1 1A	2 2A	3 3B	4 4B	5 5B	6 6B	7 7B	8 8B	9 8B	10 8B	11 1B	12 2B	13 3A	14 4A	15 5A	16 6A	17 7A	18 8A
1 **H** 1.008																	2 **He** 4.003
3 **Li** 6.941	4 **Be** 9.012											5 **B** 10.81	6 **C** 12.01	7 **N** 14.01	8 **O** 16.00	9 **F** 19.00	10 **Ne** 20.18
11 **Na** 22.99	12 **Mg** 24.31											13 **Al** 26.98	14 **Si** 28.09	15 **P** 30.97	16 **S** 32.07	17 **Cl** 35.45	18 **Ar** 39.95
19 **K** 39.10	20 **Ca** 40.08	21 **Sc** 44.96	22 **Ti** 47.88	23 **V** 50.94	24 **Cr** 52.00	25 **Mn** 54.94	26 **Fe** 55.85	27 **Co** 58.93	28 **Ni** 58.69	29 **Cu** 63.55	30 **Zn** 65.41	31 **Ga** 69.72	32 **Ge** 72.64	33 **As** 74.92	34 **Se** 78.96	35 **Br** 79.90	36 **Kr** 83.80
37 **Rb** 85.47	38 **Sr** 87.62	39 **Y** 88.91	40 **Zr** 91.22	41 **Nb** 92.91	42 **Mo** 95.94	43 **Tc** (98)	44 **Ru** 101.1	45 **Rh** 102.9	46 **Pd** 106.4	47 **Ag** 107.9	48 **Cd** 112.4	49 **In** 114.8	50 **Sn** 118.7	51 **Sb** 121.8	52 **Te** 127.6	53 **I** 126.9	54 **Xe** 131.3
55 **Cs** 132.9	56 **Ba** 137.3	57 **La** 138.9	72 **Hf** 178.5	73 **Ta** 180.9	74 **W** 183.9	75 **Re** 186.2	76 **Os** 190.2	77 **Ir** 192.2	78 **Pt** 195.1	79 **Au** 197.0	80 **Hg** 200.6	81 **Tl** 204.4	82 **Pb** 207.2	83 **Bi** 209.0	84 **Po** (209)	85 **At** (210)	86 **Rn** (222)
87 **Fr** (223)	88 **Ra** (226)	89 **Ac** (227)	104 **Rf** (261)	105 **Db** (262)	106 **Sg** (266)	107 **Bh** (262)	108 **Hs** (277)	109 **Mt** (268)	110	111	112	(113)	114	(115)	116	(117)	118

Key: 24 — Atomic number; **Cr**; 52.00 — Atomic mass

58 **Ce** 140.1	59 **Pr** 140.9	60 **Nd** 144.2	61 **Pm** (145)	62 **Sm** 150.4	63 **Eu** 152.0	64 **Gd** 157.3	65 **Tb** 158.9	66 **Dy** 162.5	67 **Ho** 164.9	68 **Er** 167.3	69 **Tm** 168.9	70 **Yb** 173.0	71 **Lu** 175.0
90 **Th** 232.0	91 **Pa** (231)	92 **U** 238.0	93 **Np** (237)	94 **Pu** (242)	95 **Am** (243)	96 **Cm** (247)	97 **Bk** (247)	98 **Cf** (251)	99 **Es** (252)	100 **Fm** (257)	101 **Md** (258)	102 **No** (259)	103 **Lr** (262)

The 1–18 group designation has been recommended by the International Union of Pure and Applied Chemistry (IUPAC) but is not yet in wide use. No names have been assigned for elements 110–112, 114, 116, and 118. Elements 113, 115, and 117 have not yet been synthesized.

Source: Chang, R. *Chemistry,* 7th ed.

INTRODUCTION TO ENVIRONMENTAL ENGINEERING

The McGraw-Hill Series in Civil and Environmental Engineering

Engineering Economy
Blank and Tarquin: *Engineering Economy*

Engineering Math and Statistics
Milton and Arnold: *Introduction to Probability and Statistics: Principles and Applications for Engineering and the Computing Sciences*
Navidi: *Statistics for Engineers and Scientists*
Reddy: *Introduction to the Finite Element Method*

Fluid Mechanics
Cengel and Cimbala: *Fluid Mechanics*
Cimbala and Cengel: *Essentials of Fluid Mechanics*
Çengel and Turner: *Fundamentals of Thermal-Fluid Sciences*
Finnemore and Franzini: *Fluid Mechanics with Engineering Applications*
White: *Fluid Mechanics*

Geotechnical Engineering
Bowles: *Foundation Analysis and Design*
Bowles: *Engineering Properties of Soils and Their Measurement*

Foundation Engineering
Salgado: *The Engineering of Foundations*

Numerical Methods
Chapra: *Applied Numerical Methods with MATLAB*
Chapra and Canale: *Numerical Methods for Engineers*
Heath: *Scientific Computing: An Introductory Survey*

Structures
Leet, Uang, Gilbert: *Fundamentals of Structural Analysis*
Nilson, Darwin, and Dolan: *Design of Concrete Structures*
Vinnakota: *Steel Structures: Behavior and LRFD*

Surveying
Anderson and Mikhail: *Surveying: Theory and Practice*
Wolf and DeWitt: *Elements of Photogrammetry (with Applications in GIS)*

Statics, Dynamics, and Mechanics of Materials
Barber: *Intermediate Mechanics of Materials*
Beer and Johnston: *Vector Mechanics for Engineers: Statics*
Beer and Johnston: *Vector Mechanics for Engineers: Dynamics*
Beer and Johnston: *Vector Mechanics for Engineers: Statics and Dynamics*
Beer and Johnston: *Mechanics of Materials*

Construction Engineering and Project Management
Raymond E. Levitt, *Stanford University,* Consulting Editor

Barrie and Paulson: *Professional Construction Management*
Bockrath: *Contracts and the Legal Environment for Engineers and Architects*
Hinze: *Construction Contracts*
Oberlender: *Project Management for Engineering and Construction*
Peurifoy and Oberlender: *Estimating Construction Costs*
Peurifoy, Schexnayder, and Shapira: *Construction Planning, Equipment, and Methods*
Schexnayder and Mayo: *Construction Management Fundamentals*

Transportation Engineering
Edward K. Morlok, *University of Pennsylvania,* Consulting Editor
Banks: *Introduction to Transportation Engineering*
Meyer and Miller: *Urban Transportation Planning*

Water Resources and Environmental Engineering
George Tchobanoglous, *University of California, Davis,* Consulting Editor

Benjamin: *Water Chemistry*
Canter: *Environmental Impact Assessment*
Chapra: *Surface Water Quality Modeling*
Chow, Maidment, Mays: *Applied Hydrology*
Crites and Tchobanoglous: *Small and Decentralized Wastewater Management Systems*
Davis and Cornwell: *Introduction to Environmental Engineering*
Davis and Masten: *Principles of Environmental Engineering and Science*
de Nevers: *Air Pollution Control Engineering*
Eckenfelder: *Industrial Water Pollution Control*
LaGrega, Buckingham, Evans: *Hazardous Waste Management*
Linsley, Franzini, Freyberg, and Tchobanoglous: *Water Resources Engineering*
Metcalf & Eddy, Inc., Tchobanoglous, Burton, and Stensel: *Wastewater Engineering: Treatment and Reuse*
Rittmann and McCarty: *Environmental Biotechnology: Principles and Applications*
Rubin: *Introduction to Engineering and the Environment*
Sawyer, McCarty, Parkin: *Chemistry for Environmental Engineering*
Sturm: *Open Channel Hydraulics*
Tchobanoglous, Theisen, Vigil: *Integrated Solid Waste Management: Engineering Principles and Management Issues*

Others Titles of Interest
Budynas: *Advanced Strength and Applied Stress Analysis*
Heerwagen: *Passive and Active Environmental Controls: Informing the Schematic Designing of Buildings*

INTRODUCTION TO ENVIRONMENTAL ENGINEERING

Fourth Edition

Mackenzie L. Davis
Michigan State University

David A. Cornwell
Environmental Engineering & Technology, Inc.

Boston Burr Ridge, IL Dubuque, IA New York San Francisco St. Louis
Bangkok Bogotá Caracas Kuala Lumpur Lisbon London Madrid Mexico City
Milan Montreal New Delhi Santiago Seoul Singapore Sydney Taipei Toronto

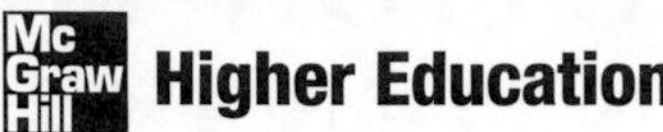

INTRODUCTION TO ENVIRONMENTAL ENGINEERING, FOURTH EDITION

Published by McGraw-Hill, a business unit of The McGraw-Hill Companies, Inc., 1221 Avenue of the Americas, New York, NY 10020.

This book is printed on acid-free paper.

1 2 3 4 5 6 7 8 9 0 DOC/DOC 0 9 8 7 6

ISBN 978–0–07–242411–9
MHID 0–07–242411–7

Senior Sponsoring Editor: *Bill Stenquist*
Developmental Editor: *Kathleen L. White*
Executive Marketing Manager: *Michael Weitz*
Project Coordinator: *Tracy L. Konrardy*
Lead Production Supervisor: *Sandy Ludovissy*
Associate Media Producer: *Christina Nelson*
Designer: *John Joran*
Cover Illustration: *Barbara Masten Cobb*
Senior Photo Research Coordinator: *John C. Leland*
Photo Research: *Jerry Marshall*
Supplement Producer: *Tracy L. Konrardy*
Compositor: *Techbooks*
Typeface: *10.5/12 Times Roman*
Printer: *R. R. Donnelley Crawfordsville, IN*

Library of Congress Cataloging-in-Publication Data

Davis, Mackenzie Leo, 1941–
Introduction to environmental engineering — 4th ed.
p. cm.
Includes bibliographical references and index.
ISBN 978–0–07–242411–9 — ISBN 0–07–242411–7
1. Environmental engineering. 2. Sanitary engineering. I. Cornwell, David A., 1948–. II. Title.

TD145.D26 2008
628—dc22

2006049134
CIP

www.mhhe.com

To our students,
who make it worthwhile

To Mack's wife Elaine, for her understanding and support these 44 years.

To Mack's daughter, Laura Safran, son-in-law, John Safran, and especially to grandsons Aaron and Zachary Safran with the hope that those who use this book may make the environment of the future a sustainable one for you and all those generations that follow.

To Nancy Cornwell for her assistance in revising the Solid Waste chapter 9 and all of her hard work in researching information for the Water Treatment chapter 4. Nancy is a recognized expert in the environmental field in her own right and her assistance and contributions are greatly appreciated.

ABOUT THE AUTHORS

Mackenzie L. Davis is an Emeritus Professor of Environmental Engineering at Michigan State University. He received all his degrees from the University of Illinois. From 1968 to 1971 he served as a Captain in the U.S. Army Medical Service Corps. During his military service he conducted air pollution surveys at Army ammunition plants. From 1971 to 1973 he was Branch Chief of the Environmental Engineering Branch at the U.S. Army Construction Engineering Research Laboratory. His responsibilities included supervision of research on air, noise, and water pollution control and solid waste management for Army facilities. In 1973 he joined the faculty at Michigan State University. He has taught and conducted research in the areas of air pollution control and hazardous waste management.

In 1987 and 1989–1992, under an intergovernmental personnel assignment with the Office of Solid Waste of the U.S. Environmental Protection Agency, Dr. Davis performed technology assessments of treatment methods used to demonstrate the regulatory requirements for the land disposal restrictions ("land ban") promulgated under the Hazardous and Solid Waste Amendments.

Dr. Davis is a member of the following professional organizations: American Chemical Society, American Institute of Chemical Engineers, American Society for Engineering Education, American Meteorological Society, American Society of Civil Engineers, American Water Works Association, Air & Waste Management Association, Association of Environmental Engineering and Science Professors, and the Water Environment Federation.

His honors and awards include the State-of-the-Art Award from the ASCE, Chapter Honor Member of Chi Epsilon, Sigma Xi, election as a Fellow in the Air & Waste Management Association, and election as a Diplomate in the American Academy of Environmental Engineers with certification in hazardous waste management. He has received teaching awards from the American Society of Civil Engineers Student Chapter, Michigan State University College of Engineering, North Central Section of the American Society for Engineering Education, Great Lakes Region of Chi Epsilon, and the Amoco Corporation. In 1998, he received the Lyman A. Ripperton Award for distinguished achievement as an educator from the Air & Waste Management Association. He is a registered professional engineer in Michigan.

In 2003, Dr. Davis retired from Michigan State University.

David A. Cornwell is the Founder and President of the consulting firm Environmental Engineering & Technology, Inc. headquartered in Newport News, VA. He attended the University of Florida in Gainesville where he received his Ph.D. in Civil/Environmental Engineering, and has remained a loyal Gator fan ever since. He was an Associate Professor in the Civil and Environmental Engineering Department at Michigan State University prior to entering the consulting field. Many of Dr. Cornwell's students now are active members of the water industry.

During his career as a consultant, Dr. Cornwell has provided design and operational services to water utilities around the world. He has lectured and written on many aspects of this field, including over 50 peer reviewed technical articles and reports. Much of his work has included the development of new and optimized water treatment processes. He is a registered professional engineer in more than 15 states.

Dr. Cornwell has an extensive record of service to the water industry. He has been an active member of AWWA since the early 1970s and has served on numerous committees in that organization. In addition, he is an active member of the Water Environment Federation, AWWA Research Foundation (subscriber), the American Chemical Society, the American Consulting Engineers Council, the Association of Environmental Engineering Professors, and is a Diplomate of the American Academy of Environmental Engineers.

In 2005, Dr. Cornwell was the recipient of the A.P. Black Research Award, given by AWWA to recognize excellence in water treatment research. He has also been the recipient of two AWWA Best Publication Awards, and was elected an Honorary Member of AWWA.

About the Cover Artist

Barbara Masten Cobb attended art school before completing an associate degree in nursing in 1983. Barbara is employed as the lead floor nurse in a New Jersey nursing home. In her spare time she pursues her avocation—painting. The water color used for the cover for this book is her third painting to grace a McGraw-Hill textbook cover.

PREFACE

Following the format of previous editions, the fourth edition of *Introduction to Environmental Engineering* is designed for use in an introductory sophomore-level engineering course with sufficient depth to allow its use in more advanced courses. The book covers the basic, traditional subject matter that forms the foundation of more advanced courses. As such, it provides the fundamental science and engineering principles that instructors in more advanced courses may assume are common knowledge for an advanced undergraduate. In the more than 60 offerings of this course, we have found that mature college students in allied fields—such as biology, chemistry, resource development, fisheries and wildlife, microbiology and soils science—have no difficulty with the material.

We have assumed the students using this text have had courses in chemistry, physics, and biology as well as sufficient mathematics to understand the concepts of differentiation and integration. Basic and environmental chemistry concepts are introduced at the beginning of the chapters in which they are relevant. This format integrates the chemistry fundamentals with their application to the subject matter of the chapter. It provides the student with the tools to analyze and understand the environmental engineering issues described in the chapter, in addition to providing an immediate feedback of the relevance of the basic chemistry. There are over 100 end-of-chapter chemistry-related problems spread throughout the text. In a similar manner, the fundamental concepts of microbiology are introduced as an introduction to biological treatment of wastewater. In the mathematical presentations, we have provided only a few derivations. In our experience, the more rigorous approach of derived mathematics may yield a result that is not more but less demonstrative—and even confusing—to the beginning engineering student.

Two themes are carried through the text. The first is an introduction to the concept of materials and energy balance as a tool for understanding environmental processes and solving environmental engineering problems. This concept is introduced in a new stand-alone chapter and then applied for conservative systems in hydrology (hydrologic cycle, development of the rational formula, and reservoir design). This theme is expanded to include sludge mass balance in Chapter 4, and the DO sag curve in Chapter 5. The design equations for a completely mixed activated sludge system and a more elaborate sludge mass balance are developed in Chapter 6. Mass balance is used to account for the production of sulfur dioxide from the combustion of coal and in the development of absorber design equations in Chapter 7. In Chapter 10, a mass balance approach is used for waste audit. There are over 100 materials and energy balance end-of-chapter problems spread throughout the text.

The second theme of the book is the concept of sustainability. First introduced in Chapter 1, the methods of waste minimization are discussed in each succeeding chapter under the topics of water conservation, sludge minimization in water treatment, land treatment of wastewater, protection of the ozone layer, global warming, resource conservation and recovery of solid waste, hazardous waste management, and reduction of the volume of radioactive waste.

Each chapter concludes with a list of review items, the traditional end-of-chapter problems, and, perhaps less traditional, discussion questions. The review items have been written in the "objective" format of the Accreditation Board for Engineering and Technology (ABET). Instructors will find this particularly helpful for directing student review for exams, for assessing continuous quality improvement for ABET and for preparing documentation for ABET curriculum review. We have found the discussion questions useful as a "minute check" or spot quiz item to see if the students understand concepts as well as number crunching.

The fourth edition has been thoroughly revised and updated. With the addition of 222 new end-of-chapter problems there are now a total of 650 problems. Sixty-six of the problems have been set up for spreadsheet solutions. The following paragraphs summarize the major changes in this edition.

- A discussion of sustainability and a discussion of the process by which laws and regulations are developed has been added to the first chapter. The discussion of ethics has been expanded.
- A new, stand-alone chapter on material and energy balances has been added.
- The hydrology chapter has been reorganized and slimmed down.
- The water treatment chapter has been revised to include a new treatment of Henry's law, new material on waterborne disease and arsenic, updated water quality standards, an updated technique for design of mixing systems, a new discussion of membrane treatment technology, and a revised and expanded discussion of ultraviolet disinfection. Two new example problems have been added.
- The water quality management chapter has been expanded to include discussions of endocrine disrupting compounds (EDCs), total maximum daily load (TMDL), water quality management in estuaries, and groundwater quality, including uncontrolled releases of contaminants and saltwater intrusion into aquifers.
- A new introduction, a new section on treatment standards, and a new section on membrane treatment have been added to the wastewater chapter. In addition, the chapter has been rearranged to place the microbiology review closer to the application of microbiology to activated sludge treatment.
- The air pollution standards have been updated and new material on mercury, lead, and $PM_{2.5}$ has been added to the air pollution chapter. In addition, the sections on origin and fate, indoor air, acid rain, ozone depletion, global warming, and control of automobile emissions have been updated. New discussions on catalytic combustion, baghouses, and mercury control have been added. Two new example problems on catalytic combustion and baghouse design have been added.
- A revised introduction to the noise pollution chapter includes the impact of hearing loss on people, as well as the economic impact of noise pollution on civil engineering projects and businesses. A new discussion of the L_{dn} concept and a

complete revision of the method of calculating airborne transmission to reflect the ISO calculation procedure are included in the noise pollution chapter.

- The solid waste chapter introduction and discussion of collection methods has been updated. A new section on bioreactor landfills has been added. All cost data have been updated.
- In the hazardous waste chapter, the section on risk assessment has been updated. The discussions on generator requirements, transporter regulations, and underground storage tanks have been slimmed down and updated. Two new sections and example problems, on retardation of uncontrolled releases into the groundwater and on pump and treat, have been added.
- The chapter on ionizing radiation has been revised to conform to the SI units of notation. Three new example problems have been added. The discussion of radioactive waste management has been updated.

As it stands in the curriculum at Michigan State University (MSU), the course bearing the title of this book provides the foundation for four follow-on senior level environmental engineering courses. The initial portions of selected chapters (hydrology, materials and energy balances, water treatment, water quality, wastewater treatment, air pollution, noise pollution, and solid waste) are included in the introductory course. Advanced material, including most of the design concepts, are covered in the upper level courses (hydrology, water and wastewater treatment plant design, solid and hazardous waste management). Some of the material is left for the students to pursue on their own (environmental legislation, ionizing radiation).

An instructor's manual and set of PowerPoint® slides are available online for qualified instructors. Please inquire with your McGraw-Hill representative for the necessary access password. The instructor's manual includes sample course outlines, solved example exams, and detailed solutions to the end-of-chapter problems. In addition, there are suggestions for using the pedagogic aids in the next.

Numerous MSU alumni have indicated that *Introduction to Environmental Engineering* is an excellent text for review and preparation for the Professional Engineers examination. It is not only readable for self-study but also provides sufficient example problems and data for practical application in the exam. Many have taken it to the exam as one of their reference resources. And they have used it!

As always, we appreciate any comments, suggestions, corrections, and contributions for future revisions.

Mackenzie L. Davis
David A. Cornwell

Acknowledgements

As with any other text, the number of individuals who have made it possible far exceeds those whose names grace the cover. At the hazard of leaving someone out, we would like to explicitly thank the following individuals for their contribution.

Over the many years of the four editions, the following students helped to solve problems, proofread text, prepare illustrations, raise embarrassing questions, and generally make sure that other students could understand the material: Shelley Agarwal, Stephanie Albert, Deb Allen, Mark Bishop, Aimee Bolen, Kristen Brandt, Jeff Brown, Amber Buhl, Nicole Chernoby, Rebecca Cline, Linda Clowater, Shauna Cohen, John Cooley, Ted Coyer, Marcia Curran, Talia Dodak, Kimberly Doherty, Bobbie Dougherty, Lisa Egleston, Karen Ellis, Craig Fricke, Elizabeth Fry, Beverly Hinds, Edith Hooten, Brad Hoos, Kathy Hulley, Geneva Hulslander, Lisa Huntington, Angela Ilieff, Alison Leach, Gary Lefko, Lynelle Marolf, Lisa McClanahan, Tim McNamara, Becky Mursch, Cheryl Oliver, Kyle Paulson, Marisa Patterson, Lynnette Payne, Jim Peters, Kristie Piner, Christine Pomeroy, Susan Quiring, Erica Rayner, Bob Reynolds, Laurene Rhyne, Sandra Risley, Carlos Sanlley, Lee Sawatzki, Stephanie Smith, Mary Stewart, Rick Wirsing, and Ya-yun Wu. To them a hearty thank you!

We would also like to thank the following reviewers for their many helpful comments and suggestions in bringing out the first three editions of the book: Wayne Chudyk, Tufts University; John Cleasby, Iowa State University; Michael J. Humenick, University of Wyoming; Tim C. Keener, University of Cincinnati; Paul King, Northeastern University; Susan Masten, Michigan State University; R. J. Murphy, University of South Florida; Thomas G. Sanders, Colorado State University; and Ron Wukasch, Purdue University. The following reviewers provided many helpful comments and useful suggestions for the fourth edition: Myron Erickson, P. E., Clean Water Plant, City of Wyoming, MI; Thomas Overcamp, Clemson University; James E. Alleman, Iowa State University; Janet Baldwin, Roger Williams University; Ernest R. Blatchley, III, Purdue University; Amy B. Chan Hilton, Florida A&M University-Florida State University; Tim Ellis, Iowa State University; Selma E. Guigard, University of Alberta; Nancy J. Hayden, University of Vermont; Jin Li, University of Wisconsin-Milwaukee; Mingming Lu, University of Cincinnati; Taha F. Marhaba, New Jersey Institute of Technology; Alexander P. Mathews, Kansas State University; William F. McTernan, Oklahoma State University; Eberhard Morgenroth, University of Illinois at Urbana-Champaign; Richard J. Schuhmann, The Pennsylvania State University; Michael S. Switzenbaum, Marquette University; Derek G. Williamson, University of Alabama.

To John Eastman, our esteemed friend and former colleague, we offer our sincere appreciation. His contribution to the initial work of Chapter 5 in the first edition, as well as constructive criticism and "independent" testing of the material was exceptionally helpful. Kristin Erickson, Radiation Safety Officer, Office of Radiation, Chemical and Biological Safety, Michigan State University, contributed to the Chapter 11 revisions for the third edition. To her we offer our hearty thanks.

And last, but certainly not least, we wish to thank our families, who have put up with the nonsense of book writing.

CONTENTS

CHAPTER 1

INTRODUCTION

1-1 WHAT IS ENVIRONMENTAL ENGINEERING?

Professions, Learned and Otherwise

Webster's dictionary defines the learned professions as law, medicine, and theology. It has been suggested that engineers may not be learned enough to rank among these because the study of law, medicine, or theology requires considerably more than four years of undergraduate work. There was a time, some hundred years ago, when the four-year engineering program was two years longer than those of the learned professions! At any rate, *Webster's* is willing to concede that engineering, along with teaching and writing, is a profession even if it is not "learned." At a minimum, a profession is an occupation that requires advanced training in the liberal arts or sciences and mental rather than manual work.

But being a professional is more than being in or of a profession. True professionals are those who pursue their learned art in a spirit of public service (ASCE, 1973). True professionalism is defined by the following seven characteristics:

1. Professional decisions are made by means of general principles, theories, or propositions that are independent of the particular case under consideration.
2. Professional decisions imply knowledge in a specific area in which the person is expert. The professional is an expert only in his or her profession and not an expert at everything.
3. The professional's relations with his or her clients are objective and independent of particular sentiments about them.
4. A professional achieves status and financial reward by accomplishment, not by inherent qualities such as birth order, race, religion, sex, or age or by membership in a union.
5. A professional's decisions are assumed to be on behalf of the client and to be independent of self-interest.
6. The professional relates to a voluntary association of professionals and accepts only the authority of those colleagues as a sanction on his or her own behavior.
7. A professional is someone who knows better what is good for clients than do the clients. The professional's expertise puts the client into a very vulnerable position. This vulnerability has necessitated the development of strong professional codes and ethics, which serve to protect the client. Such codes are enforced through the colleague peer group (Schein, 1968).

The branch of engineering called civil engineering, from which environmental engineering is primarily, but not exclusively, derived, has an established code of ethics that embodies these principles. The code is summarized in Figure 1-1.

And What Is Engineering?

Engineering is a profession that applies mathematics and science to utilize the properties of matter and sources of energy to create useful structures, machines, products, systems, and processes.

AMERICAN SOCIETY OF CIVIL ENGINEERS
CODE OF ETHICS

Fundamental Principles

Engineers uphold and advance the integrity, honor and dignity of the engineering profession by:

1. using their knowledge and skill for the enhancement of human welfare and the environment;
2. being honest and impartial and serving with fidelity the public, their employers and clients;
3. striving to increase the competence and prestige of the engineering profession; and
4. supporting the professional and technical societies of their disciplines.

Fundamental Canons

1. Engineers shall hold paramount the safety, health and welfare of the public and shall strive to comply with the principles of sustainable development in the performance of their professional duties.
2. Engineers shall perform services only in areas of their competence.
3. Engineers shall issue public statements only in an objective and truthful manner.
4. Engineers shall act in professional matters for each employer or client as faithful agents or trustees, and shall avoid conflicts of interest.
5. Engineers shall build their professional reputation on the merit of their services and shall not compete unfairly with others.
6. Engineers shall act in such a manner as to uphold and enhance the honor, integrity, and dignity of the engineering profession.
7. Engineers shall continue their professional development throughout their careers, and shall provide opportunities for the professional development of those engineers under their supervision.

FIGURE 1-1
American Society of Civil Engineers code of ethics. (ASCE, 2005)

This implies that there are fundamental differences between scientists and engineers. The key is not so much in the individual parts of the definition, but rather in the integration of the parts. It is inherent in the professional development of the engineer that he or she must attain experience, practice, and judgment under the tutelage of an experienced engineer. Engineering has at least this much in common with the learned professions!

Engineers are frequently pressed to explain why they are different from scientists. Consider the following distinction: "Scientists discover things. Engineers make them work" (MacVicar, 1983).

On to Environmental Engineering

The Environmental Engineering Division of the American Society of Civil Engineers (ASCE) has published the following statement of purpose:

> Environmental engineering is manifest by sound engineering thought and practice in the solution of problems of environmental sanitation, notably in the provision of safe, palatable, and ample public water supplies; the proper disposal of or recycle of wastewater and solid wastes; the adequate drainage of urban and rural areas for proper sanitation; and the control of water,

> soil, and atmospheric pollution, and the social and environmental impact of these solutions. Furthermore it is concerned with engineering problems in the field of public health, such as control of arthropod-borne diseases, the elimination of industrial health hazards, and the provision of adequate sanitation in urban, rural, and recreational areas, and the effect of technological advances on the environment (ASCE, 1977).

Thus, we may consider what environmental engineering is not. It is not concerned primarily with heating, ventilating, or air conditioning (HVAC), nor is it concerned primarily with landscape architecture. Neither should it be confused with the architectural and structural engineering functions associated with built environments, such as homes, offices, and other workplaces.

1-2 INTRODUCTION TO ENVIRONMENTAL ENGINEERING

Where Do We Start?

We have used the ASCE definition of an environmental engineer as a basis for this book. Given the constraints of time and space, we have limited ourselves to the following topics from the definition:

1. Provision of safe, palatable, and ample public water supplies
2. Proper disposal of or recycling of wastewater and solid wastes
3. Control of water, soil, and atmospheric pollution (including noise as an atmospheric pollutant)

A Short Outline of This Book

This short outline provides an overview of the book. It is derived from the ASCE definition of environmental engineering.

Chapter 2, Materials and Energy Balances, introduces tools that are used in environmental engineering. These tools will be applied throughout the book to develop a fundamental understanding of the subject matter and as a method for developing equations to analyze and describe the behavior of environmental processes.

Hydrology is the subject of Chapter 3. In that chapter we discuss the hydrologic cycle and the analyses used to ensure an ample supply of water from either surface water or groundwater. Because hydrology is concerned with flooding as well as with droughts, we also touch on the "adequate drainage" portion of the environmental engineering definition. The discussion of the physics of groundwater movement will give you the tools you need to understand problems of groundwater pollution.

In Chapter 4 we turn from water quantity to water quality. First, we review some basic chemistry concepts and calculations; then we examine some characteristics of water that affect its quality. Finally, we explain how to treat water for public consumption.

In Chapter 5 we consider the effects of various materials on water quality. In particular, we spend a good deal of time examining the effects of organic pollution on the levels of dissolved oxygen in the water. Dissolved oxygen is required for higher forms of aquatic life, such as fish, to survive.

Wastewater treatment is the subject of Chapter 6. Here, we look at how we can remove pollutants that reduce the quality of the lake or stream. Our emphasis is on municipal wastewater treatment.

In Chapters 7 and 8, we turn to the control of atmospheric pollution and noise control. After a brief introduction to the health effects and other environmental impacts of air pollutants and noise, we examine transport processes that carry pollutants from their source to people, as well as some methods of control.

Solid waste is the topic of Chapter 9. Collection, disposal, and recycling of solid waste are fundamental needs of our complex urban society. This chapter will present some of the tools for understanding and solving problems in solid waste management.

Hazardous waste is the topic of Chapter 10. Methods of dealing with abandoned hazardous waste sites and managing the wastes we are continually generating are discussed. We examine some alternatives for treatment of these wastes as an application of the technologies addressed in earlier chapters.

The final chapter is a brief examination of ionizing radiation. A brief introduction to health effects of radiation is followed by a discussion of management techniques for both radioactive waste and x-rays.

The appendices provide tables of the properties of air, water, and selected chemicals.

Tables at the inside front and back covers provide a list of atomic masses, the periodic table, conversion factors, and the International System of Units (SI) naming convention for factors of 10.

1-3 ENVIRONMENTAL SYSTEMS OVERVIEW

Systems as Such

Before we begin in earnest, we thought it worth taking a look at the problems to be discussed in this text in a larger perspective. Engineers like to call this the "systems approach," that is, looking at all the interrelated parts and their effects on one another. In environmental systems it is doubtful that mere mortals can ever hope to identify all the interrelated parts, to say nothing of trying to establish their effects on one another. The first thing the systems engineer does, then, is to simplify the system to a tractable size that behaves in a fashion similar to the real system. The simplified model does not behave in detail as the system does, but it gives a fair approximation of what is going on.

We have followed this pattern of simplification in our description of three environmental systems: the water resource management system, the air resource management system, and the solid waste management system. Pollution problems that are confined to one of these systems are called single-medium problems if the medium is either air, water, or soil. Many important environmental problems are not confined to one of these simple systems but cross the boundaries from one to the other. These problems are referred to as *multimedia* pollution problems.

Water Resource Management System

Water Supply Subsystem. The nature of the water source commonly determines the planning, design, and operation of the collection, purification, transmission, and distribution works. The two major sources used to supply community and industrial needs

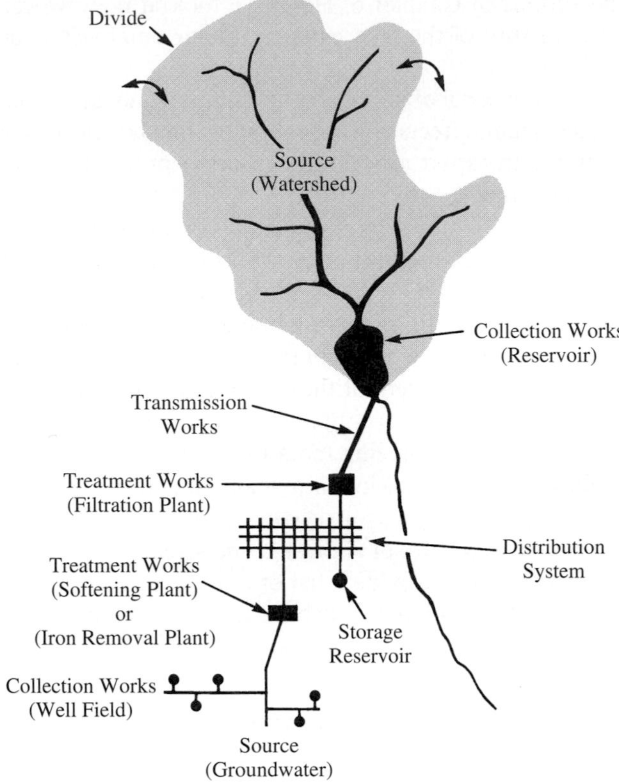

FIGURE 1-2
An extension of the water supply resource system.

are referred to as *surface water* and *groundwater.* Streams, lakes, and rivers are the surface water sources. Groundwater sources are those pumped from wells.

Figure 1-2 depicts an extension of the water resource system to serve a small community. The source in each case determines the type of collection works and the type of treatment works.* The pipe network in the city is called the distribution system. The pipes themselves are often referred to as *water mains*. Water in the mains generally is kept at a pressure between 200 and 860 kilopascals (kPa). Excess water produced by the treatment plant during periods of low *demand*† (usually the nighttime hours) is held in a storage reservoir. The storage reservoir may be elevated (the ubiquitous water tower), or it may be at ground level. The stored water is used to meet high demand during the day. Storage compensates for changes in demand and allows a smaller treatment plant to be built. It also provides emergency backup in case of a fire.

Population and water-consumption patterns are the prime factors that govern the quantity of water required and hence the source and the whole composition of the water

***Works* is a noun used in the plural to mean "engineering structures." It is used in the same sense as *art works.*

†*Demand* is the use of water by consumers. This use of the word derives from the economic term meaning "the desire for a commodity." The consumers express their desire by opening the faucet or flushing the water closet (W.C.).

resource system. One of the first steps in the selection of a suitable water-supply source is determining the demand that will be placed on it. The essential elements of water demand include average daily water consumption and peak rate of demand. Average daily water consumption must be estimated for two reasons: (1) to determine the ability of the water source to meet continuing demands over critical periods when surface flows are low or groundwater tables are at minimum elevations, and (2) for purposes of estimating quantities of stored water that would satisfy demands during these critical periods. The peak demand rates must be estimated in order to determine plumbing and pipe sizing, pressure losses, and storage requirements necessary to supply sufficient water during periods of peak water demand.

Many factors influence water use for a given system. For example, the mere fact that water under pressure is available stimulates its use, often excessively, for watering lawns and gardens, for washing automobiles, for operating air-conditioning equipment, and for performing many other activities at home and in industry. The following factors have been found to influence water consumption in a major way:

1. Climate
2. Industrial activity
3. Meterage
4. System management
5. Standard of living

The following factors also influence water consumption but to a lesser degree: extent of sewerage, system pressure, water price, and availability of private wells.

If the demand for water is measured on a *per capita** basis, climate is the most important factor influencing demand. This is shown dramatically in Table 1-1. The average annual precipitation for the "wet" states is about 100 cm per year while the average annual precipitation for the "dry" states is only about 25 cm per year. Of course, the "dry" states are also considerably warmer than the "wet" states.

The influence of industry is to increase per capita water demand. Small rural and suburban communities will use less water per person than industrialized communities.

The third most important factor in water use is whether individual consumers have water meters. Meterage imposes a sense of responsibility not found in unmetered residences and businesses. This sense of responsibility reduces per capita water consumption because customers repair leaks and make more conservative water-use decisions almost regardless of price. Because water is so inexpensive, price is not much of a factor.

Following meterage closely is the aspect called system management. If the water distribution system is well managed, per capita water consumption is less than if it is not well managed. Well-managed systems are those in which the managers know when and where leaks in the water main occur and have them repaired promptly.

---

*Per capita is a Latin term that means "by heads." Here it means "per person." This assumes that each person has one head (on the average).

TABLE 1-1
Total fresh water withdrawals for public supply[a]

State	Withdrawal (Lpcd)[b]
"Wet"	
Connecticut	471
Michigan	434
New Jersey	473
Ohio	488
Pennsylvania	449
Average	463
"Dry"	
Nevada	1,190
New Mexico	797
Utah	1,083
Average	963

[a]Compiled from Hutson et al. (2001).
[b]Lpcd = liters per capita per day.

Climate, industrial activity, meterage, and system management are more significant factors controlling water consumption than the standard of living. The rationale for the last factor is straightforward. Per capita water use increases with an increased standard of living. Highly developed countries use much more water than the less developed nations. Likewise, higher socioeconomic status implies greater per capita water use than lower socioeconomic status.

The total U.S. water withdrawal for all uses (agricultural, commercial, domestic, mining, and thermoelectric power), including both fresh and saline water, was estimated to be approximately 5,400 liters per capita per day (Lpcd) in 2000 (Hutson, et al., 2001). The amount for U.S. public supply (domestic, commercial and industrial use) was estimated to be 580 Lpcd in 2000 (Hutson, et al., 2001). The American Water Works Association estimated that the average daily household water use in the United States was 1,320 liters per day in 1999 (AWWA, 1999). For a family of three, this would amount to about 440 Lpcd. The variation in demand is normally reported as a factor of the average day. For metered dwellings the factors are as follows: maximum day = 2.2 $\times$ average day; peak hour = 5.3 $\times$ average day (Linaweaver et al., 1967). Some mid-Michigan average daily use figures and the contribution of various sectors to demand are shown in Table 1-2.

International per capita domestic water use has been estimated by the Pacific Institute for Studies in Development, Environment, and Security (Pacific Institute, 2000). For example, they report the following (all in Lpcd): Australia, 1,400; Canada, 430; China, 60; Ecuador, 85; Egypt, 130; Germany, 270; India, 30; Mexico, 130; Nigeria, 25.

Wastewater Disposal Subsystem. Safe disposal of all human wastes is necessary to protect the health of the individual, the family, and the community, and also to prevent

TABLE 1-2
Examples of variation in per capita water consumption

		Percent of per capita consumption		
Location	Lpcd	Industry	Commercial	Residential
Lansing, MI	512	14	32	54
East Lansing, MI	310	0	10	90
Michigan State University	271	0	1	99

Data from local treatment plants, 2004.

the occurrence of certain nuisances. To accomplish satisfactory results, human wastes must be disposed of so that:

1. They will not contaminate any drinking water supply.
2. They will not give rise to a public health hazard by being accessible to *vectors* (insects, rodents, or other possible carriers) that may come into contact with food or drinking water.
3. They will not give rise to a public health hazard by being accessible to children.
4. They will not cause violation of laws or regulations governing water pollution or sewage disposal.
5. They will not pollute or contaminate the waters of any bathing beach, shellfish-breeding ground, or stream used for public or domestic water-supply purposes, or for recreational purposes.
6. They will not give rise to a nuisance due to odor or unsightly appearance.

These criteria can best be met by the discharge of domestic sewage to an adequate public or community sewerage system (U.S. PHS, 1970). Where no community sewer system exists, on-site disposal by an approved method is mandatory.

In its simplest form the wastewater management subsystem is composed of six parts (Figure 1-3). The source of wastewater may be either industrial wastewater or domestic sewage or both.* Industrial wastewater may be subject to some pretreatment on site if it has the potential to upset the municipal wastewater treatment plant (WWTP). Federal regulations refer to municipal wastewater treatment systems as publicly owned treatment works, or POTWs.

The quantity of sewage flowing to the WWTP varies widely throughout the day in response to water usage. A typical daily variation is shown in Figure 1-4. Most of the water used in a community will end up in the sewer. Between 5 and 15 percent of the water is lost in lawn watering, car washing, and other consumptive uses. Consumptive use may be thought of as the difference between the average rate that water

*Domestic sewage is sometimes called sanitary sewage, although it is far from being sanitary!

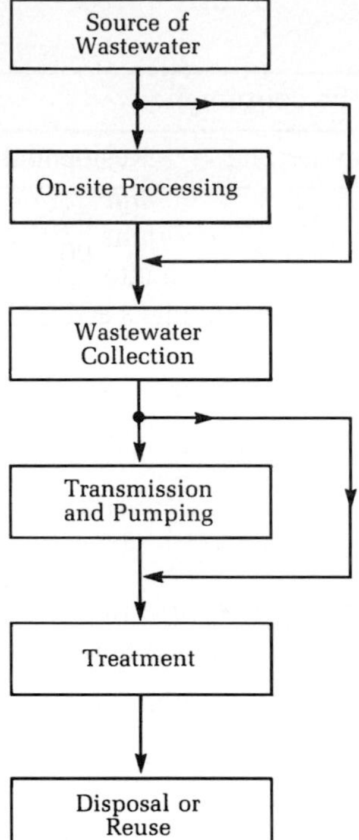

FIGURE 1-3
Wastewater management subsystem. (Linsley and Fanzini, 1979)

flows into the distribution system and the average rate that wastewater flows into the WWTP (excepting the effects of leaks in the pipes).

The quantity of wastewater, with one exception, depends on the same factors that determine the quantity of water required for supply. The major exception is that underground water (groundwater) conditions may strongly affect the quantity of water in the

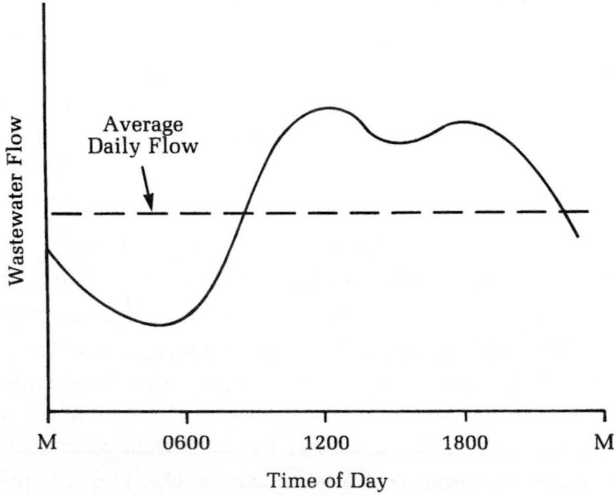

FIGURE 1-4
Typical variation in daily wastewater flow.

system because of leaks. Whereas the drinking water distribution system is under pressure and is relatively tight, the sewer system is gravity operated and is relatively open. Thus, groundwater may *infiltrate,* or leak into, the system. When manholes lie in low spots, there is the additional possibility of *inflow* through leaks in the manhole cover. Other sources of inflow include direct connections from roof gutters and downspouts, as well as sump pumps used to remove water from basement footing tiles. *Infiltration* and *inflow* (I & I) are particularly important during rainstorms. The additional water from I & I may hydraulically overload the sewer causing sewage to back up into houses as well as to reduce the efficiency of the WWTP. New construction techniques and materials have made it possible to reduce I & I to insignificant amounts.

Sewers are classified into three categories: sanitary, storm, and combined. *Sanitary sewers* are designed to carry municipal wastewater from homes and commercial establishments. With proper pretreatment, industrial wastes may also be discharged into these sewers. *Storm sewers* are designed to handle excess rainwater to prevent flooding of low areas. While sanitary sewers convey wastewater to treatment facilities, storm sewers generally discharge into rivers and streams. *Combined sewers* are expected to accommodate both municipal wastewater and stormwater. These systems are designed so that during dry periods the wastewater is carried to a treatment facility. During rain storms, the excess water is discharged directly into a river, stream or lake without treatment. Unfortunately, the storm water is mixed with untreated sewage. The U.S. Environmental Protection Agency (EPA) has estimated that 40,000 overflows occur each year. Modern design practice discourages the building of combined sewers. Many communities have already begun the process of replacing the combined sewers with separate systems for sanitary and storm flow.

When gravity flow is not possible or when sewer trenches become uneconomically deep, the wastewater may be pumped. When the sewage is pumped vertically to discharge into a higher-elevation gravity sewer, the location of the sewage pump is called a *lift station.*

Sewage treatment is performed at the WWTP to stabilize the waste material, that is, to make it less *putrescible*. The *effluent* from the WWTP may be discharged into an ocean, lake, or river (called the receiving body). Alternatively, it may be discharged onto (or into) the ground, or be processed for reuse. The by-product sludge from the WWTP also must be disposed of in an environmentally acceptable manner.

Whether the waste is discharged onto the ground or into a receiving body, care must be exercised not to overtax the assimilative capacity of the ground or receiving body. The fact that the wastewater effluent is cleaner than the river into which it flows does not justify the discharge if it turns out to be the proverbial "straw that breaks the camel's back."

In summary, water resource management is the process of managing both the quantity and the quality of the water used for human benefit without destroying its availability and purity.

Air Resource Management System

Our air resource differs from our water resource in two important aspects. The first is in regard to quantity. Whereas engineering structures are required to provide an adequate water supply, air is delivered free of charge in whatever quantity we desire. The second

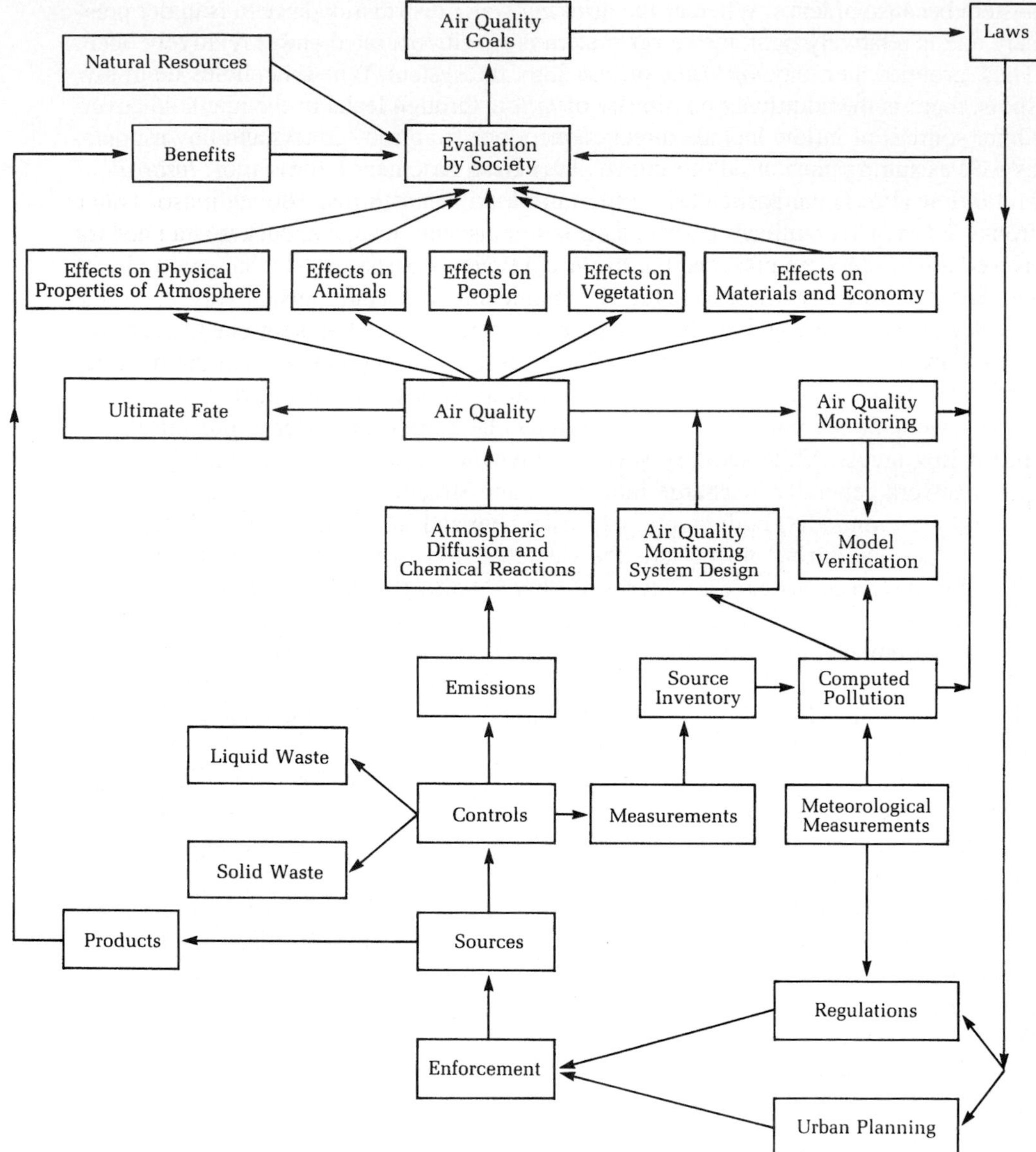

FIGURE 1-5
A simplified block diagram of an air resource management system.

aspect is in regard to quality. Unlike water, which can be treated before we use it, it is impractical to go about with a gas mask on to treat impure air and with ear plugs in to keep out the noise.

The balance of cost and benefit to obtain a desired quality of air is termed *air resource management*. Cost-benefit analyses can be problematic for at least two reasons. First is the question of what is desired air quality. The basic objective is, of course, to protect the health and welfare of people. But how much air pollution can we stand? We know the tolerable limit is something greater than zero, but tolerance varies from person to person. Second is the question of cost versus benefit. We know that we don't want to spend the entire Gross Domestic Product to ensure that no individual's health or welfare is impaired, but we do know that we want to spend some amount. Although the cost of control can be reasonably determined by standard engineering and economic means, the cost of pollution is still far from being quantitatively assessed.

Air resource management programs are instituted for a variety of reasons. The most defensible reasons are that (1) air quality has deteriorated and there is a need for correction, and (2) the potential for a future problem is strong.

In order to carry out an air resource management program effectively, all of the elements shown in Figure 1-5 must be employed. (Note that with the appropriate substitution of the word *water* for *air,* these elements apply to management of water resources as well.)

Solid Waste Management

In the past, solid waste was considered a resource, and we will examine its current potential as a resource. Generally, however, solid waste is considered a problem to be solved as cheaply as possible rather than a resource to be recovered. A simplified block diagram of a solid waste management system is shown in Figure 1-6.

While typhoid and cholera epidemics of the mid-1800s spurred water resource management efforts, and while air pollution episodes have prompted better air resource management, we have yet to feel the impact of material or energy shortages severe enough to encourage modern solid waste management. The landfill "crisis" of the 1980s appears to have abated in the early 1990s due to new or expanded landfill capacity and to many initiatives to reduce the amount of solid waste generated. By

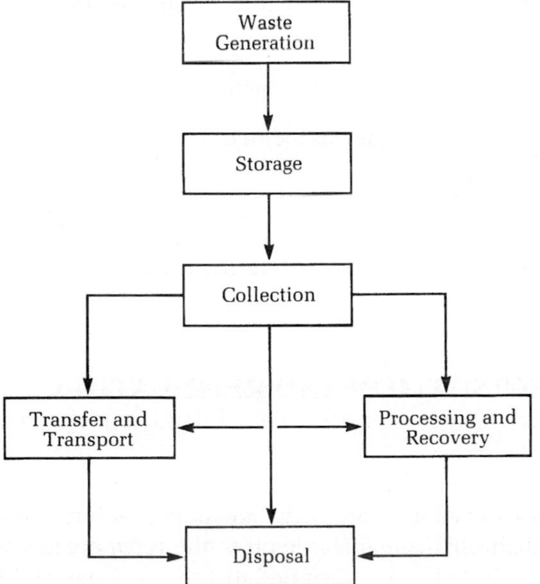

FIGURE 1-6
A simplified block diagram of a solid waste management system. (Tchobanoglous et al., 1977)

1999, more than 9,000 curbside recycling programs served roughly half of the U.S. population (U.S. EPA, 2005a).

Multimedia Systems

Many environmental problems cross the air-water-soil boundary. An example is acid rain that results from the emission of sulfur oxides and nitrogen oxides into the atmosphere. These pollutants are washed out of the atmosphere, thus cleansing it, but in turn polluting water and changing the soil chemistry, which ultimately results in the death of fish and trees. Thus, our historic reliance on the natural cleansing processes of the atmosphere in designing air-pollution-control equipment has failed to deal with the multimedia nature of the problem. Likewise, disposal of solid waste by incineration results in air pollution, which in turn is controlled by scrubbing with water, resulting in a water pollution problem.

Three lessons have come to us from our experience with multimedia problems. First, it is dangerous to develop models that are too simplistic. Second, environmental engineers must use a multimedia approach and, in particular, work with a multidisciplinary team to solve environmental problems. Third, the best solution to environmental pollution is waste minimization—if waste is not produced, it does not need to be treated or disposed of.

Sustainability

While pollution problems will remain with us for the foreseeable future, an overriding issue for the continuation of our modern living style and for the development of a similar living style for those in developing countries, is the question of *sustainability*. That is, how do we maintain our ecosystem in the light of major depletion of our natural resources. If, in our systems view, we look beyond the simple idea of controlling pollution to the larger idea of sustaining our environment, we see that there are better solutions for our pollution problems. For example:

- Pollution prevention by the minimization of waste production
- Life cycle analysis of our production techniques to include built-in features for extraction and reuse of materials
- Selection of materials and methods that have a long life
- Selection of manufacturing methods and equipment that minimize energy and water consumption

1-4 ENVIRONMENTAL LEGISLATION AND REGULATION

Acts, Laws, and Regulations

The following paragraphs provide a brief introduction to the process leading to the establishment of regulations and the terms used to identify the location of information about bills, laws, and regulations. This discussion is restricted to the federal process and nomenclature.

A proposal for a new law, called a *bill,* is introduced in either the Senate or the House of Representatives (House). The bill is given a designation, for example S. 2649 in the Senate or H.R. 5959 in the House. Bills often have "companions" in that similar bills may be started in both the Senate and House at the same time. The bill is given a title, for example, the "Safe Drinking Water Act" which implies an "act" of Congress. The act may be listed under one "Title" or it may be divided into several "Titles." References to the "Titles" of the act are given by roman numeral. For example, Title III of the Clean Air Act Amendments establishes a list of hazardous air pollutants. Frequently a bill directs some executive branch of the government such as the EPA to carry out an action such as setting limits for contaminants. On occasion, such a bill includes specific numbers for limits on contaminants. If the bills successfully pass the committee to which they are assigned, they are "reported out" to the full Senate (for example, Senate Report 99-56) or to the full House (House Report 99-168). The first digits preceding the dash refer to the session of Congress during which the bill is reported out. In this example, it is the 99th Congress. If bills pass the full Senate/House they are taken by a joint committee of senators and congressional representatives (conference committee) to form a single bill for action by both the Senate and House. If the bill is adopted by a majority of both houses, it goes to the President for approval or veto. When the President signs the bill it becomes a *law* or *statute*. It is then designated, for example, as Public Law 99-339 or PL 99-339. This means it is the 339th law passed by the 99th Congress. The law or statute approved by the President's signature may alternatively be called an act that is referred to by the title assigned the bill in Congress.

The Office of the Federal Register prepares the *United States Statutes at Large* annually. This is a compilation of the laws, concurrent resolutions, reorganization plans, and proclamations issued during each congressional session. The statutes are numbered chronologically. They are not placed in order by subject matter. The short hand reference is, for example, 104 Stat. 3000.

The *United States Code* is the compiled written set of laws in force on the day before the beginning of the current session of Congress (U.S. Code, 2005). Reference is made to the U.S. Code by "Title" and "Section" number (for example, 42 USC 6901 or 42 U.S.C. §6901). Table 1-3 gives a sample of titles and sections of environmental interest. Note that "Titles" of the U.S. Code do not match the "Titles" of the Acts of Congress.

In carrying out the directives of the Congress to develop a *regulation* or *rule,* the EPA or other executive branch of the government follows a specific set of formal procedures in a process referred to as *rule making*. The government agency (EPA, Department of Energy, Federal Aviation Agency, etc.) first publishes a *proposed rule* in the *Federal Register*. The *Federal Register* is, in essence, the government's newspaper. It is published every day that the federal government is open for business. The agency provides the logic for the rule making (called a *preamble*) as well as the proposed rule and requests comments. The preamble may be several hundred pages in length for a rule that is only a few lines long or a single page table of allowable concentrations of contaminants. Prior to the issuance of a final rule, the agency allows and considers public comment. The time period for submitting public comments varies. For rules that are not complex or controversial it may be a few weeks. For more complicated rules, the comment period may extend for as long as a year. The reference citation to *Federal Register*

TABLE 1-3
U.S. Code title and section numbers of environmental interest

Title	Sections	Statute
7	136 to 136y	Federal Insecticide, Fungicide, and Rodenticide Act
16	1531 to 1544	Endangered Species Act
33	1251 to 1387	Clean Water Act
33	2701 to 2761	Oil Pollution Act
42	300f to 300j-26	Safe Drinking Water Act
42	4321 to 4347	National Environmental Policy Act
42	4901 to 4918	Noise Control Act
42	6901 to 6922k	Solid Waste Disposal Act
42	7401 to 7671q	Clean Air Act (includes noise at §7641)
42	9601 to 9675	Comprehensive Environmental Response, Compensation, and Liability Act
42	11001 to 11050	Emergency Planning and Community Right-to-Know Act
42	13101 to 13109	Pollution Prevention Act
46	3703a	Oil Pollution Act
49	2101	Aviation Safety and Noise Abatement Act[a]
49	2202	Airport and Airway Improvement Act[a]
49	47501 to 47510	Airport Noise Abatement Act

[a]At U.S. Code Annotated (U.S.C.S.A.)

publications is in the following form: 59 FR 11863. The first number is the volume number. Volumes are numbered by year. The last number is the page number. Pages are numbered sequentially beginning with page 1 on the first day of business in January of each year. From the number shown, this rule making starts on page 11,863! Although one might assume this is late in the year, it may not be if a large number of rules have been published. This makes the date of publication very useful in searching for the rule.

Once a year, on July 1, the rules that have been finalized in the past year are *codified*. This means they are organized and published in the *Code of Federal Regulations* (CFR, 2005). Unlike the *Federal Register,* the *Code of Federal Regulations* is a compilation of the rules/regulations of the various agencies without explanation of how the government arrived at its decision. The explanation of how the rule was developed may be found only in the *Federal Register.* The notation used for *Code of Federal Regulations* is as follows: 40 CFR 280. The first number is the "Title" number. The second number in the citation refers to the part number. Unfortunately, this title number has no relation to either the title number in the Act or the United States Code title number. The CFR title numbers and subjects of environmental interest are shown in Table 1-4.

Water Quality Management

Drinking water. Under the Interstate Quarantine Act of 1893, the U.S. Public Health Service (PHS) was empowered to make and enforce regulations to prevent the spread of communicable diseases. Interstate regulations were first promulgated in 1894 and the first water-related regulation (prohibiting the use of the "common cup" on interstate carriers) was adopted in 1912. The first federal drinking water regulation was adopted

TABLE 1-4
***Code of Federal Regulations* title numbers of environmental interest**

Title number	Subject
7	Agriculture (soil conservation)
10	Energy (Nuclear Regulatory Commission)
14	Aeronautics and Space (noise)
16	Conservation
23	Highways (noise)
24	Housing and Urban Development (noise)
29	Labor (noise)
30	Mineral Resources (surface mining reclamation)
33	Navigation and Navigable Waters (wet lands and dredging)
40	Protection of the Environment (Environmental Protection Agency)
42	Public Health and Welfare
43	Public Lands: Interior
49	Transportation (transporting hazardous waste)
50	Wildlife and Fisheries

in 1914. It established limits for bacterial contamination. In 1925, still acting under the 1893 Act, the PHS tightened the bacteriological standard and added physical and chemical standards. These were reviewed and updated periodically through the 1940s. In 1962, a comprehensive update of the standards was completed. These standards were accepted by all the states but were binding on only about 2 percent of the communities, that is those that served interstate carriers.

The Safe Drinking Water Act of 1974 (SDWA) also identified as Title XIV of the U.S. Public Health Service Act, was the first congressional act focused on drinking water. It directed the newly formed EPA to revise drinking-water regulations to protect the public health. The Congress specified a two step process. First, it was to publish recommended maximum contaminant levels (RMCLs) for contaminants believed to have an adverse effect on health based on a study of health effects by the National Academy of Science. The RMCLs were to be set, with an adequate margin of safety, at a level that known or anticipated health effect would occur. The Congress specified that these levels were to be health goals and were not to be federally enforceable. EPA was then to set *maximum contaminant levels* (MCLs) as close to the RMCLs as the agency thought feasible. These became the National Primary Drinking Water Regulations. These standards applied to public water systems serving 25 or more people year-round or having 15 or more year-round service connections.

The SDWA was amended and/or reauthorized in 1977, 1979, and 1980. The 1986 revision of the SDWA resulted in significant changes. The congressional focus was on strengthening the regulation-setting process which had lagged significantly under the Reagan administration. The 1986 Act required:

1. Mandatory standards for 83 contaminants by June 1989.
2. Mandatory regulation of 25 contaminants every 3 years.

3. Designation of best available technology (BAT) for each contaminant regulated.
4. Specification of criteria for deciding when filtration of surface water supplies is required.
5. Disinfection of all public water supplies.
6. Monitoring for contaminants that were not regulated.
7. Banned lead solders, flux, and pipe in public water systems.
8. New programs for wellhead protection and protection of sole source aquifers (Pontius, 2003).

The mandate to regulate 25 contaminants every 3 years could not be met, and after 1992 regulations ceased to be issued. The 1986 SDWA amendments authorized congressional appropriations for implementation through fiscal year 1991. Reauthorization was not completed until 1996.

The 1996 SDWA amendments were signed into law by President Clinton as PL 104–182. The amendments made substantial revisions to the act. Eleven new sections were added. The amendments strengthened and expanded the protection of drinking water by providing grants for compliance and enforcement, enhanced water-system capacity, operator training, and development of solutions to source pollution. In addition, it provided for public notification of violations within 24 hours (rather than 2 weeks under the old act), and annual reporting of levels of regulated contaminants to consumers. Relief from analysis of contaminants that have never been found and are unlikely to occur was given to reduce analytical costs. EPA was funded to conduct research on health effects and treatment for arsenic, radon, and Crytosporidium. In addition, EPA was required to develop a screening program to identify the risks posed by substances that have an effect similar to that produced by naturally occurring estrogen and to screen pesticides and other chemicals for estrogenic effects. In a major shift from all preceding environmental rule making, Section 1412(b)(6) of the act requires that environmental regulations include an assessment of the costs and benefits. Furthermore, it permits the EPA administrator to "promulgate a maximum contaminant level for the contaminant that maximizes health risk reduction benefits at a cost justified by the benefits." Prior to the enactment of this legislation, cost was not to be considered in the protection of human health and the environment.

Water pollution control. The federal role in water pollution control began with the Public Health Service Act of 1912. This act established the Streams Investigation Station at Cincinnati to carry out water pollution research. The Oil Pollution Act was passed in 1924 to prevent oily discharges on coastal waters. During the 1930s and 1940s, there was a continuing debate over whether the federal government should take a greater role in controlling water pollution. This debate led to the limited expansion of federal powers expressed in the Water Pollution Control Act of 1948 (Table 1-5). The Federal Water Pollution Control Act (FWPCA) of 1956, passed by overriding President Eisehower's veto (Percival, 2003), was the cornerstone of early federal efforts to reduce pollution. Key elements of the act included a new program of subsidies for

TABLE 1-5
Federal laws controlling water pollution

Year	Title	Selected elements of legislation[a]
1948	Water Pollution Control Act	Funds for state water pollution control agencies Technical assistance to states Limited provisions for legal action against polluters
1956	Federal Water Pollution Control Act (FWPCA)	Funds for water pollution research and training Construction grants to municipalities Three-stage enforcement process
1965	Water Quality Act	States set water quality standards States prepare implementation plans
1972	FWPCA Amendments	Zero discharge of pollutants goal BPT and BAT effluent limitations NPDES permits Enforcement based on permit violations
1977	Clean Water Act	BAT requirements for toxic substances BCT requirements for conventional pollutants
1981	Municipal Waste Treatment Construction Grants Amendments	Reduced federal share in construction grants program

[a]The table entries include only the new policies and programs established by each of the laws. Often these provisions were carried forward in modified form as elements of subsequent legislation.

Legend:
BPT = Best Practical Treatment
BAT = Best Available Treatment
NPDES = National Pollution Discharge Elimination System
BCT = Best Conventional Treatment

municipal treatment plant construction and an expanded basis for federal legal action against polluters. Increased funding for state water pollution control efforts and new support for research and training activities were also provided. Each of these programs was continued in the many amendments to the Federal Water Pollution Control Acts in the 1960s and 1970s.

The Water Quality Act of 1965 carried forward many provisions of the earlier federal legislation, generally with an increase in levels of funding. The 1965 act also introduced important new requirements for states to establish ambient water quality standards and detailed plans indicating how the standards would be met. The act also shifted responsibility for administering the federal water quality program from the U.S. Public Health Service to a separate agency, the Federal Water Pollution Control Administration, within the Department of Health, Education, and Welfare (HEW). This was not a permanent change. In 1970, a presidential reorganization order placed the water pollution control activities and several other federal environmental programs in the newly created Environmental Protection Agency (EPA).

In Public Law 92-500 (Federal Water Pollution Control Act of 1972),* Congress introduced (1) national water quality goals, (2) technology-based effluent limitations, (3) a national discharge permit system, and (4) federal court actions against sources violating permit conditions.

The 1972 amendments aimed to restore and maintain "the chemical, physical and biological integrity of the nation's waters." The amendments specified, as a national goal, that the "discharge of pollutants into navigable waters be eliminated by 1985." This also included an interim goal:

> [W]herever attainable, an interim goal of water quality which provides for the protection and propagation of fish, shellfish and wildlife and provides for recreation in and on the water [should] be achieved by July 1, 1983.

The EPA administrator was required to set effluent restrictions that met the following general requirements of the 1972 amendments: By 1977, all dischargers were to achieve *"best practicable control technology currently available"* (BPT); and by 1983, all dischargers were to have the *"best available technology economically achievable"* (BAT). After delays caused by numerous legal challenges to the EPA administrator's effluent limitations guidelines, the BPT provisions were implemented. However, the BAT requirements were so heavily disputed that Congress modified them in the Clean Water Act of 1977.

The principal criticism of the original BAT effluent limitations was that the costs of the very high required percentage reductions in residuals would be much greater than the benefits. In defining BAT, costs were considered, but only in the general context of affordability by industry. Computations of the social benefits of stringent effluent controls were not a central factor. Congress presumed the benefits of eliminating water pollutants would be substantial. Congressional insistence on very strict effluent limitations can also be interpreted as an effort to guarantee the rights of Americans to high-quality waters.

In 1977, Congress responded to critics of BAT by requiring it only for toxic substances. A different requirement was introduced for "conventional pollutants," such as biochemical oxygen demand and suspended solids. The effluent limitations guidelines for these pollutants were to be based on the "*best conventional pollutant control* technology" (BCT).

The Clean Water Act of 1977 strongly endorsed the view that waterborne toxic substances must be controlled. The text of the act included a list of 65 substances, or classes of substances, to be used as the basis for defining toxics. This list resulted from a 1976 settlement of a legal action in which several environmental organizations sued the EPA administrator for failing to issue toxic pollutant standards. This list was subsequently expanded by EPA to include 127 "priority pollutants" (Table 1-6).

Effluent limitations required by the FWPCA amendments of 1972 (and later the Clean Water Act of 1977) formed the basis for issuing "*National Pollutant Discharge Elimination System*" (NPDES) permits. The permit system idea stemmed from actions taken by the Department of Justice in the late 1960s. With the support of a favorable interpretation by the Supreme Court, attorneys for the United States relied on the 1899 River and Harbor Act to prosecute industrial sources of water pollution. The 1899 act,

*Passed by override of President Nixon's veto (Percival, 2003).

TABLE 1-6
EPA's priority pollutant list

1. Antimony
2. Arsenic
3. Beryllium
4. Cadmium
5a. Chromium (III)
5b. Chromium (VI)
6. Copper
7. Lead
8. Mercury
9. Nickel
10. Selenium
11. Silver
12. Thallium
13. Zinc
14. Cyanide
15. Asbestos
16. 2,3,7,8-TCDD (Dioxin)
17. Acrolein
18. Acrylonitrile
19. Benzene
20. Bromoform
21. Carbon tetrachloride
22. Chlorobenzene
23. Chlorodibromomethane
24. Chloroethane
25. 2-Chloroethylvinyl ether
26. Chloroform
27. Dichlorobromomethane
28. 1,1-Dichloroethane
29. 1,2-Dichloroethane
30. 1,1-Dichloroethylene
31. 1,2-Dichloropropane
32. 1,3-Dichloropropylene
33. Ethylbenzene
34. Methyl bromide
35. Methyl chloride
36. Methylene chloride
37. 1,2,2,2-Tetrachloroethane
38. Tetrachloroethylene
39. Toluene
40. 1,2-trans-dichloroethylene
41. 1,1,1-Trichloroethane
42. 2,4 Dichlorophenol
43. Trichloroethylene
44. Vinyl chloride
45. 2-Chlorophenol
46. 2,4-Dichlorophenol
47. 2,4-Dimethylphenol
48. 2-Methyl-4-chlorophenol
49. 2,4-Dinitrophenol
50. 2-Nitrophenol
51. 4-Nitrophenol
52. 3-Methyl-4-chlorophenol
53. Pentachlorophenol
54. Phenol
55. 2,4,6-Trichlorophenol
56. Acenaphthene
57. Acenaphthylene
58. Anthracene
59. Benzidine
60. Benzo(a)anthracene
61. Benzo(a)pyrene
62. Benzo(a)fluoranthene
63. Benzo(ghi)perylene
64. Benzo(k)fluoranthene
65. bis(2-Chloroethoxy)methane
66. bis(2-Chloroethyl)ether
67. bis(2-Chloroisopropyl)ether
68. bis(2-Ethylhexyl)phthalate
69. 4-Bromophenyl phenyl ether
70. Butylbenzyl phthalate
71. 2-Chloronaphthalene
72. 4-Chlorophenyl phenyl ether
73. Chrysene
74. Dibenzo(a,h)anthracene
75. 1,2-Dichlorobenzene
76. 1,3-Dichlorobenzene
77. 1,4-Dichlorobenzene
78. 3,3-Dichlorobenzidine
79. Diethyl phthalate
80. Dimethyl phthalate
81. Di-n-butyl phthalate
82. 2,4-Dinitrotoluene
83. 2,6-Dinitrotoluene
84. Di-n-octyl phthalate
85. 1,2-Diphenylhydrazine
86. Fluoranthene
87. Fluorene
88. Hexachlorobenzene
89. Hexachlorobutadiene
90. Hexachlorocyclopentadiene
91. Hexachloroethane
92. Indeno(1,2,3 cd)pyrene
93. Isophorone
94. Naphthalene
95. Nitrobenzene
96. N-Nitrosodimethylamine
97. N-Nitrosodi-n-propylamine
98. N-Nitrosodiphenylamine
99. Phenanthrene
100. Pyrene
101. 1,2,4-Trichlorobenzene
102. Aldrin
103. alpha-BHC
104. beta-BHC
105. gamma-BHC
106. delta-BHC
107. Chlordane
108. 4,4′-DDT
109. 4,4′-DDE
110. 4,4′-DDD
111. Dieldrin
112. alpha-Endosulfan
113. beta-Endosulfan
114. Endosulfan sulfate
115. Endrin
116. Endrin aldehyde
117. Heptachlor
118. Heptachlor epoxide
119. PCB-1242
120. PCB-1254
121. PCB-1221
122. PCB-1232
123. PCB-1248
124. PCB-1260
125. PCB-1016
126. Toxaphene

Source: 40 CFR 131.36, July 1, 1993.

which was drafted originally to prohibit deposits of refuse in navigable waters to keep them clear for boat traffic, was interpreted in the 1960s as applying to liquid waste as well. In December 1970, the EPA administrator issued an executive order calling for a water quality management program using permits and penalties based on the River and Harbor Act of 1899. Although this program was delayed by court challenges in 1971,

Congress made it a central part of the federal strategy embodied in the FWPCA amendments of 1972.

Air Quality Management

Two factors stimulated the development of air pollution control legislation. The first was an air pollution episode at Donora, Pennsylvania, that killed 20 people and made several thousand ill. The second factor was the growing recognition of the linkage between automobile exhausts and photochemical smog. The legislative history is shown in Table 1-7.

The first federal act was the Air Pollution Control Act of 1955 (PL 84-159). It established a program of federally funded research grants to be administered by the U.S. Public Health Service. The expansion of the federal government into air pollution control was a limited one. The legislative history of the act reveals that Congress intended to limit federal involvement in deference to the states, counties, and cities.

The federal role was further extended by the Clean Air Act of 1963, which allowed direct federal intervention to reduce interstate pollution. The form of intervention followed the enforcement process in the Federal Water Pollution Control Act of 1956.

The first federal restrictions on auto emissions came with the Motor Vehicle Air Pollution Control Act of 1965. Based on earlier auto emission control efforts in California, the 1965 act gave the Secretary of the Department of Health, Education, and Welfare authority to establish permissible emission levels for new automobiles beginning with the 1968 model year. The control of emissions from older vehicles was left to individual states.

The Air Quality Act of 1967 borrowed concepts from the Water Quality Control Act of 1965 by requiring states to develop ambient air quality standards and state implementation plans (SIPs) to achieve the standards. Implementation plans were to include emission requirements for controlling air pollution and a timetable for meeting the requirements. Deadlines were set for submitting ambient standards, which were to be established on a region-wide basis.

Although the Clean Air Act Amendments of 1970 continued many of the research and state aid programs established by prior legislation, several aspects of the amendments represented dramatic changes in strategy. These involved (1) the requirement that the administrator of EPA set national ambient air quality standards (NAAQS) and emission standards for selected categories of new industrial facilities, and (2) the explicit delineation (by Congress) of auto emission standards. Another manifestation of the expanded role of the federal government was the requirement of the 1970 amendments that the EPA administrator issue *new source performance standards* (NSPS). These standards were to control new stationary sources categorized by the administrator as contributing significantly to air pollution.

The Clean Air Act Amendments of 1977 relaxed the emission requirements somewhat and extended the compliance deadlines into the early 1980s. They also defined a concept of *prevention of significant deterioration* (PSD) areas and required that an area that meets the national ambient standards for a given air pollutant be declared a PSD area for that pollutant. The amendments also defined three classes of PSD areas. For each class, numerical limits indicated the maximum permissible increment of air quality degradation from all new (or modified) stationary sources of pollution in an area.

TABLE 1-7
Federal laws controlling air pollution

Year	Title	Selected elements of legislation[a]
1955	Air Pollution Control Act	Funds for air pollution research
1960	Motor Vehicle Exhaust Act	Funds for research on vehicle emissions
1963	Clean Air Act	Three-stage enforcement process Funds for state and local air pollution control agencies
1965	Motor Vehicle Air Pollution Control Act	Emission regulations for cars beginning with 1968 models
1967	Air Quality Act	Federally issued criteria documents Federally issued control technique documents Air quality and control regions (AQCRs) defined Requirements for states to set ambient standards for AQCRs Requirements for state implementation plans
1970	Clean Air Act Amendments	National ambient air quality standards New source performance standards Technology forcing auto emission standards Transportation control plans
1977	Clean Air Act Amendments	Relaxation of previous auto emission requirements Vehicle inspection and maintenance programs Prevention of significant deterioration areas Emission offsets for nonattainment areas Study ozone depletion National emission standards for hazardous air pollutants (NESHAP)
1980	Acid Precipitation Act	Development of a long-term research plan
1986	Radon Gas and Indoor Air Quality Research Act	Research program to gather data and to coordinate and assess federal action
1990	Clean Air Act Amendments	Sets attainment dates for criteria air pollutants Imposes new requirements for auto emissions and establishes clean fuels program Identifies 189 hazardous air pollutants to be regulated Establishes SO_2 allowances for acid rain control Establishes a national permit system Sets schedule for phase-out of ozone-depleting compounds

[a]The table entries include only the new policies and programs established by each of the laws. Often these provisions were carried forward in modified form as elements of subsequent legislation.

The 1977 amendments also indicated that significant new sources of pollution could locate in areas that did not meet the NAAQS, but only if certain conditions were satisfied. The amendments required that a significant new source locating in a nonattainment area (one which has not achieved the NAAQS) had to meet strict emission-reduction requirements developed by the EPA administrator. In addition, discharges

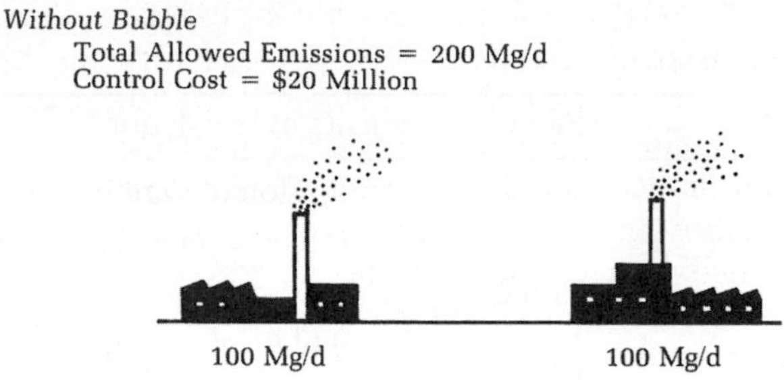

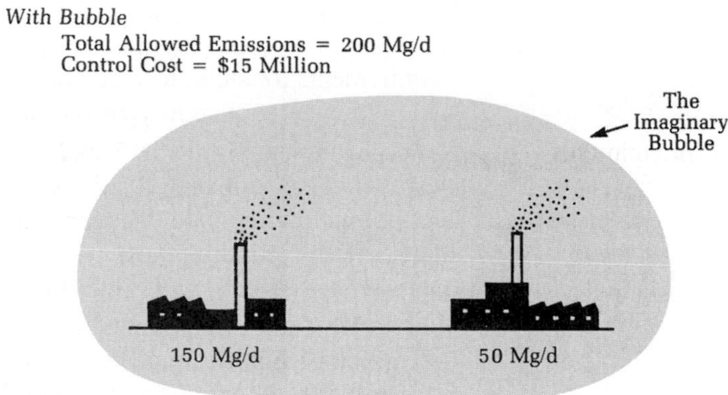

FIGURE 1-7
Illustration of bubble concept.

from the new source had to be more than offset by reductions in emissions from other sources in the region.

In 1979, the EPA extended the concept of emission offsets, as used in nonattainment areas, to a different context: multiple sources of air pollution generated at a single site. This extension, known as the *bubble policy,* is illustrated in Figure 1-7. The figure depicts a firm that must control releases from smokestacks at two adjacent plants. Before the bubble policy, the firm had to comply with emission standards that allowed only 100 Mg/d from each plant.* The total discharge was 200 Mg/d. The unit cost of emission controls for Plant A was much higher than that for Plant B, but the emission requirements were insensitive to these cost differences. Using the bubble policy, the firm is free to decide how to reduce residuals at each plant. The only restriction is that its total discharge must be no greater than 200 Mg/d. Imagine that a bubble surrounds the two plants. The policy allows the firm to make choices within the bubble, but the total discharge from the bubble is restricted. In the early 1980s, the original bubble policy was extended to include plants that were not at the same location (multiplant bubbles).

---

*Mg/d = megagram per day. 1 Mg = 1,000 kg.

The Clean Air Act Amendments of 1990 (CAAA) mandated that the EPA promulgate more than 175 new regulations, 30 guidance documents, 35 studies, and 50 new research initiatives. The Congressional mandates are categorized under eleven "*Titles*" in the Act. It has become common to refer to the requirements of the CAAA by title number.

In light of the fact that three previous deadlines for attainment had come and gone, Title I establishes 16 new deadlines. Although these are primarily aimed at ozone, there are also classifications for carbon monoxide and fine particulates.

Provisions relating to mobile sources are spelled out in Title II. Cars are required to have dashboard warning lights that signal whether or not pollution control equipment is working. These devices frequently have impregnated chemicals that react with the pollutants. The life expectancy of these devices, in terms of miles driven, is specified as 100,000 miles, rather than the previous requirement of 50,000 miles. Auto makers are required to produce some cars that use clean fuels such as alcohol and some that are powered by electricity. In addition, *inspection and maintenance* (I/M) programs for metropolitan areas have been expanded.

Because the previous legislation establishing national emission standards for hazardous pollutants (NESHAPs) based on health risk proved too cumbersome, Title III established an initial list of 189 *hazardous air pollutants* (HAPs) shown in Table 1-8 and directed EPA to establish emission standards based on technology.* These standards are to be the *maximum achievable control technology* (MACT) for a given source category.

Under Title IV, the Act outlines a new nationwide approach to the problem of acid rain. The law sets up a market-based system to lower sulfur dioxide emissions. EPA will issue emission allowances to power plants listed in the act. The allowances are set below current emission levels. Plants may meet the allowances by installing control technology or by purchasing allowances from plants that have emissions below their allowance. For example, in November of 1994, Niagara Mohawk, which serves upstate New York, and the Arizona Public Service Co. traded emission allowances for carbon monoxide and sulfur dioxide.

Unlike the Clean Water Act, no provision for permits was included in the original Clean Air Act (1963). Title V remedies this deficiency by making it unlawful to operate one of the sources listed in the Act except by compliance with a permit.

Depletion of the ozone layer is addressed in Title VI of the Act. A schedule for phasing out the production of ozone-destroying chemicals was promulgated in the Act with provision that EPA could accelerate the schedule. In 1993, EPA established an accelerated schedule that eliminated production of these chemicals by 2001.

Noise Pollution Control

The federal government's activities in noise abatement are spread over several agencies by a variety of legislative acts. The emphasis is on specific activities already regulated separately by the various agencies.

The landmark legislation in the area of occupational noise abatement was enacted in 1942 and is known as the Walsh-Healey Public Contracts Act. This act established minimum working conditions for employees of contractors who supply the federal

---

*Caprolactam was deleted from the list in 1996 (40 CFR 63.60).

TABLE 1-8
Hazardous air pollutants (HAPs)

Acetaldehyde
Acetamide
Acetonitrile
Acetophenone
2-Acetylaminofluorene
Acrolein
Acrylamide
Acrylic acid
Acrylonitrile
Allyl chloride
4-Aminobiphenyl
Aniline
o-Anisidine
Asbestos
Benzene (including benzene from gasoline)
Benzidine
Benzotrichloride
Benzyl chloride
Biphenyl
Bis(2-ethylhexyl)phthalate (DEHP)
Bis(chloromethyl)ether
Bromoform
1,3-Butadiene
Calcium cyanamide
Caprolactam (deleted 1996)
Captan
Carbaryl
Carbon disulfide
Carbon tetrachloride
Carbonyl sulfide
Catechol
Chloramben
Chlordane
Chlorine
Chloroacetic acid
2-Chloroacetophenone
Chlorobenzene
Chlorobenzilate
Chloroform
Chloromethyl methyl ether
Chloroprene
Cresols/Cresylic acid (isomers and mixture)
o-Cresol
m-Cresol
p-Cresol
Cumene
2,4-D, salts and esters
DDE
Diazomethane
Dibenzofurans
1,2-Dibromo-3-chloropropane
Dibutylphthalate
1,4-Dichlorobenzene(p)
3,3-Dichlorobenzidene
Dichloroethyl ether [Bis(2-chloroethyl)ether]
1,3-Dichloropropene
Dichlorvos
Diethanolamine
N,N-Diethyl aniline (N,N-Dimethylaniline)
Diethyl sulfate
3,3-Dimethoxybenzidine
Dimethyl aminoazobenzene
3,3′-Dimethyl benzidine
Dimethyl carbamoyl chloride
Dimethyl formamide
1,1-Dimethyl hydrazine
Dimethyl phthalate
Dimethyl sulfate
4,6-Dinitro-o-cresol, and salts
2,4-Dinitrophenol
2,4-Dinitrotoluene
1,4-Dioxane (1,4-Diethyleneoxide)
1,2-Diphenylhydrazine
Epichlorohydrin (1-chloro-2,3-epoxypropane)
1,2-Epoxybutane
Ethyl acrylate
Ethyl benzene
Ethyl carbamate (Urethane)
Ethyl chloride (Chloroethane)
Ethylene dibromide (Dibromoethane)
Ethylene dichloride (1,2-Dichloroethane)
Ethylene glycol
Ethylene imine (Aziridine)
Ethylene oxide
Ethylene thiourea
Ethylidene dichloride (1,1-Dichloroethane)
Formaldehyde
Heptachlor
Hexachlorobenzene
Hexachlorobutadiene
Hexachlorocyclopentadiene
Hexachloroethane
Hexamethylene-1,6-diisocyanate
Hexamethylphosphoramide
Hexane
Hydrazine
Hydrochloric acid
Hydrogen fluoride (Hydrofluoric acid)
Hydrogen sulfide (clerical error; deleted 1991)
Hydroquinone
Isophorone
Lindane (all isomers)
Maleic anhydride
Methanol
Methoxychlor
Methyl bromide (Bromomethane)
Methyl chloride (Chloromethane)
Methyl chloroform (1,1,1-Trichloroethane)
Methyl ethyl ketone (2-Butanone)
Methyl hydrazine
Methyl iodide (Iodomethane)
Methyl isobutyl ketone (Hexone)
Methyl isocyanate
Methyl methacrylate
Methyl tert butyl ether
4,4-Methylene bis(2-chloroaniline)
Methylene chloride (Dichloromethane)
Methylene diphenyl diisocyanate (MDI)
4,4′-Methylenedianiline
Naphthalene
Nitrobenzene
4-Nitrobiphenyl
4-Nitrophenol
2-Nitropropane
N-Nitroso-N-methylurea
N-Nitrosodimethylamine
N-Nitrosomorpholine
Parathion
Pentachloronitrobenzene (Quintobenzene)
Pentachlorophenol
Phenol
p-Phenylenediamine
Phosgene
Phosphine
Phosphorus
Phthalic anhydride
Polychlorinated biphenyls (Aroclors)
1,3-Propane sultone
beta-Propiolactone
Propionaldehyde
Propoxur (Baygon)
Propylene dichloride (1,2-Dichloropropane)
Propylene oxide
1,2-Propylenimine (2-Methyl aziridine)
Quinoline
Quinone
Styrene
Styrene oxide
2,3,7,8-Tetrachlorodibenzo-p-dioxin
1,1,2,2-Tetrachloroethane
Tetrachloroethylene (Perchloroethylene)
Titanium tetrachloride
Toluene
2,4-Toluene diamine
2,4-Toluene diisocyanate
o-Toluidine

TABLE 1-8
Hazardous air pollutants (HAPs) (*continued*)

Toxaphene (chlorinated camphene)	Vinylidene chloride (1,1-Dichloroethylene)	Coke oven emissions
1,2,4-Trichlorobenzene	Xylenes (isomers and mixture)	Cyanide compounds[1]
1,1,2-Trichloroethane	o-Xylenes	Glycol ethers[2]
Trichloroethylene	m-Xylenes	Lead compounds
2,4,5-Trichlorophenol	p-Xylenes	Manganese compounds
2,4,6-Trichlorophenol	Antimony compounds	Mercury compounds
Triethylamine	Arsenic compounds (inorganic, including arsine)	Fine mineral fibers[3]
Trifluralin		Nickel compounds
2,2,4-Trimethylpentane	Beryllium compounds	Polycylic organic matter[4]
Vinyl acetate	Cadmium compounds	Radionuclides (including radon)[5]
Vinyl bromide	Chromium compounds	Selenium compounds
Vinyl chloride	Cobalt compounds	

NOTE: For all listings above which contain the word "compounds" and for glycol ethers, the following applies: Unless otherwise specified, these listings are defined as including any unique chemical substance that contains the named chemical (i.e., antimony, arsenic, etc.) as part of that chemical's infrastructure.

[1]X′CN where X = H′ or any other group where a formal dissociation may occur. For example KCN or $Ca(CN)_2$

[2]Includes mono- and di- ethers of ethylene glycol, diethylene glycol, and triethylene glycol $R\text{-}(OCH2CH2)_n\text{-}OR'$ where
n = 1, 2, or 3
R = alkyl or aryl groups
R′ = R, H, or groups which, when removed, yield glycol ethers with the structure: $R\text{-}(OCH2CH)_n\text{-}OH$. Polymers are excluded from the glycol category. Ethylene glycol monobutyl ether and surfactant alcohol ethoxylates and derivatives delisted November 29, 2004, 69 FR 692988.

[3]Includes mineral fiber emissions from facilities manufacturing or processing glass, rock, or slag fibers (or other mineral derived fibers) of average diameter 1 micrometer or less.

[4]Includes organic compounds with more than one benzene ring, and which have a boiling point greater than or equal to 100°C.

[5]A type of atom which spontaneously undergoes radioactive decay.

Source: Public Law 101-549, Nov. 15, 1990, 40 CFR 63.60

government with materials, supplies, and equipment in excess of $10,000. However, it was not until 1969 that the Secretary of Labor interpreted this as applicable to noise! (Note: These applied only to supply contracts and not to construction contracts.)

The Occupational Safety and Health Act of 1970 (OSHA) enabled the Secretary of Labor to apply the Walsh-Healey standards with new meaning. Walsh-Healey merely excluded from bidding on federal contracts those suppliers who failed to meet minimum work condition standards. OSHA provided penalties for those suppliers, including civil and criminal law sanctions. Construction noise was brought under federal consideration in the Construction Safety Act of 1970. This act carried the Walsh-Healey provisions to the supply of construction contracts.

In response to the Housing Act of 1949, the Federal Housing Administration's 1961 appraisal guidance identified noise as an issue to be considered in property appraisals. Under the Housing and Urban Development Act of 1965, the Department issued comprehensive noise standards in the 1971 Housing and Urban Development (HUD) circular 1390.2. These rules were updated to the current standard in 1979.

Control and abatement of aircraft noise and sonic booms was the focus of the environmental component of the Federal Aviation Act of 1958. The Department of Transportation Act (1966) included provisions to promote research on noise abatement with particular attention to aircraft. This was followed by the 1968 amendments to the

Federal Aviation Administration Act that directed the Secretary of Transportation to prescribe rules for control and abatement of aircraft noise. The responsibility for noise abatement from airports was assigned to the EPA in the Noise Pollution and Abatement Act of 1970. It directed EPA to

1. Measure noise levels and exposure at airports.
2. Develop airport noise exposure maps.
3. Develop a land use noise compatibility program.
4. Develop noise standards.

Planning grant funds for noise compatibility surveys and the responsibility for noise standards for air carriers were assigned to the EPA in the Aviation Safety and Noise Abatement Act of 1979. The responsibility for airport noise abatement was assigned to the Federal Aviation Administration in the Airport Noise Abatement Act Amendments of 1994 (PL 103-s272).

In the 1962 amendments to the Federal Aid Highways Act, economic, social, and environmental impacts were included as requirements for consideration in the development of plans for construction. The Secretary of Transportation was directed to develop and promulgate standards for highway noise levels compatible with different land uses.

In 1970 Congress added Title IV to the Clean Air Act amendments. This act was entitled "Noise Pollution and Abatement Act of 1970," and it set up the Office of Noise Abatement and Control in the EPA. This was followed by the Noise Control Act of 1972 (PL 92-574). The major provisions of the act stipulated that EPA:

1. Develop and publish criteria for levels of noise requisite to the protection of public health.
2. Compile a list of noise sources, identify noise-producing products, and indicate techniques for control.
3. Set noise emission standards for products distributed in commerce including construction equipment, transportation equipment (including recreational vehicles), any motor or engine, and electrical or electronic equipment.
4. Set aircraft, railroad, and motor carrier noise standards.

In 1994, the Noise Control Act was amended to move airport noise abatement to the Federal Aviation Agency.

Solid Waste

Modern solid waste legislation dates from 1965 when the Solid Waste Disposal Act, Title II of Public Law 89-272, was enacted by Congress. The intent of this act was to

1. Promote the demonstration, construction, and application of solid waste management and resource recovery systems.
2. Provide technical and financial assistance in the planning and development of resource recovery and solid waste disposal programs.

3. Promote a national research and development program for improved management techniques.
4. Provide for the promulgation of guidelines for solid waste collection, transport, separation, recovery, and disposal systems.
5. Provide training grants in occupations involving the design, operation, and maintenance of solid waste disposal systems.

Enforcement of this act became the responsibility of the U.S. Public Health Service (USPHS) and the Bureau of Mines. The USPHS had responsibility for most of the municipal wastes. The Bureau of Mines was charged with supervision of solid wastes from mining activities and the fossil-fuel solid wastes from power plants and industrial steam plants.

The Solid Waste Disposal Act of 1965 was amended by Public Law 95-512, the Resources Recovery Act of 1970. The act directed that the emphasis of the national solid waste management program be shifted from disposal as its primary objective to that of recycling and reuse of recoverable materials in solid wastes or to the conversion of wastes to energy.

Another feature of the 1970 act was the mandate of Congress to the Secretary of Health, Education, and Welfare to prepare a report on the treatment and disposal of hazardous wastes, including radioactive, toxic chemical, biological, and other wastes of significance to the public health and welfare.

Hazardous Wastes

The Resource Conservation and Recovery Act of 1976, commonly known as RCRA (and pronounced "rick-rah") addresses the handling of hazardous waste at facilities currently operating and at those yet to be constructed. The act was designed in large part to meet disposal needs resulting from the Clean Air Act and Clean Water Act, which require industries to remove hazardous substances from their air emissions and their wastewater discharges. Neither statute, however, ensures that the ultimate disposition of waste materials will be environmentally sound. RCRA was intended to provide that ensurance. RCRA does not, however, deal directly with abandoned sites or closed facilities where hazardous wastes have been handled or disposed of in the past. These locations are covered by the Comprehensive Environmental Response, Compensation, and Liability Act (CERCLA, pronounced "sir-klah"), commonly referred to as "Superfund," enacted by Congress in 1980. Finally, RCRA also does not control the disposition of hazardous substances within the productive stream of commerce. Such substances include chemicals covered by the Toxic Substances Control Act of 1976 (PL 94-469), pesticides regulated under the Federal Insecticide, Fungicide, and Rodenticide Act of 1972 (PL 92-516), or other hazardous products subject to the 1975 Hazardous Materials Transportation Act and to other types of federal regulation.

The five major elements in the federal approach to hazardous waste management are:

1. Federal classification of hazardous waste
2. Cradle-to-grave manifest (record-keeping) system

3. Federal standards for safeguards to be followed by generators, transporters, and facilities that treat, store, or dispose of hazardous waste

4. Enforcement of federal standards for facilities through a permit program

5. Authorization of state programs to operate in lieu of the federal program

The act directs the U.S. Environmental Protection Agency to promulgate regulations necessary to put the federal program into full effect.

Unhappy with the progress in implementing RCRA, Congress in 1984 passed the Hazardous and Solid Waste Amendments (HSWA, pronounced "hiss-wah"). The scope of RCRA was significantly increased. Under the legislation:

1. Waste minimization was established as the preferred method for managing hazardous waste.

2. Untreated hazardous waste was banned from land disposal and EPA was directed to establish treatment standards for land disposal.

3. New technology standards, such as double liners, leachate collection systems, and extensive groundwater monitoring, were established for land disposal facilities.

4. New requirements were established for small quantity generators.

5. The EPA was directed to establish standards for underground storage tanks.

6. The EPA was directed to evaluate criteria for municipal solid waste landfills and upgrade monitoring requirements.

The Comprehensive Environmental Response, Compensation, and Liability Act of 1980 (CERCLA) provided authority for removal of hazardous substances from improperly constructed or operated active sites not in compliance with RCRA and from inactive disposal sites.

The most fundamental feature of CERCLA is that it provides basic operating authority to the federal government to take direct action to remove hazardous substances from dangerous inactive disposal sites and to assist with cleaning up emergency spills. This includes authority to carry out investigations, testing, and monitoring of disposal sites. It also includes authority to implement remedial measures to remove contaminants in the groundwater.

CERCLA earned its nickname, "Superfund," from the provision of a $1.6 billion Hazardous Substance Response Trust Fund. Seven-eighths of the money is to be provided by industry through taxes on crude oil, certain petroleum products, and 42 chemical feedstocks; one-eighth is to be provided by government through appropriations from general revenues.

In cases where responsibility for the wastes that cause contamination can be traced to companies with financial resources, CERCLA places financial responsibility for the cleanup on those companies. The statute establishes a set of federal laws under which liability can be imposed on such companies even when they are only indirectly involved in the ownership or operation of the facilities where the wastes were disposed. After the government has identified a site as a threat to the environment, it may call upon those liable companies to undertake the cleanup at their own cost. Alternatively, if such companies refuse to assume responsibility for the cleanup, the government can carry out the remedial program using money from the fund and then bring suit against the companies for reimbursement.

A National Contingency Plan (NCP) establishes the rules for how EPA will use its authority and spend its money. To qualify for expenditure of CERCLA funds, a site must appear on the National Priorities List (NPL). The EPA developed the Hazardous Ranking System (HRS) as a method of assigning a site to the NPL. As of 2004, 1,244 sites had been placed on the NPL (U.S. EPA, 2005b). In addition, CERCLA contains notice requirements for all releases (spills) of reportable quantities of hazardous substances and creates a Post-Closure Liability Fund for qualified disposal facilities.

The Superfund Amendments and Reauthorization Act (SARA) of 1986 extended the provisions of CERCLA. In addition to establishing an $8.5 billion fund for cleanup, SARA directs or establishes that EPA:

1. Revise the NPL and the HRS on which it is based.
2. Revise the NCP.
3. Is authorized to subpoena documents and witnesses.
4. Can spend money to investigate sites and design remedies, and can permit private parties to conduct cleanup.
5. Has broad enforcement authority to require private parties to undertake cleanup.
6. Must impose the more stringent of federal standards or state standards.
7. May use mixed funding, that is, both federal and private money.
8. Develop an administrative record of decisions.

The Toxic Substances Control Act (TSCA) is unique in hazardous waste legislation in that it requires disclosure of information about the toxicity of new materials before they enter into commercial manufacture. It deals with hazardous waste in only one instance: polychlorinated biphenyls (PCBs). At the federal level, rules for the disposal of PCBs are set under TSCA (pronounced "tos-ka") rather than RCRA or CERCLA.

Atomic Energy and Radiation

Laws and regulations to manage radioactive materials and radiation exposure began with the Atomic Energy Act of 1946. The act established the Atomic Energy Commission (AEC) and directed it to conduct research and development on peaceful applications of fissionable and radioactive materials. The Atomic Energy Act of 1954 provided for control of uranium and thorium ("source material" for nuclear reactors), plutonium and enriched uranium (classified as special nuclear material because of their potential use in atomic weapons), and other by-products of the nuclear industry. The Energy Reorganization Act of 1974 divided the developmental and regulatory functions of the AEC between two agencies: the Energy and Research and Development Administration (ERDA) and the Nuclear Regulatory Commission (NRC). In restructuring the administration of energy-related matters after the Arab oil boycott, the Energy Organization Act of 1977 replaced ERDA with the Department of Energy (DOE). The NRC was given jurisdiction over reactor construction and operation. It regulates the possession, use, transportation, handling, and disposal of radioactive materials and wastes. The DOE is responsible for research and development and will operate defense and high-level waste repositories.

The diminishing space at low-level disposal sites led to the enactment of the Low-Level Waste Policy Act (LLWPA) in 1980. Each state is responsible for providing for the availability of capacity either within or outside the state for disposal of low-level radioactive waste generated within its borders. States were encouraged to enter into *compacts* with their neighbors to more efficiently manage the waste. The law allowed the compacts to exclude wastes from other regions and allowed existing disposal sites to impose surcharges for disposal of wastes from regions without sites. The surcharge was to be used for site development. Difficulties in negotiating the compacts prompted the enactment of the Low-Level Radioactive Waste Policy Act Amendments of 1985 (LLRWPAA). It stipulated that the three existing commercial sites remain open for use by all states through 1992. Annual and total limits on the volume of waste that can be sent from reactors were established. DOE is responsible for overseeing the compact arrangements with authority to allocate additional emergency capacity to reactors. The NRC can authorize emergency access to existing sites.

The Nuclear Waste Policy Act of 1982 directed DOE to develop a plan for storage of high-level radioactive waste. Following the requirements of the law, DOE began investigation of nine sites in the west and two in the east. Under the act, the EPA established standards that specified release limits for 1,000 and 10,000 years after disposal.

Because of loudly voiced concern over the direction of the DOE's mission plan and the decision to abandon the search for a repository site in the east, Congress passed the Nuclear Waste Policy Act Amendments of 1987. The amendments restructured DOE's high-level waste program. The only site that would be considered would be Yucca Flats, Nevada. Furthermore, spent fuel would be required to be shipped in NRC-approved packages after notification of state and local governments. During the years 2007-2010, DOE is to study the need for a second repository.

For mixed wastes, that is, both hazardous and radioactive, RCRA and HSWA apply to the hazardous characteristic. As of 1987, disposal rules must comply with both NRC rules for radioactivity and EPA rules for hazardous constituents. Before then, only the NRC rules applied. Likewise, for leaking disposal sites, CERCLA and SARA rules apply as well as the NRC rules.

Radiation exposure from x-rays and medical diagnosis and treatment are regulated under the Radiation Control for Health and Safety Act of 1968.

1-5 ENVIRONMENTAL ETHICS

The birth of environmental ethics as a force is partly a result of concern for our own long-term survival, as well as our realization that humans are but one form of life, and that we share our earth with other forms of life (Vesilind, 1975).

TABLE 1-9
An Environmental Code of Ethics

1. Use knowledge and skill for the enhancement and protection of the environment.
2. Hold paramount the health, safety and welfare of the environment.
3. Perform services only in areas of personal expertise.
4. Be honest and impartial in serving the public, your employers, your clients and the environment.
5. Issue public statements only in an objective and truthful manner.

Although it seems a bit unrealistic for us to set a framework for a discussion of environmental ethics in this short introduction, we have summarized a few salient points in Table 1-9.

Although these few principles seem straightforward, real-world problems offer distinct challenges. Here is an example for each of the principles listed:

- The first principle may be threatened when it comes into conflict with the need for food for a starving population and the country is overrun with locust. Will the use of pesticides enhance and protect the environment?
- The EPA has stipulated that wastewater must be disinfected where people come into contact with the water. However, the disinfectant may also kill naturally occurring beneficial microorganisms. Is this consistent with the second principle?
- Suppose your expertise is water and wastewater chemistry. Your company has accepted a job to perform air pollution analysis and asks you to perform the work in the absence of a colleague who is the company's expert. Do you decline and risk being fired?
- The public, your employers, and your client believe that dredging a lake to remove weeds and sediment will enhance the lake. However, the dredging will destroy the habitat for muskrats. How can you be impartial to *all* these constituencies?
- You believe that a new regulation proposed by EPA is too expensive to implement but you have no data to confirm that opinion. How do you respond to a local newspaper reporter asking for your opinion? Do you violate the fifth principle even though it is "your opinion" that is being sought?

Below are two cases that are more complex. We have not attempted to provide pat solutions, but rather have left them for you and your instructor to resolve.

Case 1: To Add or Not to Add

A friend of yours has discovered that his firm is adding nitrites and nitrates to bacon to help preserve it. He also has read that these compounds are precursors to cancer-forming chemicals that are produced in the body. On the other hand, he realizes that certain disease organisms such as those that manufacture botulism toxin have been known to grow in bacon that had not been treated. He asks you whether he should (a) protest to his superiors knowing he might get fired; (b) leak the news to the press; (c) remain silent because the risk of dying from cancer is less than the absolute certainty of dying from botulism.

Note: The addition of nitrite to bacon is approved by the Food and Drug Administration. Nitrites and nitrates are, in and of themselves, not very toxic to adults. However, heating these compounds results in a reaction with the amines in proteins to form nitrosamine which is a cancer-forming compound.

Case 2: Too Close for Comfort

As the only engineer in the shop who has had any training in noise pollution, you have been asked, on your third day on the job, to review a manufacturer's bid for noise control devices on aeration blowers for a wastewater treatment plant. After reading the bid proposal and making a few calculations, you conclude that the noise silencers will protect the workers. However, your reading of *Introduction to Environmental Engineering* leads you to surmise that the noise level for nearby neighbors will be excessive at night. (The city has no noise ordinance.) You know that if you ask to go back out on bid for better noise control equipment, construction of the plant will be delayed 90 days. During these 90 days untreated sewage will enter the river. What do you recommend to your new boss?

We think it is important to point out that many environmentally related decisions such as those described above are much more difficult than the problems presented in the remaining chapters of this book. Frequently these problems are related more to ethics than to engineering. The problems arise when there are several courses of action with no *a priori* certainty as to which is best. Decisions related to safety, health, and welfare are easily resolved. Decisions as to which course of action is in the best interest of the public are much more difficult to resolve. Furthermore, decisions as to which course of action is in the best interest of the environment are at times in conflict with those that are in the best interest of the public. Whereas decisions made in the public interest are based on professional ethics, decisions made in the best interest of the environment are based on environmental ethics.

Ethos, the Greek word from which "ethic" is derived, means the character of a person as described by his or her actions. This character was developed during the evolutionary process and was influenced by the need for adapting to the natural environment. Our ethic is our way of doing things. Our ethic is a direct result of our natural environment. During the latter stages of the evolutionary process, *Homo sapiens* began to modify the environment rather than submit to what, millennia later, became known as Darwinian natural selection. As an example, consider the cave dweller who, in the chilly dawn of prehistory, realized the value of the saber-toothed tiger's coat and appropriated it for personal use. Inevitably a pattern of appropriation developed, and our ethic became more self-modified than environmentally adapted. Thus, we are no longer adapted to our natural environment but rather to our self-made environment. In the ecological context, such maladaptation results in one of two consequences: (1) the organism (*Homo sapiens*) dies out; or (2) the organism evolves to a form and character that is once again compatible with the natural environment (Vesilind, 1975). Assuming that we choose the latter course, how can this change in character (ethic) be brought about? Each individual must change his or her character or ethic, and the social system must change to become compatible with the global ecology.

The acceptable system is one in which we learn to share our exhaustible resources—to regain a balance. This requires that we reduce our needs and that the materials we use must be replenishable. We must treat all of the earth as a sacred trust to be used so that its content is neither diminished nor permanently changed; we must release no substances that cannot be reincorporated without damage to the natural system. The recognition of the need for such adaptation (as a means of survival) has developed into what we now call the *environmental ethic* (Vesilind, 1975).

1-6 CHAPTER REVIEW

When you have completed studying this chapter, you should be able to do the following without the aid of your textbook or notes:

1. Sketch and label a water resource system including (*a*) source; (*b*) collection works; (*c*) transmission works; (*d*) treatment works; and (*e*) distribution works.
2. State the proper general approach to treatment of a surface water and a groundwater (see Figure 1-2).
3. Define the word "demand" as it applies to water.
4. List the five most important factors contributing to water consumption and explain why each has an effect.
5. State the rule-of-thumb water requirement for an average city on a per-person basis and calculate the average daily water requirement for a city of a stated population.
6. Define the acronyms WWTP and POTW.
7. Explain why separate storm sewers and sanitary sewers are preferred over combined sewers.
8. Explain the purpose of a lift station.
9. Identify and explain the following acronyms and concepts found in environmental legislation: BPT, BAT, BCT, NPDES, HAP, MACT, "bubble policy," NIMLO, Walsh-Healey, OSHA, RCRA, and Superfund.

1-7 PROBLEMS

1-1. Estimate the total daily water withdrawal (in m^3/d) including both fresh and saline water for all uses for the United States in 2000. The population was 281,421,906.

Answer: 1.52×10^9 m^3/d

1-2. Estimate the per capita daily water withdrawal for public supply in the United States in 2005 (in Lpcd). Use the following population data (McGeveran, 2002) and water supply data (Hutson, 2001):

Year	Population	Public supply withdrawal, m^3/d
1950	151,325,798	5.30×10^7
1960	179,323,175	7.95×10^7
1970	203,302,031	1.02×10^8
1980	226,542,203	1.29×10^8
1990	248,709,873	1.46×10^8
2000	281,421,906	1.64×10^8

(NOTE: This problem may be worked by hand calculation and then plotted on graph paper to extrapolate to 2005, or it may be worked by using a spreadsheet to perform the calculations, plot the graph, and extrapolate to 2005.)

1-3. A residential development of 280 houses is being planned. Assume that the American Water Works Association average daily household consumption applies, and that each house has three residents. Estimate the additional average daily water production in L/d that will have to be supplied by the city.

Answer: 3.70×10^5 L/d

1-4. Repeat Problem 1-3 for 320 houses, but assume that low-flush valves reduce water consumption by 14 percent.

1-5. Using the data in Problem 1-3 and assuming that the houses are metered, determine what additional demand will be made at the peak hour.

Answer: 1.96×10^6 L/d

1-6. If a faucet is dripping at a rate of one drop per second and each drop contains 0.150 milliliters, calculate how much water (in liters) will be lost in 1 year.

1-7. Savabuck University has installed standard pressure-operated flush valves on its water closets. When flushing, these valves deliver 130.0 L/min. If the delivered water costs $0.45 per cubic meter, what is the monthly cost of not repairing a broken valve that flushes continuously?

Answer: $2,527.20, or $2,530/mo

1-8. The American Water Works Association estimates that 15% of the water that utilities process is lost each day. Assuming that the loss was from public supply withdrawal in 2000 (Problem 1-2), estimate the total value of the lost water if delivered water costs $0.45 per cubic meter.

1-9. Water delivered from a public supply in western Michigan costs $0.45 per cubic meter. A 0.5-L bottle of water purchased from a dispensing machine costs $1.00. What is the cost of the bottled water on a per cubic meter basis?

Answer: $2,000/m^3

1-10. Using U.S. Geological Survey Circular 1268 (http://usgs.gov), estimate the daily per capita domestic withdrawal of fresh water in South Carolina in Lpcd. (Note: conversion factors inside the back cover of this book may be helpful.)

1-11. Using the Pacific Institute for Studies in Development, Environment, and Security website (http://www.worldwater.org/table2.html), determine the lowest per capita domestic water withdrawal in the world in Lpcd and identify the country in which it occurs.

1-8 DISCUSSION QUESTIONS

1-1. Would you expect the demand for water to drop in half if the price (\$/L) doubled? Explain your reasoning.

1-2. The water supply for the city of Peoria, is from wells. Other than disinfection, no water treatment is provided. A filtration plant would be appropriate to improve the quality of the water. True or False? If the answer is false, revise the statement so that it is true.

1-3. The water treatment plant for the town of Gettysburg, was built 20 years ago. Over the last few years, there has been difficulty in maintaining water pressure in the system over the 4th of July weekend. In some parts of town only a trickle of water flows from the tap during early morning and late evening hours. There are no problems during the remainder of the year. Explain why the town may be having water pressure problems.

1-4. The town of West Lafayette, is considering two proposals for a new water-treatment plant. West Lafayette's average daily demand is 11,400 m^3/d. Proposal A is to build a plant that will produce 475 m^3/h and a storage reservoir to hold 2,520 m^3 of water. Proposal B is to build a plant that will produce 1,425 m^3/h but no water storage reservoir will be provided. Which proposal do you recommend? Explain why.

1-5. Homeowners in the town of Rolla, have connected their downspouts and the sump pumps from their footing drains to the sanitary sewer system. The rainwater and sump water entering the sewer is called (choose one):

(a) Infiltration

(b) Inflow

These connections to the sanitary sewer, in effect, make it a (choose one):

(c) Storm sewer

(d) Combined sewer

Explain why you have made your choices.

1-6. The Shiny Plating Company is using about 2,000.0 kg/wk of organic solvent for vapor degreasing of metal parts before they are plated. The Air Pollution Engineering and Testing Company (APET) has measured the air in the workroom and in the stack which vents the degreaser. APET has determined that 1,985.0 kg/wk is being vented up the stack and that the workroom environment is within occupational standards. The 1,985.0 kg/wk is well above the allowable emission rate of 11.28 kg/wk.

Elizabeth Fry, the plant superintendent, has asked J.R. Injuneer, the plant engineer, to review two alternative control approaches offered by APET and to recommend one of them.

The first method is to purchase a pollution control device to put on the stack. This control system will reduce the solvent emission to 1.0 kg/wk. Approximately 1,950.0 kg of the solvent which is captured each week can be recycled back to the degreaser. Approximately 34.0 kg of the solvent must be discharged to the wastewater treatment plant (WWTP). J. R. has determined that this small amount of solvent will not adversely affect the performance of the WWTP. In addition, the capital cost of the pollution control equipment will be recovered in about two years as a result of savings from recovering lost solvent.

The second method is to substitute a solvent that is not on the list of regulated emissions. The price of the substitute is about 10 percent higher than the solvent currently in use. J. R. has estimated that the substitute solvent loss will be about 100.0 kg/wk. The substitute collects moisture and loses its effectiveness in about a month's time. The substitute solvent cannot be discharged to the WWTP because it will adversely affect the WWTP performance. Consequently, about 2,000 kg must be hauled to a hazardous waste disposal site for storage each month. Because of the lack of capital funds and the high interest rate for borrowing, J. R. recommends that the substitute solvent be used. Do you agree with this recommendation? Explain your reasoning.

1-7. Ted Terrific is the manager of a leather tanning company. In part of the tanning operation a solution of chromic acid is used. It is company policy that the spent chrome solution is put in 0.20-m^3 drums and shipped to a hazardous waste disposal facility.

On Thursday the 12th, the day shift miscalculates the amount of chrome to add to a new batch and makes it too strong. Since there is not enough room in the tank to adjust the concentration, Abe Lincoln, the shift supervisor, has the tank emptied and a new one prepared and makes a note to the manager that the bad batch needs to be reworked.

On Monday the 16th, Abe Lincoln looks for the bad batch and cannot find it. He notifies Ted Terrific that it is missing. Upon investigation, Ted finds that Rip Van-Winkle, the night-shift supervisor, dumped the batch into the sanitary sewer at 3:00 a.m. on Friday the 13th. Ted makes discreet inquiries at the wastewater plant and finds that they have had no process upsets. After Ted severely disciplines Rip, he should: (Choose the correct answer and explain your reasoning.)

A. Inform the city and state authorities of the illegal discharge as required by law even though no apparent harm resulted.

B. Keep the incident quiet because it will cause trouble for the company without doing the public any good. No harm was done and the shift supervisor has been punished.
C. Advise the president and board of directors and let them decide whether to follow A or B.

1-9 REFERENCES

ASCE (1973) *Official Record,* American Society of Civil Engineers, New York.

ASCE (1977) *Official Record,* Environmental Engineering Division, Statement of Purpose, American Society of Civil Engineers, New York.

ASCE (2005) http://www.asce.org/inside/codeofethics.cfm.

AWWA (1999) American Water Works Association, "Stats on Tap," Denver, http://www.awwa.org/Advocacy/pressroom/STATS.cfm.

CFR (2005) U.S. Government Printing Office, Washington, DC, http://www.gpoaccess.gov/ecfr/ (in January 2005 this was a beta test site for searching the CFR).

Hutson, S. S., N. L. Barber, J. F. Kenny, et al. (2001) *Estimated Use of Water in the United States in 2000,* U.S. Geological Survey Circular 1268, Washington, DC. http://www.usgs.gov

Linaweaver, F. P., J. C. Geyer, and J. B. Wolff (1967) "Summary Report on the Residential Water Use Research Project," *Journal of the American Water Works Association,* vol. 59, p. 267.

Linsley, R. K. and J. B. Franzini (1979) *Water Resources Engineering*, McGraw-Hill, New York, p. 546.

MacVicar, R. (1983) Oklahoma State University, commencement address.

McGeveran, W. A. (editorial director) (2002) *The World Almanac and Book of Facts: 2002,* World Almanac Books; New York, p. 377.

Pacific Institute (2000) Pacific Institute for Studies in Development, Environment, and Security, Oakland, CA. http://www.worldwater.org/table2.html.

Percival, R. V. (2003) *Environmental Law: Statutory Supplement and Internet Guide, 2003–2004,* Aspen Publishers, New York, p. 702.

Pontius, F. W. (editor) (2003) *Drinking Water Regulation and Health,* John Wiley & Sons, New York, pp. 14–17, 73–81, 119, 125, 748.

Schein, E. H. (1968) "Organizational Socialization and the Profession of Management," 3rd Douglas Murray McGregor Memorial Lecture to the Alfred P. Sloan School of Management, Massachusetts Institute of Technology.

Tchobanoglous, G., H. Theisen, and R. Eliassen (1977) *Solid Wastes,* McGraw-Hill, New York, p. 21.

U.S. Code (2005) House of Representatives, Washington, DC, http://uscode.house.gov/.

U.S. EPA (2005a) "Municipal Solid Waste: Reduce, Reuse, Recycle," U.S. Environmental Protection Agency, Washington, DC, http://www.epa.gov/epaoswer/non-hw/muncpl/reduce.htm.

U.S. EPA (2005b) "National Priorities List," U.S. Environmental Protection Agency, Washington, DC, http://www.epa.gov/superfund/sites/npl/newfin.htm.

U.S. PHS (1970) *Manual of Septic Tank Practice,* Public Health Service Publication No. 526, Department of Health, Education and Welfare, Washington, DC.

Vesilind, P. A. (1975) *Environmental Pollution and Control,* Ann Arbor Science, Ann Arbor, MI, p. 214.

CHAPTER 2

MATERIALS AND ENERGY BALANCES

2-1 INTRODUCTION

Materials and energy balances are key tools in achieving a quantitative understanding of the behavior of environmental systems. They serve as a method of accounting for the flow of energy and materials into and out of environmental systems. Mass balances provide us with a tool for modeling the production, transport, and fate of pollutants in the environment. Energy balances likewise provide us with a tool for modeling the production, transport, and fate of energy in the environment. Examples of mass balances include prediction of rainwater runoff (Chapter 3), oxygen balance in streams (Chapter 5), and audits of hazardous waste production (Chapter 10). Energy balances predict the temperature rise in a stream from the discharge of cooling water from a power plant (Chapter 5), and the temperature rise due to global warming (Chapter 7).

2-2 UNIFYING THEORIES

Conservation of Matter

The *law of conservation of matter* states that (without nuclear reaction) matter can neither be created nor destroyed. This is a powerful theory. It means that if we observe an environmental process carefully, we should be able to account for the "matter" at any point in time. It does not mean that the form of the matter does not change nor, for that matter, the properties of the matter. Thus, if we measure the volume of a fresh glass of water on the counter on Monday and measure it again a week later and find the volume to be less, we do not presume magic has occurred but rather that matter has changed in form. The law of conservation of matter says we ought to be able to account for all the mass of the water that was originally present, that is, the mass of water remaining in the glass plus the mass of water vapor that has evaporated equals the mass of water originally present. The mathematical representation of this accounting system is called a *materials balance* or *mass balance.*

Conservation of Energy

The *law of conservation of energy* states that energy cannot be created or destroyed. Like the law of conservation of matter, this theory means that we should be able to account for the "energy" at any point in time. It also does not mean that the form of the energy does not change. Thus, we should be able to trace the energy of food through a series of organisms from green plants through animals. The mathematical representation of the accounting system we use to trace energy is called an *energy balance.*

Conservation of Matter and Energy

At the turn of the 20th century, Albert Einstein hypothesized that matter could be transformed to energy and vice versa. The birth of the nuclear age proved his hypothesis correct, so today we have a combined *law of conservation of matter and energy* that states that the total amount of energy and matter is constant. A nuclear change produces new materials by changing the identity of the atoms themselves. Significant amounts of matter are converted to energy in nuclear explosions. Exchange between mass and energy is not an issue in environmental applications. Thus, there are generally two separate balances for mass and energy.

2-3 MATERIALS BALANCES

Fundamentals

In its simplest form a materials balance or mass balance may be viewed as an accounting procedure. You perform a form of material balance each time you balance your checkbook.

$$\text{Balance} = \text{deposit} - \text{withdrawal} \tag{2-1}$$

For an environmental process, the equation would be written

$$\text{Accumulation} = \text{input} - \text{output} \tag{2-2}$$

where accumulation, input, and output refer to the mass quantities accumulating in the system or flowing into or out of the system. The "system" may be, for example, a pond, river, or a pollution control device.

The Control Volume. Using the mass balance approach, we begin solving the problem by drawing a flowchart of the process or a conceptual diagram of the environmental subsystem. All of the known inputs, outputs, and accumulation are converted to the same mass units and placed on the diagram. Unknown inputs, outputs, and accumulation are also marked on the diagram. This helps us define the problem. System boundaries (imaginary blocks around the process or part of the process) are drawn in such a way that calculations are made as simple as possible. The system within the boundaries is called the *control volume*.

We then write a materials balance equation to solve for unknown inputs, outputs, or accumulations or to demonstrate that we have accounted for all of the components by demonstrating that the materials balance "closes," that is, the accounting balances. Alternatively, when we do not have data for all inputs or outputs, we can assume that the mass balance closes and solve for the unknown quantity. The following example illustrates the technique.

Example 2-1. Mr. and Mrs. Green have no children. In an average week they purchase and bring into their house approximately 50 kg of consumer goods (food, magazines, newspapers, appliances, furniture, and associated packaging). Of this amount, 50 percent is consumed as food. Half of the food is used for biological maintenance and ultimately released as CO_2; the remainder is discharged to the sewer system. The Greens recycle approximately 25 percent of the solid waste that is generated. Approximately 1 kg accumulates in the house. Estimate the amount of solid waste they place at the curb each week.

Solution. Begin by drawing a mass balance diagram and labeling the known and unknown inputs and outputs. There are, in fact, two diagrams: one for the house and one for the people. However, the mass balance for the people is superfluous for the solution of this problem.

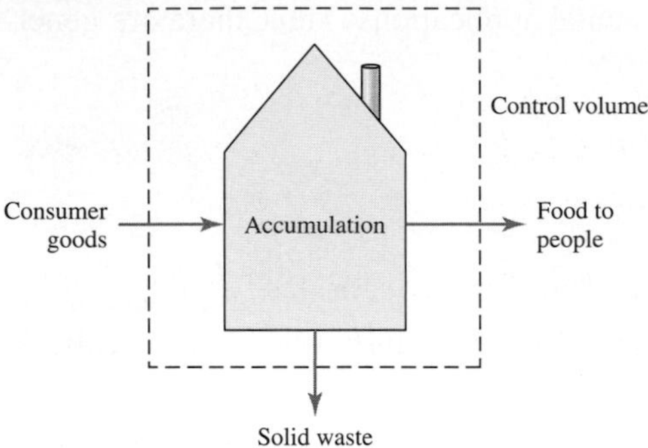

Write the mass balance equation for the house.

$$\text{Input} = \text{accumulation in house} + \text{output as food to people} + \text{output as solid waste}$$

Now we need to calculate the known inputs and outputs.

$$\text{One half of input is food} = (0.5)(50 \text{ kg}) = 25 \text{ kg}$$

This is output as food to the people. The mass balance equation is then rewritten as

$$50 \text{ kg} = 1 \text{ kg} + 25 \text{ kg} + \text{output as solid waste}$$

Solving for the mass of solid waste gives

$$\text{Output as solid waste} = 50 - 1 - 25 = 24 \text{ kg}$$

The mass balance diagram with the appropriate masses may be redrawn as shown below:

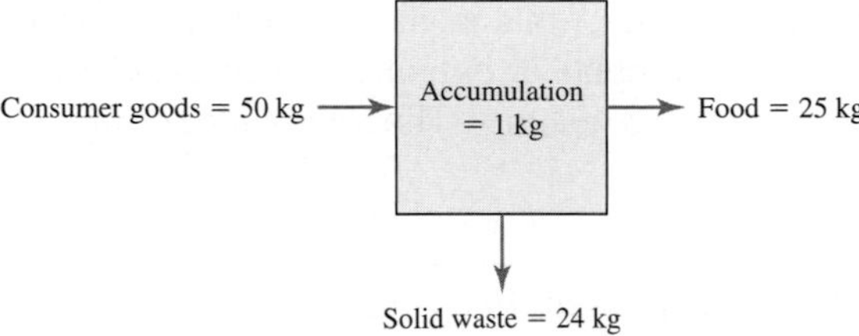

We can estimate the amount of solid waste placed at the curb by performing another mass balance around the solid waste as shown in the following diagram:

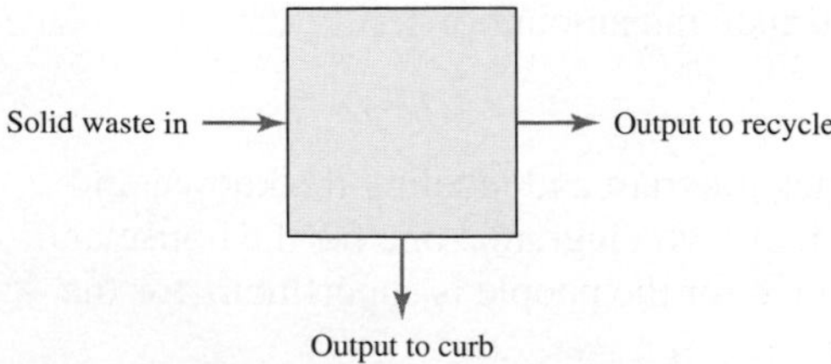

The mass balance equation is

$$\text{Solid waste in} = \text{output to recycle} + \text{output to curb}$$

Because the recycled amount is 25 percent of the solid waste

$$\text{Output to recycle} = (0.25)(24 \text{ kg}) = 6 \text{ kg}$$

Substituting into the mass balance equation for solid waste and solving for output to curb;

$$24 \text{ kg} = 6 \text{ kg} + \text{output to curb}$$

$$\text{Output to curb} = 24 - 6 = 18 \text{ kg}$$

Time as a Factor

For many environmental problems time is an important factor in establishing the degree of severity of the problem or in designing a solution. In these instances, Equation 2-2 is modified to the following form:

$$\text{Mass rate of accumulation} = \text{mass rate of input} - \text{mass rate of output} \quad (2\text{-}3)$$

where *rate* is used to mean "per unit of time." In the calculus this may be written as

$$\frac{dM}{dt} = \frac{d(\text{in})}{dt} - \frac{d(\text{out})}{dt} \quad (2\text{-}4)$$

where M refers to the mass accumulated and (in) and (out) refer to the mass flowing in or out. As part of the description of the problem, a convenient time interval that is meaningful for the system must be chosen.

Example 2-2. Truly Clearwater is filling her bathtub but she forgot to put the plug in. If the volume of water for a bath is 0.350 m^3 and the tap is flowing at 1.32 L/min and the drain is running at 0.32 L/min, how long will it take to fill the tub to bath level? Assuming Truly shuts off the water when the tub is full and does not flood the house, how much water will be wasted? Assume the density of water is 1,000 kg/m^3

Solution. The mass balance diagram is shown here.

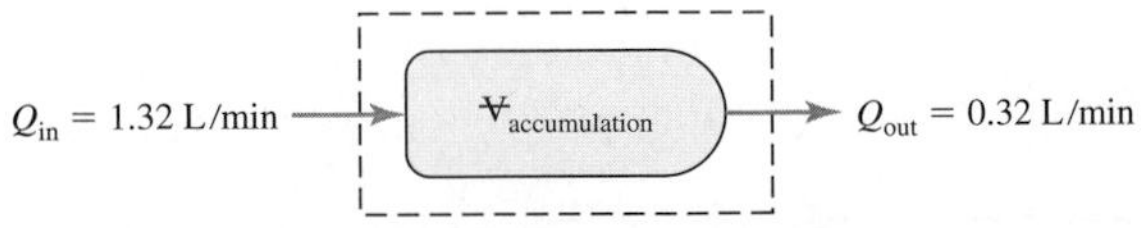

Control volume

Q_{in} = 1.32 L/min → $\forall_{accumulation}$ → Q_{out} = 0.32 L/min

Because we are working in mass units, we must convert the volumes to masses. To do this, we use the density of water.

$$\text{Mass} = (\text{volume})(\text{density}) = (\forall)(\rho)$$

where the

$$\text{Volume} = (\text{flow rate})(\text{time}) = (Q)(t)$$

So for the mass balance equation, noting that $1.0\ m^3 = 1{,}000\ L$ so $0.350\ m^3 = 350\ L$,

$$\begin{aligned}
\text{Accumulation} &= \text{mass in} - \text{mass out} \\
(\forall_{ACC})(\rho) &= (Q_{in})(\rho)(t) - (Q_{out})(\rho)(t) \\
\forall_{ACC} &= (Q_{in})(t) - (Q_{out})(t) \\
\forall_{ACC} &= 1.32t - 0.32t \\
350\ \text{L} &= (1.00\ \text{L/min})(t) \\
t &= 350\ \text{min}
\end{aligned}$$

The amount of water wasted is

$$\text{Wasted water} = (0.32\ \text{L/min})(350\ \text{min}) = 112\ \text{L}$$

More Complex Systems

A key step in the solution of mass balance problems for systems that are more complex than the previous examples is the selection of an appropriate control volume. In some instances, it may be necessary to select multiple control volumes and then solve the problem sequentially using the solution from one control volume as the input to another control volume. For some complex processes, the appropriate control volume may treat all of the steps in the process as a "black box" in which the internal process steps are not required and therefore are hidden in a black box. The following example illustrates a case of a more complex system and a method of solving the problem.

Example 2-3. A storm sewer network in a small residential subdivision is shown in the following sketch. The storm water flows by gravity in the direction shown by the pipes. Storm water only enters the storm sewer on the east–west legs of pipe. No storm water enters on the north–south legs. The flow rate for each section of pipe is also shown by each section of pipe. The capacity of each pipe is $0.120\ m^3/s$. During large rain storms, River Street floods below junction number 1 because the flow of water exceeds the capacity of the storm sewer pipe. To alleviate this problem and to provide extra capacity for expansion, it is proposed to build a retention pond to hold the storm water until the storm is over and then gradually release it. Where in the pipe network should the retention pond be built to provide approximately 50 percent extra capacity ($0.06\ m^3/s$) in the remaining system?

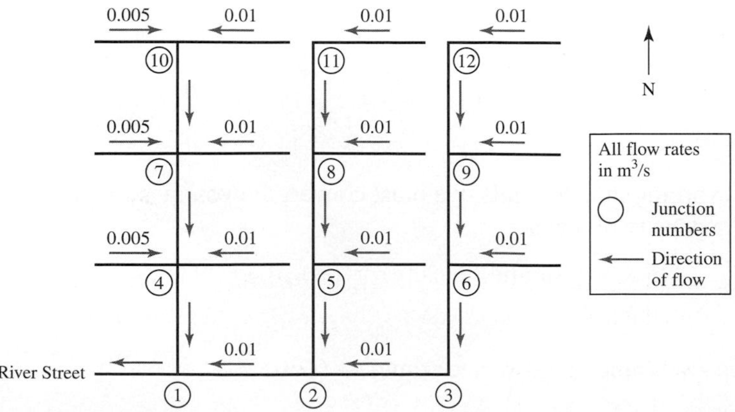

Solution. This is an example of a balanced flow problem. That is Q_{out} must equal Q_{in}. Although this problem can almost be solved by observation, we will use a sequential mass balance approach to illustrate the technique. Starting at the upper end of the system at junction number 12, we draw the following mass balance diagram:

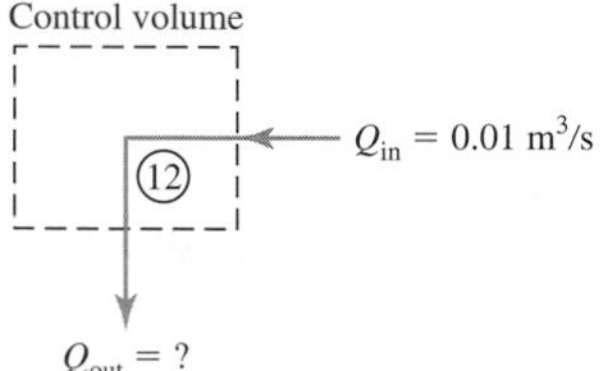

The mass balance equation is

$$\frac{dM}{dt} = \frac{d(\text{in})}{dt} - \frac{d(\text{out})}{dt}$$

Because no water accumulates at the junction

$$\frac{dM}{dt} = 0$$

and

$$\frac{d(\text{in})}{dt} = \frac{d(\text{out})}{dt}$$

$$(\rho)(Q_{in}) = (\rho)(Q_{out})$$

Because the density of water remains constant, we may treat the mass flow rate in and out as directly proportional to the flow rate in and out.

$$Q_{in} = Q_{out}$$

So the flow rate from junction 12 to junction 9 is 0.01 m^3/s.

At junction 9, we can draw the following mass balance diagram.

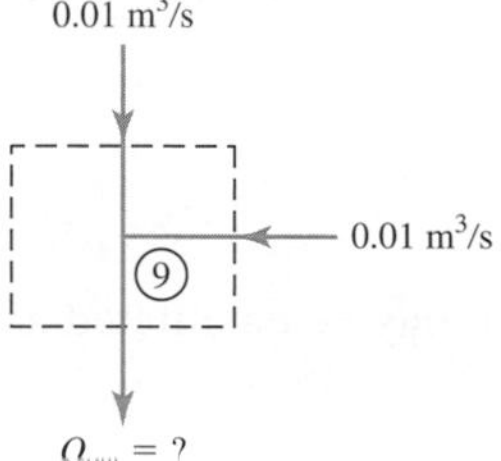

Again using our assumption that no water accumulates at the junction and recognizing that the mass balance equation may again be written in terms of flow rates,

$$Q_{\text{from junction 9}} = Q_{\text{from junction 12}} + Q_{\text{in the pipe connected to junction 9}}$$
$$= 0.01 + 0.01 = 0.02\ m^3/s$$

Similarly

$$Q_{\text{from junction 6}} = Q_{\text{from junction 9}} + Q_{\text{in the pipe connected to junction 6}}$$
$$= 0.02 + 0.01 = 0.03 \text{ m}^3/\text{s}$$

and, noting that storm water enters on the east–west legs of pipe

$$Q_{\text{from junction 3}} = Q_{\text{from junction 6}} + Q_{\text{in the pipe connecting junction 3 with junction 2}}$$
$$= 0.03 + 0.01 = 0.04 \text{ m}^3/\text{s}$$

By a similar process for all the junctions, we may label the network flows as shown in the following diagram.

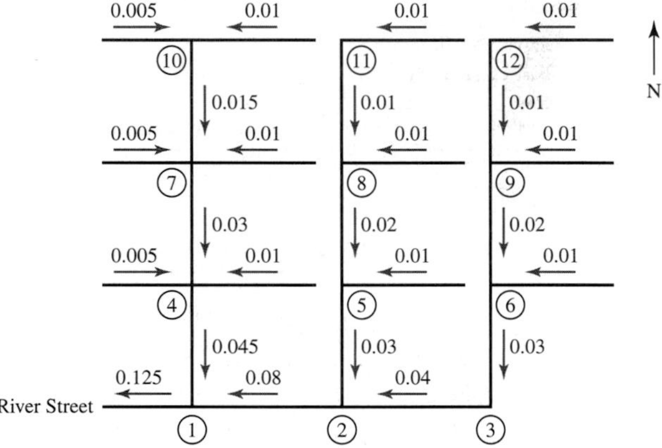

It is obvious that the pipe capacity of 0.12 m^3/s is exceeded just below junction 1. By observation we can also see that the total flow into junction 2 is 0.07 m^3/s and that a retention pond at this point would require that the pipe below junction 1 carry only 0.055 m^3/s.This meets the requirement of providing approximately 50 percent of the capacity for expansion.

Efficiency

The effectiveness of an environmental process in removing a contaminant can be determined using the mass balance technique. Starting with Equation 2-4,

$$\frac{dM}{dt} = \frac{d(\text{in})}{dt} - \frac{d(\text{out})}{dt}$$

The mass of contaminant per unit of time [$d(\text{in})/dt$ and $d(\text{out})/dt$] may be calculated as

$$\frac{\text{Mass}}{\text{Time}} = (\text{concentration})(\text{flow rate})$$

For example,

$$\frac{\text{Mass}}{\text{Time}} = (\text{mg/m}^3)(\text{m}^3/\text{s}) = \text{mg/s}$$

This is called a *mass flow rate*. In concentration and flow rate terms, the mass balance equation is

$$\frac{dM}{dt} = C_{in}Q_{in} - C_{out}Q_{out} \tag{2-5}$$

where dM/dt = rate of accumulation of contaminant in the process
C_{in}, C_{out} = concentrations of contaminant into and out of the process
Q_{in}, Q_{out} = flow rates into and out of the process

The ratio of the mass that is accumulated in the process to the incoming mass is a measure of how effective the process is in removing the contaminant

$$\frac{dM/dt}{C_{in}Q_{in}} = \frac{C_{in}Q_{in} - C_{out}Q_{out}}{C_{in}Q_{in}} \tag{2-6}$$

For convenience, the fraction is multiplied by 100%. The left-hand side of the equation is given the notation η. Efficiency (η) is then defined as

$$\eta = \frac{\text{mass in} - \text{mass out}}{\text{mass in}}(100\%) \tag{2-7}$$

If the flow rate in and the flow rate out are the same, this ratio may be simplified to

$$\eta = \frac{\text{concentration in} - \text{concentration out}}{\text{concentration in}}(100\%) \tag{2-8}$$

The following example illustrates a multistep solution using efficiency as part of the solution technique.

Example 2-4. The air pollution control equipment on a municipal waste incinerator includes a fabric filter particle collector (known as a baghouse). The baghouse contains 424 cloth bags arranged in parallel, that is 1/424 of the flow goes through each bag. The gas flow rate into and out of the baghouse is 47 m^3/s, and the concentration of particles entering the baghouse is 15 g/m^3. In normal operation the baghouse particulate discharge meets the regulatory limit of 24 mg/m^3. During preventive maintenance replacement of the bags, one bag is inadvertently not replaced, so only 423 bags are in place.

Calculate the fraction of particulate matter removed and the efficiency of particulate removal when all 424 bags are in place and the emissions comply with the regulatory requirements. Estimate the mass emission rate when one of the bags is missing and recalculate the efficiency of the baghouse. Assume the efficiency for each individual bag is the same as the overall efficiency for the baghouse.

Solution. The mass balance diagram for the baghouse in normal operation is shown here.

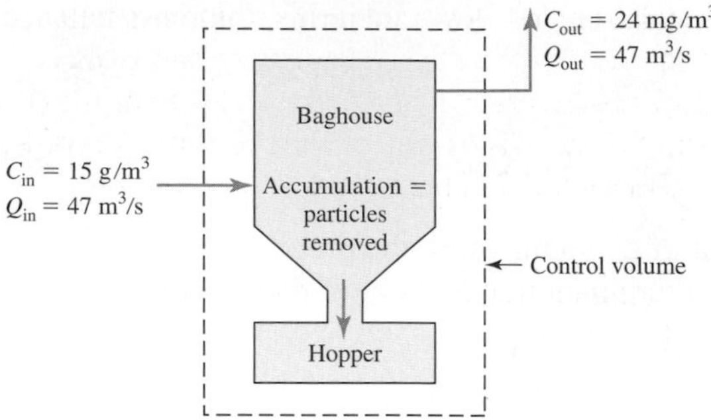

In concentration and flow rate terms, the mass balance equation is

$$\frac{dM}{dt} = C_{in}Q_{in} - C_{out}Q_{out}$$

The mass rate of accumulation in the baghouse is

$$\frac{dM}{dt} = (15{,}000 \text{ mg/m}^3)(47 \text{ m}^3\text{/s}) - (24 \text{ mg/m}^3)(47 \text{ m}^3\text{/s}) = 703{,}872 \text{ mg/s}$$

The fraction of particulates removed is

$$\frac{703{,}872 \text{ mg/s}}{(15{,}000 \text{ mg/m}^3)(47 \text{ mg/s})} = \frac{703{,}872 \text{ mg/s}}{705{,}000 \text{ mg/s}} = 0.9984$$

The efficiency of the baghouse is

$$\eta = \frac{15{,}000 \text{ mg/m}^3 - 24 \text{ mg/m}^3}{15{,}000 \text{ mg/m}^3}(100\%)$$
$$= 99.84\%$$

Note that the fraction of particulate matter removed is the decimal equivalent of the efficiency.

To determine the mass emission rate with one bag missing, we begin by drawing a mass balance diagram. Because one bag is missing, a portion of the flow (1/424 of Q_{out}) effectively bypasses the baghouse. The "Bypass" line around the baghouse is drawn to show this.

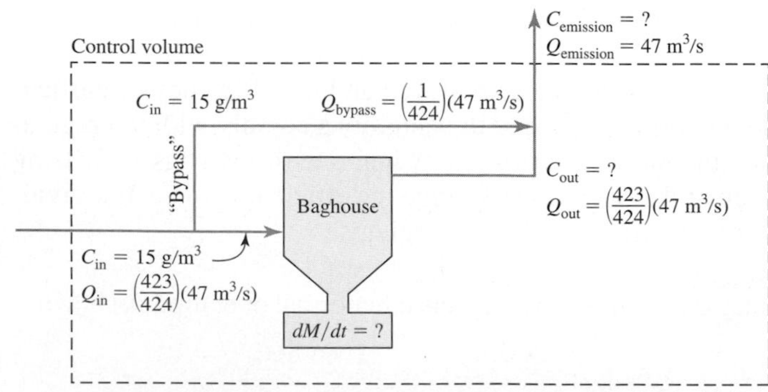

A judicious selection of the control volume aids in the solution of this problem. As shown in the diagram, a control volume around the overall baghouse and bypass flow yields three unknowns: the mass flow rate out of the baghouse, the rate of mass accumulation in the baghouse hopper, and the mass flow rate of the mixture. A control volume around the baghouse alone reduces the number of unknowns to two:

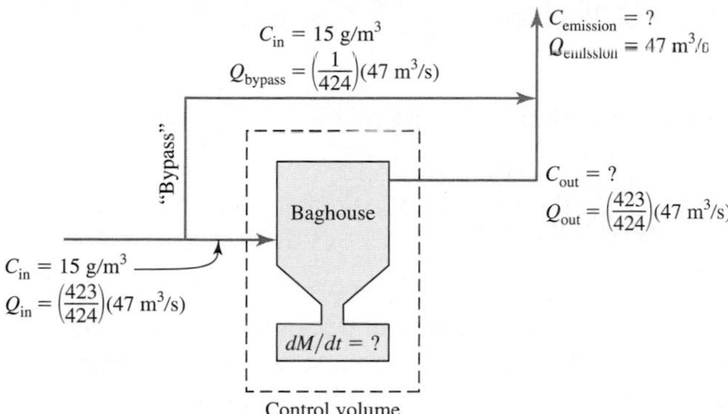

Because we know the efficiency and the influent mass flow rate, we can solve the mass balance equation for the mass flow rate out of the filter.

$$\eta = \frac{C_{in}Q_{in} - C_{out}Q_{out}}{C_{in}Q_{in}}$$

Solving for $C_{out}Q_{out}$

$$\begin{aligned} C_{out}Q_{out} &= (1 - \eta)C_{in}Q_{in} \\ &= (1 - 0.9984)(15{,}000 \text{ mg/m}^3)(47 \text{ m}^3\text{/s})(423/424) = 1{,}125 \text{ mg/s} \end{aligned}$$

This value can be used as an input for a control volume around the junction of the bypass, the effluent from the baghouse and the final effluent.

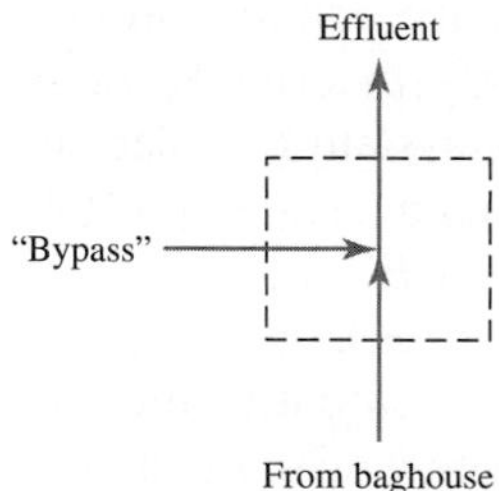

A mass balance for the control volume around the junction may be written as

$$\frac{dM}{dt} = C_{in}Q_{in \text{ from bypass}} + C_{in}Q_{in \text{ from baghouse}} - C_{emission}Q_{emission}$$

Because there is no accumulation in the junction

$$\frac{dM}{dt} = 0$$

and the mass balance equation is

$$
\begin{aligned}
C_{out}Q_{out} &= C_{in}Q_{\text{in from bypass}} + C_{in}Q_{\text{in from baghouse}} \\
&= (15{,}000 \text{ mg/m}^3)(47 \text{ m}^3\text{/s})(1/424) + 1{,}125 = 2788 \text{ mg/s}
\end{aligned}
$$

The concentration in the effluent is

$$\frac{C_{out}Q_{out}}{Q_{out}} = \frac{2{,}788 \text{ mg/s}}{47 \text{ m}^3\text{/s}} = 59 \text{ mg/m}^3$$

The overall efficiency of the baghouse with the missing bag is

$$
\begin{aligned}
\eta &= \frac{15{,}000 \text{ mg/m}^3 - 59 \text{ mg/m}^3}{15{,}000 \text{ mg/m}^3}(100\%) \\
&= 99.61\%
\end{aligned}
$$

The efficiency is still very high but the control equipment does not meet the allowable emission rate of 24 mg/m^3. It is not likely that a baghouse would ever operate with a missing bag because the unbalanced gas flows would be immediately apparent. However, many small holes in a number of bags could yield an effluent that did not meet the discharge standards but would otherwise appear to be functioning correctly. To prevent this situation, the bags undergo periodic inspection and maintenance and the effluent stream is monitored continuously.

The State of Mixing

The state of mixing in the system is an important consideration in the application of Equation 2-4. Consider a coffee cup containing approximately 200 mL of black coffee (or another beverage of your choice). If we add a dollop (about 20 mL) of cream and immediately take a sample (or a sip), we would not be surprised to find that the cream was not evenly distributed throughout the coffee. If, on the other hand, we mixed the coffee and cream vigorously and then took a sample, it would not matter if we sipped from the left or right of the cup, or, for that matter, put a valve in the bottom and sampled from there, we would expect the cream to be distributed evenly. In terms of a mass balance on the coffee cup system, the cup itself would define the system boundary for the control volume. If the coffee and cream were not mixed well, then the place we take the sample from would strongly affect the value of $d(\text{out})/dt$ in Equation 2-4. On the other hand, if the coffee and cream were instantaneously well mixed, then any place we take the sample from would yield the same result. That is, any output would look exactly like the contents of the cup. This system is called a completely mixed system. A more formal definition is that *completely mixed systems* are those in which every drop of fluid is homogeneous with every other drop, that is, every drop of fluid contains the same concentration of material or physical property (e.g., temperature). If a system is completely mixed, then we may assume that the output from the system (concentration, temperature, etc.) is the same as the contents within the system boundary. Although we frequently make use of this assumption to solve mass-balance problems, it is often very difficult to achieve in real systems. This means that solutions to mass-balance problems that make this assumption must be taken as approximations to reality.

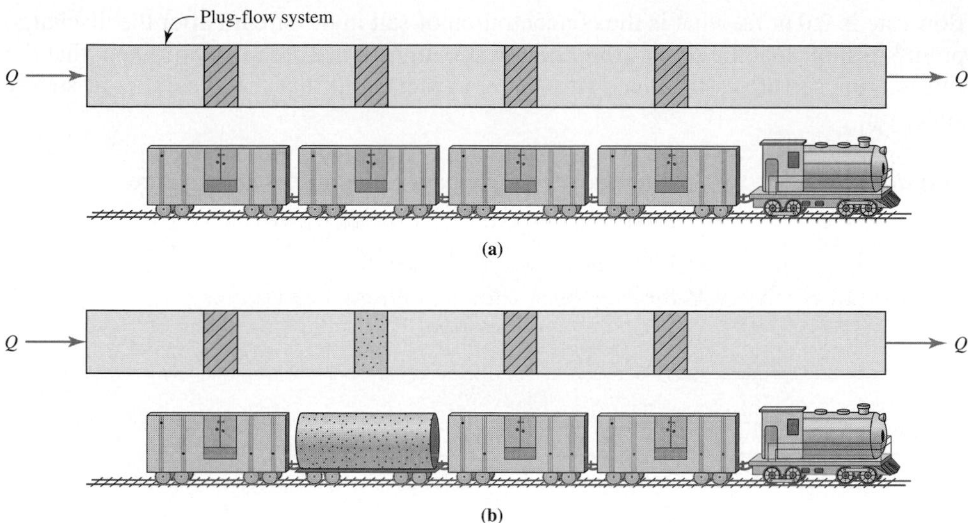

FIGURE 2-1
(*a*) Analogy of a plug-flow system and a train. (*b*) Analogy when a pulse change in influent concentration occurs.

If completely mixed systems exist, or at least systems that we can approximate as completely mixed, then it stands to reason that some systems are completely unmixed or approximately so. These systems are called *plug-flow systems*. The behavior of a plug-flow system is analogous to that of a train moving along a railroad track (Figure 2-1). Each car in the train must follow the one preceding it. If, as in Figure 2-1b, a tank car is inserted in a train of box cars, it maintains its position in the train until it arrives at its destination. The tank car may be identified at any point in time as the train travels down the track. In terms of fluid flow, each drop of fluid along the direction of flow remains unique and, if no reactions take place, contains the same concentration of material or physical property that it had when it entered the plug-flow system. Mixing may or may not occur in the radial direction. As with the completely mixed systems, ideal plug-flow systems don't happen very often in the real world.

When a system has operated in such a way that the rate of input and the rate of output are constant and equal, then, of course, the rate of accumulation is zero (i.e., $dM/dt = 0$). This condition is called *steady state*. In solving mass-balance problems, it is often convenient to make an assumption that steady-state conditions have been achieved. We should note that steady state does not imply equilibrium. For example, water running into and out of a pond at the same rate is not at equilibrium, otherwise it would not be flowing. However, if there is no accumulation in the pond, then the system is at steady state.

Example 2-5 demonstrates the use of two assumptions: complete mixing and steady state.

Example 2-5. A storm sewer is carrying snow melt containing 1.200 g/L of sodium chloride into a small stream. The stream has a naturally occurring sodium chloride concentration of 20 mg/L. If the storm sewer flow rate is 2,000 L/min and the stream flow rate is 2.0 m^3/s, what is the concentration of salt in the stream after the discharge point? Assume that the sewer flow and the stream flow are completely mixed, that the salt is a conservative substance (it does not react), and that the system is at steady state.

Solution. The first step is to draw a mass balance diagram as shown here.

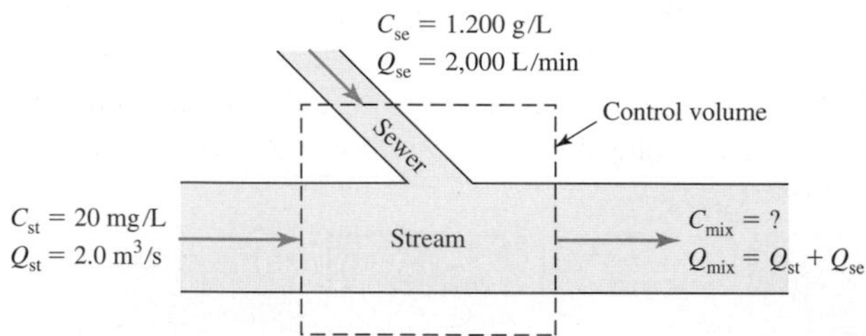

Note that the mass flow of salt may be calculated as

$$\frac{\text{Mass}}{\text{Time}} = (\text{concentration})(\text{flow rate})$$

or

$$\frac{\text{Mass}}{\text{Time}} = (\text{mg/L})(\text{L/min}) = \text{mg/min}$$

Using the notation in the diagram, where the subscript "st" refers to the stream and the subscript "se" refers to the sewer, the mass balance may be written as

$$\text{Rate of accumulation of salt} = [C_{st}Q_{st} + C_{se}Q_{se}] - C_{mix}Q_{mix}$$

where $Q_{mix} = Q_{st} + Q_{se}$.

Because we assume steady state, the rate of accumulation equals zero and

$$C_{mix}Q_{mix} = [C_{st}Q_{st} + C_{se}Q_{se}]$$

Solving for C_{mix}

$$C_{mix} = \frac{[C_{st}Q_{st} + C_{se}Q_{se}]}{Q_{st} + Q_{se}}$$

Before substituting in the values, the units are converted as follows:

$$C_{se} = (1.200 \text{ g/L})(1000 \text{ mg/g}) = 1{,}200 \text{ mg/L}$$

$$Q_{st} = (2.0 \text{ m}^3\text{/s})(1000 \text{ L/m}^3)(60 \text{ s/min}) = 120{,}000 \text{ L/min}$$

$$C_{mix} = \frac{[(20 \text{ mg/L})(120{,}000 \text{ L/min})] + [(1{,}200 \text{ mg/L})(2{,}000 \text{ L/min})]}{120{,}000 \text{ L/min} + 2{,}000 \text{ L/min}}$$

$$= 39.34 \text{ or } 39 \text{ mg/L}$$

Including Reactions

Equation 3-4 is applicable when no chemical or biological reaction takes place and no radioactive decay occurs of the substances in the mass balance. In these instances the substance is said to be *conserved.* Examples of conservative substances include salt in water and argon in air. Examples of nonconservative substances (i.e., those that react or settle out) include decomposing organic matter and particulate matter that is settling from the air.

In most systems of environmental interest, transformations occur within the system: by-products are formed (e.g., CO_2) or compounds are destroyed (e.g., ozone). Because many environmental reactions do not occur instantaneously, the time dependence of the reaction must be taken into account. Equation 2-3 may be written to account for time-dependent transformation as follows:

$$\text{Accumulation rate} = \text{input rate} - \text{output rate} \pm \text{transformation rate} \quad (2\text{-}9)$$

Time-dependent reactions are called *kinetic reactions.* The rate of transformation, or reaction rate (r), is used to describe the rate of formation or disappearance of a substance or chemical species. With reactions, Equation 2-4 may become

$$\frac{dM}{dt} = \frac{d(\text{in})}{dt} - \frac{d(\text{out})}{dt} + r \quad (2\text{-}10)$$

The reaction rate is often some complex function of temperature, pressure, the reacting components, and products of reaction.

$$r = -kC^{n} \quad (2\text{-}11)$$

where k = reaction rate constant (in s^{-1} or d^{-1})
C = concentration of substance
n = exponent or reaction order

The minus sign before reaction rate, k, indicates the disappearance of a substance or chemical species.

In many environmental problems, for example the oxidation of organic compounds by microorganisms (Chapter 6) and radioactive decay (Chapter 11), the reaction rate, r, may be assumed to be directly proportional to the amount of material remaining, that is the value of $n = 1$. This is known as a *first-order reaction.* In first-order reactions, the rate of loss of the substance is proportional to the amount of substance present at any given time, t.

$$r = -kC = \frac{dC}{dt} \quad (2\text{-}12)$$

The differential equation may be integrated to yield either

$$\ln\frac{C}{C_0} = -kt \quad (2\text{-}13)$$

or

$$C = C_0 e^{-kt} = C_0 \exp(-kt) \quad (2\text{-}14)$$

where C = concentration at any time t
C_0 = initial concentration
ln = logarithm to base e
e = exp = exponential e = 2.7183 raised to the $-kt$ power

For simple completely mixed systems with first-order reactions, the total mass of substance (M) is equal to the product of the concentration and volume ($C\mathcal{V}$) and, when $\mathcal{V}$ is a constant, the mass rate of decay of the substance is

$$\frac{dM}{dt} = \frac{d(C\mathcal{V})}{dt} = \mathcal{V}\frac{d(C)}{dt} \tag{2-15}$$

Because first-order reactions can be described by Equation 2-12, we can rewrite Equation 2-10 as

$$\frac{dM}{dt} = \frac{d(\text{in})}{dt} - \frac{d(\text{out})}{dt} - kC\mathcal{V} \tag{2-16}$$

Example 2-6. A well-mixed sewage lagoon (a shallow pond) is receiving 430 m^3/d of sewage out of a sewer pipe. The lagoon has a surface area of 10 ha (hectares) and a depth of 1.0 m. The pollutant concentration in the raw sewage discharging into the lagoon is 180 mg/L. The organic matter in the sewage degrades biologically (decays) in the lagoon according to first-order kinetics. The reaction rate constant (decay coefficient) is 0.70 d^{-1}. Assuming no other water losses or gains (evaporation, seepage, or rainfall) and that the lagoon is completely mixed, find the steady-state concentration of the pollutant in the lagoon effluent.

Solution. We begin by drawing the mass-balance diagram.

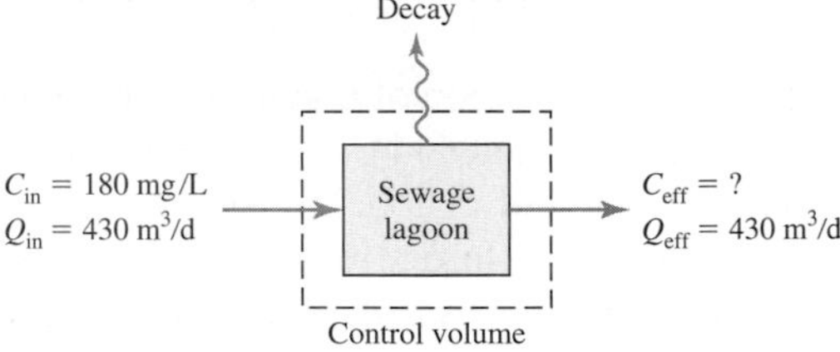

The mass-balance equation may be written as

$$\text{Accumulation} = \text{input rate} - \text{output rate} - \text{decay rate}$$

Assuming steady-state conditions, that is, accumulation = 0, then

$$\text{Input rate} = \text{output rate} + \text{decay rate}$$

This may be written in terms of the notation in the figure as

$$C_{in}Q_{in} = C_{eff}Q_{eff} + kC_{lagoon}\mathcal{V}$$

Solving for C_{eff}, we have

$$C_{eff} = \frac{C_{in}Q_{in} - kC_{lagoon}\forall}{Q_{eff}}$$

Now calculate the values for terms in the equation. The input mass rate ($C_{in}Q_{in}$) is

$$(180 \text{ mg/L})(430 \text{ m}^3/\text{d})(1{,}000 \text{ L/m}^3) = 77{,}400{,}000 \text{ mg/d}$$

With a lagoon volume of

$$(10 \text{ ha})(10^4 \text{ m}^2/\text{ha})(1 \text{ m}) = 100{,}000 \text{ m}^3$$

and the decay coefficient of 0.70 d^{-1}, the decay rate is

$$kC\forall = (0.70 \text{ d}^{-1})(100{,}000 \text{ m}^3)(1{,}000 \text{ L/m}^3)(C_{lagoon}) = (70{,}000{,}000 \text{ L/d})(C_{lagoon})$$

Now using the assumption that the lagoon is completely mixed, we assume that $C_{eff} = C_{lagoon}$. Thus,

$$kC\forall = (70{,}000{,}000 \text{ L/d})(C_{eff})$$

Substituting into the mass-balance equation

$$\text{Output rate} = 77{,}400{,}000 \text{ mg/d} - 70{,}000{,}000 \text{ L/d} \times C_{eff}$$

or

$$C_{eff}(430 \text{ m}^3/\text{d})(1{,}000 \text{ L/m}^3) = 77{,}400{,}000 \text{ mg/d} - 70{,}000{,}000 \text{ L/d} \times C_{eff}$$

Solving for C_{eff}, we have

$$C_{eff} = \frac{77{,}400{,}000 \text{ mg/d}}{70{,}430{,}000 \text{ L/d}} = 1.10 \text{ mg/L}$$

Plug-Flow with Reaction. As noted in Figure 2-1, in plug-flow systems, the tank car, or "plug" element of fluid, does not mix with the fluid ahead or behind it. However, a reaction can take place in the tank car element. Thus, even at steady state, the contents within the element can change with time as the plug moves downstream. The control volume for the mass balance is the plug or differential element of fluid. The mass balance for this moving plug may be written as

$$\frac{dM}{dt} = \frac{d(\text{in})}{dt} - \frac{d(\text{out})}{dt} + \forall\frac{d(C)}{dt} \quad (2\text{-}17)$$

Because no mass exchange occurs across the plug boundaries (in our railroad car analogy, there is no mass transfer between the box cars and the tank car), $d(\text{in})$ and $d(\text{out}) = 0$. Equation 2-17 may be rewritten as

$$\frac{dM}{dt} = 0 - 0 + \forall\frac{d(C)}{dt} \quad (2\text{-}18)$$

As noted earlier, for a first-order decay reaction, the right-hand term may be expressed as

$$\forall \frac{dC}{dt} = -kC\forall \tag{2-19}$$

The total mass of substance (M) is equal to the product of the concentration and volume ($C\forall$) and, when $\forall$ is a constant, the mass rate of decay of the substance in Equation 2-18 may be expressed as

$$\forall \frac{dC}{dt} = -kC\forall \tag{2-20}$$

where the left-hand side of the equation = dM/dt. The steady-state solution to the mass-balance equation for the plug-flow system with first-order kinetics is

$$\ln \frac{C_{\text{out}}}{C_{\text{in}}} = -k\theta \tag{2-21}$$

or

$$C_{\text{out}} = (C_{\text{in}})e^{-k\theta} \tag{2-22}$$

where k = reaction rate constant, s^{-1}, $\min^{-1}$, or d^{-1}
θ = residence time in plug-flow system, s, min, or d

In a plug-flow system of length L, each plug travels for a period = L/u, where u = the speed of flow. Alternatively, for a cross-sectional area A, the residence time is

$$\theta = \frac{(L)(A)}{(u)(A)} = \frac{\forall}{Q} \tag{2-23}$$

where $\forall$ = volume of the plug-flow system, m^3
Q = flow rate (in m^3/s)

Thus, for example, Equation 2-21, may be rewritten as

$$\ln \frac{C_{\text{out}}}{C_{\text{in}}} = -k\frac{L}{u} = -k\frac{\forall}{Q} \tag{2-24}$$

where L = length of the plug-flow segment, m
u = linear velocity, m/s

Although the concentration within a given plug changes over time as the plug moves downstream, the concentration at a fixed point in the plug-flow system remains constant with respect to time. Thus, Equation 2-24 has no time dependence.

Example 2-7 illustrates an application of plug-flow with reaction.

Example 2-7. A wastewater treatment plant must disinfect its effluent before discharging the wastewater to a near-by stream. The wastewater contains 4.5×10^5 fecal coliform colony-forming units (CFU) per liter. The maximum permissible fecal coliform concentration that may be discharged is 2,000 fecal coliform CFU/L. It is proposed that a pipe carrying the wastewater be used for disinfection process. Determine the length of pipe required if the linear velocity of the wastewater in the pipe is 0.75 m/s. Assume that the pipe behaves as a steady-state plug-flow system and that the reaction rate constant for destruction of the fecal coliforms is 0.23 mm^{-1}.

Solution. The mass balance diagram is sketched here. The control volume is the pipe itself.

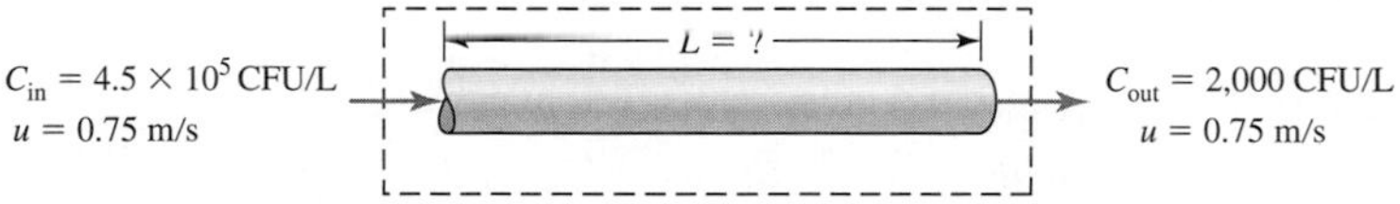

$C_{in} = 4.5 \times 10^5$ CFU/L
$u = 0.75$ m/s

$L = ?$

$C_{out} = 2{,}000$ CFU/L
$u = 0.75$ m/s

Using the steady-state solution to the mass-balance equation, we obtain

$$\ln\frac{C_{out}}{C_{in}} = -k\frac{L}{u}$$

$$\ln\frac{2{,}000\,\text{CFU/L}}{4.5 \times 10^5\,\text{CFU/L}} = -0.23\ \text{min}^{-1}\frac{L}{(0.75\ \text{m/s})(60\ \text{s/min})}$$

Solving for the length of pipe, we have

$$\ln(4.44 \times 10^{-3}) = -0.23\ \text{min}^{-1}\frac{L}{45\ \text{m/min}}$$

$$-5.42 = -0.23\ \text{min}^{-1}\frac{L}{45\ \text{m/min}}$$

$$L = 1{,}060\,\text{m}$$

A little over 1 km of pipe is needed to meet the discharge standard. For most wastewater treatment systems this would be an exceptionally long discharge and another alternative such as a mixing reactor (discussed in the following section) would be investigated.

Reactors

The tanks in which physical, chemical, and biochemical reactions occur, for example, in water softening (Chapter 4) and wastewater treatment (Chapter 6), are called *reactors*. These reactors are classified based on their flow characteristics and their mixing conditions. With appropriate selection of control volumes, ideal chemical reactor models may be used to model natural systems.

Batch reactors are of the fill-and-draw type: materials are added to the tank (Figure 2-2*a*), mixed for sufficient time to allow the reaction to occur (Figure 2-2*b*), and then drained (Figure 2-2*c*). Although the reactor is well mixed and the contents are uniform at any instant in time, the composition within the tank changes with time as the reaction proceeds. A batch reaction is unsteady. Because there is no flow into or out of a batch reactor

$$\frac{d(\text{in})}{dt} = \frac{d(\text{out})}{dt} = 0$$

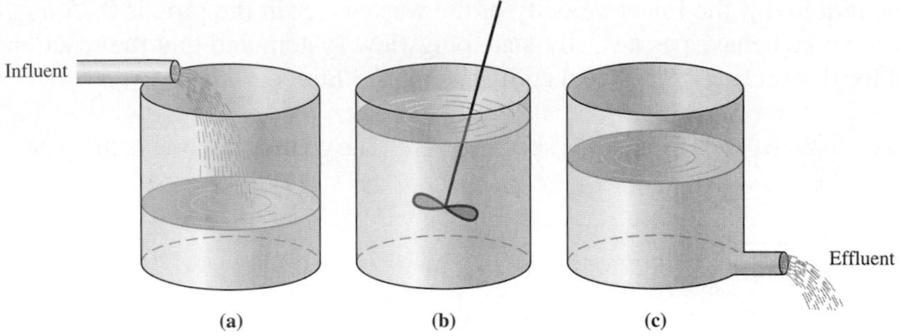

FIGURE 2-2
Batch reactor operation. (*a*) Materials added to the reactor. (*b*) Mixing and reaction. (*c*) Reactor is drained. *Note:* There is no influent or effluent during the reaction.

For a batch reactor Equation 2-16 reduces to

$$\frac{dM}{dt} = -kC\mathcal{V} \tag{2-25}$$

As we noted in Equation 2-15

$$\frac{dM}{dt} = \mathcal{V}\frac{dC}{dt}$$

So that for a first-order reaction in a batch reactor, Equation 2-25 may be simplified to

$$\frac{dC}{dt} = -kC \tag{2-26}$$

Flow reactors have a continuous type of operation: material flows into, through, and out of the reactor at all times. Flow reactors may be further classified by mixing conditions. The contents of a *completely mixed flow reactor* (CMFR), also called a continuous-flow stirred tank reactor (CSTR), ideally are uniform throughout the tank. A schematic diagram of a CMFR and the common flow diagram notation are shown in Figure 2-3. The composition of the effluent is the same as the composition in the tank.

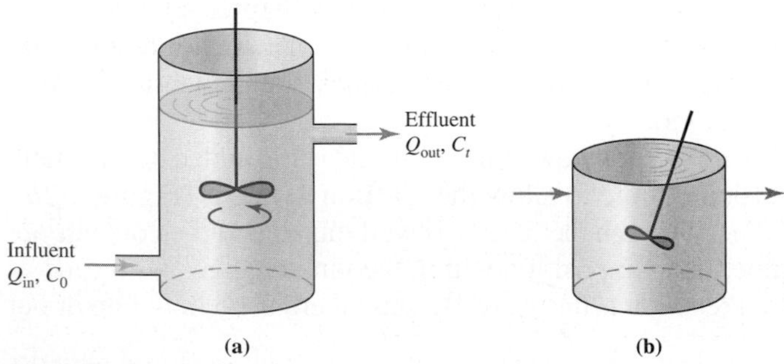

FIGURE 2-3
Schematic diagram of (*a*) completely mixed flow reactor (CMFR) and (*b*) the common diagram. The propeller indicates that the reactor is completely mixed.

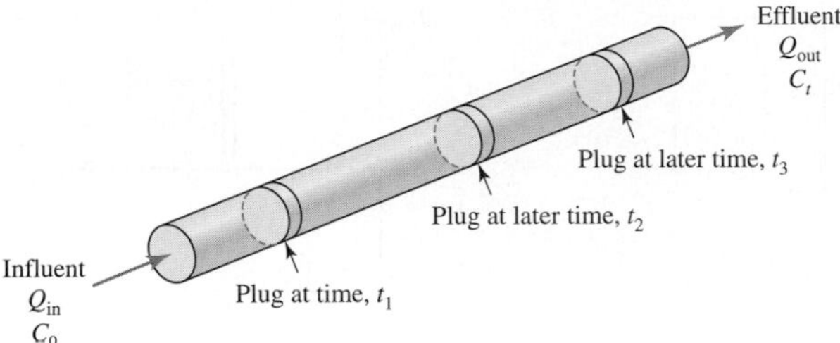

FIGURE 2-4
Schematic diagram of a plug-flow reactor (PFR). *Note:* $t_3 > t_2 > t_1$.

If the mass input rate into the tank remains constant, the composition of the effluent remains constant. The mass balance for a CMFR is described by Equation 2-16.

In *plug-flow reactors* (PFR), fluid particles pass through the tank in sequence. Those that enter first leave first. In the ideal case, it is assumed that no mixing occurs in the lateral direction. Although composition varies along the length of the tank, as long as the flow conditions remain steady, the composition of the effluent remains constant. A schematic diagram of a plug-flow reactor is shown in Figure 2-4. The mass balance for a PFR is described by Equation 2-18, where the time element (dt) is the time spent in the PFR as described by Equation 2-23. Real continuous-flow reactors are generally something in between a CMFR and PFR.

For time-dependent reactions, the time that a fluid particle remains in the reactor obviously affects the degree to which the reaction goes to completion. In ideal reactors the average time in the reactor (*detention time* or *retention time* or, for liquid systems, *hydraulic detention time* or *hydraulic retention time*) is defined as

$$\theta = \frac{V\!\!\!\!-}{Q} \tag{2-27}$$

where θ = theoretical detention time, s
$V\!\!\!\!-$ = volume of fluid in reactor, m^3
Q = flow rate into reactor, m^3/s

Real reactors do not behave as ideal reactors because of density differences due to temperature or other causes, short circuiting because of uneven inlet or outlet conditions, and local turbulence or dead spots in the tank corners. The detention time in real tanks is generally less than the theoretical detention time calculated from Equation 2-27.

Reactor Analysis

The selection of a reactor either as a treatment method or as a model for a natural process depends on the behavior desired or recognized. We will examine the behavior of batch, CMFR, and PFR reactors in several situations. Situations of particular interest are the response of the reactor to a sudden increase (Figure 2-5*a*) or decrease (Figure 2-5*b*) in the steady-state influent concentration for conservative and nonconservative species (commonly called a step increase or decrease) and the response to pulse or spike change in influent concentration (Figure 2-5*c*). We will present the plots

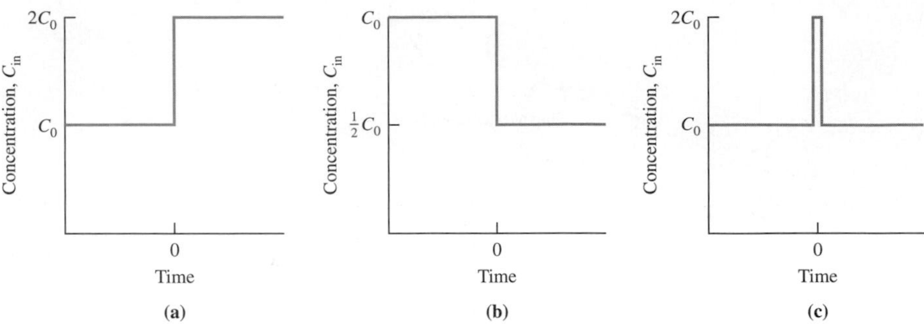

FIGURE 2-5
Example influent graphs of (*a*) step increase in influent concentration, (*b*) step decrease in influent concentration, and (*c*) a pulse or spike increase in influent concentration. *Note:* The size of the change is for illustration purposes only.

of the effluent concentration for each of the reactor types for a variety of conditions to show the response to these influent changes.

For nonconservative substances, we will present the analysis for first-order reactions. The behavior of zero-order and second-order reactions will be summarized in comparison at the conclusion of this discussion.

Batch Reactor. Laboratory experiments are often conducted in batch reactors because they are inexpensive and easy to build. Industries that generate small quantities of wastewater (less than 150 m^3/d) use batch reactors because they are easy to operate and provide an opportunity to check the wastewater for regulatory compliance before discharging it.

Because there is no influent to or effluent from a batch reactor, the introduction of a conservative substance into the reactor either as a step increase or a pulse results in an instantaneous increase in concentration of the conservative substance in the reactor. The concentration plot is shown in Figure 2-6.

Because there is no influent or effluent, for a nonconservative substance that decays as a first-order reaction, the mass balance is described by Equation 2-26. Integration yields

$$\frac{C_t}{C_0} = e^{-kt} \tag{2-28}$$

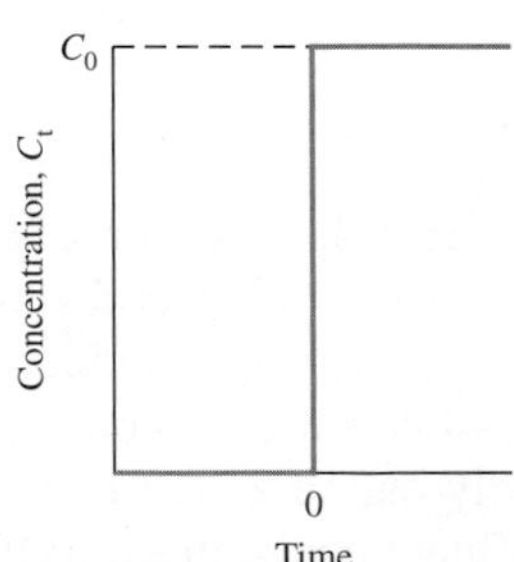

FIGURE 2-6
Batch reactor response to a step or pulse increase in concentration of a conservative substance. C_0 = mass of conservative substance/volume of reactor.

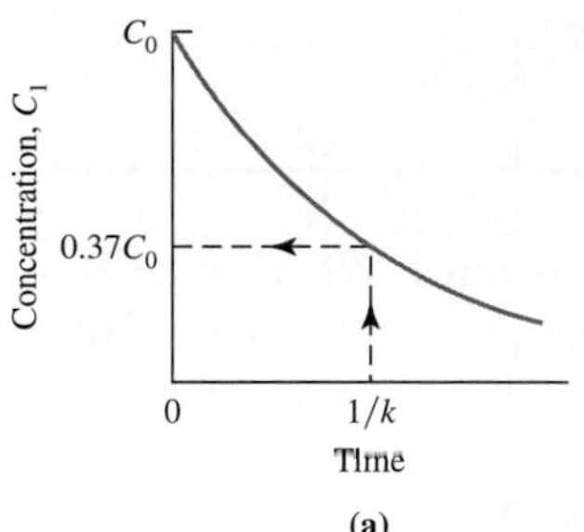

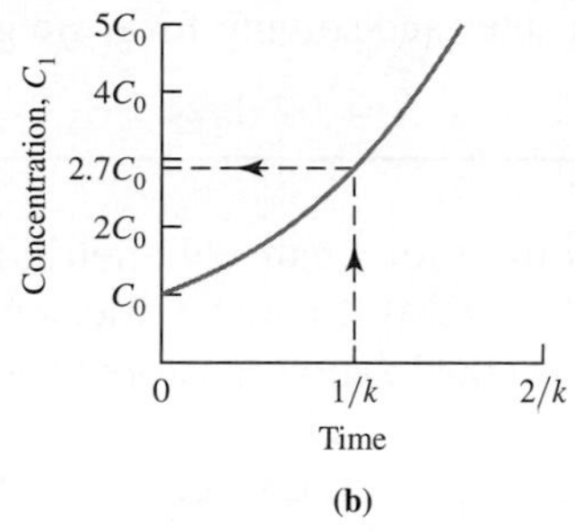

FIGURE 2-7
Batch reactor response for (*a*) decay of a non-conservative substance and (*b*) for a formation reaction.

The final concentration plot is shown in Figure 2-7*a*. For the formation reaction, where the sign in Equation 2-28 is positive, the concentration plot is shown in Figure 2-7*b*.

Example 2-8. A contaminated soil is to be excavated and treated in a completely mixed aerated lagoon at a Superfund site. To determine the time it will take to treat the contaminated soil, a laboratory completely mixed batch reactor is used to gather the following data. Assuming a first-order reaction, estimate the rate constant, k, and determine the time to achieve 99 percent reduction in the original concentration.

Time (d)	Waste Concentration (mg/L)
1	280
16	132

Solution. The rate constant may be estimated by solving Equation 2-28 for k. Using the 1st and 16th day, the time interval $t = 16 - 1 = 15$ d,

$$\frac{132 \text{ mg/L}}{280 \text{ mg/L}} = \exp[-k(15 \text{ d})]$$

$$0.4714 = \exp[-k(15)]$$

Taking the logarithm (base e) of both sides of the equation, we obtain

$$-0.7520 = -k(15)$$

Solving for k, we have

$$k = 0.0501 \text{ d}^{-1}$$

To achieve 99 percent reduction the concentration at time t must be $1 - 0.99$ of the original concentration:

$$\frac{C_t}{C_0} = 0.01$$

The estimated time is then

$$0.01 = \exp[-0.05(t)]$$

Taking the logarithm of both sides and solving for t, we get

$$t = 92 \text{ days}$$

CMFR. A batch reactor is used for small volumetric flow rates. When water flow rates are greater than 150 m^3/d, a CMFR may be selected for chemical mixing. Examples of this application include equalization reactors to adjust the pH, precipitation reactors to remove metals, and mixing tanks (called *rapid mix* or *flash mix tanks*) for water treatment. Because municipal wastewater flow rates vary over the course of a day, a CMFR (called an *equalization basin*) may be placed at the treatment plant influent point to level out the flow and concentration changes. Some natural systems such as a lake or the mixing of two streams or the air in a room or over a city may be modeled as a CMFR as an approximation of the real mixing that is taking place.

For a step increase in a conservative substance entering a CMFR, the initial level of the conservative substance in the reactor is C_0 prior to $t = 0$. At $t = 0$, the influent concentration (C_{in}) instantaneously increases to C_1 and remains at this concentration (Figure 2-8*a*). With balanced fluid flow ($Q_{in} = Q_{out}$) into the CMFR and no reaction, the mass balance equation for a step increase is

$$\frac{dM}{dt} = C_t Q_{in} - C_{out} Q_{out} \tag{2-29}$$

where $M = C\forall$. The solution is

$$C_t = C_0\left[\exp\left(-\frac{t}{\theta}\right)\right] + C_1\left[1 - \exp\left(-\frac{t}{\theta}\right)\right] \tag{2-30}$$

where C_t = concentration at any time t
C_0 = concentration in reactor prior to step change
C_1 = concentration in influent after instantaneous increase
t = time after step change
θ = theoretical detention time = $\forall/Q$
exp = exponential e such that the terms in brackets immediately following are powers of e, that is, e raised to the power of the term in the brackets, where $e = 2.7183$

Figure 2-8*b* shows the effluent concentration plot.

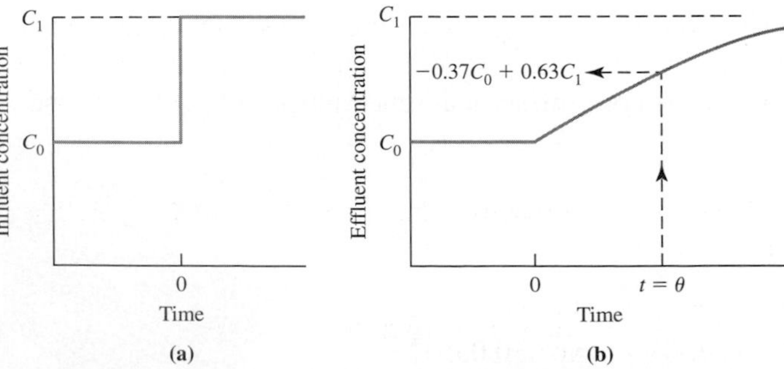

FIGURE 2-8
Response of a CMFR to (*a*) a step increase in the influent concentration of a conservative substance from concentration C_0 to a new concentration C_1. (*b*) Effluent concentration.

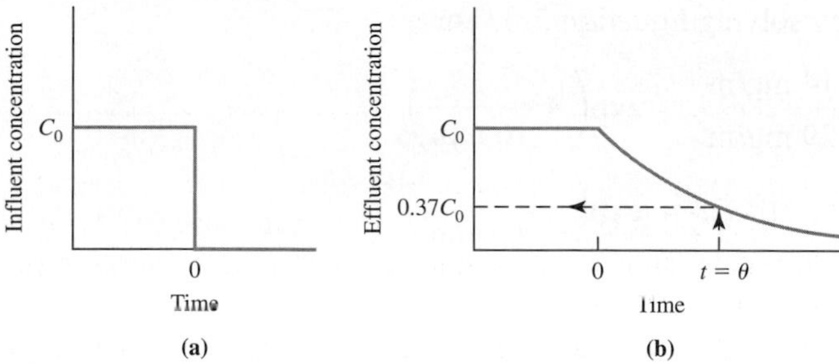

FIGURE 2-9
Flushing of CMFR resulting from (*a*) a step decrease in influent concentration of a conservative substance from C_0 to 0. (*b*) Effluent concentration.

Flushing of a nonreactive contaminant from a CMFR by a contaminant-free fluid is an example of a step change in the influent concentration (Figure 2-9*a*). Because $C_{\text{in}} = 0$ and no reaction takes place, the mass balance equation is

$$\frac{dM}{dt} = -C_{\text{out}}Q_{\text{out}} \tag{2-31}$$

where $M = C\text{V}$. The initial concentration is

$$C_0 = \frac{M}{\text{V}} \tag{2-32}$$

Solving Equation 2-31 for any time $t \geq 0$, we obtain

$$C_t = C_0 \exp\left(-\frac{t}{\theta}\right) \tag{2-33}$$

where $\theta = \text{V}/Q$ as noted in Equation 2-27. Figure 2-9*b* shows the effluent concentration plot.

Example 2-9. Before entering an underground utility vault to do repairs, a work crew analyzed the gas in the vault and found that it contained 29 mg/m^3 of hydrogen sulfide. Because the allowable exposure level is 14 mg/m^3 the work crew began ventilating the vault with a blower. If the volume of the vault is 160 m^3 and the flow rate of contaminant-free air is 10 m^3/min, how long will it take to lower the hydrogen sulfide level to a level that will allow the work crew to enter? Assume the manhole behaves as a CMFR and that hydrogen sulfide is nonreactive in the time period considered.

Solution. This is a case of flushing a nonreactive contaminant from a CMFR. The theoretical detention time is

$$\theta = \frac{\text{V}}{Q} = \frac{160 \text{ m}^3}{10 \text{ m}^3/\text{min}} = 16 \text{ min}$$

The required time is found by solving Equation 2-33 for t

$$\frac{14 \text{ mg/m}^3}{29 \text{ mg/m}^3} = \exp\left(-\frac{t}{16 \text{ min}}\right)$$

$$0.4828 = \exp\left(-\frac{t}{16 \text{ min}}\right)$$

Taking the logarithm to the base e of both sides

$$-0.7282 = -\frac{t}{16 \text{ min}}$$

$t = 11.6$ or 12 min to lower the concentration to the allowable level

Because the odor threshold for H_2S is about 0.18 mg/m^3, the vault will still have quite a strong odor after 12 min.

A precautionary note is in order here. H_2S is commonly found in confined spaces such as manholes. It is a very toxic poison and has the unfortunate property of deadening the olfactory senses. Thus, you may not smell it after a few moments even though the concentration has not decreased. Each year a few individuals in the United States die because they have entered a confined space without taking stringent safety precautions.

Because a CMFR is completely mixed, the response of a CMFR to a step change in the influent concentration of a reactive substance results in an immediate corresponding change in the effluent concentration. For this analysis we begin with the mass balance for a balanced flow ($Q_{in} = Q_{out}$) CMFR operating under steady-state conditions with first-order decay of a reactive substance.

$$\frac{dM}{dt} = C_{in}Q_{in} - C_{out}Q_{out} - kC_{out}\forall \tag{2-34}$$

where $M = C\forall$. Because the flow rate and volume are constant, we may divide through by Q and $\forall$ and simplify to obtain

$$\frac{dC}{dt} = \frac{1}{\theta}(C_{in} - C_{out}) - kC_{out} \tag{2-35}$$

where $\theta = \forall/Q$ as noted in Equation 2-27. Under steady-state conditions $dC/dt = 0$ and the solution for C_{out} is

$$C_{out} = \frac{C_0}{1 + k\theta} \tag{2-36}$$

where $C_0 = C_{in}$ immediately after the step change. Note that C_{in} can be nonzero before the step change. For a first-order reaction where material is produced, the sign of the reaction term is positive and the solution to the mass balance equations is

$$C_{out} = \frac{C_0}{1 - k\theta} \tag{2-37}$$

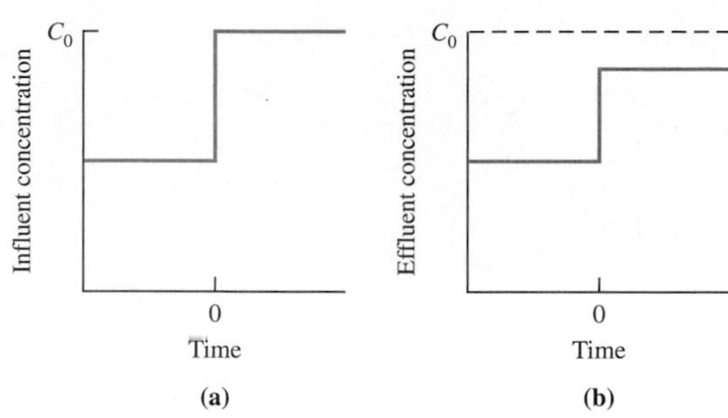

FIGURE 2-10
Steady-state response of CMFR to (*a*) a step increase in influent concentration of a reactive substance. (*b*) Effluent concentration. *Note:* Steady-state conditions exist prior to $t = 0$.

The behavior the CMFR described by Equation 2-36 is shown diagrammatically in Figure 2-10.

A step decrease in the influent concentration to zero ($C_{in} = 0$) in a balanced flow ($Q_{in} = Q_{out}$) CMFR operating under non-steady-state conditions with first-order decay of a reactive substance may be described by rewriting Equation 2-34:

$$\frac{dM}{dt} = 0 - C_{out}Q_{out} - kC_{out}\forall \tag{2-38}$$

where $M = C\forall$. Because the volume is constant, we may divide through by $\forall$ and simplify to obtain

$$\frac{dC}{dt} = \left(\frac{1}{\theta} + k\right)C_{out} \tag{2-39}$$

where $\theta = \forall/Q$ as noted in Equation 2-27. The solution for C_{out} is

$$C_{out} = C_0\exp\left[-\left(\frac{1}{\theta} + k\right)t\right] \tag{2-40}$$

where C_0 is the effluent concentration at $t = 0$.

The concentration plots are shown in Figure 2-11.

PFR. Pipes and long narrow rivers approximate the ideal conditions of a PFR. Biological treatment in municipal wastewater treatment plants is often conducted in long narrow tanks that may be modeled as a PFR.

A step change in the influent concentration of a conservative substance in a plug-flow reactor results in an identical step change in the effluent concentration at a time equal to the theoretical detention time in the reactor as shown in Figure 2-12.

Equation 2-21 is a solution to the mass-balance equation for a steady-state first-order reaction in a PFR. The concentration plot for a step change in the influent concentration is shown in Figure 2-13.

A pulse entering a PFR travels as a discrete element as illustrated in Figure 2-14 for a pulse of green dye. The passage of the pulse through the PFR and the plots of concentration with distance along the PFR are also shown.

Reactor Comparison. Although first-order reactions are common in environmental systems, other reaction orders may be more appropriate. Tables 2-1 and 2-2 compare the reactor types for zero-, first-, and second-order reactions.

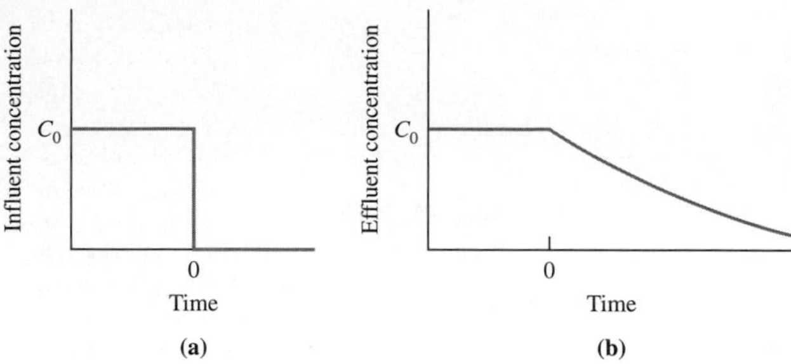

FIGURE 2-11
Non-steady-state response of CMFR to (*a*) step decrease from C_0 to 0 of influent C_0 reactive substance. (*b*) Effluent concentration.

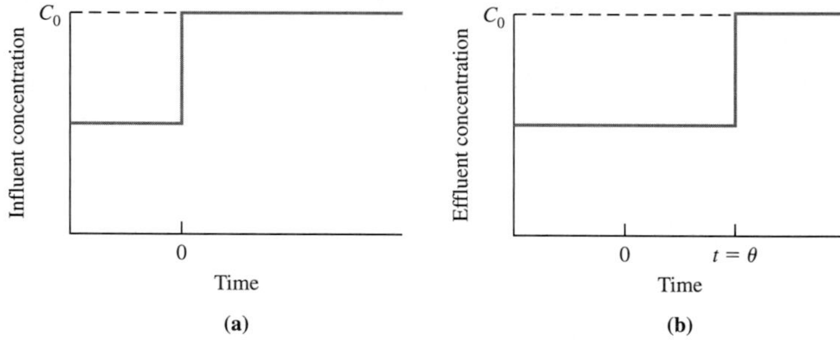

FIGURE 2-12
Response of a PFR to (*a*) a step increase in the influent concentration of a conservative substance. (*b*) Effluent concentration.

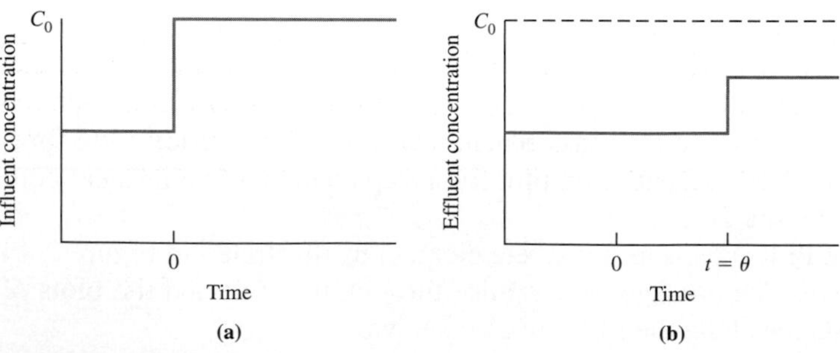

FIGURE 2-13
Response of a PFR to (*a*) a step increase in the influent concentration of a reactive substance. (*b*) Effluent concentration.

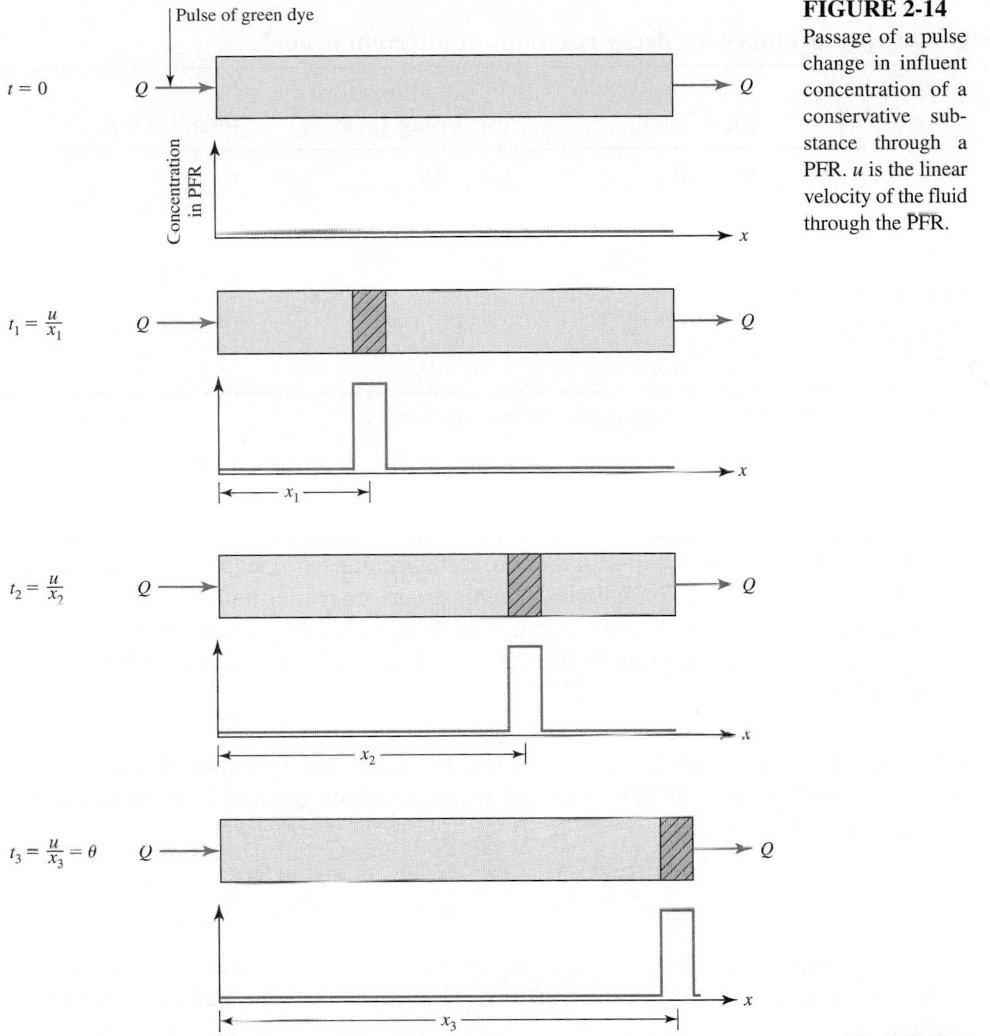

FIGURE 2-14
Passage of a pulse change in influent concentration of a conservative substance through a PFR. u is the linear velocity of the fluid through the PFR.

TABLE 2-1
Comparison of steady-state mean retention times for decay reactions of different order*

		Equations for mean retention times (θ)		
Reaction order	r	Ideal batch	Ideal plug flow	Ideal CMFR
Zero†	$-k$	$\frac{(C_0 - C_t)}{k}$	$\frac{(C_0 - C_t)}{k}$	$\frac{(C_0 - C_t)}{k}$
First	$-kC$	$\frac{\ln(C_0/C_t)}{k}$	$\frac{\ln(C_0/C_t)}{k}$	$\frac{(C_0/C_t) - 1}{k}$
Second	$-kC^2$	$\frac{(C_0/C_t) - 1}{kC_0}$	$\frac{(C_0/C_t) - 1}{kC_0}$	$\frac{(C_0/C_t) - 1}{kC_t}$

*C_0 = initial concentration or influent concentration; C_t = final condition or effluent concentration.
†Expressions are valid for $k\theta \leq C_0$; otherwise $C_t = 0$.

TABLE 2-2
Comparison of steady-state performance for decay reactions of different order*

		Equations for C_t		
Reaction order	r	Ideal batch	Ideal plug flow	Ideal CMFR
Zero† $t \leq C_0/k$	$-k$	$C_0 - kt$	$C_0 - k\theta$	$C_0 - k\theta$
$t > C_0/k$		0		
First	$-kC$	$C_0[\exp(-kt)]$	$C_0[\exp(-k\theta)]$	$\dfrac{C_0}{1 + k\theta}$
Second	$-kC^2$	$\dfrac{C_0}{1 + ktC_0}$	$\dfrac{C_0}{1 + k\theta C_0}$	$\dfrac{(4k\theta C_0 + 1)^{1/2} - 1}{2k\theta}$

*C_0 = initial concentration or influent concentration; C_t = final condition or effluent concentration.
†Time conditions are for ideal batch reactor only.

Example 2-10. A chemical degrades in a flow-balanced, steady-state CMFR according to first-order reaction kinetics. The upstream concentration of the chemical is 10 mg/L and the downstream concentration is 2 mg/L. Water is being treated at a rate of 29 m^3/min. The volume of the tank is 580 m^3. What is the rate of decay? What is the rate constant?

Solution. From Equation 2-11, we note that for a first-order reaction, the rate of decay, $r = -kC$. To find the rate of decay, we must solve Equation 2-34 for kC to determine the reaction rate.

$$\frac{dM}{dt} = C_{\text{in}}Q_{\text{in}} - C_{\text{out}}Q_{\text{out}} - kC_{\text{out}}\forall$$

For steady-state there is no mass accumulation, so $dM/dt = 0$. Because the reactor is flow-balanced, $Q_{\text{in}} = Q_{\text{out}} = 29$ m^3/min. The mass balance equation may be rewritten

$$kC_{\text{out}}\forall = C_{\text{in}}Q_{\text{in}} - C_{\text{out}}Q_{\text{out}}$$

Solving for the reaction rate, we obtain

$$r = kC = \frac{C_{\text{in}}Q_{\text{in}} - C_{\text{out}}Q_{\text{out}}}{\forall}$$

$$= kC = \frac{(10 \text{ mg/L})(29 \text{ m}^3/\text{min}) - (2 \text{ mg/L})(29 \text{ m}^3/\text{min})}{580 \text{ m}^3} = 0.4$$

The rate constant, k, can be determined by using the equations given in Table 2-1. For a first-order reaction in a CMFR

$$\theta = \frac{(C_0/C_t) - 1}{k}$$

The mean hydraulic detention time (θ) is

$$\theta = \frac{\forall}{Q} = \frac{580 \text{ m}^3}{29 \text{ m}^3/\text{min}} = 20 \text{ min}$$

Solving the equation from the table for the rate constant, k, we get

$$k = \frac{(C_0/C_t) - 1}{\theta}$$

and

$$k = \frac{(10 \text{ mg/L}/2 \text{ mg/L}) - 1}{20 \text{ min}} = 0.20 \text{ min}^{-1}$$

Reactor Design. Volume is major design parameter in reactor design. In general, the influent concentration of material, the flow rate into the reactor, and the desired effluent concentration are known. As noted in Equation 2-27, the volume is directly related to the theoretical detention time and the flow rate into the reactor. Thus, the volume can be determined if the theoretical detention time can be determined. The equations in Table 2-1 may be used to determine the theoretical detention time if the decay rate constant, k, is available. The rate constant must be determined from the literature or laboratory experiments.

2-4 ENERGY BALANCES

First Law of Thermodynamics

The *first law of thermodynamics* states that (without nuclear reaction) energy can be neither created nor destroyed. As with the law of conservation of matter, it does not mean that the form of the energy does not change. For example, the chemical energy in coal can be changed to heat and electrical power. *Energy* is defined as the capacity to do useful work. *Work* is done by a force acting on a body through a distance. One *joule* (J) is the work done by a constant force of one newton when the body on which the force is exerted moves a distance of one meter in the direction of the force. *Power* is the rate of doing work or the rate of expanding energy. The first law may be expressed as

$$Q_H = U_2 - U_1 + W \qquad (2\text{-}41)$$

where Q_H = heat absorbed, kJ
U_1, U_2 = internal energy (or thermal energy) of the system in states 1 and 2, kJ
W = work, kJ

Fundamentals

Thermal Units of Energy. Energy has many forms, among which are thermal, mechanical, kinetic, potential, electrical, and chemical. Thermal units were invented when heat was considered a substance (caloric), and the units are consistent with conservation of a quantity of substance. Subsequently, we have learned that energy is not a substance but mechanical energy of a particular form. With this in mind we will still use the common metric thermal unit of energy, the calorie.* One *calorie* (cal) is the amount of energy required to raise the temperature of one gram of water from 14.5°C to 15.5°C. In SI units 4.186 J = 1 cal.

The *specific heat* of a substance is the quantity of heat required to increase a unit mass of the substance one degree. Specific heat is expressed in metric units as kcal/kg · K and in SI units as kJ/kg · K where K is kelvins and 1 K = 1°C.

Enthalpy is a thermodynamic property of a material that depends on temperature, pressure, and the composition of the material. It is defined as

$$H = U + P\forall \qquad (2\text{-}42)$$

where H = enthalpy, kJ
U = internal energy (or thermal energy), kJ
P = pressure, kPa
$\forall$ = volume, m^3

Think of enthalpy as a combination of thermal energy (U) and flow energy ($P\forall$) Flow energy should not be confused with kinetic energy ($\frac{1}{2}Mv^2$). Historically, H has been referred to as a system's "heat content." Because *heat* is correctly defined only in terms of energy transfer across a boundary, this is not considered a precise thermodynamic description and enthalpy is the preferred term.

When a non-phase-change process† occurs without a change in volume, a change in internal energy is defined as

$$\Delta U = Mc_v\Delta T \qquad (2\text{-}43)$$

where ΔU = change in internal energy
M = mass
c_v = specific heat at constant volume
ΔT = change in temperature

When a non-phase-change process occurs without a change in pressure, a change in enthalpy is defined as

$$\Delta H = Mc_p\Delta T \qquad (2\text{-}44)$$

where ΔH = enthalpy change
c_p = specific heat at constant pressure

Equations 2-43 and 2-44 assume that the specific heat is constant over the range of temperature (ΔT). Solids and liquids are nearly incompressible and therefore do virtually no work. The change in $P\forall$ is zero, making the changes in H and U identical. Thus, for solids and liquids, we can generally assume $c_v = c_p$ and $\Delta U = \Delta H$, so the change in energy stored in a system is

$$\Delta H = Mc_v\Delta T \qquad (2\text{-}45)$$

---

*In discussing food metabolism, physiologists also use the term *Calorie*. However, the food Calorie is equivalent to a *kilocalorie* in the metric system. We will use the units cal or kcal throughout this text.

†Non-phase-change means, for example, water is not converted to steam.

TABLE 2-3
Specific heat capacities for common substances

Substance	c_p (kJ/kg · K)
Air (293.15 K)	1.00
Aluminum	0.95
Beef	3.22
Cement, portland	1.13
Concrete	0.93
Copper	0.39
Corn	3.35
Dry soil	0.84
Human being	3.47
Ice	2.11
Iron, cast	0.50
Steel	0.50
Poultry	3.35
Steam (373.15 K)	2.01
Water (288.15 K)	4.186
Wood	1.76

Adapted from Guyton (1961), Hudson (1959), Masters (1998), Salvato (1972).

Specific heat capacities for some common substances are listed in Table 2-3.

When a substance changes *phase* (i.e., it is transformed from solid to liquid or liquid to gas), energy is absorbed or released without a change in temperature. The energy required to cause a phase change of a unit mass from a solid to a liquid at constant pressure is called the *latent heat of fusion* or *enthalpy of fusion*. The energy required to cause a phase change of a unit mass from a liquid to a gas at constant pressure is called the *latent heat of vaporization* or *enthalpy of vaporization*. The same amounts of energy are released in condensing the vapor and freezing the liquid. For water the enthalpy of fusion at 0°C is 333 kJ/kg and the enthalpy of vaporization is 2257 kJ/kg at 100°C.

Example 2-11. Standard physiology texts (Guyton, 1961) report that a person weighing 70.0 kg requires approximately 2,000 kcal for simple existence, such as eating and sitting in a chair. Approximately 61% of all the energy in the foods we eat becomes heat during the process of formation of the energy-carrying molecule adenosine triphosphate (ATP) (Guyton, 1961). Still more energy becomes heat as it is transferred to functional systems of the cells. The functioning of the cells releases still more energy so that ultimately "all the energy released by metabolic processes eventually becomes heat" (Guyton, 1961). Some of this heat is used to maintain the body at a normal temperature of 37°C. What fraction of the 2,000 kcal is used to maintain the body temperature at 37°C if the room temperature is 20°C? Assume the specific heat of a human is 3.47 kJ/kg · K.

Solution. The change in energy stored in the body is

$$\Delta H = (70 \text{ kg})(3.47 \text{ kJ/kg} \cdot \text{K})(37°\text{C} - 20°\text{C}) = 4{,}129.30 \text{ kJ}$$

Converting the 2,000 kcal to kJ gives

$$(2{,}000 \text{ kcal})(4.186 \text{ kJ/kcal}) = 8{,}372.0 \text{ kJ}$$

So the fraction of energy used to maintain temperature is approximately

$$\frac{4{,}129.30 \text{ kJ}}{8{,}372.0 \text{ kJ}} = 0.49, \text{ or about } 50\%$$

The remaining energy must be removed if the body temperature is not to rise above normal. The mechanisms of removing energy by heat transfer are discussed in the following sections.

Energy Balances. If we say that the first law of thermodynamics is analogous to the law of conservation of matter, then energy is analogous to matter because it too can be "balanced." The simplest form of the energy balance equation is

$$\text{Loss of enthalpy of hot body} = \text{gain of enthalpy by cold body} \qquad (2\text{-}46)$$

Example 2-12. The Rhett Butler Peach, Co. dips peaches in boiling water (100°C) to remove the skin (a process called blanching) before canning them. The wastewater from this process is high in organic matter and it must be treated before disposal. The treatment process is a biological process that operates at 20°C. Thus, the wastewater must be cooled to 20°C before disposal. Forty cubic meters (40 m^3) of wastewater is discharged to a concrete tank at a temperature of 20°C to allow it to cool. Assuming no losses to the surroundings, and that the concrete tank has a mass of 42,000 kg and a specific heat capacity of 0.93 kJ/kg · K, what is the equilibrium temperature of the concrete tank and the wastewater?

Solution. Assuming the density of the water is 1,000 kg/m^3, the loss in enthalpy of the boiling water is

$$\Delta H = (1{,}000 \text{ kg/m}^3)(40 \text{ m}^3)(4.186 \text{ kJ/kg} \cdot \text{K})(373.15 - T) = 62{,}480{,}236 - 167{,}440T$$

where the absolute temperature is 273.15 + 100 = 373.15 K.

The gain in enthalpy of the concrete tank is

$$\Delta H = (42{,}000 \text{ kg})(0.93 \text{ kJ/kg} \cdot \text{K})(T - 293.15) = 39{,}060T - 11{,}450{,}439$$

The equilibrium temperature is found by setting the two equations equal and solving for the temperature.

$$(\Delta H)_{\text{water}} = (\Delta H)_{\text{concrete}}$$

$$62{,}480{,}236 - 167{,}440T = 39{,}060T - 11{,}450{,}439$$

$$T = 358 \text{ K or } 85°\text{C}$$

This is not very close to the desired temperature, without considering other losses to the surroundings. Convective and radiative heat losses discussed later on also play a role in reducing the temperature, but a cooling tower may be required to achieve 20°C.

For an open system, a more complete energy balance equation is

Net change in energy = energy of mass entering system − energy of mass leaving system ± energy flow into or out of system (2-47)

For many environmental systems the time dependence of the change in energy (i.e., the rate of energy change) must be taken into account. Equation 2-47 may be written to account for time dependence as follows:

$$\frac{dH}{dt} = \frac{d(H)_{\text{mass in}}}{dt} + \frac{d(H)_{\text{mass out}}}{dt} \pm \frac{d(H)_{\text{energy flow}}}{dt} \tag{2-48}$$

If we consider a region of space where a fluid flows in at a rate of dM/dt and also flows out at a rate of dM/dt, then the change in enthalpy due to this flow is

$$\frac{dH}{dt} = c_p M\frac{dT}{dt} + c_p T\frac{dM}{dt} \tag{2-49}$$

where dM/dt is the mass flow rate (e.g., in kg/s) and ΔT is the difference in temperature of the mass in the system and the mass outside of the system.

Note that Equations 2-47 and 2-48 differ from the mass-balance equation in that there is an additional term: "energy flow." This is an important difference for everything from photosynthesis (in which radiative energy from the sun is converted into plant material) to heat exchangers (in which chemical energy from fuel passes through the walls of the tubes of the heat exchanger to heat a fluid inside). The energy flow into (or out of) the system may be by conduction, convection, or radiation.

Conduction. Conduction is the transfer of heat through a material by molecular diffusion due to a temperature gradient. Fourier's law provides an expression for calculating energy flow by conduction.

$$\frac{dH}{dt} = -h_{tc}A\frac{dT}{dx} \tag{2-50}$$

where dH/dt = rate of change of enthalpy, kJ/s or kW
h_{tc} = thermal conductivity, kJ/s · m · K or kW/m · K
A = surface area, m^2
dT/dx = change in temperature through a distance, K/m

Note that 1 kJ/s = 1 kW. The average values for thermal conductivity for some common materials are given in Table 2-4.

Convection. Forced convective heat transfer is the transfer of thermal energy by means of large-scale fluid motion such as a flowing river or aquifer or the wind blowing. The convective heat transfer between a fluid at a temperature, T_f, and a solid surface at a temperature T_s, can be described by Equation 2-51.

$$\frac{dH}{dt} = h_c A(T_f - T_s) \tag{2-51}$$

TABLE 2-4
Thermal conductivity for some common materials[a]

Material	h_{tc} (W/m · K)
Air	0.023
Aluminum	221
Brick, fired clay	0.9
Concrete	2
Copper	393
Glass-wool insulation	0.0377
Steel, mild	45.3
Wood	0.126

[a]Note that the units are equivalent to J/s · m · K.
Adapted from Kuehn et al. (1998), Shortley and Williams (1955).

where h_c = convective heat transfer coefficient, kJ/s · m^2 · K
A = surface area, m^2

Radiation. Although both conduction and convection require a medium to transport energy, radiant energy is transported by electromagnetic radiation. The radiative transfer of heat involves two processes: the absorption of radiant energy by an object and the radiation of energy by that object. The change in enthalpy due to the radiative heat transfer is the energy absorbed minus the energy emitted and can be expressed as

$$\frac{dH}{dt} = E_{abs} - E_{emitted} \tag{2-52}$$

Thermal radiation is emitted when an electron moves from a higher energy state to a lower one. Radiant energy is transmitted in the form of waves. Waves are cyclical or *sinusoidal* as shown in Figure 2-15. The waves may be characterized by their wavelength (λ) or their frequency (ν). The wavelength is the distance between successive peaks or troughs. Frequency and wavelength are related by the speed of light (c).

$$c = \lambda\nu \tag{2-53}$$

Planck's law relates the energy emitted to the frequency of the emitted radiation.

$$E = h\nu \tag{2-54}$$

where h = Planck's constant = 6.63×10^{-34} J · s

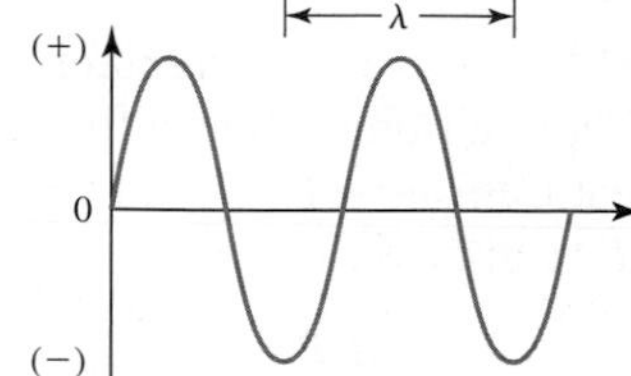

FIGURE 2-15
Sinusoidal wave. The wavelength λ is the distance between two peaks or troughs.

The electromagnetic wave emitted when an electron makes a transition between two energy levels is called a *photon.* When the frequency is high (small wavelengths), the energy emitted is high. Planck's law also applies to the absorption of a photon of energy. A molecule can only absorb radiant energy if the wavelength of radiation corresponds to the difference between two of its energy levels.

Every object emits thermal radiation. The amount of energy radiated depends on the wavelength, surface area, and the absolute temperature of the object. The maximum amount of radiation that an object can emit at a given temperature is called *blackbody radiation.* An object that radiates the maximum possible intensity for every wavelength is called a *blackbody.* The term *blackbody* has no reference to the color of the body. A blackbody can also be characterized by the fact that all radiant energy reaching its surface is absorbed.

Actual objects do not emit or absorb as much radiation as a blackbody. The ratio of the amount of radiation an object emits to that a blackbody would emit is called the emissivity (ε). The energy spectrum of the sun resembles that of a blackbody at 6,000 K. At normal atmospheric temperatures, the emissivity of dry soil and woodland is approximately 0.90. Water and snow have emissivities of about 0.95. A human body, regardless of pigmentation, has an emissivity of approximately 0.97 (Guyton, 1961). The ratio of the amount of energy an object absorbs to that which a blackbody would absorb is called absorptivity (α). For most surfaces, the absorptivity is the same value as the emissivity.

Integration of Planck's equation over all wavelengths yields the radiant energy of a blackbody.

$$E_B = \sigma T^4 \tag{2-55}$$

where E_B = blackbody emission rate, W/m^2
σ = Stephan–Boltzmann constant = 5.67×10^{-8} W/m$^2 \cdot$ K^4
T = absolute temperature, K

For other than blackbodies, the right-hand side of the equation is multiplied by the emissivity.

For a body with an emissivity ε and an absorptivity α, at a temperature T_b receiving radiation from its environment which is a blackbody of temperature $T_{environ}$, we can express the change in enthalpy as

$$\frac{dH}{dt} = A(\varepsilon\sigma T_b^4 - \alpha\sigma T_{environ}^4) \tag{2-56}$$

where A = surface area of the body, m^2.

The solution to thermal radiation problems is highly complex because of the "reradiation" of surrounding objects. In addition, the rate of radiative cooling will change with time as the difference in temperatures changes; initially the change per unit of time will be large because of the large difference in temperature. As the temperatures approach each other the rate of change will slow. In the following problem we use an arithmetic average temperature as a first approximation to the actual average temperature.

Example 2-13. As mentioned in Example 2-12, heat losses due to convection and radiation were not considered. Using the following assumptions, estimate how long it will take for the wastewater and concrete tank to come to the desired temperature (20°C) if radiative cooling and convective cooling are considered. Assume that the average temperature of the water and concrete tank while cooling between 85°C (their combined temperature from Example 2-12) and 20°C is 52.5°C. Also assume that the mean radiant temperature of the surroundings is 20°C, that both the cooling tank and the surrounding environment radiate uniformly in all directions, that their emissivities are the same (0.90), that the surface area of the concrete tank including the open water surface is 56 m^2, and that the convective heat transfer coefficient is 13 $J/s \cdot m^2 \cdot K$.

Solution. The required change in enthalpy for the wastewater is

$$\Delta H = (1{,}000 \text{ kg/m}^3)(40 \text{ m}^3)(4.186 \text{ kJ/kg} \cdot \text{K})(325.65 - 293.15) = 5{,}441{,}800 \text{ kJ}$$

where the absolute temperature of the wastewater is 273.15 + 52.5 = 325.65 K.

The required change in enthalpy of the concrete tank is

$$\Delta H = (42{,}000 \text{ kg})(0.93 \text{ kJ/kg} \cdot \text{K})(325.65 - 293.15) = 1{,}269{,}450 \text{ kJ}$$

For a total of 5,441,800 + 1,269,450 = 6,711,250 kJ, or 6,711,250,000 J

In estimating the time to cool down by radiation alone, we note that the emissivities are the same for the tank and the environment and that the net radiation is the result of the difference in absolute temperatures.

$$\begin{aligned} E_B &= \varepsilon\sigma(T_c^4 - T_{\text{environ}}^4) \\ &= \varepsilon\sigma T^4 = (0.90)(5.67 \times 10^{-8} \text{ W/m}^2 \cdot \text{K}^4)[(273.15 + 52.5)^4 - (273.15 + 20)^4] \\ &= 197 \text{ W/m}^2 \end{aligned}$$

The rate of heat loss is

$$(197 \text{ W/m}^2)(56 \text{ m}^2) = 11{,}032 \text{ W, or } 11{,}032 \text{ J/s}$$

From Equation 2-51, the convective cooling rate may be estimated.

$$\begin{aligned} \frac{dH}{dt} &= h_c A(T_f - T_s) \\ &= (13 \text{ J/s} \cdot \text{m}^2 \cdot \text{K})(56 \text{ m}^2)[(273.15 + 52.5) - (273.15 + 20)] \\ &= 23{,}660 \text{ J/s} \end{aligned}$$

The time to cool down is then

$$\frac{6{,}711{,}250{,}000 \text{ J}}{11{,}032 \text{ J/s} + 23{,}660 \text{ J/s}} = 193{,}452 \text{ s, or } 2.24 \text{ days}$$

This is quite a long time. If land is not at a premium and several tanks can be built, then the time may not be a relevant consideration. Alternatively, other options must be considered to reduce the time. One alternative is to utilize conductive heat transfer and build a heat exchanger. The heat exchanger could be used to heat the incoming water needed in the blanching process.

Overall Heat Transfer. Most practical heat transfer problems involve multiple heat transfer modes. For these cases, it is convenient to use an overall heat transfer coefficient that incorporates multiple modes. The form of the heat transfer equation then becomes

$$\frac{dH}{dt} = h_0 A(\Delta T) \tag{2-57}$$

where h_0 = overall heat transfer coefficient, kJ/s · m^2 · K
ΔT = temperature difference that drives the heat transfer, K

Among their many responsibilities, environmental scientists (often with a job title of environmental sanitarian) are responsible for checking food safety in restaurants. This includes ensuring proper refrigeration of perishable foods. The following problem is an example of one of the items, namely the electrical rating of the refrigerator, that might be investigated in a case of food poisoning at a family gathering.

Example 2-14. In evaluating a possible food "poisoning," Sam and Janet Evening evaluated the required electrical energy input to cool food purchased for a family reunion. The family purchased 12 kg of hamburger, 6 kg of chicken, 5 kg of corn, and 20 L of soda pop. They have a refrigerator in the garage in which they stored the food until the reunion. The specific heats of the food products (in kJ/kg · K) are hamburger: 3.22; chicken: 3.35; corn: 3.35; beverages: 4.186. The refrigerator dimensions are 0.70 m × 0.75 m × 1.00 m. The overall heat transfer coefficient for the refrigerator is 0.43 J/s · m^2 · K. The temperature in the garage is 30°C. The food must be kept at 4°C to prevent spoilage. Assume that it takes 2 h for the food to reach a temperature of 4°C, that the meat has risen to 20°C in the time it takes to get it home from the store, and that the soda pop and corn have risen to 30°C. What electrical energy input (in kilowatts) is required during the first 2 h the food is in the refrigerator? What is the energy input required to maintain the temperature for the second 2 h. Assume the refrigerator interior is at 4°C when the food is placed in it and that the door is not opened during the 4-h period. Ignore the energy required to heat the air in the refrigerator, and assume that all the electrical energy is used to remove heat. If the refrigerator is rated at 875 W, is poor refrigeration a part of the food poisoning problem?

Solution. The energy balance equation is of the form

$$\frac{dH}{dt} = \frac{d(H)_{\text{mass in}}}{dt} + \frac{d(H)_{\text{mass out}}}{dt} \pm \frac{d(H)_{\text{energy flow}}}{dt}$$

where dH/dt = the enthalpy change required to balance the input energy
$d(H)_{\text{mass in}}$ = change in enthalpy due to the food
$d(H)_{\text{energy flow}}$ = the change in enthalpy to maintain the temperature at 4°C

There is no $d(H)_{\text{mass out}}$.

Begin by computing the change in enthalpy for the food products.

Hamburger

$$\Delta H = (12 \text{ kg})(3.22 \text{ kJ/kg} \cdot \text{K})(20°\text{C} - 4°\text{C}) = 618.24 \text{ kJ}$$

Chicken

$$\Delta H = (6 \text{ kg})(3.35 \text{ kJ/kg} \cdot \text{K})(20°\text{C} - 4°\text{C}) = 321.6 \text{ kJ}$$

Corn

$$\Delta H = (5 \text{ kg})(3.35 \text{ kJ/kg} \cdot \text{K})(30°\text{C} - 4°\text{C}) = 435.5 \text{ kJ}$$

Beverages

Assuming that 20 L = 20 kg

$$\Delta H = (20 \text{ kg})(4.186 \text{ kJ/kg} \cdot \text{K})(30°\text{C} - 4°\text{C}) = 2{,}176.72 \text{ kJ}$$

The total change in enthalpy = 618.24 kJ + 321.6 kJ + 435.5 kJ + 2,176.72 kJ = 3,552.06 kJ

Based on a 2-h period to cool the food, the rate of enthalpy change is

$$\frac{3{,}552.06 \text{ kJ}}{(2 \text{ h})(3{,}600 \text{ s/h})} = 0.493, \quad \text{or } 0.50 \text{ kJ/s}$$

The surface area of the refrigerator is

$$0.70 \text{ m} \times 1.00 \text{ m} \times 2 = 1.40 \text{ m}^2$$
$$0.75 \text{ m} \times 1.00 \text{ m} \times 2 = 1.50 \text{ m}^2$$
$$0.75 \text{ m} \times 0.70 \text{ m} \times 2 = 1.05 \text{ m}^2$$

for a total of 3.95 m^2

The heat loss through walls of the refrigerator is then

$$\frac{dH}{dt} = (4.3 \times 10^{-4} \text{ kJ/s} \cdot \text{m}^2 \cdot \text{K})(3.95 \text{ m}^2)(30°\text{C} - 4°\text{C}) = 0.044 \text{ kJ/s}$$

In the first 2 h, the electrical energy required is 0.044 kJ/s + 0.50 kJ/s = 0.54 kJ/s Because 1 W = 1 J/s, the electrical requirement is 0.54 kW, or 540 W. It does not appear that the refrigerator is part of the food poisoning problem.

In the second 2 h, the electrical requirement drops to 0.044 kW, or 44 W.

Note that at the beginning of this example we put "poisoning" in quotation marks because illness from food spoilage may be the result of microbial infection, which is not poisoning in the same sense as, for example, that caused by arsenic.

The result of Example 2-14 is based on an assumption of 100 percent efficiency in converting electrical energy to refrigeration. This is, of course, not possible and leads us to the second law of thermodynamics.

Second Law of Thermodynamics

The second law of thermodynamics states that energy flows from a region of higher concentration to one of lesser concentration, not the reverse, and that the quality degrades as it is transformed. All natural, spontaneous processes may be studied in the light of the second law, and in all such cases a particular one-sidedness is found. Thus, heat always flows spontaneously from a hotter body to a colder one, gases seep through an opening spontaneously from a region of higher pressure to a region of lower pressure. The second law recognizes that order becomes disorder, that randomness increases, and that structure and concentrations tend to disappear. It foretells elimination of gradients, equalization of electrical and chemical potential, and leveling of contrasts in heat and molecular motion unless work is done to prevent it. Thus, gases and liquids left by themselves tend to mix, not to unmix; rocks weather and crumble; iron rusts.

The degradation of energy as it is transformed means that enthalpy is wasted in the transformation. The fractional part of the heat which is wasted is termed unavailable energy. A mathematical expression called the *change in entropy* is used to express this unavailable energy.

$$\Delta s = Mc_p \ln\frac{T_2}{T_1} \tag{2-58}$$

where Δs = change in entropy
M = mass
c_p = specific heat at constant pressure
T_1, T_2 = initial and final absolute temperature
ln = natural logarithm

By the second law, entropy increases in any transformation of energy from a region of higher concentration to a lesser one. The higher the degree of disorder, the higher the entropy. Degraded energy is entropy, dissipated as waste products and heat.

Efficiency (η) or, perhaps, lack of efficiency is another expression of the second law. Sadi Carnot (1824) was the first to approach the problem of the efficiency of a heat engine (e.g., a steam engine) in a truly fundamental manner. He described a theoretical engine, now called a Carnot engine. Figure 2-16 is a simplified representation

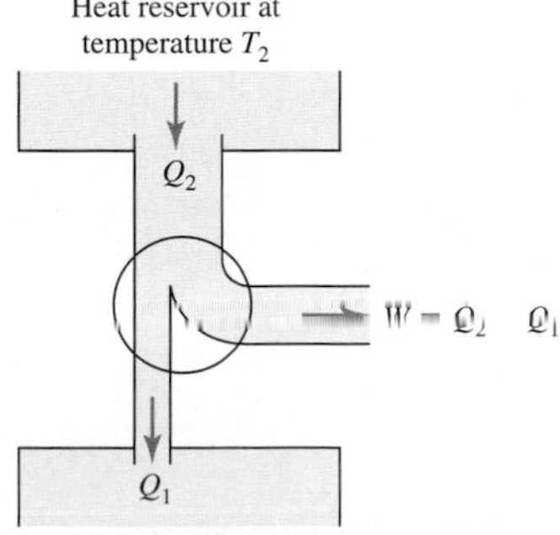

FIGURE 2-16
Schematic flow diagram of a Carnot heat engine.
(Richards et al., 1960.)

of a Carnot engine. In his engine, a material expands against a piston that is periodically brought back to its initial condition so that in any one cycle the change in internal energy of this material is zero, that is $U_2 - U_1 = 0$, and the first law of thermodynamics (Equation 2-41) reduces to

$$W = Q_2 - Q_1 \tag{2-59}$$

where Q_1 = heat rejected or exhaust heat
Q_2 = heat input

Thermal efficiency is the ratio of work output to heat input. The output is mechanical work. The exhaust heat is not considered part of the output.

$$\eta = \frac{W}{Q_2} \tag{2-60}$$

where W = work output
Q_2 = heat input

or, from Equation 2-59

$$\eta = \frac{Q_2 - Q_1}{Q_2} \tag{2-61}$$

Carnot's analysis revealed that the most efficient engine will have an efficiency of

$$\eta_{\max} = 1 - \frac{T_1}{T_2} \tag{2-62}$$

where the temperatures are absolute temperatures (in kelvins). This equation implies that maximum efficiency is achieved when the value of T_2 is as high as possible and the value for T_1 is as low as possible.

A refrigerator may be considered to be a heat engine operated in reverse (Figure 2-17). From an environmental point of view, the best refrigeration cycle is one that removes the greatest amount of heat (Q_1) from the refrigerator for the least expenditure of mechanical work. Thus, we use the *coefficient of performance* rather than efficiency.

$$\text{C.O.P} = \frac{Q}{W} = \frac{Q_1}{Q_2 - Q_1} \tag{2-63}$$

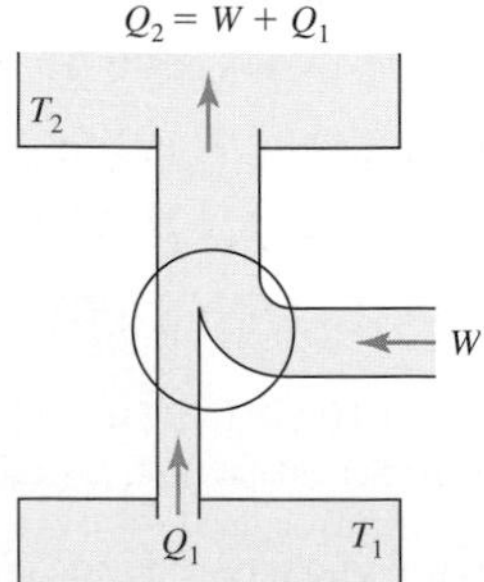

FIGURE 2-17
Schematic flow diagram of a refrigerator. (Richards et al., 1960.)

By analogy to the Carnot efficiency,

$$\text{C.O.P} = \frac{T_1}{T_2 - T_1} \tag{2-64}$$

Example 2-15. What is the coefficient of performance of the refrigerator in Example 2-14?

Solution. The C.O.P. is calculated directly from the temperatures.

$$\text{C.O.P} = \frac{273.15 + 4}{[(273.15 + 30) - (273.15 + 4)]} = 10.7$$

Note that in contrast to the heat engine, the performance increases if the temperatures are close together.

2-5 CHAPTER REVIEW

When you have completed studying the chapter, you should be able to do the following without the aid of your textbook or notes:

1. Define the law of conservation of matter (mass).
2. Explain the circumstances under which the law of conservation of matter is violated.
3. Draw a materials-balance diagram given the inputs, outputs, and accumulation or the relationship between the variables.
4. Define the following terms: rate, conservative pollutants, reactive chemicals, steady-state conditions, equilibrium, completely mixed systems, and plug-flow systems.
5. Explain why the effluent from a completely mixed system has the same concentration as the system itself.
6. Define the first law of thermodynamics and provide one example.
7. Define the second law of thermodynamics and provide one example.
8. Define energy, work, power, specific heat, phase change, enthalpy of fusion, enthalpy of evaporation, photon, and blackbody radiation.
9. Explain how the energy-balance equation differs from the materials-balance equation.
10. List the three mechanisms of heat transfer and explain how they differ.
11. Explain the relationship between energy transformation and entropy.

With the aid of this text you should be able to do the following:

1. Write and solve mass-balance equations for systems with and without transformation.
2. Write the mathematical expression for the decay of a substance by first-order kinetics with respect to the substance.
3. Solve first-order reaction problems.
4. Compute the change in enthalpy for a substance.
5. Solve heat transfer equations for conduction, convection, and radiation individually and in combination.
6. Write and solve energy balance equations.
7. Compute the change in entropy.
8. Compute the Carnot efficiency for a heat engine.
9. Compute the coefficient of performance for a refrigerator.

2-6 PROBLEMS

2-1. A sanitary landfill has available space of 16.2 ha at an average depth of 10 m. Seven hundred sixty-five (765) cubic meters of solid waste are dumped at the site 5 days per week. This waste is compacted to twice its delivered density. Draw a mass-balance diagram and estimate the expected life of the landfill in years.

Answer: 16.29 or 16 years

2-2. Each month the Speedy Dry Cleaning Company buys 1 barrel (0.160 m^3) of dry cleaning fluid. Ninety percent of the fluid is lost to the atmosphere and 10 percent remains as residue to be disposed of. The density of the dry cleaning fluid is 1.5940 g/mL. Draw a mass-balance diagram and estimate the monthly mass emission rate to the atmosphere in kg/mo.

2-3. Congress banned the production of the Speedy Dry Cleaning Company's dry cleaning fluid in 2000. Speedy is using a new cleaning fluid. The new dry cleaning fluid has one-sixth the volatility of the former dry cleaning fluid (Problem 2-2). The density of the new fluid is 1.6220 g/mL. Assume that the amount of residue is the same as that resulting from the use of the old fluid and estimate the mass emission rate to the atmosphere in kg/mo. Because the new dry cleaning fluid is less volatile, the company will have to purchase less per year. Estimate the annual volume of dry cleaning fluid saved (in m^3/y).

2-4. Gasoline vapors are vented to the atmosphere when an underground gasoline storage tank is filled. If the tanker truck discharges into the top of the tank with no vapor control (known as the splash fill method), the emission of gasoline vapors is estimated to be 2.75 kg/m^3 of gasoline delivered to the tank. If the tank is equipped with a pressure relief valve and interlocking hose connection and the tanker truck discharges into the bottom of the tank below the surface of the gasoline in the storage tank, the emission of gasoline vapors is estimated to be 0.095 kg/m^3 of gasoline delivered (Wark et al., 1998). Assume that the service station must refill the tank with 4.00 m^3 of gasoline once a week. Draw a mass-balance diagram and estimate the annual loss of gasoline vapor (in kg/y) for the splash fill method. Estimate the value of the fuel that is captured if the vapor control system is used. Assume the density of the condensed vapors is 0.800 g/mL and the cost of the gasoline is $1.06 per liter.

2-5. The Rappahannock River near Warrenton, VA, has a flow rate of 3.00 m^3/s. Tin Pot Run (a pristine stream) discharges into the Rappahannock at a flow rate of 0.05 m^3/s. To study mixing of the stream and river, a conservative tracer is to be added to Tin Pot Run. If the instruments that can measure the tracer can detect a concentration of 1.0 mg/L, what minimum concentration must be achieved in Tin Pot Run so that 1.0 mg/L of tracer can be measured after the river and stream mix? Assume that the 1.0 mg/L of tracer is to be measured after complete mixing of the stream and Rappahannock has been achieved and that no tracer is in Tin Pot Run or the Rappahannock above the point where the two streams mix. What mass rate (kg/d) of tracer must be added to Tin Pot Run?

Answer: 263.52 or 264 kg/d

2-6. The Clearwater water treatment plant uses sodium hypochlorite (NaOCl) to disinfect the treated water before it is pumped to the distribution system. The NaOCl is purchased in a concentrated solution (52,000 mg/L) that must be diluted before it is injected into the treated water. The dilution piping scheme is shown in Figure P-2-6. The NaOCl is pumped from the small tank (called a "day tank") into a small pipe carrying a portion of the clean treated water (called a "slip stream") to the main service line. The main service line carries a flow rate of 0.50 m^3/s. The slip stream flows at 4.0 L/s. At what rate of flow (in L/s) must the NaOCl from the day tank be pumped into the slip stream to achieve a concentration of 2.0 mg/L of NaOCl in the main service line? Although it is reactive, you may assume that NaOCl is not reactive for this problem.

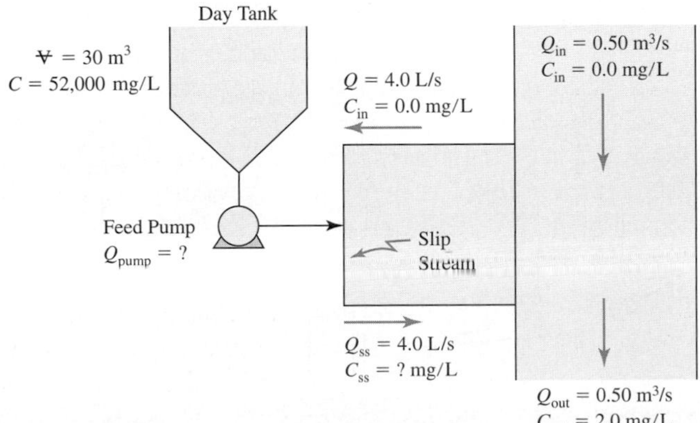

FIGURE P-2-6
Dilution piping scheme.

2-7. The Clearwater design engineer cannot find a reliable pump to move the NaOCl from the day tank into the slip stream (Problem 2-6). Therefore, she specifies in the operating instructions that the day tank be used to dilute the concentrated NaOCl solution so that a pump rated at 1.0 L/s may be used. The tank is to be filled once each shift (8 h per shift). It has a volume of 30 m^3. Determine the concentration of NaOCl that is required in the day tank if the feed rate of NaOCl must be 1,000 mg/s. Calculate the volume of concentrated solution and the volume of water that is to be added for an 8 hour operating period. Although it is reactive, you may assume that NaOCl is not reactive for this problem.

2-8. In water and wastewater treatment processes a filtration device may be used to remove water from the sludge formed by a precipitation reaction. The initial concentration of sludge from a softening reaction (Chapter 4) is 2 percent (20,000 mg/L) and the volume of sludge is 100 m^3. After filtration the sludge solids concentration is 35 percent. Assume that the sludge does not change density during filtration, and that liquid removed from the sludge contains no sludge. Using the mass balance method, determine the volume of sludge after filtration.

2-9. The U.S. EPA requires hazardous waste incinerators to meet a standard of 99.99 percent destruction and removal of organic hazardous constituents injected into the incinerator. This efficiency is referred to as "four nines DRE." For especially toxic waste the DRE must be "six nines." The efficiency is to be calculated by measuring the mass flow rate of organic constituent entering the incinerator and the mass flow rate of constituent exiting the incinerator stack. A schematic of the process is shown in Figure P-2-9. One of the

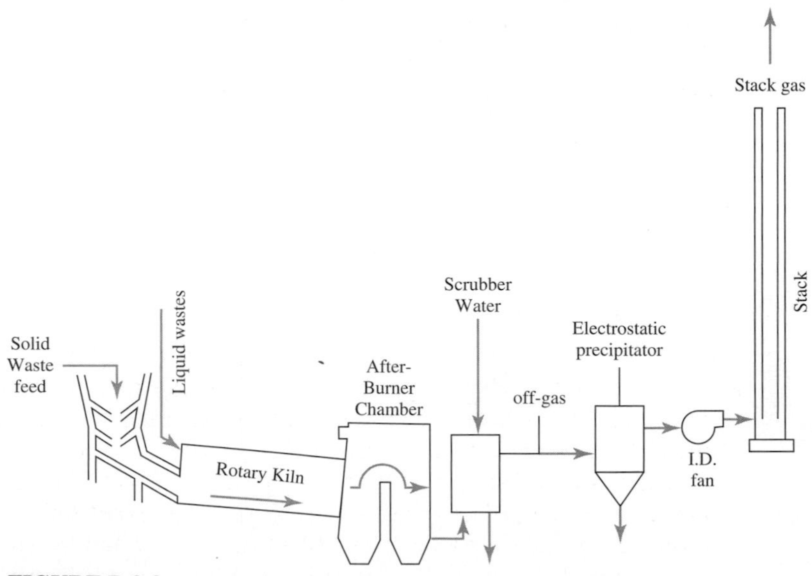

FIGURE P-2-9
Schematic of hazardous waste incinerator.

difficulties of assuring these levels of destruction is the ability to measure the contaminant in the exhaust gas. Draw a mass balance diagram for the process and determine the allowable quantity of contaminant in the exit stream if the incinerator is burning 1.0000 g/s of hazardous constituent. (*Note:* the number of significant figures is very important in this calculation.) If the incinerator is 90 percent efficient in destroying the hazardous constituent, what scrubber efficiency is required to meet the standard?

2-10. A new high-efficiency air filter has been designed to be used in a secure containment facility to do research on detection and destruction of anthrax. Before the filter is built and installed it needs to be tested. It is proposed to use ceramic microspheres of the same diameter as the anthrax spores for the test. One obstacle in the test is that the efficiency of the sampling equipment is unknown and cannot be readily tested because the rate of release of microspheres cannot be sufficiently controlled to define the number of microspheres entering the sampling device. The engineers propose the test apparatus shown in Figure P-2-10. The sampling filters capture the microspheres on a membrane filter that allows microscopic counting of the captured particles. At the end of the experiment, the number of particles on the first filter is 1,941 and the number on the second filter is 63. Assuming each filter has the same efficiency, estimate the efficiency of the sampling filters. (*Note:* this problem is easily solved by using the particle counts (C_1, C_2, C_3) and efficiency η.)

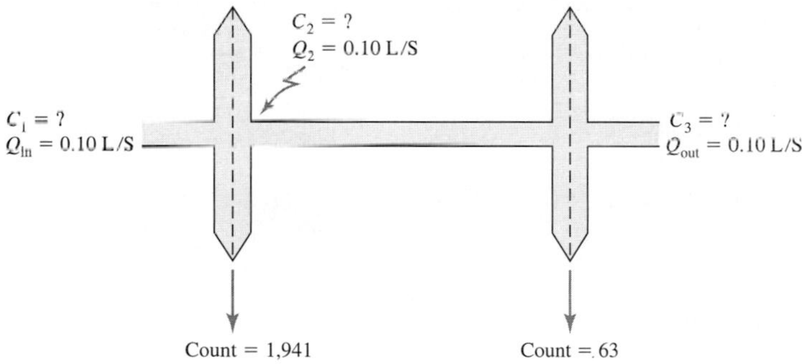

FIGURE P-2-10
Filtration test apparatus.

2-11. To remove the solution containing metal from a part after metal plating the part is commonly rinsed with water. This rinse water is contaminated with metal and must be treated before discharge. The Shiny Metal Plating Co. uses the process flow diagram shown in Figure P-2-11. The plating solution contains 85 g/L of nickel. The parts drag out 0.05 L/min of plating solution into the rinse tank. The flow of rinse water into the rinse tank is 150 L/min. Write the general mass balance equation for the rinse tank and estimate the concentration of nickel in the wastewater stream that must be

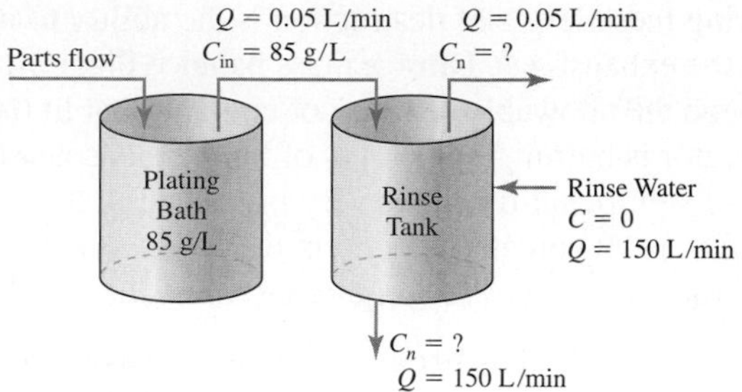

FIGURE P-2-11
Plating rinse water flow scheme.

treated. Assume that the rinse tank is completely mixed and that no reactions take place in the rinse tank.

Answer: $C_n = 28.3$ or 28 mg/L

2-12. Because the rinse water flow rate for a nickel plating bath (Problem 2-11) is quite high, it is proposed that the countercurrent rinse system shown in Figure P-2-12 be used to reduce the flow rate. Assuming that the C_n concentration remains the same at 28 mg/L, estimate the new flow rate. Assume that the rinse tank is completely mixed and that no reactions take place in the rinse tank.

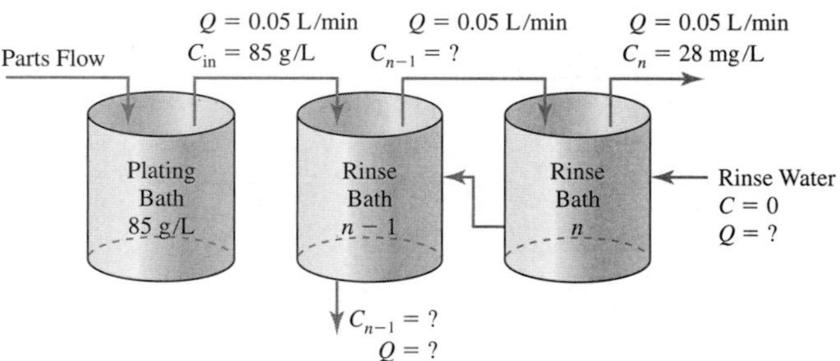

FIGURE P-2-12
Countercurrent rinse water flow scheme.

2-13. The Environmental Protection Agency (U.S. EPA, 1982) offers the following equation to estimate the flow rate for counter current rinsing (Figure P-2-12):

$$Q = \left[\left(\frac{C_{in}}{C_n}\right)^{1/n} + \frac{1}{n}\right]q$$

where Q = rinse water flow rate, L/min
C_{in} = concentration of metal in plating bath, mg/L
C_n = concentration of metal in nth rinse bath, mg/L
n = number of rinse tanks
q = flow rate of liquid dragged out of a tank by the parts, L/min

Using the EPA equation and the data from Problem 2-11, calculate the rinse water flow rate for one, two, three, four and five rinse tanks in series using a computer spread sheet you have written. Use the spreadsheet graphing function to plot a graph of rinse water flow rate versus number of rinse tanks.

2-14. If biodegradable organic matter, oxygen, and microorganisms are placed in a closed bottle, the microorganisms will use the oxygen in the process of oxidizing the organic matter. The bottle may be treated as a batch reactor and the decay of oxygen may be treated as a first-order reaction. Write the general mass balance equation for the bottle. Using a computer spreadsheet program you have written, calculate and then plot the concentration of oxygen each day for a period of 5 days starting with a concentration of 8 mg/L. Use a rate constant of 0.35 d^{-1}.

Answer: day 1 = 5.64 or 5.6 mg/L; day 2 = 3.97 or 4.0 mg/L

2-15. In 1908, H. Chick reported an experiment in which he disinfected anthrax spores with a 5 percent solution of phenol (Chick, 1908). The results of his experiment are tabulated below. Assuming the experiment was conducted in a completely mixed batch reactor, determine the decay rate constant for the die-off of anthrax.

Concentration of Survivors (number/mL)	Time (min)
398	0
251	30
158	60

2-16. A water tower containing 4,000 m^3 of water has been taken out of service for installation of a chlorine monitor. The concentration of chlorine in the water tower was 2.0 mg/L when the tower was taken out of service. If the chlorine decays by first-order kinetics with a rate constant k = 1.0 d^{-1} (Grayman and Clark, 1993), what is the chlorine concentration when the tank is put back in service 8 hours later? What mass of chlorine (in kg) must be added to the tank to raise the chlorine level back to 2.0 mg/L? Although it is not completely mixed, you may assume the tank is a completely mixed batch reactor.

2-17. The concept of "half-life" is used extensively in environmental engineering and science. For example, it is used to describe the decay of radioisotopes, elimination of poisons from people, self-cleaning of lakes, and the disappearance of pesticides from soil. Starting with the mass balance equation, develop an expression that describes the half-life ($t_{1/2}$) of a substance in terms of the reaction rate constant k, assuming the decay reaction takes place in a batch reactor.

2-18. If the initial concentration of a reactive substance in a batch reactor is 100 percent, determine the amount of substance remaining after 1, 2, 3, and 4 half-lives if the reaction rate constant is 6 mo^{-1}.

2-19. Liquid hazardous wastes are blended in a CMFR to maintain a minimum energy content before burning them in a hazardous waste incinerator. The energy content of the waste currently being fed is 8.0 MJ/kg (megajoules/kilogram). A new waste is injected in the flow line into the CMFR. It has an energy content of 10.0 MJ/kg. If the flow rate into and out of the 0.20 m^3 CMFR is 4.0 L/s, how long will it take the effluent from the CMFR to reach an energy content of 9 MJ/kg?

Answer: $t = 34.5$ or 35 s

2-20. Repeat Problem 2-19 with a new waste having an energy content of 12 MJ/kg instead of 10 MJ/kg.

2-21. An instrument is installed along a major water distribution pipe line to detect potential contamination from terrorist threats. A 2.54-cm-diameter pipe connects the instrument to the water distribution pipe. The connecting pipe is 20.0 m long. Water from the distribution pipe is pumped through the instrument and then discharged to a holding tank for verification analysis and proper disposal. If the flow rate of the water in the sample line is 1.0 L/min, how many minutes will it take a sample from the distribution pipe to reach the instrument. Use the following relationship to determine the speed of the water in the sample pipe:

$$u = \frac{Q}{A}$$

where u = speed of water in pipe, m/s
Q = flow rate of water in pipe, m^3/s
A = area of pipe, m^2

If the instrument uses 10 mL for sample analysis, how many liters of water must pass through the sampler before it detects a contaminant in the pipe?

2-22. A bankrupt chemical firm has been taken over by new management. On the property they found a 20,000 m^3 brine pond containing 25,000 mg/L of salt. The new owners propose to flush the pond into their discharge pipe leading to the Atlantic ocean, which has a salt concentration above 30,000 mg/L. What flow rate of freshwater (in m^3/s) must they use to reduce the salt concentration in the pond to 500 mg/L within one year?

Answer: $Q = 0.0025$ m^3/s

2-23. A 1,900-m^3 water tower has been cleaned with a chlorine solution. The vapors of chlorine in the tower exceed allowable concentrations for the work crew to enter and finish repairs. If the chlorine concentration is 15 mg/m^3 and the allowable concentration is 0.0015 mg/L, how long must the workers vent the tank with clean air flowing at 2.35 m^3/s? Although chlorine is a reactive substance, you may consider it nonreactive for this problem.

2-24. A railroad tank car is derailed and ruptured. It discharges 380 m^3 of pesticide into the Mud Lake drain. As shown in Figure P-2-24, the drain flows into Mud Lake which has a liquid volume of 40,000 m^3. The water in the

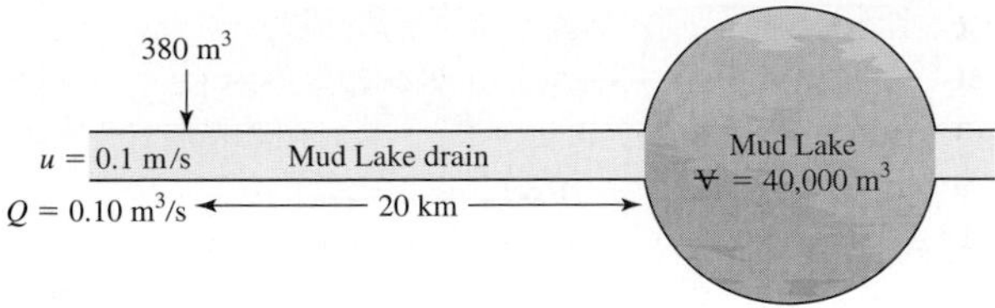

FIGURE P-2-24
Drain flow into Mud lake.

creek has a flow rate of 0.10 m^3/s, a velocity of 0.10 m/s, and the distance from the spill site to the pond is 20 km. Assume that the spill is short enough to treat the injection of the pesticide as a pulse, that the pond behaves as a flow balanced CMFR, and that the pesticide is nonreactive. Estimate the time it will take to flush 99 percent of the pesticide from the pond.

Answer: Time to reach the pond = 2.3 days;
time to flush = 21.3 or 21 days

2-25. During a snow storm the fluoride feeder in North Bend runs out of feed solution. As shown in Figure P-2-25, the rapid-mix tank is connected to a 5-km-long distribution pipe. The flow rate into the rapid-mix tank is 0.44 m^3/s and the volume of the tank is 2.50 m^3. The velocity in the pipe is 0.17 m/s. If the fluoride concentration in the rapid-mix tank is 1.0 mg/L when the feed stops, how long will it be until the concentration of fluoride is reduced to 0.01 mg/L at the end of the distribution pipe? The fluoride may be considered a nonreactive chemical.

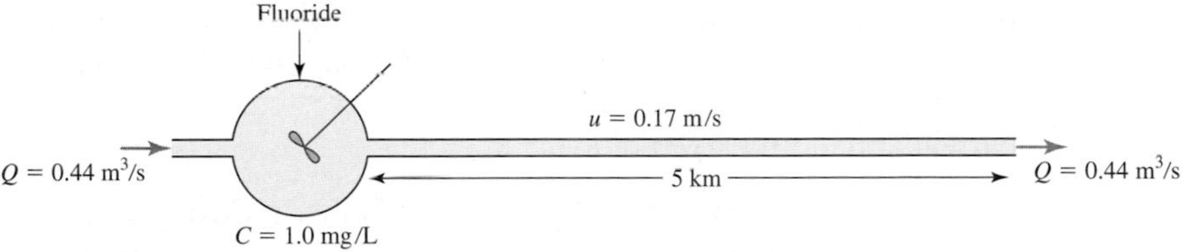

FIGURE P-2-25
Fluoride feeder.

2-26. A sewage lagoon that has a surface area of 10 ha and a depth of 1 m is receiving 8,640 m^3/d of sewage containing 100 mg/L of biodegradable contaminant. At steady state, the effluent from the lagoon must not exceed 20 mg/L of biodegradable contaminant. Assuming the lagoon is well mixed and that there are no losses or gains of water in the lagoon other than the sewage input, what biodegradation reaction rate coefficient (d^{-1}) must be achieved for a first order reaction?

Answer: $k = 0.3478$ or 0.35 d^{-1}

2-27. Repeat Problem 2-26 with two lagoons in series (see Figure P-2-27). Each lagoon has a surface area of 5 ha and a depth of 1 m.

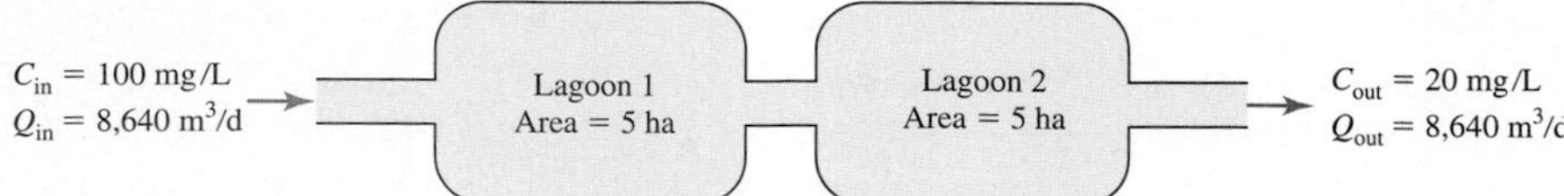

FIGURE P-2-27
Two lagoons in series.

2-28. Using a spreadsheet program you have written, determine the effluent concentration if the process producing sewage in Problem 2-26 shuts down ($C_{in} = 0$). Calculate and plot points at 1 day intervals for 10 days. Utilize the graphing function of the spreadsheet to construct your plot.

2-29. A 90-m^3 basement in a residence is found to be contaminated with radon coming from the ground through the floor drains. The concentration of radon in the room is 1.5 Bq/L under steady-state conditions. The room behaves as a CMFR and the decay of radon is a first order reaction with a decay rate constant of $2.09 \times 10^{-6}\ s^{-1}$. If the source of radon is closed off and the room is vented with radon free air at a rate of 0.14 m^3/s, how long will it take to lower the radon concentration to an acceptable level of 0.15 Bq/L?

2-30. An ocean outfall diffuser that discharges treated wastewater into the Pacific ocean is 5,000 m from a public beach. The wastewater contains 10^5 coliform bacteria per milliliter. The wastewater discharge flow rate is 0.3 m^3/s. The coliform first-order death rate in seawater is approximately 0.3 h^{-1} (Tchobanoglous and Schroeder, 1985). The current carries the wastewater plume toward the beach at a rate of 0.5 m/s. The ocean current may be approximated as a pipe carrying 600 m^3/s of seawater. Determine the coliform concentration at the beach. Assume that the current behaves as a plug-flow reactor and that the wastewater is completely mixed in the current at the discharge point.

2-31. For the following conditions, determine whether a CMFR or a PFR is more efficient in removing a reactive compound from the waste stream under steady-state conditions with a first-order reaction: reaction volume = 280 m^3, flow rate = 14 m^3/d, and reaction rate coefficient = 0.05 d^{-1}.

Answer: CMFR $\eta = 50\%$; PFR $\eta = 63\%$

2-32. Compare the reactor volume required to achieve 95 percent efficiency for a CMFR and a PFR for the following conditions: steady-state, first-order reaction, flow rate = 14 m^3/d, and reaction rate coefficient = 0.05 d^{-1}.

2-33. The discharge pipe from a sump pump in the dry well of a sewage lift station did not drain properly, and the water at the discharge end of the pipe froze. A hole has been drilled into the ice and a 200-W electric heater has been inserted in the hole. If the discharge pipe contains 2 kg of ice, how long will it take to melt the ice? Assume all the heat goes into melting the ice.

Answer: 55.5 or 56 min

2-34. As noted in Examples 2-12 and 2-13, the time to achieve the desired temperature using the cooling tank is quite long. An evaporative cooler is proposed as an alternative means of reducing the temperature. Estimate the amount of water (in m^3) that must evaporate each day to lower the temperature of the 40 m^3 of wastewater from 100°C to 20°C. (*Note:* While the solution to this problem is straightforward, the design of an evaporative cooling tower is a complex thermodynamic problem made even more complicated in this case by the contents of the wastewater that would potentially foul the cooling system.)

2-35. The water in a biological wastewater treatment system must be heated from 15°C to 40°C for the microorganisms to function. If the flow rate of the wastewater into the process is 30 m^3/d, at what rate must heat be added to the wastewater flowing into the treatment system? Assume the treatment system is completely mixed and that there are no heat losses once the wastewater is heated.

Answer: 3.14 GJ/d

2-36. The lowest flow in the Menominee River in July is about 40 m^3/s. If the river temperature is 18°C and a power plant discharges 2 m^3/s of cooling water at 80°C, what is the final river temperature after the cooling water and the river have mixed? Ignore radiative and convective losses to the atmosphere as well as conductive losses to the river bottom and banks.

2-37. The flow rate of the Seine in France is 28 m^3/s at low flow. A power plant discharges 10 m^3/s of cooling water into the Seine. In the summer the river temperature upstream of the power plant reaches 20°C. The temperature of the river after the power plant discharge mixes with the river is 27°C (Goubet, 1969). Estimate the temperature of the cooling water before it is mixed with the river water. Ignore radiative and convective losses to the atmosphere as well as conductive losses to the river bottom and banks.

2-38. An aerated lagoon (a sewage treatment pond that is mixed with air) is being proposed for a small lake community in northern Wisconsin. The lagoon must be designed for the summer population but will operate year-round. The winter population is about half of the summer population. The volume of the proposed lagoon, based on these design assumptions, is 3,420 m^3. The daily volume of sewage in the winter is estimated to be 300 m^3. In January, the temperature of the lagoon drops to 0°C but it is not yet frozen. If the temperature of the wastewater flowing into the lagoon is 15°C, estimate the temperature of the lagoon at the end of a day. Assume the lagoon is completely mixed and that there are no losses to the atmosphere or the lagoon walls or floor. Also assume that the sewage has a density of 1,000 kg/m^3 and a specific heat of 4.186 $kJ/kg \cdot K$.

2-39. Using the data in Problem 2-38 and a spreadsheet program you have written, estimate the temperature of the lagoon at the end of each day for a period of 7 days. Assume that the flow leaving the lagoon equals the flow entering the lagoon and that the lagoon is completely mixed.

2-40. A cooling water pond is to be constructed for a power plant that discharges 17.2 m^3/s of cooling water. Estimate the required surface area of the pond if the water temperature is to be lowered from 45.0°C at its inlet to 35.5°C at its outlet. Assume an overall heat transfer coefficient of 0.0412 $kJ/s \cdot m^2 \cdot K$ (Edinger et al., 1968). (*Note:* the cooling water will be mixed with river water after it is cooled. The mixture of the 35°C water and the river water will meet thermal discharge standards.)

Answer: 174.76 or 175 ha

2-41. A small building that shelters a water supply pump measures 4 m $\times$6 m $\times$ 2.4 m. It is constructed of 1-cm-thick wood having a thermal conductivity of 0.126 $W/m \cdot K$. The inside walls are to be maintained at 10°C when the outside temperature is -18°C. How much heat must be supplied each hour to maintain the desired temperature? How much heat must be supplied if the walls are lined with 10 cm of glass-wool insulation having a thermal conductivity of 0.0377 $W/m \cdot K$? Neglect the wood in the second calculation.

2-42. Because the sewage in the lagoon in Problem 2-38 is violently mixed, there is a good likelihood that the lagoon will freeze. Estimate how long it will take to freeze the lagoon if the temperature of the wastewater in the lagoon is 15°C and the air temperature is -8°C. The pond is 3 m deep. Although the aeration equipment will probably freeze before all of the wastewater in the lagoon is frozen, assume that the total volume of wastewater freezes. Use an overall heat transfer coefficient of 0.5 $kJ/s \cdot m^2 \cdot K$ (Metcalf & Eddy, 2003). Ignore the enthalpy of the influent wastewater.

2-43. Bituminous coal has a heat of combustion* of 31.4 MJ/kg. In the United States, the average coal-burning utility produces an average of 2.2 kWh of electrical energy per kilogram of bituminous coal burned. What is the average overall efficiency of this production of electricity?

2-7 DISCUSSION QUESTIONS

2-1. A piece of limestone rock ($CaCO_3$) at the bottom of Lake Superior is slowly dissolving. For the purpose of calculating a mass balance, you can assume:

(a) The system is in equilibrium.
(b) The system is at steady state.
(c) Both of the above.
(d) Neither of the above.

Explain your reasoning.

---

* Heat of combustion is the amount of energy released per unit mass when the compound reacts completely with oxygen. The mass does not include the mass of oxygen.

2-2. A can of a volatile chemical (benzene) has spilled into a small pond. List the data you would need to gather to calculate the concentration of benzene in the stream leaving the pond using the mass balance technique.

2-3. In Table 2-3, specific heat capacities for common substances, the values for c_p for beef, corn, human beings, and poultry are considerably higher than those for aluminum, copper and iron. Explain why.

2-4. If you hold a beverage glass whose contents are at 4°C, "You can feel the cold coming into your hand." Thermodynamically speaking is this statement true? Explain.

2-5. If you walk barefoot across a brick floor and a wood floor, the brick floor will feel cooler even though the room temperature is the same for both floors. Explain why.

2-8 REFERENCES

Chick, H. (1908) "An Investigation of the Laws of Disinfection," *Journal of Hygiene*, p. 698.

Edinger, J. E., D. K. Brady, and W. L. Graves (1968) "The Variation of Water Temperatures Due to Steam-Electric Cooling Operations," *Journal of Water Pollution Control Federation,* vol. 40, no. 9, pp. 1637–1639.

Gates, D. M. (1962) *Energy Exchange in the Biosphere,* Harper & Row, New York, p. 70.

Goubet, A. (1969) "The Cooling of Riverside Thermal-Power Plants," in F. L. Parker and P. A. Krenkel (eds.), *Engineering Aspects of Thermal Pollution*, Vanderbilt University Press, Nashville, p. 119.

Grayman, W. A. and R. M. Clark (1993) "Using Computer Models to Determine the Effect of Storage on Water Quality," *Journal of the American Water Works Association*, vol. 85, no. 7, pp. 67–77.

Guyton, A. C. (1961) *Textbook of Medical Physiology,* 2nd ed., W. B. Saunders, Philadelphia, pp. 920–921, 950–953.

Hudson, R. G. (1959) *The Engineers' Manual*, John Wiley & Sons, New York, p. 314.

Kuehn, T. H., J. W. Ramsey, and J. L. Threkeld (1998) *Thermal Environmental Engineering*, Prentice Hall, Upper Saddle River, NJ, pp. 425–427.

Masters, G. M. (1998) *Introduction to Environmental Engineering and Science*, Prentice Hall, Upper Saddle River, NJ, p. 30.

Metcalf & Eddy, Inc. (2003) revised by G. Tchobanoglous, F. L. Burton, and H. D. Stensel, *Wastewater Engineering, Treatment and Reuse*, McGraw-Hill, Boston, p. 844.

Richards, J. A., F. W. Sears, M. R. Wehr, and M. W. Zemansky, *Modern University Physics*, Addison-Wesley, Reading, MA, 1960, pp. 339, 344.

Salvato, Jr., J. A. (1972) *Environmental Engineering and Sanitation*, 2nd ed., Wiley-Interscience, New York, pp. 598–599.

Shortley, G., and D. Williams (1955) *Elements of Physics*, Prentice-Hall, Engelwood Cliffs, NJ, p. 290.

Tchobanoglous, G., and E. D. Schroeder (1985) *Water Quality*, Addison-Wesley, Reading, MA, p. 372.

U.S. EPA (1982) *Summary Report: Control and Treatment Technology for the Metal Finishing Industry, In-Plant Changes*, U.S. Environmental Protection Agency, Washington, DC, Report No. EPA 625/8-82-008.

Wark, K., C. F. Warner, and W. T. Davis (1998) *Air Pollution: Its Origin and Control*, 3rd ed., Addison-Wesley, Reading, MA, p. 509.

CHAPTER 3

HYDROLOGY

3-1 FUNDAMENTALS

The Hydrologic Cycle

The global system that supplies and removes water from the earth's surface is known as the hydrologic cycle (Figure 3-1). Water is transferred to the earth's atmosphere through two processes: (1) *evaporation* and (2) *transpiration.** As moist air rises, it cools. Eventually enough moisture accumulates and the mass cools sufficiently to nucleate (form small crystals) on microscopic particles. Sufficient growth causes the droplets or snowflakes to become heavy enough to fall as precipitation. As they fall on the earth's surface, the droplets either run over the ground into streams and rivers (*surface runoff,* or just *runoff*) or percolate into the ground to form groundwater.

Surface Water Hydrology

Precipitation. Surface water hydrology begins before the precipitate hits the ground. The form the precipitate takes (rain, sleet, hail, or snow) is important. For example, it takes about 10 mm of snow to make the equivalent of 1 mm of rain. Other factors of importance are the size of the area over which the precipitation falls, the intensity of the precipitation, and its duration.

Once the precipitation hits the ground, a number of things can happen. It can evaporate promptly. This is especially true if the surface is hot and impervious. If the soil is dry and/or porous, the precipitate may *infiltrate* into the ground or it may only wet the surface. This process and the process of wetting leaves and blades of grass is called *interception.* The precipitate may be trapped in small depressions or puddles. It may remain there until it evaporates or until the depressions fill and overflow. And last

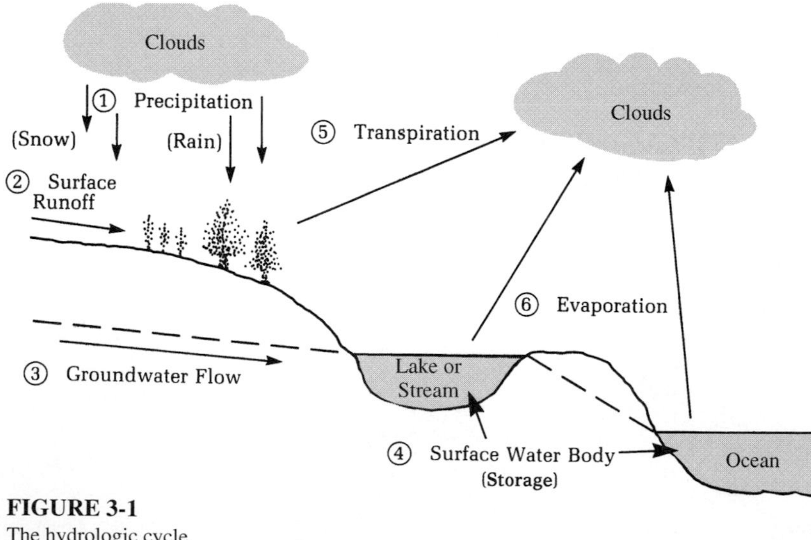

FIGURE 3-1
The hydrologic cycle.

*Transpiration is the process whereby plants give off water vapor through the pores of their leaves. The moisture comes from the roots through capillary action.

but not least, it may run off directly to the nearest stream or lake to become surface water. The four factors (evaporation, infiltration, interception, and trapping) that reduce the amount of direct runoff are called *abstractions.*

Streamflow. The water that makes up our streams and rivers is derived from two sources: *direct runoff* and groundwater *exfiltration*, or *base flow,* as it is more commonly called. Direct runoff is a consequence of precipitation. Base flow is the dry weather flow that results from the seepage of groundwater out of stream banks.

The amount of water that reaches a stream is a function of the abstractions mentioned above and the catchment area or watershed that feeds the stream. The watershed, or *basin,* is defined by the surrounding topography (Figure 3-2). The perimeter of the watershed is called a *divide.* It is the highest elevation surrounding the watershed. All of the water that falls on the inside of the divide has the potential to be shed into the streams of the basin encompassed by the divide. Water falling outside of the divide is shed to another basin.

Groundwater Hydrology

Water Table (Unconfined) Aquifer. As we mentioned earlier, part of the precipitation that falls on the soil may infiltrate. This water replenishes the soil moisture or is used by growing plants and returned to the atmosphere by transpiration. Water that drains downward below the root zone finally reaches a level at which all of the openings or voids in the earth's materials are filled with water. This zone is known as the *zone of saturation.* Water in the zone of saturation is referred to as groundwater. The geologic formation

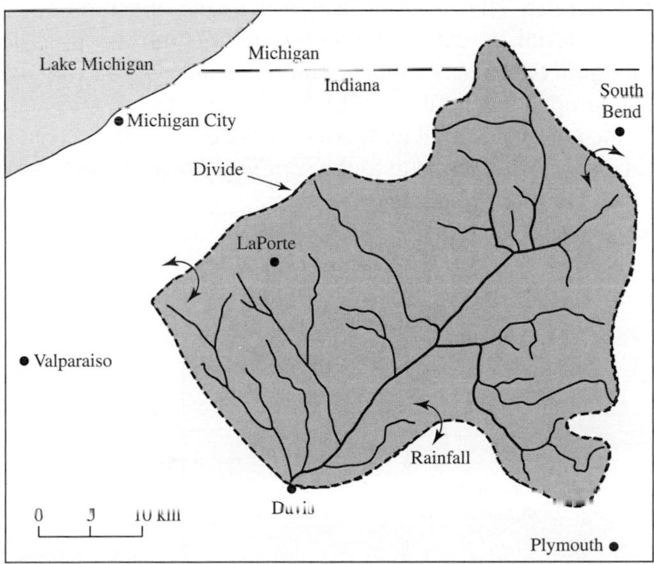

FIGURE 3-2
The Kankakee River Basin above Davis, IN. *Note:* Arrows indicate that precipitation falling inside the dashed line is in the Davis watershed, while that falling outside is in another watershed. The dashed line then "divides" the watersheds.

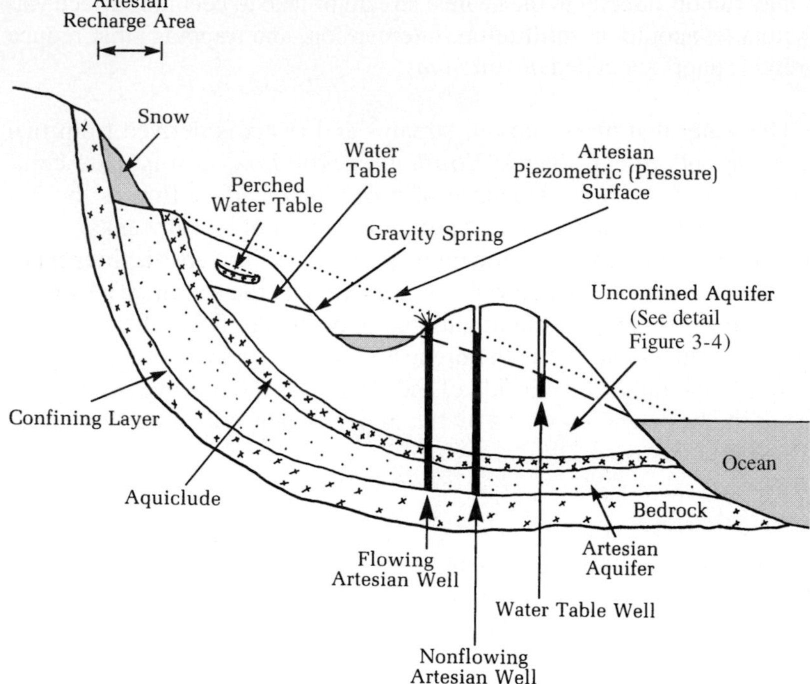

FIGURE 3-3
Schematic of groundwater aquifers.

that bears the water is called an *aquifer.* The upper surface of the zone of saturation, if not confined by impermeable material, is called the *water table* (Figure 3-3). The aquifer is called a *water table aquifer* or an *unconfined aquifer.* Water will rise to the level of the water table in an unpumped water table well.

The smaller void spaces in the porous material just above the water table may contain water as a result of capillarity. This zone is referred to as the *capillary fringe* (Figure 3-4).

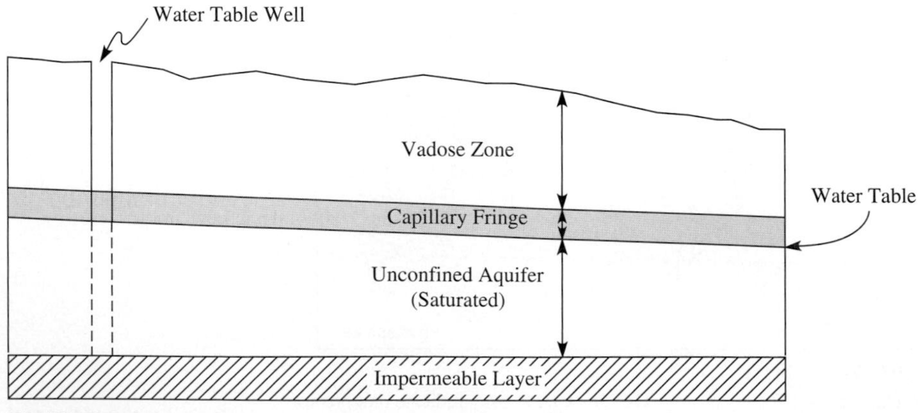

FIGURE 3-4
Detail of the unconfined aquifer.

It is not a source of supply since the water held will not drain freely by gravity. The region from the saturated zone to the surface is also called the *vadose zone.*

Springs. Because of the irregularities in underground deposits and in surface topography, the water table occasionally intersects the surface of the ground or the bed of a stream, lake, or ocean. At these points of intersection, groundwater moves out of the aquifer. The place where the water table breaks the ground surface is called a *gravity* or *seepage spring* (Figure 3-3).

Perched Water Table. A perched water table is a lens of water held above the surrounding water table by an impervious layer. It may cover an area from a few hundred square meters to several square kilometers.

Artesian (Confined) Aquifer. As water percolates into an aquifer and flows downhill, the lower layers come under pressure. This pressure is the result of the mass of water in the upper layers pressing on the water in the lower layers, much as deep sea divers are under greater and greater pressure as they go deeper and deeper into the sea. The system is analogous to a manometer (Figure 3-5). When there is no constriction in the manometer, the water level in each leg rises to the same height. If the left leg is raised, the increased water pressure in that leg pushes the water up in the right leg until the levels are equal again. If the right leg is clamped shut then, of course, the water will not rise to the same level. However, at the point where the clamp is placed, the water pressure will increase. This pressure is the result of the height of water in the left leg.

A special type of groundwater system occurs when an overlying impermeable formation and an underlying impermeable formation restrict the water, much as the walls of a manometer. The impermeable layers are called *confining layers.* Other names given to these layers are *aquicludes* if they are essentially impermeable, or *aquitards* if they are less permeable than the aquifer but not truly impermeable. An aquifer between impermeable layers is called a *confined aquifer.* If the water in the aquifer is under

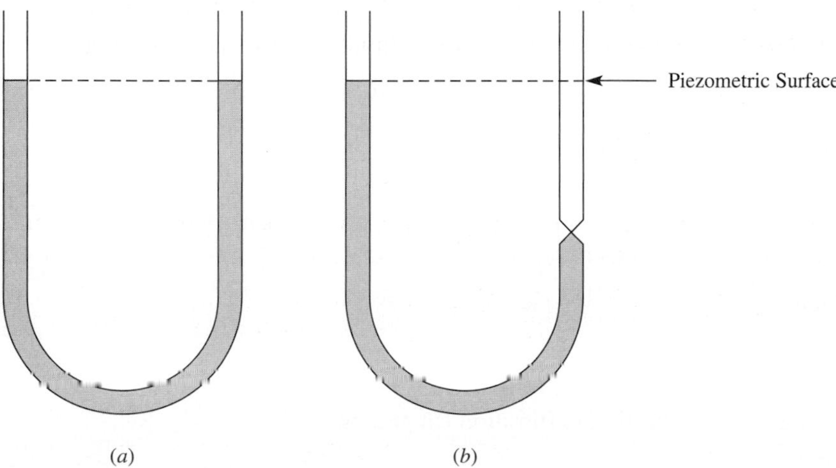

FIGURE 3-5
Manometer analogy to water in an aquifer. Manometer "a" is analogous to an unconfined aquifer. Manometer "b" is analogous to a confined aquifer.

pressure, it is called an *artesian aquifer* (Figure 3-3). The name "artesian" comes from the French province of Artois (*Artesium* in Latin) where, in the days of the Romans, water flowed to the surface of the ground from a well.

Water enters an artesian aquifer at some location where the confining layers intersect the ground surface. This is usually in an area of geological uplift. The exposed surface of the aquifer is called the *recharge area.* The artesian aquifer is under pressure for the same reason that the pinched manometer is under pressure, that is, because the recharge area is higher than the bottom of the top aquiclude and, thus, the height of the water above the aquiclude causes pressure in the aquifer. The greater the vertical distance between the recharge area and the bottom of the top aquiclude, the higher the height of the water, and the higher the pressure.

Piezometric Surfaces. If we place small tubes (*piezometers*) into an artesian aquifer along its length, the water pressure will cause water to rise in the tubes much as the water in the legs of a manometer rises to a point of equilibrium. The height of the water above the bottom of the aquifer is a measure of the pressure in the aquifer. An imaginary plane drawn through the points of equilibrium is called a *piezometric surface.* In an unconfined aquifer, the piezometric surface is the water table.

If the piezometric surface of a confined aquifer lies above the ground surface, a well penetrating into the aquifer will flow naturally without pumping. If the piezometric surface is below the ground surface, the well will not flow without pumping.

Hydrologic Mass Balance

Hydrologic problems of interest to civil and environmental engineers, such as the sizing of retention ponds and reservoirs and estimating the size of sewers for parking lots, streets, and airports, may be solved by the application of mass balance equations. The system shown in Figure 3-6*a* is an example of a small hydrologic system. A simplified mass balance diagram of this system is shown in Figure 3-6*b*. It may be described by a form of the mass balance equation given in Chapter 2 (Equation 2-3):

Mass rate of accumulation = Mass rate of input − mass rate of output

Noting that mass rate = volumetric rate × density, we may write:

$$Q_S(\rho) = Q_P(\rho) + Q_{Q_{\text{in}}}(\rho) + Q_{I_{\text{in}}}(\rho) - Q_{Q_{\text{out}}}(\rho) - Q_{I_{\text{out}}}(\rho) - Q_R(\rho) - Q_E(\rho) - Q_T(\rho) \qquad (3\text{-}1)$$

where Q refers to the volume per unit of time (m^3/s), ρ is the density of water (kg/m^3), and the subscripts are defined as follows:

S = storage
P = precipitation
Q = river flow (in and out)
I = groundwater infiltration/exfiltration (in and out)
R = runoff
E = evaporation
T = transpiration

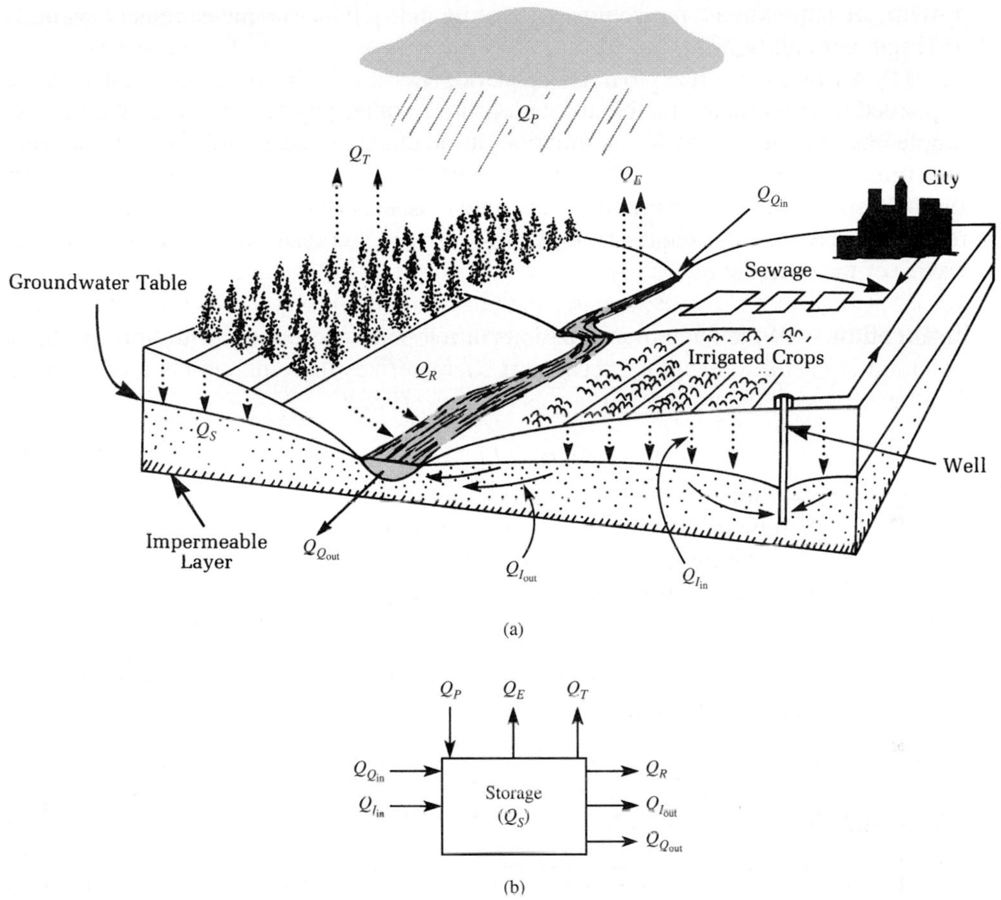

FIGURE 3-6
(*a*) Schematic diagram of a hydrologic subsystem; (*b*) mass balance diagram of a hydrologic subsystem.

We often assume that the density of the water is constant throughout the system. Thus, we divide both sides of the equation by the density to yield an equation for the volumetric rate of accumulation. Equation 3-1 may then be written as:

$$Q_S = Q_P + Q_{Q_{in}} + Q_{I_{in}} - Q_{Q_{out}} - Q_{I_{out}} - Q_R - Q_E - Q_T \tag{3-2}$$

In many hydrology texts, the equation is further simplified by writing the expression in terms of the subscripts:

$$S = P + Q_{in} + I_{in} - Q_{out} - I_{out} - R - E - T \tag{3-3}$$

The common units of expression for the measurement of these terms are not consistent with one another. For example, the common unit of measure for precipitation, infiltration, evaporation, and transpiration is mm/h while the common unit of measure for storage, river flow, and runoff is m^3/s. Because it is assumed that precipitation, infiltration, evaporation, and transpiration occur over the entire surface of the hydrologic system, we approximate the volumetric rate by multiplying the measurement (in units of length per unit time) by the surface area.

The terms of the mass balance equation for the hydrologic equation may be expanded to show their functional relationship to other physical phenomena. For example, the amount of runoff is a function of the characteristics of the surface (paved, cultivated, flat, steep-sloped). The amount of storage, for example, is a function of the type of soil or geological formation. These two aspects of the hydrologic equation are discussed in Sections 3-3 and 3-5. In the following paragraphs, we elaborate on the behavior of the other terms.

Infiltration. Of the numerous equations developed to describe infiltration, Horton's equation is useful to examine because it characterizes three phenomena of interest. Horton expressed the infiltration rate as (Horton, 1935):

$$f = f_c + (f_o - f_c)e^{-kt} \tag{3-4}$$

where f = infiltration rate, mm/h
f_c = equilibrium or final infiltration rate, mm/h
f_o = initial infiltration rate, mm/h
k = empirical constant, h^{-1}
t = time, h

This expression assumes that the rate of precipitation is greater than the rate of infiltration.

Infiltration rate is a function of the properties of the soil; thus, the values for f_o, f_c, and k are, as you might expect, a function of the soil type. Some examples are (in mm/h and h^{-1}):

	f_o	f_c	k
Dothan loamy sand	88	67	1.4
Fuquay pebbly loamy sand	159	61	4.7

Soil moisture content, vegetative cover, organic matter, and season affect these values.

The second property of interest is that the infiltration rate is an inverse exponential function of time. If the rate of precipitation exceeds the rate of infiltration, a plot of infiltration rate versus time will reveal that as rainfall continues, the rate at which the ground soaks it up decreases because the pore spaces in the soil fill up with water. Because typical values for f_o and f_c are greater than prevailing rainfall intensity, this may lead to calculated decreases in infiltration even though there is capacity to accept precipitation at higher rates.

The third property, which is directly related to hydrologic balances, is that the area under the infiltration curve represents the volume of water that infiltrates. Integration of Horton's equation yields the volume:

$$\forall = f_c t + \frac{f_o - f_c}{k}(1 - e^{-kt}) \tag{3-5}$$

Evaporation. The loss of water from the surface of a lake or other water body is a function of solar radiation, air and water temperature, wind speed, and the difference in vapor pressures at the water surface and in the overlying air. As with estimates of infiltration rate, there are numerous methods for estimating evaporation. Dalton first expressed the fundamental relationship in the form (Dalton, 1802):

$$E = (e_s - e_a)(a + bu) \tag{3-6}$$

where E = evaporation rate, mm/d
e_s = saturation vapor pressure, kPa
e_a = vapor pressure in overlying air, kPa
a, b = empirical constants
u = wind speed, m/s

Empirical studies at Lake Hefner, Oklahoma, yielded a similar relationship:

$$E = 1.22(e_s - e_a)u \tag{3-7}$$

From these expressions, it is apparent that high wind speeds and low humidities (vapor pressure in the overlying air) result in large evaporation rates. You may note that the units for these expressions do not make much sense. This is because these are empirical expressions developed from field data. The constants have implied conversion factors in them. In applying empirical expressions, care must be taken to use the same units as those used by the author of the expression.

Evapotranspiration. Water loss from plants (transpiration) is difficult to separate from losses from the soil surface or root zone. For mass balance calculations, these are often lumped together under the term evapotranspiration. The rate of evapotranspiration is a function of soil moisture, soil type, plant type, wind speed, and temperature. Plant types may affect evapotranspiration rates dramatically. For example, an oak tree may transpire as much as 160 L/d while a corn plant may transpire only about 1.9 L/d.

Example 3-1. Silk's Lake has a surface area of 70.8 ha. For the month of April the inflow was 1.5 m^3/s. The dam regulated the outflow (discharge) from Silk's Lake to be 1.25 m^3/s. If the precipitation recorded for the month was 7.62 cm and the storage volume increased by an estimated 650,000 m^3, what is the estimated evaporation in m^3 and cm? Assume that no water infiltrates out of the bottom of Silk's Lake and that the density of water is constant.

Solution. Begin by drawing the mass balance diagram:

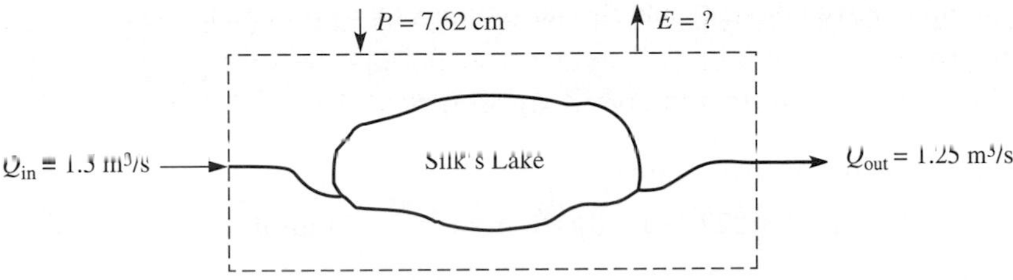

The mass balance equation is:

$$\text{Mass accumulation} = \text{Mass input} - \text{mass output}$$

Using the assumption that the density of water is constant, we may write the mass-balance equation as

$$\text{Volume accumulation} = \text{Volume in} - \text{Volume out}$$

The accumulation is given as 650,000 m^3. The input consists of the inflow and the precipitation. The product of the precipitation depth and the area on which it fell (70.8 ha) will yield a volume. The output consists of outflow plus evaporation.

$$\Delta S = [(Q_{in})(t) + (P)(\text{area})]_{input} - [(Q_{out})(t) + E]_{output}$$

where

$$\Delta S = \text{Change in storage (a volume)}$$
$$(Q)(t) = (\text{flow rate})(\text{time}) = \text{volume}$$

Noting that April has 30 days and making the appropriate units conversions:

$$\begin{aligned}650{,}000 \text{ m}^3 &= (1.5 \text{ m}^3\text{/s})(30 \text{ d})(86{,}400 \text{ s/d}) \\ &\quad + (7.62 \text{ cm})(70.8 \text{ ha})(10^4 \text{ m}^2\text{/ha})(1\text{m}/100 \text{ cm}) \\ &\quad - (1.25 \text{ m}^3\text{/s})(30 \text{ d})(86{,}400 \text{ s/d}) - E\end{aligned}$$

Solving for E:

$$E = 3.89 \times 10^6 \text{ m}^3 + 5.39 \times 10^4 \text{ m}^3 - 3.24 \times 10^6 \text{ m}^3 - 6.50 \times 10^5 \text{ m}^3$$
$$E = 5.39 \times 10^4 \text{ m}^3$$

For an area of 70.8 ha, the evaporation depth is:

$$E = \frac{5.39 \times 10^4 \text{ m}^3}{(70.8 \text{ ha})(10^4 \text{ m}^2\text{/ha})} = 0.076 \text{ m or } 7.6 \text{ cm}$$

Example 3-2. During April, the wind speed over Silk's Lake was estimated to be 4.0 m/s. The air temperature averaged 20°C and the relative humidity was 30%. The water temperature averaged 10°C. Estimate the evaporation rate using the empirical relationship in Equation 3-7.

Solution. From the water temperature and Table 3-1, the saturation vapor pressure is estimated as $e_s = 1.227$ kPa. The vapor pressure in the air may be estimated as the product of the relative humidity and the saturation vapor pressure at the air temperature:

$$e_a = (2.337 \text{ kPa})(0.30) = 0.70 \text{ kPa}$$

The daily evaporation rate is then estimated to be:

$$E = 1.22(1.227 - 0.70)(4.0 \text{ m/s}) = 2.57 \text{ mm/d}$$

The monthly evaporation would then be estimated to be:

$$E = (2.56 \text{ mm/d})(30 \text{ d}) = 76.8 \text{ mm or } 7.7 \text{ cm}$$

TABLE 3-1
Water vapor pressures at various temperatures

Temperature, °C	Vapor pressure, kPa
0	0.611
5	0.872
10	1.227
15	1.704
20	2.337
25	3.167
30	4.243
35	5.624
40	7.378
50	12.34

3-2 RAINFALL ANALYSIS

Of the many variables of rainfall that might be of interest, we are concerned primarily with four:

1. Space: the average rainfall over the area
2. Intensity: how hard it rains
3. Duration: how long it rains at any given intensity
4. Frequency: how often it rains at any given intensity and duration

Point Precipitation Analysis

Data from a single nearby rain gage are often sufficiently representative to allow their use in the design of small projects. The analysis of data from a single gage is called point precipitation analysis. Spatial analysis is much more complex and is left for more advanced courses.

Rain Gages and Rain Gage Records. There are three types of rain gage in use: the U.S. Weather Bureau standard, the weighing bucket, and the tipping bucket. The standard gage is used for manually recording 24-hour accumulations of precipitation. The weighing bucket provides a continuous strip chart record of accumulated precipitation (Figure 3-7). The tipping bucket records precipitation by logging the number of times the cup tips. The cup is designed to tip when 0.25 mm of precipitation has accumulated.

Interpreting Rain Gage Records. The analysis of standard and tipping-bucket measurements is fairly straightforward. Weighing-bucket charts require a slight degree of interpretation.

The weighing-bucket chart is wound on a drum for data collection. The drum rotates at a constant speed while the pen inscribes a line of accumulated precipitation (actually mass converted to depth). Thus, when there is no rain the line on the chart is

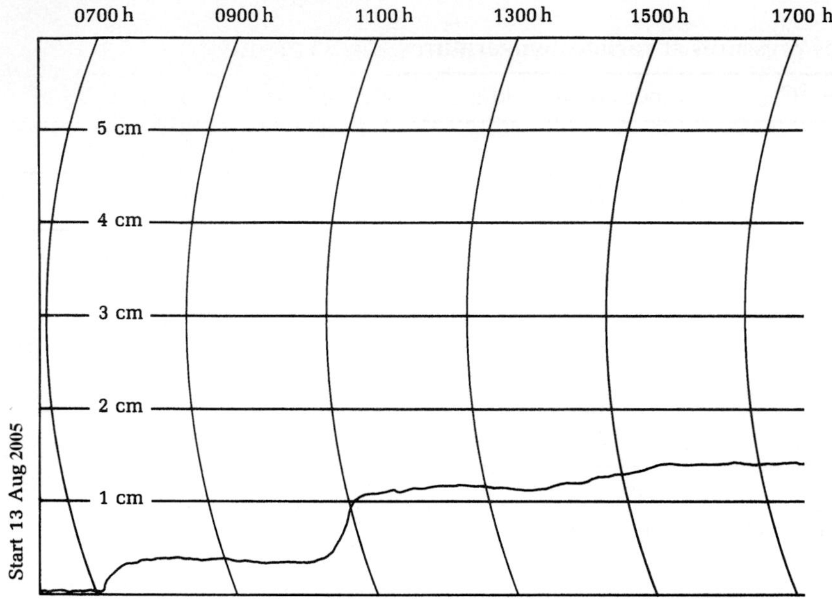

FIGURE 3-7
Strip chart record from weighing-bucket rain gage.

horizontal. When it rains, the line is sloped upward; the steeper the slope, the more intense the rain.

Intensity is computed from weighing-bucket records by determining the slope of the line (Figure 3-8):

$$\frac{\Delta p}{\Delta t} = \frac{(p_2 - p_1)}{(t_2 - t_1)} \tag{3-8}$$

where p_1 and p_2 are the accumulated precipitation at times t_1 and t_2.

The duration of precipitation for a given intensity is t. The normal procedure is to select fixed time intervals and calculate Δp at each interval. Thus, it is possible for one rainfall event to produce data for several durations.* For example, a 15-minute rain could yield the following pieces of data: one 15-minute duration event, one 10-minute duration event, and three 5-minute duration events.

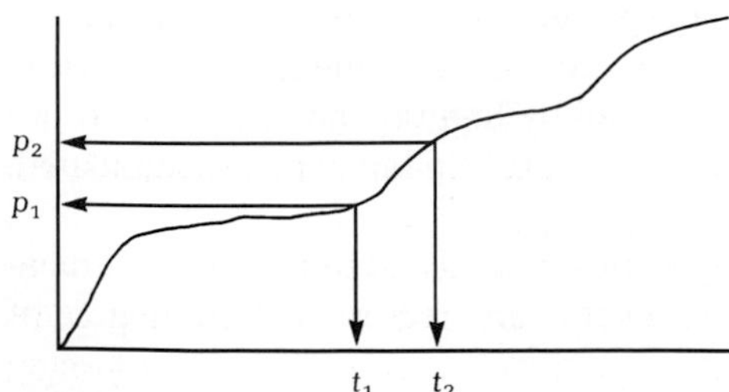

FIGURE 3-8
Computation of rainfall intensity from weighing-bucket strip chart record.

*An *event* is any continuous period of precipitation.

TABLE 3-2
Rainfall record for the Dismal Swamp (1 Oct. 1954–30 Sep. 1999)

	Number of storms of stated intensity or more										
Duration	Intensity (mm/h)										
(min)	20.0	30.0	40.0	60.0	80.0	100.0	120.0	140.0	160.0	180.0	200.0
5						245	49	16	7	3	2
10					256	64	15	7	4	1	
15				241	94	18	6	3	2		
20		240	80	36	10	4	2	1			
30	202	44	17	9	2	2	1				
40	76	31	8	1							
50	30	12	3								
60	9	2									

Intensity-Duration-Frequency Curves (IDF). The family of curves that depicts the relationship between the intensity, duration, and frequency of precipitation at a point is a fundamental part of the rational method of design for storm sewers and retention ponds (Section 3-3). "In practice, the collection and analysis of basic rainfall data by the designer are limited to extensive projects. Rainfall data compiled and processed by the U.S. Weather Bureau, the Department of Agriculture, and similar government agencies are used widely where the size of the project or lack of local records does not justify complete local statistical analysis" (ASCE, 1969).

Table 3-2 is the compilation of a partial series of rainfall events. Rather than a record of all rainfalls, it is a record of rainfall intensities above some practical minimum. It gives the frequency or number of times that a rainfall of given intensity and duration will be equaled or exceeded for the period of record. For example, looking at the first row in Table 3-2, one would expect seven rainfall events with an intensity of 160.0 mm/h or more and a duration of five minutes to occur in any 45-year period $(1999 - 1954 = 45 \text{ years})$ in the Dismal Swamp.

You should note two other facts about the table. First, the numbers in the table are also ranks. If the rainfall events for 5-minute duration storms are arranged in descending order of intensity, then the 7th storm in the sequence or "the 7th-ranked" storm has an intensity of 160.0 mm/h or more. We assume that the ranks are spaced evenly between the recorded ranks, that is, that intensity and rank are linear. Thus, for the 5th-ranked storm of 5-minute duration, by interpolation, we can estimate that it would have an intensity of 170.0 mm/h or more.

The second fact that you should note is the ranks may be used to infer the probability that a given intensity storm will be equaled or exceeded. Again, using the seventh-ranked storm, we may infer that rainfall intensities of 160.0 mm/h or greater will occur with a frequency of seven times in 45 years. An annual average probability of occurrence would be $\frac{7}{45} = 0.16$ or 16 percent. Hydrologists and engineers often use the reciprocal of annual average probability because it has some temporal significance. The reciprocal is called the average *return period* or average *recurrence interval* (T):

$$T = \frac{1}{\text{Annual average probability}} \tag{3-9}$$

For the case of our seventh-ranked storm, the average return period of a 160.0 mm/h, 5-minute storm is 6.25 years. This means that we would expect a storm of 160.0 mm/h or greater once every 6.25 years on the average.

Because the amount of reliable data available is limited,* it is customary to use Weibull's formula for calculating return period (Weibull, 1939):

$$T = \frac{n+1}{m} \tag{3-10}$$

where T = average return period in years
n = number of years of record
m = rank of storm, with most intense storm given a rank of 1

Weibull's formula allows for a small correction when the number of years of record is small. At larger values of n it closely approximates $T = n/m$.

Example 3-3. Prepare a table of plotting points for an IDF curve for a 5-year storm at the Dismal Swamp. Compute points for each duration given in Table 3-2.

Solution. Because Table 3-2 is a table of ranks, we need to determine the rank of the 5-year storm. First, we rearrange Weibull's formula:

$$m = \frac{n+1}{T}$$

where

$$n = 1999 - 1954 = 45 \text{ y}$$
$$T = 5 \text{ y}$$

thus

$$m = \frac{46}{5} = 9.2$$

Starting with the 5-minute duration, we note that the 9.2-ranked storm lies between the 16th- and 7th-ranked storm; that is,

Intensity (mm/h)		
140.0		160.0
16	9.2	7

---

*Systematic measurement of precipitation was begun by the Surgeon General of the Army in 1819, while streamflow data collection did not begin until 1888.

We also note that the ranks increase from right to left while the intensities increase from left to right. Keeping this in mind, and recalling that we assume a linear relationship between intensity and rank, we may interpolate by simple proportions:

$$\frac{9.2 - 7}{16 - 7}(160.0 - 140.0) = 4.89$$

Thus, the 9.2-ranked storm is 4.89 mm/h less than 160.0 mm/h:

$$160.0 - 4.89 = 155.11 \text{ or } 155.1 \text{ mm/h}$$

The completed table would appear as follows:

Intensity and duration values for a 5-year storm at the Dismal Swamp

Duration (min)	Intensity (mm/h)
5	155.1
10	134.5
15	114.7
20	82.7
30	59.5
40	39.5
50	33.1
60	—

Note that a similar table could be constructed for each intensity given in Table 3-2. This would give us twice as many points to use for fitting the curve.

The IDF curve for Example 3-3 and the curve for a return period of 20 years are plotted in Figure 3-9. You should note that the frequency curves join occurrences that are

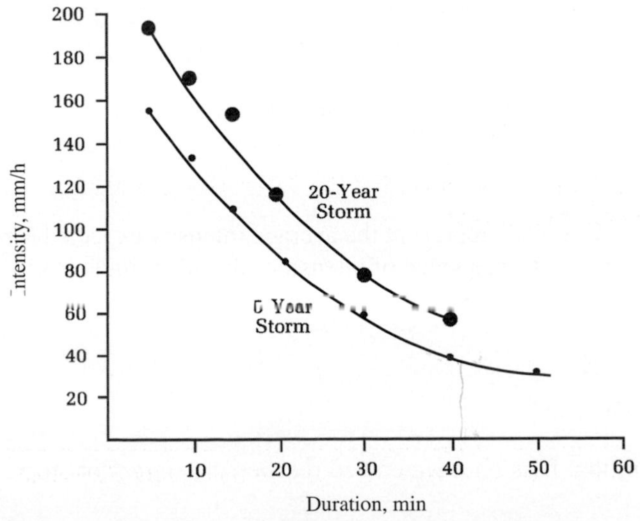

FIGURE 3-9
Intensity-duration-frequency curves for the Dismal Swamp.

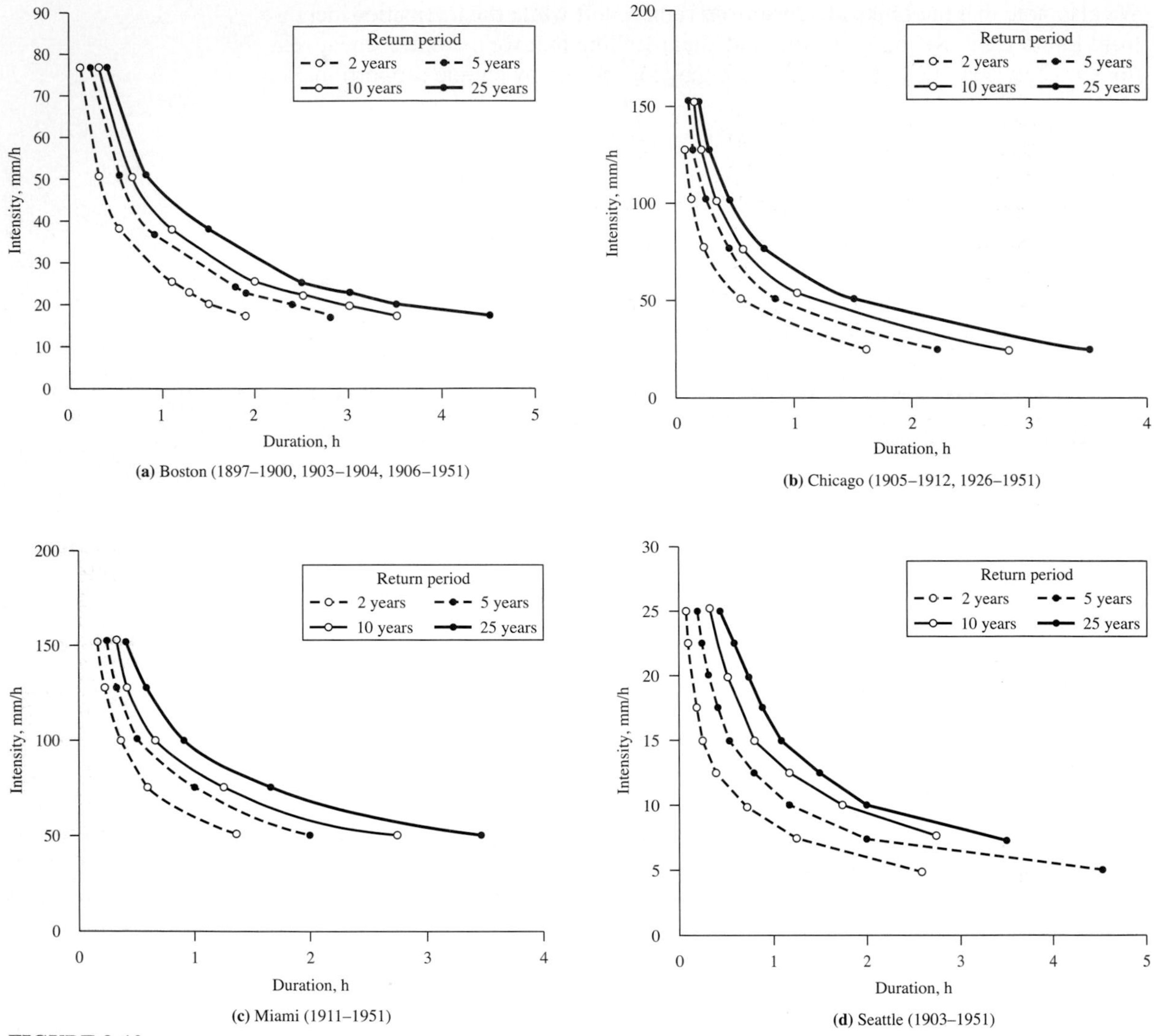

FIGURE 3-10
IDF Curves for (*a*) Boston, (*b*) Chicago, (*c*) Miami, and (*d*) Seattle. (*Source*: Gilman, 1964)

not necessarily from the same storm. They represent the average intensity expected for a given duration. They do not represent a sequence of intensities during a single storm. Example IDF curves for four U.S. cities are shown in Figure 3-10.

3-3 RUNOFF ANALYSIS

Three runoff questions are of interest:

1. How much of the rain that falls on a watershed reaches the stream or storm sewer draining it?

2. How long does it take for the runoff to reach the stream or storm sewer?
3. How often does the runoff cause a flood?

Estimation of Amount of Runoff

Stream Gages. Streamflow measurements are made by recording the height of the surface of the water above a reference datum. The elevation (*stage*) readings are calibrated in terms of streamflow (*discharge*). At manual recording stations, readings are made from a marked rod (*staff gage*) placed in the stream (Figure 3-11). At automatic recording stations, a float and cable system is used to drive a pen on a strip chart recorder (Figure 3-12). A *stilling well* (Figure 3-13) is used to minimize the effects of wave action and to protect the float from floating logs and other materials. For small streams, a dam with a weir plate (Figure 3-14) may be installed. This system increases the change in elevation for small changes in streamflow and makes readings more precise and accurate.

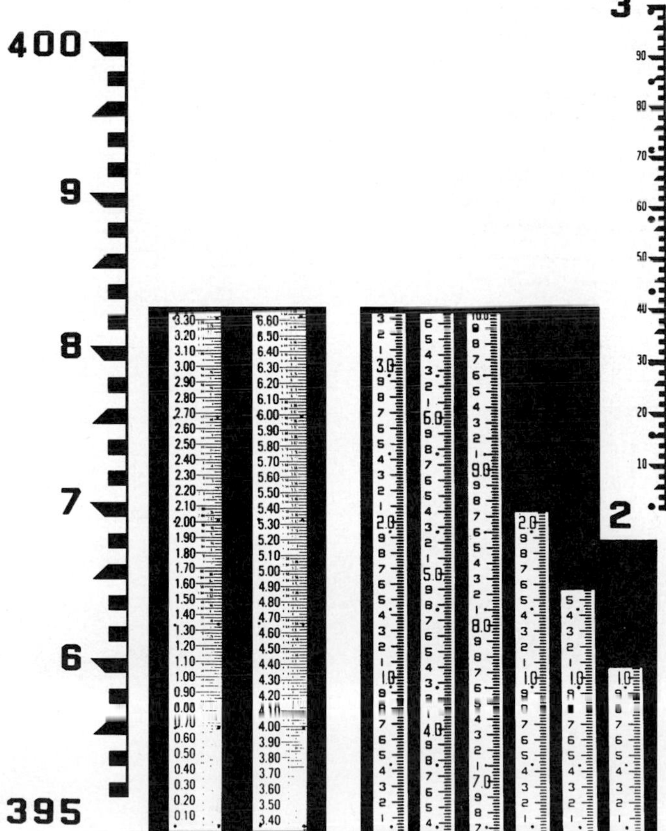

FIGURE 3-11
Staff gages for measurement of stream stage. (Courtesy of Stevens Water Monitoring Systems, Inc.)

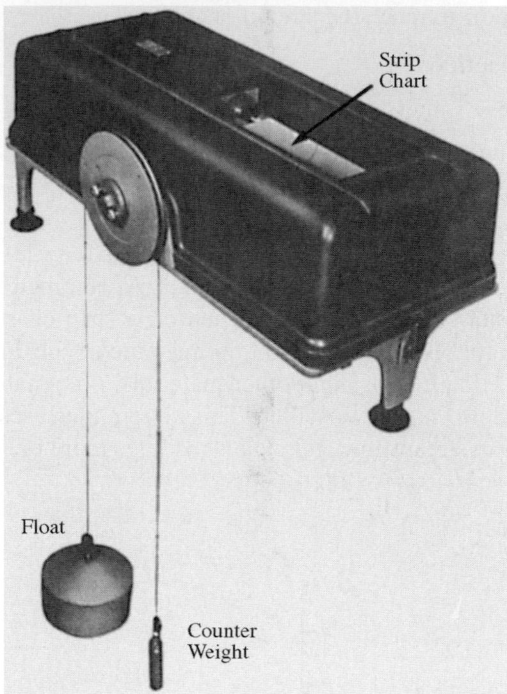

FIGURE 3-12
Float system and strip chart recorder for continuous stage measurement. (Courtesy of Stevens Water Monitoring Systems, Inc.)

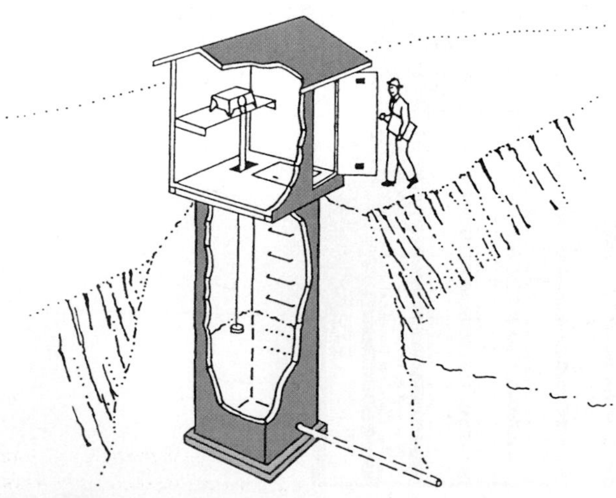

FIGURE 3-13
Stilling well. (Courtesy of Stevens Water Monitoring Systems, Inc.)

FIGURE 3-14
Dam and weir with float system for stage measurement. (Courtesy of Stevens Water Monitoring Systems, Inc.)

Hydrographs. A graphical representation of the discharge of a stream at a single gaging station is called a *hydrograph* (Figure 3-15). As we mentioned earlier, during the period between storms the base flow is a result of exfiltration of groundwater from the banks of the stream. Discharge from precipitation excess, that is, that which remains after abstractions, causes a hump in the hydrograph. This hump is called the *direct runoff hydrograph* (DRH).

Obviously, any precipitation excess that occurs at the extremities of a watershed will not be recorded at the basin outlet until some time lapse has occurred. As precipitation continues, enough time elapses for the more distant areas to add to the discharge at the gaging station. The lag time of the peak and the shape of the DRH depend on the precipitation pattern and the characteristics of the basin (size, slope, shape, and storage capacity).

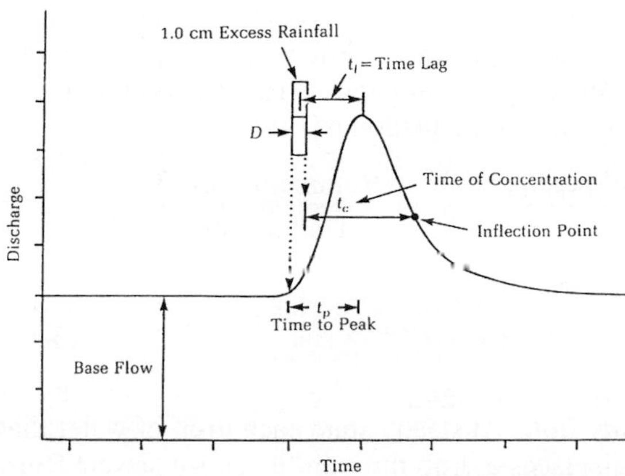

FIGURE 3-15
An idealized hydrograph showing a uniform base flow and a superimposed direct runoff hydrograph resulting from 1.0 cm of rainfall excess.

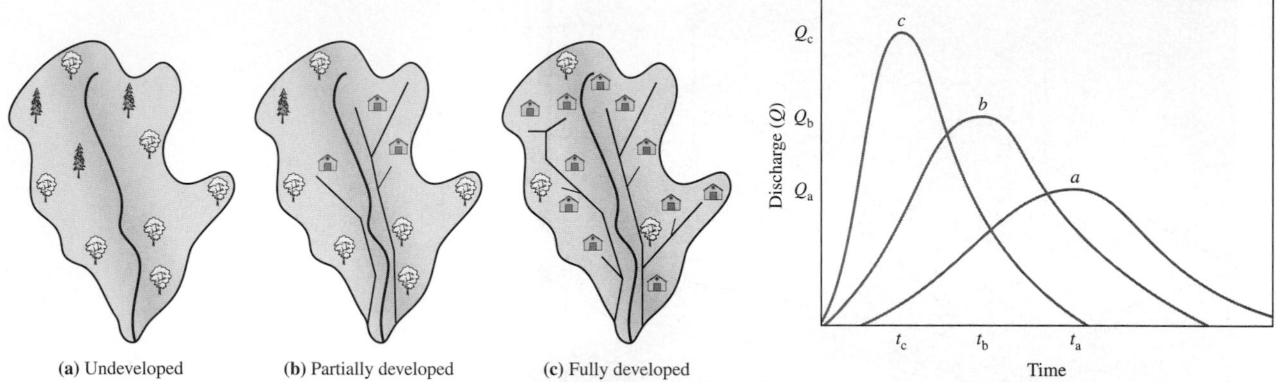

FIGURE 3-16
Effect of watershed development on a hydrograph. Note that $Q_c > Q_b > Q_a$ and that $t_c < t_a < t_a$.
(*Source:* Adapted from Davis and Master, 2004.)

The area under the DRH (volume of water discharged) will be larger for a watershed with a large surface area than for one with a small area. Because the water will flow more quickly from a steeply sloped watershed, the lag time will be smaller than for a flat watershed having the same area. It will take longer for the water to reach the gaging station from a long, narrow watershed than from a short, broad watershed of the same area and slope. The storage capacity of the watershed is dependent on a number of factors including, but not limited to, permeability of the soil, type of vegetative cover, time of year (frozen or not), and degree of development (urbanization). The effect of development on the discharge hydrograph is illustrated in Figure 3-16.

One measure of the storage capacity of a watershed is the ratio of the volume of water that is discharged past the gaging station to the volume of water that falls as precipitation. This ratio is called the *runoff coefficient*. It may be expressed as:

$$C = \frac{Q_R}{Q_P} \tag{3-11}$$

Rational Method. This method of determining runoff is one of the simplest applications of the hydrologic equation. A good example for us to look at is a paved parking lot (Figure 3-17). A mass balance equation of the form of Equation 3-2 applies in this case. The only input is precipitation. The only output is direct runoff. Assuming that the density of water is constant, the mass balance equation is:

$$\frac{\text{Storage}}{\text{Unit of time}} = \frac{\text{Volume of precipitation}}{\text{Unit of time}} - \frac{\text{Volume of runoff}}{\text{Unit of time}}$$

or

$$Q_S = Q_P - Q_R \tag{3-12}$$

If the rainfall on the lot continues for a long enough period at a constant intensity, at some time the system will reach *steady state.* At steady state each drop of water that falls on the watershed conceptually displaces a drop through the storm sewer. Thus, further rainfall at the same intensity does not increase the discharge at the storm sewer.

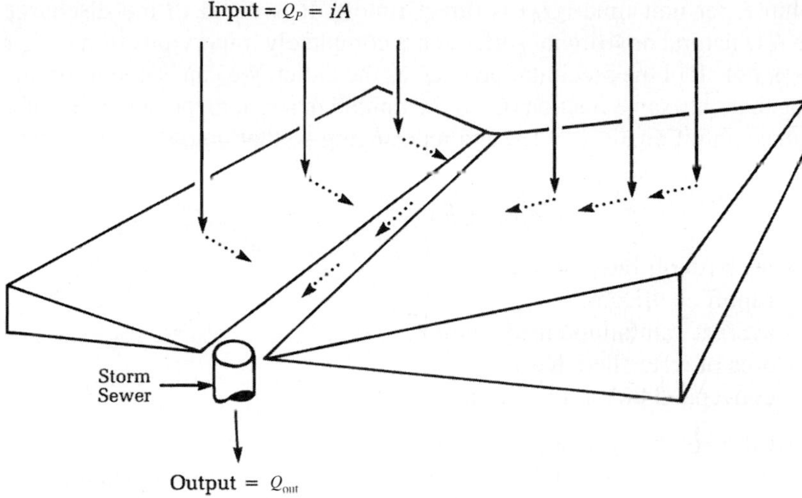

FIGURE 3-17
The application of the hydrologic equation to a parking lot having area = A.

The *hyetograph* (time versus rainfall) and corresponding DRH for this situation are shown in Figure 3-18. The time that it takes for steady state to be achieved is called the *time of concentration* (t_c). The time of concentration is primarily a function of the basin geometry, surface conditions, and slope.

At steady state the storage term (Q_S) in Equation 3-12 is equal to zero. The equation then reduces to the form

$$Q_P = Q_R \tag{3-13}$$

The input (Q_P) is a volume-per-unit time that may easily be shown to be equal to the product of the rainfall intensity (i) and the area of the watershed (A):

$$Q_P = iA \tag{3-14}$$

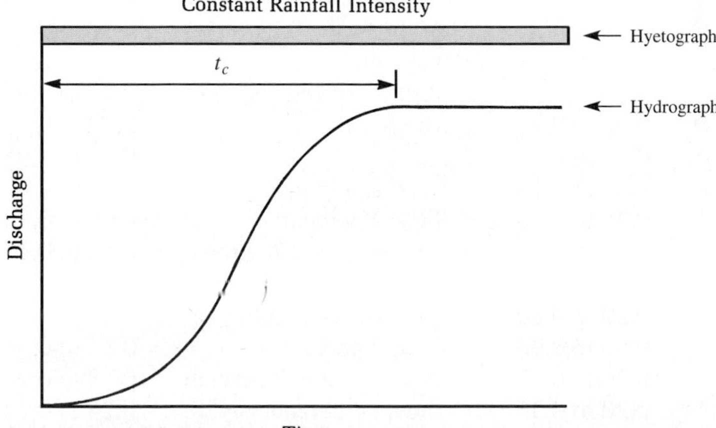

FIGURE 3-18
Hyetograph and hydrograph for a parking lot.

The output volume per unit time (Q_R) is direct runoff. It is equal to the discharge (Q_{out}). Because few natural or artificial surfaces are completely impervious (remember the chuck holes!), not all of the precipitation reaches the outlet. We can account for this loss by assuming that only some fraction (C) of the rainfall makes it to the outlet. By substituting all of these into Equation (3-13) and rearranging so that output is on the left, we obtain the *rational formula*

$$Q = 0.0028\ CiA \tag{3-15}$$

where Q = peak runoff rate, m^3/s
C = runoff coefficient
i = average rainfall intensity, mm/h
A = area of watershed, ha
0.0028 = conversion factor, $m^3 \cdot h/mm \cdot ha \cdot s$

The original derivation of the rational method was in English units. In the English system the use of intensity in inches/h and area in acres yields a runoff in ft^3/s without any conversion factor. Hence the name "rational" because the units work out rationally! Although the basic principles of the rational method are applicable to large watersheds, an upper limit of 13 square kilometers is recommended (ASCE, 1969). A selected list of runoff coefficients is given in Table 3-3.

The coefficients in Table 3-3 are applicable for storms of 5- to 10-year return period. Less frequent, higher intensity storms will require the use of higher coefficients because infiltration and other losses have a proportionally smaller effect on runoff. The coefficients are based on the assumption that the design storm does not occur when the ground is frozen.

TABLE 3-3
Selected runoff coefficients

Description of area or character of surface	Runoff coefficient	Description of area or character of surface	Runoff coefficient
Business		Railroad yard	0.20 to 0.35
Downtown	0.70 to 0.95	Unimproved	0.10 to 0.30
Neighborhood	0.50 to 0.70	Pavement	
Residential		Asphaltic and concrete	0.70 to 0.95
Single-family	0.30 to 0.50	Brick	0.70 to 0.85
Multi-units, detached	0.40 to 0.60	Roofs	0.75 to 0.95
Multi-units, attached	0.60 to 0.75	Lawns, sandy soil	
Residential (suburban)	0.25 to 0.40	Flat, 2 percent	0.05 to 0.10
Apartment	0.50 to 0.70	Average, 2 to 7 percent	0.10 to 0.15
Industrial		Steep, 7 percent	0.15 to 0.20
Light	0.50 to 0.80	Lawns, heavy soil	
Heavy	0.60 to 0.90	Flat, 2 percent	0.13 to 0.17
Parks, cemeteries	0.10 to 0.25	Average, 2 to 7 percent	0.18 to 0.22
Playgrounds	0.20 to 0.35	Steep, 7 percent	0.25 to 0.35

Extracted from ASCE, 1969.

Example 3-4. What is the peak discharge from the grounds of the Beauregard Long Ashby High School during a 5-year storm? The school grounds encompass a 16.2 ha plot that is 1.3 km east of the Dismal Swamp rain gage. Assume that the time of concentration of the grounds is 41 minutes. (Note: The method for calculating the time of concentration is illustrated in Examples 3-7 and 3-8.) The composition of the grounds is as follows:

Character of surface	Area (m^2)	Runoff coefficient
Building	10,800	0.75
Parking lot, asphaltic	11,150	0.85
Lawns, heavy soil		
2.0% slope	35,000	0.17
6.0% slope	105,050	0.20
	$\Sigma = 162{,}000$	

Solution. We begin by computing the weighted runoff coefficient, that is, the product of the fraction of the area and its runoff coefficient.

$$AC = (10{,}800)(0.75) + (11{,}150)(0.85) + (35{,}000)(0.17) + (105{,}050)(0.20)$$
$$= 44{,}537.5 \text{ m}^2 \text{ or } 4.45 \text{ ha}$$

Because the Dismal Swamp rain gage is only 1.3 km away, we shall use the IDF curve obtained in Example 3-3 to determine the intensity. By definition, the peak discharge for a watershed occurs when the duration of the storm equals the time of concentration. Thus, we select a duration of 41 minutes and read a value of 38 mm/h at the 5-year storm curve in Figure 3-9.

The peak runoff is then

$$Q = (0.0028)(4.45)(38) = 0.47 \text{ m}^3/\text{s}$$

Thus, a storm sewer large enough to handle 0.47 m^3/s of flow is required to carry storm water away from the BLAHS grounds.

Unit Hydrograph Method. A unit hydrograph (UH) is a DRH that results from a unit of precipitation excess over a watershed for a unit period of time. Although any unit depth may be selected (fathoms, furlongs, feet, hands, or cubits all would do), we have selected an excess of 1.0 cm after abstractions as a workable unit depth.* The presumption is that if you can determine an average UH, then you can approximate the DRH for any other rainfall excess over the same unit time by multiplying the UH ordinates by the amount of the rainfall excess (Sherman, 1932). For example, a 2.0-cm rainfall excess would yield a DRH with ordinates twice as large as a 1.0-cm UH. The method is limited to watersheds between 3,000- and 4,000-square kilometers in area (Viessman,

*In the original development of the UH by Sherman, a unit depth was defined as 1.0 inch of rainfall excess.

et al., 1989). Because the intent of the UH is to portray discharge caused by direct runoff so we can use it to predict the DRH for other storms, the first step in constructing a UH is to remove the groundwater contribution. This step is called hydrograph separation. There are a number of graphical procedures for hydrograph separation.

The second step in the construction of the UH is to estimate the total volume of water that occurs as direct runoff. Since the hydrograph is a plot of discharge versus time, the area under the DRH is equal to the volume of direct runoff. The volume is computed by numerical integration of the area under the curve. This is simply a summation of the products of an arbitrary unit of time (Δt) and the height of the DRH ordinate at the center of the selected time interval.

The third step is to convert the volume of direct runoff to a storm depth of runoff. This is done by dividing the volume of direct runoff by the area of the watershed in square meters and then multiplying by a conversion factor of 100 cm/m.

The fourth step is to divide the ordinates of the DRH by the storm depth computed in step three. The quotients are the ordinates of the UH. They have units of $m^3/s \cdot cm$.

The unit duration of the UH is determined from the *hyetograph* (time-rainfall graph) of the storm that was used to develop the UH. Because all of the precipitation does not result in direct runoff, an effective duration of excess precipitation must be estimated. The effective duration becomes the UH *unit duration*.

Example 3-5. Determine the unit hydrograph ordinates for the Triangle River hydrograph shown in Figure 3-19. The area of the watershed is 16.2 square kilometers.

Solution. The first step is to determine the depth of the storm precipitation spread over the watershed. The depth is equivalent to the volume of water divided by the area. The volume is equal to the area under the hydrograph. Because of the rather symmetrical shape of this particular hydrograph, it would be easy to find the area from the principles of geometry. However, in the interest of developing a technique that will also be applicable to more customary hydrographs, we will numerically integrate the area under the curve. We do this by taking a convenient slice or Δt and multiplying it by the height of the direct runoff (DRH) ordinate. The direct runoff ordinate is simply the difference between the total ordinate and the base ordinate. In this particular instance the base ordinate is, by observation, 2.0 m^3/s for all time periods. Using a convenient time interval of 1 hour, the following tabular computations are used to numerically integrate the area under the curve:

Time interval (h)	Total ordinate (m^3/s)	Base ordinate (m^3/s)	DRH ordinate (m^3/s)	Volume increment (m^3)
10–11	2.5	2.0	0.5	1,800
11–12	3.5	2.0	1.5	5,400
12–13	4.5	2.0	2.5	9,000
13–14	4.5	2.0	2.5	9,000
14–15	3.5	2.0	1.5	5,400
15–16	2.5	2.0	0.5	1,800
				$\Sigma = 32{,}400$

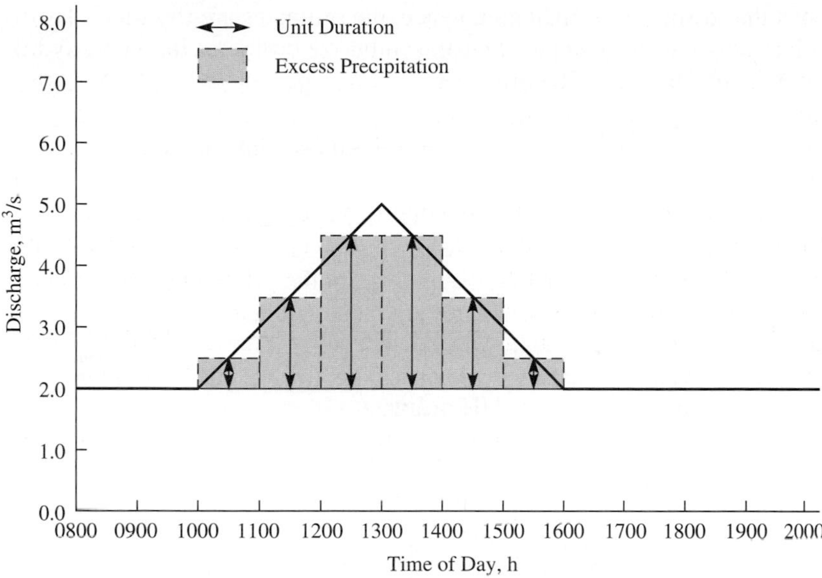

FIGURE 3-19
Triangle River hydrograph.

The volume increment is calculated as follows: First, the difference between the total ordinate and the base ordinate is found for the time increment selected. In the first row, for the time period from 10 AM to 11 AM (1000 to 1100 hours) the total ordinate is read from the hydrograph (Figure 3-19) as 2.5 m^3/s:

$$\begin{array}{ccccc} \text{Total ordinate} & - & \text{Base ordinate} & = & \text{DRH ordinate} \\ 2.5 \text{ m}^3/\text{s} & - & 2.0 \text{ m}^3/\text{s} & = & 0.5 \text{ m}^3/\text{s} \end{array}$$

To find the area (volume) represented by this slice, the flow rate is multiplied by the time interval selected (1 h) with appropriate units conversions:

$$(0.5 \text{ m}^3/\text{s})(1 \text{ h})(3{,}600 \text{ s/h}) = 1{,}800 \text{ m}^3$$

This process is continued for all the slices shown in Figure 3-19. The total volume (area under the curve) is estimated as 32,400 m^3. We can verify this by using the geometry of the triangle:

$$1/2(\text{base})(\text{height}) = (0.5)(6 \text{ h})(5.0 \text{ m}^3/\text{s} - 2.0 \text{ m}^3/\text{s})(3{,}600 \text{ s/h}) = 32{,}400 \text{ m}^3$$

Since we wish to construct a *unit hydrograph,* we need to determine whether or not this storm produced 1.0 cm of rainfall excess over the watershed. If it did, then we may use the ordinates directly. If not, then we must adjust the ordinates so that they would be equivalent to that produced by a 1.0 cm rainfall excess. We can determine whether or not this storm produced 1.0 cm by dividing the volume of rainfall by the area of the watershed (given as 16.2 km^2):

$$\text{Storm depth} = \frac{32{,}400 \text{ m}^3}{(16.2 \text{ km}^2)(1 \times 10^6 \text{ m}^2/\text{km}^2)} \times 100 \text{ cm/m} = 0.20 \text{ cm}$$

It is obvious that the storm is too small and, hence, the ordinates are too small. By dividing the ordinates by the storm depth, we can synthesize ordinates for a unit hydrograph. For example, for the first DRH ordinate:

$$\frac{\text{DRH ordinate}}{\text{Storm depth}} = \frac{0.5 \text{ m}^3/\text{s}}{0.2 \text{ cm}} = 2.5 \text{ m}^3/\text{s} \cdot \text{cm}$$

This ordinate would be located at the center of the slice that was used to establish it, i.e., halfway between 1000 and 1100 hours (see the arrows in Figure 3-19), i.e., 1030. For a generic hydrograph starting at a time equal to zero, the plotting point would be 0.5 h. The remaining unit hydrograph ordinates are tabulated below.

Triangle River plotting time (h)	Generic plotting time (h)	UH ordinate ($m^3/s \cdot cm$)
1030	0.5	2.5
1130	1.5	7.5
1230	2.5	12.5
1330	3.5	12.5
1430	4.5	7.5
1530	5.5	2.5

The unit "$m^3/s \cdot cm$" is read as

$$\frac{\text{m}^3}{(\text{s})(\text{cm})}$$

This means if we multiply a UH ordinate by the cm of excess rainfall, we will get units of m^3/s for the ordinate.

We can check our logic by calculating the area under a similar triangle using these new ordinates.

Time interval (h)	DRH ordinate (m^3/s)	Volume increment (m^3)
10–11	2.5	9,000
11–12	7.5	27,000
12–13	12.5	45,000
13–14	12.5	45,000
14–15	7.5	27,000
15–16	2.5	9,000
		$\Sigma = 162{,}000$

Recalculating our storm depth:

$$\frac{162{,}000 \text{ m}^3}{(16.2 \text{ km}^2)(1 \times 10^6 \text{ m}^2/\text{km}^2)} \times 100 \text{ cm/m} = 1.00 \text{ cm}$$

The unit hydrograph may be applied to a sequence of storms that have the same unit duration. There are two fundamental assumptions in the technique. The first is that storms of the same unit duration have ordinates that are in proportion to the unit hydrograph ordinates. Thus, simple ratios can account for differences in runoff excess. The second assumption is that a sequence of storms may be approximated by superimposing one hydrograph over another (with appropriate time lag) and adding the ordinates together. This is illustrated in the next example.

Example 3-6. Using the hyetograph in Figure 3-20, and the unit hydrograph ordinates from Example 3-5, determine the DRH ordinates and compound runoff.

Solution. The tabular computations are shown below. The explanation follows the table.

Time interval	Time (h)	Rainfall excess (cm)	DRH ordinates 1	DRH ordinates 2	DRH ordinates 3	Compound runoff (m^3/s)
1	0–1	0.5	1.25	N/A	N/A	1.25
2	1–2	2.0	3.75	5.0	N/A	8.75
3	2–3	1.0	6.25	15.0	2.5	23.75
4	3–4	0.0	6.25	25.0	7.5	38.75
5	4–5	0.0	3.75	25.0	12.5	41.25
6	5–6	0.0	1.25	15.0	12.5	28.75
7	6–7	0.0	0.0	5.0	7.5	12.5
8	7–8	0.0	0.0	0.0	2.5	2.5

The time interval is simply an enumeration of the segments. For the first hour, from the hyetograph in Figure 3-20, the rainfall excess is 0.5 cm. For the second and third hours, the rainfall excesses are 2.0 and 1.0 cm, respectively. No rain falls after the end of the third hour. The column labeled DRH 1 refers to the ordinates that are generated from the rainfall excess (0.5 cm) occurring in the first hour. Likewise, the DRH 2 refers to the ordinates resulting from the 2.0-cm rainfall excess in the second hour.

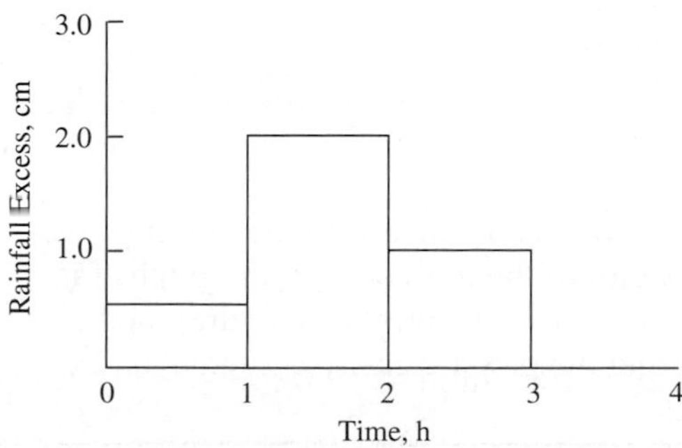

FIGURE 3-20
Hyetograph for Triangle River basin.

The first set of ordinates is obtained by multiplying the rainfall excess by each of the UH ordinates, that is:

$$(\text{Rainfall excess})(\text{UH Ordinate}) = \text{DRH ordinate}$$

Using the UH ordinates from Example 3-5:

$$(0.5 \text{ cm})(2.5 \text{ m}^3/\text{s} \cdot \text{cm}) = 1.25 \text{ m}^3/\text{s}$$
$$(0.5 \text{ cm})(7.5 \text{ m}^3/\text{s} \cdot \text{cm}) = 3.75 \text{ m}^3/\text{s}$$
$$(0.5 \text{ cm})(12.5 \text{ m}^3/\text{s} \cdot \text{cm}) = 6.25 \text{ m}^3/\text{s}$$
$$(0.5 \text{ cm})(12.5 \text{ m}^3/\text{s} \cdot \text{cm}) = 6.25 \text{ m}^3/\text{s}$$
$$(0.5 \text{ cm})(7.5 \text{ m}^3/\text{s} \cdot \text{cm}) = 3.75 \text{ m}^3/\text{s}$$
$$(0.5 \text{ cm})(2.5 \text{ m}^3/\text{s} \cdot \text{cm}) = 1.25 \text{ m}^3/\text{s}$$

The values for the second DRH start an hour later. Thus, under the column DRH 2, the first row is not applicable (N/A) since the rain that falls in the second hour (time interval 2) cannot reach the stream in the first hour. Likewise, under the column DRH 3, the first and second rows are N/A because rain that falls in the third hour cannot reach the stream in the first or second hour.

The DRH ordinates for the second hour of rainfall excess are obtained in the same fashion as those for the first, that is by multiplying the rainfall excess by each of the UH ordinates:

$$(2.0 \text{ cm})(2.5 \text{ m}^3/\text{s} \cdot \text{cm}) = 5.0 \text{ m}^3/\text{s}$$
$$(2.0 \text{ cm})(7.5 \text{ m}^3/\text{s} \cdot \text{cm}) = 15.0 \text{ m}^3/\text{s}$$
$$(2.0 \text{ cm})(12.5 \text{ m}^3/\text{s} \cdot \text{cm}) = 25.0 \text{ m}^3/\text{s}$$
$$(2.0 \text{ cm})(12.5 \text{ m}^3/\text{s} \cdot \text{cm}) = 25.0 \text{ m}^3/\text{s}$$
$$(2.0 \text{ cm})(7.5 \text{ m}^3/\text{s} \cdot \text{cm}) = 15.0 \text{ m}^3/\text{s}$$
$$(2.0 \text{ cm})(2.5 \text{ m}^3/\text{s} \cdot \text{cm}) = 5.0 \text{ m}^3/\text{s}$$

You should note that the table is carried beyond the last rainfall period in the hyetograph until all of the ordinates are used because it takes some finite length of time for the last drop of rainfall excess to reach the stream.

The compound runoff is the sum of the DRH ordinates for each of the time intervals. For example:

$$1.25 + \text{N/A} + \text{N/A} = 1.25$$
$$3.75 + 5.0 + \text{N/A} = 8.75$$
$$6.25 + 15.0 + 2.5 = 23.75$$

To plot the compound runoff hydrograph, the compound runoff ordinates are plotted at 1.0-h intervals, starting 0.5 h from time zero in accordance with the plotting position of the UH ordinates specified earlier. A plot of the individual hydrographs for each of the storms, their superposition, and the resulting compound hydrograph are shown in Figure 3-21. These computations and the plot are easily executed on a spreadsheet.

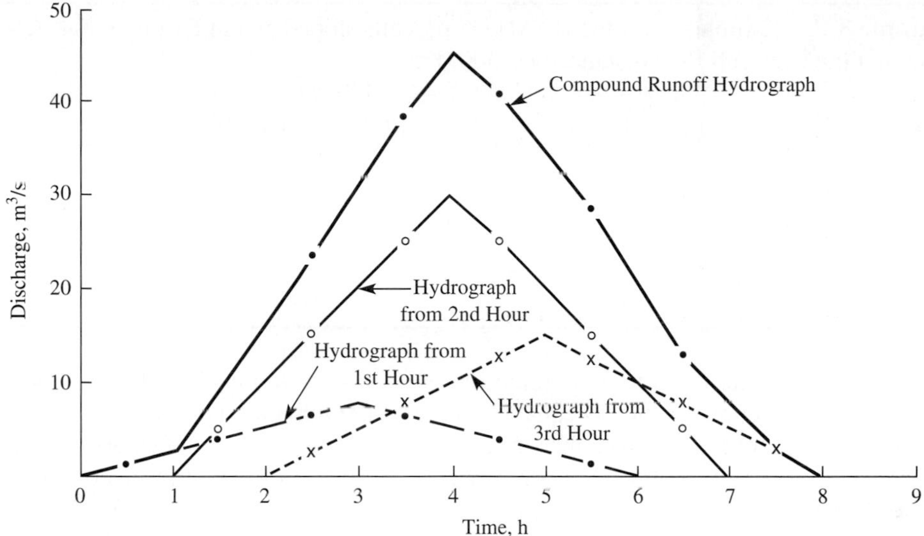

FIGURE 3-21
Compound runoff hydrograph for Triangle River. *Note:* Base flow is not shown.

Estimation of Time of Arrival

In addition to the quantity of discharge, it is often desirable to know when the peak flow will arrive at the watershed outlet or at some point along the discharge channel. This is particularly important when analyzing a series of watersheds that contribute to a river or sewer at various distances downstream from the headwater. The coincident arrival of two peaks would influence the design dramatically.

Lag Time. The time of arrival of the peak discharge is determined inherently in the UH method of estimating runoff. The lag time is the time from the midpoint of excess rainfall to the peak discharge as shown in Figure 3-15.

Time of Concentration. The time of concentration (t_c) is the time required for direct runoff to flow from the hydraulically most remote part of the drainage area to the watershed outlet. One of the major assumptions of the rational method is that the average rainfall intensity used in Equation 3-15 has continued for a period long enough to establish direct runoff and that rainfall has continued long enough to equal or exceed t_c. Thus, it is impossible to use the rational formula without being able to estimate t_c.

Although there are several methods for estimating t_c, the Federal Aviation Agency formula appears to be the easiest to use (FAA, 1970):

$$t_c = \frac{1.8(1.1 - C)\sqrt{3.28D}}{\sqrt[3]{S}} \tag{3-16}$$

where t_c = time of concentration, min
C = runoff coefficient
D = overland flow distance, m
S = slope, %

Example 3-7. Estimate t_c for the BLAHS 6-percent-slope lawn in Example 3-4. Assume that the overland flow distance was 300.0 m.

Solution. From Example 3-4 we use the same value of C, namely 0.20. Thus,

$$t_c = \frac{1.8(1.1 - 0.20)\sqrt{(3.28)(300.0)}}{\sqrt[3]{6.0}}$$

$$t_c = \frac{50.82}{1.82} = 27.97 \text{ or } 28 \text{ min}$$

For several areas that drain to a common outlet such as a drainage ditch or a storm sewer, the time of concentration is equal to the largest combination of the time of concentration for runoff to flow from the surface to the drainage inlet (t_c) plus the time of flow through the drainage ditch or storm sewer. Example 3-8 illustrates the computations.

Example 3-8. Estimate the time of concentration for BLAHS. The table below shows the estimated values for t_c for each of the areas. The building and lawns drain to a common storm sewer inlet. The storm sewer flows to another manhole inlet where the parking lot runoff is collected. The total flow from the four areas is then carried to a municipal storm sewer. The arrangement of the areas and the storm sewers is shown in the sketch below. The storm sewer on the BLAHS grounds flows at a speed of 0.6 m/s.

Character of surface	t_c (min)
Building	8
Parking lot	10
Lawns	
2.0% slope	38
6.0% slope	28

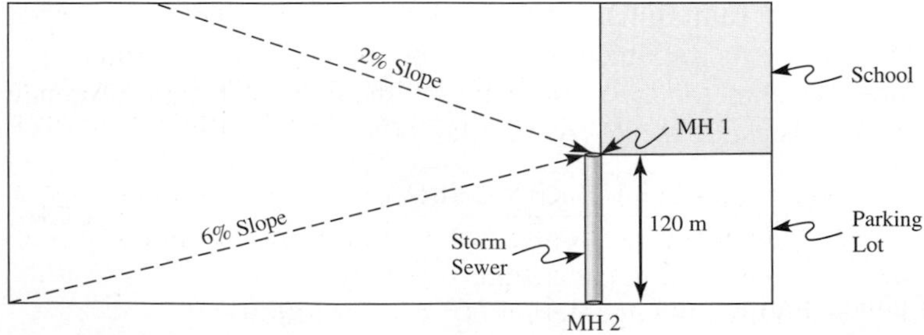

Solution. The time of concentration to be used for estimating the peak discharge is the largest combination of t_c and flow time.

For the distance marked on the sketch, the travel time of the flow from manhole 1 (MH 1) to manhole 2 (MH 2) on the school grounds is

$$\text{Travel time} = \frac{120 \text{ m}}{0.6 \text{ m/s}} = 200 \text{ s or } 3.33 \text{ min}$$

The total time of concentration (T_c) for the school building is

$$T_c = t_c + \text{travel time} = 8 + 3.33 = 11.33 \text{ or } 11 \text{ min}$$

The time of concentration for each element is summarized as follows:

Character of surface	t_c (min)	Total T_c (min)
Building	8	11
Parking lot	10	10
Lawns		
2.0% slope	38	41
6.0% slope	28	31

Note that the parking lot drains directly into MH 2. Thus, $T_c = t_c$.

According to this evaluation, the time of concentration to be used in entering the IDF curve is the maximum T_c. In this case the maximum T_c is 41 minutes. The estimate of the peak discharge using this time of concentration was shown in Example 3-4.

Estimation of Probability of Occurrence

The frequency of occurrence ($1/T$) used in the design of water supply or storm water control projects should be a function of the cost of the project and the benefits to be obtained from it. That is, the benefit-cost ratio should be greater than 1.0 to justify the project on economic grounds. While the cost of construction can be estimated in a straightforward manner, the environmental cost may be impossible to estimate. Likewise, the benefit beyond a cheaper supply of water or a reduced amount of flood damage is difficult to quantify.

The following paragraphs provide some guidance on the selection of a design frequency. They were obtained from the ASCE sewer design manual with their permission (ASCE, 1969).

In practice, benefit cost studies usually are not conducted for the ordinary urban storm drainage project. Judgment supported by records of performance in other similar areas is usually the basis of selecting the design frequency.

The range of frequencies used in engineering offices is as follows:

1. For storm sewers in residential areas, 2 to 15 years, with 5 years most commonly reported.

2. For storm sewers in commercial and high-value districts, 10 to 50 years, depending on economic justification.

3. For flood protection works and reservoir design, economics usually dictate a 50-year return period.

Other factors that may affect choice of design frequency include:

1. Use of greater return periods for design of those parts of the system not economically susceptible to future relief.

2. Use of greater recurrence intervals for design of special structures, such as expressway drainage pumping systems, where runoff exceeding capacity would seriously disrupt an important facility. Design frequencies of 50 years or more may be justified in such cases, particularly in small drainage areas, even though the project may be located in a district justifying only 5-year frequency for normal drainage.

3. Adoption of shorter return periods than normal but commensurate with available funds so that some degree of protection can be provided.

The cost of storm sewers is not directly proportional to design frequency. Studies of effects of various factors on sewer cost show that sewer systems designed for 10-year storms may cost only about 6 to 11 percent more than systems designed for 5-year storms, depending on the sewer slope (Rousculp, 1939). The lesser increase applies to steeper sewers.

If the peak flow is estimated by the rational method, it is assumed that the return period (inverse probability) of the peak flow is the same as that of the rainfall used to obtain it. If you use the UH method to determine the discharge, you must resort to some other method of estimating the probability of occurrence.

3-4 STORAGE OF RESERVOIRS

Classification of Reservoirs

For our purpose we can classify reservoirs either by size or by use. The size of the reservoir is used to establish the degree of safety to be incorporated into the design of the dam and spillway. The use or uses of the reservoir are a basis for evaluating the benefit-cost ratio.

Major dams (reservoir capacity greater than 6×10^7 m^3) are designed to withstand the maximum probable flood. Intermediate-sized dams (1×10^6 to 6×10^7 m^3) are designed to handle the discharge from the most severe storm considered to be reasonably characteristic of the watershed. For minor reservoirs (less than 1×10^6 m^3) the dams are designed to handle floods with return periods of 50 to 100 years.

Some of the benefits derived from reservoirs include the following: (1) flood control; (2) hydroelectric power; (3) irrigation; (4) water supply; (5) navigation; (6) preservation of aquatic life; and (7) recreation. The multipurpose or multiuse reservoir is the rule rather than the exception. Very seldom is it possible to justify the cost of a major reservoir on the basis of a single use.

Volume of Reservoirs

Mass Diagram. The techniques for determining the storage volume required for a reservoir are dependent both on the size and use of the reservoir. We shall discuss the simplest procedure, which is quite satisfactory for small water-supply impoundments, storm-water retention ponds, and wastewater equalization basins. It is called the mass diagram or *Rippl method* (Rippl, 1883). The main disadvantage of the Rippl method is that it assumes that the sequence of events leading to a drought or flood will be the same in the future as it was in the past. More sophisticated techniques have been developed to overcome this disadvantage, but these techniques are left for more advanced classes.

The Rippl procedure for determining the storage volume is an application of the mass balance approach (Equation 2-4). In this case it is assumed that the only input is the flow into the reservoir (Q_{in}) and that the only output is the flow out of the reservoir (Q_{out}). Therefore,

$$\frac{dS}{dt} = \frac{d(\text{In})}{dt} - \frac{d(\text{Out})}{dt}$$

becomes

$$\frac{dS}{dt} = Q_{in} - Q_{out} \qquad (3\text{-}17)$$

with the assumption that the density term cancels out because the change in density across the reservoir is negligible.

If we multiply both sides of the equation by dt, the inflow and outflow become volumes (flow rate $\times$ time = volume), that is,

$$dS = (Q_{in})(dt) - (Q_{out})(dt) \qquad (3\text{-}18)$$

By substituting finite time increments (Δt), the change in storage is then

$$(Q_{in})(\Delta t) - (Q_{out})(\Delta t) = \Delta S \qquad (3\text{-}19)$$

By cumulatively summing the storage terms, we can estimate the size of the reservoir. If the reservoir design is for water supply, then Q_{out} is the demand, and zero or positive values of storage (ΔS) indicate there is enough water to meet the demand. If the storage is negative, then the reservoir must have a capacity equal to the absolute value of cumulative storage to meet the demand. If the reservoir design is for flood protection, then Q_{out} is the capacity of the downstream river to hold water, and zero or negative values indicate the river is below flood stage. If the storage is positive, then the reservoir must have capacity equal to the cumulative storage to prevent flooding.

Example 3-9. Using the data in Table 3-4, determine the storage required to meet a demand of 2.0 m^3/s for the period from August 1976 through December 1978.

TABLE 3-4
Average monthly discharge of the Wash River at Watapitae, MI (discharge in m^3/s)

Year	J	F	M	A	M	J	J	A	S	O	N	D
1969	2.92	5.10	1.95	4.42	3.31	2.24	1.05	0.74	1.02	1.08	3.09	7.62
1970	24.3	16.7	11.5	17.2	12.6	7.28	7.53	3.03	10.2	10.9	17.6	16.7
1971	15.3	13.3	14.2	36.3	13.5	3.62	1.93	1.83	1.93	3.29	5.98	12.7
1972	11.5	4.81	8.61	27.0	4.19	2.07	1.15	2.04	2.04	2.10	3.12	2.97
1973	11.1	7.90	41.1	6.77	8.27	4.76	2.78	1.70	1.46	1.44	4.02	4.45
1974	2.92	5.10	28.7	12.2	7.22	1.98	0.91	0.67	1.33	2.38	2.69	3.03
1975	7.14	10.7	9.63	21.1	10.2	5.13	3.03	10.9	3.12	2.61	3.00	3.82
1976	7.36	47.4	29.4	14.0	14.2	4.96	2.29	1.70	1.56	1.56	2.04	2.35
1977	2.89	9.57	17.7	16.4	6.83	3.74	1.60	1.13	1.13	1.42	1.98	2.12
1978	1.78	1.95	7.25	24.7	6.26	8.92	3.57	1.98	1.95	3.09	3.94	12.7
1979	13.8	6.91	12.9	11.3	3.74	1.98	1.33	1.16	0.85	2.63	6.49	5.52
1980	4.56	8.47	59.8	9.80	6.06	5.32	2.14	1.98	2.17	3.40	8.44	11.5
1981	13.8	29.6	38.8	13.5	37.2	22.8	6.94	3.94	2.92	2.89	6.74	3.09
1982	2.51	13.1	27.9	22.9	16.1	9.77	2.44	1.42	1.56	1.83	2.58	2.27
1983	1.61	4.08	14.0	12.8	33.2	22.8	5.49	4.25	5.98	19.6	8.5	6.09
1984	21.8	8.21	45.1	6.43	6.15	10.5	3.91	1.64	1.64	1.90	3.14	3.65
1985	8.92	5.24	19.1	69.1	26.8	31.9	7.05	3.82	8.86	5.89	5.55	12.6
1986	6.20	19.1	56.6	19.5	20.8	7.73	5.75	2.95	1.49	1.69	4.45	4.22
1987	15.7	38.4	14.2	19.4	6.26	3.43	3.99	2.79	1.79	2.35	2.86	10.9
1988	21.7	19.9	40.0	40.8	11.7	13.2	4.28	3.31	9.46	7.28	14.9	26.5
1989	31.4	37.5	29.6	30.8	11.9	5.98	2.71	2.15	2.38	6.03	14.2	11.5
1990	29.2	20.5	34.9	35.3	13.5	5.47	3.29	3.14	3.20	2.11	5.98	7.62

Solution. The computations are summarized in the table below.

Month	Q_{in} (m^3/s)	$Q_{in}(\Delta t)$ (10^6 m^3)	Q_{out} (m^3/s)	$Q_{out}(\Delta t)$ (10^6 m^3)	ΔS (10^6 m^3)	$\Sigma(\Delta S)$ (10^6 m^3)
1976						
Aug	1.70	4.553	2.0	5.357	−0.8035	−0.8035
Sep	1.56	4.043	2.0	5.184	−1.140	−1.944
Oct	1.56	4.178	2.0	5.357	−1.178	−3.122
Nov	2.04	5.287	2.0	5.184	0.1036	−3.019
Dec	2.35	6.294	2.0	5.357	0.9374	−2.081
1977						
Jan	2.89	7.741	2.0	5.357	2.384	
Feb	9.57					
Mar	17.7					
Apr	16.4					
May	6.83					
Jun	3.74					
Jul	1.60	4.285	2.0	5.357	−1.071	−1.071
Aug	1.13	3.027	2.0	5.357	−2.330	−3.402
Sep	1.13	2.929	2.0	5.184	−2.255	−5.657
Oct	1.42	3.803	2.0	5.357	−1.553	−7.210
Nov	1.98	5.132	2.0	5.184	−0.052	−7.262
Dec	2.12	5.678	2.0	5.357	0.3214	−6.941
1978						
Jan	1.78	4.768	2.0	5.357	−0.5892	−7.530
Feb	1.95	4.717	2.0	4.838	−0.121	−7.651
Mar	7.25	19.418	2.0	5.357	14.061	
Apr	24.7					
May	6.26					
Jun	8.92					
Jul	3.57					
Aug	1.98	5.303	2.0	5.357	−0.0536	−0.0537
Sep	1.95	5.054	2.0	5.184	−0.1296	−0.1832
Oct	3.09	8.276	2.0	5.357	2.919	
Nov	3.94					
Dec	12.7					

The data in the first and second columns of the table were extracted from Table 3-4.

The third column is the product of the second column and the time interval for the month. For example, for August (31 d) and September (30 d), 1976:

$$(1.70 \text{ m}^3/\text{s})(31 \text{ d})(86{,}400 \text{ s/d}) = 4{,}553{,}280 \text{ m}^3$$

$$(1.56 \text{ m}^3/\text{s})(30 \text{ d})(86{,}400 \text{ s/d}) = 4{,}043{,}520 \text{ m}^3$$

The fourth column is the demand given in the problem statement.

The fifth column is the product of the demand and the time interval for the month. For example, for August and September 1976:

$$(2.0 \text{ m}^3/\text{s})(31 \text{ d})(86{,}400 \text{ s/d}) = 5{,}356{,}800 \text{ m}^3$$
$$(2.0 \text{ m}^3/\text{s})(30 \text{ d})(86{,}400 \text{ s/d}) = 5{,}184{,}000 \text{ m}^3$$

The sixth column (ΔS) is the difference between the third and fifth columns. For example, for August and September 1976:

$$4{,}553{,}280 \text{ m}^3 - 5{,}356{,}800 \text{ m}^3 = -803{,}520 \text{ m}^3$$
$$4{,}043{,}520 \text{ m}^3 - 5{,}184{,}000 \text{ m}^3 = -1{,}140{,}480 \text{ m}^3$$

The last column ($\Sigma(\Delta S)$) is the sum of the last value in that column and the value in the sixth column. For August 1976, it is $-803{,}520 \text{ m}^3$ since this is the first value. For September 1976, it is

$$(-803{,}520 \text{ m}^3) + (-1{,}140{,}480 \text{ m}^3) = -1{,}944{,}000 \text{ m}^3$$

The following logic is used in interpreting the table. From August through December 1976, the demand exceeds the flow, and storage must be provided. The maximum storage required for this interval is $3.122 \times 10^6 \text{ m}^3$. In January 1977, the storage (ΔS) exceeds the deficit ($\Sigma(\Delta S)$) from December 1976. If we view the deficit as the volume of water in a virtual reservoir with a total capacity of $3.122 \times 10^6 \text{ m}^3$, then in December 1976, the volume of water in the reservoir is $1.041 \times 10^6 \text{ m}^3$ ($3.122 \times 10^6 - 2.081 \times 10^6$). The January 1977 inflow exceeds the demand and fills the reservoir deficit of $2.081 \times 10^6 \text{ m}^3$.

Because the inflow (Q_{in}) exceeds the demand ($2.0 \text{ m}^3/\text{s}$) for the months of February through June 1977, no storage is required during this period. Hence, no computations were performed.

From July 1977 through February 1978, the demand exceeds the inflow, and storage is required. The maximum storage required is $7.651 \times 10^6 \text{ m}^3$. Note that the computations for storage did not stop in December 1977, even though the inflow exceeded the demand. This is because the storage was not sufficient to fill the reservoir deficit. The storage was sufficient to fill the reservoir deficit in March 1978.

You should note that these tabulations are particularly well suited to spreadsheet-type programs.

The storage volume determined by the Rippl method must be increased to account for water lost through evaporation and volume lost through the accumulation of sediment.

Application of the Mass Diagram. When existing storm sewers cannot handle runoff, consideration is given to providing a *retention basin* (storage pond) as an alternative to replacing the existing storm sewer. The retention basin serves two functions. First, it delays the coincidence of peak flows, and second, it can be designed to release storm water at a rate the storm sewer can handle. The procedure for applying the mass diagram to retention basins is basically the same as it is for water supply. The DRH for the design storm is substituted for the monthly discharge readings. The allowable discharge to the existing sewer is substituted for Q_{out}.

Wastewater treatment plants function best when the flow into the plant is constant. The general daily fluctuation in water use in most municipalities and industries results in an alternating pattern of high and low wastewater discharge. The peaks and valleys of flow can be evened out with an *equalization basin*. The volume of the equalization basin is chosen to shave off the peaks and fill in the valleys of the inflow mass diagram.

3-5 GROUNDWATER AND WELLS

Although the portion of the population of the United States supplied by surface water is 58 percent greater than that supplied by groundwater, the number of communities supplied by groundwater is almost 12 times that supplied by surface water (Figure 3-22). The reason for this pattern is that larger cities are supplied by surface water while many small communities use groundwater.

Groundwater has several characteristics that make it desirable as a water supply source. First, the groundwater system provides natural storage, which eliminates the cost of impoundment works. Second, since the supply frequently is available at the point of demand, the cost of transmission is reduced significantly. Third, because groundwater is filtered by the natural geologic strata, groundwater is clearer to the eye than surface water.

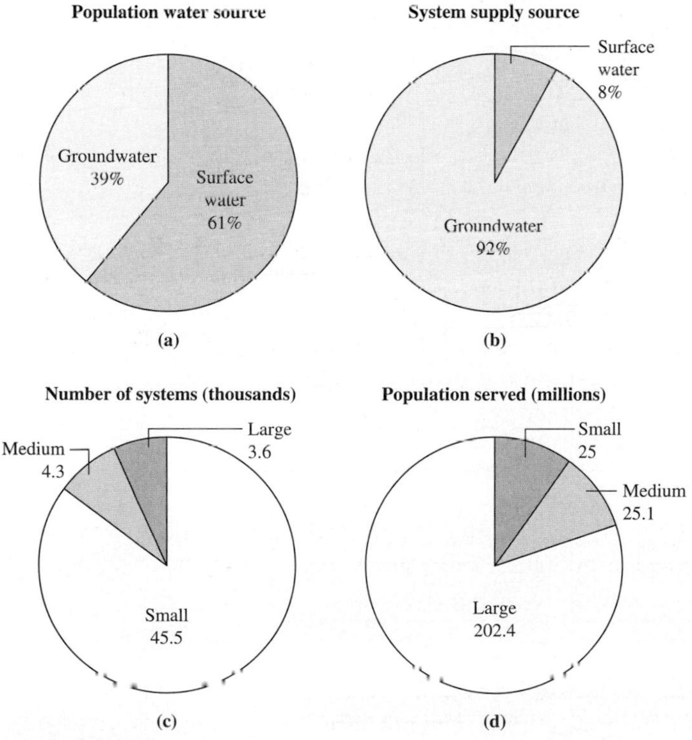

FIGURE 3-22
(*a*) Percentage of the population served by drinking-water system source. (*b*) Percentage of drinking-water systems by supply source. (*c*) Number of drinking-water systems (in thousands) by size. (*d*) Population served (in millions of people) by drinking-water system size. *Source:* 1997 National Public Water Systems Compliance Report. U.S. EPA, Office of Water. Washington, D.C. 20460. (EPA-305-R-99-002). (*Note*: Small systems serve 25-3,300 people; medium systems serve 3,301–10,000 people; large systems serve 10,000+ people.)

Construction of Wells

Modern wells consist of more than a simple hole in the ground (Figure 3-23). A steel pipe called a *casing* is placed in the well hole to maintain the integrity of the hole. The casing is sealed to the surrounding soil with a cement grout, and a screen is placed at the bottom of the casing to allow water in and to keep soil material out. Two types of pump may be used. In the diagram shown, the pump motor is at the ground surface and the pump itself is placed down in the well above the well screen. The alternative is a submersible pump. In this case both the pump and the motor are lowered into the casing; water is pumped out of the well through a discharge pipe or drop pipe.

Sanitary Considerations. The penetration of a water-bearing formation by a well provides a direct route for possible contamination of the groundwater. Although there are different types of wells and well construction, there are basic sanitary aspects that must be considered and followed (refer to Figure 3-24):

1. The annular space outside the casing should be filled with a watertight cement grout or puddled clay from the surface to the depth necessary to prevent entry of contaminated water. A minimum of 6 m is recommended.

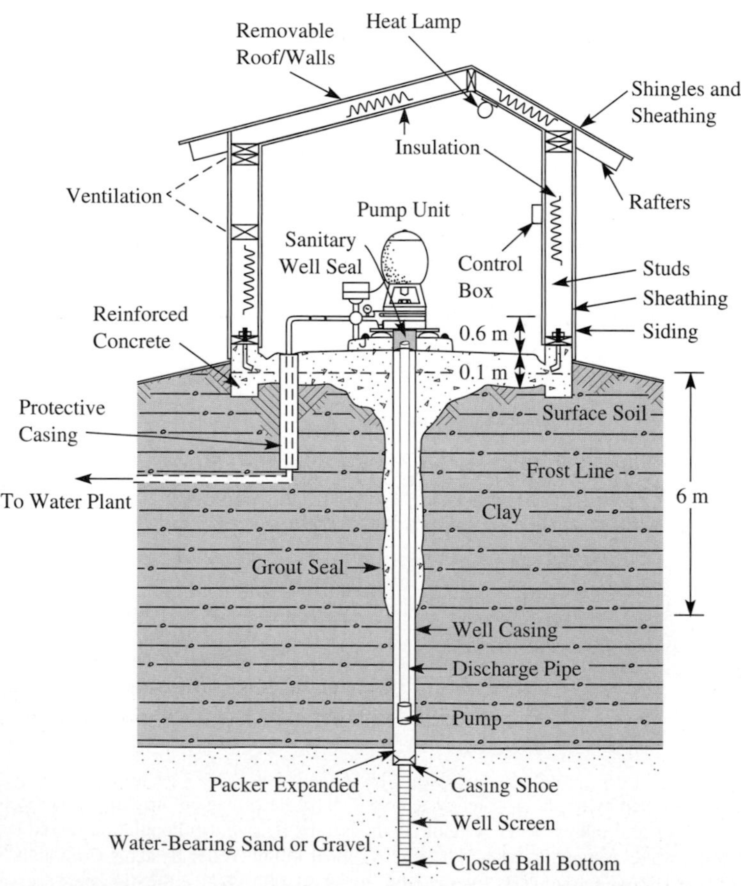

FIGURE 3-23
Pumphouse. (U.S. EPA, 1973)

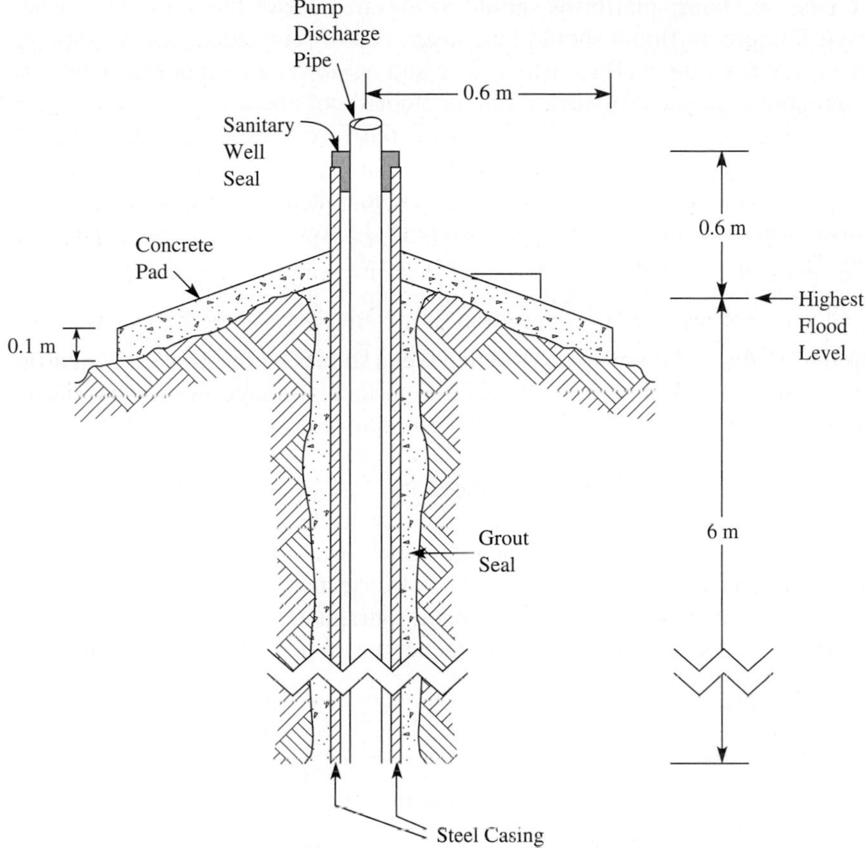

FIGURE 3-24
Sanitary considerations in well construction.

2. For artesian aquifers, the casing should be sealed into the overlying impermeable formations so as to retain the artesian pressure.
3. When a water-bearing formation containing water of poor quality is penetrated, the formation should be sealed off to prevent the infiltration of water into the well and aquifer.
4. A sanitary well-seal with an approved vent should be installed at the top of the well casing to prevent the entrance of contaminated water or other objectionable material (U.S. EPA, 1973).

Well Covers and Seals. Every well should be provided with an overlapping, tight-fitting cover at the top of the casing or pipe sleeve to prevent contaminated water or other material from entering the well.

The seal in a well that is exposed to possible flooding should be elevated at least 0.6 m above the highest known flood level. When this is not possible, the seal should be watertight and equipped with a vent line whose opening to the atmosphere is at least 0.6 m above the highest known flood level.

Well covers and pump platforms should be elevated above the adjacent finished ground level. Pumproom floors should be constructed of reinforced, watertight concrete sloped away from the well so that surface and wastewater cannot stand near the well. The minimum thickness of such a slab or floor should be 0.1 m. Concrete slabs or floors should be poured separately from the grout formation seal and, where the threat of freezing exists, insulated from it and the well casing by a plastic or mastic coating or sleeve to prevent bonding of the concrete to either.

All water wells should be readily accessible at the top for inspection, servicing, and testing. This requires that any structure over the well be easily removable to provide full, unobstructed access for well-servicing equipment.

Disinfection of Wells. All newly constructed wells should be disinfected to neutralize contamination from equipment, material, or surface drainage introduced during construction. Every well should be disinfected promptly after construction or repair.

Pumphousing. A pumphouse installed above the surface of the ground should be used. It should be unnecessary to use an underground discharge connection if an insulated, heated pumphouse is provided. For individual installations in rural areas, two 60-watt light bulbs, a thermostatically controlled electric heater, or a heating cable will generally provide adequate protection when the pumphouse is properly insulated. Because power failures may occur, an emergency gasoline-driven power supply or pump should be considered.

Cone of Depression

When a well is pumped, the level of the piezometric surface in the vicinity of the well will be lowered (Figure 3-25). This lowering, or drawdown, causes the piezometric surface to take the shape of an inverted cone called a cone of depression. Because the water level in a pumped well is lower than that in the aquifer surrounding it, the water flows from the aquifer into the well. At increasing distances from the well, the drawdown decreases until the slope of the cone merges with the static water table. The distance from the well at which this occurs is called the radius of influence. The radius of

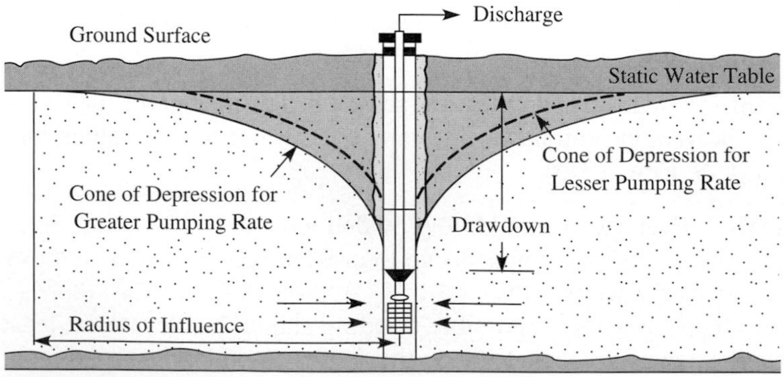

FIGURE 3-25
Effect of pumping rate on cone of depression. (U.S. EPA, 1973)

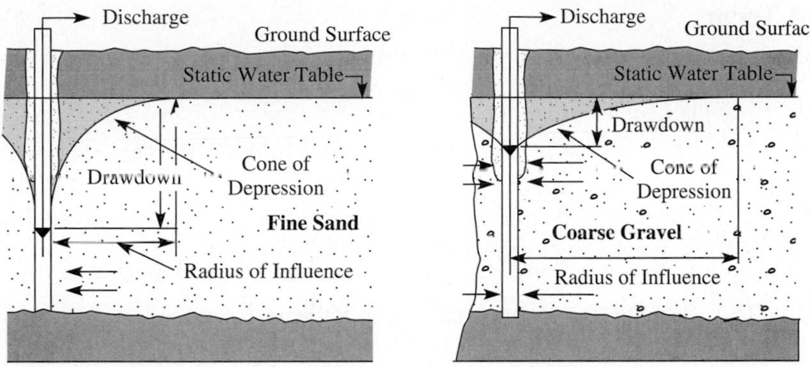

FIGURE 3-26
Effect of aquifer material on cone of depression. (U.S. EPA, 1973)

influence is not constant but tends to expand with continued pumping. At a given pumping rate, the shape of the cone of depression depends on the characteristics of the water-bearing formation. Shallow and wide cones will form in aquifers composed of coarse sands or gravel. Deeper and narrower cones will form in fine sand or sandy clay (Figure 3-26). As the pumping rate increases, the drawdown increases. Consequently the slope of the cone steepens. When other conditions are equal for two wells, it may be expected that pumping costs will be higher for the well surrounded by the finer material because of greater drawdown.

When the cones of depression overlap, the local water table will be lowered (Figure 3-27). This requires additional pumping lifts to obtain water from the interior portion of the group of wells. A wider distribution of the wells over the groundwater basin will reduce the cost of pumping and will allow the development of a larger quantity of water. One rule of thumb is that two wells should be placed no closer together than two times the thickness of the water-bearing strata. For more than two wells, they should be spaced at least 75 meters apart.

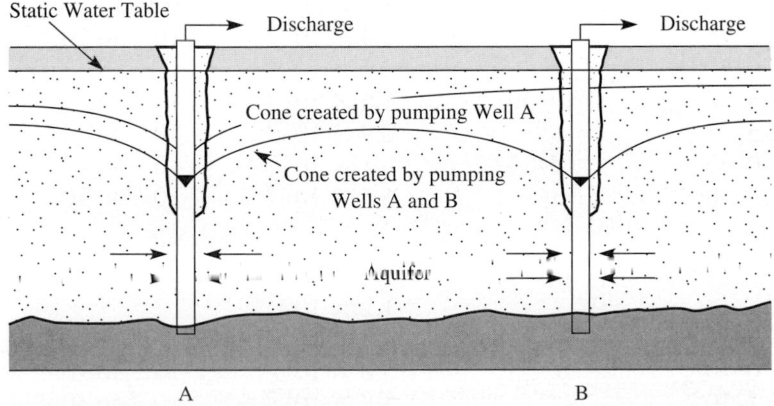

FIGURE 3-27
Effect of overlapping cones of depression. (U.S. EPA, 1973)

Definition of Terms

The aquifer parameters identified and defined in this section are those relevant to determining the available volume of water and the ease of its withdrawal.

Porosity. The ratio of the volume of voids (open spaces) in the soil to the total volume is called *porosity.* It is a measure of the amount of water that can be stored in the spaces between soil particles. It does not indicate how much of this water is available for development. Some typical values are shown in Table 3-5.

Specific Yield. The percentage of water that is free to drain from the aquifer under the influence of gravity is defined as *specific yield* (Figure 3-28). Specific yield is not equal to porosity because the molecular and surface tension forces in the pore spaces keep some of the water in the voids. Specific yield reflects the amount of water available for development. Some values are shown in Table 3-5.

Storage Coefficient (S). This parameter is akin to specific yield. The *storage coefficient* is the volume of available water resulting from a unit decline in the piezometric surface over a unit horizontal cross-sectional area. It has units of m^3 of water/m^3 of aquifer or, in essence, no units at all! Storage coefficient and specific yield may be used interchangeably for unconfined aquifers. Values of S for unconfined aquifers range from 0.01 to 0.35. For confined aquifers the values of S vary from 1×10^{-3} to 1×10^{-5}.

TABLE 3-5
Values of aquifer parameters

Aquifer material	Typical porosity (%)	Range of porosities (%)	Range of specific yield (%)	Typical hydraulic conductivity (m/s)	Range of hydraulic conductivities (m/s)
Unconsolidated					
Clay	55	50–60	1–10	1.2×10^{-6}	$0.1\text{–}2.3 \times 10^{-6}$
Loam	35	25–45		6.4×10^{-6}	10^{-6} to 10^{-5}
Fine sand	45	40–50		3.5×10^{-5}	$1.1\text{–}5.8 \times 10^{-5}$
Medium sand	37	35–40	10–30	1.5×10^{-4}	10^{-5} to 10^{-4}
Coarse sand	30	25–35		6.9×10^{-4}	10^{-4} to 10^{-3}
Sand and gravel	20	10–30	15–25	6.1×10^{-4}	10^{-5} to 10^{-3}
Gravel	25	20–30		6.4×10^{-3}	10^{-3} to 10^{-2}
Consolidated					
Shale	<5		0.5–5	1.2×10^{-12}	
Granite	<1			1.2×10^{-10}	
Sandstone	15	5–30	5–15	5.8×10^{-7}	10^{-8} to 10^{-5}
Limestone	15	10–20	0.5–5	5.8×10^{-6}	10^{-7} to 10^{-5}
Fractured rock	5	2–10		5.8×10^{-5}	10^{-8} to 10^{-4}

Adapted from Bouwer, 1978, Linsley et al., 1975, and Walton, 1970.

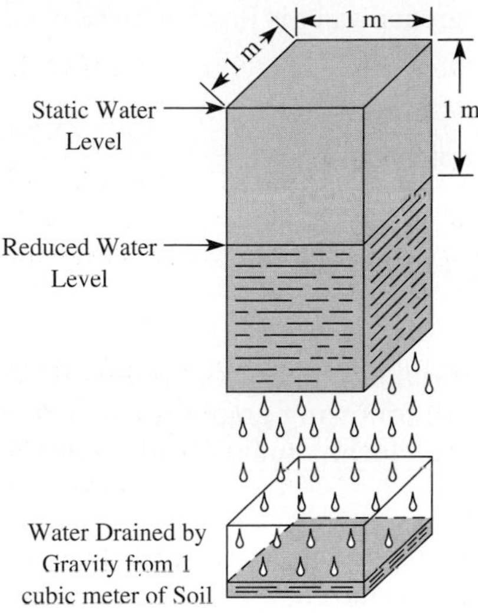

$$\text{Specific Yield} = \frac{\text{Volume Water}}{\text{Volume Soil}} (100\%)$$

FIGURE 3-28
Specific yield. (Adapted from Johnson Division, UOP, 1975)

Hydraulic Gradient and Head. The slope of the piezometric surface is called the *hydraulic gradient.* It is measured in the direction of the steepest slope of the piezometric surface. Groundwater flows in the direction of the hydraulic gradient and at a rate proportional to the slope.

The vertical distance from a reference plane (zero datum) to the bottom of a well or piezometer (Figure 3-29) is called the *elevation head.* The height to which water will rise in the piezometer is called the *pressure head.* The sum of the elevation head and the pressure head is called the *total head.*

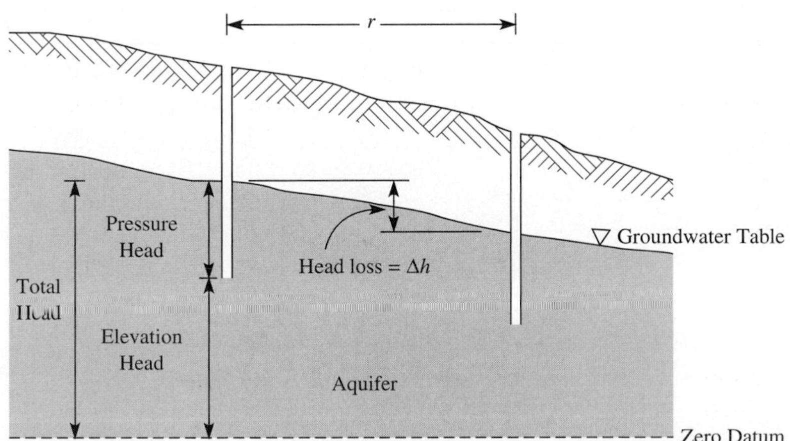

FIGURE 3-29
Geometry for definition of head and hydraulic gradient.

Using Figure 3-29 and the assumption that the groundwater is flowing in the plane of the page, we can define the gradient to be:

$$\text{Hydraulic gradient} = \frac{\text{Change in head}}{\text{Horizontal distance}} \tag{3-20}$$

or in the differential sense:

$$\text{Hydraulic gradient} = \frac{\Delta h}{r} = \frac{dh}{dr} \tag{3-21}$$

We generally require three wells, one of which is out of plane with the other two, to define the hydraulic gradient. Heath suggests the following graphical procedure for finding the hydraulic gradient between three closely spaced wells (Heath, 1983):

1. Draw a line between the two wells with the highest and lowest head and divide it into equal intervals. By interpolation, find the place on the line between these two wells where the head is equal to the head of the third well.
2. Draw a line from the third well to the point on the line between the first two wells where the head is the same as that in the third well. This line is called an *equipotential line.* This means that the head anywhere along the line should be constant. Groundwater will flow in a direction perpendicular to this line.
3. Draw a line perpendicular to the equipotential line through the well with the lowest or highest head. The groundwater flow is in a direction parallel to this line. It is called a *flow line.*
4. Calculate the gradient as the difference in head between the head on the equipotential line and the head at the lowest or highest well divided by the distance from the equipotential line to that well.

Example 3-10. For the wells shown in plan view below, determine the direction of flow and the hydraulic gradient. The total head is given for each well as follows:

Well A = 10.4 m

Well B = 9.9 m

Well C = 10.0 m

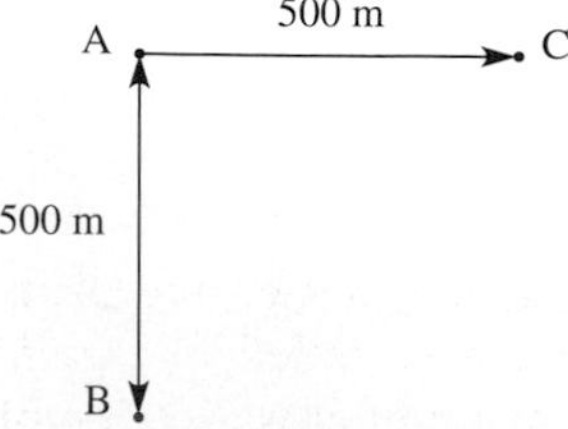

Solution. The stepwise graphical solution procedure is shown below.

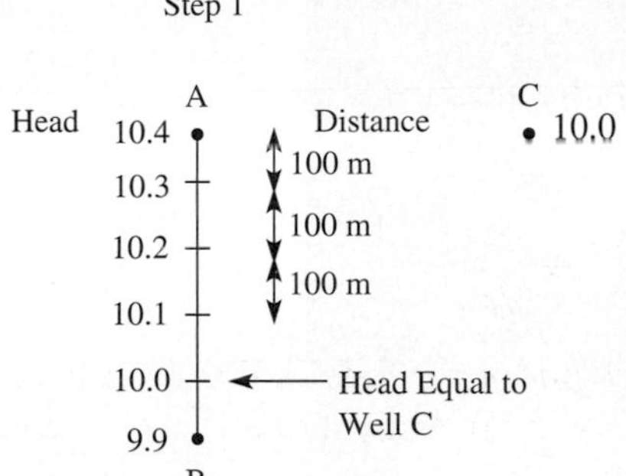

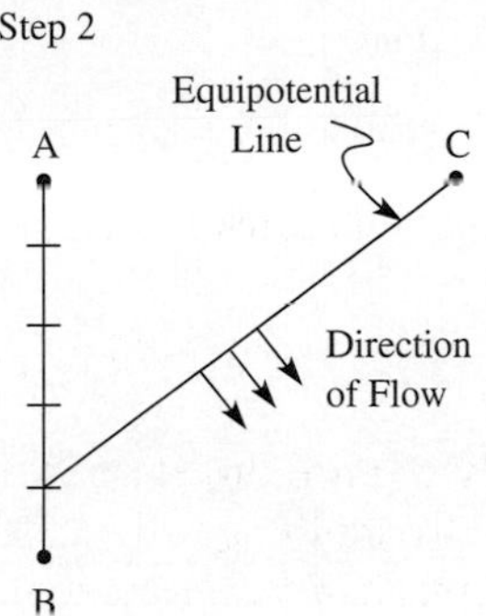

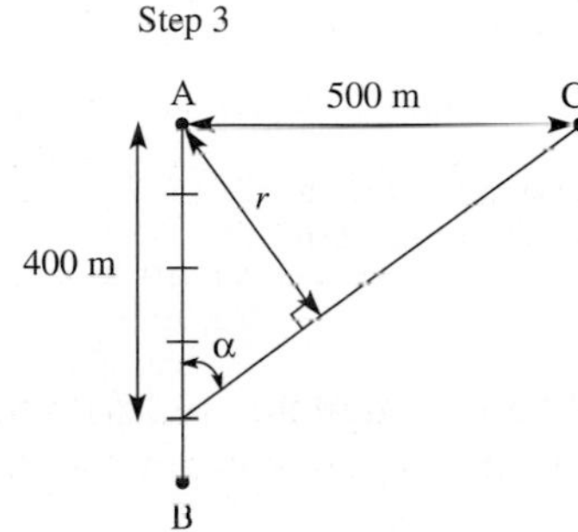

The distance r must be determined in order to calculate the hydraulic gradient. From the plan view, we may note that the wells form a right triangle with legs of 400 m and 500 m. The angle α may be computed as:

$$\tan^{-1}(\alpha) = \frac{500}{400}$$

and $\alpha = 51.34°$.

The distance r is

$$r = (400)\sin\alpha = 400 \sin 51.34 = 312.35 \text{ m}$$

The hydraulic gradient is then:

$$\text{Hydraulic gradient} = \frac{10.4 \text{ m} - 10.0 \text{ m}}{312.35 \text{ m}} = 0.00128$$

Note that the hydraulic gradient has no units.

Hydraulic Conductivity (K). The property of an aquifer that is a measure of its ability to transmit water under a sloping piezometric surface is called *hydraulic conductivity.* It is defined as the discharge that occurs through a unit cross section of

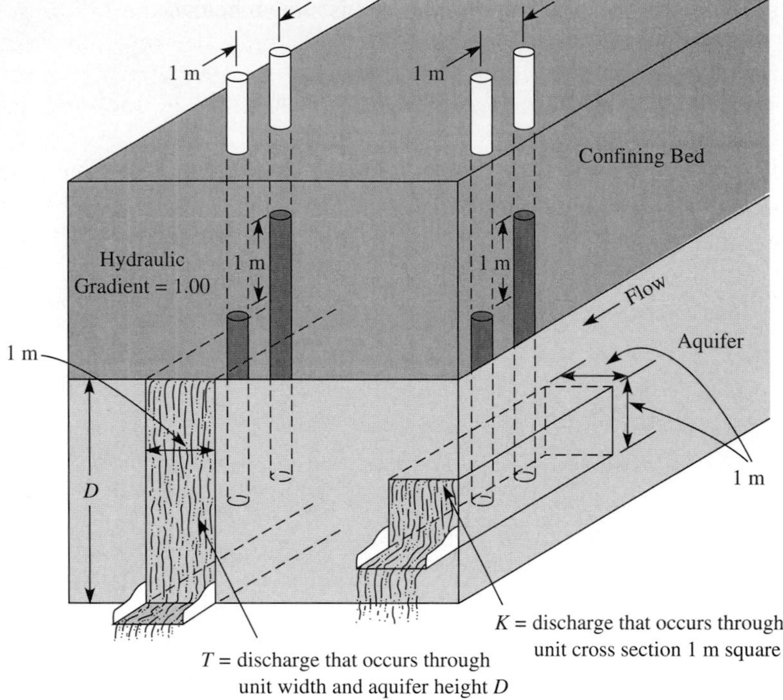

FIGURE 3-30
Illustration of definition of hydraulic conductivity (K) and transmissivity (T). (Adapted from Johnson, UOP, 1975)

aquifer (Figure 3-30) under a hydraulic gradient of 1.00. It has units of speed (m/s). Typical values are given in Table 3-5.

Transmissivity (T). The coefficient of transmissivity (T) is a measure of the rate at which water will flow through a unit width vertical strip of aquifer extending through its full saturated thickness (Figure 3-30) under a unit hydraulic gradient. It has units of m^2/s. Values of the transmissivity coefficient range from 1.0×10^{-4} to 1.5×10^{-1} m^2/s.

Well Hydraulics

Equations for calculating the discharge that results when the piezometric surface is lowered are based on the work of the French hydrologist Henri Darcy. In 1856 he discovered that the velocity of water flow in a porous aquifer was proportional to the hydraulic gradient. He proposed the following equation, which is now called Darcy's law (Darcy, 1856):

$$v = K\frac{dh}{dr} \tag{3-22}$$

where v = velocity, m/s
K = hydraulic conductivity, m/s
dh/dr = slope of the hydraulic gradient, m/m

The *Darcy velocity* is not a real velocity because it assumes that the full cross-sectional area of the aquifer is available for water to flow through. However, much of the cross-sectional area is soil material; thus, the actual area through which the water flows is much less. As a result, the actual linear velocity of the groundwater is considerably faster than the Darcy velocity.

We may determine the actual average linear velocity by taking into account the fraction of the cross-sectional area filled with soil. Let us define the flow (Q) to be the product of the total cross-sectional area and the Darcy velocity, that is $Q = Av$. The actual average linear velocity (v') through the voids times the area of the voids (A') is also equal to the flow (Q). Thus, the actual velocity is related to the Darcy velocity:

$$A'v' = Av = Q \tag{3-23}$$

If we solve this expression for the average linear velocity, v', we find:

$$v' = \frac{Av}{A'} \tag{3-24}$$

If we multiply the top and bottom by a unit length (L), then we have

$$v' = \frac{ALv}{A'L} \tag{3-25}$$

The product AL is the total volume of the soil. The product $A'L$ is the void volume. The ratio of the void volume to the total volume is the definition of porosity (η). The actual linear velocity may then be defined in terms of the porosity as

$$v' = \frac{\text{Darcy velocity}}{\text{Porosity}} = \frac{v}{\eta} \tag{3-26}$$

or in terms of the hydraulic gradient

$$v' = \frac{K(dh/dr)}{\eta} \tag{3-27}$$

The gross discharge is the product of the velocity of flow and the area (A) through which it flows.

$$Q = vA = KA\frac{dh}{dr} \tag{3-28}$$

This equation has been solved for *steady state* and nonsteady or *transient flow*. Steady state is a condition under which no changes occur with time. It will seldom, if ever, occur in practice, but may be approached after very long periods of pumping. Transient-flow equations include a factor of time. The derivation of these equations is based on the following assumptions:

1. The well is pumped at a constant rate.
2. Flow towards the well is radial and uniform.
3. Initially the piezometric surface is horizontal.

4. The well fully penetrates the aquifer and is open for the entire height of the aquifer.
5. The aquifer is homogeneous in all directions and is of infinite horizontal extent.
6. Water is released from the aquifer in immediate response to a drop in the piezometric surface.

Steady Flow in a Confined Aquifer. The equation describing steady, confined aquifer flow was first presented by Dupuit in 1863 (Dupuit, 1863) and subsequently extended by Theim in 1906 (Theim, 1906). It may be written as follows (refer to Figure 3-31 for an explanation of the notation):

$$Q = \frac{2\pi T(h_2 - h_1)}{\ln(r_2/r_1)} \tag{3-29}$$

Q = well pumping flow rate, m^3/s

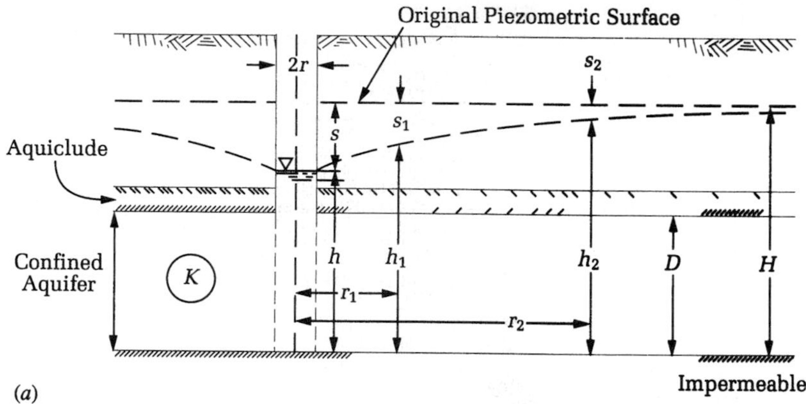

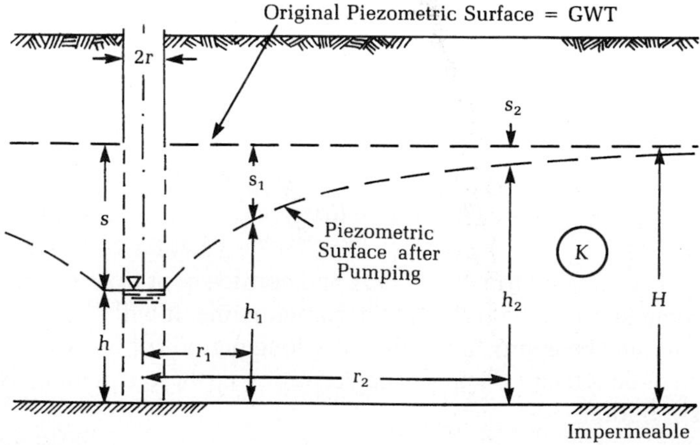

FIGURE 3-31

Geometry and symbols for a pumped well in (*a*) confined aquifer and (*b*) unconfined aquifer. (Adapted from Bouwer, 1978.)

where $T = KD$ = transmissivity, m^2/s
D = thickness of artesian aquifer, m
h_1, h_2 = height of piezometric surface above confining layer, m
r_1, r_2 = radius from pumping well, m
ln = logarithm to base e

Example 3-11. An artesian aquifer 10.0 m thick with a piezometric surface 40.0 m above the bottom confining layer is being pumped by a fully penetrating well. The aquifer is a medium sand with a hydraulic conductivity of 1.50×10^{-4} m/s. Steady state drawdowns of 5.00 m and 1.00 m are observed at two nonpumping wells located 20.0 m and 200.0 m, respectively, from the pumped well. Determine the discharge at the pumped well.

Solution. First we determine h_1 and h_2:

$$h_1 = 40.0 - 5.00 = 35.0 \text{ m}$$
$$h_2 = 40.0 - 1.00 = 39.0 \text{ m}$$

so

$$Q = \frac{(2\pi)(1.50 \times 10^{-4})(10.0)(39.0 - 35.0)}{\ln(200/20)}$$
$$Q = 0.0164 \text{ or } 0.016 \text{ m}^3/\text{s}$$

Steady Flow in an Unconfined Aquifer. For unconfined aquifers the factor D in Equation 3-29 is replaced by the height of the water table above the lower boundary of the aquifer. The equation then becomes

$$Q = \frac{\pi K(h_2^2 - h_1^2)}{\ln(r_2/r_1)} \tag{3-30}$$

Example 3-12. A 0.50 m diameter well fully penetrates an unconfined aquifer which is 30.0 m thick. The drawdown at the pumped well is 10.0 m and the hydraulic conductivity of the gravel aquifer is 6.4×10^{-3} m/s. If the flow is steady and the discharge is 0.014 m^3/s, determine the drawdown at a site 100.0 m from the well.

Solution. First we calculate h_1

$$h_1 = 30.0 - 10.0 = 20.0 \text{ m}$$

Then we apply Equation 3-30 and solve for h_2. Note that $r_1 = 0.50 \text{ m}/2 = 0.25$ m.

$$0.014 = \frac{\pi(6.4 \times 10^{-3})(h_2^2 - (20.0)^2)}{\ln(100/0.25)}$$

$$h_2^2 - 400.0 = \frac{(0.014)(5.99)}{\pi(6.4 \times 10^{-3})}$$

$$h_2 = (4.17 + 400.0)^{1/2}$$

$$h_2 = 20.10 \text{ m}$$

The drawdown is then

$$s_2 = H - h_2 = 30.0 - 20.10 = 9.90 \text{ m}$$

Unsteady Flow in a Confined Aquifer. A solution for the transient-flow problem was developed by Theis in 1935 (Theis, 1935). Using heat-flow theory as an analogy, he found the following for an infinitesimally small diameter well with radial flow:

$$s = \frac{Q}{4\pi T}\int_u^{\infty}\left(\frac{e^{-u}}{u}\right)du \tag{3-31}$$

where s = drawdown $(H - h)$, m

$u = \dfrac{r^2S}{4Tt}$

r = distance between pumping well and observation well, or radius of pumping well, m

S = storage coefficient

T = transmissivity, m^2/s

t = time since pumping began, s

Some explanations of the terms may be of use here. The lower case s refers to the drawdown at some time, t, after the start of pumping. The time does not appear explicitly in Equation 3-31 but is used to compute the value of u to be used in the integration. The transmissivity and storage coefficient also are used to calculate u. The transmissivity may be determined from the hydraulic conductivity and the thickness of the aquifer as it was for steady-state flow. Field pumping tests may also be used to define T. You should note that the r term used to calculate the value of u may take on values ranging upward from the radius of the well. Thus, you could if you wished, calculate every point on the cone of depression (i.e., value of s) by iterating the calculation with values of r from the well radius to infinity. If you wish to calculate the drawdown at a specific distance from the pumping well, then, of course, you must use that distance for r. The integral in Equation 3-31 is called the well function of u and is evaluated by the following series expansion:

$$W(u) = -0.577216 - \ln u + u - \frac{u^2}{2 \cdot 2!} + \frac{u^3}{3 \cdot 3!} - \cdots \tag{3-32}$$

A table of values of $W(u)$ was prepared by Ferris, *et al.* (Ferris et al., 1962). It is reproduced in Table 3-6.

TABLE 3-6

Values of $W(u)$

$N \backslash u$	$N \times 10^{-15}$	$N \times 10^{-14}$	$N \times 10^{-13}$	$N \times 10^{-12}$	$N \times 10^{-11}$	$N \times 10^{-10}$	$N \times 10^{-9}$	$N \times 10^{-8}$	$N \times 10^{-7}$	$N \times 10^{-6}$	$N \times 10^{-5}$	$N \times 10^{-4}$	$N \times 10^{-3}$	$N \times 10^{-2}$	$N \times 10^{-1}$	N
1.0	33.9616	31.6590	29.3564	27.0538	24.7512	22.4486	20.1460	17.8435	15.5409	13.2383	10.9357	8.6332	6.3315	4.0379	1.8229	0.2194
1.1	33.8662	31.5637	29.2611	26.9585	24.6559	22.3533	20.0507	17.7482	15.4456	13.1430	10.8404	8.5379	6.2363	3.9436	1.7371	0.1860
1.2	33.7792	31.4767	29.1741	26.8715	24.5689	22.2663	19.9637	17.6611	15.3586	13.0560	10.7534	8.4509	6.1494	3.8576	1.6595	0.1584
1.3	33.6992	31.3966	29.0940	26.7914	24.4889	22.1863	19.8837	17.5811	15.2785	12.9759	10.6734	8.3709	6.0695	3.7785	1.5889	0.1355
1.4	33.6251	31.3225	29.0199	26.7173	24.4147	22.1122	19.8096	17.5070	15.2044	12.9018	10.5993	8.2968	5.9955	3.7054	1.5241	0.1162
1.5	33.5561	31.2535	28.9509	26.6483	24.3458	22.0432	19.7406	17.4380	15.1354	12.8328	10.5303	8.2278	5.9266	3.6374	1.4645	0.1000
1.6	33.4916	31.1890	28.8864	26.5838	24.2812	21.9786	19.6760	17.3735	15.0709	12.7683	10.4657	8.1634	5.8621	3.5739	1.4092	0.08631
1.7	33.4309	31.1283	28.8258	26.5232	24.2206	21.9180	19.6154	17.3128	15.0103	12.7077	10.4051	8.1027	5.8016	3.5143	1.3578	0.07465
1.8	33.3738	31.0712	28.7686	26.4660	24.1634	21.8608	19.5583	17.2557	14.9531	12.6505	10.3479	8.0455	5.7446	3.4581	1.3089	0.06471
1.9	33.3197	31.0171	28.7145	26.4119	24.1094	21.8068	19.5042	17.2016	14.8990	12.5964	10.2939	7.9915	5.6906	3.4050	1.2649	0.05620
2.0	33.2684	30.9658	28.6632	26.3607	24.0581	21.7555	19.4529	17.1503	14.8477	12.5451	10.2426	7.9402	5.6394	3.3547	1.2227	0.04890
2.1	33.2196	30.9170	28.6145	26.3119	24.0093	21.7067	19.4041	17.1015	14.7969	12.4964	10.1938	7.8914	5.5907	3.3069	1.1829	0.04261
2.2	33.1731	30.8705	28.5679	26.2653	23.9628	21.6602	19.3576	17.0550	14.7524	12.4498	10.1473	7.8449	5.5443	3.2514	1.1454	0.03719
2.3	33.1286	30.8261	28.5235	26.2209	23.9183	21.6157	19.3131	17.0106	14.7080	12.4054	10.1028	7.8004	5.4999	3.2179	1.1099	0.03250
2.4	33.0861	30.7835	28.4809	26.1783	23.8758	21.5732	19.2706	16.9680	14.6654	12.3628	10.0603	7.7579	5.4575	3.1763	1.0762	0.02844
2.5	33.0453	30.7427	28.4401	26.1375	23.8349	21.5323	19.2298	16.9272	14.6246	12.3220	10.0194	7.7172	5.4167	3.1365	1.0443	0.02491
2.6	33.0060	30.7035	28.4009	26.0983	23.7957	21.4931	19.1905	16.8880	14.5854	12.2828	9.9802	7.6779	5.3776	3.0983	1.0139	0.02185
2.7	32.9683	30.6657	28.3631	26.0606	23.7580	21.4554	19.1528	16.8502	14.5476	12.2450	9.9425	7.6401	5.3400	3.0615	0.9849	0.01918
2.8	32.9319	30.6294	28.3268	26.0242	23.7216	21.4190	19.1164	16.8138	14.5113	12.2087	9.9061	7.6038	5.3037	3.0261	0.9573	0.01686
2.9	32.8968	30.5943	28.2917	25.9891	23.6865	21.3839	19.0813	16.7788	14.4762	12.1736	9.8710	7.5687	5.2687	2.9920	0.9309	0.01482
3.0	32.8629	30.5604	28.2578	25.9552	23.6526	21.3500	19.0474	16.7449	14.4423	12.1397	9.8371	7.5348	5.2349	2.9591	0.9057	0.01305
3.1	32.8302	30.5276	28.2250	25.9224	23.6198	21.3172	19.0146	16.7121	14.4095	12.1069	9.8043	7.5020	5.2022	2.9273	0.8815	0.01149
3.2	32.7984	30.4958	28.1932	25.8907	23.5880	21.2855	18.9829	16.6803	14.3777	12.0751	9.7726	7.4703	5.1706	2.8965	0.8583	0.01013
3.3	32.7676	30.4651	28.1625	25.8599	23.5573	21.2547	18.9521	16.6495	14.3470	12.0444	9.7418	7.4395	5.1399	2.8668	0.8361	0.008939
3.4	32.7378	30.4352	28.1326	25.8300	23.5274	21.2249	18.9223	16.6197	14.3171	12.0145	9.7120	7.4097	5.1102	2.8379	0.8147	0.007891
3.5	32.7088	30.4062	28.1036	25.8010	23.4985	21.1959	18.8933	16.5907	14.2881	11.9855	9.6830	7.3807	5.0813	2.8099	0.7942	0.006970
3.6	32.6806	30.3780	28.0755	25.7729	23.4703	21.1677	18.8651	16.5625	14.2599	11.9574	9.6548	7.3526	5.0532	2.7827	0.7745	0.006160
3.7	32.6532	30.3506	28.0481	25.7455	23.4429	21.1403	18.8377	16.5351	14.2325	11.9300	9.6274	7.3252	5.0259	2.7563	0.7554	0.005448
3.8	32.6266	30.3240	28.0214	25.7188	23.4162	21.1136	18.8110	16.5085	14.2059	11.9033	9.6007	7.2985	4.9993	2.7306	0.7371	0.004820
3.9	32.6006	30.2980	27.9954	25.6928	23.3902	21.0877	18.7851	16.4825	14.1799	11.8773	9.5748	7.2725	4.9735	2.7056	0.7194	0.004267
4.0	32.5753	30.2727	27.9701	25.6675	23.3649	21.0623	18.7598	16.4572	14.1546	11.8520	9.5495	7.2472	4.9482	2.6813	0.7024	0.003779
4.1	32.5506	30.2480	27.9454	25.6428	23.3402	21.0376	18.7351	16.4325	14.1299	11.8273	9.5248	7.2225	4.9236	2.6576	0.6859	0.003349
4.2	32.5265	30.2239	27.9213	25.6187	23.3161	21.0136	18.7110	16.4084	14.1058	11.8032	9.5007	7.1985	4.8997	2.6344	0.6700	0.002969
4.3	32.5029	30.2004	27.8978	25.5952	23.2926	20.9900	18.6874	16.3884	14.0823	11.7797	9.4771	7.1749	4.8762	2.6119	0.6546	0.002633
4.4	32.4800	30.1774	27.8748	25.5722	23.2696	20.9670	18.6644	16.3619	14.0593	11.7567	9.4541	7.1520	4.8533	2.5899	0.6397	0.002336
4.5	32.4575	30.1519	27.8523	25.5497	23.2471	20.9446	18.6420	16.3394	14.0368	11.7342	9.4317	7.1295	4.8310	2.5634	0.6253	0.002073
4.6	32.4355	30.1329	27.8303	25.5277	23.2252	20.9226	18.6200	16.3174	14.0148	11.7122	9.4097	7.1075	4.8091	2.5474	0.6114	0.001841
4.7	32.4140	30.1114	27.3088	25.5062	23.2037	20.9011	18.5985	16.2959	13.9933	11.6907	9.3882	7.0860	4.7877	2.5268	0.5979	0.001635
4.8	32.3929	30.0904	27.7878	25.4852	23.1826	20.8800	18.5774	16.2748	13.9723	11.6697	9.3671	7.0650	4.7667	2.5068	0.5848	0.001453
4.9	32.3723	30.0697	27.7672	25.4646	23.1620	20.8594	18.5568	16.2542	13.9516	11.6491	9.3465	7.0444	4.7462	2.4871	0.5721	0.001291
5.0	32.3521	30.0495	27.7470	25.4444	23.1418	20.8392	18.5366	16.2340	13.9314	11.6289	9.3263	7.0242	4.7261	2.4679	0.5598	0.001148
5.1	32.3323	30.0297	27.7271	25.4246	23.1220	20.8194	18.5168	16.2142	13.9116	11.6091	9.3065	7.0044	4.7064	2.4491	0.5478	0.001021
5.2	32.3129	30.0103	27.7077	25.4051	23.1026	20.8000	18.4974	16.1948	13.8922	11.5896	9.2871	6.9850	4.6871	2.4306	0.5362	0.0009086
5.3	32.2939	29.9913	27.6887	25.3861	23.0835	20.7809	18.4783	16.1758	13.8732	11.5706	9.2681	6.9659	4.6681	2.4126	0.5250	0.0008086
5.4	32.2752	29.9726	27.6700	25.3674	23.0648	20.7622	18.4596	16.1571	13.8545	11.5519	9.2494	6.9473	4.6495	2.3948	0.5140	0.0007198

TABLE 3-6

Values of $W(u)$ (continued)

N \ u	N × 10^{-15}	N × 10^{-14}	N × 10^{-13}	N × 10^{-12}	N × 10^{-11}	N × 10^{-10}	N × 10^{-9}	N × 10^{-8}	N × 10^{-7}	N × 10^{-6}	N × 10^{-5}	N × 10^{-4}	N × 10^{-3}	N × 10^{-2}	N × 10^{-1}	N
5.5	32.2568	29.9542	27.6516	25.3491	23.0465	20.7439	18.4413	16.1387	13.8361	11.5336	9.2310	6.9289	4.6313	2.3775	0.5034	0.0006409
5.6	32.2388	29.9362	27.6336	25.3310	23.0285	20.7259	18.4233	16.1207	13.8181	11.5155	9.2130	6.9109	4.6134	2.3604	0.4930	0.0005708
5.7	32.2211	29.9185	27.6159	25.3133	23.0108	20.7082	18.4056	16.1030	13.8004	11.4978	9.1953	6.8932	4.5958	2.3437	0.4830	0.0005085
5.8	32.2037	29.9011	27.5985	25.2959	22.9934	20.6908	18.3882	16.0856	13.7830	11.4804	9.1779	6.8758	4.5785	2.3273	0.4732	0.0004532
5.9	32.1866	29.8840	27.5814	25.2789	22.9763	20.6737	18.3711	16.0685	13.7659	11.4633	9.1608	6.8588	4.5615	2.3111	0.4637	0.0004039
6.0	32.1608	29.8672	27.5646	25.2620	22.9595	20.6569	18.3543	16.0517	13.7491	11.4465	9.1440	6.8420	4.5448	2.2953	0.4544	0.0003601
6.1	32.1533	29.8507	27.5481	25.2455	22.9429	20.6403	18.3378	16.0352	13.7326	11.4300	9.1275	6.8254	4.5283	2.2797	0.4454	0.0003211
6.2	32.1370	29.8344	27.5318	25.2293	22.9267	20.6241	18.3215	16.0189	13.7163	11.4138	9.1112	6.8092	4.5122	2.2645	0.4366	0.0002864
6.3	32.1210	29.8184	27.5158	25.2133	22.9107	20.6081	18.3055	16.0029	13.7003	11.3978	9.0952	6.7932	4.4963	2.2494	0.4280	0.0002555
6.4	32.1053	29.8027	27.5001	25.1975	22.8949	20.5923	18.2898	15.9872	13.6846	11.3820	9.0795	6.7775	4.4806	2.2346	0.4197	0.0002279
6.5	32.0898	29.7872	27.4846	25.1820	22.8794	20.5768	18.2742	15.9717	13.6691	11.3665	9.0640	6.7620	4.4652	2.2201	0.4115	0.0002034
6.6	32.0745	29.7719	27.4693	25.1667	22.8641	20.5616	18.2590	15.9564	13.6538	11.3512	9.0487	6.7467	4.4501	2.2058	0.4036	0.0001816
6.7	32.0595	29.7569	27.4543	25.1517	22.8491	20.5465	18.2439	15.9414	13.6388	11.3362	9.0337	6.7317	4.4351	2.1917	0.3959	0.0001621
6.8	32.0446	29.7421	27.4395	25.1369	22.8343	20.5317	18.2291	15.9265	13.6240	11.3214	9.0189	6.7169	4.4204	2.1779	0.3883	0.0001448
6.9	32.0300	29.7275	27.4249	25.1223	22.8197	20.5171	18.2145	15.9119	13.6094	11.3608	9.0043	6.7023	4.4059	2.1643	0.3810	0.0001293
7.0	32.0156	29.7131	27.4105	25.1079	22.8053	20.5027	18.2001	15.8976	13.5950	11.2924	8.9899	6.6879	4.3916	2.1508	0.3738	0.0001155
7.1	32.0015	29.6989	27.3963	25.0937	22.7911	20.4885	18.1860	15.8834	13.5808	11.2782	8.9757	6.6737	4.3775	2.1376	0.3668	0.0001032
7.2	31.9875	29.6849	27.3823	25.0797	22.7771	20.4746	18.1720	15.8694	13.5668	11.2642	8.9617	6.6598	4.3636	2.1246	0.3599	0.00009219
7.3	31.9737	29.6711	27.3685	25.0659	22.7633	20.4608	18.1582	15.8556	13.5530	11.2504	8.9479	6.6460	4.3500	2.1118	0.3532	0.00008239
7.4	31.9601	29.6575	27.3549	25.0523	22.7497	20.4472	18.1446	15.8420	13.5394	11.2368	8.9343	6.6324	4.3364	2.0991	0.3467	0.00007364
7.5	31.9467	29.6441	27.3415	25.0389	22.7363	20.4337	18.1311	15.8286	13.5260	11.2234	8.9209	6.6190	4.3231	2.0867	0.3403	0.00006583
7.6	31.9334	29.6308	27.3282	25.0257	22.7231	20.4205	18.1179	15.8153	13.5127	11.2102	8.9076	6.6057	4.3100	2.0744	0.3341	0.00005886
7.7	31.9203	29.6178	27.3152	25.0126	22.7100	20.4074	18.1048	15.8022	13.4997	11.1971	8.8946	6.5927	4.2970	2.0623	0.3280	0.00005263
7.8	31.9074	29.6048	27.3023	24.9997	22.6971	20.3945	18.0919	15.7893	13.4868	11.1842	8.8817	6.5798	4.2842	2.0503	0.3221	0.00004707
7.9	31.8947	29.5921	27.2895	24.9869	22.6844	20.3818	18.0792	15.7766	13.4740	11.1714	8.8689	6.5671	4.2716	2.0386	0.3163	0.00004210
8.0	31.8821	29.5795	27.2769	24.9744	22.6718	20.3692	18.0666	15.7640	13.4614	11.1589	8.8563	6.5545	4.2591	2.0269	0.3106	0.00003767
8.1	31.8697	29.5671	27.2645	24.9619	22.6594	20.3568	18.0542	15.7516	13.4490	11.1464	8.8439	6.5421	4.2468	2.0155	0.3050	0.00003370
8.2	31.8574	29.5548	27.2523	24.9497	22.6471	20.3445	18.0419	15.7393	13.4367	11.1342	8.8317	6.5298	4.2346	2.0042	0.2996	0.00003015
8.3	31.8453	29.5427	27.2401	24.9375	22.6350	20.3324	18.0298	15.7272	13.4246	11.1220	8.8195	6.5177	4.2226	1.9930	0.2943	0.00002699
8.4	31.8333	29.5307	27.2282	24.9256	22.6230	20.3204	18.0178	15.7152	13.4126	11.1101	8.8076	6.5057	4.2107	1.9820	0.2891	0.00002415
8.5	31.8215	29.5189	27.2163	24.9137	22.6112	20.3086	18.0060	15.7034	13.4008	11.0982	8.7957	6.4939	4.1990	1.9711	0.2840	0.00002162
8.6	31.8098	29.5072	27.2046	24.9020	22.5995	20.2969	17.9943	15.6917	13.3891	11.0865	8.7840	6.4822	4.1874	1.9604	0.2790	0.00001936
8.7	31.7982	29.4957	27.1931	24.8905	22.5879	20.2853	17.9827	15.6801	13.3776	11.0750	8.7725	6.4707	4.1759	1.9498	0.2742	0.00001733
8.8	31.7868	29.4842	27.1816	24.8790	22.5765	20.2739	17.9713	15.6687	13.3661	11.0635	8.7610	6.4592	4.1646	1.9393	0.2694	0.00001552
8.9	31.7755	29.4729	27.1703	24.8678	22.5652	20.2626	17.9600	15.6574	13.3548	11.0523	8.7497	6.4480	4.1534	1.9290	0.2647	0.00001390
9.0	31.7643	29.4618	27.1592	24.8566	22.5540	20.2514	17.9488	15.6462	13.3437	11.0411	8.7386	6.4368	4.1423	1.9187	0.2602	0.00001245
9.1	31.7533	29.4507	27.1481	24.8455	22.5429	20.2404	17.9378	15.6352	13.3326	11.0300	8.7275	6.4258	4.1313	1.9087	0.2557	0.00001115
9.2	31.7424	29.4398	27.1372	24.8346	22.5320	20.2294	17.9268	15.6243	13.3217	11.0191	8.7166	6.4148	4.1205	1.8987	0.2513	0.000009988
9.3	31.7315	29.4290	27.1264	24.8238	22.5212	20.2186	17.9160	15.6135	13.3109	11.0083	8.7058	6.4040	4.1098	1.8888	0.2470	0.000008948
9.4	31.7208	29.4183	27.1157	24.8131	22.5105	20.2079	17.9053	15.6028	13.3002	10.9976	8.6951	6.3934	4.0992	1.8791	0.2429	0.000008018
9.5	31.7103	29.4077	27.1051	24.8025	22.4999	20.1973	17.8948	15.5922	13.2896	10.9870	8.6845	6.3828	4.0887	1.8695	0.2387	0.000007185
9.6	31.6998	29.3972	27.0946	24.7920	22.4895	20.1869	17.8843	15.5817	13.2791	10.9765	8.6740	6.3723	4.0784	1.8599	0.2347	0.000006439
9.7	31.6894	29.3868	27.0843	24.7817	22.4791	20.1765	17.8739	15.5713	13.2688	10.9662	8.6637	6.3620	4.0681	1.8505	0.2308	0.000005771
9.8	31.6792	29.3766	27.0740	24.7714	22.4688	20.1663	17.8637	15.5611	13.2585	10.9559	8.6534	6.3517	4.0579	1.8412	0.2269	0.000005173
9.9	31.6690	29.3664	27.0639	24.7613	22.4587	20.1561	17.8535	15.5509	13.2483	10.9458	8.6433	6.3416	4.0479	1.8320	0.2231	0.000004637

Source: Ferris et al., 1962.

Example 3-13. If the storage coefficient is 2.74×10^{-4} and the transmissivity is 2.63×10^{-3} m^2/s, calculate the drawdown that will result at the end of 100 days of pumping a 0.61-m-diameter well at a rate of 2.21×10^{-2} m^3/s.

Solution. Begin by computing u. The radius is

$$r = \frac{0.61 \text{ m}}{2} = 0.305 \text{ m}$$

and

$$u = \frac{(0.305 \text{ m})^2(2.74 \times 10^{-4})}{4(2.63 \times 10^{-3} \text{ m}^2\text{/s})(100 \text{ d})(86{,}400 \text{ s/d})} = 2.80 \times 10^{-10}$$

The factor of 86,400 is to convert days to seconds.

From the table of $W(u)$ versus u find that at 2.8×10^{-10}, $W(u) = 21.4190$.

Compute s:

$$s = \frac{2.21 \times 10^{-2} \text{ m}^3\text{/s}}{4(3.14)2.63 \times 10^{-3} \text{ m}^3\text{/s}}(21.4190)$$
$$= 14.33 \text{ or } 14 \text{ m}$$

Unsteady Flow in an Unconfined Aquifer. There is no exact solution to the transient-flow problem for unconfined aquifers because T changes with time and r as the water table is lowered. Furthermore, vertical-flow components near the well invalidate the assumption of radial flow that is required to obtain an analytical solution. If the unconfined aquifer is very deep in comparison to the drawdown, the transient-flow solution for a confined aquifer may be used for an approximate solution. In general, however, numerical methods yield more satisfactory solutions.

Determining the Hydraulic Properties of a Confined Aquifer. The estimation of the transmissivity and storage coefficient of an aquifer is based on the results of a pumping test. The preferred situation is one in which one or more observation wells located at a distance from the pumping well are used to gather the data.

The transmissivity may be determined in the steady-state condition by using Equation 3-29. If we define drawdown as $s = H - h$, then the rearrangement of Equation 3-29 yields

$$T = \frac{Q\ln(r_2/r_1)}{2\pi(s_1 - s_2)} \quad (3\text{-}33)$$

where s_1 = drawdown at radius r_1, m
s_2 = drawdown at radius r_2, m

In the transient state we cannot solve for T and S directly. We have selected the Cooper and Jacob method from the several indirect methods that are available (Cooper and Jacob, 1946). For values of u less than 0.01, they found that Equation 3-31 could

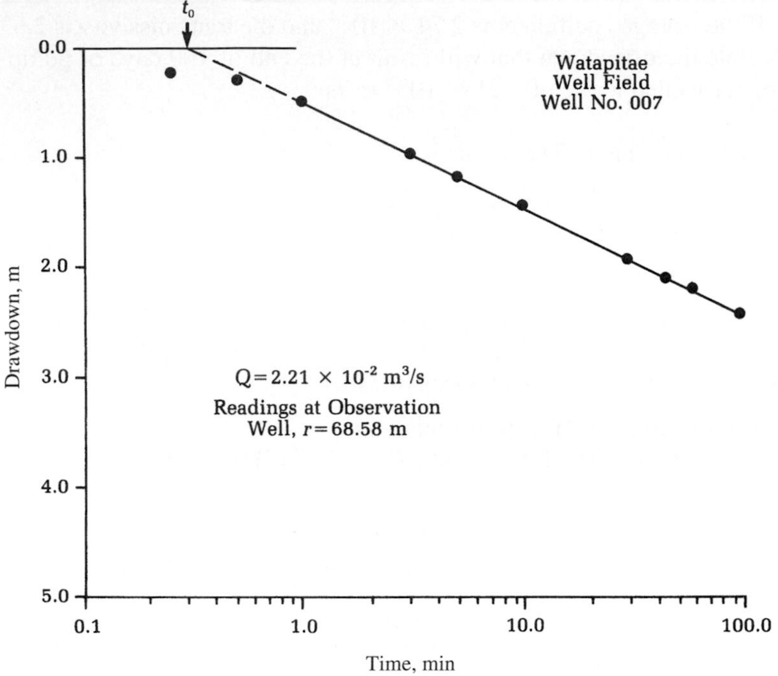

FIGURE 3-32
Pumping test results.

be rewritten as follows:

$$s = \frac{Q}{4\pi T} \ln \frac{2.25Tt}{r^2 S} \tag{3-34}$$

A semilogarithmic plot of s versus t (log scale) from the results of a pumping test (Figure 3-32) enables a direct calculation of T from the slope of the line. From Equation 3-34, the difference in drawdown at two points in time may be shown to be

$$s_2 - s_1 = \frac{Q}{4\pi T} \ln \frac{t_2}{t_1} \tag{3-35}$$

Solving for T we find

$$T = \frac{Q}{4\pi(s_2 - s_1)} \ln \frac{t_2}{t_1} \tag{3-36}$$

Cooper and Jacob showed that an extrapolation of the straight-line portion of the plot to the point where $s = 0$ yields a "virtual" (imaginary) starting time (t_0). At this virtual time, Equation 3-34 may be solved for the storage coefficient, S, as follows:

$$S = \frac{2.25Tt_0}{r^2} \tag{3-37}$$

The calculus implies that the distance to the observation well (r) may be as little as the radius of the pumping well itself. This means that drawdown measured in the pumping well may be used as a source of data.

Example 3-14. Determine the transmissivity and storage coefficient for the Watapitae Wells based on the pumping test data plotted in Figure 3-32.

Solution. Using Figure 3-32 we find $s_1 = 0.49$ m at $t_1 = 1.0$ min and at $t_2 = 10.0$ min we find $s_2 = 1.43$ m. Thus,

$$T = \frac{2.21 \times 10^{-2}}{4(3.14)(1.43 - 0.49)} \ln \frac{10.0}{1.0}$$
$$= (1.87 \times 10^{-3})(2.30) = 4.31 \times 10^{-3} \text{ m}^2/\text{s}$$

From Figure 3-32 we find that the extrapolation of the straight portion of the graph yields $t_0 = 0.30$ min. Using the distance between the pumping well and the observation ($r = 68.58$ m) we find

$$S = \frac{(2.25)(4.31 \times 10^{-3})(0.30)(60)}{(68.58)^2}$$
$$= 3.7 \times 10^{-5}$$

The factor of 60 is to convert minutes into seconds. Now we should check to see if our implicit assumption that u is less than 0.01 was true. We use $t = 10.0$ min for the check.

$$u = \frac{(68.58)^2(3.7 \times 10^{-5})}{4(4.31 \times 10^{-3})(10.0)(60)} = 0.017$$

This is a bit high. However, it is obvious that at 100 minutes it would be acceptable. Because the slope does not change, we will take this as a reasonable solution.

Determining the Hydraulic Properties of an Unconfined Aquifer. Transmissivity may be determined under steady-state conditions in a fashion similar to that for confined aquifers using Equation 3-33.

Providing that the basic assumptions can be met, the transient-state equations for a confined aquifer may be applied to an unconfined aquifer. Because the unconfined layer often does not release water immediately after a drop in the piezometric surface, the transient equations often are not valid. Thus, great care should be taken in using them.

Pumping Test Results from Nonhomogeneous Aquifers. Of course, our assumption that an aquifer is homogeneous seldom applies in real life. Under a few circumstances the inhomogeneities even out to yield average aquifer parameters. Under many other circumstances more complex techniques are required.

Two special cases of inhomogeneity are of particular interest. The first case is where the cone of depression intercepts a barrier. This is shown in Figure 3-33. The net result is that a greater pumping lift is required to achieve the same well yield as would have been obtained if the barrier had not existed.

The second case is where the cone of depression intersects a recharge area (Figure 3-34) such as a lake, stream, or underground stream. The net result is to reduce the lift required to maintain the desired yield of the well. Of course, it is possible to dry up the lake or stream if the pumping rate is large enough. This is shown by the increase in slope after the plateau.

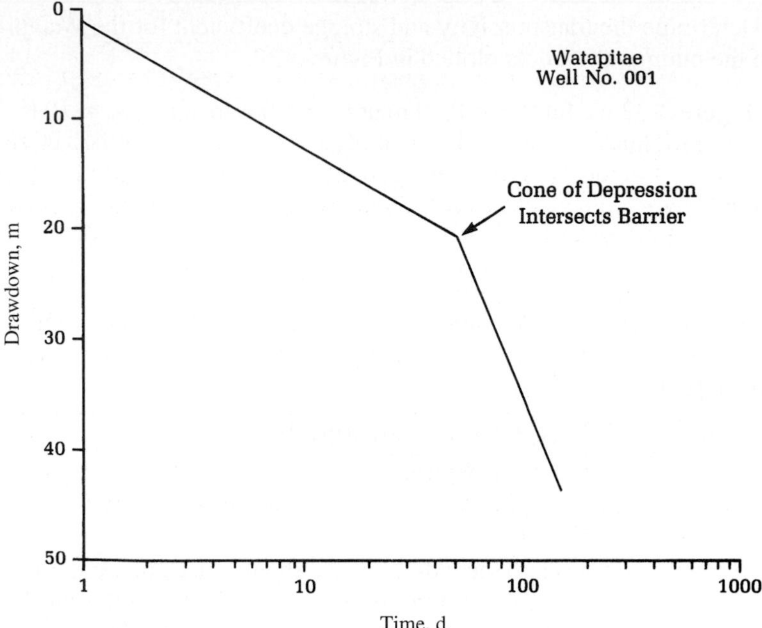

FIGURE 3-33
Pumping test curve showing effect of barrier at day 50.

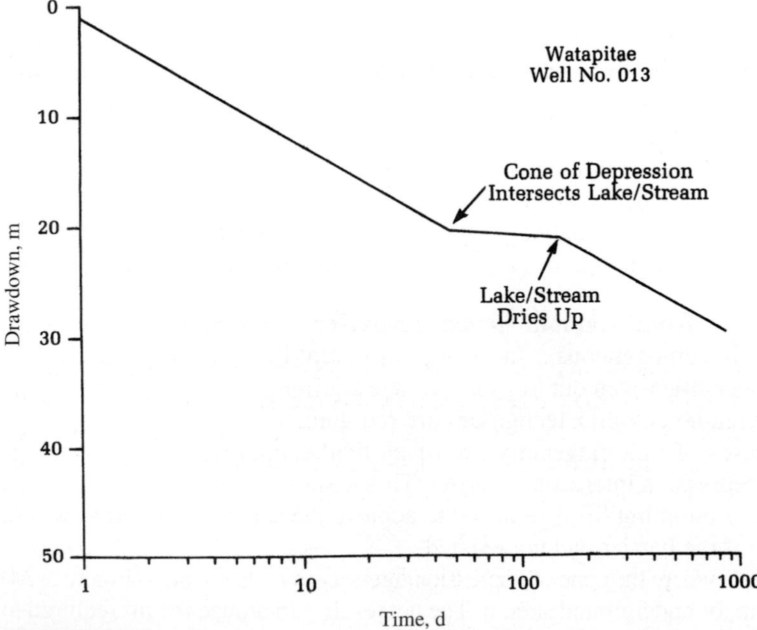

FIGURE 3-34
Pumping test curve showing effect of recharge from day 50 to day 150.

Calculating Interference. As we mentioned earlier, the cones of depression of wells located close together will overlap; this interference will reduce the potential yield of both wells. In severe circumstances, well interference can cause drawdowns that leave shallow wells dry.

A solution to the well interference problem can be achieved by the method of superposition. This method assumes that the drawdown at a particular location is equal to the sum of the drawdowns from all of the influencing wells. Mathematically this can be represented as follows:

$$s_r = \sum_{i=1}^{n} s_i \tag{3-38}$$

where s_i = individual drawdown caused by well i at location r.

Example 3-15. Three wells are located at 75-m intervals along a straight line. Each well is 0.50 m in diameter. The coefficient of transmissivity is 2.63×10^{-3} m^2/s and the storage coefficient is 2.74×10^{-4}. Determine the drawdown at each well if each well is pumped at 4.42×10^{-2} m^3/s for 10 days.

Solution. The drawdown at each well will be the sum of the drawdown of each well pumping by itself plus the interference from each of the other two wells. Because each well is the same diameter and pumps at the same rate, we may compute one value of the term $Q/(4\pi T)$ and apply it to each well.

$$\frac{Q}{4\pi T} = \frac{4.42 \times 10^{-2}}{4(3.14)(2.63 \times 10^{-3})} = 1.34$$

In addition, because each well is identical, the individual drawdowns of the wells pumping by themselves will be equal. Thus, we may compute one value of u and apply it to each well.

$$u = \frac{(0.25)^2(2.74 \times 10^{-4})}{4(2.63 \times 10^{-3})(10)(86{,}400)} = 1.88 \times 10^{-9}$$

Using $u = 1.9 \times 10^{-9}$ and referring to Table 3-6 we find $W(u) = 19.5042$. The drawdown of each individual well is then

$$s = (1.34)(19.5042) = 26.14 \text{ m}$$

Before we begin calculating interference, we should label the wells so that we can keep track of them. Let us call the two outside wells A and C and the inside well B. Let us now calculate interference of well A on well B, that is, the increase in drawdown at well B as a result of pumping well A. Because we have pumped only for ten days, we must use the transient flow equations and calculate u at 75 m.

$$u_{75} = \frac{(75)^2(2.74 \times 10^{-4})}{4(2.63 \times 10^{-3})(10)(86{,}400)} = 1.70 \times 10^{-4}$$

From Table 3-6, $W(u) = 8.1027$. The interference of well A on B is then

$$s_{\text{A on B}} = (1.34)(8.1027) = 10.86 \text{ m}$$

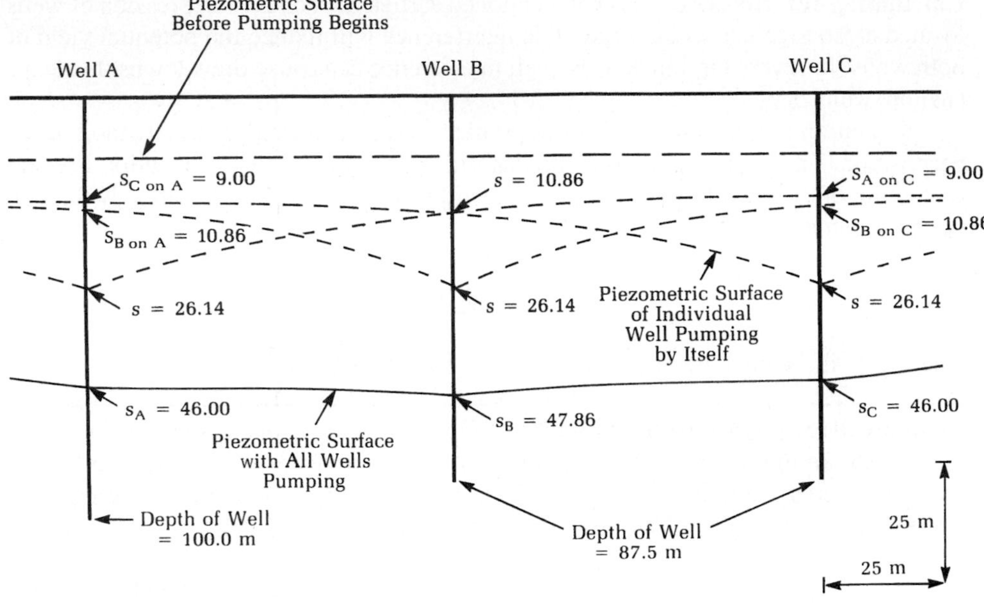

FIGURE 3-35
Interference drawdown of three wells.

In a similar fashion we calculate the interference of well A on well C.

$$u_{150} = (150)^2(3.0145 \times 10^{-8}) = 6.78 \times 10^{-4}$$

and $W(u) = 6.7169$

$$s_{A\text{ on }C} = (1.34)(6.7169) = 9.00 \text{ m}$$

Since our well arrangement is symmetrical, the following equalities may be used:

$$s_{A\text{ on }B} = s_{B\text{ on }A} = s_{B\text{ on }C} = s_{C\text{ on }B}$$

and

$$s_{A\text{ on }C} = s_{C\text{ on }A}$$

The total drawdown at each well is computed as follows:

$$s_A = s + s_{B\text{ on }A} + s_{C\text{ on }A}$$

$$s_A = 26.14 + 10.86 + 9.00 = 46.00 \text{ m}$$

$$s_B = s + s_{A\text{ on }B} + s_{C\text{ on }B}$$

$$s_B = 26.14 + 10.86 + 10.86 = 47.86 \text{ m}$$

$$s_C = s_A = 46.00 \text{ m}$$

Drawdowns are measured from the undisturbed piezometric surface. Note that if the wells are pumped at different rates, the symmetry would be destroyed and the value $Q/(4\pi T)$ would have to be calculated separately for each case. Likewise, if the distances were not symmetric, then separate u values would be required. The results of these calculations have been plotted in Figure 3-35.

Groundwater Contamination

The well hydraulics equations may be used to design purge or extraction wells to remove contaminated groundwater. The lowering of the piezometric surface will cause contaminants to flow to the well where they may be removed to the surface for treatment. In some cases, a well field may be required to intercept the contaminant plume. In these cases, the well interference effects may be put to good use in designing the distribution of the wells. The use of extraction wells is discussed in Chapter 10.

3-6 WASTE MINIMIZATION FOR SUSTAINABILITY

"Water: too much, too little, too dirty." (Loucks et al., 1981) These are the conditions that prompt water resource management. Often the incentive for increased control follows a major disaster, namely flood, drought, fish kill, or epidemic. Historically, solutions to water resource problems in general, and water supply in particular, have been structural in nature. Building reservoirs, dredging channels, and driving new and deeper wells are traditional ways that civil engineers have solved these problems. Modern water resource management also considers nonstructural means to achieve water resource goals.

A key nonstructural means to achieve water resource goals is to minimize the use of water. Water that is not used has reduced potential for being contaminated or lost through evaporation/transpiration. In addition to conserving water, a reduced municipal water demand reduces the cost of water supply works, water treatment and distribution, and wastewater collection and disposal facilities. Many examples can be cited where simple changes in habits can reduce water consumption and, in turn, pollution of natural waters. For example, washing of streets with high-pressure hoses can be replaced with wet sweeping, and spray irrigation can be replaced with trickle irrigation. The installation of water meters in Boulder, Colorado, reduced the average daily demand from 802 Lpcd to 635 Lpcd (Hanke, 1970). The use of water-saving devices such as high-pressure, low-volume shower heads and low-volume flush toilets can reduce per capita household demand by as much as 16 percent (Flack, 1980).

3-7 CHAPTER REVIEW

When you have completed studying this chapter you should be able to do the following without the aid of your textbooks or notes:

1. Sketch and explain the hydrologic cycle, labeling the parts as in Figure 3-1.
2. List and explain the four factors that reduce the amount of direct runoff.
3. Explain the difference between streamflow that results from direct runoff and that which results from baseflow.

4. Define evaporation, transpiration, runoff, baseflow, watershed, basin, and divide.
5. Sketch the groundwater hydrologic system, labeling the parts as in Figure 3-3.
6. Explain why water in an artesian aquifer is under pressure and why water may rise above the surface in some instances and not others.
7. Explain why infiltration rates decrease with time.
8. Explain return period or recurrence interval in terms of probability that an event will take place.
9. Define the rational formula and identify the units as used in this text.
10. Explain why the rational method may yield inadequate results when applied by inexperienced engineers, while experienced engineers may get adequate results.
11. Explain what a unit hydrograph is.
12. Explain the purpose of the unit hydrograph and explain how it might be applied in the analysis of a storm sewer design or a stream flood-control project.
13. Using a sketch, show how the groundwater contribution to a hydrograph is identified.
14. Define time of concentration and explain how it is used in conjunction with the rational method.
15. Sketch a subsurface cross-section from the results of a well boring log and identify pertinent hydrogeologic features.
16. Sketch a well and label the major sanitary protection features according to this text.
17. Sketch a piezometric profile for a single well pumping at a high rate, and sketch a profile for the same well pumping at a low rate.
18. Sketch a piezometric profile for two or more wells located close enough together to interfere with one another.
19. Sketch a well-pumping test curve which shows (*a*) the interception of a barrier, and (*b*) the interception of a recharge area.
20. Give two examples of methods to minimize water use.

With the use of this text you should be able to do the following:

1. Compute mass balances for open and closed hydrologic systems.
2. Compute infiltration rates by Horton's method and estimate the volume of infiltration.

3. Estimate the volume of water loss through transpiration given the air and water temperature, wind speed, and relative humidity.
4. Use Horton's equation and/or some form of Dalton's equation to solve complex mass balance problems.
5. Construct an intensity-duration-frequency curve from a compilation of intense rainfall occurrences.
6. Construct a unit hydrograph for a given stream-gaging station if you are provided with a rainfall and total flow at the gaging station.
7. Apply a given unit hydrograph to construct a compound hydrograph if you are provided with an observed rainfall and an estimate of the rainfall losses due to infiltration and evaporation.
8. Determine the peak flow (Q in the rational formula) and time of arrival (t_c) resulting from a rainfall of specified intensity and duration in a well-defined watershed.
9. Using the mass balance method, determine the volume of a reservoir or retention basin for a given demand or flood control given appropriate discharge data.
10. Calculate the drawdown at a pumped well or observation well if you are given the proper input data.
11. Calculate the transmissivity and storage coefficient for an aquifer if you are provided with the results of a pumping test.
12. Calculate the interference effects of two or more wells.

3-8 PROBLEMS

3-1. Lake Pleasant, AZ, has a surface area of approximately 2,000 ha. The inflow for the month of August is zero. The lake is dammed so that no water leaves the impoundment downstream of the dam. There is no rain in August. The evaporation is estimated to be 6.8 mm/d and the seepage is estimated to be 0.01 mm/d. Assuming that the lake shore is vertical, estimate the distance the lake elevation drops during the month of August. What distance will the shoreline recede if the lake shore slope is 5°?

Answers: Vertical drop = 211.11 or 210 mm or 21 cm
Contraction = 2,422 or 2,400 mm or 240 cm or 2.4 m

3-2. Lake Kickapoo, TX, is approximately 12 km in length by 2.5 km in width. The inflow for the month of March is 3.26 m^3/s and the outflow is 2.93 m^3/s. The total monthly precipitation is 15.2 cm and the evaporation is 10.2 cm.

The seepage is estimated to be 2.5 cm. Estimate the change in storage (in m^3) during the month of March.

3-3. A 4,000-km^2 watershed receives 102 cm of precipitation in one year. The average flow of the river draining the watershed is 34.2 m^3/s. Infiltration is estimated to be 5.5×10^{-7} cm/s and evapotranspiration is estimated to be 40 cm/y. Determine the change in storage in the watershed over one year. The ratio of runoff to precipitation (both in cm) is termed the runoff coefficient. Compute the runoff coefficient for this watershed.

3-4. Using the values of f_o, f_c, and k for a Fuquay pebbly loamy sand (see Section 3-1), find the infiltration rate at times of 12, 30, 60, and 120 minutes. Compute the total volume of infiltration over 120 minutes.

3-5. Infiltration data from an experiment yield an initial infiltration rate of 4.70 cm/h and a final equilibrium infiltration rate of 0.70 cm/h after 60 minutes of steady precipitation. The value of k was estimated to be 0.1085 h^{-1}. Determine the total volume of infiltration for the following storm sequence: 30 mm/h for 30 minutes, 53 mm/h for 30 minutes, 23 mm/h for 30 minutes.

3-6. Using the empirical equation developed for Lake Hefner, estimate the evaporation from a lake on a day that the air temperature is 30°C, the water temperature is 15°C, the wind speed is 9 m/s, and the relative humidity is 30 percent.

Answer: 4.73 or 4.7 mm/d

3-7. Using the empirical equation developed for Lake Hefner, estimate the evaporation from a lake on a day that the air temperature is 40°C, the water temperature is 25°C, the wind speed is 2.0 m/s, and the relative humidity is 5 percent.

3-8. The Dalton-type evaporation equation implies that there is a limiting relative humidity above which evaporation will be nil regardless of the wind speed. Using the Lake Hefner empirical equation, estimate the relative humidity at which evaporation will be nil if the water temperature is 10°C and the air temperature is 25°C.

3-9. Prepare an IDF curve for a 2-year storm at Dismal Swamp using the data in Table 3-2. *Hint:* Curve should intersect 98.7 mm/h at 15 minutes duration.

3-10. Prepare an IDF curve for a 10-year storm at Dismal Swamp using the data in Table 3-2.

3-11. Prepare an intensity-duration-frequency curve for a 5-year storm with the data shown in the following table.

Annual maximum intensity (mm/h)

Year	30-minute duration	60-minute duration	90-minute duration	120-minute duration
1960	122.3	100.3	81.1	55.3
1961	104.6	82.9	64.5	39.7
1962	81.0	60.8	41.5	16.5
1963	145.1	123.7	104.7	81.7
1964	83.0	61.5	40.7	16.0
1965	70.1	51.1	30.1	11.3
1966	94.7	71.0	51.7	26.7
1967	63.7	41.5	21.8	10.1
1968	57.9	35.7	17.1	9.7
1969	71.5	50.0	31.3	15.9

Hint: Curve should intersect 96.8 mm/h at 60 minutes duration.

3-12. Prepare an intensity-duration-frequency curve for a 2-year storm with the data from Problem 3-11.

3-13. A large shopping mall parking lot may be arranged in either of the two configurations shown in Figure P-3-13. The drain commissioner prefers the longest time of concentration to allow for the passage of other peak flows in the drain. Using the highest runoff coefficient for asphaltic pavement, estimate the time of concentration for each configuration and recommend the one to submit to the drain commissioner. Note: area of (a) is same as area of (b).

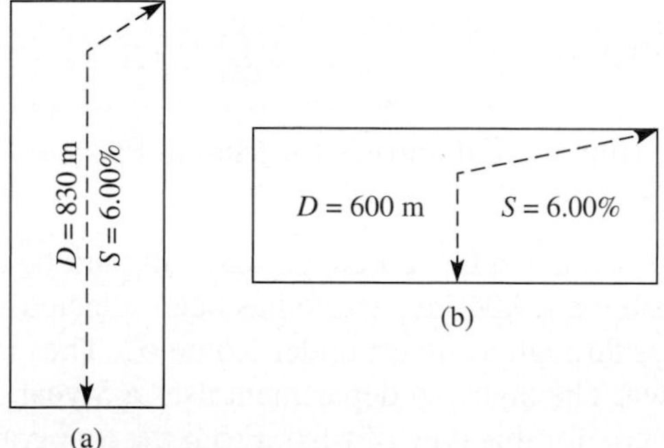

FIGURE P-3-13
Alternative parking lot configurations.

3-14. Mechanicsville has obtained a grant under the Federal Program for Urban Development of Greenspace and the Environment to build a condominium complex for the retired. The total land set aside is 74,010 m^2. Of this area 15,831 m^2 will be used for Swiss chalet condominiums with slate roofs. Public streets and drives will occupy 18,886 m^2. The remainder will be lawns. The area is flat and has a sandy soil. Using the most conservative estimates of *C*, find the peak discharge from the development for a 2-year storm. Assume an overland flow distance of 272 m and use the IDF curves given in Figure P-3-14.

Answer: $Q = 0.6097$ or $0.61\ m^3/s$

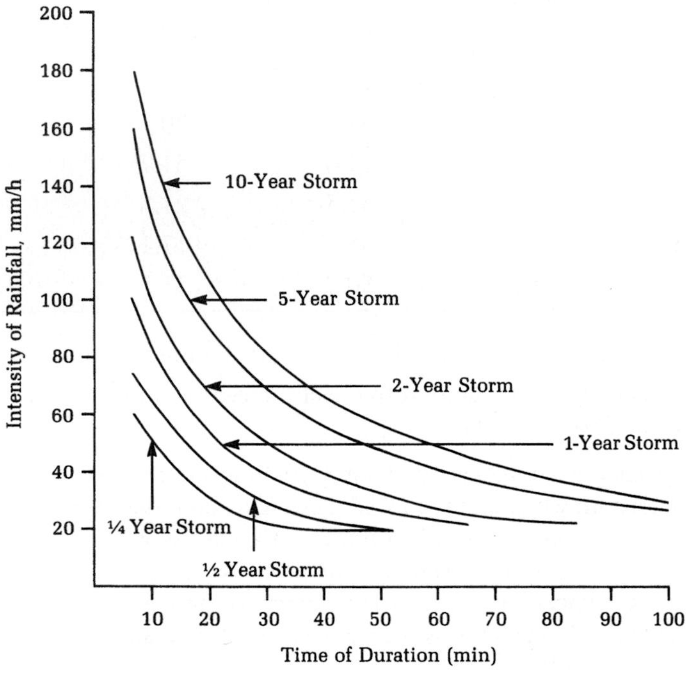

FIGURE P-3-14
IDF Curves for Mechanicsville

3-15. Repeat Problem 3-14 using the IDF curves for Miami, FL, shown in Figure 3-10c.

3-16. Paula Revere has proposed that a Little League baseball park be built near Boxborough, MA. A site on a 9.94-ha pasture has been selected. Drainage from the pasture passes through a culvert under Route 62. The capacity of the culvert is not known. The highway department uses a 5-year return period for design of culverts for this type of road. From the topographic data of the site, the overland flow distance is estimated to be 450 m, the slope of the pasture is estimated to be 2.00 percent, and the runoff coefficient is estimated to be 0.20. Using the IDF curves for Boston, MA (Figure 3-10*a*), estimate the peak flow capacity of the existing culvert in m^3/s.

Answer: $Q = 0.2115$ or $0.21\ m^3/s$

3-17. The proposed parking lot for Paula's baseball park (Problem 3-16) will occupy 2.64 ha of the site. It will be graded to a slope of 1.80 percent. The overland flow distance will be 200.0 m and the runoff coefficient is estimated to be 0.70. Ignoring the runoff from the pasture, determine whether the existing culvert has the capacity to carry the peak flow from the parking lot from a storm with a 5-year return period. Use the IDF curves for Boston, MA (Figure 3-10*a*).

3-18. A storm sewer is to be designed for a parking lot in Holland, MI, that has an area of 4.8 ha. The parking lot slope will be 1.00 percent and the estimated runoff coefficient is 0.85. The overland flow distance is 219.0 m. Using the rational method, estimate the peak discharge from the parking lot for a 5-year storm. The IDF curve for a 5-year storm at Holland, MI, can be described by the following equation:

$$i = \frac{1,193.80}{(t_c)^{0.8} + 7}$$

where i = intensity, mm/h, and t_c = time of concentration, min.

3-19. John Snow is planning to build a shopping mall and parking lot on the north side of Hindry Road as shown in Figure P-3-19. The existing culvert near the southwest corner of the proposed parking lot was designed for a 5-year storm. Determine the frequency that the capacity of the culvert will be exceeded if it is not enlarged when the mall is constructed. Also determine the peak discharge that must he handled if the design criteria are changed to specify a 10-year storm. Use the IDF curves in Figure P-3-14.

Answers: Q_{design} = 0.74 m^3/s; I that will cause flood = 37.7 mm/h; frequency of flooding ≈ 4 times/y; Q_{10y} = 2.0 m^3/s

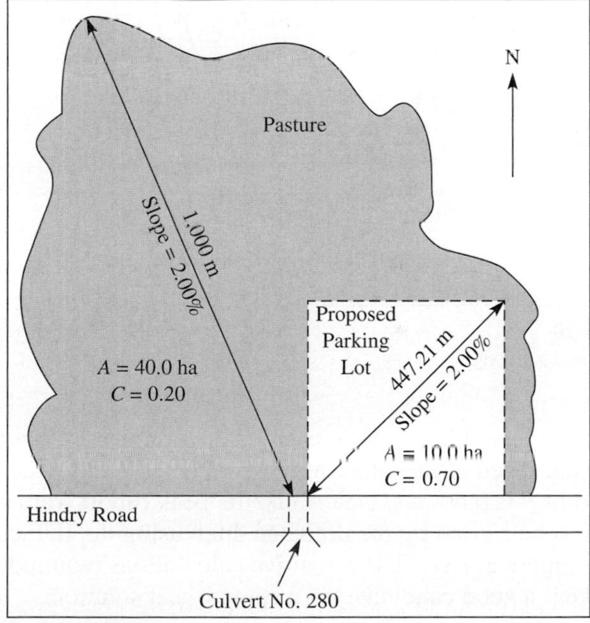

FIGURE P-3-19
Sketch of proposed shopping mall site.

3-20. Dr. William Gorgas is planning to build a clinic on the east side of Okemos Road as shown in Figure P-3-20. The existing culvert (no. 481) was designed for a 5-year storm. Determine if the capacity of the culvert will be exceeded if it is not enlarged when the clinic is built. Also determine the peak discharge that must be handled if the design criteria are changed to specify a 10-year storm. Use the IDF curves in Figure P-3-14.

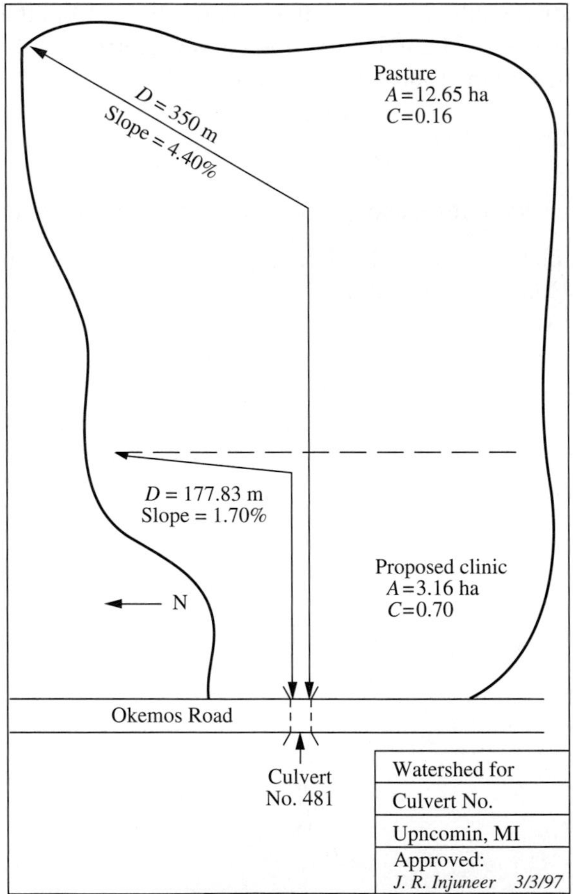

FIGURE P-3-20
Sketch of proposed clinic site.

3-21. Two adjacent parcels of land adjoin each other as shown in Figure P-3-21. Assume that the speed of the water moving to the east along the drainage ditch from point *P* is 0.60 m/s. Determine the peak runoff from a 5-year storm that must be carried by the drainage ditch using the IDF curves for Seattle, WA (Figure 3-10*d*). The repetitive calculations required in this problem make it a good candidate for a spreadsheet solution.

Answer: $Q = 0.11\ m^3/s$

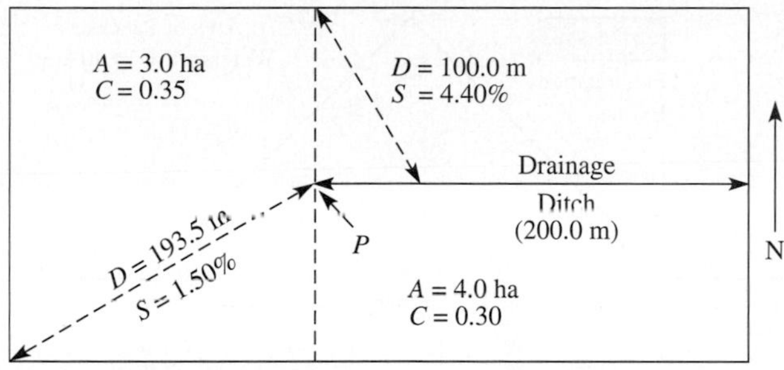

FIGURE P-3-21
Drainage from two adjacent parcels.

3-22. A parking lot is configured in three sections as shown in Figure P-3-22. There are storm water inlet manholes for each section. Storm water flows in the storm sewer from east to west. The storm water moves at a speed of 0.90 m/s once it reaches the storm sewer. Determine the peak runoff from a 5-year storm that must be carried by the storm sewer carrying storm water away from manhole no. 3. Use the IDF curves for Miami, FL (Figure 3-10*c*). The repetitive calculations required in this problem make it a good candidate for a spreadsheet solution.

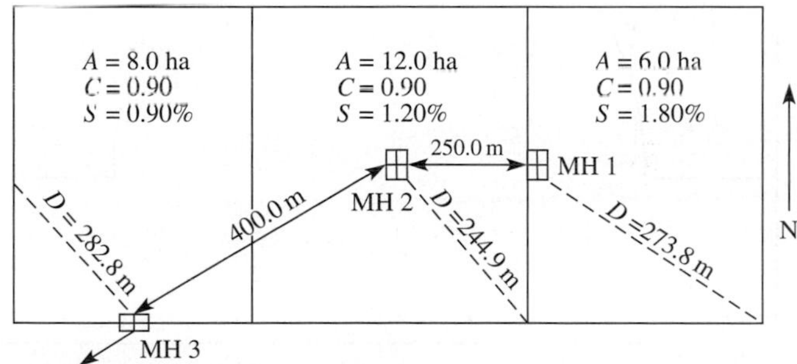

FIGURE P-3-22
Storm sewer capacity for three part parking lot.

3-23. Determine the unit hydrograph ordinates for the Isosceles River with the stream flow data shown in Figure P-3-23 on page 164. The basin area is 14.40 km^2, and the unit duration of the storm is 1 hour. For ease of computation, locate ordinates at hourly intervals, that is, 1500, 1600, and 1700 hours. Note that the time is military time; that is, 1500 h = 3 PM, 2000 h = 8 PM, etc.

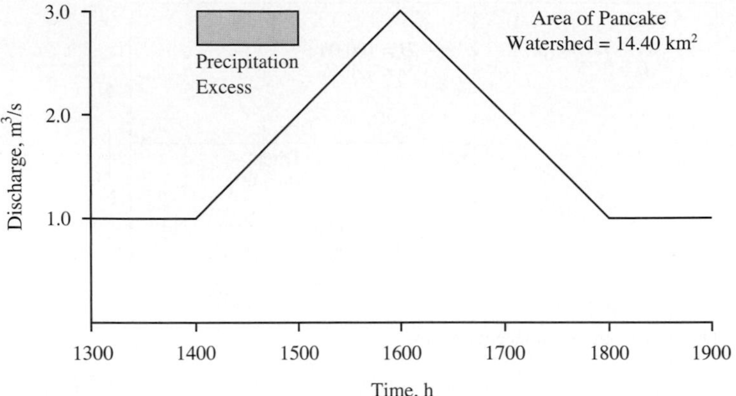

FIGURE P-3-23
Storm hydrograph for Isosceles River.

3-24. Determine the unit hydrograph ordinates for the Convex River flow data shown below. For ease of computation the total ordinates are tabulated below the hydrograph. Note that the base flow may be determined by simple graphical extrapolation. The area of the watershed is 100 ha. Note that the time is military time, that is, 1500 h = 3 PM, 2000 h = 8 PM, etc.

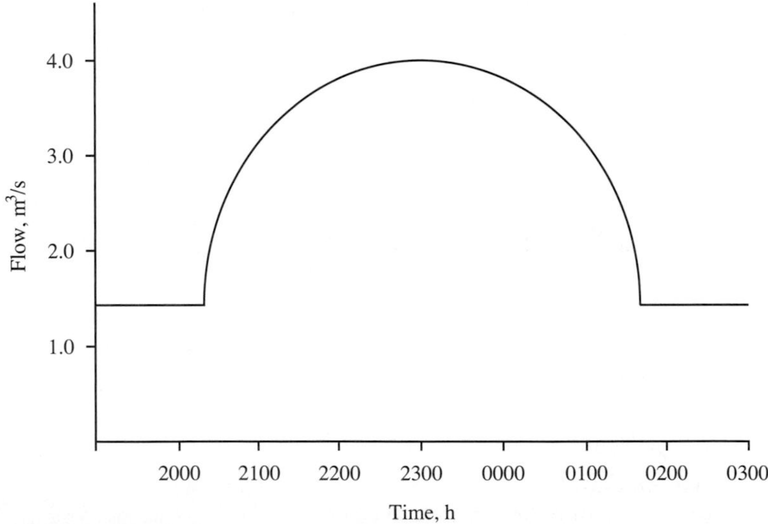

FIGURE P-3-24
Storm hydrograph for the Convex River.

Time (h)	Total stream flow (m^3/s)
2100	3.0
2200	3.8
2300	4.0
0000	3.8
0100	3.0

3-25. Determine the unit hydrograph ordinates for the Verde River with the stream flow data shown in the table below that resulted from a 5-hour storm of uniform intensity. The basin area is 64.0 km^2.

Time (h)	Flow (m^3/s)	Time (h)	Flow (m^3/s)	Time (h)	Flow (m^3/s)
0	0.55	35	5.77	65	1.64
5	0.50	40	5.02	70	1.10
10	0.45	45	4.29	75	0.79
15	1.98	50	3.50	80	0.47
20	4.82	55	2.72	85	0.25
25	6.24	60	2.19	90	0.25
30	6.86				

Answer: The plotting points of the ordinates are at midpoints of the time intervals, i.e., 2.5 h, 7.5 h, 15.0 h, 25 h, 35 h, 45 h, 55 h, 65 h. The UH ordinates in m^3/s · cm are 0.51, 2.544, 4.95, 4.55, 3.29, 1.98, 1.07, 0.42.

3-26. Determine the unit hydrograph ordinates for the Crimson River with the stream flow data shown in the table below. The basin area is 626.0 km^2. The unit duration of the storm was five hours.

Time (h)	Flow (m^3/s)	Time (h)	Flow (m^3/s)	Time (h)	Flow (m^3/s)
0	1.60	35	21.55	70	5.15
5	1.73	40	18.13	75	3.46
10	1.57	45	15.77	80	2.48
15	1.41	50	13.48	85	1.48
20	6.22	55	11.03	90	0.79
25	15.14	60	8.55	95	0.77
30	19.60	65	6.88	100	0.77

3-27. Apply the unit hydrograph distribution shown below to the following observed rainfall. Compute the compound runoff.

Day	UH ord. (m^3/s · cm)
1	0.12
2	0.75
3	0.13

Day	Rain (cm)	Abstractions (cm)
1	0.50	0.20
2	0.30	0.10
3	0.0	0.0

3-28. Using the unit hydrograph for the Isosceles River developed in Problem 3-23, determine the stream flow hydrograph resulting from the storm shown in Figure P-3-28. Note that the time is military time; that is, 1500 h = 3 PM, 2000 h = 8 PM, etc.

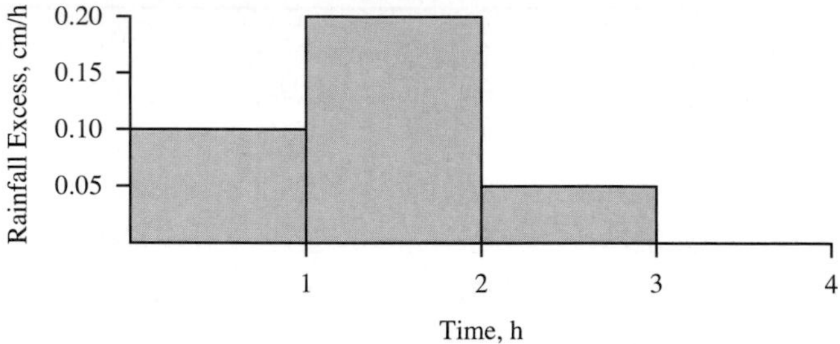

FIGURE P-3-28
Plot of rainfall on the Isosceles River watershed.

3-29. Using the unit hydrograph developed for the Verde River in Problem 3-25, determine the stream flow hydrograph resulting from the storm shown in Figure P-3-29 below.

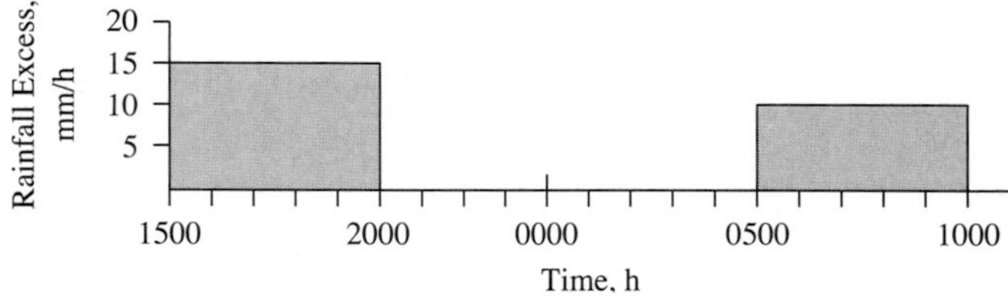

FIGURE P-3-29
Plot of rainfall on the Verde River watershed.

3-30. Using the unit hydrograph developed for the Crimson River in Problem 3-26, determine the stream flow hydrograph resulting from the storm shown in Figure P-3-30 below.

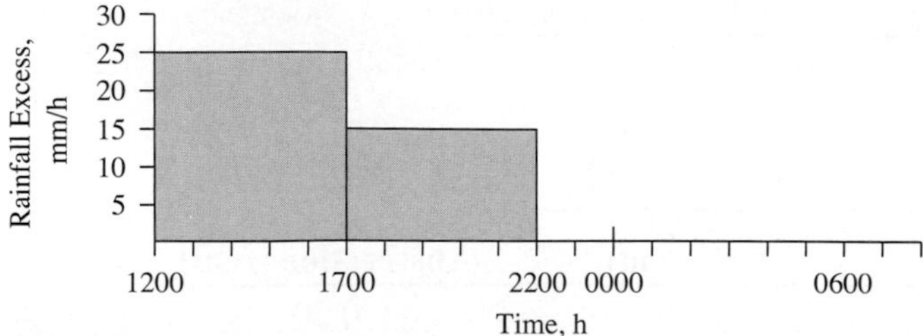

FIGURE P-3-30
Plot of rainfall on the Crimson River watershed.

3-31. A reservoir has been built to supply water during droughts. The design volume is 7.00×10^6 m^3. If the average flow into the reservoir is 3.2 m^3/s and the DNR requires that the discharge from the reservoir be at least 2.0 m^3/s, how many days will it take to achieve the design volume?

3-32. The Town of Woebegone has requested that you estimate the size of water tower (volume in liters) it should provide, on the basis of the following estimate of their demand and the fact that the town have a well that pumps 36 L/s.

Time	**Q_{out} (L/s)**
Midnight–6 AM	0.0
6 AM–noon	54.0
Noon–6 PM	54.0
6 PM–midnight	36.0

3-33. The demand for a water supply from the Clear Fork Trinity River is 0.35 m^3/s. Construct a mass balance for the period from July 1, 1951, through May 31, 1952, and determine the storage required (in m^3) to meet this demand. Data are given in the table on page 168. Assume that the reservoir is full at the beginning of the analysis period and that flows greater than the demand are discharged downstream after the reservoir is filled. Is the reservoir full at the end of May 1952? Use a spreadsheet program you have written to solve this problem.

Answer: $\forall$ = 4,838,832 or 4.8×10^6 m^3; the reservoir is full at the end of May 1952.

3-34. Construct a mass balance for the Squannacook River to determine the storage (in cubic meters) required to meet a demand of 1.76 m^3/s for the period from January 1, 1964 through December 31, 1967. Data are given in the table on page 169. Assume that the reservoir is full at the beginning of the analysis period and that flows greater than the demand are discharged downstream after the reservoir is filled. Is the reservoir full at the end of December 1967? Use a spreadsheet program you have written to solve this problem.

3-35. The demand for a water supply from the Hoko River is 0.325 m^3/s. If the Department of Environmental Quality limits the withdrawal to 10 percent of the flow, what size reservoir (in cubic meters) is required to provide the required flow during the drought that occurs between May 1 and November 30, 1972? Data are given in the table on page 170. Assume that the reservoir is full at the beginning of the analysis period and that flows greater than the demand are discharged downstream after the reservoir is filled. Is the reservoir full at the end of November 1972? Work this problem by the mass balance method, using a spreadsheet program you have written.

Clear Fork Trinity River at Fort Worth, TX (Problem 3-33)
Mean monthly discharge (m^3/s)

Year	Jan	Feb	Mar	Apr	May	June	July	Aug	Sept	Oct	Nov	Dec
1940	—	—	—	—	—	—	—	—	—	0.00	5.63	15.4
1941	4.59	23.8	7.50	6.91	10.2	17.0	2.07	2.29	0.20	1.71	0.631	0.926
1942	0.697	0.595	0.614	58.93	24.1	9.09	0.844	0.714	1.21	10.0	2.38	1.87
1943	1.33	1.00	3.99	3.71	8.38	3.77	0.140	0.00	1.33	0.014	0.00	0.139
1944	0.311	4.93	2.83	2.25	13.3	1.68	0.210	0.609	4.11	0.985	0.515	1.47
1945	3.06	30.38	35.23	28.85	5.69	21.7	2.14	0.230	0.162	0.971	0.617	0.541
1946	1.88	5.75	3.54	1.89	6.57	5.86	0.153	1.45	4.02	0.906	12.2	10.3
1947	4.64	2.62	4.87	5.13	2.27	4.25	0.292	0.054	0.535	0.371	0.331	3.51
1948	3.99	16.9	9.06	1.91	2.64	1.11	1.22	0.00	0.00	0.00	0.00	0.003
1949	0.309	4.19	9.94	4.16	55.21	11.1	1.38	0.450	0.447	4.53	0.711	0.614
1950	3.28	14.7	3.26	12.7	15.1	2.50	3.60	2.44	10.6	1.12	0.711	0.801
1951	0.708	0.994	0.719	0.527	1.37	6.20	0.980	0.00	0.00	0.00	0.006	0.090
1952	0.175	0.413	0.297	1.93	3.65	0.210	0.003	0.029	0.007	0.00	0.368	0.167
1953	0.099	0.080	0.134	0.671	0.934	0.008	0.286	0.249	0.041	0.546	0.182	0.066
1954	0.108	0.092	0.114	0.088	0.278	0.017	0.021	0.015	0.008	0.047	0.024	0.063
1955	0.091	0.153	0.317	0.145	0.464	0.640	0.049	0.050	0.119	0.104	0.055	0.058
1956	0.069	0.218	0.026	0.306	1.35	0.30	0.026	0.019	0.029	0.266	0.030	0.170
1957	0.065	0.300	0.385	12.8	23.6	59.9	6.97	1.36	0.501	0.476	0.855	1.55
1958	1.65	1.61	4.59	5.69	28.1	0.589	0.524	0.456	0.549	0.572	0.490	0.566
1959	0.759	0.776	0.120	0.261	0.097	0.685	0.379	0.668	0.473	9.03	1.64	3.65
1960	11.8	4.45	3.26	1.42	0.631	0.379	0.660	0.566	0.467	0.498	0.241	0.648
1961	2.05	1.92	3.40	1.02	0.306	2.34	0.821	0.816	1.08	0.824	0.297	0.504
1962	0.311	2.03	0.467	0.759	0.459	0.236	0.745	1.41	6.94	1.31	0.405	0.767
1963	0.345	0.268	0.379	1.74	3.79	1.48	0.527	0.586	0.331	0.277	0.249	0.266
1964	0.416	0.266	1.16	0.813	1.02	0.374	0.535	0.963	3.96	0.351	1.47	0.886
1965	2.13	14.6	4.16	2.28	20.5	2.45	1.22	0.821	0.776	0.394	0.476	0.213
1966	0.169	0.354	0.462	6.40	23.1	18.5	5.32	0.951	0.294	1.37	0.15	0.134
1967	0.244	0.244	0.198	0.688	1.04	3.65	0.354	0.068	0.697	0.473	0.394	0.558
1968	1.64	3.85	23.2	1.89	15.7	6.12	0.583	0.144	0.220	0.419	0.396	0.206
1969	0.259	0.555	2.66	12.1	21.2	0.745	0.674	0.30	1.56	0.917	0.459	1.94
1970	5.78	6.37	27.0	3.31	15.0	1.03	0.521	0.697	1.23	—	—	—

Squannacook River near West Groton, MA (Problem 3-34)

Mean monthly discharge (m^3/s)

Year	Jan	Feb	Mar	Apr	May	June	July	Aug	Sept	Oct	Nov	Dec
1951	3.48	8.18	6.63	6.63	3.20	2.38	2.40	1.49	1.06	1.82	6.60	4.47
1952	6.68	5.07	6.51	9.94	5.44	3.71	.87	1.01	.69	.45	1.05	3.37
1953	4.79	6.77	11.44	9.80	6.34	1.21	.52	.42	.29	.51	1.34	3.65
1954	2.06	3.20	4.67	4.53	9.71	2.75	1.21	1.05	6.94	2.27	6.26	7.16
1955	2.92	3.06	5.41	6.17	2.77	1.44	.46	1.63	.61	8.38	8.61	2.17
1956	9.15	3.29	3.82	14.56	5.21	2.50	.77	.40	.50	.54	1.14	2.33
1957	2.92	2.63	4.22	3.99	2.65	.87	.37	.22	.22	.29	.97	3.91
1958	5.89	3.48	6.60	12.40	5.35	1.29	.81	.49	.45	.62	1.13	1.49
1959	2.07	2.05	5.41	8.67	2.37	1.22	1.17	.59	.82	2.55	4.08	5.55
1960	3.51	3.96	3.03	14.73	5.52	2.41	1.09	1.21	2.71	2.18	3.34	2.49
1961	1.57	3.09	7.28	11.10	4.67	2.31	1.03	.80	1.23	.99	2.06	1.73
1962	3.14	1.80	5.47	10.93	3.71	1.25	.56	.69	.50	2.95	4.73	4.30
1963	2.19	1.76	6.83	7.53	2.66	.77	.38	.24	.25	.35	1.52	1.98
1964	3.77	2.57	7.33	6.57	1.85	.59	.38	.25	.21	.27	.36	.79
1965	.65	1.33	2.38	3.79	1.47	.59	.23	.20	.19	.27	.45	.64
1966	.61	1.96	5.55	2.92	2.46	.80	.26	.18	.27	.52	1.75	1.35
1967	1.68	1.53	2.64	10.62	6.29	3.17	2.22	.72	.47	.60	1.07	3.03
1968	2.02	2.14	9.60	3.79	3.82	4.79	1.92	.61	.48	.46	1.88	4.33
1969	2.21	2.17	5.81	10.70	2.80	1.01	.58	1.03	.93	.52	5.24	5.83

Hoko River near Sekiu, WA (Problem 3-35)
Mean monthly discharge (m^3/s) 1963–1973

Year	Jan	Feb	Mar	Apr	May	June	July	Aug	Sept	Oct	Nov	Dec
1963	12.1	15.0	8.55	9.09	5.78	1.28	2.59	1.11	.810	13.3	26.1	20.3
1964	27.3	12.2	18.0	8.21	4.08	3.62	4.53	2.44	4.28	7.67	13.3	14.7
1965	27.4	27.3	5.01	5.61	6.68	1.38	.705	.830	.810	7.31	16.8	19.6
1966	29.8	11.3	17.6	5.18	2.67	2.10	1.85	.986	1.54	10.3	17.0	39.0
1967	35.6	26.8	18.5	6.51	3.43	1.46	.623	.413	.937	25.7	14.2	27.8
1968	34.2	22.4	15.7	9.20	3.68	2.65	1.72	1.55	9.12	16.8	16.5	25.2
1969	17.2	18.5	12.9	12.8	3.74	2.23	1.19	.810	6.15	7.84	9.15	15.9
1970	17.3	12.1	8.50	17.7	3.85	1.32	.932	.708	4.22	7.96	13.9	25.4
1971	32.7	21.0	21.1	8.13	3.43	2.83	1.83	.932	2.22	10.7	22.7	22.0
1972	27.4	26.9	25.4	14.6	3.00	1.00	5.32	.841	2.00	1.14	11.8	37.8
1973	28.0	9.23	11.3	4.13	5.30	4.93	1.63	.736	.810	13.1	29.8	31.5

3-36. A coworker has been working on the design of a retention pond for the Bar Nunn Mall. Your boss has just informed you that your coworker has come down with chicken pox and you must finish the calculations to determine the volume of the retention basin. Your boss gives you the work shown below. Determine the volume of the retention basin in cubic meters. *Note:* each interval is 1-hour long. Use a spreadsheet program you have written to solve this problem.

Interval	Inflow (L/s)	Volume in (m^3)	Outflow (L/s)	Volume out (m^3)
1	10.0	36.0	10.0	36.0
2	20.0	72.0	10.0	36.0
3	30.0		10.0	
4	20.0		10.0	
5	15.0		10.0	
6	5.0			

3-37. The Menominee River floods when the flow rate is greater than 100 m^3/s. Using the data on page 172, determine the storage required (in cubic meters) to prevent a flood during the period from January 1, 1959, through December 31, 1960. Assume the reservoir is empty at the start of the analysis period. Also assume that, after the reservoir is full, the flow out of the reservoir is at a rate equal to Q_{out} until the reservoir is empty. Is the reservoir empty at the end of December 1960? Use a spreadsheet program you have written to solve this problem.

Answer: $\forall$ = 1,226,525,760 or 1.23×10^9 m^3; the reservoir is not empty.

3-38. The Spokane River floods when the flow rate is greater than 250.0 m^3/s. Using the data in the table on page 173, determine the storage required (in cubic meters) to prevent a flood during the period from January 1, 1957, through December 31, 1958. Assume the reservoir is empty at the start of the analysis period. Also assume that, after the reservoir is full, the flow out of the reservoir is at a rate equal to Q_{out} until the reservoir is empty. Is the reservoir empty at the end of December 1958? Use a spreadsheet program you have written to solve this problem.

3-39. The Rappahannock River floods if the flow rate exceeds 5.80 m^3/s. Using the data in the table on page 174, determine the storage required (in cubic meters) to prevent a flood during the period from January 1, 1960, through December 31,1962. Assume the reservoir is empty at the start of the analysis period. Also assume that, after the reservoir is full, the flow out of the reservoir is at a rate equal to Q_{out} until the reservoir is empty. Is the reservoir empty at the end of December 1962? Use a spreadsheet program you have written to solve this problem.

Menominee River below Koss, MI (Problem 3-37)
Mean monthly discharge (m^3/s)

Year	Jan	Feb	Mar	Apr	May	June	July	Aug	Sept	Oct	Nov	Dec
1946	76.3	69.8	149.0	106.0	85.3	132.0	89.5	62.2	58.5	56.6	76.7	56.8
1947	54.0	53.0	60.1	157.0	164.0	103.0	72.1	55.5	52.7	51.0	59.3	47.1
1948	49.2	35.4	72.5	103.0	82.3	48.5	42.2	45.8	39.6	34.8	56.1	49.2
1949	45.9	47.9	57.4	87.7	84.7	62.0	102.0	51.7	57.2	57.0	57.8	58.6
1950	59.1	57.6	61.1	199.0	243.0	101.0	69.2	65.8	49.0	43.4	48.9	48.4
1951	50.5	44.5	68.0	267.0	171.0	143.0	159.0	93.3	122.0	147.0	117.0	87.0
1952	76.9	75.6	66.5	219.0	94.0	86.7	153.0	97.0	58.6	45.6	50.2	51.9
1953	59.4	62.5	103.0	167.0	133.0	166.0	174.0	87.5	67.0	56.1	53.4	63.6
1954	57.0	66.7	68.5	182.0	184.0	138.0	73.2	61.0	91.1	123.0	84.6	68.5
1955	66.2	61.0	70.5	254.0	108.0	106.0	48.7	56.1	38.0	59.7	63.0	57.4
1956	59.0	54.8	49.7	270.0	108.0	88.4	116.0	83.4	61.5	48.8	53.1	53.6
1957	49.4	46.1	69.7	130.0	93.0	65.0	41.4	36.3	52.3	52.7	73.3	59.8
1958	54.0	51.3	61.8	123.0	65.9	62.0	132.0	47.0	58.5	47.7	68.9	48.4
1959	46.7	43.1	55.0	110.0	105.0	56.7	48.3	78.0	142.0	155.0	122.0	78.2
1960	82.3	71.0	62.4	242.0	373.0	135.0	83.4	72.1	80.8	60.5	102.0	68.2
1961	52.6	49.2	77.0	158.0	186.0	82.5	60.8	53.8	48.9	57.9	67.6	63.1
1962	53.9	52.4	69.2	168.0	168.0	107.0	59.3	52.1	73.8	65.3	54.7	51.4
1963	46.8	43.8	56.1	87.7	120.0	99.1	43.6	40.6	38.0	34.2	35.6	35.7
1964	37.2	33.8	41.8	70.2	131.0	63.2	42.1	56.9	65.1	55.6	69.8	54.2
1965	49.3	44.1	50.9	173.0	361.0	83.6	51.7	45.6	56.2	67.5	76.6	87.2
1966	78.5	66.6	131.0	157.0	117.0	113.0	44.6	63.0	43.5	62.6	63.7	59.2
1967	59.8	64.0	65.2	295.0	133.0	135.0	100.0	66.3	51.6	86.0	108.0	63.3
1968	50.5	58.0	72.6	134.0	108.0	168.0	141.0	78.1	155.0	93.8	90.2	82.7
1969	89.9	90.0	84.4	229.0	157.0	118.0	87.1	51.1	42.6	65.3	68.9	58.2
1970	62.7	51.0	58.8	114.0	109.0	157.0	56.9	46.6	49.1	64.1	122.0	97.5
1971	72.4	64.7	89.4	284.0	155.0	93.3	67.5	51.0	47.0	92.4	88.2	80.2
1972	65.9	56.7	68.5	194.0	254.0	88.6	68.2	108.0	91.0	134.0	138.0	77.2
1973	85.3	73.8	226.0	240.0	290.0	111.0	72.6	80.3	69.0	66.9	82.5	66.3
1974	62.8	65.4	73.4	142.0	107.0	111.0	61.6	86.4	77.8	58.8	106.0	75.5
1975	66.3	66.0	68.4	193.0	209.0	116.0	54.8	41.5	63.2	43.4	69.2	86.3
1976	67.5	69.2	96.9	298.0	147.0	79.1	41.2	36.9	30.3	—	—	—

Spokane River near Otis Orchards, WA (Problem 3-38)
Mean monthly discharge (m^3/s)

Year	Jan	Feb	Mar	Apr	May	June	July	Aug	Sept	Oct	Nov	Dec
1950	—	—	—	—	—	—	—	—	—	59.9	123.0	257.0
1951	217.0	492.0	200.0	422	460	156.0	36.3	4.25	8.44	75.6	98.7	145.0
1952	115.0	140.0	107.0	491	624	165.0	52.4	9.49	33.3	36.1	44.7	50.5
1953	169.0	311.0	163.0	246	525	358.0	32.3	12.8	29.8	52.2	53.0	97.8
1954	116.0	190.0	254.0	398	657	414.0	68.0	28.13	28.3	82.1	106.0	93.1
1955	60.0	76.3	62.4	257	544	468.0	82.3	19.5	26.0	86.0	166.0	354.0
1956	292.0	141.0	195.0	685	809	391.0	47.8	22.0	27.9	64.7	78.9	101.0
1957	99.0	61.0	278.0	461	792	329.0	33.6	12.5	15.7	55.5	66.9	73.0
1958	80.0	245.0	234.0	408	548	152.0	29.5	4.50	24.4	36.8	153.0	240.0
1959	356.0	233.0	192.0	465	351	410.0	35.5	11.6	45.2	84.4	224.0	172.0
1960	117.0	154.0	202.0	600	470	266.0	35.4	16.0	45.0	47.8	57.6	88.5
1961	93.0	454.0	406.0	389	559	352.0	8.86	4.96	19.1	24.5	55.9	50.7
1962	107.0	142.0	124.0	534	535	232.0	39.9	11.0	13.3	50.3	95.6	186.0
1963	152.0	268.0	208.0	353	303	92.1	26.9	5.32	10.1	21.2	63.0	71.2
1964	76.5	97.6	81.6	328	594	619.0	66.5	34.0	92.7	43.0	83.6	408.0
1965	282.0	314.0	268.0	475	602	158.0	57.3	28.9	23.2	32.5	82.7	76.7
1966	105.0	63.7	175.0	451	388	126.0	38.3	10.3	12.0	45.6	60.1	122.0
1967	205.0	282.0	191.0	272	504	413.0	40.8	13.3	30.8	49.3	68.5	93.6
1968	104.0	276.0	348.0	226	232	155.0	42.3	22.3	40.5	92.9	166.8	198.0
1969	254.0	138.0	162.0	638	651	220.0	50.7	25.2	40.5	45.8	49.0	60.8
1970	86.8	225.0	214.0	260	512	382.0	51.3	32.5	35.6	—	—	—

Rappahannock River near Warrenton, VA (Problem 3-39)
Mean monthly discharge (m^3/s)

Year	Jan	Feb	Mar	Apr	May	June	July	Aug	Sept	Oct	Nov	Dec
1943	7.08	9.23	10.5	8.55	6.12	3.31	1.18	0.294	0.269	0.057	1.38	1.06
1944	4.67	3.14	8.44	5.66	3.71	1.88	0.334	0.450	4.13	4.62	1.87	5.27
1945	4.11	5.44	4.81	3.62	3.79	2.94	4.22	9.68	13.0	3.77	4.05	8.27
1946	8.44	7.90	7.36	5.38	9.94	7.84	2.60	4.70	2.02	3.14	2.35	2.38
1947	6.94	3.79	6.77	3.57	4.84	3.74	2.92	2.39	1.40	1.04	6.60	2.74
1948	3.85	6.09	7.28	9.77	11.5	4.05	4.16	13.1	2.86	7.73	10.4	14.1
1949	13.8	11.0	8.95	11.9	13.2	5.10	4.39	3.79	1.67	2.03	2.37	2.94
1950	2.76	7.45	7.73	4.42	7.56	5.44	4.25	1.60	5.75	4.19	5.63	20.0
1951	5.72	12.8	11.0	12.1	5.44	9.40	2.89	1.19	0.447	0.453	1.93	5.13
1952	7.62	9.74	12.5	18.0	9.32	3.43	2.21	1.98	2.50	0.951	9.80	7.28
1953	11.3	7.25	12.6	8.24	10.6	4.39	1.36	0.685	0.343	0.357	0.773	2.25
1954	2.47	2.54	5.69	5.72	4.39	1.95	0.875	0.572	0.131	1.90	1.80	3.57
1955	2.42	4.11	8.21	5.07	3.85	4.16	0.801	20.3	3.54	2.47	2.19	1.62
1956	2.21	6.54	7.31	6.23	2.46	1.06	10.3	3.60	2.17	4.42	6.34	3.91
1957	4.33	8.07	8.95	8.72	3.31	2.15	0.402	0.008	0.391	1.02	1.66	6.43
1958	10.1	6.82	12.0	11.8	9.03	2.97	5.07	3.57	1.33	1.56	1.81	1.89
1959	3.45	3.06	4.76	7.08	3.28	5.04	0.804	0.513	0.759	2.68	2.05	3.88
1960	4.11	9.71	7.70	13.3	11.3	9.97	2.97	1.85	2.77	1.10	1.23	1.31
1961	3.31	15.4	9.85	15.5	11.1	6.82	3.23	2.24	1.70	1.16	1.77	4.25
1962	5.44	5.61	16.8	10.7	5.27	6.88	3.57	1.51	0.855	0.932	4.73	3.60
1963	6.51	4.19	13.6	4.45	2.55	3.20	0.496	0.136	0.160	0.121	2.28	2.21
1964	9.97	8.18	9.63	11.8	7.25	1.37	1.44	0.660	0.697	1.72	1.83	3.82
1965	6.51	13.8	15.0	6.31	3.74	1.34	5.27	0.365	3.09	0.459	0.379	0.450
1966	0.694	5.83	4.45	4.45	6.63	1.70	0.225	0.135	4.47	3.51	3.31	4.42
1967	5.63	5.86	13.6	3.82	4.53	1.57	1.04	7.42	2.27	3.82	2.57	10.0
1968	12.9	6.68	9.40	4.56	4.02	4.59	2.64	1.64	1.03	0.971	4.93	3.00
1969	4.47	4.84	5.07	3.51	1.93	1.70	1.54	1.93	2.12	1.80	3.37	6.12
1970	5.92	11.4	6.14	9.83	4.30	2.34	3.62	1.62	0.413	—	—	—

3-40. Four monitoring wells have been placed around a leaking underground storage tank. The wells are located at the corners of a 100-ha square. The total piezometric head in each of the wells is as follows: NE corner, 30.0 m; SE corner, 30.0 m; SW corner, 30.6 m: NW corner 30.6 m. Determine the magnitude and direction of the hydraulic gradient.

Answer: Hydraulic gradient $= 6.0 \times 10^{-4}$; direction = west to east

3-41. After a long wet spell, the water levels in the wells described in Problem 3-40 were measured and found to be the following distances from the ground surface: NE corner, 3.0 m; SE corner, 3.0 m; SW corner, 3.6 m; NW corner, 3.4 m. Assume that the ground surface is at the same elevation for each of the wells. Determine the magnitude and direction of the hydraulic gradient.

3-42. In preparation for a groundwater modeling study of the movement of a plume of contamination, three wells were installed to determine the hydraulic gradient. Three wells were installed in a rectangular grid at the following locations: well A at $x = 0.0$ m and $y = 0.0$ m; well B at $x = 280$ m and $y = 0.0$ m; well C at $x = 0.0$ m and $y = 500$ m. The ground surface elevation for each well is at 186.66 m above mean sea level. The depth to the groundwater table in each well is as follows: well A = 5.85 m; well B = 5.63 m; well C = 5.52 m. Determine the magnitude and direction of the hydraulic gradient.

3-43. A gravelly sand has a hydraulic conductivity of 6.9×10^{-4} m/s, a hydraulic gradient of 0.00141, and a porosity of 20 percent. Determine the Darcy velocity and the average linear velocity.

Answer: $v = 9.73 \times 10^{-7}$ m/s; $v' = 4.86 \times 10^{-6}$ m/s

3-44. A fine sand has a hydraulic conductivity of 3.5×10^{-5} m/s, a hydraulic gradient of 0.00141, and a porosity of 45 percent. Determine the Darcy velocity and the average linear velocity.

3-45. The results of a tracer study at a beach yielded an estimated average linear velocity of the groundwater of 0.60 m/d. Laboratory studies were used to determine a porosity of 30 percent and a hydraulic conductivity of 4.75×10^{-4} m/s for the sand. Estimate the Darcy velocity (in m/s) and hydraulic gradient of the groundwater.

3-46. Two piezometers have been placed along the direction of flow in a confined aquifer that is 30.0 m thick. The piezometers are 280 m apart. The difference in piezometric head between the two is 1.4 m. The aquifer hydraulic conductivity is 50 m/d and the porosity is 20 percent. Estimate the travel time for water to flow between the two piezometers.

3-47. A fully-penetrating well in a 28.0 m thick artesian aquifer pumps at a rate of 0.00380 m^3/s for 1,941 days and causes a drawdown of 64.05 m at an observation well 48.00 m from the pumping well. How much drawdown will occur at an observation well 68.00 m away? The original piezometric surface was 94.05 m above the bottom confining layer. The aquifer material is sandstone. Report your answer to two decimal places.

Answer: $s_2 = 51.08$ m

3-48. If a fully penetrating well in a 99.99 m thick artesian aquifer pumping at a rate of 0.0020 m^3/s for 1,812 days causes a drawdown of 12.73 m at an observation well 280.00 m from the pumping well, how much drawdown will occur at an observation well 1,492.0 m away? The original piezometric surface was 170.89 m above the bottom confining layer. The aquifer material is sandstone. Report your answer to two decimal places.

3-49. A 42.43 m thick fractured rock artesian aquifer supplies water to a fully penetrating well. Before pumping began, the piezometric surface in the pumping well was observed to be 70.89 m above the bottom confining layer. The pumping rate of the well is 0.0255 m^3/s. After 1,776 days of pumping an observation well is drilled 272.70 m from the pumping well. The drawdown in this well was observed to be 5.04 m. Estimate the drawdown at a distance of 64.28 m from the pumping well.

3-50. A long-term pumping test was conducted to determine the hydraulic conductivity of a 82.0 m thick confined aquifer. The nonpumping piezommetric surface was 109.5 m above the confining layer. The pumping rate was 0.0280 m^3/s. The steady-state drawdown at an observation well 41.0 m away from the pumping well was 3.55 m. The drawdown at an observation well 63.5 m away from the pumping well was 1.35 m. Determine the hydraulic conductivity to three significant figures.

3-51. It is undesirable to lower the piezometric surface of a confined aquifer below the aquiclude because this will destroy the structural integrity of the aquifer formation. Determine the maximum rate of pumping that would be permissible for the case described in Example 3-11 if the following conditions prevail:

1. The observation well at 200.0 m maintained the same drawdown.
2. Another observation well 2.0 m from the pumped well was used to observe lowering of the piezometric surface to the bottom of the aquiclude.

Report your answer to two decimal places.

3-52. An artesian aquifer 5.0 m thick with a piezometric surface 65.0 m above the bottom confining layer is being pumped by a fully penetrating well. The aquifer is a mixture of sand and gravel. A steady-state drawdown of 7.0 m is observed at a nonpumping well located 10.0 m away. If the pumping rate is 0.020 m^3/s, how far away is a second nonpumping well that has an observed drawdown of 2.0 m? Report your answer to two decimal places.

3-53. A test well was drilled to the underlying impervious stratum in an unconfined aquifer. The depth of the well was 18.3 m. Observation wells were drilled at 20.0 m and 110.0 m from the pumping test well. The static water level in each well was 4.57 m below the ground surface. The test well was pumped at a rate of 0.0347 m^3/s until steady-state conditions were achieved. The drawdown in the observation wells was 2.78 m and 0.73 m at 20.0 m and 110.0 m respectively. Determine the hydraulic conductivity in m/s. Report your answer to three significant figures.

3-54. For an unconfined aquifer, there is a possibility that drawdown will lower the piezometric surface to the bottom of the well and that the water will stop flowing. For the case described in Example 3-12, determine the maximum pumping rate that can be sustained indefinitely if the drawdown at the observation well 100.0 m from the pumped well is 9.90 m and the drawdown at the pumped well is limited by the depth of the aquifer, that is, 30.0 m. Report your answer to two decimal places.

Answer: $Q = 1.36$ m^3/s

3-55. A contractor is trying to estimate the distance to be expected of a drawdown of 4.81 m from a pumping well under the following conditions:

Pumping rate 0.0280 m^3/s

Pumping time = 1,066 d

Drawdown in observation well = 9.52 m

Observation well is located 10.00 m from the pumping well

Aquifer material = medium sand

Aquifer thickness = 14.05 m

Assume that the well is fully penetrating in an unconfined aquifer. Report your answer to two decimal places.

3-56. A fully penetrating well in a 30.0 m thick unconfined aquifer is to be used to dewater the area to be excavated for a new wastewater treatment plant. The piezometric surface must be lowered 5.25 m below the static water level 45.45 m from the pumping well. In addition, it is desired that the piezometric surface be lowered 2.50 m at a distance of 53.56 m from the pumping well. The aquifer material is loam. Assuming steady state, what pumping rate (in m^3/s) must be used to achieve the desired drawdowns. Report your answer to three significant figures.

3-57. A well with a 0.25 m diameter fully penetrates an unconfined aquifer that is 20.0 m thick. The well has a discharge of 0.0150 m^3/s and a drawdown of 8.0 m. If the flow is steady and the hydraulic conductivity is 1.5×10^{-4} m/s, what is the height of the piezometric surface above the confining layer at a site 80 m from the well?

3-58. Repeat Problem 3-57 using a well diameter of 0.50 m.

3-59. A 0.30 m diameter well fully penetrates a confined aquifer that is 28.0 m thick. The aquifer material is fractured rock. If the drawdown in the pumped well is 6.21 m after pumping for 48 hours at a rate of 0.0075 m^3/s, what will the drawdown be at the end of 48 days of pumping at this rate?

3-60. An aquifer yields the following results from pumping a 0.61 m diameter well at 0.0303 m^3/s: $s = 0.98$ m in 8 min; $s = 3.87$ m in 24 h. Determine its transmissivity. Report your answer to three significant figures.

Answer: $T = 4.33 \times 10^{-3}\ m^2/s$

3-61. Determine the transmissivity of a confined aquifer that yields the following results from a pumping test of a 0.46 m diameter well that fully penetrates the aquifer.

Pumping rate $= 0.0076\ m^3/s$

$s = 3.00$ m in 0.10 min

$s = 34.0$ m in 1.00 min

3-62. An aquifer yields a drawdown of 1.04 m at an observation well 96.93 m from a well pumping at 0.0170 m^3/s after 80 min of pumping. The virtual time is 0.6 min and the transmissivity is $5.39 \times 10^{-3}\ m^2/s$. Determine the storage coefficient.

Answer: $S = 4.647 \times 10^{-5}$ or 4.6×10^{-5}

3-63. Using the data from Problem 3-62, find the drawdown at the observation well 80 days after pumping begins.

3-64. If the transmissivity is $2.51 \times 10^{-3}\ m^2/s$ and the storage coefficient is 2.86×10^{-4}, calculate the drawdown that will result at the end of 2 days of pumping a 0.50 m diameter well at a rate of 0.0194 m^3/s.

3-65. Determine the storage coefficient for an artesian aquifer from the pumping test results shown in the table below. The measurements were made at an observation well 300.00 m away from the pumping well. The pumping rate was 0.0350 m^3/s.

Time (min)	Drawdown (m)
100.0	3.10
500.0	4.70
1,700.0	5.90

Answer: $S = 1.9 \times 10^{-5}$

3-66. Rework Problem 3-65, but assume that the data were obtained at an observation well 100.0 m away from the pumping well.

3-67. Determine the storage coefficient for an artesian aquifer from the pumping test results shown in the table below. The measurements were made at an observation well 100.00 m away from the pumping well. The pumping rate was 0.0221 m^3/s.

Time (min)	Drawdown (m)
10.0	1.35
100.0	3.65
1,440.0	6.30

3-68. Rework Problem 3-67, but assume that the data were obtained at an observation well 60.0 m away from the pumping well.

3-69. Determine the storage coefficient for an artesian aquifer from the following pumping test results on a 0.76 m diameter well that fully penetrates the aquifer. The pumping rate was 0.00350 m^3/s. The drawdowns were measured in the pumping well.

Time (min)	Drawdown (m)
0.20	2.00
1.80	3.70
10.0	5.00

3-70. Two wells located 106.68 m apart are both pumping at the same time. Well A pumps at 0.0379 m^3/s and well B pumps at 0.0252 m^3/s. The diameter of each well is 0.460 m. The transmissivity is 4.35×10^{-3} m^2/s and the storage coefficient is 4.1×10^{-5}. What is the interference of well A on well B after 365 days of pumping? Report your answer to two decimal places.

Answer: Interference of well A on B is 9.29 m.

3-71. Using the data from Problem 3-70, find the total drawdown in well B after 365 days of pumping. Report your answer to two decimal places.

3-72. If two wells, no. 12 and no. 13, located 100.0 m apart, are pumping at rates of 0.0250 m^3/s and 0.0300 m^3/s, respectively, what is the interference of well no. 12 on well no. 13 after 280 days of pumping? The diameter of each well is 0.500 m. The transmissivity is 1.766×10^{-3} m^2/s and the storage coefficient is 6.675×10^{-5}. Report your answer to two decimal places.

3-73. Using the data from Problem 3-72, find the total drawdown in well 13 after 280 days of pumping. Report your answer to two decimal places.

3-74. Wells X, Y, and Z are located equidistant at 100.0 m intervals. Their pumping rates are 0.0315 m^3/s, 0.0177 m^3/s, and 0.0252 m^3/s respectively. The diameter of each well is 0.300 m. The transmissivity is 1.77×10^{-3} m^2/s. The storage coefficient is 6.436×10^{-5}. What is the interference of well X on well Y and on well Z after 100 days of pumping? Report your answer to two decimal places.

3-75. Using the data in Problem 3-74, find the total drawdown in well X at the end of 100 days of pumping. Report your answer to two decimal places.

3-76. For the well field layout shown in Figure P-3-76, determine the effect of adding a sixth well. Is there any potential for adverse effects on the well or the aquifer? Assume all wells are pumped for 100 days and that each well is 0.300 m in diameter. Well data are given in the table below. Aquifer data are shown below the well data.

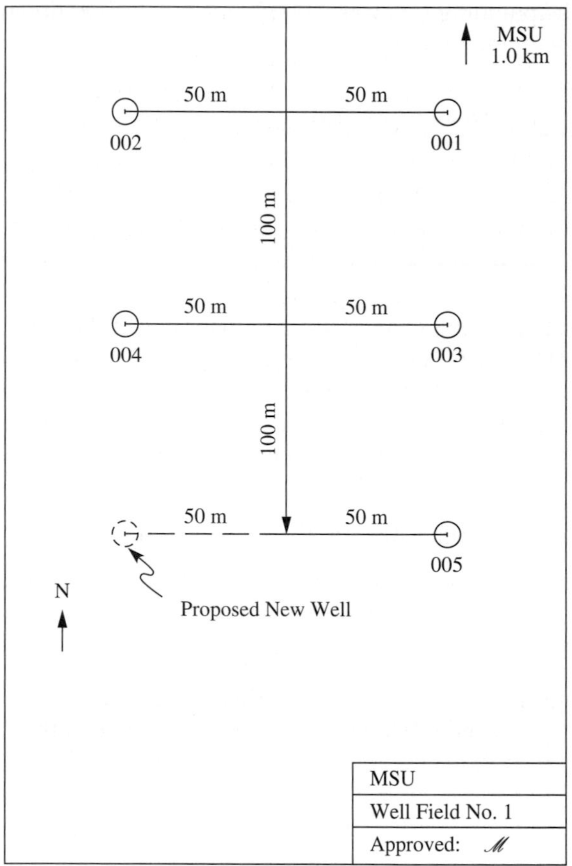

FIGURE P-3-76
Layout for MSU well field.

MSU Well Field No. 1

Well no.	Pumping rate (m^3/s)	Depth of well (m)
1	0.0221	111.0
2	0.0315	112.0
3	0.0189	110.0
4	0.0177	111.0
5	0.0284	112.0
6 (proposed)	0.0252	111.0

The aquifer characteristics are as follows:

Storage coefficient = 6.418×10^{-5}

Transmissivity = 1.761×10^{-3} m^2/s

Nonpumping water level = 6.90 m below grade

Depth to top of artesian aquifer = 87.0 m

Answer: Total drawdown for each well in numerical order: (1) 79.54 m, (2) 84.99 m, (3) 80.05 m, (4) 79.54 m, (5) 83.18 m, (6) 81.35 m. Drawdown is below the top of the aquiclude for wells 2, 5, and 6.

3-77. For the well field layout shown in Figure P-3-77, determine the effect of adding a sixth well. Is there any potential for adverse effects on the well or the aquifer? Assume all wells are pumped for 100 days and that each well is 0.300 m in diameter. Aquifer and well data are given below the Figure.

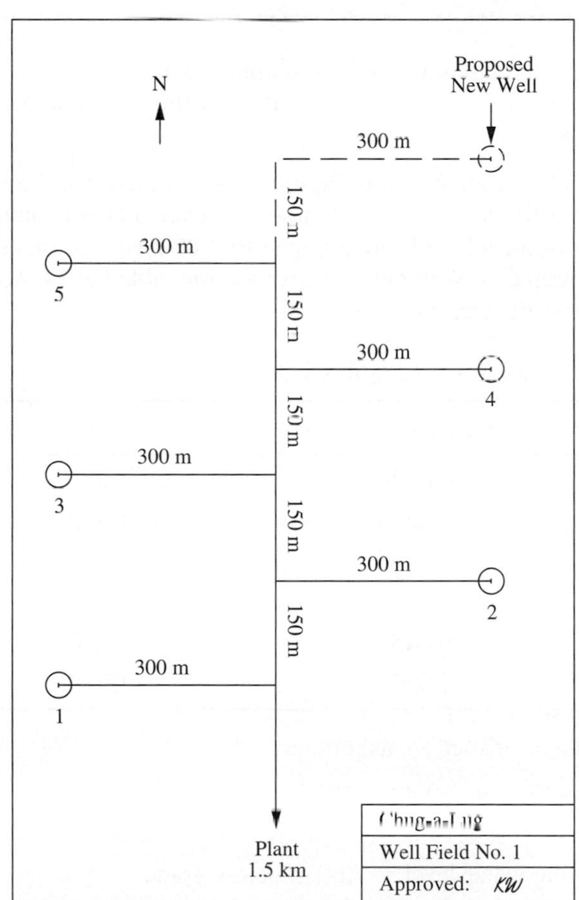

FIGURE P-3-77
Layout for brewery well field no. 1.

The aquifer characteristics are as follows:

Storage coefficient = 2.11×10^{-6}

Transmissivity = 4.02×10^{-3} m^2/s

Non-pumping water level = 9.50 m below grade
Depth to top of artesian aquifer = 50.1 m

Chug-a-Lug Brewery Well Field No. 1

Well no.	Pumping rate (m^3/s)	Depth of well (m)
1	0.020	105.7
2	0.035	112.8
3	0.020	111.2
4	0.015	108.6
5	0.030	113.3
6 (proposed)	0.025	109.7

3-78. What pumping rate, pumping time, or combination thereof can be sustained by the new well in Problem 3-77 if all of the well diameters are enlarged to 1.50 m?

3-79. For the well field layout shown in Figure P-3-79, determine the effect of adding a sixth well. Is there any potential for adverse effects on the well or the aquifer? Assume all wells are pumped for 180 days and that each well is 0.914 m in diameter. Well data are given in the table below. Aquifer data are shown below the well field data.

Chug-a-Lug Brewery Well Field No. 2

Well No.	Pumping rate (m^3/s)	Depth of well (m)
1	0.0426	169.0
2	0.0473	170.0
3	0.0426	170.0
4	0.0404	168.0
5	0.0457	170.0
6 (proposed)	0.0473	170.0

The aquifer characteristics are as follows:

Storage coefficient = 2.80×10^{-5}
Transmissivity = 1.79×10^{-3} m^2/s
Nonpumping water level = 7.60 m below grade
Depth to top of artesian aquifer = 156.50 m

3-80. What pumping rate, pumping time, or combination thereof can be sustained by the new well in Problem 3-79 if all of the well diameters are enlarged to 1.80 m?

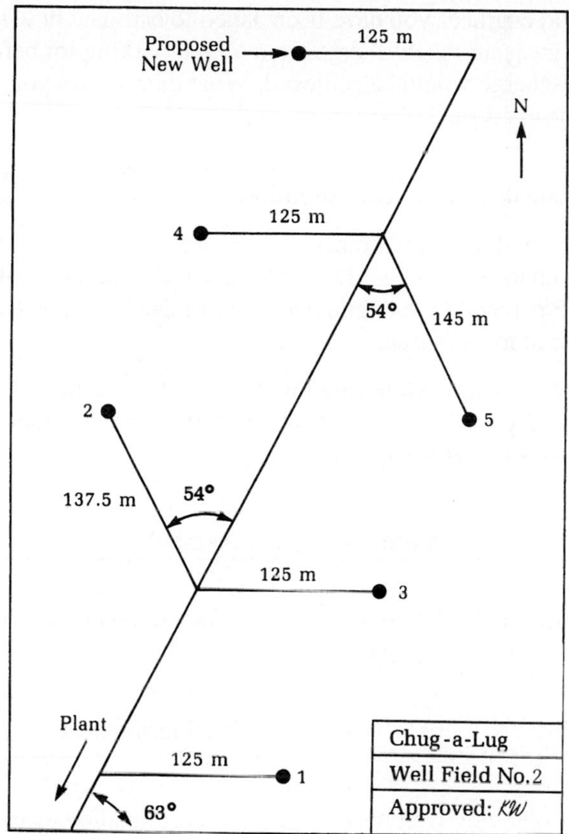

FIGURE P-3-79
Layout for brewery well field no. 2.

3-9 DISCUSSION QUESTIONS

3-1. An artesian aquifer is under pressure because of the weight of the overlying geologic strata. Is this sentence true or false? If it is false, rewrite the sentence to make it true.

3-2. Identify the base flow in the following hydrographs.

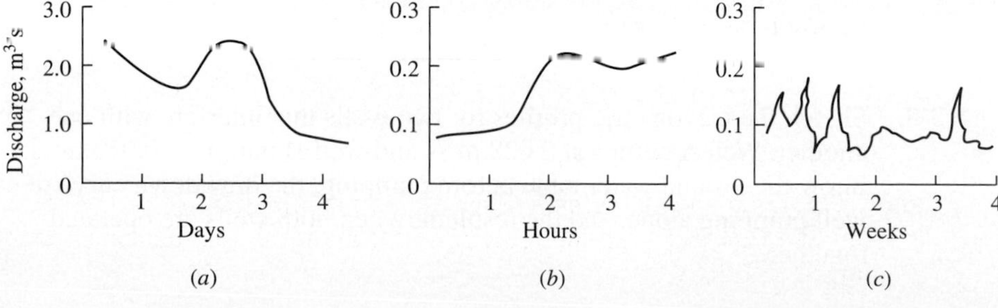

3-3. As a field engineer you have been asked to estimate how long you would have to measure the discharge from a mall parking lot before the maximum discharge would be achieved. What data would you have to gather to make the estimate?

3-4. Explain why it is impossible to use the rational formula without being able to estimate the time of concentration.

3-5. When a flood has a recurrence interval (return period) of 5 years, it means that the chance of another flood of the same or less severity occurring next year is 5 percent. Is this sentence true or false? If it is false, rewrite the sentence to make it true.

3-6. For the following well boring log, identify the pertinent hydrogeologic features. The well screen is set at 6.0–8.0 m and the static water level after drilling is 1.8 m from the ground surface.

Strata	Depth, m	Remarks
Top soil	0.0–0.5	
Sandy till	0.5–6.0	Water encountered at 1.8 m
Sand	6.0–8.0	
Clay	8.0–9.0	
Shale	9.0–10.0	Well terminated

3-7. For the following well boring log (Bracebridge, Ontario, Canada), identify the pertinent hydrogeologic features. The well screen is set at 48.0–51.8 m and the static water level after drilling is 10.2 m from the ground surface.

Strata	Depth, m	Remarks
Sand	0.0–6.1	
Gravelly clay	6.10–8.6	
Fine sand	8.6–13.7	
Clay	13.7–17.5	Casing sealed
Fine sand	17.5–51.8	
Bedrock	51.8	Well terminated

3-8. Sketch the piezometric profiles for two wells that interfere with one another. Well A pumps at 0.028 m^3/s and well B pumps at 0.052 m^3/s. Show the ground water table before pumping, the drawdown curve of each well pumping alone, and the resultant when both wells are operated together.

3-10 REFERENCES

ASCE (1969) *Design and Construction of Sanitary and Storm Sewers,* Manual of Practice No. 37 (also Water Pollution Control Federation Manual of Practice No. 9), American Society of Civil Engineers, New York, pp. 43–46.

Bouwer, H. (1978) *Groundwater Hydrology*, McGraw-Hill, New York, pp. 22, 38, 66, 68.

Cooper, H. H., and C. E. Jacob (1946) "A Generalized Graphical Method for Evaluating Formation Constants and Summarizing Well Field History," *Transactions American Geophysical Union,* vol. 27, pp. 520–534.

Dalton, J. (1802) "Experimental Essays on the Constitution of Mixed Gases; on the Force of Steam or Vapor from Waters and other Liquids, Both in a Torricellian Vacuum and in Air; on Evaporation; and on the Expansion of Gases by Heat," *Member Proceedings, Manchester Literary and Philosophical Society,* vol. 5, pp. 535–602.

Darcy, H. (1856) *Les Fontaines Publiques de la Ville de Dijon,* Victor Dalmont, Paris, pp. 570, 590–594.

Davis, M. L., and S. J. Masten, (2004) *Principles of Environmental Engineering and Science,* McGraw-Hill, New York, p. 195.

Dupuit, J. (1863) *Etudes Théoriques et Practiques sur le Mouvement des Eaux dans Les Canaux Découverts et à Travers les Terrains Perrméables*, Dunod, Paris.

FAA (1970) *Airport Drainage,* Advisory Circular A/C 150-5320-5B, Federal Aviation Agency, Department of Transportation, U.S. Government Printing Office, Washington, DC.

Ferris, J. G., D. B. Knowles, R. H. Brown, and R. W. Stallman (1962) *Theory of Aquifer Tests,* U.S. Geological Survey Water-Supply Paper 1536-E, pp. 69–174.

Flack, J. E. (1980) "Achieving Urban Water Conservation," *Water Resources Bulletin,* vol. 16, no. 1.

Gilman, C. S. (1964) "Rainfall," in V. T. Chow (ed.), *Handbook of Applied Hydrology,* McGraw-Hill, New York, p. 9–50.

Hanke, S. H. (1970) "Demand for Water Under Dynamic Conditions," *Water Resources Research*, vol. 6, No. 5.

Heath, R. C. (1983) *Basic Ground-Water Hydrology,* U.S. Geological Survey Water-Supply Paper 2220, U.S. Government Printing office, Washington, DC.

Horton, R. E. (1935) *Surface Runoff Phenomena: Part I, Analysis of the Hydrograph,* Horton Hydrologic Lab Publication 101, Edward Bros., Ann Arbor, MI.

Johnson Division, UOP (1975) *Ground Water and Wells*, Johnson Division, UOP, St. Paul, MN, pp. 37, 102.

Linsley, R. K., M. A. Kohler, and J. L. H. Paulhus (1975) *Hydrology for Engineers,* McGraw-Hill, New York, p. 200.

Loucks, D. P., J. R. Stedinger, and D. A. Haith (1981) *Water Resource Systems Planning and Analysis,* Prentice-Hall, Englewood Cliffs, NJ, p. 3

Rippl, W. (1883) "The Capacity of Storage Reservoirs for Water Supply," *Proceedings of the Institution of Civil Engineers* (London), vol. 71, p. 270.

Rousculp, J. A. (1939) "Relation of Rainfall and Runoff to Cost of Sewers," *Transactions of the American Society of Civil Engineers,* vol. 104, p. 1473.

Sherman, L. K. (1932) "Stream-Flow from Rainfall by the Unit-Graph Method," *Engineering News Record,* vol. 108, pp. 501–505.

Theim, G. (1906) *Hydrologische Methoden,* J. M. Gebhart, Leipzig, Germany.

Theis, C. V. (1935) "The Relation Between Lowering of the Piezometric Surface and the Rate and Duration of Discharge of a Well Using Ground Water Storage," *Transactions of the American Geophysical Union,* vol. 16, pp. 519–524.

U.S. EPA (1973) *Manual of Individual Water Supply Systems,* U.S. Environmental Protection Agency (EPA-430-9-73), Washington, DC, pp. 45–50, 107–109.

U.S. EPA (1997) *National Public Water Systems Compliance Report*, U.S. Environmental Protection Agency, Office of Water (EPA-305-R-99-002) Washington, D.C.

Viessman, W., J. W. Knapp, and G. L. Lewis (1989) *Introduction to Hydrology,* Harper & Row, New York, p. 186.

Walton, W. C. (1970) *Groundwater Resource Evaluation,* McGraw-Hill, New York, p. 34.

Weibull, W. (1939) "Statistical Theory of the Strength of Materials," *Ing. Vetenskapsakad. Handl,* Stockholm, vol. 151, p. 15.

CHAPTER 4

WATER TREATMENT

4-1 INTRODUCTION

Approximately 80 percent of the United States population turn their taps on every day to take a drink of publicly supplied water. They all assume that when they take a drink it is safe. They probably never even think of safety.

There are approximately 170,000 public water systems in the United States. EPA classifies these water systems according to the number of people they serve, the source of their water, and whether they serve the same customers year-round or on an occasional basis. The following statistics are based on information in the *Safe Drinking Water Information System* (SDWIS), for the fiscal year ended September 2000, as reported to EPA by the states (U.S. EPA, 2000).

Public water systems provide water for human consumption through pipes or other constructed conveyances to at least 15 service connections or serve an average of at least 25 people for at least 60 days a year. EPA has defined three types of public water systems:

- Community Water System (CWS): A public water system that supplies water to the same population year-round.
- Non-Transient Non-Community Water System (NTNCWS): A public water system that regularly supplies water to at least 25 of the same people at least six months per year, but not year-round. Some examples are schools, factories, office buildings, and hospitals that have their own water systems.
- Transient Non-Community Water System (TNCWS): A public water system that provides water in a place such as a gas station or campground where people do not remain for long periods of time.

EPA also classifies water systems according to the number of people they serve:

- Very small water systems serve 25–500 people
- Small water systems serve 501–3,300 people
- Medium water systems serve 3,301–10,000 people
- Large water systems serve 10,001–100,000 people
- Very large water systems serve 100,000+ people

The following is the number of systems and population served in the year 2000 (*Note:* populations are not summed because some people are served by multiple systems and counted more than once):

- 54,064 CWS served 263.9 million people
- 20,559 NTNCWS served 6.9 million people
- 93,210 TNCWS served 12.9 million people

In developing nations, clean water is the exception rather than the rule.

In 2005 every 15 seconds a child under the age of 5 died from a water-related illness. Seventeen percent of the earth's population (1.2 billion people) do not have

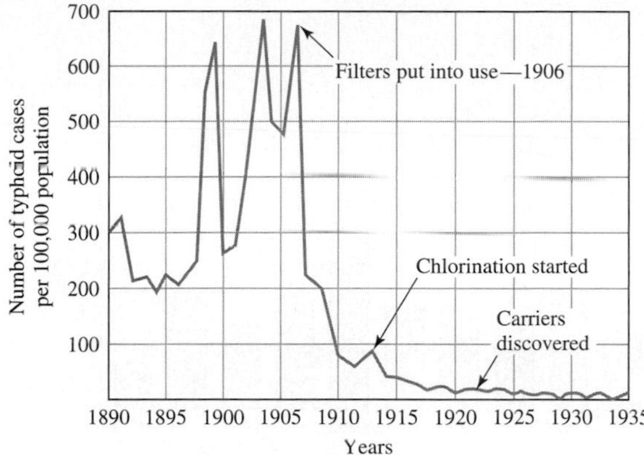

FIGURE 4-1
Typhoid fever cases per 100,000 population from 1890 to 1935, Philadelphia.

reliable drinking water and 40 percent of the population do not have access to adequate sanitation (www.waterforpeople.com).

The fact that the United States and developed countries have an outstanding water supply record is no accident. Since the early 1900s environmental engineers have been working in the United States to reduce waterborne disease. Figure 4-1 shows the incidence of typhoid cases in Philadelphia from 1890 to 1935. This is typical of the decrease in waterborne disease as public water supply treatment has increased. Philadelphia received its water from rivers and distributed the water untreated until about 1906, when slow sand filters were put into use. An immediate reduction in typhoid fever was realized. Disinfection of the water by the addition of chlorine further decreased the number of typhoid cases. A still greater decrease was accomplished after 1920 by careful control over infected persons who had become carriers. Since 1952 the death rate from typhoid fever in the United States has been less than 1 per 1,000,000. Many countries and organizations have committed their technological and financial resources to developing nations in recognition of the principle that reasonable access to safe and adequate drinking water is a fundamental right of all people.

Advances in public health tracking in the United States, and the more acute awareness of waterborne illnesses from a variety of organisms, have contributed to a better understanding of the link between water quality and illness. While many of the organisms associated with deaths or serious disease (e.g., typhoid, polio virus) have been eliminated in the United States, organisms with the potential to cause sickness and occasionally death are still being found. Between 1980 and 1996, 402 outbreaks of waterborne disease were reported in the United States (WQ&T, 1999) as show in Figure 4-2. Many of these illnesses are associated with gastrointestinal symptoms (diarrhea, fatigue, cramps). The majority of these outbreaks tend to occur in small systems. Table 4-1 shows the number of illnesses caused by various organisms. Of those 509,213 cases, 403,000 cases are attributed to one outbreak of cryptosporidiosis in Milwaukee, Wisconsin in 1993. It was estimated that 403,000 people became sick and 50 deaths occurred among severely immunocompromised individuals during this outbreak. At the time, the city complied with all water quality regulations. In 2006, EPA passed new regulations to address cryptosporidiosis.

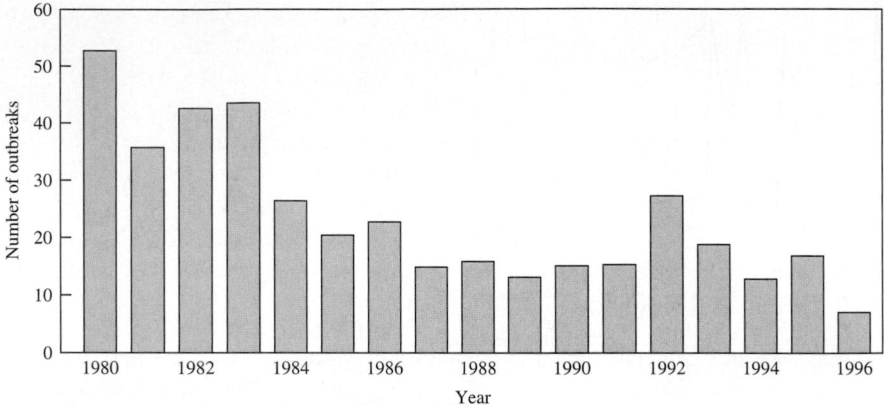

FIGURE 4-2
Waterborne disease outbreaks, 1980 to 1996. (*Source:* WQ & T, 1999.)

TABLE 4-1
Waterborne Disease Outbreaks in the United States, 1980 to 1996*†

Illness	No. of outbreaks	Cases of illness
Gastroenteritis, undefined	183	55,562
Giardiasis	84	10,262
Chemical poisoning	46	3,097
Shigellosis	19	3,864
Gastroenteritis, Norwalk virus	15	9,437
Campylobacteriosis	15	2,480
Hepatitis A	13	412
Cryptosporidiosis	10	419,939‡
Salmonellosis	5	1,845
Gastroenteritis, *E. coli* 0157:H7	3	278
Yersiniosis	2	103
Cholera	2	28
Gastroenteritis, rotavirus	1	1,761
Typhoid fever	1	60
Gastroenteritis, Plesiomonas	1	60
Amoebiasis	1	4
Cyclosporiasis	1	21
TOTAL	**402**	**509,213**

*An outbreak of waterborne disease for microorganisms is defined as: (1) two or more persons experience a similar illness after consumption or use of water intended for drinking, and (2) epidemiologic evidence implicates the water as a source of illness. A single case of chemical poisoning constitutes an outbreak if a laboratory study indicates that the water has been contaminated by the chemical.
†Data are from CDC annual surveillance summaries for 1980 through 1985 and two-year summaries for 1986 through 1994, as corrected for several missing outbreaks by G.F. Craun (personal communications).
‡Total includes 403,000 cases from a single outbreak.
Source: WQ&T, 1999.

A water that can be consumed in any desired amount without concern for adverse health effects is termed a *potable* water. Potable does not necessarily mean that the water tastes good. This is in contrast to a *palatable* water, which is one that is pleasing to drink, but not necessarily safe. We have learned that we must provide a water that is both potable and palatable, for if it is not palatable people will turn to untreated water that may not be potable. The widespread availability of potable water in the United States does not mean that there are no operational or control deficiencies in water systems, especially in smaller systems. The scientific community is continually making advances in identifying contaminants and discovering potential long-term health effects of constituents that had not been previously identified.

Water Chemistry

An understanding of the fundamentals of water chemistry is essential to the comprehension of the chapters on water quality and wastewater treatment. You should plan to spend ample time working the examples in this section.

Physical Properties of Water. The basic physical properties of water relevant to water treatment are density and viscosity. Density is a measure of the concentration of matter and is expressed in three ways:

1. Mass density, ρ. *Mass density* is mass per unit volume and is measured in units of kg/m^3. Appendix A, Table A-1, shows the variation of density with temperature for pure water free from air. Dissolved impurities change the density in direct proportion to their concentration and their own density. In environmental engineering applications, it is common to ignore the density increase due to impurities in the water. However, environmental engineers do not ignore the density of the matter when dealing with high concentrations, such as thickened sludge or commercial liquid chemicals.

2. Specific weight, γ. *Specific weight* is weight (force) per unit volume, measured in units of kN/m^3. The specific weight of a fluid is related to its density by the acceleration of gravity, g, which is 9.81 m/s^2.

$$\gamma = \rho g \tag{4-1}$$

3. Specific gravity, S. Specific gravity is given by

$$S = \rho/\rho_0 = \gamma/\gamma_0 \tag{4-2}$$

where the subscript zero denotes the density of water at 3.98°C, 1,000 kg/m^3, and the specific weight of water, 9.81 kN/m^3.

For quick approximations, the density of water at normal temperature is taken as 1,000 kg/m^3 (which is conveniently 1 kg/L) with a specific gravity = 1.

All substances, including liquids, exhibit a resistance to movement, an internal friction. The higher the friction, the harder it is to pump the liquid. A measure of the friction is viscosity. Viscosity is presented in one of two ways:

1. Dynamic viscosity, or absolute viscosity, μ, has dimensions of mass per unit length per time, with units of Pa · s.

2. Kinematic viscosity, υ, is found by

$$\upsilon = \mu/\rho \tag{4-3}$$

and has dimensions of length squared per time with the corresponding units m^2/s.

States of Solution Impurities. From an environmental engineering point of view, substances can exist in water in one of three classifications—*suspended, colloidal,* or *dissolved.*

A dissolved substance is one which is truly in solution. The substance is homogeneously dispersed in the liquid. Dissolved substances can be simple atoms or complex molecular compounds. Dissolved substances are in the liquid, that is, there is only one phase present. The substance cannot be removed from the liquid without accomplishing a phase change such as distillation, precipitation, adsorption, extraction, or passage through "ionic" pore-sized membranes. In *distillation* either the liquid or the substance itself is changed from a liquid phase to a gas phase in order to achieve separation. In *precipitation* the substance in the liquid phase combines with another chemical to form a solid phase, thus achieving separation from the water. *Adsorption* also involves a phase change, wherein the dissolved substance reacts with a solid particle to form a solid particle-substance complex. *Liquid extraction* can separate a substance from water by extracting it into another liquid, hence a phase change from water to a different liquid. A membrane with pore sizes in the ionic-size range can separate dissolved substances from the solution by a high-pressure filtering process.

Suspended solids are large enough to settle out of solution or be removed by filtration. In this case there are two phases present, the liquid water phase and the suspended-particle solid phase. The lower size range of this class is 0.1 to 1.0μm, about the size of bacteria. In environmental engineering, suspended solids are defined as those solids that can be filtered by a glass fiber filter disc and are properly called filterable solids. Suspended solids can be removed from water by physical methods such as sedimentation, filtration, and centrifugation.

Colloidal particles are in the size range between dissolved substances and suspended particles. They are in a solid state and can be removed from the liquid by physical means such as very high-force centrifugation or filtration through membranes with very small pore spaces. However, the particles are too small to be removed by sedimentation or by normal filtration processes. Colloidal particles exhibit the Tyndall effect; that is, when light passes through a liquid containing colloidal particles, the light is reflected by the particles. The degree to which a colloidal suspension reflects light at a 90° angle to the entrance beam is measured by *turbidity.* Turbidity is a relative measure, and there are various standards against which a sample is compared. The most common standard is a nephelometric turbidity unit (NTU). For our purposes we will simply refer to the measure of turbidity as a turbidity unit (TU). For a given particle size, the higher the turbidity, the higher the concentration of colloidal particles.

Another useful term in environmental engineering that is used to describe a solution state is *color*. Color is not separate from the above three categories, but rather is a combination of dissolved and colloidal materials. Color is widely used in environmental engineering because it, in itself, can be measured. However, it is very difficult to distinguish "dissolved color" from "colloidal color." Some color is caused by colloidal

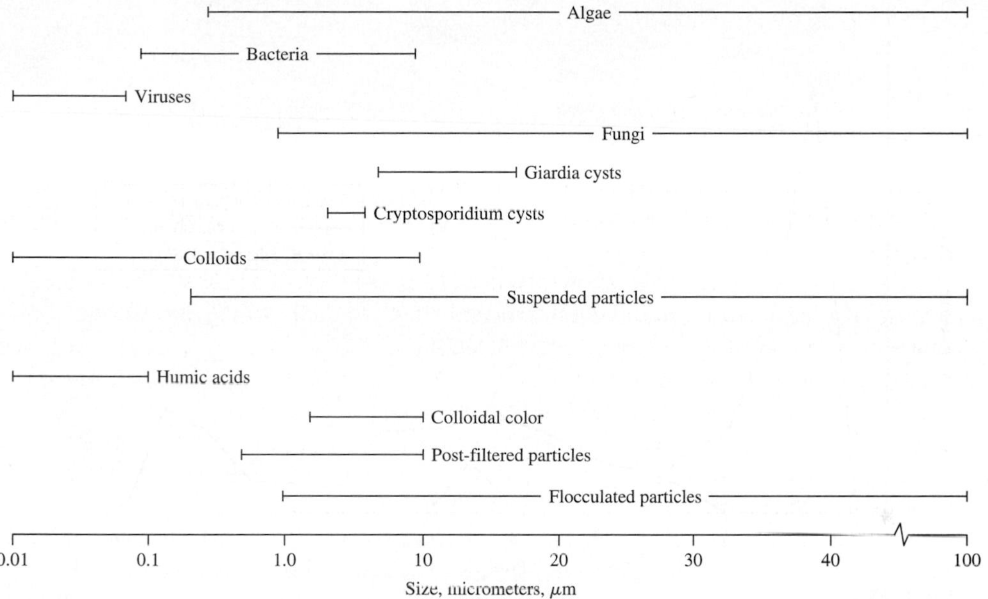

FIGURE 4-3
Particulates in water. (*Source:* McTigue and Cornwell, 1988.)

iron or manganese complexes, although the most common cause of color is from complex organic compounds that originate from the decomposition of organic matter. One common source of color is the degradation of soil humus, which produces humic acids. Humic acids impart reddish-brown color to the water. Humic acids have molecular weights between 800 and 50,000, the lower being dissolved and the greater, colloidal. Most color seems to be between 3.5 and 10μm, which is colloidal. Color is measured by the ability of the solution to absorb light. Color particles can be removed by the methods discussed for dissolved or colloidal particles, depending upon the state of the color.

Figure 4-3 presents an overview by size of the types of particles that are often dealt with in water treatment. A technique that is being used in water treatment to help evaluate water quality is *particle counting*. A particle counter counts the number of particles in a water sample and reports the results by particle size, generally from 1 to 30μm. Figure 4-4 shows a sample count comparing the distribution of particles in a raw water to that of the finished water. While particle counting does not indicate anything about the kind of particle, it can be useful in assessing overall treatment efficiency as well as characterizing water sources.

Chemical Units. Since solutes in solution are often analyzed by weight, the terms *weight percent* and *milligram per liter* are used. In order to perform stoichiometric calculations,* it is necessary to convert to common units, and the terms *molarity* and *normality* are used.

**Stoichiometry* is the part of chemistry concerned with measuring the proportions of elements involved in a reaction. Stoichiometric calculations are an application of the principle of conservation of mass to chemical reactions.

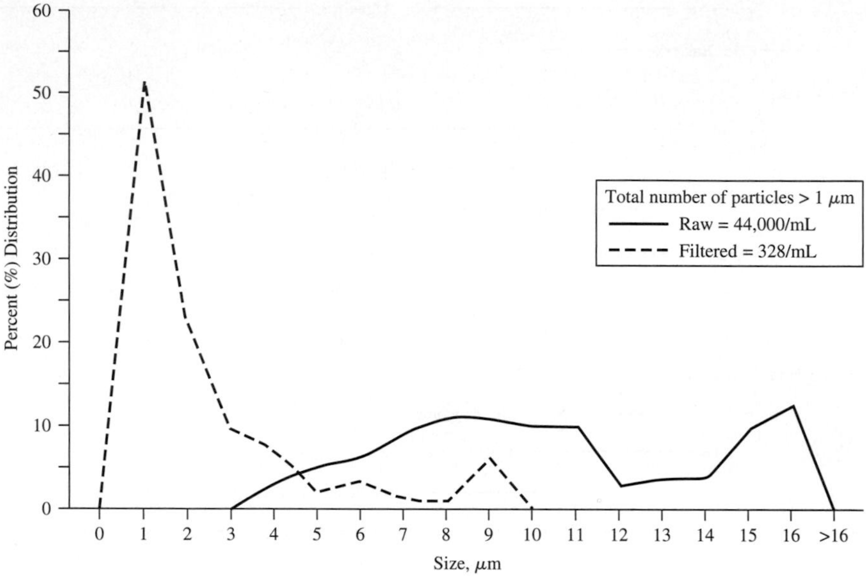

FIGURE 4-4
Particle distribution changes through treatment.

Weight percent, P, is sometimes employed to express approximate concentrations of commercial chemicals or of solid concentrations of sludges. The term specifies the grams of substance per 100 grams of solution and is mathematically expressed as

$$P = \frac{W}{W + W_0} \times 100\% \tag{4-4}$$

where P = percent of substance by weight
W = grams of substance
W_0 = grams of solution

Analysts generally give results directly in mass per volume (concentration), and the units are mg/L. In environmental engineering it is often assumed that the substance does not change the density of the water. This is generally untrue, but it does make for some useful conversions, and the assumption is not too inaccurate for dilute concentrations. If such an assumption is made and we recall that 1 mL of water weighs 1 g (again an approximation), then

$$\frac{1\text{ mg}}{\text{L}} = \frac{1\text{ mg}}{1{,}000\text{ g}} = \frac{1\text{ mg}}{10^6\text{ mg}} = 1\text{ ppm} \tag{4-5}$$

or 1 mg/L equals 1 part per million (ppm). If the same assumptions are made, then the weight percent of 1 mg/L can be determined:

$$P = \frac{W}{W + W_0} \times 100 = \frac{1\text{ mg}(100)}{1\text{ L}} = \frac{10^{-3}\text{ g}(100)}{10^3\text{ g}} = 1 \times 10^{-4}\% \tag{4-6}$$

or 1 mg/L equals $1 \times 10^{-4}\%$, which can be translated into 1% = 10,000 mg/L.

In order to work with chemical reactions it is necessary to convert weight concentrations to molarity or normality. A *mole* is 6.02×10^{23} molecules of a substance. Chemical reactions are expressed in integral numbers of moles. A mole of a substance has a relative weight called its *molecular weight* (MW). Molecular weight is the sum of the atomic weights. A table of atomic weights is given inside the front cover of this book. *Molarity* is the number of moles in a liter of solution. A 1-molar (1 M) solution has 1 mole of substance per liter of solution. Molarity is related to mg/L by

$$\begin{aligned} \text{mg/L} &= \text{molarity} \times \text{molecular weight} \times 10^3 \\ &= (\text{moles/L})(\text{g/mole})(10^3\ \text{mg/g}) \end{aligned} \tag{4-7}$$

A second unit, *equivalent weight* (EW), is frequently used in softening and redox reactions. The equivalent weight is the molecular weight divided by the number (n) of electrons transferred in redox reactions or the number of protons transferred in acid/base reactions.

The value of n depends on how the molecule reacts. In this text we are concerned with molecules that react in acid/base reactions or precipitation reactions. **In an acid/base reaction, n is the number of hydrogen ions that the molecule transfers. That is, an acid gives up an EW of hydrogen ions, and a base accepts an EW of hydrogen ions. In a precipitation reaction, n is the valence of the element in question. For compounds, n is equal to the number of hydrogen ions that would be required to replace the cation; that is, for $CaCO_3$ it would take two hydrogen ions to replace the calcium, therefore, $n = 2$. In oxidation/reduction reactions, n is equal to the change in oxidation number that the compound undergoes in the reaction.** Obviously, it is difficult to recognize reaction capacity without the context of the reaction. Common valence states of elements found in water are listed in Appendix A.

Normality (N) is the number of equivalent weights per liter of solution and is related to molarity (M) by

$$N = Mn \tag{4-8}$$

Example 4-1. Commercial sulfuric acid, H_2SO_4, is often purchased as a 93 weight percent solution. Find the mg/L of H_2SO_4 and the molarity and normality of the solution. Sulfuric acid has a specific gravity of 1.839.

Solution. Since 1 L of water weighs 1,000 g, 1 L of 100% H_2SO_4 weighs

$$1{,}000(1.839) = 1{,}839\ \text{g}$$

$(0.93)(1{,}839\ \text{g}) = 1{,}710$ g of H_2SO_4, or 1.7×10^6 mg/L of H_2SO_4 in a 93% solution. The molecular weight of H_2SO_4 is found by looking up the atomic weights on the inside cover of this book:

$$\begin{aligned} 2\text{H} &= 2(1) &&= 2 \\ \text{S} &= && 32 \\ 4\text{O} &= 4(16) &&= \underline{64} \\ & && 98\ \text{g/mole} \end{aligned}$$

The molarity is found by using Equation 4-7:

$$\frac{1{,}710 \text{ g/L}}{98 \text{ g/mole}} = 17.45 \text{ mole/L or } 17.45\ M$$

The normality is found from Equation 4-8, realizing that H_2SO_4 can give up two hydrogen ions and therefore $n = 2$ equivalents/mole:

$$N = 17.45 \text{ mole/L (2 equiv/mole)} = 34.9 \text{ equiv/L}$$

Example 4-2. Find the weight of sodium bicarbonate, $NaHCO_3$, necessary to make a 1 M solution. Find the normality of the solution.

Solution. The molecular weight of $NaHCO_3$ is 84; therefore by using Equation 4-7:

$$\text{mg/L} = (1 \text{ mole/L})(84 \text{ g/mole})(10^3 \text{ mg/g}) = 84{,}000$$

HCO_3^- is able to give or accept only one proton; therefore $n = 1$, and the normality is the same as the molarity.

Example 4-3. Find the equivalent weight of each of the following: Ca^{2+}, CO_3^{2-}, $CaCO_3$.

Solution. Equivalent weight was defined as

$$\text{EW} = \frac{\text{Atomic or molecular weight}}{n}$$

The units of EW are grams/equivalent (g/eq) or milligrams/milliequivalent (mg/meq).

For calcium, n is equal to the valence or oxidation state in water, so $n = 2$. From the table on the inside cover of the book, the atomic weight of Ca^{2+} is 40.08. The equivalent weight is then

$$\text{EW} = \frac{40.08}{2} = 20.04 \text{ g/eq or } 20.04 \text{ mg/meq}$$

For the carbonate ion (CO_3^{2-}) the oxidation state of 2^- is used for n since the base CO_3^{2-} can potentially accept 2 hydrogen ions (H^+). The molecular weight is

$$\begin{aligned} \text{C} &= & 12.01 \\ 3\text{O} &= 3(16.00) = & 48.00 \\ & & \overline{60.01} \end{aligned}$$

and the equivalent weight is

$$\text{EW} = \frac{60.01}{2} = 30.00 \text{ g/eq or } 30.00 \text{ mg/meq}$$

In $CaCO_3$, $n = 2$ since it would take two hydrogen ions to replace the cation (Ca^{2+}) to form carbonic acid, H_2CO_3. Its molecular weight is the sum of the atomic weights of Ca^{2+} and CO_3^{2-} and is, therefore, equal to 40.08 + 60.01 = 100.09. Its equivalent weight is

$$EW = \frac{100.09}{2} = 50.04 \text{ g/eq or mg/meq}$$

Chemical Reactions. There are four principal types of reactions of importance in environmental engineering: precipitation, acid/base, ion-association, and oxidation/reduction.

Dissolved ions can react with each other and form a solid compound. This phase-change reaction of dissolved to solid state is called a precipitation reaction. Typical of a precipitation reaction is the formation of calcium carbonate when a solution of calcium is mixed with a solution of carbonate:

$$Ca^{2+} + CO_3^{2-} \rightleftharpoons CaCO_3(s) \tag{4-9}$$

The (s) in the above reaction denotes that the $CaCO_3$ is in the solid state. When no symbol is used to designate state, it is assumed to be dissolved. The arrows in the reaction imply that the reaction is reversible and so could proceed to the right (that is, the ions are combining to form a solid) or to the left (that is, the solid is dissociating into the ions).

Often, out of convenience, we talk about compounds when in reality a compound does not exist in water. Take, for example, a water containing sodium chloride and calcium sulfate. We would say that the water has NaCl and $CaSO_4$ in it, but no implication is made regarding the association of Na and Cl or Ca and SO_4. The following reactions occur:

$$CaSO_4(s) \rightleftharpoons Ca^{2+} + SO_4^{2-} \tag{4-10}$$

and

$$NaCl(s) \rightleftharpoons Na^+ + Cl^- \tag{4-11}$$

such that the water consists of four unassociated ions: Na^+, Ca^{2+}, Cl^-, and SO_4^{2-}. Don't make the mistake of thinking that the sodium and chloride are together.

Acid/base reactions are a special type of ionization when a hydrogen ion is added to or removed from solution. An acid could be added to water to produce a hydrogen ion, as by the addition of hydrochloric acid to water with the reaction

$$HCl \rightleftharpoons H^+ + Cl^- \tag{4-12}$$

The above reaction is simplified in that it is assumed that water is present. The reaction is properly written

$$HCl + H_2O \rightleftharpoons H_3O^+ + Cl^- \tag{4-13}$$

A hydrogen ion could also be removed from water, as by the addition of a base:

$$NaOH + H_3O^+ \rightleftharpoons 2H_2O + Na^+ \tag{4-14}$$

In some cases, ions may exist in water complexed with other ions. Formation of dissolved complexes are ion-association reactions. In this case, the ions are "tied" together in the solution. The complex could be a neutral compound, such as soluble mercuric chloride:

$$Hg^{2+} + 2Cl^- \rightleftharpoons HgCl_2 \tag{4-15}$$

More often, the soluble complex has a charge and is itself an ion. Metal ion complexes are common examples:

$$Al^{3+} + OH^- \rightleftharpoons AlOH^{2+} \tag{4-16}$$

The $AlOH^{2+}$ is still soluble, but acts differently than did the individual species before complexation.

Oxidation/reduction reactions involve valence changes and the transfer of electrons. When iron metal corrodes, it releases electrons:

$$Fe^0 \rightleftharpoons Fe^{2+} + 2e^- \tag{4-17}$$

If one element releases electrons, then another must be available to accept the electrons. In iron pipe corrosion, hydrogen gas is often produced:

$$2H^+ + 2e^- \rightleftharpoons H_2(g) \tag{4-18}$$

where the symbol (g) indicates the hydrogen is in the gas phase.

Precipitation Reactions. All complexes are soluble in water to a certain extent. Likewise, all complexes are limited by how much can be dissolved in water. Some compounds, such as NaCl, are very soluble; other compounds, such as AgCl, are very insoluble—only a small amount will go into solution. Visualize a solid compound being placed in distilled water. Some of the compound will go into solution. At some time no more of the compound will dissolve, and equilibrium will be reached. The time to reach equilibrium may be seconds or centuries. The solubility reaction is written as follows:

$$A_aB_b(s) \rightleftharpoons aA^{b+} + bB^{a-} \tag{4-19}$$

For example,

$$Ca_3(PO_4)_2(s) \rightleftharpoons 3Ca^{2+} + 2PO_4^{3-} \tag{4-20}$$

Interestingly, the product of the activity of the ions (approximated by the molar concentration) is always a constant for a given compound at a given temperature. This constant is called the solubility constant, K_s. In the general form it is written as

$$K_s = [A]^a[B]^b \tag{4-21}$$

where, in this text, [] denotes *molar* concentrations. **Do not use mg/L!** A table of constants is shown in Table 4-2 and in Appendix A. K_s values are often reported as pK_s, where

$$pK_s = -\log K_s \tag{4-22}$$

The constant works equally well whether we are dissolving a solid (reaction going to the right) or precipitating ions (reaction going to the left). If we place $A_aB_b(s)$ in water,

TABLE 4-2
Selected solubility constants at 25°C

Substance	Equilibrium equation	pK_s	Application
Aluminum hydroxide	$Al(OH)_3(s) \rightleftharpoons Al^{3+} + 3OH^-$	32.9	Coagulation
Aluminum phosphate	$AlPO_4(s) \rightleftharpoons Al^{3+} + PO_4^{3-}$	20.0	Phosphate removal
Calcium carbonate	$CaCO_3(s) \rightleftharpoons Ca^{2+} + CO_3^{2-}$	8.305	Softening, corrosion control
Ferric hydroxide	$Fe(OH)_3(s) \rightleftharpoons Fe^{3+} + 3OH^-$	38.57	Coagulation, iron removal
Ferric phosphate	$FePO_4(s) \rightleftharpoons Fe^{3+} + PO_4^{3-}$	21.9	Phosphate removal
Magnesium hydroxide	$Mg(OH)_2(s) \rightleftharpoons Mg^{2+} + 2OH^-$	11.25	Softening

for every *a* moles of *A* that dissolve, *b* moles of *B* will dissolve until equilibrium is reached. But kinetically* it might take years to happen.

When precipitating ions, it is possible to have a higher concentration of ions in solution than dictated by the solubility product. This is called a supersaturated solution.

Example 4-4. How many mg/L of PO_4^{3-} would be in solution at equilibrium with $AlPO_4(s)$?

Solution. The pertinent reaction is

$$AlPO_4(s) \rightleftharpoons Al^{3+} + PO_4^{3-}$$

The associated pK_s is found in Table 4-2 as 20.0 and calculated as

$$K_s = 10^{-20.0} = [Al][PO_4]$$

For every mole of $AlPO_4$ that dissolves, one mole of Al^{3+} and one mole of PO_4^{3-} are released into solution. At equilibrium, the molar concentration of Al^{3+} and PO_4^{3-} in solution will be equal, so we may say

$$[Al^{3+}] = [PO_4^{3-}] = X$$

Substituting X for each compound in the K_s expression,

$$10^{-20.0} = X^2$$

Solving for X (which is equal to PO_4^{3-}), we find $PO_4^{3-} = 10^{-10}$ moles per liter in solution. The molecular weight is 95 g/mole, so the concentration in mg/L is

$$(95 \text{ g/mole})(10^3 \text{ mg/g})(10^{-10} \text{ moles/L}) = 9.5 \times 10^{-6} \text{ mg/L}$$

Example 4-5. If 50.0 mg of CO_3^{2-} and 50.0 mg of Ca^{2+} are present in 1 L of water, what will be the final (equilibrium) concentration of Ca^{2+}?

**Kinetics* is the part of chemistry concerned with rates of reactions and factors that affect them.

Solution. The molecular weight of Ca^{2+} is 40.08 and that of CO_3^{2-} is 60.01, resulting in initial molar concentrations of 1.25×10^{-3} moles/L and 8.33×10^{-4} moles/L for Ca^{2+} and CO_3^{2-} respectively.

$$K_s = 10^{-pK_s} = 10^{-8.305} = [Ca^{2+}][CO_3^{2-}]$$

For every mole of Ca^{2+} that is removed from solution, one mole of CO_3^{2-} is removed from solution. If the amount removed is given by Z, then

$$10^{-8.305} = 4.95 \times 10^{-9} = [1.25 \times 10^{-3} - Z][8.33 \times 10^{-4} - Z]$$

$$1.04 \times 10^{-6} - (2.08 \times 10^{-3})Z + Z^2 = 0$$

$$Z = \frac{-b \pm \sqrt{b^2 - 4ac}}{2a}$$

$$= \frac{2.08 \times 10^{-3} \pm \sqrt{4.34 \times 10^{-6} - 4(1.04 \times 10^{-6})}}{2}$$

$$= 8.28 \times 10^{-4}$$

so that the final Ca^{2+} concentration is

$$[Ca^{2+}] = 1.25 \times 10^{-3} - 8.28 \times 10^{-4} = 4.22 \times 10^{-4}\ M$$

or

$$(4.22 \times 10^{-4}\ \text{moles/L})(40\ \text{g/mole})(10^3\ \text{mg/g}) = 16.9\ \text{mg/L}$$

Acid/Base Reactions. For the purposes of this text, acids are defined as those compounds that release protons. Bases are those compounds that accept protons.

The simple reaction for the release of a proton is

$$HA \rightleftharpoons H^+ + A^- \tag{4-23}$$

In order for HA to release the proton (H^+), something must accept the proton. Often that something is water, that is,

$$H^+ + H_2O \rightleftharpoons H_3O^+ \tag{4-24}$$

resulting in the net reaction

$$HA + H_2O \rightleftharpoons H_3O^+ + A^- \tag{4-25}$$

It is understood that water is generally present. Hence Equation 4-23 is used in place of Equation 4-25. In the case of Equation 4-25, water is acting as the base; that is, it accepts the proton. If a base is added to water, the water can act as an acid.

$$B^- + H_2O \rightleftharpoons HB + OH^- \tag{4-26}$$

In the above reaction the base (B^-) accepts a proton from water. If a compound is a stronger acid than water, then water will act as a base. If a compound is a stronger base than water, then water will act as an acid.

You can quickly see that acid/base chemistry centers on water and that it is important to know how strong an acid water is. Water itself is ionized in water by the equation

$$H_2O \rightleftharpoons H^+ + OH^- \tag{4-27}$$

TABLE 4-3
Strong acids

Substance	Equilibrium equation	Significance
Hydrochloric acid	$HCl \rightarrow H^+ + Cl^-$	pH adjustment
Nitric acid	$HNO_3 \rightarrow H^+ + NO_3^-$	Analytical techniques
Sulfuric acid[a]	$H_2SO_4 \rightarrow 2H^+ + SO_4^{2-}$	pH adjustment, coagulation

[a]Dissociation of the second proton, $HSO_4^- \rightleftharpoons H^+ + SO_4^{2-}$, is actually a weak acid reaction with a pK_a of 1.92. As long as the pH of the solution is above 2.5, the release of both protons may be considered complete.

The degree of ionization of water is very small and can be measured by what is called the ion product of water, K_w. It is found by

$$K_w = [OH^-][H^+] \tag{4-28}$$

and has a value of 10^{-14} ($pK_w = 14$) at 25°C. A solution is said to be acidic if $[H^+]$ is greater than $[OH^-]$, neutral if equal, and basic if $[H^+]$ is less than $[OH^-]$. If the solution is neutral, then $[H^+] = [OH^-] = 10^{-7}$ *M*. If the solution is acidic, H^+ is greater than 10^{-7} *M*. A convenient expression for the hydrogen ion concentration is pH, given by

$$pH = -\log[H^+] \tag{4-29}$$

Therefore, a neutral solution at 25°C has a pH of 7 (written pH 7), an acidic solution has a pH < 7, and a basic solution has a pH > 7.

Acids are classified as strong acids or weak acids. *Strong acids* have a tendency to donate their protons to water. For example,

$$HCl \rightarrow H^+ + Cl^- \tag{4-30}$$

which we recall is the simplified form of

$$HCl + H_2O \rightarrow H_3O^+ + Cl^- \tag{4-31}$$

A list of important strong acids is in Table 4-3. Note the use of the single arrow to signify that, for practical purposes, we may assume that the reaction proceeds completely to the right.

Example 4-6. If 100 mg of H_2SO_4 (MW = 98) is added to 1 L of water, what is the final pH?

Solution. Using the molecular weight of sulfuric acid we find

$$\left(\frac{100\text{ mg}}{1\text{ L }H_2O}\right)\left(\frac{1}{98\text{ g/mole}}\right)\left(\frac{1}{10^3\text{ mg/g}}\right) = 1.02 \times 10^{-3}\text{ mole/L}$$

The reaction is

$$H_2SO_4 \rightarrow 2H^+ + SO_4^{2-}$$

and therefore $2(1.02 \times 10^{-3})M$ H^+ is produced. The pH is

$$pH = -\log(2.04 \times 10^{-3}) = 2.69$$

TABLE 4-4
Selected weak acid dissociation constants at 25°C

Substance	Equilibrium equation	pK_a	Significance
Acetic acid	$CH_3COOH \rightleftharpoons H^+ + CH_3COO^-$	4.75	Anaerobic digestion
Carbonic acid	$H_2CO_3\ (CO_2 + H_2O) \rightleftharpoons H^+ + HCO_3^-$	6.35	Corrosion, coagulation,
	$HCO_3^- \rightleftharpoons H^+ + CO_3^{2-}$	10.33	softening, pH control
Hydrogen sulfide	$H_2S \rightleftharpoons H^+ + HS^-$	7.2	Aeration, odor control,
	$HS^- \rightleftharpoons H^+ + S^{2-}$	11.89	corrosion
Hypochlorous acid	$HOCl \rightleftharpoons H^+ + OCl^-$	7.54	Disinfection
Phosphoric acid	$H_3PO_4 \rightleftharpoons H^+ + H_2PO_4^-$	2.12	Phosphate removal
	$H_2PO_4^- \rightleftharpoons H^+ + HPO_4^{2-}$	7.20	plant nutrient,
	$HPO_4^{2-} \rightleftharpoons H^+ + PO_4^{3-}$	12.32	analytical

Weak acids are acids that do not completely dissociate in water. An equilibrium exists between the dissociated ions and undissociated compound. The reaction of a weak acid is

$$HW \rightleftharpoons H^+ + W^- \tag{4-32}$$

An equilibrium constant exists that relates the degree of dissociation:

$$K_a = \frac{[H^+][W^-]}{[HW]} \tag{4-33}$$

As with other K values,

$$pK_a = -\log K_a \tag{4-34}$$

A list of important weak acids in water and in wastewater treatment is in Table 4-4. By knowing the pH of a solution (which can be easily found with a pH meter) it is possible to get a rough idea of the degree of dissociation of the acid. For example, if the pH is equal to the pK_a (that is, $[H^+] = K_a$), then from Equation 4-33, $[HW] = [W^-]$ and the acid is 50 percent dissociated. If the $[H^+]$ is two orders of magnitude (100 times) less than the K_a, then $100[H^+] = K_a$ (or pH $\gg$ pK).

$$100[H^+] = \frac{[H^+][W^-]}{[HW]}$$

or 100 $[HW] = [W^-]$. We would conclude that essentially all the acid is dissociated ($W^- \gg HW$). Correspondingly, if pH $\ll$ pK then $[HW] \gg [W^-]$, and none of the acid is dissociated.*

Example 4-7. If 15 mg/L of HOCl is added to a potable water for disinfection and the final measured pH is 7.0, what percent of the HOCl is not dissociated? Assume the temperature is 25°C.

*If $[H^+] < K_a$, then pH > pK. The symbol $\gg$ means greater by two orders of magnitude.

Solution. The reaction is

$$HOCl \rightleftharpoons H^+ + OCl^-$$

From Table 4-4, we find the pK_a is 7.54 and

$$K_a = 10^{-7.54} = 2.88 \times 10^{-8}$$

Writing the equilibrium constant expression in the form of Equation 4-33

$$K_a = \frac{[H^+][OCl^-]}{[HOCl]}$$

and substituting the values for K_a and $[H^+]$

$$2.88 \times 10^{-8} = \frac{[10^{-7}][OCl^-]}{[HOCl]}$$

Solving for the HOCl concentration

$$[HOCl] = 3.47[OCl^-]$$

Since the fraction of HOCl that has not dissociated plus the OCl^- that was formed by the dissociation must, by the law of conservation of mass, equal 100% of the original HOCl added:

$$[HOCl] + [OCl^-] = 100\% \text{ (of the total HOCl added to the solution)}$$

then

$$3.47[OCl^-] + [OCl^-] = 100\%$$

$$4.47[OCl^-] = 100\%$$

$$[OCl^-] = \frac{100\%}{4.47} = 22.37\%$$

and

$$[HOCl] = 3.47(22.37\%) = 77.6\%$$

Buffer Solutions. A solution that resists large changes in pH when an acid or base is added or when the solution is diluted is called a *buffer* solution. A solution containing a weak acid and its salt is an example of a buffer. Atmospheric carbon dioxide (CO_2) produces a natural buffer through the following reactions:

$$CO_2(g) \rightleftharpoons CO_2 + H_2O \rightleftharpoons H_2CO_3 \rightleftharpoons H^+ + HCO_3^- \rightleftharpoons 2H^+ + CO_3^{2-} \quad (4\text{-}35)$$

where H_2CO_3 = carbonic acid
HCO_3^- = bicarbonate ion
CO_3^{2-} = carbonate ion

This is perhaps the most important buffer system in water and wastewater treatment. We will be referring to it several times in this and subsequent chapters as the *carbonate buffer system.*

As depicted in Equation 4-35, the CO_2 in solution is in equilibrium with atmospheric $CO_2(g)$. Any change in the system components to the right of CO_2 causes the CO_2 either to be released from solution or to dissolve.

We can examine the character of the buffer system in resisting a change in pH by assuming the addition of an acid or a base and applying the law of mass action (Le Chatelier's principle). For example, if an acid is added to the system, it unbalances it by increasing the hydrogen ion concentration. Therefore, the carbonate combines with it to form bicarbonate. Bicarbonate reacts to form more carbonic acid, which in turn dissociates to CO_2 and water. The excess CO_2 can be released to the atmosphere in a thermodynamically open system. Alternatively, the addition of a base consumes hydrogen ions and the system moves to the right with the CO_2 being replenished from the atmosphere. When CO_2 is bubbled into the system or is removed by passing an inert gas such as nitrogen through the liquid (a process called *stripping*), the pH will change more dramatically because the atmosphere is no longer available as a source or sink for CO_2. Figure 4-5 summarizes the four general responses of the carbonate buffer system. The first two cases are common in natural settings when the reactions proceed over a relatively long period of time. In a water treatment plant, we can alter the reactions more quickly than the CO_2 can be replenished from the atmosphere. The second two cases are not common in natural settings. They are used in water treatment plants to adjust the pH.

In natural waters in equilibrium with atmospheric CO_2, the amount of CO_3^{2-} in solution is quite small in comparison to the HCO_3^- in solution. The presence of Ca^{2+} in the form of limestone rock or other naturally occurring sources of calcium results in the formation of calcium carbonate ($CaCO_3$), which is very insoluble. As a consequence, it precipitates from solution. The reaction of Ca^{2+} with CO_3^{2-} to form a precipitate is one of the fundamental reactions used to soften water.

Alkalinity. Alkalinity is defined as the sum of all titratable bases down to about pH 4.5. It is found by experimentally determining how much acid it takes to lower the pH of water to 4.5. In most waters the only significant contributions to alkalinity are the carbonate species and any free H^+ or OH^-. The total H^+ that can be taken up by a water containing primarily carbonate species is

$$\text{Alkalinity} = [HCO_3^-] + 2[CO_3^{2-}] + [OH^-] - [H^+] \tag{4-36}$$

where [] refers to concentrations in moles/L. In most natural water situations (pH 6 to 8), the OH^- and H^+ are negligible, such that

$$\text{Alkalinity} = [HCO_3^-] + 2[CO_3^{2-}] \tag{4-37}$$

Note that $[CO_3^{2-}]$ is multiplied by two because it can accept two protons. The pertinent acid/base reactions are

$$H_2CO_3 \rightleftharpoons H^+ + HCO_3^- \qquad pK_{a1} = 6.35 \text{ at } 25°C \tag{4-38}$$

$$HCO_3^- \rightleftharpoons H^+ + CO_3^{2-} \qquad pK_{a2} = 10.33 \text{ at } 25°C \tag{4-39}$$

From the pK values, some useful relationships can be found. The more important ones are as follows:

1. Below pH of 4.5, essentially all of the carbonate species are present as H_2CO_3, and the alkalinity is negative (due to the H^+).

Case I
Acid is added to carbonate buffer system[a]

Reaction shifts to the left as $H_2CO_3^*$ is formed when H^+ and HCO_3^- combine[b]

CO_2 is released to the atmosphere

pH is lowered slightly because the availability of free H^+ (amount depends on buffering capacity)

Case II
Base is added to carbonate buffer system

Reaction shifts to the right

CO_2 from the atmosphere dissolves into solution

pH is raised slightly because H^+ combines with OH^- (amount depends on buffering capacity)

Case III
CO_2 is bubbled into carbonate buffer system

Reaction shifts to the right because $H_2CO_3^*$ is formed when CO_2 and H_2O combine

CO_2 dissolves into solution

pH is lowered

Case IV
Carbonate buffer system is stripped of CO_2

Reaction shifts to the left to form more $H_2CO_3^*$ to replace that removed by stripping

CO_2 is removed from solution

pH is raised

[a]Refer to Equation 4-35
[b]The asterisk $*$ in the H_2CO_3 is used to signify the sum of CO_2 and H_2CO_3 in solution.

FIGURE 4-5
Behavior of the carbonate buffer system with the addition of acids and bases or the addition and removal of CO_2.

2. At a pH of 8.3 most of the carbonate species are present as HCO_3^-, and the alkalinity equals HCO_3^-.

3. Above a pH of 12.3, essentially all of the carbonate species are present as CO_3^{2-}, and the alkalinity equals $2[CO_3^{2-}] + [OH^-]$. The $[OH^-]$ may not be insignificant at this pH.

Figure 4-6 schematically shows the change of species described above as the pH is lowered by the addition of acid to a water containing alkalinity. Note that the pH

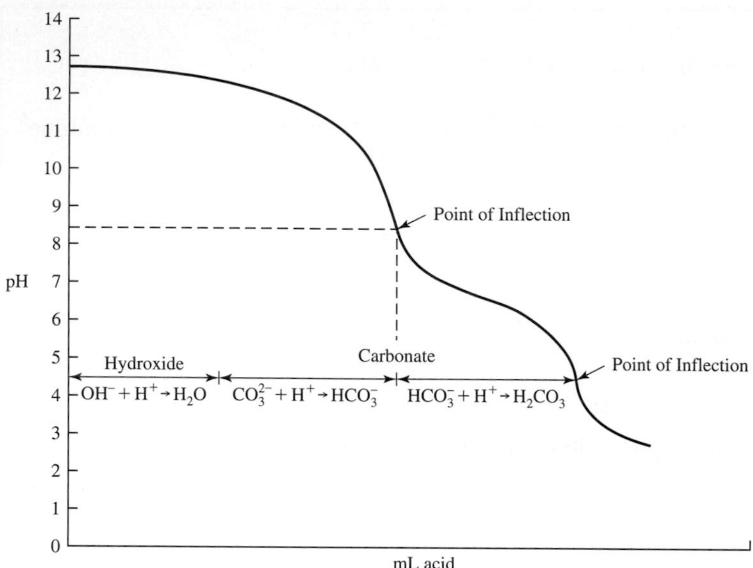

FIGURE 4-6
Titration curve for a hydroxide-carbonate mixture. (*Source:* Sawyer, McCarty, and Parkin, 1994.)

starts at above 12.3 and as acid is added the pH drops slowly as the first acid (H^+) addition is consumed by free hydroxide (OH^-), preventing a significant pH drop, and then the acid is consumed by carbonate (CO_3^{2-}) being converted to bicarbonate (HCO_3^-). At about pH 8.3 the carbonate is essentially all converted to bicarbonate, at which point there is another somewhat flat area where the acid is consumed by converting bicarbonate to carbonic acid.

From Equation 4-37 and our discussion of buffer solutions, it can be seen that alkalinity serves as a measure of buffering capacity. The greater the alkalinity, the greater the buffering capacity. In environmental engineering, then, we differentiate between alkaline water and water having high alkalinity. Alkaline water has a pH greater than 7, while a water with high alkalinity has a high buffering capacity. An alkaline water may or may not have a high buffering capacity. Likewise, a water with a high alkalinity may or may not have a high pH.

By convention, alkalinity is not expressed in molarity units as shown in the above equations, but rather in mg/L as $CaCO_3$. In order to convert species to mg/L as $CaCO_3$, multiply mg/L as the species by the ratio of the equivalent weight of $CaCO_3$ to the species equivalent weight:

$$\text{mg/L as CaCO}_3 = (\text{mg/L as species})\left(\frac{\text{EW}_{\text{CaCO}_3}}{\text{EW}_{\text{species}}}\right) \tag{4-40}$$

The alkalinity is then found by adding all the carbonate species and the hydroxide, and then subtracting the hydrogen ions. When using the units "mg/L as $CaCO_3$," the terms are added directly. The multiple of two for CO_3^{2-} has already been accounted for in the conversion.

Example 4-8. A water contains 100.0 mg/L CO_3^{2-} and 75.0 mg/L HCO_3^- at a pH of 10. Calculate the alkalinity exactly at 25°C. Approximate the alkalinity by ignoring $[OH^-]$ and $[H^+]$.

Solution. First, convert CO_3^{2-}, HCO_3^-, OH^-, and H^+ to mg/L as $CaCO_3$.

The equivalent weights are

$$\begin{aligned} CO_3^{2-}&: MW = 60, n = 2, EW = 30 \\ HCO_3^-&: MW = 61, n = 1, EW = 61 \\ H^+&: MW = 1, n = 1, EW = 1 \\ OH^-&: MW = 17, n = 1, EW = 17 \end{aligned}$$

and the concentration of H^+ and OH^- is calculated as follows: pH = 10; therefore $[H^+] = 10^{-10}$ *M*. Using Equation 4-7,

$$\text{mg/L} = (10^{-10}\ \text{moles/L})(1\ \text{g/mole})(10^3\ \text{mg/g}) = 10^{-7}$$

Using Equation 4-28,

$$[OH^-] = \frac{K_w}{[H^+]} = \frac{10^{-14}}{10^{-10}} = 10^{-4}\ \text{moles/L}$$

and

$$\text{mg/L} = (10^{-4}\ \text{moles/L})(17\ \text{g/mole})(10^3\ \text{mg/g}) = 1.7$$

Now, the mg/L as $CaCO_3$ is found by using Equation 4-40 and taking the equivalent weight of $CaCO_3$ to be 50:

$$CO_3^{2-} = 100.0\left(\frac{50}{30}\right) = 167$$

$$HCO_3^- = 75.0\left(\frac{50}{61}\right) = 61$$

$$H^+ = 10^{-7}\left(\frac{50}{1}\right) = 5 \times 10^{-6}$$

$$OH^- = 1.7\left(\frac{50}{17}\right) = 5.0$$

The exact alkalinity (in mg/L) is found by

$$\begin{aligned} \text{Alkalinity} &= 61 + 167 + 5.0 - (5 \times 10^{-6}) \\ &= 233\ \text{mg/L as } CaCO_3 \end{aligned}$$

It is approximated by 61 + 167 = 228 mg/L as $CaCO_3$. This is a 2.2 percent error.

Activity Coefficients. To this point our discussions have assumed that the solutions being analyzed were dilute. That is, the total ion concentrations were low (generally less than 10^{-2} *M*). For dilute solutions, the ions in solution can be considered to act independently from one another. As the concentration of ions in solution increases, the interaction of their electric charges affects their equilibrium relationships. This interaction is measured in terms of *ionic strength.* To account for high ionic strength, the equilibrium relationships are modified by incorporating *activity coefficients.* These are symbolized by γ(ion). Activity is then the product of the molar concentration of the species and its activity coefficient. For example, the solubility product of $CaCO_3$ would be

$$K_s = \{\gamma(Ca^{2+}) \times [Ca^{2+}]\}\{\gamma(CO_3) \times [CO_3]\} \tag{4-41}$$

Reaction Kinetics

Many reactions that occur in the environment do not reach equilibrium quickly. Some examples include disinfection of water, gas transfer into and out of water, removal of organic matter from water, and radioactive decay. The study of how these reactions proceed is called *reaction kinetics*. The *rate of reaction, r,* is used to describe the rate of formation or disappearance of a compound. Reactions that take place in a single phase (that is, liquid, gas, or solid) are called *homogeneous* reactions. Those that occur at surfaces between phases are called *heterogeneous*. For each type of reaction, the rate may be defined as follows:

For homogeneous reactions

$$r = \frac{\text{moles or milligrams}}{(\text{unit volume})(\text{unit time})} \tag{4-42}$$

For heterogeneous reactions

$$r = \frac{\text{moles or milligrams}}{(\text{unit surface})(\text{unit time})} \tag{4-43}$$

Production of a compound results in a positive sign for the reaction rate ($+r$), while disappearance of a substance yields a negative sign ($-r$). Reaction rates are a function of temperature, pressure, and the concentration of reactants. For a stoichiometric reaction of the form:

$$a\text{A} + b\text{B} \rightarrow c\text{C}$$

where a, b, and c are the proportionality coefficients for the reactants A, B, and C, the change in concentration of compound A is equal to the reaction rate equation for compound A:

$$\frac{d[\text{A}]}{dt} = r_\text{A} = -k[\text{A}]^\alpha\,[\text{B}]^\beta = k[\text{C}]^\gamma \tag{4-44}$$

where [A], [B], and [C] are the concentrations of the reactants, and α, β, and γ are empirically determined exponents. The proportionality term, k, is called the *reaction rate constant*. It is often not a constant but, rather, is dependent on the temperature and pressure. Since A and B are disappearing, the sign of the reaction rate equation is negative. It is positive for C because C is being formed.

TABLE 4-5
Example reaction orders

Reaction order	Rate Equation
Zero	$r_A = -k$
First	$r_A = -k[A]$
Second	$r_A = -k[A]^2$
Second	$r_A = -k[A][B]$

The *order of reaction* is defined as the sum of the exponents in the reaction rate equation. The exponents may be either integers or fractions. Some sample reaction orders are shown in Table 4-5.

For elementary reactions where the stoichiometric equation represents both the mass balance and the molecular scale process, the coefficients of proportionality (a, b, c) are equivalent to the exponents in the reaction rate equation:

$$r_{\mathrm{A}} = -k[\mathrm{A}]^a[\mathrm{B}]^b \tag{4-45}$$

The overall reaction rate, r, and the individual reaction rates are related:

$$r = \frac{r_{\mathrm{A}}}{a} = \frac{r_{\mathrm{B}}}{b} = \frac{r_{\mathrm{C}}}{c} \tag{4-46}$$

The reaction rate constant, k, may be determined experimentally by obtaining data on the concentrations of the reactants as a function of time and plotting on a suitable graph. The form of the graph is determined from the result of integration of the equations in Table 4-5. The integrated forms and the appropriate graphical forms are shown in Table 4-6.

Gas Transfer. An important example of time-dependent reactions is the mass transfer (dissolution or volatilization) of gas from water. In 1924 Lewis and Whitman

TABLE 4-6
Plotting procedure to determine order of reaction by method of integration for plug flow reactor and for a batch reactor[a]

Order	Rate equation	Integrated equation	Linear plot	Slope	Intercept
0	$\frac{d[A]}{dt} = -k$	$[A] - [A_0] = kt$	$[A]$ vs. t	$-k$	$[A_0]$
1	$\frac{d[A]}{dt} = -k[A]$	$\ln \frac{[A]}{[A_0]} = -kt$	$\ln [A]$ vs. t	$-k$	$\ln [A_0]$
2	$\frac{d[A]}{dt} = -k[A]^2$	$\frac{1}{[A]} - \frac{1}{[A_0]} = kt$	$\frac{1}{[A]}$ vs. t	k	$\frac{1}{[A_0]}$

[a]*Source:* J. G. Henry and G. W. Heinke, 1989.

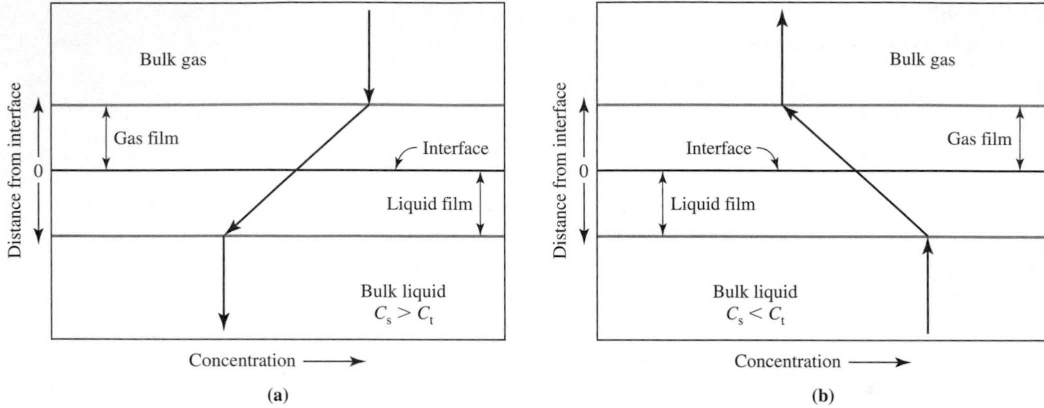

FIGURE 4-7
Two-film model of the interface between gas and liquid: (*a*) absorption mode and (*b*) desorption mode.

postulated a two-film theory to describe the mass transfer of gases. According to their theory, the boundary between the gas phase and the liquid phase (also called the *interface*) is composed of two distinct films that serve as a barrier between the bulk phases (Figure 4-7). For a molecule of gas to go into solution, it must pass through the bulk of the gas, the gas film, the liquid film, and into the bulk of the liquid (Figure 4-7*a*). To leave the liquid, the gas molecule must follow the reverse course (Figure 4-7*b*). The driving force causing the gas to move, and hence the mass transfer, is the concentration gradient: $C_s - C$. C_s is the saturation concentration of the gas in the liquid, and C is the actual concentration. When C_s is greater than C, the gas will go into solution. When C is greater than C_s, the gas will desorb.

The relationship between the equilibrium concentration of gas dissolved in solution and the partial pressure of the gas is defined by *Henry's law* (WQ&T, 1990):

$$p = \frac{Hc}{P_T} \qquad (4\text{-}47)$$

where c = the mole fraction of gas in water
p = mole fraction of gas in air
H = proportionality constant, known as Henry's constant (the slope of the straight-line portion of the distribution curve)
P_T = total pressure atm

For water treatment, P_T is usually 1 atmosphere (atm). Various units are used by different investigators for the concentrations in the two phases and, therefore, the units for Henry's constant vary. Because of units, care must be taken in using the relationship, especially when obtaining constants from different sources. Below is a discussion of the primary methods of reporting Henry's law.

Probably the most common method of expressing Henry's law is with units of c and p as mole fractions:

$$p = \frac{Hc}{P_T} \qquad (4\text{-}48)$$

where p = mol gas/mol air
c = mol gas/mol water
H = atm, actually $= \dfrac{\text{atm (mol gas/mol air)}}{\text{mol gas/mol water}}$
P_T = atm, usually = 1

Recall that according to Dalton's law, a mole of gas per mole of air is the same as the partial pressure of the gas or is also the same as the volume of gas per volume of air. A useful conversion factor when calculating c is that 1 L of water contains 55.6 mol of water:

$$\frac{1000\text{ g/L}}{18\text{ g/mol}} = 55.6\text{ mol/L}$$

Another method of reporting Henry's law is to utilize concentration units. In this case, the total pressure P_T is commonly defined as 1, and hence it is left off of the equation and atm is dropped from the units of H. In this case, any set of mass per volume or mole per volume units can be used (as long as p and c are the same), and hence it is often referred to as the dimensionless or unitless Henry's law constant:

$$p = H_u c \tag{4-49}$$

where p = concentration units, e.g., kg/m^3, mol/L, mg/L
H_u = unitless
c = same concentration units used for p

At 1 atm pressure and 0°C, 22.412 L of air is 1 mol of air. At other temperatures, 1 mol of air is $0.082T$ L [where T = temperature in kelvin (K)] of air. The following conversion between H and H_u can be made:

$$H_u = \left[H\frac{\text{atm (mol gas/mol air)}}{\text{mol gas/mol water}}\right]\left(\frac{\text{mol air}}{0.082T\text{ L air}}\right)\left(\frac{\text{L water}}{55.6\text{ mol}}\right)$$

$$= \frac{H}{4.56T} \text{ or } H_u = H \times 7.49 \times 10^{-4} \text{ at } 20°\text{C} \tag{4-50}$$

Another method for reporting Henry's constant is to use mixed units for p and c. This is very common because units of partial pressure in the air phase and concentration units in the water phase tend to be used. Different variations are available. Two are shown below:

$$p = \frac{H_m c}{P_T} \tag{4-51}$$

where p = mol gas/mol air (partial pressure)
c = mol gas/m^3 water
H_m = atm × m^3 water/mol gas

$$= \left[H\frac{\text{atm(mol gas/mol air)}}{\text{mol gas/mol water}}\right]\left(\frac{\text{m}^3\text{ water}}{55{,}600\text{ mol}}\right)$$

$$= \frac{H}{55{,}600}$$

Finally, milligram per liter units for c may be used. This is very useful in water treatment:

$$p = \frac{H_D c}{P_T} \tag{4-52}$$

where p = mol gas/mol air (partial pressure)
c = mg/L
H_D = (atm)(L)/mg
$H_D = \frac{H_m}{\text{MW}} = \frac{H}{55{,}600\ \text{MW}}$
MW = molecular weight of gas of interest

The Henry's law coefficient varies both with the temperature and the concentration of other dissolved substances. Henry's law constants are given in Table A-2, Appendix A.

The rate of mass transfer can be described by the following equation:

$$\frac{dC}{dt} = k_a(C_s - C) \tag{4-53}$$

where k_a = rate constant or mass transfer coefficient, s^{-1}.

The difference between the saturation concentration and the actual concentration $(C_s - C)$ is called the *deficit*. Since the saturation concentration is a constant for a constant temperature and pressure, this is a first-order reaction.

Example 4-9. A falling raindrop initially has no dissolved oxygen. The saturation concentration for the drop is 9.20 mg/L. If, after falling for two seconds, the droplet has an oxygen concentration of 3.20 mg/L, how long must the droplet fall (from the start of the fall) to achieve a concentration of 8.20 mg/L?

Solution. We begin by calculating the deficit after two seconds, and that at a concentration of 8.20 mg/L:

$$\text{Deficit at 2 sec} = 9.20 - 3.20 = 6.00 \text{ mg/L}$$
$$\text{Deficit at } t \text{ sec} = 9.20 - 8.20 = 1.00 \text{ mg/L}$$

Now using the integrated form of the first-order rate equation from Table 4-6, noting that the rate of change is proportional to deficit and, hence, $[A] = (C_s - C)$ and that $[A_0] = (9.20 - 0.00)$,

$$\ln \frac{6.00}{9.20} = -k(2.00 \text{ s})$$
$$k = 0.2137 \text{ s}^{-1}$$

With this value of k, we can calculate a value for t:

$$\ln \frac{(9.20 - 8.20)}{9.20} = -(0.2137)(t)$$
$$t - 10.4 \text{ s}$$

Water Quality

Precipitation in the form of rain, hail, or sleet contains very few impurities. It may contain trace amounts of mineral matter, gases, and other substances as it forms and falls through the earth's atmosphere. The precipitation, however, has virtually no bacterial content (U.S. PHS, 1962).

Once precipitation reaches the earth's surface, many opportunities are presented for the introduction of mineral and organic substances, microorganisms, and other forms of pollution (contamination).* When water runs over or through the ground surface, it may pick up particles of soil. This is noticeable in the water as cloudiness or turbidity. It also picks up particles of organic matter and bacteria. As surface water seeps downward into the soil and through the underlying material to the water table, most of the suspended particles are filtered out. This natural filtration may be partially effective in removing bacteria and other particulate materials. However, the chemical characteristics of the water may change and vary widely when it comes in contact with mineral deposits. As surface water seeps down to the water table, it dissolves some of the minerals contained in the soil and rocks. Groundwater, therefore, often contains more dissolved minerals than surface water.

The following four categories are used to describe drinking water quality:

1. Physical: Physical characteristics relate to the quality of water for domestic use and are usually associated with the appearance of water, its color or turbidity, temperature, and, in particular, taste and odor.

2. Chemical: Chemical characteristics of waters are sometimes evidenced by their observed reactions, such as the comparative performance of hard and soft waters in laundering. Most often, differences are not visible.

3. Microbiological: Microbiological agents are very important in their relation to public health and may also be significant in modifying the physical and chemical characteristics of water.

4. Radiological: Radiological factors must be considered in areas where there is a possibility that the water may have come in contact with radioactive substances. The radioactivity of the water is of public health concern in these cases.

Consequently, in the development of a water supply system, it is necessary to examine carefully all the factors that might adversely affect the intended use of a water supply source.

Physical Characteristics

Turbidity. The presence of suspended material such as clay, silt, finely divided organic material, plankton, and other particulate material in water is known as *turbidity.* The unit of measure is a Turbidity Unit (TU) or Nephlometric Turbidity Unit (NTU).

---

*Pollution as used in this text means the presence in water of any foreign substances (organic, inorganic, radiological, or biological) that tend to lower its quality to such a point that it constitutes a health hazard or impairs the usefulness of the water.

It is determined by reference to a chemical mixture that produces a reproducible refraction of light. Turbidities in excess of 5 TU are easily detectable in a glass of water and are usually objectionable for aesthetic reasons.

Clay or other inert suspended particles in drinking water may not adversely affect health, but water containing such particles may require treatment to make it suitable for its intended use. Following a rainfall, variations in the groundwater turbidity may be considered an indication of surface or other introduced pollution.

Color. Dissolved organic material from decaying vegetation and certain inorganic matter cause color in water. Occasionally, excessive blooms of algae or the growth of aquatic microorganisms may also impart color. While color itself is not usually objectionable from the standpoint of health, its presence is aesthetically objectionable and suggests that the water needs appropriate treatment.

Taste and Odor. Taste and odor in water can be caused by foreign matter such as organic compounds, inorganic salts, or dissolved gases. These materials may come from domestic, agricultural, or natural sources. Drinking water should be free from any objectionable taste or odor at point of use.

Temperature. The most desirable drinking waters are consistently cool and do not have temperature fluctuations of more than a few degrees. Groundwater and surface water from mountainous areas generally meet these criteria. Most individuals find that water having a temperature between $10^\circ - 15^\circ$C is most palatable.

Chemical Characteristics

Chloride. Most waters contain some chloride. The amount present can be caused by the leaching of marine sedimentary deposits or by pollution from sea water, brine, or industrial or domestic wastes. Chloride concentrations in excess of about 250 mg/L usually produce a noticeable taste in drinking water. Domestic water should contain less than 100 mg/L of chloride. In some areas, it may be necessary to use water with a chloride content in excess of 100 mg/L. In these cases, all of the other criteria for water purity must be met.

Fluoride. In some areas, water sources contain natural fluoride. Where the concentrations approach optimum levels, beneficial health effects have been observed. In such areas, the incidence of dental caries has been found to be below the levels observed in areas without natural fluoride. The optimum fluoride level for a given area depends upon air temperature, since temperature greatly influences the amount of water people drink. Excessive fluoride in drinking water supplies may produce fluorosis (mottling) of teeth, which increases as the optimum fluoride level is exceeded.* State or local health departments should be consulted for their recommendations, but acceptable levels are generally between 0.8 and 1.3 mg/L fluoride.

*Mottled teeth are characterized by black spots or streaks and may become brittle when exposed to large amounts of fluoride.

Iron. Small amounts of iron frequently are present in water because of the large amount of iron in the geologic materials. The presence of iron in water is considered objectionable because it imparts a brownish color to laundered goods and affects the taste of beverages such as tea and coffee.

Lead. Exposure of the body to lead, however brief, can be seriously damaging to health. Prolonged exposure to relatively small quantities may result in serious illness or death. Lead taken into the body in quantities in excess of certain relatively low "normal" limits is a cumulative poison.

Manganese. Manganese imparts a brownish color to water and to cloth that is washed in it. It flavors coffee and tea with a medicinal taste.

Sodium. The presence of sodium in water can affect persons suffering from heart, kidney, or circulatory ailments. When a strict sodium-free diet is recommended, any water should be regarded with suspicion. Home water softeners may be of particular concern because they add large quantities of sodium to the water. (See Section 4-3 for an explanation of the chemistry and operation of softeners).

Sulfate. Waters containing high concentrations of sulfate, caused by the leaching of natural deposits of magnesium sulfate (Epsom salts) or sodium sulfate (Glauber's salt), may be undesirable because of their laxative effects.

Zinc. Zinc is found in some natural waters, particularly in areas where zinc ore deposits have been mined. Zinc is not considered detrimental to health, but it will impart an undesirable taste to drinking water.

Arsenic. Arsenic occurs naturally in the environment, and it is also widely used in timber treatment, agricultural chemicals (pesticides), and manufacturing of gallium arsenide wafers, glass, and alloys. Arsenic in drinking water is associated with lung and urinary bladder cancer.

Toxic Inorganic Substances. *Nitrates* (NO_3), *cyanides* (CN), and *heavy metals* constitute the major classes of inorganic substances of health concern. Methemoglobinemia (infant cyanosis or "blue baby syndrome") has occurred in infants who have been given water or fed formula prepared with water having high concentrations of nitrate. CN ties up the hemoglobin sites that bind oxygen to red blood cells. This results in oxygen deprivation. A characteristic symptom is that the patient has a blue skin color. This condition is called cyanosis. CN causes chronic effects on the thyroid and central nervous system. The toxic heavy metals include arsenic (As), barium (Ba), cadmium (Cd), chromium (Cr), lead (Pb), mercury (Hg), selenium (Se), and silver (Ag). The heavy metals have a wide range of effects. They may be acute poisons (As and Cr^{6+} for example), or they may produce chronic disease (Pb, Cd, and Hg for example).

Toxic Organic Substances. There are over 120 toxic organic compounds listed on the U.S. Environmental Protection Agency's Priority Pollutant List (Table 1-6). These include pesticides, insecticides, and solvents. Like the inorganic substances, their effects may be acute or chronic.

Microbiological Characteristics

Water for drinking and cooking purposes must be made free from disease-producing organisms (*pathogens*). These organisms include viruses, bacteria, protozoa, and helminths (worms).

Some organisms which cause disease in people originate with the fecal discharges of infected individuals. Others are from the fecal discharge of animals.

Unfortunately, the specific disease-producing organisms present in water are not easily identified. The techniques for comprehensive bacteriological examination are complex and time-consuming. It has been necessary to develop tests that indicate the relative degree of contamination in terms of an easily defined quantity. The most widely used test estimates the number of microorganisms of the *coliform group.* This grouping includes two genera: *Escherichia coli* and *Aerobacter aerogenes.* The name of the group is derived from the word *colon.* While *E. coli* are common inhabitants of the intestinal tract, *Aerobacter* are common in the soil, on leaves, and on grain; on occasion they cause urinary tract infections. The test for these microorganisms, called the *Total Coliform Test,* was selected for the following reasons:

1. The coliform group of organisms normally inhabits the intestinal tracts of humans and other mammals. Thus, the presence of coliforms is an indication of fecal contamination of the water.
2. Even in acutely ill individuals, the number of coliform organisms excreted in the feces outnumber the disease-producing organisms by several orders of magnitude. The large numbers of coliforms make them easier to culture than disease-producing organisms.
3. The coliform group of organisms survives in natural waters for relatively long periods of time, but does not reproduce effectively in this environment. Thus, the presence of coliforms in water implies fecal contamination rather than growth of the organism because of favorable environmental conditions. These organisms also survive better in water than most of the bacterial pathogens. This means that the absence of coliforms is a reasonably safe indicator that pathogens are not present.
4. The coliform group of organisms is relatively easy to culture. Thus, laboratory technicians can perform the test without expensive equipment.

Current research indicates that testing for *Escherichia coli* specifically may be warranted. Some agencies prefer the examination for *E. coli* as a better indicator of biological contamination than total coliforms.

The two protozoa of most concern are *Giardia* cysts and *Cryptosporidium* oocysts. Both pathogens are carried by animals in the wild and on farms and make their way into the environment and water sources. Both are associated with gastrointestinal illness.

Radiological Characteristics

The development and use of atomic energy as a power source and the mining of radioactive materials, as well as naturally occurring radioactive materials, have made it necessary to establish limiting concentrations for the intake into the body of radioactive substances, including drinking water.

The effects of human exposure to radiation or radioactive materials are harmful, and any unnecessary exposure should be avoided. Humans have always been exposed to natural radiation from water, food, and air. The amount of radiation to which the individual is normally exposed varies with the amount of background radioactivity. Water with high radioactivity is not normal and is confined in great degree to areas where nuclear industries are situated.

Water Quality Standards

President Ford signed the National Safe Drinking Water Act (SDWA) into law on December 16, 1974. The Environmental Protection Agency (EPA) was directed to establish *maximum contaminant levels* (MCLs) for public water systems to prevent the occurrence of any known or anticipated adverse health effects with an adequate margin of safety. EPA defined a public water system to be any system that provides piped water for human consumption, if such a system has at least 15 service connections or regularly serves an average of at least 25 individuals daily at least 60 days out of the year. This definition includes private businesses, such as service stations, restaurants, motels, and others that serve more than 25 persons per day for greater than 60 days out of the year.

From 1975 through 1985, the EPA regulated 23 contaminants in drinking water supplied by public water systems. These regulations are known as interim primary drinking water regulations (IPDWRs). In June of 1986, the SDWA was amended. The amendments required EPA to set *maximum contaminant level goals* (MCLGs) and MCLs for 83 specific substances. This list included 22 of the IPDWRs (all except trihalomethanes). The amendments also required EPA to regulate 25 additional contaminants every three years beginning in January, 1991 and continuing for an indefinite period of time.

Table 4-7 lists each regulated contaminant and summarizes its adverse health effects. Some of these contaminant levels are being considered for revision. The notation "TT" in the table means that a treatment technique is specified rather than a contaminant level. The treatment techniques are specific processes that are used to treat the water. Some examples include coagulation and filtration, lime softening, and ion exchange. These will be discussed in the following sections.

Lead and Copper. In June 1988, EPA issued proposed regulations to define MCLs and MCLGs for lead and copper, as well as to establish a monitoring program and a treatment technique for both. The MCLG proposed for lead is zero; for copper, 1.3 mg/L. The MCL action levels, applicable to water entering the distribution system, are 0.005 mg/L for lead and 1.3 mg/L for copper.

Compliance with the regulations is also based on the quality of the water at the consumer's tap. Monitoring is required by means of collection of first-draw samples at

TABLE 4-7
Standards and potential health effects of the contaminants regulated under the SDWA

Contaminant	Maximum Contaminant Level Goal, mg/L	Maximum Contaminant Level, mg/L	BAT	Potential Health Effects
Organics				
Acrylamide	Zero	TT	PAP	Cancer, nervous system effects
Alachor	Zero	0.002	GAC	Cancer
Atrazine	0.003	0.003	GAC	Liver, kidney, lung, cardiovascular effects; possible carcinogen
Benzene	Zero	0.005	GAC, PTA	Cancer
Benzo(a)pyrene	Zero	0.0002	GAC	Cancer
Bromodichloromethane	Zero	See TTHM	GAC, NF*	Cancer
Bromoform	Zero	See TTHM	GAC, NF*	Cancer
Carbofuran	0.04	0.04	GAC	Nervous system, reproductive system effects
Carbon tetrachloride	Zero	0.005	GAC, PTA	Cancer
Chlordane	Zero	0.002	GAC	Cancer
Chloroform	0.07	See TTHM	GAC, NF*	Cancer
Chlorodibromomethane	No MCLG	See TTHM	GAC, NF*	Cancer
2,4-D	0.07	0.07	GAC	Liver, kidney effects
Dalapon	0.2	0.2	GAC	Kidney, liver effects
Di(2-ethylhexyl)adipate	0.4	0.4	GAC, PTA	Reproductive effects
Di(2-ethylhexyl)phthalate	Zero	0.006	GAC	Cancer
Dibromochloropropane (DBCP)	Zero	0.0002	GAC, PTA	Cancer
Dichloroacetic acid	No MCLG	See HAA5	GAC, PTA	Cancer
p-Dichlorobenzene	0.075	0.075	GAC, PTA	Kidney effects, possible carcinogen
o-Dichlorobenzene	0.6	0.6	GAC, PTA	Liver, kidney, blood cells effects
1,2-Dichloroethane	Zero	0.005	GAC, PTA	Cancer
1,1-Dichloroethylene	0.007	0.007	GAC, PTA	Liver, kidney effects, possible carcinogen
cis-1,2-Dichloroethylene	0.07	0.07	GAC, PTA	Liver, kidney, nervous system, circulatory effects
trans-1,2-Dichloroethylene	0.1	0.1	GAC, PTA	Liver, kidney, nervous system, circulatory effects
Dichloromethane (methylene chloride)	Zero	0.005	PTA	Cancer
1,2-Dichloropropane	Zero	0.005	GAC, PTA	Cancer
Dibromoacetic acid	No MCLG	See HAA5	GAC, NF*	Cancer
Dichloroacetic acid	No MCLG	See HAA5	GAC, NF*	Cancer
Dinoseb	0.007	0.007	GAC	Thyroid, reproductive effects
Diquat	0.02	0.02	GAC	Ocular, liver, kidney effects
Endothall	0.1	0.1	GAC	Liver, kidney, gastrointestinal effects
Endrin	0.002	0.002	GAC	Liver, kidney, nervous system effects
Epichlorohydrin	Zero	TT	PAP	Cancer
Ethylbenzene	0.7	0.7	GAC, PTA	Liver, kidney, nervous system effects
Ethylene dibromide (EDB)	Zero	0.00005	GAC, PTA	Cancer

TABLE 4-7
Standards and potential health effects of the contaminants regulated under the SDWA (*continued*)

Contaminant	Maximum Contaminant Level Goal, mg/L	Maximum Contaminant Level, mg/L	BAT	Potential Health Effects
Organics				
Glyphosate	0.7	0.7	OX	Liver, kidney effects
Haloacetic acids (sum of 5; HAA5)[1]	No MCLG	0.060	GAC, NF*	Cancer
Heptachlor	Zero	0.0004	GAC	Cancer
Heptachlor epoxide	Zero	0.0002	GAC	Cancer
Hexachlorobenzene	Zero	0.001	GAC	Cancer
Hexachlorocyclopentadiene	0.05	0.05	GAC, PTA	Kidney, stomach effects
Lindane	0.0002	0.0002	GAC	Liver, kidney, & nervous, immune, circulatory system effects
Methoxychlor	0.04	0.04	GAC	Development, liver, kidney, nervous system effects
Monochlorobenzene	0.1	0.1	GAC, PTA	Cancer
Monochloroacetic acid	0.07	See HAA5	GAC, NF*	Cancer
Monobromoacetic acid	No MCLG	See HAA5	GAC, NF*	Cancer
Oxamyl (vydate)	0.2	0.2	GAC	Kidney effects
Pentachlorophenol	Zero	0.001	GAC	Cancer
Picloram	0.5	0.5	GAC	Kidney, liver effects
Polychlorinated biphenyls (PCBs)	Zero	0.0005	GAC	Cancer
Simazine	0.004	0.004	GAC	Body weight and blood effects, possible carcinogen
Styrene	0.1	0.1	GAC, PTA	Liver, nervous system effects, possible carcinogen
2,3,7,8-TCDD (dioxin)	Zero	5×10^{-8}	GAC	Cancer
Tetrachloroethylene	Zero	0.005	GAC, PTA	Cancer
Toluene	1	1	GAC, PTA	Liver, kidney, nervous system, circulatory system effects
Toxaphene	Zero	0.003	GAC	Cancer
2,4,5-TP (silvex)	0.05	0.05	GAC	Liver, kidney effects
Trichloroacetic acid	0.02	See HAA5	GAC, NF†	Cancer
1,2,4-Trichlorobenzene	0.07	0.07	GAC, PTA	Liver, kidney effects
1,1,1-Trichloroethane	0.2	0.2	GAC, PTA	Liver, nervous system effects
1,1,2-Trichloroethene	0.003	0.005	GAC, PTA	Kidney, liver effects, possible carcinogen
Trichloroethylene	Zero	0.005	GAC, PTA	Cancer
Trihalomethanes (sum of 4; TTHM's)[2]	No MCLG	0.080	GAC, NF*	Cancer
Vinyl chloride	Zero	0.002	PTA	Cancer
Xylenes (total)	10	10	GAC, PTA	Liver, kidney, nervous system effects

TABLE 4-7
Standards and potential health effects of the contaminants regulated under the SDWA (*continued*)

Contaminant	Maximum Contaminant Level Goal, mg/L	Maximum Contaminant Level, mg/L	BAT	Potential Health Effects
Inorganics				
Antimony	0.006	0.006	C-F[3], RO	Decreased longevity, blood effects
Arsenic	Zero	0.010	IX,AA,RO, C-F,LS,ED, OX-F	Dermal, nervous system effects, cancer
Asbestos (fibers > 10 μm)	7 million (fibers/L)	7 million (fibers/L)	C-F[3], DF, DEF	Possible carcinogen by ingestion
Barium	2	2	IX,RO, LS[3]	Blood pressure effects
Beryllium	0.004	0.004	IX,RO, C-F[3] LS[3], AA,IX	Bone, lung effects, cancer
Bromate	Zero	0.010	DC	
Cadmium	0.005	0.005	C-F[3], LS[3], IX, RO	Kidney effects
Chlorite	0.8	1.0	DC	Nervous system effects
Chromium (total)	0.1	0.1	C-F[3], LS[3], (Cr III), IX, RO	Liver, kidney, circulatory system effects
Copper	1.3	TT	CC, SWT	Gastrointestinal effects
Cyanide	0.2	0.2	IX, RO, Cl_2	Thyroid, central nervous system effects
Fluoride	4	4	AA, RO	Skeletal Fluorosis
Lead	Zero	TT	CC, PE, SWT, LSLR	Cancer, kidney, central and peripheral nervous system effects
Mercury	0.002	0.002	C-F[3] (influent < 10μg/L), LS[3], GAC, RO (influent < 10 μg/L)	Kidney, central nervous system effects
Nitrate (as N)	10	10	IX,RO,ED	Methemoglobinemia (blue baby syndrome)
Nitrite (as N)	1	1	IX,RO	Methemoglobinemia (blue baby syndrome)
Nitrate + nitrite (both as N)	10	10	IX,RO	
Selenium	0.05	0.05	C-F[3] (Se IV), LS[3], AA,RO,ED	Nervous system effects
Thallium	0.0005	0.002	IX, AA	Liver, kidney, brain, intestine effects
Radionuclides				
Beta particle and photon emitters	Zero	4 mrem	C-F,IX,RO	Cancer
Alpha particles	Zero	15 pCi/L	C-F,RO	Cancer
Radium-226 + Radium-228	No MCLG	5 pCi/L	IX,LS,RO	Cancer
Uranium	Zero	30 μg/L	C-F[3], LS[3], AX	Cancer

TABLE 4-7
Standards and potential health effects of the contaminants regulated under the SDWA (*continued*)

Contaminant	Maximum Contaminant Level Goal, mg/L	Maximum Contaminant Level, mg/L	BAT	Potential Health Effects
Microbials				
Cryptosporidium	Zero	TT	NA	Gastroenteric disease
E. coli	Zero	TT[5]	NA	Gastroenteric disease
Fecal coliforms	Zero	TT[5]	NA	Gastroenteric disease
Giardia lambia	Zero	TT	NA	Gastroenteric disease
Heterotrophic bacteria	No MCLG	TT	NA	Gastroenteric disease
Legionella	Zero	TT	NA	Pneumonia-like effects
Total coliforms	Zero	TT[4]	NA	Indicator of gastroenteric infections
Turbidity		PS	NA	Interferes with disinfection, indicator of filtration performance
Viruses	Zero	TT	NA	Gastroenteric disease, respiratory disease, and other diseases, (e.g. hepatitis, myocarditis)

*Consecutive systems can use monochloramine (NH_2Cl) as BAT.

AA–activated alumina, AX–anion exchange, CC–corrosion control, C-F–coagulation and filtration, Cl_2–chlorination, DC–disinfection system control, DEF–diatomaceous earth filtration, DF–direct filtration, EF–enhanced coagulation, ED–electrodialysis, GAC–granular activated carbon, IX–ion exchange, LS–lime softening, LSLR–lead service line replacement, NA–not applicable, NF–nanofiltration, OX–oxidation, OX-F–oxidation and filtration, PAP–polymer addition practices, PE–public education, PR–precursor removal, PS–performance standard, PTA–packed-tower aeration, RO–reverse osmosis, SWT–source water treatment, TT–treatment technique.

1. Sum of the concentrations of mono-, di-, and trichloroacetic acids and mono- and dibromoacetic acids.
2. Sum of the concentrations of bromodichloromethane, dibromonochloromethane, bromoform, and chloroform.
3. Coagulation-filtration and lime-softening are not BAT for small systems for variance unless treatment is already installed.
4. No more than 5 percent of the samples per month may be positive. For systems collecting fewer than 40 samples per month, no more than 1 sample per month may be positive.
5. If a repeat total coliform sample is fecal coliform- or *E. coli*-positive, the system is in violation of the MCL for total coliforms. The system is also in violation of the MCL for total coliforms if a routine sample is fecal coliform- or *E. coli*-positive and is followed by a total coliform-positive repeat sample.

residences. The number of samples required to be collected will range from 10 per year to 50 per quarter, depending on the size of the water system.

The SDWA amendments forbid the use of pipe, solder, or flux that is not lead-free in the installation or repair of any public water system or in any plumbing system providing water for human consumption. This does not, however, apply to leaded joints necessary for the repair of cast iron pipes.

Disinfectants and Disinfectant By-Products (D-DBPs). The disinfectants used to destroy pathogens in water and the by-products of the reaction of these disinfectants with organic materials in the water are of potential health concern. One class of DBPs has been regulated since 1979. This class is known as trihalomethanes (THMs). THMs are formed when a water containing an organic precursor is chlorinated (*precursor* means forerunner). In this case it means an organic compound capable of reacting to produce a THM. The precursors are natural organic substances formed from the decay of vegetative matter, such as leaves, and aquatic organisms. THMs are of concern because they are potential carcinogens and may cause reproductive effects. The four THMs that were regulated in the 1979 rules are: chloroform ($CHCl_3$), bromodichloromethane ($CHBrCl_2$), dibromochloromethane ($CHBr_2Cl$), and bromoform ($CHBr_3$). Of these four, chloroform appears most frequently and is found in the highest concentrations.

The D-DBP rule was developed through a negotiated rule-making process, in which individuals representing major interest groups concerned with the rule (for example, public-water-system owners, state and local government officials, and environmental groups) publicly work with the EPA representatives to reach a consensus on the contents of the proposed rule.

Maximum residual disinfectant level goals (MRDLGs) and maximum residual disinfectant levels (MRDLs) were established for chlorine, chloramine, and chlorine dioxide (Table 4-8). Because ozone reacts too quickly to be detected in the distribution system, no limits on ozone were set.

The MCLGs and MCLs for the disinfection by-products are listed in Table 4-9. In addition to regulating individual compounds, the D-DBP rule set levels for two groups of compounds: HAA5 and TTHMs. These groupings were made to recognize the potential cumulative effect of several compounds. HAA5 is the sum of five haloacetic acids (monochloroacetic acid, dichloroacetic acid, trichloroacetic acid, monobromoacetic acid, and dibromoacetic acid). TTHMs (total trihalomethanes) is the sum of

TABLE 4-8
Maximum residual disinfectant goals (MRDLGs) and maximum residual disinfectant levels (MRDLs)

Disinfectant residual	MRDLGs mg/L	MRDL mg/L
Chlorine (free)	4	4.0
Chloramines (as total chlorine)	4	4.0
Chlorine dioxide	0.08	0.8

TABLE 4-9
Maximum contaminant level goals (MCLGs) and maximum contaminant levels (MCLs) for disinfection by-products (DBPs)

Contaminant	MCLG mg/L	Stage 1 MCL mg/L	Stage 2 MCL mg/L
Bromate	Zero	0.010	
Bromodichloromethane	Zero		
Bromoform	Zero		
Chloral hydrate	0.005		
Chlorite	0.3	1.0	
Chloroform	0.07		
Dibromochloromethane	0.06		
Dichloroacetic acid	Zero		
Monochloroacetic acid	0.03		
Trichloroacetic acid	0.02		
HAA5		0.060	0.060*
TTHMs		0.080	0.080*

*Calculated differently in Stage 2.

the concentrations of chloroform ($CHCl_3$), bromodichloromethane ($CHBrCl_2$), dibromochloromethane ($CHBr_2Cl$), and bromoform ($CHBr_3$).

The D-DBP rule is quite complex. In addition to the regulatory levels shown in the tables, levels are established for precursor removal. The amount of precursor required to be removed is a function of the alkalinity of the water and the amount of *total organic carbon* (TOC) present.

The D-DBP rule was implemented in stages. Stage 1 of the rule was promulgated in November 1998. Stage 2 was promulgated in 2006.

When chlorine is added to a water that contains TOC, the chlorine and TOC slowly react to form THMs and HAA5. Therefore, THM and HAA5 are continuously increasing until the point of maximum formation is reached. In order to better determine the level of DBPs consumed at the tap, compliance with the regulation is based on collecting samples in the distribution system. Although the number of samples can vary, it is in the range of four distribution samples collected quarterly for each treatment plant. In the Stage 1 rule, the sample points (say, four) are averaged over four quarters of data (so 16 data points are averaged) to determine compliance with the MCLs of Table 4-9. This method of determining compliance is called a running annual average (RAA). For the Stage 2 rule, several changes were made to the number and locations of the samples, but also the method to calculate compliance was changed. In Stage 2, the four quarters from an individual sample site are averaged (4 data points) and each site must be below the MCLs. This is referred to as a locational running annual average (LRAA). Although the MCLs in Stage 1 and Stage 2 are the same, because of the use of the LRAA, compliance is more difficult in the Stage 2 rule.

Surface Water Treatment Rule (SWTR). The Surface Water Treatment Rule (SWTR) and its companion rules, the Interim Enhanced Surface Water Treatment Rule (IESWTR) and the Long-Term Enhanced Surface Water Treatment Rules (LT1ESWTR and LT2ESWTR), set forth primary drinking water regulations requiring treatment of surface water supplies or groundwater supplies under the direct influence of surface water. The regulations require a specific treatment technique—filtration and/or disinfection—in lieu of establishing maximum contaminant levels (MCLs) for turbidity, *Cryptosporidium, Giardia,* viruses, *Legionella,* and heterotrophic bacteria, as well as many other pathogenic organisms that are removed by these treatment techniques. The regulations also establish a maximum contaminant level goal (MCLG) of zero for *Giardia, Cryptosporidium,* viruses, and *Legionella.* No MCLG is established for heterotrophic plate count or turbidity.

Turbidity Limits. Treatment by conventional or direct filtration must achieve a turbidity level of less than 0.3 NTU in at least 95 percent of the samples taken each month. Those systems using slow sand filtration must achieve a turbidity level of less than 5 NTU at all times and not more than 1 NTU in more than 5 percent of the samples taken each month. The 1 NTU limit may be increased by the state up to 5 NTU if it determines that there is no significant interference with disinfection. Other filtration technologies may be used if they meet the turbidity requirements set for slow sand filtration, provided they achieve the disinfection requirements and are approved by the state.

Turbidity measurements must be performed on representative samples of the system's filtered water every four hours or by continuous monitoring. For any system using slow sand filtration or a filtration treatment other than conventional treatment, direct filtration, or diatomaceous earth filtration, the state may reduce the monitoring requirements to once per day.

Disinfection Requirements. Filtered water supplies must achieve the same disinfection as required for unfiltered systems (that is, 99.9 or 99.99% removal, also known as 3-log and 4-log removal or inactivation, for *Giardia* and viruses respectively) through a combination of filtration and application of a disinfectant.

Giardia and viruses are both fairly well inactivated by chlorine, and hence with proper physical treatment and chlorination both can be controlled. *Cryptosporidium,* however, is resistant to chlorination. Depending on the source water concentration, EPA established levels of treatment, both additional physical barriers and disinfection techniques, that must be used to reduce the rush of illness due to *Cryptosporidium.* Ozone and ultraviolet light are effective disinfectants for *Cryptosporidium.*

Total Coliform. On June 19, 1989, the EPA promulgated the revised National Primary Drinking Water Regulations for total coliforms, including fecal coliforms and *E. coli.* These regulations apply to all public water systems.

The regulations establish a maximum contaminant level (MCL) for coliforms based on the presence or absence of coliforms. Larger systems that are required to collect at least 40 samples per month cannot obtain coliform-positive results in more than 5 percent of the samples collected each month to stay in compliance with the MCL. Smaller systems that collect fewer than 40 samples per month cannot have coliform-positive results in more than one sample per month.

The EPA will accept any one of the five analytical methods noted below for the determination of total coliforms:

Multiple-tube fermentation technique (MTF)
Membrane filter technique (MF)
Minimal media ONPG-MUG test (colilert system) (MMO-MUG)
Presence-absence coliform test (P-A)
Colisure technique

Regardless of the method used, the standard sample volume required for total coliform testing is 100 mL.

A public water system must report a violation of the total coliform regulations to the state no later than the end of the next business day. In addition to this, the system must make public notification according to the general public notification requirements of the Safe Drinking Water Act, but with special wording prescribed by the total coliform regulations.

Secondary Maximum Contaminant Levels (SMCLs). The National Safe Drinking Water Act also provided for the establishment of an additional set of standards to prescribe maximum limits for those contaminants that tend to make water disagreeable to use, but that do not have any particular adverse public health effect. These secondary maximum contaminant levels are the advisable maximum level of a contaminant in any public water supply system. The levels are shown in Table 4-10.

AWWA Goals. The primary and secondary maximum contaminant levels are the maximum allowed (or recommended) values of the various contaminants. However, a

TABLE 4-10
Secondary maximum contaminant levels

Contaminant	SMCL (mg/L)[a]
Chloride	250
Color	15 color units
Copper	1
Corrosivity	Noncorrosive
Foaming agents	0.5
Hydrogen sulfide	0.05
Iron	0.3
Manganese	0.05
Odor	3 threshold odor number
pH	6.5–8.5 units
Sulfate	250
Total dissolved solids (TDS)	500
Zinc	5

[a]All quantities are mg/L except those for which units are given.

TABLE 4-11
American Water Works Association water quality goals

Contaminant	Goal (mg/L)[a]
Turbidity	< 0.1 TU
Color	< 3 color units
Odor	None
Taste	None objectionable
Aluminum	< 0.05
Copper	< 0.2
Iron	< 0.05
Manganese	< 0.01
Total dissolved solids (TDS)	200.0
Zinc	< 1.0
Hardness	80.0

[a]All quantities are mg/L except those for which units are given.

reasonable goal may be much lower than the MCLs themselves. The American Water Works Association (AWWA) has issued its own set of goals to which its members try to adhere. These goals are shown in Table 4-11.

Water Classification and Treatment Systems

Water Classification by Source. Potable water is most conveniently classified as to its source, that is, groundwater or surface water. Generally, groundwater is uncontaminated but may contain aesthetically or economically undesirable impurities. Surface water must be considered to be contaminated with bacteria, viruses, or inorganic substances which could present a health hazard. Surface water may also have aesthetically unpleasing characteristics for a potable water. Table 4-12 shows a comparison between groundwater and surface water.

TABLE 4-12
General characteristics of groundwater and surface water

Ground	Surface
Constant composition	Varying composition
High mineralization	Low mineralization
Little turbidity	High turbidity
Low or no color	Color
Bacteriologically safe	Microorganisms present
No dissolved oxygen	Dissolved oxygen
High hardness	Low hardness
H_2S, Fe, Mn	Tastes and odors
	Possible chemical toxicity

TABLE 4-13
Raw water quality as a function of water source

Source	Turbidity (TU)	Color (Pt-Co units)	Average alum dose (mg/L)
Reservoir	11	18	16
Lake	16	28	22
River	26	44	29

Source: D. A. Cornwell and J. A. Susan, 1979.

Groundwater is further classified as to its source—deep or shallow wells. Municipal water quality factors of safety, temperature, appearance, taste and odor, and chemical balance are most easily satisfied by a deep well source. High concentrations of calcium, iron, manganese, and magnesium typify well waters. Some supplies contain hydrogen sulfide, while others may have excessive concentrations of chloride, sulfate, or carbonate.

Shallow wells are recharged by a nearby surface watercourse. They may have qualities similar to the deep wells, or they may take on the characteristics of the surface recharge water. A sand aquifer between the shallow well supply and the surface watercourse may act as an effective filter for removal of organic matter and as a heat exchanger for buffering temperature changes. To predict water quality from shallow wells, careful studies of the aquifer and nature of recharge water are necessary.

Surface water supplies are classified as to whether they come from a lake, reservoir, or river. A comparison of the three is shown in Table 4-13. Generally, a river has the lowest water quality and a reservoir the highest. Water quality in rivers depends upon the character of the watershed. River quality is largely influenced by pollution (or lack thereof) from municipalities, industries, and agricultural practices. The characteristics of a river can be highly variable. During rains or periods of runoff, turbidity may increase substantially. Many rivers will show an increase in color and taste and in odor-producing compounds. In warm months, algal blooms frequently cause taste and odor problems.

Reservoir and lake sources have much less day-to-day variation than rivers. Additionally, the quiescent conditions will reduce both the turbidity and, on occasion, the color. As in rivers, summer algal blooms can create taste and odor problems in lakes and reservoirs.

Treatment Systems. Treatment plants can be classified as simple disinfection, filter plants, or softening plants. Plants employing simple chlorination are usually groundwater sources that have a high water quality and chlorinate to ensure that the water reaching customers contains safe bacteria levels. Generally, a filtration plant is used to treat surface water and when necessary a softening plant is used to treat groundwater.

In a filtration plant, rapid mixing, flocculation, sedimentation, filtration, and disinfection are employed to remove color, turbidity, taste and odors, organic matter, and bacteria. Additional operations may include bar racks or coarse screens if floating debris and fish are a problem. Figure 4-8 shows a typical flow diagram of a filtration plant. The *raw* (untreated) *surface water* enters the plant via low-lift pumps or gravity. Usually screening has taken place prior to pumping. During mixing, chemicals called coagulants are added and rapidly dispersed through the water. The chemical reacts with the desired

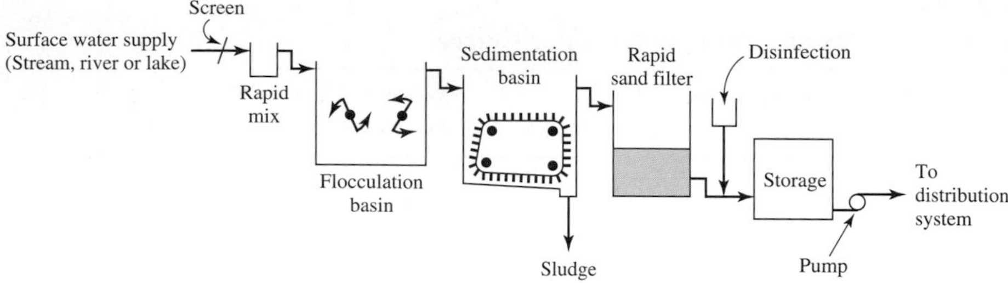

FIGURE 4-8
Flow diagram of a conventional surface water treatment plant ("filtration plant").

impurities and forms precipitates (*flocs*) that are slowly brought into contact with one another during flocculation. The objective of flocculation is to allow the flocs to collide and "grow" to a settleable size. The particles are removed by gravity (sedimentation). This is done to minimize the amount of solids that are applied to the filters. For treatment works with a high-quality raw water, it may be possible to omit sedimentation and perhaps flocculation. This modification is called *direct filtration*. Filtration is the final polishing (removal) of particles. During filtration the water is passed through sand or similar media to screen out the fine particles that will not settle. Disinfection is the addition of chemicals (usually chlorine) to kill or reduce the number of pathogenic organisms. Disinfection of the raw water is neither economical nor efficient, and may form undesirable by-products. The color and turbidity consume the disinfectant thus requiring the use of excessive amounts of chemical. In addition, the presence of turbidity may shield the pathogens from the action of the disinfectant and thereby prevent efficient destruction. Storage may be provided at the plant or located within the community to meet peak demands and to allow the plant to operate on a uniform schedule. The high-lift pumps provide sufficient pressure to convey the water to its ultimate destination. The precipitated chemicals, original turbidity, and suspended material are removed from the sedimentation basins and from the filters. These residuals must be disposed of properly.

Softening plants utilize the same unit operations as filtration plants, but use different chemicals. The primary function of a softening plant is to remove hardness (calcium and magnesium). In a softening plant (a typical flow diagram is shown in Figure 4-9), the design considerations of the various facilities are different than those

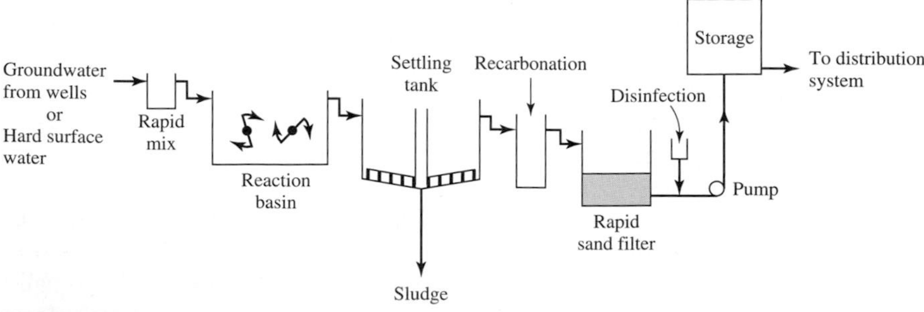

FIGURE 4-9
Flow diagram of water softening plant.

in filtration. Also the chemical doses are much higher in softening, and the corresponding sludge production is greater.

During rapid mix, chemicals are added to react with and precipitate the hardness. Precipitation occurs in the reaction basin. The other unit operations are the same as in a filtration plant except for the additional recarbonation step employed in softening to adjust the final pH.

In the next two sections of this chapter, we discuss coagulation chemistry and softening chemistry, respectively. The subsequent sections describe the physical processes themselves. These are applicable to both filtration and softening plants.

4-2 COAGULATION

Surface waters must be treated to remove turbidity, color, and bacteria. When sand filtration was developed around 1885, it became immediately apparent that filtration alone would not produce a clear water. Experience has demonstrated that direct filtration is largely ineffective in removing bacteria, viruses, soil particles, and color.

The object of coagulation (and subsequently flocculation) is to turn the small particles of color, turbidity, and bacteria into larger flocs, either as precipitates or suspended particles. These flocs are then conditioned so that they will be readily removed in subsequent processes. Technically, coagulation applies to the removal of colloidal particles. However, the term has been applied more loosely to removal of dissolved ions, which is actually precipitation. Coagulation in this chapter will refer to colloid removal only. We define *coagulation* as a method to alter the colloids so that they will be able to approach and adhere to each other to form larger floc particles.

Colloid Stability

Before discussing colloid removal, we should understand why the colloids are suspended in solution and can't be removed by sedimentation or filtration. Very simply, the particles in the colloid range are too small to settle in a reasonable time period, and too small to be trapped in the pores of a filter. For colloids to remain stable they must remain small. Most colloids are stable because they possess a negative charge that repels other colloidal particles before they collide with one another.* The colloids are continually involved in *Brownian movement,* which is merely random movement. Charges on colloids are measured by placing DC electrodes in a colloidal dispersion. The particles migrate to the pole of opposite charge at a rate proportional to the potential gradient. Generally, the larger the surface charge, the more stable the suspension.

Colloid Destabilization

Colloids are stable because of their surface charge. In order to destabilize the particles, we must neutralize this charge. Such neutralization can take place by the addition of an ion of opposite charge to the colloid. Since most colloids found in water are negatively charged, the addition of sodium ions (Na^+) should reduce the charge. Figure 4-10 shows such an effect. The plot shows surface charge as a function of

*Some colloids are stabilized by their affinity for water, but these types of colloids are of less importance.

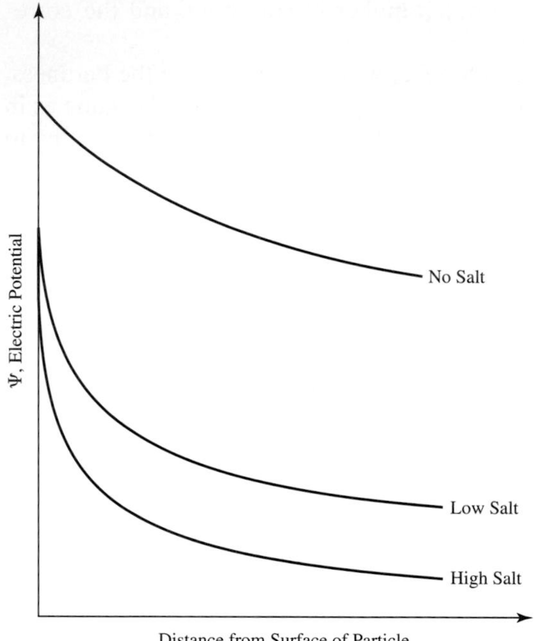

FIGURE 4-10
Effect of salt on electric potential.

distance from the colloid for no-salt (NaCl) addition, low-salt addition, and high-salt addition. As we would have predicted, the higher the concentration of sodium we add, the lower the charge, and therefore the lower the repelling forces around the colloid. If, instead of adding a monovalent ion such as sodium, we add a divalent or trivalent ion, the charge is reduced even faster, as shown in Figure 4-11. In fact, it was found by Schulze and Hardy that one mole of a trivalent ion can reduce the charge as much as 30 to 50 moles of a divalent ion and as much as 1,500 to 2,500 moles of a monovalent ion (often referred to as the *Schulze-Hardy rule*).

Coagulation

The purpose of coagulation is to alter the colloids so that they can adhere to each other. During coagulation a positive ion is added to water to reduce the surface charge to the point where the colloids are not repelled from each other. A *coagulant* is the substance (chemical) that is added to the water to accomplish coagulation. There are three key properties of a coagulant:

1. Trivalent cation. As indicated in the last section, the colloids most commonly found in natural waters are negatively charged, hence a cation is required to neutralize the charge. A trivalent cation is the most efficient cation.
2. Nontoxic. This requirement is obvious for the production of a safe water.
3. Insoluble in the neutral pH range. The coagulant that is added must precipitate out of solution so that high concentrations of the ion are not left in the water. Such precipitation greatly assists the colloid removal process.

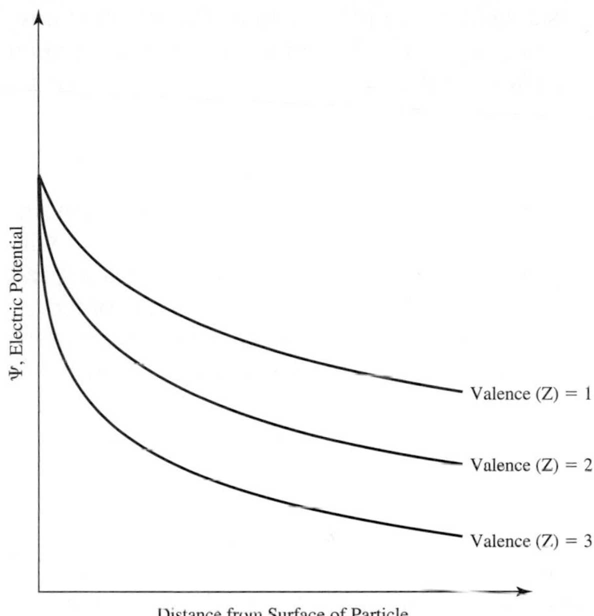

FIGURE 4-11
Effect of valence on electric potential.

The two most commonly used coagulants are aluminum (Al^{3+}) and ferric iron (Fe^{3+}). Both meet the above three requirements, and their reactions are outlined here.

Aluminum. Aluminum can be purchased as either dry or liquid alum [$Al_2(SO_4)_3 \cdot 14H_2O$]. Commercial alum has an average molecular weight of 594. Liquid alum is sold as approximately 48.8 percent alum (8.3% Al_2O_3) and 51.2 percent water. If it is sold as a more concentrated solution, there can be problems with crystallization of the alum during shipment and storage. A 48.8 percent alum solution has a crystallization point of −15.6°C. A 50.7 percent alum solution will crystallize at +18.3°C. Dry alum costs about 50 percent more than an equivalent amount of liquid alum so that only users of very small amounts of alum purchase dry alum.

When alum is added to a water containing alkalinity, the following reaction occurs:

$$Al_2(SO_4)_3 \cdot 14H_2O + 6HCO_3^- \rightleftharpoons 2Al(OH)_3 \cdot 3H_2O(s) + 6CO_2 + 8H_2O + 3SO_4^{2-} \quad (4\text{-}54)$$

such that each mole of alum added uses six moles of alkalinity and produces six moles of carbon dioxide. The above reaction shifts the carbonate equilibrium and decreases the pH. However, as long as sufficient alkalinity is present and $CO_2(g)$ is allowed to evolve, the pH is not drastically reduced and is generally not an operational problem. When sufficient alkalinity is not present to neutralize the sulfuric acid production, the pH may be greatly reduced:

$$Al_2(SO_4)_3 \cdot 14H_2O \rightleftharpoons 2Al(OH)_3 \cdot 3H_2O(s) + 3H_2SO_4 + 2H_2O \quad (4\text{-}55)$$

If the second reaction occurs, lime or sodium carbonate may be added to neutralize the acid.

Two important factors in coagulant addition are pH and dose. The optimum dose and pH must be determined from laboratory tests. The optimal pH range for alum is approximately 5.5 to 6.5 with adequate coagulation possible between pH 5 to pH 8 under some conditions.

An important aspect of coagulation is that the aluminum ion does not really exist as Al^{3+} and the final product is more complex than $Al(OH)_3$. When the alum is added to the water, it immediately dissociates, resulting in the release of an aluminum ion surrounded by six water molecules. The aluminum ion immediately starts reacting with the water, forming large $Al \cdot OH \cdot H_2O$ complexes. Some have suggested that it forms $[Al_8\,(OH)_{20} \cdot 28H_2O]^{4+}$ as the product that actually coagulates. Regardless of the actual species produced, the complex is a very large precipitate that removes many of the colloids by enmeshment as it falls through the water. This precipitate is referred to as a *floc*. Floc formation is one of the important properties of a coagulant for efficient colloid removal. The final product after coagulation has three water molecules associated with it in the solid form as indicated in the equations.

Example 4-10. One of the most common methods to evaluate coagulation efficiency is to conduct jar tests. Jar tests are performed in an apparatus such as shown in Figure 4-12. Six beakers are filled with water and then each is mixed and flocculated uniformly by a gang stirrer. The jars are square in order to prevent swirling and are called gator jars. A test is often conducted by first dosing each jar with the same alum dose and varying the pH in each jar. The test can then be repeated in a second set of jars by holding the pH constant and varying the coagulant dose.

Two sets of such jar tests were conducted on a raw water containing 15 TU and an HCO_3^- alkalinity concentration of 50 mg/L expressed as $CaCO_3$. Given the data below, find the optimal pH, coagulant dose, and the theoretical amount of alkalinity that would be consumed at the optimal dose.

Jar test I

	1	**2**	**3**	**4**	**5**	**6**
pH	5.0	5.5	6.0	6.5	7.0	7.5
Alum dose (mg/L)	10	10	10	10	10	10
Settled turbidity (TU)	11	7	5.5	5.7	8	13

Jar test II

pH	6.0	6.0	6.0	6.0	6.0	6.0
Alum dose (mg/L)	5	7	10	12	15	20
Settled turbidity (TU)	14	9.5	5	4.5	6	13

Solution. The results of the two jar tests are plotted in Figure 4-13. The optimal pH was chosen as 6.25 and the optimal alum dose was about 12.5 mg/L. The experimenter would probably try to repeat the test using a pH of 6.25 and varying the alum dose between 10 and 15 to pinpoint the optimal conditions.

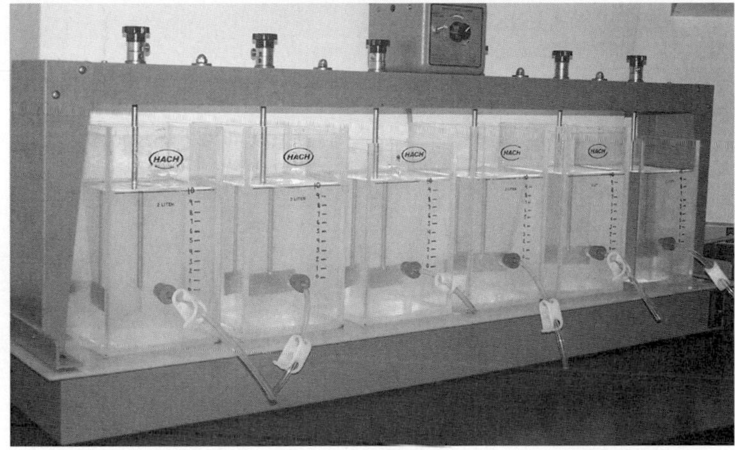

(a)

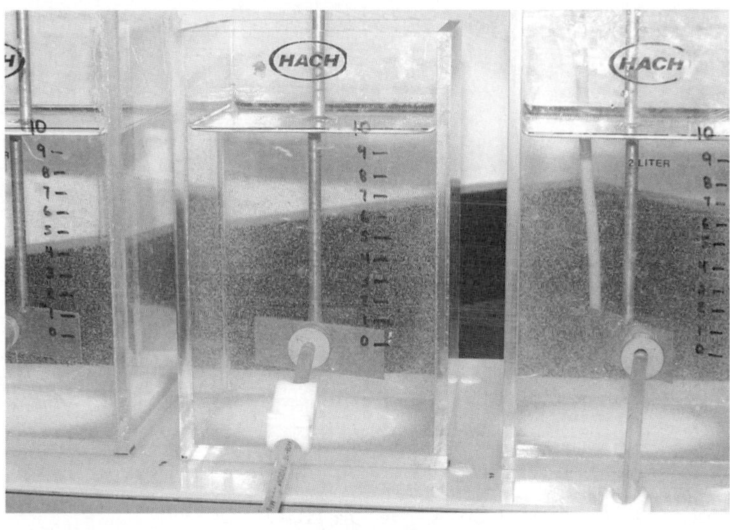

(b)

FIGURE 4-12
Jar tests: (*a*) Phipps and Bird jar tester; (*b*) close up of floc in Gator jar. (David Cornwell)

The amount of alkalinity that will be consumed is found by using Equation 4-54, which shows us that one mole of alum consumes six moles of HCO_3^-. With the molecular weight of alum equal to 594, the moles of alum added per liter is found by using Equation 4-7:

$$\frac{12.5 \times 10^{-3}\ \text{g/L}}{594\ \text{g/mole}} = 2.1 \times 10^{-5}\ \text{moles/L}$$

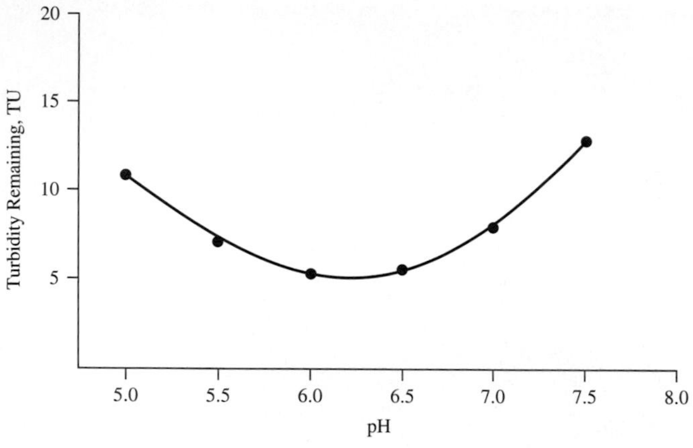

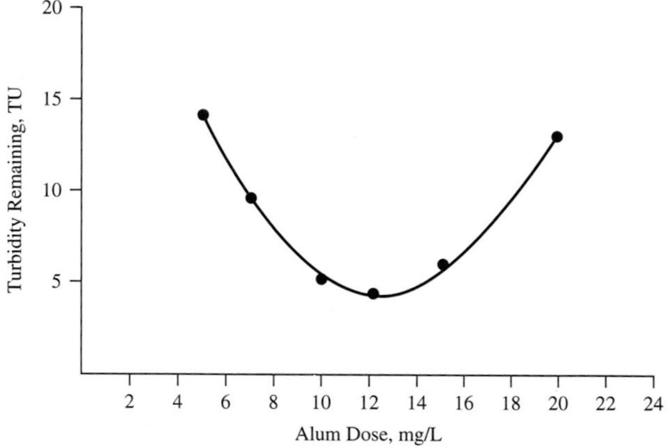

FIGURE 4-13
Results of jar test.

which will consume

$$6(2.1 \times 10^{-5}) = 1.26 \times 10^{-4}\, M\, HCO_3^-$$

The molecular weight of HCO_3^- is 61, so

$$(1.26 \times 10^{-4} \text{ moles/L})(61 \text{ g/mole})(10^3 \text{ mg/g}) = 7.7 \text{ mg/L } HCO_3$$

are consumed, which can be expressed as $CaCO_3$ by using Equation 4-40:

$$7.7 \frac{50}{61} = 6.31 \text{ mg/L } HCO_3^- \text{ as } CaCO_3$$

Iron. Iron can be purchased as either the sulfate salt ($Fe_2(SO_4)_3 \cdot xH_2O$) or the chloride salt ($FeCl_3 \cdot xH_2O$). It is available in various forms, and the individual supplier should be consulted for the specifics of the product. Dry and liquid forms are available. The properties of iron with respect to forming large complexes, dose, and pH curves are similar to those of alum. An example of the reaction of $FeCl_3$ in the presence of alkalinity is

$$FeCl_3 + 3HCO_3^- + 3H_2O \rightleftharpoons Fe(OH)_3 \cdot 3H_2O(s) + 3CO_2 + 3Cl^- \quad (4\text{-}56)$$

and without alkalinity

$$FeCl_3 + 6H_2O \rightleftharpoons Fe(OH)_3 \cdot 3H_2O(s) + 3HCL \quad (4\text{-}57)$$

forming hydrochloric acid which in turn lowers the pH. Ferric salts generally have a wider pH range for effective coagulation than aluminum, that is, pH ranges from 4 to 9.

Coagulant Aids. The four basic types of coagulant aids are pH adjusters, activated silica, clay, and polymers. Acids and alkalies are both used to adjust the pH of the water into the optimal range for coagulation. The acid most commonly used for lowering the pH is sulfuric acid. Either lime [$Ca(OH)_2$] or soda ash (Na_2CO_3) are used to raise the pH.

When activated silica is added to water, it produces a stable solution that has a negative surface charge. The activated silica can unite with the positively charged aluminum or with iron flocs, resulting in a larger, denser floc that settles faster and enhances enmeshment. The addition of activated silica is especially useful for treating highly colored, low-turbidity waters because it adds weight to the floc. However, activation of silica does require proper equipment and close operational control, and many plants are hesitant to use it.

Clays can act much like activated silica in that they have a slight negative charge and can add weight to the flocs. Clays are also most useful for colored, low-turbidity waters, but are rarely used.

Polymers can have a negative charge (*anionic*), positive charge (*cationic*), positive and negative charge (*polyamphotype*), or no charge (*nonionic*). Polymers are long-chained carbon compounds of high molecular weight that have many active sites. The active sites adhere to flocs, joining them together and producing a larger, tougher floc that settles better. This process is called *interparticle bridging*. The type of polymer, dose, and point of addition must be determined for each water, and requirements may change within a plant on a seasonal, or even daily, basis.

4-3 SOFTENING

Hardness. The term hardness is used to characterize a water that does not lather well, causes a scum in the bath tub, and leaves hard, white, crusty deposits (scale) on coffee pots, tea kettles, and hot water heaters. The failure to lather well and the formation of scum on bath tubs is the result of the reactions of calcium and magnesium with the soap. For example:

$$Ca^{2+} + (Soap)^- \rightleftharpoons Ca(Soap)_2(s) \quad (4\text{-}58)$$

As a result of this complexation reaction, soap cannot interact with the dirt on clothing, and the calcium-soap complex itself forms undesirable precipitates.

Hardness is defined as the sum of all polyvalent cations (in consistent units). The common units of expression are mg/L as $CaCO_3$ or meq/L. Qualitative terms used

TABLE 4-14
Hard water classification

Hardness range (mg/L $CaCO_3$)	Description
0–75	Soft
75–100	Moderately hard
100–300	Hard
>300	Very hard

to describe hardness are listed in Table 4-14. Because many people object to water containing hardness greater than 150 mg/L as $CaCO_3$ suppliers of public water have considered it a benefit to *soften* the water, that is, to remove some of the hardness. A common water treatment goal is to provide water with a hardness in the range of 75 to 120 mg/L as $CaCO_3$.

Although all polyvalent cations contribute to hardness, the predominant contributors are calcium and magnesium. Thus, our focus for the remainder of this discussion will be on calcium and magnesium.

The natural process by which water becomes hard is shown schematically in Figure 4-14. As rainwater enters the topsoil, the respiration of microorganisms increases the CO_2 content of the water. As shown in Equation 4-35, the CO_2 reacts with the water to form H_2CO_3. Limestone, which is made up of solid $CaCO_3$ and $MgCO_3$ reacts with the carbonic acid to form calcium bicarbonate [$Ca(HCO_3)_2$] and magnesium bicarbonate

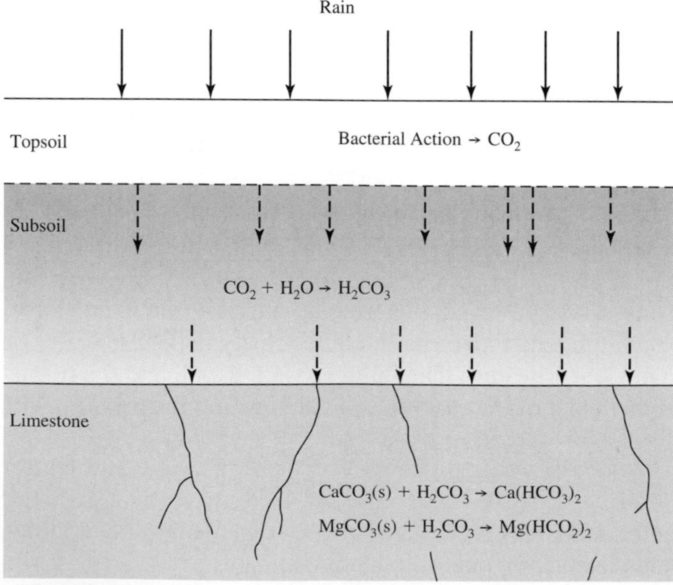

FIGURE 4-14
Natural process by which water is made hard.

$[Mg(HCO_3)_2]$. While $CaCO_3$ and $MgCO_3$ are both insoluble in water, the bicarbonates are quite soluble. Gypsum ($CaSO_4$) and $MgSO_4$ may also go into solution to contribute to the hardness.

Because calcium and magnesium predominate, it is often convenient in performing softening calculations to define the *total hardness* (TH) of a water as the sum of these elements

$$TH = Ca^{2+} + Mg^{2+} \tag{4-59}$$

where the concentrations of each element are in consistent units (mg/L as $CaCO_3$ or meq/L). Total hardness is often broken down into two components: (1) that associated with the HCO_3^- anion (called *carbonate hardness* and abbreviated CH), and (2) that associated with other anions (called *noncarbonate hardness* and abbreviated NCH).* Total hardness, then, may also be defined as

$$TH = CH + NCH \tag{4-60}$$

Carbonate hardness is defined as the amount of hardness equal to the total hardness or the total alkalinity, whichever is less. Carbonate hardness is often called temporary hardness because heating the water removes it. When the pH is less than 8.3, HCO_3^- is the dominant form of alkalinity, and total alkalinity is nominally taken to be equal to the concentration of HCO_3^-.

Noncarbonate hardness is defined as the total hardness in excess of the alkalinity. If the alkalinity is equal to or greater than the total hardness, then there is no noncarbonate hardness. Noncarbonate hardness is called permanent hardness because it is not removed when water is heated.

Bar charts of water composition are often useful in understanding the process of softening. By convention, the bar chart is constructed with cations in the upper bar and anions in the lower bar. In the upper bar, calcium is placed first and magnesium second. Other cations follow without any specified order. The lower bar is constructed with bicarbonate placed first. Other anions follow without any specified order. Construction of a bar chart is illustrated in Example 4-11.

Example 4-11. Given the following analysis of a groundwater, construct a bar chart of the constituents, expressed as $CaCO_3$.

Ion	mg/L as ion	EW $CaCO_3$/EW ion	mg/L as $CaCO_3$
Ca^{2+}	103	2.50	258
Mg^{2+}	5.5	4.12	23
Na^+	16	2.18	35
HCO_3^-	255	0.82	209
SO_4^{2-}	49	1.04	51
Cl^-	37	1.41	52

*Note that this does not imply that the compounds exist as compounds in solution. They are dissociated.

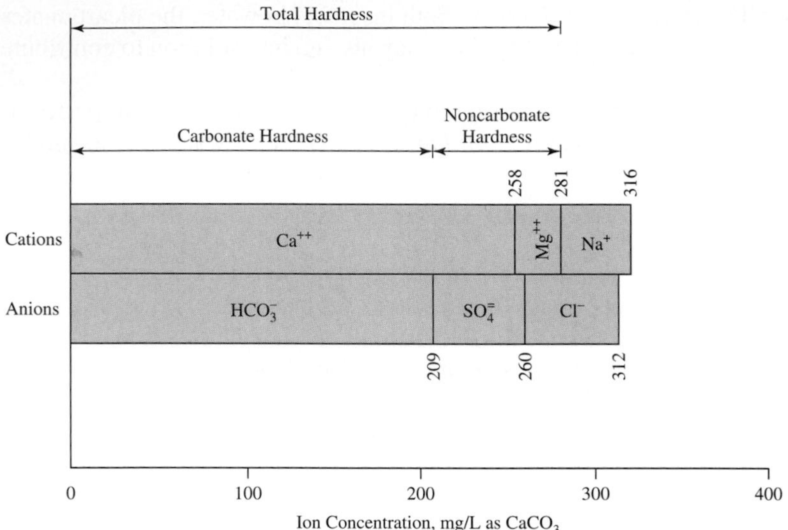

FIGURE 4-15
Bar graph of groundwater constituents.

Solution. The concentrations of the ions have been converted to $CaCO_3$ equivalents. The results are plotted in Figure 4-15.

The cations total 316 mg/L as $CaCO_3$, of which 281 mg/L as $CaCO_3$ is hardness. The anions total 312 mg/L as $CaCO_3$, of which the carbonate hardness is 209 mg/L as $CaCO_3$. There is a discrepancy between the cation and anion totals because there are other ions that were not analyzed. If a complete analysis were conducted, and no analytical error occurred, the equivalents of cations would equal exactly the equivalents of anions. Typically, a complete analysis may vary $\pm 5\%$ because of analytical errors.

The relationships between total hardness, carbonate hardness, and noncarbonate hardness are illustrated in Figure 4-16. In Figure 4-16a, the total hardness is 250 mg/L as $CaCO_3$, the carbonate hardness is equal to the alkalinity ($HCO_3^- =$ 200 mg/L as $CaCO_3$), and the noncarbonate hardness is equal to the difference between the total hardness and the carbonate hardness (NCH = TH − CH = 250 = 200 = 50 mg/L as $CaCO_3$). In Figure 4-16b, the total hardness is again 250 mg/L as $CaCO_3$. However, since the alkalinity (HCO_3^-) is greater than the total hardness, and since the carbonate hardness cannot be greater than the total hardness (see Equation 4-60), the carbonate hardness is equal to the total hardness, that is, 250 mg/L as $CaCO_3$. With the carbonate hardness equal to the total hardness, then all of the hardness is carbonate hardness and there is no noncarbonate hardness. Note that in both cases it may be assumed that the pH is less than 8.3 because HCO_3^- is the only form of alkalinity present.

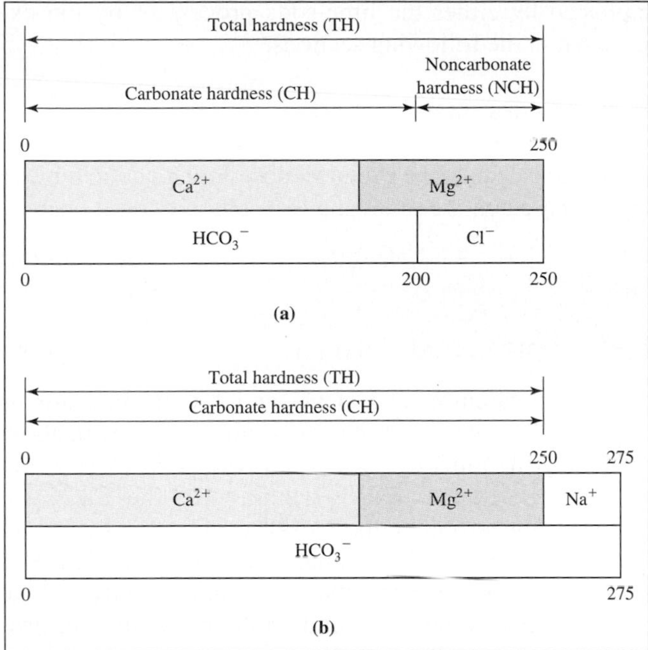

FIGURE 4-16
Relationships between total hardness, carbonate hardness, and noncarbonate hardness.

Example 4-12. A water has an alkalinity of 200 mg/L as $CaCO_3$. The Ca^{2+} concentration is 160 mg/L as the ion, and the Mg^{2+} concentration is 40 mg/L as the ion. The pH is 8.1. Find the total, carbonate, and noncarbonate hardness.

Solution. The molecular weights of calcium and magnesium are 40 and 24 respectively. Since each has a valence of 2^+, the corresponding equivalent weights are 20 and 12. Using Equation 4-40 to convert mg/L as the ion to mg/L as $CaCO_3$ and adding the two ions as shown in Equation 4-59, the total hardness is

$$\text{TH} = 160\text{ mg/L}\left(\frac{50\text{ mg/meq}}{20\text{ mg/meq}}\right) + 40\text{ mg/L}\left(\frac{50\text{ mg/meq}}{12\text{ mg/meq}}\right) = 567\text{ mg/L as CaCO}_3$$

where 50 is the equivalent weight of $CaCO_3$.

By definition, the carbonate hardness is the lesser of the total hardness or the alkalinity. Since, in this case, the alkalinity is less than the total hardness, the carbonate hardness (CH) is equal to 200 mg/L as $CaCO_3$. The noncarbonate hardness is equal to the difference

$$\text{NCH} = \text{TH} - \text{CH} = 567 - 200 = 367\text{ mg/L as CaCO}_3$$

Note that we can only add and subtract concentrations of Ca^{2+} and Mg^{2+} if they are in equivalent units, for example, moles/L or milliequivalents/L or mg/L as $CaCO_3$.

Softening can be accomplished by either the lime-soda process or by ion exchange. Both methods are discussed in the following sections.

Lime-Soda Softening

In lime-soda softening it is possible to calculate the chemical doses necessary to remove hardness. Hardness precipitation is based on the following two solubility reactions:

$$Ca^{2+} + CO_3^{2-} \rightleftharpoons CaCO_3(s) \qquad (4\text{-}61)$$

and

$$Mg^{2+} + 2OH^- \rightleftharpoons Mg(OH)_2(s) \qquad (4\text{-}62)$$

The objective is to precipitate the calcium as $CaCO_3$ and the magnesium as $Mg(OH)_2$. In order to precipitate $CaCO_3$, the pH of the water must be raised to about 10.3. To precipitate magnesium, the pH must be raised to about 11. If there is not sufficient naturally occurring bicarbonate alkalinity (HCO_3^-) for the $CaCO_3(s)$ precipitate to form (that is, there is noncarbonate hardness), we must add CO_3^{2-} to the water. Magnesium is more expensive to remove than calcium, so we leave as much Mg^{2+} in the water as possible. It is more expensive to remove noncarbonate hardness than carbonate hardness because we must add another chemical to provide the CO_3^{2-}. Therefore, we leave as much noncarbonate hardness in the water as possible.

Softening Chemistry. The chemical processes used to soften water are a direct application of the law of mass action. We increase the concentration of CO_3^{2-} and/or OH^- by the addition of chemicals, and drive the reactions given in Equations 4-61 and 4-62 to the right. Insofar as possible, we convert the naturally occurring bicarbonate alkalinity (HCO_3^-) to carbonate (CO_3^{2-}) by the addition of hydroxyl ions (OH^-). Hydroxyl ions cause the carbonate buffer system (Equation 4-35) to shift to the right and, thus, provide the carbonate for the precipitation reaction (Equation 4-61).

The common source of hydroxyl ions is calcium hydroxide [$Ca(OH)_2$]. Many water treatment plants find it more economical to buy quicklime (CaO), commonly called lime, than hydrated lime [$Ca(OH)_2$]. The quicklime is converted to hydrated lime at the water treatment plant by mixing CaO and water to produce a slurry of $Ca(OH)_2$, which is fed to the water for softening. The conversion process is called *slaking:*

$$CaO + H_2O \rightleftharpoons Ca(OH)_2 + \text{heat} \qquad (4\text{-}63)$$

The reaction is exothermic. It yields almost 1 MJ per gram mole of lime. Because of this high heat release, the reaction must be controlled carefully. All safety precautions for handling a strong base should be observed. Because the chemical is purchased as lime, it is common to speak of chemical additions as addition of "lime," when in fact we mean calcium hydroxide. When carbonate ions must be supplied, the most common chemical chosen is sodium carbonate (Na_2CO_3). Sodium carbonate is commonly referred to as *soda ash* or *soda*.

Softening Reactions. The softening reactions are regulated by controlling the pH. First, any free acids are neutralized. Then pH is raised to precipitate the $CaCO_3$; if necessary, the pH is raised further to remove $Mg(OH)_2$. Finally, if necessary, CO_3^{2-} is added to precipitate the noncarbonate hardness.

Six important softening reactions are discussed below. In each case, the chemical that has been added to the water is printed in bold type. Remember that (s) designates the solid form, and hence indicates that the substance has been removed from the water. The following reactions are presented sequentially, although in reality they occur simultaneously.

1. Neutralization of carbonic acid (H_2CO_3).

In order to raise the pH, we must first neutralize any free acids that may be present in the water. CO_2 is the principal acid present in unpolluted, naturally occurring water.* You should note that no hardness is removed in this step.

$$CO_2 + \mathbf{Ca(OH)_2} \rightleftharpoons CaCO_3(s) + H_2O \tag{4-64}$$

2. Precipitation of carbonate hardness due to calcium.

As we mentioned previously, we must raise the pH to about 10.3 to precipitate calcium carbonate. To achieve this pH we must convert all of the bicarbonate to carbonate. The carbonate then serves as the common ion for the precipitation reaction.

$$Ca^{2+} + 2HCO_3^- + \mathbf{Ca(OH)_2} \rightleftharpoons 2CaCO_3(s) + 2H_2O \tag{4-65}$$

3. Precipitation of carbonate hardness due to magnesium.

If we need to remove carbonate hardness that results from the presence of magnesium, we must add more lime to achieve a pH of about 11. The reaction may be considered to occur in two stages. The first stage occurs when we convert all of the bicarbonate in step 2 above.

$$Mg^{2+} + 2HCO_3^- + \mathbf{Ca(OH)_2} \rightleftharpoons MgCO_3 + CaCO_3(s) + 2H_2O \tag{4-66}$$

Note that the hardness of the water did not change because $MgCO_3$ is soluble. With the addition of more lime the hardness due to magnesium is removed.

$$Mg^{2+} + CO_3^{2-} + \mathbf{Ca(OH)_2} \rightleftharpoons Mg(OH)_2(s) + CaCO_3(s) \tag{4-67}$$

4. Removal of noncarbonate hardness due to calcium.

If we need to remove noncarbonate hardness due to calcium, no further increase in pH is required. Instead we must provide additional carbonate in the form of soda ash.

$$Ca^{2+} + \mathbf{Na_2CO_3} \rightleftharpoons CaCO_3(s) + 2Na^+ \tag{4-68}$$

---

*CO_2 and H_2CO_3 in water are essentially the same:

$$CO_2 + H_2O \rightleftharpoons H_2CO_3$$

Thus, the number of reaction units (n) for CO_2 is two.

5. Removal of noncarbonate hardness due to magnesium.

If we need to remove noncarbonate hardness due to magnesium, we will have to add both lime and soda. The lime provides the hydroxyl ion for precipitation of the magnesium.

$$Mg^{2+} + \mathbf{Ca(OH)_2} \rightleftharpoons Mg(OH)_2(s) + Ca^{2+} \tag{4-69}$$

Note that although the magnesium is removed, there is no change in the hardness because the calcium is still in solution. To remove the calcium we must add soda.

$$Ca^{2+} + \mathbf{Na_2CO_3} \rightleftharpoons CaCO_3(s) + 2Na^+ \tag{4-70}$$

Note that this is the same reaction as the one to remove noncarbonate hardness due to calcium.

These reactions are summarized in Figure 4-17.

Process Limitations and Empirical Considerations. Lime-soda softening cannot produce a water completely free of hardness because of the solubility of $CaCO_3$ and $Mg(OH)_2$, the physical limitations of mixing and contact, and the lack of sufficient time for the reactions to go to completion. Thus, the minimum calcium hardness that can be achieved is about 30 mg/L as $CaCO_3$, and the minimum magnesium hardness is about 10 mg/L as $CaCO_3$. Because of the slimy condition that results when soap is used with a water that is too soft, we have traditionally set a goal for final total hardness of 75 to 120 mg/L as $CaCO_3$. Because of economic

Neutralization of Carbonic Acid

$$CO_2 + \mathbf{Ca(OH)_2} \rightleftharpoons CaCO_3(s) + H_2O$$

Precipitation of Carbonate Hardness

$$Ca^{2+} + 2HCO_3^- + \mathbf{Ca(OH)_2} \rightleftharpoons 2CaCO_3(s) + 2H_2O$$

$$Mg^{2+} + 2HCO_3^- + \mathbf{Ca(OH)_2} \rightleftharpoons MgCO_3 + CaCO_3(s) + 2H_2O$$

$$MgCO_3 + \mathbf{Ca(OH)_2} = Mg(OH)_2(s) + CaCO_3(s)$$

Precipitation of Noncarbonate Hardness Due to Calcium

$$Ca^{2+} + \mathbf{Na_2CO_3} \rightleftharpoons CaCO_3(s) + 2Na^+$$

Precipitation of Noncarbonate Hardness Due to Magnesium

$$Mg^{2+} + \mathbf{Ca(OH)_2} \rightleftharpoons Mg(OH)_2(s) + Ca^{2+}$$

$$Ca^{2+} + \mathbf{Na_2CO_3} \rightleftharpoons CaCO_3(s) + 2Na^+$$

FIGURE 4-17
Summary of softening reactions. (*Note:* The chemical added is printed in bold type. The precipitate is designated by (s). The arrow indicates where a compound formed in one reaction is used in another reaction.)

constraints, many utilities will operate at total hardness goals of up to 140 to 150 mg/L.

In order to achieve reasonable removal of hardness in a reasonable time period, an excess of $Ca(OH)_2$ beyond the stoichiometric amount usually is provided. Based on our empirical experience, a minimum excess of 20 mg/L of $Ca(OH)_2$ expressed as $CaCO_3$ must be provided.

Magnesium in excess of about 40 mg/L as $CaCO_3$ forms scales on heat exchange elements in hot water heaters. Because of the expense of removing magnesium, we normally remove only that magnesium which is in excess of 40 mg/L as $CaCO_3$. For magnesium removals less than 20 mg/L as $CaCO_3$, the basic excess of lime mentioned above is sufficient to ensure good results. For magnesium removals between 20 and 40 mg/L as $CaCO_3$, we must add an excess of lime equal to the magnesium to be removed. For magnesium removals greater than 40 mg/L as $CaCO_3$, the excess lime we need to add is 40 mg/L as $CaCO_3$. Addition of excess lime in amounts greater than 40 mg/L as $CaCO_3$ does not appreciably improve the reaction kinetics.

The chemical additions (as $CaCO_3$) to soften water may be summarized as follows:

Step	Chemical addition[a]	Reason
Carbonate hardness		
1.	Lime = CO_2	Destroy H_2CO_3
2.	Lime = HCO_3^-	Raise pH; convert HCO_3^- to CO_3^{2-}
3.	Lime = Mg^{2+} to be removed	Raise pH; precipitate $Mg(OH)_2$
4.	Lime = required excess	Drive reaction
Noncarbonate hardness		
5.	Soda = noncarbonate hardness to be removed	Provide CO_3^{2-}

[a]The terms "Lime =" and "Soda =" refer to mg/L of $Ca(OH)_2$ and Na_2CO_3 as $CaCO_3$ equal to mg/L of ion (or gas in the case of CO_2) as $CaCO_3$.

These steps are diagrammed in the flow chart shown in Figure 4-18. The next three examples illustrate the technique.

Example 4-13. From the water analysis presented below, determine the amount of lime and soda (in mg/L as $CaCO_3$) necessary to soften the water to 80.00 mg/L hardness as $CaCO_3$.

Water Composition (mg/L)

Ca^{2+}:	95.20	CO_2:	19.36	HCO_3^-:	241.46
Mg^{2+}:	13.44			SO_4^{2-}:	53.77
Na^+:	25.76			Cl^-:	67.81

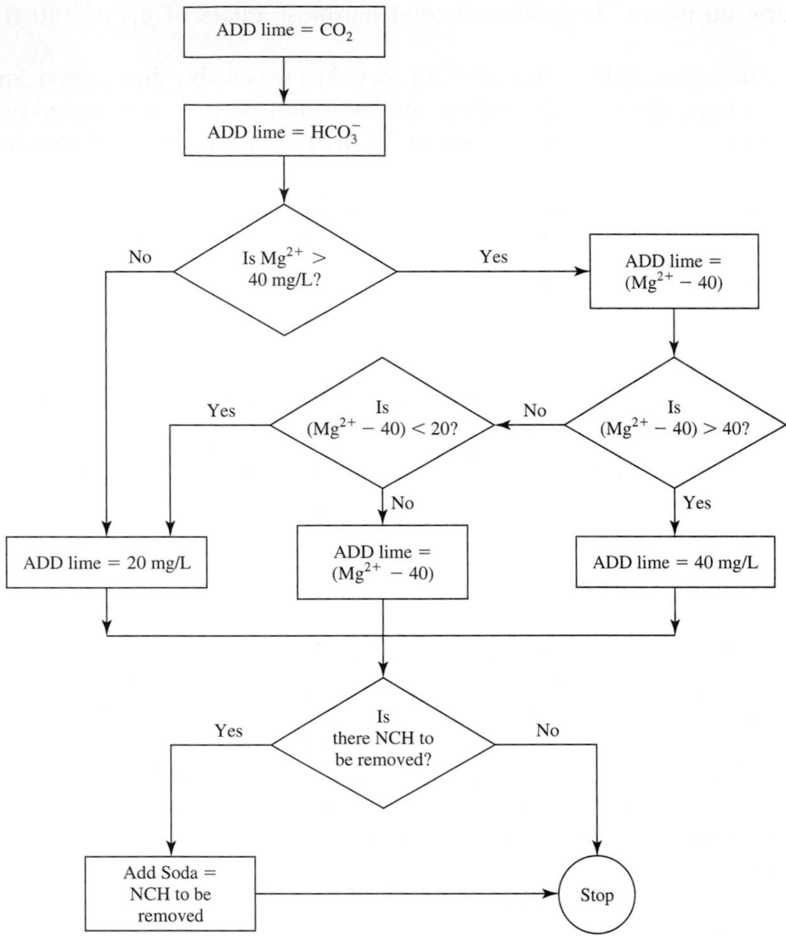

FIGURE 4-18
Flow diagram for solving softening problems. (*Note:* All additions are "as $CaCO_3$." NCH means noncarbonate hardness.)

Solution. We begin by converting the elements and compounds to $CaCO_3$ equivalents.

Ion	mg/L as ion	EW $CaCO_3$/EW ion	mg/L as $CaCO_3$
Ca^{2+}	95.20	2.50	238.00
Mg^{2+}	13.44	4.12	55.37
Na^{+}	25.76	2.18	56.16
HCO_3^-	241.46	0.820	198.00
SO_4^{2-}	53.77	1.04	55.92
Cl^-	67.81	1.41	95.61
CO_2	19.36	2.28	44.14

The resulting bar chart would appear as shown below.

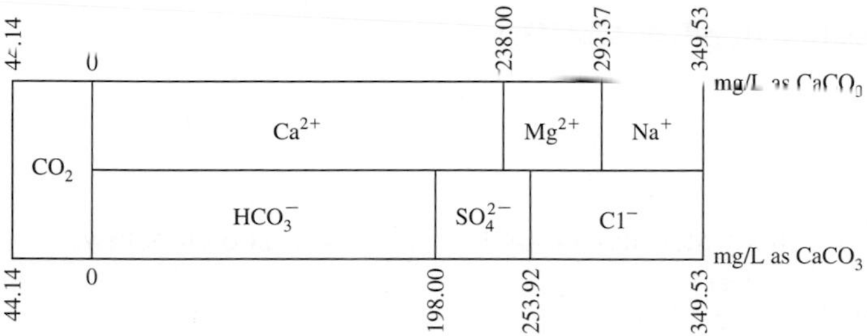

From the bar chart we note the following: CO_2 does not contribute to the hardness; the total hardness (TH) = 293.37 mg/L as $CaCO_3$; the carbonate hardness (CH) = 198.00 mg/L as $CaCO_3$; and, finally, the noncarbonate hardness (NCH) is equal to TH − CH = 95.37 mg/L as $CaCO_3$.

Using Figure 4-17 to guide our logic, we determine the lime dose as follows:

Step	Dose (mg/L as $CaCO_3$)
Lime = CO_2	44.14
Lime = HCO_3^-	198.00
Lime = $Mg^{2+} - 40 = 55.37 - 40$	15.37
Lime = excess	20.00
	277.51

The amount of lime to add is 277.51 mg/L as $CaCO_3$. The excess chosen was the minimum since ($Mg^{2+} - 40$) was less than 20.

Now we must determine if any NCH need be removed. The amount of NCH that can be left (NCH_f) is equal to the final hardness desired (80.00 mg/L) minus the CH left due to solubility and other factors (40.00 mg/L):

$$NCH_f = 80.00 - 40.00 = 40.00 \text{ mg/L}$$

Thus, 40.00 mg/L may be left. The NCH that must be removed (NCH_R) is the initial NCH_i (95.37 mg/L) minus the NCH_f:

$$NCH_R = NCH_i - NCH_f$$
$$NCH_R = 95.37 - 40.00 = 55.37 \text{ mg/L}$$

Thus, the amount of soda to be added is 55.37 mg/L as $CaCO_3$. The fact that this number is equal to the Mg^{2+} concentration is a coincidence of the numbers chosen.

Example 4-14. From the water analysis presented below, determine the amount of lime and soda (in mg/L as $CaCO_3$) necessary to soften the water to 90.00 mg/L as $CaCO_3$.

Water Composition (mg/L as $CaCO_3$)

Ca^{2+}:	149.2	CO_2:	29.3	HCO_3^-:	185.0
Mg^{2+}:	65.8			SO_4^{2-}:	29.8
Na^+:	17.4			Cl^-:	17.6

Solution. The bar chart for this water may be plotted directly as shown below.

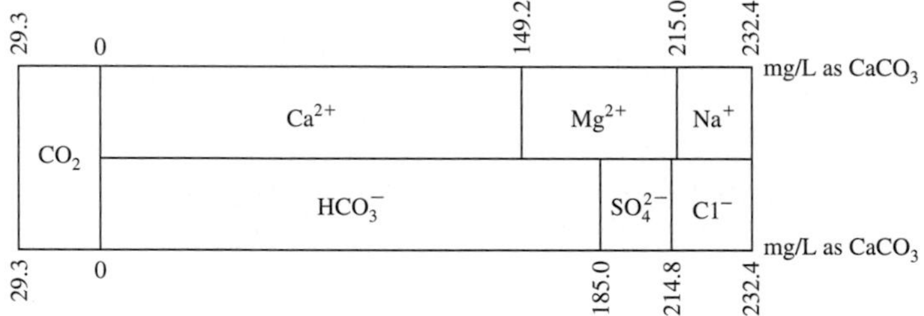

From the bar chart we note the following:

$$TH = 215.0 \text{ mg/L as } CaCO_3$$
$$CH = 185.0 \text{ mg/L as } CaCO_3$$
$$NCH = 30.0 \text{ mg/L as } CaCO_3$$

Following the logic of Figure 4-17, we calculate the lime dose as follows:

Step	Dose (mg/L as $CaCO_3$)
Lime = CO_2	29.3
Lime = HCO_3^-	185.0
Lime = $Mg^{2+} - 40 = 65.8 - 40 =$	25.8
Lime = excess	25.8
	265.9

The amount of lime to add is 265.9 mg/L as $CaCO_3$. The excess chosen was equal to the difference between the Mg^{2+} concentration and 40 since that difference was between 20 and 40, that is, $Mg^{2+} - 40 = 25.8$.

The amount of NCH_R is calculated as in Example 4-13:

$$NCH_f = 90.00 - 40.00 = 50.00$$
$$NCH_R = 30.00 - 50.00 = -20.00$$

Since NCH_R is a negative number, there is no need to remove NCH and, therefore, *no soda ash is required.*

Example 4-15. Given the following water, determine the amount (mg/L) of 90 percent purity CaO and 97 percent purity Na_2CO_3 that must be purchased to treat the water to a final hardness of 85 mg/L; 120 mg/L.

Ion	(mg/L as $CaCO_3$)
CO_2	21
HCO_3^-	209
Ca^{2+}	183
Mg^{2+}	97

Solution. First find the total hardness (TH), carbonate hardness (CH), and noncarbonate hardness (NCH):

$$TH = Ca^{2+} + Mg^{2+} = 183 + 97 = 280 \text{ mg/L}$$
$$CH = HCO_3^- = 209 \text{ mg/L}$$
$$NCH = TH - CH = 71 \text{ mg/L}$$

Use Figure 4-17 to find the lime dose as $CaCO_3$, assuming that we will leave 40 mg/L Mg^{2+} in the water:

$$Ca(OH)_2 = [21 + 209 + (97 - 40) + 40] = 327 \text{ mg/L as } CaCO_3$$

Since one mole of CaO equals one mole of $Ca(OH)_2$, we find 327 mg/L of CaO as $CaCO_3$. The molecular weight of CaO is 56 (equivalent weight = 28), and correcting for 90 percent purity:

$$CaO = 327\left(\frac{28}{50}\right)\left(\frac{1}{.9}\right) = 203 \text{ mg/L as CaO}$$

The amount of NCH that can be left in solution is equal to the final hardness desired (85 mg/L) minus the CH left behind due to solubility, inefficient mixing, etc. (40 mg/L), and is equal to 85 − 40 = 45 mg/L.

Therefore, the NCH_R is the initial NCH (71 mg/L) minus the NCH which can be left (45 mg/L) and is 71 − 45 = 26 mg/L. From Figure 4-18:

$$Na_2CO_3 = 26 \text{ mg/L as } CaCO_3$$

The equivalent weight of soda ash is 53 and the purity 97 percent:

$$Na_2CO_3 = 26\left(\frac{53}{50}\right)\left(\frac{1}{.97}\right) = 28 \text{ mg/L as } Na_2CO_3$$

If the final hardness desired is 120 mg/L, then the allowable final NCH is 120 − 40 = 80 mg/L, which is greater than the initial NCH of 71, so no soda ash is necessary. The final hardness would be about 40 mg/L (carbonate hardness solubility) plus 71 mg/L (noncarbonate hardness) or 111 mg/L.

More Advanced Concepts in Lime-Soda Softening

Estimating CO_2 Concentration. CO_2 is of importance in two instances in softening. In the first instance, it consumes lime that otherwise could be used to remove Ca and Mg. When the concentration of CO_2 exceeds 10 mg/L as CO_2 (22.7 mg/L as $CaCO_3$ or 0.45 meq/L), the economics of removal of CO_2 by aeration (stripping) are favored over removal by lime neutralization. In the second instance, CO_2 is used to neutralize the high pH of the effluent from the softening process. These reactions are an application of the concepts of the carbonate buffer system discussed in Section 4-1.

The concentration of CO_2 may be estimated by using the equilibrium expressions for the dissociation of water and carbonic acid with the definition of alkalinity (Equation 4-36). The pH and alkalinity of the water must be determined to make the estimate. Example 4-16 illustrates a simple case where one of the forms of alkalinity predominates.

Example 4-16. What is the estimated CO_2 concentration of a water with a pH of 7.65 and a total alkalinity of 310 mg/L as $CaCO_3$?

Solution. When the raw water pH is less than 8.5, we can assume that the alkalinity is predominately HCO_3^-. Thus, we can ignore the dissociation of bicarbonate to form carbonate.

With this assumption, the procedure to solve the problem is

a. Calculate the $[H^+]$ from the pH.

b. Calculate the $[HCO_3^-]$ from the alkalinity.

c. Solve the first equilibrium expression of the carbonic acid dissociation for $[H_2CO_3]$.

d. Use the assumption that $[CO_2] = [H_2CO_3]$ to estimate the CO_2 concentration.

Following this approach, the $[H^+]$ concentration is

$$[H^+] = 10^{-7.65} = 2.24 \times 10^{-8} \text{ moles/L}$$

The $[HCO_3^-]$ concentration is

$$[HCO_3^-] = 310 \text{ mg/L}\left(\frac{61 \text{ mg/meq}}{50 \text{ mg/meq}}\right)\left(\frac{1}{(61 \text{ g/mole})(10^3 \text{ mg/g})}\right)$$

$$= 6.20 \times 10^{-3} \text{ moles/L}$$

Since the alkalinity is reported as mg/L as $CaCO_3$, it must be converted to mg/L as the species using Equation 4-40 before the molar concentration may be calculated. The ratio 61/50 is the ratio of the equivalent weight of HCO_3^- to the equivalent weight of $CaCO_3$.

The equilibrium expression for the dissociation of carbonic acid is written in the form of Equation 4-33 using the reaction and pK_a given in Table 4-4.

$$K_a = \frac{[H^+][HCO_3^-]}{[H_2CO_3]}$$

where $K_a = 10^{-6.35} = 4.47 \times 10^{-7}$

Solving for $[H_2CO_3]$ gives

$$[H_2CO_3] = \frac{(2.24 \times 10^{-8} \text{ moles/L})(6.20 \times 10^{-3} \text{ moles/L})}{4.47 \times 10^{-7}}$$

$$[H_2CO_3] = 3.11 \times 10^{-4} \text{ moles/L}$$

We may assume that all the CO_2 in water forms carbonic acid. Thus, the estimated CO_2 concentration is

$$[CO_2] = 3.11 \times 10^{-4} \text{ moles/L}$$

In other units for comparison and calculation:

$$CO_2 = (3.11 \times 10^{-4} \text{ moles/L})(44 \times 10^3 \text{ mg/mole}) = 13.7 \text{ mg/L as } CO_2$$

$$CO_2 = (13.7 \text{ mg/L as } CO_2)\left(\frac{50 \text{ mg/meq}}{22 \text{ mg/meq}}\right) = 31.14 \text{ or } 31.1 \text{ mg/L as } CaCO_3$$

The equivalent weight of CO_2 is taken as 22 because it effectively behaves as carbonic acid (H_2CO_3) and thus $n = 2$.

Softening to Practical Limits. In Example 4-14, the NCH_R was found to be a negative number. This implies that the water will be softened to a hardness less than the desired value. One way to overcome this difficulty is to treat a portion of the water to the practical limits and then blend the treated water with the raw water to achieve the desired hardness. Unlike the flowchart method used in Examples 4-13 through 4-15, calculations of chemical additions to soften to the practical limits of softening (that is 0.60 meq/L or 30 mg/L as $CaCO_3$ of Ca and 0.20 meq/L or 10 mg/L as $CaCO_3$ of $Mg(OH)_2$), do not take into account the desired final hardness. Stoichiometric amounts of lime and soda are added to remove all of the Ca and Mg. Example 4-17 illustrates the technique using both mg/L as $CaCO_3$ and milliequivalents/L as units of measure.

Example 4-17. Determine the chemical dosages for softening the following water to the practical solubility limits.

Constituent	mg/L	EW	EW $CaCO_3$/EW ion	mg/L as $CaCO_3$	meq/L
CO_2	9.6	22.0	2.28	21.9	0.44
Ca^{2+}	95.2	20.0	2.50	238.0	4.76
Mg^{2+}	13.5	12.2	4.12	55.6	1.11
Na^+	25.8	23.0	2.18	56.2	1.12
Alkalinity				198	3.96
Cl^-	67.8	35.5	1.41	95.6	1.91
SO_4^{2-}	76.0	48.0	1.04	76.0	1.58

Bar chart of raw water in mg/L as $CaCO_3$:

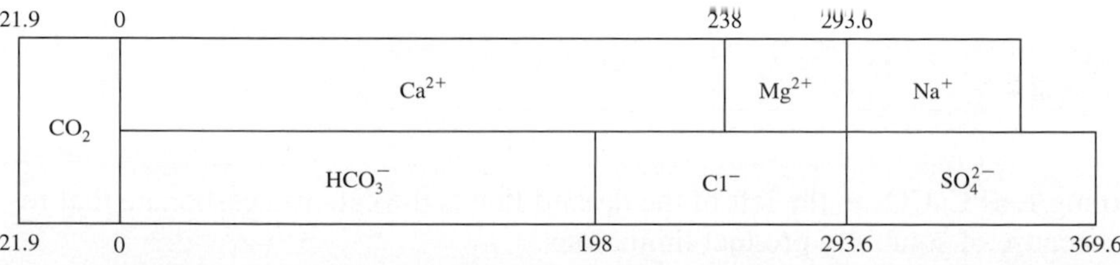

Solution. To soften to the practical solubility limits, lime and soda must be added as shown below.

Addition = to:	Lime mg/L as $CaCO_3$	Lime meq/L	Soda mg/L as $CaCO_3$	Soda meq/L
CO_2	21.9	0.44		
HCO_3^-	198.0	3.96		
Ca-HCO_3^-			40	0.80
Mg^{2+}	55.6	1.11	55.6	1.11
	275.5	5.51	95.6	1.91

Since the difference Mg − 40 = 15.6 mg/L as $CaCO_3$, the minimum excess lime of 20 mg/L as $CaCO_3$ is selected. The total lime addition is 295.5 mg/L as $CaCO_3$ or 165.5 mg/L as CaO. The soda addition is 95.6 mg/L as $CaCO_3$ or

$$95.6 \text{ mg/L as } CaCO_3(53/50) = 101.3 \text{ mg/L as } Na_2CO_3$$

Note that (53/50) is the equivalent weight of Na_2CO_3/equivalent weight of $CaCO_3$.
Reaction with CO_2:

$$CO_2 + Ca(OH)_2 \rightarrow CaCO_3(s) + H_2O$$

Bar chart after removal of CO_2:

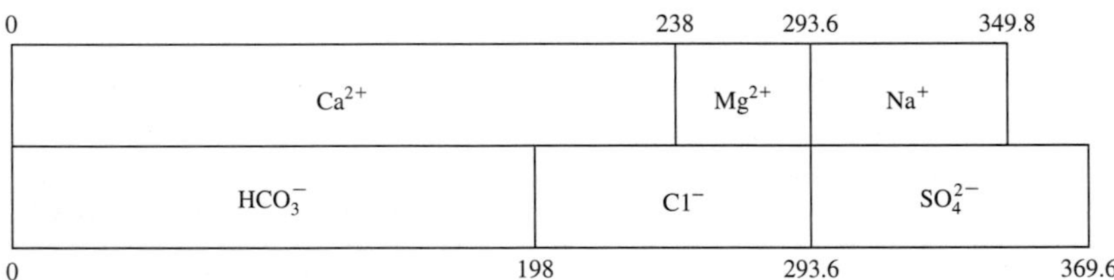

Reaction with HCO_3^-

$$Ca^{2+} + 2HCO_3^- + Ca(OH)_2 \rightarrow 2CaCO_3(s) + 2H_2O$$

Bar chart after reaction with HCO_3^-.

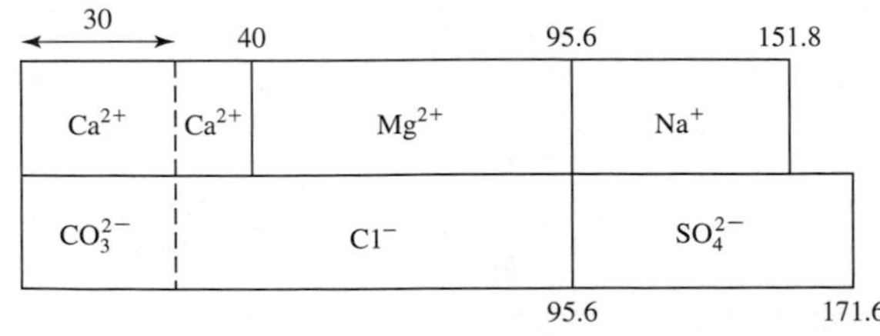

The 30 mg/L as $CaCO_3$ to the left of the dashed line is the calcium carbonate that remains because of solubility product limitations.

Reaction with calcium and soda:

$$Ca^{2+} + Na_2CO_3 \rightarrow CaCO_3(s) + 2Na^+$$

Bar chart after reaction with calcium and soda:

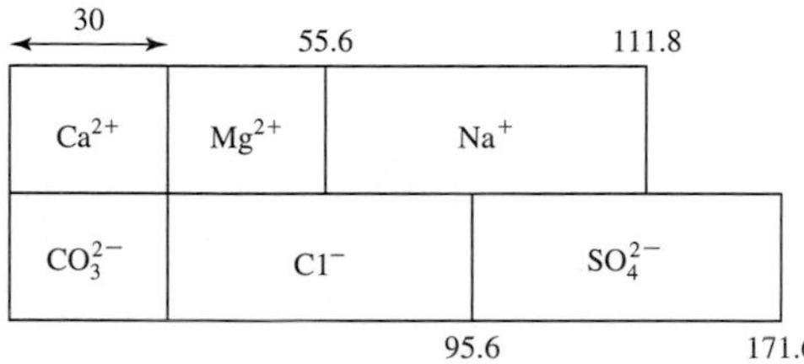

Reactions with magnesium, lime, and soda:

$$Mg^{2+} + Ca(OH)_2 \rightarrow Mg(OH)_2(s) + Ca^{2+}$$
$$Ca^{2+} + Na_2CO_3 \rightarrow CaCO_3(s) + 2Na^+$$

Bar chart of finished water:

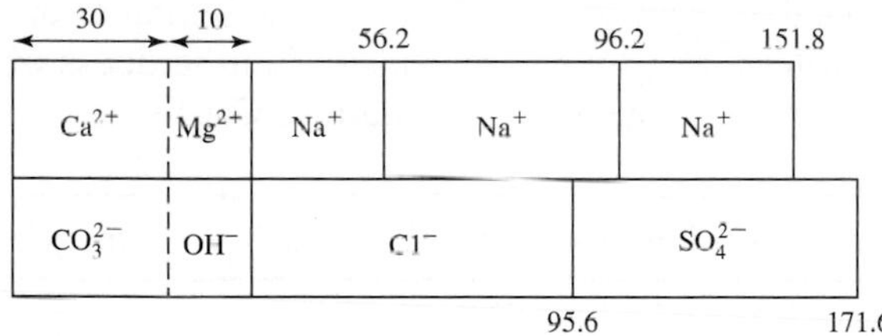

Total hardness of finished water is 30 mg/L as $CaCO_3$ + 10 mg/L as $CaCO_3$ = 40 mg/L as $CaCO_3$.

Split Treatment. As shown in Figure 4-19, in split treatment a portion of the raw water is by-passed around the softening reaction tank and the settling tank. This serves several functions. First, it allows the water to be tailored to yield a product water that has 0.80 meq/L or 40 mg/L as $CaCO_3$ of magnesium (or any other value above the solubility limit). Second, it allows for reduction in capital cost of tankage because the entire flow does not need to be treated. Third, it minimizes operating costs for chemicals by treating only a fraction of the flow. Fourth, it uses the natural alkalinity of the water to lower the pH of the product water and assist in stabilization. In many cases a second sedimentation basin is added after recarbonation and prior to filtration to reduce the solids loading onto the filters.

The fractional amount of the split is calculated as

$$X = \frac{Mg_f - Mg_i}{Mg_r - Mg_i} \qquad (4\text{-}71)$$

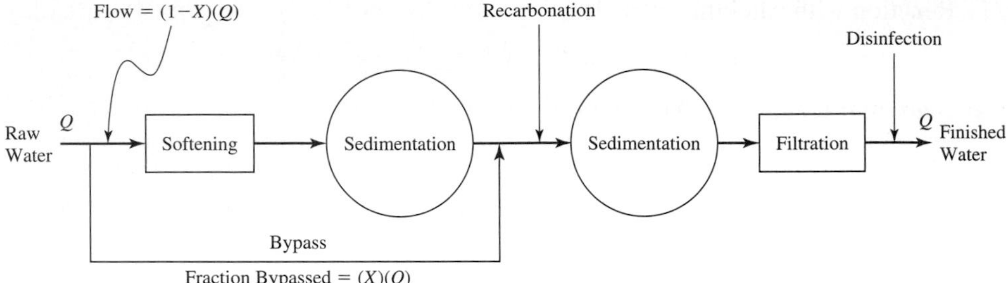

FIGURE 4-19
Split-flow treatment scheme.

where Mg_f = final magnesium concentration, mg/L as $CaCO_3$
Mg_i = magnesium concentration from first stage, mg/L as $CaCO_3$
Mg_r = raw water magnesium concentration, mg/L as $CaCO_3$

The first stage is operated to soften the water to the practical limits of softening. Thus, the value for Mg_i is commonly taken to be 10 mg/L as $CaCO_3$. Since the desired concentration of Mg is nominally set at 40 mg/L as $CaCO_3$ as noted previously, Mg_f is commonly taken as 40 mg/L as $CaCO_3$.

Example 4-18. Determine the chemical dosages for split treatment softening of the following water. The finished water criteria is a maximum magnesium hardness of 40 mg/L as $CaCO_3$ and a total hardness in the range 80 to 120 mg/L as $CaCO_3$.

Constituent	mg/L	EW	EW $CaCO_3$/EW ion	mg/L as $CaCO_3$	meq/L
CO_2	11.0	22.0	2.28	25.0	0.50
Ca^{2+}	95.2	20.0	2.50	238	4.76
Mg^{2+}	22.0	12.2	4.12	90.6	1.80
Na^+	25.8	23.0	2.18	56.2	1.12
Alkalinity				198	3.96
Cl^-	67.8	35.5	1.41	95.6	1.91
SO_4^{2-}	76.0	48.0	1.04	76.0	1.58

Solution. In the first stage the water is softened to the practical solubility limits; lime and soda must be added as shown below.

Addition = to:	Lime mg/L as $CaCO_3$	Lime meq/L	Soda mg/L as $CaCO_3$	Soda meq/L
CO_2	25.0	0.50		
HCO_3^-	198.0	3.96		
Ca-HCO_3^-			40	0.80
Mg^{2+}	90.6	1.80	90.6	1.80
	313.6	6.26	130.6	2.60

The split is calculated in terms of mg/L as $CaCO_3$:

$$X = (40 - 10)/(90.6 - 10) = 0.372$$

The fraction of water passing through the first stage is then $1 - 0.372 = 0.628$. The total hardness of the water after passing through the first stage is the practical solubility limit, that is, 40 mg/L as $CaCO_3$. Since the total hardness in the raw water is $238 + 90.6 = 328.6$ mg/L as $CaCO_3$, the mixture of the treated and bypass water has a hardness of:

$$(0.372)(328.6) + (0.628)(40) = 147.4 \text{ mg/L as } CaCO_3$$

This is above the specified finished water criteria range of 80–120 mg/L as $CaCO_3$, so further treatment is required. Since the split is designed to yield the required 40 mg/L as $CaCO_3$ of magnesium, more calcium must be removed. Removal of the calcium equivalent to the bicarbonate will leave 40 mg/L as $CaCO_3$ of calcium hardness plus the 40 mg/L as $CaCO_3$ of magnesium hardness for a total of 80 mg/L as $CaCO_3$. The additions are as follows.

Constituent	Lime mg/L as $CaCO_3$	Lime meq/L
CO_2	25.0	0.50
HCO_3^-	198.0	3.96
	223.0	4.46

Addition of lime = to CO_2 and HCO_3^- (even in second stage) is necessary to get pH high enough.

The total chemical additions are in proportion to the flows:

Lime = 0.628(313.6) + 0.372(223) = 280 mg/L as $CaCO_3$

Soda = 0.628(130.6) + 0.372(0.0) = 82 mg/L as $CaCO_3$

Cases. The selection of chemicals and their dosage depends on the raw water composition and the desired final water composition. If we use a Mg concentration of 40 mg/L as $CaCO_3$ as a product water criterion, then six cases illustrate the dosage schemes. Three of the cases occur when the Mg concentration is less than 40 mg/L as $CaCO_3$ (Figure 4-20a, b, and c) and three cases occur when Mg is greater than 40 mg/L as $CaCO_3$ (Figure 4-21a, b, and c). In the cases illustrated in Figure 4-20, no split treatment is required. Conversely, the cases illustrated in Figure 4-21 are for the first stage of a split treatment flow scheme (softening to the practical limits). In the cases illustrated in Figure 4-21, the hardness of the mixture of the treated and raw water must be checked to see if an acceptable hardness has been achieved. If the hardness after blending is above the desired concentration, then further softening in a second stage is required (Figure 4-22). Since the design of the split is to achieve a desired Mg concentration of 40 mg/L as $CaCO_3$, no further Mg removal is required. Only treatment of the Ca is required.

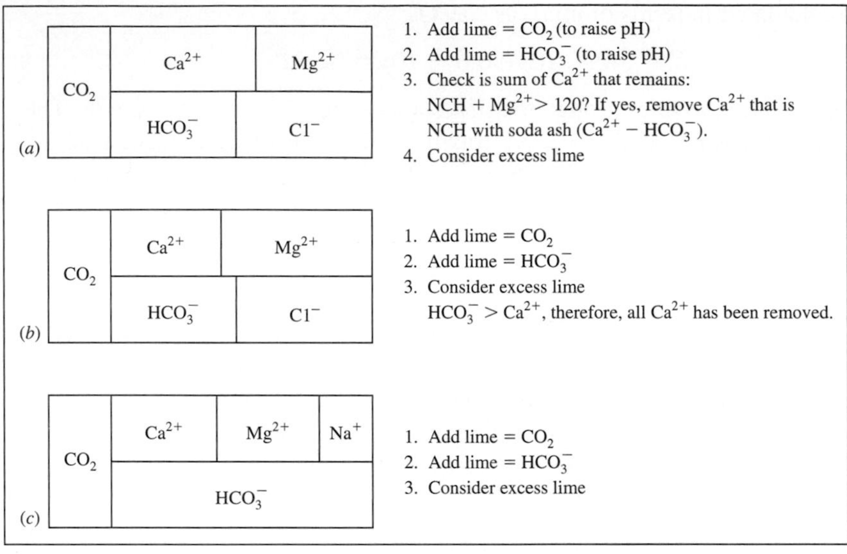

FIGURE 4-20
Dosage schemes when Mg^{2+} concentration is less than 40 mg/L as $CaCO_3$ and no split treatment is required. Note that no Mg^{2+} is removed and that reactions deal with CO_2 and Ca^{2+} only.

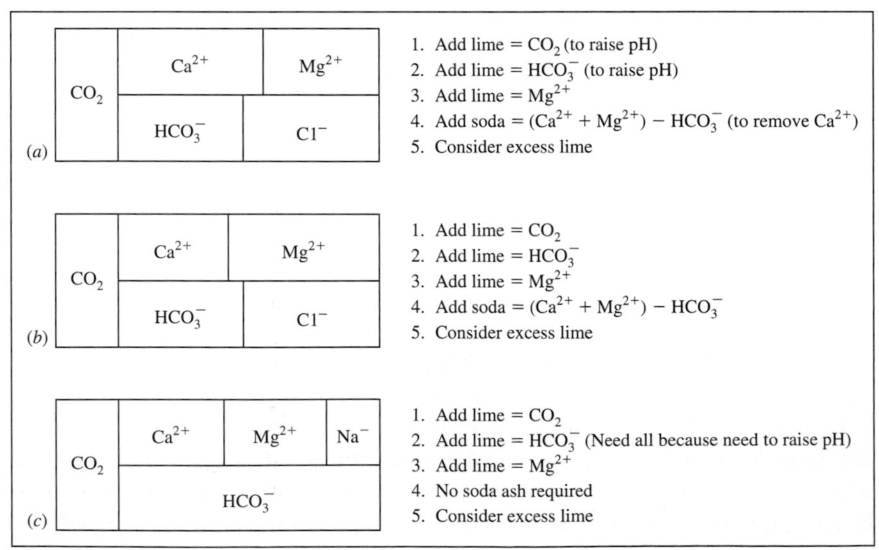

FIGURE 4-21
Cases when Mg^{2+} concentration is greater than 40 mg/L as $CaCO_3$ and split treatment is required. Note that these cases illustrate softening to the practical limits in the first stage of the split-flow scheme.

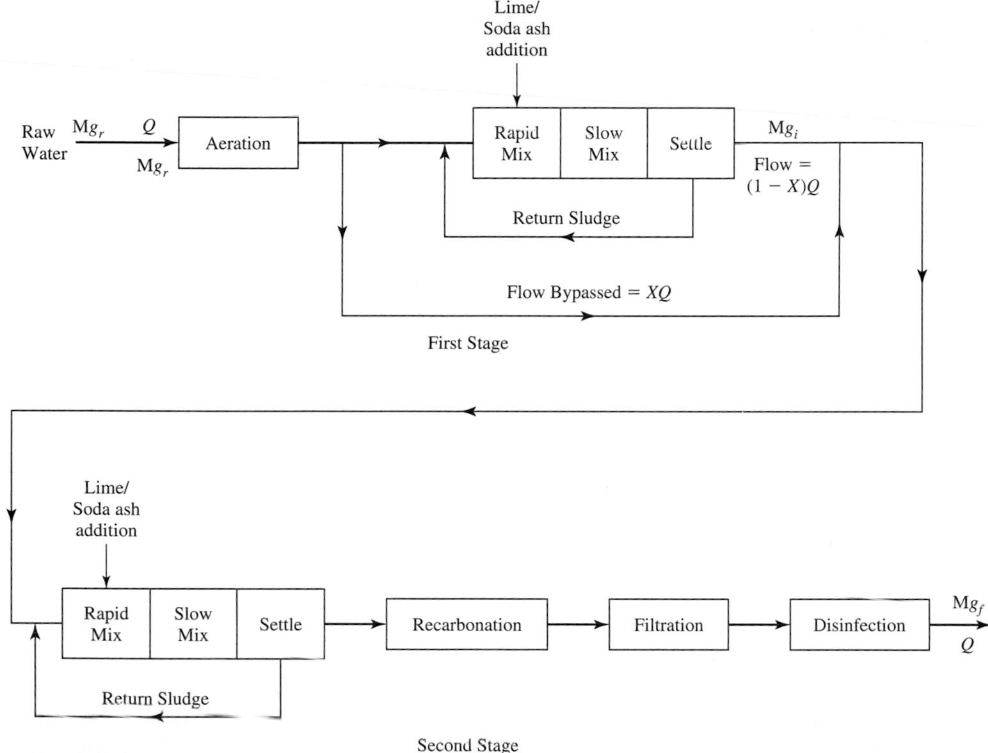

FIGURE 4-22
Flow diagram for a two stage split-treatment lime–soda ash softening plant. (Adapted from J. L. Cleasby and J. H. Dillingham, 1966.)

Ion-Exchange Softening

Ion exchange can be defined as the reversible interchange of ions between a solid and a liquid phase in which there is no permanent change in the structure of the solid. Typically, in water softening by ion exchange, the water containing the hardness is passed through a column containing the ion-exchange material. The hardness in the water exchanges with an ion from the ion-exchange material. Generally, the ion exchanged with the hardness is sodium, as illustrated in Equation 4-72:

$$Ca(HCO_3)_2 + 2NaR \rightleftharpoons CaR_2 + 2NaHCO_3 \tag{4-72}$$

where R represents the solid ion-exchange material. By the above reaction, calcium (or magnesium) has been removed from the water and replaced by an equivalent amount of sodium, that is, two sodium ions for each cation. The alkalinity remains unchanged. The exchange results in essentially 100 percent removal of the hardness from the water until the exchange capacity of the ion-exchange material is reached, as shown in Figure 4-23. When the ion-exchange material becomes saturated, no hardness will be removed. At this point *breakthrough* is said to have occurred because the hardness passes through the bed. At this point the column is taken out of service, and

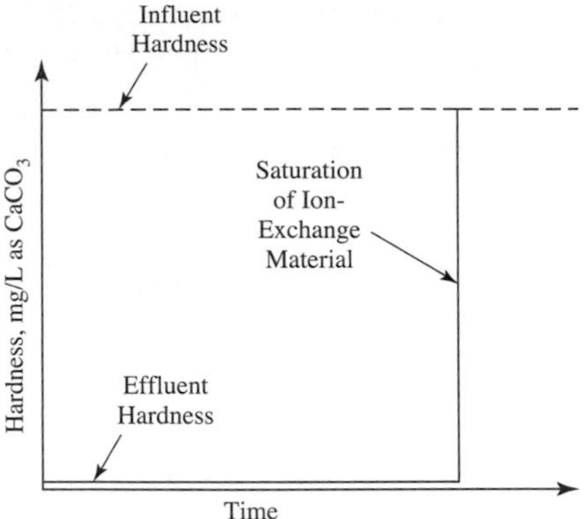

FIGURE 4-23
Hardness removal in ion-exchange column.

the ion-exchange material is regenerated. That is, the hardness is removed from the material by passing water containing a large amount of Na^+ through the column. The mass action of having so much Na^+ in the water causes the hardness of the ion-exchange material to enter the water and exchange with the sodium:

$$CaR_2 + 2NaCl \rightleftharpoons 2NaR + CaCl_2 \tag{4-73}$$

The ion-exchange material can now be used to remove more hardness. The $CaCl_2$ is a waste stream that must be disposed of.

There are some large water treatment plants that utilize ion-exchange softening, but the most common application is for residential water softeners. The ion-exchange material can either be naturally occurring clays, called *zeolites,* or synthetically made resins. There are several manufacturers of synthetic resins. The resins or zeolites are characterized by the amount of hardness that they will remove per volume of resin material and by the amount of salt required to regenerate the resin. The synthetically produced resins have a much higher exchange capacity and require less salt for regeneration. However, they also cost much more. People who work in the water softening industry often work in units of grains of hardness per gallon of water (gr/gal). It is useful to remember that 1 gr/gal equals 17.1 mg/L.

Since the resin removes virtually 100 percent of the hardness, it is necessary to bypass a portion of the water and then blend in order to obtain the desired final hardness.

$$\%\text{Bypass} = (100)\frac{\text{Hardness}_{\text{desired}}}{\text{Hardness}_{\text{initial}}} \tag{4-74}$$

Example 4-19. A home water softener has 0.1 m^3 of ion-exchange resin with an exchange capacity of 57 kg/m^3. The occupants use 2,000 L of water per day. If the water contains 280.0 mg/L of hardness as $CaCO_3$ and it is desired to soften it to 85 mg/L as $CaCO_3$, how much should be bypassed? What is the time between regeneration cycles?

Solution. The percentage of water to be bypassed is found using Equation 4-74:

$$\%\text{Bypass} = (100)\frac{85}{280} = 30.36 \text{ or } 30\%$$

The length of time between regeneration cycles is determined from the exchange capacity of the ion-exchange material (media). This is also called the "time to breakthrough," that is, the time to saturate the exchange material. If 30 percent of the water is being bypassed, then 70 percent of the water is being treated and the hardness loading rate is

$$\text{Loading rate} = (0.7)(280 \text{ mg/L})(2{,}000 \text{ L/d}) = 392{,}000 \text{ mg/d}$$

Since the bed capacity is 57 kg/m^3 and the bed contains 0.1 m^3 of ion-exchange media, the breakthrough time is approximately

$$\text{Breakthrough time} = \frac{(57 \text{ kg/m}^3)(0.1 \text{ m}^3)}{(392{,}000 \text{ mg/d})(10^{-6} \text{ kg/mg})} = 14.5 \text{ d}$$

4-4 MIXING AND FLOCCULATION

Clearly, if the chemical reactions in coagulating and softening a water are going to take place, the chemical must be mixed with the water. In this section we will begin to look at the physical methods necessary to accomplish the chemical processes of coagulation and softening.

Mixing, or *rapid mixing* as it is called, is the process whereby the chemicals are quickly and uniformly dispersed in the water. Ideally, the chemicals would be instantaneously dispersed throughout the water. During coagulation and softening the chemical reactions that take place in rapid mixing form precipitates. Either aluminum hydroxide or iron hydroxide form during coagulation, while calcium carbonate and magnesium hydroxide form during softening. The precipitates formed in these processes must be brought into contact with one another so that they can agglomerate and form larger particles, called *flocs.* This contacting process is called *flocculation* and is accomplished by slow, gentle mixing.

In the treatment of water and wastewater the degree of mixing is measured by the velocity gradient, G. The velocity gradient is best thought of as the amount of shear taking place; that is, the higher the G value, the more violent the mixing. The velocity gradient is a function of the power input into a unit volume of water. It is given by

$$G = \sqrt{\frac{P}{\mu \forall}} \tag{4-75}$$

where G = velocity gradient, s^{-1}
P = power input, W
$\forall$ = volume of water in mixing tank, m^3
μ = dynamic viscosity, Pa · s

From literature, experience, laboratory, or pilot plant work it is possible to select a G value for a particular application. The total number of particle collisions is proportional to $G\theta$, where θ is the detention time in the basin as given by Equation 2-27.

Rapid Mix

Rapid mixing is probably the most important physical operation affecting coagulant dose efficiency. The chemical reaction in coagulation is completed in less than 0.1 s;

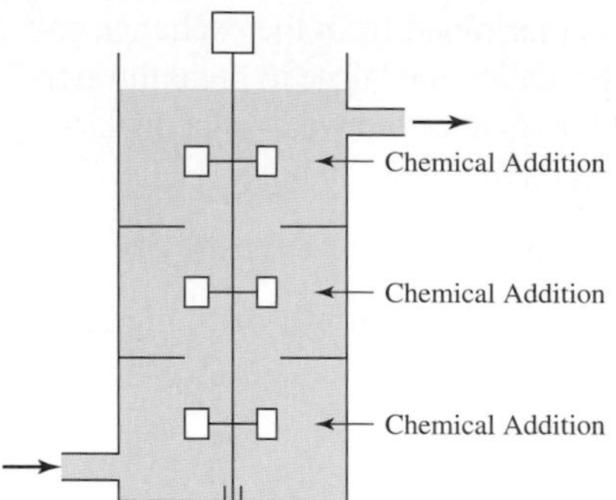

FIGURE 4-24
Rapid mix tank.

therefore, it is imperative that mixing be as instantaneous and complete as possible. Rapid mixing can be accomplished within a tank utilizing a vertical shaft mixer (Figure 4-24), within a pipe using an in-line blender (Figure 4-25), or in a pipe using a static mixer (Figure 4-26). Other methods, such as Parshall flumes, hydraulic jumps, baffled channels, or air mixing may also be used.

The selection of G and $G\theta$ values for coagulation is dependent on the mixing device, the chemicals selected, and the anticipated reactions. Coagulation occurs predominately by two mechanisms: adsorption of the soluble hydrolysis species on the colloid and destabilization or sweep coagulation where the colloid is trapped in the hydroxide precipitate. The reactions in adsorption-destabilization are extremely fast and occur within 1 second. Sweep coagulation is slower and occurs in the range of 1 to 7 seconds (Amirtharajah, 1978). Jar test data may be used to identify whether adsorption-destabilization or sweep coagulation is predominant. If charge reversal is apparent from the dose-turbidity curve (see, for example, Figure 4-13), then adsorption-destabilization is the predominant mechanism. If the dose-turbidity curve does not show charge reversal (that is, the curve is relatively flat at higher doses), then the predominant mechanism is sweep coagulation. G values in the range of 3,000 to 5,000 s^{-1} and detention times on the order of 0.5 s are recommended for adsorption-desorption

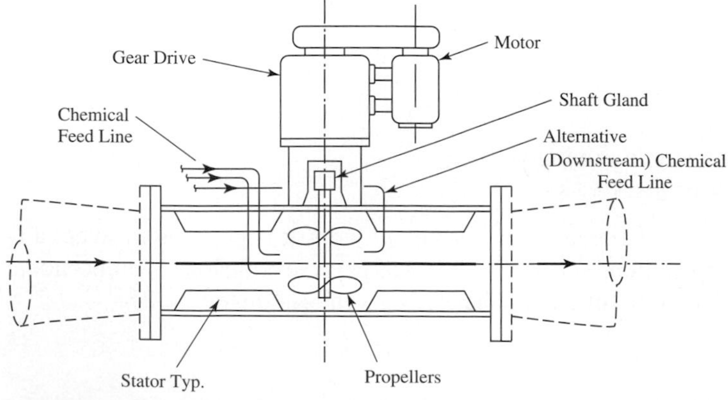

FIGURE 4-25
Typical in-line blender. (*Source:* AWWA, 1998.)

FIGURE 4-26
STATIC MIXER: A succession of reversing, flow-twisting, and flow-splitting elements provides positive dispersion proportional to number of elements. (*Source: Chemical Engineering,* March 22, 1971.)

reactions. These values are most commonly achieved in an in-line blender. For sweep coagulation, detention times of 1 to 10 s and G values in the range of 600 to 1,000 s^{-1} are recommended (Amirtharajah, 1978).

Softening. For dissolution of $CaO/Ca(OH)_2$ mixtures for softening, detention times on the order of 5 to 10 minutes may be required. G values to disperse and maintain particles in suspension may be on the order of 700 s^{-1}. In-line blenders are not used to mix softening reagents.

Rapid-Mix Tanks. The volume of a rapid-mix tank seldom exceeds 8 m^3 because of mixing equipment and geometry constraints. The mixing equipment consists of an electric motor, gear-type speed reducer, and either a radial-flow or axial-flow impeller as shown in Figure 4-27. The radial-flow impeller provides more turbulence and is preferred for rapid mixing. The tanks should be horizontally baffled into at least two and preferably three compartments in order to provide sufficient residence time. They are also baffled vertically to minimize vortexing. Chemicals should be added below the impeller, the point of the most mixing. Some unitless geometric ratios for both rapid mixing and flocculation are shown in Table 4-15. These values can be used to select the proper basin depth and surface area and the impeller diameter. For rapid mixing, in order to construct a reasonably sized basin, often more depth is required

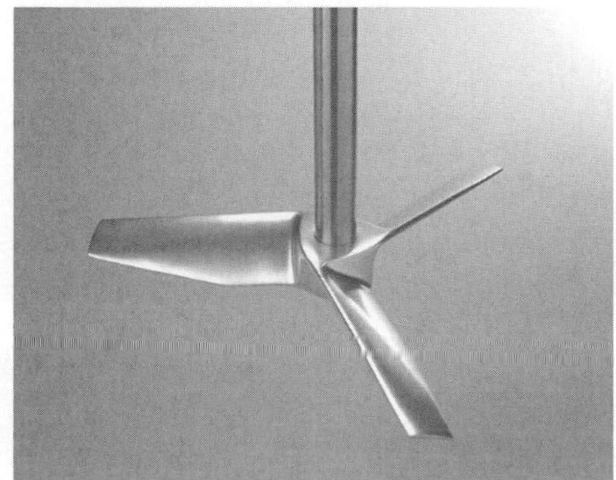

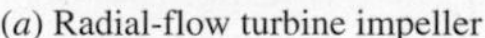

(*a*) Radial-flow turbine impeller (*b*) Axial-flow impeller

FIGURE 4-27
Basic impeller styles. (*Source:* Courtesy of SPX Process Equipment.)

TABLE 4-15
Tank and impeller geometries for mixing

Geometric Ratio	Allowed Range
D/T (radial)	0.14–0.5
D/T (axial)	0.17–0.4
H/D (either)	2–4
H/T (axial)	0.34–1.6
H/T (radial)	0.28–2
B/D (either)	0.7–1.6

D = impeller diameter
T = equivalent tank diameter
H = water depth
B = water depth below impeller

than allowed by the ratios in Table 4-15. In this case the tank is made deeper by using two impellers on the shaft. Rapid mixing is generally accomplished with a radial-flow-type turbine like that of Figure 4-27. When dual impellers are used, the top impeller is axial flow, while the bottom impeller is radial flow. Figure 4-28 shows the flow patterns of radial and axial flow impellers. When dual impellers are employed on gear-driven mixers, they are spaced approximately two impeller diameters apart. We normally assume an efficiency of transfer of motor power to water power of 0.8 for a single impeller.

Flocculation

While rapid mix is the most important physical factor affecting coagulant efficiency, flocculation is the most important factor affecting particle-removal efficiency. The

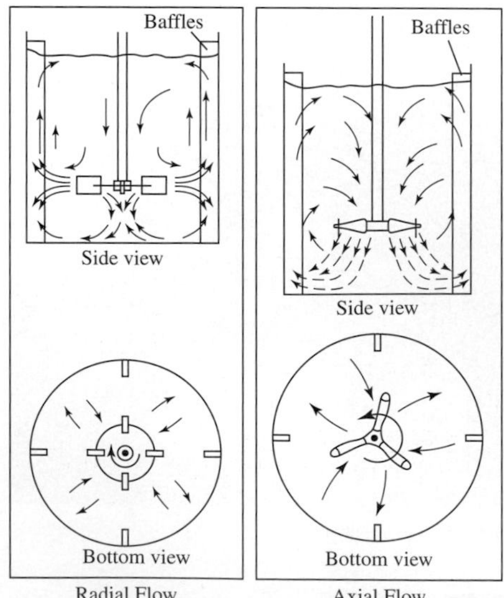

FIGURE 4-28
Basic flow patterns created by impellers. (*Source:* Cornwell and Bishop, 1983.)

TABLE 4-16
***Gθ* values for flocculation**

Type	G (s^{-1})	$G\theta$ (unitless)
Low-turbidity, color removal coagulation	20–70	60,000 to 200,000
High-turbidity, solids removal coagulation	30–80	36,000 to 96,000
Softening, 10% solids	130–200	200,000 to 250,000
Softening, 39% solids	150–300	390,000 to 400,000

objective of flocculation is to bring the particles into contact so that they will collide, stick together, and grow to a size that will readily settle. Enough mixing must be provided to bring the floc into contact and to keep the floc from settling in the flocculation basin. Too much mixing will shear the floc particles so that the floc is small and finely dispersed. Therefore, the velocity gradient must be controlled within a relatively narrow range. Flexibility should also be built into the flocculator so that the plant operator can vary the G value by a factor of two to three. The heavier the floc and the higher the suspended solids concentration, the more mixing is required to keep the floc in suspension. This is reflected in Table 4-16. Softening floc is heavier than coagulation floc and therefore requires a higher G value. An increase in the floc concentration (as measured by the suspended solids concentration) also increases the required G. With water temperatures of approximately 20°C, modern plants provide about 20 minutes of flocculation time (θ) at plant capacity. With lower temperatures, the detention time is increased. At 15°C the detention time is increased by 7 percent, at 10°C it is increased 15 percent, and at 5°C it is increased 25 percent.

Flocculation is normally accomplished with an axial-flow impeller (Figure 4-27), a paddle flocculator (Figure 4-29), or a baffled chamber (Figure 4-30). Axial-flow impellers have been recommended over the other types of flocculators because they impart a nearly constant G throughout the tank (Hudson, 1981). The flocculator basin

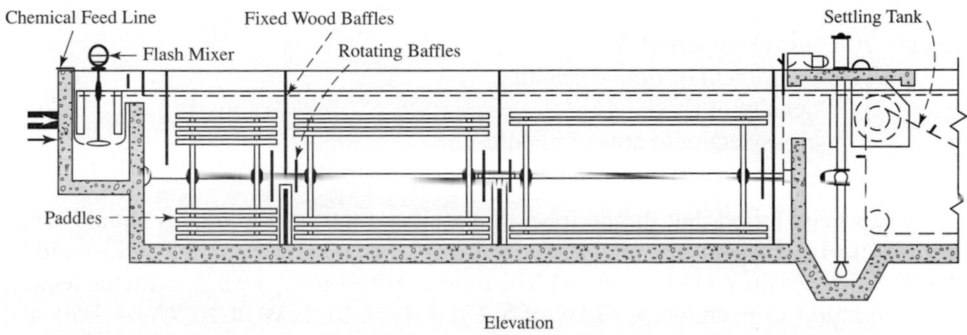

FIGURE 4-29
Paddle flocculator. (Courtesy of Envirex, Inc., a Rexnord Company.)

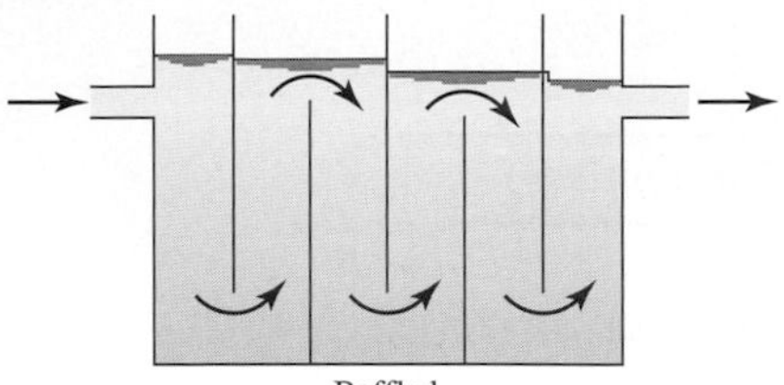

FIGURE 4-30
Baffled chamber flocculator.

should be divided into at least three compartments. The velocity gradient is tapered so that the G values decrease from the first compartment to the last and that the average of the compartments is the design value selected from Table 4-16.

Power Requirements

In the design of mixing equipment for rapid-mix and flocculation tanks, the power imparted to the liquid in a baffled tank by an impeller may be described by the following equation for fully turbulent flow developed by Rushton (1952).

$$P = N_p(n)^3(D_i)^5\rho \tag{4-76}$$

where P = power, W
N_p = impeller constant (also called power number)
n = rotational speed, revolutions/s
D_i = impeller diameter, m
ρ = density of liquid, kg/m^3

The impeller constant of a specific impeller can be obtained from the manufacturer. For the radial-flow impeller of Figure 4-27, the impeller constant is 5.7 and for the axial-flow impeller, it is 0.31. It is recommended that for flocculation, the tangential velocity (tip speed) be limited to 2.7 m/s.

The power imparted by a paddle mixer (Figure 4-29) is a function of the drag force on the paddles

$$P = \frac{C_D A\rho(v_p)^3}{2} \tag{4-77}$$

where P = power imparted, W
C_D = coefficient of drag of paddle
ρ = density of fluid, kg/m^3
A = cross-sectional area of paddles, m^2
v_p = relative velocity of paddles with respect to fluid, m/s

It has been found that the peripheral velocity of the paddle blades should range from 0.1 to 1.0 m/s and that the relative velocity of the paddles to the fluid should be 0.6 to 0.75 times the paddle-tip speed. The drag coefficient (C_D) varies with the length-to-width ratio (for example: for L:W of 5, C_D = 1.20, for L:W of 20, C_D = 1.50, and for L:W of infinity, C_D = 1.90). It is also recommended that the total paddle-blade area on a horizontal shaft not exceed 15 to 20 percent of the total basin cross-sectional area to avoid excessive rotational flow.

For pneumatic mixing, the power imparted is given by

$$P = KQ_a \ln\left(\frac{h + 10.33}{10.33}\right) \tag{4-78}$$

where P = power imparted, W
K = constant = 1.689
Q_a = air flow rate at atmospheric pressure, m^3/min
h = air pressure at the point of discharge, m

The power imparted by static-mixing devices may be computed as

$$P = \gamma Q h \tag{4-79}$$

where P = power imparted, kW
γ = specific weight of fluid, kN/m^3
Q = flow rate, m^3/s
h = head loss through the mixer, m

The specific weight of water is equal to the product of the density and the acceleration due to gravity ($\gamma = \rho g$). At normal temperatures, the specific weight of water is taken to be 9.81 kN/m^3.

Upflow Solids-Contact. Mixing, flocculation, and clarification may be conducted in a single tank such as that in Figure 4-31. The influent raw water and chemicals are mixed in the center cone-like structure. The solids flow down under the cone (sometimes called a "skirt"). As the water flows upward, the solids settle to form a *sludge blanket.* This design is called an upflow solids-contact basin. The main advantage of this unit is its reduced size. The units are best suited to treat a feed water that has a relatively constant quality. It is often favored for softening because the water quality from wells is relatively constant and the sludge blanket provides a further opportunity to drive the precipitation reactions to completion.

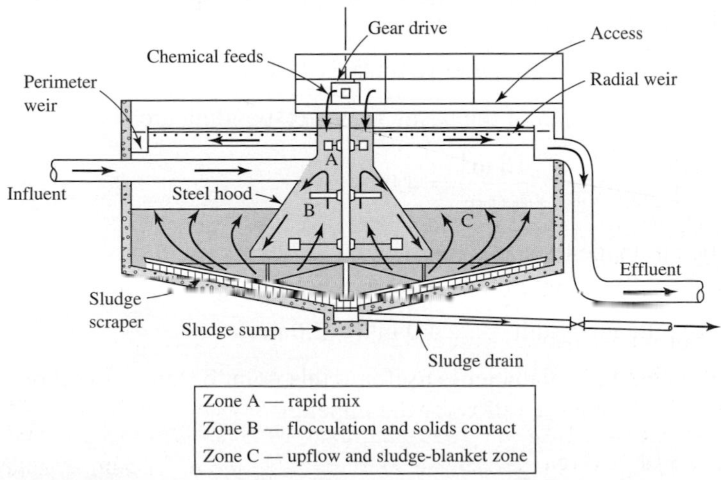

FIGURE 4-31
Typical upflow solids-contact unit. (*Source:* AWWA, 1990.)

Example 4-20. A city is planning the installation of a new water treatment plant to supply a growing population. There will be only one rapid mix basin and then the flow will be evenly split between two flocculator trains, each with three basins equal in volume. The required water depth of all basins is 4.0 m. Determine correct basin volumes, basin dimensions, tank equivalent diameter, required input power, impeller diameter from table below, and rotational speed using the following parameters:

Q = 11.5×10^3 m^3/d

Rapid mix θ = 2 min.

Rapid mix $G = 600$ s^{-1}

Total flocculation θ = 30 min

Flocculators G = 70, 50, 30 s^{-1}

Water temperature = 5°C

Place impeller at one-third the water depth

Impeller type	Impeller Diameters (m)			Power number (N_p)
Radial	0.8	1.1	1.4	5.7
Axial	0.8	1.4	2.0	0.31

Solution:

a. **Rapid-mix design**

Convert 11.5×10^3 m^3/d to m^3/min:

$$\frac{11.5 \times 10^3 \text{ m}^3}{d} \times \frac{d}{24 \text{ hr}} \times \frac{\text{h}}{60 \text{ min}} = 7.986 \text{ or } 8.0 \text{ m}^3/\text{min}$$

Now with Equation 2-27 we can determine the volume of the rapid mix basin using the flow rate and detention time.

$$\forall = (Q)(\theta) = (8.0 \text{ m}^3/\text{min})(2 \text{ min}) = 16.0 \text{ m}^3$$

Based on the required water depth of the basin, the corresponding area is:

$$\text{Area} = \frac{16 \text{ m}^3}{4.0 \text{ m}} = 4.0 \text{ m}^2$$

It is common practice to make the length and width of the mixing basins equal. Therefore, the square root of the required area will yield the side dimension.

$$\text{Length and width} = \sqrt{4.0 \text{ m}^2} = 2.0 \text{ m}$$

Calculating the equivalent tank diameter is useful at this point because it will be needed when calculating the geometric ratios for the impeller.

$$T_E = \sqrt{\frac{4 \times \text{Area}}{\pi}} = \sqrt{\frac{4 \times 4.0 \text{ m}^2}{\pi}} = 2.26 \text{ m}$$

The required input power can be calculated by using equation 4-75. Using Table A-1 in Appendix A and the temperature of the water, find the dynamic viscosity of 1.52×10^{-3} Pa · s. As noted in the table footnote, the values in the table are multiplied by 10^{-3}.

$$P = G^2\mu V = (600\ \text{s}^{-1})^2(1.52 \times 10^{-3}\ \text{Pa}\cdot\text{s})(16.0\ \text{m}^3) = 8{,}755\ \text{W}$$

By using Table 4-15, we can evaluate different size radial impellers using the geometric ratios. Below is a comparison of the ratios for the available sizes of radial impellers and the rapid mix basin.

Geometric Ratio	Allowable Range	Radial Impeller Diameter		
		0.8 m	1.1 m	1.40 m
D/T	0.14–0.5	0.35	0.49	0.62
H/D	2.0–4.0	5.00	3.64	2.85
H/T	0.28–2.0	1.77	1.77	1.77
B/D	0.7–1.6	1.67	1.21	0.95

The shaded areas indicate values that are not within the range, which leaves the 1.1-m radial impeller as best suited for this application. Finally, the required rotational speed for the impeller is calculated by using Equation 4-76.

$$n = \left[\frac{P}{N_p\rho(D_i)^5}\right]^{1/3} = \left[\frac{8{,}755\ \text{W}}{(5.7)(1{,}000\ \text{kg/m}^3)(1.1\ \text{m})^5}\right]^{1/3} = 0.98\ \text{rps} = 59\ \text{rpm}$$

***b.* Flocculator design**

Because the flow is evenly split between two flocculation trains, the flow rate from the rapid mix calculations should be divided by 2.

$$Q_{\text{per train}} = \frac{8.0\ \text{m}^3/\text{min}}{2} = 4.0\ \text{m}^3/\text{min}$$

Using Equation 2-27 again, we can find the needed basin volume for each flocculation train.

$$V_T = (Q)(\theta) = (4.0\ \text{m}^3/\text{min})(30\ \text{min}) = 120\ \text{m}^3$$

Because we need three equal flocculator tanks, the total volume per train is divided by 3.

$$V_{\text{per basin}} = \frac{V_T}{3} = \frac{120\ \text{m}^3}{3} = 40\ \text{m}^3$$

Using the same criteria as before we solve for the area, length, width, and equivalent tank diameter of the basins.

$$\text{Area} = \frac{40.0\ \text{m}^3}{4.0\ \text{m}} = 10.0\ \text{m}^2$$

$$\text{Length and width} = \sqrt{10.0 \text{ m}^2} = 3.16 \text{ m}$$

$$T_E = \sqrt{\frac{4 \times \text{Area}}{\pi}} = \sqrt{\frac{4 \times 10.0 \text{ m}^2}{\pi}} = 3.57 \text{ m}$$

The input power needed is going to be calculated three times because each flocculator has a unique velocity gradient.

$$P_{G=70} = G^2\mu V = (70 \text{ s}^{-1})^2(1.52 \times 10^{-3} \text{ Pa} \cdot \text{s})(40.0 \text{ m}^3) = 298 \text{ W}$$
$$P_{G=50} = G^2\mu V = (50 \text{ s}^{-1})^2(1.52 \times 10^{-3} \text{ Pa} \cdot \text{s})(40.0 \text{ m}^3) = 152 \text{ W}$$
$$P_{G=30} = G^2\mu V = (30 \text{ s}^{-1})^2(1.52 \times 10^{-3} \text{ Pa} \cdot \text{s})(40.0 \text{ m}^3) = 54.7 \text{ W}$$

Select an impeller that will work with the flocculator basins by using the geometric ratios for an axial impeller.

Geometric Ratio	Allowable Range	Axial Impeller Diameter		
		0.8 m	1.4 m	2.0 m
D/T	0.17–0.4	0.22	0.39	0.56
H/D	2.0–4.0	5.00	2.86	2.00
H/T	0.34–1.6	1.12	1.12	1.12
B/D	0.7–1.6	1.66	0.95	0.67

Notice the gray areas again; this table shows that a 1.4-m impeller would best fit the tank sizes.

$$n_{G=70} = \left[\frac{P}{N_p \rho (D_i)^5}\right]^{1/3} = \left[\frac{298 \text{ W}}{(0.31)(1{,}000 \text{ kg/m}^3)(1.4 \text{ m})^5}\right]^{1/3} = 0.56 \text{ rps} = 34 \text{ rpm}$$

For flocculation, the tangential velocity (tip speed) must not exceed 2.7 m/s. If this is true for the flocculator with a $G = 70 \text{ s}^{-1}$, then it is true for the others since their rotational speed is lower.

$$\text{Tip speed} = (\text{rps})(\pi \times D_i) = (0.56)(\pi \times 1.4) = 2.46 \text{ or } 2.5 \text{ m/s}$$

4-5 SEDIMENTATION

Overview

Particles that will settle within a reasonable period of time can be removed in a sedimentation basin (also called a clarifier). Sedimentation basins are usually rectangular or circular with either a radial or upward water flow pattern. Regardless of the type of basin, the design can be divided into four zones: inlet, settling, outlet, and sludge storage. While our intent is to present the concepts of sedimentation and to design a sedimentation tank, a brief discussion of all four zones is helpful in understanding the sizing of the settling zone. A schematic showing the four zones is shown in Figure 4-32.

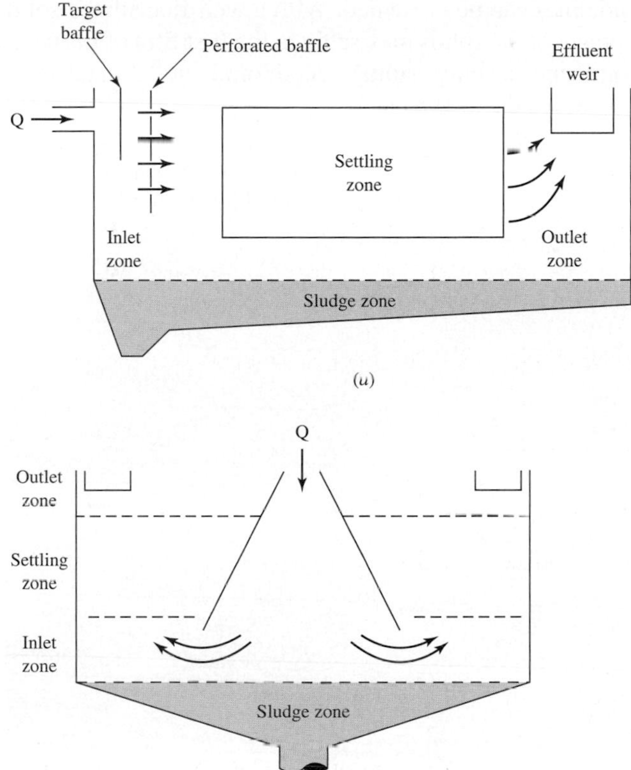

FIGURE 4-32
Zones of sedimentation: (*a*) horizontal flow clarifier; (*b*) upflow clarifier.

The purpose of the inlet zone is to evenly distribute the flow and suspended particles across the cross section of the settling zone.* The inlet zone consists of a series of inlet pipes and baffles placed about 1 m into the tank and extending the full depth of the tank. Following the baffle system, the flow takes on a flow pattern determined by the inlet structure. At some point the flow pattern is evenly distributed, and the water velocity slowed to the design velocity of the sedimentation zone. At that point the inlet zone ends and the settling zone begins. With a well-designed inlet baffle system, the inlet zone extends approximately 1.5 m down the length of the tank. Proper inlet zone design may well be the most important aspect of removal efficiency.

With improper design, the inlet velocities may never subside to the settling-zone design velocity. Typical design numbers are usually conservative enough that an inlet zone length does not have to be added to the length calculated for the settling zone. In an accurate design, the inlet and settling zones are each designed separately and their lengths added together.

The configuration and depth of the sludge storage zone depends upon the method of cleaning, the frequency of cleaning, and the quantity of sludge estimated to be produced. All these variables can be evaluated and a sludge storage zone designed. In lieu of these

*The cross section is the area through which the flow moves. For example, in Figure 4-32*a* the cross section is the settling zone width × depth and in 4-32b it is the bottom circular area.

design details, some general guidelines can be presented. With a well-flocculated solid and good inlet design, over 75 percent of the solids may settle in the first fifth of the tank. For coagulant floc, Hudson recommends a sludge storage depth of about 0.3 m near the outlet and 2 m or more near the inlet (Hudson, 1981).

If the tank is long enough, storage depth can be provided by bottom slope; if not, a sludge hopper is necessary at the inlet end. Mechanically-cleaned basins may be equipped with a bottom scraper, such as shown in Figure 4-33. The sludge is continu-

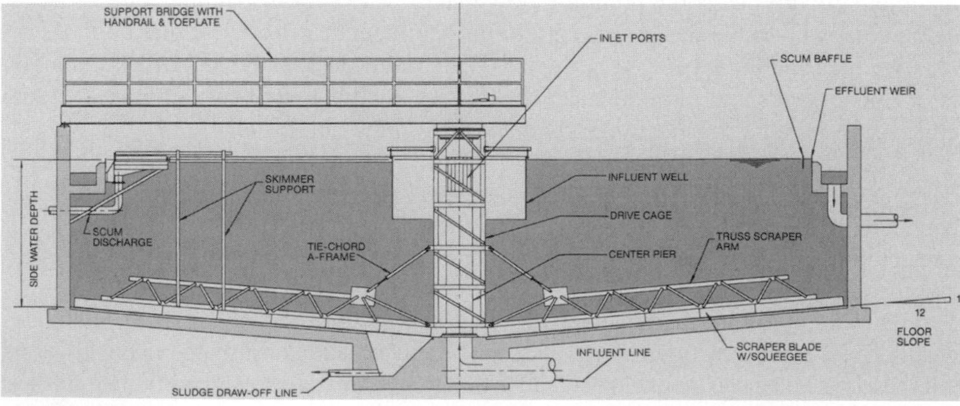

FIGURE 4-33

Photograph and schematic diagram of circular sludge scraper. (*Source:* David Cornwell/Courtesy of Siemens Water Technologies.)

ously scraped to a hopper where it is pumped out. For mechanically cleaned basins, a one percent slope toward the sludge withdrawal point is used. A sludge hopper is designed with sides sloping with a vertical to horizontal ratio of 1.2:1 to 2:1.

The outlet zone is designed so as to remove the settled water from the basin without carrying away any of the floc particles. A fundamental property of water is that the velocity of flowing water is proportional to the flow rate divided by the area through which the water flows, that is,

$$v = \frac{Q}{A_c} \tag{4-80}$$

where v = water velocity, m/s
Q = water flow, m^3/s
A_c = cross-sectional area, m^2

Within the sedimentation tank, the flow is going through a very large area (basin depth times width); consequently, the velocity is slow. To remove the water from the basin quickly, it is desirable to direct the water into a pipe or small channel for easy transport, which will produce a significantly higher velocity. If a pipe were to be placed at the end of the sedimentation basin, all the water would "rush" to the pipe. This rushing water would create high velocity profiles in the basin, which would tend to raise the settled floc from the basin and into the effluent water. This phenomenon of washing out the floc is called *scouring,* and one way to create scouring is with an improper outlet design. Rather than put a pipe at the end of the sedimentation basin, it is desirable to first put a series of troughs, called *weirs,* which provide a large area for the water to flow through and minimize the velocity in the sedimentation tank near the outlet zone. The weirs then feed into a central channel or pipe for transport of the settled water. Figure 4-34 shows various weir arrangements. The length of weir required is a function of the type of solids. The heavier the solids, the harder it is to scour them, and the higher the allowable outlet velocity. Therefore, heavier particles require a shorter length of weir than do light particles. Each state generally has a set of standards which must be followed, but Table 4-17 shows typical design values for weir loadings. The units for weir overflow rates are $m^3/d \cdot m$, which is water flow (m^3/d) per unit length of weir (m).

TABLE 4-17
Typical weir overflow rates

Type of floc	Weir overflow rate ($m^3/d \cdot m$)
Light alum floc (low-turbidity water)	143–179
Heavier alum floc (higher-turbidity water)	179–268
Heavy floc from lime softening	268–322

Source: Walker Process Equipment, Inc., 1973.

(*a*)

(*b*)

FIGURE 4-34
Weir arrangements: (*a*) rectangular; (*b*) circular. (*Source:* David Cornwell/ MacKenzie L. Davis.)

In a rectangular basin, the weirs should cover at least one-third, and preferably up to one-half, of the basin length. Spacing may be as large as 5 to 6 m on-centers but is frequently on the order of one-half this distance.

Example 4-21. The town of Urbana has a low-turbidity raw water and is designing its overflow weir at a loading rate of 150 $m^3/d \cdot m$. If its plant flow rate is 0.5 m^3/s, how many linear meters of weir are required?

Solution

$$\frac{(0.5\ m^3/s)(86{,}400\ s/d)}{150\ m^3/d \cdot m} = 288\ m$$

Sedimentation Concepts

There are two important terms to understand in sedimentation zone design. The first is the particle (floc) *settling velocity,* v_s. The second is the velocity at which the tank is designed to operate, called the *overflow rate,* v_o. The easiest way to understand these two concepts is to view an upward-flow sedimentation tank as shown in Figure 4-35. In this design, the particles fall downward and the water rises vertically. The rate at which the particle is settling downward is the particle-settling velocity, and the velocity of the liquid rising is the overflow rate. Obviously, if a particle is to be removed from the bottom of the clarifier and not go out in the settled water, then the particle-settling velocity must be greater than the liquid-rise velocity ($v_s > v_o$). If v_s is greater than v_o, one would expect 100 percent particle removal, and if v_s is less than v_o, one would expect 0 percent removal. In design, the procedure would be to determine the particle-settling velocity and set the overflow rate at some lower value. Often v_o is set at 50 to 70 percent of v_s for an *upflow clarifier.*

Let us now consider why the liquid-rise velocity is called an overflow rate and what its units are. The term *overflow rate* is used since the water is flowing over the top of the tank into the weir system. It is sometimes referred to as the *surface loading* rate because it has units of $m^3/d \cdot m^2$. The units are flow of water (m^3/d) being applied to a

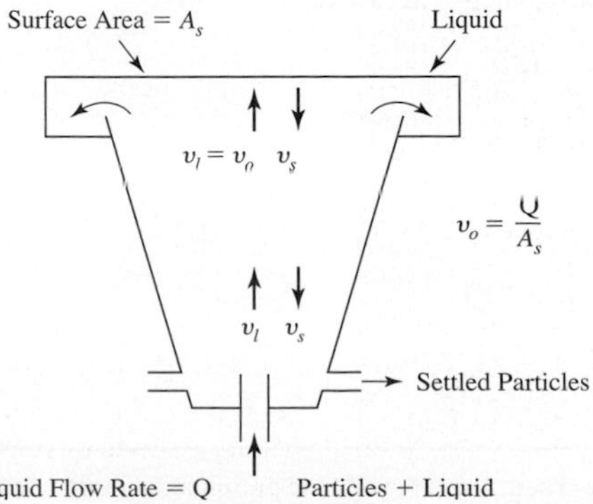

FIGURE 4-35
Settling in an upflow clarifier. (Legend: v_l = velocity of liquid; v_s = terminal settling velocity of particle.)

m^2 of surface area. This can be thought of as the amount of water that goes through each m^2 of tank surface area per day, which is similar to a loading rate. Recall from Equation 4-80 that the velocity of flow is equal to the flow rate divided by the area through which it flows. Hence an overflow rate is the same as a liquid velocity:

$$v_o = \frac{\text{Volume/Time}}{\text{Surface Area}} = \frac{(\text{Depth})(\text{Surface area})}{(\text{Time})(\text{Surface area})} = \frac{\text{Depth}}{\text{Time}} = \text{Liquid velocity} \quad (4\text{-}81)$$

$$v_o = \frac{\rlap{--}V/\theta}{A_s} = \frac{(h)(A_s)}{(\theta)(A_s)} = \frac{h}{\theta}$$

It can be seen from the above discussion that particle removal is independent of the depth of the sedimentation tank. As long as v_s is greater than v_o, the particles will settle downward and be removed from the bottom of the tank regardless of the depth. Sedimentation zones vary from a depth of a few centimeters to a depth of 6 m or greater.*

We can show that particle removal in a horizontal sedimentation tank is likewise dependent only upon the overflow rate. An ideal horizontal sedimentation tank is based upon three assumptions (Hazen, 1904, and Camp, 1946):

1. Particles and velocity vectors are evenly distributed across the tank cross section. This is the function of the inlet zone.
2. The liquid moves as an ideal slug down the length of the tank.
3. Any particle hitting the bottom of the tank is removed.

Using Figure 4-36*a* to illustrate the concept, let us consider a particle which is released at point A. In order to be removed from the water it must have a settling velocity great enough so that it reaches the bottom of the tank during the detention time (θ) of the water in the tank. We may say its settling velocity must equal the depth of the tank divided by the detention time, that is,

$$v_s = \frac{h}{\theta} \quad (4\text{-}82)$$

Now we can also show that the settling velocity of the particle must be equal to or greater than the overflow rate of the tank in order to be removed. Using the definition of detention time from Equation 2-27 and substituting into Equation 4-82:

$$v_s = \frac{h}{(\rlap{--}V/Q)} = \frac{hQ}{\rlap{--}V} \quad (4\text{-}83)$$

But we know that the tank volume is described by the product of the height, length, and width so we can rewrite this as:

$$v_s = \frac{hQ}{l \times w \times h} = \frac{Q}{l \times w} \quad (4\text{-}84)$$

*Tube settlers are designed with a very shallow settling zone. Their use is beyond the presentation of this text.

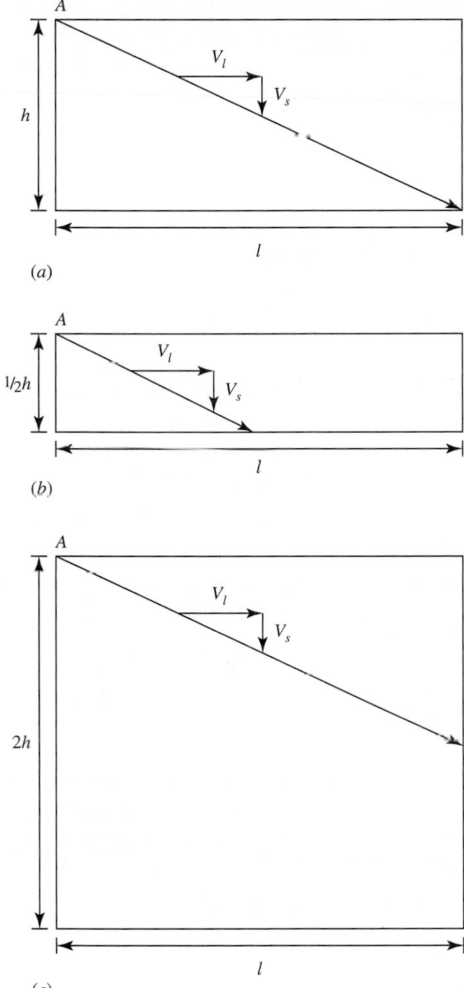

FIGURE 4-36
Ideal horizontal sedimentation tank.

And the product ($l \times w$) is the surface area (A_s) so

$$v_s = \frac{Q}{A_s} \tag{4-85}$$

which is the overflow rate (v_o). This implies that the removal of a horizontal clarifier is independent of depth! This is indeed strange and runs counter to our intuitive feeling of how the sedimentation tank should work. Why is this so? Figure 4-36*b* should help clear up this apparent ambiguity. What we see is that particles with v_s greater than or equal to v_o are removed in tanks having a depth equal to one-half the depth of Figure 4-36*a*. If the depth were greater, particles having settling velocities equal to v_o would not be completely removed (Figure 4-36*c*). But some removal would take place since those particles entering the tank at lower depths would have the correct trajectory to reach the bottom. This leads us to the concept of partial removal.

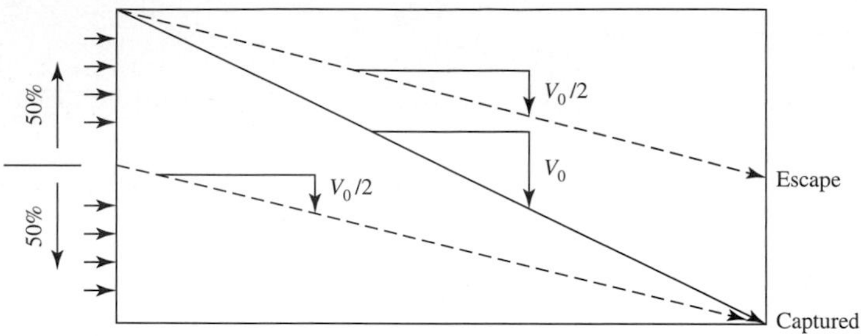

FIGURE 4-37
Partial solids removal in ideal sedimentation tank.

In a horizontal sedimentation tank, unlike an upflow clarifier, some percentage of the particles with a v_s less than v_o will be removed. For example, consider particles having a settling velocity of 0.5 v_o entering uniformly into the settling zone. Figure 4-37 shows that 50 percent of these particles (those below half the depth of the tank) will be removed. That is, they will hit the bottom of the tank before being carried out because they only have to settle one-half the tank depth. Likewise, one-fourth of the particles having a settling velocity of 0.25 v_o will be removed. The percentage of particles removed, P, with a settling velocity of v_s in a sedimentation tank designed with an overflow rate of v_o is

$$P = 100\frac{v_s}{v_o} \tag{4-86}$$

Example 4-22. The town of San Jose has an existing horizontal-flow sedimentation tank with an overflow rate of 17 $m^3/d \cdot m^2$, and it wishes to remove particles that have settling velocities of 0.1 mm/s, 0.2 mm/s, and 1 mm/s. What percentage of removal should be expected for each particle in an ideal sedimentation tank?

Solution

a. $v_s = 0.1$ mm/s

First we need to convert the overflow rate to compatible units:

$$17\frac{\text{m}^3}{\text{d}\cdot\text{m}^2} = 17\frac{\text{m}}{\text{d}}\frac{(1{,}000\text{ mm/m})}{(86{,}400\text{ s/d})} = 0.2\text{ mm/s}$$

Since $v_s < v_o$ for a v_s of 0.1 mm/s, some fraction of the particles will be removed, as given by Equation 4-86.

$$P = 100\frac{(0.1)}{(0.2)} = 50\%$$

b. $v_s = 0.2$ mm/s

These particles have $v_s = v_o$, and ideally will be 100 percent removed.

c. $v_s = 1$ mm/s

These particles have $v_s > v_o$, and 100 percent of the particles should be easily removed.

Determination of v_s

In design of an ideal sedimentation tank, one first determines the settling velocity (v_s) of the particle to be removed and then sets the overflow rate (v_o) at some value less than or equal to v_s.

Determination of the particle-settling velocity is different for different types of particles. Settling properties of particles are often categorized into one of three classes:

Type I Sedimentation. Type I sedimentation is characterized by particles that settle discretely at a constant settling velocity. They settle as individual particles and do not flocculate or stick to other particles during settling. Examples of these particles are sand and grit material. Generally speaking, the only application of Type I settling is during pre-sedimentation for sand removal prior to coagulation in a potable water plant, in settling of sand particles during cleaning of rapid sand filters, and in grit chambers (see Section 6-6).

Type II Sedimentation. Type II sedimentation is characterized by particles that flocculate during sedimentation. Because they flocculate, their size is constantly changing; therefore, the settling velocity is changing. Generally the settling velocity is increasing. These types of particles occur in alum or iron coagulation, in primary sedimentation (see Section 6-7), and in settling tanks in trickling filtration (see Section 6-8).

Type III or Zone Sedimentation. In zone sedimentation the particles are at a high concentration (greater than 1,000 mg/L) such that the particles tend to settle as a mass, and a distinct clear zone and sludge zone are present. Zone settling occurs in lime-softening sedimentation, activated-sludge sedimentation, and sludge thickeners (see Section 6-12).

Determination of v_o

There are five ways to determine effective particle-settling velocities and consequently to determine overflow rates.

Calculation. In the case of Type I sedimentation, the particle-settling velocity can be calculated and the basin designed to remove a specific size particle. In 1687, Sir Isaac Newton showed that a particle falling in a quiescent fluid accelerates until the frictional resistance, or drag, on the particle is equal to the gravitational force of the particle (Figure 4-38) (Newton, 1687). The three forces are defined as follows:

$$F_G = (\rho_s)g\mathcal{V}_p \tag{4-87}$$

$$F_B = (\rho)g\mathcal{V}_p \tag{4-88}$$

$$F_D = C_D A_p(\rho)\frac{v^2}{2} \tag{4-89}$$

where F_G = gravitational force
F_B = buoyancy force
F_D = drag force
ρ_s = density of particle, kg/m^3
ρ = density of fluid, kg/m^3
g = acceleration due to gravity, m/s^2

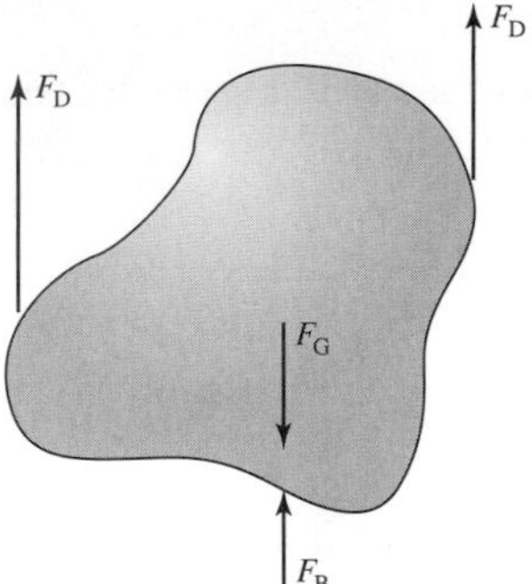

FIGURE 4-38
Forces acting on a free-falling particle in a fluid (F_D = drag force; F_G = gravitational force; F_B = buoyancy force).

$\forall_p$ = volume of particle, m^3
C_D = drag coefficient
A_p = cross sectional area of particle, m^2
v = velocity of particle, m/s

The driving force for acceleration of the particle is the difference between the gravitational and buoyant force:

$$F_G - F_B = (\rho_s - \rho)g\,\forall_p \tag{4-90}$$

When the drag force is equal to the driving force, the particle velocity reaches a constant value called the *terminal settling velocity* (v_s).

$$F_G - F_B = F_D \tag{4-91}$$

$$(\rho_s - \rho)g\forall_p = C_D A_p(\rho)\frac{v_s^2}{2} \tag{4-92}$$

For spherical particles with a diameter = d,

$$\frac{\forall_p}{A_p} = \frac{4/3\,(\pi)(d/2)^3}{(\pi)(d/2)^2} = \frac{2}{3}\,d \tag{4-93}$$

Using equation 4-29 and 4-93 to solve for the terminal settling velocity:

$$v_s = \left[\frac{4\,g(\rho_s - \rho)\,d}{3\,C_D\,\rho}\right]^{1/2} \tag{4-94}$$

The drag coefficient takes on different values depending on the flow regime surrounding the particle. The flow regime may be characterized qualitatively as laminar, turbulent, or transitional. In laminar flow, the fluid moves in layers, or laminas, one layer gliding smoothly over adjacent layers with only molecular interchange of momentum. In turbulent flow, the fluid motion is very erratic with violent transverse interchange of momentum. Osborne Reynolds (1883) developed a quantitative means of describing the different flow regimes using a dimensionless ratio that is called the *Reynolds number*. For spheres moving through a liquid this number is defined as

$$\mathbf{R} = \frac{(d)\,v_s}{\nu} \tag{4-95}$$

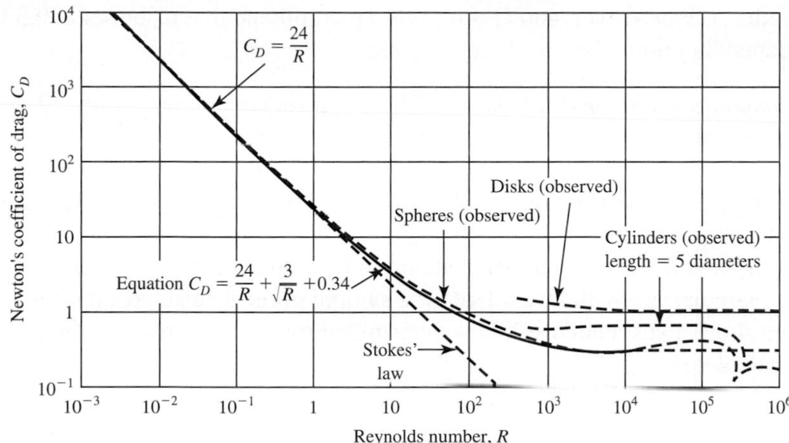

FIGURE 4-39
Newton's coefficient of drag as a function of Reynolds number. (*Source:* T. R. Camp, 1946.)

where $\mathbf{R}$ = Reynolds number
d = diameter of sphere, m
v_s = velocity of sphere, m/s
ν = kinematic viscosity, $m^2/s = \mu/\rho$
ρ = density of fluid, kg/m^3
μ = dynamic viscosity, Pa · s

Thomas Camp (1946) developed empirical data relating the drag coefficient to Reynolds number (Figure 4-39). For eddying resistance for spheres at high Reynolds numbers ($\mathbf{R} > 10^4$), C_D has a value of about 0.4. For viscous resistance at low Reynolds numbers ($\mathbf{R} < 0.5$) for spheres:

$$C_D = \frac{24}{\mathbf{R}} \tag{4-96}$$

For the transition region of $\mathbf{R}$ between 0.5 and 10^4, the drag coefficient for spheres may be approximated by the following:

$$C_D = \frac{24}{\mathbf{R}} + \frac{3}{\mathbf{R}^{1/2}} + 0.34 \tag{4-97}$$

Sir George Gabriel Stokes showed that, for spherical particles falling under laminar (quiescent) conditions, Equation 4-94 reduces to the following (Stokes, 1845):

$$v_s = \frac{g(\rho_s \quad \rho)d^2}{18\mu} \tag{4-98}$$

where μ = dynamic viscosity, Pa · s
18 = a constant

Equation 4-98 is called *Stokes' law* (Stokes, 1845). Dynamic viscosity (also called absolute viscosity) is a function of the water temperature. A table of dynamic viscosities is given in Appendix A. Fair, Geyer, and Okun (1968) recommend that v_0 be set at 0.33 to 0.7 times v_s depending upon the efficiency desired.

Flocculant Sedimentation Lab or Pilot Data. There is no adequate mathematical relationship that can be used to describe Type II settling. The Stokes equation cannot be used because the flocculating particles are continually changing in size and shape, and when water is entrapped in the floc, in specific gravity. Laboratory tests with settling columns are used to develop design data.

A settling column is filled with the suspension to be analyzed. The suspension is allowed to settle. Samples are withdrawn from the sample ports at selected time intervals. The concentration of suspended solids is determined for each sample and the percent removal is calculated:

$$R\% = 1 - \frac{C_t}{C_o}(100\%) \tag{4-99}$$

where $R\%$ = percent removal at one depth and time, %
C_t = concentration at time, t, and given depth, mg/L
C_o = initial concentration, mg/L

Percent removal versus depth is then plotted as shown in Figure 4-40. The circled numbers are the calculated percentages. Interpolations are made between these plotted points to construct curves of equal concentration at reasonable percentages, that is, 5 or 10 percent increments. Each intersection point of an isoconcentration line and the bottom of the column defines an overflow rate (v_o):

$$v_o = \frac{H}{t_i} \tag{4-100}$$

where H = height of column, m
t_i = time defined by intersection of isoconcentration line and bottom of column (x-axis) where the subscript, i, refers to the first, second, third, etc., intersection points, d

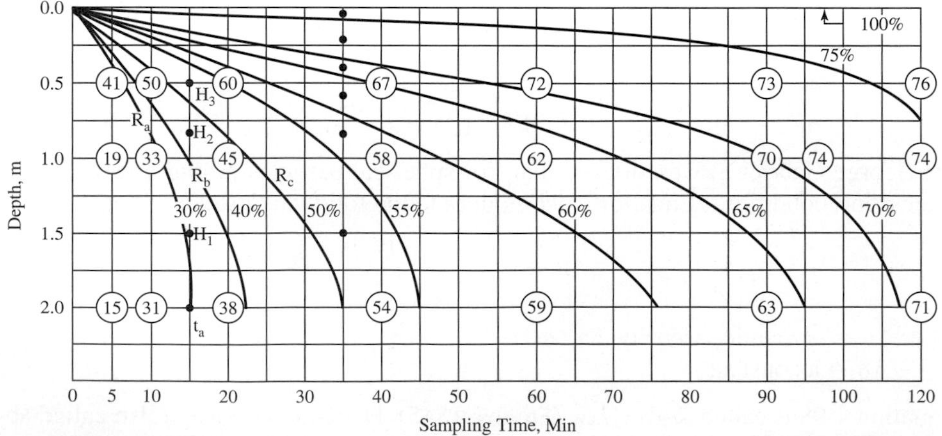

FIGURE 4-40
Isoconcentration lines for Type II settling test using a 2-m-deep column.

A vertical line is drawn from t_i to intersect all the isoconcentration lines crossing the t_i time. The midpoints between isoconcentration lines define heights H_1, H_2, H_3, etc. used to calculate the fraction of solids removed. For each time, t_i, defined by the intersection of the isoconcentration line and the bottom of the column (x-axis), you can construct a vertical line and calculate the fraction of solids removed:

$$R_{T_a} = R_a + \frac{H_1}{H}(R_b - R_a) + \frac{H_2}{H}(R_c - R_b) + \cdots \tag{4-101}$$

where R_{T_a} = total fraction removed for settling time, t_a

R_a, R_b, R_c = isoconcentration fractions a, b, c, etc.

The series of overflow rates and removal fractions are used to plot curves of suspended solids removal versus detention time and suspended solids removal versus overflow rate that can be used to size the settling tank. Eckenfelder (1980) recommends that scale-up factors of 0.65 for overflow rate and 1.75 for detention time be used to design the tank.

Example 4-23. The city of Urbana is planning to install a new water treatment plant. Design a settling tank to remove 65 percent of the influent suspended solids from their design flow of 0.5 m^3/s. A batch-settling test using a 2.0 m column and coagulated water from their existing plant yielded the following data:

Percent removal as a function of time and depth

	Sampling Time, min						
Depth, m	5	10	20	40	60	90	120
0.5	41	50	60	67	72	73	76
1.0	19	33	45	58	62	70	74
2.0	15	31	38	54	59	63	71

Solution. The plot is shown in Figure 4-40.

Calculate the overflow rate for each intersection point. For example, for the 50 percent line,

$$v_o = \frac{2.0 \text{ m}}{(35 \text{ min})}(1{,}440 \text{ min/d}) = 82.3 \text{ m/d}$$

The corresponding removal fraction is

$$\begin{aligned} R_{T50} &= 50 + \frac{1.5}{2.0}(55 - 50) + \frac{0.85}{2.0}(60 - 55) \\ &\quad + \frac{0.60}{2.0}(65 - 60) + \frac{0.40}{2.0}(70 - 65) \\ &\quad + \frac{0.20}{2.0}(75 - 70) + \frac{0.05}{2.0}(100 - 75) \\ &= 59.5 \text{ or } 60\% \end{aligned}$$

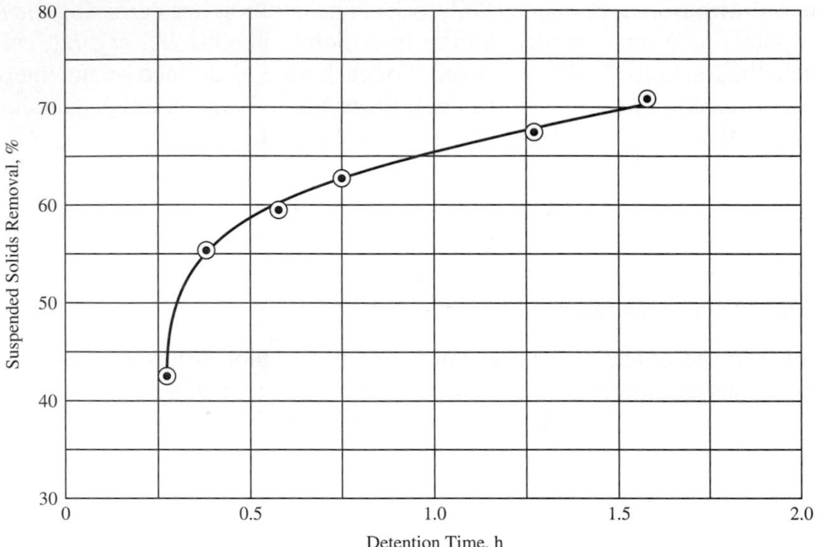

FIGURE 4-41
Suspended solids removal versus detention time.

The corresponding detention time is taken from the intersection of the isoconcentration line and the x-axis used to define the overflow rate, that is, 35 minutes for the 50 percent line.

This calculation is repeated for each isoconcentration line that intersects the x-axis except the last ones for which data are too sparse, that is, 30, 40, 50, 55, 60, and 65 percent, but not 70 or 75 percent.

Two graphs are then constructed (see Figures 4-41 and 4-42). From these graphs the bench-scale detention time and overflow rate for 65 percent removal are found to be 54 minutes and 50 m/d.

Applying the scale-up factors yields

$$t_o = (54 \text{ min})(1.75) = 94.5 \text{ or } 95 \text{ min}$$

$$v_o = (50 \text{ m/d})(0.65) = 32.5 \text{ m/d}$$

Zone Sedimentation Lab Data. For zone sedimentation, values can also be obtained from the lab. The design overflow is again set at about 0.5 to 0.7 times the lab value.

Jar Test Data. A technique has been developed to determine settling velocities of coagulant flocs from jar test data (Hudson, 1981).

Experience. Typical design numbers exist for all types of sedimentation basins. These numbers can be used in lieu of laboratory or pilot work. The applicability of the typical numbers to particles in different situations is unknown. For this reason, the typical numbers are often quite conservative. These conservative numbers also correct for

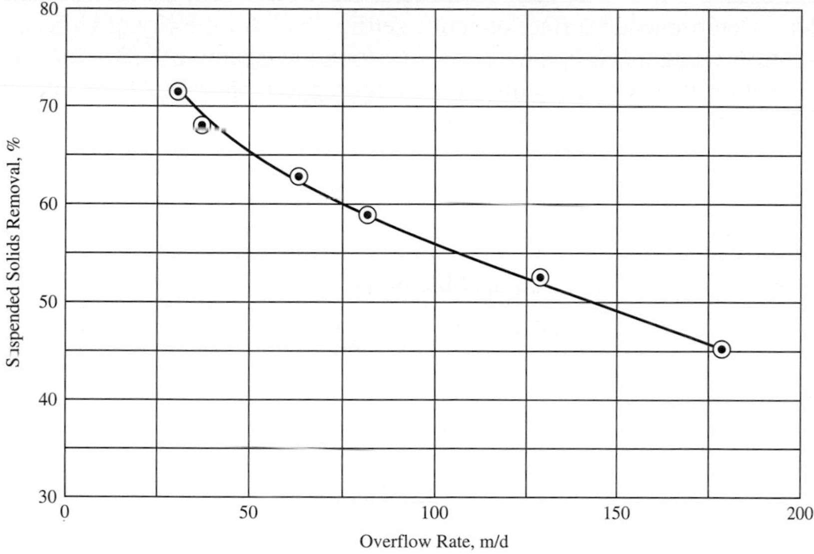

FIGURE 4-42
Suspended solids removal versus overflow rate.

ineffective inlet or outlet zone design. A design engineer with sufficient time and funds can generally save his or her client capital costs by designing a good inlet system and conducting lab tests for proper sedimentation zone design. However, clients are not always willing to expend such funds and the engineer has little choice other than to select conservative design numbers. Table 4-18 shows some overflow rates for potable water treatment.

Typical detention times for waters coagulated with alum or iron salts are on the order of 2 to 8 hours. In lime-soda softening plants, the detention times range from 4 to 8 hours (Reynolds and Richards, 1996).

TABLE 4-18
Typical sedimentation tank overflow rates

Application	Long rectangular and circular $m^3/d \cdot m^2$	Upflow solids-contact $m^3/d \cdot m^2$
Alum or iron coagulation		
Turbidity removal	40	50
Color removal	30	35
High algae	20	
Lime softening		
Low magnesium	70	130
High magnesium	57	105

Source: AWWA, 1990.

Example 4-24. Determine the surface area of a settling tank for the city of Urbana's 0.5 m^3/s design flow using the design overflow rate found in Example 4-23. Compare this surface area with that which results from assuming a typical overflow rate of 20 $m^3/d \cdot m^2$. Find the depth of the clarifier for the overflow rate and detention time found in Example 4-23.

Solution

a. Find the surface area.

First change the flow rate to compatible units:

$$\left(\frac{0.5 \text{ m}^3}{\text{s}}\right)\left(\frac{86{,}400 \text{ s}}{\text{d}}\right) = 43{,}200 \text{ m}^3/\text{d}$$

Using the overflow rate from Example 4-23, the surface area is

$$A_s = \frac{43{,}200 \text{ m}^3/\text{d}}{32.5 \text{ m}^3/\text{d} \cdot \text{m}^2} = 1{,}329.23 \text{ or } 1{,}330 \text{ m}^2$$

Using the conservative value

$$A_s = \frac{43{,}200 \text{ m}^3/\text{d}}{20 \text{ m}^3/\text{d} \cdot \text{m}^2} = 2{,}160 \text{ m}^2$$

Obviously, the use of conservative data would, in this case, result in a 60 percent overdesign of the tank area.

Common length-to-width ratios for settling are between 2:1 and 5:1, and lengths seldom exceed 100 m. A minimum of two tanks is always provided.

Assuming two tanks, each with a width of 12 m, a total surface area of 1,330 m^2 would imply a tank length of

$$\text{Length} = \frac{1{,}330 \text{ m}^2}{(2 \text{ tanks})(12 \text{ m wide})} = 55.4 \text{ or } 55 \text{ m}$$

This meets our length-to-width ratio of 5:1.

b. Find the tank depth.

First find the total tank volume from Equation 2-27 using a detention time of 95 minutes from Example 4-23:

$$\forall = (0.5 \text{ m}^3/\text{s})(95 \text{ min})(60 \text{ s/min}) = 2{,}850 \text{ m}^3$$

This would be divided into two tanks as noted above. The depth is found as the total tank volume divided by the total surface area:

$$\text{Depth} = \frac{2{,}850 \text{ m}^3}{1{,}330 \text{ m}^2} = 2.1428 \text{ or } 2 \text{ m}$$

This depth would not include the sludge storage zone.

The final design would then be two tanks, each having the following dimensions: 12 m wide × 55 m long × 2 m deep plus sludge storage depth.

4-6 FILTRATION

The water leaving the sedimentation tank still contains floc particles. The settled water turbidity is generally in the range from 1 to 10 TU with a typical value being 2 TU. In order to reduce this turbidity to less than 0.3 TU, a filtration process is normally used. Water filtration is a process for separating suspended or colloidal impurities from water by passage through a porous medium, usually a bed of sand or other medium. Water fills the pores (open spaces) between the sand particles, and the impurities are left behind, either clogged in the open spaces or attached to the sand itself.

There are several methods of classifying filters. One way is to classify them according to the type of medium used, such as sand, coal (called anthracite), dual media (coal plus sand), or mixed media (coal, sand, and garnet). Another common way to classify the filters is by allowable loading rate. *Loading rate* is the flow rate of water applied per unit area of the filter. It is the velocity of the water approaching the face of the filter:

$$v_a = \frac{Q}{A_s} \tag{4-102}$$

where v_a = face velocity, m/d
= loading rate, $m^3/d \cdot m^2$
Q = flow rate onto filter surface, m^3/d
A_s = surface area of filter, m^2

Based on loading rate, the filters are described as being slow sand filters, rapid sand filters, or high-rate sand filters.

Slow sand filters were first introduced in the 1800s. The water is applied to the sand at a loading rate of 2.9 to 7.6 $m^3/d \cdot m^2$. As the suspended or colloidal material is applied to the sand, the particles begin to collect in the top 75 mm and to clog the pore spaces. As the pores become clogged, water will no longer pass through the sand. At this point the top layer of sand is scraped off, cleaned, and replaced. Slow sand filters require large areas of land and are operator intensive.

In the early 1900s there was a need to install filtration systems in large numbers in order to prevent epidemics. Rapid sand filters were developed to meet this need. These filters have graded (layered) sand within the bed. The sand grain size distribution is selected to optimize the passage of water while minimizing the passage of particulate matter.

Rapid sand filters are cleaned in place by forcing water backwards through the sand. This operation is called *backwashing*. The washwater flow rate is such that the sand is expanded and the filtered particles are removed from the bed. After backwashing, the sand settles back into place. The largest particles settle first, resulting in a fine sand layer on top and a coarse sand layer on the bottom. Rapid sand filters are the most common type of filter used in water treatment today.

Traditionally, rapid sand filters have been designed to operate at a loading rate of 120 $m^3/d \cdot m^2$. Experiments conducted at the Chicago water treatment plant have demonstrated that satisfactory water quality can be obtained with rates as high as 235 $m^3/d \cdot m^2$ (AWWA, 1971). Filters now operate successfully at even higher loading rates through the use of proper media selection and improved pretreatment. Normally, a minimum of two filters are constructed to ensure redundancy. For larger plants

(>0.5 m^3/s), a minimum of four filters is suggested (Montgomery, 1985). The surface area of the filter tank (often called a filter box) is generally restricted in size to about 100 m^2, except for very large plants.

In the wartime era of the early 1940s, dual-media filters were developed. They are designed to utilize more of the filter depth for particle removal. In a rapid sand filter, the finest sand is on the top; hence, the smallest pore spaces are also on the top. Therefore, most of the particles will clog in the top layer of the filter. In order to use more of the filter depth for particle removal, it is necessary to have the large particles on top of the small particles. This was accomplished by placing a layer of coarse coal on top of a layer of fine sand. Coal has a lower specific gravity than sand, so, after backwash, it settles slower than the sand and ends up on top. Dual-media filters are operated up to loading rates of 300 $m^3/d \cdot m^2$.

In the mid 1980s, deep-bed monomedia filters came into use. The filters are designed to achieve higher loading rates while at the same time producing lower finished water turbidities. The filters typically consist of 1.0 mm to 1.5 mm diameter anthracite about 1.5 m to 2.5 m deep. They operate at loading rates up to 800 $m^3/d \cdot m^2$.

Example 4-25. As part of their proposed new treatment plant, Urbana is going to install rapid sand filters after their sedimentation tanks. The design loading rate to the filter is 200 $m^3/d \cdot m^2$. How much filter surface area should be provided for their design flow rate of 0.5 m^3/s? If the surface area per filter box is to be limited to 50 m^2, how many filter boxes are required?

Solution. The surface area required is the flow rate divided by the loading rate:

$$A_s = \frac{Q}{V_a} = \frac{(0.5 \text{ m}^3/\text{s})(86{,}400 \text{ s/d})}{200 \text{ m}^3/\text{d} \cdot \text{m}^2} = 216 \text{ m}^2$$

If the maximum surface area of any one tank is 50 m^2, then the number of filters required is

$$\text{Number} = \frac{216 \text{ m}^2}{50 \text{ m}^2} = 4.32$$

Since we cannot build 0.32 filter, we need to round to an integer. Normally, we build an even number of filters to make construction easier and to reduce costs. In this case we would propose to build four filters and check to see that the design loading does not exceed our guideline values. With four filters the loading would be

$$v_a = \frac{Q}{A_s} = \frac{(0.5 \text{ m}^3/\text{s})(86{,}400 \text{ s/d})}{(4 \text{ filters})(50 \text{ m}^2/\text{filter})} = 216 \text{ m/d}$$

This is less than the 235 m/d recommended maximum loading rate and would be acceptable except that many states require that the filter capacity be sufficient to handle the design flow rate with one filter out of service. Therefore, we must check the loading with three filters in service

$$v_a = \frac{(0.5 \text{ m}^3/\text{s})(86{,}400 \text{ s/d})}{(3 \text{ filters})(50 \text{ m}^2/\text{filter})} = 288 \text{ m/d}$$

This exceeds the recommended maximum loading rate. If the 50 m^2/filter cannot be altered, another filter box is required. Because the filter box may be as large as 100 m^2/filter, we would expect that four slightly larger filters would be constructed to meet the required loading with one filter out of service.

Figure 4-43 shows a cutaway drawing of a rapid sand filter. The bottom of the filter consists of a support media and collection system. The support media is designed to keep the sand in the filter and prevent it from leaving with the filtered water. Layers of graded gravel (large on bottom, small on top) traditionally have been used for the support. The underdrain blocks collect the filtered water.

On top of the support media is a layer of graded sand. The sand depth varies between 0.5 and 0.75 m. If a dual media filter is used, the sand is about 0.3 m thick and the coal about 0.45 m thick. Approximately 0.7 m to 1 m above the top of the sand are the washwater troughs. The washwater troughs collect the backwash water used to clean the filter. The troughs are placed high enough above the sand layer so that sand will not be carried out with the backwash water. Generally a total depth of 1.8 m to 3 m is allowed above the sand layer for water to build up above the filter. This depth of water provides sufficient pressure to force the water through the sand during filtration.

Figure 4-44 shows a slightly simplified version of a rapid sand filter. Water from the settling basins enters the filter and seeps through the sand and gravel bed, through a false floor, and out into a clear well that acts as a storage tank for finished water. During filtration, valves A and C are open.

During filtration the filter bed will become more and more clogged. As the filter clogs, the water level will rise above the sand as it becomes harder to force water

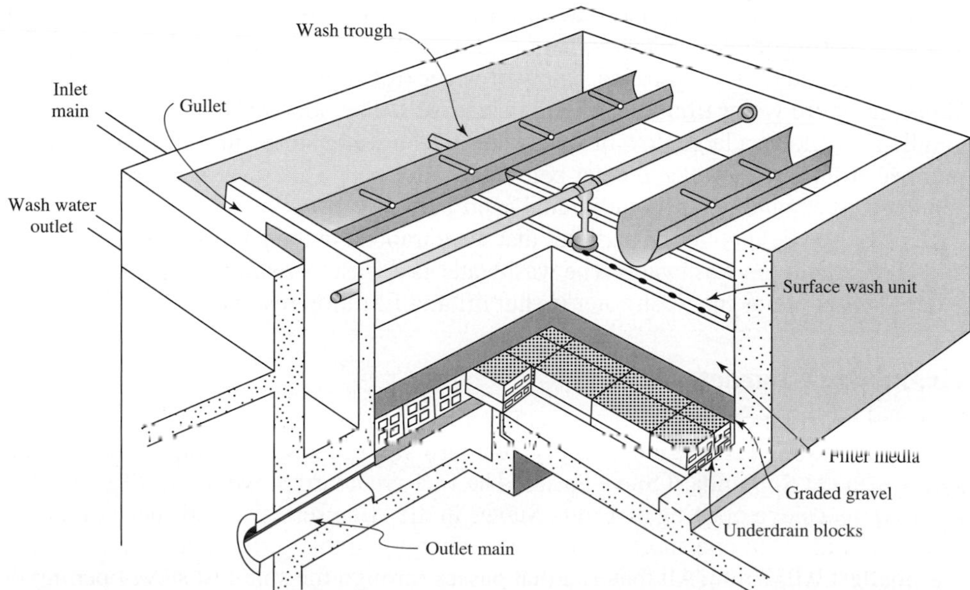

FIGURE 4-43
Typical gravity filter box. (*Source: F. B. Leopold Co.*)

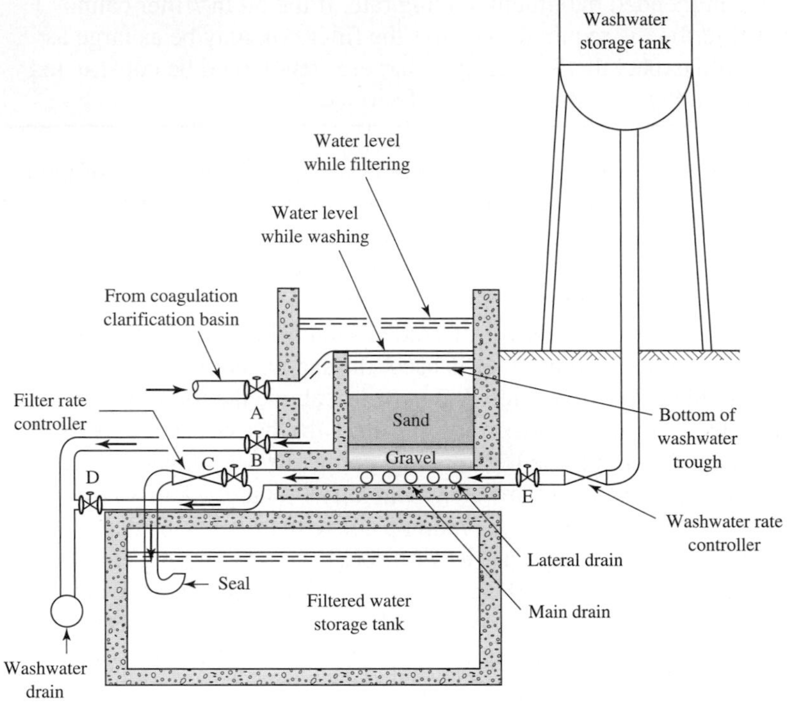

FIGURE 4-44
Operation of a rapid sand filter. (*Source:* Steel and McGhee, 1979).

through the bed. Eventually, the water level will rise to the point that the filter bed must be cleaned. This point is called *terminal head loss.* When this occurs, the operator turns off valves A and C. This stops the supply of water from the sedimentation tank and prevents any more water from entering the clear well. The operator then opens valves E and B. This allows a large flow of washwater (clean water stored in an elevated tank or pumped from a clear well) to enter below the filter bed. This rush of water forces the sand bed to expand and sets individual sand particles in motion. By rubbing against each other, the light colloidal particles that were trapped in the pore spaces are released and escape with the washwater. The washwater is a waste stream that must be treated. After a few minutes the washwater is shut off and filtration resumed.

Grain Size Characteristics

The size distribution or variation of a sample of granular material is determined by sieving the sample through a series of standard sieves (screens). One such standard series is the U.S. Standard Sieve Series. The U.S. Standard Sieve Series (Table 4-19) is based on a sieve opening of 1 mm. Sieves in the "fine series" stand successively in the ratio of $(2)^{1/4}$ to one another, the largest opening in this series being 5.66 mm and the smallest 0.037 mm. All material that passes through the smallest sieve opening in the series is caught in a pan that acts as the terminus of the series (Fair and Geyer, 1954).

TABLE 4-19
U.S. Standard Sieve Series

Sieve designation number	Size of opening (mm)	Sieve designation number	Size of opening (mm)
200	0.074	20	0.84
140	0.105	(18)	(1.00)
100	0.149	16	1.19
70	0.210	12	1.68
50	0.297	8	2.38
40	0.42	6	3.36
30	0.59	4	4.76

Source: Fair and Geyer, 1954.

The grain size analysis begins by placing the sieve screens in ascending order with the largest opening on top and the smallest opening on the bottom. A sand sample is placed on the top sieve and the stack is shaken for a prescribed amount of time. At the end of the shaking period, the mass of material retained on each sieve is determined. The cumulative mass is recorded and converted into percentages by mass equal to or less than the size of separation of the overlying sieve. Then the cumulative frequency distribution is plotted. For many natural granular materials, this curve approaches geometric normality. Logarithmic-probability paper, therefore, assures an almost straight-line plot which facilitates interpolation. The geometric mean (X_g) and geometric standard deviation (S_g) are useful parameters of central tendency and variation. Their magnitudes may be determined from the plot. The parameters most commonly used, however, are the *effective size, E*, or 10 percentile, P_{10}, and the *uniformity coefficient, U*, or ratio of the 60 percentile to the 10 percentile, P_{60}/P_{10}. Use of the 10 percentile was suggested by Allen Hazen because he had observed that resistance to the passage of water offered by a bed of sand within which the grains are distributed homogeneously remains almost the same, irrespective of size variation (up to a uniformity coefficient of about 5.0), provided that the 10 percentile remains unchanged (Hazen, 1892). Use of the ratio of the 60 percentile to the 10 percentile as a measure of uniformity was suggested by Hazen because this ratio covered the range in size of half the sand.* On the basis of logarithmic normality, the probability integral establishes the following relations between the effective size and uniformity coefficient and the geometric mean size and geometric standard deviation:

$$E = P_{10} = (X_g)(S_g)^{-1.282} \tag{4-103}$$

$$U = P_{60}/P_{10} = (S_g)^{1.535} \tag{4-104}$$

Experience has suggested that, for silica sand, the effective size should be in the range of 0.35 to 0.55 mm with a maximum of about 1.0 mm. The uniformity coefficient should range between 1.3 and 1.7. Smaller effective sizes result in a product water that

*It would be more logical to speak of this ratio as a coefficient of nonuniformity because the coefficient increases in magnitude as the nonuniformity increases.

is lower in turbidity, but they also result in higher pressure losses in the filter and shorter operating cycles between cleanings.

Example 4-26. For the size frequencies by weight and by count of the sample of sand listed below, find the effective size, E, and uniformity coefficient, U.

U. S. Standard Sieve No.	Analysis of stock sand (Cumulative mass % passing)
140	0.2
100	0.9
70	4.0
50	9.9
40	21.8
30	39.4
20	59.8
16	74.4
12	91.5
8	96.8
6	99.0

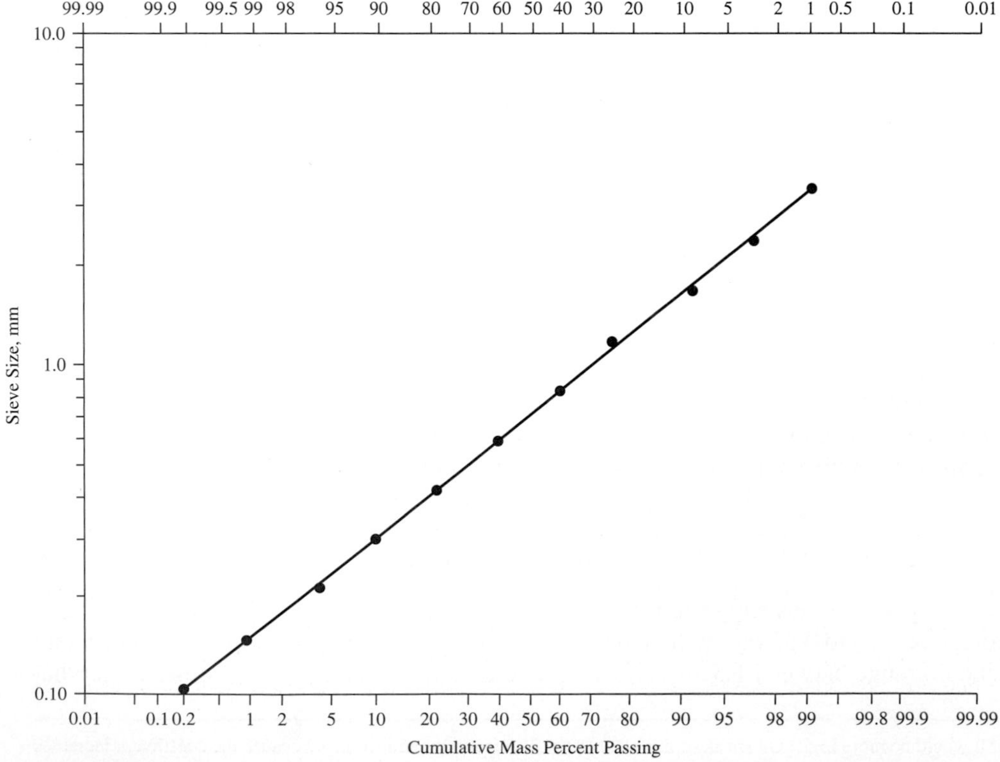

FIGURE 4-45
Grain size analysis of run-of-bank sand.

Solution. First we must plot the data on log-probability paper as shown in Figure 4-45. From this plot we then find the effective size:

$$E = P_{10} = 0.30 \text{ mm}$$

The uniformity coefficient is then

$$U = \frac{P_{60}}{P_{10}} = \frac{0.85}{0.30} = 2.8$$

Sand excavated from a natural deposit is called *run-of-bank* sand. Run-of-bank sand may be too coarse, too fine, or too nonuniform for use in filters. Within economical limits, proper sizing and uniformity are obtained by screening out coarse components and washing out fine components. In rapid sand filters, the removal of "fines" may be accomplished by stratifying the bed through backwashing and then scraping off the layer that includes the unwanted sand.

Filter Hydraulics

The loss of pressure (commonly termed *head loss*) through a clean stratified-sand filter with uniform porosity was described by Rose (1945) in the following form*:

$$h_L = \frac{1.067\,(v_a)^2(D)}{(\phi)(g)(\epsilon)^4} \sum_{i=1}^{n} \frac{(C_D)(f)}{d} \qquad (4\text{-}105)$$

where h_L = frictional head loss through the filter, m
v_a = approach velocity, m/s
D = depth of filter sand, m
C_D = drag coefficient
f = mass fraction of sand particles of diameter d
d = diameter of sand grains, m
ϕ = shape factor
g = acceleration due to gravity, m/s^2
ϵ = porosity

The drag coefficient is defined in Equations 4-96 and 4-97. The Reynolds number used to calculate the drag coefficient is multiplied by the shape factor to account for nonspherical sand grains. The summation term may be calculated using the size distribution of the sand particles found from a sieve analysis. Although the Rose equation is limited to clean filter beds, it does provide an opportunity to examine the initial stages of filtration and the effects of sand grain size distribution on head loss. As the filter clogs, the head loss will increase so that the calculated results are the minimum expected head losses. Initial head losses in excess of 0.6 m imply either that the loading rate is too high or that the sand has too large a proportion of fine grain sizes. The design of the filter must account for the additional losses that occur as the filter runs.

---

*Other headloss equations include those by Carmen-Kozeny, Fair-Hatch, and Hazen. [These equations are summarized by Cleasby and Logsdon (WQ&T, 1999) and by Metcalf & Eddy, Inc. (2003).]

Thus, the filter box must be at least as deep as the highest design head loss. As mentioned previously, this is about 3 m maximum.

The hydraulic head loss that occurs during backwashing is calculated to determine the placement of the backwash troughs above the filter bed. The trough bottom should be at least 0.15 m above the expanded bed to prevent loss of filter material. Fair and Geyer (1954) developed the following relationship to predict the depth of the expanded bed:

$$D_e = (1 - \epsilon)(D) \sum_{i=1}^{n} \frac{f}{(1 - \epsilon_e)} \tag{4-106}$$

where D_e = depth of the expanded bed, m
ϵ = porosity of the bed
ϵ_e = porosity of expanded bed
f = mass fraction of sand with expanded porosity

The porosity of the expanded bed may be calculated for a given particle by

$$\epsilon_e = \left(\frac{v_b}{v_s}\right)^{0.22} \tag{4-107}$$

where v_b = velocity of backwash, m/s
v_s = settling velocity, m/s

Strictly speaking, this form of the expanded bed porosity is applicable only for laminar conditions. Since the conditions during backwash are turbulent, a more representative model equation is that given by Richardson and Zaki (1954):

$$\epsilon_e = \left(\frac{v_b}{v_s}\right)^{0.2247\mathbf{R}^{0.1}} \tag{4-108}$$

where the Reynolds number is defined as

$$\mathbf{R} = \frac{v_s\, d_{60\%}}{\nu}$$

where $d_{60\%}$ = 60 percentile diameter, m. A more sophisticated model developed by Dharmarajah and Cleasby (1986) is also available.

The determination of D_e is not straightforward. From Equation 4-107, it is obvious that the expanded bed porosity is a function of the settling velocity. The particle settling velocity is determined by Equation 4-94. To solve Equation 4-94 you must calculate the drag coefficient (C_D). The drag coefficient is a function of the Reynolds number which, in turn, is a function of the settling velocity. Thus, you need the settling velocity to find the settling velocity! To resolve this dilemma, you must begin with an estimated settling velocity. Knowing the sand grain diameter and specific gravity, you can use Figure 4-46 to obtain a first estimate for the settling velocity to use in calculating the Reynolds number.

Backwash rates normally vary between 880 m/d and 1,200 m/d. However, the limiting factor in choosing a backwash rate is the terminal settling velocity of the smallest sand grains that are to be retained in the filter. Because the filter backwashing process is

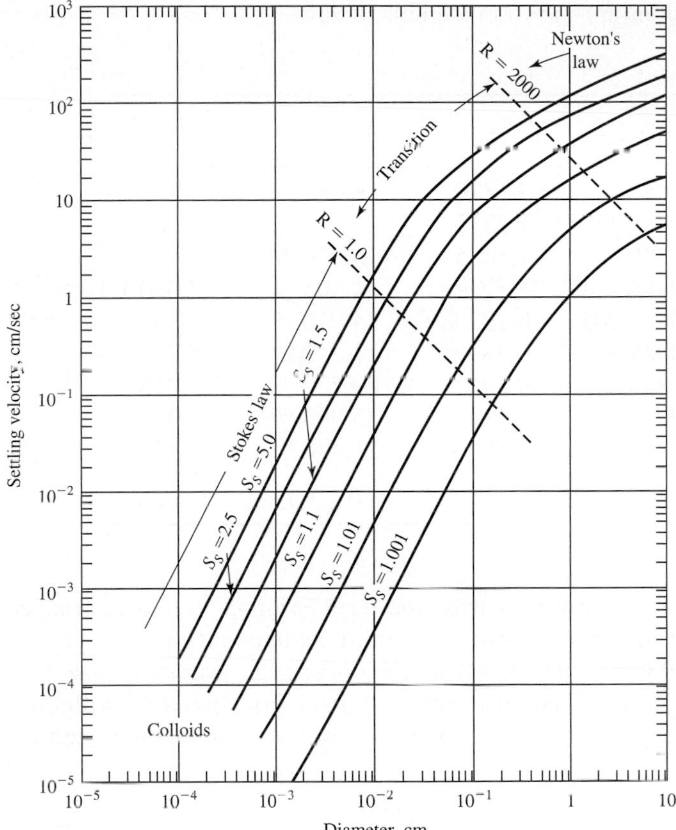

FIGURE 4-46
Settling and rising velocities of discrete spherical particles in quiescent water at 10°C. For other temperatures, multiply the Stokes values by $\nu/(1.31 \times 10^{-2})$, where ν is the kinematic viscosity at the stated temperature. (*Source:* G. M. Fair, J. C. Geyer, and D. A. Orun, 1971.)

effectively an upflow clarifier, the backwash rate becomes the overflow rate that determines whether a particle is retained in the filter or is washed out through the backwash trough.

Another design criterion that is often used is to make sure that the largest, or the 90th percentile largest, particles are fluidized. In a dual-media filter, the P_{90} of the anthracite is considered the largest-diameter particle. By fluidizing this particle during backwash it will not "sink" to the bottom of the filter and will restratify after backwash is complete.

Example 4-27. Estimate the clean filter head loss in Urbana's proposed new sand filter (Example 4-25) using the sand described in Example 4-26 and determine if it is reasonable. Use the following assumptions: specific gravity of sand is 2.65, the shape factor is 0.82, the bed porosity is 0.45, the water temperature is 10°C, and the depth of sand is 0.5 m.

Solution. The computations are shown in the table below.

Sieve No.	% Retain	d(m)	**R**	C_D	$\frac{(C_D)(f)}{d}$
8–12	5.3	0.002	3.1370	9.684551	256.64
12–16	17.1	0.00142	2.2272	13.12587	1,580.7
16–20	14.6	0.001	1.5685	18.03689	2,633.4
20–30	20.4	0.000714	1.1199	24.60549	7,030.1
30–40	17.6	0.000505	0.79208	34.01075	11,853
40–50	11.9	0.000357	0.55995	47.21035	15,737
50–70	5.9	0.000252	0.39526	60.72009	14,216
70–100	3.1	0.000178	0.27919	85.96328	14,971
100–140	.7	0.000126	0.19763	121.4402	6,746.7
				Total $(C_D)(f)/d$ =	75,025

In the first two columns, the grain size distribution from Example 4-26 is rearranged to show the fraction retained between sieves. The third column is the geometric mean diameter of the sand grain computed from the upper and lower sieve size. The fourth column is the Reynolds number computed from Equation 4-95 with the correction for nonspherical sand grains. For the first row, using the loading rate from Example 4-25,

$$\mathbf{R} = \frac{\phi\,(d)\,v_a}{\nu} = \frac{(.82)(.002\text{ m})(0.0025\text{ m/s})}{1.307 \times 10^{-6}\text{ m}^2\text{/s}} = 3.137$$

The filtration velocity is simply the conversion of the filtration rate to compatible units:

$$v_a = \frac{216\text{ m}^3\text{/d}\cdot\text{m}^2}{86{,}400\text{ s/d}} = 0.0025\text{ m/s}$$

The kinematic viscosity is determined from Appendix A using the water temperature of 10°C. The factor of 10^{-6} is to convert from μm^2/s to m^2/s.

The drag coefficient is calculated in column 5 using either Equation 4-96 or 4-97, depending on the Reynolds number. For the first row,

$$C_D = \frac{24}{\mathbf{R}} + \frac{3}{\mathbf{R}^{1/2}} + 0.34$$
$$= 7.6507 + 1.6938 + 0.34 = 9.6846$$

The final column is the product of the fractional mass retained and the drag coefficient divided by the diameter. For the first row,

$$\frac{(C_D)(f)}{d} = \frac{(9.6846)(0.053)}{0.002} = 256.64$$

The last column is summed and the head loss calculated using Equation 4-105:

$$h_L = \frac{1.067\,(0.0025)^2\,(0.5)}{(0.82)(9.8)(0.45)^4}(7.5025 \times 10^4)$$

$$= (1.0119 \times 10^{-5})(7.5025 \times 10^4) = 0.76 \text{ m}$$

This initial head loss exceeds the guideline of 0.6 m. Either the filtration rate should be lowered or the fraction of fines should be reduced.

Example 4-28. Determine the height that the backwash troughs must be placed above the filter bed for the sand filter being designed for Urbana. For ease of computation, assume that Equation 4-107 applies, even though the flow is turbulent and equation 4-108 is more applicable.

Solution. To begin, we must select a backwash rate. Assuming that we wish to retain the finest sand grains used in building the filter, the backwash rate must not wash out particles with a diameter of 0.000126 m (0.0126 cm). Using Figure 4-46, we find that, for a 0.0126 cm particle with a specific gravity of 2.65, the terminal settling velocity is approximately 1 cm/s (864 m/d). Thus, our backwash rate may not exceed 864 m/d rather than the nominal minimum of 880 m/d.

The computations are shown in the table below.

Estimated v_s (m/s)	**R**	C_D	Calculated v_s (m/s)	ϵ_e	$\frac{f}{1-\epsilon_e}$
.30	376.435	0.558380	.2778839	.4812	.10216
.20	178.179	0.699442	.2092095	.5122	.35058
.15	94.1086	0.904272	.1544058	.5476	.32275
.10	44.7957	1.32400	.1078248	.5927	.50080
.07	22.1783	2.05917	.0727132	.6463	.49762
.05	11.1989	3.37953	.0477221	.7091	.40902
.03	4.74308	6.77751	.0283125	.7954	.28831
.02	2.23351	13.0928	.0171201	.8884	.27788
0.015	1.18577	23.3350	.0107893	.9834	.42232
					$\Sigma f/(1-\epsilon_e) = 3.1715$

The estimated settling velocities in the first column were found from Figure 4-46. The Reynolds number was then computed with this estimated velocity. For the first row:

$$\mathbf{R} = \frac{\phi\,(d)\,v_s}{\nu} = \frac{(.82)(.002 \text{ m})(0.30 \text{ m/s})}{1.307 \times 10^{-6} \text{ m}^2/\text{s}} = 376.435$$

You should note that the shape factor, sand particle diameter, and viscosity are all the same as in Example 4-27. The drag coefficient (C_D) is calculated in the same fashion as Example 4-27. The settling velocity is calculated using Equation 4-94 assuming the density of water is 1,000 kg/m^3. For the first row:

$$v_s = \left[\frac{(4)(9.8 \text{ m/s}^2)(2{,}650 \text{ kg/m}^3 - 1{,}000 \text{ kg/m}^3)(0.002 \text{ m})}{(3)(0.55838)(1{,}000 \text{ kg/m}^3)}\right]^{1/2}$$

$$= 0.2778839 \text{ m/s}$$

The density of the sand grain is simply the product of the specific gravity (from Example 4-27) and the density of water:

$$\rho_s = (2.65)(1{,}000 \text{ kg/m}^3) = 2{,}650 \text{ kg/m}^3$$

The expanded bed porosity (column 5) is calculated from Equation 4-107. For the first row,

$$\epsilon_e = \left(\frac{v_b}{v_s}\right)^{0.22} = \left(\frac{0.01 \text{ m/s}}{0.2778839 \text{ m/s}}\right)^{0.22} = 0.4812$$

The first row of the last column is then

$$\frac{f}{1 - \epsilon_e} = \frac{0.053}{1 - 0.4812} = 0.10216$$

where 0.053 is taken from the first row in Example 4-27 and is the mass fraction of sand having a geometric mean diameter of 0.002 m, that is, between sieve numbers 8 and 12.

Using Equation 4-106 with a porosity of 0.45 and an undisturbed bed depth of 0.5 m from Example 4-27, the depth of the expanded bed is then

$$D_e = (1 - .45)(0.5 \text{ m})(3.1715) = 0.87 \text{ m}$$

Allowing for a safety margin of 0.15 m as mentioned above, the bottom of the backwash trough should be placed

$$(0.87 - 0.5) \text{ m} + 0.15 = 0.52 \text{ or } 0.5 \text{ m}$$

above the top of the sand surface. (Note: $(0.87 - 0.5) = D_e - D$.)

4-7 DISINFECTION

Disinfection is used in water treatment to reduce pathogens (disease-producing microorganisms) to an acceptable level. Disinfection is not the same as sterilization. Sterilization implies the destruction of all living organisms. Drinking water need not be sterile.

Three categories of human enteric pathogens are normally of consequence: bacteria, viruses, and amebic cysts. Purposeful disinfection must be capable of destroying all three.

To be of practical service, such water disinfectants must possess the following properties:

1. They must destroy the kinds and numbers of pathogens that may be introduced into water within a practicable period of time over an expected range in water temperature.

2. They must meet possible fluctuations in composition, concentration, and condition of the waters or wastewaters to be treated.

3. They must be neither toxic to humans and domestic animals nor unpalatable or otherwise objectionable in required concentrations.

4. They must be dispensable at reasonable cost and safe and easy to store, transport, handle, and apply.

5. Their strength or concentration in the treated water must be determined easily, quickly, and (preferably) automatically.

6. They must persist within disinfected water in a sufficient concentration to provide reasonable residual protection against its possible recontamination before use, or—because this is not a normally attainable property—the disappearance of residuals must be a warning that recontamination may have taken place.

Disinfection Kinetics

Under ideal conditions, when an exposed microorganism contains a single site vulnerable to a single unit of disinfectant, the rate of die-off follows *Chick's law,* which states that the number of organisms destroyed in a unit time is proportional to the number of organisms remaining (Chick, 1908):

$$-\frac{dN}{dt} = kN \tag{4-109}$$

This is a first-order reaction. Under real conditions the rate of kill may depart significantly from Chick's law. Increased rates of kill may occur because of a time lag in the disinfectant reaching vital centers in the cell. Decreased rates of kill may occur because of declining concentrations of disinfectant in solution or poor distribution of organisms and disinfectant.

Chlorine Reactions in Water

Chlorine is the most common disinfecting chemical used. The term *chlorination* is often used synonymously with disinfection. Chlorine may be used as the element (Cl_2), as sodium hypochlorite (NaOCl), or as calcium hypochlorite [$Ca(OCl)_2$].

When chlorine is added to water, a mixture of hypochlorous acid (HOCl) and hydrochloric acid (HCl) is formed:

$$Cl_2(g) + H_2O \rightleftharpoons HOCl + H^+ + Cl^- \tag{4-110}$$

This reaction is pH dependent and essentially complete within a very few milliseconds. In dilute solution and at pH levels above 1.0, the equilibrium is displaced to the right and very little Cl_2 exists in solution. Hypochlorous acid is a weak acid and dissociates poorly at levels of pH below about 6. Between pH 6.0 and 8.5 there occurs a very sharp change from undissociated HOCl to almost complete dissociation:

$$HOCl \rightleftharpoons H^+ + OCl^- \tag{4-111}$$

$$pK = 7.537 \text{ at } 25°C$$

Thus, chlorine exists predominantly as HOCl at pH levels between 4.0 and 6.0. Below pH 1.0, depending on the chloride concentration, the HOCl reverts back to Cl_2 as shown in Equation 4-110. At 20°C, above about pH 7.5, and at 0°C, above about pH 7.8, hypochlorite ions (OCl^-) predominate. Hypochlorite ions exist almost exclusively at levels of pH around 9 and above. Chlorine existing in the form of HOCl and/or OCl^- is defined as free available chlorine.

Hypochlorite salts dissociate in water to yield hypochlorite ions:

$$\text{NaOCl} \rightleftharpoons \text{Na}^+ + \text{OCl}^- \tag{4-112}$$

$$\text{Ca(OCl)}_2 \rightleftharpoons \text{Ca}^{2+} + 2\text{OCl}^- \tag{4-113}$$

The hypochlorite ions establish equilibrium with hydrogen ions (in accord with Equation 4-111), again depending on pH. Thus, the same active chlorine species and equilibrium are established in water regardless of whether elemental chlorine or hypochlorites are used. The significant difference is in the resultant pH and its influence on the relative amounts of HOCl and OCl^- existing at equilibrium. Elemental chlorine tends to decrease pH; each mg/L of chlorine added reduces the alkalinity by up to 1.4 mg/L as $CaCO_3$. Hypochlorites, on the other hand, always contain excess alkali to enhance their stability and tend to raise the pH somewhat. We seek to maintain the design pH within a range of 6.5 to 7.5 to optimize disinfecting action.

Chlorine-Disinfecting Action

Chlorine disinfection involves a very complex series of events and is influenced by the kind and extent of reactions with chlorine-reactive materials (including nitrogen), temperature, pH, the viability of test organisms, and numerous other factors. Such factors greatly complicate attempts to determine the precise mode of action of chlorine on bacteria and other microorganisms. Over the years, several theories have been advanced. One early theory held that chlorine reacts directly with water to produce nascent oxygen; another held that the action of chlorine is due to complete oxidative destruction of organisms. These theories were nullified because small concentrations of hypochlorous acid were observed to destroy bacteria whereas other oxidants (such as hydrogen peroxide or potassium permanganate) failed to do likewise. A later theory suggested that chlorine reacts with protein and amino acids of cells to alter and ultimately destroy cell protoplasm. Currently it is considered that the bactericidal action of chlorine is physiochemical, but among yet-unanswered questions are those pertaining to phenomena such as variations in resistance of bacteria, spores, cysts, and viruses, and the appearance of mutants.

Microorganism kill by disinfectants is assumed to follow the CT concept, that is, the product of disinfectant concentration (C) and time (T) yields a constant. CT is widely used in the Surface Water Treatment Rule (SWTR) as a criteria for cyst and virus disinfection. CT is an empirical expression for defining the nature of biological inactivation where:

$$CT = 0.9847C^{0.1758}\text{pH}^{2.7519}\text{temp}^{-0.1467} \tag{4-114}$$

where C = disinfectant concentration
T = contact time between the microorganism and the disinfectant
pH = $-\log\,[\text{H}^+]$
temp = temperature, °C

The relationship shown in Equation 4-114 means that the combination of concentration and time (CT) required to produce a 3-log reduction in *Giardia* cysts by free chlorine can be estimated if the free chlorine concentration, pH, and water temperature are known.

Table 4-20 shows examples of the U.S. EPA–required CT times for free chlorine under the SWTR. EPA used empirical data and a safety factor to develop the table. Therefore, the numbers in the table do not exactly match Equation 4-114. Generally, for a conventional coagulation plant, 0.5-log inactivation of *Giardia* is required and for an untreated surface water, 3-log inactivation is required.

Another pathogen that may require inactivation is *Cryptosporidium. Cryptosporidium* is not inactivated by chlorine, and either ozone or ultraviolet light (UV) is required. Those processes are discussed in later sections.

It has become common in the water industry to express the inactivation credit or to express the physical removal achieved in a plant as *log removal.* This term does not imply that the removal of physical particles is a logarithmic process, but rather that the percent removal found at a point in time can be mathematically represented by a log removal function. Log removal (LR) can be found as:

$$\text{LR} = \log\left(\frac{\text{influent concentrations}}{\text{effluent concentrations}}\right) \tag{4-115}$$

If the log removal for a series of data is to be determined, then the averages for the influent and effluent concentrations can be used. The percent removal (or inactivation) is related to the log removal or inactivation by

$$\%\ \text{removal} = 100 - \frac{100}{10^{\text{LR}}} \tag{4-116}$$

Example 4-29. A city measured the concentration of aerobic spores in its raw and finished water as an indicator of plant performance. Spores are often plentiful in water supplies and are conservative indicators of how well a plant is able to remove *cryptosporidium.* The city data are as follows:

	(spores/L)	
Day of Week	Raw	Finished
Sunday	200,000	16
Monday	145,000	4
Tuesday	170,000	2
Wednesday	150,000	8
Thursday	170,000	10
Friday	180,000	2
Saturday	180,000	3

Calculate the log removal and convert that to percent removal.

TABLE 4-20

CT values (in mg/L · min) for inactivation of *Giardia* cysts by free chlorine at 10°C

Chlorine concentration (mg/L)	pH = 6.0						pH = 7.0						pH = 8.0						pH = 9.0					
	Log inactivations						Log inactivations						Log inactivations						Log inactivations					
	0.5	1.0	1.5	2.0	2.5	3.0	0.5	1.0	1.5	2.0	2.5	3.0	0.5	1.0	1.5	2.0	2.5	3.0	0.5	1.0	1.5	2.0	2.5	3.0
≤ 0.4	12	24	37	49	61	73	17	35	52	69	87	104	25	50	75	99	124	149	35	70	105	139	174	209
0.6	13	25	38	50	63	75	18	36	54	71	89	107	26	51	77	102	128	153	36	73	109	145	182	218
0.8	13	26	39	52	65	78	18	37	55	73	92	110	26	53	79	105	132	158	38	75	113	151	188	226
1	13	26	40	53	66	79	19	37	56	75	93	112	27	54	81	108	135	162	39	78	117	156	195	234
1.2	13	27	40	53	67	80	19	38	57	76	95	114	28	55	83	111	138	166	40	80	120	160	200	240
1.4	14	27	41	55	68	82	19	39	58	77	97	116	28	57	85	113	142	170	41	82	124	165	206	247
1.6	14	28	42	55	69	83	20	40	60	79	99	119	29	58	87	116	145	174	42	84	127	169	211	253
1.8	14	29	43	57	72	86	20	41	61	81	102	122	30	60	90	119	149	179	43	86	130	173	216	259
2	15	29	44	58	73	87	21	41	62	83	103	124	30	61	91	121	152	182	44	88	133	177	221	265
2.2	15	30	45	59	74	89	21	42	64	85	106	127	31	62	93	124	155	186	45	90	136	181	226	271
2.4	15	30	45	60	75	90	22	43	65	86	108	129	32	63	95	127	158	190	46	92	138	184	230	276
2.6	15	31	46	61	77	92	22	44	66	87	109	131	32	65	97	129	162	194	47	94	141	187	234	281
2.8	16	31	47	62	78	93	22	45	67	89	112	134	33	66	99	131	164	197	48	96	144	191	239	287
3	16	32	48	63	79	95	23	46	69	91	114	137	34	67	101	134	168	201	49	97	146	195	243	292

Source: U.S. EPA, 1991.

Solution. The log removal is found by finding the average raw and finished water concentrations and then using Equation 4-115. The average raw concentration is 170,714 spores/L and the finished average concentration is 6.43. Therefore,

$$LR = \log\left(\frac{170{,}714}{6.43}\right) = 4.42$$

And the percent removal is

$$\% = 100 - \frac{100}{10^{4.42}} = 99.996$$

Chlorine/Ammonia Reactions

The reactions of chlorine with ammonia are of great significance in water chlorination processes. When chlorine is added to water that contains natural or added ammonia (ammonium ion exists in equilibrium with ammonia and hydrogen ions), the ammonia reacts with HOCl to form various *chloramines* which, like HOCl, retain the oxidizing power of the chlorine. The reactions between chlorine and ammonia may be represented as follows (AWWA, 1971):

$$NH_3 + HOCl \rightleftharpoons \underset{\text{Monochloramine}}{NH_2Cl} + H_2O \tag{4-117}$$

$$NH_2Cl + HOCl \rightleftharpoons \underset{\text{Dichloramine}}{NHCl_2} + H_2O \tag{4-118}$$

$$NHCl_2 + HOCl \rightleftharpoons NCl_3 + H_2O$$
Trichloramine or nitrogen trichloride (4-119)

The distribution of the reaction products is governed by the rates of formation of monochloramine and dichloramine, which are dependent upon pH, temperature, time, and initial Cl_2:NH_3 ratio. In general, high Cl_2:NH_3 ratios, low temperatures, and low pH levels favor dichloramine formation.

Chlorine also reacts with organic nitrogenous materials, such as proteins and amino acids, to form organic chloramine complexes. Chlorine that exists in water in chemical combination with ammonia or organic nitrogen compounds is defined as *combined available chlorine.*

The oxidizing capacity of free chlorine solutions varies with pH because of variations in the resultant HOCl:OCl^- ratios. This applies also in the case of chloramine solutions as a result of varying $NHCl_2$:NH_2Cl ratios, and where monochloramine predominates at high pH levels. The disinfecting ability of the chloramines is much lower than that of free available chlorine, indicating that the ammonia chloramines also are less reactive than free available chlorine.

Practices of Water Chlorination

Evolution. Early water chlorination practices (variously termed "plain chlorination," "simple chlorination," and "marginal chlorination") were applied for the purpose of disinfection. Chlorine-ammonia treatment was soon thereafter introduced to limit the development of objectionable tastes and odors often associated with marginal chlorine disinfection. Subsequently, "super-chlorination" was developed for the additional purpose of destroying objectionable taste- and odor-producing substances often associated with chlorine-containing organic materials. The introduction of "breakpoint chlorination" and the recognition that chlorine residuals can exist in two distinct forms established contemporary water chlorination as one of two types: combined residual chlorination or free residual chlorination.

Combined Residual Chlorination. Combined residual chlorination practice involves the application of chlorine to water in order to produce, with natural or added ammonia, a combined available chlorine residual, and to maintain that residual through part or all of a water-treatment plant or distribution system. Combined available chlorine forms have lower oxidation potentials than free available chlorine forms and, therefore, are less effective as oxidants. Moreover, they are also less effective disinfectants. In fact, about 25 times as much combined available residual chlorine as free available residual chlorine is necessary to obtain equivalent bacterial kills (*S. typhosa*) under the same conditions of pH, temperature, and contact time. And about 100 times longer contact is required to obtain equivalent bacterial kills under the same conditions and for equal amounts of combined and free available chlorine residuals.

When a combined available chlorine residual is desired, the characteristics of the water will determine how it can be accomplished:

1. If the water contains sufficient ammonia to produce with added chlorine a combined available chlorine residual of the desired magnitude, the application of chlorine alone suffices.
2. If the water contains too little or no ammonia, the addition of both chlorine and ammonia is required.
3. If the water has an existing free available chlorine residual, the addition of ammonia will convert the residual to combined available residual chlorine. A combined available chlorine residual should contain little or no free available chlorine.

The practice of combined residual chlorination is especially applicable after filtration (post treatment) for controlling certain algae and bacterial growths and for providing and maintaining a stable residual throughout the system to the point of consumer use.

Although combined chlorine residual is not a good disinfectant, it has an advantage over free chlorine residual in that it is reduced more slowly and, therefore, persists for a longer time in the distribution system. Thus, it is useful as an indicator of major contamination. Water plant personnel routinely monitor the chlorine level in the distribution system. The presence of available chlorine (either combined or free) indicates that no major contamination has occurred. If major contamination does take place, the combined chlorine residual will be depleted, albeit at a slow rate. This depletion serves as a warning that contamination may have taken place.

A chloramine residual also has the advantage of minimizing the formation of disinfection by-products (DBPs), as chloramines are not a strong enough oxidant to react with the organic matter in the water and form chlorinated hydrocarbons. An increasingly common disinfection strategy is to add chlorine in order to achieve *Giardia* and virus CT and then add ammonia to form chloramines and stop or reduce the DBP formation.

Because of its relatively poor disinfecting power, combined residual chlorination is often preceded by free residual chlorination to ensure the production of potable water.

Free Residual Chlorination. Free residual chlorination practice involves the application of chlorine to water to produce, either directly or through the destruction of ammonia, a free available chlorine residual and to maintain that residual through part or all of a water treatment plant or distribution system. Free available chlorine forms have higher oxidation potentials than combined available chlorine forms and therefore are more effective as oxidants. Moreover, as already noted, they are also the most effective disinfectants.

When free available chlorine residual is desired, the characteristics of the water will determine how it can be accomplished:

1. If the water contains no ammonia (or other nitrogenous materials), the application of chlorine will yield a free residual.
2. If the water does contain ammonia that results in the formation of a combined available chlorine residual, it must be destroyed by applying an excess of chlorine.

With molar Cl_2:NH_3 (as N), concentrations up to 1:1 (5:1 mass basis) monochloramine and dichloramine will be formed. The relative amounts of each depend on pH and other factors. Chloramine residuals generally reach a maximum at equimolar concentrations of chlorine and ammonia. Further increases in the Cl_2:NH_3 ratio result in the oxidation of ammonia and reduction of chlorine. These oxidation/reduction reactions are essentially complete when two moles of chlorine have been added for each mole of ammonia present. Sufficient time must be provided to allow the reaction to go to completion. Chloramine residuals decline to a minimum value, the breakpoint, when the molar Cl_2:NH_3 ratio is about 2:1. At this point, oxidation/reduction reactions are essentially complete. Further addition of chlorine produces free residual chlorine as illustrated in Figure 4-47.

Chlorine Dioxide

Another very strong oxidant is chlorine dioxide. Chlorine dioxide (ClO_2) is formed onsite by combining chlorine and sodium chlorite. Chlorine dioxide is often used as a primary disinfectant, inactivating the bacteria and cysts, followed by the use of chloramine as a distribution system disinfectant. Chlorine dioxide does not maintain a residual long enough to be useful as a distribution-system disinfectant. The advantage of chlorine dioxide over chlorine is that chlorine dioxide does not react with precursors to form DBPs.

When chlorine dioxide reacts in water, it forms two by-products, chlorite and, to some extent, chlorate. These compounds may be associated with a human health risk, and therefore many state regulatory agencies limit the dose of chlorine dioxide to

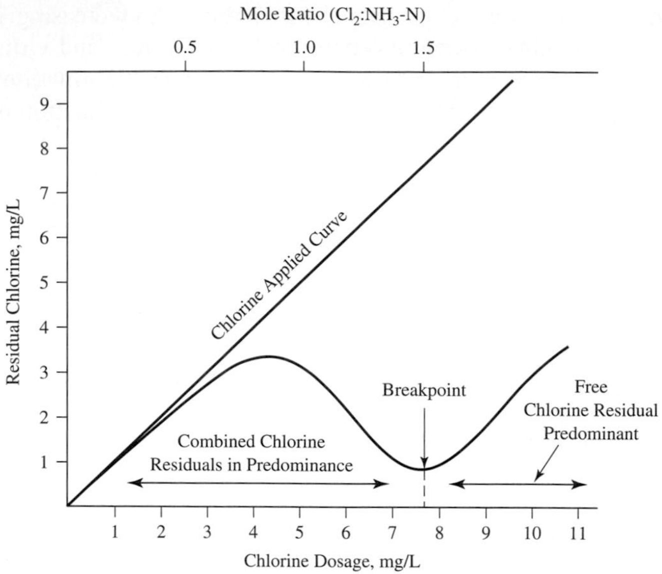

FIGURE 4-47
Breakpoint chlorination. (*Source:* AWWA, 1969.)

1.0 mg/L. In many cases this may not be a sufficient dose to provide good disinfection. Chlorine dioxide use has also been associated with tastes and odors in some communities. The health concerns, taste and odors, and relatively high cost have tended to limit the use of chlorine dioxide. However, many utilities have successfully used chlorine dioxide as a primary disinfectant.

Ozonation

Ozone is a pungent-smelling, unstable gas. It is a form of oxygen in which three atoms of oxygen are combined to form the molecule O_3. Because of its instability, it is generated at the point of use. The ozone-generating apparatus commonly is a discharge electrode. To reduce corrosion of the generating apparatus, air is passed through a drying process and then into the ozone generator. The generator consists of two plates or a wire and tube with an electric potential of 15,000 to 20,000 volts. The oxygen in the air is dissociated by the impact of electrons from the discharge electrode. The atomic oxygen then combines with atmospheric oxygen to form ozone in the following reaction:

$$O + O_2 \rightarrow O_3 \tag{4-120}$$

Approximately 0.5 to 1.0 percent by volume of the air exiting from the apparatus will be ozone. The resulting ozone-air mixture is then diffused into the water that is to be disinfected.

Ozone is widely used in drinking water treatment in Europe and is continuing to gain popularity in the United States. It is a powerful oxidant, more powerful even than hypochlorous acid. It has been reported to be more effective than chlorine in destroying viruses and cysts. Table 4-21 shows the *CT* values for ozone, chlorine dioxide, and

TABLE 4-21
CT values for Giardia and Cryptosporidium inactivation

Temperature, (°C)	Chlorine dioxide	Ozone	Chloramine
*Giardia** at 3-log inactivation			
0.5	63	2.9	3,800
5	26	1.9	2,288
10	23	1.43	1,850
15	19	0.95	1,500
20	15	0.72	1,100
25	11	0.48	750
Cryptosporidium† at 2-log inactivation			
0.5	1,275	48	N/A
5	858	32	N/A
10	553	20	N/A
15	357	12	N/A
20	232	7.8	N/A
25	150	4.9	N/A

Sources: *U.S EPA, 1991.
†U.S. Environmental Protection Agency, National Primary Drinking Water Regulations: Long Term 2 Enhanced Surface Water Treatment Rule.

chloramine for 3-log inactivation (99.90 percent removal) of *Giardia* cysts and for 2-log inactivation of *Cryptosporidium* oocysts. Table 4-21 can be compared to the values for chlorine in Table 4-20 to note how strong an oxidant ozone is. (Also the weak disinfection ability of chloramines is obvious.)

In addition to being a strong oxidant, ozone has the advantage of not forming THMs or any of the chlorinated DBPs. As with chlorine dioxide, ozone will not persist in the water, decaying back to oxygen in minutes. Therefore, a typical flow schematic would be to add ozone either to the raw water or between the sedimentation basins and filter for primary disinfection, followed by chloramine addition after the filters as the distribution system disinfectant.

Ultraviolet Radiation

Light is important to almost all life forms. We "see" only a very small fraction of the colors of light. The ultraviolet range of light falls beyond the violet end of the rainbow. Table 4-22 outlines the spectral ranges of interest in photochemistry.

Light photons with wavelengths longer than 1000 nanometers (nm) have a photon energy too small to cause chemical change when absorbed, and photons with wavelengths shorter than 100 nm have so much energy that ionization and molecular disruptions characteristic of radiation chemistry prevail.

Little photochemistry occurs in the near infrared range except in some photosynthetic bacteria. The visible range is completely active for photosynthesis in green

TABLE 4-22
Spectral ranges of interest in photochemistry

Range name	Wavelength range (nm)
Near infrared	700–1,000
Visible	400–700
Ultraviolet	
UVA	315–400
UVB	280–315
UVC	200–280
Vacuum ultraviolet (VUV)	100–200

plants and algae. Most studies in photochemistry involve the ultraviolet range. The ultraviolet range is divided into three categories connected with the human skin's sensitivity to ultraviolet light. The UVA range causes changes to the skin that lead to tanning. The UVB range can cause skin burning and is prone to induce skin cancer. The UVC range is extremely dangerous since it is absorbed by proteins and can lead to cell mutations or cell death.

Disinfection of water and wastewater using ultraviolet light has been practiced extensively in Europe and is becoming more common in the United States. Ultraviolet light disinfects water by rendering pathogenic organisms incapable of reproducing. This is accomplished by disrupting the genetic material in cells. The genetic material in cells, namely DNA, will absorb light in the ultraviolet range—primarily between 200 nm and 300 nm wavelengths (UV light is most strongly absorbed by DNA at 253.7 nm). If the DNA absorbs too much UV light, it will be damage and will be unable to replicate. It has been found that the energy required to damage the DNA is much less than that required to actually destroy the organism. The effect is the same, however, since if a microorganism cannot reproduce, it cannot cause an infection.

The inactivation of microorganisms by UV is directly related to UV dose, a concept similar to CT used for other common disinfections including chlorine and ozone. The average UV dose is calculated as follows:

$$D = It \tag{4-121}$$

where D = UV dose
I = average intensity, mW/cm^2
t = average exposure time, s

The survival fraction is calculated as follows:

$$\text{Survival fraction} = \log\left(\frac{N}{N_0}\right) \tag{4-122}$$

where N = organism concentration after inactivation
N_0 = organism concentration before inactivation

The equation for UV dose indicates that dose is directly proportional to exposure time and thus inversely proportional to system flow rate. UV intensity (I) is a function of water UV transmittance and UV reactor geometry as well as lamp age and fouling. UV intensity can be estimated by mathematical modeling and verified by bioassay. Exposure time is estimated from the UV reactor specific hydraulic characteristics and flow patterns.

The major factor affecting the performance of UV disinfection systems is the influent water quality. Particles, turbidity, and suspended solids can shield pathogens from UV light or scatter UV light to prevent it from reaching the target microorganism, thus reducing its effectiveness as a disinfectant. Some organic compounds and inorganic compounds (such as iron and permanganate) can reduce UV transmittance by absorbing UV energy, requiring higher levels of UV to achieve the same dose. Therefore, it is recommended that UV systems be installed downstream of the filters so that removal of particles and organic and inorganic compounds is maximized upstream of UV.

Water turbidity and UV transmittance are commonly used as process controls at UV facilities. The UV percent transmittance of a water sample is measured by a UV-range spectrophotometer set at a wavelength of 253.7 nm using a 1-cm-thick layer of water. The water UV transmittance is related to UV absorbance (A) at the same wavelength by the equation:

$$\text{Percent transmittance} = 100 \times 10^{-A} \tag{4-123}$$

For example, a water UV absorbance of 0.022 per cm corresponds to a water percent transmittance of 95 (i.e., at 1 cm from the UV lamp, 95 percent of lamp output is left). Similarly, a UV absorbance of 0.046 per cm is equivalent to 90 percent UV transmittance.

UV has been found to be very effective for the disinfection of *Cryptosporidium, Giardia,* and viruses. U.S. EPA has established UV dose requirements. Table 4-23 shows the U.S. EPA requirements.

TABLE 4-23
UV dose requirements for *Cryptosporidium, Giardia lamblia,* and virus

Log Credit	*Cryptosporidium* UV dose (mJ/cm^2)	*Giardia lamblia* UV dose (mJ/cm^2)	Virus UV dose (mJ/cm^2)
0.5	1.6	1.5	39
1.0	2.5	2.1	58
1.5	3.9	3.0	79
2.0	5.8	5.2	100
2.5	8.5	7.7	121
3.0	12	11	143
3.5	15	15	163
4.0	22	22	186

Although the *Cryptosporidium* log credit requirements for say 2-log inactivation are 5.8 mJ/cm^2, many design engineers will use values of 20 to 40 mJ/cm^2 for additional safety.

UV electromagnetic energy is typically generated by the flow of electrons from an electrical source through ionized mercury vapor in a lamp. Several manufactures have developed systems to align UV lamps in vessels or channels to provide UV light in the germicidal range for inactivation of bacteria, viruses, and other microorganisms. The UV lamps are similar to household fluorescent lamps, except that fluorescent lamps are coated with phosphorus, which converts the UV light to visible light. Ballasts (i.e., transformers) that control the power of the UV lamps are either electronic or electromagnetic. Electronic ballasts offer several potential advantages including lower lamp operating temperatures, higher efficiencies, and longer ballast life.

Both low pressure and medium-pressure lamps are available for disinfection applications. Low-pressure lamps emit their maximum energy output at a wavelength of 253.7 nm, while medium-pressure lamps emit energy with wavelengths ranging from 180 to 1370 nm. The intensity of medium-pressure lamps is much greater than low-pressure lamps. Thus, fewer medium-pressure lamps are required for an equivalent

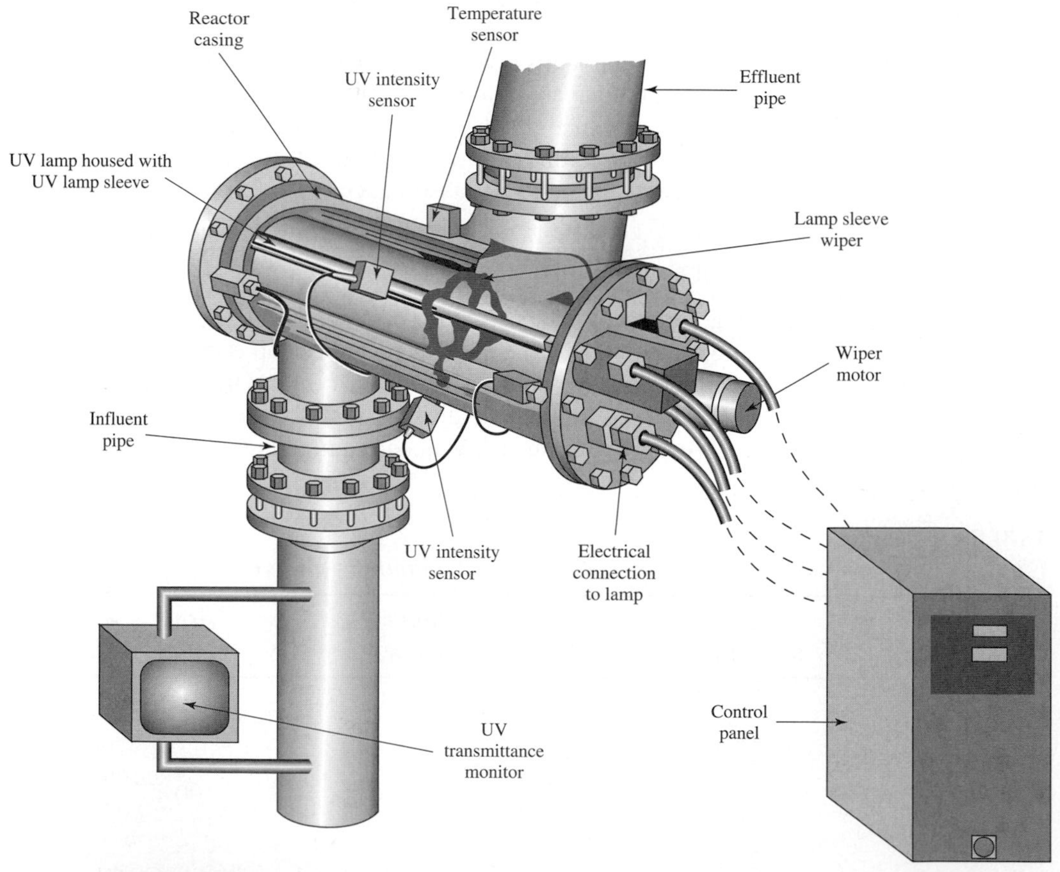

FIGURE 4-48
UV disinfection system schematic. (*Source:* Aquionics.)

dosage. For small systems, the medium-pressure system may consist of a single lamp. Although both types of lamps work equally well for inactivation of organisms, low-pressure UV lamps are recommended for small systems because of the reliability associated with multiple low-pressure lamps as opposed to a single medium-pressure lamp, and for adequate operation during cleaning cycles.

Most conventional UV reactors are available in two types: namely, closed vessel and open channel. For drinking water applications, the closed vessel is generally the preferred UV reactor for the following reasons (U.S. EPA, 1996):

- Smaller footprint
- Minimized pollution from airborne material
- Minimal personnel exposure to UV
- Modular design for installation simplicity

Figure 4-48 shows a conventional closed-vessel UV reactor.

Advanced Oxidation Processes (AOPs)

AOPs are combinations of disinfectants designed to produce hydroxyl radicals ($OH\cdot$). Hydroxyl radicals are highly reactive nonselective oxidants able to decompose many organic compounds. Most noteworthy of the AOP processes is ozone plus hydrogen peroxide.

4-8 ADSORPTION

Adsorption is a mass transfer process wherein a substance is transferred from the liquid phase to the surface of a solid where it is bound by chemical or physical forces.

Generally, in water treatment, the adsorbent (solid) is activated carbon, either granular (GAC) or powdered (PAC). PAC is fed to the raw water in a slurry and is generally used to remove taste- and odor-causing substances or to provide some removal of synthetic organic chemicals (SOCs). GAC is added to the existing filter system by replacing the anthracite with GAC, or an additional contactor is built and is placed in the flow scheme after primary filtration. The design of the GAC contactor is very similar to a filter box, although deeper.

At present, the applications of adsorption in water treatment in the United States are predominately for taste and odor removal. However, adsorption is increasingly being considered for removal of SOCs, VOCs, and naturally occurring organic matter, such as THM precursors and DBPs.

Biologically derived earthy-musty odors in water supplies are a widespread problem. Their occurrence interval and concentration vary greatly from season to season and is often unpredictable. As mentioned, one of the most popular methods for removing these compounds is the addition of PAC to the raw water. The dose is generally less than 10 mg/L. The advantage of PAC is that the capital equipment is relatively inexpensive and it can be used on an as-needed basis. The disadvantage is that the adsorption is often incomplete. Sometimes even doses of 50 mg/L are not sufficient.

As an alternative for taste and odor control, many plants have replaced the anthracite in the filters with GAC. The GAC will last from one to three years and then must be replaced. It is very effective in removing many taste and odor compounds.

Concern about SOCs in drinking water has motivated interest in adsorption as a treatment process for removal of toxic and potentially carcinogenic compounds present in minute, but significant, quantities. Few other processes can remove SOCs to the required low levels. Generally, GAC is used for SOC removal either as a filter media replacement or as a separate contactor. The data for how long the GAC will last for any given SOC are somewhat limited and must be evaluated on a case-by-case basis. If the GAC is designed to remove SOCs from a periodic "spill" into the source water, then filter media replacement may be adequate because the GAC is not being used every day and is acting as a barrier. However, if the GAC is to be used continuously for SOC removal, then a separate contactor may be warranted.

GAC has been proposed to be used to remove naturally occurring organic matter that would, in turn, reduce the formation of DBPs. Testing has shown that GAC will remove these organics. It must operate in a separate contactor since the depth of a conventional filter is inadequate. The GAC will typically last 90 to 120 days until it loses its adsorptive capacity. Because of its short life, the GAC needs to be regenerated by burning in a high-temperature furnace. This can be done on-site or can be done by returning the GAC to the manufacture.

GAC has also been considered for removal of THMs. However, the capacity is very low and the carbon may only last up to 30 days. GAC is not considered practical for THM removal.

4-9 MEMBRANES

A membrane is a thin layer of material that is capable of separating materials as a function of their physical and chemical properties when a driving force is applied across the membrane. In the case of water treatment, the driving force is supplied by using a high pressure pump and the membrane type is selected based on the constituents to be removed.

In the membrane process, the feed stream is divided into two streams, the concentrate or reject stream, and the permeate or product stream as shown in Figure 4-49. The membrane is at the heart of every membrane process and can be considered a barrier

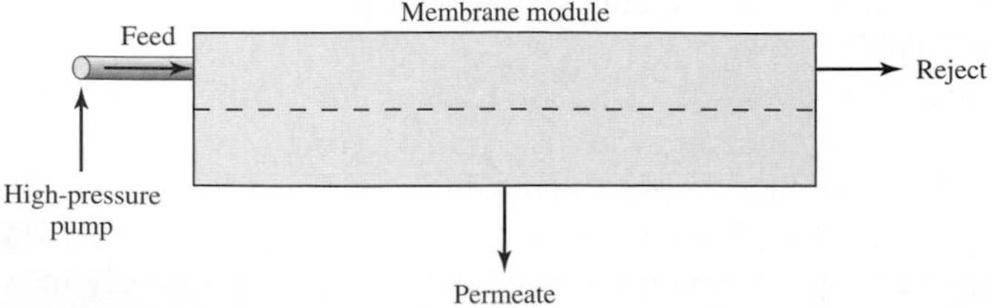

FIGURE 4-49
Schematic representation of a membrane process.

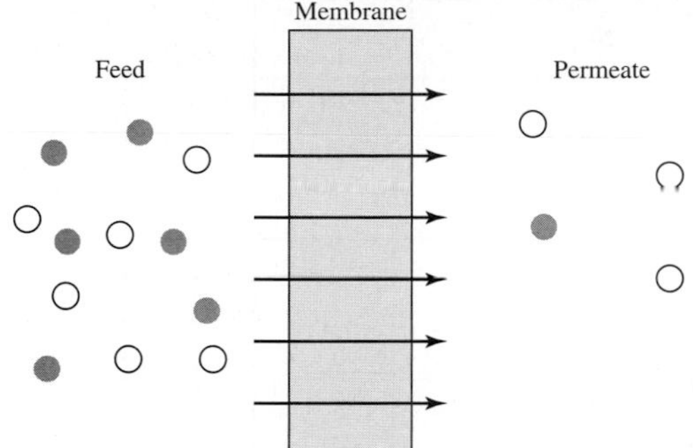

FIGURE 4-50
Schematic representation of contaminants separated by a membrane.

between the feed and product water that does not allow certain contaminants to pass, as represented by Figure 4-50. The performance or efficiency of a given membrane is determined by its selectivity and the applied flow. The efficiency is called the flux and is defined as the volume flowing through the membrane per unit area and time, $m^3/m^2 \cdot s$. The selectivity of a membrane toward a mixture is expressed by its retention, R, and is given by

$$R = \frac{c_p - c_f}{c_p} \times 100\% \qquad (4\text{-}124)$$

where c_f = contaminant concentration in feed
c_p = contaminant concentration in permeate

The value of R varies between 100 percent (complete retention) and 0 percent (no retention).

Membrane technology is becoming increasingly popular as an alternative treatment technology for drinking water. The anticipation of more stringent water quality regulations, a decrease in availability of adequate water resources, and an emphasis on water for reuse has made membrane processes more viable as a water treatment process. As advances are being made in membrane technology, capital and operation and maintenance costs continue to decline, further endorsing the use of membrane treatment techniques.

Membranes can be described by a variety of criteria including: (Jacangelo et al., 1997)

- membrane pore size
- molecular weight cutoff (MWCO)
- membrane material and geometry
- targeted materials to be removed
- type of water quality to be treated, and/or
- treated water quality

Along with these criteria, membrane processes can also be categorized broadly into pressure-driven and electrically driven processes. This discussion is limited to

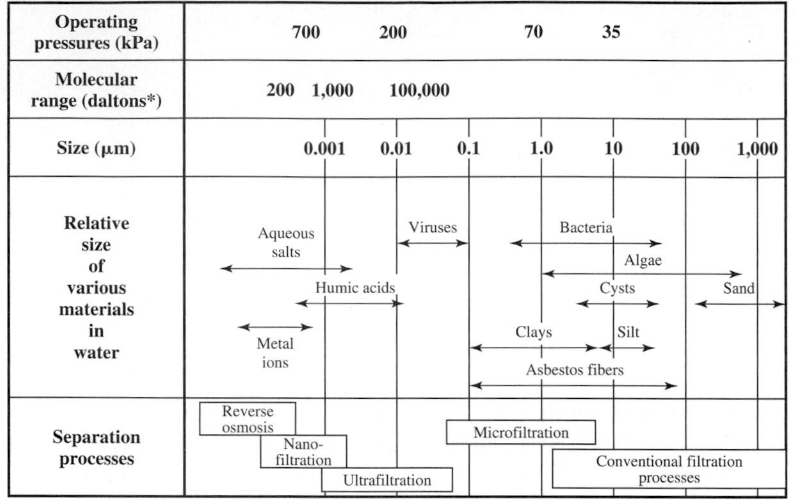

*Dalton is psi × 7 = kPa a unit of mass equal to 1/16 the mass of the lightest and most abundant isotope of oxygen.

FIGURE 4-51

Schematic comparison of selected separation processes. (*Source:* Jacangelo et. al., 1997.)

pressure-driven membrane processes. Figure 4-51 summarizes the various pressure-driven membrane processes and selected materials removed by each. A brief description of each follows.

- *Reverse osmosis* (*RO*). This process has traditionally been employed for the removal of salts from brackish water and seawater. It depends upon applying high pressures across the membrane (in the range of 1,000 to 8,000 kPa) in order to overcome the osmotic pressure differential between the saline feed and product waters.
- *Nano-filtration* (*NF*). This process, also called membrane softening, lies between RO and ultrafiltration. This membrane process employs 500 to 1,000 kPa for operation. While it provides removal of ions contributing to hardness, i.e., calcium and magnesium, the technology is also very effective for removal of color and DBP precursors.
- *Ultrafiltration* (*UF*). UF membranes cover a wide range of MWCOs and pore sizes. Operational pressures range from 70 to 700 kPa, depending on the application. "Tight" UF membranes (MWCO = 1,000 daltons) may be employed for removal of some organic materials from freshwater, while the objective of "loose" membranes (MWCO > 50,000 daltons, 70 to 200 kPa) is primarily for liquid/solid separation, i.e., particle and microbial removal.
- *Microfiltation* (*MF*). A major difference between MF and loose UF is membrane pore size; the pores of MF (= 0.1 μm or greater) are approximately an order of magnitude greater than those of UF. The primary application for this membrane process is particulate and microbial removal.

4-10 WATER PLANT RESIDUALS MANAGEMENT

The precipitated chemicals and other materials removed from raw water to make it potable and palatable are termed *residuals*. Satisfactory treatment and disposal of water treatment plant residuals can be the single most complex and costly operation in the plant.

Residuals withdrawn from coagulation and softening plants are composed largely of water, and the residuals are often referred to a sludge. As much as 98 percent of the sludge mass may be water. Thus, for example, 20 kg of solid chemical precipitate is accompanied by 980 kg of water. Assuming equal densities for the precipitate and water (a bad assumption at best), approximately 1 m^3 of sludge is produced for each 20 kg of chemicals added to the water. For even a small plant (say 0.05 m^3/s) this might mean up to 800 m^3/y of sludge—a substantial volume to say the least!

Water treatment plants and the residuals they produce can be broadly divided into four general categories. First are those treatment plants that coagulate, filter, and oxidize a surface water for removal of turbidity, color, bacteria, algae, some organic compounds, and often iron and/or manganese. These plants generally use alum or iron salts for coagulation and produce two waste streams. The majority of the waste produced from these plants is sedimentation basin (or clarifier) sludge and spent filter backwash water (SFBW). The second type of treatment plants are those that practice softening for the removal of calcium and magnesium by the addition of lime, sodium hydroxide, and/or soda ash. These plants produce clarifier basin sludges and SFBW. On occasion, plants practice both coagulation and softening. Softening plant wastes can also contain trace inorganics such as radium that could affect their proper handling. The third type of plants are those that are designed to specifically remove trace inorganics such as nitrate, fluoride, radium, arsenic, etc. These plants use processes such as ion exchange, reverse osmosis, or adsorption. They produce liquid residuals or solid residuals such as spent adsorption material. The fourth category of treatment plants are those that produce air-phase residuals during the stripping of volatile compounds. The major types of treatment plant residuals produced are shown in Table 4-24. Because 95 percent of the residuals produced are coagulants or softening sludge, they will be stressed in this section.

Hydrolyzing metal salts or synthetic organic polymers are added in the water treatment process to coagulate suspended and dissolved contaminants and yield relatively clean water suitable for filtration. Most of these coagulants and the impurities they remove settle to the bottom of the settling basin where they become part of the sludge. These sludges are referred to as alum, iron, or polymeric sludge according to which primary coagulant is used. These wastes account for approximately 70 percent of the water plant waste generated. The sludges produced in treatment plants where water softening is practiced using lime or lime and soda ash account for an additional 25 percent of the industry's waste production. It is therefore apparent that most of the waste generation involves water treatment plants using coagulation or softening processes.

The most logical sludge management program attempts to use the following approach in disposing of the sludge:

1. Minimization of sludge generation
2. Chemical recovery of precipitates
3. Sludge treatment to reduce volume
4. Ultimate disposal in an environmentally safe manner

TABLE 4-24
Major water treatment plant residuals

Solid/Liquid Residuals
1. Alum sludges
2. Iron sludges
3. Polymeric sludges
4. Softening sludges
5. SFBW
6. Spent GAC or discharge from carbon systems
7. Slow sand filter cleanings
8. Residuals from iron and manganese removal plants
9. Spent precoat filter media

Liquid-Phase Residuals
10. Ion-exchange regenerant brine
11. Pregenerant from activated alumina
12. Reverse osmosis reject

Gas-Phase Residuals
13. Air stripping off-gases

With a short digression to identify the sources of water plant sludges and their production rates, we have organized the following discussion along these lines.

Sludge Production and Characteristics

In water treatment plants, sludge is most commonly produced in the following treatment processes: presedimentation, sedimentation, and filtration (filter backwash).

Presedimentation. When surface waters are withdrawn from watercourses that contain a large quantity of suspended materials, presedimentation prior to coagulation may be practiced. The purpose of this is to reduce the accumulation of solids in subsequent units. The settled material generally consists of fine sand, silt, clays, and organic decomposition products.

Softening Sedimentation Basin. The residues from softening by precipitation with lime [$Ca(OH)_2$] and soda ash (Na_2CO_3) will vary from a nearly pure chemical to a highly variable mixture. The softening process discussed in Section 4-3 produces a sludge containing primarily $CaCO_3$ and $Mg(OH)_2$.

Theoretically, each mg/L of calcium hardness removed produces 1 mg/L of $CaCO_3$ sludge; each mg/L of magnesium hardness removed produces 0.6 mg/L of sludge; and each mg/L of lime added produces 1 mg/L of sludge. The theoretical sludge production can be calculated as:

$$M_s = 86.4\,Q(2\,\text{CaCH} + 2.6\,\text{MgCH} + \text{CaNCH} + 1.6\,\text{MgNCH} + CO_2) \tag{4-125}$$

where M_s = sludge production (kg/d)
CaCH = calcium carbonate hardness removed as $CaCO_3$ (mg/L)

MgCH = magnesium carbonate hardness removed as $CaCO_3$ (mg/L)
Q = plant flow, m^3/s
CaNCH = noncarbonate calcium hardness removed as $CaCO_3$ (mg/L)
MgNCH = noncarbonate magnesium hardness removed as $CaCO_3$ (mg/L)
CO_2 = carbon dioxide removed by lime addition, as $CaCO_3$ (mg/L)

When surface waters are softened, this equation is not valid. There will be additional sludge from coagulation of suspended materials and precipitation of metal coagulants. The solids content of lime-softening sludge in the sedimentation basin ranges between 2 and 15 percent. A nominal value of 10 percent solids is often used.

Coagulation Sedimentation Basin. Aluminum or iron salts are generally used to accomplish coagulation. (The chemistry of the two salts was discussed in Section 4-2.) The pH range of 6 to 8 is where most water treatment plants effect the coagulation process. In this range the insoluble aluminum hydroxide complex of $Al(H_2O)_3(OH)_3$ probably predominates. This species results in the production of 0.44 kg of chemical sludge for each kg of alum added. Any suspended solids present in the water will produce an equal amount of sludge. The amount of sludge produced per turbidity unit is not as obvious; however, in many waters a correlation does exist. Carbon, polymers, and clay will produce about 1 kg of sludge per kg of chemical addition. The sludge production for alum coagulation may then be approximated by:

$$M_s = 86.40\,Q(0.44\text{A} + \text{SS} + \text{M}) \tag{4-126}$$

where M_s = dry sludge produced, kg/d
Q = plant flow, m^3/s
A = alum dose, mg/L
SS = suspended solids in raw water, mg/L
M = miscellaneous chemical additions such as clay, polymer, and carbon, mg/L

Alum sludge leaving the sedimentation basin usually has a suspended solids content in the range of 0.5 to 2 percent. It is often less than 1 percent. Twenty to 40 percent of the solids are organic; the remainder are inorganic or silts. The pH of alum sludge is normally in the 5.5 to 7.5 range. Alum sludge from sedimentation basins may include large numbers of microorganisms, but it generally does not exhibit an unpleasant odor. The sludge flow rate is often in the range of 0.3 to 1 percent of the treatment plant flow.

Spent Filter Backwash Water. All water treatment plants that practice filtration produce a large volume of washwater containing a low suspended solids concentration. The volume of backwash water is usually 2 to 3 percent of the treatment plant flow. The solids in backwash water resemble those found in sedimentation units. Since filters can support biological growth, the spent filter backwash water may contain a larger fraction of organic solids than do the solids from the sedimentation basins.

Mass Balance Analysis. Clarifier sludge production can be estimated by a mass balance analysis of the sedimentation basin. Since there is no reaction taking place, the mass balance equation reduces to the form:

$$\text{Accumulation Rate} = \text{Input Rate} - \text{Output Rate} \tag{4-127}$$

The input rate of solids may be estimated from Equation 4-125 or 4-126. To estimate the mass flow (output rate) of solids leaving the clarifier through the weir, you must have an estimate of the concentration of solids and the flow rate. The mass flow out through the weir is then

$$\text{Weir output rate} = (\text{Concentration, g/m}^3)(\text{Flow Rate, m}^3\text{/s}) = \text{g/s}$$

Example 4-30. A coagulation treatment plant with a flow of 0.5 m^3/s is dosing alum at 23.0 mg/L. No other chemicals are being added. The raw-water suspended-solids concentration is 37.0 mg/L. The effluent suspended-solids concentration is measured at 12.0 mg/L. The sludge solids content is 1.00 percent and the specific gravity of the sludge solids is 3.01. What volume of sludge must be disposed of each day?

Solution. The mass-balance diagram for the sedimentation basin is

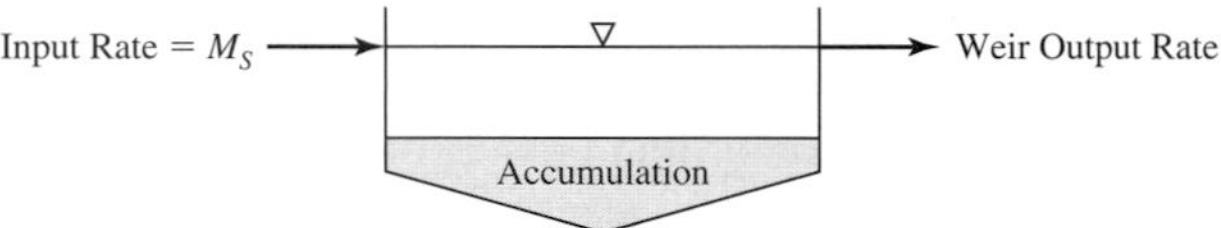

The mass of solids (sludge) flowing into the clarifier is estimated from Equation 4-126:

$$\begin{aligned} M_s &= 86.40(0.50 \text{ m}^3\text{/s})(0.44(23.0 \text{ mg/L}) + 37.0 \text{ mg/L} + 0) \\ &= 2{,}035.58 \text{ kg/d} \end{aligned}$$

Recognizing that $g/m^3 = mg/L$, the mass of solids leaving the weir is

$$\begin{aligned} \text{Weir output rate} &= (12.0 \text{ g/m}^3)(0.50 \text{ m}^3\text{/s})(86{,}400 \text{ s/d})(10^{-3} \text{ kg/g}) \\ &= 518.4 \text{ kg/d} \end{aligned}$$

The accumulation is then

$$\text{Accumulation} = 2{,}035.58 - 518.4 = 1{,}517.18 \text{ or } 1{,}517 \text{ kg/d}$$

Because this is a dry mass and the sludge has only 1.00 percent solids, we must account for the volume of water in estimating the volume to be removed each day. Using Equation 4-4, we can solve for the mass of water (W_o):

$$1.00 = \frac{1{,}517}{1{,}517 + W_o}(100)$$

$$W_o = 150{,}183 \text{ kg/d}$$

Now we use the definition of density (mass per unit volume) to find the volume of sludge and water:

$$\text{Volume} = \frac{\text{Mass}}{\text{Density}}$$

$$V_T = \text{volume of solids} + \text{volume of water}$$
$$= \frac{1{,}517 \text{ kg/d}}{(3.01)(1{,}000 \text{ kg/m}^3)} + \frac{150{,}183 \text{ kg/d}}{1{,}000 \text{ kg/m}^3}$$
$$= 0.50 + 150.18 = 150.7 \text{ m}^3\text{/d or } 150 \text{ m}^3\text{/d}$$

The specific gravity of the solids is 3.01 and the density of water is 1,000 kg/m^3.

Obviously, the solids account for only a small fraction of the total volume. This is why sludge dewatering is an important part of the water treatment process.

Minimization of Sludge Generation for Sustainability

Minimizing sludge generation can have an advantageous effect on the requirements and economics of handling, treating, and disposing of water treatment plant sludges. Minimization also results in the conservation of raw materials, energy, and labor.

There are three methods to minimize the quantity of metal hydroxide precipitates in the sludge:

1. Changing the water treatment process to direct filtration
2. Substituting other coagulants and, in particular, using polymers that are more effective at lower dosage
3. Conserving chemicals by determining optimum dosage at frequent intervals as raw water characteristics change

Sludge Treatment

The treatment of solid/liquid wastes produced in water treatment processes involves the separation of the water from the solid constituents to the degree necessary for the selected disposal method. Therefore, the required degree of treatment is a direct function of the ultimate disposal method.

There are several sludge treatment methodologies which have been practiced in the water industry. Figure 4-52 shows the most common sludge-handling options available, listed by general categories of thickening, dewatering, and disposal. In choosing a combination of the possible treatment process trains, it is probably best to first identify the available disposal options and their requirements for a final cake solids concentration. Most landfill applications will require a "handleable" sludge and this may limit the type of dewatering devices which are acceptable. Methods and costs of transportation may affect the decision "how dry is dry enough." The criteria should not be to simply reach a given solids concentration but rather to reach a solids concentration of desired properties for the handling, transport, and disposal options available.

Table 4-23 shows a generalized range of results which have been obtained for final solids concentrations from different dewatering devices for coagulant and lime sludges.

To give you an appreciation of these solids concentrations, a sludge cake with 35 percent solids would have the consistency of butter, while a 15 percent sludge would have a consistency much like rubber cement.

After removal of the sludge from the clarifier or sedimentation basin, the first treatment step is usually thickening. Thickening assists the performance of any subsequent treatment, gets rid of much water quickly, and helps to equalize flows to the subsequent

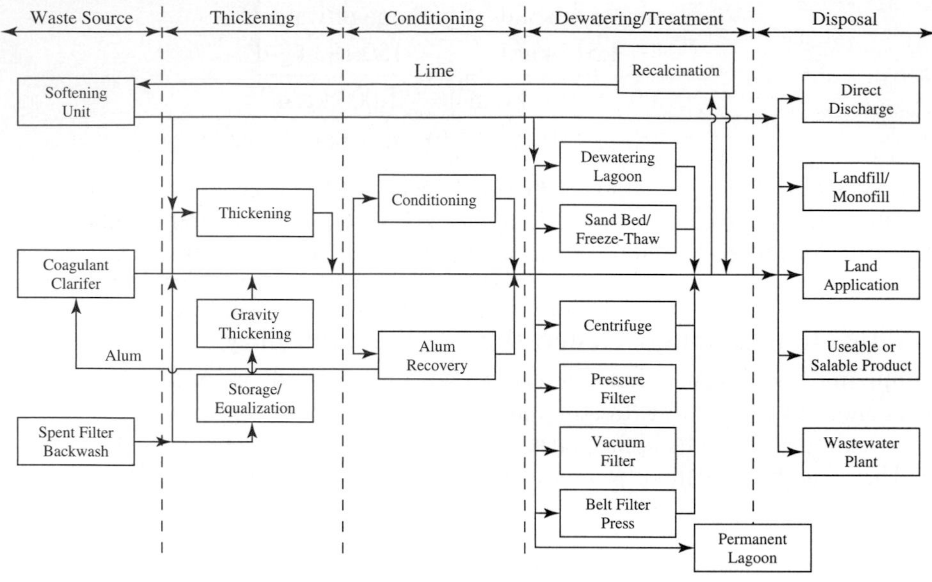

FIGURE 4-52
Sludge handling options.

treatment device. An approximation for determining sludge volume reduction via thickening is given by:

$$\frac{\forall_2}{\forall_1} = \frac{P_1}{P_2} \tag{4-128}$$

where $\forall_1$ = volume of sludge before thickening, m^3
$\forall_2$ = volume of sludge after thickening, m^3
P_1 = percent of solids concentration of sludge before thickening
P_2 = percent of solids concentration after thickening

Thickening is usually accomplished by using circular settling basins similar to a clarifier (Figure 4-53). Thickeners can be designed based on pilot evaluations or using

TABLE 4-25
Range of cake solid concentrations obtainable

	Lime sludge, %	Coagulation sludge, %
Gravity thickening	15–30	3–4
Basket centrifuge		10–15
Scroll centrifuge	55–65	20–25
Belt filter press		18–25
Vacuum filter	45–65	n/a
Pressure filter	55–70	30–45
Sand drying beds	50	20–25
Storage lagoons	50–60	7–15

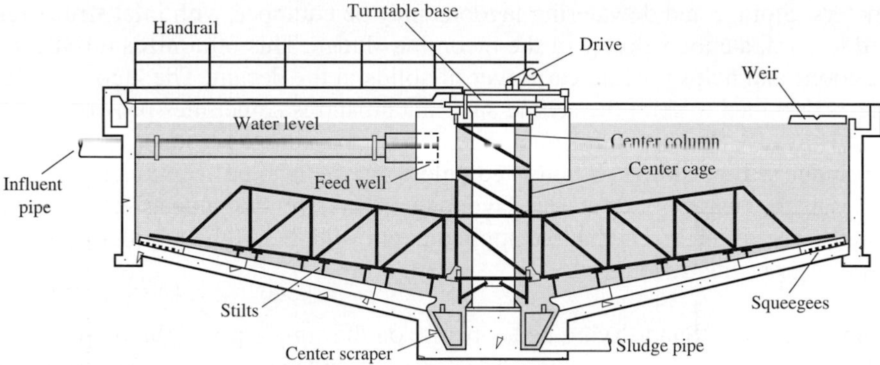

FIGURE 4-53
Continuous-flow gravity thickener. (Courtesy of Link Belt.)

data obtained from similar plants. Lime sludges are typically loaded at 100 to 200 $kg/m^2 \cdot d$, and coagulant sludge loading rates are about 15 to 25 $kg/m^2 \cdot d$.

Following thickening of the sludge, dewatering can take place by either mechanical or nonmechanical means. In nonmechanical devices, sludge is spread out with the free water draining and the remaining water evaporating. Sometimes the amount of free water available to drain is enhanced by natural freeze-thaw cycles. In mechanical dewatering, some type of device is used to force the water out of the sludge.

We begin our discussion with the nonmechanical methods and follow with the mechanical methods.

Lagoons. A lagoon is essentially a large hole that is dug out for the sludge to flow into. Lagoons can be constructed as either storage lagoons or dewatering lagoons. Storage lagoons are designed to store and collect the solids for some predetermined amount of time. They will generally have decant capabilities but no underdrain system. Storage lagoons should be equipped with sealed bottoms to protect the groundwater. Once the storage lagoon is full or decant can no longer meet discharge limitations, it must be abandoned or cleaned. To facilitate drying, the standing water may be removed by pumping, leaving a wet sludge. Coagulant sludges can only be expected to reach a 7 to 10 percent solids concentration in storage lagoons. The remaining solids must be either cleaned out wet or allowed to evaporate. Depending upon the depth of the wet solids, evaporation can take years. The top layers will often form a crust, preventing evaporation of the bottom layers of sludge.

The primary difference between a dewatering lagoon and a storage lagoon is that a dewatering lagoon has a sand and underdrain-bottom, similar to a drying bed. Dewatering lagoons can be designed to achieve a dewatered sludge cake. The advantage of a dewatering lagoon over a drying bed is that storage is built into the system to assist in meeting peak solids production or to assist in handling sludge during wet weather. The disadvantage of bottom sand layers compared to conventional drying beds is that the bottom sand layers can plug up or "blind" with multiple loadings, thereby increasing the required surface area. Polymer treatment can be useful in preventing this sand blinding.

Storage lagoons, which are generally earthen basins, have no size limitations but have been designed in areas from 2,000 to 60,000 m^2, ranging in depth from 2 to 10 or more meters. Storage and dewatering lagoons may be equipped with inlet structures designed to dissipate the velocity of the incoming sludge. This minimizes turbulence in the lagoons and helps prevent carryover of solids in the decant. The lagoon outlet structure is designed to skim the settled supernatant and is sometimes provided with flash boards to vary the draw-off depths. Any design of a storage lagoon must consider how the sludge will be ultimately removed unless the site is to be abandoned.

The basis for design of dewatering lagoons is essentially the same as that for sand drying beds. The difference is that the applied depth is high and the number of applications per year is greatly reduced.

Sand-Drying Beds. Sand-drying beds operate on the simple principle of spreading the sludge out and letting it dry. As much water as possible is removed by drainage or decant and the rest of the water must evaporate until the desired final solids concentration is reached. Sand-drying beds have been built simply by cleaning an area of land, dumping the sludge, and hoping something happens. At the other end of the spectrum sophisticated automated drying systems have been built.

Drying beds may be roughly categorized as follows:

1. Conventional rectangular beds with side walls and a layer of sand on gravel with underdrain piping to carry away the liquid. They are built either with or without provisions for mechanical removal of the dried sludge and with or without a roof or a greenhouse-type covering (see Figure 4-54).
2. Paved rectangular drying beds with a center sand drainage strip with or without heating pipes buried in the paved section and with or without covering to prevent incursion of rain.
3. "Wedge-water" drying beds that include a wedge wire septum incorporating provision for an initial flood with a thin layer of water, followed by introduction of liquid sludge on top of the water layer, controlled formation of cake, and provision for mechanical cleaning.
4. Rectangular vacuum-assisted drying beds with provision for application of vacuum to assist gravity drainage.

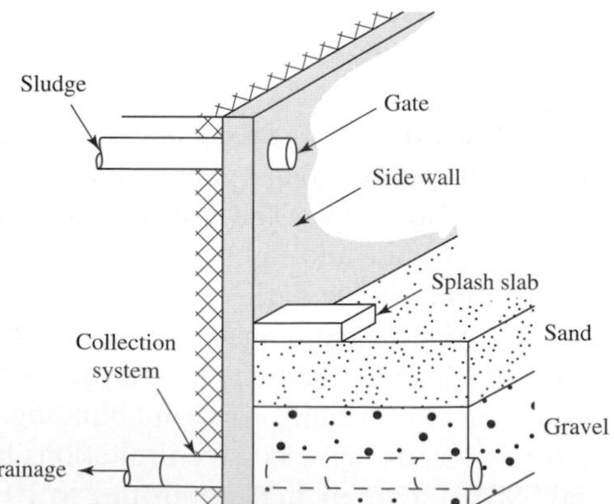

FIGURE 4-54
Typical sludge drying bed construction. (*Source:* U.S. Environmental Protection Agency, 1979.)

The dewatering of sludge on sand beds is accomplished by two major factors: drainage and evaporation. The removal of water from sludge by drainage is a two-step process. First, the water is drained from the sludge, into the sand, and out the underdrains. This process may last a few days until the sand is clogged with fine particles or until all the free water has drained away. Further drainage by decanting can occur once a supernatant layer has formed (if beds are provided with a means of removing surface water). Decanting for removal of rain water can also be particularly important with sludges that do not crack.

The water that does not drain or is not decanted must evaporate. Obviously climate plays a role here. Phoenix would be a more efficient area for a sand bed than Seattle! Much of the lower Midwest, for example, would have an annual evaporation rate of about 0.75 m, so typical annual loadings to a sand bed in that area would be 100 $kg/m^2 \cdot y$. This may be applied and cleaned in about ten cycles during the year.

The filtrate from the sand-drying beds can be either recycled, treated, or discharged to a watercourse depending on its quality. Laboratory testing of the filtrate should be performed in conjunction with sand-drying bed pilot testing before a decision is made as to what is to be done with it.

Current United States practice is to make drying beds rectangular with dimensions of 4 to 20 m by 15 to 50 m with vertical side walls. Usually 100 to 230 mm of sand is placed over 200 to 460 mm of graded gravel or stone. The sand is usually 0.3 to 1.2 mm in effective diameter and has a uniformity coefficient less than 5.0. Gravel is normally graded from 3 to 25 mm in effective diameter. Underdrain piping has normally been of vitrified clay, but plastic pipe is also becoming acceptable. The pipes should be no less than 100 mm in diameter, should be spaced 2.2 to 6 m apart, and should have a minimum slope of 1 percent. When the cost of labor is high, newly constructed beds are designed for mechanical sludge removal.

Freeze Treatment. Dewatering sludge by either of the nonmechanical methods may be enhanced by physical conditioning of the sludge through alternate natural freezing and thawing cycles. The freeze-thaw process dehydrates the sludge particles by freezing the water that is closely associated with the particles. The dewatering process takes place in two stages. The first stage reduces sludge volume by selectively freezing the water molecules. Next, the solids are dehydrated when they become frozen. When thawed, the solid mass forms granular-shaped particles. This coarse material readily settles and retains its new size and shape. This residue sludge dewaters rapidly and makes suitable landfill material.

The supernatant liquid from this process can be decanted, leaving the solids to dewater by natural drainage and evaporation. Pilot-scale systems can be utilized to evaluate this method's effectiveness and to establish design parameters. Elimination of rain and snow from the dewatering system by the provision of a roof will enhance the process considerably.

The potential advantages of a freeze-thaw system are

1. Insensitivity to variations in sludge quality
2. No conditioning required
3. Minimum operator attention

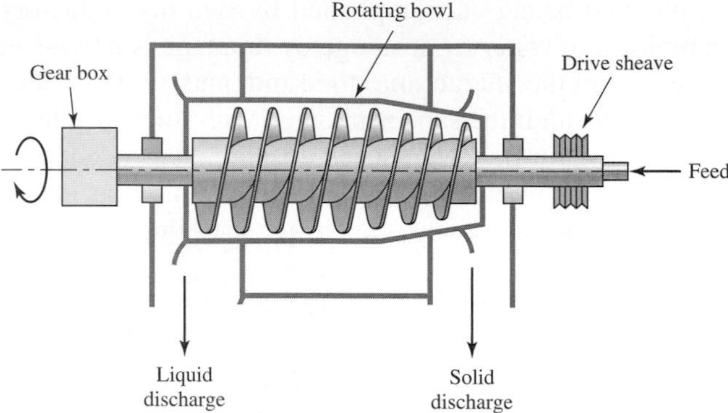

FIGURE 4-55
Solid bowl centrifuge. (Courtesy of Bird Machine Company.)

4. Natural process in cold climates (winter)
5. Solids cake more acceptable at landfills
6. Sludge easily worked with conventional equipment

Several natural freeze-thaw installations are located in New York state. At the alum coagulation plant of the Metropolitan Water Board of Oswego County, SFBW is discharged to lagoons that act as decant basins. Thickened sludge is pumped from the lagoons to special freeze-thaw basins in layers about 450 mm thick. The sludge has never been deeper than 300 mm during freezing because of additional water losses. The 300 mm sludge layer reduces to about 75 mm of dried material after freeze-thaw.

Centrifuging. A centrifuge uses centrifugal force to speed up the separation of sludge particles from the liquid. In a typical unit (Figure 4-55), sludge is pumped into a horizontal, cylindrical bowl, rotating at 800 to 2,000 rpm. Polymers used for sludge conditioning also are injected into the centrifuge. The solids are spun to the outside of the bowl where they are scraped out by a screw conveyor. The liquid, or *centrate,* is returned to the treatment plant. Two types of centrifuges are currently used for sludge dewatering: the solid bowl and the basket bowl. For dewatering water-treatment-plant sludges, the solid bowl has proven to be more successful than the basket bowl. Centrifuges are very sensitive to changes in the concentration or composition of the sludge, as well as to the amount of polymer applied.

Because of its calcium carbonate content, lime softening-sludge dewaters with relative ease in a centrifuge. A cake dryness of 20 to 25 percent can be achieved for a centrifuged alum sludge. A solids content of 50 percent or higher can be achieved with a lime sludge.

Vacuum Filtration. A vacuum filter consists of a cylindrical drum covered with a filtering material or fabric, which rotates partially submerged in a vat of conditioned sludge (Figure 4-56). A vacuum is applied inside the drum to extract water, leaving the solids, or *filter cake,* on the filter medium. As the drum completes its rotational cycles, a blade scrapes the filter cake from the filter and the cycle begins again. Two basic types of rotary-drum vacuum filters are used in water treatment: the *traveling medium* and the *precoat medium* filters. The traveling medium filter is made of fabric or stain-

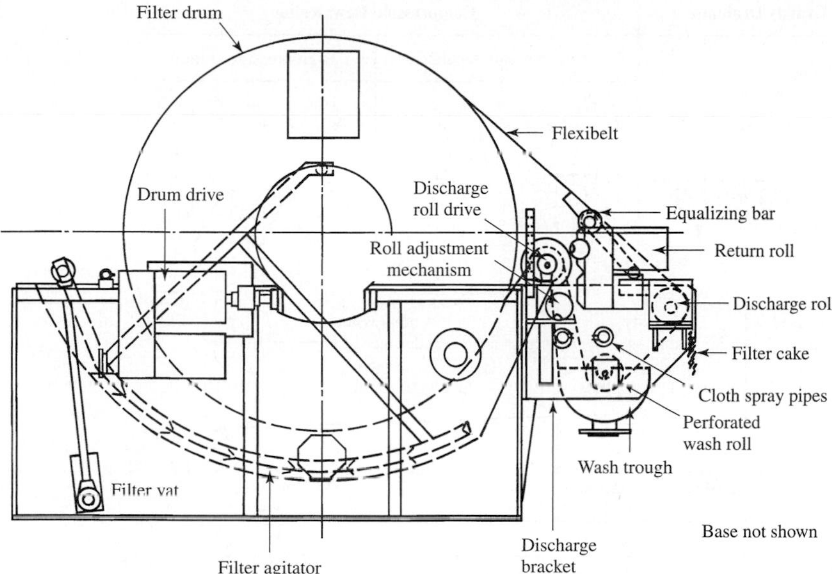

FIGURE 4-56
Vacuum filter. (Courtesy of Komline-Sanderson Engineering Corporation.)

less steel coils. This filter is continuously removed from the drum, allowing it to be washed from both sides without diluting the sludge in the sludge vat. The precoat medium filter is coated with 50 to 75 mm of inert material, which is shaved off in 0.1 mm increments as the drum moves.

Continuous Belt Filter Press (CBFP). The belt filter press operates on the principle that bending a sludge cake contained between two filter belts around a roll introduces shear and compressive forces in the cake, allowing water to work its way to the surface and out of the cake, thereby reducing the cake moisture content. The device employs double moving belts to continuously dewater sludges through one or more stages of dewatering (Figure 4-57). Typically the CBFP includes the following stages of treatment:

1. A reactor/conditioner to remove free-draining water
2. A low pressure zone of belts with the top belt being solid and the bottom belt being a sieve; here further water removal occurs and a sludge mat having significant dimensional stability is formed
3. A high pressure zone of belts with a serpentine or sinusoidal configuration to add shear to the pressure dewatering mechanisms

Plate Pressure Filters. The basic component of a plate filter is a series of recessed vertical plates. Each plate is covered with cloth to support and contain the sludge cake. The plates are mounted in a frame consisting of two head supports connected by two horizontal parallel bars. A schematic cross section is illustrated in Figure 4-58. Conditioned sludge is pumped into the pressure filter and passes through feed holes in the filter plates along the length of the filter and into the recessed chambers. As the sludge cake forms

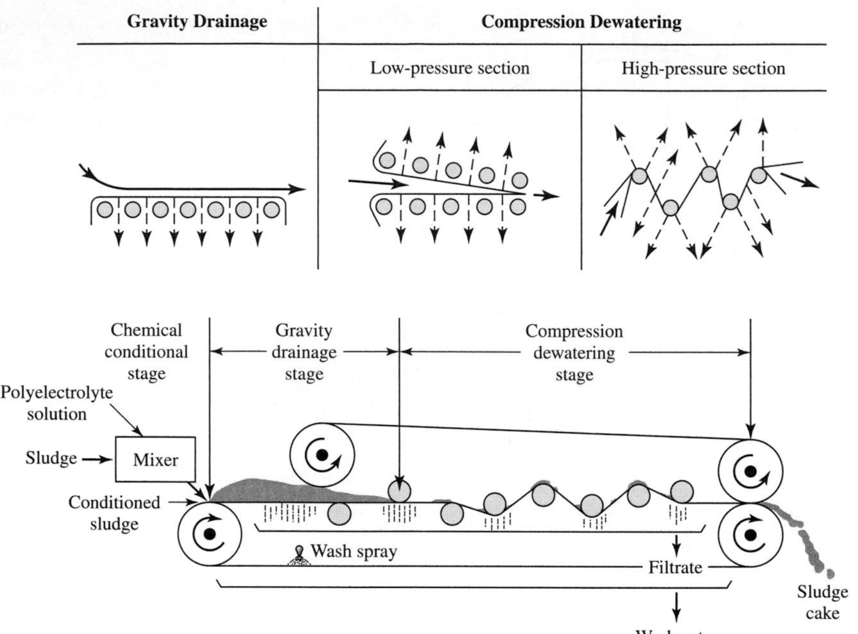

FIGURE 4-57
Continuous belt filter press. (*Source:* U.S. EPA, 1979.)

and builds up in the chamber, the pressure gradually increases to a point where further sludge injection would be counterproductive. At this time the injection ceases.

A typical pressure filtration cycle begins with the closing of the press to the position shown on Figure 4-58. Sludge is fed for a 20- to 30-minute period until the press is effectively full of cake. The pressure at this point is generally the designed maximum (700 to 1,700 kPa) and is maintained for 1 to 4 hours, during which more filtrate is removed and the desired cake solids content is achieved. The filter is then mechanically

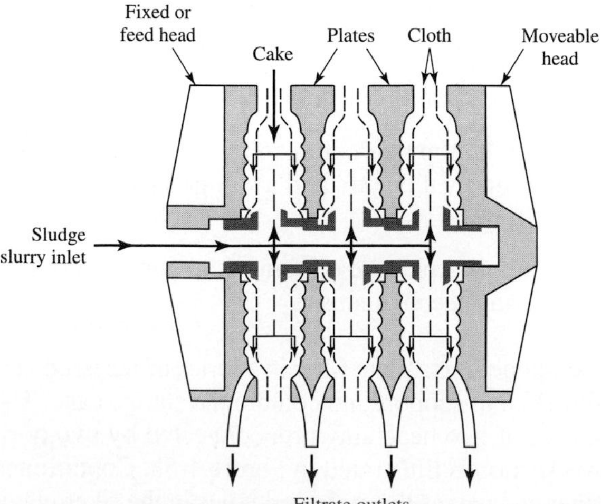

FIGURE 4-58
Schematic cross section of a fixed volume recessed plate filter assembly. (*Source:* U.S. EPA, 1979.)

opened, and the dewatered cake is dropped from the chambers onto a conveyor belt for removal. Cake breakers are usually required to break up the rigid cake into conveyable form. Because recessed-plate pressure filters operate at high pressures and because many units use lime for conditioning, the cloths require routine washing with high-pressure water, as well as periodic washing with acid.

The Erie County Water Authority's Sturgeon Point Plant in Erie County, New York, and the Monroe County Water Authority's Shoremont Plant in Rochester, New York, serve as typical examples of the application of filter presses. Both plants include gravity settling and chemical conditioning of the sludge followed by mechanical dewatering via pressure plate filtration. A typical process flow diagram of the sludge treatment system used at both plants is shown in Figure 4-59. The Sturgeon Point Plant includes a 1.6-m diameter press, while the Monroe County facility includes a 2-m by 2-m press. Under actual operating conditions, the dewatered sludge cakes produced at Sturgeon Point and at the Shoremont Plant contain between 45 and 50 percent dry solids by mass. Of the dry solids, as much as 30 percent may be conditioning chemical solids and/or fly ash. Thus, the corrected dry solids achieved is about 35 percent. This is a significantly better product than what the other mechanical methods produce.

Ultimate Disposal

After all possible sludge treatment has been accomplished, a residual sludge remains, which must be sent to ultimate disposal or used in a beneficial manner. Of

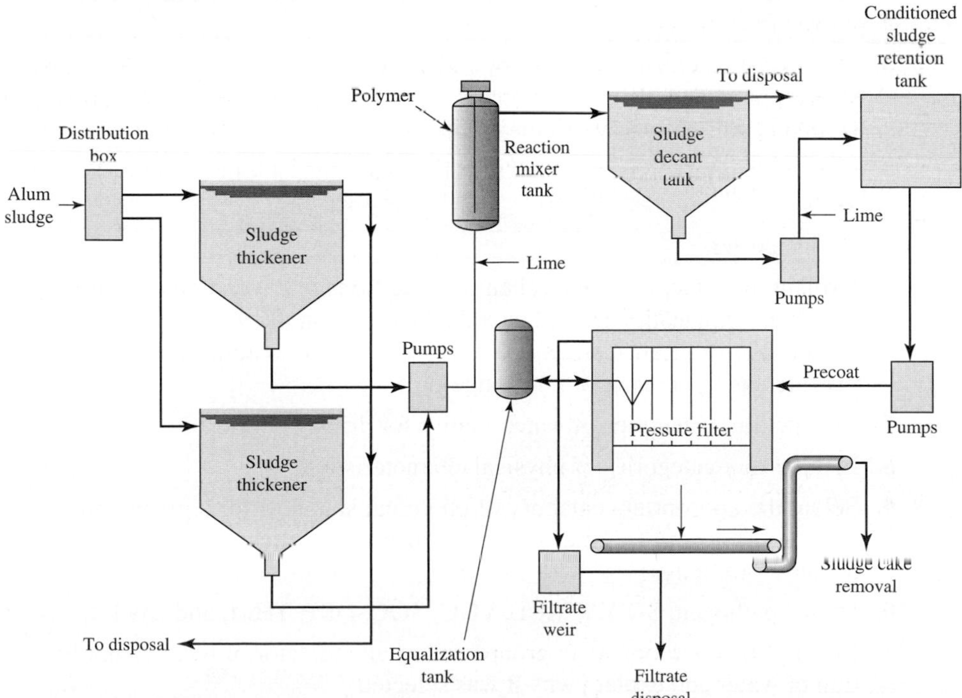

FIGURE 4-59
Pressure plate filter process for alum sludge treatment.

the many theoretical alternatives for ultimate disposal, only three are of practical interest:

1. Co-disposal with sewage sludge
2. Landfilling
3. Beneficial use

Due to the increasing costs and decreasing availability of landfills, beneficial use options are becoming more popular. Beneficial use options include land application to agricultural or forest lands, soil replacement for turf farms, additives to yard waste or biosolids compost, top soil additions, and brick or cement manufacturing. For detailed information on beneficial use options refer to: *Commercial Application and Marketing of Water Plant Residuals* by Cornwell et al. (2000)

4-11 CHAPTER REVIEW

When you have completed studying this chapter you should be able to do the following without the aid of your textbook or notes:

1. Define potable and palatable and explain why we must provide a water that is both potable and palatable.
2. Distinguish between dissolved substances, suspended solids, and colloidal substances based on their size and the mechanism by which they can be removed from water.
3. Define and calculate quantities of a given substance in water in percent by weight, parts per million (ppm), and milligrams per liter (mg/L), and convert from one unit of measure to the others.
4. Define alkalinity in terms of all the chemical species found, that is, Equation 4-36.
5. Define buffer.
6. Explain the effect of various chemical additions to the carbonate buffer system. Your explanation should include the effect on the displacement of the reaction (left or right), effect on CO_2 (into or out of solution), and effect on pH (increase, decrease, or no change).
7. List the four categories of water quality for drinking water.
8. List the four categories of physical characteristics.
9. Select the appropriate category of chemical standard for a given constituent, for example, zinc—esthetics, iron—esthetics/economics, nitrates—toxicity.
10. Define pathogen, SDWA, MCL, VOC, SOC, DBP, THM, and SWTR.
11. Identify the microorganism group used as an indicator of fecal contamination of water and explain why it was selected.
12. Sketch a water softening plant and a filtration plant, labeling all of the parts and explaining their functions.

13. Define the Schulze-Hardy rule and use it to explain the effectiveness of ions of differing valence in coagulation (Figures 4-10 and 4-11).
14. Explain the significance of alkalinity in coagulation.
15. Differentiate between coagulation and flocculation.
16. Write the reaction chemistry of alum and ferric chloride when alkalinity is present and when no alkalinity is present.
17. Explain the effect of pH on alum and ferric chloride solubility.
18. Explain how to conduct a jar test to obtain an optimum coagulant dose.
19. List the four basic types of coagulant aids; explain how each aid works and when it should be employed.
20. Define hardness in terms of the chemical constituents that cause it and in terms of the results as seen by the users of hard water.
21. Using diagrams and chemical reactions, explain how water becomes hard.
22. Given the total hardness and alkalinity, calculate the carbonate hardness and noncarbonate hardness.
23. Write the general equations for softening by ion exchange and by chemical precipitation.
24. Explain the significance of alkalinity in lime-soda softening.
25. Calculate the theoretical detention time or volume of tank if you are given the flow rate and the volume or detention time.
26. Explain how an upward-flow sedimentation tank (upflow clarifier) works, using a vector arrow diagram of a settling particle.
27. Define overflow rate in terms of liquid flow and settling-basin geometry, and state its units.
28. Using a vector arrow diagram, show how, in a horizontal-flow clarifier, a particle with a settling velocity that is less than the overflow rate may be captured (see Figure 4-36).
29. Calculate the percent of particles retained in a settling basin given the overflow rate, settling velocity, and basin flow scheme (that is, horizontal flow or upward flow). (Problems 4-88 and 4-89)
30. Explain the difference between Type I, Type II, and Type III sedimentation.
31. Compare slow sand filters, rapid sand filters, and dual media filters with respect to operating procedures and loading rates.
32. Explain how a rapid sand filter is cleaned.
33. Sketch and label a rapid sand filter identifying the following pertinent features: inlet main, outlet main, washwater outlet, collection laterals, support media (graded gravel), graded filter sand, and backwash troughs.
34. Define effective size and uniformity coefficient and explain their use in designing a rapid sand filter.

35. Explain why a disinfectant that has a residual is preferable to one that does not.

36. Write the equations for the dissolution of chlorine gas in water and the subsequent dissociation of hypochlorous acid.

37. Explain the difference between free available chlorine and combined available chlorine, and state which is the more effective disinfectant.

38. Sketch a breakpoint chlorination curve and label the axes, breakpoint, and regions of predominantly combined and predominantly free residual.

39. Define the terms thickening, conditioning, and dewatering.

40. List and describe three methods of nonmechanical dewatering of sludge.

41. List and describe four mechanical methods of dewatering of sludge.

With the aid of the text you should be able to do the following:

1. Calculate the gram equivalent weight of a chemical or compound.

2. Calculate the molarity, normality, and concentration of a given chemical compound in milligrams per liter (mg/L) and convert from one unit of measure to the others.

3. Calculate the equilibrium concentration of a compound when it is in equilibrium with its precipitate.

4. Calculate the pH of a solution containing a strong or weak acid alone (neglecting the dissociation of water).

5. Convert a concentration of a compound to calcium carbonate equivalents.

6. Calculate the reaction rate constant (k) from a set of experimental data.

7. Calculate consumption of alkalinity for a given dose of alum or ferric chloride.

8. Calculate the production of CO_2, SO_4^{2-}, or acid for a given dose of alum or ferric chloride.

9. Calculate the amount of lime (as $CaCO_3$) that must be added if insufficient alkalinity is present in order to neutralize the acid produced by the addition of alum or ferric chloride to a given water.

10. Estimate the amount of lime and soda ash required to soften water of a stated composition.

11. Calculate the fraction of the "split" for a lime-soda softening system or an ion-exchange softening system.

12. Size a rapid-mix and flocculation basin for a given type of water treatment plant and determine the required power input.

13. Design a mixer system for either a rapid-mix or flocculation basin.

14. Size a sedimentation basin and estimate the required weir length.

15. Perform a grain size analysis and determine the effective size and uniformity coefficient.

16. Size a rapid sand filter and determine the clean-sand head loss and the depth of the expanded bed during backwash.

4-12 PROBLEMS

4-1. Show that a density of 1 g/mL is the same as a density of 1,000 kg/m^3. (*Hint:* Some useful conversions are listed inside the back cover of this book.)

4-2. Show that a 4.50 percent by weight mixture contains 45.0 kg of substance in a cubic meter of water (that is, 4.50% = 45.0 kg/m^3). Assume the density of water = 1,000 kg/m^3.

4-3. What is the concentration of NH_3 (in mg/L) of household ammonia that contains 3.00 percent by weight of NH_3? Assume the density of water = 1,000 kg/m^3.

Answer: 30,000 mg/L

4-4. What is the concentration of chlorine (in mg/L) of household bleach that contains 5.25 percent by weight of Cl_2? Assume the density of water = 1,000 kg/m^3.

4-5. Show that 1 mg/L = 1 g/m^3.

4-6. In 2001 the U.S. Environmental Protection Agency promulgated a new standard MCL for arsenic in drinking water. The standard is now 10 parts per billion (ppb). What is the concentration in mg/L?

4-7. In a now antiquated and, we hope, soon to be forgotten system of measurements, it was common to consider water flows in terms of millions of gallons per day (MGD). Determine the number of MGD equivalent to the following flows in m^3/s: 0.0438; 0.05; 0.438; 0.5; 4.38; and 5. Record both your calculated answer and the answer rounded to include only significant figures.

4-8. Calculate the molarity and normality of the following:

a. 200.0 mg/L HCl
b. 150.0 mg/L H_2SO_4
c. 100.0 mg/L $Ca(HCO_3)_2$
d. 70.0 mg/L H_3PO_4

Answers:

Molarity (M)	Normality (N)
a. 0.005485	0.005485
b. 0.001529	0.003059
c. 0.0006168	0.001234
d. 0.000714	0.00214

4-9. Calculate the molarity and normality of the following:

a. 80 μg/L HNO_3
b. 135 μg/L $CaCO_3$
c. 10 μg/L $Cr(OH)_3$
d. 1000 μg/L $Ca(OH)_2$

4-10. Calculate the molarity and normality of the following:

a. 0.05 mg/L As^{3+}
b. 0.005 mg/L Cd^{2+}
c. 0.002 mg/L Hg^{2+}
d. 0.10 mg/L Ni^{2+}

4-11. Calculate the mg/L of the following:

a. 0.01000 N Ca^{2+}
b. 1.000 M HCO_3^-
c. 0.02000 N H_2SO_4
d. 0.02000 M SO_4^{2-}

Answers: Ca^{2+} = 200.4 mg/L, HCO_3^- = 61.02 mg/L, H_2SO_4 = 980.7 mg/L, SO_4^{2-} = 1,921 mg/L

4-12. Calculate the μg/L of the following:

a. 0.0500 N H_2CO_3
b. 0.0010 M $CHCl_3$
c. 0.0300 N $Ca(OH)_2$
d. 0.0080 M CO_3^{2-}

4-13. Calculate the mg/L of the following:

a. 0.2500 M NaOH
b. 0.0704 M Na_2SO_4
c. 0.0340 M $K_2Cr_2O_7$
d. 0.1342 M KCl

4-14. How many mg/L of magnesium ion will remain in solution in water that is 0.001000 M in hydroxyl ion and at 25°C?

Answer: 0.1367 mg/L

4-15. The Pherric, New Mexico, groundwater contains 1.800 mg/L of iron as Fe^{3+}. What pH is required to precipitate all but 0.300 mg/L of the iron at 25°C?

4-16. Determine the concentration, in moles per liter, to which the hydroxide concentration must be raised to produce a concentration of 0.200 mg/L of copper if the starting concentration is 2.00 mg/L. Estimate the resultant pH.

4-17. Given a saturated solution of calcium carbonate, how many moles of calcium ion will remain in solution after the addition of 3.16×10^{-4} M of Na_2CO_3 at 25°C? (Assume that the pH does not change.)

4-18. The solubility product of calcium fluoride (CaF_2) is 3.45×10^{-11} at 25°C. Will a fluoride concentration of 1.0 mg/L be soluble in a water containing 200 mg/L of calcium?

4-19. In preparation for a laboratory experiment, a technician makes up a saturated solution of $CaSO_4$. Because the container is unlabeled, a 5.00×10^{-3} M solution of Na_2SO_4 is accidently added to the container. What are the concentrations of calcium and sulfate after equilibrium is reached? Assume the pK_s of $CaSO_4$ is 4.31 at 25°C. Assume both solutions are at 25°C.

4-20. What amount of NaOH (a strong base), in mg, would be required to neutralize the acid in Example 4-6 (see Section 4-1)?

Answer: 81.568 or 81.6 mg

4-21. The pH of a finished water from a softening process is 10.74. What amount of 0.02000 N sulfuric acid, in milliliters, is required to neutralize 1.000 L of the finished water? Assume the buffering capacity is zero.

4-22. How many milliliters of 0.02000 N hydrochloric acid would be required to perform the neutralization in Problem 4-21?

4-23. Using a computer spreadsheet program you have written, plot a titration curve of the pH of a 50.0 mL solution of 0.0200 N NaOH (a strong base) being titrated with 0.0200 N HCl (a strong acid) to a pH of 7.00.

4-24. Calculate the pH of a water at 25°C that contains 0.6580 mg/L of carbonic acid. Assume that $[H^+] = [HCO_3^-]$ at equilibrium and neglect the dissociation of water.

Answer: pH = 5.66

4-25. If the pH in Problem 4-24 is adjusted to 4.50, what is the HCO_3^- concentration in moles/L?

4-26. What is the pH of a water at 25°C that contains 0.5000 mg/L of hypochlorous acid? Assume equilibrium has been achieved. Neglect the dissociation of water. Although it may not be justified by the data available to you, report the answer to two decimal places.

4-27. If the pH in Problem 4-26 is adjusted to 7.00, what would the OCl^- concentration in mg/L be at 25°C?

4-28. Convert the following from mg/L as the ion to mg/L as $CaCO_3$:

a. 83.00 mg/L Ca^{2+}
b. 27.00 mg/L Mg^{2+}
c. 48.00 mg/L CO_2 (*Hint:* See footnote on p. 241)
d. 220.00 mg/L HCO_3
e. 15.00 mg/L CO_3^{2-}

Answers:

Ca^{2+} = 207.25 or 207.3 mg/L as $CaCO_3$
Mg^{2+} = 111.20 or 111.2 mg/L as $CaCO_3$
CO_2 = 109.18 or 109.2 mg/L as $CaCO_3$
HCO_3^- = 180.41 or 180.4 mg/L as $CaCO_3$
CO_3^{2-} = 25.02 or 25.0 mg/L $CaCO_3$

4-29. Convert the following from mg/L as the ion or compound to mg/L as $CaCO_3$:

a. 200.00 mg/L HCl
b. 280.00 mg/L CaO
c. 123.45 mg/L Na_2CO_3
d. 85.05 mg/L $Ca(HCO_3)_2$
e. 19.90 mg/L Na^+

4-30. Convert the following from mg/L as $CaCO_3$ to mg/L as the ion or compound:

a. 100.00 mg/L SO_4^{2-}
b. 30.00 mg/L HCO_3^-
c. 150.00 mg/L Ca^{2+}
d. 10.00 mg/L H_2CO_3
e. 150.00 mg/L Na^+

Answers:

SO_4^{2-} = 95.98 or 96.0 mg/L
HCO_3^- = 36.58 or 36.6 mg/L
Ca^{2+} = 60.07 or 60.1 mg/L
H_2CO_3 = 6.198 or 6.20 mg/L
Na^+ = 68.91 mg/L

4-31. Convert the following from mg/L as $CaCO_3$ to mg/L as the ion or compound:

a. 10.00 mg/L CO_2
b. 13.50 mg/L $Ca(OH)_2$
c. 481.00 mg/L H_3PO_4
d. 81.00 mg/L H_2PO_4
e. 40.00 mg/L Cl^-

4-32. Convert 0.0100 N Ca^{2+} to mg/L as $CaCO_3$.

4-33. What is the "exact" alkalinity of a water that contains 0.6580 mg/L of bicarbonate, as the ion, at a pH of 5.66? No carbonate is present.

Answer: 0.4302 mg/L as $CaCO_3$

4-34. Calculate the "approximate" alkalinity (in mg/L as $CaCO_3$) of a water containing 120.0 mg/L of bicarbonate ion and 15.00 mg/L of carbonate ion.

4-35. Calculate the "exact" alkalinity of the water in Problem 4-34 if the pH is 9.43.

4-36. Calculate the "approximate" alkalinity (in mg/L as $CaCO_3$) of a water containing 15.00 mg/L of bicarbonate ion and 120.0 mg/L of carbonate ion.

4-37. Using Equations 4-28, 4-36, 4-38, and 4-39, derive two equations that allow calculation of the bicarbonate and carbonate alkalinities in mg/L as $CaCO_3$ from measurements of the total alkalinity (A) and the pH.

Answers: (in mg/L as $CaCO_3$)

$$HCO_3^- = \frac{50{,}000\left\{\left(\dfrac{A}{50{,}000}\right) + [H^+] - \left(\dfrac{K_W}{[H^+]}\right)\right\}}{1 + \left(\dfrac{2K_2}{[H^+]}\right)}$$

$$CO_3^{2-} = \left(\frac{2K_2}{[H^+]}\right)(HCO_3^-)$$

where
A = total alkalinity, mg/L as $CaCO_3$
K_2 = second dissociation constant of carbonic acid
= 4.68×10^{-11} at 25°C
K_W = ionization constant of water
= 1×10^{-14} at 25°C
HCO_3^- = bicarbonate alkalinity in mg/L as $CaCO_3$
CO_3^{2-} = carbonate alkalinity in mg/L as $CaCO_3$

4-38. Using the solution to Problem 4-37, calculate the bicarbonate and carbonate alkalinities, in mg/L as $CaCO_3$, of a water having a total alkalinity of 233.0 mg/L as $CaCO_3$ and a pH of 10.47.

4-39. Using the solution to Problem 4-37, calculate the bicarbonate and carbonate alkalinities, in mg/L as $CaCO_3$, of the water described in Problem 4-43.

4-40. If a water has a carbonate alkalinity of 120.00 mg/L as the ion and a pH of 10.30, what is the bicarbonate alkalinity in mg/L as the ion?

Answer: HCO_3^- = 130.686 or 130.7 mg/L

4-41. What is the pH of a water that contains 120.00 mg/L of bicarbonate ion and 15.00 mg/L of carbonate ion?

4-42. Calculate the alkalinity of the water in Problem 4-40, using the equations in Problem 4-37 and the "exact" method of Example 4-8 (see Section 4-1).

4-43. The following mineral analysis was reported for a water sample taken from Well No. 1 at the Eastwood Manor Subdivision near McHenry, Illnois (Woller and Sanderson, 1976a).

Well No. 1, Lab No. 02694, November 9, 1971

Iron	0.2	Silica (SiO_2)	20.0
Manganese	0.0	Fluoride	0.35
Ammonium	0.5	Boron	0.1
Sodium	4.7	Nitrate	0.0
Potassium	0.9	Chloride	4.5
Calcium	67.2	Sulfate	29.0
Magnesium	40.0	Alkalinity	284.0 as $CaCO_3$
Barium	0.5	pH	7.6 units

Note: All reported as "mg/L as the ion" unless stated otherwise.

Determine the total, carbonate, and noncarbonate hardness in mg/L as $CaCO_3$ using the predominant polyvalent cation definition in Section 4-3.

Answers: TH = 332.8 mg/L as $CaCO_3$
CH = 284.0 mg/L as $CaCO_3$
NCH = 48.8 mg/L as $CaCO_3$

4-44. Calculate the total, carbonate, and noncarbonate hardness in Problem 4-43 using all of the polyvalent cations. What is the percent error in using only the predominant cations?

4-45. The following mineral analysis was reported for a water sample taken from Well No. 1 at Magnolia, Illinois (Woller and Sanderson, 1976b). Determine the total, carbonate and noncarbonate hardness in mg/L as $CaCO_3$ using the predominant polyvalent cation definition of hardness.

Well No. 1, Lab No. B109535, April 23, 1973

Iron	0.42	Zinc	0.01
Manganese	0.04	Silica (SiO_2)	20.0
Ammonium	11.0	Fluoride	0.3
Sodium	78.0	Boron	0.3
Potassium	2.6	Nitrate	0.0
Calcium	78.0	Chloride	9.0
Magnesium	32.0	Sulfate	0.0
Barium	0.5	Alkalinity	494.0 as $CaCO_3$
Copper	0.01	pH	7.7 units

Note: All reported as "mg/L as the ion" unless stated otherwise.

4-46. The following mineral analysis was reported for Michigan State University well water (MDEQ, 1979). Determine the total, carbonate, and noncarbonate hardness in mg/L as $CaCO_3$ using the predominant polyvalent cation definition of hardness. *Note:* All units are mg/L as the ion.

Michigan State University Well Water

Fluoride	1.1	Silica (SiO_2)	3.4
Chloride	4.0	Bicarbonate	318.0
Nitrate	0.0	Sulfate	52.0
Sodium	14.0	Iron	0.5
Potassium	1.6	Manganese	0.07
Calcium	96.8	Zinc	0.27
Magnesium	30.4	Barium	0.2

4-47. An analysis of bottled water from the Kool Artesian Water Bottling Company is listed below. Determine the total, carbonate, and noncarbonate hardness in mg/L as $CaCO_3$ using the predominant polyvalent cation definition of hardness. *Note:* All units are mg/L as the ion unless otherwise stated. (*Hint:* use the solution to Problem 4-37 to find the bicarbonate concentration.)

Kool Artesian Water

Calcium	37.0	Silica	11.5
Magnesium	18.1	Sulfate	5.0
Sodium	2.1	Potassium	1.6
Fluoride	0.1	Zinc	0.02
Alkalinity	285.0 mg/L as $CaCO_3$		
pH	7.6 units		

4-48. The following data were obtained for an irreversible elementary reaction. Plot the data, determine the order of the reaction (zero, first, or second) and the rate constant *k*. Use a spreadsheet to plot the data and fit a curve.

Time, min	Reactant A Concentration, mmoles/L
0	2.80
1	2.43
2	2.12
5	1.39
10	0.69
20	0.17

4-49. Repeat Problem 4-48 for the following data.

Time, min	Reactant A Concentration, mmoles/L
0	48.0
1	6.22
2	3.32
3	2.27
5	1.39
10	0.704

4-50. Shown below are the results of water quality analyses of the Thames River in London. If the water is treated with 60.00 mg/L of alum to remove turbidity, how much alkalinity will remain? Ignore side reactions with phosphorus and assume all the alkalinity is HCO_3^-.

Thames River, London

Constituent	Expressed as	Milligrams per liter
Total hardness	$CaCO_3$	260.0
Calcium hardness	$CaCO_3$	235.0
Magnesium hardness	$CaCO_3$	25.0
Total iron	Fe	1.8
Copper	Cu	0.05
Chromium	Cr	0.01
Total alkalinity	$CaCO_3$	130.0
Chloride	Cl	52.0
Phosphate (total)	PO_4	1.0
Silica	SiO_2	14.0
Suspended solids		43.0
Total solids		495.0
pH[a]		7.4

[a]Not in mg/L.

Answer: Alkalinity remaining = 99.69 or 100 mg/L as $CaCO_3$

4-51. Shown below are the results of water quality analyses of the Mississippi River at Baton Rouge, Louisiana. If the water is treated with 30.00 mg/L of ferric chloride for turbidity coagulation, how much alkalinity will remain? Ignore the side reactions with phosphorus and assume all the alkalinity is HCO_3^-.

Mississippi River, Baton Rouge, Louisiana

Constituent	Expressed as	Milligrams per liter
Total hardness	$CaCO_3$	164.0
Calcium hardness	$CaCO_3$	108.0
Magnesium hardness	$CaCO_3$	56.0
Total iron	Fe	0.9
Copper	Cu	0.01
Chromium	Cr	0.03
Total alkalinity	$CaCO_3$	136.0
Chloride	Cl	32.0
Phosphate (total)	PO_4	3.0
Silica	SiO_2	10.0
Suspended solids		29.9
Turbidity[a]	NTU	12.0
pH[a]		7.6

[a]Not in mg/L.

4-52. Shown below are the results of water quality analyses of Crater Lake at Mount Mazama, Oregon. If the water is treated with 40.00 mg/L of alum for turbidity coagulation, how much alkalinity will remain? Assume all the alkalinity is HCO_3^-.

Crater Lake, Mount Mazama, Oregon

Constituent	Expressed as	Milligrams per liter
Total hardness	$CaCO_3$	28.0
Calcium hardness	$CaCO_3$	19.0
Magnesium hardness	$CaCO_3$	9.0
Total iron	Fe	0.02
Sodium	Na	11.0
Total alkalinity	$CaCO_3$	29.5
Chloride	Cl	12.0
Sulfate	SO_4	12.0
Silica	SiO_2	18.0
Total dissolved solids		83.0
pH[a]		7.2

[a]Not in mg/L.

4-53. Prepare a bar chart of the water described in Problem 4-43. (*Note:* Valences may be found in Appendix A.) Because all of the constituents were not analyzed, an ion balance is not achieved.

4-54. Prepare a bar chart of the water described in Problem 4-45. Because all of the constituents were not analyzed, an ion balance is not achieved.

4-55. Prepare a bar chart of the water described in Problem 4-46. Because all of the constituents were not analyzed, an ion balance is not achieved.

4-56. Prepare a bar chart of the Lake Michigan water analysis shown below. Because all of the constituents were not analyzed, an ion balance is not achieved. For the estimate of the CO_2 concentration, ignore the carbonate alkalinity.

Lake Michigan at Grand Rapids, MI Intake

Constituent	Expressed as	Milligrams per liter
Total hardness	$CaCO_3$	143.0
Calcium	Ca	38.4
Magnesium	Mg	11.4
Total iron	Fe	0.10
Sodium	Na	5.8
Total alkalinity	$CaCO_3$	119
Bicarbonate alkalinity	$CaCO_3$	115
Chloride	Cl	14.0
Sulfate	SO_4	26.0
Silica	SiO_2	1.2
Total dissolved solids		180.0
Turbidity[a]	NTU	3.70
pH[a]		8.4

[a]Not in mg/L.

4-57. Determine the lime and soda ash dose, in mg/L as $CaCO_3$, to soften the following water to a final hardness of 80.0 mg/L as $CaCO_3$. The ion concentrations reported below are all mg/L as $CaCO_3$.

$$Ca^{2+} = 120.0$$
$$Mg^{2+} = 30.0$$
$$HCO_3^- = 70.0$$
$$CO_2 = 10.0$$

Answers: Total lime addition = 100 mg/L as $CaCO_3$

Total soda ash addition = 40 mg/L as $CaCO_3$

4-58. What amount of lime and/or soda ash, in mg/L as $CaCO_3$, is required to soften the Village of Lime Ridge's water to 80.0 mg/L hardness as $CaCO_3$.

Compound	Concentration, mg/L as $CaCO_3$
CO_2	4.6
Ca^{2+}	257.9
Mg^{2+}	22.2
HCO_3^-	248.0
SO_4^{2-}	32.1

4-59. Determine the lime and soda ash dose, in mg/L as CaO and Na_2CO_3, to soften the following water to a final hardness of 80.0 mg/L as $CaCO_3$. The ion concentrations reported below are all mg/L as $CaCO_3$. Assume the lime is 90 percent pure and the soda ash is 97 percent pure.

$Ca^{2+} = 210.0$
$Mg^{2+} = 23.0$
$HCO_3^- = 165.0$
$CO_2 = 5.0$

Answers: Total lime addition = 118 mg/L as CaO
Total soda ash addition = 31 mg/L as Na_2CO_3

4-60. Determine the lime and soda ash dose, in mg/L as $CaCO_3$ to soften the following water to a final hardness of 70.0 mg/L as $CaCO_3$. The ion concentrations reported below are all as $CaCO_3$.

$Ca^{2+} = 220.0$
$Mg^{2+} = 75.0$
$HCO_3^- = 265.0$
$CO_2 = 17.0$

Answers: Total lime addition = 352 mg/L as $CaCO_3$;
Add no soda ash

4-61. What amount of lime and/or soda ash, in mg/L as $CaCO_3$, is required to soften the Village of Sarepta's water to a hardness of 80.0 mg/L as $CaCO_3$.

Compound	Concentration, mg/L as $CaCO_3$
CO_2	39.8
Ca^{2+}	167.7
Mg^{2+}	76.3
HCO_3^-	257.9
SO_4^{2-}	109.5

4-62. Determine the lime and soda ash dose, in mg/L as CaO and Na_2CO_3, to soften the following water to a final hardness of 80.0 mg/L as $CaCO_3$. The ion concentrations reported below are all as $CaCO_3$. Assume the lime is 93 percent pure and the soda ash is 95 percent pure.

$$Ca^{2+} = 137.0$$
$$Mg^{2+} = 56.0$$
$$HCO_3^- = 128.0$$
$$CO_2 = 7.0$$

4-63. What amount of lime and/or soda ash, in mg/L as $CaCO_3$, is required to soften the Village of Zap's water to 80.0 mg/L hardness as $CaCO_3$?

Compound	**Concentration, mg/L as $CaCO_3$**
CO_2	44.2
Ca^{2+}	87.4
Mg^{2+}	96.3
HCO_3^-	204.6
SO_4^{2-}	73.8

4-64. Determine the lime and soda ash dose, in mg/L as $CaCO_3$, to soften the water described in Problem 4-43 to a final hardness of 100.0 mg/L as $CaCO_3$.

Answers: CO_2 = 44.2 mg/L as $CaCO_3$
Total lime addition = 481 mg/L as $CaCO_3$
Add no soda ash

4-65. Determine the lime and soda ash dose, in mg/L as $CaCO_3$, to soften the water described in Problem 4-50 to a final hardness of 90.0 mg/L as $CaCO_3$.

4-66. The Village of Galena wants to use a softening process to remove lead from their water. The water analysis is shown next page. Determine the lime and soda ash dose, in mg/L as CaO and Na_2CO_3, to soften the following water to a final hardness of 80.0 mg/L as $CaCO_3$. The ion concentrations reported below are all as the ion. Assume the lime is 93 percent pure and the soda ash is 95 percent pure.

Village of Galena

Constituent	Expressed as	Milligrams per liter
Calcium	Ca	177.8
Magnesium	Mg	16.2
Total iron	Fe	0.20
Lead[a]	Pb	20[a]
Sodium	Na	4.9
Carbonate alkalinity	$CaCO_3$	0.0
Bicarbonate alkalinity	$CaCO_3$	276.6
Chloride	Cl	0.0
Sulfate	SO_4	276.0
Silica	SiO_2	1.2
Total dissolved solids		667
pH[b]		7.2

[a]Parts per billion.

[b]Not in mg/L.

Will the softening process remove lead?

4-67. Determine the lime and soda ash dose, in mg/L as $CaCO_3$, to soften the following water to a final hardness of 80.0 mg/L as $CaCO_3$. If the price of lime, purchased as CaO, is \$100.00 per megagram (Mg), and the price of soda ash, purchased as Na_2CO_3 is \$200.00 per Mg, what is the annual chemical cost of treating 0.500 m^3/s of this water?

Assume the lime and soda ash are 100 percent pure. The ion concentrations reported below are all mg/L as $CaCO_3$.

$$Ca^{2+} = 200.0$$
$$Mg^{2+} = 100.0$$
$$HCO_3^- = 150.0$$
$$CO_2 = 22.0$$

Answers: Lime = 272.0 mg/L as $CaCO_3$
Soda = 110.0 mg/L as $CaCO_3$
Total annual cost = \$607,703.25, or \$608,000

4-68. Determine the lime and soda ash dose, in mg/L as $CaCO_3$, to soften the following water to a final hardness of 120.0 mg/L as $CaCO_3$. If the price of lime, purchased as CaO, is \$61.70 per megagram (Mg), and the price of soda ash, purchased as Na_2CO_3, is \$172.50 per Mg, what is the annual chemical cost of treating 1.35 m^3/s of this water? Assume the lime is 87 percent pure and the soda ash is 97 percent pure. The ion concentrations reported below are all mg/L as $CaCO_3$.

$$Ca^{2+} = 293.0$$
$$Mg^{2+} = 55.0$$
$$HCO_3^- = 301.0$$
$$CO_2 = 3.0$$

4-69. Determine the lime and soda ash dose, in mg/L as $CaCO_3$, to soften the Hardin, Illinois, water to a final hardness of 95.00 mg/L as $CaCO_3$ (Woller, 1975). Using the price and purity information supplied in Problem 4-67, determine the annual chemical cost of treating 0.150 m^3/s of this water.

Well No. 2, Hardin, IL

Iron	0.10	Zinc	0.13
Manganese	0.64	Silica (SiO_2)	21.6
Ammonium	0.38	Fluoride	0.3
Sodium	21.8	Boron	0.38
Potassium	3.0	Nitrate	8.4
Calcium	102.0	Chloride	32.0
Magnesium	45.2	Sulfate	65.0
Copper	0.01	Alkalinity	344.0 as $CaCO_3$
pH	7.2 units		

Note: All reported as "mg/L as the ion" unless stated otherwise.

4-70. Determine the lime and soda ash dose, in mg/L as $CaCO_3$, to soften the following water to a final hardness of 90.0 mg/L as $CaCO_3$. If the price of lime, purchased as CaO, is \$61.70 per megagram (Mg), and the price of soda ash, purchased as Na_2CO_3, is \$172.50 per Mg. What is the annual chemical cost of treating 0.050 m^3/s of this water? Assume the lime is 90 percent pure and the soda ash is 97 percent pure. The ion concentrations reported below are all mg/L as $CaCO_3$.

$$Ca^{2+} = 137.0$$
$$Mg^{2+} = 40.0$$
$$HCO_3^- = 197.0$$
$$CO_2 = 9.0$$

4-71. Design a split treatment softening process (flow scheme/split, chemical dose in mg/L as $CaCO_3$) for the following water. The final hardness must be ≤120 mg/L as $CaCO_3$. Compounds are given in mg/L as the ion stated unless otherwise specified:

CO_2	42.7	HCO_3^-	344.0 mg/L as $CaCO_3$
Ca^{2+}	102.0	SO_4^{2-}	65.0
Mg^{2+}	45.2	Cl^-	32.0
Na^+	21.8		

4-72. Given the following water (all in meq/L), design a process to soften the water (flow scheme/split; amount of lime and/or soda required in mg/L as CaO and Na_2CO_3 respectively) and find the final hardness. The final hardness must be ≤120 mg/L as $CaCO_3$.

CO_2	0.40	Mg^{2+}	1.12
Ca^{2+}	2.16	HCO_3^-	2.72

4-73. Design a softening process for the City of What Cheer to achieve a magnesium concentration of 40 mg/L as $CaCO_3$ and a final total hardness less than 120 mg/L as $CaCO_3$. Show the flow scheme, calculated split, amount of lime and/or soda in mg/L as $CaCO_3$, and the final hardness. The water analysis is shown below.

Compound	Concentration, mg/L as $CaCO_3$
CO_2	39.8
Ca^{2+}	167.7
Mg^{2+}	76.3
HCO_3^-	257.9
SO_4^{2-}	109.5

4-74. What is the volume required for a rapid-mix basin that is to be used to treat 0.05 m^3/s of water if the detention time is 10 seconds?

Answer: 0.5 m^3

4-75. Two parallel flocculation basins are to be used to treat a water flow of 0.150 m^3/s. If the design detention time is 20 minutes, what is the volume of each tank?

4-76. Two sedimentation tanks operate in parallel. The combined flow to the two tanks is 0.1000 m^3/s. The volume of each tank is 720 m^3. What is the detention time of each tank?

4-77. Determine the power input required for the tank designed in Problem 4-74 if the water temperature is 20°C and the velocity gradient is 700 s^{-1}.

Answer: 245.49 or 250 W or 0.25 kW

4-78. The flocculation tanks in Problem 4-75 were designed for an average velocity gradient of 36 s^{-1} and a water temperature of 17°C. What power input is required?

4-79. What power input is required for Problem 4-77 if the water temperature falls to 10°C?

4-80. What power input is required for Problem 4-78 if the water temperature falls to 108°C?

4-81. Complete Example 4-20 by computing the rotational speed of the impellers in compartments 2 and 3.

4-82. The town of Eau Gaullie has requested proposals for a new coagulation water treatment plant. The design flow for the plant is 0.1065 m^3/s. The average annual water temperature is 19°C. The following design assumptions for a rapid-mix tank have been made:

1. Number of tanks = 1 (with 1 backup spare)
2. Tank configuration: circular with liquid depth = 2 × diameter
3. Detention time = 10 s
4. Velocity gradient = 800 s^{-1}
5. Impeller type: turbine, 6 flat blades, $N_p = 5.7$
6. Available impeller diameters: 0.45, 0.60, and 1.2 m
7. Assume $B = \frac{1}{3}H$

Design the rapid-mix system by providing the following:

1. Water power input in kW
2. Tank dimensions in m
3. Diameter of the impeller in m
4. Rotational speed of impeller in rpm

Answers: $P = 0.700$ kW
Diameter = 0.88 m; depth = 1.76 m
Impeller diameter = 0.45 m
Rotational speed = 399 or 400 rpm

4-83. Laramie is planning for a new softening plant. The design flow is 0.168 m^3/s The average water temperature is 5°C. The following design assumptions for a rapid-mix tank have been made:

1. Tank configuration: square plan with depth = width
2. Detention time = 5 s
3. Velocity gradient 700 s^{-1}
4. Impeller type: turbine, 6 flat blades, $N_P = 5.7$

5. Available impeller diameters: 0.45, 0.60, and 1.2 m
6. Assume $B = \frac{1}{3}H$

Design the rapid-mix system by providing the following:

1. Number of tanks
2. Water power input in kW
3. Tank dimensions in m
4. Diameter of the impeller in m
5. Rotational speed of impeller in rpm

4-84. Your boss has assigned you the job of designing a rapid-mix tank for the new water treatment plant for the town of Waffle. The design flow rate is $0.050 \text{ m}^3/\text{s}$. The average water temperature is 8°C. The following design assumptions for a rapid-mix tank have been made:

1. Number of tanks = 1 (with 1 backup)
2. Tank configuration: circular with liquid depth = 1.0 m
3. Detention time = 5 s
4. Velocity gradient = 750 s^{-1}
5. Impeller type: turbine, 6 flat blades, $N_P = 3.6$
6. Available impeller diameters: 0.25, 0.50, and 1.0 m
7. Assume $B = \frac{1}{3}H$

Design the rapid-mix system by providing the following:

1. Water power input in kW
2. Tank dimensions in m
3. Diameter of the impeller in m
4. Rotational speed of impeller in rpm

4-85. Continuing the preparation of the proposal for the Eau Gaullie treatment plant (Problem 4-82), design the flocculation tank by providing the following for the first two compartments only:

1. Water power input in kW
2. Tank dimensions in m
3. Diameter of the impeller in m
4. Rotational speed of impeller in rpm

Use the following assumptions:

1. Number of tanks = two
2. Tapered G in three compartments: 90, 60, and 30 s^{-1}

3. $G\theta = 120{,}000$
4. Compartment length = width = depth
5. Impeller type: axial-flow impeller, three blades, $N_P = 0.31$
6. Available impeller diameters: 1.0, 1.5, and 2.0 m
7. Assume $B = \frac{1}{3}H$

    Answers for first compartment only:
    $P = 295.31$, or 295 W or 0.295 kW
    $L = W = D = 3.3$ m
    Impeller diameter = 1.5 m
    Rotational speed = 30 rpm

4-86. Continuing the preparation of the proposal for the Laramie treatment plant (Problem 4-83), design the flocculation tank by providing the following for the first two compartments only:

1. Water power input in kW
2. Tank dimensions in m
3. Diameter of the impeller in m
4. Rotational speed of impeller in rpm

Use the following assumptions:

1. Number of tanks = two
2. Tapered G in three compartments: 90, 60, and 30 s^{-1}
3. $G\theta = 120{,}000$
4. Compartment length = width = depth
5. Impeller type: axial-flow impeller, three blades, $N_P = 0.40$
6. Available impeller diameters: 1.0, 1.8, and 2.4 m
7. Assume $B = \frac{1}{3}H$

4-87. Continuing the preparation of the proposal for the Waffle treatment plant (Problem 4-84), design the flocculation tank by providing the following for the first two compartments only:

1. Water power input in kW
2. Tank dimensions in m
3. Diameter of the impeller in m
4. Rotational speed of impeller in rpm

Use the following assumptions:

1. Number of tanks = 1 (with 1 backup)
2. Tapered G in three compartments: 60, 50, and 20 s^{-1}

3. Detention time = 30 min
4. Depth = 3.5 m
5. Impeller type: axial-flow impeller, three blades, $N_p = 0.43$
6. Available impeller diameters: 1.0, 1.5, and 2.0 m
7. Assume $B = \frac{1}{3}H$

4-88. If the settling velocity of a particle is 0.70 cm/s and the overflow rate of a horizontal flow clarifier is 0.80 cm/s, what percent of the particles are retained in the clarifier?

Answer: 88 percent

4-89. If the settling velocity of a particle is 2.80 mm/s and the overflow rate of an upflow clarifier is 0.560 cm/s, what percent of the particles are retained in the clarifier?

4-90. If the settling velocity of a particle is 0.30 cm/s and the overflow rate of a horizontal flow clarifier is 0.25 cm/s, what percent of the particles are retained in the clarifier?

4-91. If the flow rate of the original plant in Problem 4-88 is increased from 0.150 m^3/s to 0.200 m^3/s, what percent removal of particles would be expected?

4-92. If the flow rate of the original plant in Problem 4-89 is doubled, what percent removal of particles would be expected?

4-93. If the flow rate of the original plant in Problem 4-90 is doubled, what percent removal of particles would be expected?

4-94. If a 1.0-m^3/s flow water treatment plant uses 10 sedimentation basins with an overflow rate of 15 $m^3/d \cdot m^2$, what should be the surface area (m^2) of each tank?

Answer: 576.0 m^2

4-95. Assuming a conservative value for an overflow rate, determine the surface area (in m^2) of each of two sedimentation tanks that together must handle a flow of 0.05162 m^3/s of lime softening floc.

4-96. Repeat Problem 4-95 for an alum or iron floc.

4-97. Two sedimentation tanks operate in parallel. The combined flow to the two tanks is 0.1000 m^3/s. The depth of each tank is 2.00 m and each has a detention time of 4.00 h. What is the surface area of each tank and what is the overflow rate of each tank in $m^3/d \cdot m^2$?

4-98. Determine the detention time and overflow rate for a settling tank that will reduce the influent suspended solids concentration from 33.0 mg/L to 15.0 mg/L. The following batch settling column data are available. The data given are percent removals at the sample times and depths shown.

	Depths,* m				
Time, min	**0.5**	**1.5**	**2.5**	**3.5**	**4.5**
10	50	32	20	18	15
20	75	45	35	30	25
40	85	65	48	43	40
55	90	75	60	50	46
85	95	87	75	65	60
95	95	88	80	70	63

*Depths from top of column, column depth = 4.5 m.

4-99. The following test data were gathered to design a settling tank. The initial suspended solids concentration for the test was 20.0 mg/L. Determine the detention time and overflow rate that will yield 60 percent removal of suspended solids. The data given are suspended solids concentrations in mg/L.

	Time, min					
Depth,* m	**10**	**20**	**35**	**50**	**70**	**85**
0.5	14.0	10.0	7.0	6.2	5.0	4.0
1.0	15.0	13.0	10.6	8.2	7.0	6.0
1.5	15.4	14.2	12.0	10.0	7.8	7.0
2.0	16.0	14.6	12.6	11.0	9.0	8.0
2.5	17.0	15.0	13.0	11.4	10.0	8.8

*Depths from top of column, column depth = 2.5 m.

4-100. The following test data were gathered to design a settling tank. The initial turbidity for the test was 33.0 NTU. Determine the detention time and overflow rate that will yield 88 percent removal of suspended solids. The data given are suspended solids concentrations in NTU.

	Time, min			
Depth,* m	**30**	**60**	**90**	**120**
1.0	5.6	2.0		
2.0	11.2	5.9	2.6	
3.5	14.5	9.9	6.6	3.0

*Depths from top of column, column depth = 4.0 m.

4-101. For a flow of 0.8 m^3/s, how many rapid sand filter boxes of dimensions 10 m $\times$ 20 m are needed for a loading rate of 110 $m^3/d \cdot m^2$?

Answer: 4 filters (rounding to nearest even number)

4-102. If a dual-media filter with a loading rate of 300 $m^3/d \cdot m^2$ were built instead of the standard filter in Problem 4-101, how many filter boxes would be required?

4-103. The water flow meter at the Troublesome Creek water plant is on the blink. The plant superintendent tells you the four dual-media filters (each 5.00 m $\times$ 10.0 m) are loaded at a velocity of 280 m/d. What is the flow rate through the filters in m^3/s?

4-104. A plant expansion is planned for Urbana (Example 4-25). The new design flow rate is 1.0 m^3/s. A deep-bed monomedia filter with a design loading rate of 600 $m^3/d \cdot m^2$ of filter is to be used. If each filter box is limited to 50 m^2 of surface area, how many filter boxes will be required? Check the design loading with one filter box out of service. Propose an alternative design if the design loading rate is exceeded with one filter box out of service.

4-105. The Orono Sand and Gravel Company has made a bid to supply sand for Eau Gaullie's new sand filter. The request for bids stipulated that the sand have an effective size in the range 0.35 to 0.55 mm and a uniformity coefficient in the range 1.3 to 1.7. Orono supplied the following sieve analysis as evidence that their sand will meet the specifications. Perform a grain size analysis (semi-log plot) and determine whether or not the sand meets the specifications. Use a spreadsheet program you have written to plot the data and fit a curve.

U.S. Standard Sieve No.	Mass percent retained
8	0.0
12	0.01
16	0.39
20	5.70
30	25.90
40	44.00
50	20.20
70	3.70
100	0.10

4-106. The Lexington Sand and Gravel Company has made a bid to supply sand for Laramie's new sand filter. The request for bids stipulated that the sand have an effective size in the range 0.35 to 0.55 mm and a uniformity coefficient in the range 1.3 to 1.7. Lexington supplied the following sieve analysis (sample size = 500.00 g) as evidence that its sand will meet the specifications. Perform a grain size analysis (log-log plot) and determine whether or not the sand meets the specifications. Use a spreadsheet program you have written to plot the data and fit a curve.

U.S. Standard Sieve No.	Mass retained, g
12	0.00
16	2.00
20	65.50
30	272.50
40	151.0
50	8.925
70	0.075

4-107. Rework Example 4-26 (Section 4-6) with the 70, 100, and 140 sieve fractions removed. Assume original sample contained 100 g.

4-108. The rapid sand filter being designed for Eau Gaullie has the characteristics and sieve analysis shown below. Determine the head loss for the clean filter bed in a stratified condition.

Depth = 0.60 m
Filter loading = 120 $m^3/d \cdot m^2$
Sand specific gravity = 2.50
Shape factor = 1.00
Stratified bed porosity = 0.42
Water temperature = 19°C

Sand Analysis

U.S. Standard Sieve No.	Mass percent retained
8–12	0.01
12–16	0.39
16–20	5.70
20–30	25.90
30–40	44.00
40–50	20.20
50–70	3.70
70–100	0.10

4-109. Determine the height of the expanded bed for the sand used in Problem 4-108 if the backwash rate is 1,000 m/d. Assume Equation 4-107 applies.

4-110. The rapid sand filter being designed for Laramie has the characteristics shown below. Determine the head loss for the clean filter bed in a stratified condition.

Depth = 0.75 m
Filter loading = 230 $m^3/d \cdot m^2$
Sand specific gravity = 2.80
Shape factor = 0.91
Stratified bed porosity = 0.50
Water temperature = 5°C

Sand Analysis

U.S. Standard Sieve No.	Mass percent retained
8–12	0.00
12–16	0.40
16–20	13.10
20–30	54.50
30–40	30.20
40–50	1.785
50–70	0.015

4-111. Determine the maximum backwash rate and the height of the backwash troughs above the sand used in Problem 4-110. Assume Equation 4-107 applies.

4-112. As noted in Example 4-27 in Section 4-6, the head loss was too high. Rework the example without the 100-140 sieve fraction to see how much this would improve the head-loss characteristics.

4-113. What effect does removing the 100-140 sieve fraction have on the depth of the expanded bed in Example 4-28 (Section 4-6)?

4-114. What is the equivalent percent reduction for a 2.5 log reduction of *Giardia lambia*?

4-115. What is the log reduction of *Giardia lambia* that is equivalent to 99.96 percent reduction?

4-13 DISCUSSION QUESTIONS

4-1. Would you expect a carbonated beverage to have a pH above, below, or equal to 7.0? Explain why.

4-2. Explain the word *turbidity* in terms that the mayor of a community could understand.

4-3. Which of the chemicals added to treat a surface water aids in making the water palatable?

4-4. Microorganisms play a role in the formation of hardness in groundwater. True or false? Explain.

4-5. If there is no bicarbonate present in a well water that is to be softened to remove magnesium, which chemicals must you add?

4-6. Use a scale drawing to sketch a vector diagram of a horizontal-flow sedimentation tank that shows how 25 percent of the particles with a settling velocity one-quarter that of the overflow rate will be removed.

4-7. In the United States, chlorine is preferred as a disinfectant over ozone because it has a residual. Why is the presence of a residual important?

4-8. A new water softening plant is being designed for Lubbock, Texas. The climate is dry and land is readily available at a reasonable cost. What methods of sludge dewatering would be most appropriate? Explain your reasoning.

4-14 REFERENCES

Amirtharajah, A. (1978) "Design of Raid Mix Units," in R. L. Sanks (ed.) *Water Treatment Plant Design for the Practicing Engineer,* Ann Arbor Science, Ann Arbor, MI, pp. 132, 141, 143.

AWWA (1990) *Water Treatment Plant Design,* 2nd ed., McGraw-Hill, New York.

AWWA (1971) *Water Quality and Treatment,* 3rd ed., American Water Works Association, McGraw-Hill, New York, pp. 188, 259.

AWWA (1998) *Water Treatment Plant Design,* 3rd ed., American Water Works Association, McGraw-Hill, New York.

Camp, T. R. (1946) "Sedimentation and Design of Settling Tanks," *Transactions of the American Society of Civil Engineers,* vol. 111, p. 895.

Chick, H. (1908) "Investigation of the Law of Disinfection," *Journal of Hygiene,* vol. 8, p. 92.

Cleasby, J. L. and J. H. Dillingham (1966) "Rational Aspects of Split Treatment," *Proceedings American Society of Civil Engineers, Journal Sanitary Engineering Division,* vol. 92, SA2, pp. 1–7.

Cleasby, J. L. and G. S Logsdon, "Filtration," in Letterman, R. D. (Ed.) (1999) *Water Quality and Treatment,* 5th ed., American Water Works Association, McGraw-Hill, New York, pp. 8.11–8.16.

Cornwell, D. A., and M. Bishop (1983) "Determining Velocity Gradients in Laboratory and Full-Scale Systems," *Journal of American Water Works Association,* vol. 75, September, pp. 470–475.

Cornwell, D. A. and J. A. Susan (1979) "Characterization of Acid Treated Alum Sludges." *Journal of the American Water Works Association,* vol. 71, October, pp. 604–608.

Cornwell, D. A., R. M. Mutter, C. Vandermeyden, (2000) "Commercial Application and Marketing of Water Plant Residuals," AWWA Research Foundation and American Water Works Association, Denver, CO.

Dharmarajah, A. H., and J. L. Cleasby (1986) "Predicting the Expansion Behavior of Filter Media," *Journal of the American Water Works Association,* December, pp. 66–76.

Eckenfelder, W. W. (1980) *Industrial Water Pollution Control,* McGraw-Hill, New York, 1989, p. 61.

Fair, G. M., and J. C. Geyer (1954) *Water Supply and Wastewater Disposal,* John Wiley & Sons, New York, pp. 664–671, 678.

Fair, G. M., J. C. Geyer, and D. A. Okun (1968) *Water and Wastewater Engineering,* vol. 2, John Wiley & Sons, Inc., New York, pp. 25–14.

Fair, G. M., J. C. Geyer, and D. A. Okun (1971) *Elements of Water Supply and Wastewater Disposal,* John Wiley & Sons, New York, p. 371.

Hazen, A. (1892) *Annual Report of the Massachusetts State Board of Health,* Boston.

Hazen, A. (1904) "On Sedimentation," *Transactions of the American Society of Civil Engineers,* vol. 53, p. 45.

Henry, J. G., and C. W. Heinke (1989) *Environmental Science and Engineering,* Prentice Hall, Englewood Cliffs, NJ, p. 201.

Hudson, H. E. (1981) *Water Clarification Processes, Practical Design and Evaluation,* Van Nostrand Reinhold, New York, pp. 115–117.

Jacangelo, J. G., S. Adham, and J. Laine (1997) *Membrane Filtration for Microbial Removal,* AWWA Research Foundation and American Water Works Association.

Lewis, W. K., and W. G. Whitman (1924) "Principles of Gas Absorption," *Industrial Engineering Chemistry,* vol. 16, p. 1,215.

Lide, D. R. (2000) *CRC Handbook of Chemistry and Physics,* 81st ed., CRC Press, Boca Raton, FL, pp. 8-111–8-112.

Lightnin Mixers (2000) http://www.lightninmixers.com/sites/lightnin.

MDEQ (1979) *Annual Data Summary,* Michigan Department of Public Health, Lansing, MI.

McTigue, N., and D. Cornwell, "The Use of Particle Counting for the Evaluation of Filter Performance," AWWA Seminar Proceedings, Filtration: Meeting New Standards, AWWA Conference, 1988.

Metcalf & Eddy, Inc. (2003) *Wastewater Engineering: Treatment and Reuse,* revised by G. Tchobanoglous, F. L. Burton, and H. D. Stensel, McGraw-Hill, New York, p. 1,051.

Montgomery, J. M. (1985) *Water Treatment Principles and Design,* John Wiley & Sons, New York, p. 535.

Newton, I. (1687) *Philosophiae Naturalis Principia Mathematica.*

Reilly, W. H. (a USFilter Company) (2006) http:www.whreilly.com/USF Envirex clarifier.html

Reynolds, O. (1883) "An Experimental Investigation of the Circumstances Which Determine Whether the Motion of Water Shall Be Direct or Sinuous and the Laws of Resistance in Parallel Channels," *Transactions of the Royal Society of London,* vol. 174.

Reynolds, T. D., and P. A. Richards (1996) *Unit Operations and Processes in Environmental Engineering,* PWS-Kent, Boston, p. 256.

Richardson, J. F., and W. N. Zaki, (1954) "Sedimentation and Fluidization, Part I, *Transactions of the Institute of Chemical Engineers* (Brit.), vol. 32, pp. 35–53.

Rose, H. E. (1945) "On the Resistance Coefficient–Reynolds Number Relationship of Fluid Flow through a Bed of Granular Material," *Proceedings of the Institute of Mechanical Engineers,* vol. 153, p. 493.

Rushton, J. H. (1952) "Mixing of Liquids in Chemical Processing," *Industrial & Engineering Chemistry,* vol. 44, no. 12, p. 2,931.

Sawyer, C. N., P. L. McCarty, and G. F. Parkin (1994) *Chemistry for Environmental Engineering,* 4th ed., McGraw-Hill, Boston, p. 473.

Steele, E. W., and T. J. McGhee (1979) *Water Supply and Sewerage,* McGraw-Hill, New York.

Stokes, G. G. (1845) *Transactions of the Cambridge Philosophical Society,* vol. 8, p. 287.

U.S. EPA (1979) *Process Design Manual, Sludge Treatment and Disposal,* U. S. Environmental Protection Agency, Washington, DC.

U.S. EPA (1991) *Guidance Manual for Compliance with Filtration and Disinfection Requirements for Public Water Systems Using Surface Waters,* Compliance and Standards Division, Office of Drinking Water, U.S. Environmental Protection Agency, NTIS Pub. No. PB93-222933.

U.S. EPA (1996) *Ultraviolet Light Disinfection Technology in Drinking Water Application—An Overview,* U.S. Environmental Protection Agency Office of Ground Water and Drinking Water, Pub. No. 811-R-96-002.

U.S. EPA (2000) *Safe Drinking Water Information System,* U.S. Environmental Protection Agency, www.epa.gov/safewater/getdata.html.

U.S. PHS, (1962) *Manual of Individual Water Supply Systems,* U.S. Public Health Service Publication No. 24, U.S. Department of Health, Education and Welfare, Washington, DC.

Walker Process Equipment (1973) *Walker Process Circular Clarifiers,* Bulletin 9-w-65, Aurora, IL.

Woller, D. M. (1975) *Public Groundwater Supplies in Calhoun County,* Illinois State Water Survey, Publication No. 60-16, Urbana, IL.

Woller, D. M., and E. W. Sanderson (1976a) *Public Water Supplies in McHenry County,* Illinois State Water Survey, Publication No. 60-19, Urbana, IL.

Woller, D. M., and E. W. Sanderson (1976b) *Public Groundwater Supplies in Putnam County,* Illinois State Water Survey, Publication No. 60-15, Urbana, IL.

WQ&T (1990) *Water Quality and Treatment: A Handbook of Community Water Supplies,* F. W. Pontius (ed.), American Water Works Association, McGraw-Hill, New York, pp. 234–235.

WQ&T (1999) *Water Quality and Treatment: A Handbook of Community Water Supplies,* R. D. Letterman (ed.), American Water Works Association, McGraw-Hill, New York, pp. 23, 24. http://www.waterforpeople.org

CHAPTER 5

WATER QUALITY MANAGEMENT

5-1 INTRODUCTION

The uses we make of water in lakes, rivers, ponds, and streams is greatly influenced by the quality of the water found in them. Activities such as fishing, swimming, boating, shipping, and waste disposal have very different requirements for water quality. Water of a particularly high quality is needed for potable water supplies. In many parts of the world, the introduction of pollutants from human activity has seriously degraded water quality even to the extent of turning pristine trout streams into foul open sewers with few life forms and fewer beneficial uses.

Water quality management is concerned with the control of pollution from human activity so that the water is not degraded to the point that it is no longer suitable for intended uses. The lone frontier family, settled on the banks of the Ohio River, did not significantly degrade water quality in that mighty river even though it threw all its wastes into the river. The city of Cincinnati, however, could not discharge its untreated wastes into the Ohio River without disastrous consequences. Thus, water quality management is also the science of knowing how much is too much for a particular water body.

To know how much waste can be tolerated (the technical term is *assimilated*) by a water body, you must know the type of pollutants discharged and the manner in which they affect water quality. You must also know how water quality is affected by natural factors such as the mineral heritage of the watershed, the geometry of the terrain, and the climate of the region. A small, tumbling mountain brook will have a very different assimilative capacity than a sluggish, meandering lowland river, and lakes are different from moving waters.

Originally, the intent of water quality management was to protect the intended uses of a water body while using water as an economic means of waste disposal within the constraints of its assimilative capacity. In 1972, the Congress of the United States established that it was in the national interest to "restore and maintain the chemical, physical, and biological integrity of the nation's waters." In addition to making the water safe to drink, the Congress also established a goal of "water quality which provides for the protection and propagation of fish, shellfish, and wildlife, and provides for recreation in and on the water" (FWPCA, 1972). By understanding the impact of pollutants on water quality, the environmental engineer can properly design the treatment facilities to remove these pollutants to acceptable levels.

This chapter deals first with the major types of pollutants and their sources. In the remainder of the chapter, water quality management in rivers and in lakes is discussed, placing the emphasis on the categories of pollutants found in domestic wastewaters. For both rivers and lakes, the natural factors affecting water quality will be discussed as the basis for understanding the impact of human activities on water quality.

5-2 WATER POLLUTANTS AND THEIR SOURCES

Point Sources. The wide range of pollutants discharged to surface waters can be grouped into broad classes, as shown in Table 5-1. Domestic sewage and industrial wastes are called *point sources* because they are generally collected by a network of pipes or channels and conveyed to a single point of discharge into the receiving water.

The first edition of this chapter was written by John A. Eastman, Ph.D., of Lockwood, Jones and Beals, Inc., Kettering, OH.

TABLE 5-1
Major pollutant categories and principal sources of pollutants

Pollutant category	Point sources: Domestic sewage	Point sources: Industrial wastes	Non-point sources: Agricultural runoff	Non-point sources: Urban runoff
Oxygen-demanding material	X	X	X	X
Nutrients	X	X	X	X
Pathogens	X	X	X	X
Suspended solids/sediments	X	X	X	X
Salts		X	X	X
Toxic metals		X		X
Toxic organic chemicals		X	X	
Endocrine-disrupting chemicals	X	X	X	
Heat		X		

Domestic sewage consists of wastes from homes, schools, office buildings, and stores. The term municipal sewage is used to mean domestic sewage into which industrial wastes are also discharged.

Under the Federal Water Pollution Control Act of 1972 some *animal feeding operations* (AFOs) may be designated as a point source. AFOs are designated point sources if the feeding operation may be classified as a concentrated animal feeding operation (CAFO). The regulations define AFOs as facilities where animals are fed and confined for 45 days or more in any 12 consecutive-month period, and where crops, vegetation, forage growth, or postharvest residues are not grown or sustained in the feedlot or facility. The latter requirement is to distinguish feedlots from pastures. To qualify as a CAFO (and, thus, be a point source) an AFO must meet one of three criteria (DiMura, 2003): (1) all AFOs with 1,000 animal units* or more are CAFOs, (2) AFOs with between 300 and 1,000 animal units that discharge directly or indirectly to surface waters are CAFOs (AFOs with less than 300 animal units may not be classified as CAFOs), and (3) any AFO may be designated a CAFO if it is found to contribute significantly to pollution of surface waters.

In general, point source pollution can be reduced or eliminated through waste minimization and proper wastewater treatment prior to discharge to a natural water body.

Non-Point Sources. Urban and agricultural runoff are characterized by multiple discharge points. These are called non-point sources. Often the polluted water flows over the surface of the land or along natural drainage channels to the nearest water body. Even when urban or agricultural runoff waters are collected in pipes or channels, they are generally transported the shortest possible distance for discharge, so that wastewater treatment at each outlet is not economically feasible. Much of the non-point source pollution occurs during rainstorms or spring snowmelt resulting in large flow rates that make treatment even more difficult. Non-point pollution from urban storm water and, in particular, storm water collected in *combined sewers* that carry both storm water and municipal sewage may require major engineering work to correct. The

*Generally, one animal unit is equal to 450 kg of live animal mass.

original design of combined sewers provided a flow structure that diverted excess storm water mixed with raw sewage (above the design capacity of the wastewater treatment plant) directly to the nearest river or stream. The elimination of *combined sewer overflow* (CSO) may involve not only provision of separate storm and sanitary sewers but also the provision of storm water retention basins and expanded treatment facilities to treat the storm water. This is particularly complex and expensive because the combined sewers frequently occur in the oldest, most developed portions of the community. Thus, paved streets, utilities, and commercial activities will be disrupted. The installation of combined sewers is now prohibited in the United States.

Runoff from agricultural land is a significant non-point source. Fertilizer, whether in the form of manure or commercial fertilizer, contributes nutrients. Agricultural runoff carries toxic organic compounds in the form of pesticides. Soil erosion contributes suspended solids. Implementation of *Best Management Practices* (BMP) to reduce excess application of fertilizer and pesticides along with erosion control programs conserves the farmers economic investment while protecting the river.

Oxygen-Demanding Material. Anything that can be oxidized in the receiving water with the consumption of dissolved molecular oxygen is termed oxygen-demanding material. This material is usually biodegradable organic matter but also includes certain inorganic compounds. The consumption of *dissolved oxygen,* DO (pronounced "dee oh"), poses a threat to fish and other higher forms of aquatic life that must have oxygen to live. The critical level of DO varies greatly among species. For example, brook trout may require about 7.5 mg/L of DO, while carp may survive at 3 mg/L. As a rule, the most desirable commercial and game fish require high levels of dissolved oxygen. Oxygen-demanding materials in domestic sewage come primarily from human waste and food residue. Particularly noteworthy among the many industries that produce oxygen-demanding wastes are the food processing and paper industries. Almost any naturally occurring organic matter, such as animal droppings, crop residues, or leaves, that get into the water from non-point sources, contribute to the depletion of DO.

Nutrients. Nitrogen and phosphorus, two nutrients of primary concern, are considered pollutants because they are too much of a good thing. All living things require these nutrients for growth. Thus, they must be present in rivers and lakes to support the natural food chain.* Problems arise when nutrient levels become excessive and the food web is grossly disturbed, which causes some organisms to proliferate at the expense of others. As will be discussed in a later section, excessive nutrients often lead to large growths of algae, which in turn become oxygen-demanding material when they die and settle to the bottom. Some major sources of nutrients are phosphorus-based detergents, fertilizers, and food-processing wastes.

Pathogenic Organisms. Microorganisms found in wastewater include bacteria, viruses, and protozoa excreted by diseased persons or animals. When discharged into surface waters, they make the water unfit for drinking (that is, nonpotable). If the concentration of pathogens is sufficiently high, the water may also be unsafe for

---

*In simplistic terms, a food chain is the collection of interrelated organisms in which the lower levels are the "eatees" and the upper levels are the "eaters."

swimming and fishing. Certain shellfish can be toxic because they concentrate pathogenic organisms in their tissues, making the toxicity levels in the shellfish much greater than the levels in the surrounding water.

Cholera and typhoid are *endemic* diseases in the world with over 384,000 cases of cholera and 16 million cases of typhoid per year. These killers cause 20,000 and 600,000 deaths per year respectively (*Population Reports,* 1998). Yet, in the United States, they are unheard of outside of a historic context. The widespread disease-causing organisms in the United States are *Giardia lambia* and *Cryptosporidium parvum*. These are protozoan pathogens from both human and animal sources.

Bacteria that have developed immunity to antibiotics are now being found in natural waters (Ash et al.; 1999, Sternes, 1999, Bennett and Kramer, 1999). The resistance of the bacteria appearing in natural waters is an ominous prelude to the future effectiveness of antibiotics.

Suspended Solids. Organic and inorganic particles that are carried by the wastewater into a receiving water are termed *suspended solids* (SS). When the speed of the water is reduced by flowing into a pool or a lake, many of these particles settle to the bottom as sediment. In common usage, the word sediment also includes eroded soil particles that are being carried by water even if they have not yet settled. Colloidal particles, that do not settle readily, cause the turbidity found in many surface waters. Organic suspended solids may also exert an oxygen demand. Inorganic suspended solids are discharged by some industries but result mostly from soil erosion, which is particularly bad in areas of logging, strip mining, and construction activity. As excessive sediment loads are deposited into lakes and reservoirs, the usefulness of the water is reduced. Even in rapidly moving mountain streams, sediment from mining and logging operations has destroyed many living places (*ecological habitats*) for aquatic organisms. For example, salmon eggs can only develop and hatch in stream beds of loose gravel. As the pores between the pebbles are filled with sediment, the eggs suffocate and the salmon population is reduced.

Salts. Although most people associate salty water with oceans and salt lakes, all water contains some salt. These salts are often measured by evaporation of a filtered water sample. The salts and other things that don't evaporate are called *total dissolved solids* (TDS). A problem arises when the salt concentration in normally fresh water increases to the point where the natural population of plants and animals is threatened or the water is no longer useful for public water supplies or irrigation. High concentrations of salts are discharged by many industries, and the use of salt on roads during the winter causes high salt levels in urban runoff, especially during the spring snowmelt. Of particular concern in arid regions, where water is used extensively for irrigation, is that the water picks up salts every time it passes through the soil on its way back to the river. In addition, evapotranspiration causes the salts to be further concentrated. Thus, the salt concentration continuously increases as the water moves downstream. If the concentration gets too high, crop damage or soil poisoning can result.

Toxic Metals and Toxic Organic Compounds. Agricultural runoff often contains pesticides and herbicides that have been used on crops. Urban runoff is a major source of zinc in many water bodies. The zinc comes from tire wear. Many industrial wastewaters contain either toxic metals or toxic organic substances. If discharged in large quantities, many of these materials can render a body of water nearly useless for long periods of time. The lower James River in Virginia has been reduced to use only as a shipping channel because of a large industrial discharge of highly toxic and persistent organic compounds. Many toxic compounds are concentrated in the food chain, making fish and shellfish unsafe for human consumption. Thus, even small quantities of toxic compounds in the water can be incompatible with the natural ecosystem and many human uses.

Endocrine-Disrupting Chemicals. The class of chemicals known as *endocrine disrupters,* or EDCs, alter the normal physiological function of the endocrine system and can affect the synthesis of hormones. EDCs can also target tissues where the hormones exert their effects. EDCs can mimic estrogens, androgens, or thyroid hormones or their antagonists. They can interfere with the regulation of reproductive and developmental process in mammals, birds, reptiles, and fish (Sadik and Witt, 1999; Harries et al., 2000). Figure 5-1 shows the most frequently detected pharmaceuticals and EDCs in recent survey of U.S. waters (U.S.G.S., 2002).

Heat. Although heat is not often recognized as a pollutant, those in the electric power industry are well aware of the problems of disposing of waste heat. Also, waters released by many industrial processes are much hotter than the receiving waters. In

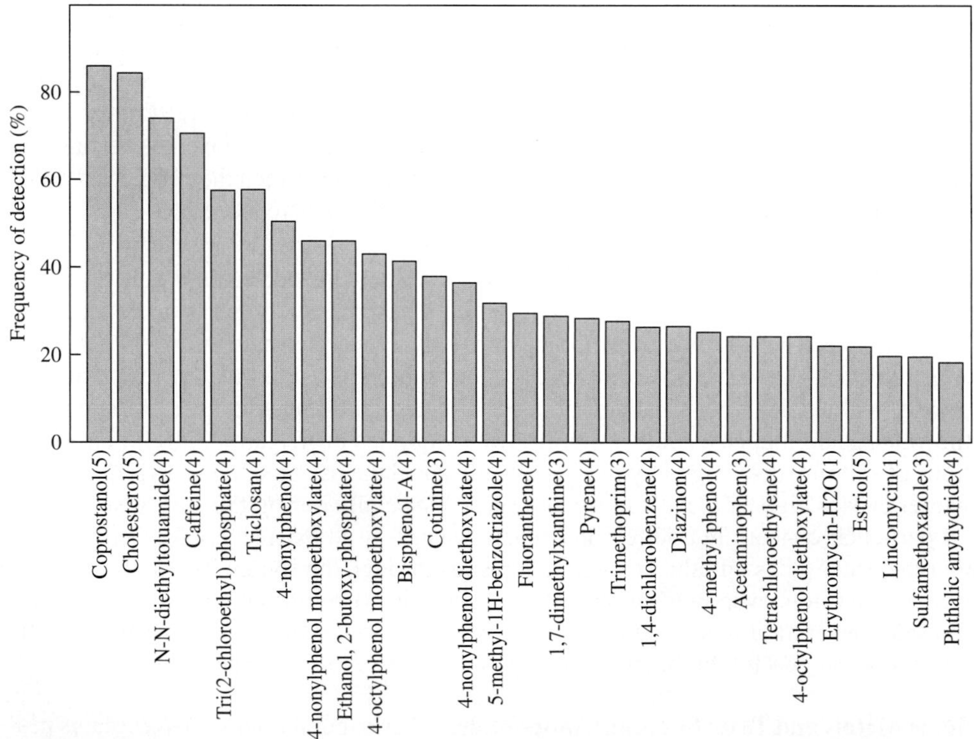

FIGURE 5-1
Most frequently detected pharmaceutical- and endocrine-disrupting compounds. *Source:* U.S.G.S., 2002.

some environments an increase of water temperature can be beneficial. For example, production of clams and oysters can be increased in some areas by warming the water. On the other hand, increases in water temperature can have negative impacts. Many important commercial and game fish, such as salmon and trout, live only in cool water. In some instances the discharge of heated water from a power plant can completely block salmon migration. Higher temperatures also increase the rate of oxygen depletion in areas where oxygen-demanding wastes are present.

5-3 WATER QUALITY MANAGEMENT IN RIVERS

The objective of water quality management is simple to state: to control the discharge of pollutants so that water quality is not degraded to an unacceptable extent below the natural background level. However, controlling waste discharges must be a quantitative endeavor. We must be able to measure the pollutants, predict the impact of the pollutant on water quality, determine the background water quality which would be present without human intervention, and decide the levels acceptable for intended uses of the water.

To most people, the tumbling mountain brook, crystal clear and icy cold, fed by the melting snow, and safe to drink is the epitome of high water quality. Certainly a stream in that condition is a treasure, but we cannot expect the Mississippi River to have the same water quality. It never did and never will. Yet both need proper management if the water is to remain usable for intended purposes. The mountain brook may serve as the spawning ground for desirable fish and must be protected from heat and sediment as well as chemical pollution. The Mississippi, however, is already warmed from hundreds of kilometers of exposure to the sun and carries the sediment from thousands of square kilometers of land. But even the Mississippi can be damaged by organic matter and toxic chemicals. Fish do live there and the river is used as a water supply for millions of people.

The impact of pollution on a river depends both on the nature of the pollutant and the unique characteristics of the individual river.* Some of the most important characteristics include the volume and speed of water flowing in the river, the river's depth, the type of bottom, and the surrounding vegetation. Other factors include the climate of the region, the mineral heritage of the watershed, land use patterns, and the types of aquatic life in the river. Water quality management for a particular river must consider all these factors. Thus, some rivers are highly susceptible to pollutants such as sediment, salt, and heat, while other rivers can tolerate large inputs of these pollutants without much damage.

Total Maximum Daily Load (TMDL)

Under Section 303(d) of the 1972 Clean Water Act, states, territories, and authorized tribes are required to develop lists of *impaired waters*. Impaired waters are those that do not meet water quality standards that the states, territories, and authorized tribes have established for them. This assessment is made after assuming that point sources of pollution have installed minimum levels of pollution control technology. The law requires that these jurisdictions establish priority rankings for waters on the lists and develop *total maximum daily loads* (TMDL) for these waters. A TMDL specifies the maximum amount of pollutant that a water body can receive and still meet water quality

*Here we will use the word "river" to include streams, brooks, creeks, and any other channel of flowing, fresh water.

standards. In addition, the TMDL allocates pollutant *loadings* (that is, the mass of pollutant) that may be contributed among point and non-point sources. The TMDL is computed on a pollutant-by-pollutant basis for a list of pollutants similar to those in Table 5-1. Additional categories include acids/bases (measured as pH), pesticides, and mercury. The TMDL computation is defined as:

$$\text{TMDL} = \Sigma\text{WLA} + \Sigma\text{LA} + \text{MOS} \tag{5-1}$$

where WLA = waste load allocations, that is, portions of the TMDL assigned to existing and future point sources

LA = load allocations, that is, portions of the TMDL assigned to existing and future non-point sources

MOS = margin of safety

The MOS is to account for uncertainty about the relationships between loads and water quality. A software system called *Better Assessment Science Integrating Point and Non-point Sources* (BASINS) that integrates a *geographic information system* (GIS), national watershed and meteorological data, and state-of-the-art environmental assessment and modeling tools is used to develop the TMDL (Ahmad, 2002; U.S EPA, 2005).

Some pollutants, particularly oxygen-demanding wastes and nutrients, are so common and have such a profound impact on almost all types of rivers that they deserve special emphasis. This is not to say that they are always the most significant pollutants in any one river, but rather that no other pollutant category has as much overall effect on our nation's rivers. For these reasons, the next sections of this chapter will be devoted to a more detailed look at how oxygen-demanding material and nutrients affect water quality in rivers.

Effect of Oxygen-Demanding Wastes on Rivers

The introduction of oxygen-demanding material, either organic or inorganic, into a river causes depletion of the dissolved oxygen in the water. This poses a threat to fish and other higher forms of aquatic life if the concentration of oxygen falls below a critical point. To predict the extent of oxygen depletion, it is necessary to know how much waste is being discharged and how much oxygen will be required to degrade the waste. However, because oxygen is continuously being replenished from the atmosphere and from photosynthesis by algae and aquatic plants, as well as being consumed by organisms, the concentration of oxygen in the river is determined by the relative rates of these competing processes. Organic oxygen-demanding materials are commonly measured by determining the amount of oxygen consumed during degradation in a manner approximating degradation in natural waters. This section begins by considering the factors affecting oxygen consumption during the degradation of organic matter, then moves on to inorganic nitrogen oxidation. Finally, the equations for predicting dissolved oxygen concentrations in rivers from degradation of organic matter are developed and discussed.

Biochemical Oxygen Demand

The amount of oxygen required to oxidize a substance to carbon dioxide and water may be calculated by stoichiometry if the chemical composition of the substance is known. This amount of oxygen is known as the *theoretical oxygen demand* (ThOD).

Example 5-1. Compute the ThOD of 108.75 mg/L of glucose ($C_6H_{12}O_6$).

Solution. We begin by writing a balanced equation for the reaction.

$$C_6H_{12}O_6 + 6O_2 \rightleftharpoons 6CO_2 + 6H_2O$$

Next, compute the gram molecular weights of the reactants using the table on the inside cover of the book.

glucose	*oxygen*
6C = 72	(6)(2)O = 192
12H = 12	
6O = 96	
180	

Thus, it takes 192 g of oxygen to oxidize 180 g of glucose to CO_2 and H_2O.

The ThOD of 108.75 mg/L of glucose is

$$(108.75 \text{ mg/L of glucose})\left(\frac{192 \text{ g } O_2}{180 \text{ g glucose}}\right) = 116 \text{ mg/L } O_2$$

In contrast to the ThOD, the *chemical oxygen demand,* COD (pronounced "see oh dee"), is a measured quantity that does not depend on knowledge of the chemical composition of the substances in the water. In the COD test, a strong chemical oxidizing agent (chromic acid) is mixed with a water sample and then boiled. The difference between the amount of oxidizing agent at the beginning of the test and that remaining at the end of the test is used to calculate the COD.

If the oxidation of an organic compound is carried out by microorganisms using the organic matter as a food source, the oxygen consumed is known as *biochemical oxygen demand,* or BOD (pronounced "bee oh dee"). The actual BOD is less than the ThOD due to the incorporation of some of the carbon into new bacterial cells. The test is a bioassay that utilizes microorganisms in conditions similar to those in natural water to measure indirectly the amount of biodegradable organic matter present. *Bioassay* means to measure by biological means. A water sample is inoculated with bacteria that consume the biodegradable organic matter to obtain energy for their life processes. Because the organisms also utilize oxygen in the process of consuming the waste, the process is called *aerobic* decomposition. This oxygen consumption is easily measured. The greater the amount of organic matter present, the greater the amount of oxygen utilized. The BOD test is an indirect measurement of organic matter because we actually measure only the change in dissolved oxygen concentration caused by the microorganisms as they degrade the organic matter. Although not all organic matter is biodegradable and the actual test procedures lack precision, the BOD test is still the most widely used method of measuring organic matter because of the direct conceptual relationship between BOD and oxygen depletion in receiving waters.

Only under rare circumstances will the ThOD, COD, and BOD be equal. If the chemical composition of all of the substances in the water is known and they are capable of being completely oxidized both chemically and biologically, then the three measures of oxygen demand will be the same.

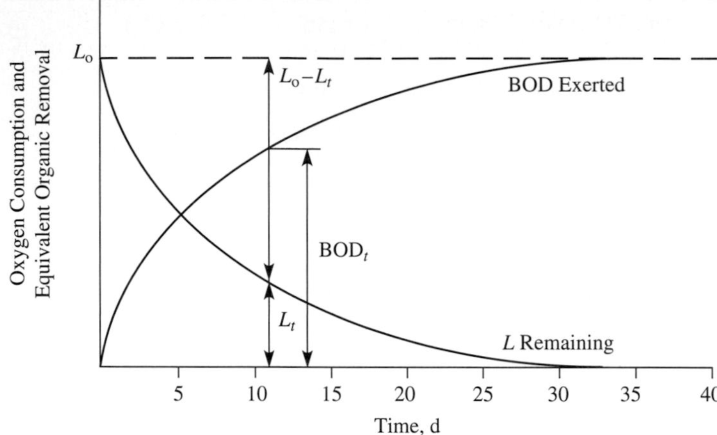

FIGURE 5-2
BOD and oxygen-equivalent relationships.

When a water sample containing degradable organic matter is placed in a closed container and inoculated with bacteria, the oxygen consumption typically follows the pattern shown in Figure 5-2. During the first few days the rate of oxygen depletion is rapid because of the high concentration of organic matter present. As the concentration of organic matter decreases, so does the rate of oxygen consumption. During the last part of the BOD curve, oxygen consumption is mostly associated with the decay of the bacteria that grew during the early part of the test. It is generally assumed that the rate at which oxygen is consumed is directly proportional to the concentration of degradable organic matter remaining at any time. As a result, the BOD curve in Figure 5-2 can be described mathematically as a first-order reaction. Using our definition of reaction rate and reaction order from Chapter 4, this may be expressed as:

$$\frac{dL_t}{dt} = -r_A \tag{5-2}$$

where L_t = oxygen equivalent of the organics remaining at time t, mg/L
$-r_A = -kL_t$
k = reaction rate constant, d^{-1}

Rearranging Equation 5-2 and integrating yields:

$$\frac{dL_t}{L_t} = -k\,dt$$

$$\int_{L_o}^{L} \frac{dL_t}{L_t} = -k\int_0^t dt$$

$$\ln\frac{L_t}{L_o} = -kt$$

or

$$L_t = L_o e^{-kt} \tag{5-3}$$

where

L_o = oxygen equivalent of organic compounds at time $t = 0$

Rather than L_t, our interest is in the amount of oxygen used in the consumption of the organics (BOD_t). From Figure 5-2, it is obvious that BOD_t is the difference between the initial value of L_o and L_t, so

$$\begin{aligned} BOD_t &= L_o - L_t \\ &= L_o - L_o e^{-kt} \\ &= L_o(1 - e^{-kt}) \end{aligned} \tag{5-4}$$

L_o is often referred to as the *ultimate BOD,* that is, the maximum oxygen consumption possible when the waste has been completely degraded. Equation 5-4 is called the *BOD rate equation* and is often written in base 10:

$$BOD_t = L_o(1 - 10^{-Kt}) \tag{5-5}$$

Note that lower case k is used for the reaction rate constant in base e and that capital K is used for the constant in base 10. They are related: $k = 2.303(K)$.

Example 5-2. If the BOD_3 of a waste is 75 mg/L and the K is 0.150 d^{-1}, what is the ultimate BOD?

Solution. Note that the rate constant is given in base 10 (K versus k), and substitute the given values into Equation 5-5 and solve for L_o:

$$75 = L_o(1 - 10^{-(.150)(3)}) = 0.645\, L_o$$

or

$$L_o = \frac{75}{0.645} = 116 \text{ mg/L}$$

In base e,

$$k = 2.303(K) = 0.345, \text{ and}$$
$$75 = L_o(1 - e^{-(.345)(3)}) = 0.645\, L_o$$

so

$$L_o = 116 \text{ mg/L}$$

You should note that the ultimate BOD (L_o) is defined as the maximum BOD exerted by the waste. It is denoted by the horizontal line in Figure 5-2. Because BOD_t approaches L_o asymptotically, it is difficult to assign an exact time to achieve ultimate BOD. Indeed, based on Equation 5-3, it is achieved only in the limit as t approaches infinity. However, from a practical point of view, we can observe that when the BOD curve is approximately horizontal, the ultimate BOD has been achieved. In Figure 5-2, we would take this to be at about 35 days. In computations, we use a rule of thumb that if BOD_t and L_o agree when rounded to three significant figures, then the time to reach ultimate BOD has been achieved. Given the vagaries of the BOD test, there are occasions when rounding to two significant figures would not be unrealistic.

While the ultimate BOD best expresses the concentration of degradable organic matter, it does not, by itself, indicate how rapidly oxygen will be depleted in a receiving water.

TABLE 5-2
Typical values for the BOD rate constant

Sample	K (20°C) (day^{-1})	k (20°C) (day^{-1})
Raw sewage	0.15–0.30	0.35–0.70
Well-treated sewage	0.05–0.10	0.12–0.23
Polluted river water	0.05–0.10	0.12–0.23

Oxygen depletion is related to both the ultimate BOD and the BOD rate constant (k). While the ultimate BOD increases in direct proportion to the concentration of degradable organic matter, the numerical value of the rate constant is dependent on the following:

1. The nature of the waste
2. The ability of the organisms in the system to utilize the waste
3. The temperature

Nature of the Waste. There are literally thousands of naturally occurring organic compounds, not all of which can be degraded with equal ease. Simple sugars and starches are rapidly degraded and will therefore have a very large BOD rate constant. Cellulose (for example, toilet paper) degrades much more slowly, and hair and fingernails are almost undegradable in the BOD test or in normal wastewater treatment. Other compounds are intermediate between these extremes. The BOD rate constant for a complex waste depends very much on the relative proportions of the various components. A summary of typical BOD rate constants is shown in Table 5-2. The lower rate constants for treated sewage compared to raw sewage result from the fact that easily degradable organic compounds are more completely removed than less readily degradable organic compounds during wastewater treatment.

Ability of Organisms to Utilize Waste. Any given microorganism is limited in its ability to utilize organic compounds. As a consequence, many organic compounds can be degraded by only a small group of microorganisms. In a natural environment receiving a continuous discharge of organic waste, that population of organisms which can most efficiently utilize this waste will predominate. However, the culture used to inoculate the BOD test may contain only a very small number of organisms that can degrade the particular organic compounds in the waste. This problem is especially common when analyzing industrial wastes. The result is that the BOD rate constant is lower in the laboratory test than in the natural water. This is an undesirable outcome. The BOD test should therefore be conducted with organisms which have been acclimated to the waste so that the rate constant determined in the laboratory can be compared to that in the river.*

Temperature. Most biological processes speed up as the temperature increases and slow down as the temperature drops. Because oxygen utilization is caused by the

*The word "acclimated" means that the organisms have had time to adapt their metabolisms to the waste or that organisms that can utilize the waste have been given the chance to predominate in the culture.

metabolism of microorganisms, the rate of utilization is similarly affected by temperature. Ideally, the BOD rate constant should be experimentally determined for the temperature of the receiving water. There are two difficulties with this ideal. Often the receiving-water temperature changes throughout the year, so a large number of tests would be required to define k. An additional difficulty is the task of comparing data from various locations having different temperatures. Laboratory testing is therefore done at a standard temperature of 20°C, and the BOD rate constant is adjusted to the receiving-water temperature using the following expression:

$$k_T = k_{20}(\theta)^{T-20} \tag{5-6}$$

where T = temperature of interest, °C
k_T = BOD rate constant at the temperature of interest, d^{-1}
k_{20} = BOD rate constant determined at 20°C, d^{-1}
θ = temperature coefficient. This has a value of 1.135 for temperatures between 4 and 20°C and 1.056 for temperatures between 20 and 30°C (Schroepfer, et al., 1964).

Example 5-3. A waste is being discharged into a river that has a temperature of 10°C. What fraction of the maximum oxygen consumption has occurred in four days if the BOD rate constant determined in the laboratory under standard conditions is 0.115 d^{-1} (base e)?

Solution. Determine the BOD rate constant for the waste at the river temperature using Equation 5-6:

$$\begin{aligned} k_{10^\circ C} &= 0.115(1.135)^{10-20} \\ &= 0.032\ d^{-1} \end{aligned}$$

Use this value of k in Equation 5-4 to find the fraction of maximum oxygen consumption occurring in four days:

$$\begin{aligned} \frac{BOD_4}{L_o} &= [1 - e^{-(.032)(4)}] \\ &= 0.12 \end{aligned}$$

Graphical Determination of BOD Constants

A variety of methods may be used to determine k and L_o from an experimental set of BOD data. The simplest and least accurate method is to plot BOD versus time. This results in a hyperbolic first-order curve of the form shown in Figure 5-2. The ultimate BOD is estimated from the asymptote of the curve. The rate equation is used to solve for k. It is often difficult to fit an accurate hyperbola to data that are frequently scattered. Methods that linearize the data are preferred. The usual graphical methods for first-order reactions cannot be used because the semilog plot requires knowledge of the initial concentration which, in this case, is one of the constants we are trying to determine, that is, L_o! One simple method around this impasse is called Thomas' graphical method (Thomas, 1950). The method relies on the similarity of the series expansion of the following two functions:

$$F_1 = 1 - e^{-kt} \tag{5-7}$$

and

$$F_2 = (kt)(1 + (1/6)\,kt)^{-3} \tag{5-8}$$

The series expansion of these functions yields:

$$F_1 = (kt)\left[1 - 0.5(kt) + \frac{1}{6}(kt)^2 - \frac{1}{24}(kt)^3 + \cdots\right] \tag{5-9}$$

$$F_2 = (kt)\left[1 - 0.5(kt) + \frac{1}{6}(kt)^2 - \frac{1}{21.9}(kt)^3 + \cdots\right] \tag{5-10}$$

The first two terms are identical and the third differs only slightly. Replacing Equation 5-9 by 5-10 in the BOD rate equation results in the following approximate equation:

$$\mathrm{BOD}_t = L_o(kt)[1 + (1/6)kt]^{-3} \tag{5-11}$$

By rearranging terms and taking the cube root of both sides, Equation 5-11 can be transformed to:

$$\left(\frac{t}{\mathrm{BOD}_t}\right)^{1/3} = \frac{1}{(kL_o)^{1/3}} + \frac{(k)^{2/3}}{6(L_o)^{1/3}}(t) \tag{5-12}$$

A plot of $(t/\mathrm{BOD}_t)^{1/3}$ versus t is linear (Figure 5-3). The intercept is defined as:

$$A = (kL_o)^{-1/3} \tag{5-13}$$

The slope is defined by:

$$B = \frac{(k)^{2/3}}{6(L_o)^{1/3}} \tag{5-14}$$

Solving for $L_o^{1/3}$ in Equation 5-13 substituting into Equation 5-14 and solving for k yields:

$$k = 6(B/A) \tag{5-15}$$

Likewise, substituting Equation 5-15 into 5-13 and solving for L yields:

$$L_o = \frac{1}{6(A)^2(B)} \tag{5-16}$$

The procedure for determining the BOD constants by this method is as follows:

1. From the experimental results of BOD for various values of t, calculate $(t/\mathrm{BOD}_t)^{1/3}$ for each day.

2. Plot $(t/\mathrm{BOD}_t)^{1/3}$ versus t on arithmetic graph paper and draw the line of best fit by eye.

3. Determine the intercept (A) and slope (B) from the plot.

4. Calculate k and L_o from Equations 5-15 and 5-16.

Example 5-4. The following data were obtained from an experiment to determine the BOD rate constant and ultimate BOD for an untreated wastewater:

Day	0	1	2	4	6	8
BOD, mg/L	0	32	57	84	106	111

Solution. First calculate values of $(t/\text{BOD}_t)^{1/3}$

Day	0	1	2	4	6	8
$(t/\text{BOD}_t)^{1/3}$	—	0.315	0.327	0.362	0.384	0.416

The plot of $(t/\text{BOD}_t)^{1/3}$ versus time for these data is shown in Figure 5-3. From the figure, $A = 0.30$ and

$$B = \frac{(0.416 - 0.300)}{(8 - 0)} = 0.0145$$

Substituting in Equations 5-15 and 5-16:

$$k = \frac{6(0.0145)}{0.30} = 0.29 \text{ d}^{-1}$$

$$L_o = \frac{1}{6(0.30)^2(0.0145)} = 128 \text{ mg/L}$$

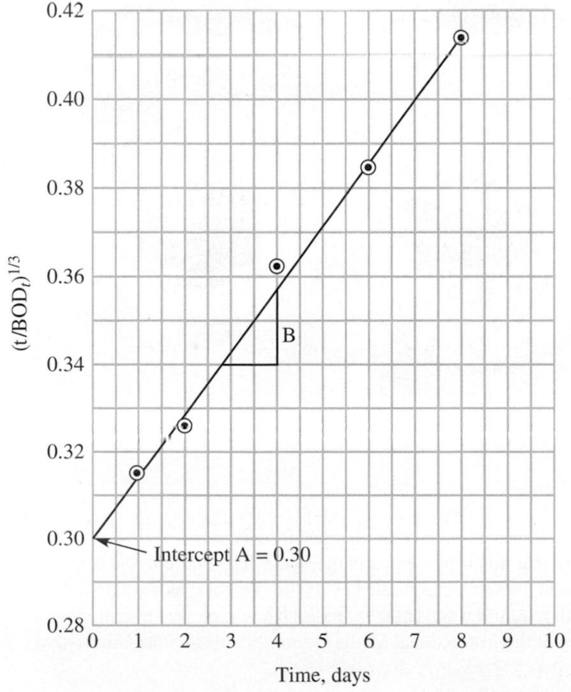

FIGURE 5-3
Plot of $(t/\text{BOD}_t)^{1/3}$ versus t for Thomas' graphical method.

Laboratory Measurement of Biochemical Oxygen Demand

In order to have as much consistency as possible, it is important to standardize testing procedures when measuring BOD. In the paragraphs that follow, the standard BOD test is outlined with emphasis placed on the reason for each step rather than the details. The detailed procedures can be found in *Standard Methods for the Examination of Water and Wastewater* (APHA, 2004) which is the authoritative reference of testing procedures in the water pollution control field.

Step 1. A special 300 mL BOD bottle (Figure 5-4) is completely filled with a sample of water that has been appropriately diluted and inoculated with microorganisms. The bottle is then stoppered to exclude air bubbles. Samples require dilution because the only oxygen available to the organisms is dissolved in the water. The most oxygen that can dissolve is about 9 mg/L, so the BOD of the diluted sample should be between 2 and 6 mg/L. Samples are diluted with a special dilution water that contains all of the trace elements required for bacterial metabolism so that degradation of the organic matter is not limited by lack of bacterial growth. The dilution water also contains an inoculum of microorganisms so that all samples tested on a given day contain approximately the same type and number of microorganisms.

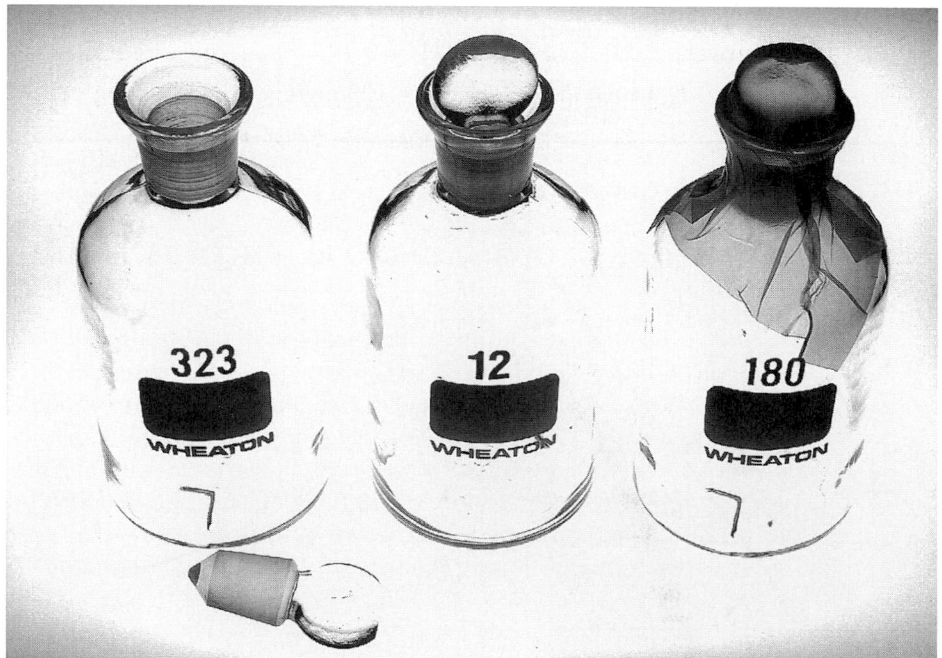

FIGURE 5-4
BOD bottles. The bottle on the left is shown with the cap removed to illustrate the shape of the glass stoppers. The point on the end of the stopper is to ensure that no air is trapped in the bottle. The bottle in the center is shown with the stopper in place. Water is placed in the small cup formed by the lip. This acts as a seal to further exclude air. The bottle on the right is shown with plastic wrap over the stopper. This is to prevent evaporation of the water seal. (Photo courtesy of Harley Seeley of the Instructional Media Center, Michigan State University.)

The ratio of undiluted to diluted sample is called the *sample size,* usually expressed as a percentage, while the inverse relationship is called the *dilution factor.* Mathematically, these are

$$\text{Sample size (\%)} = \frac{\text{vol. of undiluted sample}}{\text{vol. of diluted sample}} \times 100 \tag{5-17}$$

$$\text{Dilution factor} = \frac{\text{vol. of diluted sample}}{\text{vol. of undiluted sample}} = \frac{100}{\text{sample size (\%)}} \tag{5-18}$$

The appropriate sample size to use can be determined by dividing 4 mg/L (the midpoint of the desired range of diluted BOD) by the estimated BOD concentration in the sample being tested. A convenient volume of undiluted sample is then chosen to approximate to this sample size.

Example 5-5. The BOD of a wastewater sample is estimated to be 180 mg/L. What volume of undiluted sample should be added to a 300 mL bottle? Also, what are the sample size and dilution factor using this volume? Assume that 4 mg/L BOD can be consumed in the BOD bottle.

Solution. Estimate the sample size needed:

$$\text{Sample size} = \frac{4}{180} \times 100 = 2.22\%$$

Estimate the volume of undiluted sample needed since the volume of diluted sample is 300 mL:

$$\text{Vol. of undiluted sample} = 0.0222 \times 300 \text{ mL} = 6.66 \text{ mL}$$

Therefore a convenient sample volume would be 7.00 mL.

Compute the actual sample size and dilution factor:

$$\text{Sample size} = \frac{7.00 \text{ mL}}{300 \text{ mL}} \times 100 = 2.33\%$$

$$\text{Dilution factor} = \frac{300 \text{ mL}}{7.00 \text{ mL}} = 42.9$$

Step 2. Blank samples containing only the inoculated dilution water are also placed in BOD bottles and stoppered. Blanks are required to estimate the amount of oxygen consumed by the added inoculum of microorganisms (called *seed*) in the absence of the sample.

Step 3. The stoppered BOD bottles containing diluted samples and blanks are incubated in the dark at 20°C for the desired number of days. For most purposes, a standard time of five days is used. To determine the ultimate BOD and the BOD rate constant, additional times are used. The samples are incubated in the dark to prevent photosynthesis from adding oxygen to the water and invalidating the oxygen consumption results. As mentioned earlier, the BOD test is conducted at a standard temperature of 20°C so that the effect of temperature on the BOD rate constant is eliminated and results from different laboratories can be compared.

Step 4. After the desired number of days has elapsed, the samples and blanks are removed from the incubator and the dissolved oxygen concentration in each bottle is measured. The BOD of the undiluted sample is then calculated using the following equation:

$$\text{BOD}_t = (\text{DO}_{b,t} - \text{DO}_{s,t}) \times \text{dilution factor} \tag{5-19}$$

where $\text{DO}_{b,t}$ = dissolved oxygen concentration in blank after t days of incubation, mg/L
$\text{DO}_{s,t}$ = dissolved oxygen concentration in sample after t days of incubation, mg/L

Example 5-6. What is the BOD_5 of the wastewater sample of Example 5-5 if the DO values for the blank and diluted sample after five days are 8.7 and 4.2 mg/L, respectively?

Solution. Substitute the appropriate values into Equation 5-19:

$$\text{BOD}_5 = (8.7 - 4.2) \times 42.9 = 193 \text{ or } 190 \text{ mg/L}$$

Additional Notes on Biochemical Oxygen Demand

Although the 5-day BOD has been chosen as the standard value for most wastewater analysis and for regulatory purposes, ultimate BOD is actually a better indicator of total waste strength. For any one type of waste having a defined BOD rate constant, the ratio between ultimate BOD and BOD_5 is constant so that BOD_5 indicates relative waste strength. For different types of wastes having the same BOD_5, the ultimate BOD is the same only if, by chance, the BOD rate constants are the same. This is illustrated in Figure 5-5 for a municipal wastewater having a $K = 0.15 \text{ d}^{-1}$ and an industrial wastewater having a $K = 0.05 \text{ d}^{-1}$. Although both wastewaters have a BOD_5 of 200 mg/L, the

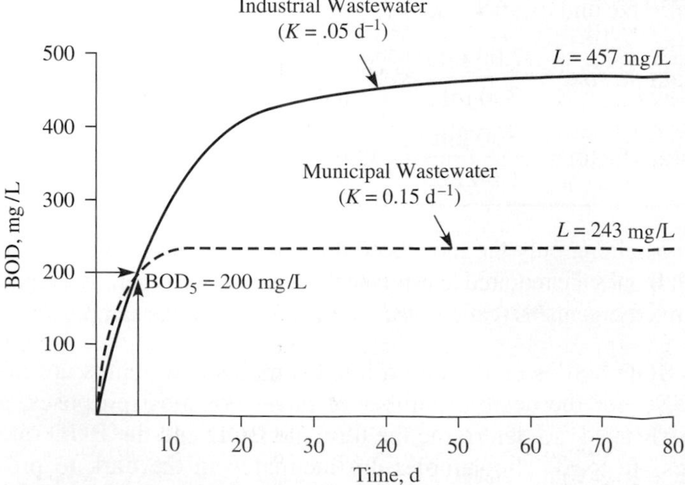

FIGURE 5-5
The effect of K on ultimate BOD for two wastewaters having the same BOD_5.

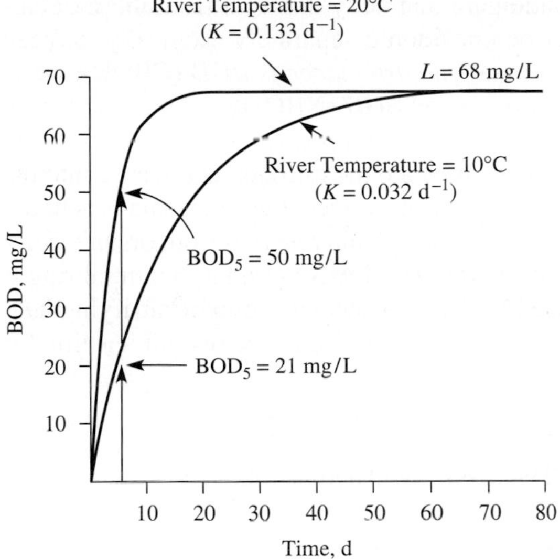

FIGURE 5-6
The effect of K on BOD_5, when the ultimate BOD is constant.

industrial wastewater has a much higher ultimate BOD and can be expected to have a greater impact on dissolved oxygen in a river. For the industrial wastewater, a smaller fraction of the BOD was exerted in the first five days due to the lower rate constant.

Proper interpretation of BOD_5 values can also be illustrated in another way. Consider a sample of polluted river water for which the following values were determined using standard laboratory techniques: $BOD_5 = 50$ mg/L, and $K = 0.115\ d^{-1}$. The ultimate BOD calculated from Equation 5-5 is, therefore, 68 mg/L. However, because the river temperature is 10°C, the K value in the river is only 0.032 d^{-1} (see Example 5-3). As shown graphically in Figure 5-6, the laboratory value of BOD_5 seriously overestimates the actual oxygen consumption in the river. Again, a smaller fraction of the BOD is exerted in 5 days when the BOD rate constant is lower.

The 5-day BOD was chosen as the standard value for most purposes because the test was devised by environmental engineers in England, where rivers have travel times to the sea of less than 5 days, so there was no need to consider oxygen demand at longer times. Since there is no other time that is any more rational than 5 days, this value has become firmly entrenched.*

Nitrogen Oxidation

Up to this point an unstated assumption has been that only the carbon in organic matter is oxidized. Actually many organic compounds, such as proteins, also contain nitrogen that can be oxidized with the consumption of molecular oxygen. However, because the

*Numerous investigations of the kinetics of the BOD test have demonstrated that the selection of a 5-day incubation period may be justified on scientific grounds regardless of the travel time of streams. For example, one explanation is that after 5 days, the microbial system in the BOD bottle is in the autodigestive phase. That is, there is no carbon source outside of the microbial cells that they can consume and that any oxygen uptake that occurs beyond 5 days is the result of consumption of carbon that they have stored (Gaudy and Gaudy, 1980).

mechanisms and rates of nitrogen oxidation are distinctly different from those of carbon oxidation, the two processes must be considered separately. Logically, oxygen consumption due to oxidation of carbon is called *carbonaceous BOD* (CBOD), while that due to nitrogen oxidation is called *nitrogenous BOD* (NBOD).

The organisms that oxidize the carbon in organic compounds to obtain energy cannot oxidize the nitrogen in these compounds. Instead, the nitrogen is released into the surrounding water as ammonia (NH_3). At normal pH values, this ammonia is actually in the form of the ammonium cation (NH_4^+). The ammonia released from organic compounds, plus that from other sources such as industrial wastes and agricultural runoff (that is, fertilizers), is oxidized to nitrate (NO_3^-) by a special group of nitrifying bacteria as their source of energy in a process called *nitrification.* The overall reaction for ammonia oxidation is

$$NH_4^+ + 2O_2 \xrightleftharpoons{\text{microorganisms}} NO_3^- + H_2O + 2H^+ \tag{5-20}$$

From this reaction the theoretical NBOD can be calculated as follows:

$$\text{NBOD} = \frac{\text{grams of oxygen used}}{\text{grams of nitrogen oxidized}} = \frac{4 \times 16}{14} = 4.57 \text{ g } O_2/\text{g N}$$

The actual nitrogenous BOD is slightly less than the theoretical value due to the incorporation of some of the nitrogen into new bacterial cells, but the difference is only a few percent.

Example 5-7. Compute the theoretical NBOD of a wastewater containing 30 mg/L of ammonia as nitrogen. (We often say "ammonia nitrogen" and write the expression as NH_3-N.) If the wastewater analysis was reported as 30 mg/L of ammonia (NH_3), what would the theoretical NBOD be?

Solution. In the first part of the problem, the amount of ammonia was reported as NH_3-N. Therefore, we can use the theoretical relationship developed from Equation 5-20.

$$\text{Theo. NBOD} = (30 \text{ mg N/L})(4.57 \text{ mg } O_2/\text{mg N}) = 137 \text{ mg } O_2/\text{L}$$

To answer the second question, we must convert mg/L of ammonia to NH_3-N by multiplying by the ratio of gram molecular weights of N to NH_3.

$$(30 \text{ mg } NH_3/\text{L})\left(\frac{14 \text{ g N}}{17 \text{ g } NH_3}\right) = 24.7 \text{ mg N/L}$$

Now we may use the relationship developed from Equation 5-20.

$$\text{Theo. NBOD} = (24.7 \text{ mg N/L})(4.57 \text{ mg } O_2/\text{mg N}) = 133 \text{ mg } O_2/\text{L}$$

The rate at which the NBOD is exerted depends heavily on the number of nitrifying organisms present. In untreated sewage, there are few of these organisms, while in a well-treated effluent, the concentration is high. When samples of untreated and

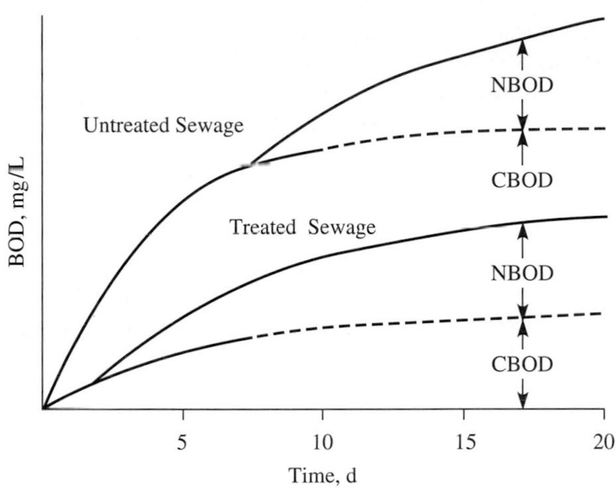

FIGURE 5-7
BOD curves showing both carbonaceous and nitrogenous BOD.

treated sewage are subjected to the BOD test, oxygen consumption follows the pattern shown in Figure 5-7. In the case of untreated sewage, the NBOD is exerted after much of the CBOD has been exerted. The lag is due to the time it takes for the nitrifying bacteria to reach a sufficient population for the amount of NBOD exertion to be significant compared with that of the CBOD. In the case of the treated sewage, a higher population of nitrifying organisms in the sample reduces the lag time. Once nitrification begins, however, the NBOD can be described by Equations 5-4 and 5-5 with a BOD rate constant comparable to that for the CBOD of a well-treated effluent (K = 0.04 to 0.10 d^{-1}). Because the lag before the nitrogenous BOD is highly variable, BOD_5 values are often difficult to interpret. When measurement of only carbonaceous BOD is desired, chemical inhibitors are added to stop the nitrification process. The rate constant for nitrification is also affected by temperature and can be adjusted using Equation 5-6.

DO Sag Curve

The concentration of dissolved oxygen in a river is an indicator of the general health of the river. All rivers have some capacity for self-purification. As long as the discharge of oxygen-demanding wastes is well within the self-purification capacity, the DO level will remain high and a diverse population of plants and animals, including game fish, can be found. As the amount of waste increases, the self-purification capacity can be exceeded, causing detrimental changes in plant and animal life. The stream loses its ability to cleanse itself and the DO level decreases. When the DO drops below about 4 to 5 mg/L, most game fish will have been driven out. If the DO is completely removed, fish and other higher animals are killed or driven out and extremely noxious conditions result. The water becomes blackish and foul smelling as the sewage and dead animal life decompose under *anaerobic* conditions (that is, without oxygen).

One of the major tools of water quality management in rivers is the ability to assess the capability of a stream to absorb a waste load. This is done by determining the profile of DO concentration downstream from a waste discharge. This profile

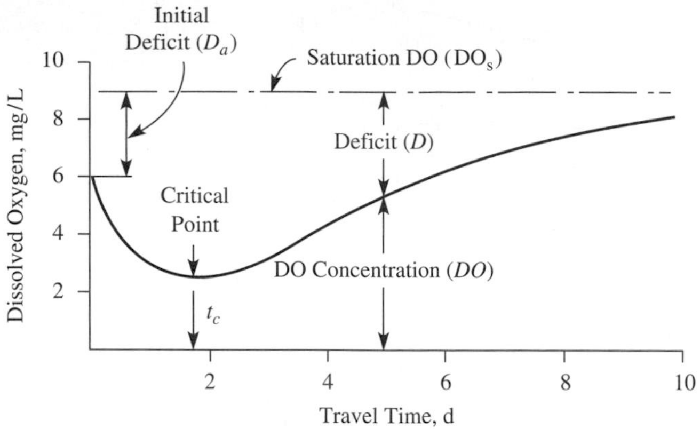

FIGURE 5-8
Typical DO sag curve.

is called the DO sag curve (see Figure 5-8) because the DO concentration dips as oxygen-demanding materials are oxidized and then rises again further downstream as the oxygen is replenished from the atmosphere. As depicted in Figure 5-9, the biota of the stream are often a reflection of the dissolved oxygen conditions in the stream.

To develop a mathematical expression for the DO sag curve, the sources of oxygen and the factors affecting oxygen depletion must be identified and quantified. The only significant sources of oxygen are reaeration from the atmosphere and photosynthesis of aquatic plants. Oxygen depletion is caused by a larger range of factors, the most important being the BOD, both carbonaceous and nitrogenous, of the waste discharge, and the BOD already in the river upstream of the waste discharge. The second most important factor is that the DO in the waste discharge is usually less than that in the river. Thus, the DO at the river is lowered as soon as the waste is added, even before any BOD is exerted. Other factors affecting dissolved oxygen depletion include nonpoint source pollution, the respiration of organisms living in the sediments (*benthic demand*), and the respiration of aquatic plants. Following the classical approach, the DO sag equation (also known as the *Streeter-Phelps* equation in recognition of the engineers who developed it) will be developed by considering only initial DO reduction, carbonaceous BOD, and reaeration from the atmosphere (Streeter-Phelps, 1925). Subsequently, the equation will be expanded to include the nitrogenous BOD. Finally, the other factors affecting DO levels will be discussed qualitatively; a quantitative discussion is beyond the scope of this book.

Mass-Balance Approach. Simplified mass balances help us understand and solve the DO sag curve problem. Three conservative (those without chemical reaction) mass balances may be used to account for initial mixing of the waste stream and the river. DO, carbonaceous BOD, and temperature all change as the result of mixing of the waste stream and the river. Once these are accounted for, the DO sag curve may be viewed as a nonconservative mass balance, that is, one with reactions.

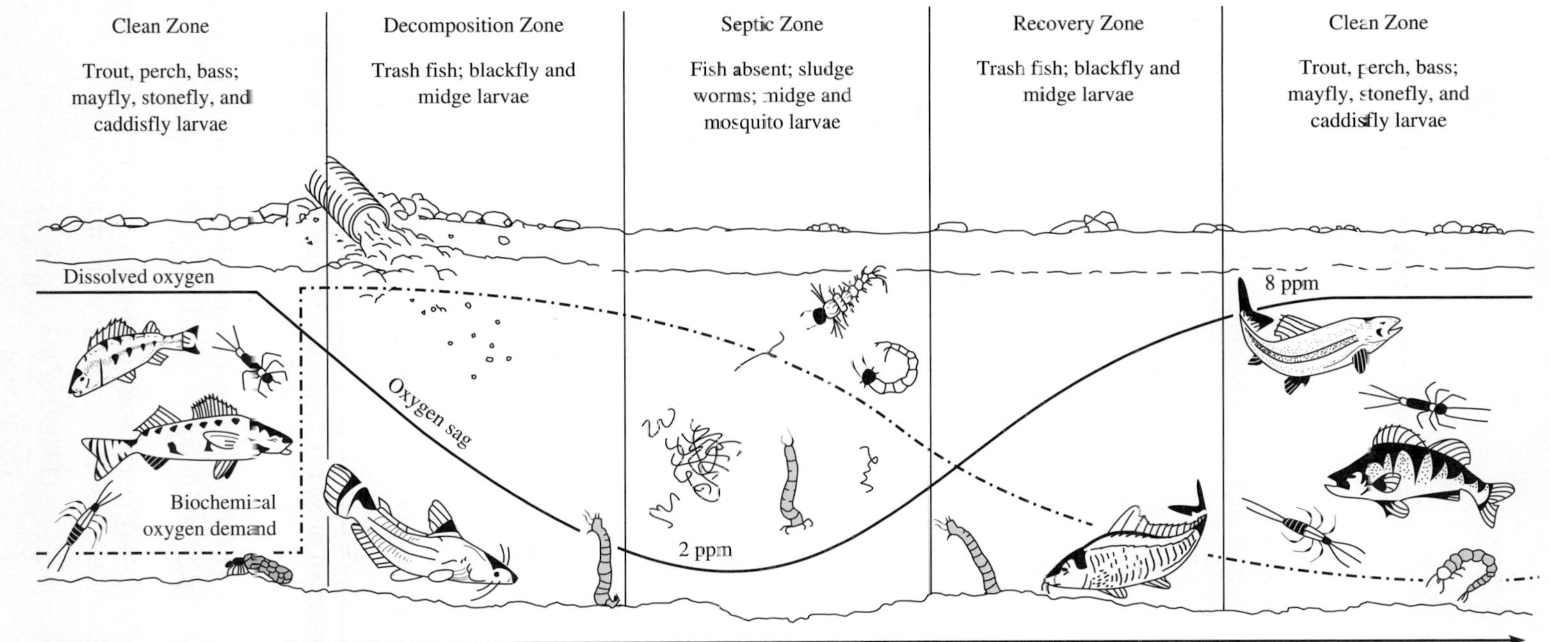

FIGURE 5-9
Oxygen sag downstream of an biodegradable organic chemical source. *Source:* U.S. EPA.

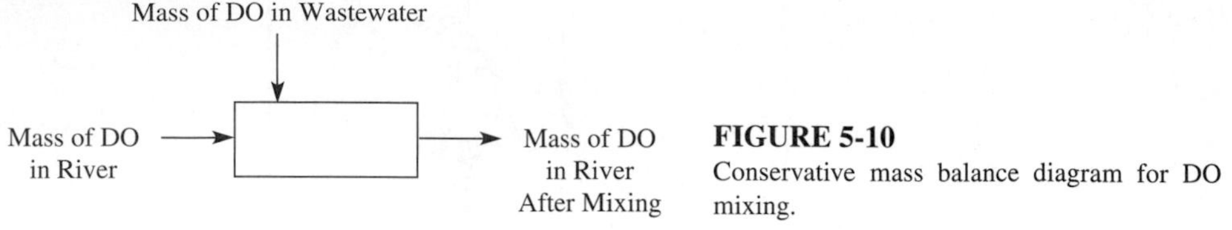

FIGURE 5-10
Conservative mass balance diagram for DO mixing.

The conservative mass balance diagram for oxygen (mixing only) is shown in Figure 5-10. The product of the water flow and the DO concentration yields a mass of oxygen per unit of time:

$$\text{Mass of DO in wastewater} = Q_w\text{DO}_w \tag{5-21}$$

$$\text{Mass of DO in river} = Q_r\text{DO}_r \tag{5-22}$$

where Q_w = volumetric flow rate of wastewater, m^3/s
Q_r = volumetric flow rate of the river, m^3/s
DO_w = dissolved oxygen concentration in the wastewater, g/m^3
DO_r = dissolved oxygen concentration in the river, g/m^3

The mass of DO in the river after mixing equals the sum of the mass flows:

$$\text{Mass of DO after mixing} = Q_w\text{DO}_w + Q_r\,\text{DO}_r \tag{5-23}$$

In a similar fashion for ultimate BOD:

$$\text{Mass of BOD after mixing} = Q_wL_w + Q_rL_r \tag{5-24}$$

where L_w = ultimate BOD of the wastewater, mg/L
L_r = ultimate BOD of the river, mg/L

The concentrations of DO and BOD after mixing are the respective masses per unit time divided by the total flow rate (that is, the sum of the wastewater and river flows):

$$\text{DO} = \frac{Q_w\text{DO}_w + Q_r\text{DO}_r}{Q_w + Q_r} \tag{5-25}$$

$$\text{L}_a = \frac{Q_wL_w + Q_rL_r}{Q_w + Q_r} \tag{5-26}$$

where L_a = initial ultimate BOD after mixing.

Example 5-8. The town of State College discharges 17,360 m^3/d of treated wastewater into the Bald Eagle Creek. The treated wastewater has a BOD_5 of 12 mg/L and a k of 0.12 d^{-1} at 20°C. Bald Eagle Creek has a flow rate of 0.43 m^3/s and an ultimate BOD of 5.0 mg/L. The DO of the river is 6.5 mg/L and the DO of the wastewater is 1.0 mg/L. Compute the DO and initial ultimate BOD after mixing.

Solution. The DO after mixing is given by Equation 5-25. To use this equation we must convert the wastewater flow to compatible units, that is, m^3/s.

$$Q_w = \frac{(17{,}360\ m^3/d)}{(86{,}400\ s/d)} = 0.20\ m^3/s$$

The DO after mixing is then

$$DO = \frac{(0.20\ m^3/s)(1.0\ mg/L) + (0.43\ m^3/s)(6.5\ mg/L)}{0.20\ m^3/s + 0.43\ m^3/s} = 4.75\ mg/L$$

Before we can determine the initial ultimate BOD after mixing, we must first determine the ultimate BOD of the wastewater. Solving Equation 5-4 for L_o:

$$L_o = \frac{BOD_5}{(1 - e^{-kt})} = \frac{12\ mg/L}{(1 - e^{-(0.12)(5)})} = \frac{12}{(1 - .55)} = 26.6\ mg/L$$

Note that we used the subscript of 5 days in BOD_5 to determine the value of t in the equation. Now setting $L_w = L_o$, we can determine the initial ultimate BOD after mixing using Equation 5-26:

$$L_a = \frac{(0.20\ m^3/s)(26.6\ mg/L) + (0.43\ m^3/s)(5.0\ mg/L)}{0.20\ m^3/s + 0.43\ m^3/s} = 11.86 \text{ or } 12\ mg/L$$

For temperature, we must consider a heat balance rather than a mass balance. As noted in Chapter 2, this is an application of a fundamental principle of physics:

$$\text{Loss of heat by hot bodies} = \text{gain of heat by cold bodies} \tag{5-27}$$

The change in *enthalpy* or "heat content" of a mass of a substance may be defined by the following equation:

$$H = mc_p\Delta T \tag{5-28}$$

where H = change in enthalpy, J
m = mass of substance, g
c_p = specific heat at constant pressure, J/g · K
ΔT = change in temperature, K

The specific heat of water varies slightly with temperature. For natural waters, a value of 4.19 will be a satisfactory approximation. Using our fundamental heat loss = heat gain equation, we may write

$$(m_w)(4.19)\Delta T_w = (m_r)(4.19)\Delta T_r \tag{5-29}$$

The temperature after mixing is found by solving this equation for the final temperature by recognizing that ΔT on each side of the equation is the difference between the final river temperature (T_f) and the starting temperature of the wastewater and the river water, respectively:

$$T_f = \frac{Q_wT_w + Q_rT_r}{Q_w + Q_r} \tag{5-30}$$

Oxygen Deficit. The DO sag equation has been developed using oxygen deficit rather than dissolved oxygen concentration, to make it easier to solve the integral equation that results from the mathematical description of the mass balance. The oxygen deficit is the amount by which the actual dissolved oxygen concentration is less than the saturation value with respect to oxygen in the air:

$$D = \mathrm{DO}_s - \mathrm{DO} \tag{5-31}$$

where D = oxygen deficit, mg/L
DO_s = saturation concentration of dissolved oxygen at the temperature of the river after mixing, mg/L
DO = actual concentration of dissolved oxygen, mg/L

The saturation value of dissolved oxygen is heavily dependent on water temperature—it decreases as the temperature increases. Values of DO_s for fresh water are given in Table A-3 of Appendix A.

Initial Deficit. The beginning of the DO sag curve is at the point where a waste discharge mixes with the river. The initial deficit is calculated as the difference between saturated DO and the concentration of the DO after mixing (Equation 5-25):

$$D_a = \mathrm{DO}_s - \frac{Q_w\mathrm{DO}_w + Q_r\mathrm{DO}_r}{Q_w + Q_r} \tag{5-32}$$

where D_a = initial deficit after river and waste have mixed, mg/L.

Example 5-9. Calculate the initial deficit of the Bald Eagle Creek after mixing with the wastewater from the town of State College (see Example 5-8 for data). The stream temperature is 10°C and the wastewater temperature is 10°C.

Solution. With the stream temperature, the saturation value of dissolved oxygen (DO_s) can be determined from the table in Appendix A. At 10°C, DO_s = 11.33 mg/L. Since we calculated the concentration of DO after mixing as 4.75 mg/L in Example 5-8, the initial deficit after mixing is

$$D_a = 11.33 \text{ mg/L} - 4.75 \text{ mg/L} = 6.58 \text{ mg/L}$$

Because wastewater commonly has a higher temperature than river water, especially during the winter, the river temperature downstream of the discharge is usually higher than that upstream. Since we are interested in downstream conditions, it is important to use the downstream temperature when determining the saturation concentration of dissolved oxygen.

DO Sag Equation. A mass balance diagram of DO in a small *reach* (stretch) of river is shown in Figure 5-11*a*. This is a comprehensive mass balance that accounts for all of the inputs and outputs. We are going to limit our development to the classical Streeter-Phelps model. The simplified mass balance diagram is shown in Figure 5-11*b*. The mass balance equation is then:

$$R\mathrm{DO}_{\mathrm{in}} + W + A - M - R\mathrm{DO}_{\mathrm{out}} = 0 \tag{5-33}$$

where $R\mathrm{DO}_{\mathrm{in}}$ = mass of DO in river flowing into reach
W = mass of DO in wastewater flowing into reach

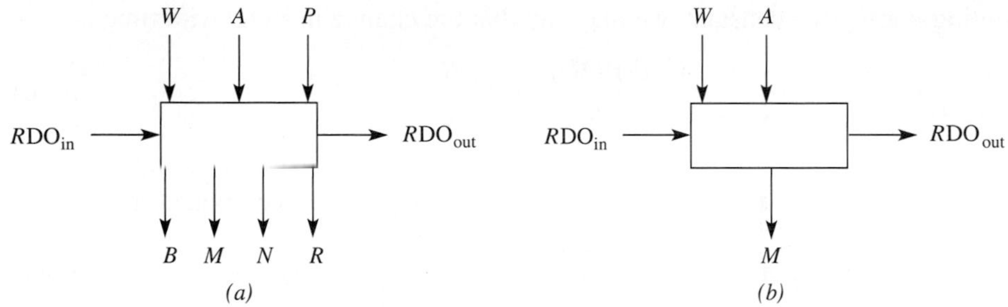

Legend

$R\mathrm{DO}_{\mathrm{in}}$, $R\mathrm{DO}_{\mathrm{out}}$ = mass of DO flowing in and out of reach
W = mass of DO in wastewater flowing into reach
A = mass of DO entering from atmosphere
P = mass of DO entering from algae photosynthetic oxygen production
B = mass of DO consumed by benthic demand
M = mass of DO removed by microbial degradation of carbonaceous BOD
N = mass of DO removed by microbial degradation of nitrogenous BOD
R = mass of DO consumed by algal respiration

FIGURE 5-11
Mass balance diagram of DO in a small reach of river (*a*) and simplified mass balance for Streeter-Phelps model (*b*).

A = mass of DO added from atmosphere
M = mass of DO removed by microbial degradation of carbonaceous BOD
$R\mathrm{DO}_{\mathrm{out}}$ = mass of DO in river flowing out of reach

In Equation 5-33 we can account for $R\mathrm{DO}_{\mathrm{in}} + W$. Our goal is to find $R\mathrm{DO}_{\mathrm{out}}$ in terms of mass per unit volume (mg/L). This leaves A and M to be accounted for before we can solve the mass balance equation.

The rate at which DO disappears from the stream as a result of microbial action (M) is exactly equal to rate of increase in the deficit. With the assumption that the saturation value for DO remains constant $[d(\mathrm{DO}_s)/dt = 0]$, differentiation of Equation 5-31 yields:

$$\frac{d(\mathrm{DO})}{dt} + \frac{dD}{dt} = 0$$

and

$$\frac{d(\mathrm{DO})}{dt} = -\frac{dD}{dt} \tag{5-34}$$

The rate at which DO disappears coincides with the rate that BOD is degraded, so

$$\frac{d(\mathrm{DO})}{dt} = -\frac{dD}{dt} = -\frac{d(\mathrm{BOD})}{dt} \tag{5-35}$$

Remembering that in Equation 5-4 BOD_t was defined as

$$\mathrm{BOD}_t = L_o - L_t$$

and noting that L_o is a constant, we may say that the change in BOD with time is

$$\frac{d(\text{BOD})}{dt} = -\frac{dL_t}{dt} \tag{5-36}$$

This leads us to see that the rate of change in deficit at time t due to BOD is a first-order reaction proportional to the oxygen equivalent of the organic compounds remaining:

$$\frac{dD}{dt} = kL_t \tag{5-37}$$

The rate constant, k, is called the *deoxygenation rate constant* and is designated k_d.

The rate of oxygen mass transfer into solution from the air (A) has been shown to be a first-order reaction proportional to the difference between the saturation value and the actual concentration:

$$\frac{d(\text{DO})}{dt} = k(\text{DO}_s - \text{DO}) \tag{5-38}$$

From Equations 5-31 and 5-34 we can see that

$$\frac{dD}{dt} = -kD \tag{5-39}$$

The rate constant is called the *reaeration rate constant,* k_r.

From Equations 5-37 and 5-39 we can see that the oxygen deficit is a function of the competition between oxygen utilization and reaeration from the atmosphere:

$$\frac{dD}{dt} = k_d L - k_r D \tag{5-40}$$

where $\frac{dD}{dt}$ = the change in oxygen deficit (D) per unit of time, mg/L · d
k_d = deoxygenation rate constant, d^{-1}
L = ultimate BOD of river water, mg/L
k_r = reaeration rate constant, d^{-1}
D = oxygen deficit in river water, mg/L

By integrating Equation 5-40, and using the initial conditions (at $t = 0$, $D = D_a$) we obtain the DO sag equation:

$$D = \frac{k_d L_a}{k_r - k_d}(e^{-k_d t} - e^{k_r t}) + D_a(e^{-k_r t}) \tag{5-41}$$

where D = oxygen deficit in river water after exertion of BOD for time, t, mg/L
L_a = initial ultimate BOD after river and wastewater have mixed (Equation 5-26), mg/L
k_d = deoxygenation rate constant, d^{-1}
k_r = reaeration rate constant, d^{-1}
t = time of travel of wastewater discharge downstream, d
D_a = initial deficit after river and wastewater have mixed (Equation 5-32), mg/L

When $k_r = k_d$, Equation 5-41 reduces to:

$$D = (k_d t L_a + D_a)(e^{-k_d t}) \tag{5-42}$$

where the terms are as previously defined.

Deoxygenation Rate Constant. The deoxygenation rate constant differs from the BOD rate constant because there are physical and biological differences between a river and a BOD bottle. In general, BOD is exerted more rapidly in a river because of turbulent mixing, larger numbers of "seed" organisms, and BOD removal by organisms on the stream bed as well as by those suspended in the water. While k rarely has a value greater than 0.7 day^{-1}, k_d may be as large as 7 day^{-1} for shallow, rapidly flowing streams. However, for deep, slowly moving rivers, the value of k_d is very close to that for k.

Bosko has developed a method of estimating k_d from k using characteristics of the stream (Bosko, 1966).

$$k_d = k + \frac{v}{H}\eta \tag{5-43}$$

where k_d = deoxygenation rate constant at 20°C, d^{-1}
v = average speed of stream flow, m/s
k = BOD rate constant determined in laboratory at 20°C, d^{-1}
H = average depth of stream, m
η = bed-activity coefficient

The bed-activity coefficient may vary from 0.1 for stagnant or deep water to 0.6 or more for rapidly flowing streams. Note that the bed-activity coefficient includes a conversion factor to make the second term dimensionally correct. After determining k_d from Equation 5-43, it should be corrected for temperature using Equation 5-6 if the stream temperature is not 20°C.

Example 5-10. Determine the deoxygenation rate constant for the reach of Bald Eagle Creek (Examples 5-8 and 5-9) below the wastewater outfall (discharge pipe). The average speed of the stream flow in the creek is 0.03 m/s. The depth is 5.0 m and the bed-activity coefficient is 0.35.

Solution. From Example 5-8, the value of k is 0.12 d^{-1}. Using Equation 5-43, the deoxygenation rate constant at 20°C is

$$k_d = 0.12\ \text{d}^{-1} + \frac{0.03\ \text{m/s}}{5.0\ \text{m}}(0.35) = 0.1221 \text{ or } 0.12\ \text{d}^{-1}$$

Note that the units are not consistent. As we have noted before, empirical expressions, such as that in Equation 5-43, may have implicit conversion factors. Thus, you must be careful to use the same units as those used by the author of the equation.

We also note that the deoxygenation rate constant of 0.1221 d^{-1} is at 20°C. In Example 5-9, we noted that the stream temperature was 10°C. Thus, we must correct the estimated k_d using Equation 5-6.

$$k_d \text{ at } 10°\text{C} = (0.1221\ \text{d}^{-1})(1.135)^{10-20} = (0.1221)(0.2819)$$
$$= 0.03442 \text{ or } 0.034\ \text{d}^{-1}$$

Reaeration. The value of k_r depends on the degree of turbulent mixing, which is related to stream velocity, and on the amount of water surface exposed to the atmosphere compared to the volume of water in the river. A narrow, deep river will have a much lower k_r than a wide, shallow river. O'Connor and Dobbins (1958) developed a generalized empirical equation to estimate the reaeration constant based on the characteristics of the stream and the molecular diffusion of oxygen into water:

$$k_r = \frac{3.9\, v^{0.5}}{H^{1.5}} \tag{5-44}$$

where k_r = reaeration rate constant at 20°C, day^{-1}
v = average stream velocity, m/s
H = average depth, m

Note that the factor of 3.9 includes a conversion factor to make the equation dimensionally correct. The reaeration rate constant is also affected by temperature and can be adjusted to the river temperature using Equation 5-6 but with a temperature coefficient (θ) of 1.024. For various streams, k_r can range from 0.05 to greater than 18 d^{-1}.

To relate travel time to a physical distance downstream, one must also know the average stream velocity. Once D has been found at any point downstream, the DO can be found from Equation 5-31. Note that it is physically impossible for the DO to be less than zero. If the deficit calculated from Equation 5-41 is greater than the saturation DO, then all the oxygen was depleted at some earlier time and the DO is zero. If the result of your calculations yields a negative DO, report it as zero because it cannot be less than zero!

The lowest point on the DO sag curve, which is called the *critical point,* is of major interest since it indicates the worst conditions in the river. The time to the critical point (t_c) can be found by differentiating Equation 5-41, setting it equal to zero, and solving for t using base e values for k_r and k_d:

$$t_c = \frac{1}{k_r - k_d} \ln\left[\frac{k_r}{k_d}\left(1 - D_a\frac{k_r - k_d}{k_d L_a}\right)\right] \tag{5-45}$$

or when $k_r = k_d$:

$$t_c = \frac{1}{k_d}\left(1 - \frac{D_a}{L_a}\right) \tag{5-46}$$

The critical deficit (D_c) is then found by using this critical time in Equation 5-41.

In some instances there may not be a sag in the DO downstream. The lowest DO may occur in the mixing zone. In these instances Equation 5-45 will not give a useful value.

Example 5-11. Determine the DO concentration at a point 5 km downstream from the State College discharge into the Bald Eagle Creek (Examples 5-8, 5-9, 5-10). Also determine the critical DO and the distance downstream at which it occurs.

Solution. All of the appropriate data are provided in the three previous examples. With the exceptions of the travel time, t, and the reaeration rate, the values needed for Equations 5-41 and 5-45 have been computed in Examples 5-8, 5-9, and 5-10. The first step then is to calculate k_r.

$$k_r \text{ at } 20°\text{C} = \frac{(3.9)(0.03 \text{ m/s})^{0.5}}{(5.0 \text{ m})^{1.5}} = 0.0604 \text{ d}^{-1}$$

Because this is at 20°C and the stream temperature is at 10°C, Equation 5-6 must be used to correct for the temperature difference.

$$k_r \text{ at } 10°\text{C} = (0.0604 \text{ d}^{-1})(1.024)^{10-20} = (0.0604)(0.7889) = 0.04766 \text{ d}^{-1}$$

Note that the temperature coefficient is the one noted in the text above rather than the ones reported with Equation 5-6.

The travel time t is computed from the distance downstream and the speed of the stream:

$$t = \frac{(5 \text{ km})(1{,}000 \text{ m/km})}{(0.03 \text{ m/s})(86{,}400 \text{ s/d})} = 1.929 \text{ d}$$

Although it is not warranted by the significant figures in the computation, we have elected to keep four significant figures because of the computational effects of truncating the value.

The deficit is estimated using Equation 5-41.

$$D = \frac{(0.03442)(11.86)}{0.04766 - 0.03442}[e^{-(0.03442)(1.929)} - e^{-(0.04766)(1.929)}] + 6.58[e^{-(0.04766)(1.929)}]$$

$$D = (30.83)(0.9358 - 0.9122) + 6.58(0.9122) = 6.729 \text{ or } 6.73 \text{ mg/L}$$

and the dissolved oxygen is

$$\text{DO} = 11.33 - 6.73 = 4.60 \text{ mg/L}$$

The critical time is computed using Equation 5-45:

$$t_c = \frac{1}{0.04766 - 0.03442} \ln\left\{\frac{0.04766}{0.03442}\left[1 - 6.58\frac{0.04766 - 0.03442}{(0.03442)(11.86)}\right]\right\}$$

$$t_c = 6.45 \text{ d}$$

Using t_c for the time in Equation 5-41, calculate the critical deficit as

$$D_c = \frac{(0.03442)(11.86)}{0.04766 - 0.03442}[e^{-(0.03442)(6.45)} - e^{-(0.04766)(6.45)}] + 6.58[e^{-(0.04766)(6.45)}]$$

$$D_c = 6.85 \text{ mg/L}$$

Solving Equation 5-31 for DO at the critical point (DO_c) and using $DO_s = 11.33$ mg/L from Table A-3 of Appendix A gives the critical DO:

$$DO_c = 11.33 - 6.85 = 4.48 \text{ mg/L}$$

The critical DO occurs downstream at a distance of

$$(6.45\text{d})(86{,}400 \text{ s/d})(0.03 \text{ m/s})\left(\frac{1}{1{,}000 \text{ m/km}}\right) = 16.7 \text{ km}$$

from the wastewater discharge point. (Note that 0.03 m/s is the speed of the stream.)

Management Strategy. The beginning point for water quality management in rivers using the DO sag curve is to determine the minimum DO concentration that will protect the aquatic life in the stream. This value, called the DO standard, is generally set to protect the most sensitive species that exist or could exist in the particular river. For a known waste discharge and a known set of river characteristics, the DO sag equation can be solved to find the DO at the critical point. If this value is higher than the standard, the stream can adequately assimilate the waste. If the DO at the critical point is less than the standard, then additional waste treatment is needed. Usually, the environmental engineer has control over just two parameters, L_a and D_a. By increasing the efficiency of the existing treatment processes or by adding additional treatment steps, the ultimate BOD of the waste discharge can be reduced, thereby reducing L_a. Often a relatively inexpensive method for improving stream quality is to reduce D_a by adding oxygen to the wastewater to bring it close to saturation prior to discharge. To determine whether a proposed improvement will be adequate, the new values for L_a and D_a are used to determine whether the DO standard will be violated at the critical point. Under unusual conditions, the engineer may artificially aerate the river with mechanical systems to increase the DO.

When using the DO sag curve to determine the adequacy of wastewater treatment, it is important to use the river conditions that will cause the lowest DO concentration. Usually these conditions occur in the late summer when river flows are low and temperatures are high. A frequently used criterion is the "10-year, 7-day low flow," which is the recurrence interval of the average low flow for a 7-day period. Low river flows reduce the dilution of the waste entering the river, causing higher values for L_a and D_a. The value of k_r is usually reduced by low river flows because of reduced velocities. In addition, higher temperatures increase k_d more than k_r and also decrease DO saturation, thus making the critical point more severe.

Example 5-12. The Pitts Canning Company is considering opening a new plant at one of two possible locations: the Green River and its twin, the White River. Among the decisions to be made are what effect the plant discharge will have on each river and which river would be impacted less. Effluent data from the Pitts A Plant and the Pitts B Plant are considered to be representative of the potential discharge characteristics. In addition, measurements from each river at summer low-flow conditions are available.

Effluent parameter	A plant	B plant
Flow, m^3/s	0.0500	0.0500
Ultimate BOD at 25°C, kg/d	129.60	129.60
DO, mg/L	0.900	0.900
Temperature, °C	25.0	25.0
K at 20°C, d^{-1}	0.0500	0.0300

River parameter	Green River	White River
Flow, m^3/s	0.500	0.500
Ultimate BOD at 25°C, mg/L	19.00	19.00
DO, mg/L	5.85	5.85
Temperature, °C	25.0	25.0
Speed, m/s	0.100	0.200
Average depth, m	4.00	4.00
Bed-activity coefficient	0.200	0.200

There are four combinations to be evaluated:

A–Green B–Green
A–White B–White

Solution. This problem will be worked in base 10. Note that the BOD rate constant (K) is given in base 10. Also note that for the purpose of explaining the calculations, the number of significant figures given for the data is greater than can probably be measured. The only difference in the combinations is the change in deoxygenation and reaeration coefficients. Thus, we need calculate only one value of L_a and one value of D_a.

We begin by converting the mass flow of ultimate BOD (kg/d) to a concentration (mg/L). Following our general approach for calculating concentration from mass flow (Chapter 2); we divide the mass discharge (kg/d) by the flow of the water carrying the waste (Q_w, Q_r, or the sum $Q_w + Q_r$):

$$\frac{\text{Mass of ultimate BOD discharged (kg/d)}}{\text{Flow of water-carrying waste (m}^3\text{/s)}}$$

The mass discharge units are then converted to mg/d and the water flow to L/d so that the days cancel.

$$\frac{(\text{kg/d}) \times (1 \times 10^6 \text{ mg/kg})}{(\text{m}^3\text{/s}) \times (86{,}400 \text{ s/d})(1 \times 10^3 \text{ L/m}^3)}$$

For either Plant A or B:

$$L_w = \frac{(129.60 \text{ kg/d})(1 \times 10^6 \text{ mg/kg})}{(0.0500 \text{ m}^3\text{/s})(86{,}400 \text{ s/d})(1 \times 10^3 \text{ L/m}^3)}$$

$$= \frac{129.60 \times 10^6 \text{ mg}}{4.320 \times 10^6 \text{ L}}$$
$$= 30.00 \text{ mg/L}$$

Now we can compute the mixed BOD using Equation 5-26.

$$L_a = \frac{(0.0500)(30.00) + (0.500)(19.00)}{0.0500 + 0.500}$$
$$= 20.0 \text{ mg/L}$$

From Table A-3 of Appendix A, we find that the DO saturation at 25°C is 8.38 mg/L. Then, using Equation 5-32, we determine the initial deficit:

$$D_a = 8.38 - \frac{(0.0500)(0.900) + (0.500)(5.85)}{0.0500 + 0.500}$$
$$= 8.38 - 5.4$$
$$= 2.98 \text{ mg/L}$$

For the combination of the A Plant discharging to the Green River, the deoxygenation coefficient and reaeration coefficient are calculated by using Equations 5-43 and 5-44, in base 10:

$$K_d = 0.0500 + \frac{0.100 \times 0.200}{2.3 \times 4.00}$$
$$= 0.05217 \text{ d}^{-1} \text{ at } 20°\text{C}$$

and

$$K_r = \frac{1.7(0.100)^{0.5}}{(4.00)^{1.5}}$$
$$= 0.067198 \text{ d}^{-1} \text{ at } 20°\text{C}$$

(Note: The factor 2.3 in the K_d equation and the 1.7 in the K_r equation are the conversions in base 10. Hence, they differ from the equations in the text.)

Since the temperature of the river is 25°C and the wastewater effluent temperature is also 25°C, we do not have to calculate a temperature after mixing. However, we will have to adjust K_d and K_r to 25°C. For K_d, we use Equation 5-6 with a value of θ of 1.056.

$$K_d = 0.05217(1.056)^{25-20}$$
$$= 0.068513 \text{ or } 0.0685 \text{ d}^{-1}$$

From the discussion that follows Equation 5-44, we note that $\theta = 1.024$ for reaeration, and thus

$$K_r = 0.067198(1.024)^{25-20}$$
$$= 0.075658 \text{ or } 0.0757 \text{ d}^{-1}$$

Although perhaps not justified by the coefficients, we round to three significant figures because we will want to calculate travel time to two decimal places.

Now we have all the information we need to calculate the time to the critical point. Using Equation 5-45 gives

$$t_c = \frac{1}{0.0757 - 0.0685} \log \left\{ \frac{0.0757}{0.0685} \left[1 - 3.0 \left(\frac{0.0757 - 0.0685}{0.0685 \times 20.0} \right) \right] \right\}$$
$$= 138.89 \log [1.105(1 - 3.0(0.005255))]$$
$$= 138.89 \log [1.0876]$$
$$= 5.07 \text{ d}$$

Using this value for t in Equation 5-41 (in base 10 because K values are in base 10), we can calculate the deficit at the critical point:

$$D_c = \frac{(0.0685)(20.0)}{0.0757 - 0.0685} [10^{-(0.0685)(5.07)} - 10^{-(0.0757)(5.07)}] + 2.98[10^{-(0.0757)(5.07)}]$$
$$= 191.76 [(0.4493) - (0.4133)] + 2.98[0.4133]$$
$$= 6.903 + 1.232$$
$$= 8.13 \text{ mg/L}$$

The DO at the critical point is determined by solving Equation 5-31 for DO and substituting the appropriate value for the DO saturation that we obtained earlier from Table A-3 of Appendix A:

$$\text{DO} = \text{DO}_s - D$$
$$= 8.38 - 8.13 = 0.25 \text{ mg/L}$$

Thus, the lowest DO for the A–Green combination is 0.25 mg/L, and it occurs at a travel time of 5.07 days downstream from the A Plant outfall. Since the Green River travels at a speed of 0.100 m/s, this would be

$$\frac{(0.100 \text{ m/s})(5.07 \text{ d})(86{,}400 \text{ s/d})}{1{,}000 \text{ m/km}} = 43.8 \text{ km}$$

downstream.

The results of the other combinations are summarized below:

	A–Green	A–White	B–Green	B–White
K_d	0.0685	0.0714	0.0422	0.0451
K_r	0.0757	0.107	0.0757	0.107
t_c	5.07	3.99	5.93	4.44
D	8.14	6.93	6.27	5.31
DO	0.25	1.47	2.13	3.08

It is obvious that the best combination is the B Plant on the White River.

Using a spreadsheet program, we have generated the deficit values for a series of times for each of the combinations and plotted the results in Figure 5-12. From this figure we can make the following general observations:

1. Increasing the reaeration rate, while holding everything else as it is, reduces the deficit and displaces the critical point upstream.

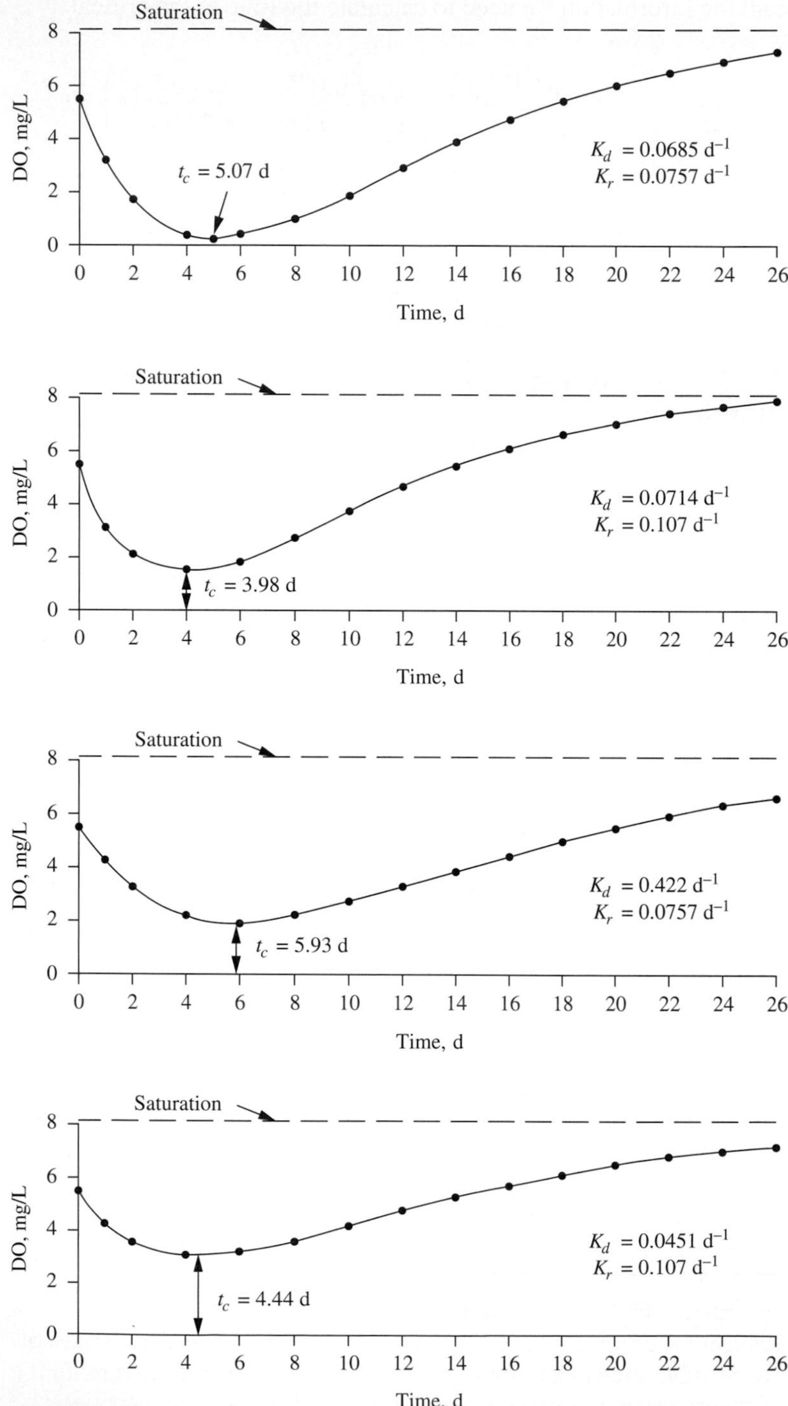

FIGURE 5-12
Effect of K_d and K_r on DO sag curve.

2. Decreasing the reaeration rate, while holding everything else as it is, increases the deficit and displaces the critical point downstream.
3. Increasing the deoxygenation rate, while holding everything else as it is, increases the deficit and displaces the critical point upstream.
4. Decreasing the deoxygenation rate, while holding everything else as it is, decreases the deficit and displaces the critical point downstream.

Nitrogenous BOD. Up to this point, only carbonaceous BOD has been considered in the DO sag curve. However, in many cases nitrogenous BOD has at least as much impact on dissolved oxygen levels. Modern wastewater treatment plants can routinely produce effluents with $CBOD_5$ of less than 30 mg/L. A typical effluent also contains approximately 30 mg/L of nitrogen, which would mean an NBOD of about 137 mg/L if it were discharged as ammonia (see Example 5-7). Nitrogenous BOD can be incorporated into the DO sag curve by adding an additional term to Equation 5-41:

$$D = \frac{k_d L_a}{k_r - k_d}(e^{-k_d t} - e^{-k_r t}) + D_a(e^{-k_r t}) + \frac{k_n L_n}{k_r - k_n}(e^{-k_n t} - e^{-k_r t}) \quad (5\text{-}47)$$

where k_n = the nitrogenous deoxygenation coefficient, d^{-1}; L_n = ultimate nitrogenous BOD after waste and river have mixed, mg/L; and the other terms are as previously defined. It is important to note that with the additional term for NBOD, it is not possible to find the critical time using Equation 5-45. Instead, it must be found by a trial and error solution of Equation 5-47.

Other Factors Affecting DO Levels in Rivers. The classical DO sag curve assumes that there is only one point-source discharge of waste into the river. In reality, this is rarely the case. Multiple point sources can be handled by dividing the river up into reaches with a point source at the head of each reach. A *reach* is a length of river specified by the engineer on the basis of its homogeneity, that is, channel shape, bottom composition, slope, etc. The oxygen deficit and residual BOD can be calculated at the end of each reach. These values are then used to determine new values of D_a and L_a at the beginning of the following reach. Non-point source pollution can also be handled this way if the reaches are made small enough. Non-point source pollution can also be incorporated directly into the DO sag equation for a more sophisticated analysis. Dividing the river into reaches is also necessary whenever the flow regime changes, since the reaeration coefficient would also change. In small rivers, rapids play a major role in maintaining high DO levels. Eliminating rapids by dredging or damming a river can have a severe impact on DO, although DO levels immediately downstream of dams are usually high because of the turbulence of the falling water.

Some rivers contain large deposits of organic matter in the sediments. These can be natural deposits of leaves and dead aquatic plants or can be sludge deposits from wastewaters receiving little or no treatment. In either case, decomposition of this organic matter places an additional burden on the stream's oxygen resources, since the oxygen demand must be supplied from the overlying water. When this benthic demand is significant, compared to the oxygen demand in the water column, it must be included quantitatively in the sag equation.

Aquatic plants can also have a substantial effect on DO levels. During the day, their photosynthetic activities produce oxygen that supplements the reaeration and can even cause oxygen supersaturation. However, plants also consume oxygen for respiration processes. Although there is a net overall production of oxygen, plant respiration can severely lower DO levels during the night. Plant growth is usually highest in the summer when flows are low and temperatures are high, so that large nighttime respiration requirements coincide with the worst cases of oxygen depletion from BOD exertion. In addition, when aquatic plants die and settle to the bottom, they increase the benthic demand. As a general rule, large growths of aquatic plants are detrimental to the maintenance of a consistently high DO level.

Effect of Nutrients on Water Quality in Rivers

Although oxygen-demanding wastes are the most important river pollutants on an overall basis, nutrients can also contribute to deteriorating water quality in rivers by causing excessive plant growth. Nutrients are those elements required by plants for their growth. They include, in order of abundance in plant tissue: carbon, nitrogen, phosphorus, and a variety of trace elements. When there are sufficient quantities of all nutrients available, plant growth is possible. By limiting the availability of any one nutrient, further plant growth is prevented.

Some plant growth is desirable, since plants form the base of the food chain and thus support the animal community. However, excessive plant growth can create a number of undesirable conditions such as thick slime layers on rocks and dense growths of aquatic weeds.

The availability of nutrients is not the only requirement for plant growth. In many rivers, the turbidity caused by eroded soil particles, bacteria, and other factors prevents light from penetrating far into the water, thereby limiting total plant growth in deep water. It is for this reason that slime growths on rocks usually occur in shallow water. Strong water currents also prevent rooted plants from taking hold, and thus limit their growth to quiet backwaters where the currents are weak and the water is shallow enough for light to penetrate.

Effects of Nitrogen. There are three reasons why nitrogen is detrimental to a receiving body:

1. In high concentrations, NH_3-N is toxic to fish.
2. NH_3, in low concentrations, and NO_3^- serve as nutrients for excessive growth of algae.
3. The conversion of NH_4^+ to NO_3^- consumes large quantities of dissolved oxygen.

Effects of Phosphorus. The major deleterious effect of phosphorus is that it serves as a vital nutrient for the growth of algae. If the phosphorus availability meets the growth demands of the algae, there is an excessive production of algae. When the algae die, they become an oxygen-demanding organic material as bacteria seek to degrade them. This oxygen demand frequently overtaxes the DO supply of the water body and, as a consequence, causes fish to die.

Management Strategy. The strategy for managing water quality problems associated with excessive nutrients is based on the sources for each nutrient. Except under rare circumstances, there is plenty of carbon available for plant growth. Plants use carbon dioxide, which is available from the bicarbonate alkalinity of the water and from the bacterial decomposition of organic matter. As carbon dioxide is removed from the water, it is replenished from the atmosphere. Generally, the major source of trace elements is the natural weathering of rock minerals, a process over which the environmental engineer has little control. However, since the acid rain caused by air pollution accelerates the weathering process, air pollution control can help reduce the supply of trace elements. Even when substantial amounts of trace elements are found in wastewater, their removal is difficult. In addition, such small amounts are needed for plant growth that nitrogen or phosphorus is more likely to be the limiting nutrient. Therefore, the practical control of nutrient-caused water-quality problems in streams is based on removal of nitrogen and/or phosphorus from wastewaters before they are discharged.

5-4 WATER QUALITY MANAGEMENT IN LAKES

Oxygen-demanding wastes can also be important lake pollutants, especially when the waste is discharged to a contained area such as a bay. Pathogens are of particular concern near bathing beaches. Again, as with rivers, there are special classes of lakes which are most seriously affected by other pollutants such as toxic chemicals from industrial discharges. However, phosphorus so dominates other pollutants in controlling water quality in the vast majority of lakes that we will give it special emphasis.

A knowledge of lake systems is essential to an understanding of the role of phosphorus in lake pollution. The study of lakes is called *limnology*. This section is essentially a short course in limnology as it relates to phosphorus pollution.

Stratification and Turnover

Nearly all lakes in the temperate climatic zone become stratified during the summer and overturn (turnover) in the fall due to changes in the water temperature that result from the annual cycle of air temperature changes. In addition, lakes in cold climates undergo winter stratification and spring overturn as well. These physical processes, which are described below, occur regardless of the water quality in the lake. Nonetheless, they do help determine the water quality.

During the summer, the surface water of a lake is heated both indirectly by contact with warm air and directly by sunlight. Warm water, being less dense than cool water, remains near the surface until mixed downward by turbulence from wind, waves, boats, and other forces. Because this turbulence extends only a limited distance below the water surface, the result is an upper layer of well-mixed, warm water (the *epilimnion*) floating on the lower water (the *hypolimnion*), which is poorly mixed and cool, as shown in Figure 5-13*a*. Because of good mixing the epilimnion will be *aerobic* (have DO). The hypolimnion will have a lower DO and may become *anaerobic* (devoid of oxygen). The boundary is called the *thermocline* because of the sharp temperature change (and therefore density change) that occurs within a relatively short

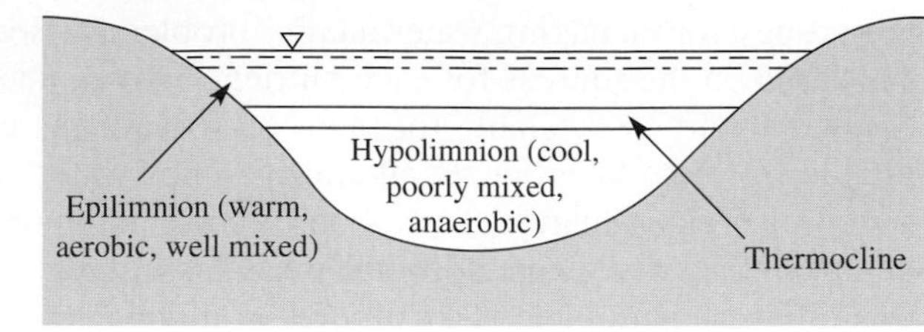

(a)

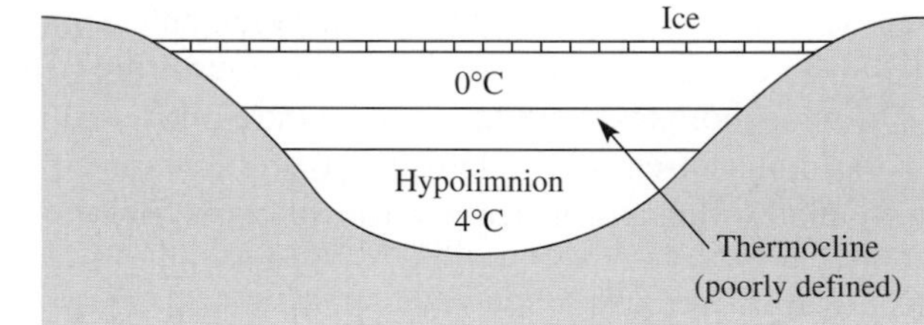

(b)

FIGURE 5-13
Stratification of a lake during (*a*) summer and (*b*) winter.

distance. The thermocline may be defined as a change in temperature with depth that is greater than 1°C/m. You may have experienced the thermocline while swimming in a small lake. As long as you are swimming horizontally, the water is warm, but as soon as you tread water or dive, the water turns cold. You have penetrated the thermocline. The depth of the epilimnion is related to the size of the lake. It is as little as one meter in small lakes and as much as 20 meters or more in large lakes. The depth of the epilimnion is also related to storm activity in the spring when stratification is developing. A major storm at the right time will mix warmer water to a substantial depth and thus create a deeper than normal epilimnion. Once formed, lake stratification is very stable. It can be broken only by exceedingly violent storms. In fact, as the summer progresses, the stability increases because the epilimnion continues to warm, while the hypolimnion remains at a fairly constant temperature.

In the fall, as temperatures drop, the epilimnion cools until it is more dense than the hypolimnion. The surface water then sinks, causing overturning. The water of the hypolimnion rises to the surface where it cools and again sinks. The lake thus becomes completely mixed. If the lake is in a cold climate, this process stops when the temperature reaches 4°C, since this is the temperature at which water is most dense. Further cooling or freezing of the surface water results in winter stratification, as shown in Figure 5-13*b*. As the water warms in the spring, it again overturns and becomes completely mixed. Thus, temperate climate lakes have at least one, if not two, cycles of stratification and turnover every year.

Biological Zones

Lakes contain several distinct zones of biological activity, largely determined by the availability of light and oxygen. The most important biological zones, shown in Figure 5-14, are the euphotic, littoral, and benthic zones.

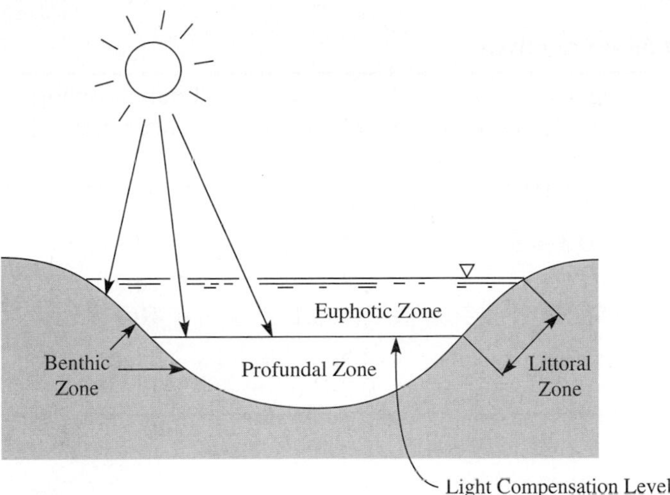

FIGURE 5-14
Biological zones in a lake.

Euphotic Zone. The upper layer of water through which sunlight can penetrate is called the *euphotic zone.* All plant growth occurs in this zone. In deep water, algae are the most important plants, while rooted plants grow in shallow water near the shore. The depth of the euphotic zone is determined by the amount of turbidity blocking sunlight penetration. In most lakes, the turbidity is due to algal growth, although color and suspended clays may substantially reduce sunlight penetration in some lakes. In the euphotic zone, plants produce more oxygen by photosynthesis than they remove by respiration. Below the euphotic zone lies the *profundal zone.* The transition between the two zones is called the *light compensation level.* The light compensation level corresponds roughly to a depth at which the light intensity is about one percent of unattenuated sunlight. It is important to note that the bottom of the euphotic zone only rarely coincides with the thermocline.

Littoral Zone. The shallow water near the shore in which rooted water plants can grow is called the *littoral zone.* The extent of the littoral zone depends on the slope of the lake bottom and the depth of the euphotic zone. The littoral zone cannot extend deeper than the euphotic zone.

Benthic Zone. The bottom sediments comprise the *benthic zone.* As organisms living in the overlying water die, they settle to the bottom where they are decomposed by organisms living in the benthic zone. Bacteria are always present. The presence of higher life forms such as worms, insects, and crustaceans depends on the availability of oxygen.

Lake Productivity

The productivity of a lake is a measure of its ability to support a food web. Algae form the base of this food web, supplying food for the higher organisms. A lake's productivity may be determined by measuring the amount of algal growth that can be supported

TABLE 5-3
Lake classification based on productivity

Lake classification		Chlorophyll *a* concentration, μg/L	Secchi depth, m	Total phosphorus concentration, μg/L
Oligotrophic	Average	1.7	9.9	8
	Range	0.3–4.5	5.4–28.3	3.0–17.7
Mesotrophic	Average	4.7	4.2	26.7
	Range	3–11	1.5–8.1	10.9–95.6
Eutrophic	Average	14.3	2.5	84.4
	Range	3–78	0.0–7.0	15–386

Source: Wetzel, 1983.

by the available nutrients. Although a more productive lake usually will have a higher fish population, the number of the most desirable fish may decline. In fact, increased productivity generally results in reduced water quality because of undesirable changes that occur as algal growth increases. Because of the important role productivity plays in determining water quality, it forms a basis for classifying lakes. Table 5-3 shows a lake classification based on productivity.

Oligotrophic Lakes. Oligotrophic lakes have a low level of productivity due to a severely limited supply of nutrients to support algal growth. As a result, the water is clear enough that the bottom can be seen at considerable depths. In this case, the euphotic zone often extends into the hypolimnion, which is aerobic. Oligotrophic lakes, therefore, support cold water game fish. Lake Tahoe on the California-Nevada border is a classic example of an oligotrophic lake.

Eutrophic Lakes. Eutrophic lakes have a high productivity because of an abundant supply of algal nutrients. The algae cause the water to be highly turbid, so the euphotic zone may extend only partially into the epilimnion. As the algae die, they settle to the lake bottom where they are decomposed by benthic organisms. In a eutrophic lake, this decomposition is sufficient to deplete the hypolimnion of oxygen during summer stratification. Because the hypolimnion is anaerobic during the summer, eutrophic lakes support only warm-water fish. In fact, most cold-water fish are driven out of the lake long before the hypolimnion becomes anaerobic because they generally require dissolved oxygen levels of at least 5 mg/L. Highly eutrophic lakes may also have large mats of floating algae that typically impart unpleasant tastes and odors to the water.

Mesotrophic Lakes. Lakes which are intermediate between oligotrophic and eutrophic are called mesotrophic. Although substantial depletion of oxygen may have occurred in the hypolimnion, it remains aerobic.

Senescent Lakes. These are very old, shallow lakes which have thick organic sediments and rooted water plants in great abundance. These lakes will eventually become marshes.

Eutrophication

Eutrophication is a natural process in which lakes gradually become shallower and more productive through the introduction and cycling of nutrients. Thus, oligotrophic lakes gradually pass through the mesotrophic, eutrophic, and senescent stages, eventually filling completely. The time for this process to occur depends on the original size of the lake and on the rate at which sediments and nutrients are introduced. In some lakes the eutrophication process is so slow that thousands of years may pass with little change in water quality. Other lakes may have been eutrophic from the day they were formed, if nutrient levels were high at that time.

Cultural eutrophication is caused when human activity speeds the processes naturally occurring by increasing the rate at which sediments and nutrients are added to the lake. Thus, lake pollution can be seen as the intensification of a natural process. This is not to say that eutrophic lakes are necessarily polluted, but that pollution contributes to eutrophication. Water quality management in lakes is primarily concerned with slowing eutrophication to at least the natural rate. To understand the factors involved in eutrophication, it is necessary to understand the factors contributing to algal growth.

Algal Growth Requirements

All algae require macronutrients, such as carbon, nitrogen, and phosphorus, and micronutrients, such as trace elements. For algae to grow, all nutrients must be available. Lack of any one nutrient will limit the total algal population. The availability of each nutrient and its natural cycle are summarized below.

Carbon. Algae obtain their carbon from carbon dioxide dissolved in the water. Since the carbon dioxide is in equilibrium with the bicarbonate buffer system (see Chapter 4), the immediately available carbon is determined by the alkalinity of the water. However, as carbon dioxide is removed from the water, it is replenished from the atmosphere. The atmosphere is, of course, a virtually inexhaustible source of this gas. When algae are either consumed by higher organisms or die and decompose, the organic carbon is oxidized back to carbon dioxide which returns either to the water or to the atmosphere to complete the carbon cycle.

Nitrogen. Nitrogen in lakes is usually in the form of nitrate (NO_3^-) and comes from external sources by way of inflowing streams or groundwater. When taken up for algal growth, the nitrogen is chemically reduced to amino-nitrogen (NH_2^-) and incorporated into organic compounds. When dead algae undergo decomposition, the organic nitrogen is released to the water as ammonia (NH_3). The ammonia is then oxidized back to nitrate by bacteria in the same nitrification process discussed earlier in river systems.

Nitrogen cycles from nitrate to organic nitrogen, to ammonia, and back to nitrate as long as the water remains aerobic. However, in anaerobic sediments, and in the hypolimnion of eutrophic lakes, when algal decomposition has depleted the oxygen supply, nitrate is reduced by anaerobic bacteria to nitrogen gas (N_2) and lost from the system in a process called *denitrification.* Denitrification reduces the average time nitrogen remains in the lake system. The denitrification reaction is

$$2NO_3^- + \text{organic carbon} \rightleftharpoons N_2 + CO_2 + H_2O \tag{5-48}$$

Some photosynthetic microorganisms can also fix nitrogen gas from the atmosphere by converting it to organic nitrogen. In lakes the most important nitrogen-fixing microorganisms are photosynthetic bacteria called cyanobacteria, formerly known as blue-green algae because of the pigments they contain. Because of their nitrogen-fixing ability, cyanobacteria have a competitive advantage over green algae when nitrate and ammonium concentrations are low but other nutrients are sufficiently abundant. These cyanobacteria are generally undesirable because of their tendency to aggregate in unsightly floating mats and because they impart unpleasant odor and taste to the water. Cyanobacteria can also produce toxins which kill fish. Fortunately, these nuisance organisms are not prevalent unless the supply of soluble fixed nitrogen is reduced to low levels.

Phosphorus. Phosphorus in lakes originates from external sources and is taken up by algae in the inorganic form (PO_4^{3-}) and incorporated into organic compounds. During algal decomposition, phosphorus is returned to the inorganic form. The release of phosphorus from dead algal cells is so rapid that only a little of it leaves the epilimnion with the settling algal cells. However, little by little, phosphorus is transferred to the sediments, some of it in undecomposed organic matter; some of it in precipitates of iron, aluminum, and calcium; and some bound to clay particles. To a large extent, the permanent removal of phosphorus from the overlying waters to the sediments depends on the amount of iron, aluminum, calcium, and clay entering the lake along with phosphorus.

Trace Elements. The quantities of trace elements required to support algal growth are so small that most fresh waters have sufficient amounts for a substantial algal population.

The Limiting Nutrient

In 1840, Justin Liebig formulated the idea that "growth of a plant is dependent on the amount of foodstuff that is presented to it in minimum quantity." This is now known as *Liebig's law of the minimum.* As applied to algae, it means that algal growth will be limited by the nutrient that is least available. Of all the nutrients, only phosphorus is not readily available from the atmosphere or the natural water supply. For this reason, phosphorus is deemed the *limiting nutrient* in lakes. The amount of phosphorus controls the quantity of algal growth and therefore the productivity of lakes. This can be seen from Figure 5-15 in which the concentration of *chlorophyll a* is plotted against phosphorus concentration. Chlorophyll *a*, one of the green pigments involved in photosynthesis, is found in all algae, so it is used to distinguish the amount of algae in the water from other organic solids such as bacteria. It has been estimated that the phosphorus concentration should be below 0.010 to 0.015 mg/L to limit algae blooms (Vollenweider, 1975).

Control of Phosphorus in Lakes

Since phosphorus is usually the limiting nutrient, control of cultural eutrophication must be accomplished by reducing the input of phosphorus to the lake. Once the input is reduced, the phosphorus concentration will gradually fall as phosphorus is buried in the

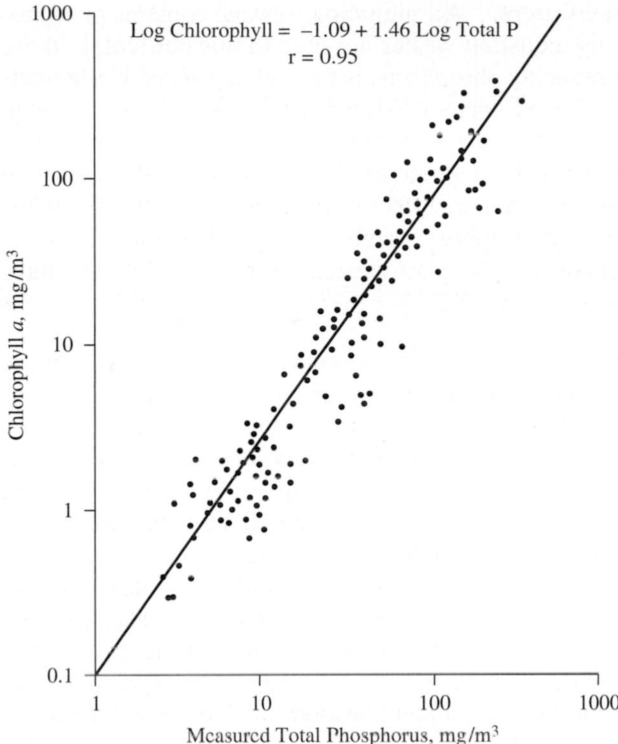

FIGURE 5-15
Relationship between summer levels of chlorophyll *a* and measured total phosphorus concentration for 143 lakes. (*Source:* Jones and Bachmann, 1976.)

sediment or flushed from the lake. Other strategies for reversing or slowing the eutrophication process, such as precipitating phosphorus with additions of aluminum (alum) or removing phosphorus-rich sediments by dredging, have been proposed. However, if the input of phosphorus is not also curtailed, the eutrophication process will continue. Thus, dredging or precipitation alone can result only in temporary improvement in water quality. In conjunction with reduced phosphorus inputs, these measures can help speed up the removal of phosphorus already in the lake system. Of course, the need to speed the recovery process must be weighed against the potential damage from inundating shoreline areas with sludge and stirring up toxic compounds buried in the sediment.

To be able to reduce phosphorus inputs, it is necessary to know the sources of phosphorus and the potential for their reduction. The natural source of phosphorus is the weathering of rock. Phosphorus released from the rock can enter the water directly, but more commonly it is taken up by plants and enters the water in the form of dead plant matter. It is exceedingly difficult to reduce the natural inputs of phosphorus. If these sources are large, the lake is generally naturally eutrophic. For many lakes the principal sources of phosphorus are the result of human activity. The most important sources are municipal and industrial wastewaters, seepage from septic tanks, and agricultural runoff that carries phosphorus fertilizers into the water.

Municipal and Industrial Wastewaters. All municipal sewage contains phosphorus from human excrement. Many industrial wastes are high in this nutrient. In these cases, the only effective way of reducing phosphorus is through advanced waste treatment processes, which are discussed in Chapter 6. Municipal wastewaters also contain large quantities of phosphorus from detergents containing polyphosphate, which is a chain of phosphate ions (usually three) linked together. The polyphosphate binds with hardness in water to make the detergent a more effective cleaning agent. By the 1970s, phosphorus loading from detergents was approximately twice that from human excrement. Today's detergents do not contain phosphorus because the manufacturers have replaced it with other chemicals.

Septic Tank Seepage. The shores of many lakes are dotted with homes and summer cottages, each with its own septic tank and tile field for waste disposal. As treated wastewater moves through the soil toward the lake, phosphorus is adsorbed by soil particles, especially clay. Thus, during the early life of the tile field, very little phosphorus gets to the lake. However, with time, the capacity of the soil to adsorb phosphorus is exceeded and any additional phosphorus will pass on into the lake, contributing to eutrophication. The time it takes for phosphorus to break through to the lake depends on the type of soil, the distance to the lake, the amount of wastewater generated, and the concentration of phosphorus in that wastewater. To prevent phosphorus from reaching the lake, it is necessary to put the tile field far enough from the lake that the adsorption capacity of the soil is not exceeded. If this is not possible, it may be necessary to replace the septic tanks and tile fields with a sewer to collect the wastewater and transport it to a treatment facility.

Agricultural Runoff. Because phosphorus is a plant nutrient, it is an important ingredient in fertilizers. As rain water washes off fertilized fields, some of the phosphorus is carried into streams and then into lakes. Most of the phosphorus not taken up by growing plants is bound to soil particles. Bound phosphorus is carried into streams and lakes through soil erosion. Waste minimization can be applied to the control of phosphorus loading to lakes from agricultural fertilization by encouraging farmers to fertilize more often with smaller amounts and to take effective action to stop soil erosion.

Acidification of Lakes

Pure rainwater is slightly acid. As we discussed in Chapter 4, CO_2 dissolves in water to form carbonic acid (H_2CO_3). The equilibrium concentration of H_2CO_3 results in a rainwater pH of approximately 5.6. Thus, acid rain is usually defined to be precipitation with a pH less than 5.6. The northeastern U.S. and Canada frequently record rainwater pH values between 4 and 5 (Figure 5-16). These low pH values have been attributed to emissions of sulfur and nitrogen oxides from the combustion of fossil fuels (see Chapter 7).

Fish, and in particular trout and Atlantic salmon, are very sensitive to low pH levels. Most are severely stressed if the pH drops below 5.5, and few are able to survive if the pH falls below 5.0. If the pH falls below 4.0, cricket frogs and spring peepers experience mortalities in excess of 85 percent.

High aluminum concentrations are often the trigger which kills fish. Aluminum is abundant in soil but it is normally bound up in the soil minerals. At normal pH values

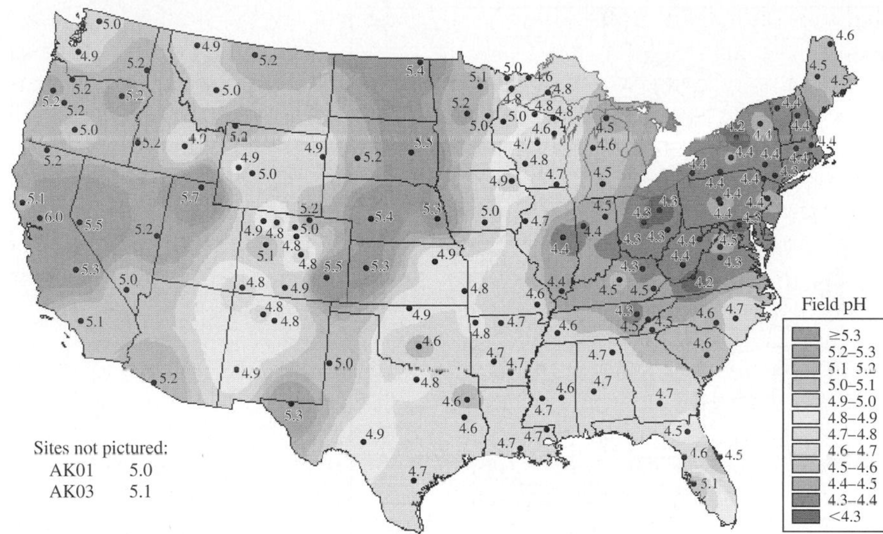

FIGURE 5-16
Hydrogen ion concentration as pH from measurements of rainwater. (*Source:* National Atmospheric Deposition Program (NRSP-3)/National Trends Network (2003). NADP Program Office, Illinois State Water Survey, 2204 Griffith Dr. Champaign, IL 61820. Data for 2001 now available. http://nadp.sws.uiuc.edu/isopleths/maps1997/phfield.gif.)

aluminum rarely occurs in solution. Acidification of the water releases highly toxic Al^{3+} to the water.

Most lakes are buffered by the carbonate buffer system (see Chapter 4). To the extent that the buffer capacity of the lake is not exceeded, the pH of the lake will not be appreciably affected by acid rain. If there is a source of carbonate to replace that consumed by the acid rain, the buffering capacity can be quite large. Calcareous soils are those containing large quantities of calcium carbonate ($CaCO_3$). As shown in Figure 4-14, carbonic acid releases bicarbonate into solution. H^+ from acid rain will also release bicarbonate. Thus, lakes formed in calcareous soils tend to be resistant to acidification.

Other factors that affect the susceptibility of a lake to acidification are the permeability and depth of the soil, the bedrock, the slope and size of the watershed, and the type of vegetation. Thin, impermeable soils provide little time for contact between the soil and the precipitation. This reduces the potential for the soil to buffer the acid precipitation. Likewise, small watersheds with steep slopes reduce the time for buffering to occur. Deciduous foliage tends to decrease acidity. Coniferous foliage tends to yield runoff that is more acid than the precipitation itself. Granite bedrock offers little potential to buffer acid rain. Galloway and Cowling (1978) used bedrock geology to predict areas where lakes are potentially most sensitive to acid rain (Figure 5-17). You may note that the predicted areas of sensitivity are also those subjected to very acid precipitation.

The control of lake acidification is related to the control of atmospheric emissions of sulfur and nitrogen oxides. The role of air pollution in acid deposition is discussed in more detail in Chapter 7.

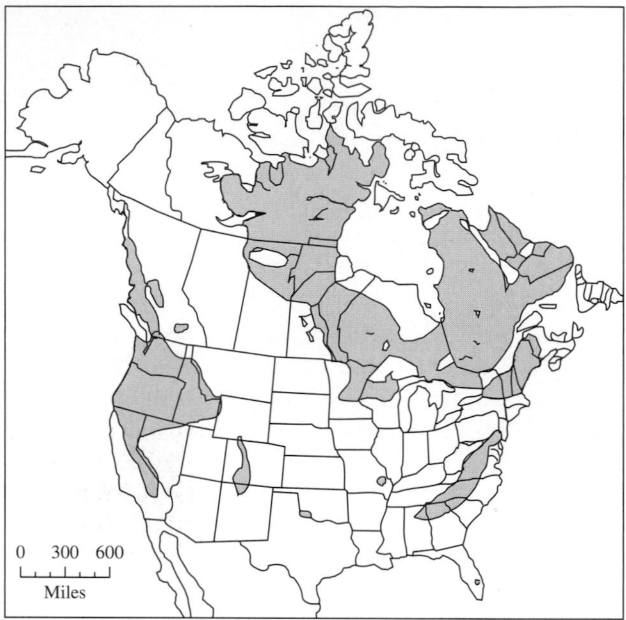

FIGURE 5-17
Regions in North America containing lakes sensitive to acidification by acid precipitation. The shaded areas have igneous or metamorphic bedrock geology; the unshaded areas have calcereous or sedimentary bedrock geology. Regions having low alkalinity lakes are concurrent with regions of igneous and metamorphic bedrock geology. (*Source*: Galloway and Cowling, 1978.)

5-5 WATER QUALITY MANAGEMENT IN ESTUARIES

An *estuary* is formed along the coastline where freshwater from rivers and streams flows into the ocean. It is a place of transition, where freshwater mixes with saltwater. Estuaries are influenced by tides. The Bay of Fundy, Boston Harbor, Chesapeake Bay, Corpus Christi Bay, Florida Bay, Hudson River estuary, Mobile Bay, Puget Sound, and San Francisco Bay are examples of estuaries.

The mix of saltwater and freshwater combined with the diurnal cycle of tidal wetting and drying and changes in salinity make the estuary ecosystem extremely complex. Numerous different habitats support an abundance of a wide variety of different animals. Each of these is sensitive to minor changes in the quality of the water.

Like other surface waters, estuaries are subject to a plethora of pollutants. Although the technologies and policies that serve to manage the water quality of lakes and rivers also apply to estuaries, the management of the water quality is complicated by the number of political jurisdictions that touch on the estuary as well as the competing needs of organisms that live there. Competing with the environmental value is the economic value of the estuary to commerce (Costanza and Voinov, 2000).

5-6 GROUNDWATER QUALITY MANAGEMENT

Groundwater, by its very location, has a large measure of protection from contaminants that are found in surface waters. However, once groundwater becomes contaminated, its location and low rate of replacement with freshwater makes it difficult to return it to a pristine state. Two major source of contaminants are of concern: uncontrolled releases of biological and chemical contaminants and saltwater intrusion from over pumping of wells.

Uncontrolled Releases

Uncontrolled releases come from a variety of sources:

- Discharge from improperly operated or located septic systems
- Leaking underground storage tanks
- Improper disposal of hazardous or other chemical wastes
- Spills from pipelines or transportation accidents
- Recharge of groundwater with contaminated surface water
- Leaking landfills
- Leaking retention ponds or lagoons

The properties of the contaminant and the aquifer material govern their migration of the contaminants. Water-soluble chemicals are likely to move vertically down through the soil to the aquifer and migrate with the water as shown in Figure 5-18. Nitrate, methyl tertiary butyl ketone (a gasoline additive) and methamidophos (a pesticide) are examples of water soluble chemicals.

Chemicals that are only sparingly soluble migrate through the aquifer as a separate nonaqueous phase. These chemicals are know as *nonaqueous phase liquids* (NAPLs). The NAPLs are divided into two categories based on their density. The light NAPLs or (LNAPLs) are less dense than water and will tend to "float" on the water table as shown in Figure 5-19. Some of the chemical will dissolve into the groundwater and some of the chemical may volatilize into the pore spaces of the unsaturated zone. Examples of LNAPLs include gasoline and aviation fuel. The dense NAPLs (DNAPLs) have a density greater than that of water. They will tend to "sink" in the aquifer until they reach an impervious barrier. The behavior of DNAPLs is shown in Figure 5-20. Trichloroethylene, tetrachloroethylene, and PCB-laden oils are examples of DNAPLs.

Management techniques for contaminant control in groundwater are discussed in Chapter 10.

Saltwater Intrusion

Freshwater aquifers near oceans or above saline aquifers may become contaminated with saltwater when water is pumped from wells that draw water that is too close to the freshwater-saltwater interface.

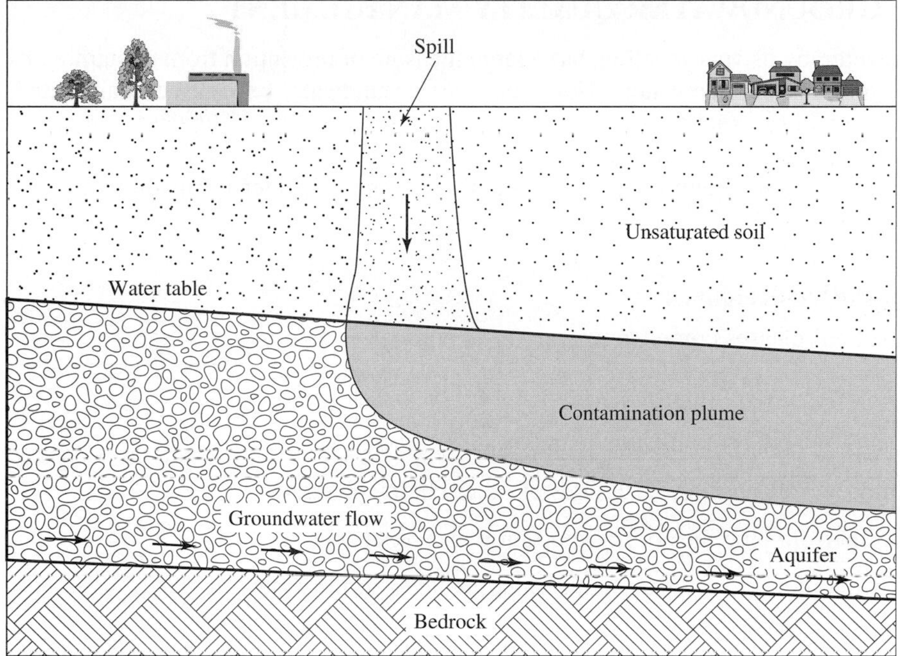

FIGURE 5-18
Dissolved contamination plume.

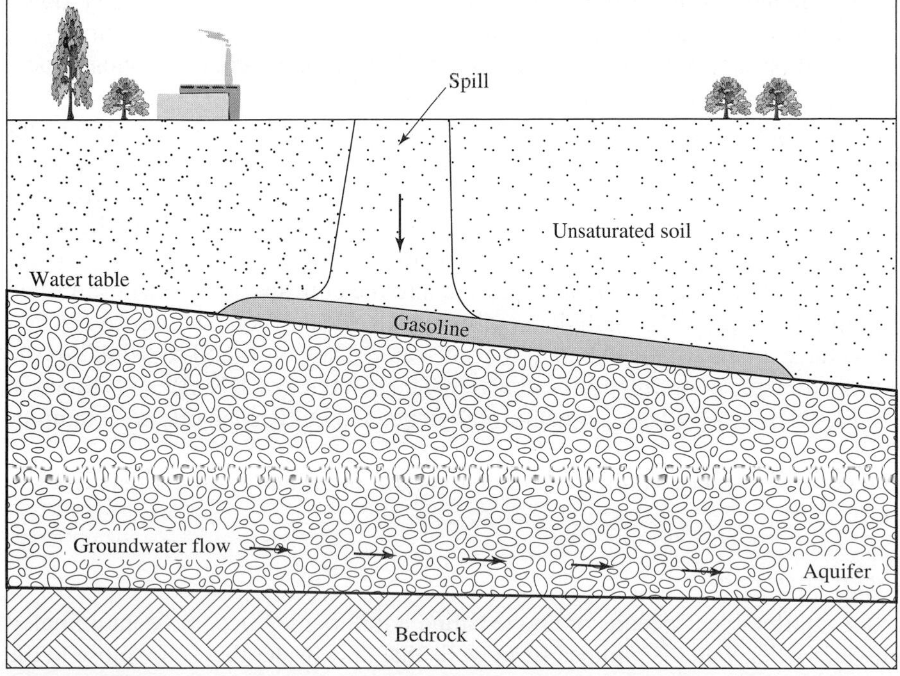

FIGURE 5-19
Immiscible plume less dense than water.

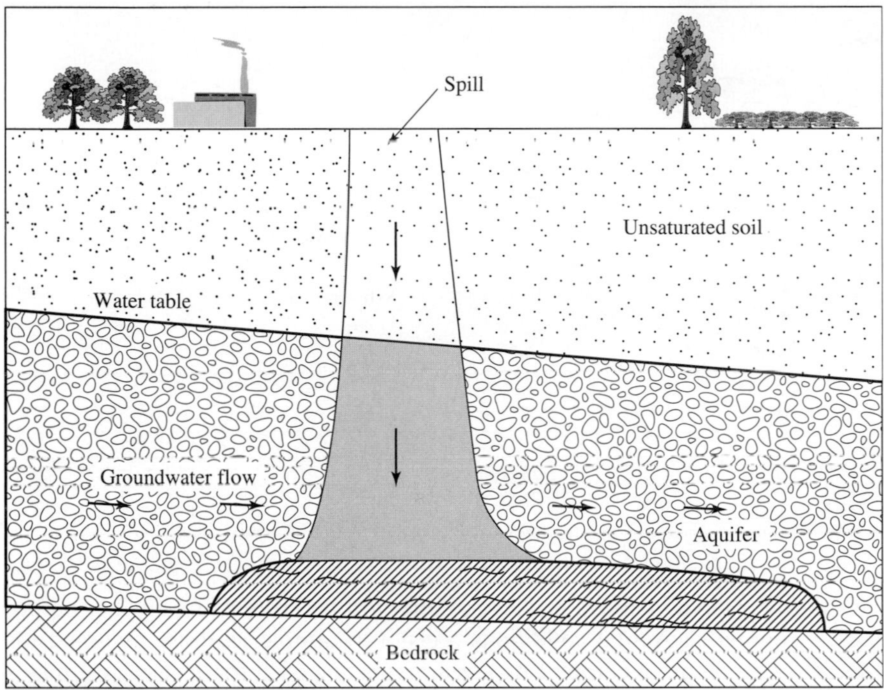

FIGURE 5-20
Immiscible plume denser than water.

Because seawater is heavier than freshwater, it will form a saltwater wedge in aquifers that drain into the ocean (Figure 5-21). Using the notation in Figure 5-21 and assuming that fresh water moves horizontally to the ocean and that the interface is abrupt and that it occurs at the shoreline, we can express the freshwater pressure at any point of the interface as

$$P_f = (h + z)\rho_f \tag{5-49}$$

where P_f = freshwater pressure, Pa
h = height of water table above sea level, m
z = distance of interface below sea level, m
ρ_f = density of fresh water, kg/m^3

This pressure must be the same as the saltwater pressure on the other side of the interface; that is, P_f is also equal to $z\rho_s$. Equating the two expressions and solving for z gives

$$z = \frac{\rho_f}{\rho_s - \rho_f} h \tag{5-50}$$

where ρ_s = density of saltwater, g/mL
ρ_f = density of freshwater, g/mL

Taking the density of fresh water as 1.000 g/mL and that of seawater as 1.025 g/mL, we can use this expression to estimate that, for coastal waters, $z \approx 40h$. Thus, for every

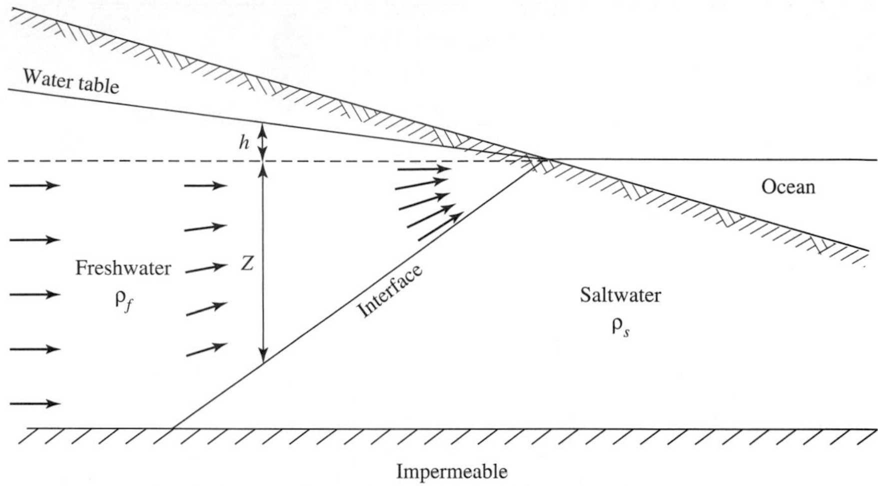

FIGURE 5-21
Freshwater-saltwater interface in coastal aquifer draining into ocean. (*Source:* Bouwer, 1978.)

meter that the water table is above sea level, there will be 40 m of freshwater before saltwater is encountered. Actual conditions are not as simple as those assumed for Figure 5-21 and a much more complex expression must be used for accurate estimates of the position of the interface (Bouwer, 1978).

When fresh groundwater is underlain by saline water, pumping a well in the freshwater will cause the interface to rise below the well as shown in Figure 5-22. This *upconing* is in response to the pressure reduction on the interface that results from the drawdown of the water table around the well. If the well screen is close to the saline water or the pumping rate is too high, saltwater may be drawn into the well.

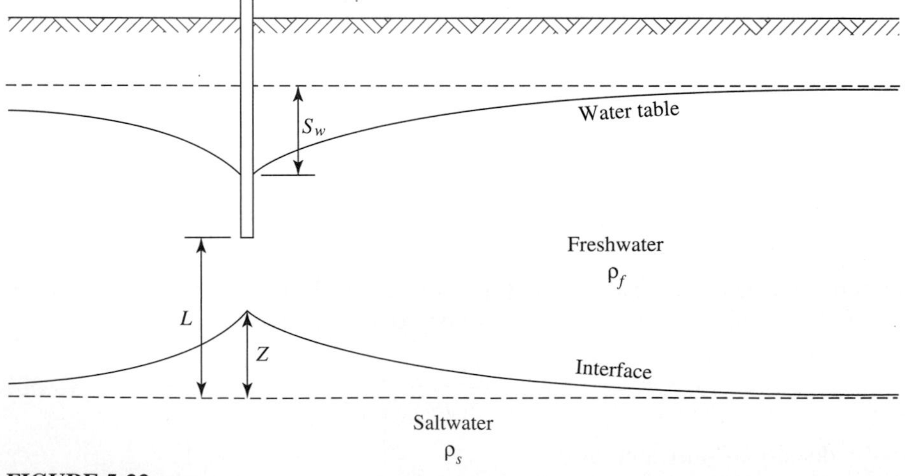

FIGURE 5-22
Geometry and symbols for upconing of saltwater beneath a pumped well (dashed lines represent static positions of water table and interface). (*Source:* Bouwer, 1978.)

The height of the upconing below the center of a well after an infinite pumping time at the freshwater well may be estimated by using the Bear and Dagan (1968) equation as modified by Schmorak and Mercado (1969):

$$z_{\infty} = \frac{\rho f Q}{2\pi(\rho_s - \rho_f)KL} \tag{5-51}$$

where z_{∞} = rise of the cone center at $t = \infty$, m
Q = well discharge, m^3/d
K = hydraulic conductivity, m/d
L = depth of freshwater-saltwater interface below well bottom prior to pumping, m

A critical point in the relationship between the upconing and the drawdown is reached when the saltwater cone height reaches approximately $0.4L$ to $0.6L$ (see Figure 5-22). When the cone height exceeds this critical height, the cone may "jump" to the bottom of the well (Bouwer, 1978). Thus, when freshwater is underlain with saline water, prediction of the upconing is important to prevent *saltwater intrusion*.

5-7 CHAPTER REVIEW

When you have completed studying this chapter; you should be able to do the following without the aid of your text or notes:

1. List the major pollutant categories (there are four) that are produced by each of the four principal sources of wastewater.
2. Explain the difference between point sources and non-point sources of pollution.
3. Explain the difference between AFOs and CAFOs and determine whether or not an animal feeding operation is a CAFO on the basis of data you are given.
4. List the two nutrients of primary concern with respect to a receiving body of water.
5. Define TMDL and explain how it is calculated.
6. Define biochemical oxygen demand (BOD).
7. Explain the procedure for determining BOD and specify the nominal values of temperature and time used in the test.
8. List three reasons why the BOD rate constant may vary.
9. Sketch a graph showing the effect of varying rate constant on 5-day BOD if the ultimate BOD is the same, and the effect on ultimate BOD if the 5-day BOD is the same.
10. Utilizing Equation 5-20 in your answer, explain what causes nitrogenous BOD.

11. Sketch a series of curves that show the deoxygenation, reaeration, and DO sag in a river. Show the effect of a change in the deoxygenation or reaeration rate on the location of the critical point and the magnitude of the DO deficit.

12. List three reasons why ammonia nitrogen is detrimental to a receiving body of water and its inhabitants.

13. Sketch and compare the epilimnion and hypolimnion with respect to the following: location in a lake, temperature, and oxygen abundance (that is, DO).

14. Describe the process of stratification and turnover in lakes.

15. Explain what determines the euphotic zone of a lake and what significance this has for biological growth.

16. Given a description of a lake that includes the productivity, clarity, and oxygen levels, classify it as oligotrophic, mesotrophic, eutrophic, or senescent.

17. Explain the process of eutrophication.

18. State Liebig's law of the minimum.

19. Name the most common "limiting nutrient" in lakes and explain why it is a limiting nutrient.

20. List three sources of phosphorus that must be controlled to reduce cultural eutrophication of lakes.

21. Explain why the pH of pure rainwater is about 5.6.

22. Define acid rain.

23. Explain why acid rain is of concern.

24. Explain the role of calcareous soils in protecting lakes from acidification.

25. Other than rainwater pH, list six variables that determine the extent of lake acidification and explain how increasing or decreasing the value of each might be expected to change the extent of acidification.

26. Identify the two components that define an estuary.

27. Describe the process of saltwater intrusion into a freshwater well.

With the aid of this text you should be able to do the following:

1. Calculate the ThOD of a compound given the balanced oxidation reaction(s).

2. Calculate the BOD_5, given the sample size and oxygen consumption, or calculate the sample size, given the allowable oxygen consumption and estimated BOD_5.

3. Calculate the ultimate BOD (L_o), given the BOD exerted (BOD_t) in time t and rate constant k, or calculate the rate constant k, given L_o and BOD_5.

4. Calculate a new k for a temperature other than 20°C, given a value at T °C.

5. Calculate the BOD rate constant k and ultimate BOD (L_o) from experimental data of BOD versus time.
6. Calculate the oxygen deficit D in a length of stream (reach), given the required input data.
7. Calculate the critical oxygen deficit DO at the DO sag point (minimum).
8. Estimate the height of upconing from a calculation of drawdown (Chapter 3).

5-8 PROBLEMS

5-1. Glutamic acid ($C_5H_9O_4N$) is used as one of the reagents for a standard to check the BOD test. Determine the theoretical oxygen demand of 63 mg/L of glutamic acid. Assume the following reactions apply:

$$C_5H_9O_4N + 4.5O_2 \rightleftharpoons 5CO_2 + 3H_2O + NH_3$$
$$NH_3 + 2O_2 \rightleftharpoons NO_3 + H^+ + H_2O$$

Answer: ThOD = 89.14 or 89 mg/L

5-2. Bacterial cells have been represented by the chemical formula $C_5H_7NO_2$. Compute the theoretical oxygen demand of 30 mg/L of bacterial cells, assuming the following reactions apply:

$$C_5H_7NO_2 + 5O_2 \rightleftharpoons 5CO_2 + 2H_2O + NH_3$$
$$NH_3 + 2O_2 \rightleftharpoons NO_3 + H^+ + H_2O$$

5-3. A step in the anaerobic decomposition of organic wastes produces acetic acid (CH_3COOH). Determine the theoretical oxygen demand of 300 mg/L of acetic acid. Assume the following reaction applies:

$$CH_3COOH + 2O_2 \rightleftharpoons 2CO_2 + H_2O$$

5-4. If the BOD_5 of a waste is 220.0 mg/L and the ultimate BOD is 320.0 mg/L, what is the rate constant (base 10)? Assume the temperature is 20 °C.

Answer: $K = 0.101\ d^{-1}$

5-5. If the BOD of a municipal wastewater at the end of 7 days is 60.0 mg/L and the ultimate BOD is 85.0 mg/L, what is the rate constant (base 10)? Assume the temperature is 20°C.

5-6. If the BOD_6 of a municipal wastewater is 213 mg/L and the ultimate BOD is 318.1 mg/L, what is the rate constant (base e)? Assume the temperature is 20°C.

5-7. Convert the rate constant found in Problem 5-4 to base e.

Answer: $k = 0.233\ d^{-1}$

5-8. Convert the rate constant found in Problem 5-5 to base e.

5-9. Convert the rate constant found in Problem 5-6 to base 10.

5-10. Assuming that the data in Problem 5-4 were taken at 20°C, compute the rate constant at a temperature of 15°C.

Answer: $K = 0.0536\ \text{d}^{-1}$

5-11. Assuming that the data in Problem 5-5 were taken at 25°C, compute the rate constant at 16°C.

5-12. What is the BOD_5 of a waste that yields an oxygen consumption of 2.00 mg/L from a 1.00 percent sample?

Answer: $BOD_5 = 200$ mg/L

5-13. What sample size (in percent) is required for a BOD_5 of 350.0 mg/L if the oxygen consumed is to be limited to 4.00 mg/L?

5-14. If the BOD_5 of a waste is 327 mg/L, what sample size (in percent) should be selected to yield an oxygen consumption of 4.8 mg/L?

5-15. If the ultimate BOD of two wastes having K values of $0.0800\ \text{d}^{-1}$ and $0.120\ \text{d}^{-1}$ is 280.0 mg/L, what would be the 5-day BOD for each?

Answer: For $K = 0.08\ \text{d}^{-1}$, $BOD_5 = 169$ mg/L; for $K = 0.12\ \text{d}^{-1}$, $BOD_5 = 210$ mg/L

5-16. If the BOD_5 of two wastes having K values of $0.0800\ \text{d}^{-1}$ and $0.120\ \text{d}^{-1}$ is 280.0 mg/L, what would be the ultimate BOD for each?

5-17. Using a computer spreadsheet program you have written, plot the BOD curves that would result for the data given in Problem 5-15. At approximately what day (± 5.0 d) does the ultimate BOD occur? Check your answer, using Equation 5-5.

5-18. Using a computer spreadsheet program you have written, plot the BOD curves that would result for the data given in Problem 5-16. At approximately what day (± 5.0 d) would the ultimate BOD occur for each waste? Check your answer, using Equation 5-5.

5-19. Using Thomas' graphical method and a computer spreadsheet program you have written, calculate the BOD rate constant in base e and the ultimate BOD from the following data:

Day	BOD, mg/L
2	70.0
5	102.4
7	111.00
8	114.0
10	118.8

Answers: $k = 0.36\ \text{d}^{-1}$; $L_0 = 129.69$ or 130 mg/L

5-20. Using Thomas' graphical method and a computer spreadsheet program you have written, calculate the BOD rate constant in base e from the following data:

Day	BOD, mg/L
2	119
5	210
10	262
20	279
35	279.98

5-21. Using Thomas' graphical method and a computer spreadsheet program you have written, calculate the BOD rate constant in base *e* from the following data:

Day	BOD, mg/L
2	86
5	169
10	236
20	273
35	279.55

5-22. Using the data from Problem 5-1, calculate the theoretical NBOD of glutamic acid.

Answer: Theo. NBOD = 27.42 or 27 mg/L

5-23. Using the data from Problem 5-2, calculate the theoretical NBOD of bacterial cells.

5-24. Calculate the NBOD of 200 mg/L of a casein ($C_8H_{12}O_3N_2$) dairy waste. Assume the following reactions:

$$C_8H_{12}O_3N_2 + 8O_2 \rightleftharpoons 8CO_2 + 3H_2O + 2NH_3$$

$$NH_3 + 2O_2 \rightleftharpoons NO_3 + H^+ + H_20$$

5-25. Derive an expression for the final temperature T_s of the mixture of wastewater flow Q_w at temperature T_w and river flow Q_r at temperature T_r. Assume that the specific heat and density of the wastewater and the river are the same.

5-26. A tannery with a wastewater flow of 0.011 m^3/s and a BOD_5 of 590 mg/L discharges into the Cattaraugus Creek (Nemerow, 1974). The creek has a 10-year, 7-day low flow of 1.7 m^3/s. Upstream of the tannery, the BOD_5 of the creek is 0.6 mg/L. The BOD rate constants k are 0.115 d^{-1} for the tannery and 3.7 d^{-1} for the creek. The temperature of both the creek and the tannery wastewater is 20°C. Calculate the initial ultimate BOD after mixing.

Answers: $L_{oTannery}$ = 1,349.2 mg/L, L_{oCreek} = 0.6 mg/L, L_a = 9.27 or 9 mg/L

5-27. The town of Pittsburgh discharges 0.126 m^3/s of treated wastewater into Cherry Creek. The BOD_5 of the wastewater is 34 mg/L. Cherry Creek has a 10-year, 7-day low flow of 0.126 m^3/s. Upstream of the wastewater outfall from Pittsburgh, the BOD_5 is 1.2 mg/L. The BOD rate constants k are 0.222 d^{-1} and 0.090 d^{-1} for the wastewater and creek respectively. The temperature of both the creek and the municipal wastewater is 20°C. Calculate the initial ultimate BOD after mixing.

5-28. A short distance downstream from the tannery in Problem 5-26, a glue factory and a municipal wastewater treatment plant also discharge into Cattaraugus Creek. The wastewater flows and ultimate BODs for these discharges are listed below. Determine the initial ultimate BOD after mixing of the creek and the three wastewater discharges.

Source	Flow, m^3/s	Ultimate BOD, mg/L	Temperature, °C
Glue factory	0.13	255	20
Municipal WWTP	0.02	75	20

5-29. Cherry Creek is joined by Peach Tree Creek and Apple Creek to form the Ambrosia River. The flows and ultimate BODs for these creeks are listed below. Determine the initial ultimate BOD after mixing of the Ambrosia River.

Source	Flow, m^3/s	Ultimate BOD, mg/L	Temperature, °C
Cherry Creek	0.252	27	20
Peach Tree Creek	0.13	8	20
Apple Creek	0.02	16	20

5-30. Compute the deoxygenation rate constant and reaeration rate constant (base e) for the following wastewater and stream conditions.

Source	k, d^{-1}	Temperature, °C	H, m	v, m/s	η
Wastewater	0.20	20			
Stream		20	1.0	0.5	0.4

5-31. As a result of snowmelt and spring flooding, the stream conditions in Problem 5-30 change as shown below. Determine the values of k_d and k_r for flood conditions.

Source	Q, m^3/s	k, d^{-1}	Temperature, °C	H, m	v, m/s	η
Wastewater	0.126	0.20	20			
Stream	0.252		10	4.0	2.5	0.6

5-32. The initial ultimate BOD after mixing of the Noir River is 50 mg/L. The DO in the river after the wastewater and river have mixed is at saturation. The river temperature is 10°C. At 10°C, the deoxygenation rate constant k_r

is 0.30 d^{-1} and the reaeration rate constant k_r is 0.30 d^{-1}. Determine the critical point t_c and the critical DO.

5-33. Repeat Problem 5-32, assuming the river temperature rises to 15°C so that k_d and k_r change.

5-34. Churchill, Elmore, and Buckingham (1962) developed the following reaeration equation based on studies of Tennessee Valley rivers:

$$k_r = \frac{5.23v}{H^{1.67}}$$

Compare this equation with the O'Connor and Dobbins equation (Equation 5-44) using a computer spreadsheet program you have written. Plot the values of k_r as a function of the speed of the stream flow for each of these equations, using four speeds (0.05, 0.10, 0.20, and 0.40 m/s) and a depth of 1.0 m. Assume the stream temperature is 20°C. Discuss the possible reasons for the results you observe.

5-35. The discharge from a sugar beet plant causes the DO at the critical point to fall to 4.0 mg/L. The stream has a negligible BOD and the initial deficit after the river and wastewater have mixed is zero. What DO will result if the concentration of the waste (L_w) is reduced by 50 percent? Assume that the flows remain the same and that the saturation value of DO is 10.83 mg/L in both cases.

5-36. The Big Bear town council has asked that you determine whether the discharge of the town's wastewater into the Salmon River will reduce the DO below the state standard of 5.00 mg/L at Alittlebit, 5.79 km downstream, or at any other point downstream. The pertinent data are as follows:

Parameter	Big Bear wastewater	Salmon River
Flow, m^3/s	0.280	0.877
Ultimate BOD at 28°C, mg/L	6.44	7.00
DO, mg/L	1.00	6.00
K_d at 28°C, d^{-1}	N/A	0.199
K_r at 28°C, d^{-1}	N/A	0.370
Speed, m/s	N/A	0.650
Temperature, °C	28°C	28°C

Answers: DO at Alittlebit = 4.75 mg/L
Critical DO = 4.72 mg/L at t_c = 0.3149 d

5-37. If the Salmon River temperature in Problem 5-36 decreases to 12°C, will the discharge from Big Bear reduce the DO in the river below 5.00 mg/L at a distance of 5.79 km downstream? *Note:* You must calculate a new temperature of the mixed river water and wastewater, then correct K_d and K_r of the river only.

5-38. Calculate the DO at a point 1.609 km downstream from a waste discharge point for the following conditions. Report your answer to two decimal places. Rate constants are already temperature-adjusted.

Parameter	**Stream**
k_d	1.911 d^{-1}
k_r	4.49 d^{-1}
Flow	2.4 m^3/s
Speed	0.100 m/s
D_a (after mixing)	0.00
Temperature	17.0°C
BOD_L (after mixing)	1,100.00 kg/d

Answer: DO = 8.69 or 8.7 mg/L

5-39. Calculate the DO at a point 2.880 km downstream from a waste discharge point for the following conditions. Report your answer to two decimal places. Rate constants are already temperature adjusted.

Parameters	**Stream**
K_d	1.830 d^{-1}
K_r	2.030 d^{-1}
Flow	0.30 m^3/s
Speed	0.100 m/s
D_a (after mixing)	0.00
Temperature	18.0°C
BOD_L (after mixing)	1,125.00 kg/d

Hint: Problems 5-40 through 5-50 require solution of a long series of sequential equations (similar to but not the same as those in Example 5-12). Some of the problems require iterative solutions. Even when an iterative solution is not required, a computer spreadsheet solution that computes the results of each of the sequential equations is recommended. This will allow you to correct small mistakes in equations you might have made early in the sequence without time-consuming recomputation of the subsequent equations. Your time is valuable, make efficient use of it!

5-40. The town of Avepitaeonmi has filed a complaint with the state Department of Natural Resources (DNR) that the City of Watapitae is restricting its use of the Wash River because of the discharge of raw sewage. The DNR water quality criterion for the Wash River is 5.00 mg/L of DO. Avepitaeonmi is 15.55 km downstream from Watapitae. What is the DO at Avepitaeonmi? What is the critical DO and where (at what distance) downstream does it occur? Is the assimilative capacity of the river restricted? The following data pertain to the 7-year, 10-day low flow at Watapitae.

Parameter	Watapitae wastewater	Wash River
Flow, m^3/s	0.1507	1.08
BOD_5 at 16°C, mg/L	128.00	N/A
BOD_u at 16°C, mg/L	N/A	11.40
DO, mg/L	1.00	7.95
Temperature, °C	16.0	16.0
k at 20°C, d^{-1}	0.4375	N/A
Speed, m/s	N/A	0.390
Depth, m	N/A	2.80
Bed-activity coefficient	N/A	0.200

5-41. Under the provisions of the Clean Water Act, the U.S. Environmental Protection Agency established a requirement that municipalities had to provide secondary treatment of their waste. This was defined to be treatment that results in an effluent BOD_5 that does not exceed 30 mg/L. The discharge from Watapitae (Problem 5-40) is clearly in violation of this standard. Given the data in Problem 5-40, rework the problem assuming that Watapitae provides treatment to lower the BOD_5 to 30.00 mg/L.

5-42. The Blue Ox Tannery has asked you to assist them in preparations to file for a NPDES permit to discharge wastewater into the Zmellsbad River. The DNR water quality criterion for the Zmellsbad River is 5.00 mg/L of DO. They have asked you the following questions: What is the critical DO and where (at what distance) downstream does it occur? They have provided you with the following data. They pertain to the 7-year, 10-day low flow at the tannery.

Parameter	Blue Ox wastewater	Zmellsbad River
Flow, m^3/s	1.148	7.222
BOD_5 at 15°C, mg/L	90.00	N/A
BOD_u at 15°C, mg/L	N/A	7.66
DO, mg/L	1.00	6.00
Temperature, °C	15.0	15.0
k at 20°C, d^{-1}	0.3685	N/A
Speed, m/s	N/A	0.300
Depth, m	N/A	2.92
Bed-activity coefficient	N/A	0.100

5-43. Under the provisions of the Clean Water Act, the U.S. Environmental Protection Agency established a requirement that municipalities had to provide secondary treatment of their waste. This was defined to be treatment that results in an effluent BOD_5 that does not exceed 30 mg/L. The discharge from the Blue Ox Tannery (Problem 5-42) is clearly in violation of this standard. Given the data in Problem 5-42, rework the problem to determine the amount of ultimate BOD (in kg/d) that the tannery may discharge to keep the DO above the DEQ water quality criteria of 5.00 mg/L at the critical point.

Hint: This problem cannot be worked backward starting with DO = 5.00 mg/L because there are two unknowns in the equation: L_a and t. Two approaches may be used. One is to set up the equations on a computer spreadsheet and decrease L_w until the DO_c becomes greater than 5.00 mg/L. An alternative is to set up the equations a computer spreadsheet and use the "solver" tool in the spreadsheet.

5-44. If the population and water use of Watapitae (Problems 5-40 and 5-41) are growing at 5 percent per year with a corresponding increase in wastewater flow, how many years' growth may be sustained before secondary treatment becomes inadequate? Assume that the treatment plant continues to maintain an effluent BOD_5 of 30.00 mg/L.

5-45. When ice covers a river, it severely limits the reaeration. There is some compensation for the reduced aeration because of the reduced water temperature. The lower temperature reduces the biological activity and, thus, the deoxygenation rate and, at the same time, the DO saturation level increases. Assuming a winter condition, rework Problem 5-40 with the reaeration reduced to zero and the river water temperature at 2°C.

5-46. What combination of BOD reduction and/or wastewater DO increase is required so the Big Bear wastewater in Problem 5-36 does not reduce the DO below 5.00 mg/L anywhere along the Salmon River? Assume that the cost of BOD reduction is 3 to 5 times that of increasing the effluent DO. Since the cost of adding extra DO is high, limit the excess above the minimum amount such that the critical DO falls between 5.00 mg/L and 5.25 mg/L.

Answer: Raising the wastewater DO to 2.7 mg/L is the most cost-effective remedy.

5-47. What amount of ultimate BOD, in kg/d, may Watapitae (Problem 5-40) discharge and still allow Avepitaeonmi 1.50 mg/L of DO above the DNR water quality criteria for assimilation of its waste?

5-48. Assuming that the mixed oxygen deficit D_a is zero and that the ultimate BOD L_r of the Looking Glass River above the wastewater outfall from Carrollville is zero, calculate the amount of ultimate BOD, in kg/d, that can be discharged if the DO must be kept at 4.00 mg/L at a point 8.05 km downstream. The stream deoxygenation rate K_d is 1.80 d^{-1} at 12°C, and the reaeration rate K_r is 2.20 d^{-1} at 12°C. The river temperature is 12°C.

The river flow is 5.95 m^3/s with a speed of 0.300 m/s. The Carrollville wastewater flow is 0.0130 m^3/s.

Answer: $Q_w L_w = 1.14 \times 10^4$ kg/d of ultimate BOD

5-49. Assume that the Carrollville wastewater (Problem 5-48) also contains 3.0 mg/L of ammonia nitrogen with a stream deoxygenation rate of 0.900 d^{-1} at 12°C. What is the amount of ultimate carbonaceous BOD, in kg/d, that Carrollville can discharge and still meet the DO level of 4.00 mg/L at a point 8.05 km downstream? Assume also that the theoretical amount of oxygen will ultimately be consumed in the nitrification process.

5-50. As part of a TMDL evaluation, you have been asked to determine the effect of the wastewater discharge from the μBrew Bottling Company on the dissolved oxygen of the Big Head River for the winter and summer conditions shown below. Use a spreadsheet to plot the DO sag curve and calculate the DO at the critical point.

		Big Head River	
Parameter	**μBrew wastewater**	**Winter**	**Summer**
Flow, m^3/s	0.200	0.483	0.241
BOD_5, mg/L	100	N/A	N/A
k at 20°C, d^{-1}	0.3685	N/A	N/A
BOD_U, mg/L	N/A	7.66	7.66
Temperature, °C	28	4	28
DO, mg/L	0.0	8.0	8.0
Speed, m/s	N/A	0.150	0.150
Depth, m	N/A	2.0	1.0
Bed Activity Coefficient	N/A	0.3	0.3

5-51. For the case of salt water intrusion, we made the statement that for coastal waters $z \approx 40h$, where z is the elevation of freshwater above sea level. Show by calculation that this is true.

5-52. A freshwater well is located in a coastal aquifer with a hydraulic conductivity of 4.63×10^{-5} m/s. The well screen is located 30 m above the freshwater-saltwater interface. To prevent saltwater from entering the well, upconing should be less than 10 m. What is the maximum permissible discharge of the well for an infinite pumping time? (After Bouwer, 1978.)

5-9 DISCUSSION QUESTIONS

5-1. Students in a graduate-level environmental engineering laboratory took samples of the influent (raw sewage) and effluent (treated sewage) of a municipal wastewater treatment plant. They used these samples to determine the BOD rate constant (k). Would you expect the rate constants to be the same or different? If different, which would be higher and why?

5-2. If it were your job to set standards for a water body and you had a choice of either BOD_5 or ultimate BOD, which would you choose and why?

5-3. A summer intern has turned in his log book for temperature measurements for a limnology survey. He was told to take the measurements in the air 1 m above the lake, 1 m deep in the lake, and at a depth of 10 m. He turned in the following results but did not record which temperatures were taken where. If the measurements were made at noon in July in Missouri, what is your best guess as to the location of the measurements (i.e. air, 1 m deep, 10-m deep)? The recorded values were: 33°C, 18°C, and 21°C.

5-4. If the critical point in a DO sag curve is found to be 18 km downstream from the discharge point of untreated wastewater, would you expect the critical point to move upstream (toward the discharge point), downstream, or remain in the same place, if the wastewater is treated?

5-5. You have been assigned to conduct an environmental study of a remote lake in Canada. Aerial photos and a ground-level survey reveal no anthropogenic waste sources are contributing to the lake. When you investigate the lake, you find a highly turbid lake with abundant mats of floating algae and a hypolimnion DO of 1.0 mg/L. What productivity class would you assign to this lake? Explain your reasoning.

5-6. The lakes in Illinois, Indiana, western Kentucky, the lower peninsula of Michigan, and Ohio do not appear to be subject to acidification even though the rainwater pH is 4.4. Based on your knowledge (or what you can discover by research) of the topography, vegetation, and bedrock, explain why the lakes in these areas are not acidic.

5-10 REFERENCES

Ahmad, R. (2002) "Watershed Assessment Power for Your PC," *Water Environment & Technology,* April, pp. 25–29.

APHA (2004) "Part 5210B: 5-day Biochemical Oxygen Demand," *Standard Methods for the Examination of Water and Wastewater,* 21st ed., American Public Health Association, Washington, DC.

Ash, R. J., B. Mauck, M. Morgan, et al. (1999) "Antibiotic Resistant Bacteria in U.S. Rivers," (Abstract Q-383), in *Abstracts of the 99th General Meeting of the American Society for Microbiology,* Chicago, May 30–June 3, p. 607.

Bear, J., and G. Dagan (1968) "Solving the Problem of Local Interfaced Upconing in a Coastal Aquifer by the Method of Small Perturbations," *Journal of Hydraulic Research,* vol. 6, no. 1, pp. 16–44.

Bennett, J., and G. Kramer (1999) "Multidrug Resistant Strains of Bacteria in the Streams of Dubuque County, Iowa," (Abstract Q-86), in *Abstracts of the 99th*

General Meeting of the American Society for Microbiology, Chicago, May 30–June 3, p. 464.

Bosko, K. (1966) "An Explanation of the Rate of BOD Progression Under Laboratory and Stream Conditions," *Advances in Water Pollution Research,* Proceedings of the Third International Conference, Munich, p. 43.

Bouwer, H. (1978) *Groundwater Hydrology,* McGraw-Hill, New York, pp. 402–406.

Churchill, M. A., H. L. Elmore, and R. A. Buckingham (1962) "Prediction of Stream Reaeration Rates," *Journal of the Sanitary Engineering Division,* American Society of Civil Engineers, vol. 88, SA4, p. 1.

Costanza, R., and A. Voinov (2000) "Integrated Ecological Economic Regional Modeling," in J. E. Hobbie (ed.), *Estuarine Science: A Synthetic Approach to Research and Practices,* Island Press, Washington, DC, pp. 461–506.

DiMura, J. (2003) "Permitting Agricultural Sources of Water Pollution," *Water Environment & Technology,* May, pp. 50–53.

FWPCA (1972) Federal Water Pollution Control Act Amendments, PL 92-500.

Galloway, J. N., and E. B. Cowling (1978) "The Effects of Precipitation on Aquatic and Terrestrial Ecosystems: A Proposed Precipitation Chemistry Network," *Journal of the Air Pollution Control Association,* vol. 28, pp. 229–235.

Gaudy, A. and E. Gaudy (1980) *Microbiology for Environmental Scientists and Engineers,* McGraw-Hill, New York, pp. 209–210, 485–492.

Harries, J. E., T. Ryunnalls, E. Hill, et al. (2000) "Development of a Reproductive Performance Test for Endocrine Disrupting Chemicals Using Pair-Breeding Fathead Minnows, *Environmental Science and Technology,* vol. 34, pp. 3003–3011.

Jones, J. R., and R. W. Bachmann (1976) "Prediction of Phosphorus and Chlorophyll Levels in Lakes," *Journal of the Water Pollution Control Federation,* vol. 48, p. 2176.

Nemerow, N. L. (1974) *Scientific Stream Pollution Analysis,* McGraw-Hill, New York, p. 272.

O'Connor, D. J., and W. E. Dobbins (1958) "Mechanisms of Reaeration in Natural Streams," *American Society of Civil Engineers Transactions,* vol. 153, p. 641.

Population Reports (1998) "Solutions for a Water-Short World," Population Information Program, Center for Communication Programs, The Johns Hopkins University School of Public Health, Baltimore, MD, vol. 26, no. 1, p. 14.

Sadik, O. A., and D. M. Witt (1999) "Monitoring Endocrine-Disrupting Chemicals," *Environmental Science and Technology,* vol. 33, pp. 368A–374A.

Schmorak, S., and A. Mercado (1969) "Upconing of Fresh Water–Sea Water Interface Below Pumping Wells, Field Study," *Water Resources Research,* vol. 5, pp. 1290–1311.

Schroepfer, G. J., M. L. Robins, and R. H. Susag (1964) "Research Program on the Mississippi River in the Vicinity of Minneapolis and St. Paul," *Advances in Water Pollution Research,* vol. 1, part 1, p. 145.

Sternes, K. L. (1999) "Presence of High-level Vancomycin Resistant Enterococci in the Upper Rio Grande," (Abstract Q-63) in *Abstracts of the 99th General Meeting of the American Society for Microbiology*, Chicago, May 30–June 3, p. 545.

Streeter, H. W., and E. B. Phelps (1925) *A Study of the Pollution and Natural Purification of the Ohio River,* U.S. Public Health Service Bulletin No. 146.

Thomas, H. A. (1950) "Graphical Determination of B.O.D. Curve Constants," *Water and Sewage Works,* pp. 123–124.

U.S. EPA (2005) http://www.epa.gov/owow/tmdl/intro.html and http://www.epa.gov/waterscience/models/allocation/def.html.

U.S.G.S. (2002) http://toxics.usgs.gov/pubs/OFR-02-94

Vollenweider, R. A. (1975) "Input-Output Models with Special Reference to the Phosphorus Loading Concept in Limnology," *Schweiz. Z. Hydro,* vol. 37, pp. 53–83.

Wetzel, R. G. (1983) *Limnology,* W. B. Saunders, Philadelphia, p. 767.

CHAPTER 6

WASTEWATER TREATMENT

6-1 INTRODUCTION

Wastewater has historically been considered a nuisance to be discarded in the cheapest, least offensive manner possible. This meant the use of on-site disposal systems such as the pit privy and direct discharge into our lakes and streams. Over the last century it has been recognized that this approach produces an undesirable impact on the environment. This led to a variety of treatment techniques that characterize the municipal treatment systems today that are the focal point of this chapter. As we look forward, it becomes obvious that in the interest of sustainability as well as fundamental economic efficiency, we must view the wastewater as a raw material to be conserved. Clean water is a scarce commodity; it should be treated as such and conserved and reused. The contents of wastewater are often viewed as pollutants. The abundance of nutrients such as phosphorus and nitrogen are, in some treatment schemes (as discussed in Section 6-11), recovered for crop growth. This approach must become more prevalent to achieve a sustainable future. The organic compounds in wastewater are a source of energy. Currently we utilize the processes described in Section 6-12 to recover some of this energy. Future efforts will focus on improving the efficiency of energy utilization in wastewater.

6-2 CHARACTERISTICS OF DOMESTIC WASTEWATER

Physical Characteristics of Domestic Wastewater

Fresh, aerobic, domestic wastewater has been said to have the odor of kerosene or freshly turned earth. Aged, septic sewage is considerably more offensive to the olfactory nerves. The characteristic rotten-egg odor of hydrogen sulfide and the mercaptans is indicative of septic sewage. Fresh sewage is typically gray in color. Septic sewage is black.

Wastewater temperatures normally range between 10 and 20°C. In general, the temperature of the wastewater will be higher than that of the water supply. This is because of the addition of warm water from households and heating within the plumbing system of the structure.

One cubic meter of wastewater weighs approximately 1,000,000 grams. It will contain about 500 grams of solids. One-half of the solids will be dissolved solids such as calcium, sodium, and soluble organic compounds. The remaining 250 grams will be insoluble. The insoluble fraction consists of about 125 grams of material that will settle out of the liquid fraction in 30 minutes under quiescent conditions. The remaining 125 grams will remain in suspension for a very long time. The result is that wastewater is highly turbid.

Chemical Characteristics of Domestic Wastewater

Because the number of chemical compounds found in wastewater is almost limitless, we normally restrict our consideration to a few general classes of compounds. These classes often are better known by the name of the test used to measure them than by what is included in the class. The biochemical oxygen demand (BOD_5) test, which we discussed in Chapter 5, is a case in point. Another closely related test is the *chemical oxygen demand* (COD) test.

The COD test is used to determine the oxygen equivalent of the organic matter that can be oxidized by a strong chemical oxidizing agent (potassium dichromate) in an acid medium. The COD of a waste, in general, will be greater than the BOD_5 because more compounds can be oxidized chemically than can be oxidized biologically, and because BOD_5 does not equal ultimate BOD.

The COD test can be conducted in about 3 hours. If it can be correlated with BOD_5, it can be used to aid in the operation and control of the wastewater treatment plant (WWTP).

Total Kjeldahl nitrogen (TKN) is a measure of the total organic and ammonia nitrogen in the wastewater.* TKN gives a measure of the availability of nitrogen for building cells, as well as the potential nitrogenous oxygen demand that will have to be satisfied.

Phosphorus may appear in many forms in wastewater. Among the forms found are the orthophosphates, polyphosphates, and organic phosphate. For our purpose, we will lump all of these together under the heading "Total Phosphorus (as P)."

Three typical compositions of untreated domestic wastewater are summarized in Table 6-1. The pH for all of these wastes will be in the range of 6.5 to 8.5, with a majority being slightly on the alkaline side of 7.0.

Characteristics of Industrial Wastewater

Industrial processes generate a wide variety of wastewater pollutants. The characteristics and levels of pollutants vary significantly from industry to industry. The Environmental Protection Agency (EPA) has grouped the pollutants into three categories: conventional pollutants, nonconventional pollutants, and priority pollutants. The conventional and nonconventional pollutants are listed in Table 6-2. The priority pollutants are listed in Table 1-6.

Because of the wide variety of industries and levels of pollutants, we can only present a snapshot view of the characteristics. A sampling of a few industries for two conventional pollutants is shown in Table 6-3.

---

*Pronounced "kell dall" after J. Kjeldahl, who developed the test in 1883.

TABLE 6-1
Typical composition of untreated domestic wastewater

Constituent	Weak	Medium	Strong
	(all mg/L except settleable solids)		
Alkalinity (as $CaCo_3$)[a]	50	100	200
BOD_5 (as O_2)	100	200	300
Chloride[a]	30	50	100
COD (as O_2)	250	500	1,000
Suspended solids (SS)	100	200	350
Settleable solids, mL/L	5	10	20
Total dissolved solids (TDS)	200	500	1,000
Total Kjeldahl nitrogen (TKN) (as N)	20	40	80
Total organic carbon (TOC) (as C)	75	150	300
Total phosphorus (as P)	5	10	20

[a]To be added to amount in domestic water supply. Chloride is exclusive of contribution from water-softener backwash.

TABLE 6-2
EPA's conventional and nonconventional pollutant categories

Conventional	*Nonconventional*
Biochemical oxygen demand (BOD_5)	Ammonia (as N)
Total suspended solids (TSS)	Chromium VI (hexavalent)
Oil and grease	Chemical oxygen demand (COD)
Oil (animal, vegetable)	COD/BOD_7
Oil (mineral)	Fluoride
pH	Manganese
	Nitrate (as N)
	Organic nitrogen (as N)
	Pesticide active ingredients (PAI)
	Phenols, total
	Phosphorus, total (as P)
	Total organic carbon (TOC)

Source: CFR, 2005a.

TABLE 6-3
Examples of industrial wastewater concentrations for BOD_5 and suspended solids

Industry	BOD_5, mg/L	Suspended solids, mg/L
Ammunition	50–300	70–1,700
Fermentation	4,500	10,000
Slaughterhouse (cattle)	400–2,500	400–1,000
Pulp and paper (kraft)	100–350	75–300
Tannery	700–7,000	4,000–20,000

TABLE 6-4
Examples of industrial wastewater concentrations for nonconventional pollutants

Industry	Pollutant	Concentration, mg/L
Coke by-product (steel mill)	Ammonia (as N)	200
	Organic nitrogen (as N)	100
	Phenol	2,000
Metal plating	Chromium VI	3–550
Nylon polymer	COD	23,000
	TOC	8,800
Plywood-plant glue waste	COD	2,000
	Phenol	200–2,000
	Phosphorus (as PO_4)	9–15

A similar sampling for nonconventional pollutants is shown in Table 6-4.

6-3 WASTEWATER TREATMENT STANDARDS

In Public Law 92-500 (see Section 1-4, Chapter 1), the Congress required municipalities and industries to provide *secondary treatment* before discharging wastewater into natural water bodies. The U.S. Environmental Protection Agency (EPA) established a definition of secondary treatment based on three wastewater characteristics: BOD_5, suspended solids and hydrogen-ion concentration (pH). The definition is summarized in Table 6-5.

TABLE 6-5
U.S. Environmental Protection Agency definition of secondary treatment [a, b]

Characteristic of discharge	Units	Average monthly concentration[c]	Average weekly concentration[c]
BOD_5	mg/L	30[d]	45
Suspended solids	mg/L	30[d]	45
Hydrogen-ion Concentration	pH units	Within the range 6.0–9.0 at all times[e]	
$CBOD_5$[f]	mg/L	25	40

[a]*Source:* CFR, 2005b.
[b]Present standards allow stabilization ponds and trickling filters to have higher 30-day average concentrations (45 mg/L) and 7-day average concentrations (65 mg/L) of BOD and suspended solids as long as the water quality of the receiving body of water is not adversely affected. Other exceptions are also permitted. The CFR and the *NPDES Permit Writers' Manual* (U.S. EPA, 1996) should be consulted for details on the exceptions.
[c]Not to be exceeded.
[d]Average removal shall not be less than 85 percent.
[e]Only enforced if caused by industrial wastewater or by in-plant inorganic chemical addition.
[f]May be substituted for BOD_5 at the option of the permitting authority.

PL 92-500 also directed that the EPA establish a permit system called the National Pollutant Discharge Elimination System (NPDES). Under the NPDES program, all facilities that discharge pollutants from any point source into waters of the United States are required to obtain a NPDES permit. Although some states elected to have EPA administer their permit system, most states administer their own program. Before a permit is granted the administering agency will model the response of the receiving body to the proposed discharge to determine if the receiving body is adversely affected (for an example of modeling, see Section 5-3). The permit may require lower concentrations than those specified in Table 6-5 to maintain the quality of the receiving body of water.

In addition, the states may impose additional conditions in the NPDES permit. For example, in Michigan, a limit of 1 mg/L of phosphorus is contained in permits for discharges to surface waters that do not have substantial problems with high levels of nutrients. More stringent limits are required for discharges to surface waters that are very sensitive to nutrients.

$CBOD_5$ limits are placed in the NPDES permits for all facilities that have the potential to contribute significant quantities of oxygen-consuming substances. The nitrogenous oxygen demand from ammonia nitrogen is typically the oxygen demand of concern from municipal discharges (see Section 5-3). It is computed separately from the $CBOD_5$ and then combined to establish a discharge limit. Ammonia is also evaluated for its potential toxicity to the stream's biota.

Bacterial effluent limits may also be included in the NPDES permit. For example, municipal wastewater treatment plants in Michigan must comply with limits of 200 fecal coliform bacteria (FC) per 100 mL of water as a monthly average and 400 FC/100 mL as a 7-day average. More stringent requirements are imposed to protect waters that are used for recreation. Total-body-contact recreation waters must meet limits of 130 *Escherichia coli* per 100 mL of water as a 30-day average and 300 *E. coli* per 100 mL at any time. Partial-body-contact recreation is permitted for water with less than 1,000 *E. coli* per 100 mL of water.

For thermal discharges such as cooling water, temperature limits may be included in the permit. Michigan rules state that the Great Lakes and connecting waters and inland lakes shall not receive a heat load that increases the temperature of the receiving water more than 1.7°C above the existing natural water temperature after mixing. For rivers, streams, and impoundments the temperature limits are 1°C for cold-water fisheries and 2.8°C for warm-water fisheries. (See Section 2-4 for a discussion of energy balances and Problems 2-36, 2-37, and 2-40 for typical problems in thermal discharge analysis.)

An example of NPDES limits is shown in Table 6-6.* Note that in addition to concentration limits, mass discharge limits are also established.

Pretreatment of Industrial Wastes

Industrial wastewaters can pose serious hazards to municipal systems because the collection and treatment systems have not been designed to carry or treat them. The

*This table outlines only the quantitative limits. The entire permit is 22 pages long.

TABLE 6-6
NPDES Limits for the city of Hailey, Idaho[a, b]

Parameter	Average monthly limit	Average weekly limit	Instantaneous maximum limit
BOD_5	30 mg/L 43 kg/d	45 mg/L 64 kg/d	N/A
Suspended solids	30 mg/L 43 kg/d	45 mg/L 64 kg/d	N/A
E. coli bacteria	126/100 mL	N/A	406/100 mL
Fecal coliform bacteria	N/A	200 colonies/100 mL	N/A
Total ammonia as N	1.9 mg/L 4.1 kg/d	2.9 mg/L 6.4 kg/d	3.3 mg/L 7.1 kg/d
Total phosphorus	6.8 kg/d	10.4 kg/d	N/A
Total Kjeldahl nitrogen	25 kg/d	35 kg/d	N/A

[a]*Source:* U.S. EPA, 2005.
[b]Renewal announcement, 7 February 2001.

wastes can damage sewers and interfere with the operation of treatment plants. They may pass through the wastewater treatment plant (WWTP) untreated or they may concentrate in the sludge, rendering it a hazardous waste.

The Clean Water Act gives the EPA the authority to establish and enforce pretreatment standards for discharge of industrial wastewaters into municipal treatment systems. Specific objectives of the pretreatment program are:

- To prevent the introduction of pollutants into WWTPs that will interfere with their operation, including interference with their use or with disposal of municipal sludge.
- To prevent the introduction of pollutants to WWTPs that will pass through the treatment works or otherwise be incompatible with such works.
- To improve opportunities to recycle and reclaim municipal and industrial wastewaters and sludge.

EPA has established "prohibited discharge standards" (40 CFR 403.5) that apply to all nondomestic discharges to the WWTP and "categorical pretreatment standards" that are applicable to specific industries (40 CFR 405-471). Congress assigned the primary responsibility for enforcing these standards to local WWTPs.

In the General Pretreatment Regulations, industrial users (IUs) are prohibited from introducing the following into a WWTP:

1. Pollutants that create a fire or explosion hazard in the municipal WWTP, including, but not limited to, waste streams with a closed-cup flash point of less than or equal to 60°C, using the test methods specified in 40 CFR 261.21.

2. Pollutants that will cause corrosive structural damage to the municipal WWTP (but in no case discharges with a pH lower than 5.0) unless the WWTP is specifically designed to accommodate such discharges.
3. Solid or viscous pollutants in amounts that will cause obstruction to the flow in the WWTP resulting in interference.
4. Any pollutant, including oxygen-demanding pollutants (such as BOD), released in a discharge at a flow rate and/or concentration that will cause interference with the WWTP.
5. Heat in amounts that will inhibit biological activity in the WWTP and result in interference, but in no case heat in such quantities that the temperature at the WWTP exceeds 40°C unless the approval authority, on request of the publicly owned treatment works (POTW), approves alternative temperature limits.
6. Petroleum oil, non-biodegradable cutting oil, or products of mineral oil origin in amounts that will cause interference or will pass through.
7. Pollutants that result in the presence of toxic gases, vapors, or fumes within the POTW in a quantity that may cause acute worker health and safety problems.
8. Any trucked or hauled pollutants, except at discharge points designated by the POTW.

6-4 ON-SITE DISPOSAL SYSTEMS

In less densely populated areas where lot sizes are large and houses are spaced widely apart, it is often more economical to treat human waste on site, rather than use a sewer system to collect the waste and treat it at a centralized location. On-site systems are generally small and may serve individual homes, small housing developments (clusters), or isolated commercial establishments, such as small hotels or restaurants. In the United States, about 25 percent of the population is served by on-site wastewater systems. In some states as much as 50 percent of the population utilize on-site systems within rural and suburban communities (U.S. EPA, 1997). As many people choose to move to rural and outer suburban areas, the number of decentralized systems is increasing. It is estimated that as much as 40 percent of new housing construction is taking place in areas that are not connected to municipal sewers.

Alternative On-Site Treatment and Disposal Systems with Water

Septic Tanks and Absorption Fields. About 85 to 90 percent of on-site wastewater disposal systems are conventional septic systems. A conventional septic system consists of three parts: the *septic tank*, a distribution box, and an absorption field (also called a leach or drain or tile field) (see Figure 6-1). The septic tank and tile field are a unit. Neither part will function as intended without the other.

The main function of the septic tank (Figure 6-2) is to remove large particles and grease that would otherwise clog the tile field. Heavy solids settle to the bottom, where

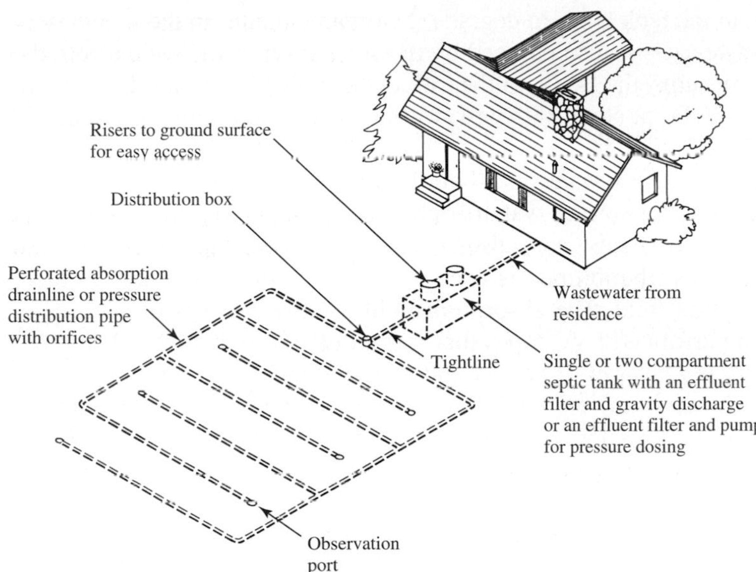

FIGURE 6-1
Schematic of a conventional septic system. (*Source:* Crites and Tchobanoglous, 1998.)

they undergo biological decomposition. Grease floats to the surface and is trapped. It is only slightly decomposed.

The size of the tank depends on the expected wastewater flow. The tank should be large enough that the holding time for water in a septic tank is a least 24 hours. For individual homes the following guidelines can be used: 3 m^3 tank minimum; 4 m^3 for a three-bedroom house; 5 m^3 for a four-bedroom house and 6 m^3 for a five-bedroom house.

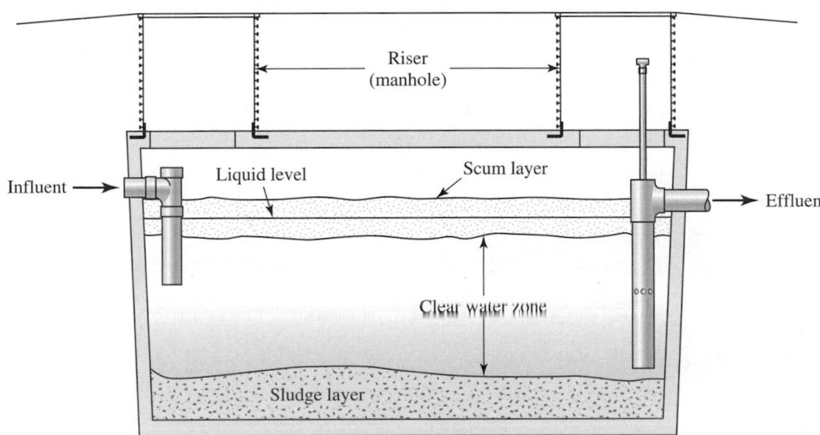

FIGURE 6-2
Definition sketch for the sludge, clear water, and scum zones that form in a septic tank. (*Source:* Crites and Tchobanoglous, 1998.)

Bacterial action in the tank helps to degrade the organic matter in the wastewater. The BOD_5 of the wastewater is also reduced by the separation of the solids from the liquid. For domestic systems, the BOD_5 of the influent is typically 210 mg/L. The septic tank effluent has a BOD_5 of about 180 mg/L without an effluent filter and about 130 mg/L if an effluent filter is used in the system. Usually the BOD_5 limit to allow wastewater to be discharged to surface waters is 20 mg/L or less, so the BOD_5 of a septic effluent is too high to allow for surface water discharge. However, as further treatment occurs in the soil, absorption fields can be used to safely dispose of the partially treated septic tank effluent. A distribution box is used to distribute the septic tank effluent throughout the absorption field. The absorption field usually consists of a series of trenches that contain perforated PVC pipes that are about 10 cm diameter. The pipes placed over a 15-cm-deep layer of drainrock and then buried with an additional layer of drainrock. The drainrock is covered building fabric or building paper (which helps to prevent the migration of fines into the drainrock) and finally the trench is filled with native soil (Figure 6-3). The trenches should be separated by a distance of at least 2 meters.

When the system is operational, bacteria produce a slime layer at the bottom of the trench. This layer is commonly called the clogging mat. The clogging mat creates a barrier that slows the movement of water into the surrounding soil. This allows the flow in the surrounding soil to remain unsaturated, which allows air to move through the soil. This maintains aerobic conditions, which are essential to obtain proper treatment of the effluent.

The size of the absorption area depends upon the wastewater flow and the permeability of the surrounding soil. The permeability of the soil is determined by a percolation (or *perc*) test. The perc test is conducted by digging a hole of a prescribed size, filling it with water and measuring the rate at which the water percolates into the soil. An alternative, and preferred, method for determining the suitability of the soil is to dig a trench in the area proposed for the tile field and visually inspect it. The inspector looks for unsuitable soil (clay, for example) and the presence of mottled (discolored) soil. Mottled soil indicates that the groundwater table has, at some point in time, risen to a level which would interfere with the operation of the tile field and, perhaps more

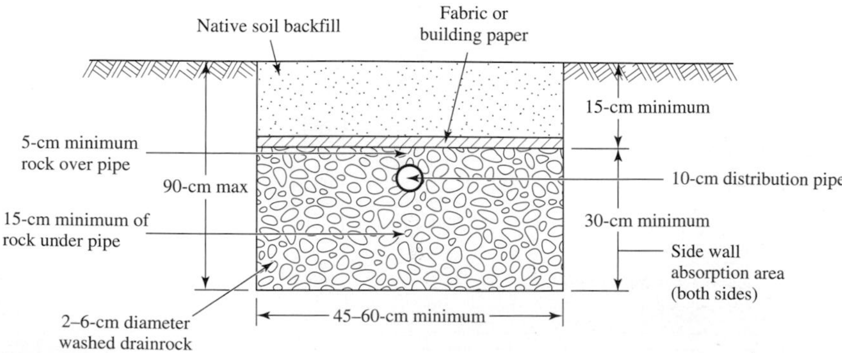

FIGURE 6-3

Typical cross section through conventional absorption trench. (*Source:* Crites and Tchobanoglous, 1998.)

TABLE 6-7
Maximum acceptable application rates for tile fields

Soil texture and structure	Percolation rate		Maximum acceptable application rate, m^3 of vol/m^2 area
	mm/h	min/mm	
Coarse and medium sand	⩾150	<0.40	0.04
Fine and loamy sand	75–150	0.40–0.80	0.03
Sandy loam	50–75	0.80–1.20	0.02
Loam and sandy clay	35–50	1.20–1.71	0.01
Loams	<35	>1.71	Not permitted
Clays, silts, muck, peat, marl	≪35	≫1.71	Not permitted

important, bring the groundwater into direct contact with sewage. The information in Table 6-7 is then used to determine the size of the tile field.

Further limitations on the use of a septic-tank tile-field system usually include the following:

1. The tile field must be located more than 30 m from any well, surface water, footing drain, or storm drain.
2. The tile field must be located at least 3 m from any property line.
3. The minimum distance between the bottom of the absorption trench and the groundwater table or any impermeable layer must not be less than 1.25 m.
4. The earth cover placed over the absorption tile must not be less than 0.3 m nor more than 0.6 m deep.
5. A clean aggregate graded between 12 and 36 mm must be placed around the tile pipe. It must be a minimum of 50 mm above the pipe and 150 mm below the pipe, with a total depth of not less than 300 mm.

Most states limit septic tank/tile field installation to facilities producing less than 40 m^3/d of wastewater. This limits their use to single-family residences, small apartments, freeway rest areas, parks, and isolated commercial establishments.

Example 6-1. Peter and Pam Piper are considering the purchase of a plot of land on which to build a retirement home. According to their water bills for the past 5 years, their average daily water consumption is about 0.4 m^3. What size septic tank and tile field should they expect to put on the lot if it perks at 1.00 min/mm?

Solution. If the septic tank must provide a detention time of 24 h, then its volume should be

$$V = 0.4 \text{ m}^3/\text{d} \times 1 \text{ d} = 0.4 \text{ m}^3$$

However, the minimum recommended volume is 4.0 m^3. Good septic tank design practice calls for length to width ($1/w$) ratios greater than 2 to 1 and a minimum liquid depth of about 1.2 m. For these criteria and a 4.0 m^3 volume, the liquid surface area would be

$$A_s = \frac{4.0 \text{ m}^3}{1.2 \text{ m}} = 3.33 \text{ m}^2$$

If we choose a width of 1.15 m and a length of 3 m, we will have a well-sized tank of 4.14 m^3 and a l/w ratio of 2.61 to 1.

From Table 6-7 we find that a perk rate of 1.00 min/mm will allow an application rate of 0.02 m^3/m^2 of trench. The bottom area of trench should then be about

$$\text{A} = \frac{0.4 \text{ m}^3}{0.02 \text{ m}^3/\text{m}^2} = 20.0 \text{ m}^2$$

One trench 1.0 m wide and 20.0 m long would meet the requirements; however, our preference is to use a 0.6-m trench width and three trenches about 12 m in length.

Most septic systems will fail eventually. The normal lifetime of an absorption field is 20 to 30 years. After that time the soil around the field becomes clogged with organic matter and the system will not operate properly. Many factors can cause the system to fail prematurely. Roots can block pipes or the pipes may be crushed if a vehicle is driven over the field. The system may also fail if the absorption field is hydraulically overloaded or if substances that are toxic to soil bacteria, such as solvents, paints, pesticides or softener salt are disposed of down the drain. However, the most common reason for premature failure is improper maintenance. Because the rate of biodegradation in septic tanks is slow, the solids that settle in the tank tend to accumulate over time. If these solids are not removed, the clear water zone between the sludge layer and the scum layer becomes too small. This leads to an increased carryover of solids to the absorption field. If too many solids reach the absorption field, then it can become clogged, resulting in premature failure of the field. To prevent the accumulation of too much sludge and scum in a septic tank, they should be periodically removed. The rate of accumulation of sludge depends on the usage of the system. It is suggested that the level of sludge in the tank be checked annually, though usually a domestic system will need to be pumped out only once every 2 or 3 years.

Septic Tank and Absorption Field Modifications. As mentioned above, often the reason for the failure of absorption fields in poorly drained soils is that the growth of the clogging mat is excessive. Reduction of the BOD of the wastewater can reduce the rate of growth of the mat. Two types of treatment systems that are commonly used to reduce the wastewater BOD are aerobic treatment systems and sand filters.

Aerobic Systems. A wide range of aerobic treatment systems are available. The common feature of these units is that they use some mechanism to inject or circulate air inside the treatment tank. If sufficient air is introduced, then aerobic conditions can be achieved in the wastewater. As aerobic degradation is rapid, good removal of BOD can be achieved under these conditions. This reduces the rate of growth of the clogging mat and extends the life of the absorption field.

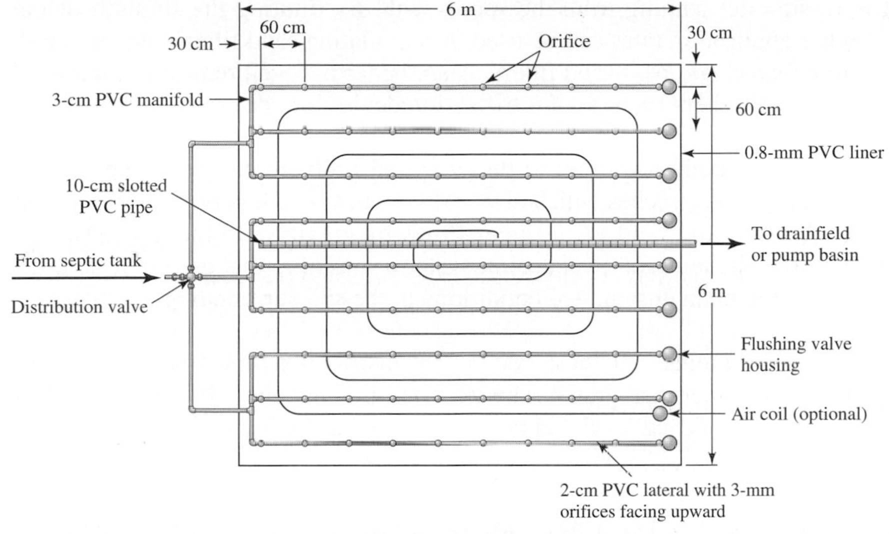

(*a*) Plan view

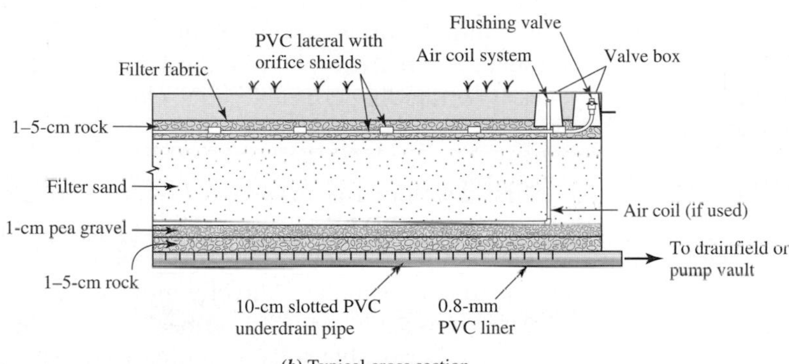

(*b*) Typical cross section

FIGURE 6-4
Schematic of modern intermittent sand filter: (*a*) plan view and (*b*) typical cross section. (Courtesy Orenco Systems, Inc.)

Sand Filter Systems. The sand filter is also an aerobic treatment system. The components of a typical sand filter are illustrated in Figure 6-4. The filter consists of a bed of granular material (usually sand, but other materials such as anthracite can be used). The surface of the bed is intermittently dosed with wastewater that percolates through the sand to the bottom of the filter. The sand bed is dosed anywhere from 12 to 72 times per day. The size of the dose should be such that the sand bed does not become saturated. This allows the wastewater to flow as a thin film around the sand particles, so good contact between the wastewater and the air can be achieved. Sand filters may be single-pass or multipass. Single-pass sand filters are commonly called intermittent sand filters (ISFs). In a single-pass system the wastewater is collected in the underdrain and passed onto an absorption field or other disposal system. In a multipass system a portion of the treated wastewater is recycled back through the sand bed. Recirculation dilutes the wastewater coming from the septic tank. By diluting the strength of the effluent, higher application rates can be used. Recirculating sand filters take up 3 to 5 times less area than single-pass sand filters. Also, better nitrogen removal is achieved in recirculating sand filters because of nitrification/denitrification (see Chapter 5).

Dosing Systems. Another solution to the clogging problem can be to replace the conventional absorption trenches with a disposal system that is less prone to failure. In a conventional absorption field the effluent flows by gravity into the trenches. The gravity system may be replaced by a dosing system, similar to that used in a sand filter. This helps to maintain unsaturated conditions in the soil surrounding the trench.

Shallow Absorption Fields. In these systems the distribution piping is often covered with large half pipe rather than gravel. The trenches are only about 0.25 m deep. Better treatment can be achieved, as the upper soil layers have a higher concentration of microbes.

On-Site Treatment/Disposal Systems for Unfavorable Site Conditions

Where site conditions are unfavorable for conventional septic systems, alternative treatment/disposal systems may be required. Among the limitations that might preclude the installation of a conventional system are: high groundwater tables, shallow limiting layers of bed rock, very slowly or rapidly permeable soils, close proximity to surface water, and small lot size.

Mound Systems. Mounds were first developed at the North Dakota Agricultural College in the 1940s. They were known as NODAK systems. The components of a typical mound system are illustrated in Figure 6-5. Mounds are both treatment and disposal systems, as the effluent from the mound percolates directly into the native soil. The design overcomes certain site restrictions such as slowly permeable soils, shallow permeable soils over porous bedrock, and permeable soils with high water tables. A mound system is a pressure-dosed absorption system that is elevated above the natural soil surface. Effluent from the septic tank is pumped or siphoned to the elevated

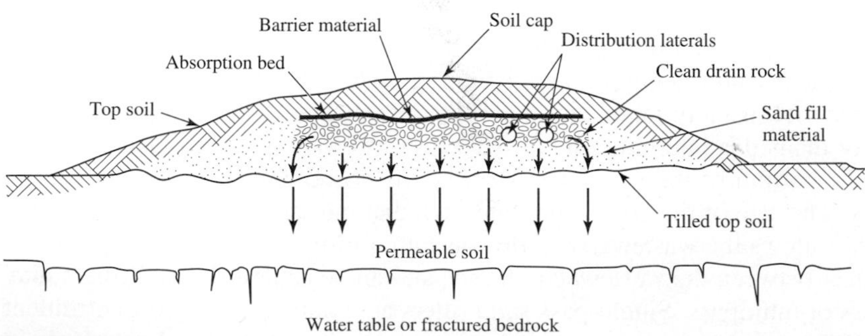

FIGURE 6-5
Typical cross section through mound effluent disposal system. (*Source:* Crites and Tchobanoglous, 1998.)

absorption area and distributed through a piping network located in the coarse aggregate at the top of the mound. The effluent then passes through the aggregate and infiltrates the sand fill. Treatment occurs in the sand and in the fill below the sand bed. As the water percolates downward, it spreads out over a large area. The size of the mound must be such that area of native soil under the mound, called the *basal area,* is large enough that wastewater does not seep out of the base or sides of the mound. During construction of mound systems, special attention should be given to ensuring that the basal area of the system is properly scarified, and that compaction of the basal area by earthmoving equipment is minimal. Compaction of the top layer of the soil can greatly reduce the rate of infiltration into the soil.

Barriered-Landscape Water-Renovation System (BLWRS). In the summer of 1969, Dr. A. Earl Erickson demonstrated the efficacy of utilizing a BLWRS (pronounced "blowers," like "flowers") to denitrify water containing 100 mg/L of nitrate. Subsequently, he and his associates demonstrated that the BLWRS could be used to renovate both dairy cow and swine feedlot wastewater (Table 6-8) (Erickson et al., 1974). The system is, of course, equally applicable to domestic wastewater.

The BLWRS differs from the NODAK mound system in that the mound of soil is underlain by an impervious water barrier (Figure 6-6*a* and *b*). As the renovated water passes beyond the edge of the barrier, it may be collected in drains or be allowed to recharge the aquifer. The mound is constructed of a fine sand. The dimensions of the BLWRS depend on the soil texture and expected wastewater application rates. A 0.15-m layer of topsoil is used to cover the sand. A water-hardy grass (quack grass or volunteer weed cover) must be established on the surface and banks to maintain the soil's permeability and stability.

TABLE 6-8
BLWRS wastewater renovation efficiencies

	Average influent concentration, mg/L	Average effluent concentration, mg/L	Efficiency, %
Swine waste[a]			
BOD_5	1,131	8.9	98.3
P	18	0.02	99.9
Suspended solids	3,000	Nil	~100.0
TKN	937	187.4	80.0
Dairy waste[b]			
BOD_5	1,637.0	18.9	98.8
P	38.5	0.22	99.4
Suspended solids	4,400.0	Nil	~100
TKN	917.0	27.5	97.0

[a]Average application rate of 15 mm/d for 503 d.
[b]Average application rate of 8.8 mm/d for 450 d.
Source: Erickson et al., 1974.

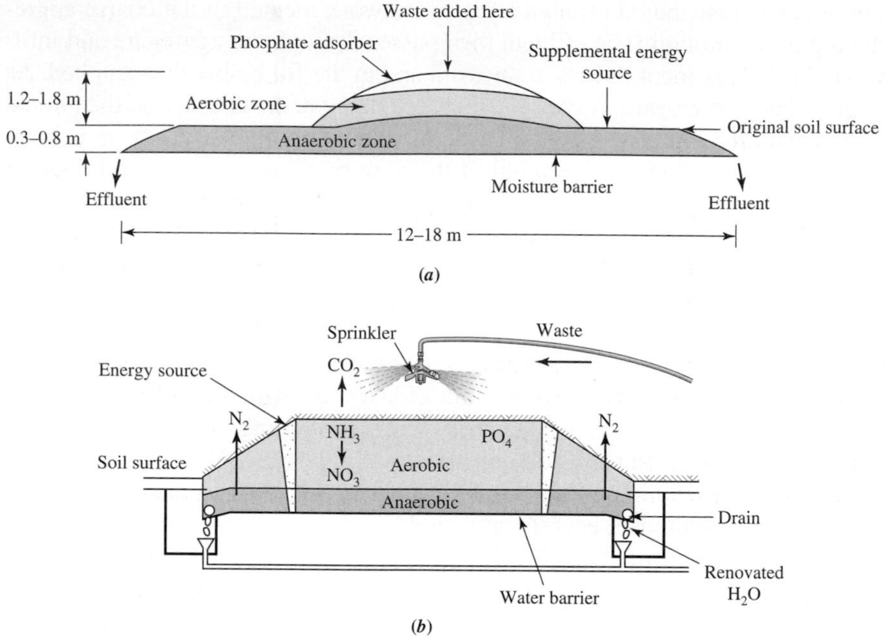

FIGURE 6-6
(*a*) Common dimensions of barriered landscape water-renovation system (BLWRS); (*b*) water chemistry change in a BLWRS.

The wastewater is spread on the top of the mound by a sprinkler. As the wastewater percolates down, the organic particles are filtered out and remain on the surface. The particles are oxidized by soil microorganisms. The soluble organic compounds and other ions move into the aerobic soil zone. Most of the soluble organic matter is oxidized by bacteria in the highly active aerobic soil. The phosphate ions are held on the clay fraction of the soil and sand bed. (Iron slag and/or limestone can be used to enhance the phosphorus adsorption capacity.) The ammonium ions are held on the soil until they are nitrified to nitrate. The downward movement of the nitrified water is stopped by the barrier. The water then is forced to move laterally through the anoxic layer. Denitrification occurs as the waste passes out of the carbon source.

The BLWRS must be operated in a cyclic fashion to allow the soil microorganisms time to degrade the waste and to maintain aerobic conditions in the soil. Application rates between 9 and 18 mm of wastewater per day may be used, provided that the BLWRS is "rested" for one-third of the time. The physical conditions of the soil govern the application rates. Ponding on the surface indicates excessive application rates.

Other On-Site Treatment/Disposal Options

Constructed wetlands can be used for on-site wastewater treatment/disposal. The use of these systems is more common in warmer climates. In arid areas, evapotranspiration (ET) beds are an alternative to conventional absorption beds. In an evapotranspiration system, water tolerant vegetation is planted in a shallow sand bed. The plant roots draw the water up through the sand and it is evaporated or transpired to the atmosphere.

Treated domestic wastewater can be reused. However, because of the risks posed by pathogens in the water, the reuse of domestic wastewater in on-site systems is not common. Some alternatives for reuse are drip irrigation and toilet flushing.

Alternative On-Site Treatment/Disposal Systems without Water

In areas away from population centers, such as national or state parks, remote roadside rest areas or vacation, there may be no reliable water supply. The absence of a water supply or water scarcity may preclude the use of flush toilets. In this case, other systems for the disposal of human waste need to be considered. Commonly used systems are the pit privy and vault, chemical and composting toilets. Vault, chemical, and composting toilets are closed systems, so there is "zero" discharge from these systems on site. The waste produced is collected and disposed of elsewhere. For this reason, these systems may also be used in environmentally sensitive areas where the discharge of wastewater may be environmentally unacceptable.

The Pit Privy. Although most modern environmental engineering and science texts would skip this subject, the mere existence of 10,000 of these or their modern equivalent in the United States is just too much for us to ignore. Furthermore, the facts of the matter are that junior engineers and environmental scientists are the most likely candidates for designing, erecting, operating, dismantling, and closing the beasts.

Figure 6-7 provides most of the information you will ever want to know about the construction of an outhouse. The slab is usually poured over flat ground on top of roofing paper. The riser hole is formed with 12-gauge galvanized iron. Once the slab has set, it is lifted into place over the pit. The concrete is a 1:2:3 mix, that is, one part portland cement, two parts sand, and three parts gravel less than 25 mm in diameter.

The principle of operation of the pit privy is that the liquid materials percolate into the soil through the cribbing and the solids "dry out." A pit of the dimensions shown in Figure 6-7 should last a family of four about 10 years. Rainwater is to be prevented from entering the pit. A cup of kerosene at weekly intervals discourages mosquito breeding, and odors can be reduced by the use of a cup of hydrated lime. Unfortunately, the lime also slows the decomposition of paper, so its use is not encouraged. Disinfectants should never be used.

The Vault Toilet. This is the modern version of the pit privy. Its construction is the same as that of the pit privy with the exception that the pit is formed as a watertight vault. A special truck (fondly called a "honey wagon") is used to pump out the vault at regular intervals. Because of the liquefying action of the bacteria and biological decomposition in the liquid (rather than in the soil as occurs with the pit privy), vault toilets are much more odiferous than the old pit privies. Many masking agents (perfumes) and disinfectants are available to mitigate the stench. Unfortunately, most of them have unpleasant odors themselves. If electricity is at hand, an ozone generator, set to vent into the gas space above the waste, will perform near-miracles in odor reduction.

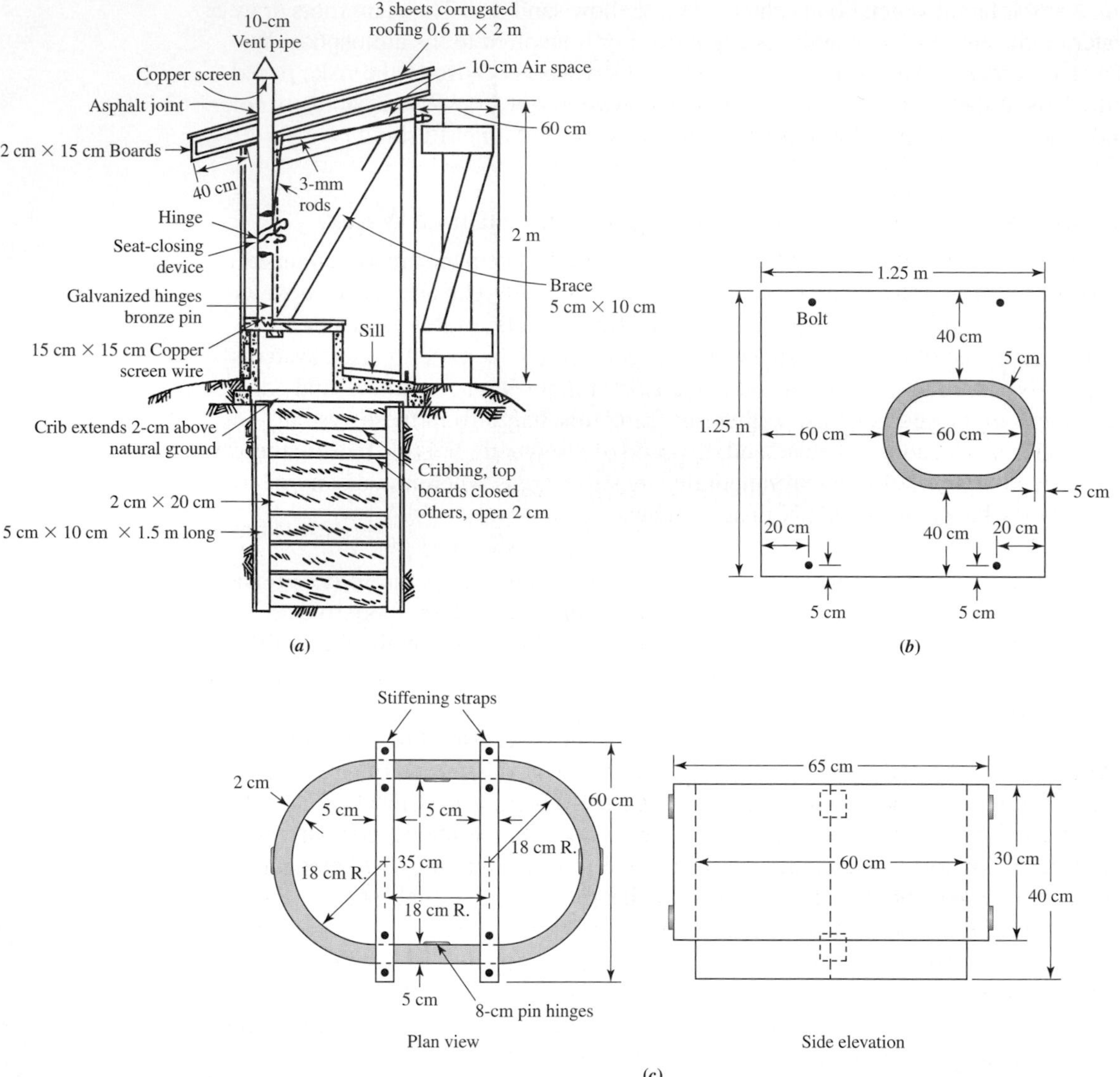

FIGURE 6-7
Construction details of the pit privy: (*a*) cross section; (*b*) plan of concrete slab; and (*c*) details of riser form. (*Source:* Ehlers, V. M., and E. W. Steel, 1943.)

The Chemical Toilet. The airplane toilet, the coach-bus toilet, and the self-contained toilets of recreation vehicles are all versions of the chemical toilet. The essence of the system is a strong disinfectant chemical used to carry the waste to a holding tank and render it inoffensive until it can be pumped from the holding tank. While these vehicular systems are quite effective, the chemical must be selected with an eye toward its impact on the treatment system that ultimately must receive it. The chemical toilet has not found wide acceptance in permanent installations. This is because of the cost of the chemical and the impracticality of maintenance.

The Composting Toilet. A composting toilet consists of a large tank located directly below the toilet room. Wastes enter the tank through a large-diameter chute that connects to the toilet. No water is used for the toilet, but a bulking agent (such as wood shavings) is added to improve liquid drainage and aeration. A small fan draws air through the tank and up the vent pipe to ensure adequate oxygen for decomposition and odorless operation. The liquid in the waste is evaporated, leaving a compost. The finished compost can be removed from the lower end of the tank about once each year. It can be used as a fertilizer. Power requirements for the system are very low, so if power from an electrical grid is not available, the electrical requirements can be met from an independent generating system, such as a photovoltaic system.

6-5 MUNICIPAL WASTEWATER TREATMENT SYSTEMS

The alternatives for municipal wastewater treatment fall into three major categories (Figure 6-8): (1) primary treatment, (2) secondary treatment, and (3) advanced treatment. It is commonly assumed that each of the "degrees of treatment" noted in Figure 6-8 includes the previous steps. For example, primary treatment is assumed to include the pretreatment processes: bar rack, grit chamber, and equalization basin. Likewise, secondary treatment is assumed to include all the processes of primary treatment: bar rack, grit chamber, equalization basin, and primary settling tank.

The purpose of pretreatment is to provide protection to the wastewater treatment plant (WWTP) equipment that follows. In some older municipal plants the equalization step may not be included.

The major goal of primary treatment is to remove from wastewater those pollutants that will either settle or float. Primary treatment will typically remove about 60 percent of the suspended solids in raw sewage and 35 percent of the BOD_5. Soluble pollutants are not removed. At one time, this was the only treatment used by many cities. Although primary treatment alone is no longer acceptable, it is still frequently used as the first treatment step in a secondary-treatment system. The major goal of secondary treatment is to remove the soluble BOD_5 that escapes the primary process and to provide added removal of suspended solids. Secondary treatment is typically achieved by using biological processes. These provide the same biological reactions that would occur in the receiving water if it had adequate capacity to assimilate the wastewater. The secondary treatment processes are designed to speed up these natural processes so that the breakdown of the degradable organic pollutants can be achieved in relatively short time periods. Although secondary treatment may remove more than 85 percent of the BOD_5 and suspended solids, it does not remove significant amounts of nitrogen, phosphorus, or heavy metals, nor does it completely remove pathogenic bacteria and viruses.

In cases where secondary levels of treatment are not adequate, additional treatment processes are applied to the secondary effluent to provide advanced wastewater treatment (AWT). These processes may involve chemical treatment and filtration of the wastewater—much like adding a typical water treatment plant to the tail end of a secondary plant—or they may involve applying the secondary effluent to the land in carefully designed irrigation systems where the pollutants are removed by a soil-crop system. Some of these processes can remove as much as 99 percent of the BOD_5,

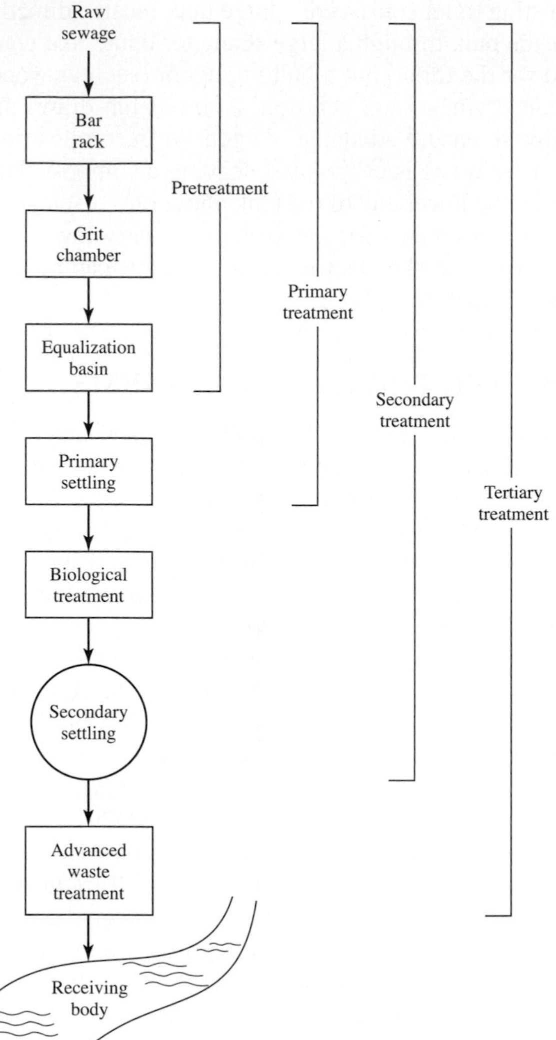

FIGURE 6-8
Degrees of treatment.

phosphorus, suspended solids and bacteria, and 95 percent of the nitrogen. They can produce a sparkling clean, colorless, odorless effluent indistinguishable in appearance from a high-quality drinking water. Although these processes and land treatment systems are often applied to secondary effluent for advanced treatment, they have also been used in place of conventional secondary treatment processes.

Most of the impurities removed from the wastewater do not simply vanish. Some organic compounds are broken down into harmless carbon dioxide and water. Most of the impurities are removed from the wastewater as a solid, that is, sludge. Because most of the impurities removed from the wastewater are present in the sludge, sludge handling and disposal must be carried out carefully to achieve satisfactory pollution control.

6-6 UNIT OPERATIONS OF PRETREATMENT

Several devices and structures are placed upstream of the primary treatment operation to provide protection to the wastewater treatment plant (WWTP) equipment. These devices and structures are classified as pretreatment because they have little effect in reducing BOD_5. In industrial WWTPs where only soluble compounds are present, bar racks and grit chambers may be absent. Equalization is frequently required in industrial WWTPs.

Bar Racks

Typically, the first device encountered by the wastewater entering the plant is a bar rack (Figure 6-9). The primary purpose of the rack is to remove large objects that would damage or foul pumps, valves, and other mechanical equipment. Rags, logs, and other objects that find their way into the sewer are removed from the wastewater on the racks. In modern WWTPs, the racks are cleaned mechanically. The solid material is stored in a hopper and removed to a sanitary landfill at regular intervals.

Bar racks (or bar screens) may be categorized as trash racks, manually cleaned racks, and mechanically cleaned racks. Trash racks are those with large openings, 40 to 150 mm, that are designed to prevent very large objects such as logs from entering the plant. These are normally followed by racks with smaller openings. Manually cleaned racks have openings that range from 25 to 50 mm. Channel approach velocities are designed to be in the range of 0.3 to 0.6 m/s. As mentioned above, manually cleaned racks are not frequently employed. They do find application in bypass channels

FIGURE 6-9
Bar rack. (Courtesy of Siemens Water Technologies.)

that are infrequently used. Mechanically cleaned racks have openings ranging from 15 to 75 mm. Maximum channel approach velocities range from 0.6 to 1.0 m/s. Minimum velocities of 0.3 to 0.5 m/s are necessary to prevent grit accumulation. Regardless of the type of rack, two channels with racks are provided to allow one to be taken out of service for cleaning and repair.

Grit Chambers

Inert dense material, such as sand, broken glass, silt, and pebbles, is called *grit*. If these materials are not removed from the wastewater, they abrade pumps and other mechanical devices, causing undue wear. In addition, they have a tendency to settle in corners and bends, reducing flow capacity and, ultimately, clogging pipes and channels.

There are three basic types of grit-removal devices: velocity controlled, aerated, and constant-level short-term sedimentation basins. We will discuss only the first two, since they are the most common.

Velocity Controlled. This type of grit chamber, also known as a *horizontal-flow* grit chamber, can be analyzed by means of the classical laws of sedimentation for discrete, nonflocculating particles (Type I sedimentation).

Stokes' law (See Section 4-5) may be used for the analysis and design of horizontal-flow grit chambers if the horizontal liquid velocity is maintained at about 0.3 m/s. Liquid velocity control is achieved by placing a specially designed weir at the end of the channel. A minimum of two channels must be employed so that one can be out of service without shutting down the treatment plant. Cleaning may be either by mechanical devices or by hand. Mechanical cleaning is favored for plants having average flows over 0.04 m^3/s. Theoretical detention times are set at about one minute for average flows. Washing facilities are normally provided to remove organic material from the grit.

Example 6-2. Will a grit particle with a radius of 0.10 mm and a specific gravity of 2.65 be collected in a horizontal grit chamber that is 13.5 m in length if the average grit-chamber flow is 0.15 m^3/s, the width of the chamber is 0.56 m, and the horizontal velocity is 0.25 m/s? The wastewater temperature is 22°C.

Solution. Before we can calculate the terminal settling velocity of the particle, we must gather some information from Table A-1 in Appendix A. At a wastewater temperature of 22°C, we find the water density to be 997.774 kg/m^3. We will use 1,000 kg/m^3 as a sufficiently close approximation. Since the particle radius is given to only two significant figures, this approximation is reasonable. From the same table, we find the viscosity to be 0.995 mPa · s. As noted in the footnote, we must multiply this by 10^{-3} to obtain the viscosity in units of Pa · s. Using a particle diameter of 0.20×10^{-3} m, we can calculate the terminal settling velocity using Equation 4-98.

$$v_s = \frac{g(\rho_s - \rho)d^2}{18\mu}$$

$$v_s = \frac{9.80(2{,}650 - 1{,}000)(0.20 \times 10^{-3})^2\,(\text{m/s}^2)(\text{kg/m}^3)(\text{m}^2)}{18(0.000995)(\text{kg} \cdot \text{m/m}^2 \cdot \text{s}^2)(\text{s})}$$

$$v_s = 3.61 \times 10^{-2} \text{ m/s or about 36 mm/s}$$

Note that the product of the specific gravity of the particle (2.65) and the density of water is the density of the particle (ρ_s). The Reynolds number for this settling velocity (3.61×10^{-2} m/s) and particle size is 7.54. This is in the transition region and Stokes' law is not valid. This value of **R** (7.54) is used as a first approximation to iteratively solve Equations 4-97 and 4-94 for v_s. From the iterative solution $v_s = 0.028$ m/s.

With a flow of 0.15 m^3/s and a horizontal velocity of 0.25 m/s, the cross-sectional area of flow may be estimated to be

$$A_c = \frac{0.15 \text{ m}^3/\text{s}}{0.25 \text{ m/s}} = 0.60 \text{ m}^2$$

The depth of flow is then estimated by dividing the cross-sectional area by the width of the channel.

$$h = \frac{0.60 \text{ m}^2}{0.56 \text{ m}} = 1.07 \text{ m}$$

If the grit particle in question enters the grit chamber at the liquid surface, it will take h/v_s seconds to reach the bottom.

$$t = \frac{1.07 \text{ m}}{0.028 \text{ m/s}} = 38.2 \text{ s}$$

Since the chamber is 13.5 m in length and the horizontal velocity is 0.25 m/s, the liquid remains in the chamber.

$$t = \frac{13.5 \text{ m}}{0.025 \text{ m/s}} = 54 \text{ s}$$

Thus, the particle will be captured in the grit chamber.

Aerated Grit Chambers. The spiral roll of the aerated grit chamber liquid "drives" the grit into a hopper which is located under the air diffuser assembly (Figure 6-10). The shearing action of the air bubbles is supposed to strip the inert grit of much of the organic material that adheres to its surface.

Aerated grit chamber performance is a function of the roll velocity and detention time. The roll velocity is controlled by adjusting the air feed rate. Nominal air flow values are in the range of 0.2 to 0.5 cubic meters per minute of air per meter of tank length ($m^3/min \cdot m$). Liquid detention times are usually set to be about three minutes at maximum flow. Length-to-width ratios range from 3:1 to 5:1 with depths on the order of 2 to 5 m.

Grit accumulation in the chamber varies greatly, depending on whether the sewer system is a combined type or a separate type, and on the efficiency of the chamber. For combined systems, 90 m^3 of grit per million cubic meters of sewage ($m^3/10^6$ m^3) is not uncommon. In separate systems you might expect something less than 40 $m^3/10^6$ m^3. Normally the grit is buried in a sanitary landfill.

Comminutors

Devices that are used to macerate wastewater solids (rags, paper, plastic, and other materials) by revolving cutting bars are called *comminutors*. Comminutors are most

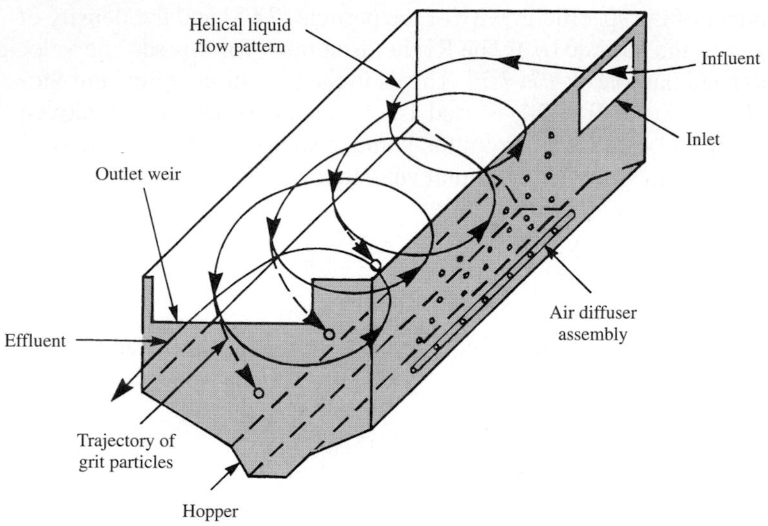

FIGURE 6-10
Aerated grit chamber. (*Source:* Metcalf & Eddy, 1979.)

commonly used in small WWTPs with flows less than 0.2 m^3/s. These devices are placed downstream of the grit chambers to protect the cutting bars from abrasion. They are used as a replacement for the downstream bar rack but must be installed with a hand-cleaned rack in parallel in case they fail. Comminutors may create a string of material such as rags that collect on downstream equipment. Because of this and a high maintenance cost, newer installations use a screen or a *macerator* (Figure 6-11).

Macerators are slow-speed grinders. One type consists of counterrotating blades. The material is chopped as it passes between the blades. The chopping action reduces the potential for producing ropes of rags or plastic.

Equalization

Flow equalization is not a treatment process in itself, but a technique that can be used to improve the effectiveness of both secondary and advanced wastewater treatment processes (U.S. EPA, 1979a). Wastewater does not flow into a municipal wastewater treatment plant at a constant rate (see Figure 1-4); the flow rate varies from hour to hour, reflecting the living habits of the area served. In most towns, the pattern of daily activities sets the pattern of sewage flow and strength. Above-average sewage flows and strength occur in mid-morning. The constantly changing amount and strength of wastewater to be treated makes efficient process operation difficult. Also, many treatment units must be designed for the maximum flow conditions encountered, which actually results in their being oversized for average conditions. The purpose of flow equalization is to dampen these variations so that the wastewater can be treated at a nearly constant flow rate. Flow equalization can significantly improve the performance of an existing plant and increase its useful capacity. In new plants, flow equalization can reduce the size and cost of the treatment units.

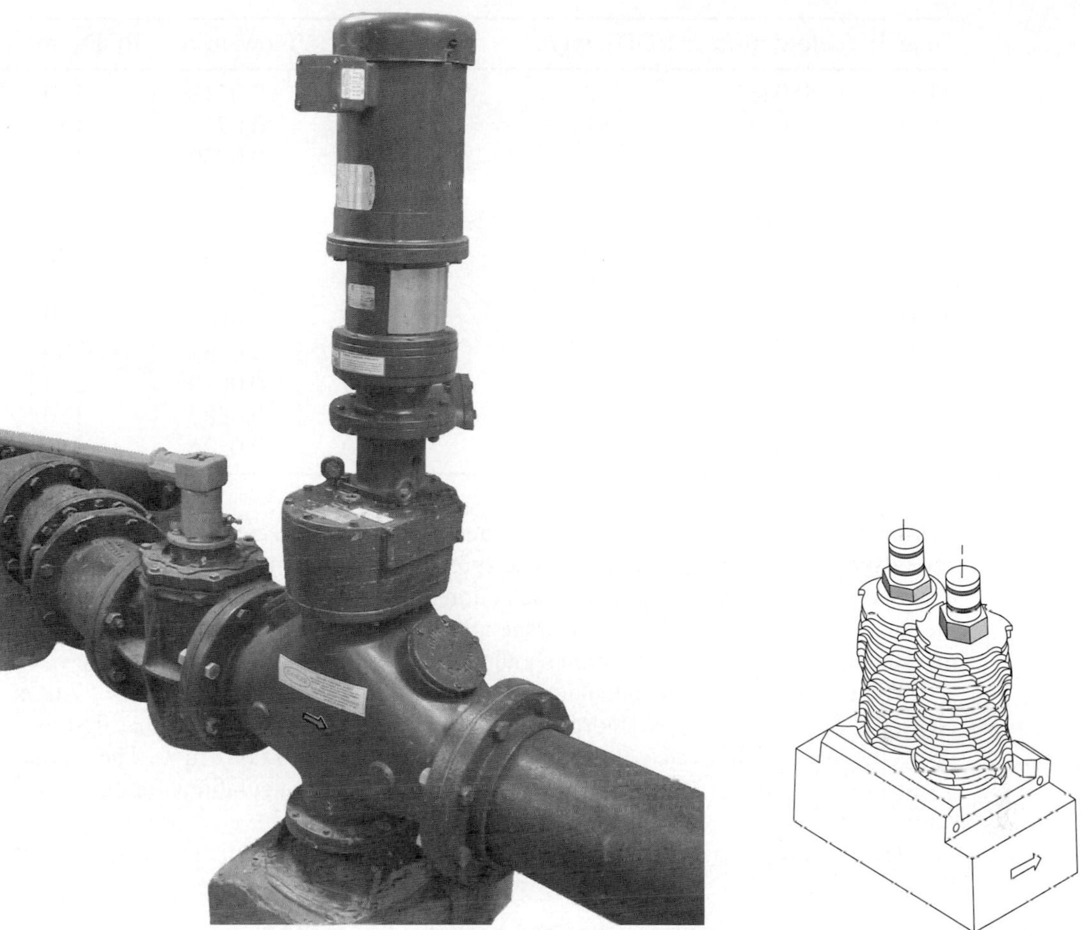

FIGURE 6-11
Photo and isometric drawing of a macerator. (*Source:* Mackenzie L. Davis.)

Flow equalization is usually achieved by constructing large basins that collect and store the wastewater flow and from which the wastewater is pumped to the treatment plant at a constant rate. These basins are normally located near the head end of the treatment works, preferably downstream of pretreatment facilities such as bar screens, comminutors, and grit chambers. Adequate aeration and mixing must be provided to prevent odors and solids deposition. The required volume of an equalization basin is estimated from a mass balance of the flow into the treatment plant with the average flow the plant is designed to treat. The theoretical basis is the same as that used to size reservoirs (see Section 3-4).

Example 6-3. Design an equalization basin for the following cyclic flow pattern. Provide a 25 percent excess capacity for equipment, unexpected flow variations, and solids accumulation. Evaluate the impact of equalization on the mass loading of BOD_5.

Time, h	Flow, m^3/s	BOD_5, mg/L	Time, h	Flow, m^3/s	BOD_5, mg/L
0000	0.0481	110	1200	0.0718	160
0100	0.0359	81	1300	0.0744	150
0200	0.0226	53	1400	0.0750	140
0300	0.0187	35	1500	0.0781	135
0400	0.0187	32	1600	0.0806	130
0500	0.0198	40	1700	0.0843	120
0600	0.0226	66	1800	0.0854	125
0700	0.0359	92	1900	0.0806	150
0800	0.0509	125	2000	0.0781	200
0900	0.0631	140	2100	0.0670	215
1000	0.0670	150	2200	0.0583	170
1100	0.0682	155	2300	0.0526	130

Solution. Because of the repetitive and tabular nature of the calculations, a computer spreadsheet is ideal for this problem. The spreadsheet solution is easy to verify if the calculations are set up with judicious selection of the initial value. If the initial value is the first flow rate greater than the average after the sequence of nighttime low flows, then the last row of the computation should result in a storage value of zero.

The first step then is to calculate the average flow. In this case it is 0.05657 m^3/s. Next, the flows are arranged in order beginning with the time and flow that first exceeds the average. In this case it is at 0900 h with a flow of 0.0631 m^3/s. The tabular arrangement is shown on page 446. An explanation of the calculations for each column follows.

The third column converts the flows to volumes using the time interval between flow measurements:

$$\forall = (0.0631 \text{ m}^3/\text{s})(1 \text{ h})(3600 \text{ s/h}) = 227.16 \text{ m}^3$$

The fourth column is the average volume that leaves the equalization basin.

$$\forall = (0.05657 \text{ m}^3/\text{s})(1 \text{ h})(3600 \text{ s/h}) = 203.655 \text{ m}^3$$

The fifth column is the difference between the inflow volume and the outflow volume.

$$dS = \forall_{\text{in}} - \forall_{\text{out}} = 227.16 \text{ m}^3 - 203.655 \text{ m}^3 = 23.505 \text{ m}^3$$

The sixth column is the cumulative sum of the difference between the inflow and outflow. For the second time interval, it is

$$\text{Storage} = \sum dS = 37.55 \text{ m}^3 + 23.51 \text{ m}^3 = 61.06 \text{ m}^3$$

Note that the last value for the cumulative storage is 0.12 m^3. It is not zero because of round-off truncation in the computations. At this point the equalization basin is empty and ready to begin the next day's cycle.

The required volume for the equalization basin is the maximum cumulative storage. With the requirement for 25 percent excess, the volume would then be

$$\text{Storage volume} = (863.74 \text{ m}^3)(1.25) = 1{,}079.68 \text{ or } 1{,}080 \text{ m}^3$$

The mass of BOD_5 into the equalization basin is the product of the inflow (Q), the concentration of BOD_5 (S_o), and the integration time (Δt):

$$M_{\text{BOD-in}} = (Q)(S_o)(\Delta t)$$

The mass of BOD_5 out of the equalization basin is the product of the average outflow (Q_{avg}), the average concentration (S_{avg}) in the basin, and the integration time (Δt):

$$M_{\text{BOD-out}} = (Q_{avg})(S_{avg})(\Delta t)$$

The average concentration is determined as

$$S_{avg} = \frac{(\forall_i)(S_o) + (\forall_s)(S_{prev})}{\forall_i + \forall_s}$$

where $\forall_i$ = volume of inflow during time interval Δt, m^3
S_o = average BOD_5 concentration during time interval Δt, g/m^3
$\forall_s$ = volume of wastewater in the basin at the end of the previous time interval, m^3
S_{prev} = concentration of BOD_5 in the basin at the end of the previous time interval
= (previous S_{avg}), g/m^3

Noting that 1 mg/L = 1 g/m^3, we find that the first row (the 0900 h time) computations are

$$M_{\text{BOD-in}} = (0.0631 \text{ m}^3/\text{s})(140 \text{ g/m}^3)(1 \text{ h})(3{,}600 \text{ s/h})(10^{-3} \text{kg/g})$$
$$= 31.8 \text{ kg}$$

$$S_{avg} = \frac{(227.16 \text{ m}^3)(140 \text{ g/m}^3) + 0}{227.16 \text{ m}^3 + 0}$$
$$= 140 \text{ mg/L}$$

$$M_{\text{BOD-out}} = (0.05657 \text{ m}^3/\text{s})(140 \text{ g/m}^3)(1 \text{ h})(3{,}600 \text{ s/h})(10^{-3} \text{ kg/g})$$
$$= 28.5 \text{ kg}$$

Note that the zero values in the computation of S_{avg} are valid only at startup of an empty basin. Also note that in this case $M_{\text{BOD-in}}$ and $M_{\text{BOD-out}}$ differ only because of the difference in flow rates. For the second row (1000 h), the computations are

$$M_{\text{BOD-in}} = (0.0670 \text{ m}^3/\text{s})(150 \text{ g/m}^3)(1 \text{ h})(3{,}600 \text{ s/h})(10^{-3} \text{ kg/g})$$
$$= 36.2 \text{ kg}$$

$$S_{avg} = \frac{(241.20 \text{ m}^3)(150 \text{ g/m}^3) + (23.51 \text{ m}^3)(140 \text{ g/m}^3)}{241.20 \text{ m}^3 + 23.51 \text{ m}^3}$$
$$= 149.11 \text{ mg/L}$$

$$M_{\text{BOD-out}} = (0.05657 \text{ m}^3/\text{s})(149.11 \text{ g/m}^3)(1 \text{ h})(3{,}600 \text{ s/h})(10^{-3} \text{ kg/g})$$
$$= 30.37 \text{ kg}$$

Note that $\forall_s$ is the volume of wastewater in the basin at the end of the previous time interval. Therefore, it equals the accumulated dS. The concentration of BOD_5

Time	Flow, m^3/s	Vol_{in}, m^3	Vol_{out}, m^3	dS, m^3	$\Sigma\, dS$, m^3	BOD_5, mg/L	$M_{BOD\text{-}in}$, kg	S, mg/L	$M_{BOD\text{-}out}$, kg
0900	0.0631	227.16	203.65	23.51	23.51	140	31.80	140.00	28.51
1000	0.067	241.2	203.65	37.55	61.06	150	36.18	149.11	30.37
1100	0.0682	245.52	203.65	41.87	102.93	155	38.06	153.83	31.33
1200	0.0718	258.48	203.65	54.83	157.76	160	41.36	158.24	32.23
1300	0.0744	267.84	203.65	64.19	221.95	150	40.18	153.06	31.17
1400	0.075	270	203.65	66.35	288.3	140	37.80	145.89	29.71
1500	0.0781	281.16	203.65	77.51	365.81	135	37.96	140.51	28.62
1600	0.0806	290.16	203.65	86.51	452.32	130	37.72	135.86	27.67
1700	0.0843	303.48	203.65	99.83	552.15	120	36.42	129.49	26.37
1800	0.0854	307.44	203.65	103.79	655.94	125	38.43	127.89	26.04
1900	0.0806	290.16	203.65	86.51	742.45	150	43.52	134.67	27.43
2000	0.0781	281.16	203.65	77.51	819.96	200	56.23	152.61	31.08
2100	0.067	241.2	203.65	37.55	857.51	215	51.86	166.79	33.97
2200	0.0583	209.88	203.65	6.23	863.74	170	35.68	167.42	34.10
2300	0.0526	189.36	203.65	−14.29	849.45	130	24.62	160.69	32.73
0000	0.0481	173.16	203.65	−30.49	818.96	110	19.05	152.11	30.98
0100	0.0359	129.24	203.65	−74.41	744.55	81	10.47	142.42	29.00
0200	0.0226	81.36	203.65	−122.29	622.26	53	4.31	133.61	27.21
0300	0.0187	67.32	203.65	−136.33	485.93	35	2.36	123.98	25.25
0400	0.0187	67.32	203.65	−136.33	349.6	32	2.15	112.79	22.97
0500	0.0198	71.28	203.65	−132.37	217.23	40	2.85	100.46	20.46
0600	0.0226	81.36	203.65	−122.29	94.94	66	5.37	91.07	18.55
0700	0.0359	129.24	203.65	−74.41	20.53	92	11.89	91.61	18.66
0800	0.0509	183.24	203.65	−20.41	0.12	125	22.91	121.64	24.77

(S_{prev}) is the average concentration at the end of previous interval (S_{avg}) and *not* the influent concentration for the previous interval (S_o).

For the third row (1100 h), the concentration of BOD_5 is

$$S_{avg} = \frac{(245.52 \text{ m}^3)(155 \text{ g/m}^3) + (61.06 \text{ m}^3)(149.11 \text{ g/m}^3)}{245.52 \text{ m}^3 + 61.06 \text{ m}^3}$$

$$= 153.83 \text{ mg/L}$$

6-7 PRIMARY TREATMENT

With the screening completed and the grit removed, the wastewater still contains light organic suspended solids, some of which can be removed from the sewage by gravity in a sedimentation tank. These tanks can be round or rectangular. The mass of settled solids is called *raw sludge.* The sludge is removed from the sedimentation tank by mechanical scrapers and pumps (Figure 6-12). Floating materials, such as grease and oil, rise to the surface of the sedimentation tank, where they are collected by a surface skimming system and removed from the tank for further processing.

Primary sedimentation basins (*primary tanks*) are characterized by Type II flocculant settling. The Stokes equation cannot be used because the flocculating particles are continually changing in size, shape, and, when water is entrapped in the floc, specific gravity. There is no adequate mathematical relationship that can be used to describe Type II settling. Laboratory tests with settling columns are used to develop design data (see Chapter 4).

Rectangular tanks with common-wall construction are frequently chosen because they are advantageous for sites with space constraints. Typically, these tanks range from 15 to 100 m in length and 3 to 24 m in width. Common length-to-width ratios for the design of new facilities range from 3:1 to 5:1. Existing plants have length-to-width ratios ranging from 1.5:1 to 15:1. The width is often controlled by the availability of

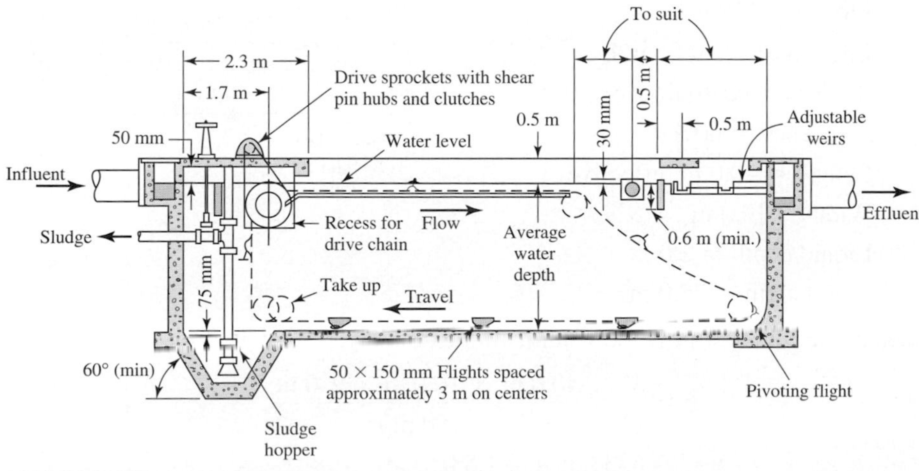

FIGURE 6-12
Primary settling tank.

sludge collection equipment. Side water depths range from 3 to 5 m. Typically the depth is about 4 m.

Circular tanks have diameters from 3 to 60 m. Side water depths range from 3 to 5 m.

As in water treatment clarifier design, overflow rate is the controlling parameter for the design of primary settling tanks. At average flow, overflow rates typically range from 25 to 60 $m^3/m^2 \cdot d$ (or 25 to 60 m/d). When waste-activated sludge is returned to the primary tank, a lower range of overflow rates is chosen (25 to 35 m/d). Under peak flow conditions, overflow rates may be in the range of 80 to 120 m/d.

Hydraulic detention time in the sedimentation basin ranges from 1.5 to 2.5 hours under average flow conditions. A 2.0-hour detention time is typical.

The Great Lakes–Upper Mississippi River Board of State Sanitary Engineers (GLUMRB) recommends that *weir loading* (hydraulic flow over the effluent weir) rates not exceed 120 m^3/d of flow per m of weir length ($m^3/d \cdot m$) for plants with average flows less than 0.04 m^3/s. For larger flows, the recommended rate is 190 $m^3/d \cdot m$ (GLUMRB, 1978). If the side water depths exceed 3.5 m, the weir loading rates have little effect on performance.

Two different approaches have been used to place the weirs. Some designers believe in the "long" approach and place the weirs to cover 33 to 50 percent of the length of the tank. Those of the "short school" (for example, see Metcalf & Eddy, Inc., 2003) assume the weir length is less important and place it across the width of the end of the tank as shown in Figure 6-12. The spacing may vary from 2.5 to 6 m between weirs.

As mentioned previously, approximately 50 to 60 percent of the raw sewage suspended solids and as much as 30 to 35 percent of the raw sewage BOD_5 may be removed in the primary tank.

Example 6-4. Evaluate the following primary tank design with respect to detention time, overflow rate, and weir loading.

Design data:

Flow = 0.150 m^3/s

Influent SS = 280 mg/L

Sludge concentration = 6.0%

Efficiency = 60%

Length = 40.0 m (effective)

Width = 10.0 m

Liquid depth = 2.0 m

Weir length = 75.0 m

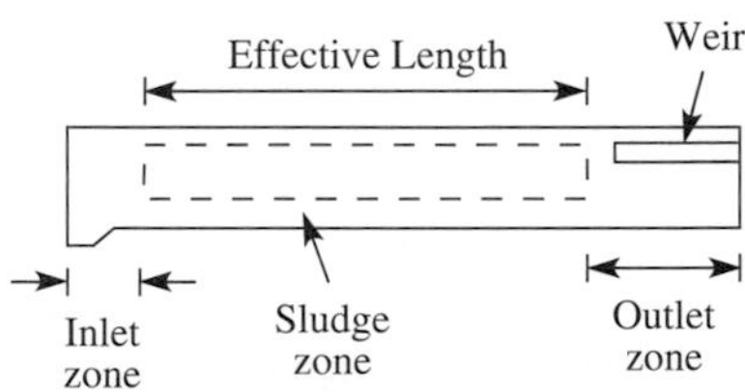

The detention time is simply the volume of the tank divided by the flow:

$$\theta = \frac{\forall}{Q} = \frac{40.0 \text{ m} \times 10.0 \text{ m} \times 2.0 \text{ m}}{0.150 \text{ m}^3/\text{s}}$$

$$= 5333.33 \text{ s or } 1.5 \text{ h}$$

This is a reasonable detention time.

The overflow rate is the flow divided by the surface area:

$$v_o = \frac{0.150 \text{ m}^3/\text{s}}{40.0 \text{ m} \times 10.0 \text{ m}}$$
$$= 3.75 \times 10^{-4} \text{ m/s} \times 86{,}400 \text{ s/d} = 32 \text{ m/d}$$

This is an acceptable overflow rate.

The weir loading is calculated in the same fashion:

$$WL = \frac{0.150 \text{ m}^3/\text{s}}{75.0 \text{ m}}$$
$$= 0.0020 \text{ m}^3/\text{s} \cdot \text{m} \times 86{,}400 \text{ s/d} = 172.8 \text{ or } 173 \text{ m}^3/\text{d} \cdot \text{m}$$

This is an acceptable weir loading.

6-8 UNIT PROCESSES OF SECONDARY TREATMENT

Overview

The major purpose of secondary treatment is to remove the soluble BOD that escapes primary treatment and to provide further removal of suspended solids. The basic ingredients needed for conventional aerobic secondary biologic treatment are the availability of many microorganisms, good contact between these organisms and the organic material, the availability of oxygen, and the maintenance of other favorable environmental conditions (for example, favorable temperature and sufficient time for the organisms to work). A variety of approaches have been used in the past to meet these basic needs. The most common approaches are, (1) activated sludge, (2) trickling filters and (3) oxidation ponds (or lagoons). A process that does not fit precisely into either the trickling filter or the activated sludge category but does employ principles common to both is the *rotating biological contactor* (RBC).

Because the secondary treatment processes are fundamentally microbiological, we begin this discussion with an introduction to wastewater microbiology.

Wastewater Microbiology

Role of Microorganisms. The stabilization of organic matter is accomplished biologically using a variety of microorganisms. The microorganisms convert the colloidal and dissolved carbonaceous organic matter into various gases and into protoplasm. Because protoplasm has a specific gravity slightly greater than that of water, it can be removed from the treated liquid by gravity settling.

It is important to note that unless the protoplasm produced from the organic matter is removed from the solution, complete treatment will not be accomplished because the protoplasm, which itself is organic, will be measured as BOD in the effluent. If the protoplasm is not removed, the only treatment that will be achieved is that associated with the bacterial conversion of a portion of the organic matter originally present to various gaseous end products (Metcalf & Eddy, Inc., 1979.)

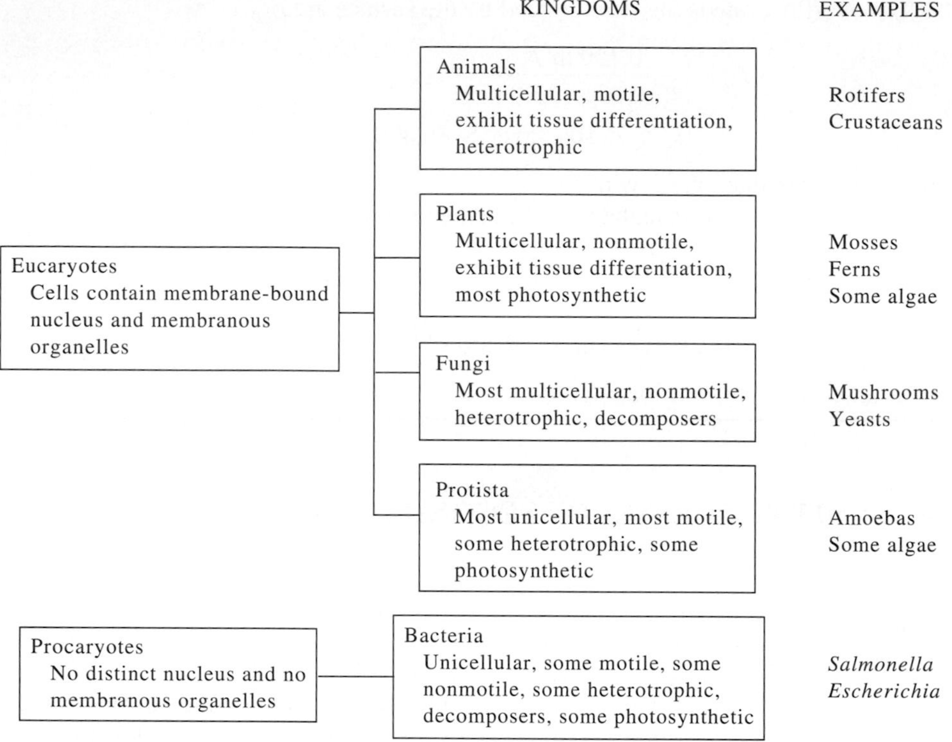

FIGURE 6-13
Classification of microorganisms by kingdom.

Classification of Microorganisms by Kingdoms. Microorganisms are organized into five broad groups based on their structural and functional differences. The groups are called *kingdoms*. The five kingdoms are *animals, plants, protista, fungi,* and *bacteria.* Representative examples and characteristics of differentiation are shown in Figure 6-13.

Classification by Energy and Carbon Source. The relationship between the source of carbon and the source of energy for the microorganism is important. Carbon is the basic building block for cell synthesis. A source of energy must be obtained from outside the cell to enable synthesis to proceed. Our goal in wastewater treatment is to convert both the carbon and the energy in the wastewater into the cells of microorganisms, which we can remove from the water by settling. Therefore, we wish to encourage the growth of organisms that use organic material for both their carbon and energy source.

If the microorganism uses organic material as a supply of carbon, it is called *heterotrophic. Autotrophs* require only CO_2 to supply their carbon needs.

Organisms that rely only on the sun for energy are called *phototrophs. Chemotrophs* extract energy from organic or inorganic oxidation/reduction reactions. *Organotrophs* use organic materials, while *lithotrophs* oxidize inorganic compounds (Bailey and Ollis, 1977).

Classification by their Relationship to Oxygen. Bacteria also are classified by their ability or inability to utilize oxygen as a terminal electron acceptor* in oxidation/reduction reactions. *Obligate aerobes* are microorganisms that must have oxygen as the terminal electron acceptor. When wastewater contains oxygen and can support obligate aerobes, it is called *aerobic.*

Obligate anaerobes are microorganisms that cannot survive in the presence of oxygen. They cannot use oxygen as a terminal electron acceptor. Wastewater that is devoid of oxygen is called *anaerobic. Facultative anaerobes* can use oxygen as the terminal electron acceptor and, under certain conditions, they can also grow in the absence of oxygen.

Under *anoxic* conditions, a group of facultative anaerobes called *denitrifiers* utilizes nitrites (NO_2^-) and nitrates (NO_3^-) as the terminal electron acceptor. Nitrate nitrogen is converted to nitrogen gas in the absence of oxygen. This process is called *anoxic denitrification.*

Classification by their Preferred Temperature Regime. Each species of bacteria reproduces best within a limited range of temperatures. Four temperature ranges are used to classify bacteria. Those that grow best at temperatures below 20°C are called *psychrophiles. Mesophiles* grow best at temperatures between 25 and 40°C. Between 45 and 60°C, the *thermophiles* grow best. Above 60°C, *stenothermophiles* grow best. The growth range of *facultative thermophiles* extends from the thermophilic range into the mesophilic range. These ranges are qualitative and somewhat subjective. You will note the gaps between 20 and 25°C and between 40 and 45°C. Don't make the mistake of saying that an organism that grows well at 20.5°C is a mesophile. The rules just aren't that hard and fast. Bacteria will grow over a range of temperatures and will survive at a very large range of temperatures. For example, *Escherichia coli,* classified as mesophiles, will grow at temperatures between 20 and 50°C and will reproduce, albeit very slowly, at temperatures down to 0°C. If frozen rapidly, they and many other microorganisms can be stored for years with no significant death rate.

Some Microbes of Interest in Wastewater Treatment

Bacteria. The highest population of microorganisms in a wastewater treatment plant will belong to the bacteria. They are single-celled organisms which use soluble food. Conditions in the treatment plant are adjusted so that chemoheterotrophs predominate. No particular species is selected as "the best."

---

*An organic substrate is not directly oxidized to carbon dioxide and water in a single chemical step because there is no energy-conserving mechanism that could trap so much energy. Thus, biological oxidation occurs in small steps. Oxidation requires the transfer of an electron from the substance being oxidized to some acceptor molecule that will subsequently be reduced. In most biological systems, each step in the oxidation process involves the removal of two electrons and the simultaneous loss of two protons (H^+). The combination of the two losses is equivalent to the molecule having lost two hydrogen atoms. The reaction is often referred to as *dehydrogenation*. The electrons and protons are not released into the cell, but are transferred to an acceptor molecule. The acceptor molecule will not accept the protons until it has accepted the electrons and thus it is referred to as an electron acceptor. Since the net result of accepting an electron and proton is the same as accepting a hydrogen atom, such acceptors are also called hydrogen acceptors. (Grady and Lim, 1980.)

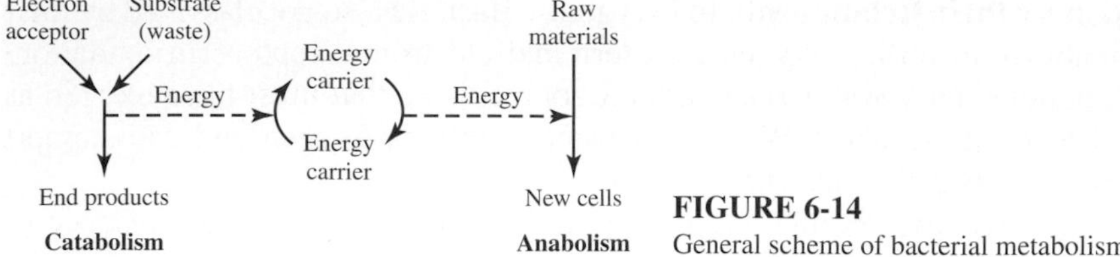

FIGURE 6-14
General scheme of bacterial metabolism.

Fungi. Fungi are multicellular, nonphotosynthetic, heterotrophic organisms. Fungi are obligate aerobes that reproduce by a variety of methods including fission, budding, and spore formation. Their cells require only half as much nitrogen as bacteria so that in a nitrogen-deficient wastewater, they predominate over the bacteria (McKinney, 1962).

Algae. This group of microorganisms are photoautotrophs and may be either unicellular or multicellular. Because of the chlorophyll contained in most species, they produce oxygen through photosynthesis. In the presence of sunlight, the photosynthetic production of oxygen is greater than the amount used in respiration. At night they use up oxygen in respiration. If the daylight hours exceed the night hours by a reasonable amount, there is a net production of oxygen.

Protozoa. Protozoa are single-celled organisms that can reproduce by *binary fission* (dividing in two). Most are aerobic chemoheterotrophs, and they often consume bacteria. They are desirable in wastewater effluents because they act as polishers in consuming the bacteria.

Rotifers and Crustaceans. Both rotifers and crustaceans are animals—aerobic, multicellular chemoheterotrophs. The rotifer derives its name from the apparent rotating motion of two sets of cilia on its head. The cilia provide mobility and a mechanism for catching food. Rotifers consume bacteria and small particles of organic matter.

Crustaceans, a group that includes shrimp, lobsters, and barnacles, are characterized by their shell structure. They are a source of food for fish and are not found in wastewater treatment systems to any extent except in underloaded lagoons. Their presence is indicative of a high level of dissolved oxygen and a very low level of organic matter.

Bacterial Biochemistry

The general term that describes all of the chemical activities performed by a cell is *metabolism.* This in turn is divided into two parts: catabolism and anabolism. *Catabolism* includes all the biochemical processes by which a substrate is degraded to end products with the release of energy.* In wastewater treatment, the substrate is oxidized. The oxidation process releases energy that is transferred to an energy carrier which stores it for future use by the bacterium (Figure 6-14). Some chemical compounds released by catabolism are used by the bacterial cell for its life functions.

Anabolism includes all the biochemical processes by which the bacterium synthesizes new chemical compounds needed by the cells to live and reproduce. The synthesis process is driven by the energy that was stored in the energy carrier.

*Substrate is food. For our application, "food" is the organic material from the human digestive tract and other biodegradable wastes.

Decomposition of Waste. The type of electron acceptor available for catabolism determines the type of decomposition (that is, aerobic, anoxic, or anaerobic) used by a mixed culture of microorganisms. Each type of decomposition has peculiar characteristics which affect its use in waste treatment.

Aerobic Decomposition. From our discussion of bacterial metabolism you will recall that molecular oxygen (O_2) must be present as the terminal electron acceptor for decomposition to proceed by aerobic oxidation. As in natural water bodies, the oxygen is measured as DO. When oxygen is present, it is the only terminal electron acceptor used. Hence, the chemical end products of decomposition are primarily carbon dioxide, water, and new cell material (Table 6-9). Odiferous gaseous end products are kept to a minimum. In healthy natural water systems, aerobic decomposition is the principal means of self-purification.

A wider spectrum of organic material can be oxidized aerobically than by any other type of decomposition. This fact, coupled with the fact that the final end products are oxidized to a very low energy level, results in a more stable end product (that is, one that can be disposed of without damage to the environment and without creating a nuisance condition) than can be achieved by the other oxidation systems.

Because of the large amount of energy released in aerobic oxidation, most aerobic organisms are capable of high growth rates. Consequently, there is a relatively large production of new cells in comparison with the other oxidation systems. This means

TABLE 6-9
Representative end products

Substrates	Aerobic and anoxic decomposition	Anaerobic decomposition
Proteins and other organic nitrogen compounds	Amino acids Ammonia → nitrites → nitrates* Alcohols → CO_2 + H_2O Organic acids	Amino acids Ammonia Hydrogen sulfide Methane Carbon dioxide Alcohols Organic acids
Carbohydrates	Alcohols → CO_2 + H_2O Fatty acids	Carbon dioxide Fatty acids Methane
Fats and related substances	Fatty acids + glycerol Alcohols → CO_2 + H_2O Lower fatty acids	Fatty acids + glycerol Carbon dioxide Alcohols Lower fatty acids Methane

*Under anoxic conditions the nitrates are converted to nitrogen gas.
Source: After Pelczar, M. J., and R. D. Reid, 1958.

that more biological sludge is generated in aerobic oxidation than in the other oxidation systems.

Aerobic decomposition is the method of choice for large quantities of dilute wastewater (BOD_5 less than 500 mg/L) because decomposition is rapid, efficient, and has a low odor potential. For high-strength wastewater (BOD_5 is greater than 1,000 mg/L), aerobic decomposition is not suitable because of the difficulty in supplying enough oxygen and because of the large amount of biological sludge produced. In small communities and in special industrial applications where aerated lagoons (see "Oxidation Ponds," below) are used, wastewaters with BOD_5 up to 3,000 mg/L may be treated satisfactorily by aerobic decomposition.

Anoxic Decomposition. Some microorganisms can use nitrate (NO_3^-) as the terminal electron acceptor in the absence of molecular oxygen. Oxidation by this route is called denitrification.

The end products from denitrification are nitrogen gas, carbon dioxide, water, and new cell material. The amount of energy made available to the cell during denitrification is about the same as that made available during aerobic decomposition. As a consequence, the rate of production of new cells, although not as high as in aerobic decomposition, is relatively high.

Denitrification is of importance in wastewater treatment where nitrogen must be removed to protect the receiving body. In this case, a special treatment step is added to the conventional process for removal of carbonaceous material. Denitrification will be discussed in detail later.

One other important aspect of denitrification is in relation to final clarification of the treated wastewater. If the environment of the final clarifier becomes anoxic, the formation of nitrogen gas will cause large globs of sludge to float to the surface and escape from the treatment plant into the receiving water. Thus, it is necessary to ensure that anoxic conditions do not develop in the final clarifier.

Anaerobic Decomposition. In order to achieve anaerobic decomposition, molecular oxygen and nitrate must not be present as terminal electron acceptors. Sulfate (SO_4^{2-}), carbon dioxide, and organic compounds that can be reduced serve as terminal electron acceptors. The reduction of sulfate results in the production of hydrogen sulfide (H_2S) and a group of equally odiferous organic sulfur compounds called *mercaptans.*

The anaerobic decomposition (fermentation) of organic matter generally is considered to be a three-step process. In the first step, waste components are hydrolysed. In the second step, complex organic compounds are fermented to low-molecular-weight fatty acids (volatile acids). In the third step, the organic acids are converted to methane. Carbon dioxide serves as the electron acceptor.

Anaerobic decomposition yields carbon dioxide, methane, and water as the major end products. Additional end products include ammonia, hydrogen sulfide, and mercaptans. As a consequence of these last three compounds, anaerobic decomposition is characterized by an unbelievably horrid stench!

Because only small amounts of energy are released during anaerobic oxidation, the amount of cell production is low. Thus, sludge production is low. We make use of this fact in wastewater treatment by using anaerobic decomposition to stabilize sludges produced during aerobic and anoxic decomposition.

Direct anaerobic decomposition of wastewater generally is not feasible for dilute waste.* The optimum growth temperature for the anaerobic bacteria is at the upper end of the mesophilic range. Thus, to get reasonable biodegradation, we must elevate the temperature of the culture. For dilute wastewater, this is not practical. For concentrated wastes (BOD_5 greater than 1,000 mg/L), anaerobic digestion is quite appropriate.

Population Dynamics. In the discussion of the behavior of bacterial cultures which follows, there is the inherent assumption that all the requirements for growth are initially present. Since these requirements are fairly extensive and stringent, it is worth taking a moment to recapitulate them. The following list summarizes the major requirements that must be satisfied:

1. A terminal electron acceptor
2. Macronutrients
   a. Carbon to build cells
   b. Nitrogen to build cells
   c. Phosphorus for ATP (energy carrier) and DNA
3. Micronutrients
   a. Trace metals
   b. Vitamins are required by some bacteria
4. Appropriate environment
   a. Moisture
   b. Temperature
   c. pH

As an illustration of growth in pure cultures, let us examine a hypothetical situation in which 1,400 bacteria of a single species are introduced into a synthetic liquid medium. Initially nothing appears to happen. The bacteria must adjust to their new environment and begin to synthesize new protoplasm. On a plot of bacterial growth versus time (Figure 6-15), this phase of growth is called the *lag phase.*

At the end of the lag phase the bacteria begin to divide. Since all of the organisms do not divide at the same time, there is a gradual increase in population. This phase is labeled *accelerated growth* on the growth plot.

At the end of the accelerated growth phase, the population of organisms is large enough and the differences in generation time are small enough that the cells appear to divide at a regular rate. Since reproduction is by binary fission (each cell divides producing two new cells), the increase in population follows in geometric progression: $1 \rightarrow 2 \rightarrow 4 \rightarrow 8 \rightarrow 16 \rightarrow 32$, and so forth. The population of bacteria (P) after the nth generation is given by the following expression:

$$P = P_0(2)^n \tag{6-1}$$

*Some researchers are exploring the use of anaerobic systems for treatment of dilute wastes, especially groundwater contaminated with hazardous waste.

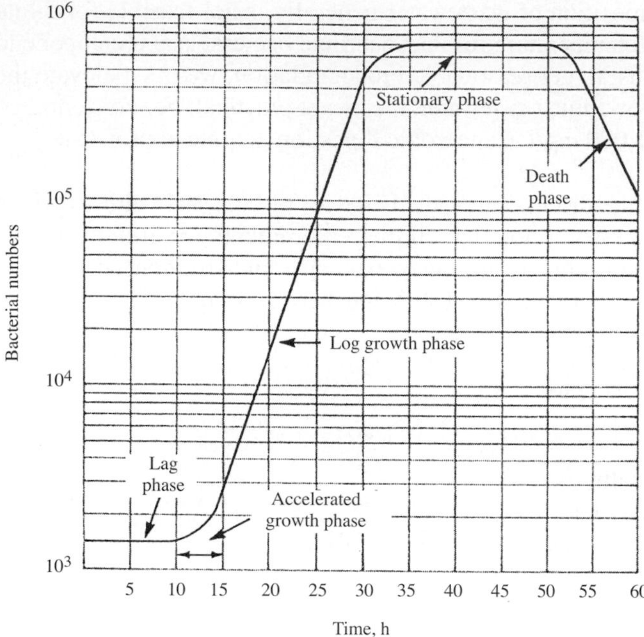

FIGURE 6-15
Bacterial growth in a pure culture: the "log-growth curve."

where P_0 is the initial population at the end of the accelerated growth phase. If we take the log of both sides of Equation 6-1, we obtain the following:

$$\log P = \log P_0 + n \log 2 \tag{6-2}$$

This means that if we plot bacterial population on a logarithmic scale, this phase of growth would plot as a straight line of slope n and intercept P_0 at t_0 equal to the end of the accelerated growth phase. Thus, this phase of growth is called the *log growth* or *exponential growth phase.*

The log growth phase tapers off as the substrate becomes exhausted or as toxic by-products build up. Thus, at some point the population becomes constant either as a result of cessation of fission or a balance in death and reproduction rates. This is depicted by the *stationary phase* on the growth curve.

Following the stationary phase, the bacteria begin to die faster than they reproduce. This death phase is due to a variety of causes that are basically an extension of those which lead to the stationary phase.

In wastewater treatment, as in nature, pure cultures of microorganisms do not exist. Rather, a mixture of species compete and survive within the limits set by the environment. *Population dynamics* is the term used to describe the time-varying success of the various species in competition. It is expressed quantitatively in terms of relative mass of microorganisms.*

*If each individual organism of species A has, on the average, twice the mass at maturity as each individual organism of species B, and both compete equally, we would expect that both would have the same total biomass, but that there would be twice as many of species B as there would be of A.

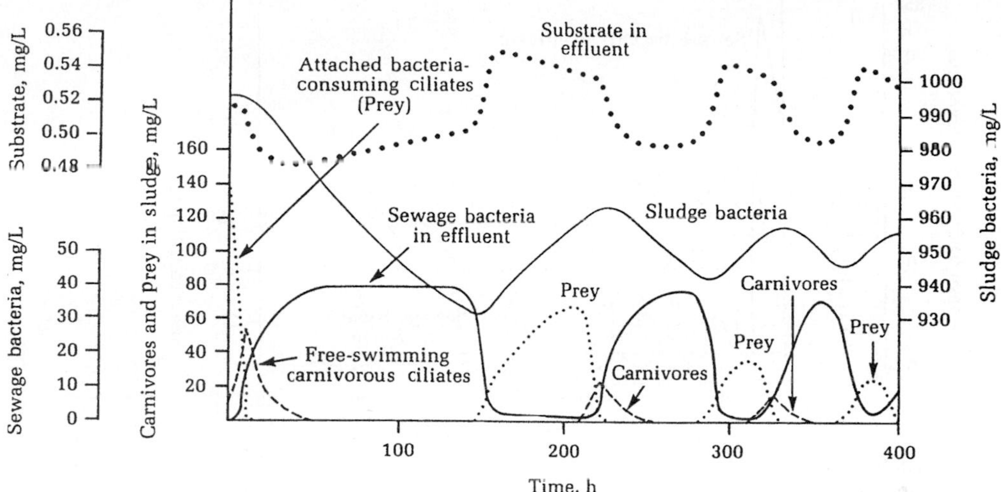

FIGURE 6-16
Population dynamics in a closed system. (*Source:* Curds, 1973.)

The prime factor governing the dynamics of the various microbial populations is the competition for food. The second most important factor is the predator-prey relationship.

The relative success of a pair of species competing for the same substrate is a function of the ability of the species to metabolize the substrate. The more successful species will be the one that metabolizes the substrate more completely. In so doing, it will obtain more energy for synthesis and consequently will achieve a greater mass.

Because of their relatively smaller size and, thus, larger surface area per unit mass, which allows a more rapid uptake of substrate, bacteria will predominate over fungi. For the same reason, the fungi predominate over the protozoa.

When the supply of soluble organic substrate becomes exhausted, the bacterial population is less successful in reproduction and the predator populations increase. In a closed system with an initial inoculum of mixed microorganisms and substrate, the populations will cycle as the bacteria give way to higher level organisms which in turn die for lack of food and are then decomposed by a different set of bacteria (Figure 6-16). In an open system, such as a wastewater treatment plant or a river, with a continuous inflow of new substrate, the predominant populations will change through the length of the plant (Figure 6-17). This condition is known as *dynamic equilibrium.* It is a highly sensitive state, and changes in influent characteristics must be regulated closely to maintain the proper balance of the various populations.

For the large numbers and mixed cultures of microorganisms found in waste treatment systems, it is convenient to measure biomass rather than numbers of organisms.*

*Frequently, this is done by measuring suspended solids or volatile suspended solids (those that burn at 550 ± 50°C). When the wastewater contains only soluble organic matter, the volatile suspended solids test is reasonably representative. The presence of organic particles (which is often the case in municipal wastewater) confuses the issue completely.

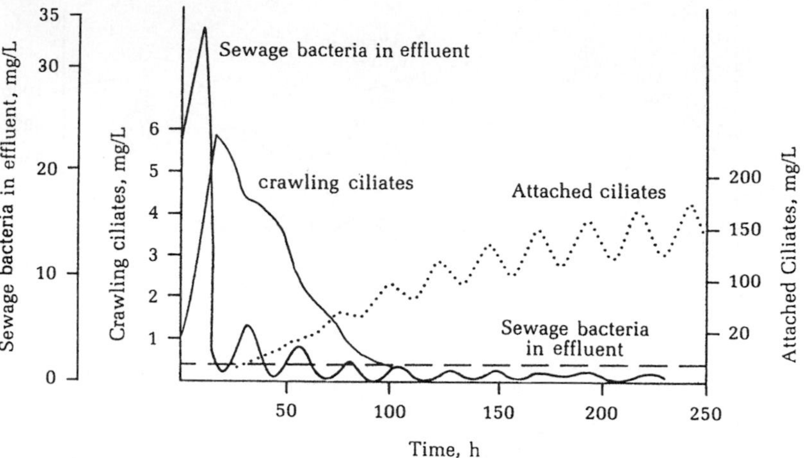

FIGURE 6-17
Population dynamics in an open system. (*Source:* Curds, 1973.)

In the log-growth phase, the rate expression for biomass increase is

$$\frac{dX}{dt} = \mu X \tag{6-3}$$

where $\frac{dX}{dt}$ = growth rate of the biomass, mg/L · t

μ = growth rate constant, t^{-1}

X = concentration of biomass, mg/L

Because of the difficulty of direct measurement of μ in mixed cultures, Monod (1949) developed a model equation that assumes that the rate of food utilization, and therefore the rate of biomass production, is limited by the rate of enzyme reactions involving the food compound that is in shortest supply relative to its need. The *Monod equation* is

$$\mu = \frac{\mu_m S}{K_s + S} \tag{6-4}$$

where μ_m = maximum growth rate constant, t^{-1}

S = concentration of limiting food in solution, mg/L

K_s = half saturation constant, mg/L

= concentration of limiting food when $\mu = 0.5\ \mu_m$

The growth rate of biomass follows a hyperbolic function as shown in Figure 6-18.

Two limiting cases are of interest in the application of Equation 6-4 to wastewater treatment systems. In those cases where there is an excess of the limiting food, then $S \gg K_s$ and the growth rate constant μ is approximately equal to μ_m. Equation 6-3 then becomes zero-order in substrate. At the other extreme, when $S \ll K_s$, the system is food-limited and the growth rate becomes first-order with respect to substrate.

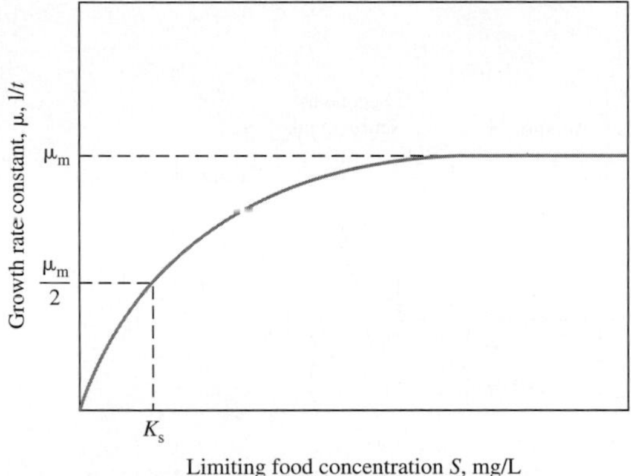

FIGURE 6-18
Monod growth rate constant as a function of limiting food concentration.

Equation 6-4 assumes only growth of microorganisms and does not take into account natural die-off. It is generally assumed that the death or decay of the microbial mass is a first-order expression in biomass and hence Equations 6-3 and 6-4 are expanded to

$$\frac{dX}{dt} = \frac{\mu_m SX}{K_s + S} - k_d X \tag{6-5}$$

where k_d = endogenous decay rate constant, t^{-1}.

If all of the food in the system were converted to biomass, the rate of food utilization (dS/dt) would equal the rate of biomass production. Because of the inefficiency of the conversion process, the rate of food utilization will be greater than the rate of biomass utilization, so

$$-\frac{dS}{dt} = \frac{1}{Y}\frac{dX}{dt} \tag{6-6}$$

where Y = decimal fraction of food mass converted to biomass

$$= \text{yield coefficient, } \frac{\text{mg/L biomass}}{\text{mg/L food utilized}}$$

Combining Equations 6-3, 6-4, and 6-6

$$-\frac{dS}{dt} = \frac{1}{Y}\frac{\mu_m SX}{K_s + S} \tag{6-7}$$

Equations 6-5 and 6-7 are a fundamental part of the development of the design equations for wastewater treatment processes.

Activated Sludge

The activated sludge process is a biological wastewater treatment technique in which a mixture of wastewater and biological sludge (microorganisms) is agitated and aerated. Because the microorganisms are suspended in the liquid wastewater, this process is known as a *suspended growth* process. The biological solids are subsequently separated from the treated wastewater and returned to the aeration process as needed.

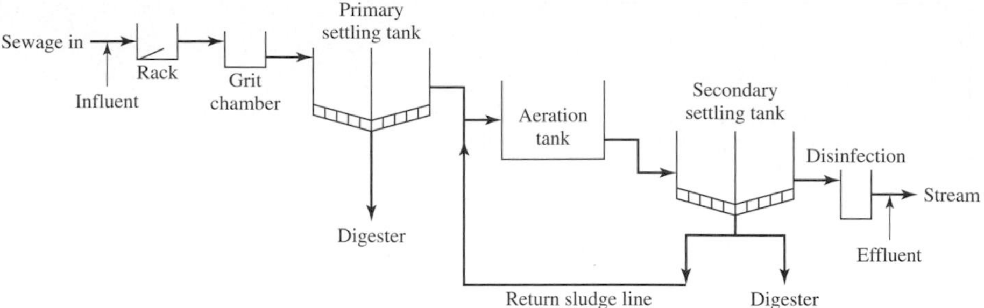

FIGURE 6-19
Conventional activated sludge plant.

The activated sludge process derives its name from the biological mass formed when air is continuously injected into the wastewater. In this process, microorganisms are mixed thoroughly with the organic compounds under conditions that stimulate their growth through use of the organic compounds as food. As the microorganisms grow and are mixed by the agitation of the air, the individual organisms clump together (flocculate) to form an active mass of microbes (biologic floc) called *activated sludge.*

In practice, wastewater flows continuously into an aeration tank (Figure 6-19) where air is injected to mix the activated sludge with the wastewater and to supply the oxygen needed for the organisms to break down the organic compounds. The mixture of

FIGURE 6-20
Activated sludge aeration tank under air. (*Source:* E. Lansing WWTP, photo courtesy of Harley Seeley of the Instructional Media Center, Michigan State University.)

activated sludge and wastewater in the aeration tank is called *mixed liquor.* The mixed liquor flows from the aeration tank to a secondary clarifier where the activated sludge is settled out. Most of the settled sludge is returned to the aeration tank (and is called *return sludge*) to maintain the high population of microbes that permits rapid breakdown of the organic compounds. Because more activated sludge is produced than is desirable in the process, some of the return sludge is diverted or wasted to the sludge handling system for treatment and disposal. In conventional activated sludge systems, the wastewater is typically aerated for six to eight hours in long, rectangular aeration basins. About 8 m^3 of air is provided for each m^3 of wastewater treated. Sufficient air is provided to keep the sludge in suspension (Figure 6-20). The air is injected near the bottom of the aeration tank through a system of diffusers (Figure 6-21). The volume of sludge returned to the aeration basin is typically 20 to 30 percent of the wastewater flow.

(*a*)

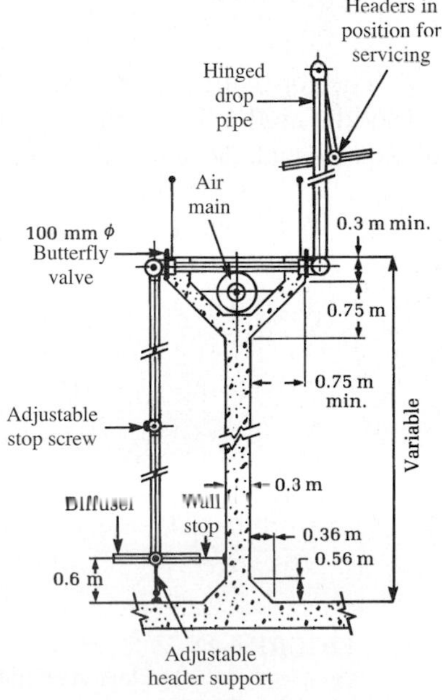

(*b*)

(*c*)

FIGURE 6-21
(*a*) Diffuser system for activated sludge tank; (*b*) cross section through tank illustrating hinged drop pipe; (*c*) close-up of a set of diffusers. (*Photo Source*: E. Lansing WWTP, courtesy of Harley Seeley of the Instructional Media Center, Michigan State University.)

The activated sludge process is controlled by wasting a portion of the microorganisms each day in order to maintain the proper amount of microorganisms to efficiently degrade the BOD_5. *Wasting* means that a portion of the microorganisms is discarded from the process. The discarded microorganisms are called *waste activated sludge* (WAS). A balance is then achieved between growth of new organisms and their removal by wasting. If too much sludge is wasted, the concentration of microorganisms in the mixed liquor will become too low for effective treatment. If too little sludge is wasted, a large concentration of microorganisms will accumulate and, ultimately, overflow the secondary tank and flow into the receiving stream.

The *mean cell residence time* θ_c, also called *solids retention time* (SRT) or *sludge age,* is defined as the average amount of time that microorganisms are kept in the system.

Many modifications of the conventional activated sludge process have been developed to address specific treatment problems. A brief description of these is given in Table 6-10. We have selected the completely mixed and conventional plug-flow processes for further discussion.

Completely Mixed Activated Sludge Process. The design formulas for the completely mixed activated sludge process are a mass balance application of the equations used to describe the kinetics of microbial growth. A mass balance diagram for the completely mixed system (CSTR) is shown in Figure 6-22. The mass balance equations are written for the system boundary shown by the dashed line. Two mass balances are required to define the design of the reactor: one for biomass and one for food (soluble BOD_5).

Under steady-state conditions, the mass balance for biomass may be written as:

$$\begin{array}{c}\text{Biomass in} \\ \text{influent}\end{array} + \begin{array}{c}\text{Net biomass} \\ \text{growth}\end{array} = \begin{array}{c}\text{Biomass in} \\ \text{effluent}\end{array} + \begin{array}{c}\text{Biomass} \\ \text{wasted}\end{array} \tag{6-8}$$

The biomass in the influent is the product of the concentration of microorganisms in the influent (X_o) and the flow rate of wastewater (Q). The concentration of microorganisms in the influent (X_o) is measured as suspended solids (mg/L). The biomass that grows in the aeration tank is the product of the volume of the tank ($\forall$) and the Monod expression for growth of microbial mass (Equation 6-5)

$$(\forall)\left(\frac{\mu_m SX}{K_s + S} - k_d X\right) \tag{6-9}$$

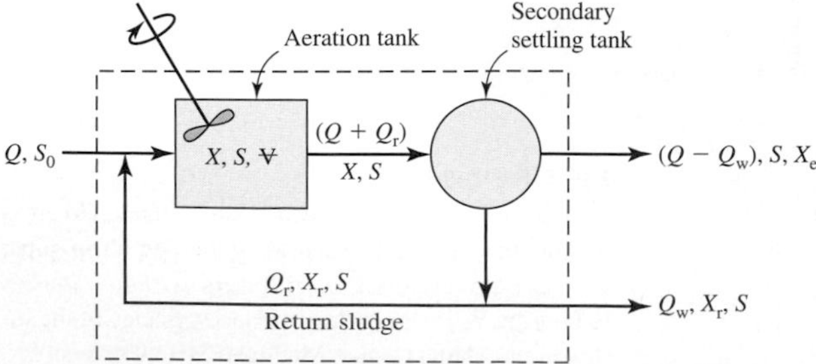

FIGURE 6-22
Completely mixed biological reactor with solids recycle.

TABLE 6-10
Description of activated sludge processes and process modifications[a]

Process or process modification	Description
Conventional plug-flow	Settled wastewater and recycled activated sludge enter the head end of the aeration tank and are mixed by diffused-air or mechanical aeration. Air application is generally uniform throughout tank length. During the aeration period, adsorption, flocculation, and oxidation of organic matter occur. Activated sludge solids are separated in a secondary settling tank.
Complete-mix	Process is an application of the flow regime of a continuous-flow stirred-tank reactor. Settled wastewater and recycled activated sludge are introduced typically at several points in the aeration tank. The organic load on the aeration tank and the oxygen demand are uniform throughout the tank length.
Tapered aeration	Tapered aeration is a modification of the conventional plug-flow process. Varying aeration rates are applied over the tank length depending on the oxygen demand. Greater amounts of air are supplied to the head end of the aeration tank, and the amounts diminish as the mixed liquor approaches the effluent end. Tapered aeration is usually achieved by using different spacing of the air diffusers over the tank length.
Step-feed aeration	Step feed is a modification of the conventional plug-flow process in which the settled wastewater is introduced at several points in the aeration tank to equalize the food-to-microorganism (F/M) ratio, thus lowering peak oxygen demand. Generally three or more parallel channels are used. Flexibility of operation is one of the important features of this process.
Modified aeration	Modified aeration is similar to the conventional plug-flow process except that shorter aeration times and higher F/M ratios are used. BOD removal efficiency is lower than other activated sludge processes.
Contact stabilization	Contact stabilization uses two separate tanks or compartments for the treatment of the wastewater and stabilization of the activated sludge. The stabilized activated sludge is mixed with the influent (either raw or settled) wastewater in a contact tank. The mixed liquor is settled in a secondary settling tank and return sludge is aerated separately in a reaeration basin to stabilize the organic matter. Aeration volume requirements are typically 50 percent less than conventional plug flow.
Extended aeration	Extended aeration process is similar to the conventional plug-flow process except that it operates in the endogenous respiration phase of the growth curve, which requires a low organic loading and long aeration time. Process is used extensively for prefabricated package plants for small communities.
High-rate aeration	High-rate aeration is a process modification in which high MLSS concentrations are combined with high volumetric loadings. This combination allows high F/M ratios and long mean cell-residence times with relatively short hydraulic detention times. Adequate mixing is very important.

(continued)

TABLE 6-10
(continued)

Process or process modification	Description
Kraus process	Kraus process is a variation of the step aeration process used to treat wastewater with low nitrogen levels. Digester supernatant is added as a nutrient source to a portion of the return sludge in a separate aeration tank designed to nitrify. The resulting mixed liquor is then added to the main plug-flow aeration system.
High-purity oxygen	High-purity oxygen is used instead of air in the activated sludge process. Oxygen is diffused into covered aeration tanks and is recirculated. A portion of the gas is wasted to reduce the concentration of carbon dioxide. pH adjustment may also be required. The amount of oxygen added is about four times greater than the amount that can be added by conventional aeration systems.
Oxidation ditch	The oxidation ditch consists of a ring- or oval-shaped channel and is equipped with mechanical aeration devices. Screened wastewater enters the ditch, is aerated, and circulates at about 0.25 to 0.35 m/s. Oxidation ditches typically operate in an extended aeration mode with long detention and solids retention times. Secondary sedimentation tanks are used for most applications.
Sequencing batch reactor	The sequencing batch reactor is a fill-and-draw type reactor system involving a single complete-mix reactor in which all steps of the activated sludge process occur. Mixed liquor remains in the reactor during all cycles, thereby eliminating the need for separate secondary sedimentation tanks.
Deep-shaft reactor	The deep vertical-shaft reactor is a form of the activated sludge process. A vertical shaft about 120 to 150 m deep replaces the primary clarifiers and aeration basin. The shaft is lined with a steel shell and fitted with a concentric pipe to form an annular reactor. Mixed liquor and air are forced down the center of the shaft and allowed to rise upward through the annulus.
Single-stage nitrification	In single-stage nitrification, both BOD and ammonia reduction occur in a single biological stage. Reactor configurations can be either a series of complete-mix reactors or plug flow.
Separate stage nitrification	In separate stage nitrification, a separate reactor is used for nitrification, operating on a feed waste from a preceding biological treatment unit. The advantage of this system is that operation can be optimized to conform to the nitrification needs.

[a]*Source:* Metcalf & Eddy, 1991.

The biomass in the effluent is the product of flow rate of treated wastewater leaving the plant ($Q - Q_w$) and the concentration of microorganisms that does not settle in the secondary clarifier (X_e). The flow rate of wastewater leaving the plant does not equal the flow rate into the plant because some of the microorganisms must be wasted. The flow rate of wasting (Q_w) is deducted from the flow exiting the plant.

The biomass that is wasted is the product of concentration of microorganisms in the WAS flow (X_r) and the WAS flow rate (Q_r). The narrative mass balance equation may be rewritten as

$$QX_o + (\mathcal{V})\left(\frac{\mu_m SX}{K_s + S} - k_d X\right) - (Q - Q_w)X_e + Q_w X_r \qquad (6\text{-}10)$$

The variables are summarized as follows:

Q = wastewater flow rate into the aeration tank, m^3/d
X_o = microorganism concentration (volatile suspended solids or VSS)* entering aeration tank, mg/L
$\mathcal{V}$ = volume of aeration tank, m^3
μ_m = maximum growth rate constant, d^{-1}
S = soluble BOD_5 in aeration tank and effluent, mg/L
X = microorganism concentration (mixed-liquor volatile suspended solids or MLVSS)** in the aeration tank, mg/L
K_s = half velocity constant
= soluble BOD_5 concentration at one-half the maximum growth rate, mg/L
k_d = decay rate of microorganisms, d^{-1}
Q_w = flow rate of liquid containing microorganisms to be wasted, m^3/d
X_e = microorganism concentration (VSS) in effluent from secondary settling tank, mg/L
X_r = microorganism concentration (VSS) in sludge being wasted, mg/L

At steady-state, the mass balance equation for food (soluble BOD_5) may be written

$$\text{Food in influent} - \text{Food consumed} = \text{Food in effluent} + \text{Food in WAS} \qquad (6\text{-}11)$$

The food in the influent is the product of the concentration of soluble BOD_5 in the influent (S_o) and the flow rate of wastewater (Q). The food that is consumed in the aeration tank is the product of the volume of the tank ($\mathcal{V}$) and the expression for rate of food utilization (Equation 6-7)

$$(\mathcal{V})\left(\frac{\mu_m SX}{Y(K_s + S)}\right) \qquad (6\text{-}12)$$

The food in the effluent is the product of flow rate of treated wastewater leaving the plant ($Q - Q_w$) and the concentration of soluble BOD_5 in the effluent (S). The concentration of soluble BOD_5 in the effluent (S) is the same as that in the aeration tank because

*Suspended solids means that the material will be retained on a filter, unlike dissolved solids such as NaCl. The amount of the suspended solids that volatilizes at 500 ± 50°C is taken to be a measure of active biomass concentration. The presence of nonliving organic particles in the influent wastewater will cause some error (usually small) in the use of volatile suspended solids as a measure of biomass.

**Mixed-liquor volatile suspended solids is a measure of the active biological mass in the aeration tank. The term "mixed liquor" implies a mixture of activated sludge and wastewater. The phrase "volatile suspended solids" has the same meaning as in the definition of X_o.

we have assumed that the aeration tank is completely mixed. Since the BOD_5 is soluble, the secondary settling tank will not change the concentration. Thus, the effluent concentration from the secondary settling tank is the same as the influent concentration.

The food in the waste activated sludge flow is the product of the concentration of soluble BOD_5 in the influent (S) and the WAS flow rate (Q_r). The narrative mass balance equation for steady-state conditions may be rewritten as

$$QS_o - (\forall)\left(\frac{\mu_m SX}{Y(K_s + S)}\right) = (Q - Q_w)S + Q_w S \tag{6-13}$$

where Y = yield coefficient (see Equation 6-6).

To develop working design equations we make the following assumptions:

1. The influent and effluent biomass concentrations are negligible compared to that in the reactor.
2. The influent food (S_o) is immediately diluted to the reactor concentration in accordance with the definition of a CSTR (see Chapter 2).
3. All reactions occur in the CSTR.

From the first assumption we may eliminate the following terms from Equation 6-10: QX_o, and $(Q - Q_w)X_e$ because X_o, and X_e, are negligible compared to X. Equation 6-10 may be simplified to

$$(\forall)\left(\frac{\mu_m SX}{K_s + S} - k_d X\right) = +Q_w X_r \tag{6-14}$$

For convenience, we may rearrange Equation 6-14 in terms of the Monod equation

$$\left(\frac{\mu_m S}{K_s + S}\right) = \frac{Q_w X_r}{\forall X} + k_d \tag{6-15}$$

Equation 6-13 may also be rearranged in terms of the Monod equation

$$\left(\frac{\mu_m S}{K_s + S}\right) = \frac{Q}{\forall}\frac{Y}{X}(S_o - S) \tag{6-16}$$

Noting that the left side of Equations 6-15 and 6-16 are the same, we set the right-hand side of these equations equal and rearrange to give:

$$\frac{Q_w X_r}{\forall X} = \frac{Q}{\forall}\frac{Y}{X}(S_o - S) - k_d \tag{6-17}$$

Two parts of this equation have physical significance in the design of a completely mixed activated sludge system. The inverse of $Q/\forall$ is the *hydraulic detention time* (θ) of the reactor:

$$\frac{\forall}{Q} = \theta \tag{6-18}$$

The inverse of the left side of Equation 6-17 defines the mean cell-residence time (θ_c):

$$\frac{\forall X}{Q_w X_r} = \theta_c \tag{6-19}$$

The mean cell-residence time expressed in Equation 6-19 must be modified if the effluent biomass concentration is not negligible. Equation 6-20 accounts for effluent losses of biomass in calculating θ_c.

$$\theta_c = \frac{\forall X}{Q_w X_r + (Q - Q_w)(X_e)} \tag{6-20}$$

From Equation 6-15, it can be seen that once θ_c is selected, the concentration of soluble BOD_5 in the effluent (S) is fixed:

$$S = \frac{K_s(1 + k_d\theta_c)}{\theta_c(\mu_m - k_d) - 1} \tag{6-21}$$

Typical values of the microbial growth constants are given in Table 6-11. Note that the concentration of soluble BOD_5 leaving the system (S) is affected only by the mean cell-residence time and not by the amount of BOD_5 entering the aeration tank or by the hydraulic detention time. It is also important to reemphasize that S is the soluble BOD_5 and not the total BOD_5. Some fraction of the suspended solids that do not settle in the secondary settling tank also contributes to the BOD_5 load to the receiving body. To achieve a desired effluent quality both the soluble and insoluble fractions of BOD_5 must be considered. Thus, to use Equation 6-21 to achieve a desired effluent quality (S) by solving for θ_c, some estimate of the BOD_5 of the suspended solids must be made first. This value is then subtracted from the total allowable BOD_5 in the effluent to find the allowable S:

$$S = \text{Total BOD}_5 \text{ allowed} - \text{BOD}_5 \text{ in suspended solids} \tag{6-22}$$

From Equation 6-17, it is also evident that the concentration of microorganisms in the aeration tank is a function of the mean cell-residence time, hydraulic detention time, and difference between the influent and effluent concentrations:

$$X = \frac{\theta_c(Y)(S_o - S)}{\theta(1 + k_d\theta_c)} \tag{6-23}$$

TABLE 6-11
Values of growth constants for domestic wastewater[a]

		Value[b]	
Parameter	Basis	Range	Typical
K_s	mg/L BOD_5	25–100	60
k_d	d^{-1}	0–0.30	0.10
μ_m	d^{-1}	1–8	3
Y	mg VSS/mg BOD_5	0.4–0.8	0.6

[a]*Sources:* Metcalf & Eddy, 2003 and Shahriari et al., 2006.
[b]Values are for 20°C.

Plug Flow with Recycle. A plug-flow reactor is one in which the fluid particles pass through the tank in sequence (see Section 2-3 for further discussion). Although it is difficult to achieve true plug flow, many long, narrow aeration tanks may be better approximated by plug flow than by completely mixed models. A kinetic model of a plug-flow system is difficult to develop from basic mass balance equations. With two simplifying assumptions, Lawrence and McCarty (1970) have developed a useful equation. The assumptions are:

1. The concentration of microorganisms in the influent to the aeration tank is approximately the same as that in the effluent from the aeration tank. This assumption applies if θ_c/θ is greater than 5.
2. The rate of soluble BOD_5 utilization as the waste passes through the aeration tank is given by

$$r_u = -\frac{\mu_m S X_{avg}}{K_s + S} \tag{6-24}$$

where X_{avg} is the average concentration of microorganisms in the aeration tank. The design equation is

$$\frac{1}{\theta_c} = \frac{Y\mu_m (S_o - S)}{(S_o - S) + (1 + \alpha)K_s \ln(S_i/S)} - k_d \tag{6-25}$$

where α = recycle ratio, Q_r/Q
ln = logarithm to base e
S_i = influent concentration to aeration tank after dilution with recycle flow, mg/L

$$= \frac{S_o + \alpha S}{1 + \alpha}$$

Other terms are the same as those defined previously.

Example 6-5. The town of Gatesville has been directed to upgrade its primary WWTP to a secondary plant that can meet an effluent standard of 30.0 mg/L BOD_5 and 30.0 mg/L suspended solids (SS). They have selected a completely mixed activated sludge system.

Assuming that the BOD_5 of the SS may be estimated as equal to 63 percent of the SS concentration, estimate the required volume of the aeration tank. The following data are available from the existing primary plant.

Existing plant effluent characteristics

Flow = 0.150 m^3/s

BOD_5 = 84.0 mg/L

Assume the following values for the growth constants: K_s = 100 mg/L BOD_5; $\mu_m = 2.5\ d^{-1}$; $k_d = 0.050\ d^{-1}$; Y = 0.50 mg VSS/mg BOD_5 removed.

Solution. Assuming that the secondary clarifier can produce an effluent with only 30.0 mg/L SS, we can estimate the allowable soluble BOD_5 in the effluent using the 63-percent assumption from above and Equation 6-22.

$$S = BOD_5 \text{ allowed} - BOD_5 \text{ in suspended solids}$$
$$S = 30.0 - (0.630)(30.0) = 11.1 \text{ mg/L}$$

The mean cell-residence time can be estimated with Equation 6-21 and the assumed values for the growth constants.

$$11.1 = \frac{100.0(1 + (0.050)\theta_c)}{\theta_c(2.5 - 0.050) - 1}$$

Solving for θ_c

$$(11.1)(2.45\theta_c - 1) = 100.0 + 5.00\theta_c$$
$$27.20\theta_c - 11.1 = 100.0 + 5.00\theta_c$$
$$\theta_c = \frac{111.1}{22.2} = 5.00 \text{ or } 5.0 \text{ d}$$

If we assume a value of 2,000 mg/L for the MLVSS, we can solve Equation 6-23 for the hydraulic detention time.

$$2{,}000 = \frac{5.00(0.50)(84.0 - 11.1)}{\theta(1 + (0.050)(5.00))}$$
$$\theta = \frac{2.50(72.9)}{2{,}000(1.25)}$$
$$\theta = 0.073 \text{ d or } 1.8 \text{ h}$$

The volume of the aeration tank is then estimated using Equation 6-18:

$$1.8 = \frac{\forall}{0.150 \times 3{,}600 \text{ s/h}}$$
$$\forall = 972 \text{ m}^3 \text{ or } 970 \text{ m}^3$$

Note: This answer results from rounding off the time. If the exact time is used, the volume us 945.63 m^3.

Another commonly used parameter in the activated sludge process is the food to microorganism ratio (F/M), which is defined as:

$$\frac{F}{M} = \frac{QS_o}{\forall X} \tag{6-26}$$

The units of the F/M ratio are

$$\frac{\text{mg BOD}_5\text{/d}}{\text{mg MLVSS}} = \frac{\text{mg}}{\text{mg} \cdot \text{d}}$$

The F/M ratio is controlled by wasting part of the microbial mass, thereby reducing the MLVSS. A high rate of wasting causes a high F/M ratio. A high F/M yields organisms that are saturated with food. The result is that efficiency of treatment is poor. A low rate of wasting causes a low F/M ratio, which yields organisms that are starved. This results in more complete degradation of the waste.

A long θ_c (low F/M) is not always used, however, because of certain trade-offs that must be considered. A long θ_c means a larger and more costly aeration tank. It also means a higher requirement for oxygen and, thus, higher power costs. Problems with poor sludge "settleability" in the final clarifier may be encountered if θ_c is too long. However, because the waste is more completely degraded to final end products and less of the waste is converted into microbial cells when the microorganisms are starved at a low F/M, there is less sludge to handle.

Because both the F/M ratio and the cell-detention time are controlled by wasting of organisms, they are interrelated. A high F/M corresponds to a short θ_c and a low F/M corresponds to a long θ_c. F/M values typically range from 0.1 to 1.0 mg/mg · d for the various modifications of the activated sludge process.

Example 6-6. Two "fill and draw," batch-operated sludge tanks are operated as follows:

Tank A is settled once each day, and half the liquid is removed with care not to disturb the sludge that settles to the bottom. This liquid is replaced with fresh settled sewage. A plot of MLVSS concentration versus time takes the shape shown below.

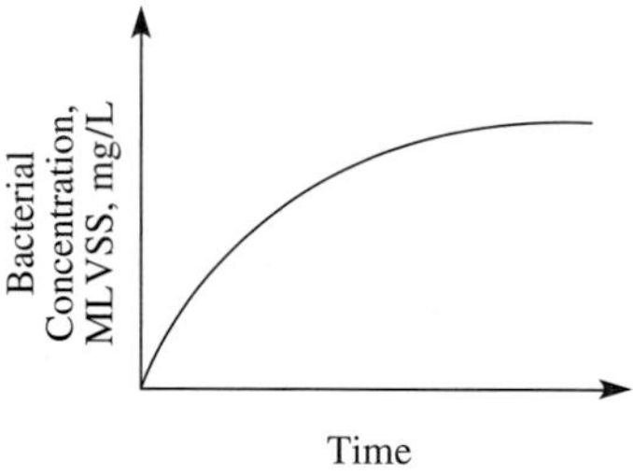

Tank B is not settled. Rather, once each day, half the mixed liquor is removed while the tank is being violently agitated. The liquid is replaced with fresh settled sewage. A plot of MLVSS concentration versus time is shown below.

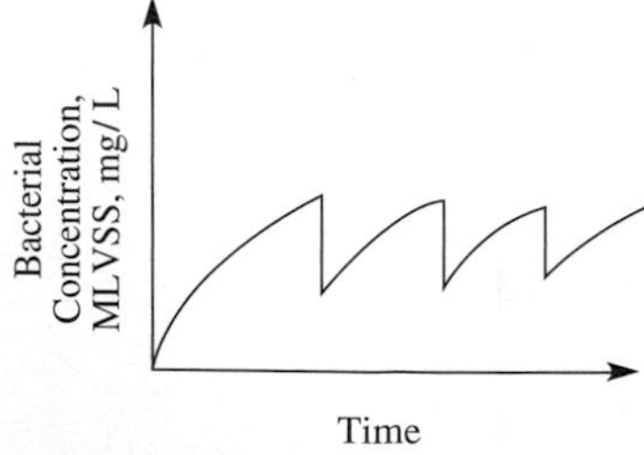

A comparison of the operating characteristics of the two systems is shown in the following table.

Parameter	Tank A	Tank B
F/M	Low	High
θ_c	Long	Short
Sludge	None	Much
Oxygen required	High	Low
Power	High	Low

The optimum choice is somewhere between these extremes. A balance must be struck between the cost of sludge disposal and the cost of power to provide oxygen (air).

Example 6-7. Compute the F/M ratio for the new activated-sludge plant at Gatesville (Example 6-5).

Solution. Using the data from Example 6-5 and Equation 6-26:

$$\frac{F}{M} = \frac{(0.150\ m^3/s)(84.0\ mg/L)(86{,}400\ s/d)}{(970\ m^3)(2{,}000\ mg/L)}$$

$$= 0.56\ mg/mg \cdot d$$

This is well within the typical range of F/M ratios.

Sludge Return. The purpose of sludge return is to maintain a sufficient concentration of activated sludge in the reactor basin. One method used to control the rate of sludge return to the reactor basin is based on the empirical measurement known as the *sludge volume index* (SVI).

SVI is determined from a standard laboratory test (APHA, 2005). The procedure involves measuring the MLSS and sludge settleability. A one-liter sample of mixed liquor is obtained from the aeration tank at the discharge end. The sludge settleability is measured by filling a standard one-liter graduated cylinder to the 1.0 liter mark, allowing undisturbed settling for 30 minutes, and then reading the volume occupied by the settled sludge. The MLSS is determined by filtering, drying, and weighing a second portion of the mixed liquor. The SVI, which is defined as the volume in milliliters occupied by 1 g of activated sludge after the aerated liquor has settled 30 min, is calculated as follows:

$$SVI = \frac{SV}{MLSS} \times 1{,}000\ mg/g \qquad (6\text{-}27)$$

where SVI = sludge volume index, mL/g
SV = volume of settled solids in one-liter graduated cylinder after 30 min settling, mL/L
MLSS = mixed liquor suspended solids, mg/L

Conceptually, SVI can be related to the quantity and solids concentration in the secondary settling tank as we have depicted in Figure 6-23. In the following discussion and mathematical relationships, the secondary tank is assumed to respond identically

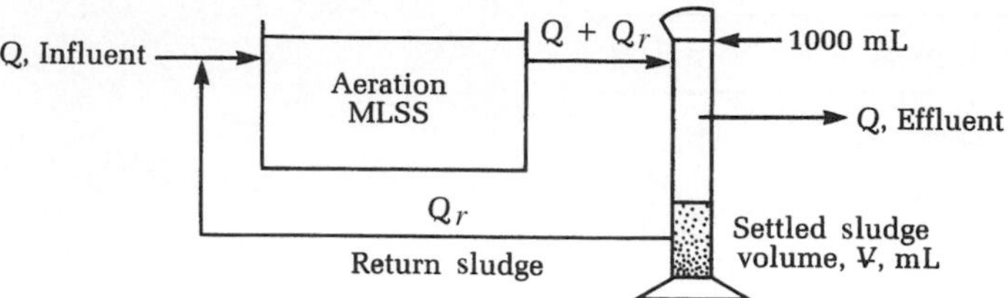

FIGURE 6-23
Hypothetical relationship between settled sludge volume from SVI test and return sludge flow. (*Source:* Hammer, 1977.)

to the graduated cylinder used in the SVI test. This assumption is extraordinary to say the least. In fact, Vesilind (1968) has shown that for MLSS concentrations of less than 5,000 mg/L, the settling rates are 10 to 20 percent greater than might be expected in a final clarifier. Nonetheless, environmental engineers have developed a large body of empirical data based on it.

The SVI can be used as an indication of the settling characteristics of the sludge, thereby impacting on return rates and MLSS. Typical values of SVI for activated sludge plants operating with an MLSS concentration of 2,000 to 3,500 mg/L range from 80 to 150 mL/g. As the sludge concentration is increased to the 3,000 to 5,000 mg/L range, there is a higher solids loading on the settling basin, and as a consequence, a lower SVI or larger settling basin is required to avoid the loss of solids caused by "washout" or hydraulic displacement.

The SVI is a key factor in the system design. Indirectly, it limits the reactor basin MLSS concentration and, in turn, the MLVSS that can be achieved, because it controls the settling tank underflow concentration. Thus, for a given SVI and return sludge rate, the maximum MLSS and MLVSS are fixed within narrow limits.

Most activated sludge plants are designed to permit variable sludge return from 10 to 100 percent of the raw wastewater flow. This range of return sludge flow gives the operator reasonable flexibility to adjust the MLSS to the desired concentration. In general, the return sludge ratio should be limited to or below 100 percent. This is particularly true if the SVI is higher than 150 mL/g and there is no provision for additional floor area in the final clarification step.

Without operating data, the Joint Task Force (WEF, 1992) suggests that MLSS be limited to 5,000 mg/L (lower at temperatures of less than 20°C), even though the SVIs may be very low. Design values over 5,000 mg/L generally will lead to inordinately low detention times that are more subject to washout unless surge control is planned. Design MLSS values should not be any higher than needed, since the final settling basin operations become critical at high MLSS levels.

The mixed liquor concentration as a function of the SVI and the return sludge ratio (Q_r/Q) is shown in Figure 6-24. The return sludge pumping rate may be determined from a mass balance around the settling tank in Figure 6-22. Assuming that the amount of sludge in the secondary settling tank remains constant (steady-state conditions) and that the effluent suspended solids (X_e) are negligible, the mass balance is

$$\text{accumulation} = \text{inflow} - \text{outflow} \tag{6-28}$$

$$0 = (Q + Q_r)(X') - (Q_r X_r' + Q_w X_r') \tag{6-29}$$

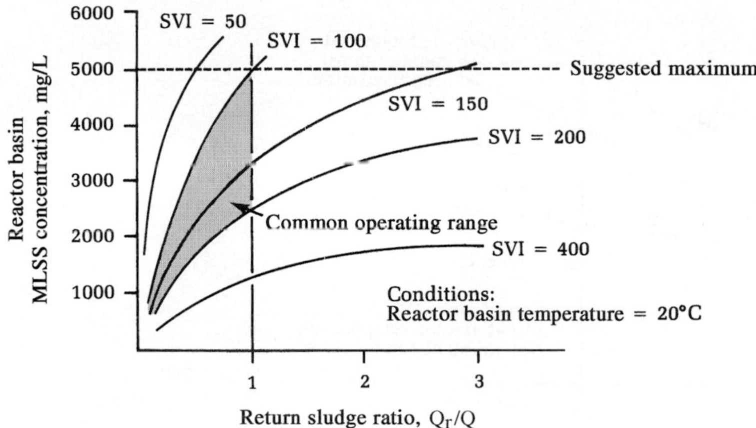

FIGURE 6-24
Design MLSS versus SVI and return sludge ratio. (*Source:* WEF, 1992.)

where Q = wastewater flow rate, m^3/d
Q_r = return sludge flow rate, m^3/d
X' = mixed liquor suspended solids (MLSS), g/m^3
X'_r = maximum return sludge concentration, g/m^3
Q_w = sludge wasting flow rate, m^3/d

Solving for the return sludge flow rate:

$$Q_r = \frac{QX' - Q_w X'_r}{X'_r - X'} \tag{6-30}$$

Frequently, the assumption that the effluent suspended solids are negligible is not valid. If the effluent suspended solids are significant, the mass balance may then be expressed as

$$0 = (Q + Q_r)(X') - (Q_r X'_r + Q_w X'_r + (Q - Q_w)X_e) \tag{6-31}$$

Solving for the return sludge flow rate gives

$$Q_r = \frac{QX' - Q_w X'_r - (Q - Q_w)X_e}{X'_r - X'} \tag{6-32}$$

Note that X'_r and X' include both the volatile and inert fractions. Thus, they differ from X_r and X by a constant factor. With the volume of the tank and the mean cell-residence time, the sludge wasting flow rate can be determined with Equation 6-19 if the maximum return sludge concentration (X'_r) can be determined. The maximum return sludge concentration is related to the SVI as follows:

$$X'_r = \frac{1{,}000 \text{ mg/g}(1{,}000 \text{ mL/L})}{\text{SVI}} = \frac{10^6}{\text{SVI}} \text{ mg/L} \tag{6-33}$$

Figure 6-24 has been constructed on the basis of rapid sludge removal and uses the concentration achieved in the 30-minute settling test as the settling basin underflow concentration. Practice has shown this to be a relatively valid approach.

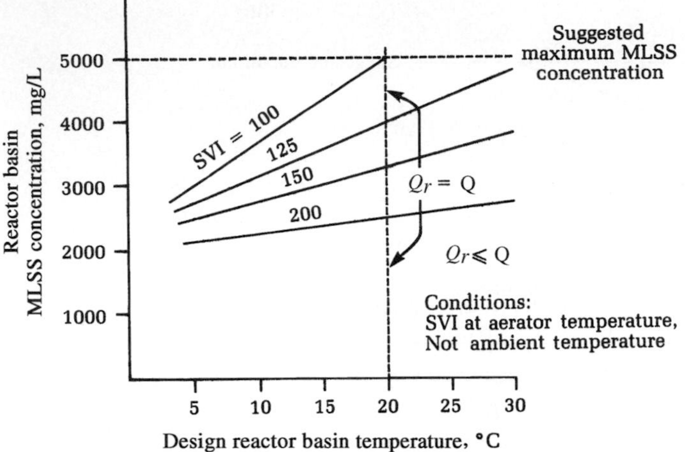

FIGURE 6-25
Recommended maximum MLSS design versus temperature and SVI. (*Source:* WEF, 1992.)

The maximum achievable underflow concentration is also a function of temperature. Temperature affects zone settling velocity, as well as the SVI. In cold weather, the SVI increases because of poor settling. The Joint Task Force's recommended mixed-liquor design concentration as a function of the minimum design reactor-basin temperature for several SVI values is shown in Figure 6-25. (The SVI is taken at the temperature of the reactor basin contents.)

Example 6-8. In the continuing saga of the Gatesville plant expansion, we now wish to consider the question of the return sludge design. Based on the aeration tank design (Example 6-5) and an informed, reliable source, we have the following data:

Design data:

Flow = 0.150 m^3/s

MLVSS(X) = 2,000 mg/L

MLSS(X') = 1.43 (MLVSS)

Effluent suspended solids = 30 mg/L

Wastewater temperature = 18.0°C

Solution. We begin by computing the anticipated concentration of the MLSS.

$$\text{MLSS} = 1.43(2{,}000)$$

$$\text{MLSS} = 2{,}860 \text{ mg/L}$$

We can't really predict SVI but Figure 6-25 gives us a reasonable range to assume a value. Alternatively, we could assume a return sludge concentration. Using Figure 6-25, we select an SVI of 175 based on the calculated MLSS and the reactor basin temperature.

Now, using Equation 6-33, we can determine the return sludge concentration.

$$X'_r = \frac{10^6}{175}$$

$$X'_r = 5{,}714.29 \text{ or } 5{,}700 \text{ mg/L}$$

The return sludge flow rate may be computed using Equations 6-19 and 6-30. Solving Equation 6-19 for the sludge wasting flow rate using the data from Example 6-5 and noting that $X_r = X'_r/1.43 = 3{,}986$ mg/L:

$$Q_w = \frac{\forall X}{\theta_c X_r} = \frac{(970 \text{ m}^3)(2{,}000 \text{ mg/L})}{(5 \text{ d})(3{,}986 \text{ mg/L})} = 97.3 \text{ m}^3/\text{d}$$

Converting Q_w to m^3/s:

$$\frac{(97.3 \text{ m}^3/\text{d})}{(86{,}400 \text{ s/d})} = 0.0011 \text{ m}^3/\text{s}$$

Noting that 1 mg/L = 1 g/m^3, if we ignore the effluent suspended solids, the estimated return sludge flow rate is

$$Q_r = \frac{(0.150 \text{ m}^3/\text{s})(2{,}860 \text{ g/m}^3) - (0.0011 \text{ m}^3/\text{s})(5{,}714 \text{ g/m}^3)}{5{,}714 \text{ g/m}^3 - 2{,}860 \text{ g/m}^3}$$

$$= 0.148 \text{ or } 0.15 \text{ m}^3/\text{s}$$

If the effluent suspended solids are not neglected, the estimated return sludge flow rate is

$$Q_r = \frac{(0.150 \text{ m}^3/\text{s})(2{,}860 \text{ g/m}^3) - (0.0011 \text{ m}^3/\text{s})(5{,}714 \text{ g/m}^3) - (0.150 - 0.0011)(30 \text{ g/m}^3)}{5{,}714 \text{ g/m}^3 - 2{,}860 \text{ g/m}^3}$$

$$= 0.146 \text{ or } 0.15 \text{ m}^3/\text{s}$$

We can check this result by using Figure 6-24. Although it appears high on first consideration, it is a valid result.

Sludge Production. The activated sludge process removes substrate, which exerts an oxygen demand by converting the food into new cell material and degrading this cell material while generating energy. This cell material ultimately becomes sludge that must be disposed of. Despite the problems in doing so, researchers have attempted to develop enough basic information on sludge production to permit a reliable design basis. Heukelekian and Sawyer both reported that a net yield of 0.5 kg MLVSS/kg BOD_5 removed could be expected for a completely soluble organic substrate (Heukelekian et al., 1951, and Sawyer, 1956). Most researchers agree that, depending on the inert solids in the system and the SRT, 0.40 to 0.60 kg MLVSS/kg BOD_5 removed will normally be observed.

The amount of sludge that must be wasted each day is the difference between the amount of increase in sludge mass and the suspended solids (SS) lost in the effluent:

$$\text{Mass to be wasted} = \text{increase in MLSS} - \text{SS lost in effluent} \qquad (6\text{-}34)$$

The net activated sludge produced each day is determined by:

$$Y_{obs} = \frac{Y}{1 + k_d\theta_c} \quad (6\text{-}35)$$

and

$$P_x = Y_{obs}Q(S_o - S)(10^{-3}\ \text{kg/g}) \quad (6\text{-}36)$$

where P_x = net waste activated sludge produced each day in terms of VSS, kg/d
Y_{obs} = observed yield, kg MLVSS/kg BOD_5 removed

Other terms are as defined previously.

The increase in MLSS may be estimated by assuming that VSS is some fraction of MLSS. It is generally assumed that VSS is 60 to 80 percent of MLVSS. Thus, the increase in MLSS in Equation 6-36 may be estimated by dividing P_x by a factor of 0.6 to 0.8 (or multiplying by 1.25 to 1.667). The mass of suspended solids lost in the effluent is the product of the flow rate $(Q - Q_w)$ and the suspended solids concentration (X_e).

Example 6-9. Estimate the mass of sludge to be wasted each day from the new activated sludge plant at Gatesville (Examples 6-5 and 6-8).

Solution. Using the data from Example 6-5, calculate Y_{obs}:

$$Y_{obs} = \frac{0.50\ \text{kg VSS/kg BOD}_5\ \text{removed}}{1 + [(0.050\ \text{d}^{-1})(5\,\text{d})]}$$
$$= 0.40\ \text{kg VSS/kg BOD}_5\ \text{removed}$$

The net waste activated sludge produced each day is

$$P_x = (0.40)(0.150\ \text{m}^3/\text{s})(84.0\ \text{g/m}^3 - 11.1\ \text{g/m}^3)(86{,}400\ \text{s/d})(10^{-3}\ \text{kg/g})$$
$$= 377.9\ \text{kg/d of VSS}$$

The total mass produced includes inert materials. Using the relationship between MLSS and MLVSS in Example 6-8,

$$\text{Increase in MLSS} = (1.43)(377.9\ \text{kg/d}) = 540.4\ \text{kg/d}$$

The mass of solids (both volatile and inert) lost in the effluent is

$$(Q - Q_w)(X_e) = (0.150\ \text{m}^3/\text{s} - 0.001\ \text{m}^3/\text{s})(30\ \text{g/m}^3)(86{,}400\ \text{s/d})(10^{-3}\ \text{kg/g})$$
$$= 385.9\ \text{kg/d}$$

The mass to be wasted is then

$$\text{Mass to be wasted} = 540.4 - 385.9 = 154.5\ \text{kg/d}$$

Note that this mass is calculated as dry solids. Because the sludge is mostly water, the actual mass will be considerably larger. This is discussed further in Section 6-12.

Oxygen Demand. Oxygen is used in those reactions required to degrade the substrate to produce the high-energy compounds required for cell synthesis and for respiration. For long SRT systems, the oxygen needed for cell maintenance can be of the same order of magnitude as substrate metabolism. A minimum residual of 0.5 to 2 mg/L DO is usually maintained in the reactor basin to prevent oxygen deficiencies from limiting the rate of substrate removal.

An estimate of the oxygen requirements may be made from the BOD_5 of the waste and amount of activated sludge wasted each day. If we assume all of the BOD_5 is converted to end products, the total oxygen demand can be computed by converting BOD_5 to BOD_L. Because a portion of waste is converted to new cells that are wasted, the BOD_L of the wasted cells must be subtracted from the total oxygen demand. An approximation of the oxygen demand of the wasted cells may be made by assuming cell oxidation can be described by the following reaction:

$$\underbrace{C_5H_7NO_2}_{\text{cells}} + 5O_2 \rightleftharpoons 5CO_2 + 2H_2O + NH_3 + \text{energy} \qquad (6\text{-}37)$$

The ratio of gram molecular weights is

$$\frac{5(32)}{113} = 1.42$$

Thus the oxygen demand of the waste activated sludge may be estimated as 1.42 (P_x).

The mass of oxygen required may be estimated as:

$$M_{O_2} = \frac{Q(S_o - S)(10^{-3}\ \text{kg/g})}{f} - 1.42(P_x) \qquad (6\text{-}38)$$

where Q = wastewater flow rate into the aeration tank, m^3/d
S_o = influent soluble BOD_5, mg/L
S = effluent soluble BOD_5, mg/L
f = conversion factor for converting BOD_5 to ultimate BOD_L
P_x = waste activated sludge produced (see Equation 6-36)

The volume of air to be supplied must take into account the percent of air that is oxygen and the transfer efficiency of the dissolution of oxygen into the wastewater.

Example 6-10. Estimate the volume of air to be supplied (m^3/d) for the new activated sludge plant at Gatesville (Examples 6-5 and 6-9). Assume that BOD_5 is 68 percent of the ultimate BOD and that the oxygen transfer efficiency is 8 percent.

Solution. Using the data from Examples 6-5 and 6-9 gives

$$M_{O_2} = \frac{(0.150\ m^3/s)(84.0\ g/m^3 - 11.1\ g/m^3)(86{,}400\ s/d)(10^{-3}\ kg/g)}{0.68}$$

$$1.42(377.9\ \text{kg/d of VSS})$$

$$= 1{,}389.4 - 536.6 = 852.8\ \text{kg/d of oxygen}$$

From Table A-5 in Appendix A, air has a density of 1.185 kg/m^3 at standard conditions. By mass, air contains about 23.2 percent oxygen. At 100 percent transfer efficiency, the volume of air required is

$$\frac{852.8\ \text{kg/d}}{(1.185\ kg/m^3)(0.232)} = 3{,}101.99 \text{ or } 3{,}100\ m^3/d$$

At 8 percent transfer efficiency the volume of air required is

$$\frac{3,101.99 \text{ m}^3/\text{d}}{0.08} = 38,774.9 \text{ or } 38,000 \text{ m}^3/\text{d}$$

Process Design Considerations. The SRT (i.e., θ_c) selected for design is a function of the degree of treatment required. A high SRT (or older sludge age) results in a higher quantity of solids being carried in the system and a higher degree of treatment being obtained. A long SRT also results in the production of less waste sludge.

SRT values for design of carbonaceous BOD_5 removal as a function of the minimum temperature at which the reactor basin will be operated are depicted in Figure 6-26. The SRT values given are those for normal domestic wastewater. It is expected that the soluble BOD_5 in the effluent from the aeration system will be 4 to 8 mg/L.

If industrial wastes are discharged to the municipal system, several additional concerns must be addressed. Municipal wastewater generally contains sufficient nitrogen and phosphorus to support biological growth. The presence of large volumes of industrial wastewater that is deficient in either of these nutrients will result in poor removal efficiencies. Addition of supplemental nitrogen and phosphorus may be required. The ratio of nitrogen to BOD_5 should be 1:32. The ratio of phosphorus to BOD_5 should be 1:150.

Although toxic metals and organic compounds may be at low enough levels that they do not interfere with the operation of the plant, two other untoward effects may result if they are not excluded in a pretreatment program. Volatile organic compounds may be stripped from solution into the atmosphere in the aeration tank. Thus, the WWTP may become a source of air pollution. The toxic metals may precipitate into the waste sludges. Thus, otherwise nonhazardous sludges may be rendered hazardous.

Oil and grease that pass through the primary treatment system will form grease balls on the surface of the aeration tank. The microorganisms cannot degrade this material because it is not in the water where they can physically come in contact with

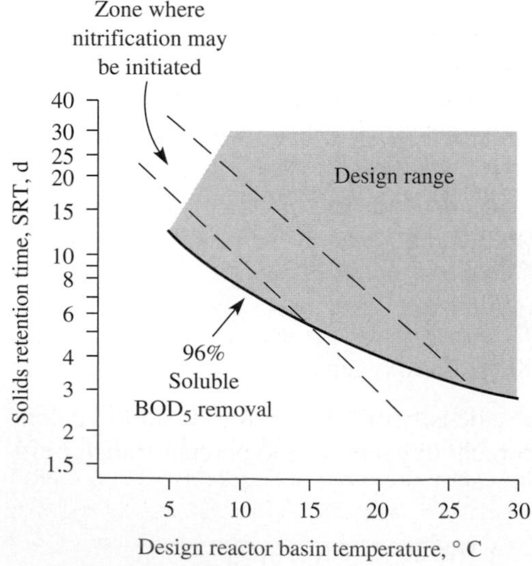

FIGURE 6-26
Design SRT for carbonaceous BOD_5 removal. (*Source:* WEF, 1992.)

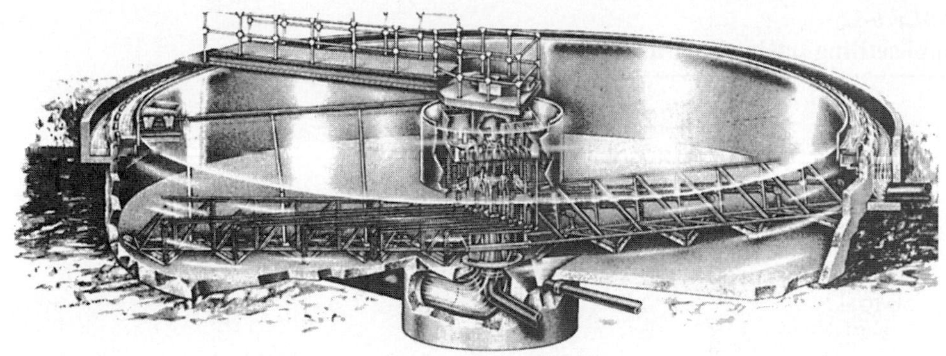

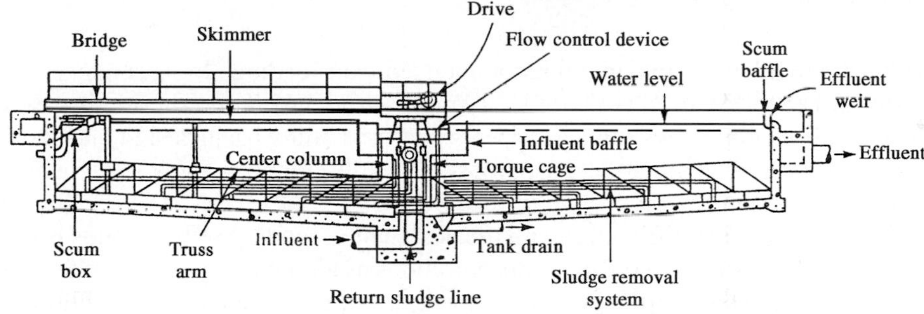

FIGURE 6-27
Rendering and cross-sectional diagram of secondary settling tank. (Courtesy of FMC Corporation.)

it. Special consideration should be given to the surface skimming equipment in the secondary clarifier to handle the grease balls.

Secondary Clarifier Design Considerations. Although the secondary settling tank (Figure 6-27) is an integral part of both the trickling filter and the activated sludge process, environmental engineers have focused particular attention on the secondary clarifier used after the activated sludge process. A secondary clarifier is important because of the high solids loading and fluffy nature of the activated sludge biological floc. Also, it is highly desirable that sludge recycle be well thickened.

Secondary settling tanks for activated sludge are generally characterized as having Type III settling. Some authors would argue that Types I and II also occur.

The following guidance has been excerpted from the Joint Task Force of the Water Pollution Control Federation and the American Society of Civil Engineers (WEF, 1992). The design factors discussed here are the result of the experiences of investigators, plant superintendents, and equipment manufacturers. The criteria primarily apply to circular (or square) center-fed basins, which comprise the majority of activated sludge secondary settling units designed in the last 40 years.

An overflow rate between 20 and 34 m/d for the average flow in a conventional process can be expected to result in good separation of liquid and SS. The design engineer also must check the peak hydraulic rates that will be imposed on the settling basin.

Suggested secondary settling tank *side water depths* (SWD) and solids loading rates are shown in Table 6-12 and Figure 6-28, respectively.

TABLE 6-12
Final settling basin side water depth

	Side water depth, m	
Tank diameter, m	Minimum	Recommended
<12	3.0	3.4
12 to 20	3.4	3.7
20 to 30	3.7	4.0
30 to 42	4.0	4.3
>42	4.3	4.6

Source: Adapted from WEF, 1992.

The GLUMRB has set maximum recommended weir loadings for secondary settling basins at 125 to 250 m^3/d per m of weir length (m^3/d · m). This criterion is based on effluent quality of operating units. It appears that the settling basin design may have had much to do with this limitation. Also, most of the observations apply to rectangular settling basins.

One of the continuing difficulties with the design of secondary settling tanks is the prediction of effluent suspended solids concentrations as a function of common design and operating parameters. Little theoretical work has been conducted, and empirical correlations have been less than satisfactory (Rex Chainbelt, Inc., 1972).

Example 6-11. The secondary clarifier for the Gatesville plant (Examples 6-5, and 6-7–6-10) must be able to handle an MLSS load of 2,860 mg/L. The flow to the secondary clarifier is 0.300 m^3/s, of which one-half is contributed by the return flow. Determine the tank diameter, depth, and weir length.

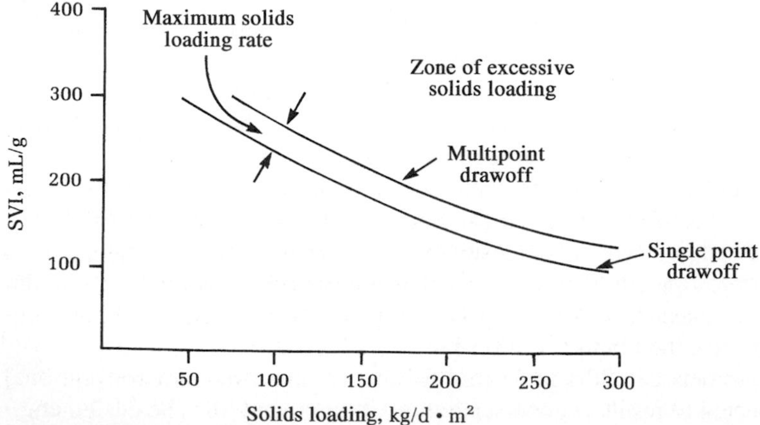

FIGURE 6-28
Design solids loading versus SVI. (*Note:* Rapid sludge removal design assumes that there will be no inventory in the settling tank. *Source:* WEF, 1992.)

Solution. Utilizing an average overflow rate of 33 m/d, we determine the diameter of the secondary tank as follows:

First, compute the surface area required:

$$A_s = \frac{0.150 \text{ m}^3\text{/s} \times 86{,}400 \text{ s/d}}{33\,\text{m/d}}$$
$$= 392.73 \text{ m}^2$$

Note that because only one-half of the flow leaves through the surface of the tank (the remainder leaves through the bottom as Q_r), only one-half of the hydraulic load is used to compute the surface area.

The diameter of the tank is then

$$392.73 = \frac{\pi D^2}{4}$$
$$D = (500.04)^{1/2} = 22.36 \text{ or } 22 \text{ m}$$

From Table 6-12 we select an SWD of 4.0 m.

Now we must check the solids loading. Using the equality 1 mg/L = 1 g/m^3:

$$\text{SL} = \frac{2{,}860 \text{ g/m}^3 \times 0.300 \text{ m}^3\text{/s}}{\dfrac{\pi(22 \text{ m})^2}{4}}$$
$$= \frac{858.00 \text{ g/s}}{380.13 \text{ m}^2} \times 10^{-3} \text{ kg/g} \times 86{,}400 \text{ s/d}$$
$$= 195\,\text{kg/d} \cdot \text{m}^2$$

Checking this rate with the maxima shown in Figure 6-28, we find that for an SVI of 175 (from Example 6-8), we have the maximum allowable loading of 200 kg/d · m^2.

The weir loading for a single weir located at the periphery is

$$\text{WL} = \frac{0.150 \times 86{,}400 \text{ s/d}}{\pi(22)}$$
$$= 187.51 \text{ or } 190 \text{ m}^3\text{/d} \cdot \text{m}$$

This is less than the prescribed loading given by GLUMRB and, therefore, is acceptable.

Sludge problems. A *bulking sludge* is one that has poor settling characteristics and poor compactability. There are two principal types of sludge bulking. The first is caused by the growth of filamentous organisms, and the second is caused by water trapped in the bacterial floc, thus reducing the density of the agglomerate and resulting in poor settling.

Filamentous bacteria have been blamed for much of the bulking problem in activated sludge. Although filamentous organisms are effective in removing organic matter, they have poor floc-forming and settling characteristics. Bulking may also be caused by a number of other factors, including long, slow-moving collection-system transport; low available ammonia nitrogen when the organic load is high; low pH, which could favor acid-favoring fungi; and the lack of macronutrients, which stimulates predomination of the filamentous actinomycetes over the normal floc-forming bacteria. The lack of nitrogen also favors slime-producing bacteria, which have a low specific gravity, even though they are not filamentous. The multicellular fungi cannot compete with the bacteria normally, but can compete under specific environmental conditions, such as low pH, low nitrogen, low oxygen, and high carbohydrates. As the pH decreases below 6.0, the fungi are less affected than the bacteria and tend to predominate. As the nitrogen concentrations drop below a $BOD_5 : N$ ratio of 20:1, the fungi, which have a lower protein level than the bacteria, are able to produce normal protoplasm, while the bacteria produce nitrogen-deficient protoplasm.

A sludge that floats to the surface after apparently good settling is called a *rising sludge.* Rising sludge results from denitrification, that is, reduction of nitrates and nitrites to nitrogen gas in the sludge blanket (layer). Much of this gas remains trapped in the sludge blanket, causing globs of sludge to rise to the surface and float over the weirs into the receiving stream.

Rising-sludge problems can be overcome by increasing the rate of return sludge flow (Q_r), by increasing the speed of the sludge-collecting mechanism, by decreasing the mean cell residence time, and, if possible, by decreasing the flow from the aeration tank to the offending tank.

Trickling Filters

A typical process flow diagram for a trickling filter plant is shown in Figure 6-29. The trickling filter itself consists of a bed of coarse material, such as stones, slats, or plastic materials (*media*), over which wastewater is applied. Because the microorganisms that biodegrade the waste form a film on the media, this process is known as an *attached growth* process. Trickling filters have been a popular biologic treatment process. The most widely used design for many years was simply a bed of stones from 1 to 3 m deep through which the wastewater passed. The wastewater is typically distributed over the surface of the rocks by a rotating arm (Figure 6-30). Rock filter diameters may range up to 60 m.

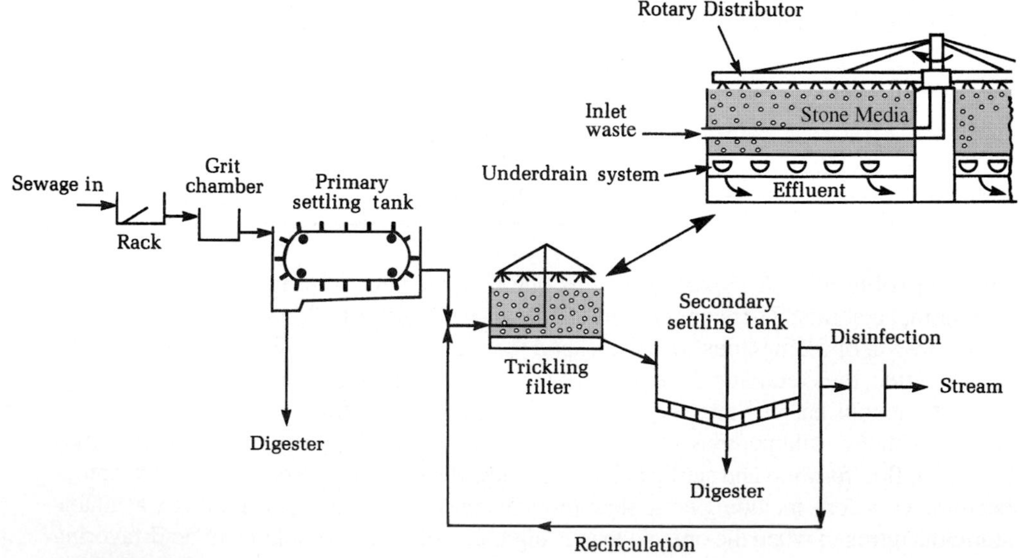

FIGURE 6-29
Trickling-filter plant with enlargement of trickling filter.

(a)

(b)

FIGURE 6-30
Trickling filters, rock media (a) and synthetic media (b). (*Sources:* (a) Wyoming, MI WWIP, photo by M. L. Davis and (b) Brentwood Industries.)

As the wastewater trickles through the bed, a microbial growth establishes itself on the surface of the stone or packing in a fixed film. The wastewater passes over the stationary microbial population, providing contact between the microorganisms and the organic contaminants.

Trickling filters are not primarily a filtering or straining process as the name implies. The rocks in a rock filter are 25 to 100 mm in diameter, and hence have openings too large to strain out solids. They are a means of providing large amounts of surface area where the microorganisms cling and grow in a slime on the rocks as they feed on the organic matter.

Excess growths of microorganisms wash from the rock media and would cause undesirably high levels of suspended solids in the plant effluent if not removed. Thus, the flow from the filter is passed through a sedimentation basin to allow these solids to settle out. As in the case of the activated sludge process this sedimentation basin is referred to as a *secondary clarifier,* or *final clarifier.*

Although rock trickling filters have performed well for years, they have certain limitations. Under high organic loadings, the slime growths can be so prolific that they plug the void spaces between the rocks, causing flooding and failure of the system. Also, the volume of void spaces is limited in a rock filter, which restricts the circulation of air and the amount of oxygen available for the microbes. This limitation, in turn, restricts the amount of wastewater that can be processed.

To overcome these limitations, other materials have become popular for filling the trickling filter. These materials include modules of corrugated plastic sheets and plastic rings. These media offer larger surface areas for slime growths (typically 90 square meters of surface area per cubic meter of bulk volume, as compared to 40 to 60 square meters per cubic meter for 75 mm rocks) and greatly increase void ratios for increased air flow. The materials are also much lighter than rock (by a factor of about 30), so that the trickling filters can be much taller without facing structural problems. While rock in filters is usually not more than 3 m deep, synthetic media depths may reach 12 m, thus reducing the overall space requirements for the trickling-filter portion of the treatment plant.

Trickling filters are classified according to the applied hydraulic and organic load. The hydraulic load may be expressed as cubic meters of wastewater applied per day per square meter of bulk filter surface area ($m^3/d \cdot m^3$) or, preferably, as the depth of water applied per unit of time (mm/s or m/d). Organic loading is expressed as kilograms of BOD_5 per day per cubic meter of bulk filter volume ($kg/d \cdot m^3$). Common hydraulic and organic loadings for the the various filter classifications are summarized in Table 6-13.

An important element in trickling filter design is the provision for return of a portion of the effluent to flow through the filter. This practice is called *recirculation.* The ratio of the returned flow to the incoming flow is called the *recirculation ratio* (r). Recirculation is practiced in stone filters for the following reasons:

1. To increase contact efficiency by bringing the waste into contact more than once with active biological material.
2. To dampen variations in loadings over a 24-hour period. The strength of the recirculated flow lags behind that of the incoming wastewater. Thus, recirculation dilutes strong influent and supplements weak influent.
3. To raise the DO of the influent.

TABLE 6-13
Comparison of different types of trickling filters[a]

	Trickling filter classification				
Design characteristics	Low or standard rate	Intermediate rate	High rate (stone media)	Super rate (plastic media)	Roughing
Hydraulic loading, m/d	1 to 4	4 to 10	10 to 40	15 to 90[b]	60 to 180[b]
Organic loading, kg $BOD_5/d \cdot m^3$	0.08 to 0.32	0.24 to 0.48	0.32 to 1.0	0.32 to 1.0	Above 1.0
Recirculation ratio	0	0 to 1	1 to 3	0 to 1	1 to 4
Filter flies	Many	Varies	Few	Few	Few
Sloughing	Intermittent	Varies	Continuous	Continuous	Continuous
Depth, m	1.5 to 3	1.5 to 2.5	1 to 2	Up to 12	1 to 6
BOD_5 removal, %	80 to 85	50 to 70	40 to 80	65 to 85	40 to 85
Effluent quality	Well nitrified	Some nitrification	Nitrites	Limited nitrification	No nitrification

[a]*Source:* Adapted from WEF, 1992.
[b]Not including recirculation.

4. To improve distribution over the surface, thus reducing the tendency to clog and also reduce filter flies.
5. To prevent the biological slimes from drying out and dying during nighttime periods when flows may be too low to keep the filter wet continuously.

Recirculation may or may not improve treatment efficiency. The more dilute the incoming wastewater, the less likely it is that recirculation will improve efficiency.

Recirculation is practiced for plastic media to provide the desired wetting rate to keep the microorganisms alive. Generally, increasing the hydraulic loading above the minimum wetting rate does not increase BOD_5 removal. The minimum wetting rate normally falls in the range of 25 to 60 m/d.

Two-stage trickling filters (Figure 6-31) provide a means of improving the performance of filters. The second stage acts as a polishing step for the effluent from the primary stage by providing additional contact time between the waste and the microorganisms. Both stages may use the same media or each stage may have different media as shown in Figure 6-31. The designer will select the types of media and their arrangement based on the desired treatment efficiencies and an economic analysis of the alternatives.

Design Formulas. Numerous investigators have attempted to correlate operating data with the bulk design parameters of trickling filters. Rather than attempt a comprehensive review of these formulations, we have selected the National Research Council (NRC, 1946) equations and Schulze's equation (Schulze, 1960) as illustrations. A thorough review of several of the more important equations is given in the Water Environment Federation's publication on wastewater treatment plant design (WEF, 1992).

During World War II, the NRC made an extensive study of the operation of trickling filters serving military installations. From this study, empirical equations were

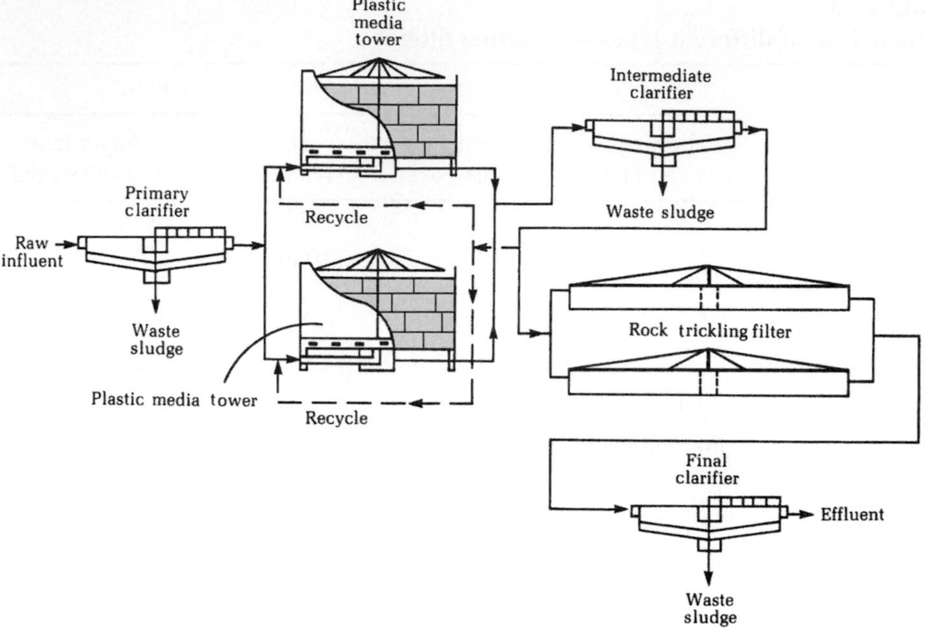

FIGURE 6-31
Two-stage trickling-filter plant. (Courtesy of Dow Chemical Company.)

developed to predict the efficiency of the filters based on the BOD load, the volume of the filter media, and the recirculation. For a single-stage filter or the first stage of a two-stage filter, the efficiency is

$$E_1 = \frac{1}{1 + 4.12\left(\dfrac{QC_{\text{in}}}{\forall F}\right)^{0.5}} \tag{6-39}$$

where E_1 = fraction of BOD_5 removal for first stage at 20°C, including recirculation and sedimentation
Q = wastewater flow rate, m^3/s
C_{in} = influent BOD_5, mg/L
$\forall$ = volume of filter media, m^3
F = recirculation factor

The recirculation factor is

$$F = \frac{1 + R}{(1 + 0.1\,R)^2} \tag{6-40}$$

where R = recirculation ratio = Q_r/Q
Q_r = recirculation flow rate, m^3/s
Q = wastewater flow rate, m^3/s

The recirculation factor represents the average number of passes of the raw wastewater BOD through the filter. The factor $0.1\,R$ is to account for the empirical observation that the biodegradability of the organic matter decreases as the number of passes increases. For the second stage filter, the efficiency is

$$E_2 = \frac{1}{1 + \dfrac{4.12}{1 - E_1}\left(\dfrac{QC_e}{\forall F}\right)^{0.5}} \tag{6-41}$$

where E_2 = fraction of BOD_5 removal for second stage filter at 20°C, including recirculation and sedimentation

E_1 = fraction of BOD_5 removed in first stage

C_e = effluent BOD_5 from first stage, mg/L

The effect of temperature on the efficiency may be estimated from the following equation:

$$E_T = E_{20}\theta^{(T-20)} \tag{6-42}$$

where a value of 1.035 is used for θ.

Some care should be used in applying the NRC equations. Military wastewater during this period (World War II) had a higher strength than domestic wastewater today. The filter media was rock. Clarifiers associated with the trickling filters were shallower and carried higher hydraulic loads than current practice would permit. The second stage filter is assumed to be preceded by an intermediate settling tank (see Figure 6-31).

Example 6-12. Using the NRC equations, determine the BOD_5 of the effluent from a single-stage, low-rate trickling filter that has a filter volume of 1,443 m^3, a hydraulic flow rate of 1,900 m^3/d, and a recirculation factor of 2.78. The influent BOD_5 is 150 mg/L.

Solution. To use the NRC equation, the flow rate must first be converted to the correct units.

$$Q = (1{,}900 \text{ m}^3/\text{d})\left(\frac{1}{86{,}400 \text{ s/d}}\right) = 0.022 \text{ m}^3/\text{s}$$

The efficiency of a single-stage filter is

$$E_1 = \frac{1}{1 + 4.12\left(\dfrac{(0.022)(150)}{(1{,}443)(2.78)}\right)^{0.5}} = 0.8943$$

The concentration of BOD_5 in the effluent is then

$$C_e = (1 - 0.8943)(150) = 15.8 \text{ mg/L}$$

Schulze (1960) proposed that the time of wastewater contact with the biological mass in the filter is directly proportional to the depth of the filter and inversely proportional to the hydraulic loading rate:

$$t = \frac{CD}{(Q/A)^n} \tag{6-43}$$

where t = contact time, d
C = mean active film per unit volume
D = filter depth, m
Q = hydraulic loading, m^3/d
A = filter area over which wastewater is applied, m^2
n = empirical constant based on filter media

The mean active film per unit volume may be approximated by

$$C \simeq \frac{1}{D^m} \tag{6-44}$$

where m is an empirical constant that is an indicator of biological slime distribution. It is normally assumed that the distribution is uniform and that $m = 0$. Thus, C is 1.0.

Schulze combined his relationship with Velz's (1948) first-order equation for BOD removal

$$\frac{S_t}{S_o} = \exp\left[-\frac{KD}{(Q/A)^n}\right] \tag{6-45}$$

where K is an empirical rate constant with the units of

$$\frac{(\text{m/d})^n}{\text{m}}$$

The values of K and n determined by Shulze were 0.69 $(\text{m/d})^n/\text{m}$ and 0.67 at 20°C. The temperature correction for K may be computed with Equation 6-42 if K_T is substituted for E_T and K_{20} is substituted for E_{20}.

Example 6-13. Determine the BOD_5 of the effluent from a low-rate trickling filter that has a diameter of 35.0 m and a depth of 1.5 m if the flow rate is 1,900 m^3/d and the influent BOD_5 is 150.0 mg/L. Assume the rate constant is 2.3 $(\text{m/d})^n/\text{m}$ and $n = 0.67$.

Solution. We begin by computing the area of the filter.

$$A = \frac{\pi(35.0)^2}{4}$$
$$= 962.11\,\text{m}^2$$

This area is then used to compute the loading rate.

$$\frac{Q}{A} = \frac{1{,}900\ \text{m}^3/\text{d}}{962.11\ \text{m}^2}$$
$$= 1.97\ \text{m}^3/\text{d}\cdot\text{m}^2$$

Now we can compute the effluent BOD using Equation 6-45

$$S_t = (150)\exp\left[\frac{-(2.3)(1.5)}{(1.97)^{0.67}}\right]$$
$$= 16.8\,\text{mg/L}$$

Oxidation Ponds

Treatment ponds have been used to treat wastewater for many years, particularly as wastewater treatment systems for small communities (Benefield and Randall, 1980). Many terms have been used to describe the different types of systems employed in wastewater treatment. For example, in recent years, *oxidation pond* has been widely used as a collective term for all types of ponds. Originally, an oxidation pond was a pond that received partially treated wastewater, whereas a pond that received raw wastewater was known as a *sewage lagoon. Waste stabilization pond* has been used as an all-inclusive term that refers to a pond or lagoon used to treat organic waste by biological and physical processes. These processes would commonly be referred to as self-purification if they took place in a stream. To avoid confusion, the classification to be employed in this discussion will be as follows (Caldwell et al., 1973):

1. *Aerobic ponds:* Shallow ponds, less than 1 m in depth, where dissolved oxygen is maintained throughout the entire depth, mainly by the action of photosynthesis.
2. *Facultative ponds:* Ponds 1 to 2.5 m deep, which have an anaerobic lower zone, a facultative middle zone, and an aerobic upper zone maintained by photosynthesis and surface reaeration.
3. *Anaerobic ponds:* Deep ponds that receive high organic loadings such that anaerobic conditions prevail throughout the entire pond depth.
4. *Maturation or tertiary ponds:* Ponds used for polishing effluents from other biological processes. Dissolved oxygen is furnished through photosynthesis and surface reaeration. This type of pond is also known as a *polishing pond.*
5. *Aerated lagoons:* Ponds oxygenated through the action of surface or diffused air aeration.

Aerobic Ponds. The aerobic pond is a shallow pond in which light penetrates to the bottom, thereby maintaining active algal photosynthesis throughout the entire system. During the daylight hours, large amounts of oxygen are supplied by the photosynthesis process; during the hours of darkness, wind mixing of the shallow water mass generally provides a high degree of surface reaeration. Stabilization of the organic material entering an aerobic pond is accomplished mainly through the action of aerobic bacteria.

Anaerobic Ponds. The magnitude of the organic loading and the availability of dissolved oxygen determine whether the biological activity in a treatment pond will occur under aerobic or anaerobic conditions. A pond may be maintained in an anaerobic condition by applying a BOD_5 load that exceeds oxygen production from photosynthesis. Photosynthesis can be reduced by decreasing the surface area and increasing the depth. Anaerobic ponds become turbid from the presence of reduced metal sulfides. This restricts light penetration to the point that algal growth becomes negligible. Anaerobic treatment of complex wastes involves three stages. In the first stage organic matter is hydrolyzed. In the second stage (known as acid fermentation), complex organic materials are broken down mainly to short-chain acids and alcohols. In the third stage (known as methane fermentation), these materials are converted to gases, primarily methane and carbon dioxide. The proper design of anaerobic ponds must result in environmental conditions favorable to methane fermentation.

Anaerobic ponds are used primarily as a pretreatment process and are particularly suited for the treatment of high-temperature, high-strength wastewaters. However, they have been used successfully to treat municipal wastewaters as well.

Facultative Ponds. Of the five general classes of lagoons and ponds, facultative ponds are by far the most common type selected as wastewater treatment systems for small communities. Approximately 25 percent of the municipal wastewater treatment plants in the United States are ponds and about 90 percent of these ponds are located in communities of 5,000 people or fewer. Facultative ponds are popular for such treatment situations because long retention times facilitate the management of large fluctuations in wastewater flow and strength with no significant effect on effluent quality. Also capital, operating, and maintenance costs are less than those of other biological systems that provide equivalent treatment.

A schematic representation of a facultative pond operation is given in Figure 6-32. Raw wastewater enters at the center of the pond. Suspended solids contained in the wastewater settle to the pond bottom, where an anaerobic layer develops. Microorganisms occupying this region do not require molecular oxygen as an electron acceptor in energy metabolism, but rather use some other chemical species. Both acid fermentation and methane fermentation occur in the bottom sludge deposits.

The facultative zone exists just above the anaerobic zone. This means that molecular oxygen will not be available in the region at all times. Generally, the zone is aerobic during the daylight hours and anaerobic during the hours of darkness.

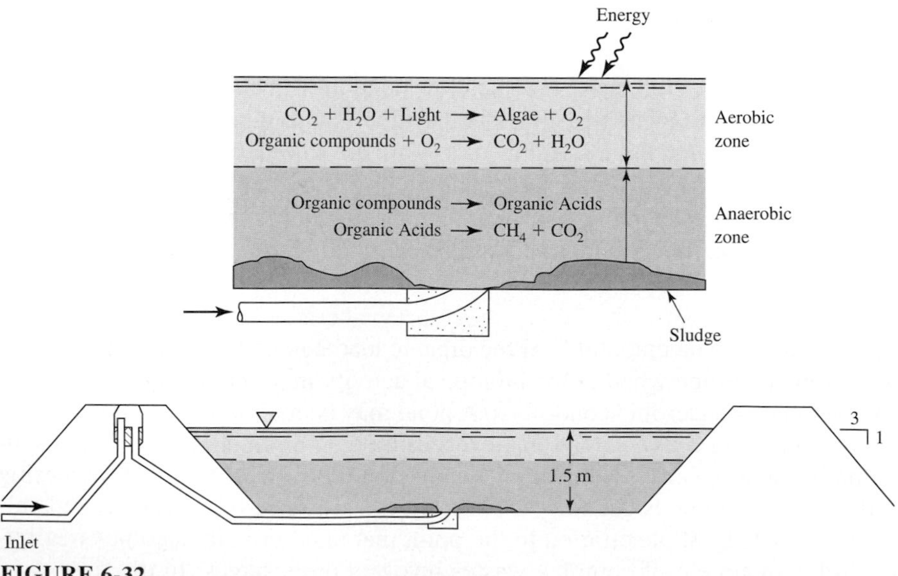

FIGURE 6-32
Schematic diagram of facultative pond relationships.

Above the facultative zone, there exists an aerobic zone that has molecular oxygen present at all times. The oxygen is supplied from two sources. A limited amount is supplied from diffusion across the pond surface. However, the majority is supplied through the action of algal photosynthesis.

Two rules of thumb commonly used in Michigan in evaluating the design of facultative lagoons are as follows:

1. The BOD_5 loading rate should not exceed 22 kg/ha · d on the smallest lagoon cell.
2. The detention time in the lagoon (considering the total volume of all cells but excluding the bottom 0.6 m in the volume calculation) should be six months.

The first criterion is to prevent the pond from becoming anaerobic. The second criterion is to provide enough storage to hold the wastewater during winter months when the receiving stream may be frozen or during the summer when the flow in the stream might be too low to absorb even a small amount of BOD.

Example 6-14. A lagoon having three cells, each 115,000 m^2 in area, a minimum operating depth of 0.6 m, and a maximum operating depth of 1.5 m, receives 1,900 m^3/d of wastewater having an average BOD_5 of 122 mg/L. What is the BOD_5 loading and what is the detention time?

Solution. To compute the BOD loading, we must first compute the mass of BOD_5 entering each day.

$$BOD_5 \text{ mass} = (122 \text{ mg/L})(1{,}900 \text{ m}^3)(1{,}000 \text{ L/m}^3)(1 \times 10^{-6} \text{ kg/mg}) = 231.8 \text{ kg/d}$$

Then, we must convert the area into hectares. Using only one cell,

$$\text{Area} = (115{,}000 \text{ m}^2)(1 \times 10^{-4} \text{ ha/m}^2)$$
$$= 11.5 \text{ ha each}$$

Now we can compute the loading.

$$BOD_5 \text{ loading} = \frac{231.8 \text{ kg/d}}{11.5 \text{ ha}} = 20.2 \text{ kg/ha} \cdot \text{d}$$

This loading rate is acceptable.

The detention time is simply the working volume between the minimum and maximum operating levels divided by the average daily flow.

$$\text{Detention time} = \frac{(115{,}000 \text{ m}^2)(3 \text{ lagoons})(1.5 - 0.6 \text{ m})}{1{,}900 \text{ m}^3/\text{d}}$$
$$= 163.4 \text{ days}$$

This is less than the desired 180 days.

We have ignored the slope of the lagoon walls in this calculation. For large lagoons, this is probably acceptable. In small lagoons, the slope should be considered.

Rotating Biological Contactors (RBCs)

The RBC process consists of a series of closely spaced discs (3 to 3.5 m in diameter) mounted on a horizontal shaft and rotated, while about one-half of their surface area is immersed in wastewater (Figure 6-33). The discs are typically constructed of lightweight plastic. The speed of rotation of the discs is adjustable.

When the process is placed in operation, the microbes in the wastewater begin to adhere to the rotating surfaces and grow there until the entire surface area of the discs is covered with a 1- to 3-mm layer of biological slime. As the discs rotate, they carry a film of wastewater into the air; this wastewater trickles down the surface of the discs, absorbing oxygen. As the discs complete their rotation, the film of water mixes with the reservoir of wastewater, adding to the oxygen in the reservoir and mixing the treated and partially treated wastewater. As the attached microbes pass through the reservoir, they absorb other organics for breakdown. The excess growth of microbes is sheared from the discs as they move through the reservoir. These

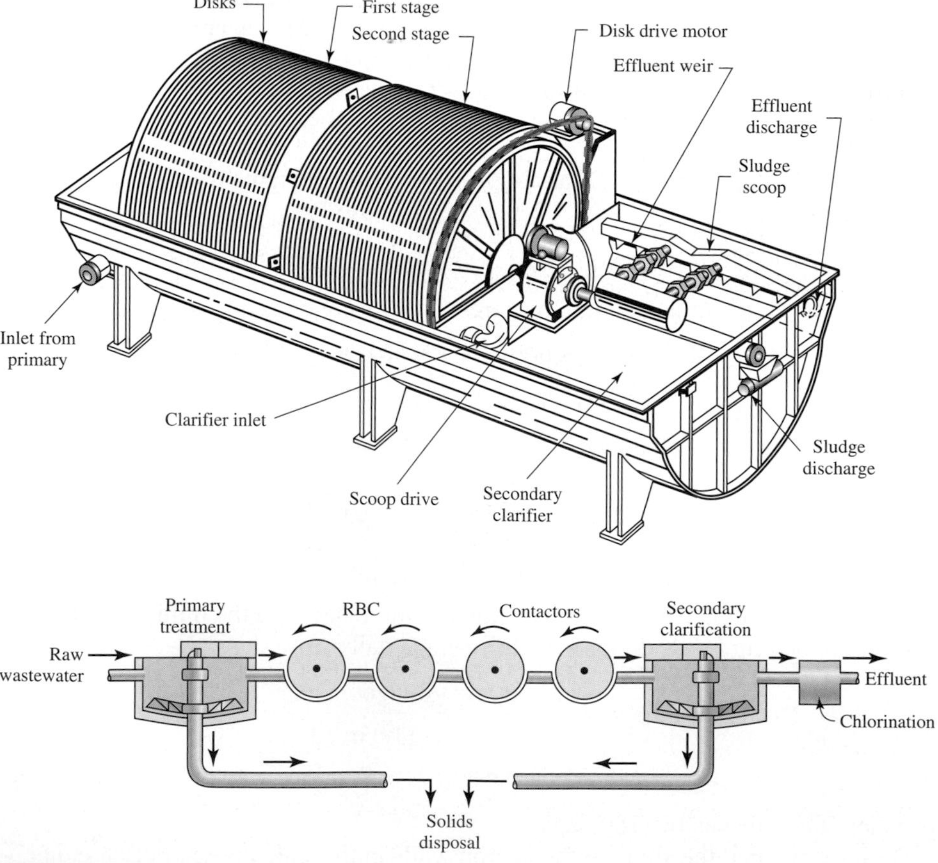

FIGURE 6-33
Rotating Biological Contactor (RBC) and process arrangement. (*Source:* U.S. EPA, 1977.)

dislodged organisms are kept in suspension by the moving discs. Thus, the discs serve several purposes:

1. They provide media for the buildup of attached microbial growth.
2. They bring the growth into contact with the wastewater.
3. They aerate the wastewater and the suspended microbial growth in the reservoir.

The attached growths are similar in concept to a trickling filter, except the microbes are passed through the wastewater rather than the wastewater passing over the microbes. Some of the advantages of both the trickling filter and activated sludge processes are realized.

As the treated wastewater flows from the reservoir below the discs, it carries the suspended growths out to a downstream settling basin for removal. The process can achieve secondary effluent quality or better. By placing several sets of discs in series, it is possible to achieve even higher degrees of treatment, including biological conversion of ammonia to nitrates.

6-9 DISINFECTION

The last treatment step in a secondary plant is the addition of a disinfectant to the treated wastewater. The addition of chlorine gas or some other form of chlorine is the process most commonly used for wastewater disinfection in the United States. Chlorine is injected into the wastewater by automated feeding systems. Wastewater then flows into a basin, where it is held for about 15 minutes to allow the chlorine to react with the pathogens.

There is concern that wastewater disinfection may do more harm than good. Early U.S. Environmental Protection Agency rules calling for disinfection to achieve 200 fecal coliforms per 100 mL of wastewater have been modified to a requirement for disinfection only during the summer season when people may come into contact with contaminated water. There were three reasons for this change. The first was that the use of chlorine and, perhaps, ozone causes the formation of organic compounds that are carcinogenic. The second was the finding that the disinfection process was more effective in killing the predators to cysts and viruses than it was in killing the pathogens themselves. The net result was that the pathogens survived longer in the natural environment because there were fewer predators. The third reason was that chlorine is toxic to fish.

6-10 ADVANCED WASTEWATER TREATMENT

Although secondary treatment processes, when coupled with disinfection, may remove over 85 percent of the BOD and suspended solids and nearly all pathogens, only minor removal of some pollutants, such as nitrogen, phosphorus, soluble COD, and heavy metals, is achieved. In some circumstances, these pollutants may be of major concern. In these cases, processes capable of removing pollutants not adequately removed by secondary treatment are used in what is called *tertiary wastewater treatment,* or *advanced wastewater treatment* (AWT). The following sections describe available AWT processes.

In addition to solving difficult pollution problems, these processes improve the effluent quality to the point that it is adequate for many reuse purposes, and may convert what was originally a wastewater into a valuable sustainable resource too good to throw away.

Filtration

Secondary treatment processes, such as the activated-sludge process, are highly efficient for removal of biodegradable colloidal and soluble organics. However, the typical effluent contains a much higher BOD_5 than one would expect from theory. The typical BOD is approximately 20 to 50 mg/L. This is principally because the secondary clarifiers are not perfectly efficient at settling out the microorganisms from the biological treatment processes. These organisms contribute both to the suspended solids and to the BOD_5 because the process of biological decay of dead cells exerts an oxygen demand.

Granular Filtration. By using a filtration process similar to that used in water treatment plants, it is possible to remove the residual suspended solids, including the unsettled microorganisms. Removing the microorganisms also reduces the residual BOD_5. Conventional sand filters identical to those used in water treatment can be used, but they often clog quickly, thus requiring frequent backwashing. To lengthen filter runs and reduce backwashing, it is desirable to have the larger filter grain sizes at the top of the filter. This arrangement allows some of the larger particles of biological floc to be trapped at the surface without plugging the filter. Multimedia filters accomplish this by using low-density coal for the large grain sizes, medium-density sand for intermediate sizes, and high-density garnet for the smallest size filter grains. Thus, during backwashing, the greater density offsets the smaller diameter so that the coal remains on top, the sand remains in the middle, and the garnet remains on the bottom.

Typically, plain filtration can reduce activated sludge effluent suspended solids from 25 to 10 mg/L. Plain filtration is not as effective on trickling filter effluents because trickling filter effluents contain more dispersed growth. However, the use of coagulation and sedimentation followed by filtration can yield suspended solids concentrations that are virtually zero. Typically, filtration can achieve 80 percent suspended solids reduction for activated sludge effluent and 70 percent reduction for trickling filter effluent.

Membrane Filtration. The alternative membrane processes have been discussed in Chapter 4. Of the 5 processes, the one most commonly used in AWT is microfiltration (MF). It is used as a replacement for granular filtration. MF processes have achieved BOD removals of 75–90 percent and total suspended solids removals of 95–98 percent. Performance is highly site-specific. Membrane fouling is of particular concern and on-site pilot testing is highly recommended (Metcalf & Eddy, 2003).

Carbon Adsorption

Even after secondary treatment, coagulation, sedimentation, and filtration, soluble organic materials that are resistant to biological breakdown will persist in the effluent. The persistent materials are often referred to as *refractory organics*. Refractory organics can be detected in the effluent as soluble COD. Secondary effluent COD values are often 30 to 60 mg/L.

The most practical available method for removing refractory organics is by *adsorbing* them on *activated carbon* (U.S. EPA, 1979a). *Adsorption* is the accumulation of materials at an *interface*. The interface, in the case of wastewater and activated carbon, is the liquid/solid boundary layer. Organic materials accumulate at the interface because of physical binding of the molecules to the solid surface. Carbon is activated by heating in the absence of oxygen. The activation process results in the formation of many pores within each carbon particle. Since adsorption is a surface phenomenon, the greater the surface area of the carbon, the greater its capacity to hold organic material. The vast areas of the walls within these pores account for most of the total surface area of the carbon, which makes it so effective in removing organics.

After the adsorption capacity of the carbon has been exhausted, it can be restored by heating it in a furnace at a temperature sufficiently high to drive off the adsorbed organics. Keeping oxygen at very low levels in the furnace prevents carbon from burning. The organics are passed through an afterburner to prevent air pollution. In small plants where the cost of an on-site regeneration furnace cannot be justified, the spent carbon is shipped to a central regeneration facility for processing.

Phosphorus Removal

All the polyphosphates (molecularly dehydrated phosphates) gradually hydrolyze in aqueous solution and revert to the ortho form (PO_4^{3-}) from which they were derived. Phosphorus is typically found as mono-hydrogen phosphate (HPO_4^{2-}) in wastewater.

The removal of phosphorus to prevent or reduce eutrophication is typically accomplished by chemical precipitation using one of three compounds. The precipitation reactions for each are shown below.

Using ferric chloride:

$$FeCl_3 + HPO_4^{2-} \rightleftharpoons FePO_4\downarrow + H^+ + 3Cl^- \tag{6-46}$$

Using alum:

$$Al_2(SO_4)_3 + 2HPO_4^{2-} \rightleftharpoons 2AlPO_4\downarrow + 2H^+ + 3SO_4^{2-} \tag{6-47}$$

Using lime:

$$5Ca(OH)_2 + 3HPO_4^{2-} \rightleftharpoons Ca_5(PO_4)_3OH\downarrow + 3H_2O + 6OH^- \tag{6-48}$$

You should note that ferric chloride and alum reduce the pH while lime increases it. The effective range of pH for alum and ferric chloride is between 5.5 and 7.0. If there is not enough naturally occurring alkalinity to buffer the system to this range, then lime must be added to counteract the formation of H^+.

The precipitation of phosphorus requires a reaction basin and a settling tank to remove the precipitate. When ferric chloride and alum are used, the chemicals may be added directly to the aeration tank in the activated sludge system. Thus, the aeration tank serves as a reaction basin. The precipitate is then removed in the secondary clarifier. This is not possible with lime since the high pH required to form the precipitate is detrimental to the activated sludge organisms. In some wastewater treatment plants, the $FeCl_3$ (or alum) is added before the wastewater enters the primary sedimentation tank. This improves the efficiency of the primary tank, but may deprive the biological processes of needed nutrients.

Example 6-15. If a wastewater has a soluble orthophosphate concentration of 4.00 mg/L as P, what *theoretical* amount of ferric chloride will be required to remove it completely?

Solution. From Equation 6-46, we see that one mole of ferric chloride is required for each mole of phosphorus to be removed. The pertinent gram molecular weights are as follows:

$$FeCl_3 = 162.2\,g$$
$$P = 30.97\,g$$

With a $PO_4{-}P$ of 4.00 mg/L, the theoretical amount of ferric chloride would be

$$4.00 \times \frac{162.2}{30.97} = 20.95 \text{ or } 21.0 \text{ mg/L}$$

Because of side reactions, solubility product limitations, and day-to-day variations, the actual amount of chemical to be added must be determined by jar tests on the wastewater. You can expect that the actual ferric chloride dose will be 1.5 to 3 times the theoretically calculated amount. Likewise, the actual alum dose will be 1.25 to 2.5 times the theoretical amount.

Nitrogen Control

Nitrogen in any soluble form (NH_3, NH_4^+, NO_2^-, and NO_3^-, but not N_2 gas) is a nutrient and may need to be removed from wastewater to help control algal growth in the receiving body. In addition, nitrogen in the form of ammonia exerts an oxygen demand and can be toxic to fish. Removal of nitrogen can be accomplished either biologically or chemically. The biological process is called *nitrification/denitrification.* The chemical process is called *ammonia stripping.*

Nitrification/Denitrification. The natural nitrification process can be forced to occur in the activated-sludge system by maintaining a cell detention time (θ_c) of 15 days in moderate climates and over 20 days in cold climates. The nitrification step is expressed in chemical terms as follows:

$$NH_4^+ + 2O_2 \rightleftharpoons NO_3^- + H_2O + 2H^+ \tag{6-49}$$

Of course, bacteria must be present to cause the reaction to occur. This step satisfies the oxygen demand of the ammonium ion. If the nitrogen level is not of concern for the receiving body, the wastewater can be discharged after settling. If nitrogen is of concern, the nitrification step must be followed by anoxic denitrification by bacteria:

$$2NO_3^- + \text{organic matter} \rightarrow N_2 + CO_2 + H_2O \tag{6-50}$$

As indicated by the chemical reaction, organic matter is required for denitrification. Organic matter serves as an energy source for the bacteria. The organic matter may be obtained from within or outside the cell. In multistage nitrogen-removal systems, because the concentration of BOD_5 in the flow to the denitrification process is usually quite low, a supplemental organic carbon source is required for rapid denitrification.

(BOD_5 concentration is low because the wastewater previously has undergone carbonaceous BOD removal and the nitrification process.) The organic matter may be either raw, settled sewage or a synthetic material such as methanol (CH_3OH). Raw, settled sewage may adversely affect the effluent quality by increasing the BOD_5 and ammonia content.

Ammonia Stripping. Nitrogen in the form of ammonia can be removed chemically from water by raising the pH to convert the ammonium ion into ammonia, which can then be stripped from the water by passing large quantities of air through the water. The process has no effect on nitrate, so the activated sludge process must be operated at a short cell-detention time to prevent nitrification. The ammonia stripping reaction is

$$NH_4^+ + OH^- \rightleftharpoons NH_3 + H_2O \qquad (6\text{-}51)$$

The hydroxide is usually supplied by adding lime. The lime also reacts with CO_2 in the air and water to form a calcium carbonate scale, which must be removed periodically. Low temperatures cause problems with icing and reduced stripping ability. The reduced stripping ability is caused by the increased solubility of ammonia in cold water.

6-11 LAND TREATMENT FOR SUSTAINABILITY

This discussion on land treatment follows two EPA publications: *Environmental Control Alternatives: Municipal Wastewater* and *Land Treatment of Municipal Wastewater Effluents, Design Factors I* (Pound et al., 1976).

An alternative to the previously discussed AWT processes for producing an extremely high-quality effluent is offered by an approach called *land treatment.* Land treatment is the application of effluents, usually following secondary treatment, on the land by one of the several available conventional irrigation methods. This approach uses wastewater, and often the nutrients it contains, as a resource rather than considering it as a disposal problem. Treatment is provided by natural processes as the effluent moves through the natural filter provided by soil and plants. Part of the wastewater is lost by evapotranspiration, while the remainder returns to the hydrologic cycle through overland flow or the groundwater system. Most of the groundwater eventually returns, directly or indirectly, to the surface water system.

Land treatment of wastewaters can provide moisture and nutrients necessary for crop growth. In semiarid areas, insufficient moisture for peak crop growth and limited water supplies make this water especially valuable. The primary nutrients (nitrogen, phosphorus, and potassium) are reduced only slightly in conventional secondary treatment processes, so that most of these elements are still present in secondary effluent. Soil nutrients that are consumed each year by crop removal and by losses through soil erosion may be replaced by the application of wastewater.

Land application is the oldest method used for treatment and disposal of wastes. Cities have used this method for more than 400 years. Several major cities, including Berlin, Melbourne, and Paris, have used "sewage farms" for at least 60 years for waste treatment and disposal. Approximately 600 communities in the United States reuse municipal wastewater treatment plant effluent in surface irrigation.

Land treatment systems use one of the three basic approaches:

1. Slow rate
2. Overland flow
3. Rapid infiltration

Each method, shown schematically in Figure 6-34, can produce renovated water of different quality, can be adapted to different site conditions, and can satisfy different overall objectives.

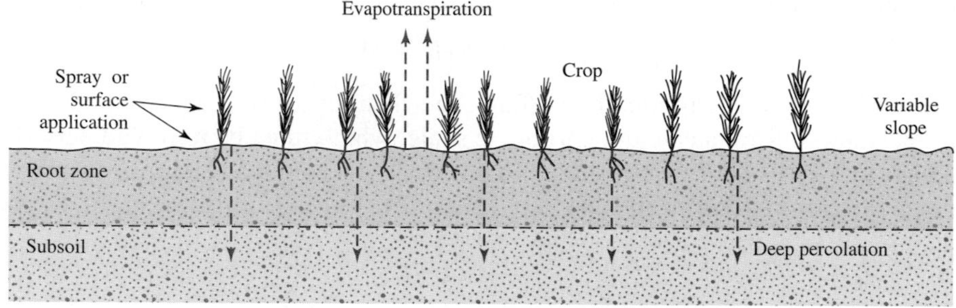

Slow Rate

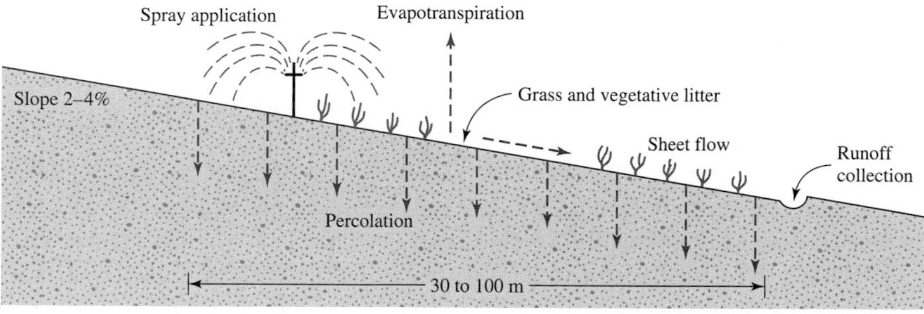

Overland Flow

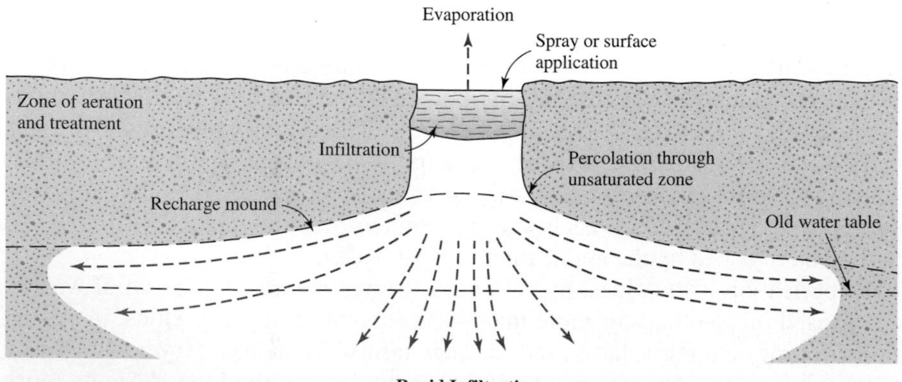

Rapid Infiltration

FIGURE 6-34
Methods of land application. (*Source:* Pound et al., 1976.)

Slow Rate

Irrigation, the predominant land application method in use today, involves the application of effluent to the land for treatment and for meeting the growth needs of plants. The applied effluent is treated by physical, chemical, and biological means as it seeps into the soil. Effluent can be applied to crops or vegetation (including forestland) either by sprinkling or by surface techniques, for purposes such as:

1. Avoidance of surface discharge of nutrients
2. Economic return from use of water and nutrients to produce marketable crops
3. Water conservation by exchange when lawns, parks, or golf courses are irrigated
4. Preservation and enlargement of greenbelts and open space

Where water for irrigation is valuable, crops can be irrigated at consumptive use rates (3.5 to 10 mm/d, depending on the crop), and the economic return from the sale of the crop can be balanced against the increased cost of the land and distribution system. On the other hand, where water for irrigation is of little value, hydraulic loadings can be maximized (provided that renovated water quality criteria are met), thereby minimizing system costs. Under high-rate irrigation (10 to 15 mm/d), water-tolerant grasses with high nutrient uptake become the crop of choice.

Overland Flow

Overland flow is essentially a biological treatment process in which wastewater is applied over the upper reaches of sloped terraces and allowed to flow across the vegetated surface to runoff collection ditches. Renovation is accomplished by physical, chemical, and biological means as the wastewater flows in a thin sheet down the relatively impervious slope.

Overland flow can be used as a secondary treatment process where discharge of a nitrified effluent low in BOD is acceptable or as an advanced wastewater treatment process. The latter objective will allow higher rates of application (18 mm/d or more), depending on the degree of advanced wastewater treatment required. Where a surface discharge is prohibited, runoff can be recycled or applied to the land in irrigation or infiltration-percolation systems.

Rapid Infiltration

In infiltration-percolation systems, effluent is applied to the soil at higher rates by spreading in basins or by sprinkling. Treatment occurs as the water passes through the soil matrix. System objectives can include:

1. Groundwater recharge
2. Natural treatment followed by pumped withdrawal or the use of underdrains for recovery
3. Natural treatment where renovated water moves vertically and laterally in the soil and recharges a surface watercourse

Where groundwater quality is being degraded by salinity intrusion, groundwater recharge can reverse the hydraulic gradient and protect the existing groundwater. Where existing groundwater quality is not compatible with expected renovated quality, or where existing water rights control the discharge location, a return of renovated water to surface water can be designed, using pumped withdrawal, underdrains, or natural drainage. At Phoenix, Arizona, for example, the native groundwater quality is poor, and the renovated water is to be withdrawn by pumping, with discharge into an irrigation canal.

6-12 SLUDGE TREATMENT

In the process of purifying the wastewater, another problem is created: sludge. The higher the degree of wastewater treatment, the larger the residue of sludge that must be handled. The exceptions to this rule are where land applications or polishing lagoons are used. Satisfactory treatment and disposal of the sludge can be the single most complex and costly operation in a municipal wastewater treatment system (U.S. EPA, 1979b). The sludge is made of materials settled from the raw wastewater and of solids generated in the wastewater treatment processes.

The quantities of sludge involved are significant. For primary treatment, they may be 0.25 to 0.35 percent by volume of wastewater treated. When treatment is upgraded to activated sludge, the quantities increase to 1.5 to 2.0 percent of this volume of water treated. Use of chemicals for phosphorus removal can add another 1.0 percent. The sludges withdrawn from the treatment processes are still largely water, as much as 97 percent. Sludge treatment processes, then, are concerned with separating the large amounts of water from the solid residues. The separated water is returned to the wastewater plant for processing.

The basic processes for sludge treatment are as follows:

1. *Thickening:* Separating as much water as possible by gravity or flotation.
2. *Stabilization:* Converting the organic solids to more refractory (inert) forms so that they can be handled or used as soil conditioners without causing a nuisance or health hazard through processes referred to as "digestion." (These are biochemical oxidation processes.)
3. *Conditioning:* Treating the sludge with chemicals or heat so that the water can be readily separated.
4. *Dewatering:* Separating water by subjecting the sludge to vacuum, pressure, or drying.
5. *Reduction:* Converting the solids to a stable form by wet oxidation or incineration. (These are chemical oxidation processes; they decrease the volume of sludge, hence the term reduction.)

Although a large number of alternative combinations of equipment and processes are used for treating sludges, the basic alternatives are fairly limited. The ultimate depository of the materials contained in the sludge must either be land, air, or water. Current policies discourage practices such as ocean dumping of sludge. Air pollution considerations necessitate air pollution control facilities as part of the sludge incineration process.

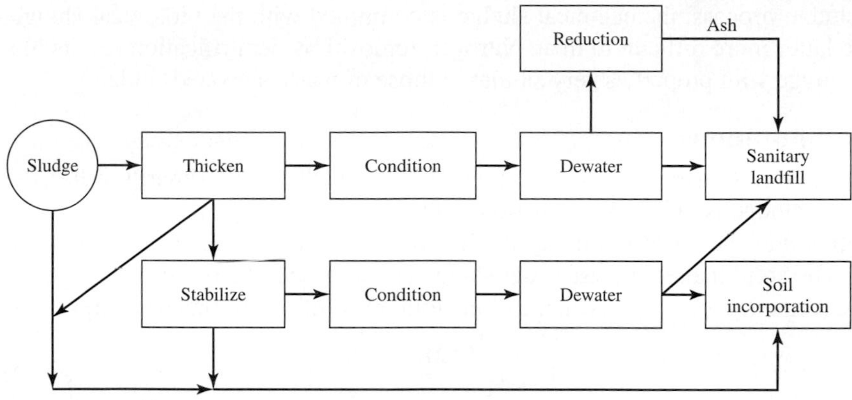

FIGURE 6-35
Basic sludge handling alternatives.

The following sections discuss the processes commonly used. The basic alternative routes by which these processes may be employed are shown in Figure 6-35.

Sources and Characteristics of Various Sludges

Before we begin the discussion of the various treatment processes, it is worthwhile to recapitulate the sources and nature of the sludges that must be treated.

Grit. The sand, broken glass, nuts, bolts, and other dense material that is collected in the grit chamber is not true sludge in the sense that it is not fluid. However, it still requires disposal. Because grit can be drained of water easily and is relatively stable in terms of biological activity (it is not biodegradable), it is generally trucked directly to a landfill without further treatment.

Primary or Raw Sludge. Sludge from the bottom of the primary clarifiers contains from 3 to 8 percent solids (1 percent solids $\simeq$ 1 g solids/100 mL sludge volume), which is approximately 70 percent organic matter. This sludge rapidly becomes anaerobic and is highly odiferous.

Secondary Sludge. This sludge consists of microorganisms and inert materials that have been wasted from the secondary treatment processes. Thus, the solids are about 90 percent organic matter. When the supply of air is removed, this sludge also becomes anaerobic, creating noxious conditions if not treated before disposal. The solids content depends on the source. Wasted activated sludge is typically 0.5 to 2 percent solids, while trickling filter sludge contains 2 to 5 percent solids. In some cases, secondary sludges contain large quantities of chemical precipitates because the aeration tank is used as the reaction basin for the addition of chemicals to remove phosphorus.

Tertiary Sludges. The characteristics of sludges from the tertiary treatment processes depend on the nature of the process. For example, phosphorus removal results in a chemical sludge that is difficult to handle and treat. When phosphorus removal occurs in the activated sludge process, the chemical sludge is combined with the biological sludge, making the latter more difficult to treat. Nitrogen removal by denitrification results in a biological sludge with properties very similar to those of waste activated sludge.

Solids Computations

Volume–Mass Relationships. Because most WWTP sludges are primarily water, the volume of the sludge is primarily a function of the water content. Thus, if we know the percent solids and the specific gravity of the solids we can estimate the volume of the sludge. The solid matter in wastewater sludge is composed of fixed (mineral) solids and volatile (organic) solids. The volume of the total mass of solids may be expressed as

$$\forall_{\text{solids}} = \frac{M_s}{S_s \rho} \tag{6-52}$$

where M_s = mass of solids, kg
S_s = specific gravity of solids
ρ = density of water = 1,000 kg/m^3

Since the total mass is composed of fixed and volatile fractions, Equation 6-52 may be rewritten as:

$$\frac{M_s}{S_s \rho} = \frac{M_f}{S_f \rho} + \frac{M_v}{S_v \rho} \tag{6-53}$$

where M_f = mass of fixed solids, kg
M_v = mass of volatile solids, kg
S_f = specific gravity of fixed solids
S_v = specific gravity of volatile solids

The specific gravity of the solids may be expressed in terms of the specific gravities of the fixed and solid fractions by solving Equation 6-53 for S_s:

$$S_s = M_s \left[\frac{S_f S_v}{M_f S_v + M_v S_f} \right] \tag{6-54}$$

The specific gravity of sludge (S_{sl}) may be estimated by recognizing that, in a similar fashion to the fractions of solids, the sludge is composed of solids and water so that

$$\frac{M_{sl}}{S_{sl} \rho} = \frac{M_s}{S_s \rho} + \frac{M_w}{S_w \rho} \tag{6-55}$$

where M_{sl} = mass of sludge, kg
M_w = mass of water, kg
S_{sl} = specific gravity of sludge
S_w = specific gravity of water

It is customary to report solids concentrations as percent solids, where the fraction of solids (P_s) is computed as

$$P_s = \frac{M_s}{M_s + M_w} \tag{6-56}$$

and the fraction of water (P_w) is computed as

$$P_w = \frac{M_w}{M_s + M_w} \tag{6-57}$$

Thus, it is more convenient to solve Equation 6-52 in terms of percent solids. If we divide each term in Equation 6-55 by ($M_s + M_w$) and recognize that $M_{sl} = M_s + M_w$, then Equation 6-55 may be expressed as

$$\frac{1}{S_{sl}\rho} = \frac{P_s}{S_s\rho} + \frac{P_w}{S_w\rho} \tag{6-58}$$

If the specific gravity of water is taken as 1.0000, as it can be without appreciable error, then solving for S_{sl} yields

$$S_{sl} = \frac{S_s}{P_s + (S_s)(P_w)} \tag{6-59}$$

With these expressions in hand, or at least where you can find them, you can calculate the volume of sludge (V_{sl}) with the following equation:

$$V_{sl} = \frac{M_s}{(\rho)(S_{sl})(P_s)} \tag{6-60}$$

Example 6-16. Using the following primary settling-tank data, determine the daily sludge production.

Operating Data:

Flow = 0.150 m^3/s

Influent SS = 280.0 mg/L = 280.0 g/m^3

Removal efficiency = 59.0%

Sludge concentration = 5.00%

Volatile solids = 60.0%

Specific gravity of volatile solids = 0.990

Fixed solids = 40.0%

Specific gravity of fixed solids = 2.65

Solution. We begin by calculating S_s. We can do this without calculating M_s, M_f, and M_v directly by recognizing that they are proportional to the percent composition. With

$$M_s = M_f + M_v$$
$$= 0.400 + 0.600 = 1.00$$

Then Equation 6-54 gives the following:

$$S_s = \frac{(2.65)(0.990)}{[(0.400)(0.990)] + [(0.600)(2.65)]}$$
$$= 1.321 \text{ or } 1.32$$

The specific gravity of the sludge is calculated with Equation 6-59:

$$S_{sl} = \frac{1.321}{0.05 + (1.321 \times 0.950)}$$
$$= 1.012 \text{ or } 1.01$$

The mass of the sludge is estimated from the incoming suspended solids concentration and the removal efficiency of the primary tank.

$$M_s = 0.59 \times 280.0 \text{ g/m}^3 \times 0.15 \text{ m}^3\text{/s} \times 86{,}400 \text{ s/d} \times 10^{-3} \text{ kg/g}$$
$$= 2.14 \times 10^3 \text{ kg/d}$$

The sludge volume is then calculated with Equation 6-60:

$$\forall_{sl} = \frac{2.14 \times 10^3 \text{ kg/d}}{1{,}000 \text{ kg/m}^3 \times 1.012 \times 0.05}$$
$$= 42.29 \text{ or } 42.3 \text{ m}^3\text{/d}$$

Mass Balance. Barring black holes* and the like, we all understand that the physical, chemical, and biological processes of wastewater treatment neither create nor destroy matter. This fact allows us to employ Equation 2-3 in a new context.

$$\frac{dS}{dt} = M_{\text{in}} - M_{\text{out}} \tag{6-61}$$

where M_{in} and M_{out} refer to the mass of dissolved chemicals, solids, or gas entering and leaving a process or group of processes. If we assume steady-state conditions, then $dS/dt = 0$ and Equation 6-61 reduces to the following:

$$M_{\text{in}} = M_{\text{out}} \tag{6-62}$$

Several interrelated processes are examined together in the flowsheet shown in Figure 6-36. When labeled with mass flows, the flowsheet may be called a *quantitative flow diagram* (QFD). The solids mass balance can be an important aid to a designer in predicting long-term average solids loadings on sludge treatment components. This allows the designer to establish such factors as operating costs and quantities of sludge for ultimate disposal. However, it does not establish the solids loading that each equipment item must be capable of processing. A particular component should be sized to handle the most rigorous loading conditions it is expected to encounter. This loading is usually not determined by applying steady-state models because of storage and plant scheduling considerations. Thus, the rate of solids reaching any particular piece of equipment does not usually rise and fall in direct proportion to the rate of solids arriving at the plant headworks.

The mass balance calculation is carried out in a step-by-step procedure:

1. Draw the flowsheet (as in Figure 6-36).
2. Identify all streams. For example, Stream A contains raw sewage solids plus chemical solids generated by dosing the sewage with chemicals. Let the *mass flow rate* of solids in Stream A be equal to *A* kg per day.

---

*We are, of course, referring to the Einsteinian black hole and not the Black Hole of Calcutta.

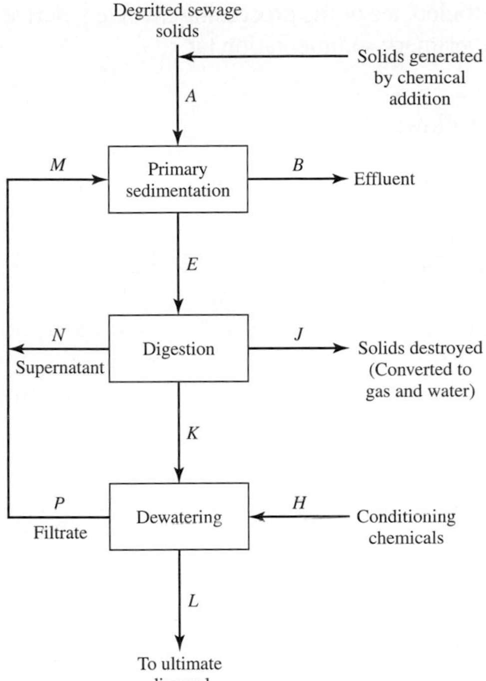

FIGURE 6-36
Primary WWTP flowsheet. (*Source:* U.S. EPA, 1979b.)

3. For each processing unit, identify the relationship of entering and leaving streams to one another in terms of mass. For example, for the primary sedimentation tank, let the ratio of solids in the tank underflow (E) to entering solids ($A + M$) be equal to η_E. η_E is actually an indicator of solids separation efficiency. The general form in which such relationships are expressed is:

$$\eta_i = \frac{\text{mass of solids in stream } i}{\text{mass of solids entering the unit}} \tag{6-63}$$

For example,

$$\eta_p = \frac{P}{K + H}; \ \eta_j = \frac{J}{E}$$

The processing unit's performance is specified when a value is assigned to η_i.

4. Combine the mass balance relationships so as to reduce them to one equation describing a specific stream in terms of given or known quantities, or ones which can be calculated from a knowledge of the process behavior.

Example 6-17. Using Figure 6-36 and assuming that A, η_E, η_j, η_N, η_P, and η_H are known or can be determined from a knowledge of water chemistry and an understanding of the general solids separation/destruction efficiencies of the processing involved, derive an expression for E, the mass flow out of the primary sedimentation tank.

Solution. The derivation is carried out as follows.

a. Define M by solids balances on streams around the primary sedimentation tank:

$$\eta_E = \frac{E}{A + M} \tag{i}$$

Therefore,

$$M = \frac{E}{\eta_E} - A \tag{ii}$$

b. Define M by balances on recycle streams:

$$M = N + P \tag{iii}$$

$$N = \eta_N E \tag{iv}$$

$$P = \eta_P (H + K) \tag{v}$$

$$H = \eta_H K \tag{vi}$$

Therefore,

$$P = \eta_P(1 + \eta_H)K \tag{vii}$$

$$K + J + N = E \tag{viii}$$

Therefore,

$$K = E - J - N = E - \eta_j E - \eta_N E = E(1 - \eta_j - \eta_N) \tag{ix}$$

and

$$P = \eta_P E\,(1 - \eta_j - \eta_N)(1 + \eta_H) \tag{x}$$

Therefore,

$$M = E\,[\eta_N + \eta_P\,(1 - \eta_j - \eta_N)(1 + \eta_H)] \tag{xi}$$

c. Equate equations (ii) and (xi) to eliminate M:

$$\frac{E}{\eta_E} - A = E[\eta_N + \eta_P(1 - \eta_j - \eta_N)(1 + \eta_H)]$$

$$E = \frac{A}{\dfrac{1}{\eta_E} - \eta_N - \eta_P\,(1 - \eta_j - \eta_N)(1 + \eta_H)}$$

E is expressed in terms of assumed or known influent solids loadings and solids separation/destruction efficiencies.

Once the equation for E is derived, equations for other streams follow rapidly; in fact, most have already been derived. These are summarized in Table 6-14.

TABLE 6-14
Mass balance equations for Figure 6-36

$$E = \frac{A}{\dfrac{1}{\eta_E} - \eta_N - \eta_P(1 - \eta_j - \eta_N)(1 + \eta_H)}$$

$$M = \frac{E}{\eta_E} - A$$

$$B = (1 - \eta_E)(A + M)$$

$$J = \eta_J E$$

$$N = \eta_N E$$

$$K = E(1 - \eta_J - \eta_N)$$

$$H = \eta_H K$$

$$P = \eta_P(1 + \eta_H)K$$

$$L = K(1 + \eta_H)(1 - \eta_P)$$

Source: U.S. EPA, 1979b.

The example just worked was relatively simple. A more complex system is illustrated in Figure 6-37. Mass balance equations for this system are summarized in Table 6-15 on page 509. For this flowsheet the following information must be specified:

A = Influent solids

X = Effluent solids, that is, overall suspended solids removal must be specified

η_F, η_G, η_J, η_N, η_R, and η_T = assumptions about the degree of solids removal, addition, or destruction

η_D = describes the net solids destruction reduction or the net solids synthesis in the biological system, and must be estimated from yield data. A positive η_D signifies net solids destruction. A negative η_D signifies net solids growth. In this example, 8 percent of the solids entering the biological process are assumed destroyed, that is, converted to gas or liquified.

Note that alternative processing schemes can be evaluated simply by manipulating appropriate variables. For example:

Filtration can be eliminated by setting η_R to zero.

Thickening can be eliminated by setting η_G to zero.

Digestion can be eliminated by setting η_J to zero.

Dewatering can be eliminated by setting η_P to zero.

A system without primary sedimentation can be simulated by setting η_E equal to approximately zero, for example, 1×10^{-8}. η_E cannot be set equal to exactly zero, since division by η_E produces indeterminate solutions when computing E.

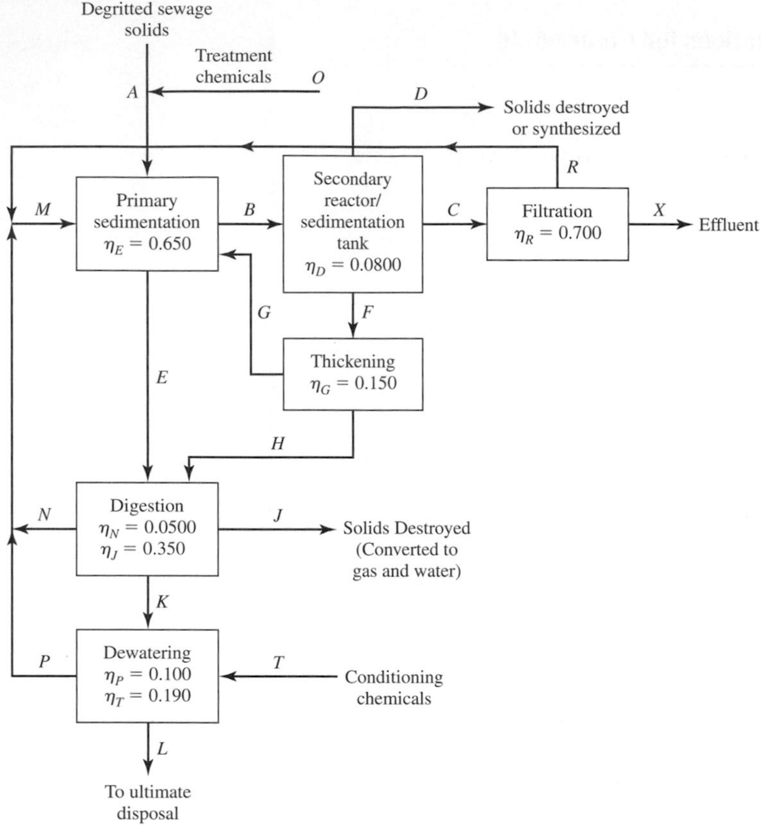

FIGURE 6-37
Flowsheet for a complex WWTP. (*Source:* U.S. EPA, 1979b.)

A set of different mass balance equations must be derived if flow paths between processing units are altered. For example, the equations of Table 6-15 do not describe operations in which the dilute stream from the thickener (Stream G) is returned to the secondary reactor instead of the primary sedimentation tank.

Thickening

Thickening is usually accomplished in one of two ways: the solids are floated to the top of the liquid (*flotation*) or are allowed to settle to the bottom (*gravity thickening*).

The goal is to remove as much water as possible before final dewatering or digestion of the sludge. The processes involved offer a low-cost means of reducing sludge volumes by a factor of two or more. The costs of thickening are usually more than offset by the resulting savings in the size and cost of downstream sludge processing equipment.

Flotation. In the flotation thickening process (Figure 6-38) air is injected into the sludge under pressure (275 to 550 kPa). Under this pressure, a large amount of air can

TABLE 6-15
Mass balance equations for Figure 6-37

$$E = \frac{A - \left(\dfrac{X}{1 - \eta_R}\right)(\gamma - \eta_R)}{\dfrac{1}{\eta_E} - \alpha - \beta(\gamma)}$$

Where $\alpha = \eta_P(1 - \eta_J - \eta_N)(1 + \eta_T) + \eta_N$

$$\beta = \frac{(1 - \eta_E)(1 - \eta_D)}{\eta_E}$$

$$\gamma = \eta_G + \alpha(1 - \eta_G)$$

$$B = \frac{(1 - \eta_E)E}{\eta_E}$$

$$C = \frac{X}{1 - \eta_R}$$

$$D = \eta_D B$$

$$F = \beta E - \frac{X}{1 - \eta_R}$$

$$G = \eta_G F$$

$$H = (1 - \eta_G)F$$

$$J = \eta_J(E + H)$$

$$K = (1 - \eta_J - \eta_N)(E + H)$$

$$L = K(1 + \eta_T)(1 - \eta_P)$$

$$M = \frac{E}{\eta_E} - G - A$$

$$N = \eta_N(E + H)$$

$$P = \eta_P(1 + \eta_T)K$$

$$R = \frac{\eta_R}{1 - \eta_R}X$$

$$T = \eta_T K$$

Source: U.S. EPA, 1979b.

be dissolved in the sludge. The sludge then flows into an open tank where, at atmospheric pressure, much of the air comes out of solution as minute bubbles. The bubbles attach themselves to sludge solids particles and float them to the surface. The sludge forms a layer at the top of the tank; this layer is removed by a skimming mechanism for further processing. The process typically increases the solids content of activated

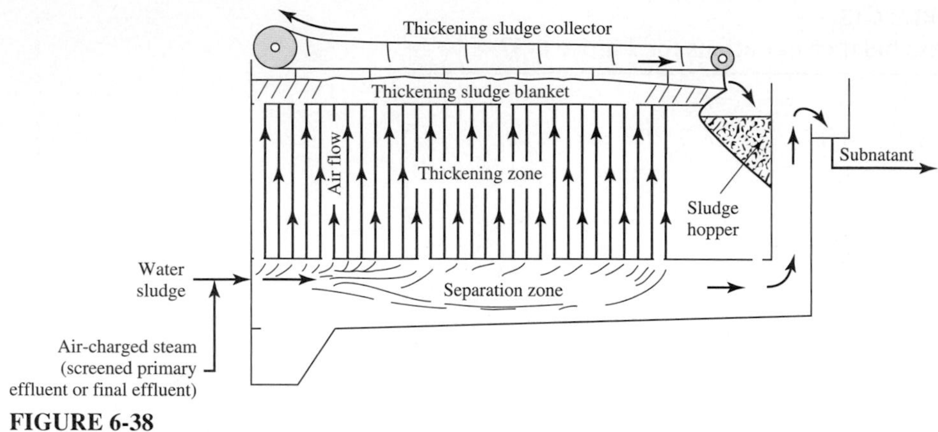

FIGURE 6-38
Air flotation thickener.

sludge from 0.5–1 percent to 3–6 percent. Flotation is especially effective on activated sludge, which is difficult to thicken by gravity.

Gravity Thickening. Gravity thickening is a simple and inexpensive process that has been used widely on primary sludges for many years. It is essentially a sedimentation process similar to that which occurs in all settling tanks. Sludge flows into a tank that is very similar in appearance to the circular clarifiers used in primary and secondary sedimentation (Figure 6-39); the solids are allowed to settle to the bottom where a heavy-duty mechanism scrapes them to a hopper from which they are withdrawn for further processing. The type of sludge being thickened has a major effect on performance. The best results are obtained with purely primary sludges. As the proportion of activated sludge increases, the thickness of settled sludge solids decreases. Purely primary sludges can be thickened from 1–3 percent to 10 percent solids. The current trend is toward using gravity thickening for primary sludges and flotation thickening for activated sludges, and then blending the thickened sludges for further processing.

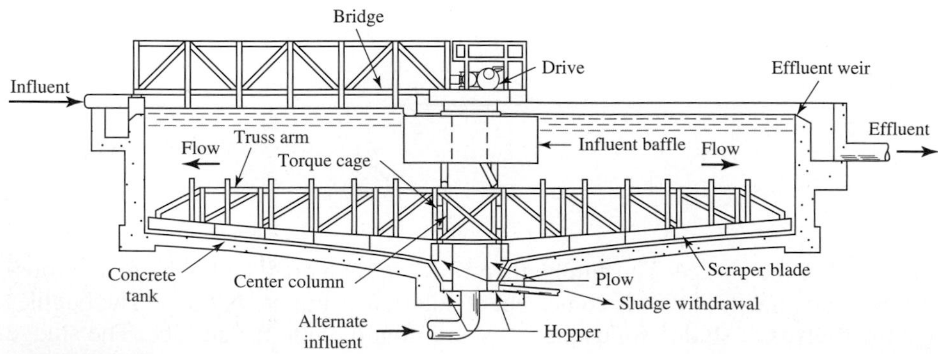

FIGURE 6-39
Gravity thickener.

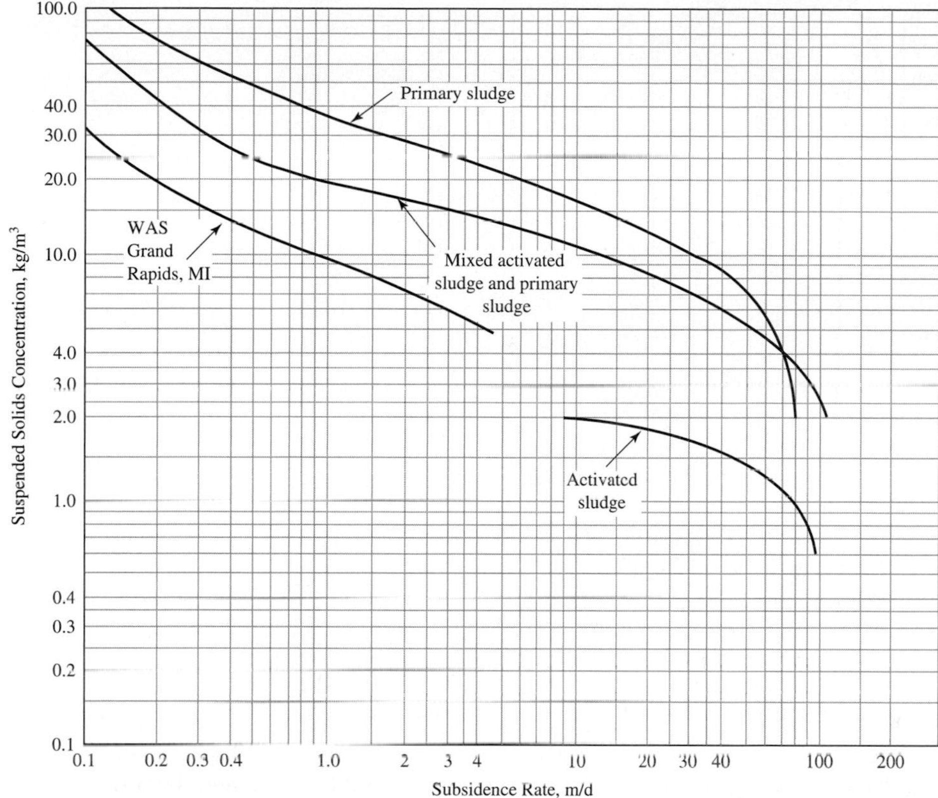

FIGURE 6-40
Batch settling curve.

Dick has described a graphical procedure for sizing gravity thickeners using a *batch flux curve** (Yoshioka et al., 1957, and Dick, 1970). *Flux* is the term used to describe the rate of settling of solids. It is defined as the mass of solids which pass through a horizontal unit area per unit of time (kg/d · m^2). This may be expressed mathematically as follows:

$$F_s = (C_u)(v)$$
$$= (C_s)(\text{zone settling velocity}) \qquad (6\text{-}64)$$

where F_s = solids flux, kg/m^2 · d
C_s = suspended solids concentration, kg/m^3
C_u = concentration of solids in underflow, that is, sludge withdrawal pipe, kg/m^3
v = underflow velocity, m/d

The sizing procedure begins with a batch settling curve such as that shown in Figure 6-40. Data from the batch settling curve are used to construct a batch flux curve (Figure 6-41). Knowing the desired underflow concentration, a line through the desired concentration and tangent to the batch flux curve is constructed. The extension of this

* The original development of this method was by N. Yoshioka and others.

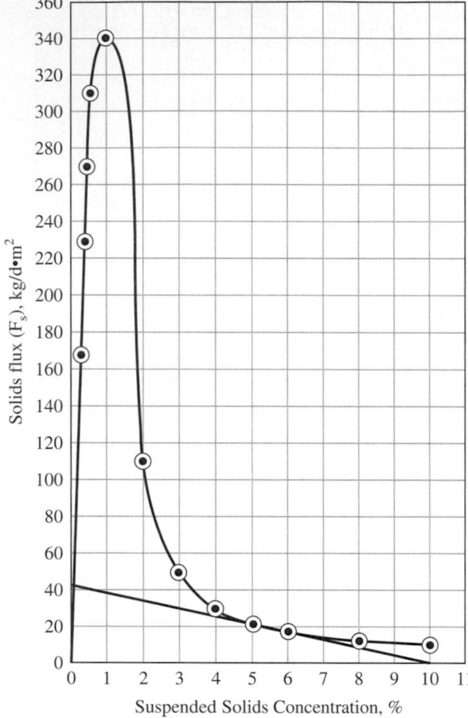

FIGURE 6-41
Batch flux curve.

line to the axis of ordinates yields the design flux. From this flux and the inflow solids concentration, the surface area may be determined.

Example 6-18. A gravity thickener is to be designed to thicken the sludge from the primary tank described in Example 6-16. The thickened sludge should have an underflow solids concentration of 10.0 percent. Assume that the sludge yields a batch settling curve such as that shown in Figure 6-40.

Solution. First we must compute the solids flux for several arbitrarily selected suspended solids concentrations.

SS, kg/m^3	v, m/d	F_s, kg/d · m^2	SS, kg/m^3	v, m/d	F_s, kg/d · m^2
100	0.125	12.5	20	5.30	106.
80	0.175	14.0	10	34.0	340.
60	0.30	18.	5	62.0	310.
50	0.44	22.	4	68.0	272.
40	0.78	31.	3	76.0	228.
30	1.70	51.	2	83.0	166.

The data in the first column were selected arbitrarily. The data in the second column were read from Figure 6-40 at the abscissa points noted in the first column. The data in the third column are the products of the first and second column, that is, $100.0 \times 0.125 = 12.5$, $80.0 \times 0.175 = 14.0$, etc.

The percent solids concentration is simply 0.10 times the SS in kg/m^3. Converting the first column to percent and plotting it versus the last column yields the batch flux curve (Figure 6-41).

The tangent line from 10.0 percent yields a solids flux of 43 $kg/d \cdot m^2$.

From Example 6-16, we find the solids mass loading to be 2.14×10^3 kg/d. Therefore, the required surface area of the thickener is

$$A_s = \frac{2.14 \times 10^3 \text{ kg/d}}{43 \text{ kg/d} \cdot \text{m}^2}$$

$$= 49.77 \text{ or } 50 \text{m}^2$$

Typical gravity-thickener design criteria are summarized in Table 6-16. Wasting to the thickener may or may not be continuous, depending upon the size of the WWTP. Frequently, smaller plants will waste intermittently because of work schedules and

TABLE 6-16
Typical gravity-thickener design criteria

Sludge Source	Influent SS, %	Expected underflow concentration, %	Mass loading kg/h · m²
Individual sludges			
PS	2–7	5–10	4–6
TF	1–4	3–6	1.5–2.0
RBC	1–3.5	2–5	1.5–2.0
WAS	0.5–1.5	2–3	0.5–1.5
Tertiary sludges			
High CaO	3–4.5	12–15	5–12
Low CaO	3–4.5	10–12	2–6
Fe	0.5–1.5	3–4	0.5–2.0
Combined sludges			
PS + WAS	0.5–4	4–7	1–3.5
PS + TF	2–6	5–9	2–4
PS + RBC	2–6	5–8	2–3
PS + Fe	2	4	1
PS + Low CaO	5	7	4
PS + High CaO	7.5	12	5
PS + (WAS + Fe)	1.5	3	1
PS + (WAS + Al)	0.2–0.4	4.5–6.5	2–3.5
(PS + Fe) + TF	0.4–0.6	6.5–8.5	3–4
(PS + Fe) + WAS	1.8	3.6	1
WAS + TF	.5–2.5	2–4	0.5–1.5

(*Source:* U.S. EPA, 1979b.)
Legend: PS = primary sedimentation; TF = trickling filter; RBC = rotating biological contactor; WAS = waste activated sludge; High CaO = high lime; Low CaO = low lime; Fe = iron; Al = alum; + = mixture of sludges from processes indicated; ( ) = chemical added to process is within parentheses.

TABLE 6-17
Reported operation results for gravity thickeners

Location	Sludge source	Influent SS, %	Mass loading, kg/h · m^2	Underflow concentration, %	Overflow SS, mg/L
Port Huron, MI	PS + WAS	0.6	1.7	4.7	2,500
Sheboygan, WI	PS + TF	0.3	2.2	8.6	400
	PS + (TF + Al)	0.5	3.6	7.8	2,400
Grand Rapids, MI	WAS	1.2	2.1	5.6	140
Lakewood, OH	PS + (WAS + Al)	0.3	2.9	5.6	1,400

(*Source:* U.S. EPA, 1979b.)
(*Note:* Values shown are average values only.)

lower volumes of sludge. Some examples of thickener performance are listed in Table 6-17. You should note that the supernatant suspended solids levels are quite high. Thus, the supernatant must be returned to the head end of the WWTP.

Example 6-19. One hundred cubic meters per day (100.0 m^3/d) of mixed sludge at 4.0 percent solids is to be thickened to 8.0 percent solids. What is the approximate volume of the sludge after thickening?

Solution. A "4.0 percent sludge" contains 4.0 percent by mass of solids and 96.0 percent by mass of water. Assuming that the specific gravity is not appreciably different from that of water, we can approximate the relationship between volume and percent solids as follows:

$$\frac{V_1}{V_2} = \frac{P_2}{P_1}$$

In this case then, the volume of sludge after thickening would be

$$\frac{100.0 \text{ m}^3/\text{d}}{V_2} = \frac{0.080}{0.040}$$

$$V_2 = 50.0 \text{ m}^3/\text{d}$$

Thus, we can see a substantial reduction in the volume that must be handled by thickening the sludge from 4 to 8 percent solids.

Stabilization

The principal purposes of sludge stabilization are to break down the organic solids biochemically so that they are more stable (less odorous and less putrescible) and more dewaterable, and to reduce the mass of sludge. If the sludge is to be dewatered and burned, stabilization is not used. There are two basic stabilization processes in use. One is carried out in closed tanks devoid of oxygen and is called *anaerobic digestion.* The other approach injects air into the sludge to accomplish *aerobic digestion.*

Aerobic Digestion. The aerobic digestion of biological sludges is nothing more than a continuation of the activated sludge process. When a culture of aerobic heterotrophs is placed in an environment containing a source of organic material, the microorganisms remove and utilize most of this material. A fraction of the organic material removed will be used for the synthesis of new biomass. The remaining material will be channeled into energy metabolism and oxidized to carbon dioxide, water, and soluble inert material to provide energy for both synthesis and maintenance (life-support) functions. Once the external source of organic material is exhausted, however, the microorganisms enter into endogenous respiration, where cellular material is oxidized to satisfy the energy of maintenance (that is, energy for life-support requirements). If this condition is continued over an extended period of time, the total quantity of biomass will be considerably reduced. Furthermore, that portion remaining will exist at such a low energy state that it can be considered biologically stable and suitable for disposal in the environment. This forms the basic principle of aerobic digestion.

Aerobic digestion is accomplished by aerating the organic sludges in an open tank resembling an activated sludge aeration tank. Like the activated sludge aeration tank, the aerobic digestor must be followed by a settling tank unless the sludge is to be disposed of on land in liquid form. Unlike the activated sludge process, the effluent (supernatant) from the clarifier is recycled back to the head end of the plant. This is because the supernatant is high in suspended solids (100 to 300 mg/L), BOD_5 (to 500 mg/L), TKN (to 200 mg/L), and total P (to 100 mg/L).

Because the fraction of volatile matter is reduced, the specific gravity of the digested sludge solids will be higher than it was before digestion. Thus, the sludge settles to a more compact mass, and the clarifier underflow concentration can be expected to reach 3 percent. Beyond this, its dewatering properties are terrible.

Anaerobic Digestion. The anaerobic treatment of complex wastes involves three distinct stages. In the first stage, complex waste components, including fats, proteins, and polysaccharides, are hydrolyzed to their component subunits. This is accomplished by a heterogeneous group of facultative and anaerobic bacteria. These bacteria then subject the products of hydrolysis (triglycerides, fatty acids, amino acids, and sugars) to fermentation and other metabolic processes leading to the formation of simple organic compounds and hydrogen in a process called *acidogenesis* or *acetogenesis*. The organic compounds are mainly short-chain (volatile) acids and alcohols. The second stage is commonly referred to as *acid fermentation*. In this stage, organic material is simply converted to organic acids, alcohols, and new bacterial cells, so that little stabilization of BOD or COD is realized. In the third stage, the end products of the first stage are converted to gases (mainly methane and carbon dioxide) by several different species of strictly anaerobic bacteria. Thus, it is here that true stabilization of the organic material occurs. This stage is generally referred to as *methane fermentation*. The stages of anaerobic waste treatment are illustrated in Figures 6-42 and 6-43. Even though the anaerobic process is presented as being sequential in nature, all stages take place simultaneously and synergistically. The primary acid produced during acid fermentation is acetic acid. The significance of this acid as a precursor for methane formation is illustrated in Figure 6-43.

The bacteria responsible for acid fermentation are relatively tolerant to changes in pH and temperature and have a much higher rate of growth than the bacteria responsible

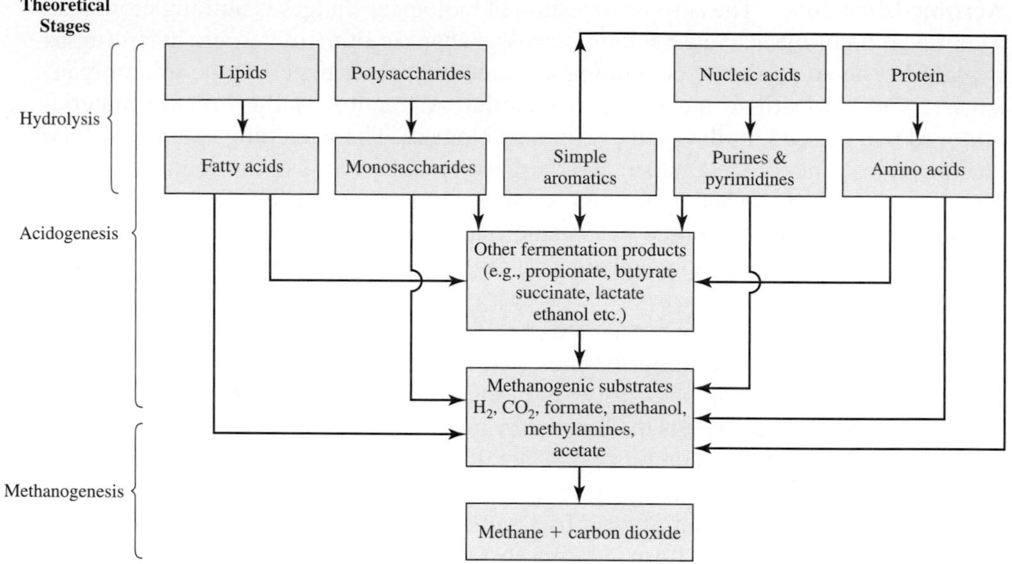

FIGURE 6-42
Schematic diagram of the patterns of carbon flow in anaerobic digestion.

for methane fermentation. As a result, methane fermentation is generally assumed to be the rate-controlling step in anaerobic waste treatment processes.

Considering 35°C as the optimum temperature for anaerobic waste treatment, Lawrence proposes that, in the range of 20 to 35°C, the kinetics of methane fermentation of long- and short-chain fatty acids will adequately describe the overall kinetics of anaerobic treatment (Lawrence and Milnes, 1971). Thus, the kinetic equations we presented to describe the completely mixed activated sludge process are equally applicable to the anaerobic process.

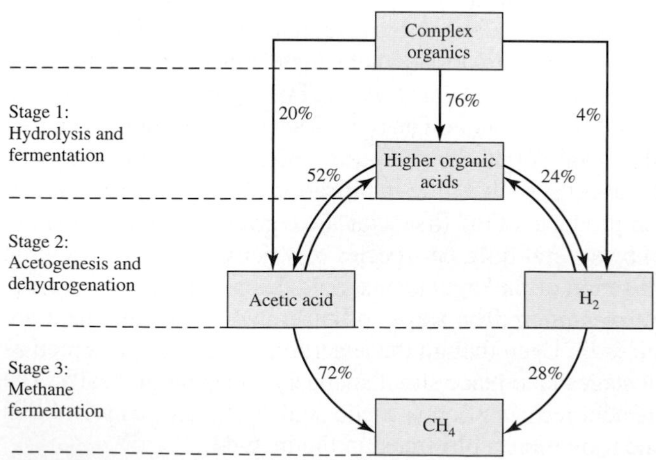

FIGURE 6-43
Steps in anaerobic digestion process with energy flow.

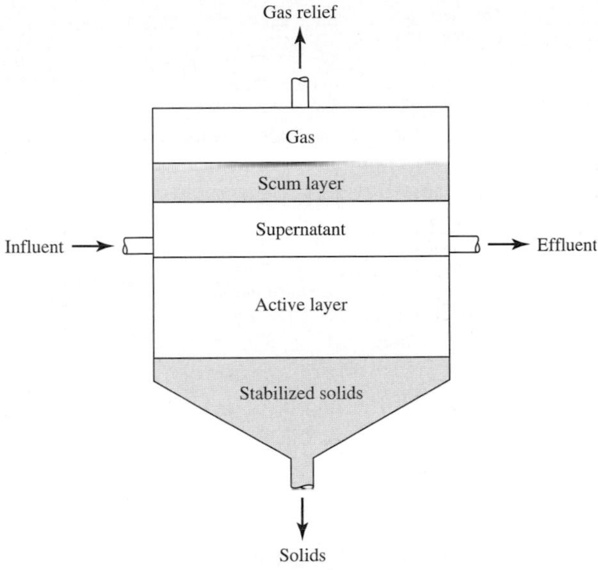

FIGURE 6-44
Schematic of a standard-rate anaerobic digester.

There are essentially two types of anaerobic digestion processes used today: the standard-rate process and the high-rate process.

The standard-rate process does not employ sludge mixing, but rather the digester contents are allowed to stratify into zones, as illustrated in Figure 6-44. Sludge feeding and withdrawal are intermittent rather than continuous. The digester is generally heated to increase the rate of fermentation and therefore decrease the required retention time. Retention time ranges between 30 and 60 days for heated digesters. The organic loading rate for a standard-rate digester is between 0.48 and 1.6 kg total volatile solids per m^3 of digester volume per day.

The major disadvantage of the standard-rate process is the large tank volume required because of long retention times, low loading rates, and thick scum-layer formation. Only about one-third of the tank volume is utilized in the digestion process. The remaining two-thirds of the tank volume contains the scum layer, stabilized solids, and the supernatant. Because of this limitation, systems of this type are generally used only at treatment plants having a capacity of 0.04 m^3/s or less.

The high-rate system evolved as a result of continuing efforts to improve the standard-rate unit. In this process, two digesters operating in series separate the functions of fermentation and solids/liquid separation (see Figure 6-45). The contents of the first-stage, high-rate unit are thoroughly mixed and the sludge is heated to increase the rate of fermentation. Because the contents are thoroughly mixed, temperature distribution is more uniform throughout the tank volume. Sludge feeding and withdrawal are continuous or nearly so. The retention time required for the first-stage unit is normally between 10 and 15 days. Organic loading rates vary between 1.6 and 8.0 kg total volatile solids per m^3 of digester per day.

The primary functions of the second-stage digester are solids/liquid separation and residual gas extraction. First-stage digesters may be equipped with fixed or floating

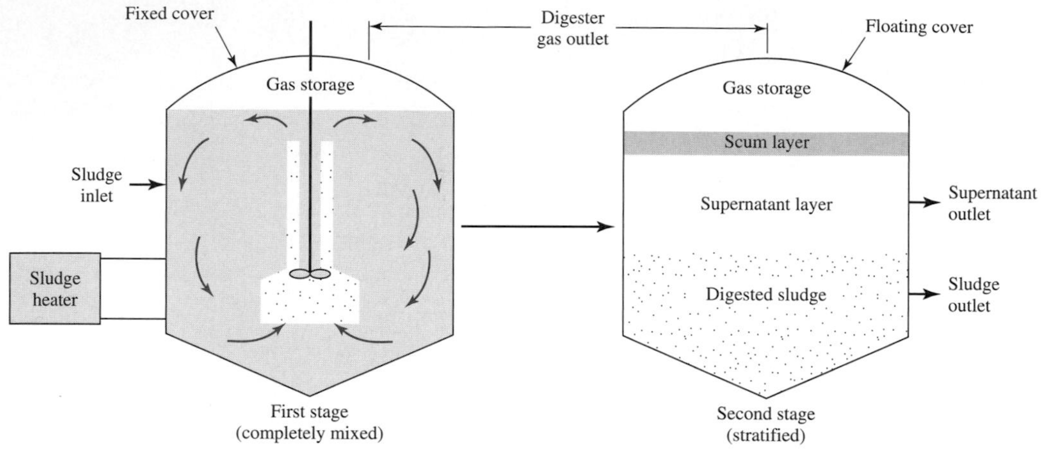

FIGURE 6-45
Schematic of a high-rate anaerobic digester.

covers. Second-stage digester covers are often of the floating type (Figure 6-46). Second-stage units are generally not heated.

The first-stage digester of a high-rate system approximates a completely mixed reactor without solids recycle. Hence, the biological solids retention time (SRT) and the hydraulic retention time are equal for this system. As with the aerobic digesters, the most important operating parameters affecting VSS reduction are solids retention time and digestion temperature.

The BOD remaining at the end of digestion is still quite high. Likewise, the suspended solids may be as high as 12,000 mg/L, while the TKN may be on the order of 1,000 mg/L. Thus, the supernatant from the secondary digester (in the high-rate process) is returned to the head end of the WWTP. The settled sludge is conditioned and dewatered for disposal.

Sludge Conditioning

Chemical Conditioning. Several methods of conditioning sludge to facilitate the separation of the liquid and solids are available. One of the most commonly used is the addition of coagulants such as ferric chloride, lime, or organic polymers. Ash from incinerated sludge has also found use as a conditioning agent. As happens when coagulants are added to turbid water, chemical coagulants act to clump the solids together so that they are more easily separated from the water. In recent years, organic polymers have become increasingly popular for sludge conditioning. Polymers are easy to handle, require little storage space, and are very effective. The conditioning chemicals are injected into the sludge just before the dewatering process and are mixed with the sludge.

Heat Treatment. Another conditioning approach is to heat the sludge at high temperatures (175 to 230°C) and pressures (1,000 to 2,000 kPa). Under these conditions, much like those of a pressure cooker, water that is bound up in the solids is released, improving the dewatering characteristics of the sludge. Heat treatment has the advantage of producing a sludge that dewaters better than chemically conditioned sludge.

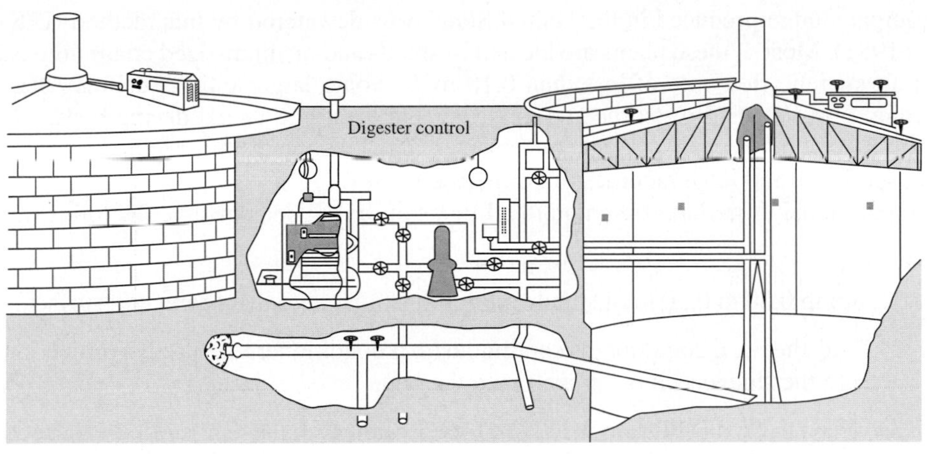

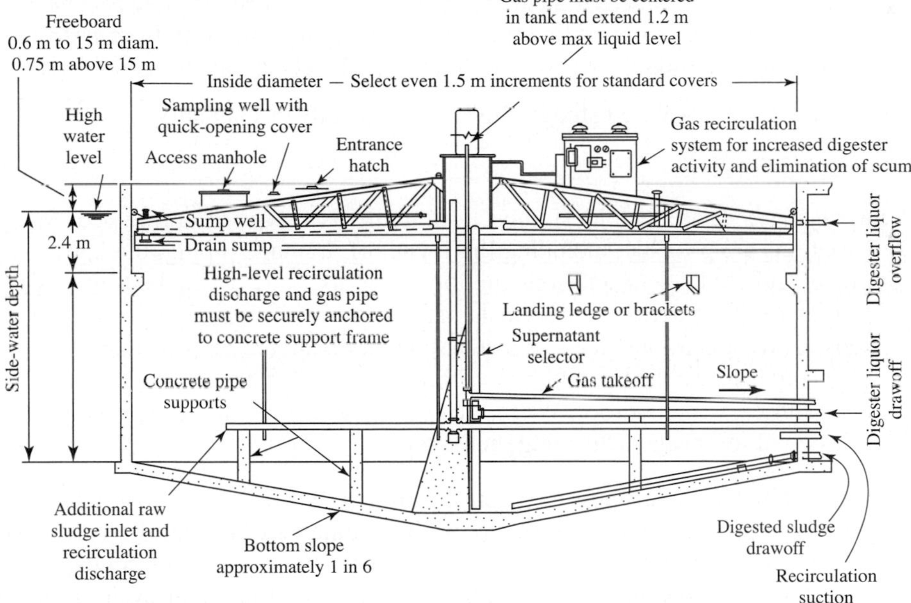

FIGURE 6-46
Phantom view of high-rate anaerobic digester and cross section of detail of floating cover. (Courtesy of Envirex.)

The process has the disadvantages of relatively complex operation and maintenance and the creation of highly polluted cooking liquors that when recycled to the treatment plant impose a significant added treatment burden.

Sludge Dewatering

Sludge Drying Beds. The most popular method of sludge dewatering in the past has been the use of sludge drying beds. These beds are especially popular in small plants because of their simplicity of operation and maintenance. Seventy-seven percent of all United States wastewater treatment plants utilized drying beds; one-half of all the municipal sludge produced in the United States was dewatered by this method (U.S. EPA, 1981). Most of these plants are located in small- and medium-sized communities, with an average flow rate of less than 0.10 m^3/s. Some larger cities, such as Albequerque, Fort Worth, Texas, Phoenix, and Salt Lake City, use sand drying beds. Although the use of drying beds might be expected in the warmer, sunny regions, they are also used in several large facilities in northern climates.

Operational procedures common to all types of drying beds involve the following steps:

1. Pump 0.20 to 0.30 m of stabilized liquid sludge onto the drying bed surface.
2. Add chemical conditioners continuously, if conditioners are used, by injection into the sludge as it is pumped onto the bed.
3. When the bed is filled to the desired level, allow the sludge to dry to the desired final solids concentration. (This concentration can vary from 18 to 60 percent, depending on several factors, including type of sludge, processing rate needed, and degree of dryness required for lifting. Nominal drying times vary from 10 to 15 d under favorable conditions, to 30 to 60 d under barely acceptable conditions.)
4. Remove the dewatered sludge either mechanically or manually.
5. Repeat the cycle.

Sand drying beds are the oldest, most commonly used type of drying bed. Many design variations are possible, including the layout of drainage piping, thickness and type of gravel and sand layers, and construction materials. Sand drying beds for wastewater sludge are constructed in the same manner as water treatment plant sludge-drying beds. Current U.S. practice was discussed and illustrated in Section 4-10.

Sand drying beds can be built with or without provision for mechanical sludge removal, and with or without a roof. When the cost of labor is high, newly constructed beds are designed for mechanical sludge removal.

Vacuum Filtration. A vacuum filter consists of a cylindrical drum covered with a filtering material or fabric, which rotates partially submerged in a vat of conditioned sludge. A vacuum is applied inside the drum to extract water, leaving the solids, or filter cake, on the filter medium. As the drum completes its rotational cycle, a blade scrapes the *filter cake* from the filter and the cycle begins again. In some systems, the filter fabric passes off the drum over small rollers to dislodge the cake. There is a wide variety of filter fabrics, ranging from Dacron to stainless-steel coils, each with its own advantages. The vacuum filter can be applied to digested sludge to produce a sludge cake dry enough (15 to 30 percent solids) to handle and dispose of by burial in a landfill or by application to the land as a relatively dry fertilizer. If the sludge is to be incinerated, it is not stabilized. In this case, the vacuum filter is applied to the raw sludge to dewater it. The sludge cake is then fed to the furnace to be incinerated. Vacuum filters are being replaced by continuous belt filter presses.

Continuous Belt Filter Presses (CBFP). The CBFP equipment used in treating wastewater sludges is the same as that used for water treatment plant sludges (Figure 6-47).

FIGURE 6-47
Continuous belt filter press. (Courtesy of Komline-Sanderson Engineering Corporation.)

The CBFP is successful with many normal mixed sludges. Typical dewatering results for digested mixed sludges with initial feed solids of 5 percent give a dewatered cake of 19 percent solids at a rate of 32.8 kg/h · m^2. In general, most of the results with these units closely parallel those achieved with rotary vacuum filters. An advantage of CBFPs is that they do not have the sludge pickup problem that sometimes occurs with rotary vacuum filters. Additionally, they have a lower energy consumption.

Reduction

Incineration. If sludge use as a soil conditioner is not practical, or if a site is not available for landfill using dewatered sludge, cities may turn to the alternative of sludge reduction. Incineration completely evaporates the moisture in the sludge and combusts the organic solids to a sterile ash. To minimize the amount of fuel used, the sludge must be dewatered as completely as possible before incineration. The exhaust gas from an incinerator must be treated carefully to avoid air pollution.

6-13 SLUDGE DISPOSAL

Ultimate Disposal

The WWTP process *residuals* (leftover sludges, either treated or untreated) are the bane of design and operating personnel. Of the five possible disposal sites for residuals, two are feasible and only one is practical. Conceivably, one could ultimately dispose of residues in the following places: in the air, in the ocean, in "outer space," on the land, or in the marketplace. Disposal in the air by burning is in reality not ultimate disposal but only temporary storage until the residue falls to the ground. If you use air pollution control devices, then the residue from these devices must be disposed of. Disposal of sewage sludge at sea by barging is now prohibited in the United States. "Outer space" is not a suitable disposal site. Thus, we are left with land disposal and utilization of the sludge to produce a product.

For ease of discussion, we have divided land disposal into three categories: land spreading, landfilling, and dedicated land disposal. We have grouped all of the utilization ideas under one category.

Land Spreading for Sustainability

The practice of applying WWTP residuals for the purposes of recovering nutrients, water, or reclaiming despoiled land such as strip mine spoils is called *land spreading.* In contrast to the other land disposal techniques, land spreading is land-use intensive. Application rates are governed by the character of the soil and the ability of the crops or forests on which the sludge is spread to accommodate it.

Landfilling

Sludge landfill can be defined as the planned burial of wastewater solids, including processed sludge, screenings, grit, and ash, at a designated site. The solids are placed into a prepared site or excavated trench and covered with a layer of soil. The soil cover must be deeper than the depth of the plow zone (about 0.20 to 0.25 m). For the most part, landfilling of screenings, grit, and ash is accomplished with methods similar to those used for sludge landfilling.

Dedicated Land Disposal (DLD)

Dedicated land disposal means the application of heavy sludge loadings to some finite land area that has limited public access and has been set aside or dedicated for all time to the disposal of wastewater sludge. Dedicated land disposal does not mean in-place utilization. No crops may be grown. Dedicated sites typically receive liquid sludges. While application of dewatered sludges is possible, it is not common. In addition, disposal of dewatered sludge in landfills is generally more cost-effective.

Utilization

Wastewater solids may sometimes be used beneficially in ways other than as a soil nutrient. Of the several methods worthy of note, composting and co-firing with municipal solid waste are two which have received increasing amounts of interest in the last few years. The recovery of lime and the use of the sludge to form activated carbon have also been in practice to a lesser extent.

Sludge Disposal Regulations

On February 19, 1993, the EPA promulgated risk-based regulations that govern the use or disposal of sewage sludge. These regulations are codified as 40 CFR Part 503 and have become known as the "503 Regulations." The regulations apply to sewage sludge generated from the treatment of domestic sewage that is land-applied, placed on a surface disposal site, or incinerated in an incinerator that accepts only sewage sludge. The regulations do not apply to sludge generated from treatment of industrial process wastes at an industrial facility, hazardous sewage sludge, sewage sludge with polychlorinated biphenyl (PCB) concentrations of 50 mg/L or greater, or drinking water sludge.

Figure 6-48 summarizes the sludge quality requirements for use or disposal. The regulation establishes two levels of sewage sludge quality with respect to heavy-metal

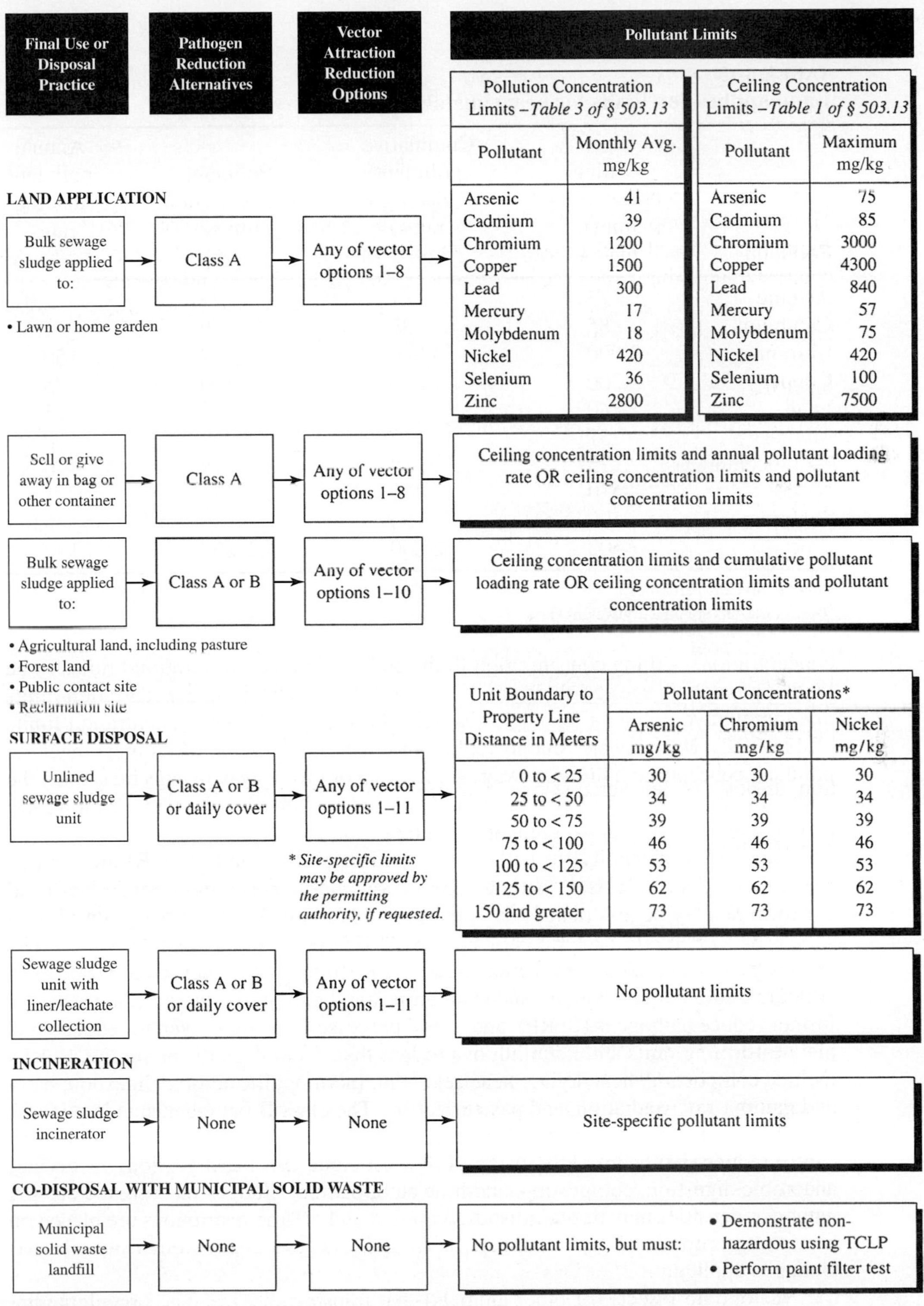

FIGURE 6-48
Sludge quality requirements for use or disposal practices.

TABLE 6-18
Land application limits for heavy metals[a,b]

Pollutant	Ceiling concentration limits, mg/kg	Cumulative pollutant loading rates, kg/ha	Pollutant concentration limits, mg/kg	Annual pollutant loading rates, kg/ha · y
Arsenic	75	41	41	2.0
Cadmium	85	39	39	1.9
Chromium	3,000	3,000	1,200	150
Copper	4,300	1,500	1,500	75
Lead	840	300	300	15
Mercury	57	17	17	0.85
Molybdenum	75	18	18	0.90
Nickel	420	420	420	21
Selenium	100	100	36	5.0
Zinc	7,500	2,800	2,800	140

[a]*Source:* 40 CFR Part 503.13
[b]Concentrations are on a dry-weight basis

concentrations: ceiling concentration limits and pollution concentration limits. To be land-applied, bulk sewage sludge must meet the pollutant ceiling concentration limits *and* cumulative pollutant loading rates (CPLR) *or* the pollutant concentration limits (Table 6-18). Bulk sewage sludge applied to lawns and home gardens must meet the pollutant concentration limits. Sewage sludge sold or given away in bags must meet the pollutant concentration limits *or* the annual sewage sludge product application rates that are based on the annual pollutant loading rates.

Two levels of quality for pathogen densities (class A and class B) are defined in the regulation. All class A pathogen reduction alternatives require that either fecal coliform density be less than 1,000 most probable number (MPN) per gram of total solids, or *Salmonella* bacteria be less than 3 MPN per 4 grams of total solids. The class A treatment alternatives include treating the sludge for a specified time and temperature combination, heat-enhanced alkaline stabilization, treatment in a process to further reduce pathogens (PFRP), and use of processes that are proven to reduce virus plaque-forming units and helminth ova to less than 1 per 4 grams of sludge. PFRPs include composting, heat drying, heat treatment, thermophilic aerobic digestion, beta- and gamma-ray irradiation, and pasteurization. The class B pathogen standard is less than 2 million fecal coliforms per gram of sludge or treatment in a process to significantly reduce pathogens (PSRP). The PSRPs include aerobic digestion, air drying, anaerobic digestion, composting, and lime stabilization. Sludges meeting the class A pathogen densities may be land-disposed immediately. Time restrictions are placed on harvesting crops, grazing of animals, and public access to sites on which class B sludge is applied.

Vectors are insects (or other animals) that transmit disease. The organic nature of sludge often attracts vectors after the sludge is land-applied. The 503 regulations provide 11 alternatives to reduce vector attraction. Some of the alternatives are: volatile solids reduction of 38 percent of more, achieving a standard oxygen uptake rate of less than 1.5 mg O_2 per hour per gram of dry solids at 20°C, aerobic treatment at greater than 40°C with an average temperature greater than 45°C for 14 days, alkaline stabilization, sludge drying, surface incorporation, and soil cover.

The 503 regulations are "self-implementing" in that permits are not required to require conformance.

6-14 CHAPTER REVIEW

When you have completed studying this chapter, you should be able to do the following without the aid of your text or notes:

1. List BOD_5 values for strong, medium, and weak domestic wastewater.
2. List and describe five on-site alternatives for treating and/or disposing of domestic sewage.
3. Choose the correct on-site treatment/disposal system on the basis of population, land use, and soil conditions.
4. Explain the difference between pretreatment, primary treatment, secondary treatment, and tertiary treatment, and show how they are related.
5. Sketch a graph showing the average variation of daily flow at a municipal wastewater treatment plant (WWTP).
6. Define and explain the purpose of equalization.
7. For each type of decomposition (aerobic, anoxic, and anaerobic), list the electron acceptor, important end products, and relative advantages and disadvantages as a waste treatment process.
8. List the growth requirements of bacteria and explain why a bacterium needs them.
9. Sketch and label the bacterial growth curve for a pure culture. Define or explain each phase labeled on the curve.
10. Define θ_c, SRT, and sludge age, and explain their use in regulating the activated sludge process.
11. Explain the purpose of the F/M ratio and define F and M in terms of BOD_5 and mixed liquor volatile suspended solids.
12. Explain the relationship between F/M and θ_c.
13. Explain how cell production is regulated using F/M and/or θ_c.
14. Compare two systems operating at two different F/M ratios.
15. Define SVI and explain its use in the design and operation of an activated sludge plant.

16. Explain the difference between bulking sludge and rising sludge and what circumstances cause each to occur.
17. Sketch, label, and explain the function of the parts of an activated sludge plant and a trickling filter plant.
18. List and explain the relationship of the five types of oxidation ponds to oxygen.
19. Explain what an RBC is and how it works.
20. Compare the positive and negative effects of disinfection of wastewater effluents.
21. List the four common advanced wastewater treatment (AWT) processes and the pollutants they remove.
22. Explain why removal of residual suspended solids effectively removes residual BOD_5.
23. Describe refractory organic compounds and the method used to remove them.
24. List three chemicals used to remove phosphorus from wastewaters.
25. Explain biological nitrification and denitrification either in words or with an equation.
26. Explain ammonia stripping either in words or with an equation.
27. Describe the three basic approaches to land treatment of wastewater.
28. State the two major purposes of sludge stabilization.
29. Explain the purpose of each of the sludge treatment steps and describe the major processes used.
30. Describe the locations for ultimate disposal of sludges and the treatment steps needed prior to ultimate disposal.

With the aid of this text, you should be able to do the following:

1. Determine the volume of a septic tank and the area of a tile field to treat wastewater from a family or institution, given the proper data.
2. Determine whether or not a grit particle of given diameter and density will be captured in a given velocity-controlled grit chamber, or determine the minimum diameter that will be captured under a given set of conditions.
3. Determine the required volume of an equalization basin to dampen a given periodic flow.
4. Determine the effect of equalization on mass loading of a pollutant.
5. Evaluate or size primary and secondary sedimentation tanks with respect to detention time, overflow rate, solids loading, and weir loading.
6. Calculate the bacterial population at a time t, given the initial population and the number of generations.

7. Estimate the soluble BOD_5 in the effluent from a completely mixed or plug-flow activated sludge plant; determine the mean cell residence time or the hydraulic detention time to achieve a desired degree of treatment; determine the "wasting" flow rate to achieve a desired mean cell residence time or F/M ratio.
8. Calculate the F/M ratio given an influent BOD_5, flow, and detention time, or calculate the volume of the aeration basin given F/M, BOD_5, and flow.
9. Calculate SVI and utilize it to determine return sludge concentration and/or flow rate.
10. Calculate the required mass of sludge to be wasted from an activated sludge process given the appropriate data.
11. Calculate the theoretical mass of oxygen required and the amount of air required to supply it given the appropriate data.
12. Use the appropriate trickling filter equation to determine one or more of the following, given the appropriate data: treatment efficiency, filter volume, filter depth, hydraulic loading rate.
13. Perform a sludge mass balance, given the separation efficiencies and appropriate mass flow rate.

6-15 PROBLEMS

6-1. Design a septic tank and tile field system for a highway rest area. Use the following assumptions:

a. Average daily traffic = 6,000 vehicles/d

b. Percent turn in = 10 percent

c. Use rate = 20.0 liters/turn in

d. Maximum use rate = 2.5 × average

e. Terrain = flat

f. GWT = average 4.2 m below grade

g. Soil percolation rate: 5 min/cm

6-2. Ginger Snap is planning to expand her Kookie Jar restaurant to a full-size restaurant to be called the Pretzel Bowl. The existing septic tank has a volume of 4.0 m^3 and the existing tile field has a trench area of 100.0 m^2. If the anticipated wastewater production from the Pretzel Bowl is 4,000 L/d, will Ms. Snap have to expand either the septic tank or the tile field or both? Assume the soil is a sandy loam.

6-3. If a particle having a 0.0170-cm radius and density of 1.95 g/cm^3 is allowed to fall into quiescent water having a temperature of 4°C, what will be the terminal settling velocity? Assume the density of water = 1,000 kg/m^3. Assume Stoke's law applies.

Answer: 3.82×10^{-2} m/s

6-4. If the terminal settling velocity of a particle falling in quiescent water having a temperature of 15°C is 0.0950 cm/s, what is its diameter? Assume a particle density of 2.05 g/cm^3 and density of water equal to 1,000 kg/m^3. Assume Stoke's law applies.

6-5. You have been asked to evaluate the ability of a horizontal-flow, gravity grit chamber to remove a 0.020 cm diameter particle under winter and summer conditions. The particle density is 1.83 g/cm^3. The winter wastewater temperature is 12°C. The summer wastewater temperature is 25°C. The depth of the grit chamber is 1.0 m. The detention time of the wastewater in the grit chamber is 60 s. Assume the density of the wastewater is 1,000 kg/m^3. Assume Stoke's law applies.

6-6. A treatment plant being designed for Cynusoidal City requires an equalization basin to even out flow and BOD variations. The average daily flow is 0.400 m^3/s. The following flows and BOD_5 have been found to be typical of the average variation over a day. What size equalization basin, in cubic meters, is required to provide for a uniform outflow equal to the average daily flow? Assume the flows are hourly averages.

Time	Flow, m^3/s	BOD_5, mg/L	Time	Flow, m^3/s	BOD_5, mg/L
0000	0.340	123	1200	0.508	268
0100	0.254	118	1300	0.526	282
0200	0.160	95	1400	0.530	280
0300	0.132	80	1500	0.552	268
0400	0.132	85	1600	0.570	250
0500	0.140	95	1700	0.596	205
0600	0.160	100	1800	0.604	168
0700	0.254	118	1900	0.570	140
0800	0.360	136	2000	0.552	130
0900	0.446	170	2100	0.474	146
1000	0.474	220	2200	0.412	158
1100	0.482	250	2300	0.372	154

$$\textit{Answer}: \text{V} = 6{,}105.6 \text{ plus } 25\% \text{ excess } = 7{,}630 \text{ m}^3$$

6-7. A treatment plant being designed for Metuchen requires an equalization basin to even out flow and BOD variations. The following flows and BOD_5 have been found to be typical of the average variation over a day. What size equalization basin, in cubic meters, is required to provide for a uniform outflow equal to the average daily flow? Assume the flows are hourly averages.

Time	Flow, m^3/s	BOD_5, mg/L	Time	Flow, m^3/s	BOD_5, mg/L
0000	0.0875	110	1200	0.135	160
0100	0.0700	81	1300	0.129	150
0200	0.0525	53	1400	0.123	140
0300	0.0414	35	1500	0.111	135
0400	0.0334	32	1600	0.103	130
0500	0.0318	42	1700	0.104	120
0600	0.0382	66	1800	0.105	125
0700	0.0653	92	1900	0.116	150
0800	0.113	125	2000	0.127	200
0900	0.131	140	2100	0.128	215
1000	0.135	150	2200	0.121	170
1100	0.137	155	2300	0.110	130

6-8. A treatment plant being designed for the village of Excel requires an equalization basin to even out flow and BOD variations. The following flows and BOD_5 have been found to be typical of the average variation over a day. What size equalization basin, in cubic meters, is required to provide for a uniform outflow equal to the average daily flow?

Time	Flow, m^3/s	BOD_5, mg/L	Time	Flow, m^3/s	BOD_5, mg/L
0000	0.0012	50	1200	0.0041	290
0100	0.0011	34	1300	0.0041	290
0200	0.0009	30	1400	0.0042	275
0300	0.0009	30	1500	0.0038	225
0400	0.0009	33	1600	0.0033	170
0500	0.0013	55	1700	0.0039	180
0600	0.0018	73	1800	0.0046	190
0700	0.0026	110	1900	0.0046	190
0800	0.0033	150	2000	0.0044	190
0900	0.0039	195	2100	0.0034	160
1000	0.0047	235	2200	0.0031	125
1100	0.0044	265	2300	0.0020	80

6-9. Compute and plot the unequalized and the equalized hourly BOD mass loadings to the Cynusoidal City WWTP (Problem 6-6). Using the plot and computations, determine the following ratios for BOD mass loading: peak to average; minimum to average; peak to minimum.

Answers :

	Unequalized	**Equalized**
P/A	1.97	1.47
M/A	0.14	0.63
P/M	14.05	2.34

6-10. Repeat Problem 6-9 using the data from Problem 6-7.

6-11. Repeat Problem 6-9 using the data from Problem 6-8.

6-12. The Ogolly Testing Company has delivered the graph of their results from a batch settling column test conducted for design of a primary settling tank (Figure P-6-12). Determine the detention time and overflow rate for a primary settling tank that will reduce the influent suspended solids concentration from 286 mg/L to 85 mg/L.

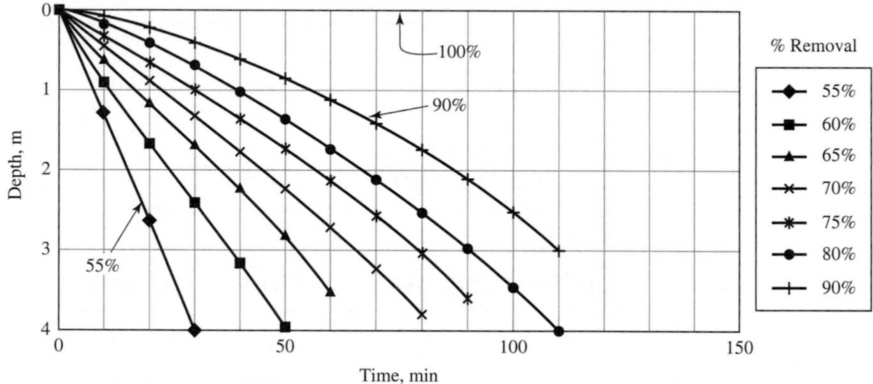

FIGURE P-6-12
Batch settling column data.

6-13. Determine the detention time and overflow rate for a primary settling tank that will reduce the influent suspended solids concentration from 330 mg/L to 150 mg/L. The following batch settling column data are available. The data given are percent removals at the sample times and depths shown.

	Depths, m				
Time (mm)	**0.5**	**1.5**	**2.5**	**3.5**	**4.5**
10	50	32	20	18	15
20	75	45	35	30	25
40	85	65	48	43	40
55	90	75	60	50	46
85	95	87	75	65	60
95	95	88	80	70	63

6-14. The following test data were gathered to design a primary settling tank for a municipal wastewater treatment plant. The initial suspended solids concentration for the test was 200 mg/L. Determine the detention time and overflow rate that will yield 60 percent removal of suspended solids. The data given are suspended solids concentrations in mg/L.

	Time, min					
Depth, m	**10**	**20**	**35**	**50**	**70**	**85**
0.5	140	100	70	62	50	40
1.0	150	130	106	82	70	60
1.5	154	142	120	100	78	70
2.0	160	146	126	110	90	80
2.5	170	150	130	114	100	88

6-15. Using an overflow rate of 26.0 m/d and a detention time of 2.0 h, size a primary sedimentation tank for the average flow at Cynusoidal City (Problem 6-6). What would the overflow rate be for the unequalized maximum flow? Assume 15 sedimentation tanks with length-to-width ratio of 4.7.

Answers: Tank dimensions = 15 tanks at 2.17 m deep by 4.34 m by 20.4 m. Maximum overflow rate 39.3 m/d

6-16. Determine the surface area of a primary settling tank sized to handle a maximum hourly flow of 0.570 m^3/s at an overflow rate of 60.0 m/d. If the effective tank depth is 3.0 m, what is the effective theoretical detention time?

Answers: Surface area = 820.80 or 821 m^2; $\theta = 1.2$ h

6-17. If an equalization basin is installed ahead of the primary tank in Problem 6-16, the average flow to the tank is reduced to 0.400 m^3/s. What is the new overflow rate and detention time?

6-18. The influent BOD_5 to a primary settling tank is 345 mg/L. The average flow rate is 0.050 m^3/s. If the BOD_5 removal efficiency is 30 percent, how many kilograms of BOD_5 are removed in the primary settling tank each day?

6-19. The influent suspended solids concentration to a primary settling tank is 435 mg/L. The average flow rate is 0.050 m^3/s. If the suspended solids removal efficiency is 60 percent, how many kilograms of suspended solids are removed in the primary settling tank each day?

6-20. If the population of microorganisms is 3.0×10^5 at time t_o and 36 hours later it is 9.0×10^8, how many generations have occurred?

Answer: $n = 11.55$ or 12 generations

6-21. The following data were gathered in a bacterial growth experiment. Plot a semilogarithm graph of the data, using a computer spreadsheet you have written. At approximately what time did log growth start and terminate? How many generations occurred during log growth?

Time, h	Bacterial count
0	1×10^3
5	1×10^3
10	1.5×10^3
15	5.4×10^3
20	2.0×10^4
25	7.5×10^4
30	2.85×10^5
35	1.05×10^5
40	1.15×10^5
45	1.15×10^5

6-22. The following data were gathered by Kajima (1923) in an *E. coli* growth experiment. Plot a semilogarithm graph of the data using a computer spreadsheet you have written. Label the following phases on the graph: log growth, stationary, and death. Note that there is no lag phase or acclimation phase. Also note the change in pH with time as a result of the accumulation of by-products of metabolism.

Time, h	Bacterial count	pH
0	50×10^3	7.2
6	175×10^6	6.9
12	320×10^6	6.8
18	538×10^6	7.2
24	609×10^6	7.6
36	559×10^6	7.9
48	493×10^6	8.2
96	330×10^6	8.3
192	53×10^6	8.5
240	7.5×10^6	8.7

6-23. Using the assumptions given in Example 6-5, the rule-of-thumb values for growth constants, and the further assumption that the influent BOD_5 was reduced by 32.0 percent in the primary tank, estimate the liquid volume of a completely mixed activated sludge aeration tank required to treat the wastewater in Problem 6-6. Assume an MLVSS of 2,000 mg/L.

Answer: Volume = 4,032 or 4,000 m^3

6-24. Repeat Problem 6-23 using the wastewater in Problem 6-7.

6-25. Repeat Problem 6-23 using the wastewater in Problem 6-8.

6-26. The town of Camp Verde has been directed to upgrade its primary WWTP to a secondary plant that can meet an effluent standard of 25.0 mg/L BOD_5 and 30 mg/L suspended solids. They have selected a completely mixed activated sludge system for the upgrade. The existing primary treatment plant has a flow rate of 0.029 m^3/s. The effluent from the primary tank has a BOD_5 of 240 mg/L. Using the following assumptions, estimate the required volume of the aeration tank:

1. BOD_5 of the effluent suspended solids is 70 percent of the allowable suspended solids concentration.
2. Growth constants values are estimated to be: $K_s = 100$ mg/L BOD_5; $k_d = 0.025\ d^{-1}$; $\mu_m = 10\ d^{-1}$; $Y = 0.8$ mg VSS/mg BOD_5 removed.
3. The design MLVSS is 3,000 mg/L.

6-27. Using a spreadsheet program you have written, rework Example 6-5 using the following MLVSS concentrations instead of the 2,000 mg/L used in the example: 1,000 mg/L; 1,500 mg/L; 2,500 mg/L; and 3,000 mg/L.

6-28. Using a spreadsheet program you have written, determine the effect of MLVSS concentration on the effluent soluble BOD_5 (S) using the data in Example 6-5. Assume the volume of the aeration tank remains constant at 970 m^3. Use the same MLVSS values used in Problem 6-27.

6-29. If the F/M of a 0.4380 m^3/s activated sludge plant is 0.200 mg/mg · d, the influent BOD_5 after primary settling is 150 mg/L and the MLVSS is 2,200 mg/L, what is the volume of the aeration tank?

Answer: Volume $= 1.29 \times 10^4\ m^3$

6-30. What sludge volume would you expect to find after settling the mixed liquor described in Example 6-8 for 30 minutes in a 1 L graduated cylinder (magna cum laude).

Answer: Volume = 500 mL

6-31. What MLVSS and SVI must be achieved to reduce the return sludge flow rate of Example 6-8 from 0.150 m^3/s to 0.0375 m^3/s? (Note that there are several combinations that will be satisfactory.)

6-32. Two activated sludge aeration tanks at Turkey Run, Indiana, are operated in series. Each tank has the following dimensions: 7.0 m wide by 30.0 m long by 4.3 m effective liquid depth. The plant operating parameters are as follows:

Flow = 0.0796 m^3/s

Soluble BOD_5 after primary settling = 130 mg/L

MLVSS = 1,500 mg/L

MLSS = 1.40 (MLVSS)

Settled sludge volume after 30 min = 230.0 mL/L

Aeration tank liquid temperature = 15°C

Determine the following: aeration period, F/M ratio, SVI, solids concentration in the return sludge.

Answers : Aeration period = 6.3 h; F/M = 0.33; SVI = 110 m/g; X = 9,130 mg/L

6-33. The 500-bed Lotta Hart Hospital has a small activated sludge plant to treat its wastewater. The average daily hospital discharge is 1,500 L per day per bed, and the average soluble BOD_5 after primary settling is 500 mg/L. The aeration tank has effective liquid dimensions of 10.0 m wide by 10.0 m long by 4.5 m deep. The plant operating parameters are as follows:

MLVSS = 2,500 mg/L

MLSS = 1.20 (MLVSS)

Settled sludge volume after 30 min = 200 mL/L

Determine the following: aeration period, F/M ratio, SVI, solids concentration in return sludge.

6-34. The Jambalaya shrimp processing plant generates 0.012 m^3/s of wastewater each day. The wastewater is treated in an activated sludge plant. The average BOD_5 of the raw wastewater before primary settling is 1,400 mg/L. The aeration tank has effective liquid dimensions of 8.0 m wide by 8.0 m long by 5.0 m deep. The plant operating parameters are as follows:

Soluble BOD_5 after primary settling = 966 mg/L

MLVSS = 2,000 mg/L

MLSS = 1.25 (MLVSS)

Settled sludge volume after 30 min = 225.0 mL/L

Aeration tank liquid temperature = 15°C

Determine the following: aeration period, F/M ratio, SVI, solids concentration in the return sludge.

6-35. Using the following assumptions, determine the sludge age, cell wastage flow rate, and the return sludge flow rate for the Turkey Run WWTP (Problem 6-32). Assume:

Suspended solids in the effluent are negligible

Wastage is from the aeration tank

Yield coefficient = 0.40 mg VSS/mg BOD_5 removed

Decay rate of microorganisms = 0.040 d^{-1}

Effluent BOD_5 = 5.0 mg/L (soluble)

Answers: $\theta_c = 11.50$ d; $Q_w = 0.00182$ m^3/s; $Q_r = 0.0214$ m^3/s

6-36. Using the following assumptions, determine the solids retention time, the cell wastage flow rate, and the return sludge flow rate for the Lotta Hart Hospital WWTP (Problem 6-33). Assume:

Allowable BOD_5 in effluent = 25.0 mg/L
Suspended solids in effluent = 25.0 mg/L
Wastage is from the return sludge line
Yield coefficient = 0.60 mg VSS/mg BOD_5 removed
Decay rate of microorganisms = 0.060 d^{-1}
Inert fraction of suspended solids = 66.67%

6-37. Using the following assumptions, determine the solids retention time, the cell wastage flow rate, and the return sludge flow rate for the Jambalaya shrimp processing plant WWTP (Problem 6-34). Assume:

Allowable BOD_5 in effluent = 25.0 mg/L
Suspended solids in effluent = 30.0 mg/L
Wastage is from the return sludge line
Yield coefficient = 0.50 mg VSS/mg BOD_5 removed
Decay rate of microorganisms = 0.075 d^{-1}
Inert fraction of suspended solids = 30.0%

6-38. The two secondary settling tanks at Turkey Run (Problem 6-32) are 16.0 m in diameter and 4.0 m deep at the side wall. The effluent weir is a single launder set on the tank wall. Evaluate the overflow rate, depth, solids loading, and weir length of this tank for conformance to standard practice.

Answers: $v_o = 17.1$ m/d < 33 m/d; OK
SWD > 3.7 m recommended depth; OK
SL $= 45.57$ kg/m$^2 \cdot$ d $\ll 250$ kg/m$^2 \cdot$ d; OK
WL $= 68.4$ m^3/d $\cdot$ m, which is acceptable

6-39. The single secondary settling tank at the Lotta Hart Hospital WWTP (Problem 6-33) is 10.0 m in diameter and 3.4 m deep at the side wall. The effluent weir is a single launder set on the tank wall. Evaluate the overflow rate, depth, solids loading, and weir length for conformance to standard practice.

6-40. The single secondary settling tank at the Jambalaya shrimp processing WWTP (Problem 6-34) is 5.0 m in diameter and 2.5 m deep at the side wall. The effluent weir is a single launder set on the tank wall. Evaluate the overflow rate, depth, solids loading, and weir length for conformance to standard practice.

6-41. Envirotech Systems markets synthetic media for use in the construction of trickling filters. Envirotech uses the following formula to determine BOD removal efficiency:

$$\frac{L_e}{L_i} = \exp\left[-\frac{k\theta D}{Q^n}\right]$$

where L_e = BOD_5 of effluent, mg/L
L_i = BOD_5 of influent, mg/L
k = treatability factor, $(m/d)^{0.5}/m$
θ = temperature correction factor
= $(1.035)^{T-20}$
T = wastewater temperature, °C
D = media depth, m
Q = hydraulic loading rate, m/d
n = 0.5

Using the following data for domestic wastewater, determine the treatability factor k.

Wastewater temperature = 13°C
Hydraulic loading rate = 41.1 m/d

% BOD remaining	Media depth, m
100.0	0.00
80.3	1.00
64.5	2.00
41.6	4.00
17.3	8.00

Answer: $k = 1.79\ (m/d)^{0.5}/m$ at 20°C

6-42. Using the Envirotech systems equation and the treatability factor from Problem 6-41, estimate the depth of filter required to achieve 82.7 percent BOD_5 removal if the wastewater temperature is 20°C and the hydraulic loading rate is 41.1 m/d.

6-43. Koon, et al. (1976), suggest that recirculation for a synthetic media filter may be considered by the following formula:

$$\frac{L_e}{L_i} = \frac{\exp\left[-\frac{k\theta D}{Q^n}\right]}{(1 + r) - r\exp\left[-\frac{k\theta D}{Q^n}\right]}$$

where r = recirculation ratio and all other terms are as described in Problem 6-41.

Use this equation to determine the efficiency of a 1.8-m-deep synthetic media filter loaded at a hydraulic loading rate of 5.00 m/d with a recirculation ratio of 2.00. The wastewater temperature is 16°C and the treatability factor is 1.79 $(m/d)^{0.5}/m$ at 20°C.

6-44. Determine the concentration of the effluent BOD_5 for the two-stage trickling filter described below. The wastewater temperature is 17°C. Assume the NRC equations apply.

Design flow = 0.0509 m^3/s

Influent BOD_5 (after primary treatment) = 260 mg/L

Diameter of each filter = 24.0 m

Depth of each filter = 1.83 m

Recirculation flow rate for each filter = 0.0594 m^3/s

6-45. Using a computer spreadsheet program you have written, plot a graph of the final effluent BOD_5 of the two-stage filter described in Problem 6-44 as a function of the influent flow rate. Assume that the ratio of recirculation flow to influent flow remains constant. Use flow rates of 0.02, 0.03, 0.04, 0.05, 0.06, 0.07, 0.08, 0.09, and 0.10 m^3/s for the influent.

6-46. Determine the diameter of a single-stage rock media filter to reduce an applied BOD_5 of 125 mg/L to 25 mg/L. Use a flow rate of 0.14 m^3/s, a recirculation ratio of 12.0, and a filter depth of 1.83 m. Assume the NRC equations apply and that the wastewater temperature is 20°C.

6-47. Using a computer spreadsheet program you have written, plot a graph of the final effluent BOD_5 of the single-stage filter described in Problem 6-46 as a function of the influent flow rate. Assume that the ratio of recirculation flow to influent flow remains constant and that the filter diameter is 35.0 m. Use hydraulic loading rates of 10, 12, 14, 16, 18, and 20 $m^3/m^2 \cdot d$.

6-48. An oxidation pond having a surface area of 90,000 m^2 is loaded with a waste flow of 500 m^3/d containing 180 kg of BOD_5. The operating depth is from 0.8 to 1.6 m. Using the Michigan rules of thumb, determine whether this design is acceptable.

Answers: Loading rate = 20.0 kg/ha · d

Detention time = 180 d

This design is acceptable.

6-49. Determine the required surface area and the loading rate for a facultative oxidation pond to treat a waste flow of 3,800 m^3/d with a BOD_5 of 100.0 mg/L.

6-50. Rework Example 6-15 using alum $[Al_2(SO_4)_3 \cdot 18\ H_2O]$ to remove the phosphorus.

Answer: 86.1 mg/L of alum

6-51. Rework Example 6-15 using lime (CaO) to remove the phosphorus.

6-52. Prepare a monthly water balance and estimate of the storage volume required (in m^3) for a spray irrigation system being designed for Wheatville, Iowa.

The design population is 1,000 and the design wastewater generation rate is 280.0 Lpcd. Based on a nitrogen balance, the allowable application rate is 27.74 mm/mo. The area available is 40.0 ha. The percolation rate during the spray season is 150 mm/mo. Assume that the runoff is contained and reapplied. Assume "spray season" is when temperature is above 0°C. In fact, spraying can continue to about −4°C but once spraying has stopped, it may not recommence until temperatures exceed +4°C. The following climatological data (from Kansas City) may be used. The water balance is a direct application of the hydrologic balance equation. The mass balance equation may be rewritten as:

$$\frac{dS}{dt} = P + \text{WW} - \text{ET} - G - R$$

where dS/dt = change in storage, mm/mo
P = precipitation, mm/mo
WW = wastewater application rate, mm/mo
ET = evapotranspiration, mm/mo
G = groundwater infiltration, mm/mo
R = runoff, mm/mo

The storage volume required may be estimated from a mass balance of the form used for determining storage volume for reservoirs in Chapter 3.

Climatological data from Kansas City, Missouri

Month	Average temperature, °C	Evapotranspiration,[a] mm	Precipitation, mm
Jan	−0.2	23	36
Feb	2.1	28	32
Mar	6.3	43	63
Apr	13.2	79	90
May	18.7	112	112
Jun	24.4	155	116
Jul	27.5	203	81
Aug	26.6	198	96
Sep	21.8	152	83
Oct	15.7	114	73
Nov	7.0	64	46
Dec	2.1	25	39

[a]Estimated.

6-53. Prepare a monthly water balance and estimate the storage volume required for Flushing Meadows. The area available for spraying is 125 ha. The design population for Flushing Meadows is 8,880. The average wastewater generation rate is 485.0 Lpcd. The percolation rate is 200 mm/mo during the spray season. Assume that runoff is to be contained and reapplied. Assume also that the following climatological data apply. Assume "spray season" is when temperature is above 0°C. In fact, spraying can continue to about −4°C but once spraying has stopped, it may not recommence until temperatures exceed +4°C. See Problem 6-52 for hints.

Climatological data from Columbus, Ohio

Month	Average temperature, °C	Evapotranspiration,[a] mm	Precipitation, mm
Jan	−1.2	15	80
Feb	−0.5	20	59
Mar	3.8	28	80
Apr	10.4	58	59
May	16.4	89	102
Jun	21.9	117	106
Jul	23.8	142	100
Aug	22.9	130	73
Sep	18.8	104	67
Oct	12.3	76	54
Nov	5.2	41	63
Dec	−0.3	15	59

[a] Estimated.

6-54. Prepare a monthly water balance for Weeping Water. The design population for Weeping Water is 10,080. The average wastewater generation rate is 385.0 Lpcd. The area available is 200.0 ha. The available lagoon volume is 300,000 m^3. The percolation rate is 150 mm/mo during the spray season. Assume that runoff is to be contained and reapplied. Assume also that the following climatological data apply. Assume "spray season" is when temperature is above 4°C. In fact, spraying can continue to about −4°C but once spraying has stopped, it may not recommence until temperatures exceed +4°C. Is the lagoon volume sufficient? If not, how much additional volume is needed? See Problem 6-52 for hints.

Climatological data from Helena, Montana[a]

Month	Average temperature, °C	Evapotranspiration,[b] mm	Precipitation, mm
Jan	−6.6	3	13
Feb	−3.1	5	10
Mar	1.7	5	16
Apr	6.7	35	23
May	11.6	102	45
Jun	16.2	178	46
Jul	19.9	180	34
Aug	19.3	163	32
Sep	13.4	76	27
Oct	8.4	5	16
Nov	4.3	5	12
Dec	3.1	3	12

[a] http//cdo.ncdc.noaa.gov/ancsum/ACS.
[b] Estimated.

6-55. Determine the daily and annual primary sludge production for a WWTP having the following operating characteristics:

Flow = $0.0500\ m^3/s$
Influent suspended solids = 155.0 mg/L
Removal efficiency = 53.0%
Volatile solids = 70.0%
Specific gravity of volatile solids = 0.970
Fixed solids = 30.0%
Specific gravity of fixed solids = 2.50
Sludge concentration = 4.50%

Answer: $\forall_{sl} = 7.83\ m^3/d$

6-56. Repeat Problem 6-55 using the following operating data:

Flow = $2.00\ m^3/s$
Influent suspended solids = 179.0 mg/L
Removal efficiency = 47.0%
Specific gravity of fixed solids = 2.50
Specific gravity of volatile solids = 0.999
Fixed solids = 32.0%

Volatile solids = 68.0%
Sludge concentration = 5.20%

6-57. Using a computer spreadsheet you have written, and the data in Problem 6-56, determine the daily and annual sludge production at the following removal efficiencies: 40, 45, 50, 55, 60, and 65 percent. Plot annual sludge production as a function of efficiency.

6-58. Using Figure 6-36, Table 6-14, and the following data, determine B, E, J, K, and L in megagrams per day (Mg/d).

$A = 185.686$ Mg/d

$\eta_E = 0.900$; $\eta_j = 0.250$; $\eta_N = 0.00$; $\eta_P = 0.150$; $\eta_H = 0.190$

Answers: $B = 21.112$ or 21.1 Mg/d
$E = 190.011$ or 190 Mg/d
$J = 47.503$ or 47.5 Mg/d
$K = 142.509$ or 143 Mg/d
$L = 144.147$ or 144 Mg/d

6-59. Rework Problem 6-58 assuming that the digestion solids are not dewatered prior to ultimate disposal; that is, $K = L$.

6-60. The value for η_E in Problem 6-58 is quite high. Rework the problem with a more realistic value of $\eta_E = 0.50$.

6-61. The flowsheet for the Doubtful WWTP is shown in Figure P-6-61. Assuming that the appropriate values of η given in Figure 6-37 may be used when needed and that $A = 7.250$ Mg/d, $X = 1.288$ Mg/d, and $N = 0.000$ Mg/d, what is the mass flow (in kg/d) of sludge to be sent to ultimate disposal?

Answer: $L = K = 3.743$ Mg/d, or 3,743 kg/d

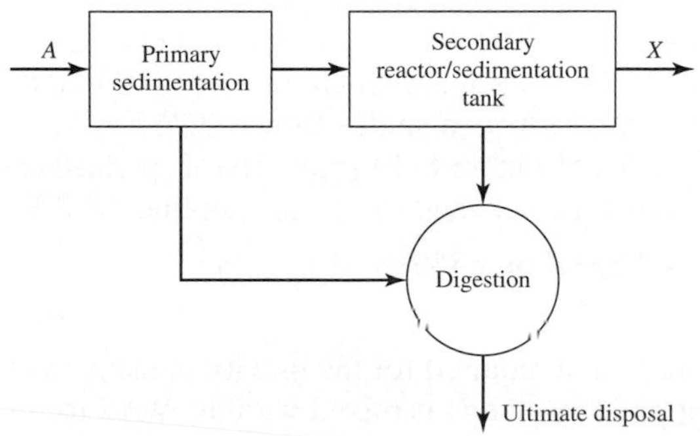

FIGURE P-6-61
Flowsheet for Doubtful WWTP.

6-62. Using the following mass flow data from the Doubtful WWTP (Problem 6-61) determine η_E, η_D, η_N, η_J, η_X. Mass flows for Doubtful WWTP in

Mg/d: $A = 7.280$, $B = 7.798$, $D = 0.390$, $E = 8.910$, $F = 6.940$, $J = 4.755$, $K = 6.422$, $N = 9.428$, $X = 0.468$

6-63. The city of Doubtful (Problem 6-61) is considering the installation of thickening and dewatering facilities. The revised flow diagram for Doubtful to include thickening and dewatering with appropriate return lines is shown in Figure P-6-63. Calculate a value for L in Mg/d. Assume that the appropriate values of η given in Figure 6-37 maybe used when needed and that $A = 7.250$ Mg/d, $X = 1.288$ Mg/d.

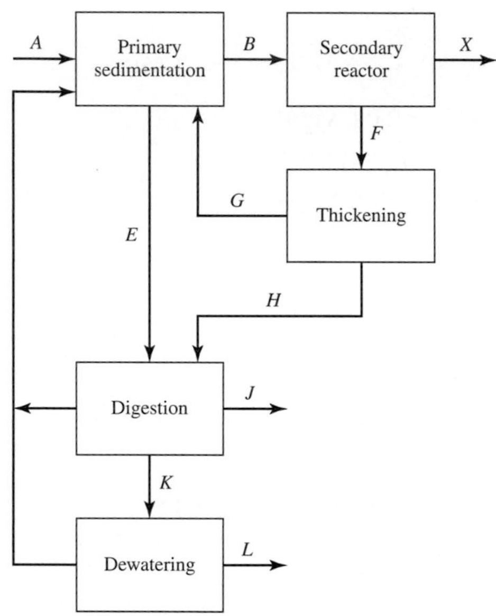

FIGURE P-6-63
Revised flowsheet for Doubtful WWTP.

6-64. Determine the surface area required for the gravity thickeners (assume that no thickener is greater than 30.0 m in diameter) to thicken the waste activated sludge (WAS) at Grand Rapids, Michigan, from 10,600 mg/L to 2.50 percent solids. The waste activated sludge flow is 3,255 m^3/d. Assume that the batch settling curves of Figure 6-40 apply. Use a spreadsheet program you have written to plot the data and fit the tangent line.

Answer: $A_s = 2{,}851.4$ or 2,850 m^2 depending on graph reading. Thus, choose four thickeners at 30 m diameter.

6-65. Determine the surface area required for the gravity thickeners of Problem 6-64 if 710 m^3/d of primary sludge is mixed with the WAS to form a sludge having 2.00 percent solids. The final sludge is to have a solids concentration of 5.00 percent. The batch settling curve for mixed WAS and PS in Figure 6-40 is assumed to apply. Because of the additional sludge, assume five thickeners will be used. Use a spreadsheet program you have written to plot the data and fit the tangent line.

6-66. Settling test data from the Little Falls WWTP are shown below. Determine the surface area for a gravity thickener for 733 m^3/d of waste activated sludge. The final sludge concentration is to be 3.6 percent. Use a spreadsheet program you have written to plot the data and fit the tangent line.

Suspended solids concentration, g/L	Initial settling velocity, m/d
4.0	58.5
6.0	36.6
8.0	24.1
14.0	8.1
29.0	2.2
41.0	0.73

6-67. The Pomdeterra wastewater treatment plant produces thickened sludge that has a suspended solids concentration of 3.8 percent. They are investigating a filter press that will yield a solids concentration of 24 percent. If they now produce 33 m^3/d of sludge, what annual volume savings will they achieve if they install the press?

6-68. Ottawa's anaerobic digester produces 13 m^3/d of sludge with a suspended solids concentration of 7.8 percent. What volume of sludge must they dispose of each year if their sand drying beds yield a solids concentration of 35 percent?

6-69. Weed Patch's digester produces 30 m^3/mo of sludge with a suspended solids concentration of 2.5 percent. What solids concentration must their drying facility achieve to reduce the volume to 3 m^3/mo?

6-16 DISCUSSION QUESTIONS

6-1. You are touring the research labs of the environmental engineers at your university. Two biological reactors are in a controlled temperature room that has a temperature of 35°C. Reactor A has a strong odor. Reactor B has virtually no odor. What electron acceptors are being used in each reactor?

6-2. If the state regulatory agency requires tertiary treatment of a municipal wastewater, what, if any, processes would you expect to find preceding the tertiary process?

6-3. What is the purpose of recirculation in a trickling filter plant and how does it differ from return sludge in an activated sludge plant?

6-4. In which of the following cases is the cost of sludge disposal higher?

a. $\theta_c = 3$ days

b. $\theta_c = 10$ days

6-5. Would an industrial wastewater containing only NH_4 at a pH of 7.00 be denitrified if pure oxygen was bubbled into it? Explain your reasoning.

6-17 REFERENCES

APHA (2005) *Standard Methods for the Examination of Waste and Wastewater,* 21st ed., American Public Health Association, Washington, DC.

Bailey, J. E., and D. F. Ollis (1977) *Biochemical Engineering Fundamentals,* McGraw-Hill, New York, NY, p. 222.

Benefield, L. D., and C. W. Randall (1980) *Biological Process Design for Wastewater Treatment,* Prentice Hall, Upper Saddle river, NJ, pp. 322–324, 338–340, 353–354.

Caldwell, D. H., D. S. Parker, and W. R. Uhte (1973) *Upgrading Lagoons,* U.S. Environmental Protection Agency Technology Transfer Publication, Washington, DC.

CFR (2005a) *Code of Federal Regulations,* 40 CFR §413.02, 464.02, 467.02, and 469.12.

CFR (2005b) *Code of Federal Regulations*, 40 CFR §133.102.

Crites, R., and G. Tchobanoglous (1998) *Small Decentralized Wastewater Management Systems,* McGraw-Hill, New York.

Curds, C. R. (1973) "A Theoretical Study of Factors Influencing the Microbial Population Dynamics of the Activated Sludge Process—I," *Water Research,* vol. 7, pp. 1269–1284.

Dick, R. I. (1970) "Thickening," in E. F. Gloyna and W. W. Eckenfelder, (eds.), *Advances in Water Quality Improvement—Physical and Chemical Processes,* University of Texas Press, Austin, p. 380.

Ehlers, V. M., and E. W. Steel (1943) *Municipal and Rural Sanitation,* McGraw-Hill, New York.

Erickson, A. E., B. G. Ellis, J. M. Tiedje, et al. (1974) *Soil Modification for Denitrification and Phosphate Reduction of Feedlot Waste,* U.S. Environmental Protection Agency, EPA Pub. No. 660/2-74-057, Washington, DC.

GLUMRB (1978) *Recommended Standards for Sewage Works,* Great Lakes–Upper Mississippi River Board of State Sanitary Engineers, Health Education Service, Inc., Albany, NY, pp. 60–63.

Grady, C. P. L., and H. C. Lim (1980) *Biological Wastewater Treatment, Theory and Applications,* Marcel Dekker, New York.

Hammer, M. J. (1977) *Water and Waste-water Technology, SI Version,* John Wiley & Sons, New York, p. 387.

Heukeleian, H., H. Orford, and R. Maganelli (1951) "Factors Affecting the Quantity of Sludge Production in the Activated Sludge Process," *Sewage & Industrial Wastes,* vol. 23, p. 8.

Kajima (1923) *Scientific Reports, Government Institute of Infectious Diseases,* Tokyo, vol. 2, p. 305.

Koon, J. H., R. F. Curran, C. E. Adams, and W. W. Eckenfelder (1976) *Evaluation and Upgrading of a Multistage Trickling Filter Facility,* U.S. Environmental Protection Agency, Publication No. EPA 600/2-76 193), Washington, DC.

Lawrence, A. W. and T. R. Milnes (1971) "Discussion Paper," *Journal of the Sanitary Engineering Division,* American Society of Civil Engineers, vol. 97, p. 121.

Lawrence, A. W. and P. L. McCarty (1970) "A Unified Basis for Biological Treatment Design and Operation," *Journal of the Sanitary Engineering Division*, American Society of Civil Engineers, vol. 96, no. SA3.

McCarty, P. L. (1968) "Anaerobic Treatment of Soluble Wastes," in E. F. Gloyna and W. W. Eckenfelder (eds.), *Advances in Water Quality Improvement—Physical and Chemical Processes,* University of Texas Press, Austin, TX.

McKinney, R. E. (1962) *Microbiology for Sanitary Engineers*, McGraw-Hill, New York, p. 40.

Metcalf & Eddy, Inc. (1979) *Wastewater Engineering: Treatment, Disposal, Reuse,* revised by G. Tchobanoglous, McGraw-Hill, New York, pp. 328, 395.

Metcalf & Eddy, Inc. (1991) *Wastewater Engineering: Treatment, Disposal, Reuse,* revised by G. Tchobanoglous and F. L. Burton, McGraw-Hill, New York, pp. 394, 540–542.

Metcalf & Eddy, Inc. (2003) *Wastewater Engineering: Treatment and Reuse*, revised by G. Tchobanoglous, F. L. Burton, and H. D. Stensel, McGraw-Hill, New York, p. 408.

Monod, J. (1949) "The Growth of Bacterial Cultures," *Annual Review of Microbiology,* vol. 3, pp. 371–394.

NRC (1946) "Sewage Treatment at Military Installations," *Sewage Works Journal,* vol. 18, p. 787.

Pelczar, M. J., and R. D. Reid (1958) *Microbiology*, McGraw-Hill, New York, p. 424.

Pound, C. E., R. W. Crites, and D. A. Griffes (1976) *Land Treatment of Municipal Wastewater Effluents, Design Factors I,* U.S. Environmental Protection Agency Technology Transfer Seminar Publication, Washington, DC.

Rex Chainbelt, Inc. (1972) *A Mathematical Model of a Final Clarifier,* U.S. Environmental Protection Agency, Report No. 17090 FJW 02/72, Washington, DC.

Sawyer, C. N. (1956) "Bacterial Nutrition and Synthesis," *Biological Treatment of Sewage and Industrial Wastes,* vol. I, p. 3.

Schulze, K. L. (1960) "Load and Efficiency of Trickling Filters," *Journal of Water Pollution Control Federation,* vol. 32, p. 245.

Shahriar, H., C. Eskicioglu, and R. L. Droste (2006) "Simulating Activated Sludge System by Simple-to-Advanced Models," *Journal of Environmental Engineering,* American Society of Civil Engineers, vol. 132, pp. 42–50.

U.S. EPA (1977) *Process Design Manual: Wastewater Treatment Facilities for Sewered Small Communities,* U.S. Environmental Protection Agency, EPA Pub. No. 625/1-77-009, Washington. DC, p. 8–12.

U.S. EPA (1979a) *Environmental Pollution Control Alternatives: Municipal Wastewater,* U.S. Environmental Protection Agency, EPA Pub. No. 625/5-79-012, Washington, DC, pp. 33–35, 52–55.

U.S. EPA (1979b) *Process Design Manual, Sludge Treatment and Disposal,* U.S. Environmental Protection Agency, Publication No. 625/1-79-011, Washington DC, pp. 3–19, 3–21, 3–25, 3–26, 5–7, 5–8.

U.S. EPA (1981) *The 1980 Needs Survey. Conveyance, Treatment, and Control of Municipal Wastewater, Combined Sewer Outflows, and Stormwater Runoff, Summaries of Technical Data,* U.S. Environmental Protection Agency, EPA-430/9-81-008, Washington, DC.

U.S. EPA (1996) *NPDES Permit Writers' Manual,* U.S. Environmental Protection Agency, EPA Pub. No. 833-B-96-003, Washington, DC, pp. 77–78.

U.S. EPA (1997) *Response to Congress on the Use of Decentralized Wastewater Treatment Systems,* U.S. Environmental Protection Agency, EPA Report No. 832-R-001b, Washington, DC.

U.S. EPA (2005) http://www.epa.gov. Search: Region 10 ⇒ Homepage ⇒ NPDES Permits ⇒ Current NPDES Permits in Pacific Northwest and Alaska ⇒ Current Individual NPDES Permits in Idaho

Velz, C. J. (1948) "A Basic Law for the Performance of Biological Filters," *Sewage Works,* vol. 20, p. 607.

Vesilind, P. A. (1968) "Discussion of Evaluation of Sludge Thickening Theories," *Journal of the Sanitary Engineering Division,* American Society of Civil Engineers, vol. 94, p. 185.

WEF (1992) *Design of Municipal Wastewater Treatment Plants,* Vol. I, Manual of Practice No. 8, Joint Task Force of the Water Environment Federation and American Society of Civil Engineers, Alexandria, VA, pp. 528, 529, 530, 586, 595, 705–714.

Yoshioka, N., et al. (1957) "Continuous Thickening of Homogenous Flocculated Slurries," *Chemical Engineering,* Tokyo (in Japanese).

CHAPTER 7

AIR POLLUTION

7-1 AIR POLLUTION PERSPECTIVE

Air pollution is of public health concern on several scales: micro, meso, and macro. Indoor air pollution results from products used in construction materials, inadequacy of general ventilation, and geophysical factors that may result in exposure to naturally occurring radioactive materials. Industrial and mobile sources contribute to meso-scale air pollution that contaminates the ambient air that surrounds us outdoors. Macro-scale effects include transport of ambient air pollutants over large distances and global impact. Examples of macro-scale impacts include acid rain and ozone pollution. Global impacts of air pollution result from sources that may potentially change the upper atmosphere. Examples include depletion of the ozone layer and global warming. While micro- and macro-scale effects are of concern, our focus will predominately be on meso-scale air pollution.

7-2 PHYSICAL AND CHEMICAL FUNDAMENTALS

Ideal Gas Law

Although polluted air may not be "ideal" from the biological point of view, we may treat its behavior with respect to temperature and pressure as if it were ideal. Thus, we assume that at the same temperature and pressure, different kinds of gases have densities proportional to their molecular masses. This may be written as

$$\rho = \frac{1}{R}\frac{PM}{T} \tag{7-1}$$

where ρ = density of gas, kg/m^3
P = absolute pressure, kPa
M = molecular mass, grams/mole
T = absolute temperature, K
R = universal gas constant = 8.3143 J/K · mole = 8.3143 Pa · m^3/K · mole

Since density is mass per unit volume, or the number of moles per unit volume, $n/\forall$, the expression may be rewritten in the general form as

$$P\forall = nRT \tag{7-2}$$

where $\forall$ is the volume occupied by n moles of gas. At 273.15 K and 101.325 kPa, one mole of an ideal gas occupies 22.414 L.

Dalton's Law of Partial Pressures

Stack and exhaust sampling measurements are made with instruments calibrated with air. Because combustion products have an entirely different composition than air, the readings must be adjusted ("corrected" in sampling parlance) to reflect this difference. Dalton's law forms the basis for the calculation of the correction factor. Dalton found that the total pressure exerted by a mixture of gases is equal to the sum of the pressures that each type of gas would exert if it alone occupied the container. In mathematical terms,

$$P_t = P_1 + P_2 + P_3 + \cdots \tag{7-3}$$

where P_t = total pressure of mixture
P_1, P_2, P_3 = pressure of each gas if it were in container alone, that is, *partial pressure*

Dalton's law also may be written in terms of the ideal gas law:

$$P_t = n_1\frac{RT}{\forall} + n_2\frac{RT}{\forall} + n_3\frac{RT}{\forall} + \cdots$$
$$= (n_1 + n_2 + n_3 + \cdots)\frac{RT}{\forall}$$

Adiabatic Expansion and Compression

Air pollution meteorology is, in part, a consequence of the thermodynamic processes of the atmosphere. One such process is adiabatic expansion and contraction. An *adiabatic* process is one that takes place with no addition or removal of heat and with sufficient slowness so that the gas can be considered to be in equilibrium at all times.

As an example, let us consider the piston and cylinder in Figure 7-1. The cylinder and piston face are assumed to be perfectly insulated. The gas is at pressure *P*. A force, *F*, equal to *PA* must be applied to the piston to maintain equilibrium. If the force is increased and the volume is compressed, the pressure will increase and work will be done on the gas by the piston. Since no heat enters or leaves the gas, the work will go into increasing the thermal energy of the gas in accordance with the first principle of thermodynamics, that is,

(Heat added to gas) = (increase in thermal energy) + (external work done by or on the gas)

Since the left side of the equation is zero (because it is an adiabatic process), the increase in thermal energy is equal to the work done. The increase in thermal energy is reflected by an increase in the temperature of the gas. If the gas is expanded adiabatically, its temperature will decrease.

Units of Measure

The three basic units of measure used in reporting air pollution data are *micrograms per cubic meter* ($\mu g/m^3$), *parts per million* (ppm), and the *micron* (μ) or, preferably, its equivalent, the *micrometer* (μm). Micrograms per cubic meter and parts per million are measures of concentration. Both $\mu g/m^3$ and ppm are used to indicate the concentration of a gaseous pollutant. However, the concentration of particulate matter may be reported only as $\mu g/m^3$. The μm is used to report particle size.

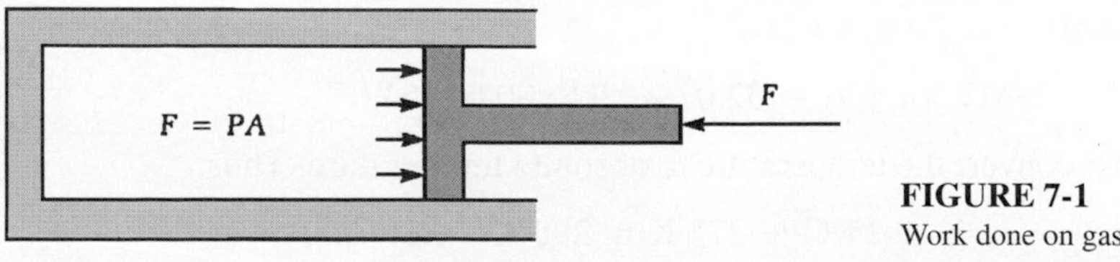

FIGURE 7-1
Work done on gas.

There is an advantage to the unit ppm that frequently makes it the unit of choice. The advantage results from the fact that ppm is a volume-to-volume ratio. (Note that this is different than ppm in water and wastewater, which is a mass-to-mass ratio.) Changes in temperature and pressure do not change the ratio of the volume of pollutant gas to the volume of air that contains it. Thus, it is possible to compare ppm readings from Denver and Washington, DC, without further conversion.

Converting $\mu g/m^3$ ppm. The conversion between $\mu g/m^3$ and ppm is based on the fact that at standard conditions (0°C and 101.325 kPa), one mole of an ideal gas occupies 22.414 L. Thus, we may write an equation that converts the mass of the pollutant M_p in grams to its equivalent volume $\forall_p$ in liters at standard temperature and pressure (STP):

$$\forall_p = \frac{M_p}{\text{GMW}} \times 22.414 \text{ L/GM} \tag{7-4}$$

where GMW is the gram molecular weight of the pollutant. For readings made at temperatures and pressures other than standard conditions, the standard volume, 22.414 L/GM, must be corrected. We use the ideal gas law to make the correction:

$$22.414 \text{ L/GM} \times \frac{T_2}{273 \text{ K}} \times \frac{101.325 \text{ kPa}}{P_2} \tag{7-5}$$

where T_2 and P_2 are the absolute temperature and absolute pressure at which the readings were made. Since ppm is a volume ratio, we may write

$$\text{ppm} = \frac{\forall_p}{\forall_a} \tag{7-6}$$

where $\forall_a$ is the volume of air in cubic meters at the temperature and pressure of the reading. We then combine Equations 7-4, 7-5, and 7-6 to form Equation 7-7.

$$\text{ppm} = \frac{\frac{M_p}{\text{GMW}} \times 22.414 \times \frac{T_2}{273 \text{ K}} \times \frac{101.325 \text{ kPa}}{P_2}}{\forall_a \times 1{,}000 \text{ L/m}^3} \tag{7-7}$$

where M_p is in μg. The factors converting μg to g and L to millions of L cancel one another. Unless otherwise stated, it is assumed that $\forall_a = 1.00 \text{ m}^3$

Example 7-1. A one-cubic-meter sample of air was found to contain 80 $\mu g/m^3$ of SO_2. The temperature and pressure were 25°C and 103.193 kPa when the air sample was taken. What was the SO_2 concentration in ppm?

Solution. First we must determine the GMW of SO_2. From the chart inside the front cover, we find

$$\text{GMW of } SO_2 = 32.07 + 2(16.00) = 64.07$$

Next we must convert the temperature to absolute temperature. Thus,

$$25°\text{C} + 273 \text{ K} = 298 \text{ K}$$

Now we may make use of Equation 7-7.

$$\text{ppm} = \frac{\frac{80\mu g}{64.07} \times 22.414 \times \frac{298}{273} \times \frac{101.325}{103.193}}{1.00\ \text{m}^3 \times 1{,}000\ \text{L/m}^3} = 0.0300\ \text{ppm of } SO_2$$

Relativity. Before we launch into the esoterics of air pollution, let's take a moment to look at the relationship of a ppm and a μm to something relevant to daily life. Four crystals of common table salt in one cup of granulated sugar is approximately equal to 1 ppm on a volume-to-volume basis. Figure 7-2 should help you visualize the size of a μm. Note that a hair has an average diameter of approximately 80 μm.

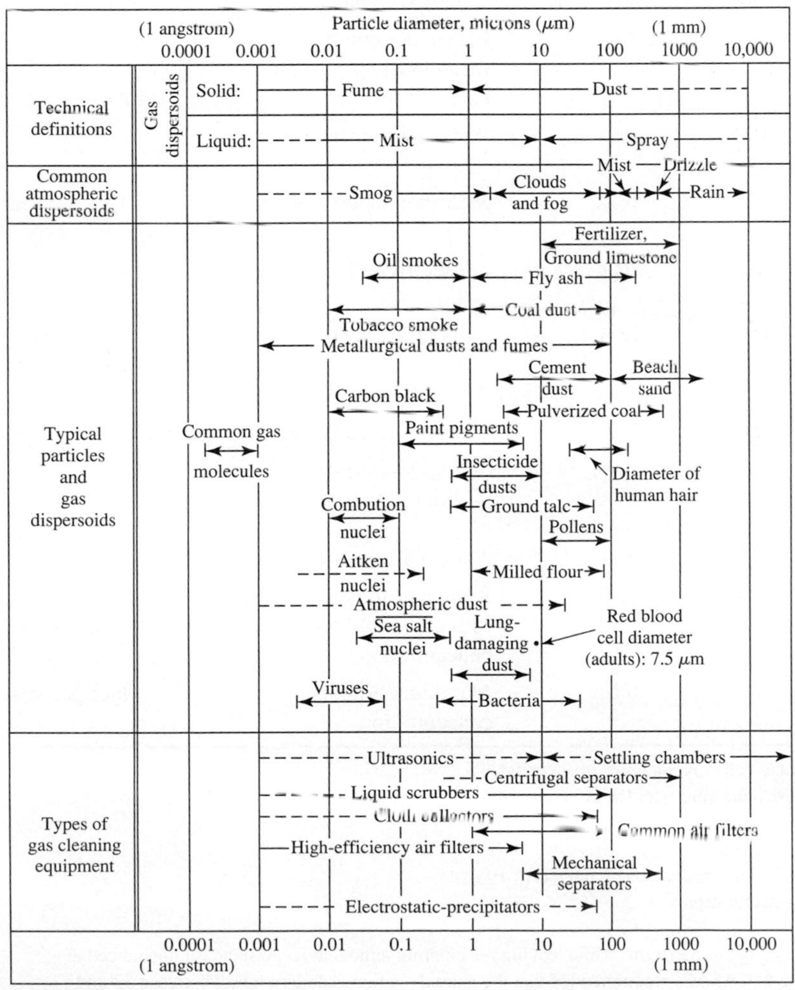

FIGURE 7-2
Characteristics of particles and particle dispersoids. (*Source:* Lapple, 1951.)

7-3 AIR POLLUTION STANDARDS

The 1970 Clean Air Act (CAA) required the U.S. Environmental Protection Agency (EPA) to investigate and describe the environmental effects of any air pollutant emitted by stationary or mobile sources that may adversely affect human health or the environment. The EPA used these studies to establish the National Ambient Air Quality

TABLE 7-1
National Ambient Air Quality Standards (NAAQS)

Criteria pollutant	Standard type	Concentration $\mu g/m^3$	Concentration ppm	Averaging period or method	Allowable exceedances[a]
CO	Primary	10,000	9	8-hour average	Once per year
	Primary	40,000	35	1-hour average	Once per year
Lead	Primary and secondary	1.5	N/A	Maximum arithmetic mean measured over a calendar quarter	
NO_2	Primary and secondary	100	0.053	Annual arithmetic mean	
Ozone	Primary and secondary	235	0.12	Maximum hourly average[b]	Once per year
Ozone	Primary and secondary	157	0.08	8-hour average	[c]
Particulate matter (PM_{10})[d]	Primary and secondary	150	N/A	24-hour average	One day per year
	Primary and secondary	50	N/A	Annual arithmetic mean	[e]
($PM_{2.5}$)	Primary and secondary	65	N/A	24-hour average	[e]
		15	N/A	Annual arithmetic mean	[f, g]
SO_2	Primary	80	0.03	Annual arithmetic mean	
	Primary	365	0.14	Maximum 24-hour concentration	Once per year
SO_2	Secondary	1,300	0.5	Maximum 3-hour concentration	Once per year

[a]Allowable exceedances may actually be an average value over a multi-year period.
[b]The 1-hour NAAQS will no longer apply to an area one year after the effective date of the designation of that area for the 8-hour ozone NAAQS. For most areas, the date of designation was June 15, 2004.
[c]Average fourth highest concentration over 3-year period.
[d]Particulate matter standard applies to particles with an aerodynamic diameter $\leq 10\ \mu m$.
[e]Three-year average of 98th percentile 24-hour concentration.
[f]Three-year average of weighted annual mean.
[g]EPA has proposed to lower the the daily average $PM_{2.5}$ to 35 $\mu g/m^3$ while leaving the existing annual average standard unchanged at 15 $\mu g/m^3$. The Clean Air Scientific Advisory Board (CASAC) recommended that the annual average standard be set between 13 and 14 $\mu g/m^3$ and that the 24-hour standard be set between 30 and 35 $\mu g/m^3$. The final rule will be issued by September 2006.
(*Source:* 40 CFR 50.4–50.12 and 69 FR 23996.)

Standards (NAAQS). These standards are for the ambient air, that is, the outdoor air that normally surrounds us. EPA calls the pollutants listed in Table 7-1 *criteria pollutants* because they were developed on health-based criteria. The *primary standard* was established to protect human health with an "adequate margin of safety." The *secondary standards* are intended to prevent environmental and property damage. In 1987, the EPA revised the NAAQS. The standard for hydrocarbons was dropped and the standard for Total Suspended Particulates (TSP) was replaced with a particulate standard based on the mass of particulate matter with an aerodynamic diameter less than or equal to 10 μm. This standard is referred to as the PM_{10} standard. It was replaced by a standard for particulate matter with an aerodynamic diameter less than or equal to 2.5 μm.

States are divided into *Air Quality Control Regions (AQRs).* An AQR that has air quality equal to or better than the primary standard is called an *attainment area.* Those areas that do not meet the primary standard are called *nonattainment areas.*

Under the 1970 CAA, the EPA was directed to establish regulations for *hazardous air pollutants (HAPs)* using a risk-based approach. These were called NESHAPs—national emission standards for hazardous air pollutants. Because EPA had difficulty defining "an ample margin of safety" as required by the law, only seven HAPs were regulated between 1970 and 1990: asbestos, arsenic, benzene, beryllium, mercury, vinyl chloride, and radionuclides. The Clean Air Act Amendments of 1990 directed EPA to establish a HAP emissions control program based on technology for 189 chemicals* (see Table 1-8 for the list). EPA will establish emission allowances based on *Maximum Achievable Control Technology (MACT)* for 174 categories of industrial sources that potentially emit 9.08 megagrams (Mg) per year of a single HAP or 22.7 Mg per year of a combination of HAPs. A MACT can include process changes, material substitutions, or air pollution control equipment.

7-4 EFFECTS OF AIR POLLUTANTS

Effects on Materials

Mechanisms of Deterioration. Five mechanisms of deterioration have been attributed to air pollution: abrasion, deposition and removal, direct chemical attack, indirect chemical attack, and electrochemical corrosion (Yocom and McCaldin, 1968).

Solid particles of large enough size and traveling at high enough speed can cause deterioration by abrasion. With the exception of soil particles in dust storms and lead particles from automatic weapons fire, most air pollutant particles either are too small or travel at too slow a speed to be abrasive.

Small liquid and solid particles that settle on exposed surfaces do not cause more than aesthetic deterioration. For certain monuments and buildings, such as the White House, this form of deterioration is in itself quite unacceptable. For most surfaces, it is the cleaning process that causes the damage. Sandblasting of buildings is an obvious case in point. Frequent washing of clothes weakens their fiber, while frequent washing of painted surfaces dulls their finish.

Solubilization and oxidation/reduction reactions typify direct chemical attack. Frequently, water must be present as a medium for these reactions to take place. Sulfur

*Subsequently modified. As of July 2005 there were 188 HAPs.

dioxide and SO_3 in the presence of water react with limestone ($CaCO_3$) to form calcium sulfate ($CaSO_4$) and gypsum ($CaSO_4 \cdot 2H_2O$). Both $CaSO_4$ and $CaSO_4 \cdot 2H_2O$ are more soluble in water than $CaCO_3$, and both are leached away when it rains. The tarnishing of silver by H_2S is a classic example of an oxidation/reduction reaction.

Indirect chemical attack occurs when pollutants are absorbed and then react with some component of the absorbent to form a destructive compound. The compound may be destructive because it forms an oxidant, reductant, or solvent. Further, a compound can be destructive by removing an active bond in some lattice structure. Leather becomes brittle after it absorbs SO_2, which reacts to form sulfuric acid because of the presence of minute quantities of iron. The iron acts as a catalyst for the formation of the acid. A similar result has been noted for paper.

Oxidation/reduction reactions cause local chemical and physical differences on metal surfaces. These differences, in turn, result in the formation of microscopic anodes and cathodes. Electrochemical corrosion results from the potential that develops in these microscopic batteries.

Factors that Influence Deterioration. Moisture, temperature, sunlight, and position of the exposed material are among the more important factors that influence the rate of deterioration.

Moisture, in the form of humidity, is essential for most of the mechanisms of deterioration to occur. Metal corrosion does not appear to occur even at relatively high SO_2 pollution levels until the relative humidity exceeds 60 percent. On the other hand, humidities above 70 to 90 percent will promote corrosion without air pollutants. Rain reduces the effects of pollutant-induced corrosion by dilution and washing away of the pollutant.

Higher air temperatures generally result in higher reaction rates. However, when low air temperatures are accompanied by cooling of surfaces to the point where moisture condenses, then the rates may be accelerated.

In addition to the oxidation effect of its ultraviolet wave lengths, sunlight stimulates air pollution damage by providing the energy for pollutant formation and cyclic reformation. The cracking of rubber and the fading of dyes have been attributed to ozone produced by these photochemical reactions.

The position of the exposed surface influences the rate of deterioration in two ways. First, whether the surface is vertical or horizontal or at some angle affects deposition and wash-off rates. Second, whether the surface is an upper or lower one may alter the rate of damage. When the humidity is sufficiently high, the lower side usually deteriorates faster because rain does not remove the pollutants as efficiently.

Effects on Vegetation

Cell and Leaf Anatomy. Because the leaf is the primary indicator of the effects of air pollution on plants, we shall define some terms and explain how the leaf functions. A typical plant cell (Figure 7-3) has three main components: the cell wall, the protoplast, and the inclusions. Much like human skin, the cell wall is thin in young plants and gradually thickens with age. Protoplast is the term used to describe the protoplasm of one cell. It consists primarily of water, but it also includes protein, fat, and carbohydrates. The nucleus contains the hereditary material (DNA), which controls the operation of the cell. The protoplasm located outside the nucleus is called cytoplasm. Within the

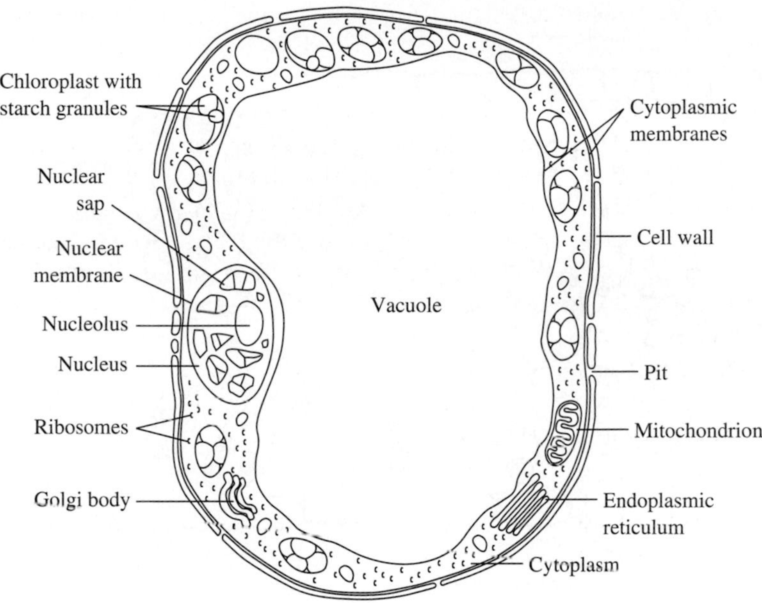

FIGURE 7-3
Typical plant cell. (*Source:* Fuller, 1960.)

cytoplasm are tiny bodies or plastids. Examples include chloroplasts, leucoplasts, chromoplasts, and mitochondria. Chloroplasts contain the chlorophyll that manufactures the plant's food through photosynthesis. Leucoplasts convert starch into starch grains. Chromoplasts are responsible for the red, yellow, and orange colors of fruit and flowers.

A cross section through a typical mature leaf (Figure 7-4) reveals three primary tissue systems: the epidermis, the mesophyll, and the vascular bundle (veins). Chloroplasts are usually not present in epidermal cells. The opening in the underside of the leaf is called a stoma. (The plural of stoma is stomata.) The mesophyll, which includes both the palisade parenchyma and the spongy parenchyma, contains chloroplasts. It is the food production center. The vascular bundles carry water, minerals, and food throughout the leaf and to and from the main stem of the plant.

The guard cells regulate the passage of gases and water vapor in and out of the leaf. When it is hot, sunny, and windy, the processes of photosynthesis and respiration are increased. The guard cells open, which allows increased removal of water vapor that otherwise would accumulate because of the increased transport of water and minerals from the roots.

Pollutant Damage. Ozone injures the palisade cells (Hindawi, 1970). The chloroplasts condense and ultimately the cell walls collapse. This results in the formation of red-brown spots that turn white after a few days. The white spots are called *fleck.* Ozone injury appears to be the greatest during midday on sunny days. The guard cells are more likely to be open under these conditions and thus allow pollutants to enter the leaf.

Plant growth may be inhibited by continuous exposure to 0.5 ppm of NO_2. Levels of NO_2 in excess of 2.5 ppm for periods of four hours or more are required to produce *necrosis* (surface spotting due to plasmolysis or loss of protoplasm).

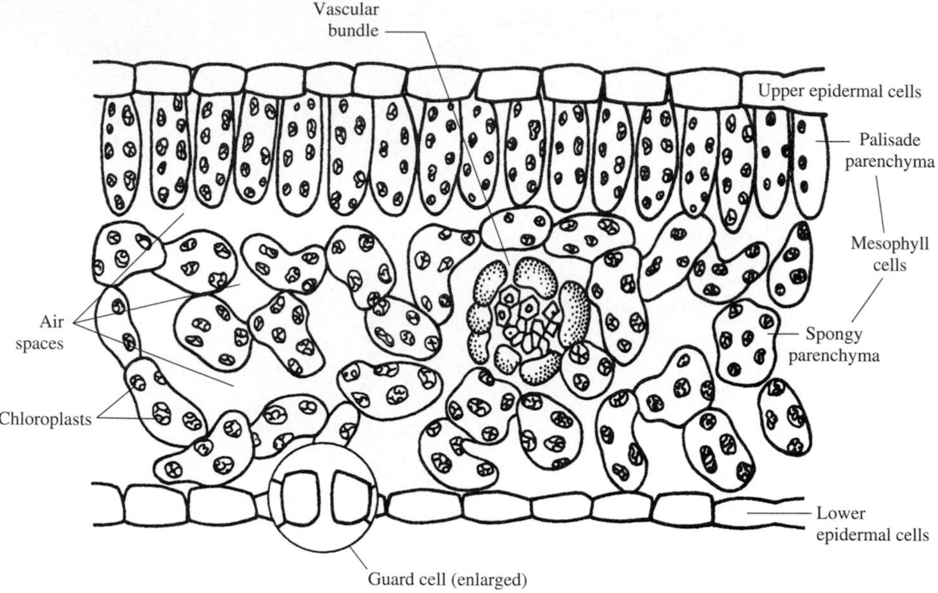

FIGURE 7-4
Cross section of intact leaf. (*Source:* Hindawi, 1970.)

Sulfur dioxide injury is also typified by necrosis, but at much lower levels. A concentration of 0.3 ppm for eight hours is sufficient (O'Gara, 1922). Lower levels for longer periods of exposure will produce a diffuse *chlorosis* (bleaching).

The net result of air pollutant damage goes beyond the apparent superficial damage to the leaves. A reduction in surface area results in less growth and small fruit. For commercial crops this results in a direct reduction in income for the farmer. For other plants the net result is likely to be an early death.

Fluoride deposition on plants not only causes them damage but may result in a second untoward effect. Grazing animals may accumulate an excess of fluoride that mottles their teeth and ultimately causes them to fall out.

Problems of Diagnosis. Various factors make it difficult to diagnose actual air pollution damage. Droughts, insects, diseases, herbicide overdoses, and nutrient deficiencies all can cause injury that resembles air pollution damage. Also, combinations of pollutants that alone cause no damage are known to produce acute effects when combined (Hindawi, 1970). This effect is known as *synergism.*

Effects on Health

Susceptible Population. It is difficult at best to assess the effects of air pollution on human health. Personal pollution from smoking results in exposure to air pollutant concentrations far higher than the low levels found in the ambient atmosphere. Occupational exposure may also result in pollution doses far above those found outdoors. Tests on rodents and other mammals are difficult to interpret and apply to human anatomy. Tests on human subjects are usually restricted to those who would be expected to survive. This leads us to a question of environmental ethics. If the allowable concentration levels (standards) are based on results from tests on rodents, they would be rather high. If the allowable concentration levels must also protect those with existing cardiorespiratory ailments, they should be lower than those resulting from the observed effects on rodents.

We noted earlier that the air quality standards were established to protect public health with an "adequate margin of safety." In the opinion of the Administrator of the EPA, the standards must protect the most sensitive responders. Thus, as you will note in the following paragraphs, the standards have been set at the lowest level of observed effect. This decision has been attacked by some theorists. They say it would make better economic sense to build more hospitals (Connolly, 1972). However, one also might apply this kind of logic in establishing speed limits for highways, that is, raise the speed limit and build more hospitals, junk yards, and cemeteries!

Anatomy of the Respiratory System. The respiratory system is the primary indicator of air pollution effects in humans. The major organs of the respiratory system are the nose, pharynx, larynx, trachea, bronchi, and lungs (Figure 7-5). The nose, pharynx,

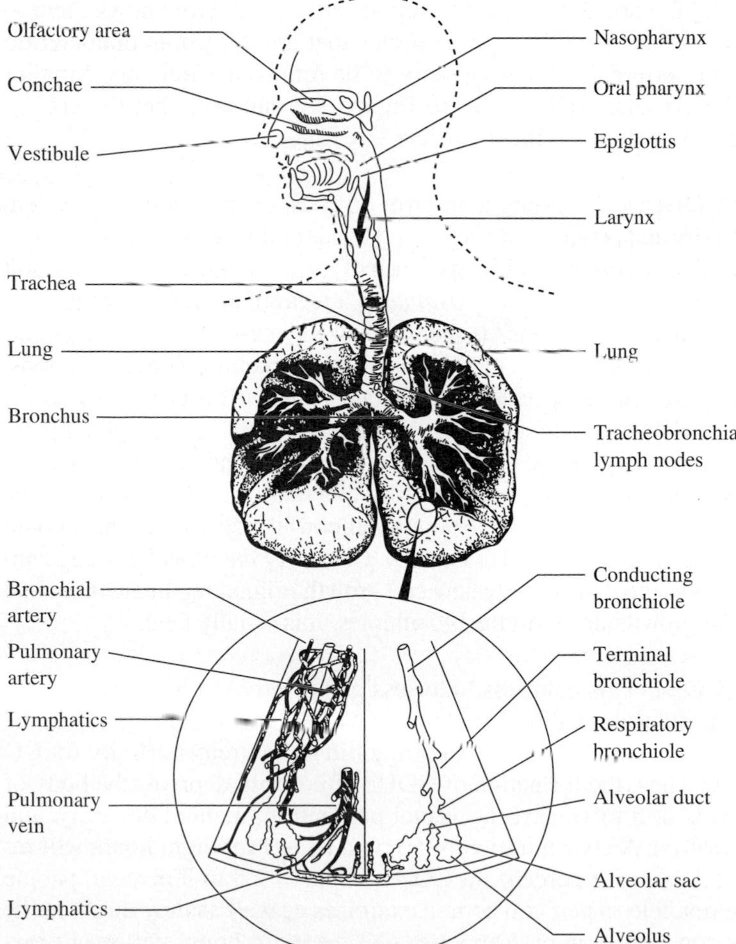

FIGURE 7-5
The Respiratory System. (*Source:* NAS, 1961.)

larynx, and trachea together are called the *upper respiratory tract* (URT). The primary effects of air pollution on the URT are aggravation of the sense of smell and inactivation of the sweeping motion of cilia, which remove mucus and entrapped particles. The *lower respiratory tract* (LRT) consists of the branching structures known as bronchi and the lung itself, which is composed of grape-like clusters of sacs called *alveoli*. The alveoli are approximately 300 μm in diameter. The walls of alveoli are lined with capillaries. Carbon dioxide diffuses through the capillary wall into the alveolus, while oxygen diffuses out of the alveolus into the blood cell. The difference in partial pressure of each of the gases causes it to move from the higher to lower partial pressure.

Inhalation and Retention of Particles. The degree of penetration of particles into the LRT is primarily a function of the size of the particles and the rate of breathing. Particles greater than 5 to 10 μm are screened out by the hairs in the nose. Sneezing also helps the screening process. Particles in the 1 to 2 μm size range penetrate to the alveoli. These particles are small enough to bypass screening and deposition in the URT, however they are big enough that their terminal settling velocity allows them to deposit where they can do the most damage. Particles that are 0.5 μm in diameter do not have a large enough terminal settling velocity to be removed efficiently. Smaller particles diffuse to the alveolar walls. Refer to Figure 7-2 and note that the size of "Lung Damaging Dust" falls in the critical particle size range.

Chronic Respiratory Disease. Several long-term diseases of the respiratory system are seriously aggravated by and perhaps may be caused by air pollution. *Airway resistance* is the narrowing of air passages because of the presence of irritating substances. The result is that breathing becomes difficult. *Bronchial asthma* is a form of airway resistance that results from an allergy. An asthma "attack" is the result of the narrowing of the bronchioles because of a swelling of the mucous membrane and a thickening of the secretions. The bronchioles return to normal after the attack. *Chronic bronchitis* is currently defined to be present in a person when excess mucus in the bronchioles results in a cough for three months a year for two consecutive years. Lung infections, tumors, and heart disease must be absent. *Pulmonary emphysema* is characterized by a breakdown of the alveoli. The small grape-like clusters become a large nonresilient balloon-like structure. The amount of surface area for gas exchange is reduced drastically. *Cancer of the bronchus* (lung cancer) is characterized by abnormal, disorderly new cell growth originating in the bronchial mucous membrane. The growth closes off the bronchioles. It is usually fatal.

Carbon Monoxide (CO). This colorless, odorless gas is lethal to humans within a few minutes at concentrations exceeding 5,000 ppm. CO reacts with hemoglobin in the blood to form carboxyhemoglobin (COHb). Hemoglobin has a greater affinity for CO than it does for oxygen. Thus, the formation of COHb effectively deprives the body of oxygen. At COHb levels of 5 to 10 percent, visual perception, manual dexterity, and ability to learn are impaired. A concentration of 50 ppm of CO for eight hours will result in a COHb level of about 7.5 percent. At COHb levels of 2.5 to 3 percent, people with heart disease are not able to perform certain exercises as well as they might in the absence of COHb. A concentration of 20 ppm of CO for eight hours will result in a COHb level of about 2.8 percent (Ferris, 1978). (We should note here that the average concentration of CO inhaled in cigarette smoke is 200 to 400 ppm!) The sensitive populations are those with heart and circulatory ailments, chronic pulmonary disease, developing fetuses, and those with conditions that cause increased oxygen demand, such as fever from an infections disease.

Hazardous Air Pollutants (HAPs). Most of the information on the direct human health effects of hazardous air pollutants (also known as air toxics) comes from studies of industrial workers. Exposure to air toxics in the workplace is generally much higher than in the ambient air. We know relatively little about the specific effects of the HAPs at the low levels normally found in ambient air.

The HAPs regulated under the NESHAP program were identified as causal agents for a variety of diseases. For example, asbestos, arsenic, benzene, coke oven emissions, and radionuclides may cause cancer. Beryllium primarily causes lung disease but also affects the liver, spleen, kidneys, and lymph glands.

Mercury has been especially targeted for regulation because it is released during the combustion of coal. Thus, it is one of the few HAPs that is widespread in the environment. Of particular concern are children who are exposed to methyl mercury prenatally. They are at increased risk of poor performance on neurobehavioral tasks such as those measuring attention, fine motor function, language skills, visual-spatial abilities and verbal memory (U.S. EPA, 1997, and U.S. EPA, 2004).

Lead (Pb). In contrast to the other criteria air pollutants, lead is a cumulative poison. A further difference is that it is ingested in food and water, as well as being inhaled. Of that portion taken by ingestion, approximately 5 to 10 percent is absorbed in the body. Between 20 and 50 percent of the inspired portion is absorbed. Those portions that are not absorbed are excreted in the feces and urine. Lead is measured in the urine and blood for diagnostic evidence of lead poisoning.

An early manifestation of acute lead poisoning is a mild anemia (deficiency of red blood cells). Fatigue, irritability, mild headache, and pallor indistinguishable from other causes of anemia occur when the blood level of lead increases to 60 to 120 μg/100 g of whole blood. Blood levels in excess of 80 μg/100 g result in constipation and abdominal cramps. When an acute exposure results in blood levels of lead greater than 120 μg/100 g, acute brain damage (encephalopathy) may result (Goyer and Chilsolm, 1972). Such acute exposure results in convulsions, coma, cardiorespiratory arrest, and death. Acute exposures may occur over a period of one to three weeks.

Chronic exposure to lead may result in brain damage characterized by seizures, mental incompetence, and highly active aggressive behavior. Weakness of extensor muscles of the hands and feet and eventual paralysis may also result. Canfield et al. (2003) found a decline in intelligence quotient (IQ) of 7.4 points for a lifetime blood lead concentration of up to 10 μg per deciliter. For a lifetime average blood lead concentration ranging from more than 10 μg per deciliter to 30 μg per deciliter, a more gradual decrease of 2.5 IQ points was observed.

Atmospheric lead occurs as a particulate. The particle size range is between 0.16 and 0.43 μm. Nonsmoking residents of suburban Philadelphia exposed to approximately 1 $\mu g/m^3$ of lead in air have blood levels averaging 11 μg/100 g. Nonsmoking residents of downtown Philadelphia exposed to approximately 2.5 $\mu g/m^3$ of lead have blood levels averaging 20 μg/100 g (U.S. PHS, 1965). In the early 1990s, 4.4 percent of U.S. children ages 1 to 5 had elevated lead levels. The percentage dropped to 1.6 percent by 2002. The U.S. Centers for Disease Control and Prevention attributed this drop to the removal of lead from gasoline as well as other efforts to screen and treat children for lead exposure (U.S. CDC, 2005).

Nitrogen Dioxide (NO_2). Exposure to NO_2 concentrations above 5 ppm for 15 minutes results in cough and irritation of the respiratory tract. Continued exposure may produce an abnormal accumulation of fluid in the lung (pulmonary edema). The gas is reddish brown in concentrated form and gives a brownish yellow tint at lower concentrations. At 5 ppm it has a pungent sweetish odor. The average NO_2 concentration in tobacco smoke is approximately 5 ppm. Slight increases in respiratory illness and decrease in pulmonary function have been associated with concentrations of about 0.10 ppm (Ferris, 1978). You should note that these concentrations are very high with respect to the NAAQS in Table 7-1.

Photochemical Oxidants. Although the photochemical oxidants include peroxyacetyl nitrate (PAN), acrolein, peroxybenzoyl nitrates (PBzN), aldehydes, and nitrogen oxides, the major oxidant is ozone (O_3). Ozone is commonly used as an indicator of the total amount of oxidant present. Oxidant concentrations above 0.1 ppm result in eye irritation. At a concentration of 0.3 ppm, cough and chest discomfort are increased. Those people who suffer from chronic respiratory disease are particularly susceptible.

PM_{10}. As noted earlier, large particles are not inhaled deeply into the lungs. This is why EPA switched from an air quality standard based on total suspended matter to one based on particles with an aerodynamic diameter less than 10 μm (PM_{10}). Studies in the United States, Brazil, and Germany have related higher levels of particulates to increased risk of respiratory, cardiovascular, and cancer-related deaths, as well as pneumonia, lung function loss, hospital admissions, and asthma (Reichhardt, 1995).

Particles 2.5 μm in aerodynamic diameter have been identified as a major contributor to elevated death rates in polluted cities (Pope et al., 1995). One hypothesized biological mechanism is pollution-induced lung damage resulting in declines in lung function, in respiratory distress, and in cardiovascular disease potentially related to hypoxemia (Pope et al., 1999).

Sulfur Oxides (SO_x) and Total Suspended Particulates (TSP). The sulfur oxides include sulfur dioxide (SO_2), sulfur trioxide (SO_3), their acids, and the salts of their acids. Rather than try to separate the effects of SO_2 and SO_3, they are usually treated together. There is speculation that a definite synergism exists whereby fine particulates carry absorbed SO_2 to the LRT. The SO_2 in the absence of particulates would be absorbed in the mucous membranes of the URT.

Patients suffering from chronic bronchitis have shown an increase in respiratory symptoms when the TSP levels exceeded 350 $\mu g/m^3$ and the SO_2 level was above 0.095 ppm. Studies made in Holland at an interval of three years showed that pulmonary function improved as SO_2 and TSP levels dropped from 0.10 ppm and 230 $\mu g/m^3$ to 0.03 ppm and 80 $\mu g/m^3$, respectively.

TABLE 7-2
Three major air pollution episodes

	Meuse Valley, 1930 (Dec. 1)	Donora, 1948 (Oct. 26–31)	London, 1952 (Dec. 5–9)
Population	No data	12,300	8,000,000
Weather	Anticyclone, inversion, and fog	Anticyclone, inversion, and fog	Anticyclone, inversion, and fog
Topography	River valley	River valley	River plain
Most probable source of pollutants	Industry (including steel and zinc plants)	Industry (including steel and zinc plants)	Household coal-burning
Nature of the illnesses	Chemical irritation of exposed membranous surfaces	Chemical irritation of exposed membranous surfaces	Chemical irritation of exposed membranous surfaces
No. of deaths	63	17	4,000
Time of deaths	Began after second day of episode	Began after second day of episode	Began on first day of episode
Suspected proximate cause of irritation	Sulfur oxides with particulates	Sulfur oxides with particulates	Sulfur oxides with particulates

(*Source:* WHO, 1961.)

Air Pollution Episodes. The word *episode* is used as a refined form of the word *disaster.** Indeed, it was the shock of these disasters that stimulated the first modern legislative action to require control of air pollutants. The characteristics of the three major episodes are summarized in Table 7-2. Careful study of the table will reveal that all of the episodes had some things in common. Comparison of these situations and others where no episode occurred (that is, where the number of dead and ill was considerably less) has revealed that four ingredients are essential for an episode. If one ingredient is omitted, fewer people will get sick and only a few people can be expected to die. The crucial ingredients are: (1) a large number of pollution sources, (2) a restricted air volume, (3) failure of officials to recognize that anything is wrong, and (4) the presence of water droplets of the "right" size (Goldsmith, 1968).

Although a sufficient quantity of any pollutant is lethal by itself, it is generally agreed that some mix is required to achieve the results seen in these episodes. Atmospheric levels of individual pollutants seldom rise to lethal levels without an explosion

*In the nuclear power business, they would call it an "incident."

or transportation accident. However, the proper combination of two or more pollutants will yield untoward symptoms at much lower levels. The sulfur oxides and particulates were the most suspect in the three major episodes.

The meteorology must be such that there is little air movement. Thus, the pollutants cannot be diluted. Although a valley is most conducive to a stagnation effect, the London episode proved that it isn't necessary. The stagnant conditions must persist for several days. Three days appears to be the minimum.

Tragically, each of these hazardous air pollution conditions became lethal because of the failure of city officials to notice anything strange. If they have no measurements of pollution levels or reports from hospitals and morgues, city authorities have no reason to alert the public, shut down factories, or restrict traffic.

The last and, perhaps, most crucial element is fog.* The fog droplets must be of the "right" size, namely, in the 1 to 2 μm diameter range or, perhaps, in the range below 0.5 μm. As mentioned earlier, these particle sizes are most likely to penetrate into the LRT. Pollutants that dissolve into the fog droplet are thus carried deep into the lungs and deposited there.

7-5 ORIGIN AND FATE OF AIR POLLUTANTS

Carbon Monoxide

Incomplete oxidation of carbon results in the production of carbon monoxide. The natural anaerobic decomposition carbonaceous material by microorganisms releases approximately 160 teragrams† (T_g) of methane (CH_4) to the atmosphere each year worldwide (IPCC, 1995). The natural formation of CO results from an intermediate step in the oxidation of the methane. The hydroxyl radical (OH·) serves as the initial oxidizing agent. It combines with CH_4 to form an alkyl radical (Wofsy, et al., 1972).

$$CH_4 + OH\cdot \rightarrow CH_3\cdot + H_2O \qquad (7\text{-}8)$$

This reaction is followed by a complex series of 39 reactions, which we have oversimplified to the following:

$$CH_3\cdot + O_2 + 2(h\upsilon) \rightarrow CO + H_2 + OH\cdot \qquad (7\text{-}9)$$

This says that $CH_3\cdot$ and O_2 are each zapped by a photon of light energy ($h\upsilon$). The symbol υ stands for the frequency of the light. The h is Planck's constant $= 6.626 \times 10^{-34}$ J/Hz.

Anthropogenic sources (those associated with the activities of human beings) include motor vehicles, fossil fuel burning for electricity and heat, industrial processes, solid waste disposal, and miscellaneous burning of such things as leaves and brush. Approximately 600–1250 Tg of CO are released by these sources (IPCC, 1995). Motor vehicles account for more than 60 percent of the emission.

*The word "smog" is a term coined by Londoners before World War I to describe the combination of smoke and fog that accounted for much of their weather. Los Angeles smog is a misnomer since little smoke and no fog is present. In fact, as we shall see later, Los Angeles smog cannot occur without a lot of sunshine. "Photochemical smog" is the correct term to describe the Los Angeles haze.

†One teragram $= 1 \times 10^{12}$ grams.

No significant change in the global atmospheric CO level has been observed over the past 20 years. Yet the worldwide anthropogenic contribution of combustion sources has doubled over the same time period. Because there is no apparent change in the atmospheric concentration, a number of mechanisms (*sinks*) have been proposed to account for the missing CO. The two most probable are

1. Reaction with hydroxyl radicals to form carbon dioxide
2. Removal by soil microorganisms

It has been estimated that these two sinks annually consume an amount of CO that just equals the production (Seinfeld, 1975).

Hazardous Air Pollutants (HAPs)

The EPA has identified 166 categories of major sources and 8 categories of area sources for the HAPs listed in Table 1-8 (57 FR 31576). The source categories represent a wide range of industrial groups: fuel combustion, metal processing, petroleum and natural gas production and refining, surface coating processes, waste treatment and disposal processes, agricultural chemicals production, and polymers and resins production. There are also a number of miscellaneous source categories, such as dry cleaning and electroplating.

In addition to these direct emissions, air toxics can result from chemical formation reactions in the atmosphere. These reactions involve chemicals emitted to the atmosphere that are not listed HAPs and may not be toxic themselves, but can undergo atmospheric transformations to generate HAPs. For organic compounds present in the gas phase, the most important transformation processes involve photolysis and chemical reactions with ozone, hydroxyl radicals (OH·), and nitrate radicals (Kao, 1994). *Photolysis* is the chemical fragmentation or rearrangement of a chemical upon the adsorption of radiation of the appropriate wavelength. Photolysis is only important during the daytime for those chemicals that absorb strongly within the solar radiation spectrum. Otherwise, reaction with OH· or O_3 is likely to predominate. The HAPs most often formed are formaldehyde and acetaldehyde.

The major removal mechanisms appear to be OH abstraction or addition. The reaction products lead to the formation of CO and CO_2. Eighty-nine of the 188 HAPs have atmospheric lifetimes of less than one day.

Lead

Volcanic activity and airborne soil are the primary natural sources of atmospheric lead. Smelters and refining processes, as well as incineration of lead-containing wastes, are major point sources of lead. Approximately 70 to 80 percent of the lead that used to be added to gasoline was discharged to the atmosphere.

Submicron lead particles, which are formed by volatilization and subsequent condensation, attach to larger particles or they form nuclei before they are removed from the atmosphere. Once they have attained a size of several microns, they either settle out or are washed out by rain.

Nitrogen Dioxide

Bacterial action in the soil releases nitrous oxide (N_2O) to the atmosphere. In the upper troposphere and stratosphere, atomic oxygen reacts with the nitrous oxide to form nitric oxide (NO).

$$N_2O + O \rightarrow 2NO \qquad (7\text{-}10)$$

The atomic oxygen results from the dissociation of ozone. The nitric oxide further reacts with ozone to form nitrogen dioxide (NO_2).

$$NO + O_3 \rightarrow NO_2 + O_2 \qquad (7\text{-}11)$$

The global formation of NO_2 by this process is estimated to be 0.45 petagrams* (Pg) annually (Seinfeld, 1975).

Combustion processes account for 96 percent of the anthropogenic sources of nitrogen oxides. Although nitrogen and oxygen coexist in our atmosphere without reaction, their relationship is much less indifferent at high temperatures and pressures. At temperatures in excess of 1,600 K, they react.

$$N_2 + O_2 \overset{\Delta}{\rightleftharpoons} 2NO \qquad (7\text{-}12)$$

If the combustion gas is rapidly cooled after the reaction by exhausting it to the atmosphere, the reaction is quenched and NO is the byproduct. The NO in turn reacts with ozone or oxygen to form NO_2. The anthropogenic contribution to global emission of NO_x amounted to 32 Tg/y (as N) in 1995 (IPCC, 1995). Between 40 and 45 percent of the NO*x* emissions in the United States come from transportation, 30 to 35 percent from power plants, and 20 percent from industrial sources (Seinfeld and Pandis, 1998).

Ultimately, the NO_2 is converted to either NO_2^- or NO_3^- in particulate form. The particulates are then washed out by precipitation. The dissolution of nitrate in a water droplet allows for the formation of nitric acid (HNO_3). This, in part, accounts for "acid" rain found downwind of industrialized areas.

Photochemical Oxidants

Unlike the other pollutants, the photochemical oxidants result entirely from atmospheric reactions and are not directly attributable to either people or nature. Thus, they are called *secondary pollutants*. They are formed through a series of reactions that are initiated by the absorption of a photon by an atom, molecule, free radical, or ion. Ozone is the principal photochemical oxidant. Its formation is usually attributed to the nitrogen dioxide photolytic cycle. Hydrocarbons modify this cycle by reacting with atomic oxygen to form free radicals (highly reactive organic species). The hydrocarbons, nitrogen oxides, and ozone react and interact to produce more nitrogen dioxide and ozone. This cycle is represented in summary form in Figure 7-6. The whole reaction sequence depends on an abundance of sunshine.

---

*One petagram = 1×10^{15} grams.

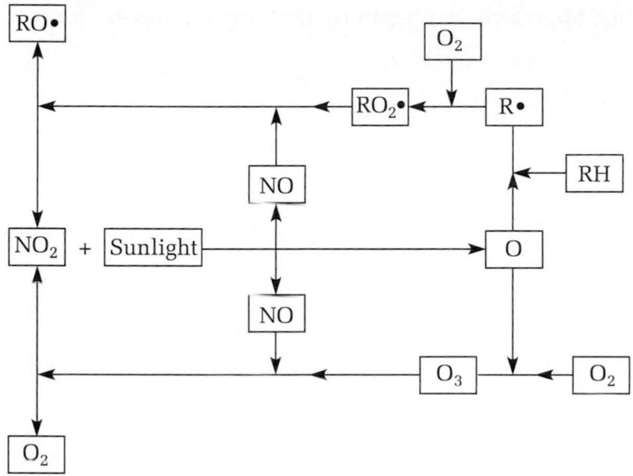

FIGURE 7-6
Interaction of hydrocarbons with atmospheric nitrogen oxide photolytic cycle. (*Source:* NAPCA, 1970.)

A result of these reactions is the photochemical "smog" for which Los Angeles is famous.

Sulfur Oxides

Sulfur oxides may be both primary and secondary pollutants. Power plants, industry, volcanoes, and the oceans emit SO_2, SO_3, and SO_4^{2-} directly as primary pollutants. In addition, biological decay processes and some industrial sources emit H_2S, which is oxidized to form the secondary pollutant SO_2. In terms of sulfur, approximately 30 Tg are emitted annually by natural sources. Approximately 75 Tg of sulfur may be attributed to anthropogenic sources each year (Seinfeld and Pandis, 1998).

The most important oxidizing reaction for H_2S appears to be one involving ozone:

$$H_2S + O_3 \rightarrow H_2O + SO_2 \tag{7-13}$$

The combustion of fossil fuels containing sulfur yields sulfur dioxide in direct proportion to the sulfur content of the fuel:

$$S + O_2 \rightarrow SO_2 \tag{7-14}$$

This reaction implies that for every gram of sulfur in the fuel, two grams of SO_2 are emitted to the atmosphere. Because the combustion process is not 100 percent efficient, we generally assume that 5 percent of the sulfur in the fuel ends up in the ash, that is, 1.90 g SO_2 per gram of sulfur in the fuel is emitted.

Example 7-2. An Illinois coal is burned at a rate of 1.00 kg per second. If the analysis of the coal reveals a sulfur content of 3.00 percent, what is the annual rate of emission of SO_2?

Solution. Using the mass balance approach, we begin by drawing a mass-balance diagram:

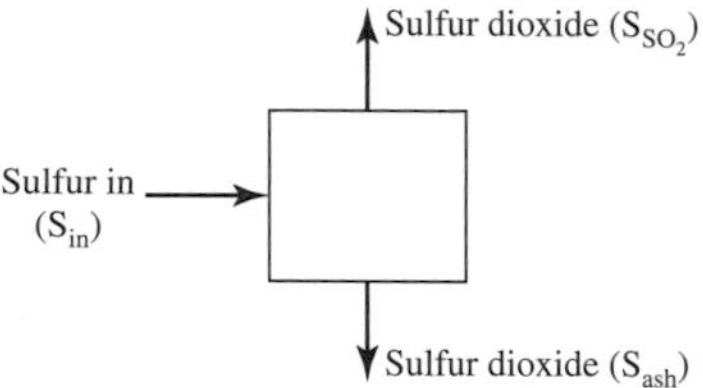

The mass balance equation may be written as

$$S_{in} = S_{ash} + S_{SO_2}$$

From the problem data, the mass of "sulfur in" is

$$S_{in} = 1.00 \text{ kg/s} \times 0.030 = 0.030 \text{ kg/s}$$

In one year,

$$S_{in} = 0.030 \text{ kg/s} \times 86{,}400 \text{ s/d} \times 365 \text{ d/y} = 9.46 \times 10^5 \text{ kg/y}$$

The sulfur in the ash is 5 percent of the input sulfur:

$$S_{ash} = (0.05)(9.46 \times 10^5 \text{ kg/y}) = 4.73 \times 10^4 \text{ kg/y}$$

The amount of sulfur available for conversion to SO_2:

$$S_{SO_2} = S_{in} - S_{ash} = 9.46 \times 10^5 - 4.73 \times 10^4 = 8.99 \times 10^5 \text{ kg/y}$$

The amount of sulfur dioxide formed is determined from the proportional weights of the oxidation reaction (Equation 7-14):

$$S + O_2 \rightarrow SO_2$$
$$\text{GMW} = 32 + 32 = 64$$

The amount of sulfur dioxide formed is then 64/32 of the sulfur available for conversion:

$$S_{SO_2} = \frac{64}{32}(8.99 \times 10^5 \text{ kg/y}) = 1.80 \times 10^6 \text{ kg/y}$$

The ultimate fate of most of the SO_2 in the atmosphere is conversion to sulfate salts, which are removed by sedimentation or by washout with precipitation. The conversion to sulfate is by either of two routes: catalytic oxidation or photochemical oxidation. The first process is most effective if water droplets containing Fe^{3+}, Mn^{2+}, or NH_3 are present:

$$2SO_2 + 2H_2O + O_2 \xrightarrow{\text{catalyst}} 2H_2SO_4 \tag{7-15}$$

At low relative humidities, the primary conversion process is photochemical oxidation. The first step is photoexcitation* of the SO_2.

$$SO_2 + h\nu \rightarrow \overset{*}{S}O_2 \qquad (7\text{-}16)$$

The excited molecule then readily reacts with O_2 to form SO_3:

$$\overset{*}{S}O_2 + O_2 \rightarrow SO_3 + O \qquad (7\text{-}17)$$

The trioxide is very hygroscopic and consequently is rapidly converted to sulfuric acid:

$$SO_3 + H_2O \rightarrow H_2SO_4 \qquad (7\text{-}18)$$

This reaction in large part accounts for acid rain (that is, precipitation with a pH value less than 5.6) found in industrialized areas. Normal precipitation has a pH of 5.6, due to the carbonate buffer system.

Particulates

Sea salt, soil dust, volcanic particles, and smoke from forest fires account for 2.9 Pg of particulate emissions each year. Anthropogenic emissions from fossil fuel burning and industrial processes account for emissions of 110 Tg per year (Kiehl and Rodhe, 1995). Secondary sources of particulates include the conversion of H_2S, SO_2, NO_x, NH_3, and hydrocarbons. H_2S and SO_2 are converted to sulfates. NO_x and NH_3 are converted to nitrates. The hydrocarbons react to form products that condense to form particles at atmospheric temperatures. Natural sources of secondary pollutants yield about 240 Tg annually. Anthropogenic sources yield about 340 Tg annually (Kiehl and Rodhe, 1995).

Dust particles that are *entrained* (picked up) by the wind and carried over long distances tend to sort themselves out to the sizes between 0.5 and 50 μm in diameter. Sea salt nuclei have sizes between 0.05 and 50 μm. Particles formed as a result of photochemical reactions tend to have very small diameters (< 0.4 μm). Smoke and fly ash particles cover a wide range of sizes from 0.05 to 200 μm or more. Particle mass distributions in urban atmospheres generally exhibit two maxima. One is between 0.1 and 1 μm in diameter. The other is between 1 and 30 μm. The smaller fraction is the result of condensation. The coarse fraction consists of fly ash and dust generated by mechanical abrasion.

Small particles are removed from the atmosphere by accretion to water droplets, which grow in size until they are large enough to precipitate. Larger particles are removed by direct washout by falling raindrops.

7-6 MICRO AND MACRO AIR POLLUTION

Air pollution problems may occur on three scales: micro, meso, and macro. Micro-scale problems range from those covering less than a centimeter to those the size of a house or slightly larger. Meso-scale air pollution problems are those of a few hectares up to the size of a city or county. Macro-scale problems extend from counties to states, nations, and in the broadest sense, the globe. Much of the remaining discussion in this

*Photoexcitation is the displacement of an electron from one shell to another, thereby storing energy in the molecule. Photoexcitation is represented in reactions by an asterisk.

chapter is focused on the meso-scale problem. In this section we will address the general micro-scale and macro-scale problems recognized today.

Indoor Air Pollution

People who live in urban, cold climates may spend more than 90 percent of their time indoors (Lewis, 2001). In the last three decades, researchers have identified sources, concentrations, and impacts of air pollutants that arise in conventional domestic residences. The startling results indicate that, in certain instances, indoor air may be substantially more polluted than outdoor air.

Carbon monoxide from improperly operating furnaces has long been a serious concern. In numerous instances, people have died from furnace malfunction. More recently, chronic low levels of CO pollution have been recognized. Gas ranges, ovens, pilot lights, gas and kerosene space heaters, and cigarette smoke all contribute (Table 7-3).

Nitrogen oxide sources are also shown in Table 7-3. NO_2 levels have been found to range from 70 $\mu g/m^3$ in air-conditioned houses with electric ranges to 182 $\mu g/m^3$ in non-air-conditioned houses with gas stoves (Hosein et al., 1985). The latter value is quite high in comparison to the national ambient air quality limits. SO_2 levels were found to be very low in all houses investigated.

TABLE 7-3
Tested combustion sources and their emission rates

	Range of emission rates,[a] mg/MJ				
Source	NO	NO_2	NO_x (as NO_2)	CO	SO_2
Range-top burner[b]	15–17	9–12	32–37	40–244	—[c]
Range oven[d]	14–29	7–13	34–53	12–19	—
Pilot light[e]	4–17	8–12	[f]	40–67	—
Gas space heaters[g]	0–15	1–15	1–37	14–64	—
Gas dryer[h]	8	8	20	69	—
Kerosene space heaters[i]	1–13	3–10	5–31	35–64	11–12
Cigarette smoke[j]	2.78	0.73	[f]	88.43	—

[a]The lowest and highest mean values of emission rates for combustion sources tested in milligrams per mega-Joule (mg/MJ). Note: It takes 4.186 Joules to raise the temperature of 1.0 g of water from 14.5°C to 15.5°C at 100 percent efficiency.
[b]Three ranges were evaluated. Reported results are for blue flame condition.
[c]Dash (–) means combustion source is not emitting the pollutant.
[d]Three ranges were evaluated. Ovens were operated for several different settings (bake, broil, self-clean cycle, etc.).
[e]One range was evaluated with all three pilot lights, two top pilots, and a bottom pilot.
[f]Emission rates not reported.
[g]Three space heaters including one convective, radiant, and catalytic were tested.
[h]One gas dryer was evaluated.
[i]Two kerosene heaters including a convective and radiant type were tested.
[j]One type of cigarette. Reported emission rates are in mg/cigarette (800 mg tobacco/cigarette).
(*Source:* D. J. Moschandreas et al., 1985.)

Over 800 volatile organic compounds (VOCs) have been identified in indoor air (Hines et al., 1993). Aldehydes, alkanes, alkenes, ethers, ketones, and polynuclear aromatic hydrocarbons (PAHs) are among them. Although they are not all present all the time, frequently there are several present at the same time. Typical sources of these compounds are listed in Table 7-4.

Between 1979 and 1987, the EPA investigated personal exposures of the general public to VOCs. These studies, titled the Total Exposure Assessment Methodology (TEAM), revealed that personal exposures exceeded median outdoor air concentrations by a factor of 2 to 5 for nearly all of the 19 VOCs investigated. Traditional sources (automobiles, industry, petrochemical plants) contributed only 20 to 25 percent of the total exposure to most of the target VOCs (Wallace, 2001).

TABLE 7-4
Common volatile organic compounds and their sources[a]

Volatile organic compounds	Major indoor sources of exposure
Acetaldehyde	Paint (water-based), sidestream smoke
Alcohols (ethanol, isopropanol)	Spirits, cleansers
Aromatic hydrocarbons (ethylbenzene toluene, xylenes, trimethylbenzenes)	Paints, adhesives, gasoline, combustion sources
Aliphatic hydrocarbons (octane, decane, undecane)	Paints, adhesives, gasoline, combustion sources
Benzene	Sidestream smoke
Butylated hydroxytoluene (BHT)	Urethane-based carpet cushions
Chloroform	Showering, washing clothes, washing dishes
p-Dichlorobenzene	Room deodorizers, moth cakes
Ethylene glycol	Paints
Formaldehyde	Sidestream smoke, pressed wood products, photocopier
Methylene chloride	Paint stripping, solvent use
Phenol	Vinyl flooring
Styrene	Smoking, photocopier
Terpenes (limonene, α-pinene)	Scented deodorizers, polishes, fabric softeners
Tetrachloroethylene	Wearing/storing dry-cleaned clothes
Tetrahydrofuran	Sealer for vinyl flooring
Toluene	Photocopier, sidestream smoke, synthetic carper fiber
1,1,1-Trichlorotheane	Aerosol sprays, solvents

[a]Compiled from Tucker, 2001, and Wallace, 2001

Formaldehyde (CH_2O) has been singled out as one of the more prevalent, as well as one of the more toxic, compounds (Hines et al., 1993). Formaldehyde may not be generated directly by the activity of the homeowner. It is emitted by a variety of consumer products and construction materials including pressed wood products, insulation materials [urea-formaldehyde foam insulation (UFFI) in trailers has been particularly suspect], textiles, and combustion sources. In a composite of several studies, CH_2O concentrations ranged from 0.01 to 5.52 ppm, with median concentration of approximately 0.18 ppm (Godish, 1989). The highest values were for manufactured homes and conventional houses in cold climates (examples included Minnesota and Indiana). For comparison, the American Society of Heating, Refrigeration and Air Conditioning Engineers (ASHRAE, 1981) set a guideline concentration of 0.1 ppm.

Unlike the other air pollution sources that continue to emit as long as there is anthropogenic activity (or in the case of radon, for geologic time), CH_2O is not regenerated unless new materials are brought into the residence. If the house is ventilated over a period of time, the concentration will drop.

The primary source of heavy metals indoors is from infiltration of outdoor air and soil and dust that is tracked into the building. Arsenic, cadmium, chromium, mercury, lead, and nickel have been measured in indoor air. Lead and mercury may be generated from indoor sources such as paint. Old lead paint is a source of particulate lead as it is abraded or during removal. Mercury vapor is emitted from latex-based paints that contain diphenyl mercury dodecenyl succinate to prevent fungus growth.

Although little or no effort has been exerted to reduce or eliminate the danger from ranges, ovens, etc., the public has come to expect that the recreational habits of smokers should not interfere with the quality of the air others breathe. The results of a general ban on cigarette smoking in one office are shown in Table 7-5. Smokers were allowed to smoke only in the designated lounge area. Period 1 was prior to the implementation of the new policy. It is obvious that the new policy had a positive effect outside of the lounge. On the other hand, respirable particulate matter (RSP) was found to increase with one smoker and to rise dramatically with two (Figure 7-7).

TABLE 7-5
Mean respirable particulates (RSP), CO, and CO_2 levels measured on the test floor

	RSP ($\mu g/m^3$)	CO (ppm)	CO_2 (ppm)
Period 1			
Floor	26	1.67	624
Lounge	51	1.98	642
Period 2			
Floor	18	1.09	569
Lounge	189	2.40	650

(*Source:* Lee et al., April 1985.)

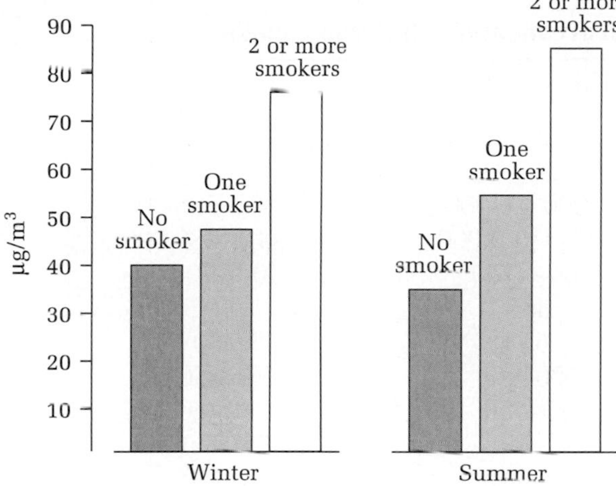

FIGURE 7-7
Respirable suspended particulate levels as a function of smoking. (*Source:* Hosein et al., 1985.)

Indoor tobacco smoking is of particular concern because of the carcinogenic properties of the smoke. While *mainstream smoking* (taking a puff) exposes the smoker to large quantities of toxic compounds, the smoldering cigarette in the ashtray (*sidestream smoke*) adds a considerable burden to the room environment. Table 7-6 illustrates the emission rates of mainstream and sidestream smoke.

In the early 1990s and in 2002, the U.S. Centers for Disease Control and Prevention (CDC) tested nonsmokers for levels of cotinine, a product of nicotine metabolism. The 2002 serum levels of cotinine were 75 percent less in adults and 68 percent less in children than 10 years previously. CDC attributed this dramatic decrease to restrictions to reduce second-hand smoke. Yet, more needs to be done. The levels in children were more than twice those of nonsmoking adults (U.S. CDC, 2005).

Bacteria, viruses, fungi, mites, and pollen are collectively referred to as *bioaerosols*. They require a reservoir (for storage), an amplifier (for reproduction), and a means of dispersal. Most bacteria and viruses in indoor air come from humans and pets. Other microorganisms and pollen are introduced from the ambient air through either natural ventilation or through the intakes of building air handling systems. Humidifiers, air-conditioning systems, and other places where water accumulates are potential reservoirs for bioaerosols.

Radon is not regulated as an ambient air pollutant but has been found in dwellings at alarmingly high concentrations. We will address the radon issue in depth in Chapter 11. Suffice it to say at this juncture that radon is a radioactive gas that emanates from natural geologic formations and, in some cases, from construction materials. It is not generated from the activities of the householder, unlike the pollutants discussed above.

It is doubtful that there will be any regulatory effort to reduce the emissions of indoor air pollutants in the near future. Thus the house or apartment dweller has little recourse other than to replace gas appliances, remove or cover formaldehyde sources, and put out the smokers.

TABLE 7-6
Emission of chemicals from mainstream and sidestream smoke

Chemicals	Mainstream (μg/cigarette)	Sidestream (μg/cigarette)
Gas and Vapor Phase		
Carbon monoxide	1,000–20,000	25,000–50,000
Carbon dioxide	20,000–60,000	160,000–480,000
Acetaldehyde	18–1,400	40–3,100
Hydrogen cyanide	430	110
Methyl chloride	650	1,300
Acetone	100–600	250–1,500
Ammonia	10–150	980–150,000
Pyridine	9–93	90–930
Acrolein	25–140	55–130
Nitric oxide	10–570	2,300
Nitrogen dioxide	0.5–30	625
Formaldehyde	20–90	1,300
Dimethylnitrosamine	10–65	520–3,300
Nitrosopyrolidine	10–35	270–945
Particulates		
Total suspended particles	36,200	25,800
Nicotine	100–2,500	2,700–6,750
Total phenols	228	603
Pyrene	50–200	180–420
Benzo (a) pyrene	20–40	68–136
Naphthalene	2.8	4.0
Methyl naphthalene	2.2	60
Aniline	0.36	16.8
Nitrosonornicotine	0.1–0.55	0.5–2.5

(*Source:* Hines et al., 1993.)

Acid Rain

Unpolluted rain is naturally acidic because CO_2 from the atmosphere dissolves to a sufficient extent to form carbonic acid (see Section 4-1). The equilibrium pH for pure rainwater is about 5.6. Measurements taken over North America and Europe have revealed lower pH values. In some cases individual readings as low as 3.0 have been recorded. The average pH in rain weighted by the amount of precipitation over the United States and lower Canada in 1997 is shown in Figure 5-16.

Chemical reactions in the atmosphere convert SO_2, NO_x, and volatile organic compounds (VOCs) to acidic compounds and associated oxidants (Figure 7-8). The primary conversion of SO_2 in the eastern United States is through the aqueous phase reaction with hydrogen peroxide (H_2O_2) in clouds. Nitric acid is formed by

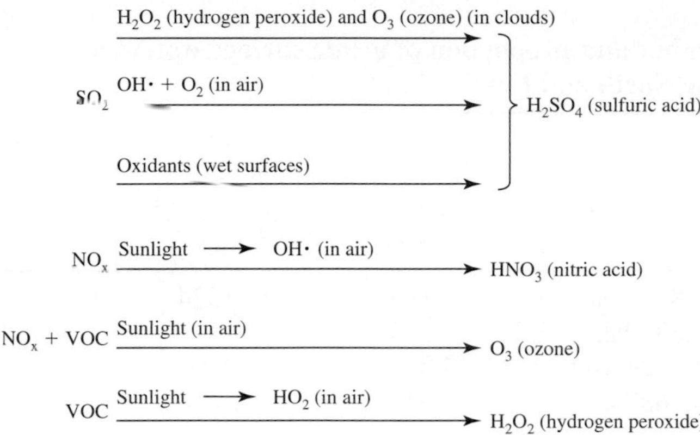

FIGURE 7-8
Acid rain precursors and products.

the reaction of NO_2 with OH radicals formed photochemically. Ozone is formed and then protected by a series of reactions involving both NO_x and VOCs.

As discussed in Chapter 5, the concern about acid rain relates to potential effects of acidity on aquatic life, damage to crops and forests, and damage to building materials. Lower pH values may affect fish directly by interfering with reproductive cycles or by releasing otherwise insoluble aluminum, which is toxic. Dramatic dieback of trees in Central Europe has stimulated concern that similar results could occur in North America. It is hypothesized that the acid rain leaches calcium and magnesium from the soil (see Figure 4-14). This lowers the molar ratio of calcium to aluminum which, in turn, favors the uptake of aluminum by the fine roots, that ultimately leads to their deterioration.

In 1980, Congress authorized a 10-year study to assess the causes and effects of acidic deposition. This study was titled the National Acid Precipitation Assessment Program (NAPAP). In September 1987, the NAPAP released an interim report that indicated that acidic precipitation appeared to have no measurable and consistent effect on crops, tree seedlings, or human health, and that a small percentage of lakes across the United States were experiencing pH values lower than 5.0 (Lefohn and Krupa, 1998). On the other hand, oxidant damage was measurable.

Approximately 70 percent of the SO_2 emissions in the United States are attributable to electric utilities. In order to decrease the SO_2 emissions, the Congress developed a two-phase control program under Title IV of the 1990 Clean Air Act Amendments. Phase I sets emission allowances for 110 of the largest emitters in the Eastern half of the U.S. Phase II will include smaller utilities. The utilities may buy or sell allowances. Each allowance is equal to about 1 Mg of SO_2 emissions. If a company does not expend its maximum allowance, it may sell it to another company. This program is called a *market-based system.* As a result of this program utility emissions have decreased by 9 Tg.

TABLE 7-7
Estimates of change in number and proportion of acidic surface waters in acid-sensitive regions of the North and East[a]

Region	Population or size	Number acidic in past surveys[b]	Estimated number currently acidic	Percent Change
New England	6,834 lakes	386	374	−3
Adirondacks	1,830 lakes	238	149	−38
Northern Appalachians	42,426 km^2	5,014	3,600	−28
Blue Ridge	32,687 km^2	1,634	1,634	0
Upper Midwest	8,574 lakes	251	800	−68

[a]Adapted from U.S. EPA, 2003a.
[b]Survey dates range from 1984 in Upper Midwest to 1993–1994 in the Northern Appalachians.

In 2003, the EPA reported on the long-term response of surface water chemistry to the 1990 Clean Air Act Amendments (U.S. EPA, 2003a). Eighty-one selected sites in the Northeast and Upper Midwest have been monitored for acidity since the early 1980s. The EPA's estimate of changes in the number and proportion of acidic surface waters is summarized in Table 7-7. In two areas, the New England lakes and the Blue Ridge Province streams, there is little evidence of reduction in acidity over the last decade. Sulfate levels have decreased significantly while nitrate levels have not changed appreciably. The widespread decrease in sulfate concentration parallels the general decrease in national emissions of sulfur dioxide since 1980. The EPA concluded from its analysis that surface waters have responded relatively rapidly to the decline in sulfate deposition and that additional reductions in deposition will result in additional declines in sulfate concentration.

In its report EPA also noted that, in many cases, sites that are not chronically acidic do undergo short-term episodic acidification during spring snow melt or during intense rain events.

Ozone Depletion

Without ozone, every living thing on the earth's surface would be incinerated. (On the other hand, as we have already noted, ozone can be lethal.) The presence of ozone in the upper atmosphere (20 to 40 km and up) provides a barrier to ultraviolet (UV) radiation. The small amounts that do seep through provide you with your summer tan. Too much UV will cause skin cancer. Although oxygen also serves as a barrier to UV radiation, it absorbs only over a narrow band centered at a wavelength of 0.2 μm. The photochemistry of these reactions is shown in Figure 7-9. The M refers to any third body (usually N_2).

In 1974, Molina and Rowland (1974) revealed a potential air pollution threat to this protective ozone shield. It is noteworthy that they, along with Paul Crutzen, jointly received the Nobel Prize in chemistry for their research. They hypothesized that chlorofluorocarbons (CF_2Cl_2 and $CFCl_3$—often abbreviated as CFC), which are used as

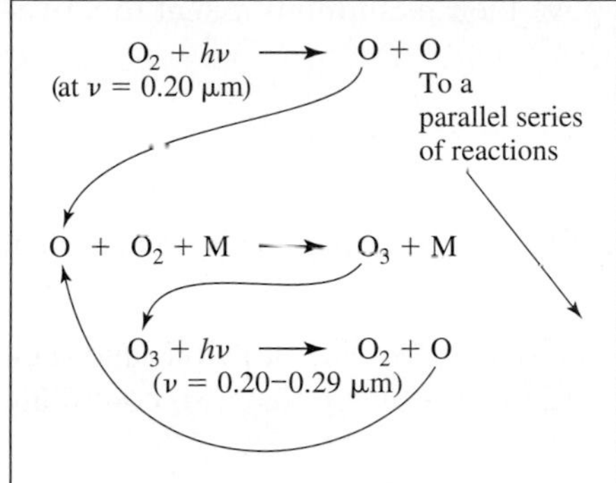

FIGURE 7-9
Photoreactions of ozone.

aerosol propellants and refrigerants, react with ozone (Figure 7-10). The frightening aspects of this series of reactions are that the chlorine atom removes ozone from the system, and that the chlorine atom is continually recycled to convert more ozone to oxygen. It has been estimated that a 5 percent reduction in ozone could result in nearly a 10 percent increase in skin cancer (ICAS, 1975). Thus, CFCs that are rather inert compounds in the lower atmosphere become a serious air pollution problem at higher elevations.

By 1987, the evidence that CFCs destroy ozone in the stratosphere above Antarctica every spring had become irrefutable. In 1987, the ozone hole was larger than ever. More than half of the total ozone column was wiped out and essentially all ozone disappeared from some regions of the stratosphere.

Research confirmed that the ozone layer, on a worldwide basis, shrunk approximately 2.5 percent in the preceding decade (Zurer, 1988). Initially, it was believed that this phenomenon was peculiar to the geography and climatology of Antarctica and that the warmer northern hemisphere was strongly protected from the processes that lead to massive ozone losses. Studies of the North Pole stratosphere in the winter of 1989 revealed that this is not the case (Zurer, 1989).

In September 1987, the Montreal Protocol on Substances That Deplete the Ozone Layer was developed. The Protocol, which was initially ratified by 36 countries and became effective in January 1989, proposed that CFC production first be frozen and then reduced 50 percent by 1998. Yet, under the terms of the Protocol, the chlorine content of the atmosphere would continue to grow because the fully halogenated CFCs have such long lifetimes in the atmosphere. CF_2Cl_2, for example, has a lifetime of 110 years (Reisch and Zurer, 1988). Eighty countries met at Helsinki, Finland, in the spring of

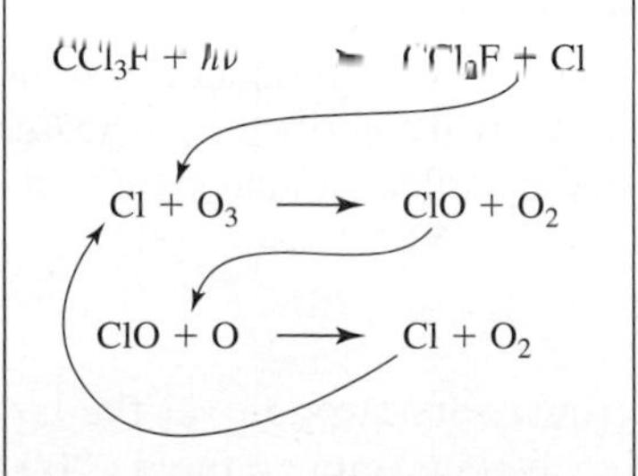

FIGURE 7-10
Ozone destruction by chlorofluoromethane.

1989 to assess new information. The delegates gave their unanimous assent to a five-point "Helsinki Declaration":

1. All join the 1985 Vienna Convention for the Protection of the Ozone Layer and the follow-up Montreal Protocol.
2. Phase out production and consumption of ozone-depleting CFCs no later than 2000.
3. Phase out production and consumption as soon as feasible of halons and such chemicals as carbon tetrachloride and methyl chloroform that also contribute to ozone depletion.
4. Commit themselves to accelerated development of environmentally acceptable alternative chemicals and technologies.
5. Make relevant scientific information, research results, and training available to developing countries (Sullivan, 1989).

The Montreal Protocol was strengthened in 1990, 1992, 1997, and 1999. The current terms of the treaty ban production of CFCs, carbon tetrachloride, and methyl chloroform as of January 1996. A ban on halon production took effect in January 1995 (Zurer, 1994). As of September 2002, 183 countries have ratified the Protocol (UNDP, 2005).

A number of alternatives to the fully chlorinated and, hence, more destructive CFCs have been developed. The two groups of compounds that emerged as significant replacements for the CFCs are hydrofluorocarbons (HFCs) and hydrochlorofluorocarbons (HCFCs). In contrast to the CFCs, HFCs and HCFCs contain one or more C-H bonds. This makes them susceptible to attack by OH radicals in the lower atmosphere. Because HFCs do not contain chlorine, they do not have the ozone depletion potential associated with the chlorine cycle shown in Figure 7-10. Although HCFCs contain chlorine, this chlorine is not transported to the stratosphere because OH scavenging in the troposphere is relatively efficient.

The implementation of the Montreal Protocol appears to be working. The use of CFCs has been reduced to one tenth of the 1990 levels (UN, 2005). Total tropospheric chlorine from the long- and short-lived chlorocarbons was about 5 percent lower in 2000 than that observed at its peak in 1992–1994. The rate of change in 2000 was about −22 parts per trillion per year. Total chlorine from CFCs is no longer increasing, in contrast to the slight increase noted in 1998. Total tropospheric bromine from halons continues to increase at about 3 percent per year, which is about two-thirds of the 1996 rate (UNEP/WHO, 2002).

The issues of ozone depletion and change are interconnected. As the atmospheric abundance of CFCs declines, their contribution to global warming will decline. On the other hand, the use of HFCs and HCFCs as substitutes for CFCs will contribute to increases in global warming. Because ozone depletion tends to cool the earth's climate system, recovery of the ozone layer will tend to warm the climate system (UNEP/WHO, 2002).

Global Warming

Scientific Basis. The case for global warming has grown very strong over the last two decades. As shown in Figure 7-11, the 5-year running average temperature in 2000

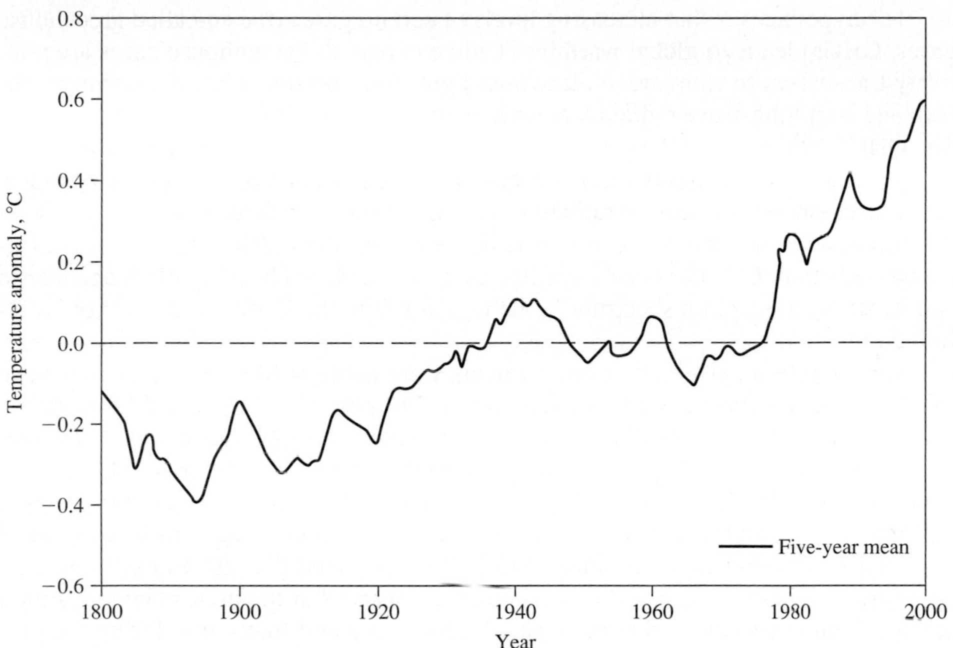

FIGURE 7-11
Global average surface temperature. Temperature anomaly is departure above and below the 1951–1980 average temperature, shown by dashed line. (*Source:* Hansen and Sato, 2004.)

was almost 0.6°C above the 1951–1980 average (Hansen and Sato, 2004). Mann and Jones (2003) have compiled proxy temperature data from sediments, ice cores, and tree-ring temperature reconstruction over the past two millennia. Their research shows (Figure 7-12) the average global surface temperature has been increasing for the last 100 years and was higher in 2000 than in any time in the past 2,000 years.

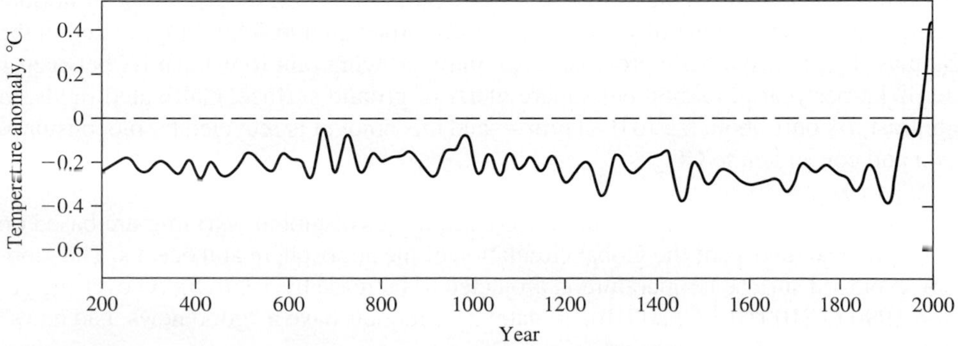

FIGURE 7-12
Global average surface temperature reconstruction. Temperature anomaly is departure from 1961–1990 instrumental reference period, for which the average is shown by dashed line (*Source:* Mann and Jones, 2003.)

The hypothesis is that increasing levels of certain gases (the so-called greenhouse gases, GHGs) leads to global warming. Unlike ozone, the greenhouse gases are relatively transparent to short-wave ultraviolet light from the sun. They do, however, absorb and emit long-wave radiation at wavelengths typical of the earth and atmosphere. The GHGs act much like the glass on a greenhouse (thus, the name *greenhouse gases*): they let in short-wave (ultraviolet) radiation from the sun that heats the ground surface, but restrict the loss of heat by radiation from the ground surface. The more GHGs in the atmosphere, the more effective it is in restricting the outflow of long-wave (infrared) radiation. CO_2 has been identified as the major GHG because of its abundance and its strong absorption spectrum in the region where the Earth emits most of its infrared radiation.

Since the first systematic measurements were made at Mauna Loa in Hawaii in 1958, CO_2 levels have risen from 316 ppm to 370 ppm (Keeling and Whorf, 2005). From analysis of air trapped in ice cores in Greenland and Antarctica, we know that preindustrial levels of CO_2 were about 280 ppm. The ice core records indicate that, over the last 160,000 years, no fluctuations of CO_2 have been larger than 70 ppm (Hileman, 1989) and that the current concentrations are higher than any level attained in the past 650,000 years (Hileman, 2005). It is estimated that the atmospheric CO_2 concentration has increased 30 percent since 1750 and that the present concentration has not been exceeded during the past 420,000 years and likely not during the past 20 million years (IPCC, 2001a). Several gases have been recognized as contributing to the greenhouse effect. Methane (CH_4), nitrous oxide (N_2O), and CFCs are similar to CO_2 in their radiative behavior. Even though their concentrations are much lower than CO_2, these gases are now estimated to trap about 60 percent as much long-wave radiation as CO_2.

In 1995, the United Nations Intergovernmental Panel on Climate Change* (IPCC) declared: "The balance of evidence suggests a discernable human influence on global climate." About three-quarters of the anthropogenic emissions of CO_2 that have been added to the atmosphere over the past 20 years is attributed to the combustion of fossil fuel (IPCC, 2001a). In the 1980s, massive deforestation was identified as a possible contribution. Both the burning of timber and the release of carbon from bacterial degradation contribute. Perhaps more important, deforestation removes a mechanism for removing CO_2 from the atmosphere (commonly referred to as a *sink*). In normal respiration, green plants utilize CO_2 much as a carbon source. This CO_2 is fixed in the biomass by photosynthetic processes. A rapidly growing rain forest can fix between 1 and 2 kg per year of carbon per square meter of ground surface. Cultivated fields, in contrast, fix only about 0.2 to 0.4 kg/m^2—and this amount is recycled by bioconsumption and conversion to CO_2.

Impacts. Attempts to understand the consequences of global warming are based on mathematical models of the global circulation of the atmosphere and oceans. The globally averaged surface temperature is projected to increase by 1.4 to 5.8°C over the period 1990 to 2100 (IPCC, 2001b). To date these models have a "good news, bad news"

*The IPCC is composed of over 673 scientists and 420 expert reviewers from around the world.

conclusion. On the the basis of the 1.4 to 5.8°C rise in global temperature, the following is predicted for North America (IPCC, 2001b):

1. A decrease in heating costs (partly offset by increased air-conditioning cost).
2. Potential increased food production in areas of Canada and an increase in warm-temperature mixed forest production with modest warming. With severe warming, crop production could possibly become a net loss.
3. Much easier navigation in the Arctic seas offset by reduced lake levels and navigational constraints for the Great Lakes and St. Lawrence sea way.
4. Drier crop conditions in the Midwest and Great Plains, requiring more irrigation.
5. Widespread melting of permanently frozen ground with adverse effects on building technology in Alaska and northern Canada.
6. A rise in sea level up to 0.9 m that would result in an increase in the severity of flooding, damage to coastal structures, destruction of wetlands, and saltwater intrusion into drinking water supplies in coastal areas particularly in Florida and much of the Atlantic coast.

On a global scale, the effects on human health may be catastrophic. Aside from heat stress, which in itself is no small thing, malaria and dengue—transmitted by mosquitoes—will increase in areas of increased precipitation as well as increased temperature (Martens, 1999). The IPCC estimates between 50 and 80 million additional cases of malaria alone will result from the forecast temperature change (IPCC, 2000).

Kyoto Protocol. The framework convention for the Protocol was signed in 1992. In 1997 the Protocol set targets for industrialized countries to reduce their GHG emissions was finalized. To become legally binding two conditions had to be fulfilled:

- Ratification by 55 countries
- Ratification by nations accounting for at least 55 percent of emissions from 38 industrialized countries plus Belarus, Turkey, and Kazakhstan

The first condition was met in 2002. Following the decision of the United States and Australia not to ratify, Russia's position became crucial to fulfill the second condition. On 18 November 2004, Russia ratified the Kyoto Protocol. It came into force 90 days later on 16 February 2005. At that time the targets for reducing emissions became binding on the countries that ratified the Protocol. The agreement set levels to reduce emissions by 5 percent from the 1990 baseline level. As of December, 2005, 157 nations had ratified the accord and the U.S. remained unwilling to make any commitments to reduce greenhouse emissions (AP, 2005a).

Russia's ratification of the treaty sounded a bell for a new international financial market in which companies buy and sell what amounts to global-warming pollution permits. The Protocol mandates emission reductions only from industrialized countries. However, it allows the industrialized countries to finance projects that reduce their emissions in developing countries and, thus, to generate credits toward their quotas. The theory is that, since global warming is global, the atmosphere doesn't care where the emissions or the emission reductions occur. Since financing an emissions-reduction project in a developing country is cheaper than in an industrialized country, there is a great incentive to put together investment funds.

Institutions including the World Bank, Japanese electric producers, the Chicago Climate Exchange (CCX), and French lender Caisse des Depots & Consignations are building funds. As of July 2005 more than $700 million had been gathered to acquire global-warming emission credits. By the time the Kyoto mandates become effective in 2008, it is estimated that $10 billion in credits will have been purchased (Noticias, 2004). It is noteworthy that the Protocol does not require Russia to decrease its emissions from the 1990 level at all, but its emission of GHGs has shrunk by nearly 40 percent because of the collapse of its economy after 1990. Russia will be able to make a lot of money selling credits when trading begins.

Although the United States did not a ratify the Protocol, 136 U.S. mayors representing more than 30 million people have signed an agreement to meet the goals spelled out in the treaty (AP, 2005b). On December 20, 2005, seven northeastern states (Connecticut, Delaware, Maine, New Hampshire, New Jersey, New York, and Vermont) signed an agreement to establish carbon dioxide emissions caps for electric utilities in their states (C&EN, 2006). In addition, the CCX has signed up more than 50 organizations including American Electric Power (Columbus, Ohio) and TECO Energy (Tampa). Several states have also indicted they will work to implement an emission credit program. It is not clear that these credits (from either the CCX or the states) will count under any government mandate.

A Rationale for Action. While there is still considerable disagreement about the potential for global warming, the consequences of ignoring these trends are sufficiently dramatic that intensive research must continue in the decades to come. Even without the risks of climate change, improvements in energy efficiency to reduce greenhouse gas emissions are amply justified from two points of view: economics and sustainability. Higher energy efficiency will yield economic benefit in reducing the cost of electricity and transportation. Higher efficiency will contribute to sustainability by extending the availability of finite energy resources. The expectation of damages from climate change provides extra incentive for pursuing these programs vigorously.

7-7 AIR POLLUTION METEOROLOGY

The Atmospheric Engine

The atmosphere is somewhat like an engine. It is continually expanding and compressing gases, exchanging heat, and generally raising chaos. The driving energy for this unwieldy machine comes from the sun. The difference in heat input between the equator and the poles provides the initial overall circulation of the earth's atmosphere. The rotation of the earth coupled with the different heat conductivities of the oceans and land produce weather.

Highs and Lows. Because air has mass, it also exerts pressure on things under it. Like water, which we intuitively understand to exert greater pressures at greater

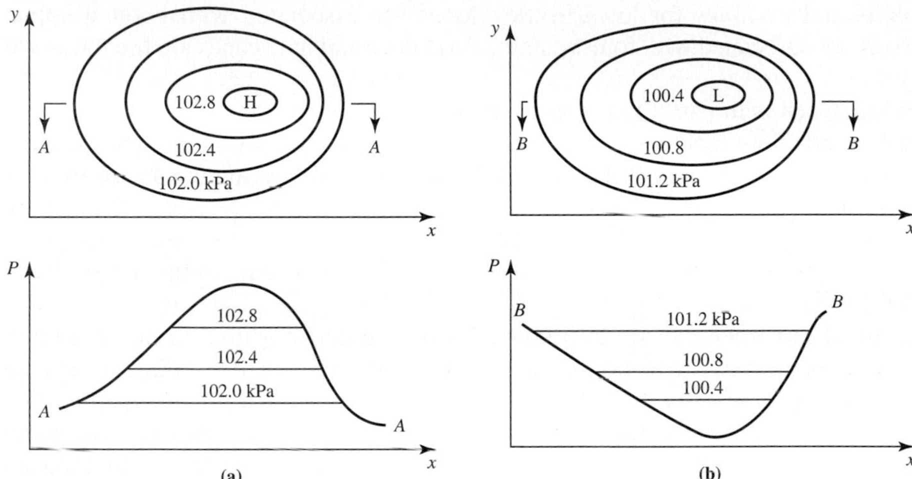

FIGURE 7-13
High (a) and low (b) pressure systems.

depths, the atmosphere exerts more pressure at the surface than it does at higher elevations. The highs and lows depicted on weather maps are simply areas of greater and lesser pressure. The elliptical lines shown on more detailed weather maps are lines of constant pressure, or *isobars.* A two-dimensional plot of pressure and distance through a high- or low-pressure system would appear as shown in Figure 7-13.

The wind flows from the higher pressure areas to the lower pressure areas. On a nonrotating planet, the wind direction would be perpendicular to the isobars (Figure 7-14*a*). However, since the earth rotates, an angular thrust called the Coriolis effect is added to this motion. The resultant wind direction in the northern hemisphere is as shown in Figure 7-14*b*. The technical names given to these systems are *anticyclones*

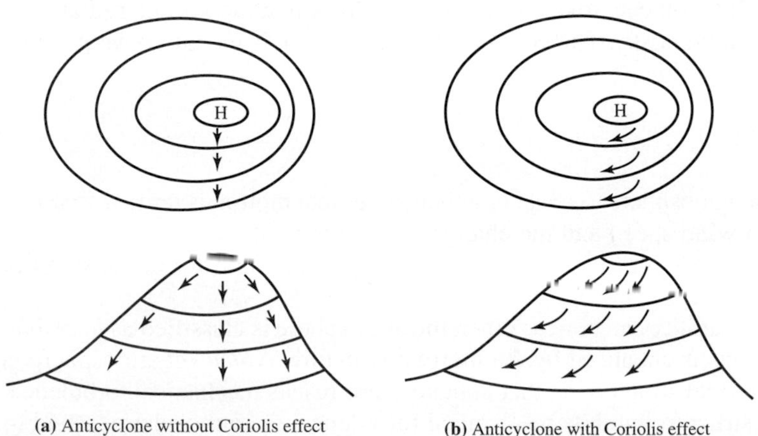

(a) Anticyclone without Coriolis effect

(b) Anticyclone with Coriolis effect

FIGURE 7-14
Wind flow due to pressure gradient.

for highs and *cyclones* for lows. Anticyclones are associated with good weather. Cyclones are associated with foul weather. Tornadoes and hurricanes are the foulest of the cyclones.

Wind speed is in part a function of the steepness of the pressure surface. When the isobars are close together, the pressure gradient (slope) is said to be steep and the wind speed relatively high. If the isobars are well spread out, the winds are light or nonexistent.

Turbulence

Mechanical Turbulence. In its simplest terms, we may consider turbulence to be the addition of random fluctuations of wind velocity (that is, speed and direction) to the overall average wind velocity. These fluctuations are caused, in part, by the fact that the atmosphere is being sheared. The shearing results from the fact that the wind speed is zero at the ground surface and rises with elevation to near the speed imposed by the pressure gradient. The shearing results in a tumbling, tearing motion as the mass just above the surface falls over the slower moving air at the surface. The swirls thus formed are called *eddies*. These small eddies feed larger ones. As you might expect, the greater the mean wind speed, the greater the mechanical turbulence. The more mechanical turbulence, the easier it is to disperse and spread atmospheric pollutants.

Thermal Turbulence. Like all other things in nature, the rather complex interaction that produces mechanical turbulence is confounded and further complicated by a third party. Heating of the ground surface causes turbulence in the same fashion that heating the bottom of a beaker full of water causes turbulence. At some point below boiling, you can see density currents rising off the bottom. Likewise, if the earth's surface is heated strongly and in turn heats the air above it, thermal turbulence will be generated. Indeed, the "thermals" sought by glider pilots and hot air balloonists are these thermal currents rising on what otherwise would be a calm day.

The converse situation can arise during clear nights when the ground radiates its heat away to the cold night sky. The cold ground, in turn, cools the air above it, causing a sinking density current.

Stability

The tendency of the atmosphere to resist or enhance vertical motion is termed *stability.* It is related to both wind speed and the change of air temperature with height (*lapse rate*). For our purpose, we may use the lapse rate alone as an indicator of the stability condition of the atmosphere.

There are three stability categories. When the atmosphere is classified as *unstable,* mechanical turbulence is enhanced by the thermal structure. A *neutral* atmosphere is one in which the thermal structure neither enhances nor resists mechanical turbulence. When the thermal structure inhibits mechanical turbulence, the atmosphere is said to be *stable.* Cyclones are associated with unstable air. Anticyclones are associated with stable air.

Neutral Stability. The lapse rate for a neutral atmosphere is defined by the rate of temperature increase (or decrease) experienced by a parcel of air that expands (or contracts) *adiabatically* (without the addition or loss of heat) as it is raised through the atmosphere. This rate of temperature decrease (dT/dz) is called the *dry adiabatic lapse rate.* It is designated by the Greek letter gamma (Γ). It has a value of approximately -1.00°C/100 m. (Note that this is not a slope in the normal sense, that is, it is not dy/dx.) In Figure 7-15*a*, the dry adiabatic lapse rate of a parcel of air is shown as a dashed line and the temperature of the atmosphere (ambient lapse rate) is shown as a solid line. Since the ambient lapse rate is the same as Γ, the atmosphere is said to have a *neutral stability.*

Unstable Atmosphere. If the temperature of the atmosphere falls at a rate greater than Γ, the lapse rate is said to be *superadiabatic,* and the atmosphere is unstable. Using Figure 7-15*b,* we can see that this is so. The actual lapse rate is shown by the solid line. If we capture a balloon full of polluted air at elevation A and adiabatically displace it 100 m vertically to elevation B, the temperature of the air inside the balloon will decrease from 21.15° to 20.15°C. At a lapse rate of -1.25°C/100 m, the temperature of the air outside the balloon will decrease from 21.15° to 19.90°C. The air inside the balloon will be warmer than the air outside; this temperature difference gives the balloon buoyancy. It will behave as a hot gas and continue to rise without any further mechanical effort. Thus, mechanical turbulence is enhanced and the atmosphere is unstable. If we adiabatically displace the balloon downward to elevation C, the temperature inside the balloon would rise at the rate of the dry adiabat. Thus, in moving 100 m, the temperature will increase from 21.15° to 22.15°C. The temperature outside the balloon will increase at the superadiabatic lapse rate to 22.40°C. The air in the balloon will be cooler than the ambient air and the balloon will have a tendency to sink. Again, mechanical turbulence (displacement) is enhanced.

Stable Atmosphere. If the temperature of the atmosphere falls at a rate less than Γ, it is called *subadiabatic,* and the atmosphere is stable. If we again capture a balloon of polluted air at elevation A (Figure 7-15*c*) and adiabatically displace it vertically to elevation B, the temperature of the polluted air will decrease at a rate equal to the dry adiabatic rate. Thus, in moving 100 m, the temperature will decrease from 21.15° to 20.15°C as before. However, since the ambient lapse rate is -0.5°C/100 m, the temperature of the air outside the balloon will have dropped to only 20.65°C. Because the air inside the balloon is cooler than the air outside the balloon, the balloon will have a tendency to sink. Thus, the mechanical displacement (turbulence) is inhibited.

In contrast, if we displace the balloon adiabatically to elevation C, the temperature inside the balloon would increase to 22.15°C, while the ambient temperature would increase to 21.65°C. In this case, the air inside the balloon would be warmer than the ambient air and the balloon would tend to rise. Again, the mechanical displacement would be inhibited.

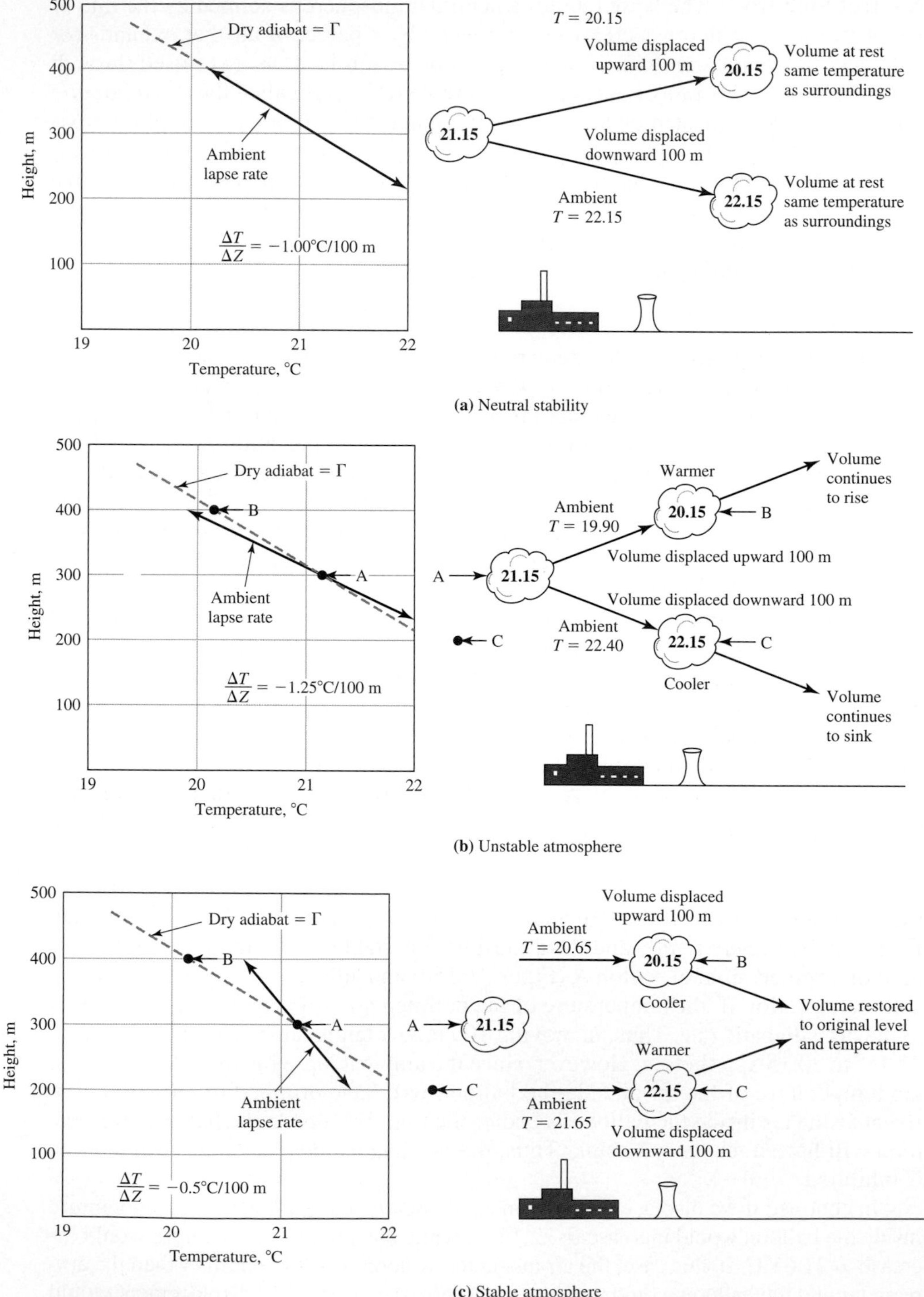

FIGURE 7-15

Lapse rate and displaced air volume. (*Source:* AEC, 1968.)

There are two special cases of subadiabatic lapse rate. When there is no change of temperature with elevation, the lapse rate is called *isothermal.* When the temperature increases with elevation, the lapse rate is called an *inversion.* The inversion is the most severe form of a stable temperature profile. It is often associated with restricted air volumes that cause air pollution episodes.

Example 7-3. Given the following temperature and elevation data, determine the stability of the atmosphere.

Elevation, m	Temperature, °C
2.00	14.35
324.00	11.13

Solution. Begin by determining the existing lapse rate:

$$\frac{\Delta T}{\Delta Z} = \frac{T_2 - T_1}{Z_2 - Z_1}$$
$$= \frac{11.13 - 14.35}{324.00 - 2.00} = \frac{-3.22}{322.00}$$
$$= -0.0100°\text{C/m} = -1.00°\text{C}/100\text{ m}$$

Now we compare this with Γ and find that they are equal. Thus, the atmospheric stability is neutral.

Plume Types. The smoke trail or plume from a tall stack located on flat terrain has been found to exhibit a characteristic shape that is dependent on the stability of the atmosphere. The six classical plumes are shown in Figure 7-16, along with the corresponding temperature profiles. In each case, Γ is given as a broken line to allow comparison with the actual lapse rate, which is given as a solid line. In the bottom three cases, particular attention should be given to the location of the inflection point with respect to the top of the stack.

Terrain Effects

Heat Islands. A heat island results from a mass of material, either natural or anthropogenic, that absorbs and reradiates heat at a greater rate than the surrounding area. This causes moderate to strong vertical convection currents above the heat island. The effect is superimposed on the prevailing meteorological conditions. It is nullified by strong winds. Large industrial complexes and small to large cities are examples of places that would have a heat island.

Because of the heat island effect, atmospheric stability will be less over a city than it is over the surrounding countryside. Depending upon the location of the pollutant sources, this can be either good news or bad news. First, the good news: For ground

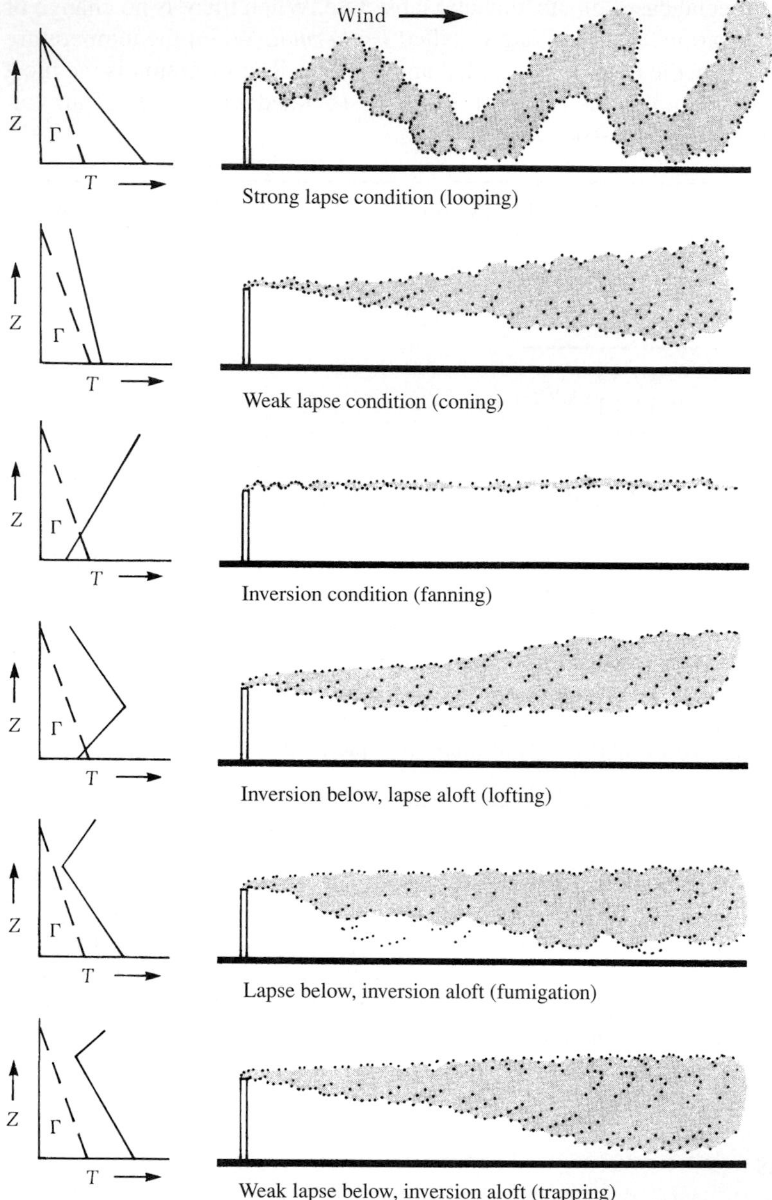

FIGURE 7-16
Six types of plume behavior. (*Source:* Church, 1949.)

level sources such as automobiles, the bowl of unstable air that forms will allow a greater air volume for dilution of the pollutants. Now the bad news: Under stable conditions, plumes from tall stacks would be carried out over the countryside without increasing ground level pollutant concentrations. Unfortunately, the instability caused by the heat island mixes these plumes to the ground level.

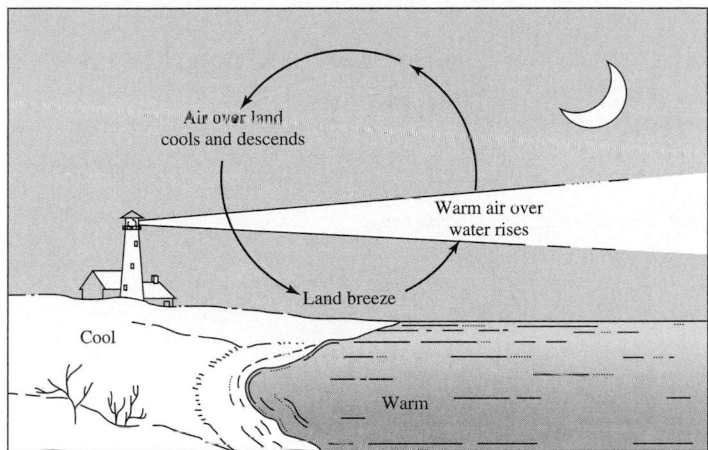

FIGURE 7-17
Land breeze during the night.

Land/Sea Breezes. Under a stagnating anticyclone, a strong local circulation pattern may develop across the shoreline of large water bodies. During the night, the land cools more rapidly than the water. The relatively cooler air over the land flows toward the water (a land breeze, Figure 7-17). During the morning the land heats faster than water. The air over the land becomes relatively warm and begins to rise. The rising air is replaced by air from over the water body (a sea or lake breeze, Figure 7-18).

The effect of the lake breeze on stability is to impose a surface-based inversion on the temperature profile. As the air moves from the water over the warm ground, it is heated from below. Thus, for stack plumes originating near the shoreline, the stable

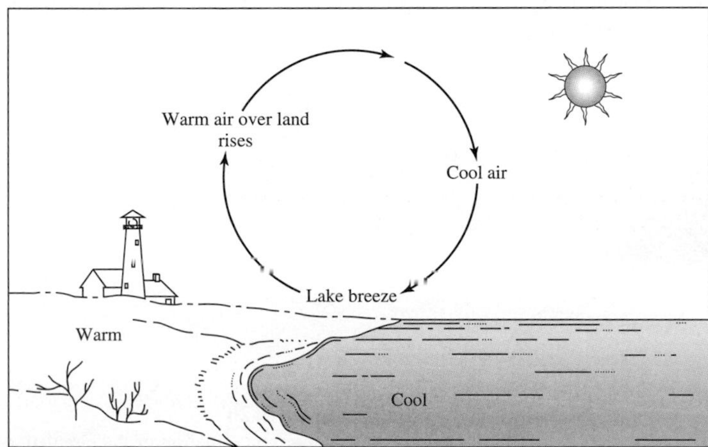

FIGURE 7-18
Lake breeze during the day.

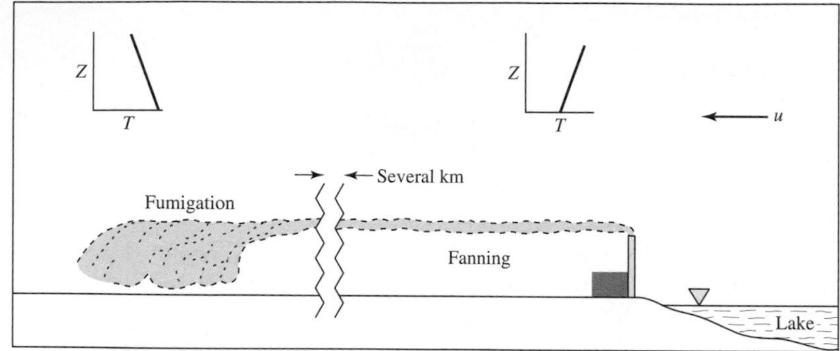

FIGURE 7-19
Effect of lake breeze on plume dispersion.

lapse rate causes a fanning plume close to the stack (Figure 7-19). The lapse condition grows to the height of the stack as the air moves inland. At some point inland, a fumigation plume results.

Valleys. When the general circulation imposes moderate to strong winds, valleys that are oriented at an acute angle to the wind direction channel the wind. The valley effectively peels off part of the wind and forces it to follow the direction of the valley floor (Figure 7-20).

Under a stagnating anticyclone, the valley will set up its own circulation. Warming of the valley walls will cause the valley air to be warmed. It will become more buoyant and flow up the valley. At night the cooling process will cause the wind to flow down the valley.

Valleys oriented in the north-south direction are more susceptible to inversions than level terrain. The valley walls protect the floor from radiative heating by the sun. Yet the walls and floor are free to radiate heat away to the cold night sky. Thus, under weak winds, the ground cannot heat the air rapidly enough during the day to dissipate the inversion that formed during the night.

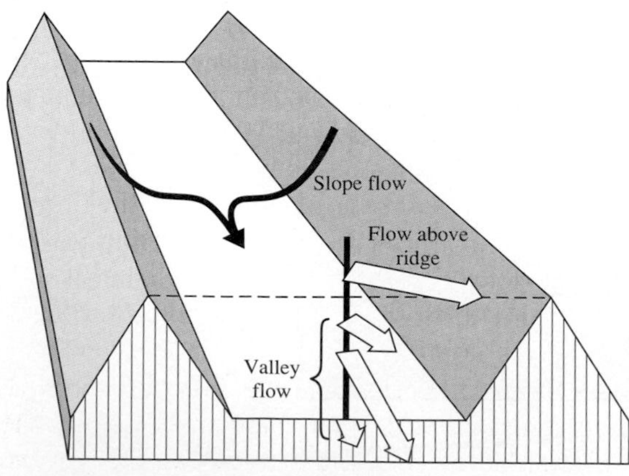

FIGURE 7-20
Idealized representation of the circulation that might be expected in a typical valley on a clear night. (*Source:* AEC, 1968.)

7-8 ATMOSPHERIC DISPERSION

Factors Affecting Dispersion of Air Pollutants

The factors that affect the transport, dilution, and dispersion of air pollutants can generally be categorized in terms of the emission point characteristics, the nature of the pollutant material, meteorological conditions, and effects of terrain and anthropogenic structures. We have discussed all of these except the source conditions. Now we wish to integrate the first and third factors to describe the qualitative aspects of calculating pollutant concentrations. We shall follow this with a simple quantitative model for a point source. More complex models for point sources (in rough terrain, in industrial settings, or for long time periods), area sources, and mobile sources are left for more advanced texts.

Source Characteristics. Most industrial effluents are discharged vertically into the open air through a stack or duct. As the contaminated gas stream leaves the discharge point, the plume tends to expand and mix with the ambient air. Horizontal air movement will tend to bend the discharge plume toward the downwind direction. At some point between 300 and 3,000 m downwind, the effluent plume will level off. While the effluent plume is rising, bending, and beginning to move in a horizontal direction, the gaseous effluents are being diluted by the ambient air surrounding the plume. As the contaminated gases are diluted by larger and larger volumes of ambient air, they are eventually dispersed toward the ground.

The plume rise is affected by both the upward inertia of the discharge gas stream and by its buoyancy. The vertical inertia is related to the exit gas velocity and mass. The plume's buoyancy is related to the exit gas mass relative to the surrounding air mass. Increasing the exit velocity or the exit gas temperature will generally increase the plume rise. The plume rise, together with the physical stack height, is called the *effective stack height.*

The additional rise of the plume above the discharge point as the plume bends and levels off is a factor in the resultant downwind ground level concentrations. The higher the plume rises initially, the greater distance there is for diluting the contaminated gases as they expand and mix downward.

For a specific discharge height and a specific set of plume dilution conditions, the ground level concentration is proportional to the amount of contaminant materials discharged from the stack outlet for a specific period of time. Thus, when all other conditions are constant, an increase in the pollutant discharge rate will cause a proportional increase in the downwind ground level concentrations.

Downwind Distance. The greater the distance between the point of discharge and a ground level receptor downwind, the greater will be the volume of air available for diluting the contaminant discharge before it reaches the receptor.

Wind Speed and Direction. The wind direction determines the direction in which the contaminated gas stream will move across local terrain. Wind speed affects the plume rise and the rate of mixing or dilution of the contaminated gases as they leave the discharge point. An increase in wind speed will decrease the plume rise by bending the plume over more rapidly. The decrease in plume rise tends to increase the pollutant's ground level concentration. On the other hand, an increase in wind speed will increase the rate of dilution of the effluent plume, tending to lower the downwind concentrations. Under different conditions, one or the other of the two wind speed effects becomes the predominant effect. These effects, in turn, affect the distance downwind of the source at which the maximum ground level concentration will occur.

Stability. The turbulence of the atmosphere follows no other factor in power of dilution. The more unstable the atmosphere, the greater the diluting power. Inversions that are not ground based, but begin at some height above the stack exit, act as a lid to restrict vertical dilution.

Dispersion Modeling

General Considerations and Use of Models. A dispersion model is a mathematical description of the meteorological transport and dispersion process that is quantified in terms of source and meteorologic parameters during a particular time. The resultant numerical calculations yield estimates of concentrations of the particular pollutant for specific locations and times.

To verify the numerical results of such a model, actual measured concentrations of the particular atmospheric pollutant must be obtained and compared with the calculated values by means of statistical techniques. The meteorological parameters required for use of the models include wind direction, wind speed, and atmospheric stability. In some models, provisions may be made for including lapse rate and vertical mixing height. Most models will require data about the physical stack height, the diameter of the stack at the emission discharge point, the exit gas temperature and velocity, and the mass rate of emission of pollutants.

Models are usually classified as either short-term or climatological models. Short-term models are generally used under the following circumstances: (1) to estimate ambient concentrations where it is impractical to sample, such as over rivers or lakes, or at great distances above the ground; (2) to estimate the required emergency source reductions associated with periods of air stagnations under air pollution episode alert conditions; and (3) to estimate the most probable locations of high, short-term, ground-level concentrations as part of a site selection evaluation for the location of air monitoring equipment.

Climatological models are used to estimate mean concentrations over a long period of time or to estimate mean concentrations that exist at particular times of the day for each season over a long period of time. Long-term models are used as an aid for developing emissions standards. We will be concerned only with short-term models in their most simple application.

Basic Point Source Gaussian Dispersion Model. The basic Gaussian diffusion equation assumes that atmospheric stability is uniform throughout the layer into which the contaminated gas stream is discharged. The model assumes that turbulent diffusion is a random activity and hence the dilution of the contaminated gas stream in both the horizontal and vertical direction can be described by the Gaussian or normal equation.

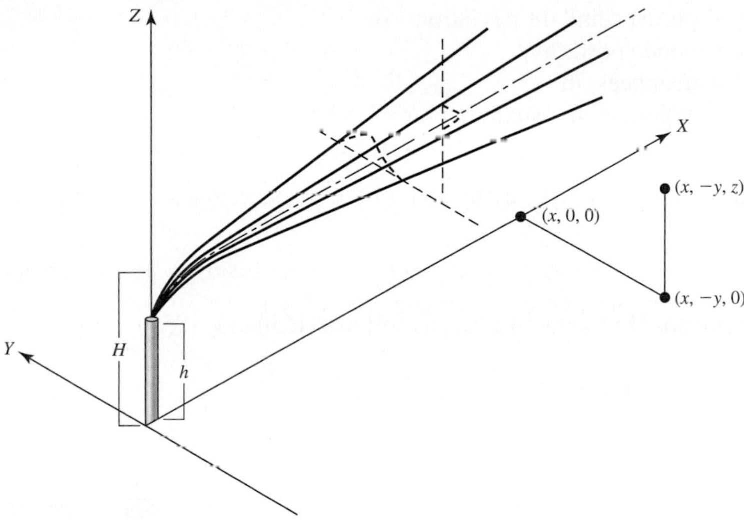

FIGURE 7-21
Plume dispersion coordinate system. (*Source:* Turner, 1967.)

The model further assumes that the contaminated gas stream is released into the atmosphere at a distance above ground level that is equal to the physical stack height plus the plume rise (ΔH). The model assumes that the degree of dilution of the effluent plume is inversely proportional to the wind speed (u). The model also assumes that pollutant material that reaches ground level is totally reflected back into the atmosphere like a beam of light striking a mirror at an angle. Mathematically, this ground reflection is accounted for by assuming a virtual or imaginary source located at a distance of $-H$ with respect to ground level, and emitting an imaginary plume with the same source strength as the real source being modeled. The same general idea can be used to establish other boundary layer conditions for the equations, such as limiting horizontal or vertical mixing.

The Model. We have selected the model* equation in the form presented by D. B. Turner (1967). It gives the ground level concentration (χ) of pollutant at a point (coordinates x and y) downwind from a stack with an effective height (H) (Figure 7-21). The standard deviation of the plume in the horizontal and vertical directions is designated by s_y and s_z, respectively. The standard deviations are functions of the downward distance from the source and the stability of the atmosphere. The equation is as follows:

$$\chi_{(x,y,0,H)} = \left[\frac{F}{\pi s_y s_z u}\right]\left[\exp\left[-\frac{1}{2}\left(\frac{y}{s_y}\right)^2\right]\right]\left[\exp\left[-\frac{1}{2}\left(\frac{H}{s_z}\right)^2\right]\right] \tag{7-19}$$

where $\chi_{(x,y,0,H)}$ = downwind concentration at ground level, g/m^3

E = emission rate of pollutant, g/s

Note: Turner provides guidelines on the accuracy of this model. It is an estimating tool and not a definitive model to be used indiscriminately.

s_y, s_z = plume standard deviations, m
u = wind speed, m/s
$x, y, z,$ and H = distances, m
exp = exponential e such that terms in brackets immediately following are powers of e, that is, $e^{[\ ]}$ where e = 2.7182

The value for the effective stack height is the sum of the physical stack height (h) and the plume rise ΔH:

$$H = h + \Delta H \tag{7-20}$$

ΔH may be computed from Holland's formula as follows (Holland, 1953):

$$\Delta H = \frac{v_s d}{u}\left[1.5 + \left(2.68 \times 10^{-2}(P)\left(\frac{T_s - T_a}{T_s}\right)d\right)\right] \tag{7-21}$$

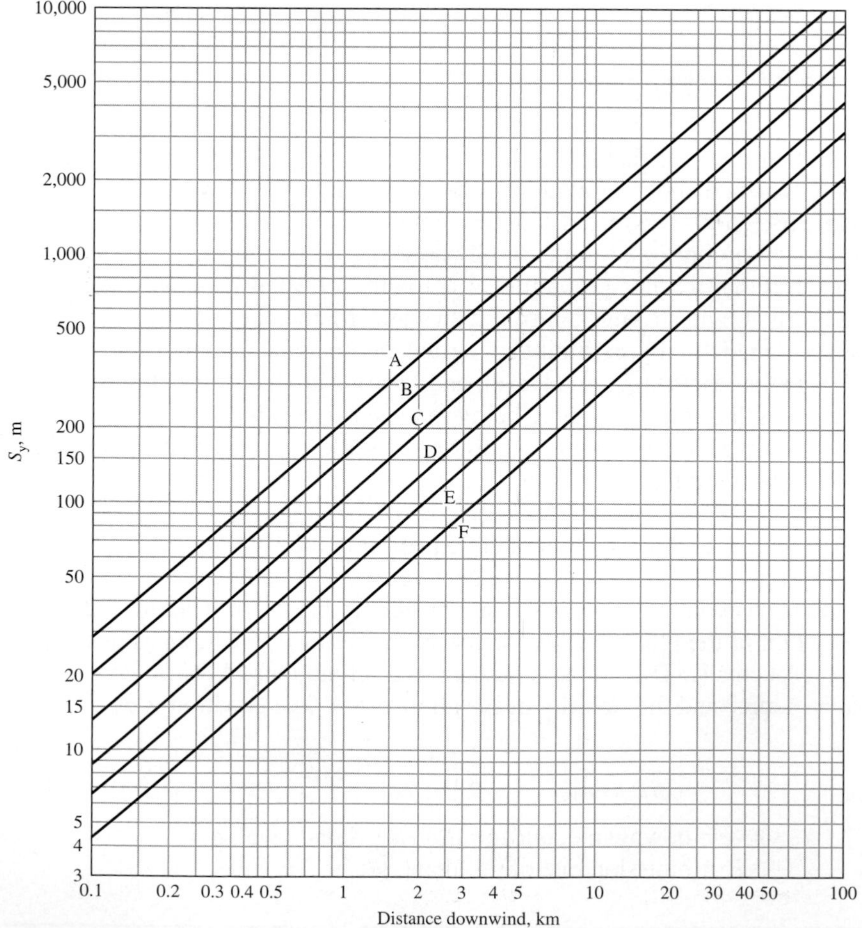

FIGURE 7-22
Horizontal dispersion coefficient. (*Source:* Turner, 1967).

where υ_s = stack velocity, m/s
d = stack diameter, m
u = wind speed, m/s
P = pressure, kPa
T_s = stack temperature, K
T_a = air temperature, K

The values of s_y and s_z depend upon the turbulent structure or stability of the atmosphere. Figures 7-22 and 7-23 provide graphical relationships between the downwind distance x in kilometers and values of s_y and s_z in meters. The curves on the two figures are labeled A through F. The label A refers to very unstable atmospheric conditions, B to unstable atmospheric conditions, C to slightly unstable conditions, D to neutral conditions, E to stable atmospheric conditions, and F to very stable atmospheric

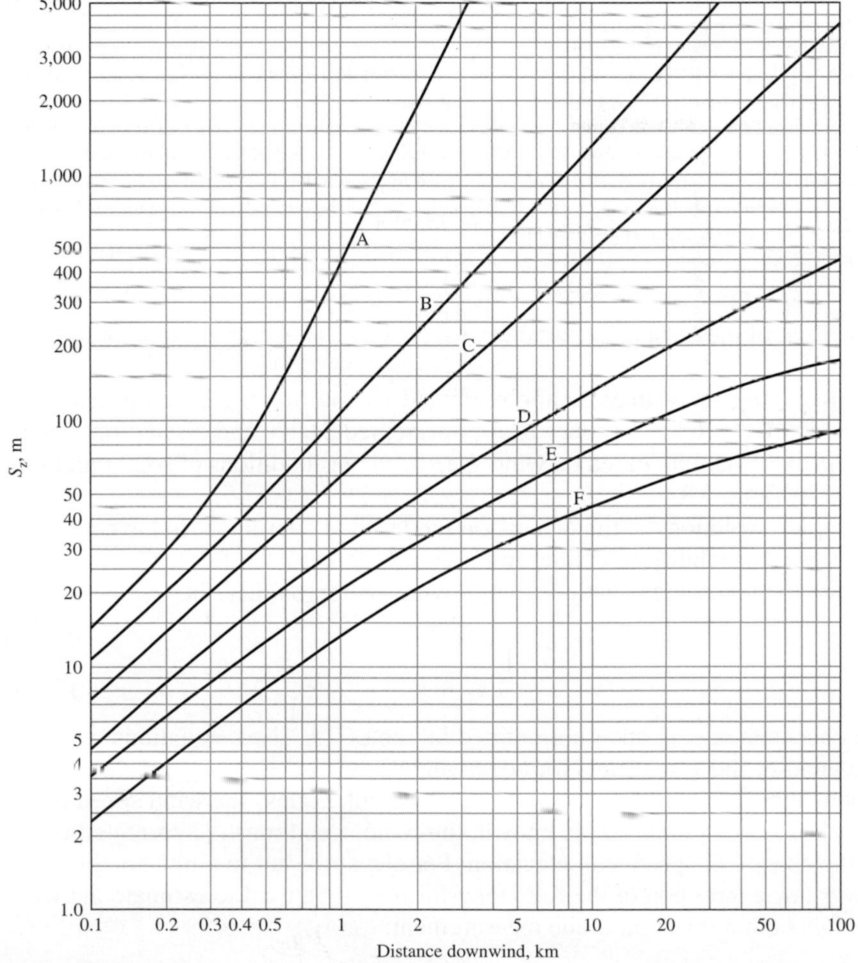

FIGURE 7-23
Vertical dispersion coefficient. (*Source:* Turner, 1967.)

TABLE 7-8
Key to stability categories

Surface wind speed (at 10 m) (m/s)	Day[a]			Night[a]	
	Incoming solar radiation				
	Strong	Moderate	Slight	Thinly overcast or ≥ 1/2 Low cloud	≤ 3/8 Cloud
<2	A	A–B	B	—	—
2–3	A–B	B	C	E	F
3–5	B	B–C	C	D	E
5–6	C	C–D	D	D	D
>6	C	D	D	D	D

[a]The neutral class, D, should be assumed for overcast conditions during day or night. Note that "thinly overcast" is not equivalent to "overcast."

Notes: Class A is the most unstable and class F is the most stable class considered here. Night refers to the period from one hour before sunset to one hour after sunrise. Note that the neutral class, D, can be assumed for overcast conditions during day or night, regardless of wind speed.

"Strong" incoming solar radiation corresponds to a solar altitude greater than 60° with clear skies; "slight" insolation corresponds to a solar altitude from 15° to 35° with clear skies. Table 170, Solar Altitude and Azimuth, in the Smithsonian Meteorological Tables, can be used in determining solar radiation. Incoming radiation that would be strong with clear skies can be expected to be reduced to moderate with broken (5/8 to 7/8 cloud cover) middle clouds and to slight with broken low clouds.

(*Source:* Turner, 1967.)

conditions. Each of these stability parameters represents an averaging time of approximately 3 to 15 min.

Other averaging times may be approximated by multiplying by empirical constants, for example, 0.36 for 24 hours. Turner presented a table and discussion that allows an estimate of stability based on wind speed and the conditions of solar radiation. This is given in Table 7-8.

For computer solutions of the dispersion model, it is convenient to have an algorithm to express the stability class lines in Figures 7-22 and 7-23. D. O. Martin (1976) has developed the following equations that provide an approximate fit.

$$s_y = ax^{0.894} \tag{7-22}$$

$$s_z = cx^d + f \tag{7-23}$$

where the constants a, c, d, and f are defined in Table 7-9. These equations were developed to yield s_y and s_z in meters for downwind distance x in kilometers.

As noted above, the wind speed varies with height. Unless the wind speed at the effective height of the plume (H) is known, the wind speed must be corrected to account for the change in speed with elevation. For elevations up to a few hundred meters, a power law expression of the following form may be used to estimate the wind speed at heights other than that of the measurement:

$$u_2 = u_1\left(\frac{z_2}{z_1}\right)^p \tag{7-24}$$

TABLE 7-9
Values of a, c, d, and f for calculating s_y and s_z

Stability class	a	$x \leq 1$ km c	d	f	$x > 1$ km c	d	f
A	213	440.8	1.941	9.27	459.7	2.094	−9.6
B	156	100.6	1.149	3.3	108.2	1.098	2
C	104	61	0.911	0	61	0.911	0
D	68	33.2	0.725	−1.7	44.5	0.516	−13.0
E	50.5	22.8	0.678	−1.3	55.4	0.305	−34.0
F	34	14.35	0.74.0	−0.35	62.6	0.18	−48.6

(*Source:* Martin, 1976.)

TABLE 7-10
Exponent p values for rural and urban regimes

Stability class	Rural	Urban
A	0.07	0.15
B	0.07	0.15
C	0.10	0.20
D	0.15	0.25
E	0.35	0.30
F	0.55	0.30

(*Source:* U.S. EPA, 1995.)

where u_2 is the windspeed at elevation z_2 and u_1 is the windspeed at elevation z_1. The exponent p is a function of the terrain roughness and the stability. EPA's recommended values for p are shown in Table 7-10.

Example 7-4. It has been estimated that the emission of SO_2 from a coal-fired power plant is 1,656.2 g/s. At 3 km downwind on an overcast summer afternoon, what is the centerline concentration of SO_2 if the wind speed is 4.50 m/s? (Note: "centerline" implies $y = 0$.)

Stack parameters:

Height = 120.0 m

Diameter = 1.20 m

Exit velocity = 10.0 m/s

Temperature = 315°C

Atmospheric conditions:

Pressure = 95.0 kPa

Temperature = 25.0°C

Solution. We begin by determining the effective stack height (H).

$$\Delta H = \frac{(10.0)(1.20)}{4.50}\left[1.5 + \left(2.68 \times 10^{-2}(95.0)\frac{588 - 298}{588}1.20\right)\right]$$

$$\Delta H = 8.0 \text{ m}$$

$$H = 120.0 + 8.0 = 128.0 \text{ m}$$

Next, we must determine the atmospheric stability class. The footnote to Table 7-8 indicates that the D class should be used for overcast conditions.

From Equations 7-22 and 7-23 we can determine that, at 3 km downwind with a D stability, the plume standard deviations are as follows:

$$s_y = 68(3)^{0.894} = 181.6 \text{ m}$$

$$s_z = 44.5(3)^{0.516} + (-13) = 65.4 \text{ m}$$

Thus,

$$\chi = \left[\frac{1{,}656.2}{\pi(181.6)(65.4)(4.50)}\right]\left\{\exp\left[-\frac{1}{2}\left(\frac{0}{181.5}\right)^2\right]\right\}\left\{\exp\left[-\frac{1}{2}\left(\frac{128.0}{65.4}\right)^2\right]\right\}$$

$$= 1.45 \times 10^{-3} \text{ g/m}^3\text{, or } 1.5 \times 10^{-3} \text{ g/m}^3\text{, of } SO_2$$

Inversion Aloft. When an inversion is present, the basic diffusion equation must be modified to take into account the fact that the plume cannot disperse vertically once it reaches the inversion layer. The plume will begin to mix downward when it reaches the base of the inversion layer (Figure 7-24). The downward mixing will begin at a distance x_L downwind from the stack. The x_L distance is a function of the stability in the layer below the inversion. It has been determined empirically that the vertical standard deviation of the plume can be calculated with the following formula at the distance x_L:

$$s_z = 0.47(L - H) \tag{7-24}$$

where L = height to bottom of inversion layer, m
H = effective stack height, m

When the plume reaches twice the distance to initial contact with the inversion base, the plume is said to be completely mixed throughout the layer below the inversion.

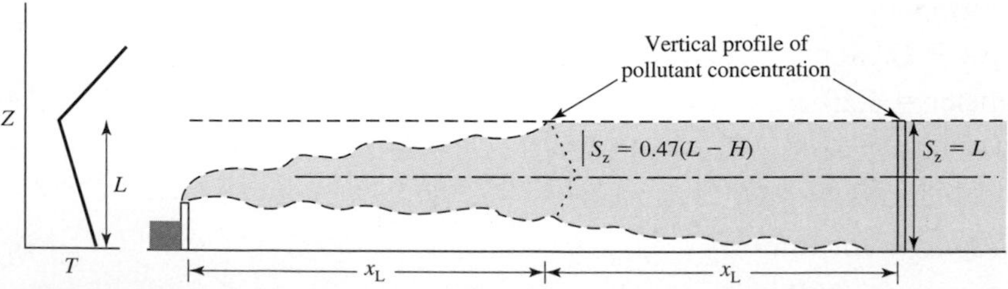

FIGURE 7-24
Effect of elevated inversion on dispersion.

Beyond a distance equal to $2x_L$ the centerline concentration of pollutants may be estimated by using the following equation:

$$\chi = \frac{E}{(2\pi)^{1/2} s_y (u)(L)} \tag{7-25}$$

Note that s_y is determined by the stability of the layer below the inversion and the distance to the receptor. We call this the "inversion" or "short form" of the dispersion equation.

Example 7-5. Determine the distance downwind from a stack at which we must switch to the "inversion form" of the dispersion model given the following meteorologic situation:

Effective stack height: 50 m

Inversion base: 350 m

Wind speed: 7.3 m/s

Cloud cover: none

Time: 1130 h

Season: summer

Solution. Determine the stability class using Table 7-8. At > 6 m/s with strong radiation, the stability class is C.

Calculate the value of s_z.

$$s_z = 0.47(350 \text{ m} - 50 \text{ m}) = 141 \text{ m}$$

Using Figure 7-23, find x_L. With $s_z = 141$, draw a horizontal line to stability class C. Drop a vertical line to the "distance downwind." Find $x_L = 2.5$ km.

Therefore, at any distance equal to or greater than 5 km downwind ($2x_L$), use the "inversion form" of the equation (Equation 7-25).

For distances less than 5 km, we use Equation 7-19 with s_z determined from the distance to the point of interest and the stability. Thus, in no case do we use s_z computed from Equation 7-24 to calculate χ.

7-9 INDOOR AIR QUALITY MODEL

If we envision a house or room in a house or other enclosed space as a simple box (Figure 7-25), then we can construct a simple mass balance model to explore the behavior of the indoor air quality as a function of infiltration of outdoor, indoor sources and sinks, and leakage to the outdoor air. If we assume the contents of the box are well mixed, then

Rate of pollutant increase in box	=	Rate of pollutant entering box from outdoors	+	Rate of pollutant entering box from indoor emissions	−	Rate of pollutant leaving box by leakage to outdoors	−	Rate of pollutant leaving box by decay	(7-26)

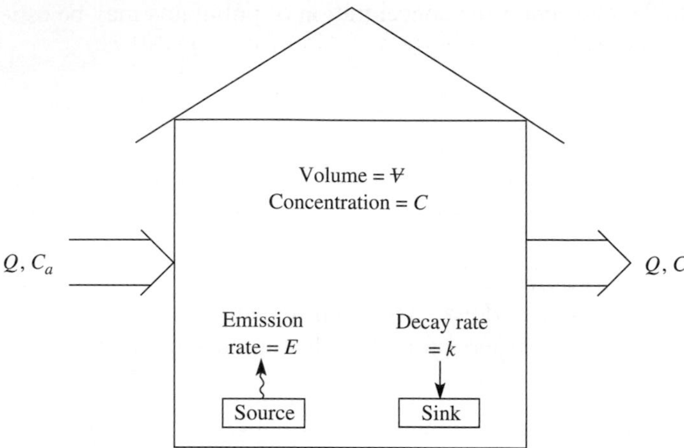

FIGURE 7-25
Mass balance model for indoor air pollution.

or

$$\forall \frac{dC}{dt} = QC_a + E - QC - kC\forall \qquad (7\text{-}27)$$

where $\forall$ = volume of box, m^3
C = concentration of pollutant, g/m^3
Q = rate of infiltration of air into and out of box, m^3/s
C_a = concentration of pollutant in outdoor air, g/m^3
E = emission rate of pollutant into box from indoor source, g/s
k = pollutant decay rate or reaction rate coefficient, s^{-1}

Reaction rate coefficients for a selected list of pollutants are given in Table 7-11. Emission factors for selected indoor air pollution sources are listed in Table 7-12.

TABLE 7-11
Reaction rate coefficients for selected pollutants

Pollutant	k, s^{-1}
CO	0.0
CH_2O	1.11×10^{-4}
NO	0.0
NO_x (as N)	4.17×10^{-5}
Particulates ($< 0.5\ \mu m$)	1.33×10^{-4}
Radon	2.11×10^{-6}
SO_2	6.39×10^{-5}

(*Source:* Traynor et al., 1982.)

TABLE 7-12
Emission factors for selected indoor air pollution sources and pollutants[a]

Pollutant	Emission factor, $\mu g/h \cdot m^2$, at various times after being put into use				
	1 h	1 day	1 week	1 month	1 year
	Floor materials: carpet, synthetic fiber				
Formaldehyde	15	10	5	2	1
Styrene	50	20	6	3	2
Toluene	300	40	20	10	1
TVOC[b]	600	80	20	10	5
	Paints and coatings: solvent-based paint				
Decane	200,000	2,000	0	0	0
Nonane	100,000	100	0	0	0
Pentylcyclohexane	10,000	3,000	0	0	0
Undecane	100,000	10,000	0	0	0
m,p Xylenes	50,000	5	0	0	0
TVOC	3×10^6	200,000	0	0	0
	Paints and coatings: water-based paints				
Acetaldehyde	100	10	2	1	0
Ethylene glycol	20,000	20,000	15,000	4,000	0
Formaldehyde	40	100	2	1	0
TVOC	50,000	40,000	20,000	200	20
	Photocopiers: dry-process, $\mu g/h$ per machine				
	Machine in standby mode		Machine making copies		
Ethylebenzene	10		30,000		
Styrene	500		7,000		
m,p Xylenes	200		20,000		
	Formaldehyde emission factors, $\mu g/d \cdot m^2$				
Medium density fiberboard		17,600–55,000			
Particleboard		2,000–25,000			
Paper products		260–280			
Fiberglass products		400–470			
Clothing		35–570			

[a]Compiled from Godish, 2001, and Tucker, 2001.
[b]TVOC = total volatile organic compounds.

The general solution for Equation 7-27 is

$$C_t = \frac{\dfrac{E}{\forall} + C_a \dfrac{Q}{\forall}}{\dfrac{Q}{\forall} + k}\left[1 - \exp\left(-\left(\frac{Q}{\forall} + k\right)t\right)\right] + C_o \exp\left[-\left(\frac{Q}{\forall} + k\right)t\right] \tag{7-28}$$

The steady-state solution for Equation 7-27 may be found by setting $dC/dt = 0$ and solving for C:

$$C = \frac{QC_a + E}{Q + k\forall} \tag{7-29}$$

When the pollutant is conservative and does not decay with time or have a significant reactivity, $k = 0$. In the special case when the pollutant is conservative and the ambient concentration is negligible and the initial indoor concentration is zero, Equation 7-27 reduces to:

$$C_t = \frac{E}{Q}\left[1 - \exp\left(-\left(\frac{Q}{\forall}\right)t\right)\right] \tag{7-30}$$

Example 7-6. An unvented kerosene heater is operated for one hour in an apartment having a volume of 200 m^3. The heater emits SO_2 at a rate of 50 μg/s. The ambient air concentration (C_a) and the initial indoor air concentration (C_0) of SO_2 are 100 $\mu g/m^3$. If the rate of ventilation is 50 L/s, and the apartment is assumed to be well mixed, what is the indoor air concentration of SO_2 at the end of one hour?

Solution. The concentration may be determined using the general solution form of the indoor air quality model (Equation 7-28). The decay rate for SO_2 from Table 7-11 is $6.39 \times 10^{-5}\ s^{-1}$ and 50 L/s is equivalent to 0.050 m^3/s.

$$C_t = \frac{\dfrac{50\ \mu g/s}{(200\ m^3)} + 100\ \mu g/m^3\ \dfrac{0.050\ m^3/s}{200\ m^3}}{\dfrac{0.050\ m^3/s}{200\ m^3} + 6.39 \times 10^{-5}\ s^{-1}}$$

$$\times \left[1 - \exp\left(-\left(\frac{0.050\ m^3/s}{200\ m^3} + 6.39 \times 10^{-5}\ s^{-1}\right)(3600\ s)\right)\right]$$

$$+(100\ \mu g/m^3)\exp\left[-\left(\frac{0.050\ m^3/s}{200\ m^3} + 6.39 \times 10^{-5}\ s^{-1}\right)(3600\ s)\right]$$

$$= 876.08(1 - \exp(-1.13)) + 100\exp(-1.13) = 876.08(1 - 0.323) + 100(0.323)$$

$$= 593.09 + 32.3 = 625.39 \text{ or } 630\ \mu g/m^3$$

In addition to the mass balance model, statistical and *computational fluid dynamics* (CFD) models have been developed. Sparks (2001) provides an overview of the different types, their advantages and disadvantages, and a list of complex computer models that are available.

7-10 AIR POLLUTION CONTROL OF STATIONARY SOURCES

Gaseous Pollutants

Absorption. Control devices based on the principle of absorption attempt to transfer the pollutant from a gas phase to a liquid phase. This is a *mass transfer* process in which the gas dissolves in the liquid (see Section 4-1). The dissolution may or may not be accompanied by a reaction with an ingredient of the liquid. Mass transfer is a diffusion process wherein the pollutant gas moves from points of higher concentration to points of lower concentration. The removal of the pollutant gas takes place in three steps:

1. Diffusion of the pollutant gas to the surface of the liquid
2. Transfer across the gas/liquid interface (dissolution)
3. Diffusion of the dissolved gas away from the interface into the liquid

Structures such as *spray chambers* (Figure 7-26) and *towers* or *columns* (Figure 7-27) are two classes of devices employed to absorb pollutant gases. In scrubbers, which are a type of spray chamber, liquid droplets are used to absorb the gas. In towers, a thin film of liquid is used as the absorption medium. Regardless of the type of

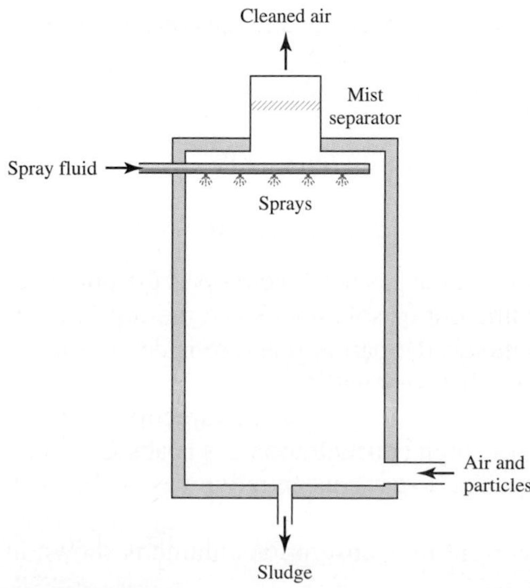

FIGURE 7-26
Spray chamber.

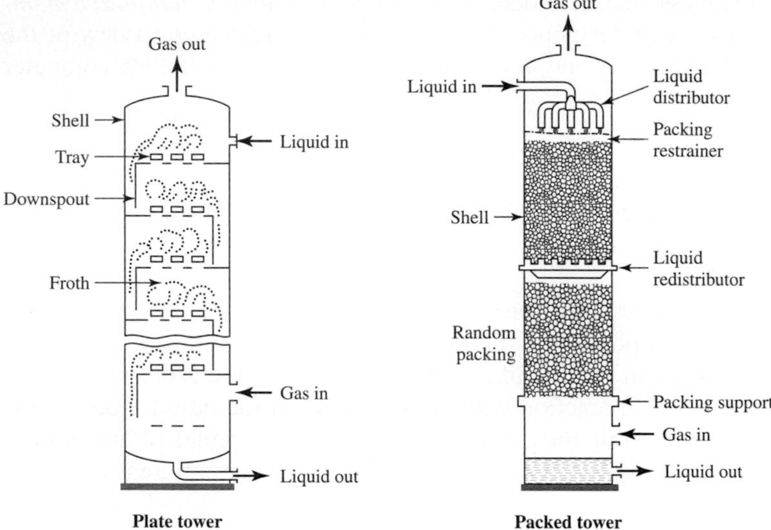

FIGURE 7-27
Absorption systems.

device, the solubility of the pollutant in the liquid must be relatively high. If water is the solute, this generally limits the application to a few inorganic gases such as NH_3, Cl_2, and SO_2. Scrubbers are relatively inefficient absorbers but have the advantage of being able to simultaneously remove particulates. Towers are much more efficient absorbers but they become plugged by particulate matter.

The amount of absorption that can take place for a nonreactive solution is governed by the partial pressure of the pollutant. For dilute solutions, as we have in pollution control systems, the relationship between partial pressure and the concentration of the gas in solution is given by *Henry's law:*

$$P_g = K_H C_{equil} \tag{7-31}$$

where P_g = partial pressure of gas in equilibrium with liquid, kPa
K_H = Henry's law constant, kPa · m^3/g
C_{equil} = concentration of pollutant gas in the liquid phase, g/m^3

Equation 7-31 implies that the partial pressure of the gas must increase as the liquid accumulates more pollutant or else it will come out of solution. Since the liquid is removing pollutant from the gas phase, this means the partial pressure is decreasing as the gas is cleaned. This is just the reverse of what we want to happen. The easiest way to get around this problem is to run the gas and liquid in opposite directions. This is called *countercurrent flow.* In this manner, the high concentration gas is absorbed into a liquid with a high pollutant concentration. The lower concentration gas is absorbed by liquid with no pollutants in it.

A mass balance diagram of a countercurrent flow absorption column is shown in Figure 7-28. The mass balance equation is

$$(G_{m1})(y_1) - (G_{m2})(y_2) = (L_{m1})(x_1) - (L_{m2})(x_2) \tag{7-32}$$

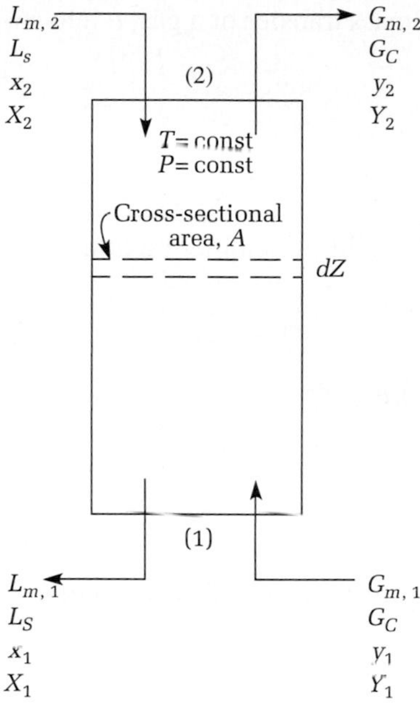

FIGURE 7-28
Notation for a countercurrent flow packed absorption tower.

where G_{m1}, G_{m2} = total gas flow (air plus pollutant) into and out of the column respectively, kg · mole/h
y_1, y_2 = mole fraction of pollutant in the gas phase at inlet and outlet of column, respectively*
L_{m1}, L_{m2} = total liquid flow (solvent plus absorbed pollutant) out of and into the column respectively, kg · mole/h
x_1, x_2 = mole fraction of pollutant in the liquid phase out of and into the column, respectively

Three variables of interest in the design of a packed tower are the gas flow rate, the liquid flow rate, and the height of the tower. As you might expect, the three are related. If we consider a differential height of the absorber, dZ, as shown in Figure 7-28, the total interfacial area open to mass transfer is defined as

$$\text{area for mass transfer} = (a)(A)(dZ) \tag{7-33}$$

where a = area per unit volume of packing
A = cross sectional area of column

---

*A mole fraction is defined as follows:

$$y = \frac{P}{P_t}$$

P = partial pressure of gas
P_t = total pressure of gas
$y^* = (P^*/P_t)$

As we did in Section 4-1, we may describe the rate of mass transfer of a gas, i, into solution (N_i) by the following differential equation:

$$N_i = \frac{dC}{dt} = K_y(y - y^*) \tag{7-34}$$

where K_y = overall mass transfer coefficient for gas
y, y^* = mole fraction of gaseous pollutant and equilibrium mole fraction, respectively

The rate of transfer of species i then is

$$\text{rate of mass transfer} = (N_i)(A)(a)(dZ) \tag{7-35}$$

This mass is equal to the mass loss from the gas phase as it passes through the differential height dZ:

$$\text{mass loss} = d(G_m y) \tag{7-36}$$

We may expand this expression by defining two new terms: the mass flow rate per unit area G'_m and the mole ratio:

$$Y = \frac{y}{1 - y} \tag{7-37}$$

and noting that

$$G_c Y = G_m y \tag{7-38}$$

where G_c is the mass flow of the carrier gas without the pollutant. Equating the mass transfer (Equation 7-34) with the mass loss (Equation 7-36) and making substitutions from Equations 7-35, 7-37, and 7-38 yields

$$K_y a(y - y^*)\,dZ = \frac{G'_m dy}{1 - y} \tag{7-39}$$

or

$$dZ = \frac{G'_m dy}{K_y a(y - y^*)(1 - y)} \tag{7-40}$$

The overall driving force ($y - y^*$) at any location in the tower may be written in the form

$$y - y^* = (1 - y^*) - (1 - y) \tag{7-41}$$

It is convenient then to define the log-mean value of $(1 - y^*)$ and $(1 - y)$:

$$(1 - y)_{LM} = \frac{(1 - y^*) - (1 - y)}{\ln[(1 - y^*)/(1 - y)]} \tag{7-42}$$

Multiplying the numerator and denominator of Equation 7-40 by $(1 - y)_{LM}$, we obtain

$$dZ = \left(\frac{G'_m}{K_y a(1 - y)_{LM}}\right)\left(\frac{(1 - y)_{LM} dy}{(y - y^*)(1 - y)}\right) \tag{7-43}$$

Although G'_m, K_ya, and $(1 - y)_{LM}$ vary along the absorption column, the first term of this equation is reasonably constant. This quantity is called the *overall height of a transfer unit* (H_{og}). As a first approximation to the height of the column, Equation 7-43 may be rewritten as

$$Z = (H_{og}) \int_{y_2}^{y_1} \frac{(1 - y)_{LM} dy}{(y - y^*)(1 - y)} \tag{7-44}$$

The integral is called the *number of transfer units* (N_{og}). The height of the tower is computed from the following equation:

$$Z_t = (H_{og})(N_{og}) \tag{7-45}$$

For dilute solutions that obey Henry's law, the number of overall gas transfer units may be calculated as follows (Treybal, 1968):

$$N_{og} = \frac{\ln\left[\left(\frac{y_1 - mx_2}{y_2 - mx_2}\right)(1 - A) + A\right]}{1 - A} \tag{7-46}$$

where y_1, y_2 = mole fraction of pollutant in the gas phase at inlet and outlet of tower, respectively
m = slope of equilibrium curve defined by Henry's law = y^*/x^* in mole fraction units (m has no units)
x_2 = mole fraction of pollutant in the liquid phase entering the tower
$A = mQ_g/Q_l$
Q_l = liquid flow rate, kg · mole/h · m^2
Q_g = gas flow rate, kg · mole/h · m^2

The height of a single overall mass transfer unit (HTU) may also be expressed as the sum of the gas and liquid HTUs.

$$H_{og} = H_g + AH_l \tag{7-47}$$

where H_g and H_l are complex functions of the flow rate, surface area of the packing, viscosity of the liquid and air, and the diffusivity of the pollutant gas.

Example 7-7. Determine the height of a packed tower that is to reduce NH_3 in air from a concentration of 0.10 kg/m^3 to a concentration of 0.0005 kg/m^3 given the following data:

Column diameter = 3.00 m

Operating temperature = 20.0°C

Operating pressure = 101.325 kPa

H_g = 0.438 m

H_l = 0.250 m

$Q_g = Q_l$ = 10.0 kg/s

Incoming liquid is water free of NH_3

Solution. We begin by converting to mole fractions. NH_3 has a GMW of 17.03. For air we assume a GMW of 28.970 and a density of 1.185 kg/m^3 at 25°C. Since the operating temperature is 20°C, we correct the density of the air:

$$1.185 \times \frac{298}{293} = 1.205 \text{ kg/m}^3$$

Now we compute the mole fractions at the inlet (y_1) and outlet (y_2):

$$y_1 = \frac{\dfrac{0.10 \text{ kg/m}^3}{17.03 \text{ GMW NH}_3}}{\dfrac{1.205 \text{ kg/m}^3}{28.970 \text{ GMW air}}} = \frac{0.005872}{0.04159} = 0.14118$$

In a like manner, $y_2 = 0.000706$. Since the incoming liquid has no NH_3, the mole fraction is zero, that is, $x_2 = 0.0$.

The Henry's law constant in mole fraction units must be determined from experimental data. From the *Chemical Engineers' Handbook* we find the following data (Perry and Chilton, 1973):

P_{NH_3}, kPa	kg NH_3 per 100 kg H_2
15.199	15
9.319	10
4.266	5
1.600	1

If we convert each value to mole fractions and plot x^* versus y^* (the asterisk refers to the steady-state condition), the slope of the line will be m. An example calculation is shown for the first value of x^* and y^*. The total pressure is taken to be 101.325 kPa. The GMW of H_2O is 18.015. For 15 kg NH_3 per 100 kg H_2O:

$$X^* = \frac{\dfrac{15 \text{ kg}}{17.030 \text{ GMW NH}_3}}{\dfrac{15 \text{ kg}}{17.03 \text{ GMW NH}_3} + \dfrac{100 \text{ kg}}{18.015 \text{ GMW H}_2\text{O}}}$$

$$x^* = 0.1369$$

$$y^* = \frac{15.199 \text{ kPa}}{101.325 \text{ kPa}}$$

$$y^* = 0.1500$$

The value of m is then found by a least squares linear regression fit of a line through the four pairs of x^* and y^* values. The slope of the line is m.

$$m = 1.068$$

The value of A is computed in mole units as follows:

$$A = \frac{1.068\left[\dfrac{10.0 \text{ kg/s of air}}{28.97 \text{ GMW of air}}\right]}{\dfrac{10.0 \text{ kg/s of } H_2O}{18.015 \text{ GMW of } H_2O}}$$
$$= 0.6641$$

The number of gas transfer units is then

$$N_{og} = \frac{\ln\left[\dfrac{0.14118 - 1.068(0)}{0.000706 - 1.068(0)}(1 - 0.6641) + 0.6641\right]}{1 - 0.6641}$$
$$= 12.5545$$

The height of an individual gas transfer unit is

$$H_{og} = 0.438 + 0.6641(0.250) = 0.6040$$

The height of the tower is then

$$Z_t = (0.6040)(12.5545) = 7.5832$$

Since the limiting concentration data were given to only one significant figure, the answer would be

$$Z_t = 8 \text{ m}$$

Before we leave this example, we should look back and see what we have wrought. Since the absorption tower neither creates nor destroys matter, the mass of NH_3 entering and leaving the column must be the same. If we assume isothermal, steady-state conditions (that is, gas and liquid rates in and out are equal), we can solve the mass balance equation (Equation 7-32) for x_1. After some calculations we find $x_1 =$ 0.08734. This is 90,300 mg/L of NH_3. This is a classic example of a multimedia problem. In solving an air pollution problem, we have created a serious water pollution problem. Catch-22!

Adsorption. This is a *mass-transfer* process in which the gas is bonded to a solid. It is a surface phenomenon. The gas (the *adsorbate*) penetrates into the pores of the solid (the *adsorbent*) but not into the lattice itself. The bond may be physical or chemical. Electrostatic forces hold the pollutant gas when physical bonding is significant. Chemical bonding is by reaction with the surface. Pressure vessels having a fixed bed are used to hold the adsorbent (Figure 7-29). Active carbon (activated charcoal), molecular sieves, silica gel, and activated alumina are the most common adsorbents. Active carbon is manufactured from nut shells (coconuts are great) or coal subjected to heat treatment in a reducing atmosphere. Molecular sieves are dehydrated zeolites (alkali-metal silicates). Sodium silicate is reacted with sulfuric acid to make silica gel. Activated alumina is a porous hydrated aluminum oxide. The common property of these adsorbents is a large "active" surface area per unit volume after treatment. They are

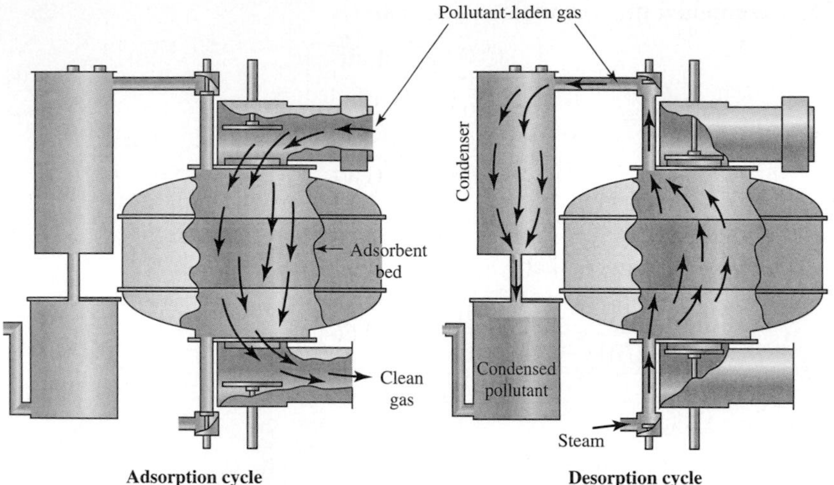

FIGURE 7-29
Adsorption system.

very effective for hydrocarbon pollutants. In addition, they can capture H_2S and SO_2. One special form of molecular sieve can also capture NO_2. With the exception of the active carbons, adsorbents have the drawback that they preferentially select water before any of the pollutants. Thus, water must be removed from the gas before it is treated. All of the adsorbents are subject to destruction at moderately high temperatures (150°C for active carbon, 600°C for molecular sieves, 400°C for silica gel, and 500°C for activated alumina). They are very inefficient at these high temperatures. In fact, their activity is regenerated at these temperatures!

The relation between the amount of pollutant adsorbed and the equilibrium pressure at constant temperature is called an *adsorption isotherm.* The equation that best describes this relation for gases is the one derived by Langmuir (Buonicore and Theodore, 1975).

$$W = \frac{aC_R^*}{1 + bC_g^*} \tag{7-48}$$

where W = amount of gas per unit mass of adsorbent, kg/kg
a, b = constants determined by experiment
C_g^* = equilibrium concentration of gaseous pollutant, g/m^3

In the analysis of experimental data, Equation 7-48 is rewritten as follows:

$$\frac{C_g^*}{W} = \frac{1}{a} + \frac{b}{a}C_g^* \tag{7-49}$$

In this arrangement, a plot of (C_g^*/W) versus C_g^* should yield a straight line with a slope of (b/a) and an intercept equal to $(1/a)$.

In contrast to absorption towers where the collected pollutant is continuously removed by flowing liquid, the collected pollutant remains in the adsorption bed. Thus, while the bed has sufficient capacity, no pollutants are emitted. At some point in time, the bed will become saturated with pollutant. As saturation is approached, pollutant will begin to leak out of the bed. This is called *breakthrough.* When the bed capacity is exhausted, the influent and effluent concentration will be equal. A typical breakthrough curve is shown in Figure 7-30. In order to allow for continuous operation, two beds are provided (Figure 7-29). While one is collecting pollutant, the other is being regenerated. The concentrated gas released during regeneration is usually returned to the process as recovered product. The critical factor in the operation of the bed is the length of time it can operate before breakthrough occurs. The time to breakthrough may be calculated from the following (Crawford, 1976):

$$t_B = \frac{Z_t - \delta}{v_f} \tag{7-50}$$

where Z_t = height of bed, m
δ = width of adsorption zone, m
v_f = velocity of adsorption zone as defined by Equation 7-52, m/s

The height of the adsorption bed (Z_t) can be determined in the same manner as it was for absorption towers, with a few exceptions. The value of N_{og} must be determined by

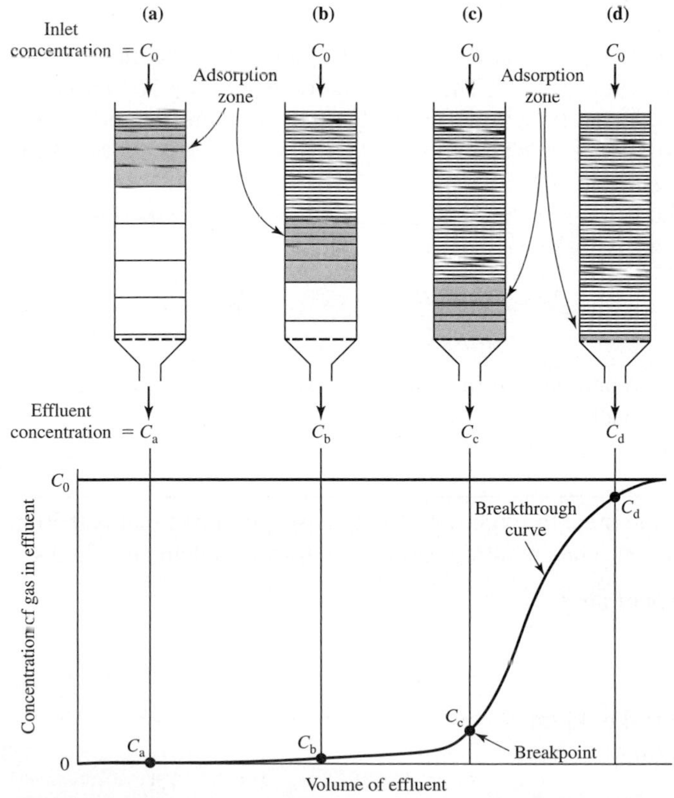

FIGURE 7-30
Adsorption wave and breakthrough curve. (*Source:* Treybal, 1968.)

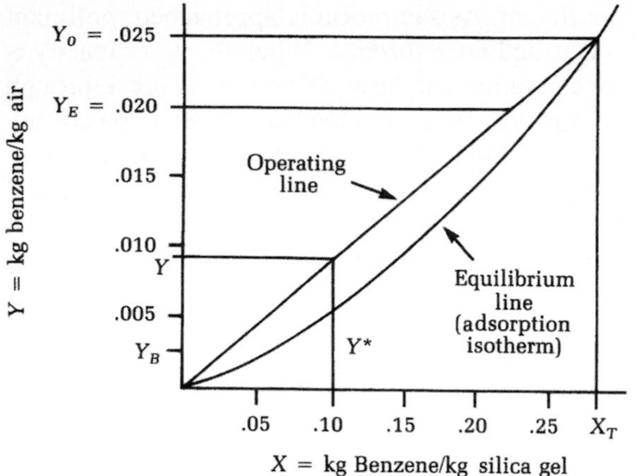

FIGURE 7-31
Equilibrium and operating lines for adsorption of benzene on silica gel. (*Source:* Seinfeld, 1975.)

integration of the following expression (Treybal, 1968):

$$N_{og} = \int_{C_2}^{C} \frac{dC}{C - C_g^*} \tag{7-51}$$

where C_g^* = the equilibrium partial pressure described by Equation 7-48 and C is described by the operating line (Figure 7-31). The H_{og} equation is modified by replacing H_l with H_s. The value of slope m in Equation 7-46 is determined from Equation 7-49.

The width of the adsorption zone is shown in Figure 7-30. It is a function of the shape of the adsorption isotherm.

The velocity of the adsorption zone may be calculated from the properties of the system:

$$v_f = \frac{(Q_g)(1 + bC_g^*)}{a\rho_s\rho_g A_c} \tag{7-52}$$

where ρ_s, ρ_g = density of solid and gas, kg/m^3 (Note that ρ_s is the density of the absorbent "as packed.")

A_c = cross-sectional area of bed, m^2

Example 7-8. Determine the breakthrough time for an adsorption bed that is 0.50 m thick and 10 m^2 in cross section. The operating parameters for the bed are as follows:

Gas flow rate = 1.3 kg/s of air

Gas temperature = 25°C

Gas pressure = 101.325 kPa

Bed density as packed = 420 kg/m^3

Inlet pollutant concentration = 0.0020 kg/m^3

Langmuir parameters: a = 18; b = 124

Width of adsorption zone = 0.03 m

Solution. Using Table A-5 in Appendix A, we find $\rho_g = 1.185\ \text{kg/m}^3$. Then the face velocity of the adsorption wave is

$$v_f = \frac{(1.3\ \text{kg/s})[1 + 124(0.0020\ \text{kg/m}^3)]}{(18)(420\ \text{kg/m}^3)(1.185\ \text{kg/m}^3)(10\ \text{m}^2)}$$
$$= 1.8 \times 10^{-5}\ \text{m/s}$$

The breakthrough time is calculated directly from Equation 7-50:

$$t_B = \frac{0.50\ \text{m} - 0.03\ \text{m}}{1.8 \times 10^{-5}\ \text{m/s}}$$
$$= 2.6 \times 10^4\ \text{s or } 7.2\ \text{h}$$

Combustion. When the contaminant in the gas stream is oxidizable to an inert gas, combustion is a possible alternative method of control. Typically, CO and hydrocarbons fall into this category. Both direct flame combustion by afterburners (Figure 7-32) and catalytic combustion (Figure 7-33) have been used in commercial applications.

Direct flame incineration is the method of choice if two criteria are satisfied. First, the gas stream must have a *net heating value* (NHV) greater than 3.7 MJ/m^3. At this NHV, the gas flame will be *autogenous* (self-supporting after ignition). Below this point, supplementary fuel is required. The second requirement is that none of the by-products of combustion be toxic. In some cases the combustion by-product may be more toxic than the original pollutant gas. For example, the combustion of trichloroethylene produces phosgene, which was used as a poison gas in World War I. Direct flame incineration has been successfully applied to varnish-cooking, meat-smokehouse, and paint bake-oven emissions.

Some catalytic materials enable oxidation to be carried out in gases that have an NHV of less than 3.7 MJ/m^3. Conventionally, the catalyst is placed in beds similar to adsorption beds. Frequently, the active catalyst is a platinum or palladium compound. The supporting lattice is usually a ceramic. Aside from expense, a major drawback of

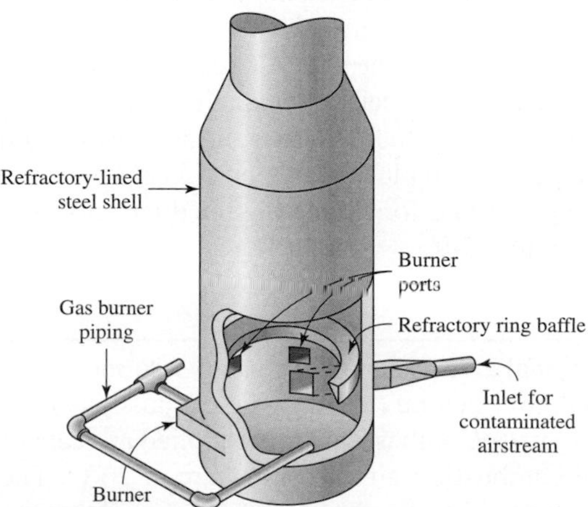

FIGURE 7-32
Direct flame incineration.

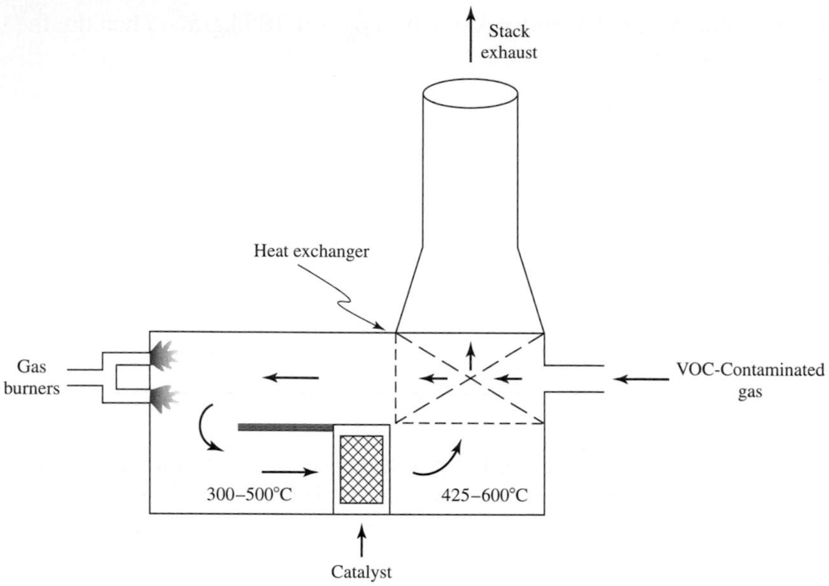

FIGURE 7-33
Catalytic incinerator.

the catalysts is their susceptibility to poisoning by sulfur and lead compounds in trace amounts. Catalytic combustion has successfully been applied to printing-press, varnish-cooking, and asphalt-oxidation emissions.

The fundamental problem in the design of a catalytic reactor is to determine the volume and dimensions of the catalyst bed for a given conversion and flow rate. The catalyst increases the rate of reaction at lower temperatures than are required in direct flame incineration. The reaction is assumed to be a first-order reaction (Equation 2-14). While the reaction rate constant k may be estimated from the Arrhenius equation for flame incineration (Beard et al., 1980), the reaction rate constant for catalytic incineration is highly dependent on the catalyst. Thus, manufacturers' data or pilot plant data must be used to estimate the required retention time in the catalyst bed. Gas velocities in the range of 6 to 12 m/s are commonly used. Typical catalyst operating temperatures are in the range of 250–550°C (Noll, 1999). The actual residence time is estimated from the total gas flow rate (contaminated gas stream plus the combustion gases) at the operating temperature of the catalyst. The procedure for estimating the dimension and volume of the catalyst bed is shown in Example 7-9.

Example 7-9. Determine the cross-sectional area and depth of catalyst to meet a 95 percent destruction efficiency for methyl ethyl ketone (MEK). The exhaust gas flow rate entering the control equipment is 5.66 m^3/s. Combustion air is supplied at a rate of 0.60 m^3/s. Both the exhaust gas and the combustion air flow rates are at 20°C. The manufacturers' specifications require that the catalyst be operated at 480°C and that the bed gas velocity be limited to 6.0 m/s. Assume that the MEK combustion reaction follows first order kinetics and that the rate constant at 480°C is 100 s^{-1}.

Solution.

1. Calculate the volumetric gas flow rate at the catalyst operating temperature.

$$(5.66 \text{ m}^3/\text{s} + 0.60 \text{ m}^3/\text{s})\left(\frac{480 + 273}{20 + 273}\right) = 16.09 \text{ m}^3/\text{s}$$

2. The cross-sectional area is computed from the bed gas velocity and the gas flow rate:

$$\text{Area} = \frac{16.09 \text{ m}^3/\text{s}}{6.0 \text{ m/s}} = 2.68 \text{ m}^2$$

3. Base on first order kinetics and 95 percent destruction efficiency, the desired retention time in the bed is calculated by using Equation 2-14:

$$C_t = C_0 \exp(-kt)$$

with $C_t/C_0 = 0.05$ for 95 percent destruction

$$0.05 = \exp(-100t)$$

Taking the natural logarithm of both sides of this equation and solving for t gives

$$-2.9957 = -100t$$

$$t = 0.030 \text{ s}$$

4. The depth of the catalyst is then

$$\text{Depth} = (6.0 \text{ m/s})(0.030 \text{ s}) = 0.18 \text{ m}$$

Flue Gas Desulfurization (FGD)

Flue gas desulfurization systems fall into two broad categories: nonregenerative and regenerative. Nonregenerative means that the reagent used to remove the sulfur oxides from the gas stream is used and discarded. Regenerative means that the reagent is recovered and reused. In terms of the number and size of systems installed, nonregenerative systems dominate.

Nonregenerative Systems. There are nine commercial nonregenerative systems (Hance and Kelly, 1991). All have reaction chemistries based on lime (CaO), caustic soda (NaOH), soda ash (Na_2CO_3), or ammonia (NH_3).

The SO_2 removed in a lime/limestone-based FGD system is converted to sulfite. The overall reactions are generally represented by (Karlsson and Rosenberg, 1980):

$$SO_2 + CaCO_3 \rightarrow CaSO_3 + CO_2 \quad (7\text{-}53)$$

$$SO_2 + Ca(OH)_2 \rightarrow CaSO_3 + H_2O \quad (7\text{-}54)$$

when using limestone and lime, respectively. Part of the sulfite is oxidized with the oxygen content in the flue gas to form sulfate:

$$CaSO_3 + \frac{1}{2}O_2 \rightarrow CaSO_4 \tag{7-55}$$

Although the overall reactions are simple, the chemistry is quite complex and not well defined. The choice between lime and limestone, the type of limestone, and method of calcining and slaking can influence the gas-liquid-solid reactions taking place in the absorber.

The principal types of absorbers used in the wet scrubbing systems include venturi scrubber/absorbers, static packed scrubbers, moving-bed absorbers, tray towers, and spray towers (Black & Veatch, 1983).

Spray dryer-based FGD systems consist of one or more spray dryers and a particulate collector.* The reagent material is typically a slaked lime slurry or a slurry of lime and recycled material. Although lime is the most common reagent, soda ash has also been used. The reagent is injected in droplet form into the flue gas in the spray dryer. The reagent droplets absorb SO_2 while simultaneously being dried. Ideally, the slurry or solution droplets are completely dried before they impact the wall of the dryer vessel. The flue gas stream becomes more humidified in the process of evaporation of the reagent droplets, but it does not become saturated with water vapor. This is the single most significant difference between spray dryer FGD and wet scrubber FGD. The humidified gas stream and a significant portion of the particulate matter (fly ash, FGD reaction products, and unreacted reagent) are carried by the flue gas to the particulate collector located downstream of the spray dryer vessel (Cannel and Meadows, 1985). Generally, larger units firing high-sulfur coals use wet FGD. Smaller units use spray dryers.

Control Technologies for Nitrogen Oxides

Almost all nitrogen oxide (NO_x) air pollution results from combustion processes. They are produced from the oxidation of nitrogen bound in the fuel, from the reaction of molecular oxygen and nitrogen in the combustion air at temperatures above 1,600 K (see Equation 7-12), and from the reaction of nitrogen in the combustion air with hydrocarbon radicals. Control technologies for NO_x are grouped into two categories: those that prevent the formation of NO_x during the combustion process and those that convert the NO_x formed during combustion into nitrogen and oxygen (Prasad, 1995).

Prevention. The processes in this category employ the fact that reduction of the peak flame temperature in the combustion zone reduces NO_x formation. Nine alternatives have been developed to reduce flame temperature: (1) minimizing operating temperatures, (2) fuel switching, (3) low excess air, (4) flue gas recirculation, (5) lean combustion, (6) staged combustion, (7) low NO_x burners, (8) secondary combustion, and (9) water/steam injection.

---

*Historically, from a mass transfer point of view, spray drying refers to the evaporation of a solvent from an atomized spray. Simultaneous diffusion of a gaseous species into the evaporating droplet is not true spray drying. Nonetheless, many authors have adopted the term "spray drying" as synonymous with dry scrubbing.

Routine burner tune-ups and operation with combustion zone temperatures at minimum values reduce the fuel consumption and NO_x formation. Converting to a fuel with a lower nitrogen content or one that burns at a lower temperature will reduce NO_x formation. For example, petroleum coke has a lower nitrogen content and burns with a lower flame temperature than coal. On the other hand, natural gas has no nitrogen content but burns at a relatively high flame temperature and, thus, produces more NO_x than coal.

Low excess air and flue gas recirculation work on the principle that reduced oxygen concentrations lower the peak flame temperatures. In contrast, in lean combustion, additional air is introduced to cool the flame.

In staged combustion and low NO_x burners, initial combustion takes place in a fuel-rich zone that is followed by the injection of air downstream of the primary combustion zone. The downstream combustion is completed under fuel-lean conditions at a lower temperature.

Staged combustion consists of injecting part of the fuel and all of the combustion air into the primary combustion zone. Thermal NO_x production is limited by the low flame temperatures that result from high excess air levels.

Water/steam injection reduces thermal NO_x emissions by lowering the flame temperature.

Post-Combustion. Three processes may be used to convert NO_x to nitrogen gas: selective catalytic reduction (SCR), selective noncatalytic reduction (SNCR), and nonselective catalytic reduction (NSCR).

The SCR process uses a catalyst bed (usually vanadium-titanium, or platinum-based and zeolite) and anhydrous ammonia (NH_3). After the combustion process, ammonia is injected upstream of the catalyst bed. The NO_x reacts with the ammonia in the catalyst bed to form N_2 and water.

In the SNCR process ammonia or urea is injected into the flue gas at an appropriate temperature (870 to 1,090°C). The urea is converted to ammonia, which reacts to reduce the NO_x to N_2 and water.

NSCR uses a three-way catalyst similar to that used in automotive applications. In addition to NO_x control, hydrocarbons and carbon monoxide are converted to CO_2 and water. These systems require a reducing agent similar to CO and hydrocarbons upstream of the catalyst. Larger boilers that have post combustion NO_x controls are generally equipped with SCR.

Typical reduction capabilities of the NO_x techniques range from 30 to 60 percent for the "prevention" methods, 30 to 50 percent for SNCR, and 70 to 90 percent for the SCR systems (Srivastava et al., 2005).

Particulate Pollutants

Cyclones. For particle sizes greater than about 10 μm in diameter, the collector of choice is the cyclone (Figure 7-34). This is an inertial collector with no moving parts. The particulate-laden gas is accelerated through a spiral motion, which imparts a centrifugal force to the particles. The particles are hurled out of the spinning gas and impact on the cylinder wall of the cyclone. They then slide to the bottom of the cone. Here they are removed through an airtight valving system. The standard single-barrel cyclone will have dimensions proportioned as shown in Figure 7-35.

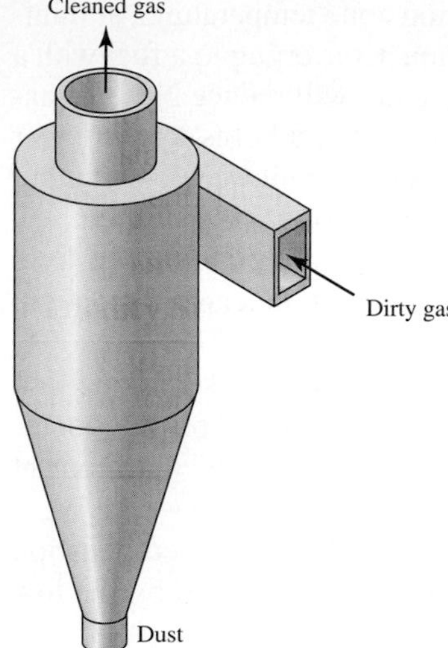

FIGURE 7-34
Reverse flow cyclone. (*Source:* Crawford, 1976.)

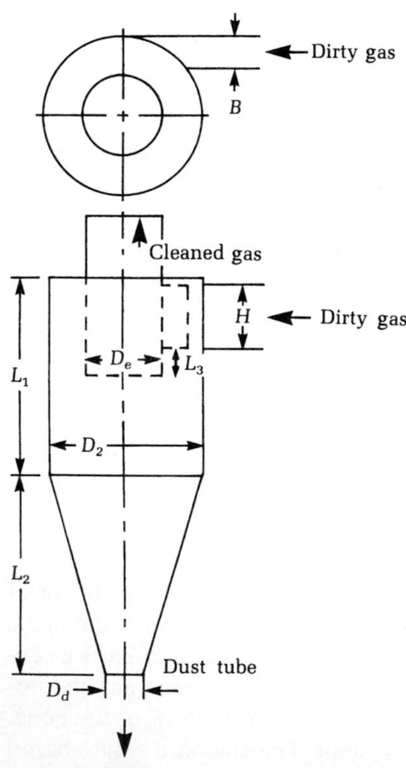

FIGURE 7-35
Standard reverse flow cyclone proportions.
Note: Standard cyclone proportions are as follows:

Length of cylinder, $L_1 = 2D_2$
Length of cone, $L_2 = 2D_2$
Diameter of exit, $D_e = 0.5D_2$
Height of entrance, $H = 0.5D_2$
Width of entrance, $B = 0.25D_2$
Diameter of dust exit, $D_d = 0.25D_2$
Length of exit duct, $L_3 = 0.125D_2$

(*Source:* Crawford, 1976.)

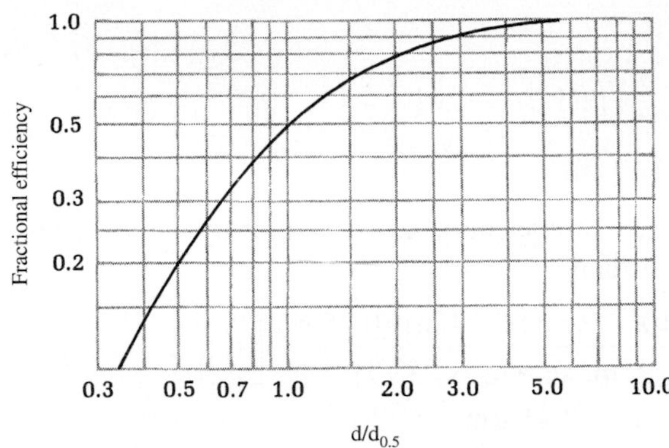

FIGURE 7-36
Empirical cyclone collection efficiency. (*Source:* Lapple, 1951.)

The efficiency of collection of various particle sizes (η) can be determined from an empirical expression and graph (Figure 7-36) developed by Lapple (1951):

$$d_{0.5} = \left[\frac{9\mu B^2}{\rho_p Q_g}\frac{H}{\theta}\right]^{1/2} \tag{7-56}$$

where $d_{0.5}$ = cut diameter, the particle size for which the collection efficiency is 50 percent
μ = dynamic viscosity of gas, Pa · s
B = width of entrance, m
H = height of entrance, m
ρ_p = particle density, kg/m^{3}s
Q_g = gas flow rate, m^3/s
θ = effective number of turns made in traversing the cyclone as defined in Equation 7-57.

The value of θ may be determined approximately by the following:

$$\theta = \frac{\pi}{H}(2L_1 + L_2) \tag{7-57}$$

where L_1 and L_2 are the length of the cylinder and cone, respectively.

Example 7-10. Determine the efficiency of a "standard" cyclone having the following characteristics for particles 10 μm in diameter with a density of 800 kg/m^3:

Cyclone barrel diameter = 0.50 m
Gas flow rate = 4.0 m^3/s
Gas temperature = 25°C

Solution. From the standard cyclone dimensions we can calculate the following:

$$B = (0.25)(0.50 \text{ m}) = 0.13 \text{ m}$$
$$H = (0.50)(0.50 \text{ m}) = 0.25 \text{ m}$$
$$L_1 = L_2 = (2.00)(0.50 \text{ m}) = 1.0 \text{ m}$$

The number of turns i is then

$$\theta = \frac{\pi}{0.25}[2(1.0) + 1.0]$$
$$= 37.7$$

From the gas temperature and Table A-4 of Appendix A, we find the dynamic viscosity is 18.5 μPa · s. The cut diameter is then

$$d_{0.5} = \left[\frac{9(18.5 \times 10^{-6}\ \text{Pa} \cdot \text{s})(0.13\ \text{m})^2(0.25\ \text{m})}{(800\ \text{kg/m}^3)(4.0\ \text{m}^3/\text{s})(37.7)}\right]^{1/2}$$
$$= 2.41 \times 10^{-6}\ \text{m} = 2.41\ \mu\text{m}$$

The ratio of particle sizes is

$$\frac{d}{d_{0.5}} = \frac{10\ \mu\text{m}}{2.41\ \mu\text{m}} = 4.15$$

From Figure 7-36 we find that the collection efficiency is about 95 percent.

As the diameter of the cyclone is reduced, the efficiency of collection is increased. However, the pressure drop also increases. This increases the power requirements for moving the gas through the collector. Since an efficiency increase will result, even if the tangential velocity remains constant, the efficiency may be increased without increasing the power consumption by using multiple cyclones in parallel (*multiclones*).

From the example, you can see that cyclones are quite efficient for particles larger than 10 μm. Conversely, you should note that cyclones are not very efficient for particles 1 μm or less in diameter. Thus, they are employed only for coarse dusts. Some applications include controlling emissions of wood dust, paper fibers, and buffing fibers. Multiclones are frequently used as precleaners for fly-ash control devices in power plants.

Filters. When high efficiency control of particles smaller than 5 μm is desired, a filter may be selected as the control method. Two types are in use: (1) the deep bed filter, and (2) the baghouse (Figure 7-37). The deep bed filter resembles a furnace filter. A packing of fibers is used to intercept particles in the gas stream. For relatively clean gases and low volumes, such as air conditioning systems, these are quite effective. For dirty industrial gas with high volumes, the baghouse is preferable.

The fundamental mechanisms of collection include screening or sieving (where the particles are larger than the openings between the fibers), interception by the fibers themselves, and electrostatic attraction (because of the difference in static charge on the particle and fiber). Once a dust cake begins to form on the fabric, sieving is probably the dominant mechanism. As particulate matter collects on the bag, the collection efficiency increases. The buildup of the dust cake also increases the resistance to gas flow.

At some point the pressure drop across the filter bags reduces the gas flow rate to an unacceptable level and the filter bags must be cleaned. The three methods used to clean the bags are mechanical shaking, reverse air flow, and pulse-jet cleaning.

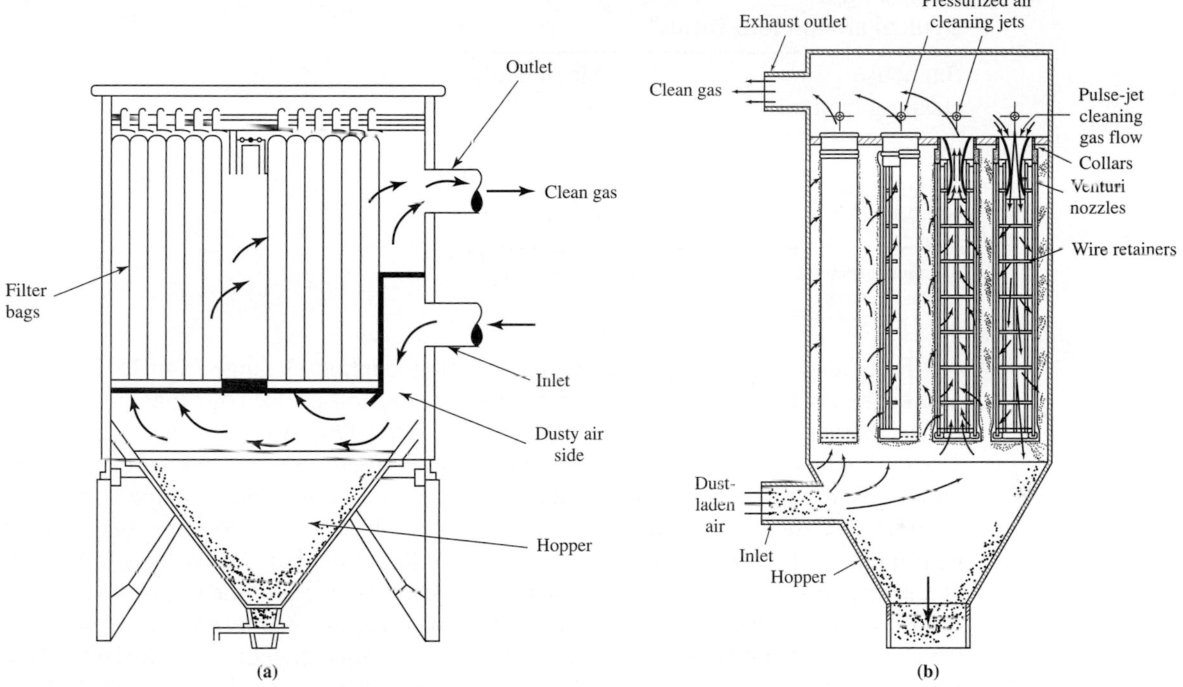

FIGURE 7-37
Mechanically cleaned (shaker) baghouse (*a*) and pulse-jet-cleaned baghouse (*b*). Pulse-jet baghouse shows normal operation for three left-hand-side bags and pulse-jet cleaning for the bag on the right-hand side. (*Source:* Walsh, 1967.)

Mechanically cleaned baghouses operate by directing the dirty gas into the inside of the bag. The particulate matter is collected on the inside of the bag much in the same manner as a vacuum cleaner bag. The bags are hung on a frame that oscillates. They are shaken at periodic intervals, ranging from 30 minutes to more than 2 hours. The bags are arranged in groups in separate compartments that are taken off line during cleaning.

In reverse air flow cleaning, a compartment is isolated and a large volume of gas flow is forced countercurrent to normal operation. The dust cake is removed by collapsing or flexing the bag. The reverse flow combined with the inward collapse of the bag causes the collected dust cake to fall into the hopper below.

Pulse-jet baghouses are designed with frame structures, called cages, that support the bags. In contrast to the other two cleaning methods, the particulate matter is collected on the outside of the bag instead of the inside of the bag. The dust cake is removed by directing a pulsed jet of compressed air into the bag. This causes a sudden expansion of the bag. Dust is removed primarily by inertial forces as the bag reaches maximum expansion. The pulse of cleaning air is at such a high pressure drop and short duration that cleaning is normally accomplished with the baghouse on line. Cleaning occurs at 2- to 15-minute intervals. Extra bags, which are normally provided to compensate for the bags that are required in the other cleaning schemes, are not required in pulse-jet baghouses (Noll, 1999).

TABLE 7-13
Typical air-to-cloth ratios[a]

Baghouse cleaning method	Air-to-cloth ratio, m/s
Shaking	0.010 to 0.017
Reverse air	0.010 to 0.020
Pulse jet	0.033 to 0.083

[a]Compiled from Davis, 2000; Noll, 1999; Wark, Warner, and Davis, 1998.

Bag diameters for shaker and reverse air flow baghouses range from 15 to 45 cm. The bags may be up to 12 m in length. Pulse-jet baghouses use bags that are 10 to 15 cm in diameter with lengths less than 5 m (Noll, 1999; Wark, Warner, and Davis, 1998). The bags are made of either natural or synthetic fibers. Synthetic fibers are widely used as filtration fabrics because of their low cost, better temperature- and chemical-resistance characteristics, and small fiber diameter. Cotton fiber bags cannot be used for sustained temperatures above 90°C. Glass fiber bags, however, can be used at temperatures up to 260°C (McKenna et al., 2000). Because of the stress produced in cleaning, only woven fibers are used when the bags are cleaned mechanically or by reverse air flow. Felted fabrics are used in pulse-jet-cleaned baghouse (Noll, 1999). Bag life varies between 1 and 5 years. Two years is considered normal.

The fundamental design parameter for baghouses is the ratio of the volumetric flow rate of the gas to be cleaned to the area of filter fabric. This ratio is termed the *air-to-cloth ratio*.* It has units of $m^3/s \cdot m^2$ or m/s. Typical air-to-cloth ratios are shown in Table 7-13.

Example 7-11. An aggregate plant at Lime Ridge has been found to be in violation of particulate discharge standards. A mechanical shaker baghouse has been selected for particulate control. Estimate the number of bags required for a gas flow rate of 20 m^3/s if each bag is 15 cm in diameter and 12 m in length. One-eighth of the bags are taken off line for cleaning. The manufacturer's recommended air-to-cloth ratio for aggregate plants is 0.010 m/s.

Solution.

1. Noting that the air-to-cloth ratio units of m/s are equivalent to $m^3/s \cdot m^2$, calculate the net cloth area required with one compartment off line for cleaning:

$$A = \frac{Q}{V} = \frac{20 \text{ m}^3/\text{s}}{0.010 \text{ m}^3/\text{s} \cdot \text{m}^2} = 2{,}000 \text{ m}^2$$

2. The net number of bags is the total area divided by the area of one bag:

$$\frac{2{,}000 \text{ m}^2}{(\pi)(0.15 \text{ m})(12 \text{ m})} = 353.67 \text{ or } 354 \text{ bags}$$

---

*It may also be called the *gas-to-cloth ratio, filtration velocity,* or the *face velocity.*

3. With one-eighth of the bags off line, an additional one-eighth of the net number is required:

$$\frac{354 \text{ bags}}{8} = 44.25 \text{ or } 44 \text{ bags}$$

4. The total number of bags is 354 + 44 = 398.

In order to have an equal number of bags in each compartment, the total number of bags will have to be slightly larger (398 bags/8 compartments = 49.75 bags per compartment). With 50 bags per compartment, the total will be 400 bags.

Baghouses have found a wide variety of applications. Examples include the carbon black and gypsum industries, cement crushing, feed and grain handling, limestone crushing, sanding machines, and coal-fired utility boilers. Of all of the particulate control devices, filtration is the only technology that has the potential to include the addition of adsorption media to facilitate concurrent removal of gas phase contaminants.

Liquid Scrubbing. When the particulate matter to be collected is wet, corrosive, or very hot, the fabric filter may not work. Liquid scrubbing might. Typical scrubbing applications include control of emission of talc dust, phosphoric acid mist, foundry cupola dust, and open hearth steel furnace fumes.

Liquid scrubbers vary in complexity. Simple spray chambers are used for relatively coarse particle sizes. For high efficiency removal of fine particles, the combination of a venturi scrubber followed by a cyclone would be selected (Figure 7-38). The underlying principle of operation of the liquid scrubbers is that a differential velocity between the droplets of collecting liquid and the particulate pollutant allows the particle to impinge onto the droplet. Since the droplet-particle combination is still suspended in the gas stream, an inertial collection device is placed downstream to remove it. Because the droplet enhances the size of the particle, the collection efficiency of the inertial device is higher than it would be for the original particle without the liquid drop.

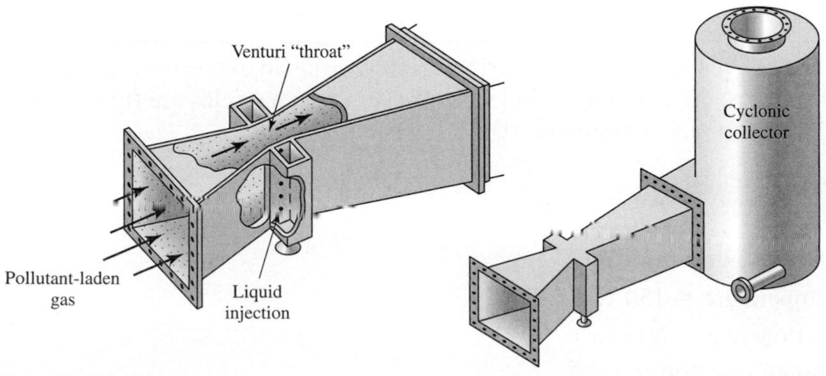

FIGURE 7-38
Venturi scrubber.

The most popular collection efficiency equation is that proposed by Johnstone, Field, and Tassler (1954):

$$\eta = 1 - \exp\left(-\kappa R \sqrt{\psi}\right) \tag{7-58}$$

where η = efficiency
exp = exponential to base e
κ = correlation coefficient, m^3 of gas/m^3 of liquid
R = liquid flow rate, m^3/m^3 of gas
ψ = inertial impaction parameter defined by Equation 7-59

The inertial impaction parameter (ψ) relates the particle and droplet sizes and relative velocities:

$$\psi = \frac{C\rho_p v_g (d_p)^2}{18 d_d \mu} \tag{7-59}$$

where C = Cunningham correction factor defined by Equation 7-60, unitless
ρ_p = particle density, kg/m^3
v_g = speed of gas at throat, m/s
d_p = diameter of particle, m
d_d = diameter of droplet, m
μ = dynamic viscosity of gas, Pa · s

The Cunningham correction factor accounts for the fact that very small particles do not obey Stokes' settling equation. They tend to "slip" between the gas molecules. Thus, the drag coefficient (C_D) is reduced and the particles fall faster than otherwise would be expected. This is particularly true for particles less than 1 μm in diameter. The Cunningham factor may be approximated with the following equation (Hesketh, 1977):

$$C = 1 + \frac{6.21 \times 10^{-4}(T)}{d_p} \tag{7-60}$$

where T = absolute temperature, K
d_p = diameter of particle, μm

Example 7-12. Given the scrubber described below, write an expression for collection efficiency that is a function of particle size. Assume the particles are fly ash with a density of 700 kg/m^3 and a minimum size of 10 μm diameter.

Venturi characteristics:

Throat area = 1.00 m^2

Gas flow rate = 94.40 m^3/s

Gas temperature = 150°C

Liquid flow rate = 0.13 m^3/s

Coefficient κ = 200

Droplet diameter = 100 μm

Solution. We begin by determining the value of the Cunningham correction factor for the smallest particle to see if the d_p term in the denominator must be retained.

$$C = 1 + \frac{6.21 \times 10^{-4}(423\text{ K})}{10\ \mu\text{m}}$$
$$= 1 + 0.0263$$

For this we can see that the term containing d_p will be small for all particles greater than 10 μm and we can use the approximation:

$$C = 1$$

Before we can proceed to calculate a value for ψ, we must determine the gas velocity at the throat:

$$v_g = \frac{Q_g}{A_t}$$

where A_t = cross-sectional area of throat

$$v_g = \frac{94.40\text{ m}^3/\text{s}}{1.00\text{ m}^2} = 94.40\text{ m/s}$$

The dynamic viscosity of the gas is determined from Table A-4 of Appendix A and from the temperature of the gas (150°C). It is 25.2 μPa · s.

Now we can calculate in terms of d_p in µm. Note that $C = 1$ and that 18 is a constant.

$$\psi = \frac{(1)(700\text{ kg/m}^3)(94.40\text{ m}^3/\text{s})(1 \times 10^{-12}\ \mu\text{m}^2/\text{m}^2)(d_p)^2}{(18)(100 \times 10^{-6}\text{ m})(25.2 \times 10^{-6}\text{ Pa} \cdot \text{s})}$$
$$= (1.46)(d_p)^2$$

Taking the square root of ψ and computing R as 0.13/94.40, the expression for efficiency as a function of diameter is then

$$\eta = 1 - \exp[-(200)(1.38 \times 10^{-3})(1.21)d_p]$$
$$= 1 - \exp[-0.33(d_p)]$$

Electrostatic Precipitation (ESP). High efficiency, dry collection of particles from hot gas streams can be obtained by electrostatic precipitation of the particles. The ESP is usually constructed of alternating plates and wires (Figure 7-39). A large direct current potential (30–75 kV) is established between the plates and wires. This results in the creation of an ion field between the wire and plate (Figure 7-40*a*). As the particle-laden gas stream passes between the wire and the plate, ions attach to the particles, giving them a net negative charge (Figure 7-40*b*). The particles then migrate toward the positively charged plate where they stick (Figure 7-40*c*). The plates are rapped at frequent intervals and the agglomerated sheet of particles falls to a hopper.

Unlike the baghouse, the gas flow between the plates is not stopped during cleaning. The gas velocity through the ESP is kept low (less than 1.5 m/s) to allow particle

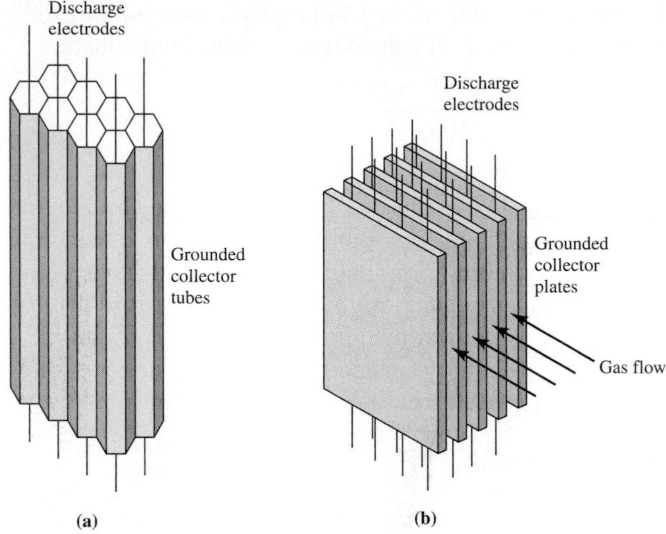

FIGURE 7-39
Electrostatic precipitator with (*a*) wire in tube, (*b*) wire and plate. (*Source:* EPA Training Manual.)

migration. Thus, the terminal settling velocity of the sheet is sufficient to carry it to the hopper before it exits the precipitator.

The classic ESP efficiency equation is the one proposed by Deutsch (1922).

$$\eta = 1 - \exp\left(-\frac{Aw}{Q_g}\right) \tag{7-61}$$

where A = collection area of plates, m^2
w = migration velocity of particles, m/s
Q_g = gas flow rate, m^3/s

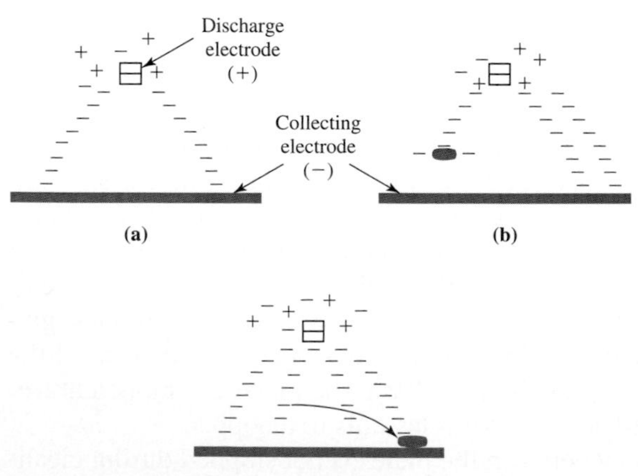

FIGURE 7-40
Particle charging and collection in ESP. (*Source:* EPA Training Manual.)

The migration velocity of the particles is a function of the electrostatic force. The migration velocity is described by the following equation:

$$w = \frac{qE_pC}{6\pi r\mu} \tag{7-62}$$

where q = charge, coulombs (C)
E_p = collection field intensity, volts/m
r = particle radius, m
μ = dynamic viscosity of gas, Pa · s
C = Cunningham correction factor

Example 7-13. Determine the collection efficiency of the electrostatic precipitator described below for a particle 154 μm in diameter having a drift velocity of 0.184 m/s. What is the effect of reducing the plate spacing to one-half of its current value and doubling the number of plates?

ESP specifications:

Height = 7.32 m

Length = 6.10 m

Number of passages = 5

Plate spacing = 0.28 m

Gas flow rate = 19.73 m^3/s

Solution. First we calculate the area of the plates. For a single plate,

$$A = 7.32 \times 6.1 = 44.65 \text{ m}^2$$

Since there are eight collecting surfaces (two for each plate, 4 plates form 5 passages):

$$A = 44.65 \text{ m}^2 \times 8 = 357.2 \text{ m}^2$$

The efficiency is then calculated in a straightforward manner using Equation 7-61.

$$\eta = 1 - \exp\left[-\frac{(357.2)(0.184)}{19.73}\right]$$
$$= 0.964$$

Therefore the efficiency is 96.4 percent. Now what is the effect of reducing the plate spacing? The spacing enters into the efficiency equation through the calculation of the collection field intensity (E_p). Treating everything else in Equation 7-62 as a constant, we can write the following equation:

$$w = KE_p$$

where E_p is measured in volts per meter. If the distance between the plates is reduced, the collection field intensity is proportionately increased:

$$w = KE_p\frac{0.28}{0.14} = 2KE_p$$

Thus, w increases by a factor of two. In order to maintain the same gas velocity, the number of plates and, hence, the surface area (A) must double. The new efficiency would then be:

$$\eta = 1 - \exp\left[-\frac{(714.4)(0.368)}{19.73}\right]$$
$$= 1 - 0.0000016$$
$$= 0.999998 \text{ or } 1.00$$

Thus, the efficiency would be increased to 100 percent. This plate spacing may not be feasible because of sparkover problems.

One operational problem of ESPs is of particular note. *Fly ash* is a generic term used to describe the particulate matter carried in the effluent gases from furnaces burning fossil fuels. ESPs often are used to collect fly ash. The strongest force holding fly ash to the collection plate is electrostatic and is caused by the flow of current through the fly ash. The fly ash acts like a resistor and, hence, resists the flow of current. This resistance to current flow is called the resistivity of the fly ash. It is measured in units of ohm · cm. If the resistivity is too low (less than 10^4 ohm · cm), not enough charge will be retained to produce a strong force and the particles will not "stick" to the plate. Conversely, and often more importantly, if the resistivity is too high (greater than 10^{10} ohm · cm), there is an insulating effect. The layer of fly ash breaks down locally and a local discharge of current (*back corona*) from the normally passive collection electrode occurs. This discharge lowers the sparkover voltage and produces positive ions that decrease particle charging and, hence, collection efficiency.

The presence of SO_2 in the gas stream reduces the resistivity of the fly ash. This makes particle collection relatively easy. However, the mandate to reduce SO_2 emissions has frequently been satisfied by switching to low sulfur coal. The result has been increased particulate emissions. This problem can be resolved by adding conditioners such as SO_3 or NH_3 to reduce the resistivity or by building larger precipitators.

Electrostatic precipitators have been used to control air pollution from electric power plants, Portland cement kilns, blast furnace gas, kilns and roasters for metallurgical processes, and mist from acid production facilities.

Control Technologies for Mercury

During combustion, the mercury in coal is volatilized and converted to Hg^0 vapor. As the flue gas cools, a series of complex reactions convert Hg^0 to Hg^{2+} and particulate Hg compounds (Hg_p). The presence of chlorine favors the formation of mercuric chloride. In general, the majority of gaseous mercury in bituminous coal-fired boilers is Hg^{2+}. The majority of gaseous mercury in subbituminous- and lignite-fired boilers is Hg^0.

Existing boiler control equipment achieves some ancillary removal mercury compounds. Hg_p is collected in particulate control equipment. Soluble Hg^{2+} compounds are collected in FGD systems. Particulate control equipment has achieved a range of emission reductions from 0 to 90 percent. Of these units, fabric filters obtained the highest levels of control. Dry scrubbers achieve average total mercury (particulate plus compounds) ranging from 0 to 98 percent. Wet FGD scrubber efficiencies were similar. Higher efficiencies were achieved at boilers using bituminous coal than at those using subbituminous and lignite coal. EPA estimates that existing controls remove about 36 percent of the 75 Mg of mercury input with coal in U.S. coal-fired boilers (Srivastava et al., 2005).

There are two broad approaches being developed to control mercury emissions: powdered activated carbon (PAC) injection and enhancement of existing control devices. The leading candidates for top efficiency (90 percent) are PAC with pulse-jet fabric filters and FGD (wet or dry) with fabric filters (U.S. EPA, 2003b).

7-11 AIR POLLUTION CONTROL OF MOBILE SOURCES

Engine Fundamentals

Before we examine some cures for the pollution from the common gasoline auto engine, it may be useful to compare the three familiar types of engines: the gasoline engine, the diesel engine, and the jet engine.

The Gasoline Engine. Each of the four strokes of the engine is diagrammed in Figure 7-41. In the typical automobile engine with no air pollution controls, a mixture of fuel and air is fed into a cylinder and is compressed and ignited by a spark from the spark plug. The explosive energy of the burning mixture moves the pistons. The pistons' motion is transmitted to the crankshaft that drives the car. The burnt, spent mixture passes out of the engine and out through the tail pipe.

One kg of gasoline can burn completely when mixed with about 15 kg of air. For maximum power, however, the proportion of air to fuel must be less. Most driving takes place at less than the 15-to-1 *air-to-fuel* ratio. Combustion is incomplete, and substantial amounts of material other than carbon dioxide and water are discharged through the tail pipe. One result of having an inadequate supply of air is the emission of carbon monoxide instead of carbon dioxide. Other by-products are unburned gasoline and hydrocarbons.

Because of the high temperatures and pressures that exist in the cylinder, copious amounts of NO_x are formed (see Equation 7-12).

The Diesel Engine. As shown in Figure 7-42, the diesel engine differs from the four-stroke engine in two respects.

First, the air supply is unthrottled; that is, its flow into the engine is unrestricted. Thus, a diesel normally operates at a higher air-to-fuel ratio than does a gasoline engine.

Second, there is no spark ignition system. The air is heated by compression. That is, the air in the engine cylinder is squeezed until it exerts a pressure high enough to raise the air temperature to about 540°C, which is enough to ignite the fuel oil as it is injected into the cylinder.

A well-designed, well-maintained, and properly adjusted diesel engine will emit less CO and hydrocarbons than the four-stroke engine because of the diesel's high air-to-fuel ratio. However, the higher operating temperatures lead to substantially higher NO_x emissions. In addition, when the engine is overloaded during acceleration from a

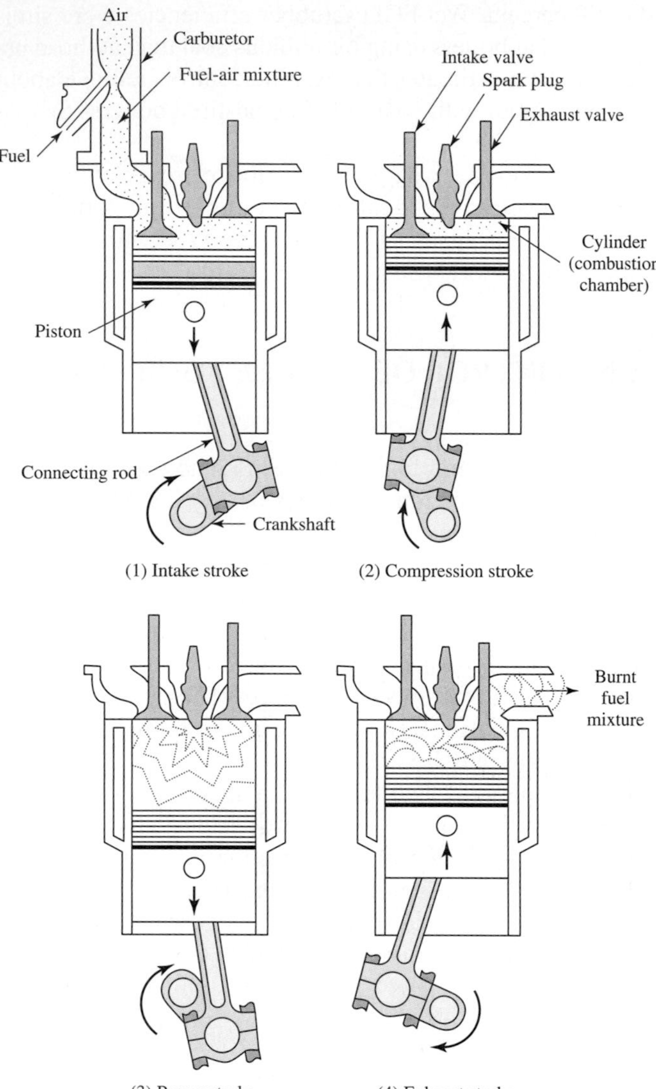

FIGURE 7-41

Combustion in an automobile engine. On the intake stroke (*1*), the piston moves down and a mixture of fuel and air is drawn into the cylinder past the open intake valve. With the compression stroke (*2*), the intake valve closes and the piston moves and compresses the air-fuel mixture. On the power stroke (*3*), a spark from the spark plug ignites the heated, compressed mixture, which begins to burn, expands, and pushes the piston down. For the exhaust stroke (*4*), the exhaust valve opens, the spent, burned mixture exits with its pollutants, and the piston returns to the top of the cylinder. (*Source:* NTRDA, 1969.)

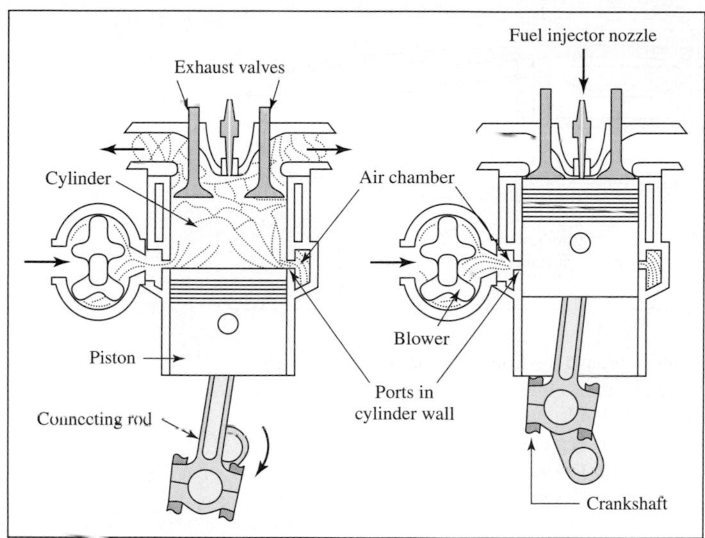

FIGURE 7-42
Combustion in a two-stroke diesel engine. With the piston at the bottom of the cylinder (shown left), and exhaust valves and ports open, fresh air is forced into the cylinder by the blower, and the used air-fuel mixture—along with any polluting byproducts—left from the previous stroke is forced out. On the second stroke (shown right), the exhaust valves close, the piston rises—shutting off the ports—and compresses the air. When the piston reaches a position near the top of the cylinder, fuel is injected into the now highly compressed, heated air. This heated air ignites the fuel without a spark, and the resulting combustion forces the piston down to its first position. (*Source:* NTRDA, 1969.)

stop, CO, VOCs, odors and particulate matter (smoke) may be emitted in large quantities (Cooper and Alley, 2002).

The Jet Engine. Large commercial aircraft that utilize the thrust of compressed gases for propulsion may contribute significant amounts of particulates and NO_x to urban atmospheres. Their largest emission rate is on takeoff and climb-out. However, on an annual basis, emissions from jet engines are small relative to those from highway vehicles.

As shown in Figure 7-43, air drawn into the front of the engine is compressed and then heated by burning fuel. The expanding gas passes through turbine blades, which drive the compressor. The gas then exits the engine through an exhaust nozzle.

Effect of Design and Operating Variables on Emissions. The list of variables that affect internal combustion (automobile) emissions includes the following (Patterson and Henein, 1972):

1. Air-to-fuel ratio
2. Load or power level
3. Speed
4. Spark timing

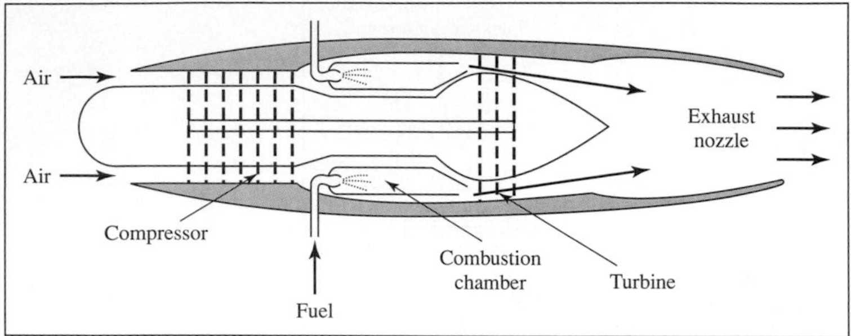

FIGURE 7-43
Combustion in a jet engine. Air enters through the front and goes to a compressor, where it is increasingly compressed and forced into combustion chambers that are arranged in circles around the engine. Fuel is sprayed into the front end of the combustion chamber in a steady stream so that it ignites and burns continuously. The burning air-fuel mixture expands and pushes toward the rear. (On the way, it hits turbine wheel blades and forces them to rotate. This rotation drives the compressor.) As the expanded mixture moves toward the tailpipe, the areaway narrows and the stream of burning air-fuel mixture is compressed into the exceedingly strong jet stream that shoots out of the rear of the plane. (*Source:* NTRDA, 1969.)

5. Exhaust back pressure
6. Valve overlap
7. Intake manifold pressure
8. Combustion chamber deposit buildup
9. Surface temperature
10. Surface-to-volume ratio
11. Combustion chamber design
12. Stroke-to-bore ratio
13. Displacement per cylinder
14. Compression ratio

A discussion of all of these items is beyond the scope of an introductory text such as this. Therefore, we shall restrict ourselves to a few of the variables that serve to illustrate the kinds of problems encountered in trying to design pollution out of an internal combustion engine.

The *air-to-fuel* ratio (A/F) is fairly easy to regulate. As we noted previously, it has a direct effect on all three emissions. As shown in Figure 7-44*a*, the A/F of 14.6 is the *stoichiometric* mixture for complete combustion.* At lower ratios, both CO and HC emissions increase. At higher ratios, to about 15.5, NO_x emissions increase. At very lean mixtures (high A/F ratios), the NO_x emission begins to decrease.

*Note that stoichiometric (stoi-chio-met-ric) means "combined in exactly the proper proportions according to their molecular weight."

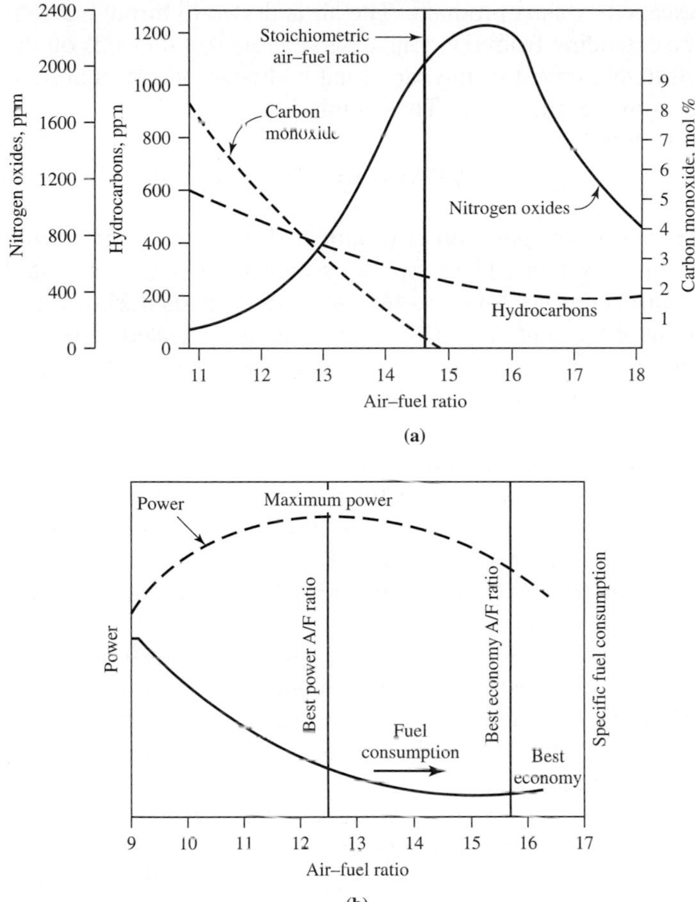

FIGURE 7-44
Effect of air-to-fuel ratio (*a*) on emissions, and (*b*) on power and economy. (*Source:* Seinfeld, 1975.)

Combustion chamber design has changed dramatically over the last three decades. For example, the *extra-lean-burn engine* cylinder has an A/F ratio that is richer at the spark plug and leaner elsewhere. In this engine the A/F ratio is as high as 25. The result is a very low emission of CO and VOCs. Coupled with a higher drive ratio, this engine also gives improved gasoline efficiency. As with all lean-burn engine modifications, this one has the drawback of higher NO formation (Cooper and Alley, 2002).

Retarding the timing of the spark relative to the stroke of the piston decreases the hydrocarbon emissions by reducing the amount of unburned fuel. NO_x emissions also decrease with increased retarding. Little or no change occurs in CO emissions.

Control of Automobile Emissions

Blowby. The flow of air past the moving vehicle is directed through the crankcase in order to rid it of any gas-air mixture that has blown past the pistons, any evaporated lubricating oil, and any escaped exhaust products. The air is drawn in through a vent and emitted through a tube extending from the crankcase at a rate that depends on the speed of the car. About 20 to 40 percent of the car's total hydrocarbon emissions are sent into the atmosphere from the crankcase. These emissions are called *crankcase blowby.* All vehicles manufactured after 1963 are required to have a positive crankcase ventilation (PCV) valve to eliminate blowby emissions.

Fuel Tank Evaporation Losses. Evaporation of volatile hydrocarbons from the fuel tank is controlled by one of two systems. The simplest system is to place an activated charcoal adsorber in the tank vent line. Thus, as the gasoline expands during warm weather and forces vapor out of the vent, the HC is trapped on the activated carbon.

An alternative system is to vent the tank to the crankcase. With this method, it is more difficult to achieve 100 percent control than with the activated charcoal system.

Engine Exhaust. Because engine modifications alone are not sufficient to meet stringent emission standards, an external catalytic reactor (commonly referred to as the *catalytic converter*) is placed on the exhaust system. The function of the catalytic converter is to promote reactions that convert NO_x to N_2, CO to CO_2, and hydrocarbons to CO_2 and H_2O. A *three-way catalyst* (TWC) that simultaneously oxidizes the hydrocarbons and CO and reduces the NO_x is employed. The catalyst is a precious metal (for example platinum/rhodium) on an alumina support structure. The catalyst must operate in a narrow band of A/F ratios that is centered about the stoichiometric point (Figure 7-44). In addition, the gases entering the catalyst bed must have a specific composition and the catalyst temperature must be carefully controlled. A sophisticated computerized electronic control system maintains the correct A/F and temperature (Figure 7-45).

The major problems with the catalysts are their susceptibility to "poisoning" by lead, phosphorus, and sulfur, and their poor wear characteristics under thermal cycling. The poisoning problem is solved by removing the lead, phosphorus, and sulfur from the fuel.

Another approach being implemented is fuel modification. The use of lead in fuels was completely phased out by January 1996. In addition, diesel fuel refining is being changed so that it will contain less sulfur and emit 20 percent less VOC's. Lowering the gasoline vapor pressure (called the *Reid vapor pressure*) reduces hydrocarbon emissions. *Oxyfuel* is yet another alternative. Oxyfuel is one with more oxygen

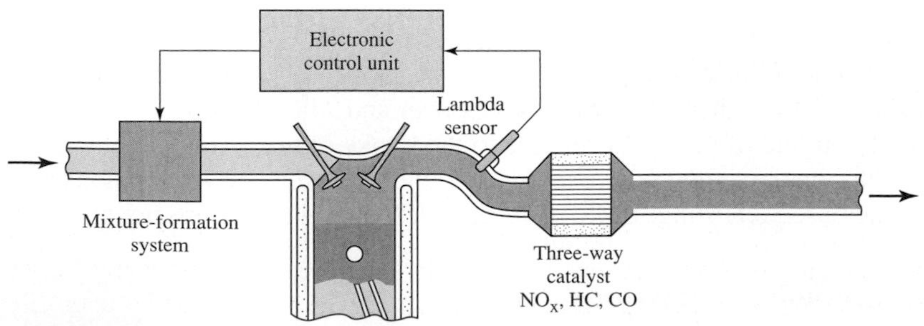

FIGURE 7-45
Single-bed, three-way catalyst with electronic control systems. (*Source:* Bosch, 1988.)

to allow the fuel to burn more efficiently. Other alternatives include alcohols, liquified petroleum gas, and natural gas.

Inspection/Maintenance (I/M) Programs. The devices installed by automobile manufacturers are extremely successful in minimizing the pollution from the exhaust and from evaporating fuel. However, as with other aspects of running an automobile, these devices wear out and fail. Since their failure does not inhibit the operation of the automobile, they are not likely to be repaired by the owner. In those areas that have exceeded the NAAQS (nonattainment areas), inspection/maintenance programs have been implemented to ensure that the control devices are in good working order. These programs require periodic checks of the exhaust and, in some instances, the evaporative controls. If the vehicle fails the inspection, the owner is required to provide the required maintenance and have the vehicle reinspected. Failure to pass the inspection may be cause to deny the issuance of license plates or tags.

7-12 WASTE MINIMIZATION FOR SUSTAINABILITY

The best and first step in any air pollution control strategy should be to minimize the production of pollutants in the first place. Since a large proportion of air pollutants results from the combustion of fossil fuels, an obvious approach to waste minimization is to conserve energy. Modern technology has yielded more efficient furnaces that improve fuel use, but simple measures such as turning off the lights in unoccupied rooms, turning down the heat at night and, in factories, during weekends and holidays, can have a dramatic impact. Because of the interrelationship between energy consumption and water supply, water conservation also reduces air pollution. In a similar manner, building smaller, lighter automobiles reduces air pollution because less fuel is burned to propel them. The introduction of the hybrid automobile is a major step in improving fuel economy. Alternatives such as mass transit, walking, and bicycles can contribute significantly to reduced fuel consumption. Alternative sources of energy such as solar, wind, and nuclear also reduce air pollution emissions. (Nuclear power, of course, has a series of pollution problems that may outweigh the benefits of reduced air pollution.)

The chlorofluorocarbon destruction of the ozone layer can only be resolved by waste minimization. Preventing the escape of CFCs from refrigeration systems, the use of alternative propellants for spray cans, and similar measures are the only ones that will be successful, since control devices make no sense. Waste minimization is, in fact, the method of control specified by the Montreal Protocol (see Section 7-6). In a similar fashion, the production of ozone in the lower atmosphere can only be reduced by minimizing the release of precursor hydrocarbons and the production of NO_x. Reduced use of solvents and the substitution of water-based paints for solvent-based paints are examples of methods to reduce hydrocarbon release.

The fundamental method for minimizing anthropogenic inputs to global warming is to minimize the anthropogenic emissions of GHGs. As noted earlier, more efficient use of energy is a major method of reducing GHGs. In and of itself, efficient energy use makes economic sense. The fact that efficient energy use will contribute immensely to the reduction of air pollution and GHGs while enhancing sustainability makes it an ideal candidate for waste minimization.

7-13 CHAPTER REVIEW

When you have completed studying this chapter, you should be able to do the following without the aid of your textbook or notes:

1. List the six criteria air pollutants for which the U.S. Environmental Protection Agency has designated National Ambient Air Quality Standards (NAAQS).
2. List and define three units of measure used to report air pollution data (that is, ppm, $\mu g/m^2$, and μm).
3. Explain the difference between ppm in air pollution and ppm in water pollution.
4. Explain the effect of temperature and pressure on readings made in ppm.
5. Explain the influence of moisture, temperature, and sunlight on the severity of air pollution effects on materials.
6. Differentiate between acute and chronic health effects from air pollution.
7. State which particle sizes are more important with respect to alveolar deposition and explain why.
8. Explain why it is difficult to define a causal relationship between air pollution and health effects.
9. List three potential chronic health effects of air pollution.
10. List four common features of air pollution episodes and identify the locations of three "killer" episodes.
11. Discuss the natural and anthropogenic origin of the six criteria air pollutants and identify the likely mechanisms for their removal from the atmosphere.
12. Identify one indoor air pollution source for each of the following pollutants: CH_2O, CO, NO_x, Rn, respirable particulates, and SO_x.
13. Define the term "acid rain" and explain how it occurs.
14. Discuss the photochemistry of ozone in the upper atmosphere using the pertinent chemical reactions. Discuss the hypothesized effect of chlorofluorocarbons on these reactions.
15. Explain the term "greenhouse effect", its hypothesized cause, and why it is being debated, pro and con.
16. Determine the stability (ability to dissipate pollutants) of the atmosphere from vertical temperature readings.
17. Explain why valleys are more susceptible to inversions than is flat terrain.
18. Explain why lake breezes and land breezes occur.
19. Explain how a lake breeze adversely affects the dispersion of pollutants.

20. State the theoretical principle on which each of the following air pollution control devices operates: (*a*) absorption column (either a packed tower or plate tower), (*b*) adsorption column, (*c*) either afterburner or catalytic combustor, (*d*) cyclone, (*e*) baghouse, (*f*) venturi scrubber, and (*g*) electrostatic precipitator.
21. Select the correct air pollution control device for a given pollutant and source.
22. Discuss the pros and cons of FGD and the problem of fly ash resistivity.
23. Explain the difference between prevention and post-combustion techniques for reduction of nitrogen oxide emissions and give one example of each.
24. Graph the relationship between air-to-fuel ratio and emission of CO, HC, and NO_x from automobiles.
25. Explain how evaporative emissions are commonly controlled.
26. Explain how exhaust emissions are commonly controlled and the role of computerized control systems in making them work.

With the aid of this text, you should be able to do the following:

1. Solve gas law problems.
2. Convert parts per million (ppm) to micrograms per cubic meter ($\mu g/m^3$) and vice versa.
3. Calculate the amount of SO_2 that will be released from burning coal or fuel oil with a given sulfur content in percent.
4. Calculate the ground level concentration of air pollutants released from a stationary elevated source or the emission rate (QE) for a given ground level concentration.
5. Use the air pollution control equations to analyze the performance and modify the design of absorption and adsorption control devices, cyclones, scrubbers, and ESPs.

7-14 PROBLEMS

7-1. What is the density of oxygen at a temperature of 273.0 K and a pressure of 98.0 kPa?

Answer: 1.382 kg/m^3

7-2. Determine the density of carbon monoxide gas at a pressure of 102.0 kPa and a temperature of 298.0 K.

7-3. Calculate the density of methane at a temperature of 273.0 K and at a pressure of 101.325 kPa?

7-4. Show that 1 mole of any ideal gas will occupy 22.414 L at standard temperature and pressure (STP). (STP is 273.16 K and 101.325 kPa.)

7-5. What volume would 1 mole of an ideal gas occupy at 20°C and 101.325 kPa?

7-6. A sample of air contains 8.583 moles/m^3 of oxygen and 15.93 moles/m^3 of nitrogen at STP. Determine the partial pressures of oxygen and nitrogen in 1.0 m^3 of the air.

Answer: 19.45 kPa; 36.18 kPa

7-7. A 1.000 m^3 volume tank contains a gas mixture of 8.32 moles of oxygen, 16.40 moles of nitrogen, and 16.15 moles of carbon dioxide. What is the partial pressure of each component in the gas mixture at 25.0°C?

7-8. A 1.000 m^3 volume tank contains a gas mixture of oxygen, nitrogen, and carbon dioxide. How many moles are there of each of these components of the gas mixture at 25.0°C if the partial pressures of each gas are as shown below?

$P_{O_2} = 45.39$ kPa

$P_{N_2} = 40.63$ kPA

$P_{CO_2} = 15.24$ kPa

7-9. Calculate the volume occupied by 5.2 kg of carbon dioxide at 152.0 kPa and 315.0 K.

Answer: 2,036 L

7-10. Determine the mass of oxygen contained in a 5.0 m^3 volume under a pressure of 568.0 Pa and at a temperature of 263.0 K.

7-11. Calculate the volume occupied by 235 μg of O_3 at STP. If this volume is contained in 1.00 m^3 of air, what is the volumetric ratio (that is volume of O_3 per volume of air)?

7-12. A gas mixture at 0°C and 108.26 kPa contains 250 mg/L of H_2S gas. What is the partial pressure exerted by this gas?

Answer: 16.7 kPa/L

7-13. A 28-L volume of gas at 300.0 K contains 11 g of methane, 1.5 g of nitrogen, and 16 g of carbon dioxide. Determine the partial pressure exerted by each gas.

7-14. Given the gas mixture of Problem 7-13, how many moles of each gas are present in the 28-L volume?

Answer: 0.688 moles of CH_4; 0.054 moles of N_2; 0.364 moles of CO_2

7-15. The partial pressures of the gases in a 22.414 L volume of air at STP are: oxygen, 21.224 kPa; nitrogen, 79.119 kPa; argon 0.946 kPa; and carbon dioxide, 0.036 kPa. Determine the gram-molecular weight of air.

7-16. Using the data in Problem 7–15, determine the gram–molecular weight of air at 500°C and 101.325 kPa.

7-17. Convert 80 $\mu g/m^3$ of SO_2 to ppm at 25°C and 101.325 kPa pressure.

Answer: 0.031 ppm

7-18. Convert 0.55 ppm of NO_2 to $\mu g/m^3$ at -17.7°C and 100.0 kPa pressure.

7-19. Convert 370 ppm of CO_2 to $\mu g/m^3$ at 20°C and 101.325 kPa.

7-20. Given the following temperature profiles, determine whether the atmosphere is unstable, neutral, or stable. Show all work and explain choices.

a.

Z, m	*T*, °C
2	−3.05
318	−6.21

b.

Z, m	T, °C
10	6.00
202	3.09

c.

Z, m	T, °C
18	14.03
286	16.71

Answers: (a) neutral; (b) unstable; (c) stable (inversion)

7-21. Determine the atmospheric stability for each of the following temperature profiles. Show all work and explain choices.

a.

Z, m	*T*, °C
1.5	−4.49
339	0.10

b.

Z, m	*T*, °C
12	28.05
279	19.67

c.

Z, m	*T*, °C
8	19.55
339	18.93

7-22. Determine the atmospheric stability for each of the following temperature profiles. Show all work and explain choices.

a.

Z, m	*T*, °C
2.00	5.00
50.00	4.52

b.

Z, m	*T*, °C
2.00	5.00
50.00	5.00

c.

Z, m	*T*, °C
2.00	−21.01
50.00	−25.17

7-23. Given the following observations, use the key to stability categories (Table 7-8) to determine the stability.

a. Clear winter morning at 9:00 A.M.; wind speed of 5.5 m/s

b. Overcast summer afternoon at 1:30 P.M.; wind speed of 2.8 m/s

c. Clear winter night at 2:00 A.M.; wind speed of 2.8 m/s
d. Summer morning at 11:30 A.M.; wind speed of 4.1 m/s

Answers: (a) D; (b) D; (c) F; (d) B

7-24. Determine the atmospheric stability category of the following observations.
a. Clear summer afternoon at 1:00 P.M.; wind speed of 1.6 m/s
b. Overcast summer night at 1:30 A.M.; wind speed of 2.1 m/s
c. Clear winter morning at 9:30 A.M.; wind speed of 6.6 m/s
d. Thinly overcast winter night at 8:00 P.M.; wind speed of 2.4 m/s

7-25. Determine the atmospheric stability category of the following observations.
a. Clear summer afternoon at 1:00 P.M.; wind speed of 5.6 m/s
b. Clear summer night at 1:30 A.M.; wind speed of 2.1 m/s
c. Overcast winter afternoon at 2:30 P.M.; wind speed of 6.6 m/s
d. Summer afternoon at 1:00 P.M. with broken low clouds; wind speed of 5.2 m/s

7-26. A power plant in a college town is burning coal on a cold, clear winter morning at 8:00 A.M. with a wind speed of 2.6 m/s and an inversion layer with its base at a height of 697 m. The effective stack height is 30 m. Calculate the distance downwind x_L at which the plume released will reach the inversion layer and begin to mix downward.

Answer: 5.8 km

7-27. A factory releases a plume into the atmosphere on an overcast summer afternoon. At what distance downwind will the plume begin mixing downward if an inversion layer exists at a base height of 414 m and the wind speed is 1.8 m/s? The effective stack height is 45 m.

7-28. At what distance downwind will the plume from a stack begin mixing downward if an inversion layer exists at a base height of 265 m and the wind speed is 4.0 m/s on an overcast summer afternoon? The effective stack height is 85 m.

7-29. Given the same power plant and conditions that were found in Example 7-4, determine the concentration of SO_2 at a point 4 km downwind and 0.2 km perpendicular to the plume centerline ($y = 0.2$ km) if there is an inversion with a base height of 328 m.

Answer: 1.16×10^{-3} g/m^3

7-30. On a clear summer afternoon with a wind speed of 3.20 m/s, the particulate concentration was found to be 1,520 μg/m^3 at a point 2 km downwind and 0.5 km perpendicular to the plume centerline from a coal-fired power plant. Given the following parameters and conditions, determine the particulate emission rate of the power plant:

Stack parameters:
Height = 75.0 m
Diameter = 1.50 m
Exit velocity = 12.0 m/s
Temperature – 322°C

Atmospheric conditions:
Pressure = 100.0 kPa
Temperature = 28.0°C

7-31. Calculate the downwind concentration at 30 km ($y = 0$) in g/m^3 resulting from an emission of 1,976 g/s of SO_2 into a 2.5 m/s wind at 1:00 A.M. on a clear winter night. Assume an effective stack height of 85 m and an inversion layer at 185 m. Identify the stability class and show all work.

7-32. Using a computer spreadsheet program you have written determine the maximum concentration of SO_2 (in ppm) and the distance downwind of a power plant stack that the maximum concentration of SO_2 occurs for the following conditions:

Power plant data:

Coal specifications
Bituminous (Saginaw No. 1, Belmont, OH)
Sulfur: 2.80 percent
Ash: 9.8 percent

Burning rate: 28.82 megagrams per hour
Stack conditions:
Height: 40.0 m
Inside diameter: 1.8 m
Exit velocity: 10.5 m/s
Exit temperature: 297°C

Meteorological conditions to be investigated:

a. Wind speed: 3.8 m/s
b. Inversion base: 170.0 m above ground surface
c. Ambient temperature: −11°C
d. Ambient pressure: 103.285 kPa
e. Thinly overcast winter night (midnight to 4:00 A.M.)

Your solution should include the microcomputer spreadsheet and a graph of the ground level concentration of SO_2 in g/m^3 versus distance from the stack. Provide a table of values for distances from 0.1 km to 100 km in the following steps:

From 0.1 to 1.0 km in 0.1-km steps
From 1.0 to 10.0 km in 1.0-km steps
From 10.0 to 100.0 km in 10.0-km steps

HINTS: Some of the initial and final values may be very small and may be ignored in the plot. Plot enough values to identify the maximum point and to show the effect of the elevated inversion. A log–log graph will be required to include enough points on a reasonable scale. You will have to determine the distance downwind where the short form of the dispersion equation is to be used (i.e., 2 x_L) and switch equations at that point in your spreadsheet calculations.

7-33. Anna Lytical purchased a mobile home last year. She has been suffering severe allergic symptoms and has detected a strong odor of formaldehyde. An analysis of the mobile home air has revealed that the formaldehyde concentration is 0.28 ppm. Anna's friend, Sybil Injuneer, has measured the air flow in the ventilation system and found that the ventilation rate is 0.56 air changes per hour (ach). She has recommended increasing the ventilation rate to reduce the formaldehyde concentration below the threshold odor level of 0.05 ppm (Lee et al., 2001). Assuming the moile home volume is 148 m^3 and that the outdoor air concentration is 0.0 ppm, estimate the ach required to achieve the threshold odor level.

7-34. A manufacturer of carbon monoxide detector/alarms has asked you to perform an analysis of the time to achieve various levels of CO in a standard house so it can set the detection level of the monitor. The standard house has a volume of 540 m^3 and a ventilation rate of 100 m^3/h. The manufacturer uses an assumption of a furnace flue malfunction that results in an emission of 3.0 mg/s of CO into the house. They use the following World Health Organization guidelines to prevent excess levels of COHb (WHO, 1987): 10 mg/m^3 for 8 h, 30 mg/m^3 for 1 h, 60 mg/m^3 for 30 min, and 100 mg/m^3 for 15 min. Using a computer spreadsheet program you have written, estimate the time it will take to achieve each of these concentrations. What safety factors (time to achieve a given level divided by the allowable time) will be achieved if the alarm is set at each of these levels. Assume that the outdoor and indoor air concentration equals the 8-hour NAAQS for CO and the starting concentration is 1.0 mg/m^3.

7-35. Determine the slope of the equilibrium curve defined by Henry's law for HCl gas at 20°C from the following data:

P_{HCl}, kPa	kg HCl per 100 kg H_2O
0.6533	38.9
0.0871	31.6
0.02733	25.0

Answer: Henry's law is not followed well. By linear regression, $m = 0.120$.

7-36. Find the slope of the equilibrium curve defined by Henry's law for SO_2 gas at 30°C from the following data:

P_{SO_2}, kPa	Kg SO_2 per 100 kg H_2O
10.532	1.0
6.933	0.7
4.800	0.5
2.626	0.3

7-37. Determine the height of a packed tower that is to reduce the concentration in air of H_2S from 0.100 kg/m^3 to 0.005 kg/m^3 given the following data:

Incoming liquid is water free of H_2S
Operating temperature = 25.0°C
Operating pressure = 101.325 kPa
Henry's law constant, m = 5.522 mole fraction units
$H_g = 0.444$ m
$H_l = 0.325$ m
Liquid flow rate = 20.0 kg/s
Gas flow rate = 5.0 kg/s

Answer: 7 m

7-38. Neighbors are complaining about odor from the facility that produces the H_2S in Problem 7-37. Using the data in Problem 7-37, determine the height of a packed tower that is to reduce the concentration in air of H_2S from 0.100 kg/m^3 to the threshold odor concentration of 0.0002 mg/L of air. Is the tower height practical?

7-39. The concentration of chlorine gas in air must be reduced from 10.0 mg/m^3 to 2.95 mg/m^3. Determine the height of the packed tower that should be used if the following parameters apply:

Incoming liquid is water free of Cl_2
Operating temperature = 20.0°C
Operating pressure 101.325 kPa
$H_g = 0.662$ m
$H_l = 0.285$ m
Henry's law constant, m = 6.820 mole fraction units
Liquid flow rate = 15.0 kg/s
Gas flow rate = 3.00 kg/s

7-40. Determine the Langmuir constants a and b for the following isotherm data for adsorption of H_2S on a molecular sieve. Use a computer spreadsheet program to plot the data.

P_{H_2S}, kPa	W, g H_2S/g sieve
0.840	0.082
1.667	0.1065
2.666	0.118
3.333	0.122

Answer: $a = 20$; $b = 135$

7-41. Determine the Langmuir constants a and b for the adsorption of benzene on activated carbon, given the following isotherm data.

$P_{C_6H_6}$ kPa	W, kg C_6H_6/kg carbon
0.027	0.129
0.067	0.170
0.133	0.204
0.266	0.240

7-42. When the isotherm data are nonlinear, the Freundlich model may be used to fit the data:

$$q_e = KP^n$$

where q_e = mg of pollutant/g of adsorbent
K, n = curve fitting constants
P = pollutant partial pressure

The following adsoption data were obtained using beaded activated carbon to remove tetrachloroethylene (Noll et al., 1992).

q_e, mg/g of carbon	C_e, ppm
520	70
550	170
640	700
690	1,750
740	4,000
780	7,000

Use a computer spreadsheet to plot the data and fit a curve. Use the computer spreadsheet "trendline" to determine the Freundlich curve-fitting constants.

7-43. Determine the breakthrough time for toluene on an adsorption bed of activated carbon that is 0.75 m thick and 5.0 m^2 in cross section. The operating parameters for the bed are as follows:

Gas flow rate = 1.185 kg/s

Gas temperature = 25°C

Bed density = 450 kg/m^3

Inlet pollutant concentration = 0.00350 kg/m^3

Langmuir parameters: $a = 465$; $b = 3{,}000$

Width of adsorption zone = 0.045 m

Answer: 17.8 h

7-44. What thickness of molecular sieve adsorption bed is required for the following system to ensure an SO_2 breakthrough time (t_B) of not less than 8.00 h?

Gas flow rate = 2.36 m^3/s of air

Gas temperature = 25.0°C

Gas pressure = 105.0 kPa

Bed density as packed = 390 kg/m^3

Inlet pollutant concentration = 3,000 ppm

Langmuir parameters: $a = 400$; $b = 900$

Width of adsorption zone = 0.028 m

Bed diameter = 3.00 m

7-45. Determine the cross-sectional area and depth of catalyst to reduce an inlet concentration of toluene from 1.87 g/m^3 to 0.00187 g/m^3. The exhaust gas flow rate entering the control equipment is 16.33 m^3/s. Combustion air is supplied at a rate of 1.80 m^3/s. Both the exhaust gas and the combustion air flow rates are at 20°C. The manufacturer's specifications require that the catalyst be operated at 510°C and that the bed gas velocity be limited to 7.5 m/s. Assume that the toluene combustion reaction follows first-order kinetics and that the rate constant at 510°C is 120 s^{-1}.

Answers: area = 6.5 m^2, depth = 0.43 m

7-46. Hexane (C_6H_{14}) is emitted from a baking oven at a rate of 454 g/min. The exhaust gas flow rate is 7.1 m^3/s at a temperature of 315°C. Determine the cross-sectional area and depth of catalyst to produce an exhaust concentration of 100 ppm at STP. Combustion air is supplied at 0.70 m^3/s at 20°C. The manufacturer's specifications require that the catalyst be operated at 550°C and that the bed gas velocity be limited to 9.5 m/s. Assume that the hexane combustion reaction follows first order kinetics and that the rate constant at 550°C is 55 s^{-1}.

7-47. Calculate the efficiency of removal of a 2.50-μm-diameter particle having a density of 1,250 kg/m^3 for a cyclone barrel diameter of 1.0 m. The gas flow rate is 2.80 m^3/s and the gas temperature is 25°C.

Answer: $\eta \approx 14\%$

7-48. Because the efficiency of the large-barrel-diameter cyclone in Problem 7-47 is low for fine particles, a multiclone consisting of 10 barrels has been proposed as an alternative. Each barrel is to be 0.10 m in diameter. Calculate the efficiency of the multiclone using the particle and gas data given Problem 7-47.

7-49. Determine the efficiency of the cyclone in Example 7-10 for particles having a density of 1,000 kg/m^3 and radii of 1.00, 5.00, 10.00, and 25.00 μm. Using a computer spreadsheet, plot the efficiency as a function of particle diameter for the specified cyclone and gas conditions.

7-50. A consultant has proposed that a pulse-jet baghouse with bags that are 15 cm in diameter and 5 m in length be used instead of the mechanical shaker bag system proposed in Example 7-11. Estimate the net number of bags required if the manufacturer's recommended air-to-cloth ratio for aggregate plants is 0.050 m/s.

7-51. A green coffee bean screening and handling operation emits 0.75 g/m^3 of fine particulate matter. A reverse-air baghouse is being proposed for controlling the particulate emissions. The gas handling system has an exhaust flow rate of 3.3 m^3/s. A manufacturer has supplied the following data:

Bag diameter = 20 cm

Bag length = 12 m

Air-to-cloth ratio = 0.010

Bag cleaning = 0.5

Estimate the number of bags required and the mass of particulate matter collected each day if the efficiency is 99 percent. Assume 24-hour operation.

7-52. Calculate the overall mass efficiency (η) of the venturi described in Example 7-12 for the following particle size distribution

Average diameter, μm	% of total mass
2.5	25
7.5	20
15.0	15
25.0	15
35.0	10
50.0	15

Answer: overall η = 87.73 or 88 percent

7-53. Calculate the venturi throat area required to achieve 99.0 percent removal of a 1.25-μm-radius particle having a density of 1,400 kg/m^3 for the following gas stream and venturi characteristics.

Gas flow rate = $10.0\ m^3/s$
Gas temperature = 180°C
Droplet diameter = $100\ \mu m$
Liquid flow rate = $0.100\ m^3/s$
Coefficient $\kappa = 200$

7-54. Using a spreadsheet progrm you have written, calculate the overall mass efficiency (η) of the venturi described in Problem 7-53 for a throat velocity of 26.3 m/s and the following fly ash particle size distribution (after Noll, 1999).

Average diameter, μm	% of total mass
0.05	0.01
0.3	0.21
0.8	0.78
3.0	13.0
8.0	16.0
13.0	12.0
18.0	8.0
80.0	50.0

7-55. Determine the collection efficiency of an electrostatic precipitator (ESP) tube that is 0.300 m in diameter and 2.00 m in length for particles that are 1.00 μm in diameter. The flow rate is $0.150\ m^3/s$, the collection field intensity is 100,000 V/m, the particle charge is 0.300 femtocoulombs (fC), and the gas temperature is 25°C.

Answer: 92.4 percent

7-56. Rework Problem 7-55 with the gas flow rate reduced to $0.075\ m^3/s$.

7-15 DISCUSSION QUESTIONS

7-1. A gas sample is collected in a special gas sampling bag that does not react with the pollutants collected but is free to expand and contract. When the sample was collected, the atmospheric pressure was 103.0 kPa. At the time the sample was analyzed the atmospheric pressure was 100.0 kPa. The bag was found to contain 0.020 ppm of SO_2. Would the original concentration of SO_2 be more, less, or the same? Explain.

7-2. Under which of the following conditions would you expect the strongest inversion (largest positive lapse rate) to form?

a. Foggy day in the fall after the leaves have fallen
b. Clear winter night with fresh snow on the ground
c. Clear summer morning just before sunrise

Explain why.

7-3. Cement dust is characterized by very fine particulates. The exhaust gas temperatures from a cement kiln are very hot. Which of the following air pollution control devices would appear to be appropriate? Explain the reasoning for your selection.
a. Venturi scrubber
b. Baghouse
c. Electrostatic precipitator

7-4. Photochemical oxidants are not directly attributable to either people or natural sources. Why, then, are automobiles singled out as the major cause of the formation of ozone?

7-5. Explain why the $PM_{2.5}$ standard is more appropriate than a "Total Suspended Particulate" for protection of human health.

7-16 REFERENCES

AEC (1968) *Meteorology and Atomic Energy,* 1968, U.S. Atomic Energy Commission, USAEC Division of Technical Information Extension, Oak Ridge, TN.

AP (2005a) Associated Press, *Lansing State Journal,* December 18, p. 14A.

AP (2005b) Associated Press, www.myrtlebeachonline.com.

ASHRAE (1981) *Ventilation for Acceptable Air Quality,* American Society of Heating, Refrigerating, and Air Conditioning Engineers, Standard 62-1981, Atlanta.

Beard, J., F. Lachetta, and L. Lilleleht (1980) *Combustion Evaluation,* U.S. Environmental Protection Agency Report No. 4500/2-80-063.

Betts, K. S. (2005) "The Changing Chemistry of Office Cubicles," *Environmental Science and Technology,* vol. 39, pp. 319A–320A.

Black & Veatch Consulting Engineers (1983) *Lime FGD Systems Data Book,* 2nd ed., EPRI Publication No. CS-2781.

Bosch, R. (1988) *Automotive Electric/Electronic Systems,* Society of Automotive Engineers, Warrendale, PA.

Buonicore, A. J., and L. Theodore (1975) *Industrial Control Equipment for Gaseous Pollutants,* Vol. I, CRC Press, Cleveland, pp. 149–1500.

Canfield, R. L., C. R. Henderson, D. A. Cory-Slechta, et al. (2003) "Intellectual Impairment in Children with Blood Lead Concentrations Below 10 μg per Deciliter," *The New England Journal of Medicine,* vol. 348, pp. 1517–1526.

Cannell, A. L., and M. L. Meadows (1985) "Effects of Recent Operating Experience on the Design of Spray Dryer FGD Systems," *Journal of the Air Pollution Control Association*, vol. 35 (7), pp. 782–789.

C&EN (2006) "Government Concentrates: Seven States Agree to Cut CO_2 Emissions," January 2, p. 16.

CDC (2005) *Third National Report on Human Exposure to Environmental Chemicals,* National Center for Environmental Health Publication No. 05-0570, Centers for Disease Control and Prevention, Atlanta, pp. 41, 74–75.

Church, P. E. (1949) "Dilution of Waste Stack Gases in the Atmosphere," *Industrial Engineering Chemistry,* vol. 41, pp. 3753–3756.

Connolly, C. H. (1972) *Air Pollution and Public Health,* Holt, Rinehart & Winston, New York, p. 7.

Cooper, C. D., and F. C. Alley (2002) *Air Pollution Control: A Design Approach,* Waveland Press, Long Grove, IL, p. 547.

Crawford, M. (1976) *Air Pollution Control Theory,* McGraw-Hill, New York, p. 516.

Deutsch, W. (1922) "Motion and Charge of a Charged particle in the Cylindrical Condenser," *Annals of Physics,* vol. 68, pp. 335–344s.

Ferris, B. G. (1978) "Health Effects of Exposure to Low Levels of Regulated Air Pollutants," *Journal of the Air Pollution Control Association,* vol. 28, pp. 482–497.

Fuller, H. J. (1960) *The Plant World,* Henry Holt, New York.

Godish, T. (1989) *Indoor Air Pollution Control,* Lewis Publishers, Chelsea, MI.

Godish, T. (2001) "Aldehydes," in J. D. Spengler, J. M. Samet, and J. F. McCarthy (eds.), *Indoor Air Quality Handbook,* McGraw-Hill, New York, pp. 32.1–32.22.

Goldsmith, J. R. (1968) "Effects of Air Pollution on Human Health," in A. C. Stern (ed.), *Air Pollution,* Academic Press, New York, pp. 554–557.

Goyer, R. A., and J. J. Chilsolm (1972) "Lead," in D. K. K. Lee (ed.), *Metallic Contaminants and Human Health,* Academic Press, New York, pp. 57–95.

Hance, S. B., and J. L. Kelly (1991) "Status of Flue Gas Desulfurization Systems," Paper No. 91-157.3, 84th Annual Meeting of the Air and Waste Management Association.

Hansen, J., and M. Sato (2004) "Temperature Trends: 2004 Summation," http://www.giss.nasa.gov/data/update/gistemp/2004/.

Hesketh, H. E. (1977) *Fine Particles in Gaseous Media,* Ann Arbor Science, Ann Arbor, MI, p. 19.

Hileman, B. (1989) "Global Warming," *C&E News,* March 13, pp. 25–44.

Hileman, B. (2005) "Ice Core Record Extended," *C&E News,* November 28, p. 7.

Hindawi, I. (1970) *Air Pollution Injury to Plants,* U.S. Department of Health, Education, and Welfare, National Air Pollution Control Administration Publication No. AP-71, Washington, DC, p. 13.

Hines, A. L., T. K. Ghosh, S. K. Loyalka, and R. C. Warder (1993) *Indoor Air Quality & Control,* PTR Prentice Hall, Englewood Cliffs, NJ, pp. 21, 22 , 34.

Holland, J. Z. (1953) *A Meteorological Survey of the Oak Ridge Area,* U.S Atomic Energy Commission Report No. ORO-99, Washington, DC, p. 540.

Hosein, R., F. Silverman, P. Coreg, et al. (1985) "The Relationship Between Pollutant Levels in Homes and Potential Sources," *Transactions, Indoor Air Quality in Cold Climates, Hazards and Abatement Measures,* Air Pollution Control Association, Pittsburgh, pp. 250–260.

ICAS (1975) *The Possible Impact of Fluorocarbons and Hydrocarbons on Ozone,* Interdepartmental Committee for Atmospheric Sciences, Federal Council for Science and Technology, National Science Foundation Publication No. ICAS 18a-FY 75, Washington, DC, p. 3.

IPCC (1995) *Climate Change1994: Radiative Forcing of Climate Change and an Evaluation of the IPCC 1992 Emission Scenarios,* Intergovernmental Panel on Climate Change, Cambridge University Press, Cambridge, U.K.

IPCC (2000) Intergovernmental Panel on Climate Change, http://www.gcrio.org/edu.html, April 2000.

IPCC (2001a) *Climate Change 2001: The Scientific Basis, Summary for Policymakers,* Intergovernmental Panel on Climate Change, Cambridge University Press, Cambridge, U.K., pp. 1–18.

IPCC (2001b) *Climate Change 2001: Impacts, Adaptation and Vulnerability, Summary for Policymakers,* Intergovernmental Panel on Climate Change, Cambridge University Press, Cambridge, U.K., pp. 1–17.

Johnstone, H. F., R. B. Field, and M. C. Tassler (1954) "Gas Absorption and Aerosol Collection in a Venturi Atomizer," *Industrial Engineering Chemistry,* vol. 46, pp. 1601–1608.

Karlsson, H. T., and H. S. Rosenberg, (1980) "Technical Aspects of Lime/Limestone Scrubbers for Coal fired Power Plants, Part I, Process Chemistry and Scrubber Systems," *Journal of the Air Pollution Control Association,* vol. 30(6), pp. 710–714.

Kao, A.S. (1994) "Formation and Removal Reactions of Hazardous Air Pollutants," *Journal of the Air Pollution Control Association,* vol. 44, pp. 683–696.

Keeling, C. M., and T. P. Whorf (2005) "Atmospheric Carbon Dioxide Record From Mauna Loa," http://www.mlo.noaa.gov.

Kiehl, J. T., and H. Rodhe (1995) "Modeling Geographic and Seasonal Forcing Due to Aerosols," in R. J. Charlson and J. Heintzenberg (eds.), *Aerosol Forcing of Climate,* John Wiley & Sons, New York, pp. 281–296.

Lapple, C. E. (1951) "Processes Use Many Collection Types," *Chemical Engineer,* vol. 58, pp. 144–151.

Lee, H. K., T. K. McKenna, L. N. Renton, et al. (1985) "Impact of a New smoking Policy on Office Air Quality," *Indoor Air Quality in Cold Climates, Transactions of the Air Pollution Control Association,* Pittsburgh, pp. 307–322.

Lee, W. G., H. Chen, and C. Wu (2001) "Emission of VOCs from Wooden Building Materials in Indoor Environment," *Proceedings of the Air & Waste Management Association 94th Annual Conference & Exhibition,* Orlando, June 24–28, 2001.

Lefohn, A. S., and S. V. Krupa (1988) "Acidic Precipitation, A Technical Amplification of NAPAP's Findings," *Proceedings of APCA International Conference,* Pittsburgh, p. 1.

Lewis, R. G. (2001) "Pesticides," in J. D. Spengler, J. M. Samet, and J. F. McCarthy (eds.) *Indoor Air Quality Handbook,* McGraw-Hill, New York, p. 35.14.

Mann, M. E., and P. D. Jones (2003) "Global Surface Temperatures Over the Past Two Millennia," *Geophysical Research Letters,* vol. 30, no. 15, pp. CLM 5-1–CLM 5-4.

Martin, D. O. (1976) "Comment on the Change of Concentration Standard Deviations with Distance," *Journal of the Air Pollution Control Association,* vol. 26, pp. 145–146.

Martens, P. (1999) "How Will Climate Change Affect Human Health," *American Scientist,* vol. 87, pp. 534–541.

McKenna, J. D., A. B. Nunn, and D. A. Furlong (2000) "Fabric Filters," in W. T. Davis (ed.), *Air Pollution Engineering Manual,* 2nd ed., Air Pollution Control Association and John Wiley & Sons, New York, p. 104.

Molina, M. J., and F. S. Rowland (1974) "Stratospheric Sink for Chlorofloromethanes; Chlorine Atom Catalysed Destruction of Ozone," *Nature,* vol. 248, pp. 810–812.

Moschandreas, D. J., J. D. Zabpansky and S. D. Pelta, (1985) *Characteristics of Emissions from Indoor Combustion Sources,* Gas Research Institute Report No. 85/0075, Chicago, IL.

NAPCA (1970) *Air Quality Criteria for Photochemical Oxidants,* U.S. Department of Health, Education and Welfare, National Air Pollution Control Administration Publication No. AP-63, Washington, DC.

NAS (1961) *Effects of Inhaled Radioactive Particles,* National Academy of Sciences Publication 848, Washington, DC.

Noll, K. E., V. Gounaris, and Wain-Sun Hou (1992) Adsorption for *Air and Water Pollution Control,* Lewis Publishers, Chelsea, MI, pp. 74–79.

Noll, K. (1999) *Fundamentals of Air Quality Systems: Design of Air Pollution Control Devices,* American Academy of Environmental Engineers, Annapolis, MD, pp. 228, 402–403.

Noticias (2004) Noticias.info, Agencia International de Noticias, http://www.noticias.info/Archivo 08/11/2004 notas_de_prensa_archivo.

NTRDA (1969) *Air Pollution Primer,* National Tuberculosis and Respiratory disease Association.

O'Gara, P. J. (1922) "Sulfur Dioxide and Fume Problems and Their Solutions," *Industrial Engineering Chemistry,* vol. 14, p. 744.

Patterson, D. J., and N. A. Henein (1972) *Emissions from Combustion Engines and Their Control,* Ann Arbor Science, Ann Arbor, MI, p. 143.

Pope, C. A., M. J. Thun, M. M. Namboodri, et al. (1995) "Particulate Air Pollution as a Predictor of Mortality in a Prospective Study of U.S. Adults," *American Journal of Respiratory and Critical Care Medicine,* vol. 151, pp. 669–674.

Pope, C. A., D. W. Dockery, R. E. Kanner, G. M. Villegas and J. Schwartz (1999) "Oxygen Saturation, Pulse Rate, and Particulate Air Pollution: A Daily Time-series Panel Study," *American Journal of Respiratory and Critical Care Medicine,* vol. 159, pp. 365–372.

Perry, R. H., and C. H. Chilton (eds.) (1973) *Chemical Engineers Handbook,* 5th ed., McGraw-Hill, New York, pp. 3–96.

Prasad, A. (1995) "Air Pollution Control Technologies for Nitrogen Oxides," *The National Environmental Journal,* May/June, pp. 46–50.

Reichhardt, T. (1995) "Weighing the Health Risks of Airborne Particulates," *Environmental Science and Technology,* vol. 29, pp. 360A–364A.

Reisch, M., and P. S. Zurer (1988) "CFC production: DuPont Seeks Total Phaseout," *C&E News,* April 4, p. 4.

Seinfeld, J. H. (1975) *Air Pollution, Physical and Chemical Fundamentals,* McGraw-Hill, New York, p. 71.

Seinfeld, J. H., and S. N. Pandis (1998) *Atmospheric Chemistry and Physics,* John Wiley & Sons, New York, pp. 59, 71.

Sparks, L. E. (2001) "Indoor Air Quality Modeling," in J. D. Spengler, J. M. Samet, and J. F. McCarthy (eds.), *Indoor Air Quality Handbook,* McGraw-Hill, New York, pp. 58.1–58.28.

Srivastava, R. K., J. E. Staudt, and W. Josewicz (2005) "Preliminary Estimates of Performance and Cost of Mercury Emission Control Technology Applications on electric Utility Boilers: An Update," *Environmental Progress,* vol. 24, no. 2, pp. 198–213.

Sullivan, D. A. (1989) "International Gathering Plans Ways to Safeguard Atmospheric Ozone," *C&E News,* June 26, pp. 33–36.

Traynor, G. W., J. R. Allen, and M. G. Apte (1982) *Indoor Air Pollution from Portable Kerosene-fired Space Heaters, Woodburning Stoves and Woodburning Furnaces,* Lawrence Berkeley Laboratory Report No. LBL-14027.

Treybal, R. E. (1968) *Mass Transfer Operations,* McGraw-Hill, New York, pp. 253, 535.

Tucker, W. G. (2001) "Volatile Organic Compounds," in J. D. Spengler, J. M. Samet, and J. F. McCarthy (eds.), *Indoor Air Quality Handbook,* McGraw-Hill, New York, pp. 31.1–31.20.

Turner, D. B. (1967) *Workbook of Atmospheric Dispersion Estimates,* U.S. Department of Health, Education, and Welfare, U.S. Public Health Service Publication No. 999-AP-28, Washington, DC.

UN (2005) *The Millennium Development Goals Report:* 2005, United Nations, New York, p. 32.

UNDP (2005) *The Montreal Protocol,* http://www.undp.org/seed/eap/montreal/montreal.htm.

UNEP/WHO (2002) "Executive Summary," *Scientific Assessment of Ozone Depletion: 2002,* United Nations Environmental Programme/World Health Organization, New York.

U.S. CDC (2005) *Third National Report on Human Exposure to Environmental Chemicals,* U.S. Centers for Disease Control and Prevention, National Center for Environmental Health, NCEH Pub. No. 05-0570, pp. 41, 74–75.

U.S. EPA (1995) *User's Guide for ISC3 Dispersion Models,* Vol. II, EPA-454/B-95-003b, U.S. Environmental Protection Agency, Research Triangle Park, NC.

U.S. EPA (1997) *1997 Mercury Study Report To Congress,* U.S. Environmental Protection Agency, http://www.epa.gov.

U.S. EPA (2003a) *Response of Surface Water Chemistry to the Clean Air Act Amendments of 1990,* U.S. Environmental Protection Agency, Report No. 620/R-03/001, Research Triangle Park, NC, pp. 59–62.

U.S. EPA (2003b) *Performance and Cost of Mercury Emission Control Technology Applications on Electric Utility Boilers,* Report No. 600/R-03/1100, Research Triangle Park, NC.

U.S. EPA (2004) *EPA Fact Sheet,* U.S. Environmental Protection Agency, http://www.epa.gov.

U.S. PHS (1965) *Survey of Lead in the Atmosphere of Three Urban Communities,* U.S. Department of Health, Education, and Welfare, U.S. Public Health Service Publication No. 999-AP-12, Washington, DC.

Wallace, L. A. (2001) "Assessing Human Exposure to Volatile Organic Compounds," in J. D. Spengler, J. M. Samet, and J. F. McCarthy (eds.), *Indoor Air Quality Handbook,* McGraw-Hill, New York, p. 33.1–33.35.

Walsh, G.W. (1967) "Fabric Filtration," in *Control of Particulate Emissions Training Course Manual in Air Pollution,* National Center for Air Pollution Control, U.S. Public Health Service, Cincinnati, p. 9.

Wark, K., C. F. Warner, and W. T. Davis (1998) *Air Pollution. Its Origin and Control,* 3rd ed., Addison-Wesley, Menlo Park, CA, pp. 250–251.

Wofsy, S. C., J. C. McConnell, and M. B. McElroy (1972) "Atmospheric CH_4, CO, and CO_2," *Journal of Geophysical Research,* vol. 67, pp. 4477–4493.

WHO (1961) *Air Pollution,* World Health Organization, Geneva, Switzerland, p. 180.

WHO (1987) "Carbon Monoxide," in *Air Quality Guidelines for Europe,* World Health Organization Regional Office for Europe, European Series 23, Copenhagen, pp. 210–220.

Yocom, J. E. and R. O. McCaldin (1968) "Effects on Materials and the Economy," in A. C. Stern (ed.), *Air **Pollution,*** vol. I, 2nd ed., Academic Press, New York, pp. 617–654.

Zurer, P. S. (1988) "Studies on Ozone Destruction Expand Beyond Antarctic," *C&E News,* May 30, pp. 18–25.

Zurer, P. S. (1989) "Scientists Find Arctic May Face Ozone Hole," *C&E News,* February 27, p. 5.

Zurer, P. S. (1994) "Scientists Expect Ozone Loss to Peak About 1998," *C&E, News,* September 12, p. 5.

CHAPTER 8

NOISE POLLUTION

8-1 INTRODUCTION

Noise, commonly defined as unwanted sound, is an environmental phenomenon to which we are exposed before birth and throughout life. Noise is an environmental pollutant, a waste product generated in conjunction with various anthropogenic activities. Under the latter definition, noise is any sound—independent of loudness—that can produce an undesired physiological or psychological effect in an individual, and that may interfere with the social ends of an individual or group. These social ends include all of our activities—communication, work, rest, recreation, and sleep.

As waste products of our way of life, we produce two general types of pollutants. The general public has become well aware of the first type—the mass residuals associated with air and water pollution—that remain in the environment for extended periods of time. However, only recently has attention been focused on the second general type of pollution, the energy residuals such as the waste heat from manufacturing processes that creates thermal pollution of our streams. Energy in the form of sound waves constitutes yet another kind of energy residual, but, fortunately, one that does not remain in the environment for extended periods of time. The total amount of energy dissipated as sound throughout the earth is not large when compared with other forms of energy; it is only the extraordinary sensitivity of the ear that permits such a relatively small amount of energy to adversely affect us and other biological species.

It has long been known that noise of sufficient intensity and duration can induce temporary or permanent hearing loss, ranging from slight impairment to nearly total deafness. In general, a pattern of exposure to any source of sound that produces high enough levels can result in temporary hearing loss. If the exposure persists over a period of time, this can lead to permanent hearing impairment. It has been estimated that 1.7 million workers in the United States between 50 and 59 years of age have enough hearing loss to be awarded compensation. The potential cost to U.S. industry could be in excess of $1 billion* (Olishifshi and Harford, 1975). Short–term, but frequently serious, effects include interference with speech communication and the perception of other auditory signals, disturbance of sleep and relaxation, annoyance, interference with an individual's ability to perform complicated tasks, and general diminution of the quality of life.

Beginning with the technological expansion of the Industrial Revolution and continuing through a post-World War II acceleration, environmental noise in the United States and other industrialized nations has been gradually and steadily increasing, with more geographic areas becoming exposed to significant levels of noise. Where once noise levels sufficient to induce some degree of hearing loss were confined to factories and occupational situations, noise levels approaching such intensity and duration are today being recorded on city streets and, in some cases, in and around the home.

There are valid reasons why widespread recognition of noise as a significant environmental pollutant and potential hazard or, as a minimum, a detractor from the

*In 2005 dollars.

quality of life, has been slow in coming. In the first place, noise, if defined as unwanted sound, is a subjective experience. What is considered noise by one listener may be considered desirable by another.

Secondly, noise has a short decay time and thus does not remain in the environment for extended periods, as do air and water pollution. By the time the average individual is spurred to action to abate, control, or, at least, complain about sporadic environmental noise, the noise may no longer exist.

Thirdly, the physiological and psychological effects of noise on us are often subtle and insidious, appearing so gradually that it becomes difficult to associate cause with effect. Indeed, to those persons whose hearing may already have been affected by noise, it may not be considered a problem at all.

Further, the typical citizen is proud of this nation's technological progress and is generally happy with the things that technology delivers, such as rapid transportation, labor-saving devices, and new recreational devices. Unfortunately, many technological advances have been associated with increased environmental noise, and large segments of the population have tended to accept the additional noise as part of the price of progress.

In the last three decades, the public has begun to demand that the price of progress not fall to them. They have demanded that the environmental impact of noise be mitigated. The cost of mitigation is not trivial. The average cost of soundproofing each of 600 suburban houses around the Chicago O'Hare airport was about $27,500 in 1997 (Sylvan, 2000). Through 2001, the Boston Logan airport had spent about $99 million and the Los Angeles International airport had allocated about $119 million for soundproofing and land acquisition. At the end of 2001, the total amount spent in the United States for noise mitigation exceeded $5.2 billion (de Neufville and Odoni, 2003). The cost to retrofit and replace airplanes to reduce noise probably exceeds $3.6 billion* (Achitoff, 1973). Traffic noise reduction programs have been in place since the first noise barrier was built in 1963. As of 2001, departments of transportation in 44 states and the Commonwealth of Puerto Rico had constructed more than 2,900 linear kilometers of noise barriers at a cost of more than $2.8 billion* (FHWA, 2005).

The engineering and scientific community has already accumulated considerable knowledge concerning noise, its effects, and its abatement and control. In that regard, noise differs from most other environmental pollutants. Generally, the technology exists to control most indoor and outdoor noise. As a matter of fact, this is one instance in which knowledge of control techniques exceeds the knowledge of biological and physical effects of the pollutant.

Properties of Sound Waves

Sound waves result from the vibration of solid objects or the separation of fluids as they pass over, around, or through holes in solid objects. The vibration and/or separation causes the surrounding air to undergo alternating compression and rarefaction, much in the same manner as a piston vibrating in a tube (Figure 8-1). The compression

*In 2005 dollars.

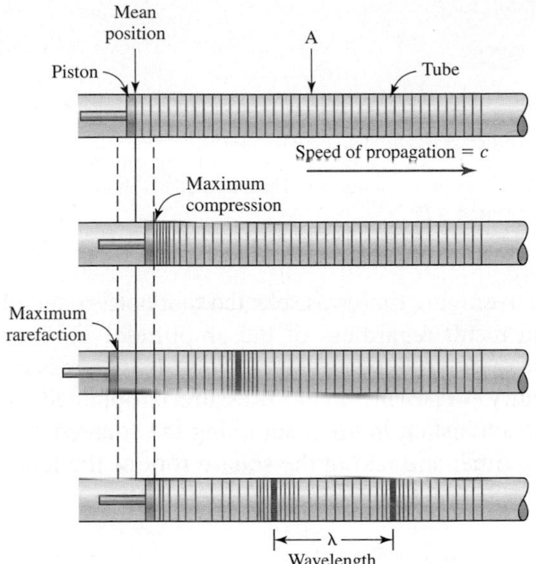

FIGURE 8-1
Alternating compression and rarefaction of air molecules resulting from a vibrating piston.

of the air molecules causes a local increase in air density and pressure. Conversely, the rarefaction causes a local decrease in density and pressure. These alternating pressure changes are the sound detected by the human ear.

Let us assume that you could stand at Point A in Figure 8-1. Also let us assume that you have an instrument that will measure the air pressure every 0.000010 seconds and plot the value on a graph. If the piston vibrates at a constant rate, the condensations and rarefactions will move down the tube at a constant speed. That speed is the *speed of sound* (c). The rise and fall of pressure at point A will follow a cyclic or wave pattern over a "period" of time (Figure 8-2). The wave pattern is called *sinusoidal.* The time between successive peaks or between successive troughs of the oscillation is called the *period (P).* The inverse of this, that is, the number of times a peak arrives in one second of oscillations, is called the *frequency (f).* Period and frequency are then related as follows:

$$P = \frac{1}{f} \tag{8-1}$$

Since the pressure wave moves down the tube at a constant speed, you would find that the distance between equal pressure readings would remain constant. The distance between adjacent crests or troughs of pressure is called the *wavelength* (λ). Wavelength and frequency are then related as follows:

$$\lambda = \frac{c}{f} \tag{8-2}$$

The *amplitude (A)* of the wave is the height of the peak or depth of the trough measured from the zero pressure line (Figure 8-2). From Figure 8-2, we can also note that

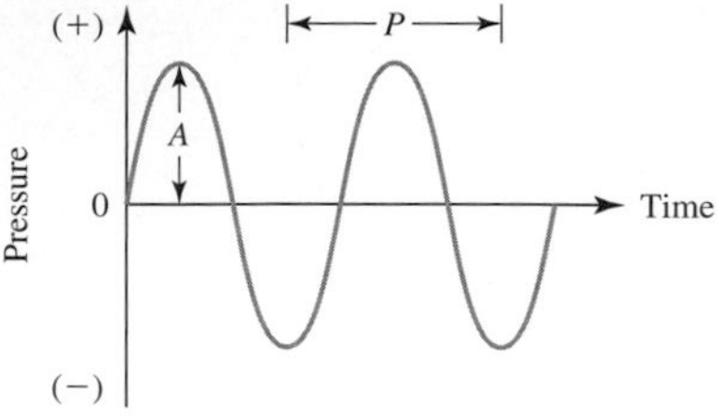

FIGURE 8-2
Sinusoidal wave that results from alternating compression and rarefaction of air molecules. The amplitude is shown as A and the period is P.

the average pressure could be zero if an averaging time was selected that corresponded to the period of the wave. This would result regardless of the amplitude! This, of course, is not an acceptable state of affairs. The root mean square (rms) sound pressure (p_{rms}) is used to overcome this difficulty.* The rms sound pressure is obtained by squaring the value of the amplitude at each instant in time; summing the squared values; dividing the total by the averaging time; and taking the square root of the total. The equation for rms is

$$p_{\text{rms}} = \left(\overline{p^2}\right)^{1/2} = \left[\frac{1}{T}\int_0^T P^2(t)dt\right]^{1/2} \tag{8-3}$$

where the overbar refers to the time-weighted average and T is the time period of the measurement.

Sound Power and Intensity

Work is defined as the product of the magnitude of the displacement of a body and the component of force in the direction of the displacement. Thus, traveling waves of sound pressure transmit energy in the direction of propagation of the wave. The rate at which this work is done is defined as the *sound power* (W).

Sound intensity (I) is defined as the time-weighted average sound power per unit area normal to the direction of propagation of the sound wave. Intensity and power are related as follows:

$$I = \frac{W}{A} \tag{8-4}$$

where A is a unit area perpendicular to the direction of wave motion. Intensity, and hence, sound power, is related to sound pressure in the following manner:

$$I = \frac{(p_{\text{rms}})^2}{\rho c} \tag{8-5}$$

where I = intensity, W/m^2
p_{rms} = root mean square sound pressure, Pa
ρ = density of medium, kg/m^3
c = speed of sound in medium, m/s

*Sound pressure = (total atmospheric pressure) − (barometric pressure).

Both the density of air and speed of sound are a function of temperature. Given the temperature and pressure, the density of air (1.185 kg/m^3 at 101.325 kPa and 298 K) may be determined using the gas laws. The speed of sound in air at 101.325 kPa may be determined from the following equation:

$$c = 20.05\sqrt{T} \tag{8-6}$$

where T is the absolute temperature in kelvins (K) and c is in m/s.

Levels and the Decibel

The sound pressure of the faintest sound that a normal healthy individual can hear is about 0.00002 pascal. The sound pressure produced by a Saturn rocket at liftoff is greater than 200 pascal. Even in scientific notation this is an "astronomical" range of numbers.

In order to cope with this problem, a scale based on the logarithm of the ratios of the measured quantities is used. Measurements on this scale are called *levels*. The unit for these types of measurement scales is the *bel*, which was named after Alexander Graham Bell:

$$L' = \log \frac{Q}{Q_o} \tag{8-7}$$

where L' = level, bels
Q = measured quantity
Q_o = reference quantity
log = logarithm in base10

A bel turns out to be a rather large unit, so for convenience it is divided into 10 subunits called *decibels* (dB). Levels in dB are computed as follows:

$$L = 10 \log \frac{Q}{Q_o} \tag{8-8}$$

The dB does not represent any physical unit. It merely indicates that a logarithmic transformation has been performed.

Sound Power Level. If the reference quantity (Q_o) is specified, then the dB takes on physical significance. For noise measurements, the reference power level has been established as 10^{-12} watts. Thus, sound power level may be expressed as

$$L_w = 10 \log \frac{W}{10^{-12}} \tag{8-9}$$

Sound power levels computed with Equation 8-9 are reported as dB re: 10^{-12} W.

Sound Intensity Level. For noise measurements, the reference sound intensity (Equation 8-4) is 10^{-12} W/m^2. Thus, the sound intensity level is given as

$$L_I = 10 \log \frac{I}{10^{-12}} \tag{8-10}$$

Sound Pressure Level. Because sound-measuring instruments measure the root mean square pressure, the sound pressure level is computed as follows:

$$L_P = 10 \log \frac{(p_{\text{rms}})^2}{(p_{\text{rms}})_o^2} \tag{8-11}$$

which, after extraction of the squaring term, is given as

$$L_P = 20 \log \frac{p_{\text{rms}}}{(p_{\text{rms}})_o} \tag{8-12}$$

The reference pressure has been established as 20 micropascals (μPa). A scale showing some common sound pressure levels is shown in Figure 8-3.

Combining Sound Pressure Levels. Because of their logarithmic heritage, decibels don't add and subtract the way apples and oranges do. Remember: adding the logarithms of numbers is the same as multiplying them. If you take a 60-decibel noise (re: 20 μPa) and add another 60-decibel noise (re: 20 μPa) to it, you get a 63-decibel noise (re: 20 μPa). If you're strictly an apple-and-orange mathematician, you may take this on faith. For skeptics, this can be demonstrated by converting the dB to sound power

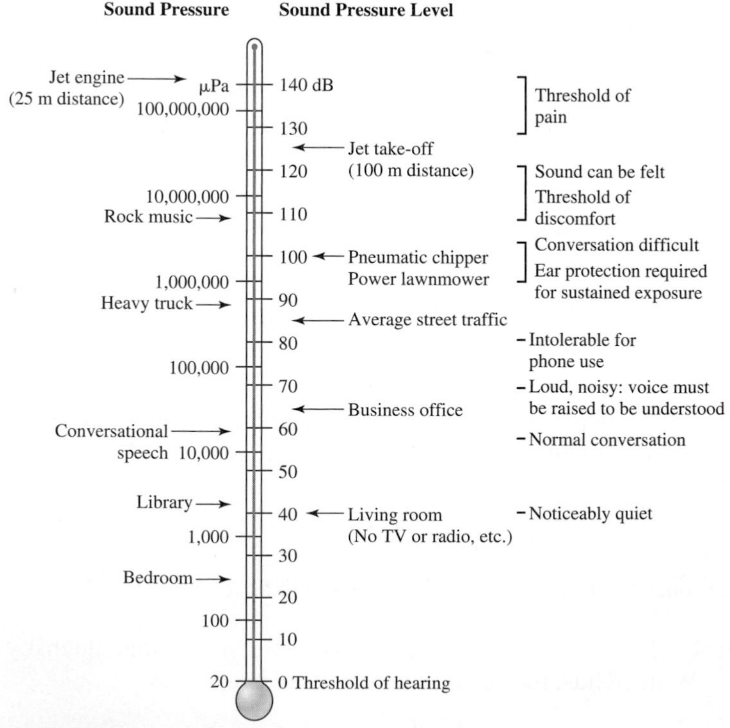

FIGURE 8-3
Relative scale of sound pressure levels.

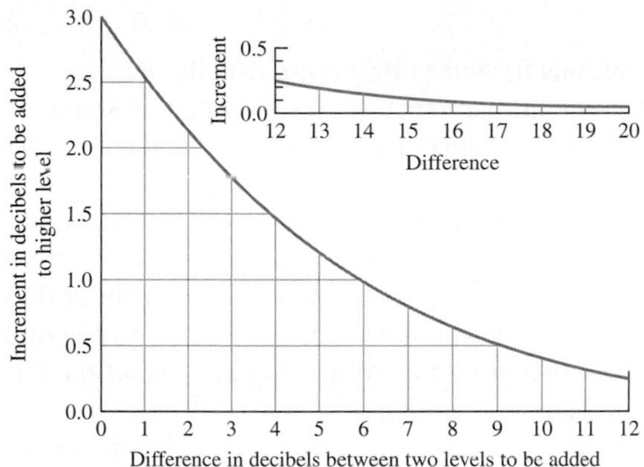

FIGURE 8-4
Graph for solving decibel addition problems.

level, adding them, and converting back to dB. Figure 8-4 provides a graphical solution for this type of problem. For noise pollution work, results should be reported to the nearest whole number. When there are several levels to be combined, they should be combined two at a time, starting with lower-valued levels and continuing two at a time with each successive pair until one number remains. Henceforth, in this chapter we will assume levels are all "re: 20 μPa" unless stated otherwise.

Example 8-1. What sound power level results from combining the following three levels: 68 dB, 79 dB, and 75 dB?

Solution. This problem can be worked by converting the readings to sound power level, adding them, and converting back to dB.

$$\begin{aligned} L_P &= 10 \log \Sigma 10^{(68/10)} + 10^{(75/10)} + 10^{(79/10)} \\ &= 10 \log(117{,}365{,}173) \\ &= 80.7 \text{ dB} \end{aligned}$$

Rounding off to the nearest whole number yields an answer of 81 dB re: 20 μPa.

An alternative solution technique using Figure 8-4 begins by selecting the two l-owest levels: 68 dB and 75 dB. The difference between the values is $75 - 68 = 7.00$. Using Figure 8-4, draw a vertical line from 7.00 on the abscissa to intersect the curve. A horizontal line from the intersection to the ordinate yields about 0.8 dB. Adding this value to the highest value, the combination of 68 dB and 75 dB results in a level of 75.8 dB. This, and the remainder of the computation, is shown diagrammatically below.

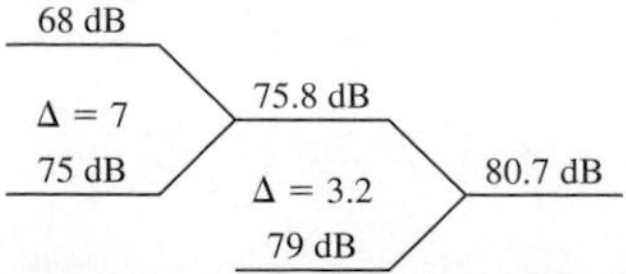

Characterization of Noise

Weighting Networks. Because our reasons for measuring noise usually involve people, we are ultimately more interested in the human reaction to sound than in sound as a physical phenomenon. Sound pressure level, for instance, can't be taken at face value as an indication of loudness because the frequency (or pitch) of a sound has quite a bit to do with how loud it sounds. For this and other reasons, it often helps to know something about the frequency of the noise you're measuring. Weighting networks are used to account for the frequency of a sound. They are electronic filtering circuits built into the meter to attenuate certain frequencies. They permit the sound level meter to respond more to some frequencies than to others with a prejudice something like that of the human ear. Writers of the acoustical standards have established three weighting characteristics: A, B, and C. The chief difference among them is that very low frequencies are filtered quite severely by the A network, moderately by the B network, and hardly at all by the C network. Therefore, if the measured sound level of a noise is much higher on C weighting than on A weighting, much of the noise is probably of low frequency. If you really want to know the frequency distribution of a noise (and most serious noise measurers do), it is necessary to use a *sound analyzer*. But if you are unable to justify the expense of an analyzer, you can still find out something about the frequency of a noise by shrewd use of the weighting networks of a sound level meter.

Figure 8-5 shows the response characteristics of the three basic networks as prescribed by the American National Standards Institute (ANSI) specification number S1.4–1971. When a weighting network is used, the sound level meter electronically subtracts or adds the number of dB shown at each frequency shown in Table 8-1 from or to the actual sound pressure level at that frequency. It then sums all the resultant numbers by logarithmic addition to give a single reading. Readings taken when a network is in use are said to be "sound levels" rather than "sound pressure levels." The readings taken are designated in decibels in one of the following forms: dB(A); dBa; dBA; dB(B); dBb; dBB; and so on. Tabular notations may refer to L_A, L_B, L_C.

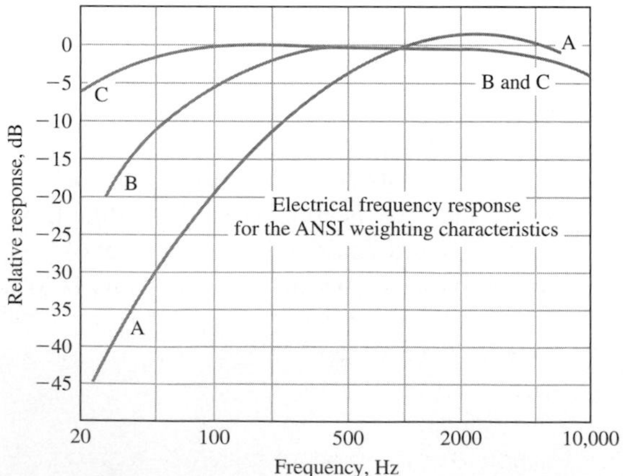

FIGURE 8-5
Response characteristics of the three basic weighting networks.

TABLE 8-1
Sound level meter network weighting values

Frequency (Hz)	Curve A (dB)	Curve B (dB)	Curve C (dB)
10	−70.4	−38.2	−14.3
12.5	−63.4	−33.2	−11.2
16	−56.7	−28.5	−8.5
20	−50.5	−24.2	−6.2
25	−44.7	−20.4	−4.4
31.5	−39.4	−17.1	−3.0
40	−34.6	−14.2	−2.0
50	−30.2	−11.6	−1.3
63	−26.2	−9.3	−0.8
80	−22.5	−7.4	−0.5
100	−19.1	−5.6	−0.3
125	−16.1	−4.2	−0.2
160	−13.4	−3.0	−0.1
200	−10.9	−2.0	0
250	−8.6	−1.3	0
315	−6.6	−0.8	0
400	−4.8	−0.5	0
500	−3.2	−0.3	0
630	−1.9	−0.1	0
800	−0.8	0	0
1,000	0	0	0
1,250	0.6	0	0
1,600	1.0	0	−0.1
2,000	1.2	−0.1	−0.2
2,500	1.3	−0.2	−0.3
3,150	1.2	−0.4	−0.5
4,000	1.0	−0.7	−0.8
5,000	0.5	−1.2	−1.3
6,300	−0.1	−1.9	−2.0
8,000	−1.1	−2.9	−3.0
10,000	−2.5	−4.3	−4.4
12,500	−4.3	−6.1	−6.2
16,000	−6.6	−8.4	−8.5
20,000	−9.3	−11.1	−11.2

Example 8-2. A new Type 2 sound level meter is to be tested with two pure tone sources that emit 90 dB. The pure tones are at 1,000 Hz and 100 Hz. Estimate the expected readings on the A, B, and C weighting networks.

Solution. From Table 8-1 at 1,000 Hz, we note that the relative response (correction factor) for each of the weighting networks is zero. Thus for the pure tone at 1,000 Hz we would expect the readings on the A, B, and C networks to be 90 dB.

From Table 8-1 at 100 Hz, the relative response for each weighting network differs. For the A network, the meter will subtract 19.1 dB from the actual reading, for the B network, the meter will subtract 5.6 dB from the actual reading, and for the C network, the meter will subtract 0.3 dB. Thus, the anticipated readings would be:

A network: $90 - 19.1 = 70.9$ or 71 dB(A)

B network: $90 - 5.6 = 84.4$ or 84 dB(B)

C network: $90 - 0.3 = 89.7$ or 90 dB(C)

Example 8-3. The following sound levels were measured on the A, B, and C weighting networks:

Source 1: 94 dB(A), 95 dB(B), and 96 dB(C)

Source 2: 74 dB(A), 83 dB(B), and 90 dB(C)

Characterize the sources as "low frequency" or "mid/high frequency."

Solution. From Figure 8-5, we can see that readings on the A, B, and C networks will be close together if the source emits noise in the frequency range above about 500 Hz. This range may be classified "mid/high frequency" since we cannot distinguish between "mid" and "high" frequency using a Type 2 sound level meter. Likewise, we can see that below 200 Hz (low frequency), readings on the A, B, and C scale will be substantially different. The readings from the A network will be lower than the readings from the B network, and readings from both the A and B networks will be lower than those from the C network.

Source 1: Note that the sound levels on each of the weighting networks differ by 1 dB. From Figure 8-5, it appears that the sound level will be in the mid/high frequency range.

Source 2: Note that the sound levels on each of the weighting networks differ by several dB and that the reading from the A network is lower than that from the B network and both are below that from the C network. From Figure 8-5, it appears that the sound level will be in the low frequency range.

Octave Bands. To completely characterize a noise, it is necessary to break it down into its frequency components or spectra. Normal practice is to consider 8 to 11 octave bands.* The standard octave bands and their geometric mean frequencies (center band frequencies) are given in Table 8-2. Octave analysis is performed with a combination precision sound level meter and an octave filter set.

While octave band analysis is frequently satisfactory for community noise control (that is, identifying violators), more refined analysis is required for corrective action and design. One-third octave band analysis provides a slightly more refined picture of the noise source than the full octave band analysis (Figure 8-6a). This improved

*An octave is the frequency interval between a given frequency and twice that frequency. For example, given the frequency 22 Hz, the octave band is from 22 to 44 Hz. A second octave band would then be from 44 to 88 Hz.

TABLE 8-2
Octave bands

Octave frequency range (Hz)	Geometric mean frequency (Hz)
22–44	31.5
44–88	63
88–175	125
175–350	250
350–700	500
700–1,400	1,000
1,400–2,800	2,000
2,800–5,600	4,000
5,600–11,200	8,000
11,200–22,400	16,000
22,400–44,800	31,500

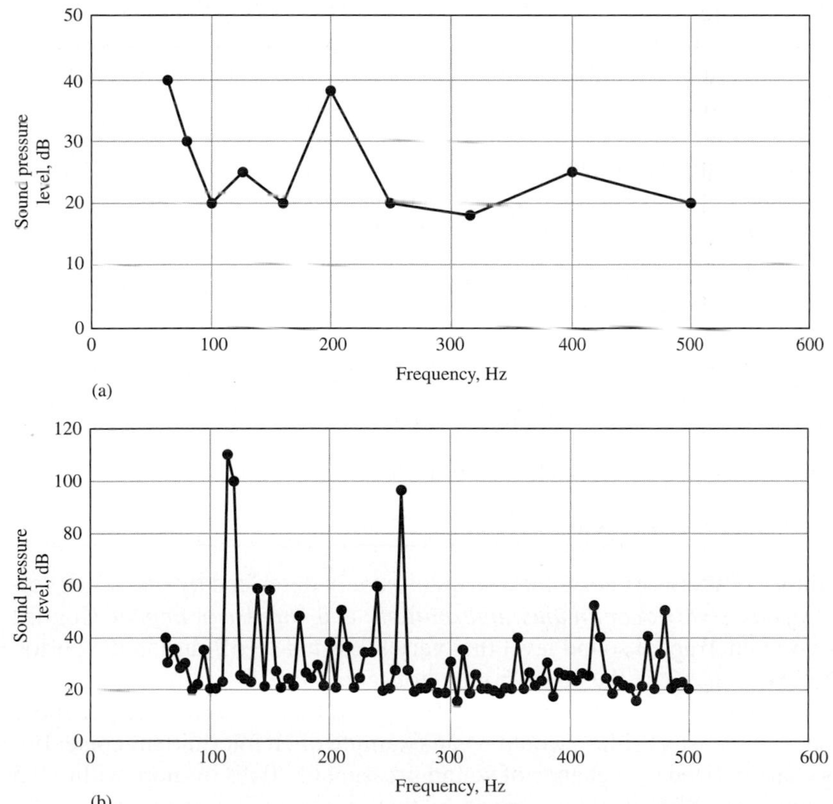

FIGURE 8-6
(*a*) One-third octave band analysis of a small electric motor. (*b*) Narrowband analysis of a small electric motor.

resolution is usually sufficient for determining corrective action for community noise problems. Narrow band analysis is highly refined and may imply band widths down to 2 Hz (Figure 8-6b). This degree of refinement is only justified in product design and testing or in troubleshooting industrial machine noise and vibration.

Averaging Sound Pressure Levels. Because of the logarithmic nature of the dB, the average value of a collection of sound pressure level measurements cannot be computed in the normal fashion. Instead, the following equation must be used:

$$\overline{L}_p = 20 \log \frac{1}{N} \sum_{j=1}^{N} 10^{(L_j/20)} \tag{8-13}$$

where L_p = average sound pressure level, dB re: 20 μPa
N = number of measurements
L_j = the jth sound pressure level, dB re: 20 μPa
$j = 1, 2, 3 \ldots, N$

This equation is equally applicable to sound levels in dBA. It may also be used to compute average sound power levels if the factors of 20 are replaced with 10s.

Example 8-4. Compute the mean sound level from the following four readings (all dBA): 38, 51, 68, and 78.

Solution. First we compute the sum:

$$\sum_{j=1}^{4} = 10^{(38/20)} + 10^{(51/20)} + 10^{(68/20)} + 10^{(78/20)}$$

$$= 1.09 \times 10^4$$

Now we complete the computation:

$$\overline{L}_p = 20 \log \frac{1.09 \times 10^4}{4}$$

$$= 68.7 \text{ or } 69 \text{ dBA}$$

Straight arithmetic averaging would yield 58.7 or 59 dB.

Types of Sounds. Patterns of noise may be qualitatively described by one of the following terms: *steady-state* or *continuous; intermittent;* and *impulse* or *impact.* Continuous noise is an uninterrupted sound level that varies less than 5 dB during the period of observation. An example is the noise from a household fan. Intermittent noise is a continuous noise that persists for more than one second that is interrupted for more than one second. A dentist's drilling would be an example of an intermittent noise. Impulse noise is characterized by a change of sound pressure of 40 dB or more within 0.5 second with a duration of less than one second.* The noise from firing a weapon would be an example of an impulsive noise.

*The Occupational Safety and Health Administration (OSHA) classifies repetitive events, including impulses, as steady noise if the interval between events is less than 0.5 seconds.

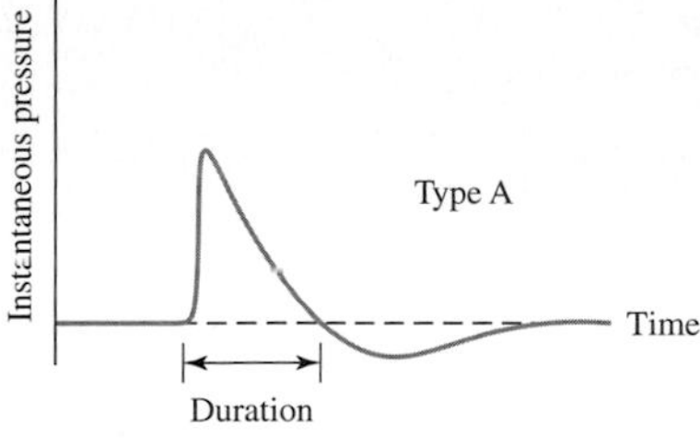

FIGURE 8-7
Type A impulse noise.

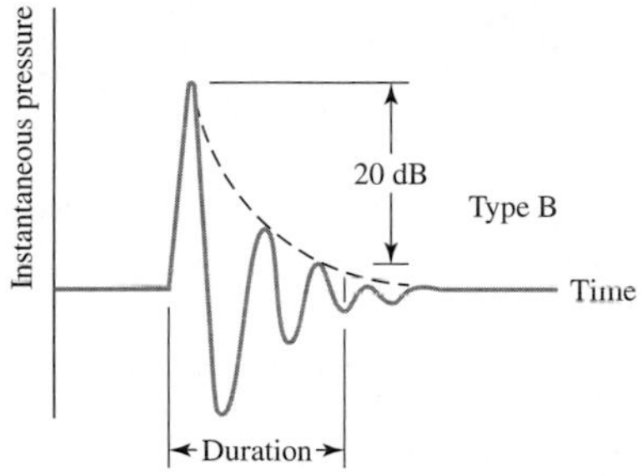

FIGURE 8-8
Type B impulse noise.

Two types of impulse noise generally are recognized. The type A impulse is characterized by a rapid rise to a peak sound pressure level followed by a small negative pressure wave or by decay to the background level (Figure 8-7). The type B impulse is characterized by a damped (oscillatory) decay (Figure 8-8). Where the duration of the type A impulse is simply the duration of the initial peak, the duration of the type B impulse is the time required for the envelope to decay to 20 dB below the peak. Because of the short duration of the impulse, a special sound-level meter must be employed to measure impulse noise. You should note that the peak sound pressure level is different than the impulse sound level because of the time-averaging used in the latter.

8-2 EFFECTS OF NOISE ON PEOPLE

For the purpose of our discussion, we have classified the effects of noise on people into the following two categories: auditory effects and psychological/sociological effects. Auditory effects include both hearing loss and speech interference. Psychological/sociological effects include annoyance, sleep interference, effects on performance, and acoustical privacy.

The Hearing Mechanism

Before we can discuss hearing loss, it is important to outline the general structure of the ear and how it works.

Anatomically, the ear is separated into three sections: the outer ear, the middle ear, and the inner ear (Figure 8-9). The outer and middle ear serve to convert sound pressure to vibrations. In addition, they perform the protective role of keeping debris and objects from reaching the inner ear. The Eustachian tube extends from the middle ear space to the upper part of the throat behind the soft palate. The tube is normally closed.

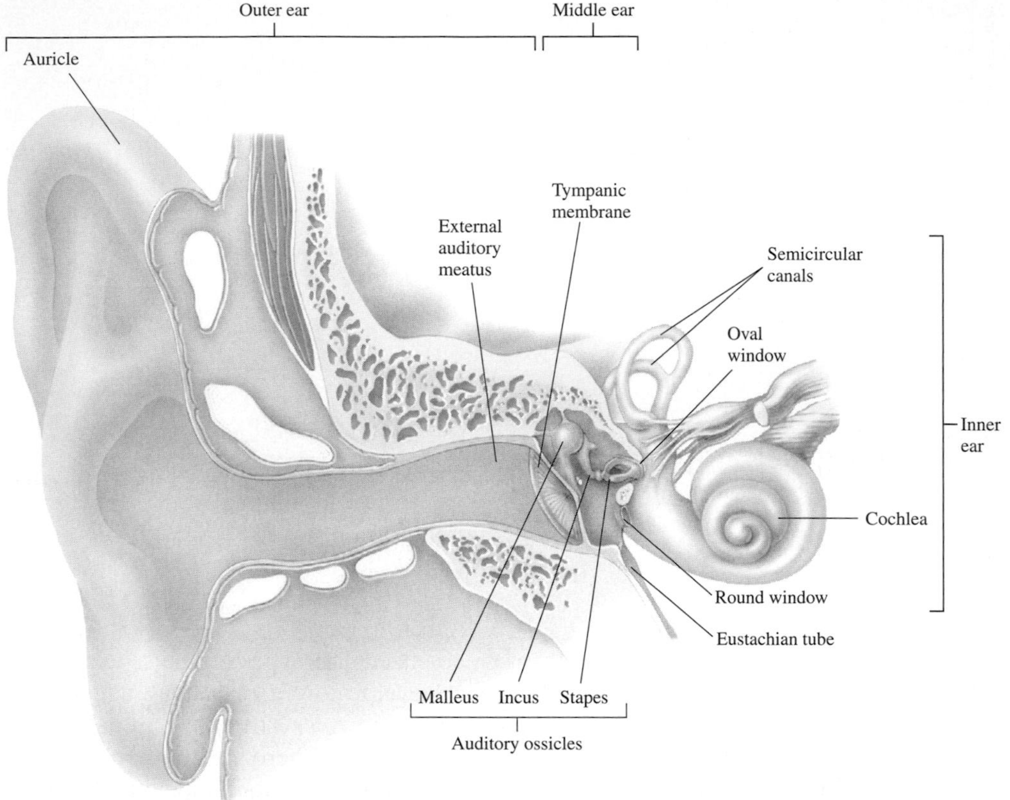

FIGURE 8-9
Anatomical divisions of the ear. (*Source:* Seeley et al., 2003.)

Contraction of the palate muscles during yawning, chewing, or swallowing opens the tubes. This allows the middle ear to ventilate and equalize pressure. If external air pressure changes rapidly, for example, by a sudden change in elevation, the tube is opened by involuntary swallowing or yawning to equalize the pressure.

The sound transducer mechanism is housed in the middle ear.* It consists of the *tympanic membrane* (eardrum) and three *ossicles* (bones) (Figure 8-10). The ossicles are supported by ligaments and may be moved by two muscles or by deflection of the tympanic membrane. The muscle movement is involuntary. Loud sounds cause these muscles to contract. This stiffens and diminishes the movement of the ossicular chain (Borg and Counter, 1989). The discussion on the middle ear that follows is excerpted from Clemis (1975).

> The primary function of the middle ear in the hearing process is to transfer sound energy from the outer to the inner ear. As the eardrum vibrates, it transfers its motion to the malleus. Since

*A transducer is a device that transmits power from one system to another. In this case, sound power is converted to mechanical displacement, which is later measured and interpreted by the brain.

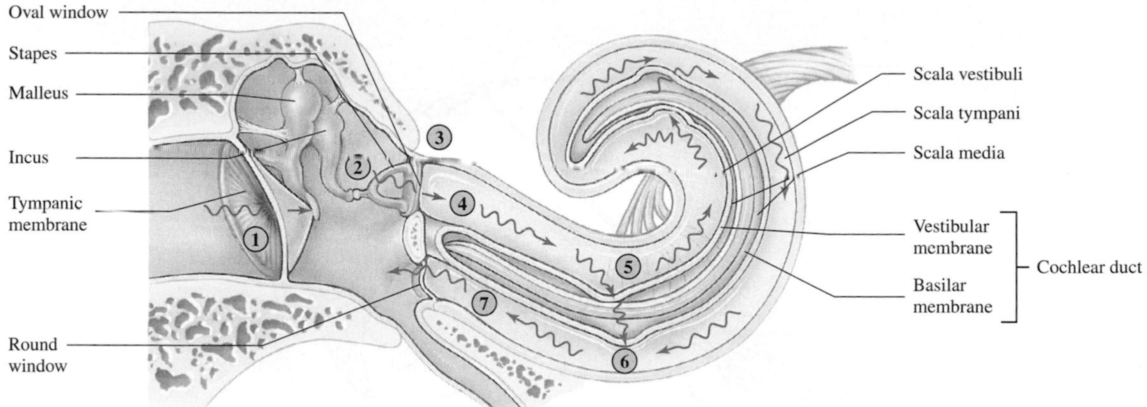

1. Sound waves strike the tympanic membrane and cause it to vibrate.
2. Vibration of the tympanic membrane causes the three bones of the middle ear to vibrate.
3. The foot plate of the stapes vibrates in the oval window.
4. Vibration of the foot plate causes the perilymph in the scala vestibuli to vibrate.
5. Vibration of the perilymph causes displacement of the basilar membrane. Short waves (high pitch) cause displacement of the basilar membrane near the oval window, and longer waves (low pitch) cause displacement of the basilar membrane some distance from the oval window. Movement of the basilar membrane is detected in the hair cells of the spiral organ, which are attached to the basilar membrane.
6. Vibrations of the perilymph in the scala vestibuli and of the endolymph in the cochlear duct are transferred to the perilymph of the scala tympani.
7. Vibrations in the perilymph of the scala tympani are transferred to the round window, where they are dampened.

FIGURE 8-10
The sound transducer mechanism housed in the middle ear. (*Source:* Seeley et al., 2003.)

the bones of the ossicular chain are connected to one another, the movements of the malleus are passed on to the incus, and finally to the stapes, which is imbedded in the oval window.

As the stapes moves back and forth in a rocking motion, it passes the vibrations into the inner ear through the oval window. Thus, the mechanical motion of the eardrum is effectively transmitted through the middle ear and into the fluid of the inner ear.

The sound-conducting transducer amplifies sound by two main mechanisms. First, the large surface area of the drum as compared to the small surface area of the base of the stapes (footplate) results in a hydraulic effect. The eardrum has about 25 times as much surface area as the oval window. All of the sound pressure collected on the eardrum is transmitted through the ossicular chain and is concentrated on the much smaller area of the oval window. This produces a significant increase in pressure.

The bones of the ossicular chain are arranged in such a way that they act as a series of levers. The long arms are nearest the eardrum, and the shorter arms are toward the oval window. The fulcrums are located where the individual bones meet. A small pressure on the long arm of the lever produces a much stronger pressure on the shorter arm. Since the longer arm is attached to the eardrum and the shorter arm is attached to the oval window, the ossicular chain acts as an amplifier of sound pressure. The magnification effect of the entire sound-conducting mechanism is about 22-to-1.

The inner ear houses both the balance receptors and the auditory receptors. The auditory receptors are in the *cochlea.* It is a bone shaped like a snail coiled two and one-half times around its own axis (Figure 8-9). A cross section through the cochlea

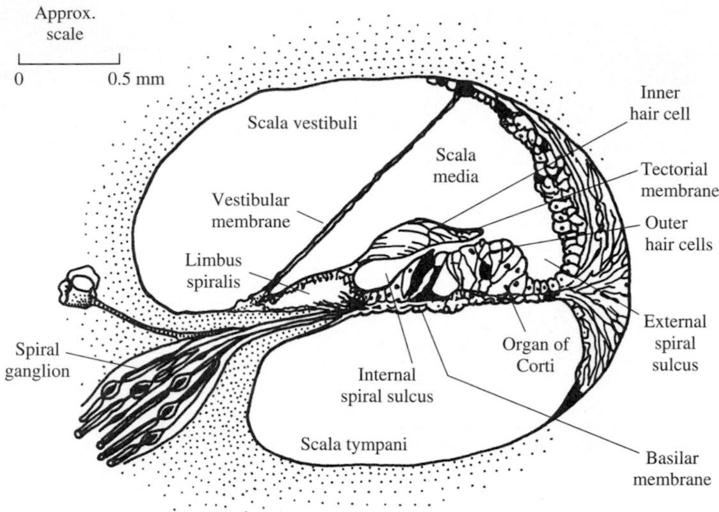

FIGURE 8-11
Cross section through the cochlea.

(Figure 8-11) reveals three compartments: the *scala vestibuli;* the *scala media;* and the *scala tympani.* The scala vestibuli and the scala tympani are connected at the apex of the cochlea. They are filled with a fluid called *perilymph,* in which the scala media floats. The hearing organ, the *organ of Corti,* is housed in the scala media. The scala media contains a different fluid, *endolymph,* which bathes the organ of Corti.

The scala media is triangular in shape and is about 34 mm in length. As shown in Figure 8-11, there are cells growing up from the *basilar membrane.* They have a tuft of hair at one end and are attached to the hearing nerve at the other end. A gelatinous membrane (*tectorial membrane*) extends over the hair cells and is attached to the *limbus spiralis.* The hair cells are embedded in the tectorial membrane.

Vibration of the oval window by the stapes causes the fluids of the three scalae to develop a wave-like motion. The movement of the basilar membrane and the tectorial membrane in opposite directions causes a shearing motion on the hair cells. The dragging of the hair cells sets up electrical impulses in the auditory nerves, which are transmitted to the brain.

The nerve endings near the oval and round windows are sensitive to high frequencies. Those near the apex of the cochlea are sensitive to low frequencies.

Normal Hearing

Frequency Range and Sensitivity. The ear of the young, audiometrically healthy, adult male responds to sound waves in the frequency range of 20 to 16,000 Hz. Young children and women often have the capacity to respond to frequencies up to 20,000 Hz. The speech zone lies in the frequency range of 500 to 2,000 Hz. The ear is most sensitive

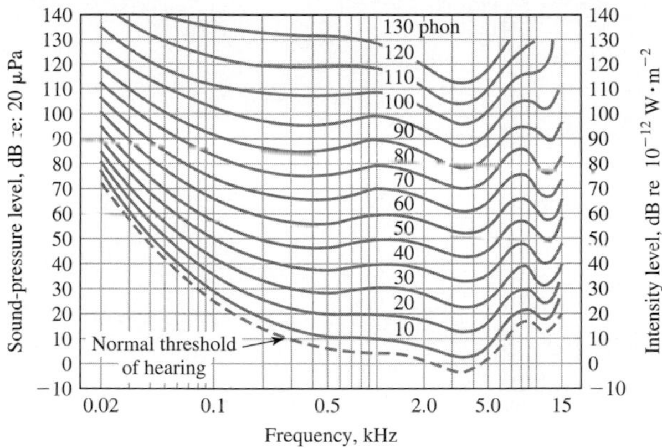

FIGURE 8-12
Fletcher-Munson equal loudness contours. (*Source:* Magrab, 1975).

in the frequency range from 2,000 to 5,000 Hz. The smallest perceptible sound pressure in this frequency range is 20 μPa.

A sound pressure of 20 μPa at 1,000 Hz in air corresponds to a 1.0 nm displacement of the air molecules. The thermal motion of the air molecules corresponds to a sound pressure of about 1 μPa. If the ear were much more sensitive, you would hear the air molecules crashing against your ear like waves on the beach!

Loudness. In general, two pure tones having different frequencies but the same sound pressure level will be heard as different loudness levels. Loudness level is a psychoacoustic quantity.

Fletcher and Munson (1935) conducted a series of experiments to determine the relationship between frequency and loudness. A reference tone and a test tone were presented alternately to the test subjects. They were asked to adjust the sound level of the test tone until it sounded as loud as the reference. The results were plotted as sound pressure level in dB versus the test tone frequency (Figure 8-12). The curves are called the Fletcher-Munson or *equal loudness contours*. The reference frequency is 1,000 Hz. The curves are labeled in *phons,* which are the sound pressure levels of the 1,000 Hz pure tone in dB. The lowest contour (dashed line) represents the "threshold of hearing." The actual threshold may vary by as much as ± 10 dB between individuals with normal hearing.

Audiometry Hearing tests are conducted with a device known as an *audiometer.* Basically, it consists of a source of pure tones with variable sound pressure level output into a pair of earphones. If the instrument also automatically prepares a graph of the test results (an *audiogram*), then it will include a weighting network called the *hearing threshold level* (HTL) scale.

The HTL scale is one in which the loudness of each pure tone is adjusted by frequency such that "0" dB is the level just audible for the average normal young ear. Two reference standards are in use: ASA–1951 and ANSI–1969. The ANSI reference

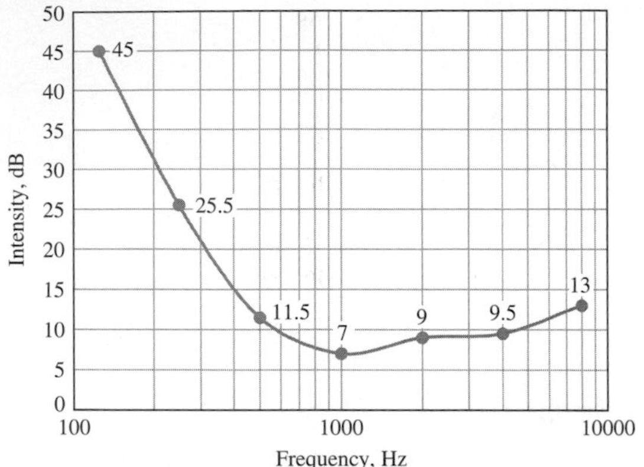

FIGURE 8-13
The ANSI reference values for hearing threshold level.

values are shown in Figure 8-13. Note the similarity to the Fletcher-Munson contours. The initial audiogram prepared for an individual may be referred to as the baseline HTL or simply as the HTL.

The audiogram shown in Figure 8-14 reflects excellent hearing response. The average normal response may vary ± 10 dB from the "0" dB value. As noted on the audiogram, this test was conducted with the ANSI–1969 weighting network.

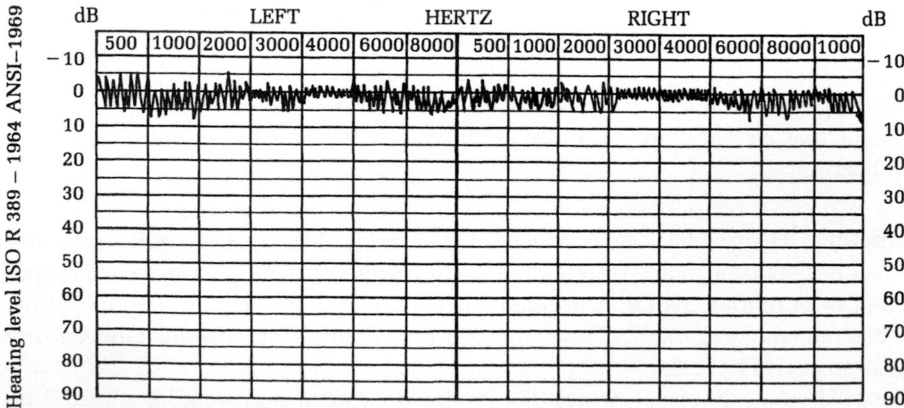

FIGURE 8-14
An audiogram illustrating excellent hearing response.

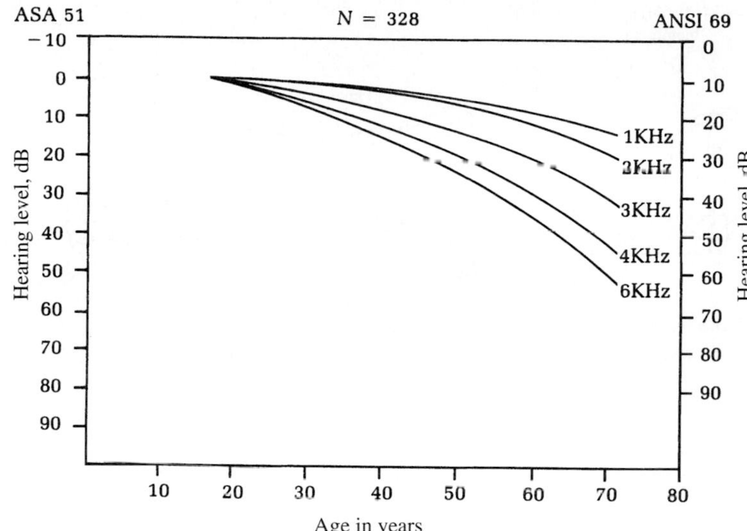

FIGURE 8-15
Hearing loss as a result of presbycusis. (*Source:* Olishifski and Harford, 1975.)

You may have noted that we keep stressing young in our references to normal hearing. This is because there is hearing loss due to the aging process. This type of loss is called *presbycusis.* The average amount of loss as a function of age is shown in Figure 8-15.

Hearing Impairment

Mechanism. With the exception of eardrum rupture from intense explosive noise, the outer and middle ear rarely are damaged by noise. More commonly, hearing loss is a result of neural damage involving injury to the hair cells (Figure 8-16). Two theories are offered to explain noise-induced injury. The first is that excessive shearing forces mechanically damage the hair cells. The second is that intense noise stimulation forces the hair cells into high metabolic activity, which overdrives them to the point of metabolic failure and consequent cell death. Once destroyed, hair cells are not capable of regeneration.

Measurement. Since direct observation of the organ of Corti in persons having potential hearing loss is impossible, injury is inferred from losses in their HTL. The increased sound pressure level required to achieve a new HTL is called *threshold shift.* Obviously, any measurement of threshold shift is dependent upon having a baseline audiogram taken before the noise exposure.

Hearing losses may be either temporary or permanent. Noise-induced losses must be separated from other causes of hearing loss such as age (presbycusis), drugs, disease, and blows on the head. *Temporary threshold shift* (TTS) is distinguished *from permanent threshold shift* (PTS) by the fact that in TTS removal of the noise overstimulation will result in a gradual return to baseline hearing thresholds.

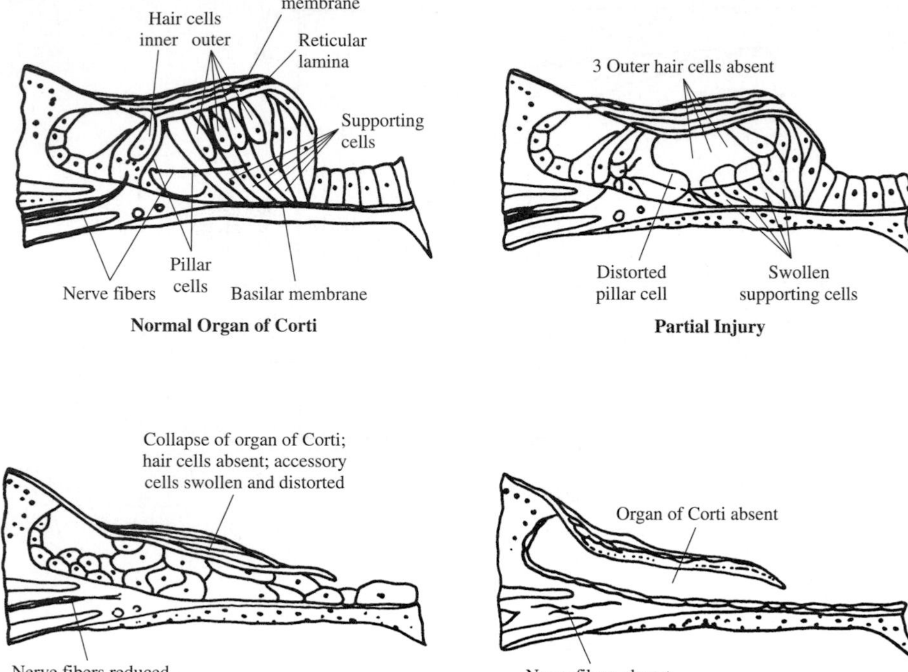

FIGURE 8-16
Various degrees of injury to the hair cells.

Factors Affecting Threshold Shift. Important variables in the development of temporary and permanent hearing threshold changes include the following (NIOSH, 1972).

1. Sound level: Sound levels must exceed 60 to 80 dBA before the typical person will experience TTS.
2. Frequency distribution of sound: Sounds having most of their energy in the speech frequencies are more potent in causing a threshold shift than are sounds having most of their energy below the speech frequencies.
3. Duration of sound: The longer the sound lasts, the greater the amount of threshold shift.
4. Temporal distribution of sound exposure: The number and length of quiet periods between periods of sound influences the potentiality of threshold shift.
5. Individual differences in tolerance of sound may vary greatly among individuals.
6. Type of sound—steady-state, intermittent, impulse, or impact: The tolerance to peak sound pressure is greatly reduced by increasing the duration of the sound.

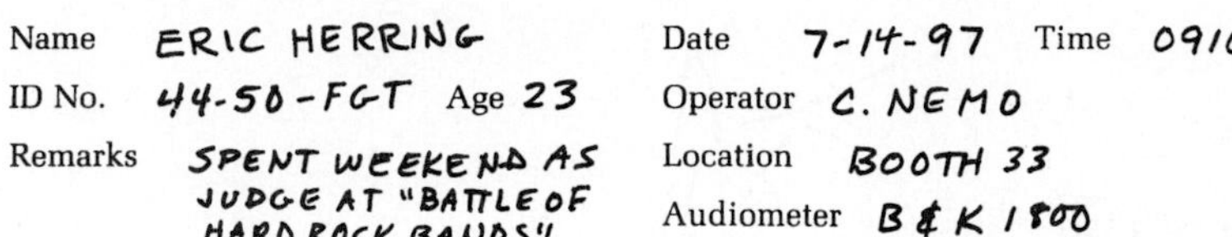

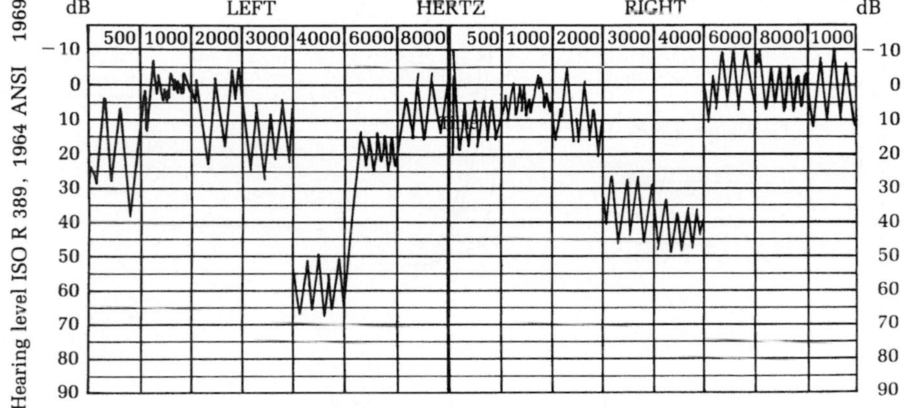

FIGURE 8-17
An audiogram illustrating hearing loss at the high frequency notch.

Temporary Threshold Shift (TTS). TTS is often accompanied by a ringing in the ear, muffling of sound, or discomfort of the ears. Most of the TTS occurs during the first two hours of exposure. Recovery to the baseline HTL after TTS begins within the first hour or two after exposure. Most of the recovery that is going to be attained occurs within 16 to 24 hours after exposure.

Permanent Threshold Shift (PTS). There appears to be a direct relationship between TTS and PTS. Noise levels that do not produce TTS after two to eight hours of exposure will not produce PTS if continued beyond this time. The shape of the TTS audiogram will resemble the shape of the PTS audiogram.

Noise-induced hearing loss generally is first characterized by a sharply localized dip in the HTL curve at the frequencies between 3,000 and 6,000 Hz. This dip commonly occurs at 4,000 Hz (Figure 8-17). This is the *high frequency notch.* The progress from TTS to PTS with continued noise exposure follows a fairly regular pattern. First, the high frequency notch broadens and spreads in both directions. While substantial losses may occur above 3,000 Hz, the individual will not notice any change in hearing. In fact, the individual will not notice any hearing loss until the speech frequencies between 500 and 2,000 Hz average more than a 25 dB increase in HTL on the ANSI—1969 scale. The onset and progress of noise-induced permanent hearing loss is slow and insidious. The exposed individual is unlikely to notice it. Total hearing loss from noise exposure has not been observed.

Acoustic Trauma. The outer and middle ear rarely are damaged by intense noise. However, explosive sounds can rupture the tympanic membrane or dislocate the ossicular chain. The permanent hearing loss that results from very brief exposure to a very

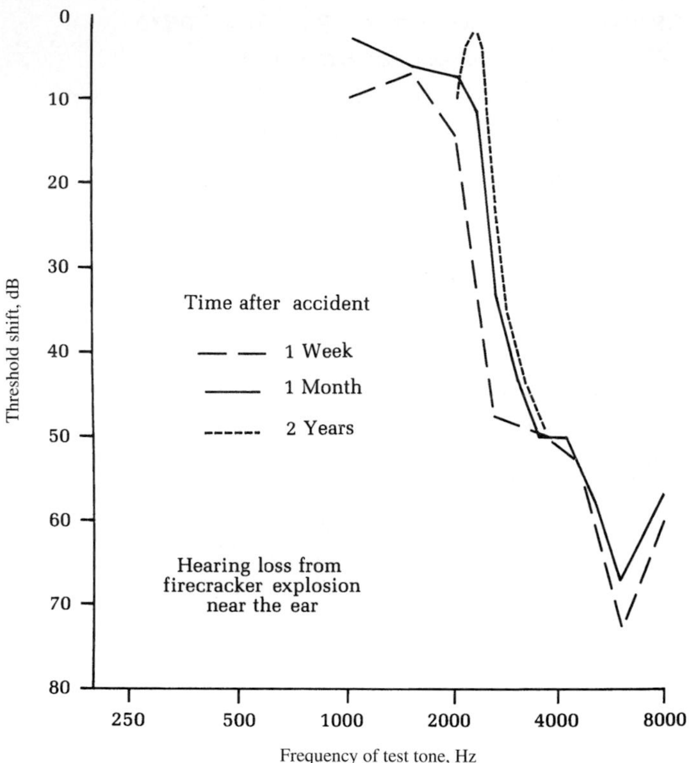

FIGURE 8-18
An example audiogram illustrating acoustic trauma. (*Source:* Ward and Glorig, 1961.)

loud noise is termed *acoustic trauma* (Davis, 1958). Damage to the outer and middle ear may or may not accompany acoustic trauma. Figure 8-18 is an example of an audiogram that illustrates acoustic trauma.

Protective Mechanisms. Although the extent and mechanisms are not clear, it appears that the structures of the middle ear offer some protection to the delicate sensory organs of the inner ear (Borg and Counter, 1989). One mechanism of protection is a change in the mode of vibration of the stapes. As noted earlier, there is evidence that the muscles of the middle ear contract reflexively in response to loud noise. This contraction results in a reduction in the amplification that this series of levers normally produces. Changes in transmission may be on the order of 20 dB. However, the reaction time of the muscle/bone structure is on the order of 100–200 milliseconds. Thus, this protection is not effective against steep acoustic wave fronts that are characteristic of impact or impulsive noise.

Damage-Risk Criteria

A damage-risk criterion specifies the maximum allowable exposure to which a person may be exposed if risk of hearing impairment is to be avoided. The American

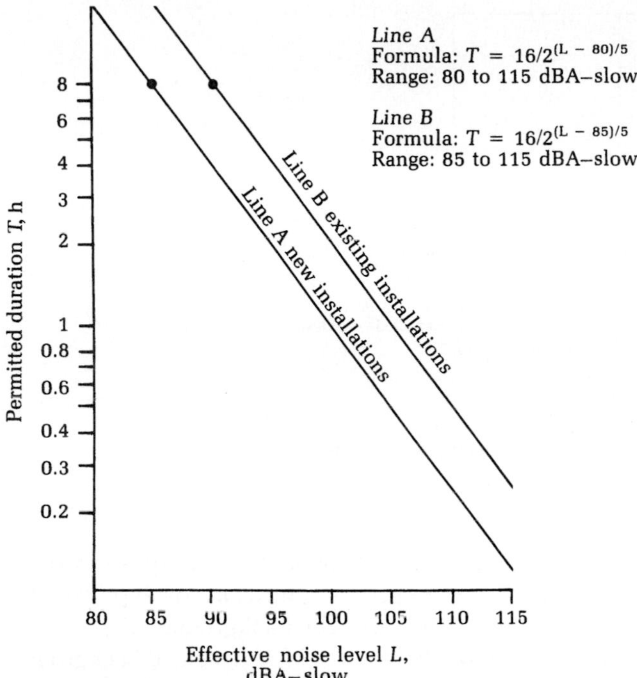

FIGURE 8-19
NIOSH occupational noise exposure limits for continuous or intermittent noise exposure.

Academy of Ophthalmology and Otolaryngology has defined hearing impairment as an average HTL in excess of 25 dB (ANSI–1969) at 500, 1,000, and 2,000 Hz. This is called the *low fence.* Total impairment is said to occur when the average HTL exceeds 92 dB. Presbycusis is included in setting the 25 dB ANSI low fence. Two criteria have been set to provide conditions under which nearly all workers may be repeatedly exposed without adverse effect on their ability to hear and understand normal speech.

Continuous or Intermittent Exposure. The National Institute for Occupational Safety and Health (NIOSH) has recommended that occupational noise exposure be controlled so that no worker is exposed in excess of the limits defined by line B in Figure 8-19. In addition, NIOSH recommends that new installations be designed to hold noise exposure below the limits defined by line A in Figure 8-19. The Walsh-Healey Act, which was enacted by Congress in 1969 to protect workers, used a damage-risk criterion equivalent to the line A criterion.

Speech Interference

As we all know, noise can interfere with our ability to communicate. Many noises that are not intense enough to cause hearing impairment can interfere with speech communication. The interference, or *masking,* effect is a complicated function of the distance

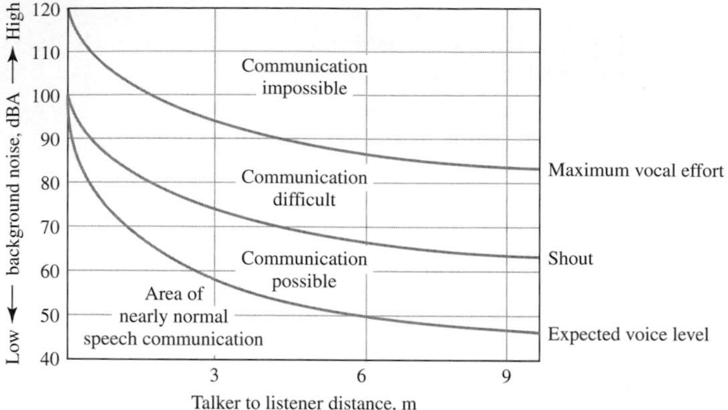

FIGURE 8-20
Quality of speech communication as a function of sound level and distance. (*Source:* Miller, 1971.)

between the speaker and listener and the frequency components of the spoken words. The Speech Interference Level (SIL) was developed as a measure of the difficulty in communication that could be expected with different background noise levels (Beranek, 1954). It is now more convenient to talk in terms of A-weighted background noise levels and the quality of speech communication (Figure 8-20).

Example 8-5. Consider the problem of a speaker in a quiet zone who wishes to speak to a listener operating a 4.5 Mg (megagram) truck 6.0 m away. The sound level in the truck cab is about 73 dBA.

Solution. Using Figure 8-20, we can see that she is going to have to shout very loudly to be heard. However, if she moved to within about 1.0 m, she would be able to use her "expected" voice level, that is, the unconscious slight rise in voice level that one would normally use in a noisy situation.

It can be seen that at distances not uncommon in living rooms or classrooms (4.5 to 6.0 m), the A-weighted background level must be below about 50 dB for normal conversation.

Annoyance

Annoyance by noise is a response to auditory experience. Annoyance has its base in the unpleasant nature of some sounds, in the activities that are disturbed or disrupted by noise, in the physiological reactions to noise, and in the responses to the meaning of "messages" carried by the noise (Miller, 1971). For example, a sound heard at night may be more annoying than one heard by day, just as one that fluctuates may be more annoying than one that does not. A sound that resembles another sound that we already dislike and that perhaps threatens us may be especially annoying. A sound that we know is mindlessly inflicted and will not be removed soon may be more annoying than one that is temporarily and regretfully inflicted. A sound, the source of which is visible, may be more annoying than one with an invisible source. A sound that is new may be less annoying. A sound that is locally a political issue may have a particularly high or low annoyance (May, 1978).

The degree of annoyance and whether that annoyance leads to complaints, product rejection, or action against an existing or anticipated noise source depend upon many factors. Some of these factors have been identified, and their relative importance has been assessed. Responses to aircraft noise have received the greatest attention. There is less information available concerning responses to other noises, such as those of surface transportation and industry, and those from recreational activities (Miller, 1971). Many of the noise rating or forecasting systems that are now in existence were developed in an effort to predict annoyance reactions.

Sonic Booms. One noise of special interest with respect to annoyance is called *sonic boom* or, more correctly as we shall see, sonic booms.

> The flow of air around an aircraft or other object whose speed exceeds the speed of sound (supersonic) is characterized by the existence of discontinuities in the air known as *shock wave*. These discontinuities result from the sudden encounter of an impenetrable body with air. At subsonic speeds, the air seems to be forewarned; thus, it begins its outward flow before the arrival of the leading edge. At supersonic speeds, however, the air in front of the aircraft is undisturbed, and the sudden impulse at the leading edge creates a region of overpressure (Figure 8-21) where the pressure is higher than atmospheric pressure. This overpressure region travels outward with the speed of sound, creating a conically shaped shock wave called the *bow wave* that changes the direction of airflow. A second shock wave, the *tail wave,* is produced by the tail of the aircraft and is associated with a region where the pressure is lower than normal. This underpressure discontinuity causes the air behind the aircraft to move sideways.

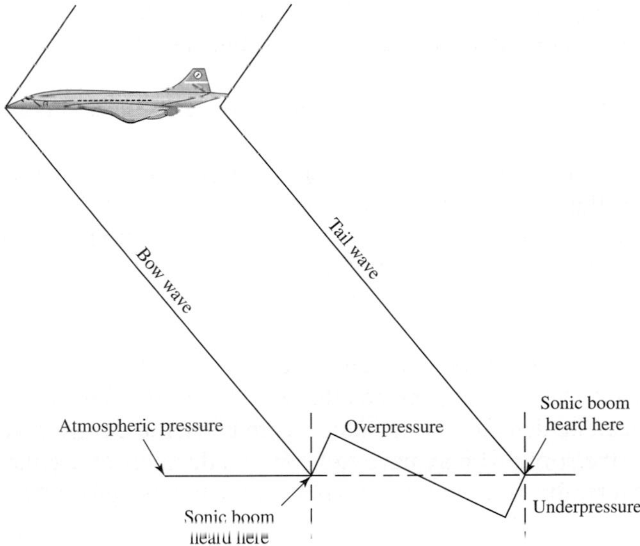

FIGURE 8-21
Sonic booms resulting from bow wave and tail wave set in motion by supersonic flight.

> Major pressure changes are experienced at the ear as the bow and tail shock waves reach an observer. Each of these pressure deviations produces the sensation of an explosive sound (Minnix, 1978).

You should note that the pressure wave and, hence, the sonic boom exist whenever the aircraft is at supersonic speed and not "just when it breaks the sound barrier."

Both the loudness of the noise and the startling effect of the impulse (it makes us "jump") are found to be very annoying. Apparently we can never get used to this kind of noise. Supersonic flight by commercial aircraft is forbidden in the airspace above the United States. Supersonic flight by military aircraft is restricted to sparsely inhabited areas.

Sleep Interference

Sleep interference is a special category of annoyance that has received a great deal of attention and study. Almost all of us have been wakened or kept from falling asleep by loud, strange, frightening, or annoying sounds. It is commonplace to be wakened by an alarm clock or clock radio. But it also appears that one can get used to sounds and sleep through them. Possibly, environmental sounds only disturb sleep when they are unfamiliar. If so, disturbance of sleep would depend only on the frequency of unusual or novel sounds. Everyday experience also suggests that sound can help to induce sleep and, perhaps, to maintain it. The soothing lullaby, the steady hum of a fan, or the rhythmic sound of the surf can serve to induce relaxation. Certain steady sounds can serve as an acoustical shade and mask disturbing transient sounds.

Common anecdotes about sleep disturbance suggest an even greater complexity. A rural person may have difficulty sleeping in a noisy urban area. An urban person may be disturbed by the quiet when sleeping in a rural area. And how is it that a parent may wake to a slight stirring of his or her child, yet sleep through a thunderstorm? These observations all suggest that the relations between exposure to sound and the quality of a night's sleep are complicated.

The effects of relatively brief noises (about three minutes or less) on a person sleeping in a quiet environment have been studied the most thoroughly. Typically, presentations of the sounds are widely spaced throughout a sleep period of 5 to 7 hours. A summary of some of these observations is presented in Figure 8-22. The dashed lines are hypothetical curves that represent the percent of awakenings under conditions in which the subject is a normally rested young adult male who has been adapted for several nights to the procedures of a quiet sleep laboratory. He has been instructed to press an easily reached button to indicate that he has awakened, and had been moderately motivated to awake and respond to the noise.

While in light sleep, subjects can awake to sounds that are about 30–40 decibels above the level at which they can be detected when subjects are conscious, alert, and attentive. While in deep sleep, the stimulus may have to be 50–80 decibels above the level at which they can be detected by conscious, alert, attentive subjects before they will awaken the sleeping subject.

The solid lines in Figure 8-22 are data from questionnaire studies of persons who live near airports. The percentage of respondents who claim that flyovers wake them or keep them from falling asleep is plotted against the A-weighted sound level of a single

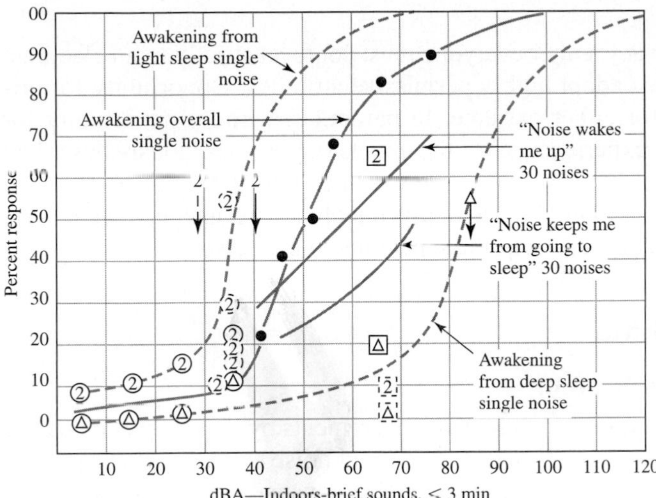

FIGURE 8-22
Effects of brief noise on sleep (*Source:* Miller, 1971.)

flyover. These curves are for the case of approximately 30 flyovers spaced over the normal sleep period of six to eight hours. The filled circles represent the percentage of sleepers that awake to a three minute sound at each A-weighted sound level (dBA) or lower. This curve is based on data from 350 persons, each tested in his or her own bedroom. These measures were made between 2:00 and 7:00 A.M. It is reasonable to assume that most of the subjects were roused from a light sleep.

Effects on Performance

When a task requires the use of auditory signals, speech or nonspeech, then noise at any intensity level sufficient to mask or interfere with the perception of these signals will interfere with the performance of the task.

Where mental or motor tasks do not involve auditory signals, the effects of noise on their performance have been difficult to assess. Human behavior is complicated, and it has been difficult to discover exactly how different kinds of noises might influence different kinds of people doing different kinds of tasks. Nonetheless, the following general conclusions have emerged. Steady noises without special meaning do not seem to interfere with human performance unless the A-weighted noise level exceeds about 90 decibels. Irregular bursts of noise (intrusive noise) are more disruptive than steady noises. Even when the A-weighted sound levels of irregular bursts are below 90 decibels, they may sometimes interfere with performance of a task. High-frequency components of noise, above about 1,000–2,000 hertz, may produce more interference with performance than low-frequency components of noise. Noise does not seem to influence the overall rate of work, but high levels of noise may increase the variability of the rate of work. There may be "noise pauses" followed by compensating increases in work rate. Noise is more likely to reduce the accuracy of work than to reduce the total quantity of work. Complex tasks are more likely to be adversely influenced by noise than are simple tasks.

Acoustic Privacy

Without opportunity for privacy, either everyone must conform strictly to an elaborate social code or everyone must adopt highly permissive attitudes. Opportunity for privacy avoids the necessity for either extreme. In particular, without opportunity for acoustical privacy, one may experience all of the effects of noise previously described and, in addition, one is constrained because one's own activities may disturb others. Without acoustical privacy, sound, like a faulty telephone exchange, reaches the "wrong number." The result disturbs both the sender and the receiver.

8-3 RATING SYSTEMS

Goals of a Noise-Rating System

An ideal noise-rating system is one that allows measurements by sound level meters or analyzers to be summarized succinctly and yet represent noise exposure in a meaningful way. In our previous discussions on loudness and annoyance, we noted that our response to sound is strongly dependent on the frequency of the sound. Furthermore, we noted that the type of noise (continuous, intermittent, or impulsive) and the time of day that it occurred (night being worse than day) were significant factors in annoyance.

Thus, the ideal system must take frequency into account. It should differentiate between daytime and nighttime noise. And, finally, it must be capable of describing the cumulative noise exposure. A statistical system can satisfy these requirements.

The practical difficulty with a statistical rating system is that it would yield a large set of parameters for each measuring location. A much larger array of numbers would be required to characterize a neighborhood. It is literally impossible for such an array of numbers to be used effectively in enforcement. Thus, there has been a considerable effort to define a single number measure of noise exposure. The following paragraphs describe two of the systems now being used.

The L_N Concept

The parameter L_N is a statistical measure that indicates how frequently a particular sound level is exceeded. If, for example, we write $L_{30} = 67$ dBA, then we know that

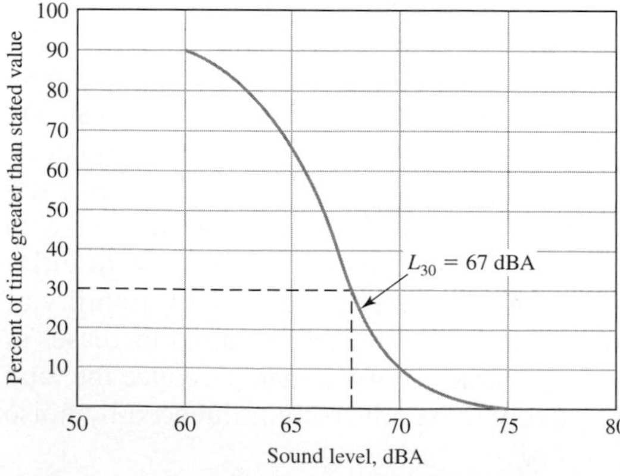

FIGURE 8-23
Cumulative distribution curve.

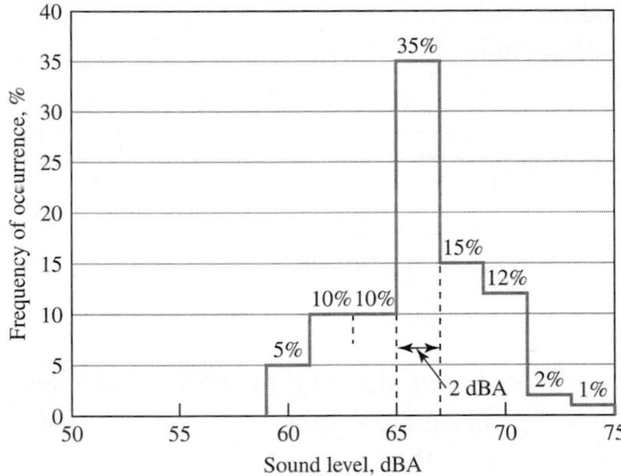

FIGURE 8-24
Probability distribution plot.

67 dB(A) was exceeded for 30 percent of the measuring time. A plot of L_N against N where $N = 1$ percent, 2 percent, 3 percent, and so forth, would look like the cumulative distribution curve shown in Figure 8-23.

Allied to the cumulative distribution curve is the probability distribution curve. A plot of this will show how often the noise levels fall into certain class intervals. In Figure 8-24 we can see that 35 percent of the time the measured noise levels ranged between 65 and 67 dBA; for 15 percent of the time they ranged between 67 and 69 dBA; and so on. The relationship between this picture and the one for L_N is really quite simple. By adding the percentages given in successive class intervals from right to left, we can arrive at a corresponding L_N where N is the sum of the percentages and L is the lower limit of the left-most class interval added, thus, L_{30}

$$L\,(1 + 2 + 12 + 15) = 67 \text{ dBA}$$

The L_{eq} Concept

The equivalent continuous equal energy level (L_{eq}) can be applied to any fluctuating noise level. It is that constant noise level that, over a given time, expends the same amount of energy as the fluctuating level over the same time period. It is expressed as follows:

$$L_{eq} = 10 \log \frac{1}{t} \int_0^t 10^{L(t)/10}\, dt \qquad (8\text{-}14)$$

where t = the time over which L_{eq} is determined
$L(t)$ = the time varying noise level in dBA

Generally speaking, there is no well-defined relationship between $L(t)$ and time, so a series of discrete samples of $L(t)$ have to be taken. This modifies the expression to:

$$L_{eq} = 10 \log \sum_{i=t}^{i=n} (10^{L_i/10})(t_i) \qquad (8\text{-}15)$$

where n = the total number of samples taken
L_i = the noise level in dBA of the ith sample
t_i = fraction of total sample time

Example 8-6. Consider the case where a noise level of 90 dBA exists for 10 minutes and is followed by a reduced noise level of 70 dBA for 30 minutes. What is the equivalent continuous equal energy level for the 40-minute period? Assume a five-minute sampling interval.

Solution. If the sampling interval is five minutes, then the total number of samples (n) is 8, and the fraction of total sample time (t_i) for each sample is 1/8 = 0.125. With these preliminary calculations, we may now compute the sum:

$$\sum_{t=1}^{2} = (10^{90/10})(0.250) + (10^{70/10})(0.750)$$

$$= (2.50 \times 10^8) + (7.50 \times 10^6) = 2.58 \times 10^8$$

And finally, we take the log to find

$$L_{eq} = 10 \log(2.58 \times 10^8) = 84.11, \text{ or } 84, \text{ dBA}$$

The example calculation is depicted graphically in Figure 8-25. From this you may note that great emphasis is put on occasional high noise levels.

The equivalent noise level was introduced in 1965 in Germany as a rating specifically to evaluate the impact of aircraft noise upon the neighbors of airports (Burck et al., 1965). It was almost immediately recognized in Austria as appropriate for evaluating the impact of street traffic noise in dwellings and schoolrooms. It has been embodied in the National Test Standards of Germany for rating the subjective effects of fluctuating noises of all kinds, such as from street and road traffic, rail traffic, canal and river ship traffic, aircraft, industrial operations (including the noise from individual machines), sports stadiums, playgrounds, and the like.

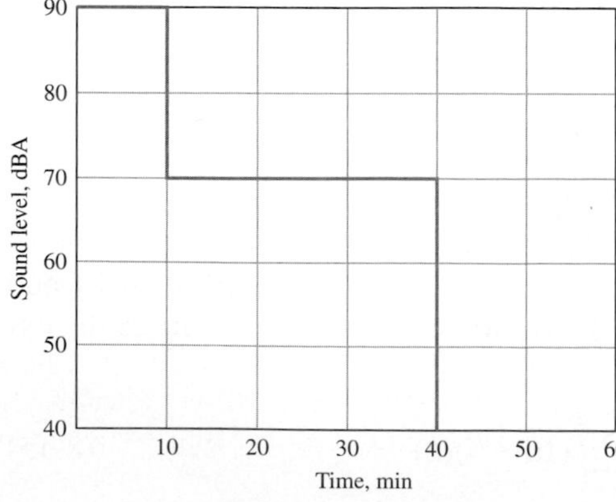

FIGURE 8-25
Graphical illustration of L_{eq} computation given in Example 8-6.

The L_{dn} Concept

The L_{dn} is the L_{eq} computed over a 24-hour period with a "penalty" of 10 dBA for a designated nighttime period. Thus, it is a day-night average and the subscript "dn" is assigned instead of "eq." In applications to airport noise, the L_{dn} may be referred to as DNL. The nighttime period is from 10 PM. to 7 AM. The L_{dn} equation is derived from the L_{eq} equation with the time increment specified as 1 second. Because the time over which the L_{dn} is computed is a day, the total time period is 86,400 seconds. Equation 8–15 is then written as

$$L_{dn} = 10 \log \left[\frac{1}{86{,}400} \sum 10^{Li/10} t_i + \sum 10^{(L_j + 10)/10} t_i \right] \tag{8-16}$$

Because 10(log 86,400) ≈ 49.4, the day-night average sound level may be written as

$$L_{dn} = 10 \log \left[\sum 10^{\mathrm{L}i/10} t_i + \sum 10^{(L_j + 10)/10} t_i \right] - 49.4 \tag{8-17}$$

8-4 COMMUNITY NOISE SOURCES AND CRITERIA

It is not our intent to provide a detailed discussion of the noise characteristics of all community noise sources. Likewise, we have not attempted to provide a comprehensive list of noise criteria. Rather, we have selected a few examples to provide you with a feeling for the magnitude and range of the numbers.

Transportation Noise

Aircraft Noise. The noise spectra of a wide body fan jet (for example, the Boeing 747) reveal that sound pressure levels are higher on takeoff than during the approach to land. This is typical of all aircraft. With the notable exception of the turbojets, smaller aircraft have lower sound pressure levels.

The annoyance criteria for aircraft operations are based on extensive field measurements and opinion surveys. The results of annoyance surveys at nine airports in the United States and Great Britain are summarized in Figure 8-26.

Highway Vehicle Noise. For most automobiles, exhaust noise constitutes the predominant source for normal operation below about 55 km/h (Figure 8-27). Although tire noise is much less of a problem in automobiles than in trucks, it is the dominant noise source at speeds above 80 km/h. While not as noisy as trucks, the total contribution of automobiles to the noise environment is significant because of the very large number in operation.

Diesel trucks are 8 to 10 dB noisier than gasoline-powered ones. At speeds above 80 km/h, tire noise often becomes the dominant noise source on the truck. The "crossbar" tread is the noisiest.

Motorcycle noise is highly dependent on the speed of the vehicle. The primary source of noise is the exhaust. The noise spectra of two-cycle and four-cycle engines are of somewhat different character. The two-cycle engines exhibit more high frequency spectra energy content.

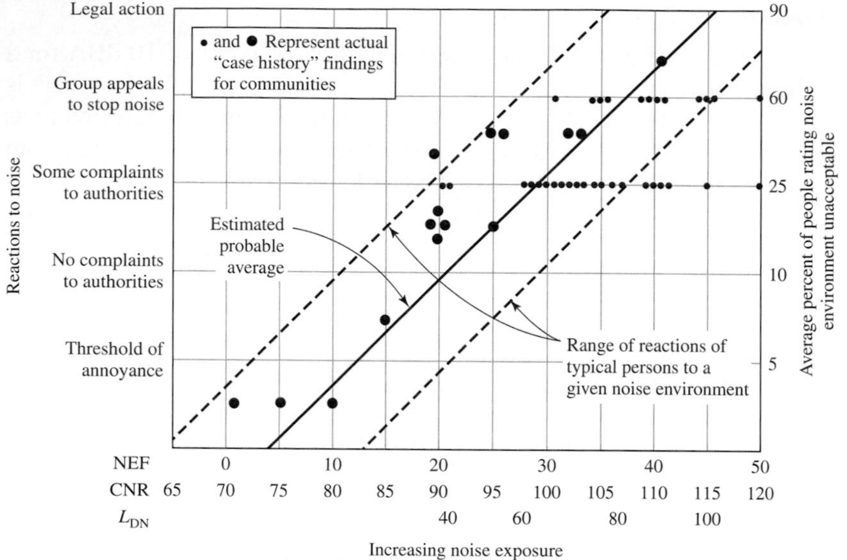

FIGURE 8-26
Relationship between exposure to aircraft noise and annoyance. (*Source:* Kryter, et al., 1971.)

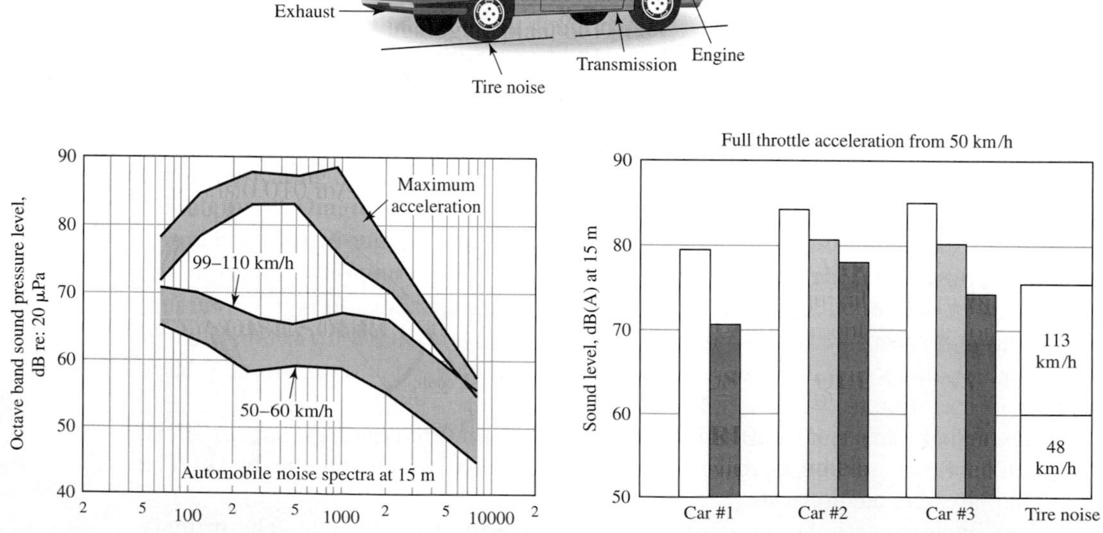

FIGURE 8-27
Typical noise spectra of automobiles. (*Source:* U.S. EPA, 1971.)

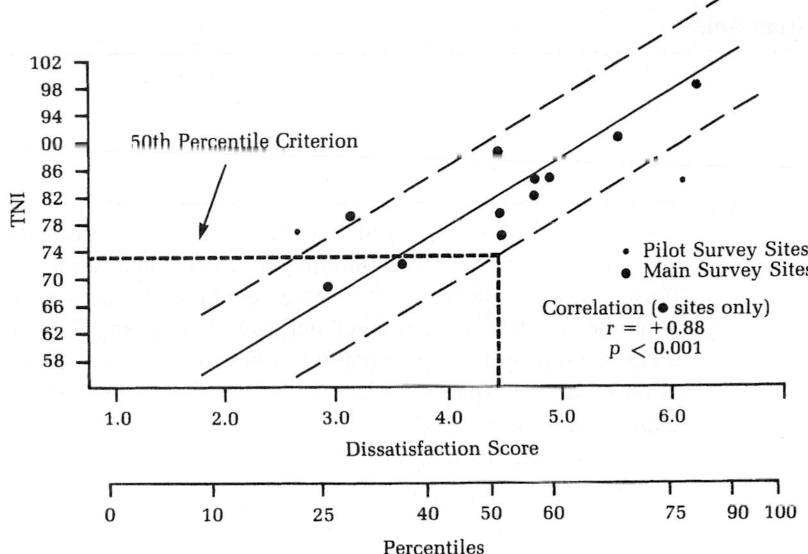

FIGURE 8-28
Annoyance as a function of the Traffic Noise Index (TNI). (*Source:* Alexandre et al., 1975.)

In 1968, Griffiths and Langdon (1968) reported on the results of an extensive attitude survey on traffic noise. They correlated their results with the Traffic Noise Index rating system (Figure 8-28). The U.S. Federal Highway Administration has developed the standards shown in Table 8-3. The levels are above those that would be expected to yield no problems but are below those of many existing highways.

Other Internal Combustion Engines

Because of their ubiquitous nature and the general interest they stimulate, the combustion engines listed in Table 8-4 are included at this point. "In general, these devices are not significant contributors to average residential noise levels in urban areas. However, the relative annoyance of most of the equipment tends to be high" (U.S. EPA, 1971). The eight-hour exposure level is in reference to the equipment operator.

Construction Noise

The range of sound levels found for 19 common types of construction equipment is shown in Figure 8-29 on page 687. Although the sample was limited, the data appear to be reasonably accurate. The noise produced by the interaction of the machine and the material on which it acts often contributes greatly to the sound level.

It is difficult, at best, to quantify the annoyance that results from construction noise. The following generalizations appear to hold:

1. Single house construction in suburban communities will generate sporadic complaints if the boundary line eight-hour L_{eq} exceeds 70 dBA.

TABLE 8-3
FHA noise standards for new construction[a]

Land use category	Exterior design noise level dBA[b] L_{eq}	L_{10}	Description of land use category
A	57	60	Tracts of lands in which serenity and quiet are of extraordinary significance and serve an important public need, and where the preservation of those qualities is essential if the area is to continue to serve its intended purpose. For example, such areas could include amphitheaters, particular parks or portions of parks, or open spaces, which are dedicated or recognized by appropriate local officials for activities requiring special qualities of serenity and quiet.
B	67	70	Residences, motels, hotels, public meeting rooms, schools, churches, libraries, hospitals, picnic areas, recreation areas, playgrounds, active sports areas, and parks.
C	72	75	Developed lands, properties, or activities not included in categories A and B above.
D	Unlimited	Unlimited	Undeveloped lands.
E	52 (Interior)	55 (Interior)	Public meeting rooms, schools, churches, libraries, hospitals, and other such public buildings.

[a]FHWA, 1973.
[b]Either L_{eq} or L_{10} may be used, but not both. The levels are to be based on a 1-hour sample.

TABLE 8-4
Summary of noise characteristics of internal combustion engines

Source	A-weighted noise energy (kw · h/d)[a]	Typical A-weighted noise level at 15.2 m [dB(A)]	8-hr exposure level [db(A)][b] Average	Maximum	Typical exposure time (h)
Lawn mowers	63	74	74	82	1.5
Garden tractors	63	78	N/A	N/A	N/A
Chain saws	40	82	85	95	1
Snow blowers	40	84	61	75	1
Lawn edgers	16	78	67	75	0.5
Model aircraft	12	78	70[c]	79[c]	0.25
Leaf blowers	3.2	76	67	75	0.25
Generators	0.8	71	—	—	—
Tillers	0.4	70	72	80	1

[a]Based on estimates of the total number of units in operation per day.
[b]Equivalent level for evaluation of relative hearing damage risk.
[c]During engine trimming operation.
(*Source:* U.S. EPA, 1971.)

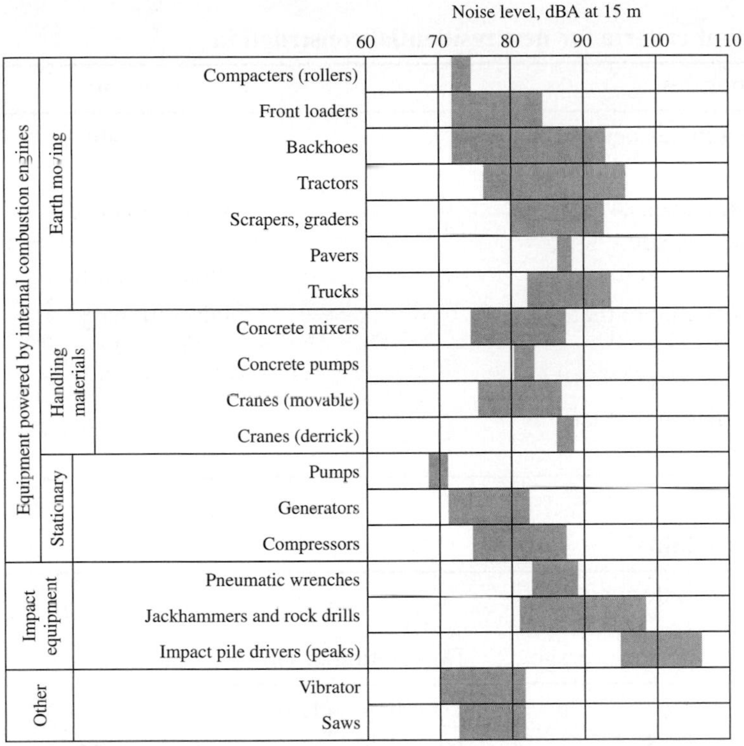

Noise level, dBA at 15 m

60 70 80 90 100 110

Equipment powered by internal combustion engines

Earth moving: Compacters (rollers); Front loaders; Backhoes; Tractors; Scrapers, graders; Pavers; Trucks

Handling materials: Concrete mixers; Concrete pumps; Cranes (movable); Cranes (derrick)

Stationary: Pumps; Generators; Compressors

Impact equipment: Pneumatic wrenches; Jackhammers and rock drills; Impact pile drivers (peaks)

Other: Vibrator; Saws

FIGURE 8-29
Range of sound levels from various types of construction equipment (based on limited available data samples). (*Source:* U.S. EPA, 1972.)

2. Major excavation and construction in a normal suburban community will generate threats of legal action if the boundary line eight-hour L_{eq} exceeds 85 dBA.

Zoning and Siting Considerations

The U.S. Department of Housing and Urban Development (HUD) set out guideline criteria for noise exposure at residential sites for new construction (Table 8-5). The Federal Aviation Administration (FAA) specifies the L_{dn} for various types of land use compatibility (Table 8-6). These guidelines, and those given on page 686 for traffic noise (Table 8-3), if followed in zoning and siting, will minimize annoyance and complaints.

Levels to Protect Health and Welfare

In accordance with the directive from Congress, the U.S. Environmental Protection Agency published noise criteria levels that it deemed necessary to protect the health and welfare of U.S. citizens (Table 8-7) (U.S. EPA, 1974). The EPA maintained that a quiet residential environment is necessary in both urban and rural areas to prevent

TABLE 8-5
HUD noise assessment criteria for new residential construction

General external exposures	Assessment
Exceeds 89 dBA 60 minutes per 24 hours Exceeds 75 dBA 8 hours per 24 hours	Unacceptable
Exceeds 65 dBA 8 hours per 24 hours Loud repetitive sounds on site	Discretionary: normally unacceptable
Does not exceed 65 dBA more than 8 hours per 24 hours	Discretionary: normally acceptable
Does not exceed 45 dBA more than 30 minutes per 24 hours	Acceptable

TABLE 8-6
FAA land use compatability[a]

Land use category	Exterior L_{dn}, dBA yearly average	Description of land use category
Residential	<65	Dwellings schools
Public	<65	Hospitals, nursing homes, churches and auditoriums[b]
Public	65–70	Government services[c]
Commercial	65–70	Offices, retail trade, communication[c]
Commercial	80–85	Wholesale and retail equipment, utilities
Manufacturing and Production	60–75	Photographic and optical[c]
Manufacturing and Production	70–75	Livestock farming and breeding
Manufacturing and Production	80–85	General manufacturing
Manufacturing and Production	>85	Agriculture, forestry, mining, and fishing
Recreational	<65	Outdoor amphitheaters
Recreational	65–70	Nature exhibits and zoos
Recreational	65–70	Golf course, riding stables[c]
Recreational	70–75	Outdoor sports arena
Recreational	>85	Amusement parks and camps

[a]Adapted from FAA Advisory Circular AC150/5020-1.
[b]L_{dn} of 65–70 if 25 dB reduction indoors is provided; 70–75 if 30 dB reduction is provided.
[c]L_{dn} of 70–75 if 25 dB reduction indoors is provided; 75–80 if 30 dB reduction is provided.

TABLE 8-7

Yearly energy average L_{eq} identified as requisite to protect the public health and welfare with an adequate margin of safety

	Measure	Indoor			Outdoor		
		Activity interference	Hearing loss consideration	To protect against both effects (*b*)	Activity interference	Hearing loss consideration	To protect against both effects (*b*)
Residential with outside	L_{dn}	45		45	55		55
space and farm residences	$L_{eq(24)}$		70			70	
Residential with no	L_{dn}	45		45			
outside space	$L_{eq(24)}$		70				
Commercial	$L_{eq(24)}$	(*a*)	70	70(*c*)	(*a*)	70	70(*c*)
Inside transportation	$L_{eq(24)}$	(*a*)	70	(*a*)			
Industrial	$L_{eq(24)(d)}$	(*a*)	70	70(*c*)	(*a*)	70	70(*c*)
Hospitals	L_{dn}	45		45	55		55
	$L_{eq(24)}$		70			70	
Educational	$L_{eq(24)}$	45		45	55		55
	$L_{eq(24)(d)}$		70			70	
Recreational areas	$L_{eq(24)}$	(*a*)	70	70(*c*)	(*a*)	70	70(*c*)
Farm land and general unpopulated land	$L_{eq(24)}$				(*a*)	70	70(*c*)

Code:

(*a*) Since different types of activities appear to be associated with different levels, identification of a maximum level for activity interference may be difficult except in those circumstances where speech communication is a critical activity.

(*b*) Based on lowest level.

(*c*) Based only on hearing loss.

(*d*) An $L_{eq(8)}$ of 75 dB may be identified in these situations so long as the exposure over the remaining 16 hours per day is low enough to result in negligible contribution to the 24-hour average, that is, no greater than an L_{eq} of 60 dB.

Note: Explanation of identified level for hearing loss: The exposure period that results in hearing loss at the identified level is a period of 40 years.

(*Source:* U.S. EPA, 1974).

activity interference and annoyance and to permit the hearing mechanism an opportunity to recuperate if it is exposed to high levels during the day. The L_{dn} of 45 provides a fair margin of safety.

8-5 TRANSMISSION OF SOUND OUTDOORS

Inverse Square Law

If a sphere of radius δ vibrates with a uniform radial expansion and contraction, sound waves radiate uniformly from its surface. If the sphere is placed such that no sound waves are reflected back in the direction of the source, and if the product $\kappa\delta$, where κ is the wave number, is much less than 1, then the sound intensity at any radial distance r from the sphere is inversely proportional to the square of distance, that is*:

$$I = \frac{W}{4\pi r^2} \tag{8-18}$$

where I = sound intensity, watts/m^2
W = sound power of source, watts

This is the *inverse square law.* It explains that portion of the reduction of sound intensity with distance that is due to wave divergence (Figure 8-30). For a line source such as a roadway or a railroad, the reduction of sound intensity is inversely proportional to r rather than r^2. If we measure sound power level (L_w, re: 10^{-12} W) rather than sound power (W), we can rewrite Equation 8-18 in terms of sound pressure level†:

$$L_P \cong L_w - 20 \log r - 11 \tag{8-19}$$

where L_P = sound pressure level, dB re: 20 μPa
L_w = sound power level, dB re: 10^{-12} W
r = distance between source and receiver, m
$20 \log r$ = decibel transform = $10 \log r^2$
11 = decibel transform $\cong$ [$10 \log (4\pi) = 10.99$]

The tilde (~), indicating "approximately," results from the assumptions used above. L_w should be computed for all frequency bands of interest.

From a practical point of view it is difficult, if not impossible, to measure the sound power of the source. In such instances we measure the sound pressure level at some known distance from the source and then use the inverse square law or radial dependence relationships to estimate the sound pressure level at some other distance. For

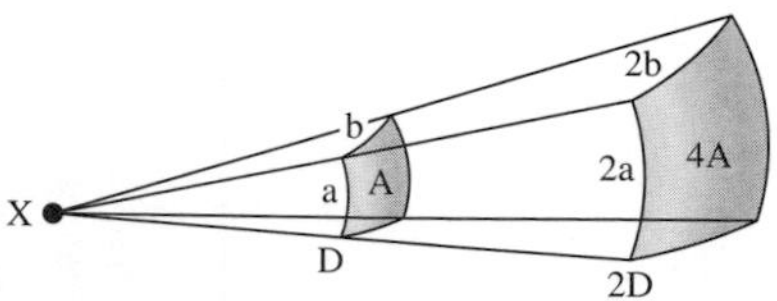

FIGURE 8-30
Illustration of inverse square law.

---

*$\kappa = 2\pi/\lambda$, where λ = wavelength, κ has units of reciprocal length, m^{-1}.
†This can be proved by using Equations 8-4, 8-5, 8-9, 8-10, and 8-11, and the assumption that $\rho c = 400$ kg/m$^2 \cdot$ s.

example, using the inverse square law, the sound pressure level L_{p2} at a distance r_2 from the source may be determined if the sound pressure level L_{p1} at some closer point r_1 is known:

$$L_{p2} = L_{p1} - 10\log\left(\frac{r_2}{r_1}\right)^2 \tag{8-20}$$

For a line source, the sound pressure level L_{p2} at a distance r_2 from the source may be determined at some closer point r_1 by a similar equation:

$$L_{p2} = L_{p1} - 10\log\left(\frac{r_2}{r_1}\right) \tag{8-21}$$

Radiation Fields of a Sound Source

The character of the wave radiation from a noise source will vary with distance from the source (Figure 8-31). At locations close to the source, the *near field,* the particle velocity is not in phase with the sound pressure. In this area, L_p fluctuates with distance and does not follow the inverse square law. When the particle velocity and sound pressure are in phase, the location of the sound measurement is said to be in the *far field.* If the sound source is in free space, that is, there are no reflecting surfaces, then measurements in the far field are also *free field measurements.* If the sound source is in a highly reflective space, for example, a room with steel walls, ceiling, and floor, then measurements in the far field are also *reverberant field measurements.* The shaded area in the far field of Figure 8-31 shows that L_p does not follow the inverse square law in the reverberant field.

Directivity

Most real sources do not radiate sound uniformly in all directions. If you were to measure the sound pressure level in a given frequency band at a fixed distance from a real source, you would find different levels for different directions. If you plotted these data in polar coordinates, you would obtain the directivity pattern of the source.

The *directivity factor* is the numerical measure of the directivity of a sound source. In logarithmic form the directivity factor is called the *directivity index.* For a spherical source it is defined as follows:

$$DI_\theta = L_{p\theta} - L_{ps} \tag{8-22}$$

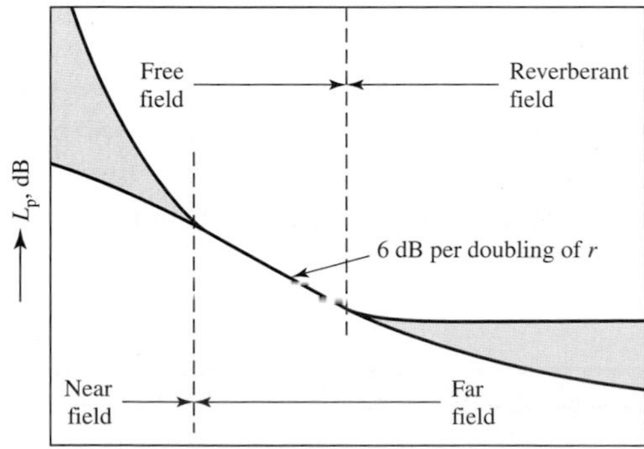

FIGURE 8-31
Variation of sound-pressure level in an enclosure along radius r from a noise source. (*Source:* Beranek, 1971.)

where $L_{p\theta}$ = sound pressure level measured at distance r' and angle 0° from a directive source radiating power W into an echo-free (*anechoic*) space, dB
L_{ps} = sound pressure level measured at distance r' from a nondirective source radiating power W into anechoic space,* dB

For a source located on or near a hard, flat surface, the directivity index takes the following form:

$$DI_\theta = L_{p\theta} - L_{ps} + 3 \tag{8-23}$$

The 3 dB addition is made because the measurement is made over a hemisphere instead of a sphere. That is, the intensity at a radius, r, is twice as large if a source radiates into a hemisphere rather than the ideal sphere we have used up to this point. Each directivity index is applicable only to the angle at which $L_{p\theta}$ was measured and only for the frequency at which it was measured.

We assume that the directivity pattern does not change its shape regardless of the distance from the source. This allows us to apply the inverse square law to directive sources simply by adding the directivity index:

$$L_{p\theta} \cong L_w + DI_\theta - 20 \log r - 11 \tag{8-24}$$

You should note that it is not possible to reduce the equation by using the equality given in Equation 8-22. The values of $L_{p\theta}$ are at a distance r, which is different than the r' in Equation 8-22.

Favorable propagation conditions are those shown in Figure 8-32a. They are environmentally relevant. These conditions are also stable in the sense that they are suitable for reproducible measurements. It has become standard practice to restrict prediction of airborne transmission of noise to conditions favorable for propagation. These are specified as (ISO, 1989, 1990):

1. The wind direction is within an angle of 45° of the direction connecting the center of the sound source and the center of the specified area, with the wind blowing from the source to the receiver.
2. The wind speed is between approximately 1 and 5 m/s measured at a height of 3–11 m.
3. Propagation in any near horizontal direction is under a well developed ground-based inversion.

Airborne Transmission

Effects of Atmospheric Conditions. Sound energy is absorbed in quiet isotropic air by molecular excitation and relaxation of oxygen molecules and, at very low temperatures, by heat conduction and viscosity in the air. Molecular excitation is a complex function of the frequency of noise, humidity, and temperature. In general, we may say that as the humidity decreases, sound absorption increases. As the

*This is the same source as the directive source, but acting in the ideal fashion that we assumed in developing the inverse square law.

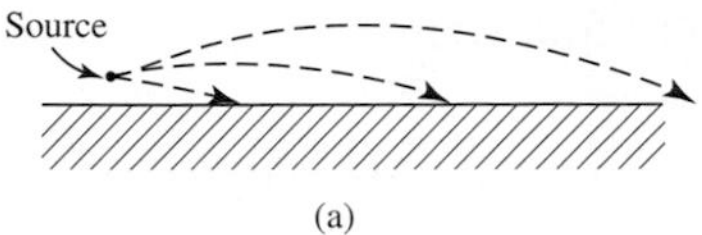

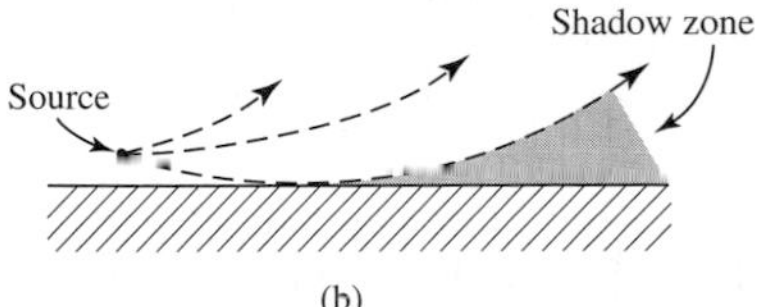

FIGURE 8-32
Refraction of sound (*a*) when the propagation is downwind or under conditions of a temperature inversion and (*b*) when the propagation is upwind or under temperature lapse conditions. (*Source:* Piercy and Daigle, 1991.)

temperature increases to about 10 to 20°C (depending upon the noise frequency), absorption increases. Above 25°C, absorption decreases. Sound absorption is higher at higher frequencies.

The vertical temperature profile greatly alters the propagation paths of sound. If a superadiabatic lapse rate exists, sound rays bend upward and noise shadow zones are formed (Figure 8-32b). If an inversion exists, sound rays are bent back toward the ground (Figure 8-32a). This results in an increase in the sound level. These effects are negligible for short distances but may exceed 20 dB at distances over 800 m.

In a similar fashion, wind speed gradients alter the way noise propagates. Sound traveling with the wind is bent down, while sound traveling against the wind is bent upward. When sound waves are bent down, there is little or no increase in sound levels. But when sound waves are bent upward, there can be a noticeable reduction in sound levels.

Basic Point Source Model. A point source is one for which $\kappa\delta \ll 1$ and for which Equation 8-18 holds. According to Magrab (1975),

> In practice most noise sources cannot be classified as simple point sources. However, the sound field of a complicated sound source will look as if it were a point source if the following two conditions are met: (1) $r/\delta \gg 1$, that is, the distance from the source is large compared to its characteristic dimension, and (2) $\delta/\lambda \ll r/\delta$, that is, the ratio of the size of the source to the wavelength of sound in the medium is small compared to the ratio of the distance from the source to its characteristic dimension. Recall that $r/\delta \gg 1$ from the first condition. A value of $r/\delta > 3$ is a sufficient approximation; therefore, $\delta\lambda \ll 3$.

The basic point source equation is

$$L_p \cong L_w - 20 \log r - 11 - A_e \tag{8-25}$$

where L_p = the desired SPL (re: (20 μPa) at angle θ and distance r from source, dB
L_w = the measured sound power level (re: 10^{-12} W) at angle θ, dB
A_e = attenuation for the distance r, dB

With the exception of the last term (A_e), it is the inverse square law (Equations 8–18 and 8–19). The A_e term is the excess attenuation beyond wave divergence. It is caused by environmental conditions and has units of dB.

The A_e term may be further divided into five terms as follows:

A_{e1} = attenuation by absorption in the air, dB

A_{e2} = attenuation by the ground, dB

A_{e3} = attenuation by barriers, dB

A_{e4} = attenuation by foliage, dB

A_{e5} = attenuation by houses, dB

Because of the introductory nature of this text, we have chosen to limit the following discussion to the first two terms, A_{e1} and A_{e2}. In addition, we will consider only the case of ground attenuation ranges greater than 100 m. For detailed examination of the the other cases we recommend that you consult Piercy and Daigle (1991).

TABLE 8-8

Air attenuation coefficient, dB/km, for an ambient pressure of 101.3 kPa (one standard sea-level atmosphere) for sound propagation in open air

		Frequency, Hz					
Temperature	Relative humidity, %	125	250	500	1,000	2,000	4,000
30°C	10	0.96	1.8	3.4	8.7	29	96
	20	0.73	1.9	3.4	6.0	15	47
	30	0.54	1.7	3.7	6.2	12	33
	50	0.35	1.3	3.6	7.0	12	25
	70	0.26	0.96	3.1	7.4	13	23
	90	0.20	0.78	2.7	7.3	14	24
20°C	10	0.78	1.6	4.3	14	45	109
	20	0.71	1.4	2.6	6.5	22	74
	30	0.62	1.4	2.5	5.0	14	49
	50	0.45	1.3	2.7	4.7	9.9	29
	70	0.34	1.1	2.8	5.0	9.0	23
	90	0.27	0.97	2.7	5.3	9.1	20
10°C	10	0.79	2.3	7.5	22	42	57
	20	0.58	1.2	3.3	11	36	92
	30	0.55	1.1	2.3	6.8	24	77
	50	0.49	1.1	1.9	4.3	13	47
	70	0.41	1.0	1.9	3.7	9.7	33
	90	0.35	1.0	2.0	3.5	8.1	26
0°C	10	1.3	4.0	9.3	14	17	19
	20	0.61	1.9	6.2	18	35	47
	30	0.47	1.2	3.7	13	36	69
	50	0.41	0.82	2.1	6.8	24	71
	70	0.39	0.76	1.6	4.6	16	56
	90	0.38	0.76	1.5	3.7	12	43

(*Source:* ISO, 1990.)

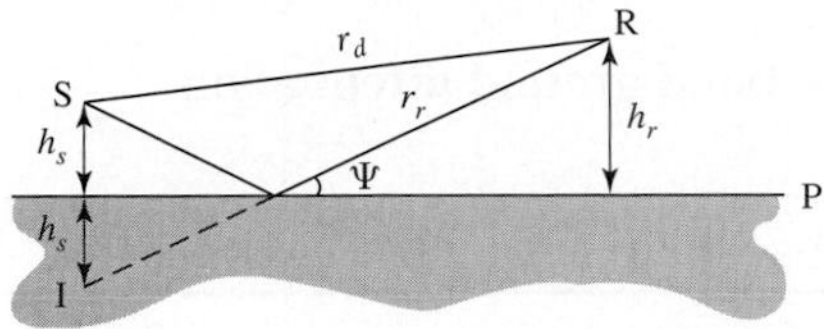

FIGURE 8-33
Paths for propagation from source S to receiver R. The direct ray is r_d and the ray reflected from plane P (which effectively comes from image source I) is r_r. (*Source:* Piercy and Daigle, 1991.)

The attenuation of sound by air absorption is given as (Piercy and Daigle, 1991):

$$A_{e1} = \frac{\alpha d}{1000 \text{ m/km}} \qquad (8\text{-}26)$$

where α = air attenuation coefficient, dB/km
d = distance, m

The air attenuation coefficient α as a function of temperature and humidity is listed in Table 8-8 on page 694.

The sound above a reflecting ground surface arrives at a receiver R from a source S by two paths (Figure 8-33). The path r_d is a direct ray. The path r_r is the reflection of the sound from the ground. The attenuation A_{e2} is a result of interference between the reflected ray and the direct ray. It strongly depends on the type of ground surface, the grazing angle ψ, the path length difference $(r_r - r_d)$, and the frequency of the sound. The ground surface may be classified as follows:

- Hard: asphalt or concrete pavement, water, and all surfaces with low porosity.
- Soft: ground covered by grass or other vegetation and all surfaces suitable for growth of vegetation.
- Mixed: surface that includes both hard and soft conditions.

Very soft ground such as snow cover is not considered in the following analysis.

As noted above, for long range propagation (>100 m) the ground attenuation (A_{e2}) is calculated for atmospheric conditions favorable to propagation. At distances less than 100 m the results obtained by the following method will differ slightly from the short range technique (Piercy and Daigle, 1991).

Figure 8-34 is used to define the zones and terms used in calculating A_{e2}. Each zone is assigned a *ground factor* according to the following rules:

1. The *source zone* (A_s) extends from the source S toward the receiver R a distance of $30h_s$, with a maximum of *r.*

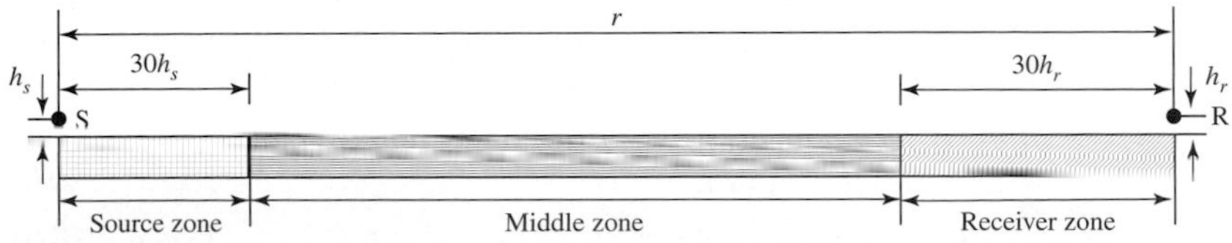

FIGURE 8-34
Three zones between a source S and receiver R separated by distance *r,* used in determining the ground attenuation A_{ground} at long ranges. (*Source:* Piercy and Daigle, 1991.)

TABLE 8-9
Expressions to be used in calculating the octave-band ground attenuation (A_{ground}) in decibels at long range

Octave-band frequency, Hz	A_s or A_r, dB	A_m, dB
63	−1.5	$-3e$
125	(a · G) − 1.5	$-3e(1 - G)$
250	(b · G) − 1.5	$-3e(1 - G)$
500	(c · G) − 1.5	$-3e(1 - G)$
1,000	(d · G) − 1.5	$-3e(1 - G)$
2,000	(1 − G) − 1.5	$-3e(1 - G)$
4,000	(1 − G) − 1.5	$-3e(1 - G)$
8,000	(1 − G) − 1.5	$-3e(1 - G)$

	Source or receiver height, m				
Distance, m	0.5	1.5	3.0	6.0	>10.0
	Factor a				
50	1.7	2.0	2.7	3.2	1.6
100	1.9	2.2	3.2	3.8	1.6
200	2.3	2.7	3.6	4.1	1.6
500	4.6	4.5	4.6	4.3	1.6
>1,000	7.0	6.6	5.7	4.4	1.7
	Factor b				
50	6.8	5.9	3.9	1.7	1.5
100	8.8	7.6	4.8	1.8	1.5
>200	9.8	8.4	5.3	1.8	1.5
	Factor c				
50	9.4	4.6	1.6	1.5	1.5
100	12.3	5.8	1.7	1.5	1.5
>200	13.8	6.5	1.7	1.5	1.5
	Factor d				
50	4.0	1.9	1.5	1.5	1.5
>100	5.0	2.1	1.5	1.5	1.5

(*Source:* ISO, 1989.)

2. The *receiver zone* (A_r) extends from the receiver R toward the source S a distance of $30h_r$ with a maximum of r.

3. The *middle zone* (A_m) lies between the source and receiver zones. If $r < 30(h_s + h_r)$, then the source and receiver zones overlap and there is no middle zone.

The ground factor G for each zone is based on the surface characteristics:

- Hard ground: $G = 0$
- Soft ground: $G = 1$
- Mixed ground: G equals the fraction of the ground that is soft [for example if 25 percent of the ground is soft, then $G = (0.25)(1) = 0.25$]

The ground attenuation for an octave band is calculated for each zone by using the equations and data in Table 8-9 on page 696. The value for e for the middle zone calculation in the table is determined from the following equation:

$$e = 1 - \left[\frac{30(h_s + h_r)}{r}\right] \tag{8-27}$$

The total ground attenuation is then

$$A_{e2} = A_s + A_r + A_m \tag{8-28}$$

For a complete analysis, the attenuation must computed for each relevant octave band.

Example 8-7. The sound power level (re: 10^{-12} W) of a compressor is 124.5 dB at 1,000 Hz. Determine the SPL 200 m downwind on a clear summer afternoon if the wind speed is 5 m/s, the temperature is 20°C, the relative humidity is 50 percent, and the barometric pressure is 101.325 kPa. The heights of the compressor and the receiver are 1.2 m. The ground surface characteristics are shown in the sketch below.

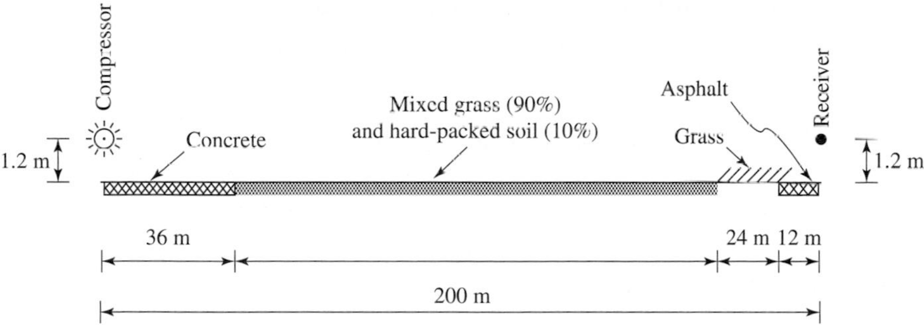

Solution. The attenuation by air absorption (A_{e1}) is calculated directly from Table 8-8 with the distance being 200 m.

$$A_{e1} = (4.7 \text{ dB/km})\left(\frac{200 \text{ m}}{1{,}000 \text{ m/km}}\right) = 0.94 \text{ dB}$$

Calculate the ground attenuation in three steps.

1. The source zone attenuation extends a distance

$$30h_s = (30)(1.2 \text{ m}) = 36 \text{ m}$$

From the sketch we note that the source zone is 100 percent hard and that $G = 0$. From Table 8-9, the equation for the source zone at 1,000 Hz is

$$A_s = [(d)(\mathrm{G})] - 1.5$$

From Table 8-9, for distances >100 m and a source height of 1.5 m, select $d = 2.1$. The ground attenuation is then

$$A_s = [(2.1)(0)] - 1.5 = -1.5 \text{ dB}$$

(Note that we did not interpolate for the height because the product would obviously be zero.)

2. The receiver zone extends a distance

$$30h_r = (30)(1.2 \text{ m}) = 36 \text{ m}$$

From the sketch we note that 12 m is hard and 36 − 12 = 24 m is soft. The fraction that is soft is then 12/36 = 0.33. The G value for "soft" is 1.0. From Table 8-9, the equation for receiver attenuation at 1,000 Hz is

$$A_r = [(d)(G)] - 1.5$$

The d value is the same as that for the source zone. The G value must be multiplied by the fraction of receiver zone that is soft. The receiver zone attenuation is

$$A_r = [(2.97)(0.33)(1.0)] - 1.5 = 0.98 - 1.5 = -0.52 \text{ dB}$$

(Note that in this case we did interpolate the receiver height to obtain 2.97 for the factor d.)

3. The middle zone is 90 percent covered in grass, so $G = (0.90)(1.0) = 0.90$. From Table 8-9, the equation for the middle zone attenuation is

$$A_m = -3e(1 - G)$$

The value for e is calculated by using Equation 8-27:

$$e = 1 - \left[\frac{30(1.2 + 1.2)}{200}\right] = 1 - 0.36 = 0.64$$

The attenuation in the middle zone is:

$$A_m = -3(0.64)(1 - 0.90) = -0.19 \text{ dB}$$

4. The total ground attenuation is

$$A_{e2} = -1.5 - 0.52 - 0.19 = -2.21 \text{ dB}$$

Note that the attenuation is negative. Thus, the ground surface reflection actually increases the SPL.

Using the basic point source model (Equation 8-25) gives the SPL at the receiver as

$$\begin{aligned} L_p &= 124.5 - 20 \log(200) - 11 - 0.94 - (-2.21) \\ &= 124.5 - 46 - 11 - 0.94 + 2.21 = 68.77 \text{ or } 69 \text{ dB at } 1{,}000 \text{ Hz} \end{aligned}$$

8-6 TRAFFIC NOISE PREDICTION

National Cooperative Highway Research Program 174

The National Cooperative Highway Research Program has developed a series of documents (NCHRP 117, NCHRP 144, and NCHRP 174) that provide design guidance for the prediction and control of highway noise (Kugler et al., 1976). These documents have been used widely because of their simplicity and relatively high success in making accurate noise predictions. The NCHRP 174 procedure is the last revision in the series. It contains a four-step procedure for the prediction and control of highway noise. We have limited ourselves to the first prediction step, that is, the "short method." The Federal Highway Administration (FHWA) has developed sophisticated computer models to replace the NCHRP 174 manual technique used in this text for illustration purposes. The FHWA released Version 2.5 of the Traffic Noise Model—TNM in 2004.*

The objective of the "short method" is to obtain a quick and gross (always overpredicting) prediction of the expected noise levels. This is necessary because the prediction of true highway noise levels is a rather complicated subject. In many instances it is desirable to first obtain a rough idea of the potential problem areas before full knowledge of the horizontal and vertical roadway design parameters has been gained. Such is the case, for example, of a location study where a number of alignments must be considered. Also, this first step helps to eliminate areas that do not represent a problem in terms of noise levels, thus simplifying further evaluation.

The "short method" prediction can be performed quickly through use of two *nomographs* and knowledge of a few traffic and roadway parameters.† By its design, the "short method" requires many assumptions and approximations and should not be used as a final tool.

The second step (the "complete method") utilizes a microcomputer program to refine the predictions made in the first step. The third step is the selection of a noise control design. The fourth step is to redo the second step and check the design solution. In the following paragraphs we have reproduced the short method as it appears in NCHRP 174, with the addition of clarifying comments and modification to SI units.

Methodology. The flow diagram that illustrates the methodology of the short method is shown in Figure 8-35. The method assumes that the roadway can be approximated by one infinite element with constant traffic parameters and roadway characteristics.

The initial step in using the short method consists of defining an infinite straight-line approximation to the real highway configuration. On-ramps, off-ramps, and interchange ramps are omitted from the short method analysis.

---

*TNM 2.5.

†A nomograph is a graph that provides the solution to an equation or series of equations containing three or more variables (see Figure 8-39).

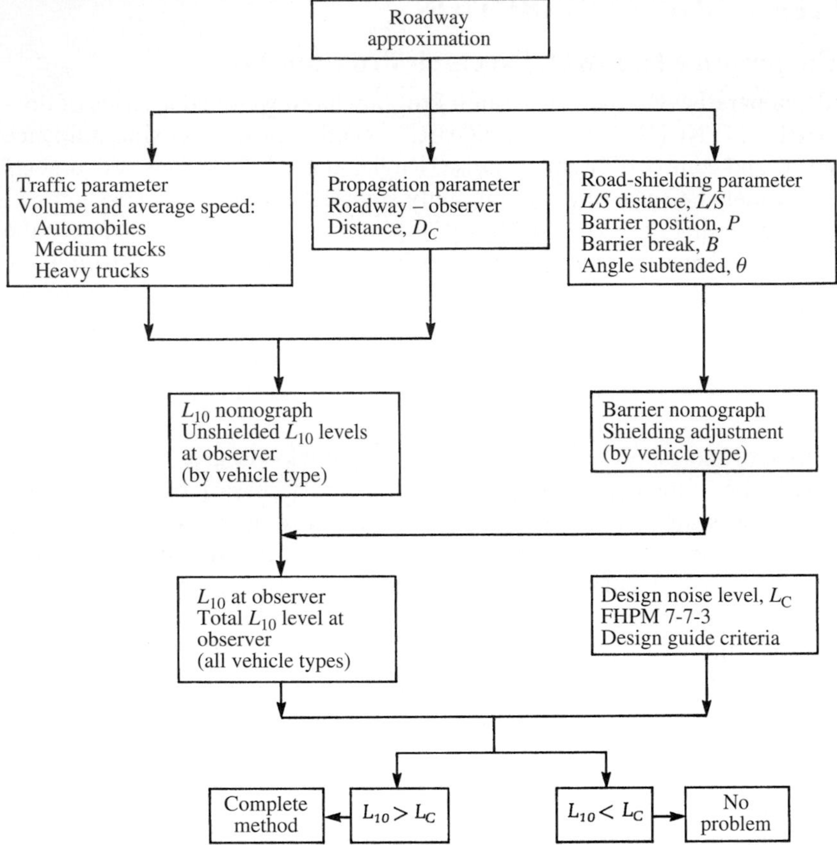

FIGURE 8-35
Flow diagram of methodology for applying NCHRP 174 method for estimating L_{10} from traffic. (*Source:* Kugler et al., 1976.)

Once the approximate roadway has been chosen, the following parameters must be computed or estimated: (*a*) the traffic parameters, which include the speed and volume of each class of vehicles; (*b*) the propagation characteristics, which describe the location of the receiver relative to the roadway; and (*c*) the roadway-shielding parameters, which describe the shielding provided by the roadway, if any. [Only barriers located within the right-of-way (ROW) may be considered.]

These parameters are used in two operations. First, the traffic and propagation parameters are combined in the L_{10} nomograph to determine, for each type of source, the unshielded L_{10} level at the observer.

The final result is then compared to the criteria level, L_c, at the observer (see, for example, Table 8-3) to define a "no problem" or "potential problem" condition. If a potential problem is identified, the observer location in question should be evaluated using the complete method.

Procedure. The step-by-step procedures necessary to calculate noise levels by the short method are presented in the following numbered paragraphs. In addition to the

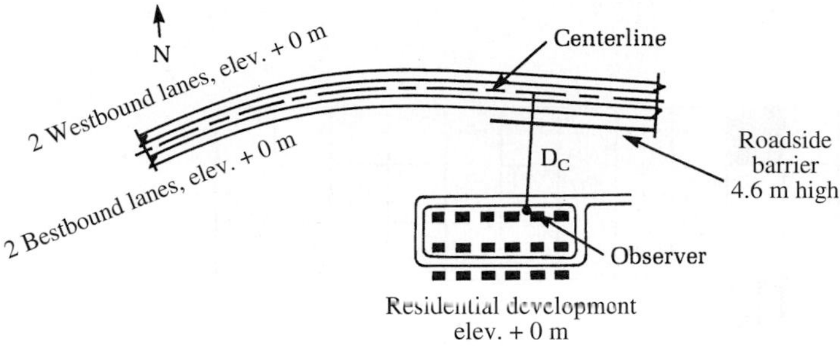

(a) Route map showing observer location and observer - roadway centerline distance, D_C

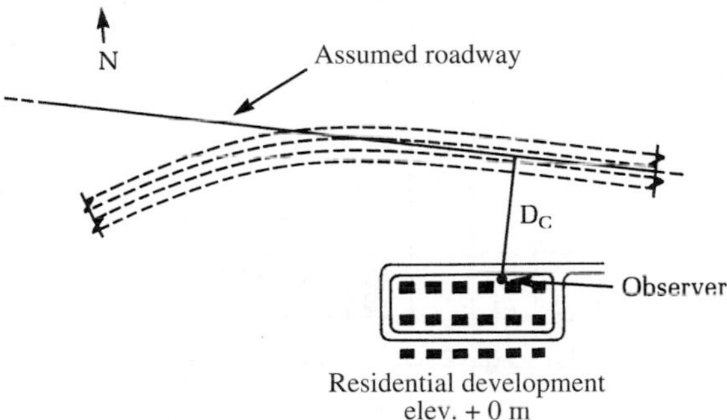

(b) Model of assumed roadway alignment

FIGURE 8-36
(*a*) Roadway and (*b*) roadway approximation. (*Source:* Kugler et al., 1976.)

nomographs, the method uses a noise prediction worksheet to aid the user in the sequential steps. A blank worksheet and larger scale nomographs are included in Appendix B.

1. Observer identification: On a route map of convenient scale, identify all observer locations at which analysis is desired.

2. Roadway approximation: Approximate the roadway alignment by a straight, infinite line. The procedure is as follows: Determine and measure the nearest perpendicular distance, D_c, between the roadway centerline and observer, as shown in Figure 8-36*a*. Enter on line 4 on the noise prediction worksheet (Figure 8-37). Note that the infinite roadway approximation automatically assumes a line perpendicular to the centerline distance, D_c. There is no need to draw this line on the route map for computation reasons. An illustration of this assumption is shown in Figure 8-36*b*. Note that for each observer location, a

Project BRISTOL HWY. Date 13 AUG. 1980 Engineer I. THOMPSON

Step			Ex. 9-10a			Ex. 9-10b														
			A	T_M	T_H	A	T_M	T_H	A	T_M	T_H	A	T_M	T_H	A	T_M	T_H	A	T_M	T_H
1	Traffic	Vehicle Volume, V(Vph)	2000	100	100	2000	100	100												
2		Vehicle Av. Speed, S(km/h)	80.5	80.5	80.5	80.5	80.5	80.5												
3		Combined Veh. Vol.*, V_C(Vph)	3000			3000														
4	Prop.	Observer-Roadway Dist., D_C(m)	60			60														
5	Shielding	Line-of-Sight Dist., L/S(m)	—			60														
6		Barrier Position Dist., P(m)	—			15.2														
7		Break in Barrier., B(m)	—			4.6		2.7												
8		Angle Subtended, θ (deg)				170														
9	Prediction**	Unshield L_{10} Level (dBA)	66	-	68	66	-	68												
10		Shielding Adjust. (dBA)	0		0	13		10												
11		L_{10} at Observer (By Veh. Class)	66	-	68	53	-	58												
12		L_{10} at Observer – Total	70 dBA			59.25 or 59 dBA														

Code:

A = Automobiles, T_M = Medium Trucks, T_H = Heavy Trucks

* Applies only when automobile and medium truck average speeds are equal. $V_C = V_A + (10)V_{T_M}$

** If automobile-medium truck volume V_C is combined, use L_{10} nomograph prediction only once for these two vehicle classes

FIGURE 8-37
Noise prediction worksheet. (*Source:* Kugler et al., 1976.)

different roadway approximation might result. Note also that the noise prediction worksheet (NPWS) allows for computations for six different observer locations by entering an observer location identification at the head of each column. Similarly, the NPWS can be used to calculate six different traffic conditions for the same observer location.

3. Traffic parameters: Determine the vehicle operating conditions by using the traffic parameters at the roadway point nearest the observer (if these parameters vary along the roadway). The procedure is as follows:

a. Determine the automobile volume (vph) and average speed (km/h) and enter them on lines 1 and 2 under automobiles (A).

b. Determine the medium truck volume (vph) and average speed (km/h) and enter them on lines 1 and 2 under medium trucks (T_M).

c. Determine the heavy truck volume (vph) and average speed (km/h) and enter them on lines 1 and 2 under heavy trucks (T_H).

d. If the automobile and medium truck speeds are the same, multiply the medium truck volume by 10 and add to the automobile volume. Enter combined volume V_c on line 3 of the NPWS. If the automobile and medium truck volumes are combined, in subsequent operations consider the two vehicle classes as one source. If the speed of the medium trucks differs from that of the automobiles, the volumes are *not* combined *but* the medium truck volume is still multiplied by 10 for the determination of the L_{10} value for the reason noted in the footnote below.*

4. Roadway-shielding parameters: If the roadway cross section at the nearest point is not at grade (either elevated or depressed), or if a roadside barrier (on the roadway right-of-way) is present, determine the roadway-shielding parameters. If the elevation, depression, or roadside barrier is less than 1.5 m high (compared to the surrounding terrain), disregard it. The procedure is as follows: Determine the barrier parameters and enter on lines 5 through 8 of the NPWS. Use Figure 8-38 for definitions of parameters. The parameters that must be measured are: (*a*) line-of-sight distance, *L/S*, (*b*) break in line of sight, (*c*) barrier position distance, and (*d*) angle subtended, θ (degree). The angle subtended is measured from the ends of the barrier with respect to the position of the observer. For an observer placed equidistant from the ends of the barrier, the angle may be determined from simple trigonometric principles. Graphical methods may be more appropriate for other configurations. Note that as the barrier length increases, the angle approaches 180 degrees.

5. Unshielded L_{10} level at observer location: Determine the unshielded L_{10} level at the observer location for all three traffic sources (automobiles, medium trucks, and heavy trucks) using the L_{10} nomograph (Figure 8-39 on page 705).

---

*The medium truck volume is multiplied by 10 because this traffic noise source behaves similarly to automobile noise, but the overall level is 10 dBA higher than for automobiles.

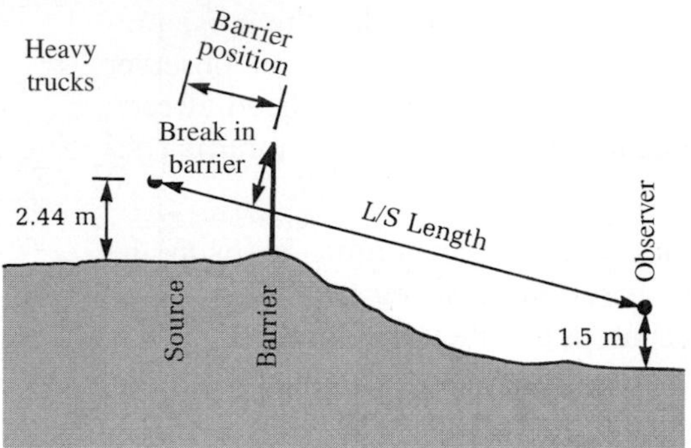

(*a*) Barrier parameters for simple barrier, section view

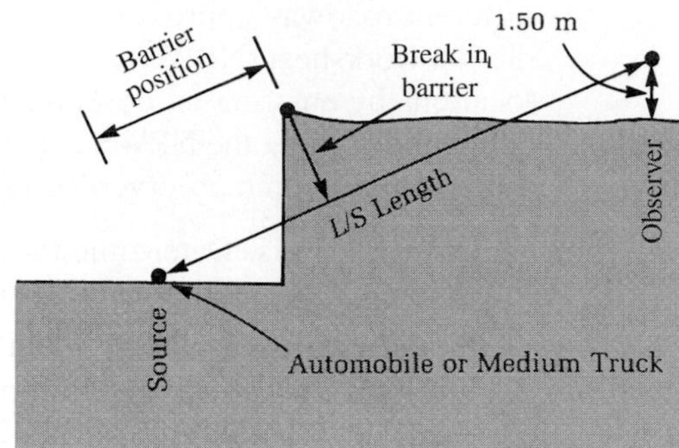

(*b*) Barrier parameters for depressed raodway, section view

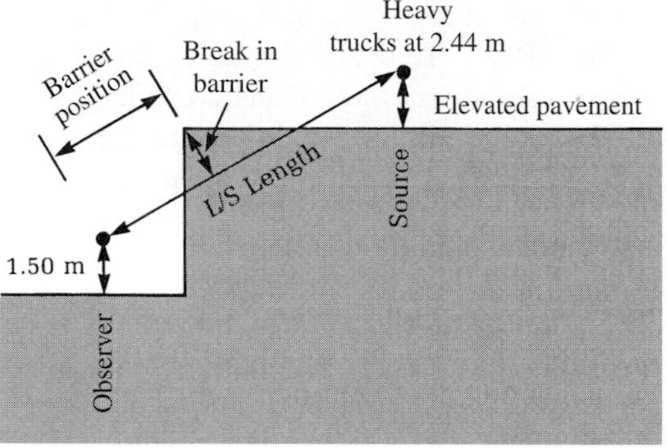

(*c*) Barrier parameters for elevated roadway, section view

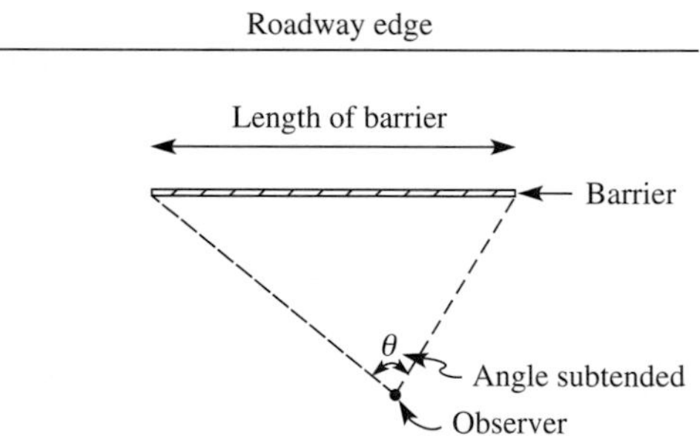

(*d*) Barrier parameters, plan view

FIGURE 8-38
Definitions of barrier parameters. (*Source:* Kugler et al., 1976.)

Note that if the automobile and medium truck speeds are equal, these two sources may be evaluated together using the combined volume, V_c, and average speed, S_A or S_M on lines 3 and 2 of NPWS. The procedure is as follows:

a. Automobiles (and medium trucks): Using the vehicle volume, V_A (this corresponds to V_c, the combined auto and medium truck volumes, when the speeds of these two populations are equal), and the average speed, S_A (or S_M), enter the L_{10} nomograph and determine the unshielded L_{10} noise level at the observer. Enter on line 9 of the NPWS.

b. Medium trucks: Using the vehicle volume, V_M, *multiplied by ten,* and the average speed, S_M, enter the L_{10} nomograph and determine the unshielded L_{10} noise level at the observer. Enter on line 9 of the NPWS. If automobiles and medium trucks were combined in step a, this step should be omitted.

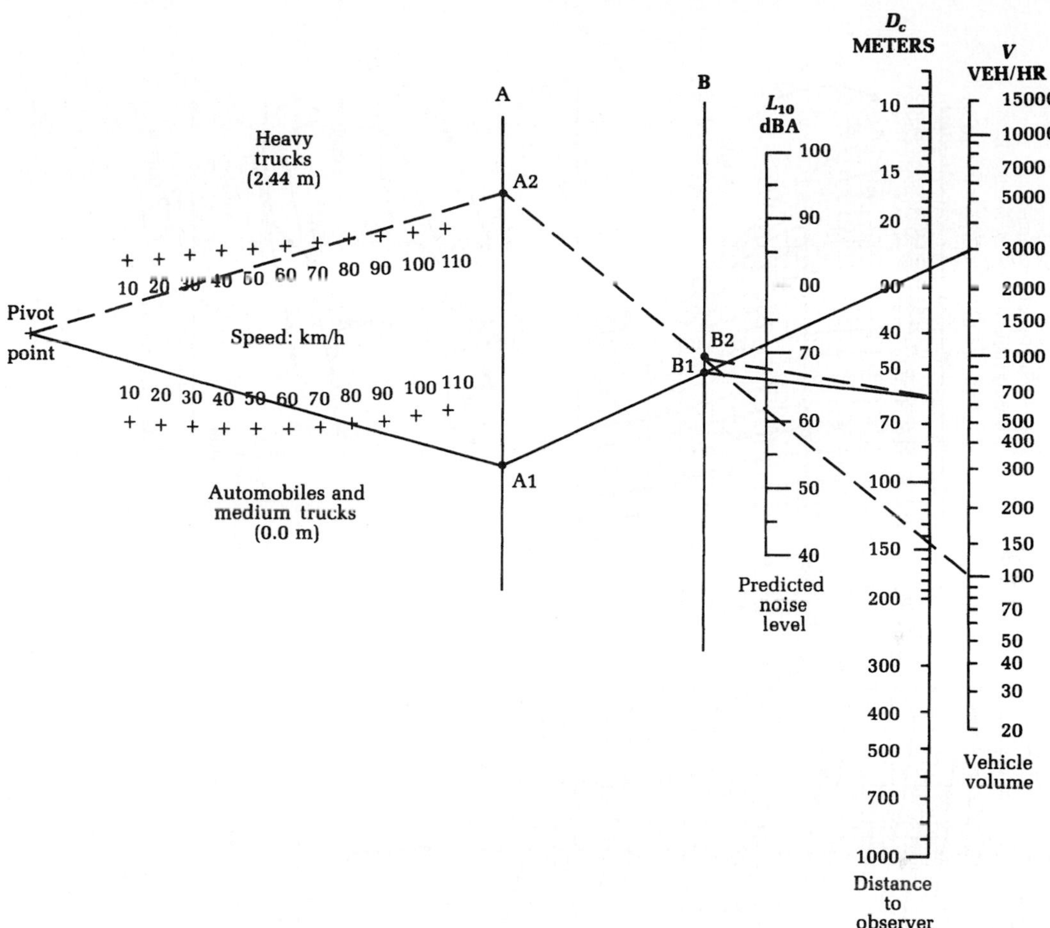

FIGURE 8-39
L_{10} nomograph.

c. Heavy trucks: Using the vehicle volume, V_T, and the average speed, S_T, enter the L_{10} nomograph and determine the unshielded L_{10} noise level at the observer. Enter on line 9 of the NPWS.

6. Shielding adjustment: Determine the noise reduction afforded by the roadway geometry using the barrier nomograph (Figure 8-40) and the roadway parameters listed in the NPWS. This procedure must be performed twice: once for the 0 m source elevation (automobiles and medium trucks) and once for the 2.44 m elevation (heavy trucks).

   a. Low sources (0 m): Using the line-of-sight distance, *L/S*, barrier position distance, *P*, break in *L/S* distance, *B* (for sources at ground level), and the angle subtended, θ, enter the barrier nomograph and calculate the shielding adjustment. Enter on line 10 of the NPWS under automobile and medium trucks.

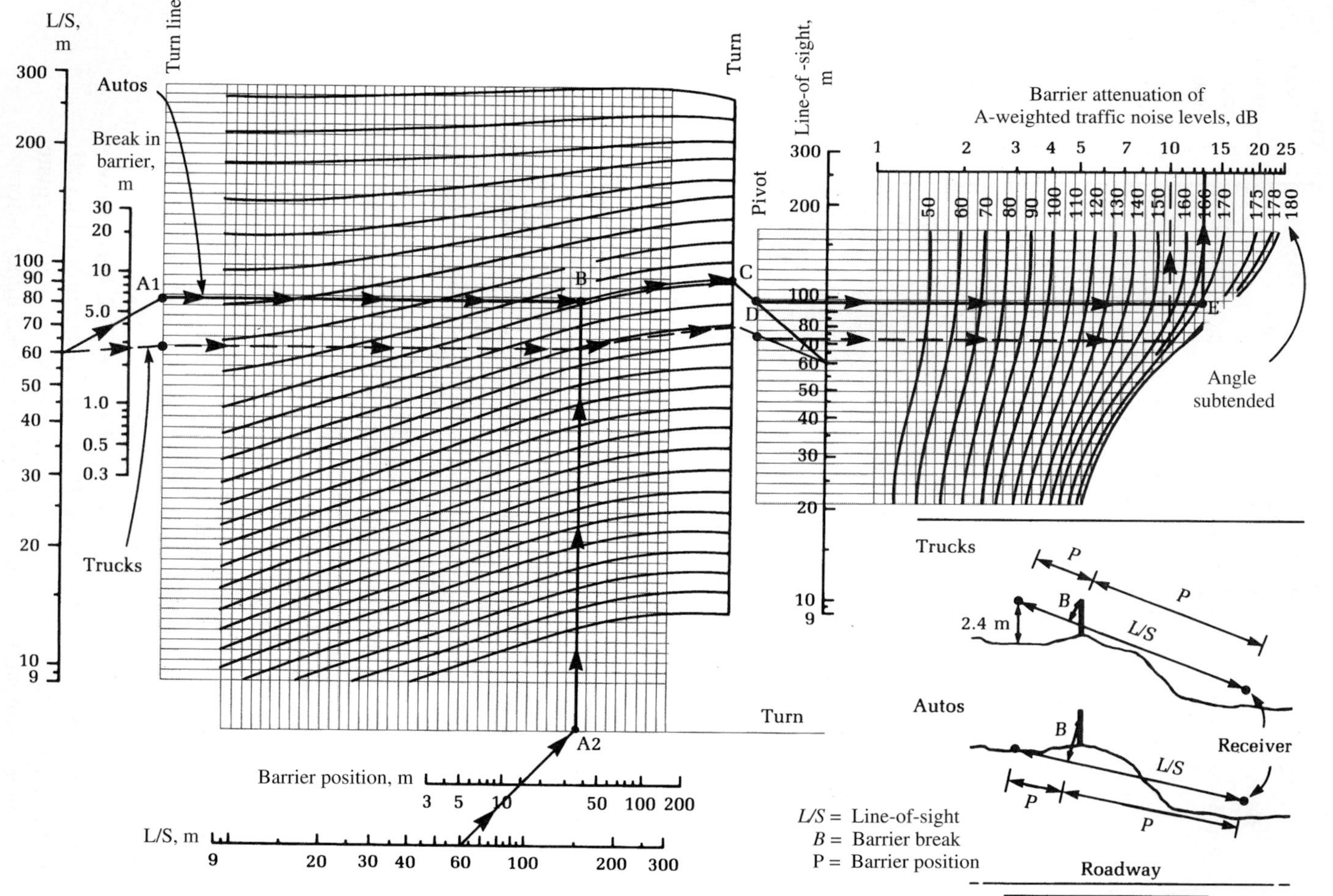

FIGURE 8-40
Barrier nomograph. (*Source:* Kugler et al., 1976.)

b. High sources (2.44 m): Using the line-of-sight distance, *L/S*, barrier position distance, *P*, break in *L/S* distance, *B* (for 2.44 m sources), and the angle subtended, θ, enter the barrier nomograph and calculate the shielding adjustment. Enter on line 10 of the NPWS under heavy trucks.

7. L_{10} at observer, by vehicle type: Calculate the L_{10} noise level at the observer for each individual source by subtracting the shielding adjustment (line 10) from the unshielded L_{10} level at the observer (line 9), and enter the result in line 11. Note that the shielding adjustment is always negative and can be subtracted algebraically from line 9.

8. Total L_{10} level at observer: Determine the total L_{10} noise level at the observer and enter on line 12. This is done by logarithmically adding (decibel addition) the contributions from automobiles, medium trucks, and heavy trucks computed in line 11. Taking two L_{10} levels at a time, find the difference between them and enter the addition scale provided in Figure 8-4 to find the "adjustment." The "adjustment" should be added to the higher of the two L_{10} levels. The operation is then repeated, two levels at a time, until only one level remains. The lowest levels should be added first for maximum accuracy.

Example 8-8. The county road commissioner has requested a noise evaluation of a proposed highway near Bristol. The proposed highway is to be routed such that the centerline of the roadway will be 60 m from a school. Using the following data, determine whether or not the FHA criterion will be met:

Average vehicle speed = 80.5 km/h for all vehicles

Automobiles = 2,000/h

Medium trucks = 100/h

Heavy trucks = 100/h

The terrain is level and no shielding is present.

Solution. According to step 3d, when the speed of the medium trucks and the cars is the same, the volumes can be combined by multiplying the medium truck volume by ten, and then combining it with the car volume. The combined volume, V_c, for cars and medium trucks is 2,000 + 10 (100) = 3,000 vph. Using the L_{10} nomograph shown in Figure 8-39, proceed as follows:

1. Draw a straight line from the left pivot point through the 80.5 km/h point on the automobile speed scale. Extend the straight line to turn line A. The intersection is marked A1.
2. Draw a second straight line from the intersection point A1 to the 3,000 vph point on the volume scale on the far right of the figure. The intersection of this straight line with turn line B is marked B1.
3. Draw a third straight line from point B1 to the point on the D_c scale (60 m). The intersection of this third line with the L_{10} scale gives the predicted A-weighted L_{10} level at the observer. For this example, the predicted L_{10} level is 66 dBA. This value is entered on line 9 of the NPWS (Figure 8-37).

Now repeat the procedure for the heavy trucks. It is shown by the dashed line in Figure 8-39. The predicted L_{10} for heavy trucks is 68 dBA.

The combined level of the automobiles and trucks is found by "decibel addition" (Example 8-1) to be 70 dBA. This just meets the FHA design level for land use category B (Table 8-3). Let us see what effect a barrier will have. We have arbitrarily selected the following characteristics for the barrier:

Height = 5.0 m, which yields a "Break in Barrier" of 4.6 m for automobiles and 2.6 m for heavy trucks

Position = 15.2 m from centerline of roadway

Subtended angle = 170°

Note: The "Break in Barrier" is determined by constructing a scale drawing of the roadway and receiver as shown in Figure 8-38. Because the automobile noise is assumed to originate at an elevation of 0.0 m, the "Break in Barrier" is greater than for the heavy trucks, where the noise is asssumed to originate at an elevation of 2.44 m.

Using the barrier nomograph shown in Figure 8-40, proceed as follows:

1. Starting from the vertical *L/S* scale on the left: From the 60-m point on the *L/S* scale, draw a straight line going through the 4.6-m point on the "Break in barrier" to the "Turn line". Note that the scale is logarithmic. From the intersection, called A_1, draw a straight horizontal line.
2. Starting from the horizontal *L/S* scale on the bottom: Draw a straight line through the 60-m point on the *L/S* scale, and the 15.2-m point on the "Barrier position" scale to the "Turn line". From the intersection, called A_2, draw a straight vertical line until it intersects (at B) with the horizontal drawn from A_1.
3. From B, move to the right upward following the nearest curve to turn line C. (If B is on one of the curves, simply follow it; if B is between curves, follow parallel to the nearest curve upward to the right until it intersects with the turn line.)
4. From C, draw a straight line to the line-of-sight (*L/S*) distance (60 m) point on the vertical *L/S* scale. It will intersect with the pivot line at D.
5. From D, draw a horizontal line to the right until it intersects the curve corresponding to the subtended angle (170°) at E.
6. Finally, draw a vertical line from E upward until it intersects the barrier attenuation scale. The attenuation can now be read, that is, 13 dBA for automobiles and medium trucks.

The same procedure is followed for heavy trucks using 2.6 m for the "Break in Barrier" because of the higher noise emission elevation from the heavy trucks. Note from Figure 8-38 that this distance is perpendicular to *L/S* and not perpendicular to the horizontal. The attenuation is about 10 dBA.

The revised overall combined L_{10} would then be 60 dBA. This is very acceptable for Class B land use category. All of the tabulations are summarized in Figure 8-37.

L_{eq} Prediction

At about the same time that the NCHRP 174 report was being finalized, the Ontario Ministry of Transportation and Communications completed development of a predictive equation based on the L_{eq} concept (Hajek, 1977). The empirical equation they developed is as follows:

$$L_{eq} = 42.3 + 10.2 \log(V_c + 6V_t) - 13.9 \log D + 0.13S \qquad (8\text{-}29)$$

where L_{eq} = energy equivalent sound level during one hour, dBA
V_c = volume of automobiles (four tires only), veh/h
V_t = volume of trucks (six or more tires), veh/h
D = distance from edge of pavement to receiver, m
S = average speed of traffic flow during one hour, km/h

The simplicity of the equation is an obvious advantage over the NCHRP method. It does have the restriction that it does not account for barriers. A nomograph technique similar to the NCHRP method is available to take barriers into account.

L_{dn} Prediction

As a direct extension of the L_{eq} methodology, the Ontario method was extended to enable the calculation of L_{dn}. The modified model has the following form:

$$L_{dn} = 31.0 + 10.2 \log [\text{AADT} + (\text{T\% AADT}/20)] - 13.9 \log D + 0.13\ S \qquad (8\text{-}30)$$

where L_{dn} = equivalent A-weighted sound level during 24-hour time period with 10 dBA weighting applied to 2200–0700 h, dBA
AADT = annual average daily traffic, veh/d
$\%T$ = average percentage of trucks during a typical day, %

Equation 8-30 has the same advantages and disadvantages as Equation 8-29.

8-7 NOISE CONTROL

Source-Path-Receiver Concept

If you have a noise problem and want to solve it, you have to find out something about what the noise is doing, where it comes from, how it travels, and what can be done about it. A straightforward approach is to examine the problem in terms of its three basic elements: that is, sound arises from a source, travels over a path, and affects a receiver or listener.*

The source may be one or any number of mechanical devices that radiate noise or vibratory energy. Such a situation occurs when several appliances or machines are in operation at a given time in a home or office.

The most obvious transmission path by which noise travels is simply a direct line-of-sight air path between the source and the listener. For example, aircraft flyover noise

*This discussion in large part was taken from Berendt, Corliss and Ojalvo, 1976.

reaches an observer on the ground by the direct line-of-sight air path. Noise also travels along structural paths. Noise can travel from one point to another via any one path or a combination of several paths. Noise from a washing machine operating in one apartment may be transmitted to another apartment along air passages such as open windows, doorways, corridors, or duct work. Direct physical contact of the washing machine with the floor or walls sets these building components into vibration. This vibration is transmitted structurally throughout the building, causing walls in other areas to vibrate and to radiate noise.

The receiver may be, for example, a single person, a classroom of students, or a suburban community.

Solution of a given noise problem might require alteration or modification of any or all of these three basic elements:

1. Modifying the source to reduce its noise output
2. Altering or controlling the transmission path and the environment to reduce the noise level reaching the listener
3. Providing the receiver with personal protective equipment

Control of Noise Source by Design

Reduce Impact Forces. Many machines and items of equipment are designed with parts that strike forcefully against other parts, producing noise. Often, this striking action or impact is essential to the machine's function. A familiar example is the typewriter—its keys must strike the ribbon and paper in order to leave an inked impression. But the force of the key also produces noise as the impact falls on the ribbon, paper, and platen.

Several steps can be taken to reduce noise from impact forces. The particular remedy to be applied will be determined by the nature of the machine in question. Not all of the steps listed below are practical for every machine and for every impact-produced noise. But application of even one suggested measure can often reduce the noise appreciably.

Some of the more obvious design modifications are as follows:

1. Reduce the weight, size, or height of fall of the impacting mass.
2. Cushion the impact by inserting a layer of shock-absorbing material between the impacting surfaces. (For example, insert several sheets of paper in the typewriter behind the top sheet to absorb some of the noise-producing impact of the keys.) In some situations, you could insert a layer of shock-absorbing material behind each of the impacting heads or objects to reduce the transmission of impact energy to other parts of the machine.
3. Whenever practical, one of the impact heads or surfaces should be made of nonmetallic material to reduce resonance (ringing) of the heads.
4. Substitute the application of a small impact force over a long time period for a large force over a short period to achieve the same result.
5. Smooth out acceleration of moving parts by applying accelerating forces gradually. Avoid high, jerky acceleration or jerky motion.

6. Minimize overshoot, backlash, and loose play in cams, followers, gears, linkages, and other parts. This can be achieved by reducing the operational speed of the machine, better adjustment, or by using spring-loaded restraints or guides. Machines that are well made, with parts machined to close tolerances, generally produce a minimum of such impact noise.

Reduce Speeds and Pressures. Reducing the speed of rotating and moving parts in machines and mechanical systems results in smoother operation and lower noise output. Likewise, reducing pressure and flow velocities in air, gas, and liquid circulation systems lessens turbulence, resulting in decreased noise radiation. Some specific suggestions that may be incorporated in design are the following:

1. Fans, impellers, rotors, turbines, and blowers should be operated at the lowest bladetip speeds that will still meet job needs. Use large-diameter, low-speed fans rather than small-diameter, high-speed units for quiet operation. In short, maximize diameter and minimize tip speed.
2. All other factors being equal, centrifugal squirrel-cage type fans are less noisy than vane axial or propeller type fans.
3. In air ventilation systems, a 50 percent reduction in the speed of the air flow may lower the noise output by 10 to 20 dB, or roughly one-quarter to one-half of the original loudness. Air speeds less than 3 m/s measured at a supply or return grille produce a level of noise that usually is unnoticeable in residential or office areas. In a given system, reduction of air speed can be achieved by operating at lower motor or blower speeds, installing a greater number of ventilating grilles, or increasing the cross-sectional area of the existing grilles.

Reduce Frictional Resistance. Reducing friction between rotating, sliding, or moving parts in mechanical systems frequently results in smoother operation and lower noise output. Similarly, reducing flow resistance in fluid distribution systems results in less noise radiation.

Four of the more important factors that should be checked to reduce frictional resistance in moving parts are the following:

1. Alignment: Proper alignment of all rotating, moving, or contacting parts results in less noise output. Good axial and directional alignment in pulley systems, gear trains, shaft couplings, power transmission systems, and bearing and axle alignment are fundamental requirements for low noise output.
2. Polish: Highly polished and smooth surfaces between sliding, meshing, or contacting parts are required for quiet operation, particularly where bearings, gears, cams, rails, and guides are concerned.
3. Balance: Static and dynamic balancing of rotating parts reduces frictional resistance and vibration, resulting in lower noise output.
4. Eccentricity (out-of-roundness): Off-centering of rotating parts such as pulleys, gears, rotors, and shaft/bearing alignment causes vibration and noise.

Likewise, out-of-roundness of wheels, rollers, and gears causes uneven wear, resulting in flat spots that generate vibration and noise.

The key to effective noise control in fluid systems is *streamline flow.* This holds true regardless of whether one is concerned with air flow in ducts or vacuum cleaners, or with water flow in plumbing systems. Streamline flow is simply smooth, nonturbulent, low-friction flow.

The two most important factors that determine whether flow will be streamline or turbulent are the speed of the fluid and the cross-sectional area of the flow path, that is, the pipe or duct diameter. The rule of thumb for quiet operation is to use a low-speed, large-diameter system to meet a specified flow capacity requirement. However, even such a system can inadvertently generate noise if certain aerodynamic design features are overlooked or ignored. A system designed for quiet operation will employ the following features:

1. Low fluid speed: Low fluid speeds avoid turbulence, which is one of the main causes of noise.
2. Smooth boundary surfaces: Duct or pipe systems with smooth interior walls, edges, and joints generate less turbulence and noise than systems with rough or jagged walls or joints.
3. Simple layout: A well-designed duct or pipe system with a minimum of branches, turns, fittings, and connectors is substantially less noisy than a complicated layout.
4. Long-radius turns: Changes in flow direction should be made gradually and smoothly. It has been suggested that turns should be made with a curve radius equal to about five times the pipe diameter or major cross-sectional dimension of the duct.
5. Flared sections: Flaring of intake and exhaust openings, particularly in a duct system, tends to reduce flow speeds at these locations, often with substantial reductions in noise output.
6. Streamline transition in flow path: Changes in flow path dimensions or cross-sectional areas should be made gradually and smoothly with tapered or flared transition sections to avoid turbulence. A good rule of thumb is to keep the cross-sectional area of the flow path as large and as uniform as possible throughout the system.
7. Remove unnecessary obstacles: The greater the number of obstacles in the flow path, the more tortuous, turbulent, and hence noisier, the flow. All other required and functional devices in the path, such as structural supports, deflectors, and control dampers, should be made as small and as streamlined as possible to smooth out the flow patterns.

Reduce Radiating Area. Generally speaking, the larger the vibrating part or surface, the greater the noise output. The rule of thumb for quiet machine design is to minimize the effective radiating surface areas of the parts without impairing their operation or structural strength. This can be done by making parts smaller, removing excess material, or by cutting openings, slots, or perforations in the parts. For example, replacing a large, vibrating sheet-metal safety guard on a machine with a guard made of wire mesh or metal webbing might result in a substantial reduction in noise because of the drastic reduction in surface area of the part.

Reduce Noise Leakage. In many cases, machine cabinets can be made into rather effective soundproof enclosures through simple design changes and the application of some sound absorbing treatment. Substantial reductions in noise output may be achieved by adopting some of the following recommendations:

1. All unnecessary holes or cracks, particularly at joints, should be caulked.
2. All electrical or plumbing penetrations of the housing or cabinet should be sealed with rubber gaskets or a suitable nonsetting caulk.
3. If practical, all other functional or required openings or ports that radiate noise should be covered with lids or shields edged with soft rubber gaskets to effect an airtight seal.
4. Other openings required for exhaust, cooling, or ventilation purposes should be equipped with mufflers or acoustically lined ducts.
5. Openings should be directed away from the operator and other people.

Isolate and Dampen Vibrating Elements. In all but the simplest machines, the vibrational energy from a specific moving part is transmitted through the machine structure, forcing other component parts and surfaces to vibrate and radiate sound—often with greater intensity than that generated by the originating source itself.

Generally, vibration problems can be considered in two parts. First, we must prevent energy transmission between the source and surfaces that radiate the energy. Second, we must dissipate or attenuate the energy somewhere in the structure. The first part of the problem is solved by *isolation.* The second part is solved by *damping.*

The most effective method of vibration isolation involves the resilient mounting of the vibrating component on the most massive and structurally rigid part of the machine. All attachments or connections to the vibrating part, in the form of pipes, conduits, and shaft couplers, must be made with flexible or resilient connectors or couplers. For example, pipe connections to a pump that is resiliently mounted on the structural frame of a machine should be made of resilient tubing and be mounted as close to the pump as possible. Resilient pipe supports or hangers may also be required to avoid bypassing the isolated system (Figure 8-41).

Damping material or structures are those that have some viscous properties. They tend to bend or distort slightly, thus consuming part of the noise energy in molecular motion. The use of spring mounts on motors and laminated galvanized steel and plastic in air-conditioning ducts are two examples.

When the vibrating noise source is not amenable to isolation, as, for example, in ventilation ducts, cabinet panels, and covers, then damping materials can be used to reduce the noise.

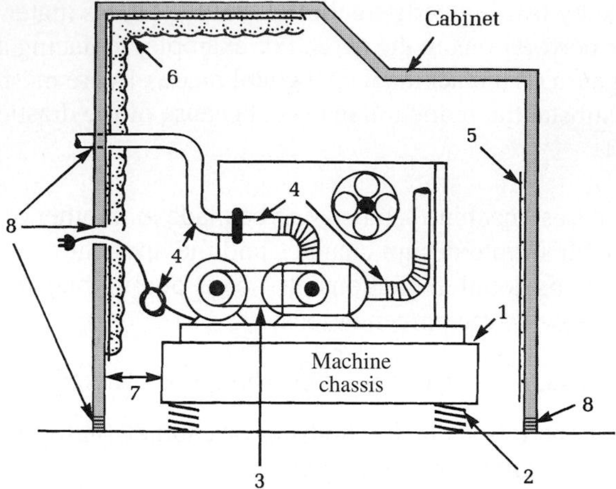

FIGURE 8-41
Examples of vibration isolation.

Code:
1. Motors, pumps, and fans installed on most massive part of the machine
2. Resilient mounts or vibration isolators used for the installation
3. Belt-drive or roller-drive systems used in place of gear trains
4. Flexible hoses and wiring used instead of rigid piping and stiff wiring
5. Vibration-damping materials applied to surfaces undergoing most vibration
6. Acoustical lining installed to reduce noise buildup inside machine
7. Mechanical contact minimized between the cabinet and the machine chassis
8. Openings at the base and other parts of the cabinet scaled to prevent noise leakage

(*Source:* Berendt et al., 1976.)

The type of material best suited for a particular vibration problem depends on factors such as size, mass, vibrational frequency, and operational function of the vibrating structure. Generally speaking, the following guidelines should be observed in the selection and use of such materials to maximize vibration damping efficiency:

1. Damping materials should be applied to those sections of a vibrating surface where the most flexing, bending, or motion occurs. These usually are the thinnest sections.

2. For a single layer of damping material, the stiffness and mass of the material should be comparable to that of the vibrating surface to which it is applied. This means that single-layer damping materials should be about two or three times as thick as the vibrating surface to which they are applied.

3. Sandwich materials (*laminates*) made up of metal sheets bonded to mastic (sheet metal viscoelastic composites) are much more effective vibration dampers than single-layer materials; the thickness of the sheet-metal constraining layer and the viscoelastic layer should each be about one-third the thickness of the vibrating surface to which they are applied. Ducts and panels can be purchased already fabricated as laminates.

Provide Mufflers/silencers. There is no real distinction between mufflers and silencers. They are often used interchangeably. They are, in effect, acoustical filters and are used when fluid flow noise is to be reduced. The devices can be classified into two fundamental groups: *absorptive mufflers* and *reactive mufflers.* An absorptive muffler is one whose noise reduction is determined mainly by the presence of fibrous or porous materials, which absorb the sound. A reactive muffler is one whose noise reduction is determined mainly by geometry. It is shaped to reflect or expand the sound waves with resultant self-destruction.

Although there are several terms used to describe the performance of mufflers, the most frequently used appears to be *insertion loss* (IL). Insertion loss is the difference between two sound pressure levels that are measured at the same point in space before and after a muffler has been inserted. Since each muffler's IL is highly dependent on the manufacturer's selection of materials and configuration, we will not present general IL prediction equations.

Noise Control in the Transmission Path

After you have tried all possible ways of controlling the noise at the source, your next line of defense is to set up devices in the transmission path to block or reduce the flow of sound energy before it reaches your ears. This can be done in several ways: (*a*) absorb the sound along the path, (*b*) deflect the sound in some other direction by placing a reflecting barrier in its path, or (*c*) contain the sound by placing the source inside a sound-insulating box or enclosure.

Selection of the most effective technique will depend upon various factors, such as the size and type of source, intensity and frequency range of the noise, and the nature and type of environment.

Separation. We can make use of the absorptive capacity of the atmosphere, as well as divergence, as a simple, economical method of reducing the noise level. Air absorbs high-frequency sounds more effectively than it absorbs low-frequency sounds. However, if enough distance is available, even low-frequency sounds will be absorbed appreciably.

If you can double your distance from a point source, you will have succeeded in lowering the sound pressure level by 6 dB. It takes about a 10 dB drop to halve the loudness. If you have to contend with a line source such as a railroad train, the noise level drops by only 3 dB for each doubling of distance from the source. The main reason for this lower rate of attenuation is that line sources radiate sound waves that are cylindrical in shape. The surface area of such waves only increases two-fold for each doubling of distance from the source. However, when the distance from the train becomes comparable to its length, the noise level will begin to drop at a rate of 6 dB for each subsequent doubling of distance.

Indoors, the noise level generally drops only from 3 to 5 dB for each doubling of distance in the near vicinity of the source. However, further from the source, reductions of only 1 or 2 dB occur for each doubling of distance due to the reflections of sound off hard walls and ceiling surfaces.

Absorbing Materials. Noise, like light, will bounce from one hard surface to another. In noise control work, this is called *reverberation.* If a soft, spongy material is placed on the walls, floors, and ceiling, the reflected sound will be diffused and soaked up (absorbed). Sound-absorbing materials are rated either by their *Sabin absorption coefficients* (α_{SAB}) at 125, 500, 1,000, 2,000, and 4,000 Hz or by a single number rating called the *noise reduction coefficient* (NRC). If a unit area of open window is assumed to transmit all and reflect none of the acoustical energy that reaches it, it is assumed to be 100 percent absorbent. This unit area of totally absorbent surface is called a "sabin" (Sabin, 1942). The absorptive properties of acoustical materials are then compared with this standard. The performance is expressed as a fraction or percentage of the sabin (α_{SAB}). The NRC is the average of the α_{SAB}s at 250, 500, 1,000, and 2,000 Hz rounded to the nearest multiple of 0.05. The NRC has no physical meaning. It is a useful means of comparing similar materials.

Sound-absorbing materials such as acoustical tile, carpets, and drapes placed on ceiling, floor, or wall surfaces can reduce the noise level in most rooms by about 5 to 10 dB for high-frequency sounds, but only by 2 or 3 dB for low-frequency sounds. Unfortunately, such treatment provides no protection to an operator of a noisy machine who is in the midst of the direct noise field. For greatest effectiveness, sound-absorbing materials should be installed as close to the noise source as possible.

If you have a small or limited amount of sound-absorbing material and wish to make the most effective use of it in a noisy room, the best place to put it is in the upper trihedral corners of the room, formed by the ceiling and two walls. Due to the process of reflection, the concentration of sound is greatest in the trihedral corners of a room. Additionally, the upper corner locations also protect the lightweight fragile materials from damage.

Because of their light weight and porous nature, acoustical materials are ineffectual in preventing the transmission of either airborne or structure-borne sound from one room to another. In other words, if you can hear people walking or talking in the room or apartment above, installing acoustical tile on your ceiling will not reduce the noise transmission.

Acoustical Lining. Noise transmitted through ducts, pipe chases, or electrical channels can be reduced effectively by lining the inside surfaces of such passageways with sound-absorbing materials. In typical duct installations, noise reductions on the order of 10 dB/m for an acoustical lining 2.5 cm thick are well within reason for high-frequency noise. A comparable degree of noise reduction for the lower frequency sounds is considerably more difficult to achieve because it usually requires at least a doubling of the thickness and/or length of acoustical treatment.

Barriers and Panels. Placing barriers, screens, or deflectors in the noise path can be an effective way of reducing noise transmission, provided that the barriers are large enough in size, and depending upon whether the noise is high frequency or low frequency. High-frequency noise is reduced more effectively than low-frequency noise.

The effectiveness of a barrier depends on its location, its height, and its length. Referring to Figure 8-42, we can see that the noise can follow five different paths.

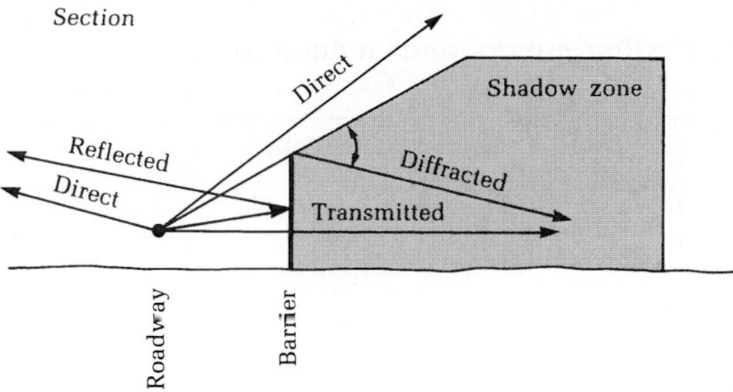

FIGURE 8-42
Noise paths from a source to a receiver. (*Source:* Kugler et al., 1976.)

First, the noise follows a direct path to receivers who can see the source well over the top of the barrier. The barrier does not block their line of sight (*L/S*) and therefore provides no attenuation. No matter how absorptive the barrier is, it cannot pull the sound downward and absorb it.

Second, the noise follows a diffracted path to receivers in the shadow zone of the barrier. The noise that passes just over the top edge of the barrier is diffracted (bent) down into the apparent shadow shown in the figure. The larger the angle of diffraction, the more the barrier attenuates the noise in this shadow zone. In other words, less energy is diffracted through large angles than through smaller angles.

Third, in the shadow zone, the noise transmitted directly through the barrier may be significant in some cases. For example, with extremely large angles of diffraction, the diffracted noise may be less than the transmitted noise. In this case, the transmitted noise compromises the performance of the barrier. It can be reduced by constructing a heavier barrier. The allowable amount of transmitted noise depends on the total barrier attenuation desired. More is said about this transmitted noise later.

The fourth path shown in Figure 8-42 is the reflected path. After reflection, the noise is of concern only to a receiver on the opposite side of the source. For this reason, acoustical absorption on the face of the barrier may sometimes be considered to reduce this reflected noise; however, this treatment will not benefit any receivers in the shadow zone. It should be noted that in most practical cases the reflected noise does not play an important role in barrier design. If the source of noise is represented by a line of noise, another short-circuit path is possible. Part of the source may be unshielded by the barrier. For example, the receiver might see the source beyond the ends of the barrier if the barrier is not long enough. This noise from around the ends may compromise, or short-circuit, barrier attenuation. The required barrier length depends on the total net attenuation desired. When 10 to 15 dB attenuation is desired, barriers must, in general, be very long. Therefore, to be effective, barriers must not only break the line of sight to the nearest section of the source, but also to the source far up and down the line.

Of these four paths, the noise diffracted over the barrier into the shadow zone represents the most important parameter from the barrier design point of view. Generally, the determination of barrier attenuation or barrier noise reduction involves only calculation

TABLE 8-10
Relation between sound level reduction, energy, and loudness for line sources

To reduce A-level by dB	Remove portion of energy (%)	Divide loudness by
3	50	1.2
6	75	1.5
10	90	2
20	90	4
30	99.9	8
40	99.99	16

(*Source:* Kugler et al., 1976.)

of the amount of energy diffracted into the shadow zone. The procedures presented in the barrier nomograph used to predict highway noise are based on this concept.

Another general principle of barrier noise reduction that is worth reviewing at this point is the relation between noise attenuation expressed in (1) decibels, (2) energy terms, and (3) subjective loudness. Table 8-10 gives these relationships for line sources. As indicated in the loudness column, a barrier attenuation of 3 dB will be barely discerned by the receiver. However, to attain this reduction, 50 percent of the acoustical energy must be removed. To cut the loudness of the source in half, a reduction of 10 dB is necessary. That is equivalent to eliminating 90 percent of the energy initially directed toward the receiver. As indicated previously, this drastic reduction in energy requires very long and high barriers. In summary, when designing barriers, you can expect the complexity of the design to be about as follows:

Attenuation (dB)	Complexity
5	Simple
10	Attainable
15	Very difficult
20	Nearly impossible

Roadside barriers can be designed using the barrier nomograph in reverse order. A set of typical solutions is summarized in Table 8-11. The noise reduction at 152 m is less than that at 30 m because the barrier does not cast as large a shadow at a distance. The effectiveness of the barrier is reduced for trucks because of the elevated nature of the source.

Transmission Loss. When the position of the noise source is very close to the barrier, the diffracted noise is less important than the transmitted noise. If the barrier is in fact a wall panel that is sealed at the edges, the transmitted noise is the only one of concern.

TABLE 8-11
Noise reductions for various highway configurations

Highway configuration[a]		Height or depth (m)	Truck mix (%)	Noise reduction[b] at distance from ROW (dBA)	
Sketch	Description			30 m	152 m
	Roadside barriers 7.6 m from edge of shoulders; ROW = 78 m wide	6.1	0	13.9	13.3
			5	13.0	12.1
			10	12.6	11.7
			20	12.3	11.3
	Depressed roadway w/2:1 slopes; ROW = 102 m	6.1	0	9.9	11.4
			5	8.8	10.3
			10	8.4	9.8
			20	8.1	9.4
	Fill elevated roadway w/2:1 slopes; ROW = 102 m	6.1	0	9.0	6.3
			5	7.6	2.7
			10	7.1	1.8
			20	6.7	1.1
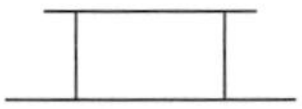	Elevated structure; ROW = 78 m	7.3	0	9.8	6.0
			5	9.6	2.4
			10	9.3	1.5
			20	8.8	0.8

[a]Assumes divided 8 lanes with 9.1 m median.
[b]Based on observed 1.5 m above grade.
(*Source:* Kugler et al., 1976.)

The ratio of the sound energy incident on one surface of a panel to the energy radiated from the opposite surface is called the *sound transmission loss* (TL). The actual energy loss is partially reflected and partially absorbed. Since TL is frequency-dependent, only a complete octave or one-third octave band curve provides a full description of the performance of the barrier.

Enclosures. Sometimes it is much more practical and economical to enclose a noisy machine in a separate room or box than to quiet it by altering its design, operation, or component parts. The walls of the enclosure should be massive and airtight to contain the sound. Absorbent lining on the interior surfaces of the enclosure will reduce the reverberant buildup of noise within it. Structural contact between the noise source and the enclosure must be avoided, so that the source vibration is not transmitted to the enclosure walls, thus short-circuiting the isolation. For maximum effective noise control, all of the techniques illustrated in Figure 8-43 must be employed.

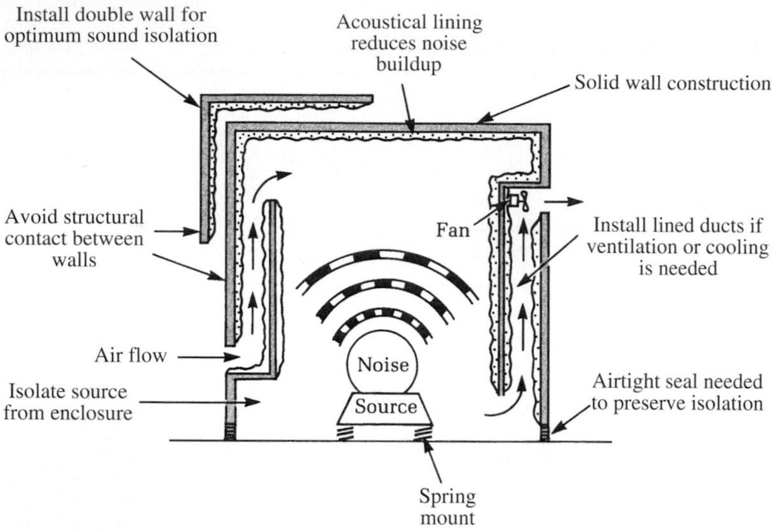

FIGURE 8-43
Enclosures for controlling noise. (*Source:* Berendt et al., 1976.)

Control of Noise Source by Redress

The best way to solve noise problems is to design them out of the source. However, we are frequently faced with an existing source that, either because of age, abuse, or poor design, is a noise problem. The result is that we must redress, or correct, the problem as it currently exists. The following sections identify some measures that might apply if you are allowed to tinker with the source.

Balance Rotating Parts. One of the main sources of machinery noise is structural vibration caused by the rotation of poorly balanced parts, such as fans, fly wheels, pulleys, cams, shafts, and so on. Measures used to correct this condition involve the addition of counterweights to the rotating unit or the removal of some weight from the unit. You are probably familiar with noise caused by imbalance in the high-speed spin cycle of washing machines. The imbalance results from clothes not being distributed evenly in the tub. By redistributing the clothes, balance is achieved and the noise ceases. This same principle of balance can be applied to furnace fans and other common sources of such noise.

Reduce Frictional Resistance. A well-designed machine that has been poorly maintained can become a serious source of noise. General cleaning and lubrication of all rotating, sliding, or meshing parts at contact points should go a long way toward fixing the problem.

Apply Damping Materials. Since a vibrating body or surface radiates noise, the application of any material that reduces or restrains the vibrational motion of that body will decrease its noise output. Three basic types of redress vibration damping materials are available:

1. Liquid mastics, which are applied with a spray gun and harden into relatively solid materials, the most common being automobile "undercoating"

2. Pads of rubber, felt, plastic foam, leaded vinyls, adhesive tapes, or fibrous blankets, which are glued to the vibrating surface
3. Sheet metal viscoelastic laminates or composites, which are bonded to the vibrating surface

Seal Noise Leaks. Small holes in an otherwise noise-tight structure can reduce the effectiveness of the noise control measures. As you can see in Figure 8-44, if the designed transmission loss of an acoustical enclosure is 40 dB, an opening that comprises only 0.1 percent of the surface area will reduce the effectiveness of the enclosure by 10 dB.

Perform Routine Maintenance. We all recognize the noise of a worn muffler. Likewise, studies of automobile tire noise in relation to pavement roughness show that maintenance of the pavement surface is essential to keep noise at minimum levels. Normal road wear can yield noise increases on the order of 6 dBA.

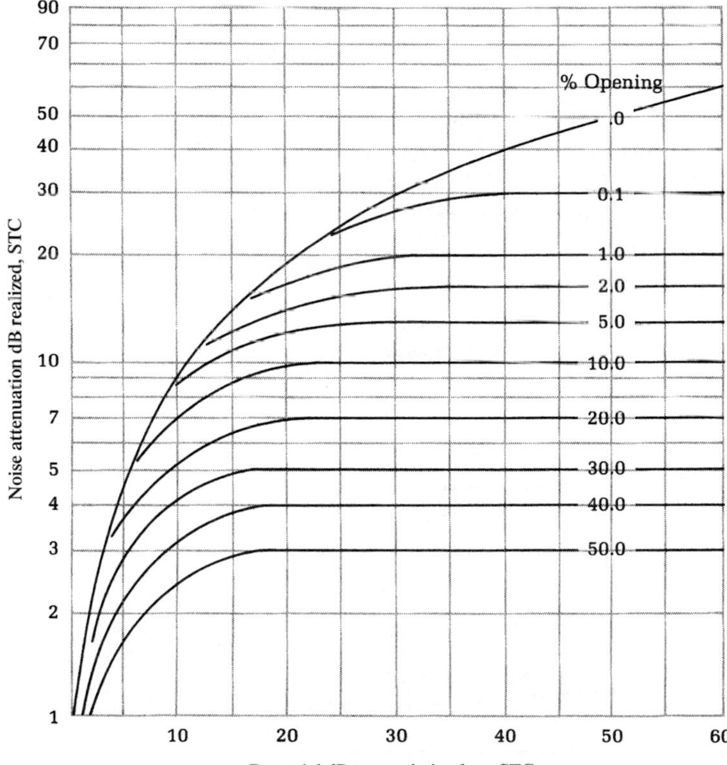

FIGURE 8-44
Transmission loss potential versus transmission loss realized for various opening sizes as a percent of total wall area. STC = sound transmission coefficient (*Source:* Adapted from Warnock and Quirt, 1991.)

Protect the Receiver

When All Else Fails. When exposure to intense noise fields is required and none of the measures discussed so far is practical, as, for example, for the operator of a chain saw or pavement breaker, then measures must be taken to protect the receiver. The following two techniques are commonly employed.

Alter Work Schedule. Limit the amount of continuous exposure to high noise levels. In terms of hearing protection, it is preferable to schedule an intensely noisy operation for a short interval of time each day over a period of several days rather than a continuous eight-hour run for a day or two.

In industrial or construction operations, an intermittent work schedule would benefit not only the operator of the noisy equipment, but also other workers in the vicinity. If an intermittent schedule is not possible, then workers should be given relief time during the day. They should take their relief time at a low-noise-level location, and should be discouraged from trading relief time for dollars, paid vacation, or an "early out" at the end of the day!

Inherently noisy operations, such as street repair, municipal trash collection, factory operation, and aircraft traffic, should be curtailed at night and early morning to avoid disturbing the sleep of the community. Remember: operations between 10 P.M. and 7 A.M. are effectively 10 dBA higher than the measured value.

Ear Protection. Molded and pliable earplugs, cup-type protectors, and helmets are commercially available as hearing protectors. Such devices may provide noise reductions ranging from 15 to 35 dB (Figure 8-45). Earplugs are effective only if they are properly fitted by medical personnel. As shown in Figure 8-45, maximum protection

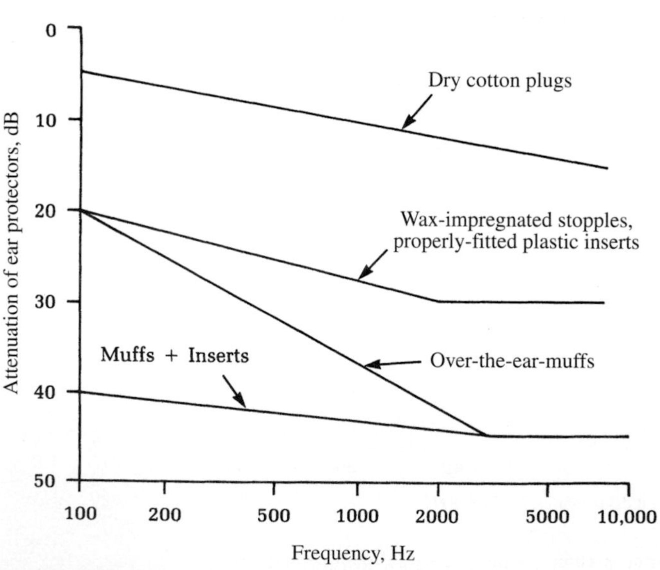

FIGURE 8-45
Attenuation of ear protectors at various frequencies. (*Source:* Berendt et al., 1976.)

can be obtained when both plugs and muffs are employed. Only muffs that have a certification stipulating the attenuation should be used.

These devices should be used only as a last resort, after all other methods have failed to lower the noise level to acceptable limits. Ear protection devices should be used while operating lawn mowers, mulchers, and chippers, and while firing weapons at target ranges. It should be noted that protective ear devices do interfere with speech communication and can be a hazard in some situations where warning calls may be a routine part of the operation (for example, TIMBERRRR!). A modern ear-destructive device is a portable digital music player that uses earphones. In this "reverse" muff, high noise levels are directed at the ear without attenuation. If you can hear someone else's music player, that person is subjecting him or herself to noise levels in excess of 90–95 dBA!

8-8 CHAPTER REVIEW

When you have completed studying this chapter, you should be able to do the following without the aid of your textbook or notes:

1. Define frequency, based on a sketch of a harmonic wave you have drawn, and state its units of measure (namely, hertz, Hz).
2. State the basic unit of measure used in measuring sound energy (namely, the decibel) and explain why it is used.
3. Define sound pressure level in mathematical terms, that is,

$$\text{SPL} = 20 \log \frac{P_{\text{rms}}}{(P_{\text{rms}})_0}$$

4. Explain why a weighting network is used in a sound level meter.
5. List the three common weighting networks and sketch their relative frequency response curves. (Label frequencies, that is, 20, 1,000, and 10,000 Hz; and relative response, that is, 0, −5, −20, and −45 dB, as in Figure 8-5.)
6. Differentiate between a mid/high-frequency noise source and a low-frequency noise source on the basis of A, B, and C scale readings.
7. Explain the purpose of octave band analysis.
8. Differentiate between continuous, intermittent, and impulsive noise.
9. Sketch the curves and label the axes of the two typical types of impulsive noise.
10. Sketch a Fletcher-Munson curve, label the axes, and explain what the curve depicts.
11. Define "phon."
12. Explain the mechanism by which hearing damage occurs.
13. Explain what hearing threshold level (HTL) is.

14. Define presbycusis and explain why it occurs.
15. Distinguish between temporary threshold shift (TTS), permanent threshold shift (PTS), and acoustic trauma with respect to cause of hearing loss, duration of exposure, and potential for recovery.
16. Explain why impulsive noise is more dangerous than steady state noise.
17. Explain the relationship between the allowable duration of noise exposure and the allowable level for hearing protection, that is, damage-risk criteria.
18. List five effects of noise other than hearing damage.
19. List the three basic elements that might require alteration or modification to solve a noise problem.
20. Describe two techniques to protect the receiver when design and/or redress are not practical, that is, when all else fails.

With the aid of this text, you should be able to do the following:

1. Calculate the resultant sound pressure level from a combination of two or more sound pressure levels.
2. Determine the A-, B-, and C-weighted sound levels from octave band readings.
3. Compute the mean sound level from a series of sound level readings.
4. Compute the following noise statistics if you are provided the appropriate data: L_N and/or L_{eq}; L_{dn}.
5. Determine whether or not a noise level will be acceptable given a series of measurements and the criteria listed in Tables 8-3, 8-5, 8-6, 8-7, and/or Figures 8-26 and 8-28.
6. Calculate the sound level at a receptor site after transmission through the atmosphere.
7. Estimate the noise level L_{10} that might be expected for a given roadway configuration and traffic pattern.

8-9 PROBLEMS

8-1. A building located near a road is 6.92 m high. How high is the building in terms of wavelengths of a 50.0-Hz sound? Assume that the speed of sound is 346.12 m/s.

Answer: One wavelength

8-2. Repeat Problem 8-1 for a 500-Hz sound if the temperature is 25.0°C.

8-3. Determine the sum of the following sound levels (all in dB): 68, 82, 76, 68, 74, and 81.

Answer: 85.5 or 86 dB

8-4. A motorcyclist is warming up his racing cycle at a racetrack approximately 200 m from a sound level meter. The meter reading is 56 dBA. What meter reading would you expect if 15 of the motorcyclist's friends join him with motorcycles having exactly the same sound emission characteristics? You may assume that the sources may be treated as ideal point sources located at the same point.

8-5. A sound power level reading of 127 dB was taken near a construction site where chippers were being used. When all but one of the chippers stopped working, the sound power level reading was 120 dB. Estimate the number of chippers in operation when the reading of 127 was obtained. You may assume that the sources may be treated as ideal point sources located at the same point.

8-6. A law enforcement officer has taken the following readings with her sound level meter. Is the noise source a predominantly low- or middle-frequency emitter? Readings: 80 dBA, 84 dBB, and 90 dBC.

Answer: Predominantly low frequency

8-7. The following readings have been made outside the open stage door of the opera house: 109 dBA, 110 dBB, and 111 dBC. Is the singer a bass or a soprano? Explain how you arrived at your answer.

8-8. Convert the following octave band measurements to an equivalent A-weighted sound level.

Band center frequency (Hz)	Band level (dB)
31.5	78
63	76
125	78
250	82
500	81
1,000	80
2,000	80
4,000	73
8,000	65

Answer: 85.5 or 86 dBA

8-9. The following noise spectrum was obtained from a jet aircraft flying overhead at an altitude of 250 m. Compute the equivalent A-weighted sound level using sound power level addition in a spreadsheet program you have written.

Band center frequency (Hz)	Band level (dB)
125	85
250	88
500	96
1,000	100
2,000	104
4,000	101

8-10. Using the typical noise spectrum for automobiles traveling at 50 to 60 km/h, determine the equivalent A-weighted level using sound power level addition in a spreadsheet program you have written. The following band levels were estimated from Figure 8-27.

Band center frequency (Hz)	Band level (dB)
63	67
125	64
250	58
500	59
1,000	59
2,000	55
4,000	51
8,000	45

8-11. You have been asked to evaluate the A-weighted sound level of a new model lawn mower and make a recommendation on an acceptable noise spectrum to achieve 74 dBA. Three approaches are being considered by the manufacturer: (1) an improved muffler that will reduce the sound level 3 dB in each frequency band, (2) a reduction in the speed of the mower which will reduce the sound level 5 dB in each frequency band, and (3) an engine redesign that will reduce the sound level 15 dB in the five highest frequency bands. Using a computer spreadsheet program you have written, compute the A-weighted sound level for the sound spectrum shown on the following page and develop a recommended noise spectrum based on the manufacturer's alternatives that results in a sound level of less than 74 dBA. Assume that each of the alternative reductions may be added together (by decibel addition) in each frequency band in which it is applicable.

Band center frequency (Hz)	Band level (dB)
63	78
125	76
250	76
500	77
1,000	79
2,000	80
4,000	78
8,000	70

8-12. Compute the average sound pressure level of the following readings by simple arithmetic averaging and by logarithmic averaging (Equation 8-13) (all readings in dB): 42, 50, 65, 71, and 47. Does arithmetic averaging underestimate or overestimate the sound pressure level?

Answers: $\bar{x} = 55.00$ or 55 dB $L_p = 61.57$ or 62 dB

8-13. Repeat Problem 8-12 for the following data (all in dB): 76, 59, 35, 69, and 72.

8-14. The following noise record was obtained in the front yard of a home. Is this a relatively quiet or a relatively noisy neighborhood? Determine the equivalent continuous equal energy level.

Time (h)	Sound level (dBA)
0000–0600	42
0600–0800	45
0800–0900	50
0900–1500	47
1500–1700	50
1700–1800	47
1800–0000	45

Answers: It is a quiet neighborhood. $L_{eq} = 46.2$ or 46 dBA

8-15. A developer has proposed putting a small shopping mall next to a very quiet residential area in Nontroppo, Michigan. Based on the measurements given on the following page, which were taken at a similar size mall in a similar setting, should the developer expect complaints or legal action? Calculate L_{eq}.

Time (h)	Sound level (dBA)
0000–0600	42
0600–0800	55
0800–1000	65
1000–2000	70
2000–2200	68
2200–0000	57
1800–0000	45

8-16. The U.S. EPA (1974) estimated that the following was a typical noise exposure pattern for a factory worker living in an urban area. Estimate the L_{dn} for the exposure shown.

Time (h)	Sound level (dBA)
0000–0500	52
0500–0700	78
0700–1130	90
1130–1200	70
1200–1530	90
1530–1800	52
1800–2200	60
2200–0000	52

8-17. The U.S. EPA (1974) estimated that the following was a typical noise exposure pattern for a middle school student living in an urban area. Estimate the L_{dn} for the exposure shown.

Time (h)	Sound level (dBA)
0000–0700	52
0700–0900	82
0900–1200	60
1200–1300	65
1300–1500	60
1500–1700	75
1700–1800	90
1800–2100	60
2100–0000	52

8-18. Two oil-fired boilers for a 600 megawatt (MW) power plant produce a sound power level of 139 dB (re: 10^{-12} W) at 4,000 Hz, from the induced draft fans. Determine the sound pressure level 408.0 m downwind on a clear winter night when the wind speed is 4.50 m/s, the temperature is 0.0°C, the relative humidity is 30.0 percent, and the barometric pressure is 101.3 kPa. The height of the boiler is 12 m. The height of the receiver is 1.5 m. The ground surface characteristics are shown in Figure P-8-18.

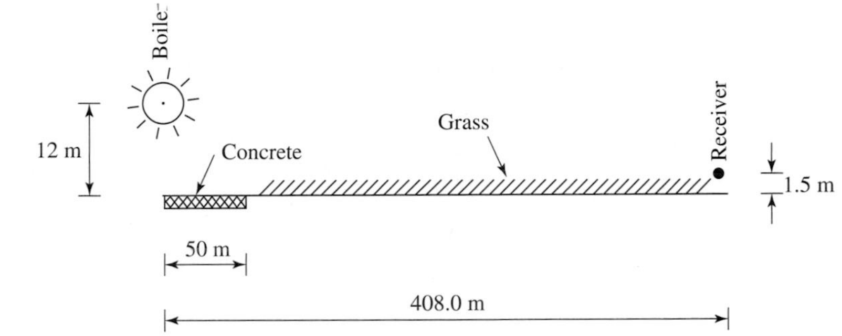

FIGURE P-8-18
Ground surface characteristics for Problem 8-18.

Answer: SPL at 408.0 m = 50.50 or 50 dB at 4,000 Hz

8-19. The 125-Hz sound power level (re: 10^{-12} W) from a jet engine test cell is 149 dB. What is the 125-Hz sound pressure level 1,200 m downwind on a clear summer morning during an inversion when the wind speed is 1.50 m/s, the temperature is 25.0°C, the relative humidity is 70.0 percent, and the barometric pressure is 101.3 kPa? The height of the engine and the receiver are 1.5 m. The ground surface characteristics are shown in Figure P-8-19.

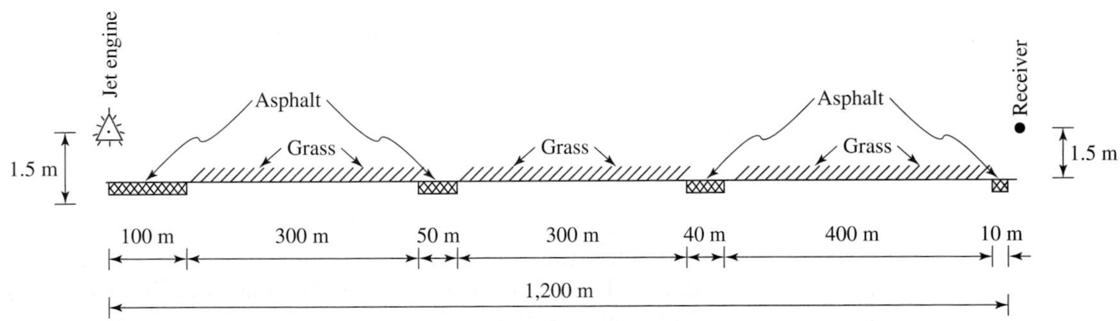

FIGURE P-8-19
Ground surface characteristics for Problem 8-19.

8-20. Using a computer spreadsheet you have written, calculate the A-weighted sound level produced by a jet airplane that is 2,000 m from the receiver.

The sound power level in each octave band for the jet (at the source) is tabulated below. The meteorological conditions are as follows: clear summer morning during an inversion when the wind speed is 2.50 m/s, the temperature is 20.0°C, the relative humidity is 70.0 percent, and the barometric pressure is 101.3 kPa. The source and the receiver heights are 1.5 m. The ground surface characteristics are shown in Figure P-8-20.

Band center frequency (Hz)	Band level, re: 10^{-12} W
125	144
250	148
500	155
1,000	160
2,000	165
4,000	168

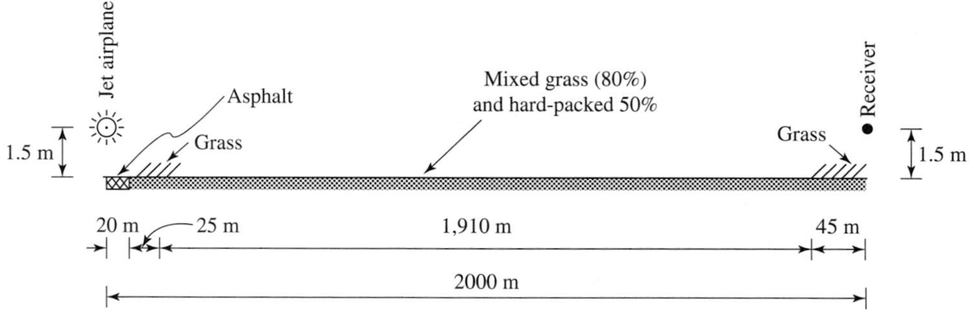

FIGURE P-8-20
Ground surface characteristics for Problem 8-20.

8-21. Consider an ideal single lane of road that carries 1,200 vehicles per hour uniformly spaced along the road and determine the following:

a. The average center-to-center spacing of the vehicles for an average traffic speed of 40.0 km/h.

b. The number of vehicles in a 1 km length of the lane when the average speed is 40.0 km/h.

c. The sound level (dBA) 60.0 m from a 1 km length of this roadway with automobiles emitting 71 dBA at the edge of an 8.0 m-wide roadway. Assume that the autos travel at a speed of 40.0 km/h, that the sound radiates ideally from a hemisphere, and that contributions of less than 0.3 dBA may be ignored.

Answers:

a. Average center-to-center spacing = 33.3 m

b. Number of vehicles in a 1 km length = 30 vehicles/km

c. L_p = 47.47 or 48 dBA

8-22. Repeat Problem 8-21 if the vehicle speed is increased to 80.0 km/h and the spacing is decreased or increased appropriately to maintain 1,200 vehicles per hour.

8-23. Determine the L_{10} noise level for the following highway traffic situation:

Autos = 1,500 per hour Speed = 60 km/h

Medium trucks = 150 per hour Speed = 60 km/h

Heavy trucks = 0

Observer-roadway distance = 40.0 m

Line-of-sight distance = 40.0 m

Barrier position = 3.0 m

Break in line-of-sight distance = 6.0 m

Angle subtended = 170°

Answer: L_{10} = 51 dBA

8-24. Determine the L_{10} noise level for the following highway traffic situation:

Autos = 2,000 per hour Speed = 70 km/h

Medium trucks = 200 per hour Speed = 70 km/h

Heavy trucks = 0

Observer-roadway distance = 60.0 m

Line-of-sight distance = 60.0 m

Barrier position = 5.0 m

Break in line-of-sight distance = 1.0 m

Angle subtended = 160°

8-25. In preparation for a public hearing on a proposed interstate bypass at Non-troppo, Michigan (Figure P-8-25a), the County Road Commission has requested that you prepare an estimate of the potential for violation of FHA noise standards 75 m from the interstate. The city engineer has supplied sketch maps (see Figure P-8-25b) and data summary for your use.

Data for I-481 at Pianissimo Avenue

Estimated traffic:
Automobiles: 7,800 per hour at 88.5 km/h
Medium trucks: 520 per hour at 80.5 km/h
Heavy trucks: 650 per hour at 80.5 km/h

Roadway configuration: Depressed

Section length: 857.25 m east and 857.25 m west of center line of Pianissimo Avenue

Assume that the receiver is located on the center line of Pianissimo Avenue 75.00 m from the center line of I-481.

Answer: L_{10} at observer = 68.6 or 69 dBA

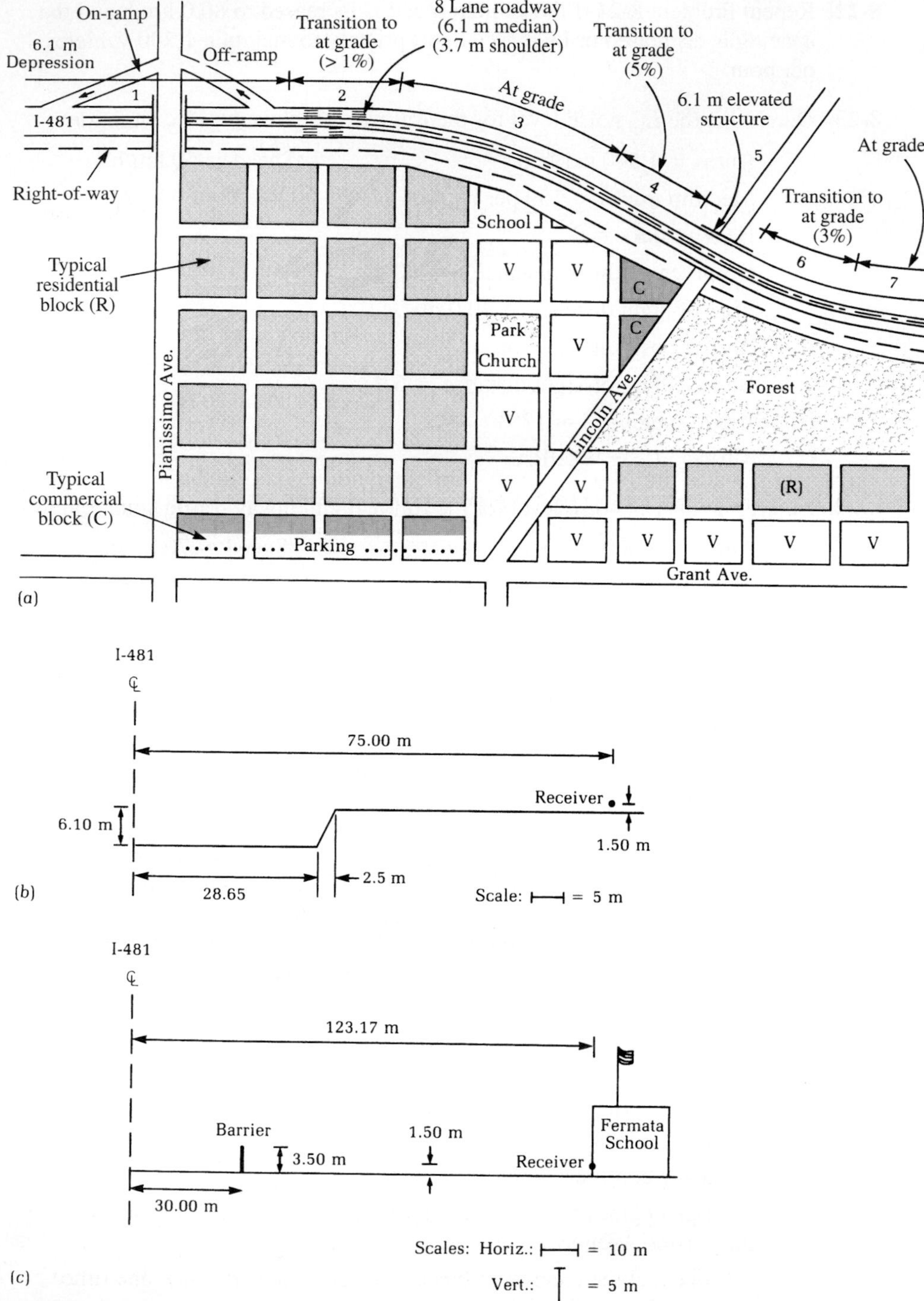

FIGURE P-8-25

Sketch maps for proposed bypass. (*a*) plan view, (*b*) cross section along Pianissimo Avenue, (*c*) cross section at Fermata School. (See Problems 8-25 and 8-26.)

8-26. Determine the potential for violation of FHA noise standards at the north side of Fermata School. The city engineer has supplied sketch maps (see Figure P-8-25c) and data for your use.

Data for I-481 at Fermata School

Estimated traffic: Same as at Pianissimo Avenue

Roadway configuration: At grade

Barrier length: 199.80 m east and 199.80 m west of Fermata School

Assume that the receiver is located just outside of the north side of the school, 123.17 m from the center line of I-481, and is 1.5 m above the ground.

8-27. Using the data from Problem 8-25, compute the *unattenuated* L_{eq}, at the receiver for autos only. Assume the edge of the roadway is at the "toe" of the road cut.

Answer: $L_{eq} = 70$ dBA

8-28. Rework Problem 8-27 using the data from Problem 8-26. Assume the edge of the roadway is at the barrier.

8-10 DISCUSSION QUESTIONS

8-1. Classify each of the following noise sources by "type," that is, continuous, intermittent or impulse. (Not all sources fit these three classifications.)
(a) electric saw
(b) air conditioner
(c) alarm clock (bell type)
(d) punch press

8-2. Is the following statement true or false? If it is false, correct it in a nontrivial manner.

"A sonic boom occurs when an aircraft breaks the sound barrier."

8-3. Is the following statement true or false? If it is false, correct it in a nontrivial manner.

"Excessive continuous noise causes hearing damage by breaking the stapes."

8-4. As the safety officer of your company, you have been asked to determine the feasibility of reducing exposure time as a method of reducing hearing damage for the following situation:

The worker is operating a high speed grinder on steel girders for a high rise building. The effective noise level at the operator's ear is 100 dBA. She cannot wear protective ear devices because she must communicate with others.

What amount of exposure time would you set as the limit?

8-5. In Figure 8-43, identify where the following noise-control techniques are applied: isolation and/or damping, reduction in noise leakage, use of absorbing materials, use of acoustical lining, enclosure.

8-11 REFERENCES

Achitoff, L. (1973) "Aircraft Noise—A Threat to Aviation," *Journal of Water, Air and Soil Pollution*, vol. 2, no. 3, pp. 357–363.

Alexandre, Barde, Lamure and Langdon (1975) *Road Traffic Noise*, Applied Science Publishers.

Berendt, R. D., E. L. R. Corliss, and M. S. Ojalro, (1976) A Practical Guide to Noise Control, *National Bureau of Standards Handbook* 119, U.S. Department of Commerce, pp. 16-41.

Beranek, L. L. (1954) *Acoustics,* McGraw-Hill, New York.

Borg, E., and S. A. Counter (1989) "The Middle-Ear Muscles," *Scientific American,* August, pp. 74–79.

Burck, W., et al. (1965.) "Gutachten erstatet im Auftrag des Bundesministers für Gesundheits wesen," *Flugärm.*

Clemis, J. D. (1975) "Anatomy, Physiology, and Pathology of the Ear," in J. B Olishifski and E. R. Harford (eds.), *Industrial Noise and Hearing Conservation*, National Safety Council, Chicago, p. 213.

Davis, H. (1957) "The Hearing Mechanism," in C. M. Harris (ed.), *Handbook of Noise Control,* McGraw–Hill, New York, pp. 4–6.

Davis, H. (1958) "Effects of High Intensity Noise on Navy Personnel," *U.S. Armed Forces Medical Journal*, vol. 9, pp. 1027–1047.

De Neufville, R., and A. R. Odoni (2003) *Airport Systems: Planning, Design, and Management*, McGraw-Hill, New York, p. 198.

FHWA (1973) *Policy and Procedure Memorandum 90-2, Noise Standards and Procedures,* U.S. Department of Transportation, Washington, DC, http://www.fhwa.dot.gove/environment.

FHWA (2005) "Priority, Market-Ready Technologies and Innovations, FHWA Traffic Noise Model®, Version 2.1," Federal Highway Administration, U.S. Department of Transportation.

Fletcher, H., and W. A. Munson (1935) "Loudness, Its Definition, Measurement and Calculation," *Journal of Acoustic Society of America,* vol. 5, October, pp. 82–105.

Griffiths, I. D., and F. J. Langdon (1968) "Subjective Response to Road Noise," *Journal of Sound & Vibration,* vol. 8, pp. 16–32.

Hajek, J. (1977) "L_{eq} Traffic Noise Prediction Method," *Environmental and Conservation Concerns in Transportation: Energy, Noise and Air Quality,* (Transportation Research Record No. 648), Transportation Research Board, National Academy of Sciences, pp. 48–53.

ISO (1989) *Acoustics—Attenuation of Sound During Propagation Outdoors, Part 2, A General Calculation*, International Organization for Standardization, ISO/DIS 9613-2, Geneva.

ISO (1990) *Acoustics—Attenuation of Sound During Propagation Outdoors, Part 1, Calculation of Absorption of Sound by the Atmosphere*, International Organization for Standardization, ISO/DIS 9613-1, Geneva.

Kryter, K. D. et al. (1971) *Non-auditory Effects of Noise,* Report WG-63, National Academy of Science, Washington, DC.

Kugler, B. A., D. E. Commins, and W. J. Galloway (1976) *Highway Noise: A Design Guide for Prediction and Control,* National Cooperative Highway Research Program Report.

Magrab, E. B. (1975) *Environmental Noise Control,* John Wiley & Sons, New York.

May, D. (1978) *Handbook of Noise Assessment,* Van Nostrand Reinhold, New York, p. 5.

Miller, J. D. (1971) *Effects of Noise on People,* U.S. Environmental Protection Agency Publication No. NTID 300.7, Washington, DC, p. 93.

Minnix, R. B. (1978) "The Nature of Sound," D. M. Lipscomb and A. C. Taylor (eds.), *Noise Control Handbook of Principles and Practices*, pp. 29–30.

NIOSH (1972) *Criteria for a Recommended Standard: Occupational Exposure to Noise,* National Institute for Occupational Safety and Health, U.S. Department of Health Education and Welfare, Washington, DC.

Olishifski, J. B., and E. R. Harford (eds.), (1975) *Industrial Noise and Hearing Conservation,* National Safety Council, Chicago, pp. 7, 340.

Piercy, J. E., and G. A. Daigle (1991) "Sound Propagation in the Open Air," in C. M. Harris (ed.), *Handbook of Acoustical Measurements and Noise Control,* McGraw-Hill, New York, pp. 3.1–3.26.

Sabin, H. J. (1942) "Notes on Acoustic Impedance Measurement," *Journal of the Acoustical Society of America,* vol. 14, p. 143.

Seeley, R., T. Stephens, and P. Tate (2003) *Anatomy and Physiology,* 6th ed., McGraw-Hill, New York.

Sylvan, S. (2000) *Best Environmental Practices in Europe and North America,* County Administration of Vastra Gotaland, Sweden.

U.S. EPA (1971) *Transportation Noise and Noise from Equipment Powered by Internal Combustion Engines,* U.S. Environmental Protection Agency Publication No. NTID 300.13, Washington, DC, p. 230.

U.S. EPA (1972) Report to the President and Congress on Noise, U.S. Environmental Protection Agency, Washington, DC.

U.S. EPA (1974) *Information on Levels of Environmental Noise Requisite to Protect Public Health and Welfare With an Adequate Margin of Safety,* U.S. Environmental Protection Agency, Publication No. 550/9-74-004, Washington, DC, pp. 29, B9, B10.

Ward, W. D., and A. Glorig (1961) "A Case of Firecracker-induced Hearing Loss," *Laryngoscope*, vol. 71, pp. 1590–1596.

Warnock, A. C. C. and J. D. Quirt (1991) "Noise Control in Buildings," in C. M. Harris, (ed.), *Handbook of Acoustical Measurements and Noise Control,* McGraw-Hill, New York, p. 33.13.

CHAPTER 9

SOLID WASTE MANAGEMENT

9-1 PERSPECTIVE

Solid waste is a generic term used to describe the things we throw away. It includes things we commonly describe as garbage, refuse, and trash. The U.S. Environmental Protection Agency's (EPA) regulatory definition is broader in scope. It includes any discarded item; things destined for reuse, recycle, or reclamation; sludges; and hazardous wastes. The regulatory definition specifically excludes radioactive wastes and *in situ* mining wastes.

We have limited the discussion in this chapter to solid wastes generated from residential and commercial sources. Sludges were discussed in Chapters 4 and 6. Hazardous waste will be discussed in Chapter 10, and radioactive waste will be discussed in Chapter 11.

Magnitude of the Problem

Solid waste disposal creates a problem primarily in highly populated areas. The more concentrated the population, the greater the problem becomes. Various estimates have been made of the quantity of solid waste generated and collected per person per day. In 2003, the EPA estimated that the national average rate of solid waste generated was 2.04 kg/capita · day (U.S. EPA, 2003). On this basis, in 2003, the U.S. produced 214 teragrams (Tg) of solid waste.* This is a 56 percent increase over the 1980 estimate of 137.8 Tg and a nearly 170 percent increase over the 1960 estimate of 80.1 Tg. The EPA estimates that 60 percent of the waste stream comes from residential sources, and the remainder is from commercial sources. Individual cities may vary greatly from these estimates. For example, Los Angeles, California, generates about 3.18 kg/capita · day while the rural community of Wilson, Wisconsin, generates about 1.0 kg/capita · day.

Figure 9-1 shows solid waste production rates. Averages are subject to adjustment depending on many local factors. Studies show there are wide differences in amounts

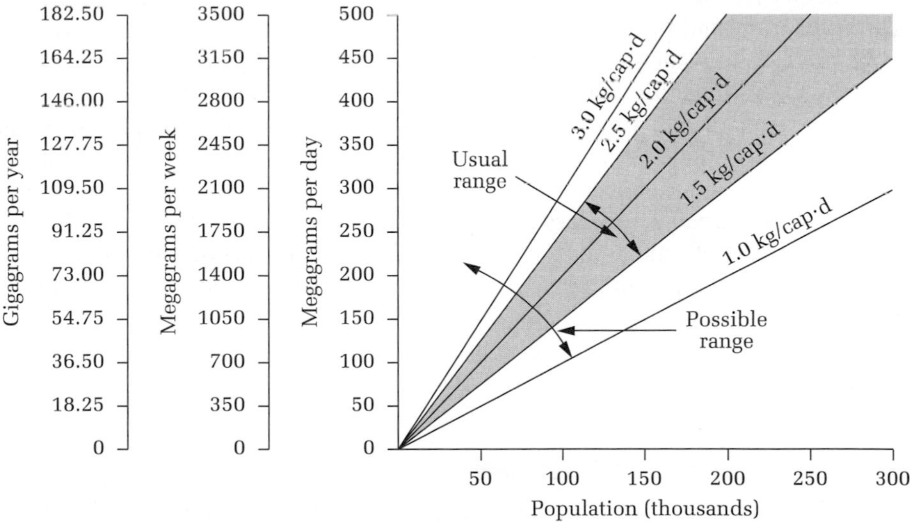

FIGURE 9-1
Solid waste produced: varying per capita figures.

---

*In keeping with correct SI notation, we use teragrams (1×10^{12} grams). One Tg is equivalent to 1×10^{9} kilograms (kg) or 1×10^{6} megagrams (Mg). The megagram is often referred to as the "metric ton."

collected by municipalities because of differences in climate, living standards, time of year, education, location, and collection and disposal practices.

Characteristics of Solid Waste

The terms *refuse* and *solid waste* are used more or less synonymously, although the latter term is preferred. The common materials of solid waste can be classified in several different ways. The point of origin is important in some cases, so classification as domestic, institutional, commercial, industrial, street, demolition, or construction may be useful. The nature of the material may be important, so classification can be made on the basis of organic, inorganic, combustible, noncombustible, putrescible, and nonputrescible fractions. One of the most useful classifications is based on the kinds of materials as shown in Table 9-1. Another classification system that is similar to this is the one used by the Incinerator Institute of America (Table 9-2). This is based primarily on the heat content of the waste.

Garbage is the animal and vegetable waste resulting from the handling, preparation, cooking, and serving of food. It is composed largely of putrescible organic matter and moisture; it includes a minimum of free liquids. The term does not include food processing wastes from canneries, slaughterhouses, packing plants, and similar facilities, or large quantities of condemned food products. Garbage originates primarily in home kitchens, stores, markets, restaurants, and other places where food is stored, prepared, or served. Garbage decomposes rapidly, particularly in warm weather, and may quickly produce disagreeable odors. There is some commercial value in garbage as animal food and as a base for commercial feeds. However, this use may be precluded by health considerations.

Rubbish consists of a variety of both combustible and noncombustible solid wastes from homes, stores, and institutions, but does not include garbage. Trash is synonymous with rubbish in some parts of the country, but trash is technically a subcomponent of rubbish. Combustible rubbish (the "trash" component of rubbish) consists of paper, rags, cartons, boxes, wood, furniture, tree branches, yard trimmings, and so on. Some cities have separate designations for yard wastes. Combustible rubbish is not putrescible and may be stored for long periods of time. Noncombustible rubbish is material that cannot be burned at ordinary incinerator temperatures of 700 to 1,100°C. It is the inorganic portion of refuse, such as tin cans, heavy metals, glass, ashes, and so on.

The average municipal solid waste composition in the United States in 2003 is shown in Figure 9-2 on page 741.

The density of loose combustible refuse is approximately 115 kg/m^3, while the density of collected solid waste is 235 to 300 kg/m^3.

Solid Waste Management Overview

The first objective of solid waste management is to remove discarded materials from inhabited places in a timely manner to prevent the spread of disease, to minimize the likelihood of fires, and to reduce aesthetic insults arising from putrifying organic matter. The second objective, which is equally important, is to dispose of the discarded materials in a manner that is environmentally responsible.

TABLE 9-1
Refuse materials by kind, composition, and sources

Kind	Composition	Sources
Garbage	Wastes from preparation, cooking, and serving of food; market wastes; wastes from handling, storage, and sale of produce	Households, restaurants, institutions, stores, markets
Rubbish	Combustible: paper, cartons, boxes, barrels, wood, excelsior, tree branches, yard trimmings, wood furniture, bedding, dunnage	
	Noncombustible: metals, tin cans, metal furniture, dirt, glass, crockery, minerals	
Ashes	Residue from fires used for cooking and heating and from on-site incineration	
Street refuse	Sweepings, dirt, leaves, catch basin dirt, contents of litter receptacles	Streets, sidewalks, alleys, vacant lots
Dead animals	Cats, dogs, squirrels, deer	
Abandoned vehicles	Unwanted cars and trucks left on public property	
Industrial wastes	Food-processing wastes, boiler house cinders, lumber scraps, metal scraps, shavings	Factories, power plants
Demolition wastes	Lumber, pipes, brick, masonry, and other construction materials from razed buildings and other structures	Demolition sites to be used for new buildings, renewal projects, expressways
Construction wastes	Scrap lumber, pipe, other construction materials	New construction, remodeling
Special wastes	Hazardous solids and liquids; explosives, pathological wastes, radioactive materials	Households, hotels, hospitals, institutions, stores, industry
Sewage treatment residue	Solids from coarse screening and from grit chambers; septic tank sludge	Sewage treatment plants, septic tanks

(*Source:* ISW, 1970.)

Table 9-2
Incinerator Institute of America waste classification

Classification of Wastes		Classification of wastes to be incinerated						
Type	Description	Principal components	Approximate composition % by weight	Moisture content %	Incombustible solids %	MJ heat value/kg of refuse as fired	MJ of aux. fuel per kg of waste to be included in combustion calculations	Recommended min. MJ burner input per kg waste
[a]0	Trash	Highly combustible waste, paper, wood, cardboard cartons, including up to 10% treated papers, plastic or rubber scraps; commercial and industrial sources	Trash 100%	10%	5%	19.8	0	0
[a]1	Rubbish	Combustible waste, paper, cartons, rags, wood scraps, combustible floor sweepings; domestic, commercial, and industrial sources	Rubbish 80% Garbage 20%	25%	10%	15.1	0	0
[a]2	Refuse	Rubbish and garbage; residential sources	Rubbish 50% Garbage 50%	50%	7%	10.0	0	3.5
[a]3	Garbage	Animal and vegetable wastes; restaurants, hotels, markets; institutional, commercial, and club sources	Garbage 65% Rubbish 35%	70%	5%	5.8	3.5	7.0
4	Animal solids and organic wastes	Carcasses, organs, solid organic wastes; hospital, laboratory, abattoirs, animal pounds, and similar sources	100% Animal and human tissue	85%	5%	2.3	7.0	18.6 (11.6 Primary) (7.0 Secondary)
5	Gaseous, liquid, or semi-liquid wastes	Industrial process wastes	Variable	Dependent on predominant components	Variable according to wastes survey	Variable according to wastes survey	Variable according to wastes survey	Variable according to wastes survey
6	Semi-solid and solid wastes	Combustibles requiring hearth, retort, or grate burning equipment	Variable	Dependent on predominant components	Variable according to wastes survey	Variable according to wastes survey	Variable according to wastes survey	Variable according to wastes survey

[a]The above figures on moisture content, ash, and MJ as fired have been determined by analysis of many samples. They are recommended for use in computing heat release, burning rate, velocity, and other details of incinerator designs. Any design based on these calculations can accomodate minor variations.
(*Source:* IIA, 1968.)

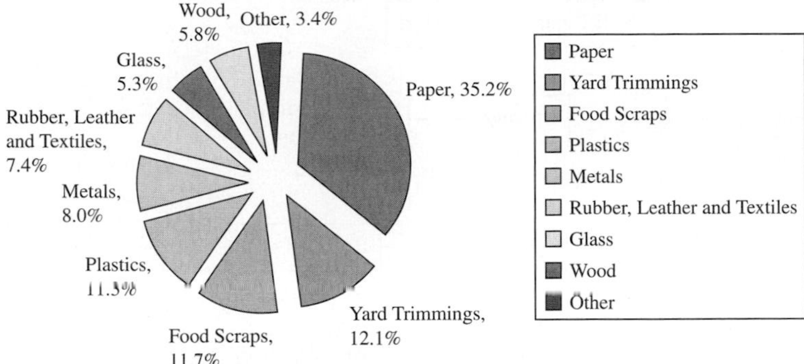

FIGURE 9-2
Materials generated in municipal solid waste (percent by mass), 2003. (*Source:* U.S. EPA, 2003.)

Policy making. Solid waste system policy making is primarily a function of the public sector rather than the private sector. The goal of a private firm is to minimize a well-defined cost function or to maximize profits. These are generally not the only, or even the primary, constraints of the public sector. The public objective function is more vague and difficult to express formally.

Constraints on the public sector, especially those of a political or a social nature, are difficult to measure, and criteria of effectiveness may not exist in units that can be quantified. Criteria of effectiveness against which public efficiency might be measured include such things as the frequency of collection, types of waste collected, location from which waste is collected, method of disposal, location of disposal site, environmental acceptability of disposal system, and the level of satisfaction of the customers. Public receptivity of a solid waste management system also depends on even less quantifiable parameters, which we group under the term *institutional factors.* Institutional factors include such things as political feasibility of the system, legislative constraints, and administrative simplicity.

Additional constraints on decision making in the public sector are environmental factors and resource conservation. Environmental factors are most important in the areas of waste storage and disposal because these functions represent prolonged exposure of wastes to the environment. Resource conservation is considered seriously by local governments as we become increasingly conscious of the limits of our natural resources.

Decisions in solid waste management policy formulation must be made in four basic areas: collection, transport, processing, and disposal. The flowchart in Figure 9-3 illustrates the decisions that must be made from the point of generation to the ultimate disposal of residential solid waste.

In designing a solid waste collection system, one of the first decisions to be made is where the waste will be picked up: the curb or the backyard. This is an important decision because it affects many other collection variables, including choice of storage containers, crew size, and the selection of collection trucks. Backyard service, once the predominant method of pickup, is still used by some communities. It is generally more costly, but it eliminates the need for scheduled pickups.

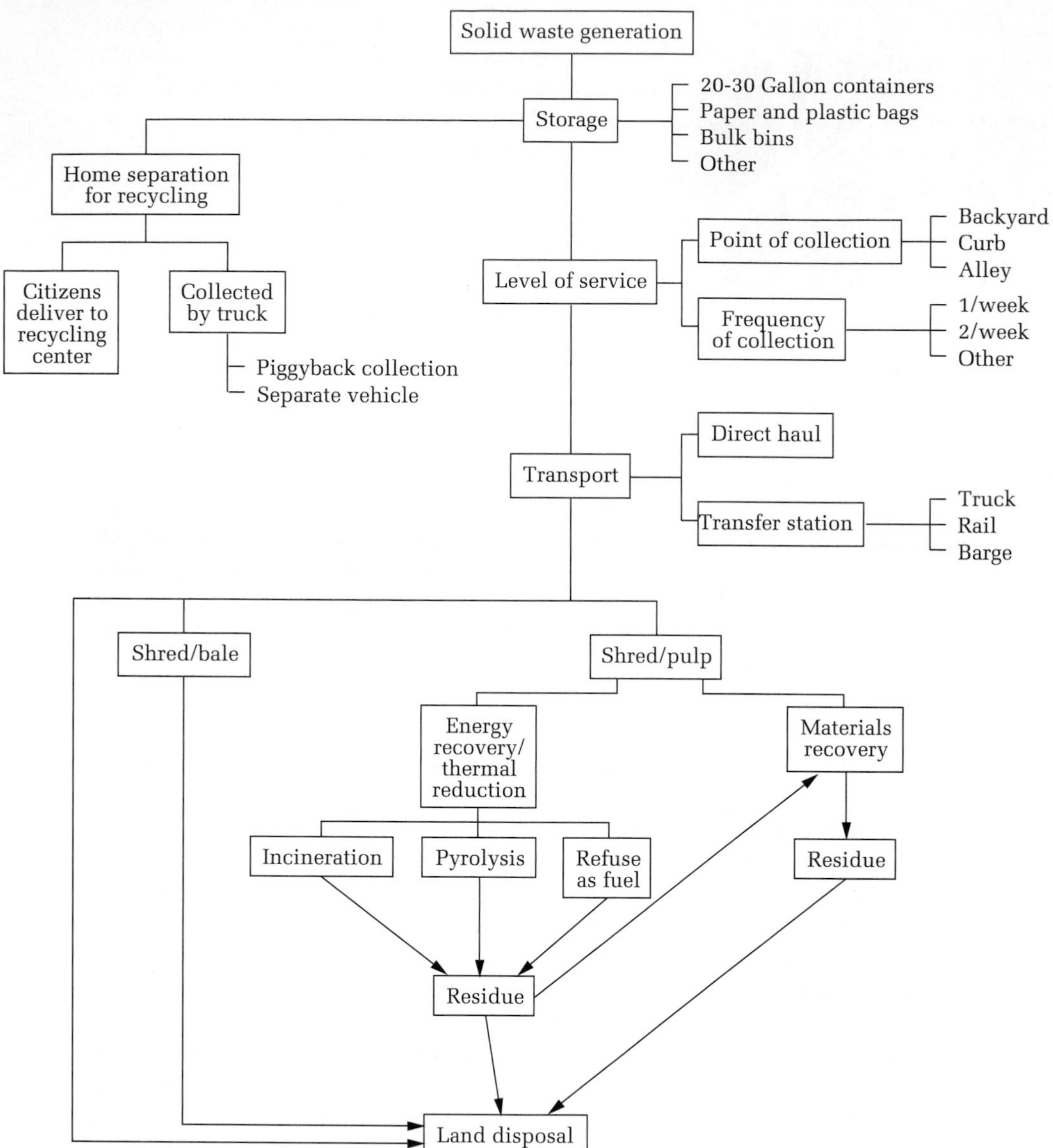

FIGURE 9-3
Solid waste management decision alternatives. (*Source:* U.S. EPA, 1974.)

Another key decision is frequency of collection. Both point of collection and frequency of collection should be evaluated in terms of their impact on collection costs. Since collection costs generally account for 70 to 85 percent of total solid waste management costs, and labor represents 60 to 75 percent of collection costs, increases in the productivity of collection personnel can dramatically reduce overall costs. Most communities offer collection once or twice a week, with once per week being the most common schedule (U.S. EPA, 1995).

Systems with once-a-week curbside collection help maximize labor productivity and result in significantly lower costs than systems with more frequent collection and/or backyard pickup. The main reason many communities retain twice-a-week backyard service is that the citizens demand this convenience and are willing to pay for it. In warmer regions of the country, twice-a-week service may be deemed essential to prevent gross odors and to break the fly-breeding cycle. The egg-larvae-adult cycle is about 4–5 days.

The choice of solid waste storage containers must be evaluated in terms of both environmental effects and costs. From the environmental standpoint, some storage containers can present health and safety problems to the collectors, as well as to the general public. Therefore, the decision facing a community is which storage system is both environmentally sound and most economical, given the collection system characteristics. For example, paper and plastic bags are superior to many other containers from a health and esthetic standpoint and can increase productivity when used in conjunction with curbside collection. However, with backyard collection systems, bags have little effect on productivity.

The type of container used may also be dictated by the type of collection. If solid waste is collected manually, then plastic bags or cans can be used. Some communities have recently begun to sell special plastic bags or stickers to put on plastic bags that include the cost of the bag as well as the disposal fee. If the system is automated or semiautomated, then the container must be specifically designed to fit the truck-mounted loading system. The containers typically hold from 1 to 20 cubic meters of waste.

Another factor to be considered in examining storage alternatives is home separation of various materials for recovery. The collection of materials for recovery/recycle is a growing practice that many cities are implementing. The technique of greatest interest to municipal decision makers is home separation and collection by either the regular collection truck equipped with special bins or by separate trucks.

One of the primary factors to consider in implementing a separate collection system is whether the benefits of recovery outweigh the costs involved. The economic viability of separate collection depends primarily on the local market price for the material and the degree of participation by the citizens. If these factors are positive, it may be possible to implement a recovery system with no increase, and possibly a savings, in collection operating costs; often no additional capital expenditure is required. Another factor to be considered is the expectation of the community that the municipality be actively involved in recycling. Most people perceive recycling as an environmentally friendly practice and so expect municipalities to provide the opportunity.

The distance between the disposal site and the center of the city will determine the advisability of including a *transfer station* in the transport system.* In addition to distance traveled to the disposal site, the time required for the transport is a key factor, especially in traffic-congested large cities.

The tradeoffs involved in transfer station operations are the capital and operating costs of the transfer station as compared to the cost (mostly labor) of having route collection vehicles travel excessive distances to the disposal site. These tradeoffs can be computed to find the point at which transfer becomes economically advantageous.

---

*A transfer station is a place where trucks dump their loads into a larger vehicle where it is compacted. By combining loads, the cost per Mg · km for transport to the landfill is reduced.

The sheer quantities of solid waste to be disposed of daily makes the problem of what to do with the waste, once it has been collected, among the most difficult problems confronting community officials. A crisis situation can develop very quickly, for example, in the case of an incinerator or land disposal site forced to shut down because of failure to meet newly passed environmental regulations. Alternatively, a crisis can build gradually over a period of time if needed new facilities are not properly planned for and put into service.

There are three basic alternatives for disposal. Some have subalternatives. The major alternatives are: (1) direct disposal of unprocessed waste in a municipal solid waste landfill, (2) processing of waste followed by land disposal, and (3) processing of waste to recover resources (materials and/or energy) with subsequent disposal of the residues. Most municipal solid waste is landfilled, but the amount landfilled declined from 73 percent in 1988 to 56 percent in 2003. Fourteen percent of the waste was incinerated in 2003 and 30 percent was recycled or composted. EPA projects the increase in waste incineration and recycling to continue.

Direct haul to a sanitary landfill (with or without transfer and long haul) is usually the cheapest disposal alternative in terms of both operating and capital costs. In 1988, it was estimated that about 8,000 landfills were in operation, but by 2002 the number had decreased to about 1800. Many were closed as a result of regulatory restrictions. Municipalities own 75 percent of the sites (Wolpin, 1994). With rising *tipping fees* (the cost to dump solid waste at a disposal facility), a surplus of disposal capacity has replaced the late 1980s predictions of lack of landfill space.

With the second alternative, processing prior to land disposal, the primary objective is to reduce the volume of wastes. Such volume reduction has definite advantages since it reduces hauling costs and ultimate disposal cost, both of which are, to some extent, a function of waste volume. However, the capital and operating cost to achieve this volume reduction are significant and must be balanced against the savings achieved.

An additional consideration is the environmental benefit that might be derived from the volume reduction process. In some cases, shredding and baling may reduce the chances for water pollution from leachate. This alternative is more conserving of land than sanitary landfilling of unprocessed wastes, but by itself provides no opportunity for material or energy recovery.

The third category of disposal alternatives includes those processes that recover energy or materials from solid waste and leave only a residue for ultimate land disposal. There are significant capital and operating costs associated with all these energy and/or materials recovery systems. However, if markets are available, both energy and materials can be sold to reduce the net costs of recovery.

While resource recovery techniques may be more costly than other disposal alternatives, they do achieve the goal of resource conservation while enhancing sustainability and the residuals of the processes require much less space for land disposal than unprocessed wastes.

Affecting all four major functions are basic decisions regarding how the solid waste system will be managed and operated. This includes how the system will be financed, which level of government will administer it, and whether a public agency or private firm will operate the collection, transport, processing, and disposal functions. The criteria most relevant for making these decisions are the institutional factors of political feasibility and legislative constraints.

Integrated Solid Waste Management (ISWM). The selection of a combination of techniques, technologies, and management programs to achieve waste management objectives is called *integrated solid waste management* (ISWM). This approach has made major strides in recent years. The EPA proposed a hierarchy of actions to implement ISWM: source reduction (including reuse and waste reduction), recycling and composting, and disposal in combustion facilities and landfills (U.S. EPA, 1995). The most obvious effect of the integrated approach is to reduce the size of the incineration facility. This reduces the capital cost of the incineration facility. Although the energy output is also reduced, the waste that remains has a higher energy content so that the reduction in energy output is less than the reduction in plant size. Recycling also reduces waste elements that can damage the boilers and removes those components that slag in the furnace and foul it (Shortsleeve and Roche, 1990).

9-2 COLLECTION

The solid waste collection policies of a city begin with decisions made by elected representatives about whether collection is to be made by: (1) city employees (municipal collection), (2) private firms that contract with city government (contract collection), or (3) private firms that contract with private residents (private collection). Many communities have moved away from exclusive municipal collection and toward a combined system. More and more communities are moving toward mandatory recycling of materials such as paper, plastic, and glass. In these situations, separation of waste is required.

Elected officials may also determine what type of solid wastes are to be collected and from whom. In some municipalities broad classes of solid wastes (such as rubbish) are not accepted for collection. In others, certain materials (such as tires, grass trimmings, furniture, or dead animals) may be excluded. Hazardous wastes are generally excluded from regular collections because of disposal and collection dangers. The nature of the service may be governed by limitations of disposal facilities or by the opinion of the legislative body as to what service should be performed. A city may collect garbage only or it may collect everything but garbage. Almost all municipal systems collect residential waste, but only about one-third collect industrial waste.

The final decision concerning collection, which is made by the elected officials, is the frequency of collection. The proper frequency for the most satisfactory and economical service is governed by the amount of solid waste that must be collected and by climate, cost, and public requests. For the collection of solid waste that contains garbage, the maximum period should not be greater than

1. The normal time for the accumulation of the amount that can be placed in containers of reasonable size.
2. The time it takes for fresh garbage to putrefy and emit foul odors under average storage conditions.
3. The length of the fly-breeding cycle, which, during the hot summer months, is less than seven days.

In the last three decades the prevailing frequency of collection has changed from twice a week pickup to once a week. The increased use of once per week service is due to two factors. First, unit costs are reduced when frequency is cut from twice to once per week. Second, the increased percentage of paper and decreased percentage of garbage in the solid waste permit longer periods of acceptable storage.

Once policy has been set, the actual method of collection is determined by engineers or managers. Major considerations include how the solid waste will be collected, how the crews will be managed, how the trucks will be routed, and the type of equipment to be used.

Collection Methods

The first decision to be made is how the solid waste container will get from the residence to the collection vehicle. The three basic methods are: (1) *curbside* or alley pickup, (2) *set-out, set-back collection,* and (3) *backyard pickup,* or the tote barrel method. Most urban and suburban areas utilize curbside pickup, but a few communities still use backyard pickup. In some less populated areas, municipal waste collection is sometimes accomplished by requiring residents to transport waste to a specified point. This point may be a transfer station or the disposal site. This is the least expensive method for a municipality, but it is the least convenient method for the homeowner.

The quickest and most economical point of collection is from curbs or alleys using standard containers. It is the most common type of collection used. It costs only about one-half as much as backyard collection. Usually the city designates what type of containers are to be used. The crews simply empty the containers into the collection vehicles. Whenever possible the crews collect from both sides of the street at the same time. Municipal ordinances or administrative regulations usually specify when the containers must be placed at the curb or in the alley for pickup and also how long they may remain after pickup. Common limits are out by 7 A.M. and back by 7 P.M. When solid wastes are loaded from curbs or alleys, work progresses rapidly. A typical crew consists of a driver and two collectors. Some crews still have three or even four collectors, but the trend is toward fewer collectors. Recent studies indicate that small crews are more efficient than larger ones, since labor costs are a major element of the total cost. Aside from the cost advantage of this method, it also eliminates the need for the collectors to enter private property, and the amount of service given each homeowner is relatively uniform. However, many citizens dislike having to set their solid wastes out at certain times and object to the unsightly appearance on the streets. Some surveys have shown that many homeowners would prefer to pay more in order to receive backyard service.

When curbside removal is chosen, automatic and semiautomatic collection vehicles can be utilized. In an automated system, residents are provided with large specialized containers (approximately 90 gallons), which they roll to the curb. These containers are then lifted by powerful hydraulic arms that empty the contents of the container into the truck's hopper. The crew, or often just the driver, performs the operation from inside the cab of the collection vehicle. A typical side-loading vehicle with a hydraulic arm is shown in Figure 9-4. A fully automated system can be the most economical for a community, particularly if the community also uses this single truck to collect recyclables. The city of Los Angeles converted to such a system and in 2000 collected 712 Gg of refuse, recyclables, and yard waste with automated sideloading trucks. The waste is then transported to a waste processing facility where the materials are separated.

FIGURE 9-4
Side-loading refuse collection vehicle with hydraulic lift arm. In this model, the tractor-trailer configuration allows for additional maneuverability. (*Source:* Heil Environmental, 2006.)

However, many communities cannot accommodate these large vehicles in their existing residential neighborhoods. They therefore use some combination of automatic and semiautomatic vehicles. In a semiautomatic system, the crew wheels the cart to the collection vehicle, lines the cart up with the lifting device and activates the lifter. A hydraulic device lifts the cart and tips the car, allowing the contents to fall into the hopper of the truck.

The existence of cul de sacs, alleys, and narrow streets as well as low-hanging utility lines may dictate the type of vehicle selected. For example, the city of Houston uses three different types of vehicles in its fleet of 200 vehicles. The city uses automated sideloaders to pick up curbside trash as well as recyclables, semiautomatic rear loaders to pick up yard waste, and a combination of rear loaders and a one-operator heavy-duty vehicle equipped with a grapple to pick up heavy trash to deposit in the rear loader (Bader, 2001, and Luken and Bush, 2002).

The set-out, set-back method eliminates most of the disadvantages of the curb method, but it does require the collector to enter private property. This method consists of the following operations: (1) the set-out crew carries the full containers from the residential storage location to the curb or alley before the collection vehicle arrives, (2) the collection crew loads the refuse in the same manner as the curb method, and (3) the set-back crew returns the empty cans. Any of the crew may be required to do more than one step or the homeowner may be required to do one of the steps. This method has not been shown to be more economical or advantageous than the backyard method, and it is more costly and time-consuming than curbside pickup.

Backyard pickup is usually accomplished by the use of tote barrels. In this method, the collector enters the resident's property, dumps the container into a tote barrel,

FIGURE 9-5
Typical rear-loading refuse collection vehicle. (*Source:* Heil Environmental, 2006.)

carries it to the truck, and dumps it. The collector may collect refuse from more than one house before returning to the truck to dump. The primary advantage of this system is in the convenience to the homeowner. The major disadvantage is the high cost. Many homeowners object to having the collectors enter their private property. With this collection method, a rear-loading vehicle, such as the one shown in Figure 9-5, is used.

Cost analyses have revealed that 70 to 85 percent of the cost of solid waste collection and disposal can be attributed to the collection phase. For this reason, it would seem that a great deal of municipal effort should be directed to studying collection alternatives to determine the most efficient system. However, many analyses begin their studies assuming that waste loads are already collected and waiting for disposal. There are two major reasons why the collection system is not studied more often. First, the collection system is a complex and expensive system to analyze. The primary reasons for this are that it involves people, equipment, and levels of service, plus the possibility of numerous variations in secondary factors such as collection methodology; quantity, nature, and the method of storage of refuse; location of pickup point; equipment type and characteristics of operation; road factors; service density; route topography; climatic factors; and human factors. Human factors would include morale, incentive, fatigue, and other variables that influence the time required to complete a given task. Secondly, most cities are already collecting refuse in some manner, and the cliche "leave well enough alone" often prevails. It is generally on the disposal system that the public is placing pressure for improvement, rather than the collection system.

Most changes in collection systems will require a great deal of investigation and testing. Even if the change is an obvious one, often "proof" of some sort is needed to convince the elected officials. The most important thing to realize about the solid waste collection system is that it is too big, complex, and vital to allow actual experimentation except on a very small scale. Coupled with this are all the other problems peculiar to studying large-scale public systems. A relevant data base is probably nonexistent. The political implications of control of the system and cost distribution may override an otherwise practical solution. A large investment will have already been made in the existing system and the designer is not allowed the luxury of starting at the beginning, but must start with a system that may be founded on a pyramid of errors.

EPA suggests a method that can be used to estimate the time requirements of a waste collection system in order to evaluate and subsequently optimize the system (U. S. EPA, 1995). The steps included in a time study are shown in Table 9-3.

Waste Collection System Design Calculations

Often, it is desirable to calculate "quick and dirty" estimates of such things as crew size, desired truck capacity, and labor and capital costs. Simple formulas have been developed that enable such calculations. The formulas are based on crude averages regarding collection times, and they make broad assumptions. An example of a not-always-justifiable assumption is that if one collector can collect a house in one minute, then two can do it in one-half minute. Several such equations follow.

Estimating Truck Capacity. Given that you are able to estimate a large number of factors, the following equation will allow you to estimate the volume of solid waste a truck must be able to carry.

$$\forall_T = \frac{\forall_p}{rt_p}\left[\frac{H}{N_d} - \frac{2x}{s} - 2t_d - t_u - \frac{B}{N_d}\right] \tag{9-1}$$

where $\forall_T$ = volume of solid waste carried per trip by truck at a mean density, D_T, m^3

$\forall_p$ = volume of solid waste per pickup location or stop, m^3/stop

r = compaction ratio

t_p = mean time per collection stop plus the mean time to reach the next stop, h

H = length of working day,* h

N_d = number of trips to the disposal site per day

x = one-way distance to disposal site, km

s = average haul speed to and from disposal site, km/h

t_d = one-way delay time, h/trip

t_u = unloading time at disposal site, h/trip

B = off route time per day, h

*We should note that it is standard practice to allow two fifteen-minute breaks during the day. Since the crew is paid for this, the number of hours in the workday (H) are unchanged. However, some allowance must be made for it. Hence the off route time (B) is included in the equation.

TABLE 9-3
Steps for conducting a time study

1. Select crew(s) representative of average level and skill level.
2. Determine the best method (series of movements) for conducting the work.
3. Set up a data sheet that can be used to record the following information: date, name of crew members and time recorder, type of collection method and equipment (including loading mechanism), specific area of municipality, and distance between collection points.
4. Divide loading activity into elements that are appropriate for the type of collection service. For example, the following elements might be appropriate for a study of residential collection loading times:
   - Time to travel from last loading point to next one
   - Time to get out of vehicle and carry container to the loading area
   - Time to load vehicle
   - Time to return container to the collection point and return to the vehicle.
5. Using a stop watch, record the time required to complete each element for a representative number of repetitions. Time may be measured using one of the following two methods:
   - *Snapback method:* The time recorder records the time after each element and then resets watch to zero for measurement of the next element.
   - *Continuous method:* The time recorder records the time after each element but does not reset the watch so that it moves continuously until the last elements is completed.

   Because the continuous method requires the time recorder to perform fewer movements and no time is lost for watch resetting, the continuous method is usually recommended.

   The number of repetitions that will be representative depends on the time required to complete the overall activity (cycle). The following numbers of repetitions have been suggested as sufficient:

Number of Repetitions	Minutes Per Cycle	Number of Repetitions	Minutes Per Cycle
60	0.50	20	2.0
40	0.75	15	5.0
30	1.00	10	10.5

6. Determine the average time recorded (T_o) and adjust it for "normal" conditions.

   In the case of waste collection, adjustments should be made for delays and for crew fatigue. These adjustments are typically in terms of the percent of time spent in a workday. The delay allowance (D) should include time for traffic conditions, equipment failures and other uncontrollable delays. Crew fatigue allowance (F) should include adequate rest time for recovery from heavy lifting, extreme hot and cold weather conditions, and other circumstances encountered in waste collection. The allowance factors (D and F) along with the average observed time (T_o), can be used to estimate the "normal" time (T_n):

$$T_n = (T_o) \times [1 + (F + D)/100]$$

   This "normal" time is the loading time required for the particular area, and collection system.

   For other activities, adjustments are also made for personal time (bathroom breaks). In this case, adjustment for personal time is made when calculating the number of loads/crew/day.

(*Source:* U.S. EPA, 1995.)

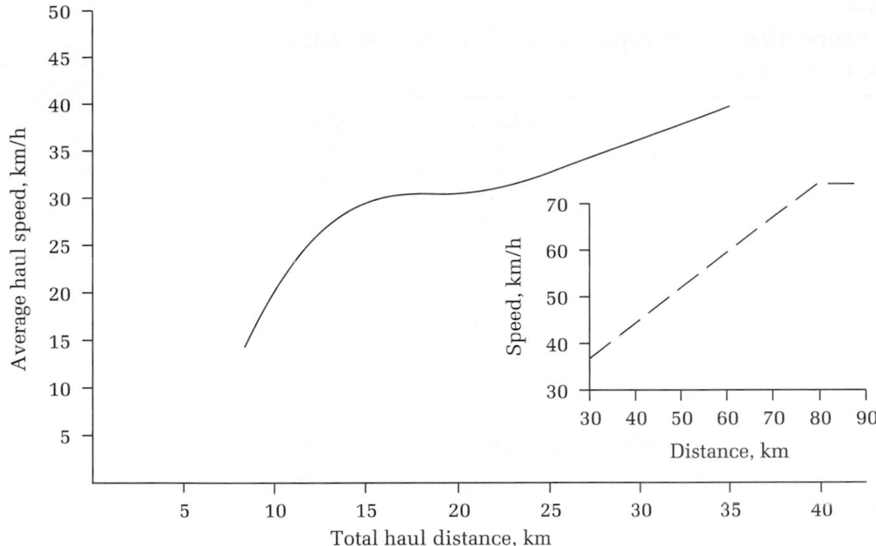

FIGURE 9-6
Effect of haul distance on average haul speed. (Adapted from U. Calif., 1952.)

The factor of two in Equation 9-1 accounts for travel both to and from the disposal site. The average haul speed is a function of the total round trip distance to the disposal site (Figure 9-6). As noted in the definitions, the volume carried presumes a mean density, D_T. This is the density that results after the waste has been compacted in the truck. The compaction ratio (r) is the ratio of the density after compaction to that before compaction. Typical densities "as discarded" are given for several solid waste components in Table 9-4. If, for example, paper waste was compacted to a density of 163.4 kg/m^3, the compaction ratio would be two to one. Compactor trucks can achieve densities ranging from 300 to 600 kg/m^3.

A value for t_p can be estimated from empirical data (U. Calif., 1952; Stone, 1969). The data may be approximated by linear equations of the following form:

$$t'_p = t_{b_p} + a(C_n) + b(PRH) \tag{9-2}$$

where t'_p = mean time per collection stop plus mean time to reach next stop, min/stop
t_{b_p} = mean time between collection stops, min/stop
a, b = coefficients of regression fit to data points
C_n = mean number of containers at each pickup location
PRH = rear of house pickup locations, %

To convert t'_p to t_p, we must divide by 60 min/h.

The number of pickup locations that can be handled by a given crew is simply the available time after haul divided by the mean pickup time:

$$N_p = \frac{\frac{H}{N_d} - \frac{2x}{s} - 2t_d - t_u - \frac{B}{N_d}}{t_p} \tag{9-3}$$

where N_p = number of pickup locations per load

TABLE 9-4
Typical properties of uncompacted solid waste as discarded in Davis, California

Component	Mass (kg)	Density (kg/m^3)	Volume (m^3)
Food wastes	4.3	288	0.0149
Paper	19.6	81.7	0.240
Cardboard[a]	2.95	99.3	0.0297
Plastics	0.82	64	0.013
Textiles	0.091	64	0.0014
Rubber	—	128	—
Leather	0.68	160	0.0043
Garden trimmings	6.5	104	0.063
Wood	1.59	240	0.00663
Glass	3.4	194	0.018
Tin cans	2.36	88.1	0.0268
Nonferrous metals	0.68	160	0.0043
Ferrous metals	1.95	320	0.00609
Dirt, ashes, brick	0.50	480	0.0010
Total	45.4		0.429

[a]Cardboard partially compressed by hand before being placed in container.
(*Source:* Tchobanoglous et al., 1977.)

Example 9-1. The solid waste collection vehicle of Watapitae, Michigan is about to expire, and city officials are in need of advice on the size of truck they should purchase. The compactor trucks available from a local supplier are rated to achieve a density (D_T) of 400 kg/m^3 and a dump time of 6.0 minutes. In order to ensure once-a-week pickup the truck must service 250 locations per day. The disposal site is 6.4 km away from the collection route. From past experience, a delay time of 13 minutes can be expected. The data given in Table 9-4 have been found to be typical for the entire city. Each stop typically has three cans containing 4 kg each. About 10 percent of the stops are backyard pickups. Assume that two trips per day will be made to the disposal site. Also assume that the crew size will be two and that the empirical equation of Tchobanoglous, Theisen, and Eliassen for a two-person crew applies (1977). That equation is given as follows:

$$t'_p = 0.72 + 0.18(C_n) + 0.014(PRH)$$

$$t'_p = 0.72 + 0.54 + 0.14 = 1.40 \text{ min/stop}$$

$$t_p = \frac{1.40 \text{ min}}{60 \text{ min/h}} = 0.0233 \text{ h}$$

Solution. Using Table 9-4 we determine the mean density of the uncompacted solid waste to be

$$D_u = \frac{\text{Total Mass}}{\text{Total Volume}} = \frac{45.4 \text{ kg}}{0.429 \text{ m}^3} = 105.83 \text{ or } 106 \text{ kg/m}^3$$

The volume per pickup is then

$$\forall_p = \frac{(3 \text{ cans})(4 \text{ kg/can})}{106 \text{ kg/m}^3} = 0.11 \text{ m}^3$$

The compaction ratio is determined from the densities:

$$r = \frac{D_T}{D_u} = \frac{400 \text{ kg/m}^3}{106 \text{ kg/m}^3} = 3.77$$

The average haul speed is determined from Figure 9-6. Since the graph is for total haul distance, we enter with (2)(6.4) = 12.8 km and determine that $s = 27$ km/h. All of the other required data were given; thus, we can now use Equation 9-1. The factor of 60 is to convert minutes to hours. For two 15-minute breaks, $B = 0.50$.

$$\forall_t = \frac{0.11}{(3.77)(0.0233)}\left[\frac{8}{2} - \frac{(2)(6.4)}{27} - 2\frac{13 \text{ min}}{60 \text{ min/h}} - \frac{6 \text{ min}}{60 \text{ min/h}} - \frac{0.50}{2}\right]$$

$$= (1.25)(2.74) = 3.43 \text{ m}^3$$

The number of stops that can be handled is given by Equation 9-3:

$$N_p = \frac{2.74}{0.0233} = 117.60 \text{ or } 118 \text{ pickups per load}$$

The smallest compactor truck available is one that will hold 4.0 m^3. Obviously, this will be satisfactory. However, the crew will not be able to reach the required 250 stops per day. Thus, some other alternative must be considered. One would be to extend the workday by 30 minutes.

Estimating Costs. Most of the decisions involved in the collection of solid waste are based on economic considerations rather than technical ones. The costs are considered on the basis of a unit mass of solid waste to facilitate comparison between different size vehicles, crews, and the like. Furthermore, truck costs are considered separately from labor costs.

Truck costs include depreciation of the initial capital investment plus the *operating and maintenance (O & M)* costs.*

---

* Government-operated collection systems, by the nature of their operation, do not actually depreciate purchases. First of all, they get no tax credit for doing so and, secondly, they do not save or put aside money in a bank and therefore cannot draw interest. In spite of all this, good engineering economics demands that capital costs be depreciated in order to allow valid comparisons between alternatives.

The following equation may be used to estimate the annual cost per Mg (U. Calif., 1952):

$$A_T = \frac{1{,}000(F)}{\forall_T D_T N_T Y}\left[1 + \frac{i(Y+1)}{2}\right] + \frac{1{,}000(X_t)(OM)}{\forall_T D_T} \tag{9-4}$$

where A_T = annual truck cost, \$/Mg
F = initial (first) cost of truck, \$
D_T = mean density of solid waste in truck, kg/m^3
N_T = number of trips per year
Y = useful life of truck, y
i = interest rate on capital
X_t = distance per trip, pickup plus haul, km
OM = operating and maintenance cost, \$/km

The factor of 1,000 is to convert kg to Mg.

Labor costs consist of direct wages plus some overhead costs for such things as supervision, secretarial support, phone, utilities, insurance, and fringe benefits. Equation 9-5 can be used to estimate the annual labor cost per Mg:

$$A_L = \frac{1{,}000(CS)(W)(H)}{\forall_T D_T N_d}[1 + (OH)] \tag{9-5}$$

where A_L = annual labor cost, \$/Mg
CS = average crew size
W = average hourly wage rate, \$/h
OH = overhead as a fraction of wages

Again, the factor of 1,000 is to convert kg to Mg.

Example 9-2. Estimate the customer service charge for the situation of Example 9-1. The initial truck cost of a 4.0 m^3 compactor truck is \$104,000, and the average O & M cost over the five-year life of the truck is expected to be \$5.50/km. The interest rate is 8.25 percent. The average route length is 6.3 km. The average hourly wage rate is \$13.50 per hour with time and a half for overtime. The overhead rate is 125 percent of the hourly wage rate.

Solution. Assuming a five-day work week and ignoring holidays, the number of trips per year would be

$$N_t = N_d(5)(52) = 2(5)(52) = 520$$

Since the average route length is 6.3 km and the average haul distance from Example 9-1 is 2(6.4) = 12.8 km, then

$$X_t = 6.3 + 12.8 = 19.1 \text{ km}$$

For the extended workday proposed at the end of Example 9-1, the volume of solid waste per trip would be

$$V_T = (1.25)(2.74 + 1/2(0.5)) = 3.74 \text{ m}^3$$

The factor of one-half times the extra half hour was selected because we assumed the time to be equally divided between each of the two trips. Note that we do not use the actual volume of the truck, which is somewhat larger than V_T. (The truck size is the nearest standard size.) Now we may compute the annualized truck cost.

$$A_T = \frac{1{,}000(104{,}000)}{(3.74)(400)(520)(5)}\left[1 + \frac{0.0825(5 + 1)}{2}\right] + \frac{1{,}000(19.1)(5.50)}{(3.74)(400)}$$
$$= (26.74)(1.25) + 70.22 = \$103.65/\text{Mg}$$

Since we have planned for an extra half hour of work each workday, we must adjust the hourly wage rate accordingly before we can use Equation 9-5. The adjustment is simply a determination of the weighted average rate.

$$W = \frac{(\text{reg. shift hours})(\text{wage}) + (\text{overtime hours})(\text{OT rate})(\text{wage})}{\text{total hours}}$$
$$= \frac{8(13.50) + 0.5(1.5)(13.50)}{8.5} = \$13.90/\text{h}$$

Now we may apply Equation 9-5 directly.

$$A_L = \frac{(1{,}000)(2)(13.90)(8.5)}{(3.74)(400)(2)}[1 + 1.25] = \$177.70/\text{Mg}$$

The total annual cost is then

$$A_{\text{tot}} = \$103.65 + \$177.70 = \$281.35/\text{Mg}$$

From Example 9-1, we know that each service stop averages three cans per week at 4 kg per can. Thus, each service stop contributes 3(4)(52) = 624 kg or 0.624 Mg per year. The annual cost per service stop should be (\$281.35/Mg)(0.624 Mg) = \$175.56. For 52 pickups per year, this is an average cost of about \$3.38 per week (that is, \$175.56/52).

Truck Routing

The routing of trucks may follow one of four methods. The first possibility is the daily route method. In this method the crew has a definite route that must be finished before going home. When the route is finished the crew can leave, but if necessary, they must work overtime to finish the route. This is the simplest method and the most common. The advantages of this method are as follows:

1. The homeowner knows when the refuse will be picked up.
2. The route sizes can be adjusted for the load to maximize crew and truck utilization.
3. The crew likes the method because it provides an incentive to get done early.

The disadvantages include:

1. If the route is not finished, the crew will work overtime, which will increase the expense.

2. The crew may have a tendency to become careless as they try to finish the job sooner.

3. Frequently the result is underutilization of the crew and equipment due to the increased incentive of the crew.

4. A breakdown seriously affects operations.

5. It is hard to plan routes if the load is variable, because of the disposal of yard wastes and the like.

The next method is the large route method. In this scheme the crew has enough work to last the entire week. The route must be completed in one week. The crew is left on its own to decide when to pick up the route. Usually some time off at the end of the week is the goal of the crew. This method is only good for backyard pickup since the residents don't know when pickup will be. The same advantages and disadvantages apply to this method as to the daily route method.

In the single load method, the routes are planned to get a full truck load. Each crew is assigned as many loads as it can collect per day. The biggest advantage of this method is that it can minimize travel time. The method must consider size of crew, capacity of truck, length of travel, refuse generated, and similar variables. Other advantages include:

1. A full day's work can be provided for maximum utilization of the crew and equipment.

2. It can be used for any type of pickup.

The major disadvantage is that it is hard to predict the number of homes that can be serviced before the truck is filled.

The last method is the definite working day method. As its name implies, the crew works for its assigned number of hours and quits. This method predominates in areas where unions are strong. With this method, the crew and the equipment get maximum utilization. Regularity is sacrificed with this method, and residents have little idea when pickup will occur.

Having determined the method by which the trucks will be managed, it is still necessary to find the actual route the truck will follow through the city. The purpose of routing and districting is to subdivide the community into units that will permit collection crews to work efficiently. No matter what the size of the community, it can be divided into districts, with each district constituting one day's work for the crew. The route is the detailed path of travel for the collection vehicle. The size of each route depends upon various factors as discussed earlier. The Office of Solid Waste Management Programs of the U.S. Environmental Protection Agency has developed a simple, noncomputerized "heuristic" (rule-of-thumb) approach to routing based on logical principles. The goal is to minimize deadheading, delay, and left turns. This method relies on developing, recognizing, and using certain patterns that repeat themselves in every municipality. Routing skills can be quickly acquired by applying the rules and developing experience. The following rules are taken from an EPA publication (Shuster and Schur, 1974).

1. Routes should not be fragmented or overlapped. Each route should be compact, consisting of street segments clustered in the same geographical area.
2. Total collection plus haul times should be reasonably constant for each route in the community (equalized workloads).
3. The collection route should be started as close to the garage or motor pool as possible, taking into account heavily traveled and one-way streets. (See rules 4 and 5.)
4. Heavily traveled streets should not be collected during rush hours.
5. In the case of one-way streets, it is best to start the route near the upper end of the street, working down it through the looping process.
6. Services on dead end streets can be considered as services on the street segment that they intersect, since they can only be collected by passing down that street segment. To keep left turns at a minimum, collect the dead end streets when they are to the right of the truck. They must be collected by walking down, backing down, or making a U-turn.
7. When practical, service stops on steep hills should be collected on both sides of the street while the vehicle is moving downhill for safety, ease, speed of collection, wear on vehicle, and conservation of gas and oil.
8. Higher elevations should be at the start of the route.
9. For collection from one side of the street at a time, it is generally best to route with many clockwise turns around blocks. (Authors' note: Heuristic rules 8 and 9 emphasize the development of a series of clockwise loops in order to minimize left turns, which generally are more difficult and time-consuming than right turns. Especially for right-hand-drive vehicles, right turns are safer.)
10. For collection from both sides of the street at the same time, it is generally best to route with long, straight paths across the grid before looping clockwise.
11. For certain block configurations within the route, specific routing patterns should be applied.

See Figure 9-7 for an example of the heuristic routing procedure.

Crew Integration

Another area of consideration is the integration of several crews. There are four ways of managing crews; usually some combination of the four is employed by any given city.

The swing crew method utilizes an extra crew as standby for heavy pickups, breakdown, or illness. Many times this crew will not report until noon to begin its day.

Crew sizes may be varied because of heavy loads, rain, different route sizes, and other factors. This is referred to as the variable crew method.

With the interroute relay method, when a crew member finishes one job, he or she is put on another route that needs additional help. This method requires more

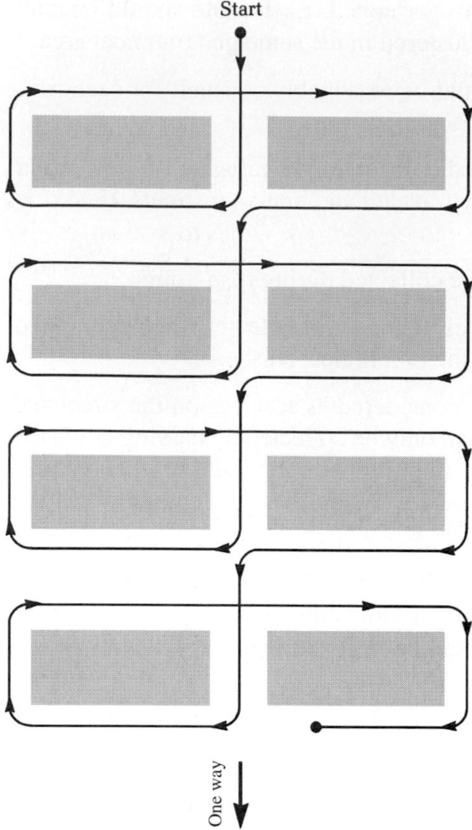

FIGURE 9-7
Arrows show heuristic routing pattern developed for a north-south, one-way street combined with east-west, two-way streets. If both sides of the one-way street cannot be collected in one pass, it is necessary to loop back to the upper end and make a straight pass down the other side. (*Source:* Shuster and Schur, 1974.)

administration to operate, but results in better utilization of personnel and helps ensure that all routes will be completed during the day. Some form of this method has found wide acceptance with good results. Management must be sure that the work load is being balanced fairly and that a faster worker doesn't have to carry the load for others.

The last possibility is the reservoir route method. In this method, the crews work around a central core. When they have finished the route, the crews go to the core and begin picking up there. The core is usually an every day pickup, such as a park or a downtown area.

9-3 INTERROUTE TRANSFER

It is not always economical, or even possible, to haul the solid waste directly to the disposal site in the collection vehicle. In these cases, the solid waste is transferred from several collection vehicles to a larger vehicle, which then carries it to the disposal site. The larger vehicle (*transfer vehicle*) may be a tractor-trailer, railroad car, or barge. A special facility, called a *transfer station,* must be constructed to permit this exchange in a rapid and sanitary fashion.

Among the more important considerations in planning and designing a transfer station are location, type of station, sanitation, access, and accessories such as weighing scales and fences. The use of a transfer station may also provide for present or future resource recovery facilities.

Maximum Haul Time

As in estimating collection times, it is possible to use average values to evaluate trade-offs in transfer station effectiveness. One such method is to compute the travel time available to the crew to travel to the disposal site and still collect the appointed route. This can be done by rearranging Equation 9-3:

$$T_H = \frac{H}{N_d} - t_p N_p - 2t_d - t_u - \frac{B}{N_d} \tag{9-6}$$

where T_H = maximum available haul time, h.

If the maximum available haul time is less than the round trip distance divided by the average route speed ($2x/s$), then you have a problem. Up to a point, changes in t_d, t_u, B, and/or H may alleviate the situation.

Economical Haul Time

The travel time in and of itself is not usually the prime consideration. Cost is usually the prime consideration. Costs are saved when a transfer operation is used because

1. The nonproductive time of collectors is reduced, since they no longer ride to and from the disposal site. It may be possible to reduce the number of collection crews needed because of increased productive collection time.
2. Any reduction in mileage traveled by the collection trucks results in a savings in operating costs.
3. The maintenance requirements for collection trucks can be reduced when these vehicles are no longer required to drive into the landfill site. Much of the damage to suspensions, drive trains, and tires occurs at landfills.
4. The capital cost of collection equipment may be reduced; since the trucks will be traveling only on improved roads, lighter duty, less expensive models can be used (U.S. EPA, 1995).

In order to compare "direct haul" with "transfer" costs, the costs are computed on the basis of \$/Mg · km or, preferably, \$/Mg · min. The time-based comparison is preferred because the average haul speed of the collection vehicle will often be greater than that of the transfer vehicle. Since it is time, not distance, that costs money, this gives a fairer comparison. In addition to the travel cost of operating the transfer vehicle, there are fixed costs for the construction and operation of the transfer station and for maneuvering and unloading the transfer vehicle. Figure 9-8 may be used to estimate the cost of the transfer station.

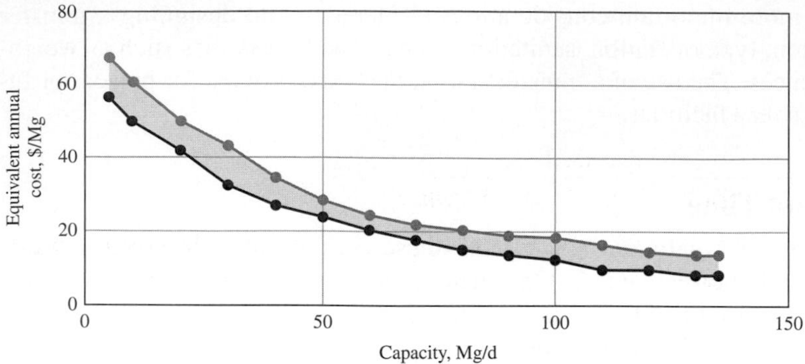

FIGURE 9-8
Transfer station equivalent annual cost as a function of capacity. Costs adjusted to 2006. (*Source:* Zuena, 1987.)

Example 9-3. The disposal site for Watapitae will be closed in two years because of the lack of capacity. An alternative disposal site will be available when the present site is closed. It will be a county-wide regional system that will be 32.5 km from the collection route. Using the data from Examples 9-1 and 9-2 and the following assumptions, determine the maximum haul time for the collection vehicle and the cost for collection vehicle and transfer vehicle haul: $N_d = 1$, $B = 0.50$ h, and the amortized capital cost and operating cost for the transfer station is approximately \$37/Mg.

Solution. First we must determine whether or not the collection vehicle has the time to get to the disposal site while still making all of its pickups.

$$T_H = \frac{8.5}{1} - (0.0233)(250) - 2\frac{13}{60} - \frac{6}{60} - \frac{0.5}{1}$$
$$= 1.64 \text{ h or } 98.5 \text{ min}$$

We now note that the round trip distance is two times the distance from the collection route. The average haul speed can be determined from Figure 9-6. The average haul speed is 64 km/h. Thus, we find the round trip travel time to the regional facility to be

$$\frac{2(32.5 \text{ km})}{64 \text{ km/h}} = 1.02 \text{ h or } 61 \text{ min}$$

The collection vehicle can make it to the disposal site. However, since we have reduced the number of trips to the disposal site, we must either provide an additional vehicle of the same size or replace the existing one with one that is twice as large. Since the existing crew size can handle the 250 pickups per day, the more logical choice would seem to be to choose the larger vehicle. (This is especially true since the existing one is about to expire.) Let us assume the new vehicle will have a capacity of 10.0 m^3.

Now let us examine the comparative haul costs. First we will look at the collection vehicle. We will take the annual cost for a new vehicle exclusive of O & M to be \$29,851. Assuming eight hours of operation per day for five days a week for 52 weeks per year, the annual cost per minute of operation is

$$\frac{\$29{,}851}{(8 \text{ h/d})(60 \text{ min/h})(5 \text{ d/w})(52 \text{ wk/y})} = \$0.2392/\text{min}$$

With the effective wage rate of \$13.90 per hour from Example 9-2, the cost of wages and 125 percent overhead is

$$\frac{(\$13.90 \times 2.25)}{60 \text{ min/h}} = \$0.5213/\text{min}$$

per worker or \$1.0425/min for the crew. The operating cost will be about \$5.50 per kilometer. For travel to the disposal site, the cost per minute would be

$$\frac{(\$5.50/\text{km})(32.5 \text{ km})(2)}{61 \text{ min}} = \$5.8607/\text{min}$$

The factor of two is for the round trip to the disposal site. The total haul cost per trip would be

$$61[(\$0.2392) + (\$1.0425) + (\$5.8607)] = \$435.69$$

The mass of solid waste hauled per trip is

$$(V_T)(D_T) = \text{mass}$$
$$(7.48 \text{ m}^3)(400 \text{ kg/m}^3) = 2{,}992 \text{ kg or } 3.0 \text{ Mg}$$

Note that the volume is twice that of a single trip (Example 9-2), but is considerably less than the capacity of the new vehicle. The unit cost of the haul would then be

$$\frac{\$435.69}{3.0 \text{ Mg}} = 145.23 \text{ or } \$145/\text{Mg}$$

Now let us look at the transfer vehicle. Assume that a tractor-trailer rig having a capacity of 46 m^3 has an annual cost exclusive of O & M of \$37,601. The cost per minute is then

$$\frac{\$37{,}601}{(8 \text{ h/d})(60 \text{ min/h})(5 \text{ d/wk})(52 \text{ wk/y})} = \$0.3013/\text{min}$$

Since the tractor-trailer rig requires an operator with higher skill, the wage rate will be higher. Using a rate of \$19.85 per hour and an overhead rate of 125 percent of wages, the cost per minute is

$$\frac{(\$19.85 \times 2.25)}{60 \text{ min/h}} = \$0.7444/\text{min}$$

In contrast to the collection vehicle, the crew is comprised of only the operator. Thus, the crew cost is \$0.7444/min.

The operating cost will be about \$6.50 per kilometer. The time for the rig to travel to the disposal site will be about 25 percent more than the collection vehicle. The travel cost would then be

$$\frac{(\$6.50)(32.5)(2)}{61 \times 1.25} = \$5.541/\text{min}$$

The total haul cost per trip would be

$$(1.25)(61)[(\$0.3013) + (\$0.7444) + (\$5.541)] = \$502.23$$

Since the capacity of the rig is four times that of the collection vehicle, the mass hauled per trip is

$$4(3.0) = 12 \text{ Mg}$$

The unit cost of the haul, including the cost of building and operating the transfer station (approximately \$37/Mg), would be

$$\frac{\$502.23}{12} + \$37 = 78.83 \text{ or } \$79/\text{Mg}$$

Obviously, consideration should be given to the construction and operation of a transfer station as an alternative to direct haul.

9-4 DISPOSAL BY MUNICIPAL SOLID WASTE LANDFILL

A municipal solid waste (MSW) landfill is defined as a land disposal site employing an engineered method of disposing of solid wastes on land in a manner that minimizes environmental hazards by spreading the solid wastes to the smallest practical volume, and applying and compacting cover material at the end of each day.

Site Selection

Site location is perhaps the most difficult obstacle to overcome in the development of a MSW landfill. Opposition by local citizens eliminates many potential sites. In choosing a location for a landfill, consideration should be given to the following variables:

1. Public opposition
2. Proximity of major roadways
3. Speed limits
4. Load limits on roadways
5. Bridge capacities
6. Underpass limitations
7. Traffic patterns and congestion
8. Haul distance (in time)

9. Detours
10. Hydrology
11. Availability of cover material
12. Climate (for example, floods, mud slides, snow)
13. Zoning requirements
14. Buffer areas around the site (for example, high trees on the site perimeter)
15. Historic buildings, endangered species, wetlands, and similar environmental factors.

In October of 1991, under Subtitle D of the Resource Conservation and Recovery Act (RCRA), the EPA promulgated new federal regulations for landfills. These regulations are known as the Criteria for Municipal Solid Waste Landfills (MSWLF Criteria). EPA also published a companion document to assist owners and municipalities comply with these criteria (U.S. EPA, 1998). These included siting criteria that specify restrictions on distances from airports, flood plains, and fault areas, as well as limitations on construction in wetlands, seismic impact areas, and other areas of unstable geology such as landslide areas and those susceptible to sink holes. Other restrictions may apply. For example, a landfill should be more than:

30 m from streams,

160 m from drinking water wells,

65 m from houses, schools, and parks, and

3,000 m from airport runways.

Site Preparation

The plans and specifications for a MSW landfill should require that certain steps be carried out before operations begin. These steps include grading the site area, constructing access roads and fences, and installing signs, utilities, and operating facilities.

On-site access roads should be of all-weather construction and wide enough to permit two-way truck travel (7.3 m). Grades should not exceed equipment limitations. For loaded vehicles, most uphill grades should be less than 7 percent, and downhill grades should be less than 10 percent.

All MSW landfill sites should have electric, water, and sanitary services. Remote sites may have to use acceptable substitutes, for example, portable chemical toilets, trucked-in drinking water, and electric generators. Water should be available for drinking, fire-fighting, dust control, and sanitation. Telephone or radio communications are desirable.

A small MSW landfill operation will usually require only a small building for storing hand tools and equipment parts and a shelter with sanitary facilities. A single building may serve both purposes. Buildings may be temporary and preferably movable.

Equipment

The size, type, and amount of equipment required at an MSW landfill depends on the size and method of operation, quantities and time of solid waste deliveries, and, to a degree, the experience and preference of the designer and equipment operators. Another factor to be considered is the availability and dependability of service from the equipment.

The most common equipment used on MSW landfills is the crawler or rubber-tired tractor (Figure 9-9). The tractor can be used with a dozer blade, trash blade, or a front-end loader. A tractor is versatile and can perform a variety of operations: spreading, compacting, covering, trenching, and even hauling the cover material. The decision on whether to select a rubber-tired or a crawler-type tractor, and a dozer blade, trash blade, or front-end loader must be based on the conditions at each individual site (see Table 9-5).

The crawler dozer is excellent for grading and can be economically used for dozing solid waste or soil over distances up to 100 m. The larger trash or landfill blade can be used in lieu of a straight dozer blade, thereby increasing the volume of solid waste

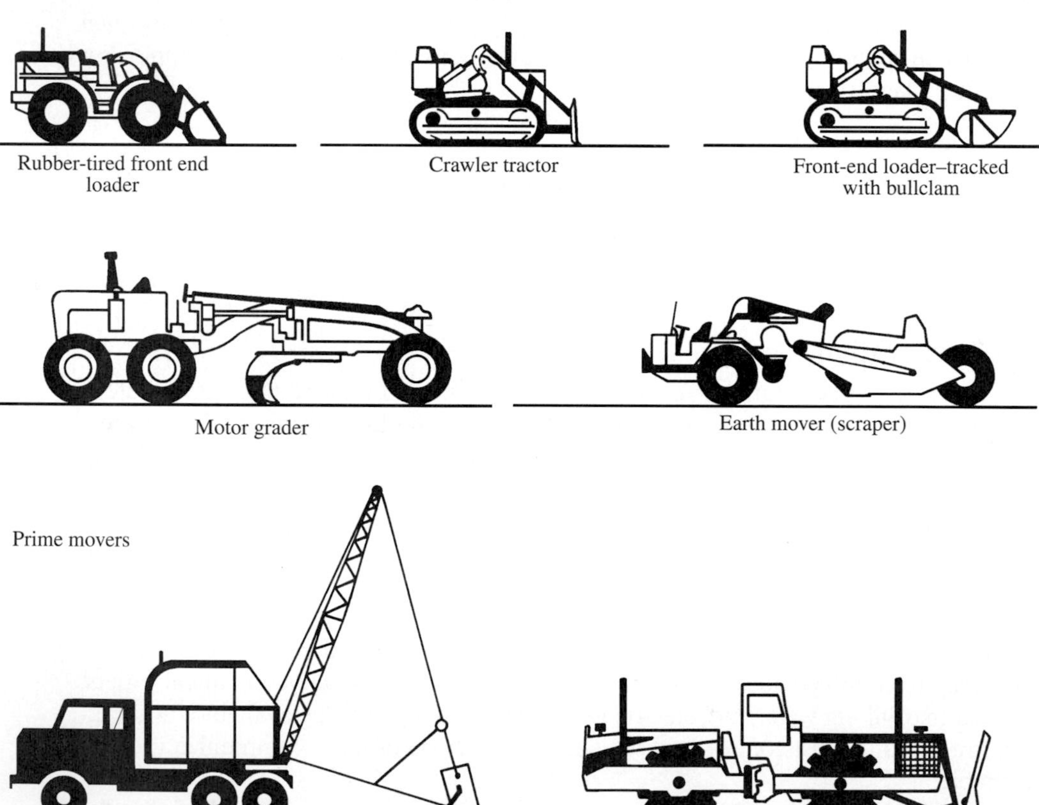

FIGURE 9-9
Municipal solid waste landfill equipment.

TABLE 9-5
Performance characteristics of landfill equipment[a]

Equipment	Spreading	Compacting	Excavating	Spreading	Compacting	Hauling	Density of compacted solid waste (kg/m^3)
Crawler dozer	E	G	E	E	G	NA	750
Crawler loader	G	G	E	G	G	NA	—
Rubber-tired dozer	E	G	F	G	G	G	733
Rubber-tired loader	G	G	F	G	G	G	—
Steel-wheeled compactor	E	E	P	G	E	NA	809
Scraper	NA	NA	G	E	NA	E	NA
Dragline	NA	NA	E	F	NA	NA	NA

[a]*Basis of evaluation:* Easily workable soil and cover material haul distance greater than 300 m.

Rating key: E, excellent; G, good; F, fair; P, poor; NA, not applicable.

Note: Density of "well-compacted" solid waste resulting from four passes over each square meter. Density measured after daily soil cover emplaced but not including soil in volume and weight measurements.

(*Source:* Data from Stone and Conrad, 1969, and O'Leary and Walsh, 2002.)

that can be dozed. The crawler loader has the capability to lift materials off the ground for carrying. It is an excellent excavator, well suited for trench operations.

Rubber-tired machines are generally faster than crawler machines. Because their loads are concentrated more, rubber-tired machines have less flotation and traction than crawler machines. Rubber-tired machines can be economically operated at distances of up to 200 m.

Steel-wheeled compactors are finding increased application at MSW landfills. In basic design, compactors are similar to rubber-tired tractors. The unique feature of compactors is the design of their wheels, which are steel and equipped with teeth or lugs of varying shape and configuration. This design is employed to impart greater crushing and demolition forces to the solid waste. Use of compactors should be restricted to solid waste, because their design does not lend them to application of a smooth layer of compacted cover material. Thus, compactors are best used in conjunction with tracked or rubber-tired machines that can be used for cover material application.

Other equipment used at MSW landfills are scrapers, water wagons, drag-lines, dump trucks, and graders. This type of equipment is normally found only at large solid waste landfills where specialized equipment increases the overall efficiency.

Equipment size depends on the size of the operation. Small landfills for communities of 15,000 or less, or landfills handling 50 Mg of solid wastes per day or less, can operate successfully with one tractor in the 20 to 30 Mg range. Heavier equipment in the 30 to 45 Mg range, or larger, can handle more waste and achieve better compaction. Heavy equipment is recommended for MSW landfill sites serving more than 15,000 people or handling more than 50 Mg per day. MSW landfills serving 50,000 people or less or handling no more than about 150 Mg of solid waste per day normally can manage well with one piece of heavy equipment (30 to 45 Mg range).

Operation

Although various titles are used to describe the operating methods employed at MSW landfills, only two basic techniques are involved. They are termed the *area method* (Figure 9-10) and the *trench method* (Figure 9-11). At many sites, both methods are used, either simultaneously or sequentially.

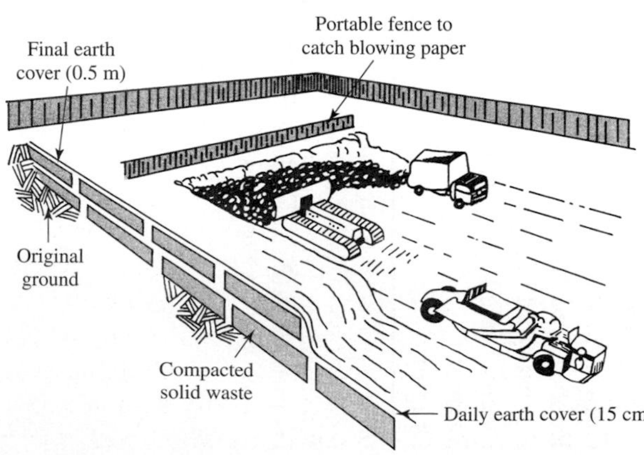

FIGURE 9-10
The area method.

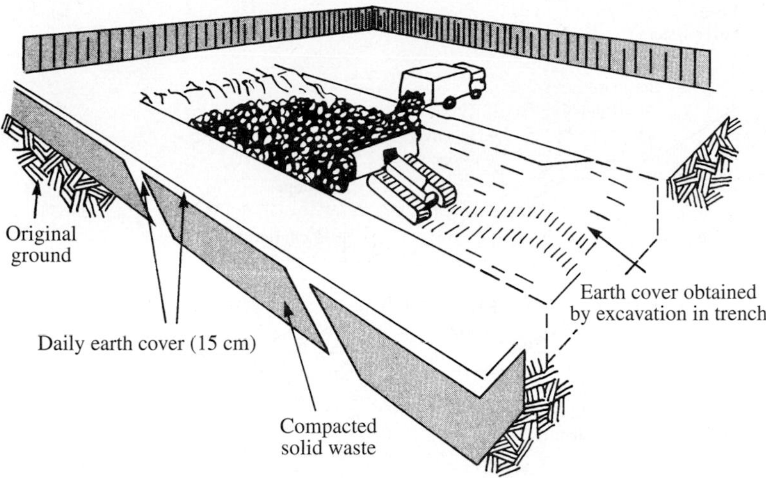

FIGURE 9-11
The trench method.

In the area method, the solid waste is deposited on the surface, compacted, then covered with a layer of compacted soil at the end of the working day. Use of the area method is seldom restricted by topography; flat or rolling terrain, canyons, and other types of depressions are all acceptable. The cover material may come from on or off site.

The trench method is used on level or gently sloping land where the water table is low. In this method a trench is excavated; the solid waste is placed in it and compacted; and the soil that was taken from the trench is then laid on the waste and compacted. The advantage of the trench method is that cover material is readily available as a result of trench excavation. Stockpiles can be created by excavating long trenches, or the material can be dug up daily. The depth depends on the location of the groundwater and/or the character of the soil. Trenches should be at least twice as wide as the compacting equipment so that the treads or wheels can compact all the material on the working area.

A MSW landfill does not need to be operated by using only the area or trench method. Combinations of the two are possible. The methods used can be varied according to the constraints of the particular site.

A profile view of a typical landfill is shown in Figure 9-12. The waste and the daily cover placed in a landfill during one operational period form a *cell*. The operational period is usually one day. The waste is dumped by the collection and transfer vehicles onto the working *face*. It is spread in 0.4 to 0.6 m layers and compacted by driving a crawler tractor or other compaction equipment over it. At the end of each day *cover* material is placed over the cell. The cover material may be native soil or other approved materials. Its purpose is to prevent fires, odors, blowing litter, and scavenging. The federal regulations also permit the state regulatory authority to allow the use of alternative daily covers (ADC) if the owner of the landfill can demonstrate that the alternative material functions as well as the earthen cover without presenting a threat to human health or the environment. Some landfills have successfully demonstrated that diverted wastes such as chipped tires, yard waste, shredded wood waste, and

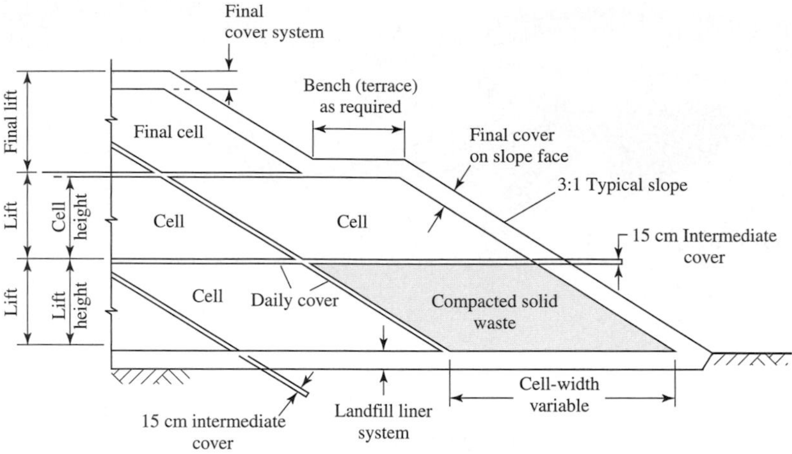

FIGURE 9-12
Sectional view through a MSW landfill. (*Source:* Tchobanoglous et al., 1993.)

petroleum-contaminated soils can be used effectively as ADCs. Using these waste products as ADCs presents a cost savings for the landfill and also increases the landfill's available space. The use of manufactured ADCs such as colored tarps is also being accepted in some localities. Recommended depths of cover for various exposure periods are given in Table 9-6. The dimensions of a cell are determined by the amount of waste and the operational period.

A *lift* may refer to the placement of a layer of waste or the completion of the horizontal active area of the landfill. In Figure 9-12 a lift is shown as the completion of the active area of the landfill. An extra layer of intermediate cover may be provided if the lift is exposed for long periods. The active area may be up to 300 m in length and width. The side slopes typically range from 1.5:1 to 2:1. Trenches vary in length from 30 to 300 m with widths of 5 to 15 m. The trench depth may be 3 to 9 m (Tchobanoglous et al., 1993).

Benches are used where the height of the landfill exceeds 15 to 20 m. They are used to maintain the slope stability of the landfill, for the placement of surface water drainage channels, and for the location of landfill gas collection piping.

Final cover is applied to the entire landfill site after all landfilling operations are complete. A modern final cover will contain several different layers of material to perform different functions. These are discussed more fully in the landfill design section of this chapter.

TABLE 9-6
Recommended depths of cover

Type of cover	Minimum depth (m)	Exposure time (d)
Daily	0.15	<7
Intermediate	0.30	7 to 365
Final	0.60	>365

Additional considerations in the operation of the landfill are those required by the 1991 Subtitle D regulations promulgated by EPA. These require exclusion of hazardous waste, use of cover materials, disease vector control, explosive gas control, air quality measurements, access control, runoff and run-on controls, surface water and liquids restrictions, and groundwater monitoring, as well as record keeping (40 CFR 257 and 258; FR 9 OCT 1991).

Environmental Considerations

Vectors (carriers of disease) and water and air pollution should not be a problem in a properly operated and maintained landfill. Good compaction of the waste, daily covering of the solid waste with good compaction of the cover, and good housekeeping are musts for control of flies, rodents, and fires.

Burning, which may cause air pollution, is never permitted at a MSW landfill. If accidental fires should occur, they should be extinguished immediately using soil, water, or chemicals. Odors can be controlled by covering the wastes quickly and carefully, and by sealing any cracks that may develop in the cover.

Landfill Gases. The principal gaseous products emitted from a landfill (methane and carbon dioxide) are the result of microbial decomposition. Typical concentrations of landfill gases and their characteristics are summarized in Table 9-7. During the early life of the landfill, the predominant gas is carbon dioxide. As the landfill matures, the gas is composed almost equally of carbon dioxide and methane. Because the methane is explosive, its movement must be controlled. The heat content of this landfill gas mixture (16,000 to 20,000 kJ/m^3), although not as substantial as methane alone

TABLE 9-7
Typical constituents found in MSW landfill gas

Component	Percent (dry volume basis)
Methane	45–60
Carbon dioxide	40–60
Nitrogen	2–5
Oxygen	0.1–1.0
Sulfides, disulfides, mercaptans, etc.	0–1.0
Ammonia	0.1–1.0
Hydrogen	0–0.2
Carbon monoxide	0–0.2
Trace constituents	0.01–0.06
Characteristic	**Value**
Temperature, °C	35–50
Specific gravity	1.02–1.05
Moisture content	Saturated
High heating value, kJ/m^3	16,000–20,000

(*Source:* G. Tchobanoglous et al., 1993.)

(37,000 kJ/m^3), has sufficient economic value that many landfills have been tapped with wells to collect it. At the end of 2004, there were 378 landfill gas (LFG) recovery projects in the United States. This is a four-fold increase over the 86 LFG projects operating in 1990.

Because of their toxicity, trace gas emissions from landfills are of concern. More than 150 compounds have been measured at various landfills. Many of these may be classified as volatile organic compounds (VOCs). The occurrence of significant VOC concentrations is often associated with older landfills that previously accepted industrial and commercial wastes containing these compounds. The concentrations of 10 compounds measured in landfill gases from several California sites are shown in Table 9-8.

Leachate

Liquid that passes through the landfill and that has extracted dissolved and suspended matter from it is called *leachate.* The liquid enters the landfill from external sources such as rainfall, surface drainage, groundwater, and the liquid in and produced from the decomposition of the waste.

Leachate Quantity. The amount of leachate generated from a landfill site may be estimated using a hydrologic mass balance for the landfill. Those portions of the global hydrologic cycle (see Chapter 3) that typically apply to a landfill site include precipitation, surface runoff, evaporation, transpiration (when the landfill cover is completed), infiltration, and storage. Precipitation may be estimated in the conventional fashion from climatological records. Surface runoff or run-on may be estimated using the rational formula (Equation 3-15). Evaporation and transpiration are often lumped together as *evapotranspiration.* It may be estimated from regional data such as that provided by the U.S. Geologic Service *Water Atlas.* Infiltration (and exfiltration) may be estimated using Darcy's law (Equation 3-22). Until the landfill becomes saturated, some of the water infiltration will be stored in both the cover material and the waste. The quantity of water that can be held against the pull of gravity is referred to as *field capacity* (Figure 9-13 on page 772). Theoretically, when the landfill reaches its field capacity, leachate will begin to be produced. Then, the potential quantity of leachate is the amount of moisture within the landfill in excess of the field capacity. In reality, leachate will begin to be produced almost immediately because of channeling in the waste. The following equation may be used to estimate the field capacity of the waste (Tchobanoglous et al., 1993):

$$\mathrm{FC} = 0.6 - 0.55\left(\frac{2.205W}{10{,}000 + 2.205W}\right) \tag{9-7}$$

where FC = field capacity (fraction of water in the waste based on dry weight of the waste)

W = overburden mass of waste calculated at midheight of the lift in question, kg

The EPA and the Waterways Experiment Station of the U.S. Army Corps of Engineers developed a microcomputer model of the hydrologic balance called the Hydrologic Evaluation of Landfill Performance (HELP) (Schroeder et al., 1984). The

TABLE 9-8
Concentrations of specified air contaminants measured in landfill gases (in parts per billion)

	Landfill Site						
Compound	Yolo Co.	City of Sacramento	Yuba Co.	El Dorado Co.	L.A.-Pacific (Ukiah)	City of Clovis	City of Willits
Vinyl chloride	6,900	1,850	4,690	2,200	<2	66,000	7.5
Benzene	1,860	289	963	328	<2	895	<18
Ethylene dibromide	1,270	<10	<50	<1	<1	<1	<0.5
Ethylene dichloride	nr	nr	nr	<20	0.2	<20	4
Methylene chloride	1,400	54	4,500	12,900	<1	41,000	<1
Perchloroethylene	5,150	92	140	233	<0.2	2,850	8.1
Carbon tetrachloride	13	<5	<7	<5	<0.2	<5	<0.2
1,1,1-TCA[1]	1,180	6.8	<60	3,270	0.52	113	0.8
TCE[2]	1,200	470	65	900	<0.6	895	8
Chloroform	350	<10	<5	120	<0.8	1,200	<0.8
Methane	nr	nr	nr	nr	0.11%	17%	0.14%
Carbon dioxide	nr	nr	nr	nr	0.12%	24%	<0.1%
Oxygen	nr	nr	nr	nr	nr	10%	21%

nr: Not reported by operator

[1]1,1,1-TCA: 1,1,1-trichloroethane, methyl chloroform

[2]TCE: Trichloroethene, trichloroethylene

(*Source:* CARB, 1988.)

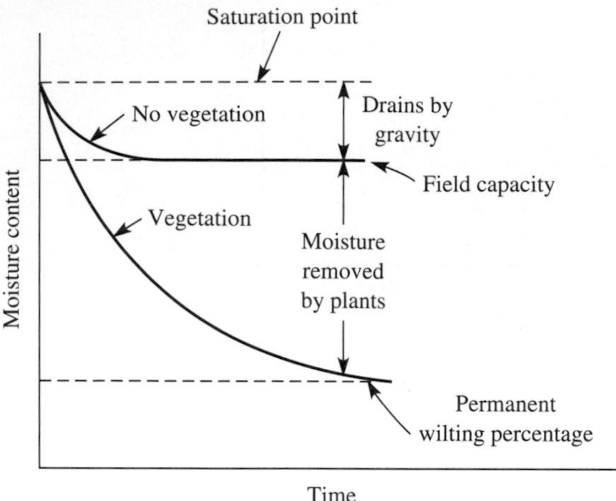

FIGURE 9-13
Moisture relationships in soil. (*Source:* Pfeffer, 1992.)

program contains extensive data on the characteristics of various soil types, precipitation patterns, and evapotranspiration-temperature relationships as well as the algorithms to perform a routing of the moisture flow through the landfill.

Leachate Composition. Solid wastes placed in a sanitary landfill may undergo a number of biological, chemical, and physical changes. Aerobic and anaerobic decomposition of the organic matter results in both gaseous and liquid end products. Some materials are chemically oxidized. Some solids are dissolved in water percolating through the fill. A range of leachate compositions is listed in Table 9-9. The VOCs in the landfill gas often contribute to contamination of groundwater because they dissolve in the leachate as it passes through the landfill. Henry's law (see Chapter 4) may be used to estimate the VOC concentrations that might occur in the leachate. Because of the differential heads (slope of the piezometric surface), the water containing dissolved substances moves into the groundwater system. The result is gross pollution of the groundwater.

Bioreactor Landfills

The implementation of RCRA Subtitle D resulted in more stringent protection of the environment, particularly the groundwater resources. The future trend in landfill design appears to be the development of engineered systems that optimize waste degradation and so minimize the amount of land needed for waste disposal. One technology that shows a lot of promise is bioreactor landfills. EPA has initiated a number of studies and partnerships with waste management companies to fully investigate the potential of this technology.

In traditional municipal solid waste landfills, organic waste eventually decomposes and stabilizes. These processes are controlled by microorganisms. In bioreactor landfills, biological decomposition is accelerated by enhancing the conditions necessary for these microorganisms to flourish. This is accomplished by the controlled addition of supplemental air and water. The degradation and stabilization of organic waste is then accelerated.

EPA defines a bioreactor landfill as "any permitted Subtitle D landfill (under RCRA) or landfill cell where liquid or air is injected in a controlled fashion into the waste mass in

TABLE 9-9
Typical data of the composition of leachate from new and mature landfills

Constituent	Value, mg/L		
	New landfill (less than 2 years)		Mature landfill (greater than 10 years)
	Range	Typical	
BOD_5 (5-day biochemical oxygen demand)	2,000–30,000	10,000	100–200
TOC (total organic carbon)	1,500–20,000	6,000	80–160
COD (chemical oxygen demand)	3,000–60,000	18,000	100–500
Total suspended solids	200–2,000	500	100–400
Organic nitrogen	10–800	200	80–120
Ammonia nitrogen	10–800	200	20–40
Nitrate	5–40	25	5–10
Total phosphorus	5–100	30	5–10
Ortho phosphorus	4–80	20	4–8
Alkalinity as $CaCO_3$	1,000–10,000	3,000	200–1,000
pH (no units)	4.5–7.5	6	6.6–7.5
Total hardness as $CaCO_3$	300–10,000	3,500	200–500
Calcium	200–3,000	1,000	100–400
Magnesium	50–1,500	250	50–200
Potassium	200–1,000	300	50–400
Sodium	200–2,500	500	100–200
Chloride	200–3,000	500	100–400
Sulfate	50–1,000	300	20–50
Total iron	50–1,200	60	20–200

(*Source:* Tchobanoglous et al., 1993.)

order to accelerate or enhance biostabilization of the waste" (40 CFR 257 and 258, FR, 9 OCT 1991.) In these landfills additional moisture is introduced to the waste, typically by recirculating the leachate and adding additional moisture such as stormwater, wastewater, and wastewater treatment plant sludge. The goal is to provide enough moisture to the waste to maintain the optimal moisture content for microbial decomposition, typically 35 to 65 percent moisture.

One of the benefits of this system is that the decomposition rate is increased, so complete decomposition can occur in years instead of decades. The waste density is increased, so over the life of the landfill, 15 to 30 percent additional space is available. Also, the cost of the leachate disposal is reduced, since it is recirculated. And there is a significant increase in the landfill gas that is generated. If this is captured on site, then it can be used to produce energy.

These systems have a higher initial cost to build and operate, since extensive recirculation and monitoring is required. Bioreactor landfills can be designed to use aerobic, anaerobic, or facultative microorganisms.

Phases of Bioreaction. Five more or less sequential phases of bioreaction are thought to occur in a landfill. In the *initial adjustment phase,* the organic biodegradable components in the MSW undergo aerobic biodegradation because some air is trapped when the waste is placed in the landfill. In a conventional landfill, the principle source of microorganisms is the soil material that is used as daily and final cover. Digested wastewater treatment plant sludge as well as recycled leachate are also sources of microorganisms. In a bioreactor landfill, the latter sources provide a means of accelerating the decomposition process.

The second phase is called the *transitional phase.* Oxygen is depleted and anoxic and anaerobic conditions begin to develop. As the landfill becomes anaerobic, nitrate and sulfate serve as electron acceptors. Nitrogen, hydrogen, and hydrogen sulfide are products of the decomposition process. As the conversion process proceeds, the microbial community responsible for conversion or organic material to methane and carbon dioxide begin the three-step process described in Chapter 6 (Figure 6-43).

In the *acid phase,* the anaerobic microbial activity initiated in the second phase accelerates. Significant amounts of organic acids are produced and the production of hydrogen decreases. Carbon dioxide is the principle gas produced in this phase. The pH of the leachate will often drop to 5 or lower (Tchobanoglous et al., 1993).

The fourth phase is called the *methane fermentation phase. Methanogens* convert the acetic acid and hydrogen gas produced by the acid formers into methane (CH_4) and CO_2. The pH of the leachate rises to more neutral values in the range 6 to 8.

The *maturation phase* begins after the readily available biodegradable organic matter has been converted to CH_4 and CO_2. The rate of landfill gas generation decreases dramatically.

Volume of Gas Produced. Cossu et al. (1996) present the following reaction representing the overall methane fermentation process:

$$C_aH_bO_cN_d + nH_2O \rightarrow x\,CH_4 + y\,CO_2 + w\,NH_3 = z\,C_sH_7O_2N + \text{energy} \tag{9-8}$$

where $C_aH_bO_cN_d$ is the empirical formula for the biodegradable organic matter and $C_5H_7O_2N$ is the empirical chemical formula of bacterial cells.

The maximum theoretical landfill gas yield (neglecting bacterial cell conversion) may be estimated as (Tchobanoglous et al., 1993):

$$C_aH_bO_cN_d + \left(\frac{4a - b - 2c + 3d}{4}\right)H_2O \rightarrow \left(\frac{4a + b - 2c - 3d}{8}\right)CH_4 + \left(\frac{4a - b + 2c + 3d}{8}\right)CO_2 + dNH_3 \tag{9-9}$$

For the purpose of analysis, the MSW may be divided into two classes: rapidly biodegradable and slowly biodegradable. Food waste, newspaper, office paper, cardboard, leaves, and leafy yard trimmings fall into the first category. Textiles, rubber, leather, tree branches, and wood fall into the second category.

Tchobanoglous et al. (1993) developed empirical chemical formulas for typical U.S. MSW as collected in 1990 for each of these categories:

- Rapidly decomposable = $C_{68}H_{111}O_{50}N$
- Slowly decomposable = $C_{20}H_{29}O_9N$

These formulas may be used to estimate the maximum theoretical gas production. Actual quantities of gas will be lower because (1) all of the biodegradable organic matter is not available for decomposition, (2) the biodegradability is less for organic wastes with high lignin content, and (3) moisture may be limiting. The construction and operation of a bioreactor landfill is designed to minimize these limitations. Actual gas production rates from typical MSW landfills ranges from 40 to 400 m^3/Mg of MSW.

The rate of decomposition that is reflected in gas production of MSW is highly variable. Most models use first-order equations in two stages to describe the gas production as it rises to some peak value and then falls (Cossu et al., 1996).

Landfill Design

The design of the landfill has many components including site preparation, buildings, monitoring wells, size, liners, leachate collection system, final cover, and gas collection system. Figure 9-14 shows a schematic of a typical municipal solid waste landfill with all of these components shown. In the following discussion we will limit ourselves to introductory consideration of the design of the size of the landfill, the

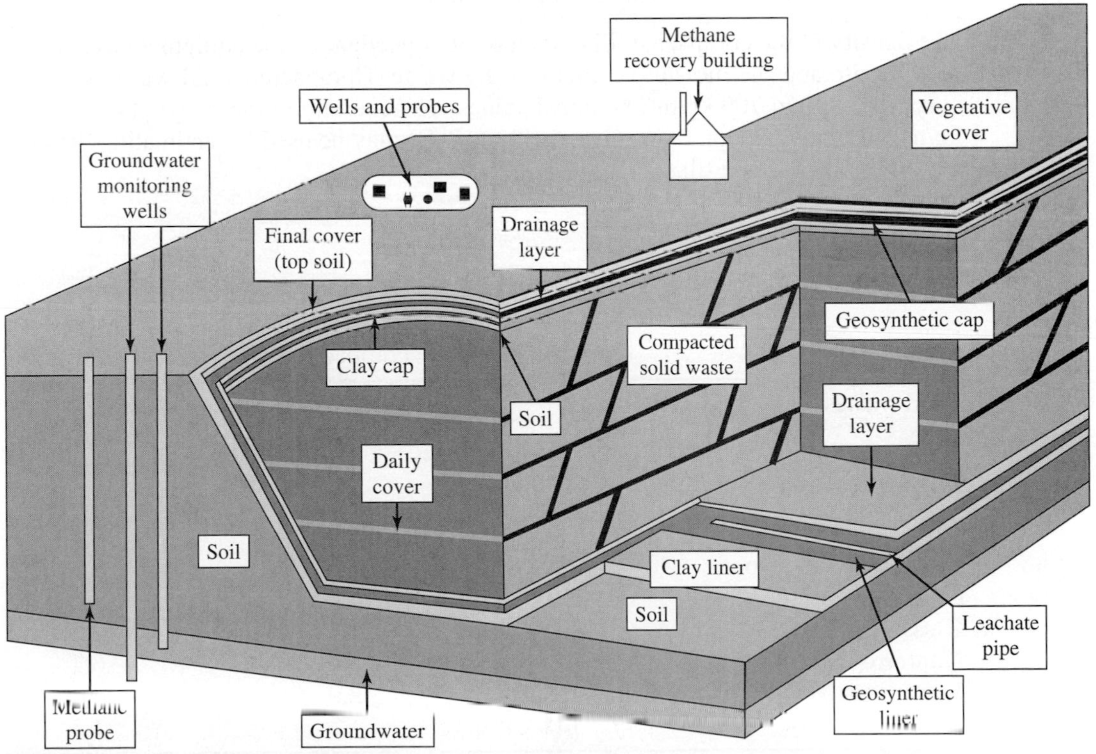

FIGURE 9-14
Schematic of a typical municipal solid waste landfill. (*Source:* U.S. EPA, 1995.)

selection of a liner system, the design of a leachate collection system, and a discussion of the final cover system.

Volume Required. To estimate the volume required for a landfill, it is necessary to know the amount of refuse being produced and the density of the in-place, compacted refuse. The volume of refuse differs markedly from one city to another because of local conditions.

Salvato recommends a formula of the following form for estimating the annual volume required (Salvato, 1972).

$$\mathcal{V}_{LF} = \frac{PEC}{D_c} \tag{9-10}$$

where $\mathcal{V}_{LF}$ = volume of landfill, m^3
P = population
E = ratio of cover (soil) to compacted fill

$$= \frac{\mathcal{V}_{SW} + \mathcal{V}_c}{\mathcal{V}_{SW}}$$

$\mathcal{V}_{SW}$ = volume of solid waste, m^3
$\mathcal{V}_c$ = volume of cover, m^3
C = average mass of solid waste collected per capita per year, kg/person
D_c = density of compacted fill, kg/m3

The density of the compacted fill is somewhat dependent on the equipment used at the landfill site and the moisture content of the waste. Compacted solid waste densities vary from 300 to 700 kg/m^3. Nominal values are generally in the range of 475 to 600 kg/m^3. The compaction ratios given in Table 9-10 may be used for estimating the density of the compacted fill.

TABLE 9-10
Typical compaction ratios[a]

Component	Poorly compacted	Normal compaction	Well-compacted
Food wastes	2.0	2.8	3.0
Paper	2.5	5.0	6.7
Cardboard	2.5	4.0	5.8
Plastics	5.0	6.7	10.0
Textiles	2.5	5.8	6.7
Rubber, leather, wood	2.5	3.3	3.3
Garden trimmings	2.0	4.0	5.0
Glass	1.1	1.7	2.5
Nonferrous metal	3.3	5.6	6.7
Ferrous metal	1.7	2.9	3.3
Ashes, masonry	1.0	1.2	1.3

[a] The ratio of the density after compaction to that as discarded, that is, before pickup by collection vehicle.
(*Source:* Tchobanoglous et al., 1977)

Example 9-4. How much landfill space does Watapitae require for 20 years of operation? Assume that the village will use a cell height of 2.4 m and that it will follow normal practice and use 0.15 m of soil for daily cover; 0.3 m to complete the cell; and a final cover of 0.6 m for every stack of three cells. Assume that compaction will be "normal."

Solution. Although we do not know the population or per capita waste generation rate, we can estimate the mass generated per year from other data. From Example 9-1 we know that 1,250 service stops must be collected each week. From Example 9-2 we know that each service stop contributes an average of 0.624 Mg per year. Then the annual mass generation rate is

$$\text{Mass} = (1{,}250 \text{ stops}) \times (0.624 \text{ Mg/y stop}) = 780 \text{ Mg/y}$$

This is equivalent to the product $(P)(C)$ in Equation 9-10.

In Example 9-1 we determined that the mean density of the uncompacted solid waste was 106 kg/m^3. Using the fractional mass composition of the waste as given in Table 9-4 and the "normal" compaction ratios in Table 9-10, we can determine the weighted compaction ratio by multiplying the fractional mass by the compaction ratio (Table 9-11).

With a compaction ratio of 4.18, the density of the compacted fill is estimated to be

$$D_c = (106 \text{ kg/m}^3) \times (4.18) = 443 \text{ kg/m}^3 \text{ or } 0.443 \text{ Mg/m}^3$$

Note that this implies that waste dumped at the face of the fill in a 1.25-m layer would have to be compressed to a depth of 0.3 m, that is,

$$\left(\frac{1}{4.18}\right)(1.25 \text{ m})$$

TABLE 9-11
Weighted compaction ratios for Example 9-4

Component	Mass fraction	Weighted compaction ratio
Food wastes	0.0947	0.27
Paper	0.4317	2.16
Cardboard	0.0650	0.26
Plastics	0.0181	0.12
Textiles	0.0020	0.01
Rubber	—	—
Leather	0.0150	0.05
Garden trimmings	0.1432	0.57
Wood	0.0350	0.12
Glass	0.0749	0.12
Tin cans	0.0520	0.29
Nonferrous metals	0.0150	0.08
Ferrous metals	0.0430	0.12
Dirt, ashes, brick	0.0110	0.01
Total	1.0006	4.18

Before we can estimate E, we must determine the daily volume of solid waste and the area over which it will be spread. For a five-day week, the daily volume is determined as follows:

$$\forall = \frac{780 \text{ Mg/y}}{0.443 \text{ Mg/m}^3} \times \frac{1}{52 \text{ wk/y}} \times \frac{1}{5 \text{ d/wk}} = 6.77 \text{ m}^3\text{/d}$$

If this is spread in a 0.3-m layer, then the area would be

$$\frac{6.77 \text{ m}^3}{0.3 \text{ m}} = 22.57 \text{ m}^2\text{/d}$$

This is equivalent to a square 4.75 m on each side. This seems reasonable for a small community.

If 0.15 m of soil is used as cover each day, then 0.45 m will be placed each day and it will take

$$\frac{2.4 \text{ m} - 0.15 \text{ m}}{0.45 \text{ m/day}} = 5.00 \text{ days}$$

to complete the cell. (The 0.15 m is the addition to daily cover to complete the cell with 0.3 m of cover.) At this rate we will complete a stack of three cells every three weeks (15 working days).

The soil volume separating a stack of three cells will be about

$$0.3 \text{ m thick} \times 2.4 \text{ m high} \times 4.75 \text{ m long} \times 3 \text{ cells} = 10.26 \text{ m}^3$$

To account for two sides of the cell, this number needs to be multiplied by two.

$$10.26 \text{ m}^3 \times 2 = 20.52 \text{ m}^3$$

If we ignore this volume, E can be calculated as

$$E = \frac{0.3 + [0.15 + 0.03 + 0.02]}{0.3} = 1.67$$

The terms in the brackets account for the daily cover of 0.15 m; the cell cover of an additional 0.15 m each five days or 0.03 m per day; and the final stack cover of an additional 0.3 m to the three-cell cover each 15 days or 0.02 m per day.

If we do not ignore the soil separating the cells, then the soil volume per stack of three cells as shown in Figure 9-15 is calculated as follows:

$$(3 \text{ cells/stack})(5 \text{ lifts/cell})(22.57 \text{ m}^2)(0.15 \text{ m}) = 50.78 \text{ m}^3$$

plus the 0.15 m of additional soil to bring the weekly cell cover to 0.30 m is

$$(3 \text{ cells/stack})(22.57 \text{ m}^2)(0.15 \text{ m}) = 10.16 \text{ m}^3$$

plus the additional 0.3 m to bring the final cover to 0.6 m,

$$(22.57 \text{ m}^2)(0.3 \text{ m}) = 6.77 \text{ m}^3$$

The total soil volume, including the 20.52 m^3 for the sides of the stack, is

$$50.78 + 10.16 + 6.77 + 20.52 = 88.23 \text{ m}^3$$

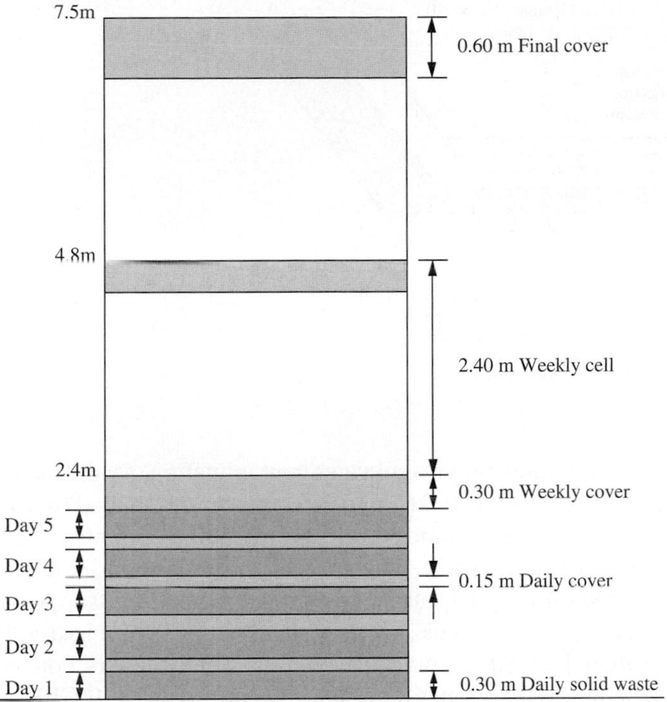

FIGURE 9-15
Schematic diagram of MSW landfill stack of three cells (Example 9-4).

The value for $\forall_{SW}$ would then be

$$\forall_{SW} = (6.77\ \text{m}^3/\text{d})(15\ \text{d/stack}) = 101.55\ \text{m}^3/\text{stack}$$

The value for E would then be

$$E = \frac{101.55 + 88.23}{101.55} = 1.87$$

Thus, for this landfill, the separation wall will increase the volume by about 12 percent. This is not insignificant!

The estimated volume requirement for 20 years would be

$$\forall_{LF} = \frac{(780\ \text{Mg/y})(1.87)}{0.443\ \text{Mg/m}^3} \times 20\ \text{y} = 6.59 \times 10^4\ \text{m}^3$$

Since the average landfill depth will be three 2.4 m cells plus an additional 0.3 m final cover, the area will be

$$A_{LF} = \frac{6.59 \times 10^4}{(3)(2.4) + 0.3} = 8.78 \times 10^3\ \text{m}^2$$

An area approximately 100 m on a side would do very nicely.

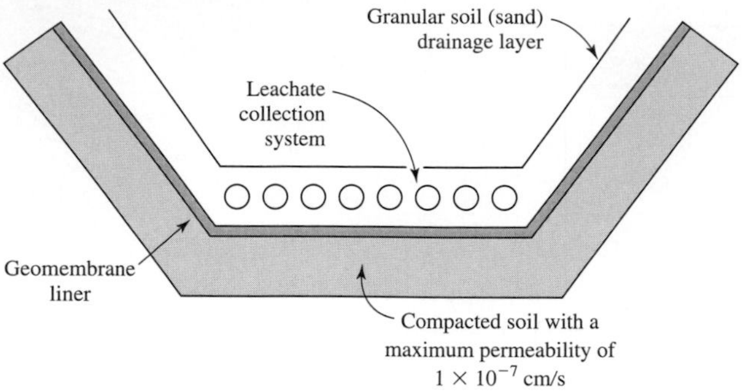

FIGURE 9-16
A composite liner and leachate collection system.

Liner Selection. In order to prevent groundwater contamination, strict leachate control measures are required. Under the 1991 Subtitle D rules promulgated by EPA, new landfills must be lined in a specific manner or meet maximum contaminant levels for the groundwater at the landfill boundary. The specified liner system includes a synthetic membrane (*geomembrane*) at least 30 mils (0.76 mm) thick supported by a compacted soil liner at least 0.6 m thick. The soil liner must have a hydraulic conductivity of no more than 1×10^{-7} cm/s. Flexible membrane liners consisting of high-density polyethylene (HDPE) must be at least 60 mils thick (40 CFR 257 and 258, and FR 9 OCT 1991). A schematic of the EPA specified liner system is shown in Figure 9-16.

Several geomembrane materials are available. Some examples include polyvinyl chloride (PVC), high-density polyethylene (HDPE), chlorinated polyethylene (CPE), and ethylene propylene diene monomer (EPDM). Designers show a strong preference for PVC and especially for HDPE. Although the geomembranes are highly impermeable (hydraulic conductivities are often less than 1×10^{-12} cm/s), they can be easily damaged or improperly installed. Damage may occur during construction by construction equipment, by failure due to tensile stress generated by the overburden, tearing as a result of differential settling of the supporting soil, puncture from sharp objects in the overburden, puncture from coarse aggregate in the supporting soil, and tearing by landfill equipment during operation. Installation errors primarily occur during seaming when two pieces of geomembrane must be attached or when piping must pass through the liner. A liner placed with adequate quality control should have less than 3 to 5 defects per hectare.

The soil layer under the geomembrane acts as a foundation for the geomembrane and as a backup for control of leachate flow to the groundwater. Compacted clay generally meets the requirement for a hydraulic conductivity of less than 1×10^{-7} cm/s. In addition to having a low permeability, it should be: free of sharp objects greater than 1 cm in diameter, graded evenly without pockets or hillocks, compacted to prevent differential settlement, and free of cracks.

Leachate Collection. Under the 1991 Subtitle D rules promulgated by EPA, the leachate collection system must be designed so that the depth of leachate above the liner does not exceed 0.3 m. The leachate collection system is designed by sloping the

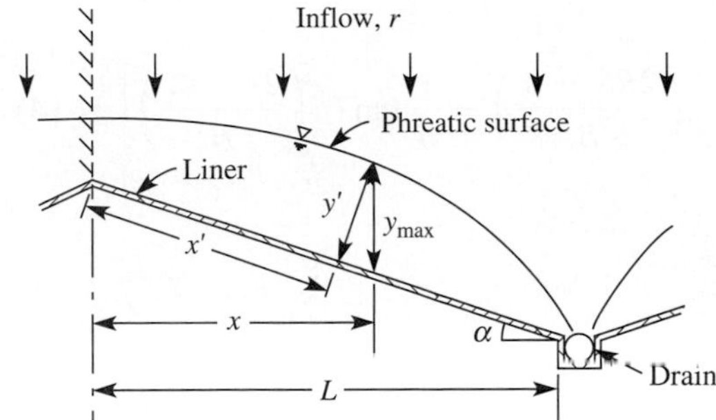

FIGURE 9-17
Geometry and symbols for calculating Y_{max}.
(*Source:* McEnroe, 1993.)

floor of the landfill to a grid of underdrain pipes* that are placed above the geomembrane. A 0.3-m-deep layer of granular material (for example, sand) with a high hydraulic conductivity (EPA recommends greater than 1×10^{-2} cm/s) is placed over the geomembrane to conduct the leachate to the underdrains. In addition to carrying the leachate, this layer also protects the geomembrane from mechanical damage from equipment and solid waste. In some instances a geonet (a synthetic matrix that resembles a miniature chain link fence), with a geofabric (an open-weave cloth) protective layer to keep out the sand, is placed under the sand and above the geomembrane to increase the flow of leachate to the pipe system.

Several different methods for estimating the steady-state maximum leachate depth have been proposed. EPA has proposed the following formula (refer to Figure 9-17 for an explanation of the notation) (U.S. EPA, 1989):

$$y_{max} = L\left(\frac{r}{2K}\right)^{0.5}\left[\frac{KS^2}{r} + 1 - \frac{KS}{r}\left(S^2 + \frac{r}{K}\right)^{0.5}\right] \tag{9-11}$$

where y_{max} = maximum saturated depth, m
L = drainage distance, measured horizontal, m
r = vertical flow rate per unit horizontal area, $m^3/s \cdot m^2$
K = hydraulic conductivity of drainage layer, m/s
S = slope of liner (= tan α)

This formula may overestimate the value of y_{max} where the underdrain system has free drainage, that is, it is not undersized or clogged. Since this is commonly the case, McEnroe has proposed the following equations as a better approximation. (McEnroe, 1993):

$$Y_{max} = (R - RS + R^2S^2)^{0.5}\left[\frac{(1 - A - 2R)(1 + A - 2RS)}{(1 + A - 2R)(1 - A - 2RS)}\right]^{0.5A} \tag{9-12}$$

for $R < 1/4$;

$$Y_{max} = \frac{R(1 - 2RS)}{1 - 2R}\exp\left[\frac{2R(S - 1)}{(1 - 2RS)(1 - 2R)}\right] \tag{9-13}$$

for $R = 1/4$;

*Underdrain pipes are perforated pipes designed to collect the leachate.

and

$$Y_{\max} = (R - RS + R^2S^2)^{0.5} \exp\left[\frac{1}{B}\tan^{-1}\left(\frac{2RS - 1}{B}\right) - \frac{1}{B}\tan^{-1}\left(\frac{2R - 1}{B}\right)\right] \quad (9\text{-}14)$$

for $R > 1/4$;

where $Y_{\max} = y_{\max}/(L \tan \alpha)$
$R = r/(K \sin^2 \alpha)$
S = slope of liner (= $\tan \alpha$)
$A = (1 - 4R)^{0.5}$
$B = (4R - 1)^{0.5}$

The collected leachate must be treated because of the high concentration of pollutants it contains. In some instances on-site treatment is provided. This frequently is a biological treatment system. In other cases, the leachate may be pumped to a municipal treatment plant. In some recent designs, the leachate is recirculated through the landfilled waste. This provides moisture for the microbial population and accelerates the stabilization process. It also promotes the production of methane and provides some treatment for the biodegradable fraction of the constituents in the leachate.

Final Cover. The major function of the final cover is to prevent moisture from entering the finished landfill. If no moisture enters, then at some point in time the leachate production will reach minimal proportions and the chance of groundwater contamination will be minimized.

Modern final cover design consists of a surface layer, biotic barrier, drainage layer, hydraulic barrier, foundation layer, and gas control. The surface layer is to provide suitable soil for plants to grow. This minimizes erosion. A soil depth of about 0.3 m is appropriate for grass. The biotic barrier is to prevent the roots of the plants from penetrating the hydraulic barrier. At this time, there does not seem to be a suitable material for this barrier. The drainage layer serves the same function here as in the leachate collection system—that is, it provides an easy flow path to a grid of perforated pipes. This collection piping system is subject to differential settling and may fail because of this settling. Some designers do not recommend installing it as they prefer to use the funds to develop a thicker hydraulic barrier. The hydraulic barrier serves the same function as the liner in that it prevents movement of water into the landfill. The EPA recommends a composite liner consisting of a geomembrane and a low hydraulic conductivity soil that also serves as the foundation for the geomembrane. This soil also protects the geomembrane from the rough aggregate in the gas control layer. The gas control layer is constructed of coarse gravel that acts as a vent to carry the gases to the surface. If the gas is to be collected for its energy value, a series of gas recovery wells is installed. A negative pressure is placed on these wells to draw the gas into the system.

Completed MSW Landfills

Completed landfills generally require maintenance because of uneven settling. Maintenance consists primarily of regrading the surface to maintain good drainage and filling in small depressions to prevent ponding and possible subsequent groundwater pollution. The final soil cover should be about 0.6 m deep.

Completed landfills have been used for recreational purposes such as parks, playgrounds, or golf courses. Parking and storage areas or botanical gardens are other final uses. Because of the characteristic uneven settling and gas evolution from landfills, construction of buildings on completed landfills should be avoided.

On occasion, one-story buildings and runways for light aircraft might be constructed. In such cases, it is important to avoid concentrated foundation loading, which can result in uneven settling and cracking of the structure. The designer must provide the means for the gas to dissipate into the atmosphere and not into the structure.

9-5 WASTE TO ENERGY

Utilization of the organic fraction of solid waste for fuel, while simultaneously reducing the volume, may be an important part of an integrated waste management plan. Specially designed power plants known as waste-to-energy facilities can produce energy through the combustion of municipal solid waste. In these facilities, trash volume is reduced by 90 percent and its weight by 75 percent. The remaining residue is disposed of in a MSW landfill. According to a 2004 Integrated Waste Services Association publication, 89 waste-to-energy facilities were in operation as of that time, disposing of 86 Gg of waste each day (IWSA, 2004). This waste was converted to approximately 2,500 megawatts of electric power.

Heating Value of Waste

The heating value of waste is measured in kilojoules per kilogram (kJ/kg), and is determined experimentally using a bomb calorimeter. A dry sample is placed in a chamber and burned. The heat released at a constant temperature of 25°C is calculated from a heat balance. Because the combustion chamber is maintained at 25°C, combustion water produced in the oxidation reaction remains in the liquid state. This condition produces the maximum heat release and is defined as the *higher heating value* (HHV).

In actual combustion processes, the temperature of the combustion gas remains above 100°C until the gas is discharged into the atmosphere. Consequently, the water from actual combustion processes is always in the vapor state. The heating value for actual combustion is termed the *lower heating value* (LHV). The following equation gives the relationship between HHV and LHV:

$$\text{LHV} = \text{HHV} - [(\Delta H_v)(9\text{ H})] \tag{9-15}$$

where ΔH_v = heat of vaporization of water
= 2,420 kJ/kg
H = hydrogen content of combusted material

The factor of 9 results because one gram mole of hydrogen will produce 9 gram moles of water (that is, 18/2). Note that this water is only that resulting from the combustion reaction. If the waste is wet, the free water must also be evaporated. The energy required to evaporate this water may be substantial. This results in a very inefficient combustion process from the point of view of energy recovery. The ash content also reduces the energy yield because it reduces the proportion of dry organic matter per kilogram of fuel and because it retains some heat when it is removed from the furnace.

Fundamentals of Combustion

Combustion is a chemical reaction where the elements in the fuel are oxidized. In waste-to-energy (WTE) plants, the fuel is, of course, the solid waste. The major oxidizable elements in the fuel are carbon and hydrogen. To a lesser extent sulfur and nitrogen are also present. With complete oxidation, carbon is oxidized to carbon dioxide, hydrogen to water, and sulfur to sulfur dioxide. Some fraction of the nitrogen may be oxidized to nitrogen oxides.

The combustion reactions are a function of oxygen, time, temperature, and turbulence (O, T, T, T). There must be a sufficient excess of oxygen to drive the reaction to completion in a short period of time. The oxygen is most frequently supplied by forcing air into the combustion chamber. Over 100 percent excess air may be provided to ensure a sufficient excess. Sufficient time must be provided for the combustion reactions to proceed. The amount of time is a function of the combustion temperature and the turbulence in the combustion chamber. Some minimum temperature must be exceeded to initiate the combustion reaction (that is, to ignite the waste). Higher temperatures also yield higher quantities of nitrogen oxide emissions, so there is a tradeoff in destroying the solid waste and forming air pollutants. Mixing of the combustion air and the combustion gases is essential for completion of the reaction.

As the solid waste enters the combustion chamber and its temperature increases, volatile materials are driven off as gases. Rising temperatures cause the organic components to thermally "crack" and form gases. When the volatile compounds are driven off, fixed carbon remains. When the temperature reaches the ignition temperature of carbon (700°C), it is ignited. To achieve destruction of all the combustible material (*burnout*), it is necessary to achieve 700°C throughout the bed of waste and ash (Pfeffer, 1992).

The flame zone is that area where the hot volatilized gases mix with oxygen. This reaction is very rapid. It goes to completion within 1 or 2 seconds if there is sufficient excess air and turbulence.

The evolution of solid waste combustion has led to higher temperatures both to destroy toxic compounds and to increase the opportunity to utilize the waste as an energy source by producing steam.

Conventional Incineration

The basic arrangement of the conventional incinerator is shown in Figure 9-18. Although the solid waste may have some heat value, it is normally quite wet and is not *autogenous* (self-sustaining in combustion) until it is dried. Conventionally, auxiliary fuel is provided for the initial drying stages. Because of the large amount of particulate matter generated in the combustion process, some form of air pollution control device is required. Normally, electrostatic precipitators or scrubbers are chosen. Bulk volume reduction in incinerators is about 90 percent. Thus, about 10 percent of the material still must be carried to a landfill.

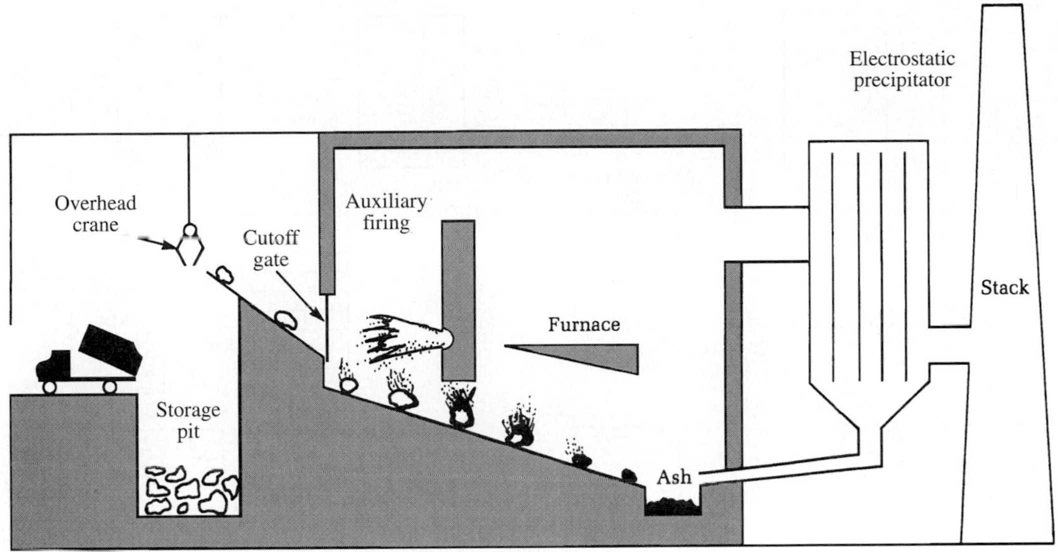

FIGURE 9-18
Schematic of a conventional traveling grate incinerator.

Recovering Energy from Waste

In order to utilize the heat value of solid waste, most modern combustion devices are designed to recover the energy. The concept is more than 100 years old. The first refuse-to-electricity system was built in Hamburg, Germany, in 1896. In 1903, the first of several solid waste-fired electricity generating plants in the United States was installed in New York City.

There are now many WTE plants operating in the United States. They burn solid waste in a specially designed incinerator furnace jacketed with water-filled tubes to recover the heat as steam. The steam may be used directly for heating or to produce electricity.

Many states require public utilities to buy the electricity produced at these plants. With efficient heat recovery and electric generators, WTE plants can produce about 600 kWh per mg of waste.

Refuse-Derived Fuel (RDF). Refuse-derived fuel is the combustible portion of solid waste that has been separated from the noncombustible portion through processes such as shredding, screening, and air classifying (Vence and Powers, 1980). By processing municipal solid waste (MSW), refuse-derived fuel containing 12 to 16 MJ/kg can be produced from between 55 and 85 percent of the refuse received. This system is also called a supplemental fuel system because the combustible fraction is typically marketed as a fuel to outside users (utilities or industries) as a supplement to coal or other solid fuels in their existing boilers.

In a typical system, MSW is fed into a trommel or rotating screen to remove glass and dirt, and the remaining fraction is conveyed to a shredder for size reduction. Shredded wastes may then pass through an air classifier to separate the "light fraction" (plastics, paper, wood, textiles, food wastes, and smaller amounts of light metals) from the "heavy fraction" (metals, aluminum, and small amounts of glass and ceramics).

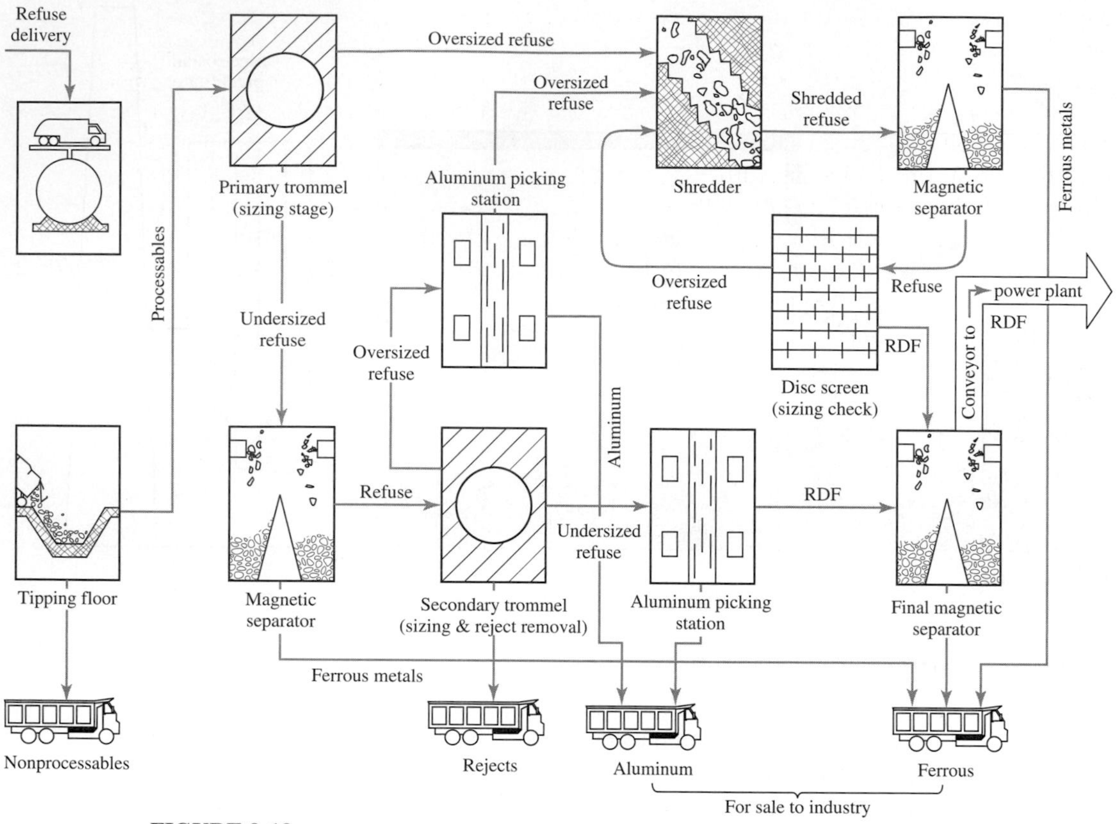

FIGURE 9-19
Southeastern Virginia Public Service Authority's refuse-derived fuel (RDF) plant.

The light fraction, after being routed through a magnetic system to remove ferrous metals, is ready for fuel use. The heavy fraction is conveyed to another magnetic removal system for recovery of ferrous metals. Aluminum may also be recovered. The remaining glass, ceramics, and other nonmagnetic materials from the heavy fraction are then sent to the landfill.

The first full-scale plant to prepare RDF has been in operation in Ames, Iowa, since 1975. Subsequently, other plants using similar technology have been designed and constructed. Figure 9-19 shows the process flow diagram for the Southeastern Virginia Public Service Authority's RDF plant.

Although there are a number of RDF production systems operating or starting up, they are still developmental in terms of process, equipment, and application. Data still are being gathered for prediction of performance and maintenance requirements.

Modular Incinerators. These units are available in various sizes. Their modularity enables them to be coupled with similar units to process available tonnage.

Most modular incinerators that produce energy incorporate a controlled air principle, use unprocessed MSW, and require a small amount of auxiliary fuel for startup. The waste is fed into a primary chamber where it is burned in the absence of sufficient oxygen for complete combustion. The resulting combustible gas passes through a second chamber, where excess air is injected, completing combustion. Auxiliary fuel may also be required in minimal quantities to maintain proper combustion temperatures.

After most of the particulate matter burns off, the hot effluent passes through a waste heat boiler to produce steam. The ash is water-quenched and disposed of at a landfill. The steam can be used directly or can be converted to electricity with the addition of a turbine generator.

The newer waste-to energy plants are not without their problems. Serious concern has been raised about emissions of dioxins that result from the combustion process. Two approaches are used to reduce the dioxin emission. Because the dioxin is formed as a combustion by-product from chlorinated plastics, it can be minimized by reducing the plastic in the feed stream. The second approach is to utilize sophisticated air pollution control equipment.

A second problem is associated with the ash from the combustion process. There are two categories of ash generated: fly ash from the air pollution control equipment and bottom ash from the furnace. Fly ash is of greater concern because the metals are adsorbed on particulates and are easily leached with water. When fly ash is mixed with bottom ash, the leachability of the metals is reduced. In 1994, the Supreme Court ruled that ash from municipal incinerators is not excluded from being considered as a hazardous waste (Chicago vs. EDF, 1994). It must be tested before it can be landfilled and must be treated if it fails the tests.

9-6 RESOURCE CONSERVATION AND RECOVERY FOR SUSTAINABILITY

Background and Perspective

The earth's prime mineral deposits are limited. As high-quality ores are depleted, lower-grade ores must be used. Lower-grade ores require proportionately greater amounts of energy and capital investment to extract. In a broad economic context, we should view with concern the long-term reasonableness of a market-accounting system that applies only current development costs to our use of depletable, nonrenewable natural resources such as aluminum, copper, iron, and petroleum. High rates of solid waste production imply high rates of virgin raw material extraction. In the United States, blatant mispricing—including the "depletion allowance" on minerals and unreasonably low rail rate fares on ores in contrast to scrap—is in no small way responsible for this state of affairs. Furthermore, our high-waste, low-recycle lifestyle is inherently wasteful of a bountiful endowment of natural resources.

Our renewable resources, primarily timber, are also under siege. Our pre-packaged society, in combination with a wanton lack of care in our forests, has strained nature's capacity for growth and replenishment. Europe, India, and Japan have long been faced with a want of timber. We in the United States should learn from their predicaments.

The prevention of waste generation (resource conservation) and the productive use of waste material (resource recovery) represent means of alleviating some of the problems of solid waste management. At one time in our history, resource recovery played an important role in our industrial production. Until the mid-twentieth century, salvage (recovery and recycling) from household wastes was an important source of

TABLE 9-12
Generation, materials recovery, composting, and discards of municipal solid waste, 1960–2003[a, b]

	1960	1970	1980	1990	2000	2003
Generation	80.1	110.1	137.8	186.6	212.8	214.8
Recovery for recycling	5.1	7.3	13.2	26.4	47.6	50.4
Recovery for composting[c]				3.8	15.0	15.4
Total Materials Recovery	**5.1**	**7.3**	**13.2**	**30.2**	**62.6**	**65.7**
Discards after recovery	75.0	102.7	124.7	156.4	150.1	149.0

[a]*Source:* U.S. EPA, 2005.
[b]In teragrams (Tg).
[c]Composting of yard trimmings, food scraps, and other MSW organic material. Does not include backyard composting. Details may not add because of rounding.

materials. In the five years preceding 1939, recycled copper, lead, aluminum, and paper supplied 44, 39, 28, and 30 percent, respectively, of the total raw materials shipments to fabricators in the United States (NCRR, 1974). Ultimately, it became more economical to process virgin materials than to use recovered materials.

In principle, processable municipal solid waste could provide 95 percent and 73 percent of our nation's needs in glass and paper, respectively. EPA estimates that overall, 30 percent of municipal solid waste was recovered in 2003. This represents an increasing trend. Table 9-12 shows the trend in recycling and reuse from 1960 to 2003 in millions of tons of waste. In 2003, 65.7 Tg million tons of waste were diverted from landfills by recycling and composting.

Table 9-13 shows a breakdown of recovered waste by 1=1 product in 2003. EPA estimates that during 2003, nearly 39 percent of containers and packaging were recycled. About 44 percent of aluminum beverage cans were recycled, as well as 48 percent of paper and paperboard, 22 percent of glass containers, and 8 percent of plastic packaging and containers. Newspapers, the most recycled product, were recycled at a rate of about 82 percent, while used telephone books were recycled at a rate of only 16 percent.

Recycling of municipal solid waste for profit or for energy recovery is rarely cost-effective. However, many communities have initiated recycling programs as a means of protecting the environment. Citizens have become increasingly aware of their role in protecting the natural environment, and so demand that communities offer recycling services. EPA has also set national goals to encourage active resource conservation and recovery programs.

Most states and the District of Columbia have enacted laws on recycling ranging from purchasing preferences to comprehensive recycling goals. Over 8,000 curbside recycling programs, 3,000 composting programs, and 200 municipal recycling facilities are in operation (Wolpin, 1994, and U.S. EPA, 2003). The recyclable market continues to fluctuate dramatically. For example, the price of old newsprint fell from \$50/Mg in 1988 to less than \$10/Mg in 1993 (Rogoff and Williams, 1995). It rose to over \$100/Mg in 1995 (Paul, 1995).

The remainder of our discussion will be devoted to the technical details of several of the more promising resource conservation and recovery (RC & R) techniques. We

TABLE 9-13
Generation and recovery of products in MSW by material 2003[a,b]

	Mass generated[c]	Mass recovered[c]	Recovery as a percent of generation
Durable goods			
Steel	10.16	3.06	30.2
Aluminum	0.96	Neg.[f]	Neg.
Other nonferrous metals[d]	1.44	0.96	66.7
Total metals	12.52	4.02	32.1
Glass	1.61	Neg.	Neg.
Plastics	7.61	0.30	3.9
Rubber and leather	5.36	1.00	18.6
Wood	4.78	Neg.	Neg.
Textiles	2.75	0.29	10.6
Other materials	1.18	0.89	75.4
Total durable goods	35.83	6.50	18.1
Nondurable goods			
Paper and paperboard	40.19	16.42	40.8
Plastics	5.76	Neg.	Neg.
Rubber and leather	0.80	Neg.	Neg.
Textiles	6.69	1.09	16.3
Other materials	2.96	Neg.	Neg.
Total nondurable goods	56.34	17.51	31.0
Containers and packaging			
Steel	2.58	1.56	60.6
Aluminum	1.76	0.63	35.6
Total metals	4.34	2.19	50.4
Glass	9.71	2.13	22.0
Paper and paperboard	35.20	19.87	56.4
Plastics	10.80	0.96	8.9
Wood	7.58	1.16	15.3
Other materials	0.20	Neg.	Neg.
Total containers and packaging	67.86	26.31	38.8
Other wastes			
Food, other[e]	25.04	0.68	2.7
Yard trimmings	25.95	14.61	56.3
Miscellaneous inorganic wastes	3.28	Neg.	Neg.
Total other wastes	54.25	15.33	28.2
Total MSW	214.28	65.59	30.6

[a]Source: U.S. EPA, 2003
[b]Includes waste from residential, commercial, and institutional sources.
[c]In teragrams (Tg).
[d]Includes lead from lead-acid batteries.
[e]Includes recovery of other MSW organic material for composting.
[f]Neg. = negligible.

have divided these into three broad categories entitled low technology, medium technology, and high technology. These categories refer to increasing degrees of sophistication in terms of implementation, equipment, and capital investment. No municipal government should be enticed into any one of these schemes with the hope of making money. The best that can be hoped for is defraying the additional costs over conventional landfilling and extending the life of the landfill by some modest amount. In some cases, even these modest goals may not be achieved.

Low Technology RC & R

Returnable Beverage Containers. The substitution of reusable products for single-use "disposable" products is a workable means of conserving natural resources. Legislation requiring mandatory refunds and/or deposits on both returnable and nonreturnable beverage containers has been and will continue to be hotly contested by the beverage and beverage container industries. States that have enacted mandatory refund and/or deposit legislation include California, Connecticut, Delaware, Hawaii, Maine, Massachusetts, Michigan, New York, Oregon, and Vermont as of 2002. The programs are successful in encouraging recycling of containers. Between 90 and 95 percent of the bottles are returned and between 80 and 85 percent of the cans are returned. In Oregon, a reduction in total roadside litter of 39 percent by item count and 47 percent by volume was reported after the second year of implementation of its law. Furthermore, for glass containers there is a significant energy savings in that a glass bottle reused 10 times consumes less than one-third of the energy of a single-use container. Average reuse cycles vary from 10 to 20 times per container.

Recycling. The reprocessing of wastes to recover an original raw material was formerly called *salvage* and is now called *recycling*. At its lowest and most appropriate technological level, the materials are separated at the source by the consumer (*source-separation*). This is the most appropriate level because it requires the minimum expenditure of energy. With stringent goals for recycling, municipalities are looking at detailed recycling options.

Generally, the recycling options available to a municipality for residential use include:

Curbside collection

Drop-off centers

Material processing facility

Material transfer stations

Leaf/yard waste compost

Bulky waste collection and processing

Tire recovery

The primary method of recycling in the United States today is curbside collection. This has the advantage of being easier on the resident than having to drive to a recycling center. There are two basic types of curbside collection for recycling. In the first, the homeowner is given a number of bins or bags. The homeowner separates the refuse as it is used, placing it in the appropriate bin. On collection day the container is placed on the curb. The primary disadvantage of supplying home storage containers is the cost, which can represent a significant investment. A second method of curbside recycling is to provide the homeowner with only one bin, into which is placed all the recyclable materials. Curbside personnel then separate material as it is being picked up, placing each type of material into a separate compartment in the vehicle.

A second alternative is a drop-off center. Because recycling is a community-specific operation, a drop-off system must be designed around and in consideration of conditions particular to the area of involvement. To evaluate and select the most appropriate drop-off system, we must consider critical factors such as location, materials handled, population, number of centers, operation, and public information. When drop-offs are used to supplement curbside programs, fewer and smaller drop-off sites may be required. When drop-off sites are the only, or primary, recycling system in a community, the system must provide for increased capacity. Careful planning to accommodate traffic flow, as well as storage and collection of materials, must be part of the siting activity.

The convenience of a drop-off center will directly affect the amount of citizen participation. Strategically locating a drop-off center in an area of high traffic flow, where the center is highly visible, will encourage a greater level of participation. Even rural areas with widely scattered populations provide good locations for drop-offs. Rural homeowners have certain common travel patterns that bring them to a few locations at regular intervals—to a grocery store, church, or post office, to name a few. Figure 9-20 shows an example of a drive-through material recycling center.

A third major type of recycling is a materials recovery facility. In this case the recyclable material is taken by the municipality to a central facility where the material is separated via mechanical and labor-intensive means. Figure 9-21 shows an example layout of a separation facility and Figure 9-22 shows a mass balance of what can be expected at such a facility.

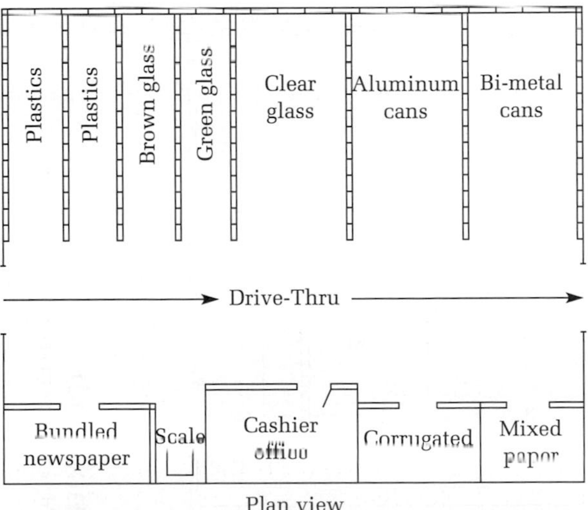

FIGURE 9-20
Enclosed drive-through drop-off center.

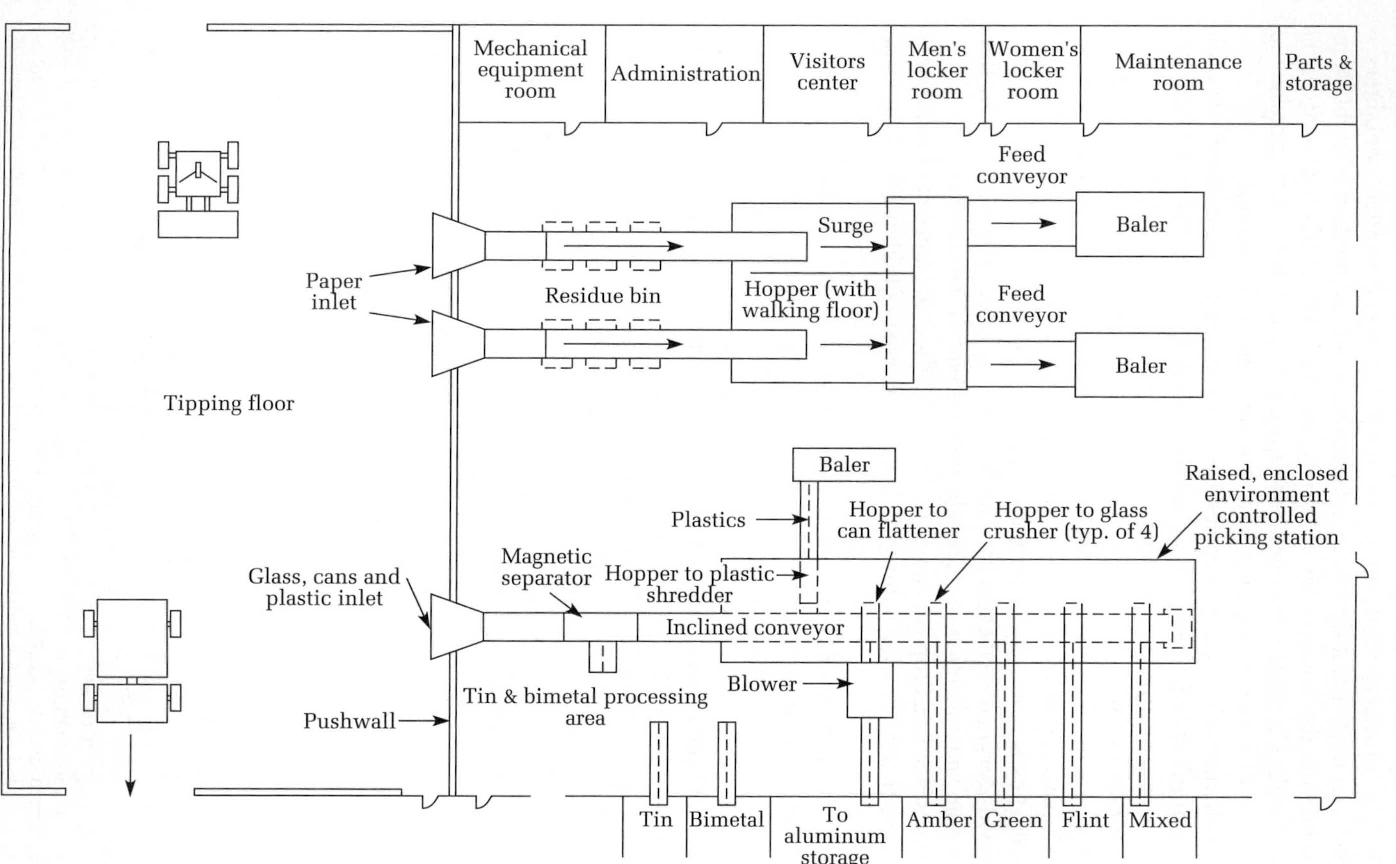

FIGURE 9-21
Material processing conceptual floor plan.

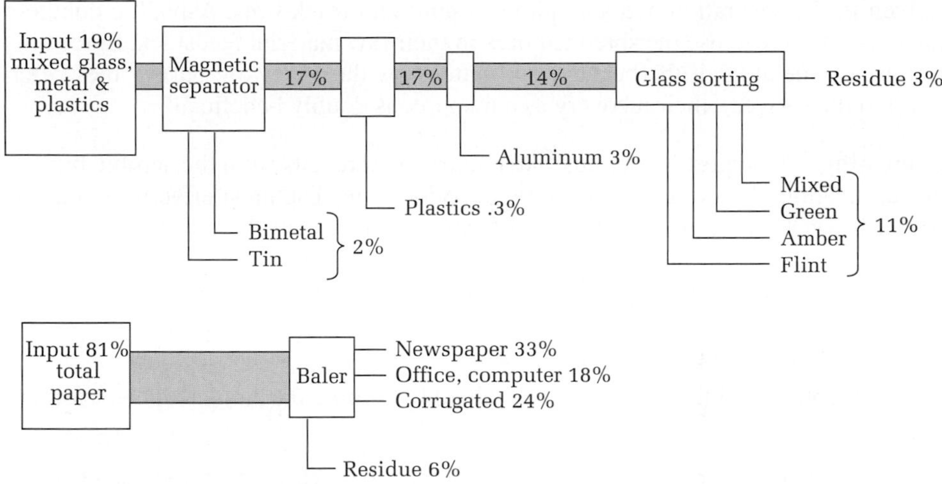

FIGURE 9-22
Material recovery facility process mass flow.

Medium Technology RC & R

Product Design. Simple changes in product configuration or packaging can result in conservation of resources. Three examples will suffice to illustrate the concept. In the mid-1970s several newspapers (for example, *Los Angeles Times, Washington Post,* and *New York Times*) switched from a traditional eight-column format to a new six-column format for news and nine-column format for advertising. This shift resulted in a 5 percent reduction in the amount of newsprint consumed. A large retail grocery store found that it could eliminate the custom of double bagging groceries by using a slightly heavier-weight bag with a reinforced bottom. This resulted in a 30 percent savings in the amount of fiber consumed. Many fast food snacks eliminated styrofoam containers for their sandwiches and now use paper wrapping, which is more readily biodegraded.

These kinds of changes are generally beyond the scope of the environmental engineer. However, their use can be encouraged, and purchases can be made that support those who use environmentally conservative packages and products.

Shredding and Separation. As a first step in a medium technology system or as an add-on to a landfill volume enhancement program, some materials may be reclaimed at a central processing point. The most likely candidates for recycling are paper, nonferrous metals (for example, aluminum), and ferrous metals. Paper generally is removed by hand as the MSW passes along on a conveyor belt.* After passing through a shredder, ferrous metals can be removed using a magnetic separator. In large communities, where more than 1,000 Mg/wk of MSW is collected, some consideration may

*Depending upon the economy hand sorting may be a losing proposition. An average worker can pick about 2.0 Mg of newspaper in an eight-hour day. At a wage of \$5.50/h, a day's wages amount to \$44.00, exclusive of overhead and fringe benefits. Using an overhead rate of 100 percent, the cost of sorting is \$44.00/Mg. If the price for No. 6 newsprint (a grade of paper) is \$22/Mg as it was in 1994, this is a loss of \$22/Mg before transportation costs are deducted. Of course, in 1995, when the price was \$116/Mg, it was a winning proposition.

be given to the separation and shredding of auto and truck tires. Asphaltic concrete plants may be able to use the shredded tires in their raw material feedstock. Since tires are troublesome at landfills (because no matter how deep they are buried, they often pop up to the surface), their recovery as a resource is doubly beneficial.

Composting. Compost is a humus-like material that results from the aerobic biological stabilization of the organic materials in solid waste. The most effective composting occurs when the waste stream is free of inorganic materials. Frequently, this makes source-separated yard waste ideal. For the biological process to be effective, the following conditions must be met (Tchobanoglous et al., 1993).

1. Particle size must be small (< 5 cm).
2. Aerobic conditions must be maintained by turning the compost pile or forcing air through it.
3. Adequate, but not excessive, moisture must be present (50 to 60 percent).
4. An adequate population of acclimated microorganisms must be present.
5. The carbon-to-nitrogen ratio must be in the range of 20–25 to 1.

The biodegradation process is exothermic and a well-operating compost will have a temperature between 55 and 60°C during the period of active degradation. These temperatures are effective in destroying pathogens. The processing cycle for composting is about 20 to 25 days with active degradation taking place over a 10- to 15-day period. One of the major drawbacks of composting is odors. Maintenance of aerobic conditions and a proper cure time minimize odor problems.

Compost is useful as a soil conditioner. In this role compost will: (1) improve soil structure, (2) increase moisture-holding capacity, (3) reduce leaching of soluble nitrogen, and (4) increase the buffer capacity of the soil. It should be emphasized that compost is not a valuable fertilizer. It contains only 1 percent or less of the major nutrients, such as nitrogen, phosphorus, and potash.

Composting is one of the fastest-growing aspects of ISWM. The driving force is legislation enacted to extend the life of landfills by removing yard waste from the waste stream. According to the EPA, recovery by composting was negligible in 1988. By 1990, EPA estimated that 2 percent of the nation's solid waste was being composted. The 2000 estimate was that 7 percent of the solid waste was being composted. In 1994, over 3,000 composting facilities were operating in the United States. Sludge composting facilities numbered over 180, and municipal solid waste composting was being practiced by 21 cities (Monk, 1994).

Methane Recovery. Methane is produced in sanitary landfills as a result of anaerobic decomposition of the organic fraction of the waste. In addition to gas extraction wells and a collection system, some gas processing equipment is employed. The minimum processing consists of dehydration, gas cooling, and, perhaps, removal of heavy hydrocarbons. The gas produced is a low-Joule gas having heating value of 18.6 MJ/m^3. In high-Joule processing systems, carbon dioxide and some hydrocarbons are removed to yield essentially pure methane. The resulting gas is of pipeline quality and has a heating value of approximately 37.3 MJ/m^3. The anticipated quantity of landfill gas (LFG) varies between 0.6 and 8.7 liters per kilogram of solid waste present per year (L/kg · y). The average production rate is 5 L/kg · y.

Although landfill sites as small as 11 ha have yielded substantial quantities of recoverable methane, the capital investment and complexity of the gas processing equipment will limit this technique to the larger sites (>65 ha). Otherwise, the technology is readily available and can make use of a resource that otherwise would dissipate into the atmosphere. According to EPA data, in 1999, 360 LFG-recovery projects nationwide produced the equivalent of 1,200 MW of power (Skinner, 1999).

High Technology RC & R

In the mid-1970s, under the auspices of the U.S. Environmental Protection Agency and with federal financing, several innovative high technologies for resource recovery were examined. At the end of the decade, a few workable systems and a large number of unworkable systems were identified.

Since the successful high technology systems depend, to a large measure, on the recovery of energy for their success, we will consider the worth of solid waste as a fuel. As illustrated in Table 9-14, MSW is not a very good fuel. On the other hand, its cost of \$0.00/Mg may seem quite attractive. This is especially so when the price of anthracite coal may be \$50/Mg and the price of No. 2 fuel oil is \$250/Mg. Unfortunately, solid waste, as

TABLE 9-14
Net heating value of various materials

Material	Net heating value (MJ/kg)
Charcoal	26.3
Coal, anthracite	25.8
Coal, bituminous (hi volatile B)	28.5
Fuel oil, no. 2 (home heating)	45.5
Fuel oil, no. 6 (bunker C)	42.5
Garbage	4.2
Gasoline (regular, 84 octane)	48.1
Methane[a]	55.5
Municipal solid waste (MSW)	10.5
Natural gas[a]	53.0
Newsprint	18.6
Refuse derived fuel (RDF)	18.3
Rubber	25.6
Sewage gas[a]	21.3 to 26.6
Sewage sludge (dry solids)	23.3
Trash	19.8
Wood, oak	13.3 to 19.3
Wood, pine	14.9 to 22.3

[a]Densities taken as follows (all in kg/m^3): CH_4 = 0.680; natural gas = 0.756; sewage gas = 1.05.

a fuel, has a hidden cost. Unless the physical characteristics are upgraded by removing metals and glass and by reducing the particle size, MSW cannot be burned in conventional coal-fired power plants. The alternative is the construction of a special power plant that can handle the MSW as it is received. In either case, some cost is imposed.

It appears that if a high technology resource recovery facility is to be successful, it must meet the following criteria (Serper, 1980):

1. High technology resource recovery can only be economical in large metropolitan areas where landfill sites are unavailable or are very expensive, above $25/Mg, or in geographic locations where the water table makes safe landfilling impossible, as, for example, the city of New Orleans and its surrounding suburbs.
2. There must be an adequate refuse supply committed to the facility (a minimum of 1.8 Gg/d is needed). In general, this implies a population of 250,000 or more.
3. A customer must be obtained for the steam or the power generated by the plant and must be located close by. Firm contracts must be obtained for both the refuse supply and the sale of energy.
4. If the customer is totally dependent on the energy supplied by the facility, the combustion facility must be designed with the capacity to burn fossil fuel when refuse is unavailable or when the plant cannot process the raw refuse due to malfunctions of the processing equipment.
5. The logistics of delivering refuse to the resource recovery facility should be planned long in advance. It may be necessary to establish transfer stations and storage locations that will operate in conjunction with the resource recovery plant.
6. Systems that can dispose of both municipal refuse and sewage sludge will have economic advantages over systems that dispose of refuse only. With the ban of ocean dumping now in effect, local sewage districts are being forced to spend astronomical amounts of money to incinerate sludge. A co-disposal plant should reduce both the refuse and sludge disposal costs. In order to be economically competitive, sewage sludge must be dewatered to the maximum practical extent. A number of co-disposal plants are now in operation in Europe. Except for large installations, there will not be sufficient excess energy to warrant exporting it.

Many of the high technology systems have, as a common starting point, the medium technology materials recovery systems as their first process steps. These were discussed in a previous section.

9-7 CHAPTER REVIEW

When you have completed studying this chapter you should be able to do the following without the aid of your textbooks or notes:

1. State the average mass of solid waste produced per capita per day in the United States in 2003.

2. Differentiate between garbage, rubbish, refuse, and trash, based on their composition and source.
3. Compare the advantages and disadvantages of public and private solid waste collection systems.
4. List the three pickup methods (backyard, set-out/set-back, and curbside) and explain the advantages and disadvantages of each.
5. List the components of a time study for a waste collection system.
6. Compare the advantages and disadvantages of the four methods of collection truck routing.
7. Explain the four methods of integrating several crews.
8. Explain what a transfer station is and what purpose it serves.
9. List and discuss the factors pertinent to the selection of a landfill site.
10. Describe the two methods of constructing a MSW landfill.
11. Explain the purpose of daily cover in a MSW landfill and state the minimum desirable depth of daily cover.
12. Define leachate and explain why it occurs.
13. Sketch a MSW landfill that includes proper cover and a leachate collection system.
14. Define or explain the following terms: WTE, autogenous, HHV, LHV, RDF, source-separation.
15. Explain the relationship between oxygen, time, temperature, and turbulence in establishing efficient combustion reactions.
16. Explain the effect of source-separation on the heating value of solid waste and on the potential for hazardous air pollution emissions.
17. List two highly feasible methods of resource conservation and/or recovery in low technology and medium technology RC & R.
18. Describe and explain, in a basic manner, each of the two methods listed in number 17 above such that the average citizen could understand the method.

With the aid of this text you should be able to do the following:

1. Determine the volume and mass of solid waste from various establishments.
2. Determine the required volume capacity of a solid waste collection truck, or conversely, determine the number of stops possible for a given truck volume, or the allowable mean time per collection.
3. Estimate the annual truck and labor cost for solid waste collection and the cost per service stop.

4. Lay out a truck route using the heuristic routing technique.
5. Determine the necessity and/or advisability of constructing a transfer station.
6. Estimate the volume and area requirements for a landfill.
7. Compute the LHV given the HHV and the chemical formula for a compound to be burned.

9-8 PROBLEMS

9-1. The student population of Metuchen High School is 881. The school has 30 standard classrooms. Assuming a 5-day school week with solid waste pickups on Wednesday and Friday before school starts in the morning, determine the size of storage container (dumpster) required. Assume waste is generated at a rate of 0.11 kg/cap · d plus 3.6 kg per room and that the density of uncompacted solid waste is 120.0 kg/m^3. Standard container sizes are as follows (all in m^3): 1.5, 2.3, 3.0, and 4.6.

Answer: Select one 1.5-m^3 and one 4.6-m^3 container.

9-2. The Bailey Stone Works employs six people. Assuming that the density of uncompacted waste is 480 kg/m^3, determine the annual volume of solid waste produced by the stone works assuming a waste generation rate of 1 kg/cap · d.

9-3. As the supply of high-grade ores is used up, lower grade ores are used to produce minerals. Assuming that you are producing 100 kg of metal, use the mass balance method to calculate the kilograms of waste rock per kilogram of metal for ore containing 50, 25, 10, 5, and 2.5 percent metal.

9-4. Professor Green has made measurements of her household solid waste, shown in the table below. If the container volume is 0.0757 m^3, what is the average density of the solid waste produced in her household? Assume that the mass of each empty container is 3.63 kg.

Date	Can no.	Gross mass[a] (kg)
March 18	1	7.26
	2	7.72
March 25	1	10.89
	2	7.26
	3	8.17
April 8	1	6.35
	2	8.17
	3	8.62

[a]Container plus solid waste.

Answer: Average density = 58.4 kg/m^3

9-5. The collection vehicle compacts the household solid waste in Problem 9-4 to 37 percent of its original volume. Estimate the density of the compacted waste in kg/m^3.

9-6. The typical composition of solid waste from Davis, California, is shown in Table 9-4. Calculate the density of this waste in kg/m^3 if the paper, cardboard, plastic, glass, and tin cans are removed.

9-7. Early Collection Systems is considering bidding on a solid waste management contract to collect all of the residential solid waste generated by Midden (population 44,000). The average solid waste generation rate is 1.17 kg/cap · d and the average uncompacted density is 144.7 kg/m^3. The request for bids specifies that each residence must have a minimum of two pickups per week (maximum of 4 days between pickups) and that there will be no rear-of-house pickups. Using the following assumptions, determine how many trucks of what size Early Collection Systems should plan on using. Assumptions for Midden:

Average residential occupancy = 4/residence

Average number of cans per stop = 3/wk at 0.0757 m^3/can

Side loader compactor truck with a crew of one

Truck compactor density rating = 475 kg/m^3

Truck dump time = 7.50 min

Delay time = 20.0 min

Distance to disposal site = 24.0 km

Number of trips to disposal site = 2/d

Time between pickup stops = 18.00 s

Dump time (regression coefficient a) = 12.60 s/can

Standard side-loading compactor truck capacities (all in m^3): 9.0, 12.0, 15.0, 18.0, 19.0, 21, and 27

Answer: Should have 12 trucks of 9.0 m^3 capacity.

9-8. The City of Forty Two (population 361,564) has requested your assistance in evaluating its solid waste collection system. Determine the mean time per collection stop plus the mean time to reach the next stop, the number of pickup locations per load, and the minimum number of trucks the city must own. Forty Two collection data:

Average truck capacity = 18.0 m^3

Average observed compaction ratio = 3.97

Crew size = 2

Number of pickups = 1 /wk (no rear-of-house service)

Average number of cans per stop = 2.53/wk at 0.1136 m^3/can

Average number of residents per stop = 4

Average uncompacted density = 100.76 kg/m^3

Average transport time to disposal site including delays and dumping = 1.00 h/trip

Average number of trips to disposal site = 2/d

Rest breaks = 2 at 15.0 min

Average maintenance downtime = 24.0 min/d

Average work day = 8.00 h

Average percent of trucks out of service for major repairs = 15.0%

9-9. The City of Bon Chance (population 161,565) has requested your assistance in evaluating its solid waste collection system. Determine the mean time per collection stop plus the mean time to reach the next stop, the number of pickup locations per load, and the minimum number of trucks the city must own. Bon Chance collection data:

Average truck capacity = 18.0 m^3

Average observed compaction ratio = 3.28

Crew size = 2

Number of pickups = 1 /wk (no rear-of-house service)

Average number of cans per stop = 2.95/wk at 0.0911 m^3/can

Average number of residents per stop = 2.5

Average uncompacted density = 122.0 kg/m^3

Average transport time to disposal site including delays and dumping = 1.50 h/trip

Average number of trips to disposal site = 2/d

Rest breaks = 2 at 15.0 min

Average maintenance downtime = 36.0 min/d

Average work day = 8.00 h

Average percent of trucks out of service for major repairs = 15.0%

9-10. Rework Example 9-3 assuming no rear-of-yard pickup and only one trip per day to the disposal site.

9-11. Rework Problem 9-8 using a time between pickup stops of 28.20 s and a dump time (regression coefficient a) of 12.80 s/can for a side-loading truck and a crew of one. Assume that the truck size remains the same but the number of trips to the disposal site is reduced to one per day.

9-12. Mr. Midas, owner and manager of Early Collection Systems, would like to make a 20 percent profit (before taxes) on the Midden collection system work (Problem 9-7). Using the data provided by Mr. Midas, shown in the table below, determine the annual cost per megagram (Mg) and the average weekly charge to each household in order for Mr. Midas to make a 20 percent profit before taxes.

Labor costs for Midden

Employee title	Number	Wage rate, $/h
Route supervisor[a]	1	29.60
Secretary/bookkeeper[a]	1	16.20
Mechanic[b]	1	20.61
Driver/collector	12	17.74
General laborer[a]	2	7.40

[a]paid by overhead.
[b]Mechanic is included in 0 & M cost.

Average work week 40.0 h/wk, 5 d/wk

Overhead rate = 101.38% of total driver/collector wages

Truck data:
- Size = 9.0 m^3
- Capital cost = $117,000
- O&M cost = $6.46/km
- Anticipated life of truck = 5 y
- Interest rate = 8.75%
- Average annual distance for each truck = 16,412 km

9-13. Determine the annual cost per megagram and the average weekly charge per household for a system using a crew of two and for a system using a crew of one for the city of Nosleep (population 361, 564). Nosleep collection data:

Number of pickups = 1 /wk (no rear-of-house service)

Average number of cans per stop = 2.53/wk at 0.1136 m^3/can

Average number of residents per stop = 4

Mean time per collection stop plus mean time to reach next stop:
- For crew of one = 0.01180 h
- For crew of two = 0.00883 h

Average uncompacted density = 100.76 kg/m^3

Average observed compaction ratio = 3.97

Average transport time to disposal site including delays and dumping = 1.00 h/trip

Average number of trips to disposal site = 1/d

Rest breaks = 2 at 15.0 min

Average maintenance downtime = 24.0 min/d

Average work day = 8.00 h

Average percent of trucks out of service for major repairs = 15.0%

Labor costs for Nosleep

Employee title	Number	Wage rate, \$/h
Director[a]	1	48.95
Secretary[a]	1	13.44
Bookkeeper[a]	1	23.02
Route supervisors	4	33.50
Senior mechanic[b]	1	37.68
Mechanic[b]	3	25.11
Crew of two:		
Driver (1/truck)	*c*	15.25
Collector (1/truck)	*c*	14.70
Crew of one:		
Driver/collector	*c*	16.00
General laborers[a]	4	7.40

[a]Paid by overhead.

[b]Included in O & M cost.

[c]Dependent on number and type of trucks required.

Average work week = 40 h/wk, 5 d/wk

Overhead rate = 75.04% of total crew wages

Truck data:

Capital cost of 15.0-m^3 truck = \$122,000
O & M for 15.0-m^3 truck = \$5.75/km
Capital cost of 21.0-m^3 truck = \$141,000
O & M cost for 21.0-m^3 truck = \$6.55/km
Truck compactor density rating = 400 kg/m^3
Anticipated life = 5 y
Interest rate = 6.75%
Average annual distance for each truck = 11,797 km

9-14. The city manager of Bon Chance (population 161,565) has requested your services in analyzing three alternative city-managed schemes for collection of the city's solid waste. The three schemes are: (1) a system with crew-of-one trucks, (2) a system with crew-of-two trucks, and (3) a system with crew-of-three trucks. Using a spreadsheet program you have written, prepare an estimate of the annual cost per megagram (Mg) of each of these systems for the city manager. Bon Chance collection data:

Average observed compaction ratio = 3.28

Number of pickups = 1 /wk (no rear-of-house service)

Average number of cans per stop = 2.95/wk at 0.0911 m^3/can

Average number of people per stop = 2.5

Average uncompacted density = 122.0 kg/m^3

Mean time per colection stop plus mean time to reach next stop:

For crew of one = 0.88 min
For crew of two = 0.57 min
For crew of three = 0.37 min

Average transport time to disposal site including delays and dumping = 1.50 h/trip

Average number of trips to disposal site = 1/d

Rest breaks = 2 at 15.0 min

Average maintenance downtime = 36.0 min/d

Average work day = 8.00 h

Average percent of trucks out of service for major repairs = 15.0%

Bon Chance labor costs

Employee title	Number	Wage rate, $/h
Director[a]	1	48.95
Secretary[a]	1	13.44
Bookkeeper[a]	1	23.02
Route supervisors	4	33.50
Senior mechanic[b]	1	37.68
Mechanic[b]	3	25.11
Crew of one:		
Driver/collector	*c*	16.00
Crew of two:		
Driver (1/trusck)	*c*	15.25
Collector (1/truck)	*c*	14.70
Crew of three:		
Driver (1/truck)	*c*	15.25
Collector (2/truck)	*c*	14.70
General laborers[a]	4	7.40

[a]Paid by overhead.
[b]Included in O & M cost.
[c]Dependent on number and type of trucks required.

Average work week = 40 h/wk, 5 d/wk

Overhead rate = 75.04% of total crew wages for crew of one.

Truck data:

Capital cost of 15.0 m^3 truck = \$122,000
O & M for 15.0 m^3 truck = \$5.75/km
Capital cost of 18.0 m^3 truck = \$131,500
O & M cost for 18.0 m^3 truck = \$6.55/km
Capital cost of 21.0 m^3 truck = \$141,000
O & M cost for 21.0 m^3 truck = \$7.60/km
Truck compactor density rating = 400 kg/m^3
Anticipated life = 5 y
Interest rate = 6.75%
Average annual distance for each truck = 15,260 km

9-15. Using the rules for heuristic routing, plan a collection route for the section of Redbud shown in Figure P-9-15. Assume that all streets are two-way and that the pattern is bounded by two-way streets on all four sides. Also assume that collection is on one side of the street at a time.

Answer: The solution has no dead distance and two left turns. Both left turns occur at the intersection of Simons and Garson.

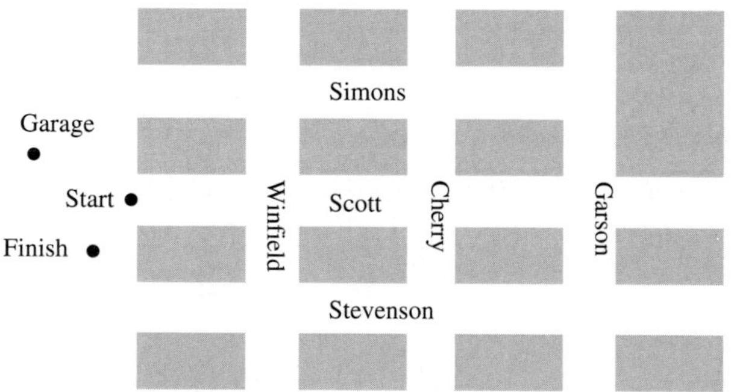

FIGURE P-9-15
Sketch map no. 1, Redbud.

9-16. Rework Problem 9-15 for the section of Mundy shown in Figure P-9-16. All of the streets are two-way and the pattern is bounded by two-way streets on all four sides. Collection is from one side of the street at a time.

9-17. Rework Problem 9-16 but assuming that West Zacks is one-way going north and that East Zacks is one-way going south.

Answer: The solution has three dead distances and 14 left turns. The dead distances are in the middle block of North and South Avenues. The left turns occur at the traffic signals.

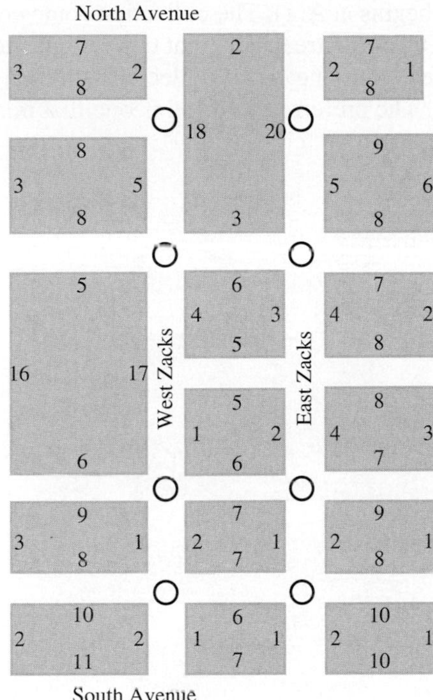

FIGURE P-9-16
Sketch map no. 2, Mundy. The numbers refer to the number of stops in a block. The circles denote traffic signals.

9-18. Using the rules for heuristic routing, plan a collection route for the section of Travail shown in Figure P-9-18. Assume collection is on one side of the street at a time and that all streets are two-way.

FIGURE P-9-18
Sketch map no. 3, Travail.

9-19. Divide the collection area shown in Figure P-9-19 into two approximately equal collection routes with starting points at A(1) and A(2). The difference in the number of stops for each route should not exceed 25. Lay out the collection route that begins at A(1). The collection route constraints are that there are to be no U turns in streets and that collection is to be made from each side of the street with one driver/collector using a right-hand-drive collection vehicles. The preferred solution is one that minimizes overlaps.

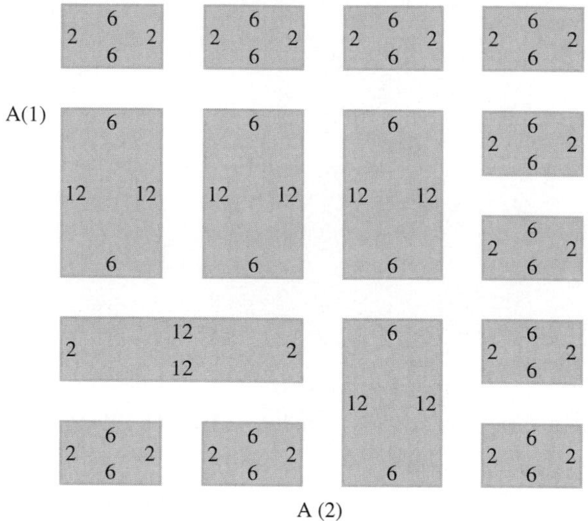

FIGURE P-9-19

Sketch map of Troublesome Creek collection area. 2, 6, 12 = number of residences along each block. (*Source:* Tchobanoglow, et al., 1993)

9-20. Divide the Olson collection area shown in Figure P-9-20 into two approximately equal collection routes with starting points at A(1) and A(2). The difference in the number of stops for each route should not exceed 25. Assuming that both sides of the street can be collected in one pass, lay out the collection route that begins at A(2). *HINTS:* $N_p \approx 500$, Huntington Rd. divides the two routes.

9-21. Repeat Problem 9-20 for the Masters collection area shown in Figure P-9-21 for the collection route that begins at A(l). *HINT:* $N_p = 488$, The route is roughly bounded by Highland Ave. and Concord Ave.

The following equations may be used to determine the speed as a function of the haul distance x for Problems 9-22, 9-23, and 9-24:

From 7.5 to 22 km: $s = -17.76 + \ln 2x$

From 22 to 40 km: $s = 10.36 + 0.86(2x)$

From 40 to 80 km: $s = 4.75 + 0.925(2x)$

Beyond 80 km: $s = 80$

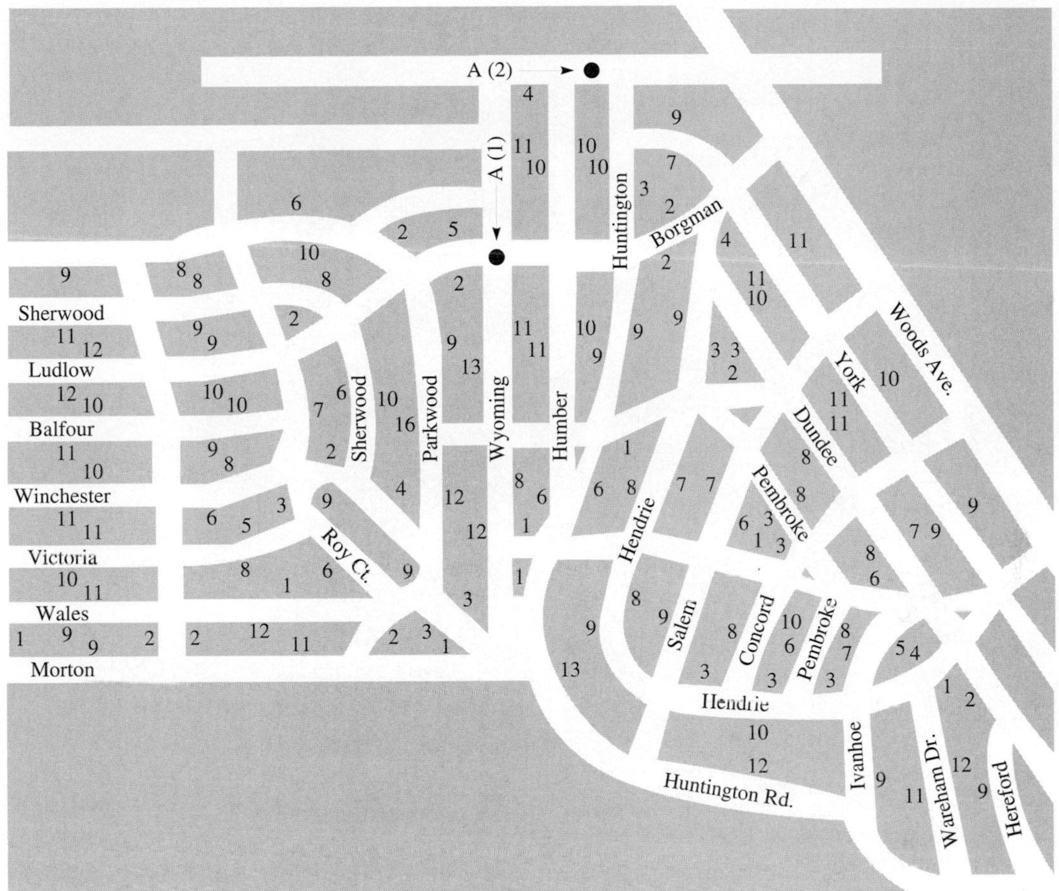

FIGURE P-9-20
Sketch map no. 4, Olson. The numbers refer to the number of stops in a block.

9-22. Write an equation for the relationship between the cost per megagram for hauling solid waste to a disposal site at a distance x, and the round trip time H_t it takes to travel to the disposal site for a crew of one in a 9.0-m^3 compactor truck (see Problems 9-7 and 9-12 for data). The haul speed is determined from the equations noted above.

Answer: $TC = 13.29 + 1.51x + 4.18H_t$

9-23. Repeat Problem 9-22 using a crew of one with 18-m^3 compactor trucks. (See Problem 9-14 data.)

9-24. Repeat Problem 9-22 using a crew of two with 18-m^3 compactor trucks. (See Problems 9-9 and 9-14 for data.)

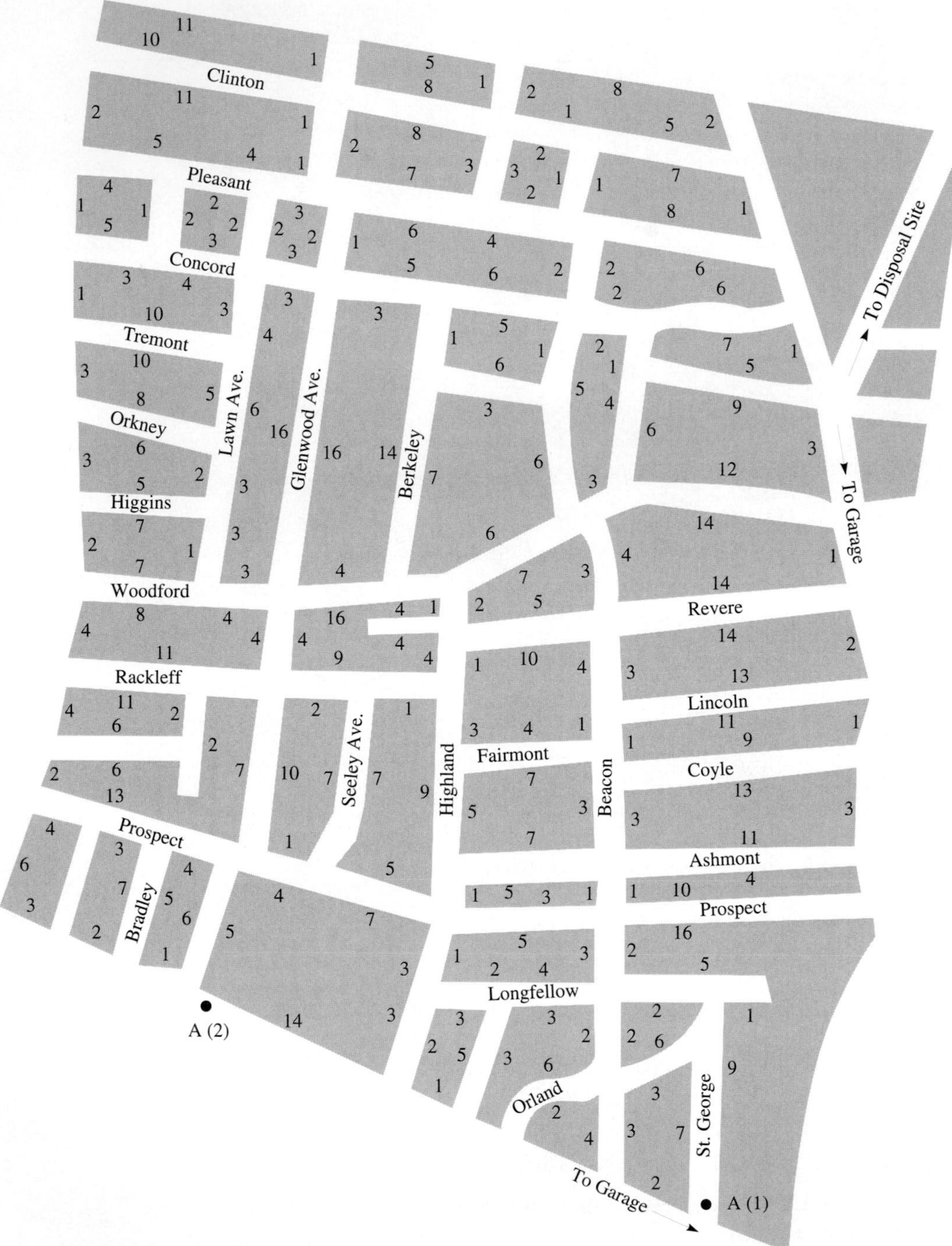

FIGURE P-9-21
Sketch map no. 5, Masters. The numbers refer to the number of stops in a block.

9-25. The town of Trooper (population 8,500) is closing its open dump and will transport its solid waste (9.53 Mg/wk) to the regional landfill at Tuppance Junction. If the one-way distance to the disposal site is 64.0 km and the crew size is one, should Trooper consider using a transfer station? Assume that the capital cost of the transfer station is $25,000 amortized at 6.0 percent over 5 years, and that the annual cost of owning and operating the transfer vehicle and station is $20,000. *Note: A/P* (6.0%, 5y) = 0.2374 and

$$V_T = 11.321 - 3.827\left(\frac{2x}{s}\right)$$

$$H_t = \frac{2x}{s}$$

Answer: No.

9-26. Calamity (population 35,000) generates 48,800 m^3 of solid waste at a density of 425.0 kg/m^3 each year. Four crews of three each are now averaging 1.08 h/d in haul time to the disposal site. Using the data given below, and the crew-of-three cost curves determine whether or not a transfer station should be considered based on an economical analysis.

Transfer station data:

Capital cost = $1,200,000 at 6.00% over 8 years
Transfer vehicle amortization = $55,000/y
Operator cost (including overhead) = $64,960/y
O & M = $6.55/km

Round trip travel and dump time to disposal site = 1.35 h
Number of trips = 5/d
Distance to disposal site = 46.7 km
$T_C = 13.22 + 0.6319(x) + 3.869(\text{Hz})$

Savings from transfer station:

Number of collection crews will be reduced to two
Average daily round trip haul time to transfer station for each vehicle = 20 min

9-27. Estimate the area and volume of landfill to handle the solid waste from Midden (Problem 9-7) for 20 years. The Science Club at Midden High School has furnished the following data based on a 12-month survey. (One sample having a mass of 1.000 Mg was taken at the existing landfill during normal off-loading operations 1 day each month.) Assume a cell height of 2.40 m and that the recommended depths of cover will be used and that compaction will be normal.

Characterization of Midden solid waste

Component	Mass fraction
Food waste	0.0926
Paper	0.4954
Plastics, rubber, leather	0.0438
Textiles	0.0379
Metals	0.0741
Glass	0.1668
Miscellaneous	0.0894
Total	1.0000

9-28. Rework Problem 9-27 assuming that 50 percent of the paper is recycled.

9-29. A MSW landfill is being designed to handle solid waste generated by Binford at a rate of 50 Mg/d. It is expected that the waste will be delivered by compactor truck on a 5 d/week basis. The density as spread is 122 kg/m^3. It will be spread in 0.50 m layers and compacted to 0.25 m. Assuming three such lifts per day and a daily cover of 0.15 m, determine the following: (a) annual volume of landfill consumed in m^3, and (b) daily horizontal area covered by the solid waste. Ignore the soil volume between stacks.

9-30. Estimate the theoretical production of landfill gas (CH_4 only) from the degradation of 20.3 kg of rapidly decomposable MSW. Assume the density of methane is 0.7177 kg/m^3.

9-31. Estimate the theoretical production of landfill gas (CH_4 plus CO_2) from the degradation of 3.3 kg of slowly decomposable MSW. Assume the density of methane is 0.7167 kg/m^3 and that of carbon dioxide is 1.9768 kg/m^3 at STP.

9-32. The city of Nosleep (Problem 9-13) is considering instituting a recycling program in which the residents presort the solid waste into four components: (1) mixed waste, (2) paper, (3) glass, and (4) metallics. From a research study report, we find that the mean time per collection stop plus the mean time to reach the next stop (t_p) can be estimated from the following equation (Tichenor, 1980):

$$t_p = 22.6 + 3.80R + 5.50S$$

where t = mean collection time, s
R = number of units of mixed waste per stop
S = sum of the number of units of separated paper, glass, and metallics per stop

Assuming that $S = 3.00$ and $R = 1.53$, rework Problem 9-13 to determine what savings in disposal cost is needed to offset the additional cost of collection for a crew of one.

9-33. Rework Example 9-4 assuming that 50 percent of the paper and 80 percent of the glass and metal are separated at the source and recycled.

9-34. Using the EPA method, estimate the maximum drainage distance L for a solid waste landfill with a rainfall of 4.0 cm/mo. Assume the hydraulic conductivity of the drainage layer is 2×10^{-2} cm/s and the slope of the liner is 1.0 percent.

9-35. Using a spreadsheet program you have written, rework Problem 9-34 with slopes of 0.5, 1.0, 2.0, and 3.0 percent. Estimate the maximum depth of leachate for wet season rainfall of 40.0 cm/mo for the case of the 1.0 percent slope.

9-36. The higher heating value for cellulose ($C_6H_{10}O_5$) is 32,600 kJ/kg. Compute the lower heating value.

9-37. The higher heating value for methane (CH_4) is 888,500 kJ/kg. Compute the lower heating value.

9-38. Typical residential food waste has a higher heating value of 4,500 kJ/kg on a dry mass basis. Compute the lower heating value if 6.0 percent of the waste by mass is hydrogen.

9-9 DISCUSSION QUESTIONS

9-1. What is the effect of crew size on the mean time per collection stop (t'_p)? How does the container location affect the mean time per collection stop?

9-2. Under what conditions would you recommend consideration of a transfer station?

9-3. Which of the following soil types would be suitable for (a) composite liner, (b) drainage layer, (c) gas venting:

1. Gravel (>2.5 cm diameter)
2. Glacial till
3. Clay ($K = 1 \times 10^{-9}$ cm/s)
4. Clay ($K = 1 \times 10^{6}$ cm/s)
5. Sand ($K = 0.1$ cm/s)
6. Sand ($K = 0.001$ cm/s)

9-4. A WTE plant is being proposed as part of an ISWM plan. The proponents of the WTE argue that recycling is not necessary and will have no effect on the performance of the plant. Do you agree or disagree? Explain.

9-5. Although the market value of compost is negligible, many communities have implemented yard waste composting systems. Explain why.

9-10 REFERENCES

Bader, C. (2001) "Where are Collection Trucks Going?", *MSW Management: The Journal for Municipal Solid Waste Professionals,* vol. 12, no. 6, September/October.

CARB (1988) *The Landfill Gas Testing Program: A Report to the Legislature*, State of California Air Resources Board.

Chicago vs. EDF (1994) *City of Chicago, et al. Vs. Environmental Defense Fund, et al.*, No. 92-1639, May.

Cossu, R., G. Andreottola, and A. Muntoni (1996) "Modeling Landfill Gas Production," in T. H. Christensen, R. Cossu, and R. Stegmann (eds.), E&FN Spon, Landfilling of Waste: Biogas London, pp. 237–250.

Heil (2006) at http://www.heil.com/products/starr.asp. and http://www.heil.com/products/pt1000.asp.

IIA (1968) *I.I.A. Standards*, Incinerator Institute of America, New York.

ISW (1970) *Municipal Refuse Disposal*, Institute for Solid Waste, American Public Works Association, Chicago.

IWSA (2004) *Waste-to-Energy, Clean, Reliable, Renewable Power,* Integrated Waste Serrvices Association, Washington, DC.

Luken, K., and S. Bush (2002) "Automated Collection: Getting the Biggest Bang for Your Buck," *MSW Management: The Journal for Municipal Solid Waste Professionals*, vol. 12, no. 6 September/October.

McEnroe, B. M. (1993) "Maximum Saturated Depth Over Landfill Liner," *Journal of Environmental Engineering Division*, American Society of Civil Engineers, vol. 119, pp. 262–270.

Monk, R. B. (1994) "Digging in the Dirt, Unearthing Potential," *World Wastes*, vol. 37, no. 4 (April), cs1-cs-14.

NCRR (1974) *Resource Recovery from Municipal Solid Waste,* National Center for Resurce Recovery, Lexington Books, Lexington, MA.

O'Leary, P., and P. Walsh (2002) "Landfill Equipment and Operating Procedures," *Waste Age,* September, pp. 53–59.

Paul, S. (1995) "Reaching equilibrium in Recycling Marketables," *World Wastes,* vol. 38, no. 8 (August), p. 52.

Pfeffer, J. T. (1992) *Solid Waste Management Engineering,* Prentice Hall, Upper Saddle River, NJ, p. 172.

Rogoff, M. J. and J. F. Williams (1995) "Marketing Efforts to Close Loop," *World Wastes,* vol. 38, no. 5 (May), p. 28.

Salvato, J. A. (1972) *Environmental Engineering and Sanitation,* Wiley-Interscience, New York, p. 427.

Schroeder, P. R., et al. (1984) *The Hydrologic Evaluation of Landfill Performance (HELP) Model Documentation, User's Guide*, U.S. Environmental Protection Agency Publication No. EPA 530 SW-84-009, Washington, DC.

Serper, A. (1980) "Resource Recovery Field Stands Poised Between Problems, Solutions," *Solid Waste Management/Resource Recovery Journal,* May, p. 86.

Shortsleeve, J., and R. Roche (1990) "Analyzing the Integrated Approach," *Waste Age*, March 1990, pp. 92–94.

Shuster, K. A., and D. A. Schur (1974) *Heuristic Routing for Solid Waste Collection Vehicles,* U.S. Environmental Protection Agency Publication No. SW–113.

Skinner, J. M. (1999) "Advancements in Reduction and Recovery," *MSW Management.*

Stone, R. (1969) *A Study of Solid Waste Collection systems: Comparing One Man with Multi-man Crews,* U.S. Department of Health, Education and Welfare, Report No. SW-9C, Washington, DC, pp. 96–98.

Stone, R., and E. T. Conrad (1969) "Landfill Compaction Equipment Efficency," *Public Works,* May, pp. 111–113 and 160.

Tchobanoglous, G., H. Theisen, and R. Eliassen (1977) *Solid Wastes: Engineering Principles and Management Issues,* McGraw-Hill, New York, p. 95.

Tchobanoglous, G., H. Theisen, and S. Vigil (1993) *Integrated Solid Waste Management: Engineering Principles and Management Issues,* McGraw-Hill, New York, pp. 49, 214, 374, 388–391, 424, 686–695, 932–935.

Tichenor, Richard (1980) "Designing a Vehicle to Collect Source-Separated Recyclables," *Compost Science/Land Utilization,* vol. 21(l), pp. 36–4l, January/February.

U. Calif. (1952) *An Analysis of Refuse Collection and Sanitary Landfill Disposal,* University of California Technical Bulletin 8, Series 73, University of California Press, Berkeley, CA, p. 22.

U.S. EPA (1974) *Decision Makers Guide to Solid Waste Management,* U.S. Environmental Protection Agency, Washington, D.C.

U.S. EPA (1989) *Requirements for Hazardous Waste Landfill Design, Construction and Closure*, U.S. Environmental Protection Agency Publication No. EPA 625/4-89/022, Washington, DC, p. 89.

U.S. EPA (1995) *Decision Makers Guide to Solid Waste Management, Vol. II,* U.S. Environmental Protection Agency Publication No. EPA 530-R-95-023, Washington, DC.

U.S. EPA (1998) *Solid Waste Disposal Facility Criteria: Technical Manual,* U.S Environmental Protection Agency Publication No. EPA 530-R-93-017, Washington, DC.

U.S. EPA (2003) *Municipal Solid Waste in the United States: Facts and Figures*, U.S. Environmental Protection Agency, Washington, DC, found at http://www.epa.gov/msw/facts.htm and http://www.epa.gov/msw/msw99.htm.

Vence, T. D., and D. L. Powers (1980) "Resource Recovery systems, Part 1, Technological Comparison," *Solid Waste Management/Resource Recovery Journal,* May, pp. 26–28, 32, 34, 72, 92, 93.

Wolpin, B. (1994) "Go Figure," *World Wastes*, vol. 37, no. 10, October 1994, p. 4.

Zuena, A. J. (1987) "Snapshot of Small Transfer Station Costs," *Waste Age.*

CHAPTER 10

HAZARDOUS WASTE MANAGEMENT

10-1 THE HAZARD

A hazardous waste, in short, is any waste or combination of wastes that poses a substantial danger, now or in the future, to human, plant, or animal life, and which therefore cannot be handled or disposed of without special precautions. The following examples illustrate the potential problems that may arise when special precautions are not taken in the disposal of hazardous waste.

1. Judy Piatt hired Russell Bliss to spray oil around her stables at Moscow Mills, Missouri, to control the dust. A few days later, hundreds of birds fell to the ground and died. Within the next three-and-a-half years, 20 of her cats went bald and died. Sixty-two of her horses died. Bliss's oil was waste from a defunct hexachlorophene plant that had paid him to dispose of it. That same waste oil was used to settle the dust in the streets of Times Beach, Missouri. It contained dioxin as a contaminant. It was 1971 and virtually no one knew that this waste oil was a hazardous waste.
2. The lagoon was once a licensed disposal site for toxic wastes. The owner was Berlin & Farro Liquid Incineration, Inc. At the bottom of the mysterious blue liquid in the lagoon were some barrels. They were thought to contain hydrochloric acid. The blue liquid in the lagoon was a cyanide waste. The combination of the two chemicals would result in a lethal cloud of cyanide gas. The citizens of nearby Swartz Creek, Michigan, were evacuated while the State Department of Natural Resources oversaw the cleanup.

These actual case histories epitomize our concern with hazardous wastes. That which is common practice today may be the seed of disaster for tomorrow. What is considered good disposal practice may become a nightmare if the operators are not responsible and/or are not good enough business people to make money within the rules, and are therefore tempted to stray beyond them.

Dioxins and PCBs

We would like to elaborate on two particular hazardous wastes that have achieved national prominence: dioxins and PCBs. Because of their newsworthiness, we provide you with a brief summary of what these compounds are, where they come from, and their environmental impact.

Dioxins are found as over twenty different isomers of a basic chlorodioxin structure (Figure 10-1). The most common form, 2,3,7,8-tetrachlorodibenzo-p-dioxin (TCDD), has become recognized as probably the most poisonous of all synthetic chemicals. Dioxins are a contaminant by-product that may be thermally generated during the manufacture or burning of chlorophenols; pesticides such as 2,4,5-T; Agent Orange, a defoliant made of a 50/50 mix of 2,4-D and 2,4,5-T; algae-controlling herbicides; insecticides; and preservatives. Dioxins are not manufactured for any commercial purpose. They occur only as a contaminant by-product. To date no dioxin has been found to be formed naturally in the environment. Widespread TCDD contamination has been reported in particulate matter from commercial and domestic combustion processes. Additional background dioxin contamination (0.1 to 10 parts

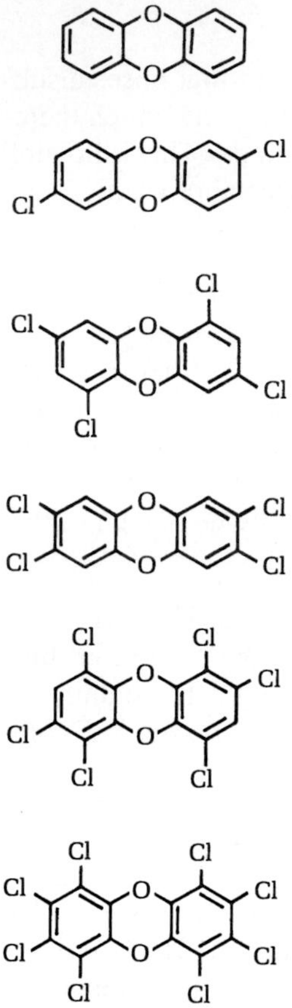

Unsubstituted dioxin

2, 7-DCDD

1, 3, 6, 8-TCDD

2, 3, 7, 8-TCDD

1, 2, 4, 6, 7, 9-HEXA-CDD

OCDD

FIGURE 10-1
Some examples of dioxins.

per million, ppm) may persist and bioaccumulate following the field application of herbicides.

TCDD is a crystalline solid at room temperature. It is only slightly soluble in water (0.2 to 0.6 parts per billion, ppb). TCDD is considered to be a highly stable compound. It is thermally degraded at temperatures over 700°C. It is photochemically degraded under ultraviolet light in the presence of a hydrogen-donating solvent such as a solution of olive oil in cyclohexanone.

TCDD contamination was found at ppm levels in 2,4,5-T and 2,4-D used for weed control in the United States and as a defoliant in Vietnam; in wastes at the Love Canal disposal sites; in orthochlorophenol crude spill residues in the Sturgeon, Missouri, train derailment; and in fallout from an explosion at a chlorophenol manufacturing plant spill in Seveso, Italy. It is at this last site that engineers and scientists were challenged to develop environmentally safe control strategies.

The environmental health effects of dioxin in people are not well documented. However, alleged birth defects in newborns in South Vietnam caused researchers to begin animal toxicological investigations. TCDD is known to cause severe skin disorders, such as chloracne. In test animals it is a carcinogen, teratogen, mutagen, and embryo-toxin, and is known to affect immune responses in mammals. It is considered persistent, and it bioaccumulates in aquatic organisms and people (U.S. EPA, 2005a). At this date (2005) no deaths have been directly correlated with low-level TCDD exposure. Nor have epidemiological findings shown any increased incidence of carcinogenesis, teratogenesis, mutagenesis or newborn defects, miscarriages, or similar adverse health effects in people. In 1994, the U.S. Environmental Protection Agency (EPA) released a report compiled by more than 100 scientists, including many not affiliated with EPA, that presents evidence that dioxin, even in trace amounts, may cause adverse human health effects (Hileman, 1994). EPA believes that dioxins are carcinogens and may cause a wide range of other effects including disruption of regulatory hormones, reproductive and immune system disorders, and abnormal fetal development (U.S. EPA, 2001 and 2005a). Levels of dioxins in the environment were negligible until about 1930, peaked about 1970, and have been declining since then. Concentrations of dioxins in human lipid tissue have declined since 1980.

The term PCB (polychlorinated biphenyls) refers to a class of organic chemicals produced by the chlorination of a biphenyl molecule. It is composed of ten possible forms and, theoretically, more than 200 isomers. These forms arise from a specified number of chlorine substitutions on the biphenyl molecule and correspond to the chemical nomenclatures monochlorobiphenyl, dichlorobiphenyl, trichlorobiphenyl, and so on. Several isomers for each PCB molecule are possible, the number depending on available substitution sites on each biphenyl portion (2–6, $2'-6'$) of the molecule. However, not all possible isomers are likely to be formed during the manufacturing processes. In general, the most common ones are those that have either an equal number of chlorine atoms on both rings or a difference of only one chlorine atom between rings. Some examples are shown in Figure 10-2.

Commercial PCB mixtures were manufactured under a variety of trade names. The chlorine content of any product varied from 18 to 79 percent, depending on the extent of chlorination during the manufacturing process or on the amount of isomeric mixing engaged in by individual producers. Each company had a specific system for identifying the chlorine content of its product. For example, Aroclor 1248, 1254, and 1260 indicate 48 percent, 54 percent, and 60 percent chlorine, respectively; Clophen A60, Phenochlor DP6, and Kaneclor 600 designate that these products contain mixtures of hexachlorobiphenyls.

The only important U.S. producer of PCBs was Monsanto Industrial Chemicals Co., which had plants at Anniston, Alabama, where production of PCBs ended in 1970; and Sauget, Illinois, where production ceased in 1977. Sold under Monsanto's registered trademark of Aroclors, mixtures of PCBs had been used originally as a coolant/dielectric for transformers and capacitors, as heat transfer fluids, and as protective coatings for woods when low flammability was essential or desirable. Producers and users alike, apparently unaware of any potential hazards from exposure to PCBs, initially operated in accordance with earlier results of toxicity tests that indicated no effects (Penning, 1930). The expansion of open-ended applications between 1930 and 1960, incorporating PCBs into such commodities as paints, inks, dedusting

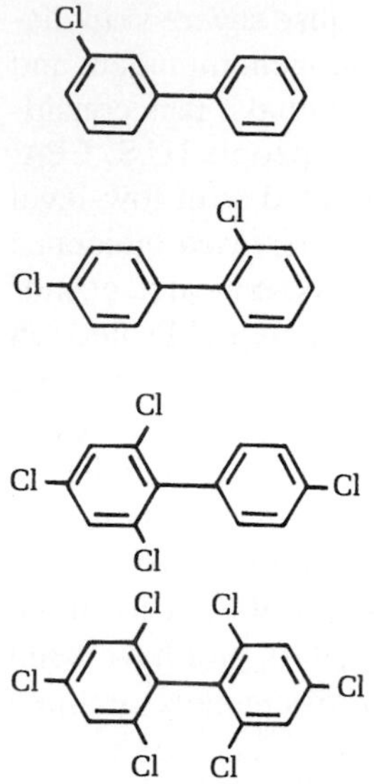

3-chlorobiphenyl

2,4′-dichlorobiphenyl

2,4,4′,6-tetrachlorobiphenyl

2,2′,4,4′,6,6′-hexachlorobiphenyl

FIGURE 10-2
Molecular structure and names of a few selected polychlorinated biphenyls.

agents, and pesticides, led to the widespread dissemination of which we are now aware. By 1937, toxic effects were noted in occupationally-exposed workers, and threshold limit values were imposed at manufacturing sites.

The general pattern of release of PCBs to the environment changed significantly during the early 1970s. Until then, essentially no restrictions were imposed either on the use or on the disposal of PCBs. After evidence became available in 1969 and 1970 that chronic exposure could result in hazards to human health and the environment, Monsanto voluntarily banned sales of PCBs, and the release rate from industrial use was reduced through stringent control measures. However, significant reservoirs of mobile PCBs (those available for transport among environmental media and biota) still exist along with even larger, currently immobile reservoirs. The latter include those materials containing PCBs that are still in service and those deposited in landfills and dumps. The major factor affecting future release of PCBs from these sources will be government regulations controlling storage and disposal of the chemical.

10-2 RISK

The concepts of risk and hazard are inextricably intertwined. *Hazard* implies a probability of adverse effects in a particular situation. *Risk* is a measure of the probability. In some instances the measure is subjective, or *perceived risk.* Scientists and engineers use models to calculate an estimated risk. In some instances actual data may be used to estimate the risk. Today this process is called *quantitative risk assessment,* or more simply *risk assessment.* The use of the results of a risk assessment to make policy decisions is called *risk management.*

Risk Perception

There is an old political saying: "Perception is reality." This is no less true for environmental concerns than it is for politics. People respond to the hazards they perceive. If their perceptions are faulty, risk management efforts to improve environmental protection may be misdirected.

Some risks are well quantified. For example, the frequency and severity of automobile accidents are well documented. In contrast, other hazardous activities such as the use of alcohol and tobacco are more difficult to document. Their assessment requires complex epidemiological studies (Slovic et al., 1979).

When lay people (and some experts for that matter) are asked to evaluate risk, they seldom have ready access to the statistics. In most cases, they rely on inferences based on their experience. People are likely to judge an event as likely or frequent if instances of it are easy to imagine or recall. Also, it is evident that acceptable risk is inversely related to the number of people participating in the activity. In addition, recent events such as a disaster can seriously distort risk judgments.

Table 10-1 and Figure 10-3 illustrate different perceptions of risk. Four different groups were asked to rate thirty activities and technologies according to the present risk of death from each. Three of the groups were from Eugene, Oregon. They included 30 college students, 40 members of the League of Women Voters (LOWV), and 25 business and professional members of the "Active Club." The fourth group was composed of 15 people selected from across the United States because of their professional involvement in risk assessment. Table 10-1 shows how the various groups ranked the risk of the activities or technologies. The same groups were asked to estimate the mean fatality for the same group of activities and technologies given the fact that the annual death toll from motor vehicle accidents in the U.S. was 50,000. The results are plotted in Figure 10-3. The dashed line is the line of best fit to the results. If the dashed line was at 45 degrees, the estimate would be perfect. The steeper slope of the line for the experts' risk judgments shows that they are more closely associated with the actual annual fatality rates than those of the lay groups.

Putting risk perception in perspective, we can calculate the risk of death from some familiar causes. To begin, we recognize that we will all die at some time. So, as a trivial example, the lifetime risk of death from all causes is 100 percent, or 1.0. In 2001, there were about 3.9 million deaths per year. Of these, about 541,532 were cancer related. Without considering age factors, the risk of dying from cancer in a lifetime was about

$$\frac{541{,}532}{3.9 \times 10^6} = 0.14$$

The annual risk (assuming a 70-year life expectancy and again neglecting age factors) is about

$$\frac{0.14}{70} = 0.002$$

TABLE 10-1
Ordering of perceived risk for 30 activities and technologies[a]

	Group 1: LOWV	Group 2: College Students	Group 3: Active Club Members	Group 4: Experts
Nuclear power	1	1	8	20
Motor vehicles	2	5	3	1
Handguns	3	2	1	4
Smoking	4	3	4	2
Motorcycles	5	6	2	6
Alcoholic beverages	6	7	5	3
General (private) aviation	7	15	11	12
Police work	8	8	7	17
Pesticides	9	4	15	8
Surgery	10	11	9	5
Firefighting	11	10	6	18
Large construction	12	14	13	13
Hunting	13	18	10	23
Spray cans	14	13	23	26
Mountain climbing	15	22	12	29
Bicycles	16	24	14	15
Commercial aviation	17	16	18	16
Electric power	18	19	19	9
Swimming	19	30	17	10
Contraceptives	20	9	22	11
Skiing	21	25	16	30
X-rays	22	17	24	7
High school college football	23	26	21	27
Railroads	24	23	20	19
Food preservatives	25	12	28	14
Food coloring	26	20	30	21
Power mowers	27	28	25	28
Prescription antibiotics	28	21	26	24
Home appliances	29	27	27	22
Vaccinations	30	29	29	25

[a]The ordering is based on the geometric mean risk ratings within each group. Rank 1 represents the most risky activity or technology. (*Source:* Slovic et al., 1979.)

For comparison, Table 10-2 summarizes the risk of dying from various causes of death.

In developing standards for environmental protection, the EPA often selects a lifetime risk in the range of 10^{-7} to 10^{-4} as acceptable. Table 10-3 shows a comparison of other activities that, based on statistical evidence, yield a risk of 10^{-6}.

Of course, if the risk of dying in one year is increased, the risk of dying from another cause in a later year is decreased. Since accidents often occur early in life, a

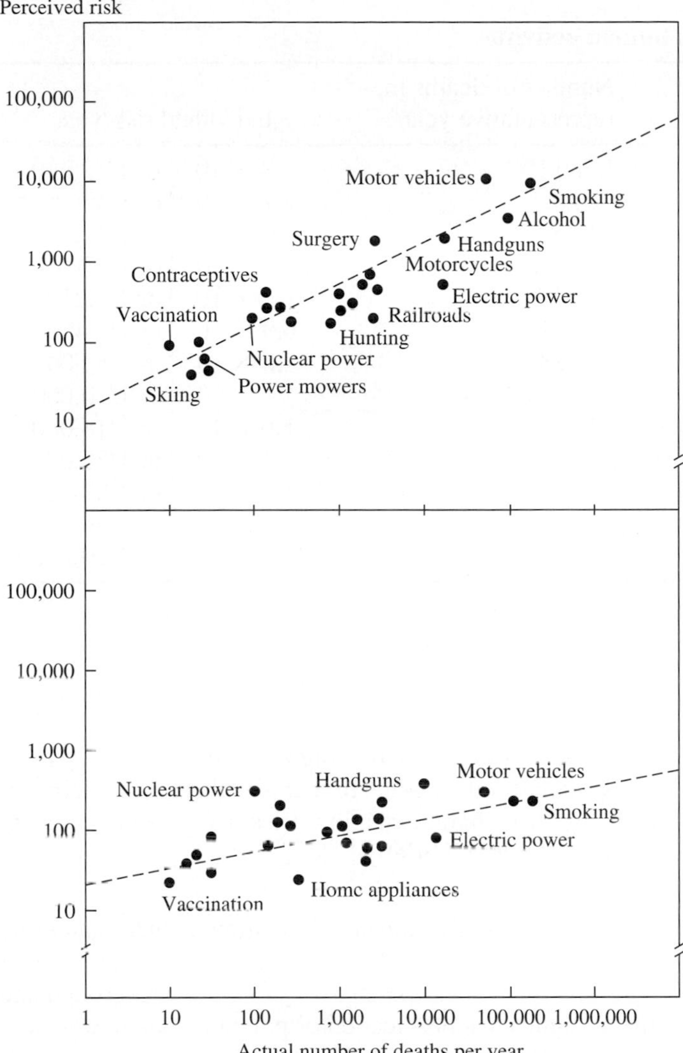

FIGURE 10-3
Judgments of perceived risk for experts (top) and lay people (bottom) plotted against the best technical estimates of annual fatalities for 25 technologies and activities. Each point represents the average responses of the participants. The dashed lines are the straight lines that best fit the points. The experts' risk judgments are seen to be more closely associated with annual fatality rates than are the lay judgments. (*Source:* Slovic et al., 1979.)

typical accident may shorten life by 30 years. In contrast, diseases, such as cancer, cause death later in life, and life is shortened by about 15 years. Therefore, a risk of 10^{-6} shortens life on the average of 30×10^{-6} years, or 15 minutes for an accident. The same risk for a fatal illness shortens life by about 8 minutes. It has been noted that smoking a cigarette takes 10 minutes and shortens life by 5 minutes (Wilson, 1979).

TABLE 10-2
Annual risk of death from selected common human activities

Cause of death	Number of deaths in representative year	Individual risk/year
Black lung disease (coal mining)	1,135	8×10^{-3} or 1/125
Heart attack	724,859	2.7×10^{-3} or 1/370
Cancer	541,532	2.0×10^{-3} or 1/500
Coal mining accident	180	1.3×10^{-3} or 1/770
Firefighting		3×10^{-4} or 1/1,250
Motorcycle driving	3,714	6.9×10^{-4} or 1/1,450
Motor vehicle	42,884	1.8×10^{-4} or 1/5,500
Truck driving	761	10^{-4} or 1/10,000
Falls	16,257	5.6×10^{-5} or 1/18,000
Football (averaged over participants)		4×10^{-5} or 1/25,000
Home accidents	25,000	1.2×10^{-5} or 1/83,000
Bicycling (assuming one person per bicycle)	500	2×10^{-6} or 1/500,000
Air travel: one transcontinental trip/year		2×10^{-6} or 1/500,000

(*Sources:* CDC, 2004; NHTSA, 2005; Hutt, 1978, and Rodricks and Taylor, 1983.)

Risk Assessment

In 1989, the EPA adopted a formal process for conducting a baseline risk assessment (U.S. EPA, 1989a). This process includes data collection and evaluation, toxicity assessment, exposure assessment, and risk characterization. Risk assessment is considered to be site-specific. Each step is described briefly below.

Data Collection and Evaluation. Data collection and evaluation includes gathering and analyzing site-specific data relevant to human health concerns for the purpose of identifying substances of major interest. This step includes gathering background and site information as well as the preliminary identification of potential human exposure through sampling, and development of a sample collection strategy.

When collecting background information, it is important to identify the following:

1. Possible contaminants on the site
2. Concentrations of the contaminants in key sources and media of interest, characteristics of sources, and information related to the chemical's release potential
3. Characteristics of the environmental setting that could affect the fate, transport, and persistence of the contaminants

The review of the available site information determines basic site characteristics such as groundwater movement or soil characteristics. With these data, it is possible to initially identify potential exposure pathways and exposure points important for assessing exposure. A conceptual model of pathways and exposure points can be formed

TABLE 10-3
Risks that increase chance of death by 0.000001[a]

Smoking 1.4 cigarettes	Cancer, heart disease
Drinking 1/2 liter of wine	Cirrhosis of the liver
Spending 1 hour in a coal mine	Black lung disease
Spending 3 hours in a coal mine	Accident
Living 2 days in New York or Boston	Air pollution
Travelling 6 minutes by canoe	Accident
Travelling 10 miles by bicycle	Accident
Travelling 300 miles by car	Accident
Flying 1,000 miles by jet	Accident
Flying 6,000 miles by jet	Cancer caused by cosmic radiation
Living 2 months in Denver on vacation from New York	Cancer caused by cosmic radiation
Living 2 months in average stone or brick building	Cancer caused by natural radioactivity
One chest X-ray taken in a good hospital	Cancer caused by radiation
Living two months with a cigarette smoker	Cancer, heart disease
Eating 40 tablespoons of peanut butter	Liver cancer caused by aflatoxin B
Drinking Miami drinking water for a year	Cancer caused by chloroform
Living 5 years at a site boundary of a typical nuclear power plant in the open	Cancer caused by radiation
Drinking 1,000 24 oz. soft drinks from banned plastic bottles	Cancer from acrylonitrile monomer
Living 20 years near PVC plant	Cancer caused by vinyl chloride (1976 standard)
Living 150 years within 20 miles of a nuclear power plant	Cancer caused by radiation
Eating 100 charcoal broiled steaks	Cancer from benzopyrene
Risk of accident by living within 5 miles of a nuclear reactor for 50 years	Cancer caused by radiation

[a](1 part in 1 million)
(*Source:* Wilson, 1979.)

from the background data and site information. This conceptual model can then be used to help refine data needs.

Toxicity Assessment. Toxicity assessment is the process of determining the relationship between the exposure to a contaminant and the increased likelihood of the occurrence or severity of adverse effects to people. This procedure includes hazard identification and dose-response evaluation. *Hazard identification* determines whether exposure to a contaminant causes increased adverse effects towards humans and to what level of severity. *Dose-response* evaluation uses quantitative information on the dose of the contaminant and relates it to the incidence of adverse health in an exposed population. Toxicity values can be determined from this quantitative relationship and used in the risk characterization step to estimate different occurrences of adverse health effects based on various exposure levels.

The single factor that determines the degree of harmfulness of a compound is the dose of that compound (Loomis, 1978). *Dose* is defined as the mass of chemical received by the animal or exposed individual. Dose is usually expressed in units of milligrams per kilogram of body mass (mg/kg). Some authors use parts per million (ppm) instead of mg/kg. Where the dose is administered over time, the units may be mg/kg · d. It should be noted that dose differs from the concentration of the compound in the medium (air, water, or soil) to which the animal or individual is exposed.

For toxicologists to establish the "degree of harmfulness" of a compound, they must be able to observe a quantitative effect. The ultimate effect manifested is death of the organism. Much more subtle effects may also be observed. Effects on body weight, blood chemistry, and enzyme inhibition or induction are examples of *graded responses.* Mortality and tumor formation are examples of *quantal* (all-or-nothing) responses. If a dose is sufficient to alter a biological mechanism, a harmful consequence will result. The experimental determination of the range of changes in a biologic mechanism to a range of doses is the basis of the dose-response relationship.

The statistical variability of organism response to dose is commonly expressed as a cumulative-frequency distribution known as a dose-response curve. Figure 10-4 illustrates the method by which a common toxicological measure, namely the LD_{50}, or lethal dose for 50 percent of the animals, is obtained. The assumption inherent in the plot of the dose-response curve is that the test population variability follows a Gaussian distribution and, hence, that the dose-response curve has the statistical properties of a Gaussian cumulative-frequency curve.

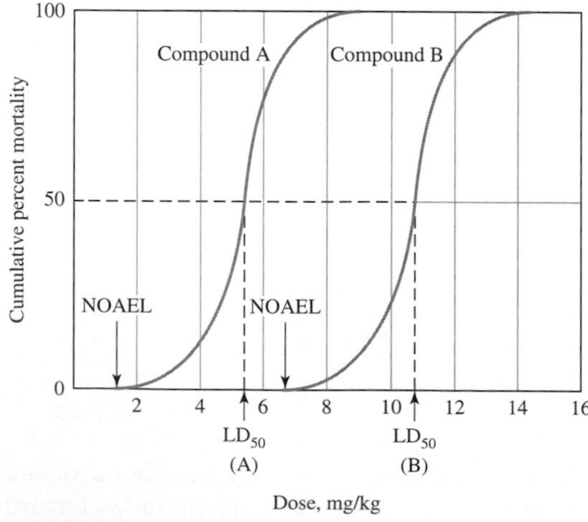

FIGURE 10-4
Hypothetical dose-response curves for two chemical agents (A and B) administered to a uniform population. NOAEL = no observed adverse effect level. (*Source:* Davis and Masten, 2004.)

Toxicity is a relative term. That is, there is no absolute scale for establishing toxicity; one may only specify that one chemical is more or less toxic than another. Comparison of different chemicals is uninformative unless the organism or biologic mechanism is the same and the quantitative effect used for comparison is the same. Figure 10-4 serves to illustrate how a toxicity scale might be developed. Of the two curves in the figure, the LD_{50} for compound B is greater than that for compound A. Thus, for the test animal represented by the graph, compound A is more toxic than compound B as measured by lethality. There are many difficulties in establishing toxicity relationships. Species respond differently to toxicants so that the LD_{50} for a mouse may be very different than that for a human. The shape (slope) of the dose-response curve may differ for different compounds so that a high LD_{50} may be associated with a low "no observed adverse effect level" (NOAEL) and vice versa.

The nature of a statistically obtained value, such as the LD_{50}, tends to obscure a fundamental concept of toxicology: that there is no fixed dose that can be relied on to produce a given biologic effect in every member of a population. In Figure 10-4, the mean value for each test group is plotted. If, in addition, the extremes of the data are plotted as in Figure 10-5, it is apparent that the response of individual members of the population may vary widely from the mean. This implies not only that single point comparisons, such as the LD_{50}, may be misleading, but that even knowing the slope of the average dose-response curve may not be sufficient if one wishes to protect hypersensitive individuals.

Organ toxicity is frequently classified as an acute or subacute effect. Carcinogenesis, teratogenesis, reproductive toxicity, and mutagenesis have been classified as chronic effects.* It is self-evident that an organ may exhibit acute, subacute, and chronic effects and that this system of classification is not well bounded.

Virtually all of the data used in hazard identification and, in particular, hazard quantification, is derived from animal studies. Aside from the difficulty of extrapolating from one species to another, the testing of animals to estimate low-dose response is difficult. Example 10-1 illustrates the problem.

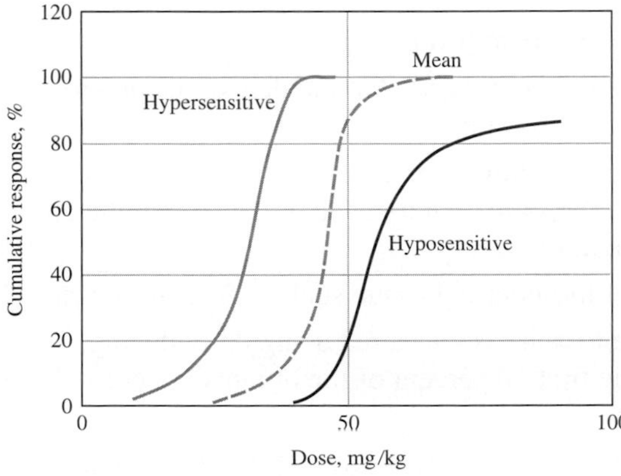

FIGURE 10-5
Hypothetical dose-response relationships for a chemical agent administered to a uniform population. (*Source:* Davis and Masten, 2004.)

*A glossary of these toxicology terms is given in Table 10-4.

TABLE 10-4
Glossary of toxicological terms

Acute toxicity	An adverse effect that has a rapid onset, short course, and pronounced symptoms.
Cancer	An abnormal growth process in which cells begin a phase of uncontrolled growth and spread.
Carcinogen	A cancer-producing substance.
Carcinomas	Cancers of epithelial tissues. Lung cancer and skin cancer are examples of carcinomas.
Chronic toxicity	An adverse effect that frequently takes a long time to run its course and initial onset of symptoms may go undetected.
Genotoxic	Toxic to the genetic material (DNA).
Initiator	A chemical that starts the change in a cell that irreversibly converts the cell into a cancerous or precancerous state. Needs to have a promoter to develop cancer.
Leukemias	Cancers of white blood cells and the tissue from which they are derived.
Lymphomas	Cancers of the lymphatic system. An example is Hodgkin's disease.
Metastasis	Process of spreading/migration of cancer cells throughout the body.
Mutagenesis	Mutagens cause changes in the genetic materal of cells. The mutations may occur either in somatic (body) cells or germ (reproductive) cells.
Neoplasm	A new growth. Usually an abnormally fast-growing tissue.
Oncogenic	Causing cancers to form.
Promoter	A chemical that increases the incidence to a previous carcinogen exposure.
Reproductive toxicity	Decreases in fertility, increases in miscarriages, and fetal or embryonic toxicity as manifested in reduced birth weight or size.
Sarcoma	Cancer of mesodermal tissue such as fat and muscle.
Subacute toxicity	Subacute toxicity is measured using daily dosing during the first 10 percent of the organism's normal life expectancy and checking for effects throughout the normal lifetime.
Teratogenesis	Production of a birth defect in the offspring after maternal or paternal exposure.

Example 10-1. An experiment was developed to ascertain whether a compound has a 5 percent probability of causing a tumor. The same dose of the compound was administered to 10 groups of 100 test animals. A control group of 100 animals was, with the exception of the test compound, exposed to the same environmental conditions for the same period of time. The following results were obtained:

Group	No. of Tumors
A	6
B	4
C	10
D	1
E	2
F	9
G	5
H	1
I	4
J	7

No tumors were detected in the controls (not likely in reality).

Solution. The average number of excess tumors is 4.9 percent. These results tend to confirm that the probability of causing a tumor is 5 percent.

If, instead of using 1,000 animals (10 groups × 100 animals), only 100 animals were used, it is fairly evident from the data that, statistically speaking, some very anomalous results might be achieved. That is, we might find a risk from 1 percent to 10 percent.

Note that a 5 percent risk (probability of 0.05) is very high in comparison to the EPA's objective of achieving an environmental contaminant risk of 10^{-7} to 10^{-4}.

Animal studies are only capable of detecting risks on the order of 1 percent. To extrapolate the data taken from animals exposed to high doses to humans who will be exposed to doses several orders of magnitude lower, toxicologists employ mathematical models.

One of the most controversial aspects of toxicological assessment is the method chosen to extrapolate the carcinogenic dose-response curve from the high doses actually administered to test animals to the low doses that humans actually experience in the environment. The conservative worst-case assessment is that one event capable of altering DNA will lead to tumor formation. This is called the *one-hit hypothesis.* From this hypothesis, it is assumed that there is no threshold dose below which the risk is zero, so that for carcinogens, there is no NOAEL and the dose-response curve passes through the origin.

Many models have been proposed for extrapolation to low doses. The selection of an appropriate model is more a policy decision than a scientific one since there are no data to confirm or refute any model. The *one-hit model* is frequently used:

$$P(d) = 1 - \exp(-q_0 - q_1 d) \tag{10-1}$$

where $P(d)$ = lifetime risk (probability) of cancer
d = dose
q_0 and q_1 = parameter to fit data

This model corresponds to the simplest mechanistic model of carcinogenesis, namely that a single chemical hit will induce a tumor.

The background rate of cancer incidence, $P(0)$, may be represented by expanding the exponential as

$$\exp(x) = 1 + x + \frac{x^2}{2!} + \cdots + \frac{x^n}{n!} \tag{10-2}$$

For small values of x, this expansion is approximately

$$\exp(x) \simeq 1 + x \tag{10-3}$$

Assuming the background rate for cancer is small, then

$$P(0) = 1 - \exp(-q_0) \simeq 1 - [1 + (-q_0)] = q_0 \tag{10-4}$$

This implies that q_0 corresponds to the background cancer incidence. For small dose rates, the one-hit model can then be expressed as

$$P(d) \simeq 1 - [1 - (q_0 + q_1 d)] = q_0 + q_1 d = P(0) + q_1 d \tag{10-5}$$

For low doses, the additional cancer risk above the background level may be estimated as

$$A(d) = P(d) - P(0) = (P(0) + q_1 d) - P(0) \tag{10-6}$$

or

$$A(d) = q_1 d \tag{10-7}$$

This model, therefore, assumes that the excess lifetime probability of cancer is linearly related to dose.

Some authors prefer a model that is based on an assumption that tumors are formed as a result of a sequence of biological events. This model is called the *multistage model:*

$$P(d) = 1-\exp[-(q_0 + q_1 d + q_2 d^2 + \cdots + q_n d^n)] \tag{10-8}$$

where q_i values are selected to fit the data. The one-hit model is a special case of the multistage model.

EPA has selected a modification of the multistage model for toxicological assessment. It is called the *linearized multistage model.* This model assumes that we can extrapolate from high doses to low doses with a straight line. At low doses, the slope of the dose-response curve is represented by a *slope factor* (SF). It has units of *risk per unit dose* or *risk* (kg · d/mg).

The EPA maintains a toxicological data base called IRIS (*Integrated Risk Information System*) that provides background information on potential carcinogens. IRIS includes suggested values for the slope factor. A list of slope factors for several compounds is shown in Table 10-5.

TABLE 10-5
Slope factors for potential carcinogens[a]

Chemical	CPS_0, kg · d/mg	RfC per $\mu g/m^3$	CPS_i, kg · d/mg
Arsenic	1.5	4.3×10^{-3}	15.1
Benzene	0.015	2.2×10^{-6}	0.029
Benzo(a)pyrene	7.3	Not available	
Cadmium	N/A	1.8×10^{-3}	6.3
Carbon tetrachloride	0.13	1.5×10^{-5}	0.0525
Chloroform	0.0061	2.3×10^{-5}	0.08
Chromium VI	N/A	1.2×10^{-2}	42.0
DDT	0.34	9.7×10^{-5}	0.34
1,1-Dichloroethylene	0.6	2×10^{-1}	0.175
Dieldrin	16.0	4.6×10^{-3}	16.1
Heptachlor	4.5	1.3×10^{-3}	4.55
Hexachloroethane	0.014	4.0×10^{-6}	0.014
Methylene chloride	0.0075	4.7×10^{-7}	0.00164
Pentachlorophenol	0.02	Not available	
Polychlorinated biphenyls	0.04	1×10^{-4}	
2,3,7,8-TCDD[b]	1.5×10^5		1.16×10^5
Tetrachloroethylene[b]	0.052		0.002
Trichloroethylene[c]			0.006
Vinyl chloride[b]	0.072	4.4×10^{-6}	

CPS_0 = cancer potency slope, oral; RfC = reference air concentration unit risk; CPS_i = cancer potency slope, inhalation, derived from RfC.
[a]Values are frequently updated. Refer to IRIS for current data.
[b]From Health Effects Assessment Summary Tables (HEAST), 1994.
[c]From U.S. EPA–NCEA Regional Support—provisional value, http://www.epa.gov/ncea.
(*Source:* with exceptions noted above: U.S. Environmental Protection Agency, IRIS database, September 2005.)

In contrast to the carcinogens, it is assumed that for noncarcinogens there is a dose below which there is no adverse effect; that is, there is an NOAEL. The EPA has estimated the acceptable daily intake, or *reference dose* (RfD), that is likely to be without appreciable risk. The RfD is obtained by dividing the NOAEL by safety factors to account for the transfer from animals to humans, sensitivity, and other uncertainties in developing the data. A list of several compounds and their RfD values is given in Table 10-6.

Limitations of Animal Studies. No species provides an exact duplicate of human response. Certain effects that occur in common lab animals generally occur in people. Many effects produced in people can, in retrospect, be produced in some species. Notable exceptions are toxicities dependent on immunogenic mechanisms. Most sensitization reactions are difficult if not impossible to induce in lab animals. The procedure in transferring animal data to people is then to find the "proper" species and study it in context. Observed differences are then often quantitative rather than qualitative.

Carcinogenicity as a result of application or administration to lab animals is often assumed to be transposable to people because of the seriousness of the consequence of

TABLE 10-6
RfDs for chronic noncarcinogenic effects for selected chemicals[a]

Chemical	Oral RfD, mg/kg · d
Acetone	0.9
Barium	0.2
Cadmium	0.0005
Chloroform	0.01
Cyanide	0.02
1,1-Dichloroethylene	0.05
Hydrogen cyanide	0.02
Methylene chloride	0.06
Pentachlorophenol	0.03
Phenol	0.3
PCB	
Aroclor 1016	7.0×10^{-5}
Aroclor 1254	2.0×10^{-5}
Silver	0.005
Tetrachloroethylene	0.01
Toluene	0.2
1,2,4-Trichlorobenzene	0.01
Xylenes	0.2

[a]Values are frequently updated. Refer to IRIS for current data.
(*Source:* U.S. Environmental Protection Agency IRIS database, 2005.)

ignoring such evidence. However, slowly induced, subtle toxicity—because of the effects of ancillary factors (environment, age, etc.)—is difficult at best to transfer. This becomes even more difficult when the incidence of toxicity is restricted to a small hypersensitive subset of the population.

Limitations of Epidemiological Studies. There are four difficulties in epidemiological studies of toxicity in human populations. The first is that large populations are required to detect a low frequency of occurrence of a toxicological effect. The second difficulty is that there may be a long or highly variable latency period between the exposure to the toxicant and a measurable effect. Competing causes of the observed toxicological response make it difficult to attribute a direct cause and effect. For example, cigarette smoking, the use of alcohol or drugs, and personal characteristics such as sex, race, age, and prior disease states tend to mask environmental exposures. The fourth difficulty is that epidemiological studies are often based on data collected in specific political boundaries that do not necessarily coincide with environmental boundaries such as those defined by an aquifer or the prevailing wind patterns.

Exposure Assessment. The objective of this step is to estimate the magnitude of exposure to chemicals of potential concern. The magnitude of exposure is based on

TABLE 10-7
Potential contaminated media and corresponding routes of exposure

Media	Routes of potential exposure
Groundwater	Ingestion, dermal contact, inhalation during showering
Surface water	Ingestion, dermal contact, inhalation during showering
Sediment	Ingestion, dermal contact
Air	Inhalation of airborne (vapor phase) chemicals (indoor and outdoor)
	Inhalation of particulates (indoor and outdoor)
Soil/dust	Incidental ingestion, dermal contact
Food	Ingestion

chemical intake and pathways of exposure. The most important route (or pathway) of exposure may not always be clearly established. Arbitrarily eliminating one or more routes of exposure is not scientifically sound. The more reasonable approach is to consider an individual's potential contact with all contaminated media through all possible routes of entry. These are summarized in Table 10-7.

The evaluation of all major sources of exposure is known as total exposure assessment (Butler et al., 1993). After reviewing the available data, it may be possible to decrease or increase the level of concern for a particular route of entry to the body. Elimination of a pathway of entry can be justified if:

1. The exposure from a particular pathway is less than that of exposure through another pathway involving the same media at the same exposure point.
2. The magnitude of exposure from the pathway is low.
3. The probability of exposure is low and incidental risk is not high.

There are two methods of quantifying exposure: point estimate methods and probabilistic methods. The EPA utilizes the point estimate procedure by estimating the *reasonable maximum exposure* (RME). Because this method results in very conservative estimates, some scientists believe probabilistic methods are more realistic (Finley and Paustenbach, 1994).

RME is defined as the highest exposure that is reasonably expected to occur and is intended to be a conservative estimate of exposure within the range of possible exposures. Two steps are involved in estimating RME: first, exposure concentrations are predicted using a transport model such as the Gaussian plume model for atmospheric dispersion (Section 7-8), then pathway-specific intakes are calculated using these exposure concentration estimates. The following equation is a generic intake equation*:

$$\text{CDI} = C\left[\frac{(\text{CR})(\text{EDF})}{\text{BW}}\right]\left(\frac{1}{\text{AT}}\right) \tag{10-9}$$

*The notation in Equation 10-9 and subsequent equations follows that used in EPA guidance documents. The abbreviation CDI does not imply multiplication of three variables C, D, and I. CDI, CR, EFD, BW, and so on are the notation for the variables. They do not refer to product of terms.

where CDI = chronic daily intake, (mg/kg body weight · day)
C = chemical concentration, contacted over the exposure period (e.g., mg/L water)
CR = contact rate, the amount of contaminated medium contacted per unit time or event (e.g., L/day)
EFD = exposure frequency and duration, describes how long and how often exposure occurs. Often calculated using two terms (EF and ED):
EF = exposure frequency (days/year)
ED = exposure duration (years)
BW = body weight, the average body weight over the exposure period (kg)
AT = averaging time, period over which exposure is averaged (days)

For each different media and corresponding route of exposure, it is important to note that additional variables are used to estimate intake. For example, when calculating intake for the inhalation of airborne chemicals, an inhalation rate and exposure time are required. Specific equations for media and routes of exposure are given in Table 10-8. Standard values for use in the intake equations are shown in Table 10-9.

Example 10-2. Estimate the lifetime average chronic daily intake of benzene from exposure to a city water supply that contains a benzene concentration equal to the drinking water standard. Assume the exposed individual is an adult male who consumes water at the adult rate for 63 years, that he is an avid swimmer and swims in a local pool (supplied with city water) 3 days a week for 30 minutes and has been doing so since he was 30 years old. He has taken a long shower every day for 63 years. Assume that the average air concentration of benzene during the shower is 5 $\mu g/m^3$ (McKone, 1987). From the literature, it is estimated that the dermal uptake from water is 0.0020 $m^3/m^2 \cdot h$ (This is PC in Table 10-8. PC also has units of m/h or cm/h.) and that direct dermal absorption during showering is no more than 1 percent of the available benzene because most of the water does not stay in contact with skin long enough (Byard, 1989).

Solution. From Table 10-8, we note that five routes of exposure are possible from the drinking water medium: (1) ingestion, dermal contact while (2) showering and (3) swimming, (4) inhalation of vapor while showering, and (5) ingestion while swimming. The allowable drinking water concentration is determined to be 0.005 mg/L from Chapter 4, Table 4-7.

We begin by calculating the CDI for ingestion (Equation 10-10):

$$\text{CDI} = \frac{(0.005 \text{ mg/L})(2.3 \text{ L/d})(365 \text{ d/y})(63 \text{ y})}{(78 \text{ kg})(75 \text{ y})(365 \text{ d/y})}$$

$$= 1.24 \times 10^{-4} \text{ mg/kg} \cdot \text{d}$$

The ingestion rate (IR) and body weight (BW) were selected from Table 10-9. Although the actual ingestion is over 63 y, the lifetime average (Table 10.9) is over 75 y

TABLE 10-8
Residential exposure equations for various pathways[a]

Ingestion in drinking water

$$\text{CDI} = \frac{(\text{CW})(\text{IR})(\text{EF})(\text{ED})}{(\text{BW})(\text{AT})} \tag{10-10}$$

Ingestion while swimming

$$\text{CDI} = \frac{(\text{CW})(\text{CR})(\text{ET})(\text{EF})(\text{ED})}{(\text{BW})(\text{AT})} \tag{10-11}$$

Dermal contact with water

$$\text{AD} = \frac{(\text{CW})(\text{SA})(\text{PC})(\text{ET})(\text{EF})(\text{ED})(\text{CF})}{(\text{BW})(\text{AT})} \tag{10-12}$$

Ingestion of chemicals in soil

$$\text{CDI} = \frac{(\text{CS})(\text{IR})(\text{CF})(\text{FI})(\text{EF})(\text{ED})}{(\text{BW})(\text{AT})} \tag{10-13}$$

Dermal contact with soil

$$\text{AD} = \frac{(\text{CS})(\text{CF})(\text{SA})(\text{AF})(\text{ABS})(\text{EF})(\text{ED})}{(\text{BW})(\text{AT})} \tag{10-14}$$

Inhalation of airborn (vapor phase) chemicals

$$\text{CDI} = \frac{(\text{CA})(\text{IR})(\text{ET})(\text{EF})(\text{ED})}{(\text{BW})(\text{AT})} \tag{10-15}$$

Ingestion of contaminated fruits, vegetables, fish and shellfish

$$\text{CDI} = \frac{(\text{CF})(\text{IR})(\text{FI})(\text{EF})(\text{ED})}{(\text{BW})(\text{AT})} \tag{10-16}$$

where ABS = absorption factor for soil contaminant, unitless
AD = absorbed dose, mg/kg · d
AF = soil-to-skin adherence factor, mg/cm^2
AT = averaging time, d
BW = body weight, kg
CA = contaminant concentration in air, mg/m^3
CDI = chronic daily intake, mg/kg · d
CF = volumetric conversion factor for water = 1 L/1,000 cm^3
= conversion factor for soil = 10^{-6} kg/mg
CR = contact rate, L/h
CS = chemical concentration in soil, mg/kg
CW = chemical concentration in water, mg/L
ED = exposure duration, y
EF = exposure frequency, d/y or events/y
ET = exposure time, h/d or h/event
FI = fraction ingested, unitless
IR = ingestion rate, L/d or mg soil/d or kg/meal
= inhalation rate, m^3/h
PC = chemical-specific dermal permeability constant, cm/h
SA = skin surface area available for contact, cm^2

([a]*Source:* U.S. EPA, 1989a.)

TABLE 10-9
EPA recommended values for estimating intake[a, b]

Parameter	Standard value
Body weight, adult female	65.4 kg
Body weight, adult male	78 kg
Body weight, child	
6–11 months	9 kg
1–5 y	16 kg
6–12 y	33 kg
Amount of water ingested daily, adult[c]	2.3 L
Amount of water ingested daily, child[c]	1.5 L
Amount of air breathed daily, adult female	11.3 m^3
Amount of air breathed daily, adult male	15.2 m^3
Amount of air breathed daily, child (3–5 y)	8.3 m^3
Amount of fish consumed daily, adult	6 g/d
Water swallowing rate, swimming	50 mL/h
Skin surface available, adult female	1.69 m^2
Skin surface available, adult male	1.94 m^2
Skin surface available, child	
3–6 y (avg for male and female)	0.720 m^2
6–9 y (avg for male and female)	0.925 m^2
9–12 y (avg for male and female)	1.16 m^2
12–15 y (avg for male and female)	1.49 m^2
15–18 y (female)	1.60 m^2
15–18 y (male)	1.75 m^2
Soil ingestion rate, children 1 to 6 y	100 mg/d
Soil ingestion rate, persons $>$ 6 y	50 mg/d
Skin adherence factor, gardeners	0.07 mg/cm^2
Skin adherence factor, wet soil	0.2 mg/cm^2
Exposure duration	
Lifetime	75 y
At one residence, 90th percentile	30 y
National median	5 y
Averaging time	(ED)(365 d/y)
Exposure frequency (EF)	
Swimming	7 d/y
Eating fish and shell fish	48 d/y
Exposure time (ET)	
Bath or shower, 90th percentile	30 min
Bath or shower, 50th percentile	15 min

([a]*Sources:* U.S. EPA, 1989a; U.S. EPA, 1997; U.S. EPA, 2004b.)
[b]Average value unless otherwise noted.
[c]90th percentile.

Equation 10-12 may be used to estimate absorbed dose while showering:

$$AD = \frac{(0.005\ \text{mg/L})(1.94\ \text{m}^2)(0.0020\ \text{m/h})(0.5\ \text{h/event})}{(78\ \text{kg})(75\ \text{y})}$$

$$\times \frac{(1\ \text{event/d})(365\ \text{d/y})(63\ \text{y})(10^3\ \text{L/m}^3)}{(365\ \text{d/y})}$$

$$= 1.04 \times 10^{-4}\ \text{mg/kg} \cdot \text{d}$$

But only about 1 percent of this amount is available for adsorption in a shower because of the limited contact time, so the actual adsorbed dose by dermal contact is

$$AD = (0.01)(1.04 \times 10^{-4}\ \text{mg/kg} \cdot \text{d}) = 1.04 \times 10^{-6}\ \text{mg/kg} \cdot \text{d}$$

The surface area (SA) and exposure time were obtained from Table 10-9. The permeability constant was given in the problem statement. The exposure time is estimated by converting a long shower of 30 minutes to hours ($30/60 = 0.5$).

The adsorbed dose for swimming is calculated in the same fashion:

$$AD = \frac{(0.005\ \text{mg/L})(1.94\ \text{m}^2)(0.0020\ \text{m/h})(0.5\ \text{h/event})}{(78\ \text{kg})(75\ \text{y})}$$

$$\times \frac{(3\ \text{events/w})(52\ \text{w/y})(45\ \text{y})(10^3\ \text{L/m}^3)}{(365\ \text{d/y})}$$

$$= 3.19 \times 10^{-5}\ \text{mg/kg} \cdot \text{d}$$

In this case, since there is virtually total body immersion for the entire contact period and since there is virtually an unlimited supply of water for contact, there is no reduction for availability. The value of ET is computed from the swimming time (30 minutes = 0.5 h/event). The exposure frequency is computed from the number of swimming events per week and the number of weeks in a year. The exposure duration (ED) is calculated from the lifetime and beginning time of swimming = 75 y – 30 y = 45 y.

The inhalation rate from showering is estimated from Equation 10-15:

$$CDI = \frac{(5\ \mu\text{g/m}^3)(10^{-3}\ \text{mg}/\mu\text{g})(0.833\ \text{m}^3\text{/h})(0.5\ \text{h/event})(1\ \text{event/d})(365\ \text{d/y})(63\ \text{y})}{(78\ \text{kg})(75\ \text{y})(365\ \text{d/y})}$$

$$= 2.24 \times 10^{-5}\ \text{mg/kg} \cdot \text{d}$$

The inhalation rate (IR) is taken from Table 10-9 and converted to an hourly basis.

For ingestion while swimming, we apply Equation 10-11:

$$CDI = \frac{(0.005\ \text{mg/L})(50\,\text{mL/h})(10^{-3}\,\text{L/mL})(0.5\ \text{h/event})(3\ \text{events/w})(52\,\text{w/y})(45\,\text{y})}{(78\ \text{kg})(75\ \text{y})(365\ \text{d/y})}$$

$$= 4.11 \times 10^{-7}\ \text{mg/kg} \cdot \text{d}$$

The contact rate (CR) was determined from Table 10-9. Other values were obtained in the same fashion as those for dermal contact while swimming.

The total exposure would be estimated as:

$$CDI_T = 1.04 \times 10^{-4} + 1.04 \times 10^{-6} + 3.19 \times 10^{-5} + 2.24 \times 10^{-5} + 4.11 \times 10^{-7}$$

$$= 1.60 \times 10^{-4}\ \text{mg/kg} \cdot \text{d}$$

From these calculations, it becomes readily apparent that, in this case, drinking the water dominates the intake of benzene.

Risk Characterization. In the risk characterization step, all data collected from exposure and toxicity assessments are reviewed to corroborate qualitative and quantitative conclusions about risk. The risk for each media source and route of entry is calculated. This includes the evaluation of compounding effects due to the presence of more than one chemical contaminant and the combination of risk across all routes of entry.

For low-dose cancer risk (risk below 0.01), the quantitative risk assessment for a single compound by a single route is calculated as:

$$\text{Risk} = (\text{Intake})(\text{Slope Factor}) \tag{10-17}$$

where intake is calculated from one of the equations in Table 10-8 or a similar relationship. The slope factor is obtained from IRIS (see, for example, Table 10-5). For high carcinogenic risk levels (risk above 0.01), the one-hit equation is used:

$$\text{Risk} = 1 - \exp[-(\text{Intake})(\text{Slope Factor})] \tag{10-18}$$

The measure used to describe the potential for noncarcinogenic toxicity to occur in an individual is not expressed as a probability. Instead, EPA uses the noncancer hazard quotient, or hazard index (HI):

$$\text{HI} = \frac{\text{Intake}}{\text{RfD}} \tag{10-19}$$

These ratios are not to be interpreted as statistical probabilities. A ratio of 0.001 does *not* mean that there is a one in one thousand chance of an effect occurring. If the HI exceeds unity, there may be concern for potential noncancer effects. As a rule, the greater the value above unity, the greater the level of concern.

To account for multiple substances in one pathway, EPA sums the risks for each constituent:

$$\text{Risk}_T = \sum \text{Risk}_i \tag{10-20}$$

For multiple pathways

$$\text{Total Exposure Risk} = \sum \text{Risk}_{ij} \tag{10-21}$$

where i = the compounds and j = pathways.

In a like manner, the hazard index for multiple substances and pathways is estimated as

$$\text{Hazard Index}_T = \sum \text{HI}_{ij} \tag{10-22}$$

In EPA's guidance documents, they recommend segregation of the hazard index into chronic, subchronic, and short-term exposure.

Example 10-3. Using the results from Example 10-2, estimate the risk from exposure to drinking water containing the MCL for benzene.

Solution. Equation 10-21 in the form

$$\text{Total Exposure Risk} = \sum \text{Risk}_i$$

may be used to estimate the risk. Since the problem is only to consider one compound, namely benzene, i = 1 and others do not need to be considered. Since the total exposure from Example 10-2 included each of the routes of concern for drinking water, that is all j's, the final sum may be used to compute risk. The slope factor is obtained from Table 10-5. The risk is

$$\text{Risk} = (1.60 \times 10^{-4} \text{ mg/kg} \cdot \text{d})(1.5 \times 10^{-2} (\text{mg/kg} \cdot \text{d})^{-1})$$
$$= 2.40 \times 10^{-6}$$

This is the total lifetime risk (70 years) for benzene in drinking water at the MCL. Another way of viewing this is to estimate the number of people that might develop cancer. For example, in a population of 2 million,

$$(2 \times 10^6)(2.40 \times 10^{-6}) = 5 \text{ people might develop cancer.}$$

This risk falls within the EPA guidelines of 10^{-4} to 10^{-7} risk. It, of course, does not account for all sources of benzene by all routes. None the less, the risk, compared to some other risks in daily life, appears to be quite small.

Risk Management

Though some might wish it, it is clear that establishment of zero risk cannot be achieved. There are risks in all societal decisions from driving a car to drinking water with benzene at the MCL concentration. Even banning the production of chemicals, as was done for PCBs, for example, does not remove those that already permeate our environment. Risk management is performed to decide the magnitude of risk that is tolerable in specific circumstances (NRC, 1983). This is a policy decision that weighs the results of the risk assessment against costs and benefits as well as the public acceptance. The risk manager recognizes that if a very high certainty in avoiding risk (that is, a very low risk, for example, 10^{-7}) is required, the costs in achieving low concentrations of the contaminant are likely to be high.

Unfortunately, there is very little guidance that can be provided to the risk manager. We know that people are willing to accept a higher risk for things that they expose themselves to voluntarily than for involuntary exposures, and, hence, insist on lower levels of risk, regardless of cost, for involuntary exposure. We also know that people are willing to accept risk if it approaches that for disease, that is, a fatality rate of 10^{-6} people per person-hour of exposure (Starr, 1969).

10-3 DEFINITION AND CLASSIFICATION OF HAZARDOUS WASTE

There are two ways a waste material is found to be hazardous (40 CFR 260): (1) by its presence on the EPA-developed lists, or (2) by evidence that the waste exhibits ignitable, corrosive, reactive, or toxic characteristics.

EPA's Hazardous Waste Designation System

The list of hazardous wastes includes spent halogenated and nonhalogenated solvents; electroplating baths; wastewater treatment sludges from many individual production processes; and heavy ends, light ends, bottom tars, and side-cuts from various distillation processes.

Some commercial chemical products are also listed as being hazardous wastes when discarded. These include "acutely hazardous" wastes such as arsenic acid, cyanides, and many pesticides, as well as "toxic" wastes such as benzene, toluene, and phenols.

EPA has designated five hazardous waste categories. Each hazardous waste is given an EPA Hazardous Waste Number. This is often referred to as the *Hazardous Waste Code*. Each of the five categories may be identified by the prefix letter assigned by EPA. The five categories may be described as follows:

1. Specific types of wastes from nonspecific sources; examples include halogenated solvents, nonhalogenated solvents, electroplating sludges, and cyanide solutions from plating batches. (There are 28 listings in this category. See 40 CFR 261.31.) These wastes have a waste code prefix letter F.
2. Specific types of wastes from specific sources; examples include oven residue from the production of chrome oxide green pigments and brine purification muds from the mercury cell process in chlorine production where separated, prepurified brine is not used. (There are 111 listings in this category. See 40 CFR 261.32.) These wastes have a waste code prefix letter K.
3. Any commercial chemical product or intermediate, off-specification product, or residue that has been identified as an acute hazardous waste. Examples include potassium silver cyanide, toxaphene, and arsenic oxide. (There are approximately 203 listings in this category. See 40 CFR 261.33.) These wastes have a waste code prefix letter P.
4. Any commercial chemical product or intermediate, off-specification product, or residue that has been identified as hazardous waste. Examples include xylene, DDT, and carbon tetrachloride. (There are approximately 450 listings in this category. See 40 CFR 261.33.) These wastes have a waste code prefix letter U.
5. Characteristic wastes (40 CFR 261.21 through 40 CFR 261.27), which are wastes not specifically identified elsewhere, that exhibit properties of ignitability, corrosivity, reactivity, or toxicity. These wastes have a waste code prefix letter D.

The wastes that appear on one of the lists specified in items one through four are called *listed wastes*. The current list may be found at www.gpoaccess.gov/cfs.* Those wastes that are declared hazardous because of their general properties are called *characteristic wastes*. The characteristics of ignitability, corrosivity, and reactivity may be referred to as *ICR*. The toxicity characteristic may be referred to as *TC*.

---

*In 2005, the **gpoaccess** search engine would locate only 40 CFR 261. To find a subparagraph such as 40 CFR 261.31, first search for 40 CFR 261, then scroll to the subparagraph of interest.

Ignitability

A solid waste is said to exhibit the characteristic of ignitability if a representative sample of the waste has any of the following properties:

1. It is a liquid, other than an aqueous solution containing less than 24 percent alcohol by volume, and has a flash point less than 60°C*.
2. It is not a liquid and is capable, under standard temperature and pressure, of causing fire through friction, absorption of moisture, or spontaneous chemical changes; and, when ignited, burns so vigorously and persistently that it creates a hazard.
3. It is an ignitable, compressed gas.
4. It is an oxidizer.

A solid waste that exhibits the characteristic of ignitability is given an EPA Hazardous Waste Number of D001.

Corrosivity

A solid waste is said to exhibit the characteristic of corrosivity if a representative sample of the waste has either of the following properties:

1. It is aqueous and has a pH less than or equal to 2 or greater than or equal to 12.5.
2. It is a liquid that corrodes steel at a rate greater than 6.35 mm per year at a test temperature of 55°C.

A solid waste that exhibits the characteristic of corrosivity is given an EPA Hazardous Waste Number of D002.

Reactivity

A solid waste is said to exhibit the characteristic of reactivity if a representative sample of the waste has any of the following properties:

1. It is normally unstable and readily undergoes violent change without detonating.
2. It reacts violently with water.
3. It forms potentially explosive mixtures with water.
4. When mixed with water, it generates toxic gases, vapors, or fumes in a quantity sufficient to present a danger to human health or the environment.

*Although it would seem to be a contradiction in terms, that is, calling a solid waste a liquid, Congress has done what was once only the province of the gods. In Section 1004 (27) of the Resource Conservation and Recovery Act of 1976, they saw fit to violate the laws of physics and make all of the physical states (liquids, gases, and solids) one and the same, that is, solid waste. By their definition, almost any discarded material is solid waste.

5. It is a cyanide or sulfide-bearing waste that, when exposed to pH between 2 and 12.5, can generate toxic gases, vapors, or fumes in a quantity sufficient to present a danger to human health or the environment.
6. It is capable of detonation or explosive reaction if it is subjected to a strong initiating source or if heated under confinement.
7. It is readily capable of detonation or explosive decomposition or reaction at standard temperature and pressure.
8. It is a forbidden explosive, as defined in Department of Transportation regulations (49 CFR 173.51, 173.53, and 173.88).

A solid waste that exhibits the characteristic of reactivity is given an EPA Hazardous Waste Number of D003.

Toxicity

A solid waste is said to exhibit the characteristic of extraction procedure (EP) toxicity if, using the test methods described in Appendix II of the *Federal Register* (55 FR 11863 and 55 FR 26986), the extract from a representative sample of the waste contains any of the contaminants listed in Table 10-10 at a concentration equal to or greater than the respective value given in the table.

Figure 10-6 shows a generalized flow scheme for determining if a waste is hazardous according to EPA definitions. Of particular importance in the scheme are those things that are not included in the RCRA regulations. For example, domestic sewage, certain nuclear material, household wastes, including toxic and hazardous materials, and small quantities (less than 100 kg/mo) are excluded from the RCRA regulations. This does not mean that these wastes are not regulated at all. In fact they are regulated under other statutes and, thus, do not need to be regulated under RCRA.

Four important but controversial parts of the definition of a hazardous waste are the mixture rule, "contained in" policy, "derived from" rule, and waste-code carry through principle.

The *mixture rule* prevents dilution of waste for the purpose of escaping RCRA Subtitle C regulation. Under 40 CFR 261.3(a)(2), mixtures of a listed hazardous waste and other solid wastes become hazardous wastes. In certain instances for characteristic wastes and in those cases where the listed waste is not to be land disposed, dilution is permitted. The *dilution rules* are summarized in 56 FR 3875, 31 JAN 1991. When the dilution rules apply, the mixture of a hazardous waste with the diluent does not cause the diluent to become hazardous and may render the hazardous waste nonhazardous.

A corollary to the mixture rule is the "*contained in*" policy. Under this policy, media such as soil and water are treated as hazardous wastes if they "contain" listed hazardous wastes.

Any solid waste generated from the treatment, storage, or disposal of a hazardous waste, including any sludge, spill residue, ash, emission-control dust, or leachate (but not precipitation run-off) is a hazardous waste (40 CFR 261.3(c)). This is known as the "*derived from*" rule.

TABLE 10-10
Toxicity characteristic constituents and regulatory levels

EPA HW No.[a]	Constituent	Regulatory level (mg/L)
D004	Arsenic	5.0
D005	Barium	100.0
D018	Benzene	0.5
D006	Cadmium	1.0
D019	Carbon tetrachloride	0.5
D020	Chlordane	0.03
D021	Chlorobenzene	100.0
D022	Chloroform	6.0
D007	Chromium	5.0
D023	o-Cresol	200.0[b]
D024	m-Cresol	200.0[b]
D025	p-Cresol	200.0[b]
D026	Cresol	200.0[b]
D016	2,4-D	10.0
D027	1,4-Dichlorobenzene	7.5
D028	1,2-Dichloroethane	0.5
D029	1,1-Dichloroethylene	0.7
D030	2,4-Dinitrotoluene	0.13[c]
D012	Endrin	0.02
D031	Heptachlor (and its epoxide)	0.008
D032	Hexachlorobenzene	0.13[c]
D033	Hexachloro-1,3-butadiene	0.5
D034	Hexachloroethane	3.0
D008	Lead	5.0
D013	Lindane	0.4
D009	Mercury	0.2
D014	Methoxychlor	10.0
D035	Methyl ethyl ketone	200.0
D036	Nitrobenzene	2.0
D037	Pentachlorophenol	100.0
D038	Pyridine	5.0[c]
D010	Selenium	1.0
D011	Silver	5.0
D039	Tetrachloroethylene	0.7
D015	Toxaphene	0.5
D040	Trichloroethylene	0.5
D041	2,4,5-Trichlorophenol	400.0
D042	2,4,6-Trichlorophenol	2.0
D017	2,4,5-TP (Silvex)	1.0
D043	Vinyl chloride	0.2

[a] Hazardous waste number.
[b] If o-, m-, and p-cresol concentrations cannot be differentiated, the total cresol (D026) concentration is used. The regulatory level for total cresol is 200 mg/L.
[c] Quantitation limit is greater than the calculated regulatory level. The quantitation limit therefore becomes the regulatory level.

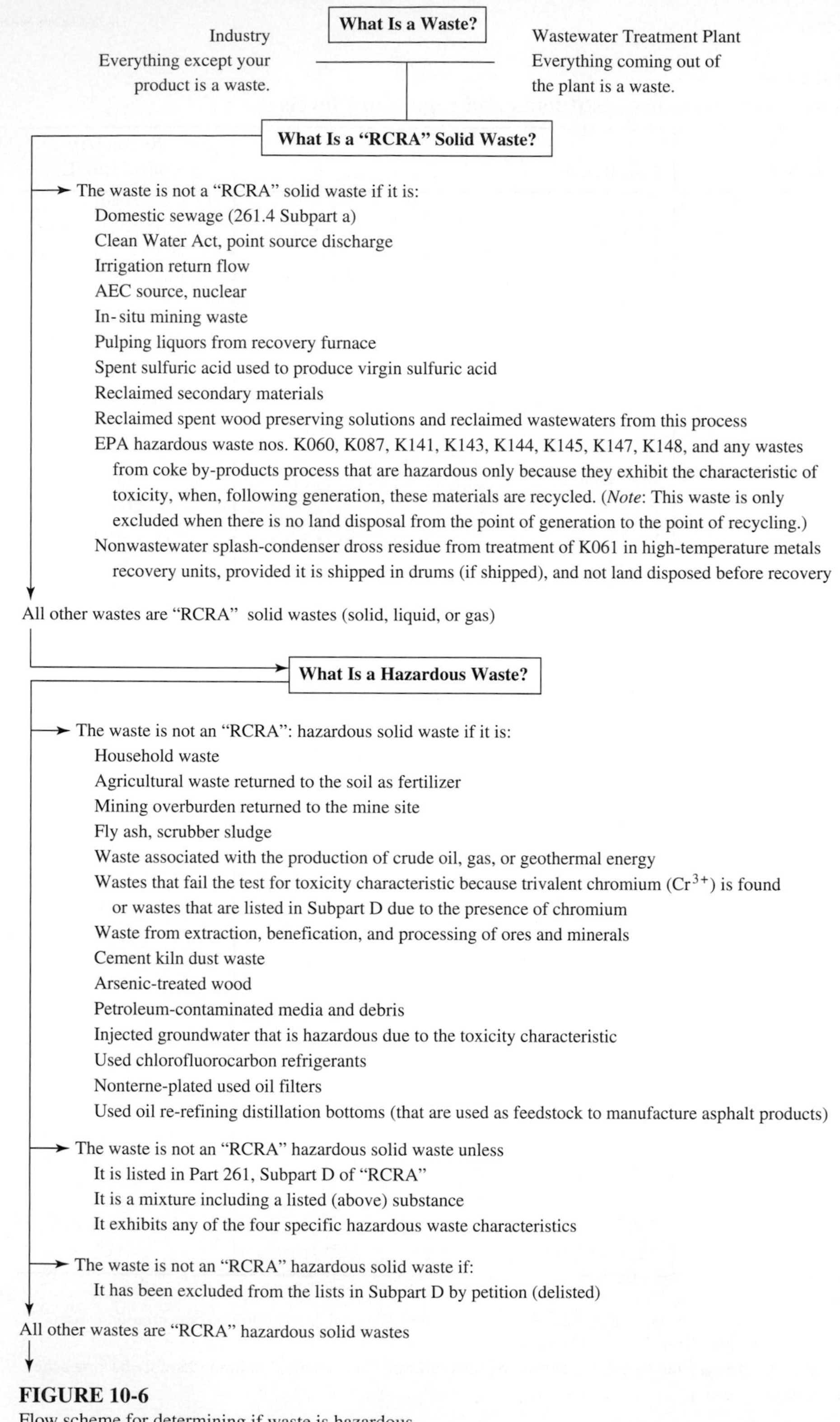

What Is a Waste?
Industry
Everything except your product is a waste.
Wastewater Treatment Plant
Everything coming out of the plant is a waste.
What Is a "RCRA" Solid Waste?
The waste is not a "RCRA" solid waste if it is:
Domestic sewage (261.4 Subpart a)
Clean Water Act, point source discharge
Irrigation return flow
AEC source, nuclear
In-situ mining waste
Pulping liquors from recovery furnace
Spent sulfuric acid used to produce virgin sulfuric acid
Reclaimed secondary materials
Reclaimed spent wood preserving solutions and reclaimed wastewaters from this process
EPA hazardous waste nos. K060, K087, K141, K143, K144, K145, K147, K148, and any wastes from coke by-products process that are hazardous only because they exhibit the characteristic of toxicity, when, following generation, these materials are recycled. (Note: This waste is only excluded when there is no land disposal from the point of generation to the point of recycling.)
Nonwastewater splash-condenser dross residue from treatment of K061 in high-temperature metals recovery units, provided it is shipped in drums (if shipped), and not land disposed before recovery
All other wastes are "RCRA" solid wastes (solid, liquid, or gas)
What Is a Hazardous Waste?
The waste is not an "RCRA": hazardous solid waste if it is:
Household waste
Agricultural waste returned to the soil as fertilizer
Mining overburden returned to the mine site
Fly ash, scrubber sludge
Waste associated with the production of crude oil, gas, or geothermal energy
Wastes that fail the test for toxicity characteristic because trivalent chromium (Cr^{3+}) is found or wastes that are listed in Subpart D due to the presence of chromium
Waste from extraction, benefication, and processing of ores and minerals
Cement kiln dust waste
Arsenic-treated wood
Petroleum-contaminated media and debris
Injected groundwater that is hazardous due to the toxicity characteristic
Used chlorofluorocarbon refrigerants
Nonterne-plated used oil filters
Used oil re-refining distillation bottoms (that are used as feedstock to manufacture asphalt products)
The waste is not an "RCRA" hazardous solid waste unless
It is listed in Part 261, Subpart D of "RCRA"
It is a mixture including a listed (above) substance
It exhibits any of the four specific hazardous waste characteristics
The waste is not an "RCRA" hazardous solid waste if:
It has been excluded from the lists in Subpart D by petition (delisted)
All other wastes are "RCRA" hazardous solid wastes

FIGURE 10-6
Flow scheme for determining if waste is hazardous.

What Wastes Are Subject to Regulation?

The "RCRA" hazardous solid waste is currently not subject to Subtitle C regulations if:

The total combined "RCRA" hazardous waste generated at the site is less than 100 kg/month

It is intended to be legitimately reclaimed or reused, (261.6). However it is subject to RCRA reporting requirements regarding storage and transportion if it is a sludge or contains a Part 261 listed substance.

The "RCRA" hazardous solid waste is temporarily exempt from certain regulations if:

It is a hazardous waste that is generated in a product or raw material storage tank, a product or raw material transport vehicle or vessel, a product or raw material pipeline, or in a manufacturing process unit or an associated nonwaste-treatment-manufacturing unit.

All other "RCRA" hazardous solid wastes are subject to Subtitle C of RCRA regulation with respect to disposal, transport, and storage.

Requirements for Recyclable Materials

Hazardous wastes that are not subject to requirements for generator, transporters, and storage facilities:

Regulated under Subparts C through H (261.6):

Recyclable materials used in a manner constituting disposal

Hazardous wastes burned for energy recovery in boilers and industrial furnaces

Recyclable materials from which precious metals are reclaimed

Spent lead-acid batteries that are being reclaimed

Not subject to regulation or to the notification requirements of RCRA:

Industrial ethyl alcohol that is reclaimed

Used batteries returned to a battery manufacturer for regeneration

Scrap metal

Fuels produced from the refining of oil bearing hazardous wastes

Oil reclaimed from hazardous waste resulting from normal petroleum refining, production, and transportation practices

Hazardous waste fuel produced from oil-bearing hazardous wastes

Petroleum coke produced from petroleum refinery hazardous wastes

Used oil that is recycled and is also a hazardous waste solely because it exhibits a hazardous characteristic is not subject to the requirements of Parts 260–268 of this chapter, but is regulated under Part 279 of this chapter.

FIGURE 10-6
(*continued*)

A corollary to the derived from and mixture rules is the "*waste-code carry through*" principle. The principle states that a solid waste derived from a hazardous waste or a mixture of hazardous and nonhazardous waste contains all of the same waste codes as the original waste (53 FR 31138, 31148).

Because of a legal suit and Court order regarding the mixture rule, contained in policy, derived from rule, and waste-code carry through principle, EPA developed and proposed the *Hazardous Waste Identification Rule* (HWIR) (60 FR 66344, 12 DEC 1995). The proposed HWIR establishes exit levels for low-risk solid wastes that are designated as hazardous because they are listed, or have been mixed with, derived from, or contain a listed hazardous waste. If the constituents of the waste are below the exit levels, the waste may be disposed of as a Subtitle D waste.

The *Universal Waste Regulations* (40 CFR 273) were developed by EPA to streamline hazardous waste management standards for federal universal wastes (batteries, pesticides, thermostats, and lamps) that are widely generated. Batteries such as nickel-cadmium and small sealed lead-acid types, unused or banned pesticides, thermostats that contain mercury, and lamps such as fluorescent lights, neon, and mercury vapor that contain mercury or lead are considered *universal wastes*. The streamlining includes, for example, provisions to extend the amount of time that businesses can accumulate wastes, to increase the amount of waste that can be accumulated, and to eliminate the need for a manifest.

RCRA provides a petition mechanism (40 CFR 260.20 and 260.22) for excluding a waste from nonspecific sources and at a particular generating facility. Those wastes that successfully pass the petition process are *delisted*. The list of delisted wastes appears in Appendix IX of 40 CFR 261.

Some waste streams do not come under the purview of RCRA but are, nonetheless, considered hazardous. These special wastes include, for example, polychlorinated biphenyls (PCBs) and asbestos. PCBs and asbestos are regulated under the Toxic Substances Control Act (abbreviated TSCA and pronounced "tas-kah").

10-4 RCRA AND HSWA

Congressional Actions on Hazardous Waste

In 1976 Congress passed the Resource Conservation and Recovery Act (abbreviated RCRA and pronounced "rick-rah") directing the U.S. Environmental Protection Agency to establish hazardous waste regulations. RCRA was amended in 1984 by the Hazardous and Solid Waste Amendments (abbreviated HSWA and pronounced "hiss-wah"). RCRA and HSWA were enacted to regulate the generation and disposal of hazardous wastes. These acts did not address abandoned or closed waste disposal sites or spills. The Comprehensive Environmental Response, Compensation, and Liability Act (abbreviated by CERCLA and pronounced "sir-klah"), commonly referred to as "Superfund," was enacted in 1980 to address these problems. SARA, the Superfund Amendments and Reauthorization Act of 1986, extended the provisions of CERCLA. In the following sections, we shall attempt to tell you about the who, what, where, and how of RCRA, HSWA, CERCLA, and SARA.

Cradle-to-Grave Concept

The EPA's cradle-to-grave hazardous waste management system is an attempt to track hazardous waste from its generation point (the "cradle") to its ultimate disposal point (the "grave"). The system requires generators to attach a manifest (itemized list describing the contents) form to their hazardous waste shipments. This procedure is designed to ensure that wastes are directed to, and actually reach, a permitted disposal site.

Generator Requirements

Generators of hazardous waste are the first link in the cradle-to-grave chain of hazardous waste management established under RCRA. Generators of more than 100 kilograms of hazardous waste or 1 kilogram of acutely hazardous waste per month must (with a few exceptions) comply with all of the generator regulations.

The regulatory requirements for hazardous waste generators include (U.S. EPA, 1986):

1. Obtaining an EPA ID number
2. Handling of hazardous waste before transport
3. Manifesting of hazardous waste
4. Recordkeeping and reporting

EPA assigns each generator a unique identification number. Without this number the generator is barred from treating, storing, disposing of, transporting, or offering for transportation any hazardous waste. Furthermore, the generator is forbidden from offering the hazardous waste to any transporter, or treatment, storage, or disposal (TSD) facility that does not also have an EPA ID number.

Pretransport regulations are designed to ensure safe transportation of a hazardous waste from origin to ultimate disposal. In developing these regulations, EPA adopted those used by the Department of Transportation (DOT) for transporting hazardous wastes (49 CFR Parts 172, 173, 178, and 179). These DOT regulations require:

1. Proper packaging to prevent leakage of hazardous waste during both normal transport conditions and in potentially dangerous situations, such as when a drum falls out of a truck.
2. Identification of the characteristics and dangers associated with the wastes being transported through labeling, marking, and placarding of the packaged waste.

These pretransport regulations only apply to generators shipping waste off site.

In addition to adopting the DOT regulations outlined above, EPA also developed pretransport regulations that cover the accumulation of waste prior to transport. A generator may accumulate hazardous waste on site for 90 days or less as long as the following requirements are met:

1. *Proper storage:* The waste is properly stored in containers or tanks marked with the words "Hazardous Waste" and the date on which accumulation began.
2. *Emergency plan:* There is a contingency plan and emergency procedures for use in an emergency.
3. *Personnel training:* Facility personnel are trained in the proper handling of hazardous waste.

The 90-day period allows a generator to collect enough waste to make transportation more cost effective, that is, instead of paying to haul several small shipments of waste, the generator can accumulate waste until there is enough for one big shipment.

If the generator accumulates hazardous waste on site for more than 90 days, it is considered an operator of a storage facility and must comply with requirements for such facilities. Under temporary, unforeseen, and uncontrollable circumstances, the 90-day period may be extended for up to 30 days by the EPA Regional Administrator on a case-by-case basis.

There is an exception to this 90-day accumulation period that applies to generators of between 100 and 1,000 kg/mo of hazardous waste who ship their waste off site. People who fall in this category are called *small quantity generators* (SQG). HSWA required, and the EPA developed, regulations that allow such generators to accumulate waste for 180 days (or 270 days if the waste must be shipped over 320 km) before they are considered the operator of a storage facility.

The Uniform Hazardous Waste Manifest (the *manifest*) is the key to cradle-to-grave waste management (see Figure 10-7). Through the use of a manifest, generators can track the movement of hazardous waste from the point of generation to the point of ultimate treatment, storage, or disposal.

HSWA requires that each manifest certify that the generator has in place a program to reduce the volume and toxicity of the waste to the degree that is economically practicable, as determined by the generator, and that the treatment, storage, or disposal method chosen by the generator is the best practicable method currently available that minimizes the risk to human health and the environment.

It is especially important for the generators to prepare the manifest properly since they are responsible for the hazardous waste they produce and its ultimate disposition.

The manifest is part of a controlled tracking system. Each time the waste is transferred, that is, from a transporter to the designated facility or from a transporter to another transporter, the manifest must be signed to acknowledge receipt of the waste. A copy of the manifest is retained by each link in the transportation chain. Once the waste is delivered to the designated facility, the owner or operator of the facility must send a copy of the manifest back to the generator. This system ensures that the generator has documentation that the hazardous waste has made it to its ultimate destination.

If 35 days pass from the date on which the waste was accepted by the initial transporter and the generator has not received a copy of the manifest from the designated facility, the generator must contact the transporter and/or the designated facility to determine the whereabouts of the waste. If 45 days pass and the manifest still has not been received, the generator must submit an exception report.

The recordkeeping and reporting requirements for generators provide EPA and states with a method for tracking the quantities of waste generated and the movement of hazardous wastes.

Transporter Regulations

Transporters of hazardous waste are the critical link between the generator and the ultimate off-site treatment, storage, or disposal of hazardous waste. The transporter regulations were developed jointly by EPA and the DOT to avoid contradictory requirements coming from the two agencies (U.S. EPA, 1986). Although

Please print or type. *(Form designed for use on elite (12-pitch) typewriter.)* *Form Approved. OMB No. 2000-0404. Expires 7-31-86*

UNIFORM HAZARDOUS WASTE MANIFEST

1. Generator's US EPA ID No. — Manifest Document No. — 2. Page 1 of — Information in the shaded areas is not required by Federal law.

GENERATOR

3. Generator's Name and Mailing Address
A. State Manifest Document Number
B. State Generator's ID
4. Generator's Phone ( )
5. Transporter 1 Company Name — 6. US EPA ID Number
C. State Transporter's ID
D. Transporter's Phone
7. Transporter 2 Company Name — 8. US EPA ID Number
E. State Transporter's ID
F. Transporter's Phone
9. Designated Facility Name and Site Address — 10. US EPA ID Number
G. State Facility's ID
H. Facility's Phone

11. US DOT Description *(Including Proper Shipping Name, Hazard Class, and ID Number)*	12. Containers No.	12. Containers Type	13. Total Quantity	14. Unit Wt/Vol	I. Waste No.
a.					
b.					
c.					
d.					

J. Additional Descriptions for Materials Listed Above
K. Handling Codes for Wastes Listed Above

15. Special Handling Instructions and Additional Information

16. GENERATOR'S CERTIFICATION: I hereby declare that the contents of this consignment are fully and accurately described above by proper shipping name and are classed, packed, marked, and labeled, and are in all respects in proper condition for transport by highway according to applicable international and national government regulations.

Unless I am a small quantity generator who has been exempted by statute or regulation from the duty to make a waste minimization certification under Section 3002(b) of RCRA, I also certify that I have a program in place to reduce the volume and toxicity of waste generated to the degree I have determined to be economically practicable and I have selected the method of treatment, storage, or disposal currently available to me which minimizes the present and future threat to human health and the environment.

Printed/Typed Name — Signature — Month Day Year

TRANSPORTER

17. Transporter 1 Acknowledgement of Receipt of Materials
Printed/Typed Name — Signature — Month Day Year

18. Transporter 2 Acknowledgement of Receipt of Materials
Printed/Typed Name — Signature — Month Day Year

FACILITY

19. Discrepancy Indication Space

20. Facility Owner or Operator: Certification of receipt of hazardous materials covered by this manifest except as noted in Item 19.
Printed/Typed Name — Signature — Month Day Year

EPA Form 8700-22 (Rev. 4-85) Previous edition is obsolete.

FIGURE 10-7
Uniform hazardous waste manifest.

the regulations are integrated, they are not contained under the same act. A transporter must comply with the regulations under 49 CFR 171-179 (The Hazardous Materials Transportation Act) as well as those under 40 CFR Part 263 (Subtitle C of RCRA).

Even if generators and transporters of hazardous waste comply with all appropriate regulations, transporting hazardous waste can still be dangerous. There is always the possibility that an accident will occur. To deal with this possibility, the regulations require transporters to take immediate action to protect health and the environment if a release occurs by notifying local authorities and/or diking off the discharge area.

The regulations also give officials special authority to deal with transportation accidents. Specifically, if a federal, state, or local official, with appropriate authority, determines that the immediate removal of the waste is necessary to protect human health or the environment, the official can authorize waste removal by a transporter who lacks an EPA ID and without the use of a manifest.

Treatment, Storage, and Disposal Requirements

Treatment, storage, and disposal facilities (TSDs) are the last link in the cradle-to-grave hazardous waste management system. All TSDs handling hazardous waste must obtain an operating permit and abide by the treatment, storage, and disposal regulations. The TSD regulations establish performance standards that owners and operators must apply to minimize the release of hazardous waste into the environment.

A TSD facility may perform one or more of the following functions (U.S. EPA, 1986):

1. *Treatment:* Any method, technique, or process, including neutralization, designed to change the physical, chemical, or biological character or composition of any hazardous waste so as to neutralize it or render it nonhazardous or less hazardous; to recover it; make it safer to transport, store, or dispose of; or make it amenable for recovery, storage, or volume reduction.

2. *Storage:* The holding of hazardous waste for a temporary period, at the end of which the hazardous waste is treated, disposed, or stored elsewhere.

3. *Disposal:* The discharge, deposit, injection, dumping, spilling, leaking, or placing of any solid waste or hazardous waste into or on any land or water so that any constituent thereof may enter the environment or be emitted into the air or discharged into any waters, including groundwaters.

The act establishes standards that consist of administrative-nontechnical requirements and technical requirements.

The purpose of the administrative/nontechnical requirements is to ensure that owners and operators of TSDs establish the necessary procedures and plans to operate a facility properly and to handle any emergencies or accidents. They cover the subject areas shown below:

Subpart	Subject
A	Who is subject to the regulations
B	General facility standards
	Waste analysis, security, inspections, training
	Ignitable, reactive, or incompatible wastes
	Location standards (permitted facilities)
C	Preparedness and prevention
D	Contingency plans and emergency procedures
E	Manifest system, recordkeeping, and reporting

The objective of the interim status technical requirements is to minimize the potential for threats resulting from hazardous waste treatment, storage, and disposal at existing facilities waiting to receive an operating permit. There are two groups of interim status requirements: general standards that apply to several types of facilities and specific standards that apply to a waste management method.

The general standards cover three areas:

1. Groundwater monitoring requirements
2. Closure, postclosure requirements
3. Financial requirements

Groundwater monitoring is only required of owners or operators of a surface impoundment, landfill, land treatment facility, or some waste piles used to manage hazardous waste. The purpose of these requirements is to assess the impact of a facility on the groundwater beneath it. Monitoring must be conducted for the life of the facility except at land disposal facilities, which must continue monitoring for up to 30 years after the facility has closed.

The groundwater monitoring program outlined in the regulations requires a monitoring system of four wells to be installed: one upgradient from the waste management unit and three downgradient. The downgradient wells must be placed so as to intercept any waste migrating from the unit, should such a release occur. The upgradient wells must provide data on groundwater that is not influenced by waste coming from the waste management unit (called background data). If the wells are properly located, comparison of data from upgradient and downgradient wells should indicate if contamination is occurring.

Once the wells have been installed, the owner or operator monitors them for one year to establish background concentrations for selected chemicals. These data form the basis for all future data comparisons. There are three sets of parameters for which background concentrations are established: drinking water parameters, groundwater quality parameters, and groundwater contamination parameters.

Closure is the period when wastes are no longer accepted, during which owners or operators of TSD facilities complete treatment, storage, and disposal operations, apply final covers to or cap landfills, and dispose of or decontaminate equipment, structures, and soil. Postclosure, which applies only to disposal facilities, is the 30-year period after closure during which owners or operators of disposal facilities conduct monitoring and maintenance activities to preserve and look after the integrity of the disposal system.

Financial requirements were established to ensure that funds are available to pay for closing a facility, for rendering post-closure care at disposal facilities, and to compensate third parties for bodily injury and property damage caused by sudden and non-sudden accidents related to the facility's operation (states and federal governments are exempted from abiding by these requirements). There are two kinds of financial requirements: financial assurance for closure/postclosure and liability coverage for injury and property damage.

Land Ban. The Hazardous and Solid Waste Amendments (HSWA) of 1984 significantly expanded the scope of the Resource Conservation and Recovery Act (RCRA). HSWA was created, in large part, in response to strongly voiced citizen concerns that existing methods of hazardous waste disposal, particularly land disposal, were not safe. Section 3004 of the act sets restrictions on land disposal of specific wastes. This is commonly called the "land ban," or *land disposal restrictions* (LDR). As specifically required by Section 3004(m), the agency established levels or methods of treatment, if any, which substantially reduce the likelihood of migration of hazardous constituents from waste so that short-term and long-term threats to human health and the environment are minimized. Congress established a stringent timetable for development of treatment standards. After the effective date of the promulgated standards, listed and characteristic wastes must be treated to meet the standards before the wastes can be placed in any form of land disposal facility. The only exception is where a special variance is approved based on a showing of no migration of hazardous constituents from the land disposal site for as long as the waste remains hazardous. The last set of Congressionally mandated standards was promulgated in accordance with the timetable on May 8, 1990. EPA has subsequently published revisions to clarify and streamline the standards. The *Universal Treatment Standards* (UTS) are of particular note in this respect.

Prior to 1994, treatment facilities managing hazardous waste often had to meet LDR treatment standards established for many different listed and characteristic wastes. In some cases, a constituent regulated to a given concentration level for one waste was also regulated in another waste at a different concentration level. On 18 SEP 1994, EPA published the UTS to eliminate these differences (59 FR 47980, 18 SEP 1994, and 60 FR 242, 3 JAN 1995).

Underground Storage Tanks (UST)

A "UST system"* includes an underground storage tank, connected piping, underground ancillary equipment, and containment system, if any. On September 23, 1988, the EPA promulgated the final rules for underground storage tanks.

There are a number of exclusions to the new regulations, including:

Hazardous waste UST systems

Regulated wastewater treatment facilities

---

* You can imagine the chagrin of regulators and others when the acronym for Leaking Underground Storage Tanks appears on meeting agendas and technical symposia!

Any equipment or machinery that contains regulated substances for operational purposes such as hydraulic lift tanks and electrical equipment tanks

Any UST system of less than 415 liters

Any UST system containing a *de minimis* (negligible) concentration of regulated substances

Any emergency spill or overflow containment system that is expeditiously emptied after use

All UST systems must have corrosion protection. There are three ways to obtain corrosion protection for tanks: (1) construction of fiberglass-reinforced plastic, (2) steel- and fiberglass-reinforced plastic composite, or (3) a coated steel tank with cathodic protection. Cathodic protection systems must be regularly tested and inspected. All owners and operators must also provide spill and overfill prevention equipment and a certificate of installation to ensure that the methods of installation were in compliance with the regulations.

Release (leak) detection must be instituted for all UST systems. Several different methods are allowed for petroleum UST systems. However, some systems have specific requirements, for instance, a pressurized delivery system must be equipped with an automatic line leak detector and have an annual line tightness test. All new or upgraded UST systems storing hazardous substances must have secondary containment with interstitial monitoring.

When release is confirmed, owners and operators must begin corrective action. Immediate corrective action measures include mitigation of safety and fire hazards, removal of saturated soils and floating free product, and an assessment of further corrective action needed. As with any remediation situation, a corrective action plan may be required for long-term cleanups of contaminated soil and groundwater.

10-5 CERCLA AND SARA

The Superfund Law

The Comprehensive Environmental Response, Compensation, and Liability Act (CERCLA) of 1980, better known as "Superfund," became law "to provide for liability, compensation, cleanup and emergency response for hazardous substances released into the environment and the cleanup of inactive hazardous waste disposal sites." CERCLA was generally intended to give EPA authority and funds to clean up abandoned waste sites and to respond to emergencies related to hazardous waste. The law provides for both response and enforcement mechanisms. The four major provisions of the law establish:

1. A fund (the "superfund") to pay for investigations and remedies at sites where the responsible people cannot be found or will not voluntarily pay;
2. A priority list of abandoned or inactive hazardous waste sites for cleanup (the National Priority List);
3. The mechanism for action at abandoned or inactive sites (the National Contingency Plan);
4. Liability for those responsible for cleaning up.

Initially the trust fund was supported by taxes on producers and importers of petroleum and 42 basic chemicals. In its first five-year period, Superfund collected about 1.6 billion dollars, with 86 percent of that money coming from industry and the remainder from federal government appropriations. In 1986 the Superfund Amendments and Reauthorization Act (SARA) greatly expanded the money available to remediate Superfund sites. The fund was raised to $8.6 billion for a five-year period by taxing petroleum products ($2.75 billion), business income ($2.5 billion), and chemical feedstocks ($1.4 billion). The remainder is from general revenues.

The National Priority List (NPL)

The NPL serves as a tool for the EPA to use in identifying sites that appear to present a significant risk to public health or the environment and that may merit use of Superfund money. First published in 1982, it is updated three times a year. In September 2005 the list contained 1,239 sites (U.S. EPA, 2005b). The first NPL was formulated from notification procedures and existing information sources. Subsequently, a numeric ranking system known as the *Hazard Ranking System* (HRS) was developed. Sites with high HRS scores may be added to the list. Sites on the NPL are eligible for Superfund money. Those with lower scores are not likely to be eligible.

The Hazard Ranking System (HRS)

The HRS is a procedure for ranking uncontrolled hazardous waste sites in terms of the potential threat based upon containment of the hazardous substances, route of release, characteristics and amount of the substances, and likely targets (40 CFR 300, Appendix A). The methodology of the HRS provides a quantitative estimate that represents the relative hazards posed by a site and takes into account the potential for human and environmental exposure to hazardous substances. The HRS score is based on the probability of contamination from four pathways—groundwater, surface water, soil, and air—on the site in question. The groundwater and air migration pathways are evaluated for ingestion and inhalation respectively. The surface water migration and soil exposure pathways are evaluated for multiple intake routes. Surface water is evaluated for (1) drinking water, (2) human food chain, and (3) environmental (contact) exposures. These exposures are evaluated for two separate migration components—overland/flood migration and groundwater to surface water migration. Soil is evaluated for potential exposure to the (1) resident population and (2) nearby population.

Use of the HRS requires considerable information about the site and its surroundings, the hazardous substances present, and the geology of the aquifers and the intervening strata. The factors that most affect an HRS site score are the proximity to a densely populated area or source of drinking water, the quantity of hazardous substances present, and the toxicity of those hazardous substances. The HRS methodology has been criticized for the following reasons:

1. There is a strong bias toward human health effects, with only a slight chance of a site in question receiving a high score if it represents only a threat or hazard to the environment.
2. Because of the human health bias, there is an even stronger bias in favor of highly populated affected areas.

3. The air emission migration route must be documented by an actual release, while groundwater and surface water routes have no such documentation requirement.

4. The scoring for toxicity and persistence of chemicals may be based on site containment, which is not necessarily related to a known or potential release of the toxic chemicals.

5. A high score for one migration route can be more than offset by low scores for the other migration routes.

6. Averaging of the route scores creates a bias against a site that has only one hazard, even though that one hazard may pose extreme threat to human health and the environment.

The HRS scores range from 0 to 100, with a score of 100 representing the most hazardous site. Occasional exceptions have been made in the HRS priority ranking to meet the CERCLA requirement that a site designated by a state as its top priority be included on the NPL.

The National Contingency Plan (NCP)

The *National Contingency Plan* (NCP) provides detailed direction on the action to be taken at a hazardous waste site, including initial assessment to determine if an emergency or imminent threat exists, emergency response actions, and a method to rank sites (the HRS) and establish priority for future action. When there is sufficient indication that a site poses a potential risk to the environment, a detailed study is required.

The NCP describes the steps to be taken for the detailed evaluation of the risks associated with a site. Such an evaluation is termed a *remedial investigation* (RI). The process of selecting an appropriate remedy is termed the *feasibility study* (FS). The remedial investigation and the feasibility study are often combined into a single measure, known popularly as a remedial investigation/feasibility study (RI/FS). The requirements of the RI/FS are usually outlined in a written work plan, which must be approved by the relevant federal and state agencies before it may be implemented.

A remedial investigation includes the development of detailed plans that address the following items (40 CFR 300.400):

1. *Site characterization:* A description of the hydrogeological and geophysical sampling and analytical procedures to be applied in order to discover the nature and extent of the waste materials, the physical characteristics of the site, and any receptors that could be affected by the wastes at the site.

2. *Quality control:* The guidelines to be enforced to ensure that all the data collected from the characterization program are valid and satisfactorily accurate.

3. *Health and safety:* The procedures to be employed to protect the safety of the individuals who will work at the site and perform the site characterization.

The RI activities and subsequent evaluation of the data gathered are termed a *risk assessment* or an endangerment assessment. The remedial investigation report documents the evaluation.

The remedial investigation report serves as a basis for the feasibility study, which evaluates various remedial alternatives. The review criteria include: overall protection of human health and the environment; compliance with applicable or relevant and appropriate regulations; long-term effectiveness; reduction in toxicity, mobility, or volume; short-term effectiveness; technical and administrative implementability; cost; state acceptance; and community acceptance. All the remedies selected must be capable of reducing the risk at the hazardous waste site to an acceptable level. And, in general, the lowest-cost alternative that achieves this objective is chosen as the course of action. The results of the feasibility study are presented in a written report, called the *record of decision* (ROD). This document serves as a preliminary basis for the design of the selected alternative.

One of the keys to the National Contingency Plan is that it specifies that the degree of cleanup be selected in accordance with several criteria, including the degree of hazard to the "public health, welfare and the environment." Therefore, there is no predetermined level of remediation that can be required or that must be achieved at any site. Rather, the degree of correction is established on a site-by-site basis. What is acceptable in one location may not necessarily be acceptable in another.

On completion and approval of the RI/FS, the next step is the preparation of plans and specifications for the selected remedy—the *remedial design* (RD). To complete the process, the actual construction and other activities are undertaken in accordance with the plans and specifications.

Liability

Perhaps the most far reaching provision of CERCLA that has stood the test of the courts was the establishment of *strict, joint, and several liability* for cleanup of an NPL site. Those identified by EPA as *potentially responsible parties* (PRPs) may include generators, present owners, or former owners of facilities or real property where hazardous wastes have been stored, treated, or disposed of, as well as those who accepted hazardous waste for transport and selected the facility. PRPs have strict liability; that is, liability without fault. Neither care nor negligence, neither good nor bad faith, neither knowledge nor ignorance, can be claimed as a defense. Congress correctly predicted that there would be instances where the PRPs would contest their contribution to the problem and would, then, be unwilling to share the costs or the responsibility. The strict liability provision orders that the PRP is liable even if the method of disposal was in accordance with prevailing standards, laws, and practice at the time of disposal. In other words, CERCLA is a "pay now, argue later" statute (O'Brien & Gere, 1988).

Although the language specific to "joint and several" liability was removed from CERCLA, the courts have interpreted the law as though the language were included. This means that if a PRP contributed any wastes to a site, that PRP can be held accountable for all costs associated with the cleanup. This concept was strongly reaffirmed in SARA. If the PRP refuses to pay, the federal government can sue to recover costs. These actions have been successful. In certain instances where those liable fail, without sufficient cause, to properly provide for cleanup, they may be liable for treble damages!

Superfund Amendments and Reauthorization Act (SARA)

SARA reaffirmed and strengthened many of the provisions and concepts of the CERCLA program. In SARA, Congress clearly expressed a preference, but not a require ment, for remedies such as incineration or chemical treatment that render a waste nonhazardous rather than transport to another disposal site or simple containment on site.

SARA directs that the level of cleanup should achieve compliance with *Applicable or Relevant and Appropriate Requirements* (ARARs). ARARs are environmental standards from programs other than CERCLA and SARA. For example, if a state has a regulation regarding atmospheric emissions from incinerators, then a Superfund cleanup using incineration must meet those *applicable* standards. Furthermore, if a similar standard appears to be *relevant* and *appropriate,* then EPA may elect to apply it. For example, if drums of waste found on an uncontrolled hazardous waste site have contents that appear to have the same constituents as F001-F005 spent solvent, then the UTS standards for RCRA waste may be considered relevant and appropriate even though there is no specific evidence to identify the origin of the waste.

SARA significantly strengthens the requirement to consider damages to natural resources, especially those off site.

Title III. SARA includes a major addition to the provisions of CERCLA, namely Title III—Emergency Planning and Community Right-to-Know. Under the Emergency Planning provisions, facilities must notify the State Emergency Response Commission if they have quantities of extremely hazardous substances that exceed EPA specified *Threshold Planning Quantities* (TPQ). In addition, communities must establish Local Emergency Planning Committees (LEPCs) to develop a chemical emergency response plan. This plan must include identification of regulated facilities, emergency response and notification procedures, training programs, and evacuation plans in case of a chemical release.

If a facility accidentally releases chemicals that are on one of two lists [that is, EPA's Extremely Hazardous Substance list or the CERCLA Section 103(a) list], in regulated quantities (RQ), and the release has the potential for exposure off-site, they must notify the LEPC immediately. The law also requires a report on response actions taken, known or anticipated health risks, and advice on medical attention for exposed individuals.

Perhaps the most revolutionary provision of Title III is the establishment of the *Community's Right-to-Know* amounts of chemicals and their location in facilities in their community. Thus, information about potential hazards from chemicals is available to the public. In addition, each year those facilities that release chemicals above specified threshold amounts must submit a *Toxic Release Inventory* (TRI) on an EPA-specified form ("Form R"). This inventory includes both accidental and routine releases, as well as off-site shipments of waste. The publication of these data has resulted in strenuous efforts by industry to control their previously unregulated and, hence, uncontrolled emissions because of the public outcry at the large quantities of materials being dumped into their environment.

10-6 HAZARDOUS WASTE MANAGEMENT

A logical priority in managing hazardous waste would be to:

1. Reduce the amount of hazardous wastes generated in the first place.
2. Stimulate "waste exchange." (One factory's hazardous wastes can become another's feedstock; for instance, acid and solvent wastes from some industries can be utilized by others without processing.)
3. Recycle metals, the energy content, and other useful resources contained in hazardous wastes.
4. Detoxify and neutralize liquid hazardous waste streams by chemical and biological treatment.
5. Reduce the volume of waste sludges generated in number four, above, by dewatering.
6. Destroy combustible hazardous wastes in special high-temperature incinerators equipped with proper pollution-control and monitoring systems.
7. Stabilize/solidify sludges and ash from numbers five and six to reduce leachability of metals.
8. Dispose of remaining treated residues in specially designed landfills.

Waste Minimization

The key elements necessary to the success of a waste-minimization program include (Fromm et al., 1986):

Top-level organizational commitment

Financial resources

Technical resources

Appropriate organization, goals, and strategy

The commitment of senior management is the first element that must be in place. Efforts to establish the other elements can follow. The organizational structure adopted should promote communication and feedback from participants. Often, the best ideas come from line operators who work with the processes day in and day out.

Some firms set quantitative waste-minimization goals. Other firms are more qualitative in their goal setting.

Waste Audit. An important first step in establishing a strategy for waste minimization is to conduct a waste audit. The audit should proceed stepwise:

1. Identify waste streams
2. Identify sources
3. Establish priority of waste streams for waste-minimization activity

4. Screen alternatives
5. Implement
6. Track
7. Evaluate progress

The key question that must be asked at the outset of a waste audit is "why is this waste being generated?" You must first establish the primary cause(s) of waste generation before attempting to find solutions. The audit should be waste stream–oriented in order to produce a list of specific waste-minimization options for additional evaluation or implementation. Once the causes are understood, solution options can be formulated. An efficient materials and waste tracking system that allows computation of mass balances is useful in establishing priorities. Knowing how much material is going in and how much of it is ending up as waste allows you to decide which process and which waste to address first.

Example 10-4. A manufacturing company has, as part of their first audit, gathered the following data. Estimate the potential annual air emissions in kg of VOCs from the company.

Purchasing department records

Material	Purchase Quantity (barrels)
Methylene chloride (CH_2Cl_2)	228
Trichloroethylene (C_2HCl_3)	505

Wastewater treatment plant influent

Material	Average Concentration (mg/L)
CH_2Cl_2	4.04
C_2HCl_3	3.25

(Average flow into treatment plant is 0.076 m^3/s.)

Hazardous waste manifests

Material	Barrels	Concentration (%)
CH_2Cl_2	228	25
C_2HCl_3	505	80

Unused barrels at end of year

CH_2Cl_2	8
C_2HCl_3	13

Solution. The materials balance diagram will be the same for each waste.

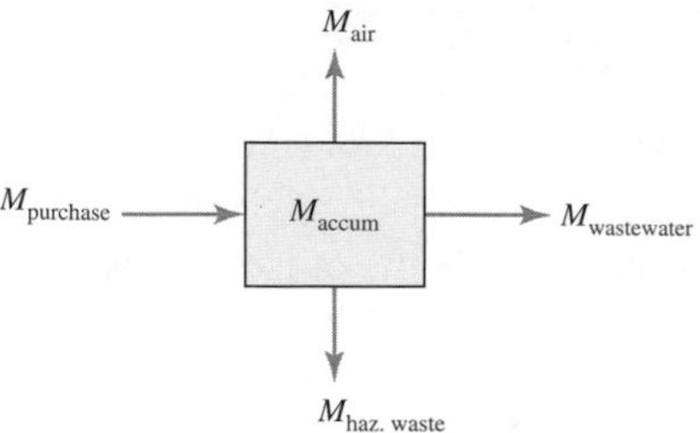

The mass balance equation would be

$$M_{purchase} = M_{air} + M_{ww} + M_{hw} + M_{accum}$$

Solving this equation for M_{air} gives us the estimated VOC emission.

First, we calculate the mass purchased. The density of each compound is found in Appendix A.

Mass Purchased

$$M(CH_2Cl_2) = (228 \text{ barrels/y})(0.12 \text{ m}^3\text{/barrel})(1{,}326 \text{ kg/m}^3)$$
$$= 36{,}279.36 \text{ kg/y}$$

$$M(C_2HCl_3) = (505 \text{ barrels/y})(0.12 \text{ m}^3\text{/barrel})(1{,}476 \text{ kg/m}^3)$$
$$= 89{,}445.60 \text{ kg/y}$$

Now we calculate the mass received at the wastewater treatment plant. (Note that mg/L = g/m^3.)

$$M(CH_2Cl_2) = (4.04 \text{ g/m}^3)(0.076 \text{ m}^3\text{/s})(86{,}400)(365)(10^{-3})$$
$$= 9{,}682.81 \text{ kg/y}$$

$$M(C_2HCl_3) = (3.25)(0.076)(86{,}400)(365)(10^{-3})$$
$$= 7{,}789.39 \text{ kg/y}$$

The mass shipped to the hazardous waste disposal facility is calculated next.

$$M(CH_2Cl_2) = (228)(0.12)(1{,}326)(0.25) = 9{,}069.84 \text{ kg/y}$$
$$M(C_2HCl_3) = (505)(0.12)(1{,}476)(0.80) = 71{,}556.48 \text{ kg/y}$$

Accumulated

$$M(CH_2Cl_2) = (8)(0.12)(1{,}326) = 1{,}272.96 \text{ kg/y}$$
$$M(C_2HCL_3) = (13)(0.12)(1{,}476) = 2{,}302.56 \text{ kg/y}$$

The estimated air emission for each compound is then

$$M(CH_2Cl_2) = 36{,}279.36 - 9{,}682.81 - 9{,}069.84 - 1{,}272.96$$
$$= 16{,}253.75 \text{ or } 16{,}000 \text{ kg/y}$$

$$M(C_2HCL_3) = 89{,}445.60 - 7{,}789.39 - 71{,}556.48 - 2{,}302.56$$
$$= 7{,}797.17 \text{ or } 7{,}800 \text{ kg/y}$$

Note that we round to two significant figures because the volume of the barrels is known to only two significant figures. From this analysis, to reduce the mass of air pollutants emitted, the company should attack the methylene chloride source first. We should also point out that simply counting "barrels in" from the purchasing record and "barrels out" from the hazardous waste manifest would give a highly erroneous picture of the environmental impact of this company's emissions. From a waste minimization point of view, it is also apparent that C_2HCl_3, at 80 percent concentration in barrels going to hazardous waste disposal, is a candidate for recycling.

The first four steps of the waste audit allow you to generate a comprehensive set of waste management options following the hierarchy of source reduction first, waste exchange second, recycling third, and treatment last.

The screening of options begins with source control. The source control investigation should focus on (1) changes in input materials, (2) changes in process technology, and (3) changes in the human aspect of production. Input material changes can be classified into three separate elements: purification, substitution, and dilution.

Purification of input materials is performed in order to avoid the introduction of inerts or impurities into the production process. Such an introduction results in waste because the process inventory must be purged in order to prevent the undesirable accumulation of impurities. Examples of purification of feed materials to lower waste generation include the use of deionized rinse water in electroplating or the use of oxygen instead of air in oxychlorination reactors for production of ethylene dichloride.

Substitution is the replacement of a toxic material with one characterized by lower toxicity or higher environmental desirability. Examples include using phosphates in place of dichromates as cooling water corrosion inhibitors or the use of alkaline cleaners in place of chlorinated solvents for degreasing.

Dilution is a minor component of input material changes and is exemplified by use of more dilute plating solutions to minimize *dragout* (material carried out of one tank into another).

Technology changes are those made to the physical plant. Examples include process changes, equipment, piping or layout changes, changes to process operational settings, additional automation, energy conservation, and water conservation.

Procedural and/or institutional changes consist of improvements in the ways people affect the production process. Also referred to as "good operating practices" or "good housekeeping," these include operating procedures, loss prevention, waste segregation, and material handling improvements.

Waste Exchange

Waste minimization by consignment of excess unused materials to an independent party for resale to a third party, saves both in waste production and in the cost (environmental and financial) of production from new raw materials. In essence "one person's trash becomes another person's treasure." The difference between a manufacturing by-product, which is costly to treat or dispose, and a usable or salable by-product involves opportunity, knowledge of processes outside the generator's immediate production line, and comparative pricing of virgin material. Waste exchanges serve as information clearinghouses through which the availability and need for various types of materials can be established.

Recycling

Under RCRA and HSWA, EPA has carefully defined recycling to prohibit bogus recyclers that are really TSDs from taking advantage of more lenient rules for recycling. The definition says that a material is *recycled* if it is used, reused, or reclaimed (40 CFR 261.1 (c)(7)). A material is "used or reused" if it is either (1) employed as an ingredient (including its use as an intermediate) to make a product (however, a material will not satisfy this condition if distinct components of the material are recovered as separate end products, as when metals are recovered from metal-containing secondary materials); or (2) employed in a particular function as an effective substitute for a commercial product (40 CFR 261.1 (c)(5)). A material is *reclaimed* if it is processed to recover a useful product or if it is regenerated. Examples include the recovery of lead from spent batteries and the regeneration of spent solvents (40 CFR 261.1 (c)(4)) (U.S. EPA, 1988a).

Distillation processes can be utilized to recover spent solvent. The principal characteristics that determine the potential for recovery are the boiling points of the various useful constituents and the water content. The more dilute the waste solvent, the less economical it is to recover. Recovered solvents can be reused by the generator or sold for at least a substantial fraction of the cost of virgin material, and the credit for recovered solvent can more than offset the cost of recovery.

There are several technologies for recovery of metals from metal-plating rinse water. Most are applicable only to waste streams containing a single metal constituent. Examples include ion exchange, electrodialysis, evaporation, and reverse osmosis.

In October 1988 a federal appeals court struck down an EPA policy not to list used oil collected for recycling as a hazardous waste. Prior to that ruling, PCB-contaminated oils, petroleum industry sludges, and leaded tank bottoms were the only oils regulated. The majority of oil and oily wastes generated were not classified as hazardous under EPA regulations. These wastes are amenable either to recovery for use as fuel or to refinement for use as lubricants. Although all waste oil is now deemed hazardous, those oils that were formerly recovered may still be recovered, but the requirements for tracking them are more stringent.

10-7 TREATMENT TECHNOLOGIES

The wastes that remain after the implementation of waste minimization must be detoxified and neutralized. There are a large number of treatment technologies available to accomplish this. Many of these are applications of processes we have discussed in earlier chapters. Examples include: biological oxidation (Chapter 6), chemical precipitation, ion exchange, and oxidation-reduction (Chapter 4), and carbon adsorption (Chapter 7). Here we will discuss these as they apply to hazardous waste treatment, and we will introduce some new technologies.

Biological Treatment

In contrast to naturally occurring compounds, *anthropogenic compounds* (those created by human beings) are relatively resistant to biodegradation. One reason is that the organisms that are naturally present often cannot produce the enzymes necessary to bring about transformation of the original compound to a point at which the resultant intermediates can enter into common metabolic pathways and be completely mineralized.

Many environmentally important anthropogenic compounds are halogenated, and halogenation is often implicated as a reason for their persistence. The list of halogenated organic compounds includes pesticides, plasticizers, plastics, solvents, and trihalomethanes. Chlorinated compounds are the best known and most studied because of the highly publicized problems associated with DDT and other pesticides and numerous industrial solvents. Hence, chlorinated compounds serve as the basis for most of the information available on halogenated compounds.

Some of the characteristics that appear to confer persistence to halogenated compounds are the location of the halogen atom, the halide involved, and the extent of halogenation (Kobayashi and Rittman, 1982). The first step in biodegradation, then, is sometimes dehalogenation, for which there are several biological mechanisms.

Simple generalizations do not appear to be applicable. For example, until recently, oxidative pathways were mostly believed to be the typical means by which halogenated compounds were dehalogenated. Anaerobic, reductive dehalogenation, either biological or nonbiological, is now recognized as the critical factor in the transformation or biodegradation of certain classes of compounds. Compounds that require reductive dechlorination are common among the pesticides, as well as halogenated one- and two-carbon aliphatic compounds.

Reductive dehalogenation involves the removal of a halogen atom by oxidation-reduction. In essence, the mechanism involves the transfer of electrons from reduced organic substances via microorganisms or a nonliving (abiotic) mediator, such as inorganic ions (for example, Fe^{3+}) and biological products (for example, NAD(P), flavin, flavoproteins, hemoproteins, porphyrins, chlorophyll, cytochromes, and glutathione). The mediators are responsible for accepting electrons from reduced organic substances and transferring them to the halogenated compounds. The major requirements for the process are believed to be available free electrons and direct contact between the donor, mediator, and acceptor of electrons. Significant reductive dechlorination usually occurs only when the oxidation-reduction potential of the environment is 0.35 V and lower; the exact requirements appear to depend upon the compound involved (Kobayashi and Rittman, 1982).

Although simple studies using pure cultures of microorganisms and single substrates are valuable, if not essential, for determining biochemical pathways, they cannot always be used to predict biodegradability or transformation in more natural situations. The interactions among environmental factors, such as dissolved oxygen, oxidation-reduction potential, temperature, pH, availability of other compounds, salinity, particulate matter, competing organisms, and concentrations of compounds and organisms, often control the feasibility of biodegradation. The compound's physical or chemical characteristics, such as solubility, volatility, hydrophobicity, and octanol-water partition coefficient, contribute to the compound's availability in solution. Often compounds not soluble in the water are not readily available to organisms for biodegradation. There are some exceptions. For example, DDT, which is only slightly soluble in water, may be degraded by the white rot fungus found on decaying trees. This is because the enzymes involved in the white rot reaction are secreted from the cell (Kobayashi and Rittman, 1982).

Simple culture studies are similarly inadequate for predicting the fate of substances in the environment if there are many interactions between different organisms. First, substances that cannot be changed significantly in pure culture studies often will be degraded or transformed under mixed culture conditions. A good example of this type of interaction is *cometabolism,* in which a compound, the nongrowth substrate, is not metabolized as a source of carbon or energy, but is incidentally transformed by organisms using other compounds as growth substrates. The growth substrates provide the energy needed to cometabolize the nongrowth substrates. Second, products of the initial transformation by one organism may subsequently be broken down by a series of different organisms until compounds that can be metabolized by normal metabolic pathways are formed. An example is the degradation of DDT, which is reportedly mineralized directly by only one organism, a fungus; other organisms studied appear to degrade DDT only through cometabolism, resulting in numerous transformation products that subsequently can be used by other organisms. For example, *Hydrogenomonas* can metabolize DDT only as far as p-chlorophenylacetic acid (PCPA), while *Arthrobacter* species can then remove the PCPA (Kobayashi and Rittman, 1982).

Table 10-11 demonstrates that members of almost every class of anthropogenic compound can be degraded by some microorganism. The table also illustrates the wide variety of microorganisms that participate in environmentally significant biodegradation.

The metabolic capabilities of many microorganisms, in particular algae and oligotrophic bacteria, are not well understood. Such knowledge is necessary if limiting reactions are to be determined and the proper types of organisms selected for specific applications. More information about appropriate types of microorganisms to be selected and maintained in "real-world" treatment systems is needed, especially for the more novel microbial cultures. In order to develop special-purpose organisms by genetic manipulation, major advances in the understanding of the genetic structure of the many different types of organisms in nature are needed.

Conventional biological treatment processes such as activated sludge and trickling filters have been used to treat hazardous wastes. The major modification to the activated sludge processes has been to extend the mean cell residence time from the conventional values of 4 to 15 days to much longer periods of 3 to 6 months. In a similar fashion, trickling filter loading rates are much lower than those employed in municipal treatment systems. One innovation that has been adopted by TSD facilities is the *sequencing batch reactor* (SBR). The SBR is a periodically operated, fill-and-draw reactor (Herzbron et al., 1985). Each reactor in an SBR system has five discrete periods in each cycle: fill, react, settle, draw, and idle. Biological reactions are initiated as the raw wastewater fills the tank. During the fill and react phase, the waste is aerated in the same fashion as an activated sludge unit. After the react phase, the mixed liquor suspended solids (MLSS) are allowed to settle. The treated supernatant is discharged during the draw phase. The idle stage, the time between the draw and fill, may be zero or may be a few days depending on wastewater flow demand. The SBR has a major advantage in that wastes may be tested for completeness of treatment before discharge.

TABLE 10-11
Examples of anthropogenic compounds and microorganisms that can degrade them

Compound	Organism
Aliphatic (nonhalogenated)	
Acrylonitrile	Mixed culture of yeast mold, protozoan bacteria
Aliphatic (halogenated)	
Trichloroethane, trichloroethylene, methyl chloride, methylene chloride	Marine bacteria, soil bacteria, sewage sludge
Aromatic compounds (nonhalogenated)	
Benzene, 2,6-dinitrotoluene, creosol, phenol	*Pseudomonas* sp., sewage sludge
Aromatic compounds (halogenated)	Sewage sludge
1,2-; 2,3-; 1,4-dichlorobenzene, hexachlorobenzene, trichlorobenzene	
Pentachlorophenol	Soil microbes
Polycyclic aromatics (nonhalogenated)	
Benzo(a)pyrene, naphthalene	*Cunninghamella elegans*
Benzo(a)anthracene	*Pseudomonas*
Polycyclic aromatics (halogenated)	
PCBs	*Pseudomonas, Flavobacterium*
4-Chlorobiphenyl	Fungi
Pesticides	
Toxaphene	*Corynebacterium pyrogenes*
Dieldrin	Anacystic nidulans
DDT	Sewage sludge, soil bacteria
Kepone	Treatment lagoon sludge
Nitrosamines	
Dimethylnitrosamines	*Rhodopseudomonas*
Phthalate esters	Micrococcus 12B

(*Source:* Extracted from Table 1 of Kobayashi and Rittman, 1982.)

Chemical Treatment

Chemical detoxification is a treatment technology, either employed as the sole treatment procedure or used to reduce the hazard of a particular waste prior to transport, incineration, and burial.

It is important to remember that a chemical procedure cannot magically make a toxic chemical disappear from the *matrix* (wastewater, sludge, etc.) in which it is found, but can only convert it to another form. Thus, it is vital to ensure that the products of a chemical detoxification step are less of a problem than the starting material. It is equally important to remember that the reagents for such a reaction can be hazardous.

The spectrum of chemical methods includes complexation, neutralization, oxidation, precipitation, and reduction. An optimum method would be fast, quantitative, inexpensive, and leave no residual reagent, which itself would be a pollution problem. The following paragraphs describe a few of these techniques.

Neutralization. Solutions are neutralized by a simple application of the law of mass balance to bring about an acceptable pH. Sulfuric or hydrochloric acid is added to basic solutions, while caustic (NaOH) or slaked lime [$Ca(OH)_2$] is added to acidic solutions. Though a waste is hazardous at pH values less than 2 or greater than 12.5, and it would seem that simply bringing the pH into the range 2 to 12.5 would be adequate, good treatment practice requires that final pH values be in the range 6 to 8 to protect natural biota.

Oxidation. The cyanide molecule is destroyed by oxidation. Chlorine is the oxidizing agent most frequently used. Oxidation must be conducted under alkaline conditions to avoide the generation of hydrogen cyanide gas. This process is often referred to as alkaline chlorination. In chlorine oxidation, the reaction is carried out in two steps:

$$\text{NaCN} + 2\text{NaOH} + \text{Cl}_2 \rightleftharpoons \text{NaCNO} + 2\text{NaCl} + \text{H}_2\text{O} \quad (10\text{-}23)$$

$$2\text{NaCNO} + 5\text{NaOH} + 3\text{Cl}_2 \rightleftharpoons 6\text{NaCl} + \text{CO}_2 + \text{N}_2 + \text{NaHCO}_3 + 2\text{H}_2\text{O} \quad (10\text{-}24)$$

In the first step, the pH is maintained above 10 and the reaction proceeds in a matter of minutes. In this step, great care must be taken to maintain relatively high pH values, because at lower pHs there is a potential for the evolution of highly toxic hydrogen cyanide gas. The second reaction step proceeds most rapidly around a pH of 8, but it is not as rapid as the first step. Higher pH values may be selected for the second step to reduce chemical consumption in the following precipitation steps. This increases the reaction time. Often the second reaction is not carried out because the CNO is considered non-toxic by current regulations.

Ozone also may be used as the oxidizing agent. Ozone has a higher redox potential than chlorine, thus there is a higher driving force toward the oxidized state. When ozone is used, the pH considerations are similar to those discussed for chlorine. Ozone cannot be purchased. It must be made on site as part of the process.

This technology can be applied to a wide range of cyanide wastes: copper, zinc, and brass plating solutions; cyanide from cyanide salt heating baths; and passivating solutions. The process has been practiced on an industrial scale since the early 1940s. For extremely high cyanide concentrations ($>$1 percent), oxidation may not be desirable. Cyanide complexes of metals, particularly iron and to some extent nickel, cannot be decomposed easily by cyanide oxidation techniques.

Electrolytic oxidation of cyanide is carried out by anodic electrolysis at high temperatures. The theoretical basis of the process is that cyanide reacts with oxygen in solution in the presence of an electric potential to produce carbon dioxide and nitrogen gas. Normally, the destruction is carried out in a closed cell. Two electrodes are suspended in the solution and a DC current is applied to drive the reaction. The bath temperature must be maintained in the range of 50 to 95°C.

This technology is used for the destruction of cyanide in concentrated spent stripping solutions; in plating solutions for copper, zinc, and brass; in alkaline descalers; and in passivating solutions. It has been more successful for wastes containing high concentrations of cyanide (50,000 to 100,000 mg/L), but it has also been successfully used for concentrations as low as 500 mg/L.

Chemical oxidation methods for organic compounds in wastewater have received extensive study. In general they apply only to dilute solutions and often are considered expensive in comparison to the biological methods. Some examples include wet air oxidation, hydrogen peroxide, permanganate, chlorine dioxide, chlorine, and ozone oxidation. Of these, wet air oxidation and ozonation have shown promise as a pretreatment step for biological processes.

Wet air oxidation, also known as the Zimmerman process, operates on the principle that most organic compounds can be oxidized by oxygen given sufficient temperature and pressure. Wet air oxidation may be described as the aqueous phase oxidation of dissolved or suspended organic particles at temperatures of 175 to 325°C and sufficiently high pressure to prevent excessive evaporation. Air is bubbled through the liquid. The process is fuel efficient; once the oxidation reaction has started, it is usually self-sustaining. As this method is not limited by reagent cost, it is potentially the most widely applicable of all chemical oxidation methods. The method has been shown to be of use in destroying a wide range of organic compounds, including some pesticides. Although wet oxidation can provide acceptable levels of destruction for many hazardous compounds, it generally is not as complete as incineration. In many instances, the addition of metal salt catalysts can increase the destruction efficiency or allow the process to be run at lower temperature and/or pressure.

Precipitation. Metals are often removed from plating rinse waters by precipitation. This is a direct application of the solubility product principle (see Section 4-1). By raising the pH with lime or caustic, the solubility of the metal is reduced (Figure 10-8) and the metal hydroxide precipitates. Optimum removal is achieved by selecting the optimum pH as shown in Figure 10-8. Though there is an optimum for each metal, in many cases, the metals are mixed and the lowest value for an individual metal may not be achievable for the mixture.

Example 10-5. A metal plating firm is installing a precipitation system to remove zinc. They plan to use a pH meter to control the feed of hydroxide solution to the mixing tank. What pH should the controller be set at to achieve a zinc effluent concentration of 0.80 mg/L? The K_{sp} of $Zn(OH)_2$ is 7.68×10^{-17}.

Solution. From Table A-9 in Appendix A we find that the zinc hydroxide reaction is

$$Zn^{2+} + 2OH \rightleftharpoons Zn(OH)_2$$

As shown in Section 4-1, we can write the solubility product equation as

$$K_{sp} = [Zn^{2+}][OH^-]^2$$

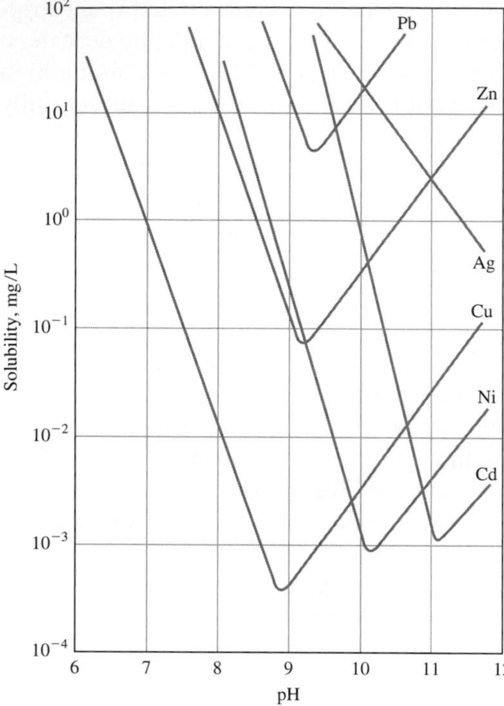

FIGURE 10-8
Solubilities of metal hydroxides as a function of pH. (*Source:* U.S. EPA, 1981b.)

Because we want the zinc concentration to be no greater than 0.80 mg/L, we calculate the moles per liter of zinc.

$$[Zn^{2+}] = \frac{0.80 \text{ mg/L}}{(65.41 \text{ g/mol})(1{,}000 \text{ mg/g})} = 1.223 \times 10^{-5} \text{ moles/L}$$

Now we solve for the hydroxide concentration.

$$[OH^-]^2 = \frac{7.68 \times 10^{-17}}{1.223 \times 10^{-5}} = 6.28 \times 10^{-12}$$

$$= (6.28 \times 10^{-12})^{\frac{1}{2}} = 2.51 \times 10^{-6}$$

The pOH is

$$\text{pOH} = -\log(2.505 \times 10^{-6}) = 5.60$$

And the pH set point for the controller is

$$\text{pH} = 14 - \text{pOH}$$

$$= 14 - 5.60 = 8.4$$

Reduction. Although most heavy metals readily precipitate as hydroxides, hexavalent chromium used in plating solutions must be reduced to trivalent chromium before it will precipitate. Reduction is usually done with sulfur dioxide (SO_2) or sodium bisulfite

($NaHSO_3$). With SO_2 the reaction is

$$3SO_2 + 2H_2CrO_4 + 3H_2O \rightleftharpoons Cr_2(SO_4)_3 + 5H_2O \qquad (10\text{-}25)$$

Because the reaction proceeds rapidly at low pH, an acid is added to control the pH between 2 and 3.

Physical/Chemical Treatment

Several treatment processes are used to separate hazardous waste from aqueous solution. The waste is not detoxified but only concentrated for further treatment or recovery.

Carbon Adsorption. Adsorption is a mass transfer process in which gas vapors or chemicals in solution are held to a solid by intermolecular forces (for example, hydrogen bonding and van der Waals' interactions). It is a surface phenomenon. Pressure vessels having a fixed bed are used to hold the adsorbent (see Section 7-10). Activated carbon, molecular sieves, silica gel, and activated alumina are the most common adsorbents. The active sites become saturated at some point in time. When the organic material has commercial value, the bed is then regenerated by passing steam through it. The vapor-laden steam is condensed and the organic fraction is separated from the water. If the organic compounds have no commercial value, the carbon may be either incinerated or shipped to the manufacturer for regeneration. Carbon systems for recovery of vapor from degreasers and for polishing wastewater effluents have been in commercial application for over 20 years.

Distillation. The separation of more volatile materials from less volatile materials by a process of vaporization and condensation is called *distillation*. When a liquid mixture of two or more components is brought to the boiling point of the mixture, a vapor phase is created above the liquid phase. If the vapor pressures of the pure components are different (which is usually the case), then the constituent(s) having the higher vapor pressure will be more concentrated in the vapor phase than the constituent(s) having the lower vapor pressure. If the vapor phase is cooled to yield a liquid, a partial separation of the constituents will result. The degree of separation depends on the relative differences in the vapor pressures. The larger the differences, the more efficient the separation. If the difference is large enough, a single separation cycle of vaporization and condensation is sufficient to separate the components. If the difference is not large enough, multiple cycles (stages) are required. Four types of distillation may be used: batch distillation, fractionation, steam stripping, and thin film evaporation.*

Both batch distillation and fractionation are well proven technologies for recovery of solvents. Batch distillation is particularly applicable for wastes with high solids concentrations. Fractionation is applicable where multiple constituents must be separated and where the waste contains minimal suspended solids.

When the volatility of the organic compound is relatively high and the concentration relatively low, then some form of stripping may be appropriate. *Air stripping* has

*Air stripping, though not strictly a distillation process because the condensation step is omitted, employs the same general principles of volatilization and, hence, is included in the discussion.

been used to purge large quantities of contaminated groundwater of small concentrations of volatile organic matter. The behavior of the process is the inverse of absorption discussed in Section 7-10. Air and contaminated liquid are passed countercurrently through a packed tower. The volatiles evaporate into the air, leaving a clean liquid stream. The contaminated air stream must then be treated to avoid an air pollution problem. Frequently this is accomplished by passing the air through an activated carbon column. The carbon is then incinerated. Air stripping has been used to remove tetrachloroethylene, trichloroethylene, and toluene from water (Gross and TerMaath, 1985; U.S. EPA, 1987).

The air stripper design equation may be developed in the same fashion as the absorber equation in Chapter 7. It is given here without that development:

$$Z_T = \frac{L}{A}\,\frac{\ln\left[\dfrac{C_1}{C_2} - \dfrac{\mathrm{LRT}_g}{\mathrm{GH}_c}\left(\dfrac{C_1}{C_2} - 1\right)\right]}{K_L a\left(1 - \dfrac{\mathrm{LRT}_g}{\mathrm{GH}_c}\right)} \tag{10-26}$$

where Z_T = depth of packing in tower, m
L = water flow, $\mathrm{m^3/min}$
A = cross-sectional area of tower, $\mathrm{m^2}$
G = air flow, $\mathrm{m^3/min}$
H_c = Henry's constant, $\mathrm{atm \cdot m^3/mol}$
R = universal gas constant = 8.206×10^{-5} $\mathrm{atm \cdot m^3/mole \cdot K}$
T_g = temperature of air, K
C_1, C_2 = influent and effluent organic concentration in the water, $\mathrm{mol/m^3}$
K_L = overall mass transfer coefficient, $\mathrm{mol/min \cdot m^2 \cdot mol/m^3}$
a = effective interfacial area of packing per unit volume for mass transfer, $\mathrm{m^2/m^3}$

Realistic values of the air-to-water ratio (G/L) range from 5 to several hundred. In an actual design, a safety factor of 20 percent would be added to Z_t. The column holding the packing would be somewhat larger to accommodate support structures and distribution piping (LaGrega et al., 2001).

Example 10-6. Well 12A at the City of Tacoma, WA, is contaminated with 350 μg/L of 1,1,2,2-tetrachloroethane. The water must be cleaned to the detection limit of 1.0 μg/L. Design a packed tower stripping column to meet this requirement using the following design parameters.

Henry's law constant = 5.0×10^{-4} $\mathrm{atm \cdot m^3/mol}$

$K_L a = 10 \times 10^{-3}\ \mathrm{s^{-1}}$

Air flow rate = 13.7 $\mathrm{m^3/s}$

Liquid flow rate = 0.044 $\mathrm{m^3/s}$

Temperature = 25°C

Column diameter may not exceed 4.0 m

Column height may not exceed 6.0 m

Solution. The Henry's law constants given in Appendix A are in kPa · m^3/moles. To convert these to atm · m^3/mole, divide by the atmospheric pressure at standard conditions, that is, 101.325 kPa/atm.

The stripper equation is then solved for $Z_T A$, the column volume.

$$Z_T A = (0.044)\frac{\ln\left[\frac{350}{1} - \frac{(0.044)(8.206 \times 10^{-5})(298)}{(13.7)(5.0 \times 10^{-4})}\left(\frac{350}{1} - 1\right)\right]}{10 \times 10^{-3}\left[1 - \frac{(0.044)(8.206 \times 10^{-5})(298)}{(13.7)(5.0 \times 10^{-4})}\right]}$$

$$= (0.044)(6.75 \times 10^2)$$

$$= 29.7 \text{ m}^3$$

Any number of solutions are now possible within the boundary conditions of 4 m diameter and 6 m height. For example, with the 20 percent safety factor and rounding,

Diameter (m)	Z_T (m)	Tower height (m)
4.00	2.36	3
3.34	3.39	5

For gases of lower volatility or higher concentration (>100 ppm) *steam stripping* may be employed. The physical arrangement of the process is much like that of an air stripper, except that steam is introduced instead of air. The addition of steam enhances the stripping process by decreasing the solubility of the organic in the aqueous phase and by increasing the vapor pressure. Steam stripping has been used to treat aqueous waste contaminated with chlorinated hydrocarbons, xylenes, acetone, methyl ethyl ketone, methanol, and pentachlorophenol. Concentrations treated range from 100 ppm to 10 percent organic compound (U.S. EPA, 1987).

Recovery of metals by evaporation is accomplished by boiling off sufficient water from the collected rinse stream to allow the concentrate to be returned to the plating bath. The condensed steam is recycled for use as rinse water. The boil-off rate, or evaporator duty, is set to maintain the water balance of the plating bath. Evaporation is usually performed under vacuum to prevent thermal degradation of additives and to reduce the amount of energy required for evaporation of the water.

There are four types of evaporators: rising film, flash evaporators using waste heat, submerged tube, and atmospheric pressure. Rising film evaporators are built so that the evaporative heating surface is covered by a wastewater film and does not lie in a pool of boiling wastewater. Flash evaporators are of similar configuration, but the plating solution is continuously recirculated through the evaporator along with the wastewater. This allows the use of waste heat in the plating bath to augment the evaporation process. In the submerged tube design, the heating coils are submerged in the wastewater. Atmospheric evaporators do not recover the distillate for reuse and they do not operate under vacuum.

Ion Exchange. Metals and ionized organic chemicals can be recovered by ion exchange. Ion exchange chemistry was discussed in Section 4-3. In ion exchange, the

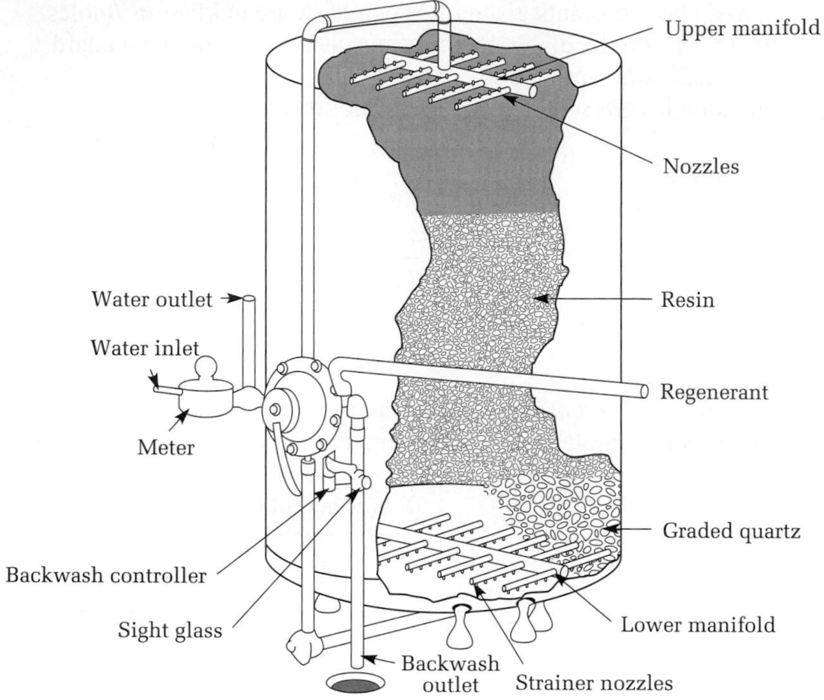

FIGURE 10-9
Typical ion exchange resin column. (*Source:* U.S. EPA, 1981a)

waste stream containing the ion to be removed is passed through a bed of resin. The resin is selected to remove either cations or anions. In the exchange process, ions of like charge are removed from the resin surface in exchange for ions in solution. Typically, either hydrogen or sodium is exchanged for cations (metal) in solution. When the bed becomes saturated with the exchanged ion, it is shut down and the resin is regenerated by passing a concentrated solution containing the original ion (hydrogen or sodium) back through the bed. The exchanged pollutant is forced off the bed in a concentrated form that may be recycled. A typical ion exchange column is shown in Figure 10-9. A prefilter is required to remove suspended material that would hydraulically foul the column. It also removes organic contaminants and oils that would foul the resin.

As a rule, ion exchange systems are suitable for chemical recovery applications where the rinse water feed has a relatively dilute concentration ($< 1{,}000$ mg/L) and a relatively low concentration is required for recycle. Ion exchange has been demonstrated commercially for recovery of plating chemicals from acid-copper, acid-zinc, nickel, tin, cobalt, and chromium plating baths.

The breakthrough curves for an ion-exchange column and an adsorption column (Section 7-10) are similar. Thomas (1948) proposed a kinetic equation to describe the removal of the contaminant in the column:

$$\ln\left(\frac{C_o}{C} - 1\right) = \frac{(k)(q_o)(M)}{Q} - \frac{(k)(C_o)(\forall)}{Q} \tag{10-27}$$

where C_o = influent solute concentration, mg/L or milliequivalents/L (meq/L)
C = effluent solute concentration, mg/L or milliequivalents/L (meq/L)
k = rate constant, L/d · equivalent
q_o = maximum solid phase concentration of exchanged solute, equivalents/kg of resin
M = mass of resin, kg
$\forall$ = volume of solution passed through column, L
Q = flow rate, L/d

This equation is of the form $y = mx + b$

where $y = \ln\left(\frac{C_o}{C} - 1\right)$
$x = \forall$

This allows us to determine the rate constant and the maximum solid phase concentration from a plot of $\ln (C_o/C - 1)$ versus $\forall$ as shown in Figure 10-10.

The slope of the line is equal to

$$\frac{kC_o}{Q}$$

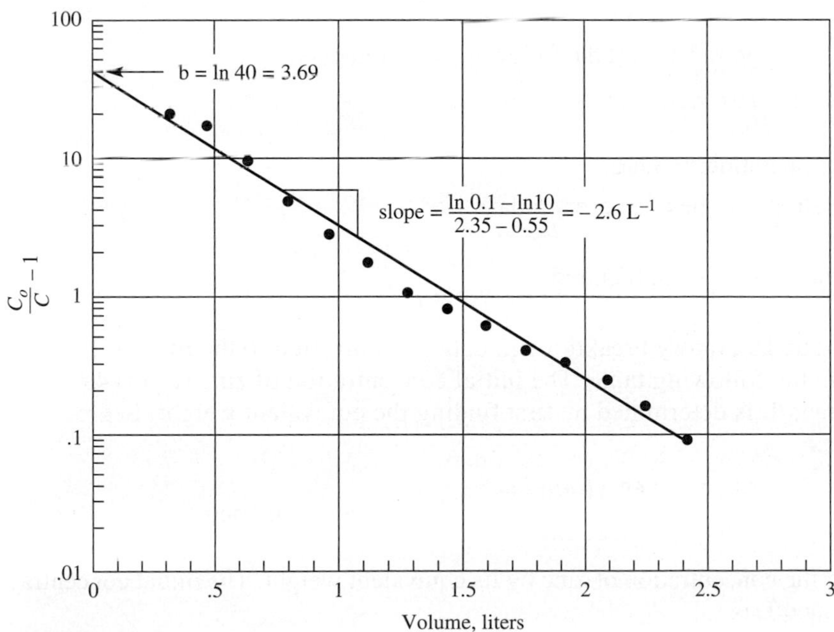

FIGURE 10-10
Plot of breakthrough data to estimate kinetic equation constants. (Note: ordinate scale is logarithm to the base *e*.)

and the intercept is equal to

$$\frac{(k)(q_o)(M)}{Q}$$

Data from a laboratory or pilot scale breakthrough curve are required to obtain the plot. The same flow rate, in terms of bed volumes per unit time, should be used for both the pilot studies and the full scale column.

Example 10-7. An electroplating rinse water containing 49 mg/L of zinc is to be treated by an ion exchange column to meet an allowable effluent concentration of 2.6 mg/L. A laboratory scale column has provided the breakthrough data shown in the first two columns of the table on page 873. The laboratory column data are as follows:

Inside diameter = 1.0 cm

Length = 10.0 cm

Mass of resin (moist basis) = 5.2 g

Water content = 17%

Density of dry resin = 0.65 g/cm^3

Liquid flow rate = 7.87 L/d

Initial concentration of zinc = 49 mg/L

The full scale design must meet the following requirements:

Flow rate = 36,000 L/d

Hours of operation = 8 h/d

Regeneration is to be once every 5 days

Determine the mass of resin required.

Solution. The laboratory breakthrough data are converted to the form of Equation 10-27 in the following table. The initial concentration of zinc (C_o) is 49 mg/L. The meq/L is determined by first finding the equivalent weight (See Chapter 4) as

$$\frac{\text{GMW}}{\text{n}} = \frac{65.41 \text{ g/mole}}{2 \text{ eq/mole}} = 32.71 \text{ g/eq or mg/meq}$$

and dividing the concentration of zinc by its equivalent weight. The initial concentration (C_o) in meq/L is

$$\frac{49 \text{ mg/L}}{32.71 \text{ mg/meq}} = 1.50 \text{ meq/L}$$

Breakthrough Data

$\mathcal{V}$, L	C, mg/L	C, meq/L	$\frac{C_0}{C} - 1$
0.32	2.25	0.06826	20.973
0.48	2.74	0.08313	17.044
0.64	4.56	0.13835	9.8421
0.80	8.32	0.25243	4.9423
0.96	12.74	0.38653	2.8807
1.12	17.70	0.53701	1.7932
1.28	23.54	0.71420	1.1003
1.44	27.48	0.83374	0.7991
1.60	30.58	0.92779	0.6167
1.76	35.34	1.07221	0.3990
1.92	37.02	1.12317	0.3355
2.08	39.38	1.19478	0.2555
2.24	42.50	1.28944	0.1632
2.40	45.10	1.36833	0.0962
2.56	44.10	1.33799	0.1211

The plot of these data is shown in Figure 10-10.
From the plot

$$k = (\text{slope})\left(\frac{Q}{C_o}\right) = (2.6\ \text{L}^{-1})\left(\frac{7.87\ \text{L/d}}{1.50\ \text{meq/L}}\right)$$
$$= 13.64\ \text{L/d} \cdot \text{meq}$$

and

$$q_o = \frac{(b)(Q)}{(k)(M)} = \frac{(3.69)(7.87\ \text{L/d})}{(13.64\ \text{L/d} \cdot \text{meq})(4.316\ \text{g})}$$
$$= 0.4933\ \text{meq/g}$$

Note that the mass of the test column resin (M) is corrected for moisture, that is, the dry weight is

$$(5.2\ \text{g})(1 - 0.17) = 4.316\ \text{g}$$

Using these values of k and q_o and reapplying Equation 10-27 we can determine the mass of resin for the full scale column. Since the effluent concentration must not exceed 2.6 mg/L, we can solve the left-hand side of the equation as:

$$\ln\left(\frac{49}{2.6} - 1\right) = 2.882$$

The first term on the right-hand side contains the unknown (M). Using the constants determined above, and the daily flow rate, it may be simplified to

$$\frac{(13.64\ \text{L/d} \cdot \text{meq})(0.4933\ \text{meq/g})(M)}{36{,}000\ \text{L/d}} = 1.87 \times 10^{-4}\,(M)$$

Using a flow rate of 36,000 L/d and a 5 day operating cycle, the volume to treat ($\mathcal{V}$) is

$$(36{,}000\ \text{L/d})(5\ \text{d}) = 180{,}000\ \text{L}$$

The second term on the right-hand side of the equation is then

$$\frac{(13.64\ \text{L/d} \cdot \text{meq})(1.50\ \text{meq/L})(180{,}000\ \text{L})}{36{,}000\ \text{L/d}} = 102.30$$

Setting the left-hand side of the equation equal to the right-hand side and solving for M yields

$$2.882 = 1.87 \times 10^{-4}(M) - 102.30$$
$$M = 5.6 \times 10^{5}\ \text{g or } 560\ \text{kg}$$

In full-scale operation, the resin bed is not allowed to reach saturation because the concentration of the solute will exceed most discharge standards before this occurs. Normal operation then requires either an operating cycle that will allow regeneration of the spent resin during nonworking hours or, in the case of 24 hour, 7 day per week schedules, multiple beds so that one may be taken off-line.

The diameters of ion exchange columns may vary from centimeters to 6 m. Resin bed depths range from 1 to 3 m. Bed height-to-diameter ratios range from 1.5:1 to 1:3. The column shell is designed to allow for 100 percent expansion of the resin bed during backwashing (regeneration). Columns are normally prefabricated and shipped by truck. Column height generally does not exceed 4 m. Multiple columns in series are provided where the design height exceeds 4 m. The maximum column diameter is often controlled by the clearance under bridges passing over the highway.

During ion exchange, the normal flow pattern is downward through the bed. The hydraulic loading may range from 25 to 600 $m^3/d \cdot m^2$. Lower hydraulic loadings result in longer contact periods and better exchange efficiency. Because the surface of the bed acts like a filter, regeneration is often countercurrent, that is, the regenerating solution is pumped into the bottom of the column. This results in a cleansing of the column much like the backwashing of a rapid sand filter cleans it. Regeneration hydraulic loadings range from 60 to 120 $m^3/d \cdot m^2$.

Electrodialysis. The electrodialysis unit uses a membrane to selectively retain or transmit specific molecules. The membranes are thin sheets of ion exchange resin reinforced by a synthetic fiber backing. The construction of the unit is such that anion membranes are alternated with cation membranes in stacks of cells in series (Figure 10-11). An electric potential is applied across the membrane to provide the motive force for ion migration. Cation membranes permit passage of only positively charged ions, while anion membranes permit passage of only negatively charged ions. The flow is directed through the membrane in two hydraulic circuits (Figure 10-12). One circuit is ion-depleted and the other is ion-concentrated. The degree of purification achieved in the dilute circuit is set by the electric potential. The ability to pass the charge is proportional to the concentration of the ionic species in the dilute stream. Because ion

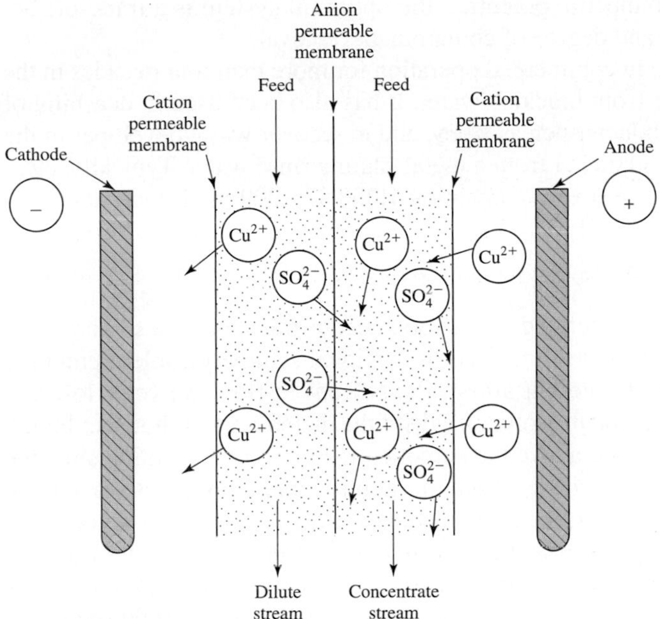

FIGURE 10-11

Electrodialysis. (Cations in the feed water show the same behavior as copper (Cu^{2+}) and anions show the same behavior as sulfate (SO_4^{2-}). Under the action of an electric field, cation-exchange membranes permit passage only of positive ions, while anion-exchange membranes permit passage only of negatively charged ions.) (*Source:* Davis and Masten, 2004.)

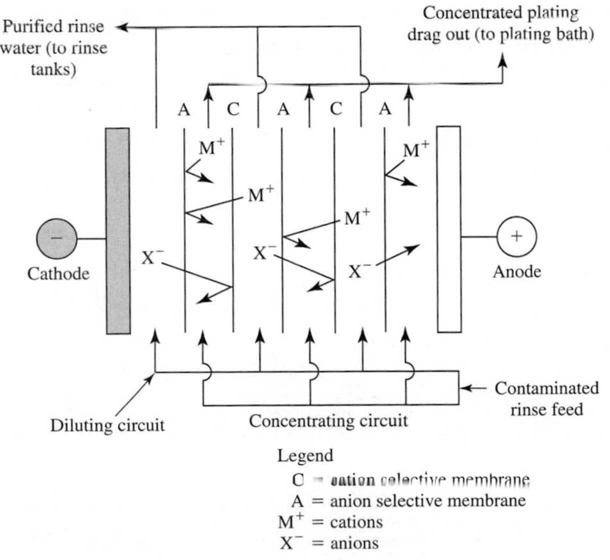

FIGURE 10-12

Electrodialysis unit flow schematic.

migration is proportional to electric potential, the optimum system is a trade-off between energy requirements and degree of contaminant removal.

Electrodialysis has been in commercial operation for more than four decades in the production of potable water from brackish water. It has also been used in deashing of sugars, desalting of food products such as whey, and to recover waste developer in the photo processing industry and nickel from a metal-plating rinse water. Typically, electrodialysis can separate a waste stream containing 1,000 to 5,000 mg/L inorganic salts into a dilute stream that contains 100 to 500 mg/L salt and a concentrated stream that contains up to 10,000 mg/L salt.

Reverse Osmosis. Osmosis is defined as the spontaneous transport of a solvent from a dilute solution to a concentrated solution across an ideal semipermeable membrane that impedes passage of the solute but allows the solvent to flow. Solvent flow can be reduced by exerting pressure on the solution side of the membrane, as shown in Figure 10-13. If the pressure is increased above the osmotic pressure on the solution side, the flow reverses. Pure solvent will then pass from the solution into the solvent. As applied to metal finishing wastewater, the solute is the metal and the solvent is pure water.

Many configurations of the membrane are possible. The driving pressure is on the order of 1,000 to 5,500 kPa. No commercially available membrane polymer has demonstrated tolerance to all extreme chemical factors such as pH, strong oxidizing agents, and aromatic hydrocarbons. However, selected membranes have been demonstrated on nickel, copper, zinc, and chrome baths.

Solvent Extraction. Solvent extraction is also called *liquid extraction* and *liquid-liquid extraction.* Contaminants can be removed from a waste stream using liquid-liquid extraction if the wastewater is contacted with a solvent having a greater solubility for the target contaminants than the wastewater. The contaminants will tend to migrate from the wastewater into the solvent. Although predominately a method for separating organic materials, it may also be applied to remove metals if the solvent contains a material that will react with the metal. *Liquid ion exchange* is one kind of these reactions.

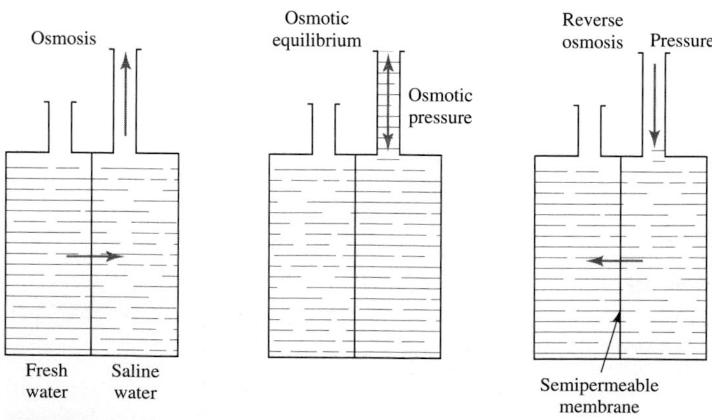

FIGURE 10-13
Direct and reverse osmosis.

In the solvent extraction process, the solvent and the waste stream are mixed to allow mass transfer of the constituent(s) from the waste to the solvent. The solvent, immiscible in water, is then allowed to separate from the water by gravity. The solvent solution containing the extracted contaminant is called the *extract.* The extracted waste stream with the contaminants removed is called the *raffinate.* As in distillation, the separation may need to be done in one or more stages. In general, more stages result in a cleaner raffinate. The degree of complexity of the apparatus varies from simple mixer/settlers to more exotic contacting devices. If the extract is sufficiently enriched, it may be possible to recover useful material. Distillation is often employed to recover the solvent and reusable organic chemicals. For metal recovery, the ion exchange material is regenerated by the addition of an acid or alkali. The process has found wide application in the ore processing industry, in food processing, in pharmaceuticals, and in the petroleum industry.

Incineration

In an incinerator, chemicals are decomposed by oxidation at high temperatures (800°C and greater). The waste, or at least its hazardous components, must be combustible in order to be destroyed. The primary products from combustion of organic wastes are carbon dioxide, water vapor, and inert ash. However, there are a multitude of other products that can be formed.

Products of Combustion. The percentages of carbon, hydrogen, oxygen, nitrogen, sulfur, halogens, and phosphorus in the waste, as well as the moisture content, need to be known to determine stoichiometric combustion air requirements and to predict combustion gas flow and composition. Actual incineration conditions generally require excess oxygen to maximize the formation of *products of complete combustion* (POCs) and minimize the formation of *products of incomplete combustion* (PICs).

The incineration of halogenated organics results in the formation of halogenated acids, which require further treatment to ensure environmentally acceptable air emissions from the incineration process. Chlorinated organics are the most common halogenated hydrocarbons found in hazardous waste. The incineration of chlorinated hydrocarbons with excess air results in the formation of carbon dioxide, water, and hydrogen chloride. An example is the following reaction for the incineration of dichlorethane (Wentz, 1989):

$$2C_2H_4Cl_2 + 5O_2 \rightarrow 4CO_2 + 2H_2O + 4HCl \qquad (10\text{-}28)$$

The hydrogen chloride must be removed before the carbon dioxide and steam can be safely exhausted into the atmosphere.

Hazardous waste may contain either organic or inorganic sulfur compounds. When these wastes are incinerated, sulfur dioxide is produced. For example, the destruction of ethyl mercaptan results in the following reaction:

$$2C_2H_5SH + 9O_2 \rightarrow 4CO_2 + 6H_2O + 2SO_2 \qquad (10\text{-}29)$$

The sulfur dioxide produced by the incineration of sulfur-containing wastes must not exceed air quality standards.

Excess air must be provided to ensure complete combustion. However, the amount of the excess can only be determined empirically. For example, a highly volatile, clean, hydrocarbon waste would probably require much less excess air than would a heavy hydrocarbon sludge within a high solids content. Incineration of sludges and solids may require as much as two to three times excess air above stoichiometric equivalents. Too much excess air should be avoided because it increases the fuel required to heat the waste to destruction temperatures, reduces residence time for the hazardous wastes to be oxidized, and increases the volume of air emissions to be handled by the air pollution control equipment.

By-products from the incineration of hazardous wastes may also result from incomplete combustion as well as from the products of combustion. Products of incomplete combustion (PICs) include carbon monoxide, hydrocarbons, aldehydes, ketones, amines, organic acids, and polycyclic aromatic hydrocarbons (PAHs). In a well-designed incinerator, these products are insignificant in amount. However, in poorly designed or overloaded incinerators, PICs may pose environmental concerns. Polychlorinated biphenyls, for instance, decompose under such conditions into highly toxic chlorinated dibenzo furans (CDBF). The hazardous material, hexachlorocyclopentadiene (HCCPD), found in many hazardous wastes, is known to decompose into the even more hazardous compound hexachlorobenzene (HCB) (Oppelt, 1981).

Suspended particulate emissions are also produced during incineration. These include particles of mineral oxides and salts from the mineral constituents in the waste material, as well as fragments of incompletely burned combustibles.

Last, but not least, ash is a product of combustion. The ash is considered a hazardous waste. Metals not volatilized end up in the ash. Unburned organic compounds may also be found in the ash. When organic compounds remain, the ash may simply be incinerated. The metals must be treated prior to land disposal.

Design Considerations. The most important factors for proper incinerator design and operation are combustion temperature, combustion gas residence time, and the efficiency of mixing the waste with combustion air and auxiliary fuel.

Chemical and thermal dynamic properties of the waste that are important in determining its time/temperature requirements for destruction are its elemental composition, net heating value, and any special properties (for example, explosive properties) that may interfere with incineration or require special design considerations.

In general, higher heating values are required for solids versus liquids or gases, for higher operating temperatures, and for higher excess air rates if combustion is to be sustained without auxiliary fuel consumption. While sustained combustion (*autogenous combustion*) is possible with heating values as low as 9.3 MJ/kg, in the hazardous waste incineration industry it is common practice to blend wastes (and fuel oil, if necessary) to obtain an overall heating value of 18.6 MJ/kg or greater (Davis et al, 2000).

Blending is also used to limit the net chlorine content of chlorinated hazardous waste to a maximum of roughly 30 percent by weight to reduce chlorine concentrations in the combustion gas. The chlorine and, especially, hydrogen chloride that forms from the chlorine, are very corrosive. They oxidize the fire brick in the incinerator which causes it to fail.

Hazardous waste incinerators must be designed to achieve a 99.99 percent *destruction and removal efficiency* (DRE) of the *principal organic hazardous components* (POHCs) in the waste. This is commonly referred to as "four 9s DRE"; higher DREs may be referred to as five 9s, six 9s, that is, 99.999 and 99.9999 percent DRE, respectively. Because of the complexity of the wastes being burned, little success has been achieved in predicting the time and temperature requirements for achieving the 99.99 percent DRE. Empirical tests (*trial burns*) are required to demonstrate compliance. Experience has demonstrated that highly halogenated materials are more difficult to destroy than those with low halogen content.

Incinerator Types. Two technologies dominate the incineration field: liquid injection and rotary kiln incinerators. Over 90 percent of all incineration facilities use one of these technologies. Of these, more than 90 percent are liquid injection units. Less commonly used incinerators include fluidized beds and starved air/pyrolysis systems.

Horizontal, vertical, and tangential liquid injection units are used. The majority of the incinerators for hazardous wastes inject liquid hazardous waste at 350 to 700 kPa through an atomizing nozzle into the combustion chamber. These liquid incinerators vary in size from 300,000 to 90 million Joules of heat released per second. An auxiliary fuel such as natural gas or fuel oil is often used when the waste is not autogenous. The liquid wastes are atomized into fine droplets as they are injected. A droplet size in the range 40 to 100 μm is obtained with atomizers or nozzles. The droplet volatilizes in the hot gas stream and the gas is oxidized. Efficient destruction of liquid hazardous wastes requires minimizing unevaporated droplets and unreacted vapors.

Residence time, temperature, and turbulence (often referred to as the "three Ts") are optimized to increase destruction efficiencies. Typical residence times are 0.5 to 2 seconds. Incinerator temperatures usually range between 800 and 1600°C. A high degree of turbulence is desirable for achieving effective destruction of the organic chemicals in the waste. Depending on whether the liquid incinerator flow is axial, radial, or tangential, additional fuel burners and separate waste injection nozzles can be arranged to achieve the desired temperature, turbulence, and residence time. Vertical units are less likely to experience ash buildup. Tangential units have a much higher heat release and generally superior mixing.

The rotary kiln is often used in hazardous waste disposal systems because of its versatility in processing solid, liquid, and containerized wastes. Waste is incinerated in a refractory-lined rotary kiln, as shown in Figure 10-14. The shell is mounted at a slight incline from the horizontal plane to facilitate mixing the waste materials with circulating air. Solid wastes and drummed wastes are usually fed by a conveyor system or a ram. Liquids and pumpable sludges are injected through a nozzle. Noncombustible metal and other residues are discharged as ash at the end of the kiln.

Rotary kilns are typically 1.5 to 4 m in diameter and range in length from 3 to 10 m. Rotary kiln incinerators usually have a length-to-diameter ratio (L/D) of between two and eight. Rotational speeds range from 0.5 to 2.5 cm/s, depending on kiln periphery. High L/D ratios, along with slower rotational speeds, are used for wastes requiring longer residence times. The feed end of the kiln has airtight seals to adequately control the initial incineration reactions.

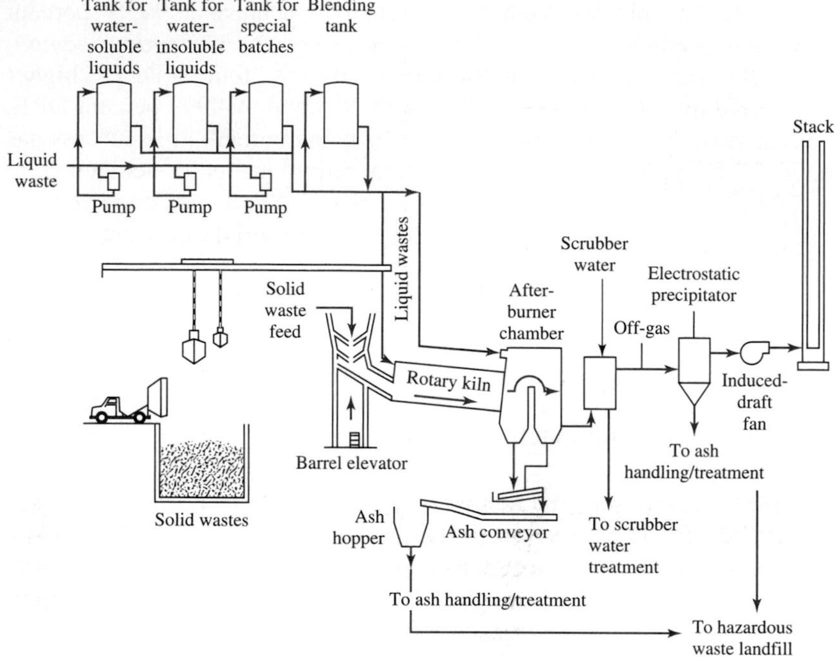

FIGURE 10-14
Rotary kiln incinerator.

Residence times for solid wastes are based on the rotational speed of the kiln and its angle. The residence time to volatilize waste is controlled by the gas velocity. The retention time of solids in the incinerator can be estimated from the following, where the coefficient 0.19 is based on limited experimental data:

$$\theta = \frac{0.19\, L}{NDS} \tag{10-30}$$

where θ = retention time, min
L = kiln length, m
N = kiln rotational speed, rev/min
D = kiln diameter, m
S = kiln slope, m/m

Rotary kiln systems typically include secondary combustion chambers or afterburners to ensure complete destruction of the hazardous waste. Kiln operating temperatures range from 800 to 1,600°C. Afterburner temperatures range from 1,000 to 1,600°C. Liquid wastes are often injected into the secondary combustion chamber. The volatilized and combusted wastes leave the kiln and enter the secondary chamber, where additional oxygen is available and high heating value liquid wastes or fuel may be introduced. Both the secondary combustion chamber and the kiln are usually equipped with an auxiliary fuel firing system for startup.

Cement kilns are very efficient at destroying hazardous waste. Their long residence times and high operating temperatures exceed the requirements for destruction of most wastes. Hydrochloric acid generated from chlorinated hydrocarbon wastes is neutralized by the lime in the kiln while slightly lowering the alkalinity of the cement products. While cement plants can save energy by incinerating liquid wastes, the expense of obtaining permits and public resistance have inhibited use of this process.

Air Pollution Control (APC). Typical APC equipment on an incinerator will include an afterburner, liquid scrubber, demister, and fine particulate control device. Afterburners are used to control emission of unburned organic by-products by providing additional combustion volume at an elevated temperature. Scrubbers are used to physically remove particulate matter, acid gases, and residual organics from the combustion gas stream. Metals, of course, are not destroyed in the incineration process. Some are volatilized and then collected in the air pollution control device. The large liquid droplets that escape from the scrubber are captured in a mist collector. The final stage in gas cleaning is to remove the fine particles that remain. Electrostatic precipitators have been used for this step. Scrubber water and residues from other APC devices are still considered hazardous and must be treated before ultimate land disposal.

Permitting of Hazardous Waste Incinerators. The permitting of hazardous waste incinerators is a complex, multifaceted program conducted simultaneously on federal, state, and local levels. Because of the variety of state and local regulations for the handling, transportation, treatment, and disposal of hazardous wastes, as well as those concerning the operation of incinerators, each startup has a unique set of permit requirements.

Generally speaking, hazardous waste incinerators require at least the following permits: federal RCRA, state RCRA, for PCBs—the Toxic Substances and Control Act (TSCA), state and federal wastewater discharge, and state and federal air pollution control. A variety of local permits may also be necessary. Each of these require data substantiating an incinerator's operation at or above performance levels determined by environmental legislation. Each requires a public hearing and discussion of environmental impacts as well.

Hazardous waste incinerators must meet three performance standards (Theodore and Reynolds, 1987):

1. *Principal Organic Hazardous Constituents* (POHC). The DRE for a given POHC is defined as the mass percentage of the POHC removed from the waste. The POHC performance standard requires that the DRE for each POHC *designated* in the permit be 99.99 percent or higher. The DRE performance standard implicitly requires sampling and analysis to measure the amounts of the designated POHC(s) in both the waste stream and the stack effluent gas during a trial burn. (The term *designated POHC* is described in more detail later in this section.) The DRE is determined for each designated

POHC from a mass balance of the waste introduced into the incinerator and in the stack gas*:

$$DRE = \frac{(W_{in} - W_{out})}{W_{in}} \times 100\% \tag{10-31}$$

where W_{in} = mass feed rate of one POHC in the waste stream
W_{out} = mass emission rate of the same POHC present in exhaust emissions prior to release to the atmosphere

2. *Hydrochloric acid.* An incinerator burning hazardous waste and producing stack emissions of more than 1.8 kg/h of hydrogen chloride (HCl) must control HCl emissions such that the rate of emission is no greater than the larger of either 1.8 kg/h or 1 percent of the HCl in the stack gas prior to entering any pollution control equipment.

3. *Particulates.* Stack emissions of particulate matter are limited to 180 milligrams per dry standard cubic meter (mg/dscm) for the stack gas corrected to 7 percent oxygen. This adjustment is made by calculating a corrected concentration:

$$P_c = P_m \frac{14}{21 - Y} \tag{10-32}$$

where P_c = corrected concentration of particulate, mg/dscm
P_m = measured concentration of particulate, mg/dscm
Y = percent oxygen in the dry flue gas

In this way, a decrease in the particulate concentration due solely to increasing air flow in the stack is not rewarded, and an increase in the particulate concentration due solely to reduction in the air flow in the stack is not penalized. Special rules for this calculation are being developed for oxygen-enriched combustion systems where the oxygen content is greater than the 21 percent found in the atmosphere.

Compliance with these performance standards is documented by a trial burn of the facility's waste streams. As part of the RCRA permit application, a trial burn plan detailing waste analysis, an engineering description of the incinerator, sampling and monitoring procedures, test schedule and protocol, as well as control information, must be developed. If EPA determines that the design is adequate, a temporary or draft permit is issued. This allows the owner or operator to build the incinerator and initiate the trial burn procedure.

The temporary permit covers four phases of operation. During the first phase, immediately following construction, the unit is operated for *shake-down* purposes to identify possible mechanical deficiencies and to ensure its readiness for the trial burn procedures. This phase of the permit is limited to 720 h of operation using hazardous

*Note that this is not a mass balance around the incinerator. Hazardous waste that ends up in the scrubber water, APC residue, and ash are not counted. Hence, the oxidation can be very poor and the incinerator can still meet the 99.99 percent rule if the scrubber is efficient and/or the waste ends up in the ash. This is one reason that residues are considered hazardous and must be treated before land disposal.

waste feed. The trial burn is conducted during the second phase. This is the most critical component of the permitting process, since it demonstrates the incinerator's ability to meet the three performance standards. In addition, performance data collected during the trial burn phase are reviewed by the permitting official and become the basis for setting the conditions of the facility permit. These conditions are: (1) allowable waste analysis procedures, (2) allowable waste feed composition (including acceptable variations in the physical or chemical properties of the waste feed), (3) acceptable operating limits for carbon monoxide in the stack, (4) waste feed rate, (5) combustion temperature, (6) combustion gas flow rate, and (7) allowable variations in incinerator design and operating procedures (including a requirement for shutoff of waste feed during startup, shutdown, and at any time when conditions of the permit are violated).

To verify compliance with the POHC performance standard during the trial burn, it is not required that the incinerator DRE for every POHC identified in the waste be measured. The POHCs with the greatest potential for a low DRE, based on the expected difficulty of thermal degradation (incinerability) and the concentration of the POHC in the waste, become the *designated POHCs* for the trial burn. The EPA permit review personnel work with the owners/operators of the incinerator facility in determining which POHCs in a given waste should be designated for sampling and analysis during the trial burn.

If a wide variety of wastes are to be treated, a difficult-to-incinerate POHC at high concentration may be proposed for the trial burn. The substitute POHC is referred to as a *surrogate POHC*. The surrogate POHC does not have to be actually present in the normal waste. It does, however, have to be considered more difficult to incinerate than any POHC found in the waste.

The third phase consists of completing the trial burn and submitting the results. This phase can last several weeks to several months, during which the incinerator is allowed to operate under specified conditions. The data to be reported to regulatory agencies after the burn are: (1) a quantitative analysis of the POHCs in the waste feed, (2) a determination of the concentration of the particulates, POHCs, oxygen, and HCl in the exhaust gas, (3) a quantitative analysis of any scrubber water, ash residues, and other residues to determine the fate of the POHCs, (4) a computation of the DRE for the POHCs, (5) a computation of the HCl removal efficiency if the HCl emission rate exceeds 1.8 kg/h, (6) a computation of particulate emissions, (7) the identification of sources of fugitive emissions and their means of control, (8) a measurement of average, maximum, and minimum temperatures and combustion gas velocities (gas flows), (9) a continuous measurement of carbon monoxide (CO) in the exhaust gas, and (10) any other information EPA may require to determine compliance.

Provided that performance standards are met in the trial burn, the facility can begin its *fourth* and final phase, which continues through the duration of the permit. In the event that the trial burn results do not demonstrate compliance with standards, the temporary permit must be modified to allow for a second trial burn.

Example 10-8. A test burn waste mixture consisting of three designated POHCs (chlorobenzene, toluene, and xylene) is incinerated at 1,000°C. The waste feed rate and the stack discharge are shown on page 884. The stack gas flow rate is 375.24 dscm/min (dry standard cubic meters per minute). Is the unit in compliance?

Compound	Inlet (kg/h)	Outlet (kg/h)
Chlorobenzene (C_6H_5Cl)	153	0.010
Toluene (C_7H_8)	432	0.037
Xylene (C_8H_{10})	435	0.070
HCl	—	1.2
Particulates at 7% O_2	—	3.615

Outlet concentrations were measured in the stack after APC equipment.

Solution. We begin by calculating the DRE for each of the POHCs.

$$\text{DRE} = \frac{(W_{\text{in}}) - (W_{\text{out}})}{(W_{\text{in}})} \times 100$$

$$\text{DRE}_{\text{chlorobenzene}} = \frac{153 - 0.010}{153} \times 100 = 99.993\%$$

$$\text{DRE}_{\text{toluene}} = \frac{432 - 0.037}{432} \times 100 = 99.991\%$$

$$\text{DRE}_{\text{xylene}} = \frac{435 - 0.070}{435} \times 100 = 99.984\%$$

The DRE for each designated POHC must be at least 99.99 percent. In this case, the designated POHC xylene fails to meet the standard. The other POHCs exhibit a DRE of greater than 99.99 percent.

Now we check compliance for the HCl emission. The HCl emission may not exceed 1.8 kg/h or 1 percent of the HCl prior to the control equipment, whichever is greater. It is obvious that the 1.2 kg/h emission meets the 1.8 kg/h limit. This would be sufficient to demonstrate compliance, but we will calculate the mass emission rate prior to control for the purpose of comparison. To do this we assume all the chlorine in the feed is converted to HCl. The molar feed rate of chlorobenzene (M_{CB}) is

$$M_{\text{CB}} = \frac{W_{\text{CB}}}{(MW)_{\text{CB}}} = \frac{(153 \text{ kg/h})(1{,}000 \text{ g/kg})}{112.5 \text{ g/mole}}$$
$$= 1{,}360 \text{ mole/h}$$

where M_{CB} = molar flow rate of chlorobenzene
$(MW)_{\text{CB}}$ = molecular weight of chlorobenzene

Each molecule of chlorobenzene contains one atom of chlorine. Therefore,

$$M_{\text{HCl}} = M_{\text{CB}}$$
$$= 1{,}360 \text{ mole/h}$$
$$W_{\text{HCl}} = (\text{GMW of HCl})(\text{mole/h})$$
$$= (36.5 \text{ g/mole})(1{,}360 \text{ mole/h})$$
$$= 49{,}640 \text{ g/h or } 49.64 \text{ kg/h}$$

This is the HCl emission prior to control. The emission of 1.2 kg/h is greater than 1 percent of the uncontrolled emission, that is,

$$1\% \text{ of uncontrolled} = (0.01)(49.64)$$
$$= 0.4964 \text{ kg/h.}$$

However, the incinerator passes the HCl limits because the HCl emission is less than 1.8 kg/h.

The particulate concentration was measured at 7 percent O_2 and, therefore, does not need to be corrected. The outlet loading (W_{out}) of the particulates is

$$W_{out} = \frac{(3.615 \text{ kg/h})(10^6 \text{ mg/kg})}{(375.24 \text{ dscm/min})(60 \text{ min/h})}$$
$$= 160 \text{ mg/dscm}$$

This is less than the standard of 180 mg/dscm and is, therefore, in compliance with regard to particulates. However, because the incinerator fails the DRE for xylene, the unit is out of compliance.

Regulations for PCBs. Incineration of PCBs is regulated under the Toxic Substances Control Act (TSCA) rather than RCRA. Thus, some of the permit conditions for incineration of PCBs are different from other RCRA hazardous wastes.

The conditions for incineration of liquid PCBs may be summarized as follows (Wentz, 1989):

1. *Time and temperature.* Either of two conditions must be met. The residence time of the PCBs in the furnace must be 2 seconds at 1200°C ± 100°C with 3 percent excess oxygen in the stack gas or, alternatively, the furnace residence time must be 1.5 seconds at 1600°C ± 100°C with 2 percent excess oxygen in the stack gas.

   The EPA has interpreted these conditions to require a liquid PCB DRE ≥ 99.9999 percent.

2. *Combustion efficiency.* The combustion efficiency shall be at least 99.99 percent, computed as follows:

$$\text{Combustion efficiency} = \frac{C_{co_2}}{C_{co_2} + C_{co}} \times 100\% \qquad (10\text{-}33)$$

   where C_{co_2} = concentration of carbon dioxide in stack gas
   C_{co} = concentration of carbon monoxide in stack gas

3. *Monitoring and controls.* In addition to these permitted limits, owners or operators of incinerators are required to monitor and control the variables that affect performance. The rate and quantity of PCBs fed to the combustion system must be measured and recorded at regular intervals of no longer than 15 minutes. The temperatures of the incineration process must be continuously measured and recorded. The flow of PCBs to the incinerator must stop automatically whenever one of the following occurs: the combustion temperature

drops below the temperatures specified, that is, 1200 or 1600°C; when there is a failure of monitoring operations; when the PCB rate and quantity measuring and recording equipment fails; or when excess oxygen falls below the percentage specified. Scrubbers must be used for HCl removal during PCB incineration.

In addition, a trial burn must be conducted and the following exhaust emissions must be monitored:

Oxygen (O_2)

Carbon monoxide (CO)

Oxides of nitrogen (NO_x)

Hydrogen chloride (HCl)

Total chlorinated organic content

PCBs

Total particulate matter

An incinerator used for incinerating nonliquid PCBs, PCB articles, PCB equipment, or PCB containers must comply with the same rules as those for liquid PCBs, and the mass air emissions from the incinerator must be no greater than 0.001 g PCB per kilogram of the PCB introduced into the incinerator, that is, a DRE of 99.9999 percent.

Stabilization/Solidification

Because of their elemental composition, some wastes, such as nickel, cannot be destroyed or detoxified by physical or chemical means. Thus, once they have been separated from aqueous solution and concentrated in ash or sludge, the hazardous constituents must be bound up in stable compounds that meet the LDR restrictions for leachability.

The terminology for this treatment technology has evolved in the last decade. In the early to mid 1980s "chemical fixation," "encapsulation," and "binding" were often used interchangeably with solidification and stabilization. With the promulgation of the LDR restrictions, the EPA established a more precise definition for solidification/stabilization and discouraged the use of the other terms to describe the technology (U.S. EPA, 1988b). EPA linked solidification and stabilization because the resultant material from the treatment must be both stable and solid. "Stability" is determined by the degree of resistance of the mixture of the hazardous waste and additive chemical to leaching in the *Toxicity Characteristic Leaching Procedure* (TCLP) (55 FR 26986, JUN 29, 1990). In the EPA definition, then, solidification/stabilization refers to chemical treatment processes that chemically reduce the mobility of the hazardous constituent.

Reduced leachability is accomplished by the formation of a lattice structure and/or chemical bonds that bind the hazardous constituent and thereby limit the amount of constituent that can be leached when water or a mild acid solution comes into contact with the waste matrix. There are two principal solidification/stabilization processes: cement based and lime based. The cement or lime additive is mixed with the ash or sludge and water. It is then allowed to cure to form a solid. The correct mix proportions are determined by trial-and-error experiments on waste samples. In both techniques the stabilizing agent may be modified by other additives such as silicates. In general, this technology is applicable to wastes containing metals with little or no organic contamination, oil, or grease.

10-8 LAND DISPOSAL

Deep Well Injection

Deep well injection consists of pumping wastes into geologically secure formations. Pumping of wastes into these formations has been practiced primarily in Louisiana and Texas. In promulgating the final third of the LDR restrictions (55 FR 22530, 1 JUN 1990), the EPA allowed disposal of waste in Class I injection wells for wastes disposed under clean water act regulations.

Land Treatment

Land treatment is sometimes called "land farming" of the waste. In this practice, waste was incorporated with soil material in the manner that fertilizer or manure might be. Microorganisms in the soil degraded the organic fraction of the waste. Under the LDR restrictions, this practice is prohibited.

The Secure Landfill

Although far from ideal, the use of land for the disposal of hazardous wastes is a major option for the foreseeable future. Furthermore, we recognize that incinerator ash, scrubber bottoms, and the results of biological, chemical, and physical treatment leave residues of up to 20 percent of the original mass. These residues must be secured in an economical fashion. At this juncture, the secure landfill is the only option.

The basic physical problem with land disposal of hazardous waste stems from the movement of water. The dissolution of waste material results in contaminants being transported from the waste site to larger regions of the soil zone and, too often, to an underlying aquifer. Problems of groundwater pollution frequently lead to the condemnation of wells and to the contamination of surface water bodies fed by the associated aquifer. In many instances, well contamination is not detected until years after land disposal of waste has begun, because of the slow movement of the conveying ground water (Wood et al., 1984).

Water pollution, caused by a hazardous waste facility, may evolve in a variety of ways. Leachate from landfills may drain out of the side of the landfill and appear as surface runoff. It may seep down slowly through the unsaturated zone and enter an underlying aquifer. Fissures in liners lead to a downward migration of contaminants toward the water table.

Without the institution of remedial measures, buried waste usually acts as a continuing source of pollution. The waste constituents continue to be transported in the subsurface by infiltrating precipitation. Thus, sites that handle hazardous wastes are located above a natural barrier, as well as an applied liner. Moreover, the site is instrumented to continuously monitor the condition of any associated aquifers. In addition, a system for the collection and treatment of the leachate is required.

The technology of the secure landfill may be divided into two phases: siting and construction. The following discussion on siting is drawn primarily from E. F. Wood et al. (1984).

Landfill Siting. In siting a hazardous waste landfill, the four main considerations are air quality, groundwater quality, surface water quality, and subsurface migration of gases and leachates. Aside from the sociopolitical aspects, the last three components are the major factors to be considered in siting the landfill.

Air quality must be considered to prevent adverse effects to the air caused by volatilization, gas generation, gas migration, and wind dispersal of landfilled hazardous wastes. Generally, these can be controlled by proper construction techniques and do not inhibit the siting.

The hydrogeologic siting problem can be divided into four main areas: geology, soil, hydrology, and climate. Bedrock geology determines the structural framework that surfaces as landforms and the structural integrity of the landfill site.

Structural integrity of host rock is important in terms of seismic risk zones, dipping, and cleavage. Seismic risk zones indicate the presence of geologic faults and fractures. Faults and fractures provide a natural pathway for the flow of contaminants, even in low-permeability and low-porosity rock.

Transport capacity refers to a soil's ability to allow migration of contaminants. A soil with low permeability and porosity can lengthen the flow period and act as a natural defense by retarding the movement of contaminants. Glacial outwash plains and deltaic sands are both well-sorted sand and gravel beds with high permeability. Thus, they allow wastes to move faster and further. Clays and silts have lower permeabilities and, thus, inhibit the movement of wastes.

Most contaminants will move either at the same rate or slower than the water. The relative speeds of the water and contaminant are a function of the contaminant and water characteristics. For example, organic contaminants that are relatively insoluble in water will be *retarded* more by the soil than organic contaminants that are relatively soluble in water. The pH of the water will also affect retardation. For example, at low pH (and in the absence of oxygen), iron will be present predominantly as ferrous iron (Fe^{2+}). This iron is quite soluble and will move with the water. If the pH is high (>6) and oxygen is present, the iron will be in the ferric (Fe^{3+}) form, which is much less soluble in water. The Fe^{3+} will precipitate and, therefore, will not move with the groundwater. The extent to which the chemicals are retarded is defined by the *retardation coefficient:*

$$R = \frac{v'_{\text{water}}}{v'_{\text{contaminant}}} \tag{10-34}$$

where v'_{water} = linear speed of the water
$v'_{\text{contaminant}}$ = linear speed of the contaminant

The retardation coefficient is a function of the hydrophobicity of the contaminant on a specified soil. For neutral organic chemicals, R is defined as:

$$R = 1 + \left(\frac{\rho_b}{\eta}\right) K_{oc} f_{oc} \tag{10-35}$$

TABLE 10-12
Retardation coefficients of typical groundwater contaminants[a]

Compound	Soil A[b]	Soil B[b]	Soil C[b]
Benzene	1.2	1.7	5.3
Toluene	1.7	3.6	17.0
Aniline	1.1	1.2	2.2
Di-n-propyl phthalate	5.6	19.0	110.0
Fluorene	23.0	86.0	500.0
n-Pentane	7.0	24.0	140.0

[a]Data and formulas used for computation of K_{oc} are from Schwarzenback et al., 1993.
[b]For soil A: $\rho_b = 1.4$ g/cm^3; $\eta = 0.40$; $f_{oc} = 0.002$; for soil B: $\rho_b = 1.6$ g/cm^3; $\eta = 0.30$; $f_{oc} = 0.005$; for soil C: $\rho_b = 1.75$ g/cm^3; $\eta = 0.55$; $f_{oc} = 0.05$.

where ρ_b = bulk density of the soil
η = porosity of soil as a fraction
K_{oc} = partition coefficient into the organic carbon fraction of the soil
f_{oc} = fraction of organic carbon in the soil

Some retardation coefficients for typical groundwater contaminants are given in Table 10–12.

Example 10-9. An illegally buried drum of toluene has begun to leak into an unconfined drinking water aquifer. A homeowner's well is located 60 m down-gradient from the leaking drum. If this is a type C soil and the linear speed of the water in the aquifer is 4.7×10^{-6} m/s, how many days will it take for the toluene to reach the well?

Solution. Using the value of R from Table 10–12, compute the linear speed of the contaminant as

$$v'_{\text{contaminant}} = \frac{v'_{\text{water}}}{R}$$

$$= \frac{4.7 \times 10^{-6} \text{ m/s}}{17.0} = 2.76 \times 10^{-7} \text{ m/s}$$

The travel time is then

$$\frac{\text{Distance}}{v'_{\text{contaminant}}} = \left(\frac{60 \text{ m}}{2.76 \times 10^{-7} \text{ m/s}}\right)\left(\frac{1}{86{,}400 \text{ s/d}}\right)\left(\frac{1}{365 \text{ d/y}}\right) = 6.88 \text{ or about 7 years}$$

We should point out two important notes: (1) this is an extreme simplification of a very complex problem that, in reality, may result in a very different answer from this computation, and (2) the concentration of the toluene is not addressed in this problem. The actual travel time may vary by an order of magnitude depending on the pumping rate of the well, precipitation patterns, and other undetermined hydrogeologic parameters. The concentration depends on the mass of toluene released, the quantity of water that dilutes it, the solubility of toluene in this water, and undetermined reactions in the soil.

Sorption capacity depends on the organic content, predominant minerals, pH, and soil. Sorption includes both absorption and adsorption of contaminants. Sorption is important in limiting the movement of metals, phosphorus, and organic chemicals. *Cation exchange capacity* (CEC) is a measure of the ability of the soil to trade cations in the soil for those in waste. The higher the CEC, the more metal will be retained. The capacity of soil to retard contaminant migration also depends on the presence of numerous hydrous oxides, particularly iron oxides, and other compounds such as phosphates and carbonates. These compounds precipitate heavy metals out of solution, making them unable to travel further.

The hydrogen-ion concentration (pH) of soil influences the dominant removal mechanism for metal cations. The dominant removal mechanism for metal cations when $pH < 5$ is exchange or adsorption; when $pH > 6$, it is precipitation.

Hydrologic considerations in locating a hazardous waste landfill include distance to the groundwater table, the hydraulic gradient, the proximity of wells, and the proximity to surface waters.

When distance from the surface to the groundwater table is short, contaminant travel time is also short, allowing for little attenuation before pollutants disperse laterally in the saturated zone. It is desirable to have the average distance to the groundwater table large enough so that contaminants may be significantly attenuated. This also facilitates monitoring of the saturated zone. This will permit remedial action to be undertaken, if necessary.

A hydraulic gradient that slopes away from local groundwater supplies is desired. The steeper the hydraulic gradient, the lower the attenuation time and the faster the water movement. Therefore, a moderate hydraulic gradient may be most acceptable.

The distance from the disposal site to water-supply wells and surface waters must be as large as possible to protect them from potential contamination in case the landfill leaks. Furthermore, the proximity to surface waters must take into account the potential for flooding. Site flooding will weaken the structure of a land emplacement facility, causing it to fail and leak wastes. Therefore, it is essential that the facility not be built on a floodplain or area subject to local flooding. The facility should be designed so that it will not be flooded.

Climate is considered a driving force in contaminant migration, but we may exclude it when considering potential sites within the same region, where climate is unlikely to vary significantly.

Landfill Construction. A secure landfill means, in essence, that no leachate or other contaminant can escape from the fill and cause adverse impacts on the surface water or groundwater. Leakage from the site is not acceptable during or after operations. Neither is any external or internal displacement, which could be brought about by slumping, sliding, and flooding. Wastes must not be allowed to migrate from the site.

It is nearly impossible to create an impervious burial vault for hazardous wastes and guarantee its integrity forever. Landfill design and operation is regulated to minimize migration of wastes from the site. The current EPA rules (40 CFR 264.300) for hazardous waste landfills require a minimum of (1) two or

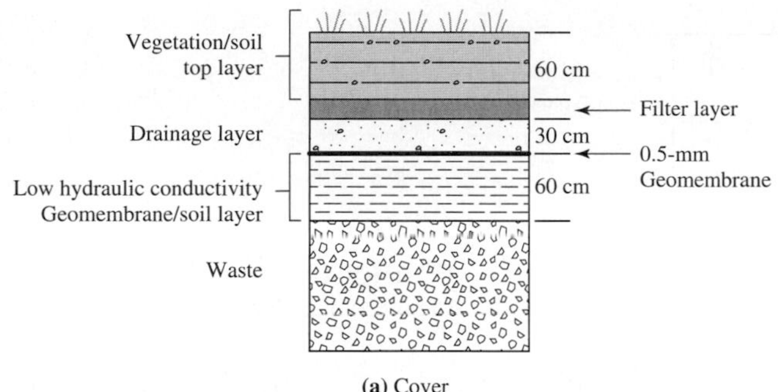

(a) Cover

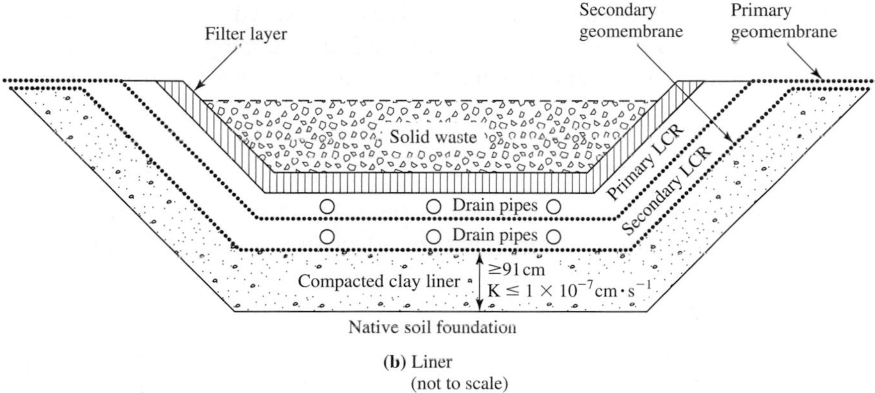

(b) Liner
(not to scale)

FIGURE 10-15
Minimum technology landfill liner, design and recommended final cover design. (*Source:* U.S. EPA, 1989b and 1991.)

more liners, (2) a leachate collection system above and between the liners, (3) surface run-on and run-off control to collect and control at least the water volume resulting from a 24-hour, 25-year storm, (4) monitoring wells, and (5) a "cap" (Figure 10-15).

The liner system must include (57 FR 3462, 29 JAN 1992):

1. A top liner designed and constructed of materials (for example, a geomembrane) to prevent migration of hazardous constituents into the liner during the active life and post-closure care period;

2. A composite bottom liner consisting of at least two components. The upper component must be designed and constructed of materials (for example, a geomembrane) to prevent migration of hazardous constituents into the liner during the active life and post-closure care period. The lower component must be designed and constructed of materials to minimize migration of hazardous constituents if a breach in the upper component were to occur. The lower component must be constructed of at least 91 cm of compacted soil material with a hydraulic conductivity of no more than 1×10^{-7} cm/s.

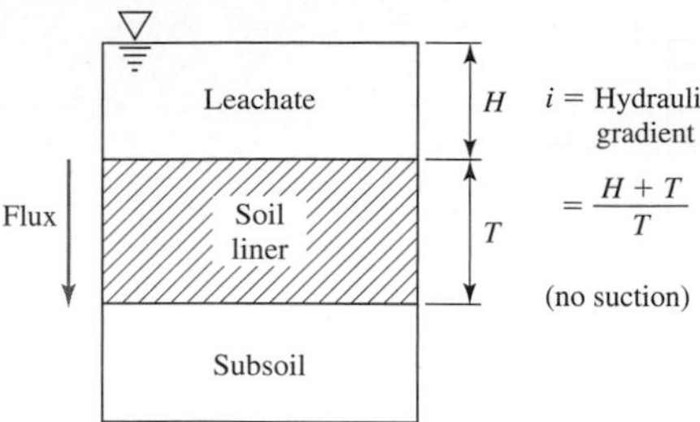

FIGURE 10-16
Definition of hydraulic gradient for landfill liner.

The leachate collection and removal system (LCR) immediately above the top liner must be designed, constructed, operated, and maintained to collect and remove leachate so that the leachate depth over the liner does not exceed 30 cm. The leachate collection and removal system between the liners and immediately above the bottom liner is also a leak detection system. The leachate collection system must, at a minimum, be:

1. Constructed with a bottom slope of one percent or more;
2. Constructed of a granular drainage material with a hydraulic conductivity of 1×10^{-2} cm/s or more and a thickness of 30 cm or more; or be constructed of synthetic or geonet drainage materials with a transmissibility of 3×10^{-5} m^2/s or more;
3. Constructed of sufficient strength to prevent collapse and be designed to prevent clogging.

The design equations for the leachate collection system are the same as those used for a municipal landfill (Section 9-4). The leachate collection system must include pumps of sufficient size to remove the liquids to prevent leachate from backing up into the drainage layer. The leachate must be treated to meet discharge limits. The treated leachate may be discharged into the municipal wastewater treatment system or into a waterway.

The amount of leachate may be estimated using Darcy's law (Equation 3-22). The hydraulic gradient for a liner is defined as shown in Figure 10-16. The flow rate cannot exceed the amount of water available, that is the product of the precipitation rate and the area of the landfill. The travel time of a contaminant through a soil layer may be estimated as the linear length of the flow path (T) divided by the seepage velocity (Equation 3-26).

Example 10-10. How long will it take for leachate to migrate through a 0.9 m clay liner with a hydraulic conductivity of 1×10^{-7} cm/s if the depth of leachate above the clay layer is 30 cm and the porosity of the clay is 55 percent?

Solution. The Darcy velocity is found using Equation 3-22.

$$v = K\left(\frac{dh}{dr}\right)$$

where the hydraulic gradient (dh/dr) is defined as in Figure 10-16:

$$\frac{dh}{dr} = \frac{0.30 \text{ m} + 0.9 \text{ m}}{0.9 \text{ m}} = 1.33$$

The Darcy velocity is then

$$v = (1 \times 10^{-7} \text{ cm/s})(1.33) = 1.33 \times 10^{-7} \text{ cm/s}$$

From Equation 3-27, the seepage velocity is

$$v' = \frac{K(dh/dr)}{\eta} = \frac{1.33 \times 10^{-7} \text{ cm/s}}{0.55} = 2.42 \times 10^{-7} \text{ cm/s}$$

The travel time is then

$$t = \frac{T}{v'} = \frac{(0.9 \text{ m})(100 \text{ cm/m})}{2.42 \times 10^{-7} \text{ cm/s}} = 3.71 \times 10^{8} \text{ s or about 12 years}$$

The site operator must keep careful records of the location and dimensions of each cell and must depict each cell on a map keyed to permanently surveyed vertical and horizontal markers. Records must show the contents of each cell and the approximate location of each hazardous waste type within the cell.

The purpose of groundwater monitoring is to ensure that programs for managing runon, runoff, and leachates are functioning properly so that groundwater remains uncontaminated. If contamination is occurring, early warning can be given and countermeasures taken. The site owner/operator has to place a sufficient number of monitoring wells around the limits of the facility to be able to describe the background (upgradient) and downgradient water quality. The regulations set forth, in detail, how the monitoring wells must be sunk, screened, sealed, sampled, and located, with special emphasis on location of the downgradient wells.

General groundwater quality, especially the suitability of the uppermost aquifer for use as a drinking water source, must meet EPA's primary drinking water standards. The flow rate for each sump must be calculated weekly during the active life and closure period, and monthly during the post-closure care period. If the landfill is leaking to the groundwater, the site operator must file an assessment plan with the EPA that shows how the problem is to be remedied.

10-9 GROUNDWATER CONTAMINATION AND REMEDIATION

The Process of Contamination

Hazardous waste landfills are, of course, not the only source of groundwater contamination. Other sources include municipal landfills, septic tanks, mining and agricultural activities, "midnight dumping," and leaking underground storage tanks. It has been estimated that more than 35,000 underground storage tanks are leaking (U.S. EPA, 2004a).

The threat of contamination to groundwater depends on the specific geologic and hydrologic conditions of the site. The following paragraphs describe general considerations. All conditions may not exist at every site.

Leaking chemicals pass through several different hydrologic zones as they migrate through the soil to the groundwater system. The pore spaces in the unsaturated zone in the top soil layers are occupied by both air and water. Flow in this zone for liquid contaminants is downward by gravity. The upper region of the unsaturated zone is important for pollutant attenuation. Some chemicals are retained by adsorption onto organic material and chemically active soil particles. Some are trapped in the pore spaces and held by surface tension. These adsorbed and trapped chemicals may decompose through abiotic processes such as oxidation, reduction, and hydrolysis, as well as microbial activity, or they may simply remain sorbed onto the particles. Migration of precipitation may leach this sorbed and trapped material and carry it to the underlying aquifer for long periods of time after the source of contamination has been removed (Wentz, 1989).

In the capillary zone just above the saturated zone that marks the groundwater table, spaces between soil particles may be saturated by water rising from the water table by capillary action. Chemicals that are lighter than water will "float" on top of the water table in this zone and move in different directions and rates than dissolved contaminants.

The pore spaces between soil particles below the water table are saturated. Generally, the saturated zone is devoid of oxygen. The lack of dissolved oxygen limits the oxidation of chemicals.

Groundwater flow is laminar, with minimal mixing occurring as the groundwater moves. Dissolved chemicals will flow with groundwater and form distinct plumes. The shape and size of a contaminant plume depends upon the local hydrogeological setting, groundwater flow, the characteristics of the contaminants, and geochemistry. Solubility, adsorption characteristics, and degradation affect mobility. The density of the contaminant is important in determining the shape and movement of the plume. Lighter, less soluble chemicals, like gasoline, will tend to flow on top of the aquifer (see Figure 5-19). Water soluble contaminants tend to dissolve in and then flow with groundwater (see Figure 5-18). Dense, insoluble contaminants will sink to the bottom (see Figure 5-20) of the aquifer. *Volatile organic chemicals* (VOCs) in groundwater are extremely mobile. Polyvalent metal contaminants tend to adsorb onto clays and, hence, are not very mobile.

EPA's Groundwater Remediation Procedure

The federal program for cleanup of contaminated sites follows a procedural sequence as shown in Figure 10-17. Each of these steps is discussed in the following paragraphs.

Preliminary Assessment. EPA involvement usually begins with the identification of a potential hazardous waste site. The initial information can come from a variety of sources, including local citizens and officials, state environmental agencies, the site owners themselves, or simply from awareness of potential problems associated with particular industries.

EPA has developed an inventory system called the Comprehensive Environmental Response, Compensation, and Liability Information System (CERCLIS) to document

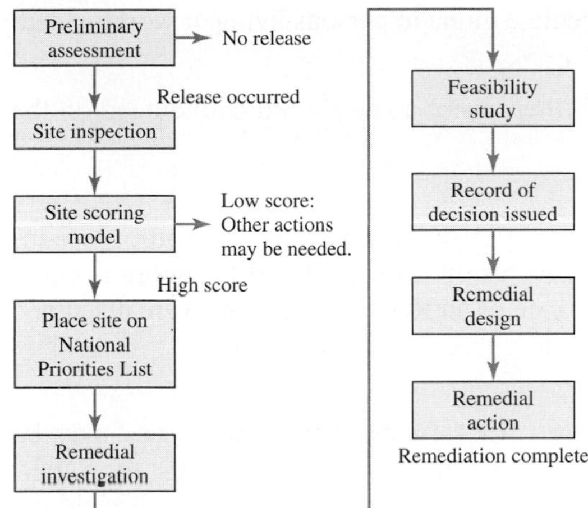

FIGURE 10-17
Steps involved in the Superfund cleanup process.

all of the sites in the United States that may be candidates for remedial action. This is a continuing program that identifies sites as information about them becomes available. The growth in the number of CERCLIS sites has been dramatic and is expected to continue for the foreseeable future as additional abandoned and contaminated sites are discovered. As of September 2005, 12,031 sites were in the inventory. This list did not include on estimated 130,000 leaking underground storage tanks (U.S. EPA, 2004a and 2005b).

A *preliminary assessment* (PA) is the first step in identifying the potential for contamination from a particular site. The primary objectives of the PA are to determine if there has been a release of contaminant to the environment, if there is immediate danger to persons living or working near the site, and whether a site inspection is necessary. Samples for environmental analysis are generally not taken during the PA. Following the preliminary assessment, EPA or the designated state agency might determine that an immediate threat to residents or employees at the site requires an immediate removal action. Otherwise, on the basis of the preliminary assessment, the site is classified by EPA into one of the three following categories:

1. There is no further action needed, since there is no threat to human health or the environment.
2. Additional information is required to complete the preliminary assessment.
3. Inspection of the site is necessary.

Site Inspection. Site inspection requires sampling to determine the types of hazardous substances that are present and to identify the extent of contamination and its migration. The actual site inspection includes preparation of a work plan and an on-site safety plan. The site assessment has three objectives:

1. To determine which releases pose no threat to public health and the environment;

2. To determine if there is any immediate threat to persons living or working near the release;

3. To collect data to determine whether or not a site should be included on the National Priorities List (NPL).

HRS, NPL, RI/FS, and ROD. The next series of steps in the EPA's procedure include calculations to complete the HRS; inclusion on the NPL if the score is sufficiently high; conduct of a RI/FS; and issuance of an ROD. These steps were discussed in detail in Section 10-5.

Remedial Design and Remedial Action. EPA-funded remedial actions may be taken only at those sites that are on the NPL. This ranking helps ensure that the Superfund dollars are used in the most cost-effective manner and where they will yield the greatest benefit.

Before a remedial action can be taken at a site, a number of questions must be answered. These can be classified as problem definition, design alternatives, and policy.

1. Problem definition questions: What are the contaminants and how much contamination is present? How large is the surface area of the contaminated site? What is the size of the contaminated groundwater plume? Where is the exact location of the plume and in what direction is it moving?

2. Design questions: Based on the alternatives available, what is the best way to clean up the site? How should these alternatives be implemented? What products will be produced during treatment? How long will it take to complete the remediation and what will it cost?

3. Policy questions: What level of protection is adequate? In other words, how clean is clean?

The answers to the first two sets of questions require scientific and engineering background that is supported by extensive sampling of the contaminated site area. The last question cannot be answered objectively; rather it is a subjective and oftentimes political question.

The NCP defines three types of responses for incidents involving hazardous substances. In these responses *removal* is differentiated from *remediation.* Removal is, as its name suggests, the physical relocation of the waste—usually to a secure hazardous waste landfill. Remediation means that the waste is to be treated to make it less toxic and/or less mobile or the site is to be contained to minimize further release. Remediation can take place on site or at a TSD facility. The three types of responses are:

1. *Immediate removal* is a prompt response to prevent immediate and significant harm to human health or the environment. Immediate removals must be completed within six months.

2. *Planned removal* is an expedited removal when some response, not necessarily an emergency response, is required. The same six-month limitation also applies to planned removal.

3. *Remedial response* is intended to achieve a site solution that is a permanent remedy for the particular problem involved.

Immediate removals are done to prevent an emergency involving hazardous substances. These emergencies might include fires; explosions; direct human contact with a hazardous substance; human, animal, or food-chain exposure; or contamination of drinking water sources. An immediate removal involves cleaning up the hazardous site to protect human health and life, containing the hazardous release, and minimizing the potential for damage to the environment. For example, a truck, train, or barge spill could involve an immediate removal determination by EPA to get the spill cleaned up.

Immediate removal responses may include activities such as sample collection and analysis, containment or control of the release, removal of the hazardous substances from the site, provision of alternate water supplies, installation of security fences, evacuation of threatened citizens, or general deterrent of the spread of the hazardous contaminants.

A planned removal involves a hazardous site that does not present an immediate emergency. Under Superfund, EPA may initiate a planned removal if the action will minimize the damage or risk and is consistent with a more effective long-term solution to the problem. Planned removals are carried out by EPA if the responsible party is either unknown or cannot or will not take timely and appropriate action. The state in which the cleanup is located must be willing to match at least 10 percent of the costs of the removal action, as well as agree to nominate the site in question for the National Priority List.

Mitigation and Treatment

Because the spread of contaminants is usually confined to a plume, only localized areas of an aquifer need to be reclaimed and restored. Cleanup of a contaminated aquifer, however, is often time-consuming and costly. The original source of contamination can be eliminated, but the complete restoration of the groundwater is fraught with additional problems, such as defining the site's subsurface soil and geologic composition, locating contamination sources, defining contaminant transport pathways, determining the extent and concentration of the contaminants, and choosing and implementing an effective remedial process (Griffin, 1988).

Cleanup methods for contaminated aquifers range from containment to destruction of the contaminants. Because, in the long run, containment does not really solve the problem, destruction of the contaminants is the preferred objective of a cleanup program. Examples of remedial methods include (LaGrega et al., 2001):

- Installing pumping wells to remove the contaminated water and then treating it with one of the technologies described in Section 10-7 (called *pump and treat*)
- Air sparging
- Soil vapor extraction
- Reactive treatment walls (also known as *permeable reactive barriers*) that allow the contaminated plume to pass through a reactive chemical or a an acclimated biomass.

In the next few paragraphs we will discuss the first of these systems. The other systems are discussed in detail in LaGrega et al. (2001).

Certainly, combinations of barriers and treatment methods should be considered. Source control (removal or remediation of the source), physical control, and treatment methods all will have their part in mitigating groundwater contamination problems. Legal implications may also dictate strategies that may be utilized (Griffin, 1988).

Pump and Treat. The objectives of a pump-and-treat system include hydraulic containment of the contaminated plume and removal of the contaminant from the groundwater. The design of the well system for a pump-and-treat remediation is an application of well hydraulics described in Chapter 3.

The *capture zone* of an extraction well is that portion of the groundwater that will discharge into the well. The capture zone is not necessarily coincident with the cone of depression (Section 3-5) because the groundwater flow lines can be diverted by the influence of the pumping well without being captured by the well (Figure 10-18). Under steady-state conditions, the extent of the cone of depression largely depends on the

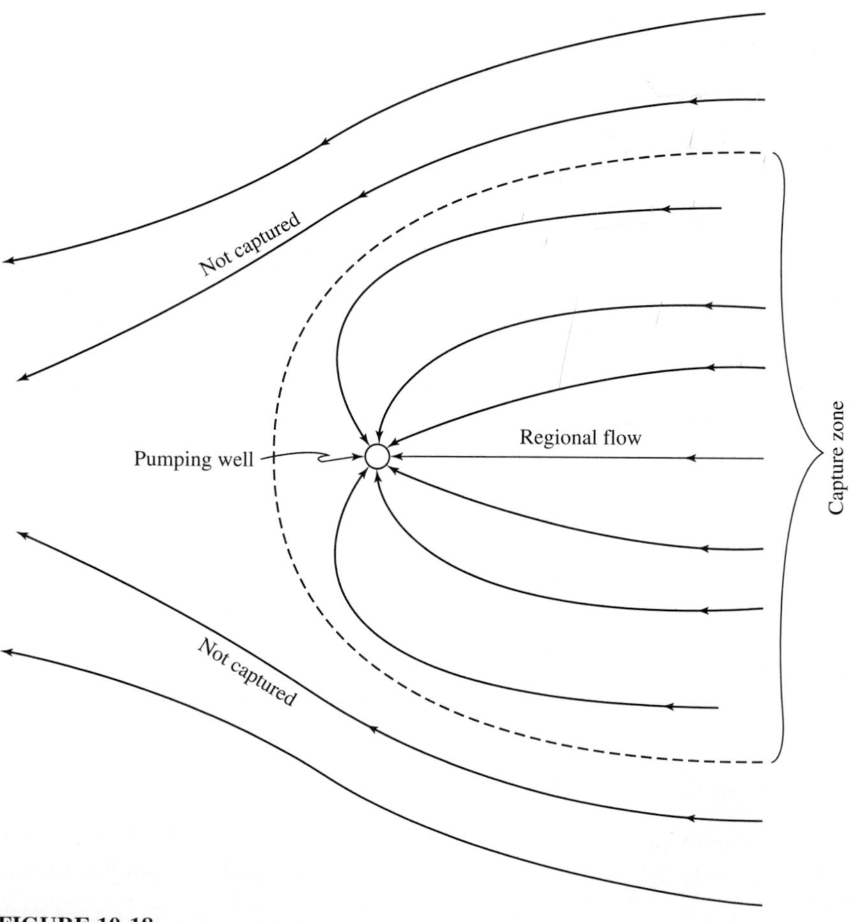

FIGURE 10-18
Groundwater flow lines influenced by pumping well.

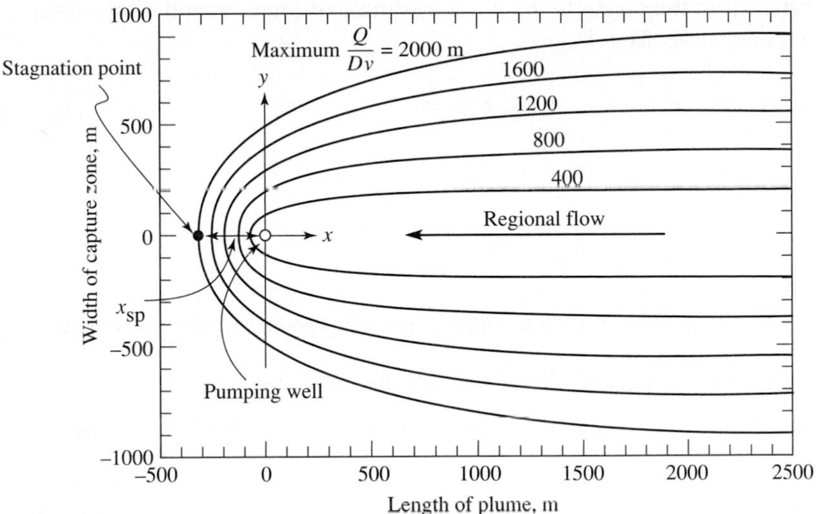

FIGURE 10-19
Type curve for analytical solution to capture zone analysis for a single extraction well. (*Source:* Javandel and Tsang, 1986.)

transmissivity and pumping rate. The extent of the capture zone depends on the regional hydraulic gradient as well as the transmissivity and pumping rate.

Three parameters are used to delineate the capture zone: (1) the width of the capture zone at an infinite distance up-gradient from the pumping well, (2) the width of the capture zone at the location of the pumping well, and (3) the location of the down-gradient distance of the capture zone from the pumping well (called the *stagnation point*). These parameters are shown in Figure 10-19.

Javendel and Tsang (1986) developed a highly idealized model of the capture zone that can be used to examine the relationship between some of the important variables. The model assumes a homogeneous, isotropic aquifer uniform in cross section and infinite in width. The aquifer may be either confined or unconfined. However, in the case of the unconfined aquifer, drawdown must be insignificant with respect to the total thickness of the aquifer. The extraction wells are assumed to be fully penetrating.

With a single well located at the origin of the coordinate system shown in Figure 10-19, Javendel and Tsang (1986) developed the following equation to describe the y coordinate of the capture zone envelope:

$$y = \pm\frac{Q}{2Dv} - \frac{Q}{2\pi Dv}\tan^{-1}\frac{y}{x} \tag{10-36}$$

where x, y = distances from the origin m
Q = well pumping rate, m^3/s
D = aquifer thickness, m
v = Darcy velocity, m/s

Note that the $\pm$ allows computation of the y coordinate above and below the x axis. Masters (1998) has shown that this equation may be rewritten in terms of the angle ϕ

(in radians) drawn from the origin to the x, y coordinate of interest on the line describing the capture zone curve. That is,

$$\tan\phi = \frac{y}{x} \tag{10-37}$$

so that, for $0 \leq \phi \geq 2\pi$, Equation 10-36 may be rewritten as

$$y = \pm\frac{Q}{2Dv} - \left(1 - \frac{\phi}{\pi}\right) \tag{10-38}$$

This equation allows us to examine some important fundamental relationships:

- The width of the capture zone is directly proportional to the pumping rate.
- The width of the capture zone is inversely proportional to the Darcy velocity.
- As x approaches infinity, $\phi = 0$ and $y = Q/(2Dv)$. This sets the maximum total width of the capture zone at $2\,[Q/(2Dv)] = Q/(Dv)$ as shown in Figure 10-19.
- For $\phi = \pi/2$, $x = 0$ and y is equal to $Q/(4Dv)$. Thus, the width of the capture zone at $x = 0$ is $2[Q/(4Dv)] = Q/(2Dv)$.

The distance to the stagnation point down-gradient of the extraction well (x_{sp}) may be estimated from the following equation (Legrega et al., 2001):

$$x_{sp} = \frac{Q}{2\pi Dv} \tag{10-39}$$

Javandel and Tsang prepared a series of "type" curves for various well configurations (one to four wells) and several widths of the capture zone at $x = \infty$. The suggested approach to using the capture zone technique is summarized as follows:

1. Prepare a site map with the plume shape at the same scale as the type curves.
2. Superimpose the site map on the one-well type curve with the direction of regional flow parallel to the x axis. Place the leading edge of the plume just beyond the location of the extraction well. Select the capture zone curve that completely captures the plume. This defines the required value of Q/Dv at $x = \infty$.
3. Determine the required pumping rate by multiplying Q/Dv by Dv. If the required pumping rate can be achieved by the use of one well, then the problem is solved. If one well does not produce the required pumping rate, then go to step 4.
4. Repeat step 2 using the two, three, or four well–type curves as required to achieve an acceptable pumping rate. Each well in the multiple-well scenarios is assumed to pump at the same rate.

The capture zone of multiple extraction wells must overlap to prevent the groundwater flow from passing between them. If the distance between the extraction wells is less than or qual to $Q/\pi Dv$, the capture zones will overlap. Assuming that the wells are located symmetrically around the x axis, the optimum spacing may be calculated by using the following:

- For two wells space at $Q/\pi Dv$
- For three wells space at $1.26Q/\pi Dv$
- For four wells space at $1.2Q/\pi Dv$

The question of whether or not the required pumping rate can be achieved is determined, in a confined aquifer, by the available drawdown that will not lower the piezometric surface into the aquifer. This can be calculated using the methods discussed in Section 3-5. For an unconfined aquifer, the restriction noted above, that the drawdown must be insignificant with respect to the total thickness of the aquifer, requires a judgement decision.

Example 10-11. The drinking water well at village of Oh Six is threatened by a contaminant plume in the aquifer. The confined aquifer is 28.7 m thick. It has a hydraulic conductivity of 1.5×10^{-4} m/s, a storage coefficient of 3.7×10^{-5}, and a regional hydraulic gradient of 0.003. The contaminant plume is 300 m wide at its widest point. The maximum allowable pumping rate based on the allowable drawdown is 0.006 m^3/s. Locate a single extraction well so that the stagnation point is 100 m from the drinking water well and so that the capture zone encompasses the plume. A sketch map of the drinking water well and the plume relationship is shown below.

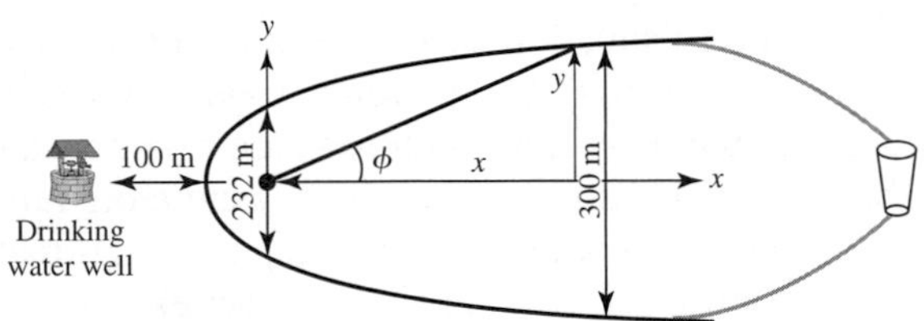

Solution. Determine the Darcy velocity, using Equation 3-22:

$$v = K\frac{dh}{dr} = (1.5 \times 10^{-4}\ \text{m/s})(0.003) = 4.5 \times 10^{-7}\ \text{m/s}$$

The width of the capture zone at an infinite distance up-gradient is

$$\frac{Q}{Dv} = \frac{0.006\ \text{m}^3/\text{s}}{(28.7\ \text{m})(4.5 \times 10^{-7}\ \text{m/s})} = 464.58 \text{ or } 465\ \text{m}$$

The width of the capture zone at the extraction well will be

$$\frac{Q}{2Dv} = \frac{0.006 \text{ m}^3/\text{s}}{2(28.7 \text{ m})(4.5 \times 10^{-7} \text{ m/s})} = 232.29 \text{ or } 232 \text{ m}$$

The stagnation point will be

$$x_{sp} = \frac{0.006 \text{ m}^3/\text{s}}{2\pi(28.7 \text{ m})(4.5 \times 10^{-7} \text{ m/s})} = 73.94 \text{ or } 74 \text{ m}$$

down-gradient from the extraction well.

The distance down-gradient from the lead edge of the plume that the extraction well must be placed is determined by using Equation 10-38. With $y = 150$ m,

$$150 \text{ m} = (232.29 \text{ m})\left(1 - \frac{\phi}{\pi}\right)$$

Solving for the angle (in radians) from the extraction well to the point where the plume just touches the capture zone is

$$\phi = 0.35\ \pi \text{ rad}$$

Using the geometry shown in the sketch map, the distance is therefore

$$x = \frac{y}{\tan\phi} = \frac{150 \text{ m}}{\tan(0.35\ \pi)} = \frac{150 \text{ m}}{1.96} = 76.4 \text{ or } 76 \text{ m}$$

This solution is, of course, highly idealized. The lead edge of the plume is conveniently of a geometry that allows us to locate it by using Equation 10-38. A more ellipsoidal plume geometry would project the lead edge in advance of the tangent point. This technique would then lead to a very erroneous positioning of the extraction well.

While the highly idealized situation used by Javendel and Tsang is useful in understanding the behavior of a well system to control the movement of a contaminant plume, in actual field sites the boundary conditions are rarely met. Computer models that have been calibrated to the local conditions will yield more reliable, although not perfect, understanding of the behavior of a proposed pump-and-treat system.

Because the rate of removal of contaminant decreases exponentially over time, and because the concentrations in the water rebound over time because of diffusion and desorption from the soil, pump-and-treat systems are limited in their value as a mass removal technology.

Non-Aqueous Phase Liquids (NAPLs) such as gasoline are referred to as "product" because their recovery may have some commercial value. When the NAPL floats on the groundwater table, special recovery techniques may be employed to recover it. Product recovery systems to recover NAPL use wells that terminate in the NAPL plume rather than in the aquifer. Because all hydrocarbons are slightly soluble in water, the product recovery system is usually accompanied by a groundwater pumping system to remove and treat the contaminated groundwater. A typical pump-and-treat system is shown in Figure 10-20.

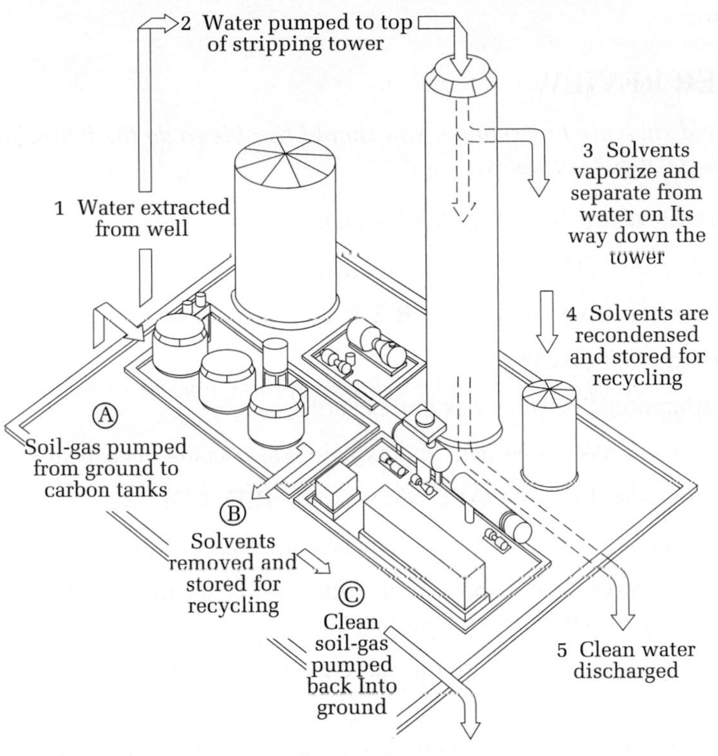
2 Water pumped to top
of stripping tower
1 Water extracted
from well
3 Solvents
vaporize and
separate from
water on Its
way down the
tower
4 Solvents are
recondensed
and stored for
recycling
Ⓐ
Soil-gas pumped
from ground to
carbon tanks
Ⓑ
Solvents
removed and
stored for
recycling
Ⓒ
Clean
soil-gas
pumped
back Into
ground
5 Clean water
discharged

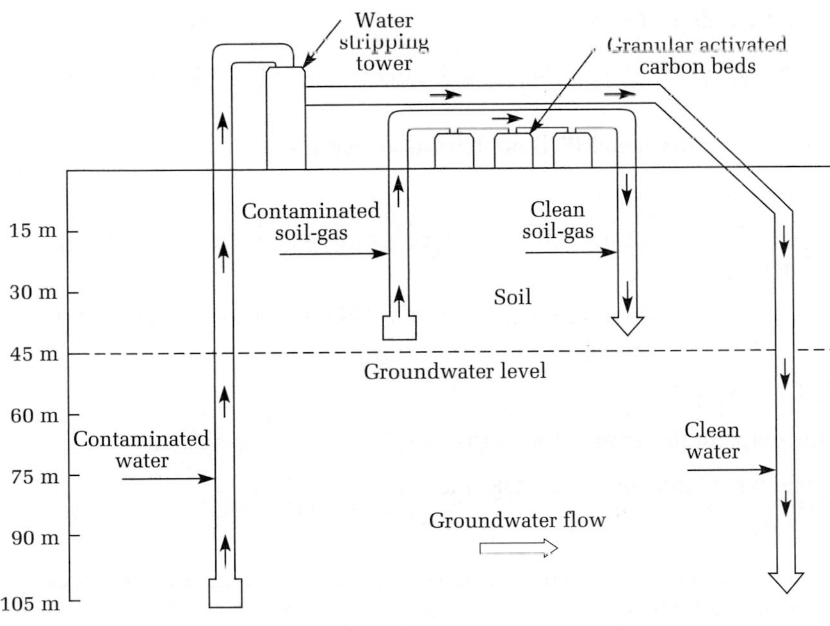
Water
stripping
tower
Granular activated
carbon beds
15 m
30 m
45 m
60 m
75 m
90 m
105 m
Contaminated
soil-gas
Clean
soil-gas
Soil
Groundwater level
Contaminated
water
Clean
water
Groundwater flow
Underground view

FIGURE 10-20
Lockheed Aeronautical Systems Company's Aqua-Detox groundwater treatment system. (*Source: Hazmat World,* November 1989.)

10-10 CHAPTER REVIEW

When you have finished studying this chapter, you should be able to do the following without the aid of your textbooks or notes:

1. Sketch the chemical structure of 2,3,7,8-TCDD.
2. Explain how 2,3,7,8-TCDD occurs and/or when it is found in nature.
3. Sketch the chemical structure of the PCB 2,4′-dichlorobiphenyl.
4. Explain the origin of PCBs.
5. Define and differentiate between risk and hazard.
6. List the four steps in risk assessment and explain what occurs in each step.
7. Define the terms dose, LD_{50}, NOAEL, slope factor, RfD, CDI, IRIS.
8. Explain why it is not possible to establish an absolute scale of toxicity.
9. Explain why an average dose-response curve may not be an appropriate model to develop environmental protection standards.
10. Identify routes of potential exposure for the release of contaminants in multiple media.
11. Explain how risk management differs from risk assessment and the role of risk perception in risk management.
12. Define hazardous waste.
13. List the five ways a waste can be found to be hazardous and briefly explain each.
14. Explain why dioxin and PCB are hazardous wastes.
15. State how long generators may store their waste.
16. Explain what defines a small quantity generator and what "break" the rules give them.
17. Define the abbreviations CFR, FR, RCRA, HSWA, CERCLA, and SARA.
18. Explain the major difference (objective) between RCRA/HSWA and CERCLA/SARA.
19. Define/explain the terms "cradle-to-grave" and manifest system.
20. Explain what "land ban," or LDR, means.
21. Define the abbreviations TSD and UST.
22. Describe the three ways to meet corrosion protection standards for underground storage tanks.
23. List the four major provisions of CERCLA.
24. Define/explain the following abbreviations: NCP, NPL, HRS, RI, FS, ROD, and PRP.

25. Explain why it is important for a site to be placed on the NPL.
26. Explain the concept of "joint and several liability" and the implications to those with wastes found in an abandoned hazardous waste site.
27. List and explain four hazardous waste management techniques.
28. List the objectives of a waste audit.
29. Differentiate between waste minimization, waste exchange, and recycling.
30. List six disposal technologies for hazardous wastes.
31. Explain why seismic risk is important in landfill siting.
32. Explain how permeability, porosity, and sorption capacity of soil limit the migration of hazardous wastes.
33. Explain what hydrologic features are important in siting a landfill.
34. List the minimum EPA requirements for a hazardous waste landfill and sketch a landfill that meets these.
35. Explain the difference between deep well injection and land treatment.
36. Define the following acronyms: PIC, POC, POHC, and DRE, as they apply to incineration.
37. List the most important factors for proper incinerator design and operation.
38. List the two types of incinerators most commonly used for destroying hazardous waste.
39. Explain the terms "designated POHC" and "surrogate" as they apply to a trial burn.
40. Outline the steps in EPA's remediation procedures.
41. Differentiate between "remediation" and "removal" as they pertain to a CERCLA/SARA cleanup.
42. Explain by pump-and-treat remediation systems may take a very long time to clean up groundwater.

With the aid of this text, you should be able to do the following:

1. Calculate lifetime risk using the one-hit or multistage model.
2. Calculate chronic daily intake or other variables given the media and values for remaining variables.
3. Perform a risk characterization calculation for carcinogenic and noncarcinogenic threats by multiple contaminants and multiple pathways.
4. Determine whether or not a waste is an EPA hazardous waste based on its composition, source, or characteristics.
5. Perform a mass balance to identify waste sources or waste-minimization opportunities.

6. Write the reactions for oxidation or reduction of chemical contaminants to mineralized form.
7. Perform solubility product calculations to estimate treatment doses for precipitation or the concentration of contaminants that remain in solution.
8. Determine the dimensions of an air stripping column, air or liquid flow rate given the values for remaining variables.
9. Determine the mass of resin and column dimensions for an ion exchange column given laboratory or pilot breakthrough data.
10. Evaluate a chemical feed to an incinerator to determine whether or not the chlorine content is acceptable and design a mix of waste feeds to achieve a desired chlorine feed rate.
11. Evaluate the operating variables for an incinerator to determine regulatory compliance for DRE, HCl emissions, and particulate emissions.
12. Estimate the hydraulic conductivity of a liner material based on laboratory measurements.
13. Estimate the quantity of leachate given the precipitation rate, area, hydraulic gradient, and hydraulic conductivity.
14. Estimate the seepage velocity and travel time of a contaminant through a soil given the hydraulic gradient, hydraulic conductivity, porosity, and length of the flow path.
15. Locate one or more extraction walls in a contaminant plume for a specified pumping rate, aquifer thickness, hydraulic conductivity, and hydraulic gradient.
16. Estimate the required pumping rate for a single extraction well for a specified aquifer thickness, hydraulic conductivity, hydraulic gradient and capture zone width.

10-11 PROBLEMS

10-1. The recommended time weighted average air concentration for occupational exposure to water-soluble hexavalent chromium (Cr VI) is 0.05 mg/m^3. This concentration is based on an assumption that the individual is generally healthy and is exposed for 8 hours per day over a working lifetime (that is from age 18 to 65 years). Assuming a body weight of 70 kg and an inhalation rate of 20 m^3/h over the working life of the individual, what is the lifetime (70 y) CDI?

Answer: 2.2×10^{-3} mg/kg · d

10-2. The National Ambient Air Quality Standard for sulfur dioxide is 80 $\mu g/m^3$. Assuming a lifetime exposure (24 h/d, 365 d/y) for an adult male of average body weight, what is the estimated CDI for this concentration?

10-3. Children are one of the major concerns of environmental exposure. Compare the CDIs for a 1-year-old child and an adult female drinking a water contaminated with 10 mg/L of nitrate (as N). Assume a 1-year averaging time.

10-4. Agricultural chemicals such as 2,4-D (2,4-dichlorophenoxyacetic acid) may be ingested by routes other than food. Compare the CDIs for ingestion of a soil contaminated with 10 mg/kg of 2,4-D by a 3-year-old child and an adult. Assume a 1-year averaging time.

10-5. Estimate the chronic daily intake of toluene from exposure to a city water supply that contains a toluene concentration equal to the drinking water standard of 1 mg/L. Assume the exposed individual is an adult female who consumes water at the adult rate for 70 years, that she abhors swimming, and that she takes a long (20-min) bath every day. Assume that the average air concentration of toluene during the bath is 1 $\mu g/m^3$. Assume the dermal uptake from water (PC) is 9.0×10^{-6} m/h and that direct dermal absorption during bathing is no more than 80 percent of the available toluene because she is not completely submerged.

Answer: 3.5×10^{-2} mg/kg · d

10-6. Estimate the chronic daily intake of 1,1,1-trichloroethane from exposure to a city water supply that contains a 1,1,1-trichloroethane concentration equal to the drinking water standard of 0.2 mg/L. Assume the exposed individual is a 1 to 5 year old child who consumes water at the child rate for 5 years, that she swims once a week for 30 min, and that she takes a short (10-min) bath every day. Assume her average age over the exposure period is 8. Assume that the average air concentration of 1,1,1-trichloroethane during the bath is 1 $\mu g/m^3$. Assume the dermal uptake from water (PC) is 0.0060 m/h and that direct dermal absorption during bathing is no more than 50 percent of the available 1,1,1-trichloroethane because she is not completely submerged.

10-7. Estimate the risk from occupational inhalation exposure to hexavalent chromium. (See Problem 10-1 for assumptions.)

Answer: Risk $= 8.83 \times 10^{-2}$ or 0.09

10-8. In its rule making for burning hazardous waste in boilers and industrial furnaces, EPA calculated doses of various contaminants that would result in a risk of 10^{-5} (56 FR 7233, 21 FEB 1991). Using the standard assumptions in Table 10-9 for an adult male, estimate the dose of hexavalent chromium that results in an inhalation risk of 10^{-5}.

10-9. Characterize the risk for a chronic daily exposure by the water pathway (oral) of 0.03 mg/kg · d of toluene, 0.06 mg/kg · d of barium, and 0.3 mg/kg · d of xylenes.

Answer: HI = 1.5

10-10. Characterize the risk for a chronic daily exposure by the water pathway (oral) of 1.34×10^{-4} mg/kg · d of tetrachloroethylene, 1.43×10^{-3} mg/kg · d of arsenic, and 2.34×10^{-4} mg/kg · d of dichloromethane (methylene chloride).

10-11. Determine whether the following is a RCRA hazardous waste: Municipal wastewater containing 2.0 mg/L of selenium.

10-12. Determine whether the following is a RCRA hazardous waste: An empty pesticide container that a homeowner wishes to discard.

10-13. The town of What Cheer has set up a recycling center to collect old fluorescent light bulbs. They anticipate collecting about 250 kg/mo of fluorescent bulbs. What is the maximum time the fluorescent bulbs can be stored before they must be disposed? (*Hint:* use the internet to access the appropriate CFR)

10-14. A vapor degreaser uses 590 kg/week of trichloroethylene (TCE). It is never dumped. The incoming parts have no TCE on them and the exiting parts drag out 3.8 L/h of TCE. The sludge removed from the bottom of the degreaser each week has 1.0% of the incoming TCE in it. The plant operates 8 h/d for 5 d/week. Draw the mass balance diagram for the degreaser and estimate the loss due to evaporation (in kg/week). The density of TCE is 1.460 kg/L.

Answer: $M_{evap} = 362.18$ or 360 kg/week

10-15. Using the following data and Figure P-10-15, use the mass balance technique to determine the mass flow rate (kg/d) of organic compounds to the condensate collection tank (sample location 4 in Figure P-10-15).

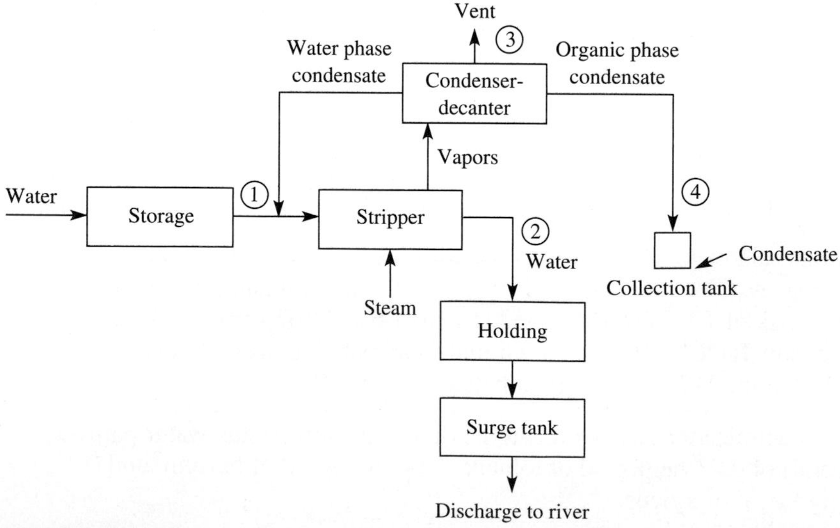

FIGURE P-10-15

Sample location	Flow rate, L/min	Total volatile organic	Temperature, °C
1	40.5	5,858. mg/L	25
2	44.8	0.037 mg/L	80
3	57.0 (vapor)	44.13%	20

Notes:

% is volume percent.

Vapor flow rate is corrected to 1 atm and 20°C.

Liquid organic density may be assumed to be 0.95 kg/L.

Assume the molecular weight of the organic vapor is equal to that of methylene chloride.

Steam mass flow rate is 252 kg/h at 106°C.

10-16. What is the efficiency of the condenser-decanter in Problem 10-15?

10-17. Given the waste constituent and concentration shown below, determine the quantity (in kg/d) of hydrated lime ($Ca(OH)_2$) required to neutralize the waste. Estimate the total dissolved solids (TDS) after neutralization. Report your answer in mg/L.

Constituent	Concentration, mg/L	Flow, L/min
HCl	100	5

Answers: Lime = 0.730 kg/d, TDS =152 mg/L

10-18. Given the constituents and concentrations shown below, determine the quantities (in kg/d) of sulfuric acid required to neutralize the waste. Estimate the total dissolved solids (TDS) after neutralization. Report your answer in mg/L.

Constituent	Concentration, mg/L	Flow, L/min
NaOH	15	200

10-19. It has been proposed to mix a 1500 L bath containing 5.00 percent by volume of H_2SO_4 with a 1,500 L bath containing 5.00 percent by weight of NaOH. The specific gravity of the acid added is 1.841 and its purity is 96 percent. The base added is 100 percent pure. Estimate the final pH (to two decimal places) and the final TDS (in mg/L) of the mixture of the two baths. (*Note:* the pH is very low.)

10-20. Write the reaction equation to oxidize sodium cyanide using sodium hypochlorite (NaOCl).

10-21. Write the reaction equation to oxidize sodium cyanide using ozone (O_3).

10-22. Write the reaction equation to reduce hexavalent chromium in chromic acid ($H_2Cr_2O_7$) to trivalent chromium using $NaHSO_3$.

10-23. A metal plating solution contains 50.00 mg/L of copper. Determine the concentration, in moles/L, to which the hydroxide concentration must be raised to precipitate all but 1.3 mg/L of the copper using lime. The K_{sp} of copper hydroxide is 2.00×10^{-19}. Estimate the final pH (report your answer to two decimal places).

10-24. A plating rinse water flowing at 100 L/min contains 50.0 mg/L of Zn. Calculate the theoretical pH required to achieve the EPA's pretreatment standard for existing dischargers of 2.6 mg/L and estimate the theoretical dose rate (g/min) of hydrated lime to remove only the required amount of Zn to achieve the standard (i.e., 50 mg/L minus the standard). Assume the lime is 100 percent pure.

10-25. A metal plating sludge as removed from a clarifier has a solids concentration of 4 percent. If the volume of sludge is 1.0 m^3/d, what volume will result if the sludge is processed in a filter press to a solids concentration of 30 percent? If the pressed sludge is dried to 80 percent solids, what volume will result?

Answer: $\maltese_1 = 0.133$ m^3/d,
$\maltese_2 = 0.05$ m^3/d

10-26. In Problem 10-25, ferrocyanide is found in the clarifier sludge at a concentration of 400 mg/kg (4 percent solids). Assuming that the ferrocyanide is part of the precipitate and that none escapes from the filter press, what concentration would be expected in filter cake? (*Hint: Set this up as a mass balance problem.*)

10-27. A drinking water supply at Oscoda, Michigan, has been contaminated by trichloroethylene. The average concentration in the water is estimated to be 6,000 μg/L. Using the following design parameters, design a packed-tower stripping column to reduce the water concentration to the state of Michigan discharge limit of 1.5 μg/L. Note that more than one column in series may be required for reasonable tower heights.

Henry's law constant $= 6.74 \times 10^{-3}$ m$^3 \cdot$ atm/mole

$K_La = 0.720$ min^{-1}

Air flow rate = 60 m^3/min

G/L = 18

Temperature = 25°C

Column diameter may not exceed 4.0 m

Column height may not exceed 6.0 m

Answer: With an assumed height of 6 m, the diameter is 3.15 m

10-28. Well 13 at Watapitae is contaminated with 340 μg/L of tetrachloroethylene (perchloroethylene). The water must be remediated to achieve a concentration of 0.2 μg/L (the detection limit). Using a spreadsheet program you have written, design a packed-tower stripping column to meet this requirement using the following design parameters. (*Note:* more than one column in series may be required for reasonable tower heights.)

Henry's law constant $= 100 \times 10^{-4}$ m$^3 \cdot$ atm/mole
$K_La = 14.5 \times 10^{-3}$ s^{-1}
Air flow rate = 15 m^3/s
Liquid flow rate = 0.22 m^3/s
Temperature = 20°C
Column diameter may not exceed 4.0 m
Column height may not exceed 6.0 m

10-29. An alternative form of Equation 10-26 uses the transfer unit concept discussed in Section 7-10. The relevant equations are (LaGrega, 2001):

1. Dimensionless Henry's law constant

$$H' = \frac{H_c}{RT_g}$$

2. Stripping factor

$$R_{\text{sf}} = \frac{(H')(G)}{L}$$

3. Height of transfer unit

$$\text{HTU} = \frac{L}{(A)(M_W)(K_La)}$$

4. Number of transfer units

$$\text{NTU} = \left(\frac{R_{\text{sf}}}{R_{\text{sf}} - 1}\right) \ln\left[\frac{(C_1/C_2)(R_{\text{sf}} - 1) + 1}{R_{\text{sf}}}\right]$$

5. Height of packing in column

$$Z = (\text{NTU})(\text{HTU})$$

where G = molar flow rate of air, moles/s
L = molar flow rate of water, moles/s
A = cross-sectional area of column, m^2
M_W = molar density of water
= 55,600 moles/m^3

Other terms are as defined for Equation 10-26.

Using the stripping factor equations, determine the height of packing for an air stripping column to reduce the concentration of ethylbenzene from 1.0 mg/L to 35 μg/L, using the following design parameters.

Henry's law constant = $6.44 \times 10^{-3}\ m^3 \cdot atm/mole$

$K_La = 1.6 \times 10^{-2}\ s^{-1}$

Liquid flow rate = 7.14 L/s

Temperature = 20°C

Column diameter may not exceed 4.0 m

Column height may not exceed 6.0 m

Because the air flow rate and diameter are not given, a trial and error solution is required. Use a spreadsheet program you have written to perform the trial-and-error solution. Use a 20 percent safety factor to estimate the final packing height.

10-30. An electroplating rinse water containing 55 mg/L of nickel is to be treated by an ion exchange column to meet an allowable effluent concentration of 2.6 mg/L. A laboratory-scale column has provided the breakthrough data shown in the table below. The laboratory column data are as follows:

Inside diameter = 1.0 cm

Length = 7.0 cm

Mass of resin (moist basis) = 5.2 g

Water content = 17%

Density of resin = 0.65 g/cm^3

Liquid flow rate = 7.68 L/d

Initial concentration = 55 mg/L

Breakthrough data

¥, L	C, mg/L	¥, L	C, mg/L
0.160	4.23	1.280	39.04
0.320	5.14	1.440	44.04
0.480	10.03	1.600	49.54
0.640	16.65	1.760	53.32
0.800	23.62	1.920	54.14
0.960	29.54	2.080	53.22
1.120	35.46		

The full scale design must meet the following requirements:

Flow rate = 36,000 L/d

Hours of operation = 8 h/d

Regeneration is to be once every 5 days

Use a spreadsheet to plot the breakthrough data and determine the mass of resin required. (*Hint:* Some initial and final data points may be hard to plot and may need to be ignored to achieve a straight line on the semi-log plot.)

Answer: Mass of resin = 8.38×10^5 g or 840 kg

10-31. An electroplating rinse water containing 10 mg/L of silver is to be treated by an ion exchange column to meet an allowable effluent concentration of 0.24 mg/L. A laboratory scale column has provided the breakthrough data shown in the table below. The laboratory column data are as follows:

Inside diameter = 1.0 cm

Length = 14.85 cm

Mass of resin (moist basis) = 7.58 g

Water content = 34%

Density of resin = 0.65 g/cm^3

Liquid flow rate = 4.523 L/d

Initial concentration = 10 mg/L

Breakthrough data

V, L	*C*, mg/L	*V*, L	*C*, mg/L
0.1	0.00	1.1	2.00
0.2	0.00	1.2	3.33
0.3	0.01	1.3	5.00
0.4	0.02	1.4	6.67
0.5	0.04	1.5	8.00
0.6	0.08	1.6	8.89
0.7	0.16	1.7	9.41
0.8	0.31	1.8	9.69
0.9	0.61	1.9	9.84
1.0	1.15	2.0	9.92

The full scale design must meet the following requirements:

Flow rate = 3,600 L/d

Hours of operation = 8 h/d

Regeneration is to be once every 5 days

Use a spreadsheet to plot the breakthrough data and determine the mass of resin required. (*Hint:* Some initial and final data points may be hard to plot and may need to be ignored to achieve a straight line on the semi-log plot.)

10-32. A very hard water is to be softened for use as an electroplating rinse water. The raw water analysis shows a calcium concentration of 107 mg/L and a magnesium concentration of 18 mg/L. The desired final hardness is 10 mg/L as $CaCO_3$. An ion exchange column has been selected to achieve this hardness. A pilot scale column has provided the breakthrough data shown in the table below. The laboratory column data are as follows (after Reynolds and Richards, 1996):

Inside diameter = 10.0 cm

Length = 91.5 cm

Mass of resin (moist basis) = 5.0 kg

Water content = 34%

Density of resin = 0.7 g/cm^3

Liquid flow rate = 2.25 L/h

Initial concentrations:

Ca = 107 mg/L as ion

Mg = 18 mg/L as ion

Breakthrough data

$\mathcal{V}$, m^3	C, meq/L
2.35	0.21
2.9	0.48
3.1	1.10
3.26	1.64
3.39	2.47
3.49	3.22
3.56	3.56
3.71	4.52
3.81	5.07
4.03	5.96
4.62	6.78

The full scale design must meet the following requirements:

Flow rate = 570 m^3/d

Regeneration is to be once every 60 days

Determine the mass of resin required. (*Hint:* see Sections 4-1 and 4-3 for hardness equivalent weight and mg/L as $CaCO_3$ calculations and conversions.) Use a spreadsheet to plot the breakthrough data and determine the mass of resin required. (*Hint:* Some initial and final data points may be hard to plot and may need to be ignored to achieve a straight line on the semi-log plot.)

10-33. An incinerator operator receives the following shipments of waste for incineration. Can the operator mix these wastes to achieve 30 percent by mass of chlorine in the feed?

Trichloroethylene = 18.9 m^3

1,1,1 Trichloroethane = 5.3 m^3

Toluene = 213 m^3

o-Xylene = 4.8 m^3

10-34. An incinerator operator receives the following shipments of waste for incineration. What volume of methanol (CH_3OH) must the operator mix to achieve 30 percent by mass of chlorine in the feed? Assume the density of methanol is 0.7913 g/mL.

Carbon tetrachloride = 12.2 m^3

Hexachlorobenzene = 153 m^3

Pentachlorophenol = 2.5 m^3

10-35. A hazardous waste incinerator is being fed methylene chloride at a concentration of 5,858 mg/L in an aqueous stream at a rate of 40.5 L/min. Calculate the mass flow rate of the feed in units of g/min.

10-36. Methylene chloride was measured in the flue gas of a hazardous waste incinerator at a concentration of 211.86 $\mu g/m^3$. If the flow rate of gas from the incinerator was 597.55 m^3/min, what was the mass flow rate of methylene chloride in g/min?

10-37. Assuming that the same incinerator is being evaluated in Problems 10-35 and 10-36, what is the DRE for the incinerator?

10-38. Xylene is fed into an incinerator at a rate of 481 kg/h. If the mass flow rate at the stack is 72.2 g/h, is the unit in compliance with the EPA rules?

10-39. 1,2-Dichlorobenzene is being burned in an incinerator under the following conditions:

Operating temperature = 1,150°C

Feed flow rate = 173.0 L/min

Feed concentration = 13.0 g/L

Residence time = 2.4 s

Oxygen in stack gas = 7.0%

Stack gas flow rate = 6.70 m^3/s at standard conditions

Stack gas concentrations after APC equipment

Dichlorobenzene = 338.8 μg/dscm

HCl = 77.2 mg/dscm

Particulates = 181.6 mg/dscm

Assume all of the chlorine in the feed is converted to HCl. Does the incinerator comply with the EPA rules?

10-40. The POHCs from a trial burn are shown in the table below. The incinerator was operated at a temperature of 1,100°C. The stack gas flow rate was 5.90 dscm/s with 10.0 percent oxygen. Assuming that all the chlorine in the feed is converted to HCl, is the unit in compliance if the emissions are measured downstream of the APC equipment?

Compound	Inlet kg/h	Outlet kg/h
Benzene	913.98	0.2436
Chlorobenzene	521.63	0.0494
Xylenes	1,378.91	0.5670
HCl	n/a	4.85
Particulates	n/a	10.61

n/a = not applicable.

10-41. During a trial burn, an incinerator was fed a mixed feed containing trichloroethylene, 1,1,1-trichloroethane, and toluene in a aqueous solution. Each component accounted for 5.0 percent of the feed solution on a volume basis. The feed rate was 40 L/min. The incinerator was operated at a temperature of 1,200°C. The stack gas flow rate was 9.0 dscm/s with 7 percent oxygen. Assuming that all the chlorine in the feed is converted to HCl, is the unit in compliance with the following emissions measured after the APC equipment?

Trichloroethylene = 170 μg/dscm

1,1,1-Trichloroethane = 353 μg/dscm

Toluene = 28 μg/dscm

HCl = 83.2 mg/dscm

Particulates = 123.4 mg/dscm

10-42. During a trial burn, an incinerator was fed a mixed feed containing hexachlorobenzene (HCB), pentachlorophenol (PCP), and acetone (ACET) in a aqueous solution. Each component accounted for 9.3 percent of the feed solution on a volume basis, that is, HCB = 9.3 percent, PCP = 9.3 percent, and ACET = 9.3 percent. The feed rate was 140 L/min. The incinerator was operated at a temperature of 1,200°C. The stack gas flow rate was 28.32 dscm/s with 14 percent oxygen. Assuming that all the chlorine in the feed is converted to HCl, is the unit in compliance if the following emissions are measured downstream of the APC equipment?

Hexachlorobenzene = 170 μg/dscm

Pentachlorophenol = 353 μg/dscm

Acetone = 28 μg/dscm

HCl = 83.2 μg/dscm

Particulates = 123.4 mg/dscm

10-43. The permit for a rotary kiln hazardous waste incinerator specifies that the retention time for solids is 1 hour. The proposed dimensions and operating condition for the incinerator are:

Diameter = 3.00 m

Length = 6.00 m

Slope = 2.00%

Peripheral speed = 1.5 m/min

Determine if the permit requirement will be met.

10-44. A standard permeameter is being considered for testing a clay for a hazardous waste landfill base. If the clay must have a hydraulic conductivity of 10^{-7} cm/s and the dimensions of the permeameter are as shown below, how long will the test take if a minimum of 100.0 milliliter of liquid must be collected for an accurate measurement? See Figure P-10-44 for notation and permeameter equation. Dimensions are given on page 918.

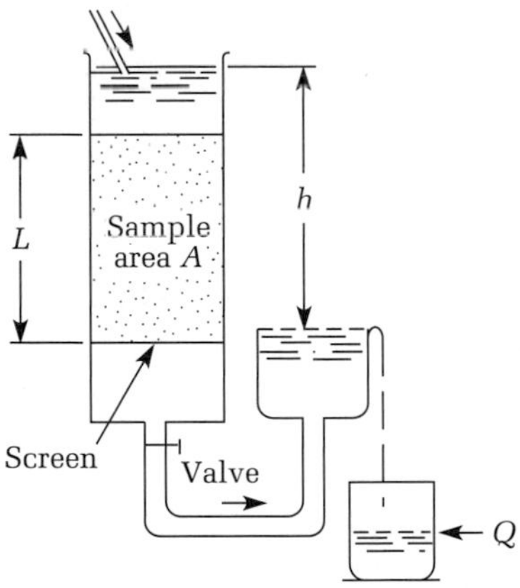

Standard constant head permeameter equation:

$$K = \frac{QL}{htA}$$

where K = hydraulic conductivity
Q = quantity of discharge
L = length of sample
h = hydraulic head
A = cross-sectioned area of sample
t = time

FIGURE P-10-44

$L = 10$ cm

$h = 1$ m

Diameter of sample = 5.0 cm

Answer: $t = 58.95$ or 60 d

10-45. A standard permeameter is being considered for testing a clay for a hazardous waste landfill base. If the clay must have a hydraulic conductivity of 10^{-7} cm/s and the dimensions of the permeameter are as shown in Figure P-10-44, the test will take 60 days if a minimum of 100.0 mL of liquid must be collected for an accurate measurement. You need the results in 30 days. What change in the design of the permeameter would you make to obtain results in 30 days? Show by calculation that your redesign would work.

10-46. A soil sample has been tested to determine permeability using a falling-head permeameter. (See Figure P-10-46.) The data on page 919 were recorded:

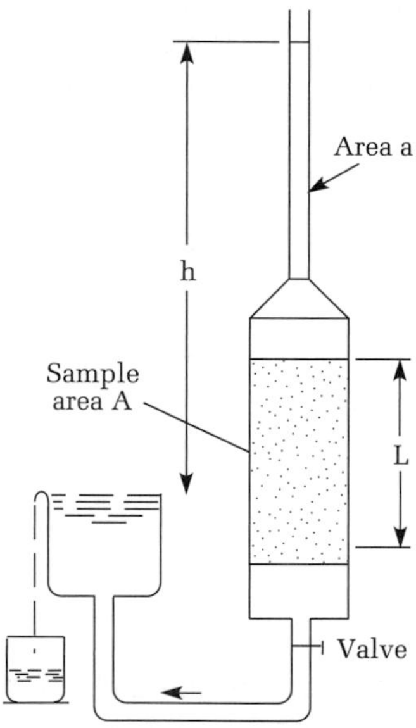

Falling head permeameter equation:

$$K = 2.3 \frac{a\,L}{A\,t} \log\left(\frac{h_0}{h_1}\right)$$

where K = hydraulic conductivity
a = cross-sectional area of stand pipe
A = cross-sectional area of sample
L = length of sample
t = time
h_0, h_1 = head at beginning of test and at time t, respectively

FIGURE P-10-46

Diameter of a = 1 mm

Diameter of A = 10 cm

Length, L = 25 cm

Initial head = 1.0 m

Final head = 25 cm

Duration of test = 14 days

From these data, calculate the hydraulic conductivity of the sample. Assuming the sample is representative of the landfill site, is this a good soil for a hazardous waste landfill base?

10-47. An old hazardous waste landfill was built on a 10-m-deep clay liner. An aquifer lies immediately below the clay layer. The clay layer through which the leachate must pass has a hydraulic conductivity of 1×10^{-7} cm/s. If the liquid level (leachate) is 1.0 m deep above the clay layer, how much leachate (in m^3/d) will reach the aquifer when the clay layer becomes saturated? Assume Darcy's law applies.

10-48. The three soil layers described below lie between the bottom of a hazardous waste landfill and the underlying aquifer. The depth of leachate above the top soil layer is 0.3 m. How long will it take (in years) for the leachate to migrate to an aquifer located at the bottom of soil C?

Soil A
Depth = 3.0 m
Hydraulic conductivity = 1.8×10^{-7} cm/s
Porosity = 55%

Soil B
Depth = 10 m
Hydraulic conductivity = 2.2×10^{-5} m/s
Porosity = 25%

Soil C
Depth = 12.0 m
Hydraulic conductivity = 5.3×10^{-5} mm/s
Porosity = 35%

10-49. The practical quantitation limit (PQL) for the solvent trichloroethylene is 5 μg/L. If a barrel (approximately 0.12 m^3) of spent solvent leaked into an aquifer, approximately how many cubic meters of water would be contaminated at the PQL?

10-50. An aquifer has a hydraulic gradient of 8.6×10^{-4}, a hydraulic conductivity of 200 m/d, and a porosity of 0.23. A chemical with a retardation factor of 2.3 contaminates the aquifer. What is the linear velocity of the contaminant? How long will it take to travel 100 m in the aquifer?

10-51. An extraction well must be installed at the site of a leaking gasoline storage tank. The depth of the unconfined aquifer is 60.00 m and the hydraulic conductivity is 6.4×10^{-3} m/s. Measurements show that the plume does not extend more than 150 m from the center of the leak. At 130 m from the center of the leak, the plume is 0.1 m in depth. If the extraction well is 28 cm in diameter, what size pump (in m^3/s) is required so that the plume does not migrate any farther? (*Note:* this is an application of the well equations in Chapter 3.)

10-52. A drum (0.12 m^3) of carbon tetrachloride has leaked into a sandy soil. The soil has a hydraulic conductivity of 7×10^{-4} m/s and a porosity of 0.38. The groundwater table is 3 m below grade and has a hydraulic gradient of 0.002. The aquifer is 28 m thick. A single well intercept system is proposed, using a well pumping at 0.014 m^3/s. Estimate the width of the capture zone at the well.

10-53. If the leading edge of the plume in Problem 10-52 has spread to a width of 200 m, how far ahead of the plume must the well be located to intercept it?

10-12 DISCUSSION QUESTIONS

10-1. What was the outcome of the hazardous waste episode at Times Beach? (*Hint:* You will need to do an internet search.)

10-2. It has been stated that, on the basis of LD_{50}, 2,3,7,8-TCDD is the most toxic chemical known. Why might this statement be misleading? How would you rephrase the statement to make it more scientifically correct?

10-3. Which of the following individuals is at greater risk from inhalation of an airborne contaminant: a 1-year-old child; an adult female; an adult male? Explain your reasoning.

10-4. Which of the following individuals is at greater risk from ingestion of a soil contaminant: a 1-year-old child; an adult female; an adult male? Explain your reasoning.

10-5. A hazard index of 0.001 implies:

a. Risk = 10^{-3}

b. The probability of hazard is 0.001

c. The RfD is small compared to the CDI

d. There is little concern for potential health effects

10-6. A dry cleaner accumulates 10 kg per month of perchloroethylene (a hazardous waste solvent). To save shipping cost he would like to accumulate 6 months' worth before he ships it to a TSD facility. Can he do this? Explain. (*Hint*: search the applicable regulations in the CFR.)

10-7. Does the "land ban" actually ban the disposal of hazardous waste on the land? Explain.

10-8. A multimillion dollar company has just learned that one drum out of several hundred found at an abandoned waste disposal site has been identified as its property. Their attorney explains that the company may potentially be responsible for cleanup of all the drums at the site if no other former owners of the drums can be identified. Is this correct? Why or why not?

10-9. Your boss has proposed that your company institute a recycling program to minimize the generation of waste. Is recycling the best first step to investigate in a waste minimization program? If not, what others would you suggest and in what order?

10-10. A metal plater is proposing to treat waste sludge to recover the nickel from it. Would this be

a. recycling?

b. reusing,

c. reclaiming.

State the correct answer(s) and explain why you made your choice(s).

10-11. It is not necessary to measure every POHC in an incinerator trial burn. True or false? Explain your answer.

10-13 REFERENCES

Butler, J. P., A. Greenberg, P. J. Lioy, G. B. Post, and J. M. Waldman (1993) "Assessment of Carcinogenic Risk from Personal Exposure to Benzo(a)pyrene in the Total Human Environmental Exposure Study (THEES)," *Journal of the Air & Waste Management Association,* vol. 43, pp. 970–977.

Byard, J. L. (1989) "Hazard Assessment of 1,1,1-Trichloroethane in Groundwater," in D. J. Paustenbach (ed.), *The Risk Assessment of Environmental Hazards,* John Wiley & Sons, New York, pp. 331–344.

CDC (2004) "Deaths: Final Data for 2002," *National Vital Statistics Reports,* vol. 53, no. 5, pp. 5–10, 77.

Copeland, R., et al. (1994) "Use of Probabilistic Methods to Understand the Conservativism in California's Approach to Assessing Health Risks Posed by Air Contaminants," *Journal of Air & Waste Management Association,* vol. 44, pp. 1399–1413.

Davis, M. L., C. R. Dempsey and E. T. Oppelt (2000) "Waste Incineration Sources: Hazardous Waste," in W. T. Davis, *Air Pollution Engineering Manual*, Air Pollution Control Association, Pittsburgh, and John Wiley & Sons, New York, pp. 268–274.

Davis, M. L. and S. J. Masten (2004) *Principles of Environmental Engineering and Science,* McGraw-Hill, Boston, MA, p. 175.

Finley, B., and D. Paustenbach (1994) "The Benefits of Probabilistic Exposure Assessment; Three Case Studies Involving Contaminated Air, Water, and Soil." *Risk Analysis,* vol. 14, no. 1, pp. 53–73.

Fromm, C. H., A. Bachrach, and M. S. Callahan (1986) "Overview of Waste Minimization Issues, Approaches and Techniques," in E. T. Oppelt, B. L. Blaney, and W. F. Kemner, (eds.), *Transactions of an APCA International Specialty Conference on Peformance and Costs of Alternatives to Land Disposal of Hazardous Waste*, Air Pollution Control Association, Pittsburgh, pp. 6–20.

Griffin R. D. (1988) Principles of Hazardous Materials Management, Lewis Publishers; Ann Arbor, MI.

Gross, R. L., and S. G. TerMaath, (1985) "Packed Tower Aeration Strips Trichloroethylene from Groundwater," *Environmental Progress,* vol. 4, pp. 119–124.

Herzbron, P. A., R. L. Irvine, and K. C. Malinowski, (1985) "Biological Treatment of Hazardous Waste in Sequencing Batch Reactors," *Journal of the Water Pollution Control Federation*, vol. 57, pp. 1163–1167.

Hileman, B. (1994) "EPA Reassesses Dioxins," *C&E News*, September 19, p. 6.

Hutt, P. B. (1978) "Legal Considerrtions in Risk Assessment," *Food, Drugs, Cosmetic Law Journal,* vol. 33, pp. 558–559.

Javandel, I., and C. Tsang (1986) "Capture Zone Type Curves: A Tool for Cleanup," *Ground Water,* vol. 24, no. 5, pp. 616–625.

Kobayashi, H. and B. P. Rittman (1982) "Microbial Removal of Hazardous Organic Compounds," *Environmental Science and Technology,* vol. 16. pp. 170A–172A.

LaGrega, M. D., P. L. Buckingham, and J. C. Evans (2001) *Hazardous Waste Management,* McGraw–Hill, Boston, pp. 471–473, 899–903, 1014–1016.

Loomis, T. A. (1978) *Essentials of Toxicology,* Lea & Febiger, Philadelphia, p. 2.

Masters, G. M. (1998) *Introduction to Environmental Engineering and Science,* Prentice Hall, Upper Saddle River, NJ, p. 240.

McKone, T. E. (1987) "Human Exposure to Volatile Organic Compounds in Household Tap Water: The Indoor Inhalation Pathway," *Environmental Science & Technology,* vol. 21, no. 12, pp. 1194–1201.

NHTSA (2005) "2004 Annual Assessment," National Highway Traffic Safety Administration, Washington, DC.

NRC (1983) *Risk Assessment in the Federal Government: Managing the Process,* National Research Council, National Academy Press, Washington, DC, pp. 18–19.

O'Brien & Gere Engineers Inc. (1988) *Hazardous Waste Site Remediation*, Van Nostrand Reinhold, New York, pp. 11–13.

Oppelt, E. T. (1981) "Thermal Destruction Options for Controlling Hazardous Wastes," *Civil Engineering ASCE,* pp. 72–75, September.

Penning, C. H. (1930) "Physical Characteristics and Commercial Possibility of Chlorinated Diphenyl," *Industrial & Engineering Chemistry,* vol. 22, pp. 1180–1183.

Reynolds, T. D. and P. A. Richards (1996) *Unit Operations and Processes in Environmental Engineering,* PWS Publishing, Boston, pp. 392–393.

Rodricks, I., and M. R. Taylor (1983) "Application of Risk Assessment to Good Safety Decision Making," *Regulatory Toxicology and Pharmacology*, vol. 3, pp. 275–284.

Schwarzebach, R. P., P. M. Gschwend, and D. M. Imboden (1993) *Environmental Organic Chemistry,* John wiley and sons, New York, p. 274.

Slovic, P., B. Fischoff, and S. Lichenstein (1979) "Rating the Risks," *Environment*, vol. 21, no. 3, pp. 14–20.

Starr, C. (1969) "Social Benefit Versus Technological Risk," *Science*, vol. 165, pp. 1232–1238.

Theodore, L., and J. Reynolds (1987) *Introduction to Hazardous Waste Incineration*, John Wiley & Sons, New York, pp. 76–85.

Thomas, H. C. (1948) "Chromatography: A Problem of Kinetics," *Annals of the New York Academy of Science*, vol. 49, p. 161.

U.S. EPA (1981a) *Summary Report: Control and Treatment Technology for the Metal Finishing Industry—Ion Exchange*, U.S. Environmental Protection Agency Publication No. EPA 625/8-81-007.

U.S. EPA (1981b) *Development Document for Effluent Limitations: Guideline and Standards for the Metal Finishing Point Source Category*, U.S. Environmental Protection Agency Publication No. EPA/440/1-83-091, Washington, DC.

U.S. EPA (1986) *RCRA Orientation Manual*, U.S. Environmental Protection Agency, Publication No. EPA/530-SW-86-001, Washington, DC.

U.S. EPA (1987) *A Compendium of Technologies Used in the Treatment of Hazardous Waste*, U.S. Environmental Protection Agency Publication No. EPA/625/8-87/014), Cincinnati.

U.S. EPA (1988a) *Waste Minimization Opportunity Assessment Manual*, U.S. Environmental Protection Agency Publication No. EPA/625/7-88/003), Cincinnati, p. 2.

U.S. EPA (1988b) *Best Demonstrated Available Technology (BDAT) Background Document for FOO6*, U.S. Environmental Protection Agency Publication No. EPA/530-SW-88-009-I, Washington, DC.

U.S. EPA (1989a) *Risk Assessment Guidance for Superfund, Volume 1, Human Health Evaluation Manual (Part A)*, U.S. Environmental Protection Agency, EPA/540/1-89/002.

U.S. EPA (1989b) *U.S. EPA Seminar Publication: Requirements for Hazardous Waste Landfill Design, Construction and Closure*, U.S. Environmental Protection Agency Publication No. EPA 625/4-89/022.

U.S. EPA (1991) *Design and Construction of RCRA/CERCLA Final Covers*, U.S. Environmental Protection Agency Publication No. EPA 625/4-89/022, Washington, DC.

U.S. EPA (1997) *Exposure Factor Handbook*, U.S. Environmental Protection Agency National Center for Environmental Assessment, Washington, DC.

U.S. EPA (2001) *Ninth Report on Carcinogens*, U.S. Environmental Protection Agency Report to Congress, Washington, DC.

U.S. EPA (2004a) *Underground Storage Tanks: Building on the Past to Protect the Future*, U.S. Environmental Protection Agency Publication No. EPA 510-R-04-001, Washington, DC.

U.S. EPA (2004b) *Risk Assessment Guidance Manual for Superfund, Volume I: Human Health Evaluation Manual*, U.S. Environmental Protection Agency Publication No. EPA/540/R/99/005, Washington, DC.

U.S. EPA (2005a) *Eleventh Report on Carcinogens*, U.S. Environmental Protection Agency Report to Congress, Washington, DC.

U.S EPA (2005b) CERCLIS Database, at http://cfpub.epa.gov/superrcpad/cursites.

Wentz, C. A. (1989) *Hazardous Waste Management,* McGraw-Hill, New York, pp. 206–207.

Wilson, R. (1979) "Annalyzing the Daily Risks of Life," *Technology Review,* vol. 81, pp. 41–46.

Wood, E. F., R. A. Ferrara, W. G. Gray, and G. F. Pinder (1984) *Groundwater Contamination from Hazardous Waste,* Prentice Hall, Englewood Cliffs, NJ, pp. 2–4, 145–158.

CHAPTER 11

IONIZING RADIATION*

*Ms. Kristin Erickson, Radiation Safety Officer, Office of Radiation. Chemical, and Biological Safety, Michigan State University, contributed to this chapter.

11-1 FUNDAMENTALS*

Atomic Structure

We assume that you are familiar with the Bohr model of atomic structure. In this model the atom is described as consisting of a central nucleus surrounded by a number of electrons in closed orbits about the nucleus. The orbital electrons are grouped in shells.

The nucleus itself can be considered as composed of two distinct kinds of particles: protons, which carry a positive unit charge, e^+, and neutrons, which are uncharged. In a particular atom there are Z electrons, each carrying a charge e^-, orbiting around the nucleus, and a nucleus composed of N neutrons and P protons. The condition of electrical neutrality for the atom as a whole yields $Pe - Ze = 0$, that is, the number of protons in the nucleus is equal to the number of orbital electrons.

The number Z is the atomic charge or atomic number of the atom, and $Z + N$ is the atomic mass number, usually denoted by A. The parameters A and Z completely define a particular atomic species, this being known as a *nuclide*.

The masses of nuclides are measured in terms of the *unified atomic mass unit,* with the symbol u. This is defined as the unit of mass equal to one-twelfth the mass of an atom of carbon of atomic mass number 12. This gives 1 u as 1.6606×10^{-27} kg. On this scale, the mass of the neutron is 1.0088665 u, the mass of the proton 1.0088925 u, and the mass of the electron 0.0005486 u.

From the definition of the mass scale, giving proton and neutron masses of the order unity, it is clear that the *atomic mass number* will be a whole number approximation to the nuclidic mass in u. For example, a nuclide of magnesium which contains 12 protons and 12 neutrons has $A = 24$ and a nuclidic mass of 23.985045 u. The difference between the nuclidic mass and the atomic mass number is called the *mass excess*.

The chemical properties of the atom, and hence its designation as a particular element, depend on the number of orbital electrons, that is, on the atomic number Z. Given Z, the element is uniquely defined. As an example, if a given atom has two orbital electrons, it must be helium (assuming that the atom is not ionized or in some similar nonequilibrium state). Similarly an atom with eight electrons must be oxygen.

A particular nuclide is denoted by ${}^{A}_{Z}\mathrm{X}$, where X takes the place of the element symbol. But as Z determines the element, Z and X denote the same thing. Thus, the shorthand can be amended to ${}^{A}\mathrm{X}$. For example, carbon has six neutrons and six protons. Therefore, this nuclide can be written ${}^{12}\mathrm{C}$, or carbon-12.

For each element (determined only by Z) several nuclides (determined by Z and A) have the same Z value but different values of A. These different nuclides of the same element are called *isotopes*. Hydrogen with $Z = 1$ has three isotopes with atomic mass numbers of 1, 2, and 3. As Z must remain constant at 1, this means that they have zero, one, and two neutrons, respectively. This is illustrated in Figure 11-1. These isotopes all act chemically as hydrogen, but their nuclidic masses are different. The nuclidic mass of ${}^{1}\mathrm{H}$ is 1.007825 u, that of ${}^{2}\mathrm{H}$ (known as deuterium) is 2.014102 u, and that of ${}^{3}\mathrm{H}$ (known as tritium) is 3.016049 u.

*This discussion follows R. A. Coombe, 1968.

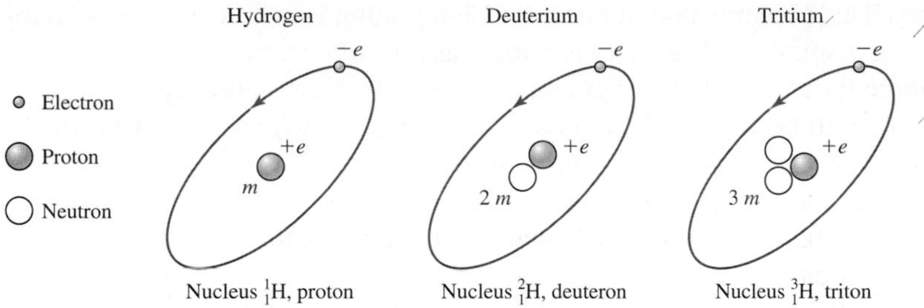

FIGURE 11-1
Three isotopes of hydrogen. (*Source:* R. A. Coombe, 1968.)

The atomic weight of an element is defined as the combined nuclidic masses of all the isotopes, weighted according to their natural relative abundances. It is denoted by A. In the case of hydrogen it follows that the atomic weight is

$$1.007825(0.9844) + 2.014102(0.0156) + 3.016049(0) = 1.00797$$

The masses of the hydrogen isotopes are not obtained by simple addition of neutron masses. For example, the nuclidic mass of ^{1}H plus a neutron is 2.016490 u, whereas the mass of deuterium is 2.014102 u. This difference of 0.002388 u is called the *mass defect*. This is because when a proton and a neutron are brought together to form a deuteron (the nucleus of deuterium), energy is released to bind them together. Conversely, energy must be supplied to split them apart. This required energy, the *binding energy,* is obtained from Einstein's equation for the conversion of mass into energy,

$$E = \Delta mc^2 \tag{11-1}$$

where Δm is the mass defect and c is the speed of light.

All energies of emitted radiation and particles, as well as the various atomic and nuclear energy levels, are quoted in terms of the *electron volt,* eV. This is the energy that would be acquired by an electron in falling through a potential difference of one volt. From this definition the following equivalent units of energy can be established:

$$1 \text{ eV} = 1.602 \times 10^{-12} \text{ erg} = 1.602 \times 10^{-19} \text{ J}$$

For nuclear energy levels and radiation energies, the electron volt is usually an inconveniently small unit. The units MeV and KeV are then used for 10^6 eV and 10^3 eV, respectively. Using Equation 11-1, with the information that $c = 2.99793 \times 10^8$ m/s, and 1 u $= 1.6606 \times 10^{-27}$ kg, then the energy equivalent of 1 u is 931.634 MeV. In other words, this means that if an electron of mass 0.0005486 u were completely annihilated, the energy released would be approximately 0.511 MeV.

Radioactivity and Radiation

By definition, isotopes have different ratios of neutrons to protons in the nucleus. Some ratios give rise to unstable conditions. This is usually because the neutron-to-proton ratio is too large. Because of this instability, the nucleus changes its state to attain equilibrium, and in so doing emits either a particle or electromagnetic radiation to carry off the excess energy. This phenomenon of nuclear disintegration is known as *radioactivity,* and an isotope that displays such activity is known as a *radioisotope.*

There are three types of isotopes. Some are stable, others are naturally radioactive, and the third group can be artificially produced and are also radioactive. These artificially produced radioisotopes are the isotopes most used in industrial application.

Three major types of decay product carry off the surplus energy when a radioisotope decays: alpha particles, beta particles, and gamma radiation.

Alpha-Particle Emission. Conceptually, the source of the instability of the heavy elements is their size; their nuclei are too large. How can they become smaller? One method would be to eject protons or neutrons. Rather than doing this singly, the heavy-element atoms expel them in "packages" containing two protons and two neutrons. This "package" is called an *alpha particle* (α). An alpha particle is equivalent to the nucleus of the helium-4 atom, that is, it is simply a body consisting of two protons and two neutrons bound together. Consequently, if an alpha particle is emitted, the nucleus must change to one that has a charge $2e$ less and a mass approximately 4 u less. The general expression is

$$^{A}_{Z}X \rightleftharpoons {}^{A-4}_{Z-2}X + {}^{4}_{2}He \qquad (11\text{-}2)$$

Atoms that eject the helium "package" are said to decay through emission of alpha radiation. Alpha-particle emission occurs mainly in radioisotopes whose atomic number is greater than 82. With increasing atomic number, the occurrence of alpha-particle decay increases rapidly, and it is a characteristic of the very heavy elements. It is particularly in evidence in the main decay chains of the natural radioactive isotopes.

You should note that an atom undergoing alpha-particle decay changes into a new element. It is a new element because the product nucleus (often called a *daughter*) contains two fewer protons than the parent atom. Through emission of an alpha particle, uranium becomes thorium. Similarly, radium becomes radon.

Beta-Particle Emission. The instability that is the cause of beta-particle emission arises from the fact that the neutron-to-proton ratio in the nucleus is too high (there are too many neutrons in the nucleus). To achieve stability, a neutron in the nucleus can decay into a proton and an electron. The proton remains in the nucleus so that the neutron-to-proton ratio is decreased, and the electron is ejected. This ejected electron is known as a *beta particle* (β). The general expression for the decay is

$$^{A}_{Z}X \rightleftharpoons {}_{Z+1}^{A}X + \beta^{-} \qquad (11\text{-}3)$$

Note that we use the β^{-} to represent an electron of nuclear origin to differentiate it from electrons from other sources. The negative sign is used with the β if there is any chance of ambiguity, because a similar particle, called a positron, also exists that carries a positive charge.

Again, as in alpha radiation, emission of a beta particle changes the parent atom into a new element because the number of protons in the nucleus increases by one. If the daughter product also is radioactive, it will, in turn, emit a beta particle, becoming another new element, and so on, until finally a stable neutron-to-proton ratio is reached. Through such a series of changes, for example, the fission-product element krypton becomes rubidium which, in turn, becomes strontium, which finally converts to stable yttrium.

Example 11-1. Identify the particles that are emitted in each step of the decay chain represented by

$$^{86}_{36}\mathrm{Kr} \rightarrow {}^{89}_{37}\mathrm{Rb} \rightarrow {}^{89}_{38}\mathrm{Sr} \rightarrow {}^{89}_{39}\mathrm{Y}$$

Solution. Because z increases by 1 in each case, the particle emitted in each step is a beta particle.

Gamma-Ray Emission. Either alpha or beta particles may be accompanied by gamma radiation. Whereas alpha or beta radiation brings about a change in the size of the nucleus or the number of a particular type of particle therein, the emission of gamma radiation represents only a release of energy. This is the energy that remains in the newly formed nucleus after emission of the alpha or beta particle. Electromagnetic radiation in the form of gamma rays is emitted when a nucleus in an excited state transfers to a more stable state. The nucleus thus retains its original composition, the excess energy being radiated away. If the frequency of the radiation is v, and the nucleus changes from a state of energy E_1 to a state of energy E_2, then the two energies are related by the equation

$$E_1 - E_2 = hv \qquad (11\text{-}4)$$

where h is Planck's constant, having a value of 6.624×10^{-27} ergs. The energy of the emitted gamma ray is thus hv. In equations, the gamma ray is represented by the Greek letter gamma (γ).

X-Rays. Gamma rays are similar to x-rays. Their difference lies only in their source. Gamma rays originate from a nucleus transferring from one nuclear excited state to another, whereas x-rays originate from electrons transferring from a higher to a lower atomic energy state. As atomic energy levels are in general spaced much closer in terms of energy than nuclear levels, it follows from Equation 11-4 that the frequencies of x-rays are much less than those of gamma rays. As far as industrial applications are concerned, the only difference between them is the penetrating power. Because penetrating power increases with frequency, gamma rays have more penetrating power than x-rays.

Multiple Emissions. In the preceding discussions, only single emission has been considered. In practice, two or more different types of emission are possible, and in a great many cases several particles of the same type but of different energies are emitted. This latter effect is due to the multiplicity of nuclear energy levels both in the original isotope nucleus and in the nucleus formed by particle emission.

Radioactive Decay

Each unstable (radioactive) atom will eventually achieve stability by ejecting an alpha or beta particle. This shift to a more stable form is called *decay*. Each radioactive decay process is characterized by the fact that only a fraction of the unstable nuclei in a given sample will decay in a given time. The probability that a particular nucleus in the sample will decay during a time interval dt is $\lambda\, dt$, where λ is the radioactive decay constant. It is defined as the probability that any particular nucleus will disintegrate in unit time.

For a large number of like nuclei together, we make the assumption that λ is independent of the age of a nucleus and is the same for all nuclei. This means that λ is a constant. If N is the number of nuclei present at a time t, then the number of decays occurring in a time dt can be written $\lambda N dt$. As the number of nuclei decreases by dN in this time, we can write

$$dN = -\lambda N dt \tag{11-5}$$

The negative sign denotes that N is decreasing with time. Equation 11-5 shows that the rate of decay is proportional to the number of nuclei present, that is, it is a first-order reaction.

Equation 11-5 can be rearranged and integrated.

$$\int_{N_0}^{N} \frac{dN}{N} = \int_0^1 \lambda\, dt$$

$$\ln \frac{N}{N_0} = -\lambda t$$

or

$$N = N_0 \exp(-\lambda t) \tag{11-6}$$

where N_0 is the number of radioactive nuclei present at time $t = 0$. Equation 11-6 shows that radioactive decay follows an exponential form. In particular, the time taken for a given number of nuclei to decay to half that number, $T_{1/2}$, is obtained from Equation 11-6 as

$$\ln \frac{N/2}{N_0} = -\lambda T_{1/2}$$

Solving for $T_{1/2}$ yields

$$T_{1/2} = \frac{\ln 2}{\lambda} = \frac{0.693}{\lambda} \tag{11-7}$$

This equation relates two important parameters of a radioactive species: λ and the half-life, $T_{1/2}$. These quantities are characteristic properties of a particular species. Half-lives of radioisotopes cover an enormous range of values, from microseconds to millions of years. To illustrate this, some values are given in Table 11-1.

Example 11-2. Kal Karbonate has a vial containing 2.0 μCi/L of ^{45}Ca that must be disposed of. How long must the radioisotope be held to meet an allowable sewer discharge standard of 2.0×10^{-4} μCi/mL?

Solution. From Table 11-1 find the half-life of ^{45}Ca is 165 d. Calculate the value of λ, using Equation 11-7:

$$\lambda = \frac{0.693}{165 \text{ d}} = 4.20 \times 10^{-3} \text{ d}^{-1}$$

TABLE 11-1
Some radioisotope half-lives

Radioisotope	Half-Life	Radioisotope	Half-Life
Polonium-212	3.04×10^{-7} s	Calcium-45	165 days
Carbon-10	19.3 s	Cobalt-60	5.27 years
Oxygen-15	2.05 min	Tritium	12.5 years
Carbon-11	20.4 min	Strontium-90	28 years
Radon-222	3.825 days	Cesium-137	30 years
Iodine-131	8.06 days	Radium-226	1622 years
Phosphorus-32	14.3 days	Carbon-14	5570 years
Polonium-210	138.4 days	Potassium-40	1.4×10^9 years

Calculate the holding time, using Equation 11-6:

$$2.0 \times 10^{-4}\ \mu\text{Ci/mL} = (2.0\ \mu\text{Ci/L})(10^{-3}\ \text{L/mL}) \exp[(-4.20 \times 10^{-3})(t)]$$
$$0.10 = \exp[(-4.20 \times 10^{-3})(t)]$$

Taking the logarithm of both sides of the equation:

$$\ln(0.10) = \ln\{\exp[(-4.20 \times 10^{-3})(t)]\}$$
$$-2.30 = (-4.20 \times 10^{-3})(t)$$
$$t = 548.23 \text{ or } 550 \text{ days}$$

Specific Activity and the Becquerel. The quantity N is called the *activity* of a sample. In SI units the becquerel (Bq) is the unit used for activity. One *becquerel* of radioactive material is that quantity of unstable atoms whose frequency of decay is one disintegration per second. This definition covers all modes of disintegration for both single isotopes and mixtures.

For many years the unit used for activity was the curie. One *curie* of radioactive material is that quantity of unstable atoms whose frequency of decay is 3.700×10^{10} disintegrations per second. One becquerel is equal to 2.7×10^{-11} Ci. The curie is quite a large unit for a lot of purposes. *Millicuries* (1 mCi = 10^{-3} Ci) or *microcuries* (1 μCi = 10^{-6} Ci) and even *picocuries* (1 pCi = 10^{-12} Ci) were chosen as more manageable units to work with.

The specific activity of a radioisotope is the activity per gram of the pure radioisotope. The number of atoms of a pure radioisotope in one gram is given by

$$N = \frac{N_A}{A} \qquad (11\text{-}8)$$

where N_A is Avogadro's number (6.0248×10^{23}) and A is the nuclidic mass. The specific activity S of a particular radioisotope is an intrinsic property of that radioisotope.

$$S = \frac{\lambda N_A}{A} \text{ disintegrations} \cdot \text{s}^{-1} \qquad (11\text{-}9)$$

Growth of Subsidiary Products. In the process of decay, a new nuclide is formed, the daughter product. If the daughter product is stable, its concentration will gradually increase as the parent decays. On the other hand, if the daughter product is itself radioactive, the variation in concentrations of parent, daughter, and granddaughter products will very much depend on the relative rates of decay.

In several cases a radioactive isotope decays into another nuclide that is itself radioactive. This can continue for a large number of nuclides, resulting in a decay chain. The characteristics of any particular chain depend largely on the relative decay constants of its various members.

The simplest case is the growth of a radioactive daughter product from the parent atoms. Let us assume we begin with N_1 parent atoms of decay constant λ_1, and N_2 daughter atoms of decay constant λ_2. The rate at which the daughter product is increasing is then the difference between the rate at which it is produced by its parent and the rate at which it decays. This can be written as

$$\frac{dN_2}{dt} = \lambda_1 N_1 - \lambda_2 N_2 \tag{11-10}$$

The rate of production of the daughter is simply the decay rate of the parent.

Using Equation 11-6 with the notation that N_1 is the number of nuclei of the parent and N_{10} is the initial number gives

$$N_1 = N_{10} \exp(-\lambda_1 t)$$

Substituting in Equation 11-10, we obtain

$$\frac{dN_2}{dt} = \lambda_1 N_{10} \exp(-\lambda_1 t) - \lambda_2 N_2 \tag{11-11}$$

Rearranging, we get

$$\frac{dN_2}{dt} + \lambda_2 N_2 = \lambda_1 N_{10} \exp(-\lambda_1 t) \tag{11-12}$$

This equation can readily be solved by multiplying throughout by the factor $e^{\lambda 2t}$. Thus,

$$\exp(\lambda_2 t)\frac{dN_2}{dt} + \exp(\lambda_2 t)\lambda_2(N_2) = \lambda_1 N_{10} \exp(-\lambda_1 t) \exp(\lambda_2 t) \tag{11-13}$$

and

$$\frac{dN_2 e^{\lambda_2 t}}{dt} = \lambda_1 N_{10} \exp[(\lambda_2 - \lambda_{10})t] \tag{11-14}$$

On integration this yields

$$N_2 e^{\lambda_2 t} = \frac{\lambda_1 N_{10}}{\lambda_2 - \lambda_1} \exp[(\lambda_2 - \lambda_1)t] + C \tag{11-15}$$

The integration constant C is determined from the boundary conditions. For this case, at $t = 0$, there was no daughter product present, that is, $N_2 = 0$ at $t = 0$. Using these boundary conditions, Equation 11-15 reduces to

$$N_2 = \frac{\lambda_1 N_{10}}{\lambda_2 - \lambda_1}(e^{-\lambda_1 t} - e^{-\lambda_2 t}) \tag{11-16}$$

Characteristics of Daughter Products. In the derivation of Equation 11-16, it was assumed that N_2 was zero at zero time. Because the daughter nuclide itself decays, then at an infinite time, N_2 will again be zero. Between these two times when $N_2 = 0$, there will be a time, say t', when N_2 will reach a maximum. At this time, the rate of increase will be passing through a turning point, that is, $dN_2/dt = 0$. Using this fact, together with Equation 11-16, it can be shown that

$$t' = \frac{\ln\lambda_2 - \ln\lambda_1}{\lambda_2 - \lambda_1} \tag{11-17}$$

Secular Equilibrium. A limiting case of radioactive equilibrium in which $\lambda_1 \ll \lambda_2$ and in which the parent activity does not decrease measurably during many daughter half-lives is known as *secular equilibrium.* An example of this is ^{238}U decaying to ^{234}Th. In this case, a useful approximation of the value of N_2 after a large number of half-lives is

$$N_2 = N_{10}\frac{\lambda_1}{\lambda_2} \tag{11-18}$$

Continuous Production of Parent. The previous calculations assumed that at zero time a certain number of parent atoms were present and then decayed. In many cases of interest the parent is continuously replenished. Such cases occur for instance in nuclear reactors, where the parent nuclides are continuously being created by neutron bombardment. Another case is the continuous production of carbon-14 by cosmic rays incident on the nuclei present in the upper atmosphere.

End Products. Any radioactive decay chain must finally arrive at a nuclide that is stable. The relevant equations can readily be obtained, for any stable nuclide has $\lambda = 0$. For example, consider the case of a radioisotope whose daughter is stable. For this, Equation 11-16 can be used with $\lambda_2 = 0$. Thus,

$$N_2 = N_{10}(1 - e^{-\lambda_1 t}) \tag{11-19}$$

Similar modifications can be made to other equations concerned with longer decay chains.

Radioisotopes

Naturally Occurring Radioisotopes. Most of the 50 naturally occurring radioisotopes are associated with three distinct series: the thorium series, the uranium series, and the actinium series. Each one of these series starts with an element of high atomic mass (uranium-238, thorium-232, and uranium-235, respectively) and then decays by a long series of alpha- and beta-particle emissions to reach a stable nuclide (lead-206, lead-208, and lead-207, respectively). The three chains are associated with the heavy elements, and very few naturally occurring radioisotopes are found with atomic masses less than 82.

The half-lives of the naturally occurring radioisotopes are very long. Presumably they were constituents of the earth at its formation and their activity has not yet died away beyond detection.

Two other important isotopes that occur in the natural environment but that are not strictly naturally occurring are hydrogen-3 (tritium) and carbon-14. These radioisotopes are artificially produced by cosmic rays bombarding the upper atmosphere of the earth. At present the quantities of these isotopes are in equilibrium, their production rate by cosmic radiation being balanced by their natural decay rate. Because of this phenomenon these isotopes are of particular use in archaeological dating.

Artificially Produced Radioisotopes. The artificial production of radioisotopes is mainly carried out either by nuclear reactors or by particle accelerators. The cyclotron is the accelerator in most general use because the required bombarding particle energies are easily obtained and the output is reasonably high. The transmutation of a stable isotope to a radioactive one is effected by bombarding a target nucleus with a suitable projectile, either electromagnetic or a particle, to produce the required isotope from the resultant nuclear reaction.

When an accelerator is used, the bombarding particles are usually protons, deuterons, or alpha particles. In the nuclear reaction brought about by the bombardment of zinc-64 with energetic deuterons from a cyclotron, the deuteron and zinc-64 nucleus combine to form a new element. The new element has a charge of $30e + e$ and an atomic mass number of $64 + 2$. This compound nucleus is then ^{66}Ga, gallium-66. This intermediate nucleus disintegrates almost immediately by one of several possible modes of decay. If a proton is emitted, for example, the final nucleus must be left with a nuclear charge of $32e$ and an atomic mass number of 65, so it is ^{65}Zn. This isotope does not occur in nature.

For the production of radioisotopes for industrial application, the most common nuclear reactions used are those from thermal neutrons. A target sample, in a suitable container, is inserted into the core of a reactor and left there for varying amounts of time. In the core of a reactor there is a copious supply of thermal neutrons. These interact with the target nucleus to produce the required radioisotope, a process known as *neutron activation.*

Fission

A *nuclear reactor* is an assembly of fissionable material (such as uranium-235, plutonium-239, or uranium-233) arranged in such a way that a self-sustaining *chain reaction* is maintained. When these nuclei are bombarded with neutrons of the appropriate energy, they split up, or *fission,* into fission fragments and neutrons. For the nuclear reaction to continue, at least one of the neutrons produced must be available to produce another fission instead of escaping from the assembly or being used up in some other nuclear reaction. Thus, there is a minimum (*critical*) mass below which the reaction cannot be self-sustaining. Actual reactors are built with an excess mass to make large amounts of neutrons available. The excess neutron production is controlled by the use of *moderators*. The moderators are made of materials with large neutron-capture cross sections, such as boron, cadmium, or hafnium. These are formed into *control rods* that are moved in and out of the reactor to moderate the excess neutrons.

The fission chain reaction is characterized by an enormous release of heat. This heat must be carried away by an efficient cooling system to prevent mechanical failure of the reactor assembly (*meltdown*) and, ultimately, an uncontrolled fission. The ultimate uncontrolled reaction is, of course, an atomic explosion.

The fission fragments are simply lower mass elements. There are, most commonly, two fission fragments from each nucleus with an energy of the order of 200 MeV shared between them. The uranium nucleus does not split into the same two fragments each time. The breakup is far from symmetrical and can occur in more than 30 different ways. The most commonly produced isotopes are grouped around the mass numbers 95 and 139.

The fragments produced from the fission process have very large neutron-to-proton ratios so that they are highly unstable. Many transitions have to occur before a stable nucleus is finally achieved. These successive decays give rise to a decay chain.

Fission fragments, because of their high mass and very high initial charge, have extremely short ranges in matter. Hence, they are contained within the fuel element when a uranium nucleus fissions. The spent nuclear reactor fuel elements thus provide a very intense radioactive source that presents many problems in the subsequent handling and processing. Fission fragments themselves can sometimes be used as a radioactive source for industrial application.

The Production of X-Rays*

X-rays were discovered in 1895 by Wilhelm Conrad Roentgen. During the course of some studies, he covered a cathode ray tube with a black cardboard box and observed fluorescence on a screen coated with barium platinocyanide near the tube. After further investigation of this phenomenon, he concluded that the effect was caused by the generation of new invisible rays capable of penetrating opaque materials and producing visible fluorescence in certain chemicals. He called these new invisible rays *x-rays*. Because of their discoverer, x-rays are also sometimes referred to as Roentgen rays.

As pointed out previously, x-rays are electromagnetic waves and occupy the same portion of the electromagnetic spectrum as gamma rays. Like gamma rays, x-rays can pass through solid material. The mode of interaction of x-rays with matter is the same, as are the biological and photographic effects.

Whereas gamma rays come from within the nucleus of the atom, x-rays are generated outside the nucleus by the interaction of high-speed electrons with the atom. For this reason, there is a difference in the energy distribution of x- and gamma rays. Gamma rays from any single radionuclide consist only of rays of one or several discrete energies. X-rays consist of a broad, continuous spectrum of energies. The continuous spectrum will be discussed in detail later.

The X-Ray Tube. X-rays are produced whenever a stream of high-speed electrons strikes a substance. This is caused by their sudden stoppage or deflection by atoms within the target material. The x-ray tube (Figure 11-2) is designed to provide the high-speed electrons and the interacting material. Essential components of a x-ray tube are (1) a highly evacuated glass envelope containing the cathode and anode; (2) a source of electrons proceeding from a cathode; and (3) a target (or anode) placed in the path of the electron stream.

---

*This discussion follows U.S. PHS, 1968.

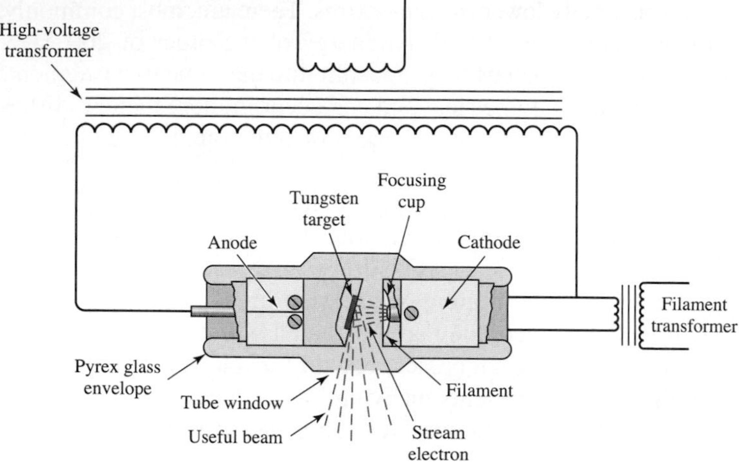

FIGURE 11-2
Typical x-ray tube in self-rectified circuit.

The development of the hot filament tube by William D. Coolidge in 1913 was a major advance. Most x-ray tubes in use today are of this type. Here, the free electrons are "boiled out" of an incandescent filament within an evacuated tube and given their velocity by accelerating them through an electric field. In the hot filament tube, the quality and intensity of radiation can be controlled independently by simple electrical means. The intensity of radiation is directly proportional to the current and is proportional to the square of the voltage. This allows a much wider range of wavelengths and intensities, while the characteristics of the tube remain reasonably constant throughout its useful life.

The high voltages required for x-ray tube operation are best obtained by step-up transformers, whose output is always alternating current. Because the electrons must flow only from cathode to anode within the tube some means of rectification is necessary. A self-rectified tube acts as its own rectifier. When an alternating voltage is applied to such a tube, electrons flow only from the cathode to anode as long as the anode remains cool. If the anode becomes hot, the flow of electrons reverses during the second half-cycle and the cathode is damaged. Thus the self-rectified tube is limited to low currents and short periods of operation. The use of "valves" (rectifiers) in the power supply circuit eliminates the inverse voltage on the x-ray tube. Thus, more power can be handled by the x-ray tube, the radiation output is increased, and the time of exposure is shortened.

X-Ray Production Efficiency. On average, the fraction of the electron energy emitted as electromagnetic radiation increases with the atomic number of the atoms of the target and the velocity of the electrons. This fraction is very small and can be represented by the following empirical equation:

$$F = 1.1 \times 10^{-9} ZV \tag{11-20}$$

where F = fraction of the energy of the electrons converted into x-rays
Z = atomic number of the target
V = energy of the electrons (in volts)*

Typically, less than 1 percent of the electrical power supplied is converted into x-ray energy. The remaining energy (over 99 percent) appears as heat produced at the target (largely through ionization and excitation). As a result, electron bombardment of the target raises it to a high temperature and, if the heat produced is not dissipated fast enough, the target will melt. This heat production is a serious factor in limiting the capacity of a x-ray tube.

A suitable target must have the following characteristics:

1. A high atomic number because efficiency is directly proportional to Z
2. A high melting point because of the high temperatures involved
3. A high thermal conductivity to dissipate the heat
4. Low vapor pressure at high temperatures to prevent target evaporation

The Continuous Spectrum. When high-speed electrons are stopped by a target, the radiation produced has a continuous distribution of energies (wavelengths). As the fast-moving electrons enter the surface layers of the target, they are abruptly slowed down by collision with the strong Coulomb field of the nucleus and are diverted from their original direction of motion. Each time the electron suffers an abrupt change of speed, a change in direction, or both, energy in the form of x-rays is radiated. The energy of the x-ray photon emitted depends on the degree of deceleration. If the electron is brought to rest in a single collision, the energy of the resulting photon corresponds to the kinetic energy of the electron stopped and will be a maximum. If the electron suffers a less drastic collision, a lower energy photon is produced. Because a variety of types of collisions will be occurring, photons of all energies up to the maximum will be produced. This accounts for the continuous distribution of a x-ray spectrum. The maximum intensity (peak of the curve) occurs at a wavelength about 1.5 times the minimum wavelength. The total intensity of radiation from a given x-ray tube is represented by the area under the spectral curve. The intensity has been found, as might be expected, to be directly proportional to the electron current (number of electrons striking the target).

Radiation Dose†

Fundamentally, the harmful consequences of ionizing radiations to a living organism are due to the energy absorbed by the cells and tissues of the organism. This absorbed energy (or dose) produces chemical decomposition of the molecules present in the living cells. The mechanism of the decomposition appears to be related to ionization and excitation interactions between the radiation and atoms within the tissue. The amount of ionization or number of ion pairs produced by ionizing radiations in the cells or

*The electron energy is generally expressed in terms of the voltage applied across the tube.
†This discussion follows U.S. PHS, 1968.

tissues provides some measure of the amount of decomposition or physiological damage that might be expected from a given quantity or dose. The ideal basis for radiation dose measurement could be, therefore, the number of ion pairs (or ionizations) taking place within the medium of interest. For certain practical reasons, the medium chosen for defining exposure dose is air.

Exposure Dose—the Roentgen. The exposure dose of x- or gamma radiation within a specific volume is a measure of the radiation based on its ability to produce ionization in air. The unit used for expressing the exposure to x- or gamma radiation is the roentgen (R). Its merit lies in the fact that the magnitude of the exposure dose in roentgens can usually be related to the absorbed dose, which is of importance in predicting or quantifying the expected biological effect (or injury) resulting from the radiation.

The *roentgen* is an exposure dose of x- or gamma radiation such that the associated corpuscular emission per 0.001293 g of air* produces, in air, ions carrying one electrostatic unit (esu) of quantity of electricity of either sign. Because the ionizing property of radiation provides the basis for several types of detection instruments and methods, such devices may be used to quantify the exposure dose. Note that this is a unit of exposure dose based on ionization of air; it is not a unit of ionization, nor is it an absorbed dose in air.

Absorbed Dose—the Gray. The absorbed dose of any ionizing radiation is the energy imparted to matter by ionizing radiations per unit mass of irradiated material at the place of interest. The SI unit of absorbed dose is the gray (Gy). One *gray* is equivalent to the absorption of 1 J/kg (joule per kilogram). The former unit of absorbed dose was the rad. One *rad* is equivalent to the absorption of 100 ergs/g. One Gy = 100 rads. It should be emphasized that although the roentgen unit is strictly applicable only to x- or gamma radiation, the gray may be used regardless of the type of ionizing radiation or the type of absorbing medium.

To make a conversion from roentgens to grays two things must be known: the energy of the incident radiation and the mass absorption coefficient of the absorbing material.

Example 11-3. A dose of 1.0 R of gamma radiation was measured in air. From empirical studies it is known that, on the average, 34 eV of energy is transferred (or absorbed) in the process of forming each ion pair in air. What is the equivalent absorbed dose in 1.0 cm^3 of air?

Solution. To form 1 esu per 0.001293 g of air (mass of 1 cm^3 at STP), the radiation must produce 1.61×10^{12} ion pairs when absorbed in air. Thus, using the empirical estimate, we find that the total energy absorbed is

$$(34 \text{ eV/ion pair})(1.61 \times 10^{12} \text{ ion pairs/g}) = 5.48 \times 10^{13} \text{ eV/g}$$

---

*One cubic centimeter of air at STP has a mass of 0.001293 g.

In ergs rather than electron volts,

$$(5.48 \times 10^{13}\ \text{eV/g})(1.602 \times 10^{-12}\ \text{erg/eV}) = 87\ \text{ergs/g}$$

Because 1 erg $= 1 \times 10^{-7}$ J, 1 R of exposure dose to 1.0 cm^3 of air at standard conditions results in the absorbed dose of

$$(87\ \text{erg/g})(10^{-7}\ \text{J/erg})(10^{3}\ \text{g/kg}) = 8.7 \times 10^{-3}\ \text{J/kg} = 8.7 \times 10^{-3}\ \text{Gy}$$

Relative Biological Effectiveness (Quality Factor). Although all ionizing radiations are capable of producing similar biological effects, the absorbed dose, measured in grays, that will produce a certain effect may vary appreciably from one type of radiation to another. The difference in behavior, in this connection, is expressed by means of a quantity called the *relative biological effectiveness* (RBE) of the particular radiation. The RBE of a given radiation may be defined as the ratio of the absorbed dose (grays) of gamma radiation (of a specified energy) to the absorbed dose of the given radiation required to produce the same biological effect. Thus, if an absorbed dose of 0.2 Gy of slow neutron radiation produces the same biological effect as 1 Gy of gamma radiation, the RBE for slow neutrons would be

$$\text{RBE} = \frac{1\ \text{Gy}}{0.2\ \text{Gy}} = 5$$

The value of the RBE for a particular type of nuclear radiation depends on several factors, such as the energy of the radiation, the kind and degree of the biological damage, and the nature of the organisms or tissue under consideration.

Tissue Weighting Factor (W_T). The *tissue weighting factor* (W_T) is a modifying factor used in dose calculations to correct for the fact that different tissues and organs have varying degrees of radiosensitivity depending on the radioisotope and the chemical form of the radioisotope. Some tissues and organs are very sensitive; others are not radiosensitive at all. For example, because iodine is easily incorporated in thyroid tissue, the thyroid gland is very sensitive to the radioiodines. The W_T is, therefore, high for the radioiodines. When the tissue or organ is not radiosensitive, the value of W_T may be very small or zero for that tissue.

The Sievert. With the concept of the RBE in mind, it is now useful to introduce another SI unit, known as the sievert (Sv). One *sievert* equals the radiation dose having the same biological effect as a gray of gamma radiation. This was formerly know as the *rem,* an abbreviation of "roentgen equivalent man" (1 Sv = 100 rem). The gray is a convenient unit for expressing energy absorption, but it does not take into account the biological effect of the particular nuclear radiation absorbed. The sievert, however, which is defined by

$$\text{Dose in Sv} = \text{RBE} \times \text{dose in grays} \times W_T$$

provides an indication of the extent of biological injury (of a given type) that would result from the absorption of nuclear radiation. Thus, the sievert is a unit of biological dose.

11-2 BIOLOGICAL EFFECTS OF IONIZING RADIATION*

The fact that ionizing radiation produces biological damage has been known for many years. The first case of human injury was reported in the literature just a few months following Roentgen's original paper in 1895 announcing the discovery of x-rays. As early as 1902, the first case of x-ray induced cancer was reported in the literature.

Early human evidence for harmful effects as a result of exposure to radiation in large amounts existed in the 1920s and 1930s based on the experience of early radiologists, persons working in the radium industry, and other special occupational groups. The long-term biological significance of smaller, chronic doses of radiation, however, was not widely appreciated until the 1950s, and most of our current knowledge of the biological effects of radiation has been accumulated since World War II.

Sequential Pattern of Biological Effects

The sequence of events following radiation exposure may be classified into three periods: a latent period, a period of demonstrable effect, and a recovery period.

Latent Period. Following the initial radiation event, and often before the first detectable effect occurs, there is a time lag referred to as the *latent period.* There is a vast time range possible in the latent period. In fact, the biological effects of radiation are arbitrarily divided into short-term, or acute, and long-term, or delayed, effects on this basis. Those effects that appear within a matter of minutes, days, or weeks are called *acute effects* and those which appear years, decades, and sometimes generations later are called *delayed effects.*

Demonstrable Effects Period. During or immediately following the latent period, certain discrete effects can be observed. One of the phenomena seen most frequently in growing tissues exposed to radiation is the cessation of mitosis or cell division. This may be temporary or permanent, depending on the radiation dosage. Other effects observed are chromosome breaks, clumping of chromatin, formation of giant cells or other abnormal mitoses, increased granularity of cytoplasm, changes in staining characteristics, changes in motility or ciliary activity, cytolysis, vacuolization, altered viscosity of protoplasm, arid altered permeability of the cell wall. Many of these effects can be duplicated individually with other types of stimuli. The entire gamut of effects however, cannot be reproduced by any single chemical agent.

Recovery Period. Following exposure to radiation, recovery can and does take place to a certain extent. This is particularly manifest in the case of the acute effects, that is, those appearing within a matter of days or weeks after exposure. There is, however, a residual damage from which no recovery occurs, and it is this irreparable injury which can give rise to later delayed effects.

Determinants of Biological Effects

The Dose-Response Curve. For any biologically harmful agent, it is useful to correlate the dosage administered with the response or damage produced. "Amount of dam-

---

*This discussion follows U.S. PHS, 1968.

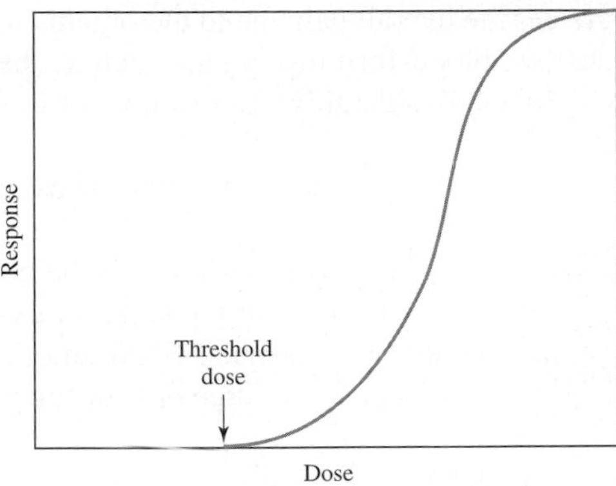

FIGURE 11-3
Dose-response curve depicting "threshold" dose.

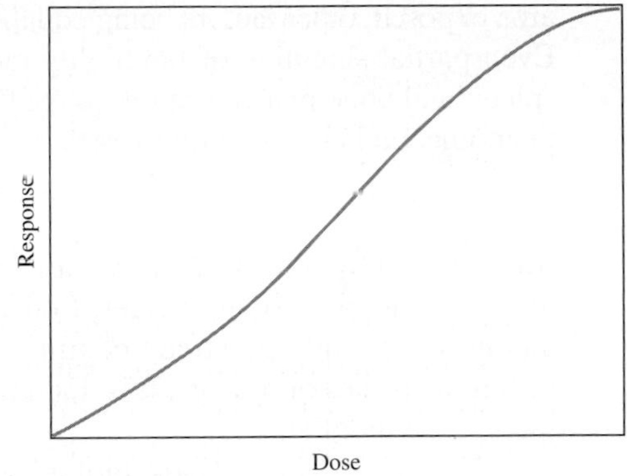

FIGURE 11-4
Dose-response curve depicting "no threshold" dose.

age" in the case of radiation might be the frequency of a given abnormality in the cells of an irradiated animal, or the incidence of some chronic disease in an irradiated human population. In plotting these two variables, a dose-response curve is produced. With radiation, an important question has been the nature and shape of this curve. Two possibilities are illustrated in Figures 11-3 and 11-4.

Figure 11-3 is a typical "threshold" curve. The point at which the curve intersects the abscissa is the threshold dose, that is, the dose below which there is no response. If an acute and easily observable radiation effect, such as reddening of the skin, is taken as "response," then this type of curve is applicable. The first evidence of the effect does not occur until a certain minimum dose is reached.

Figure 11-4 represents a linear, or nonthreshold, relationship, in which the curve intersects the abscissa at the origin. Here any dose, no matter how small, involves some degree of response. There is some evidence that the genetic effects of radiation constitute a nonthreshold phenomenon, and one of the underlying (and prudent) assumptions in the establishment of radiation protection guidelines and in radiation control activities in public health programs has been the assumption of a nonthreshold effect. Thus, some degree of risk is assumed when large populations of people are exposed to even very small amounts of radiation. This assumption often makes the establishment of guidelines for acceptable radiation exposure an enormously complex task, because the concept of "acceptable risk" comes into play, in which the benefit to be accrued from a given radiation exposure must be weighed against its hazard.

Rate of Absorption. The rate at which the radiation is administered or absorbed is most important in the determination of what effects will occur. Because a considerable degree of recovery occurs from the radiation damage, a given dose will produce less of an effect if divided (thus allowing time for recovery between dose increments) than if it were given in a single exposure.

Area Exposed. Generally when an external radiation exposure is referred to without qualification as to the area of the body involved, whole-body irradiation is assumed. The portion of the body irradiated is an important exposure parameter because the larger the area exposed, other factors being equal, the greater the overall damage to the organism. Even partial shielding of the highly radiosensitive blood-forming organs such as the spleen and bone marrow can mitigate the total effect considerably. An example of this phenomenon is in radiation therapy, in which doses that would be lethal if delivered to the whole body are commonly delivered to very limited areas, such as to tumor sites.

Variation in Species and Individual Sensitivity. There is a wide variation in the radiosensitivity of various species. Lethal doses for plants and microorganisms, for example, are usually hundreds of times larger than those for mammals. Even among different species of rodents, it is not unusual for one to demonstrate three or four times the sensitivity of another.

Within the same species, biological variability accounts for a difference in sensitivity among individuals. For this reason the lethal dose for each species is expressed in statistical terms. The LD_{50} for that species, or the dose required to kill 50 percent of the individuals in a large population, is the standard statistical measure. For people, the LD_{50} is estimated to be approximately 450 R.

Variation in Cell Sensitivity. Within the same individual, a wide variation in susceptibility to radiation damage exists among different types of cells and tissues. In general, cells that are rapidly dividing or have a potential for rapid division are more sensitive than those that do not divide are. Furthermore, nondifferentiated cells (i.e., primitive, or nonspecialized) are more sensitive than highly specialized cells. Within the same cell families then, the immature forms, which are generally primitive and rapidly dividing, are more radiosensitive than the older, mature cells, which have specialized in structure and function arid have ceased to divide.

Acute Effects

An acute dose of radiation is one delivered to a large portion of the body during a very short time. If the amount of radiation involved is large enough, acute doses may result in effects that can manifest themselves within a period of hours or days. Here the latent period, or time elapsed between the radiation insult and the onset of effects, is relatively short and grows progressively shorter as the dose level is raised. These short-term radiation effects are composed of signs and symptoms collectively known as *acute radiation syndrome*.

The stages in acute radiation syndrome may be described as follows:

1. *Prodrome.* This is the initial phase of the syndrome and is usually characterized by nausea, vomiting, and malaise. It may be considered analogous to the prodrome state in acute viral infections in which the individual is subject to nonspecific systemic reactions.
2. *Latent stage.* During this phase, which may be likened to the incubation period of a viral infection, the subjective symptoms of illness may subside, and the individual may feel well. Changes, however, may be taking place within the blood-forming organs and elsewhere that will subsequently give rise to the next aspect of the syndrome.

3. *Manifest illness stage.* This phase reflects the clinical picture specifically associated with the radiation injury. Among the possible signs and symptoms are fever, infection, hemorrhage, severe diarrhea, prostration, disorientation, and cardiovascular collapse. Which, if any, of the foregoing phenomena are observed in a given individual largely depend on the radiation dose received.

4. *Recovery or death.*

Relation of Dose to Type of Acute Radiation Syndrome

As mentioned earlier, each kind of cell has a different sensitivity to radiation. At relatively low doses, for example, the most likely cells to be injured are those with greatest sensitivity, such as the immature white blood cells of lymph nodes and bone marrow. At low doses the observable effects during the manifest illness stage would be in these cells. Thus, you would expect to observe fever, infection, and hemorrhage. This is known as the *hematopoietic form* of the acute radiation syndrome.

At higher doses, usually over 6 Gy cells of somewhat lower sensitivity will be injured. Of particular importance are the epithelial cells lining the gastrointestinal tract, for when these are destroyed a vital biological barrier is broken down. As a result, fluid loss may occur, as well as overwhelming infection, and severe diarrhea in the *gastrointestinal form* of the acute radiation syndrome.

In the *cerebral form,* which may result from doses of 100 Gy or more, the relatively resistant cells of the central nervous system are damaged, and the affected individual undergoes a rapid illness, characterized by disorientation and shock.

Considering the large degree of individual variation that exists in the manifestation of radiation injury, it is difficult to assign a precise dose range to each of these forms of the syndrome. The following generalizations, however, may serve to provide a rough indication of the kinds of doses involved. At 0.5 Gy or less, ordinary laboratory or clinical methods will show no indications of injury. At 1 Gy, most individuals show no symptoms, although a small percentage may show mild blood changes. At 2 Gy, most persons show definite signs of injury; this dose level may prove fatal to those individuals most sensitive to the effects of radiation. At 4.5 Gy, the mean lethal dose has been reached, and 50 percent of exposed individuals will succumb. Approximately 6 Gy usually marks the threshold of the gastrointestinal form of the acute radiation syndrome, with a very poor prognosis for all individuals involved. A fatal outcome may well be certain at 8–10 Gy.

Delayed Effects

Long term effects of radiation are those that may manifest themselves years after the original exposure. The latent period, then, is much longer than that associated with acute radiation syndrome. Delayed radiation effects may result from previous acute, high-dose exposures or from chronic, low-level exposures over a period of years.

No unique disease is associated with the long-term effects of radiation. These effects manifest themselves in human populations simply as a statistical increase in the incidence of certain already existing conditions. Because of the low incidence of these conditions, it is usually necessary to observe large populations of irradiated persons to measure these effects. Biostatistical and epidemiological methods are then used to indicate relationships between exposure and effect. In addition to the large numbers of people needed for human studies of delayed radiation effects, the situation is further complicated by the latent period. In some cases, a radiation-induced increase in a disease may go unrecorded unless the study is continued for many years.

Also note that although it is possible to perform true experiments with animal populations, in which all factors with the exception of radiation exposure are kept identical in study populations, human data are limited to "secondhand" information accrued from populations irradiated for reasons other than radiobiological information. Often a special characteristic of irradiated human populations is the presence of some preexisting disease that makes it extremely difficult to draw meaningful conclusions when these groups are compared with nonirradiated ones.

Despite these difficulties, many epidemiologic investigations of irradiated human beings have provided convincing evidence that ionizing radiation may indeed result in an increased risk of certain diseases long after the initial exposure. This information supplements and corroborates that gained from animal experimentation that demonstrates these same effects.

Among the delayed effects thus far observed have been somatic damage, which may result in an increased incidence of cancer, embryological defects, cataracts, lifespan shortening, and genetic mutations. With proper selection of animal species and strains, and of dose, ionizing radiation may be shown to exert an almost universal carcinogenic action, resulting in tumors in a great variety of organs and tissues. There is human evidence as well that radiation may contribute to the induction of various kinds of neoplastic diseases (cancers).

Human Evidence. Both empirical observations and epidemiologic studies of irradiated individuals have more or less consistently demonstrated the carcinogenic properties of radiation. Some of these findings are summarized here.

Early in the 1900s, when delayed radiation effects were little recognized, luminous numerals on watches and clocks were painted by hand with fine sable brushes, dipped first in radium-containing paint and then often "tipped" on the lips or tongue. Young women commonly were employed in this occupation. Years later, studies of these individuals who had ingested radium paint have disclosed an increased incidence of bone sarcomas and other malignancies resulting from the burdens of radium that had accumulated in their bones.

Some early medical and dental users of x-rays, largely unaware of the hazards involved, accumulated considerable doses of radiation. As early as the year 1910, there were reports of cancer deaths among physicians, presumably attributable to x-ray exposure. Skin cancer was a notable finding among these early practitioners. Dentists, for example, developed lesions on the fingers with which they repeatedly held dental films in their patients' mouths.

Early in the 1900s, certain large mines in Europe were worked for pitchblende, a uranium ore. Lung cancer was highly prevalent among the miners as a result of the inhalation of large quantities of airborne radioactive materials. It was estimated that the risk of lung cancer in the pitchblende miners was at least 50 percent higher than that of the general population.

One of the strongest supports for the concept that radiation is a leukemogenic agent in people comes from the epidemiologic studies of the survivors of the atomic bombing in Hiroshima, Japan. Survivors exposed to radiation above an estimated dose of approximately 1 Sv showed a significant increase in the incidence of leukemia. In addition, leukemia incidence correlated well with the estimated dose (expressed as distance from the detonation point), thus strengthening the hypothesis that the excess leukemia cases were indeed attributable to the radiation exposure. There is also some indication of an increase in thyroid cancer among the heavily irradiated survivors.

A pioneering study of children of mothers irradiated during pregnancy purported to show an increased risk of leukemia among young children if they had been irradiated in utero as a result of pelvic x-ray examination of the mother. Mothers of leukemic children were questioned as to their radiation histories during pregnancy with the child in question, and these responses were compared with those of a control group, consisting of mothers of healthy playmates of the leukemic children. Originally this work received much criticism, based partly on the questionnaire technique used to elicit the information concerning radiation history. It was believed that differences in recall between the two groups of mothers might have biased the results. A larger subsequent study designed to correct for the objections to the first one corroborated its essential findings and established the leukemogenic effect on the fetus of prenatal x-rays.

Considering the fact that immature, undifferentiated, and rapidly dividing cells are highly sensitive to radiation, it is not surprising that embryonic and fetal tissues are readily damaged by relatively low doses of radiation. It has been shown in animal experiments that deleterious effects may be produced with doses of only 0.10 Gy delivered to the embryo. There is no reason to doubt that the human embryo is equally susceptible.

The majority of the anomalies produced by prenatal irradiation involve the central nervous system, although the specific type of damage is related to the dose and to the stage of pregnancy during which irradiation takes place. In terms of embryonic death, the very earliest stages of pregnancy, perhaps the first few weeks in human beings, are most radiosensitive. From the standpoint of practical radiation protection, this very early sensitivity is of great significance, because it involves a stage in human embryonic development in which pregnancy may well be unsuspected. For this reason, the International Committee on Radiological Protection has recommended that routine nonemergency diagnostic irradiation involving the pelvic area of women in the childbearing years be limited to the 10-day interval following the onset of menstruation. Such precautions would virtually eliminate the possibility of inadvertently exposing a fertilized egg.

The period from approximately the second through the sixth week of human gestation, when pregnancy could still be unsuspected, is the most sensitive for the production of congenital anomalies in the newborn. During this period, embryonic death is less likely than in the extremely early stage, but the production of morphological defects in the newborn is a major consideration.

During later stages of pregnancy, embryonic tissue is more resistant to gross and easily observable damage. However, functional changes, particularly those involving the central nervous system, may result from such late exposures. These would be difficult to measure or evaluate at birth. They usually involve subtle alterations in such phenomena as learning patterns and development and may have a considerable latent period before they manifest themselves. There is some evidence that the decreasing sensitivity of the fetus to gross radiation damage as pregnancy progresses may not apply for the leukemogenic effects of prenatal irradiation. Another important factor to be considered in evaluating the radiation hazard during late pregnancy is that irradiation may produce true genetic mutations in the immature germ cells of the fetus for which no threshold dose has been established.

Lifespan Shortening. In a number of animal experiments, radiation has been demonstrated to have a lifespan-shortening effect. The aging process is complex and largely obscure, and the exact mechanisms involved in aging are as yet uncertain. Irradiated animals in these investigations appear to die of the same diseases as the nonirradiated controls, but they do so at an earlier age. How much of the total effect is due to premature aging and how much to an increased incidence of radiation-induced diseases is still unresolved.

Genetic Effects

Background. The fertilized egg is a single cell resulting from the union of sperm and egg; millions of cell divisions develop it into a complete new organism. The information that produces the characteristics of the new individual is carried in the nucleus of the fertilized egg on rod-shaped structures called chromosomes, arranged in 23 pairs. In each pair, one member is contributed by the mother and the other by the father. With each cell division that the rapidly developing embryonic tissue undergoes, all of this information is faithfully duplicated, so that the nucleus in each new cell of the developing organism contains essentially all of the information. This, of course, includes the germ cells in the new organism, which are destined to become sperm or eggs, and thus the information is transmitted from one generation to the next. This hereditary information is often likened to a template or to a code, which is reproduced millions of times over with remarkable accuracy. It is possible to damage the hereditary material in the cell nucleus by means of external influences, and when this is done the garbled or distorted genetic information will be reproduced just as faithfully when the cell divides as was the original message. When this kind of alteration occurs in those cells of the testes or ovaries that will become mature sperm or eggs, it is referred to as *genetic mutation;* if the damaged sperm or egg cell is then used in conception, the defect is reproduced in all of the cells of the new organism that results from this conception, including those that will become sperm or eggs, and thus whatever defect resulted from the original mutation can be passed on for many generations.

Most geneticists agree that the great preponderance of genetic mutations are harmful. By virtue of their damaging effects, they can be gradually eliminated from a population by natural means because individuals afflicted with this damage are themselves less likely to reproduce successfully than are normal individuals. The more severe the defect produced by a given mutation, the more rapidly it will be eliminated and vice versa; mildly damaging mutations may require a great many generations before they gradually disappear.

As a balance to this natural elimination of harmful mutations, fresh ones are constantly occurring. A large number of agents have mutagenic properties, and it is probable that our current knowledge includes just a fraction of these. In addition, mutations can arise within the germ cells of an organism without external insult. Among the various external influences found to be mutagenic are a wide variety of chemicals, certain drugs, and physical factors such as elevated temperatures and ionizing radiation. Natural background radiation probably accounts for a small proportion of naturally occurring mutations. For people, it has been estimated that background radiation probably produces less than 10 percent of these. Anthropogenic radiation, of course, if delivered to the gonads, can also produce mutations over and above those that occur spontaneously. Radiation, it should be noted, is not unique in this respect and is probably one of a number of environmental influences capable of increasing the mutation rate.

Animal Evidence. The mutagenic properties of ionizing radiation were first discovered in 1927, using the fruit fly as the experimental animal. Since that time, experiments have been extended to include other species, and many investigations have been carried out on mice. Animal experimentation remains our chief source of information concerning the genetic effects of radiation, and as a result of the intensive experimentation, certain generalizations may be made. Among those of health significance are (1) there is no indication of a threshold dose for the genetic effects of radiation, that is, a dose below which genetic damage does not occur; and (2) the degree of mutational damage that results from radiation exposure seems to be dose-rate dependent, so that a given dose is less effective in producing damage if it is protracted or fractionated over a long period.

Human Evidence. A major human study on genetic effects was made with the Japanese who survived the atomic bomb in 1945. As the index of a possible increase of the mutation rate, the sex ratio in the offspring of certain irradiated groups (families, for example, in which the mother had been irradiated but the father had not) was observed. Assuming that some of the mutational damage in the mothers would be recessive, lethal, and sex-linked, a shift in the sex ratio among these families might be expected in the direction of fewer male births than in completely nonirradiated groups, and this seemed to be the case in early reports. Later evaluation of more complete data, however, did not bear out the original suggestion of an effect on the sex ratio.

The preconception radiation histories of the parents of leukemic children compared with those of normal children was a part of the subject of another investigation. From the results, it would appear that there is a statistically significant increase in leukemia risk among children whose mothers had received diagnostic x-rays during this period. The effect here is apparently a genetic rather than an embryonic one because the irradiation occurred prior to the conception of the child.

A somewhat similar study ascertained the radiation exposure histories of the parents of children with Down syndrome. Most of this exposure was prior to the conception of the child. A significantly greater number of the mothers of children with Down syndrome reported receiving fluoroscopy and x-ray therapy than did mothers of the normal children in the control group.

The findings of these two studies may provide evidence that ionizing radiation is a mutational agent in people. However, the findings should be viewed with some reservations because there could be significant differences to begin with between populations of people requiring x-rays and those who do not. These differences alone might account for a slightly higher incidence of leukemia or Down syndrome in the offspring of the former group, irrespective of the radiation received. To date, there has been no incontrovertible evidence found of genetic effects in humans from radiation exposure.

11-3 RADIATION STANDARDS

Two population groups are given distinctly different treatment in the establishment of exposure— dose guidelines and rules. Standards are set for those occupationally engaged in work requiring ionizing radiation and for the general public. Although there are many standard-setting bodies, in general, the limits are consistent between groups. The Nuclear Regulatory Commission (NRC) has published guidelines in the *Code of Federal Regulations* (10 CFR 20) that serve as the standard in the United States. The dose guidelines are in addition to the natural background dose.

The allowable dose for occupational exposure is predicated on the following assumptions: the exposure group is under surveillance and control; it is adult; it is knowledgeable of its work and the associated risks; its exposure is at work, that is, 40 h/week; and it is in good health. On this basis, no individual is to receive more than 0.05 Sv per year of radiation exposure.

For the population at large, the allowable whole body dose in one calendar year is 0.001 Sv. This dose does not include medical and dental doses that, for diagnostic and therapeutic reasons, may far exceed this amount.

In addition to these dose rules, the NRC has set standards for the discharge of radionuclides into the environment. Table 11-2 is an extract from that list. These concentrations are measured above the existing background concentration and are annual averages. Discharges must be limited such that the amounts shown are not exceeded in ambient air or natural waters. If a mixture of isotopes is released into an unrestricted area, the concentrations shall be limited so that the following relationship exists:

$$\frac{C_A}{\text{MPC}_A} + \frac{C_B}{\text{MPC}_B} + \frac{C_C}{\text{MPC}_C} \leq 1 \tag{11-21}$$

where C_A, C_B, C_C = concentrations of radionuclides A, B, and C, respectively (in μCi/mL)

$\text{MPC}_A, \text{MPC}_B, \text{MPC}_C$ = maximum permissible concentrations of radionuclides A, B, and C from Table II of Appendix B, Part 20 of the *CFR* (10 CFR 20)

Radon. Unlike the standards for exposure and releases to the environment, those for radon in indoor air are established by the EPA. This is because radon is not the result of anthropogenic activity but rather occurs naturally. The EPA guidelines suggest that the annual average radon exposure be limited to 4 pCi/L of air.

TABLE 11-2

Selected maximum permissible concentrations of radionuclides in air and water above background

		Occupational values			Effluent concentrations		Releases to sewers
Radionuclide	Class	Oral Ingestion ALI (μCi)[b]	Inhalation ALI (μCi)	Inhalation DAC (μCi)[c]	Air (μCi/mL)	Water (μCi/mL)	Monthly average conc. (μCi/mL)
Barium-131	D[a], all compounds	3×10^3	8×10^3	3×10^{-6}	1×10^{-8}	4×10^{-5}	4×10^{-4}
Beryllium-7	W, all compounds except those given for Y	4×10^4	2×10^4	9×10^{-6}	3×10^{-8}	6×10^{-4}	6×10^{-3}
	Y, oxides, halides and nitrates	—	2×10^4	8×10^{-6}	3×10^{-8}	—	—
Calcium-45	W, all compounds	2×10^3	8×10^2	4×10^{-7}	1×10^{-9}	2×10^{-5}	2×10^{-4}
Carbon-14	Monoxide	—	2×10^6	7×10^{-4}	2×10^{-6}	—	—
	Dioxide	—	2×10^5	9×10^{-5}	3×10^{-7}	—	—
	Compounds	2×10^3	2×10^3	1×10^{-6}	3×10^{-9}	3×10^{-5}	3×10^{-4}
Cesium-137	D, all compounds	1×10^2	2×10^2	6×10^{-8}	2×10^{-10}	1×10^{-6}	1×10^{-5}
Iodine-131	D, all compounds	3×10^1	5×10^1	2×10^{-8}	—	—	—
		Thyroid (9×10^1)	Thyroid (2×10^3)		2×10^{-10}	1×10^{-6}	1×10^{-5}
Iron-55	D, all compounds except those given for W	9×10^3	2×10^3	8×10^{-7}	3×10^{-9}	1×10^{-4}	1×10^{-3}
	W, oxides, hydroxides and halides	—	4×10^3	2×10^{-6}	6×10^{-9}	—	—
Phosphorus-32	D, all compounds except those given for W	6×10^2	9×10^2	4×10^{-7}	1×10^{-9}	9×10^{-6}	9×10^{-5}
	W, phosphates of Zn^{2+}, S^{3+}, Mg^{2+}, Fe^{3+}, Bi^{3+} and lanthanides	—	4×10^2	2×10^{-7}	5×10^{-10}	—	—
Radon-222	With daughters removed	—	1×10^4	4×10^{-6}	1×10^{-8}	—	—
	With daughters present	—	1×10^2 (or 4 working level months)	3×10^{-8} (or 0.33 working level)	1×10^{-10}	—	—
Strontium-90	D, all soluble compounds except $SrTiO_3$	3×10^1	2×10^1	8×10^{-9}			
		Bone surface 4×10^1	Bone surface 2×10^1		3×10^{-11}	5×10^{-7}	5×10^{-6}
	Y, all insoluble compounds and $SrTiO_3$	—	4	2×10^{-9}	6×10^{-12}	—	—
Zinc-65	Y, all compounds	4×10^2	3×10^2	1×10^{-7}	4×10^{-10}	5×10^{-6}	5×10^{-5}

[a]D, W, and Y are classes denoting the time of retention in the body, days, weeks and years, respectively.

[b]ALI is the annual limit of intake.

[c]DAC is the derived air concentration.

(*Source:* Excerpted from title 10, *CFR*, part 20, Appendix B.)

11-4 RADIATION EXPOSURE

External and Internal Radiation Hazards

External radiation hazards result from exposure to sources of ionizing radiation of sufficient energy to penetrate the body and cause harm. Generally speaking, it requires an alpha particle of at least 7.5 MeV to penetrate the 0.07 mm protective layer of the skin. A beta particle requires 70 keV to penetrate the same layer (U.S. PHS, 1970). Unless the sources of alpha or beta radiation are quite close to the skin, they pose only a small external radiation hazard. X-rays and gamma rays constitute the most common type of external hazard. When of sufficient energy, both are capable of deep penetration into the body. As a result no radiosensitive organ is beyond the range of their damaging power.

Radioactive materials may gain access to the body by ingestion, by inhalation of air containing radioactive materials, by absorption of a solution of radioactive materials through the skin, and by absorption of radioactive material into the tissue through a cut or break in the skin. The danger of ingesting radioactive materials is not necessarily from swallowing a large amount at one time, but rather from the accumulation of small amounts on the hands, on cigarettes, on foodstuffs, and other objects that bring the material into the mouth.

Any radioactive material that gains entry into the body is an internal hazard. The extent of the hazard depends on the type of radiation emitted, its energy, the physical and biological half-life of the material, and the radiosensitivity of the organ where the isotope localizes. Alpha and beta emitters are the most dangerous radionuclides from the standpoint internal hazard because their specific ionization is very high. Radionuclides with half-lives of intermediate length are the most dangerous because they combine fairly high activity with a half-life sufficiently long to cause considerable damage. Polonium is an example of a potentially very serious internal hazard. It emits a highly ionizing alpha particle of energy 5.3 MeV and has a half-life of 138 days.

Natural Background

People are exposed to natural radiation from cosmic, terrestrial, and internal sources. Typical gonadal exposures from natural background are summarized in Figure 11-5. Cosmic radiation is that originating outside of our atmosphere. This radiation consists predominately, if not entirely, of protons whose energy spectrum peaks in the range of 1 to 2 GeV. Heavy nuclei are also present. The impact of primary and very high energy secondary cosmic rays produces violent nuclear reactions in which many neutrons, protons, alpha particles, and other fragments are emitted. Most of the neutrons produced by cosmic rays are slowed to thermal energies and, by *n*, *p* (neutron–proton) reaction with ^{14}N, produce ^{14}C. The lifetime of carbon-14 is long enough that it becomes thoroughly mixed with the exchangeable carbon at the earth's surface (carbon dioxide, dissolved bicarbonate it the oceans, living organisms, etc.). Some of the cosmic radiation penetrates to the earth's surface and contributes directly to our whole body dose. Terrestrial radiation exposure comes from the 50 naturally occurring radionuclides found in the earth's crust. Of these, radon has come to have the most significance as a common environmental hazard to the general public.

Radon is the product of the radioactive decay of its parent, radium. Radium is produced from each of the three major series: ^{235}U, ^{238}U, and ^{232}U. The radon iso-

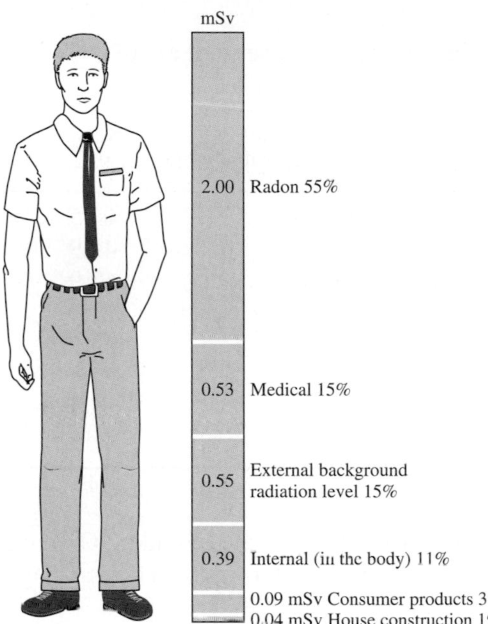

FIGURE 11-5
Average dose per year to person living in the United States. (*Source: U.S. Department of Energy.*)

topes produced are ^{222}Rn, ^{220}Rn, and ^{219}Rn. These have half-lives of 3.8 days, 55.6 s, and 3.92 s, respectively. ^{222}Rn, because of its longer half-life and the abundance of its parent uranium in geologic materials, is generally more abundant and, hence, is considered the greater environmental hazard. Because the half-life of radium and its parents is so long, the source is essentially undiminished over human time scales.

The hazard of radon does not come from radon itself but from its radioactive decay products (^{218}Po, ^{214}Po, ^{214}Bi). The decay products are charged atoms of heavy metals that readily attach themselves to airborne particulates. The main health problem stems from the inhalation of unattached decay products and these particulates. The decay products and particulates become lodged in the lung. As they continue to decay, they release small bursts of energy in the form of alpha, beta, and gamma radiation that damage the lung tissue and could ultimately lead to lung cancer (Kuennen and Roth, 1989).

Radon is a gas. It is colorless, odorless, and generally chemically inert like other noble gases such as helium, neon, krypton, and argon. It does not sorb, hydrolyze, oxidize, or precipitate. Thus, its movement through the ground is not inhibited by chemical interaction with the soil.

Migration of radon occurs by two mechanisms: diffusion as a gas through the pore spaces in the soil and by dissolution and transport in the groundwater. The rate of diffusion or transport is a function of the emanation rate, porosity, structural channels, moisture content, and hydrologic conditions. These migration routes lead to two mechanisms of effect on people. Buildings constructed in areas of high radon emanation may have radon gas penetrate the structure through natural construction openings such as floor drains or joints (Table 11-3) or through structural failures such as cracks that develop from foundation settlement. In areas where the public water supply is drawn

TABLE 11-3
Radon gas measurements in the floor drains and in the basement air of seven houses

House No.	Radon Concentration in Floor Drain (pCi/L)	Radon Concentration in Basement Air (pCi/L)	Ratio Drain/Basement
1	169.3	2.51	67.5
2	98.4	2.24	43.9
3	91.4	1.43	63.9
4	413.3	1.87	221
5	255.4	3.95	64.7
6	173.4	3.02	57.4
7	52.1	9.63	5.4
	Average 179.0	3.52	

from an aquifer that has radon emanation, shower water may release radon. One rule of thumb is that a radon concentration of 10,000 pCi/L of water, when heated and agitated, will produce about 1 pCi/L of air (Murane and Spears, 1987).

X-Rays

X-ray machine use is widespread in industry, medicine, and research. All such uses are potential sources of exposure.

Medical and Dental Use. In addition to the 300,000–400,000 medical-technical personnel that are occupationally exposed to radiation in the use of these machines, a considerable portion of the general population is also exposed. A large portion of the 2,500,000 persons seen daily by physicians have some x-ray diagnostic procedure performed on them.

Industrial Uses. Industrial x-ray devices include radiographic and fluoroscopic units used for the determination of defects in castings, fabricated structures, and welds, and fluoroscopic units used for the detection of foreign material in, for example, airline luggage. Use of these units may result in whole body exposure to the operators and people who are nearby.

Research Use. High-voltage x-ray machines are becoming familiar features of research laboratories in universities and similar institutions. Other x-ray equipment used in research includes x-ray diffraction units used for crystal analysis, electron microscopes, and particle accelerators.

Radionuclides

Naturally Occurring. Thousands of becquerels of radium is in use in the medical field. In this use, many individuals besides the patient, including other patients, nurses, technicians, radiologists, and physicians, are potentially exposed to radiation.

Static eliminators, employing polonium or radium as the radioactive source, have been widely used in industry. Typical industries where they may be found are the textile and paper trades, printing, photographic processing, and telephone companies.

Artificially Produced. Over 6,000 universities, hospitals, and research laboratories in the United States are using radionuclides for medical, biological, industrial, agricultural, and scientific research and for medical diagnosis and therapy. Over a million people in the United States receive radiotherapy treatment each year. Possible exposure from such radionuclides is involved with their preparation, handling, application, and transportation. Exposures, internal and external, might also arise through contamination of the environment by wastes originating from the use of these materials.

Nuclear Reactor Operations

Sources of radiation exposure associated with nuclear reactor operations include the reactor itself; its ventilation and cooling wastes; procedures associated with the removal and reprocessing of its "spent" fuel and the resulting fission product wastes; and procedures associated with the mining, milling, and fabrication of new fuels.

Radioactive Wastes

There are three principal sources of radioactive wastes: reactors and chemical processing plants, research facilities, and medical facilities. Regulations for the handling and disposal of radioactive wastes are designed to minimize exposure to the general public, but the regulations obviously provide less protection to those handling the waste.

11-5 RADIATION PROTECTION*

The principles discussed here are generally applicable to all types or energies of radiation. Their application will vary however, depending on the type, intensity, and energy of the source. For example, beta particles from radioactive materials require different shielding from that for high-speed electrons from an accelerator. Ideally, we would like to provide protection that results in a radiation exposure of zero. In actuality, technical and economic limitations force us to compromise so that the risks are small compared with the benefits obtained. The radiation standards set the limit above which the risk is deemed to be too great.

Reduction of External Radiation Hazards

Three fundamental methods are employed to reduce external radiation hazards: distance, shielding, and reduction of exposure time.

Distance. Distance is not only very effective, but also in many instances the most easily applied principle of radiation protection. Beta particles of a single energy have

---

*This discussion follows U.S. PHS, 1968.

a finite range in air. Sometimes the distance afforded by the use of remote control handling devices will supply complete protection.

The inverse square law for reduction of radiation intensity applies for point sources of x-, gamma, and neutron radiation. The inverse square law states that radiation intensity from a point varies inversely as the square of the distance from the source.

$$\frac{I_1}{I_2} = \frac{(R_2)^2}{(R_1)^2} \tag{11-22}$$

where I_1 is the radiation intensity at distance R_1 from the source, and I_2 is the radiation intensity at distance R_2 from the source. Inspection of this formula will show that increasing the distance by a factor of 3, for example, reduces the radiation intensity to one-ninth of its value. The inverse square law does not apply to extended sources or to radiation fields from multiple sources.

X-ray tubes act sufficiently like point sources so that reduction calculations by this law are valid. Gamma ray sources whose dimensions are small in comparison with the distances involved may also be considered point sources, as can capsule neutron sources.

Shielding. Shielding is one of the most important methods for radiation protection. It is accomplished by placing some absorbing material between the source and the person to be protected. Radiation is attenuated in the absorbing medium. When so used, "absorption," does not imply an occurrence such as a sponge soaking up water, but rather absorption here refers to the process of transferring the energy of the radiation to the atoms of the material through which the radiation passes. X- and gamma radiation energy is lost by three methods: photoelectric effect, Compton effect, and pair production.

The *photoelectric effect* is an all-or-none energy loss. The x-ray, or photon, imparts all of its energy to an orbital electron of some atom. This photon, because it consisted only of energy in the first place, simply vanishes. The energy is imparted to the orbital electron in the form of kinetic energy of motion, and this greatly increased energy overcomes the attractive force of the nucleus for the electron and causes the electron to fly from its orbit with considerable velocity. Thus, an ion pair results. The high-velocity electron (which is called a *photoelectron*) has sufficient energy to knock other electrons from the orbits of other atoms, and it goes on its way producing secondary ion pairs until all of its energy is expended.

The *Compton effect* provides a means of partial energy loss for the incoming x- or gamma ray. Again the ray appears to interact with an orbital electron of some atom, but in the case of Compton interactions, only a part of the energy is transferred to the electron, and the x- or gamma ray "staggers on" in a weakened condition. The high-velocity electron, now referred to as a Compton electron, produces secondary ionization in the same manner as does the photoelectron, and the weakened x-ray continues on until it loses more energy in another Compton interaction or disappears completely via the photoelectric effect. The unfortunate aspect of Compton interaction is that the direction of flight of the weakened x- or gamma ray is different from that of the original. In fact, the weakened x- or gamma ray is frequently referred to as a "scattered" photon, and the entire process is known as Compton scattering. By this mechanism of interaction, the direction of photons in a beam may be randomized, so that scattered radiation may appear around corners and behind shields although at a lesser intensity.

Pair production, the third type of interaction, is much rarer than either the photoelectric or Compton effect. In fact, pair production is impossible unless the x- or gamma ray possesses at least 1 MeV of energy. (Practically speaking, it does not become important until it possesses 2 MeV of energy.) *Pair production* may be thought of as the lifting of an electron from a negative to a positive energy state. The pair is a positron–electron pair that results from the photon ejecting an electron and leaving a "hole" the positron. If there is any excess energy in the photon above the 1 MeV required to create two electron masses, it is simply shared between the two electrons as kinetic energy of motion, and they fly out of the atom with great velocity. The negative electron behaves in exactly the ordinary way, producing secondary ion pairs until it loses all of its energy of motion. The positron also produces secondary ionization so long as it is in motion, but when it has lost its energy and slowed almost to a stop, it encounters a free negative electron somewhere in the material. The two are attracted by their opposite charges, and, upon contact, annihilate each other, converting both their masses into pure energy. Thus, two gamma rays of 0.51 MeV arise at the site of the annihilation. The ultimate fate of the annihilation gammas is either photoelectric absorption or Compton scattering followed by photoelectric absorption.

Because the energy of the photon must be greater than 1 MeV for pair production to occur, this process is not a factor in the absorption of x-rays used in dental and medical radiography. The energies of x-rays used in this type of radiography are rarely more than 0.1 MeV.

The predominating mechanism of interaction with the shielding material depends on the energy of the radiation and the absorbing material. The photoelectric effect is most important at low energies, the Compton effect at intermediate energies, and pair production at high energies. As x- and gamma ray photons travel through an absorber, their decrease in number caused by the above-mentioned absorption processes is governed by the energy of radiation, the specific absorber medium, and the thickness of the absorber traversed. The general attenuation may be expressed as follows:

$$\frac{dI}{dx} = -uI_0 \tag{11-23}$$

where dI = reduction of radiation
I_0 = incident radiation
u = proportionality constant
dx = thickness of absorber traversed

Integrating yields

$$I = I_0 \exp(-ux) \tag{11-24}$$

Using this formula it is easy to calculate the radiation intensity behind a shield of thickness x, or to calculate the thickness of absorber necessary to reduce radiation intensity

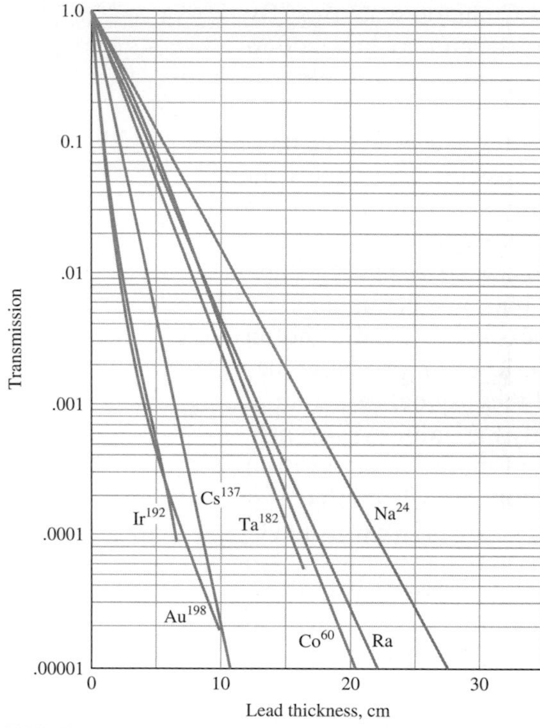

FIGURE 11-6
Transmission through lead of gamma rays from radium; cobalt-60; cesium-137; gold-198; iridium-192; tantalum-182; and sodium-24.

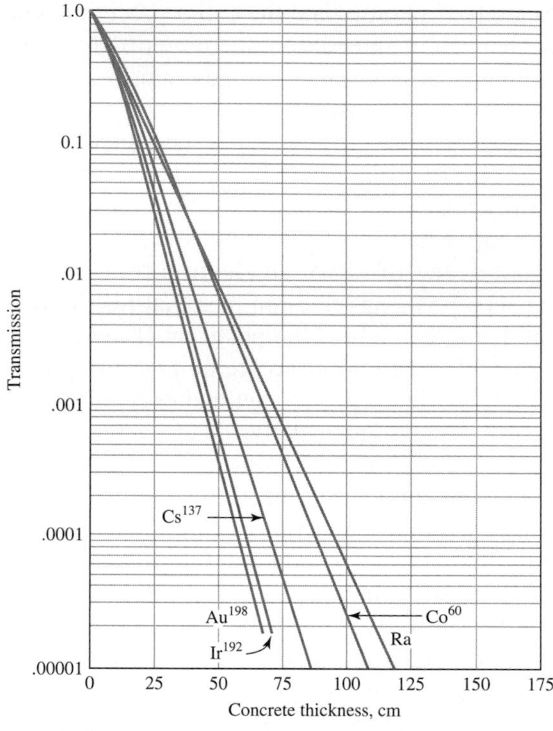

FIGURE 11-7
Transmission through concrete (density 2.35 Mg/m^3) of gamma rays from radium; cobalt-60, cesium-137, gold-198; iridium-192.

to a desired level, if the factor u is known. This factor is called the *linear absorption coefficient* when x is a linear dimension. The value of u depends on the energy of the radiation and the absorbing medium. The ratio I/I_0 is sometimes called the *transmission*. Tables and graphs are available that give values of u determined experimentally or that give transmission values for varying thickness or different shielding materials (Figures 11-6 through 11-9).

If the radiation being attenuated does not meet narrow-beam conditions, or thick absorbers are involved, the absorption equation becomes

$$I = BI_0 \exp(-ux) \tag{11-25}$$

where B is the buildup factor that takes into account an increasing radiation intensity due to scattered radiation within the absorber.

For alpha and beta emissions from radionuclides (not accelerators), substantial attenuation can be achieved with modest shielding. The amount of shielding required is, of course, a function of the particle energy. For example, a 10-MeV alpha particle has a range of 1.14 m in air, whereas a 1-MeV particle has a range of 2.28 cm. Virtually any solid material of any substance can be used to shield alpha particles. Beta particles can also be shielded relatively easily. For example a ^{32}P

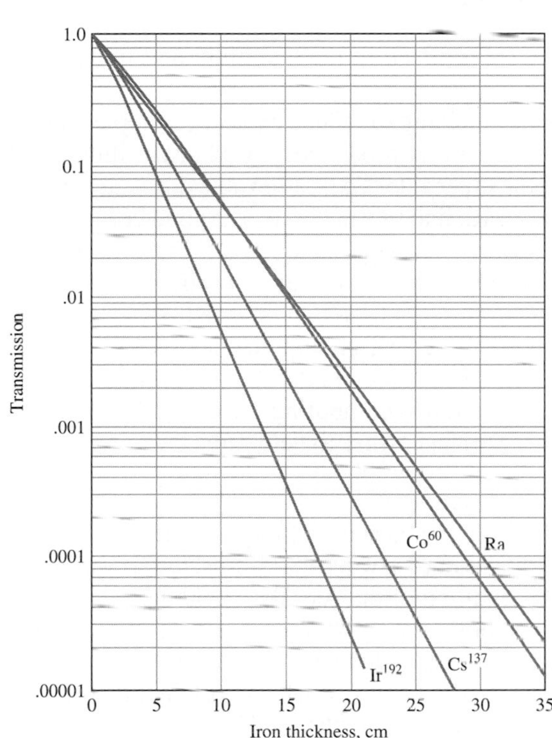

FIGURE 11-8
Transmission through iron of gamma rays from radium; cobalt-60, cesium-137; iridium-192.

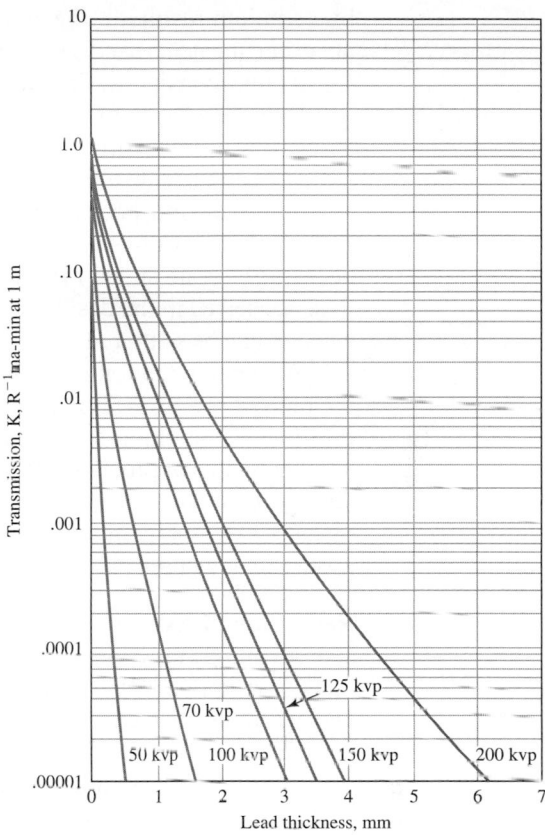

FIGURE 11-9
Transmission through lead of x-rays.

beta at 1.71 MeV can be attenuated 99.8% by 0.25 cm of aluminum. However, materials with high atomic numbers, such as metals, should not be used for high-energy beta shielding due to the production of *Bremsstrahlung radiation* (radiation produced by stopping another kind of radiation). In materials with high atomic numbers, the beta particle is absorbed, but the excess "trapped" energy is released in the form of an x-ray. For this reason, Plexiglas or Lucite, typically 6–12 mm thick, is often used.

Fast neutrons are poorly absorbed by most materials. Therefore, it is necessary to slow them down for efficient absorption. Because the greatest transfer of energy takes place in collisions between particles of equal mass, hydrogenous materials are most effective for slowing down fast neutrons. Water, paraffin, and concrete are all rich in hydrogen, and thus, important in neutron shielding. Once the neutrons have been reduced in energy, they may be absorbed by either boron or cadmium. When a boron atom captures a neutron, it emits an alpha particle, but because of the extremely short range of alpha particles, no additional hazard results. Neutron capture by cadmium results in the emission of gamma radiation. Lead or a similar gamma absorber must be used as a shield. A complete shield for a capsule-type neutron source may consist of, first, a thick layer of paraffin to slow down the neutrons, then a surrounding layer of cadmium to absorb the slow neutrons, and finally, an outer layer of lead to absorb both the gammas produced in the cadmium and those emanating from the capsule.

Some care must be exercised in using shielding to reduce exposure. People outside the "shadow" cast by the shield are not necessarily protected. A wall or partition is not necessarily a safe shield for persons on the other side. Their allowable dose may be less than conceived in the design of the barrier. Radiation can "bounce around corners" because it can be scattered.

Scattered radiation is present to some extent whenever an absorbing medium is in the path of radiation. The absorber then acts as a new source of radiation. Frequently, room walls, the floor, and other solid objects are near enough to a source of radiation to make scatter appreciable. When a point source is used under these conditions, the inverse square law is no longer completely valid for computing radiation intensity at a distance. Measurement of the radiation is then necessary to determine the potential exposure at any point.

Reduction of Exposure Time. By limiting the duration of exposure to all radiation sources and by providing ample recuperative time between exposures, the untoward effects of radiation can be minimized. Recognition of the zero threshold theory of damage warrants that exposures, no matter how small, be minimized. The standards established by the NRC are upper bounds to be avoided and not goals to be achieved.

In emergency situations it may occasionally be necessary to work in areas of very high dose rates. This can be done with safety by limiting the total exposure time so that the average permissible value for a day based on the radiation protection guide dose of 1×10^{-3} Gy (0.1 rad) per week is not exceeded. This does not imply that a worker should be allowed to extend this practice beyond receiving 1×10^{-3} Gy in a short period of time, that is, a dose of 1×10^{-3} Gy one day and no dose for 6 days would comply with the rule but would be considered excessive. Repetitions of this cycle would be unacceptable. Emergency situations may require that work be done in relays of several people in the same job so that the value of the radiation protection guide is not exceeded by any one person.

Reduction of Internal Radiation Hazards

Occupational. The prevention and control of contamination is the most effective way to reduce internal hazards in the workplace. The use of protective devices and good handling techniques affords a large measure of protection. Dust should be kept to a minimum by elimination of dry sweeping. Laboratory operations should be carried out in a hood. The exhaust air from the hood must be filtered with a high-efficiency filter. The filter must be replaced regularly in an approved manner. Protective clothing should be worn so that normal street clothes do not become contaminated. Respirators should be worn during emergency operations or when dust is generated. Eating must not be permitted in areas where radioactive materials are handled. Proper training in the care and handling of radioactive materials is, perhaps, the most

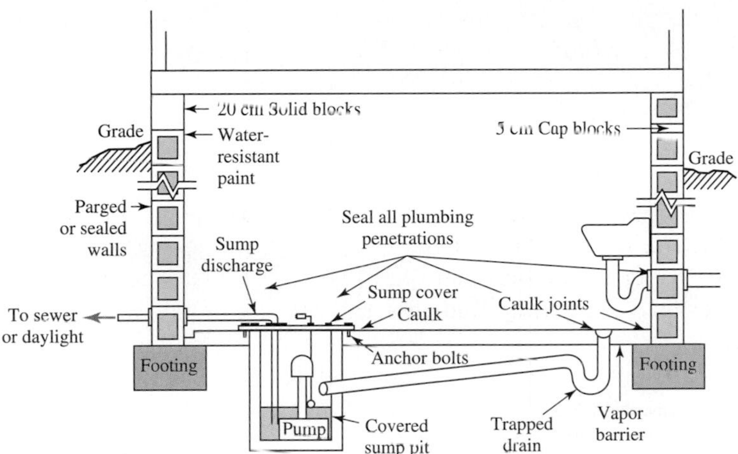

FIGURE 11-10
Methods to reduce pathways for radon entry.

important method for reducing the potential for internal radiation exposure in the workplace.

Radon. The most likely nonoccupational internal radiation hazard is from radon in private dwellings. Because the radon primarily originates in the soil beneath the house, control efforts are aimed at the basement or crawl space.

The EPA suggests two major approaches for new construction: reduction of the pathways for radon entry and reduction of the draft of the house on surrounding and underlying soil. The methods to reduce the pathways for entry are summarized in Figure 11-10. Of particular concern are penetrations into the foundation such as floor drains (see Table 11-3) and cracks in the floor. The use of a polyethylene sheet below the slab is particularly effective for controlling leaks that result from slab cracks that develop as the house settles. Because heat in the upper floors tends to rise, creating a draft much like a chimney, the house has a tendency to create a negative pressure on the basement and, hence, "suck in" radon from the soil pore spaces. Figure 11-11 shows some techniques to minimize the draft effect (Murane and Spears, 1987).

For existing structures, the remedies are more difficult to install, will be expensive, and may not yield satisfactory results. If drain tiles are present around the outside or inside of the perimeter footings, these are ideally located to permit vacuum to be drawn near some of the major soil gas entry routes (the joint between the slab and the foundation wall and the footing region where the radon can enter the voids in the block walls). Other efforts have included drilling holes in the slab itself and creating a vacuum system beneath the whole slab. Several suction points (three to seven) are required for this technique to work (Henschel and Scott, 1987). One demonstration project showed that jacking the house off the foundation and sealing the block walls was effective. In addition, a proprietary epoxy coating was applied to the floor and walls (Figure 11-12) (Ibach and Gallagher, 1987).

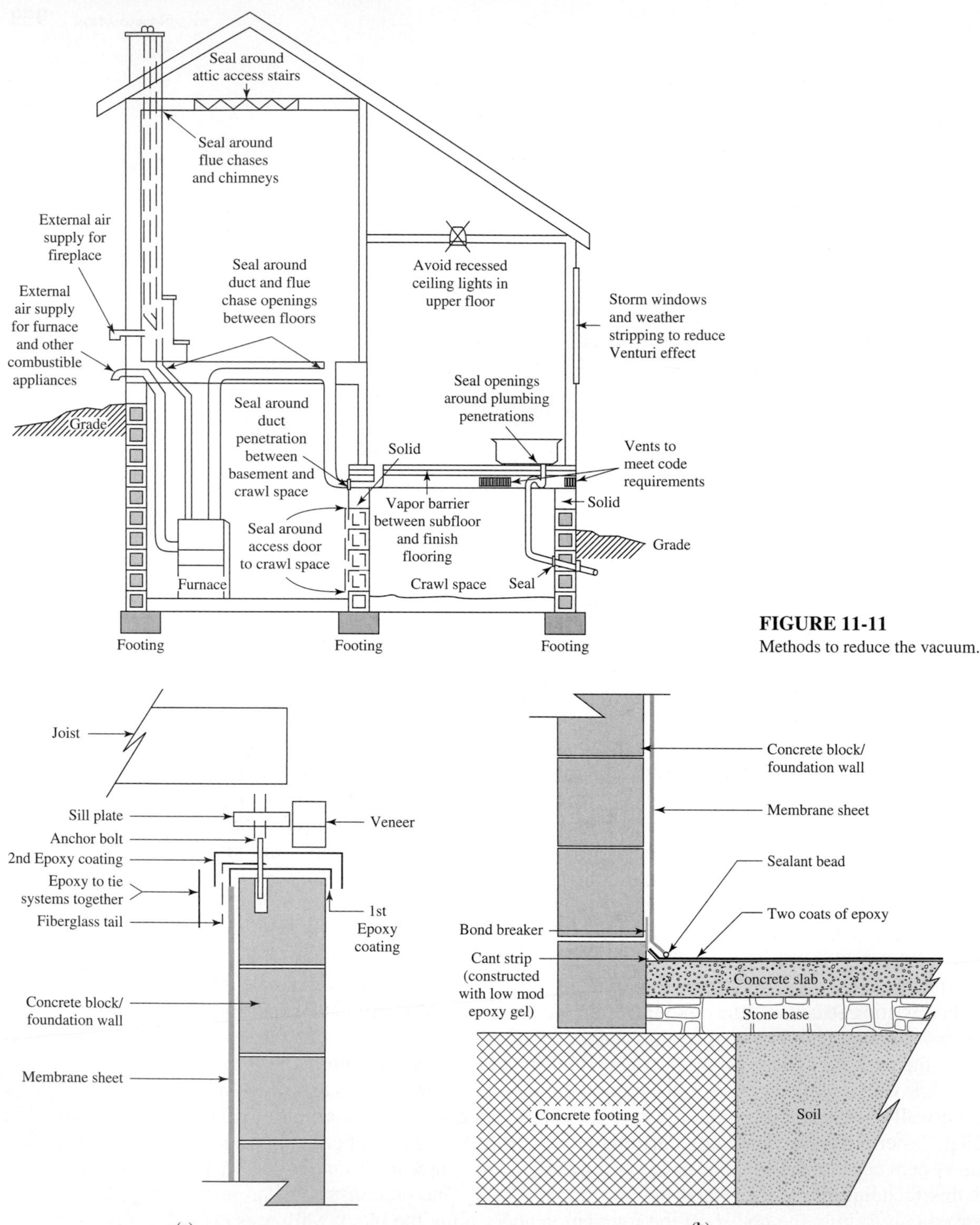

FIGURE 11-11
Methods to reduce the vacuum.

FIGURE 11-12
Interior membrane linings and sealants to prevent radon gas infiltration. (*Source:* Ibach and Gallagher, 1987.)

11-6 RADIOACTIVE WASTE

Types of Waste

No single scheme is satisfactory for classifying radioactive waste in a quantitative way. Usage has led us to categorize wastes into "levels." *High-level wastes* are those with activities measured in curies per liter; *intermediate-level wastes* have activities measured in millicuries per liter; *low-level wastes* have activities measured in microcuries per liter. Other classifications skip the intermediate-level wastes and use the terms high-level, *transuranic,* and low-level. The high-level wastes (HLW) are those resulting from reprocessing of spent fuel or the spent fuel itself from nuclear reactors. Transuranic wastes are those containing isotopes above uranium in the periodic table. They are the by-products of fuel assembly, weapons fabrication, and reprocessing. In general their radioactivity is low but they contain long-lived isotopes (those with half-lives greater than 20 years). The bulk of low-level wastes (LLW) has relatively little radioactivity. Most require little or no shielding and may be handled by direct contact.

Management of High-Level Radioactive Waste

In 2005, there were about 104 operating reactors in the United States (EIA, 2005). Roughly 10 m^3 of spent fuel is generated annually from each of these reactors. The construction of the fuel assembly results in considerably less fission product waste. Approximately 0.1 m^3 of the 10 m^3 is fission product waste. Of course, it is evenly distributed throughout the assembly and cannot be easily separated. The management choices are (1) store it indefinitely in the form in which it was removed from the reactor, (2) reprocess it to extract the fission products and recycle the other materials, or (3) dispose of it by burial or other isolation technique.

Under the Nuclear Waste Policy Act of 1987, Congress has prescribed that a storage facility be constructed that will not become permanent. President George W. Bush, on July 23, 2002, designated the monitored retrievable storage facility to be sited at Yucca Mountain, Nevada. The NRC has detailed the rules for the site in the *Code of Federal Regulations* (10 CFR 60.113). Some of the important provisions are summarized here (Murray, 1989).

1. The design and operation of the facility should not pose an unreasonable risk to the health and safety of the public. The radiation dose limit is a small fraction of that due to natural background.
2. A multiple barrier is to be used.
3. A thorough site study must be made. Geologic and hydrologic characteristics of the site must be favorable.
4. The repository must be located where there are no attractive resources, be far from population centers, and be under federal control.
5. High-level wastes are to be retrievable for up to 50 years from the start of operations.

6. The waste package must be designed to take into account all of the possible effects from earthquakes to accidental mishandling.
7. The package is to have a design life of 300 hundred years.
8. Groundwater travel time from the repository to the source of public water is to be at least 1,000 years.
9. The annual release of radionuclides must be less than one part in 100,000 per year of the amount of the radioactivity that is present 1,000 years after the repository is closed.

Waste Isolation Pilot Plant

The waste isolation pilot plant (WIPP) Project was authorized by Congress in 1979. After much political negotiation, the WIPP was authorized as a military transuranic waste facility exempt from licensing by the NRC. The facility consists of 16 km of shafts and tunnels 650 m below ground in southeast New Mexico. The geologic material is a Permian salt basin. It began accepting waste in March, 1999.

Management of Low-Level Radioactive Waste

Historical Perspective. Between 1962 and 1971, six commercial waste disposal sites were licensed. Three were subsequently closed because they failed. The three sites (Maxey Flats, Kentucky; Sheffield, Illinois; and West Valley, New York) all experienced similar problems. They used shallow land burial to dispose of the waste. This was accomplished by excavating a trench about 3–6 m deep and placing the drums and other containers (often cardboard boxes) of radionuclides in the trench and covering them with excavated soil. The completed trench was covered with a mound of earth and seeded.

Water seeped through the cover material and animals burrowed through it. The heavy clay sites chosen precisely to limit passage to the groundwater system served as holding ponds for the rainwater and ultimately accelerated the corrosion of the drums. At West Valley, when increased radioactivity called attention to this phenomenon, the trenches were opened and pumped to the nearest stream! Concurrently, it was discovered that the drums were often 30–50% empty. This, combined with the fact that the backfill material was heavy clay that did not completely fill the void spaces between the drums, allowed significant settlement of the cover material. This enhanced the collection of precipitation that contributed to the corrosion and failure of the drums.

These episodes led to a major rethinking at how we should manage our radioactive wastes. One result was that in 1980, Congress enacted the Low-Level Waste Policy Act. It says that each state is responsible for providing for the availability of capacity either within or outside of the state for disposal of low-level radioactive waste generated within its borders. The law provided for the formation of *compacts* between states to allow a regional approach to management. As of December 1995, the compact organization was as shown in Figure 11-13. The compacts decide what facilities are required and which states will serve as hosts. Although the compacts were supposed to begin accepting waste in 1986, the negotiation process has taken longer than expected

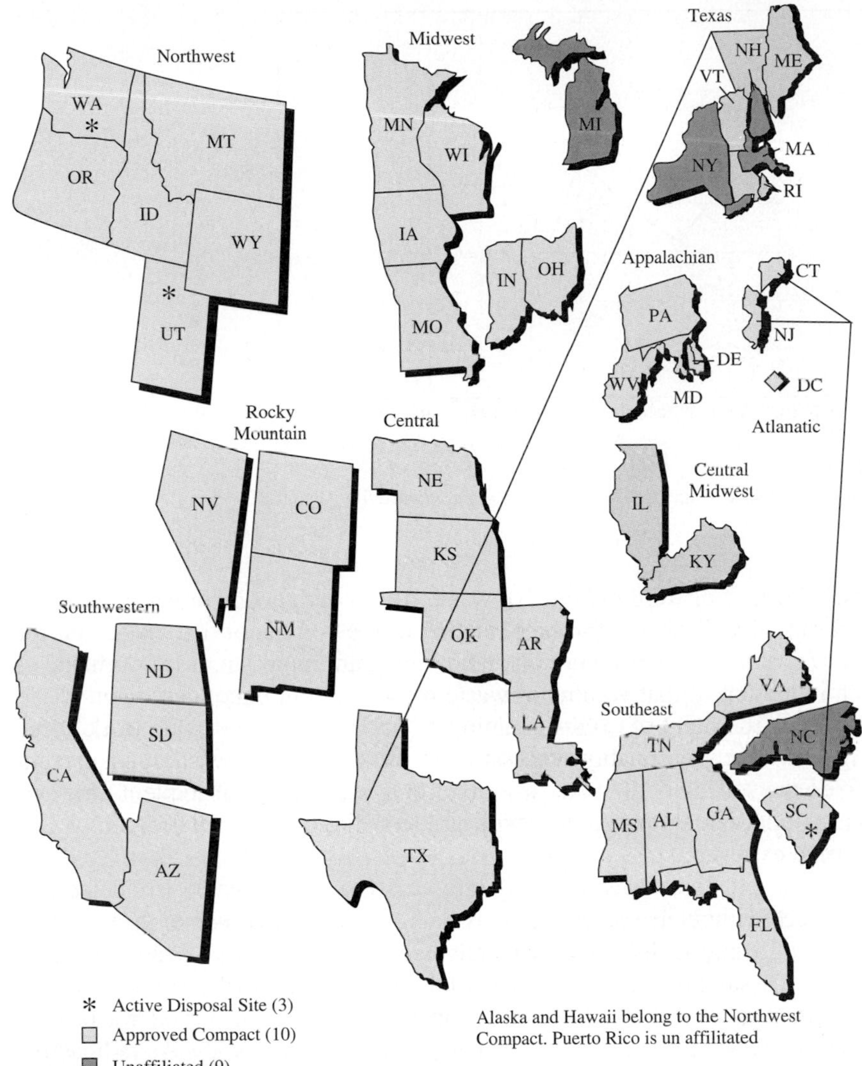

FIGURE 11-13
Low-level radioactive waste compacts. Data as of March 2004. (*Source:* Nuclear Regulatory Commission.)

and the deadline has been extended to beyond 2010. Many compacts have yet to select sites, let alone begin construction. The three currently available sites will soon run out of capacity, so there is some urgency to solve the problem.

Waste Minimization. As with all waste problems we have dealt with in this text, the first step in managing low-level radioactive waste is to minimize its production. Since 1980, considerable strides have been made in reducing the volume of LLRW (Figure 11-14). A number of procedures can be effectively employed.

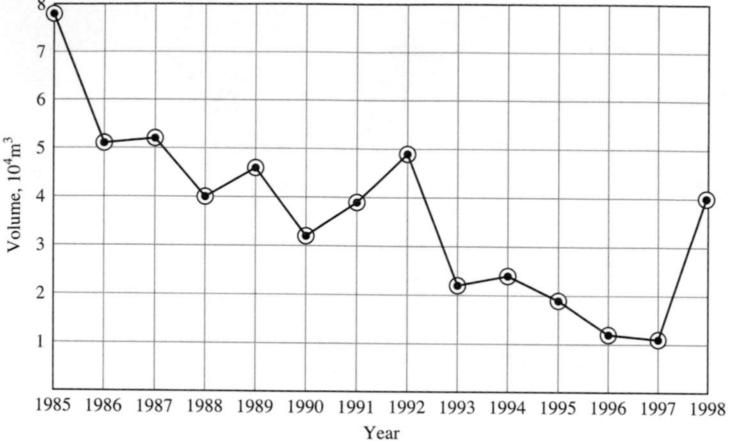

FIGURE 11-14
Low-level radioactive waste disposal. (*Source:* www.nrc.gov/waste/llw_disposal/statistics.html.)

Immediate sorting of solid radioactive waste from nonradioactive waste is an essential initial step in any scheme for the reduction of the volume of that waste and for the recovery of radionuclides from uranium and transuranium waste. It is optimistic to expect much reduction of that volume of waste by sorting out uncontaminated waste unless it is done at the point of origin. Training plant personnel to do this work at the point of origin has been reasonably successful. To ask radioactive waste management personnel to do the sorting of an unknown mixture of wastes at a subsequent time and place creates an unacceptable hazard of exposure to radiation by inhalation, injury, or ambient external exposure.

Often material only suspected of being radioactively contaminated is labeled and disposed of as such without necessarily actually being radioactive. Much of the so-called radioactive waste fits into such a category merely because of the place where it was generated. The cost of assaying such suspected low-level solid wastes to determine their true radioactive content is such that it is often cheaper to combine suspected waste with known radioactive waste than to separate it. This suspicious but not always radioactive waste takes up burial space unnecessarily. Time, effort, and money are needlessly expended in putting these nonradioactive wastes in the special radioactive waste landfills.

It has been a general practice to assume that all waste is radioactive if it has been generated in a laboratory using radioactive materials or by a radiochemical or similar processing activity. It is termed "radiation zone" or "contaminated area" wastes. Thus, waste that is suitable for disposal in a municipal landfill is mixed with contaminated waste. The burden of proof that the waste is not radioactive is on the person certifying or releasing the waste. Testing of the waste is time-consuming and is often omitted.

Probably the method most likely to succeed in reducing the amount of nonradioactive waste is a careful delineation and reduction of the so-called radiation zone and contamination areas. It is now common to define such areas rather broadly and to include certain zones and areas from which it should be obvious that the waste would not be radioactive. An example would be the office and administrative areas within a radiation zone. Such areas produce much nonradioactive waste that is often included for convenience in the low-level solid radioactive waste from the technical areas. In laboratory situations where nonradioactive wastes are generated alongside the radioactive waste, point source segregation can result in minimal radioactive waste generation.

Separation of combustible or compactible waste at the point of origin both improves waste handling and reduces volume. By sorting, wastes that are not compatible for incineration do not have to be handled at the incinerator. Because the volume reduction in an incinerator is greater than that in a compactor, the more wastes that are capable of incineration that reach the incinerator rather than the compactor, the greater the volume reduction.

Volume Reduction by Compression. Compression of solid low-level radioactive waste is suitable for about half the waste generated. There are three kinds of compression devices: compactors, balers, and baggers.

Compactors force material into the final storage, shipping, or disposal container. A favorite container is the 0.21 m^3 drum. Some space saving is possible. A variant of the compactor is called the *packer*. In this device, the material is compressed into a reusable container. At the burial grounds, the compacted material is dumped directly without any effort to retain its compacted form. Space saving is minimal with packer systems.

Balers compress the waste into bales that are wrapped, tied, or banded and then stored, shipped, or disposed of in burial grounds. Considerable space saving is possible with balers.

Baggers compress waste into a predetermined shape that is injected into round or rectangular bags, boxes, or drums before storage, shipment, or disposal. Some space saving is possible with this method of compaction.

These three techniques may be suited to general and sometimes even to unique situations. Unfortunately, such treatment does not reduce the possibility of burning while in storage, and only certain materials are suitable for compaction. These include paper, cloth, rubber, plastics, wood, glass, and small light metal objects. Large, rigid metal objects must be excluded because they are usually relatively incompressible and can damage the container and compressing machinery. Moisture (free or absorbed in large quantities by blotting paper or rags) has to be avoided because of its potential forcible release under high pressure, creating a great hazard to operators. Obviously, corrosive, pyrophoric, and explosive waste must be excluded from such processing, whether it is organic or inorganic.

The compression machinery must be economical, reliable, and easy to operate. Many commercial devices are available, but all must be modified by providing air containment, off-gas ventilation, often filtration, and, if necessary, shielding.

Volume Reduction by Incineration. Reduction of volumes of solid radioactive waste by incineration has interested managers of low-level radioactive waste, particularly in those parts of the world where land area is at a premium and costs are high. Under these conditions, the advantages of volume reduction are so great that the drawbacks seem only obstacles to be surmounted. In Europe, where land is scarce and more revered, the incineration of solid combustible radioactive waste is a common and apparently satisfactory method of pretreatment before final disposal.

There are certain advantages, such as volume reductions of 80–90 percent, reported for selected burnable waste. This may be a high estimate if such factors as residues from off-gas treatment and refractory changes are considered. This would represent a considerable saving in land used for burial, in transportation, and in long-term monitoring. In addition, it would free us from the nagging worry about the possible problem of long-burning subterranean fires. Special attention should be given to the problems of burning organic matter (solvents, ion-exchange resins, etc.) and putrescible biological material (animal cadavers, excreta, etc.). Incineration of radioactive waste must be carried out under controlled conditions to prevent the formation of radioactive aerosols and must comply with both RCRA and NRC rules if the wastes are RCRA wastes as well as being radioactive.

Long-Term Management and Containment

Site Selection. One concern in the burial of radioactive waste is that groundwater or infiltrating surface water will leach the waste and mobilize the radioactive materials. The radionuclides would be carried by this water back to the surface as a part of natural groundwater discharge or through a water well. Because of this concern, hydrogeologic and hydrochemical considerations in site selection become paramount.

The types of hydrogeologic and hydrochemical data that may be needed to determine whether or not a site is adequate include (Papadopulos and Winograd, 1974).

1. Depth to water table, including perched water tables, if present
2. Distance to nearest points of groundwater, spring water, or surface water usage (including well and spring inventory, and, particularly, wells available to the public)
3. Ratio of pan evaporation to precipitation minus runoff (by month for a period of at least 2 years)
4. Water table contour map
5. Magnitude of annual water table fluctuation
6. Stratigraphy and structure to base of shallowest confined aquifer
7. Baseflow data on perennial streams traversing or adjacent to storage site
8. Chemistry of water in aquifers and aquitards and of leachate from the waste trenches
9. Laboratory measurements of hydraulic conductivity, effective porosity, and mineralogy of core and grab samples (from trenches) of each lithology in unsaturated and saturated (to base of shallowest confined aquifer) zone-hydraulic conductivity to be measured at different water contents and tensions
10. Neutron moisture meter measurements of moisture content of unsaturated zone measurements to be made in specially constructed holes (at least 2 years' record needed)

11. In situ measurements of soil moisture tension in upper 4.5–9 m of unsaturated zone (at least 2 years' record needed)

12. Three-dimensional distribution of head in all saturated hydrostratigraphic units to base of shallowest confined aquifer

13. Pumping, bailing, or slug tests to determine transmissivity and storage coefficients

14. Definition of recharge and discharge areas for unconfined and shallowest confined aquifers

15. Field measurements of dispersivity coefficients

16. Laboratory and field determination of the distribution coefficient for movement of critical nuclides through all hydrostratigraphic units

17. Rates of denudation or slope retreat

These data are necessary for a complete definition of flow and nuclide transport through both the unsaturated and saturated zones.

It is not possible to immobilize a radioactive contaminant in a burial site for long periods of geologic time (i.e., for millions of years) with complete certainty. However, there appear to be hydrogeologic environments in which these contaminants can be kept below the surface and away from people until they have decayed to acceptable levels.

The problem is not merely a matter of ensuring optimum confinement, but also one of ensuring confinement for a minimum but specified time or describing and predicting the performance of these radioactive contaminants in the subsurface until this period has elapsed. For this reason, burial sites having complex hydrogeology in which such predictions are difficult or impossible are probably not suitable for storing radioactive waste.

From a geological standpoint, there appear to be two basic approaches to the long-term control of buried radioactive waste. The simplest approach is to prevent water from reaching the waste and thereby to eliminate the possibility of contaminants in the waste being mobilized. In arid climates, where there is little or no infiltration, this appears to be feasible.

In humid climates, where there is infiltration, some sort of engineered container or facilities that would isolate the waste from the water for hundreds of years is necessary. Whether or not such a facility can be designed, constructed, and demonstrated remains to be seen.

The second approach to long-term control involves burying the waste in a hydrogeologic environment that can be demonstrated to be safe despite the fact that radioactive contaminants can and will be mobilized. Demonstrating that such sites are, in fact, safe requires a quantitative evaluation of the factors influencing contaminant movement. Such an evaluation may be quite difficult but appears to be our only option if we wish to bury radioactive waste in humid climates or in climates where infiltration is capable of mobilizing or leaching the buried waste.

It is also important to give attention to the possible biological and microbiological environment of a burial site. Soil microorganisms, earthworms, larger burrowing animals, and the deep taproots of plants seeking water and nourishment (particularly in desert areas) can all be factors in moving components of waste out of a burial place into the biosphere. Some organisms can release organic compounds into the soil that can serve as complexing agents to mobilize otherwise insoluble contaminants. Some organisms can concentrate radionuclides by surprisingly high factors from their environment and so can change both the biochemical availability and the distribution of a radionuclide.

Site Selection Criteria. Michigan's site selection criteria serve to illustrate the factors that need to be considered in selecting disposal sites.

The first objective is to avoid population centers and conflicts with human activities. Michigan established an isolation distance of 1 km and required that projected population growth must not infringe to the extent that it would interfere with health and safety performance objectives of environmental monitoring.

Areas within 1.6 km of a fault where tectonic movement has occurred within the last 10,000 years are excluded as candidate sites. Likewise excluded are areas where significant earthquake intensity has been measured and flood plains exist. Mass wasting, erosion, and similar geologic processes are to be evaluated for possible damage to the facility.

Areas where groundwater flows from sites more than 30 m in 100 years or where groundwater could reach an aquifer in less than 500 years are excluded. The criteria also exclude areas over sole source aquifers and areas where groundwater discharges to the surface within 1 km. The facility may not be built within 16 km of the Great Lakes.

The criteria specify that the safest transportation net will be used. Highways with low accident rates located away from population centers are favored.

The site must have no complex meteorological characteristics and must avoid resource development conflicts. Likewise, environmentally sensitive areas such as wetlands and shorelands must be avoided. Areas that have formally proposed or approved development plans as of January 1, 1988, are excluded.

These criteria are extremely rigorous. Because of these constraints and the more serious problem of public opposition, no new sites have been finalized in the United States. Some compacts have had severe problems and conflicts that have resulted in the expulsion of one of the states. For example, after being selected as the host state, Michigan failed to identify an acceptable site and was expelled from the Midwest Compact.

A few compacts are proceeding quite well. These compacts are involving the public, community officials, regulators, and generators in joint efforts to identify sites, complete licensing applications, secure contractors, and construct the site. The most successful approach appears to be one of identification of actual candidate sites followed by a volunteer applicant.

The two currently operating sites in the United States are at Hanford, Washington, and Barnwell, South Carolina. These sites are accepting low-level radioactive waste from across the United States. There is a tremendous financial advantage to them in doing so. In November 1995, the total cost of disposing of a 0.21 m^3 drum was about \$3,000. Many generators have been storing this waste for several years (e.g., in Michigan, 55 generators have been storing waste for 5 years) and they are willing to pay these prices because of the lack of space.

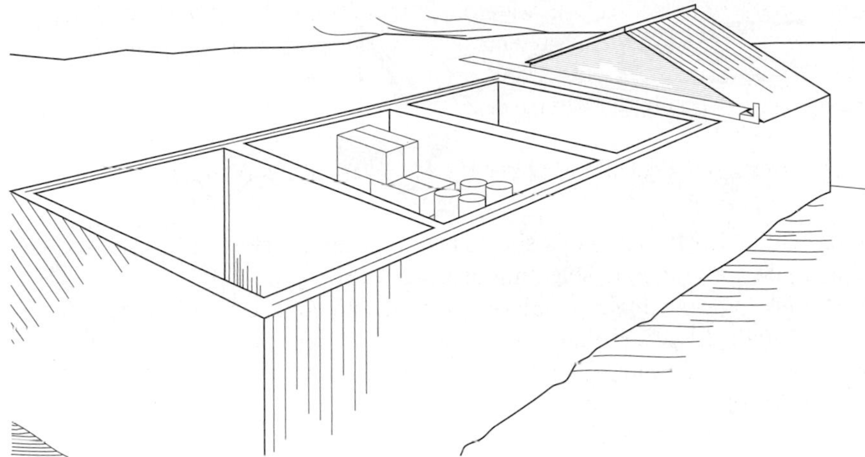

FIGURE 11-15
Aboveground vault for low-level radioactive waste disposal/storage. (*Source:* Midwest Compact.)

Engineered Containment Structures. As mentioned above, in humid climates, land burial is not acceptable. Michigan, as an example, has passed a statute that prohibits it. The alternative is an engineered structure. Engineered structures for the containment of waste must be designed with the intent to keep water, which can mobilize the contaminants, out of the facility. The Michigan statute specifies that each technology considered fulfills three requirements:

1. Maximum containment until the waste naturally decays to nonhazardous levels;
2. Capability to identify and retrieve wastes if necessary; and
3. Comprehensive monitoring of the facility and its environment.

There are four conceptual designs being considered by the Midwest Compact that would typify the various possible approaches. They are (1) aboveground vault, (2) below-ground vault, (3) aboveground modular concrete canister, and (4) below-ground concrete canister.

An aboveground vault is a large, reinforced concrete structure with access through the top or side walls for placing the waste inside (Figure 11-15). When a cell is filled, the vault will be sealed with a roof of concrete or some other suitable material. It must be designed to withstand earthquakes, tornadoes, floods, and fire.

The below-ground vault is similar to the aboveground vault except that it is located below ground (Figure 11-16). A compacted clay cover serves as part of the seal.

The aboveground concrete canister method consists of placing the low-level radioactive waste in large, precast concrete containers that are then stacked in an engineered structure (Figure 11-17).

The below-ground concrete canister method follows the same principles as the aboveground system, but the canisters are placed in a vault below ground (Figure 11-18).

FIGURE 11-16
Below-ground vault for low-level radioactive waste disposal/storage. (*Source:* Midwest Compact.)

Monitoring Systems. A monitoring system must operate both at permanent burial sites and at storage sites so that surface or air contamination will be detected quickly. Ground and surface water beneath or very near to the burial facilities should be monitored sufficiently often to give the earliest practical warning of failure of any facilities. "Failure" is defined as significant contamination of the ground or surface water in excess of standards that have been set for the disposal site.

Early detection of contamination is most important. Unlike surface water, groundwater usually moves slowly, and if contaminants move unexpectedly, we must know about it before significant amounts have left the disposal site. Interception of the

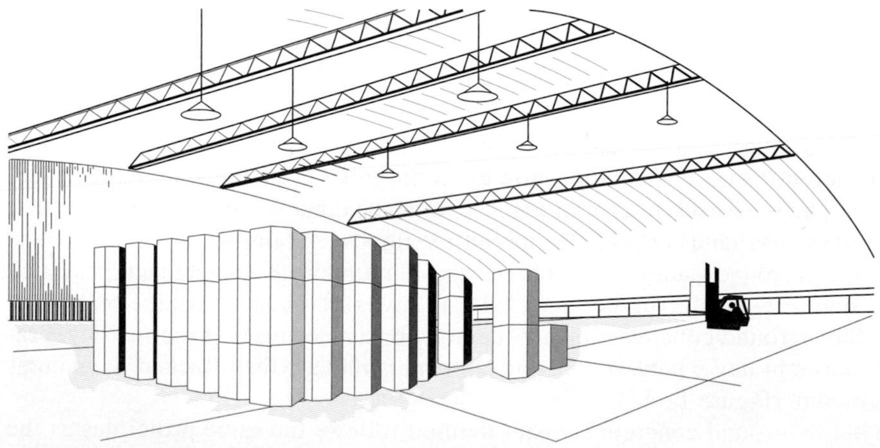

FIGURE 11-17
Aboveground canister storage for low-level radioactive waste. (*Source:* Midwest Compact.)

FIGURE 11-18
Below-ground canister storage for low-level radioactive waste. (*Source:* Midwest Compact.)

contaminants is not likely to be simple or prompt if this has not been considered in selection of the site or the design of facilities.

Should it be necessary to take remedial measures to eliminate further discharges, the smaller the amount of waste involved, the simpler these measures are likely to be. Early detection of contaminants generally requires that monitoring points be placed as close as possible to the waste.

Air monitoring should be provided around the site. Likewise, monitoring should also include adequate biological and ecological sampling to detect entrance of radionuclides into the local biosphere.

Contingency Plan. Contingency plans must be made to cover all foreseeable accidents or failures. They must include plans for corrective action in the event that monitoring shows a hazardous spread of contamination. These plans should include natural disaster precautions as well as more chronic types of failures.

Records Management. Duplicate records of the types, quantities, and concentrations of radioactive waste nuclides delivered to a burial site must be made and filed with more than one record bank. Reports on monitoring results and significant incidents, such as spills or unanticipated release of waste, must be filed with more than one record bank. These records should show the real (that is, observed, not calculated) level of contamination of the environment (including the ground area). These records must be in such a form that they will be useful and available for the effective length of time that the waste burial facility will require human attention.

Nonexhumation of Radioactive Wastes. Exhumation of waste originally buried without any intent of later retrieval is potentially a very hazardous operation. The

National Academy of Science recommends that exhumation not be made unless there is a credible reason to believe that a significant radiation hazard could arise from leaving the waste where it is and that the wastes can be exhumed safely (National Research Council, 1976). As a corollary to this recommendation, radioactive waste should not be exhumed and put into temporary engineered storage where the material must await a final decision on permanent disposal. Experience has shown that "temporary" storage may in reality be permanent storage because of the political realities in being able to relocate it.

11-7 CHAPTER REVIEW

When you have completed studying this chapter you should be able to do the following without the aid of your textbook or notes:

1. Explain what an isotope is.
2. Explain why some isotopes are radioactive and others are not.
3. Explain how alpha, beta, x-ray, and gamma ray emissions occur and how they differ.
4. Define the unit becquerel.
5. Explain the process of fission in a nuclear reactor.
6. Explain how x-rays are produced in an x-ray machine.
7. Define the concept of radiation dose and the units of roentgen, rad, Gy, Sv, and rem.
8. Explain the concepts of RBE and WT.
9. List the pattern of biological effects of radiation.
10. Discuss the determinants of biological effects.
11. Discuss the difference between acute and delayed biological effects of radiation.
12. List three possible delayed effects of radiation exposure.
13. State the acceptable occupational and nonoccupational dose of radiation as established by the NRC.
14. Explain the difference between internal and external radiation hazard.
15. Select a material and its thickness to protect against alpha or beta radiation.
16. Describe the sources of background radiation.
17. Explain why radon is a hazard and the mechanism by which the hazard is realized.
18. List three fundamental methods of reducing external radiation hazard.
19. Explain how to reduce occupational exposure to internal radiation hazards.

20. Describe how radon enters a house and give some techniques that may be used to inhibit radon entry.
21. List and describe the three types of radioactive waste (HLW, transuranic, and LLW).
22. Describe how each type of radioactive waste is to be disposed of.
23. Discuss waste minimization practice in reducing the volume of LLW.

With the aid of this text, you should be able to do the following:

1. Determine what particles are emitted in a given decay chain.
2. Determine the activity of a radioisotope given the original activity and the time interval.
3. Determine the activity resulting from the growth of a daughter product from a parent radionuclide.
4. Determine the time to achieve maximum activity of a daughter product.
5. Apply the inverse square law to determine radiation intensity.
6. Determine whether a combination of radionuclides exceeds the permissible concentrations.
7. Calculate the radiation intensity behind a shielding material or the desired thickness of a shielding material to achieve a reduction of radiation intensity.

11-8 PROBLEMS

11-1. What are the elements $^{40}_{18}X$ and $^{14}_{7}X$?

Answer: Argon and nitrogen

11-2. What are the elements $^{8}_{4}X$ and $^{238}_{92}X$?

11-3. What particle is emitted in the decay chain represented by

$$^{14}_{6}C \rightarrow ^{14}_{7}N$$

Answer: Beta

11-4. What particle is emitted in the decay chain represented by

$$^{32}_{15}P \rightarrow ^{32}_{16}S$$

11-5. What particles are emitted in each step in the decay chain represented by

$$^{226}_{88}Ra \rightarrow ^{222}_{86}Rn \rightarrow ^{218}_{84}Po \rightarrow ^{214}_{82}Pb$$

Answer: Alpha particle in each case

11-6. What particles are emitted in each step in the decay chain represented by

$$^{214}_{82}Pb \rightarrow ^{214}_{83}Bi \rightarrow ^{214}_{84}Po \rightarrow ^{210}_{82}Pb \rightarrow ^{210}_{83}Bi \rightarrow ^{210}_{84}Po \rightarrow ^{206}_{82}Pb$$

11-7. What particles are emitted in each step in the decay chain represented by

$$^{238}_{92}U \rightarrow ^{234}_{90}Th \rightarrow ^{234}_{91}Pa \rightarrow ^{234}_{92}U$$

11-8. Show that if a positron and electron are annihilated, then an energy of 1.02 MeV is released.

11-9. A laboratory solution containing 0.5 μCi/L of ^{32}P is to be disposed of. How long must the radioisotope be held to meet the allowable discharge activity?

11-10. An accident has contaminated a laboratory with ^{45}Ca. The radiation level is 10 times the tolerance level. How long must the room be isolated before the tolerance level is reached?

11-11. A hospital waste containing 100 μCi/L of ^{131}I is to be disposed of. How long must the radioisotope be held to meet the allowable discharge activity?

11-12. If in August 1911, Mme. Curie prepared an international standard containing 20.00 mg of $RaCl_2$, what was the radium content of this standard in August 2010?

11-13. What is the mass of a 50 μCi sample of pure ^{131}I?

Answer: 4.04×10^{-10} g

11-14. By emitting an alpha particle, ^{210}Po decays to ^{206}Pb. If the half-life of ^{210}Po is 138.4 d, what volume of ^{4}He will be produced in 1 year from 50 Ci of ^{210}Po? Assume the gas is at standard temperature and pressure.

11-15. Using a spreadsheet program you have written, calculate and plot the growth curve of ^{222}Rn from an initially pure sample of ^{226}Ra. Assume no ^{222}Rn is present initially.

11-16. When an x-ray unit is operated at 70 kV and 5 mA, it produces an intensity of D R/min at 1.0 m from the source. What intensity will it produce 2.0 m from the source?

11-17. If the source of x-rays in Problem 11-16 is operated at 15 mA, what intensity will be produced 2.0 m from the source?

Answer: 0.75 D

11-18. What thickness (in cm) of lead is required to shield a ^{60}Co source so that the transmission is reduced 99.6 percent?

11-19. What is the equivalent thickness (in cm) of concrete to accomplish the same attenuation as the lead in Problem 11-18?

Answer: ~55 cm

11-20. An existing concrete wall that is 25 cm thick is to be used to shield a ^{60}Co source so that the transmission is reduced 99.6 percent. What additional thickness (in cm) of lead is required to achieve this transmission reduction?

11-21. Determine the proportionality constant u for lead when it is used to shield ^{137}Cs.

11-22. Determine the proportionality constant u for iron when it is used to shield ^{137}Cs.

Answer: $u = 0.391$

11-9 DISCUSSION QUESTIONS

11-1. Explain why an archaeological artifact such as wood or bone may be dated by measuring its concentration of carbon-14.

11-2. Would you expect the tissue weighting factor (W_T) for x-rays to the big toe to be greater than, less than, or the same as that for radioiodine to the thyroid? Explain your choice.

11-3. What kind of radionuclide emitter (alpha, beta, gamma, or x-ray) is most dangerous from an internal hazard point of view? Explain why.

11-4. A laboratory worker has requested your advice on a shield for work she is doing with high-energy beta particles. What would you recommend?

11-5. You have an opportunity to purchase an older home with a basement that is serviced by a floor drain. What measures might you request to limit the migration of radon into the basement?

11-6. What is the status of the proposed Yucca Mountain disposal site? How much money has been spent to determine if this site is acceptable.

11-10 REFERENCES

Coombe, R. A. (1968) *An Introduction to Radioactivity for Engineers,* Macmillan/St. Martin's Press, New York, pp. 1–37.

EIA (2005) Energy Information Administration web site http://www.eia.doe.gov/cneaf/nuclear/page/nuc_reactors/reactsum.html.

Henschel, D. B. and A. G. Scott (1987) "Testing of Indoor Radon Reduction Techniques in Eastern Pennsylvania: An Update, "*Indoor Radon II, Proceedings of the Second APCA International Specialty Conference,* Cherry Hill, NJ, Air Pollution Control Association, Pittsburgh, pp. 146–159.

Ibach, M. T. and J. H. Gallagher (1987) "Retrofit and Preoccupancy Radon Mitigation Program for Homes," *Indoor Radon IL Proceedings of the Second APCA International Specialty Conference,* Cherry Hill, NJ, Air Pollution Control Association, Pittsburgh, pp. 172–182.

Kuennen, W., and R. C. Roth (1989) "Reduction of Radon Working Level by a Room Air Cleaner," presented at the 82nd Annual Meeting of the Air & Waste Management Association, Anaheim, CA, June 1989.

Murane, D. M., and J. Spears (1987) "Radon Reduction in New Construction," *Indoor Radon II Proceedings of the Second APCA International Specialty Conference*, Cherry Hill, NJ, Air Pollution Control Association, Pittsburgh, pp. 183–194.

Murray, R. L. (1989) Understanding Radioactive Waste, Battelle Press, Richland, WA, pp. 137–138.

National Research Council (1976) *The Shallow Land Burial of Low-Level Radioactively Contaminated Solid Waste,* National Academy of Science, Washington, DC.

Papadopulos, S. S., and I. J. Winograd (1974) *Storage of Low-Level Radioactive Wastes in the Ground: Hydrogeological and Hydrochemical Factors,* U.S. Environmental Protection Agency Report No. 520/3-74-009, Washington, DC.

U.S. PHS (1968) *Introduction to Medical X-Ray Protection, Training and Manpower Development Program,* U.S. Public Health Service, Rockville, MD.

U.S. PHS (1970) *Radiological Health Handbook,* PHS Publication No. 2016, U.S. Public Health Service, Rockville, MD, p. 204.

APPENDIX A

PROPERTIES OF AIR, WATER, AND SELECTED CHEMICALS

TABLE A-1
Physical properties of water at 1 atm

Temperature (°C)	Density, ρ (kg/m^3)	Specific weight, γ (kN/m^3)	Dynamic viscosity, μ (m(Pa · s))*	Kinematic viscosity, ν (μ(m^2/s))*
0	999.842	9.805	1.787	1.787
3.98	1,000.000	9.807	1.567	1.567
5	999.967	9.807	1.519	1.519
10	999.703	9.804	1.307	1.307
12	999.500	9.802	1.235	1.236
15	999.103	9.798	1.139	1.140
17	998.778	9.795	1.081	1.082
18	998.599	9.793	1.053	1.054
19	998.408	9.791	1.027	1.029
20	998.207	9.789	1.002	1.004
21	997.996	9.787	0.998	1.000
22	997.774	9.785	0.955	0.957
23	997.542	9.783	0.932	0.934
24	997.300	9.781	0.911	0.913
25	997.048	9.778	0.890	0.893
26	996.787	9.775	0.870	0.873
27	996.516	9.773	0.851	0.854
28	996.236	9.770	0.833	0.836
29	995.948	9.767	0.815	0.818
30	995.650	9.764	0.798	0.801
35	994.035	9.749	0.719	0.723
40	992.219	9.731	0.653	0.658
45	990.216	9.711	0.596	0.602
50	988.039	9.690	0.547	0.554
60	983.202	9.642	0.466	0.474
70	977.773	9.589	0.404	0.413
80	971.801	9.530	0.355	0.365
90	965.323	9.467	0.315	0.326
100	958.366	9.399	0.282	0.294

*Pa · s = (mPa · s) × 10^{-3}
*m^2/s = (μm^2/s) × 10^{-6}

TABLE A-2
Henry's law constants at 20°C

	H* (atm)	H_u† (dimensionless)	H_D† (atm · L/mg)	H_m† (atm · m^3/mol)
Oxygen	4.3×10^4	3.21×10	2.42×10^{-2}	7.73×10^{-1}
Methane	3.8×10^4	2.84×10	9.71×10^{-2}	6.38×10^{-1}
Carbon dioxide	1.51×10^2	1.13×10^{-1}	6.17×10^{-5}	2.72×10^{-3}
Hydrogen sulfide	5.15×10^2	3.84×10^{-1}	2.72×10^{-4}	9.26×10^{-3}
Vinyl chloride	3.55×10^5	2.65×10^2	1.02×10^{-1}	6.38
Carbon tetrachloride	1.29×10^3	9.63×10^{-1}	1.51×10^{-4}	2.32×10^{-2}
Trichloroethylene	5.5×10^2	4.1×10^{-1}	7.46×10^{-5}	9.89×10^{-3}
Benzene	2.4×10^2	1.8×10^{-1}	5.52×10^{-5}	4.31×10^{-3}
Chloroform	1.7×10^2	1.27×10^{-1}	2.55×10^{-5}	3.06×10^{-3}
Bromoform	3.5×10	2.61×10^{-2}	2.40×10^{-6}	6.29×10^{-4}
Ozone	5.0×10^3	3.71	1.87×10^{-3}	8.99×10^{-2}

*H values from Montgomery, 1985.
†H_u, H_D, and H_m calculated via Eqs. 4-50 to 4-52.

TABLE A-3

Saturation values of dissolved oxygen in freshwater exposed to a saturated atmosphere containing 20.9% oxygen under a pressure of 101.325 kPa[a]

Temperature (°C)	Dissolved oxygen (mg/L)	Saturated vapor pressure (kPa)
0	14.62	0.6108
1	14.23	0.6566
2	13.84	0.7055
3	13.48	0.7575
4	13.13	0.8129
5	12.80	0.8719
6	12.48	0.9347
7	12.17	1.0013
8	11.87	1.0722
9	11.59	1.1474
10	11.33	1.2272
11	11.08	1.3119
12	10.83	1.4017
13	10.60	1.4969
14	10.37	1.5977
15	10.15	1.7044
16	9.95	1.8173
17	9.74	1.9367
18	9.54	2.0630
19	9.35	2.1964
20	9.17	2.3373
21	8.99	2.4861
22	8.83	2.6430
23	8.68	2.8086
24	8.53	2.9831
25	8.38	3.1671
26	8.22	3.3608
27	8.07	3.5649
28	7.92	3.7796
29	7.77	4.0055
30	7.63	4.2430
31	7.51	4.4927
32	7.42	4.7551
33	7.28	5.0307
34	7.17	5.3200
35	7.07	5.6236
36	6.96	5.9422
37	6.86	6.2762
38	6.75	6.6264

[a]For other barometric pressures, the solubilities vary approximately in proportion to the ratios of these pressures to the standard pressures.

(*Source:* Calculated by G. C. Whipple and M. C. Whipple from measurements of C. J. J. Fox, *Journal of the American Chemical Society,* vol. 33, p. 362, 1911.)

TABLE A-4
Viscosity of dry air at approximately 100 kPa[a]

Temperature (°C)	Dynamic viscosity (μPa · s)
0	17.1
5	17.4
10	17.7
15	17.9
20	18.2
25	18.5
30	18.7
35	19.0
40	19.3
45	19.5
50	19.8
55	20.1
60	20.3
65	20.6
70	20.9
75	21.1
80	21.4
85	21.7
90	21.9
95	22.2
100	22.5
150	25.2

$\mu = 17.11 + 0.0536\ T + (P/8280)$ where T is in °C and P is in kPa.

TABLE A-5
Properties of air at standard conditions[a]

Molecular weight	M	28.97
Gas constant	R	287 J/kg · K
Specific heat at constant pressure	c_p	1,005 J/kg · K
Specific heat at constant volume	c_v	718 J/kg · K
Density	ρ	1.185 kg/m^3
Dynamic viscosity	μ	1.8515×10^{-5} Pa · s
Kinematic viscosity	ν	1.5624×10^{-5} m^2/s
Thermal conductivity	k	0.0257 W/m · K
Ratio of specific heats, c_p/c_v	k	1.3997
Prandtl number	Pr	0.720

[a] Measured at 101.325 kPa pressure and 298 K temperature.

TABLE A-6
Properties of saturated water at 298 K

Molecular weight	M	18.02
Gas constant	R	461.4 J/kg · K
Specific heat	c	4,181 J/kg · K
Prandtl number	Pr	6.395
Thermal conductivity	k	0.604 W/m · K

TABLE A-7
Frequently used constants

Standard atmospheric pressure	P_{atm}	101.325 kPa
Standard gravitational acceleration	g	9.8067 m/s^2
Universal gas constant	R_u	8,314.3 J/kg · mol · K
Electrical permittivity constant	ϵ_0	8.85×10^{-12} C/V · m
Electron charge	q_e	1.60×10^{-19} C
Boltzmann's constant	k	1.38×10^{-23} J/K

TABLE A-8
Properties of selected organic compounds

Name	Formula	M.W.	Density, g/mL	Vapor pressure, mm Hg	Henry's law constant kPa · m^3/mol
Acetone	CH_3COCH_3	58.08	0.79	184	0.01
Benzene	C_6H_6	78.11	0.879	95	0.6
Bromodichloromethane	$CHBrCl_2$	163.8	1.971		0.2
Bromoform	$CHBr_3$	252.75	2.8899	5	0.06
Bromomethane	CH_3Br	94.94	1.6755	1,300	0.5
Carbon tetrachloride	CCl_4	153.82	1.594	90	3
Chlorobenzene	C_6H_5Cl	112.56	1.107	12	0.4
Chlorodibromomethane	$CHBr_2Cl$	208.29	2.451	50	0.09
Chloroethane	C_2H_5Cl	64.52	0.8978	700	0.2
Chloroethylene	C_2H_3Cl	62.5	0.912	2,550	4
Chloroform	$CHCl_3$	119.39	1.4892	190	0.4
Chloromethane	CH_3Cl	50.49	0.9159	3,750	1.0
1,2-Dibromoethane	$C_2H_2Br_2$	187.87	2.18	10	0.06
1,2-Dichlorobenzene	1,2-Cl_2-C_6H_4	147.01	1.3048	1.5	0.2
1,3-Dichlorobenzene	1,3-Cl_2-C_6H_4	147.01	1.2884	2	0.4
1,4-Dichlorobenzene	1,4-Cl_2-C_6H_4	147.01	1.2475	0.7	0.2
1,1-Dichloroethylene	$CH_2{=}CCl_2$	96.94	1.218	500	15
1,2-Dichloroethane	$ClCH_2CH_2Cl$	98.96	1.2351	60	0.1
1,1-Dichloroethane	CH_3CHCl_2	98.96	1.1757	180	0.6
Trans-1,2-Dichloroethylene	$CHCl{=}CHCl$	96.94	1.2565	300	0.6
Dichloromethane	CH_2Cl_2	84.93	1.327	350	0.3
1,2-Dichloropropane	$CH_3CHClCH_2Cl$	112.99	1.1560	50	0.4
Cis-1,3-Dichloropropylene	$ClCH_2CH{=}CHCl$	110.97	1.217	40	0.2
Ethyl benzene	$C_6H_5CH_2CH_3$	106.17	0.8670	9	0.8
Formaldehyde	$HCHO$	30.05	0.815		
Hexachlorobenzene	C_6Cl_6	284.79	1.5691		
Pentachlorophenol	Cl_5C_6OH	266.34	1.978		
Phenol	C_6H_5OH	94.11	1.0576		
1,1,2,2-Tetrachloroethane	$CHCl_2CHCl_2$	167.85	1.5953	5	0.05
Tetrachloroethylene	$Cl_2C{=}CCl_2$	165.83	1.6227	15	3
Toluene	$C_6H_5CH_3$	92.14	0.8669	28	0.7
1,1,1-Trichloroethane	CH_3CCl_3	133.41	1.3390	100	3.0
1,1,2-Trichloroethane	$CH_2ClCHCl_2$	133.41	1.4397	25	0.1
Trichloroethylene	$ClHC{=}CCl_2$	131.29	1.476	50	0.9
Vinyl chloride	$H_2C{=}CHCl$	62.50	0.9106	2,200	50
o-Xylene	1,2-$(CH_3)_2C_6H_4$	106.17	0.8802	6	0.5
m-Xylene	1,3-$(CH_3)_2C_6H_4$	106.17	0.8642	8	0.7
p-Xylene	1,4-$(CH_3)_2C_6H_4$	106.17	0.8611	8	0.7

Note: Ethene = ethylene; ethyl chloride = chloroethane; ethylene chloride = 1,2-dichloroethane; ethylidene chloride = 1,1-dichloroethane; methyl benzene = toluene; methyl chloride = chloromethane; methyl chloroform = 1,1,1-trichloroethane; methylene chloride = dichloromethane; tetrachloromethane = carbon tetrachloride; tribromomethane = bromoform.

TABLE A-9
Typical solubility product constants

Equilibrium equation	K_{sp} at 25°C
$AgCl \rightleftharpoons Ag^{+} + Cl^{-}$	1.76×10^{-10}
$Al(OH)_3 \rightleftharpoons Al^{3+} + 3OH^{-}$	1.26×10^{-33}
$AlPO_4 \rightleftharpoons Al^{3+} + PO_4^{3-}$	9.84×10^{-21}
$BaSO_4 \rightleftharpoons Ba^{2+} + SO_4^{2-}$	1.05×10^{-10}
$Cd(OH)_2 \rightleftharpoons Cd^{2+} + 2OH^{-}$	5.33×10^{-15}
$CdS \rightleftharpoons Cd^{2+} + S^{2-}$	1.40×10^{-29}
$CdCO_3 \rightleftharpoons Ca^{2+} + CO_3^{2-}$	6.20×10^{-12}
$CaCO_3 \rightleftharpoons Ca^{2+} + CO_3^{2-}$	4.95×10^{-9}
$CaF_2 \rightleftharpoons Ca^{2+} + 2F^{-}$	3.45×10^{-11}
$Ca(OH)_2 \rightleftharpoons Ca^{2+} + 2OH^{-}$	7.88×10^{-6}
$Ca_3(PO_4)_2 \rightleftharpoons 3Ca^{2+} + 2PO_4^{3-}$	2.02×10^{-33}
$CaSO_4 \rightleftharpoons Ca^{2+} + SO_4^{2-}$	4.93×10^{-5}
$Cr(OH)_3 \rightleftharpoons Cr^{3+} + 3OH^{-}$	6.0×10^{-31}
$Cu(OH)_2 \rightleftharpoons Cu^{2+} + 2OH^{-}$	2.0×10^{-19}
$CuS \rightleftharpoons Cu^{2+} + S^{2-}$	1.0×10^{-36}
$Fe(OH)_3 \rightleftharpoons Fe^{3+} + 3OH^{-}$	2.67×10^{-39}
$FePO_4 \rightleftharpoons Fe^{3+} + PO_4^{3-}$	1.3×10^{-22}
$FeCO_3 \rightleftharpoons Fe^{2+} + CO_3^{2-}$	3.13×10^{-11}
$Fe(OH) \rightleftharpoons Fe^{2+} + 2OH^{-}$	4.79×10^{-17}
$FeS \rightleftharpoons Fe^{2+} + S^{2-}$	1.57×10^{-19}
$PbCO_3 \rightleftharpoons Pb^{2+} + CO_3^{2-}$	1.48×10^{-13}
$Pb(OH)_2 \rightleftharpoons Pb^{2+} + 2OH^{-}$	1.40×10^{-20}
$PbS \rightleftharpoons Pb^{2+} + S^{2-}$	8.81×10^{-29}
$Mg(OH)_2 \rightleftharpoons Mg^{2+} + 2OH^{-}$	5.66×10^{-12}
$MgCO_3 \rightleftharpoons Mg^{2+} + CO_3^{2-}$	1.15×10^{-5}
$MnCO_3 \rightleftharpoons Mn^{2+} + CO_3^{2-}$	2.23×10^{-11}
$Mn(OH)_2 \rightleftharpoons Mn^{2+} + 2OH^{-}$	2.04×10^{-13}
$NiCO_3 \rightleftharpoons Ni^{2+} + CO_3^{2-}$	1.45×10^{-7}
$Ni(OH)_2 \rightleftharpoons Ni^{2+} + 2OH^{-}$	5.54×10^{-16}
$NiS \rightleftharpoons Ni^{2+} + S^{2-}$	1.08×10^{-21}
$SrCO_3 \rightleftharpoons Sr^{2+} + CO_3^{2-}$	5.60×10^{-10}
$Zn(OH)_2 \rightleftharpoons Zn^{2+} + 2OH^{-}$	7.68×10^{-17}
$ZnS \rightleftharpoons Zn^{2+} + S^{2-}$	2.91×10^{-25}

(*Sources:* Linde, 2000; Sawyer, McCarty, and Parkin, 2003; Weast, 1983.)

TABLE A-10
Typical valences of elements and compounds in water

Element or compound	Valence
Aluminum	3^+
Ammonium (NH_4^+)	1^+
Barium	2^+
Boron	3^+
Cadmium	2^+
Calcium	2^+
Carbonate (CO_3^{2-})	2^-
Carbon dioxide (CO_2)	[a]
Chloride (*not* chlorine)	1^-
Chromium	$3^+, 6^+$
Copper	2^+
Fluoride (*not* fluorine)	1^-
Hydrogen	1^+
Hydroxide (OH^-)	1^-
Iron	$2^+, 3^+$
Lead	2^+
Magnesium	2^+
Manganese	2^+
Nickel	2^+
Oxygen	2^-
Nitrogen	$3^+, 5^+, 3^-$
Nitrate (NO_3^-)	1^-
Nitrite (NO_2^-)	1^-
Phosphorus	$5^+, 3^-$
Phosphate (PO_4^{3-})	3^-
Potassium	1^+
Silver	1^+
Silica	[b]
Silicate (SiO_4^{4-})	4^-
Sodium	1^+
Sulfate (SO_4^{2-})	2^-
Sulfide (S^{2-})	2^-
Zinc	2^+

[a]Carbon dioxide in water is essentially carbonic acid:

$$CO_2 + H_2O \rightleftharpoons H_2CO_3$$

As such, the equivalent weight = GMW/2.

[b]Silica in water is reported as SiO_2. The equivalent weight is equal to the gram molecular weight.

SOURCES

Linde, D. R. (2000) *CRC Handbook of Chemistry and Physics,* 81st ed., CRC Press, Boca Raton, FL, pp. 8-111–8-112.

Montgomery, J. M. (1985) *Water Treatment Principles and Design,* John Wiley & Sons, New York, p. 236.

Sawyer, C. N., P. L. McCarty, and G. F. Parkin (2003) *Chemistry for Environmental Engineering and Science,* 5th ed., McGraw-Hill, Boston, pp. 39–40

Weast, R. C. (1983) *CRC Handbook of Chemistry and Physics,* 64th ed., CRC Press, Boca Raton, FL, pp. B-219–B-220.

APPENDIX B

NOISE COMPUTATION TABLES AND NOMOGRAPHS

Project ______________________ Date ______________ Engineer ______________

Step			A	T_M	T_H	A	T_M	T_H	A	T_M	T_H	A	T_M	T_H	A	T_M	T_H	A	T_M	T_H
1	Traffic	Vehicle Volume, V(Vph)																		
2		Vehicle Av. Speed, S(km/h)																		
3		Combined Veh. Vol.*, V_C(Vph)																		
4	Prop.	Observer-Roadway Dist., D_C(m)																		
5	Shielding	Line-of-Sight Dist., L/S(m)																		
6		Barrier Position Dist., P(m)																		
7		Break in Barrier, B(m)																		
8		Angle Subtended, θ (deg)																		
9	Prediction**	Unshield L_{10} Level (dBA)																		
10		Shielding Adjust. (dBA)																		
11		L_{10} at Observer (By Veh. Class)																		
12		L_{10} at Observer–Total																		

Code:

A = Automobiles, T_M = Medium Trucks, T_H = Heavy Trucks

* Applies only when automobile and medium truck average speeds are equal. $V_C = V_A + (10)V_{T_M}$. Otherwise, multiply medium truck volume by 10 and use this volume to compute the Unshielded L_{10} level.

** If automobile-medium truck volume V_C is combined, use L_{10}. Nomograph prediction only once for these two vehicle classes.

FIGURE B-1

Blank noise prediction worksheet. (*Source*: *NCHRP 174*, 1976.)

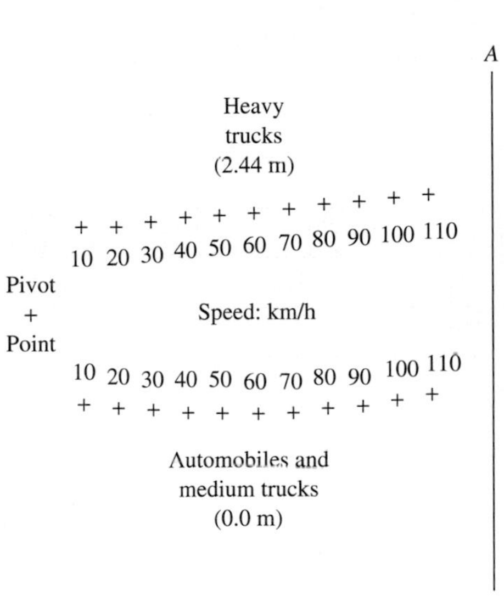

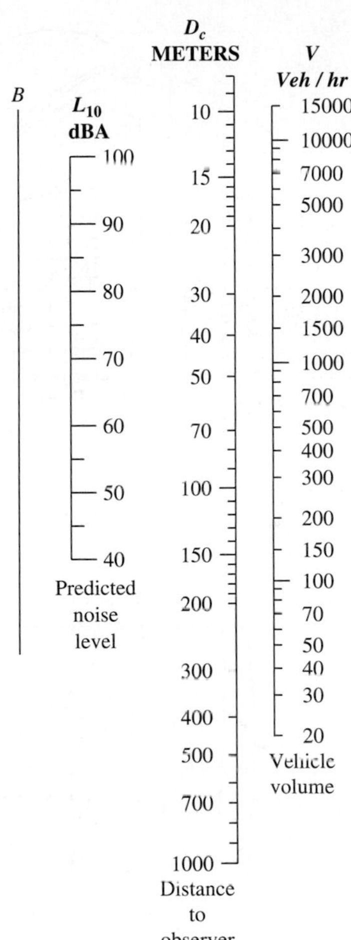

FIGURE B-2
Blank L_{10} nomograph. (*Source*: *NCHRP 174*,1976.)

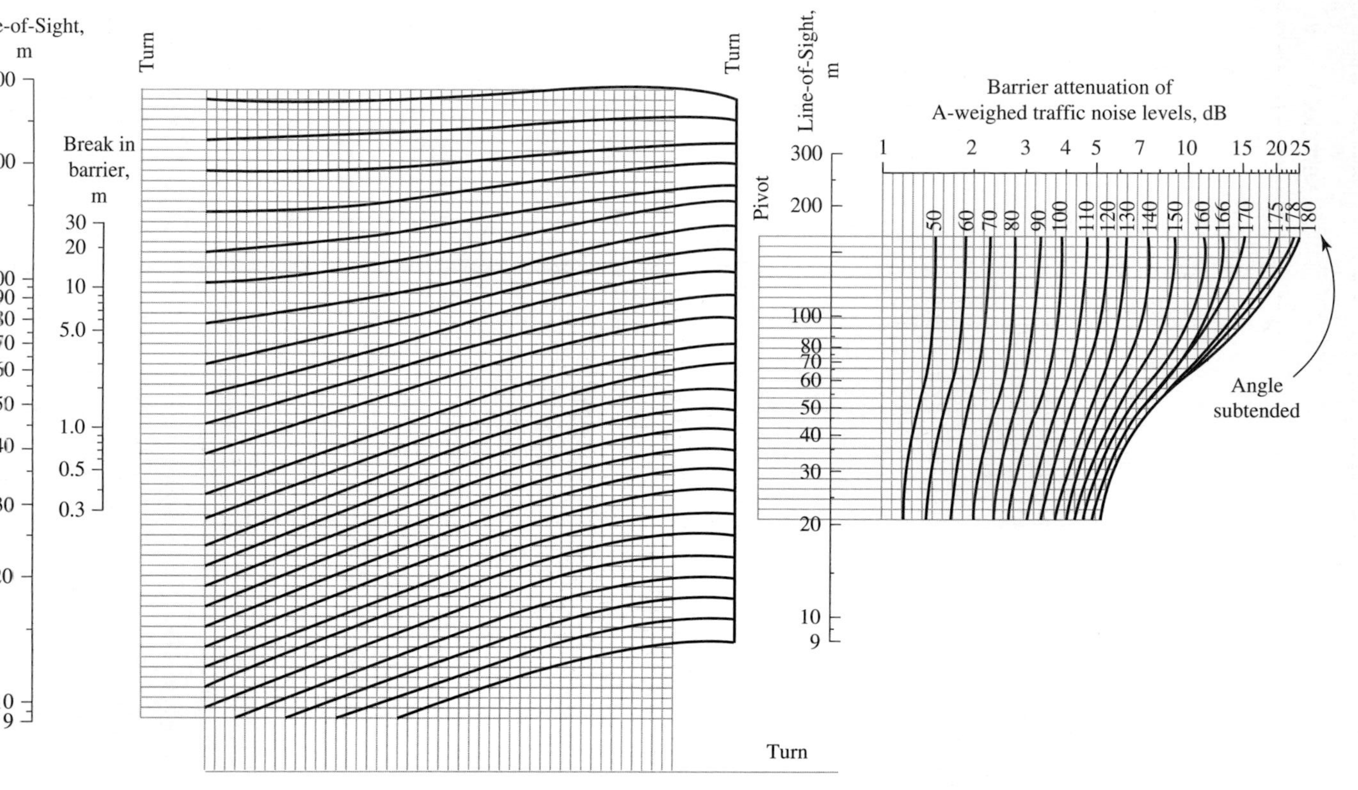

FIGURE B-3
Blank barrier nomograph. (*Source*: *NCHRP 174*, 1976.)

INDEX

Useful conversion factors

Multiply	By	To Obtain
atmosphere (atm)	101.325	kilopascal (kPa)
Calorie (international)	4.1868	Joules (J)
centipoise	10^{-3}	Pa · s
centistoke	10^{-6}	m^2/s
cubic meter (m^3)	35.31	cubic feet (ft^3)
cubic meter	1.308	cubic yard (yd^3)
cubic meter	1,000.00	liter (L)
cubic meter/s	15,850.0	gallons/min (gpm)
cubic meter/s	22.8245	million gal/d (MGD)
cubic meter/m^2	24.545	gallons/sq ft (gal/ft^2)
cubic meter/d · m	80.52	gal/d · ft (gpd/ft)
cubic meter/d · m^2	24.545	gal/d · ft^2 (gpd/ft^2)
cubic meter/d · m^2	1.0	meters/d (m/d)
days (d)	24.00	hours (h)
days (d)	1,440.00	minutes (min)
days (d)	86,400.00	seconds (s)
dyne	10^{-5}	Newtons (N)
erg	10^{-7}	Joules (J)
grains (gr)	6.480×10^{-2}	grams (g)
grains/U.S. gallon	17.118	mg/L
grams (g)	2.205×10^{-3}	pounds mass (lb_m)
hectare (ha)	10^4	m^2
Hertz (Hz)	1	cycle/s
Joule (J)	1	N · m
J/m^3	2.684×10^{-5}	Btu/ft^3
kilogram/m^3 (kg/m^3)	8.346×10^{-3}	lb_m/gal
kilogram/m^3	1.6855	lb_m/yd^3
kilogram/ha (kg/ha)	8.922×10^{-1}	lb_m/acre
kilogram/m^2 (kg/m^2)	2.0482×10^{-1}	lb_m/ft^2
kilometers (km)	6.2150×10^{-1}	miles (mi)
kilowatt (kW)	1.3410	horsepower (hp)
kilowatt-hour	3.600	megajoules (MJ)
liters (L)	10^{-3}	cubic meters (m^3)
liters	1,000.00	milliliters (mL)
liters	2.642×10^{-1}	U.S. gallons
megagrams (Mg)	1.1023	U.S. short tons
meters (m)	3.281	feet (ft)
meters/d (m/d)	2.2785×10^{-3}	ft/min
meters/d	3.7975×10^{-5}	meters/s (m/s)
meters/s (m/s)	196.85	ft/min
meters/s	3.600	km/h
meters/s	2.237	miles/h (mph)
micron (μ)	10^{-6}	meters
milligrams (mg)	10^{-3}	grams (g)
milligrams/L	1	g/m^3
milligrams/L	10^{-3}	kg/m^3
Newton (N)	1	kg · m/s^2
Pascal (Pa)	1	N/m^2
Poise (P)	10^{-1}	Pa · s
square meter (m^2)	2.471×10^{-4}	acres
square meter (m^2)	10.7639	sq ft (ft^2)
square meter/s	6.9589×10^6	gpd/ft
Stoke (St)	10^{-4}	m^2/s
Watt (W)	1	J/s
Watt/cu meter (W/m^3)	3.7978×10^{-2}	hp/1,000 ft^3
Watt/sq meter · °C (W/m^2 · °C)	1.761×10^{-1}	Btu/h · ft^2 · °F

D0605568

Webster's New Explorer Encyclopedic Dictionary

Webster's New Explorer Encyclopedic Dictionary

Created in Cooperation with the Editors of

MERRIAM-WEBSTER

A Division of Merriam-Webster, Incorporated
Springfield, Massachusetts

This 2006 edition published by
Federal Street Press
a Division of Merriam-Webster, Incorporated
P.O. Box 281
Springfield, MA 01102

Federal Street Press books are available for bulk purchase for sales promotion and premium use. For details write the manager of special sales, Federal Street Press, P.O. Box 281, Springfield, MA 01102

ISBN-13 978-1-59695-007-8

ISBN-10 1-59695-007-2

Printed in the United States of America

06 07 08 09 10 5 4 3 2 1

Contents

Preface

This dictionary has been specially created to serve the needs of a broad audience: everyone who uses English to communicate. It is produced in cooperation with the editors of Merriam-Webster, and it draws on the scholarship and reputation of a company that is the oldest and most respected publisher of dictionaries in America.

The book you are holding incorporates all of the essential features of a general purpose dictionary and a good deal more. The heart and soul of any dictionary is its main vocabulary and its ever-expanding stock of new words and uses. New words and senses come into use far more rapidly than older uses die out, and in the front of the book is a section of recently established words and senses, set aside to highlight these new additions to the language.

Entries known to be trademarks or service marks are so labeled and are treated in accordance with a formula approved by the United States Trademark Association. No entry in this dictionary, however, should be regarded as affecting the validity of any trademark or service mark.

Of particular note in this dictionary is the easy-to-read type, the abundance of pictorial illustrations, and the many set-off paragraphs with discussions of the shades of difference among synonyms, pointed discussions of some matters of confused or disputed usage, and brief stories of the interesting histories of certain words, giving more information than can be presented in the etymologies. Another feature that has proved popular with people interested in the histories of words is the date of first recorded use of a word, given before the first definition. Similarly, each letter section of the main vocabulary is introduced by a short history of the letter itself.

Other features meant to make this dictionary more inviting to the eye than dictionaries commonly are include the placing of numbered definitions on a new line, to make them easier to locate, and the spelling out of words typically abbreviated in other dictionaries.

The explantions in the front of this book establish a context for understanding what this dictionary is and how it may be used most effectively. The Explanatory Notes answer the user's questions about the conventions, devices, and techniques by which the editors have been able to present large amounts of information in a single volume. All users of the dictionary are urged to read this section through and then consult it for special information as the need arises.

In order to fit as much information as possible into a convenient size, the dictionary contains less-often referenced material in smaller sections in the back of the book including Abbreviations and Symbols, Biographical Names, Geographical Names, and a Handbook of Style, in which various stylistic conventions are explained.

The publisher is happy to offer this work in the belief that this single-volume reference will serve the user as a reliable resource for many years.

Explanatory Notes

Entries

MAIN ENTRIES

A boldface letter or a combination of such letters, including punctuation marks and diacritics where needed, that is set flush with the left-hand margin of each column of type is a main entry or entry word. The main entry may consist of letters set solid, of letters joined by a hyphen or a diagonal, or of letters separated by one or more spaces:

[1]alone . . . *adjective*

au·to–da–fé . . . *noun*

and/or . . . *conjunction*

automatic pilot *noun*

The material in lightface type that follows each main entry on the same line and on succeeding indented lines explains and justifies its inclusion in the dictionary.

Variation in the styling of compound words in English is frequent and widespread. It is often completely acceptable to choose freely among open, hyphenated, and closed alternatives (as *lifestyle, life-style,* or *life style*). However, to show all the stylings that are found for English compounds would require space that can be better used for other information. So this dictionary limits itself to a single styling for a compound:

peace·mak·er

pell–mell

boom box

When a compound is widely used and one styling predominates, that styling is shown. When a compound is uncommon or when the evidence indicates that two or three stylings are approximately equal in frequency, the styling shown is based on the analogy of parallel compounds.

ORDER OF MAIN ENTRIES

The main entries follow one another in alphabetical order letter by letter without regard to intervening spaces or hyphens: *battle royal* follows *battlement* and *earth-shattering* follows *earthshaking.* Those containing an Arabic numeral are alphabetized as if the numeral were spelled out: *3-D* comes between *three-color* and *three-decker.* Those that often begin with the abbreviation *St.* in common usage have the abbreviation spelled out: *Saint Anthony's fire.*

Full words come before parts of words made up of the same letters. Solid compounds come first and are followed by hyphenated compounds and then open compounds. Lowercase entries come before entries that begin with a capital letter:

semi . . . *noun*

semi- . . . *prefix*

take·out . . . *noun*

take–out . . . *adjective*

take out . . . *transitive verb*

tim·o·thy . . . *noun*

Tim·o·thy . . . *noun*

HOMOGRAPHS

When one main entry has exactly the same written form as another, the two are distinguished by superscript numerals preceding each word:

[1]melt . . . *verb* **[1]pine** . . . *noun*

[2]melt *noun* **[2]pine** *intransitive verb*

Sometimes such homographs are related: the two entries *melt* are derived from the same root. Sometimes there is no relationship: the two entries *pine* are unrelated beyond the accident of spelling. The order of homographs is usually historical: the one first used in English is entered first. A homograph derived from an earlier homograph by functional shift, however, follows its parent immediately, with the result that occasionally one homograph appears ahead of another that is older in usage. For example, of the three entries *kennel* the second (a verb) is derived from the first (a noun). Even though the unrelated third entry *kennel* was used in English many years before the second, it follows the two related entries.

GUIDE WORDS

A pair of guide words is printed at the top of each page. The entries that fall alphabetically between the guide words are found on that page.

It is important to remember that alphabetical order rather than position of an entry on the page determines the selection of guide words. The first guide word is the alphabetically first entry on the page. The second guide word is usually the alphabetically last entry on the page:

game **gamut**

The entry need not be a main entry. Another boldface word—a variant, an inflected form, or a defined or undefined run-on—may be selected as a guide word. For this reason the last printed main entry on a page is not always the last entry alphabetically:

Gentoo **geomagnetism**

On the page where these guide words are used, *geomagnetic storm* is the last printed entry, but *geomagnetism,* an undefined run-on at *geomagnetic,* is the last entry alphabetically and so has been chosen as the second guide word.

All guide words must themselves be in alphabetical order from page to page throughout the dictionary; thus, the alphabetically last entry on a page is not used if it follows alphabetically the first guide word on the next page:

Giemsa stain **gillied**

On the page where these guide words are found, *gillying,* an inflected form at the entry [2]*gillie,* is the last entry alphabetically, but it is not used as the second guide word because it follows alphabetically the entry *gillnet,* which is the first guide word on the next page. To use *gillying* would violate the alphabetical order of guide words from page to page, and so the inflected form *gillied* is the second guide word instead.

END-OF-LINE DIVISION

The centered dots within entry words indicate division points at which a hyphen may be put at the end of a line of print or writing. Thus the noun *pos·si·bil·i·ty* may be ended on one line with:

pos-
possi-
possibil-
possibili-

and continued on the next with:

sibility
bility
ity
ty

Centered dots are not shown after a single initial letter or before a single terminal letter because printers seldom cut off a single letter:

aswirl . . . *adjective*

mouthy . . . *adjective*

idea . . . *noun*

Nor are they shown at second and succeeding homographs unless these differ among themselves:

[1]**re·form** . . . *verb*	[1]**min·ute** . . . *noun*
[2]**reform** *noun*	[2]**minute** *transitive verb*
[3]**reform** *adjective*	[3]**mi·nute** . . . *adjective*

There are acceptable alternative end-of-line divisions just as there are acceptable variant spellings and pronunciations. It is, for example, all but impossible to produce a convincing argument that either of the divisions *aus·ter·i·ty, au·ster·i·ty* is better than the other. But space cannot be taken for entries like *aus·ter·i·ty or au·ster·i·ty,* and *au·s·ter·i·ty* would likely be confusing to many. No more than one division is, therefore, shown for an entry in this dictionary.

Many words have two or more common pronunciation variants, and the same end-of-line division is not always appropriate for each of them. The division *pi·an·ist,* for example, best fits the variant \pē-ˈa-nist\ whereas the division *pi·a·nist* best fits the variant \ˈpē-ə-nist\. In instances like this, the division falling farther to the left is used, regardless of the order of the pronunciations:

pi·a·nist \pē-ˈa-nist, ˈpē-ə-\

A double hyphen ⸗ at the end of a line in this dictionary (as in the etymology at *dally*) stands for a hyphen that belongs at that point in a hyphenated word and that is retained when the word is written as a unit on one line.

VARIANTS

When a main entry is followed by the word *or* and another spelling, the two spellings are equal variants. Both are standard, and either one may be used according to personal inclination:

ocher *or* **ochre**

If two variants joined by *or* are out of alphabetical order, they remain equal variants. The one printed first is, however, slightly more common than the second:

plow *or* **plough**

When another spelling is joined to the main entry by the word *also,* the spelling after *also* is a secondary variant and occurs less frequently than the first:

can·cel·la·tion *also* **can·cel·ation**

Secondary variants belong to standard usage and may be used according to personal inclination. If there are two secondary variants, the second is joined to the first by *or.* Once the word *also* is used to signal a secondary variant, all following variants are joined by *or:*

[1]**Shake·spear·ean** *or* **Shake·spear·ian**
also **Shak·sper·ean** *or* **Shak·sper·ian**

Variants whose spelling places them alphabetically more than a column away from the main entry are entered at their own alphabetical places and usually not at the main entry:

[1]**jibe** . . . *variant of* GIBE

[3]**rime, rimester** *variant of* RHYME, RHYMESTER

Variants having a usage label appear only at their own alphabetical places:

metre . . . *chiefly British variant of* METER

agin . . . *dialect variant of* AGAINST

RUN-ON ENTRIES

A main entry may be followed by one or more derivatives or by a homograph with a different functional label. These are run-on entries. Each is introduced on a new line by a boldface dash and each has a functional label. They are not defined, however, since their meanings are readily derivable from the meaning of the root word:

slay . . . *verb* . . .
— slay·er *noun*

spir·it·ed . . . *adjective* . . .
— spir·it·ed·ly *adverb*
— spir·it·ed·ness *noun*

stac·ca·to . . . *adjective* . . .
— staccato *adverb*
— staccato *noun*

A main entry may be followed by one or more phrases containing the entry word or an inflected form of it. These are also run-on entries. Each is introduced on a new line by a boldface dash but there is no functional label. They are, however, defined since their meanings are more than the sum of the meanings of their elements:

[1]**hole** . . . *noun* . . .
— in the hole : . . .

[1]**live** . . . *verb* . . .
— live it up : . . .

Defined phrases of this sort are run on at the entry constituting the first major element in the phrase. The first major element is ordinarily a verb or a noun, but when these are absent another part of speech may serve instead:

[1]**but** . . . *conjunction* . . .
— but what : . . .

When there are variants, however, the run-on appears at the entry constituting the first major invariable element in the phrase:

[1]**clock** . . . *noun* . . .
— kill the clock *or* **run out the clock**
: . . .

[1]**hand** . . . *noun* . . .
— on all hands *or* **on every hand**
: . . .

A run-on entry is an independent entry with respect to function and status. Labels at the main entry do not apply unless they are repeated.

Attention is called to the definition of *vocabulary entry* in this book. The term *dictionary entry* includes all vocabulary entries as well as all boldface entries in the separate sections of the back matter headed "Abbreviations and Symbols for Chemical Elements," "Foreign Words and Phrases," "Biographical Names," and "Geographical Names."

Pronunciation

Pronunciation is indicated between a pair of reversed virgules \ \ following the entry word. The symbols used are listed in the chart printed on the page facing the first page of the dictionary proper. An abbreviated list appears at the bottom of the third column of each right-hand page of the vocabulary. Explanations of the symbols are given in the Guide to Pronunciation.

SYLLABLES

A hyphen is used in the pronunciation to show syllabic division. These hyphens sometimes coincide with the centered dots in the entry word that indicate end-of-line division; sometimes they do not:

ab•sen•tee \ˌab-sən-ˈtē\

[1]**met•ric** \ˈme-trik\

STRESS

A high-set mark \ˈ\ indicates primary (strongest) stress or accent; a low-set mark \ˌ\ indicates secondary (medium) stress or accent:

heart•beat \ˈhärt-ˌbēt\

The stress mark stands at the beginning of the syllable that receives the stress.

Stress marks are an indication of the relative prominence of the syllables in a word. In running speech the primary stress can vary in English words for several contextual and semantic reasons. Because the variation is so great, this book shows the primary stress of a word in its pronunciation as a single word out of context.

VARIANT PRONUNCIATIONS

The presence of variant pronunciations indicates that not all educated speakers pronounce words the same way. A second-place variant is not to be regarded as less acceptable than the pronunciation that is given first. It may, in fact, be used by as many educated speakers as the first variant, but the requirements of the printed page make one precede the other:

apri•cot \ˈa-prə-ˌkät, ˈā-\

for•eign \ˈfȯr-ən, ˈfär-\

A variant that is appreciably less common than the preceding variant is preceded by the word *also* :

[1]**al•loy** \ˈa-ˌlȯi *also* ə-ˈlȯi\

A variant preceded by *sometimes* is even less common, though it does occur in educated speech:

in•vei•gle \in-ˈvā-gəl *sometimes* -ˈvē-\

Sometimes a regional label precedes a variant:

[1]**great** \ˈgrāt, *Southern also* ˈgre(ə)t\

The label *dial* precedes a variant that is noteworthy or common in a dialect or dialects of American English, but that is not considered to be a standard pronunciation:

ask \ˈask, ˈȧsk; *dialect* ˈaks\

The symbol \÷\ is placed before a pronunciation variant that occurs in educated speech but that is considered by some to be unacceptable:

cu•po•la \ˈkyü-pə-lə, ÷-ˌlō\

This symbol refers only to the immediately following variant and not to subsequent variants separated from it by a comma or a semicolon.

PARENTHESES IN PRONUNCIATIONS

Symbols enclosed by parentheses represent elements that are present in the pronunciation of some speakers but are absent from the pronunciation of other speakers, or elements that are present in some but absent from other utterances of the same speaker:

[1]**twin•kle** \ˈtwiŋ-kəl\ *verb* . . . **twin•kling** \-k(ə-)liŋ\

sat•is•fac•to•ry \ˌsa-təs-ˈfak-t(ə-)rē\

re•sponse \ri-ˈspän(t)s\

Thus, the parentheses at *twinkling* mean that there are some who pronounce the \ə\ between \k\ and \l\ and others who do not pronounce it.

PARTIAL AND ABSENT PRONUNCIATIONS

When a main entry has less than a full pronunciation, the missing part is to be supplied from a pronunciation in a preceding entry or within the same pair of reversed virgules:

cham•pi•on•ship \-ˌship\

Ma•dei•ra \mə-ˈdir-ə, -ˈder-\

The pronunciation of the first three syllables of *championship* is found at the main entry *champion* :

[1]**cham•pi•on** \ˈcham-pē-ən\

The hyphens before and after \ˈder\ in the pronunciation of *Madeira* indicate that both the first and the last parts of the pronunciation are to be taken from the immediately preceding pronunciation.

Partial pronunciations are usually shown when two or more variants have a part in common. When a variation of stress is involved, a partial pronunciation may be terminated at the stress mark which stands at the beginning of a syllable not shown:

di•verse \dī-ˈvərs, də-ˈ, ˈdī-ˌ\

an•cho•vy \ˈan-ˌchō-vē, an-ˈ\

In general, no pronunciation is indicated for open compounds consisting of two or more English words that have own-place entry:

witch doctor *noun*

A pronunciation is shown, however, for any element of an open compound that does not have entry at its own alphabetical place:

Oc·cam's razor \'ä-kəmz-\

sieve of Er·a·tos·the·nes \-ˌer-ə-'täs-thə-ˌnēz\

Only the first entry in a sequence of numbered homographs is given a pronunciation if their pronunciations are the same:

¹re·ward \ri-'wȯrd\

²reward

Pronunciations are shown for obsolete words only if they occur in Shakespeare:

clois·tress \'klȯi-strəs\ *noun* . . . *obsolete*

The pronunciation of unpronounced derivatives and compounds run on at a main entry is a combination of the pronunciation at the main entry and the pronunciation of the suffix or final element as given at its alphabetical place in the vocabulary:

— oval·ness *noun*

— shot in the dark

Thus, the pronunciation of *ovalness* is the sum of the pronunciations given at *oval* and *-ness* ; that of *shot in the dark,* the sum of the pronunciation of the four elements that make up the phrase.

Functional Labels

An italic label indicating a part of speech or some other functional classification follows the pronunciation or, if no pronunciation is given, the main entry. The eight traditional parts of speech are indicated as follows:

¹bold . . . *adjective*

hand·some·ly . . . *adverb*

¹but . . . *conjunction*

oops . . . *interjection*

bo·le·ro . . . *noun*

²under . . . *preposition*

some·one . . . *pronoun*

¹shrink . . . *verb*

If a verb is both transitive and intransitive, the labels *transitive verb* and *intransitive verb* introduce the subdivisions:

flat·ten . . . *verb* . . .
transitive verb . . .
intransitive verb . . .

If there is no subdivision, *transitive verb* or *intransitive verb* takes the place of *verb*:

²fleece *transitive verb*

ap·per·tain . . . *intransitive verb*

Labeling a verb as transitive, however, does not preclude occasional intransitive use (as in absolute constructions).

Other italicized labels used to indicate functional classifications that are not traditional parts of speech are:

poly- *combining form*

-logy *noun combining form*

-iferous *adjective combining form*

super- *prefix*

Gram·my . . . *service mark*

¹-ic *adjective suffix*

²-ward *or* **-wards** *adverb suffix*

-itis *noun suffix*

-ize *verb suffix*

NC-17 . . . *certification mark*

Ly·cra . . . *trademark*

-nd *symbol*

¹may . . . *verbal auxiliary*

gid·dap . . . *verb imperative*

me·thinks . . . *impersonal verb*

Two functional labels are sometimes combined:

zilch . . . *adjective or noun*

afloat . . . *adjective or adverb*

Inflected Forms

In comparison with some other languages English does not have many inflected forms. Of those which it has, several are inflected forms of words belonging to small, closed groups (as the personal pronouns or the demonstratives). These forms can readily be found at their own alphabetical places with a full entry (as *whom,* the objective case form of *who*) or with a cross-reference in small capital letters to another entry (as *those,* the plural form of *that*).

Most other inflected forms, however, are covered explicitly or by implication at the main entry for the base form. These are the plurals of nouns, the principal parts of verbs (the past tense, the past participle when it differs from the past tense, and the present participle), and the comparative and superlative forms of adjectives and adverbs. In general, it may be said that when these inflected forms are created in a manner considered regular in English (as by adding *-s* or *-es* to nouns, *-ed* and *-ing* to verbs, and *-er* and *-est* to adjectives and adverbs) and when it seems that there is nothing about the formation likely to give the dictionary user doubts, the inflected form is not shown in order to save space for information more likely to be sought. Inflected forms are also not shown at undefined run-ons or at some entries bearing a limiting label:

gour·mand . . . *noun* . . .
— gour·man·dize . . . *intransitive verb*

¹fem·i·nine . . . *adjective* . . .
— fem·i·nine·ness . . . *noun*

²lake *noun* . . .
— laky . . . *adjective*

²cote . . . *transitive verb* . . . *obsolete* **:** to pass by

crouse . . . *adjective* . . .
chiefly Scottish **:** BRISK, LIVELY

On the other hand, if the inflected form is created in an irregular way or if the dictionary user is likely to have doubts about it (even though it is formed regularly), the inflected form is shown in boldface, either in full or cut back to a convenient and easily recognizable point. Full details about the kinds of entries at which inflected forms are shown and the kinds at which they are not shown are given in the three following sections.

NOUNS

The plurals of nouns are shown in this dictionary when suffixation brings about a change of final *-y* to *-i-,* when the noun ends in a consonant plus *-o* or in *-ey,* when the noun ends in *-oo,* when the noun has an irregular plural or a zero plural or a foreign plural, when the noun is a compound that pluralizes any element but the last, when a final consonant is doubled, when the noun has variant plurals, and when it is believed that the dictionary user might have reasonable doubts about the spelling of the plural or when the plural is spelled in a way contrary to expectations:

[2]**spy** *noun, plural* **spies**

si·lo . . . *noun, plural* **silos**

[1]**ro·deo** . . . *noun, plural* **ro·de·os**

[2]**shampoo** *noun, plural* **shampoos**

[1]**mouse** . . . *noun, plural* **mice**

moose . . . *noun, plural* **moose**

cri·te·ri·on . . . *noun, plural* **-ria**

son–in–law . . . *noun, plural* **sons–in–law**

[1]**quiz** . . . *noun, plural* **quiz·zes**

[1]**fish** . . . *noun, plural* **fish** *or* **fish·es**

cor·gi . . . *noun, plural* **corgis**

[3]**dry** *noun, plural* **drys**

Cutback inflected forms are used when the noun has three or more syllables:

ame·ni·ty . . . *noun, plural* **-ties**

The plurals of nouns are usually not shown when the base word is unchanged by suffixation, when the noun is a compound whose second element is readily recognizable as a regular free form entered at its own place, or when the noun is unlikely to occur in the plural:

[1]**night** . . . *noun*

[2]**crunch** *noun*

fore·foot . . . *noun*

mo·nog·a·my . . . *noun*

Nouns that are plural in form and that regularly occur in plural construction are labeled *noun plural*:

munch·ies . . . *noun plural*

Nouns that are plural in form but that are not always construed as plurals are appropriately labeled:

ro·bot·ics . . . *noun plural but singular in construction*

two bits *noun plural but singular or plural in construction*

A noun that is singular in construction takes a singular verb when it is used as a subject; a noun that is plural in construction takes a plural verb when it is used as a subject.

VERBS

The principal parts of verbs are shown in this dictionary when suffixation brings about a doubling of a final consonant or an elision of a final *-e* or a change of final *-y* to *-i-*, when final *-c* changes to *-ck* in suffixation, when the verb ends in *-ey*, when the inflection is irregular, when there are variant inflected forms, and when it is believed that the dictionary user might have reasonable doubts about the spelling of an inflected form or when the inflected form is spelled in a way contrary to expectations:

[2]**snag** *transitive verb* **snagged; snag·ging**

[1]**move** . . . *verb* **moved; mov·ing**

[1]**cry** . . . *verb* **cried; cry·ing**

[2]**frolic** *intransitive verb* **frol·icked; frol·ick·ing**

[1]**sur·vey** . . . *verb* **sur·veyed; sur·vey·ing**

[1]**drive** . . . *verb* **drove** . . . ; **driv·en** . . . ; **driv·ing**

[2]**bus** *verb* **bused** *or* **bussed; bus·ing** *or* **bus·sing**

[2]**visa** *transitive verb* **vi·saed** . . . ; **vi·sa·ing**

[2]**chagrin** *transitive verb* **cha·grined** . . . ; **cha·grin·ing**

The principal parts of a regularly inflected verb are shown when it is desirable to indicate the pronunciation of one of the inflected forms:

learn . . . *verb* **learned** \'lərnd, 'lərnt\; **learn·ing**

rip·en . . . *verb* **rip·ened; rip·en·ing** \'rī-pə-niŋ, 'rīp-niŋ\

Cutback inflected forms are often used when the verb has three or more syllables, when it is a disyllable that ends in *-l* and has variant spellings, and when it is a compound whose second element is readily recognized as an irregular verb:

elim·i·nate . . . *transitive verb* **-nat·ed; -nat·ing**

[3]**quarrel** *intransitive verb* **-reled** *or* **-relled; -rel·ing** *or* **-rel·ling**

[1]**re·take** . . . *transitive verb* **-took** . . . ; **-tak·en** . . . ; **-tak·ing**

The principal parts of verbs are usually not shown when the base word is unchanged by suffixation or when the verb is a compound whose second element is readily recognizable as a regular free form entered at its own place:

[1]**jump** . . . *verb*

pre·judge . . . *transitive verb*

Another inflected form of English verbs is the third person singular of the present tense, which is regularly formed by the addition of *-s* or *-es* to the base form of the verb. This inflected form is not shown except at a handful of entries (as *have* and *do*) for which it is in some way anomalous.

ADJECTIVES & ADVERBS

The comparative and superlative forms of adjectives and adverbs are shown in this dictionary when suffixation brings about a doubling of a final consonant or an elision of a final *-e* or a change of final *-y* to *-i-*, when the word ends in *-ey*, when the inflection is irregular, and when there are variant inflected forms:

[1]**red** . . . *adjective* **red·der; red·dest**

[1]**tame** . . . *adjective* **tam·er; tam·est**

[1]**kind·ly** . . . *adjective* **kind·li·er; -est**

[1]**ear·ly** . . . *adverb* **ear·li·er; -est**

dic·ey . . . *adjective* **dic·i·er; -est**

[1]**good** . . . *adjective* **bet·ter** . . . ; **best**

[1]**bad** . . . *adjective* **worse** . . . ; **worst**

[1]**far** . . . *adverb* **far·ther** . . . *or* **fur·ther** . . . ; **far·thest** *or* **fur·thest**

The superlative forms of adjectives and adverbs of two or more syllables are usually cut back:

[3]**fancy** *adjective* **fan•ci•er; -est**

The comparative and superlative forms of regularly inflected adjectives and adverbs are shown when it is desirable to indicate the pronunciation of the inflected forms:

[1]**young** . . . *adjective* **youn•ger** \ˈyəŋ-gər\; **youn•gest** \ˈyəŋ-gəst\

The inclusion of inflected forms in *-er* and *-est* at adjective and adverb entries means nothing more about the use of *more* and *most* with these adjectives and adverbs than that their comparative and superlative degrees may be expressed in either way; *lazier* or *more lazy; laziest* or *most lazy.*

At a few adjective entries only the superlative form is shown:

[3]**mere** . . . *adjective, superlative* **mer•est**

The absence of the comparative form indicates that there is no evidence of its use.

The comparative and superlative forms of adjectives and adverbs are not shown when the base word is unchanged by suffixation or when the word is a compound whose second element is readily recognizable as a regular free form entered at its own place:

[1]**near** . . . *adverb*

un•wary . . . *adjective*

The comparative and superlative forms of adverbs are not shown when they are identical with the inflected forms of a preceding adjective homograph:

[1]**hot** . . . *adjective* **hot•ter; hot•test**

[2]**hot** *adverb*

Capitalization

Most entries in this dictionary begin with a lowercase letter. A few of these have an italicized label *often capitalized,* which indicates that the word is as likely to be capitalized as not, that it is as acceptable with an uppercase initial as it is with one in lowercase. Some entries begin with an uppercase letter, which indicates that the word is usually capitalized. The absence of an initial capital or of an *often capitalized* label indicates that the word is not ordinarily capitalized:

lunk•head . . . *noun*

gar•gan•tuan . . . *adjective, often capitalized*

Mo•hawk . . . *noun*

The capitalization of entries that are open or hyphenated compounds is similarly indicated by the form of the entry or by an italicized label:

obstacle course *noun*

neo-Dar•win•ian . . . *adjective, often N capitalized*

off-off-Broadway *noun, often both Os capitalized*

un-Amer•i•can . . . *adjective*

Dutch oven *noun*

Old Glory *noun*

A word that is capitalized in some senses and lowercase in others shows variations from the form of the main entry by the use of italicized labels at the appropriate senses:

re•nais•sance . . . *noun* . . .
1 *capitalized* . . .
2 *often capitalized*

Shet•land . . . *noun* . . .
2 *often not capitalized*

Trin•i•ty . . . *noun* . . .
2 *not capitalized*

Attributive Nouns

The italicized label *often attributive* placed after the functional label *noun* indicates that the noun is often used as an adjective equivalent in attributive position before another noun:

gold . . . *noun, often attributive*

busi•ness . . . *noun, often attributive*

Examples of the attributive use of these nouns are *gold chain* and *business ethics.*

While any noun may occasionally be used attributively, the label *often attributive* is limited to those having broad attributive use. This label is not used when an adjective homograph (as *iron* or *paper*) is entered. And it is not used at open compounds (as *X ray*) that may be used attributively with an inserted hyphen (as in *X-ray therapy*).

Etymology

The matter in square brackets preceding the definition is the etymology. Meanings given in roman type within these brackets are not definitions of the entry, but are meanings of the Middle English, Old English, or non-English words within the brackets.

The etymology traces a vocabulary entry as far back as possible in English (as to Old English), tells from what language and in what form it came into English, and (except in the case of such words outside the general vocabulary of English as *bascule* and *zloty*) traces the pre-English source as far back as possible if the source is an Indo-European language. These etyma are printed in italics.

OLD, MIDDLE, AND MODERN ENGLISH

The etymology usually gives the Middle English and the Old English forms of words in the following style:

[1]**nap** . . . *intransitive verb* . . . [Middle English *nappen,* from Old English *hnappian* . . .]

[1]**old** . . . *adjective* [Middle English, from Old English *eald* . . .]

An etymology in which a word is traced back to Middle English but not to Old English indicates that the word is found in Middle English but not in those texts that have survived from the Old English period:

[1]**slab** . . . *noun* [Middle English *slabbe*]

[1]**stale** . . . *adjective* . . . [Middle English, aged (of ale), not fresh; akin to Middle Dutch *stel* stale]

An etymology in which a word is traced back directly to Old English with no intervening mention of Middle English indicates that the word has not survived continuously from Old English times to the present. Rather, it died out after the Old English period and has been revived in modern times:

ge•mot . . . *noun* [Old English *gemōt* . . .]

thegn . . . *noun* [Old English . . .]

An etymology is not usually given for a word created in English by the combination of existing constituents or by functional shift. This indicates that the identity of the constituents is expected to be self-evident to the user.

book•shelf . . . *noun* . . .
: an open shelf for holding books

[1]**fire•proof** . . . *adjective* . . .
: proof against or resistant to fire

off–put•ting . . . *adjective* . . .
: that puts off **:** REPELLENT, DISCONCERTING

penal code *noun* . . . **:** a code of laws concerning crimes and offenses and their punishment

[3]**stalk** *noun* . . .
1 : the act of stalking

In the case of a family of words obviously related to a common English word but differing from it by containing various easily recognizable suffixes, an etymology is usually given only at the base word, even though some of the derivatives may have been formed in a language other than English:

[1]**equal** . . . *adjective* [Middle English, from Latin *aequalis*, from *aequus* level, equal]

1 a (1) **:** of the same measure, quantity, amount, or number as another

equal•i•ty . . . *noun* . . .
1 : the quality or state of being equal

equal•ize . . . *transitive verb*
1 : to make equal

While *equalize* was formed in Modern English, *equality* was actually borrowed into Middle English (via Middle French) from Latin *aequalitas*.

LANGUAGES OTHER THAN ENGLISH

The etymology gives the language from which words borrowed into English have come. It also gives the form or a transliteration of the word in that language if the form differs from that in English:

[1]**mar•ble** . . . *noun* [Middle English, from Old French *marbre*, from Latin *marmor*, from Greek *marmaros*]

pome•gran•ate . . . *noun* [Middle English *poumgrenet*, from Middle French *pomme grenate*, literally, seedy apple]

souk . . . *noun* [Arabic *sūq* market]

In a few cases the expression "ultimately from" replaces the more usual "from." This expression indicates that one or more intermediate steps have been omitted in tracing the derivation of the form preceding the expression from the form following it:

tri•lo•bite . . . *noun* [ultimately from Greek *trilobos* three-lobed, from *tri-* + *lobos* lobe]

Words cited from certain American Indian languages and from some other languages that are infrequently printed have been rendered with the phonetic symbols used by scholars of those languages. These symbols include the following: a raised dot to the right of a vowel letter to mark vowel length; a hook below a vowel letter to mark nasality; an apostrophe over a consonant letter to mark glottal release; a superscript *w* to the right of a consonant letter to mark labialization; the symbol ɔ to render \ȯ\; the symbol ɨ to render a high central vowel; the Greek letters β, δ, and γ to render voiced labial, dental, and velar fricatives; the symbol *x* to render \k̲\; the symbol ʔ to render a glottal stop; and the symbol ƛ ("crossed lambda") for a voiceless lateral affricate. Examples of these symbols can be found at etymologies for the words *Athabascan, babassu, coho, geoduck, muskellunge, obeah, potlatch,* and *sego lily.*

WORDS OF UNKNOWN ORIGIN

When the source of a word appearing as a main entry is unknown, the expression "origin unknown" is usually used. Only in exceptional circumstances (as with some ethnic names) does the absence of an etymology mean that it has not been possible to furnish an informative etymology. More often, it means that no etymology is believed to be necessary. This is the case, for instance, with most of the entries identified as variants and with many derivatives.

ETYMOLOGIES OF TECHNICAL WORDS

Much of the technical vocabulary of the sciences and other specialized studies consists of words or word elements that are current in two or more languages, with only such slight modifications as are necessary to adapt them to the structure of the individual language in each case. Many words and word elements of this kind have become sufficiently a part of the general vocabulary of English as to require entry in an abridged dictionary. Because of the vast extent of the relevant published material in many languages and in many scientific and other specialized fields, it is impracticable to ascertain the language of origin of every such term. Yet it would not be accurate to formulate a statement about the origin of any such term in a way that could be interpreted as implying that it was coined in English. Accordingly, whenever a term that is entered in this dictionary belongs recognizably to this class of internationally current terms and whenever no positive evidence is at hand to show that it was coined in English, the etymology recognizes its international status and the possibility that it originated elsewhere than in English by use of the label International Scientific Vocabulary:

mega•watt . . . *noun* [International Scientific Vocabulary]

phy•lo•ge•net•ic . . . *adjective* [International Scientific Vocabulary, from New Latin *phylogenesis* . . .]

[1]**-ol** *noun suffix* [International Scientific Vocabulary, from *alcohol*]

COMPRESSION OF INFORMATION

An etymology giving the name of a language (including Middle English or Old English) and not giving the foreign (or Middle English or Old English) form indicates that this form is the same as that of the entry word:

ka•pok . . . *noun* [Malay]

[1]**po•grom** . . . *noun* [Yiddish, from Russian . . .]

> **[1]dumb** . . . *adjective* [Middle English, from Old English . . .]

An etymology giving the name of a language (including Middle English or Old English) and the form in that language but not giving the foreign (or Middle English or Old English) meaning indicates that this meaning is the same as that expressed in the first definition in the entry:

> **[1]wea•ry** . . . *adjective* . . . [Middle English *wery,* from Old English *wērig* . . .] . . .
> **1 :** exhausted in strength . . .

When a word from a foreign language (or Middle English or Old English) is a key element in the etymologies of several related entries that are found close together, the meaning of the word is usually given at only one of the entries:

> **ve•lo•ce** . . . *adverb or adjective* [Italian, from Latin *veloc-, velox*]

> **ve•loc•i•pede** . . . *noun* [French *vélocipède,* from Latin *veloc-, velox* + *ped-, pes* foot — more at FOOT]

> **ve•loc•i•ty** . . . *noun* . . . [Middle French *velocité,* from Latin *velocitat-, velocitas,* from *veloc-, velox* quick; probably akin to Latin *vegēre* to enliven — more at WAKE]

When an etymology includes the expression "by alteration" and the altered form is not cited, the form is the term given in small capital letters as the definition:

> **crit•ter** . . . *noun* [by alteration] . . .
> **:** CREATURE

When the origin of a word is traced to the name of a person or place not further identified, additional information may be found in the Biographical Names or Geographical Names section in the back matter:

> **far•ad** . . . *noun* [Michael *Faraday*]

> **jodh•pur** . . . *noun* [*Jodhpur,* India]

RELATED WORDS

When a word of Indo-European origin has been traced back to the earliest language in which it is attested, words descended from the same Indo-European base in other languages (especially Old High German, Latin, Greek, and Sanskrit) are usually given:

> **na•vel** . . . *noun* [Middle English, from Old English *nafela;* akin to Old High German *nabalo* navel, Latin *umbilicus,* Greek *omphalos*]

> **[1]wind** . . . *noun* . . . [Middle English, from Old English; akin to Old High German *wint* wind, Latin *ventus,* Greek *aēnai* to blow, Sanskrit *vāti* it blows]

Sometimes, however, to avoid space-consuming repetition, the expression "more at" directs the user to another entry where the cognates are given:

> **ho•ly** . . . *adjective* . . . [Middle English, from Old English *hālig;* akin to Old English *hāl* whole — more at WHOLE]

Besides the use of "akin to" to denote relatedness, some etymologies make special use of "akin to" as part of a longer formula "of — origin; akin to —." This formula indicates that a word was borrowed from some language belonging to a group of languages whose name is inserted in the blank before the word *origin,* that it is impossible to say that the word in question is a borrowing of a particular attested word in a particular language of the source group, and that the form cited in the blank after the expression *akin to* is related to the word in question as attested within the source group:

> **ba•nana** . . . *noun* . . . [Spanish or Portuguese; Spanish, from Portuguese, of African origin; akin to Wolof *banăna* banana]

> **[2]brier** *noun* [French *bruyère* heath, from Middle French *bruiere,* from (assumed) Vulgar Latin *brucaria,* from Late Latin *brucus* heather, of Celtic origin; akin to Old Irish *froech* heather; akin to Greek *ereikē* heather]

This last example shows the two contrasting uses of "akin to." The word cited immediately after "of Celtic origin; akin to" is an attested Celtic word descended from the same etymon as the unattested Celtic source of the Latin word. The word cited after the second "akin to" is evidence that the Celtic etymon has deeper relations within Indo-European.

Word History Paragraphs

Word history paragraphs are provided for a number of words with especially interesting origins. Such paragraphs explain word origins in greater detail than is possible in etymologies. A brown-shaded diamond after the last definition of a word signals that a word history paragraph discussing that word can be found in the right-hand column on that page or on the facing page to the right:

> **den•im** . . . *noun* [French *(serge) de Nîmes* serge of Nîmes, France] (1965)
> **1 a :** a firm durable twilled usually cotton fabric woven with colored warp and white filling threads **b :** a similar fabric woven in colored stripes
> **2** *plural* **:** overalls or trousers usually of blue denim ◆

The paragraph itself is set on a light-brown background:

> ◇ WORD HISTORY
> **denim** It is not unusual for fabrics to be named after the places where they were originally made. Nîmes is a city in France known traditionally for its textiles. A heavy serge produced there became known as *serge de Nîmes,* meaning literally, "serge from Nîmes." The "s" in *Nîmes* is not pronounced in French, so when the name of the fabric came into English in the late 17th century, it was often written *serge de Nim* and later *serge denim.* In time the latter was shortened to *denim.*

When a word is included in the word history paragraph of a different entry, the main entry for that word is followed by a run-on "***word history*** see ______" which refers to the entry where the word history paragraph appears.

Dates

At most main entries a date will be found enclosed in parentheses before the boldface colon or number that introduces the first sense:

> **som•bre•ro** . . . *noun* . . . [Spanish, from *sombra* shade] (1599)

: a high-crowned hat of felt or straw with a very wide brim worn especially in the Southwest and Mexico

This is the date of the earliest recorded use in English, as far as it could be determined, of the sense which the date precedes. Several caveats are appropriate at this point. First, a few classes of main entries that are not complete words (as prefixes, suffixes, and combining forms) or are not generic words (as trademarks and names of figures from mythology) are not given dates. Second, the date given applies only to the first sense of the word entered in this dictionary and not necessarily to the word's very earliest meaning in English. Many words, especially those with long histories, have obsolete, archaic, or uncommon senses that are not entered in this dictionary, and such senses have been excluded from consideration in determining the date:

> [1]**slur** . . . *noun* [obsolete English dialect *slur* thin mud, from Middle English *sloor;* akin to Middle High German *slier* mud] (1609)
> **1 a :** an insulting or disparaging remark or innuendo **:** ASPERSION

The 1609 date is for a sense of *slur* synonymous with *aspersion. Slur* also has an obsolete sense, "thin mud," that was recorded as early as the fifteenth century; but since this sense is not entered, it is ignored for purposes of dating. Third, the printed date should not be taken to mark the very first time that the word—or even the sense—was used in English. Many words were certainly in spoken use for decades or even longer before they passed into the written language. The date is for the earliest written or printed use that the editors have been able to discover. This fact means further that any date is subject to change as evidence of still earlier use may emerge, and many dates given now can confidently be expected to yield to others in future printings and editions.

A date will appear in one of three different styles:

> **put·tee** . . . *noun* [Hindi *paṭṭī* strip of cloth, from Sanskrit *paṭṭikā*] (1886)
> **1 :** a cloth strip wrapped around the leg from ankle to knee

> [1]**moon·light** . . . *noun* (14th century) **:** the light of the moon

> [1]**thrall** . . . *noun* [Middle English *thral;* from Old English *thræl,* from Old Norse *thræll*] (before 12th century)
> **1 a :** a servant slave **:** BONDMAN; *also* **:** SERF

The style that names a year (as 1886) is the one used for the period from the sixteenth century to the present. The style that names only a century (as 14th century) is the one used for the period from the twelfth century through the fifteenth century, a span that roughly approximates the period of Middle English. The style (before 12th century) is used for the period before the twelfth century back to the earliest records of English, a span that approximates the period of Old English. Words first attested after 1500 can usually be dated to a single year because the precise dates of publication of modern printed texts are known. If a word must be dated from a modern text of uncertain chronology, it will be assigned the latest possible date of the text's publication prefixed by the word *circa*. For words from the Old and Middle English periods the examples of use on which the dates depend very often occur in manuscripts which are themselves of uncertain date and which may record a text whose date of composition is highly conjectural. To date words from these periods by year would frequently give a quite misleading impression of the state of our knowledge, and so the broader formulas involving centuries are used instead.

Each date reflects a particular instance of the use of a word, most often within a continuous text. In cases where the earliest appearance of a word dated by year is not from continuous text but from a source (as a dictionary or glossary) that defines or explains the word instead of simply using it, the year is preceded by *circa*:

> **magnesium sulfate** *noun* (circa 1890)
> **:** a sulfate of magnesium . . .

In such instances, *circa* indicates that while the source providing the date attests that the word was in use in the relevant sense at that time, it does not offer an example of the normal use of the word and thus gives no better than an approximate date for such use. For the example above no use has so far been found that is earlier than its appearance (spelled *magnesium sulphate*) as an entry in Webster's International Dictionary, published in 1890, so the date is given with the qualifying abbreviation.

Usage

USAGE LABELS

Three types of status labels are used in this dictionary—temporal, regional, and stylistic—to signal that a word or a sense of a word is not part of the standard vocabulary of English.

The temporal label *obsolete* means that there is no evidence of use since 1755:

> [1]**per·du** . . . *noun* . . .
> *obsolete*

> **gov·ern·ment** . . . *noun* . . .
> **2** *obsolete*

The label *obsolete* is a comment on the word being defined. When a thing, as distinguished from the word used to designate it, is obsolete, appropriate orientation is usually given in the definition:

> [1]**cat·a·pult** . . . *noun* . . .
> **1 :** an ancient military device for hurling missiles

> **far·thin·gale** . . . *noun* . . .
> **:** a support (as of hoops) worn especially in the 16th century beneath a skirt to expand it at the hipline

The temporal label *archaic* means that a word or sense once in common use is found today only sporadically or in special contexts:

> [1]**goody** . . . *noun* . . .
> *archaic*

> **lon·gi·tude** . . . *noun* . . .
> **2** *archaic*

A word or sense limited in use to a specific region of the U.S. has a regional label. Some regional labels correspond loosely to areas defined in Hans Kurath's *Word Geography of the Eastern United States.* The adverb *chiefly* precedes a label when the word has some currency outside the specified region, and a double label is used to indicate considerable currency in each of two specific regions:

> **pung** . . . *noun* . . .
> *New England*

> **ban·quette** . . . *noun* . . .
> **1** . . . **b** *Southern*

> [3]**pas·tor** . . . *noun* . . .
> *chiefly Southwest*

> **do·gie** . . . *noun* . . .
> *chiefly West*

> **gal·lery** . . . *noun* . . .
> **2** . . . **b** *Southern & Midland*

[1]pot·latch . . . *noun* . . .
2 *Northwest*

smear·case . . . *noun* . . .
chiefly Midland

crul·ler . . . *noun* . . .
2 *Northern & Midland*

Words current in all regions of the U.S. have no label.

A word or sense limited in use to one of the other countries of the English-speaking world has an appropriate regional label:

cutty sark . . . *noun* . . .
chiefly Scottish

lar·ri·kin . . . *noun* . . .
chiefly Australian

in·da·ba . . . *noun* . . .
chiefly South African

spal·peen . . . *noun* . . .
chiefly Irish

[1]bon·net . . . *noun* . . .
2 a *British*

book off *intransitive verb* . . .
Canadian

[1]din·kum . . . *adjective* . . .
Australian & New Zealand

gar·ron . . . *noun* . . .
Scottish & Irish

The label *dialect* indicates that the pattern of use of a word or sense is too complex for summary labeling: it usually includes several regional varieties of American English or of American and British English:

least·ways . . . *adverb* . . .
dialect

The label *dialect British* indicates currency in several dialects of the Commonwealth; *dialect English* indicates currency in one or more provincial dialects of England:

bo·gle . . . *noun* . . .
dialect British

[1]hob . . . *noun* . . .
1 *dialect English*

The stylistic label *slang* is used with words or senses that are especially appropriate in contexts of extreme informality, that usually have a currency not limited to a particular region or area of interest, and that are composed typically of shortened or altered forms or extravagant or facetious figures of speech:

[4]barb *noun* . . .
slang **:** BARBITURATE

[2]skin·ny *noun* . . .
slang **:** inside information **:** DOPE

main squeeze *noun* . . .
slang **:** one's principal romantic partner

There is no satisfactory objective test for slang, especially with reference to a word out of context. No word, in fact, is invariably slang, and many standard words can be given slang applications.

The stylistic label *nonstandard* is used for a few words or senses that are disapproved by many but that have some currency in reputable contexts:

learn . . . *verb* . . .
2 a *nonstandard*

ir·re·gard·less . . . *adverb* . . .
nonstandard

The stylistic label *substandard* is used for those words or senses that conform to a widespread pattern of usage that differs in choice of word or form from that of the prestige group of the community:

is . . . *present 3d singular of* BE, *dialect present 1st & 2d singular of* BE, *substandard present plural of* BE

A subject label or guide phrase is sometimes used to indicate the specific application of a word or sense:

knock·about . . . *adjective* . . .
3 *of a sailing vessel*

[2]break *noun* . . .
5 . . . **d** *mining*

[2]up *adjective* . . .
5 *of a quark*

In general, however, subject orientation is given in the definition:

Di·do . . . *noun* . . .
: a legendary queen of Carthage in Virgil's *Aeneid* who kills herself when Aeneas leaves her

je·té . . . *noun* . . .
: a springing jump in ballet made from one foot to the other in any direction

ILLUSTRATIONS OF USAGE

Definitions are sometimes followed by verbal illustrations that show a typical use of the word in context. These illustrations are enclosed in angle brackets:

[1]key . . . *noun* . . .
3 a . . . <the *key* to a riddle>

nary . . . *adjective*
. . . <*nary* a person wanted to go>

[2]plummet *intransitive verb* . . .
2 . . . <prices *plummeted*>

[1]true . . . *adjective* . . .
8 . . . <in the *truest* sense>

turn off *transitive verb* . . .
4 . . . <*turn* the water *off*>

Illustrative quotations are also used to show words in typical contexts:

con·flict·ed . . . *adjective*
. . . <this unhappy and *conflicted* modern woman —John Updike>

Omissions in quotations are indicated by suspension points:

alien·ation . . . *noun*
. . . <*alienation* . . . from the values of one's society and family —S.L. Halleck>

USAGE NOTES

Definitions are sometimes followed by usage notes that give supplementary information about such matters as idiom, syntax, semantic relationship, and status. A usage note is introduced by a lightface dash:

[2]cry *noun* . . .
11 : . . . — usually used in the phrase *a far cry*

[2]drum . . . *transitive verb* . . .
2 : . . . — usually used with *out*

¹so . . . *adverb* . . .
1 a : . . . — often used as a substitute for a preceding clause

¹sfor·zan·do . . . *adjective or adverb* . . . **:** . . . — used as a direction in music

grin·go . . . *noun* . . .
: . . . — often used disparagingly

pissed . . . *adjective* . . .
1 . . . **:** . . . — sometimes considered vul gar

hajji . . . *noun* . . .
: . . . — often used as a title

Two or more usage notes are separated by a semicolon:

²thine *pronoun* . . .
: that which belongs to thee — used without a following noun as a pronoun equivalent in meaning to the adjective *thy*; used especially in ecclesiastical or literary language and still surviving in the speech of Friends especially among themselves

Sometimes a usage note calls attention to one or more terms with the same denotation as the main entry:

water moccasin *noun* . . .
1 : a venomous semiaquatic pit viper (*Agkistrodon piscivorus*) chiefly of the southeastern U.S. that is closely related to the copperhead — called also *cottonmouth, cottonmouth moccasin*

The called-also terms are shown in italic type. If such a term falls alphabetically more than a column away from the main entry, it is entered at its own place with the sole definition being a synonymous cross-reference to the entry where it appears in the usage note:

cotton·mouth . . . *noun* . . .
: WATER MOCCASIN

cottonmouth moccasin *noun* . . .
: WATER MOCCASIN

Sometimes a usage note is used in place of a definition. Some function words (as conjunctions and prepositions) have little or no semantic content; most interjections express feelings but are otherwise untranslatable into meaning; and some other words (as oaths and honorific titles) are more amenable to comment than to definition:

¹of . . . *preposition* . . .
1 — used as a function word to indicate a point of reckoning

¹oyez . . . *verb imperative* . . .
— used by a court or public crier to gain attention before a proclamation

¹or . . . *conjunction* . . .
1 — used as a function word to indicate an alternative

gol·ly . . . *interjection* . . .
— used as a mild oath or to express surprise

sir . . . *noun* . . .
2 a — used as a usually respectful form of address

USAGE PARAGRAPHS

Brief usage paragraphs are provided for a number of entries that are considered to present problems of confused or disputed usage. A usage paragraph typically summarizes the historical background of the item and its associated body of opinion, compares these with available evidence of current usage, and often adds a few words of suitable advice for the dictionary user.

A small green-shaded square after the last definition of a word signals that a usage paragraph discussing that word can be found in the right-hand column on that page or on the facing page to the right:

ag·gra·vate . . . *transitive verb* . . .
1 *obsolete* **a :** to make heavy **:** BURDEN **b :** INCREASE
2 : to make worse, more serious, or more severe **:** intensify unpleasantly <problems have been *aggravated* by neglect>
3 a : to rouse to displeasure or anger by usually persistent and often petty goading **b :** to produce inflammation in ■

The paragraph itself is set on a green background. Where appropriate, discussion in the paragraph is keyed by sense number to the definition of the meaning in question. Most paragraphs incorporate appropriate verbal illustrations and illustrative quotations to clarify and exemplify the points being made:

> □ USAGE
> **Aggravate** Although *aggravate* has been used in sense 3a since the 17th century, it has been the object of disapproval only since about 1870. It is used in expository prose <when his silly conceit . . . about his not-very-good early work has begun to *aggravate* us —William Styron> but seems to be more common in speech and casual writing <a good profession for him, because bus drivers get *aggravated* —Jackie Gleason (interview, 1986)> <& now this letter comes to *aggravate* me a thousand times worse —Mark Twain (letter, 1864)>. Sense 2 is far more common than sense 3a in published prose. Such is not the case, however, with *aggravation* and *aggravating. Aggravation* is used in sense 3 somewhat more than in its earlier senses; *aggravating* has practically no use other than to express annoyance.

When a second word is also discussed in a paragraph, the main entry for that word is followed by a run-on "***usage*** see______" which refers to the entry where the paragraph may be found:

²af·fect . . . *verb*
. . . ***usage*** see EFFECT

When a word that has its own usage paragraph is also discussed in a usage paragraph at another entry, the word's usage paragraph is followed by a cross-reference to the other entry.

Definitions

DIVISION OF SENSES

A boldface colon is used in this dictionary to introduce a definition:

¹coo·per . . . *noun* . . .
: one that makes or repairs wooden casks or tubs

It is also used to separate two or more definitions of a single sense:

un·cage . . . *transitive verb* . . .
: to release from or as if from a cage **:** free from restraint

Boldface Arabic numerals separate the senses of a word that has more than one sense:

> **[2]savage** *noun* . . .
> **1 :** a person belonging to a primitive society
> **2 :** a brutal person
> **3 :** a rude or unmannerly person

Boldface lowercase letters separate the subsenses of a word:

> **[1]grand** . . . *adjective* . . .
> **5 a :** LAVISH, SUMPTUOUS . . . **b :** marked by a regal form and dignity **c :** fine or imposing in appearance or impression **d :** LOFTY, SUBLIME

Lightface numerals in parentheses indicate a further division of subsenses:

> **take out** . . . *transitive verb* . . .
> **1 a** (1) **:** DEDUCT, SEPARATE (2) **:** EXCLUDE, OMIT (3) **:** WITHDRAW, WITHHOLD

A lightface colon following a definition and immediately preceding two or more subsenses indicates that the subsenses are subsumed by the preceding definition:

> **[2]crunch** *noun* . . .
> **3 :** a tight or critical situation: as **a :** a critical point in the buildup of pressure between opposing elements . . . **b :** a severe economic squeeze . . . **c :** SHORTAGE

> **se•quoia** . . . *noun* . . .
> **:** either of two huge coniferous California trees of the bald cypress family that may reach a height of over 300 feet (90 meters): **a :** BIG TREE **b :** REDWOOD 3a

The word *as* may or may not follow the lightface colon. Its presence (as at [2]*crunch*) indicates that the following subsenses are typical or significant examples. Its absence (as at *sequoia*) indicates that the subsenses which follow are exhaustive.

The system of separating the various senses of a word by numerals and letters is a lexical convenience. It reflects something of their semantic relationship, but it does not evaluate senses or set up a hierarchy of importance among them.

Sometimes a particular semantic relationship between senses is suggested by the use of one of four italic sense dividers: *especially, specifically, also,* or *broadly.*

The sense divider *especially* is used to introduce the most common meaning subsumed in the more general preceding definition:

> **[2]slick** *adjective* . . .
> **3 a :** characterized by subtlety or nimble wit **:** CLEVER; *especially* **:** WILY

The sense divider *specifically* is used to introduce a common but highly restricted meaning subsumed in the more general preceding definition:

> **pon•tiff** . . . *noun* . . .
> **2 :** BISHOP; *specifically* **:** POPE

The sense divider *also* is used to introduce a meaning that is closely related to but may be considered less important than the preceding sense:

> **chi•na** . . . *noun* . . .
> **1 :** PORCELAIN; *also* **:** vitreous porcelain wares (as dishes, vases, or ornaments) for domestic use

The sense divider *broadly* is used to introduce an extended or wider meaning of the preceding definition:

> **flot•sam** . . . *noun* . . .
> **1 :** floating wreckage of a ship or its cargo; *broadly* **:** floating debris

ORDER OF SENSES

The order of senses within an entry is historical: the sense known to have been first used in English is entered first. This is not to be taken to mean, however, that each sense of a multisense word developed from the immediately preceding sense. It is altogether possible that sense 1 of a word has given rise to sense 2 and sense 2 to sense 3, but frequently sense 2 and sense 3 may have arisen independently of one another from sense 1.

When a numbered sense is further subdivided into lettered subsenses, the inclusion of particular subsenses within a sense is based upon their semantic relationship to one another, but their order is likewise historical: subsense 1a is earlier than 1b, 1b is earlier than 1c, and so forth. Divisions of subsenses indicated by lightface numerals in parentheses are also in historical order with respect to one another. Subsenses may be out of historical order, however, with respect to the broader numbered senses:

> **[1]job** . . . *noun* . . . (circa 1627)
> **1 a :** a piece of work; *especially* **:** a small miscellaneous piece of work undertaken on order at a stated rate **b :** the object or material on which work is being done **c :** something produced by or as if by work <do a better *job* next time> **d :** an example of a usually specified type **:** ITEM <this *job* is round-necked and sleeveless —Lois Long>
> **2 a :** something done for private advantage <suspected the whole incident was a put-up *job*> **b :** a criminal enterprise; *specifically* **:** ROBBERY **c :** a damaging or destructive bit of work <did a *job* on him>
> **3 a** (1) **:** something that has to be done **:** TASK (2) **:** an undertaking requiring unusual exertion <it was a real *job* to talk over that noise> **b :** a specific duty, role, or function **c :** a regular remunerative position **d** *chiefly British* **:** state of affairs — used with *bad* or *good* <it was a good *job* you didn't hit the old man —E. L. Thomas>

At *job* the date indicates that the earliest unit of meaning, sense 1a, was born in the seventeenth century, and it is readily apparent how the following subsenses are linked to it and to each other by the idea of work. Even subsense 1d is so linked, because while it does not apply exclusively to manufactured items, it often does so, as the illustrative quotation suggests. Yet 1d did not exist before the 1920s, while 2a and 3a(1) both belong to the seventeenth century, although they are later than 1a. Even the very last subsense, 3d, is earlier than 1d, as it is found in the works of Dickens.

Historical order also determines whether transitive or intransitive senses are given first at verbs which have both kinds. If the earliest sense is transitive, all the transitive senses precede all the intransitive senses.

OMISSION OF A SENSE

Occasionally the dictionary user, having turned to an entry, may not find a particular sense that was expected or hoped for. This usually means no more than that the editors judged the sense insufficiently common or otherwise important to include in a dictionary of this scope. Such a sense will frequently be found at the appropriate entry in a dictionary (as Webster's Third New International Dictionary) that has room for less common words and meanings. One special case is worth noting, however.

At times it would be possible to include the definition of a meaning at more than one entry (as at a simple verb and a verb-adverb collocation or at a verb and an adjective derived from a participle of that verb). To save space for other information such double coverage is avoided, and the meaning is generally defined only at the base form. For the derivative term the meaning is then considered to be essentially self-explanatory and is not defined. For example *cast off* has a sense "to get rid of" in such typical contexts as "cast off all restraint," and so has the simple verb *cast* in contexts like "cast all restraint to the winds." This meaning is defined as sense 1e(2) of *cast* and is omitted from the entry *cast off,* where the dictionary user will find a number of senses that cannot be considered self-explanatory in relation to the entries for *cast* and *off.* Likewise, the entry for the adjective *picked* gives only one sense—"CHOICE, PRIME"—which is not the meaning of *picked* in such a context as "the picked fruit lay stacked in boxes awaiting shipment." A definition suitable for this use is not given at *picked* because one is given at the first homograph *pick,* the verb from which the adjective *picked* is derived, as sense 3a—"to gather by plucking."

INFORMATION AT INDIVIDUAL SENSES

Information coming between the entry word and the first definition of a multisense word applies to all senses and subsenses. Information applicable only to some senses or subsenses is given between the appropriate boldface numeral or letter and the symbolic colon. A variety of kinds of information is offered in this way:

[2]**palm** *noun* . . .
3 [Latin *palmus,* from *palma*]

[2]**rally** *noun* . . .
4 *also* **ral·lye**

[1]**disk** *or* **disc** . . . *noun* . . .
4 . . . **b** *usually disc*

cru·ci·fix·ion . . . *noun* . . .
1 a *capitalized*

[1]**tile** . . . *noun* . . .
1 *plural* **tiles** *or* **tile a** . . .

del·i·ca·tes·sen . . . *noun plural* . . .
1 . . .
2 *singular, plural* **delicatessens**

fix·ing . . . *noun* . . .
2 *plural*

[2]**die** . . . *noun, plural* **dice** . . . *or* **dies**
. . .
1 *plural dice* . . .
2 *plural dies* . . .
3 *plural dies*

[1]**folk** . . . *noun, plural* **folk** *or* **folks** . . .
4 *folks plural*

At *palm* the subetymology indicates that the third sense, while ultimately derived from the same source (Latin *palma*) as the other senses of the word, has a different immediate etymon (Latin *palmus*), from which it receives its meaning. At *rally* one is told that in the fourth sense the word has a variant spelling not used for other senses and that this variant is a secondary or less common one. At *disk* the italic label of sense 4b indicates that, while the spelling *disk* is overall somewhat the more common (since it precedes *disc* out of alphabetical order at the beginning of the entry), *disc* is the usual spelling for this particular sense. At *crucifixion* the label *capitalized* points out the one meaning of the word in which it is capitalized. At the first homograph *tile* no plural is shown at the beginning of the entry because the usual plural, *tiles,* is regular. The subsenses of sense 1, however, have a zero plural as well as the usual one, and so both plurals appear in boldface at sense 1. At *delicatessen* the situation is different: the entry as a whole is labeled a plural noun, but sense 2 is used as a singular. In this sense *delicatessen* can take the plural ending *-s* when needed, a fact that is indicated by the appearance of the plural in boldface at the sense. At *fixing* the italic abbreviation simply means that when used in this sense the word is always written in its plural form, *fixings.* At the second homograph *die* the actual distribution of the variant plurals can be given sense by sense in italic type because both variants are shown in boldface earlier in the entry. At the first homograph *folk* a singular noun is shown with variant plurals of nearly equal frequency, when all senses are taken into account. The fourth sense, however, is unique in being always plural in form and construction. The form of the plural for this sense is *folks,* as shown, and the placement of the form before the label instead of after it (as at the senses of *die*) means that this sense is always plural.

When an italicized label or guide phrase follows a boldface numeral, the label or phrase applies only to that specific numbered sense and its subsenses. It does not apply to any other boldface numbered senses:

[1]**boot** . . . *noun* . . .
1 *archaic* . . .
2 *chiefly dialect* . . .
3 *obsolete* . . .

[1]**fa·vor** . . . *noun* . . .
2 *archaic* **a** . . . **b** (1) . . . (2) . . .
3

At *boot* the *archaic* label applies only to sense 1, the *chiefly dialect* label only to sense 2, and the *obsolete* label only to sense 3. At *favor* the *archaic* label applies to all the subsenses of sense 2 but not to sense 3.

When an italicized label or guide phrase follows a boldface letter, the label or phrase applies only to that specific lettered sense and its subsenses. It does not apply to any other boldface lettered senses:

[2]**stour** *noun* . . .
1 a *archaic* . . . **b** *dialect British*

The *archaic* label applies to sense 1a but not to sense 1b. The *dialect British* label applies to sense 1b but not to sense 1a.

When an italicized label or guide phrase follows a parenthesized numeral, the label or phrase applies only to that specific numbered sense:

in·car·na·tion . . . *noun* . . .
1a (1) . . . (2) *capitalized*

The label *capitalized* applies to sense 1a(2) and to no other subsenses of the word.

Names of Plants & Animals

The entries that define the common or vernacular names (as *peach* and *lion*) of plants and animals or sometimes a related term (as *streptomycin*), if a common name is rare or does not exist, employ in part the formal, codified, New Latin vocabulary of biological systematics. This vocabulary has been developed and used by biologists in accordance with international codes of botanical and zoological nomenclature for the purpose of identifying and indicating the relationships of plants and animals. This system of names classifies organisms into a hierarchy of groups—taxa—with each kind of organism having one—and only one—correct name and belonging to one—and only one—taxon at each level of classification in the hierarchy.

This contrasts with the system of common or vernacular names which is determined by popular usage and in which one organism may have several names (as *mountain lion, cougar,* and *painter*), different organisms may have the same name (as *dolphin*), and there may be variation in meaning or overlapping of the categories denoted by the names (as *whale, dolphin,* and *porpoise*).

The fundamental taxon is the genus. It includes a group of closely related kinds of plants (as *Prunus,* which includes the wild and cultivated cherries, apricots, peaches, and almonds) or animals (as *Canis,* which includes domestic dogs, coyotes, jackals, and wolves). The genus name is a capitalized singular noun.

The unique name of each kind of organism or species—the binomial or species name—consists of a singular capitalized genus name combined with an uncapitalized specific epithet. The name for a variety or subspecies—the trinomial, variety name, or subspecies name—adds a similar varietal or subspecific epithet. The cultivated cabbage (*Brassica oleracea capitata*), the cauliflower (*Brassica oleracea botrytis*), and brussels sprouts (*Brassica oleracea gemmifera*) belong to the same species (*Brassica oleracea*) of cole.

Names of taxa higher than the genus (as family, order, and class) are capitalized plural nouns that are often used with singular verbs and that are not abbreviated in normal use. No two genera of animals in good standing are permitted to have the same name, nor are any two genera of plants in good standing. Since the botanical and zoological codes are independent, however, a plant genus and an animal genus may have the same name. Thus, a number of cabbage butterflies (as *Pieris rapae*) are placed in a genus of animals with the same name as the plant genus to which the Japanese andromeda (*Pieris japonica*) belongs. Although no two higher taxa of plants are permitted to have the same name, the rules of zoological nomenclature do not apply to taxa above the family so that rarely some widely separated groups may receive the same taxonomic name (as the ordinal name Decapoda in the entry *decapod*) from different specialists.

The taxonomic names of biological nomenclature used in this dictionary are enclosed in parentheses and usually come immediately after the primary orienting noun. Genus names as well as binomials and trinomials are italicized, but names of taxa above the genus are not italicized:

> [1]**bee·tle** . . . *noun* . . .
> **1 :** any of an order (Coleoptera) of insects having four wings of which the outer pair are modified into stiff elytra that protect the inner pair when at rest
>
> **rob·in** . . . *noun* . . .
> **1 a :** a small chiefly European thrush (*Erithacus rubecula*) resembling a warbler and having a brownish olive back and orangish face and breast **b :** any of various Old World songbirds that are related to or resemble the European robin
> **2 :** a large North American thrush (*Turdus migratorius*) with olivaceous to slate gray upperparts, blackish head and tail, black and whitish streaked throat, and dull reddish breast and underparts
>
> **strep·to·my·cin** . . . *noun* . . .
> **:** an antibiotic organic base $C_{21}H_{39}N_7O_{12}$ produced by a soil actinomycete (*Streptomyces griseus*), active against many bacteria, and used especially in the treatment of infections (as tuberculosis) by gram-negative bacteria

Sometimes two or more different New Latin names can be found used in current literature for the same organism or group. This happens when older monographs and field guides are kept in print after name changes occur, when there are legitimate differences of opinion about the validity of names, and when the rules of priority do not apply or are not applied. To help the reader in recognizing an organism or group, some entries in this dictionary give two taxonomic names connected by the word *synonym*:

> **li·on** . . . *noun* . . .
> **1 a** . . . **:** a large heavily-built social cat (*Panthera leo* synonym *Leo leo*) of open or rocky areas chiefly of sub-Saharan Africa . . .

Taxonomic names are used in this dictionary to provide precise technical identifications through which defined terms may be pursued in technical writings. Because of their specialized nature taxonomic names do not have separate entry. However, many common names are derived directly from the names of taxa and especially genera with little or no modification. It is particularly important to distinguish between a common name and the genus name from which it is derived without a change in spelling, as these names appear in print. The common name (as clostridium or drosophila) is not usually capitalized or italicized but does have a plural (as clostridia or drosophilas) which often has an ending different from that of the singular. In contrast the genus name (as *Clostridium* or *Drosophila*) is capitalized and italicized but never takes a plural. A common name in plural form (as coleoptera) may sometimes be spelled like the name of a taxon, but it is not usually capitalized.

The entries defining the names of plants and animals are usually oriented to higher taxa by other vernaculars (as by *alga* at *seaweed* or *thrush* at *robin*) or by technical adjectives (as by *composite* at *daisy, leguminous* at *pea,* or *oscine* at *warbler*) so that the name of a higher taxon may often be found by consulting an entry defining a more inclusive common name or a term related to it. Among the higher plants, except for the composites and legumes and a few tropical groups, such orientation is by a vernacular family name linked at the corresponding taxonomic entry to its technical equivalent:

> [2]**rose** . . . *noun* . . .
> **1 a :** any of a genus (*Rosa* of the family Rosaceae, the rose family) of usually prickly shrubs with pinnate leaves and showy flowers . . .
>
> **ap·ple** . . . *noun, often attributive* . . .
> **1 :** the fleshy usually rounded and red, yellow, or green edible pome fruit of a tree (genus *Malus*) of the rose family; *also* **:** an apple tree

A genus name may be abbreviated to its initial letter when it is used more than once in senses not separated by a boldface number:

> **nas·tur·tium** . . . *noun* . . .
> **:** any of a genus (*Tropaeolum* of the family Tropaeolaceae, the nasturtium family) of herbs of Central and South America with showy spurred flowers and pungent seeds; *especially* **:** either of two widely cultivated ornamentals (*T. majus* and *T. minus*)

Cross-Reference

Four different kinds of cross-references are used in this dictionary: directional, synonymous, cognate, and inflectional. In each instance the cross-reference is readily recognized by the lightface small capitals in which it is printed.

A cross-reference following a lightface dash and beginning with *see* or *compare* is a directional cross-reference. It directs the dictionary user to look elsewhere for further information. A *compare* cross-reference is regularly appended to a definition; a *see* cross-reference may stand alone:

> **wel·ter·weight** . . . *noun*
> . . . — compare LIGHTWEIGHT, MIDDLEWEIGHT
>
> [1]**ri·al** . . . *noun*
> . . . — see MONEY table

A cross-reference immediately following a boldface colon is a synonymous cross-reference. It may stand alone as the only definitional matter for an entry or for a sense or subsense of an entry; it may follow an analytical definition; it may be one of two synonymous cross-references separated by a comma:

> **gar·ban·zo** . . . *noun* . . .
> **:** CHICKPEA
>
> [1]**ne·glect** . . . *transitive verb* . . .
> **1 :** to give little attention or respect to **:** DISREGARD

[2]main *adjective* . . .
1 : CHIEF, PRINCIPAL

A synonymous cross-reference indicates that a definition at the entry cross-referred to can be substituted as a definition for the entry or the sense or subsense in which the cross-reference appears.

A cross-reference following an italic *variant of* is a cognate cross-reference:

kaf·tan *variant of* CAFTAN

Sometimes a cognate cross-reference has a limiting label preceding *variant of* as a specific indication that the variant is not standard English:

haul·ier . . . *British variant of* HAULER

[2]hist . . . *dialect variant of* HOIST

sher·ris . . . *archaic variant of* SHERRY

A cross-reference following an italic label that identifies an entry as an inflected form of a noun, of an adjective or adverb, or of a verb is an inflectional cross-reference. Inflectional cross-references appear only when the inflected form falls at least a column away from the entry cross-referred to:

calves *plural of* CALF

[3]wound . . . *past and past participle of* WIND

When guidance seems needed as to which one of several homographs or which sense of a multisense word is being referred to, a superscript numeral may precede the cross-reference or a sense number may follow it or both:

[1]toss . . . *transitive verb* . . .
3 . . . **c :** MATCH 5a

Synonyms

Brief paragraphs discriminating words of closely associated meaning from one another have been placed at a number of entries. A small blue-shaded star after the last definition of a word signals that a synonym paragraph discussing that word can be found in the right-hand column on that page or on the facing page to the right:

[2]answer . . . *intransitive verb* . . .
6 : to offer a solution for; *especially* **:** SOLVE ☆

The paragraph itself is set on a blue background and begins with a list of the words to be discussed in it, followed by a concise statement of the element of meaning that the words have in common. The discriminations are amplified with verbal illustrations:

☆ SYNONYMS
Answer, respond, reply, rejoin, retort mean to say, write, or do something in return. ANSWER implies the satisfying of a question, demand, call, or need <*answered* all the questions>. RESPOND may suggest an immediate or quick reaction <*responded* eagerly to a call for volunteers>. REPLY implies making a return commensurate with the original question or demand <an invitation that requires you to *reply*>. REJOIN often implies sharpness or quickness in answering <"who asked you?" she *rejoined*>. RETORT suggests responding to an explicit charge or criticism by way of retaliation <he *retorted* to the attack with biting sarcasm>.

When a word is included in a synonym paragraph, the main entry for that word is followed by a run-on "***synonym*** see ______" which refers to the entry where the synonym paragraph appears:

[2]respond *verb* . . . ***synonym*** see ANSWER

When a word is a main entry at which there is a synonym paragraph and is also included in another paragraph elsewhere, the paragraph at the main entry is followed by a run-on "See in addition ______" which refers to the entry where the other paragraph may be found:

[1]nat·u·ral . . . *adjective* . . .
15 : of an off-white or beige color ☆

☆ SYNONYMS
Natural, ingenuous, naive, unsophisticated, artless mean free from pretension or calculation . . . See in addition REGULAR.

[1]reg·u·lar . . . *adjective* . . .
5 : of, relating to, or constituting the permanent standing military force of a state <*regular* army> <*regular* soldiers> ☆

☆ SYNONYMS
Regular, normal, typical, natural mean being of the sort or kind that is expected as usual, ordinary, or average . . .

Combining Forms, Prefixes & Suffixes

An entry that begins or ends with a hyphen is a word element that forms part of an English compound:

mega- *or* **meg-** *combining form* . . .
1 . . . **b** . . .<*mega*hit>

-logy *noun combining form* . . .
1 . . . <phraseo*logy*>

-lyze *verb combining form*
. . . <electro*lyze*>

-like *adjective combining form*
. . . <bell-*like*> <lady*like*>

pre- *prefix* . . .
1 a (1) . . . <*pre*historic>

[1]-ory . . . *noun suffix* . . .
1 . . . <observat*ory*>

[1]-ic *adjective suffix* . . .
2 a . . . <alderman*ic*>

[2]-ly *adverb suffix* . . .
1 a . . . <slow*ly*>

-ize *verb suffix* . . .
2 a . . . <crystall*ize*>

Combining forms, prefixes, and suffixes are entered in this dictionary for three reasons: to make easier the writing of etymologies of words in which these word elements occur over and over again; to make understandable the meaning of many undefined run-ons which for reasons of space would be omitted if they had to be given etymologies and definitions; and to make recognizable the meaningful elements of new words that are not well enough established in the language to warrant dictionary entry.

Lists of Undefined Words

Lists of undefined words occur after the entries of these prefixes and combining forms:

anti-	**multi-**	**re-**
co-	**non-**	**self-**
counter-	**out-**	**sub-**
hyper-	**over-**	**super-**
inter-	**post-**	**ultra-**
mis-	**pre-**	**un-**

These words are not defined because they are self-explanatory; their meanings are simply the sum of a meaning of the prefix or combining form and a meaning of the root word. Centered dots are shown to save the dictionary user the trouble of consulting another entry. The lists are not exhaustive of all the words that might be, or actually have been, formed with these prefixes and combining forms. The dictionary has room for only the most common or important examples.

New Words and Senses

Cross-references to other entries refer you to the main body of the book unless otherwise specified. Entries here which contain the date of first recorded use are new words; those without dates shown are new meanings of words already entered in the main vocabulary section.

accelerator *noun*
: an item of computer hardware that increases the speed at which a program or function operates ⟨a graphics *accelerator*⟩

achoo *also* **ah·choo** \ä-ˈchü\ *interjection* (1882)
— used to represent the sound of a sneeze

acid snow *noun* (1981)
: acid precipitation in the form of snow

ac·tion·er \ˈak-sh(ə-)nər\ *noun* (1973)
: a film dominated by a high degree of exciting action

action figure *noun* (1987)
: a small-scale figure usually of a superhero used especially as a toy

active–matrix \ˈak-tiv-ˌmā-triks\ *adjective* (1980)
: of, relating to, or being an LCD in which each pixel is individually controlled

adren·al·ized \ə-ˈdre-nᵊl-ˌīzd\ *adjective* (1973)
: filled with a sudden rush of energy : EXCITED

adult–on·set diabetes \ə-ˈdəlt-ˈän-ˌset-\ *noun* (1975)
: TYPE 2 DIABETES

advance directive *noun* (1984)
: a legal document (as a living will) signed by a competent person in order to provide guidance for medical and health-care decisions (as the termination of life support or organ donation) in the event the person becomes incompetent to make such decisions

aerial *noun*
: an acrobatic maneuver performed (as by skiers and gymnasts) in the air; *also, plural* : a ski event featuring aerials

aer·o·bi·cize \ˌa(-ə)r-ˈō-bə-ˌsīz, ˌe(-ə)r-\ *verb* **aer·o·bi·cized; aer·o·bi·ciz·ing**
transitive verb (1981)
: to bring to good physical condition through aerobics
intransitive verb
: to engage in aerobics

agent *noun*
: a computer application designed to automate certain tasks (as gathering information online)

agita \ˈa-jə-tə\ *noun* [southern Italian dialect pronunciation of Italian *acido,* literally, heartburn, acid, from Latin *acidus*] (1982)
: a feeling of agitation or anxiety

agree *transitive verb*
: to consent to as a course of action ⟨*agreed* to sell him the house⟩

ag·ri·tour·ism \ˌa-gri-ˈtu̇r-ˌi-zəm\ *noun* [*agriculture* + *tourism*] (1979)
: the practice of touring agricultural areas to see farms and often to participate in farm activities

Ahab \ˈā-ˌhab\ *noun* [Hebrew *Aḥĕʼābh*] (1540)
: a king of Israel in the ninth century B.C. and husband of Jezebel

AHA \ˌā-(ˌ)āch-ˈā\ *noun* (1991)
: ALPHA HYDROXY ACID (in this list)

ahchoo *var of* ACHOO

air marshal *noun*
: SKY MARSHAL

air rage *noun* (1996)
: an airline passenger's uncontrolled anger that is usually expressed in aggressive or violent behavior

all–ter·rain vehicle *noun* (1969)
: a small motor vehicle with three or four wheels that is designed for use on various types of terrain — called also *ATV*

alpha hydroxy acid *noun* (1986)
: any of various carboxylic acids with a hydroxyl group attached at the alpha position; *specifically* : one (as malic acid or lactic acid) that occurs in natural products (as fruits, sugar cane, or yogurt) and is used in cosmetics for its exfoliating effect on the surface layer of skin — called also *AHA*

al·ter·i·ty \ȯl-ˈter-ə-tē, -ˈte-rə-\ *noun* [Late Latin *alteritat-, alteritas,* from *alter*] (1642)
: OTHERNESS; *specifically* : the quality or state of being radically alien to the conscious self or a particular cultural orientation

al·ter·na·tive *adjective*
: of, relating to, or being rock music that is regarded as an alternative to conventional rock and is typically influenced by punk rock, hard rock, hip-hop, or folk music

alternative medicine *noun* (1977)
: any of various systems of healing or treating disease (as chiropractic, homeopathy, or faith healing) not included in the traditional medical curricula taught in the U.S. and Britain

aluminum oxide *noun* (1907)
: ALUMINA

ama·ro·ne \ˌä-mä-ˈrō-nā\ *noun* [Italian, from Italian dialect (Veneto), literally, tart, very dry, augmentative of *amaro* tart, bitter, from Latin *amarus* bitter] (1973)
: a robust dry red Italian wine with a high alcohol content

amuse–bouche \ˈä-ˌmüz-ˈbüsh, -ˌmœz-\ *noun* [French, literally, (it) entertains (the) mouth] (1984)
: a small complimentary appetizer offered at some restaurants

an·al·ge·sic \ˌa-nᵊl-ˈjē-zik, -sik\ *noun* (1875)
: an agent for producing analgesia
— **analgesic** *adjective*

an·cho \ˈän-chō\ *noun, plural* **anchos** [American Spanish *chile ancho* literally, wide chile] (1902)
: a poblano chile pepper especially when mature and dried to a reddish black

an·i·ma·tron·ic \ˌa-nə-mə-ˈträ-nik\ *adjective* [short for *audio-animatronic*] (1978)
: of, relating to, or being a puppet or similar figure that is animated by means of electromechanical devices
— **an·i·ma·tron·i·cal·ly** \-ni-k(ə-)lē\ *adverb*

an·ti·choice \ˌan-ti-ˈchȯis, ˌan-ˌtī-\ *adjective* (1978)
: ANTIABORTION
— **an·ti·choic·er** \-ˈchȯi-sər\ *noun*

an·ti·re·tro·vi·ral \-ˈre-trō-ˌvī-rəl\ *adjective* (1987)
: acting, used, or effective against retroviruses ⟨*antiretroviral* drugs⟩ ⟨*antiretroviral* therapy⟩
— **antiretroviral** *noun*

ap·o·pto·sis \ˌa-pəp-ˈtō-səs, -pə-ˈtō-\ *noun, plural* **ap·o·pto·ses** \-ˌsēz\ [New Latin, from Greek *apoptōsis* a falling off, from *apopiptein* to fall off, from *apo-* + *piptein* to fall — more at FEATHER] (1972)
: a genetically determined destruction of cells from within due to activation of a stimulus or removal of a suppressing agent or stimulus that is postulated to exist to explain the orderly elimination of superfluous cells — called also *programmed cell death*
— **ap·o·pto·tic** \-ˈtä-tik\ *adjective*

app \ˈap\ *noun* (1987)
: APPLICATION 1a(3)

ap·plet \ˈa-plət, -(ˌ)plet\ *noun* [*appl*ication + ¹*-et*] (1990)
: a short application program especially for performing a simple specific task

aqua·scape \ˈä-kwə-ˌskāp, ˈa-\ *noun* (1954)
1 : a scenic view of a body of water
2 : an area having a natural or constructed aquatic feature (as a pond or fountain)

ar·chaea \är-ˈkē-ə\ *noun plural* [New Latin, from Greek *archaios*] (1990)
: microorganisms of a domain (Archaea) comprising the archaebacteria when considered as equal in taxonomic rank to the other prokaryotes and the eukaryotes
— **ar·chae·al** \-əl\ *adjective*
— **ar·chae·an** \-ən\ *adjective or noun*

arena football *noun* (1986)
: a game resembling American football that is played on a shorter indoor field between two teams of eight players each

arm candy *noun* (1992)
: a young attractive person who accompanies a usually older person at social events

ar·mored \ˈär-mərd\ *adjective* (1862)
1 a : equipped or protected with armor **b :** equipped with armored fighting vehicles ⟨an *armored* division⟩
2 : marked by the use of armor ⟨*armored* combat⟩

assault weapon *noun* (1973)
: any of various automatic or semiautomatic firearms; *especially* : ASSAULT RIFLE

as·sis·ted living \ə-ˈsis-təd-\ *noun* (1981)
: a system of housing and limited care that is designed for senior citizens who need some assistance with daily activities but do not require care in a nursing home — usually hyphenated when used attributively ⟨an *assisted-living* facility⟩

assisted suicide *noun* (1976)
: suicide committed by someone with assistance from another person; *especially* : PHYSICIAN-ASSISTED SUICIDE (in this list)
ator•va•stat•in \ə-ˌtȯr-və-ˈsta-tᵊn, -ˈtȯr-və-ˌsta-\ *noun* [*ator-* (perhaps alteration of lip*id* con*trol*) + *-vastatin* (as in *lovastatin*)] (1994)
: a statin administered orally in the form of its hydrated calcium salt $(C_{33}H_{34}FN_2O_5)_2 \cdot Ca_3H_2O$ to lower lipid levels in the blood
at sign *noun* (1982)
: the symbol @ especially when used as part of an Internet user's e-mail address
attention deficit/hyperactivity disorder *noun* (1987)
: ATTENTION DEFICIT DISORDER — abbreviation *ADHD*
ATV \ˌā-(ˌ)tē-ˈvē\ *noun* (1969)
: ALL-TERRAIN VEHICLE (in this list)
au•dio•book \ˈȯ-dē-ō-ˌbu̇k\ *noun* (1988)
: TALKING BOOK
avatar *noun*
: an electronic image that represents and is manipulated by a computer user (as in a computer game or an online shopping site)
avian influenza *noun* (1980)
: a highly variable mild to fulminant influenza of birds that is caused by strains of the influenza A virus which may mutate and be transmitted to other vertebrates — called also *bird flu*
Azer•bai•ja•ni \ˌa-zər-ˌbī-ˈjä-nē, ˌä-\ *noun, plural* **Azerbaijanis** *also* **Azerbaijani** [Persian *āzarbāyjānī*, from *Āzarbāyjān* Azerbaijan] (circa 1909)
1 : a member of a Turkic-speaking people of Azerbaijan and northwest Iran
2 : the Turkic language of the Azerbaijanis
— **Azerbaijani** *adjective*
backbone *noun*
: the primary high-speed hardware and transmission lines of a telecommunication network (as the Internet)
backslash \ˈbak-ˌslash\ (1982)
: a mark \ used especially in computer programming
bad cholesterol *noun* (1980)
: LDL
baf•fle•gab \ˈba-fəl-ˌgab\ *noun* (1952)
: GOBBLEDYGOOK
ba•ris•ta \bə-ˈrēs-tə, bä-ˈrēs-tä\ *noun* [Italian, person working behind a bar, from *bar* bar (from English) + *-ista* -ist] (1982)
: a person who makes and serves coffee (as espresso) to the public
battle dress uniform *noun* (1982)
: a military uniform for field service
be•lon \bā-ˈlōn, -ˈlōⁿ\ *noun, often capitalized* [French, from *Bélon*, river in Brittany] (1940)
: EUROPEAN FLAT (in this list); *specifically* : a European flat oyster of coastal waters of northwestern France
bench•mark•ing \-ˌmär-kiŋ\ *noun* (1976)
: the study of a competitor's product or business practices in order to improve the performance of one's own company
benign prostatic hyperplasia *noun* (1968)
: enlargement of the prostate gland caused by a benign overgrowth of chiefly glandular tissue that occurs especially in men over 50 years old and that tends to obstruct urination by constricting the urethra — abbreviation *BPH*; called also *benign prostatic hypertrophy*
be•ta–am•y•loid \-ˈa-mə-ˌlȯid\ *noun* (1987)
: an amyloid that is derived from a larger precursor protein and is a component of the neurofibrillary tangles and plaques characteristic of Alzheimer's disease
beta test *noun* (1978)
: a field test of a prototype version of a product (as software) especially by testers outside the company developing it that is conducted prior to commercial release
— **beta test** *transitive verb*
— **beta tester** *noun*
big–box \ˈbig-ˌbäks\ *adjective* (1990)
: of, relating to, or being a large chain store having a boxlike structure
— **big box** *noun*
big crunch *noun, often capitalized B&C* (1979)
: a hypothetical cosmological event in which all matter in the universe collapses to a singularity and which is posited to be a possible fate of the universe if the density of matter in it is sufficiently high — compare BIG BANG
big–time \ˈbig-ˌtīm\ *adjective* (1930)
: relating to or involved in the big time ⟨*big-time* sports⟩; *also* : MAJOR 4 ⟨*big-time* operators⟩
big–time *adverb* (1980)
: in a major or large-scale manner : to a great extent or degree ⟨owes me *big-time*⟩
bikini wax *noun* (1985)
: a procedure for removing pubic hair from the skin near the edge of the bottom half of a bikini by applying hot wax, covering the wax with a cloth to which the wax and hair adhere, and then peeling it off quickly
bio•die•sel \-ˈdē-zəl, -səl\ *noun* (1986)
: a fuel that is similar to diesel fuel and is derived from usually vegetable sources (as soybean oil)
bio•in•for•mat•ics \ˌbī-ō-in-fər-ˈma-tiks\ *noun plural but singular in construction* (1993)
: the collection, classification, storage, and analysis of biochemical information using computers especially as applied to molecular genetics and genomics
— **bio•in•for•mat•ic** \-tik\ *adjective*
bio•mark•er \ˈbī-ō-ˌmär-kər\ *noun* (1982)
: a distinctive usually biochemical indicator (as a metabolite) of a biological or geochemical process or event (as aging, poisoning, fossilization, or oil formation)
bio•mi•met•ics \ˌbī-ō-mə-ˈme-tiks, -mī-\ *noun plural but singular in construction* (1974)
: the study of the formation, structure, or function of biologically produced substances and materials (as enzymes or silk) and biological mechanisms and processes (as protein synthesis or photosynthesis) especially for the purpose of synthesizing similar products by artificial mechanisms which mimic natural ones
— **bio•mi•met•ic** \-tik\ *adjective*
bio•re•gion•al•ism \ˌbī-ō-ˈrēj-nə-ˌli-zəm, -ˈrē-jə-nᵊl-i-\ *noun* (1981)
: an environmentalist movement to make political boundaries coincide with bioregions
— **bio•re•gion•al•ist** \-list, -ist\ *noun or adjective*
bio•re•gion \ˈbī-ō-ˌrē-jən\ *noun* (1978)
: a region whose limits are naturally defined by topographic and biological features (as mountains and ecosystems)
— **bio•re•gion•al** \ˌbī-ō-ˈrēj-nəl, -ˈrē-jə-nᵊl\ *adjective*
bio•re•me•di•a•tion \-ri-ˌmē-dē-ˈā-shən\ *noun* (1986)
: the treatment of pollutants or waste (as in an oil spill, contaminated groundwater, or industrial process) by the use of microorganisms (as bacteria) that break down the undesirable substances
bio•sol•id \ˈbī-ō-ˌsä-ləd\ *noun* (1990)
: solid organic matter recovered from a sewage treatment process and used especially as fertilizer — usually used in plural
bio•ter•ror•ism \-ˈter-ər-ˌi-zəm, -ə-ˌri-\ *noun* (1991)
: terrorism involving the use of biological weapons
— **bio•ter•ror•ist** \-ər-ist, -ə-rist\ *adjective or noun*
bipolar disorder *noun* (1980)
: any of several psychological disorders of mood characterized usually by alternating episodes of depression and mania — called also *manic depression, manic-depressive illness*
bis•cot•to \bi-ˈskät-ō\ *noun, plural* **bis•cot•ti** \-ē\ [Italian, biscuit, cookie, from (*pane*) *biscotto*, literally, bread baked twice] (1973)
: a crisp cookie or biscuit of Italian origin that is flavored usually with anise and filberts or almonds
bitch•in' \ˈbi-chən\ *adjective* [probably short for *sonofabitching*, from *son of a bitch* + *-ing*] (1957)
1 *slang* : remarkably bad : DETESTABLE ⟨of all of the *bitchin'* luck⟩
2 *slang* : remarkable good or cool ⟨a *bitchin'* car⟩
bit–map \ˈbit-ˌmap\ *noun* (1982)
: an array of binary data representing a bit-mapped image or display; *also* : the image or display itself
bit–mapped \ˈbit-ˌmapt\ *adjective* (1978)
: of, relating to, or being a digital image or display for which an array of binary data specifies the value of each pixel ⟨*bit-mapped* graphics⟩
bling–bling \ˈbliŋ-ˌbliŋ\ *also* **bling** \ˈbliŋ\ *noun* [imitative] (1999)
: flashy jewelry worn especially as an indication of wealth; *broadly* : expensive and ostentatious possessions
blog \ˈblȯg, ˈbläg\ *noun* [short for *Weblog*] (1999)
: a Web site that contains an online personal journal with reflections, comments, and often hyperlinks provided by the writer
— **blog•ger** *noun*
— **blog•ging** *noun*
blo•vi•ate \ˈblō-vē-ˌāt\ *intransitive verb* **blo•vi•at•ed; blo•vi•at•ing** [perhaps irregularly from *blow*] (circa 1897)
: to speak or write verbosely and windily
— **blo•vi•a•tion** \ˌblō-vē-ˈā-shən\ *noun*
blow off *transitive verb*
1 : to outperform in a contest
2 : to end a relationship with
blue screen *noun* (1977)
: a photographic technique in which a subject is filmed in front of a blue background so as to allow compositing with other footage; *also* : the blue background
bobblehead doll \ˈbä-bəl-ˌhed-\ *noun* (1964)
: a doll having a head that makes repeated bobbing movements
body•board \ˈbä-dē-ˌbȯrd\ *noun* (1982)
: a short surfboard on which the rider lies prone
— **bodyboard** *intransitive verb*
— **body•board•er** *noun*
body wrap *noun* (1974)
: a body treatment involving the application of usually oils or gels followed by a wrapping of the body with a sheet
bohr•i•um \ˈbȯr-ē-əm, ˈbȯr-\ *noun* [New Latin, from Niels *Bohr*] (1994)
: a short-lived radioactive element that is artificially produced
bollocks \ˈbä-ləks\ *noun plural* [alteration of *ballocks*, plural of *ballock* testis, from Middle English, from Old English *bealluc* — more at BALL] (1774)
1 *chiefly British, usually vulgar* : TESTICLES
2 *chiefly British, usually vulgar* : NONSENSE
bo•lus *noun*
: a dose of a substance (as a drug) that is administered intravenously; *specifically* : a large dose given intravenously so that the desired therapeutic concentration in the blood is reached rapidly
bo•no•bo \bə-ˈnō-bō; ˈbä-nə-bō, -nō-\ *noun* [origin unknown] (1954)
: PYGMY CHIMPANZEE
book *intransitive verb*
slang : LEAVE, GO; *especially* : to depart quickly
book•mark *noun*
: a menu entry or icon on a computer that is most often created by the user and that serves as a shortcut to a previously viewed location (as an Internet address)
boot *intransitive verb*
: to become ready for use especially by booting a program ⟨the computer *boots* quickly⟩ — often used with *up*
Boston marriage *noun* (1980)

: a long-term loving relationship between two women
Botox \'bō-ˌtäks\ *trademark*
— used of a preparation of botulinum toxin
bot·tom–feed·er \'bä-təm-ˌfē-dər\ *noun* (1920)
1 : a fish that feeds at the bottom
2 : one that is of the lowest status or rank
3 : an opportunist who seeks quick profit usually at the expense of others or from their misfortune
bovine spon·gi·form encephalopathy \-'spən-ji-ˌfȯrm-\ *noun* (1987)
: a fatal disease of cattle affecting the nervous system, resembling or identical with scrapie of sheep and goats, and probably caused by a prion transmitted by infected tissue in food — abbreviation *BSE*; called also *mad cow disease*
box cutter *noun* (1977)
: a small cutting tool that is designed for opening cardboard boxes and typically consists of a retractable razor blade in a thin metal sheath
brain attack *noun* (1990)
: STROKE 5
brain freeze *noun* (1991)
: a sudden shooting pain in the head caused by ingesting very cold food (as ice cream) or drink
brain·i·ac \'brā-nē-ˌak\ *noun* [probably from *Brainiac,* superintelligent villain in the *Superman* comic-book series] (1982)
: a person characterized by unusual brainpower
breakout *adjective* (1977)
: being or relating to a sudden or smashing success especially in comparison to previous efforts ⟨a *breakout* book⟩
brew·ski \'brü-skē\ *noun* [*brew* + *-ski,* suffix in Slavic surnames] (1978)
slang **:** a drink or serving of beer
brick–and–mortar *or* **bricks–and–mortar** *adjective* (1992)
: relating or being a traditional business serving customers in a building as contrasted to an online business ⟨a *brick-and-mortar* store⟩
bright–line \'brīt-ˌlīn\ *adjective* (1982)
: providing an unambiguous criterion or guideline especially in a law ⟨a *bright-line* distinction⟩
brown·field \'brau̇n-ˌfē(ə)ld\ *noun, often attributive* (1977)
: a tract of land that has been developed for industrial purposes, polluted, and then abandoned
browse *transitive verb*
: to access (a network) by means of a browser
brows·er \'brau̇-zər\ *noun* (1845)
1 : one that browses
2 : a computer program used for accessing sites or information on a network (as the World Wide Web)
bru·schet·ta \brü-'she-tə, -'ske-\ *noun* [Italian, from Italian dialect (Tuscany), from *bruscare* to toast, burn, probably from Vulgar Latin **brusicare,* frequentative of **brusare, *brusiare* to burn] (1954)
: thick slices of bread grilled, rubbed with garlic, drizzled with olive oil, and often topped with tomatoes and herbs and served as an appetizer
bub·ble *noun*
: a state of booming economic activity (as in a stock market) that often ends in a sudden collapse
bub·kes *also* **bup·kes** *or* **bup·kus** \'bəp-kəs, 'bu̇p-\ *noun plural but singular in construction* [Yiddish (probably short for *kozebubkes,* literally, goat droppings), plural of *bubke, bobke,* diminutive of *bub, bob* bean, of Slavic origin; akin to Polish *bób* bean] (1942)
: the least amount **:** BEANS ⟨won't win *bubkes* this year — Ivan Maisel⟩; *also* **:** NOTHING ⟨received *bubkes* at nomination time — Lewis Beale⟩
bucky·tube \-ˌtüb, -ˌtyüb\ *noun* [*bucky*ball + *tube*] (1991)
: a nanotube composed of pure carbon with a molecular arrangement similar to that of fullerene
buff *or* **buffed** *adjective*
: having a physique enhanced by bodybuilding exercises
buffalo wing *noun* [*Buffalo,* New York] (1984)
: a deep-fried chicken wing coated with a spicy sauce and usually served with a blue cheese dressing
bur·ka *or* **bur·qa** \'bu̇r-kə\ *noun* [Urdu, Persian & Arabic; Urdu *burqa',* from Persian *burqa', burqu',* from Arabic *burqu'*] (1836)
: a loose enveloping garment that covers the face and body and is worn in public by certain Muslim women
burn *transitive verb*
: to record digital data or music on (an optical disk) using a laser ⟨*burn* a CD⟩; *also* **:** to record (data or music) in this way ⟨*burns* songs onto a disk⟩
bust·er *noun*
chiefly Midland **:** someone or something extraordinary ⟨a *buster* of a breakfast —Harriet B. Stowe⟩
button *noun, often attributive*
: a usually box-shaped computer icon that initiates a specific software function
caf·fein·at·ed \'ka-fə-ˌnā-təd, -fē-ə\ *adjective* (1970)
1 : stimulated by or as if by caffeine ⟨*caffeinated* workers⟩
2 : containing caffeine
camp·out \'kam-ˌpau̇t\ *noun* (1879)
: an occasion on which a group camps out
ca·noo·dle \kə-'nü-dᵊl\ *intransitive verb* **ca·noo·dled; ca·noo·dling** \-'nü-dliŋ, -'nü-dᵊl-iŋ\ [perhaps from English dialect *canoodle,* noun, donkey, fool, foolish lover] (1859)
: PET, FONDLE ⟨lovers *canoodling* in the park⟩
cap·i·tat·ed \'ka-pə-ˌtā-təd\ *adjective* [back-formation from *capitation*] (1983)
: of, relating to, participating in, or being a health-care system in which a medical provider is given a set fee per patient (as by an HMO) regardless of treatment required
car·ne asa·da \ˌkär-nā-ə-'sä-də\ *noun* [Spanish, grilled meat] (1977)
: a grilled Mexican dish of spicy marinated steak strips sometimes served in a burrito or taco
car seat *noun* (1968)
: a portable seat for a small child that attaches to an automobile seat and holds the child safely
cellophane noodle *noun* (1977)
: a translucent noodle made from mung beans
cell phone *noun* (1984)
: a cellular telephone
cer·ve·za \sər-'vā-sə; ser-'bā-sä, ther-'bā-thä\ *noun* [Spanish, from Latin *cervesia,* a kind of beer] (1949)
1 : BEER 1
2 : a drink or serving of beer
chai \'chī\ *noun* [Turkish *çay* & Russian, Persian, Hindi, & Urdu *chay* tea] (1974)
: a beverage that is a blend of black tea, honey, spices, and milk
challenged *adjective* (1983)
: having a disability or deficiency
charter school *noun* (1992)
: a tax-supported school established by a charter between a granting body (as a school board) and an outside group (as of teachers and parents) which operates the school without most local and state educational regulations so as to achieve set goals
chat room *noun* (1986)
: a real-time on-line interactive discussion group
chat·ter *intransitive verb*
: to vibrate especially audibly as a consequence of repeated sticking and slipping ⟨the brakes were *chattering*⟩
cher·ry–pick \'cher-ē-ˌpik\ *intransitive verb* (1965)
: to select the best or most desirable
transitive verb
: to select as being the best or most desirable
Chet·nik \chet-'nēk, 'chet-nik\ *noun* [Serbian *četnik,* from *četa* band, troop] (1909)
1 : an irregular Slav soldier in the Balkans; *especially* **:** a member of various irregular Serbian military forces that in periods of disorder (as during World War II and following the breakup of Yugoslavia in 1991) pursued ultranational aims
2 : an ultranationalist Serb
chick flick *noun* (1988)
: a motion picture intended to appeal especially to women
chin music *noun*
: a usually high inside pitch in baseball intended to intimidate the batter
chip *noun*
: a small piece of food ⟨chocolate *chips*⟩
cho·lo \'chō-(ˌ)lō\ *noun, plural* **cholos** [American Spanish, Europeanized Indian, mestizo]
1 *often disparaging* **:** a man or boy of Mexican descent
2 : a Mexican-American youth who belongs to a street gang
chronic fatigue syndrome *noun* (1986)
: a group of symptoms of unknown cause that include fatigue, cognitive dysfunction, and sometimes fever and affect especially people between the ages of 20 and 40 — abbreviation *CFS*
Cine·plex \'si-nə-ˌpleks\ *service mark*
— used for a movie theater containing several auditoriums in one building
cip·ro·flox·a·cin \ˌsi-prə-'fläk-sə-sən, -prō-\ *noun* [probably from International Scientific Vocabulary *ci-* (alteration of *cycl-*) + *pro*pyl + *fluor-* + *ox-* + *az-* + *-mycin*] (1983)
: a synthetic broad-spectrum antibiotic $C_{17}H_{18}FN_3O_3$ that is a fluorinated derivative of quinolone and is often administered in the form of its hydrochloride
civil union *noun* (1992)
: the legal status that ensures to same-sex couples specified rights and responsibilities of married couples
clad·dagh \'kla-də, 'klä-\ *noun* [*Claddagh,* former village across the river Corrib from Galway city, Ireland] (1922)
: an Irish design (as on a ring) of two hands holding a crowned heart that symbolizes friendship, loyalty, and love
cla·fou·ti *also* **cla·fou·tis** \ˌklä-fü-'tē\ *noun* [French] (circa 1968)
: a dessert consisting of a layer of fruit (as cherries) topped with batter and baked
click *transitive verb*
: to select especially on a computer by pressing a button on a control device (as a mouse)
intransitive verb
: to select something by clicking — often used with *on* ⟨*click* on the icon⟩
click *noun*
: an instance of clicking ⟨a mouse *click*⟩
click·er \'kli-kər\ *noun* (1985)
: REMOTE CONTROL 2
cli·ent *noun*
: a computer in a network that uses the services (as access to files or shared peripherals) provided by a server
climacteric *noun*
: a period in the life of a male corresponding to female menopause and usually occurring with less well defined physiological and psychological changes
clip art *noun* (1968)
: ready-made usually copyright-free illustrations sold in books or as part of a software package from which they may be cut and pasted or inserted as artwork
clip·board *noun*
: a section of computer memory that temporarily stores data (as text) especially to facilitate its movement or duplication
clot–bust·er \-ˌbəs-tər\ *noun* (1984)
: a drug (as streptokinase) used to dissolve blood clots
— clot–bust·ing \-tiŋ\ *adjective*
cochlear implant *noun* (1980)

: an electronic prosthetic device that enables individuals with sensorineural hearing loss to recognize sounds and consists of an external microphone and speech processor and one or more electrodes implanted in the cochlea

co·in·ci·den·tal·ly \-'dent-lē, -'den-tᵊl-ē\ *adverb* (1837)
1 : in a coincidental manner **:** by coincidence ⟨lonely singles who meet *coincidentally* and click —*People*⟩
2 : it is or seems coincidental that ⟨*coincidentally*, the dog died exactly one year after his owner did⟩

collateral damage *noun* (1972)
: injury inflicted on something other than the intended target; *specifically* **:** civilian casualties of a military operation

comb–over \'kōm-ˌō-vər\ *noun* (1980)
: an arrangement of hair on a balding man in which hair from the side of the head is combed over the bald spot

comfort food *noun* (1977)
: food prepared in a traditional style having a usually nostalgic or sentimental appeal

com·mod·i·ty *noun*
: one that is subject to ready exchange or exploitation within a market ⟨the sensitive female singer-songwriter as a viable pop *commodity* —Elysa Gardner⟩

com·pa·dre \kəm-'pä-drā, -drē\ *noun* [Spanish, literally, godfather, from Medieval Latin *compater* — more at COMPEER] (1834)
: a close friend **:** BUDDY

convergence *noun*
: the merging of distinct technologies, industries, or devices into a unified whole

cook·ie *noun*
: a small file or part of a file stored on a World Wide Web user's computer, created and subsequently read by a Web site server, and containing personal information (as a user identification code, customized preferences, or a record of pages visited)

cop *intransitive verb*
slang **:** ADMIT 2b — used with *to* ⟨these small-timers would . . . *cop* to the smallest offense their attorney could negotiate — Tom Clancy⟩

co·qui \kō-'kē\ *noun* [American Spanish *coquí*] (circa 1903)
: a small chiefly nocturnal arboreal frog (*Eleutherodactylus coqui*) native to Puerto Rico that has a high-pitched call and has been introduced into Hawaii and southern Florida

co·ro·na·vi·rus \kə-'rō-nə-ˌvī-rəs\ *noun* [New Latin, from *corona* + *virus*] (1968)
: any of a family (*Coronaviridae* and especially genus *Coronavirus*) of single-stranded RNA viruses having a lipid envelope studded with club-shaped projections and including pathogens of animals and humans

COX \'käks\ *noun* [by shortening & alteration] (1990)
: CYCLOOXYGENASE (in this list) — often used with the number 1 or 2 to indicate one of the two variants of the enzyme

COX–2 inhibitor \'käks-'tü-\ *noun* (1994)
: any of a class of drugs used to treat the pain and inflammation of arthritis by selectively blocking the variant of cyclooxygenase causing these symptoms

creative *noun* (1962)
1 : one (as an artist or writer) that is creative; *especially* **:** one directly involved in the creation of advertisements
2 : creative activity or the material produced by it especially in advertising

cred \'kred\ *noun* (1981)
: CREDIBILITY; *specifically* **:** the quality of being deserving of acceptance as a member of a particular group or class ⟨used . . . his new street *cred* to develop contacts —Dale Keiger⟩

cre·mi·ni \kri-'mē-nē\ *noun* [Italian, plural of *cremino*, from *crema* cream, from Middle French *cresme;* probably from their color — more at CREAM] (1984)
: a meaty cultivated brown or tan mushroom belonging to the same variety of button mushroom as the larger and more mature portobello

cri·mi·ni \kri-'mē-nē\ *noun* (1986)
: CREMINI (in this list)

cryp·to·spo·rid·i·o·sis \ˌkrip-tō-spȯr-ˌid-ē-'ō-səs\ *noun, plural* **cryp·to·spo·rid·i·o·ses** \-ˌsēz\ [New Latin] (1982)
: a disease caused by cryptosporidia

cryp·to·spo·rid·i·um \ˌkrip-tō-spȯr-'id-ē-əm\ *noun, plural* **cryp·to·spo·rid·ia** \-ē-ə\ [New Latin, from *crypto-* + *spora* spore + *-idium*] (1982)
: a protozoan (genus *Cryptosporidium* of the order Coccidia) that is parasitic in the gut of vertebrates including humans and sometimes causes diarrhea

cut *noun*
: an edited version of a film ⟨a director's *cut*⟩

cu·vée *also* **cu·vee** \kü-'vā, kyü-; kᵫ-vā\ *noun* [French, from *cuve*] (1833)
1 : bulk wine; *especially* **:** wine in casks or vats so blended as to ensure uniformity and marketability
2 : a blend of still wines used in the production of champagne

cyber- *combining form* [*cybernetic*]
: computer **:** computer network ⟨*cyber*space⟩

cy·ber·se·cu·ri·ty \-si-ˌkyu̇r-ə-tē\ *noun* (1994)
: measures taken to protect a computer or computer system (as on the Internet) against unauthorized access or attack

cy·ber·sex \-ˌseks\ *noun* (1991)
1 : on-line sex-oriented conversations and exchanges
2 : sex-oriented material available on the Internet and on CD-ROMs

cy·brar·ian \sī-'brer-ē-ən, -'bre-rē-\ *noun* [blend of *cyber-* and *librarian*] (1992)
: a person whose job is to find, collect, and manage information that is available on the World Wide Web

cy·clo·ox·y·gen·ase \-'äk-si-jə-ˌnās\ *noun* [*cycl-* + *oxygenase,* an enzyme, fr. *oxygen* + *-ase*] (1975)
: an enzyme that catalyzes the conversion of arachidonic acid to prostaglandins and that has two isoforms of which one is involved in the creation of prostaglandins which mediate inflammation and pain

darm·stadt·i·um \ˌdärm-'sta-tē-əm\ *noun* [New Latin, from *Darmstadt,* Germany] (2003)
: a short-lived radioactive element produced artificially

date *intransitive verb*
: to go out on usually romantic dates

date rape drug *noun* (1995)
: a drug (as GHB) administered surreptitiously (as in a drink) to induce an unconscious state or sedated state in a potential date rape victim

day job *noun* (1977)
: one's regular employment as contrasted with an occasional, secondary, or coveted job

day trader *noun* (1953)
: a speculator who seeks profit from the intraday fluctuation in the price of a security or commodity by completing double trades of buying and selling or selling and covering during a single session of the market
— day–trade \'dā-ˌtrād\ *noun or verb*

dead–cat bounce *noun* [from the facetious notion that even a dead cat would bounce slightly if dropped from a sufficient height] (1985)
: a brief and insignificant recovery (as of stock prices) after a steep decline

dead–tree \'ded-'trē\ *adjective* (1993)
: being the print version of a work available in both print and electronic formats ⟨the *dead-tree* edition of the paper⟩

dead presidents *noun plural* (1944)
slang **:** U.S. money in the form of bills; *specifically* **:** DOLLARS

dear *adverb* (before 12th century)
1 : DEARLY 3 (in this list) ⟨the effort cost them *dear*⟩
2 : DEARLY 1 (in this list) ⟨so *dear* I loved the man —Shak.⟩

dear·ly *adverb* (13th century)
1 : with affection **:** FONDLY
2 : HEARTILY, EARNESTLY ⟨prayed so *dearly* for peace⟩
3 : at a high rate or price ⟨paid *dearly* for the error⟩

death·care \'deth-ˌker\ *adjective* (1987)
: of, relating to, or providing products or services for the burial or cremation of the dead ⟨the *deathcare* industry⟩

DEF·CON \'def-ˌkän\ *noun* [*def*ense *con*dition] (1962)
: any of five levels of U.S. military defense readiness ranked according to the perceived threat to national security

def \'def\ *adjective* **def·fer; def·fest** [probably alter of *death* (from the phrase *to death* excessively)] (1979)
slang **:** COOL 7

de·frag \dē-'frag\ *transitive verb* **de·fragged; de·frag·ging** (1988)
: DEFRAGMENT (in this list)

de·frag·ment \(ˌ)dē-'frag-mənt\ *transitive verb* (1985)
: to reorganize separated fragments of related data on (a computer disk) into a contiguous arrangement

de·gen·der·ize \dē-'jen-də-ˌrīz\ *transitive verb* **-ized; -iz·ing** (1987)
: to eliminate any reference to a specific gender in (as a word, text, or act)

deke \'dēk\ *verb* **deked; deke·ing** [short for *decoy*]
transitive verb (1960)
: to fake (an opponent) out of position (as in ice hockey)
intransitive verb
: to deke an opponent
— deke *noun*

delt \'delt\ *noun* (1980)
: DELTOID — usually used in plural

de min·i·mis \dē-'mi-nə-məs, dā-'mē-ni-mis\ *adjective* [New Latin, concerning trifles] (1952)
: lacking significance or importance **:** so minor as to merit disregard ⟨*de minimis* fringe benefits⟩

desktop *noun*
: an area or window on a computer screen in which icons are arranged in a manner analogous to objects on top of a desk

dex·fen·flur·a·mine \ˌdeks-'fen-flu̇r-ə-ˌmēn\ *noun* [*dex*trorotatory] (1987)
: the dextrorotatory form of fenfluramine formerly used to treat obesity but withdrawn due to its association with heart valve disease — compare FEN-PHEN (in this list)

d4T \ˌdē-(ˌ)fȯr-'tē\ *noun* [*d*ideoxy-*4*-*t*hymidine] (1988)
: a synthetic antiretroviral nucleoside $C_{10}H_{12}N_2O_4$ that is an analog of thymidine and is administered orally in the treatment of HIV infection — called also *stavudine*

dialog box *noun* (1984)
: a window on a computer screen for choosing options or inputting information

di·ge·ra·ti \ˌdi-jə-'rä-(ˌ)tē\ *noun plural* [*dig*ital + *-erati* (as in *literati*)] (1992)
: persons well versed in computer use and technology

digital subscriber line *noun* (1984)
: a high-speed communications connection used for accessing the Internet and carrying short-range transmissions over ordinary telephone lines

digital versatile disc *or* **digital video disc** *noun* (1982)
: DVD (in this list)

diminishing returns *noun*
: benefits that beyond a certain point fail to increase in proportion to extended efforts

disconnect *noun* (1976)
: a lack of or break in connection, consistency, or agreement

dis *also* **diss** \'dis\ *noun* (1990)
1 *slang* **:** a disparaging remark or act **:** INSULT ⟨was meant as a tribute, not a *dis* — *Vibe*⟩
2 *slang* **:** DISRESPECT

distance learning *noun* (1972)

: learning that takes place via electronic media linking instructors and students who are not together in a classroom

DNA virus *noun* (1963)
: a virus whose genome consists of DNA

do·main *noun*
: the highest taxonomic category in biological classification ranking above the kingdom

domestic partner *noun* (1975)
1 : a company especially in a developing country that joins in a commercial venture with an international company
2 : either one of an unmarried heterosexual or homosexual cohabiting couple especially when considered as to eligibility for spousal benefits

dope *adjective* (1981)
slang **:** EXCELLENT — used as a generalized term of approval

dot–com \'dät-ˌkäm\ *noun, often attributive* [from the use of *.com* in the URLs of such companies] (1994)
: a company that markets its products or services on-line via a Web site

dot–com·mer \-ˌkä-mər\ *noun* (1997)
: a person who owns or works for a dot-com

down *adjective*
1 *slang* **:** COOL 7
2 *slang* **:** understanding or supportive of something or someone — usually used with *with* ⟨trying to prove that they were *down* with hip-hop culture — J. E. White⟩

download *noun* (1985)
: the act or an instance of downloading something; *also* **:** the item downloaded

downtown *adjective*
: HIP, TRENDY ⟨*downtown* music⟩

drag–and–drop *adjective* (1989)
: of, relating to, or allowing movement of items on a computer screen by dragging them and fixing their new location by releasing the mouse button ⟨a *drag-and-drop* interface⟩

drama queen *noun* (1979)
: a person given to often excessively emotional performances or reactions

dream catcher *noun* (1989)
: a circular framed net with a hole in the center that is used by some American Indian peoples to help block bad dreams and catch good ones

dream·scape \'drēm-ˌskāp\ *noun* (1948)
: a dreamlike usually surrealistic scene; *also* **:** a painting of a dreamscape

DRG \ˌdē-(ˌ)är-'jē\ *noun* (1980)
: any of the payment categories that are used to classify patients and especially Medicare patients for the purpose of reimbursing hospitals for each case in a given category with a fixed fee regardless of actual costs incurred — called also *diagnosis related group*

dub·ni·um \'düb-nē-əm, 'dəb-\ *noun* [New Latin, from *Dubna,* city in Russia where a center for investigation of heavy elements is located] (1994)
: a short-lived radioactive element that is artificially produced

duh \'də, *usu with prolonged* ə\ *interjection* (1974)
1 — used to express actual or feigned ignorance or stupidity
2 — used derisively to indicate that something just stated is all too obvious or self-evident

DVD \ˌdē-(ˌ)vē-'dē\ *noun* [*d*igital *v*ideo *d*isc] (1993)
: a high-capacity optical disk format; *also* **:** an optical disk using such a format and containing especially a video recording (as a movie) or computer data

e- *combining form* [*e-mail*]
: electronic ⟨*e*-commerce⟩

ear·bud \'ir-ˌbəd\ *noun* (1984)
: a small earphone inserted into the ear

Earl Grey *noun* [Charles Grey, 2d *Earl Grey* †1845 English statesman] (1958)
: a black-tea blend flavored with bergamot oil

eating disorder *noun* (1984)
: any of several psychological disorders (as anorexia nervosa or bulimia) characterized by serious disturbances of eating behavior

eau de toi·lette \ˌō-də-twä-'let\ *noun, plural* **eaux de toilette** \ˌō(z)-\ *or* **eaux de toilettes** \ˌō(z)-də-twä-'let(s)\ *or* **eau de toilettes** \ˌō-də-twä-'let(s)\ [French, literally, water for washing and dressing] (1907)
: a perfumed liquid containing a lower percentage of fragrant oils than does ordinary perfume — called also *toilet water*

Ebon·ics \ē-'bä-niks, i-, e-\ *noun plural but singular in construction* [blend of *ebony* and *phonics*] (1973)
: BLACK ENGLISH

ech·i·na·cea \ˌe-ki-'nā-sē-ə, -shə\ *noun* [New Latin, genus name, from *echin-* + *-acea* (feminine of *-aceus* -aceous)] (1921)
: the dried rhizome, roots, or other parts of any of three composite herbs (*Echinacea angustifolia, E. pallida,* and *E. purpurea*) used in folk medicine and some patent medicines especially for a supposed beneficial effect on the immune system

eco·fem·i·nism \ˌē-kō-'fe-mə-ˌni-zəm, ˌe-kō-\ *noun* (1987)
: a movement or theory that applies feminist principles and ideas to ecological issues
— eco·fem·i·nist \-nist\ *noun or adjective*

E. coli \ˌē-'kō-ˌlī\ *noun, plural* **E. coli** (1926)
: a straight rod-shaped gram-negative bacterium (*Escherichia coli* of the family Enterobacteriaceae) occurring in various strains that are used in medical and genetic research, live as harmless inhabitants of the human lower intestine, are used in public health as indicators of fecal pollution (as of water or food), or produce a toxin causing intestinal illness

eco·ter·ror·ism \ˌē-kō-'ter-ər-i-zəm, ˌe-kō-\ *noun* (1987)
1 : sabotage intended to hinder activities that are considered damaging to the environment
2 : political terrorism intended to damage an enemy's natural environment
— eco·ter·ror·ist \-ər-ist\ *noun or adjective*

eco·tour \-ˌtu̇r\ *noun* (1987)
: a trip conducted by the principles of ecotourism

ecphrasis *var of* EKPHRASIS

edge city *noun* (1988)
: a suburb that has developed its own political, economic, and commercial base independent of the central city

ek·phra·sis *also* **ec·phra·sis** \'ek-frə-səs\ *noun, plural* **ek·phra·ses** *also* **ec·phra·ses** \-ˌsēz\ [Greek *ekphrasis,* literally, description, from *ekphrazein* to recount, describe, from *ex-* out + *phrazein* to point out, explain] (1715)
: a literary description of or commentary on a visual work of art

elec·tron·i·ca \i-ˌlek-'trä-ni-kə\ *noun* [probably from *New Electronica,* recording label of the British firm Beechwood Music Ltd.] (1994)
: dance music featuring extensive use of synthesizers, electronic percussion, and samples of recorded music or sound

elephant garlic *noun* (circa 1923)
: an Old World herb (*Alium ampeloprasum*) of the lily family that is related to the leek and has a bulb that resembles that of garlic but is much larger; *also* **:** the mildly flavored bulb of the elephant garlic used especially as a seasoning

Elo·him \ˌe-lō-'hēm, e-'lō-ˌhim\ *noun* [Hebrew *ĕlōhīm*] (1617)
: GOD 1a — used especially in the Hebrew Bible

e–mail \'ē-ˌmāl\ *noun* (1982)
: a means or system for transmitting messages electronically (as between terminals linked by telephone lines or microwave relays)
— e–mail *verb*
— e–mail·er \-ˌmā-lər\ *noun*

emo·ti·con \i-'mō-ti-ˌkän\ *noun* [*emot*ion + *icon*] (1990)
: a group of keyboard characters (as :-)) typically representing a facial expression or an emotion or otherwise conveying tone or attitude that is used especially in computerized communications (as e-mail)

empty suit *noun* (1950)
: an ineffectual executive

en·abler \i-'nā-blər, -bᵊl-ər\ *noun* (1615)
: one that enables another to achieve an end; *especially* **:** one who enables another to persist in self-destructive behavior (as substance abuse) by providing excuses or by helping that individual avoid the consequences of such behavior

en·cul·tur·a·tion \in-ˌkəl-chə-'rā-shən, (ˌ)en-\ *noun* (1948)
: the process by which an individual learns the traditional content of a culture and assimilates its practices and values
— en·cul·tur·ate \-'kəl-chə-ˌrāt\ *verb*

end·cap \'end-ˌkap\ *noun* (1983)
: a display of products placed at the end of an aisle in a store

end–stage \'end-ˌstāj\ *adjective* (1977)
: being or occurring in the final stages of a terminal disease or condition ⟨*end-stage* renal failure⟩

ephe·dra *noun*
: an extract of ma huang containing ephedrine and related alkaloids and used as a dietary supplement

equine encephalitis *noun* (1946)
: any of three forms of encephalitis that attack chiefly equines and humans in various parts of North and South America and are caused by three related single-stranded RNA viruses (genus *Alphavirus* of the family Togaviridae)

era *noun*
: a stage in development (as of a person or thing)

es·tan·cia \e-'stän(t)s-(ˌ)yä\ *noun* [American Spanish, from Spanish, stay, room, from Vulgar Latin **stantia* — more at STANCE] (1704)
: a South American cattle ranch or stock farm

estate–bottled *adjective* (1940)
of a wine **:** entirely produced or bottled by a single winery

estrogen replacement therapy *noun* (1967)
: hormone replacement therapy involving the administration of estrogen without progestin

euro *noun, plural* **euros** *also* **euro** [short for the equivalent of *Europe* or *European* in the languages of the European Union] (1981)
: the basic monetary unit shared by countries of the European Union since 1999

Eu·ro \'yu̇r-(ˌ)ō\ *adjective* (1963)
: EUROPEAN
— Euro *noun*

European flat *noun* (1981)
: a flat-shelled European oyster (*Ostrea edulis*)

evangelist *noun*
: an enthusiastic advocate ⟨an *evangelist* for physical fitness⟩

ex·fo·liant \(ˌ)eks-'fō-lē-ənt, -'fōl-yənt\ *noun* (1983)
: a mechanical or chemical agent (as an abrasive skin wash or salicylic acid) that is applied to the skin to remove dead cells from the surface

expansion card *noun* (1982)
: a circuit board connecting to a motherboard which expands the capabilities of a computer

expansion slot *noun* (1980)
: a socket on the motherboard of a computer into which an expansion card may be inserted

ex·tra–vir·gin \'ek-strə-'vər-jən\ *adjective* (1980)
: being a virgin olive oil that is lowest in acidity and highest in quality

ex·tra·net \'ek-strə-ˌnet\ *noun* (1995)
: a network (as of a company) similar to an intranet that also allows access by certain others (as customers or suppliers)

ex·treme *adjective*
1 : of, relating to, or being an outdoor activity or a form of a sport (as skiing) that involves an unusually high degree of physical risk ⟨*extreme* mountain biking down steep slopes⟩

2 : involved in an extreme sport ⟨an *extreme* snowboarder⟩
eyeballs *noun plural*
: people who view something (as an advertisement) ⟨Web sites competing for *eyeballs*⟩
eye candy *noun* (1981)
: something superficially attractive to look at
face time *noun* (1978)
1 : the amount of time one spends appearing on television
2 : time spent in a face-to-face meeting with someone
3 : time spent at one's place of employment especially beyond normal work hours
fade *noun*
: a haircut similar to a crew cut in which the hair on top of the head stands high
family leave *noun* (1981)
: a usually unpaid leave of absence for an employee to attend to family concerns (as care of an infant or a serious illness)
far·fal·le \fär-ˈfä-(ˌ)lā, -lē\ *noun* [Italian, plural of *farfalla,* literally, butterfly] (1941)
: butterfly-shaped pasta
fash·ion·is·ta \ˌfa-shə-ˈnēs-tə\ *noun* [*fashion* + *-ista* (as in *Sandinista*)] (1993)
: a designer, promoter, or follower of the latest fashions
fast–track \ˈfast-ˌtrak\ *adjective*
: of or relating to authority granted to the President of the U.S. by Congress that allows the President to negotiate trade agreements which Congress must confirm or reject in their entirety
fault–tol·er·ant \ˈfȯlt-ˌtä-lə-rənt\ *adjective* (1975)
: relating to or being a computer or program with a self-contained back-up system that allows continued operation when major components fail
— fault tolerance *noun*
fee–for–service *noun, often attributive* (1945)
: separate payment to a health-care provider for each medical service rendered to a patient ⟨a *fee-for-service* health plan⟩
fen·flur·amine \ˌfen-ˈflu̇r-ə-ˌmēn\ *noun* [*fen-* (alteration of *phen-*) + *-flur-* (alteration of *fluor-*) + *amine*] (1965)
: an anorectic amphetamine derivative $C_{12}H_{16}F_3N$ with little stimulant effect on the central nervous system formerly used in the form of its hydrochloride to treat obesity but no longer used due to its associations with heart valve disease
feng shui \ˈfəŋ-ˈshwē, -ˈshwā\ *noun* [Chinese (Beijing) *fēngshuǐ* geomantic omen, literally, wind-water] (1797)
: a Chinese geomantic practice in which a structure or site is chosen or configured so as to harmonize with the spiritual forces that inhabit it; *also* **:** auspicious orientation, placement, or arrangement as determined by feng shui
fen–phen \ˈfen-ˌfen\ *noun* [*fen*fluramine + *phen*termine] (1994)
: a former drug combination of phentermine with either fenfluramine or dexfenfluramine — called also *phen fen*
Fer·mat's last theorem \fer-ˈmäz-\ *noun* [Pierre de *Fermat*] (1847)
: a theorem in number theory: the equation $x^n + y^n = z^n$ has no solutions when x, y, z and n are all positive integers and n is greater than 2
fi·bro·my·al·gia \ˌfī-ˌbrō-ˌmī-ˈal-j(ē)ə\ *noun* [New Latin] (1983)
: a painful rheumatic condition of uncertain cause that is characterized by diffuse or localized pain, tenderness, and stiffness of skeletal muscles and associated connective tissue and that is usually accompanied by fatigue
field *noun* **— from the field**
: in field goals as opposed to free throws ⟨made 40 percent of his shots *from the field*⟩
fi·nas·te·ride \fə-ˈnas-tə-ˌrīd\ *noun* [*fina-* (of unknown origin) + testo*ster*one + ami*de*] (1989)
: a nitrogenous steroid derivative $C_{23}H_{36}N_2O_2$ that is used especially to treat symptoms of benign prostatic hyperplasia and to increase hair growth in male-pattern baldness
fin·ca \ˈfiŋ-kə\ *noun* [Spanish] (1878)
: a rural property, ranch, or farm in Spain or Spanish America
firewall \ˈfīr-ˌwȯl\ *noun*
: a computer or computer software that prevents unauthorized access to private data (as on a company's local area network or intranet) by outside computer users (as of the Internet)
fitting *noun*
: something used in fitting up **:** ACCESSORY
five *noun*
: a slapping of extended right hands by two people (as in greeting or celebration) — usually used in phrases with *give* or *slap* ⟨so I slapped him *five* and hugged him — J. R. Burke⟩
flare *noun*
1 : a short pass in football thrown to a back who is running toward the sideline
2 : a weakly hit fly ball in baseball ⟨a *flare* into short right field⟩
flat·line \ˈflat-ˌlīn\ *intransitive verb* (1980)
1 a : to register on an electronic monitor as having no brain waves or heartbeat **b :** DIE
2 a : to be in a state of no progress or advancement **b :** to come to an end
— flat·lin·er *noun*
flat pan·el \ˈflat-ˈpa-nᵊl\ *adjective* (1977)
: relating to or being a thin flat video display (as for a portable computer)
flau·ta \ˈflau̇-tə\ *noun* [American Spanish, literally, flute]
: a usually corn tortilla rolled tightly around a filling (as of meat) and deep-fried
float·er \ˈflō-tər\ *noun*
: a bit of optical debris (as a dead cell or cell fragment) in the vitreous body or lens that may be perceived as a spot before the eye; *also* **:** a spot in the visual field due to such debris — usually used in plural
floor *noun* **— from the floor**
: in field goals as opposed to free throws ⟨made 16 of 18 shots *from the floor*⟩
flu·ox·e·tine \flü-ˈäk-sə-ˌtēn\ *noun* [International Scientific Vocabulary, probably from *flu*or- + *ox*y + m*e*thyl + ami*ne*] (1976)
: an antidepressant drug $C_{17}H_{18}F_3NO$ that enhances serotonin activity
focus group *noun* (1985)
: a small group of people whose response to something (as a new product or a politician's image) is studied to determine the response that can be expected from a larger population
folder *noun*
: an organizational element of a computer operating system used to group files or other folders together
foos·ball \ˈfüz-ˌbȯl\ *noun, often capitalized* [probably modification of German *Tischfussball,* from *Tisch* table + *Fussball* soccer, from *Fuss* foot + *Ball* ball] (1977)
: a table game resembling soccer in which the ball is moved by manipulating rods to which small figures of players are attached
foreground *noun*
: a level of computer processing at which the processor responds immediately to input to a designated high priority task
format *noun*
: a method of organizing data (as for storage) ⟨various file *formats*⟩
format *transitive verb*
: to prepare (as a computer disk) for storing in a particular format
forward air controller *noun* (1952)
: a military officer who directs from a forward position on the ground or in the air the action of combat aircraft engaged in close air support of land forces
fos·car·net \fäs-ˈkär-nət\ *noun* [probably from International Scientific Vocabulary *fos-* (alteration of *phosph-*) + *car*b- + *-net* (of unknown origin)] (1981)
: a hydrated sodium salt $Na_3CO_5 \cdot 6H_2O$ that is administered intravenously to individuals infected with HIV to treat retinitis caused by a cytomegalovirus
401(k) \ˌfōr-(ˌ)ō-(ˌ)wən-ˈkā, ˌfȯr-\ *noun* [from the section of the Internal Revenue Code that established it] (1989)
: a retirement account to which employee and employer contribute, on which taxes are deferred until withdrawal, and for which the employee selects the types of investments
411 \ˈfȯr-ˈwən-ˈwən\ *noun* [from the telephone number *411* used to reach directory assistance] (1985)
slang **:** relevant information **:** SKINNY ⟨fiber-optic *411* fed 24/7 in satellite real time — Jeff MacGregor⟩
four–peat \ˈfȯr-ˌpēt\ *noun* [*four* + three-*peat*] (1989)
: a fourth consecutive championship
— four–peat *intransitive verb*
fragile X syndrome *noun* (1979)
: an X-linked inherited disorder that is characterized especially by moderate to severe mental retardation, by a long face and large ears, and by large testes in males and that often has limited or no effect in heterozygous females — called also *fragile X*
Fran·ken·food \ˈfraŋ-kən-ˌfüd\ *noun* [*Franken-* (as in *Frankenstein*) + *food*] (1992)
: genetically engineered food
free·ware \ˈfrē-ˌwar, -ˌwer\ *noun* (1983)
: software that is available for use at no cost or for a nominal usually voluntary fee
French press *noun* (1986)
: a coffeepot in which ground beans are infused and then pressed to the bottom by means off a plunger
fric·tion *noun*
: sound produced by the movement of air through a narrow constriction in the mouth or glottis
friendly *adjective*
: compatible, accommodating ⟨environmentally *friendly* packaging⟩ — often used in combination ⟨a kid-*friendly* restaurant⟩
fri·sée *also* **fri·sé** \frē-ˈzā\ *noun* [French, short for *chicorée frisée* curly chicory] (1982)
: curly endive leaves that have finely dissected edges and are used in salads — called also *frisée lettuce*
front end *noun*
: a software interface (as a graphical user interface) designed to enable user-friendly interaction with a computer
fry *transitive verb*
: to damage or destroy (an electrical device or its circuitry) by overheating especially as a result of unusually high voltage
ful·fill *transitive verb*
: to meet the requirements of (a business order)
full bore *adverb* (1927)
: with maximum effort or speed
full–bore \ˈfu̇l-ˈbōr, -ˈbȯr\ *adjective* (1974)
1 : FULL-BLOWN 2
2 : made with maximum effort
function key *noun* (1964)
: any of a set of keys on a computer keyboard that have or can be programmed to have special functions
fun·plex \ˈfən-ˌpleks\ *noun* (1986)
: an entertainment complex that includes facilities for various sports and games and often restaurants
fusion cuisine *noun* (1986)
: food prepared using techniques and ingredients of various ethnic or regional cuisines
F/X *or* **FX** \ˌef-ˈeks\ *noun plural* [letter names *ef* and *ex* representing *effects*] (1985)
: SPECIAL EFFECTS
ga·lette \gə-ˈlet\ *noun* [French, from Old French, from *galet* rounded pebble, from Old French dialect (Picard), diminutive of *gal* pebble] (1775)
1 : a flat round cake or pastry often topped with fruit
2 : food prepared and served in the shape of a flat round cake ⟨a *galette* of potatoes⟩
gamete in·tra·fal·lo·pi·an transfer \-ˌin-trə-fə-ˈlō-pē-ən-\ *noun* (1984)
: a method of assisting reproduction in cases of infertility that involves obtaining eggs from

an ovary, mixing them with sperm, and inserting them into a fallopian tube by a laparoscope — abbreviation *GIFT*; called also *gamete intrafallopian tube transfer*

gamma interferon *noun* (1980)
: an interferon that is produced by T cells, regulates the immune response, and in a form produced by recombinant DNA technology is used especially to control infections due to inability of white blood cells to destroy certain bacteria and fungi

gan·ci·clo·vir \gan-'sī-klə-(,)vir\ *noun* [perhaps from *guanosine* + *-ciclovir,* alteration of *-cyclovir* (as in *acyclovir*)] (1986)
: an antiviral drug $C_9H_{13}N_5O_4$ related to acyclovir and used especially to treat cytomegalovirus retinitis in immunocompromised individuals

gang·sta \'gaŋ(k)-stə\ *noun, often attributive* [alteration of *gangster*] (1988)
1 : a member of an urban street gang
2 : a performer of gangsta rap

gang·sta rap \'gaŋ(k)-stə-\ *noun* [*gangsta* gang member, alteration of *gangster*] (1991)
: rap music with lyrics explicitly relating to urban gangs, violence, drug use, and the degradation of women

garage band *noun* (1972)
: an amateur rock band holding its rehearsals in a garage and usually having only a local audience

gastroesophageal reflux *noun* (1966)
: backward flow of the contents of the stomach into the esophagus that is due to improper functioning of a sphincter at the lower end of the esophagus and results especially in heartburn

gated community *noun* (1981)
: a residential area protected by a private security force, enclosed by physical barriers, and entered through a controlled gate

gay·dar \'gā-,där\ *noun* [blend of *gay* and *radar*] (1982)
slang **:** the ability to recognize homosexuals through observation or intuition

ga·zil·lion \gə-'zil-yən\ *noun* [alteration of *zillion*] (1978)
: an indeterminately large number
— gazillion *adjective*

ga·zoo \gə-'zü\ *noun, plural* **gazoos** [origin unknown] (1965)
slang **:** WAZOO (in this list)

geek *noun*
: an enthusiast or expert especially in a technological field or activity ⟨computer *geek*⟩
— geek·dom \'gēk-dəm\ *noun*
— geek·i·ness *noun*

gel·cap \'jel-,kap\ *noun* (1986)
: a capsule-shaped tablet coated with gelatin for easy swallowing

gel ice *noun* (1983)
: a gel matrix in a solution that is frozen in a pack especially for use in preserving perishable products

Generation X *noun* (1989)
: the generation of Americans born in the 1960s and 1970s
— Generation Xer \-'ek-sər\ *noun*

genetic fingerprinting *noun* (1984)
: DNA FINGERPRINTING
— genetic fingerprint *noun*

ge·no·mics \jē-'nō-miks, jə-\ *noun plural but singular in construction* (1987)
: a branch of biotechnology concerned with applying the techniques of genetics and molecular biology to the genetic mapping and DNA sequencing of sets of genes or the complete genomes of selected organisms, with organizing the results in a database, and with applications of the data (as in medicine or biology)

Gen X \'jen-'eks\ *noun* (1992)
: GENERATION X (in this list) — often hyphenated in attributive use ⟨a *Gen-X* celebrity⟩
— Gen Xer \-'ek-sər\ *noun*

GHB \,jē-(,)āch-'bē\ *noun* (1964)
: a metabolite $C_4H_8O_3$ of gamma-aminobutyric acid that is a depressant of the central nervous system and is used illicitly in the form of its synthetic sodium salt to produce sedative and euphoric effects or to stimulate release of growth hormone to increase muscle mass — called also *gamma hydroxybutyrate*

GIF \'gif, 'jif\ *noun* [graphic interchange format] (1987)
: a computer file format for the compression and storage of digital video images; *also* **:** such an image itself

giga·flop \-,fläp\ *noun* [*f*loating-point *op*eration] (1976)
: a unit of measure for the speed of calculation of a computer equal to one billion floating-point operations per second

gig \'gig\ *noun* (1987)
: GIGABYTE

gimme cap *noun* (1978)
: an adjustable visored cap that often features a corporate logo or slogan

ginkgo bi·lo·ba \-,bī-'lō-bə\ *noun* [New Latin, literally, bilobed ginkgo] (1980)
: an extract of the leaves of gingko that is held to enhance mental functioning by increasing blood circulation to the brain

giv·en *adjective*
: assumed as actual or hypothetical **:** GRANTED ⟨*given* that all are equal before the law⟩

Global Positioning System *noun* (1975)
: GPS

global warm·ing \-'wȯr-miŋ\ *noun* (1969)
: an increase in the earth's atmospheric and oceanic temperatures widely predicted to occur due to an increase in the greenhouse effect resulting especially from pollution

Gloomy Gus \-'gəs\ *noun, plural* **Gloomy Gus·es** *also* **Gloomy Gus·ses** *often not capitalized 1st G* [from a comic-strip character created by Frederick Burr Opper †1937 American cartoonist]
: a person who is habitually gloomy

gly·ce·mia \glī-'sē-mē-ə\ *noun* [New Latin] (1901)
: the presence of glucose in the blood
— gly·ce·mic \-'sē-mik\ *adjective*

glycemic index *noun* (1981)
: a measure of the rate at which ingested food causes the level of glucose in the blood to rise; *also* **:** a ranking of foods according to the glycemic index

gnarly *adjective*
1 *slang* **:** DIFFICULT, HAIRY ⟨skidded around some *gnarly* hairpin turns — Austin Murphy⟩
2 *slang* **:** BAD, NASTY ⟨has some pretty *gnarly* karma coming — Drew Barrymore⟩
3 *slang* **:** COOL, GOOD

golden handcuffs *noun plural* (1976)
: special benefits offered to an employee as an inducement to continue service

go–to \'gō-,tü\ *adjective* (1985)
: relied on for expert knowledge or skill ⟨the company's *go-to* guy⟩

gon·zo *adjective*
: freewheeling or unconventional especially to the point of outrageousness ⟨a *gonzo* comedian⟩

good cholesterol *noun* (1980)
: HDL

goo·gle \'gü-gəl\ *transitive verb* **goo·gled; goo·gling** \-g(ə-)liŋ\ *often capitalized* [*Google,* trademark for a search engine] (2001)
: to use the Google search engine to obtain information about (as a person) on the World Wide Web

goom·bah \'güm-,bä\ *noun* [Italian dialect (Campania) *cumbà,* vocative form of *cumbare* respected older man, literally, godfather, from Medieval Latin *compater* — more at COMPEER] (1968)
1 : a close friend or associate — used especially among Italian-American men
2 : a member of a secret chiefly Italian-American crime organization **:** MAFIOSA; *broadly* **:** GANGSTER
3 : a macho Italian-American man

got·cha \'gä-chə\ *noun* [alteration of *got you*] (1974)
: an unexpected usually disconcerting challenge, revelation, or catch

Goth *noun*
1 *often not capitalized* **a :** rock music marked by dark and morbid lyrics **b :** a fan or performer of goth
2 : a person who wears mostly black clothing, uses dark dramatic makeup, and often has dyed black hair

GPS \,jē-(,)pē-'es\ *noun* [*G*lobal *P*ositioning *S*ystem] (1975)
: a navigational system using satellite signals to fix the location of a radio receiver on or above the earth's surface; *also* **:** the radio receiver so used

graphical user interface *noun* (1988)
: a computer program designed to allow a computer user to interact easily with the computer typically by using a mouse to make choices from menus or groups of icons

gray water *noun* (1977)
: household wastewater (as from a sink or bath) that does not contain serious contaminants (as from toilets or diapers)

green *adjective*
: tending to preserve environmental quality (as by being recyclable, biodegradable, or nonpolluting)

green·field \'grēn-,fēld\ *noun, often attributive* (1972)
: undeveloped land especially when unpolluted ⟨a *greenfield* site⟩

gross domestic product *noun* (1970)
: the gross national product excluding the value of net income earned abroad

group·ware \'grüp-,wer\ *noun* (1980)
: software that enables users to work collaboratively on projects or files via a network

Gulf War syndrome *noun* (1992)
: a syndrome of uncertain cause including fatigue, joint pain, memory loss, skin rash, and headache that has been reported in veterans of the war fought in the Persian Gulf in 1991

gut check *noun* (1972)
: a test or assessment of courage, character, or determination

ha·ba·ne·ro *also* **ha·ba·ñe·ro** \,(h)ä-bə-'n(y)er-ō\ *noun* [American Spanish (*chile*) *habanero,* literally, Havanan chili] (1988)
: a very hot roundish chili pepper (*Capsicum chinense*) that is usually orange when mature

hack *noun*
: a usually creative solution to a computer hardware or programming problem or limitation

Hail Mary *noun*
: a long forward pass in football thrown into or near the end zone in a last-ditch attempt to score as time runs out

half–pipe \'haf-,pīp, 'háf-\ *noun* (1984)
: a V-shaped high-sided ramp or runway used in snowboarding, skateboarding, or in-line skating

half·tone *noun*
: an image (as one printed on an offset press or laser printer) that renders smooth variations of color in an original by means of dots assigned to areas of the image electronically

Han·a·fi \'ha-nə-(,)fē\ *adjective* [Arabic *ḥanafī,* from Abū *Ḥanīfa* †767 Muslim jurist] (1913)
: of or relating to an orthodox school of Sunni Muslim jurisprudence followed especially in southern and central Asia

han·ta·vi·rus \'han-tə-,vī-rəs, 'hən-, 'hän-\ *noun* [New Latin, from *hanta-* (from *Hantaan,* river in South Korea near where rodents carrying the virus were collected 1974–78) + *virus*] (1984)
: any of a genus (*Hantavirus* of the family Bunyaviridae) of RNA viruses that are transmitted by rodent feces and urine and cause acute respiratory illness and hemorrhagic fever marked by renal necrosis

hardwired *adjective*
1 : connected or incorporated by or as if by permanent electrical connection ⟨a *hardwired* phone⟩ ⟨concepts of attractiveness may be universal and *hardwired* into the human brain — Jane E. Brody⟩

2 : genetically or innately determined : INBORN ⟨creature whose every action is a reflexive, *hardwired* response — Natalie Angier⟩; *also* : genetically or innately predisposed ⟨a human being who is *hardwired* to be sociable — *Forbes*⟩

has·si·um \'ha-sē-əm\ *noun* [New Latin, from *Hassia* Hesse (German state), location of the laboratory that first produced the element] (1992)
: a short-lived radioactive element that is artificially produced

haz·mat \'haz-,mat\ *noun, often attributive* [*haz*ardous *mat*erial] (1980)
: a material (as flammable or poisonous material) that would be a danger to life or to the environment if released without precautions

head·bang·er \'hed-,baŋ-ər\ *noun* (1979)
: a musician who performs hard rock; *also* : a fan of hard rock

head·hunt \'hed-,hənt\ *transitive verb* (1969)
: to recruit (personnel and especially executives) for top-level jobs ⟨was *headhunted* by three different firms⟩
intransitive verb
: to recruit personnel for top-level jobs

heart–healthy \'härt-,hel-thē\ *adjective* (1980)
: conducive to a healthy heart and circulatory system ⟨*heart-healthy* exercise⟩ ⟨a *heart-healthy* diet⟩

he·li·o·sphere \'hē-lē-ə-,sfir, -ō-\ *noun* [International Scientific Vocabulary] (1976)
: the region in space influenced by the sun or solar wind

hepatitis C *noun* (1980)
: a hepatitis caused by a single-stranded RNA virus (family Flaviviridae) that is usually transmitted by parenteral means (as injection of an illicit drug, blood transfusion, or exposure to blood or blood products)

hep·tath·lete \hep-'tath-,lēt\ *noun* [blend of *heptathlon* and *athlete*] (1981)
: an athlete who competes in a heptathlon

hep·tath·lon \hep-'tath-lən, -,län, ÷-'ta-thə-\ *noun* [*hept-* + *-athlon* (as in *decathlon*)] (1977)
: a 7-event athletic contest; *specifically* : a composite contest for female athletes that consists of the 100-meter hurdles, the high jump, the shot put, the 200-meter dash, the long jump, the javelin throw, and the 800-meter run

het·ero·so·cial \,he-tə-rō-'sō-shəl\ *adjective* (1965)
: of, relating to, or involving social relationships between persons of the opposite sex

Hib \'hib\ *noun, often attributive* [*Hi-* (from the species name *Haemophilus influenzae* + type *B*] (1984)
: a bacterial serotype (*Haemophilus influenzae* type B) that causes bacterial meningitis and pneumonia especially in children ⟨*Hib* disease⟩ ⟨*Hib* vaccine⟩

hic·cup *noun*
1 : a slight irregularity, error, or malfunction ⟨a few *hiccups* in the computer system⟩
2 : a usually minor and short-lived interruption, disruption, or change ⟨a *hiccup* in the stock market⟩

high–definition *adjective* (1981)
: being or relating to a television system that has twice as many scan lines per frame as a conventional system, a proportionally sharper image, and a wide-screen format

him·bo \'him-(,)bō\ *noun, plural* **himbos** [blend of *him* and *bimbo*] (1988)
: an attractive but vacuous man

hin·ky \'hiŋ-kē\ *adjective* **hin·ki·er; hin·ki·est** [alteration of argot *hincty* suspicious] (1956)
1 *slang* : NERVOUS, JITTERY
2 *slang* : SUSPICIOUS

hip–hop·per \'hip-,hä-pər\ *noun* (1983)
: a devotee of hip-hop and culture; *also* : a performer of hip-hop

hissy fit *noun* (circa 1970)
: TANTRUM

hit *noun*
: an instance of connecting to a particular Web site ⟨a million *hits* per day⟩

Hmong \'məŋ\ *noun, plural* **Hmong** [Hmong *hmoŋ* (with high level tone), a self-designation] (1977)
1 : a member of a mountain-dwelling people inhabiting southeastern China and the northern parts of Vietnam, Laos, and Thailand
2 : the language of the Hmong people

ho \'hō\ *noun, plural* **hos** *or* **hoes** [alteration of *whore*] (1965)
slang : WHORE 1

home·girl \'hōm-,gərl\ *noun* (1934)
1 : a girl or woman from one's neighborhood, hometown, or region
2 : a girl or woman who is a member of one's peer group

home page *noun* (1992)
: the page typically encountered first at a World Wide Web site that usually contains links to the other pages of the site or to other sites

home·school·er \-,skü-lər\ *noun* (1981)
1 : one that homeschools
2 : a child who is homeschooled

ho·mo·so·cial \-'sō-shəl\ *adjective* (1968)
: of, relating to, or involving social relationships between persons of the same sex and especially between men
— **ho·mo·so·ci·al·i·ty** \-,sō-shē-'a-lə-tē\ *noun*

hooch·ie \'hü-chē\ *noun* [perhaps from *hootchy* (as in *hootchy-kootchy* exotic dance)] (1991)
slang : a sexually promiscuous young woman

hood·ie \'hù-dē\ *noun* [*hood* + *-ie*] (1992)
: a hooded sweatshirt

hormone replacement therapy *noun* (1966)
: the administration of estrogen often along with synthetic progestin to ameliorate the symptoms of menopause and reduce the risk of postmenopausal osteoporosis

hos·pi·tal·ist \'häs-(,)pi-tə-list\ *noun* (1996)
: a physician who specializes in treating hospitalized patients of other physicians in order to minimize the number of hospital visits by other physicians

hot key *noun* (1983)
: a key or combination of keys on a computer keyboard programmed to perform a specific function when pressed

hot link *noun* (1989)
: HYPERLINK (in this list)

HTML \,āch-(,)tē-(,)em-'el\ *noun* [*h*yper*t*ext *m*arkup *l*anguage] (1992)
: a markup language that is a subset of SGML and is used to create hypertext and hypermedia documents on the World Wide Web incorporating text, graphics, sound, video, and hyperlinks

hub *noun*
: a central device that connects multiple computers on a single network

hy·dro·mas·sage \,hī-drō-mə-'säzh, -'säj\ *noun* (1940)
1 : a massage using jets of water
2 : SPA 5

hype *adjective* (1989)
slang : EXCELLENT, COOL

hy·per·link \'hī-pər-,liŋk\ *noun* (1989)
: an electronic link providing direct access from one distinctively marked place in a hypertext or hypermedia document to another in the same or a different document
— **hyperlink** *transitive verb*

hy·per·me·dia \-,mē-dē-ə\ *noun* (1965)
: a database format similar to hypertext in which text, sound, or video images related to that on a display can be accessed directly from the display

hypertext transfer protocol *noun* (1992)
: a communications protocol governing the exchange of data (as HTML) files especially on the World Wide Web — called also *hypertext transport protocol*

ice–cream headache *noun* (1964)
: BRAIN FREEZE (in this list)

identity theft *noun* (1991)
: the illegal use of someone else's personal information (as a Social Security number) in order to obtain money or credit

im·pel·ler *noun*
: a rotor located in a conduit to impart motion to a fluid

import *transitive verb*
: to transfer (as files or data) from one format to another usually within a new file

in·ar·gu·ably \-blē\ *adverb* (1925)
: it cannot be argued ⟨*inarguably*, December is the best month for retailers⟩

in·cent \in-'sent\ *transitive verb* [back-formation from *incentive*] (1981)
: INCENTIVIZE (in this list)

in·cen·tiv·ize \in-'sen-tə-,vīz\ *transitive verb* **in·cen·tiv·ized; in·cen·tiv·iz·ing** (1970)
: to provide with an incentive

indemnity *noun*
: FEE-FOR-SERVICE (in this list) — usually used attributively ⟨an *indemnity* plan⟩

information technology *noun* (1978)
: the technology involving the development, maintenance, and use of computer systems, software, and networks for the processing and distribution of data

in·fo·tech \'in-fō-,tek\ *noun* (1981)
: INFORMATION TECHNOLOGY (in this list)

infuse *transitive verb*
: to administer or inject by infusion ⟨stem cells were *infused* into the patient⟩

in·stal·la·tion *noun*
: a work of art that usually consists of multiple components often in mixed media and that is exhibited in a usually large space in an arrangement specified by the artist

insulin–dependent diabetes *noun* (1980)
: TYPE I DIABETES (in this list)

insulin–dependent diabetes mellitus *noun* (1980)
: TYPE I DIABETES (in this list) — abbreviation *IDDM*

insulin resistance syndrome *noun* (1992)
: METABOLIC SYNDROME (in this list)

intelligent design *noun* (1847)
: the theory that matter, the various forms of life, and the world were created by a designing intelligence

in·ter·est *noun*
: the profit in goods or money that is made on invested capital

In·ter·net \'in-tər-,net\ *noun* (1986)
: an electronic communications network that connects computer networks and organizational computer facilities around the world

in·ter·tex·tu·al·i·ty \-,teks-chə-'wa-lə-tē\ *noun, plural* **in·ter·tex·tu·al·i·ties** [French *intertextualité*, from *inter-* + *textuel* textual + *-ité* -ity] (1975)
: the complex interrelationship between a text and other texts taken as basic to the creation or interpretation of the text
— **in·ter·tex·tu·al** \-'teks-chə-wəl\ *adjective*
— **in·ter·tex·tu·al·ly** *adverb*

in·ti·fa·da \,in-tə-'fä-də\ *noun* [Arabic *intifāḍa*, literally, the act of shaking off] (1985)
: UPRISING, REBELLION; *specifically* : an armed uprising of Palestinians against Israeli occupation of the West Bank and Gaza Strip

in·tra·net \'in-trə-,net\ *noun* (1983)
: a network operating like the World Wide Web but having access restricted to a limited group of authorized users (as employees of a company)

investigational *adjective*
: relating to or being a drug or medical procedure that is not approved for general use but is under investigation in clinical trials regarding its safety and efficacy ⟨an *investigational* new drug⟩

IP address \'ī-'pē-ə-'dres\ *noun* [*I*nternet *p*rotocol] (1988)
: the numeric address of a computer on the Internet

irritable bowel syndrome *noun* (1943)
: a chronic functional disorder of the colon that is characterized especially by constipation

or diarrhea, cramping abdominal pain, and the passage of mucus in the stool — abbreviation *IBS*

Is·lam·ism *noun*
: a popular reform movement advocating the reordering of government and society in accordance with laws prescribed by Islam

iso·thio·cy·a·nate \ˌī-sō-ˌthī-ō-ˈsī-ə-ˌnāt\ *noun* (1891)
: a compound containing the monovalent group −NCS

it·er·a·tion *noun*
: VERSION, INCARNATION ⟨the latest *iteration* of the operating system⟩

jar·gon·ish \ˈjär-gə-nish\ *adjective* (1816)
: JARGONISTIC (in this list)

jar·gon·is·tic \ˌjär-gə-ˈnis-tik\ *adjective* (1929)
: characterized by the use of jargon **:** phrased in jargon

Jersey barrier *noun* [New *Jersey,* U.S.] (1981)
: a concrete slab 32 inches high with slanted sides that is used in tandem with others to block or reroute traffic or to divide highways

Jet·way \ˈjet-ˌwā\ *trademark*
— used for a telescoping passenger ramp between an aircraft and a terminal building

ji·had *noun*
: a personal struggle in devotion to Islam that may include spiritual discipline, verbal defense of beliefs, and physical combat

ji·had·ist \ji-ˈhä-dist, *chiefly British* -ˈha-\ *noun* (1989)
: a Muslim who advocates or participates in a jihad
— jihadist *adjective*

job *noun*
: plastic surgery for cosmetic purposes ⟨a nose *job*⟩

jock *noun*
: a person devoted to a single pursuit or interest ⟨computer *jocks*⟩

Joe Six-Pack *noun* [from the stereotype of a six-pack of beer as a workingman's drink] (1975)
: an ordinary man; *specifically* **:** a blue-collar worker

joke·ster \ˈjōk stər\ *noun* (1877)
: JOKER 1

jones *intransitive verb* (1974)
slang **:** to have a strong desire or craving for something ⟨he was *jonesing* for a drink⟩

jones \ˈjōnz\ *noun* [origin unknown] (1965)
1 *slang* **:** HABIT, ADDICTION; *especially* **:** addiction to heroin
2 *slang* **:** HEROIN
3 *slang* **:** an avid desire or appetite for something **:** CRAVING

JPEG \ˈjā-ˌpeg\ *noun* [Joint Photographic Experts Group] (1988)
: a computer file format for the compression and storage of usually high-quality photographic digital images

juiced *adjective*
: full of energy and motivation **:** EXCITED

June·teenth \ˌjün-ˈtēn(t)th\ *noun* [blend of *June* and *nineteenth*] (1940)
: June 19 celebrated especially in Texas to commemorate the announcement of the Emancipation Proclamation in 1865

junk DNA *noun* (1972)
: a region of DNA that usually consists of a repeating DNA sequence, does not code for protein, and has no known function

jury nullification *noun* (1982)
: the acquitting of a defendant by a jury in disregard of the judge's instructions and contrary to the jury's finding of fact

ka·hu·na *noun*
: a preeminent person or thing **:** BIG GUN ⟨the industry's big *kahuna,* with . . . 57 percent of the market — A.E. Serwer⟩

Ka·li \ˈkä-ˌlē\ *noun* [Sanskrit *Kālī*] (1798)
: the Hindu goddess of death and destruction

kazoo *noun, plural* **kazoos** [origin unknown] (1968)
slang **:** WAZOO (in this list)

Ke·gel exercises \ˈkā-gəl-, ˈkē-\ *noun plural* [Arnold H. *Kegel* †1976 American gynecologist] (1975)
: repetitive contractions by a woman of the pelvic muscles that control urination in order to strengthen these muscles especially to control or prevent incontinence or to enhance sexual responsiveness during intercourse — called also *Kegels*

kei·ret·su \kā-ˈret-(ˌ)sü, ke-\ *noun, plural* **keiretsu** *also* **keiretsus** [Japanese, literally, system, series, from *kei* system + *retsu* row, line] (1975)
: a powerful alliance of Japanese businesses often linked by cross-shareholding

ke·ta·mine \ˈkē-tə-ˌmēn\ *noun* [*ket-* + *amine*] (1966)
: a general anesthetic administered intravenously and intramuscularly in the form of its hydrochloride $C_{13}H_{16}ClNO{\cdot}HCl$ — compare SPECIAL K (in this list)

killer app *noun* (1988)
: a computer application of such great value or popularity that it assures the success of the technology with which it is associated; *broadly* **:** a feature or component that in itself makes something worth having

ki·net·o·chore *noun*
: a specialized structure on the centromere to which the microtubular spindle fibers attach during mitosis and meiosis

ki·osk *noun*
: a small stand-alone device providing information and services on a computer screen ⟨a museum with interactive *kiosks*⟩

kung pao \ˈkəŋ-ˈpau̇, ˈküŋ-, ˈku̇ŋ-\ *adjective* [Chinese (Beijing) *gōng báo,* literally, palace guardian] (1976)
: being stir-fried or sometimes deep-fried and served in spicy hot sauce with peanuts ⟨*kung pao* chicken⟩

la·bel·mate \ˈlā-bəl-ˌmāt\ *noun* (1981)
: a singer or musician who records for the same company as another

lac·to–ovo–vegetarian \ˈlak-tō-ˈō-vō-\ *noun* (1952)
: a vegetarian whose diet includes dairy products, eggs, fruits, grains, and nuts — called also *ovo-lacto-vegetarian*; compare LACTO-VEGETARIAN

lac·to–vegetarian \ˌlak-tō-\ *noun* (1971)
: a vegetarian whose diet includes dairy products, vegetables, fruits, grains, and nuts — compare LACTO-OVO-VEGETARIAN

landscape *adjective* (1932)
: of, relating to, or being a document having the horizontal dimension longer than the vertical dimension

La Ni·ña \lä-ˈnē-nyə, -nyä\ *noun, plural* **La Niñas** [Spanish, the (female) child] (1988)
: an irregularly recurring upwelling of unusually cold water to the surface along the western coast of South America that often occurs following an El Niño and that disrupts typical regional and global weather patterns especially in a manner opposite to that of El Niño

lao·gai \ˈlau̇-ˈgī\ *noun, plural* **laogai** [Chinese (Beijing) *láogái* reform through labor, short for *láodòng gáizào,* from *láodòng* work, labor + *gáizào* transform, reform] (1983)
: the penal system of China consisting of a network of labor camps; *also* **:** such a labor camp

laptop *noun* (1984)
: a portable microcomputer having its main components (as processor, keyboard, and display screen) integrated into a single unit capable of battery-powered operation

la·ser \ˈlā-zər\ *noun, often attributive* [*l*ight *a*mplification by *s*timulated *e*mission of *r*adiation] (1957)
1 : a device that utilizes the natural oscillations of atoms or molecules between energy levels for generating a beam of coherent electromagnetic radiation usually in the ultraviolet, visible, or infrared regions of the spectrum
2 : something resembling a laser beam in accuracy, speed, or intensity ⟨threw a *laser* into the end zone⟩ ⟨a *laser* stare⟩

laser *transitive verb* (1978)
: to subject to the action of a laser **:** treat with a laser

LA·SIK \ˈlā-sik\ *noun* [*las*er-assisted *i*n situ *k*eratomileusis] (1994)
: a surgical operation to reshape the cornea for correction of myopia, farsightedness, or astigmatism in which the surface layer of the cornea is separated to create a hinged flap providing access to the inner cornea where varying amounts of tissue are removed by an excimer laser

La·ti·na \lə-ˈtē-nə\ *noun* [American Spanish, feminine of *latino* Latino] (1983)
1 : a woman who is a native or inhabitant of Latin America
2 : a woman of Latin-American origin living in the U.S.
— Latina *adjective*

launch *transitive verb*
: to load into a computer's memory and run ⟨*launch* a program⟩

lead·er·board \-ˌbōrd, -ˌbȯrd\ *noun* (1963)
: a large board for displaying the ranking of the leaders in a competitive event (as a golf tournament)

learning curve *noun*
: the course of progress while learning something

legacy *adjective* (1990)
: of, relating to, or being a previous or outmoded computer system ⟨transfer the *legacy* data⟩ ⟨a *legacy* system⟩

lemon verbena *noun* (1869)
: a small shrub (*Aloysia triphylla*) of Chile and Argentina that has narrow lemon-scented leaves

letter bomb *noun* (1973)
: an explosive device concealed in an envelope and mailed to the intended victim

lib \ˈlib\ *noun* (1970)
: LIBERATION 2

line dance *noun* (1948)
1 : CONTREDANSE 1
2 : a dance performed by a group usually in single file
3 : a dance in which the dancers stand in ranks while performing a particular set of steps in unison
— line dancer *noun*
— line dancing *noun*

line–item veto *noun* (1979)
: the power of government executive to veto specific items in an appropriations bill without vetoing the bill altogether

Lip·i·tor \ˈli-pə-ˌtȯr\ *trademark*
— used for a preparation of atorvastatin

List·serv \ˈlist-ˌsərv\ *trademark*
— used for software for managing e-mail transmissions to and from a list of subscribers

Little Ice Age *noun* (1951)
: an episode of glacial expansion whose maximum extension occurred in the 17th and 18th centuries

load *transitive verb*
: to copy or transfer (as a program or data) into a computer's memory especially from an external sources (as a disk drive or the Internet)
intransitive verb
: to become loaded into a computer's memory ⟨the program *loads* quickly⟩

logic bomb *noun* (1978)
: a computer program often hidden within another seemingly innocuous program that is designed to perform usually malicious actions (as deleting files) when certain conditions have been met

long ball *noun* (1938)
: HOME RUN

long·neck \ˈlȯŋ-ˌnek\ *noun* (1978)
: beer served in a bottle that has a long neck

look·ism \ˈlu̇-ki-zəm\ *also* **looks·ism** \ˈlu̇k-si-\ *noun* (1978)

: prejudice or discrimination based on physical appearance and especially physical appearance believed to fall short of societal notions of beauty

lose *transitive verb*
slang : REGURGITATE, VOMIT — often used in such phrases as *lose one's lunch*

loss·less \'lȯs-ləs\ *adjective* (circa 1934)
: done or being without loss (as of power or data) ⟨*lossless* data compression⟩ ⟨*lossless* power transmission⟩

lude \'lüd\ *noun* [short for *Quaalude,* a proprietary name for methaqualone] (1973)
: a pill of methaqualone — usually used in plural

lurk *intransitive verb*
: to read messages on an Internet discussion forum (as a newsgroup or a chat room) without contributing

Mac·Guf·fin *or* **Mc·Guf·fin** \mə-'gə-fən\ *noun* [coined by Alfred Hitchcock] (circa 1939)
: an object, event, or character in a film or story that serves to set and keep the plot in motion despite usually lacking intrinsic importance

mack daddy \'mak-\ *noun* [argot *mac, mack* pimp, probably short for obsolete argot *mackerel,* from Middle English *makerel,* from Anglo-French *makerelle* procuress, ultimately from Middle Dutch *mākelaer* broker] (1989)
1 *slang* : a conspicuously successful pimp
2 *slang* : a slick womanizer
3 *slang* : one that is the best

mac·ro·bi·ot·ic \-bī-'ä-tik, -bē-\ *adjective* (1965)
: of, relating to, or being a diet based on the Chinese cosmological principles of yin and yang that consists of whole cereals and grains supplemented especially with beans and vegetables and that in its especially former more restrictive forms has been linked to nutritional deficiencies
— **macrobiotics** *noun plural but singular in construction*

macular degeneration *noun* (1918)
: a gradual loss of the central part of the field of vision usually affecting both eyes that occurs especially in the elderly and that in a slowly progressing form is marked especially by accumulation of yellow deposits in and thinning of the macula lutea and in a rapidly progressing form by scarring produced by bleeding and fluid leakage below the macula lutea

mad cow disease *noun* (1988)
: BOVINE SPONGIFORM ENCEPHALOPATHY (in this list)

ma·dras·sa *or* **ma·dra·sa** *also* **ma·dras·sah** *or* **ma·dra·sah** \mə-'dra-sə, -'drä-\ *noun* [Arabic *madrasa*] (1662)
: a Muslim school, college, or university that is often part of a mosque

magic mushroom *noun* (1966)
: a fungus (as genus *Psilocybe*) containing hallucinogenic alkaloids (as psilocybin)

mag·ne·to·en·ceph·a·log·ra·phy *noun* (1968)
: a noninvasive technique that detects and records the magnetic field associated with electrical activity in the brain

ma huang \'mä-'hwäŋ\ *noun* [Chinese (Beijing) *máhuáng*] (1926)
1 : any of several eastern Asian ephedras (especially *Ephedra sinica*) having stems and roots yielding ephedrine
2 : EPHEDRA (in this list)

Mail·lard reaction \mə-'lärd-, -'yär-\ *noun* [Louis-Camille *Maillard* †1936 French biochemist] (1929)
: a nonenzymatic reaction between sugars and proteins that occurs upon heating and that produces browning of some foods (as meat and bread)

main·stream *transitive verb*
: to incorporate in the mainstream

mai tai \'mī-,tī\ *noun* [Tahitian *maitai* good] (1961)
: a cocktail made with rum, curaçao, orgeat, lime and fruit juices

major depressive disorder *noun* (1978)
: a mood disorder having a clinical course involving one or more episodes of serious psychological depression lasting two or more weeks each with no intervening episodes of mania

major *adjective*
: prominent or significant in size, amount, or degree ⟨earned some *major* cash⟩

major *noun*
: any of several high-level tournaments in professional golf

male menopause *noun* (1949)
: CLIMACTERIC (in this list)

male–pattern baldness *noun* (1966)
: typical hereditary baldness in the male characterized by loss of hair on the crown and temples

malto·dex·trin \,mȯl-tō-'dek-strən\ *noun* [*malt*ose + *-o-* + *dextrin*] (1885)
: any of various complex carbohydrates derived from the partial hydrolysis of starch (as of corn or potatoes) and used in prepared foods especially as a filler and to enhance texture and flavor

mal·va·sia \,mal-və-'zē-ə, -'sē-\ *noun, often capitalized* [Italian, a sweet wine, from Modern Greek *Monobasia* Monemvasia, village in Greece] (1882)
: a medium to large cultivated grape of Mediterranean regions that is often blended with other grape varieties to produce aromatic dry or sweet wines

managed care *noun* (1982)
: a system of health care (as by an HMO) that controls costs by placing limits on physicians' fees and by restricting the patient's choice of physicians

Man·del·brot set \'man-dəl-,brät-, -,brȯt-\ *noun* [Benoit *Mandelbrot*] (1984)
: a fractal that when plotted on a computer screen roughly resembles a series of heart-shaped disks to which smaller disks are attached and that consists of a connected set of all points *c* in the complex plane for which the recursive expression $z_{n+1} = z_n^2 + c$ for $n = 0, 1, 2, 3, \ldots$ with the starting value $z_0 = 0$ remains bounded as *n* approaches infinity

man·ga \'mäŋ-gə\ *noun* [Japanese, comic, cartoon, from *man-* involuntary, aimless + *-ga* picture] (circa 1951)
: a Japanese comic book or graphic novel

Manhattan clam chowder *noun* (1910)
: chowder made with chopped clams, tomatoes, vegetables, and seasonings

ma·nia *noun*
: excessive or unreasonable enthusiasm ⟨a *mania* for saving things⟩ — often used in combination; *also* : the object of such enthusiasm

manic depression *noun* (1911)
: BIPOLAR DISORDER (in this list)

ma·nip·u·la·tives \mə-'ni-pyə-,lā-tivz, -lə-\ *noun plural* (1965)
: objects (as blocks) that a student is instructed to use in a way that teaches or reinforces a lesson

ma·no a ma·no \,mä-nō-ä-'mä-nō\ *adverb or adjective* [Spanish, literally, hand to hand] (1950)
: in direct competition or conflict especially between one person and another

MAOI *noun*
: monoamine oxidase inhibitor

map *intransitive verb*
: to be assigned in a relation or connection ⟨the major problems confronting humankind . . . do not *map* well onto the traditional disciplines — P.W. Porter⟩

mar·gin *noun*
: an area, state, or condition excluded from or existing outside the mainstream ⟨the *margins* of critical discourse — Barbara L. Packer⟩ ⟨living in society's *margins*⟩

market capitalization *noun* (1975)
: CAPITALIZATION 1d

market economy *noun* (1951)
: an economy in which most goods and services are produced and distributed through free markets

mar·ket·i·za·tion \,mär-kə-tə-'zā-shən\ *noun* (1961)
: the act or process of entering into, participating in, or introducing a free market economy

market maker *noun* (1962)
: an intermediary in a stock exchange who controls buy and sell orders (as by purchase and resale) for a particular stock or group of stocks

markup language *noun* (1980)
: a system (as HTML or SGML) for marking or tagging a document that indicates its logical structure (as paragraphs) and gives instructions for its layout on the page for electronic transmission and display

mar·mite \'mär-,mīt, mär-'mēt\ *noun* [Middle French] (1581)
: a usually tall covered pot

Mar·mite \'mär-,mīt\ *trademark*
— used for an edible yeast extract

marquee *adjective*
: having or associated with the name recognition and attraction of one whose name appears on a marquee : BIG-NAME, STAR ⟨*marquee* athletes⟩ ⟨*marquee* events⟩

mar·riage *noun*
: the state of being united to a person of the same sex in a relationship like that of a traditional marriage ⟨same-sex *marriage*⟩

married name *noun* (1903)
: a surname acquired by a woman through marriage

masa ha·ri·na \-ä-'rē-nä, -nə\ *noun* [Mexican Spanish, probably literally, flour masa (masa in the form of flour)] (1972)
: a flour made from dried masa

ma·sa \'mä-sə\ *noun* [Spanish, mash, dough] (circa 1896)
: a dough used in Mexican cuisine (as for tortillas and tamales) that is made from ground corn soaked in a lime and water solution; *also* : MASA HARINA (in this list)

ma·sa·la \mä-'sä-lä, -lə, mə-\ *noun* [Hindi & Urdu *masālā* materials, ingredients, spices] (1780)
: a varying blend of spices used in Indian cooking

mas·cu·lin·ist \'mas-kyə-lə-(,)nist\ *noun* (1918)
: an advocate of male superiority or dominance
— **masculinist** *adjective*

mas·ters \'mas-tərz\ *adjective* (1971)
: competing in, relating to, or being a competition for athletes over a specified age (as 40) ⟨a *masters* runner⟩

mat·su·ta·ke \,mät-sü-'tä-kē, -kā\ *noun, plural* **matsutake** *also* **matsutakes** [Japanese *matsu-take, matsudake,* from *matsu* pine tree + *take* mushroom] (1883)
: a large brownish edible Japanese mushroom (*Tricholoma matsutake*) having firm flesh and a spicy aroma; *also* : a large whitish mushroom (*Tricholoma magnivelere* synonym *Armillaria ponderosa*) of northern North America that is similar to the Japanese matsutake

ma·ture *adjective*
: of, relating to, or being an older adult : ELDERLY ⟨airline discounts for *mature* travelers⟩

max out *intransitive verb* (1967)
: to be at the upper limit ⟨the car *maxed out* at 85 mph⟩
transitive verb
: to use up all the available credit on ⟨*maxed out* the credit cards⟩

mba·qan·ga \,ùm-bä-'käŋ-gä\ *noun* [Zulu *umbaqanga,* literally, steamed cornmeal bread] (1987)
: a South African dance music that combines traditional elements (as chanting and drumming) with elements of modern music (as jazz)

Mc- \mək; mə *before forms beginning with* k *or* g\ *prefix* [*McDonald's,* chain of fast-food restaurants]

— used to indicate an inexpensive, convenient, or easy but usually low-quality or commercialized version of something specified ⟨*Mc*Book⟩ ⟨*Mc*Doctor⟩

Mc·Job \mək-ˈjäb\ *noun* (1986)
: a low-paying job that requires little skill and provides little opportunity for advancement

MDMA \ˌem-(ˌ)dē-(ˌ)em-ˈā\ *noun* [*m*ethylene + *d*i- + *m*eth*a*mphetamine]
: ECSTASY 4

med *noun* (1982)
: MEDICATION 2 — usually used in plural

med·i·cal·ize \ˈme-di-kə-ˌlīz\ *transitive verb* **-ized; -iz·ing** (1970)
: to view or treat as a medical concern, problem, or disorder ⟨those who seek to dispose of social problems by *medicalizing* them —Liam Hudson⟩
— **med·i·cal·i·za·tion** \ˌme-di-kə-lə-ˈzā-shən\ *noun*

me·di·e·val *adjective*
: having a quality (as cruelty) associated with the Middle Ages

med·i·gap \ˈme-də-ˌgap\ *noun, often capitalized, often attributive* [*Medi*care + *gap*] (1975)
: supplemental health insurance that covers costs (as of medical care or a hospital stay) not covered by Medicare

mega \ˈme-gə\ *adjective* [*mega-*] (1968)
1 : VAST ⟨a *mega* electronics store⟩
2 : of the highest level of rank, excellence, or importance ⟨a number one hit made her *mega*⟩

mega·merg·er \ˈme-gə-ˌmər-jər\ *noun* (1980)
: a merger of megacorporations

mega·plex \-ˌpleks\ *noun* (1986)
: a large multiplex typically housing 16 or more movie theaters

meg \ˈmeg\ *noun* (1975)
: MEGABYTE

mei·ster \ˈmīs-tər\ *noun* [Yiddish *mayster* & German *Meister* master, from Middle High German *meister,* from Old High German *meistar,* from Latin *magister* — more at MASTER] (1979)
: one who is knowledgeable about something specified — often used in combination ⟨puzzle-*meister*⟩

meit·ner·i·um \mīt-ˈnir-ē-əm, -ˈner-\ *noun* [New Latin, from Lise *Meitner*] (1992)
: a short-lived radioactive element that is artificially produced

melon baller *noun* (1950)
: a spoonlike utensil with a sharp edge used especially for cutting ball-shaped pieces from the pulp of a fruit

melt·down *noun*
: a breakdown of self-control (as from fatigue or overstimulation)

melt down *intransitive verb*
: to suffer a meltdown : COLLAPSE

meme \ˈmēm\ *noun* [alteration of *mimeme,* from *mim-* (as in *mimesis*) + *-eme*] (1976)
: an idea, behavior, style, or usage that spreads from person to person within a culture

menstrual cycle *noun* (1912)
: the complete cycle of physiological changes from the beginning of one menstrual period to the beginning of the next

me·nu·do \mə-ˈnü-dō, me-ˈnü-thō\ *noun, plural* **-dos** [Mexican Spanish, from Spanish *menudos* innards, giblets] (1929)
: a tripe stew seasoned with chili peppers

merch \ˈmərch\ *noun* (1982)
: MERCHANDISE 2

mes·clun \ˈmes-klən\ *noun* [French, from Provençal, literally, mixture, from *mescla* to mix, from Old Provençal *mesclar,* from (assumed) Vulgar Latin *misculare* — more at MEDDLE] (1980)
: a salad consisting of a mixture of young tender greens (as lettuces, arugula, and chicory)

message board *noun* (1973)
: BULLETIN BOARD 2

mess *noun*
: a large quantity or number ⟨a *mess* of problems⟩

Messrs. \ˈme-sərz\ *plural of* MR.
⟨*Messrs.* Jones, Brown, and Robinson⟩

metabolic syndrome *noun* (1991)
: a syndrome marked by the presence of usually three or more of a group of factors (as high blood pressure, abdominal obesity, high triglyceride levels, low HDL levels, and high fasting levels of blood sugar) that are linked to increased risk of cardiovascular disease and type 2 diabetes — called also *insulin resistance syndrome*

meta·cog·ni·tion \-käg-ˈni-shən\ *noun* (1977)
: awareness or analysis of one's own learning or thinking process

meta·da·ta \-ˈdā-tə, -ˈda- *also* -ˈdä-\ *noun plural but singular or plural in construction* (1983)
: data that provides information about other data

met·al·de·hyde \ˌmet-ˈal-də-ˌhīd\ *noun* [probably from *meta-* + *aldehyde*] (1841)
: a crystalline compound $(CH_3CHO)_4$ that is a polymer of acetaldehyde and is used as a lure and poison for snails and slugs

met·al·head \ˈme-tᵊl-ˌhed\ *noun* (1982)
: a fan or performer of heavy metal music

meta–anal·y·sis \ˌme-tə-ə-ˈna-lə-səs\ *noun* (1976)
: a quantitative statistical analysis of several separate but similar experiments or studies in order to test the pooled data for statistical significance

meta·noia \ˌme-tə-ˈnȯi-ə\ *noun* [Greek, from *metanoiein* to change one's mind, repent, from *meta-* + *noein* to think, from *nous* mind] (1577)
: a transformative change of heart; *especially* : a spiritual conversion

met·for·min \met-ˈfȯr-mən\ *noun* [*met*hyl + *-formin* (as in *phenformin,* an earlier antidiabetic drug)] (1961)
: a drug $C_4C_{11}N_5$ used in the form of its hydrochloride to treat type 2 diabetes

meth \ˈmeth\ *noun* (1966)
: METHAMPHETAMINE

met·ro·sex·u·al \ˌme-trə-ˈsek-sh(ə-)wəl, -ˈsek-shəl\ *noun* [*metro*politan + *-sexual* (as in *heterosexual*)] (1994)
: a usually urban heterosexual male given to enhancing his personal appearance by fastidious grooming, beauty treatments, and fashionable clothes
— **metrosexual** *adjective*
— **met·ro·sex·u·al·i·ty** \-ˌsek-shə-ˈwa-lə-tē\ *noun*

mic \ˈmīk\ *noun* (1961)
: MICROPHONE

mi·con·a·zole \mī-ˈkä-nə-ˌzōl\ *noun* [*micon-* (perhaps part blend, part alteration of *myc-* and New Latin *Monilia,* a genus of fungi) + imid*azole*] (1970)
: an antifungal agent $C_{18}H_{14}Cl_4N_2O$ administered especially in the form of its nitrate

mi·cro·ar·ray \ˌmī-krō-ə-ˈrā\ *noun* (1995)
: a supporting material (as a glass or plastic slide) onto which numerous molecules or fragments usually of DNA or protein are attached in a regular pattern for use in biochemical or genetic analysis

mi·cro·bi·cide \mī-ˈkrō-bə-ˌsīd\ *noun* (1887)
: an agent that destroys microbes
— **mi·cro·bi·ci·dal** \mī-ˌkrō-bə-ˈsī-dᵊl\ *adjective*

mi·cro·con·trol·ler \ˈmī-krō-kən-ˌtrō-lər\ *noun* (1971)
: a microprocessor that controls some or all of the functions of an electronic device (as a home appliance) or system

mi·cro·fi·ber \ˈmī-krō-ˌfī-bər\ *noun* (1966)
: a fine usually soft polyester fiber; *also* : a fabric made from such fibers

mi·cro·pol·i·tan \ˌmī-krō-ˈpä-lə-tən\ *adjective* [*micr-* + metro*politan*] (1982)
: of, relating to, or being a population area that includes a city with 10,000 to 50,000 residents and its surrounding communities

mi·cro·sat·el·lite \ˌmī-krō-ˈsa-tə-ˌlīt\ *noun* (1989)
: any of numerous short segments of DNA that are distributed throughout the genome, that consist of repeated sequences of usually two to five nucleotides, and that tend to vary from one individual to another

mi·cro·tech·nol·o·gy \-tek-ˈnä-lə-jē\ *noun* (1970)
: technology on a small or microscopic scale

Middle Passage *noun* (1788)
: the forced voyage of enslaved Africans across the Atlantic Ocean to the Americas

MIDI \ˈmi-dē\ *noun* [*m*usical *i*nstrument *d*igital *i*nterface] (1983)
: an electronic standard used for the transmission of digitally encoded music

mid–ocean ridge \ˈmid-ˈō-shən-\ *noun* (1961)
: an elevated region with a central valley on an ocean floor at the boundary between two diverging tectonic plates where new crust forms from upwelling magma

mif·e·pris·tone \ˌmi-fə-ˈpris-ˌtōn\ *noun* [*mife-* (perhaps alteration of *dimethyl* and *phenyl* + *-pristone* (alteration of *progesterone*)] (1985)
: RU 486

mi·grain·eur \ˌmē-gre-ˈnər\ *noun* [probably from *migraine* + *-eur* (as in *entrepreneur*)] (1970)
: an individual who experiences migraines

mind–al·ter·ing \ˈmīn(d)-ˌȯl-t(ə-)riŋ\ *adjective* (1962)
: PSYCHOACTIVE

mind–numb·ing \ˈmīn(d)-ˌnə-miŋ\ *adjective* (1898)
: relentlessly tedious : DULL
— **mind–numb·ing·ly** *adverb*

miner's lettuce *noun* (1897)
: a glossy green herb (*Montia perfoliata* synonym *Claytonia perfoliata*) of the purslane family especially of western North America that produces racemes of pink to white flowers subtended by a disk of two leaves united around the stem and is used as a salad green

mini·bar \ˈmi-nē-ˌbär\ *noun* (1984)
: a small refrigerator in a hotel room that is stocked with especially alcoholic beverages and snacks for guests

mini·disc \ˈmi-nē-ˌdisk\ *noun* (1989)
: a miniature optical disc

mini·dress \ˈmi-nē-ˌdres\ *noun* (1965)
: a short close-fitting dress

mini–mart \ˈmi-nē-ˌmärt\ *noun* (1981)
: CONVENIENCE STORE

mini·pill \ˈmi-nē-ˌpil\ *noun* (1968)
: a birth control pill that contains a very low dose of progesterone but no estrogen, is taken daily, and is intended to minimize side effects

mini·stroke \ˈmi-nē-ˌstrōk\ *noun* (1980)
: TRANSIENT ISCHEMIC ATTACK

mini·tow·er \ˈmi-nē-ˌtau̇(-ə)r\ *noun* (1987)
: a midsize personal computer case that usually stands upright

Mi·ran·dize \mə-ˈran-ˌdīz\ *transitive verb* **-dized; -diz·ing** (1984)
: to recite the Miranda warnings to (a person under arrest)

mire·poix \mir-ˈpwä\ *noun, plural* **mirepoix** [French, probably from Charles de Lévis, duc de *Mirepoix* †1757 French general, or one of his successors] (1877)
: a sautéed mixture of diced vegetables (as carrots, celery, and onions), herbs, and sometimes ham or bacon used especially as a basis for soups, stews, and sauces

mir·in \ˈmir-in\ *noun* [Japanese] (1880)
: a sweet Japanese cooking wine made from fermented rice

mis·an·dry \ˈmi-ˌsan-drē\ *noun* (circa 1909)
: a hatred of men
— **mis·an·drist** *noun or adjective*

misery index *noun* (1975)
: the sum of the rate of unemployment and the rate of inflation used as an economic indicator

mi·so·pros·tol \ˌmī-sō-ˈpräs-ˌtōl, -ˌtȯl\ *noun* [*miso-* (perhaps from *methyl* + *iso-*) + *prost*aglandin + *-ol*] (1982)
: a synthetic prostaglandin analog $C_{22}H_{38}O_5$ used to prevent stomach ulcers associated with NSAID use and to induce abortion in conjunction with RU 486

mitochondrial DNA *noun* (1964)

: an extranuclear double-stranded DNA found exclusively in mitochondria that in most eukaryotes is a circular molecule and is maternally inherited — abbreviation *mtDNA*
mixed–use \'mikst-'yüs\ *adjective* (1972)
: used or suitable for several different functions ⟨a *mixed-use* building⟩
mi•zu•na \mi-'zü-nə\ *noun* [Japanese, from *mizu* water + *na* greens] (1976)
: a Japanese mustard (*Brassica rapa nipposinica* synonym *B rapa japonica*) having mild tasting deeply dissected leaves used especially in salads; *also* : its leaves
mobile *adjective*
: CELLULAR 3 ⟨a *mobile* phone⟩
mock turtleneck *noun* (1966)
1 : a collar that is lower and usually looser than a turtleneck and is not turned over
2 : a garment with a mock turtleneck
mock•u•men•ta•ry \ˌmä-kyə-'men-tə-rē, -'men-trē\ *noun* [blend of *mock* and *documentary*] (1984)
: a facetious or satirical work (as a film) presented in the style of a documentary
modem *transitive verb* (1984)
: to send (as data) via a modem
molecular genetics *noun plural but singular in construction* (1963)
: a branch of genetics dealing with the structure and activity of genetic material at the molecular level
mon•key•pox \'məŋ-kē-ˌpäks\ *noun* (1959)
: a rare virus disease especially of central and western Africa that is caused by a poxvirus (species *Monkeypox virus* of the genus *Orthopoxvirus*), occurs chiefly in wild rodents and primates, and when transmitted to humans resembles smallpox but is milder
mono•pod \'mä-nə-ˌpäd\ *noun* [*mon-* + *-pod* (as in *tripod*)]
: a one-legged support (as for a camera)
monstrous *adverb* (1663)
chiefly dialect : VERY, EXTREMELY ⟨a *monstrous* long raft—Mark Twain⟩
mood disorder *noun* (1969)
: any of several psychological disorders (as major depressive disorder or bipolar disorder) characterized by abnormalities of emotional state — called also *affective disorder*
mook \'mük\ *noun* [perhaps alteration of *moke*] (1930)
slang : a foolish, insignificant, or contemptible person
moon•roof \'mün-ˌrüf, -ˌru̇f\ *noun* (1973)
: a glass sunroof
moon suit *noun* (1980)
: a sealed garment worn especially for protection from hazardous material (as toxic waste or infectious disease)
moon•walk \'mün-ˌwȯk\ *intransitive verb* (1984)
: to dance by gliding backwards while appearing to make forward walking motions
— moonwalk *noun*
Moore's law \'mȯrz-ˌlȯ, 'mu̇rz-\ *noun, often capitalized L* [Gordon E. *Moore* b1929 American computer industry executive] (1980)
: an axiom of microprocessor development usually holding that processing power doubles about every 18 months especially relative to cost or size
morning breath *noun* (1986)
: halitosis upon awakening from sleep that is caused by the buildup of bacteria in the mouth from decreased saliva production
morph *transitive verb* [short for *metamorphose*] (1975)
: to change the form or character of : TRANSFORM
intransitive verb
: to undergo transformation
mosh \'mäsh\ *intransitive verb* [perhaps alteration of *mash* or *mush*] (1987)
: to engage in uninhibited often frenzied activities (as intentional collision) with others near the stage at a rock concert
— mosh•er *noun*
mosh pit *noun* (1988)
: an area in front of a stage where very physical and rough dancing takes place at a rock concert
mouse pad *noun* (1983)
: a thin flat pad (as of rubber) on which a computer mouse is used
mouse potato *noun* [after *couch potato*] (1993)
slang : a person who spends a great deal of time using a computer
mouth•feel \'mau̇th-ˌfēl\ *noun* (1951)
: the sensation created by food or drink in the mouth
moxa \'mäk-sə\ *noun* [New Latin, from Japanese *mogusa*] (1675)
: a soft woolly mass prepared from the ground young leaves of an Eurasian artemisia (especially *Artemisia vulgaris*) that is used in traditional Chinese and Japanese medicine typically in the form of sticks or cones which are ignited and placed on or close to the skin or used to heat acupuncture needles
mox•i•bus•tion \ˌmäk-si-'bəs-chən\ *noun* [*moxa* + *-i-* + *-bustion* (as in *combustion*)] (1910)
: the therapeutic use of moxa
MP3 \ˌem-(ˌ)pē-'thrē\ *noun* [from the file extension *.mp3* used for such files, short for *MPEG Audio Layer 3*] (1996)
1 : a computer file format for the compression and storage of digital audio data
2 : a computer file (as of a song) in the MP3 format
MPEG \'em-ˌpeg\ *noun* [*M*oving *P*ictures *E*xperts *G*roup] (1988)
1 : any of a group of computer file formats for the compression and storage of digital video and audio data
2 : a computer file (as of a movie) in an MPEG format
MRI \ˌem-(ˌ)är-'ī\ *noun* (1982)
: MAGNETIC RESONANCE IMAGING; *also* : the procedure in which magnetic resonance imaging is used
muf•fu•let•ta *also* **muf•fa•let•ta** \ˌmə-fə-'le-tə\ *noun* [probably from Italian dialect, from Italian *muffoletta* little muff, diminutive of *muffola* muff, from French *moufle*, from Middle French] (1973)
: a sandwich made with round Italian bread and filled usually with cold cuts, cheese, and olive salad
mul•ti•eth•nic \-'eth-nik\ *adjective* (1966)
: made up of people of various ethnicities ⟨a *multiethnic* country⟩; *also* : of, relating to, reflecting, or adapted to diverse ethnicities ⟨*multiethnic* literature⟩
— mul•ti•eth•nic•i•ty \-ˌeth-'ni-sə-tē\ *noun*
mul•ti•task•ing \-ˌtas-kiŋ\ *noun*
: the performance of multiple tasks at one time
— mul•ti•task \-ˌtask\ *intransitive verb*
— mul•ti•task•er \-ˌtas-kər\ *noun*
mus•cu•lar *adjective*
: FULL-BODIED ⟨*muscular* wines⟩
must–have \'məst-ˌhav\ *noun* (1980)
: something that is essential to have or obtain
— must–have *adjective*
nan•dro•lone \'nan-drə-ˌlōn\ *noun* [perhaps from *nor-* + *andr-* + *-ol* + testoster*one*] (1963)
: a semisynthetic anabolic steroid $C_{18}H_{26}O_2$ derived from testosterone and used chiefly in the form of various ester derivatives
nano•crys•tal \'na-nə-ˌkris-tᵊl\ *noun* (1984)
: a nanoscale crystal
— nano•crys•tal•line \-'kris-tə-lən *also* -ˌlīn, -ˌlēn\ *adjective*
nano•ma•chine \-mə-ˌshēn\ *noun* (1986)
: a microscopic machine constructed by the use of nanotechnology
nano•par•ti•cle \-ˌpär-ti-kəl\ *noun* (1983)
: a microscopic particle whose size is measured in nanometers
nano•scale \-ˌskāl\ *adjective* (1986)
: having dimensions measured in nanometers
nano•struc•ture \'na-nə-ˌstrək-chər\ *noun* (1978)
: a nanoscale structure; *especially* : an arrangement, structure, or part of something of molecular dimensions
— nano•struc•tured \-chərd\ *adjective*
nano•tech \'na-nō-ˌtek\ *noun* (1991)
: NANOTECHNOLOGY
nano•tube \-ˌtüb\ *noun* (1992)
: a microscopic tube whose diameter is measured in nanometers ⟨carbon *nanotubes*⟩
nar•co•ter•ror•ism \'när-ˌkō-'ter-ər-ˌi-zəm\ *noun* (1982)
: terrorism financed by profits from illegal drug trafficking
— nar•co•ter•ror•ist *noun*
navel–gaz•ing \'nā-vəl-'gā-ziŋ\ *noun* (1963)
: useless or excessive self-contemplation
neat•nik \'nēt-nik\ *noun* (1959)
: a person who is compulsively neat
nee•dle•stick \'nē-dᵊl-ˌstik\ *noun* (1976)
: an accidental puncture of the skin with an unsterilized instrument (as a syringe) — called also *needlestick injury*
neo–pagan \-'pā-gən\ *noun* (1869)
: a person who practices a contemporary form of paganism (as Wicca)
— neo–pa•gan•ism \-'pā-gə-ˌni-zəm\ *noun*
net *noun, often capitalized*
: INTERNET (in this list)
net•i•quette \'ne-ti-kət, -ˌket\ *noun* [blend of *net* and *etiquette*] (1988)
: etiquette governing communication on the Internet
net•i•zen \'ne-tə-zən *also* -sən\ *noun* [blend of *net* and *citizen*] (1994)
: an active participant in the on-line community of the Internet
neu•ro•im•ag•ing \ˌn(y)u̇r-ō-'i-mə-jiŋ\ *noun* (1983)
: a clinical speciality concerned with producing images of the brain by noninvasive techniques (as computed tomography and magnetic resonance imaging); *also* : imaging of the brain by these techniques
neu•ter *transitive verb*
: to remove the force or effectiveness of
ne•vi•ra•pine \nə-'vir-ə-ˌpēn, -'vī-rə-\ *noun* [perhaps from *ne-* (by reversal of *en*zyme) + *vir*al + dipyridodi*azepin*one, class of drugs to which nevirapine belongs] (1991)
: an antiretroviral drug $C_{15}H_{14}N_4O$ that inhibits reverse transcriptase and is administered orally in combination with at least one other antiretroviral in the treatment of infection by HIV-1 and AIDS
new•bie \'nü-bē, 'nyü-\ *noun* [irregular from *new*] (1985)
: BEGINNER, NOVICE; *especially* : a newcomer to cyberspace
New Historicism *noun* (1986)
: a method of literary criticism that emphasizes the historicity of a text by relating it to the configurations of power, society, or ideology in a given time
— New Historicist *adjective or noun*
new jack *adjective, often capitalized N&J* [*jack* (man, guy)] (1987)
1 : of, relating to, or consisting of new jack swing ⟨*new jack* grooves⟩
2 : of, relating to, or being urban, hip, and usually black ⟨the *new jack* generation⟩
new jack swing *noun, often capitalized N&J&S* (1989)
: pop music usually performed by black musicians that combines elements of jazz, funk, rap, and rhythm and blues
news•group \-ˌgrüp\ *noun* (1993)
: an electronic bulletin board on the Internet that is devoted to a particular topic
New York minute *noun* (1980)
: INSTANT, FLASH
nick•el *noun*
slang : a packet containing five dollars worth of an illicit drug (as marijuana) — called also *nickel bag*
NIMBY \'nim-bē\ *noun* [*n*ot *i*n *m*y *b*ack*y*ard] (1980)
: opposition to the locating of something considered undesirable (as a prison or incinerator) in one's neighborhood
— NIMBY•ism \-ˌi-zəm\ *noun*
Nim•rod *noun, not capitalized*
slang : IDIOT, JERK

nine–to–fiv·er \ˈnīn-tə-ˈfī-vər\ *noun* (1959)
: one who works at a job with regular daytime hours

non·com·e·do·gen·ic \(ˈ)nän-ˌkä-mə-dō-ˈje-nik\ *adjective* [*non-* + *comedo* + *-genic*] (1983)
: not tending to clog pores (as by the formation of blackheads) ⟨a *noncomedogenic* cosmetic⟩

no–good·nik \ˌnō-ˈgu̇d-nik\ *noun* (1936)
: NO-GOOD, LOWLIFE

no–hit *transitive verb* (1967)
: to give up no base hits to ⟨*no-hit* them for five innings⟩

no–name \ˈnō-ˌnām\ *adjective* (1942)
: having a name that is not readily recognized by the public ⟨a *no-name* product⟩ ⟨a *no-name* baseball team⟩
— no–name *noun*

no–no *noun*
: NO-HITTER

no–show *adjective* (1975)
: of, relating to, or being a job for which the holder is paid although performing few duties or rarely being present for work

non–insulin–dependent diabetes *noun* (1984)
: TYPE II DIABETES (in this list)

non–insulin–dependent diabetes mellitus *noun* (1979)
: TYPE II DIABETES (in this list) — abbreviation *NIDDM*

non·in·ter·laced \ˌnän-ˌin-tər-ˈlāst\ *adjective* (1980)
: not interlaced; *specifically* **:** of, relating to, or using a method of video scanning (as for a television or computer screen) in which the horizontal lines of each frame are drawn consecutively in a single pass

non·is·sue \ˈnän-ˈi-(ˌ)shü\ *noun* (1964)
: an issue of little importance, validity, or concern

nose·bleed *adjective* (1978)
: extremely or excessively high ⟨seats in the *nosebleed* section⟩ ⟨*nosebleed* stock prices⟩

no·wheres·ville \ˈnō-ˌ(h)werz-ˌvil, -(h)wərz-\ *noun* (1965)
: NOWHERE: as **a :** a location lacking identifying or individualizing qualities **b :** a place or state denoting failure or relative obscurity

nuoc mam \nü-ˈäk-ˈmäm\ *noun* [Vietnamese *nước mắm*, literally, salted fish sauce] (1919)
: a sauce made of fish (as anchovies) fermented in brine

nu·tra·ceu·ti·cal \ˌnü-trə-ˈsü-ti-kəl\ *noun* [*nutri*tive + pharm*aceutical*] (1990)
: a foodstuff (as a fortified food or dietary supplement) that provides health benefits

Nu·yo·ri·can \ˌnü-yȯr-ˈrē-kən\ *noun* [blend of American Spanish *nuyorquino* New Yorker and English *Puerto Rican*] (1975)
: a person of Puerto Rican birth or descent who is a current or former resident of New York City
— Nuyorican *adjective*

object code *noun* (1961)
: a computer program after translation from source code usually into machine language by a compiler

object–oriented *adjective* (1982)
: relating to, used in, or implemented by object-oriented programming ⟨an *object-oriented* language⟩ ⟨*object-oriented* databases⟩

object–oriented programming *noun* (1984)
: a type of computer programming that uses programming objects as basic building blocks

ob–gyn \ˌō-(ˌ)bē-ˈjin, -jē-ˌwī-ˈen\ *noun, plural* **ob–gyns** [*ob*stetrician *gyn*ecologist] (circa 1960)
: a physician who specializes in obstetrics and gynecology

off–gas·sing \ˈȯf-ˌga-siŋ\ *noun* (1966)
: the emission of especially noxious gases (as from building material)

off–la·bel \ˈȯf-ˌlā-bəl\ *adjective* (1988)
: of, relating, to or being a drug used to treat a condition for which it has not been officially approved

oles·tra \ō-ˈles-trə\ *noun* [probably by shortening & alteration from (*sucrose*) *polyester*] (1987)
: a noncaloric fat substitute consisting of a series of compounds that are sucrose esters of six to eight fatty acids resistant to absorption by the digestive system because of their large size

ol·lie \ˈä-lē\ *noun* [*Ollie,* nickname of Alan Gelfand *b*1963 U.S. skateboarder] (1979)
1 : a maneuver in skateboarding in which the skater kicks the tail of the board down while jumping in order to make the board pop into the air
2 : a maneuver in snowboarding in which the rider transfers weight from the front to the back foot to snap the board up off the ground
— ollie *verb*

open mike *noun* (1978)
: an event in which amatuers may perform (as at a coffeehouse) usually without auditioning first

open–faced \-ˈfāst\ *also* **open–face** \-ˈfās\ *adjective* (1918)
: served without a covering layer (as of bread or pastry) ⟨an *open-faced* sandwich⟩

orec·chi·et·te \ō-ˌrä-kē-ˈe-tā\ *noun* [Italian, plural of *orecchietta,* diminutive of *orecchia* ear, from Latin *auricula* — more at AURICLE] (1973)
: small oval pasta

organic brain syndrome *noun* (1966)
: an acute or chronic mental dysfunction (as Alzheimer's disease) resulting chiefly from physical changes in brain structure and characterized especially by impaired cognition

orgasm *intransitive verb* (1972)
: to experience orgasm

or·tho·myxo·vi·rus \ˌȯr-thō-ˈmik-sə-ˌvī-rəs\ *noun* [New Latin, from *orth-* + *myxovirus*] (1973)
: any of a family (*Orthomyxoviridae*) of single-stranded RNA viruses that have their RNA divided into six to eight segments and that include the causative agents of influenza in vertebrates

otol·o·gy \ō-ˈtä-lə-jē\ *noun* [*ot-* + *-logy*] (1842)
: a science that deals with the ear and its diseases
— oto·log·ic \ˌō-tə-ˈlä-jik\ *also* **oto·log·i·cal** \-ji-kəl\ *adjective*
— otol·o·gist \ō-ˈtä-lə-jist\ *noun*

out *transitive verb*
: to identify publicly as being such secretly ⟨wanted to *out* pot smokers⟩; *especially* **:** to identify as being a closet homosexual

out *adjective*
: publicly known or identified as a homosexual

out·er·course \ˈau̇-tər-ˌkȯrs\ *noun* [*outer* + inter*course*] (1986)
: sexual activity between individuals that does not involve vaginal or anal intercourse

out·li·er \-ˌlī(-ə)r\ *noun*
: a statistical observation that is markedly different in value from the others of the sample

out–there \ˈau̇t-ˈther\ *adjective* (1991)
: UNCONVENTIONAL ⟨*out-there* styles⟩

over·class \ˈō-vər-ˌklas\ *noun* (1982)
: the highest social stratum **:** the segment of a society usually having the most wealth, influence, education, and prestige — compare UNDERCLASS

over·di·ag·no·sis \-ˌdī-ig-ˈnō-səs, -əg-\ *noun* (1978)
: the diagnosis of a condition or disease more often than it is actually present
— over·di·ag·nose \-ˈdī-ig-ˌnōs, -ˌnōz, -ˌdī-ig-ˈ, -əg-\ *transitive verb*

over·heat·ed *adjective*
: characterized by marked inflation from an increase in demand and a decrease in supply ⟨an *overheated* economy⟩

oxy·co·done \ˌäk-sē-ˈkō-ˌdōn\ *noun* [*oxy* + *co*deine + *-one*] (1966)
: a narcotic analgesic $C_{18}H_{21}NO_4$ used especially in the form of its hydrochloride

ox·y·gen *noun*
: something that sustains or fuels ⟨disagreement is the true *oxygen* of these magazines — Joseph Epstein⟩

oy \ˈȯi\ *interjection* [Yiddish] (1892)
— used especially to express exasperation or dismay ⟨*oy*, what a mess⟩

Oz \ˈäz\ *noun* [from *Oz* mythical land in a series of books by L. Frank Baum] (1936)
: an ideal or fantastical place

p53 \ˌpē-ˌfif-tē-ˈthrē\ *noun* [from *p53*, the protein made by the gene, from *p*rotein + *53*, the gene's molecular weight] (1990)
: a gene that acts to inhibit unrestrained cell division and when inactivated (as by mutation) places the cell at increased risk for malignant proliferation (as in certain cancers)

pac·li·tax·el \ˌpa-kli-ˈtak-səl\ *noun* [*Pac*ific yew + *-litax-* (perhaps from *Taxus brevifolia*) + *-el* (alteration of *-ol* or *-ol*)] (1992)
: an antineoplastic drug $C_{47}H_{51}NO_{14}$ derived especially from a yew tree of the western U.S. (*Taxus brevifolia*) and used to treat ovarian cancer

pad thai \ˈpäd-ˈtī, ˈpad-\ *noun, often capitalized T* [Thai *phàd thaj,* literally, Thai stir-fried mixture] (1978)
: a Thai dish consisting of rice noodles stir-fried usually with any of various additional ingredients (as bean sprouts, peanuts, chicken, shrimp, and egg)

page *noun*
: the block of information found at a single World Wide Web address

paint·ball \ˈpānt-ˌbȯl\ *noun* (1987)
: a game in which two teams try to capture each other's flag while defending their own using compressed-air guns that shoot paint-filled pellets

paint *noun*
: computer-generated color design ⟨a *paint* program⟩

pa·leo·con·ser·va·tive \-kən-ˈsər-və-tiv\ *noun* (1981)
: a conservative espousing traditional principles and policies
— paleoconservative *adjective*

pal·li·dot·o·my \ˌpa-li-ˈdä-tə-mē\ *noun, plural* **-mies** [New Latin (*globus*) *pallidus* structure within the corpus striatum, literally, pale globe + International Scientific Vocabulary *-tomy*] (1951)
: the surgical inactivation of a part of the basal ganglia in the treatment of involuntary movements (as in Parkinson's disease)

palm·top \ˈpä(l)m-ˌtäp, ˈpȯ(l)m-\ *noun* [*palm* + lap*top*] (1987)
: a small portable computer easily held in the palm of the hand

parachute pants *noun plural* (1977)
: baggy casual pants of lightweight fabric often with an elastic or drawstring at the waist and cuffs

para·glid·er \ˈpa-rə-ˌglī-dər\ *noun* (1944)
1 : a modified parachute used for paragliding
2 : a person who paraglides

para·glid·ing \ˈpa-rə-ˌglī-diŋ\ *noun* (1978)
: the recreational sport of soaring from a slope or a cliff using a modified parachute
— para·glide \-ˌglīd\ *intransitive verb*

Para·lym·pics \ˌpar-ə-ˈlim-piks\ *noun plural* [*para*plegic + *Olympics*] (1953)
: a series of international contests for athletes with disabilities that are associated with and held following the summer and winter Olympic Games — called also *Paralympic Games*
— Para·lym·pi·an \-pē-ən\ *noun*
— Para·lym·pic \-pik\ *adjective*

party animal *noun* (1982)
: a person known for frequent often wild partying

pash·mi·na \ˌpəsh-ˈmē-nə\ *noun* [Persian, from *pashmin* woolen, from *pashm* wool] (1885)
: a fine wool similar to cashmere made from the undercoat of domestic Himalayan goats; *also* **:** a shawl made from this wool

passive–aggressive *adjective* (1946)
: being, marked by, or displaying behavior characterized by expression of negative

feelings, resentment, and aggression in an unassertive passive way (as through procrastination and stubbornness)
— **passive-aggressive** *noun*
pas·ti·tsio \päs-ˈtēt-sē-(ˌ)ō\ *also* **pastitso** \-ˈtēt-(ˌ)sō\ *noun* [Modern Greek, from Italian *pasticcio*] (circa 1950)
: a Greek baked dish made of ground meat layered with pasta and usually topped with white sauce and cheese
patch *noun*
: a usually disk-shaped piece of material that is worn on the skin and contains a substance (as a drug) that is absorbed at a constant rate through the skin into the bloodstream ⟨a nicotine *patch*⟩
patch *transitive verb*
: to connect (as a person or message) to a communication system especially temporarily ⟨they *patched* him into the conference call⟩
pa·tho·o·gize \pə-ˈthä-lə-ˌjīz\ *transitive verb* **-ized; -iz·ing** (1649)
: to view or characterize as medically or psychologically abnormal ⟨natural hormonal shifts have been *pathologized* — Joyce C. Mills⟩
PDA \ˌpē-(ˌ)dē-ˈā\ *noun* [*p*ersonal *d*igital *a*ssistant] (1992)
: a small hand-held device equipped with a microprocessor that is used especially for storing and organizing personal information (as addresses, schedules, and notes)
peck *noun*
: a quick light kiss ⟨a *peck* on the cheek⟩
peel *noun*
: a cosmetic operation for the removal of skin blemishes by the application of a caustic chemical and especially an acid to the skin
peeps \ˈpēps\ *noun plural* [by shortening & alteration] (1992) *slang* **:** PEOPLE 1
pel·o·ton \ˌpe-lə-ˈtän, ˈpe-lə-ˌtän\ *noun* [French, literally, ball — more at PLATOON] (1951)
: the main body of riders in a bicycle race
pepper *noun* [probably from *pepper*] (1943)
: a baseball practice or warm-up game in which usually several fielders toss the ball a short distance to a single batter who hits it back
pepper spray *noun* (1989)
: a temporarily disabling aerosol that is composed partly of capsicum oleoresin and causes irritation and blinding of the eyes and inflammation of the nose, throat, and skin
peri·men·o·pause \ˌper-ē-ˈme-nə-ˌpȯz, -ˈmē-\ *noun* (1962)
: the period around the onset of menopause that is often marked by various physical signs (as hot flashes and menstrual irregularity)
— **peri·men·o·paus·al** \-ˌme-nə-ˈpȯ-zəl, -ˌmē-\ *adjective*
per·ma·cul·ture \ˈpər-mə-ˌkəl-chər\ *noun* [*perma*nent + agri*culture*] (1978)
: an agricultural system or method that seeks to integrate human activity with natural surroundings so as to create highly efficient self-sustaining ecosystems
perp \ˈpərp\ *noun* (1981)
: a perpetrator especially of a crime
petite si·rah *also* **petite sy·rah** \-sə-ˈrä\ *noun, often capitalized P&S* [French *petite syrah,* literally, little syrah (a grape variety)] (1948)
: a dry red wine of spicy fruitiness made from a grape grown chiefly in California; *also* **:** the grape
PET scan *noun* (1980)
: a sectional view of the body constructed by positron-emission tomography
— **PET scanner** *noun*
— **PET scanning** *noun*
phal·lo·crat·ic \ˌfa-lə-ˈkra-tik, -(ˌ)lō-\ *adjective* (1977)
: relating to, resulting from, or advocating masculine power and dominance
phat \ˈfat\ *adjective* **phat·ter; phat·test** [probably alteration of *fat*] (1963)
slang **:** highly attractive or gratifying **:** EXCELLENT ⟨a *phat* beat moving through my body — Tara Roberts⟩
phen–fen \ˈfen-ˌfen\ *noun* (1984)
: FEN-PHEN (in this list)
phen·ter·mine \ˈfen-tər-ˌmēn\ *noun* [probably from *phen*yl + *tert-* (from *tertiary*) + *amine*] (1962)
: an anorectic drug $C_{10}H_{15}N$ used in the form of its hydrochloride to treat obesity
phish·ing \ˈfi-shiŋ\ *noun* [alteration (influenced by *phreaking*) of *fishing*] (1997)
: a scam by which an e-mail user is duped into revealing personal or confidential information which the scammer can use illicitly
— **phish·er** \-shər\ *noun*
phone card *noun* (1982)
: a prepaid card used for making telephone calls
phone sex *noun* (1982)
1 : prerecorded sex-oriented telephone messages available by calling a commercial service
2 : sex-oriented conversations with an operator employed by a commercial service
phreak·er \ˈfrē-kər\ *noun* (1984)
: one who gains illegal access to the telephone system
— **phreak·ing** \-kiŋ\ *noun*
phreak \ˈfrēk\ *noun* [alteration of *freak*] (1972)
: PHREAKER
physician–assisted suicide *noun* (1987)
: suicide by a patient facilitated by means (as a drug prescription) or by information (as an indication of a lethal dosage) provided by a physician aware of the patient's intent
pi·cho·line \ˌpē-shō-ˈlēn\ *noun* [French, from Occitan *pichoulino*] (circa 1959)
: a medium-sized brine-cured green olive of French origin
pie·hole \ˈpī-ˌhōl\ *noun* (1993)
slang **:** MOUTH 1a
pierc·ing *noun* (1977)
: a piece of jewelry (as a ring or stud) that is attached to pierced flesh
— **pierc·ing·ly** \ˈpir-siŋ-lē\ *adverb*
Pi·la·tes \pə-ˈlä-tēz\ *trademark*
— used for an exercise regimen typically performed with the use of specialized apparatus
pi·not blanc \ˈpē-(ˌ)nō-ˈblä^{n}, pē-ˈnō-\ *noun, often capitalized P&B* [French, literally, white Pinot (a grape variety)] (circa 1948)
: a dry white wine similar to chardonnay
pinot gri·gio \-ˈgrē-j(ē-)ō, -zh(ē-)ō\ *noun, often capitalized P&G* [Italian, literally, gray Pinot] (1976)
: a dry white wine that is produced in Italy
plasma \ˈplaz-mə\ *noun*
: a display (as a television screen) consisting of discrete cells of plasma sandwiched between two layers of glass and electrodes such that each cell emits light when it receives an electric current
plat·form *noun*
: a computer architecture that uses a particular operating system
pleath·er \ˈple-t͟hər\ *noun* [blend of *plastic* and *leather*] (1982)
: a plastic fabric made to look like leather
plug and play *noun* (1994)
: a feature of a computer system by which peripherals are automatically detected and configured by the operating system
— **plug–and–play** *adjective*
plugged–in \ˈpləgd-ˈin\ *adjective* (1968)
: technologically or socially informed and connected ⟨*plugged-in* teenagers⟩
plug–in *noun*
: a small piece of software that supplements a larger program (as a browser)
plyo·met·rics \ˌplī-ə-ˈme-triks\ *noun plural but singular or plural in construction* [perhaps irregular from *plio-* + *-metrics* (as in *isometrics*)] (1981)
: exercise involving repeated rapid stretching and contracting of muscles (as by jumping and rebounding) to increase muscle power
— **plyo·met·ric** \-trik\ *adjective*
po·bla·no \pō-ˈblä-nō\ *noun, plural* **-nos** [Mexican Spanish *(chile) poblano,* literally, chili pepper of Puebla (Mexico)] (1972)
: a large usually mild heart-shaped chili pepper especially when fresh and dark green — compare ANCHO (in this list)
point–and–click *adjective* (1983)
: of, relating to, or being a computer interface that allows the activation of a file or function by selection with a pointing device (as a mouse)
point–and–shoot *adjective* (1975)
: having or using preset or automatically adjusted controls (as for focus or shutter speed) ⟨a *point-and-shoot* camera⟩
point–of–purchase *adjective* (1939)
: of or relating to the place (as a supermarket aisle) where a decision to purchase is made ⟨*point-of-purchase* displays⟩
point–of–service *adjective*
: of, relating to, or being a health-care insurance plan that allows enrollees to seek care from a physician affiliated with the service provider at a fixed co-payment or to choose a nonaffiliated physician and pay more — abbreviation *POS*
point–shav·ing \ˈpoint-ˌshā-viŋ\ *noun, often attributive* (1975)
: an attempt (as by a member of the team favored to win) to influence the final score of a game so that the predicted winner wins by less than the point spread
point spread *noun* (circa 1949)
: the number of points by which an oddsmaker expects a favorite to defeat an underdog
polarity therapy *noun* (1964)
: a holistic discipline that seeks to achieve physical and emotional health through a system of touch, diet, exercise, and self-awareness designed to balance energy flows in the body
poly·am·ory \ˌpä-lē-ˈa-mə-rē\ *noun, plural* **-ories** [*polyamor*ous (from *poly-* + *amorous*) + *-y*] (1994)
: the state or practice of having more than one open romantic relationship at a time
— **poly·am·or·ist** \-rist\ *noun*
— **poly·am·o·rous** \-ˈa-mə-rəs, -ˈam-rəs\ *adjective*
pon·zu \ˈpän-(ˌ)zü\ *noun* [Japanese *ponsu, ponzu* juice squeezed from sour oranges, from Dutch *pons,* literally, punch, from English *punch*] (1972)
: a tangy sauce made with citrus juice, rice wine vinegar, and soy sauce and used especially on seafood
pop *noun*
: power to hit a baseball hard ⟨a hitter with some *pop* in his bat⟩
pop–up \ˈpäp-ˌəp\ *adjective* (1934)
: of, relating to, or having a component or device that pops up ⟨a *pop-up* book⟩ ⟨a *pop-up* menu⟩
port *transitive verb* [perhaps from *port* (hardware interface)] (1984)
: to translate (software) into a version for another computer or operating system
por·ta·bil·i·ty \ˌpȯr-tə-ˈbi-lə-tē\ *noun*
: the transferability of a worker's benefits from one pension fund to another when the worker changes jobs
portal *noun*
: a site serving as a guide or point of entry to the World Wide Web and usually including a search engine or a collection of links to other sites arranged especially by topic
por·to·bel·lo \ˌpȯr-tə-ˈbe-(ˌ)lō\ *also* **por·ta·bel·la** \-lə\ *or* **por·ta·bel·lo** \ˌpȯr-tə-ˈbe-(ˌ)lō\ *noun, plural* **por·to·bel·los** *also* **por·ta·bel·las** *or* **por·ta·bel·los** [perhaps alteration of Italian *prataiolo, prataiuolo* or Italian dialect *pratarolo* meadow mushroom, from *prato* meadow, from Latin *pratum*] (1986)
: a large dark mature cultivated mushroom noted for its meaty texture and belonging to a variety of the button mushroom — compare CREMINI (in this list)

por·trait \'pȯr-trət, -ˌtrāt\ *noun* [Middle French, from past participle of *portraire*] (1570)
1 : PICTURE; *especially* **:** a pictorial representation of a person usually showing the face
2 : a sculptured figure **:** BUST
3 : a graphic portrayal in words

portrait *adjective* (1932)
: of, relating to, or being a document having the vertical dimension longer than the horizontal dimension

po·so·le *or* **po·zo·le** \pō-'sō-(ˌ)lā\ *noun* [Mexican Spanish, from Nahuatl *pozolli,* from the base of *pozōn-* boil, be covered with foam] (1931)
: a thick soup chiefly of Mexico and the Southwest U.S. made with pork, hominy, garlic, and chili

post·al *adjective*
: insanely or murderously violent — usually used in the phrase *go postal*

post·con·sum·er \-kən-'sü-mər\ *adjective* (1984)
1 : discarded by an end consumer ⟨*postconsumer* waste⟩
2 : having been used and recycled for reuse in another consumer product ⟨*postconsumer* plastics⟩

poster boy *noun* (1978)
: a male poster child

poster child *noun* (1969)
1 : a child afflicted with a disease who is pictured in posters to solicit funds for combating the disease
2 : a person having a public image that is identified with something (as a cause or group)

poster girl *noun* (1969)
: a female poster child

post *noun*
: something (as a message) that is published online

post *transitive verb*
: to publish (as a message) in an online forum (as an electronic bulletin board)

post–polio syndrome *noun* (1985)
: a condition that affects poliomyelitis patients long after recovery from the disease and that is characterized by muscle weakness, joint and muscle pain, and fatigue

post–struc·tur·al·ism \ˌpōs(t)-'strək-chə-rəˌli-zəm, -'strək-shrə-\ *noun* (1980)
: a movement or theory (as in literary theory or psychoanalysis) that sees inquiry not as the objective exploration of stable structures and categories (as the self) but rather as a relative undertaking shaped by discursive and interpretive practices
— post–struc·tur·al·ist \-list\ *adjective or noun*

post up *intransitive verb* (1974)
: to take up a position against a defender in the post in basketball while standing with one's back to the basket
transitive verb
: to post up against (a defender) in basketball

pot sticker *noun* (1975)
: a crescent-shaped dumpling filled usually with pork, steamed, and then fried

pot·ty–mouthed \-ˌmau̇t͟hd, -ˌmau̇tht\ *adjective* (1987)
: given to the use of vulgar language
— potty mouth *noun*

power forward *noun* (1973)
: a basketball forward whose size and strength are used primarily in controlling play near the basket

power walk *intransitive verb* (1984)
: to walk quickly for exercise especially while carrying or wearing weights

prav·a·stat·in \'pra-və-ˌsta-tᵊn\ *noun* [International Scientific Vocabulary *pra-* (of unknown origin) + *-vastatin* (as in *lovastatin*)] (1987)
: a drug $C_{23}H_{35}NaO_7$ that inhibits the production of cholesterol in the body and is used to treat hypercholesterolemia

preferred provider *noun* (1983)
1 : PPO — usually used attributively
2 : a health-care provider (as a doctor or hospital) that is part of a PPO

pre·in·stall \(ˌ)prē-in-'stȯl\ *transitive verb* (1983)
: to install (as software) on a computer prior to sale
— pre·in·stal·la·tion *noun*

preload \'prē-ˌlōd\ *transitive verb* (1945)
: to load in advance and especially at a time removed from that of use ⟨*preloaded* software⟩

prenuptial agreement *noun* (1978)
: an agreement made between a man and a woman before marrying in which they give up future rights to each other's property in the event of divorce or death — called also *prenup, prenuptial*

pre·press \'prē-ˌpres\ *adjective* (1965)
: of or relating to the processing of copy preparatory to printing ⟨*prepress* costs⟩ ⟨*prepress* equipment⟩

primary care *noun* (1970)
: health care provided by a medical professional (as a general practitioner, pediatrician, or nurse) with whom a patient has initial contact and by whom the patient may be referred to a specialist — often used attributively ⟨a *primary care* physician⟩; — called also *primary health care*; compare SECONDARY CARE (in this list), TERTIARY CARE (in this list)

pri·ma·vera \ˌprē-mə-'ver-ə\ *adjective* [Italian *(alla) primavera* in the style of springtime] (1976)
: served with a mixture of fresh vegetables (as zucchini, snow peas, and broccoli) ⟨pasta *primavera*⟩

primordial soup *noun* (1969)
: a mixture of organic molecules in evolutionary theory from which life on earth originated

programmed cell death *noun* (1982)
: APOPTOSIS (in this list)

pro·pel·ler–head *noun* [from cartoon images of science fiction fans wearing caps with a propeller protruding from the top] (1982)
often disparaging **:** an enthusiast of technology and especially of computers **:** TECHNOPHILE

props \'präps\ *noun plural but usually singular in construction* [short for *proper dues*] (1992)
1 *slang* **:** DUE a ⟨takes pains to give the man his *props* — Dan Epstein⟩
2 *slang* **:** RESPECT 3b ⟨teachers have to earn their *props* just like everybody else — Greg Donaldson⟩
3 *slang* **:** CREDIT 6b ⟨at least deserves *props* for writing a song about something that rings true — Jim Abbott⟩

prostate–specific antigen *noun* (1981)
: a protease secreted by epithelial cells of the prostate that is used in the diagnosis of prostate cancer since its concentration in blood serum tends to be proportional to the clinical stage of the disease — abbreviation *PSA*

protease inhibitor *noun* (1981)
: a substance that inhibits the action of a protease; *specifically* **:** any of a group of drugs that inhibit the protease of HIV so that the cleavage of viral proteins into mature infectious particles is prevented and that are used especially in combination with other antiviral agents in the treatment of HIV infection

pro·te·o·mics \ˌprō-tē-'ō-miks\ *noun plural but singular in construction* (1997)
: a branch of biotechnology concerned with applying the techniques of molecular biology, biochemistry, and genetics to analyzing the structure, function, and interactions of the proteins produced by the genes of a particular cell, tissue, or organism, with organizing the information in databases, and with applications of the data — compare GENOMICS (in this list)
— pro·te·o·mic \-mik\ *adjective*

pseu·do·ephed·rine \ˌsü-dō-i-'fed-rən, *British also* -'e-fə-drən\ *noun* (1974)
: an isomer of ephedrine used in the form of its hydrochloride or sulfate especially to relieve nasal congestion

psy·ops \'sī-ˌäps\ *noun plural, often attributive* [*psy*chological *op*eration*s*] (1966)
: military operations usually aimed at influencing the enemy state of mind through noncombative means (as distribution of leaflets)

pub·lic–key \'pə-blik-'kē\ *noun, often attributive* (1977)
: a cryptographic element that is the publicly shared half of an encryption code and that can be used only to encode messages

pull–down *adjective* (1984)
: being or appearing below a selected item (as an icon) in a window overlaying the original view on a computer display ⟨a *pull-down* menu⟩

pumped \'pəm(p)t\ *adjective* (1984)
: filled with energetic excitement and enthusiasm ⟨*pumped* for the football game⟩

pump fake *noun* (1977)
: a fake in which a player simulates throwing a pass (as in football) or taking a shot (as in basketball)
— pump–fake *verb*

pun·dit·oc·ra·cy \ˌpən-də-'tä-krə-sē\ *noun, plural* **-cies** [*pundit* + *-cracy*] (1987)
: a group of powerful and influential political commentators

pu·pu \'pü-ˌpü\ *noun* [Hawaiian *pūpū* appetizer] (1956)
: an Asian dish served as an appetizer or main course and consisting of a variety of foods (as egg rolls, spareribs, and fried shrimp) ⟨*pupu* platter⟩

pus·sy \'pu̇-sē\ *noun* [short for *pussycat*] (circa 1942)
slang **:** a weak or cowardly man or boy **:** WIMP, SISSY

put·ta·nes·ca \ˌpü-tä-'nes-kä\ *adjective* [Italian, short for *alla puttanesca,* literally, in the style of a prostitute] (1969)
: served with or being a pungent tomato sauce typically containing olives, garlic, capers, hot pepper, and sometimes anchovies — usually used postpositively ⟨pasta *puttanesca*⟩

putt–putt \'pət-ˌpət\ *noun* (1972)
: MINIATURE GOLF

pylon *noun*
: one of the flexible upright markers positioned on a football field at the corners of the end zone

pyramid scheme *noun* (1975)
: a usually illegal operation in which participants pay to join and profit mainly from payments made by subsequent participants

qi·gong \'chē-'gu̇ŋ\ *noun, often capitalized* [Chinese (Beijing) *qìgōng,* from *qì* chi + *gōng* achievement, skill] (1974)
: an ancient Chinese healing art involving meditation, controlled breathing, and movement exercises

Q rating *noun* [*q*uotient] (1977)
: a scale measuring the popularity of a person or thing typically based on dividing an assessment of familiarity or recognition by an assessment of favorable opinion; *also* **:** position on such a scale

quar·tier \kär-'tyā\ *noun* [French, literally, quarter] (1828)
: a district or neighborhood especially in a French city

qua·si·crys·tal \'kwā-ˌzī-ˌkris-tᵊl, -sī-, 'kwä-zē-, -sē-\ *noun* (1984)
: a body of solid material that resembles a crystal in being composed of repeating structural units but that incorporates two or more unit cells into a quasiperiodic structure
— qua·si·crys·tal·line \-'kris-tə-lən *also* -ˌlīn, -ˌlēn\ *adjective*

queer theory *noun* (1988)
: an approach to literary and cultural study that rejects traditional categories of gender and sexuality

Ra·chel \'rā-chəl\ *noun* (13c)
: a wife of Jacob and the mother of Joseph and Benjamin

ra·dar *noun*
: range of notice ⟨fell off the *radar* after losing their first three games⟩

ral·ox·i·fene \ˌra-'läk-sə-ˌfēn\ *noun* [*ral-* (of unknown origin) + *-oxifene,* alteration of *-oxifen* (as in *tamoxifen*)] (1993)

: a drug used orally in the form of its hydrochloride $C_{28}H_{27}NO_4S \cdot HCl$ as prophylaxis against osteoporosis after menopause

ra·men \'rä-mən\ *noun* [Japanese *rāmen*] (1974)
: quick-cooking egg noodles usually served in a broth with bits of meat and vegetables

recreational drug *noun* (1976)
: a drug (as cocaine, marijuana, or methamphetamine) used without medical justification for its psychoactive effects often in the belief that occasional use of such a substance is not habit-forming or addictive

red–eye *noun*
1 : the phenomenon of a subject's eyes appearing red in a color photograph taken with a flash
2 : a late night or overnight flight

red zone *noun* (1983)
: the area of a football field inside an opponent's 20-yard line

refresh *transitive verb*
: to update or renew (as an image, a display screen, or the contents of a computer memory) especially by sending a new signal

Rei·ki \'rā-ˌkē\ *noun* [Japanese, literally, spirit, from *rei* spirit, soul + *ki* vital force, mind] (1985)
: a system of touching with the hands based on the belief that such touching by an experienced practitioner produces beneficial effects by strengthening and normalizing certain vital energy fields held to exist within the body

release *intransitive verb*
: to move from one's normal position (as in football or basketball) in order to assume another position or to perform a second assignment

release *noun*
: the action or manner of throwing a ball ⟨has a quick *release*⟩

re·mix \'rē-ˌmiks\ *noun* (1980)
: a variant of an original recording (as of a song) made by rearranging or adding to the original

rent–a–cop *noun* (1971)
often disparaging **:** a security worker (as a guard) who is not a police officer

repetitive strain injury *noun* (1983)
: any of various painful musculoskeletal disorders (as carpal tunnel syndrome or tendinitis) caused by cumulative damage to muscles, tendons, ligaments, nerves, or joints (as of the hand or shoulder) from highly repetitive movements — called also *repetitive stress injury*

rep·re·sent *intransitive verb*
slang **:** to perform a task or duty admirably **:** serve as an outstanding example

re·pur·pose \(ˌ)rē-'pər-pəs\ *transitive verb* (1984)
: to give a new purpose or use to ⟨*repurpose* the company's Web site⟩ ⟨*repurpose* the archived material⟩

restless legs syndrome *noun* (1976)
: a nervous disorder characterized by aching, crawling, or creeping sensations of the legs that occur especially at night usually when lying down (as before sleep) and cause a compelling urge to move the legs — called also *restless legs, restless leg syndrome*

ret·ro·nym \'re-trō-ˌnim\ *noun* [*retro-* + *-onym*] (1980)
: a term consisting of a noun and a modifier which specifies the original meaning of the noun ⟨"film camera" is a *retronym*⟩

reverse engineer *transitive verb* (1973)
: to disassemble and examine or analyze in detail (as a product or device) to discover the concepts involved in manufacture usually in order to produce something similar
— reverse engineering *noun*

right·size \'rīt-ˌsīz\ *transitive verb* (1989)
: to reduce (as a workforce) to an optimal size
intransitive verb
: to undergo a reduction to an optimal size

rig·or *noun*
: RIGOR MORTIS

ring·ette \riŋ-'et\ *noun* (1974)
: a game of Canadian origin for women and girls that is played on ice with two teams of six players on skates whose object is to drive a rubber or plastic ring into the opponents' goal with a straight stick

ring·tone \-ˌtōn\ *noun* (1983)
: the sound made by a cell phone to signal an incoming call

RNA virus *noun* (1963)
: a virus (as a paramyxovirus or a retrovirus) whose genome consists of RNA

roam *intransitive verb*
: to use a cellular phone outside one's local calling area ⟨*roaming* charges⟩

rock *intransitive verb*
slang **:** to be extremely enjoyable, pleasing, or effective ⟨her new car *rocks*⟩

rock *noun*
: the ball used in basketball

rogue *adjective*
: of or being a nation whose leaders defy international law or norms of international behavior ⟨*rogue* states⟩

roof·ie \'rü-fē\ *noun* [probably by shortening & alteration from *Rohypnol* a trademark name for the drug] (1994)
slang **:** a tablet of a powerful benzodiazepine sedative and hypnotic drug $C_{16}H_{12}FN_3O_3$ that is not licensed for medical use in the U.S. but is used illicitly

rosy periwinkle *noun* (1982)
: PERIWINKLE 1b

ro·tis·ser·ie *adjective, often capitalized* [from La *Rotisserie* Française, restaurant in New York City where a group of fans began the organization of a fantasy baseball league in 1979] (1980)
: of, relating to, or being a sports league consisting of imaginary teams whose performance is based on the statistics of actual players ⟨a *rotisserie* baseball league⟩

router *noun*
: a device that mediates the transmission routes of data packets over an electronic communications network (as the Internet)

royal antler *noun* (circa 1727)
: the third tine above the base of a stag's antler

rub·ber–chick·en \'rə-bər-'chi-kən\ *adjective* [from the low quality of the food stereotypically served at such events] (1972)
: of, relating to, or being a series of social gatherings (as fund-raising dinners) at which speeches are given

rug rat *noun* (1975)
slang **:** a child not yet old enough for school

rule *intransitive verb*
slang **:** to be extremely cool or popular — used as a generalized term of praise or approval ⟨for a little attitude at the right price, sneakers *rule* — Tish Hamilton⟩

run–and–gun *adjective* (1977)
: relating to or being a fast, freewheeling style of play in basketball that de-emphasizes set plays and defense

run–and–shoot *noun, often attributive* (1977)
: a freewheeling style of offense in football that emphasizes passing

runner's high *noun* (1978)
: a feeling of euphoria that is experienced by some individuals engaged in strenuous running and that is held to be associated with a release of endorphins by the brain

ruth·er·ford·ium \ˌrə-t͟hə(r)-'fȯr-dē-əm\ *noun* [New Latin, from Ernest *Rutherford* †1937 British physicist] (1969)
: a short-lived radioactive element that is artificially produced

sab·bat *noun*
: any of eight neo-pagan religious festivals commemorating phases of the changing seasons

Saint–John's–wort *noun*
: the dried aerial parts of a Saint-John's-wort (*Hypericum perforatum*) that are held to relieve depression and are used in herbal remedies and dietary supplements

sa·mo·sa \sə-'mō-sə\ *noun* [Hindi *samosā* & Urdu *samosa, sambūsa,* from Persian *sambūsa*] (1932)
: a small triangular pastry filled with spiced meat or vegetables and fried in ghee or oil

sample *noun*
: an excerpt from a musical recording that is used in another artist's recording

sample *transitive verb*
: to use a segment of (another's musical recording) as part of one's own recording

sandwich generation *noun* (1987)
: a generation of people who are caring for their aging parents while supporting their own children

San·gio·vese \ˌsän-jō-'vā-zā, -'vēz, -'vēs\ *noun* [It] (1943)
: a dry red Italian wine made from a single variety of red grape; *also* **:** a similar wine made elsewhere

SARS \'särz\ *noun* [*s*evere *a*cute *r*espiratory *s*yndrome] (2003)
: a severe respiratory illness that is caused by a coronavirus (genus *Coronavirus*), is transmitted especially by contact with infectious material (as respiratory droplets), and is marked by fever, headache, body aches, a dry cough, hypoxia, and usually pneumonia

save *transitive verb*
: to store (data) in a computer or on a storage device (as a floppy disk or CD)

scalable *adjective*
: capable of being easily expanded or upgraded on demand ⟨a *scalable* computer network⟩
— scal·abil·i·ty \ˌskā-lə-'bi-lə-tē\ *noun*

sca·lop·pi·ne *also* **scal·lo·pi·ni** \ˌskä-lə-'pē-nē, ˌska-\ *noun* [Italian *scaloppine,* ultimately from French *escalope* thin slice of meat, probably from Middle French, shell] (1946)
: thin slices of meat (as veal) sautéed or coated with flour and fried

scanner *noun*
: a device that scans an image (as a photograph) or document (as a page of text) especially for use or storage on a computer

schlub *also* **shlub** \'shləb\ *noun* [Yiddish *zhlob, zhlub* yokel, boor] (1950)
: a stupid, worthless, or unattractive person

schlump \'shləmp\ *noun* [Yiddish *shlump* sloppy or dowdy person] (1948)
slang **:** SCHLUB
— schlumpy \'shləm-pē\ *adjective*

schnoz *or* **schnozz** \'shnäz\ *noun* (1940)
slang **:** NOSE; *specifically* **:** a large nose

scientific creationism *noun* (1979)
: a doctrine holding that the biblical account of creation is supported by scientific evidence

Scotch bonnet *noun* (1986)
: a small roundish very hot chili pepper especially of the Caribbean that is usually red or yellow when mature

scratch *intransitive verb*
: to produce a rhythmic scratching sound by moving a phonograph record back and forth under a phonograph needle

screen saver *noun* (1982)
: a computer program that usually displays various images on the screen of a computer that is on but not in use so as to prevent damage to the screen's phosphors

scrunch·ie *or* **scrunchy** \'skrən-chē, 'skrůn-\ *noun, plural* **scrunchies** (1988)
: a fabric-covered elastic used for holding back hair (as in a ponytail)

scuzz·ball \'skəz-ˌbȯl\ *noun* [*scuzz-,* back-formation from *scuzzy*] (1981)
slang **:** an unpleasant, dirty, or dangerous person **:** CREEP

sea·borg·i·um \sē-'bȯr-gē-əm\ *noun* [New Latin, from Glenn T. *Seaborg* †1999 American chemist] (1994)
: a short-lived radioactive element that is artificially produced

search engine *noun* (1984)
: computer software used to search data (as text or a database) for specified information

secondary care *noun* (1976)
: medical care provided by a specialist or facility upon referral by a primary care physician — compare PRIMARY CARE (in this list), TERTIARY CARE (in this list)

sei·tan \'sā-ˌtan, -ˌtän\ *noun* [origin unknown] (1974)
: flavored wheat gluten often used as a meat analogue

selective serotonin reuptake inhibitor *noun* (1987)
: SSRI

self–help \'self-'help; *Southern also* -'hep\ *noun* (1831)
: the action or process of bettering oneself or overcoming one's problems without the aid of others; *especially* : the copying with one's personal or emotional problems without professional help
— **self–help** *adjective*

self–tan·ner \'self-'ta-nər\ *noun* (1980)
: a product (as one containing dihydroxyacetone) that when applied to the skin reacts chemically with its surface layer to give the appearance of a suntan

sell *noun*
: something to be sold or cause to be accepted ⟨the new mystery novel was an easy *sell*⟩; *also* : someone to whom something is sold ⟨the new purchasing agent was a tough *sell*⟩

se·nior·i·tis \ˌsē-nyər-'ī-təs\ *noun* (1957)
: an ebbing of motivation and effort by school seniors as evidenced by tardiness, absences, and lower grades

senior moment *noun* (1996)
: an instance of momentary forgetfulness or confusion that is attributed to the aging process

sen·sei \'sen-ˌsā\ *noun, plural* **sensei** *or* **senseis** [Japanese, teacher, master] (1968)
: a teacher of martial arts (as karate or judo)

severe acute respiratory syndrome *noun* (2003)
: SARS (in this list)

SGML \ˌes-(ˌ)jē-(ˌ)em-'el\ *noun* [*s*tandard *g*eneralized *m*arkup *l*anguage] (1983)
: a markup language used to define the structure of and manage documents in electronic form — compare HTML (in this list)

sha·bu–sha·bu \'shä-bü-'shä-bü\ *noun* [Japanese, of imitative origin] (1967)
: a Japanese dish consisting of thinly sliced beef and vegetables cooked briefly in simmering broth at the table

shag *noun* (1932)
: a dance step executed by hopping livelily on each foot in turn

sheesh \'shēsh\ *interjection* (1972)
— used to express disappointment, annoyance, or surprise

Shi'i \shē-'ē, 'shē-ē\ *noun* [Arabic *shī'ī*, from *shī'a*] (1728)
: SHIITE
— **Shi'i** *adjective*

shit·less \'shit-ləs\ *adverb* (1936)
usually considered vulgar : to an extreme degree — used as an intensive especially with *scare*

shit·load \'shit-ˌlōd\ *noun* (1973)
usually considered vulgar : a very large amount : LOT

Sho·ah \'shō-ə, -ˌä\ *noun* [Modern Hebrew *shō'āh*, literally, catastrophe, from Hebrew] (1967)
: HOLOCAUST 3a

shock jock *noun* (1986)
: a radio personality noted for provocative or inflammatory commentary

shoot·around \'shüt-ə-ˌrau̇nd\ *noun* (1978)
: a usually informal basketball practice session

shooting guard *noun* (1977)
: a guard in basketball whose chief role is as an outside shooter

shop·a·hol·ic \ˌshä-pə-'hȯ-lik, -'hä-\ *noun* (1983)
: one who is extremely or excessively fond of shopping

shot clock *noun* (1976)
: a clock in basketball that displays a countdown of the time within which shooting the ball is required

shout–out \'shau̇t-ˌau̇t\ *noun* (1990)
: a brief expression of greeting or praise given especially on a broadcast or audio recording

shovel pass *noun* (1940)
: a short underhand pass (as in football)

sick building syndrome *noun* (1983)
: a set of symptoms (as headache, fatigue, and eye irritation) typically affecting workers in modern airtight office buildings that is believed to be caused by indoor pollutants (as formaldehyde fumes or microorganisms)

SIG·INT \'sig-ˌint\ *noun* [*sig*nals *int*elligence] (1969)
: intelligence obtained through the interception of transmission signals

sil·den·a·fil \sil-'de-nə-ˌfil\ *noun* [perhaps by alteration and recombination of letters from *sulfonyl, phenyl,* and *pyrimidine*] (1995)
: a drug used in the form of its citrate $C_{22}H_{30}N_6O_4S$ to treat erectile dysfunction

simian immunodeficiency virus *noun* (1986)
: SIV (in this list)

sim·va·stat·in \'sim-və-ˌsta-tᵊn, ˌsim-və-'sta-\ *noun* [*sim-* (probably alteration of *syn*thetic) + lo*vastatin*] (1987)
: a semisynthetic drug $C_{25}H_{38}O_5$ that decreases the level of cholesterol in the bloodstream and is derived from a compound produced by a mold (*Aspergillus terreus*)

sin·gle–pay·er \-'pā-ər\ *adjective* (1987)
: of, relating to, or being a system in which health-care providers are paid for their services by the government rather than by private insurers

site *noun*
: one or more Internet addresses at which an individual or organization provides information to others often including links to other locations where related information may be found

SIV \ˌes-ˌī-'vē\ *noun* (1987)
: a lentivirus (species *Simian immunodeficiency virus*) that causes a disease in monkeys similar to AIDS and that is closely related to HIV-2

skank \'skaŋk\ *noun* [origin unknown] (1964)
slang : a person and especially a woman of low or sleazy character

skank *noun* [Jamaican English] (1964)
: a rhythmic dance performed by swinging the arms while bending the knees especially to reggae or ska; *also* : the music for this dance

skank *intransitive verb* (1976)
: to dance the skank

skank *transitive verb* [probably alteration of *shank*] (1990)
slang : to bungle (a stroke) in golf

skanky \'skaŋ-kē\ *adjective* [origin unknown] (1982)
1 *slang* : repugnantly filthy or squalid
2 *slang* : of low or sleazy character

skort \'skȯrt\ *noun* [blend of *skirt* and *shorts*] (1951)
: a pair of shorts made to resemble a skirt (as with an overlapping front panel)

slack·er *noun*
: a person and especially a young person who is perceived to be disaffected, apathetic, cynical, or lacking ambition
— **slacker** *adjective*

sleep apnea *noun* (1975)
: apnea that recurs during sleep and is caused especially by obstruction of the airway or a disturbance in the brain's respiratory center

slick *noun*
slang : a military helicopter without armaments that is used to transport troops or light cargo

slo–mo \'slō-ˌmō\ *adjective* (1972)
: SLOW-MOTION
— **slo–mo** *noun*

slot *noun*
: the area on a hockey rink in front of the crease and between the face-off circles

slotting fee *noun* (1984)
: a fee charged by a vendor in exchange for carrying a manufacturer's product — called also *slotting allowance*

slow–wave sleep *noun* (1967)
: a state of deep usually dreamless sleep that occurs regularly during a normal period of sleep with intervening periods of REM sleep and that is characterized by delta waves and a low level of autonomic physiological activity

slurb \'slərb\ *noun* [*sl-* (as in *slovenly, sleazy*) + sub*urb*] (1962)
: a suburb of wearisomely uniform and usually poorly constructed houses

slurve \'slərv\ *noun* [*sl*ider + *curve*] (1973)
: a baseball pitch having the characteristics of both a slider and a curve

small forward *noun* (1977)
: a basketball forward who is usually smaller than a power forward and whose play is characterized by quickness and scoring ability

smart–mouthed \'smärt-ˌmau̇tht, -ˌmau̇t͟hd\ *adjective* (1976)
: annoyingly cocky or sarcastic in speech

smash·mouth \'smash-ˌmau̇th\ *adjective* (1984)
: characterized by brute force without finesse ⟨*smashmouth* football⟩

smiley *noun* [short for *smiley face*] (1987)
: EMOTICON

smiley face *noun* (1972)
: a line drawing of a smiling face

smoke *transitive verb*
1 *slang* : KILL 1a
2 *slang* : to defeat or surpass decisively

smooth·ie \'smü-t͟hē\ *noun*
: a creamy beverage made of fruit blended with juice, milk, or yogurt

s'more \'smȯr\ *noun* [alteration of *some more*] (1974)
: a dessert consisting usually of toasted marshmallow and pieces of chocolate bar between two graham crackers

snail mail *noun* (1983)
1 : mail delivered by a postal system
2 : a postal system

snow·cat \'snō-ˌkat\ *noun* (1955)
: a tracklaying vehicle for travel on snow

so·ba \'sō-bə\ *noun* [Japanese] (circa 1896)
: a Japanese noodle made from buckwheat flour

so·ca \'sō-kə, -kä\ *noun* [*so*ul music + *ca*lypso] (1973)
: a blend of soul and calypso music

soccer mom *noun* (1987)
: a typically suburban mother who accompanies her children to their soccer games and is considered as part of a significant voting bloc or demographic group

social promotion *noun* (1960)
: the practice of promoting a student from one grade level to the next one the basis of age rather than academic achievement

soft·ball *noun*
: a question requiring only an easy or simple response

so·man \'sō-mən\ *noun, often capitalized* [German] (1951)
: an extremely toxic chemical warfare agent $C_7H_{16}FO_2P$ similar to sarin in action

so·ma·ti·za·tion \ˌsō-mə-tə-'zā-shən\ *noun* (1925)
: conversion of a mental state (as depression or anxiety) into physical symptoms; *also* : the existence of physical bodily complaints in the absence of a known medical condition

some·thing *pronoun*
: some indeterminate amount more than a specified number — used in combination ⟨twenty-*something* years old⟩ ⟨a group of fifty-*somethings*⟩

sophomore *adjective* (1953)
: being the second in a series ⟨their *sophomore* album⟩

sou·kous \'sü-ˌküs\ *noun* [African French (of Brazzaville and Kinshasa) *soucous, soukous,* a dance popular in the late 1960s, alteration of French *secousse* jolt, jerk] (1982)
: popular guitar-driven dance music created in the Democratic Republic of the Congo under the influence of Cuban rhumba

soul patch *noun* (1991)
: a small growth of beard under a man's lower lip

sound card *noun* (1983)

: a circuit board in a computer system designed to produce or reproduce sound
sound·scape \'saůn(d)-,skāp\ *noun* (1964)
: a mélange of musical and sometimes nonmusical sounds
source code *noun* (1970)
: a computer program in its original programming language (as FORTRAN or C) before translation into object code usually by a compiler
spaghetti strap *noun* (1972)
: a very slender fabric shoulder strap
spam \'spam\ *noun* [from a skit on the British television series *Monty Python's Flying Circus* in which chanting of the word *Spam* (trademark for a canned meat product) overrides the other dialogue] (1994)
: unsolicited usually commercial e-mail sent to a large number of addresses
spam *verb* **spammed; spam·ming**
transitive verb (1994)
: to send spam to
intransitive verb
: to send spam
— spam·mer *noun*
spastic colon *noun* (1973)
: IRRITABLE BOWEL SYNDROME (in this list); *also* **:** a colon affected with spasms
Special K *noun* (1987)
: the anesthetic ketamine used illicitly usually by being inhaled in powdered form especially for the dreamlike or hallucinogenic state it produces
-specific *combining form*
: relating or applying specifically to or intended specifically for ⟨gender-*specific*⟩
speed dial *noun* (1983)
: a telephone function by which a selected stored number can be dialed by pressing only one key
— speed–dial *verb*
spelling checker *noun* (1981)
: a computer program that identifies possible misspellings in a text file by comparing the contents of the file with a database of accepted spellings — called also *spell check, spell checker*
spider vein *noun* (1976)
: a telangiectasia (as of the legs or face) often appearing as a central area with outward radiations resembling the legs of a spider
spin·meis·ter \'spin-,mīs-tər\ *also* **spin·mas·ter** *noun* (1986)
: SPIN DOCTOR
spi·ru·lina \,spī-rə-'lī-nə\ *noun* [New Latin, from *spirula* small coil, diminutive of Latin *spira* coil] (1977)
: a microscopic filamentous aquatic cyanobacterium (genus *Spriulina* especially *S platensis* synonym *Arthospira platensis*) that is sometimes cultivated for use as food especially as a dietary supplement
split end *noun*
: a hair tip that has become frayed (as from dryness) — usually used in plural
spokes·mod·el \-,mä-dᵊl\ *noun* (1984)
: a model who is a spokesman or spokeswoman
spon·gi·form encephalopathy \'spən-ji-,fȯrm-\ *noun* [*spongiform* resembling a sponge, from Latin *spongia* + English *-iform*] (1960)
: any of a group of degenerative diseases of the brain that are characterized by development of brain tissue having a structure like that of a porous sponge and by deterioration in neurological functioning — compare BOVINE SPONGIFORM ENCEPHALOPATHY (in this list)
spool·er \'spü-lər\ *noun* (1971)
: a computer utility that regulates data flow by receiving data (as from a word processor), queuing the data in a buffer, and then transmitting it (as to a printer) with increased efficiency
sports bar *noun* (1975)
: a bar catering especially to sports fans and typically containing several televisions and often sports memorabilia
sports medicine *noun* (1961)
: a field of medicine concerned with the prevention and treatment of injuries and disorders that are related to participation in sports
sport–util·i·ty vehicle \'spōrt-yü-'ti-lə-tē-, 'spȯrt-\ *noun* (1987)
: a rugged automotive vehicle similar to a station wagon but built on a light-truck chassis
spot–on \'spät-'än\ *adjective* (1949)
: exactly correct **:** ACCURATE, PERFECT ⟨a *spot-on* forecast⟩ ⟨a *spot-on* impersonation⟩
spy·ware \'spī-,wer\ *noun* (1994)
: software that is installed in a computer without the user's knowledge and transmits information about the user's computer activities over the Internet
squamous cell carcinoma *noun* (1907)
: a carcinoma that is made up of or arises from squamous cells and usually occurs in areas of the body exposed to strong sunlight over many years
SSRI \,es-(,)es-(,)är-'ī\ *noun* (1991)
: any of a class of antidepressants (as fluoxetine) that inhibit the inactivation of serotonin by blocking its reuptake by presynaptic neuron endings — called also *selective serotonin reuptake inhibitor*
stair–climb·er \'ster-'klī-mər\ *noun* (1986)
: an exercise apparatus that simulates the act of climbing stairs
— stair–climb·ing \-miŋ\ *noun*
start·er *adjective* (1946)
: of, relating to, or being an item acquired with the expectation that a more elaborate or sophisticated model will be acquired in the future ⟨a *starter* home⟩
stat·in \'sta-tᵊn\ *noun* [from *-statin* (as in *lovastatin*)] (1986)
: any of a group of drugs (as lovastatin and simvastatin) that inhibit the synthesis of cholesterol and promote the production of LDL-binding receptors in the liver resulting in a usually marked decrease in the level of LDL and a modest increase in the level of HDL circulating in blood plasma
stav·u·dine \'stav-yü-,dēn\ *noun* [*sta-* (of unknown origin) + *vudine* (as in *zidovudine*)] (1992)
: D4T (in this list)
stay *noun*
: a thin firm strip (as of plastic) used for stiffening a garment or part (as a shirt collar)
steg·a·nog·ra·phy \,ste-gə-'nä-grə-fē\ *noun* [New Latin *steganographia,* from Greek *steganos* covered, reticent (from *stegein* to cover) + Latin *-graphia* -graphy — more at THATCH] (1985)
1 *archaic* **:** CRYPTOGRAPHY
2 : the art or practice of concealing a message, image, or file within another message, image, or file
— steg·a·no·graph·ic \-nə-'gra-fik\ *adjective*
stent \'stent\ *noun* [Charles Thomas *Stent* †1885 English dentist] (1961)
: a short narrow metal or plastic tube that is inserted into the lumen of an anatomical vessel (as an artery or a bile duct) especially to keep a formerly blocked passageway open
step aerobics *noun plural but singular or plural in construction* (1985)
: aerobics that involves repeatedly stepping on and off a raised platform — called also *step training*
stick *transitive verb*
: to execute (a landing) flawlessly in gymnastics
sticky bun *noun* (1974)
: a spiral-shaped cinnamon roll topped with melted brown sugar and butter
sticky note *noun* (1984)
: a slip of notepaper having an adhesive strip on the back that allows attachment to and removal from a surface
stiff *adverb*
: close enough to the hole for an easy putt in golf ⟨hit it *stiff* and tapped it in for an easy birdie⟩
Stock·holm syndrome \'stäk-,hō(l)m-\ *noun* [from a 1973 robbery attempt in *Stockholm* Sweden, during which the bank employees held hostage developed sympathetic feelings toward their captors] (1978)
: the psychological tendency of a hostage to bond with, identify with, or sympathize with his or her captor
stop–ac·tion \'stäp-'ak-shən\ *noun, often attributive* (1946)
: STOP-MOTION
stop–mo·tion \'stäp-'mō-shən\ *noun, often attributive* (1912)
: a filming technique in which successive positions of objects (as clay models) are photographed to produce the appearance of movement
strange *noun, often attributive* (1974)
: a fundamental quark that has an electric charge of -1/3 and a measured energy of approximately 150 MeV; *also* **:** the flavor characterizing this particle
straw mushroom *noun* [from their cultivation on rice straw compost] (1961)
: a mushroom (*Volvariella volvacea*) that has a conical cap and is cultivated in southeastern Asia and used especially in Chinese cooking
streaming *adjective* (1980)
: relating to or being the transfer of data (as audio or video material) in a continuous stream especially for immediate processing or playback
street hockey *noun* (1964)
: a game resembling ice hockey played on a hard surface by players wearing shoes or roller skates and using hockey sticks and a small ball
stress incontinence *noun* (1935)
: involuntary leakage of urine from the bladder accompanying physical activity (as laughing or coughing) — compare URGE INCONTINENCE (in this list)
stretch *adjective*
: longer than the standard size ⟨*stretch* limousine⟩
strike–slip \'strīk-,slip\ *noun, often attributive* (1964)
1 : a fault about which movement is predominately horizontal
2 : slip along the strike of a fault ⟨*strike-slip* earthquakes⟩
string cheese *noun* (1974)
: cheese formed usually into sticks that can be pulled apart in narrow strips
string theory *noun* (1975)
: a theory in physics: all elementary particles are manifestations of the vibrations of one-dimensional strings
strip *transitive verb*
: to remove (a subcutaneous vein) by means of a surgical instrument ⟨*stripping* a varicose saphenous vein⟩
strip mall *noun* (1980)
: a long usually one-story building or group of buildings housing several adjacent retail stores or service establishments
stutter step *noun* (1966)
: a momentary hesitation or false step by a runner (as in football) done to fake a defender out of position
— stutter–step *verb*
sub·dwarf \'səb-,dwȯrf\ *noun* (1939)
: a small hot star containing few elements heavier than helium and having lower luminosity than a main-sequence star of similar temperature
sub·note·book \'səb-'nōt-,bůk\ *noun* (1990)
: a portable microcomputer similar to but smaller and lighter than a notebook computer
sub·woof·er \'səb-'wů-fər\ *noun* (1978)
: a loudspeaker responsive only to the lowest acoustic frequencies
sucky \'sə-kē\ *adjective* **suck·i·er; suck·i·est** (1984)
slang **:** AWFUL 3
suicide *adjective* (1773)
: of or relating to suicide; *especially* **:** being or performing a deliberate act resulting in the voluntary death of the person who does it ⟨a *suicide* mission⟩ ⟨a *suicide* bomber⟩

sul·fa·meth·ox·a·zole \ˌsəl-fə-ˌme-ˈthäk-sə-ˌzōl\ *noun* [*sulfa* + *methyl* + *oxazole*, a compound C_3H_3NO (from International Scientific Vocabulary *ox-* + *azole*)] (1960)
: an antibacterial sulfonamide $C_{10}H_{11}N_3O_3S$ used alone or in combination with trimethoprim (as in the treatment of urinary tract infections or acute otitis media)

sul·fo·raph·ane \ˌsəl-fō-ˈra-ˌfan, -ˈrā-\ *noun* [*sulforaphen*, a chemically similar substance (from *sulfo-* + *-raphen*, perhaps alteration of *raphanin*, an alternate name, from New Latin *Raphanus*, a cruciferous plant genus) + *-ane*] (1992)
: an anticarcinogenic isothiocyanate $C_6H_{11}NOS_2$ found in cruciferous vegetables (as broccoli and cauliflower) that is thought to function by stimulating the production of enzymes in the body that detoxify cancer-causing substances

su·ma·trip·tan \ˌsü-mə-ˈtrip-ˌtan, -tən\ *noun* [perhaps from *suma-* (by shortening & alteration from *sulfonamide*) + *triptan* (by shortening & alteration from *tryptamine*)] (1989)
: a triptan $C_{14}H_{21}N_3O_2S$ that is administered as a nasal spray or in the form of its succinate by mouth or by injection and is used in the treatment of migraine attacks

sun protection factor *noun* (1978)
: a number assigned to a sunscreen that is the factor by which the time required for unprotected skin to become sunburned is increased when the sunscreen is used — abbreviation *SPF*

su·per·bug \ˈsü-pər-ˌbəg\ *noun* (1985)
: a pathogenic microorganism and especially a bacterium that has developed resistance to the medications normally used against it

su·per·cen·ter \ˈsü-pər-ˌsen-tər\ *noun* (1977)
: a very large discount department store that also sells a complete line of grocery merchandise

su·per·cool *adjective* (1970)
: extremely cool: as **a :** showing extraordinary reserve and self-control **b :** being the latest style or fashion ⟨*supercool* sunglasses⟩

su·per·ma·jor·i·ty \ˈsü-pər-mə-ˌjȯr-ə-tē, -jär-\ *noun* (1977)
: a majority (as two-thirds or three-fifths) greater than a simple majority

su·per·mod·el \ˈsü-pər-ˌmä-dᵊl\ *noun* (1977)
: a famous and successful fashion model

su·per·mom \ˈsü-pər-ˌmäm\ *noun* (1974)
: an exemplary mother; *also* **:** a woman who performs the traditional duties of housekeeping and childrearing while also having a full-time job

su·per·size \ˈsü-pər-ˌsīz\ *transitive verb* **-ized; -iz·ing** (1994)
: to increase considerably the size, amount, or extent of

su·per·ti·tle \ˈsü-pər-ˌtī-tᵊl\ *noun* (1984)
: a translation of foreign-language dialogue displayed above a screen or performance ⟨an opera with *supertitles*⟩ — compare SUBTITLE

surf *intransitive verb*
: to scan a wide range of offerings for something of interest
transitive verb
: to scan the offerings of (as television or the Internet) for something that is interesting or fills a need

surround sound *noun* (1969)
: sound reproduction that often uses three or more transmission channels to enhance the illusion of a live hearing

sur·viv·al·ism \sər-ˈvī-və-ˌli-zəm\ *noun* (1980)
: an attitude, policy, or practice based on the primacy of survival as a value

sus·tained–re·lease *adjective* (1956)
: designed to release a drug in the body slowly over an extended period of time ⟨*sustained-release* capsules⟩

sus·tain *noun* (1972)
: a musical effect that prolongs a note's resonance ⟨utilizing heavy *sustain* on his guitar — Bill Dahl⟩

SWAT \ˈswät\ *noun, often attributive* [*s*pecial *w*eapons *a*nd *t*actics] (1968)
: a police or military unit specially trained and equipped to handle unusually hazardous situations or missions

swimmer's ear *noun* (1961)
: inflammation of the canal in the outer ear that is characterized by itching, redness, swelling, pain, and discharge and that typically occurs when water trapped in the outer ear during swimming becomes infected usually with a bacterium

swipe *transitive verb*
: to slide (a card with a magnetic strip or bar-code) through a slot in a reading device so that information stored on the strip can be processed (as in making a purchase)

switch–hit·ter *noun*
: one that is flexible or adaptable; *especially* **:** a person who can work equally well in either of two jobs or capacities

syn·er·gy *noun*
: a mutually advantageous conjunction or compatibility of distinct business participants or elements (as resources or efforts)

syn·the·size *transitive verb*
: to produce (as music) by an electronic synthesizer

Sy·rah \sē-ˈrä\ *noun* [French *syrah, syrac*] (1974)
1 : a grape whose skin has a dark blue to bluish-black color that was originally grown in the northern valley of the Rhone and is now widely grown elsewhere (as in California and Australia); *also* **:** a vine producing Syrah grapes
2 : a red wine made from Syrah grapes

tac·rine \ˈta-ˌkrēn\ *noun* [*t*etra- + *acridine*] (1965)
: an anticholinesterase $C_{13}H_{14}N_2$ used especially in the palliative treatment of cognitive deficits occurring early in Alzheimer's disease

tag *noun*
1 : an element of code in a computer document used especially to control format and layout or to establish a hyperlink
2 : a graffito in the form of an identifying name or symbol

tag *transitive verb*
: to deface with a graffito usually in the form of the defacer's nickname

ta·leg·gio \tä-ˈle-j(ē-)ō\ *noun, usually capitalized* [Italian, from *Taleggio* commune and valley in Italy] (1952)
: a soft creamy cheese made from the whole milk of cows

talk therapy *noun* (1979)
: psychotherapy emphasizing conversation between therapist and patient

T & A *noun* [*t*its & *a*ss] (1979)
: curvaceous and often scantily clothed women; *also* **:** entertainment featuring such women

tandem repeat *noun* (1973)
: any of several identical DNA segments lying one after the other in a sequence

tank *intransitive verb*
: to suffer rapid decline, failure, or collapse ⟨bought a stock that quickly *tanked*⟩

tank·i·ni \taŋ-ˈkē-nē\ *noun* [blend of *tank* (*top*) and *bikini*] (1985)
: a woman's two-piece swimsuit consisting of bikini briefs and a tank top

ta·pe·nade \ˌtä-pə-ˈnäd\ *noun* [French *tapénade*, from Provençal *tapenado*, from *tapeno* caper, ultimately from Latin *capparis* — more at CAPER] (1952)
: a seasoned spread made chiefly with mashed black olives, capers, and anchovies

ta·que·ria *also* **ta·que·ría** \ˌtä-kə-ˈrē-ə\ *noun* [Mexican Spanish, from *taco* taco] (1982)
: a Mexican restaurant specializing especially in tacos and burritos

tarte ta·tin \ˌtär(t)-tä-ˈtan, -ˈtaⁿ\ *noun, plural* **tarte tatins** *also* **tartes tatin** *often capitalized 2d T* [French, Tatin tart, after the *Tatin* sisters of Lamotte-Beuvron, France] (1979)
: a caramelized apple tart that is baked with pastry on top and then inverted for serving

Ta·ser \ˈtā-zər\ *trademark*
— used for a gun that fires electrified darts to stun and immobilize a person

ta–ta \tä-ˈtä\ *interjection, chiefly British* [baby talk] (1823)
— used to express farewell

tat·soi \ˈtät-ˈsȯi\ *noun* [Chinese (Guangdong) *daat-choi*, from *daat-* sink, fall flat + *choi* vegetable (or from a cognate compound in another Chinese dialect)] (1987)
: an Asian mustard (*Brassica rapa rosularis*) having a rosette of edible dark green spoon-shaped glossy leaves; *also* **:** the leaves

Tax·ol \ˈtax-ˌsȯl\ *trademark*
— used for a preparation of paclitaxel

TCP/IP \ˌtē-(ˌ)sē-ˈpē-ˌī-ˈpē\ *noun* [*t*ransmission-*c*ontrol *p*rotocol/*I*nternet *p*rotocol] (1980)
: a set of communications protocols used for the exchange of information over networks and especially over the Internet

tech·no \ˈtek-nō\ *noun* [*techno-* (as in *techno-pop* or *techno-rock*, styles of popular music utilizing electronically created sounds)] (1987)
: electronic dance music that features a fast beat and synthesized sounds usually without vocals or a conventional popular song structure

tech·no·pre·neur \ˌtek-nō-prə-ˈnər, -ˈn(y)u̇r\ *noun* [*techno-* + entre*preneur*] (1987)
: an entrepreneur whose business involves high technology

teens·ploi·ta·tion \ˌtēn-(ˌ)splȯi-ˈtā-shən\ *noun* [*teen* + *-sploitation* (from *blaxploitation*)] (1982)
: the exploitation of teenagers by producers of teen-oriented films

Te·ja·no \tā-ˈhä-(ˌ)nō\ *noun, plural* **-nos** *often attributive* [Mexican Spanish, from *Tejas* Texas] (1976)
1 : a Texan of Hispanic descent
2 [probably short for *conjunto Tejano* literally, Texan ensemble] **:** Tex-Mex popular music combining elements of traditional, rock, and country music and often featuring an accordion

tele·cine \ˈte-li-ˌsi-nē, ˌte-li-ˈsi-\ *noun* [probably short for *telecinema*, from French *télécinéma*, from *télé-* tele- + *cinéma* cinema] (1935)
: the equipment used in the process of transferring a motion picture to videotape or converting it into television images; *also* **:** such a process

tel·e·com \ˈte-li-ˌkäm\ *noun* (1948)
1 : TELECOMMUNICATION
2 : the telecommunications industry

tele·med·i·cine \ˌte-lə-ˈme-də-sən, -ˈmed-sən\ *noun* (1970)
: the practice of medicine when the doctor and patient are widely separated using two-way voice and visual communication (as by satellite, computer, or closed-circuit television)

Telestrator \ˈte-lə-ˌstrā-tər\ *trademark*
— used for an electronic device that generates drawn video images over a background image

tel·net \ˈtel-ˌnet\ *noun* [*tel*etype *net*work] (1971)
: a telecommunications protocol providing specifications for emulating a remote computer terminal so that one can access a distant computer and function on-line using an interface that appears to be part of the user's local system
— telnet *intransitive verb*

ten·der *noun* [probably short for *tenderloin*] (1983)
: an often breaded strip of usually breast meat ⟨chicken *tenders*⟩; *also* **:** the tenderloin of a chicken

TENS \ˈtenz\ *noun* (1980)
1 [*t*ranscutaneous *e*lectrical *n*erve *s*timulation] **:** electrical stimulation of the skin to relieve pain by interfering with the neural transmission of signals from underlying pain receptors
2 [*t*ranscutaneous *e*lectrical *n*erve *s*timulator] **:** a device used for TENS

tension headache *noun* (1953)

: bilateral headache marked by mild to moderate pain of variable duration that typically is accompanied by contraction of the neck and scalp muscles

tep·pan·ya·ki \ˌte-pän-ˈyä-kē\ *noun* [Japanese, from *teppan* griddle + *yaki* broiling] (circa 1970)
: a Japanese dish of meat, fish, or vegtables cooked on a large griddle usually built into the diner's table; *also* : this style of cooking

tera·byte \ˈter-ə-ˌbīt\ *noun* (1982)
: a unit of computer information storage capacity equal to about one trillion bytes; *specifically* : 1,099,511,627,776 bytes

tertiary care *noun* (1979)
: highly specialized medical care usually over an extended period of time that involves advanced and complex procedures and treatments performed by medical specialists in state-of-the-art facilities — compare PRIMARY CARE (in this list), SECONDARY CARE (in this list)

tes·tos·ter·one *noun*
: qualities (as brawn and aggressiveness) usually associated with males : MANLINESS

text messaging *noun* (1982)
: the sending of short text messages electronically especially from one cell phone to another

tex·tu·al·ize \ˈteks-ch(ə-w)ə-ˌlīz\ *transitive verb* **-ized; -iz·ing** (1981)
: to put into text : set down as concrete and unchanging ⟨the novel *textualizes* complex emotions⟩
— **tex·tu·al·i·za·tion** \ˌteks-ch(ə-w)ə-lə-ˈzā-shən\ *noun*

tha·las·so·ther·a·py \thə-ˌla-sō-ˈthe-rə-pē\ *noun* [Greek *thalassa* + English *therapy*] (1899)
: exposure to seawater (as in a hot tub) or application of sea products (as seaweed or sea salt) to the body for health or beauty benefits

third–party *adjective*
: of, relating to, or being software that is created by a vendor to be compatible with the products of another vendor

thong *noun*
: a garment consisting of a narrow strip of fabric fitting between the legs and a waistband that is worn as underwear or the bottom piece of a bikini

thread *noun*
: a series of newsgroup messages following a single topic

three–pointer *noun* (1977)
: a basketball shot or field goal from beyond the three-point line

three–point line *noun* (1977)
: a line on a basketball court forming an arc at a set distance (as 22 feet) from the basket beyond which a field goal counts for three points

three–peat \ˈthrē-ˌpēt\ *noun* [blend of *three* and *repeat*] (1988)
: a third consecutive championship

throw *transitive verb*
: to perform (as a stunt) successfully ⟨*throwing* tricks on a skateboard⟩

thumbnail *noun*
: a miniature computer graphic sometimes hyperlinked to a full-size version

tide pool *noun* (1853)
: a pool of salt water left (as in a rock basin) by an ebbing tide — called also *tidal pool*

tiki bar *noun* (1980)
: a restaurant or bar decorated in a simulated Polynesian theme that usually serves exotic cocktails

tik·ka \ˈti-kə\ *noun* [Hindi & Urdu *tikkā* small piece of meat, from Persian *tikka*] (1955)
: an Indian dish of marinated meat cooked on a skewer

tinea pe·dis \-ˈpe-dəs\ *noun* [New Latin, tinea of the foot] (1948)
: ATHLETE'S FOOT

Tin·sel·town \ˈtin(t)-səl-ˌtau̇n *also* ˈtin-zəl-\ *noun* (1978)
: HOLLYWOOD ⟨seduced by *Tinseltown*⟩
— **Tinseltown** *adjective*

T–bone *transitive verb*
: BROADSIDE

toast *noun*
slang : one that is finished or done for ⟨soon their relationship was *toast* —Rick Reilly⟩

tobacco mosaic virus *noun* (1938)
: a single-stranded RNA virus (genus *Tobamovirus*) that occurs worldwide and causes mosaic disease in tobacco and numerous other plants

toe loop *noun* (circa 1964)
: a backward jump in figure skating with a takeoff from the outside edge of one skate followed by a full turn in the air and a landing on the outside edge of the same skate

toggle *intransitive verb*
: to switch between two options especially of an electronic device usually by pressing a single button or a simple key combination

toll–free \ˈtōl-ˈfrē\ *adjective or adverb* (1970)
: having or using a direct telephone line or number (as an 800 number) for a long-distance call that is not charged to the caller ⟨a *toll-free* number⟩ ⟨called *toll-free*⟩

toot *transitive verb*
slang : to take in (as cocaine) by inhalation : SNORT

top *noun*
: a fundamental quark that has an electric charge of +2/3 and a measured energy of approximately 175 GeV — often used attributively; *also* : the flavor characterizing this particle

tough love *noun* (1968)
: love or affectionate concern expressed in a stern or unsentimental manner (as through discipline) especially to promote responsible behavior

tower *noun*
: a personal computer case that stands in an upright position

tracking stock *noun* (1989)
: a stock the value of which is linked to the performance of a company division but which does not confer ownership in the company or the division

trade show *noun* (1895)
: a large exposition to promote awareness and sales of especially new products within an industry ⟨a computer *trade show*⟩

transcription factor *noun* (1972)
: any of various proteins that bind to DNA and play a role in the regulation of gene expression by promoting transcription

trans fat *noun* (1978)
: a fat containing trans-fatty acids

trans–fat·ty acid \ˈtran(t)s-ˈfa-tē-, ˈtranz-\ *noun* (1953)
: an unsaturated fatty acid characterized by a trans arrangement of alkyl chains that is formed especially during the hydrogenation of vegetable oils and has been linked to an increase in blood cholesterol

transfer station *noun* (1969)
: a site where recyclables and refuse are collected and sorted in preparation for processing or landfill

trans·gen·der \-ˈjen-dər\ *or* **trans·gen·dered** \-dərd\ *adjective* (1979)
: exhibiting the appearance and behavioral characteristics of the opposite sex

trans·gene \ˈtran(t)s-ˌjēn, ˈtranz-\ *noun* (1984)
: a gene that is taken from the genome of one organism and introduced into the genome of another organism by artificial techniques

tran·si·tion *intransitive verb* (1946)
: to make a transition ⟨while they *transition* to a new, career-oriented lifestyle — Sarah Bright⟩

trash talk *noun* (1981)
: disparaging, taunting, or boastful comments especially between opponents trying to intimidate each other
— **trash–talk** \ˈtrash-ˌtȯk\ *verb*

traveler's diarrhea *noun* (1968)
: intestinal sickness and diarrhea affecting a traveler that is typically caused by ingestion of pathogenic microorganisms (as some E. coli)

traz·o·done \ˈtra-zə-ˌdōn\ *noun* [perhaps from International Scientific Vocabulary *triazo* azido + pyridine + *-one*] (1971)
: an antidepressant drug $C_{19}H_{22}ClN_5O$ administered in the form of its hydrochloride

treb·bia·no \tre-ˈbyä-(ˌ)nō\ *noun* [Italian] (1860)
: a widely cultivated Italian white grape used especially in making white wine and brandy

tren·doid \ˈtren-ˌdȯid\ *noun* (1985)
: a trendy person

tricho·til·lo·ma·nia \ˌtri-kə-ˌti-lə-ˈmā-nē-ə\ *noun* [New Latin, from *trich-* + Greek *tillein* to pull, pluck + New Latin *mania*] (circa 1903)
: an abnormal desire to pull out one's hair

tri·clo·san \trī-ˈklō-ˌsan\ *noun* [*tri-* + *chlor-* + *-san* (of unknown origin)] (1973)
: a whitish crystalline powder $C_{12}H_7C1_3O_2$ that is a phenyl ether derivative used especially as a broad-spectrum antibacterial agent (as in soaps, deodorants, and mouthwash)

trigger point *noun* (circa 1891)
: a localized usually tender or painful area of the body and especially of a muscle that when stimulated gives rise to pain elsewhere in the body

trip *intransitive verb*
slang : FREAK 3b ⟨she was *tripping* over the news⟩

tri·ple–dou·ble \-ˈdə-bəl\ *noun* (1982)
: an instance of a player accumulating 10 or more points, assists, and rebounds in a single game

trip–hop \ˈtrip-ˌhäp\ *noun* [probably blend of *trip* (high from a psychedelic drug) + *hip-hop*] (1989)
: electronic dance music usually based on a slow hip-hop beat and incorporating hypnotic synthesized and prerecorded sounds

trip·tan \ˈtrip-ˌtan, -tən\ *noun* [*-triptan* (as in *sumatriptan*)] (1997)
: any of a class of drugs (as sumatriptan) that bind to and are agonists of serotonin receptors and are used to treat migraine attacks

tro·phy *noun*
: one that is prized for qualities that enhance prestige or social status — usually used attributively ⟨a *trophy* wife⟩ ⟨a *trophy* house⟩

tsk *a dental click; often read as* ˈtisk\ *interjection* (1937)
— used to express disapproval

tsk–tsk \ˈtisk-ˌtisk\ *transitive verb* (1943)
: to express disapproval of by or as if by uttering tsk
intransitive verb
: to tsk-tsk someone or something

tsu·ris *also* **tsou·ris** \ˈtsü-ris, ˈtsu̇r-is\ *noun* [Yiddish *tsures,* plural of *tsure, tsore* trouble, distress, from Hebrew *ṣārāh*] (1941)
: TROUBLE, DISTRESS

tubal pregnancy *noun* (circa 1834)
: ectopic pregnancy in a fallopian tube

'tude \ˈtüd, ˈtyüd\ *noun* [short for *attitude*] (1976)
slang : a cocky or arrogant attitude

tumm·ler \ˈtu̇m-lər\ *noun* [Yiddish *tumler,* literally, one who makes a racket] (1965)
: a comic entertainer or social director at a Jewish resort

tumor suppressor gene *noun* (1985)
: any of class of genes (as p53) that act in normal cells to inhibit unrestrained cell division and that when inactivated (as by mutation) place the cell at increased risk for malignant proliferation

turf toe *noun* [from the occurrence of the injury among athletes who play on artificial turf] (1981)
: a minor but painful usually sports-related injury typically involving hyperextension of the big toe that results in spraining or tearing of the ligaments at the joint between the metatarsal and basal phalanx

turn·around \-ˌrau̇nd\ *noun*
: a jump shot by a player facing away from the basket who turns toward the basket while shooting — often used attributively ⟨a *turnaround* jumper⟩

TVP \ˌtē-(ˌ)vē-ˈpē\ *trademark*
— used for textured vegetable protein

tweak *transitive verb*

1a : ANNOY, BOTHER ⟨*tweaking* the establishment⟩ **b :** to criticize especially in a sly or sharp manner **c :** to poke fun at
2 : to injure slightly ⟨*tweak* an ankle⟩

tween·er \ˈtwē-nər\ *noun* [be*tween* + ¹*-er*] (1978)
: a player who has some but not all of the necessary characteristics for each of two or more positions (as in football or basketball)

12–step \ˈtwelv-ˌstep\ *adjective* (1983)
: of, relating to, characteristic of, or being a program that is designed especially to help an individual overcome an addiction, compulsion, serious shortcoming, or traumatic experience by adherence to 12 tenets emphasizing personal growth and dependence on a higher spiritual being

24–7 *or* **24/7** \ˈtwen-tē-ˈfōr-ˈse-vən, ˈtwən-, -ˈfȯr-\ *adverb or adjective* (1986)
: for twenty-four hours seven days a week ⟨can now shop *24-7*⟩

twen·ty·some·thing \ˈtwen-tē-ˌsəm(p)-thiŋ, -ˈtwən-\ *adjective* (1990)
: of, relating to, or being a person who is in his or her twenties ⟨a *twentysomething* professional⟩
— twentysomething *noun*

twin primes *noun plural* (1930)
: a pair of prime numbers (as 3 and 5 or 11 and 13) differing by two

twist tie *noun* (1975)
: a tie used for closing or securing (as a plastic bag) by twisting the ends together

two's complement *noun* (1958)
: the negative of a binary number represented by switching all ones to zeros and all zeros to ones and then adding one to the result

type B *adjective* (1976)
: relating to, characteristic of, having, or being a personality that is marked by a lack of aggressiveness and tension and that has been implicated by some studies as a factor decreasing the risk of cardiovascular disease ⟨*type B* behavior⟩

type I diabetes \ˈtīp-ˈwən-\ *noun* (1982)
: a form of diabetes mellitus that usually develops during childhood or adolescence and is characterized by a severe deficiency in insulin secretion resulting from atrophy of the islets of Langerhans and causing hyperglycemia and a marked tendency toward ketoacidosis — called also *insulin-dependent diabetes, insulin-dependent diabetes mellitus*

type II diabetes \-ˈtü-\ *noun* (1982)
: a common form of diabetes mellitus that develops especially in adults and most often in obese individuals and that is characterized by hyperglycemia resulting from impaired insulin utilization coupled with the body's inability to compensate with increased insulin production — called also *non-insulin-dependent diabetes, non-insulin-dependent diabetes mellitus*

ul·tra–pas·teur·ized \-ˈpas-chə-ˌrīzd, -ˈpas-tyə-, -tə-\ *adjective* (1953)
: subjected to pasteurization at higher than normal temperatures especially to extend shelf life ⟨*ultra-pasteurized* cream⟩

uma·mi \ü-ˈmä-mē\ *noun* [Japanese, savoriness, flavor] (1979)
: a taste sensation that is meaty or savory and is produced by several amino acids and nucleotides (as glutamate and aspartate)

un·ar·gu·able \ˌən-ˈär-gyə-wə-bəl\ *adjective* (1881)
: not arguable

un·ar·gu·ably \-blē\ *adverb* (1929)
: it cannot be argued **:** UNQUESTIONABLY

un·der·di·ag·nose \ˌən-dər-ˈdī-ig-ˌnōs, -ˌnōz; -ˌdī-ig-ˈ, -əg-\ *transitive verb* (1974)
: to diagnose (a condition or disease) less often than it is actually present
— un·der·di·ag·no·sis \-ˌdī-ig-ˈnō-səs, -əg-\ *noun*

un·der·dress \ˌən-dər-ˈdres\ *intransitive verb* (circa 1784)
: to dress in overly or inappropriately simple or informal clothing
transitive verb
: to dress (as oneself) more simply or informally than is appropriate

un·der·throw \ˈən-dər-ˌthrō\ *transitive verb* (1963)
: to throw (a ball or pass) short of the intended receiver in football; *also* **:** to throw a pass short of ⟨*underthrow* a receiver⟩

un·der·treat \ˌən-dər-ˈtrēt\ *transitive verb* (1908)
: to treat inadequately ⟨*undertreat* a disease⟩

un·der·wire \ˈən-dər-ˌwī(-ə)r\ *noun* (1973)
: a wire running through the bottom edge of a brassiere to aid in support

uni·brow \ˈyü-nə-ˌbrau̇\ *noun* (1988)
: a single continuous brow resulting from the growing together of eyebrows

uniform resource locator *noun* (1993)
: URL (in this list)

uni·lat·er·al·ism \ˌyü-ni-ˈla-t(ə-)rə-ˌli-zəm\ *noun* (1926)
: a policy of taking unilateral action (as in international affairs) regardless of outside support or reciprocity; *also* **:** advocacy of such a policy
— uni·lat·er·al·ist \-list\ *noun or adjective*

un·in·stall \ˌən-in-ˈstȯl\ *transitive verb* (1985)
: to remove (software) from a computer system especially by using a specially designed program

union territory *noun* (1979)
: a centrally administered subdivision of India

uni·po·lar \ˌyü-ni-ˈpō-lər\ *adjective* (1965)
: relating to, affected with, or being a manic-depressive disorder in which there is only a depressive phase ⟨*unipolar* depression⟩

universal resource locator *noun* (1993)
: URL (in this list)

Universal Serial Bus *noun* (1994)
: USB (in this list)

un·plugged \-ˈpləgd\ *adjective* (1990)
: ACOUSTIC 2 ⟨an *unplugged* performance⟩

un·re·tire \ˌən-ri-ˈtī(-ə)r\ *intransitive verb* (1966)
: to leave retirement **:** rejoin the workforce

unstable angina *noun* (1972)
: angina pectoris characterized by sudden changes (as an increase in the severity or length of anginal attacks or a decrease in the exertion required to precipitate an attack) especially when symptoms were previously stable

un·track \ˌən-ˈtrak\ *transitive verb* (1939)
: to cause to escape from a slump ⟨couldn't get *untracked* and played poorly throughout the game⟩

up *noun*
: a fundamental quark that has an electric charge of +2/3 and that is one of the constituents of a nucleon

urban legend *noun* (1979)
: an often lurid story or anecdote that is based on hearsay and widely circulated as true ⟨the *urban legend* of alligators living in the sewers⟩ — called also *urban myth*

urge incontinence *noun* (1980)
: involuntary leakage of urine from the bladder when a sudden strong need to urinate is felt — compare STRESS INCONTINENCE (in this list)

URL \ˌyü-(ˌ)är-ˈel, ˈər(-ə)l\ *noun* (1992)
: the address of a computer or a document on the Internet that consists of a communications protocol followed by a colon and two slashes (as http://), the identifier of a computer (as www.m-w.com), and usually a path through a directory to a file — called also *uniform resource locator, universal resource locator*

uro·gy·ne·col·o·gy \ˌyu̇r-ō-ˌgī-nə-ˈkä-lə-jē, -ˌji-\ *noun* (1989)
: a branch of medicine concerned with urological problems affecting women
— uro·gy·ne·col·o·gist \-jəst\ *noun*

USB \ˌyü-(ˌ)es-ˈbē\ *noun* (1995)
: a standardized serial computer interface that allows simplified attachment of peripherals especially in a daisy chain

Use·net \ˈyüz-ˌnet\ *noun* [probably from *Use*nix, an association of computer programmers using the operating system Unix (from *us*ers of *U*nix) + *net* (network)] (1980)
: the aggregation of all the newsgroups on the Internet

über- *also* **uber-** \ˈü-bər, ˈue-bər\ *prefix* [German, from *über* over, beyond, from Old High German *ubar* — more at OVER]
1 : being a superlative example of its kind or class **:** SUPER- ⟨*über*nerd⟩
2 : to an extreme or excessive degree **:** SUPER- ⟨*über*cool⟩

UVA \ˌyü-(ˌ)vē-ˈā\ *noun* (1975)
: radiation that is in the region of the ultraviolet spectrum which extends from about 320 to 400 nm in wavelength and that causes tanning and contributes to the aging of the skin

UVB \ˌyü-(ˌ)vē-ˈbē\ *noun* (1975)
: radiation that is in the region of the ultraviolet spectrum which extends from about 280 to 320 nm in wavelength and that is primarily responsible for sunburn, aging of the skin, and the development of skin cancer

UVC \ˌyü-(ˌ)vē-ˈsē\ *noun* (1978)
: radiation that is in the region of the ultraviolet spectrum which extends from about 200 to 280 nm in wavelength and that is more hazardous than UVB but is mostly absorbed by the earth's upper atmosphere

Valley girl *noun, often capitalized G* (1982)
: an adolescent girl from the San Fernando Valley; *also* **:** one whose values, mannerisms, and especially speech patterns resemble those of such a girl

val·ue–add·ed \ˈval-ˌyü-ˈa-dəd\ *adjective* (1935)
: of, relating to, or being a product whose value has been increased especially by special manufacturing, marketing, or processing ⟨*value-added* goods⟩

veggie burger *noun* (1972)
: a patty chiefly of vegetable-derived protein used as a meat substitute; *also* **:** a sandwich containing such a patty

ve·loc·i·rap·tor \və-ˈlä-sə-ˌrap-tər\ *noun* [New Latin, from Latin *veloc-, velox* + *raptor* plunderer, predator — more at RAPTOR] (1990)
: any of a genus (*Velociraptor*) of theropod dinosaurs of the late Cretaceous having a long head with a flat snout and a large sickle-shaped claw on the second toe of each foot

ventricular assist device *noun* (1970)
: a device implanted in the chest or upper abdomen to assist a damaged or weakened heart in pumping blood

ver·mi·cul·ture \ˈvər-mə-ˌkəl-chər\ *noun* (1976)
: the cultivation of annelid worms (as earthworms or bloodworms) especially for use as bait or in composting

Vi·ag·ra \vī-ˈa-grə\ *trademark*
— used for a preparation of the citrate of sildenafil

vi·at·i·cal settlement \vī-ˈa-ti-kəl-\ *noun* [probably from *viaticum*] (1991)
: an agreement by which the owner of a life insurance policy that covers a person (as the owner) who has a catastrophic or life-threatening illness receives compensation for less than the expected death benefit of the policy in return for a turning over (as by sale or bequest) of the death benefit or ownership of the policy to the other party (as a company specializing in such transfers) — called also *viatical*

Vi·da·lia \və-ˈdāl-yə\ *certification mark*
— used for certain mild sweet yellow onions grown in Georgia

video card *noun* (1982)
: a circuit board in a computer system designed to generate output for the system's video display screen

vir·gin *adjective*
: containing no alcohol ⟨a *virgin* daiquiri⟩

virtual *adjective*
: being on or simulated on a computer or computer network ⟨print or *virtual* books⟩ ⟨a *virtual* keyboard⟩: as **a :** occurring or existing primarily online ⟨a *virtual* library⟩ ⟨*virtual*

shopping⟩ **b :** of, relating to, or existing within a virtual reality ⟨a *virtual* world⟩ ⟨a *virtual* tour⟩
vision quest *noun* (1922)
: a solitary vigil by an adolescent American Indian boy to seek spiritual power and learn through a vision the identity of his usually animal or bird guardian spirit
vi·tel·lo·gen·in \vī-ˌte-lō-ˈje-nən\ *noun* [probably from *vitellogenesis* + *-in*] (1971)
: a precursor protein of egg yolk normally in the blood or hemolymph only of females that is used as a biomarker in vertebrates of exposure to environmental estrogens which stimulate elevated levels in males as well as females
V–chip \ˈvē-ˌchip\ *noun* [*v*iolence] (1993)
: a computer chip in a television set that can prevent the viewing of certain programs or channels especially on the basis of content
voo·doo *adjective*
: based on highly improbable suppositions **:** extremely implausible or unrealistic
vouch·er *noun*
: a coupon issued by government to a parent or guardian to be used to fund a child's education in either a public or private school
VX \ˈvē-ˈeks\ *noun* [American or British government code designation for the gas] (1965)
: an extremely toxic chemical weapon $C_{11}H_{26}NO_2PS$ similar to sarin and tabun in action
wack \ˈwak\ *adjective* [probably alteration of *wacky*] (1984)
slang **:** not up to the mark **:** LOUSY, LAME ⟨while there are skilled moments, there are *wack* ones as well — Danyel Smith⟩
wait·ron \ˈwā-ˌträn, -trən\ *noun* [blend of *waiter* or *waitress* and *-tron* (suggesting the machinelike impersonality of such work), later (perhaps influenced by *neutron*) taken as a gender-neutral term] (1980)
: WAITPERSON
wait·staff \ˈwāt-ˌstaf\ *noun* (1983)
: the staff of servers at a restaurant
wa·ka·me \wä-ˈkä-me\ *noun* [Japanese] (1950)
: an edible brown seaweed (*Undaria pinnatifida*) native to Asia
wake·board \ˈwāk-ˌbȯrd\ *noun* (1991)
: a short board with foot bindings on which a rider is towed by a motorboat across its wake and especially up off the crest for aerial maneuvers
— wake·board·er *noun*
— wake·board·ing *noun*
walk *intransitive verb*
: to avoid criminal prosecution or conviction
WAN \ˈwan\ *noun* (1983)
: WIDE AREA NETWORK (in this list)
warp speed *noun* [from the use in science fiction of space-time warps to allow faster-than-light travel] (1980)
: the highest possible speed
washboard *noun*
: an abdominal area characterized by prominent and well-defined musculature ⟨a bodybuilder with *washboard* abs⟩ ⟨a *washboard* stomach⟩
wax *transitive verb*
slang **:** to defeat decisively (as in an athletic contest)
wa·zoo \(ˌ)wä-ˈzü\ *noun* [origin unknown] (1983)
slang **:** the posterior opening of the alimentary canal **:** ANUS
— up the wazoo *also* **out the wazoo**
: in excess ⟨we've got lawyers *up the wazoo* —Steven Bochco⟩
weap·on·ize \ˈwe-pə-ˌnīz\ *transitive verb* **-ized; -iz·ing** (1957)
: to adapt for use as a weapon of war
— weap·on·i·za·tion *noun*
Web *noun*
: WORLD WIDE WEB (in this list)
Web·log \ˈweb-ˌlȯg, -ˌläg\ *noun* (1997)
: BLOG (in this list)
Web·mas·ter \ˈweb-ˌmas-tər\ *noun* (1994)
: a person responsible for the creation or maintenance of a World Wide Web site especially for a company or organization
weekend warrior *noun* (1981)
: a person who participates in a usually physically strenuous activity only on weekends or part-time
weird out *transitive verb* (1973)
: to make uneasy, bewildered, or disquieted by something considered very strange ⟨that movie *weirds* me *out*⟩
Wens·ley·dale \ˈwenz-lē-ˌdāl\ *noun* [from *Wensleydale,* valley in North Yorkshire] (1896)
: a mild white friable cheese of English origin
Wer·nick·e's area \ˈver-nə-kəz-, -kēz-\ *noun* [Karl *Wernicke* †1905 German neurologist] (1950)
: an area of the brain that is located in the posterior left temporal lobe and is associated with comprehension of language
West Nile encephalitis *noun* (1949)
: severe West Nile fever marked by encephalitis
West Nile fever *noun* (1943)
: illness caused by the West Nile virus
West Nile virus *noun* [from *West Nile* province of Uganda, where the virus was isolated in 1937] (1940)
: a flavivirus (species *West Nile virus* of the genus *Flavivirus*) that causes an illness marked by fever, headache, muscle ache, skin rash, and sometimes encephalitis or meningitis and that is spread chiefly by mosquitoes; *also* **:** WEST NILE FEVER
wet·ware \ˈwet-ˌwar, -ˌwer\ *noun* [*wet* + *software*] (1977)
: the human brain or a human being considered especially with respect to human logical and computational capabilities
whack *transitive verb*
slang **:** MURDER, KILL
wheat berry *noun* (1848)
: an unprocessed whole kernel of wheat
wheat·grass \ˈhwēt-ˌgras, ˈwēt-\ *noun* (1668)
: any of a genus (*Agropyron*) of perennial grasses including some which are important pasture, hay, or turf grasses
white·board \ˈhwīt-ˌbȯrd, ˈwīt-\ *noun* (1951)
: a hard smooth white surface used for writing or drawing on with markers
white–coat hypertension *noun* [from the white laboratory coats worn by physicians] (1986)
: a temporary elevation in a patient's blood pressure that occurs when measured in a medical setting (as a physician's office) and that is usually due to anxiety on the part of the patient
whole food *noun* (1970)
: a natural food and especially an unprocessed one (as a vegetable or fruit)
whole language *noun* (1984)
: a method of teaching reading and writing that emphasizes learning whole words and phrases by encountering them in meaningful contexts rather than by phonics exercises
whup \ˈhwu̇p, ˈwu̇p\ *transitive verb* **whupped; whup·ping** [alteration of *whip*] (1852)
1 : to administer a beating to especially as punishment
2 : to defeat decisively
Wic·ca \ˈwi-kə\ *noun* [probably from Old English *wicca* wizard — more at WITCH] (1959)
: a religion influenced by pre-Christian beliefs and practices of western Europe that affirms the existence of supernatural power (as magic) and of both male and female deities who inhere in nature, and that emphasizes ritual observance of seasonal and life cycles
— Wic·can \ˈwi-kən\ *adjective or noun*
wide area network *noun* (1982)
: a network of computers (as the Internet) in a large area (as a country or the globe) for sharing resources or exchanging data
wife·beat·er \ˈwīf-ˌbē-tər\ *noun* (1994)
slang **:** a man's white tank top
Wi–Fi \ˈwī-ˈfī\ *certification mark*
— used to certify the interoperability of wireless computer networking devices
wild card *noun*
usually **wildcard :** a symbol (as ? or *) used in a keyword database search to represent the presence of zero, one, or more than one unspecified characters
Wil·liams syndrome \ˈwil-yəmz-\ *noun* [J. C. P. *Williams* *b*1922 New Zealand physician] (1981)
: a rare genetic disorder marked especially by hypercalcemia of infants, heart defects, characteristic facial abnormalities, and mild to moderate mental retardation but a high verbal aptitude
wil·ly *also* **wil·lie** \ˈwi-lē\ *noun* [from the name *Willy*] (circa 1905)
slang **:** PENIS
wine·sap \ˈwīn-ˌsap\ *noun, often attributive* (1826)
: an apple with deep red skin and juicy somewhat tart flesh
win–win \ˈwin-ˈwin, -ˌwin\ *adjective* (1977)
: advantageous or satisfactory to all parties involved ⟨a *win-win* situation⟩ ⟨a *win-win* deal⟩
wired *adjective*
1 : connected to a telecommunications network and especially to the Internet
2 : characterized by a connection to the Internet ⟨the *wired* world⟩
wire *transitive verb*
: to predispose, determine, or establish genetically or innately ⟨controversy over the extent to which human violence is *wired* biologically⟩
wireless *adjective*
: of or relating to data communications using radio waves ⟨*wireless* Internet access⟩
— wirelessly *adverb*
witch *noun*
: a practitioner of Wicca
witch·craft *noun*
: WICCA (in this list)
wom·an·ism \ˈwu̇-mə-ˌni-zəm\ *noun* (1984)
: a form of feminism focused especially on the conditions and concerns of black women
— wom·an·ist \-nist\ *noun or adjective*
women's studies *noun plural but singular or plural in construction* (1972)
: the multidisciplinary study of the social status and societal contributions of women and the relationship between power and gender
word wrap *noun* (1977)
: a word processing feature that automatically transfers a word for which there is insufficient space from the end of one line of text to the beginning of the next
work–around \ˈwərk-ə-ˌrau̇nd\ *noun* (1971)
: a plan or method to circumvent a problem (as in computer software) without eliminating it
world beat *noun* (1984)
: WORLD MUSIC — usually hyphenated when used attributively
world music *noun* (1982)
: popular music originating from or influenced by non-Western musical traditions and often having a danceable rhythm — usually hyphenated when used attributively
World Wide Web *noun* (1992)
: a part of the Internet designed to allow easier navigation of the network through the use of graphical user interfaces and hypertext links between different addresses — called also *Web*
worry line *noun* (1972)
: a crease or wrinkly on the forehead or between the eyebrows
wrap *noun*
: a treatment for the care of the skin in which material (as hot wet cloth or seaweed) is wrapped around the entire body; *also* **:** this material
writable *adjective*
: being an electronic storage medium capable of having new data written on it ⟨a *writable* DVD⟩
wrongful birth *noun* (1979)

: a malpractice claim brought by the parents of a child born with a birth defect against a physician or health-care provider whose alleged negligence (as in diagnosis) effectively deprived the parents of the opportunity to make an informed decision whether to avoid or terminate the pregnancy; *also* : the birth or injury at issue in such a claim

wrongful death *noun* (1952)
: a death caused by the negligent, willful, or wrongful act, neglect, omission, or default of another

wu·shu \'wü-'shü\ *noun* [Chinese (Beijing) *wǔshù*, from *wǔ* martial, military + *shù* art] (1971)
: Chinese martial arts

xe·no·trans·plan·ta·tion \ˌze-nə-ˌtran(t)s-ˌplan-'tā-shən, ˌzē-\ *noun* (1969)
: transplantation of an organ, tissue, or cells between two different species
— **xe·no·trans·plant** \-'tran(t)s-ˌplant\ *noun*

Xer \'ek-sər\ *noun* (1991)
: a member of Generation X

XML \ˌeks-(ˌ)em-'el\ *noun* [*X* (from *extensible*) + *m*arkup *l*anguage] (1989)
: a markup language with use and design similar to HTML but employing tags that indicate the logical structure in addition to the display specifications of the coded data

X–linked \'eks-ˌliŋkt\ *adjective* (1989)
: located on an X chromosome ⟨an *X-linked* gene⟩; *also* : transmitted by an X-linked gene ⟨an *X-linked* disease⟩

yada yada *or* **yadda yadda** \'yä-də-'yä-də\ *noun* [alteration of earlier *yatata* idle chatter, probably ultimately from British dialect and argot *yatter-yatter* to chatter, of imitative origin] (1980)
: boring or empty talk ⟨listening to a lot of *yada yada* about the economy⟩ — often used interjectionally especially in recounting words regarded as too dull or predictable to be worth repeating

yeast infection *noun* (1979)
: infection of the female genital tract that is caused by a yeast fungus (*Candida albicans*) and is characterized by a discharge and inflammation; *broadly* : an infection (as thrush) caused by a yeast fungus

yoc·to- *combining form* [International Scientific Vocabulary, blend of *yotta-* and *octo-*]
: one septillionth (10^{-24}) part of ⟨*yocto*second⟩

yoc·to·sec·ond \'yäk-tə-ˌse-kənd, -kənt\ *noun* (1996)
: one septillionth of a second

yot·ta- *combining form* [International Scientific Vocabulary, alteration of Greek *iōta* (representing *y*, next-to-last letter of Latin alphabet)]
: septillion ⟨*yotta*byte⟩

yot·ta·byte \'yä-tə-ˌbīt\ *noun* (1993)
: one septillion bytes

yuk *or* **yuck** \'yək\ *intransitive verb* **yukked** *or* **yucked; yuk·king** *or* **yuck·ing** (1964) *slang* : LAUGH, JOKE — usually used in the phrase *yuk it up*

yuppie flu *noun* (1987)
: CHRONIC FATIGUE SYNDROME (in this list)

yup·pi·fy \'yə-pə-ˌfī\ *transitive verb* **yup·pi·fied; yup·pi·fy·ing** (1984)
: to make appealing to yuppies; *also* : to infuse with the qualities or values of yuppies
— **yup·pi·fi·ca·tion** \ˌyə-pə-fə-'kā-shən\ *noun*

zai·bat·su \ˌzī-'bät-ˌsü\ *noun* [Japanese *zai* money, wealth + *batsu* clique, clan] (1947)
: a powerful financial and industrial conglomerate of Japan

Zam·bo·ni \zam-'bō-nē\ *trademark*
— used for an ice resurfacing machine

z distribution *noun* (1968)
: a probability density function and especially a normal distribution that has a mean equal to zero and a standard deviation equal to one and that is used especially in testing hypotheses about means or proportions of samples drawn from populations whose population standard deviations are known — compare Z-TEST (in this list)

ze·a·xan·thin \ˌzē-ə-'zan-thən\ *noun* [International Scientific Vocabulary *zea-* (from New Latin *Zea*) + *xanthin* carotenoid pigment, from *xanth-* + [1]*-in*] (1929)
: an isomer of lutein occurring especially in fruits and vegetables (as spinach and corn)

zep·to- *combining form* [International Scientific Vocabulary, blend of *zetta-* and *hepta-*]
: one sextillionth (10^{-21}) part of ⟨*zepto*second⟩

zep·to·sec·ond \'zep-tə-ˌse-kənd, -kənt\ *noun* (1994)
: one sextillionth of a second

zet·ta- *combining form* [International Scientific Vocabulary, alteration of Greek *zēta* (representing *z*, last letter of Latin alphabet)]
: sextillion ⟨*zetta*byte⟩

zet·ta·byte \'ze-tə-ˌbīt\ *noun* (1993)
: one sextillion bytes

zine \'zēn\ *noun* [*-zine* (as in *fanzine*)] (1965)
: a magazine; *especially* : a noncommercial often homemade or online publication usually devoted to specialized and often unconventional subject matter ⟨a punk *zine*⟩ ⟨a feminist *zine*⟩

zip·lock \'zip-ˌläk\ *adjective* (1980)
: having an interlocking groove and ridge that form a tight seal when pressed together ⟨a *ziplock* plastic bag⟩

zone *noun*
: a temporary state of heightened concentration experienced by a performing athlete that enables peak performance ⟨players in the *zone*⟩

zouk \'zük\ *noun* [Lesser Antillean French Creole, literally, dance party, dance, probably alteration of *mazouk* French Caribbean ballroom and club dance of the earlier 20th century, alteration of French *mazurka* mazurka] (1986)
: a form of French West Indian music blending African rhythms, reggae, calypso, and electronic dance music

z–score \'zē-ˌskȯr\ *noun* (1966)
: STANDARD SCORE

z–test \'zē-ˌtest\ *noun* (1951)
: any of several statistical tests that use a random variable having a z distribution to test hypotheses about the mean of a population based on a single sample or about the difference between the means of two populations based on a sample from each when the standard deviations of the populations are known or to test hypotheses about the proportion of successes in a single sample or the difference between the proportion of successes in two samples when the standard deviations are estimated from the sample data

zygote intrafallopian transfer *noun* (1986)
: a method of assisting reproduction in cases of infertility that is similar to gamete intrafallopian transfer but in which eggs are fertilized in vitro and some of the resulting fertilized eggs are inserted into a fallopian tube — abbreviation *ZIFT*

Guide to Pronunciation

ə ... b**a**nan**a**, c**o**llide, **a**but

ˈə, ˌə ... hum**drum**, ab**u**t

ə ... immediately preceding \l\, \n\, \m\, \ŋ\, as in batt**le**, mitt**en**, eat**en**, and sometimes op**en** \ˈō-p^{ə}m\, lock **and** key \-^{ə}ŋ-\; immediately following \l\, \m\, \r\, as often in French tab**le**, pris**me**, tit**re**

ər ... f**ur**th**er**, m**er**g**er**, b**ir**d

ˈər-, ˈə-r ... as in two different pronunciations of h**urry** \ˈhər-ē, ˈhə-rē\

a ... m**a**t, m**a**p, m**a**d, g**a**g, sn**a**p, p**a**tch

ā ... d**ay**, f**a**de, d**a**te, **a**orta, dr**a**pe, c**a**pe

ä ... b**o**ther, c**o**t, and, with most American speakers, f**a**ther, c**a**rt

ȧ ... f**a**ther as pronounced by speakers who do not rhyme it with *bother*; French p**a**tte

au̇ ... n**ow**, l**ou**d, **ou**t

b ... **b**a**b**y, ri**b**

ch ... **ch**in, na**t**ure \ˈnā-chər\

d ... **d**i**d**, a**dd**er

e ... b**e**t, b**e**d, p**e**ck

ˈē, ˌē ... b**ea**t, nosebl**ee**d, **e**venly, **ea**sy

ē ... eas**y**, meal**y**

f ... **f**i**f**ty, cu**ff**

g ... **g**o, bi**g**, **g**ift

h ... **h**at, a**h**ead

hw ... **wh**ale as pronounced by those who do not have the same pronunciation for both *whale* and *wail*

i ... t**i**p, ban**i**sh, act**i**ve

ī ... s**i**te, s**i**de, b**uy**, tr**i**pe

j ... **j**ob, **g**em, e**dge**, **j**oin, **judge**

k ... **k**in, **c**oo**k**, a**ch**e

k̲ ... German i**ch**, Bu**ch**; one pronunciation of lo**ch**

l ... **l**i**l**y, poo**l**

m ... **m**ur**m**ur, di**m**, ny**m**ph

n ... **n**o, ow**n**

^{n} ... indicates that a preceding vowel or diphthong is pronounced with the nasal passages open, as in French *un bon vin blanc* \œ^{n}-bō^{n}-va^{n}-blä^{n}\

ŋ ... si**ng** \ˈsiŋ\, si**ng**er \ˈsiŋ-ər\, fi**ng**er \ˈfiŋ-gər\, i**n**k \ˈiŋk\

ō ... b**o**ne, kn**ow**, b**eau**

ȯ ... s**aw**, **a**ll, gn**aw**, c**augh**t

œ ... French b**oeu**f, German H**ö**lle

œ̄ ... French f**eu**, German H**öh**le

ȯi ... c**oi**n, destr**oy**

p ... **p**e**pp**er, li**p**

r ... **r**ed, ca**r**, **r**a**r**ity

s ... **s**our**c**e, le**ss**

sh ... as in **sh**y, mi**ss**ion, ma**ch**ine, spe**ci**al (actually, this is a single sound, not two); with a hyphen between, two sounds as in *grasshopper* \ˈgras-ˌhä-pər\

t ... **t**ie, a**tt**ack, la**te**, la**t**er, la**tt**er

th ... as in **th**in, e**th**er (actually, this is a single sound, not two); with a hyphen between, two sounds as in *knighthood* \ˈnīt-ˌhu̇d\

t̲h̲ ... **th**en, ei**th**er, **th**is (actually, this is a single sound, not two)

ü ... r**u**le, y**ou**th, union \ˈyün-yən\, few \ˈfyü\

u̇ ... p**u**ll, w**oo**d, b**oo**k, curable \ˈkyu̇r-ə-bəl\, fury \ˈfyu̇r-ē\

ue ... German f**ü**llen, h**ü**bsch

ūe ... French r**ue**, German f**üh**len

v ... **v**i**v**id, gi**ve**

w ... **w**e, a**w**ay

y ... **y**ard, **y**oung, cue \ˈkyü\, mute \ˈmyüt\, union \ˈyün-yən\

^{y} ... indicates that during the articulation of the sound represented by the preceding character the front of the tongue has substantially the position it has for the articulation of the first sound of *yard,* as in French *digne* \dēn^{y}\

z ... **z**one, rai**se**

zh ... as in vi**si**on, a**z**ure \ˈa-zhər\ (actually, this is a single sound, not two); with a hyphen between, two sounds as in *hogshead* \ˈhȯgz-ˌhed, ˈhägz-\

\ ... slant line used in pairs to mark the beginning and end of a transcription: \ˈpen\

ˈ ... mark preceding a syllable with primary (strongest) stress: \ˈpen-mən-ˌship\

ˌ ... mark preceding a syllable with secondary (medium) stress: \ˈpen-mən-ˌship\

\- ... mark of syllable division

, ... mark separating variant pronunciations

; ... mark separating groups of variant pronunciations

( ) ... indicate that what is symbolized between is present in some utterances but not in others: *factory* \ˈfak-t(ə-)rē\

÷ ... indicates that many regard as unacceptable the pronunciation variant immediately following: *cupola* \ˈkyü-pə-lə, ÷-ˌlō\

A is the first letter of the English Alphabet and of most alphabets that are closely related. It comes from Latin A, *which was itself descended, via Etruscan, from Greek* A (alpha). *The Greek letter, in turn, was borrowed from the first letter of the Phoenician alphabet,* ʾ aleph. *That letter represented a consonant sound called the glottal stop, pronounced something like a cough. Since the Greeks did not use the glottal stop, they assigned the letter to their own equivalent of the vowel sound (pronounced "ah") following the glottal stop in Phoenician* ʾ aleph. *In modern English,* A *represents various sounds, including the short* a *in* bad *and the long* a *in* fate. *The two small forms of* a *developed through gradual transformation of the capital form. First, the horizontal bar of* A *dropped out; next, it was replaced by a line slanting upward from the bottom of the left leg to the center of the right leg; then, the triangle thus formed was rounded to the left to form a loop; and, eventually, the right upward stroke was curved leftward or shortened and straightened to a vertical position.*

[1]a \ˈā\ *noun, plural* **a's** *or* **as** \ˈāz\ *often capitalized, often attributive* (before 12th century)
1 a : the 1st letter of the English alphabet **b :** a graphic representation of this letter **c :** a speech counterpart of orthographic *a*
2 : the 6th tone of a C-major scale
3 : a graphic device for reproducing the letter *a*
4 : one designated *a* especially as the 1st in order or class
5 a : a grade rating a student's work as superior or in quality **b :** one graded or rated with an A
6 : something shaped like the letter A

[2]a \ə, (ˈ)ā, *Canadian* ˈa\ *indefinite article* [Middle English, from Old English *ān* one — more at ONE] (before 12th century)
1 — used as a function word before singular nouns when the referent is unspecified ⟨*a* man overboard⟩ and before number collectives and some numbers ⟨*a* dozen⟩
2 : the same ⟨birds of *a* feather⟩ ⟨swords all of *a* length⟩
3 a — used as a function word before a singular noun followed by a restrictive modifier ⟨*a* man who was here yesterday⟩ **b :** ANY ⟨*a* man who is sick can't work⟩ **c** — used as a function word before a mass noun to denote a particular type or instance ⟨*a* bronze made in ancient times⟩ ⟨glucose is *a* simple sugar⟩ **d** — used as a function word before a proper noun representing an example or type ⟨the attractions of *a* Boston or *a* Cleveland⟩
4 — used as a function word with nouns to form adverbial phrases of quantity, amount, or degree ⟨felt *a* little tired⟩ ▫

[3]a \ə *also* (ˈ)ā\ *preposition* [Middle English, from Old English *a-, an, on*] (before 12th century)
1 *chiefly dialect* **:** ON, IN, AT
2 : in, to, or for each ⟨twice *a* week⟩ ⟨five dollars *a* dozen⟩
usage see [2]A

[4]a \ə, (ˌ)a\ *verb* [Middle English, contraction of *have*] (14th century)
archaic **:** HAVE ⟨I might *a* had husbands afore now —John Bunyan⟩

[5]a \ə\ *preposition* [Middle English, by contraction] (15th century)
: OF — often attached to the preceding word ⟨kind*a*⟩ ⟨lott*a*⟩

[1]a- \ə\ *prefix* [Middle English, from Old English]
1 : on **:** in **:** at ⟨*a*bed⟩
2 : in (such) a state or condition ⟨*a*fire⟩
3 : in (such) a manner ⟨*a*loud⟩
4 : in the act or process of ⟨gone *a*-hunting⟩ ⟨*a*tingle⟩

[2]a- \(ˈ)ā *also* (ˈ)a *or* (ˈ)ä\ *or* **an-** \(ˈ)an\ *prefix* [Latin & Greek; Latin, from Greek — more at UN-]
: not **:** without ⟨*a*sexual⟩ — *a-* before consonants other than *h* and sometimes even before *h, an-* before vowels and usually before *h* ⟨*a*chromatic⟩ ⟨*a*historical⟩ ⟨*an*astigmatic⟩ ⟨*an*hydrous⟩

-a- *combining form* [International Scientific Vocabulary]
: replacing carbon especially in a ring ⟨az*a*-⟩

-a \ə\ *noun suffix* [New Latin, from *-a* (as in *magnesia*)]
: OXIDE ⟨silic*a*⟩

aah \ˈä, *often prolonged and/or followed by* ə\ *intransitive verb* (1953)
: to exclaim in amazement, joy, or surprise ⟨one finds oneself oohing and *aahing* over the exciting new TV commercials —Walter Goodman⟩
— aah *noun*

aard·vark \ˈärd-ˌvärk\ *noun* [obsolete Afrikaans, from Afrikaans *aard* earth + *vark* pig] (1827)
: a large burrowing nocturnal ungulate mammal (*Orycteropus afer*) of sub-Saharan Africa that has a long snout, extensile tongue, powerful claws, large ears, and heavy tail and feeds especially on termites and ants

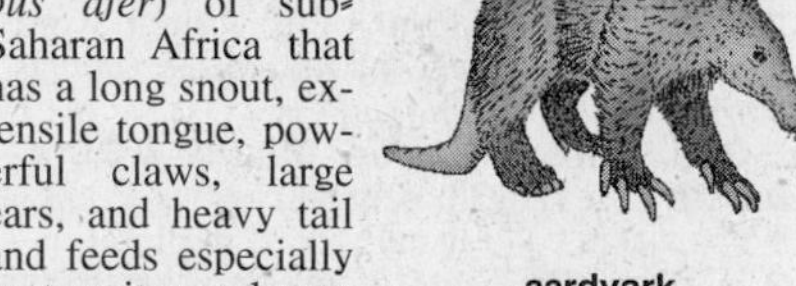
aardvark

aard·wolf \-ˌwu̇lf\ *noun* [Afrikaans, from *aard* + *wolf*] (1833)
: a maned striped nocturnal mammal (*Proteles cristatus*) of southern and eastern Africa that resembles the related hyenas and feeds chiefly on insects and especially termites

aardwolf

Aar·on \ˈar-ən, ˈer-\ *noun* [Late Latin, from Greek *Aarōn,* from Hebrew *Ahărōn*]
: a brother of Moses and high priest of the Hebrews

Aa·ron·ic \a-ˈrä-nik, e-\ *adjective* (circa 1828)
1 : of or stemming from Aaron
2 : of or relating to the lower order of the Mormon priesthood

ab \ˈab\ *noun* (1983)
: an abdominal muscle — usually used in plural

Ab \ˈäb, ˈäv, ˈóv\ *noun* [Hebrew *Ābh*] (circa 1771)
: the 11th month of the civil year or the 5th month of the ecclesiastical year in the Jewish calendar — see MONTH table

ab- \(ˌ)ab, əb\ *prefix* [Middle English, from Old French & Latin; Old French, from Latin *ab-, abs-, a-,* from *ab, a* — more at OF]
: from **:** away **:** off ⟨*ab*axial⟩

aba \ə-ˈbä, ä-ˈba\ *noun* [Arabic *ʿabāʾ*] (1811)
1 : a loose sleeveless outer garment worn as traditional dress by men in the Middle East
2 : a fabric woven from the hair of camels or goats

ab·a·ca \ˌa-bə-ˈkä, ˈa-bə-ˌ\ *noun* [Spanish *abacá,* from Tagalog *abaká*] (circa 1818)
1 : a strong fiber obtained from the leafstalk of a banana (*Musa textilis*) native to the Philippines — called also *Manila hemp*
2 : the plant that yields abaca

aback \ə-ˈbak\ *adverb* (before 12th century)
1 *archaic* **:** BACKWARD, BACK
2 : in a position to catch the wind upon the forward surface of a square sail
3 : by surprise **:** UNAWARES ⟨was taken *aback* by her sharp retort⟩

abac·te·ri·al \ˌā-(ˌ)bak-ˈtir-ē-əl\ *adjective* (circa 1935)
: not caused by or characterized by the presence of bacteria ⟨*abacterial* prostatitis⟩

ab·a·cus \ˈa-bə-kəs, ə-ˈba-\ *noun, plural* **aba·ci** \ˈa-bə-ˌsī, -ˌkē; ə-ˈba-ˌkī\ *or* **ab·a·cus·es** [Latin, from Greek *abak-, abax,* literally, slab] (14th century)
1 : an instrument for performing calculations by sliding counters along rods or in grooves
2 : a slab that forms the uppermost member or division of the capital of a column

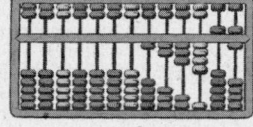

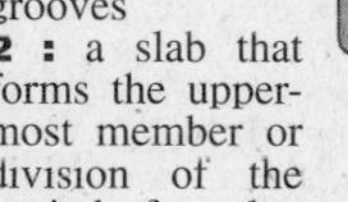
abacus 1

[1]abaft \ə-ˈbaft\ *preposition* [[1]*a-* + *baft* (aft)] (1594)
: to the rear of; *specifically* **:** toward the stern from

[2]abaft *adverb* (1628)
: toward or at the stern **:** AFT

ab·a·lo·ne \ˌa-bə-ˈlō-nē, ˈa-bə-ˌ\ *noun* [American Spanish *abulón,* from Rumsen (American

□ USAGE

a In speech and writing *a* is used before a consonant sound ⟨*a* door⟩ ⟨*a* human⟩. Before a vowel sound *an* is usual ⟨*an* icicle⟩ ⟨*an* honor⟩ but especially in speech *a* is used occasionally, more often in some dialects than in others ⟨*a* apple⟩ ⟨*a* hour⟩ ⟨*a* obligation⟩. Before a consonant sound represented by a vowel letter *a* is usual ⟨*a* one⟩ ⟨*a* union⟩ but *an* also occurs though less frequently now than formerly ⟨*an* unique⟩ ⟨such *an* one⟩. Before unstressed or weakly stressed syllables with initial *h* both *a* and *an* are used in writing ⟨*a* historic⟩ ⟨*an* historic⟩ but in speech *an* is more frequent whether \h\ is pronounced or not. In the King James Version of the Old Testament and occasionally in writing and speech *an* is used before *h* in a stressed syllable ⟨*an* huntress⟩ ⟨*an* hundred⟩ ⟨children are *an* heritage of the Lord —Psalms 127:3 (Authorized Version)⟩.

Indian language of Monterey Bay, Calif.) *aulon*] (1850)
: any of a genus (*Haliotis*) of edible rock-clinging gastropod mollusks that have a flattened shell slightly spiral in form, lined with mother-of-pearl, and with a row of apertures along its outer edge

^{1}**aban•don** \ə-ˈban-dən\ *transitive verb* [Middle English *abandounen,* from Middle French *abandoner,* from *abandon,* noun, surrender, from *a bandon* in one's power] (14th century)
1 a : to give up to the control or influence of another person or agent **b :** to give up with the intent of never again claiming a right or interest in
2 : to withdraw from often in the face of danger or encroachment ⟨*abandon* ship⟩
3 : to withdraw protection, support, or help from
4 : to give (oneself) over unrestrainedly
5 a : to cease from maintaining, practicing, or using ⟨*abandoned* their native language⟩ **b :** to cease intending or attempting to perform ⟨*abandoned* the escape⟩ ☆
— aban•don•er *noun*
— aban•don•ment \-dən-mənt\ *noun*

^{2}**abandon** *noun* (1822)
: a thorough yielding to natural impulses; *especially* **:** ENTHUSIASM, EXUBERANCE ⟨with reckless *abandon*⟩

aban•doned \ə-ˈban-dənd\ *adjective* (14th century)
1 : wholly free from restraint
2 : given up **:** FORSAKEN

à bas \ä-ˈbä\ [French] (circa 1897)
: down with ⟨*à bas* the profiteers⟩

abase \ə-ˈbās\ *transitive verb* **abased; abas•ing** [Middle English *abassen,* from Middle French *abaisser,* from *a-* (from Latin *ad-*) + (assumed) Vulgar Latin *bassiare* to lower] (15th century)
1 *archaic* **:** to lower physically
2 : to lower in rank, office, prestige, or esteem
— abase•ment \-ˈbā-smənt\ *noun*

abash \ə-ˈbash\ *transitive verb* [Middle English *abaishen,* from (assumed) Middle French *abaiss-, abair* to astonish, alteration of Middle French *esbair,* from *ex-* + *baer* to yawn — more at ABEYANCE] (14th century)
: to destroy the self-possession or self-confidence of **:** DISCONCERT
synonym see EMBARRASS
— abash•ment \-mənt\ *noun*

abate \ə-ˈbāt\ *verb* **abat•ed; abat•ing** [Middle English, from Old French *abattre* to beat down — more at REBATE] (13th century)
transitive verb
1 a : to put an end to ⟨*abate* a nuisance⟩ **b :** NULLIFY ⟨*abate* a writ⟩
2 a : to reduce in degree or intensity **:** MODERATE **b :** to reduce in value or amount **:** make less especially by way of relief ⟨*abate* a tax⟩
3 : DEDUCT, OMIT ⟨*abate* part of the price⟩
4 a : to beat down or cut away so as to leave a figure in relief **b** *obsolete* **:** BLUNT
5 : DEPRIVE
intransitive verb
1 : to decrease in force or intensity
2 a : to become defeated or become null or void **b :** to decrease in amount or value ☆
— abat•er *noun*

abate•ment \ə-ˈbāt-mənt\ *noun* (14th century)
1 : the act or process of abating **:** the state of being abated
2 : an amount abated; *especially* **:** a deduction from the full amount of a tax

ab•a•tis \ˈa-bə-ˌtē, ˈa-bə-təs\ *noun, plural* **ab•a•tis** \ˈa-bə-ˌtēz\ *or* **ab•a•tis•es** \ˈa-bə-tə-səz\ [French, from *abattre*] (1766)
: a defensive obstacle formed by felled trees with sharpened branches facing the enemy

ab•at•toir \ˈa-bə-ˌtwär, -ˌtwȯr, -ˌtȯr\ *noun* [French, from *abattre*] (1820)
: SLAUGHTERHOUSE

ab•ax•i•al \(ˌ)a-ˈbak-sē-əl\ *adjective* (1857)
: situated out of or directed away from the axis ⟨the *abaxial* or lower surface of a leaf⟩

ab•ba•cy \ˈa-bə-sē\ *noun, plural* **-cies** [Middle English *abbatie,* from Late Latin *abbatia*] (15th century)
: the office, dignity, jurisdiction, or tenure of an abbot

Ab•bas•id \ə-ˈba-səd, ˈa-bə-səd\ *noun* (1788)
: a member of a dynasty of caliphs (750–1258) ruling the Islamic empire especially from their capital Baghdad and claiming descent from Abbas the uncle of Muhammad

ab•ba•tial \ə-ˈbā-shəl, a-\ *adjective* (circa 1642)
: of or relating to an abbot, abbess, or abbey

ab•bé \a-ˈbā, ˈa-ˌbā\ *noun* [French, from Late Latin *abbat-, abbas*] (1530)
: a member of the French secular clergy in major or minor orders — used as a title

ab•bess \ˈa-bəs\ *noun* [Middle English *abbesse,* from Old French, from Late Latin *abbatissa,* feminine of *abbat-, abbas*] (13th century)
: a woman who is the superior of a convent of nuns

Abbe•vil•li•an \ˌab-ˈvi-lē-ən, ˌa-bə-\ *adjective* [*Abbeville,* France] (circa 1934)
: of or relating to an early Lower Paleolithic culture of Europe characterized by bifacial stone hand axes

ab•bey \ˈa-bē\ *noun, plural* **abbeys** [Middle English, from Old French *abaïe,* from Late Latin *abbatia* abbey, from *abbat-, abbas*] (13th century)
1 a : a monastery ruled by an abbot **b :** a convent ruled by an abbess
2 : an abbey church

ab•bot \ˈa-bət\ *noun* [Middle English *abbod,* from Old English, from Late Latin *abbat-, abbas,* from Late Greek *abbas,* from Aramaic *abbā* father] (before 12th century)
: the superior of a monastery for men

ab•bre•vi•ate \ə-ˈbrē-vē-ˌāt\ *transitive verb* **-at•ed; -at•ing** [Middle English, from Late Latin *abbreviatus,* past participle of *abbreviare* — more at ABRIDGE] (15th century)
: to make briefer; *especially* **:** to reduce to a shorter form intended to stand for the whole
synonym see SHORTEN
— ab•bre•vi•a•tor \-ˌā-tər\ *noun*

ab•bre•vi•a•tion \ə-ˌbrē-vē-ˈā-shən\ *noun* (15th century)
1 : the act or result of abbreviating **:** ABRIDGMENT
2 : a shortened form of a written word or phrase used in place of the whole ⟨*amt* is an *abbreviation* for *amount*⟩

ABC \ˌā-(ˌ)bē-ˈsē\ *noun, plural* **ABC's** *or* **ABCs** \-ˈsēz\ (13th century)
1 : ALPHABET — usually used in plural
2 a : the rudiments of reading, writing, and spelling — usually used in plural **b :** the rudiments of a subject

ABD \ˌā-(ˌ)bē-ˈdē\ *noun, plural* **ABDs** *also* **ABD's** [*a*ll *b*ut *d*issertation] (1965)
: a doctoral candidate who has completed required courses and examinations but not a dissertation

Ab•di•as \ab-ˈdī-əs\ *noun* [Late Latin, from Greek]
: OBADIAH

ab•di•cate \ˈab-di-ˌkāt\ *verb* **-cat•ed; -cat•ing** [Latin *abdicatus,* past participle of *abdicare,* from *ab-* + *dicare* to proclaim — more at DICTION] (1541)
transitive verb
1 : to cast off **:** DISCARD
2 : to relinquish (as sovereign power) formally
intransitive verb
: to renounce a throne, high office, dignity, or function ☆
— ab•di•ca•ble \-kə-bəl\ *adjective*
— ab•di•ca•tion \ˌab-di-ˈkā-shən\ *noun*
— ab•di•ca•tor \ˈab-di-ˌkā-tər\ *noun*

ab•do•men \ˈab-də-mən, -ˌdō-; ab-ˈdō-mən, ab-\ *noun* [Middle French & Latin; Middle French, from Latin] (1615)
1 : the part of the body between the thorax and the pelvis; *also* **:** the cavity of this part of the trunk containing the chief viscera
2 : the posterior section of the body behind the thorax in an arthropod — see INSECT illustration
— ab•dom•i•nal \ab-ˈdä-mə-n^{ə}l, əb-, -ˈdäm-n^{ə}l\ *adjective*
— ab•dom•i•nal•ly \-ē\ *adverb*

ab•du•cens nerve \ab-ˈdü-ˌsenz-, -ˈdyü-\ *noun* [New Latin *abducent-, abducens,* from Latin, present participle of *abducere*] (1947)
: either of the 6th pair of cranial nerves that are motor nerves supplying the rectus on the outer and lateral side of each eye — called also *abducens*

ab•du•cent nerve \ab-ˈdü-s^{ə}nt-, -ˈdyü-\ *noun* (1875)
: ABDUCENS NERVE

ab•duct \ab-ˈdəkt, əb-; *2 also* ˈab-ˌ\ *transitive verb* [Latin *abductus,* past participle of *abducere,* literally, to lead away, from *ab-* + *ducere* to lead — more at TOW] (1834)
1 : to carry off (as a person) by force
2 : to draw or spread away (as a limb or the fingers) from a position near or parallel to the median axis of the body or from the axis of a limb
— ab•duc•tor \-ˈdək-tər\ *noun*

ab•duc•tion \ab-ˈdək-shən, əb-\ *noun* (1666)
1 : the action of abducting **:** the condition of being abducted
2 : the unlawful carrying away of a woman for marriage or intercourse

abeam \ə-ˈbēm\ *adverb or adjective* (circa 1836)
: off to the side of a ship or plane especially at a right angle to the middle of the ship or plane's length

^{1}**abe•ce•dar•i•an** \ˌā-bē-(ˌ)sē-ˈder-ē-ən\ *noun* [Middle English *abecedary,* from Medieval Latin *abecedarium* alphabet, from Late Latin,

☆ SYNONYMS

Abandon, desert, forsake mean to leave without intending to return. ABANDON suggests that the thing or person left may be helpless without protection ⟨*abandoned* children⟩. DESERT implies that the object left may be weakened but not destroyed by one's absence ⟨a *deserted* town⟩. FORSAKE suggests an action more likely to bring impoverishment or bereavement to that which is forsaken than its exposure to physical dangers ⟨a *forsaken* lover⟩. See in addition RELINQUISH.

Abate, subside, wane, ebb mean to die down in force or intensity. ABATE stresses the idea of progressive diminishing ⟨the storm *abated*⟩. SUBSIDE implies the ceasing of turbulence or agitation ⟨the protests *subsided* after a few days⟩. WANE suggests the fading or weakening of something good or impressive ⟨*waning* enthusiasm⟩. EBB suggests the receding of something (as the tide) that commonly comes and goes ⟨the *ebbing* of daylight⟩. See in addition DECREASE.

Abdicate, renounce, resign mean to give up a position with no possibility of resuming it. ABDICATE implies a giving up of sovereign power or sometimes an evading of responsibility such as that of a parent ⟨*abdicated* the throne⟩. RENOUNCE may replace it but often implies additionally a sacrifice for a greater end ⟨*renounced* her inheritance by marrying a commoner⟩. RESIGN applies to the giving up of an unexpired office or trust ⟨*resigned* from the board⟩.

neuter of *abecedarius* of the alphabet, from the letters *a* + *b* + *c* + *d*] (1603)
: one learning the rudiments of something (as the alphabet)

[2]abecedarian *adjective* (1665)
1 a : of or relating to the alphabet **b :** alphabetically arranged
2 : RUDIMENTARY

abed \ə-'bed\ *adverb or adjective* (13th century)
: in bed

Abel \'ā-bəl\ *noun* [Late Latin, from Greek, from Hebrew *Hebhel*]
: a son of Adam and Eve killed by his brother Cain

abe·lia \ə-'bēl-yə\ *noun* [New Latin, from Clarke *Abel* (died 1826) English botanist] (circa 1899)
: any of a genus (*Abelia*) of shrubs of Asian or Mexican origin having opposite leaves and white, red, or pink flowers

abe·li·an \ə-'bē-lē-ən\ *adjective, often capitalized* [Niels *Abel* (died 1829) Norwegian mathematician] (circa 1909)
: COMMUTATIVE 2 ⟨*abelian* group⟩ ⟨*abelian* ring⟩

Abe·na·ki \,a-bə-'nä-kē\ *noun, plural* **Abenaki** *or* **Abenakis** (1721)
1 : a member of a group of American Indian peoples of northern New England and adjoining parts of Quebec
2 : either of the two Algonquian languages spoken by the Abenaki peoples

Ab·er·deen An·gus \'a-bər-,dēn-'aŋ-gəs\ *noun* [*Aberdeen* & *Angus,* counties in Scotland] (1862)
: ANGUS

[1]ab·er·rant \a-'ber-ənt, ə-; 'a-bə-rənt, -,ber-ənt\ *adjective* [Latin *aberrant-, aberrans,* present participle of *aberrare* to go astray, from *ab-* + *errare* to wander, err] (circa 1780)
1 : straying from the right or normal way
2 : deviating from the usual or natural type **:** ATYPICAL
— ab·er·rance \-ən(t)s\ *noun*
— ab·er·ran·cy \-ən(t)-sē\ *noun*
— ab·er·rant·ly *adverb*

[2]aberrant *noun* (1938)
1 : an aberrant group, individual, or structure
2 : a person whose behavior departs substantially from the standard

ab·er·rat·ed \'a-bə-,rā-təd\ *adjective* [Latin *aberratus,* past participle of *aberrare*] (1893)
: ABERRANT

ab·er·ra·tion \,a-bə-'rā-shən\ *noun* [Latin *aberrare*] (1594)
1 : the fact or an instance of being aberrant especially from a moral standard or normal state
2 : failure of a mirror, refracting surface, or lens to produce exact point-to-point correspondence between an object and its image
3 : unsoundness or disorder of the mind
4 : a small periodic change of apparent position in celestial bodies due to the combined effect of the motion of light and the motion of the observer
5 : an aberrant individual
— ab·er·ra·tion·al \-shnəl, -shə-nᵊl\ *adjective*

abet \ə-'bet\ *transitive verb* **abet·ted; abet·ting** [Middle English *abetten,* from Middle French *abeter,* from Old French, from *a-* (from Latin *ad-*) + *beter* to bait, of Germanic origin; akin to Old English *bǣtan* to bait] (14th century)
1 : to actively second and encourage (as an activity or plan) **:** FORWARD
2 : to assist or support in the achievement of a purpose ⟨*abetted* the thief in his getaway⟩
synonym see INCITE
— abet·ment \-mənt\ *noun*
— abet·tor *or* **abet·ter** \ə-'be-tər\ *noun*

abey·ance \ə-'bā-ən(t)s\ *noun* [Middle French *abeance* expectation, from *abaer* to desire, from *a-* + *baer* to yawn, from Medieval Latin *batare*] (1660)
1 : temporary inactivity **:** SUSPENSION
2 : a lapse in succession during which there is no person in whom a title is vested

abey·ant \-ənt\ *adjective* [back-formation from *abeyance*] (circa 1859)
: being in abeyance

ab·hor \əb-'hȯr, ab-\ *transitive verb* **ab·horred; ab·hor·ring** [Middle English *abhorren,* from Latin *abhorrēre,* from *ab-* + *horrēre* to shudder — more at HORROR] (15th century)
: to regard with extreme repugnance **:** LOATHE
synonym see HATE
— ab·hor·rer \-'hȯr-ər\ *noun*

ab·hor·rence \əb-'hȯr-ən(t)s, -'här-\ *noun* (1660)
1 a : the act or state of abhorring **b :** the feeling of one who abhors
2 : one that is abhorred

ab·hor·rent \-ənt\ *adjective* [Latin *abhorrent-, abhorrens,* present participle of *abhorrēre*] (1599)
1 a *archaic* **:** strongly opposed **b :** feeling or showing abhorrence
2 : not agreeable **:** CONTRARY ⟨a notion *abhorrent* to their philosophy⟩
3 : being so repugnant as to stir up positive antagonism ⟨acts *abhorrent* to every right-minded person⟩
— ab·hor·rent·ly *adverb*

Abib \ä-'vēv\ *noun* [Hebrew *Ābhībh,* literally, ear of grain] (1535)
: the 1st month of the ancient Hebrew calendar corresponding to Nisan — see MONTH table

abid·ance \ə-'bī-dᵊn(t)s\ *noun* (1647)
1 : an act or state of abiding **:** CONTINUANCE
2 : COMPLIANCE ⟨*abidance* by the rules⟩

abide \ə-'bīd\ *verb* **abode** \-'bōd\ *or* **abid·ed; abid·ing** [Middle English, from Old English *ābīdan,* from *ā-,* perfective prefix + *bīdan* to bide; akin to Old High German *ir-,* perfective prefix] (before 12th century)
transitive verb
1 : to wait for **:** AWAIT
2 a : to endure without yielding **:** WITHSTAND **b :** to bear patiently **:** TOLERATE ⟨cannot *abide* such bigots⟩
3 : to accept without objection ⟨will *abide* your decision⟩
intransitive verb
1 : to remain stable or fixed in a state
2 : to continue in a place **:** SOJOURN
synonym see BEAR, CONTINUE
— abid·er *noun*
— abide by 1 : to conform to **2 :** to acquiesce in

abid·ing \ə-'bī-diŋ\ *adjective* (14th century)
: ENDURING, CONTINUING ⟨an *abiding* interest in nature⟩
— abid·ing·ly *adverb*

ab·i·gail \'a-bə-,gāl\ *noun* [*Abigail,* servant in *The Scornful Lady,* a play by Francis Beaumont & John Fletcher] (1671)
: a lady's personal maid

abil·i·ty \ə-'bi-lə-tē\ *noun, plural* **-ties** [Middle English *abilite,* from Middle French *habilité,* from Latin *habilitat-, habilitas,* from *habilis* apt, skillful — more at ABLE] (14th century)
1 a : the quality or state of being able ⟨*ability* of the soil to hold water⟩; *especially* **:** physical, mental, or legal power to perform **b :** competence in doing **:** SKILL
2 : natural aptitude or acquired proficiency ⟨children whose *abilities* warrant higher education⟩

-ability *also* **-ibility** *noun suffix* [Middle English *-abilite, -ibilite,* from Middle French *-abilité, -ibilité,* from Latin *-abilitas, -ibilitas,* from *-abilis, -ibilis* -able + *-tas* -ty]
: capacity, fitness, or tendency to act or be acted on in a (specified) way ⟨agglutin*ability*⟩

ab in·i·tio \,ab-ə-'ni-shē-,ō\ *adverb* [Latin] (1599)
: from the beginning

abi·o·gen·e·sis \,ā-,bī-ō-'je-nə-səs\ *noun* [New Latin, from [2]*a-* + *bio-* + Latin *genesis*] (1870)
: the supposed spontaneous origination of living organisms directly from lifeless matter
— abi·og·e·nist \,ā-(,)bī-'ä-jə-nist\ *noun*

abi·o·gen·ic \,ā-,bī-ō-'je-nik\ *adjective* (1891)
: not produced by the action of living organisms
— abi·o·gen·i·cal·ly \-ni-k(ə-)lē\ *adverb*

abi·o·log·i·cal \,ā-,bī-ə-'lä-ji-kəl\ *adjective* (1868)
: not biological; *especially* **:** not involving or produced by organisms ⟨*abiological* synthesis of amino acids⟩

abi·ot·ic \,ā-(,)bī-'ä-tik\ *adjective* (circa 1893)
: not biotic **:** ABIOLOGICAL ⟨the *abiotic* environment⟩
— abi·ot·i·cal·ly \-ti-k(ə-)lē\ *adverb*

ab·ject \'ab-,jekt\ *adjective* [Middle English, from Latin *abjectus,* from past participle of *abicere* to cast off, from *ab-* + *jacere* to throw — more at JET] (15th century)
1 : sunk to or existing in a low state or condition ⟨to lowest pitch of *abject* fortune thou art fallen —John Milton⟩
2 a : cast down in spirit **:** SERVILE, SPIRITLESS ⟨a man made *abject* by suffering⟩ **b :** showing utter hopelessness or resignation ⟨*abject* surrender⟩
3 : expressing or offered in a humble and often ingratiating spirit ⟨*abject* flattery⟩ ⟨an *abject* apology⟩
synonym see MEAN
— ab·ject·ly \'ab-,jek(t)-lē, ab-'\ *adverb*
— ab·ject·ness \-,jek(t)-nəs, -'jek(t)-\ *noun*

ab·jec·tion \ab-'jek-shən\ *noun* (14th century)
1 : a low or downcast state **:** DEGRADATION
2 : the act of making abject **:** HUMBLING, REJECTION ⟨I protest . . . this vile *abjection* of youth to age —G. B. Shaw⟩

ab·ju·ra·tion \,ab-jə-'rā-shən\ *noun* (15th century)
1 : the act or process of abjuring
2 : an oath of abjuring

ab·jure \ab-'ju̇r\ *transitive verb* **ab·jured; ab·jur·ing** [Middle English, from Middle French or Latin; Middle French *abjurer,* from Latin *abjurare,* from *ab-* + *jurare* to swear — more at JURY] (15th century)
1 a : to renounce upon oath **b :** to reject solemnly
2 : to abstain from **:** AVOID ⟨*abjure* extravagance⟩ ☆
— ab·jur·er *noun*

ab·late \a-'blāt\ *verb* **ab·lat·ed; ab·lat·ing** [Latin *ablatus* (past participle of *auferre* to remove), from *ab-* + *latus,* past participle of

☆ SYNONYMS

Abjure, renounce, forswear, recant, retract mean to withdraw one's word or professed belief. ABJURE implies a firm and final rejecting or abandoning often made under oath ⟨*abjured* the errors of his former faith⟩. RENOUNCE often equals ABJURE but may carry the meaning of disclaim or disown ⟨*renounced* abstract art and turned to portrait painting⟩. FORSWEAR may add to ABJURE an implication of perjury or betrayal ⟨I cannot *forswear* my principles⟩. RECANT stresses the withdrawing or denying of something professed or taught ⟨if they *recant* they will be spared⟩. RETRACT applies to the withdrawing of a promise, an offer, or an accusation ⟨the newspaper had to *retract* its allegations against the mayor⟩.

\ə\ **abut** \ᵊ\ **kitten** \ər\ **further** \a\ **ash** \ā\ **ace** \ä\ **mop, mar** \au̇\ **out** \ch\ **chin** \e\ **bet** \ē\ **easy** \g\ **go** \i\ **hit** \ī\ **ice** \j\ **job** \ŋ\ **sing** \ō\ **go** \ȯ\ **law** \ȯi\ **boy** \th\ **thin** \t͟h\ **the** \ü\ **loot** \u̇\ **foot** \y\ **yet** \zh\ **vision** *see also* Guide to Pronunciation

ferre — more at UKASE, BEAR, TOLERATE] (1542)
transitive verb
: to remove especially by cutting, abrading, or evaporating
intransitive verb
: to become ablated; *especially* **:** VAPORIZE 1
ab·la·tion \a-ˈblā-shən\ *noun* (15th century)
: the process of ablating: as **a :** surgical removal **b :** loss of a part (as ice from a glacier or the outside of a nose cone) by melting or vaporization
[1]**ab·la·tive** \ˈa-blə-tiv\ *adjective* (15th century)
: of, relating to, or constituting a grammatical case expressing typically the relations of separation and source and also frequently such relations as cause or instrument
— ablative *noun*
[2]**ab·la·tive** \a-ˈblā-tiv\ *adjective* (circa 1569)
1 : of or relating to ablation
2 : tending to ablate ⟨*ablative* material on a nose cone⟩
— ab·la·tive·ly *adverb*
ablative absolute \ˈa-blə-tiv-\ *noun* (circa 1828)
: a construction in Latin in which a noun or pronoun and its adjunct both in the ablative case form together an adverbial phrase expressing generally the time, cause, or an attendant circumstance of an action
ab·laut \ˈä-blau̇t, ˈa-; ˈäp-ˌlau̇t\ *noun* [German, from *ab* away from + *Laut* sound] (1849)
: a systematic variation of vowels in the same root or affix or in related roots or affixes especially in the Indo-European languages that is usually paralleled by differences in use or meaning (as in *sing, sang, sung, song*)
ablaze \ə-ˈblāz\ *adjective or adverb* (1801)
1 : being on fire
2 : radiant with light or emotion ⟨his face all *ablaze* with excitement —Bram Stoker⟩
able \ˈā-bəl\ *adjective* **abler** \-b(ə-)lər\; **ablest** \-b(ə-)ləst\ [Middle English, from Middle French, from Latin *habilis* apt, from *habēre* to have — more at HABIT] (14th century)
1 a : having sufficient power, skill, or resources to accomplish an object **b :** susceptible to action or treatment
2 : marked by intelligence, knowledge, skill, or competence
— ably \ˈā-b(ə-)lē\ *adverb*
-able *also* **-ible** *adjective suffix* [Middle English, from Old French, from Latin *-abilis, -ibilis,* from *-a-, -i-,* verb stem vowels + *-bilis* capable or worthy of]
1 : capable of, fit for, or worthy of (being so acted upon or toward) — chiefly in adjectives derived from verbs ⟨break*able*⟩ ⟨collect*ible*⟩
2 : tending, given, or liable to ⟨agree*able*⟩ ⟨perish*able*⟩
— -ably *also* **-ibly** *adverb suffix*
able–bod·ied \ˌā-bəl-ˈbä-dēd\ *adjective* (circa 1622)
: having a sound strong body
able–bodied seaman *noun* (circa 1909)
: ABLE SEAMAN
able seaman *noun* (1702)
: an experienced deck-department seaman qualified to perform routine duties at sea
abloom \ə-ˈblüm\ *adjective* (1855)
: abounding with blooms **:** BLOOMING ⟨parks *abloom* with roses⟩
ab·lut·ed \ə-ˈblü-təd, a-\ *adjective* [back-formation from *ablution*] (1650)
: washed clean
ab·lu·tion \ə-ˈblü-shən, a-\ *noun* [Middle English, from Middle French or Latin; Middle French, from Latin *ablution-, ablutio,* from *abluere* to wash away, from *ab-* + *luere* to wash; akin to Latin *lavere* to wash — more at LYE] (1533)
1 a : the washing of one's body or part of it (as in a religious rite) **b** *plural* **:** the act or action of bathing
2 *plural, British* **:** a building housing bathing and toilet facilities on a military base
— ab·lu·tion·ary \-shə-ˌner-ē\ *adjective*
ABM \ˌā-(ˌ)bē-ˈem\ *noun, plural* **ABM's** *or* **ABMs** \-ˈemz\ (1963)
: ANTIBALLISTIC MISSILE
Ab·na·ki \ab-ˈnä-kē\ *variant of* ABENAKI
ab·ne·gate \ˈab-ni-ˌgāt\ *transitive verb* **-gat·ed; -gat·ing** [back-formation from *abnegation*] (1623)
1 : DENY, RENOUNCE ⟨*abnegated* their God⟩
2 : SURRENDER, RELINQUISH ⟨*abnegated* her powers⟩
— ab·ne·ga·tor \-ˌgā-tər\ *noun*
ab·ne·ga·tion \ˌab-ni-ˈgā-shən\ *noun* [Late Latin *abnegation-, abnegatio,* from Latin *abnegare* to refute, from *ab-* + *negare* to deny — more at NEGATE] (14th century)
: DENIAL; *especially* **:** SELF-DENIAL
[1]**ab·nor·mal** \(ˌ)ab-ˈnȯr-məl, əb-\ *adjective* [alteration of French *anormal,* from Medieval Latin *anormalis,* from Latin *a-* + Late Latin *normalis* normal] (circa 1836)
: deviating from the normal or average **:** UNUSUAL, EXCEPTIONAL ⟨*abnormal* behavior⟩
— ab·nor·mal·ly \-mə-lē\ *adverb*
[2]**abnormal** *noun* (1912)
: an abnormal person
ab·nor·mal·i·ty \ˌab-nər-ˈma-lə-tē, -(ˌ)nȯr-\ *noun, plural* **-ties** (1854)
1 : the quality or state of being abnormal
2 : something abnormal
abnormal psychology *noun* (circa 1903)
: a branch of psychology concerned with mental and emotional disorders (as neuroses, psychoses, and mental deficiency) and with certain incompletely understood normal phenomena (as dreams and hypnosis)
abo \ˈa-(ˌ)bō\ *noun, plural* **ab·os** (1908)
Australian **:** ABORIGINE — often used disparagingly
[1]**aboard** \ə-ˈbōrd, -ˈbȯrd\ *adverb or adjective* (14th century)
1 : ALONGSIDE
2 a : on, onto, or within a vehicle (as a car or ship) **b :** in or into a group, association, or organization ⟨her second promotion since coming *aboard*⟩
3 *baseball* **:** on base
[2]**aboard** *preposition* (15th century)
: ON, ONTO, WITHIN ⟨go *aboard* ship⟩ ⟨*aboard* a plane⟩
ABO blood group \ˌā-(ˌ)bē-ˈō-\ *noun* (1949)
: any of the four blood groups A, B, AB, and O comprising the ABO system
abode \ə-ˈbōd\ *noun* [Middle English *abod,* from *abiden* to abide] (13th century)
1 *obsolete* **:** WAIT, DELAY
2 : a temporary stay **:** SOJOURN
3 : the place where one abides **:** HOME
aboil \ə-ˈbȯi(ə)l\ *adjective or adverb* (1810)
1 : being at the boiling point **:** BOILING
2 : intensely excited or stirred up ⟨the meeting was *aboil* with controversy⟩
abol·ish \ə-ˈbä-lish\ *transitive verb* [Middle English *abolisshen,* from Middle French *aboliss-,* stem of *abolir,* from Latin *abolēre;* probably akin to *adolescere* to grow up — more at ADULT] (15th century)
1 : to end the observance or effect of **:** ANNUL
2 : DESTROY
— abol·ish·able \-li-shə-bəl\ *adjective*
— abol·ish·er *noun*
— abol·ish·ment \-mənt\ *noun*
ab·o·li·tion \ˌa-bə-ˈli-shᵊn\ *noun* [Middle French, from Latin *abolition-, abolitio,* from *abolēre*] (1529)
1 : the act of abolishing **:** the state of being abolished
2 : the abolishing of slavery
— ab·o·li·tion·ary \-ˈli-shᵊn-ˌer-ē\ *adjective*
ab·o·li·tion·ism \-ˈli-shᵊn-ˌi-zəm\ *noun* (1808)
: principles or measures fostering abolition especially of slavery
— ab·o·li·tion·ist \-ist\ *noun or adjective*
ab·oma·sum \ˌa-bō-ˈmā-səm\ *noun, plural* **-sa** \-sə\ [New Latin, from Latin *ab-* + *omasum* ox's tripe] (circa 1706)
: the fourth compartment of the ruminant stomach that follows the omasum and has a true digestive function — compare RUMEN, RETICULUM
— ab·oma·sal \-səl\ *adjective*
A–bomb \ˈā-ˌbäm\ *noun* (1945)
: ATOMIC BOMB 1
abom·i·na·ble \ə-ˈbäm-nə-bəl, -ˈbä-mə-\ *adjective* (14th century)
1 : worthy of or causing disgust or hatred **:** DETESTABLE ⟨the *abominable* treatment of the poor⟩
2 : quite disagreeable or unpleasant ⟨*abominable* weather⟩
— abom·i·na·bly \-blē\ *adverb*
abominable snow·man \-ˈsnō-mən, -ˌman\ *noun, often A&S capitalized* (1921)
: a mysterious creature with human or apelike characteristics reported to exist in the high Himalayas — called also *yeti*
abom·i·nate \ə-ˈbä-mə-ˌnāt\ *transitive verb* **-nat·ed; -nat·ing** [Latin *abominatus,* past participle of *abominari,* literally, to deprecate as an ill omen, from *ab-* + *omin-, omen* omen] (1644)
: to hate or loathe intensely **:** ABHOR
synonym see HATE
— abom·i·na·tor \-ˌnā-tər\ *noun*
abom·i·na·tion \ə-ˌbä-mə-ˈnā-shən\ *noun* (14th century)
1 : something abominable
2 : extreme disgust and hatred **:** LOATHING
ab·oral \(ˌ)a-ˈbōr-əl, -ˈbȯr-\ *adjective* (1857)
: situated opposite to or away from the mouth ⟨a sea urchin's *aboral* surface⟩
— ab·oral·ly \-ə-lē\ *adverb*
[1]**ab·orig·i·nal** \ˌa-bə-ˈrij-nəl, -ˈri-jə-nᵊl\ *adjective* (1667)
1 : being the first or earliest known of its kind present in a region ⟨*aboriginal* forests⟩ ⟨*aboriginal* rocks⟩
2 a : of or relating to aborigines **b** *often capitalized* **:** of or relating to the indigenous peoples of Australia
synonym see NATIVE
— ab·orig·i·nal·ly *adverb*
[2]**aboriginal** *noun* (1767)
1 : ABORIGINE 1
2 *often capitalized* **:** ABORIGINE 2
ab·orig·i·ne \ˌa-bə-ˈrij-(ˌ)nē, -ˈri-jə-\ *noun* [Latin *aborigines,* plural, from *ab origine* from the beginning] (1533)
1 : an aboriginal inhabitant especially as contrasted with an invading or colonizing people
2 *often capitalized* **:** a member of any of the indigenous peoples of Australia
[1]**aborn·ing** \ə-ˈbȯr-niŋ\ *adverb* [[1]*a-* + English dialect *borning* (birth)] (1916)
: while being born or produced ⟨a resolution that died *aborning*⟩
[2]**aborning** *adjective* (1943)
: being born or produced ⟨the *aborning* fiasco⟩
[1]**abort** \ə-ˈbȯrt\ *verb* [Latin *abortus,* past participle of *aboriri* to miscarry, from *ab-* + *oriri* to rise, be born — more at ORIENT] (1580)
intransitive verb
1 : to bring forth premature or stillborn offspring
2 : to become checked in development so as to degenerate or remain rudimentary
transitive verb
1 a : to induce the abortion of or give birth to prematurely **b :** to terminate the pregnancy of before term
2 a : to terminate prematurely **:** CANCEL ⟨*abort* a project⟩ ⟨*abort* a spaceflight⟩ **b :** to stop in the early stages ⟨*abort* a disease⟩
— abort·er *noun*
[2]**abort** *noun* (1944)
: the premature termination of a flight (as of

an aircraft or spacecraft), a mission, or an action or procedure relating to a flight ⟨a launch *abort*⟩

abor·ti·fa·cient \ə-ˌbȯr-tə-ˈfā-shənt\ *noun* (1873)
: an agent (as a drug) that induces abortion
— **abortifacient** *adjective*

abor·tion \ə-ˈbȯr-shən\ *noun* (1547)
1 : the termination of a pregnancy after, accompanied by, resulting in, or closely followed by the death of the embryo or fetus: as **a :** spontaneous expulsion of a human fetus during the first 12 weeks of gestation — compare MISCARRIAGE **b :** induced expulsion of a human fetus **c :** expulsion of a fetus by a domestic animal often due to infection at any time before completion of pregnancy — compare CONTAGIOUS ABORTION
2 : MONSTROSITY
3 : arrest of development (as of a part or process) resulting in imperfection; *also* **:** a result of such arrest

abor·tion·ist \-sh(ə-)nist\ *noun* (1871)
: one who induces abortions

abor·tive \ə-ˈbȯr-tiv\ *adjective* (14th century)
1 *obsolete* **:** prematurely born
2 : FRUITLESS, UNSUCCESSFUL
3 : imperfectly formed or developed
4 : tending to cut short ■
— **abor·tive·ly** *adverb*
— **abor·tive·ness** *noun*

ABO system \ˌā-(ˌ)bē-ˈō-\ *noun* (1944)
: the basic system of antigens of human blood behaving in heredity as an allelic unit to produce any of the ABO blood groups

abound \ə-ˈbau̇nd\ *intransitive verb* [Middle English, from Middle French *abonder,* from Latin *abundare,* from *ab-* + *unda* wave — more at WATER] (14th century)
1 : to be present in large numbers or in great quantity **:** be prevalent
2 : to be copiously supplied ⟨life *abounded* in mysteries —Norman Mailer⟩ ⟨institutions *abound* with evidence of his success —*Johns Hopkins Magazine*⟩

^{1}**about** \ə-ˈbau̇t\ *adverb* [Middle English, from Old English *abūtan,* from ^{1}*a-* + *būtan* outside — more at BUT] (before 12th century)
1 a : reasonably close to ⟨*about* a year ago⟩ **b :** ALMOST ⟨*about* starved⟩ **c :** on the verge of — usually used with *be* and a following infinitive ⟨is *about* to join the army⟩; used with a negative to express intention or determination ⟨not *about* to quit⟩
2 : on all sides **:** AROUND
3 a : in rotation **b :** around the outside
4 : HERE AND THERE
5 : in the vicinity **:** NEAR
6 : in succession **:** ALTERNATELY ⟨turn *about* is fair play⟩
7 : in the opposite direction ⟨face *about*⟩ ⟨the other way *about*⟩

^{2}**about** *preposition* (before 12th century)
1 : in a circle around **:** on every side of **:** AROUND
2 a : in the immediate neighborhood of **:** NEAR **b :** on or near the person of **c :** in the makeup of ⟨a mature wisdom *about* him⟩ **d :** at the command of ⟨has his wits *about* him⟩
3 : engaged in ⟨act as if they know what they're *about* —T. S. Matthews⟩
4 a : with regard to **:** CONCERNING **b :** concerned with **c :** fundamentally concerned with or directed toward ⟨poker is *about* money —David Mamet⟩
5 : over or in different parts of

^{3}**about** *adjective* (1815)
1 : moving from place to place; *specifically* **:** being out of bed
2 : AROUND 2

about–face \ə-ˈbau̇t-ˈfās\ *noun* [from the imperative phrase *about face*] (1861)
1 : a 180° turn to the right from the position of attention
2 : a reversal of direction
3 : a reversal of attitude, behavior, or point of view
— **about–face** *intransitive verb*

about–turn \-ˈtərn\ *noun* (1893)
British **:** ABOUT-FACE

^{1}**above** \ə-ˈbəv\ *adverb* [Middle English, from Old English *abufan,* from *a-* + *bufan* above, from *be-* + *ufan* above; akin to Old English *ofer* over] (before 12th century)
1 a : in the sky **:** OVERHEAD **b :** in or to heaven
2 a : in or to a higher place **b :** higher on the same page or on a preceding page **c :** UPSTAIRS **d :** above zero ⟨10 degrees *above*⟩
3 : in or to a higher rank or number ⟨30 and *above*⟩
4 *archaic* **:** in addition **:** BESIDES
5 : UPSTAGE

^{2}**above** *preposition* (before 12th century)
1 : in or to a higher place than **:** OVER
2 a : superior to (as in rank, quality, or degree) **b :** out of reach of **c :** in preference to **d :** too proud or honorable to stoop to
3 : exceeding in number, quantity, or size **:** more than
4 : as distinct from and in addition to ⟨heard the whistle *above* the roar of the crowd⟩

^{3}**above** *noun, plural* **above** (13th century)
1 a : something that is above **b :** a person whose name is written above
2 a : a higher authority **b :** HEAVEN ■

^{4}**above** *adjective* (1604)
: written or discussed higher on the same page or on a preceding page
usage see ^{3}ABOVE

above all *adverb* (14th century)
: before every other consideration **:** ESPECIALLY

^{1}**above·board** \ə-ˈbəv-ˌbōrd, -ˌbȯrd\ *adverb* [from the difficulty of cheating at cards when the hands are above the table] (1594)
: in a straightforward manner **:** OPENLY

^{2}**aboveboard** *adjective* (1648)
: free from all traces of deceit or duplicity

above·ground \ə-ˈbəv-ˌgrau̇nd\ *adjective* (1878)
1 : located or occurring on or above the surface of the ground
2 : existing, produced, or published by or within the establishment ⟨*aboveground* movies⟩

ab ovo \ab-ˈō-(ˌ)vō\ *adverb* [Latin, literally, from the egg] (circa 1586)
: from the beginning

ab·ra·ca·dab·ra \ˌa-brə-kə-ˈda-brə\ *noun* [Late Latin] (1565)
1 : a magical charm or incantation
2 : unintelligible language

abrade \ə-ˈbrād\ *verb* **abrad·ed; abrad·ing** [Latin *abradere* to scrape off, from *ab-* + *radere* to scrape — more at RODENT] (1677)
transitive verb
1 a : to rub or wear away especially by friction **:** ERODE **b :** to irritate or roughen by rubbing
2 : to wear down in spirit **:** IRRITATE, WEARY
intransitive verb
: to undergo abrasion
— **abrad·able** \-ˈbrā-də-bəl\ *adjective*
— **abrad·er** *noun*

Abra·ham \ˈā-brə-ˌham\ *noun* [Late Latin, from Greek *Abraam,* from Hebrew *ʼAbhrāhām*]
: an Old Testament patriarch regarded by Jews as the founder of the Hebrew people through his son Isaac and by Muslims as the founder of the Arab peoples through his son Ishmael

abra·sion \ə-ˈbrā-zhən\ *noun* [Medieval Latin *abrasion-, abrasio,* from Latin *abradere*] (circa 1656)
1 a : a wearing, grinding, or rubbing away by friction **b :** IRRITATION
2 : an abraded area of the skin or mucous membrane

^{1}**abra·sive** \ə-ˈbrā-siv, -ziv\ *noun* (1853)
: a substance (as emery or pumice) used for abrading, smoothing, or polishing

^{2}**abrasive** *adjective* (1875)
1 : tending to abrade
2 : causing irritation ⟨*abrasive* manners⟩
— **abra·sive·ly** *adverb*
— **abra·sive·ness** *noun*

ab·re·ac·tion \ˌa-brē-ˈak-shən\ *noun* [part translation of German *Abreagierung* catharsis, from *ab* off, away (from Old High German *aba*) + *Reagierung* reaction, from Latin *reagere* to react — more at OF, REACT] (1912)
: the expression and emotional discharge of unconscious material (as a repressed idea or emotion) by verbalization especially in the presence of a therapist
— **ab·re·act** \-ˈakt\ *verb*

abreast \ə-ˈbrest\ *adverb or adjective* (15th century)
1 : beside one another with bodies in line ⟨columns of men five *abreast*⟩
2 : up to a particular standard or level especially of knowledge of recent developments ⟨keeps *abreast* of the latest trends⟩

abridge \ə-ˈbrij\ *transitive verb* **abridged; abridg·ing** [Middle English *abregen,* from Middle French *abregier,* from Late Latin *abbreviare,* from Latin *ad-* + *brevis* short — more at BRIEF] (14th century)
1 a *archaic* **:** DEPRIVE **b :** to reduce in scope **:** DIMINISH ⟨attempts to *abridge* the right of free speech⟩
2 : to shorten in duration or extent ⟨modern transportation that *abridges* distance⟩
3 : to shorten by omission of words without sacrifice of sense **:** CONDENSE
synonym see SHORTEN
— **abridg·er** *noun*

abridg·ment *or* **abridge·ment** \ə-ˈbrij-mənt\ *noun* (15th century)
1 : the action of abridging **:** the state of being abridged
2 : a shortened form of a work retaining the general sense and unity of the original

abroach \ə-ˈbrōch\ *adverb or adjective* (14th century)
1 : in a condition for letting out a liquid (as wine) ⟨a cask set *abroach*⟩
2 : in action or agitation **:** ASTIR ⟨mischiefs that I set *abroach* —Shakespeare⟩

abroad \ə-ˈbrȯd\ *adverb or adjective* (13th century)
1 : over a wide area **:** WIDELY

☐ USAGE

abortive Sense 2 has been objected to occasionally since the 1880s. These objections depend upon an etymological argument that the suffix *-ive* implies continuation, repetition, or tendency, rather than completion; therefore, something unsuccessful must be *aborted,* rather than *abortive.* This argument ignores the whole history of the word in English; from Shakespeare's time and earlier *abortive* has meant, literally and figuratively, "aborted." One may safely ignore this objection, as other writers have ⟨our first design, my friend, has prov'd *abortive* —Joseph Addison⟩ ⟨the *abortive* coup against him in October —Elizabeth Drew⟩.

above Although still objected to by some, the use of *above* as a noun in sense 1a ⟨none of the *above*⟩ ⟨the *above* is Theseus's opinion —William Blake⟩ and as an adjective ⟨without the *above* reserve —O. W. Holmes (died 1935)⟩ ⟨I was brought up on the *above* words —Viscount Montgomery⟩ has been long established as standard.

2 : away from one's home
3 : beyond the boundaries of one's country
4 : in wide circulation **:** ABOUT
5 : wide of the mark **:** ASTRAY

ab·ro·gate \ˈa-brə-ˌgāt\ *transitive verb* **-gat·ed; -gat·ing** [Latin *abrogatus,* past participle of *abrogare,* from *ab-* + *rogare* to ask, propose a law — more at RIGHT] (1526)
1 : to abolish by authoritative action **:** ANNUL
2 : to treat as nonexistent ⟨*abrogating* their responsibilities⟩
synonym see NULLIFY
— **ab·ro·ga·tion** \ˌa-brə-ˈgā-shən\ *noun*

abrupt \ə-ˈbrəpt\ *adjective* [Latin *abruptus,* from past participle of *abrumpere* to break off, from *ab-* + *rumpere* to break — more at REAVE] (1591)
1 a : characterized by or involving action or change without preparation or warning **:** UNEXPECTED ⟨came to an *abrupt* stop⟩ ⟨an *abrupt* turn to the left⟩ ⟨an *abrupt* decision to retire⟩ **b :** unceremoniously curt ⟨an *abrupt* manner⟩ **c :** lacking smoothness or continuity ⟨an *abrupt* transition⟩
2 : giving the impression of being cut or broken off; *especially* **:** involving a sudden steep rise or drop ⟨*abrupt* hills⟩ ⟨a high *abrupt* bank bounded the stream⟩
synonym see PRECIPITATE, STEEP
— **abrupt·ly** \ə-ˈbrəp(t)-lē\ *adverb*
— **abrupt·ness** \ə-ˈbrəp(t)-nəs\ *noun*

abrup·tion \ə-ˈbrəp-shən\ *noun* (1606)
: a sudden breaking off or away

ABS \ˌā-(ˌ)bē-ˈes\ *noun* [*a*crylonitrile-*b*utadiene-*s*tyrene] (1966)
: a tough rigid plastic used especially for automobile parts and building materials

ab·scess \ˈab-ˌses\ *noun, plural* **ab·scess·es** \ˈab-ˌse-səz, -ˌsēz; -sə-səz\ [Latin *abscessus,* literally, act of going away, from *abscedere* to go away, from *abs-, ab-* + *cedere* to go] (1615)
: a localized collection of pus surrounded by inflamed tissue
— **ab·scessed** \-ˌsest\ *adjective*

ab·scise \ab-ˈsīz\ *verb* **ab·scised; ab·scis·ing** [Latin *abscisus,* past participle of *abscidere,* from *abs-* + *caedere* to cut] (1612)
transitive verb
: to cut off by abscission
intransitive verb
: to separate by abscission

ab·scis·ic acid \ˌab-ˈsi-zik-, -sik-\ *noun* [*ab*scis*ion* (variant of *abscission*) + *-ic*] (1968)
: a plant hormone $C_{15}H_{20}O_4$ that is a sesquiterpene widespread in nature and that typically promotes leaf abscission and dormancy and has an inhibitory effect on cell elongation

ab·sci·sin \ˈab-sə-sən, ab-ˈsi-sᵊn\ *noun* [*abscision* + *-in*] (1961)
: ABSCISIC ACID

ab·scis·sa \ab-ˈsi-sə\ *noun, plural* **abscis·sas** *also* **ab·scis·sae** \-ˈsi-(ˌ)sē\ [New Latin, from Latin, feminine of *abscissus,* past participle of *abscindere* to cut off, from *ab-* + *scindere* to cut — more at SHED] (1694)
: the horizontal coordinate of a point in a plane Cartesian coordinate system obtained by measuring parallel to the x-axis — compare ORDINATE

Y
P A
X

AP **abscissa of point *P***

ab·scis·sion \ab-ˈsi-zhən\ *noun* [Latin *abscission-, abscissio,* from *abscindere*] (15th century)
1 : the act or process of cutting off **:** REMOVAL
2 : the natural separation of flowers, fruit, or leaves from plants at a special separation layer

ab·scond \ab-ˈskänd, əb-\ *intransitive verb* [Latin *abscondere* to hide away, from *abs-* + *condere* to store up, conceal — more at CONDIMENT] (circa 1578)
: to depart secretly and hide oneself
— **ab·scond·er** *noun*

ab·seil \ˈab-ˌsāl, -ˌsīl\ *intransitive verb* [German *abseilen,* from *ab* down, off + *Seil* rope] (1941)
chiefly British **:** RAPPEL

ab·sence \ˈab-sən(t)s\ *noun* (14th century)
1 : the state of being absent
2 : the period of time that one is absent
3 : WANT, LACK ⟨an *absence* of detail⟩
4 : inattention to present surroundings or occurrences ⟨*absence* of mind⟩

¹ab·sent \ˈab-sənt\ *adjective* [Middle English, from Middle French, from Latin *absent-, absens,* present participle of *abesse* to be absent, from *ab-* + *esse* to be — more at IS] (14th century)
1 : not present or attending **:** MISSING
2 : not existing **:** LACKING ⟨danger in a situation where power is *absent* —M. H. Trytten⟩
3 : INATTENTIVE, PREOCCUPIED
— **ab·sent·ly** *adverb*

²ab·sent \ab-ˈsent, ˈab-ˌ\ *transitive verb* (15th century)
: to keep (oneself) away

³ab·sent \ˈab-sənt\ *preposition* (1945)
: in the absence of **:** WITHOUT

ab·sen·tee \ˌab-sən-ˈtē\ *noun* (1605)
: one that is absent: as **a :** a proprietor that lives away from his or her estate or business **b :** one missing from work or school
— **absentee** *adjective*

absentee ballot *noun* (1932)
: a ballot submitted (as by mail) in advance of an election by a voter who is unable to be present at the polls

ab·sen·tee·ism \ˌab-sən-ˈtē-ˌi-zəm\ *noun* (1829)
1 : prolonged absence of an owner from his or her property
2 : chronic absence (as from work or school); *also* **:** the rate of such absence

ab·sent·mind·ed \ˌab-sənt-ˈmīn-dəd\ *adjective* (1854)
1 : lost in thought and unaware of one's surroundings or actions **:** PREOCCUPIED; *also* **:** given to absence of mind
2 : indicative of or resulting from preoccupation or absence of mind
— **ab·sent·mind·ed·ly** *adverb*
— **ab·sent·mind·ed·ness** *noun*

absent without leave *adjective* (circa 1919)
: absent without authority from one's place of duty in the armed forces

ab·sinthe *also* **ab·sinth** \ˈab-(ˌ)sin(t)th\ *noun* [French *absinthe,* from Latin *absinthium,* from Greek *apsinthion*] (1612)
1 : WORMWOOD 1; *especially* **:** a common European wormwood (*Artemisia absinthium*)
2 : a green liqueur flavored with wormwood or a substitute, anise, and other aromatics

ab·so·lute \ˈab-sə-ˌlüt, ˌab-sə-ˈ\ *adjective* [Middle English *absolut,* from Latin *absolutus,* from past participle of *absolvere* to set free, absolve] (14th century)
1 a : free from imperfection **:** PERFECT **b :** free or relatively free from mixture **:** PURE ⟨*absolute* alcohol⟩ **c :** OUTRIGHT, UNMITIGATED ⟨an *absolute* lie⟩
2 : being, governed by, or characteristic of a ruler or authority completely free from constitutional or other restraint
3 a : standing apart from a normal or usual syntactical relation with other words or sentence elements ⟨the *absolute* construction *this being the case* in the sentence "this being the case, let us go"⟩ **b** *of an adjective or possessive pronoun* **:** standing alone without a modified substantive ⟨*blind* in "help the blind" and *ours* in "your work and ours" are *absolute*⟩ **c** *of a verb* **:** having no object in the particular construction under consideration though normally transitive ⟨*kill* in "if looks could kill" is an *absolute* verb⟩
4 : having no restriction, exception, or qualification ⟨an *absolute* requirement⟩ ⟨*absolute* freedom⟩
5 : POSITIVE, UNQUESTIONABLE ⟨*absolute* proof⟩
6 a : independent of arbitrary standards of measurement **b :** relating to or derived in the simplest manner from the fundamental units of length, mass, and time ⟨*absolute* electric units⟩ **c :** relating to, measured on, or being a temperature scale based on absolute zero ⟨*absolute* temperature⟩; *specifically* **:** KELVIN ⟨10° *absolute*⟩
7 : FUNDAMENTAL, ULTIMATE ⟨*absolute* knowledge⟩
8 : perfectly embodying the nature of a thing ⟨*absolute* justice⟩
9 : being self-sufficient and free of external references or relationships ⟨an *absolute* term in logic⟩ ⟨*absolute* music⟩
10 : being the true distance from an aircraft to the earth's surface ⟨*absolute* altitude⟩
— **absolute** *noun*
— **ab·so·lute·ness** *noun*

absolute ceiling *noun* (circa 1920)
: the maximum height above sea level at which a particular airplane can maintain horizontal flight under standard air conditions — called also *ceiling*

absolute convergence *noun* (circa 1909)
: convergence of a mathematical series when the absolute values of the terms are taken

absolute humidity *noun* (1867)
: the amount of water vapor present in a unit volume of air — compare RELATIVE HUMIDITY

ab·so·lute·ly \ˈab-sə-ˌlüt-lē, ˌab-sə-ˈ\ *adverb* (14th century)
1 : in an absolute manner or condition — often used as an intensive ⟨*absolutely* brilliant⟩
2 : with respect to absolute values ⟨an *absolutely* convergent series⟩

absolute magnitude *noun* (1902)
: the intrinsic luminosity of a celestial body (as a star) if viewed from a distance of 10 parsecs — compare APPARENT MAGNITUDE

absolute pitch *noun* (1864)
1 : the position of a tone in a standard scale independently determined by its rate of vibration
2 : the ability to recognize or sing a given isolated note

absolute space *noun* (circa 1889)
: SPACE 4b

absolute value *noun* (1907)
1 : a nonnegative number equal in numerical value to a given real number
2 : the positive square root of the sum of the squares of the real and imaginary parts of a complex number

absolute zero *noun* (1848)
: a theoretical temperature characterized by complete absence of heat and equivalent to exactly -273.15°C or -459.67°F

ab·so·lu·tion \ˌab-sə-ˈlü-shən\ *noun* (13th century)
: the act of absolving; *specifically* **:** a remission of sins pronounced by a priest (as in the sacrament of reconciliation)

ab·so·lut·ism \ˈab-sə-ˌlü-ˌti-zəm\ *noun* (1830)
1 a : a political theory that absolute power should be vested in one or more rulers **b :** government by an absolute ruler or authority **:** DESPOTISM
2 : advocacy of a rule by absolute standards or principles
3 : an absolute standard or principle
— **ab·so·lut·ist** \-ˌlü-tist\ *noun or adjective*
— **ab·so·lu·tis·tic** \ˌab-sə-(ˌ)lü-ˈtis-tik\ *adjective*

ab·so·lu·tive \ˌab-sə-ˈlü-tiv\ *adjective* (1952)
: of, being, or relating to an inflectional morpheme that typically marks the subject of an intransitive verb or the direct object of a transitive verb in an ergative language

ab·so·lut·ize \ˈab-sə-ˌlü-ˌtīz\ *transitive verb* **-ized; -iz·ing** (1919)
: to make absolute **:** convert into an absolute

ab·solve \əb-ˈzälv, -ˈsälv, -ˈzȯlv, -ˈsȯlv *also without* l\ *transitive verb* **ab·solved; ab-**

solv·ing [Middle English, from Latin *absolvere,* from *ab-* + *solvere* to loosen — more at SOLVE] (15th century)
1 : to set free from an obligation or the consequences of guilt
2 : to remit (a sin) by absolution
synonym see EXCULPATE
— **ab·solv·er** *noun*

ab·sorb \əb-ˈsȯrb, -ˈzȯrb\ *transitive verb* [Middle French *absorber,* from Latin *absorbēre,* from *ab-* + *sorbēre* to suck up; akin to Lithuanian *surbti* to sip, Greek *rophein* to gulp down] (15th century)
1 : to take in and make part of an existent whole ⟨the capacity of China to *absorb* invaders⟩
2 a : to suck up or take up ⟨a sponge *absorbs* water⟩ ⟨charcoal *absorbs* gas⟩ ⟨plant roots *absorb* water⟩ **b :** to take in **:** ACQUIRE, LEARN ⟨convictions *absorbed* in youth —M. R. Cohen⟩ **c :** USE UP, CONSUME ⟨the fever *absorbed* her strength⟩
3 : to engage or engross wholly ⟨*absorbed* in thought⟩
4 a (1) **:** to receive without recoil or echo ⟨provided with a sound-*absorbing* surface⟩ (2) **:** ENDURE, SUSTAIN ⟨*absorbing* hardships⟩ (3) **:** ASSUME, BEAR ⟨the expenses were *absorbed* by the company⟩ **b :** to transform (radiant energy) into a different form usually with a resulting rise in temperature ⟨the earth *absorbs* the sun's rays⟩
— **ab·sorb·abil·i·ty** \əb-ˌsȯr-bə-ˈbi-lə-tē, -ˌzȯr-\ *noun*
— **ab·sorb·able** \əb-ˈsȯr-bə-bəl, -ˈzȯr-\ *adjective*
— **ab·sorb·er** *noun*

ab·sor·bance \əb-ˈsȯr-bən(t)s, -ˈzȯr-\ *noun* (1947)
: the ability of a layer of a substance to absorb radiation expressed mathematically as the negative common logarithm of transmittance

ab·sor·ben·cy \əb-ˈsȯr-bən(t)-sē, -ˈzȯr-\ *noun, plural* **-cies** (1859)
1 : the quality or state of being absorbent
2 *or* **ab·sor·ban·cy :** ABSORBANCE

ab·sor·bent *also* **ab·sor·bant** \-bənt\ *adjective* [Latin *absorbent-, absorbens,* present participle of *absorbēre*] (1718)
: able to absorb ⟨as *absorbent* as a sponge⟩
— **absorbent** *also* **absorbant** *noun*

ab·sorb·ing *adjective* (1876)
: fully taking one's attention **:** ENGROSSING ⟨an *absorbing* novel⟩
— **ab·sorb·ing·ly** \-biŋ-lē\ *adverb*

ab·sorp·tance \əb-ˈsȯrp-tən(t)s, -ˈzȯrp-\ *noun* [*absorp*tion + *-ance*] (circa 1931)
: the ratio of the radiant energy absorbed by a body to that incident upon it

ab·sorp·tion \əb-ˈsȯrp-shən, -ˈzȯrp-\ *noun* [French & Latin; French, from Latin *absorption-, absorptio,* from *absorbēre*] (1741)
1 a : the process of absorbing or of being absorbed — compare ADSORPTION **b :** interception of radiant energy or sound waves
2 : entire occupation of the mind ⟨*absorption* in his work⟩
— **ab·sorp·tive** \-tiv\ *adjective*

absorption band *noun* (1867)
: a dark band in an absorption spectrum

absorption line *noun* (circa 1889)
: a dark line in an absorption spectrum

absorption spectrum *noun* (1879)
: an electromagnetic spectrum in which a decrease in intensity of radiation at specific wavelengths or ranges of wavelengths characteristic of an absorbing substance is manifested especially as a pattern of dark lines or bands

ab·sorp·tiv·i·ty \əb-ˌsȯrp-ˈti-və-tē, -ˌzȯrp-\ *noun, plural* **-ties** (circa 1859)
: the property of a body that determines the fraction of incident radiation absorbed by the body

ab·stain \əb-ˈstān, ab-\ *intransitive verb* [Middle English *absteinen,* from Middle French *abstenir,* from Latin *abstinēre,* from *abs-, ab-* + *tenēre* to hold — more at THIN] (14th century)
: to refrain deliberately and often with an effort of self-denial from an action or practice
— **ab·stain·er** *noun*

ab·ste·mi·ous \ab-ˈstē-mē-əs\ *adjective* [Latin *abstemius,* from *abs-* + *-temius;* akin to Latin *temetum* intoxicating drink] (1609)
: marked by restraint especially in the consumption of food or alcohol; *also* **:** reflecting such restraint ⟨an *abstemious* diet⟩
— **ab·ste·mi·ous·ly** *adverb*
— **ab·ste·mious·ness** *noun*

ab·sten·tion \əb-ˈsten(t)-shən, ab-\ *noun* [Late Latin *abstention-, abstentio,* from Latin *abstinēre*] (1521)
: the act or practice of abstaining
— **ab·sten·tious** \-shəs\ *adjective*

ab·sti·nence \ˈab-stə-nən(t)s\ *noun* [Middle English, from Middle French, from Latin *abstinentia,* from *abstinent-, abstinens,* present participle of *abstinēre*] (14th century)
1 : voluntary forbearance especially from indulgence of an appetite or craving or from eating some foods **:** ABSTENTION
2 : habitual abstaining from intoxicating beverages
— **ab·sti·nent** \-nənt\ *adjective*
— **ab·sti·nent·ly** *adverb*

[1]ab·stract \ab-ˈstrakt, ˈab-ˌ\ *adjective* [Medieval Latin *abstractus,* from Latin, past participle of *abstrahere* to drag away, from *abs-, ab-* + *trahere* to pull, draw] (14th century)
1 a : disassociated from any specific instance ⟨*abstract* entity⟩ **b :** difficult to understand **:** ABSTRUSE ⟨*abstract* problems⟩ **c :** insufficiently factual **:** FORMAL ⟨possessed only an *abstract* right⟩
2 : expressing a quality apart from an object ⟨the word *poem* is concrete, *poetry* is *abstract*⟩
3 a : dealing with a subject in its abstract aspects **:** THEORETICAL ⟨*abstract* science⟩ **b :** IMPERSONAL, DETACHED ⟨the *abstract* compassion of a surgeon —*Time*⟩
4 : having only intrinsic form with little or no attempt at pictorial representation or narrative content ⟨*abstract* painting⟩
— **ab·stract·ly** \ab-ˈstrak(t)-lē, ˈab-ˌ\ *adverb*
— **ab·stract·ness** \ab-ˈstrak(t)-nəs, ˈab-ˌ\ *noun*

[2]ab·stract \ˈab-ˌstrakt, *in sense 2 also* ab-ˈ\ *noun* [Middle English, from Latin *abstractus*] (15th century)
1 : a summary of points (as of a writing) usually presented in skeletal form; *also* **:** something that summarizes or concentrates the essentials of a larger thing or several things
2 : an abstract thing or state
3 : ABSTRACTION 4a

[3]ab·stract \ab-ˈstrakt, ˈab-ˌ, *in sense 3 usually* ˈab-ˌ\ (1542)
transitive verb
1 : REMOVE, SEPARATE
2 : to consider apart from application to or association with a particular instance
3 : to make an abstract of **:** SUMMARIZE
4 : to draw away the attention of
5 : STEAL, PURLOIN
intransitive verb
: to make an abstraction
— **ab·stract·able** \-ˈstrak-tə-bəl, -ˌstrak-\ *adjective*
— **ab·strac·tor** *or* **ab·stract·er** \-tər\ *noun*

ab·stract·ed \ab-ˈstrak-təd, ˈab-ˌ\ *adjective* (1643)
1 : PREOCCUPIED, ABSENTMINDED ⟨the *abstracted* look of a professor⟩
2 : ABSTRACT 4 ⟨*abstracted* geometric shapes⟩
— **ab·stract·ed·ly** *adverb*
— **ab·stract·ed·ness** *noun*

abstract expressionism *noun* (1951)
: an artistic movement of the mid-20th century comprising diverse styles and techniques and emphasizing especially an artist's liberty to convey attitudes and emotions through nontraditional and usually nonrepresentational means
— **abstract expressionist** *noun or adjective*

ab·strac·tion \ab-ˈstrak-shən, əb-\ *noun* (1549)
1 a : the act or process of abstracting **:** the state of being abstracted **b :** an abstract idea or term
2 : absence of mind or preoccupation
3 : abstract quality or character
4 a : an abstract composition or creation in art **b :** ABSTRACTIONISM
— **ab·strac·tion·al** \-shnəl, -shə-nᵊl\ *adjective*
— **ab·strac·tive** \ab-ˈstrak-tiv, ˈab-ˌ\ *adjective*

ab·strac·tion·ism \ab-ˈstrak-shə-ˌni-zəm, əb-\ *noun* (1926)
: the principles or practice of creating abstract art
— **ab·strac·tion·ist** \-sh(ə-)nist\ *adjective or noun*

abstract of title (1858)
: a summary statement of the successive conveyances and other facts on which a title to a piece of land rests

ab·struse \əb-ˈstrüs, ab-\ *adjective* [Latin *abstrusus,* from past participle of *abstrudere* to conceal, from *abs-, ab-* + *trudere* to push — more at THREAT] (1599)
: difficult to comprehend **:** RECONDITE ⟨the *abstruse* calculations of mathematicians⟩
— **ab·struse·ly** *adverb*
— **ab·struse·ness** *noun*

ab·stru·si·ty \-ˈstrü-sə-tē\ *noun, plural* **-ties** (1646)
1 : the quality or state of being abstruse **:** ABSTRUSENESS
2 : something that is abstruse

[1]ab·surd \əb-ˈsərd, -ˈzərd\ *adjective* [Middle French *absurde,* from Latin *absurdus,* from *ab-* + *surdus* deaf, stupid] (1557)
1 : ridiculously unreasonable, unsound, or incongruous
2 : having no rational or orderly relationship to human life **:** MEANINGLESS; *also* **:** lacking order or value
3 : dealing with the absurd or with absurdism
— **ab·surd·ly** *adverb*
— **ab·surd·ness** *noun*

[2]absurd *noun* (1946)
: the state or condition in which human beings exist in an irrational and meaningless universe and in which human life has no ultimate meaning — usually used with *the*

ab·surd·ism \əb-ˈsər-ˌdi-zəm, -ˈzər-\ *noun* (1946)
: a philosophy based on the belief that the universe is irrational and meaningless and that the search for order brings the individual into conflict with the universe — compare EXISTENTIALISM
— **ab·surd·ist** \-dist\ *noun or adjective*

ab·sur·di·ty \əb-ˈsər-də-tē, -ˈzər-\ *noun, plural* **-ties** (1528)
1 : the quality or state of being absurd **:** ABSURDNESS
2 : something that is absurd

abub·ble \ə-ˈbə-bəl\ *adjective* (circa 1869)
1 : being in the process of bubbling
2 : being in a state of agitated activity or motion **:** ASTIR

abuild·ing \ə-ˈbil-diŋ\ *adjective* (1535)
: being in the process of building or of being built

abu·lia \ā-ˈbü-lē-ə, -ˈbyü-, ə-\ *noun* [New Latin, from ²*a-* + Greek *boulē* will] (circa 1864)
: abnormal lack of ability to act or to make decisions
— **abu·lic** \-lik\ *adjective*

abun·dance \ə-ˈbən-dən(t)s\ *noun* (14th century)
1 : an ample quantity **:** PROFUSION
2 : AFFLUENCE, WEALTH
3 : relative degree of plentifulness ⟨low *abundances* of uranium and thorium —H. C. Urey⟩

abun·dant \-dənt\ *adjective* [Middle English, from Middle French, from Latin *abundant-*, *abundans,* present participle of *abundare* to abound] (14th century)
1 a : marked by great plenty (as of resources) ⟨a fair and *abundant* land⟩ **b :** amply supplied **:** ABOUNDING ⟨an area *abundant* with bird life⟩
2 : occurring in abundance **:** AMPLE ⟨*abundant* rainfall⟩
synonym see PLENTIFUL
— **abun·dant·ly** *adverb*

¹abuse \ə-ˈbyüs\ *noun* [Middle English, from Middle French *abus,* from Latin *abusus,* from *abuti* to consume, from *ab-* + *uti* to use] (15th century)
1 : a corrupt practice or custom
2 : improper or excessive use or treatment **:** MISUSE ⟨drug *abuse*⟩
3 *obsolete* **:** a deceitful act **:** DECEPTION
4 : language that condemns or vilifies usually unjustly, intemperately, and angrily
5 : physical maltreatment ☆

²abuse \ə-ˈbyüz\ *transitive verb* **abused; abus·ing** (15th century)
1 : to put to a wrong or improper use ⟨*abuse* a privilege⟩
2 *obsolete* **:** DECEIVE
3 : to use so as to injure or damage **:** MALTREAT
4 : to attack in words **:** REVILE
— **abus·able** \-ˈbyü-zə-bəl\ *adjective*
— **abus·er** *noun*

abu·sive \ə-ˈbyü-siv, -ziv\ *adjective* (1583)
1 : characterized by wrong or improper use or action; *especially* **:** CORRUPT ⟨*abusive* financial practices⟩
2 a : using harsh insulting language **:** characterized by or serving for abuse **b :** physically injurious ⟨*abusive* behavior⟩
— **abu·sive·ly** *adverb*
— **abu·sive·ness** *noun*

abut \ə-ˈbət\ *verb* **abut·ted; abut·ting** [Middle English *abutten,* partly from Old French *aboter* to border on, from *a-* (from Latin *ad-*) + *bout* blow, end, from *boter* to strike; partly from Old French *abuter* to come to an end, from *a-* + *but* end, aim — more at ¹BUTT, ⁴BUTT] (15th century)
intransitive verb
1 : to touch along a border or with a projecting part ⟨land *abuts* on the road⟩
2 a : to terminate at a point of contact **b :** to lean for support
transitive verb
1 : to border on
2 : to cause to abut

abu·ti·lon \ə-ˈbyü-tᵊl-ˌän, -tᵊl-ən\ *noun* [New Latin, genus name, from Arabic *awbūtīlūn* abutilon] (circa 1578)
: any of a genus (*Abutilon*) of plants of the mallow family often having lobed leaves and solitary bell-shaped flowers

abut·ment \ə-ˈbət-mənt\ *noun* (1644)
1 : the place at which abutting occurs
2 : the part of a structure (as an arch or a bridge) that directly receives thrust or pressure

abut·tals \ə-ˈbə-tᵊlz\ *noun plural* (1630)
: the boundaries of lands with respect to adjacent lands

abut·ter \ə-ˈbə-tər\ *noun* (1673)
: one that abuts; *specifically* **:** the owner of a contiguous property

abut·ting *adjective* (1599)
: that abuts or serves as an abutment **:** ADJOINING, BORDERING

abuzz \ə-ˈbəz\ *adjective* (1859)
: filled or resounding with or as if with a buzzing sound ⟨a lake *abuzz* with outboards⟩ ⟨a town *abuzz* with excitement⟩

aby *or* **abye** \ə-ˈbī\ *transitive verb* [Middle English *abien,* from Old English *ābycgan,* from *ā-* + *bycgan* to buy — more at ABIDE, BUY] (before 12th century)
archaic **:** to suffer a penalty for

abysm \ə-ˈbi-zəm\ *noun* [Middle English *abime,* from Middle French *abisme,* modification of Late Latin *abyssus*] (14th century)
: ABYSS ⟨the dark backward and *abysm* of time —Shakespeare⟩

abys·mal \ə-ˈbiz-məl, a-\ *adjective* (circa 1656)
1 a : having immense or fathomless extension downward, backward, or inward ⟨an *abysmal* cliff⟩ **b :** immeasurably great **:** PROFOUND ⟨*abysmal* ignorance⟩ **c :** immeasurably low or wretched ⟨*abysmal* living conditions of the poor⟩
2 : ABYSSAL
— **abys·mal·ly** \-mə-lē\ *adverb*

abyss \ə-ˈbis, a- *also* ˈa-(ˌ)bis\ *noun* [Middle English *abissus,* from Late Latin *abyssus,* from Greek *abyssos,* from *abyssos,* adjective, bottomless, from *a-* + *byssos* depth; perhaps akin to Greek *bathys* deep] (14th century)
1 : the bottomless gulf, pit, or chaos of the old cosmogonies
2 a : an immeasurably deep gulf or great space **b :** intellectual or moral depths

abys·sal \ə-ˈbi-səl\ *adjective* (1691)
1 : UNFATHOMABLE a
2 : of or relating to the bottom waters of the ocean depths

abyssal plain *noun* (1954)
: any of the great flat areas of ocean floor

Ab·ys·sin·i·an cat \ˌa-bə-ˈsi-nē-ən-, -ˈsin-yən-\ *noun* [*Abyssinia,* kingdom in Africa] (1876)
: any of a breed of small slender cats of African origin with short brownish hair ticked with bands of darker color

Abyssinian cat

ac- — see AD-

-ac *noun suffix* [New Latin *-acus,* from Greek *-akos,* variant of *-ikos* -ic after stems ending in *-i-*]
: one affected with ⟨hemophili*ac*⟩

aca·cia \ə-ˈkā-shə\ *noun* [New Latin, genus name, from Latin, acacia tree, from Greek *akakia*] (14th century)
1 : GUM ARABIC
2 : any of a large genus (*Acacia*) of leguminous shrubs and trees of warm regions with leaves pinnate or reduced to phyllodes and white or yellow flower clusters

ac·a·deme \ˈa-kə-ˌdēm, ˌa-kə-ˈ\ *noun* [Latin *Academus* (in the phrase *inter silvas Academi* among the groves of Academus), from Greek *Akadēmos* — more at ACADEMY] (1588)
1 a : a place of instruction **:** SCHOOL **b :** the academic life, community, or world ⟨in the halls of *academe*⟩
2 : ACADEMIC; *especially* **:** PEDANT

ac·a·de·mia \ˌa-kə-ˈdē-mē-ə\ *noun* [New Latin, from Latin, academy] (1946)
: ACADEME 1b

¹ac·a·dem·ic \ˌa-kə-ˈde-mik\ *noun* (1587)
1 : a member of an institution of learning
2 : one who is academic in background, outlook, or methods
3 *plural* **:** academic subjects

²academic *also* **ac·a·dem·i·cal** \-mi-kəl\ *adjective* (1588)
1 a : of, relating to, or associated with an academy or school especially of higher learning **b :** of or relating to performance in academic courses ⟨*academic* excellence⟩ **c :** very learned but inexperienced in practical matters ⟨*academic* thinkers⟩ **d :** based on formal study especially at an institution of higher learning
2 : of or relating to literary or artistic rather than technical or professional studies
3 a : THEORETICAL, SPECULATIVE ⟨an *academic* question⟩ **b :** having no practical or useful significance
4 : conforming to the traditions or rules of a school (as of literature or art) or an official academy **:** CONVENTIONAL
— **ac·a·dem·i·cal·ly** \-mi-k(ə-)lē\ *adverb*

academic freedom *noun* (1901)
: freedom to teach or to learn without interference (as by government officials)

ac·a·de·mi·cian \ˌa-kə-də-ˈmi-shən, ə-ˌka-də-\ *noun* (1748)
1 a : a member of an academy for promoting science, art, or literature **b :** a follower of an artistic or philosophical tradition or a promoter of its ideas
2 : ACADEMIC

ac·a·dem·i·cism \ˌa-kə-ˈde-mə-ˌsi-zəm\ *also* **acad·e·mism** \ə-ˈka-də-ˌmi-zəm\ *noun* (1610)
1 : the doctrines of Plato's Academy; *specifically* **:** the skeptical doctrines of the later Academy holding that nothing can be known — compare PYRRHONISM
2 : a formal academic quality (as in art or music)
3 : purely speculative thoughts and attitudes

academic year *noun* (circa 1934)
: the annual period of sessions of an educational institution usually beginning in September and ending in June

acad·e·my \ə-ˈka-də-mē\ *noun, plural* **-mies** [Latin *academia,* from Greek *Akadēmeia,* from *Akadēmeia,* gymnasium where Plato taught, from *Akadēmos* Attic mythological hero] (1549)
1 a : a school usually above the elementary level; *especially* **:** a private high school **b :** a high school or college in which special subjects or skills are taught **c :** higher education — used with *the* ⟨the functions of the *academy* in modern society⟩
2 *capitalized* **a :** the school for advanced education founded by Plato **b :** the philosophical doctrines associated with Plato's Academy
3 : a society of learned persons organized to advance art, science, or literature
4 : a body of established opinion widely accepted as authoritative in a particular field ◆

Aca·di·an \ə-ˈkā-dē-ən, a-\ *noun* (1705)
1 : a native or inhabitant of Acadia
2 : a descendant of the French-speaking inhabitants of Acadia expelled after the French loss of the colony in 1755; *especially* **:** CAJUN
— **Acadian** *adjective*

acan·tho·ceph·a·lan \ə-ˌkan(t)-thə-ˈse-fə-lən\ *noun* [ultimately from Greek *akantha* thorn, spine + *kephalē* head — more at CEPHALIC] (circa 1909)
: SPINY-HEADED WORM

☆ SYNONYMS
Abuse, vituperation, invective, obloquy, billingsgate mean vehemently expressed condemnation or disapproval. ABUSE, the most general term, usually implies the anger of the speaker and stresses the harshness of the language ⟨scathing verbal *abuse*⟩. VITUPERATION implies fluent and sustained abuse ⟨a torrent of *vituperation*⟩. INVECTIVE implies a comparable vehemence but suggests greater verbal and rhetorical skill and may apply to a public denunciation ⟨blistering political *invective*⟩. OBLOQUY suggests defamation and consequent shame and disgrace ⟨subjected to *obloquy* and derision⟩. BILLINGSGATE implies practiced fluency and variety of profane or obscene abuse ⟨directed a stream of *billingsgate* at the cabdriver⟩.

— **acanthocephalan** *adjective*

acan·thus \ə-ˈkan(t)-thəs\ *noun, plural* **acanthus** [New Latin, genus name, from Greek *akanthos,* an acanthus, from *akantha* thorn] (1616)

1 : any of a genus (*Acanthus* of the family Acanthaceae, the acanthus family) of prickly herbs of the Mediterranean region

2 : an ornamentation (as in a Corinthian capital) representing or suggesting the leaves of the acanthus

acanthus 2

a cap·pel·la *also* **a ca·pel·la** \ˌä-kə-ˈpe-lə\ *adverb or adjective* [Italian *a cappella* in chapel style] (circa 1864)
: without instrumental accompaniment

ac·a·ri·a·sis \ˌa-kə-ˈrī-ə-səs\ *noun* (1828)
: infestation with or disease caused by mites

acar·i·cide \ə-ˈkar-ə-ˌsīd\ *noun* [*acarus* + *-i-* + *-cide*] (circa 1879)
: a pesticide that kills mites and ticks
— **acar·i·cid·al** \ə-ˌkar-ə-ˈsī-dᵊl\ *adjective*

ac·a·rid \ˈa-kə-rəd\ *noun* (1881)
: any of an order (Acarina) of arachnids including the mites and ticks; *especially* **:** a typical mite (family Acaridae)
— **acarid** *adjective*

ac·a·rus \ˈa-kə-rəs\ *noun, plural* **-ri** \-ˌrī\ [New Latin, genus name, from Greek *akari,* a mite] (1658)
: MITE; *especially* **:** one of a formerly extensive genus (*Acarus*)

acat·a·lec·tic \(ˌ)ā-ˌka-tə-ˈlek-tik\ *adjective* [Late Latin *acatalecticus,* from *acatalectus,* from Greek *akatalēktos,* from *a-* + *katalēgein* to leave off — more at CATALECTIC] (1589)
: not catalectic ⟨*acatalectic* verse⟩
— **acatalectic** *noun*

acau·les·cent \ˌā-kȯ-ˈle-sᵊnt\ *adjective* [*a-* + Latin *caulis* stem — more at HOLE] (1854)
: having no stem or appearing to have none

ac·cede \ak-ˈsēd, ik-\ *intransitive verb* **ac·ced·ed; ac·ced·ing** [Middle English, from Latin *accedere* to go to, be added, from *ad-* + *cedere* to go] (15th century)
1 a : to become a party (as to an agreement) **b :** to express approval or give consent **:** give in to a request or demand
2 *archaic* **:** APPROACH
3 : to enter upon an office or position
synonym see ASSENT

¹ac·ce·le·ran·do \(ˌ)ä-ˌche-lə-ˈrän-(ˌ)dō; ik-ˌse-, (ˌ)ak-\ *adverb or adjective* [Italian, literally, accelerating, from Latin *accelerandum,* gerund of *accelerare*] (circa 1842)
: gradually faster — used as a direction in music

²accelerando *noun, plural* **-dos** (1889)
: a gradual increase in tempo

ac·cel·er·ant \ik-ˈse-lə-rənt, ak-\ *noun* (1916)
: a substance used to accelerate a process (as the spreading of a fire)

ac·cel·er·ate \-lə-ˌrāt\ *verb* **-at·ed; -at·ing** [Latin *acceleratus,* past participle of *accelerare,* from *ad-* + *celer* swift — more at HOLD] (circa 1530)
transitive verb
1 : to bring about at an earlier time
2 : to cause to move faster; *also* **:** to cause to undergo acceleration
3 a : to hasten the progress or development of **b :** INCREASE ⟨*accelerate* food production⟩
4 a : to enable (a student) to complete a course in less than usual time **b :** to speed up (as a course of study)
intransitive verb
1 a : to move faster **:** gain speed **b :** GROW, INCREASE ⟨inflation was *accelerating*⟩
2 : to follow an accelerated educational program
— **ac·cel·er·at·ing·ly** \-ˌrā-tiŋ-lē\ *adverb*

ac·cel·er·a·tion \ik-ˌse-lə-ˈrā-shən, (ˌ)ak-\ *noun* (1531)
1 : the act or process of accelerating **:** the state of being accelerated
2 : the rate of change of velocity with respect to time; *broadly* **:** change of velocity

acceleration of gravity (circa 1889)
: the acceleration of a body in free fall under the influence of earth's gravity expressed as the rate of increase of velocity per unit of time and assigned the standard value of 980.665 centimeters per second per second

acceleration principle *noun* (circa 1941)
: a theory in economics: an increase or decrease in income induces a corresponding but magnified change in investment

ac·cel·er·a·tive \ik-ˈse-lə-ˌrā-tiv, ak-\ *adjective* (1751)
: of, relating to, or tending to cause acceleration **:** ACCELERATING

ac·cel·er·a·tor \ik-ˈse-lə-ˌrā-tər, ak-\ *noun* (1611)
: one that accelerates: as **a :** a muscle or nerve that speeds the performance of an action **b :** a device (as a pedal) for controlling the speed of a motor vehicle engine **c :** a substance that speeds a chemical reaction **d :** an apparatus for imparting high velocities to charged particles (as electrons)

ac·cel·er·om·e·ter \ik-ˌse-lə-ˈrä-mə-tər, ak-\ *noun* [International Scientific Vocabulary *acceleration* + *-o-* + *-meter*] (circa 1890)
: an instrument for measuring acceleration or for detecting and measuring vibrations

¹ac·cent \ˈak-ˌsent, ak-ˈ\ *transitive verb* [Middle French *accenter,* from *accent* intonation, from Latin *accentus,* from *ad-* + *cantus* song — more at CHANT] (1530)
1 a : to pronounce with accent **:** STRESS **b :** to mark with a written or printed accent
2 : to give prominence to **:** make more prominent

²ac·cent \ˈak-ˌsent, *chiefly British* -sənt\ *noun* (1538)
1 : a distinctive manner of expression: as **a :** an individual's distinctive or characteristic inflection, tone, or choice of words — usually used in plural **b :** a way of speaking typical of a particular group of people and especially of the natives or residents of a region
2 : an articulative effort giving prominence to one syllable over adjacent syllables; *also* **:** the prominence thus given a syllable
3 : rhythmically significant stress on the syllables of a verse usually at regular intervals
4 *archaic* **:** UTTERANCE
5 a : a mark (as ´, `, ˆ) used in writing or printing to indicate a specific sound value, stress, or pitch, to distinguish words otherwise identically spelled, or to indicate that an ordinarily mute vowel should be pronounced **b :** an accented letter
6 a : greater stress given to one musical tone than to its neighbors **b :** ACCENT MARK 2
7 a : emphasis laid on a part of an artistic design or composition **b :** an emphasized detail or area; *especially* **:** a small detail in sharp contrast with its surroundings **c :** a substance or object used for emphasis
8 : a mark placed to the right of a letter or number and usually slightly above it: as **a :** a double prime **b :** PRIME
9 : special concern or attention **:** EMPHASIS ⟨an *accent* on youth⟩
— **ac·cent·less** \-ləs\ *adjective*

accent mark *noun* (circa 1889)
1 : ACCENT 5a, 8
2 a : a symbol used to indicate musical stress **b :** a mark placed after a letter designating a note of music to indicate in which octave the note occurs

ac·cen·tu·al \ak-ˈsen(t)-sh(ə-)wəl, ik-\ *adjective* [Latin *accentus*] (1610)
: of, relating to, or characterized by accent; *specifically* **:** based on accent rather than on quantity or syllabic recurrence ⟨*accentual* poetry⟩
— **ac·cen·tu·al·ly** *adverb*

ac·cen·tu·ate \ik-ˈsen(t)-shə-ˌwāt, ak-\ *transitive verb* **-at·ed; -at·ing** [Medieval Latin *accentuatus,* past participle of *accentuare,* from Latin *accentus*] (circa 1731)
: ACCENT, EMPHASIZE; *also* **:** INTENSIFY ⟨*accentuates* the feeling of despair⟩
— **ac·cen·tu·a·tion** \ik-ˌsen(t)-shə-ˈwā-shən, (ˌ)ak-\ *noun*

ac·cept \ik-ˈsept, ak- *also* ek-\ *verb* [Middle English, from Middle French *accepter,* from Latin *acceptare,* frequentative of *accipere* to receive, from *ad-* + *capere* to take — more at HEAVE] (14th century)
transitive verb
1 a : to receive willingly ⟨*accept* a gift⟩ **b :** to be able or designed to take or hold (something applied or added) ⟨a surface that will not *accept* ink⟩
2 : to give admittance or approval to ⟨*accept* her as one of the group⟩
3 a : to endure without protest or reaction ⟨*accept* poor living conditions⟩ **b :** to regard as proper, normal, or inevitable ⟨the idea is widely *accepted*⟩ **c :** to recognize as true **:** BELIEVE ⟨refused to *accept* the explanation⟩
4 a : to make a favorable response to ⟨*accept* an offer⟩ **b :** to agree to undertake (a responsibility) ⟨*accept* a job⟩
5 : to assume an obligation to pay; *also* **:** to take in payment ⟨we don't *accept* personal checks⟩
6 : to receive (a legislative report) officially
intransitive verb
: to receive favorably something offered — usually used with *of*
— **ac·cept·ing·ly** \-ˈsep-tiŋ-lē\ *adverb*
— **ac·cept·ing·ness** \-tiŋ-nəs\ *noun*

ac·cept·able \ik-ˈsep-tə-bəl, ak- *also* ek-\ *adjective* (14th century)
1 : capable or worthy of being accepted ⟨no compromise would be *acceptable*⟩
2 a : WELCOME, PLEASING ⟨compliments are always *acceptable*⟩ **b :** barely satisfactory or adequate ⟨performances varied from excellent to *acceptable*⟩
— **ac·cept·abil·i·ty** \ik-ˌsep-tə-ˈbi-lə-tē, (ˌ)ak-, ek-\ *noun*
— **ac·cept·able·ness** \ik-ˈsep-tə-bəl-nəs, ak-\ *noun*
— **ac·cept·ably** \-blē\ *adverb*

ac·cep·tance \ik-ˈsep-tən(t)s, ak-\ *noun* (1574)
1 : an agreeing either expressly or by conduct to the act or offer of another so that a contract is concluded and the parties become legally bound

◇ WORD HISTORY
academy Our word *academy* derives from the Greek word *Akadēmeia,* the name of a grove and gymnasium located outside of ancient Athens. It was there that the philosopher Plato established his celebrated school. Just as public places often bear the names of local heroes today, the *Akadēmeia* had been named in honor of *Akadēmos,* a mythic figure largely remembered as the one who revealed the place where Helen of Troy was being held captive by Theseus. When *academy* first appeared in English, in the 15th century, it was used with specific reference to Plato's school. Later, under the influence of French *académie, academy* came to be used of any school above the elementary level.

2 : the quality or state of being accepted or acceptable
3 : the act of accepting **:** the fact of being accepted **:** APPROVAL
4 a : the act of accepting a time draft or bill of exchange for payment when due according to the specified terms **b :** an accepted draft or bill of exchange
5 : ACCEPTATION 2

ac·cep·tant \-tənt\ *adjective* (1851)
: willing to accept **:** RECEPTIVE

ac·cep·ta·tion \ˌak-ˌsep-ˈtā-shən\ *noun* (15th century)
1 : ACCEPTANCE; *especially* **:** favorable reception or approval
2 : a generally accepted meaning of a word or understanding of a concept

ac·cept·ed *adjective* (15th century)
: generally approved or used
— ac·cept·ed·ly *adverb*

ac·cept·er \ik-ˈsep-tər, ak-\ *noun* (1585)
1 : one that accepts
2 : ACCEPTOR 2

ac·cep·tive \ak-ˈsep-tiv\ *adjective* (1596)
1 : ACCEPTABLE
2 : RECEPTIVE

ac·cep·tor \ik-ˈsep-tər, ak-\ *noun* (14th century)
1 : ACCEPTER 1
2 : one that accepts an order or a bill of exchange
3 : an atom, molecule, or subatomic particle capable of receiving another entity (as an electron) especially to form a compound — compare DONOR 3a

¹ac·cess \ˈak-ˌses *also* ik-ˈses\ *noun* [Middle English, from Middle French & Latin; Middle French *acces* arrival, from Latin *accessus* approach, from *accedere* to approach — more at ACCEDE] (14th century)
1 a : ONSET 2 **b :** a fit of intense feeling **:** OUTBURST
2 a : permission, liberty, or ability to enter, approach, communicate with, or pass to and from **b :** freedom or ability to obtain or make use of **c :** a way or means of access **d :** the act or an instance of accessing
3 : an increase by addition ⟨a sudden *access* of wealth⟩

²access *transitive verb* (1962)
: to get at **:** gain access to ⟨*accessed* the computer by phone⟩

ac·ces·si·ble \ik-ˈse-sə-bəl, ak-, ek-\ *adjective* (15th century)
1 : providing access
2 a : capable of being reached ⟨*accessible* by rail⟩; *also* **:** being within reach ⟨fashions at *accessible* prices⟩ **b :** easy to speak or deal with ⟨*accessible* people⟩
3 : capable of being influenced **:** OPEN
4 : capable of being used or seen **:** AVAILABLE
5 : capable of being understood or appreciated ⟨the author's most *accessible* stories⟩ ⟨an *accessible* film⟩
— ac·ces·si·bil·i·ty \-ˌse-sə-ˈbi-lə-tē\ *noun*
— ac·ces·si·ble·ness \-ˈse-sə-bəl-nəs\ *noun*
— ac·ces·si·bly \-blē\ *adverb*

¹ac·ces·sion \ik-ˈse-shən, ak-\ *noun* (1588)
1 a : increase by something added **b :** acquisition of additional property (as by growth or increase of existing property)
2 : something added **:** ACQUISITION
3 : the act of assenting or agreeing
4 a : the act of becoming joined **:** ADHERENCE **b :** the act by which one nation becomes party to an agreement already in force between other powers
5 a : an act of coming near or to **:** APPROACH, ADMITTANCE **b :** the act of coming to high office or a position of honor or power
6 : a sudden fit or outburst **:** ACCESS
— ac·ces·sion·al \-ˈsesh-nəl, -ˈse-shə-nᵊl\ *adjective*

²accession *transitive verb* (1892)
: to record in order of acquisition

ac·ces·so·ri·al \ˌak-sə-ˈsōr-ē-əl, -ˈsȯr-\ *adjective* (1726)
1 : of or relating to an accessory ⟨*accessorial* liability⟩
2 : of, relating to, or constituting an accession **:** SUPPLEMENTARY ⟨*accessorial* services⟩

ac·ces·so·rise *British variant of* ACCESSORIZE

ac·ces·so·rize \ik-ˈse-sə-ˌrīz, ak-\ *verb* **-rized; -riz·ing** (1939)
transitive verb
: to furnish with accessories
intransitive verb
: to wear clothing accessories

¹ac·ces·so·ry *also* **ac·ces·sa·ry** \ik-ˈse-sə-rē, ak-, ek-, -ˈses-rē, *also* ə-ˈse-\ *noun, plural* **-ries** (15th century)
1 a : a person not actually or constructively present but contributing as an assistant or instigator to the commission of an offense — called also *accessory before the fact* **b :** a person who knowing that a crime has been committed aids or shelters the offender with intent to defeat justice — called also *accessory after the fact*
2 a : a thing of secondary or subordinate importance **:** ADJUNCT **b :** an object or device not essential in itself but adding to the beauty, convenience, or effectiveness of something else ⟨auto *accessories*⟩ ⟨clothing *accessories*⟩

²accessory *adjective* (1607)
1 : assisting as a subordinate; *especially* **:** contributing to a crime but not as the chief agent
2 : aiding or contributing in a secondary way **:** SUPPLEMENTARY
3 : present in a minor amount and not essential as a constituent ⟨an *accessory* mineral in a rock⟩

accessory fruit *noun* (circa 1900)
: a fruit (as the apple) of which a conspicuous part consists of tissue other than that of the ripened ovary

accessory nerve *noun* (circa 1842)
: either of a pair of motor nerves that are the 11th cranial nerves of higher vertebrates, arise from the medulla and the upper part of the spinal cord, and supply chiefly the pharynx and muscles of the upper chest, back, and shoulders

access time *noun* (1950)
: the time lag between the time stored information (as in a computer) is requested and the time it is delivered

ac·ciac·ca·tu·ra \(ˌ)ä-ˌchä-kə-ˈtu̇r-ə\ *noun* [Italian, literally, crushing] (circa 1819)
: a discordant note sounded with a principal note or chord and immediately released

ac·ci·dence \ˈak-sə-dən(t)s, -ˌden(t)s\ *noun* [Middle English, from Middle French, from Latin *accidentia* inflections of words, nonessential qualities, plural of *accident-, accidens,* noun] (15th century)
: a part of grammar that deals with inflections

ac·ci·dent \ˈak-sə-dənt, -ˌdent; ˈaks-dənt\ *noun* [Middle English, from Middle French, from Latin *accident-, accidens* nonessential quality, chance, from present participle of *accidere* to happen, from *ad-* + *cadere* to fall — more at CHANCE] (14th century)
1 a : an unforeseen and unplanned event or circumstance **b :** lack of intention or necessity **:** CHANCE ⟨met by *accident* rather than by design⟩
2 a : an unfortunate event resulting especially from carelessness or ignorance **b :** an unexpected and medically important bodily event especially when injurious ⟨a cerebrovascular *accident*⟩ **c :** an unexpected happening causing loss or injury which is not due to any fault or misconduct on the part of the person injured but for which legal relief may be sought
3 : a nonessential property or quality of an entity or circumstance ⟨the *accident* of nationality⟩

¹ac·ci·den·tal \ˌak-sə-ˈden-tᵊl\ *adjective* (14th century)
1 : arising from extrinsic causes **:** INCIDENTAL, NONESSENTIAL
2 a : occurring unexpectedly or by chance **b :** happening without intent or through carelessness and often with unfortunate results ☆
— ac·ci·den·tal·ly \-ˈdent-lē, -ˈden-tᵊl-ē\ *also* **ac·ci·dent·ly** \-ˈdent-lē\ *adverb*
— ac·ci·den·tal·ness \-ˈden-tᵊl-nəs\ *noun*

²accidental *noun* (1651)
1 : a nonessential property
2 a : a note foreign to a key indicated by a signature **b :** a prefixed sign indicating an accidental

accident insurance *noun* (1866)
: insurance against loss through accidental bodily injury to the insured

accident–prone *adjective* (1926)
1 : having a greater than average number of accidents
2 : having personality traits that predispose to accidents

ac·cid·ie \ˈak-sə-dē\ *noun* (13th century)
: ACEDIA

ac·cip·i·ter \ak-ˈsi-pə-tər, ik-\ *noun* [New Latin, genus name, from Latin, hawk] (circa 1828)
: any of a genus (*Accipiter*) of medium-sized forest-inhabiting hawks that have short broad wings and a long tail and a characteristic flight pattern of several quick flaps and a glide
— ac·cip·i·trine \-ˈsi-pə-ˌtrīn\ *adjective or noun*

¹ac·claim \ə-ˈklām\ *verb* [Latin *acclamare,* literally, to shout at, from *ad-* + *clamare* to shout — more at CLAIM] (1633)
transitive verb
1 : APPLAUD, PRAISE
2 : to declare by acclamation
intransitive verb
: to shout praise or applause
— ac·claim·er *noun*

²acclaim *noun* (1667)
1 : the act of acclaiming
2 : PRAISE, APPLAUSE

ac·cla·ma·tion \ˌa-klə-ˈmā-shən\ *noun* [Latin *acclamation-, acclamatio,* from *acclamare*] (1585)
1 : a loud eager expression of approval, praise, or assent
2 : an overwhelming affirmative vote by cheers, shouts, or applause rather than by ballot

ac·cli·mate \ˈa-klə-ˌmāt; ə-ˈklī-mət, -ˌmāt\ *verb* **-mat·ed; -mat·ing** [French *acclimater,* from *a-* (from Latin *ad-*) + *climat* climate] (1792)
: ACCLIMATIZE

ac·cli·ma·tion \ˌa-klə-ˈmā-shən, -ˌklī-\ *noun* (1826)
: ACCLIMATIZATION; *especially* **:** physiological adjustment by an organism to environmental change

ac·cli·ma·tise *British variant of* ACCLIMATIZE

ac·cli·ma·ti·za·tion \ə-ˌklī-mə-tə-ˈzā-shən\ *noun* (1830)
: the process or result of acclimatizing

ac·cli·ma·tize \ə-ˈklī-mə-ˌtīz\ *verb* **-tized; -tiz·ing** (1836)
transitive verb

☆ SYNONYMS
Accidental, fortuitous, casual, contingent mean not amenable to planning or prediction. ACCIDENTAL stresses chance ⟨any resemblance to actual persons is entirely *accidental*⟩. FORTUITOUS so strongly suggests chance that it often connotes entire absence of cause ⟨a series of *fortuitous* events⟩. CASUAL stresses lack of real or apparent premeditation or intent ⟨a *casual* encounter with a stranger⟩. CONTINGENT suggests possibility of happening but stresses uncertainty and dependence on other future events for existence or occurrence ⟨the *contingent* effects of the proposed law⟩.

: to adapt to a new temperature, altitude, climate, environment, or situation
intransitive verb
: to become acclimatized
— **ac·cli·ma·tiz·er** *noun*

ac·cliv·i·ty \ə-ˈkli-və-tē, a-\ *noun, plural* **-ties** [Latin *acclivitas,* from *acclivis* ascending, from *ad-* + *clivus* slope — more at DECLIVITY] (1614)
: an ascending slope (as of a hill)

ac·co·lade \ˈa-kə-ˌlād, -ˌläd\ *noun* [French, from *accoler* to embrace, from (assumed) Vulgar Latin *accollare,* from Latin *ad-* + *collum* neck — more at COLLAR] (1623)
1 a : a ceremonial embrace **b :** a ceremony or salute conferring knighthood
2 a : a mark of acknowledgment **:** AWARD **b :** an expression of praise
3 : a brace or a line used in music to join two or more staffs carrying simultaneous parts

ac·com·mo·date \ə-ˈkä-mə-ˌdāt\ *verb* **-dat·ed; -dat·ing** [Latin *accommodatus,* past participle of *accommodare,* from *ad-* + *commodare* to make fit, from *commodus* suitable — more at COMMODE] (1550)
transitive verb
1 : to make fit, suitable, or congruous
2 : to bring into agreement or concord **:** RECONCILE
3 : to provide with something desired, needed, or suited (as a helpful service, a loan, or lodgings)
4 a : to make room for **b :** to hold without crowding or inconvenience
5 : to give consideration to **:** allow for ⟨*accommodate* the special interests of various groups⟩
intransitive verb
: to adapt oneself; *also* **:** to undergo visual accommodation
synonym see ADAPT, CONTAIN
— **ac·com·mo·da·tive** \-ˌdā-tiv\ *adjective*
— **ac·com·mo·da·tive·ness** *noun*

ac·com·mo·dat·ing *adjective* (1775)
: HELPFUL, OBLIGING
— **ac·com·mo·dat·ing·ly** \-ˌdā-tiŋ-lē\ *adverb*

ac·com·mo·da·tion \ə-ˌkä-mə-ˈdā-shən\ *noun* (1603)
1 : something supplied for convenience or to satisfy a need: as **a :** lodging, food, and services or traveling space and related services — usually used in plural ⟨tourist *accommodations* on the boat⟩ ⟨overnight *accommodations*⟩ **b :** a public conveyance (as a train) that stops at all or nearly all points **c :** LOAN
2 : the act of accommodating **:** the state of being accommodated: as **a :** the providing of what is needed or desired for convenience **b :** ADAPTATION, ADJUSTMENT **c :** a reconciliation of differences **:** SETTLEMENT **d :** the automatic adjustment of the eye for seeing at different distances effected chiefly by changes in the convexity of the crystalline lens; *also* **:** the range over which such adjustment is possible
— **ac·com·mo·da·tion·al** \-shnəl, -shənᵊl\ *adjective*

[1]**ac·com·mo·da·tion·ist** \-ˈdā-sh(ə-)nist\ *noun* (1964)
: a black who adapts to the ideals or attitudes of whites ⟨making Uncle Toms, compromisers, and *accommodationists* . . . thoroughly ashamed —Ossie Davis⟩

[2]**accommodationist** *adjective* (1964)
: favoring or practicing accommodation or compromise

accommodation ladder *noun* (1769)
: a light ladder or stairway hung over the side of a ship for ascending from or descending to small boats

ac·com·mo·da·tor \ə-ˈkä-mə-ˌdā-tər\ *noun* (circa 1630)
: one that accommodates; *especially* **:** a part-time or special-occasion domestic worker

ac·com·pa·ni·ment \ə-ˈkəm-pə-nē-mənt, -ˈkəmp-nē-\ *noun* (circa 1744)
1 : an instrumental or vocal part designed to support or complement a melody
2 a : an addition (as an ornament) intended to give completeness or symmetry **:** COMPLEMENT **b :** an accompanying situation or occurrence **:** CONCOMITANT

ac·com·pa·nist \ə-ˈkəmp-nist, -ˈkəm-pə-\ *noun* (circa 1828)
: one (as a pianist) who plays an accompaniment

ac·com·pa·ny \ə-ˈkəmp-nē, -ˈkämp-; -ˈkəm-pə-, -ˈkäm-\ *verb* **-nied; -ny·ing** [Middle English *accompanien,* from Middle French *acompaignier,* from *a-* (from Latin *ad-*) + *compaing* companion — more at COMPANION] (15th century)
transitive verb
1 : to go with as an associate or companion
2 : to perform an accompaniment to or for
3 a : to cause to be in association ⟨*accompanied* their advice with a warning⟩ **b :** to be in association with ⟨the pictures that *accompany* the text⟩
intransitive verb
: to perform an accompaniment

ac·com·plice \ə-ˈkäm-pləs, -ˈkəm-\ *noun* [alteration (from incorrect division of *a complice*) of *complice*] (1589)
: one associated with another especially in wrongdoing

ac·com·plish \ə-ˈkäm-plish, -ˈkəm-\ *transitive verb* [Middle English *accomplisshen,* from Middle French *acompliss-,* stem of *acomplir,* from (assumed) Vulgar Latin *accomplēre,* from Latin *ad-* + *complēre* to fill up — more at COMPLETE] (14th century)
1 : to bring about (a result) by effort ⟨have much to *accomplish* today⟩ ⟨regretted that he had never *accomplished* a marriage for her —Francis Hackett⟩
2 : to bring to completion **:** FULFILL ⟨we can *accomplish* the job in an hour⟩
3 : to succeed in reaching (a stage in a progression) ⟨would starve before *accomplishing* half the distance —W. H. Hudson (died 1922)⟩
4 *archaic* **a :** to equip thoroughly **b :** PERFECT
synonym see PERFORM
— **ac·com·plish·able** \-pli-shə-bəl\ *adjective*
— **ac·com·plish·er** *noun*

ac·com·plished *adjective* (15th century)
1 a : proficient as the result of practice or training ⟨an *accomplished* dancer⟩ **b :** having many social accomplishments
2 : established beyond doubt or dispute ⟨an *accomplished* fact⟩

ac·com·plish·ment \ə-ˈkäm-plish-mənt, -ˈkəm-\ *noun* (15th century)
1 : the act of accomplishing **:** COMPLETION
2 : something that has been accomplished **:** ACHIEVEMENT
3 a : a quality or ability equipping one for society **b :** a special skill or ability acquired by training or practice

[1]**ac·cord** \ə-ˈkȯrd\ *verb* [Middle English, from Old French *acorder,* from (assumed) Vulgar Latin *accordare,* from Latin *ad-* + *cord-, cor* heart — more at HEART] (12th century)
transitive verb
1 : to bring into agreement **:** RECONCILE
2 : to grant or give especially as appropriate, due, or earned
intransitive verb
1 *archaic* **:** to arrive at an agreement
2 *obsolete* **:** to give consent
3 : to be consistent or in harmony **:** AGREE
synonym see GRANT

[2]**accord** *noun* [Middle English, from Old French *acort,* from *acorder*] (14th century)
1 a : AGREEMENT, CONFORMITY ⟨acted in *accord* with the company's policy⟩ **b :** a formal reaching of agreement **:** COMPACT, TREATY
2 : balanced interrelationship **:** HARMONY
3 *obsolete* **:** ASSENT
4 : voluntary or spontaneous impulse to act ⟨gave generously of their own *accord*⟩

ac·cor·dance \ə-ˈkȯr-dᵊn(t)s\ *noun* (14th century)
1 : AGREEMENT, CONFORMITY ⟨in *accordance* with a rule⟩
2 : the act of granting

ac·cor·dant \-dᵊnt\ *adjective* (14th century)
1 : CONSONANT, AGREEING
2 : HARMONIOUS, CORRESPONDENT
— **ac·cor·dant·ly** *adverb*

according as *conjunction* (1500)
1 : in accord with the way in which
2 a : depending on how **b :** depending on whether **:** IF

ac·cord·ing·ly \ə-ˈkȯr-diŋ-lē\ *adverb* (14th century)
1 : in accordance **:** CORRESPONDINGLY
2 : CONSEQUENTLY, SO

according to *preposition* (14th century)
1 : in conformity with
2 : as stated or attested by
3 : depending on

[1]**ac·cor·di·on** \ə-ˈkȯr-dē-ən\ *noun* [German *Akkordion,* from *Akkord* chord, from French *accord,* from Old French *acort*] (1831)
: a portable keyboard wind instrument in which the wind is forced past free reeds by means of a hand-operated bellows
— **ac·cor·di·on·ist** \-dē-ə-nist\ *noun*

accordion

[2]**accordion** *adjective* (1885)
: folding or creased or hinged to fold like an accordion ⟨an *accordion* pleat⟩ ⟨an *accordion* door⟩

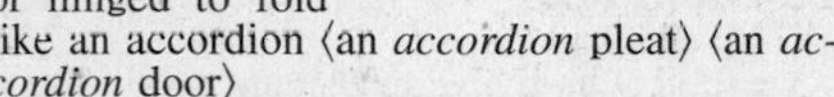

ac·cost \ə-ˈkȯst, -ˈkäst\ *transitive verb* [Middle French *accoster,* ultimately from Latin *ad-* + *costa* rib, side — more at COAST] (1612)
: to approach and speak to often in a challenging or aggressive way

ac·couche·ment \ˌa-ˌküsh-ˈmäⁿ, ə-ˈküsh-ˌ\ *noun* [French] (1803)
: the time or act of giving birth

ac·cou·cheur \ˌa-ˌkü-ˈshər\ *noun* [French] (1759)
: one that assists at a birth; *especially* **:** OBSTETRICIAN

[1]**ac·count** \ə-ˈkau̇nt\ *noun* (14th century)
1 *archaic* **:** RECKONING, COMPUTATION
2 a : a record of debit and credit entries to cover transactions involving a particular item or a particular person or concern **b :** a statement of transactions during a fiscal period and the resulting balance
3 a : a statement explaining one's conduct **b :** a statement or exposition of reasons, causes, or motives ⟨no satisfactory *account* of these phenomena⟩ **c :** a reason for an action **:** BASIS ⟨on that *account* I must refuse⟩
4 a : a formal business arrangement providing for regular dealings or services (as banking, advertising, or store credit) and involving the establishment and maintenance of an account; *also* **:** CLIENT, CUSTOMER **b :** money deposited in a bank account and subject to withdrawal by the depositor
5 a : VALUE, IMPORTANCE ⟨it's of no *account* to me⟩ **b :** ESTEEM ⟨stood high in their *account*⟩
6 : ADVANTAGE ⟨turned her wit to good *account*⟩
7 a : careful thought **:** CONSIDERATION ⟨have to take many things into *account*⟩ **b :** a usually mental record **:** TRACK ⟨keep *account* of all you do⟩

8 : a description of facts, conditions, or events **:** REPORT, NARRATIVE ⟨the newspaper *account* of the fire⟩ ⟨by all *accounts* they're well-off⟩; *also* **:** PERFORMANCE ⟨a straightforward *account* of the sonata⟩
— on account : with the price charged to one's account
— on account of : for the sake of **:** by reason of
— on no account : under no circumstances
— on one's own account 1 : on one's own behalf **2 :** at one's own risk **3 :** by oneself **:** on one's own

²account *verb* [Middle English, from Middle French *acompter,* from *a-* (from Latin *ad-*) + *compter* to count] (14th century)
transitive verb
1 : to think of as **:** CONSIDER ⟨*accounts* himself lucky⟩
2 : to probe into **:** ANALYZE
intransitive verb
1 : to furnish a justifying analysis or explanation — used with *for* ⟨couldn't *account* for the loss⟩
2 a : to be the sole or primary factor — used with *for* ⟨the pitcher *accounted* for all three putouts⟩ **b :** to bring about the capture, death, or destruction of something ⟨*accounted* for two rabbits⟩

ac·count·abil·i·ty \ə-ˌkau̇n-tə-ˈbi-lə-tē\ *noun* (1794)
: the quality or state of being accountable; *especially* **:** an obligation or willingness to accept responsibility or to account for one's actions ⟨public officials lacking *accountability*⟩

ac·count·able \ə-ˈkau̇n-tə-bəl\ *adjective* (14th century)
1 : subject to giving an account **:** ANSWERABLE
2 : capable of being accounted for **:** EXPLAINABLE
synonym see RESPONSIBLE
— ac·count·able·ness \-ˈkau̇n-tə-bəl-nəs\ *noun*
— ac·count·ably \-blē\ *adverb*

ac·coun·tan·cy \ə-ˈkau̇n-tᵊn(t)-sē\ *noun* (1854)
: the profession or practice of accounting

¹ac·coun·tant \ə-ˈkau̇n-tᵊnt\ *noun* (15th century)
1 : one that gives an account or is accountable
2 : one who is skilled in the practice of accounting or who is in charge of public or private accounts
— ac·coun·tant·ship \-tᵊn(t)-ˌship\ *noun*

²accountant *adjective* (15th century)
obsolete **:** ACCOUNTABLE, ANSWERABLE ⟨I stand *accountant* for as great a sin —Shakespeare⟩

account executive *noun* (1931)
: a business executive (as in an advertising agency) responsible for dealing with a client's account

ac·count·ing \ə-ˈkau̇n-tiŋ\ *noun* (circa 1716)
1 : the system of recording and summarizing business and financial transactions and analyzing, verifying, and reporting the results; *also* **:** the principles and procedures of accounting
2 a : work done in accounting or by accountants **b :** an instance of applied accounting or of the settling or presenting of accounts

account payable *noun, plural* **accounts payable** (circa 1936)
: the balance due to a creditor on a current account

account receivable *noun, plural* **accounts receivable** (1936)
: a balance due from a debtor on a current account

ac·cou·tre *or* **ac·cou·ter** \ə-ˈkü-tər\ *transitive verb* **-cou·tred** *or* **-cou·tered; -cou·tring** *or* **-cou·ter·ing** \-ˈkü-tə-riŋ, -ˈkü-triŋ\ [French *accoutrer,* from Middle French *acoustrer,* from *a-* + *costure* seam, from (assumed) Vulgar Latin *consutura* — more at COUTURE] (1596)
: to provide with equipment or furnishings **:** OUTFIT
synonym see FURNISH

ac·cou·tre·ment *or* **ac·cou·ter·ment** \ə-ˈkü-trə-mənt, -ˈkü-tər-mənt\ *noun* (1549)
1 a : EQUIPMENT, TRAPPINGS; *specifically* **:** a soldier's outfit usually not including clothes and weapons — usually used in plural **b :** an accessory item of clothing or equipment — usually used in plural
2 *archaic* **:** the act of accoutring
3 : an identifying and often superficial characteristic or device — usually used in plural ⟨*accoutrements* of power that define our diplomacy —Elizabeth Drew⟩

ac·cred·it \ə-ˈkre-dət\ *transitive verb* [Latin *accreditus,* past participle of *accredere* to give credence to, from *ad-* + *credere* to believe — more at CREED] (1535)
1 : to give official authorization to or approval of: **a :** to provide with credentials; *especially* **:** to send (an envoy) with letters of authorization **b :** to recognize or vouch for as conforming with a standard **c :** to recognize (an educational institution) as maintaining standards that qualify the graduates for admission to higher or more specialized institutions or for professional practice
2 : to consider or recognize as outstanding
3 : ATTRIBUTE, CREDIT
synonym see APPROVE
— ac·cred·i·table \-də-tə-bəl\ *adjective*
— ac·cred·i·ta·tion \ə-ˌkre-də-ˈtā-shən, -ˈdā-\ *noun*

ac·crete \ə-ˈkrēt\ *verb* **ac·cret·ed; ac·cret·ing** [back-formation from *accretion*] (1784)
intransitive verb
: to grow or become attached by accretion
transitive verb
: to cause to adhere or become attached; *also* **:** ACCUMULATE

ac·cre·tion \ə-ˈkrē-shən\ *noun* [Latin *accretion-, accretio,* from *accrescere* — more at ACCRUE] (1615)
1 : the process of growth or enlargement by a gradual buildup: as **a :** increase by external addition or accumulation (as by adhesion of external parts or particles) **b :** the increase of land by the action of natural forces
2 : a product of accretion; *especially* **:** an extraneous addition ⟨*accretions* of grime⟩
— ac·cre·tion·ary \-shə-ˌner-ē\ *adjective*
— ac·cre·tive \ə-ˈkrē-tiv\ *adjective*

¹ac·cru·al \ə-ˈkrü-əl\ *noun* (1880)
1 : the action or process of accruing
2 : something that accrues or has accrued

²accrual *adjective* (1917)
: being a method of accounting that recognizes income when earned and expenses when incurred regardless of when cash is received or disbursed

ac·crue \ə-ˈkrü\ *verb* **ac·crued; ac·cru·ing** [Middle English *acreuen,* probably from Middle French *acreue* increase, from *acreistre* to increase, from Latin *accrescere,* from *ad-* + *crescere* to grow — more at CRESCENT] (15th century)
intransitive verb
1 : to come into existence as a legally enforceable claim
2 a : to come about as a natural growth, increase, or advantage ⟨the wisdom that *accrues* with age⟩ **b :** to come as a direct result of some state or action ⟨rewards due to the feminine will *accrue* to me —Germaine Greer⟩
3 : to accumulate or be added periodically ⟨interest *accrues* on a daily basis⟩
transitive verb
: to accumulate or have due after a period of time ⟨*accrue* vacation time⟩
— ac·cru·able \-ˈkrü-ə-bəl\ *adjective*
— ac·crue·ment \-ˈkrü-mənt\ *noun*

ac·cul·tur·ate \ə-ˈkəl-chə-ˌrāt, a-\ *transitive verb* **-at·ed; -at·ing** [back-formation from *acculturation*] (1930)
: to change through acculturation

ac·cul·tur·a·tion \ə-ˌkəl-chə-ˈrā-shən, a-\ *noun* (1880)
1 : cultural modification of an individual, group, or people by adapting to or borrowing traits from another culture; *also* **:** a merging of cultures as a result of prolonged contact
2 : the process by which a human being acquires the culture of a particular society from infancy
— ac·cul·tur·a·tion·al \-shnəl, -shə-nᵊl\ *adjective*
— ac·cul·tur·a·tive \ə-ˈkəl-chə-ˌrā-tiv, a-\ *adjective*

ac·cu·mu·late \ə-ˈkyü-m(y)ə-ˌlāt\ *verb* **-lat·ed; -lat·ing** [Latin *accumulatus,* past participle of *accumulare,* from *ad-* + *cumulare* to heap up — more at CUMULATE] (15th century)
transitive verb
: to gather or pile up especially little by little **:** AMASS ⟨*accumulate* a fortune⟩
intransitive verb
: to increase gradually in quantity or number

ac·cu·mu·la·tion \ə-ˌkyü-m(y)ə-ˈlā-shən\ *noun* (15th century)
1 : something that has accumulated or has been accumulated
2 : the action or process of accumulating **:** the state of being or having accumulated
3 : increase or growth by addition especially when continuous or repeated ⟨*accumulation* of interest⟩

ac·cu·mu·la·tive \ə-ˈkyü-m(y)ə-ˌlā-tiv, -lə-\ *adjective* (circa 1651)
1 : CUMULATIVE ⟨an age of rapid and *accumulative* change⟩
2 : tending or given to accumulation
— ac·cu·mu·la·tive·ly *adverb*
— ac·cu·mu·la·tive·ness *noun*

ac·cu·mu·la·tor \ə-ˈkyü-m(y)ə-ˌlā-tər\ *noun* (1748)
: one that accumulates: as **a :** a device (as in a hydraulic system) in which a fluid is collected and especially in which it is kept under pressure as a means of storing energy **b** *British* **:** STORAGE BATTERY **c :** a part (as in a computer) where numbers are totaled or stored

ac·cu·ra·cy \ˈa-kyə-rə-sē, ˈa-k(ə-)rə-\ *noun, plural* **-cies** (1662)
1 : freedom from mistake or error **:** CORRECTNESS
2 a : conformity to truth or to a standard or model **:** EXACTNESS **b :** degree of conformity of a measure to a standard or a true value — compare PRECISION 2a

ac·cu·rate \ˈa-kyə-rət, ˈa-k(ə-)rət\ *adjective* [Latin *accuratus,* from past participle of *accurare* to take care of, from *ad-* + *cura* care] (1596)
1 : free from error especially as the result of care ⟨an *accurate* diagnosis⟩
2 : conforming exactly to truth or to a standard **:** EXACT ⟨providing *accurate* color⟩
3 : able to give an accurate result ⟨an *accurate* gauge⟩
synonym see CORRECT
— ac·cu·rate·ly \ˈa-kyə-rət-lē, ˈa-k(ə-)rət-, ˈa-k(y)ərt-\ *adverb*
— ac·cu·rate·ness \-kyə-rət-nəs, -k(ə-)rət-nəs\ *noun*

ac·cursed \ə-ˈkərst, -ˈkər-səd\ *or* **ac·curst** \ə-ˈkərst\ *adjective* [Middle English *acursed,* from past participle of *acursen* to consign to destruction with a curse, from *a-* (from Old English *ā,* perfective prefix) + *cursen* to curse — more at ABIDE] (13th century)
1 : being under or as if under a curse
2 : DAMNABLE
— ac·curs·ed·ly \-ˈkər-səd-lē\ *adverb*
— ac·curs·ed·ness \-ˈkər-səd-nəs\ *noun*

ac·cus·al \ə-ˈkyü-zəl\ *noun* (1594)
: ACCUSATION

ac·cu·sa·tion \ˌa-kyə-ˈzā-shən, -(ˌ)kyü-\ *noun* (14th century)
1 : the act of accusing **:** the state or fact of being accused
2 : a charge of wrongdoing

[1]**ac·cu·sa·tive** \ə-ˈkyü-zə-tiv\ *adjective* [Middle English, from Middle French or Latin; Middle French *accusatif,* from Latin *accusativus,* from *accusatus,* past participle of *accusare*] (15th century)
1 : of, relating to, or being the grammatical case that marks the direct object of a verb or the object of any of several prepositions
2 : ACCUSATORY

[2]**accusative** *noun* (circa 1620)
: the accusative case of a language **:** a form in the accusative case

ac·cu·sa·to·ry \ə-ˈkyü-zə-ˌtōr-ē, -ˌtȯr-\ *adjective* (14th century)
: containing or expressing accusation **:** ACCUSING

ac·cuse \ə-ˈkyüz\ *verb* **ac·cused; ac·cus·ing** [Middle English, from Old French *acuser,* from Latin *accusare* to call to account, from *ad-* + *causa* lawsuit] (14th century)
transitive verb
1 : to charge with a fault or offense **:** BLAME
2 : to charge with an offense judicially or by a public process
intransitive verb
: to bring an accusation
— ac·cus·er \ə-ˈkyü-zər\ *noun*
— ac·cus·ing·ly \-ˈkyü-ziŋ-lē\ *adverb*

ac·cused *noun, plural* **accused** (1593)
: one charged with an offense; *especially* **:** the defendant in a criminal case

ac·cus·tom \ə-ˈkəs-təm\ *transitive verb* [Middle English, from Middle French *acostumer,* from *a-* (from Latin *ad-*) + *costume* custom] (15th century)
: to make familiar with something through use or experience
— ac·cus·tom·ation \-ˌkəs-tə-ˈmā-shən\ *noun*

ac·cus·tomed \ə-ˈkəs-təmd\ *adjective* (15th century)
1 : often used or practiced **:** CUSTOMARY ⟨her *accustomed* cheerfulness⟩
2 : adapted to existing conditions ⟨eyes *accustomed* to the dark⟩
3 : being in the habit or custom ⟨*accustomed* to making decisions⟩
synonym see USUAL
— ac·cus·tomed·ness \-təm(d)-nəs\ *noun*

AC/DC \ˈā-(ˌ)sē-ˈdē-(ˌ)sē\ *adjective* [from the likening of a bisexual person to an electrical appliance which can operate on either alternating or direct current] (circa 1960)
: BISEXUAL 1b

[1]**ace** \ˈās\ *noun* [Middle English *as,* from Middle French, from Latin, unit, a copper coin] (14th century)
1 a : a die face marked with one spot **b :** a playing card marked in its center with one pip **c :** a domino end marked with one spot
2 : a very small amount or degree **:** PARTICLE
3 : a point scored especially on a service (as in tennis or handball) that an opponent fails to touch
4 : a golf score of one stroke on a hole; *also* **:** a hole made in one stroke
5 : a combat pilot who has brought down at least five enemy airplanes
6 : one that excels at something
— ace in the hole 1 : an ace dealt face down to a player (as in stud poker) and not exposed until the showdown **2 :** an effective and decisive argument or resource held in reserve
— within an ace of : on the point of **:** very near to ⟨came *within an ace of* winning⟩

[2]**ace** *transitive verb* **aced; ac·ing** (1923)
1 : to score an ace against (an opponent)
2 : to make (a hole in golf) in one stroke
3 : to gain a decisive advantage over **:** DEFEAT — usually used with *out*
4 : to earn a high grade on (as an examination); *especially* **:** to get an A on

[3]**ace** *adjective* (1926)
: of first or high rank or quality

-aceae *noun plural suffix* [New Latin, from Latin, feminine plural of *-aceus* -aceous]
: plants of the nature of ⟨Ros*aceae*⟩ — in names of families of plants

ace·dia \ə-ˈsē-dē-ə\ *noun* [Late Latin, from Greek *akēdeia,* from *a-* + *kēdos* care, grief — more at HATE] (1607)
: APATHY, BOREDOM

ACE inhibitor \ˌā-(ˌ)sē-ˈē-; ˈās-\ *noun* [*a*ngiotensin *c*onverting *e*nzyme] (1985)
: any of a group of antihypertensive drugs (as captopril) that relax arteries and promote renal excretion of salt and water by inhibiting the activity of angiotensin converting enzyme

Acel·da·ma \ə-ˈsel-də-mə\ *noun* [Greek *Akeldama,* from Aramaic *ḥăqēl dĕmā,* literally, field of blood]
: the potter's field bought with the money Judas had been paid for betraying Christ

acel·lu·lar \(ˌ)ā-ˈsel-yə-lər\ *adjective* (1940)
1 : containing no cells ⟨*acellular* vaccines⟩
2 : not divided into cells **:** consisting of a single complex cell — used especially of protozoa and ciliates

acen·tric \(ˌ)ā-ˈsen-trik\ *adjective* (1937)
: lacking a centromere ⟨*acentric* chromosomes⟩

-aceous *adjective suffix* [Latin *-aceus*]
1 a : characterized by **:** full of ⟨set*aceous*⟩ **b :** consisting of ⟨diatom*aceous*⟩ **:** having the nature or form of ⟨tuff*aceous*⟩
2 a : of or relating to a group of animals typified by (such) a form ⟨cet*aceous*⟩ or characterized by (such) a feature ⟨crust*aceous*⟩ **b :** of or relating to a plant family ⟨solan*aceous*⟩

aceph·a·lous \(ˌ)ā-ˈse-fə-ləs, ə-ˈse-\ *adjective* [Greek *akephalos,* from *a-* + *kephalē* head — more at CEPHALIC] (circa 1731)
1 : lacking a head or having the head reduced
2 : lacking a governing head or chief

ace·quia \ə-ˈsā-kē-ə, ä-\ *noun* [Spanish, from Arabic *as-sāqiyah* the irrigation stream] (1844)
Southwest **:** an irrigation ditch or canal

acerb \ə-ˈsərb, a-\ *adjective* [French or Latin; French *acerbe,* from Latin *acerbus;* akin to Latin *acer* sharp — more at EDGE] (1622)
: ACERBIC

ac·er·bate \ˈa-sər-ˌbāt\ *transitive verb* **-bat·ed; -bat·ing** (circa 1731)
: IRRITATE, EXASPERATE

acer·bic \ə-ˈsər-bik, a-\ *adjective* (1865)
: acid in temper, mood, or tone
— acer·bi·cal·ly \-bi-k(ə-)lē\ *adverb*

acer·bi·ty \-bə-tē\ *noun, plural* **-ties** (1572)
: the quality of being acerbic

ac·er·o·la \ˌa-sə-ˈrō-lə\ *noun* [American Spanish, from Spanish, fruit of a shrub (*Crataegus azarolus*), from Arabic *az-zu'rūr*] (1945)
: any of various West Indian shrubs (genus *Malpighia*) with mildly acid cherrylike fruits very rich in vitamin C

acet- *or* **aceto-** *combining form* [French & Latin; French *acét-,* from Latin *acet-,* from *acetum* vinegar; akin to *acēre* to be sour, *acer* sharp — more at EDGE]
: acetic acid **:** acetic ⟨*acet*yl⟩

ac·e·tab·u·lum \-ˈta-byə-ləm\ *noun, plural* **-lums** *or* **-la** \-lə\ [Latin, literally, vinegar cup, from *acetum* vinegar] (1661)
1 : a ventral sucker of a trematode
2 : the cup-shaped socket in the hipbone
— ac·e·tab·u·lar \-lər\ *adjective*

ac·e·tal \ˈa-sə-ˌtal\ *noun* [German *Azetal,* from *azet-* acet- + *Al*kohol alcohol] (1853)
: any of various compounds characterized by the grouping $C(OR)_2$ and obtained especially by heating aldehydes or ketones with alcohols

ac·et·al·de·hyde \ˌa-sə-ˈtal-də-ˌhīd\ *noun* [International Scientific Vocabulary] (1877)
: a colorless volatile water-soluble liquid aldehyde C_2H_4O used chiefly in organic synthesis

acet·amide \ə-ˈse-tə-ˌmīd, ˌa-sə-ˈta-ˌmīd\ *noun* [International Scientific Vocabulary] (1873)
: a white crystalline amide C_2H_5NO of acetic acid used especially as a solvent and in organic synthesis

acet·amin·o·phen \ə-ˌsē-tə-ˈmi-nə-fən, ˌa-sə-tə-\ *noun* [*acet-* + *amino* + *phen*ol] (1958)
: a crystalline compound $C_8H_9NO_2$ that is a hydroxy derivative of acetanilide and is used in chemical synthesis and in medicine to relieve pain and fever

ac·et·an·i·lide *or* **ac·et·an·i·lid** \ˌa-sə-ˈta-nə-ˌlīd, -ləd\ *noun* [International Scientific Vocabulary] (circa 1864)
: a white crystalline compound C_8H_9NO that is derived from aniline and acetic acid and is used especially to relieve pain or fever

ac·e·tate \ˈa-sə-ˌtāt\ *noun* (1827)
1 : a salt or ester of acetic acid
2 : CELLULOSE ACETATE; *also* **:** something (as a textile fiber) made from cellulose acetate
3 : a phonograph recording disk made of an acetate or coated with cellulose acetate

ac·et·azol·amide \ˌa-sə-tə-ˈzō-lə-ˌmīd, -ˈzä-, -məd\ *noun* [*acet-* + *azole* + *amide*] (1954)
: a diuretic drug $C_4H_6N_4O_3S_2$ used especially in the treatment of edema associated with congestive heart failure and of glaucoma

ace·tic acid \ə-ˈsē-tik-\ *noun* [probably from French *acétique,* from Latin *acetum* vinegar] (1808)
: a colorless pungent liquid acid $C_2H_4O_2$ that is the chief acid of vinegar and that is used especially in synthesis (as of plastics)

acetic anhydride *noun* (1876)
: a colorless liquid $C_4H_6O_3$ with a pungent odor used in organic synthesis (as of cellulose acetate and aspirin)

ace·ti·fy \ə-ˈsē-tə-ˌfī, -ˈse-\ *transitive verb* **-fied; -fy·ing** (circa 1828)
: to turn into acetic acid or vinegar
— ace·ti·fi·ca·tion \-ˌsē-tə-fə-ˈkā-shən, -ˌse-\ *noun*

ace·to·ace·tic acid \ˌa-sə-(ˌ)tō-ə-ˌsē-tik-, ə-ˌsē-tō-\ *noun* [part translation of German *Azetessigsäure,* from *azet-* acet- + *Essigsaure* acetic acid] (circa 1900)
: an unstable acid $C_4H_6O_3$ that is a ketone body found in abnormal quantities in the blood and urine in some conditions (as diabetes)

ac·e·tone \ˈa-sə-ˌtōn\ *noun* [German *Azeton,* from Latin *acetum*] (circa 1839)
: a volatile fragrant flammable liquid ketone C_3H_6O used chiefly as a solvent and in organic synthesis and found in abnormal quantities in diabetic urine
— ac·e·ton·ic \ˌa-sə-ˈtä-nik\ *adjective*

ace·to·ni·trile \ə-ˌsē-tō-ˈnī-trəl, ˌa-sə-tō-, -ˌtrīl\ *noun* (circa 1869)
: the colorless liquid nitrile CH_3CN of acetic acid used chiefly in organic synthesis and as a solvent

ace·to·phe·net·i·din \ˌa-sə-(ˌ)tō-fə-ˈne-tə-dən, ə-ˌsē-tō-\ *noun* [International Scientific Vocabulary] (1910)
: PHENACETIN

ace·tous \ə-ˈsē-təs, ˈa-sə-təs\ *adjective* (1778)
: relating to or producing vinegar ⟨*acetous* fermentation⟩; *also* **:** SOUR, VINEGARY

ace·tyl \ə-ˈsē-tᵊl, ˈa-sə-; ˈa-sə-ˌtēl\ *noun* (circa 1864)
: the radical CH_3CO- of acetic acid — often used in combination

acet·y·late \ə-ˈse-tᵊl-ˌāt\ *transitive verb* **-lat·ed; -lat·ing** (circa 1900)
: to introduce the acetyl radical into (a compound)
— acet·y·la·tion \-ˌse-tᵊl-ˈā-shən\ *noun*

\ə\ **abut** \ᵊ\ **kitten** \ər\ **further** \a\ **ash** \ā\ **ace** \ä\ **mop, mar** \au̇\ **out** \ch\ **chin** \e\ **bet** \ē\ **easy** \g\ **go** \i\ **hit** \ī\ **ice** \j\ **job** \ŋ\ **sing** \ō\ **go** \ȯ\ **law** \ȯi\ **boy** \th\ **thin** \t͟h\ **the** \ü\ **loot** \u̇\ **foot** \y\ **yet** \zh\ **vision** *see also* Guide to Pronunciation

— **acet·y·la·tive** \-ˈset-ᵊl-ˌā-tiv\ *adjective*
ace·tyl·cho·line \ə-ˌse-tᵊl-ˈkō-ˌlēn, -ˌsē-; ˈa-sə-ˌtēl-\ *noun* [International Scientific Vocabulary] (1906)
: a neurotransmitter $C_7H_{17}NO_3$ released at autonomic synapses and neuromuscular junctions and formed enzymatically in the tissues from choline
ace·tyl·cho·lin·es·ter·ase \-ˌkō-lə-ˈnes-tə-ˌrās, -ˌrāz\ *noun* (1947)
: an enzyme that occurs especially in some nerve endings and in the blood and promotes the hydrolysis of acetylcholine
acetyl CoA \-ˌkō-ˈā\ *noun* (circa 1959)
: ACETYL COENZYME A
acetyl coenzyme A *noun* (1952)
: a compound $C_{25}H_{38}N_7O_{17}P_3S$ formed as an intermediate in metabolism and active as a coenzyme in biological acetylations
acet·y·lene \ə-ˈse-tᵊl-ən, -tᵊl-ˌēn\ *noun* (1864)
: a colorless gaseous hydrocarbon HC≡CH used chiefly in organic synthesis and as a fuel (as in welding and soldering)
— **acet·y·le·nic** \ə-ˌse-tᵊl-ˈē-nik, -ˈe-nik\ *adjective*
ace·tyl·sa·lic·y·late \ə-ˌsē-tᵊl-sə-ˈli-sə-ˌlāt\ *noun* (circa 1960)
: a salt or ester of acetylsalicylic acid
ace·tyl·sal·i·cyl·ic acid \ə-ˈsē-tᵊl-ˌsa-lə-ˌsi-lik-\ *noun* [International Scientific Vocabulary] (1897)
: ASPIRIN 1
ac·ey–deuc·ey *also* **ac·ey–deu·cy** \ˌā-sē-ˈdü-sē, -ˈdyü-\ *noun* (1925)
: a variation of backgammon in which a throw of a 1-2 wins extra turns
[1]**Achae·an** \ə-ˈkē-ən\ *or* **Achai·an** \-ˈkī-ən, -ˈkā-ən\ *adjective* (1567)
: of, relating to, or characteristic of Achaea; *broadly* **:** of or relating to Greece
[2]**Achaean** *or* **Achaian** *noun* (1607)
: a native or inhabitant of Achaea; *broadly* **:** GREEK
Ach·ae·me·ni·an \ˌa-kə-ˈmē-nē-ən\ *adjective* (1717)
: of or relating to the Achaemenids
Achae·me·nid \ə-ˈkē-mə-nəd\ *noun, plural* **-menids** *also* **-men·i·dae** \ˌa-kə-ˈme-nə-ˌdē\ [Greek *Achaimenides,* from *Achaimenes,* 7th century B.C. Persian king, founder of the dynasty + *-ides* (patronymic suffix)] (1889)
: a member of the ruling house of ancient Persia generally considered historically important from the assumption of power by Cyrus the Great (559 B.C.) to the overthrow of Darius III (330 B.C.)
acha·la·sia \ˌā-kə-ˈlā-zh(ē-)ə\ *noun* [New Latin, from *a-* + Greek *chalasis* slackening + New Latin *-ia*] (1914)
: failure of a ring of muscle (as the anal sphincter or one of the esophagus) to relax
Acha·tes \ə-ˈkā-tēz\ *noun* [Latin]
: a faithful companion of Aeneas in Virgil's *Aeneid*
[1]**ache** \ˈāk\ *intransitive verb* **ached; ach·ing** [Middle English *aken,* from Old English *acan*] (before 12th century)
1 a : to suffer a usually dull persistent pain **b :** to become distressed or disturbed (as with anxiety or regret) **c :** to feel compassion
2 : to experience a painful eagerness or yearning
[2]**ache** *noun* (before 12th century)
1 : a usually dull persistent pain
2 : a condition marked by aching
achene \ə-ˈkēn\ *noun* [New Latin *achaenium,* from *a-* + Greek *chainein* to yawn — more at YAWN] (1855)
: a small dry indehiscent one-seeded fruit (as of a sunflower) developing from a simple ovary and usually having a thin pericarp attached to the seed at only one point
Ach·er·on \ˈa-kə-ˌrän, -rən\ *noun* [Greek *Acherōn*]
: a river in Hades
Acheu·le·an *or* **Acheu·li·an** \ə-ˈshü-lē-ən\ *adjective* [French *acheuléen,* from Saint *Acheul,* near Amiens, France] (circa 1909)
: of or relating to a Lower Paleolithic culture typified by bifacial tools with round cutting edges
à che·val \ˌä-shə-ˈväl\ *adverb* [French, literally, on horseback] (1832)
1 : with a leg on each side **:** ASTRIDE
2 : in such a way as to be played or chanced simultaneously on two numbers or events (as in roulette)
achieve \ə-ˈchēv\ *verb* **achieved; achiev·ing** [Middle English *acheven,* from Middle French *achever* to finish, from *a-* (from Latin *ad-*) + *chief* end, head — more at CHIEF] (14th century)
transitive verb
1 : to carry out successfully **:** ACCOMPLISH ⟨*achieve* a gradual increase in production⟩
2 : to get or attain as the result of exertion **:** REACH ⟨*achieved* a high degree of skill⟩ ⟨*achieved* greatness⟩
intransitive verb
: to attain a desired end or aim **:** become successful
synonym see PERFORM
— **achiev·able** \-ˈchē-və-bəl\ *adjective*
— **achiev·er** *noun*
achieved *adjective* (1918)
: brought to or marked by a high degree of development or refinement **:** FINISHED ⟨fully *achieved* poems⟩
achieve·ment \ə-ˈchēv-mənt\ *noun* (15th century)
1 : the act of achieving **:** ACCOMPLISHMENT
2 a : a result gained by effort **b :** a great or heroic deed
3 : the quality and quantity of a student's work
synonym see FEAT
Achil·les \ə-ˈki-lēz\ *noun* [Latin, from Greek *Achilleus*]
: the greatest warrior among the Greeks at Troy and slayer of Hector
Achilles' heel *noun* [from the story that Achilles was vulnerable only in the heel] (1864)
: a vulnerable point
Achilles tendon *noun* (circa 1879)
: the strong tendon joining the muscles in the calf of the leg to the bone of the heel
ach·ing \ˈā-kiŋ\ *adjective* (13th century)
1 : that aches ⟨an *aching* back⟩
2 : causing or reflecting distress, deep emotion, or longing ⟨*aching* country ballads⟩
— **ach·ing·ly** *adverb*
achlor·hyd·ria \ˌā-ˌklōr-ˈhi-drē-ə, -ˌklór-\ *noun* [New Latin, from *a-* + International Scientific Vocabulary *chlor-* + *hydr-* + New Latin *-ia*] (1898)
: absence of hydrochloric acid from the gastric juice
— **achlor·hy·dric** \-ˈhi-drik, -ˈhī-\ *adjective*
achon·drite \(ˌ)ā-ˈkän-ˌdrīt\ *noun* (circa 1904)
: a stony meteorite without rounded grains
— **achon·drit·ic** \ˌā-ˌkän-ˈdri-tik\ *adjective*
achon·dro·pla·sia \ˌā-ˌkän-drə-ˈplā-zh(ē-)ə\ *noun* [New Latin, from *a-* + *chondr-* + *-plasia*] (circa 1893)
: a genetic disorder disturbing normal growth of cartilage, resulting in a form of dwarfism characterized by a usually normal torso and shortened limbs, and usually inherited as an autosomal dominant
— **achon·dro·plas·tic** \-ˈplas-tik\ *adjective*
ach·ro·mat \ˈa-krə-ˌmat\ *noun* (1900)
: ACHROMATIC LENS
ach·ro·mat·ic \ˌa-krə-ˈma-tik, (ˌ)ā-\ *adjective* (1766)
1 : refracting light without dispersing it into its constituent colors **:** giving images practically free from extraneous colors ⟨an *achromatic* telescope⟩
2 : not readily colored by the usual staining agents
3 : possessing no hue **:** being or involving black, gray, or white **:** NEUTRAL ⟨*achromatic* visual sensations⟩
4 : being without accidentals or modulation **:** DIATONIC
— **ach·ro·mat·i·cal·ly** \-ti-k(ə-)lē\ *adverb*
— **achro·ma·tism** \(ˌ)ā-ˈkrō-mə-ˌti-zəm, a-\ *noun*
— **achro·ma·tize** \(ˌ)ā-ˈkrō-mə-ˌtīz, a-\ *transitive verb*
achromatic lens *noun* (circa 1864)
: a lens made by combining lenses of different glasses having different focal powers so that the light emerging from the lens forms an image practically free from unwanted colors
achy \ˈā-kē\ *adjective* **ach·i·er; ach·i·est** (1875)
: afflicted with aches
— **ach·i·ness** *noun*
acic·u·lar \ə-ˈsi-kyə-lər\ *adjective* [Late Latin *acicula* (diminutive of Latin *acus* needle) + English *-ar* — more at ACUTE] (1794)
: shaped like a needle ⟨*acicular* leaves⟩ ⟨*acicular* crystals⟩
[1]**ac·id** \ˈa-səd\ *adjective* [French or Latin; French *acide,* from Latin *acidus,* from *acēre* to be sour — more at ACET-] (1626)
1 a : sour, sharp, or biting to the taste **b :** sharp, biting, or sour in manner, disposition, or nature ⟨an *acid* individual⟩ **c :** sharply clear, discerning, or pointed ⟨an *acid* wit⟩ **d :** piercingly intense and often jarring ⟨*acid* yellow⟩
2 a : of, relating to, or being an acid; *also* **:** having the reactions or characteristics of an acid ⟨*acid* soil⟩ ⟨an *acid* solution⟩ **b** *of salts and esters* **:** derived by partial exchange of replaceable hydrogen ⟨*acid* sodium carbonate $NaHCO_3$⟩ **c :** containing or involving the use of an acid (as in manufacture) **d :** marked by or resulting from an abnormally high concentration of acid ⟨*acid* indigestion⟩
3 : relating to or made by a process (as in making steel) in which the furnace is lined with acidic material and an acidic slag is used
4 : rich in silica ⟨*acid* rocks⟩
— **ac·id·ly** *adverb*
— **ac·id·ness** *noun*
[2]**acid** *noun* (1696)
1 : a sour substance; *specifically* **:** any of various typically water-soluble and sour compounds that in solution are capable of reacting with a base to form a salt, redden litmus, and have a pH less than 7, that are hydrogen-containing molecules or ions able to give up a proton to a base, or that are substances able to accept an unshared pair of electrons from a base
2 : something incisive, biting, or sarcastic ⟨a social satire dripping with *acid*⟩
3 : LSD
— **ac·idy** \ˈa-sə-dē\ *adjective*
ac·id–fast \ˈa-səd-ˌfast\ *adjective* (1903)
: not easily decolorized by acids
ac·id·head \-ˌhed\ *noun* (1966)
: an individual who uses LSD
acid·ic \ə-ˈsi-dik, a-\ *adjective* (1880)
1 : acid-forming
2 : ACID
acid·i·fi·er \ə-ˈsi-də-ˌfī(-ə)r, a-\ *noun* (circa 1828)
: one that acidifies; *especially* **:** a substance used to increase soil acidity
acid·i·fy \-ˌfī\ *transitive verb* **-fied; -fy·ing** (1797)
1 : to make acid
2 : to convert into an acid
— **acid·i·fi·ca·tion** \-ˌsi-də-fə-ˈkā-shən\ *noun*
acid·i·met·ric \ə-ˌsi-də-ˈme-trik\ *adjective* (1900)
: of or relating to the precise determination of the amount of acid present in a solution
— **ac·i·dim·e·ter** \ˌa-sə-ˈdi-mə-tər\ *noun*
— **ac·i·dim·e·try** \-ˈdi-mə-trē\ *noun*

acid·i·ty \ə-ˈsi-də-tē, a-\ *noun, plural* **-ties** (1620)
1 : the quality, state, or degree of being acid
2 : the state of being excessively acid

acid·o·phil \ə-ˈsi-də-ˌfil, a-\ *also* **acid·o·phile** \-ˌfīl\ *noun* (circa 1900)
: a substance, tissue, or organism that stains readily with acid stains
— acidophil *also* **acidophile** *adjective*

ac·i·do·phil·ic \ˌa-sə-dō-ˈfi-lik\ *adjective* (circa 1900)
1 : staining readily with acid stains **:** ACIDOPHIL
2 : preferring or thriving in a relatively acid environment

ac·i·doph·i·lus milk \ˌa-sə-ˈdä-f(ə-)ləs-\ *noun* [New Latin *Lactobacillus acidophilus,* literally, acidophilic lactobacillus] (1921)
: milk fermented by any of several bacteria and used therapeutically to change the intestinal flora

ac·i·do·sis \ˌa-sə-ˈdō-səs\ *noun* [New Latin] (1900)
: an abnormal condition characterized by reduced alkalinity of the blood and of the body tissues
— ac·i·dot·ic \-ˈdä-tik\ *adjective*

acid phosphatase *noun* (1949)
: a phosphatase (as the phosphomonoesterase from the prostate gland) optimally active in acid medium

acid precipitation *noun* (1979)
: precipitation (as rain or snow) having increased acidity caused by environmental factors (as atmospheric pollutants)

acid rain *noun* (1858)
: acid precipitation in the form of rain

acid rock *noun* (1966)
: rock music with lyrics and sound relating to or suggestive of drug-induced experiences

acid test *noun* (1912)
: a severe or crucial test

acid·u·late \ə-ˈsi-jə-ˌlāt\ *transitive verb* **-lat·ed; -lat·ing** [Latin *acidulus*] (1732)
: to make acid or slightly acid
— acid·u·la·tion \-ˌsi-jə-ˈlā-shən\ *noun*

acid·u·lent \ə-ˈsi-jə-lənt\ *adjective* [French *acidulant,* from present participle of *aciduler* to acidulate, from Latin *acidulus*] (1834)
: ACIDULOUS

acid·u·lous \ə-ˈsi-jə-ləs\ *adjective* [Latin *acidulus,* from *acidus*] (1769)
: somewhat acid or harsh in taste or manner

ac·i·nar \ˈa-sə-nər, -ˌnär\ *adjective* (1936)
: of, relating to, or comprising an acinus ⟨pancreatic *acinar* cells⟩

ac·i·nus \ˈa-sə-nəs\ *noun, plural* **-ni** \-ˌnī\ [New Latin, from Latin, berry, berry seed] (circa 1751)
: any of the small sacs terminating the ducts of some exocrine glands and lined with secretory cells
— ac·i·nous \-nəs\ *adjective*

ack–ack \ˈak-ˌak\ *noun* [British signalmen's former telephone pronunciation of *AA,* abbreviation of *antiaircraft*] (1926)
: an antiaircraft gun; *also* **:** antiaircraft fire

ackee *variant of* AKEE

ac·knowl·edge \ik-ˈnä-lij, ak-\ *transitive verb* **-edged; -edg·ing** [*ac-* (as in *accord*) + *knowledge*] (15th century)
1 : to recognize the rights, authority, or status of
2 : to disclose knowledge of or agreement with
3 a : to express gratitude or obligation for **b :** to take notice of **c :** to make known the receipt of
4 : to recognize as genuine or valid ⟨*acknowledge* a debt⟩ ☆

ac·knowl·edged \-lijd\ *adjective* (1598)
: generally recognized, accepted, or admitted
— ac·knowl·edged·ly \-lijd-lē, -li-jəd-\ *adverb*

ac·knowl·edg·ment *or* **ac·knowl·edge·ment** \ik-ˈnä-lij-mənt, ak-\ *noun* (1594)
1 a : the act of acknowledging **b :** recognition or favorable notice of an act or achievement
2 : a thing done or given in recognition of something received
3 : a declaration or avowal of one's act or of a fact to give it legal validity

ac·me \ˈak-mē\ *noun* [Greek *akmē* point, highest point — more at EDGE] (1620)
: the highest point or stage; *also* **:** one that represents perfection of the thing expressed
synonym see SUMMIT

ac·ne \ˈak-nē\ *noun* [Greek *aknē* eruption of the face, manuscript variant of *akmē,* literally, point] (circa 1828)
: a disorder of the skin caused by inflammation of the skin glands and hair follicles; *specifically* **:** a form found chiefly in adolescents and marked by pimples especially on the face
— ac·ned \-nēd\ *adjective*

acock \ə-ˈkäk\ *adjective or adverb* (1846)
: being in a cocked position

acoe·lo·mate \(ˌ)ā-ˈsē-lə-ˌmāt\ *noun* (circa 1889)
: an invertebrate lacking a coelom; *especially* **:** one belonging to the group comprising the flatworms and nemerteans and characterized by bilateral symmetry and a digestive cavity that is the only internal cavity
— acoelomate *adjective*

acold \ə-ˈkōld\ *adjective* (14th century)
archaic **:** COLD, CHILLED ⟨the owl, for all his feathers, was *acold* —John Keats⟩

ac·o·lyte \ˈa-kə-ˌlīt\ *noun* [Middle English, from Middle French & Medieval Latin; Middle French, from Medieval Latin *acoluthus,* from Middle Greek *akolouthos,* from Greek, adjective, following, from *a-, ha-* together (akin to Greek *homos* same) + *keleuthos* path] (14th century)
1 : one who assists the clergyman in a liturgical service by performing minor duties
2 : one who attends or assists **:** FOLLOWER

ac·o·nite \ˈa-kə-ˌnīt\ *noun* [Middle French or Latin; from Latin *aconitum,* from Greek *akoniton*] (1578)
1 : MONKSHOOD
2 : the dried tuberous root of a common monkshood (*Aconitum napellus*) formerly used as a sedative and anodyne

acorn \ˈā-ˌkȯrn, -kərn\ *noun* [Middle English *akern,* from Old English *æcern;* akin to Old English *æcer* field, Middle High German *ackeran* acorns collectively, Old Irish *áirne* sloe, Lithuanian *uoga* berry] (before 12th century)
: the nut of the oak usually seated in or surrounded by a hard woody cupule of indurated bracts

acorn

acorn squash *noun* (1937)
: an acorn-shaped dark green winter squash with a ridged surface and sweet yellow to orange flesh

acorn squash

acorn worm *noun* (circa 1889)
: any of a class (Enteropneusta) of burrowing wormlike marine animals having an acorn-shaped proboscis and classified with the hemichordates

acous·tic \ə-ˈkü-stik\ *or* **acous·ti·cal** \-sti-kəl\ *adjective* [Greek *akoustikos* of hearing, from *akouein* to hear — more at HEAR] (1605)
1 : of or relating to the sense or organs of hearing, to sound, or to the science of sounds ⟨*acoustic* apparatus of the ear⟩ ⟨*acoustic* energy⟩: as **a :** deadening or absorbing sound ⟨*acoustic* tile⟩ **b :** operated by or utilizing sound waves
2 : of, relating to, or being a musical instrument whose sound is not electronically modified
— acous·ti·cal·ly \-k(ə-)lē\ *adverb*

ac·ous·ti·cian \ˌa-ˌkü-ˈsti-shən, ə-ˌkü-\ *noun* (1859)
: a specialist in acoustics

acous·tics \ə-ˈkü-stiks\ *noun plural* (1683)
1 *singular in construction* **:** a science that deals with the production, control, transmission, reception, and effects of sound
2 *also* **acoustic :** the qualities that determine the ability of an enclosure (as an auditorium) to reflect sound waves in such a way as to produce distinct hearing

ac·quaint \ə-ˈkwānt\ *transitive verb* [Middle English, from Old French *acointier,* from *acointe* familiar, from Latin *accognitus,* past participle of *accognoscere* to recognize, from *ad-* + *cognoscere* to know — more at COGNITION] (14th century)
1 : to cause to know personally ⟨was *acquainted* with the mayor⟩
2 : to make familiar **:** cause to know firsthand
synonym see INFORM

ac·quain·tance \ə-ˈkwān-tᵊn(t)s\ *noun* (14th century)
1 a : the state of being acquainted **b :** personal knowledge **:** FAMILIARITY
2 a : the persons with whom one is acquainted ⟨should auld *acquaintance* be forgot —Robert Burns⟩ **b :** a person whom one knows but who is not a particularly close friend
— ac·quain·tance·ship \-ˌship\ *noun*

ac·qui·esce \ˌa-kwē-ˈes\ *intransitive verb* **-esced; -esc·ing** [French *acquiescer,* from Latin *acquiescere,* from *ad-* + *quiescere* to be quiet — more at QUIESCENT] (circa 1620)
: to accept, comply, or submit tacitly or passively — often used with *in* and sometimes with *to*
synonym see ASSENT

ac·qui·es·cence \-ˈe-sᵊn(t)s\ *noun* (1612)
1 : the act of acquiescing **:** the state of being acquiescent
2 : an instance of acquiescing

ac·qui·es·cent \-ˈe-sᵊnt\ *adjective* [Latin *acquiescent-, acquiescens,* present participle of *acquiescere*] (1753)
: inclined to acquiesce
— ac·qui·es·cent·ly *adverb*

ac·quir·able \ə-ˈkwī-rə-bəl\ *adjective* (1646)
: capable of being acquired

ac·quire \ə-ˈkwīr\ *transitive verb* **ac·quired; ac·quir·ing** [Middle English *aqueren,* from Middle French *aquerre,* from Latin *acquirere,* from *ad-* + *quaerere* to seek, obtain] (15th century)

☆ SYNONYMS

Acknowledge, admit, own, avow, confess mean to disclose against one's will or inclination. ACKNOWLEDGE implies the disclosing of something that has been or might be concealed ⟨*acknowledged* an earlier peccadillo⟩. ADMIT implies reluctance to disclose, grant, or concede and refers usually to facts rather than their implications ⟨*admitted* the project was over budget⟩. OWN implies acknowledging something in close relation to oneself ⟨must *own* I know little about computers⟩. AVOW implies boldly declaring, often in the face of hostility, what one might be expected to be silent about ⟨*avowed* that he was a revolutionary⟩. CONFESS may apply to an admission of a weakness, failure, omission, or guilt ⟨*confessed* a weakness for sweets⟩.

1 : to get as one's own: **a :** to come into possession or control of often by unspecified means **b :** to come to have as a new or added characteristic, trait, or ability (as by sustained effort or natural selection) ⟨*acquire* fluency in French⟩ ⟨bacteria that *acquire* tolerance to antibiotics⟩
2 : to locate and hold (a desired object) in a detector ⟨*acquire* a target by radar⟩

acquired immune deficiency syndrome *noun* (1982)
: AIDS

acquired immunodeficiency syndrome *noun* (1982)
: AIDS

acquired taste *noun* (1858)
: one that is not easily or immediately liked or appreciated

ac•quire•ment \ə-'kwīr-mənt\ *noun* (1630)
1 : a skill of mind or body usually resulting from continued endeavor
2 : the act of acquiring

ac•qui•si•tion \ˌa-kwə-'zi-shən\ *noun* [Middle English *acquisicioun,* from Middle French or Latin; Middle French *acquisition,* from Latin *acquisition-, acquisitio,* from *acquirere*] (14th century)
1 : the act of acquiring
2 : something acquired or gained
— ac•qui•si•tion•al \-shnəl, -shə-nᵊl\ *adjective*
— ac•quis•i•tor \ə-'kwi-zə-tər\ *noun*

ac•quis•i•tive \ə-'kwi-zə-tiv\ *adjective* (1846)
: strongly desirous of acquiring and possessing
synonym see COVETOUS
— ac•quis•i•tive•ly *adverb*
— ac•quis•i•tive•ness *noun*

ac•quit \ə-'kwit\ *transitive verb* **ac•quit•ted; ac•quit•ting** [Middle English *aquiten,* from Old French *aquiter,* from *a-* (from Latin *ad-*) + *quite* free of — more at QUIT] (13th century)
1 a *archaic* **:** to pay off (as a claim or debt) **b** *obsolete* **:** REPAY, REQUITE
2 : to discharge completely (as from an obligation or accusation) ⟨the court *acquitted* the prisoner⟩
3 : to conduct (oneself) usually satisfactorily especially under stress ⟨the recruits *acquitted* themselves like veterans⟩
synonym see BEHAVE, EXCULPATE
— ac•quit•ter *noun*

ac•quit•tal \ə-'kwi-tᵊl\ *noun* (15th century)
: a setting free from the charge of an offense by verdict, sentence, or other legal process

ac•quit•tance \ə-'kwi-tᵊn(t)s\ *noun* (14th century)
: a document evidencing a discharge from an obligation; *especially* **:** a receipt in full

acr- *or* **acro-** *combining form* [Middle French or Greek; Middle French *acro-,* from Greek *akr-, akro-,* from *akros* topmost, extreme; akin to Greek *akmē* point — more at EDGE]
1 : beginning **:** end **:** tip ⟨*acro*nym⟩
2 a : top **:** peak **:** summit ⟨*acro*petal⟩ **b :** height ⟨*acro*phobia⟩

acre \'ā-kər\ *noun* [Middle English, from Old English *æcer;* akin to Old High German *ackar* field, Latin *ager,* Greek *agros,* and perhaps to Latin *agere* to drive — more at AGENT] (before 12th century)
1 a *archaic* **:** a field especially of arable land or pastureland **b** *plural* **:** LANDS, ESTATE
2 : any of various units of area; *specifically* **:** a unit in the U.S. and England equal to 43,560 square feet (4047 square meters) — see WEIGHT table
3 : a broad expanse or great quantity ⟨*acres* of free publicity⟩

acre•age \'ā-k(ə-)rij\ *noun* (1859)
: area in acres **:** ACRES

acre–foot \'ā-kər-'fu̇t\ *noun* (1900)
: the volume (as of irrigation water) that would cover one acre to a depth of one foot

acre–inch \'ā-kər-'inch\ *noun* (circa 1909)
: one twelfth of an acre-foot

ac•rid \'a-krəd\ *adjective* [modification of Latin *acr-, acer* sharp — more at EDGE] (1712)
1 : sharp and harsh or unpleasantly pungent in taste or odor **:** IRRITATING
2 : deeply or violently bitter **:** ACRIMONIOUS ⟨an *acrid* denunciation⟩
synonym see CAUSTIC
— acrid•i•ty \a-'kri-də-tē, ə-\ *noun*
— ac•rid•ly \'a-krəd-lē\ *adverb*
— ac•rid•ness *noun*

ac•ri•dine \'a-krə-ˌdēn\ *noun* (circa 1877)
: a colorless crystalline compound $C_{13}H_9N$ occurring in coal tar and important as the parent compound of dyes and pharmaceuticals

acridine orange *noun* (circa 1909)
: a basic orange dye structurally related to acridine and used especially to stain nucleic acids

ac•ri•fla•vine \ˌa-krə-'flā-ˌvēn, -vən\ *noun* [*acri*dine + *flavine*] (1917)
: a yellow dye $C_{14}H_{14}N_3Cl$ used as an antiseptic especially for wounds

Ac•ri•lan \'a-krə-ˌlan, -lən\ *trademark*
— used for an acrylic fiber

ac•ri•mo•ni•ous \ˌa-krə-'mō-nē-əs\ *adjective* (1775)
: caustic, biting, or rancorous especially in feeling, language, or manner ⟨an *acrimonious* dispute⟩
— ac•ri•mo•ni•ous•ly *adverb*
— ac•ri•mo•ni•ous•ness *noun*

ac•ri•mo•ny \'a-krə-ˌmō-nē\ *noun, plural* **-nies** [Middle French or Latin; Middle French *acrimonie,* from Latin *acrimonia,* from *acr-, acer*] (1542)
: harsh or biting sharpness especially of words, manner, or disposition

ac•ri•tarch \'a-kri-ˌtärk\ *noun* [Greek *akritos* uncertain (from *a-* + *kritos,* verbal of *krinein* to decide) + *archē* beginning — more at CERTAIN, ARCH-] (1963)
: any of a group of fossil one-celled marine planktonic organisms of uncertain and possibly various taxonomic affinities held to represent the earliest known eukaryotes

ac•ro•bat \'a-krə-ˌbat\ *noun* [French & Greek; French *acrobate,* from Greek *akrobatēs,* from *akros* + *bainein* to go — more at COME] (1825)
1 : one that performs gymnastic feats requiring skillful control of the body
2 a : one skillful at exercises of intellectual or artistic dexterity **b :** one adept at swiftly changing or adapting a position or viewpoint ⟨a political *acrobat*⟩
— ac•ro•bat•ic \ˌa-krə-'ba-tik\ *adjective*
— ac•ro•bat•i•cal•ly \-ti-k(ə-)lē\ *adverb*

ac•ro•bat•ics \ˌa-krə-'ba-tiks\ *noun plural but singular or plural in construction* (1882)
1 : the art, performance, or activity of an acrobat
2 : a spectacular, showy, or startling performance or demonstration involving great agility or complexity

ac•ro•cen•tric \ˌa-krō-'sen-trik\ *adjective* (1945)
: having the centromere situated so that one chromosomal arm is much shorter than the other
— acrocentric *noun*

ac•ro•lect \'a-krə-ˌlekt\ *noun* [*acr-* + *-lect* (as in *dialect*)] (1964)
: the language variety of a speech community closest to the standard or prestige form of a language

acro•le•in \ə-'krō-lē-ən\ *noun* [International Scientific Vocabulary *acr-* (from Latin *acr-, acer*) + Latin *olēre* to smell — more at ODOR] (circa 1857)
: a colorless irritant pungent liquid aldehyde C_3H_4O used chiefly in organic synthesis

ac•ro•meg•a•ly \ˌa-krō-'me-gə-lē\ *noun* [International Scientific Vocabulary] (1889)
: chronic hyperpituitarism marked by progressive enlargement of hands, feet, and face
— ac•ro•me•gal•ic \-mə-'ga-lik\ *adjective or noun*

ac•ro•nym \'a-krə-ˌnim\ *noun* [*acr-* + *-onym*] (1943)
: a word (as *radar* or *snafu*) formed from the initial letter or letters of each of the successive parts or major parts of a compound term
— ac•ro•nym•ic \ˌa-krə-'ni-mik\ *adjective*
— ac•ro•nym•i•cal•ly \-mi-k(ə-)lē\ *adverb*

acrop•e•tal \ə-'krä-pə-tᵊl, a-\ *adjective* [*acr-* + *-petal* (as in *centripetal*)] (1875)
: proceeding from the base toward the apex or from below upward ⟨*acropetal* development of floral buds⟩
— acrop•e•tal•ly \-tᵊl-ē\ *adverb*

ac•ro•pho•bia \ˌa-krə-'fō-bē-ə\ *noun* [New Latin] (circa 1892)
: abnormal dread of being at a great height
— ac•ro•phobe \'a-krə-ˌfōb\ *noun*

acrop•o•lis \ə-'krä-pə-ləs\ *noun* [Greek *akropolis,* from *akr-* acr- + *polis* city — more at POLICE] (1662)
: the upper fortified part of an ancient Greek city (as Athens); *also* **:** a usually fortified height of a city or district elsewhere (as in Central America)

ac•ro•some \'a-krə-ˌsōm\ *noun* [International Scientific Vocabulary] (1899)
: an anterior prolongation of a spermatozoon that releases egg-penetrating enzymes
— ac•ro•so•mal \ˌa-krə-'sō-məl\ *adjective*

[1]across \ə-'krȯs, *chiefly dialect* -'krȯst\ *adverb* [Middle English *acros,* from Anglo-French *an crois,* from *an* in (from Latin *in*) + *crois* cross, from Latin *crux*] (14th century)
1 : in a position reaching from one side to the other **:** CROSSWISE
2 : to or on the opposite side
3 : so as to be understandable, acceptable, or successful ⟨get an argument *across*⟩

[2]across *preposition* (1591)
1 a : from one side to the opposite side of **:** OVER, THROUGH ⟨swam *across* the river⟩ **b :** on the opposite side of ⟨lives *across* the street from us⟩
2 : so as to intersect or pass through at an angle ⟨sawed *across* the grain of the wood⟩
3 : so as to find or meet ⟨came *across* your football in the hall closet⟩
4 a : THROUGHOUT ⟨obvious interest *across* the nation —Robert Goralski⟩ **b :** so as to include or take into consideration all classes or categories ⟨*across* differences, they insist, there can be no rational dialogue —Huston Smith⟩

[3]across *adjective* (1646)
: being in a crossed position

across–the–board *adjective* (1945)
1 : placed to win if a competitor wins, places, or shows ⟨an *across-the-board* racing bet⟩
2 : embracing or affecting all classes or categories **:** BLANKET ⟨an *across-the-board* price increase⟩

acros•tic \ə-'krȯs-tik, -'kräs-\ *noun* [Middle French & Greek; Middle French *acrostiche,* from Greek *akrostichis,* from *akr-* acr- + *stichos* line; akin to *steichein* to go — more at STAIR] (1530)
1 : a composition usually in verse in which sets of letters (as the initial or final letters of the lines) taken in order form a word or phrase or a regular sequence of letters of the alphabet
2 : ACRONYM
— acrostic *also* **acros•ti•cal** \-ti-kəl\ *adjective*
— acros•ti•cal•ly \-ti-k(ə-)lē\ *adverb*

ac•ryl•am•ide \ˌa-krᵊl-'a-ˌmīd, ə-'kri-lə-\ *noun* [*acryl*ic + *amide*] (1946)
: an amide C_3H_5NO that is derived from acrylic acid, that polymerizes readily, and that is used in the manufacture of synthetic textile fibers

ac•ry•late \'a-krə-ˌlāt\ *noun* (1873)
1 : a salt or ester of acrylic acid
2 : ACRYLIC RESIN

[1]acryl·ic \ə-ˈkri-lik\ *adjective* [International Scientific Vocabulary *acro*lein + *-yl* + [1]*-ic*] (1855)
1 : of or relating to acrylic acid or its derivatives ⟨*acrylic* polymers⟩
2 : made or consisting of an acrylic ⟨an *acrylic* window⟩

[2]acrylic *noun* (1942)
1 a : ACRYLIC RESIN **b :** a paint in which the vehicle is an acrylic resin **c :** a painting done in an acrylic resin
2 : ACRYLIC FIBER

acrylic acid *noun* (circa 1855)
: an unsaturated liquid acid $C_3H_4O_2$ that polymerizes readily to form useful products (as constituents for varnishes and lacquers)

acrylic fiber *noun* (1951)
: a quick-drying synthetic textile fiber made by polymerization of acrylonitrile usually with other monomers

acrylic resin *noun* (1936)
: a glassy thermoplastic made by polymerizing acrylic or methacrylic acid or a derivative of either and used for cast and molded parts or as coatings and adhesives

ac·ry·lo·ni·trile \ˌa-krə-lō-ˈnī-trəl, -ˌtrēl\ *noun* (1893)
: a colorless volatile flammable liquid nitrile C_3H_3N used chiefly in organic synthesis and for polymerization

[1]act \ˈakt\ *noun* [Middle English, partly from Latin *actus* doing, act, from *agere* to drive, do; partly from Latin *actum* thing done, record, from neuter of *actus,* past participle of *agere* — more at AGENT] (14th century)
1 a : the doing of a thing **:** DEED **b :** something done voluntarily
2 : a state of real existence rather than possibility
3 : the formal product of a legislative body **:** STATUTE; *also* **:** a decision or determination of a sovereign, a legislative council, or a court of justice
4 : the process of doing **:** ACTION ⟨caught in the *act*⟩
5 *often capitalized* **:** a formal record of something done or transacted
6 : one of the principal divisions of a theatrical work (as a play or opera)
7 a : one of successive parts or performances (as in a variety show or circus) **b :** the performer or performers in such an act **c :** a performance or presentation identified with a particular individual or group **d :** the sum of a person's actions or effects that serve to create an impression or set an example ⟨a hard *act* to follow⟩
8 : a display of affected behavior **:** PRETENSE

[2]act (1594)
transitive verb
1 a : to represent or perform by action especially on the stage **b :** FEIGN, SIMULATE **c :** IMPERSONATE
2 *obsolete* **:** ACTUATE, ANIMATE
3 : to play the part of as if in a play ⟨*act* the man of the world⟩
4 : to behave in a manner suitable to ⟨*act* your age⟩
intransitive verb
1 a : to perform on the stage **b :** to behave as if performing on the stage **:** PRETEND
2 : to take action **:** MOVE ⟨think before *acting*⟩ ⟨*acted* favorably on the recommendation⟩
3 : to conduct oneself **:** BEHAVE ⟨*act* like a fool⟩
4 : to perform a specified function **:** SERVE ⟨trees *acting* as a windbreak⟩
5 : to produce an effect **:** WORK ⟨wait for a medicine to *act*⟩
6 *of a play* **:** to be capable of being performed ⟨the play *acts* well⟩
7 : to give a decision or award ⟨adjourned without *acting* on the bill⟩
— act·abil·i·ty \ˌak-tə-ˈbi-lə-tē\ *noun*
— act·able \ˈak-tə-bəl\ *adjective*

Ac·tae·on \ak-ˈtē-ən\ *noun* [Latin, from Greek *Aktaiōn*]
: a hunter turned into a stag and killed by his own hounds for having seen Artemis bathing

ACTH \ˌā-(ˌ)sē-(ˌ)tē-ˈāch\ *noun* [*a*dreno*c*ortico*t*ropic *h*ormone] (1944)
: a protein hormone of the anterior lobe of the pituitary gland that stimulates the adrenal cortex — called also *adrenocorticotropic hormone*

ac·tin \ˈak-tən\ *noun* [International Scientific Vocabulary, from Latin *actus*] (1942)
: a cellular protein found especially in microfilaments (as those comprising myofibrils) and active in muscular contraction, cellular movement, and maintenance of cell shape

actin- *or* **actini-** *or* **actino-** *combining form* [New Latin, ray, from Greek *aktin-, aktino-,* from *aktin-, aktis;* perhaps akin to Old English *ūhte* morning twilight, Latin *noct-, nox* night — more at NIGHT]
1 : having a radiate form ⟨*actino*lite⟩
2 : actinic radiation (as X rays) ⟨*actino*meter⟩

[1]act·ing \ˈak-tiŋ\ *noun* (1664)
: the art or practice of representing a character on a stage or before cameras

[2]acting *adjective* (1797)
1 : holding a temporary rank or position **:** performing services temporarily ⟨*acting* president⟩
2 a : suitable for stage performance ⟨an *acting* play⟩ **b :** prepared with directions for actors ⟨an *acting* text of a play⟩

ac·tin·i·an \ak-ˈti-nē-ən\ *noun* [New Latin *actinia,* from Greek *aktin-, aktis*] (1888)
: SEA ANEMONE

ac·tin·ic \ak-ˈti-nik\ *adjective* (1844)
: of, relating to, resulting from, or exhibiting actinism
— ac·tin·i·cal·ly \-ni-k(ə-)lē\ *adverb*

ac·ti·nide \ˈak-tə-ˌnīd\ *noun* [International Scientific Vocabulary] (1945)
: any of the members of the series of elements that begins with actinium or thorium and ends with lawrencium — see PERIODIC TABLE table

ac·ti·nism \ˈak-tə-ˌni-zəm\ *noun* (1844)
: the property of radiant energy especially in the visible and ultraviolet spectral regions by which chemical changes are produced

ac·tin·i·um \ak-ˈti-nē-əm\ *noun* [New Latin] (1900)
: a radioactive trivalent metallic element that resembles lanthanum in chemical properties and that is found especially in pitchblende — see ELEMENT table

ac·tin·o·lite \ak-ˈti-nᵊl-ˌīt\ *noun* (circa 1828)
: a bright or grayish green amphibole occurring in fibrous, radiate, or columnar forms

ac·ti·nom·e·ter \ˌak-tə-ˈnä-mə-tər\ *noun* (1833)
: any of various instruments for measuring the intensity of incident radiation; *especially* **:** one in which the intensity of radiation is measured by the speed of a photochemical reaction
— ac·ti·no·me·tric \-nō-ˈme-trik\ *adjective*
— ac·ti·nom·e·try \-ˈnä-mə-trē\ *noun*

ac·ti·no·mor·phic \ˌak-(ˌ)ti-nō-ˈmȯr-fik, -tə-nō-; ak-ˌti-nō-\ *adjective* [International Scientific Vocabulary] (1900)
: being radially symmetrical and capable of division by any longitudinal plane into essentially symmetrical halves ⟨an *actinomorphic* tulip flower⟩
— ac·ti·no·mor·phy \ˈak-tə-nō-ˌmȯr-fē, ak-ˈti-nō-\ *noun*

ac·ti·no·my·ces \ˌak-(ˌ)ti-nō-ˈmī-ˌsēz, -tə-nō-; ak-ˌti-nō-\ *noun, plural* **actinomyces** [New Latin, genus name, from *actin-* + Greek *mykēt-, mykēs* fungus; akin to Greek *myxa* mucus — more at MUCUS] (1882)
: any of a genus (*Actinomyces*) of filamentous or rod-shaped bacteria that includes usually commensal and sometimes pathogenic forms inhabiting mucosal surfaces especially of the oral cavity of warm-blooded vertebrates

ac·ti·no·my·cete \-ˈmī-ˌsēt, -mī-ˈsēt\ *noun* [ultimately from Greek *aktin-, aktis* + *mykēt-, mykēs*] (1911)
: any of an order (Actinomycetales) of filamentous or rod-shaped bacteria (as the actinomyces and streptomyces)
— ac·ti·no·my·ce·tous \-mī-ˈsē-təs\ *adjective*

ac·ti·no·my·cin \-ˈmī-sᵊn\ *noun* (1940)
: any of various red or yellow-red mostly toxic polypeptide antibiotics isolated from soil bacteria (especially *Streptomyces antibioticus*); *specifically* **:** one used to inhibit DNA or RNA synthesis

ac·ti·no·my·co·sis \-mī-ˈkō-səs\ *noun* [New Latin] (1882)
: infection with or disease caused by actinomycetes; *especially* **:** a chronic disease of cattle, swine, and humans characterized by hard granulomatous masses usually in the mouth and jaw
— ac·ti·no·my·cot·ic \-ˈkä-tik\ *adjective*

ac·ti·non \ˈak-tə-ˌnän\ *noun* [New Latin, from *actinium*] (1926)
: a gaseous radioactive isotope of radon that has a half-life of about 4 seconds

ac·tion \ˈak-shən\ *noun* (14th century)
1 : the initiating of a proceeding in a court of justice by which one demands or enforces one's right; *also* **:** the proceeding itself
2 : the bringing about of an alteration by force or through a natural agency
3 : the manner or method of performing: **a :** the deportment of an actor or speaker or his expression by means of attitude, voice, and gesture **b :** the style of movement of the feet and legs (as of a horse) **c :** a function of the body or one of its parts
4 : an act of will
5 a : a thing done **:** DEED **b :** the accomplishment of a thing usually over a period of time, in stages, or with the possibility of repetition **c** *plural* **:** BEHAVIOR, CONDUCT ⟨unscrupulous *actions*⟩ **d :** INITIATIVE, ENTERPRISE ⟨a man of *action*⟩
6 a (1) **:** an engagement between troops or ships (2) **:** combat in war ⟨gallantry in *action*⟩ **b** (1) **:** an event or series of events forming a literary composition (2) **:** the unfolding of the events of a drama or work of fiction **:** PLOT (3) **:** the movement of incidents in a plot **c :** the combination of circumstances that constitute the subject matter of a painting or sculpture
7 a : an operating mechanism **b :** the manner in which a mechanism or instrument operates
8 a : the price movement and trading volume of a commodity, security, or market **b :** the process of betting including the offering and acceptance of a bet and determination of a winner **c :** an opportunity for financial gain ⟨a piece of the *action*⟩
9 : the most vigorous, productive, or exciting activity in a particular field, area, or group ⟨they itch to go where the *action* is —D. J. Henahan⟩

ac·tion·able \ˈak-sh(ə-)nə-bəl\ *adjective* (1591)
: subject to or affording ground for an action or suit at law
— ac·tion·ably \-blē\ *adverb*

ac·tion·less \ˈak-shən-ləs\ *adjective* (circa 1817)
: marked by inaction **:** IMMOBILE

action painting *noun* (1952)
: abstract expressionism marked especially by the use of spontaneous techniques (as dribbling, splattering, or smearing)
— action painter *noun*

action potential *noun* (1926)
: a momentary change in electrical potential

(as between the inside of a nerve cell and the extracellular medium) that occurs when a cell or tissue has been activated by a stimulus

ac·ti·vate \'ak-tə-ˌvāt\ *verb* **-vat·ed; -vat·ing** (1626)
transitive verb
: to make active or more active: as **a** (1) **:** to make (as molecules) reactive or more reactive (2) **:** to convert (as a provitamin) into a biologically active derivative **b :** to make (a substance) radioactive **c :** to treat (as carbon or alumina) so as to improve adsorptive properties **d** (1) **:** to set up or formally institute (as a military unit) with the necessary personnel and equipment (2) **:** to put (an individual or unit) on active duty
intransitive verb
: to become active
— **ac·ti·va·tion** \ˌak-tə-'vā-shən\ *noun*
— **ac·ti·va·tor** \'ak-tə-ˌvā-tər\ *noun*

activated carbon *noun* (1921)
: a highly adsorbent powdered or granular carbon made usually by carbonization and chemical activation and used chiefly for purifying by adsorption — called also *activated charcoal*

activation analysis *noun* (1949)
: NEUTRON ACTIVATION ANALYSIS

activation energy *noun* (1940)
: the minimum amount of energy required to convert a normal stable molecule into a reactive molecule

ac·tive \'ak-tiv\ *adjective* [Middle English, from Middle French or Latin; Middle French *actif,* from Latin *activus,* from *actus,* past participle of *agere* to drive, do — more at AGENT] (14th century)
1 : characterized by action rather than by contemplation or speculation
2 : producing or involving action or movement
3 a *of a verb form or voice* **:** asserting that the person or thing represented by the grammatical subject performs the action represented by the verb ⟨*hits* in "he hits the ball" is *active*⟩ **b :** expressing action as distinct from mere existence or state
4 : quick in physical movement **:** LIVELY
5 : marked by vigorous activity **:** BUSY ⟨the stock market was *active*⟩
6 : requiring vigorous action or exertion ⟨*active* sports⟩
7 : having practical operation or results **:** EFFECTIVE ⟨an *active* law⟩
8 a : disposed to action **:** ENERGETIC ⟨took an *active* interest⟩ **b :** engaged in an action or activity ⟨an *active* club member⟩ **c** *of a volcano* **:** currently erupting or likely to erupt — compare DORMANT 2a, EXTINCT 1b **d :** characterized by emission of large amounts of electromagnetic energy ⟨an *active* galactic nucleus⟩
9 : engaged in full-time service especially in the armed forces ⟨*active* duty⟩
10 : marked by present operation, transaction, movement, or use ⟨*active* account⟩
11 a : capable of acting or reacting **:** reacting readily ⟨*active* nitrogen⟩ **b :** tending to progress or to cause degeneration ⟨*active* tuberculosis⟩ **c** *of an electronic circuit element* **:** capable of controlling voltages or currents **d** (1) **:** requiring the expenditure of energy ⟨*active* calcium ion uptake⟩ (2) **:** functioning by the emission of radiant energy ⟨radar is an *active* sensor⟩
12 : still eligible to win the pot in poker
13 : moving down the line **:** visiting in the set — used of couples in contredanses or square dances
— **active** *noun*
— **ac·tive·ly** *adverb*
— **ac·tive·ness** *noun*

active immunity *noun* (circa 1903)
: usually long-lasting immunity that is acquired through production of antibodies within the organism in response to the presence of antigens — compare PASSIVE IMMUNITY

active transport *noun* (1963)
: movement of a chemical substance by the expenditure of energy through a gradient (as across a cell membrane) in concentration or electrical potential and opposite to the direction of normal diffusion

ac·tiv·ism \'ak-ti-ˌvi-zəm\ *noun* (1915)
: a doctrine or practice that emphasizes direct vigorous action especially in support of or opposition to one side of a controversial issue
— **ac·tiv·ist** \-vist\ *noun or adjective*
— **ac·tiv·is·tic** \ˌak-ti-'vis-tik\ *adjective*

ac·tiv·i·ty \ak-'ti-və-tē\ *noun, plural* **-ties** (1530)
1 : the quality or state of being active
2 : vigorous or energetic action **:** LIVELINESS
3 : natural or normal function: as **a :** a process (as digestion) that an organism carries on or participates in by virtue of being alive **b :** a similar process actually or potentially involving mental function; *specifically* **:** an educational procedure designed to stimulate learning by firsthand experience
4 : an active force
5 a : a pursuit in which a person is active **b :** a form of organized, supervised, often extracurricular recreation
6 : an organizational unit for performing a specific function; *also* **:** its function or duties

act of God (circa 1859)
: an extraordinary interruption by a natural cause (as a flood or earthquake) of the usual course of events that experience, prescience, or care cannot reasonably foresee or prevent

ac·to·my·o·sin \ˌak-tə-'mī-ə-sən\ *noun* [International Scientific Vocabulary *act*in + *-o-* + *myosin*] (1942)
: a viscous contractile complex of actin and myosin concerned together with ATP in muscular contraction

ac·tor \'ak-tər *also* -ˌtȯr\ *noun* (15th century)
1 : one that acts **:** DOER
2 a : one who represents a character in a dramatic production **b :** a theatrical performer **c :** one that behaves as if acting a part
3 : one that takes part in any affair
— **ac·tor·ish** \-tə-rish\ *adjective*

act out *transitive verb* (1611)
1 a : to represent in action ⟨children *act out* what they read⟩ **b :** to translate into action ⟨unwilling to *act out* their beliefs⟩
2 : to express (as an impulse or a fantasy) directly in overt behavior without modification to comply with social norms

ac·tress \'ak-trəs\ *noun* (1676)
: a woman who is an actor
— **ac·tressy** \-trə-sē\ *adjective*
usage see -ESS

Acts \'akts\ *noun plural but singular in construction*
: a book in the New Testament narrating the beginnings of the Christian Church — called also *Acts of the Apostles;* see BIBLE table

ac·tu·al \'ak-ch(ə-w)əl, -sh(ə-w)əl\ *adjective* [Middle English *actuel,* from Middle French, from Late Latin *actualis,* from Latin *actus* act] (14th century)
1 *obsolete* **:** ACTIVE
2 a : existing in act and not merely potentially **b :** existing in fact or reality ⟨*actual* and imagined conditions⟩ **c :** not false or apparent ⟨*actual* costs⟩
3 : existing or occurring at the time **:** CURRENT ⟨caught in the *actual* commission of a crime⟩

actual cash value *noun* (circa 1946)
: money equal to the cost of replacing lost, stolen, or damaged property after depreciation

ac·tu·al·i·ty \ˌak-chə-'wa-lə-tē, ˌak-shə-\ *noun, plural* **-ties** (1652)
1 : the quality or state of being actual
2 : something that is actual **:** FACT, REALITY ⟨possible risks which have been seized upon as *actualities* —T. S. Eliot⟩

ac·tu·al·ize \'ak-ch(ə-w)ə-ˌlīz, -sh(ə-w)ə-ˌlīz\ *verb* **-ized; -iz·ing** (1701)
transitive verb
: to make actual **:** REALIZE
intransitive verb
: to become actual
— **ac·tu·al·iza·tion** \ˌak-ch(ə-w)ə-lə-'zā-shən, -sh(ə-w)ə-lə-\ *noun*

ac·tu·al·ly \'ak-ch(ə-w)ə-lē, -sh(ə-w)ə-lē; 'aksh-lē, 'aks-\ *adverb* (15th century)
1 : in act or in fact **:** REALLY ⟨nominally but not *actually* independent —Karl Loewenstein⟩ ⟨don't know how old they *actually* are⟩ ⟨*actually,* they just arrived⟩
2 : in point of fact **:** in truth — used to suggest something unexpected ⟨I have *actually* been invited⟩ ⟨he could *actually* read the Greek⟩

ac·tu·ar·i·al \ˌak-chə-'wer-ē-əl, -shə-\ *adjective* (1869)
1 : of or relating to actuaries
2 : relating to statistical calculation especially of life expectancy
— **ac·tu·ar·i·al·ly** \-ē-ə-lē\ *adverb*

ac·tu·ary \'ak-chə-ˌwer-ē, -shə-\ *noun, plural* **-ar·ies** [Latin *actuarius* shorthand writer, alteration of *actarius,* from *actum* record — more at ACTIVE] (1553)
1 *obsolete* **:** CLERK, REGISTRAR
2 : one who calculates insurance and annuity premiums, reserves, and dividends

ac·tu·ate \'ak-chə-ˌwāt, -shə-\ *transitive verb* **-at·ed; -at·ing** [Medieval Latin *actuatus,* past participle of *actuare* to execute, from Latin *actus* act] (1645)
1 : to put into mechanical action or motion
2 : to move to action
synonym see MOVE
— **ac·tu·a·tion** \ˌak-chə-'wā-shən, -shə-\ *noun*

ac·tu·a·tor \'ak-chə-ˌwā-tər, -shə-\ *noun* (circa 1864)
: one that actuates; *specifically* **:** a mechanical device for moving or controlling something

act up *intransitive verb* (1903)
1 : to act in a way different from that which is normal or expected: as **a :** to behave in an unruly, recalcitrant, or capricious manner **b :** SHOW OFF **c :** to function improperly ⟨this typewriter is *acting up* again⟩
2 : to become active or acute after being quiescent ⟨her rheumatism started to *act up*⟩

acu·ity \ə-'kyü-ə-tē, a-\ *noun, plural* **-ities** [Middle French *acuité,* from Late Latin *acuitat-, acuitas,* from Latin *acuere*] (1543)
: keenness of perception **:** SHARPNESS

acu·le·ate \ə-'kyü-lē-ət\ *adjective* [Latin *aculeatus* having stings, from *aculeus* sting, from *acus*] (1875)
: relating to or being hymenopterans (as bees, ants, and many wasps) of a division (Aculeata) typically having the ovipositor modified into a sting

acu·men \ə-'kyü-mən, 'a-kyə-mən\ *noun* [Latin *acumin-, acumen,* literally, point, from *acuere*] (circa 1580)
: keenness and depth of perception, discernment, or discrimination especially in practical matters **:** SHREWDNESS
synonym see DISCERNMENT

acu·mi·nate \ə-'kyü-mə-nət\ *adjective* (1646)
: tapering to a slender point

acu·pres·sure \'a-kyə-ˌpre-shər, 'a-kə-\ *noun* (1958)
: SHIATSU

acu·punc·ture \-ˌpəŋ(k)-chər\ *noun* [Latin *acus* + English *puncture*] (1684)
: an originally Chinese practice of puncturing the body (as with needles) at specific points to cure disease or relieve pain (as in surgery)
— **acu·punc·tur·ist** \-ˌpəŋ(k)-chə-rist\ *noun*

acute \ə-'kyüt\ *adjective* **acut·er; acut·est** [Latin *acutus,* past participle of *acuere* to sharpen, from *acus* needle; akin to Latin *acer* sharp — more at EDGE] (14th century)

1 a (1) **:** characterized by sharpness or severity ⟨*acute* pain⟩ (2) **:** having a sudden onset, sharp rise, and short course ⟨*acute* disease⟩ **b :** lasting a short time ⟨*acute* experiments⟩
2 : ending in a sharp point: as **a :** being or forming an angle measuring less than 90 degrees ⟨*acute* angle⟩ **b :** composed of acute angles ⟨*acute* triangle⟩
3 a *of an accent mark* **:** having the form ´ **b :** marked with an acute accent **c :** of the variety indicated by an acute accent
4 a : marked by keen discernment or intellectual perception especially of subtle distinctions **:** PENETRATING ⟨an *acute* thinker⟩ **b :** responsive to slight impressions or stimuli ⟨*acute* hearing⟩
5 : felt, perceived, or experienced intensely ⟨*acute* distress⟩
6 : seriously demanding urgent attention ☆
— **acute·ly** *adverb*
— **acute·ness** *noun*

acy·clic \(ˌ)ā-ˈsī-klik, -ˈsi-\ *adjective* (1878)
: not cyclic: as **a :** not disposed in whorls or cycles **b :** having an open-chain structure **:** ALIPHATIC ⟨an *acyclic* compound⟩

acy·clo·vir \(ˌ)ā-ˈsī-klō-ˌvir\ *noun* [[2]*a-* + *cycl-* + *virus*] (1979)
: a cyclic nucleoside $C_8H_{11}N_5O_3$ used especially to treat the symptoms of the genital form of herpes simplex

ac·yl \ˈa-səl\ *noun, often attributive* [International Scientific Vocabulary, from *acid*] (1899)
: a radical RCO— derived usually from an organic acid by removal of the hydroxyl from all acid groups — often used in combination

ac·yl·ate \ˈa-sə-ˌlāt\ *transitive verb* **-at·ed; -at·ing** (1907)
: to introduce an acyl group into
— **ac·yl·a·tion** \ˌa-sə-ˈlā-shən\ *noun*

[1]**ad** \ˈad\ *noun, often attributive* (1841)
1 : ADVERTISEMENT 2
2 : ADVERTISING

[2]**ad** *noun* (1947)
: ADVANTAGE 4

ad- *or* **ac-** *or* **af-** *or* **ag-** *or* **al-** *or* **ap-** *or* **as-** *or* **at-** *prefix* [Middle English, from Middle French, Old French & Latin; Middle French, from Old French, from Latin, from *ad* — more at AT]
1 : to **:** toward — usually *ac-* before *c, k,* or *q* ⟨*ac*culturation⟩ and *af-* before *f* ⟨*af*fluent⟩ and *ag-* before *g* ⟨*ag*gradation⟩ and *al-* before *l* ⟨*al*literation⟩ and *ap-* before *p* ⟨*ap*portion⟩ and *as-* before *s* ⟨*as*suasive⟩ and *at-* before *t* ⟨*at*tune⟩ and *ad-* before other sounds but sometimes *ad-* even before one of the listed consonants ⟨*ad*sorb⟩
2 : near **:** adjacent to — in this sense always in the form *ad-* ⟨*ad*renal⟩

[1]**-ad** \ˌad, əd\ *adverb suffix* [Latin *ad*]
: in the direction of **:** toward ⟨cephal*ad*⟩

[2]**-ad** *noun suffix* [probably from New Latin *-ad-, -as,* from Greek, suffix denoting descent from or connection with]
: member of a botanical group ⟨bromeli*ad*⟩

Ada \ˈā-də\ *trademark*
— used for a structured computer programming language

ad·age \ˈa-dij\ *noun* [Middle French, from Latin *adagium,* from *ad-* + *-agium* (akin to *aio* I say); akin to Greek *ē* he said] (1548)
: a saying often in metaphorical form that embodies a common observation ■

[1]**ada·gio** \ə-ˈdä-j(ē-ˌ)ō, ä-, -zh(ē-ˌ)ō\ *adverb or adjective* [Italian, from *ad* to + *agio* ease] (1724)
: at a slow tempo — used chiefly as a direction in music

[2]**adagio** *noun, plural* **-gios** (1754)
1 : a musical composition or movement in adagio tempo
2 : a ballet duet by a man and woman or a mixed trio displaying difficult feats of balance, lifting, or spinning

[1]**Ad·am** \ˈa-dəm\ *noun* [Middle English, from Late Latin, from Greek, from Hebrew *Ādhām*]
1 : the first man and father by Eve of Cain and Abel
2 : the unregenerate nature of man — used especially in the phrase *the old Adam*
— **Adam·ic** \ə-ˈda-mik\ *or* **Adam·i·cal** \-mi-kəl\ *adjective*

[2]**Adam** *adjective* [Robert *Adam* & James *Adam*] (1872)
: of, relating to, or being an 18th century decorative style (as of furniture) characterized by straight lines, surface decoration, and conventional designs (as festooned garlands and medallions)

ad·a·mance \ˈa-də-mən(t)s\ *noun* (1954)
: ADAMANCY

ad·a·man·cy \-mən(t)-sē\ *noun* [[2]*adamant* + *-cy*] (1937)
: OBSTINACY

adam–and–eve \ˌa-də-mən-ˈ(d)ēv\ *noun* (1807)
: PUTTYROOT

[1]**ad·a·mant** \ˈa-də-mənt, -ˌmant\ *noun* [Middle English, from Middle French, from Latin *adamant-, adamas* hardest metal, diamond, from Greek] (14th century)
1 : a stone (as a diamond) formerly believed to be of impenetrable hardness
2 : an unbreakable or extremely hard substance

[2]**adamant** *adjective* (1923)
: unshakable or immovable especially in opposition **:** UNYIELDING
synonym see INFLEXIBLE
— **ad·a·mant·ly** *adverb*

ad·a·man·tine \ˌa-də-ˈman-ˌtēn, -ˌtīn, -ˈman-tᵊn\ *adjective* [Middle English, from Latin *adamantinus,* from Greek *adamantinos,* from *adamant-, adamas*] (13th century)
1 : made of or having the quality of adamant
2 : rigidly firm **:** UNYIELDING
3 : resembling the diamond in hardness or luster

Adam's apple *noun* (circa 1775)
: the projection in the front of the neck formed by the largest cartilage of the larynx

Adam's needle *noun* (circa 1760)
: an often cultivated yucca (*Yucca filamentosa*) of coastal pine barrens of the eastern U.S. with a basal rosette of sharp-tipped leaves having loose threads along the margins

adapt \ə-ˈdapt, a-\ *verb* [French or Latin; French *adapter,* from Latin *adaptare,* from *ad-* + *aptare* to fit, from *aptus* apt, fit] (15th century)
transitive verb
: to make fit (as for a specific or new use or situation) often by modification
intransitive verb
: to become adapted ☆
— **adapt·ed·ness** *noun*

adapt·able \ə-ˈdap-tə-bəl, a-\ *adjective* (1800)
: capable of being adapted **:** SUITABLE
synonym see PLASTIC
— **adapt·abil·i·ty** \-ˌdap-tə-ˈbi-lə-tē\ *noun*

ad·ap·ta·tion \ˌa-ˌdap-ˈtā-shən, -dəp-\ *noun* (1610)
1 : the act or process of adapting **:** the state of being adapted
2 : adjustment to environmental conditions: as **a :** adjustment of a sense organ to the intensity or quality of stimulation **b :** modification of an organism or its parts that makes it more fit for existence under the conditions of its environment
3 : something that is adapted; *specifically* **:** a composition rewritten into a new form
— **ad·ap·ta·tion·al** \-shnəl, -shə-nᵊl\ *adjective*
— **ad·ap·ta·tion·al·ly** *adverb*

adapt·er *also* **adap·tor** \ə-ˈdap-tər, a-\ *noun* (1801)
1 : one that adapts
2 a : a device for connecting two parts (as of different diameters) of an apparatus **b :** an attachment for adapting apparatus for uses not originally intended

adap·tion \ə-ˈdap-shən, a-\ *noun* (1704)
: ADAPTATION

adap·tive \ə-ˈdap-tiv, a-\ *adjective* (1824)
: showing or having a capacity for or tendency toward adaptation
— **adap·tive·ly** *adverb*
— **adap·tive·ness** *noun*
— **ad·ap·tiv·i·ty** \ˌa-ˌdap-ˈti-və-tē\ *noun*

adaptive radiation *noun* (1902)
: evolutionary diversification of a generalized ancestral form with production of a number of adaptively specialized forms

Adar \ä-ˈdär, ˈä-\ *noun* [Middle English, from Hebrew *Ădhār*] (14th century)
: the 6th month of the civil year or the 12th month of the ecclesiastical year in the Jewish calendar — see MONTH table

Adar She·ni \ä-ˌdär-shā-ˈnē\ *noun* [Hebrew *Ădhār Shēnī* second Adar] (circa 1901)
: VEADAR

ad·ax·i·al \(ˌ)a-ˈdak-sē-əl\ *adjective* (circa 1900)

☆ SYNONYMS

Acute, critical, crucial mean of uncertain outcome. ACUTE stresses intensification of conditions leading to a culmination or breaking point ⟨an *acute* housing shortage⟩. CRITICAL adds to ACUTE implications of imminent change, of attendant suspense, and of decisiveness in the outcome ⟨the war has entered a *critical* phase⟩. CRUCIAL suggests a dividing of the ways and often a test or trial involving the determination of a future course or direction ⟨a *crucial* vote⟩. See in addition SHARP.

Adapt, adjust, accommodate, conform, reconcile mean to bring one thing into correspondence with another. ADAPT implies a modification according to changing circumstances ⟨*adapted* themselves to the warmer climate⟩. ADJUST suggests bringing into a close and exact correspondence or harmony such as exists between parts of a mechanism ⟨*adjusted* the budget to allow for inflation⟩. ACCOMMODATE may suggest yielding or compromising to effect a correspondence ⟨*accommodated* his political beliefs in order to win⟩. CONFORM applies to bringing into accordance with a pattern, example, or principle ⟨refused to *conform* to society's idea of morality⟩. RECONCILE implies the demonstration of the underlying compatibility of things that seem to be incompatible ⟨tried to *reconcile* what they said with what I knew⟩.

□ USAGE

adage Quite a few commentators call *old adage* redundant because they say *adage* means "an old saying," a phrase Noah Webster did use in his definition of 1828. But *old* has been absent from Merriam-Webster definitions for more than 50 years—adages simply are not always old ⟨the popular *adage* is that 12 years is the ideal median age —Michael Jackson⟩. Even so, *old* has gone with *adage* from the beginning ⟨that old *ad age* 'Much courtesy, much subtlety' —Thomas Nashe (1594)⟩. It is sometimes used actually to impute age ⟨racegoers remembered the old *adage,* "Second in the Trial, first in the Derby" —Audax Minor⟩ and is sometimes merely casual ⟨the old *adage* "misery loves company" —Natalie Babbitt⟩. It may be used where it seems apt.

: situated on the same side as or facing the axis (as of an organ) ⟨the *adaxial* or upper surface of a leaf⟩

add \ˈad\ *verb* [Middle English, from Latin *addere,* from *ad-* + *-dere* to put — more at DO] (14th century)

transitive verb

1 : to join or unite so as to bring about an increase or improvement ⟨*adds* 60 acres to his land⟩ ⟨wine *adds* a creative touch to cooking⟩

2 : to say further **:** APPEND

3 : to combine (numbers) into an equivalent simple quantity or number

4 : to include as a member of a group ⟨don't forget to *add* me in⟩

intransitive verb

1 a : to perform addition **b :** to come together or unite by addition

2 a : to serve as an addition ⟨the movie will *add* to his fame⟩ **b :** to make an addition ⟨*added* to her savings⟩

— add·able *or* **add·ible** \ˈa-də-bəl\ *adjective*

ad·dax \ˈa-ˌdaks\ *noun, plural* **ad·dax·es** [Latin] (1693)

: a large light-colored Saharan antelope (*Addax nasomaculatus*) that has long spiralling horns

ad·dend \ˈa-ˌdend, ə-ˈdend\ *noun* [short for *addendum*] (1674)

: a number to be added to another

ad·den·dum \ə-ˈden-dəm\ *noun, plural* **-den·da** \-ˈden-də\ [Latin, neuter of *addendus,* gerundive of *addere*] (1684)

1 : a thing added **:** ADDITION

2 : a supplement to a book — often used in plural but singular in construction

[1]ad·der \ˈa-dər\ *noun* [Middle English, alteration (by false division of *a naddre*) of *naddre,* from Old English *nǣdre;* akin to Old High German *nātara* adder, Latin *natrix* water snake] (14th century)

1 : the common venomous viper (*Vipera berus*) of Europe; *broadly* **:** a terrestrial viper (family Viperidae)

2 : any of several North American snakes (as the hognose snakes) that are harmless but are popularly believed to be venomous

adder 1

[2]add·er \ˈa-dər\ *noun* (1580)

: one that adds; *especially* **:** a device (as in a computer) that performs addition

ad·der's–tongue \ˈa-dərz-ˌtəŋ\ *noun* (1578)

1 : any of a genus (*Ophioglossum,* family Ophioglossaceae) of small ferns having a spore-bearing stalk resembling a serpent's tongue

2 : DOGTOOTH VIOLET

[1]ad·dict \ə-ˈdikt\ *transitive verb* [Latin *addictus,* past participle of *addicere* to favor, from *ad-* + *dicere* to say — more at DICTION] (1534)

1 : to devote or surrender (oneself) to something habitually or obsessively ⟨*addicted* to gambling⟩

2 : to cause addiction to a substance in

[2]ad·dict \ˈa-(ˌ)dikt\ *noun* (1909)

1 : one who is addicted to a substance

2 : DEVOTEE ⟨a detective novel *addict*⟩

ad·dic·tion \ə-ˈdik-shən, a-\ *noun* (1599)

1 : the quality or state of being addicted ⟨*addiction* to reading⟩

2 : compulsive need for and use of a habit-forming substance (as heroin, nicotine, or alcohol) characterized by tolerance and by well-defined physiological symptoms upon withdrawal; *broadly* **:** persistent compulsive use of a substance known by the user to be harmful

ad·dic·tive \-ˈdik-tiv\ *adjective* (1939)

: causing or characterized by addiction

Ad·di·son's disease \ˈa-də-sənz-\ *noun* [Thomas *Addison* (died 1860) English physician] (circa 1856)

: a destructive disease marked by deficient adrenocortical secretion and characterized by extreme weakness, loss of weight, low blood pressure, gastrointestinal disturbances, and brownish pigmentation of the skin and mucous membranes

ad·di·tion \ə-ˈdi-shən, a-\ *noun* [Middle English, from Middle French, from Latin *addition-, additio,* from *addere*] (14th century)

1 : a part added (as to a building or residential section)

2 : the result of adding **:** INCREASE

3 : the act or process of adding; *especially* **:** the operation of combining numbers so as to obtain an equivalent simple quantity

4 : direct chemical combination of substances into a single product

— in addition : [2]BESIDES, ALSO

— in addition to : combined or associated with **:** [1]BESIDES 2

ad·di·tion·al \-ˈdish-nəl, -ˈdi-shə-nᵊl\ *adjective* (1646)

: existing by way of addition **:** ADDED

ad·di·tion·al·ly \-ˈdish-nə-lē, -ˈdi-shən-lē, ˈdi-shə-nᵊl-ē\ *adverb* (circa 1665)

: in or by way of addition **:** FURTHERMORE

[1]ad·di·tive \ˈa-də-tiv\ *adjective* (1699)

1 : of, relating to, or characterized by addition

2 : produced by addition

3 : characterized by, being, or producing effects (as drug responses or gene products) that when the causative factors act together are the sum of their individual effects

— ad·di·tive·ly *adverb*

— ad·di·tiv·i·ty \ˌa-də-ˈti-və-tē\ *noun*

[2]additive *noun* (1945)

: a substance added to another in relatively small amounts to effect a desired change in properties ⟨food *additives*⟩

additive identity *noun* (1960)

: an identity element (as 0 in the group of whole numbers under the operation of addition) that in a given mathematical system leaves unchanged any element to which it is added

additive inverse *noun* (1958)

: a number that when added to a given number gives zero ⟨the *additive inverse* of 4 is −4⟩ — compare OPPOSITE 3

[1]ad·dle \ˈa-dᵊl\ *adjective* [Middle English *adel* filth, from Old English *adela;* akin to Middle Low German *adele* liquid manure] (1592)

1 *of an egg* **:** ROTTEN

2 : CONFUSED

[2]addle *verb* **ad·dled; ad·dling** \ˈad-liŋ, ˈa-dᵊl-iŋ\ (circa 1712)

transitive verb

: to throw into confusion **:** CONFOUND

intransitive verb

1 : to become rotten **:** SPOIL

2 : to become confused

ad·dle·pat·ed \ˈa-dᵊl-ˌpā-təd\ *adjective* (1630)

1 : being mixed up **:** CONFUSED

2 : ECCENTRIC

[1]add–on *noun* (1946)

: something added on: as **a :** a sum or amount added on **b :** something (as an accessory or added feature) that enhances the thing it is added to

[2]add–on \ˈa-ˌdän, -ˌdȯn\ *adjective* (1955)

1 : being or able to be added on

2 : able to be added to ⟨*add-on* certificates of deposit⟩

[1]ad·dress \ə-ˈdres, a- *also* ˈa-ˌdres\ *verb* [Middle English *adressen,* from Middle French *adresser,* from *a-* (from Latin *ad-*) + *dresser* to arrange — more at DRESS] (14th century)

transitive verb

1 *archaic* **a :** DIRECT, AIM **b :** to direct to go **:** SEND

2 a : to direct the efforts or attention of (oneself) ⟨will *address* himself to the problem⟩ **b :** to deal with **:** TREAT ⟨intrigued by the chance to *address* important issues —I. L. Horowitz⟩

3 *archaic* **:** to make ready; *especially* **:** DRESS

4 a : to communicate directly ⟨*addresses* his thanks to his host⟩ **b :** to speak or write directly to; *especially* **:** to deliver a formal speech to

5 a : to mark directions for delivery on ⟨*address* a letter⟩ **b :** to consign to the care of another (as an agent or factor)

6 : to greet by a prescribed form

7 : to adjust the club preparatory to hitting (a golf ball)

8 : to identify (as a peripheral or memory location) by an address or a name for information transfer

intransitive verb

obsolete **:** to direct one's speech or attentions

— ad·dress·er *noun*

[2]ad·dress \ə-ˈdres, *for 5 & 7 & 4 also* ˈa-ˌdres\ *noun* (1539)

1 : dutiful and courteous attention especially in courtship — usually used in plural

2 a : readiness and capability for dealing (as with a person or problem) skillfully and smoothly **:** ADROITNESS **b** *obsolete* **:** a making ready; *also* **:** a state of preparedness

3 a : manner of bearing oneself ⟨a man of rude *address*⟩ **b :** manner of speaking or singing **:** DELIVERY

4 : a formal communication; *especially* **:** a prepared speech delivered to a special audience or on a special occasion

5 a : a place where a person or organization may be communicated with **b :** directions for delivery on the outside of an object (as a letter or package) **c :** the designation of place of delivery placed between the heading and salutation on a business letter

6 : a preparatory position of the player and club in golf

7 : a location (as in the memory of a computer) where particular information is stored; *also* **:** the digits that identify such a location

synonym see TACT

ad·dress·able \ə-ˈdre-sə-bəl\ *adjective* (1953)

1 : able to be addressed **:** directly accessible ⟨*addressable* registers in a computer⟩

2 : of or relating to a subscription television system that uses decoders addressable by the system operator

— ad·dress·abil·i·ty \ə-ˌdre-sə-ˈbi-lə-tē\ *noun*

ad·dress·ee \ˌa-ˌdre-ˈsē, ə-ˌdre-ˈsē\ *noun* (1810)

: one to whom something is addressed

ad·duce \ə-ˈdüs *also* -ˈdyüs\ *transitive verb* **ad·duced; ad·duc·ing** [Latin *adducere,* literally, to lead to, from *ad-* + *ducere* to lead — more at TOW] (15th century)

: to offer as example, reason, or proof in discussion or analysis

— ad·duc·er *noun*

[1]ad·duct \ə-ˈdəkt, a-\ *transitive verb* [Latin *adductus,* past participle of *adducere*] (circa 1839)

: to draw (as a limb) toward or past the median axis of the body; *also* **:** to bring together (similar parts) ⟨*adduct* the fingers⟩

— ad·duc·tive \-ˈdək-tiv\ *adjective*

[2]ad·duct \ˈa-ˌdəkt\ *noun* [German *Addukt,* from Latin *adductus*] (1941)

: a chemical addition product

ad·duc·tion \ə-ˈdək-shən, a-\ *noun* (14th century)

1 : the action of adducting **:** the state of being adducted

2 : the act or action of adducing

ad·duc·tor \-ˈdək-tər\ *noun* [New Latin, from Latin, one that draws to, from *adductus*] (1615)

1 : a muscle that draws a part toward the median line of the body or toward the axis of an extremity

2 : a muscle that closes the valves of a bivalve mollusk

add up (1850)
intransitive verb
1 a : to come to the expected total ⟨the bill doesn't *add up*⟩ **b :** to form an intelligible pattern **:** make sense ⟨her story just doesn't *add up*⟩
2 a : AMOUNT 1b — used with *to* ⟨the play *adds up* to a lot of laughs⟩ **b :** to amount to a lot ⟨just a little each time, but it all *adds up*⟩
transitive verb
: to form an opinion of ⟨*added* him *up* at a glance⟩

-ade *noun suffix* [Middle English, from Middle French, from Old Provençal *-ada,* from Late Latin *-ata,* from Latin, feminine of *-atus* -ate]
1 : act **:** action ⟨block*ade*⟩
2 : product; *especially* **:** sweet drink ⟨lime*ade*⟩

Adé·lie penguin \ə-ˈdā-lē-\ *noun* [*Adélie* Coast, Antarctica] (1907)
: a small antarctic penguin (*Pygoscelis adeliae*) — called also *Adélie*

Adélie penguin

-adelphous *adjective combining form* [probably from New Latin *-adelphus,* from Greek *adelphos* brother, from *ha-, a-* together (akin to *homos* same) + *delphys* womb — more at SAME, DOLPHIN]
: having (such or so many) stamen fascicles ⟨mon*adelphous*⟩

aden- *or* **adeno-** *combining form* [New Latin, from Greek, from *aden-, adēn;* akin to Latin *inguen* groin]
: gland ⟨*aden*ine⟩ **:** adenoid ⟨*adeno*virus⟩

ad·e·nine \ˈa-dᵊn-ˌēn\ *noun* [International Scientific Vocabulary, from its presence in glandular tissue] (1885)
: a purine base $C_5H_5N_5$ that codes hereditary information in the genetic code in DNA and RNA — compare CYTOSINE, GUANINE, THYMINE, URACIL

ad·e·ni·tis \ˌa-dᵊn-ˈī-təs\ *noun* [New Latin] (circa 1848)
: inflammation of a gland; *especially* **:** LYMPHADENITIS

ad·e·no·car·ci·no·ma \ˌa-dᵊn-(ˌ)ō-ˌkär-sᵊn-ˈō-mə\ *noun* [New Latin] (circa 1889)
: a malignant tumor originating in glandular epithelium
— ad·e·no·car·ci·no·ma·tous \-mə-təs\ *adjective*

ad·e·no·hy·poph·y·sis \-hī-ˈpä-fə-səs\ *noun, plural* **-y·ses** \-fə-ˌsēz\ [New Latin] (1935)
: the anterior glandular lobe of the pituitary gland
— ad·e·no·hy·poph·y·se·al \-(ˌ)hī-ˌpä-fə-ˈsē-əl\ *or* **ad·e·no·hy·po·phys·i·al** \-ˌhī-pə-ˈfi-zē-əl\ *adjective*

¹ad·e·noid \ˈa-dᵊn-ˌȯid, ˈad-ˌnȯid\ *noun* [Greek *adenoeidēs* glandular, from *adēn*] (circa 1890)
: an enlarged mass of lymphoid tissue at the back of the pharynx characteristically obstructing breathing — usually used in plural

²adenoid *adjective* (circa 1947)
1 : of or relating to the adenoids
2 : relating to, affected with, or associated with abnormally enlarged adenoids ⟨a severe adenoid condition⟩ ⟨*adenoid* facies⟩

ad·e·noi·dal \ˌa-dᵊn-ˈȯi-dᵊl\ *adjective* (1919)
: exhibiting the characteristics (as snoring, mouth breathing, and voice nasality) of one affected with abnormally enlarged adenoids **:** ADENOID ⟨an *adenoidal* tenor⟩ — not usually used technically

ad·e·no·ma \ˌa-dᵊn-ˈō-mə\ *noun, plural* **-mas** *also* **-ma·ta** \-mə-tə\ [New Latin *adenomat-, adenoma*] (1870)
: a benign tumor of a glandular structure or of glandular origin
— ad·e·no·ma·tous \-mə-təs\ *adjective*

aden·o·sine \ə-ˈde-nə-ˌsēn, -sən\ *noun* [International Scientific Vocabulary, blend of *adenine* and *ribose*] (circa 1909)
: a nucleoside $C_{10}H_{13}N_5O_4$ that is a constituent of RNA yielding adenine and ribose on hydrolysis

adenosine diphosphate *noun* (1938)
: ADP

adenosine mo·no·phos·phate \-ˌmä-nə-ˈfäs-ˌfāt, -ˌmō-\ *noun* (1950)
: AMP

adenosine 3′,5′-monophosphate \-ˌthrē-ˌfīv-\ *noun* (1970)
: CYCLIC AMP

adenosine tri·phos·pha·tase \-trī-ˈfäs-fə-ˌtās, -ˌtāz\ *noun* (1943)
: ATPASE

adenosine tri·phos·phate \-trī-ˈfäs-ˌfāt\ *noun* (1938)
: ATP

ad·e·no·vi·rus \ˌa-dᵊn-ō-ˈvī-rəs\ *noun* (1956)
: any of a group of DNA-containing viruses originally identified in human adenoid tissue, causing respiratory diseases (as catarrh), and including some capable of inducing malignant tumors in experimental animals
— ad·e·no·vi·ral \-rəl\ *adjective*

ad·e·nyl·ate cy·clase \ə-ˌde-nᵊl-ət-ˈsī-ˌklās, -ˌāt-, -ˌklāz; ˌa-dᵊn-ˌi-lət-, -ə-ˌlāt-\ *noun* (1968)
: an enzyme that catalyzes the formation of cyclic AMP from ATP

ad·e·nyl cyclase \ˈa-dᵊn-ˌil-\ *noun* [*aden*ine + *-yl*] (1968)
: ADENYLATE CYCLASE

ad·e·nyl·ic acid \ˈa-dᵊn-ˌi-lik-\ *noun* (1894)
: AMP

¹ad·ept \ˈa-ˌdept, ə-ˈdept, a-ˈ\ *noun* [New Latin *adeptus* alchemist who has attained the knowledge of how to change base metals into gold, from Latin, past participle of *adipisci* to attain, from *ad-* + *apisci* to reach — more at APT] (1709)
: a highly skilled or well-trained individual **:** EXPERT ⟨an *adept* at chess⟩

²adept \ə-ˈdept *also* ˈa-ˌdept\ *adjective* (circa 1691)
: thoroughly proficient **:** EXPERT
synonym see PROFICIENT
— adept·ly \ə-ˈdep-(t)lē, a-\ *adverb*
— adept·ness \-ˈdep(t)-nəs\ *noun*

ad·e·qua·cy \ˈa-di-kwə-sē\ *noun, plural* **-cies** (1808)
: the quality or state of being adequate

ad·e·quate \-kwət\ *adjective* [Latin *adaequatus,* past participle of *adaequare* to make equal, from *ad-* + *aequare* to equal — more at EQUABLE] (circa 1617)
1 : sufficient for a specific requirement ⟨*adequate* taxation of goods⟩; *also* **:** barely sufficient or satisfactory ⟨her first performance was merely *adequate*⟩
2 : lawfully and reasonably sufficient
synonym see SUFFICIENT
— ad·e·quate·ly *adverb*
— ad·e·quate·ness *noun*

ad eun·dem \ˌa-dē-ˈən-dəm\ *or* **ad eundem gra·dum** \-ˈgrā-dəm\ *adverb or adjective* [New Latin *ad eundem gradum*] (1711)
: to, in, or of the same rank — used especially of the honorary granting of academic standing or a degree by a university to one whose actual work was done elsewhere

¹à deux \(ˌ)a-ˈdə(r), (ˌ)à-dœ\ *adjective* [French] (1886)
: involving two people especially in private ⟨a cozy evening *à deux*⟩

²à deux *adverb* (1927)
: privately or intimately with only two present ⟨dined *à deux*⟩

ad·here \ad-ˈhir, əd-\ *verb* **ad·hered; ad·her·ing** [Middle French or Latin; Middle French *adhérer,* from Latin *adhaerēre,* from *ad-* + *haerēre* to stick] (1536)
intransitive verb
1 : to give support or maintain loyalty
2 *obsolete* **:** ACCORD 3
3 : to hold fast or stick by or as if by gluing, suction, grasping, or fusing
4 : to bind oneself to observance
transitive verb
: to cause to stick fast
synonym see STICK

ad·her·ence \-ˈhir-ən(t)s\ *noun* (1531)
1 : the act, action, or quality of adhering
2 : steady or faithful attachment **:** FIDELITY

¹ad·her·ent \ad-ˈhir-ənt, əd-\ *adjective* [Middle English, from Middle French or Latin; Middle French *adhérent,* from Latin *adhaerent-, adhaerens,* present participle of *adhaerēre*] (15th century)
1 : able or tending to adhere
2 : connected or associated with especially by contract
3 : ADNATE
— ad·her·ent·ly *adverb*

²adherent *noun* (15th century)
: one that adheres: as **a :** a follower of a leader, party, or profession **b :** a believer in or advocate especially of a particular idea or church
synonym see FOLLOWER

ad·he·sion \ad-ˈhē-zhən, əd-\ *noun* [French or Latin; French *adhésion,* from Latin *adhaesion-, adhaesio,* from *adhaerēre*] (1624)
1 : steady or firm attachment **:** ADHERENCE
2 : the action or state of adhering
3 : the abnormal union of separate tissue surfaces by new fibrous tissue resulting from an inflammatory process; *also* **:** the newly formed uniting tissue
4 : agreement to join ⟨*adhesion* of all nations to a copyright convention⟩
5 : the molecular attraction exerted between the surfaces of bodies in contact
— ad·he·sion·al \-ˈhēzh-nəl, -ˈhē-zhə-nᵊl\ *adjective*

¹ad·he·sive \-ˈhē-siv, -ziv\ *adjective* (1670)
1 : tending to remain in association or memory
2 : tending to adhere or cause adherence
3 : prepared for adhering
— ad·he·sive·ly *adverb*
— ad·he·sive·ness *noun*

²adhesive *noun* (1912)
1 : an adhesive substance (as glue or cement)
2 : a postage stamp with a gummed back

adhesive binding *noun* (1955)
: PERFECT BINDING
— ad·he·sive–bound \-ˌbau̇nd\ *adjective*

adhesive tape *noun* (1928)
: tape coated on one side with an adhesive mixture; *especially* **:** one used for covering wounds

¹ad hoc \ˈad-ˈhäk, -ˈhōk; ˈäd-ˈhōk\ *adverb* [Latin, for this] (1659)
: for the particular end or case at hand without consideration of wider application

²ad hoc *adjective* (1879)
1 a : concerned with a particular end or purpose ⟨an *ad hoc* investigating committee⟩ **b :** formed or used for specific or immediate problems or needs ⟨*ad hoc* solutions⟩
2 : fashioned from whatever is immediately available **:** IMPROVISED ⟨large *ad hoc* parades and demonstrations —Nat Hentoff⟩

¹ad ho·mi·nem \(ˈ)ad-ˈhä-mə-ˌnem, -nəm\ *adjective* [New Latin, literally, to the person] (1598)
1 : appealing to feelings or prejudices rather than intellect
2 : marked by an attack on an opponent's character rather than by an answer to the contentions made

²ad hominem *adverb* (1962)
: in an ad hominem manner ⟨was arguing *ad hominem*⟩

adi·a·bat·ic \ˌa-dē-ə-ˈba-tik, ˌā-ˌdī-ə-\ *adjective* [Greek *adiabatos* impassable, from *a-* + *diabatos* passable, from *diabainein* to go across, from *dia-* + *bainein* to go — more at COME] (1870)
: occurring without loss or gain of heat ⟨*adiabatic* expansion of a body of air⟩
— adi·a·bat·i·cal·ly \-ti-k(ə-)lē\ *adverb*

adieu \ə-ˈdü, a-, -ˈdyü\ *noun, plural* **adieus** *or* **adieux** \-ˈdüz, -ˈdyüz\ [Middle English, from Middle French, from *a* (from Latin *ad*) + *Dieu* God, from Latin *Deus* — more at DEITY] (14th century)
: FAREWELL — often used interjectionally

ad in·fi·ni·tum \ˌad-ˌin-fə-ˈnī-təm *also* ˌäd-\ *adverb or adjective* [Latin] (1610)
: without end or limit

¹ad in·ter·im \ˈad-ˈin-tə-rəm, -ˌrim *also* ˈäd-\ *adverb* [Latin] (1787)
: for the intervening time **:** TEMPORARILY

²ad interim *adjective* (1818)
: made or serving ad interim

adi·os \ˌä-dē-ˈōs, ˌa-\ *interjection* [Spanish *adiós,* from *a* (from Latin *ad*) + *Dios* God, from Latin *Deus*] (1837)
— used to express farewell

adip- *or* **adipo-** *combining form* [Latin *adip-, adeps,* probably from Greek *aleipha* fat, oil, from *aleiphein* to rub with oil — more at ALIPHATIC]
: fat ⟨*adipo*cyte⟩

adip·ic acid \ə-ˈdi-pik-\ *noun* [International Scientific Vocabulary] (1877)
: a white crystalline dicarboxylic acid $C_6H_{10}O_4$ formed by oxidation of various fats and also made synthetically for use especially in the manufacture of nylon

ad·i·po·cyte \ˈa-di-pō-ˌsīt\ *noun* (1959)
: FAT CELL

ad·i·pose \ˈa-də-ˌpōs\ *adjective* [New Latin *adiposus,* from Latin *adip-, adeps*] (1743)
: of or relating to animal fat; *broadly* **:** FAT
— ad·i·pos·i·ty \ˌa-də-ˈpä-sə-tē\ *noun*

adipose tissue *noun* (1854)
: connective tissue in which fat is stored and which has the cells distended by droplets of fat

ad·it \ˈa-dət\ *noun* [Latin *aditus* approach, from *adire* to go to, from *ad-* + *ire* to go — more at ISSUE] (1602)
: a nearly horizontal passage from the surface in a mine

ad·ja·cen·cy \ə-ˈjā-sᵊn(t)-sē\ *noun, plural* **-cies** (1646)
1 : something that is adjacent
2 : the quality or state of being adjacent **:** CONTIGUITY

ad·ja·cent \ə-ˈjā-sᵊnt\ *adjective* [Middle English, from Middle French or Latin; Middle French, from Latin *adjacent-, adjacens,* present participle of *adjacēre* to lie near, from *ad-* + *jacēre* to lie; akin to Latin *jacere* to throw — more at JET] (15th century)
1 a : not distant **:** NEARBY ⟨the city and *adjacent* suburbs⟩ **b :** having a common endpoint or border ⟨*adjacent* lots⟩ ⟨*adjacent* sides of a triangle⟩ **c :** immediately preceding or following
2 *of two angles* **:** having the vertex and one side in common ☆
— ad·ja·cent·ly *adverb*

ad·jec·ti·val \ˌa-jik-ˈtī-vəl\ *adjective* (1797)
1 : ADJECTIVE
2 : characterized by the use of adjectives
— ad·jec·ti·val·ly \-və-lē\ *adverb*

¹ad·jec·tive \ˈa-jik-tiv *also* ˈa-jə-tiv\ *adjective* [Middle English, from Middle French or Late Latin; Middle French *adjectif,* from Late Latin *adjectivus,* from Latin *adjectus,* past participle of *adjicere* to throw to, from *ad-* + *jacere* to throw — more at JET] (14th century)
1 : of, relating to, or functioning as an adjective ⟨an *adjective* clause⟩
2 : not standing by itself **:** DEPENDENT
3 : requiring or employing a mordant ⟨*adjective* dyes⟩
4 : PROCEDURAL ⟨*adjective* law⟩
— ad·jec·tive·ly *adverb*

²adjective *noun* (14th century)
: a word belonging to one of the major form classes in any of numerous languages and typically serving as a modifier of a noun to denote a quality of the thing named, to indicate its quantity or extent, or to specify a thing as distinct from something else

ad·join \ə-ˈjȯin, a-\ *verb* [Middle English, from Middle French *adjoindre,* from Latin *adjungere,* from *ad-* + *jungere* to join — more at YOKE] (14th century)
transitive verb
1 : to add or attach by joining
2 : to lie next to or in contact with
intransitive verb
: to be close to or in contact with one another

ad·join·ing *adjective* (15th century)
: touching or bounding at a point or line
synonym see ADJACENT

ad·joint \ˈa-ˌjȯint\ *noun* [French, from past participle of *adjoindre* to adjoin] (1907)
: the transpose of a matrix in which each element is replaced by its cofactor

ad·journ \ə-ˈjərn\ *verb* [Middle English *ajournen,* from Middle French *ajourner,* from *a-* (from Latin *ad-*) + *jour* day — more at JOURNEY] (15th century)
transitive verb
: to suspend indefinitely or until a later stated time
intransitive verb
1 : to suspend a session indefinitely or to another time or place
2 : to move to another place

ad·journ·ment \-mənt\ *noun* (1607)
1 : the act of adjourning
2 : the state or interval of being adjourned

ad·judge \ə-ˈjəj\ *transitive verb* **ad·judged; ad·judg·ing** [Middle English *ajugen,* from Middle French *ajugier,* from Latin *adjudicare,* from *ad-* + *judicare* to judge — more at JUDGE] (14th century)
1 a : to decide or rule upon as a judge **:** ADJUDICATE **b :** to pronounce judicially **:** RULE
2 *archaic* **:** SENTENCE, CONDEMN
3 : to hold or pronounce to be **:** DEEM ⟨*adjudge* the book a success⟩
4 : to award or grant judicially in a case of controversy

ad·ju·di·cate \ə-ˈjü-di-ˌkāt\ *verb* **-cat·ed; -cat·ing** (1775)
transitive verb
: to settle judicially
intransitive verb
: to act as judge
— ad·ju·di·ca·tive \-ˌkā-tiv, -kə-\ *adjective*
— ad·ju·di·ca·tor \-ˌkā-tər\ *noun*

ad·ju·di·ca·tion \ə-ˌjü-di-ˈkā-shən\ *noun* [French or Late Latin; French, from Late Latin *adjudicatio,* from Latin *adjudicare*] (1691)
1 : the act or process of adjudicating
2 a : a judicial decision or sentence **b :** a decree in bankruptcy
— ad·ju·di·ca·to·ry \-ˈjü-di-kə-ˌtōr-ē, -ˌtȯr-\ *adjective*

¹ad·junct \ˈa-ˌjəŋ(k)t\ *noun* [Latin *adjunctum,* from neuter of *adjunctus,* past participle of *adjungere*] (1588)
1 : something joined or added to another thing but not essentially a part of it
2 a : a word or word group that qualifies or completes the meaning of another word or other words and is not itself a main structural element in its sentence **b :** an adverb or adverbial (as *heartily* in "Most children eat heartily" or *at noon* in "We will leave at noon") attached to the verb of a clause especially to express a relation of time, place, frequency, degree, or manner — compare DISJUNCT 2
3 : an associate or assistant of another
— ad·junc·tive \a-ˈjəŋ(k)-tiv, ə-\ *adjective*

²adjunct *adjective* (1595)
1 : added or joined as an accompanying object or circumstance
2 : attached in a subordinate or temporary capacity to a staff ⟨an *adjunct* psychiatrist⟩
— ad·junct·ly \ˈa-ˌjəŋ(k)-tlē, -ˌjəŋ-klē\ *adverb*

ad·junc·tion \a-ˈjəŋ(k)-shən\ *noun* (1618)
: the act or process of adjoining

ad·ju·ra·tion \ˌa-jə-ˈrā-shən\ *noun* (1611)
1 : a solemn oath
2 : an earnest urging or advising
— ad·jur·a·to·ry \ə-ˈju̇r-ə-ˌtōr-ē, -ˌtȯr-\ *adjective*

ad·jure \ə-ˈju̇r\ *transitive verb* **ad·jured; ad·jur·ing** [Middle English, from Middle French & Latin; Middle French *ajurer,* from Latin *adjurare,* from *ad-* + *jurare* to swear — more at JURY] (14th century)
1 : to command solemnly under or as if under oath or penalty of a curse
2 : to urge or advise earnestly
synonym see BEG

ad·just \ə-ˈjəst\ *verb* [Middle English *ajusten,* from Middle French *ajuster* to gauge, adjust, from *a-* (from Latin *ad-*) + *juste* right, exact — more at JUST] (14th century)
transitive verb
1 a : to bring to a more satisfactory state: (1) **:** SETTLE, RESOLVE (2) **:** RECTIFY **b :** to make correspondent or conformable **:** ADAPT **c :** to bring the parts of to a true or more effective relative position ⟨*adjust* a carburetor⟩
2 : to reduce to a system **:** REGULATE
3 : to determine the amount to be paid under an insurance policy in settlement of (a loss)
intransitive verb
1 : to adapt or conform oneself (as to new conditions)
2 : to achieve mental and behavioral balance between one's own needs and the demands of others
synonym see ADAPT
— ad·just·abil·i·ty \-ˌjəs-tə-ˈbi-lə-tē\ *noun*
— ad·just·able \-ˈjəs-tə-bəl\ *adjective*
— ad·jus·tive \-ˈjəs-tiv\ *adjective*

adjustable rate mortgage *noun* (1981)
: a mortgage having an interest rate which is usually initially lower than that of a mortgage with a fixed rate but is adjusted periodically according to the cost of funds to the lender

ad·just·ed *adjective* (circa 1674)
1 : accommodated to suit a particular set of circumstances or requirements
2 : having achieved an often specified and usually harmonious relationship with the environment or with other individuals ⟨a well-*adjusted* schoolchild⟩

ad·just·er *also* **ad·jus·tor** \ə-ˈjəs-tər\ *noun* (1673)
: one that adjusts; *especially* **:** an insurance agent who investigates personal or property damage and makes estimates for effecting settlements

ad·just·ment \ə-ˈjəs(t)-mənt\ *noun* (1644)
1 : the act or process of adjusting
2 : a settlement of a claim or debt in a case in which the amount involved is uncertain or full payment is not made
3 : the state of being adjusted
4 : a means (as a mechanism) by which things are adjusted one to another

☆ SYNONYMS
Adjacent, adjoining, contiguous, juxtaposed mean being in close proximity. ADJACENT may or may not imply contact but always implies absence of anything of the same kind in between ⟨a house with an *adjacent* garage⟩. ADJOINING definitely implies meeting and touching at some point or line ⟨had *adjoining* rooms at the hotel⟩. CONTIGUOUS implies having contact on all or most of one side ⟨offices in all 48 *contiguous* states⟩. JUXTAPOSED means placed side by side especially so as to permit comparison and contrast ⟨a skyscraper *juxtaposed* to a church⟩.

5 : a correction or modification to reflect actual conditions
— **ad·just·men·tal** \ə-ˌjəs(t)-ˈmen-tᵊl, ˌa-ˌjəs(t)-\ *adjective*

ad·ju·tan·cy \ˈa-jə-tən(t)-sē\ *noun* (1775)
: the office or rank of an adjutant

ad·ju·tant \ˈa-jə-tənt\ *noun* [Latin *adjutant-, adjutans,* present participle of *adjutare* to help — more at AID] (1539)
1 : a staff officer in the army, air force, or marine corps who assists the commanding officer and is responsible especially for correspondence
2 : one who helps **:** ASSISTANT

adjutant general *noun, plural* **adjutants general** (1645)
1 : the chief administrative officer of an army who is responsible especially for the administration and preservation of personnel records
2 : the chief administrative officer of a major military unit (as a division or corps)

¹ad·ju·vant \ˈa-jə-vənt\ *adjective* [French or Latin; French, from Latin *adjuvant-, adjuvans,* present participle of *adjuvare* to aid — more at AID] (1574)
1 : serving to aid or contribute **:** AUXILIARY
2 : assisting in the prevention, amelioration, or cure of disease ⟨*adjuvant* chemotherapy following surgery⟩

²adjuvant *noun* (1609)
: one that helps or facilitates: as **a :** an ingredient (as in a prescription or a solution) that modifies the action of the principal ingredient **b :** something (as a drug or method) that enhances the effectiveness of medical treatment **c :** a substance enhancing the immune response to an antigen

Ad·le·ri·an \ad-ˈlir-ē-ən, ad-\ *adjective* [Alfred *Adler*] (1924)
: of, relating to, or being a theory and technique of psychotherapy emphasizing the importance of feelings of inferiority, a will to power, and overcompensation in neurotic processes

¹ad–lib \ˈad-ˈlib\ *verb* **ad–libbed; ad–lib·bing** [*ad lib*] (1919)
transitive verb
: to deliver spontaneously
intransitive verb
: to improvise especially lines or a speech
— **ad–lib** *noun*

²ad–lib *adjective* (1935)
: spoken, composed, or performed without preparation

ad lib *adverb* [New Latin *ad libitum*] (circa 1811)
1 : in accordance with one's wishes
2 : without restraint or limit

¹ad li·bi·tum \(ˌ)ad-ˈli-bə-təm\ *adverb* [New Latin, in accordance with desire] (1610)
: AD LIB ⟨rats fed *ad libitum*⟩

²ad libitum *adjective* (circa 1801)
: omissible according to a performer's wishes — used as a direction in music; compare OBBLIGATO

ad·man \ˈad-ˌman\ *noun* (1909)
: a person who writes, solicits, or places advertisements

ad·mass \ˈad-ˌmas\ *noun, often attributive* [advertising + *mass*] (1955)
chiefly British **:** mass-media advertising; *also* **:** the society influenced by it

ad·mea·sure \ad-ˈme-zhər, -ˈmā-\ *transitive verb* **-sured; -sur·ing** [Middle English *amesuren,* from Middle French *amesurer,* from *a-* (from Latin *ad-*) + *mesurer* to measure] (1641)
: to determine the proper share of **:** APPORTION

ad·mea·sure·ment \-ˈme-zhər-mənt, -ˈmā-\ *noun* (1523)
1 : determination and apportionment of shares
2 : determination or comparison of dimensions
3 : DIMENSIONS, SIZE

Ad·me·tus \ad-ˈmē-təs\ *noun* [Latin, from Greek *Admētos*]
: a king of Pherae who is saved by Apollo from his fated death when his wife Alcestis offers to die in his place

ad·min·is·ter \əd-ˈmi-nə-stər\ *verb* **-is·tered; -is·ter·ing** \-st(ə-)riŋ\ [Middle English *administren,* from Middle French *administrer,* from Latin *administrare,* from *ad-* + *ministrare* to serve, from *minister* servant — more at MINISTER] (14th century)
transitive verb
1 : to manage or supervise the execution, use, or conduct of ⟨*administer* a trust fund⟩
2 a : to mete out **:** DISPENSE ⟨*administer* punishment⟩ **b :** to give ritually ⟨*administer* the last rites⟩ **c :** to give remedially ⟨*administer* a dose of medicine⟩
intransitive verb
1 : to perform the office of administrator
2 : to furnish a benefit **:** MINISTER ⟨*administer* to an ailing friend⟩
3 : to manage affairs
— **ad·min·is·tra·ble** \-strə-bəl\ *adjective*
— **ad·min·is·trant** \-strənt\ *noun*

ad·min·is·trate \-ˌstrāt\ *verb* **-trat·ed; -trat·ing** [Latin *administratus,* past participle of *administrare*] (circa 1617)
: ADMINISTER

ad·min·is·tra·tion \əd-ˌmi-nə-ˈstrā-shən, (ˌ)ad-\ *noun* (14th century)
1 : performance of executive duties **:** MANAGEMENT
2 : the act or process of administering
3 : the execution of public affairs as distinguished from policy-making
4 a : a body of persons who administer **b** *often capitalized* **:** a group constituting the political executive in a presidential government **c :** a governmental agency or board
5 : the term of office of an administrative officer or body

ad·min·is·tra·tive \əd-ˈmi-nə-ˌstrā-tiv, -strə-\ *adjective* (circa 1731)
: of or relating to administration or an administration **:** EXECUTIVE
— **ad·min·is·tra·tive·ly** *adverb*

administrative county *noun* (1949)
: a British local administrative unit often not coincident with an older county

administrative law *noun* (1896)
: law dealing with the establishment, duties, and powers of and available remedies against authorized agencies in the executive branch of the government

ad·min·is·tra·tor \əd-ˈmi-nə-ˌstrā-tər, -ˌstrā-ˌtȯr\ *noun* (15th century)
1 : a person legally vested with the right of administration of an estate
2 a : one that administers especially business, school, or governmental affairs **b :** a priest appointed to administer a diocese or parish temporarily

ad·min·is·tra·trix \-ˌmi-nə-ˈstrā-triks\ *noun, plural* **-tra·tri·ces** \-ˈstrā-trə-ˌsēz\ [New Latin] (circa 1623)
: a woman administrator especially of an estate

ad·mi·ra·ble \ˈad-m(ə-)rə-bəl\ *adjective* (15th century)
1 : deserving the highest esteem **:** EXCELLENT
2 *obsolete* **:** exciting wonder **:** SURPRISING
— **ad·mi·ra·bil·i·ty** \ˌad-m(ə-)rə-ˈbi-lə-tē\ *noun*
— **ad·mi·ra·ble·ness** \ˈad-m(ə-)rə-bəl-nəs\ *noun*
— **ad·mi·ra·bly** \-blē\ *adverb*

ad·mi·ral \ˈad-m(ə-)rəl\ *noun* [Middle English, from Middle French *amiral* admiral & Medieval Latin *admiralis* emir, *admirallus* admiral, from Arabic *amīr-al-* commander of the (as in *amīr-al-baḥr* commander of the sea)] (15th century)
1 *archaic* **:** the commander in chief of a navy
2 a : FLAG OFFICER **b :** a commissioned officer in the navy or coast guard who ranks above a vice admiral and whose insignia is four stars — compare GENERAL
3 *archaic* **:** FLAGSHIP
4 : any of several brightly colored butterflies (family Nymphalidae) ◆

admiral of the fleet (1660)
: the highest-ranking officer of the British navy

ad·mi·ral·ty \ˈad-m(ə-)rəl-tē\ *noun* (15th century)
1 *capitalized* **:** the executive department or officers formerly having general authority over British naval affairs
2 : the court having jurisdiction over questions of maritime law; *also* **:** the system of law administered by admiralty courts

ad·mi·ra·tion \ˌad-mə-ˈrā-shən\ *noun* (15th century)
1 *archaic* **:** WONDER
2 : an object of esteem
3 : delighted or astonished approbation

ad·mire \əd-ˈmīr\ *verb* **ad·mired; ad·mir·ing** [Middle French *admirer,* from Latin *admirari,* from *ad-* + *mirari* to wonder, from *mirus* astonishing] (1579)
transitive verb
1 *archaic* **:** to marvel at
2 : to regard with admiration
intransitive verb
dialect **:** to like very much ⟨I would *admire* to know why not —A. H. Lewis⟩
synonym see REGARD
— **ad·mir·er** *noun*
— **ad·mir·ing·ly** \-ˈmī-riŋ-lē\ *adverb*

ad·mis·si·ble \əd-ˈmi-sə-bəl, ad-\ *adjective* [French, from Medieval Latin *admissibilis,* from Latin *admissus,* past participle of *admittere*] (circa 1611)
1 : capable of being allowed or conceded **:** PERMISSIBLE ⟨evidence legally *admissible* in court⟩
2 : capable or worthy of being admitted ⟨*admissible* to the university⟩
— **ad·mis·si·bil·i·ty** \-ˌmi-sə-ˈbi-lə-tē\ *noun*

ad·mis·sion \əd-ˈmi-shən, ad-\ *noun* (15th century)
1 a : the act or process of admitting **b :** the state or privilege of being admitted **c :** a fee paid at or for admission
2 a : the granting of an argument or position not fully proved **b :** acknowledgment that a fact or statement is true
synonym see ADMITTANCE

◇ WORD HISTORY

admiral It is a curiosity of history that *admiral,* a word meaning "naval commander," ultimately has its source in Arabic, the language of a desert people who acquired their seafaring skills largely from the Mediterranean peoples they dominated after the great Islamic expansion of the 7th century. As a name for a Muslim chieftain, the Arabic word *amīr* "commander" appears as early as the 9th century in Medieval Latin documents, in forms such as *amiratus, admirandus,* and *admirallus.* These display a variety of suffixes and an extraneous *d* through confusion with the Latin verb *admirari* "to admire." The ending *-allus* is probably from the Arabic definite article *al,* which actually belongs to the following word in phrases such as *amīr al-'alī* "supreme commander." The specific application of *admirallus* to a commander of a fleet originated in 12th century Sicily, when it was ruled by the Normans. This usage was acquired by the Genoese and then spread to the rest of western Europe, including France and England.

— **ad•mis•sive** \-'mi-siv\ *adjective*

ad•mit \əd-'mit, ad-\ *verb* **ad•mit•ted; ad•mit•ting** [Middle English *admitten,* from Latin *admittere,* from *ad-* + *mittere* to send] (15th century)
transitive verb
1 a : to allow scope for **:** PERMIT ⟨*admits* no possibility of misunderstanding⟩ **b :** to concede as true or valid ⟨*admitted* making a mistake⟩
2 : to allow entry (as to a place, fellowship, or privilege) ⟨an open window had *admitted* rain⟩ ⟨*admitted* to the club⟩
intransitive verb
1 : to give entrance or access
2 a : ALLOW, PERMIT ⟨*admits* of two interpretations⟩ **b :** to make acknowledgment — used with *to*
synonym see ACKNOWLEDGE

ad•mit•tance \əd-'mi-tᵊn(t)s, ad-\ *noun* (1536)
1 : permission to enter (as a place or office) **:** ENTRANCE
2 : the reciprocal of the impedance of a circuit ☆

ad•mit•ted•ly \əd-'mi-təd-lē, ad-\ *adverb* (1804)
1 : as has been or must be admitted ⟨an *admittedly* inadequate treatment⟩
2 : it must be admitted ⟨*admittedly,* we took a chance⟩

ad•mix \ad-'miks\ *transitive verb* [back-formation from obsolete *admixt* mingled (with), from Middle English, from Latin *admixtus*] (1533)
: to mix in

ad•mix•ture \ad-'miks-chər\ *noun* [Latin *admixtus,* past participle of *admiscēre* to mix with, from *ad-* + *miscēre* to mix — more at MIX] (1605)
1 a : the action of mixing **b :** the fact of being mixed
2 a : something added by mixing **b :** a product of mixing **:** MIXTURE

ad•mon•ish \ad-'mä-nish\ *transitive verb* [Middle English *admonesten,* from Middle French *admonester,* from (assumed) Vulgar Latin *admonestare,* alteration of Latin *admonēre* to warn, from *ad-* + *monēre* to warn — more at MIND] (14th century)
1 a : to indicate duties or obligations to **b :** to express warning or disapproval to especially in a gentle, earnest, or solicitous manner
2 : to give friendly earnest advice or encouragement to
synonym see REPROVE
— **ad•mon•ish•er** *noun*
— **ad•mon•ish•ing•ly** \-ni-shiŋ-lē\ *adverb*
— **ad•mon•ish•ment** \-mənt\ *noun*

ad•mo•ni•tion \,ad-mə-'ni-shən\ *noun* [Middle English *amonicioun,* from Middle French *amonition,* from Latin *admonition-, admonitio,* from *admonēre*] (14th century)
1 : gentle or friendly reproof
2 : counsel or warning against fault or oversight

ad•mon•i•to•ry \əd-'mä-nə-,tōr-ē, -,tȯr-\ *adjective* (1594)
: expressing admonition **:** WARNING
— **ad•mon•i•to•ri•ly** \-,mä-nə-'tōr-ə-lē, -'tȯr-\ *adverb*

ad•nate \'ad-,nāt\ *adjective* [Latin *adnatus, adgnatus,* past participle of *adgnasci* to be born in addition, grow later — more at AGNATE] (1661)
: grown to a usually unlike part especially along a margin ⟨a calyx *adnate* to the ovary⟩
— **ad•na•tion** \ad-'nā-shən\ *noun*

ad nau•se•am \ad-'nȯ-zē-əm *also* -,am\ *adverb* [Latin] (1647)
: to a sickening or excessive degree

ad•nexa \ad-'nek-sə\ *noun plural* [New Latin, from Latin *annexa,* neuter plural of *annexus,* past participle of *annectere* to bind to — more at ANNEX] (1899)
: conjoined, subordinate, or associated anatomic parts
— **ad•nex•al** \-səl\ *adjective*

ado \ə-'dü\ *noun* [Middle English, from *at do,* from *at* + *don, do* to do] (14th century)
1 : fussy bustling excitement **:** TO-DO
2 : time-wasting bother over trivial details ⟨wrote the paper without further *ado*⟩
3 : TROUBLE, DIFFICULTY

ado•be \ə-'dō-bē\ *noun* [Spanish, from Arabic *aṭ-ṭūb* the brick, from Coptic *tōbe* brick] (1748)
1 : a brick or building material of sun-dried earth and straw
2 : a structure made of adobe bricks
3 : a heavy clay used in making adobe bricks; *broadly* **:** alluvial or playa clay in desert or arid regions
— **ado•be•like** \-,līk\ *adjective*

adobe 2

ado•bo \ə-'dō-bō, ä-'t͟hō-bō\ *noun, plural* **-bos** [Spanish] (circa 1951)
: a Philippine dish of fish or meat marinated in a sauce usually containing vinegar and garlic, browned in fat, and simmered in the marinade

ad•o•les•cence \,a-dᵊl-'e-sᵊn(t)s\ *noun* (15th century)
1 : the state or process of growing up
2 : the period of life from puberty to maturity terminating legally at the age of majority
3 : a stage of development (as of a language or culture) prior to maturity

¹ad•o•les•cent \-sᵊnt\ *noun* [French, from Latin *adolescent-, adolescens,* present participle of *adolescere* to grow up — more at ADULT] (15th century)
: one that is in the state of adolescence

²adolescent *adjective* (1785)
1 : of, relating to, or being in adolescence
2 : emotionally or intellectually immature
— **ad•o•les•cent•ly** *adverb*

Ado•nai \,ä-də-'nȯi, -'nī\ *noun* [Hebrew *'ădhōnāy*] (14th century)
— used as a name of the God of the Hebrews

Ado•nis \ə-'dä-nəs, -'dō-\ *noun* [Latin, from Greek *Adōnis*]
1 : a youth loved by Aphrodite who is killed at hunting by a wild boar and restored to Aphrodite from Hades for a part of each year
2 : a very handsome young man

adopt \ə-'däpt\ *transitive verb* [Middle English, from Middle French or Latin; Middle French *adopter,* from Latin *adoptare,* from *ad-* + *optare* to choose] (1500)
1 : to take by choice into a relationship; *especially* **:** to take voluntarily (a child of other parents) as one's own child
2 : to take up and practice or use ⟨*adopted* a moderate tone⟩
3 : to accept formally and put into effect ⟨*adopt* a constitutional amendment⟩
4 : to choose (a textbook) for required study in a course ☆
— **adopt•abil•i•ty** \-,däp-tə-'bi-lə-tē\ *noun*
— **adopt•able** \-'däp-tə-bəl\ *adjective*
— **adopt•er** *noun*

adopt•ee \ə-,däp-'tē\ *noun* (1892)
: one that is adopted

adop•tion \ə-'däp-shən\ *noun* (14th century)
: the act of adopting **:** the state of being adopted

adop•tion•ism *or* **adop•tian•ism** \-shə-,ni-zəm\ *noun, often capitalized* (1874)
: the doctrine that Jesus of Nazareth became the Son of God by adoption
— **adop•tion•ist** \-sh(ə-)nist\ *noun, often capitalized*

adop•tive \ə-'däp-tiv\ *adjective* (15th century)
1 : made or acquired by adoption ⟨the *adoptive* father⟩
2 : tending to adopt
3 : of or relating to adoption
— **adop•tive•ly** *adverb*

ador•able \ə-'dōr-ə-bəl, -'dȯr-\ *adjective* (1611)
1 : worthy of being adored
2 : extremely charming ⟨an *adorable* child⟩
— **ador•abil•i•ty** \-,dōr-ə-'bi-lə-tē, -,dȯr-\ *noun*
— **ador•able•ness** \-'dōr-ə-bəl-nəs, -'dȯr-\ *noun*
— **ador•ably** \-blē\ *adverb*

ad•o•ra•tion \,a-də-'rā-shən\ *noun* (1528)
: the act of adoring **:** the state of being adored

adore \ə-'dōr, -'dȯr\ *transitive verb* **adored; ador•ing** [Middle English *adouren,* from Middle French *adorer,* from Latin *adorare,* from *ad-* + *orare* to speak, pray — more at ORATION] (14th century)
1 : to worship or honor as a deity or as divine
2 : to regard with loving admiration and devotion ⟨*adored* his daughter⟩
3 : to be extremely fond of ⟨*adores* pecan pie⟩
synonym see REVERE
— **ador•er** *noun*
— **ador•ing•ly** *adverb*

adorn \ə-'dȯrn\ *transitive verb* [Middle English, from Middle French *adorner,* from Latin *adornare,* from *ad-* + *ornare* to furnish — more at ORNATE] (14th century)
1 : to enhance the appearance of especially with beautiful objects
2 : to enliven or decorate as if with ornaments ⟨people of fashion who *adorned* the Court⟩ ☆

☆ SYNONYMS
Admittance, admission mean permitted entrance. ADMITTANCE is usually applied to mere physical entrance to a locality or a building ⟨members must show their cards upon *admittance* to the club⟩. ADMISSION applies to entrance or formal acceptance (as into a club) that carries with it rights, privileges, standing, or membership ⟨two recommendations are required for *admission* to the club⟩.

Adopt, embrace, espouse mean to take an opinion, policy, or practice as one's own. ADOPT implies accepting something created by another or foreign to one's nature ⟨forced to *adopt* new policies⟩. EMBRACE implies a ready or happy acceptance ⟨*embraced* the customs of their new homeland⟩. ESPOUSE adds an implication of close attachment to a cause and a sharing of its fortunes ⟨*espoused* the cause of women's rights⟩.

Adorn, decorate, ornament, embellish, beautify, deck, garnish mean to enhance the appearance of something by adding something unessential. ADORN implies an enhancing by something beautiful in itself ⟨a diamond necklace *adorned* her neck⟩. DECORATE suggests relieving plainness or monotony by adding beauty of color or design ⟨*decorate* a birthday cake⟩. ORNAMENT and EMBELLISH imply the adding of something extraneous, ORNAMENT stressing the heightening or setting off of the original ⟨a white house *ornamented* with green shutters⟩, EMBELLISH often stressing the adding of superfluous or adventitious ornament ⟨*embellish* a page with floral borders⟩. BEAUTIFY adds to EMBELLISH a suggestion of counterbalancing plainness or ugliness ⟨will *beautify* the grounds with flower beds⟩. DECK implies the addition of something that contributes to gaiety, splendor, or showiness ⟨a house all *decked* out for Christmas⟩. GARNISH suggests decorating with a small final touch and is used especially in referring to the serving of food ⟨an entrée *garnished* with parsley⟩.

adorn·ment \-mənt\ *noun* (14th century)
1 : the action of adorning **:** the state of being adorned
2 : something that adorns

ADP \ˌā-(ˌ)dē-ˈpē\ *noun* [*a*denosine *d*i*p*hosphate] (1944)
: an ester of adenosine that is reversibly converted to ATP for the storing of energy by the addition of a high-energy phosphate group — called also *adenosine diphosphate*

ad rem \(ˌ)ad-ˈrem\ *adverb or adjective* [Latin, to the thing] (1599)
: to the point or purpose **:** RELEVANTLY

adren- *or* **adreno-** *combining form* [*adrenal*]
1 : adrenal glands ⟨*adreno*cortical⟩
2 : adrenaline ⟨*adren*ergic⟩

[1]ad·re·nal \ə-ˈdrē-nᵊl\ *adjective* [*ad-* + *renal*] (1875)
: of, relating to, or derived from the adrenal glands or their secretions ⟨*adrenal* steroids⟩

[2]adrenal *noun* (1882)
: ADRENAL GLAND

ad·re·nal·ec·to·my \ə-ˌdrē-nᵊl-ˈek-tə-mē\ *noun* (circa 1910)
: surgical removal of one or both adrenal glands
— ad·re·nal·ec·to·mized \-ˌmīzd\ *adjective*

adrenal gland *noun* (1875)
: either of a pair of complex endocrine organs near the anterior medial border of the kidney consisting of a mesodermal cortex that produces glucocorticoid, mineralocorticoid, and androgenic hormones and an ectodermal medulla that produces epinephrine and norepinephrine — called also *adrenal, suprarenal gland*

Adren·a·lin \ə-ˈdre-nᵊl-ən\ *trademark*
— used for a preparation of levorotatory epinephrine

adren·a·line \ə-ˈdre-nᵊl-ən\ *noun* (1901)
: EPINEPHRINE — often used in nontechnical contexts ⟨the fans were jubilant, raucous, their *adrenaline* running high —W. P. Kinsella⟩

ad·ren·er·gic \ˌa-drə-ˈnər-jik\ *adjective* [*adren-* + *ergic*] (1934)
1 : liberating or activated by adrenaline or a substance like adrenaline ⟨an *adrenergic* nerve⟩
2 : resembling adrenaline especially in physiological action ⟨*adrenergic* drugs⟩
— ad·ren·er·gi·cal·ly \-ji-k(ə-)lē\ *adverb*

ad·re·no·chrome \ə-ˈdrē-nō-ˌkrōm\ *noun* (circa 1913)
: a red-colored mixture of quinones derived from epinephrine by oxidation

ad·re·no·cor·ti·cal \ə-ˌdrē-nō-ˈkȯr-ti-kəl\ *adjective* (1936)
: of, relating to, or derived from the cortex of the adrenal glands

ad·re·no·cor·ti·co·ste·roid \-ˌkȯr-ti-kō-ˈstir-ˌȯid *also* -ˈster-\ *noun* (1960)
: a steroid obtained from, resembling, or having physiological effects like those of the adrenal cortex

ad·re·no·cor·ti·co·tro·pic \ə-ˈdrē-nō-ˌkȯr-ti-kō-ˈtrō-pik\ *also* **ad·re·no·cor·ti·co·tro·phic** \-ˈtrō-fik\ *adjective* (1936)
: acting on or stimulating the adrenal cortex ⟨*adrenocorticotropic* activity⟩

adrenocorticotropic hormone *noun* (1937)
: ACTH

ad·re·no·cor·ti·co·tro·pin \-ˈtrō-pən\ *also* **ad·re·no·cor·ti·co·tro·phin** \-ˈtrō-fən\ *noun* (1952)
: ACTH

Adria·my·cin \ˌā-drē-ə-ˈmī-sᵊn, ˌa-drē-\ *trademark*
— used for a preparation of the hydrochloride of doxorubicin

adrift \ə-ˈdrift\ *adverb or adjective* (1624)
1 : without motive power and without anchor or mooring
2 : without ties, guidance, or security
3 : free from restraint or support

adroit \ə-ˈdrȯit\ *adjective* [French, from Old French, from *a-* (from Latin *ad-*) + *droit* right, droit] (1652)
: having or showing skill, cleverness, or resourcefulness in handling situations ⟨an *adroit* leader⟩
synonym see CLEVER, DEXTEROUS
— adroit·ly *adverb*
— adroit·ness *noun*

ad·sci·ti·tious \ˌad-sə-ˈti-shəs\ *adjective* [Latin *adscitus,* from past participle of *adsciscere* to admit, adopt, from *ad-* + *sciscere* to get to know, from *scire* to know — more at SCIENCE] (1620)
: derived or acquired from something extrinsic **:** ADVENTITIOUS

ad·sorb \ad-ˈsȯrb, -ˈzȯrb\ *verb* [*ad-* + ab*sorb*] (1882)
transitive verb
: to take up and hold by adsorption
intransitive verb
: to become adsorbed
— ad·sorb·able \-ˈsȯr-bə-bəl, -ˈzȯr-\ *adjective*
— ad·sorb·er \-ˈsȯr-bər, -ˈzȯr-\ *noun*

ad·sor·bate \ad-ˈsȯr-bət, -ˈzȯr-, -ˌbāt\ *noun* (1928)
: an adsorbed substance

ad·sor·bent \-bənt\ *noun* (1917)
: a usually solid substance that adsorbs another substance
— adsorbent *adjective*

ad·sorp·tion \ad-ˈsȯrp-shən, -ˈzȯrp-\ *noun* [*ad-* + ab*sorption*] (1882)
: the adhesion in an extremely thin layer of molecules (as of gases, solutes, or liquids) to the surfaces of solid bodies or liquids with which they are in contact — compare ABSORPTION
— ad·sorp·tive \-ˈsȯrp-tiv, -ˈzȯrp-\ *adjective*

ad·u·lar·ia \ˌa-jə-ˈlar-ē-ə, ˌa-dyə-, -ˈler-\ *noun* [Italian, from French *adulaire,* from *Adula,* Swiss mountain group] (1798)
: a transparent or translucent orthoclase

ad·u·late \ˈa-jə-ˌlāt, ˈa-dyə-; ˈa-dᵊl-ˌāt\ *transitive verb* **-lat·ed; -lat·ing** [back-formation from *adulation,* from Middle English, from Middle French, from Latin *adulation-, adulatio,* from *adulari* to fawn on (of dogs), flatter] (1777)
: to flatter or admire excessively or slavishly
— ad·u·la·tion \ˌa-jə-ˈlā-shən, ˌa-dyə-; ˌa-dᵊl-ˈā-\ *noun*
— ad·u·la·tor \ˈa-jə-ˌlā-tər, ˈa-dyə-; ˈa-dᵊl-ˌā-\ *noun*
— ad·u·la·to·ry \-lə-ˌtȯr-ē, -ˌtȯr-\ *adjective*

[1]adult \ə-ˈdəlt, ˈa-ˌdəlt\ *adjective* [Latin *adultus,* past participle of *adolescere* to grow up, from *ad-* + *-olescere* (from *alescere* to grow) — more at OLD] (1531)
1 : fully developed and mature **:** GROWN-UP
2 : of, relating to, intended for, or befitting adults ⟨an *adult* approach to a problem⟩
3 : dealing in or with explicitly sexual material ⟨*adult* bookstores⟩ ⟨*adult* movies⟩
— adult·hood \ə-ˈdəlt-ˌhu̇d\ *noun*
— adult·ly \ə-ˈdəlt-lē, ˈa-ˌdəlt-\ *adverb*
— adult·ness \ə-ˈdəlt-nəs, ˈa-ˌdəlt-\ *noun*

[2]adult *noun* (1658)
: one that is adult; *especially* **:** a human being after an age (as 21) specified by law
— adult·like \ə-ˈdəlt-ˌlīk\ *adjective*

adult education *noun* (1851)
: CONTINUING EDUCATION

adul·ter·ant \ə-ˈdəl-t(ə-)rənt\ *noun* (circa 1755)
: an adulterating substance or agent
— adulterant *adjective*

[1]adul·ter·ate \ə-ˈdəl-tə-ˌrāt\ *transitive verb* **-at·ed; -at·ing** [Latin *adulteratus,* past participle of *adulterare,* from *ad-* + *alter* other — more at ELSE] (1531)
: to corrupt, debase, or make impure by the addition of a foreign or inferior substance or element; *especially* **:** to prepare for sale by replacing more valuable with less valuable or inert ingredients
— adul·ter·a·tor \-ˌrā-tər\ *noun*

[2]adul·ter·ate \ə-ˈdəl-t(ə-)rət\ *adjective* (1590)
1 : tainted with adultery **:** ADULTEROUS
2 : being adulterated **:** SPURIOUS

adul·ter·a·tion \ə-ˌdəl-tə-ˈrā-shən\ *noun* (1506)
1 : the process of adulterating **:** the condition of being adulterated
2 : an adulterated product

adul·ter·er \ə-ˈdəl-tər-ər\ *noun* (1513)
: a person who commits adultery; *especially* **:** a man who commits adultery

adul·ter·ess \ə-ˈdəl-t(ə-)rəs\ *noun* (1611)
: a woman who commits adultery

adul·ter·ine \ə-ˈdəl-tə-ˌrīn, -ˌrēn\ *adjective* (1542)
1 a : marked by adulteration **:** SPURIOUS **b :** ILLEGAL
2 : born of adultery

adul·ter·ous \ə-ˈdəl-t(ə-)rəs\ *adjective* (1606)
: relating to, characterized by, or given to adultery
— adul·ter·ous·ly *adverb*

adul·tery \ə-ˈdəl-t(ə-)rē\ *noun, plural* **-ter·ies** [Middle English, alteration of *avoutrie,* from Middle French, from Latin *adulterium,* from *adulter* adulterer, back-formation from *adulterare*] (15th century)
: voluntary sexual intercourse between a married man and someone other than his wife or between a married woman and someone other than her husband; *also* **:** an act of adultery

ad·um·brate \ˈa-dəm-ˌbrāt, a-ˈdəm-\ *transitive verb* **-brat·ed; -brat·ing** [Latin *adumbratus,* past participle of *adumbrare,* from *ad-* + *umbra* shadow — more at UMBRAGE] (1581)
1 : to foreshadow vaguely **:** INTIMATE
2 a : to give a sketchy representation or outline of **b :** to suggest or disclose partially
3 : OVERSHADOW, OBSCURE
— ad·um·bra·tion \ˌa-(ˌ)dəm-ˈbrā-shən\ *noun*
— ad·um·bra·tive \a-ˈdəm-brə-tiv\ *adjective*
— ad·um·bra·tive·ly *adverb*

adust \ə-ˈdəst\ *adjective* [Middle English, from Latin *adustus,* past participle of *adurere* to set fire to, from *ad-* + *urere* to burn — more at EMBER] (15th century)
1 : SCORCHED, BURNED
2 *archaic* **:** of a sunburned appearance
3 *archaic* **:** of a gloomy appearance or disposition

ad va·lo·rem \ˌad-və-ˈlōr-əm, -ˈlȯr-\ *adjective* [Latin, according to the value] (1698)
: imposed at a rate percent of value ⟨*ad valorem* tax on goods⟩

[1]ad·vance \əd-ˈvan(t)s\ *verb* **ad·vanced; ad·vanc·ing** [Middle English *advauncen,* from Old French *avancier,* from (assumed) Vulgar Latin *abantiare,* from Late Latin *abante* in front, from Latin *ab-* + *ante* before — more at ANTE-] (15th century)
transitive verb
1 : to accelerate the growth or progress of
2 : to bring or move forward
3 : to raise to a higher rank
4 *archaic* **:** to lift up **:** RAISE
5 a : to bring forward in time; *especially* **:** to make earlier ⟨*advance* the date of the meeting⟩ **b :** to place later in time
6 : to bring forward for notice, consideration, or acceptance **:** PROPOSE
7 : to supply or furnish in expectation of repayment
8 : to raise in rate **:** INCREASE ⟨*advance* the rent⟩

intransitive verb
1 : to move forward **:** PROCEED
2 : to make progress **:** INCREASE ⟨*advance* in age⟩
3 : to rise in rank, position, or importance
4 : to rise in rate or price ☆
— ad•vanc•er *noun*

²advance *noun* (1668)
1 : a moving forward
2 a : progress in development ⟨mistaking material *advance* for spiritual enrichment —H. J. Laski⟩ **b :** a progressive step **:** IMPROVEMENT ⟨an *advance* in medical technique⟩
3 : a rise in price, value, or amount
4 : a first step or approach made ⟨her attitude discouraged all *advances*⟩
5 : a provision of something (as money or goods) before a return is received; *also* **:** the money or goods supplied
— in advance : before a deadline or an anticipated event
— in advance of : AHEAD OF

³advance *adjective* (1701)
1 : made, sent, or furnished ahead of time ⟨*advance* sales⟩
2 : going or situated before ⟨an *advance* party of soldiers⟩

ad•vanced *adjective* (1534)
1 : far on in time or course ⟨a man *advanced* in years⟩
2 a : being beyond others in progress or ideas ⟨tastes a bit too *advanced* for the times⟩ **b :** being beyond the elementary or introductory ⟨*advanced* chemistry⟩ **c :** greatly developed beyond an initial stage ⟨the most *advanced* scientific methods⟩ ⟨*advanced* weapons systems⟩

advanced degree *noun* (1928)
: a university degree (as a master's or doctor's degree) higher than a bachelor's

Advanced level *noun* (1947)
: A LEVEL

advance man *noun* (1906)
: an employee who makes arrangements and handles publicity in advance of an appearance or engagement by the employer (as a political candidate or a circus)

ad•vance•ment \əd-ˈvan(t)-smənt\ *noun* (1599)
1 : the action of advancing **:** the state of being advanced: **a :** promotion or elevation to a higher rank or position **b :** progression to a higher stage of development
2 : an improved feature **:** IMPROVEMENT

¹ad•van•tage \əd-ˈvan-tij\ *noun* [Middle English *avantage,* from Middle French, from *avant* before, from Late Latin *abante*] (1523)
1 : superiority of position or condition ⟨higher ground gave the enemy the *advantage*⟩
2 : a factor or circumstance of benefit to its possessor ⟨lacked the *advantages* of an education⟩
3 a : BENEFIT, GAIN; *especially* **:** benefit resulting from some course of action ⟨a mistake which turned out to our *advantage*⟩ **b** *obsolete* **:** INTEREST 2a
4 : the first point won in tennis after deuce
— to advantage : so as to produce a favorable impression or effect

²advantage *transitive verb* **-taged; -tag•ing** (1549)
: to give an advantage to **:** BENEFIT

ad•van•ta•geous \ˌad-ˌvan-ˈtā-jəs, -vən-\ *adjective* (1598)
: giving an advantage **:** FAVORABLE
— ad•van•ta•geous•ly *adverb*
— ad•van•ta•geous•ness *noun*

ad•vec•tion \ad-ˈvek-shən\ *noun* [Latin *advection-, advectio* act of bringing, from *advehere* to carry to, from *ad-* + *vehere* to carry — more at WAY] (1910)
: the usually horizontal movement of a mass of fluid (as air or an ocean current); *also* **:** transport (as of pollutants or plankton) by such movement
— ad•vect \-ˈvekt\ *transitive verb*
— ad•vec•tive \-ˈvek-tiv\ *adjective*

Ad•vent \ˈad-ˌvent, *chiefly British* -vənt\ *noun* [Middle English, from Medieval Latin *adventus,* from Latin, arrival, from *advenire*] (12th century)
1 : the period beginning four Sundays before Christmas and observed by some Christians as a season of prayer and fasting
2 a : the coming of Christ at the Incarnation **b :** SECOND COMING
3 *not capitalized* **:** a coming into being or use ⟨the *advent* of spring⟩ ⟨the *advent* of pasteurization⟩ ⟨the *advent* of personal computers⟩

Ad•vent•ism \ˈad-ˌven-ˌti-zəm\ *noun* (1874)
1 : the doctrine that the second coming of Christ and the end of the world are near at hand
2 : the principles and practices of Seventh-Day Adventists
— Ad•vent•ist \əd-ˈven-tist, ad-ˈ, ˈad-ˌ\ *adjective or noun*

ad•ven•ti•tia \ˌad-vən-ˈti-shə, -(ˌ)ven-\ *noun* [New Latin, alteration of Latin *adventicia,* neuter plural of *adventicius* coming from outside, from *adventus,* past participle] (1876)
: an external chiefly connective tissue covering of an organ; *especially* **:** the external coat of a blood vessel
— ad•ven•ti•tial \-shəl\ *adjective*

ad•ven•ti•tious \ˌad-(ˌ)ven-ˈti-shəs, -vən-\ *adjective* [Latin *adventicius*] (1603)
1 : coming from another source and not inherent or innate
2 : arising or occurring sporadically or in other than the usual location ⟨*adventitious* roots⟩
— ad•ven•ti•tious•ly *adverb*

ad•ven•tive \ad-ˈven-tiv\ *adjective* (circa 1859)
1 : introduced but not fully naturalized
2 : ADVENTITIOUS 2
— adventive *noun*

Advent Sunday *noun* (15th century)
: the first Sunday in Advent

¹ad•ven•ture \əd-ˈven-chər\ *noun* [Middle English *aventure,* from Old French, from (assumed) Vulgar Latin *adventura,* from Latin *adventus,* past participle of *advenire* to arrive, from *ad-* + *venire* to come — more at COME] (14th century)
1 a : an undertaking usually involving danger and unknown risks **b :** the encountering of risks ⟨the spirit of *adventure*⟩
2 : an exciting or remarkable experience ⟨an *adventure* in exotic dining⟩
3 : an enterprise involving financial risk

²adventure *verb* **ad•ven•tured; ad•ven•tur•ing** \-ˈven-ch(ə-)riŋ\ (14th century)
transitive verb
1 : to expose to danger or loss **:** VENTURE
2 : to venture upon **:** TRY
intransitive verb
1 : to proceed despite risk
2 : to take the risk

ad•ven•tur•er \əd-ˈven-ch(ə-)rər\ *noun* (1539)
1 : one that adventures: as **a :** SOLDIER OF FORTUNE **b :** one that engages in risky commercial enterprises for profit
2 : one who seeks unmerited wealth or position especially by playing on the credulity or prejudice of others

ad•ven•ture•some \əd-ˈven-chər-səm\ *adjective* (circa 1731)
: inclined to take risks **:** VENTURESOME
— ad•ven•ture•some•ness *noun*

ad•ven•tur•ess \əd-ˈven-ch(ə-)rəs\ *noun* (1754)
: a female adventurer; *especially* **:** one who seeks position or livelihood by questionable means

ad•ven•tur•ism \əd-ˈven-chə-ˌri-zəm\ *noun* (1932)
: improvisation or experimentation (as in politics or military or foreign affairs) in the absence or in defiance of accepted plans or principles
— ad•ven•tur•ist \-ˈven-ch(ə-)rist\ *noun*
— ad•ven•tur•is•tic \-ˌven-chə-ˈris-tik\ *adjective*

ad•ven•tur•ous \əd-ˈven-ch(ə-)rəs\ *adjective* (14th century)
1 a : disposed to seek adventure or to cope with the new and unknown ⟨an *adventurous* explorer⟩ **b :** INNOVATIVE ⟨an *adventurous* artistic style⟩
2 : characterized by unknown dangers and risks ⟨an *adventurous* journey⟩ ☆
— ad•ven•tur•ous•ly *adverb*
— ad•ven•tur•ous•ness *noun*

¹ad•verb \ˈad-ˌvərb\ *noun* [Middle English *adverbe,* from Middle French, from Latin *adverbium,* from *ad-* + *verbum* word — more at WORD] (14th century)
: a word belonging to one of the major form classes in any of numerous languages, typically serving as a modifier of a verb, an adjective, another adverb, a preposition, a phrase, a clause, or a sentence, expressing some relation of manner or quality, place, time, degree, number, cause, opposition, affirmation, or denial, and in English also serving to connect and to express comment on clause content — compare ADJUNCT, CONJUNCT, DISJUNCT

²adverb *adjective* (1879)
: ADVERBIAL

ad•ver•bi•al \ad-ˈvər-bē-əl\ *adjective* (1611)
: of, relating to, or having the function of an adverb
— adverbial *noun*
— ad•ver•bi•al•ly \-ə-lē\ *adverb*

ad ver•bum \(ˌ)ad-ˈvər-bəm\ *adverb* [Latin] (circa 1580)
: to a word **:** VERBATIM

ad•ver•sar•i•al \ˌad-və(r)-ˈser-ē-əl\ *adjective* (1926)
: of, relating to, or characteristic of an adversary or adversary procedures **:** ADVERSARY

¹ad•ver•sary \ˈad-və(r)-ˌser-ē\ *noun, plural* **-sar•ies** (14th century)
: one that contends with, opposes, or resists **:** ENEMY
— ad•ver•sari•ness *noun*

☆ SYNONYMS
Advance, promote, forward, further mean to help (someone or something) to move ahead. ADVANCE stresses effective assisting in hastening a process or bringing about a desired end ⟨*advance* the cause of peace⟩. PROMOTE suggests an encouraging or fostering and may denote an increase in status or rank ⟨a campaign to *promote* better health⟩. FORWARD implies an impetus forcing something ahead ⟨a wage increase would *forward* productivity⟩. FURTHER suggests a removing of obstacles in the way of a desired advance ⟨used the marriage to *further* his career⟩.

Adventurous, venturesome, daring, daredevil, rash, reckless, foolhardy mean exposing oneself to danger more than required by good sense. ADVENTUROUS implies a willingness to accept risks but not necessarily imprudence ⟨*adventurous* pioneers⟩. VENTURESOME implies a jaunty eagerness for perilous undertakings ⟨*venturesome* stunt pilots⟩. DARING heightens the implication of fearlessness in courting danger ⟨*daring* mountain climbers⟩. DAREDEVIL stresses ostentation in daring ⟨*daredevil* motorcyclists⟩. RASH suggests imprudence and lack of forethought ⟨a *rash* decision⟩. RECKLESS implies heedlessness of probable consequences ⟨a *reckless* driver⟩. FOOLHARDY suggests a recklessness that is inconsistent with good sense ⟨only a *foolhardy* sailor would venture into this storm⟩.

²adversary *adjective* (14th century)
1 : of, relating to, or involving an adversary
2 : having or involving antagonistic parties or opposing interests ⟨divorce can be an *adversary* proceeding⟩

ad·ver·sa·tive \əd-'vər-sə-tiv, ad-\ *adjective* (15th century)
: expressing antithesis, opposition, or adverse circumstance ⟨the *adversative* conjunction *but*⟩
— adversative *noun*
— ad·ver·sa·tive·ly *adverb*

ad·verse \ad-'vərs, 'ad-,\ *adjective* [Middle English, from Middle French *advers,* from Latin *adversus,* past participle of *advertere*] (14th century)
1 : acting against or in a contrary direction **:** HOSTILE ⟨hindered by *adverse* winds⟩
2 a : opposed to one's interests ⟨an *adverse* verdict⟩ ⟨heard testimony *adverse* to their position⟩; *especially* **:** UNFAVORABLE ⟨*adverse* criticism⟩ **b :** causing harm **:** HARMFUL ⟨*adverse* drug effects⟩
3 *archaic* **:** opposite in position ■
— ad·verse·ly *adverb*
— ad·verse·ness *noun*

ad·ver·si·ty \ad-'vər-sə-tē\ *noun, plural* **-ties** (13th century)
1 : a state or condition contrary to one of well-being
2 : an instance of adversity
synonym see MISFORTUNE

¹ad·vert \ad-'vərt\ *intransitive verb* [Middle English *adverten,* from Middle French & Latin; Middle French *advertir,* from Latin *advertere,* from *ad-* + *vertere* to turn — more at WORTH] (15th century)
1 : to turn the mind or attention — used with *to* ⟨*adverted* to the speaker⟩
2 : to call attention in the course of speaking or writing **:** make reference — used with *to* ⟨*adverted* to foreign-language sources⟩

²ad·vert \'ad-,vərt\ *noun* (1860)
chiefly British **:** ADVERTISEMENT

ad·ver·tence \ad-'vər-tᵊn(t)s\ *noun* (14th century)
1 : the action or process of adverting **:** ATTENTION
2 : ADVERTENCY 1

ad·ver·ten·cy \-tᵊn(t)-sē\ *noun, plural* **-cies** (1646)
1 : the quality or state of being advertent **:** HEEDFULNESS
2 : ADVERTENCE 1

ad·ver·tent \-tᵊnt\ *adjective* [Latin *advertent-, advertens,* present participle of *advertere*] (1671)
: giving attention **:** HEEDFUL
— ad·ver·tent·ly *adverb*

ad·ver·tise \'ad-vər-,tīz\ *verb* **-tised; -tis·ing** [Middle English, from Middle French *advertiss-,* stem of *advertir*] (15th century)
transitive verb
1 : to make something known to **:** NOTIFY
2 a : to make publicly and generally known ⟨*advertising* their readiness to make concessions⟩ **b :** to announce publicly especially by a printed notice or a broadcast **c :** to call public attention to especially by emphasizing desirable qualities so as to arouse a desire to buy or patronize **:** PROMOTE
intransitive verb
: to issue or sponsor advertising ⟨*advertise* for a secretary⟩
— ad·ver·tis·er *noun*

ad·ver·tise·ment \,ad-vər-'tīz-mənt; əd-'vər-təz-mənt, -tə-smənt\ *noun* (15th century)
1 : the act or process of advertising
2 : a public notice; *especially* **:** one published in the press or broadcast over the air

ad·ver·tis·ing *noun* (1762)
1 : the action of calling something to the attention of the public especially by paid announcements
2 : ADVERTISEMENTS ⟨the magazine contains much *advertising*⟩
3 : the business of preparing advertisements for publication or broadcast

ad·ver·tize, ad·ver·tize·ment *British variant of* ADVERTISE, ADVERTISEMENT

ad·ver·to·ri·al \,ad-vər-'tōr-ē-əl, -'tȯr-\ *noun* [blend of *advertisement* and *editorial*] (1946)
: an advertisement that imitates editorial format

ad·vice \əd-'vīs\ *noun* [Middle English, from Middle French *avis* opinion, probably from the phrase *ce m'est a vis* that appears to me, part translation of Latin *mihi visum est* it seemed so to me, I decided] (14th century)
1 : recommendation regarding a decision or course of conduct **:** COUNSEL ⟨he shall have power, by and with the *advice* and consent of the Senate, to make treaties —*U.S. Constitution*⟩
2 : information or notice given — usually used in plural
3 : an official notice concerning a business transaction

ad·vis·able \əd-'vī-zə-bəl\ *adjective* (1582)
: fit to be advised or done **:** PRUDENT
synonym see EXPEDIENT
— ad·vis·abil·i·ty \-,vī-zə-'bi-lə-tē\ *noun*
— ad·vis·able·ness \-'vī-zə-bəl-nəs\ *noun*
— ad·vis·ably \-blē\ *adverb*

ad·vise \əd-'vīz\ *verb* **ad·vised; ad·vis·ing** [Middle English, from Middle French *aviser,* from *avis*] (14th century)
transitive verb
1 a : to give advice to **:** COUNSEL ⟨*advise* her to try a drier climate⟩ **b :** CAUTION, WARN ⟨*advise* them of the consequences⟩ **c :** RECOMMEND ⟨*advise* prudence⟩
2 : to give information or notice to **:** INFORM ⟨*advise* them of their rights⟩
intransitive verb
1 : to give advice ⟨*advise* on legal matters⟩
2 : to take counsel **:** CONSULT ⟨*advise* with your parents⟩
— ad·vis·er *also* **ad·vi·sor** \-'vī-zər\ *noun*

ad·vised \əd-'vīzd\ *adjective* (14th century)
: thought out **:** CONSIDERED — often used in combination ⟨ill-*advised* plans⟩
— ad·vis·ed·ly \-'vī-zəd-lē\ *adverb*

ad·vis·ee \əd-,vī-'zē\ *noun* (1824)
: one that is advised

ad·vise·ment \əd-'vīz-mənt\ *noun* (14th century)
1 : careful consideration **:** DELIBERATION
2 : the act or process of advising (as a college student)

¹ad·vi·so·ry \əd-'vīz-rē, -'vī-zə-\ *adjective* (1778)
1 : having or exercising power to advise
2 : containing or giving advice

²advisory *noun, plural* **-ries** (1936)
: a report giving information (as on the weather) and often recommending action to be taken

ad·vo·ca·cy \'ad-və-kə-sē\ *noun* (15th century)
: the act or process of advocating or supporting a cause or proposal

advocacy journalism *noun* (1970)
: journalism that advocates a cause or expresses a viewpoint
— advocacy journalist *noun*

¹ad·vo·cate \'ad-və-kət, -,kāt\ *noun* [Middle English *advocat,* from Middle French, from Latin *advocatus,* from past participle of *advocare* to summon, from *ad-* + *vocare* to call, from *voc-, vox* voice — more at VOICE] (14th century)
1 : one that pleads the cause of another; *specifically* **:** one that pleads the cause of another before a tribunal or judicial court
2 : one that defends or maintains a cause or proposal

²ad·vo·cate \-,kāt\ *transitive verb* **-cat·ed; -cat·ing** (1599)
: to plead in favor of
synonym see SUPPORT
— ad·vo·ca·tion \,ad-və-'kā-shən\ *noun*
— ad·vo·ca·tive \'ad-və-,kā-tiv\ *adjective*
— ad·vo·ca·tor \-,kā-tər\ *noun*

ad·vow·son \əd-'vau̇-zᵊn\ *noun* [Middle English, from Old French *avoueson,* from Medieval Latin *advocation-, advocatio,* from Latin, act of calling, from *advocare*] (14th century)
: the right in English law of presenting a nominee to an ecclesiastical benefice

ady·nam·ic \,ā-(,)dī-'na-mik, ,a-də-'na-\ *adjective* [Greek *adynamia* lack of strength, from *a-* + *dynamis* power, from *dynasthai* to be able] (1829)
: characterized by or causing a loss of strength or function ⟨*adynamic* ileus⟩

ad·y·tum \'a-də-təm\ *noun, plural* **-ta** \-tə\ [Latin, from Greek *adyton,* neuter of *adytos* not to be entered, from *a-* + *dyein* to enter] (1611)
: the innermost sanctuary in an ancient temple open only to priests **:** SANCTUM

adze *also* **adz** \'adz\ *noun* [Middle English *adse,* from Old English *adesa*] (before 12th century)
: a cutting tool that has a thin arched blade set at right angles to the handle and is used chiefly for shaping wood

adze

ad·zu·ki bean \ad-'zü-kē-\ *noun* [Japanese *azuki*] (1795)
: an annual bushy leguminous plant (*Vignis angularis* synonym *Phaseolus angularis*) widely grown in Japan and China for its seeds which are used as food and to produce a flour; *also* **:** its seed — called also *adzuki*

ae \'ā\ *adjective* [Middle English (northern dialect) *a,* alteration of *an*] (1737)
chiefly Scottish **:** ONE

Ae·a·cus \'ē-ə-kəs\ *noun* [Latin, from Greek *Aiakos*]
: a son of Zeus who is given the Myrmidons as followers and becomes on his death a judge of the underworld

ae·cio·spore \'ē-shə-,spōr, 'ē-sē-ə-, -,spȯr\ *noun* (1905)
: one of the spores arranged within an aecium in a series like a chain

ae·ci·um \'ē-shē-əm, 'ē-sē-\ *noun, plural* **-cia** \-shē-ə, -sē-\ [New Latin, from Greek *aikia* outrage, assault, from *aikēs, aeikēs* unseemly, from *a-* + *-eikēs,* from *eikenai* to seem] (1905)

□ USAGE
adverse Here are a few observations to help with the use of *adverse* and *averse. Adverse* is often used attributively ⟨under *adverse* circumstances⟩ ⟨an *adverse* reaction⟩ while *averse* is rarely so used. It is more often a thing than a person that is adverse ⟨the whole Parliamentary tradition . . . is *adverse* to it —Sir Winston Churchill⟩. When used with *to* and of people, *adverse* and *averse* are pretty much synonymous, but *adverse* generally refers to opinion or intention ⟨was *adverse* to his union with this young lady —George Meredith⟩ and *averse* to feeling or inclination ⟨I am not *averse* to pillorying the innocent —John Barth⟩. But this is a subtle distinction and is not observed universally, even by respected writers ⟨liked to settle herself and was *adverse* to complications —Carson McCullers⟩ ⟨Her Majesty . . . was by no means *averse* to reforms —Edith Sitwell⟩.

: the fruiting body of a rust fungus in which the first binucleate spores are usually produced
— **ae·cial** \-sh(ē-)əl\ *adjective*

ae·des \ā-'ē-(,)dēz\ *noun, plural* **aedes** [New Latin, genus name, from Greek *aēdēs* unpleasant, from *a-* + *ēdos* pleasure; akin to Greek *hēdys* sweet — more at SWEET] (circa 1909)
: any of a genus (*Aedes*) of mosquitoes including the vector of yellow fever, dengue, and other diseases
— **ae·dine** \-'ē-,dīn\ *adjective*

ae·dile \'ē-,dīl, 'ē-dᵊl\ *noun* [Latin *aedilis,* from *aedes* temple — more at EDIFY] (1540)
: an official in ancient Rome in charge of public works and games, police, and the grain supply

Ae·ge·an \i-'jē-ən\ *adjective* [Latin *Aegaeus,* from Greek *Aigaios*] (1550)
1 : of or relating to the arm of the Mediterranean Sea east of Greece
2 : of or relating to the chiefly Bronze Age civilization of the islands of the Aegean Sea and the countries adjacent to it

ae·gis \'ē-jəs *also* 'ā-\ *noun* [Latin, from Greek *aigis,* literally, goatskin, from *aig-, aix* goat; akin to Armenian *ayc* goat] (1611)
1 : a shield or breastplate emblematic of majesty that was associated with Zeus and Athena
2 a : PROTECTION ⟨under the *aegis* of the constitution⟩ **b :** controlling or conditioning influence ⟨many American mothers, under the *aegis* of benevolent permissiveness . . . actually neglect their children —*Time*⟩
3 a : AUSPICES, SPONSORSHIP ⟨under the *aegis* of the museum⟩ **b :** control or guidance especially by an individual, group, or system ⟨acted under the court's *aegis*⟩ ◊

Ae·gis·thus \i-'jis-thəs\ *noun* [Latin, from Greek *Aigisthos*]
: a lover of Clytemnestra slain with her by her son Orestes

-aemia *chiefly British variant of* -EMIA

Ae·ne·as \i-'nē-əs\ *noun* [Latin, from Greek *Aineias*]
: a son of Anchises and Aphrodite, defender of Troy, and hero of Virgil's *Aeneid*

Ae·ne·o·lith·ic \ā-,ē-nē-ō-'li-thik\ *adjective* [Latin *aeneus* of copper or bronze, from *aes* copper, bronze — more at ORE] (1901)
: of or relating to a transitional period between the Neolithic and Bronze ages in which some copper was used

¹ae·o·lian \ē-'ō-lē-ən, ē-'ōl-yən\ *adjective* (1605)
1 *often capitalized* **:** of or relating to Aeolus
2 : giving forth or marked by a moaning or sighing sound or musical tone produced by or as if by the wind

²aeolian *variant of* EOLIAN

¹Ae·o·lian \ē-'ō-lē-ən, ā-, -'ōl-yən\ *adjective* (1589)
: of or relating to Aeolis or its inhabitants

²Aeolian *noun* (circa 1889)
1 : a member of a group of Greek peoples of Thessaly and Boeotia that colonized Lesbos and the adjacent coast of Asia Minor
2 : AEOLIC

aeolian harp *noun* (1791)
: a box-shaped musical instrument having stretched strings usually tuned in unison on which the wind produces varying harmonics over the same fundamental tone

¹Ae·ol·ic \ē-'ä-lik\ *adjective* (1674)
: AEOLIAN

²Aeolic *noun* (1902)
: a group of ancient Greek dialects used by the Aeolians

Ae·o·lus \'ē-ə-ləs\ *noun* [Latin, from Greek *Aiolos*]
: the Greek god of the winds

ae·on *or* **eon** \'ē-ən, 'ē-,än\ *noun* [Latin, from Greek *aiōn* — more at AYE] (1647)
1 : an immeasurably or indefinitely long period of time **:** AGE
2 a *usually eon* **:** a very large division of geologic time usually longer than an era **b :** a unit of geologic time equal to one billion years

ae·o·ni·an \ē-'ō-nē-ən\ *or* **ae·on·ic** \-'ä-nik\ *adjective* (1765)
: lasting for an immeasurably or indefinitely long period of time

ae·py·or·nis \,ē-pē-'ȯr-nəs\ *noun* [New Latin, genus name, from Greek *aipys* high + *ornis* bird — more at ERNE] (1851)
: any of a group (genus *Aepyornis* or order Aepyornithiformes) of gigantic ratite birds known only from remains found in Madagascar — called also *elephant bird*

aer- *or* **aero-** *combining form* [Middle English *aero-,* from Middle French, from Latin, from Greek *aer-, aero-,* from *aēr*]
1 a : air **:** atmosphere ⟨*aer*ate⟩ ⟨*aero*biology⟩ **b :** air and ⟨*aero*space⟩
2 : gas ⟨*aero*sol⟩
3 : aviation ⟨*aero*drome⟩

aer·ate \'a(-ə)r-,āt, 'e(-ə)r-\ *transitive verb* **aer·at·ed; aer·at·ing** (1794)
1 : to supply (the blood) with oxygen by respiration
2 : to supply or impregnate (as the soil or a liquid) with air
3 a *British* **:** CARBONATE 2 **b :** to make light or sparkling
— **aer·a·tion** \,a(-ə)r-'ā-shən, ,e(-ə)r-\ *noun*

aer·a·tor \'a(-ə)r-,ā-tər, 'e(-ə)r-\ *noun* (1861)
: one that aerates; *especially* **:** an apparatus for aerating something (as sewage)

aer·en·chy·ma \,ar-'eŋ-kə-mə, ,er-\ *noun* [New Latin] (circa 1893)
: the spongy modified cork tissue of many aquatic plants that facilitates gaseous exchange and maintains buoyancy

¹ae·ri·al \'ar-ē-əl, 'er-; ā-'ir-ē-əl\ *adjective* [Latin *aerius,* from Greek *aerios,* from *aēr*] (1604)
1 a : of, relating to, or occurring in the air or atmosphere **b :** existing or growing in the air rather than in the ground or in water **c :** high in the air ⟨*aerial* spires⟩ **d :** operating or operated overhead on elevated cables or rails ⟨an *aerial* tram⟩
2 : suggestive of air: as **a :** lacking substance **b :** FANCIFUL, ETHEREAL ⟨visions of *aerial* joy —P. B. Shelley⟩
3 a : of or relating to aircraft ⟨*aerial* navigation⟩ **b :** designed for use in, taken from, or operating from or against aircraft **c :** effected by means of aircraft ⟨*aerial* transportation⟩
4 : of, relating to, or gained by the forward pass in football
— **ae·ri·al·ly** \-ə-lē\ *adverb*

²aer·i·al \'ar-ē-əl, 'er-\ *noun* (1902)
1 : ANTENNA 2
2 : FORWARD PASS

ae·ri·al·ist \'ar-ē-ə-list, 'er-, ā-'ir-\ *noun* (1905)
: one that performs feats in the air or above the ground especially on the trapeze

aerial ladder *noun* (1904)
: a mechanically operated extensible ladder usually mounted on a fire truck

aerial perspective *noun* (1720)
: the expression of space in painting by gradation of color and distinctness

ae·rie \'ar-ē, 'er-, 'ir-, 'ā-(ə-)rē\ *noun* [Medieval Latin *aerea,* from Old French *aire,* probably from (assumed) Vulgar Latin *agrum* origin, nest, lair, from Latin *ager* field — more at ACRE] (1581)
1 : the nest of a bird on a cliff or a mountaintop
2 *obsolete* **:** a brood of birds of prey
3 : an elevated often secluded dwelling, structure, or position

aero \'a(-ə)r-(,)ō, 'e(-ə)r-\ *adjective* [*aero-*] (1874)
: of or relating to aircraft or aeronautics ⟨an *aero* engine⟩

aer·o·bat·ics \,ar-ə-'ba-tiks, ,er-\ *noun plural but singular or plural in construction* [*aer-* + acro*batics*] (circa 1911)
: spectacular flying feats and maneuvers (as rolls and dives)
— **aer·o·bat·ic** \-tik\ *adjective*

aer·obe \'a(-ə)r-,ōb, 'e(-ə)r-\ *noun* [French *aérobie,* from *aéro-* aer- + *-bie* (from Greek *-bion,* from *bios* life) — more at QUICK] (1886)
: an organism (as a bacterium) that lives only in the presence of oxygen

aer·o·bic \,a(-ə)r-'ō-bik, ,e(-ə)r-\ *adjective* (1884)
1 : living, active, or occurring only in the presence of oxygen ⟨*aerobic* respiration⟩
2 : of, relating to, or induced by aerobes ⟨*aerobic* fermentation⟩
3 : involving, utilizing, or used in aerobics ⟨an *aerobic* workout⟩ ⟨*aerobic* fitness⟩
— **aer·o·bi·cal·ly** \-bi-k(ə-)lē\ *adverb*

aer·o·bics \-biks\ *noun plural* (1967)
1 *singular or plural in construction* **:** a system of physical conditioning involving exercises (as running, walking, swimming, or calisthenics) strenuously performed so as to cause marked temporary increase in respiration and heart rate
2 : aerobic exercises

aero·bi·ol·o·gy \,ar-ō-bī-'ä-lə-jē\ *noun* [*aer-* + *biology*] (circa 1937)
: the science dealing with the occurrence, transportation, and effects of airborne materials (as viruses, pollen, or pollutants)
— **aero·bi·o·log·i·cal** \-,bī-ə-'lä-ji-kəl\ *adjective*

aero·bi·o·sis \,ar-ō-bī-'ō-səs, ,er-, -bē-\ *noun, plural* **-o·ses** \-,sēz\ [New Latin] (circa 1900)
: life in the presence of air or oxygen

aero·brake \'ar-ō-,brāk, 'er-\ *transitive verb* (1965)
: to decelerate (as a spacecraft) by passage through a planetary atmosphere

aero·drome \'ar-ə-,drōm, 'er-\ *noun* (1908) *chiefly British* **:** AIRFIELD

aero·dy·nam·i·cist \,ar-ō-dī-'na-mə-sist, ,er-\ *noun* (1926)
: one who specializes in aerodynamics

aero·dy·nam·ics \-'na-miks\ *noun plural but singular or plural in construction* (circa 1837)
: a branch of dynamics that deals with the motion of air and other gaseous fluids and with the forces acting on bodies in motion relative to such fluids
— **aero·dy·nam·ic** \-mik\ *also* **aero·dy·nam·i·cal** \-mi-kəl\ *adjective*
— **aero·dy·nam·i·cal·ly** \-mi-k(ə-)lē\ *adverb*

aero·dyne \'ar-ə-,dīn, 'er-\ *noun* [*aerodynamic*] (circa 1906)
: a heavier-than-air aircraft (as an airplane, helicopter, or glider) — compare AEROSTAT

aero·elas·tic·i·ty \,ar-ō-,ē-,las-'ti-sə-tē, ,er-, -i-,las-\ *noun* (1935)

◇ WORD HISTORY

aegis When we speak of something as being "under the aegis of" another, we mean that it is under the authority, sponsorship, or control of another. The basic idea is that whatever is in the subordinate position is being protected. In ancient Greece the word *aigis* meant literally "goatskin," but literal usage was uncommon. More frequently, *aigis* referred to a protective mantle, presumably of goat hide, or perhaps a shield, brandished by the god Zeus. Sometimes he entrusted the aegis to other deities, especially his daughter Athena. In its center the aegis bore an image of the Gorgon Medusa's head, and its border was decorated with golden fringe according to Homer. The aegis became a familiar classical reference in English literature, and by the 18th century the word *aegis* was being used figuratively for other kinds of seemingly impregnable protection.

: distortion (as from bending or flexing) in a structure (as an airplane wing or a building) caused by aerodynamic forces
— **aero·elas·tic** \-'las-tik\ *adjective*
aero·em·bo·lism \ˌar-ō-'em-bə-ˌli-zəm, ˌer-\ *noun* (circa 1939)
: a condition equivalent to bends caused by rapid ascent to high altitudes and resulting exposure to rapidly lowered air pressure
aero·foil \'ar-ə-ˌfȯil, 'er-\ *noun* (1907)
chiefly British : AIRFOIL
aero·gram *or* **aero·gramme** \'ar-ə-ˌgram, 'er-\ *noun* (1899)
: AIR LETTER 2
aer·og·ra·pher's mate \ˌa-'rä-grə-fərz-, ˌe-\ *noun* (circa 1952)
: a navy petty officer specializing in meteorology
aer·o·lite \'ar-ə-ˌlīt, 'er-\ *noun* (circa 1815)
: a stony meteorite
aero·mag·net·ic \ˌar-ō-mag-'ne-tik, ˌer-\ *adjective* (1948)
: of, relating to, or derived from a study of the earth's magnetic field especially from the air ⟨an *aeromagnetic* survey⟩
aero·me·chan·ics \-mə-'ka-niks\ *noun plural but singular or plural in construction* (circa 1909)
: mechanics that deals with the equilibrium and motion of gases and of solid bodies immersed in them
aero·med·i·cine \-'me-də-sən\ *noun* (1942)
: a branch of medicine that deals with the diseases and disturbances arising from flying and the associated physiological and psychological problems
— **aero·med·i·cal** \-'me-di-kəl\ *adjective*
aer·om·e·ter \ˌa(-ə)r-'ä-mə-tər, ˌe(-ə)r-\ *noun* [probably from French *aéromètre,* from *aér-* + *-mètre* -meter] (1794)
: an instrument for ascertaining the weight or density of air or other gases
aero·naut \'ar-ə-ˌnȯt, 'er-, -ˌnät\ *noun* [French *aéronaute,* from *aér-* aer- + Greek *nautēs* sailor — more at NAUTICAL] (1784)
: one that operates or travels in an airship or balloon
aero·nau·tics \ˌar-ə-'nȯ-tiks, ˌer-, -'nä-\ *noun plural but singular in construction* (circa 1824)
1 : a science dealing with the operation of aircraft
2 : the art or science of flight
— **aero·nau·ti·cal** \-ti-kəl\ *also* **aero·nau·tic** \-tik\ *adjective*
— **aero·nau·ti·cal·ly** \-ti-k(ə-)lē\ *adverb*
aer·on·o·my \ˌa(-ə)r-'ä-nə-mē, ˌe(-ə)r-\ *noun* (1957)
: a science that deals with the physics and chemistry of the upper atmosphere of planets
— **aer·on·o·mer** \-mər\ *noun*
— **aer·o·nom·ic** \ˌar-ə-'nä-mik, ˌer-\ *or* **aer·o·nom·i·cal** \-mi-kəl\ *adjective*
— **aer·on·o·mist** \ˌa(-ə)r-'ä-nə-mist, ˌe(-ə)r-\ *noun*
aero·plane \'ar-ə-ˌplān, 'er-\ *noun* [French *aéroplane,* from *aéro-* aer- + *-plane,* probably from feminine of *plan* flat, level, from Latin *planus* — more at FLOOR] (1873)
chiefly British : AIRPLANE
aero·sol \'ar-ə-ˌsäl, 'er-, -ˌsȯl\ *noun* (1923)
1 : a suspension of fine solid or liquid particles in gas ⟨smoke, fog, and mist are *aerosols*⟩
2 : a substance (as an insecticide or cosmetic) dispensed from a pressurized container especially as an aerosol; *also* **:** the container for this
aero·sol·ize \-ˌsä-ˌlīz, -ˌsȯ-, -sə-\ *transitive verb* **-ized; -iz·ing** (1944)
: to disperse as an aerosol
— **aero·sol·iza·tion** \ˌar-ə-ˌsä-lə-'zā-shən, -ˌsȯ-, -sə-\ *noun*
[1]**aero·space** \'ar-ō-ˌspās, 'er-\ *noun* (circa 1958)
1 : space comprising the earth's atmosphere and the space beyond
2 : a physical science that deals with aerospace
3 : the aerospace industry
[2]**aerospace** *adjective* (1958)
: of or relating to aerospace, to vehicles used in aerospace or the manufacture of such vehicles, or to travel in aerospace ⟨*aerospace* research⟩ ⟨*aerospace* profits⟩ ⟨*aerospace* medicine⟩
aerospace plane *noun* (1960)
: an airplane designed to be able to reach low earth orbit
aero·stat \-ˌstat\ *noun* [French *aérostat,* from *aér-* + *-stat*] (1784)
: a lighter-than-air aircraft (as a balloon or blimp) — compare AERODYNE
aero·stat·ics \ˌar-ō-'sta-tiks, ˌer-\ *noun plural but singular or plural in construction* [modification of New Latin *aerostatica,* from *aer-* + *statica* statics] (1784)
: a branch of statics that deals with the equilibrium of gaseous fluids and of solid bodies immersed in them
aero·ther·mo·dy·nam·ics \-ˌthər-mə-(ˌ)dī-'na-miks\ *noun plural but singular or plural in construction* (1949)
: the thermodynamics of gases and especially of air
— **aero·ther·mo·dy·nam·ic** \-mik\ *adjective*
[1]**aery** \'ar-ē, 'er-ē, 'ā-ə-rē\ *adjective* **aer·i·er; -est** [Latin *aerius* — more at AERIAL] (14th century)
: having an aerial quality : ETHEREAL ⟨*aery* visions⟩
— **aer·i·ly** \'ar-ə-lē, 'er-\ *adverb*
[2]**aery** *like* AERIE\ *variant of* AERIE
Aes·cu·la·pi·an \ˌes-k(y)ə-'lā-pē-ən\ *adjective* [*Aesculapius,* Greco-Roman god of medicine, from Latin, from Greek *Asklēpios*] (1605)
: of or relating to Aesculapius or the healing art ◆
Ae·sir \'ā-ˌzir, -ˌsir\ *noun plural* [Old Norse *Æsir,* plural of *āss* god]
: the principal race of Norse gods
Ae·so·pi·an \ē-'sō-pē-ən, -'sä-\ *also* **Ae·sop·ic** \-'sä-pik\ *adjective* (1728)
1 : of, relating to, or characteristic of Aesop or his fables
2 : conveying an innocent meaning to an outsider but a hidden meaning to a member of a conspiracy or underground movement ⟨*Aesopian* language⟩
aes·thete \'es-ˌthēt, *British usually* 'ēs-\ *noun* [back-formation from *aesthetic*] (1881)
: one having or affecting sensitivity to the beautiful especially in art
[1]**aes·thet·ic** \es-'the-tik, is-, *British usually* ēs-\ *or* **aes·thet·i·cal** \-ti-kəl\ *adjective* [German *ästhetisch,* from New Latin *aestheticus,* from Greek *aisthētikos* of sense perception, from *aisthanesthai* to perceive — more at AUDIBLE] (1798)
1 a : of, relating to, or dealing with aesthetics or the beautiful ⟨*aesthetic* theories⟩ **b :** ARTISTIC ⟨a work of *aesthetic* value⟩ **c :** pleasing in appearance : ATTRACTIVE ⟨easy-to-use keyboards, clear graphics, and other ergonomic and *aesthetic* features —Mark Mehler⟩
2 : appreciative of, responsive to, or zealous about the beautiful; *also* **:** responsive to or appreciative of what is pleasurable to the senses
— **aes·thet·i·cal·ly** \-ti-k(ə-)lē\ *adverb*
[2]**aesthetic** *noun* (1822)
1 *plural but singular or plural in construction* **:** a branch of philosophy dealing with the nature of beauty, art, and taste and with the creation and appreciation of beauty
2 : a particular theory or conception of beauty or art : a particular taste for or approach to what is pleasing to the senses and especially sight ⟨modernist *aesthetics*⟩ ⟨staging new ballets which reflected the *aesthetic* of the new nation —Mary Clarke & Clement Crisp⟩
3 *plural* **:** a pleasing appearance or effect : BEAUTY ⟨appreciated the *aesthetics* of the gemstones⟩
aesthetic distance *noun* (1938)
: the frame of reference that an artist creates by the use of technical devices in and around the work of art to differentiate it psychologically from reality
aes·the·ti·cian \ˌes-thə-'ti-shən\ *noun* (1829)
: a specialist in aesthetics
aes·thet·i·cism \es-'the-tə-ˌsi-zəm, is-\ *noun* (1855)
1 : a doctrine that the principles of beauty are basic to other and especially moral principles
2 : devotion to or emphasis on beauty or the cultivation of the arts
aes·thet·i·cize \-ˌsīz\ *transitive verb* **-cized; -ciz·ing** (1864)
: to make aesthetic
aes·ti·val \'es-tə-vəl\ *variant of* ESTIVAL
aes·ti·vate, aes·ti·va·tion *variant of* ESTIVATE, ESTIVATION
aether *variant of* ETHER 2a
ae·ti·ol·o·gy *chiefly British variant of* ETIOLOGY
af- — see AD-
[1]**afar** \ə-'fär\ *adverb* [Middle English *afer,* from *on fer* at a distance and *of fer* from a distance] (14th century)
: from, to, or at a great distance ⟨roamed *afar*⟩
: far away
[2]**afar** *noun* (14th century)
: a great distance ⟨saw him from *afar*⟩
afeard *or* **afeared** \ə-'fird\ *adjective* [Middle English *afered,* from Old English *āfǣred,* past participle of *āfǣran* to frighten, from *ā-,* perfective prefix + *fǣran* to frighten — more at ABIDE, FEAR] (before 12th century)
dialect : AFRAID
afe·brile \(ˌ)ā-'fe-ˌbrīl *also* -'fē-\ *adjective* (1875)
: not marked by fever
af·fa·ble \'a-fə-bəl\ *adjective* [Middle French, from Latin *affabilis,* from *affari* to speak to, from *ad-* + *fari* to speak — more at BAN] (15th century)
1 : being pleasant and at ease in talking to others
2 : characterized by ease and friendliness
synonym see GRACIOUS
— **af·fa·bil·i·ty** \ˌa-fə-'bi-lə-tē\ *noun*
— **af·fa·bly** \-blē\ *adverb*
af·fair \ə-'far, -'fer\ *noun* [Middle English & Middle French; Middle English *affaire,* from Middle French, from *a faire* to do] (14th century)
1 a *plural* **:** commercial, professional, public, or personal business **b :** MATTER, CONCERN
2 : a procedure, action, or occasion only

◇ WORD HISTORY
Aesculapian Aesculapius or Asclepius was the god of healing in Greco-Roman religious belief. The son of Apollo and the nymph Coronis, he was raised by the Centaur Chiron, who taught him the arts of medicine. According to one myth, Aesculapius was struck dead by one of Zeus's thunderbolts for bringing Hippolytus back to life. Snakes were the god's sacred emblems, and he was believed to be incarnate in them. It was in the figure of a snake that he was introduced into Rome during a time of pestilence. His branched staff with a snake coiled around it became the traditional symbol for medicine and is today the official insignia of the American Medical Association.

vaguely specified; *also* **:** an object or collection of objects only vaguely specified ⟨their house was a 2-story *affair*⟩
3 *also* **af·faire a :** a romantic or passionate attachment typically of limited duration **:** LIAISON 2b **b :** a matter occasioning public anxiety, controversy, or scandal **:** CASE

¹af·fect \'a-ˌfekt\ *noun* [Middle English, from Latin *affectus,* from *afficere*] (14th century)
1 *obsolete* **:** FEELING, AFFECTION
2 : the conscious subjective aspect of an emotion considered apart from bodily changes
usage see EFFECT

²af·fect \ə-'fekt, a-\ *verb* [Middle English, from Middle French & Latin; Middle French *affecter,* from Latin *affectare,* frequentative of *afficere* to influence, from *ad-* + *facere* to do — more at DO] (15th century)
transitive verb
1 *archaic* **:** to aim at
2 a *archaic* **:** to have affection for **b :** to be given to **:** FANCY ⟨*affect* flashy clothes⟩
3 : to make a display of liking or using **:** CULTIVATE ⟨*affect* a worldly manner⟩
4 : to put on a pretense of **:** FEIGN ⟨*affect* indifference, though deeply hurt⟩
5 : to tend toward ⟨drops of water *affect* roundness⟩
6 : FREQUENT
intransitive verb
obsolete **:** INCLINE 2
synonym see ASSUME
usage see EFFECT

³affect *transitive verb* [Middle English, from *affectus,* past participle of *afficere*] (15th century)
: to produce an effect upon: as **a :** to produce a material influence upon or alteration in ⟨paralysis *affected* his limbs⟩ **b :** to act upon (as a person or a person's mind or feelings) so as to effect a response **:** INFLUENCE ☆
usage see EFFECT
— af·fect·abil·i·ty \-ˌfek-tə-'bi-lə-tē\ *noun*
— af·fect·able \-'fek-tə-bəl\ *adjective*

af·fec·ta·tion \ˌa-ˌfek-'tā-shən\ *noun* (1548)
1 a : the act of taking on or displaying an attitude or mode of behavior not natural to oneself or not genuinely felt **b :** speech or conduct not natural to oneself **:** ARTIFICIALITY
2 *obsolete* **:** a striving after
synonym see POSE

af·fect·ed \ə-'fek-təd, a-\ *adjective* (1587)
1 : INCLINED, DISPOSED ⟨was well *affected* toward her⟩
2 a : given to affectation **b :** assumed artificially or falsely **:** PRETENDED ⟨an *affected* interest in art⟩
— af·fect·ed·ly *adverb*
— af·fect·ed·ness *noun*

af·fect·ing \ə-'fek-tiŋ, a-\ *adjective* (1720)
: evoking a strong emotional response
synonym see MOVING
— af·fect·ing·ly \-tiŋ-lē\ *adverb*

af·fec·tion \ə-'fek-shən\ *noun* [Middle English, from Old French *affection,* from Latin *affection-, affectio,* from *afficere*] (13th century)
1 : a moderate feeling or emotion
2 : tender attachment **:** FONDNESS ⟨she had a deep *affection* for her parents⟩
3 a (1) **:** a bodily condition (2) **:** DISEASE, MALADY **b :** ATTRIBUTE ⟨shape and weight are *affections* of bodies⟩
4 *obsolete* **:** PARTIALITY, PREJUDICE
5 : the feeling aspect (as in pleasure) of consciousness
6 a : PROPENSITY, DISPOSITION **b** *archaic* **:** AFFECTATION 1
7 : the action of affecting **:** the state of being affected
synonym see FEELING
— af·fec·tion·less \-ləs\ *adjective*

af·fec·tion·al \ə-'fek-shnəl, -shə-nᵊl\ *adjective* (1859)
: of or relating to the affections
— af·fec·tion·al·ly *adverb*

af·fec·tion·ate \ə-'fek-sh(ə-)nət\ *adjective* (15th century)
1 *obsolete* **:** INCLINED, DISPOSED
2 : having affection or warm regard **:** LOVING
3 : proceeding from affection **:** TENDER ⟨*affectionate* care⟩
— af·fec·tion·ate·ly *adverb*

af·fec·tioned \-shənd\ *adjective* (1555)
archaic **:** having a tendency, disposition, or inclination **:** DISPOSED

af·fec·tive \a-'fek-tiv\ *adjective* (1623)
1 : relating to, arising from, or influencing feelings or emotions **:** EMOTIONAL ⟨*affective* disorders⟩
2 : expressing emotion ⟨*affective* language⟩
— af·fec·tive·ly *adverb*
— af·fec·tiv·i·ty \ˌa-ˌfek-'ti-və-tē\ *noun*

af·fect·less \'a-ˌfekt-ləs, a-'fekt-\ *adjective* (1967)
: showing or expressing no emotion; *also* **:** UNFEELING ⟨a ruthless *affectless* society⟩
— af·fect·less·ness *noun*

af·fen·pin·scher \'a-fən-ˌpin-chər\ *noun* [German, from *Affe* ape + *Pinscher,* a breed of hunting dog] (1903)
: any of a breed of toy dogs with a wiry black, red, tan, or gray coat, erect ears, large round eyes, and bushy eyebrows, chin tuft, and mustache

¹af·fer·ent \'a-fə-rənt, -ˌfer-ənt\ *adjective* [Latin *afferent-, afferens,* present participle of *afferre* to bring to, from *ad-* + *ferre* to bear — more at BEAR] (circa 1847)
: bearing or conducting inward; *specifically* **:** conveying impulses toward a nerve center (as the brain or spinal cord) — compare EFFERENT
— af·fer·ent·ly *adverb*

²afferent *noun* (1949)
: an afferent anatomical part (as a nerve)

¹af·fi·ance \ə-'fī-ən(t)s\ *noun* [Middle English, from Middle French, from *affier* to pledge, trust, from Medieval Latin *affidare* to pledge, from Latin *ad-* + (assumed) Vulgar Latin *fidare* to trust — more at FIANCÉ] (14th century)
archaic **:** TRUST, CONFIDENCE

²affiance *transitive verb* **-anced; -anc·ing** (1555)
: to solemnly promise (oneself or another) in marriage **:** BETROTH

af·fi·ant \ə-'fī-ənt\ *noun* [Middle French, from present participle of *affier*] (1807)
: one that swears to an affidavit; *broadly* **:** DEPONENT

af·fi·cio·na·do *variant of* AFICIONADO

af·fi·da·vit \ˌa-fə-'dā-vət\ *noun* [Medieval Latin, he has made an oath, from *affidare*] (1593)
: a sworn statement in writing made especially under oath or on affirmation before an authorized magistrate or officer

¹af·fil·i·ate \ə-'fi-lē-ˌāt\ *verb* **-at·ed; -at·ing** [Medieval Latin *affiliatus,* past participle of *affiliare* to adopt as a son, from Latin *ad-* + *filius* son — more at FEMININE] (1761)
transitive verb
1 a : to bring or receive into close connection as a member or branch **b :** to associate as a member ⟨*affiliates* herself with the local club⟩
2 : to trace the origin of
intransitive verb
: to connect or associate oneself **:** COMBINE
— af·fil·i·a·tion \-ˌfi-lē-'ā-shən\ *noun*

²af·fil·i·ate \ə-'fi-lē-ət, -ˌāt\ *noun* (1879)
: an affiliated person or organization

af·fil·i·at·ed \-lē-ˌā-təd\ *adjective* (1795)
: closely associated with another typically in a dependent or subordinate position ⟨the university and its *affiliated* medical school⟩

¹af·fine \a-'fīn, ə-\ *noun* [Middle French *affin,* from Latin *affinis,* from *affinis* related] (circa 1509)
: a relative by marriage **:** IN-LAW

²affine *adjective* [Latin *affinis,* adjective] (1918)
: of, relating to, or being a transformation (as a translation, a rotation, or a uniform stretching) that carries straight lines into straight lines and parallel lines into parallel lines but may alter distance between points and angles between lines ⟨*affine* geometry⟩
— af·fine·ly *adverb*

af·fined \a-'fīnd, ə-\ *adjective* (1597)
1 : joined in a close relationship **:** CONNECTED
2 : bound by obligation

af·fin·i·ty \ə-'fi-nə-tē\ *noun, plural* **-ties** [Middle English *affinite,* from Middle French or Latin; Middle French *afinité,* from Latin *affinitas,* from *affinis* bordering on, related by marriage, from *ad-* + *finis* end, border] (14th century)
1 : relationship by marriage
2 a : sympathy marked by community of interest **:** KINSHIP **b** (1) **:** an attraction to or liking for something ⟨people with an *affinity* to darkness —Mark Twain⟩ ⟨pork and fennel have a natural *affinity* for each other —Abby Mandel⟩ (2) **:** an attractive force between substances or particles that causes them to enter into and remain in chemical combination **c :** a person especially of the opposite sex having a particular attraction for one
3 a : likeness based on relationship or causal connection ⟨found an *affinity* between the teller of a tale and the craftsman —Mary McCarthy⟩ ⟨this investigation, with *affinities* to a case history, a psychoanalysis, a detective story and a novel — Oliver Sacks⟩ **b :** a relation between biological groups involving resemblance in structural plan and indicating a common origin
synonym see ATTRACTION

affinity chromatography *noun* (1970)
: chromatography in which a macromolecule (as a protein) is isolated and purified by passing it in solution through a column treated with a substance having a ligand for which the macromolecule has an affinity that causes it to be retained on the column

affinity group *noun* (1970)
: a group of people having a common interest or goal or acting together for a specific purpose (as for a chartered tour)

af·firm \ə-'fərm\ *verb* [Middle English *affermen,* from Middle French *afermer,* from Latin *affirmare,* from *ad-* + *firmare* to make firm, from *firmus* firm — more at FIRM] (14th century)
transitive verb
1 a : VALIDATE, CONFIRM **b :** to state positively
2 : to assert (as a judgment or decree) as valid or confirmed
3 : to express dedication to

☆ SYNONYMS
Affect, influence, touch, impress, strike, sway mean to produce or have an effect upon. AFFECT implies the action of a stimulus that can produce a response or reaction ⟨the sight *affected* her to tears⟩. INFLUENCE implies a force that brings about a change (as in nature or behavior) ⟨our beliefs are *influenced* by our upbringing⟩ ⟨a drug that *influences* growth rates⟩. TOUCH may carry a vivid suggestion of close contact and may connote stirring, arousing, or harming ⟨plants *touched* by frost⟩ ⟨his emotions were *touched* by her distress⟩. IMPRESS stresses the depth and persistence of the effect ⟨only one of the plans *impressed* him⟩. STRIKE, similar to but weaker than *impress,* may convey the notion of sudden sharp perception or appreciation ⟨*struck* by the solemnity of the occasion⟩. SWAY implies the acting of influences that are not resisted or are irresistible, with resulting change in character or course of action ⟨politicians who are *swayed* by popular opinion⟩.

intransitive verb
1 : to testify or declare by affirmation as distinguished from swearing an oath
2 : to uphold a judgment or decree of a lower court
synonym see ASSERT
— af·firm·able \ə-ˈfər-mə-bəl\ *adjective*
— af·fir·mance \ə-ˈfər-mən(t)s\ *noun*

af·fir·ma·tion \ˌa-fər-ˈmā-shən\ *noun* (15th century)
1 a : the act of affirming **b :** something affirmed **:** a positive assertion
2 : a solemn declaration made under the penalties of perjury by a person who conscientiously declines taking an oath

¹af·fir·ma·tive \ə-ˈfər-mə-tiv\ *adjective* (15th century)
1 : asserting a predicate of a subject
2 : asserting that the fact is so
3 : POSITIVE ⟨*affirmative* approach⟩
4 : favoring or supporting a proposition or motion
— af·fir·ma·tive·ly *adverb*

²affirmative *noun* (15th century)
1 : an expression (as the word *yes*) of affirmation or assent
2 : an affirmative proposition
3 : the side that upholds the proposition stated in a debate

affirmative action *noun* (1965)
: an active effort to improve the employment or educational opportunities of members of minority groups and women

¹af·fix \ə-ˈfiks, a-\ *transitive verb* [Latin *affixus,* past participle of *affigere* to fasten to, from *ad-* + *figere* to fasten — more at FIX] (1533)
1 : to attach physically ⟨*affix* a stamp to a letter⟩
2 : to attach in any way **:** ADD, APPEND ⟨*affix* a signature to a document⟩
3 : IMPRESS ⟨*affixed* my seal⟩
synonym see FASTEN
— af·fix·able \-ˈfik-sə-bəl\ *adjective*
— af·fix·a·tion \ˌa-ˌfik-ˈsā-shən\ *noun*
— af·fix·ment \ə-ˈfik-smənt, a-\ *noun*

²af·fix \ˈa-ˌfiks\ *noun* (1612)
1 : one or more sounds or letters occurring as a bound form attached to the beginning or end of a word, base, or phrase or inserted within a word or base and serving to produce a derivative word or an inflectional form
2 : APPENDAGE
— af·fix·al \-ˌfik-səl\ *or* **af·fix·i·al** \a-ˈfik-sē-əl\ *adjective*

af·fla·tus \ə-ˈflā-təs, a-\ *noun* [Latin, act of blowing or breathing on, from *afflare* to blow on, from *ad-* + *flare* to blow — more at BLOW] (1660)
: a divine imparting of knowledge or power **:** INSPIRATION

af·flict \ə-ˈflikt\ *transitive verb* [Middle English, from Latin *afflictus,* past participle of *affligere* to cast down, from *ad-* + *fligere* to strike — more at PROFLIGATE] (14th century)
1 *obsolete* **a :** HUMBLE **b :** OVERTHROW
2 a : to distress so severely as to cause persistent suffering or anguish **b :** TROUBLE, INJURE
☆

af·flic·tion \ə-ˈflik-shən\ *noun* (14th century)
1 : the state of being afflicted
2 : the cause of persistent pain or distress
3 : great suffering

af·flic·tive \ə-ˈflik-tiv\ *adjective* (circa 1611)
: causing affliction **:** DISTRESSING, TROUBLESOME
— af·flic·tive·ly *adverb*

af·flu·ence \ˈa-(ˌ)flü-ən(t)s *also* a-ˈflü- *or* ə-\ *noun* (14th century)
1 a : an abundant flow or supply **:** PROFUSION **b :** abundance of property **:** WEALTH
2 : a flowing to or toward a point **:** INFLUX

af·flu·en·cy \-ən(t)-sē\ *noun, plural* **-cies** (1664)
: AFFLUENCE

¹af·flu·ent \-ənt\ *adjective* [Middle English, from Middle French, from Latin *affluent-, affluens,* present participle of *affluere* to flow to, flow abundantly, from *ad-* + *fluere* to flow — more at FLUID] (15th century)
1 : flowing in abundance ⟨*affluent* streams⟩ ⟨*affluent* creativity⟩
2 : having a generously sufficient and typically increasing supply of material possessions ⟨our *affluent* society⟩
synonym see RICH
— af·flu·ent·ly *adverb*

²affluent *noun* (1828)
1 : a tributary stream
2 : an affluent person

af·ford \ə-ˈfōrd, -ˈfȯrd\ *transitive verb* [Middle English *aforthen,* from Old English *geforthian* to carry out, from *ge-,* perfective prefix + *forthian* to carry out, from *forth* — more at CO-, FORTH] (14th century)
1 a : to manage to bear without serious detriment ⟨you can't *afford* to neglect your health⟩ **b :** to be able to bear the cost of ⟨can't *afford* to be out of work long⟩ ⟨able to *afford* a new car⟩
2 : to make available, give forth, or provide naturally or inevitably ⟨the sun *affords* warmth to the earth⟩ ⟨the roof *afforded* a fine view⟩
synonym see GIVE
— af·ford·abil·i·ty \-ˌfōr-də-ˈbi-lə-tē, -ˌfȯr-\ *noun*
— af·ford·able \-ˈfōr-də-bəl, -ˈfȯr-\ *adjective*
— af·ford·ably \ə-ˈfōr-də-blē, -ˈfȯr-\ *adverb*

af·for·es·ta·tion \(ˌ)a-ˌfȯr-ə-ˈstā-shən, ə-, -ˌfär-\ *noun* [Medieval Latin *afforestation-, afforestatio,* from *afforestare* to put under forest laws, from Latin *ad-* + Medieval Latin *foresta, forestis* forest] (1615)
: the act or process of establishing a forest especially on land not previously forested
— af·for·est \a-ˈfȯr-əst, -ˈfär-\ *transitive verb*

¹af·fray \ə-ˈfrā\ *noun* [Middle English, from Middle French, from *affreer* to startle] (14th century)
: FRAY, BRAWL

²affray *transitive verb* [Middle English *affraien,* from Middle French *affreer*] (14th century)
archaic **:** STARTLE, FRIGHTEN

af·fri·cate \ˈa-fri-kət\ *noun* [probably from German *Affrikata,* from Latin *affricata,* feminine of *affricatus,* past participle of *affricare* to rub against, from *ad-* + *fricare* to rub — more at FRICTION] (1880)
: a stop and its immediately following release into a fricative that are considered to constitute a single phoneme (as the \t\ and \sh\ of \ch\ in *choose*)
— af·fric·a·tive \a-ˈfri-kə-tiv, ə-\ *noun or adjective*

¹af·fright \ə-ˈfrīt\ *transitive verb* [Middle English *afyrht, afright* frightened, from Old English *āfyrht,* past participle of *āfyrhtan* to frighten, from *ā-,* perfective prefix + *fyrhtan* to fear; akin to Old English *fyrhto* fright — more at ABIDE, FRIGHT] (before 12th century)
: FRIGHTEN, ALARM

²affright *noun* (1596)
: sudden and great fear **:** TERROR

¹af·front \ə-ˈfrənt\ *transitive verb* [Middle English *afronten,* from Middle French *afronter* to defy, from (assumed) Vulgar Latin *affrontare,* from Latin *ad-* + *front-, frons* forehead] (14th century)
1 a : to insult especially to the face by behavior or language **b :** to cause offense to ⟨a system of law about both family and marriage that *affronted* lay society —J. H. Mundy⟩
2 a : to face in defiance **:** CONFRONT ⟨*affront* death⟩ **b** *obsolete* **:** to encounter face-to-face
3 : to appear directly before
synonym see OFFEND

²affront *noun* (1533)
1 *obsolete* **:** a hostile encounter
2 : a deliberate offense **:** INSULT ⟨an *affront* to his dignity⟩

af·fu·sion \a-ˈfyü-zhən\ *noun* [Late Latin *affusion-, affusio,* from Latin *affundere* to pour on, from *ad-* + *fundere* to pour — more at FOUND] (1615)
: an act of pouring a liquid on (as in baptism)

Af·ghan \ˈaf-ˌgan *also* -gən\ *noun* [Persian *afghān* Pashtun] (1767)
1 a (1) **:** PASHTUN (2) **:** PASHTO **b :** a native or inhabitant of Afghanistan
2 *not capitalized* **:** a blanket or shawl of colored wool knitted or crocheted in strips or squares
3 *not capitalized* **:** a Turkoman carpet of large size and long pile woven in geometric designs
4 : AFGHAN HOUND
— Afghan *adjective*

Afghan hound *noun* (1925)
: any of a breed of tall slim swift hunting dogs of Near Eastern origin with a coat of silky thick hair and a long silky topknot

Afghan hound

af·ghani \af-ˈga-nē, -ˈgä-\ *noun* [Persian *afghānī,* literally, of the Pashtuns] (1927)
— see MONEY table

afi·cio·na·da \ə-ˌfi-sh(ē-)ə-ˈnä-də, -fē-, -sē-ə-, -ˌdä\ *noun* [Spanish, feminine of *aficionado*] (1952)
: a female aficionado

afi·cio·na·do \-ˈnä-(ˌ)dō\ *noun, plural* **-dos** [Spanish, from past participle of *aficionar* to inspire affection, from *afición* affection, from Latin *affection-, affectio* — more at AFFECTION] (1845)
: a person who likes, knows about, and appreciates a usually fervently pursued interest or activity **:** DEVOTEE ⟨*aficionados* of the bullfight⟩ ⟨movie *aficionados*⟩

afield \ə-ˈfē(ə)ld\ *adverb or adjective* (before 12th century)
1 : to, in, or on the field ⟨was weak at bat but strong *afield*⟩
2 : away from home **:** ABROAD
3 : out of the way **:** ASTRAY ⟨irrelevant remarks that carried us far *afield*⟩

afire \ə-ˈfīr\ *adjective or adverb* (13th century)
: being on fire **:** BLAZING

aflame \ə-ˈflām\ *adjective or adverb* (1555)
: AFIRE

☆ SYNONYMS
Afflict, try, torment, torture, rack mean to inflict on a person something that is hard to bear. AFFLICT is a general term and applies to the causing of pain or suffering or of acute annoyance, embarrassment, or any distress ⟨ills that *afflict* the elderly⟩. TRY suggests imposing something that strains the powers of endurance or of self-control ⟨children often *try* their parents' patience⟩. TORMENT suggests persecution or the repeated inflicting of suffering or annoyance ⟨a horse *tormented* by flies⟩. TORTURE adds the implication of causing unbearable pain or suffering ⟨*tortured* by a sense of guilt⟩. RACK stresses straining or wrenching ⟨a body *racked* by pain⟩.

af·la·tox·in \ˌa-flə-ˈtäk-sən\ *noun* [New Latin *Aspergillus flavus,* species of mold + English *toxin*] (1962)
: any of several carcinogenic mycotoxins that are produced especially in stored agricultural crops (as peanuts) by molds (as *Aspergillus flavus*)

afloat \ə-ˈflōt\ *adjective or adverb* [Middle English *aflot,* from Old English *on flot,* from *on* + *flot,* from *flot* deep water, sea; akin to Old English *flēotan* to float — more at FLEET] (before 12th century)
1 a : borne on or as if on the water **b :** being at sea
2 : free of difficulties **:** SELF-SUFFICIENT ⟨the inheritance kept them *afloat* for years⟩
3 a : circulating about ⟨nasty stories were *afloat*⟩ **b :** ADRIFT

aflut·ter \ə-ˈflə-tər\ *adjective* (1830)
1 : being in a flutter **:** FLUTTERING
2 : nervously excited
3 : filled with or marked by the presence of fluttering things ⟨roofs *aflutter* with flags⟩

afoot \ə-ˈfu̇t\ *adverb or adjective* (13th century)
1 : on foot
2 : in the process of development **:** UNDER WAY ⟨something out of the ordinary was *afoot* —Hamilton Basso⟩

afore \ə-ˈfōr, -ˈfȯr\ *adverb or conjunction or preposition* [Middle English, from Old English *onforan,* from *on* + *foran* before — more at BEFORE] (before 12th century)
chiefly dialect **:** BEFORE

afore·men·tioned \-ˈmen(t)-shənd\ *adjective* (1587)
: mentioned previously

afore·said \-ˌsed\ *adjective* (14th century)
: said or named before or above

afore·thought \-ˌthȯt\ *adjective* (1581)
: previously in mind **:** PREMEDITATED, DELIBERATE ⟨with malice *aforethought*⟩

a for·ti·o·ri \ˌā-ˌfȯr-shē-ˈōr-ˌī, ˌä-ˌfȯr-shē-ˈōr-ē, -ˌfȯr-tē-, -ˈȯr-\ *adverb* [New Latin, literally, from the stronger (argument)] (1588)
: with greater reason or more convincing force — used in drawing a conclusion that is inferred to be even more certain than another ⟨the man of prejudice is, *a fortiori,* a man of limited mental vision⟩

afoul of \ə-ˈfau̇-ləv\ *preposition* (1824)
1 : in or into conflict with
2 : in or into collision or entanglement with

Afr- *or* **Afro-** *combining form* [Latin *Afr-, Afer*]
: African ⟨*Afro*-American⟩ **:** African and ⟨*Afro*-Asiatic⟩

afraid \ə-ˈfrād, *Southern also* ə-ˈfred\ *adjective* [Middle English *affraied,* from past participle of *affraien* to frighten — more at AFFRAY] (14th century)
1 : filled with fear or apprehension ⟨*afraid* of machines⟩ ⟨*afraid* for his job⟩
2 : filled with concern or regret over an unwanted situation ⟨I'm *afraid* I won't be able to go⟩
3 : having a dislike for something ⟨*afraid* of hard work⟩
synonym see FEARFUL

A–frame \ˈā-ˌfrām\ *noun* (circa 1909)
1 : a support structure shaped like the letter A
2 : a building typically having triangular front and rear walls and a roof reaching to the ground

afreet *or* **afrit** \ˈa-ˌfrēt, ə-ˈfrēt\ *noun* [Arabic *ʽifrīt*] (1786)
: a powerful evil jinni, demon, or monstrous giant in Arabic mythology

afresh \ə-ˈfresh\ *adverb* (15th century)
: from a fresh beginning **:** ANEW, AGAIN

¹Af·ri·can \ˈa-fri-kən *also* ˈä-\ *noun* (before 12th century)
1 : a native or inhabitant of Africa
2 : a person and especially a black person of African ancestry

²African *adjective* (1564)
: of, relating to, or characteristic of the continent of Africa or its people
— Af·ri·can·ness \-kə(n)-nəs\ *noun*

Af·ri·ca·na \ˌa-fri-ˈka-nə, -ˈkä-, -ˈkā- *also* ˌä-\ *noun plural* (1908)
: materials (as books, documents, or artifacts) relating to African history and culture

Af·ri·can–Amer·i·can \ˌa-fri-kə-nə-ˈmer-ə-kən *also* ˌä-\ *noun* (1984)
: AFRO-AMERICAN
— African–American *adjective*

African buffalo *noun* (1902)
: CAPE BUFFALO

African daisy *noun* (circa 1889)
: any of a genus (*Arctotis*) of widely cultivated composite herbs

Af·ri·can·der *or* **Af·ri·kan·der** \ˌa-fri-ˈkan-dər\ *noun* [Afrikaans *Afrikaner, Afrikaander,* literally, Afrikaner] (1852)
: any of a breed of tall red large-horned humped southern African cattle used chiefly for meat or draft

African elephant *noun* (1607)
: ELEPHANT 1a

Af·ri·can·ise *British variant of* AFRICANIZE

Af·ri·can·ism \ˈa-fri-kə-ˌni-zəm *also* ˈä-\ *noun* (1641)
1 : a characteristic feature (as a custom or belief) of African culture
2 : a characteristic feature of an African language occurring in a non-African language
3 : allegiance to the traditions, interests, or ideals of Africa

Af·ri·can·ist \-nist\ *noun* (1895)
: a specialist in African languages or cultures

Af·ri·can·ize \-ˌnīz\ *transitive verb* **-ized; -iz·ing** (1853)
1 : to cause to acquire a distinctively African trait
2 : to bring under the influence, control, or cultural or civil supremacy of Africans and especially black Africans
— Af·ri·can·iza·tion \ˌa-fri-kə-nə-ˈzā-shən *also* ˌä-\ *noun*

Africanized bee *noun* (1974)
: a honeybee that originated in Brazil as an accidental hybrid between an aggressive African subspecies (*Apis mellifera scutellata*) and previously established European honeybees and has spread to Mexico and the southernmost U.S. by breeding with local bees producing populations retaining most of the African bee's traits — called also *Africanized honeybee, killer bee*

African mahogany *noun* (1842)
: MAHOGANY 1b

African violet *noun* (1902)
: any of several tropical African gesneriads (especially *Saintpaulia ionantha*) widely grown as houseplants for their velvety fleshy leaves and showy purple, pink, or white flowers

¹Af·ri·kaans \ˌa-fri-ˈkän(t)s, ˌä-, -ˈkänz, ˈa-fri-ˌ, ˈä-\ *noun* [Afrikaans, from *afrikaans,* adjective, African, from obsolete Afrikaans *afrikanisch,* from Latin *africanus*] (1908)
: a language developed from 17th century Dutch that is one of the official languages of the Republic of South Africa

²Afrikaans *adjective* (1923)
: of or relating to Afrikaners or Afrikaans

Af·ri·ka·ner \ˌa-fri-ˈkä-nər\ *noun, often attributive* [Afrikaans, literally, African, from Latin *africanus*] (1824)
: a South African of European descent whose native language is Afrikaans
— Af·ri·ka·ner·dom \-dəm\ *noun*

¹Af·ro \ˈa-(ˌ)frō\ *adjective* [probably from *Afro-American*] (1938)
: characterized by or being a hairstyle of tight curls in a full evenly rounded shape

²Afro *noun, plural* **Afros** (1968)
: an Afro hairstyle

Af·ro–Amer·i·can \ˌa-frō-ə-ˈmer-ə-kən\ *noun* (1853)
: an American of African and especially of black African descent
— Afro–American *adjective*

Afro

Af·ro–Asi·at·ic \ˌa-frō-ˌā-zhē-ˈa-tik, -zē- *also* -shē-\ *adjective* (1953)
: of, relating to, or being a family of languages widely distributed over southwestern Asia and Africa comprising the Semitic, Egyptian, Berber, Cushitic, and Chadic subfamilies

Af·ro·cen·tric \ˌa-frō-ˈsen-trik\ *adjective* (1966)
: centered on or derived from Africa or the Africans
— Af·ro·cen·trism \-ˌtri-zəm\ *noun*

¹aft \ˈaft\ *adverb* [Middle English *afte* back, from Old English *æftan* from behind, behind; akin to Old English *æfter*] (1628)
: near, toward, or in the stern of a ship or the tail of an aircraft **:** ABAFT ⟨called all hands *aft*⟩

²aft *adjective* (1816)
: REARWARD, AFTER 2 ⟨the *aft* decks⟩

³aft *Scottish variant of* OFT

¹af·ter \ˈaf-tər\ *adverb* [Middle English, from Old English *æfter;* akin to Old High German *aftar* after, and probably to Old English *of* of] (before 12th century)
: following in time or place **:** AFTERWARD, BEHIND, LATER ⟨we arrived shortly *after*⟩ ⟨returned 20 years *after*⟩

²after *preposition* (before 12th century)
1 a : behind in place **b** (1) **:** subsequent to in time or order (2) **:** subsequent to and in view of ⟨*after* all our advice⟩
2 — used as a function word to indicate the object of a stated or implied action ⟨go *after* gold⟩ ⟨was asking *after* you⟩
3 : so as to resemble: as **a :** in accordance with **b :** with the name of or a name derived from that of **c :** in the characteristic manner of **d :** in imitation of

³after *conjunction* (before 12th century)
: subsequently to the time when

⁴after *adjective* (before 12th century)
1 : later in time ⟨in *after* years⟩
2 : located toward the rear and especially toward the stern of a ship or tail of an aircraft

⁵af·ter \ˈäf-tər\ *verbal auxiliary* (1800)
chiefly Irish — used with a present participle to indicate action completed and especially just completed ⟨the poor old man is *after* dying on me —J. M. Synge⟩

⁶after *noun* (circa 1902)
: AFTERNOON

after all *adverb* (1846)
1 : in spite of considerations or expectations to the contrary **:** NEVERTHELESS ⟨decided to take the train *after all*⟩ ⟨didn't rain *after all*⟩
2 — used as a sentence modifier to emphasize something to be taken into consideration ⟨literature which is *after all* only a special department of reading —W. W. Watt⟩

af·ter·birth \ˈaf-tər-ˌbərth\ *noun* (1587)
: the placenta and fetal membranes that are expelled after delivery

af·ter·burn·er \-ˌbər-nər\ *noun* (1947)
1 : a device incorporated into the tail pipe of a turbojet engine for injecting fuel into the hot exhaust gases and burning it to provide extra thrust
2 : a device for burning or catalytically destroying unburned or partially burned carbon compounds in exhaust (as from an automobile)

af·ter·care \-ˌker, -ˌkar\ *noun* (1894)
: the care, treatment, help, or supervision given to persons discharged from an institution (as a hospital)

af·ter·clap \-ˌklap\ *noun* (14th century)

: an unexpected damaging or unsettling event following a supposedly closed affair

af·ter·deck \-ˌdek\ *noun* (1897)
: the part of a deck abaft amidships

af·ter·ef·fect \'af-tə-rə-ˌfekt\ *noun* (1817)
: an effect that follows its cause after an interval

af·ter·glow \'af-tər-ˌglō\ *noun* (1871)
1 : a reflection of past splendor, success, or emotion
2 : a glow remaining where a light has disappeared

af·ter–hours \ˌaf-tər-'au̇(-ə)rz\ *adjective* (1929)
: engaged in or operating after a legal or conventional closing time ⟨*after-hours* drinking⟩ ⟨an *after-hours* club⟩

af·ter·im·age \'af-tər-ˌi-mij\ *noun* (1874)
: a usually visual sensation occurring after stimulation by its external cause has ceased

af·ter·life \'af-tər-ˌlīf\ *noun* (circa 1593)
1 : an existence after death
2 : a later period in one's life

af·ter·mar·ket \-ˌmär-kət\ *noun* (1940)
1 : the market for parts and accessories used in the repair or enhancement of a product (as an automobile)
2 : a secondary market available after sales in the original market are finished ⟨a movie in the videocassette *aftermarket*⟩

af·ter·math \-ˌmath\ *noun* [⁴*after* + *math* (mowing, crop)] (1523)
1 : a second-growth crop — called also *rowen*
2 : CONSEQUENCE, RESULT ⟨stricken with guilt as an *aftermath* of the accident⟩
3 : the period immediately following a usually ruinous event ⟨in the *aftermath* of the war⟩

af·ter·most \-ˌmōst\ *adjective* (1773)
: farthest aft

af·ter·noon \ˌaf-tər-'nün\ *noun* (13th century)
1 : the part of day between noon and sunset
2 : a relatively late period (as of time or life) ⟨in the *afternoon* of the 19th century⟩
— afternoon *adjective*

af·ter·noons \-'nünz\ *adverb* (1896)
: in the afternoon repeatedly : on any afternoon

af·ter·piece \'af-tər-ˌpēs\ *noun* (1779)
: a short usually comic entertainment performed after a play

af·ters \'af-tərz\ *noun plural* (circa 1909)
British : DESSERT

af·ter·shave \'af-tər-ˌshāv\ *noun* (1946)
: a usually scented lotion for use on the face after shaving

af·ter·shock \-ˌshäk\ *noun* (1894)
1 : a minor shock following the main shock of an earthquake
2 : an aftereffect of a distressing or traumatic event

af·ter·taste \-ˌtāst\ *noun* (circa 1798)
: persistence of a sensation (as of flavor or an emotion) after the stimulating agent or experience has gone

af·ter–tax \-'taks\ *adjective* (1954)
: remaining after payment of taxes and especially of income tax ⟨an *after-tax* profit⟩

af·ter·thought \-ˌthȯt\ *noun* (circa 1661)
1 : an idea occurring later
2 : a part, feature, or device not thought of originally

af·ter·time \-ˌtīm\ *noun* (1597)
: FUTURE

af·ter·ward \'af-tə(r)-wərd\ *or* **af·ter·wards** \-wərdz\ *adverb* (13th century)
: at a later or succeeding time : SUBSEQUENTLY, THEREAFTER

af·ter·word \-ˌwərd\ *noun* (1890)
: EPILOGUE 1

af·ter·world \-ˌwərld\ *noun* (1596)
: a future world : a world after death

ag \'ag\ *adjective* (circa 1918)
: of or relating to agriculture ⟨*ag* schools⟩

ag- — see AD-

Aga·da \ə-'gä-də, -'gō-\ *variant of* HAGGADAH

again \ə-'gen, -'gin, -'gān\ *adverb* [Middle English, opposite, again, from Old English *ongēan* opposite, back, from *on* + *gēn, gēan* still, again; akin to Old English *gēan-* against, Old High German *gegin* against, toward] (13th century)
1 : in return : BACK ⟨swore he would pay him *again* when he was able —Shakespeare⟩
2 : another time : once more : ANEW ⟨I shall not look upon his like *again* —Shakespeare⟩
3 : on the other hand ⟨he might go, and *again* he might not⟩
4 : in addition : BESIDES ⟨*again*, there is another matter to consider⟩

again and again *adverb* (1604)
: OFTEN, REPEATEDLY

¹against \ə-'gen(t)st, -'gin(t)st, -'gān(t)st\ *preposition* [Middle English, alteration of *againes*, from *again*] (13th century)
1 a : in opposition or hostility to **b :** contrary to ⟨*against* the law⟩ **c :** in competition with **d :** as a basis for disapproval of ⟨had nothing *against* him⟩
2 a : directly opposite : FACING **b** *obsolete* : exposed to
3 : compared or contrasted with
4 a : in preparation or provision for **b :** as a defense or protection from
5 a : in the direction of and into contact with **b :** in contact with
6 : in a direction opposite to the motion or course of : counter to
7 a : as a counterbalance to **b :** in exchange for **c :** as a charge on
8 : before the background of

²against *conjunction* (14th century)
archaic : in preparation for the time when ⟨throw on another log of wood *against* father comes home —Charles Dickens⟩

Ag·a·mem·non \ˌa-gə-'mem-ˌnän, -nən\ *noun* [Latin, from Greek *Agamemnōn*]
: a king of Mycenae and leader of the Greeks in the Trojan War

aga·mete \(ˌ)ā-'ga-ˌmet *also* -gə-'mēt\ *noun* [International Scientific Vocabulary, from Greek *agametos* unmarried, from *a-* + *gamein* to marry, from *gamos* marriage] (circa 1920)
: an asexual reproductive cell (as a spore)

agam·ic \(ˌ)ā-'ga-mik\ *adjective* [Greek *agamos* unmarried, from *a-* + *gamos*] (1850)
: ASEXUAL, PARTHENOGENETIC

agam·ma·glob·u·lin·emia \ˌā-ˌga-mə-ˌglä-byə-lə-'nē-mē-ə\ *noun* [New Latin, from *a-* + International Scientific Vocabulary *gamma globulin* + New Latin *-emia*] (circa 1952)
: a condition in which the body forms few or no gamma globulins or antibodies
— agam·ma·glob·u·lin·emic \-'nē-mik\ *adjective*

aga·mo·sper·my \(ˌ)ā-'ga-mə-ˌspər-mē, 'a-gə-mō-ˌspər-\ *noun* [Greek *agamos* + English *-spermy*] (1944)
: APOGAMY; *specifically* : apogamy in which sexual union is not completed and the embryo is produced from the innermost layer of the integument of the female gametophyte

ag·a·pan·thus \ˌa-gə-'pan(t)-thəs\ *noun, plural* **-thus** *also* **-thuses** [New Latin, genus name, from Greek *agapē* + *anthos* flower — more at ANTHOLOGY] (circa 1789)
: any of several African plants (genus *Agapanthus*) of the lily family cultivated for their umbels of showy blue or purple flowers

¹aga·pe \ä-'gä-(ˌ)pā, 'ä-gə-ˌpā\ *noun* [Late Latin, from Greek *agapē*, literally, love] (1607)
1 : LOVE FEAST
2 : LOVE 4a

²agape \ə-'gāp *also* -'gap\ *adjective or adverb* (1667)
1 : wide open : GAPING
2 : being in a state of wonder

agar \'ä-gər\ *noun* [Malay *agar-agar*] (1889)
1 : a gelatinous colloidal extractive of a red alga (as of the genera *Gelidium, Gracilaria,* and *Eucheuma*) used especially in culture media or as a gelling and stabilizing agent in foods
2 : a culture medium containing agar

agar–agar \ˌä-gər-'ä-gər\ *noun* [Malay] (1820)
: AGAR

aga·ric \'a-gə-rik, ə-'gar-ik\ *noun* [Latin *agaricum*, a fungus, from Greek *agarikon*] (15th century)
1 : the dried fruiting body of a fungus (*Fomes officinalis* synonym *Polyporus officinalis*) formerly used in medicine
2 : any of a family (Agaricaceae) of fungi with the sporophore usually resembling an umbrella and with numerous gills on the underside of the cap

aga·rose \'a-gə-ˌrōs, 'ä-, -ˌrōz\ *noun* (1964)
: a polysaccharide obtained from agar and used especially as a supporting medium in gel electrophoresis

ag·ate \'a-gət\ *noun, often attributive* [Middle French, from Latin *achates,* from Greek *achatēs*] (1570)
1 : a fine-grained variegated chalcedony having its colors arranged in stripes, blended in clouds, or showing mosslike forms
2 : something made of or fitted with agate: as **a :** a drawplate used by gold-wire drawers **b :** a playing marble of agate
3 : a size of type approximately 5½ point

agate line *noun* (circa 1935)
: a space one column wide and 1/14 inch deep used as a unit of measurement in classified advertising

agate ware *noun* (1857)
1 : pottery veined and mottled to resemble agate
2 : an enameled iron or steel ware for household utensils

aga·ve \ə-'gä-vē\ *noun* [New Latin *Agave,* genus name, from Latin, a daughter of Cadmus, from Greek *Agauē*] (circa 1797)
: any of a genus (*Agave* of the family Agavaceae, the agave family) of plants having spiny-margined leaves and flowers in tall spreading panicles and including some cultivated for their fiber or sap or for ornament

agave

agaze \ə-'gāz\ *adjective* (circa 1902)
: engaged in the act of gazing

¹age \'āj\ *noun* [Middle English, from Old French *aage,* from (assumed) Vulgar Latin *aetaticum,* from Latin *aetat-, aetas,* from *aevum* lifetime — more at AYE] (13th century)
1 a : the time of life at which some particular qualification, power, or capacity arises or rests ⟨the voting *age* is 18⟩; *specifically* : MAJORITY **b :** one of the stages of life **c :** the length of an existence extending from the beginning to any given time ⟨a boy 10 years of *age*⟩ **d :** LIFETIME **e :** an advanced stage of life
2 : a period of time dominated by a central figure or prominent feature ⟨the *age* of Pericles⟩: as **a :** a period in history or human progress ⟨the *age* of reptiles⟩ ⟨the *age* of exploration⟩ **b :** a cultural period marked by the prominence of a particular item ⟨entering the atomic *age*⟩ **c :** a division of geologic time that is usually shorter than an epoch
3 a : the period contemporary with a person's lifetime or with his or her active life **b :** a long time — usually used in plural ⟨haven't seen him in *ages*⟩ **c :** GENERATION
4 : an individual's development measured in terms of the years requisite for like development of an average individual
synonym see PERIOD

²age *verb* **aged; ag·ing** *or* **age·ing** (14th century)
intransitive verb
1 : to become old **:** show the effects or the characteristics of increasing age
2 : to acquire a desirable quality (as mellowness or ripeness) by standing undisturbed for some time ⟨letting cheese *age*⟩
transitive verb
1 : to cause to become old
2 : to bring to a state fit for use or to maturity
— ag·er \ˈā-jər\ *noun*

-age *noun suffix* [Middle English, from Old French, from Latin *-aticum*]
1 : aggregate **:** collection ⟨track*age*⟩
2 a : action **:** process ⟨haul*age*⟩ **b :** cumulative result of ⟨break*age*⟩ **c :** rate of ⟨dos*age*⟩
3 : house or place of ⟨orphan*age*⟩
4 : state **:** rank ⟨peon*age*⟩
5 : charge ⟨post*age*⟩

aged \ˈā-jəd, ˈājd; ˈājd *for 1b*\ *adjective* (15th century)
1 : grown old: as **a :** of an advanced age **b :** having attained a specified age ⟨a man *aged* 40 years⟩
2 : typical of old age
— aged·ness \ˈā-jəd-nəs\ *noun*

age–group \ˈāj-ˌgrüp\ *noun* (1904)
: a segment of a population that is of approximately the same age or is within a specified range of ages

age·ism *also* **ag·ism** \ˈā-(ˌ)ji-zəm\ *noun* (1969)
: prejudice or discrimination against a particular age-group and especially the elderly
— age·ist *also* **ag·ist** \-jist\ *adjective*

age·less \ˈāj-ləs\ *adjective* (1651)
1 : not growing old or showing the effects of age
2 : TIMELESS, ETERNAL ⟨*ageless* truths⟩
— age·less·ly *adverb*
— age·less·ness *noun*

age·long \ˈāj-ˌlȯŋ\ *adjective* (1810)
: lasting for an age **:** EVERLASTING

age–mate \-ˌmāt\ *noun* (1583)
: one who is of about the same age as another

agen·cy \ˈā-jən(t)-sē\ *noun, plural* **-cies** (1658)
1 : the capacity, condition, or state of acting or of exerting power **:** OPERATION
2 : a person or thing through which power is exerted or an end is achieved **:** INSTRUMENTALITY ⟨communicated through the *agency* of the ambassador⟩
3 a : the office or function of an agent **b :** the relationship between a principal and his agent
4 : an establishment engaged in doing business for another ⟨an advertising *agency*⟩
5 : an administrative division (as of a government) ⟨the *agency* for consumer protection⟩

agency shop *noun* (circa 1946)
: a shop in which the union serves as the agent for and receives dues and assessments from all employees in the bargaining unit regardless of union membership

agen·da \ə-ˈjen-də\ *noun* [Latin, neuter plural of *agendum,* gerundive of *agere*] (1871)
1 : a list or outline of things to be considered or done ⟨*agendas* of faculty meetings⟩
2 : an underlying often ideological plan or program ⟨a political *agenda*⟩
— agen·da·less \-də-ləs\ *adjective*

agen·dum \-dəm\ *noun, plural* **-da** \-də\ *or* **-dums** [Latin] (circa 1847)
1 : AGENDA
2 : an item on an agenda

agen·e·sis \(ˌ)ā-ˈje-nə-səs\ *noun* [New Latin] (circa 1879)
: lack or failure of development (as of a body part)

agent \ˈā-jənt\ *noun* [Middle English, from Medieval Latin *agent-, agens,* from Latin, present participle of *agere* to drive, lead, act, do; akin to Old Norse *aka* to travel in a vehicle, Greek *agein* to drive, lead] (15th century)
1 : one that acts or exerts power
2 a : something that produces or is capable of producing an effect **:** an active or efficient cause **b :** a chemically, physically, or biologically active principle
3 : a means or instrument by which a guiding intelligence achieves a result
4 : one who is authorized to act for or in the place of another: as **a :** a representative, emissary, or official of a government ⟨crown *agent*⟩ ⟨federal *agent*⟩ **b :** one engaged in undercover activities (as espionage) **:** SPY ⟨secret *agent*⟩ **c :** a business representative (as of an athlete or entertainer) ⟨a theatrical *agent*⟩

agent–general *noun, plural* **agents–general** (1833)
: a chief agent; *specifically* **:** the representative in England of a British dominion

agent·ing \ˈā-jən-tiŋ\ *noun* (1681)
: the business or activities of an agent

Agent Orange *noun* [so called from the identifying color stripe on its container] (1970)
: an herbicide widely used as a defoliant in the Vietnam War that is composed of 2,4-D and 2,4,5-T and contains dioxin as a contaminant

agent pro·vo·ca·teur \ˈä-ˌzhäⁿ-prō-ˌvä-kə-ˈtər, ˈā-jənt-\ *noun, plural* **agents provocateurs** \ˈä-ˌzhäⁿ-prō-ˌvä-kə-ˈtər, ˈā-jən(t)s-prō-\ [French, literally, provoking agent] (1877)
: one employed to associate with suspected persons and by pretending sympathy with their aims to incite them to some incriminating action

agent·ry \ˈā-jən-trē\ *noun, plural* **-ries** (1925)
: the office, duties, or activities of an agent

age of consent (1504)
: the age at which one is legally competent to give consent especially to marriage or to sexual intercourse

age of reason (circa 1794)
1 : a period characterized by a prevailing belief in the use of reason; *especially* **:** the 18th century in England and France
2 : the time of life when one begins to be able to distinguish right from wrong

age–old \ˈā-ˈjōld\ *adjective* (1904)
: having existed for ages **:** ANCIENT

ag·er·a·tum \ˌa-jə-ˈrā-təm\ *noun, plural* **-tum** *also* **-tums** [New Latin, genus name, from Greek *agēratos* ageless, from *a-* + *gēras* old age — more at GERONT-] (1866)
: any of a genus (*Ageratum*) of tropical American composite herbs often cultivated for their small showy heads of blue or white flowers; *also* **:** a related blue-flowered plant (*Eupatorium coelestinum*)

Ag·ga·dah \ə-ˈgä-də, -ˈgȯ-\ *variant of* HAGGADAH

Ag·ge·us \a-ˈgē-əs\ *noun* [Late Latin *Aggaeus,* from Greek *Aggaios,* from Hebrew *Ḥaggai*]
: HAGGAI

¹ag·gie \ˈa-gē\ *noun, often capitalized* [*agri*cultural + *-ie*] (1902)
: an agricultural school or college; *also* **:** a student at such an institution

²aggie *noun* [agate + *-ie*] (1915)
: a playing marble; *specifically* **:** AGATE 2b

ag·gior·na·men·to \ə-ˌjȯr-nə-ˈmen-(ˌ)tō\ *noun, plural* **-tos** [Italian, from *aggiornare* to bring up to date, from *a* to (from Latin *ad-*) + *giorno* day, from Late Latin *diurnum* day — more at JOURNEY] (1963)
: a bringing up to date **:** MODERNIZATION ⟨dedicated to the *aggiornamento* of the church⟩

¹ag·glom·er·ate \ə-ˈglä-mə-ˌrāt\ *transitive verb* **-at·ed; -at·ing** [Latin *agglomeratus,* past participle of *agglomerare* to heap up, join, from *ad-* + *glomer-, glomus* ball — more at CLAM] (1684)
: to gather into a ball, mass, or cluster

²ag·glom·er·ate \-rət\ *adjective* (1828)
: gathered into a ball, mass, or cluster; *specifically* **:** clustered or growing together but not coherent ⟨an *agglomerate* flower head⟩

³ag·glom·er·ate \-rət\ *noun* (1830)
1 : a rock composed of volcanic fragments of various sizes and degrees of angularity
2 : a jumbled mass or collection **:** AGGLOMERATION

ag·glom·er·a·tion \ə-ˌglä-mə-ˈrā-shən\ *noun* (1774)
1 : the action or process of collecting in a mass
2 : a heap or cluster of usually disparate elements ⟨urban *agglomerations* knit together by the new railways —*Times Literary Supplement*⟩
— ag·glom·er·a·tive \-ˈglä-mə-ˌrā-tiv\ *adjective*

ag·glu·ti·na·bil·i·ty \ə-ˌglü-tᵊn-ə-ˈbi-lə-tē\ *noun* (1901)
: capacity (as of red blood cells) to be agglutinated
— ag·glu·ti·na·ble \ə-ˈglü-tᵊn-ə-bəl\ *adjective*

¹ag·glu·ti·nate \ə-ˈglü-tᵊn-ˌāt\ *verb* **-nat·ed; -nat·ing** [Latin *agglutinatus,* past participle of *agglutinare* to glue to, from *ad-* + *glutinare* to glue, from *glutin-, gluten* glue — more at CLAY] (1586)
transitive verb
1 : to cause to adhere **:** FASTEN
2 : to combine into a compound **:** attach to a base as an affix
3 : to cause to undergo agglutination
intransitive verb
1 : to unite or combine into a group or mass
2 : to form words by agglutination

²ag·glu·ti·nate \-tᵊn-ət, -tᵊn-ˌāt\ *noun* (1952)
: a clump of agglutinated material (as blood cells or mineral particles in soil)

ag·glu·ti·na·tion \ə-ˌglü-tᵊn-ˈā-shən\ *noun* (1541)
1 : the action or process of agglutinating
2 : a mass or group formed by the union of separate elements
3 : the formation of derivational or inflectional words by putting together constituents of which each expresses a single definite meaning
4 : a reaction in which particles (as red blood cells or bacteria) suspended in a liquid collect into clumps and which occurs especially as a serologic response to a specific antibody

ag·glu·ti·na·tive \ə-ˈglü-tᵊn-ˌā-tiv, -ə-tiv\ *adjective* (1634)
1 : ADHESIVE
2 : characterized by linguistic agglutination

ag·glu·ti·nin \ə-ˈglü-tᵊn-ən\ *noun* [International Scientific Vocabulary *agglutin*ation + *-in*] (1902)
: a substance (as an antibody) producing agglutination

ag·glu·ti·no·gen \ə-ˈglü-tᵊn-ə-jən\ *noun* [*agglutin*in + *-o-* + *-gen*] (1904)
: an antigen whose presence results in the formation of an agglutinin
— ag·glu·ti·no·gen·ic \-ˌglü-tᵊn-ə-ˈje-nik\ *adjective*

ag·gra·da·tion \ˌa-grə-ˈdā-shən\ *noun* [*ad-* + *gradation*] (1898)
: a modification of the earth's surface in the direction of uniformity of grade by deposition

ag·gran·dise *British variant of* AGGRANDIZE

ag·gran·dize \ə-ˈgran-ˌdīz *also* ˈa-grən-\ *transitive verb* **-dized; -diz·ing** [French *agrandiss-,* stem of *agrandir,* from *a-* (from Latin *ad-*) + *grandir* to increase, from Latin *grandire,* from *grandis* great] (1634)
1 : to make great or greater **:** INCREASE, ENLARGE
2 : to make appear great or greater **:** praise highly
3 : to enhance the power, wealth, position, or reputation of ⟨exploited the situation to *aggrandize* himself⟩
— ag·gran·dize·ment \ə-ˈgran-dəz-mənt, -ˌdīz- *also* ˌa-grən-ˈdīz-\ *noun*
— ag·gran·diz·er \ə-ˈgran-ˌdī-zər *also* ˈa-grən-\ *noun*

ag·gra·vate \'a-grə-ˌvāt\ *transitive verb* **-vat·ed; -vat·ing** [Latin *aggravatus,* past participle of *aggravare* to make heavier, from *ad-* + *gravare* to burden, from *gravis* heavy — more at GRIEVE] (1530)
1 *obsolete* **a :** to make heavy **:** BURDEN **b :** INCREASE
2 : to make worse, more serious, or more severe **:** intensify unpleasantly ⟨problems have been *aggravated* by neglect⟩
3 a : to rouse to displeasure or anger by usually persistent and often petty goading **b :** to produce inflammation in ■

aggravated assault *noun* (1925)
: an assault that is more serious than a common assault: as **a :** an assault combined with an intent to commit a crime **b :** any of various assaults so defined by statute

aggravating *adjective* (1775)
: arousing displeasure, impatience, or anger
usage see AGGRAVATE

ag·gra·va·tion \ˌa-grə-'vā-shən\ *noun* (circa 1555)
1 : an act or circumstance that intensifies or makes worse
2 : the act, action, or result of aggravating; *especially* **:** an increasing in seriousness or severity
3 : IRRITATION, PROVOCATION
usage see AGGRAVATE

[1]**ag·gre·gate** \'a-gri-gət\ *adjective* [Middle English *aggregat,* from Latin *aggregatus,* past participle of *aggregare* to add to, from *ad-* + *greg-, grex* flock] (15th century)
: formed by the collection of units or particles into a body, mass, or amount **:** COLLECTIVE: as **a** (1) **:** clustered in a dense mass or head ⟨an *aggregate* flower⟩ (2) **:** formed from several separate ovaries of a single flower ⟨*aggregate* fruit⟩ **b :** composed of mineral crystals of one or more kinds or of mineral rock fragments **c :** taking all units as a whole ⟨*aggregate* sales⟩
— ag·gre·gate·ly *adverb*
— ag·gre·gate·ness *noun*

[2]**ag·gre·gate** \-ˌgāt\ *transitive verb* **-gat·ed; -gat·ing** (15th century)
1 : to collect or gather into a mass or whole
2 : to amount in the aggregate to **:** TOTAL

[3]**ag·gre·gate** \-gət\ *noun* (15th century)
1 : a mass or body of units or parts somewhat loosely associated with one another
2 : the whole sum or amount **:** SUM TOTAL
3 a : an aggregate rock **b :** any of several hard inert materials (as sand, gravel, or slag) used for mixing with a cementing material to form concrete, mortar, or plaster **c :** a clustered mass of individual soil particles varied in shape, ranging in size from a microscopic granule to a small crumb, and considered the basic structural unit of soil
4 : SET 21
5 : MONETARY AGGREGATE
— in the aggregate : considered as a whole **:** COLLECTIVELY ⟨dividends for the year amounted *in the aggregate* to 25 million dollars⟩

ag·gre·ga·tion \ˌa-gri-'gā-shən\ *noun* (1547)
1 : a group, body, or mass composed of many distinct parts or individuals
2 a : the collecting of units or parts into a mass or whole **b :** the condition of being so collected
— ag·gre·ga·tion·al \-shnəl, -shə-nᵊl\ *adjective*

ag·gre·ga·tive \'a-gri-ˌgā-tiv\ *adjective* (1644)
1 : of or relating to an aggregate
2 : tending to aggregate
— ag·gre·ga·tive·ly *adverb*

ag·gress \ə-'gres\ *intransitive verb* (circa 1714)
: to commit aggression **:** act aggressively

ag·gres·sion \ə-'gre-shən\ *noun* [Latin *aggression-, aggressio* attack, from *aggredi* to attack, from *ad-* + *gradi* to step, go — more at GRADE] (1611)
1 : a forceful action or procedure (as an unprovoked attack) especially when intended to dominate or master
2 : the practice of making attacks or encroachments; *especially* **:** unprovoked violation by one country of the territorial integrity of another
3 : hostile, injurious, or destructive behavior or outlook especially when caused by frustration

ag·gres·sive \ə-'gre-siv\ *adjective* (1824)
1 a : tending toward or exhibiting aggression ⟨*aggressive* behavior⟩ **b :** marked by combative readiness ⟨an *aggressive* fighter⟩
2 a : marked by obtrusive energy **b :** marked by driving forceful energy or initiative **:** ENTERPRISING ⟨an *aggressive* salesman⟩
3 : strong or emphatic in effect or intent ⟨*aggressive* colors⟩ ⟨*aggressive* flavors⟩
4 : more severe, intensive, or comprehensive than usual especially in dosage or extent ⟨*aggressive* chemotherapy⟩ ☆
— ag·gres·sive·ly *adverb*
— ag·gres·sive·ness *noun*
— ag·gres·siv·i·ty \ˌa-ˌgre-'si-və-tē\ *noun*

ag·gres·sor \ə-'gre-sər\ *noun* (1646)
: one that commits or practices aggression

ag·grieve \ə-'grēv\ *transitive verb* **ag·grieved; ag·griev·ing** [Middle English *agreven,* from Middle French *agrever,* from Latin *aggravare* to make heavier] (14th century)
1 : to give pain or trouble to **:** DISTRESS
2 : to inflict injury on
synonym see WRONG

ag·grieved \ə-'grēvd\ *adjective* (14th century)
1 : troubled or distressed in spirit
2 a : suffering from an infringement or denial of legal rights ⟨*aggrieved* minority groups⟩ **b :** showing or expressing grief, injury, or offense ⟨an *aggrieved* plea⟩
— ag·griev·ed·ly \-'grē-vəd-lē\ *adverb*

ag·grieve·ment \ə-'grēv-mənt\ *noun* (1847)
: the quality or state of being aggrieved

ag·gro \'a-(ˌ)grō\ *noun, plural* **aggros** [probably *aggravation* + [1]*-o*] (1969)
1 *British* **:** deliberately aggressive, provoking, or violent behavior
2 *British* **:** EXASPERATION, IRRITATION

aghast \ə-'gast\ *adjective* [Middle English *agast,* from past participle of *agasten* to frighten, from *a-* (perfective prefix) + *gasten* to frighten — more at ABIDE, GAST] (13th century)
: struck with terror, amazement, or horror **:** SHOCKED

ag·ile \'a-jəl, -ˌjīl\ *adjective* [Middle French, from Latin *agilis,* from *agere* to drive, act — more at AGENT] (1581)
1 : marked by ready ability to move with quick easy grace
2 : having a quick resourceful and adaptable character ⟨an *agile* mind⟩
— ag·ile·ly \-jə(l)-lē, -ˌjī(l)-lē\ *adverb*

agil·i·ty \ə-'ji-lə-tē\ *noun, plural* **-ties** (15th century)
: the quality or state of being agile **:** NIMBLENESS, DEXTERITY ⟨played with increasing *agility*⟩

agin \ə-'gin\ *dialect variant of* AGAINST

aging *present participle of* AGE

ag·ism *variant of* AGEISM

ag·i·tate \'a-jə-ˌtāt\ *verb* **-tat·ed; -tat·ing** [Latin *agitatus,* past participle of *agitare,* frequentative of *agere* to drive — more at AGENT] (15th century)
transitive verb
1 a *obsolete* **:** to give motion to **b :** to move with an irregular, rapid, or violent action ⟨the storm *agitated* the sea⟩
2 : to excite and often trouble the mind or feelings of **:** DISTURB
3 a : to discuss excitedly and earnestly **b :** to stir up public discussion of
intransitive verb
: to attempt to arouse public feeling ⟨*agitated* for better schools⟩
synonym see SHAKE, DISCOMPOSE
— ag·i·tat·ed·ly *adverb*
— ag·i·ta·tion \ˌa-jə-'tā-shən\ *noun*
— ag·i·ta·tion·al \-shnəl, -shə-nᵊl\ *adjective*

ag·i·ta·tive \'a-jə-ˌtā-tiv\ *adjective* (1687)
: causing or tending to cause agitation

ag·i·ta·to \ˌa-jə-'tä-(ˌ)tō\ *adverb or adjective* [Italian, literally, agitated, from Latin *agitatus*] (circa 1801)
: in a restless and agitated manner — used as a direction in music

ag·i·ta·tor \'a-jə-ˌtā-tər\ *noun* (circa 1734)
: one that agitates: as **a :** one who stirs up public feeling on controversial issues ⟨political *agitators*⟩ **b :** a device or an apparatus for stirring or shaking

ag·it·prop \'a-jət-ˌpräp\ *noun* [Russian, from *agitatsiya* agitation + *propaganda*] (1935)
: PROPAGANDA; *especially* **:** political propaganda promulgated chiefly in literature, drama, music, or art
— agitprop *adjective*

Aglaia \ə-'glī-ə, -'glā-ə\ *noun* [Latin, from Greek]
: one of the three Graces

aglare \ə-'glar, -'gler\ *adjective* (1872)
: GLARING ⟨his eyes *aglare* with fury⟩

agleam \ə-'glēm\ *adjective* (1870)
: gleaming especially with reflected light

☆ SYNONYMS
Aggressive, militant, assertive, self-assertive mean obtrusively energetic especially in pursuing particular goals. AGGRESSIVE implies a disposition to dominate often in disregard of others' rights or in determined and energetic pursuit of one's ends ⟨was taught to be *aggressive* in his business dealings⟩. MILITANT also implies a fighting disposition but suggests not self-seeking but devotion to a cause, movement, or principle ⟨*militant* protesters held a rally against racism⟩. ASSERTIVE suggests bold self-confidence in expression of opinion ⟨the more *assertive* speakers dominated the forum⟩. SELF-ASSERTIVE connotes forwardness or brash self-confidence ⟨a *self-assertive* young executive climbing the corporate ladder⟩.

□ USAGE
aggravate Although *aggravate* has been used in sense 3a since the 17th century, it has been the object of disapproval only since about 1870. It is used in expository prose ⟨when his silly conceit . . . about his not-very-good early work has begun to *aggravate* us —William Styron⟩ but seems to be more common in speech and casual writing ⟨a good profession for him, because bus drivers get *aggravated* —Jackie Gleason (interview, 1986)⟩ ⟨& now this letter comes to *aggravate* me a thousand times worse —Mark Twain (letter, 1864)⟩. Sense 2 is far more common than sense 3a in published prose. Such is not the case, however, with *aggravation* and *aggravating. Aggravation* is used in sense 3 somewhat more than in its earlier senses; *aggravating* has practically no use other than to express annoyance.

ag·let \'a-glət\ *noun* [Middle English, from Middle French *aguillette, aiguillette,* diminutive of *aguille, aiguille* needle, from Late Latin *acicula, acucula* ornamental pin, diminutive of Latin *acus* needle, pin — more at ACUTE] (15th century)
1 : the plain or ornamental tag covering the ends of a lace or point
2 : any of various ornamental studs, cords, or pins worn on clothing

agley \ə-'glā, -'glē, -'glī\ *adverb* [Scots, from 1*a-* + *gley* to squint] (1785)
chiefly Scottish **:** AWRY, WRONG ⟨the best-laid schemes o' mice an' men gang aft *agley* —Robert Burns⟩

aglit·ter \ə-'gli-tər\ *adjective* (1865)
: glittering especially with reflected light

aglow \ə-'glō\ *adjective* (1817)
: glowing especially with warmth or excitement

agly·cone \a-'glī-ˌkōn\ *also* **agly·con** \-ˌkän\ *noun* [International Scientific Vocabulary *a-* (from Greek *ha-, a-* together) + *glyc-* + *-one, -on*] (1925)
: an organic compound (as a phenol or alcohol) combined with the sugar portion of a glycoside

1ag·nate \'ag-ˌnāt\ *noun* [Latin *agnatus,* from past participle of *agnasci* to be born in addition to, from *ad-* + *nasci* to be born — more at NATION] (1534)
1 : a relative whose kinship is traceable exclusively through males
2 : a paternal kinsman

2agnate *adjective* (1782)
1 : ALLIED, AKIN
2 : related through male descent or on the father's side
— ag·nat·ic \ag-'na-tik\ *adjective*

Ag·ne·an \'äg-nē-ən\ *noun* [*Agni,* ancient kingdom in Turkestan] (1939)
: TOCHARIAN A

ag·nize \ag-'nīz\ *transitive verb* **ag·nized; ag·niz·ing** [Latin *agno*scere to acknowledge (from *ad-* + *noscere* to know) + English *-ize* (as in *recognize*) — more at KNOW] (1535)
archaic **:** RECOGNIZE, ACKNOWLEDGE

ag·no·men \ag-'nō-mən\ *noun, plural* **-nom·i·na** \-'nä-mə-nə\ *or* **-no·mens** [Latin, irregular from *ad-* + *nomen* name — more at NAME] (1665)
: an additional cognomen given to a person by the ancient Romans (as in honor of some achievement)

ag·no·sia \ag-'nō-zhə, -shə\ *noun* [New Latin, from Greek *agnōsia* ignorance, from *a-* + *gnōsis* knowledge, from *gignōskein*] (circa 1900)
: loss or diminution of the ability to recognize familiar objects or stimuli usually as a result of brain damage

1ag·nos·tic \ag-'näs-tik, əg-\ *noun* [Greek *agnōstos* unknown, unknowable, from *a-* + *gnōstos* known, from *gignōskein* to know — more at KNOW] (1869)
: a person who holds the view that any ultimate reality (as God) is unknown and probably unknowable; *broadly* **:** one who is not committed to believing in either the existence or the nonexistence of God or a god ◆
— ag·nos·ti·cism \-tə-ˌsi-zəm\ *noun*

2agnostic *adjective* (1873)
1 : of, relating to, or being an agnostic or the beliefs of agnostics
2 : NONCOMMITTAL, UNDOGMATIC

Ag·nus Dei \ˌäg-ˌnüs-'dā(-ˌē), -ˌnüs-; ˌän-yüs-; ˌag-nəs-\ *noun* [Middle English, from Late Latin, lamb of God; from its opening words] (14th century)
1 : a liturgical prayer addressed to Christ as Savior
2 : an image of a lamb often with a halo and a banner and cross that is used as a symbol of Christ

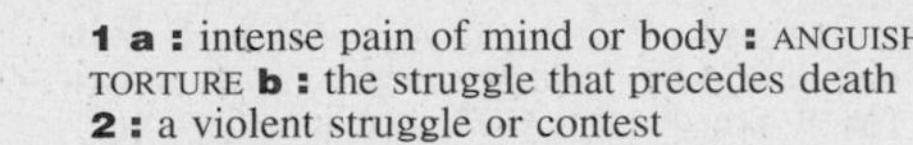
Agnus Dei

ago \ə-'gō\ *adjective or adverb* [Middle English *agon, ago,* from past participle of *agon* to pass away, from Old English *āgān,* from *ā-* (perfective prefix) + *gān* to go — more at ABIDE, GO] (14th century)
: earlier than the present time ⟨10 years *ago*⟩

agog \ə-'gäg\ *adjective* [Middle French *en gogues* in mirth] (1559)
: full of intense interest or excitement **:** EAGER ⟨kids all *agog* over new toys⟩

1a–go–go \ä-'gō-(ˌ)gō, ə-\ *noun* [*Whisky à Gogo,* café and discotheque in Paris, France, from French *à gogo* galore, from Middle French] (1965)
: a nightclub for dancing to pop music **:** DISCO

2a–go–go *adjective* (1965)
1 : GO-GO 1
2 : being in a whirl of motion
3 : being up-to-date — often used postpositively

-agogue *noun combining form* [French & New Latin; French, from Late Latin *-agogus* promoting the expulsion of, from Greek *-agōgos,* from *agein* to lead; New Latin *-agogon,* from Greek, neuter of *-agōgos* — more at AGENT]
: substance that promotes the secretion or expulsion of ⟨emmen*agogue*⟩

agon \'ä-ˌgän, ä-'gōn\ *noun* [Greek *agōn*] (1600)
: CONTEST, CONFLICT; *especially* **:** the dramatic conflict between the chief characters in a literary work

ag·o·nal \'a-gə-nᵊl\ *adjective* (1901)
: of, relating to, or associated with agony and especially the death agony

agone \ə-'gȯn *also* -'gän\ *adjective or adverb* (14th century)
archaic **:** AGO

ag·o·nise, agonised, agonising *British variant of* AGONIZE, AGONIZED, AGONIZING

ag·o·nist \'a-gə-nist\ *noun* [Late Latin *agonista* competitor, from Greek *agōnistēs,* from *agōnizesthai* to contend, from *agōn*] (circa 1623)
1 : one that is engaged in a struggle
2 [from *antagonist*] **a :** a muscle that is controlled by the action of an antagonist with which it is paired **b :** a chemical substance capable of combining with a receptor on a cell and initiating a reaction or activity — compare ANTAGONIST 2b

ag·o·nis·tic \ˌa-gə-'nis-tik\ *adjective* (1648)
1 : of or relating to the athletic contests of ancient Greece
2 : ARGUMENTATIVE
3 : striving for effect **:** STRAINED
4 : of, relating to, or being aggressive or defensive social interaction (as fighting, fleeing, or submitting) between individuals usually of the same species
— ag·o·nis·ti·cal·ly \-ti-k(ə-)lē\ *adverb*

ag·o·nize \'a-gə-ˌnīz\ *verb* **-nized; -niz·ing** (1583)
transitive verb
: to cause to suffer agony **:** TORTURE
intransitive verb
1 : to suffer agony, torture, or anguish ⟨*agonizes* over every decision⟩
2 : STRUGGLE

agonized *adjective* (1583)
: characterized by, suffering, or expressing agony

agonizing *adjective* (1593)
: causing agony **:** PAINFUL
— ag·o·niz·ing·ly *adverb*

ag·o·ny \'a-gə-nē\ *noun, plural* **-nies** [Middle English *agonie,* from Late Latin *agonia,* from Greek *agōnia* struggle, anguish, from *agōn* gathering, contest for a prize, from *agein* to lead, celebrate — more at AGENT] (14th century)
1 a : intense pain of mind or body **:** ANGUISH, TORTURE **b :** the struggle that precedes death
2 : a violent struggle or contest
3 : a strong sudden display (as of joy or delight) **:** OUTBURST ◆
synonym see DISTRESS

agony column *noun* (1863)
: a newspaper column of personal advertisements relating especially to missing relatives or friends

1ag·o·ra \'a-gə-rə\ *noun, plural* **-ras** *or* **-rae** \-ˌrē, -ˌrī\ [Greek, from *ageirein* to gather] (1589)
: a gathering place; *especially* **:** the marketplace in ancient Greece

2ago·ra \ˌä-gə-'rä\ *noun, plural* **ago·rot** \-'rōt\ [New Hebrew *ăgōrāh,* from Hebrew, a small coin] (1963)
— see *shekel* at MONEY table

ag·o·ra·pho·bia \ˌa-g(ə-)rə-'fō-bē-ə\ *noun* [New Latin, from Greek *agora* + New Latin *-phobia*] (1873)
: abnormal fear of being helpless in an embarrassing or unescapable situation that is characterized especially by the avoidance of open or public places
— ag·o·ra·phobe \'a-g(ə-)rə-ˌfōb\ *noun*
— ag·o·ra·pho·bic \-'fō-bik\ *adjective or noun*

agou·ti \ə-'gü-tē\ *noun* [French, from Spanish *agutí,* from Tupi] (1625)
1 : any of a genus (*Dasyprocta*) of tropical American rodents about the size of a rabbit
2 : a grizzled color of fur resulting from the barring of each hair in several alternate dark and light bands

agouti 1

agrafe *or* **agraffe** \ə-'graf\ *noun* [French *agrafe*] (1643)
: a hook-and-loop fastening; *especially* **:** an ornamental clasp used on armor or costumes

◇ WORD HISTORY

agnostic The word *agnostic* owes its origin to a particular occasion on the evening of April 21, 1869, when it was coined by the English biologist Thomas Henry Huxley at an organizational meeting of the Metaphysical Society. The precise etymological route by which Huxley arrived at his coinage has been a matter of debate. Two decades after the fact, the coiner himself recalled that the word seemed "suggestively antithetic to the 'gnostic' of church history" and so he chose it on that basis. In any event, *agnostic* is undoubtedly rooted in *agnōstos,* a Greek word meaning "unknown" or "unknowable." The ending *-ic* was probably suggested by the previously established English word *gnostic.*

agony In ancient Greece the word *agōn* denoted any large public assembly. As the Greeks placed a high value on athletic skills, assemblies on festival days featured athletic contests, the prototype of our modern Olympic Games. The word *agōn* by association with public games came to denote any competitive event. The struggle to win, as well as the pain accompanying the struggle, was called *agōnia,* which began also to be applied to any intense pain, physical or mental. The irony is that English *agony,* a word synonymous with intense pain of mind or body, ultimately comes from a word for a festive game.

agran·u·lo·cyte \(ˌ)ā-ˈgran-yə-lō-ˌsīt\ *noun* (circa 1923)
: a white blood cell without cytoplasmic granules

agran·u·lo·cy·to·sis \ˌā-ˌgran-yə-lō-ˌsī-ˈtō-səs\ *noun, plural* **-to·ses** \-ˈtō-ˌsēz\ [New Latin] (1927)
: an acute febrile condition marked by severe decrease in blood granulocytes and often associated with the use of certain drugs

ag·ra·pha \ˈa-grə-fə\ *noun plural* [Greek, neuter plural of *agraphos* unwritten, from *a-* + *graphein* to write — more at CARVE] (1890)
: sayings of Jesus not in the canonical gospels but found in other New Testament or early Christian writings

agraph·ia \(ˌ)ā-ˈgra-fē-ə\ *noun* [New Latin, from [2]*a-* + Greek *graphein* to write] (1871)
: the pathologic loss of the ability to write

[1]**agrar·i·an** \ə-ˈgrer-ē-ən, -ˈgrar-\ *adjective* [Latin *agrarius,* from *agr-, ager* field — more at ACRE] (1618)
1 : of or relating to fields or lands or their tenure
2 a : of, relating to, or characteristic of farmers or their way of life **b :** organized or designed to promote agricultural interests ⟨an *agrarian* political party⟩ ⟨*agrarian* reforms⟩

[2]**agrarian** *noun* (1818)
: a member of an agrarian party or movement

agrar·i·an·ism \-ē-ə-ˌni-zəm\ *noun* (1830)
: a social or political movement designed to bring about land reforms or to improve the economic status of the farmer

agree \ə-ˈgrē\ *verb* **agreed; agree·ing** [Middle English, from Middle French *agreer,* from *a gre* at will, from *a* (from Latin *ad*) + *gre* will, pleasure, from Latin *gratum,* neuter of *gratus* pleasing, agreeable — more at GRACE] (15th century)
transitive verb
1 : ADMIT, CONCEDE ⟨*agrees* that he is right⟩
2 *chiefly British* **:** to settle on by common consent **:** ARRANGE ⟨I *agreed* rental terms with him —Eric Bennett⟩
intransitive verb
1 : to accept or concede something (as the views or wishes of another) ⟨*agree* to a plan⟩
2 a : to achieve or be in harmony (as of opinion, feeling, or purpose) ⟨we *agree* in our taste in music⟩ **b :** to get along together **c :** to come to terms
3 a : to be similar **:** CORRESPOND ⟨both copies *agree*⟩ **b :** to be consistent ⟨the story *agrees* with the facts⟩
4 : to be fitting, pleasing, or healthful **:** SUIT ⟨this climate *agrees* with him⟩
5 : to have an inflectional form denoting identity or other regular correspondence in a grammatical category (as gender, number, case, or person) ☆

agree·able \ə-ˈgrē-ə-bəl\ *adjective* (14th century)
1 : pleasing to the mind or senses especially as according well with one's tastes or needs ⟨an *agreeable* companion⟩ ⟨an *agreeable* change⟩
2 : ready or willing to agree or consent
3 : being in harmony **:** CONSONANT
— agree·abil·i·ty \-ˌgrē-ə-ˈbi-lə-tē\ *noun*
— agree·able·ness \-ˈgrē-ə-bəl-nəs\ *noun*
— agree·ably \-blē\ *adverb*

agree·ment \ə-ˈgrē-mənt\ *noun* (15th century)
1 a : the act or fact of agreeing **b :** harmony of opinion, action, or character **:** CONCORD
2 a : an arrangement as to a course of action **b :** COMPACT, TREATY
3 a : a contract duly executed and legally binding **b :** the language or instrument embodying such a contract

ag·ri·busi·ness \ˈa-grə-ˌbiz-nəs, -nəz\ *noun* [*agriculture* + *business*] (circa 1955)
: an industry engaged in the producing operations of a farm, the manufacture and distribution of farm equipment and supplies, and the processing, storage, and distribution of farm commodities

ag·ri·busi·ness·man \ˌa-grə-ˈbiz-nəs-ˌman, -mən, -nəz-\ *noun* (1961)
: a person who works in or manages an agribusiness

ag·ri·cul·tur·al \ˌa-gri-ˈkəl-ch(ə-)rəl\ *adjective* (1776)
: of, relating to, used in, or concerned with agriculture
— ag·ri·cul·tur·al·ly *adverb*

ag·ri·cul·ture \ˈa-gri-ˌkəl-chər\ *noun* [Middle English, from Middle French, from Latin *agricultura,* from *ager* field + *cultura* cultivation — more at ACRE, CULTURE] (15th century)
: the science, art, or practice of cultivating the soil, producing crops, and raising livestock and in varying degrees the preparation and marketing of the resulting products **:** FARMING
— ag·ri·cul·tur·ist \ˌa-gri-ˈkəl-ch(ə-)rist\ *or* **ag·ri·cul·tur·al·ist** \-ch(ə-)rə-list\ *noun*

ag·ri·mo·ny \ˈa-grə-ˌmō-nē\ *noun, plural* **-nies** [Middle English, from Middle French & Latin; Middle French *aigremoine,* from Latin *agrimonia,* manuscript variant of *argemonia,* from Greek *argemōnē*] (14th century)
: any of a genus (*Agrimonia* and especially *A. eupatoria*) of herbs of the rose family having compound leaves, slender spikes of small yellow flowers, and fruits like burs

agro- *combining form* [French, from Greek, from *agros* field — more at ACRE]
1 : of or belonging to fields or soil **:** agricultural ⟨*agro*chemical⟩
2 : agricultural and ⟨*agro*-industrial⟩

ag·ro·chem·i·cal \ˌa-grō-ˈke-mi-kəl\ *also* **ag·ri·chem·i·cal** \ˌa-gri-\ *noun* (1956)
: an agricultural chemical (as an herbicide or an insecticide)

ag·ro·for·est·ry \ˌa-grō-ˈfȯr-ə-strē, -ˈfär-\ *noun* (1977)
: land management for the simultaneous production of food, crops, and trees; *also* **:** the science of agroforestry
— ag·ro·for·est·er \-stər\ *noun*

ag·ro–in·dus·tri·al \ˌa-grō-in-ˈdəs-trē-əl\ *adjective* (1940)
: of or relating to production (as of power for industry and water for irrigation) for both industrial and agricultural purposes

agron·o·my \ə-ˈgrä-nə-mē\ *noun* [probably from French *agronomie,* from *agro-* + *-nomie* -nomy] (1814)
: a branch of agriculture dealing with field-crop production and soil management
— ag·ro·nom·ic \ˌa-grə-ˈnä-mik\ *adjective*
— ag·ro·nom·i·cal·ly \-mi-k(ə-)lē\ *adverb*
— agron·o·mist \ə-ˈgrä-nə-mist\ *noun*

aground \ə-ˈgrau̇nd\ *adverb or adjective* (14th century)
1 : on the ground ⟨planes aloft and *aground*⟩
2 : on or onto the shore or the bottom of a body of water ⟨a ship run *aground*⟩

ague \ˈā-(ˌ)gyü\ *noun* [Middle English, from Middle French *aguë,* from Medieval Latin (*febris*) *acuta,* literally, sharp fever, from Latin, feminine of *acutus* sharp — more at ACUTE] (14th century)
1 : a fever (as malaria) marked by paroxysms of chills, fever, and sweating that recur at regular intervals
2 : a fit of shivering **:** CHILL
— agu·ish \ˈā-ˌgyü-ish\ *adjective*

ah \ˈä\ *interjection* [Middle English] (13th century)
— used to express delight, relief, regret, or contempt

aha \ä-ˈhä\ *interjection* [Middle English] (14th century)
— used to express surprise, triumph, or derision

ahead \ə-ˈhed\ *adverb or adjective* (1596)
1 a : in a forward direction or position **:** FORWARD **b :** in front
2 : in, into, or for the future ⟨plan *ahead*⟩
3 : in or toward a more advantageous position ⟨helped others to get *ahead*⟩
4 : at or to an earlier time **:** in advance ⟨make payments *ahead*⟩

ahead of *preposition* (1748)
1 : in front or advance of
2 : in excess of

ahem *a throat-clearing sound; often read as* ə-ˈhem\ *interjection* [imitative] (1763)
— used especially to attract attention

ahim·sa \ə-ˈhim-ˌsä\ *noun* [Sanskrit *ahiṁsā* noninjury] (1875)
: the Hindu and Buddhist doctrine of refraining from harming any living being

ahis·tor·i·cal \ˌā-his-ˈtȯr-i-kəl, -ˈtär-\ *or* **ahis·tor·ic** \-ik\ *adjective* (1945)
: not concerned with or related to history, historical development, or tradition ⟨the *ahistorical* attitudes of the radicals⟩

ahold \ə-ˈhōld\ *noun* [probably from the phrase *a hold*] (1872)
: HOLD ⟨if you could get *ahold* of a representative —Norman Mailer⟩

A horizon \ˈā-\ *noun* (1936)
: the uppermost dark-colored layer of a soil consisting largely of partly disintegrated organic debris

ahoy \ə-ˈhȯi\ *interjection* [*a-* (as in *aha*) + *hoy*] (1751)
— used in hailing ⟨ship *ahoy*⟩

Ah·ri·man \ˈär-i-mən, -ˌmän\ *noun* [Persian, modification of Avestan *aṅrō mainyuš* hostile spirit]
: Ahura Mazda's antagonist who is a spirit of darkness and evil in Zoroastrianism

Ahu·ra Maz·da \ə-ˌhu̇r-ə-ˈmaz-də, ä-ˌhu̇r-\ *noun* [Avestan *Ahuramazda,* literally, wise god]
: the Supreme Being represented as a deity of goodness and light in Zoroastrianism

Ai·as \ˈī-əs\ *noun* [Greek]
: AJAX

ai·blins \ˈā-blənz\ *adverb* [*able* + *-lings, -lins* -lings] (circa 1605)
chiefly Scottish **:** PERHAPS

[1]**aid** \ˈād\ *verb* [Middle English *eyden,* from Middle French *aider,* from Latin *adjutare,* frequentative of *adjuvare,* from *ad-* + *juvare* to help] (15th century)
transitive verb
: to provide with what is useful or necessary in achieving an end
intransitive verb
: to give assistance
— aid·er *noun*

[2]**aid** *noun* (15th century)
1 : a subsidy granted to the king by the English parliament until the 18th century for an extraordinary purpose
2 a : the act of helping **b :** help given **:** ASSISTANCE; *specifically* **:** tangible means of assistance (as money or supplies)

☆ SYNONYMS
Agree, concur, coincide mean to come into or be in harmony regarding a matter of opinion. AGREE implies complete accord usually attained by discussion and adjustment of differences ⟨on some points we all can *agree*⟩. CONCUR tends to suggest cooperative thinking or acting toward an end but sometimes implies no more than approval (as of a decision reached by others) ⟨if my wife *concurs,* it's a deal⟩. COINCIDE, used more often of opinions, judgments, wishes, or interests than of people, implies an agreement amounting to identity ⟨their wishes *coincide* exactly with my desire⟩. See in addition ASSENT.

3 a : an assisting person or group — compare AIDE **b :** something by which assistance is given **:** an assisting device ⟨an *aid* to understanding⟩ ⟨a visual *aid*⟩; *especially* **:** HEARING AID
4 : a tribute paid by a vassal to his lord

aide \ˈād\ *noun* [short for *aide-de-camp*] (1777)
: a person who acts as an assistant; *specifically* **:** a military officer acting as assistant to a superior

aide–de–camp \ˌād-di-ˈkamp, -ˈkä^{n}\ *noun, plural* **aides–de–camp** \ˌād(z)-di-\ [French *aide de camp,* literally, camp assistant] (1670)
: a military aid; *also* **:** a civilian aide usually to an executive

aide–mé·moire \ˌād-mām-ˈwär\ *noun, plural* **aide–mémoire** [French, from *aider* to aid + *mémoire* memory] (1846)
1 : an aid to the memory; *especially* **:** a mnemonic device
2 : a written summary or outline of important items of a proposed agreement or diplomatic communication

aid·man \ˈād-ˌman\ *noun* (1944)
: an army medical corpsman attached to a field unit

AIDS \ˈādz\ *noun* [*a*cquired *i*mmuno*d*eficiency *s*yndrome] (1982)
: a disease of the human immune system that is caused by infection with HIV, that is characterized cytologically especially by severe reduction in the numbers of helper T cells, that in modern industrialized nations occurs especially in homosexual and bisexual men and in intravenous users of illicit drugs, that is commonly transmitted in blood and bodily secretions (as semen), and that renders the subject highly vulnerable to life-threatening conditions (as Pneumocystis carinii pneumonia) and to some that become life-threatening (as Kaposi's sarcoma)

AIDS–related complex *noun* (1984)
: a group of symptoms (as fever, weight loss, and lymphadenopathy) that is associated with the presence of antibodies to HIV and is followed by the development of AIDS in a certain proportion of cases

AIDS virus *noun* (1985)
: HIV

ai·grette \ā-ˈgret, ˈā-ˌ\ *noun* [French, plume, egret, from Middle French — more at EGRET] (1630)
1 : a spray of feathers (as of the egret) for the head
2 : a spray of gems worn on a hat or in the hair

ai·guille \ā-ˈgwē(ə)l, -ˈgwē\ *noun* [French, literally, needle — more at AGLET] (1816)
: a sharp-pointed pinnacle of rock

ai·guil·lette \ˌā-gwi-ˈlet\ *noun* [French — more at AGLET] (1816)
: AGLET; *specifically* **:** a shoulder cord worn by designated military aides — compare FOURRAGÈRE

ai·ki·do \ˌī-ki-ˈdō, ī-ˈkē-(ˌ)dō\ *noun* [Japanese *aikidō,* from *ai-* match, coordinate + *ki* breath, spirit + *dō* art, way] (1956)
: a Japanese art of self-defense employing locks and holds and utilizing the principle of nonresistance to cause an opponent's own momentum to work against him

[1]ail \ˈā(ə)l\ *verb* [Middle English *eilen,* from Old English *eglan;* akin to Gothic *agljan* to harm] (before 12th century)
transitive verb
: to give physical or emotional pain, discomfort, or trouble to
intransitive verb
: to have something the matter; *especially* **:** to suffer ill health

[2]ail *noun* (13th century)
: AILMENT

ai·lan·thus \ā-ˈlan(t)-thəs\ *noun* [New Latin, from Ambonese *ai lanito,* literally, tree (of) heaven] (1807)
: any of a small Asian genus (*Ailanthus* of the family Simaroubaceae, the ailanthus family) of chiefly tropical trees and shrubs with bitter bark, pinnate leaves, and terminal panicles of ill-scented greenish flowers

ai·le·ron \ˈā-lə-ˌrän\ *noun* [French, from diminutive of *aile* wing — more at AISLE] (1909)
: a movable airfoil at the trailing edge of an airplane wing that is used for imparting a rolling motion especially in banking for turns — see AIRPLANE illustration

ail·ment \ˈā(ə)l-mənt\ *noun* (circa 1706)
1 : a bodily disorder or chronic disease
2 : UNREST, UNEASINESS

ai·lu·ro·phile \ī-ˈlu̇r-ə-ˌfīl, ā-\ *noun* [Greek *ailouros* cat] (1927)
: a cat fancier **:** a lover of cats

ai·lu·ro·phobe \-ˌfōb\ *noun* (1905)
: a person who hates or fears cats

[1]aim \ˈām\ *verb* [Middle English, from Middle French *aesmer & esmer;* Middle French *aesmer,* from Old French, from *a-* (from Latin *ad-*) + *esmer* to estimate, from Latin *aestimare*] (14th century)
intransitive verb
1 : to direct a course; *specifically* **:** to point a weapon at an object
2 : ASPIRE, INTEND ⟨*aims* to reform the government⟩
transitive verb
1 *obsolete* **:** GUESS, CONJECTURE
2 a : POINT **b :** to direct to or toward a specified object or goal ⟨a program *aimed* at reducing pollution⟩

[2]aim *noun* (14th century)
1 *obsolete* **:** MARK, TARGET
2 a : the pointing of a weapon at a mark **b :** the ability to hit a target **c :** a weapon's accuracy or effectiveness
3 *obsolete* **a :** CONJECTURE, GUESS **b :** the directing of effort toward a goal
4 : a clearly directed intent or purpose
synonym see INTENTION
— aim·less \-ləs\ *adjective*
— aim·less·ly *adverb*
— aim·less·ness *noun*

ain \ˈān\ *adjective* [probably from Old Norse *eiginn*] (1721)
Scottish **:** OWN

ain't \ˈānt\ [contraction of *are not*] (1778)
1 : am not **:** are not **:** is not
2 : have not **:** has not
3 : do not **:** does not **:** did not — used in some varieties of Black English ■

Ai·nu \ˈī-(ˌ)nü\ *noun, plural* **Ainu** *or* **Ainus** [Ainu *aynu* person] (1819)
1 : a member of an indigenous people of the Japanese archipelago, the Kuril Islands, and part of Sakhalin Island
2 : the language of the Ainu people

ai·o·li \(ˌ)ī-ˈō-lē, (ˌ)ā-\ *noun* [Provençal, from *ai* garlic + *oli* oil] (circa 1900)
: a sauce made of crushed garlic, egg yolks, olive oil, and lemon juice and sometimes potato **:** garlic mayonnaise

[1]air \ˈar, ˈer\ *noun, often attributive* [Middle English, from Middle French, from Latin *aer,* from Greek *aēr*] (14th century)
1 a *archaic* **:** BREATH **b :** the mixture of invisible odorless tasteless gases (as nitrogen and oxygen) that surrounds the earth **c :** a light breeze
2 a : empty space **b :** NOTHINGNESS ⟨vanished into thin *air*⟩ **c :** a sudden severance of relations ⟨she gave me the *air*⟩
3 [probably translation of Italian *aria*] **a :** TUNE, MELODY **b** *Elizabethan & Jacobean music* **:** an accompanied song or melody in usually strophic form **c :** the chief voice part or melody in choral music
4 a : outward appearance of a thing ⟨an *air* of luxury⟩ **b :** a surrounding or pervading influence **:** ATMOSPHERE ⟨an *air* of mystery⟩ **c :** the look, appearance, or bearing of a person especially as expressive of some personal quality or emotion **:** DEMEANOR ⟨an *air* of dignity⟩ **d :** an artificial or affected manner ⟨put on *airs*⟩
5 : public utterance ⟨he gave *air* to his opinion⟩
6 : COMPRESSED AIR
7 a (1) **:** AIRCRAFT ⟨go by *air*⟩ (2) **:** AVIATION ⟨*air* safety⟩ ⟨*air* rights⟩ (3) **:** AIR FORCE ⟨*air* headquarters⟩ **b :** the medium of transmission of radio waves; *also* **:** RADIO, TELEVISION ⟨went on the *air*⟩
8 : a football offense utilizing primarily the forward pass ⟨trailing by 20 points, the team took to the *air*⟩
9 : an air-conditioning system
synonym see POSE
— air·less \-ləs\ *adjective*
— air·less·ness *noun*
— in the air : in wide circulation **:** ABOUT
— up in the air : not yet settled

[2]air (1530)
transitive verb
1 : to expose to the air for drying, purifying, or refreshing **:** VENTILATE — often used with *out*
2 : to expose to public view or bring to public notice
3 : to transmit by radio or television ⟨*air* a program⟩
intransitive verb
1 : to become exposed to the open air
2 : to become broadcast ⟨the program *airs* daily⟩
synonym see EXPRESS

air bag *noun* (1969)
: an automobile safety device consisting of a bag designed to inflate automatically in front of an occupant in case of collision

air ball *noun* (1981)
: a missed shot in basketball that fails to touch the rim and backboard

air base *noun* (1915)
: a military base intended chiefly for the operation of aircraft

air bladder *noun* (1731)
: a sac containing gas and especially air; *especially* **:** a hydrostatic organ present in most fishes that serves as an accessory respiratory organ

air·boat \ˈar-ˌbōt, ˈer-\ *noun* (1946)
: a shallow-draft boat driven by an airplane propeller and steered by an airplane rudder

air·borne \-ˌbōrn, -ˌbȯrn\ *adjective* (1641)
1 : done or being in the air **:** being off the ground: as **a :** carried through the air (as by an

□ USAGE
ain't Although widely disapproved as nonstandard and more common in the habitual speech of the less educated, *ain't* in senses 1 and 2 is flourishing in American English. It is used in both speech and writing to catch attention and to gain emphasis ⟨the wackiness of movies, once so deliciously amusing, *ain't* funny anymore —Richard Schickel⟩ ⟨I am telling you—there *ain't* going to be any blackmail —R. M. Nixon⟩. It is used especially in journalistic prose as part of a consistently informal style ⟨the creative process *ain't* easy —Mike Royko⟩. This informal *ain't* is commonly distinguished from habitual *ain't* by its frequent occurrence in fixed constructions and phrases ⟨well—class it *ain't* —Cleveland Amory⟩ ⟨for money? say it *ain't* so, Jimmy! —Andy Rooney⟩ ⟨you *ain't* seen nothing yet⟩ ⟨that *ain't* hay⟩ ⟨two out of three *ain't* bad⟩ ⟨if it *ain't* broke, don't fix it⟩. In fiction *ain't* is used for purposes of characterization; in familiar correspondence it tends to be the mark of a warm personal friendship. It is also used for metrical reasons in popular songs ⟨*Ain't* She Sweet⟩ ⟨It *Ain't* Necessarily So⟩. Our evidence shows British use to be much the same as American.

aircraft) **b :** supported especially by aerodynamic forces or propelled through the air by force
2 : trained for deployment by air and especially by parachute

air brake *noun* (1871)
1 : a brake operated by a piston driven by compressed air
2 : a surface that may be projected into the airstream for increasing drag and lowering the speed of an airplane

[1]air·brush \-ˌbrəsh\ *noun* (circa 1889)
: an atomizer for applying by compressed air a fine spray (as of paint or liquid color)

[2]airbrush *transitive verb* (1938)
: to paint, treat, or alter with an airbrush

air·burst \-ˌbərst\ *noun* (1917)
: the burst of a shell or bomb in the air

air·bus \-ˌbəs\ *noun* (1945)
: a short-range or medium-range subsonic jet passenger airplane

air chief marshal *noun* (1919)
: a commissioned officer in the British air force who ranks with a general in the army

air commodore *noun* (1919)
: a commissioned officer in the British air force who ranks with a brigadier in the army

air–con·di·tion \ˌar-kən-ˈdi-shən, ˌer-\ *transitive verb* [back-formation from *air conditioning*] (1933)
: to equip (as a building) with an apparatus for washing air and controlling its humidity and temperature; *also* **:** to subject (air) to these processes
— **air con·di·tion·er** \-ˈdi-sh(ə-)nər\ *noun*
— **air–con·di·tion·ing** \-ˈdi-sh(ə-)niŋ\ *noun*

air·craft \ˈar-ˌkraft, ˈer-\ *noun, plural* **aircraft** *often attributive* (1850)
: a vehicle (as an airplane or balloon) for traveling through the air

aircraft carrier *noun* (1919)
: a warship with a flight deck on which airplanes can be launched and landed

air·crew \ˈar-ˌkrü, ˈer-\ *noun* (1921)
: the crew manning an airplane

air–cush·ion vehicle *noun* (circa 1962)
: HOVERCRAFT

air dam *noun* (1965)
: a device attached to the underside of the front of an automobile to improve stability, aerodynamic performance, and engine cooling by redirecting the flow of air

air·date \-ˌdāt\ *noun* (1971)
: the scheduled date of a broadcast

air·drome \ˈar-ˌdrōm, ˈer-\ *noun* [alteration of *aerodrome*] (1917)
: AIRPORT

air·drop \-ˌdräp\ *noun* (circa 1945)
: delivery of cargo or personnel by parachute from an airplane in flight
— **air–drop** *transitive verb*
— **air–drop·pa·ble** \-ˌdrä-pə-bəl\ *adjective*

air–dry \-ˈdrī\ *adjective* (1856)
: dry to such a degree that no further moisture is given up on exposure to air

Aire·dale terrier \ˌar-ˌdāl-, ˌer-\ *noun* [*Airedale,* valley of the Aire river, England] (1880)
: any of a breed of large terriers with a hard, wiry, black-and-tan coat — called also *Airedale*

air·er \ˈar-ər, ˈer-\ *noun* (circa 1847)
British **:** a frame on which clothes are aired or dried

air·fare \ˈar-ˌfar, ˈer-ˌfer\ *noun* (1918)
: fare for travel by airplane

air·field \-ˌfēld\ *noun* (1927)
: an area of land from which aircraft operate: as **a :** AIRPORT **b :** AIR BASE

air·flow \-ˌflō\ *noun* (circa 1911)
: a flow of air; *specifically* **:** the motion of air (as around parts of an airplane in flight) relative to the surface of a body immersed in it

air·foil \-ˌfȯil\ *noun* (circa 1922)
: a body (as an airplane wing or propeller blade) designed to provide a desired reaction force when in motion relative to the surrounding air

air force *noun* (1917)
1 : the military organization of a nation for air warfare
2 : a unit of the U.S. Air Force higher than a division and lower than a command

air·frame \-ˌfrām\ *noun* (1931)
: the structure of an aircraft, rocket vehicle, or missile without the power plant

air·freight \-ˈfrāt\ *noun* (1929)
: freight transport by air in volume; *also* **:** the charge for this service
— **airfreight** *transitive verb*

air·glow \-ˌglō\ *noun* (circa 1950)
: light that is observed especially during the night, that originates in the high atmosphere of a planet (as the earth), and that is associated with photochemical reactions of gases caused by solar radiation

air gun *noun* (circa 1753)
1 : a gun from which a projectile is propelled by compressed air
2 : any of various hand tools that work by compressed air; *especially* **:** AIRBRUSH

[1]air·head \-ˌhed\ *noun* [[1]*air* + *-head* (as in *beachhead*)] (circa 1944)
: an area in hostile territory secured usually by airborne troops for further use in bringing in troops and materiel by air

[2]airhead *noun* (1972)
: a mindless or stupid person
— **air·head·ed** \-ˌhe-dəd\ *adjective*

air·hole \-ˌhōl\ *noun* (1766)
: a hole to admit or discharge air

air·ing \ˈar-iŋ, ˈer-\ *noun* (circa 1606)
1 : exposure to air or heat for drying or freshening
2 : exposure to or exercise in the open air especially to promote health or fitness
3 : exposure to public view or notice
4 : a radio or television broadcast

air lane *noun* (circa 1910)
: a path customarily followed by airplanes

air letter *noun* (1920)
1 : an airmail letter
2 : a sheet of airmail stationery that can be folded and sealed with the message inside and the address outside

air·lift \ˈar-ˌlift, ˈer-\ *noun* (1945)
: a system of transporting cargo or passengers by aircraft usually to or from an otherwise inaccessible area
— **airlift** *transitive verb*

air·line \-ˌlīn\ *noun* (1910)
: an air transportation system including its equipment, routes, operating personnel, and management

air line *noun* (1813)
: a straight line through the air between two points

air·lin·er \-ˌlī-nər\ *noun* (1908)
: an airplane operated by an airline

air lock *noun* (1857)
1 : an intermediate chamber with two airtight doors or openings to permit passage between two dissimilar spaces (as two places of unequal atmospheric pressure)
2 : a stoppage of flow caused by air being in a part where liquid ought to circulate

air·mail \ˈar-ˌmā(ə)l, ˈer \ *noun* (1913)
: the system of transporting mail by aircraft; *also* **:** the mail thus transported
— **airmail** *transitive verb*

air·man \-mən\ *noun* (1873)
1 : a civilian or military pilot, aviator, or aviation technician
2 : an enlisted man in the air force: as **a :** an enlisted man of one of the three ranks below sergeant **b :** an enlisted man ranking above an airman basic and below an airman first class

airman basic *noun* (circa 1961)
: an enlisted man of the lowest rank in the air force

airman first class *noun* (1952)
: an enlisted man in the air force ranking above an airman and below a sergeant

air·man·ship \ˈar-mən-ˌship, ˈer-\ *noun* (1908)
: skill in piloting or navigating airplanes

air marshal *noun* (1919)
: a commissioned officer in the British air force who ranks with a lieutenant general in the army

air mass *noun* (1893)
: a body of air extending hundreds or thousands of miles horizontally and sometimes as high as the stratosphere and maintaining as it travels nearly uniform conditions of temperature and humidity at any given level

air mattress *noun* (1926)
: MATTRESS 1b

Air Medal *noun* (1942)
: a U.S. military decoration awarded for meritorious achievement while participating in an aerial flight

air mile *noun* (1919)
: a mile in air travel

air–mind·ed \ˈar-ˈmīn-dəd, ˈer-\ *adjective* (1924)
: interested in aviation or in air travel
— **air–mind·ed·ness** *noun*

air·mo·bile \-ˌmō-bəl, -ˌbēl, -ˌbīl\ *adjective* (1965)
: of, relating to, or being a military unit whose members are transported to combat areas usually by helicopter

air·park \-ˌpärk\ *noun* (1929)
: a small airport usually near an industrial area

air piracy *noun* (1948)
: the hijacking of a flying airplane **:** SKYJACKING

air·plane \ˈar-ˌplān, ˈer-\ *noun* [alteration of *aeroplane*] (1907)
: a powered heavier-than-air aircraft that has fixed wings from which it derives most of its lift

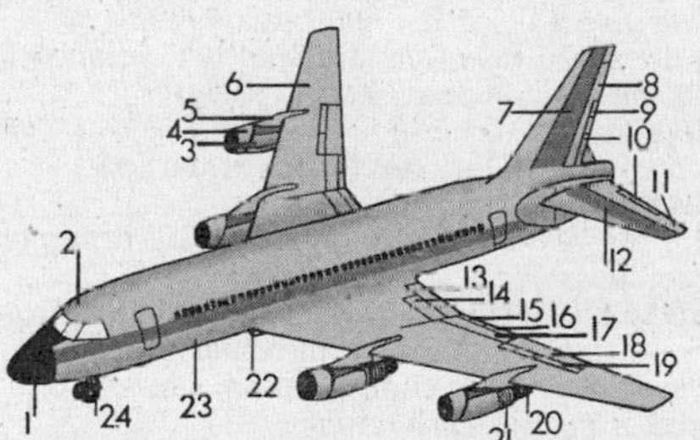

airplane: *1* **weather radar,** *2* **cockpit,** *3* **jet engine,** *4* **engine pod,** *5* **pylon,** *6* **wing,** *7* **vertical stabilizer,** *8* **rudder,** *9, 10* **tabs,** *11* **elevator,** *12* **horizontal stabilizer,** *13* **inboard flap,** *14* **inboard spoiler,** *15, 16* **tabs,** *17* **aileron,** *18* **outboard flap,** *19* **outboard spoiler,** *20* **sound suppressor,** *21* **thrust reverser,** *22* **cabin air intake,** *23* **fuselage,** *24* **nose landing gear**

air plant *noun* (1841)
1 : EPIPHYTE
2 : BRYOPHYLLUM

air·play \-ˌplā\ *noun* (1966)
: the playing of a musical recording on the air by a radio station

air pocket *noun* (1912)
: a condition of the atmosphere (as a local down current) that causes an airplane to drop suddenly

air police *noun* (1944)
: the military police of an air force

air·port \ˈar-ˌpōrt, ˈer-, -ˌpȯrt\ *noun* (1919)
: a place from which aircraft operate that usually has paved runways and maintenance facilities and often serves as a terminal

\ə\ abut \ᵊ\ kitten \ər\ further \a\ ash \ā\ ace \ä\ mop, mar \au̇\ out \ch\ chin \e\ bet \ē\ easy \g\ go \i\ hit \ī\ ice \j\ job \ŋ\ sing \ō\ go \ȯ\ law \ȯi\ boy \th\ thin \t͟h\ the \ü\ loot \u̇\ foot \y\ yet \zh\ vision *see also* Guide to Pronunciation

air·post \-ˈpōst\ *noun* (circa 1927)
: AIRMAIL
air·pow·er \-ˌpau̇(-ə)r\ *noun* (1908)
: the military strength of a nation's air force
air pump *noun* (1660)
: a pump for exhausting air from a closed space or for compressing air or forcing it through other apparatus
air raid *noun* (1914)
: an attack by armed airplanes on a surface target — usually hyphened when used attributively ⟨*air-raid* shelter⟩
air rifle *noun* (1886)
: a rifle whose projectile (as a BB or pellet) is propelled by compressed air or carbon dioxide
air right *noun* (1922)
: a property right to the space above a surface area or object
air sac *noun* (circa 1828)
1 : one of the air-filled spaces in the body of a bird connected with the air passages of the lungs
2 : ALVEOLUS 1b
3 : a thin-walled dilation of a trachea occurring in many insects
air·screw \ˈar-ˌskrü, ˈer-\ *noun* (1894)
chiefly British **:** an airplane propeller
air·ship \-ˌship\ *noun* (1819)
: a lighter-than-air aircraft having propulsion and steering systems
air show *noun* (1950)
: an exhibition of aircraft and aviation skills
air·sick \-ˌsik\ *adjective* (1785)
: affected with motion sickness associated with flying
— air·sick·ness *noun*
air·space \-ˌspās\ *noun* (1911)
: the space lying above the earth or above a certain area of land or water; *especially* **:** the space lying above a nation and coming under its jurisdiction
air·speed \-ˌspēd\ *noun* (circa 1909)
: the speed (as of an airplane) with relation to the air — compare GROUND SPEED
air·stream \-ˌstrēm\ *noun* (1869)
: a current of air; *specifically* **:** AIRFLOW
air·strip \-ˌstrip\ *noun* (1942)
: a runway without normal air base or airport facilities
[1]airt \ˈärt, ˈert\ *noun* [Middle English *art,* from Scottish Gaelic *àirt*] (15th century)
chiefly Scottish **:** compass point **:** DIRECTION
[2]airt *transitive verb* (circa 1782)
chiefly Scottish **:** DIRECT, GUIDE
air taxi *noun* (1920)
: a small commercial airplane used for short flights between localities not served by scheduled airlines
air·tight \ˈar-ˌtīt, ˈer-\ *adjective* (1760)
1 : impermeable to air or nearly so
2 a : having no noticeable weakness, flaw, or loophole ⟨an *airtight* argument⟩ **b :** permitting no opportunity for an opponent to score ⟨an *airtight* defense⟩
— air·tight·ness *noun*
air·time \-ˌtīm\ *noun* (1942)
1 : the time at which a radio or television broadcast is scheduled to begin
2 : the time or any part thereof that a radio or television station is on the air
air–to–air \ˌar-tə-ˈ(w)ar, ˌer-tə-ˈ(w)er\ *adjective* (1941)
: launched from one airplane in flight at another ⟨*air-to-air* missiles⟩; *also* **:** involving aircraft in flight ⟨*air-to-air* combat⟩
air vice–marshal *noun* (1919)
: a commissioned officer in the British air force who ranks with a major general in the army
air·wave \ˈar-ˌwāv, ˈer-\ *adjective* (1944)
: of, created for, or heard on the airwaves
air·waves \-ˌwāvz\ *noun plural* (1928)
: the medium of radio and television transmission — not used technically
air·way \-ˌwā\ *noun* (1849)
1 : a passage for a current of air (as in a mine or to the lungs)
2 : a designated route along which airplanes fly from airport to airport; *especially* **:** such a route equipped with navigational aids
3 : AIRLINE
4 : a channel of a designated radio frequency for broadcasting or other radio communication
air·wor·thy \-ˌwər-t͟hē\ *adjective* (1829)
: fit for operation in the air ⟨an *airworthy* airplane⟩
— air·wor·thi·ness *noun*
airy \ˈar-ē, ˈer-\ *adjective* **air·i·er; -est** (14th century)
1 a : of or relating to air **:** ATMOSPHERIC **b :** high in the air **:** LOFTY ⟨*airy* perches⟩ **c :** performed in air **:** AERIAL ⟨*airy* leaps⟩
2 : UNREAL, ILLUSORY ⟨*airy* romances⟩
3 a : being light and graceful in movement or manner **:** SPRIGHTLY, VIVACIOUS ⟨an *airy* laugh⟩ **b :** ETHEREAL ⟨*airy* fantasy⟩
4 a : open to the free circulation of air ⟨an *airy* front porch⟩ **b :** having openings or spaces ⟨*airy* lacework⟩
5 : AFFECTED, PROUD ⟨*airy* condescension⟩
— air·i·ly \ˈar-ə-lē, ˈer-\ *adverb*
— air·i·ness \-ē-nəs\ *noun*
airy–fairy \-ˈfar-ē, -ˈfer-ē\ *adjective* (1869)
1 *chiefly British* **:** DELICATE, FAIRYLIKE
2 *chiefly British* **:** lacking substance or purpose ⟨in . . . an *airy-fairy,* unserious, insufficiently careful fashion —*Times Literary Supplement*⟩
aisle \ˈī(ə)l\ *noun* [Middle English *ile,* alteration of *ele,* from Middle French, wing, from Latin *ala;* akin to Old English *eaxl* shoulder, Latin *axis* axletree — more at AXIS] (15th century)
1 : the side of a church nave separated by piers from the nave proper
2 a : a passage (as in a theater or railroad passenger car) separating sections of seats **b :** a passage (as in a store or warehouse) for inside traffic
aisle·way \-ˌwā\ *noun* (1926)
: AISLE 2b
ait \ˈāt\ *noun* [Middle English *eyt,* alteration of Old English *īggoth,* from *īg* island — more at ISLAND] (before 12th century)
British **:** a little island
aitch \ˈāch\ *noun* [French *hache,* from (assumed) Vulgar Latin *hacca*] (circa 1580)
: the letter *h*
aitch·bone \ˈāch-ˌbōn\ *noun* [Middle English *hachbon,* alteration (resulting from incorrect division of *a nachebon*) of (assumed) Middle English *nachebon,* from Middle English *nache* buttock (from Middle French, from Late Latin *natica,* from Latin *natis*) + *bon* bone] (15th century)
1 : the hipbone especially of cattle
2 : the cut of beef containing the aitchbone
ajar \ə-ˈjär\ *adjective or adverb* [earlier *on char,* from *on* + *char* turn — more at CHARE] (15th century)
: slightly open ⟨a door *ajar*⟩
Ajax \ˈā-ˌjaks\ *noun* [Latin, from Greek *Aias*]
1 : a Greek hero in the Trojan War who kills himself because the armor of Achilles is awarded to Odysseus
2 : a fleet-footed Greek hero in the Trojan War
aju·ga \ˈa-jə-gə\ *noun, plural* **-ga** *or* **-gas** [New Latin, from [2]*a-* + Latin *jugum* yoke — more at YOKE] (circa 1899)
: [1]BUGLE
Akan \ˈä-ˌkän\ *noun, plural* **Akan** *or* **Akans** (1694)
1 : a member of any of the Akan-speaking peoples (as the Ashanti)
2 : a Kwa language of southern Ghana and the southeast Ivory Coast
akee \ˈa-ˌkē, a-ˈkē\ *noun* [origin unknown] (1794)
: the fleshy fruit of an African tree (*Blighia sapida*) of the soapberry family grown in the Caribbean area, Florida, and Hawaii that is edible when ripe but has a toxic pink raphe attaching the aril to the seed and toxic arils when immature or overripe; *also* **:** the tree
AK–47 \ˈā-ˈkā-ˌfȯr-tē-ˈse-vᵊn, -ˌfȯr-\ *noun* [Russian *avtomat Kalashnikova 1947* Kalashnikov automatic rifle of 1947] (1968)
: a Soviet-designed 7.62 mm (.30 cal.) gas-operated magazine-fed rifle for automatic or semiautomatic fire
akim·bo \ə-ˈkim-(ˌ)bō\ *adjective or adverb* [Middle English *in kenebowe*] (15th century)
1 : having the hand on the hip and the elbow turned outward
2 : set in a bent position ⟨a tailor sitting with legs *akimbo*⟩
akin \ə-ˈkin\ *adjective* (1586)
1 : related by blood **:** descended from a common ancestor or prototype
2 : essentially similar, related, or compatible
Aki·ta \ə-ˈkē-tə, ä-\ *noun* [from *Akita,* Japan] (1928)
: any of a breed of large muscular dogs of Japanese origin
Ak·ka·di·an \ə-ˈkā-dē-ən\ *noun* (circa 1855)
1 : an extinct Semitic language of ancient Mesopotamia
2 : a Semitic inhabitant of central Mesopotamia before 2000 B.C.
— Akkadian *adjective*
ak·va·vit \ˈä-kwə-ˌvēt, ˈäk-vä-\ *variant of* AQUAVIT
al- — see AD-
[1]-al *adjective suffix* [Middle English, from Old French & Latin; Old French, from Latin *-alis*] **:** of, relating to, or characterized by ⟨direction*al*⟩ ⟨fiction*al*⟩
[2]-al *noun suffix* [Middle English *-aille,* from Old French, from Latin *-alia,* neuter plural of *-alis*]
: action **:** process ⟨rehears*al*⟩
[3]-al *noun suffix* [French, from *alcool* alcohol, from Medieval Latin *alcohol*]
: aldehyde ⟨furfur*al*⟩
ala \ˈā-lə\ *noun, plural* **alae** \-ˌlē\ [Latin — more at AISLE] (1738)
: a wing or a winglike anatomic process or part
— alar \ˈā-lər\ *adjective*
— ala·ry \-lə-rē\ *adjective*
à la *also* **a la** \ˌä-(ˌ)lä, ˌä-lə, ˌa-lə\ *preposition* [French *à la*] (1589)
: in the manner of
al·a·bas·ter \ˈa-lə-ˌbas-tər\ *noun* [Middle English *alabastre,* from Middle French, from Latin *alabaster* vase of alabaster, from Greek *alabastros*] (14th century)
1 : a compact fine-textured usually white and translucent gypsum often carved into vases and ornaments
2 : a hard compact calcite or aragonite that is translucent and sometimes banded
— alabaster *or* **al·a·bas·trine** \ˌa-lə-ˈbas-trən\ *adjective*
à la carte *also* **a la carte** \ˌä-lə-ˈkärt, ˌa-lə-\ *adverb or adjective* [French, by the bill of fare] (1826)
: according to a menu that prices each item separately
alack \ə-ˈlak\ *interjection* [Middle English] (15th century)
— used to express sorrow or regret
alac·ri·ty \ə-ˈla-krə-tē\ *noun* [Latin *alacritas,* from *alacr-, alacer* lively, eager] (15th century)
: promptness in response **:** cheerful readiness ⟨accepted the invitation of the president with *alacrity*⟩
— alac·ri·tous \-krə-təs\ *adjective*
Alad·din \ə-ˈla-dᵊn\ *noun*
: a youth in the *Arabian Nights' Entertainments* who comes into possession of a magic lamp
à la grecque \ˌä-lə-ˈgrek, ˌa-lə-\ *adjective, often G capitalized* [French, in the Greek manner] (circa 1925)

: served in a sauce made of olive oil, lemon juice, and several seasonings (as fennel, coriander, sage, and thyme)

à la king \ˌä-lə-ˈkiŋ, ˌa-lə-\ *adjective* (1919)
: served in a cream sauce with mushrooms and pimiento or green peppers ⟨chicken *à la king*⟩

al·a·me·da \ˌa-lə-ˈmē-də, -ˈmā-\ *noun* [Spanish, from *álamo* poplar] (1797)
: a public promenade bordered with trees

à la mode *also* **a la mode** \ˌä-lə-ˈmōd, ˌa-lə-\ *adjective* [French, according to the fashion] (1650)
1 : FASHIONABLE, STYLISH
2 : topped with ice cream

al·a·nine \ˈa-lə-ˌnēn\ *noun* [German *Alanin,* irregular from *Aldehyd* aldehyde] (circa 1879)
: a simple nonessential crystalline amino acid $C_3H_7NO_2$

al·a·nyl \ˈa-lə-ˌnil\ *noun* [International Scientific Vocabulary *alan*ine + *-yl*] (circa 1928)
: an acyl radical of alanine

[1]alarm \ə-ˈlärm\ *also* **ala·rum** \ə-ˈlär-əm, -ˈlar-, *British also* -ˈler-\ *noun* [Middle English *alarme, alarom,* from Middle French *alarme,* from Old Italian *all'arme,* literally, to the arms] (14th century)
1 *usually alarum, archaic* **:** a call to arms ⟨the angry trumpet sounds *alarum* —Shakespeare⟩
2 : a signal (as a loud noise or flashing light) that warns or alerts; *also* **:** a device that signals ⟨set the *alarm* to wake me at seven⟩
3 : sudden sharp apprehension and fear resulting from the perception of imminent danger
4 : a warning notice
synonym see FEAR

[2]alarm *also* **alarum** *transitive verb* (1605)
1 : DISTURB, EXCITE
2 : to give warning to
3 : to strike with fear
— **alarm·ing·ly** \-ˈlär-miŋ-lē\ *adverb*

alarm clock *noun* (1697)
: a clock that can be set to sound an alarm at a desired time

alarm·ism \ə-ˈlär-ˌmi-zəm\ *noun* (1867)
: the often unwarranted exciting of fears or warning of danger
— **alarm·ist** \-mist\ *noun or adjective*

alarm reaction *noun* (1936)
: the initial reaction of an organism (as increased hormonal activity) to stress

alarums and excursions *noun plural* (1592)
1 : martial sounds and the movement of soldiers across the stage — used as a stage direction in Elizabethan drama
2 : clamor, excitement, and feverish or disordered activity

alas \ə-ˈlas\ *interjection* [Middle English, from Old French, from *a* ah + *las* weary, from Latin *lassus* — more at LASSITUDE] (13th century)
— used to express unhappiness, pity, or concern

Alas·kan malamute \ə-ˈlas-kən-\ *noun* (1938)
: any of a breed of powerful heavy-coated deep-chested dogs of Alaskan origin that have erect ears, heavily cushioned feet, and a plumy tail

Alas·ka time \ə-ˈlas-kə-\ *noun* (1945)
: the time of the 9th time zone west of Greenwich that includes most of Alaska

[1]alate \ˈā-ˌlāt\ *adjective* [Latin *alatus,* from *ala*] (1668)
: having wings or a winglike part

[2]alate *noun* (1941)
: a winged insect (as an aphid) of a kind having winged and wingless forms

alb \ˈalb\ *noun* [Middle English *albe,* from Old English, from Medieval Latin *alba,* from Latin, feminine of *albus* white; akin to Greek *alphos* white leprous spot] (before 12th century)
: a full-length white linen ecclesiastical vestment with long sleeves that is gathered at the waist with a cincture

al·ba·core \ˈal-bə-ˌkōr, -ˌkȯr\ *noun, plural* **-core** *or* **-cores** [Portuguese *albacor,* from Arabic *al-bakūrah* the albacore] (1579)
: a large pelagic tuna (*Thunnus alalunga*) with long pectoral fins that is a source of canned tuna; *broadly* **:** any of various tunas (as a bonito)

albacore

Al·ba·nian \al-ˈbā-nē-ən, -nyən *also* ȯl-\ *noun* (1579)
1 : a native or inhabitant of Albania
2 : the Indo-European language of the Albanian people — see INDO-EUROPEAN LANGUAGES table
— **Albanian** *adjective*

al·ba·tross \ˈal-bə-ˌtrȯs, -ˌträs\ *noun, plural* **-tross** *or* **-tross·es** [probably alteration of obsolete *alcatrace* frigate bird, from Spanish or Portuguese *alcatraz* pelican, from Arabic *al-ghaṭṭās,* a kind of sea eagle] (1672)
1 : any of a family (Diomedeidae) of large web-footed seabirds that have long slender wings, are excellent gliders, and include the largest seabirds
2 a : something that causes persistent deep concern or anxiety **b :** something that greatly hinders accomplishment **:** ENCUMBRANCE

albatross 1

al·be·do \al-ˈbē-(ˌ)dō\ *noun, plural* **-dos** [Late Latin, whiteness, from Latin *albus*] (circa 1859)
: reflective power; *specifically* **:** the fraction of incident radiation (as light) that is reflected by a surface or body (as the moon or a cloud)

al·be·it \ȯl-ˈbē-ət, al-\ *conjunction* [Middle English, literally, all though it be] (14th century)
: conceding the fact that **:** even though

Al·bi·gen·ses \ˌal-bə-ˈjen-ˌsēz\ *noun plural* [Medieval Latin, plural of *Albigensis,* literally, inhabitant of Albi, from *Albiga* (Albi), France] (1625)
: members of a Catharistic sect of southern France flourishing primarily in the 12th and 13th centuries
— **Al·bi·gen·sian** \-ˈjen(t)-shən, -ˈjen(t)-sē-ən\ *adjective or noun*
— **Al·bi·gen·sian·ism** \-shə-ˌni-zəm, -sē-ə-ˌ\ *noun*

al·bi·nism \ˈal-bə-ˌni-zəm, al-ˈbī-\ *noun* (1836)
: the condition of an albino
— **al·bi·nis·tic** \ˌal-bə-ˈnis-tik\ *adjective*

al·bi·no \al-ˈbī-(ˌ)nō\ *noun, plural* **-nos** [Portuguese, from Spanish, from *albo* white, from Latin *albus*] (1777)
: an organism exhibiting deficient pigmentation; *especially* **:** a human being or nonhuman mammal that is congenitally deficient in pigment and usually has a milky or translucent skin, white or colorless hair, and eyes with pink or blue iris and deep-red pupil

al·bi·not·ic \ˌal-bə-ˈnä-tik\ *adjective* [*albino* + *-tic* (as in *melanotic*)] (1872)
1 : of, relating to, or affected with albinism
2 : tending toward albinism

Al·bi·on \ˈal-bē-ən\ *noun* [Latin] (before 12th century)
1 : Great Britain
2 : England

al·bite \ˈal-ˌbīt\ *noun* [Swedish *albit,* from Latin *albus*] (circa 1843)
: a triclinic usually white feldspar consisting of a sodium aluminum silicate
— **al·bit·ic** \al-ˈbit-ik\ *adjective*

al·bum \ˈal-bəm\ *noun* [Latin, a white tablet, from neuter of *albus*] (1612)
1 a : a book with blank pages used for making a collection (as of autographs, stamps, or photographs) **b :** a paperboard container for a phonograph record **:** JACKET **c :** one or more recordings (as on tape or disc) produced as a single unit ⟨a 2-record *album*⟩
2 : a collection usually in book form of literary selections, musical compositions, or pictures **:** ANTHOLOGY

al·bu·men \al-ˈbyü-mən; ˈal-ˌbyü-, -byə-\ *noun* [Late Latin, from *albus*] (1599)
1 : the white of an egg — see EGG illustration
2 : ALBUMIN

al·bu·min \al-ˈbyü-mən; ˈal-ˌbyü-, -byə-\ *noun* [International Scientific Vocabulary *albumen* + *-in*] (1869)
: any of numerous simple heat-coagulable water-soluble proteins that occur in blood plasma or serum, muscle, the whites of eggs, milk, and other animal substances and in many plant tissues and fluids

al·bu·min·ous \al-ˈbyü-mə-nəs\ *adjective* (1791)
: relating to, containing, or having the properties of albumen or albumin

al·bu·min·uria \(ˌ)al-ˌbyü-mə-ˈnu̇r-ē-ə, -ˈnyu̇r-\ *noun* [New Latin] (circa 1842)
: the presence of albumin in the urine often symptomatic of kidney disease
— **al·bu·min·uric** \-ˈnu̇r-ik, -ˈnyu̇r-\ *adjective*

al·ca·ic \al-ˈkā-ik\ *adjective, often capitalized* [Late Latin *Alcaicus* of Alcaeus, from Greek *Alkaïkos,* from *Alkaios* Alcaeus, (flourished about 600 B.C.) Greek poet] (circa 1637)
: relating to or written in a verse or strophe marked by complicated variation of a dominant iambic pattern
— **alcaic** *noun, often capitalized*

al·cai·de *or* **al·cay·de** \al-ˈkī-dē\ *noun* [Spanish *alcaide,* from Arabic *al-qāʼid* the captain] (1502)
: a commander of a castle or fortress (as among Spaniards, Portuguese, or Moors)

al·cal·de \al-ˈkäl-dē\ *noun* [Spanish, from Arabic *al-qāḍī* the judge] (1565)
: the chief administrative and judicial officer of a Spanish town

al·ca·zar \al-ˈkä-zər, -ˈka-\ *noun* [Spanish *alcázar,* from Arabic *al-qaṣr* the castle] (circa 1615)
: a Spanish fortress or palace

Al·ces·tis \al-ˈses-təs\ *noun* [Latin, from Greek *Alkēstis*]
: the wife of Admetus who dies for her husband and is restored to him by Hercules

al·che·mist \ˈal-kə-mist\ *noun* (15th century)
: a person who studies or practices alchemy
— **al·che·mis·tic** \ˌal-kə-ˈmis-tik\ *also* **al·che·mis·ti·cal** \-ti-kəl\ *adjective*

al·che·mize \ˈal-kə-ˌmīz\ *transitive verb* **-mized; -miz·ing** (1597)
: to change by alchemy **:** TRANSMUTE

al·che·my \ˈal-kə-mē\ *noun* [Middle English *alkamie, alquemie,* from Middle French or Medieval Latin; Middle French *alquemie,* from Medieval Latin *alchymia,* from Arabic *al-kīmiyāʼ,* from *al* the + *kīmiyāʼ* alchemy, from Late Greek *chēmeia*] (14th century)
1 : a medieval chemical science and speculative philosophy aiming to achieve the transmutation of the base metals into gold, the discovery of a universal cure for disease, and the discovery of a means of indefinitely prolonging life
2 : a power or process of transforming something common into something special

\ə\ **abut** \ᵊ\ **kitten** \ər\ **further** \a\ **ash** \ā\ **ace** \ä\ **mop, mar** \au̇\ **out** \ch\ **chin** \e\ **bet** \ē\ **easy** \g\ **go** \i\ **hit** \ī\ **ice** \j\ **job** \ŋ\ **sing** \ō\ **go** \ȯ\ **law** \ȯi\ **boy** \th\ **thin** \t͟h\ **the** \ü\ **loot** \u̇\ **foot** \y\ **yet** \zh\ **vision** *see also* Guide to Pronunciation

3 : an inexplicable or mysterious transmuting
— **al·chem·i·cal** \-mi-kəl\ *also* **al·chem·ic** \al-ˈke-mik\ *adjective*
— **al·chem·i·cal·ly** \-mi-k(ə-)lē\ *adverb*

Alc·me·ne \alk-ˈmē-nē\ *noun* [Greek *Alkmēnē*]
: the mother of Hercules by Zeus in the form of her husband Amphitryon

al·co·hol \ˈal-kə-ˌhȯl\ *noun* [New Latin, from Medieval Latin, powdered antimony, from Old Spanish, from Arabic *al-kuḥul* the powdered antimony, from *kolḥ* kohl] (1672)
1 a : ethanol especially when considered as the intoxicating agent in fermented and distilled liquors **b :** drink (as whiskey or beer) containing ethanol **c :** a mixture of ethanol and water that is usually 95 percent ethanol
2 : any of various compounds that are analogous to ethanol in constitution and that are hydroxyl derivatives of hydrocarbons

[1]al·co·hol·ic \ˌal-kə-ˈhȯ-lik, -ˈhä-\ *adjective* (1790)
1 a : of, relating to, or caused by alcohol **b :** containing alcohol
2 : affected with alcoholism
— **al·co·hol·i·cal·ly** \-li-k(ə-)lē\ *adverb*

[2]alcoholic *noun* (circa 1890)
: a person affected with alcoholism

al·co·hol·ism \ˈal-kə-ˌhȯ-ˌli-zəm, -kə-hə-\ *noun* (1860)
1 : continued excessive or compulsive use of alcoholic drinks
2 : poisoning by alcohol; *especially* **:** a complex chronic psychological and nutritional disorder associated with excessive and usually compulsive drinking

Al·co·ran \ˌal-kə-ˈran\ *noun* [Middle English, from Middle French or Medieval Latin; Middle French & Medieval Latin, from Arabic *al-qur'ān,* literally, the reading] (14th century)
archaic **:** KORAN

al·cove \ˈal-ˌkōv\ *noun* [French *alcôve,* from Spanish *alcoba,* from Arabic *al-qubbah* the arch] (1676)
1 a : a small recessed section of a room **:** NOOK **b :** an arched opening (as in a wall) **:** NICHE
2 : SUMMERHOUSE 2
— **al·coved** \-ˌkōvd\ *adjective*

al·cy·o·nar·i·an \ˌal-sē-ə-ˈnar-ē-ən\ *noun* [ultimately from Greek *alkyoneion,* a zoophyte, from neuter of *alkyoneios* of a kingfisher, from *alkyon* kingfisher; from its resemblance to a kingfisher's nest] (1878)
: any of a subclass (Alcyonaria) of colonial anthozoans (as the sea pen) having polyps with eight branched tentacles and eight septa

Al·cy·o·ne \al-ˈsī-ə-(ˌ)nē\ *noun* [Latin, from Greek *Alkyonē*]
: the brightest star in the Pleiades

Al·deb·a·ran \al-ˈde-bə-rən\ *noun* [Arabic *al-dabarān,* literally, the follower]
: a red star of the first magnitude that is seen in the eye of Taurus and is the brightest star in the Hyades

al·de·hyde \ˈal-də-ˌhīd\ *noun* [German *Aldehyd,* from New Latin *al. dehyd.,* abbreviation of *alcohol dehydrogenatum* dehydrogenated alcohol] (circa 1846)
: ACETALDEHYDE; *broadly* **:** any of various highly reactive compounds typified by acetaldehyde and characterized by the group CHO
— **al·de·hy·dic** \ˌal-də-ˈhī-dik\ *adjective*

al den·te \äl-ˈden-(ˌ)tā, al-\ *adjective* [Italian; literally, to the tooth] (1935)
: cooked just enough to retain a somewhat firm texture

al·der \ˈȯl-dər\ *noun* [Middle English, from Old English *alor;* akin to Old High German *elira* alder, Latin *alnus*] (before 12th century)
: any of a genus (*Alnus*) of toothed-leaved trees or shrubs of the birch family that have catkins which become woody, that grow in moist ground, and that have wood used in turnery

al·der·man \ˈȯl-dər-mən\ *noun* [Middle English, from Old English *ealdorman,* from *ealdor* parent (from *eald* old) + *man* — more at OLD] (before 12th century)
1 : a person governing a kingdom, district, or shire as viceroy for an Anglo-Saxon king
2 a : a magistrate formerly ranking next below the mayor in an English or Irish city or borough **b :** a high-ranking member of a borough or county council in Ireland or formerly in England chosen by elected members
3 : a member of a city legislative body
— **al·der·man·ic** \ˌȯl-dər-ˈma-nik\ *adjective*

al·der·wom·an \ˈȯl-dər-ˌwu̇-mən\ *noun* [*alder-* (as in *alderman*) + *woman*] (1768)
: a female member of a city legislative body

al·dol \ˈal-ˌdȯl, -ˌdōl\ *noun* [International Scientific Vocabulary *ald*ehyde + [1]*-ol*] (1874)
: a colorless beta-hydroxy aldehyde $C_4H_8O_2$ used especially in organic synthesis; *broadly* **:** any of various similar aldehydes
— **al·dol·iza·tion** \ˌal-ˌdȯ-lə-ˈzā-shən, -ˌdō-\ *noun*

al·dol·ase \ˈal-də-ˌlās, -ˌlāz\ *noun* [*aldol* + *-ase*] (1940)
: a crystalline enzyme that occurs widely in living systems and catalyzes reversibly the cleavage of a phosphorylated fructose into triose sugars

al·dose \ˈal-ˌdōs, -ˌdōz\ *noun* [International Scientific Vocabulary *ald*ehyde + *-ose*] (1894)
: a sugar containing one aldehyde group per molecule

al·do·ste·rone \al-ˈdäs-tə-ˌrōn; ˌal-dō-ˈster-ˌōn, -ˈstir-; ˈal-dō-stə-ˌrōn\ *noun* [*ald*ehyde + *-o-* + *-sterone*] (1954)
: a steroid hormone $C_{21}H_{28}O_5$ of the adrenal cortex that functions in the regulation of the salt and water balance of the body

al·do·ste·ron·ism \al-ˈdäs-tə-rə-ˌni-zəm; ˌal-dō-ˈster-(ˌ)ō-, -ˈstir-; ˌal-dō-stə-ˈrō-\ *noun* (1955)
: a condition that is characterized by excessive secretion of aldosterone and typically by loss of body potassium, muscular weakness, and elevated blood pressure

al·drin \ˈȯl-drən, ˈal-\ *noun* [Kurt *Alder* (died 1958) German chemist + English [1]*-in*] (1949)
: an exceedingly poisonous cyclodiene insecticide $C_{12}H_8Cl_6$

ale \ˈā(ə)l\ *noun* [Middle English, from Old English *ealu;* akin to Old Norse *ǫl* ale, Lithuanian *alus*] (before 12th century)
1 : an alcoholic beverage brewed especially by rapid fermentation from an infusion of malt with the addition of hops
2 : an English country festival at which ale is the principal beverage

ale·a·tor·ic \ˌā-lē-ə-ˈtȯr-ik, -ˈtär-\ *adjective* [Latin *aleatorius*] (1961)
: characterized by chance or indeterminate elements ⟨*aleatoric* music⟩

ale·a·to·ry \ˈā-lē-ə-ˌtōr-ē, -ˌtȯr-\ *adjective* [Latin *aleatorius* of a gambler, from *aleator* gambler, from *alea* a dice game] (1693)
1 : depending on an uncertain event or contingency as to both profit and loss ⟨an *aleatory* contract⟩
2 : relating to luck and especially to bad luck
3 : ALEATORIC

alee \ə-ˈlē\ *adverb* (14th century)
: on or toward the lee — compare AWEATHER

ale·house \ˈā(ə)l-ˌhau̇s\ *noun* (before 12th century)
: a place where ale is sold to be drunk on the premises

Al·e·man·nic \ˌa-lə-ˈma-nik\ *noun* [Late Latin *alemanni,* of Germanic origin; akin to Gothic *alamans* totality of people] (circa 1797)
: the group of dialects of German spoken in Alsace, Switzerland, and southwestern Germany

alem·bic \ə-ˈlem-bik\ *noun* [Middle English, from Middle French & Medieval Latin; Middle French *alambic* & Medieval Latin *alembicum,* from Arabic *al-anbīq,* from *al* the + *anbīq* still, from Late Greek *ambik-, ambix* alembic, from Greek, cap of a still] (14th century)
1 : an apparatus used in distillation
2 : something that refines or transmutes as if by distillation ⟨philosophy . . . filtered through the *alembic* of Plato's mind —B. T. Shropshire⟩

alembic 1

alen·çon \ə-ˈlen-ˌsän, -ˈlen(t)-sən\ *noun, often capitalized* [*Alençon,* France] (1865)
: a delicate needlepoint lace

aleph \ˈä-ˌlef, -ləf\ *noun* [Hebrew *āleph,* probably from *eleph* ox] (14th century)
: the 1st letter of the Hebrew alphabet — see ALPHABET table

aleph–null \-ˈnəl\ *noun* (circa 1909)
: the number of elements in the set of all integers which is the smallest transfinite cardinal number

[1]alert \ə-ˈlərt\ *adjective* [Italian *all'erta,* literally, on the ascent] (1618)
1 a : watchful and prompt to meet danger or emergency **b :** quick to perceive and act
2 : ACTIVE, BRISK
synonym see WATCHFUL, INTELLIGENT
— **alert·ly** *adverb*
— **alert·ness** *noun*

[2]alert *noun* (1796)
1 : the state of readiness of those warned by an alert
2 : an alarm or other signal of danger
3 : the period during which an alert is in effect
— **on the alert :** on the lookout especially for danger or opportunity

[3]alert *transitive verb* (circa 1868)
1 : to call to a state of readiness **:** WARN
2 : to make aware of ⟨*alerted* the public to the dangers of pesticides⟩

-ales *noun plural suffix* [New Latin, from Latin, plural of *-alis* -al]
: plants consisting of or related to — in the names of orders of plants ⟨Conifer*ales*⟩

al·eu·rone \ˈal-yə-ˌrōn\ *noun* [German *Aleuron,* from Greek, flour; akin to Armenian *ałam* I grind] (1869)
: protein matter in the form of minute granules or grains occurring in seeds in endosperm or in a special peripheral layer

Aleut \ˌa-lē-ˈüt, ˈa-lē-ˌ; ə-ˈlüt\ *noun* [Russian] (1780)
1 : a member of a people of the Aleutian and Shumagin islands and the western part of Alaska Peninsula
2 : the language of the Aleuts

A level *noun* (1951)
1 : the second of three standardized British examinations in a secondary school subject used as a qualification for university entrance; *also* **:** successful completion of an A-level examination in a particular subject — called also *Advanced level;* compare O LEVEL, S LEVEL
2 a : the level of education required to pass an A-level examination **b :** a course leading to an A-level examination

al·e·vin \ˈa-lə-vən\ *noun* [French, from Old French, from *alever* to lift up, rear (offspring), from Latin *allevare,* from *ad-* + *levare* to raise — more at LEVER] (1868)
: a young fish; *especially* **:** a newly hatched salmon when still attached to the yolk sac

[1]ale·wife \ˈā(ə)l-ˌwīf\ *noun* (15th century)
: a woman who keeps an alehouse

[2]alewife *noun* [perhaps alteration of obsolete *allowes,* a kind of shad, from French *alose* shad, from Old French, from Late Latin *alausa*] (1633)
: a clupeid food fish (*Alosa pseudoharengus*) very abundant along the Atlantic coast; *also* **:** any of several related fishes (as the menhaden)

al•ex•an•der \ˌa-lig-ˈzan-dər, ˌe-\ *noun, often capitalized* (1928)
: an iced cocktail made from crème de cacao, sweet cream, and gin or brandy

Al•ex•an•dri•an \ˌa-lig-ˈzan-drē-ən, ˌe-\ *adjective* (circa 1860)
1 : of or relating to Alexander the Great
2 : HELLENISTIC

al•ex•an•drine \-ˈzan-ˌdrēn, -drən, -ˌdrīn\ *noun, often capitalized* [Middle French *alexandrin,* adjective, from *Alexandre* Alexander the Great; from its use in a poem on Alexander] (1667)
: a line of verse of 12 syllables consisting regularly of 6 iambs with a caesura after the 3d iamb
— **alexandrine** *adjective*

al•ex•an•drite \-ˈzan-ˌdrīt\ *noun* [German *Alexandrit,* from *Alexander I* Russian emperor] (circa 1880)
: a grass-green chrysoberyl that shows a red color by transmitted or artificial light

alex•ia \ə-ˈlek-sē-ə\ *noun* [New Latin, from *a-* + Greek *lexis* speech, from *legein* to speak — more at LEGEND] (1878)
: aphasia marked by loss of ability to read

Al•fa \ˈal-fə\ (1952)
— a communications code word for the letter *a*

al•fal•fa \al-ˈfal-fə\ *noun* [Spanish, modification of Arabic dialect *al-faṣfaṣah* the alfalfa] (1845)
: a deep-rooted European leguminous plant (*Medicago sativa*) widely grown for hay and forage

alfalfa weevil *noun* (1912)
: a small dark brown European weevil (*Hypera postica*) that is now a widespread pest of alfalfa in North America

al•fil•a•ria \(ˌ)al-ˌfi-lə-ˈrē-ə\ *noun* [American Spanish *alfilerillo,* from Spanish, diminutive of *alfiler* pin, modification of Arabic *al-khilāl* the thorn] (1868)
: a European weed (*Erodium cicutarium*) of the geranium family grown for forage in the western U.S.

al•for•ja \al-ˈfȯr-(ˌ)hä\ *noun* [Spanish, from Arabic *al khurj*] (1611)
West **:** SADDLEBAG

al•fres•co \al-ˈfres-(ˌ)kō\ *adjective or adverb* [Italian] (1753)
: taking place or located in the open air **:** OUTDOOR, OUTDOORS ⟨an *alfresco* lunch⟩ ⟨an *alfresco* restaurant⟩
word history see FRESCO

al•ga \ˈal-gə\ *noun, plural* **al•gae** \ˈal-(ˌ)jē\ *also* **algas** [Latin, seaweed] (1551)
: a plant or plantlike organism of any of several phyla, divisions, or classes of chiefly aquatic usually chlorophyll-containing nonvascular organisms of polyphyletic origin that usually include the green, yellow-green, brown, and red algae in the eukaryotes and the blue-green algae in the prokaryotes
— **al•gal** \-gəl\ *adjective*

al•ga•ro•ba *also* **al•gar•ro•ba** \ˌal-gə-ˈrō-bə\ *noun* [Spanish *algarroba,* from Arabic *al-kharrūbah* the carob] (1577)
1 : CAROB 1
2 [Mexican Spanish, from Spanish] **:** MESQUITE; *also* **:** its pods

al•ge•bra \ˈal-jə-brə\ *noun* [Medieval Latin, from Arabic *al-jabr,* literally, the reduction] (1551)
1 : a generalization of arithmetic in which letters representing numbers are combined according to the rules of arithmetic
2 : any of various systems or branches of mathematics or logic concerned with the properties and relationships of abstract entities (as complex numbers, matrices, sets, vectors, groups, rings, or fields) manipulated in symbolic form under operations often analogous to those of arithmetic — compare BOOLEAN ALGEBRA
— **al•ge•bra•ist** \-ˌbrā-ist\ *noun*

al•ge•bra•ic \ˌal-jə-ˈbrā-ik\ *adjective* (1662)
1 : relating to, involving, or according to the laws of algebra
2 : involving only a finite number of repetitions of addition, subtraction, multiplication, division, extraction of roots, and raising to powers ⟨*algebraic* equation⟩ — compare TRANSCENDENTAL
— **al•ge•bra•i•cal•ly** \-ˈbrā-ə-k(ə-)lē\ *adverb*

algebraic number *noun* (1904)
: a root of an algebraic equation with rational coefficients

-algia *noun combining form* [Greek, from *algos* pain]
: pain ⟨neur*algia*⟩

al•gi•cide *or* **al•gae•cide** \ˈal-jə-ˌsīd\ *noun* (1904)
: an agent used to kill algae
— **al•gi•cid•al** \ˌal-jə-ˈsī-dᵊl\ *adjective*

al•gid \ˈal-jəd\ *adjective* [Latin *algidus,* from *algēre* to feel cold] (circa 1623)
: COLD

al•gin \ˈal-jən\ *noun* (1883)
: any of various colloidal substances (as an alginate or alginic acid) derived from marine brown algae and used especially as emulsifiers or thickeners

al•gi•nate \ˈal-jə-ˌnāt\ *noun* (circa 1909)
: a salt or ester of alginic acid

al•gin•ic acid \(ˌ)al-ˈji-nik-\ *noun* [International Scientific Vocabulary *algin* + [1]*-ic*] (1885)
: an insoluble colloidal acid $(C_6H_8O_6)_n$ that in the form of its salts is a constituent of the cell walls of brown algae

Al•gol \ˈal-ˌgäl, -ˌgȯl\ *noun* [Arabic *al-ghūl,* literally, the ghoul]
: a binary star in the constellation Perseus whose larger member orbits and eclipses the smaller brighter star causing periodic variation in brightness

AL•GOL *or* **Al•gol** \ˈal-ˌgäl, -ˌgȯl\ *noun* [*algo*rithmic *l*anguage] (1959)
: an algebraic computer programming language used especially in mathematical and scientific applications

al•go•lag•nia \ˌal-gō-ˈlag-nē-ə\ *noun* [New Latin, from Greek *algos* pain + Greek *lagneia* lust, from *lagnos* lustful — more at SLACK] (circa 1900)
: a perversion (as sadism or masochism) characterized by pleasure and especially sexual gratification in inflicting or suffering pain
— **al•go•lag•ni•ac** \-ˈlag-nē-ˌak\ *noun*

al•gol•o•gy \al-ˈgä-lə-jē\ *noun* (1849)
: the study or science of algae — called also *phycology*
— **al•go•log•i•cal** \ˌal-gə-ˈlä-ji-kəl\ *adjective*
— **al•gol•o•gist** \al-ˈgä-lə-jist\ *noun*

Al•gon•qui•an \al-ˈgän-kwē-ən, -ˈgäŋ-\ *or* **Al•gon•quin** \-kwən\ *or* **Al•gon•ki•an** \-ˈgäŋ-kē-ən\ *also* **Al•gon•kin** \-ˈgäŋ-kən\ *noun* [Canadian French *Algonquin*] (1625)
1 *usually Algonquin* **a :** an American Indian people of the Ottawa river valley **b :** the dialect of Ojibwa spoken by these people
2 *usually Algonquian* **a :** a family of American Indian languages spoken by peoples from Labrador to Carolina and westward into the Great Plains **b :** a member of the peoples speaking Algonquian languages

al•go•rithm \ˈal-gə-ˌri-t͟həm\ *noun* [alteration of Middle English *algorisme,* from Old French & Medieval Latin; Old French, from Medieval Latin *algorismus,* from Arabic *al-khuwārizmi,* from *al-Khuwārizmi* (flourished A.D. 825) Arab mathematician] (circa 1894)
: a procedure for solving a mathematical problem (as of finding the greatest common divisor) in a finite number of steps that frequently involves repetition of an operation; *broadly* **:** a step-by-step procedure for solving a problem or accomplishing some end especially by a computer
— **al•go•rith•mic** \ˌal-gə-ˈrit͟h-mik\ *adjective*
— **al•go•rith•mi•cal•ly** \-mi-k(ə-)lē\ *adverb*

Al•ham•bra \al-ˈham-brə\ *noun* [Spanish, from Arabic *al-ḥamrā'* the red house] (1612)
: the palace of the Moorish kings at Granada, Spain

[1]**alias** \ˈā-lē-əs, ˈāl-yəs\ *adverb* [Latin, otherwise, from *alius* other — more at ELSE] (15th century)
: otherwise called **:** otherwise known as

[2]**alias** *noun* (1605)
: an assumed or additional name

Ali Ba•ba \ˌa-lē-ˈbä-bə, ˌä-lē-\ *noun*
: a woodcutter in the *Arabian Nights' Entertainments* who enters the cave of the Forty Thieves by using the password *Sesame*

[1]**al•i•bi** \ˈa-lə-ˌbī\ *noun* [Latin, elsewhere, from *alius*] (1743)
1 : the plea of having been at the time of the commission of an act elsewhere than at the place of commission; *also* **:** the fact or state of having been elsewhere at the time
2 : an excuse usually intended to avert blame or punishment (as for failure or negligence) ◆
synonym see APOLOGY

[2]**alibi** *verb* **-bied; -bi•ing** (1909)
transitive verb
: to exonerate by an alibi **:** furnish an excuse for
intransitive verb
: to offer an excuse

Al•ice–in–Won•der•land \ˈa-ləs-ən-ˈwən-dər-ˌland\ *adjective* [from *Alice's Adventures in Wonderland* (1865) by Lewis Carroll] (1925)
: suitable to a world of fantasy or illusion **:** UNREAL

ali•cy•clic \ˌa-lə-ˈsī-klik, -ˈsi-klik\ *adjective* [International Scientific Vocabulary *ali*phatic + *cyclic*] (1891)
: of, relating to, or being an organic compound that contains a ring but is not aromatic — compare ALIPHATIC

al•i•dade \ˈa-lə-ˌdād\ *noun* [Middle English *alidatha,* from Medieval Latin *alhidada,* from Arabic *al-'iḍādah* the revolving radius of a circle] (15th century)
: a rule equipped with simple or telescopic sights and used for determination of direction: as **a :** a part of an astrolabe **b :** a part of a surveying instrument consisting of the telescope and its attachments

[1]**alien** \ˈā-lē-ən, ˈāl-yən\ *adjective* [Middle English, from Middle French, from Latin *alienus,* from *alius*] (14th century)

◇ WORD HISTORY
alibi In Latin *alibi* was an adverb that meant "elsewhere." When *alibi* was first adopted into English in the 18th century, it was still limited to adverbial use. A person on trial might be said to try to prove himself alibi when the crime was committed. By the last quarter of that century, however, *alibi* had acquired the status of a noun and was used in legal contexts for "the plea of having been at the time of the commission of an act elsewhere than at the place of commission." The meaning of the word was then extended to apply to the fact or state of having been elsewhere when a crime was committed. Though still disapproved by some, a generalized sense, "an excuse, especially for failure or negligence," dates back at least to 1912 and has become firmly established as a meaning of *alibi.*

1 a : belonging or relating to another person, place, or thing **:** STRANGE **b :** relating, belonging, or owing allegiance to another country or government **:** FOREIGN
2 : differing in nature or character typically to the point of incompatibility
synonym see EXTRINSIC
— alienly *adverb*
— alien•ness \-lē-ən-nəs, -yən-nəs\ *noun*

²alien *noun* (14th century)
1 : a person of another family, race, or nation
2 : a foreign-born resident who has not been naturalized and is still a subject or citizen of a foreign country; *broadly* **:** a foreign-born citizen
3 : EXTRATERRESTRIAL

³alien *transitive verb* (14th century)
1 : ALIENATE, ESTRANGE
2 : to make over (as property)

alien•able \'āl-yə-nə-bəl, 'ā-lē-ə-nə-\ *adjective* (1611)
: transferable to another's ownership
— alien•abil•i•ty \'āl-yə-nə-'bi-lə-tē, ˌā-lē-ə-nə-\ *noun*

alien•age \'āl-yə-nij, 'ā-lē-ə-nij\ *noun* (1809)
: the status of an alien

alien•ate \'ā-lē-ə-ˌnāt, 'āl-yə-\ *transitive verb* **-at•ed; -at•ing** (circa 1509)
1 : to make unfriendly, hostile, or indifferent where attachment formerly existed
2 : to convey or transfer (as property or a right) usually by a specific act rather than the due course of law
3 : to cause to be withdrawn or diverted
synonym see ESTRANGE
— alien•ator \-ˌnā-tər\ *noun*

alien•ation \ˌā-lē-ə-'nā-shən, ˌāl-yə-\ *noun* (14th century)
1 : a withdrawing or separation of a person or a person's affections from an object or position of former attachment **:** ESTRANGEMENT ⟨*alienation* . . . from the values of one's society and family —S. L. Halleck⟩
2 : a conveyance of property to another

alien•ee \-'nē\ *noun* (1531)
: one to whom property is transferred

alien•ism \'ā-lē-ə-ˌni-zəm, 'āl-yə-\ *noun* (1808)
: ALIENAGE

alien•ist \-nist\ *noun* [French *aliéniste,* from *aliéné* insane, from Latin *alienatus,* past participle of *alienare* to estrange, from *alienus*] (1864)
: PSYCHIATRIST

alien•or \ˌā-lē-ə-'nȯr, ˌāl-yə-\ *noun* (circa 1552)
: one who transfers property to another

¹alight \ə-'līt\ *intransitive verb* **alight•ed** *also* **alit** \ə-'lit\; **alight•ing** [Middle English, from Old English *ālīhtan,* from *ā-* (perfective prefix) + *līhtan* to alight — more at ABIDE, LIGHT] (before 12th century)
1 : to come down from something (as a vehicle): as **a :** DISMOUNT **b :** DEPLANE
2 : to descend from or as if from the air and come to rest **:** LAND, SETTLE
3 *archaic* **:** to come by chance
— alight•ment *noun*

²alight *adjective* (15th century)
1 *chiefly British* **:** being on fire
2 : lighted up

align *also* **aline** \ə-'līn\ *verb* [French *aligner,* from Old French, from *a-* (from Latin *ad-*) + *ligne* line, from Latin *linea*] (circa 1693)
transitive verb
1 : to bring into line or alignment
2 : to array on the side of or against a party or cause
intransitive verb
1 : to get or fall into line
2 : to be in or come into precise adjustment or correct relative position
— align•er *noun*

align•ment *also* **aline•ment** \ə-'līn-mənt\ *noun* (1790)
1 : the act of aligning or state of being aligned; *especially* **:** the proper positioning or state of adjustment of parts (as of a mechanical or electronic device) in relation to each other
2 a : a forming in line **b :** the line thus formed
3 : the ground plan (as of a railroad or highway) in distinction from the profile
4 : an arrangement of groups or forces in relation to one another ⟨new *alignments* within the political party⟩

¹alike \ə-'līk\ *adverb* (14th century)
: in the same manner, form, or degree **:** EQUALLY ⟨was denounced by teachers and students *alike*⟩

²alike *adjective* [Middle English *ilik, ilich* (from Old English *gelīc*) & *alik,* alteration of Old English *onlīc,* from *on* + *līc* body — more at LIKE] (15th century)
: exhibiting close resemblance without being identical ⟨*alike* in their beliefs⟩
— alike•ness *noun*

¹al•i•ment \'a-lə-mənt\ *noun* [Middle English, from Latin *alimentum,* from *alere* to nourish — more at OLD] (15th century)
: FOOD, NUTRIMENT; *also* **:** SUSTENANCE ⟨there was nothing there of conversational *aliment* —Kingsley Amis⟩

²al•i•ment \-ˌment\ *transitive verb* (15th century)
: to give aliment to **:** NOURISH, SUSTAIN

al•i•men•ta•ry \ˌa-lə-'men-t(ə-)rē\ *adjective* (1615)
1 : of or relating to nourishment or nutrition
2 : furnishing sustenance or maintenance

alimentary canal *noun* (1764)
: the tubular passage that extends from mouth to anus and functions in digestion and absorption of food and elimination of residual waste

al•i•men•ta•tion \ˌa-lə-mən-'tā-shən, -ˌmen-\ *noun* (circa 1656)
: the act or process of affording nutriment or nourishment ⟨intravenous *alimentation*⟩

al•i•mo•ny \'a-lə-ˌmō-nē\ *noun, plural* **-nies** [Latin *alimonia* sustenance, from *alere*] (1656)
1 : an allowance made to one spouse by the other for support pending or after legal separation or divorce
2 : the means of living **:** MAINTENANCE

A–line \'ā-ˌlīn\ *adjective* (1964)
: having a flared bottom and a close-fitting top — used of a garment ⟨an *A-line* skirt⟩

al•i•phat•ic \ˌa-lə-'fa-tik\ *adjective* [International Scientific Vocabulary, from Greek *aleiphat-, aleiphar* oil, from *aleiphein* to smear; perhaps akin to Greek *lipos* fat — more at LEAVE] (1889)
: of, relating to, or being an organic compound having an open-chain structure (as an alkane) — compare ALICYCLIC, AROMATIC 2

al•i•quot \'a-lə-ˌkwät, -kwət\ *adjective* [Medieval Latin *aliquotus,* from Latin *aliquot* some, several, from *alius* other + *quot* how many — more at ELSE, QUOTE] (1570)
1 : contained an exact number of times in something else — used of a divisor or part ⟨5 is an *aliquot* part of 15⟩ ⟨an *aliquot* portion of a solution⟩
2 : FRACTIONAL ⟨an *aliquot* part of invested capital⟩
— aliquot *noun*

A–list \'ā-ˌlist\ *noun* (1980)
: a list or group of individuals of the highest level of society, excellence, or eminence

alit•er•a•cy \ˌā-'li-tər-ə-ˌsē, ə-'\ *noun* (1984)
: the quality or state of being able to read but uninterested in doing so
— alit•er•ate \-'li-tər-ət\ *adjective or noun*

alive \ə-'līv\ *adjective* [Middle English, from Old English *on life,* from *on* + *līf* life] (before 12th century)
1 : having life **:** not dead or inanimate
2 : still in existence, force, or operation **:** ACTIVE ⟨kept hope *alive*⟩
3 : knowing or realizing the existence of **:** SENSITIVE ⟨*alive* to the danger⟩
4 : marked by alertness, energy, or briskness
5 : marked by much life, animation, or activity **:** SWARMING ⟨streets *alive* with traffic⟩
6 — used as an intensive following the noun ⟨the proudest boy *alive*⟩
synonym see AWARE
— alive•ness *noun*

ali•yah *or* **ali•ya** \ä-'lē-(ˌ)yä, ˌä-lē-'yä\ *noun* [New Hebrew *'alīyāh,* from Hebrew, ascent] (circa 1934)
: the immigration of Jews to Israel

aliz•a•rin \ə-'li-zə-rən\ *noun* [probably from French *alizarine*] (circa 1835)
1 : an orange or red crystalline compound $C_{14}H_8O_4$ formerly prepared from madder and now made synthetically and used especially to dye Turkey reds and in making red pigments
2 : any of various acid, mordant, and solvent dyes derived like alizarin proper from anthraquinone

al•ka•hest \'al-kə-ˌhest\ *noun* [New Latin *alchahest*] (1641)
: the universal solvent believed by alchemists to exist
— al•ka•hes•tic \ˌal-kə-'hes-tik\ *adjective*

al•ka•li \'al-kə-ˌlī\ *noun, plural* **-lies** *or* **-lis** [Middle English, from Medieval Latin, from Arabic *al-qili* the ashes of the plant saltwort] (14th century)
1 : a soluble salt obtained from the ashes of plants and consisting largely of potassium or sodium carbonate; *broadly* **:** a substance (as a hydroxide or carbonate of an alkali metal) having marked basic properties — compare BASE 7a
2 : ALKALI METAL
3 : a soluble salt or a mixture of soluble salts present in some soils of arid regions in quantity detrimental to agriculture

alkali metal *noun* (circa 1885)
: any of the univalent mostly basic metals of group I of the periodic table comprising lithium, sodium, potassium, rubidium, cesium, and francium — see PERIODIC TABLE table

al•ka•lim•e•ter \ˌal-kə-'li-mə-tər\ *noun* [French *alcalimètre,* from *alcali* alkali + *-mètre* -meter] (circa 1828)
: an apparatus for measuring the strength or the amount of alkali in a mixture or solution
— al•ka•lim•e•try \-'lim-ə-trē\ *noun*

al•ka•line \'al-kə-lən, -ˌlīn\ *adjective* (1677)
: of, relating to, containing, or having the properties of an alkali or alkali metal **:** BASIC; *especially, of a solution* **:** having a pH of more than 7
— al•ka•lin•i•ty \ˌal-kə-'li-nə-tē\ *noun*

alkaline battery *noun* (1941)
: a long-lived dry cell that has an alkaline electrolyte which decreases corrosion of the cell — called also *alkaline cell*

alkaline earth metal *noun* (circa 1903)
: any of the bivalent strongly basic metals of group II of the periodic table comprising beryllium, magnesium, calcium, strontium, barium, and radium — called also *alkaline earth;* see PERIODIC TABLE table

alkaline phosphatase *noun* (1949)
: any of the phosphatases that are optimally active in alkaline medium and occur in especially high concentrations in bone, the liver, the kidneys, and the placenta

al•ka•lin•ize \'al-kə-lə-ˌnīz\ *transitive verb* **-ized; -iz•ing** (1800)
: to make alkaline
— al•ka•lin•iza•tion \ˌal-kə-ˌli-nə-'zā-shən, -lə-nə-\ *noun*

al•ka•loid \'al-kə-ˌlȯid\ *noun* (circa 1831)
: any of numerous usually colorless, complex, and bitter organic bases (as morphine or codeine) containing nitrogen and usually oxygen that occur especially in seed plants
— al•ka•loi•dal \ˌal-kə-'lȯi-dᵊl\ *adjective*

al•ka•lo•sis \ˌal-kə-'lō-səs\ *noun* (1911)
: an abnormal condition of increased alkalinity of the blood and tissues
— al•ka•lot•ic \-'lä-tik\ *adjective*

al•kane \'al-ˌkān\ *noun* [*alk*yl + *-ane*] (1899)

: any of numerous saturated hydrocarbons; *specifically* : any of a series of open-chain hydrocarbons C_nH_{2n+2} (as methane and butane) — called also *paraffin*

al·ka·net \ˈal-kə-ˌnet\ *noun* [Middle English, from Old Spanish *alcaneta,* diminutive of *alcana* henna shrub, from Medieval Latin *alchanna,* from Arabic *al-ḥinnā'* the henna] (14th century)
1 a : a European plant (*Alkanna tinctoria*) of the borage family; *also* **:** its root **b :** a red dyestuff prepared from the root
2 : a plant (*Anchusa officinalis*) of the borage family with delicate usually blue flowers

al·kene \ˈal-ˌkēn\ *noun* [International Scientific Vocabulary *alkyl* + *-ene*] (1899)
: any of numerous unsaturated hydrocarbons having one double bond; *specifically* **:** any of a series of open-chain hydrocarbons C_nH_{2n} (as ethylene)

alk·ox·ide \al-ˈkäk-ˌsīd, -səd\ *noun* [*alkyl* + *oxide*] (circa 1889)
: a basic salt derived from an alcohol by the replacement of the hydroxyl hydrogen with a metal

alk·oxy \ˈal-ˌkäk-sē\ *adjective* [International Scientific Vocabulary *alkyl* + *oxygen*] (circa 1925)
: of, relating to, or containing a univalent radical composed of an alkyl group united with oxygen — often used in combination

al·kyd \ˈal-kəd\ *noun* [blend of *alkyl* and *acid*] (1929)
1 : any of numerous synthetic resins that are used especially for protective coatings and in paint
2 : a paint in which the vehicle is an alkyd resin

¹al·kyl \ˈal-kəl\ *adjective* (1882)
: having a monovalent organic group and especially one C_nH_{2n+1} (as methyl) derived from an alkane (as methane)

²alkyl *noun* [probably from German, from *Alkohol* alcohol] (1952)
: a compound of one or more alkyl groups with a metal ⟨mercury *alkyls*⟩

alkylating agent *noun* (1900)
: a substance that causes replacement of hydrogen by an alkyl group especially in a biologically important molecule; *specifically* **:** one with mutagenic activity that inhibits cell division and growth and is used to treat some cancers

al·kyl·ation \ˌal-kə-ˈlā-shən\ *noun* (1900)
: the act or process of introducing one or more alkyl groups into a compound (as to increase octane number in a motor fuel)
— al·kyl·ate \ˈal-kə-ˌlāt\ *transitive verb*

al·kyne \ˈal-ˌkīn\ *noun* [*alkyl* + *-yne,* alteration of *-ine*] (circa 1909)
: any of a series of open-chain hydrocarbons C_nH_{2n-2} (as acetylene) having one triple bond

¹all \ˈȯl\ *adjective* [Middle English *all, al,* from Old English *eall;* akin to Old High German *all* all] (before 12th century)
1 a : the whole amount or quantity of ⟨needed *all* the courage they had⟩ ⟨sat up *all* night⟩ **b :** as much as possible ⟨spoke in *all* seriousness⟩
2 : every member or individual component of ⟨*all* men will go⟩ ⟨*all* five children were present⟩
3 : the whole number or sum of ⟨*all* the angles of a triangle are equal to two right angles⟩
4 : EVERY ⟨*all* manner of hardship⟩
5 : any whatever ⟨beyond *all* doubt⟩
6 : nothing but **:** ONLY: **a :** completely taken up with, given to, or absorbed by ⟨became *all* attention⟩ **b :** having or seeming to have (some physical feature) in conspicuous excess or prominence ⟨*all* legs⟩ **c :** paying full attention with ⟨*all* ears⟩
7 *dialect* **:** used up **:** entirely consumed — used especially of food and drink
8 : being more than one person or thing ⟨who *all* is coming⟩
synonym see WHOLE
— all the : as much of . . . as **:** as much of a . . . as ⟨*all the* home I ever had⟩

²all *adverb* (before 12th century)
1 a : WHOLLY, QUITE ⟨sat *all* alone⟩ — often used as an intensive ⟨*all* out of proportion⟩ ⟨*all* over the yard⟩ **b :** selected as the best (as at a sport) within an area or organization — used in combination ⟨*all*-league halfback⟩
2 *obsolete* **:** ONLY, EXCLUSIVELY
3 *archaic* **:** JUST
4 : so much ⟨*all* the better for it⟩
5 : for each side **:** APIECE ⟨the score is two *all*⟩

³all *pronoun* (before 12th century)
1 : the whole number, quantity, or amount **:** TOTALITY ⟨*all* that I have⟩ ⟨*all* of us⟩ ⟨*all* of the books⟩
2 : EVERYBODY, EVERYTHING ⟨gave equal attention to *all*⟩ ⟨that is *all*⟩
— all in all : on the whole **:** GENERALLY ⟨*all in all,* things might have been worse⟩
— and all : and everything else especially of a kind suggested by a previous context ⟨cards to fill out with . . . numbers *and all* —Sally Quinn⟩

⁴all *noun* (1593)
: the whole of one's possessions, resources, or energy ⟨gave his *all* for the cause⟩

all- *or* **allo-** *combining form* [Greek, from *allos* other — more at ELSE]
1 : other **:** different **:** atypical ⟨*allo*gamous⟩ ⟨*allo*tropy⟩
2 *allo-* **:** isomeric form or variety of (a specified chemical compound) ⟨*allo*purinol⟩
3 *allo-* **:** being one of a group whose members together constitute a structural unit especially of a language ⟨*allo*phone⟩

¹al·la breve \ˌa-lə-ˈbrev, ˌä-lə-ˈbre-(ˌ)vā\ *noun* [Italian, literally, according to the breve] (circa 1740)
: the sign marking a piece or passage to be played alla breve; *also* **:** a passage so marked

alla breve

²alla breve *adverb or adjective* (circa 1823)
: in duple or quadruple time with the beat represented by the half note

Al·lah \ˈä-lə, ˈa-lə, ˈä-ˌlä, ä ˈlä\ *noun* [Arabic *allāh*] (1584)
: GOD 1a — used in Islam

all along *adverb* (1670)
: all the time ⟨knew the truth *all along*⟩

¹all–Amer·i·can \ˌȯ-lə-ˈmer-ə-kən\ *adjective* (1888)
1 a *also* **all–Amer·i·ca :** selected (as by a poll of journalists) as one of the best in the U.S. in a particular category at a particular time ⟨an *all-American* quarterback⟩ **b :** having only all-American participants ⟨an *all-American* basketball team⟩
2 : composed wholly of American elements
3 : representative or typical of the U.S. or its ideals ⟨an *all-American* boy⟩ ⟨her *all-American* optimism⟩
4 : of or relating to the American nations as a group

²all–American *noun* (1920)
: one (as an athlete) that is voted all-American

al·lan·to·in \ə-ˈlan-tə-wən\ *noun* [probably from German, from New Latin *allantois* + German *-in*] (circa 1845)
: a crystalline oxidation product $C_4H_6N_4O_3$ of uric acid used to promote healing of local wounds and infections

al·lan·to·is \ə-ˈlan-tə-wəs\ *noun, plural* **al·lan·to·ides** \ˌa-lən-ˈtō-ə-ˌdēz, ˌa-ˌlan-\ [New Latin, ultimately from Greek *allant-, allas* sausage] (1646)
: a vascular fetal membrane of reptiles, birds, and mammals that is formed as a pouch from the hindgut and that in placental mammals is intimately associated with the chorion in formation of the placenta
— al·lan·to·ic \ˌa-lən-ˈtō-ik, ˌa-ˌlan-\ *adjective*

al·lar·gan·do \ˌä-lär-ˈgän-(ˌ)dō\ *adjective or adverb* [Italian, widening, verbal of *allargare* to widen, from *al-* (from Latin *ad-*) + *largare* to widen] (circa 1893)
: becoming gradually slower and more stately — used as a direction in music

all–around \ˌȯ-lə-ˈraủnd\ *adjective* (1867)
1 : considered in or encompassing all aspects **:** COMPREHENSIVE ⟨the best *all-around* performance so far⟩
2 : competent in many fields ⟨an *all-around* performer⟩
3 : having general utility or merit

al·lay \a-ˈlā, ə-\ *verb* [Middle English *alayen,* from Old English *ālecgan,* from *ā-* (perfective prefix) + *lecgan* to lay — more at ABIDE, LAY] (14th century)
transitive verb
1 : to subdue or reduce in intensity or severity **:** ALLEVIATE ⟨expect a breeze to *allay* the heat⟩
2 : to make quiet **:** CALM
intransitive verb
obsolete **:** to diminish in strength **:** SUBSIDE
synonym see RELIEVE

all but *adverb* (1593)
: very nearly **:** ALMOST ⟨would be *all but* impossible⟩

all clear *noun* (1902)
: a signal that a danger has passed

all–day \ˈȯl-ˈdā\ *adjective* (circa 1870)
: lasting for, occupying, or appearing throughout an entire day ⟨an *all-day* trip⟩

al·lée \ä-ˈlā, a-\ *noun* [French, from Middle French *alee* — more at ALLEY] (1759)
: a walkway lined with trees or tall shrubs

al·le·ga·tion \ˌa-li-ˈgā-shən\ *noun* (15th century)
1 : the act of alleging
2 : a positive assertion; *specifically* **:** a statement by a party to a legal action of what the party undertakes to prove
3 : an assertion unsupported and by implication regarded as unsupportable ⟨vague *allegations* of misconduct⟩

al·lege \ə ˈlej\ *transitive verb* **al·leged; al·leg·ing** [Middle English *alleggen,* from Middle French *alleguer,* from Latin *allegare* to dispatch, cite, from *ad-* + *legare* to depute — more at LEGATE] (14th century)
1 *archaic* **:** to adduce or bring forward as a source or authority
2 : to assert without proof or before proving ⟨the newspaper *alleges* the mayor's guilt⟩
3 : to bring forward as a reason or excuse

al·leged \ə-ˈlejd, -ˈle-jəd\ *adjective* (1509)
1 : asserted to be true or to exist ⟨an *alleged* miracle⟩
2 : questionably true or of a specified kind **:** SUPPOSED, SO-CALLED ⟨bought an *alleged* antique vase⟩
3 : accused but not proven or convicted ⟨an *alleged* gangster⟩
— al·leg·ed·ly \-ˈle-jəd-lē\ *adverb*

Al·le·ghe·ny spurge \ˈa-lə-ˌgā-nē- *also* -ˌge-nē-\ *noun* [*Allegheny* Mountains, U.S.A.] (circa 1936)
: a low herb or subshrub (*Pachysandra procumbens*) of the box family widely grown as a ground cover

al·le·giance \ə-ˈlē-jən(t)s\ *noun* [Middle English *allegeaunce,* modification of Middle French *ligeance,* from Old French, from *lige* liege] (14th century)
1 a : the obligation of a feudal vassal to his liege lord **b** (1) **:** the fidelity owed by a subject or citizen to a sovereign or government (2) **:** the obligation of an alien to the government under which the alien resides

2 : devotion or loyalty to a person, group, or cause
synonym see FIDELITY
— al·le·giant \-jənt\ *adjective*
al·le·gor·i·cal \ˌa-lə-ˈgȯr-i-kəl, -ˈgär-\ *adjective* (1528)
1 : of, relating to, or having the characteristics of allegory
2 : having hidden spiritual meaning that transcends the literal sense of a sacred text
— al·le·gor·i·cal·ly \-k(ə-)lē\ *adverb*
— al·le·gor·i·cal·ness \-kəl-nəs\ *noun*
al·le·go·rise *British variant of* ALLEGORIZE
al·le·go·rist \ˈa-lə-ˌgōr-əst, -ˌgȯr-\ *noun* (1684)
: a creator of allegory
al·le·go·ri·za·tion \ˌa-lə-ˌgōr-ə-ˈzā-shən, -ˌgȯr-, -gər-\ *noun* (1847)
: allegorical representation or interpretation
al·le·go·rize \ˈa-lə-ˌgōr-ˌīz, -ˌgȯr-, -gər-\ *verb* **-rized; -riz·ing** (1581)
intransitive verb
1 : to give allegorical explanations
2 : to compose or use allegory
transitive verb
1 : to treat or explain as an allegory
2 : to make into allegory
— al·le·go·riz·er *noun*
al·le·go·ry \ˈa-lə-ˌgōr-ē, -ˌgȯr-\ *noun, plural* **-ries** [Middle English *allegorie,* from Latin *allegoria,* from Greek *allēgoria,* from *allēgorein* to speak figuratively, from *allos* other + *-ēgorein* to speak publicly, from *agora* assembly — more at ELSE, AGORA] (14th century)
1 : the expression by means of symbolic fictional figures and actions of truths or generalizations about human existence; *also* **:** an instance (as in a story or painting) of such expression
2 : a symbolic representation **:** EMBLEM 2
[1]al·le·gret·to \ˌa-lə-ˈgre-(ˌ)tō, ˌä-lə-\ *adverb or adjective* [Italian, diminutive of *allegro*] (circa 1740)
: faster than andante but not so fast as allegro — used as a direction in music
[2]allegretto *noun, plural* **-tos** (circa 1846)
: a musical composition or movement in allegretto tempo
[1]al·le·gro \ə-ˈle-(ˌ)grō, -ˈlā-\ *noun, plural* **-gros** (1683)
: a musical composition or movement in allegro tempo
[2]allegro *adverb or adjective* [Italian, merry, from (assumed) Vulgar Latin *alecrus* lively, alteration of Latin *alacr-, alacer*] (circa 1721)
: at a brisk lively tempo — used as a direction in music
al·lele \ə-ˈlē(ə)l\ *noun* [German *Allel,* short for *Allelomorph*] (1928)
1 : any of the alternative forms of a gene that may occur at a given locus
2 : either of a pair of alternative Mendelian characters (as smooth and wrinkled seed in the pea)
— al·le·lic \-ˈlē-lik, -ˈle-\ *adjective*
— al·lel·ism \-ˈlē(ə)-ˌli-zəm, -ˈle-\ *noun*
allelo- *combining form* [Greek *allēlōn* of each other, from *allos . . . allos* one . . . the other, from *allos* other — more at ELSE]
1 : alternative ⟨*allelo*morph⟩
2 : reciprocal ⟨*allelo*pathy⟩
al·le·lo·morph \ə-ˈlē-lə-ˌmȯrf, -ˈle-lə-\ *noun* (1902)
: ALLELE
— al·le·lo·mor·phic \ə-ˌlē-lə-ˈmȯr-fik, -ˌle-lə-\ *adjective*
— al·le·lo·mor·phism \ə-ˈlē-lə-ˌmȯr-ˌfi-zəm, -ˈle-lə-\ *noun*
al·le·lop·a·thy \ə-ˈlē-lə-ˌpa-thē, -ˈle-lə-; *also* ˌa-lə-ˈlä-pə-thē\ *noun* [International Scientific Vocabulary] (1948)
: the suppression of growth of one plant species by another due to the release of toxic substances
— al·le·lo·path·ic \ə-ˌlē-lə-ˈpa-thik, -ˌlə-lə-\ *adjective*
al·le·lu·ia \ˌa-lə-ˈlü-yə\ *interjection* [Middle English, from Late Latin, from Greek *allēlouia,* from Hebrew *hallălūyāh* praise ye Jehovah] (14th century)
: HALLELUJAH
al·le·mande \ˈa-lə-ˌman(d), -mən, -ˌmänd, *1a & 2 also* ˌa-lə-ˈ\ *noun, often capitalized* [French, from feminine of *allemand* German] (1685)
1 : a musical composition or movement (as in a baroque suite) in moderate tempo and duple or quadruple time
2 a : a 17th and 18th century court dance developed in France from a German folk dance **b :** a dance step with arms interlaced
all–em·brac·ing \ˌȯ-lim-ˈbrā-siŋ\ *adjective* (circa 1649)
: COMPLETE, SWEEPING
Al·len wrench \ˈa-lən-\ *noun* [*Allen* Manufacturing Co., Hartford, Conn.] (1943)
: an L-shaped hexagonal metal bar either end of which fits the socket of a screw or bolt
al·ler·gen \ˈa-lər-jən\ *noun* [International Scientific Vocabulary *allergy* + *-gen*] (1910)
: a substance that induces allergy
— al·ler·gen·ic \ˌa-lər-ˈje-nik\ *adjective*
— al·ler·ge·nic·i·ty \ˌa-lər-jə-ˈni-sə-tē\ *noun*
al·ler·gic \ə-ˈlər-jik\ *adjective* (1911)
1 : of, relating to, inducing, or affected by allergy
2 : having an aversion ⟨*allergic* to work⟩
al·ler·gist \ˈa-lər-jist\ *noun* (1928)
: a specialist in allergy
al·ler·gy \ˈa-lər-jē\ *noun, plural* **-gies** [German *Allergie,* from *all-* + Greek *ergon* work — more at WORK] (1910)
1 : altered bodily reactivity (as hypersensitivity) to an antigen in response to a first exposure ⟨a bee venom *allergy* so severe that a second sting may be fatal⟩
2 : exaggerated or pathological reaction (as by sneezing, respiratory embarrassment, itching, or skin rashes) to substances, situations, or physical states that are without comparable effect on the average individual
3 : medical practice concerned with allergies
4 : a feeling of antipathy or repugnance
al·le·thrin \ˈa-lə-thrən\ *noun* [*all*yl + *pyrethrin*] (1950)
: a light yellow viscous oily synthetic insecticide $C_{19}H_{26}O_3$ used especially in household aerosols
al·le·vi·ate \ə-ˈlē-vē-ˌāt\ *transitive verb* **-at·ed; -at·ing** [Late Latin *alleviatus,* past participle of *alleviare,* from Latin *ad-* + *levis* light — more at LIGHT] (15th century)
: RELIEVE, LESSEN: as **a :** to make (as suffering) more bearable ⟨her sympathy *alleviated* his distress⟩ **b :** to partially remove or correct
synonym see RELIEVE
— al·le·vi·a·tion \-ˌlē-vē-ˈā-shən\ *noun*
— al·le·vi·a·tor \-ˈlē-vē-ˌā-tər\ *noun*
[1]al·ley \ˈa-lē\ *noun, plural* **alleys** [Middle English, from Middle French *alee,* from Old French, from *aler* to go] (14th century)
1 : a garden or park walk bordered by trees or bushes
2 a (1) **:** a grassed enclosure for bowling or skittles (2) **:** a hardwood lane for bowling; *also* **:** a room or building housing a group of such lanes **b :** the space on each side of a tennis doubles court between the sideline and the service sideline **c :** an area in a baseball outfield between two outfielders when they are in normal positions
3 : a narrow street; *especially* **:** a thoroughfare through the middle of a block giving access to the rear of lots or buildings
— up one's alley *also* **down one's alley :** suited to one's own tastes or abilities
[2]alley *noun, plural* **alleys** [by shortening and alteration from *alabaster*] (1720)
: a playing marble; *especially* **:** one of superior quality
al·ley–oop \ˌa-lē-ˈüp\ *noun* [alteration of *allez-oop,* cry of a circus acrobat about to leap, probably from French *allez,* 2d person plural imperative of *aller* to go + English *-oop,* perhaps alteration of *up*] (1967)
: a basketball play in which a leaping player catches a pass above the basket and immediately dunks the ball; *also* **:** the usually looping pass thrown on such a play
al·ley·way \ˈa-lē-ˌwā\ *noun* (1788)
1 : a narrow passageway
2 : ALLEY 3
All Fools' Day *noun* (1712)
: APRIL FOOLS' DAY
all fours *noun plural* (1563)
1 a : all four legs of a quadruped **b :** the two legs and two arms of a person when used to support the body
2 *singular in construction* **:** any of various card games in which points are scored for the high trump, low trump, jack of trumps, and game
all get–out \ˌȯl-ˈget-ˌau̇t, -ˈgit-\ *noun* (1884)
: the utmost conceivable degree — used in comparisons to suggest something superlative ⟨is handsome as *all get-out* and has a deft way with the ladies —John McCarten⟩
all hail *interjection* (14th century)
— used to express greeting, welcome, or acclamation
All·hal·lows \ȯl-ˈha-(ˌ)lōz, -ləz\ *noun, plural* **Allhallows** [short for *All Hallows' Day*] (1503)
: ALL SAINTS' DAY
al·li·a·ceous \ˌa-lē-ˈā-shəs\ *adjective* [Latin *allium* garlic] (1792)
: resembling garlic or onion especially in smell or taste
al·li·ance \ə-ˈlī-ən(t)s\ *noun* (13th century)
1 a : the state of being allied **:** the action of allying **b :** a bond or connection between families, states, parties, or individuals ⟨a closer *alliance* between government and industry⟩
2 : an association to further the common interests of the members; *specifically* **:** a confederation of nations by treaty
3 : union by relationship in qualities **:** AFFINITY
4 : a treaty of alliance
al·lied \ə-ˈlīd, ˈa-ˌlīd\ *adjective* (14th century)
1 : having or being in close association **:** CONNECTED ⟨a strong personal pride *allied* with the utmost probity⟩ ⟨two families *allied* by marriage⟩
2 : joined in alliance by compact or treaty; *specifically, capitalized* **:** of or relating to the nations united against Germany and its allies in World War I or those united against the Axis powers in World War II
3 a : related especially by common properties or qualities ⟨heraldry and *allied* subjects⟩ **b :** related genetically
allies *plural of* ALLY
al·li·ga·tor \ˈa-lə-ˌgā-tər\ *noun* [Spanish *el lagarto* the lizard, from *el* the (from Latin *ille* that) + *lagarto* lizard, from (assumed) Vulgar Latin *lacartus,* from Latin *lacertus, lacerta* — more at LIZARD] (1579)
1 a : either of two crocodilians (genus *Alligator*) having broad heads not tapering to the snout and a special pocket in the upper jaw for reception of the enlarged lower fourth tooth **b :** CROCODILIAN
2 : leather made from alligator hide

alligator 1a

alligator clip *noun* (circa 1941)

: a spring-loaded clip that has jaws resembling an alligator's and is used for making temporary electrical connections

alligator clip

alligator pear *noun* [by folk etymology from Spanish *aguacate* — more at AVOCADO] (1763)
: AVOCADO

alligator snapper *noun* (1884)
: a snapping turtle (*Macrochelys temminckii*) of the rivers of the Gulf states that may reach nearly 150 pounds (68 kilograms) in weight and 5 feet (1.5 meters) in length

all–im·por·tant \ˌȯ-lim-ˈpȯr-t^{ə}nt, -tənt\ *adjective* (1839)
: of very great or greatest importance ⟨an *all-important* question⟩

all–in \ˌȯ-ˈlin\ *adjective* (1890)
1 *chiefly British* **:** ALL-INCLUSIVE
2 *chiefly British* **:** being almost without restrictions ⟨*all-in* wrestling⟩

all in *adjective* (1903)
: TIRED, EXHAUSTED ⟨after a day of wood-splitting he was *all in*⟩

all–in·clu·sive \ˌȯ-lin-ˈklü-siv, -ziv\ *adjective* (circa 1855)
: including everything ⟨a broader and more nearly *all-inclusive* view⟩
— all–in·clu·sive·ness *noun*

al·lit·er·ate \ə-ˈli-tə-ˌrāt\ *verb* **-at·ed; -at·ing** [back-formation from *alliteration*] (1816)
intransitive verb
1 : to form an alliteration
2 : to write or speak alliteratively
transitive verb
: to arrange or place so as to make alliteration ⟨*alliterate* syllables in a sentence⟩

al·lit·er·a·tion \ə-ˌli-tə-ˈrā-shən\ *noun* [*ad-* + Latin *littera* letter] (circa 1656)
: the repetition of usually initial consonant sounds in two or more neighboring words or syllables (as *w*ild and *w*oolly, *thr*eatening *thr*ongs) — called also *head rhyme, initial rhyme*

al·lit·er·a·tive \ə-ˈli-tər-ə-tiv, -tə-ˌrā-tiv\ *adjective* (1764)
: of, relating to, or marked by alliteration
— al·lit·er·a·tive·ly *adverb*

al·li·um \ˈa-lē-əm\ *noun* [New Latin, genus name, from Latin, garlic] (1807)
: any of a large genus (*Allium*) of bulbous herbs of the lily family including the onion, garlic, chive, leek, and shallot

all–night \ˈȯl-ˈnīt\ *adjective* (1888)
1 : lasting throughout the night ⟨an *all-night* poker game⟩
2 : open throughout the night ⟨an *all-night* diner⟩

all–night·er \ˈȯl-ˈnī-tər\ *noun* (1967)
: something that lasts all night; *specifically* **:** an all-night study session

allo- — see ALL-

al·lo·an·ti·body \ˌa-lō-ˈan-ti-ˌbä-dē\ *noun* (1964)
: an antibody produced following introduction of an alloantigen into the system of an individual of a species lacking that particular antigen — called also *isoantibody*

al·lo·an·ti·gen \ˌa-lō-ˈan-ti-jən\ *noun* (1964)
: an antigen present only in some individuals (as of a particular blood group) of a species and capable of inducing the production of an isoantibody by individuals which lack it — called also *isoantigen*

al·lo·ca·ble \ˈa-lə-kə-bəl\ *adjective* (1916)
: capable of being allocated

al·lo·cate \ˈa-lə-ˌkāt\ *transitive verb* **-cat·ed; -cat·ing** [Medieval Latin *allocatus,* past participle of *allocare,* from Latin *ad-* + *locare* to place, from *locus* place — more at STALL] (circa 1641)
1 : to apportion for a specific purpose or to particular persons or things **:** DISTRIBUTE ⟨*allocate* tasks among human and automated components⟩
2 : to set apart or earmark **:** DESIGNATE ⟨*allocate* a section of the building for special research purposes⟩
— al·lo·cat·able \-ˌkā-tə-bəl\ *adjective*
— al·lo·ca·tion \ˌa-lə-ˈkā-shən\ *noun*
— al·lo·ca·tor \ˈa-lə-ˌkā-tər\ *noun*

al·lo·cu·tion \ˌa-lə-ˈkyü-shən\ *noun* [Latin *allocution-, allocutio,* from *alloqui* to speak to, from *ad-* + *loqui* to speak] (1615)
: a formal speech; *especially* **:** an authoritative or hortatory address

all of *adverb* (1829)
: FULLY ⟨she's *all of* 20 years old⟩

al·log·a·mous \ə-ˈlä-gə-məs\ *adjective* (circa 1890)
: reproducing by cross-fertilization
— al·log·a·my \-mē\ *noun*

al·lo·ge·ne·ic \ˌa-lō-jə-ˈnē-ik\ *also* **al·lo·gen·ic** \-ˈje-nik\ *adjective* [*all-* + *-geneic* (as in *syngeneic*)] (1961)
: involving, derived from, or being individuals of the same species that are sufficiently unlike genetically to interact antigenically

al·lo·graft \ˈa-lə-ˌgraft\ *noun* (1961)
: a homograft between allogeneic individuals
— allograft *transitive verb*

al·lo·graph \ˈa-lə-ˌgraf\ *noun* (1951)
1 : a letter of an alphabet in a particular shape (as A or a)
2 : a letter or combination of letters that is one of several ways of representing one phoneme (as *pp* in *hopping* representing the phoneme \p\)
— al·lo·graph·ic \ˌa-lə-ˈgra-fik\ *adjective*

al·lom·e·try \ə-ˈlä-mə-trē\ *noun* (1936)
: relative growth of a part in relation to an entire organism or to a standard; *also* **:** the measure and study of such growth
— al·lo·me·tric \ˌa-lə-ˈme-trik\ *adjective*

allomorph *noun* [*allo-* + *morph*eme] (1945)
: one of a set of forms that a morpheme may take in different contexts ⟨the *-s* of *cats,* the *-en* of *oxen,* and the zero suffix of *sheep* are *allomorphs* of the English plural morpheme⟩
— al·lo·mor·phic \ˌa-lə-ˈmȯr-fik\ *adjective*
— al·lo·mor·phism \ˈa-lə-ˌmȯr-ˌfi-zəm\ *noun*

al·longe \a-ˈlō^{n}zh\ *noun* [French, literally, lengthening] (circa 1859)
: RIDER 2a

al·lo·pat·ric \ˌa-lə-ˈpa-trik\ *adjective* [*all-* + Greek *patra* fatherland, from *patēr* father — more at FATHER] (1942)
: occurring in different geographical areas or in isolation ⟨*allopatric* speciation⟩ — compare SYMPATRIC
— al·lo·pat·ri·cal·ly \-tri-k(ə-)lē\ *adverb*
— al·lop·a·try \ə-ˈlä-pə-trē\ *noun*

al·lo·phane \ˈa-lə-ˌfān\ *noun* [Greek *allophanēs* appearing otherwise, from *all-* + *phainesthai* to appear, middle voice of *phainein* to show — more at FANCY] (circa 1821)
: an amorphous translucent mineral of various colors often occurring in incrustations or stalactite forms and consisting of a hydrous aluminum silicate

al·lo·phone \ˈa-lə-ˌfōn\ *noun* [*allo-* + *phone*] (1938)
: one of two or more variants of the same phoneme ⟨the aspirated \p\ of *pin* and the unaspirated \p\ of *spin* are *allophones* of the phoneme \p\⟩
— al·lo·phon·ic \ˌa-lə-ˈfä-nik\ *adjective*

al·lo·poly·ploid \ˌa-lō-ˈpä-li-ˌplȯid\ *noun* (1928)
: a polyploid individual or strain having a chromosome set composed of two or more chromosome sets derived more or less complete from different species
— allopolyploid *adjective*
— al·lo·poly·ploi·dy \-ˌplȯi-dē\ *noun*

al·lo·pu·ri·nol \ˌa-lō-ˈpyu̇r-ə-ˌnȯl, -ˌnōl\ *noun* [*all-* + *purine* + [1]*-ol*] (1964)
: a drug $C_5H_4N_4O$ used to promote excretion of uric acid

all–or–none \ˌȯ-lər-ˈnən\ *adjective* (1900)
: marked either by entire or complete operation or effect or by none at all ⟨*all-or-none* response of a nerve cell⟩

all–or–noth·ing \-ˈnə-thiŋ\ *adjective* (1765)
1 : ALL-OR-NONE
2 a : accepting no less than everything ⟨he's an *all-or-nothing* perfectionist⟩ **b :** risking everything ⟨an *all-or-nothing* combat strategy⟩

al·lo·sau·rus \ˌa-lə-ˈsȯr-əs\ *noun* [New Latin, from Greek *all-* + *sauros* lizard] (1899)
: any of a genus (*Allosaurus*) of huge carnivorous North American theropod dinosaurs of the Upper Jurassic period

al·lo·ste·ric \ˌa-lō-ˈster-ik, -ˈstir-\ *adjective* [*all-* + *steric*] (1962)
: of, relating to, or being a change in the shape and activity of a protein (as an enzyme) that results from combination with another substance at a point other than the chemically active site
— al·lo·ste·ri·cal·ly \-i-k(ə-)lē\ *adverb*
— al·lo·ste·ry \ˈa-lō-ˌster-ē, -ˌstir-\ *noun*

al·lot \ə-ˈlät\ *transitive verb* **al·lot·ted; al·lot·ting** [Middle English *alotten,* from Middle French *aloter,* from *a-* (from Latin *ad-*) + *lot,* of Germanic origin; akin to Old English *hlot* lot] (15th century)
1 : to assign as a share or portion ⟨*allot* 10 minutes for the speech⟩
2 : to distribute by or as if by lot ⟨*allot* seats to the press⟩
— al·lot·ter *noun*

al·lo·te·tra·ploid \ˌa-lō-ˈte-trə-ˌplȯid\ *noun* (1930)
: AMPHIDIPLOID
— al·lo·te·tra·ploi·dy \-ˌplȯi-dē\ *noun*

al·lot·ment \ə-ˈlät-mənt\ *noun* (1574)
1 : the act of allotting **:** APPORTIONMENT
2 : something that is allotted; *especially, chiefly British* **:** a plot of land let to an individual for cultivation

al·lo·trope \ˈa-lə-ˌtrōp\ *noun* [International Scientific Vocabulary, back-formation from *allotropy*] (circa 1889)
: a form showing allotropy

al·lot·ro·py \ə-ˈlä-trə-pē\ *noun, plural* **-pies** (1850)
: the existence of a substance and especially an element in two or more different forms (as of crystals) usually in the same phase
— al·lo·trop·ic \ˌa-lə-ˈträ-pik\ *adjective*

all' ot·ta·va \ˌa-lə-ˈtä-və, ˌä-lō-\ *adverb or adjective* [Italian, at the octave] (circa 1823)
: OTTAVA

al·lot·tee \ə-ˌlä-ˈtē\ *noun* (1846)
: one to whom an allotment is made

al·lo·type \ˈa-lə-ˌtīp\ *noun* (1960)
: an alloantigen that is part of a plasma protein (as an antibody)
— al·lo·typ·ic \ˌa-lə-ˈti-pik\ *adjective*
— al·lo·typ·i·cal·ly \-pi-k(ə-)lē\ *adverb*
— al·lo·typy \ˈa-lə-ˌtī-pē\ *noun*

all–out \ˈȯ-ˈlau̇t\ *adjective* (1908)
1 : made with maximum effort **:** THOROUGHGOING ⟨an *all-out* effort to win the contest⟩
2 : FULL-BLOWN 2

all out *adverb* (1895)
: with full determination or enthusiasm **:** with maximum effort — used chiefly in the phrase *go all out*

[1]all·over \ˈȯ-ˌlō-vər\ *adjective* (1859)
: covering the whole extent or surface ⟨a sweater with an *allover* pattern⟩

[2]allover *noun* (1838)
1 : an embroidered, printed, or lace fabric with a design covering most of the surface

2 : a pattern or design in which a single unit is repeated so as to cover an entire surface

all over *adverb* (1577)
1 : over the whole extent ⟨decorated *all over* with a flower pattern⟩
2 : EVERYWHERE ⟨looked *all over* for the book⟩
3 : in every respect **:** THOROUGHLY ⟨she is her mother *all over*⟩

al·low \ə-ˈlau̇\ *verb* [Middle English, from Middle French *alouer* to place, (from Medieval Latin *allocare*) & *allouer* to approve, from Latin *allaudare* to extol, from *ad-* + *laudare* to praise — more at ALLOCATE] (14th century)
transitive verb
1 a : to assign as a share or suitable amount (as of time or money) ⟨*allow* an hour for lunch⟩ **b :** to reckon as a deduction or an addition ⟨*allow* a gallon for leakage⟩
2 a *chiefly Southern & Midland* **:** to be of the opinion **:** THINK **b** *dialect* **:** SAY, STATE **c :** to express an opinion — usually used with *as how* or *that*
3 *chiefly Southern & Midland* **:** INTEND, PLAN
4 : ADMIT, CONCEDE ⟨must *allow* that money causes problems in marriage⟩
5 a : PERMIT ⟨doesn't *allow* people to smoke in his home⟩ **b :** to forbear or neglect to restrain or prevent ⟨*allow* the dog to roam⟩
intransitive verb
1 : to make a possibility **:** ADMIT — used with *of* ⟨evidence that *allows* of only one conclusion⟩
2 : to give consideration to circumstances or contingencies — used with *for* ⟨*allow* for expansion⟩

al·low·able \ə-ˈlau̇-ə-bəl\ *adjective* (15th century)
: PERMISSIBLE
— al·low·ably \-blē\ *adverb*

¹al·low·ance \ə-ˈlau̇-ən(t)s\ *noun* (14th century)
1 a : a share or portion allotted or granted **b :** a sum granted as a reimbursement or bounty or for expenses ⟨salary includes cost-of-living *allowance*⟩; *especially* **:** a sum regularly provided for personal or household expenses ⟨each child has an *allowance*⟩ **c :** a fixed or available amount ⟨provide an *allowance* of time for recreation⟩ **d :** a reduction from a list price or stated price ⟨a trade-in *allowance*⟩
2 : an imposed handicap (as in a horse race)
3 : an allowed dimensional difference between mating parts of a machine
4 : the act of allowing **:** PERMISSION
5 : a taking into account of mitigating circumstances or contingencies

²allowance *transitive verb* **-anced; -ancing** (circa 1828)
1 *archaic* **:** to put on a fixed allowance (as of food and drink)
2 *archaic* **:** to supply in a fixed or regular quantity

al·low·ed·ly \ə-ˈlau̇-əd-lē\ *adverb* (1602)
: by allowance **:** ADMITTEDLY

al·lox·an \ə-ˈläk-sən\ *noun* [German, from *Al*lantoin + *Ox*alsäure oxalic acid + *-an*] (1853)
: a crystalline compound $C_4H_2N_2O_4$ causing diabetes mellitus when injected into experimental animals

¹al·loy \ˈa-ˌlȯi *also* ə-ˈlȯi\ *noun* [French *aloi,* from Old French *alei,* from *aleir* to combine, from Latin *alligare* to bind — more at ALLY] (1604)
1 : the degree of mixture with base metals **:** FINENESS
2 : a substance composed of two or more metals or of a metal and a nonmetal intimately united usually by being fused together and dissolving in each other when molten; *also* **:** the state of union of the components
3 a : an admixture that lessens value **b :** an impairing alien element
4 : a compound, mixture, or union of different things ⟨an ethnic *alloy* of many peoples⟩
5 *archaic* **:** a metal mixed with a more valuable metal to give durability or some other desired quality

²al·loy \ə-ˈlȯi *also* ˈa-ˌlȯi\ (1661)
transitive verb
1 a : TEMPER, MODERATE **b :** to impair or debase by admixture
2 : to reduce the purity of by mixing with a less valuable metal
3 : to mix so as to form an alloy
intransitive verb
: to lend itself to being alloyed ⟨iron *alloys* well⟩

all–pow·er·ful \ˈȯl-ˈpau̇(-ə)r-fəl\ *adjective* (1667)
: having complete or sole power

all–pur·pose \-ˈpər-pəs\ *adjective* (1928)
: suited for many purposes or uses

¹all right *adjective* (1701)
1 : SATISFACTORY, AGREEABLE ⟨whatever you decide is *all right* with me⟩
2 : SAFE, WELL ⟨he was ill but he's *all right* now⟩
3 : GOOD, PLEASING — often used as a generalized term of approval ⟨an *all right* guy⟩
usage see ALRIGHT

²all right *adverb* (1837)
1 : very well ⟨*all right,* let's go⟩
2 : beyond doubt **:** CERTAINLY ⟨she has pneumonia *all right*⟩
3 : well enough **:** SATISFACTORILY ⟨does *all right* in school⟩
usage see ALRIGHT

all–round \ˈȯl-ˈrau̇nd\ *variant of* ALL-AROUND

all–round·er \ˈȯl-ˈrau̇n-dər\ *noun* (1875)
British **:** one that is all-around

All Saints' Day *noun* (circa 1798)
: November 1 observed in Western liturgical churches as a Christian feast in honor of all the saints

All Souls' Day *noun* (14th century)
: November 2 observed in some Christian churches as a day of prayer for the souls of the faithful departed

all·spice \ˈȯl-ˌspīs\ *noun* (1621)
1 : the berry of a West Indian tree (*Pimenta dioica*) of the myrtle family; *also* **:** the allspice tree
2 : a mildly pungent and aromatic spice prepared from allspice berries

¹all–star \ˈȯl-ˈstär\ *adjective* (1889)
: composed wholly or chiefly of stars or of outstanding performers or participants ⟨an *all-star* cast⟩

²all–star \ˈȯl-ˌstär\ *noun* (circa 1934)
: a member of an all-star team

all that *adverb* (1945)
: to an indicated or suggested extent or degree **:** SO ⟨didn't take his threats *all that* seriously⟩

all–time \ˈȯl-ˌtīm\ *adjective* (1914)
1 : FULL-TIME 1
2 : being for or of all time up to and including the present; *especially* **:** exceeding all others of all time ⟨an *all-time* best-seller⟩

all told *adverb* (1850)
: with everything taken into account **:** in all

al·lude \ə-ˈlüd\ *intransitive verb* **al·lud·ed; al·lud·ing** [Latin *alludere,* literally, to play with, from *ad-* + *ludere* to play — more at LUDICROUS] (1533)
: to make indirect reference ⟨*alluded* vaguely to personal problems⟩; *broadly* **:** REFER

¹al·lure \ə-ˈlu̇r\ *transitive verb* **al·lured; al·lur·ing** [Middle English *aluren,* from Middle French *alurer,* from Old French, from *a-* (from Latin *ad-*) + *loire* lure — more at LURE] (15th century)
: to entice by charm or attraction
synonym see ATTRACT
— al·lure·ment \-ˈlu̇r-mənt\ *noun*
— al·lur·ing·ly *adverb*

²allure *noun* (1548)
: power of attraction or fascination **:** CHARM

al·lu·sion \ə-ˈlü-zhən\ *noun* [Late Latin *allusion-, allusio,* from Latin *alludere*] (1548)
1 : an implied or indirect reference especially in literature; *also* **:** the use of such references
2 : the act of alluding or hinting at
— al·lu·sive \-ˈlü-siv, -ziv\ *adjective*
— al·lu·sive·ly *adverb*
— al·lu·sive·ness *noun*

¹al·lu·vi·al \ə-ˈlü-vē-əl\ *adjective* (1802)
: relating to, composed of, or found in alluvium ⟨*alluvial* soil⟩ ⟨*alluvial* diamonds⟩

²alluvial *noun* (1866)
: an alluvial deposit

alluvial fan *noun* (1873)
: the alluvial deposit of a stream where it issues from a gorge upon a plain or of a tributary stream at its junction with the main stream

alluvial fan

al·lu·vi·on \ə-ˈlü-vē-ən\ *noun* [Latin *alluvion-, alluvio,* from *alluere* to flow past, deposit (of water), from *ad-* + *lavere* to wash — more at LYE] (1536)
1 : the wash or flow of water against a shore
2 : FLOOD, INUNDATION
3 : ALLUVIUM
4 : an accession to land by the gradual addition of matter (as by deposit of alluvium) that then belongs to the owner of the land to which it is added; *also* **:** the land so added

al·lu·vi·um \-vē-əm\ *noun, plural* **-vi·ums** *or* **-via** \-vē-ə\ [Medieval Latin, alteration of Latin *alluvio*] (circa 1656)
: clay, silt, sand, gravel, or similar detrital material deposited by running water

¹al·ly \ə-ˈlī, ˈa-ˌlī\ *verb* **al·lied; al·ly·ing** [Middle English *allien,* from Old French *alier,* from Latin *alligare* to bind to, from *ad-* + *ligare* to bind — more at LIGATURE] (14th century)
transitive verb
1 : to unite or form a connection between **:** ASSOCIATE ⟨*allied* himself with a wealthy family by marriage⟩
2 : to connect or form a relation between (as by likeness or compatibility) **:** RELATE
intransitive verb
: to form or enter into an alliance

²al·ly \ˈa-ˌlī, ə-ˈlī\ *noun, plural* **allies** (1598)
1 : a sovereign or state associated with another by treaty or league
2 : a plant or animal linked to another by genetic or taxonomic proximity
3 : one that is associated with another as a helper **:** AUXILIARY

-ally *adverb suffix* [¹*-al* + *-ly*]
: ²-LY ⟨terrific*ally*⟩ — in adverbs formed from adjectives in *-ic* with no alternative form in *-ical*

al·lyl \ˈa-ləl\ *adjective* [International Scientific Vocabulary, from Latin *allium* garlic] (1854)
: being or containing the unsaturated univalent group C_3H_5-

al·lyl·ic \ə-ˈli-lik, a-\ *adjective* (1857)
: involving or characteristic of an allyl group

al·ma·gest \ˈal-mə-ˌjest\ *noun* [Middle English *almageste,* from Middle French & Medieval Latin, from Arabic *al-majusti,* the Arabic version of Ptolemy's astronomy treatise, from *al* the + Greek *megistē* (*syntaxis*), literally, greatest (composition)] (14th century)
: any of several early medieval treatises on a branch of knowledge

al·ma ma·ter \ˌal-mə-ˈmä-tər\ *noun* [Latin, fostering mother] (1696)
1 : a school, college, or university which one has attended or from which one has graduated
2 : the song or hymn of a school, college, or university

al·ma·nac \ˈȯl-mə-ˌnak, ˈal-\ *noun* [Middle

English *almenak,* from Medieval Latin *almanach,* probably from Arabic *al-manākh* the almanac] (14th century)
1 : a publication containing astronomical and meteorological data for a given year and often including a miscellany of other information
2 : a usually annual publication containing statistical, tabular, and general information

al•man•dine \'al-mən-ˌdēn, -ˌdīn\ *noun* [Middle English *alemaundine,* from Middle French *alemandine,* alteration of *alabandine,* from Medieval Latin *alabandina,* from *Alabanda,* ancient city in Asia Minor] (15th century)
: ALMANDITE

al•man•dite \'al-mən-ˌdīt\ *noun* [alteration of *almandine*] (circa 1868)
: a deep red garnet consisting of an iron aluminum silicate

¹al•mighty \ȯl-'mī-tē\ *adjective* [Middle English, from Old English *ealmihtig,* from *eall* all + *mihtig* mighty] (before 12th century)
1 *often capitalized* **:** having absolute power over all ⟨*Almighty* God⟩
2 : relatively unlimited in power
3 : great in magnitude or seriousness
— al•mighti•ness *noun*

²almighty *adverb* (1833)
: to a great degree **:** EXTREMELY ⟨although he did not precisely starve, he was *almighty* hungry —W. A. Swanberg⟩

Almighty *noun* (before 12th century)
: GOD 1 — used with *the*

al•mond \'ä-mənd, 'a-, 'äl-, 'al-\ *noun* [Middle English *almande,* from Middle French, from Late Latin *amandula,* alteration of Latin *amygdala,* from Greek *amygdalē*] (14th century)
1 a : the drupaceous fruit of a small tree (*Prunus dulcis* synonym *P. amygdalus*) of the rose family with flowers and young fruit resembling those of the peach; *especially* **:** its ellipsoidal edible kernel used as a nut **b :** any of several similar fruits
2 : a tree that produces almonds

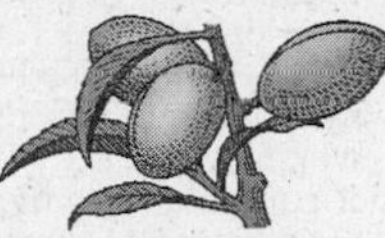
almond 1a

al•mond–eyed \-ˌīd\ *adjective* (1870)
: having narrow slant almond-shaped eyes

al•mo•ner \'al-mə-nər, 'ä-mə-\ *noun* [Middle English *almoiner,* from Middle French *almosnier,* from *almosne* alms, from Late Latin *eleemosyna*] (15th century)
1 : one who distributes alms
2 *British* **:** a social-service worker in a hospital

¹al•most \'ȯl-ˌmōst, ȯl-'\ *adverb* [Middle English, from Old English *ealmǣst,* from *eall* + *mǣst* most] (before 12th century)
: very nearly but not exactly or entirely

²almost *adjective* (1709)
: very near but not quite ⟨an *almost* failure⟩

alms \'ä(l)mz, *New England also* 'àmz\ *noun, plural* **alms** [Middle English *almesse, almes,* from Old English *ælmesse, ælmes,* from Late Latin *eleemosyna* alms, from Greek *eleēmosynē* pity, alms, from *eleēmōn* merciful, from *eleos* pity] (before 12th century)
1 *archaic* **:** CHARITY
2 : something (as money or food) given freely to relieve the poor
— alms•giv•er \-ˌgi-vər\ *noun*
— alms•giv•ing \-ˌgi-viŋ\ *noun*

alms•house \-ˌhau̇s\ *noun* (14th century)
1 *British* **:** a privately financed home for the poor
2 : POORHOUSE

alms•man \-mən\ *noun* (before 12th century)
: a recipient of alms

al•ni•co \'al-ni-ˌkō\ *noun* [*al*uminum + *ni*ckel + *co*balt] (1935)
: a powerful permanent-magnet alloy containing iron, nickel, aluminum, and one or more of the elements cobalt, copper, and titanium

al•oe \'a-(ˌ)lō\ *noun* [Middle English, from Late Latin, from Latin, dried juice of aloe leaves, from Greek *aloē*] (before 12th century)
1 *plural* **:** the fragrant wood of an East Indian tree (*Aquilaria agallocha*) of the mezereon family
2 a : any of a large genus (*Aloe*) of succulent chiefly southern African plants of the lily family with basal leaves and spicate flowers **b :** the dried juice of the leaves of various aloes used especially formerly as a purgative — usually used in plural but singular in construction

aloe vera \-'ver-ə, -'vir-\ *noun* [New Latin, species name, from *Aloe* + Latin *vera,* feminine of *verus* true — more at VERY] (circa 1936)
: an aloe (*Aloe barbadensis* synonym *A. vera*) whose leaves furnish an emollient extract used especially in cosmetics and skin creams; *also* **:** such a preparation

¹aloft \ə-'lȯft\ *adverb* [Middle English, from Old Norse *ā lopt,* from *ā* on, in + *lopt* air — more at ON, LOFT] (13th century)
1 : at or to a great height
2 : in the air; *especially* **:** in flight (as in an airplane) ⟨meals served *aloft*⟩
3 : at, on, or to the masthead or the higher rigging

²aloft *preposition* (14th century)
: on top of **:** ABOVE ⟨bright signs *aloft* hotels⟩

alog•i•cal \(ˌ)ā-'lä-ji-kəl\ *adjective* (1694)
: being outside the bounds of that to which logic can apply
— alog•i•cal•ly \-k(ə-)lē\ *adverb*

alo•ha \ə-'lō-ˌ(h)ä, ä-, -(h)ə\ *interjection* [Hawaiian, from *aloha* love] (1820)
— used as a greeting or farewell

aloha shirt *noun* (1940)
: HAWAIIAN SHIRT

¹alone \ə-'lōn\ *adjective* [Middle English, from *al* all + *one* one] (13th century)
1 : separated from others **:** ISOLATED
2 : exclusive of anyone or anything else **:** ONLY
3 a : considered without reference to any other ⟨the children *alone* would eat that much⟩ **b :** INCOMPARABLE, UNIQUE ⟨*alone* among their contemporaries in this respect⟩ ☆
— alone•ness \-'lōn-nəs\ *noun*

²alone *adverb* (13th century)
1 : SOLELY, EXCLUSIVELY
2 : without aid or support

¹along \ə-'lȯŋ\ *preposition* [Middle English, from Old English *andlang,* from *and-* against + *lang* long — more at ANTE-] (before 12th century)
1 : in a line parallel with the length or direction of
2 : in the course of
3 : in accordance with **:** IN

²along *adverb* (14th century)
1 : FORWARD, ON ⟨move *along*⟩
2 : from one to another ⟨word was passed *along*⟩
3 a : in company **:** as a companion ⟨brought his wife *along*⟩ — often used with *with* ⟨walked to school *along* with her friends⟩ **b :** in association — used with *with* ⟨work *along* with colleagues⟩
4 a : sometime within a specified or implied extent of time — usually used with *about* ⟨*along* about July 17⟩ **b :** at or to an advanced point ⟨plans are far *along*⟩
5 : in addition **:** ALSO — often used with *with* ⟨a bill came *along* with the package⟩
6 : at hand **:** as a necessary or useful item ⟨brought an extra one *along*⟩ ⟨had his gun *along*⟩
7 : on hand **:** THERE ⟨tell him I'll be *along* to see him⟩

along of *preposition* [Middle English *ilong on,* from Old English *gelang on,* from *ge-,* associative prefix + *lang* — more at CO-] (before 12th century)
dialect **:** BECAUSE OF

along•shore \ə-'lȯŋ-'shōr, -'shȯr\ *adverb or adjective* (1779)
: along the shore or coast ⟨walked *alongshore*⟩ ⟨*alongshore* currents⟩

¹along•side \-'sīd\ *adverb* (1707)
1 : along the side **:** in parallel position
2 : at the side **:** close by ⟨a guard with a prisoner *alongside*⟩

²alongside *preposition* (1793)
1 a : along the side of **b :** BESIDE 1
2 a : in company with ⟨men she has been working *alongside* —Richard Halloran⟩ **b :** in addition to ⟨a special category *alongside* the awards it annually presents —*Horizon*⟩

alongside of *preposition* (1781)
: ALONGSIDE

¹aloof \ə-'lüf\ *adverb* [obsolete *aloof* to windward, from ¹*a-* + *louf, luf* luff] (circa 1540)
: at a distance

²aloof *adjective* (1608)
: removed or distant either physically or emotionally ⟨the *aloof* composer neither worried nor cared about public opinion —Mary Jane Matz⟩ ⟨he stood *aloof* from worldly success —John Buchan⟩
synonym see INDIFFERENT
— aloof•ly *adverb*
— aloof•ness *noun*

al•o•pe•cia \ˌa-lə-'pē-sh(ē-)ə\ *noun* [Middle English *allopicia,* from Latin *alopecia,* from Greek *alōpekia,* from *alōpek-, alōpēx* fox; akin to Armenian *ałuēs* fox, Sanskrit *lopāśa*] (14th century)
: loss of hair, wool, or feathers **:** BALDNESS
— al•o•pe•cic \-'pē-sik\ *adjective*

aloud \ə-'lau̇d\ *adverb* [Middle English, from ¹*a-* + *loud*] (13th century)
1 *archaic* **:** in a loud manner **:** LOUDLY
2 : with the speaking voice

alow \ə-'lō\ *adverb* [Middle English, from ¹*a-* + *low*] (13th century)
: BELOW ⟨*alow* in the ship's hold⟩

alp \'alp\ *noun* [back-formation from *Alps,* mountain system of Europe] (15th century)
1 : a high rugged mountain
2 : something suggesting an alp in height, size, or ruggedness

al•paca \al-'pa-kə\ *noun* [Spanish, from Aymara *allpaqa*] (1811)
1 : a domesticated mammal (*Lama pacos*) especially of Peru that is probably descended from the guanaco
2 a : wool of the alpaca **b** (1) **:** a thin cloth

☆ SYNONYMS
Alone, solitary, lonely, lonesome, lone, forlorn, desolate mean isolated from others. ALONE stresses the objective fact of being by oneself with slighter notion of emotional involvement than most of the remaining terms ⟨everyone needs to be *alone* sometimes⟩. SOLITARY may indicate isolation as a chosen course ⟨glorying in the calm of her *solitary* life⟩ but more often it suggests sadness and a sense of loss ⟨left *solitary* by the death of his wife⟩. LONELY adds to SOLITARY a suggestion of longing for companionship ⟨felt *lonely* and forsaken⟩. LONESOME heightens the suggestion of sadness and poignancy ⟨an only child often leads a *lonesome* life⟩. LONE may replace LONELY or LONESOME but typically is as objective as ALONE ⟨a *lone* robin pecking at the lawn⟩. FORLORN stresses dejection, woe, and listlessness at separation from one held dear ⟨a *forlorn* lost child⟩. DESOLATE implies inconsolable grief at loss or bereavement ⟨*desolate* after her brother's death⟩.

made of or containing this wool (2) **:** a rayon or cotton imitation of this cloth

alpaca 1

al·pen·glow \'al-pən-ˌglō\ *noun* [part translation of German *Alpenglühen,* from *Alpen* Alps + *Glühen* glow] (1871)
: a reddish glow seen near sunset or sunrise on the summits of mountains

al·pen·horn \'al-pən-ˌhȯrn\ *or* **alp·horn** \'alp-ˌhȯrn\ *noun* [German, from *Alpen* + *Horn* horn] (1864)
: a straight wooden horn 5 to 14 feet (about 1.5 to 4.3 meters) in length used chiefly by Swiss herdsmen

al·pen·stock \'al-pən-ˌstäk\ *noun* [German, from *Alpen* + *Stock* staff] (1829)
: a long iron-pointed staff used in mountain climbing

¹al·pha \'al-fə\ *noun* [Middle English, from Latin, from Greek, of Semitic origin; akin to Hebrew *āleph* aleph] (13th century)
1 : the 1st letter of the Greek alphabet — see ALPHABET table
2 : something that is first **:** BEGINNING
3 : ALPHA WAVE
4 : ALPHA PARTICLE

²alpha *adjective* (1863)
1 : closest in the structure of an organic molecule to a particular group or atom — symbol α ⟨α-substitution⟩
2 : socially dominant especially in a group of animals
3 : ALPHABETIC

al·pha–ad·ren·er·gic \'al-fə-ˌa-drə-'nər-jik\ *adjective* (1966)
: of, relating to, or being an alpha-receptor ⟨*alpha-adrenergic* blocking action⟩

alpha and omega *noun* [from the fact that alpha and omega are respectively the first and last letters of the Greek alphabet] (1526)
1 : the beginning and ending
2 : the principal element

al·pha·bet \'al-fə-ˌbet, -bət\ *noun* [Middle English *alphabete,* from Late Latin *alphabetum,* from Greek *alphabētos,* from *alpha* + *bēta* beta] (1513)
1 a : a set of letters or other characters with which one or more languages are written especially if arranged in a customary order **b :** a system of signs or signals that serve as equivalents for letters
2 : RUDIMENTS, ELEMENTS

al·pha·bet·ic \ˌal-fə-'be-tik\ *or* **al·pha·bet·i·cal** \-ti-kəl\ *adjective* (1567)
1 : arranged in the order of the letters of the alphabet
2 : of, relating to, or employing an alphabet
— al·pha·bet·i·cal·ly \-ti-k(ə-)lē\ *adverb*

al·pha·bet·iza·tion \ˌal-fə-ˌbe-tə-'zā-shən\ *noun* (1889)
1 : the act or process of alphabetizing
2 : an alphabetically arranged series, list, or file

al·pha·bet·ize \'al-fə-bə-ˌtīz\ *transitive verb* **-ized; -iz·ing** (1796)
1 : to arrange alphabetically
2 : to furnish with an alphabet
— al·pha·bet·iz·er *noun*

alphabet soup *noun* (1934)
: a hodgepodge especially of initials (as of the names of organizations)

al·pha–fe·to·pro·tein \ˌal-fə-ˌfē-tō-'prō-ˌtēn, -'prō-tē-ən\ *noun* (1968)
: a fetal blood protein present abnormally in adults with some cancers (as of the liver) and normally in the amniotic fluid of pregnant women with high or low levels tending to be associated with certain birth defects (as spina bifida or Down's syndrome)

alpha globulin *noun* [International Scientific Vocabulary] (1923)
: any of several globulins of plasma or serum that have at alkaline pH the greatest electrophoretic mobility next to albumin — compare BETA GLOBULIN, GAMMA GLOBULIN

al·pha–he·lix \ˌal-fə-'hē-liks\ *noun* (1955)
: the coiled structural arrangement of many proteins consisting of a single chain of amino acids stabilized by hydrogen bonds
— al·pha–he·li·cal \-'he-li-kəl, -'hē-li-\ *adjective*

alpha iron *noun* (1902)
: the form of iron stable below 910°C

al·pha·mer·ic \ˌal-fə-'mer-ik\ *adjective* [*alphabetic* + *numeric*] (circa 1952)
: ALPHANUMERIC

al·pha·nu·mer·ic \-nü-'mer-ik, -nyü-\ *also* **al·pha·nu·mer·i·cal** \-i-kəl\ *adjective* [*alphabetic* + *numeric, numerical*] (1950)
1 : consisting of both letters and numbers and often other symbols (as punctuation marks and mathematical symbols) ⟨an *alphanumeric* code⟩; *also* **:** being a character in an alphanumeric system
2 : capable of using or displaying alphanumeric characters
— al·pha·nu·mer·i·cal·ly \-i-k(ə-)lē\ *adverb*
— al·pha·nu·mer·ics \-iks\ *noun plural*

alpha particle *noun* (1903)
: a positively charged nuclear particle identical with the nucleus of a helium atom that consists of two protons and two neutrons and is ejected at high speed in certain radioactive transformations — called also *alpha, alpha ray*

alpha privative *noun* (1590)
: the prefix *a-* or *an-* expressing negation in Greek and in English

al·pha–re·cep·tor \'al-fə-ri-ˌsep-tər\ *noun* (1961)
: any of a group of receptors postulated to exist on nerve cell membranes of the sympathetic nervous system to explain the specificity of certain adrenergic agents in affecting only some sympathetic activities (as vasoconstriction, relaxation of intestinal muscle, and contraction of most smooth muscle)

alpha wave *noun* (1936)
: an electrical rhythm of the brain with a frequency of 8 to 13 cycles per second that is often associated with a state of wakeful relaxation — called also *alpha, alpha rhythm*

Al·phe·us \al-'fē-əs\ *noun* [Latin, from Greek *Alpheios*]
: a Greek river-god who pursues the nymph Arethusa and is finally united with her

al·pine \'al-ˌpīn\ *noun* (circa 1828)
1 : a plant native to alpine or boreal regions that is often grown for ornament
2 *capitalized* **:** a person possessing Alpine physical characteristics

Alpine *adjective* (15th century)
1 *often not capitalized* **:** of, relating to, or resembling the Alps or any mountains
2 *often not capitalized* **:** of, relating to, or growing in the biogeographic zone including the elevated slopes above timberline
3 : of or relating to a physical type characterized by a broad head, stockiness, medium height, and brown hair or eyes often regarded as constituting a branch of the Caucasian race
4 : of or relating to competitive ski events consisting of slalom and downhill racing — compare NORDIC

al·pin·ism \'al-pə-ˌni-zəm\ *noun, often capitalized* (1884)
: mountain climbing in the Alps or other high mountains
— al·pin·ist \-nist\ *noun*

al·ready \ȯl-'re-dē, 'ȯl-ˌ\ *adverb* [Middle English *al redy,* from *al redy,* adjective, wholly ready, from *al* all + *redy* ready] (14th century)
1 : prior to a specified or implied past, present, or future time **:** by this time **:** PREVIOUSLY ⟨he had *already* left when I called⟩
2 — used as an intensive ⟨all right *already*⟩ ⟨enough *already*⟩

al·right \(ˌ)ȯl-'rīt, 'ȯl-ˌ\ *adverb or adjective* (1887)
: ALL RIGHT ■

Al·sa·tian \al-'sā-shən\ *noun* [Medieval Latin *Alsatia* Alsace] (1917)
: GERMAN SHEPHERD

al·sike clover \'al-ˌsak-, -ˌsīk-\ *noun* [*Alsike,* Sweden] (1852)
: a European perennial clover (*Trifolium hybridum*) much used as a forage plant

al·so \'ȯl(t)-(ˌ)sō, 'ȯ-\ *adverb* [Middle English, from Old English *eallswā,* from *eall* all + *swā* so — more at SO] (before 12th century)
1 : LIKEWISE 1
2 : in addition **:** BESIDES, TOO

al·so–ran \-ˌran\ *noun* (1896)
1 : a horse or dog that finishes out of the money in a race
2 : a contestant that does not win
3 : one that is of little importance especially competitively ⟨was just an *also-ran* in the scramble for . . . privileges —C. A. Buss⟩

Al·ta·ic \al-'tā-ik\ *adjective* (circa 1828)
1 : of or relating to the Altai Mountains
2 : of, relating to, or constituting the Turkic, Tungusic, and Mongolian language families collectively

Al·tair \al-'tīr, -'tar, -'ter, 'al-ˌ\ *noun* [Arabic *al-ṭā'ir,* literally, the flier]
: the brightest star in the constellation Aquila

al·tar \'ȯl-tər\ *noun, often attributive* [Middle English *alter,* from Old English *altar,* from Latin *altare;* probably akin to Latin *adolēre* to burn up] (before 12th century)
1 : a usually raised structure or place on which sacrifices are offered or incense is burned in worship
2 : a table on which the eucharistic elements are consecrated or which serves as a center of worship or ritual

altar boy *noun* (1772)
: a boy who assists the celebrant in a liturgical service

altar call *noun* (1946)
: an appeal by an evangelist to worshipers to come forward to signify their decision to commit their lives to Christ

altar of repose *often A&R capitalized* (circa 1872)
: REPOSITORY 2

al·tar·piece \'ȯl-tər-ˌpēs\ *noun* (1644)
: a work of art that decorates the space above and behind an altar

altar rail *noun* (1860)
: a railing in front of an altar separating the chancel from the body of the church

altar stone *noun* (14th century)
: a stone slab with a compartment containing the relics of martyrs that forms an essential part of a Roman Catholic altar

alt·az·i·muth \(ˌ)al-'taz-məth, -'ta-zə-\ *noun, often attributive* [International Scientific Vocabulary *al*titude + *azimuth*] (1860)
: a telescope mounted so that it can swing horizontally and vertically; *also* **:** any of several other similarly mounted instruments

☐ USAGE
alright The one-word spelling *alright* appeared some 75 years after *all right* itself had reappeared from a 400-year-long absence. Since the early 20th century some critics have insisted *alright* is wrong, but it has its defenders and its users. It is less frequent than *all right* but remains in common use especially in journalistic and business publications. It is quite common in fictional dialogue, and is used occasionally in other writing ⟨the first two years of medical school were *alright* —Gertrude Stein⟩.

ALPHABET TABLE

Showing the letters of five non-Roman alphabets and the transliterations used in the etymologies

HEBREW[1,4]

Letter	Name	Transliteration
א	aleph	' [2]
ב	beth	b, bh
ג	gimel	g, gh
ד	daleth	d, dh
ה	he	h
ו	waw	w
ז	zayin	z
ח	heth	ḥ
ט	teth	ṭ
י	yod	y
כ ך	kaph	k, kh
ל	lamed	l
מ ם	mem	m
נ ן	nun	n
ס	samekh	s
ע	ayin	'
פ ף	pe	p, ph
צ ץ	sadhe	ṣ
ק	qoph	q
ר	resh	r
שׂ	sin	ś
שׁ	shin	sh
ת	taw	t, th

ARABIC[3,4]

				Name	Transliteration
ا	ـا			alif	[5]
ب	ـب	ـبـ	بـ	bā	b
ت	ـت	ـتـ	تـ	tā	t
ث	ـث	ـثـ	ثـ	thā	th
ج	ـج	ـجـ	جـ	jīm	j
ح	ـح	ـحـ	حـ	ḥā	ḥ
خ	ـخ	ـخـ	خـ	khā	kh
د	ـد			dāl	d
ذ	ـذ			dhāl	dh
ر	ـر			rā	r
ز	ـز			zāy	z
س	ـس	ـسـ	سـ	sīn	s
ش	ـش	ـشـ	شـ	shīn	sh
ص	ـص	ـصـ	صـ	ṣād	ṣ
ض	ـض	ـضـ	ضـ	ḍād	ḍ
ط	ـط	ـطـ	طـ	ṭā	ṭ
ظ	ـظ	ـظـ	ظـ	ẓā	ẓ
ع	ـع	ـعـ	عـ	'ayn	'
غ	ـغ	ـغـ	غـ	ghayn	gh
ف	ـف	ـفـ	فـ	fā	f
ق	ـق	ـقـ	قـ	qāf	q
ك	ـك	ـكـ	كـ	kāf	k
ل	ـل	ـلـ	لـ	lām	l
م	ـم	ـمـ	مـ	mīm	m
ن	ـن	ـنـ	نـ	nūn	n
ه	ـه	ـهـ	هـ	hā	h[6]
و	ـو			wāw	w
ى	ـى	ـيـ	يـ	yā	y

GREEK[7]

Letter	Name	Transliteration
Α α	alpha	a
Β β	beta	b
Γ γ	gamma	g, n
Δ δ	delta	d
Ε ε	epsilon	e
Ζ ζ	zeta	z
Η η	eta	ē
Θ θ	theta	th
Ι ι	iota	i
Κ κ	kappa	k
Λ λ	lambda	l
Μ μ	mu	m
Ν ν	nu	n
Ξ ξ	xi	x
Ο ο	omicron	o
Π π	pi	p
Ρ ρ	rho	r, rh
Σ σ ς	sigma	s
Τ τ	tau	t
Υ υ	upsilon	y, u
Φ φ	phi	ph
Χ χ	chi	ch
Ψ ψ	psi	ps
Ω ω	omega	ō

RUSSIAN[8]

Letter	Transliteration
А а	a
Б б	b
В в	v
Г г	g
Д д	d
Е е	e
Ж ж	zh
З з	z
И и Й й	i, ĭ
К к	k
Л л	l
М м	m
Н н	n
О о	o
П п	p
Р р	r
С с	s
Т т	t
У у	u
Ф ф	f
Х х	kh
Ц ц	ts
Ч ч	ch
Ш ш	sh
Щ щ	shch
Ъ ъ[9]	"
Ы ы	y
Ь ь[10]	'
Э э	e
Ю ю	yu
Я я	ya

SANSKRIT[11]

Letter	Transliteration	Letter	Transliteration
अ	a	ञ	ñ
आ	ā	ट	ṭ
इ	i	ठ	ṭh
ई	ī	ड	ḍ
उ	u	ढ	ḍh
ऊ	ū	ण	ṇ
ऋ	ṛ	त	t
ॠ	ṝ	थ	th
ऌ	ḷ	द	d
ॡ	ḹ	ध	dh
ए	e	न	n
ऐ	ai	प	p
ओ	o	फ	ph
औ	au	ब	b
ं	ṁ	भ	bh
ः	ḥ	म	m
क	k	य	y
ख	kh	र	r
ग	g	ल	l
घ	gh	व	v
ङ	ṅ	श	ś
च	c	ष	ṣ
छ	ch	स	s
ज	j	ह	h
झ	jh		

1 See ALEPH, BETH, etc., in the vocabulary. Where two forms of a letter are given, the one at the right is the form used at the end of a word. 2 Not represented in transliteration when initial. 3 The left column shows the form of each Arabic letter that is used when it stands alone, the second column its form when it is joined to the preceding letter, the third column its form when it is joined to both the preceding and the following letter, and the right column its form when it is joined to the following letter only. In the names of the Arabic letters, *ā*, *ī*, and *ū* respectively are pronounced like *a* in *father*, *i* in *machine*, *u* in *rude*. 4 Hebrew and Arabic are written from right to left. The Hebrew and Arabic letters are all primarily consonants; a few of them are also used secondarily to represent certain vowels, but full indication of vowels, when provided at all, is by means of a system of dots or strokes adjacent to the consonantal characters. 5 Alif represents no sound in itself, but is used principally as an indicator of the presence of a glottal stop (transliterated ' medially and finally; not represented in transliteration when initial) and as the sign of a long *a*. 6 When ه has two dots above it (ة), it is called *tā marbūta* and, if it immediately precedes a vowel, is transliterated *t* instead of *h*. 7 See ALPHA, BETA, GAMMA, etc., in the vocabulary. The letter gamma is transliterated *n* only before velars; the letter upsilon is transliterated *u* only as the final element in diphthongs. 8 See CYRILLIC in the vocabulary. 9 This sign indicates that the immediately preceding consonant is not palatalized even though immediately followed by a palatal vowel. 10 This sign indicates that the immediately preceding consonant is palatalized even though not immediately followed by a palatal vowel. 11 The alphabet shown here is the Devanagari. When vowels are combined with preceding consonants they are indicated by various strokes or hooks instead of by the signs here given, or, in the case of short *a*, not written at all. Thus the character क represents *ka;* the character का, *kā;* the character कि, *ki;* the character की, *kī;* the character कु, *ku;* the character कू, *kū;* the character कृ, *kṛ;* the character कॄ, *kṝ;* the character के, *ke;* the character कै, *kai;* the character को, *ko;* the character कौ, *kau;* and the character क्, *k* without any following vowel. There are also many compound characters representing combinations of two or more consonants.

al·ter \'ȯl-tər\ *verb* **al·tered; al·ter·ing** \-t(ə-)riŋ\ [Middle English, from Middle French *alterer,* from Medieval Latin *alterare,* from Latin *alter* other (of two); akin to Latin *alius* other — more at ELSE] (14th century)
transitive verb
1 : to make different without changing into something else
2 : CASTRATE, SPAY
intransitive verb
: to become different
synonym see CHANGE
— al·ter·abil·i·ty \ˌȯl-t(ə-)rə-'bi-lə-tē\ *noun*
— al·ter·able \'ȯl-t(ə-)rə-bəl\ *adjective*
— al·ter·ably \-blē\ *adverb*
— al·ter·er \-tər-ər\ *noun*

al·ter·ation \ˌȯl-tə-'rā-shən\ *noun* (14th century)
1 : the act or process of altering **:** the state of being altered
2 : the result of altering **:** MODIFICATION

al·ter·cate \'ȯl-tər-ˌkāt\ *intransitive verb* **-cat·ed; -cat·ing** [Latin *altercatus,* past participle of *altercari,* from *alter*] (1530)
: to dispute angrily or noisily **:** WRANGLE

al·ter·ca·tion \ˌȯl-tər-'kā-shən\ *noun* (14th century)
: a noisy heated angry dispute; *also* **:** noisy controversy

al·ter ego \ˌȯl-tər-'ē-(ˌ)gō *also* -'e-(ˌ)gō\ *noun* [Latin, literally, second I] (1537)
: a second self: as **a :** a trusted friend **b :** the opposite side of a personality **c :** COUNTERPART 3

¹al·ter·nate *US & Canadian* 'ȯl-tər-nət *also* 'al-; *chiefly British* ȯl-'tər-\ *adjective* [Latin *alternatus,* past participle of *alternare,* from *alternus* alternate, from *alter*] (1513)
1 : occurring or succeeding by turns ⟨a day of *alternate* sunshine and rain⟩
2 a : arranged first on one side and then on the other at different levels or points along an axial line ⟨*alternate* leaves⟩ — compare OPPOSITE **b :** arranged one above or alongside the other
3 : every other **:** every second ⟨he works on *alternate* days⟩
4 : constituting an alternative ⟨took the *alternate* route home⟩
5 : ALTERNATIVE 3
— al·ter·nate·ly *adverb*

²al·ter·nate \'ȯl-tər-ˌnāt *also* 'al-\ *verb* **-nat·ed; -nat·ing** (1599)
transitive verb
1 : to perform by turns or in succession
2 : to cause to alternate
intransitive verb
: to change from one to another repeatedly ⟨rain *alternated* with sun⟩

³al·ter·nate *same as* ¹\ *noun* (1717)
1 : ALTERNATIVE
2 : one that substitutes for or alternates with another

alternate angle *noun* (1660)
: one of a pair of angles with different vertices and on opposite sides of a transversal at its intersection with two other lines: **a :** one of a pair of angles inside the two intersected lines — called also *alternate interior angle* **b :** one of a pair of angles outside the two intersected lines — called also *alternate exterior angle*

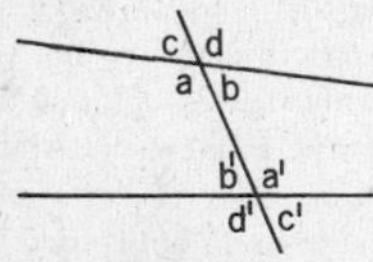

alternate interior angles *a, a', b, b'*; alternate exterior angles *c, c', d, d'*

alternating current *noun* (1839)
: an electric current that reverses its direction at regularly recurring intervals — abbreviation *AC*

alternating group *noun* (1904)
: a permutation group whose elements comprise those permutations of *n* objects which can be formed from the original order by making consecutively an even number of interchanges of pairs of objects

alternating series *noun* (circa 1909)
: a mathematical series in which consecutive terms are alternatively positive and negative

al·ter·na·tion \ˌȯl-tər-'nā-shən *also* ˌal-\ *noun* (15th century)
1 a : the act or process of alternating or causing to alternate **b :** alternating occurrence **:** SUCCESSION
2 : INCLUSIVE DISJUNCTION
3 : the occurrence of different allomorphs or allophones

alternation of generations (1858)
: the occurrence of two or more forms differently produced in the life cycle of a plant or animal usually involving the regular alternation of a sexual with an asexual generation

¹al·ter·na·tive \ȯl-'tər-nə-tiv, al-\ *adjective* (1540)
1 : ALTERNATE 1
2 : offering or expressing a choice ⟨several *alternative* plans⟩
3 : existing or functioning outside the established cultural, social, or economic system ⟨*alternative* newspaper⟩ ⟨*alternative* lifestyles⟩; *also* **:** different from the usual or conventional ⟨*alternative* fuels⟩
— al·ter·na·tive·ly *adverb*
— al·ter·na·tive·ness *noun*

²alternative *noun* (1624)
1 a : a proposition or situation offering a choice between two or more things only one of which may be chosen **b :** an opportunity for deciding between two or more courses or propositions
2 a : one of two or more things, courses, or propositions to be chosen **b :** something which can be chosen instead ⟨the only *alternative* to intervention⟩ ■
synonym see CHOICE

alternative school *noun* (1972)
: an elementary or secondary school with a nontraditional curriculum

al·ter·na·tor \'ȯl-tər-ˌnā-tər *also* 'al-\ *noun* (1892)
: an electric generator for producing alternating current

alt·horn \'alt-ˌhȯrn\ *noun* [German, from *alt* alto + *Horn* horn] (1859)
: an alto saxhorn

al·though *also* **al·tho** \ȯl-'t͟hō\ *conjunction* [Middle English *although,* from *al* all + *though*] (14th century)
: in spite of the fact that **:** even though

al·tim·e·ter \al-'ti-mə-tər, 'al-tə-ˌmē-tər\ *noun* [Latin *altus* + English *-meter*] (circa 1828)
: an instrument for measuring altitude; *especially* **:** an aneroid barometer designed to register changes in atmospheric pressure accompanying changes in altitude
— al·tim·e·try \al-'ti-mə-trē\ *noun*

al·ti·pla·no \ˌal-ti-'plä-(ˌ)nō\ *noun, plural* **-nos** [American Spanish, from Latin *altus* + *planum* plain] (1919)
: a high plateau or plain **:** TABLELAND

al·ti·tude \'al-tə-ˌtüd *also* -ˌtyüd\ *noun* [Middle English, from Latin *altitudo* height, depth, from *altus* high, deep — more at OLD] (14th century)
1 a : the angular elevation of a celestial object above the horizon **b :** the vertical elevation of an object above a surface (as sea level or land) of a planet or natural satellite **c** (1) **:** a perpendicular line segment from a vertex of a geometric figure (as a triangle or a pyramid) to the opposite side or the opposite side extended or from a side or face to a parallel side or face or the side or face extended (2) **:** the length of an altitude
2 : a high level (as of quality or feeling) ⟨the *altitudes* of his anger⟩
3 a : vertical distance or extent **b :** position at a height **c :** an elevated region **:** EMINENCE — usually used in plural
synonym see HEIGHT
— al·ti·tu·di·nal \ˌal-tə-'tü-dᵊn-əl, -'tyü-\ *adjective*
— al·ti·tu·di·nous \-dᵊn-əs\ *adjective*

altitude sickness *noun* (1920)
: the effects (as nosebleed or nausea) of oxygen deficiency in the blood and tissues developed in rarefied air at high altitudes

¹al·to \'al-(ˌ)tō\ *noun, plural* **altos** [Italian, literally, high, from Latin *altus*] (circa 1724)
1 a : COUNTERTENOR **b :** CONTRALTO
2 : the second highest voice part in a 4-part chorus
3 : a member of a family of instruments having a range lower than that of the treble or soprano; *especially* **:** an alto saxophone

²alto *adjective* (circa 1724)
: relating to or having the range or part of an alto

al·to·cu·mu·lus \ˌal-tō-'kyü-myə-ləs\ *noun, plural* **-li** \-ˌlī, -ˌlē\ [New Latin, from Latin *altus* + New Latin *-o-* + *cumulus*] (1894)
: a fleecy cloud formation consisting of large whitish globular cloudlets with shaded portions — see CLOUD illustration

¹al·to·geth·er \ˌȯl-tə-'ge-t͟hər\ *adverb* [Middle English *altogedere,* from *al* all + *togedere* together] (13th century)
1 : WHOLLY, COMPLETELY ⟨an *altogether* different problem⟩ ⟨stopped crying *altogether*⟩
2 : in all **:** ALL TOLD ⟨spent a hundred dollars *altogether*⟩
3 : on the whole **:** in the main ⟨*altogether* their efforts were successful⟩

²altogether *noun* (1894)
: NUDE — used with *the* ⟨posed in the *altogether*⟩

al·to–re·lie·vo *or* **al·to–ri·lie·vo** \ˌal-(ˌ)tō-ri-'lē-(ˌ)vō, ˌäl-(ˌ)tō-rēl-'yā-(ˌ)vō\ *noun, plural* **alto–relievos** *or* **al·to–ri·lie·vi** \ˌäl-(ˌ)tō-rēl-'yā-(ˌ)vē\ [Italian *altorilievo*] (1664)
1 : HIGH RELIEF
2 : a sculpture in high relief

al·to·stra·tus \ˌal-tō-'strā-təs, -'stra-\ *noun, plural* **-ti** \-ˌtī\ [New Latin, from Latin *altus* + New Latin *-o-* + *stratus*] (1894)
: a cloud formation similar to cirrostratus but darker and at a lower level — see CLOUD illustration

al·tri·cial \al-'tri-shəl\ *adjective* [Latin *altric-, altrix,* feminine of *altor* one who nourishes, from *alere* to nourish — more at OLD] (1872)
: being hatched or born or having young that are hatched or born in a very immature and helpless condition so as to require care for some time ⟨*altricial* birds⟩ — compare PRECOCIAL

al·tru·ism \'al-trü-ˌi-zəm\ [French *altruisme,* from *autrui* other people, from Old French, oblique case form of *autre* other, from Latin *alter*] (1853)
1 : unselfish regard for or devotion to the welfare of others
2 : behavior by an animal that is not beneficial to or may be harmful to itself but that benefits others of its species

□ USAGE
alternative Sometime during the 19th century the notion arose that *alternative* could not properly be applied to a choice of more than two. This notion, based on the word's derivation from Latin *alter* "other of two," has come down to modern times ⟨*alternatives,* like the horns of a dilemma, come only in pairs —Heywood Hale Broun⟩. But modern usage largely ignores the stricture ⟨anyone who doesn't like the other *alternatives —Reader's Digest Success with Words*⟩ and many modern authorities dismiss the etymological objection as pedantry.

— **al·tru·ist** \-trü-ist\ *noun*
— **al·tru·is·tic** \ˌal-trü-ˈis-tik\ *adjective*
— **al·tru·is·ti·cal·ly** \-ti-k(ə-)lē\ *adverb*

al·u·la \ˈal-yə-lə\ *noun, plural* **-lae** \-ˌlē, -ˌlī\ [New Latin, from Latin, diminutive of *ala* wing — more at AISLE] (1772)
: the process of a bird's wing corresponding to the thumb and bearing a few short quills — called also *bastard wing*

¹al·um \ˈa-ləm\ *noun* [Middle English, from Middle French *alum, alun,* from Latin *alumen*] (14th century)
1 : a potassium aluminum sulfate $KAl(SO_4)_2 \cdot 12H_2O$ or an ammonium aluminum sulfate $NH_4Al(SO_4)_2 \cdot 12H_2O$ used especially as an emetic and as an astringent and styptic
2 : any of various double salts isomorphous with potassium aluminum sulfate
3 : ALUMINUM SULFATE

²alum \ə-ˈləm\ *noun* [by shortening] (1930)
: ALUMNUS, ALUMNA

alu·mi·na \ə-ˈlü-mə-nə\ *noun* [New Latin, from Latin *alumin-, alumen* alum] (1801)
: aluminum oxide Al_2O_3 occuring native as corundum and in hydrated forms (as in bauxite)

alu·mi·nate \-nət\ *noun* (1841)
: a compound of alumina with a metallic oxide

al·u·min·i·um \ˌal-yə-ˈmi-nē-əm\ *noun* [New Latin, from *alumina*] (1812)
chiefly British **:** ALUMINUM

alu·mi·nize \ə-ˈlü-mə-ˌnīz\ *transitive verb* **-nized; -niz·ing** (1934)
: to treat or coat with aluminum

alu·mi·no·sil·i·cate \ə-ˌlü-mə-nō-ˈsi-lə-ˌkāt, -ˈsi-li-kət\ *noun* [Latin *alumin-, alumen* + *-o-* + International Scientific Vocabulary *silicate*] (1907)
: a combined silicate and aluminate

alu·mi·nous \ə-ˈlü-mə-nəs\ *adjective* (15th century)
: of, relating to, or containing alum or aluminum

alu·mi·num \ə-ˈlü-mə nəm\ *noun, often attributive* [New Latin, from *alumina*] (1812)
: a bluish silver-white malleable ductile light trivalent metallic element that has good electrical and thermal conductivity, high reflectivity, and resistance to oxidation and is the most abundant metal in the earth's crust where it always occurs in combination — see ELEMENT table

aluminum sulfate *noun* (1873)
: a white salt $Al_2(SO_4)_3$ usually made by treating bauxite with sulfuric acid and used in making paper, in water purification, and in tanning

alum·na \ə-ˈləm-nə\ *noun, plural* **-nae** \-(ˌ)nē *also* -ˌnī\ [Latin, feminine of *alumnus*] (1879)
1 : a girl or woman who has attended or has graduated from a particular school, college, or university
2 : a girl or woman who is a former member, employee, contributor, or inmate

alum·nus \ə-ˈləm-nəs\ *noun, plural* **-ni** \-ˌnī\ [Latin, foster son, pupil, from *alere* to nourish — more at OLD] (1645)
1 : one who has attended or has graduated from a particular school, college, or university
2 : one who is a former member, employee, contributor, or inmate

al·um·root \ˈa-ləm-ˌrüt, -ˌrůt\ *noun* (1813)
: any of a genus (*Heuchera*) of North American herbs of the saxifrage family having basal rounded or lobed toothed leaves; *especially* **:** one (*H. americana*) of eastern North America

al·u·nite \ˈal-yə-ˌnīt, ˈa-lə-\ *noun* [French, from *alun* alum] (1868)
: a mineral that consists of a hydrous potassium aluminum sulfate and occurs in massive form or in rhombohedral crystals

al·ve·o·lar \al-ˈvē-(ə-)lər\ *adjective* (1799)
1 : of, relating to, resembling, or having alveoli; *especially* **:** of, relating to, or constituting the part of the jaws where the teeth arise, the air-containing cells of the lungs, or glands with secretory cells about a central space
2 : articulated with the tip of the tongue touching or near the teethridge
— **al·ve·o·lar·ly** *adverb*

al·ve·o·late \-lət\ *adjective* (circa 1823)
: pitted like a honeycomb ⟨*alveolate* pollen⟩

al·ve·o·lus \al-ˈvē-ə-ləs\ *noun, plural* **-li** \-ˌlī, -(ˌ)lē\ [New Latin, from Latin, diminutive of *alveus* cavity, hollow, from *alvus* belly, beehive; akin to Lithuanian *aulys* beehive, Greek *aulos* tube, flute] (circa 1706)
1 : a small cavity or pit: as **a :** a socket for a tooth **b :** an air-containing cell of the lungs **c :** an acinus of a compound gland **d :** a cell or compartment of a honeycomb
2 : TEETHRIDGE

al·way \ˈól-(ˌ)wā\ *adverb* [Middle English *alwey, alneway,* from Old English *ealne weg,* literally, all the way, from *ealne* (accusative of *eall* all) + *weg* (accusative) way — more at WAY] (14th century)
archaic **:** ALWAYS

al·ways \ˈól-wēz, -wəz, -(ˌ)wāz *also* ˈó-\ *adverb* [Middle English *alwayes,* from *alwey*] (14th century)
1 : at all times **:** INVARIABLY
2 : FOREVER, PERPETUALLY
3 : at any rate **:** in any event ⟨as a last resort one can *always* work⟩

Al·yce clover \ˈa-ləs-\ *noun* [probably by folk etymology from New Latin *Alysicarpus,* genus name, from Greek *halysis* chain + *karpos* fruit] (1941)
: a low spreading annual Old World legume (*Alysicarpus vaginalis*) used in the southern U.S. as a cover crop and for hay and forage

alys·sum \ə-ˈli-səm\ *noun* [New Latin, from Greek *alysson,* plant believed to cure rabies, from neuter of *alyssos* curing rabies, from *a-* + *lyssa* rabies] (1548)
1 : any of a genus (*Alyssum*) of Old World herbs of the mustard family with small usually yellow racemose flowers
2 : SWEET ALYSSUM

Alz·hei·mer's disease \ˈälts-ˌhī-mərz-, ˈalts-\ *noun* [Alois *Alzheimer* (died 1915) German physician] (1912)
: a degenerative disease of the central nervous system characterized especially by premature senile mental deterioration — called also *Alzheimer's* ◆

am [Middle English, from Old English *eom;* akin to Old Norse *em* am, Latin *sum,* Greek *eimi*] *present 1st singular of* BE

AM \ˈā-ˌem\ *noun, often attributive* [amplitude modulation] (1940)
: a broadcasting system using amplitude modulation; *also* **:** a radio receiver of such a system

ama \ˈä-(ˌ)mä\ *noun, plural* **amas** *or* **ama** [Japanese] (1946)
: a Japanese diver especially for pearls

amah \ˈä-(ˌ)mä\ *noun* [Portuguese *ama* wet nurse, from Medieval Latin *amma*] (1839)
: an Oriental female servant; *especially* **:** a Chinese nurse

amain \ə-ˈmān\ *adverb* (1540)
1 : with all one's might ⟨attacking a huge rack of beef with a cleaver, she flailed away *amain* —Jay Jacobs⟩
2 *archaic* **a :** at full speed **b :** in great haste
3 *archaic* **:** to a high degree **:** EXCEEDINGLY ⟨they whom I favour thrive in wealth *amain* —John Milton⟩

Ama·le·kite \ˈa-mə-ˌle-ˌkīt, ə-ˈma-lə-ˌkīt\ *noun* [Hebrew *ʿĂmālēqī,* plural from *ʿĂmālēq* Amalek, grandson of Esau] (1560)
: a member of an ancient nomadic people living south of Canaan

amal·gam \ə-ˈmal-gəm\ *noun* [Middle English *amalgame,* from Middle French, from Medieval Latin *amalgama*] (15th century)
1 : an alloy of mercury with another metal that is solid or liquid at room temperature according to the proportion of mercury present and is used especially in making tooth cements
2 : a mixture of different elements **:** COMBINATION

amal·gam·ate \-gə-ˌmāt\ *transitive verb* **-at·ed; -at·ing** (1617)
: to unite in or as if in an amalgam; *especially* **:** to merge into a single body
synonym see MIX
— **amal·gam·ator** \-ˌmā-tər\ *noun*

amal·gam·ation \ə-ˌmal-gə-ˈmā-shən\ *noun* (1612)
1 a : the action or process of amalgamating **:** UNITING **b :** the state of being amalgamated
2 : the result of amalgamating **:** AMALGAM
3 : CONSOLIDATION, MERGER ⟨*amalgamation* of two corporations⟩

aman·dine \ˌä-ˌmän-ˈdēn\ *adjective* [French] (1945)
: prepared or served with almonds

am·a·ni·ta \ˌa-mə-ˈnī-tə, -ˈnē-\ *noun* [New Latin, genus name, from Greek *amanitai,* plural, a kind of fungus] (1899)
: any of a genus (*Amanita*) of white-spored fungi that typically have a volva and an annulus about the stipe and that includes some deadly poisonous forms

amanita

am·a·ni·tin \-ˈnī-tᵊn, -ˈnē-\ *noun* [*amanita* + *-in*] (circa 1847)
: a highly toxic peptide that is produced by the death cap and that selectively inhibits mammalian RNA polymerase

aman·ta·dine \ə-ˈman-tə-ˌdēn\ *noun* [International Scientific Vocabulary *amantad-* (alteration of *adamantane,* $C_{10}H_{16}$) + am*ine*] (1964)
: a drug used especially as the hydrochloride $C_{10}H_{17}N \cdot HCl$ to prevent infection (as by an influenza virus) by interfering with virus penetration into host cells

aman·u·en·sis \ə-ˌman-yə-ˈwen(t)-səs\ *noun, plural* **-en·ses** \-(ˌ)sēz\ [Latin, from (*servus*) *a manu* slave with secretarial duties] (1619)
: one employed to write from dictation or to copy manuscript

am·a·ranth \ˈa-mə-ˌran(t)th\ *noun* [Latin *amarantus,* a flower, from Greek *amaranton,* from neuter of *amarantos* unfading, from *a-* + *marainein* to waste away] (1548)

◇ WORD HISTORY

Alzheimer's disease Alzheimer's disease is named after the German neurologist Alois Alzheimer (1864–1915). The majority of his medical contributions centered on the structure of nerve tissue. In 1906 he undertook the autopsy of a 55-year-old patient who had died with severe dementia. In examining the brain he noted the abnormal presence of neuritic plaques, structures that had previously been found only in the brains of quite elderly patients. He also noted a neurofibrillary tangle (an abnormal fibrous structure within the neurons) that had not been previously described and whose presence signaled the discovery of a new disease entity. In 1907 Alzheimer published his classic description of premature senile dementia. The disease was later named *Alzheimersche Krankheit* (German for "Alzheimer's disease") by the psychiatrist Emil Kraepelin.

1 : any of a large genus (*Amaranthus* of the family Amaranthaceae, the amaranth family) of coarse herbs including forms cultivated as food crops and various pigweeds
2 : a flower that never fades
3 : a red azo dye

am·a·ran·thine \ˌa-mə-ˈran(t)-thən, -ˈran-ˌthīn\ *adjective* (1667)
1 a : of or relating to an amaranth **b :** UNDYING
2 : of the color amaranth

am·a·ret·to \ˌa-mə-ˈre-(ˌ)tō, ˌä-\ *noun* [Italian, diminutive of *amaro* bitter, from Latin *amarus*] (1945)
1 am·a·ret·ti \-(ˌ)tē\ *plural* **:** macaroons made with bitter almonds
2 *often capitalized* **:** an almond-flavored liqueur

am·a·ryl·lis \ˌa-mə-ˈri-ləs\ *noun* [New Latin, genus name, probably from Latin, name of a shepherdess in Virgil's *Eclogues*] (circa 1794) **:** an autumn-flowering South African bulbous herb (*Amaryllis belladonna* of the family Amaryllidaceae, the amaryllis family) widely grown for its deep red to whitish umbellate flowers; *also* **:** a plant of any of several related genera (as *Hippeastrum* or *Sprekelia*)

amass \ə-ˈmas\ *verb* [Middle French *amasser,* from Old French, from *a-* (from Latin *ad-*) + *masser* to gather into a mass, from *masse* mass] (15th century)
transitive verb
1 : to collect for oneself **:** ACCUMULATE ⟨*amass* a great fortune⟩
2 : to collect into a mass **:** GATHER ⟨must select rather than simply *amass* details⟩
intransitive verb
: to come together **:** ASSEMBLE
— amass·er *noun*
— amass·ment \-mənt\ *noun*

am·a·teur \ˈa-mə-(ˌ)tər, -ˌtu̇r, -ˌtyu̇r, -ˌchu̇r, -chər\ *noun, often attributive* [French, from Latin *amator* lover, from *amare* to love] (1784)
1 : DEVOTEE, ADMIRER
2 : one who engages in a pursuit, study, science, or sport as a pastime rather than as a profession
3 : one lacking in experience and competence in an art or science ☆
— am·a·teur·ish \ˌa-mə-ˈtər-ish, -ˈt(y)u̇r-, -ˈchu̇r-, -ˈchər-\ *adjective*
— am·a·teur·ish·ly *adverb*
— am·a·teur·ish·ness *noun*
— am·a·teur·ism \ˈa-mə-ˌtər-ˌi-zəm, -ˌt(y)u̇r-, -ˌchu̇r-; -ˌtə-ˌri-, -ˌchə-ˌri-\ *noun*

Ama·ti \ä-ˈmä-tē, ə-\ *noun, plural* **Amatis** (1833)
: a violin made by a member of the Amati family of Cremona

am·a·tive \ˈa-mə-tiv\ *adjective* [Medieval Latin *amativus,* from Latin *amatus,* past participle of *amare*] (1636)
: AMOROUS 1, 3
— am·a·tive·ly *adverb*
— am·a·tive·ness *noun*

am·a·to·ry \ˈa-mə-ˌtōr-ē, -ˌtȯr-\ *adjective* (1599)
: of, relating to, or expressing sexual love

am·au·ro·sis \ˌa-mȯ-ˈrō-səs\ *noun, plural* **-ro·ses** \-ˌsēz\ [New Latin, from Greek *amaurōsis,* literally, dimming, from *amauroun* to dim, from *amauros* dim] (circa 1657)
: partial or complete loss of sight occurring especially without an externally perceptible change in the eye
— am·au·rot·ic \-ˈrä-tik\ *adjective*

amaurotic idiocy *noun* (1896)
: any of several recessive genetic conditions characterized by the accumulation of lipid-containing cells in the viscera and nervous system, mental retardation, and impaired vision or blindness; *especially* **:** TAY-SACHS DISEASE

¹amaze \ə-ˈmāz\ *verb* **amazed; amaz·ing** [Middle English *amasen,* from Old English *āmasian,* from *ā-* (perfective prefix) + (assumed) *masian* to confuse — more at ABIDE] (before 12th century)
transitive verb
1 *obsolete* **:** BEWILDER, PERPLEX
2 : to fill with wonder **:** ASTOUND
intransitive verb
: to show or cause astonishment
synonym see SURPRISE
— amaz·ed·ly \-ˈmā-zəd-lē\ *adverb*

²amaze *noun* (15th century)
: AMAZEMENT

amaze·ment \ə-ˈmāz-mənt\ *noun* (1595)
1 *obsolete* **:** CONSTERNATION, BEWILDERMENT
2 : the quality or state of being amazed
3 : something that amazes

amaz·ing·ly \ə-ˈmā-ziŋ-lē\ *adverb* (1673)
1 : to an amazing degree ⟨*amazingly* low prices⟩
2 : what is amazing ⟨*amazingly,* the runnerup was only two seconds behind —Jim Doherty⟩

am·a·zon \ˈa-mə-ˌzän, -zən\ *noun* [Middle English, from Latin, from Greek *Amazōn*] (14th century)
1 *capitalized* **:** a member of a race of female warriors of Greek mythology
2 : a tall strong often masculine woman

Am·a·zo·nian \ˌa-mə-ˈzō-nē-ən, -nyən\ *adjective* (1594)
1 : relating to, resembling, or befitting an Amazon or an amazon
2 : of or relating to the Amazon River or its valley

am·a·zon·ite \ˈa-mə-zə-ˌnīt\ *noun* [*Amazon* River] (circa 1879)
: an apple-green or bluish-green microcline

am·a·zon·stone \-zən-ˌstōn\ *noun* (1836)
: AMAZONITE

am·bage \ˈam-bij\ *noun, plural* **am·ba·ges** \am-ˈbā-(ˌ)jēz, ˈam-bi-jəz\ [back-formation from Middle English *ambages,* from Middle French or Latin; Middle French, from Latin, from *ambi-* + *agere* to drive — more at AGENT] (14th century)
1 *archaic* **:** AMBIGUITY, CIRCUMLOCUTION — usually used in plural
2 *plural, archaic* **:** indirect ways or proceedings

am·bas·sa·dor \am-ˈba-sə-dər, əm-, im-, -ˌdȯr, -ˈbas-dər\ *noun* [Middle English *ambassadour,* from Middle French *ambassadeur,* ultimately of Germanic origin; akin to Old High German *ambaht* service] (14th century)
1 : an official envoy; *especially* **:** a diplomatic agent of the highest rank accredited to a foreign government or sovereign as the resident representative of his own government or sovereign or appointed for a special and often temporary diplomatic assignment
2 a : an authorized representative or messenger **b :** an unofficial representative ⟨traveling abroad as *ambassadors* of goodwill⟩
— am·bas·sa·do·ri·al \-ˌba-sə-ˈdōr-ē-əl, -ˈdȯr-\ *adjective*
— am·bas·sa·dor·ship \-ˈba-sə-dər-ˌship\ *noun*

ambassador–at–large *noun, plural* **ambassadors–at–large** (1908)
: a minister of the highest rank not accredited to a particular foreign government or sovereign

am·bas·sa·dress \am-ˈba-sə-drəs, əm-, im-\ *noun* (1594)
1 : a woman who is an ambassador
2 : the wife of an ambassador

am·beer \ˈam-ˌbir\ *noun* [probably alteration of *amber;* from its color] (1848)
chiefly Southern & southern Midland **:** TOBACCO JUICE

¹am·ber \ˈam-bər\ *noun* [Middle English *ambre,* from Middle French, from Medieval Latin *ambra,* from Arabic *ʽanbar* ambergris] (14th century)
1 : a hard yellowish to brownish translucent fossil resin that takes a fine polish and is used chiefly in making ornamental objects (as beads)
2 : a variable color averaging a dark orange yellow

²amber *adjective* (15th century)
1 : consisting of amber
2 : resembling amber; *especially* **:** having the color amber

am·ber·gris \ˈam-bər-ˌgris, -ˌgrē(s)\ *noun* [Middle English *ambregris,* from Middle French *ambre gris,* from *ambre* + *gris* gray — more at GRIZZLE] (15th century)
: a waxy substance found floating in or on the shores of tropical waters, believed to originate in the intestines of the sperm whale, and used in perfumery as a fixative

am·ber·i·na \ˌam-bə-ˈrē-nə\ *noun* [from *Amberina,* a trademark] (1885)
: a late 19th century American clear glassware of a graduated color that shades from ruby to amber

am·ber·jack \-ˌjak\ *noun* [from its color] (circa 1893)
: any of several carangid fishes (genus *Seriola*); *especially* **:** a large vigorous sport fish (*S. dumerili*) of the western Atlantic

ambi- *prefix* [Latin *ambi-*, *amb-* both, around; akin to Latin *ambo* both, Greek *amphō* both, *amphi* around — more at BY]
: both ⟨*ambi*valence⟩

am·bi·dex·ter·i·ty \ˌam-bi-(ˌ)dek-ˈster-ə-tē\ *noun* (1593)
: the quality or state of being ambidextrous

am·bi·dex·trous \ˌam-bi-ˈdek-strəs\ *adjective* [Late Latin *ambidexter,* from Latin *ambi-* + *dexter* right-hand — more at DEXTER] (1646)
1 : using both hands with equal ease
2 : unusually skillful **:** VERSATILE
3 : characterized by duplicity **:** DOUBLE-DEALING
— am·bi·dex·trous·ly *adverb*

am·bi·ence *or* **am·bi·ance** \ˈam-bē-ən(t)s, ˈäm-bē-än(t)s; äⁿ-byäⁿs\ *noun* [French *ambiance,* from *ambiant* ambient] (1889)
: a feeling or mood associated with a particular place, person, or thing **:** ATMOSPHERE

¹am·bi·ent \ˈam-bē-ənt\ *adjective* [Latin *ambient-, ambiens,* present participle of *ambire* to go around, from *ambi-* + *ire* to go — more at ISSUE] (1596)
: existing or present on all sides **:** ENCOMPASSING

²ambient *noun* (1624)
: an encompassing atmosphere **:** ENVIRONMENT

am·bi·gu·i·ty \ˌam-bə-ˈgyü-ə-tē\ *noun, plural* **-ties** (15th century)
1 a : the quality or state of being ambiguous especially in meaning **b :** an ambiguous word or expression
2 : UNCERTAINTY

am·big·u·ous \am-ˈbi-gyə-wəs\ *adjective* [Latin *ambiguus,* from *ambigere* to be undecided, from *ambi-* + *agere* to drive — more at AGENT] (1528)

☆ SYNONYMS
Amateur, dilettante, dabbler, tyro mean a person who follows a pursuit without attaining proficiency or professional status. AMATEUR often applies to one practicing an art without mastery of its essentials ⟨a painting obviously done by an *amateur*⟩; in sports it may also suggest not so much lack of skill but avoidance of direct remuneration ⟨remained an *amateur* despite lucrative offers⟩. DILETTANTE may apply to the lover of an art rather than its skilled practitioner but usually implies elegant trifling in the arts and an absence of serious commitment ⟨had no patience for *dilettantes*⟩. DABBLER suggests desultory habits of work and lack of persistence ⟨a *dabbler* who started novels but never finished them⟩. TYRO implies inexperience often combined with audacity with resulting crudeness or blundering ⟨shows talent but is still a mere *tyro*⟩.

1 a : doubtful or uncertain especially from obscurity or indistinctness ⟨eyes of an *ambiguous* color⟩ **b :** INEXPLICABLE
2 : capable of being understood in two or more possible senses or ways
synonym see OBSCURE
— am·big·u·ous·ly *adverb*
— am·big·u·ous·ness *noun*

am·bi·sex·u·al \ˌam-bi-ˈsek-sh(ə-)wəl, -ˈsek-shəl\ *adjective* (1939)
: BISEXUAL
— ambisexual *noun*
— am·bi·sex·u·al·i·ty \-ˌsek-shə-ˈwa-lə-tē\ *noun*

am·bit \ˈam-bət\ *noun* [Middle English, from Latin *ambitus,* from *ambire*] (1597)
1 : CIRCUIT, COMPASS
2 : the bounds or limits of a place or district
3 : a sphere of action, expression, or influence **:** SCOPE

¹am·bi·tion \am-ˈbi-shən\ *noun* [Middle English, from Middle French or Latin; Middle French, from Latin *ambition-, ambitio,* literally, act of soliciting for votes, from *ambire*] (14th century)
1 a : an ardent desire for rank, fame, or power **b :** desire to achieve a particular end
2 : the object of ambition
3 : a desire for activity or exertion ⟨felt sick and had no *ambition*⟩ ☆ ◆
— am·bi·tion·less \-ləs\ *adjective*

²ambition *transitive verb* (1664)
: to have as one's ambition **:** DESIRE

am·bi·tious \am-ˈbi-shəs\ *adjective* (14th century)
1 a : having or controlled by ambition **b :** having a desire to achieve a particular goal **:** ASPIRING
2 : resulting from, characterized by, or showing ambition
— am·bi·tious·ly *adverb*
— am·bi·tious·ness *noun*

am·biv·a·lence \am-ˈbi-və-lən(t)s\ *noun* [International Scientific Vocabulary] (1918)
1 : simultaneous and contradictory attitudes or feelings (as attraction and repulsion) toward an object, person, or action
2 a : continual fluctuation (as between one thing and its opposite) **b :** uncertainty as to which approach to follow
— am·biv·a·lent \-lənt\ *adjective*
— am·biv·a·lent·ly *adverb*

am·bi·ver·sion \ˌam-bi-ˈvər-zhən, -shən\ *noun* [*ambi-* + *-version* (as in *introversion*)] (1927)
: the personality configuration of an ambivert

am·bi·vert \ˈam-bi-ˌvərt\ *noun* [*ambi-* + *-vert* (as in *introvert*)] (1927)
: a person having characteristics of both extrovert and introvert

¹am·ble \ˈam-bəl\ *intransitive verb* **am·bled; am·bling** \-b(ə-)liŋ\ [Middle English, from Middle French *ambler,* from Latin *ambulare* to walk, from *ambi-* + *-ulare* (verb base akin to Middle Welsh *el* he may go, Greek *ēlythe* he went) — more at ELASTIC] (14th century)
: to go at or as if at an amble **:** SAUNTER
— am·bler \-b(ə-)lər\ *noun*

²amble *noun* (14th century)
1 a : an easy gait of a horse in which the legs on the same side of the body move together **b :** ¹RACK b
2 : an easy gait
3 : a leisurely walk

am·blyg·o·nite \am-ˈbli-gə-ˌnīt\ *noun* [German *Amblygonit,* from Greek *amblygōnios* obtuse-angled, from *amblys* blunt, dull + *gōnia* angle — more at MOLLIFY, -GON] (circa 1828)
: a mineral that consists of basic lithium aluminum phosphate commonly containing sodium and fluorine and occurs in white cleavable masses

am·bly·opia \ˌam-blē-ˈō-pē-ə\ *noun* [New Latin, from Greek *amblyōpia,* from *amblys* + *-ōpia* -opia] (circa 1706)
: dimness of sight especially in one eye without apparent change in the eye structures — called also *lazy eye*
— am·bly·opic \-ˈō-pik, -ˈä-pik\ *adjective*

Am·boi·nese \ˌam-ˌbȯi-ˈnēz, -ˈnēs, am-ˈbȯi-ˌ\ *or* **Am·bo·nese** \ˌam-bə-ˈnēz, -ˈnēs\ *noun, plural* **Amboinese** *or* **Ambonese** (circa 1864)
1 : a native or inhabitant of Ambon
2 : the group of closely related Austronesian languages spoken on Ambon
— Amboinese *or* **Ambonese** *adjective*

am·boy·na *or* **am·boi·na** \am-ˈbȯi-nə\ *noun* [*Amboina,* Moluccas, Indonesia] (circa 1859)
: a mottled curly-grained wood of a leguminous tree (*Pterocarpus indicus*) of southeastern Asia

am·bro·sia \am-ˈbrō-zh(ē-)ə\ *noun* [Latin, from Greek, literally, immortality, from *ambrotos* immortal, from *a-* + *-mbrotos* (akin to *brotos* mortal) — more at MURDER] (15th century)
1 a : the food of the Greek and Roman gods **b :** the ointment or perfume of the gods
2 : something extremely pleasing to taste or smell
3 : a dessert made of oranges and shredded coconut ◆
— am·bro·sial \-zh(ē-)əl\ *adjective*
— am·bro·sial·ly \-zh(ē-)ə-lē\ *adverb*

ambrosia beetle *noun* (circa 1900)
: any of various small wood-boring beetles (family Scolytidae) that cultivate a fungus on which they feed and raise their larvae

am·bro·type \ˈam-brə-ˌtīp\ *noun* [Greek *ambrotos* + English *type*] (1858)
: a positive picture made of a photographic negative on glass backed by a dark surface

am·bry \ˈam-brē; ˈäm-rē, ˈȯm-\ *noun, plural* **ambries** [Middle English *almery,* from Middle French *almarie, armarie,* from Latin *armarium,* from *arma* weapons — more at ARM] (14th century)
1 *dialect chiefly British* **:** PANTRY
2 : a recess in a church wall (as for holding sacramental vessels)

ambs·ace \ˈām-ˌzās\ *noun* [Middle English *ambes as,* from Old French, from *ambes* both + *as* aces] (13th century)
archaic **:** the lowest throw at dice; *also* **:** something worthless or unlucky

am·bu·la·cral \ˌam-byə-ˈla-krəl, -ˈlā-\ *adjective* (1836)
: of, relating to, or being any of the radial areas of echinoderms along which run the principal nerves, blood vessels, and elements of the water-vascular system ⟨*ambulacral* grooves⟩

am·bu·la·crum \-krəm\ *noun, plural* **-cra** \-krə\ [New Latin, from Latin, alley, from *ambulare* to walk — more at AMBLE] (1837)
: an ambulacral area or part

am·bu·lance \ˈam-byə-lən(t)s, -bə- *also* -ˌlan(t)s\ *noun* [French, from (*hôpital*) *ambulant,* literally, ambulant field hospital, from *ambulant* itinerant, from Latin *ambulant-, ambulans,* present participle of *ambulare*] (1809)
: a vehicle equipped for transporting the injured or sick

ambulance chaser *noun* (1897)
: a lawyer or lawyer's agent who incites accident victims to sue for damages
— ambulance chasing *noun*

am·bu·lant \ˈam-byə-lənt\ *adjective* (1619)
: moving about **:** AMBULATORY

am·bu·late \-ˌlāt\ *intransitive verb* **-lat·ed; -lat·ing** [Latin *ambulatus,* past participle of *ambulare*] (circa 1623)
: to move from place to place **:** WALK
— am·bu·la·tion \ˌam-byə-ˈlā-shən\ *noun*

¹am·bu·la·to·ry \ˈam-byə-lə-ˌtōr-ē, -ˌtȯr-\ *adjective* (1622)
1 : of, relating to, or adapted to walking; *also* **:** occurring during a walk
2 : moving from place to place **:** ITINERANT
3 : capable of being altered ⟨a will is *ambulatory* until the testator's death⟩
4 a : able to walk about and not bedridden **b :** performed on or involving an ambulatory patient or an outpatient ⟨*ambulatory* medical care⟩ ⟨an *ambulatory* electrocardiogram⟩
— am·bu·la·to·ri·ly \ˌam-byə-lə-ˈtōr-ə-lē, -ˈtȯr-\ *adverb*

²ambulatory *noun, plural* **-ries** (circa 1616)
: a sheltered place (as in a cloister or church) for walking

am·bus·cade \ˈam-bə-ˌskād, ˌam-bə-ˈ\ *noun* [Middle French *embuscade,* modification of Old Italian *imboscata,* from *imboscare* to place in ambush, from *in* (from Latin) + *bosco* forest, of Germanic origin; akin to Old High German *busc* forest — more at IN, BUSH] (circa 1588)
: AMBUSH
— ambuscade *verb*
— am·bus·cad·er *noun*

☆ SYNONYMS
Ambition, aspiration, pretension mean strong desire for advancement. AMBITION applies to the desire for personal advancement or preferment and may suggest equally a praiseworthy or an inordinate desire ⟨driven by *ambition*⟩. ASPIRATION implies a striving after something higher than oneself and usually implies that the striver is thereby ennobled ⟨an *aspiration* to become president someday⟩. PRETENSION suggests ardent desire for recognition of accomplishment often without actual possession of the necessary ability and therefore may imply presumption ⟨has literary *pretensions*⟩.

◇ WORD HISTORY
ambition When candidates for public office in ancient Rome wanted to be elected, they had to make the rounds of their districts urging citizens to vote for them, just as modern political candidates must do. The Latin word for this effort was *ambitio,* which came from *ambire,* a verb meaning literally "to go around." Because the act of canvassing votes was motivated by a desire for honor or power, the word came to denote this desire itself. *Ambitio* entered French and English as *ambition* in the late Middle Ages. Later its meaning broadened to include any desire for advancement or improvement and still later the object of this desire.

ambrosia The Greek and Roman gods all too often behaved like mere mortals, but what set them apart was the gift of immortality. This they enjoyed as a result of their peculiar diet. Their food, ambrosia, and their drink, nectar, had the power to prevent death. In Greek *ambrosia* literally means "immortality" and is a derivative of *ambrotos* "immortal." The notion of ambrosia as something fit for the gods has resulted in the name being given to various things whose taste or smell is especially pleasing. The ancient Greeks and Romans called several aromatic-leaved plants *ambrosia,* but in modern botany *ambrosia* is the scientific name for the genus of plants commonly known as the ragweeds—which have neither flavor nor fragrance to recommend them. Indeed, to allergy sufferers they are among the most unappealing of plants.

[1]am·bush \'am-ˌbu̇sh\ *verb* [Middle English *embushen,* from Middle French *embuschier,* from *en* in (from Latin *in*) + *busche* stick of firewood] (14th century)
transitive verb
1 : to station in ambush
2 : to attack from an ambush **:** WAYLAY
intransitive verb
: to lie in wait **:** LURK
— am·bush·er *noun*
— am·bush·ment \-mənt\ *noun*

[2]ambush *noun* (15th century)
1 : a trap in which concealed persons lie in wait to attack by surprise
2 : the persons stationed in ambush; *also* **:** their concealed position
3 : an attack especially from an ambush

ame·ba, ame·boid *variant of* AMOEBA, AMOEBOID

am·e·bi·a·sis \ˌa-mi-'bī-ə-səs\ *noun, plural* **-a·ses** \-sēz\ (1905)
: infection with or disease caused by amoebas (especially *Entamoeba histolytica*)

ame·bic dysentery \ə-'mē-bik-\ *noun* (1891)
: acute human intestinal amebiasis caused by an amoeba (*Entamoeba histolytica*) and marked by dysentery, gripes, and erosion of the intestinal wall

ame·bo·cyte *variant of* AMOEBOCYTE

ameer *variant of* EMIR

ame·lio·rate \ə-'mēl-yə-ˌrāt, -'mē-lē-ə-\ *verb* **-rat·ed; -rat·ing** [alteration of *meliorate*] (1767)
transitive verb
: to make better or more tolerable
intransitive verb
: to grow better
synonym see IMPROVE
— ame·lio·ra·tion \-ˌmēl-yə-'rā-shən, -ˌmē-lē-ə-\ *noun*
— ame·lio·ra·tive \-'mēl-yə-ˌrā-tiv, -'mē-lē-ə-\ *adjective*
— ame·lio·ra·tor \-ˌrā-tər\ *noun*
— ame·lio·ra·to·ry \-rə-ˌtōr-ē, -ˌtȯr-\ *adjective*

am·e·lo·blast \'a-mə-lō-ˌblast\ *noun* [obsolete *amel* enamel (Middle English, ultimately from Old French *esmail*) + *-o-* + *-blast* — more at ENAMEL] (1882)
: one of a group of columnar cells that produce and deposit enamel on the surface of a developing vertebrate tooth

amen \(ˈ)ä-'men, (ˈ)ā-; 'ä- *when sung*\ *interjection* [Middle English, from Old English, from Late Latin, from Greek *amēn,* from Hebrew *āmēn*] (before 12th century)
— used to express solemn ratification (as of an expression of faith) or hearty approval (as of an assertion)

ame·na·ble \ə-'mē-nə-bəl, -'me-\ *adjective* [probably from (assumed) Anglo-French, from Middle French *amener* to lead up, from Old French, from *a-* (from Latin *ad-*) + *mener* to lead, from Late Latin *minare* to drive, from Latin *minari* to threaten — more at MOUNT] (1596)
1 : liable to be brought to account **:** ANSWERABLE ⟨citizens *amenable* to the law⟩
2 a : capable of submission (as to judgment or test) **:** SUITED ⟨the data is *amenable* to analysis⟩ **b :** readily brought to yield, submit, or cooperate **c :** WILLING 1 ⟨was *amenable* to spending more time at home⟩
synonym see RESPONSIBLE, OBEDIENT
— ame·na·bil·i·ty \-ˌmē-nə-'bil-ət-ē, -ˌme-\ *noun*
— ame·na·bly \-'mē-nə-blē, -'me-\ *adverb*

amen corner \'ā-ˌmen-\ *noun* (1860)
: a conspicuous corner in a church occupied by fervent worshipers

amend \ə-'mend\ *verb* [Middle English, from Old French *amender,* modification of Latin *emendare,* from *e, ex* out + *menda* fault; akin to Latin *mendax* lying, *mendicus* beggar, and perhaps to Sanskrit *mindā* physical defect] (13th century)
transitive verb
1 : to put right; *especially* **:** to make emendations in (as a text)
2 a : to change or modify for the better **:** IMPROVE ⟨*amend* the situation⟩ **b :** to alter especially in phraseology; *especially* **:** to alter formally by modification, deletion, or addition ⟨*amend* the constitution⟩
intransitive verb
: to reform oneself
synonym see CORRECT
— amend·able \-'men-də-bəl\ *adjective*
— amend·er *noun*

amen·da·to·ry \ə-'men-də-ˌtōr-ē, -ˌtȯr-\ *adjective* [*amend* + *-atory* (as in *emendatory*)] (circa 1828)
: CORRECTIVE

amend·ment \ə-'men(d)-mənt\ *noun* (13th century)
1 : the act of amending **:** CORRECTION
2 : a material (as compost or sand) that aids plant growth indirectly by improving the condition of the soil
3 a : the process of amending by parliamentary or constitutional procedure **b :** an alteration proposed or effected by this process ⟨the 18th *amendment*⟩

amends \ə-'men(d)z\ *noun plural but singular or plural in construction* [Middle English *amendes,* from Middle French, plural of *amende* reparation, from *amender*] (14th century)
: compensation for a loss or injury **:** RECOMPENSE ⟨make *amends*⟩

ame·ni·ty \ə-'me-nə-tē, -'mē-\ *noun, plural* **-ties** [Middle English *amenite,* from Latin *amoenitat-, amoenitas,* from *amoenus* pleasant] (14th century)
1 a : the quality of being pleasant or agreeable **b** (1) **:** the attractiveness and value of real estate or of a residential structure (2) **:** a feature conducive to such attractiveness and value
2 : something (as a conventional social gesture) that conduces to smoothness or pleasantness of social relationships
3 : something that conduces to comfort, convenience, or enjoyment

amen·or·rhea \ˌā-ˌme-nə-'rē-ə, ˌä-\ *noun* [New Latin, from *a-* + Greek *mēn* month + New Latin *-o-* + *-rrhea* — more at MOON] (1804)
: abnormal absence or suppression of menses
— amen·or·rhe·ic \-'rē-ik\ *adjective*

ament \'a-mənt, 'ā-\ *noun* [New Latin *amentum,* from Latin, thong, strap] (1791)
: CATKIN
— amen·tif·er·ous \-mən-'ti-f(ə-)rəs\ *adjective*

amen·tia \(ˌ)ā-'men(t)-sh(ē-)ə, (ˌ)ä-\ *noun* [New Latin, from Latin, madness, from *ament-, amens* mad, from *a-* (from *ab-*) + *ment-, mens* mind — more at MIND] (14th century)
: MENTAL RETARDATION; *specifically* **:** a condition of lack of development of intellectual capacity

Am·er·asian \ˌa-mə-'rā-zhən, -shən\ *noun* [*American* + *Asian*] (1953)
: a person of mixed American and Asian descent; *especially* **:** one fathered by an American and especially an American serviceman in Asia

amerce \ə-'mərs\ *transitive verb* **amerced; amerc·ing** [Middle English *amercien,* from Anglo-French *amercier,* from Old French *a merci* at (one's) mercy] (15th century)
: to punish by a fine whose amount is fixed by the court; *broadly* **:** PUNISH
— amerce·ment \-'mər-smənt\ *noun*
— amer·cia·ble \-'mər-sē-ə-bəl, -'mər-shə-bəl\ *adjective*

[1]Amer·i·can \ə-'mer-ə-kən, -'mər-, -'mar-, -i-kən\ *noun* (1578)
1 : an American Indian of North America or South America
2 : a native or inhabitant of North America or South America
3 : a citizen of the U.S.
4 : AMERICAN ENGLISH

[2]American *adjective* (1598)
1 : of or relating to America
2 : of or relating to the U.S. or its possessions or original territory
— Amer·i·can·ness \-kə(n)-nəs\ *noun*

Amer·i·ca·na \ə-ˌmer-ə-'kä-nə, -ˌmər-, -ˌmar-, -'ka-nə\ *noun plural* (1841)
1 : materials concerning or characteristic of America, its civilization, or its culture; *broadly* **:** things typical of America
2 : American culture

American chameleon *noun* (1881)
: a lizard (*Anolis carolinensis*) of the southeastern U.S. that can vary its skin color from green to brown and is often kept as a pet

American cheese *noun* (1804)
: a process cheese made from American cheddar

American dog tick *noun* (1927)
: a common North American ixodid tick (*Dermacentor variabilis*) especially of dogs and humans that is an important vector of Rocky Mountain spotted fever and tularemia — called also *dog tick*

American dream *noun, often D capitalized* (1931)
: an American social ideal that stresses egalitarianism and especially material prosperity

American eel *noun* (circa 1949)
: a yellow to greenish-brown eel (*Anguilla rostrata*) that is lighter below, has 103 to 111 vertebrae, is found in fresh and coastal waters along the Atlantic coasts of North America, and is held to spawn in or near the Sargasso Sea

American elm *noun* (1813)
: a large elm (*Ulmus americana*) with gradually spreading branches and pendulous branchlets that is common in eastern North America

American English *noun* (1806)
: the English language as spoken in the U.S. — used especially with the implication that it is clearly distinguishable from British English yet not so divergent as to be a separate language

Amer·i·ca·nese \ə-ˌmer-ə-kə-'nēz, -ˌmər-, -ˌmar-, -i-kə-, -'nēs\ *noun* (1882)
: AMERICAN ENGLISH

American foxhound *noun* (circa 1891)
: any of an American breed of foxhounds smaller than the English foxhound but with longer ears and a short glossy coat usually of black, tan, and white

American Indian *noun* (1732)
: a member of any of the aboriginal peoples of the western hemisphere except usually the Eskimos; *especially* **:** an American Indian of North America and especially the U.S.

Amer·i·can·isa·tion, Amer·i·can·ise *British variant of* AMERICANIZATION, AMERICANIZE

Amer·i·can·ism \ə-'mer-ə-kə-ˌni-zəm, -'mər-, -'mar-\ *noun* (1781)
1 : a characteristic feature of American English especially as contrasted with British English
2 : attachment or allegiance to the traditions, interests, or ideals of the U.S.
3 a : a custom or trait peculiar to America **b :** the political principles and practices essential to American culture

Amer·i·can·ist \-kə-nist\ *noun* (1881)
1 : a specialist in American culture or history
2 : a specialist in the languages or cultures of the aboriginal inhabitants of America

Amer·i·can·iza·tion \ə-ˌmer-ə-kə-nə-'zā-shən, -ˌmər-, -ˌmar-\ *noun* (1858)
1 : the act or process of Americanizing

2 : instruction of foreigners (as immigrants) in English and in U.S. history, government, and culture

Amer·i·can·ize \ə-'mer-ə-kə-ˌnīz, -'mər-, -'mar-\ *verb* **-ized; -iz·ing** (1797)
transitive verb
1 : to cause to acquire or conform to American characteristics
2 : to bring (as an area) under the political, cultural, or commercial influence of the U.S.
intransitive verb
: to acquire or conform to American traits **:** assimilate to American life or culture

American pit bull terrier *noun* (1950)
: any of a breed of dogs developed to combine the traits of terriers and bulldogs that have extremely powerful jaws and great strength and tenacity — called also *pit bull terrier*

American plan *noun* (1856)
: a hotel plan whereby the daily rates cover the costs of the room and three meals — compare EUROPEAN PLAN

American saddlebred *noun* (1948)
: any of a breed of 3-gaited or 5-gaited saddle horses developed chiefly in Kentucky from Thoroughbreds and smooth-gaited stock — called also *American saddle horse*

American saddlebred

American shorthair *noun* (1974)
: any of a breed of cats with a usually solid-colored or tabby coat; *broadly* **:** SHORTHAIR

American Sign Language *noun* (1960)
: a sign language for the deaf in which meaning is conveyed by a system of articulated hand gestures and their placement relative to the upper body

American Staffordshire terrier *noun* (1971)
: a strong stocky terrier of a breed originally developed for dogfighting

American Standard Version *noun* (1901)
: an American version of the Bible that is based on the Revised Version and was published in 1901 — called also *American Revised Version*

American trotter *noun* (1894)
: STANDARDBRED

American water spaniel *noun* (1947)
: any of a breed of medium-sized spaniels of American origin that have a thick curly chocolate or liver-colored coat

am·er·i·ci·um \ˌa-mə-'ri-shē-əm, -sē-\ *noun* [New Latin, from *America* + New Latin *-ium*] (1946)
: a radioactive metallic element produced by bombardment of plutonium with high-energy neutrons — see ELEMENT table

Am·er·in·di·an \ˌa-mə-'rin-dē-ən\ *noun* [*American* + *Indian*] (circa 1898)
: AMERICAN INDIAN
— Am·er·ind \'a-mə-ˌrind\ *noun or adjective*
— Amerindian *adjective*

Ame·slan \'a-məs-ˌlan, 'am-ˌslan\ *noun* (1972)
: AMERICAN SIGN LANGUAGE

Ames test \'āmz-\ *noun* [Bruce N. *Ames* (born 1928) American biochemist] (1977)
: a test for identifying potential carcinogens by studying their mutagenic effect on bacteria

am·e·thyst \'a-mə-thəst, -(ˌ)thist\ *noun* [Middle English *amatiste,* from Old French & Latin; Old French, from Latin *amethystus,* from Greek *amethystos,* literally, remedy against drunkenness, from *a-* + *methyein* to be drunk, from *methy* wine — more at MEAD] (13th century)
1 a : a clear purple or bluish violet variety of crystallized quartz that is much used as a jeweler's stone **b :** a deep purple variety of corundum
2 : a moderate purple ◇
— am·e·thys·tine \ˌa-mə-'this-tən\ *adjective*

am·e·tro·pia \ˌa-mə-'trō-pē-ə\ *noun* [New Latin, from Greek *ametros* without measure (from *a-* + *metron* measure) + New Latin *-opia* — more at MEASURE] (1875)
: an abnormal refractive condition (as myopia, hyperopia, or astigmatism) of the eye in which images fail to focus upon the retina
— am·e·tro·pic \-'trō-pik, -'trä-\ *adjective*

Am·har·ic \am-'har-ik; am-'här-, äm-\ *noun* (1813)
: a Semitic language that is the official language of Ethiopia
— Amharic *adjective*

ami·a·ble \'ā-mē-ə-bəl\ *adjective* [Middle English, from Middle French, from Late Latin *amicabilis* friendly, from Latin *amicus* friend; akin to Latin *amare* to love] (14th century)
1 *archaic* **:** PLEASING, ADMIRABLE
2 a : generally agreeable ⟨an *amiable* comedy⟩ **b :** being friendly, sociable, and congenial ☆
— ami·a·bil·i·ty \ˌā-mē-ə-'bi-lə-tē\ *noun*
— ami·a·ble·ness \'ā-mē-ə-bəl-nəs\ *noun*
— ami·a·bly \-blē\ *adverb*

am·i·ca·ble \'a-mi-kə-bəl\ *adjective* [Middle English, from Late Latin *amicabilis*] (15th century)
: characterized by friendly goodwill **:** PEACEABLE ☆
— am·i·ca·bil·i·ty \ˌa-mi-kə-'bi-lə-tē\ *noun*
— am·i·ca·ble·ness \'a-mi-kə-bəl-nəs\ *noun*
— am·i·ca·bly \-blē\ *adverb*

am·ice \'a-məs\ *noun* [Middle English *amis,* probably from Middle French, plural of *amit,* from Medieval Latin *amictus,* from Latin, cloak, from *amicire* to wrap around, from *am-, amb-* around + *jacere* to throw — more at AMBI-, JET] (13th century)
: a liturgical vestment made of an oblong piece of cloth usually of white linen and worn about the neck and shoulders and partly under the alb

ami·cus \ə-'mē-kəs, -'mī-\ *noun, plural* **ami·ci** \-'mē-ˌkē, -'mī-ˌsī\ (1951)
: AMICUS CURIAE

amicus cu·ri·ae \-'kyu̇r-ē-ˌī, -'ku̇r-, -i-ˌē\ *noun, plural* **amici curiae** [New Latin, literally, friend of the court] (1612)
: one (as a professional person or organization) that is not a party to a particular litigation but that is permitted by the court to advise it in respect to some matter of law that directly affects the case in question

amid \ə-'mid\ *or* **amidst** \-'midst, -'mitst\ *preposition* [*amid* from Middle English *amidde,* from Old English *onmiddan,* from *on* + *middan,* dative of *midde* mid; *amidst* from Middle English *amiddes,* from *amidde* + *-es* -s] (before 12th century)
1 : in or into the middle of **:** surrounded by **:** AMONG
2 a : DURING **b :** with the accompaniment of ⟨resigned *amid* rumors of misconduct⟩

am·i·dase \'a-mə-ˌdās, -ˌdāz\ *noun* [International Scientific Vocabulary] (1921)
: an enzyme that hydrolyzes acid amides usually with the liberation of ammonia

am·ide \'a-ˌmīd, -məd\ *noun* [International Scientific Vocabulary, from New Latin *ammonia*] (circa 1847)
1 : an inorganic compound derived from ammonia by replacement of an atom of hydrogen with another element (as a metal)
2 : an organic compound derived from ammonia or an amine by replacement of ammonical hydrogen with an acyl group — compare AMINE, IMIDE

ami·do \ə-'mē-(ˌ)dō, 'a-mə-ˌdō\ *adjective* [International Scientific Vocabulary *amide* + *-o-*] (1877)
: relating to or containing an organic amide group — often used in combination

am·i·dol \'a-mə-ˌdȯl, -ˌdōl\ *noun* [German, from *Amidol,* a trademark] (1892)
: a colorless crystalline salt $C_6H_8N_2O{\cdot}2HCl$ used chiefly as a photographic developer

amid·ships \ə-'mid-ˌships\ *adverb* (1692)
1 : in or toward the part of a ship midway between bow and stern
2 : in or toward the middle

ami·go \ə-'mē-(ˌ)gō, ä-\ *noun, plural* **-gos** [Spanish, from Latin *amicus* — more at AMIABLE] (1837)
: FRIEND

amine \ə-'mēn, 'a-ˌmēn\ *noun* [International Scientific Vocabulary, from New Latin *ammonia*] (1863)
: any of a class of organic compounds derived from ammonia by replacement of hydrogen with one or more alkyl groups — compare AMIDE 2

ami·no \ə-'mē-(ˌ)nō\ *adjective* [International Scientific Vocabulary *amine* + *-o-*] (1904)
: relating to, being, or containing an amine group — often used in combination

☆ SYNONYMS

Amiable, good-natured, obliging, complaisant mean having the desire or disposition to please. AMIABLE implies having qualities that make one liked and easy to deal with ⟨an *amiable* teacher not easily annoyed⟩. GOOD-NATURED implies cheerfulness or helpfulness and sometimes a willingness to be imposed upon ⟨a *good-natured* girl who was always willing to pitch in⟩. OBLIGING stresses a friendly readiness to be helpful ⟨our *obliging* innkeeper found us a bigger room⟩. COMPLAISANT often implies passivity or a yielding to others because of weakness ⟨was too *complaisant* to protest a decision he thought unfair⟩.

Amicable, neighborly, friendly mean exhibiting goodwill and an absence of antagonism. AMICABLE implies a state of peace and a desire on the part of the parties not to quarrel ⟨maintained *amicable* relations⟩. NEIGHBORLY implies a disposition to live on good terms with others and to be helpful on principle ⟨*neighborly* concern⟩. FRIENDLY stresses cordiality and often warmth or intimacy of personal relations ⟨sought *friendly* advice⟩.

◇ WORD HISTORY

amethyst For the ancient Greeks and Romans, gemstones had medical as well as ornamental value. The violet-colored variety of quartz known as amethyst, for example, was believed to protect against the effects of overdrinking, and this quality is reflected in its name. The Greek word *amethystos* means literally "not inducing drunkenness" and is a derivative of the verb *methyein* "to be drunk," itself from the noun *methy* "wine" (which is related to the English word *mead*). The Roman author Pliny associated the stone's supposed magical property with its color, for to him it resembled wine that had been heavily diluted with water and was hence not very intoxicating.

amino acid *noun* (1898)
: an amphoteric organic acid containing the amino group NH_2; *especially* : any of the alpha-amino acids that are the chief components of proteins and are synthesized by living cells or are obtained as essential components of the diet

ami•no•ac•id•uria \ə-ˌmē-nō-ˌa-sə-ˈdu̇r-ē-ə,-ˈdyu̇r-\ *noun* [New Latin] (circa 1923)
: a condition in which one or more amino acids are excreted in excessive amounts

ami•no•ben•zo•ic acid \ə-ˌmē-nō-ben-ˈzō-ik-\ *noun* [International Scientific Vocabulary] (1904)
: any of three crystalline derivatives $C_7H_7NO_2$ of benzoic acid; *especially* : PARA-AMINOBENZOIC ACID

ami•no•pep•ti•dase \ə-ˌmē-nō-ˈpep-tə-ˌdās, -ˌdāz\ *noun* (1943)
: an enzyme that hydrolyzes peptides by acting on the peptide bond next to a terminal amino acid containing a free amino group

am•i•noph•yl•line \ˌa-mə-ˈnä-fə-lən\ *noun* [*amino* + theo*phylline*] (1934)
: a theophylline derivative $C_{16}H_{24}N_{10}O_4$ used especially to stimulate the heart in congestive heart failure and to dilate the air passages in respiratory disorders

am•i•nop•ter•in \ˌa-mə-ˈnäp-tə-rən\ *noun* [*amino* + *pter*oylglutamic acid + *-in*] (1948)
: a derivative of glutamic acid $C_{19}H_{20}N_8O_5$ used especially as a rodenticide

ami•no•py•rine \ə-ˌmē-nō-ˈpī(ə)r-ˌēn\ *noun* [International Scientific Vocabulary, from *amino* + anti*pyrine*] (circa 1936)
: a crystalline compound $C_{13}H_{17}N_3O$ formerly used to relieve pain and fever but now largely abandoned for this purpose because of the occurrence of fatal agranulocytosis as a side effect in some users

ami•no•sal•i•cyl•ic acid \ə-ˌmē-nō-ˌsa-lə-ˈsi-lik-\ *noun* (1925)
: any of four isomeric derivatives $C_7H_7NO_3$ of salicylic acid that have a single amino group; *especially* : PARA-AMINOSALICYLIC ACID

ami•no•trans•fer•ase \-ˈtran(t)s-fə-ˌrās, -ˌrāz\ *noun* (circa 1965)
: TRANSAMINASE

amir *variant of* EMIR

Amish \ˈä-mish, ˈa-, ˈā-\ *adjective* [probably from German *amisch,* from Jacob *Amman* or *Amen* (flourished 1693) Swiss Mennonite bishop] (1844)
: of or relating to a strict sect of Mennonite followers of Amman that settled in America chiefly in the 18th century
— **Amish** *noun*

¹amiss \ə-ˈmis\ *adverb* (13th century)
1 a : in a mistaken way : WRONGLY ⟨if you think he is guilty, you judge *amiss*⟩ **b :** ASTRAY ⟨something had gone *amiss*⟩
2 : in a faulty way : IMPERFECTLY

²amiss *adjective* (14th century)
1 : not being in accordance with right order
2 : FAULTY, IMPERFECT
3 : out of place in given circumstances — usually used with a negative ⟨a few remarks may not be *amiss* here⟩

ami•to•sis \ˌā-mī-ˈtō-səs\ *noun* [New Latin, from ²*a-* + *mitosis*] (1894)
: cell division by simple cleavage of the nucleus and division of the cytoplasm without spindle formation or appearance of chromosomes
— **ami•tot•ic** \-ˈtä-tik\ *adjective*
— **ami•tot•i•cal•ly** \-ti-k(ə-)lē\ *adverb*

am•i•trip•ty•line \ˌa-mə-ˈtrip-tə-ˌlēn\ *noun* [*amino* + *trypt*ophan + *-yl* + ²*-ine*] (1961)
: a tricyclic antidepressant drug $C_{20}H_{23}N$

am•i•trole \ˈa-mə-ˌtrōl\ *noun* [*amino* + *triazole*] (circa 1960)
: a systemic herbicide $C_2H_4N_4$ used in areas other than food croplands

am•i•ty \ˈa-mə-tē\ *noun, plural* **-ties** [Middle English *amite,* from Middle French *amité,* from Medieval Latin *amicitas,* from Latin *amicus* friend — more at AMIABLE] (15th century)
: FRIENDSHIP; *especially* : friendly relations between nations

am•me•ter \ˈa-ˌmē-tər\ *noun* [*ampere* + *-meter*] (1882)
: an instrument for measuring electric current in amperes

am•mine \ˈa-ˌmēn, a-ˈmēn\ *noun* [International Scientific Vocabulary *ammonia* + ²*-ine*] (1897)
1 : a molecule of ammonia as it exists in a coordination complex ⟨hex-*ammine*-cobalt chloride $Co(NH_3)_6Cl_3$⟩
2 : a compound that contains an ammine

am•mo \ˈa-(ˌ)mō\ *noun* [by shortening & alteration] (1911)
: AMMUNITION

am•mo•nia \ə-ˈmō-nyə\ *noun* [New Latin, from Latin *sal ammoniacus* sal ammoniac, literally, salt of Ammon, from Greek *ammōniakos* of Ammon, from *Ammōn* Ammon, Amen, an Egyptian god near one of whose temples it was prepared] (circa 1799)
1 : a pungent colorless gaseous alkaline compound of nitrogen and hydrogen NH_3 that is very soluble in water and can easily be condensed to a liquid by cold and pressure
2 : AMMONIA WATER

am•mo•ni•ac \ə-ˈmō-nē-ˌak\ *noun* [Middle English & Latin; Middle English, from Latin *ammoniacum,* from Greek *ammōniakon,* from neuter of *ammōniakos* of Ammon] (15th century)
: the aromatic gum resin of a Persian herb (*Dorema ammoniacum*) of the carrot family used as an expectorant and stimulant and in plasters

am•mo•ni•a•cal \ˌa-mə-ˈnī-ə-kəl\ *also* **am•mo•ni•ac** \ə-ˈmō-nē-ˌak\ *adjective* (1798)
: of, relating to, containing, or having the properties of ammonia

am•mo•ni•ate \ə-ˈmō-nē-ˌāt\ *transitive verb* **-at•ed; -at•ing** (circa 1928)
1 : to combine or impregnate with ammonia or an ammonium compound
2 : to subject to ammonification
— **am•mo•ni•a•tion** \-ˌmō-nē-ˈā-shən\ *noun*

ammonia water *noun* (circa 1928)
: a water solution of ammonia

am•mo•ni•fi•ca•tion \ə-ˌmä-nə-fə-ˈkā-shən, -ˌmō-nə-\ *noun* (1886)
1 : the act or process of ammoniating
2 : decomposition with production of ammonia or ammonium compounds especially by the action of bacteria on nitrogenous organic matter
— **am•mo•ni•fy** \-ˌfī\ *verb*

am•mo•nite \ˈa-mə-ˌnīt\ *noun* [New Latin *ammonites,* from Latin *cornu Ammonis,* literally, horn of Ammon] (1758)
: any of a subclass (Ammonoidea) of extinct cephalopods with flat spiral shells that were especially abundant in the Mesozoic age
— **am•mo•nit•ic** \ˌa-mə-ˈni-tik\ *adjective*

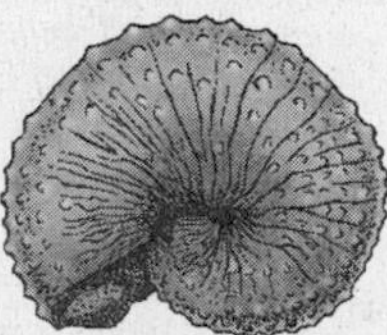
ammonite

Am•mon•ite \ˈa-mə-ˌnīt\ *noun* [Late Latin *Ammonites,* from Hebrew *ʿAmmōn* Ammon (son of Lot), descendant of Ammon] (1537)
: a member of a Semitic people who in Old Testament times lived east of the Jordan between the Jabbok and the Arnon
— **Ammonite** *adjective*

am•mo•ni•um \ə-ˈmō-nē-əm\ *noun* [New Latin, from *ammonia*] (1808)
: an ion NH_4^+ derived from ammonia by combination with a hydrogen ion and known in compounds (as salts) that resemble in properties the compounds of the alkali metals

ammonium carbonate *noun* (circa 1881)
: a carbonate of ammonium; *specifically* : the commercial mixture of the bicarbonate and carbamate used especially in smelling salts

ammonium chloride *noun* (1869)
: a white crystalline volatile salt NH_4Cl that is used in dry cells and as an expectorant — called also *sal ammoniac*

ammonium cyanate *noun* (circa 1881)
: an inorganic white crystalline salt NH_4CNO that can be converted into organic urea

ammonium hydroxide *noun* (1904)
: a weakly basic compound NH_4OH that is formed when ammonia dissolves in water and that exists only in solution

ammonium nitrate *noun* (circa 1881)
: a colorless crystalline salt NH_4NO_3 used in explosives and fertilizers and in veterinary medicine

ammonium phosphate *noun* (circa 1881)
: a phosphate of ammonium; *especially* : DIAMMONIUM PHOSPHATE

ammonium sulfate *noun* (circa 1881)
: a colorless crystalline salt $(NH_4)_2SO_4$ used chiefly as a fertilizer

am•mo•noid \ˈa-mə-ˌnȯid\ *noun* (1884)
: AMMONITE

am•mu•ni•tion \ˌam-yə-ˈni-shən\ *noun* [obsolete French *amunition,* from Middle French, alteration of *munition*] (circa 1626)
1 a : the projectiles with their fuses, propelling charges, or primers fired from guns **b :** CARTRIDGES **c :** explosive military items (as grenades or bombs)
2 : material for use in attacking or defending a position ⟨*ammunition* for the defense lawyers⟩

am•ne•sia \am-ˈnē-zhə\ *noun* [New Latin, from Greek *amnēsia* forgetfulness, alteration of *amnēstia*] (1786)
1 : loss of memory due usually to brain injury, shock, fatigue, repression, or illness
2 : a gap in one's memory
3 : the selective overlooking or ignoring of those events or acts that are not favorable or useful to one's purpose or position
— **am•ne•si•ac** \-zhē-ˌak, -zē-\ *or* **am•ne•sic** \-zik, -sik\ *adjective or noun*

am•nes•ty \ˈam-nə-stē\ *noun, plural* **-ties** [Greek *amnēstia* forgetfulness, from *amnēstos* forgotten, from *a-* + *mnasthai* to remember — more at MIND] (1580)
: the act of an authority (as a government) by which pardon is granted to a large group of individuals
— **amnesty** *transitive verb*

am•nio•cen•te•sis \ˌam-nē-ō-(ˌ)sen-ˈtē-səs\ *noun, plural* **-te•ses** \-ˌsēz\ [New Latin, from *amnion* + *centesis* puncture, from Greek *kentēsis,* from *kentein* to prick — more at CENTER] (1957)
: the surgical insertion of a hollow needle through the abdominal wall and into the uterus to obtain amniotic fluid especially for the determination of fetal sex or chromosomal abnormality

am•ni•on \ˈam-nē-ˌän, -ən\ *noun, plural* **amnions** *or* **am•nia** \-nē-ə\ [New Latin, from Greek, caul, from *amnos* lamb — more at YEAN] (1667)
1 : a thin membrane forming a closed sac about the embryos of reptiles, birds, and mammals and containing a serous fluid in which the embryo is immersed
2 : a membrane analogous to the amnion and occurring in various invertebrates
— **am•ni•ot•ic** \ˌam-nē-ˈä-tik\ *adjective*

am•ni•ote \ˈam-nē-ˌōt\ *noun* [New Latin *Amniota,* from *amnion*] (circa 1909)
: any of a group (Amniota) of vertebrates that undergo embryonic development within an amnion and include the birds, reptiles, and mammals
— **amniote** *adjective*

amniotic fluid *noun* (1855)
: the serous fluid in which the embryo is suspended within the amnion

amniotic sac *noun* (circa 1881)

: AMNION

amo·bar·bi·tal \ˌa-mō-ˈbär-bə-ˌtȯl\ *noun* [*amyl* + *-o-* + *barbital*] (1949)
: a barbiturate $C_{11}H_{18}N_2O_3$ used as a hypnotic and sedative; *also* : its sodium salt

amoe·ba \ə-ˈmē-bə\ *noun, plural* **-bas** *or* **-bae** \-(ˌ)bē\ [New Latin, genus name, from Greek *amoibē* change, from *ameibein* to change — more at MIGRATE] (1855)
: any of a large genus (*Amoeba*) of naked rhizopod protozoans with lobed and never anastomosing pseudopodia, without permanent organelles or supporting structures, and of wide distribution in fresh and salt water and moist terrestrial environments; *broadly* : a naked rhizopod or other amoeboid protozoan
— **amoe·bic** \-bik\ *adjective*

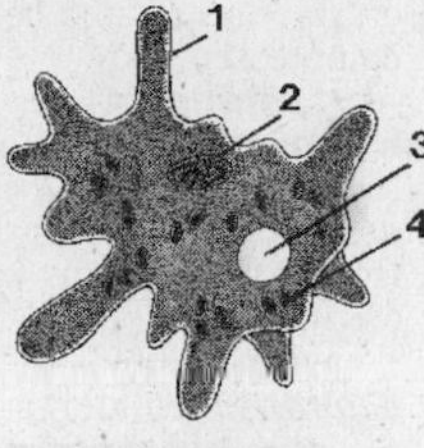

amoeba: *1* pseudopodium, *2* nucleus, *3* contractile vacuole, *4* food vacuole

am·oe·bi·a·sis *variant of* AMEBIASIS

amoe·bo·cyte \ə-ˈmē-bə-ˌsīt\ *noun* (1892)
: a cell (as a phagocyte) having amoeboid form or movements

amoe·boid \ə-ˈmē-ˌbȯid\ *adjective* (1856)
: resembling an amoeba specifically in moving or changing in shape by means of protoplasmic flow

[1]amok \ə-ˈmək, -ˈmäk\ *noun* [Malay *amok*] (1665)
: a murderous frenzy that occurs chiefly among Malays ◆

[2]amok *adverb* (1672)
1 : in a murderously frenzied state
2 a : in a violently raging manner ⟨a virus that had run *amok*⟩ **b** : in an undisciplined, uncontrolled, or faulty manner ⟨films . . . about computers run *amok* —*People*⟩

[3]amok *adjective* (1944)
: possessed with or motivated by a murderous or violently uncontrollable frenzy

amo·le \ə-ˈmo-lē\ *noun* [American Spanish, from Nahuatl *ahmōlli* soap] (1831)
: a plant part (as a root) possessing detergent properties and serving as a substitute for soap; *also* : a plant (as a yucca or agave) so used

among \ə-ˈməŋ\ *also* **amongst** \-ˈməŋ(k)st\ *preposition* [*among* from Middle English, from Old English *on gemonge*, from *on* + *gemonge*, dative of *gemong* crowd, from *ge-* (associative prefix) + *-mong* (akin to Old English *mengan* to mix); *amongst* from Middle English *amonges*, from *among* + *-es* -s — more at CO-, MINGLE] (before 12th century)
1 : in or through the midst of : surrounded by
2 : in company or association with ⟨living *among* artists⟩
3 : by or through the aggregate of ⟨discontent *among* the poor⟩
4 : in the number or class of ⟨wittiest *among* poets⟩ ⟨*among* other things she was president of her college class⟩
5 : in shares to each of ⟨divided *among* the heirs⟩
6 a : through the reciprocal acts of ⟨quarrel *among* themselves⟩ **b** : through the joint action of ⟨made a fortune *among* themselves⟩
usage see BETWEEN

amon·til·la·do \ə-ˌmän-tə-ˈlä-(ˌ)dō, -ti(l)-ˈyä-(ˌ)thō\ *noun, plural* **-dos** [Spanish, literally, done in the manner of *Montilla*, town in Andalusia] (1825)
: a medium dry sherry

amor·al \(ˌ)ā-ˈmȯr-əl, (ˌ)a-, -ˈmär-\ *adjective* (1882)
1 a : being neither moral nor immoral; *specifically* : lying outside the sphere to which moral judgments apply ⟨science as such is completely *amoral* —W. S. Thompson⟩ **b** : lacking moral sensibility ⟨infants are *amoral*⟩
2 : being outside or beyond the moral order or a particular code of morals ⟨*amoral* customs⟩
— **amor·al·ism** \-ə-ˌli-zəm\ *noun*
— **amo·ral·i·ty** \ˌā-mə-ˈra-lə-tē, ˌa-, -(ˌ)mȯ-\ *noun*
— **amor·al·ly** \ˌā-ˈmȯr-ə-lē, (ˌ)a-, -ˈmär-\ *adverb*

amo·ret·to \ˌa-mə-ˈre-(ˌ)tō, ˌä-\ *noun, plural* **-ti** \-(ˌ)tē\ *or* **-tos** [Italian, diminutive of *amore* love, cupid, from Latin *amor*] (1622)
: CUPID, CHERUB 2

am·or·ist \ˈa-mə-rist\ *noun* (1581)
1 : a devotee of love and especially sexual love : GALLANT
2 : one that writes about romantic love
— **am·or·is·tic** \ˌa-mə-ˈris-tik\ *adjective*

Am·o·rite \ˈa-mə-ˌrīt\ *noun* [Hebrew *Ĕmōrī*] (1535)
: a member of one of various Semitic peoples living in Mesopotamia, Syria, and Palestine during the 3d and 2d millennia B.C.
— **Amorite** *adjective*

am·o·rous \ˈa-mə-rəs, ˈam-rəs\ *adjective* [Middle English, from Middle French, from Medieval Latin *amorosus*, from Latin *amor* love, from *amare* to love] (14th century)
1 : strongly moved by love and especially sexual love ⟨*amorous* women⟩
2 : being in love : ENAMORED — usually used with *of* ⟨*amorous* of the girl⟩
3 a : indicative of love ⟨received *amorous* glances from her partner⟩ **b** : of or relating to love ⟨an *amorous* novel⟩
— **am·o·rous·ly** *adverb*
— **am·o·rous·ness** *noun*

amor·phous \ə-ˈmȯr-fəs\ *adjective* [Greek *amorphos*, from *a-* + *morphē* form] (circa 1731)
1 a : having no definite form : SHAPELESS ⟨an *amorphous* cloud mass⟩ **b** : being without definite character or nature : UNCLASSIFIABLE ⟨an *amorphous* segment of society⟩ **c** : lacking organization or unity ⟨an *amorphous* style⟩
2 : having no real or apparent crystalline form : UNCRYSTALLIZED ⟨an *amorphous* mineral⟩
— **amor·phous·ly** *adverb*
— **amor·phous·ness** *noun*

amort \ə-ˈmȯrt\ *adjective* [short for *all-a-mort*, by folk etymology from Middle French *a la mort* to the death] (1590)
archaic : being at the point of death

am·or·ti·za·tion \ˌa-mər-tə-ˈzā-shən *also* ə-ˌmȯr-\ *noun* (circa 1859)
1 : the act or process of amortizing
2 : the result of amortizing

am·or·tize \ˈa-mər-ˌtīz *also* ə-ˈmȯr-\ *transitive verb* **-tized; -tiz·ing** [Middle English *amortisen* to deaden, alienate in mortmain, modification of Middle French *amortiss-*, stem of *amortir*, from (assumed) Vulgar Latin *admortire* to deaden, from Latin *ad-* + *mort-*, *mors* death — more at MURDER] (1882)
1 : to provide for the gradual extinguishment of (as a mortgage) usually by contribution to a sinking fund at the time of each periodic interest payment
2 : to amortize an expenditure for ⟨*amortize* intangibles⟩ ⟨*amortize* the new factory⟩
— **am·or·tiz·able** \-ˌtī-zə-bəl\ *adjective*

Amos \ˈā-məs\ *noun* [Hebrew *ʽĀmōs*]
1 : a Hebrew prophet of the 8th century B.C.
2 : a prophetic book of canonical Jewish and Christian Scripture — see BIBLE table

am·o·site \ˈa-mə-ˌsīt, -ˌzīt\ *noun* [*Amosa* (from *A*sbestos *M*ines *o*f *S*outh *A*frica) + [1]*-ite*] (circa 1918)
: an iron-rich amphibole that is a variety of asbestos

[1]amount \ə-ˈmau̇nt\ *intransitive verb* [Middle English, from Middle French *amonter*, from *amont* upward, from *a-* (from Latin *ad-*) + *mont* mountain — more at MOUNT] (14th century)
1 a : to be equivalent ⟨acts that *amount* to treason⟩ **b** : to reach in kind or quality ⟨wants her son to *amount* to something⟩ ⟨doesn't *amount* to much⟩
2 : to reach a total : add up ⟨the bill *amounts* to $10⟩

[2]amount *noun* (1710)
1 a : the total number or quantity : AGGREGATE **b** : the quantity at hand or under consideration ⟨has an enormous *amount* of energy⟩
2 : the whole effect, significance, or import
3 : a principal sum and the interest on it ◻

amour \ə-ˈmu̇r, ä-, a-\ *noun* [Middle English, love, affection, from Middle French, from Old Provençal *amor*, from Latin, from *amare* to love] (14th century)
: a usually illicit love affair; *also* : LOVER

amour pro·pre \ˌa-ˌmu̇r-ˈprōprᵊ, ˌä-, -ˈprȯprᵊ\ *noun* [French *amour-propre*, literally, love of oneself] (1775)
: SELF-ESTEEM

amox·i·cil·lin \ə-ˌmäk-sē-ˈsi-lən\ *noun* [*am*ino + *ox-* + peni*cillin*] (1973)
: a semisynthetic penicillin $C_{16}H_{19}N_3O_5S$ derived from ampicillin

amox·y·cil·lin *British variant of* AMOXICILLIN

Amoy \ä-ˈmȯi, a-, ə-\ *noun* (1904)
: the dialect of Chinese spoken in and near Amoy in southeastern China

amp \ˈamp\ *noun* [by shortening] (1886)
1 : AMPERE
2 : AMPLIFIER; *also* : a unit consisting of an electronic amplifier and a loudspeaker

AMP \ˌā-ˌem-ˈpē\ *noun* [*a*denosine *m*ono*p*hosphate] (1951)
: a mononucleotide of adenine $C_{10}H_{12}N_5O_3H_2PO_4$ that was originally isolated from mammalian muscle and is reversibly convert-

◻ USAGE

amount *Number* is regularly used with count nouns ⟨a large *number* of mistakes⟩ ⟨any *number* of times⟩ while *amount* is mainly used with mass nouns ⟨annual *amount* of rainfall⟩ ⟨a substantial *amount* of money⟩. The use of *amount* with count nouns has been frequently criticized; it usually occurs when the number of things is thought of as a mass or collection ⟨glad to furnish any *amount* of black pebbles —*New Yorker*⟩ ⟨a substantial *amount* of film offers —Lily Tomlin⟩ or when money is involved ⟨a substantial *amount* of loans —E. R. Black⟩.

◇ WORD HISTORY

amok From the earliest Portuguese contacts with Southeast Asia in the sixteenth century, travelers have reported on a psychiatric disorder peculiar to the region that afflicts only men and is known in Malay as *amok*. In its typical modern manifestation, a man affected by amok first goes through a period of brooding and then explodes into homicidal frenzy, seizing a parang (a short sword) and cutting down everyone in his path, whether relative, friend, or stranger. If the man is not first killed by efforts to stop him, he eventually collapses in exhaustion, with no recollection of the episode. The phrase *run amuck* (translating the Malay verb *mengamok*) was being used in English to describe any undisciplined behavior by the late 17th century, and it occurs in the poetry of Alexander Pope ("Satire's my weapon, but I'm too discreet/To run a muck, and tilt at all I meet"). Most English speakers are probably no longer aware of the literal violence that the Malay word describes.

ible to ADP and ATP in metabolic reactions — called also *adenosine monophosphate, adenylic acid;* compare CYCLIC AMP

am·per·age \'am-p(ə-)rij, -ˌpir-ij\ *noun* (1893)
: the strength of a current of electricity expressed in amperes

am·pere \'am-ˌpir *also* -ˌper\ *noun* [André-Marie *Ampère*] (1881)
1 : the practical meter-kilogram-second unit of electric current that is equivalent to a flow of one coulomb per second or to the steady current produced by one volt applied across a resistance of one ohm
2 : the base unit of electric current in the International System of Units that is equal to a constant current which when maintained in two straight parallel conductors of infinite length and negligible circular sections one meter apart in a vacuum produces between the conductors a force equal to 2×10^{-7} newton per meter of length ◆

ampere–hour *noun* (1885)
: a unit quantity of electricity equal to the quantity carried past any point of a circuit in one hour by a steady current of one ampere

ampere–turn *noun* (1884)
: the meter-kilogram-second unit of magnetomotive force equal to the magnetomotive force around a path that links with one turn of wire carrying an electric current of one ampere

am·per·o·met·ric \ˌam-pir-ə-'me-trik\ *adjective [ampere + -o- + -metric]* (1940)
: relating to or being a chemical titration in which the measurement of the electric current flowing under an applied potential difference between two electrodes in a solution is used for detecting the end point

am·per·sand \'am-pər-ˌsand\ *noun* [alteration of *and* (&) *per se and,* literally, (the character) & by itself (is the word) *and*] (1835)
: a character typically & standing for the word *and* ◆

am·phet·amine \am-'fe-tə-ˌmēn, -mən\ *noun* [International Scientific Vocabulary *a*lpha + *m*ethyl + *ph*en- + *et*hyl + *amine*] (1938)
: a racemic compound $C_9H_{13}N$ or one of its derivatives (as dextroamphetamine or methamphetamine) frequently abused as a stimulant of the central nervous system but used clinically especially as the sulfate or hydrochloride salt to treat hyperactive children and the symptoms of narcolepsy and as a short-term appetite suppressant in dieting

amphi- *or* **amph-** *prefix* [Latin *amphi-* around, on both sides, from Greek *amphi-, amph-,* from *amphi* — more at AMBI-]
: on both sides **:** of both kinds **:** both ⟨*amphi*brach⟩ ⟨*amphi*diploid⟩

am·phib·ia \am-'fi-bē-ə\ *noun plural* (1607)
: AMPHIBIANS

am·phib·i·an \-bē-ən\ *noun* [ultimately from Greek *amphibion* amphibious being, from neuter of *amphibios*] (1835)
1 : an amphibious organism; *especially* **:** any of a class (Amphibia) of cold-blooded vertebrates (as frogs, toads, or salamanders) intermediate in many characters between fishes and reptiles and having gilled aquatic larvae and air-breathing adults
2 : an airplane designed to take off from and land on either land or water
— amphibian *adjective*

am·phib·i·ous \am-'fi-bē-əs\ *adjective* [Greek *amphibios,* literally, living a double life, from *amphi-* + *bios* mode of life — more at QUICK] (1643)
1 : combining two characteristics
2 a : relating to or adapted for both land and water ⟨*amphibious* vehicles⟩ **b :** executed by coordinated action of land, sea, and air forces organized for invasion; *also* **:** trained or organized for such action ⟨*amphibious* forces⟩
3 : able to live both on land and in water ⟨*amphibious* plants⟩
— am·phib·i·ous·ly *adverb*
— am·phib·i·ous·ness *noun*

am·phi·bole \'am(p)-fə-ˌbōl\ *noun* [French, from Late Latin *amphibolus,* from Greek *amphibolos* ambiguous, from *amphiballein* to throw round, doubt, from *amphi-* + *ballein* to throw — more at DEVIL] (circa 1823)
1 : HORNBLENDE
2 : any of a group of complex silicate minerals with like crystal structures that contain calcium, sodium, magnesium, aluminum, and iron ions or a combination of them

am·phib·o·lite \am-'fi-bə-ˌlīt\ *noun* (1833)
: a usually metamorphic rock consisting essentially of amphibole

am·phi·bol·o·gy \ˌam(p)-fə-'bä-lə-jē\ *noun, plural* **-gies** [Middle English *amphibologie,* from Late Latin *amphibologia,* alteration of Latin *amphibolia,* from Greek, from *amphibolos*] (14th century)
: a sentence or phrase (as "nothing is good enough for you") susceptible of more than one interpretation

am·phib·o·ly \am-'fi-bə-lē\ *noun, plural* **-lies** [Late Latin *amphibolia*] (circa 1588)
: AMPHIBOLOGY

am·phi·brach \'am(p)-fə-ˌbrak\ *noun* [Latin *amphibrachys,* from Greek, literally, short at both ends, from *amphi-* + *brachys* short — more at BRIEF] (1589)
: a metrical foot consisting of a long syllable between two short syllables in quantitative verse or of a stressed syllable between two unstressed syllables in accentual verse ⟨*romantic* is an accentual *amphibrach*⟩
— am·phi·brach·ic \ˌam(p)-fə-'bra-kik\ *adjective*

am·phic·ty·o·ny \am-'fik-tē-ə-nē\ *noun, plural* **-nies** [Greek *amphiktyonia*] (1835)
: an association of neighboring states in ancient Greece to defend a common religious center; *broadly* **:** an association of neighboring states for their common interest
— am·phic·ty·on·ic \(ˌ)am-ˌfik-tē-'ä-nik\ *adjective*

am·phi·dip·loid \ˌam(p)-fi-'di-ˌplȯid\ *noun* (1930)
: an interspecific hybrid having a complete diploid chromosome set from each parent form — called also *allotetraploid*
— amphidiploid *adjective*
— am·phi·dip·loi·dy \-ˌplȯi-dē\ *noun*

am·phim·a·cer \am-'fi-mə-sər\ *noun* [Latin *amphimacrus,* from Greek *amphimakros,* literally, long at both ends, from *amphi-* + *makros* long — more at MEAGER] (1589)
: a metrical foot consisting of a short syllable between two long syllables in quantitative verse or of an unstressed syllable between two stressed syllables in accentual verse ⟨*twenty-two* is an accentual *amphimacer*⟩

am·phi·mix·is \ˌam(p)-fi-'mik-səs\ *noun, plural* **-mix·es** \-ˌsēz\ [New Latin, from *amphi-* + Greek *mixis* mingling, from *mignynai* to mix — more at MIX] (1893)
: the union of gametes in sexual reproduction

Am·phi·on \am-'fī-ən\ *noun* [Latin, from Greek *Amphiōn*]
: a musician of Greek mythology who builds the walls of Thebes by charming the stones into place with his lyre

am·phi·ox·us \ˌam(p)-fē-'äk-səs\ *noun, plural* **-oxi** \-ˌsī\ *or* **-ox·us·es** [New Latin, from *amphi-* + Greek *oxys* sharp] (1847)
: any of a genus (*Branchiostoma*) of lancelets; *broadly* **:** LANCELET

am·phi·path·ic \ˌam(p)-fə-'pa-thik\ *adjective [amphi- + -pathic* (as in *empathic*)] (1945)
: AMPHIPHILIC

am·phi·phil·ic \-'fi-lik\ *adjective* (1950)
: of, relating to, or being a compound (as a surfactant) consisting of molecules having a polar water-soluble group attached to a water-insoluble hydrocarbon chain; *also* **:** being a molecule of such a compound
— am·phi·phile \'am(p)-fə-ˌfīl\ *noun*

am·phi·ploid \'am(p)-fi-ˌplȯid\ *adjective* (1945)
of an interspecific hybrid **:** having at least one complete diploid set of chromosomes derived from each parent species
— amphiploid *noun*
— am·phi·ploi·dy \-ˌplȯi-dē\ *noun*

am·phi·pod \-ˌpäd\ *noun* [ultimately from Greek *amphi-* + *pod-, pous* foot — more at FOOT] (1835)
: any of a large order (Amphipoda) of small crustaceans (as the beach flea) with a laterally compressed body
— amphipod *adjective*

am·phi·pro·style \ˌam(p)-fi-'prō-ˌstīl\ *adjective* [Latin *amphiprostylos,* from Greek, from *amphi-* + *prostylos* having pillars in front, from *pro-* + *stylos* pillar — more at STEER] (1850)
: having columns at each end only ⟨an *amphiprostyle* building⟩
— amphiprostyle *noun*

am·phis·bae·na \ˌam(p)-fəs-'bē-nə\ *noun* [Latin, from Greek *amphisbaina,* from *amphis* on both sides (from *amphi* around) + *bainein* to walk, go — more at BY, COME] (14th century)
: a serpent in classical mythology having a head at each end and capable of moving in either direction
— am·phis·bae·nic \-nik\ *adjective*

am·phi·the·ater \'am(p)-fə-ˌthē-ə-tər *also* 'am-pə-ˌthē-\ *noun* [Latin *amphitheatrum,* from Greek *amphitheatron,* from *amphi-* + *theatron* theater] (14th century)
1 : an oval or circular building with rising tiers of seats ranged about an open space and used in ancient Rome especially for contests and spectacles
2 a : a very large auditorium **b :** a room with a gallery from which doctors and students may observe surgical operations **c :** a rising gallery in a modern theater **d :** a flat or gently sloping area surrounded by abrupt slopes
3 : a place of public games or contests
— am·phi·the·at·ric \ˌam(p)-fə-thē-'a-trik *also* ˌam-pə-thē-\ *or* **am·phi·the·at·ri·cal** \-tri-kəl\ *adjective*
— am·phi·the·at·ri·cal·ly \-tri-k(ə-)lē\ *adverb*

◇ WORD HISTORY

ampere The *ampere* was named in honor of the French physicist André-Marie Ampère (1775–1836). Ampère is credited with founding, naming, and developing the science of electrodynamics. He was the formulator of two laws in electromagnetism relating magnetic fields to electric currents. The first person to develop techniques for measuring electricity, he invented an instrument that was a forerunner of the galvanometer. In 1881 an international congress on electricity adopted *ampere* as a term for the standard unit of electric current.

ampersand For pupils studying Latin in the grammar schools of late medieval and Tudor England, the traditional practice was to spell words by syllables. When a single letter formed a whole word (like *I*) or a complete syllable (like the first *i-* in *iris*), it was followed by the words *per se,* which in Latin means "by itself." Thus in the schoolroom *I* became *I per se, I.* Because the character &, a symbol for *and,* was placed after *z* in primers and treated like any other letter of the alphabet, it too was spelled out orally as "and per se, and." Frequent use of the phrase and diminishing recognition of its literal meaning shrank it to *ampersand* by the 19th century. The character *&* is a development of &, a ligature of *et,* the Latin word for "and."

Am·phit·ry·on \am-'fi-trē-ən\ *noun* [Greek *Amphitryōn*]
: the husband of Alcmene

am·pho·ra \'am(p)-fə-rə\ *noun, plural* **-rae** \-,rē, -,rī\ *or* **-ras** [Latin, modification of Greek *amphoreus, amphiphoreus,* from *amphi-* + *phoreus* bearer, from *pherein* to bear — more at BEAR] (14th century)
1 : an ancient Greek jar or vase with a large oval body, narrow cylindrical neck, and two handles that rise almost to the level of the mouth
2 : a 2-handled vessel shaped like an amphora

amphora 1

am·pho·ter·ic \,am(p)-fə-'ter-ik\ *adjective* [International Scientific Vocabulary, from Greek *amphoteros* each of two, from *amphō* both — more at AMBI-] (circa 1849)
: partly one and partly the other; *specifically* **:** capable of reacting chemically either as an acid or as a base

am·pho·ter·i·cin B \,am(p)-fə-'ter-ə-sən-'bē\ *noun* [*amphoteric* + [1]*-in*] (1955)
: an antifungal antibiotic obtained from a soil actinomycete (*Streptomyces nodosus*) and used especially to treat systemic fungal infections

am·pi·cil·lin \,am-pə-'si-lən\ *noun* [*am*ino + *penicillin*] (1961)
: a penicillin $C_{16}H_{19}N_3O_4S$ that is effective against gram-negative and gram-positive bacteria and is used to treat various infections of the urinary, respiratory, and intestinal tracts

am·ple \'am-pəl\ *adjective* **am·pler** \-p(ə-)lər\; **am·plest** \-p(ə-)ləst\ [Middle French, from Latin *amplus*] (15th century)
1 : generous or more than adequate in size, scope, or capacity ⟨there was room for an *ample* garden⟩
2 : generously sufficient to satisfy a requirement or need ⟨they had *ample* money for the trip⟩
3 : BUXOM, PORTLY ⟨an *ample* figure⟩
synonym see SPACIOUS, PLENTIFUL
— am·ple·ness \-pəl-nəs\ *noun*
— am·ply \-plē\ *adverb*

am·plex·us \am-'plek-səs\ *noun* [New Latin, from Latin, embrace, from *amplecti* to embrace, from *am-, amb-* around + *plectere* to braid — more at AMBI-, PLY] (circa 1927)
: the mating embrace of a frog or toad during which eggs are shed into the water and there fertilized

am·pli·dyne \'am-plə-,dīn\ *noun* [*ampli*fier + Greek *dyn*amis power — more at DYNAMIC] (1940)
: a direct-current generator that precisely controls a large power output whenever a small power input is varied in the field winding of the generator

am·pli·fi·ca·tion \,am-plə-fə-'kā-shən\ *noun* (1546)
1 a : an act, example, or product of amplifying **b :** a usually massive replication especially of a gene or DNA sequence (as in a polymerase chain reaction)
2 a : the particulars by which a statement is expanded **b :** an expanded statement

am·pli·fi·er \'am-plə-,fī(-ə)r\ *noun* (1542)
: one that amplifies; *specifically* **:** an electronic device (as in a stereo system) for amplifying voltage, current, or power

am·pli·fy \-,fī\ *verb* **-fied; -fy·ing** [Middle English *amplifien,* from Middle French *amplifier,* from Latin *amplificare,* from *amplus*] (15th century)
transitive verb
1 : to expand (as a statement) by the use of detail or illustration or by closer analysis
2 a : to make larger or greater (as in amount, importance, or intensity) **:** INCREASE **b :** to increase the strength or amount of; *especially* **:** to make louder **c :** to cause (a gene or DNA sequence) to undergo amplification
intransitive verb
: to expand one's remarks or ideas
synonym see EXPAND

am·pli·tude \-,tüd, -,tyüd\ *noun* (1542)
1 : the quality or state of being ample **:** FULLNESS
2 : the extent or range of a quality, property, process, or phenomenon: as **a :** the extent of a vibratory movement (as of a pendulum) measured from the mean position to an extreme **b :** the maximum departure of the value of an alternating current or wave from the average value
3 : ARGUMENT 6b

amplitude modulation *noun* (1921)
: modulation of the amplitude of a radio carrier wave in accordance with the strength of the audio or other signal; *also* **:** a broadcasting system using such modulation — compare FREQUENCY MODULATION

am·poule *or* **am·pule** *also* **am·pul** \'am-,pyü(ə)l, -pül\ *noun* [Middle English *ampulle* flask, from Old English & Old French; Old English *ampulle* & Old French *ampoule,* from Latin *ampulla*] (1886)
1 : a hermetically sealed small bulbous glass vessel that is used to hold a solution for hypodermic injection
2 : a vial resembling an ampoule

am·pul·la \am-'pu̇-lə, 'am-,pyü-lə\ *noun, plural* **-lae** \-(,)lē, -,lī\ [Middle English, from Old English, from Latin, diminutive of *amphora*] (before 12th century)
1 : a glass or earthenware flask with a globular body and two handles used especially by the ancient Romans to hold ointment, perfume, or wine
2 : a saccular anatomic swelling or pouch
— am·pul·la·ry \am-'pu̇-lər-ē, 'am-pyə-,ler-ē\ *adjective*

am·pu·tate \'am-pyə-,tāt\ *transitive verb* **-tat·ed; -tat·ing** [Latin *amputatus,* past participle of *amputare,* from *am-, amb-* around + *putare* to cut, prune — more at AMBI-] (1638)
: to remove by or as if by cutting; *especially* **:** to cut (as a limb) from the body
— am·pu·ta·tion \,am-pyə-'tā-shən\ *noun*

am·pu·tee \,am-pyə-'tē\ *noun* (1910)
: one that has had a limb amputated

am·trac *or* **am·track** \'am-,trak\ *noun* [*am*phibious + *trac*tor] (1944)
: a flat-bottomed military vehicle that moves on tracks on land or water

amuck \ə-'mək\ *variant of* AMOK

am·u·let \'am-yə-lət\ *noun* [Latin *amuletum*] (1584)
: a charm (as an ornament) often inscribed with a magic incantation or symbol to protect the wearer against evil (as disease or witchcraft) or to aid him

amuse \ə-'myüz\ *verb* **amused; amus·ing** [Middle French *amuser,* from Old French, from *a-* (from Latin *ad-*) + *muser* to muse] (15th century)
transitive verb
1 a *archaic* **:** to divert the attention of so as to deceive **b** *obsolete* **:** to occupy the attention of **c** *obsolete* **:** DISTRACT, BEWILDER
2 a : to entertain or occupy in a light, playful, or pleasant manner ⟨*amuse* the child with a story⟩ **b :** to appeal to the sense of humor of ⟨the joke doesn't *amuse* me⟩
intransitive verb
obsolete **:** MUSE ☆
— amus·ed·ly \-'myü-zəd-lē\ *adverb*
— amus·er *noun*
— amus·ive \ə-'myü-zəv, -səv\ *adjective*

amuse·ment \ə-'myüz-mənt\ *noun* (1603)
1 : a means of amusing or entertaining ⟨what are her favorite *amusements*⟩
2 : the condition of being amused ⟨his *amusement* knew no bounds⟩
3 : pleasurable diversion **:** ENTERTAINMENT ⟨plays the piano for *amusement*⟩

amusement park *noun* (1909)
: a commercially operated park having various devices for entertainment (as a merry-go-round and roller coaster) and usually booths for the sale of food and drink

amus·ing \ə-'myü-ziŋ\ *adjective* (1712)
: giving amusement **:** DIVERTING
— amus·ing·ly \-ziŋ-lē\ *adverb*
— amus·ing·ness *noun*

amyg·da·la \ə-'mig-də-lə\ *noun, plural* **-lae** \-,lē, -,lī\ [New Latin, from Latin, almond, from Greek *amygdalē*] (circa 1860)
: the one of the four basal ganglia in each cerebral hemisphere that is part of the limbic system and consists of an almond-shaped mass of gray matter in the anterior extremity of the temporal lobe — called also *amygdaloid nucleus*

amyg·da·lin \-lən\ *noun* [Latin *amygdala*] (1651)
: a white crystalline cyanogenetic glucoside $C_{20}H_{27}NO_{11}$ found especially in the seeds of the apricot, peach, and bitter almond

amyg·da·loid \-,lȯid\ *adjective* [Greek *amygdaloeidēs,* from *amygdalē*] (1836)
1 : almond-shaped
2 : of, relating to, or affecting an amygdala

amyg·da·loi·dal \ə-,mig-də-'lȯi-dᵊl\ *adjective* (1813)
: of, being, or containing small cavities in igneous rock that are filled with deposits of different minerals (as chalcedony or calcite)
— amyg·da·loid \ə-'mig-də-,lȯid\ *noun*

am·yl \'a-məl\ *noun* [Latin *amy*lum + English *-yl*] (1850)
: a univalent hydrocarbon radical $C_5H_{11}-$ that occurs in various isomeric forms and is derived from pentane

amyl- *or* **amylo-** *combining form* [Latin *amylum,* from Greek *amylon,* from neuter of *amylos* unmilled (of grain), from *a-* + *mylē* mill — more at MEAL]
: starch ⟨*amylo*plast⟩

amyl acetate *noun* (circa 1881)
: BANANA OIL

amyl alcohol *noun* (1863)
: any of eight isomeric alcohols $C_5H_{12}O$ used especially as solvents and in making esters; *also* **:** a commercially produced mixture of amyl alcohols used especially as a solvent

am·y·lase \'a-mə-,lās, -,lāz\ *noun* (1893)
: any of a group of enzymes (as amylopsin) that catalyze the hydrolysis of starch and glycogen or their intermediate hydrolysis products

amyl nitrite *noun* (circa 1881)
: a pale yellow pungent flammable liquid ester $C_5H_{11}NO_2$ of commercial amyl alcohol and nitrous acid — compare POPPER 2

am·y·loid \-,lȯid\ *noun* (1872)
: a waxy translucent substance consisting of protein in combination with polysaccharides

☆ SYNONYMS
Amuse, divert, entertain mean to pass or cause to pass the time pleasantly. AMUSE suggests that one's attention is engaged lightly or frivolously ⟨*amuse* yourselves while I prepare dinner⟩. DIVERT implies the distracting of the attention from worry or routine occupation especially by something funny ⟨a light comedy to *divert* the tired businessman⟩. ENTERTAIN suggests supplying amusement or diversion by specially prepared or contrived methods ⟨a magician *entertaining* children at a party⟩.

that is deposited in some animal organs and tissues under abnormal conditions (as Alzheimer's disease)
— **amyloid** *adjective*
am·y·loid·osis \ˌa-mə-ˌlȯi-ˈdō-səs\ *noun* [New Latin] (circa 1900)
: a disorder characterized by the deposition of amyloid in bodily organs and tissues
am·y·lo·lyt·ic \ˌa-mə-lō-ˈli-tik\ *adjective* [New Latin *amylolysis,* from *amyl-* + *-lysis*] (1876)
: characterized by or capable of the enzymatic splitting of starch into soluble products ⟨*amylolytic* enzymes⟩ ⟨*amylolytic* activity⟩
am·y·lo·pec·tin \ˌa-mə-lō-ˈpek-tən\ *noun* (1905)
: a component of starch that has a high molecular weight and branched structure and does not tend to gel in aqueous solutions
am·y·lo·plast \ˈa-mə-(ˌ)lō-ˌplast\ *noun* (1886)
: a colorless plastid that forms and stores starch
am·y·lop·sin \ˌa-mə-ˈläp-sən\ *noun* [*amyl-* + *-psin* (as in *trypsin*)] (circa 1881)
: the amylase of the pancreatic juice
am·y·lose \ˈa-mə-ˌlōs, -ˌlōz\ *noun* (1877)
: a component of starch characterized by its straight chains of glucose units
amyo·to·nia \ˌā-ˌmī-ə-ˈtō-nē-ə\ *noun* [New Latin] (circa 1919)
: deficiency of muscle tone
amyo·tro·phic lateral sclerosis \ˌā-ˌmī-ə-ˈtrō-fik-, -ˈträ-\ *noun* [²*a-* + *my-* + *-trophic*] (circa 1889)
: a rare progressive degenerative fatal disease affecting the spinal cord, usually beginning in middle age, and characterized especially by increasing and spreading muscular weakness — called also *Lou Gehrig's disease*
Am·y·tal \ˈa-mə-ˌtȯl\ *trademark*
— used for amobarbital
¹an \ən, (ˈ)an\ *indefinite article* [Middle English, from Old English *ān* one — more at ONE] (before 12th century)
: ²A
usage see ²A
²an \ən, an\ *preposition* (before 12th century)
: ³A 2
usage see ²A
³an *or* **an'** *conjunction* (12th century)
1 *see* AND\ **:** AND
2 \ˈan\ *archaic* **:** IF
an- — see ²A-
¹-an *or* **-ian** *also* **-ean** *noun suffix* [*-an* & *-ian* from Middle English *-an, -ian,* from Old French & Latin; Old French *-ien,* from Latin *-ianus,* from *-i-* + *-anus,* from *-anus,* adjective suffix; *-ean* from such words as *Mediterranean, European*]
1 : one that is of or relating to ⟨Americ*an*⟩ ⟨Boston*ian*⟩
2 : one skilled in or specializing in ⟨phoneti*cian*⟩
²-an *or* **-ian** *also* **-ean** *adjective suffix*
1 : of or belonging to ⟨Americ*an*⟩ ⟨Florid*ian*⟩
2 : characteristic of **:** resembling ⟨Mozart*ean*⟩
³-an *noun suffix* [International Scientific Vocabulary *-an, -ane,* alteration of *-ene, -ine,* & *-one*]
1 : unsaturated organic compound ⟨fur*an*⟩
2 : anhydride of a carbohydrate ⟨dextr*an*⟩
¹ana \ˈa-nə\ *adverb* [Middle English, from Medieval Latin, from Greek, at the rate of, literally, up] (14th century)
: of each an equal quantity — used in prescriptions
²ana \ˈa-nə, ˈä-, ˈā-\ *noun, plural* **ana** *or* **anas** [*-ana*] (circa 1751)
1 : a collection of the memorable sayings of a person
2 : a collection of anecdotes or interesting information about a person or a place
ana- *or* **an-** *prefix* [Latin, from Greek, up, back, again, from *ana* up — more at ON]
1 : up **:** upward ⟨*ana*bolism⟩
2 : back **:** backward ⟨*ana*tropous⟩

-ana *or* **-iana** *noun plural suffix* [New Latin, from Latin, neuter plural of *-anus* -an & *-ianus* -ian]
: collected items of information especially anecdotal or bibliographical concerning ⟨Ameri*cana*⟩
ana·bap·tism \ˌa-nə-ˈbap-ˌti-zəm\ *noun* [New Latin *anabaptismus,* from Late Greek *anabaptismos* rebaptism, from *anabaptizein* to rebaptize, from Greek *ana-* again + *baptizein* to baptize] (1577)
1 *capitalized* **a :** the doctrine or practices of the Anabaptists **b :** the Anabaptist movement
2 : the baptism of one previously baptized
Ana·bap·tist \-ˈbap-tist\ *noun* (1532)
: a Protestant sectarian of a radical movement arising in the 16th century and advocating the baptism and church membership of adult believers only, nonresistance, and the separation of church and state
— **Anabaptist** *adjective*
anab·a·sis \ə-ˈna-bə-səs\ *noun, plural* **-a·ses** \-ˌsēz\ [Greek, inland march, from *anabainein* to go up or inland, from *ana-* + *bainein* to go — more at COME] (circa 1706)
1 : a going or marching up **:** ADVANCE; *especially* **:** a military advance
2 [from the retreat of Greek mercenaries in Asia Minor described in the *Anabasis* of Xenophon] **:** a difficult and dangerous military retreat
an·a·bat·ic \ˌa-nə-ˈba-tik\ *adjective* [Greek *anabatos,* verbal of *anabainein*] (1853)
: moving upward **:** RISING ⟨an *anabatic* wind⟩
anabolic steroid *noun* (1961)
: any of a group of usually synthetic hormones that increase constructive metabolism and are sometimes abused by athletes in training to increase temporarily the size of their muscles
anab·o·lism \ə-ˈna-bə-ˌli-zəm\ *noun* [International Scientific Vocabulary *ana-* + meta*bolism*] (1886)
: the constructive part of metabolism concerned especially with macromolecular synthesis — compare CATABOLISM
— **an·a·bol·ic** \ˌa-nə-ˈbä-lik\ *adjective*
anach·ro·nism \ə-ˈna-krə-ˌni-zəm\ *noun* [probably from Middle Greek *anachronismos,* from *anachronizesthai* to be an anachronism, from Late Greek *anachronizein* to be late, from Greek *ana-* + *chronos* time] (circa 1646)
1 : an error in chronology; *especially* **:** a chronological misplacing of persons, events, objects, or customs in regard to each other
2 : a person or a thing that is chronologically out of place; *especially* **:** one from a former age that is incongruous in the present
— **anach·ro·nis·tic** \ə-ˌna-krə-ˈnis-tik\ *also* **ana·chron·ic** \ˌa-nə-ˈkrä-nik\ *adjective*
— **anach·ro·nis·ti·cal·ly** \ə-ˌna-krə-ˈnis-ti-k(ə-)lē\ *adverb*
— **anach·ro·nous** \ə-ˈna-krə-nəs\ *adjective*
— **anach·ro·nous·ly** *adverb*
an·a·clit·ic \ˌa-nə-ˈkli-tik\ *adjective* [Greek *anaklitos,* verbal of *anaklinein* to lean upon, from *ana-* + *klinein* to lean — more at LEAN] (1922)
: of, relating to, or characterized by the direction of love toward an object (as the mother) that satisfies nonsexual needs (as hunger)
an·a·co·lu·thon \ˌa-nə-kə-ˈlü-ˌthän\ *noun, plural* **-tha** \-thə\ *also* **-thons** [Late Latin, from Late Greek *anakolouthon* inconsistency in logic, from Greek, neuter of *anakolouthos* inconsistent, from *an-* + *akolouthos* following, from *ha-, a-* together + *keleuthos* path] (circa 1706)
: syntactical inconsistency or incoherence within a sentence; *especially* **:** a shift in an unfinished sentence from one syntactic construction to another (as in "you really ought—well, do it your own way")
— **an·a·co·lu·thic** \-thik\ *adjective*
— **an·a·co·lu·thi·cal·ly** \-thi-k(ə-)lē\ *adverb*

an·a·con·da \ˌa-nə-ˈkän-də\ *noun* [probably modification of Sinhalese *henakandayā,* a slender green snake] (1768)
: a large semiaquatic constricting snake (*Eunectes murinus*) of the boa family of tropical South America that may reach a length of 30 feet (9.1 meters); *broadly* **:** any of the large constricting snakes

anaconda

anac·re·on·tic \ə-ˌna-krē-ˈän-tik\ *noun* (1656)
: a poem in the manner of Anacreon; *especially* **:** a drinking song or light lyric ◆
Anacreontic *adjective* [Latin *anacreonticus,* from *Anacreont-, Anacreon* Anacreon, from Greek *Anakreont-, Anakreōn*] (1611)
1 : of, relating to, or resembling the poetry of Anacreon
2 : convivial or amatory in tone or theme
an·a·cru·sis \ˌa-nə-ˈkrü-səs\ *noun, plural* **-cru·ses** \-ˌsēz\ [New Latin, from Greek *anakrousis* beginning of a song, from *anakrouein* to begin a song, from *ana-* + *krouein* to strike, beat; akin to Lithuanian *kraušyti* to strike] (1830)
1 : one or more syllables at the beginning of a line of poetry that are regarded as preliminary to and not a part of the metrical pattern
2 : UPBEAT; *specifically* **:** one or more notes or tones preceding the first downbeat of a musical phrase
an·a·dama bread \ˌa-nə-ˈda-mə-\ *noun* [origin unknown] (1954)
: a leavened bread made with flour, cornmeal, and molasses
an·a·dem \ˈa-nə-ˌdem\ *noun* [Latin *anadema,* from Greek *anadēma,* from *anadein* to wreathe, from *ana-* + *dein* to bind — more at DIADEM] (1604)
archaic **:** a wreath for the head **:** GARLAND
ana·di·plo·sis \ˌa-nə-də-ˈplō-səs, ˌa-nə-(ˌ)dī-ˈplō-\ *noun, plural* **-plo·ses** \-ˌsēz\ [Late Latin, from Greek *anadiplōsis,* literally, repetition, from *anadiploun* to double, from *ana-* + *diploun* to double — more at DIPLOMA] (circa 1550)
: repetition of a prominent and usually the last word in one phrase or clause at the beginning of the next (as in "rely on his honor—honor such as his?")
anad·ro·mous \ə-ˈna-drə-məs\ *adjective* [Greek *anadromos* running upward, from

◇ WORD HISTORY
anacreontic Although the Greek lyric poet Anacreon (ca. 582–ca. 485 B.C.) wrote a number of different kinds of poetry (now mostly lost), he was remembered and quoted by later writers for his lyrics celebrating love and wine. According to ancient critics well versed in his work, his songs often made claims of having been composed in moments of drunken revelry. The Roman author Valerius Maximus relates that Anacreon died at an advanced age by choking on a grape seed. While the story may be apocryphal, it has served to enhance Anacreon's reputation as one devoted to the pleasure of the senses. From antiquity onward, Anacreon had many imitators. Since the 17th century, a poem or song, and especially one celebrating the pleasures of love and drinking, has been called an *anacreontic.* The tune of one such song, "To Anacreon in Heaven," was later used as the music for Francis Scott Key's lyric "The Star-Spangled Banner."

anadramein to run upward, from *ana-* + *dramein* to run — more at DROMEDARY] (circa 1753)
: ascending rivers from the sea for breeding ⟨shad are *anadromous*⟩ — compare CATADROMOUS

anae·mia, anae·mic *chiefly British variant of* ANEMIA, ANEMIC

an·aer·obe \'a-nə-ˌrōb; (ˌ)an-'a(-ə)r-ˌōb, -'e(-ə)r-\ *noun* [International Scientific Vocabulary] (1884)
: an anaerobic organism

an·aer·o·bic \ˌa-nə-'rō-bik; ˌan-ˌa(-ə)-, -ˌe(-ə)-\ *adjective* (circa 1881)
1 a : living, active, occurring, or existing in the absence of free oxygen ⟨*anaerobic* respiration⟩ **b :** of, relating to, or being activity in which the body incurs an oxygen debt ⟨*anaerobic* exercise⟩
2 : relating to or induced by anaerobes
— an·aer·o·bi·cal·ly \-bi-k(ə-)lē\ *adverb*

an·aero·bi·o·sis \ˌa-nə-rō-(ˌ)bī-'ō-səs, -bē-; ˌan-ˌa(-ə)-, -ˌe(-ə)-\ *noun, plural* **-o·ses** \-'ō-ˌsēz\ [New Latin] (circa 1889)
: life in the absence of air or free oxygen

an·aes·the·sia, an·aes·thet·ic *chiefly British variant of* ANESTHESIA, ANESTHETIC

ana·gen·e·sis \ˌa-nə-'je-nə-səs\ *noun* [New Latin] (1889)
: evolutionary change producing a single lineage in which one taxon replaces another without branching — compare CLADOGENESIS

ana·glyph \'a-nə-ˌglif\ *noun* [Late Latin *anaglyphus* embossed, from Greek *anaglyphos,* from *anaglyphein* to emboss, from *ana-* + *glyphein* to carve — more at CLEAVE] (1651)
1 : a sculptured, chased, or embossed ornament worked in low relief
2 : a stereoscopic motion or still picture in which the right component of a composite image usually red in color is superposed on the left component in a contrasting color to produce a three-dimensional effect when viewed through correspondingly colored filters in the form of spectacles
— ana·glyph·ic \ˌa-nə-'gli-fik\ *adjective*

an·ag·no·ri·sis \ˌa-ˌnag-'nȯr-ə-səs\ *noun, plural* **-ri·ses** \-ˌsēz\ [Greek *anagnōrisis,* from *anagnōrizein* to recognize, from *ana-* + *gnōrizein* to make known; akin to Greek *gnōrimos* well-known, *gignōskein* to come to know — more at KNOW] (circa 1800)
: the point in the plot especially of a tragedy at which the protagonist recognizes his or her or some other character's true identity or discovers the true nature of his or her own situation

an·a·go·ge *or* **an·a·go·gy** \'a-nə-ˌgō-jē\ *noun, plural* **-ges** *or* **-gies** [Late Latin *anagoge,* from Late Greek *anagōgē,* from Greek, reference, from *anagein* to refer, from *ana-* + *agein* to lead — more at AGENT] (15th century)
: interpretation of a word, passage, or text (as of Scripture or poetry) that finds beyond the literal, allegorical, and moral senses a fourth and ultimate spiritual or mystical sense
— an·a·gog·ic \ˌa-nə-'gä-jik\ *or* **an·a·gog·i·cal** \-ji-kəl\ *adjective*
— an·a·gog·i·cal·ly \-ji-k(ə-)lē\ *adverb*

[1]**an·a·gram** \'a-nə-ˌgram\ *noun* [probably from Middle French *anagramme,* from New Latin *anagrammat-, anagramma,* modification of Greek *anagrammatismos,* from *anagrammatizein* to transpose letters, from *ana-* + *grammat-, gramma* letter — more at GRAM] (1589)
1 : a word or phrase made by transposing the letters of another word or phrase
2 *plural but singular in construction* **:** a game in which words are formed by rearranging the letters of other words or by arranging letters taken (as from a stock of cards or blocks) at random
— an·a·gram·mat·ic \ˌa-nə-grə-'ma-tik\ *also* **an·a·gram·mat·i·cal** \-ti-kəl\ *adjective*
— an·a·gram·mat·i·cal·ly \-ti-k(ə-)lē\ *adverb*

[2]**anagram** *transitive verb* **-grammed; -gram·ming** (1630)
1 : ANAGRAMMATIZE
2 : to rearrange (the letters of a text) in order to discover a hidden message

an·a·gram·ma·tize \ˌa-nə-'gra-mə-ˌtīz\ *transitive verb* **-tized; -tiz·ing** (1588)
: to transpose (as letters in a word) so as to form an anagram
— an·a·gram·ma·ti·za·tion \-ˌgra-mə-tə-'zā-shən\ *noun*

anal \'ā-nᵊl\ *adjective* (1769)
1 : of, relating to, or situated near the anus ⟨*anal* fin⟩
2 a : of, relating to, characterized by, or being the stage of psychosexual development in psychoanalytic theory during which the child is concerned especially with its feces **b :** of, relating to, characterized by, or being personality traits (as parsimony, meticulousness, and ill humor) considered typical of fixation at the anal stage of development ⟨*anal* disposition⟩ ⟨*anal* neatness⟩
— anal·ly \-nᵊl-ē\ *adverb*

anal·cime \ə-'nal-ˌsēm\ *noun* [French, from Greek *analkimos* weak, from *an-* + *alkimos* strong, from *alkē* strength] (1803)
: a white or slightly colored mineral that consists of hydrated silicate of sodium and aluminum and occurs in various igneous rocks in massive form or in crystals

anal·cite \ə-'nal-ˌsīt\ *noun* (1868)
: ANALCIME

an·a·lects \'a-nᵊl-ˌek(t)s\ *also* **an·a·lec·ta** \ˌa-nᵊl-'ek-tə\ *noun plural* [New Latin *analecta,* from Greek *analekta,* neuter plural of *analektos,* verbal of *analegein* to collect, from *ana-* + *legein* to gather — more at LEGEND] (1652)
: selected miscellaneous written passages

an·a·lem·ma \ˌa-nə-'le-mə\ *noun* [Latin, sundial on a pedestal, from Greek *analēmma,* lofty structure, sundial, from *analambanein* to take up, restore, from *ana-* + *lambanein* to take — more at LATCH] (1832)
: a plot or graph of the position of the sun in the sky at a certain time of day (as noon) at one locale measured at regular intervals throughout the year that has the shape of a figure 8; *also* **:** a scale (as on a globe or sundial) based on such a plot that shows the sun's position for each day of the year or that allows local mean time to be determined
— an·a·lem·mat·ic \ˌa-nə-le-'ma-tik, -lə-\ *adjective*

an·a·lep·tic \ˌa-nə-'lep-tik\ *noun* [Greek *analēptikos,* from *analambanein*] (1671)
: a drug that stimulates the central nervous system
— analeptic *adjective*

an·al·ge·sia \ˌa-nᵊl-'jē-zh(ē-)ə, -zē-ə\ *noun* [New Latin, from Greek *analgēsia,* from *an-* + *algēsis* sense of pain, from *algein* to suffer pain, from *algos* pain] (circa 1706)
: insensibility to pain without loss of consciousness
— an·al·ge·sic \-'jē-zik, -sik\ *adjective or noun*
— an·al·ge·tic \-'je-tik\ *adjective or noun*

anal·i·ty \ā-'na-lə-tē\ *noun, plural* **-ties** (1939)
: the psychological state or quality of being anal

an·a·log \'a-nᵊl-ˌȯg, -ˌäg\ *adjective* (1948)
1 : of, relating to, or being an analogue
2 a : of, relating to, or being a mechanism in which data is represented by continuously variable physical quantities **b :** of or relating to an analog computer **c :** being a timepiece having hour and minute hands

analog computer *noun* (1948)
: a computer that operates with numbers represented by directly measurable quantities (as voltages or rotations) — compare DIGITAL COMPUTER, HYBRID COMPUTER

an·a·log·i·cal \ˌa-nᵊl-'ä-ji-kəl\ *also* **an·a·log·ic** \-jik\ *adjective* (1609)
1 : of, relating to, or based on analogy
2 : expressing or implying analogy
— an·a·log·i·cal·ly \-ji-k(ə-)lē\ *adverb*

anal·o·gist \ə-'na-lə-jist\ *noun* (circa 1828)
: one who searches for or reasons from analogies

anal·o·gize \-ˌjīz\ *verb* **-gized; -giz·ing** (1655)
intransitive verb
: to use or exhibit analogy
transitive verb
: to compare by analogy

anal·o·gous \ə-'na-lə-gəs\ *adjective* [Latin *analogus,* from Greek *analogos,* literally, proportionate, from *ana-* + *logos* reason, ratio, from *legein* to gather, speak — more at LEGEND] (1646)
1 : showing an analogy or a likeness that permits one to draw an analogy
2 : being or related to as an analogue
synonym see SIMILAR
— anal·o·gous·ly *adverb*
— anal·o·gous·ness *noun*

[1]**an·a·logue** *or* **an·a·log** \'a-nᵊl-ˌȯg, -ˌäg\ *noun* [French *analogue,* from *analogue* analogous, from Greek *analogos*] (1826)
1 : something that is analogous or similar to something else
2 : an organ similar in function to an organ of another animal or plant but different in structure and origin
3 *usually analog* **:** a chemical compound that is structurally similar to another but differs slightly in composition (as in the replacement of one atom by an atom of a different element or in the presence of a particular functional group)
4 : a food product made by combining a less expensive food (as soybeans or whitefish) with additives to give the appearance and taste of a more expensive food (as beef or crab)

[2]**an·a·logue** *chiefly British variant of* ANALOG

anal·o·gy \ə-'na-lə-jē\ *noun, plural* **-gies** (15th century)
1 : inference that if two or more things agree with one another in some respects they will probably agree in others
2 a : resemblance in some particulars between things otherwise unlike **:** SIMILARITY **b :** comparison based on such resemblance
3 : correspondence between the members of pairs or sets of linguistic forms that serves as a basis for the creation of another form
4 : correspondence in function between anatomical parts of different structure and origin — compare HOMOLOGY
synonym see LIKENESS

an·al·pha·bet \(ˌ)an-'al-fə-ˌbet, -bət\ *noun* [Greek *analphabētos* not knowing the alphabet, from *an-* + *alphabētos* alphabet] (circa 1889)
: one who cannot read **:** ILLITERATE
— an·al·pha·bet·ic \ˌan-ˌal-fə-'be-tik\ *adjective or noun*
— an·al·pha·bet·ism \(ˌ)an-'al-fə-bə-ˌti-zəm\ *noun*

anal·y·sand \ə-'na-lə-ˌsand\ *noun* [*analyse* + *-and* (as in *multiplicand*)] (1917)
: one who is undergoing psychoanalysis

an·a·lyse *chiefly British variant of* ANALYZE

anal·y·sis \ə-'na-lə-səs\ *noun, plural* **-y·ses** \-ˌsēz\ [New Latin, from Greek, from *analyein* to break up, from *ana-* + *lyein* to loosen — more at LOSE] (1581)

1 : separation of a whole into its component parts
2 a : the identification or separation of ingredients of a substance **b :** a statement of the constituents of a mixture
3 a : proof of a mathematical proposition by assuming the result and deducing a valid statement by a series of reversible steps **b** (1) **:** a branch of mathematics concerned mainly with functions and limits (2) **:** CALCULUS 1b
4 a : an examination of a complex, its elements, and their relations **b :** a statement of such an analysis
5 a : a method in philosophy of resolving complex expressions into simpler or more basic ones **b :** clarification of an expression by an elucidation of its use in discourse
6 : the use of function words instead of inflectional forms as a characteristic device of a language
7 : PSYCHOANALYSIS

analysis of variance (circa 1939)
: analysis of variation in an experimental outcome and especially of a statistical variance in order to determine the contributions of given factors or variables to the variance

analysis si·tus \-ˈsī-təs, -ˈsē-, -ˌtüs\ *noun* [New Latin, literally, analysis of situation] (circa 1909)
: TOPOLOGY 2a(1)

an·a·lyst \ˈa-nᵊl-əst\ *noun* [French *analyste,* from *analyse* analysis] (1656)
1 : a person who analyzes or who is skilled in analysis
2 : PSYCHOANALYST

an·a·lyt·ic \ˌa-nᵊl-ˈi-tik\ *or* **an·a·lyt·i·cal** \-ti-kəl\ *adjective* [Late Latin *analyticus,* from Greek *analytikos,* from *analyein*] (1601)
1 : of or relating to analysis or analytics; *especially* **:** separating something into component parts or constituent elements
2 : being a proposition (as "no bachelor is married") whose truth is evident from the meaning of the words it contains — compare SYNTHETIC
3 : skilled in or using analysis especially in thinking or reasoning ⟨a keenly *analytic* person⟩
4 : characterized by analysis rather than inflection ⟨*analytic* languages⟩
5 : PSYCHOANALYTIC
6 : treated or treatable by or using the methods of algebra and calculus
7 a *of a function of a real variable* **:** capable of being expanded in a Taylor's series in powers of $x - h$ in some neighborhood of the point h **b** *of a function of a complex variable* **:** differentiable at every point in some neighborhood of a given point or points
— **an·a·lyt·i·cal·ly** \-ti-k(ə-)lē\ *adverb*
— **an·a·ly·tic·i·ty** \ˌa-nᵊl-ə-ˈti-sə-tē\ *noun*

analytic geometry *noun* (circa 1886)
: the study of geometric properties by means of algebraic operations upon symbols defined in terms of a coordinate system — called also *coordinate geometry*

analytic philosophy *noun* (1936)
: a philosophical movement that seeks the solution of philosophical problems in the analysis of propositions or sentences — called also *philosophical analysis; compare* ORDINARY-LANGUAGE PHILOSOPHY

an·a·lyt·ics \ˌa-nᵊl-ˈi-tiks\ *noun plural but singular or plural in construction* (circa 1590)
: the method of logical analysis

an·a·ly·za·tion \ˌa-nᵊl-ə-ˈzā-shən\ *noun* (1742)
: ANALYSIS

an·a·lyze \ˈa-nᵊl-ˌīz\ *transitive verb* **-lyzed; -lyz·ing** [probably irregular from *analysis*] (1587)
1 : to study or determine the nature and relationship of the parts of by analysis ⟨*analyze* a traffic pattern⟩
2 : to subject to scientific or grammatical analysis
3 : PSYCHOANALYZE ☆
— **an·a·lyz·abil·i·ty** \ˌa-nᵊl-ˌī-zə-ˈbi-lə-tē\ *noun*
— **an·a·lyz·able** \ˈa-nᵊl-ˌī-zə-bəl\ *adjective*
— **an·a·lyz·er** \-ˌī-zər\ *noun*

an·am·ne·sis \ˌa-ˌnam-ˈnē-səs\ *noun, plural* **-ne·ses** \-ˌsēz\ [New Latin, from Greek *anamnēsis,* from *anamimnēskesthai* to remember, from *ana-* + *mimnēskesthai* to remember — more at MIND] (circa 1593)
1 : a recalling to mind **:** REMINISCENCE
2 : a preliminary case history of a medical or psychiatric patient

an·am·nes·tic \-ˈnes-tik\ *adjective* [Greek *anamnēstikos* easily recalled, from *anamimnēskesthai*] (circa 1753)
1 : of or relating to an anamnesis
2 : of or relating to a secondary response to an immunogenic substance after serum antibodies can no longer be detected in the blood

ana·mor·phic \ˌa-nə-ˈmȯr-fik\ *adjective* [New Latin *anamorphosis* distorted optical image] (circa 1925)
: producing, relating to, or marked by intentional distortion (as by unequal magnification along perpendicular axes) of an image ⟨an *anamorphic* lens⟩

An·a·ni·as \ˌa-nə-ˈnī-əs\ *noun* [Greek, probably from Hebrew *Ḥănanyāh*]
1 : an early Christian struck dead for lying
2 : LIAR

an·a·pest \ˈa-nə-ˌpest\ *noun* [Latin *anapaestus,* from Greek *anapaistos,* literally, struck back (a dactyl reversed), from *ana-* + *-paistos,* verbal of *paiein* to strike] (circa 1678)
: a metrical foot consisting of two short syllables followed by one long syllable or of two unstressed syllables followed by one stressed syllable (as *unabridged*)
— **an·a·pes·tic** \ˌa-nə-ˈpes-tik\ *adjective or noun*

ana·phase \ˈa-nə-ˌfāz\ *noun* [International Scientific Vocabulary] (1887)
: the stage of mitosis and meiosis in which the chromosomes move toward the poles of the spindle
— **ana·pha·sic** \ˌa-nə-ˈfā-zik\ *adjective*

ana·phor \ˈa-nə-ˌfȯr\ *noun, plural* **anaphors** *also* **anaph·o·ra** \ə-ˈna-f(ə-)rə\ [back-formation from *anaphoric*] (1975)
: a word or phrase with an anaphoric function

anaph·o·ra \ə-ˈna-f(ə-)rə\ *noun* [Late Latin, from Late Greek, from Greek, act of carrying back, reference, from *anapherein* to carry back, refer, from *ana-* + *pherein* to carry — more at BEAR] (circa 1589)
1 : repetition of a word or expression at the beginning of successive phrases, clauses, sentences, or verses especially for rhetorical or poetic effect ⟨Lincoln's "we cannot dedicate—we cannot consecrate—we cannot hallow—this ground" is an example of *anaphora*⟩ — compare EPISTROPHE
2 : use of a grammatical substitute (as a pronoun or a pro-verb) to refer to the denotation of a preceding word or group of words; *also* **:** the relation between a grammatical substitute and its antecedent

an·a·phor·ic \ˌa-nə-ˈfȯr-ik, -ˈfär-\ *adjective* (1904)
: of or relating to anaphora ⟨an *anaphoric* usage⟩; *especially* **:** being a word or phrase that takes its reference from another word or phrase and especially from a preceding word or phrase — compare CATAPHORIC
— **ana·phor·i·cal·ly** \-i-k(ə-)lē\ *adverb*

an·aph·ro·di·si·ac \ˌa-ˌna-frə-ˈdē-zē-ˌak, -zhē-, -ˈdi-zē-\ *adjective* (1823)
: inhibiting or discouraging sexual desire
— **anaphrodisiac** *noun*

ana·phy·lac·tic \ˌa-nə-fə-ˈlak-tik\ *adjective* (1907)
: of, relating to, affected by, or causing anaphylaxis or anaphylactic shock
— **ana·phy·lac·ti·cal·ly** \-ti-k(ə-)lē\ *adverb*
— **ana·phy·lac·toid** \-ˈlak-ˌtȯid\ *adjective*

anaphylactic shock *noun* (1910)
: an often severe and sometimes fatal systemic reaction in a susceptible individual upon exposure to a specific antigen (as wasp venom or penicillin) after previous sensitization that is characterized especially by respiratory symptoms, fainting, itching, and urticaria

ana·phy·lax·is \-ˈlak-səs\ *noun, plural* **-lax·es** \-ˌsēz\ [New Latin, from *ana-* + *prophylaxis*] (1907)
1 : hypersensitivity (as to foreign proteins or drugs) resulting from sensitization following prior contact with the causative agent
2 : ANAPHYLACTIC SHOCK

an·a·pla·sia \ˌa-nə-ˈplā-zh(ē-)ə\ *noun* [New Latin] (circa 1909)
: reversion of cells to a more primitive or undifferentiated form
— **an·a·plas·tic** \-ˈplas-tik\ *adjective*

an·a·plas·mo·sis \ˌa-nə-ˌplaz-ˈmō-səs\ *noun, plural* **-mo·ses** \-ˌsēz\ [New Latin, from *Anaplasma,* genus name, from *ana-* + *plasma* (protoplasm)] (1920)
: a tick-borne disease of cattle and sheep caused by a bacterium (*Anaplasma marginale*) and characterized especially by anemia without blood-tinged urine and by jaundice

an·arch \ˈa-ˌnärk\ *noun* [back-formation from *anarchy*] (1667)
: a leader or advocate of revolt or anarchy

an·ar·chic \a-ˈnär-kik, ə-\ *also* **an·ar·chi·cal** \-ki-kəl\ *adjective* (1649)
1 a : of, relating to, or advocating anarchy **b :** likely to bring about anarchy ⟨*anarchic* violence⟩
2 : lacking order, regularity, or definiteness ⟨*anarchic* art forms⟩
— **an·ar·chi·cal·ly** \-k(ə-)lē\ *adverb*

an·ar·chism \ˈa-nər-ˌki-zəm, -ˌnär-\ *noun* (1642)
1 : a political theory holding all forms of governmental authority to be unnecessary and undesirable and advocating a society based on voluntary cooperation and free association of individuals and groups
2 : the advocacy or practice of anarchistic principles

an·ar·chist \ˈa-nər-kist, -ˌnär-\ *noun* (1678)
1 : one who rebels against any authority, established order, or ruling power
2 : one who believes in, advocates, or promotes anarchism or anarchy; *especially* **:** one who uses violent means to overthrow the established order
— **anarchist** *or* **an·ar·chis·tic** \ˌa-nər-ˈkis-tik, -(ˌ)när-\ *adjective*

an·ar·cho–syn·di·cal·ism \a-ˌnär-kō-ˈsin-di-kə-ˌli-zəm, ˌa-nər-kō-\ *noun* (circa 1928)
: SYNDICALISM
— **an·ar·cho–syn·di·cal·ist** \-kə-list\ *noun or adjective*

an·ar·chy \ˈa-nər-kē, -ˌnär-\ *noun* [Medieval Latin *anarchia,* from Greek, from *anarchos* having no ruler, from *an-* + *archos* ruler — more at ARCH-] (1539)

☆ SYNONYMS
Analyze, dissect, break down mean to divide a complex whole into its parts or elements. ANALYZE suggests separating or distinguishing the component parts of something (as a substance, a process, a situation) so as to discover its true nature or inner relationships ⟨*analyzed* the problem of the trade deficit⟩. DISSECT suggests a searching analysis by laying bare parts or pieces for individual scrutiny ⟨commentators *dissected* every word of the President's statement⟩. BREAK DOWN implies a reducing to simpler parts or divisions ⟨*break down* the budget⟩.

1 a : absence of government **b :** a state of lawlessness or political disorder due to the absence of governmental authority **c :** a utopian society of individuals who enjoy complete freedom without government
2 a : absence or denial of any authority or established order **b :** absence of order **:** DISORDER ⟨not manicured plots but a wild *anarchy* of nature —Israel Shenker⟩
3 : ANARCHISM
ana·sar·ca \ˌa-nə-ˈsär-kə\ *noun* [New Latin, from *ana-* + Greek *sark-, sarx* flesh — more at SARCASM] (14th century)
: generalized edema with accumulation of serum in the connective tissue
— ana·sar·cous \-kəs\ *adjective*
An·a·sa·zi \ˌä-nə-ˈsä-zē\ *noun, plural* **Anasazi** [Navajo *anaasází,* literally, enemy ancestors] (1938)
: a prehistoric American Indian inhabitant of the canyons of northern Arizona and New Mexico and southwestern Colorado
an·as·tig·mat \a-ˈnas-tig-ˌmat, ˌa-nə-ˈstig-\ *noun* [German, back-formation from *anastigmatisch* anastigmatic] (1890)
: an anastigmatic lens
an·a·stig·mat·ic \ˌa-nə-(ˌ)stig-ˈma-tik, ˌa-ˌnas-tig-\ *adjective* [International Scientific Vocabulary] (1890)
: not astigmatic — used especially of lenses that are able to form approximately point images of object points
anas·to·mose \ə-ˈnas-tə-ˌmōz, -ˌmōs\ *verb* **-mosed; -mos·ing** [probably back-formation from *anastomosis*] (1697)
transitive verb
: to connect or join by anastomosis
intransitive verb
: to communicate by anastomosis
anas·to·mo·sis \ə-ˌnas-tə-ˈmō-səs\ *noun, plural* **-mo·ses** \-ˌsēz\ [Late Latin, from Greek *anastomōsis,* from *anastomoun* to provide with an outlet, from *ana-* + *stoma* mouth, opening — more at STOMACH] (1541)
1 : the union of parts or branches (as of streams, blood vessels, or leaf veins) so as to intercommunicate **:** INOSCULATION
2 : a product of anastomosis **:** NETWORK
— anas·to·mot·ic \-ˈmä-tik\ *adjective*
anas·tro·phe \ə-ˈnas-trə-(ˌ)fē\ *noun* [Medieval Latin, from Greek *anastrophē,* literally, turning back, from *anastrephein* to turn back, from *ana-* + *strephein* to turn] (circa 1550)
: inversion of the usual syntactical order of words for rhetorical effect — compare HYSTERON PROTERON
an·a·tase \ˈa-nə-ˌtās, -ˌtāz\ *noun* [French, from Greek *anatasis* extension, from *anateinein* to extend, from *ana-* + *teinein* to stretch — more at THIN] (circa 1828)
: a tetragonal form of titanium dioxide used especially as a white pigment
anath·e·ma \ə-ˈna-thə-mə\ *noun* [Late Latin *anathemat-, anathema,* from Greek, thing devoted to evil, curse, from *anatithenai* to set up, dedicate, from *ana-* + *tithenai* to place, set — more at DO] (1526)
1 a : one that is cursed by ecclesiastical authority **b :** someone or something intensely disliked or loathed — usually used as a predicate nominative ⟨this notion was *anathema* to most of his countrymen —S. J. Gould⟩
2 a : a ban or curse solemnly pronounced by ecclesiastical authority and accompanied by excommunication **b :** the denunciation of something as accursed **c :** a vigorous denunciation **:** CURSE ◇
anath·e·ma·tize \-ˌtīz\ *transitive verb* **-tized; -tiz·ing** (1566)
: to pronounce an anathema upon
synonym see EXECRATE
An·a·to·lian \ˌa-nə-ˈtō-lē-ən, -ˈtōl-yən\ *noun* (1590)
1 : a native or inhabitant of Anatolia and specifically of the western plateau lands of Turkey in Asia
2 : a branch of the Indo-European language family that includes a group of extinct languages of ancient Anatolia — see INDO-EUROPEAN LANGUAGES table
— Anatolian *adjective*
anat·o·mise *British variant of* ANATOMIZE
anat·o·mist \ə-ˈna-tə-mist\ *noun* (1543)
1 : a student of anatomy; *especially* **:** one skilled in dissection
2 : one who analyzes minutely and critically ⟨an *anatomist* of urban society⟩
anat·o·mize \-ˌmīz\ *transitive verb* **-mized; -miz·ing** (15th century)
1 : to cut in pieces in order to display or examine the structure and use of the parts **:** DISSECT
2 : ANALYZE
anat·o·my \ə-ˈna-tə-mē\ *noun, plural* **-mies** [Late Latin *anatomia* dissection, from Greek *anatomē,* from *anatemnein* to dissect, from *ana-* + *temnein* to cut] (14th century)
1 : a branch of morphology that deals with the structure of organisms
2 : a treatise on anatomic science or art
3 : the art of separating the parts of an organism in order to ascertain their position, relations, structure, and function **:** DISSECTION
4 *obsolete* **:** a body dissected or to be dissected
5 : structural makeup especially of an organism or any of its parts
6 : a separating or dividing into parts for detailed examination **:** ANALYSIS
7 a (1) **:** SKELETON (2) **:** MUMMY **b :** the human body
— an·a·tom·ic \ˌa-nə-ˈtä-mik\ *or* **an·a·tom·i·cal** \-mi-kəl\ *adjective*
— an·a·tom·i·cal·ly \-mi-k(ə-)lē\ *adverb*
anat·ro·pous \ə-ˈna-trə-pəs\ *adjective* (circa 1846)
: having or being an ovule inverted so that the micropyle is bent down to the funiculus to which the body of the ovule is united
-ance *noun suffix* [Middle English, from Old French, from Latin *-antia,* from *-ant-, -ans* -ant + *-ia* -y]
1 : action or process ⟨further*ance*⟩ **:** instance of an action or process ⟨perform*ance*⟩
2 : quality or state **:** instance of a quality or state ⟨protuber*ance*⟩
3 : amount or degree ⟨conduct*ance*⟩
an·ces·tor \ˈan-ˌses-tər *also* -səs-\ *noun* [Middle English *ancestre,* from Old French, from Latin *antecessor* predecessor, from *antecedere* to go before, from *ante-* + *cedere* to go — more at CEDE] (13th century)
1 a : one from whom a person is descended and who is usually more remote in the line of descent than a grandparent **b :** FOREFATHER 2
2 : FORERUNNER, PROTOTYPE
3 : a progenitor of a more recent or existing species or group
ancestor worship *noun* (1854)
: the custom of venerating deceased ancestors who are considered still a part of the family and whose spirits are believed to have the power to intervene in the affairs of the living
an·ces·tral \an-ˈses-trəl\ *adjective* (15th century)
: of, relating to, or inherited from an ancestor ⟨*ancestral* estates⟩
— an·ces·tral·ly \-trə-lē\ *adverb*
an·ces·tress \ˈan-ˌses-trəs\ *noun* (1580)
: a female ancestor
an·ces·try \ˈan-ˌses-trē\ *noun* (14th century)
1 : line of descent **:** LINEAGE; *especially* **:** honorable, noble, or aristocratic descent
2 : persons initiating or comprising a line of descent **:** ANCESTORS
An·chi·ses \an-ˈkī-(ˌ)sēz, aŋ-\ *noun* [Latin, from Greek *Anchisēs*]
: the father of Aeneas rescued by his son from the burning city of Troy
¹an·chor \ˈaŋ-kər\ *noun, often attributive* [Middle English *ancre,* from Old English *ancor,* from Latin *anchora,* from Greek *ankyra;* akin to Old English *anga* hook — more at ANGLE] (before 12th century)
1 : a device usually of metal attached to a ship or boat by a cable and cast overboard to hold it in a particular place by means of a fluke that digs into the bottom
2 : a reliable or principal support **:** MAINSTAY
3 : something that serves to hold an object firmly
4 : an object shaped like a ship's anchor
5 : an anchorman or anchorwoman
6 : the member of a team (as a relay team) that competes last
7 : a large business (as a department store) that attracts customers and other businesses to a shopping center or mall
— an·chor·less \-ləs\ *adjective*
— at anchor : being anchored

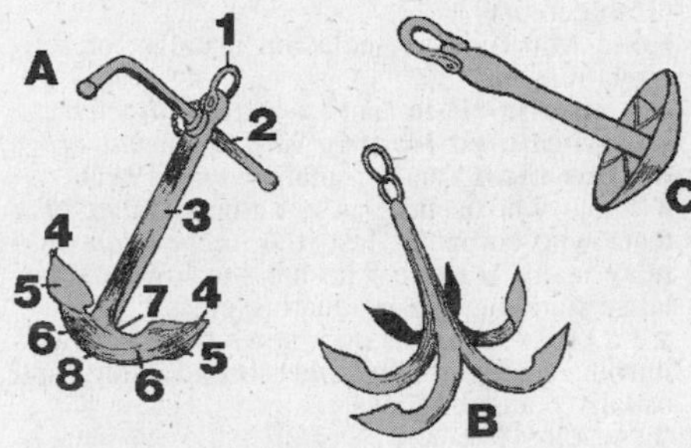

anchor 1: ***A*** **yachtsman's:** ***1*** **ring,** ***2*** **stock,** ***3*** **shank,** ***4*** **bill,** ***5*** **fluke,** ***6*** **arm,** ***7*** **throat,** ***8*** **crown;** ***B*** **grapnel;** ***C*** **mushroom**

²anchor *verb* **an·chored; an·chor·ing** \-k(ə-)riŋ\ (13th century)
transitive verb
1 : to hold in place in the water by an anchor
2 : to secure firmly **:** FIX
3 : to act or serve as an anchor for ⟨it is she who is *anchoring* the rebuilding campaign —G. D. Boone⟩ ⟨*anchoring* the evening news⟩
intransitive verb
1 : to cast anchor
2 : to become fixed
an·chor·age \ˈaŋ-k(ə-)rij\ *noun* (15th century)

◇ WORD HISTORY
anathema The word *anathema* is of Greek origin, a derivative of the verb *anatithenai,* which means "to set up, dedicate." In the Septuagint, an influential Greek translation of the Hebrew scriptures, *anathema* was used to mean "something offered to God," and could refer either to a revered object or an object representing destruction brought about in the name of the Lord. Since the objects symbolic of consecrated destruction were often the spoils of war stripped from the bodies of despised enemies, *anathema* came to signify something odious or accursed. In the New Testament, Paul uses the term in the sense of a curse and the forced exclusion of a person from the Christian community. In Medieval Latin *anathema* was used interchangeably with *excommunicatio* and was pronounced chiefly against unrepentant heretics. In English the word *anathema* has been and continues to be used in its ecclesiastical sense. By the 18th century, however, its use in a broadened sense for "the denunciation of anything as accursed" is attested. That secular use continues to be common, as does the transferred sense, "someone or something intensely disliked."

1 a : a place where vessels anchor **:** a place suitable for anchoring **b :** the act of anchoring **:** the condition of being anchored
2 : a means of securing **:** a source of reassurance ⟨this *anchorage* of Christian hope —T. O. Wedel⟩
3 : something that provides a secure hold
an·cho·ress \ˈaŋ-k(ə-)rəs\ *or* **an·cress** \-krəs\ *noun* [Middle English *ankeresse,* from *anker* hermit, from Old English *ancor,* from Old Irish *anchara,* from Late Latin *anachoreta*] (14th century)
: a woman who is an anchorite
an·cho·rite \ˈaŋ-kə-ˌrīt\ *also* **an·cho·ret** \-ˌret\ *noun* [Middle English, from Medieval Latin *anchorita,* alteration of Late Latin *anachoreta,* from Late Greek *anachōrētēs,* from Greek *anachōrein* to withdraw, from *ana-* + *chōrein* to make room, from *chōros* place] (15th century)
: one who lives in seclusion usually for religious reasons
— an·cho·rit·ic \ˌaŋ-kə-ˈri-tik\ *adjective*
— an·cho·rit·i·cal·ly \-ti-k(ə-)lē\ *adverb*
an·chor·man \ˈaŋ-kər-ˌman\ *noun* (1911)
1 : one who is last: as **a :** the member of a team who competes last ⟨the *anchorman* on a relay team⟩ **b :** one who has the lowest scholastic standing in a graduating class
2 : a broadcaster (as on a news program) who introduces reports by other broadcasters and usually reads the news
3 : MODERATOR 2c
an·chor·peo·ple \-ˌpē-pəl\ *noun plural* (1974)
: ANCHORPERSONS
an·chor·per·son \-ˌpər-sᵊn\ *noun* (1973)
: an anchorman or anchorwoman
an·chor·wom·an \-ˌwu̇-mən\ *noun* (1973)
: a woman who anchors a broadcast
an·cho·veta *also* **an·cho·vet·ta** \ˌan-chō-ˈve-tə\ *noun* [Spanish *anchoveta,* diminutive of *anchova*] (1940)
: a small anchovy (*Cetengraulis mysticetus*) of the Pacific coast of North America
an·cho·vy \ˈan-ˌchō-vē, an-ˈ\ *noun, plural* **-vies** *or* **-vy** [Spanish *anchova*] (1596)
: any of a family (Engraulidae) of small fishes resembling herrings; *especially* **:** a common Mediterranean fish (*Engraulis encrasicholus*) used especially in appetizers, as a garnish, and for making sauces and relishes
an·cien ré·gime \äⁿs-yaⁿ-rā-ˈzhēm\ *noun* [French, literally, old regime] (1794)
1 : the political and social system of France before the Revolution of 1789
2 : a system or mode no longer prevailing
¹an·cient \ˈān(t)-shənt, ˈāŋ(k)-shənt\ *adjective* [Middle English *ancien,* from Middle French, from (assumed) Vulgar Latin *anteanus,* from Latin *ante* before — more at ANTE-] (14th century)
1 : having had an existence of many years
2 : of or relating to a remote period, to a time early in history, or to those living in such a period or time; *especially* **:** of or relating to the historical period beginning with the earliest known civilizations and extending to the fall of the western Roman Empire in A.D. 476
3 : having the qualities of age or long existence: **a :** VENERABLE **b :** OLD-FASHIONED, ANTIQUE
synonym see OLD
— an·cient·ness *noun*
²ancient *noun* (1502)
1 : an aged living being ⟨a penniless *ancient*⟩
2 : one who lived in ancient times: **a** *plural* **:** the civilized people of antiquity; *especially* **:** those of the classical nations **b :** one of the classical authors ⟨Plutarch and other *ancients*⟩
3 : an ancient coin
³ancient *noun* [alteration of *ensign*] (1554)
1 *archaic* **:** ENSIGN, STANDARD, FLAG
2 *obsolete* **:** the bearer of an ensign
ancient history *noun* (1595)
1 : the history of ancient times
2 : knowledge or information that is widespread and has lost its initial freshness or importance **:** common knowledge
an·cient·ly *adverb* (1502)
: in ancient times **:** long ago
an·cient·ry \-shən-trē\ *noun* (1580)
: ANTIQUITY, ANCIENTNESS
an·cil·la \an-ˈsi-lə\ *noun, plural* **-lae** \-(ˌ)lē\ [Latin, female servant] (1902)
: an aid to achieving or mastering something difficult
an·cil·lary \ˈan(t)-sə-ˌler-ē, *especially British* an-ˈsi-lə-rē\ *adjective* (1667)
1 : SUBORDINATE, SUBSIDIARY ⟨the main factory and its *ancillary* plants⟩
2 : AUXILIARY, SUPPLEMENTARY ⟨the need for *ancillary* evidence⟩
— ancillary *noun*
an·con \ˈaŋ-ˌkän\ *noun, plural* **an·co·nes** \aŋ-ˈkō-nēz\ [Latin, from Greek *ankōn* elbow; akin to Old English *anga* hook — more at ANGLE] (circa 1706)
: a bracket, elbow, or console used as an architectural support
-ancy *noun suffix* [Latin *-antia* — more at -ANCE]
: quality or state ⟨piqu*ancy*⟩
an·cy·lo·sto·mi·a·sis \ˌaŋ-ki-ˌlō-stə-ˈmī-ə-səs, ˌan(t)-sə-\ *noun, plural* **-a·ses** \-ˌsēz\ [New Latin, from *Ancylostoma,* genus of hookworms, from Greek *ankylos* hooked (akin to Old English *anga* hook) + *stoma* mouth — more at ANGLE, STOMACH] (1887)
: infestation with or disease caused by hookworms; *especially* **:** a lethargic anemic state in humans due to blood loss from hookworms feeding in the small intestine
and \ən(d), (ˈ)an(d), *usually* ᵊn(d) *after* t, d, s *or* z, *often* ᵊm *after* p *or* b, *sometimes* ᵊŋ *after* k *or* g\ *conjunction* [Middle English, from Old English; akin to Old High German *unti* and] (before 12th century)
1 — used as a function word to indicate connection or addition especially of items within the same class or type; used to join sentence elements of the same grammatical rank or function
2 a — used as a function word to express logical modification, consequence, antithesis, or supplementary explanation **b —** used as a function word to join one finite verb (as *go, come, try*) to another so that together they are logically equivalent to an infinitive of purpose ⟨come *and* see me⟩
3 *obsolete* **:** IF
4 — used in logic to form a conjunction
— and so forth \ən-ˈsō-ˌfōrth, -ˌfȯrth\ **1 :** and others or more of the same or similar kind **2 :** further in the same or similar manner **3 :** and the rest **4 :** and other things
— and so on \ən-ˈsō-ˌȯn, -ˌän\ **:** and so forth
AND \ˈand\ *noun* (1949)
: a logical operator that requires both of two inputs to be present or two conditions to be met for an output to be made or a statement to be executed
An·da·lu·sian \ˌan-də-ˈlü-zhən\ *noun* [*Andalusia,* Spain] (1966)
: any of a breed of horses of Spanish origin that have a high-stepping gait
an·da·lu·site \ˌan-də-ˈlü-ˌsīt\ *noun* [French *andalousite,* from *Andalousie* Andalusia, region in Spain] (circa 1828)
: a mineral consisting of a silicate of aluminum usually in thick orthorhombic prisms of various colors
¹an·dan·te \än-ˈdän-(ˌ)tā, -ˈdän-tē; an-ˈdan-tē\ *adverb or adjective* [Italian, literally, going, present participle of *andare* to go] (1724)
: moderately slow — usually used as a direction in music
²andante *noun* (1784)
: a musical composition or movement in andante tempo
¹an·dan·ti·no \ˌän-ˌdän-ˈtē-(ˌ)nō\ *adverb or adjective* [Italian, diminutive of *andante*] (1819)
: slightly faster than andante — used as a direction in music
²andantino *noun, plural* **-nos** (1845)
: a musical composition or movement in andantino tempo
an·des·ite \ˈan-di-ˌzīt\ *noun* [German *Andesit,* from *Andes*] (1850)
: an extrusive usually dark grayish rock consisting essentially of oligoclase or feldspar
— an·des·it·ic \ˌan-di-ˈzi-tik\ *adjective*
and how *adverb* (1865)
— used to emphasize the preceding idea ⟨having a great time—*and how!*⟩
and·iron \ˈan-ˌdī(-ə)rn\ *noun* [Middle English *aundiren,* modification of Middle French *andier*] (14th century)
: either of a pair of metal supports for firewood used on a hearth and made of a horizontal bar mounted on short legs with usually a vertical shaft surmounting the front end
and/or \ˈan-ˈdȯr\ *conjunction* (1853)
— used as a function word to indicate that two words or expressions are to be taken together or individually ⟨language comprehension *and/or* production —David Crystal⟩
an·dou·ille \än-ˈdü-ē, ˈän-dü-ē\ *noun* [French, from Old French *andoille,* from (assumed) Vulgar Latin *inductilia,* neuter plural of *inductilis* made by insertion, from Latin *inductus,* past participle of *inducere* to insert, bring in — more at INDUCE] (1605)
: a highly spiced smoked pork sausage
an·douill·ette \ˌän-dü-ˈyet\ *noun* [French, diminutive of *andouille*] (1611)
: a fresh pork sausage made with tripe or chitterlings
andr- *or* **andro-** *combining form* [Latin, from Greek, from *andr-, anēr;* akin to Oscan *nerman,* Sanskrit *nar-,* Old Irish *nert* strength]
1 : male human being ⟨*andro*centric⟩
2 : male ⟨*andro*ecium⟩
an·dra·dite \an-ˈdrä-ˌdīt, ˈan-drə-ˌdīt\ *noun* [José B. de *Andrada* e Silva (died 1838) Brazilian geologist] (1868)
: a garnet of any of various colors ranging from yellow and green to brown and black
an·dro·cen·tric \ˌan-drə-ˈsen-trik\ *adjective* (1903)
: dominated by or emphasizing masculine interests or a masculine point of view
An·dro·cles \ˈan-drə-ˌklēz\ *noun* [Latin, from Greek *Androklēs*]
: a fabled Roman slave spared in the arena by a lion from whose foot he had years before extracted a thorn
an·droe·ci·um \an-ˈdrē-shē-əm, -sē-əm\ *noun, plural* **-cia** \-shē-ə, -sē-ə\ [New Latin, from *andr-* + Greek *oikion,* diminutive of *oikos* house — more at VICINITY] (circa 1839)
: the aggregate of stamens in the flower of a seed plant
an·dro·gen \ˈan-drə-jən\ *noun* [International Scientific Vocabulary] (1936)
: a male sex hormone (as testosterone)
— an·dro·gen·ic \ˌan-drə-ˈje-nik\ *adjective*
an·dro·gen·e·sis \ˌan-drō-ˈje-nə-səs\ *noun* [New Latin] (circa 1900)
: development of an embryo containing only paternal chromosomes due to failure of the egg to participate in fertilization
— an·dro·ge·net·ic \-jə-ˈne-tik\ *adjective*
an·dro·gyne \ˈan-drə-ˌjīn\ *noun* [Middle French, from Latin *androgynus*] (1552)
: one that is androgynous
an·drog·y·nous \an-ˈdrä-jə-nəs\ *adjective* [Latin *androgynus* hermaphrodite, from Greek *androgynos,* from *andr-* + *gynē* woman — more at QUEEN] (1651)
1 : having the characteristics or nature of both male and female
2 a : neither specifically feminine nor masculine ⟨the *androgynous* pronoun *them*⟩ **b :** suitable to or for either sex ⟨*androgynous* clothing⟩

3 : having traditional male and female roles obscured or reversed ⟨an *androgynous* marriage⟩
— an·drog·y·ny \-nē\ *noun*
an·droid \'an-ˌdrȯid\ *noun* [Late Greek *androeidēs* manlike, from Greek *andr-* + *-oeidēs* -oid] (circa 1751)
: a mobile robot usually with a human form
An·drom·a·che \an-'drä-mə-(ˌ)kē\ *noun* [Latin, from Greek *Andromachē*]
: the wife of Hector
an·drom·eda \an-'drä-mə-də\ *noun* [New Latin *Andromeda,* genus name, from Latin] (circa 1760)
: any of several evergreen shrubs (genera *Pieris* and *Andromeda*) of the heath family; *especially* **:** JAPANESE ANDROMEDA
An·drom·e·da \an-'drä-mə-də\ *noun* [Latin, from Greek *Andromedē*]
1 : an Ethiopian princess of Greek mythology rescued from a monster by her future husband Perseus
2 [Latin (genitive *Andromedae*)] **:** a northern constellation directly south of Cassiopeia between Pegasus and Perseus
an·dros·ter·one \an-'dräs-tə-ˌrōn\ *noun* [International Scientific Vocabulary] (1934)
: an androgenic hormone that is a hydroxy ketone $C_{19}H_{30}O_2$ found in human male and female urine
ane \'ān\ *adjective or noun or pronoun* (before 12th century)
chiefly Scottish **:** ONE
-ane *noun suffix* [International Scientific Vocabulary *-an, -ane,* alteration of *-ene, -ine,* & *-one*]
1 : [3]-AN 1 ⟨fur*ane*⟩
2 : saturated hydrocarbon ⟨alk*ane*⟩ ⟨meth*ane*⟩
an·ec·dot·age \'a-nik-ˌdō-tij\ *noun* (1823)
1 : the telling of anecdotes; *also* **:** ANECDOTES
2 : garrulous old age
an·ec·dot·al \ˌa-nik-'dō-tᵊl\ *adjective* (1836)
1 a : of, relating to, or consisting of anecdotes ⟨an *anecdotal* biography⟩ **b :** ANECDOTIC 2 ⟨my *anecdotal* uncle⟩
2 : based on or consisting of reports or observations of usually unscientific observers ⟨*anecdotal* evidence⟩
3 : of, relating to, or being the depiction of a scene suggesting a story ⟨*anecdotal* painting⟩ ⟨*anecdotal* detail⟩
— an·ec·dot·al·ly \-tᵊl-ē\ *adverb*
an·ec·dot·al·ist \ˌa-nik-'dō-tᵊl-ist\ *or* **an·ec·dot·ist** \'a-nik-ˌdō-tist\ *noun* (1837)
: one who is given to or is skilled in telling anecdotes
— an·ec·dot·al·ism \-tᵊl-ˌi-zəm\ *noun*
an·ec·dote \'a-nik-ˌdōt\ *noun, plural* **anec·dotes** *also* **an·ec·dota** \ˌa-nik-'dō-tə\ [French, from Greek *anekdota* unpublished items, from neuter plural of *anekdotos* unpublished, from *a-* + *ekdidonai* to publish, from *ex* out + *didonai* to give — more at EX-, DATE] (circa 1721)
: a usually short narrative of an interesting, amusing, or biographical incident ◆
an·ec·dot·ic \ˌa-nik-'dä-tik\ *or* **an·ec·dot·i·cal** \-'dä-ti-kəl\ *adjective* (circa 1744)
1 : ANECDOTAL 1a
2 : given to or skilled in telling anecdotes
— an·ec·dot·i·cal·ly \-'dä-ti-k(ə-)lē\ *adverb*
an·echo·ic \ˌa-ni-'kō-ik\ *adjective* (1946)
: free from echoes and reverberations ⟨an *anechoic* chamber⟩
an·elas·tic \ˌa-nᵊl-'as-tik\ *adjective* (1947)
: relating to the property of a substance in which there is no definite relation between stress and strain
— an·elas·tic·i·ty \-nᵊl-ˌas-'ti-sə-tē, -'tis-tē\ *noun*
anem- *or* **anemo-** *combining form* [Greek, from *anemos* — more at ANIMATE]
: wind ⟨*anemo*meter⟩
ane·mia \ə-'nē-mē-ə\ *noun* [New Latin, from Greek *anaimia* bloodlessness, from *a-* + *-aimia* -emia] (1824)
1 a : a condition in which the blood is deficient in red blood cells, in hemoglobin, or in total volume **b :** ISCHEMIA
2 : lack of vitality
ane·mic \ə-'nē-mik\ *adjective* (1858)
1 : relating to or affected with anemia
2 a : lacking force, vitality, or spirit ⟨an *anemic* rendition of the song⟩ ⟨*anemic* efforts at enforcement⟩ **b :** lacking interest or savor **:** INSIPID ⟨*anemic* wines⟩ **c :** lacking in substance or quantity ⟨*anemic* returns on an investment⟩ ⟨*anemic* attendance⟩
— ane·mi·cal·ly \-mi-k(ə-)lē\ *adverb*
anemo·graph \ə-'ne-mə-ˌgraf\ *noun* (1865)
: a recording anemometer
an·e·mom·e·ter \ˌa-nə-'mä-mə-tər\ *noun* (circa 1751)
: an instrument for measuring and indicating the force or speed of the wind
an·e·mom·e·try \ˌa-nə-'mä-mə-trē\ *noun* (1847)
: the process of ascertaining the force, speed, and direction of wind or an airflow
anem·o·ne \ə-'ne-mə-nē\ *noun* [Latin, from Greek *anemōnē*] (1548)
1 : any of a large genus (*Anemone*) of the buttercup family having lobed or divided leaves and showy flowers without petals but with conspicuous often colored sepals — called also *windflower*
2 : SEA ANEMONE
an·e·moph·i·lous \ˌa-nə-'mä-fə-ləs\ *adjective* (1874)
: pollinated by wind
an·en·ceph·a·ly \ˌan-(ˌ)en-'se-fə-lē\ *noun, plural* **-lies** [[2]*a-* + *encephal-* + [2]-y] (circa 1889)
: congenital absence of all or a major part of the brain
— an·en·ce·phal·ic \-ˌen(t)-sə-'fa-lik\ *adjective*
anent \ə-'nent\ *preposition* [Middle English *onevent, anent,* from Old English *on efen* alongside, from *on* + *efen* even] (13th century)
: ABOUT, CONCERNING
an·er·oid \'a-nə-ˌrȯid\ *adjective* [French *anéroïde,* from Greek *a-* + Late Greek *nēron* water, from Greek, neuter of *nearos, nēros* fresh; akin to Greek *neos* new — more at NEW] (circa 1848)
: using no liquid; *specifically* **:** operating by the effect of outside air pressure on a diaphragm forming one wall of an evacuated container ⟨*aneroid* barometer⟩
an·es·the·sia \ˌa-nəs-'thē-zhə\ *noun* [New Latin, from Greek *anaisthēsia* insensibility, from *a-* + *aisthēsis* perception, from *aisthanesthai* to perceive — more at AUDIBLE] (circa 1721)
: loss of sensation with or without loss of consciousness
an·es·the·si·ol·o·gist \-ˌthē-zē-'ä-lə-jist\ *noun* (1942)
: ANESTHETIST; *specifically* **:** a physician specializing in anesthesiology
an·es·the·si·ol·o·gy \-jē\ *noun* (circa 1914)
: a branch of medical science dealing with anesthesia and anesthetics
[1]an·es·thet·ic \ˌa-nəs-'the-tik\ *adjective* (1846)
1 : of, relating to, or capable of producing anesthesia
2 : lacking awareness or sensitivity ⟨unmoved and quite *anesthetic* to his presence —S. J. Perelman⟩
— an·es·thet·i·cal·ly \-ti-k(ə-)lē\ *adverb*
[2]anesthetic *noun* (1848)
1 : a substance that produces anesthesia
2 : something that brings relief **:** PALLIATIVE
anes·the·tist \ə-'nes-thə-tist, *British* -'nēs-\ *noun* (1882)
: one who administers anesthetics
anes·the·tize \-thə-ˌtīz\ *transitive verb* **-tized; -tiz·ing** (1848)
: to subject to anesthesia
an·es·trous \(ˌ)an-'es-trəs\ *adjective* (circa 1909)
1 : not exhibiting estrus
2 : of or relating to anestrus
an·es·trus \-trəs\ *noun* [New Latin] (1927)
: the period of sexual quiescence between two periods of sexual activity in cyclically breeding mammals
an·eu·ploid \'an-yu̇-ˌplȯid\ *adjective* (1926)
: having or being a chromosome number that is not an exact multiple of the usually haploid number — compare EUPLOID
— aneuploid *noun*
— an·eu·ploi·dy \-ˌplȯi-dē\ *noun*
an·eu·rysm *also* **an·eu·rism** \'an-yə-ˌri-zəm\ *noun* [Greek *aneurysma,* from *aneurynein* to dilate, from *ana-* + *eurynein* to stretch, from *eurys* wide — more at EURY-] (15th century)
: an abnormal blood-filled dilatation of a blood vessel and especially an artery resulting from disease of the vessel wall
— an·eu·rys·mal \ˌan-yə-'riz-məl\ *adjective*
anew \ə-'nü, -'nyü\ *adverb* [Middle English *of newe,* from Old English *of nīwe,* from *of* + *nīwe* new] (before 12th century)
1 : for an additional time **:** AFRESH
2 : in a new or different form
an·frac·tu·os·i·ty \(ˌ)an-ˌfrak-chə-'wä-sə-tē, -shə-\ *noun, plural* **-ties** (1596)
1 : the quality or state of being anfractuous
2 : a winding channel or course; *especially* **:** an intricate path or process (as of the mind)
an·frac·tu·ous \an-'frak-chə-wəs, -shə-\ *adjective* [French *anfractueux,* from Late Latin *anfractuosus,* from Latin *anfractus* coil, bend, from *an-* (from *ambi-* around) + *-fractus,* from *frangere* to break — more at AMBI-, BREAK] (1621)
: full of windings and intricate turnings **:** TORTUOUS
an·gel \'ān-jəl\ *noun* [Middle English, from Old English *engel* & Old French *angele;* both from Late Latin *angelus,* from Greek *angelos,* literally, messenger] (before 12th century)
1 a : a spiritual being superior to man in power and intelligence; *especially* **:** one in the lowest rank in the celestial hierarchy **b** *plural* **:** an order of angels — see CELESTIAL HIERARCHY

◇ WORD HISTORY
anecdote Procopius, a 6th century historian and official at the court of the eastern Roman emperor Justinian, published two important historical works in his lifetime. A third work, however, did not begin to circulate until after his death. Entitled *Anekdota,* "unpublished things," it contained bitter attacks on the emperor Justinian and his wife, Theodora, as well as on many other noted officials in Constantinople. Some of these attacks are in the form of juicy bits of scandalous gossip. Procopius's work is often referred to as *Historia Arcana* or *Secret History.* It was the original title that became a byword, however. Procopius's title first appeared in English in its Medieval Latin form, *anecdota,* and then as *anecdotes,* borrowed from French. When first used in the latter half of the 17th century, these plurals meant "unpublished secret history." By the 18th century, the word was being used in the singular, *anecdote,* and in the sense familiar to us today.

2 : an attendant spirit or guardian
3 : a white-robed winged figure of human form in fine art
4 : MESSENGER, HARBINGER ⟨*angel* of death⟩
5 : a person like an angel (as in looks or behavior)
6 *Christian Science* **:** inspiration from God
7 : one (as a backer of a theatrical venture) who aids or supports with money or influence
8 : ANGELFISH
— an·gel·ic \an-ˈje-lik\ *or* **an·gel·i·cal** \-li-kəl\ *adjective*
— an·gel·i·cal·ly \-li-k(ə-)lē\ *adverb*

angel dust *noun* (1973)
: PHENCYCLIDINE

An·ge·le·no \ˌan-jə-ˈlē-(ˌ)nō\ *noun, plural* **-nos** [American Spanish *angeleño,* from Los *Angeles,* Calif.] (1888)
: a native or resident of Los Angeles, Calif.

an·gel·fish \ˈān-jəl-ˌfish\ *noun* (1668)
1 : any of several compressed bright-colored bony fishes (family Pomacanthidae) of warm seas
2 : SCALARE

angel food cake *noun* (1920)
: a usually white sponge cake made of flour, sugar, and whites of eggs

angel–hair pasta \ˈan-jəl-ˌher-, -ˌhar-\ *noun* (1981)
: pasta smaller in diameter than vermicelli

an·gel·i·ca \an-ˈje-li-kə\ *noun* [New Latin, genus name, from Medieval Latin, from Late Latin, feminine of *angelicus* angelic, from Late Greek *angelikos,* from Greek, of a messenger, from *angelos*] (1527)
1 a : any of a genus (*Angelica*) of herbs of the carrot family; *especially* **:** a Eurasian biennial or perennial (*A. archangelica*) whose roots and fruit yield a flavoring oil **b :** a confection prepared from angelica
2 *capitalized* **:** a sweet fortified wine produced in California

angelica tree *noun* (1785)
: HERCULES'-CLUB 1

an·gel·ol·o·gy \ˌān-jə-ˈlä-lə-jē\ *noun, often capitalized* (circa 1828)
: the theological doctrine of angels or its study
— an·gel·ol·o·gist \-jist\ *noun*

An·ge·lus \ˈan-jə-ləs\ *noun* [Medieval Latin, from Late Latin, angel; from the first word of the opening versicle] (1658)
1 : a devotion of the Western church that commemorates the Incarnation and is said in the morning, at noon, and in the evening
2 : a bell announcing the time for the Angelus

[1]anger *verb* **an·gered; an·ger·ing** \-g(ə-)riŋ\ (13th century)
transitive verb
: to make angry
intransitive verb
: to become angry

[2]an·ger \ˈaŋ-gər\ *noun* [Middle English, affliction, anger, from Old Norse *angr* grief; akin to Old English *enge* narrow, Latin *angere* to strangle, Greek *anchein*] (14th century)
1 : a strong feeling of displeasure and usually of antagonism
2 : RAGE 2 ☆
— an·ger·less \-ləs\ *adjective*

An·ge·vin \ˈan-jə-vən\ *adjective* [French, from Old French, from Medieval Latin *andegavinus,* from *Andegavia* Anjou] (1769)
: of, relating to, or characteristic of Anjou or the Plantagenets
— Angevin *noun*

angi- *or* **angio-** *combining form* [New Latin, from Greek *angei-, angeio-,* from *angeion* vessel, blood vessel, diminutive of *angos* vessel]
1 : blood or lymph vessel **:** blood vessels and ⟨*angi*oma⟩ ⟨*angio*cardiography⟩
2 : pericarp ⟨*angio*sperm⟩

an·gi·na \an-ˈjī-nə, ˈan-jə-\ *noun* [Latin, throat inflammation, from Greek *anchonē* strangling, from *anchein* to strangle] (1578)
: a disease marked by spasmodic attacks of intense suffocative pain: as **a :** a severe inflammatory or ulcerated condition of the mouth or throat **b :** ANGINA PECTORIS
— an·gi·nal \an-ˈjī-nᵊl, ˈan-jə-\ *adjective*

angina pec·to·ris \-ˈpek-t(ə-)rəs\ *noun* [New Latin, literally, angina of the chest] (1744)
: a disease marked by brief paroxysmal attacks of chest pain precipitated by deficient oxygenation of the heart muscles

an·gio·car·di·og·ra·phy \ˌan-jē-ō-ˌkär-dē-ˈä-grə-fē\ *noun* (1938)
: the roentgenographic visualization of the heart and its blood vessels after injection of a radiopaque substance
— an·gio·car·dio·graph·ic \-dē-ə-ˈgraf-ik\ *adjective*

an·gio·gen·e·sis \ˌan-jē-ō-ˈje-nə-səs\ *noun* [New Latin] (1899)
: the formation and differentiation of blood vessels
— an·gio·gen·ic \-ˈje-nik\ *adjective*

an·gio·gram \ˈan-jē-ə-ˌgram\ *noun* (1933)
: a roentgenogram made by angiography

an·gi·og·ra·phy \ˌan-jē-ˈä-grə-fē\ *noun* (1933)
: the roentgenographic visualization of the blood vessels after injection of a radiopaque substance
— an·gio·graph·ic \ˌan-jē-ə-ˈgra-fik\ *adjective*

an·gi·o·ma \ˌan-jē-ˈō-mə\ *noun* [New Latin] (1871)
: a tumor composed chiefly of blood vessels or lymph vessels
— an·gi·o·ma·tous \-mə-təs\ *adjective*

an·gio·plas·ty \ˈan-jē-ə-ˌplas-tē\ *noun, plural* **-ties** (circa 1919)
: surgical repair of a blood vessel; *especially* **:** BALLOON ANGIOPLASTY

an·gio·sperm \ˈan-jē-ə-ˌspərm\ *noun* [ultimately from New Latin *angi-* + Greek *sperma* seed — more at SPERM] (circa 1828)
: any of a class (Angiospermae) or division (Magnoliophyta) of vascular plants (as magnolias, grasses, oaks, roses, and daisies) that have the ovules and seeds enclosed in an ovary, form the embryo and endosperm by double fertilization, and typically have each flower surrounded by a perianth composed of two sets of floral envelopes comprising the calyx and corolla — called also *flowering plant*
— an·gio·sper·mous \ˌan-jē-ə-ˈspər-məs\ *adjective*

an·gio·ten·sin \ˌan-jē-ō-ˈten(t)-sən\ *noun* [*angi-* + hyper*tension* + *-in*] (circa 1961)
: either of two forms of a kinin of which one has marked vasoconstrictive action; *also* **:** a synthetic amide derivative of the physiologically active form used to treat some forms of hypotension

angiotensin converting enzyme *noun* (1970)
: a proteolytic enzyme that converts the physiologically inactive form of angiotensin to the active vasoconstrictive form

[1]an·gle \ˈaŋ-gəl\ *noun* [Middle English, from Middle French, from Latin *angulus*] (14th century)
1 : a corner whether constituting a projecting part or a partially enclosed space ⟨they sheltered in an *angle* of the building⟩
2 a : the figure formed by two lines extending from the same point; *also* **:** DIHEDRAL ANGLE **b :** a measure of an angle or of the amount of turning necessary to bring one line or plane into coincidence with or parallel to another
3 a : the precise viewpoint from which something is observed or considered; *also* **:** the aspect seen from such an angle **b** (1) **:** a special approach, point of attack, or technique for accomplishing an objective (2) **:** an often improper or illicit method of obtaining advantage ⟨he always had an *angle* to beat the other fellow⟩
4 : a sharply divergent course ⟨the road went off at an *angle*⟩
5 : a position to the side of an opponent in football from which a player may block his opponent more effectively or without penalty — usually used in the phrases *get an angle* or *have an angle*
— an·gled \-gəld\ *adjective*

[2]angle *verb* **an·gled; an·gling** \-g(ə-)liŋ\ (1741)
intransitive verb
: to turn or proceed at an angle
transitive verb
1 : to turn, move, or direct at an angle
2 : to present (as a news story) from a particular or prejudiced point of view **:** SLANT

[3]angle *intransitive verb* **an·gled; an·gling** \-g(ə-)liŋ\ [Middle English *angelen,* from *angel* fishhook, from Old English, from *anga* hook; akin to Old High German *ango* hook, Latin *uncus,* Greek *onkos* barbed hook, *ankos* glen] (15th century)
1 : to fish with a hook
2 : to use artful means to attain an objective ⟨*angled* for an invitation⟩

An·gle \ˈaŋ-gəl\ *noun* [Latin *Angli,* plural, of Germanic origin; akin to Old English *Engle* Angles] (before 12th century)
: a member of a Germanic people that invaded England along with the Saxons and Jutes in the 5th century A.D. and merged with them to form the Anglo-Saxon peoples

angle bracket *noun* (circa 1956)
: BRACKET 3b

angle iron *noun* (circa 1853)
1 : an iron cleat for joining parts of a structure at an angle
2 : a piece of structural steel rolled with an L-shaped section

angle of attack (1908)
: the acute angle between the direction of the relative wind and the chord of an airfoil

angle of depression (1790)
: the angle formed by the line of sight and the horizontal plane for an object below the horizontal

angle of elevation (circa 1737)
: the angle formed by the line of sight and the horizontal plane for an object above the horizontal

angle of incidence (1628)
: the angle that a line (as a ray of light) falling on a surface or interface makes with the normal drawn at the point of incidence — see CRITICAL ANGLE illustration

angle of reflection (1638)
: the angle between a reflected ray and the normal drawn at the point of incidence to a reflecting surface

angle of refraction (circa 1737)
: the angle between a refracted ray and the normal drawn at the point of incidence to the interface at which refraction occurs

☆ SYNONYMS

Anger, ire, rage, fury, indignation, wrath mean an intense emotional state induced by displeasure. ANGER, the most general term, names the reaction but in itself conveys nothing about intensity or justification or manifestation of the emotional state ⟨tried to hide his *anger*⟩. IRE, more frequent in literary contexts, may suggest greater intensity than *anger,* often with an evident display of feeling ⟨cheeks flushed dark with *ire*⟩. RAGE suggests loss of self-control from violence of emotion ⟨screaming with *rage*⟩. FURY is overmastering destructive rage that can verge on madness ⟨in her *fury* she accused everyone around her of betrayal⟩. INDIGNATION stresses righteous anger at what one considers unfair, mean, or shameful ⟨a refusal to listen that caused general *indignation*⟩. WRATH is likely to suggest a desire or intent to revenge or punish ⟨rose in his *wrath* and struck his tormentor to the floor⟩.

an·gler \'aŋ-glər\ *noun* (15th century)
1 : one that angles
2 : ANGLERFISH

an·gler·fish \-,fish\ *noun* (circa 1889)
: any of several pediculate fishes (as the goosefishes); *especially* **:** MONKFISH

angle shot *noun* (circa 1922)
: a picture taken with the camera pointed at an angle from the horizontal

an·gle·site \'aŋ-gəl-,sīt, -glə-\ *noun* [French *anglésite,* from *Anglesey* Island, Wales] (circa 1841)
: a mineral consisting of lead sulfate formed by the oxidation of galena

an·gle·worm \'aŋ-gəl-,wərm\ *noun* (1832)
: EARTHWORM

An·gli·an \'aŋ-glē-ən\ *noun* (1726)
1 : a member of the Angles
2 : the Old English dialects of Mercia and Northumbria
— Anglian *adjective*

An·gli·can \'aŋ-gli-kən\ *adjective* [Medieval Latin *anglicanus,* from *anglicus* English, from Latin *Angli* Angles] (1635)
1 : of or relating to the established episcopal Church of England and churches of similar faith and order in communion with it
2 : of or relating to England or the English nation
— Anglican *noun*
— An·gli·can·ism \-kə-,ni-zəm\ *noun*

an·gli·ce \'aŋ-glə-(,)sē\ *adverb, often capitalized* [Medieval Latin, adverb of *anglicus*] (1602)
: in English; *especially* **:** in readily understood English ⟨the city of Napoli, *anglice* Naples⟩

an·gli·cise *often capitalized, variant of* ANGLICIZE

an·gli·cism \'aŋ-glə-,si-zəm\ *noun, often capitalized* [Medieval Latin *anglicus* English] (1642)
1 : a characteristic feature of English occurring in another language
2 : adherence or attachment to English customs or ideas

An·gli·cist \'aŋ-glə-sist\ *noun* (1930)
: a specialist in English linguistics

an·gli·cize \'aŋ-glə-,sīz\ *transitive verb* **-cized; -ciz·ing** *often capitalized* (1710)
1 : to make English in quality or characteristics
2 : to adapt (a foreign word or phrase) to English usage; *especially* **:** to borrow into English without alteration of form or spelling and with or without change in pronunciation
— an·gli·ci·za·tion \,aŋ-glə-sə-'zā-shən\ *noun, often capitalized*

an·gling \'aŋ-gliŋ\ *noun* (15th century)
: the action of one who angles; *especially* **:** the action or sport of fishing with hook and line

An·glist \'aŋ-glist\ *noun* (1888)
: ANGLICIST

An·glo \'aŋ-(,)glō\ *noun, plural* **Anglos** [in sense 2, from American Spanish, short for Spanish *angloamericano* Anglo-American] (1800)
1 : ANGLO-AMERICAN
2 : a white inhabitant of the U.S. of non-Hispanic descent
— Anglo *adjective*

An·glo- *combining form* [New Latin, from Late Latin *Angli*]
1 \'aŋ-(,)glō, -glə\ **:** English ⟨*Anglo*-Norman⟩
2 \-(,)glō\ **:** English and ⟨*Anglo*-Japanese⟩

An·glo–Amer·i·can \,aŋ-glō-ə-'mer-ə-kən\ *noun* (circa 1782)
: a North American whose native language is English; *especially* **:** an inhabitant of the U.S. of English origin or descent
— Anglo–American *adjective*

An·glo–Cath·o·lic \-'kath-lik, -'ka-thə-\ *adjective* (1838)
: of or relating to a High Church movement in Anglicanism emphasizing its continuity with historic Catholicism and fostering Catholic dogmatic and liturgical traditions
— Anglo–Catholic *noun*
— An·glo–Ca·thol·i·cism \-kə-'thä-lə-,si-zəm\ *noun*

An·glo–French \-'french\ *noun* (circa 1884)
: the French language used in medieval England

An·glo·ma·nia \-'mā-nē-ə, -nyə\ *noun* (1787)
: an absorbing or pervasive interest in England or things English

An·glo–Nor·man \-'nȯr-mən\ *noun* (1735)
1 : any of the Normans living in England after the Norman conquest of 1066
2 : the form of Anglo-French used by Anglo-Normans
— Anglo–Norman *adjective*

An·glo·phile \'aŋ-glə-,fīl\ *also* **An·glo·phil** \-,fil\ *noun* [French, from *anglo-* + *-phile*] (1883)
: one who greatly admires or favors England and things English
— Anglophile *or* **An·glo·phil·ic** \,aŋ-glə-'fi-lik\ *adjective*

An·glo·phil·ia \,aŋ-glə-'fi-lē-ə\ *noun* (1896)
: unusual admiration or partiality for England, English ways, or things English
— An·glo·phil·i·ac \-lē-,ak\ *adjective*

An·glo·phobe \'aŋ-glə-,fōb\ *noun* [probably from French, from *anglo-* + *-phobe*] (1866)
: one who is averse to or dislikes England and things English
— An·glo·pho·bia \,aŋ-glə-'fō-bē-ə\ *noun*
— An·glo·pho·bic \-bik\ *adjective*

an·glo·phone \'aŋ-glə-,fōn\ *adjective, often capitalized* (1965)
: consisting of or belonging to an English-speaking population especially in a country where two or more languages are spoken
— Anglophone *noun*

An·glo–Sax·on \,aŋ-glō-'sak-sən\ *noun* [New Latin *Anglo-Saxones,* plural, alteration of Medieval Latin *Angli Saxones,* from Latin *Angli* Angles + Late Latin *Saxones* Saxons] (before 12th century)
1 : a member of the Germanic peoples conquering England in the 5th century A.D. and forming the ruling class until the Norman conquest — compare ANGLE, JUTE, SAXON
2 a : ENGLISHMAN; *specifically* **:** a person descended from the Anglo-Saxons **b :** a white gentile of an English-speaking nation
3 : OLD ENGLISH 1
4 : direct plain English; *especially* **:** English using words considered crude or vulgar
— Anglo–Saxon *adjective*

an·go·ra \aŋ-'gōr-ə, an-, -'gȯr-\ *noun* (1852)
1 : the hair of the Angora rabbit or Angora goat — called also *angora wool*
2 : a yarn of Angora rabbit hair used especially for knitting
3 *capitalized* **a :** ANGORA CAT **b :** ANGORA GOAT **c :** ANGORA RABBIT

Angora cat *noun* [*Angora* (Ankara), Turkey] (1819)
: a long-haired domestic cat; *specifically* **:** any of a breed that differs from the Persian in having a narrower head and slighter body

Angora goat *noun* (1833)
: any of a breed or variety of the domestic goat raised for its long silky hair which is the true mohair

Angora goat

Angora rabbit *noun* (1849)
: any of a breed of long-haired usually white rabbits with red eyes that is raised for fine wool

an·gry \'aŋ-grē\ *adjective* **an·gri·er; -est** (14th century)
1 : feeling or showing anger **:** WRATHFUL
2 a : indicative of or proceeding from anger ⟨*angry* words⟩ **b :** seeming to show anger or to threaten in an angry manner ⟨an *angry* sky⟩
3 : painfully inflamed ⟨an *angry* rash⟩
— an·gri·ly \-grə-lē\ *adverb*
— an·gri·ness \-grē-nəs\ *noun*

Angora rabbit

angry young man *noun* (1941)
1 : an outspoken critic of or protester against an economic condition or social injustice
2 : one of a group of mid-20th century British authors whose works express the bitterness of the lower classes toward the established sociopolitical system and toward the mediocrity and hypocrisy of the middle and upper classes

angst \'äŋ(k)st, 'aŋ(k)st\ *noun* [Danish & German; Danish, from German] (circa 1942)
: a feeling of anxiety, apprehension, or insecurity

ang·strom \'aŋ-strəm *also* 'ȯŋ-\ *noun* [Anders J. *Ångström*] (1892)
: a unit of length equal to one ten-billionth of a meter

¹an·guish \'aŋ-gwish\ *noun* [Middle English *angwisshe,* from Old French *angoisse,* from Latin *angustiae,* plural, straits, distress, from *angustus* narrow; akin to Old English *enge* narrow — more at ANGER] (13th century)
: extreme pain, distress, or anxiety
synonym see SORROW

²anguish (14th century)
intransitive verb
: to suffer anguish
transitive verb
: to cause to suffer anguish

an·guished *adjective* (14th century)
1 : suffering anguish **:** TORMENTED ⟨the *anguished* martyrs⟩
2 : expressing anguish **:** AGONIZED ⟨*anguished* cries⟩

an·gu·lar \'aŋ-gyə-lər\ *adjective* [Middle French or Latin; Middle French *angulaire,* from Latin *angularis,* from *angulus* angle] (15th century)
1 a : forming an angle **:** sharp-cornered **b :** having one or more angles
2 : measured by an angle ⟨*angular* distance⟩
3 a : stiff in character or manner **:** lacking smoothness or grace **b :** lean and having prominent bone structure
— an·gu·lar·ly *adverb*

angular acceleration *noun* (1883)
: the rate of change per unit time of angular velocity

an·gu·lar·i·ty \,aŋ-gyə-'lar-ə-tē\ *noun, plural* **-ties** (1642)
1 : the quality of being angular
2 *plural* **:** angular outlines or characteristics

angular momentum *noun* (1870)
: a vector quantity that is a measure of the rotational momentum of a rotating body or system, that is equal in classical physics to the product of the angular velocity of the body or system and its moment of inertia with respect to the rotation axis, and that is directed along the rotation axis

angular velocity *noun* (1819)
: the rate of rotation around an axis usually expressed in radians or revolutions per second or per minute

an·gu·la·tion \,aŋ-gyə-'lā-shən\ *noun* (1869)
1 : the action of making angular
2 : an angular position, formation, or shape

An·gus \'aŋ-gəs\ *noun* [*Angus,* county in Scotland] (1842)
: any of a breed of usually black hornless beef cattle originating in Scotland

an·he·do·nia \ˌan-(ˌ)hē-ˈdō-nē-ə, -nyə\ *noun* [New Latin, from ²*a-* + Greek *hēdonē* pleasure — more at HEDONISM] (1897)
: a psychological condition characterized by inability to experience pleasure in normally pleasurable acts
— an·he·don·ic \-ˈdä-nik\ *adjective*

an·hin·ga \an-ˈhiŋ-gə\ *noun* [Portuguese, from Tupi] (1769)
: any of a genus (*Anhinga*) of fish-eating birds related to the cormorants but distinguished by a longer neck and sharply pointed rather than hooked bill; *especially* **:** one (*A. anhinga*) occurring from the southern U.S. to Argentina

an·hy·dride \(ˌ)an-ˈhī-ˌdrīd\ *noun* (1863)
: a compound derived from another (as an acid) by removal of the elements of water

an·hy·drite \-ˌdrīt\ *noun* [German *Anhydrit,* from Greek *anydros*] (circa 1823)
: a mineral consisting of an anhydrous calcium sulfate that is usually massive and white or slightly colored

an·hy·drous \-drəs\ *adjective* [Greek *anydros,* from *a-* + *hydōr* water — more at WATER] (1819)
: free from water and especially water of crystallization

ani \ä-ˈnē\ *noun* [Spanish *aní,* or Portuguese *ani,* from Tupi *ani*] (circa 1823)
: any of a genus (*Crotophaga*) of black cuckoos of the warmer parts of America

anile \ˈa-ˌnīl, ˈā-\ *adjective* [Latin *anilis,* from *anus* old woman] (1652)
: of or resembling a doddering old woman; *especially* **:** SENILE
— anil·i·ty \a-ˈni-lə-tē, ā-, ə-\ *noun*

an·i·line \ˈa-nᵊl-ən\ *noun* [German *Anilin,* from *Anil* indigo, from French, from Portuguese, from Arabic *an-nīl* the indigo plant, from Sanskrit *nīlī* indigo, from feminine of *nīla* dark blue] (1850)
: an oily liquid poisonous amine $C_6H_5NH_2$ obtained especially by the reduction of nitrobenzene and used chiefly in organic synthesis (as of dyes)

aniline dye *noun* (1864)
: a dye made by the use of aniline or one chemically related to such a dye; *broadly* **:** a synthetic organic dye

ani·lin·gus \ˌā-ni-ˈliŋ-gəs\ *or* **ani·linc·tus** \-ˈliŋ(k)-təs\ *noun* [New Latin, from *anus* + *-i-* + *-lingus, -linctus* (as in *cunnilingus, cunnilinctus*)] (1949)
: erotic stimulation achieved by contact between mouth and anus

an·i·ma \ˈa-nə-mə\ *noun* [New Latin, from Latin, soul] (1923)
: an individual's true inner self that in the analytic psychology of C. G. Jung reflects archetypal ideals of conduct; *also* **:** an inner feminine part of the male personality — compare ANIMUS, PERSONA

an·i·mad·ver·sion \ˌa-nə-ˌmad-ˈvər-zhən, -məd-, -ˈvər-shən\ *noun* [Latin *animadversion-, animadversio,* from *animadvertere*] (1599)
1 : a critical and usually censorious remark — often used with *on*
2 : adverse criticism

an·i·mad·vert \-ˈvərt\ *verb* [Latin *animadvertere* to pay attention to, censure, from *animum advertere,* literally, to turn the mind to] (15th century)
transitive verb
archaic **:** NOTICE, OBSERVE
intransitive verb
: to make an animadversion

¹an·i·mal \ˈa-nə-məl\ *noun* [Latin, from *animale,* neuter of *animalis* animate, from *anima* soul — more at ANIMATE] (14th century)
1 : any of a kingdom (Animalia) of living things including many-celled organisms and often many of the single-celled ones (as protozoans) that typically differ from plants in having cells without cellulose walls, in lacking chlorophyll and the capacity for photosynthesis, in requiring more complex food materials (as proteins), in being organized to a greater degree of complexity, and in having the capacity for spontaneous movement and rapid motor responses to stimulation
2 a : one of the lower animals as distinguished from human beings **b :** MAMMAL; *broadly* **:** VERTEBRATE
3 : a human being considered chiefly as physical or nonrational; *also* **:** this nature
4 : an individual with a particular interest or aptitude ⟨a political *animal*⟩
5 : MATTER, THING ⟨the theater . . . is an entirely different *animal* —Arthur Miller⟩; *also* **:** CREATURE 1c ◆
— an·i·mal·like \-mə(l)-ˌlīk\ *adjective*

²animal *adjective* (1615)
1 : of, relating to, resembling, or derived from animals
2 a : of or relating to the physical or sentient as contrasted with the intellectual or rational **b :** SENSUAL, FLESHLY
3 : of or relating to the animal pole of an egg or to the part from which ectoderm normally develops
synonym see CARNAL
— an·i·mal·ly \-mə-lē\ *adverb*

animal control *noun* (1957)
: an office or department responsible for enforcing ordinances relating to the control, impoundment, and disposition of animals

animal cracker *noun* (1898)
: a small cookie in the shape of an animal

an·i·mal·cule \ˌa-nə-ˈmal-(ˌ)kyü(ə)l\ *also* **an·i·mal·cu·lum** \-ˈmal-kyə-ləm\ *noun, plural* **-cules** *also* **-cu·la** \-kyə-lə\ [New Latin *animalculum,* diminutive of Latin *animal*] (1662)
: a minute usually microscopic organism

animal heat *noun* (1779)
: heat produced in the body of a living animal by functional chemical and physical activities

animal husbandry *noun* (1919)
: a branch of agriculture concerned with the production and care of domestic animals

an·i·mal·ier \ˌa-nə-mə-ˈlir\ *noun* [French, from *animal* animal, from Latin] (1912)
: a sculptor or painter of animal subjects

an·i·mal·ism \ˈa-nə-mə-ˌli-zəm\ *noun* (1831)
: ANIMALITY
— an·i·mal·is·tic \ˌa-nə-mə-ˈlis-tik\ *adjective*

an·i·mal·i·ty \ˌa-nə-ˈma-lə-tē\ *noun* (1615)
1 : qualities associated with animals: **a :** VITALITY **b :** a natural unrestrained unreasoned response to physical drives or stimuli
2 : the animal nature of human beings

an·i·mal·ize \ˈa-nə-mə-ˌlīz\ *transitive verb* **-ized; -iz·ing** (1741)
1 : to represent in animal form
2 : to cause to be or act like an animal
— an·i·mal·iza·tion \ˌa-nə-mə-lə-ˈzā-shən\ *noun*

animal kingdom *noun* (1776)
: a basic group of natural objects that includes all living and extinct animals — compare MINERAL KINGDOM, PLANT KINGDOM

animal magnetism *noun* (1784)
1 : a mysterious force claimed by Mesmer to enable him to hypnotize patients
2 : a magnetic charm or appeal; *especially* **:** SEX APPEAL

animal model *noun* (1976)
: an animal sufficiently like humans in its anatomy, physiology, or response to a pathogen to be used in medical research in order to obtain results that can be extrapolated to human medicine

animal pole *noun* (1887)
: the point on the surface of an egg that is diametrically opposite to the vegetal pole and usually marks the most active part of the protoplasm or the part containing least yolk

animal rights *noun plural but singular in construction* (1879)
: fair and humane treatment of animals — often used attributively

animal spirits *noun plural* (1543)
1 *sometimes* **animal spirit** *obsolete* **:** the nervous energy that is the source of physical sensation and movement
2 : vivacity arising from physical health and energy

animal starch *noun* (circa 1860)
: GLYCOGEN

¹an·i·mate \ˈa-nə-mət\ *adjective* [Middle English, from Latin *animatus,* past participle of *animare* to give life to, from *anima* breath, soul; akin to Old English *ōthian* to breathe, Latin *animus* spirit, Greek *anemos* wind, Sanskrit *aniti* he breathes] (15th century)
1 : possessing or characterized by life **:** ALIVE
2 : full of life **:** ANIMATED
3 : of or relating to animal life as opposed to plant life
4 : referring to a living thing ⟨an *animate* noun⟩
— an·i·mate·ly *adverb*
— an·i·mate·ness *noun*

²an·i·mate \-ˌmāt\ *transitive verb* **-mat·ed; -mat·ing** (15th century)
1 : to give spirit and support to **:** ENCOURAGE
2 a : to give life to **b :** to give vigor and zest to
3 : to move to action
4 a : to make or design in such a way as to create apparently spontaneous lifelike movement **b :** to produce in the form of an animated cartoon
synonym see QUICKEN

an·i·mat·ed \-ˌmā-təd\ *adjective* (1534)
1 a : endowed with life or the qualities of life **:** ALIVE ⟨viruses that can behave as *animated* bodies or inert crystals⟩ **b :** full of movement and activity **c :** full of vigor and spirit **:** LIVELY ⟨an *animated* discussion⟩
2 : having the appearance of something alive
3 : made in the form of an animated cartoon
synonym see LIVELY
— an·i·mat·ed·ly *adverb*

animated cartoon *noun* (1915)
1 : a motion picture made from a series of drawings simulating motion by means of slight progressive changes in the drawings
2 : ANIMATION 2a

an·i·ma·tion \ˌa-nə-ˈmā-shən\ *noun* (1597)
1 : the act of animating **:** the state of being animate or animated
2 a : a motion picture made by photographing successive positions of inanimate objects (as puppets or mechanical parts) **b :** ANIMATED CARTOON 1
3 : the preparation of animated cartoons

an·i·ma·to \ˌä-nə-ˈmä-(ˌ)tō, ˌa-\ *adverb or adjective* [Italian, from Latin *animatus*] (circa 1724)
: with animation — used as a direction in music

an·i·ma·tor \ˈa-nə-ˌmā-tər\ *noun* (1611)
1 : one that animates ⟨the prime *animator* of the movement⟩
2 : an artist who creates drawings for an animated cartoon

an·i·mism \ˈa-nə-ˌmi-zəm\ *noun* [German *Animismus,* from Latin *anima* soul] (1832)
1 : a doctrine that the vital principle of organic development is immaterial spirit

◇ WORD HISTORY
animal Latin *anima* means "breath," the vital principle that distinguishes living things from dead ones, and the adjective derivative *animalis* means "having breath, living." Though plants may be said to breathe insofar as they take in certain gases from the atmosphere and release others, this is not a process seen by the naked eye. Thus the Latin noun *animal,* derived from *animalis,* is used to designate those living beings that breathe perceptibly, and so is our word, borrowed from Latin.

2 : attribution of conscious life to objects in and phenomena of nature or to inanimate objects
3 : belief in the existence of spirits separable from bodies
— **an·i·mist** \-mist\ *noun*
— **an·i·mis·tic** \ˌa-nə-ˈmis-tik\ *adjective*

an·i·mos·i·ty \ˌa-nə-ˈmä-sə-tē\ *noun, plural* **-ties** [Middle English *animosite,* from Middle French or Late Latin; Middle French *animosité,* from Late Latin *animositat-, animositas,* from Latin *animosus* spirited, from *animus*] (1605)
: ill will or resentment tending toward active hostility **:** an antagonistic attitude
synonym see ENMITY

an·i·mus \ˈa-nə-məs\ *noun* [Latin, spirit, mind, courage, anger] (1816)
1 : basic attitude or governing spirit **:** DISPOSITION, INTENTION
2 : a usually prejudiced and often spiteful or malevolent ill will
3 : an inner masculine part of the female personality in the analytic psychology of C. G. Jung — compare ANIMA
synonym see ENMITY

an·ion \ˈa-ˌnī-ən\ *noun* [Greek, neuter of *aniōn,* present participle of *anienai* to go up, from *ana-* + *ienai* to go — more at ISSUE] (1834)
: the ion in an electrolyzed solution that migrates to the anode; *broadly* **:** a negatively charged ion

an·ion·ic \ˌa-(ˌ)nī-ˈä-nik\ *adjective* (circa 1920)
1 : of or relating to anions
2 : characterized by an active and especially surface-active anion

anis- *or* **aniso-** *combining form* [New Latin, from Greek, from *anisos,* from *a-* + *isos* equal]
: unequal ⟨*aniso*tropic⟩

an·ise \ˈa-nəs\ *noun* [Middle English *anis,* from Middle French, from Latin *anisum,* from Greek *annēson, anison*] (14th century)
: an herb (*Pimpinella anisum*) of the carrot family having carminative and aromatic seeds; *also* **:** ANISEED

ani·seed \ˈa-nə(s)-ˌsēd\ *noun* [Middle English *anis seed,* from *anis* + *seed*] (14th century)
: the seed of anise often used as a flavoring in liqueurs and in cooking

an·is·ei·ko·nia \ˌa-ˌnī-ˌsī-ˈkō-nē-ə\ *noun* [New Latin, from *anis-* + Greek *eikōn* image — more at ICON] (1934)
: a defect of binocular vision in which the two retinal images of an object differ in size
— **an·is·ei·kon·ic** \-ˈkä-nik\ *adjective*

an·is·ette \ˌa-nə-ˈset, -ˈzet\ *noun* [French, from *anis*] (1837)
: a usually colorless sweet liqueur flavored with aniseed

an·isog·a·mous \ˌa-(ˌ)nī-ˈsä-gə-məs\ *adjective* (1891)
: characterized by fusion of heterogamous gametes or of individuals that usually differ chiefly in size ⟨*anisogamous* reproduction⟩
— **an·isog·a·my** \-(ˌ)nī-ˈsä-gə-mē\ *noun*

an·iso·me·tro·pia \ˌa-ˌnī-sə-mə-ˈtrō-pē-ə\ *noun* [New Latin, from Greek *anisometros* of unequal measure (from *anis-* + *metron* measure) + New Latin *-opia* — more at MEASURE] (circa 1880)
: unequal refractive power in the two eyes
— **an·iso·me·tro·pic** \-ˈträ-pik, -ˈtrō-\ *adjective*

an·iso·trop·ic \ˌa-ˌnī-sə-ˈträ-pik\ *adjective* (1879)
: exhibiting properties with different values when measured in different directions ⟨an *anisotropic* crystal⟩
— **an·iso·trop·i·cal·ly** \-pi-k(ə-)lē\ *adverb*
— **an·isot·ro·py** \-(ˌ)nī-ˈsä-trə-pē\ *also* **an·isot·ro·pism** \-ˌpi-zəm\ *noun*

an·ker·ite \ˈaŋ-kə-ˌrīt\ *noun* [German *Ankerit,* from M. J. *Anker* (died 1843) Austrian mineralogist] (circa 1843)
: a dolomitic iron-containing mineral

ankh \ˈäŋk\ *noun* [Egyptian *ʿnḫ* live] (1888)
: a cross having a loop for its upper vertical arm and serving especially in ancient Egypt as an emblem of life

ankh

an·kle \ˈaŋ-kəl\ *noun* [Middle English *ankel,* from Old English *anclēow;* akin to Old High German *anchlāo* ankle] (before 12th century)
1 : the joint between the foot and the leg; *also* **:** the region of this joint
2 : the joint between the cannon bone and pastern (as in the horse)

an·kle·bone \ˈaŋ-kəl-ˌbōn, ˌaŋ-kəl-ˈ\ *noun* (14th century)
: [2]TALUS 1

an·klet \ˈaŋ-klət\ *noun* (1819)
1 : something (as an ornament) worn around the ankle
2 : a short sock reaching slightly above the ankle

an·ky·lo·saur \ˈaŋ-kə-lō-ˌsȯr\ *noun* [New Latin *Ankylosauria,* from *Ankylosaurus*] (1949)
: any of a suborder (Ankylosauria) of herbivorous Cretaceous dinosaurs having a long low-lying thickset body covered dorsally with bony plates

an·ky·lo·sau·rus \ˌaŋ-kə-lō-ˈsȯr-əs\ *noun* [New Latin, genus name, from Greek *ankylos* + *sauros* lizard] (1908)
: any of a North American genus (*Ankylosaurus*) of large ankylosaurs having a bony club at the end of the tail

an·ky·lose \ˈaŋ-ki-ˌlōs, -ˌlōz\ *verb* **-losed; -los·ing** [back-formation from *ankylosis*] (1787)
transitive verb
: to unite or stiffen by ankylosis
intransitive verb
: to undergo ankylosis

an·ky·lo·sis \ˌaŋ-ki-ˈlō-səs\ *noun, plural* **-lo·ses** \-ˌsēz\ [New Latin, from Greek *ankylōsis,* from *ankyloun* to make crooked, from *ankylos* crooked — more at ANGLE] (1713)
1 : stiffness or fixation of a joint by disease or surgery
2 : union of separate bones or hard parts to form a single bone or part
— **an·ky·lot·ic** \-ˈlä-tik\ *adjective*

an·ky·lo·sto·mi·a·sis \ˌaŋ-ki-lō-stə-ˈmī-ə-səs\ *variant of* ANCYLOSTOMIASIS

an·la·ge \ˈän-ˌlä-gə\ *noun, plural* **-gen** \-gən\ *also* **-ges** \-gəz\ [German, literally, act of laying on] (1892)
: the foundation of a subsequent development; *especially* **:** PRIMORDIUM

an·na \ˈä-nə\ *noun* [Hindi *ānā*] (1708)
1 : a former monetary unit of Burma, India, and Pakistan equal to 1/16 rupee
2 : a coin representing one anna

an·nal·ist \ˈa-nᵊl-ist\ *noun* (circa 1611)
: a writer of annals **:** CHRONICLER
— **an·nal·is·tic** \ˌa-nᵊl-ˈis-tik\ *adjective*

an·nals \ˈa-nᵊlz\ *noun plural* [Latin *annales,* from plural of *annalis* yearly — more at ANNUAL] (1542)
1 : a record of events arranged in yearly sequence
2 : historical records **:** CHRONICLES
3 : records of the activities of an organization

An·nam·ese \ˌa-nə-ˈmēz, -ˈmēs\ *noun, plural* **Annamese** [*Annam,* region of Vietnam] (1826)
1 *or* **An·nam·ite** \ˈa-nə-ˌmīt\ **:** a native or inhabitant of Annam
2 : VIETNAMESE 2
— **Annamese** *adjective*
— **Annamite** *adjective*

an·nat·to \ə-ˈnä-(ˌ)tō\ *noun* [Carib *annoto* tree producing annatto] (1629)
: a yellowish red dyestuff made from the pulp around the seeds of a tropical tree (*Bixa orellana,* family Bixaceae); *also* **:** the tree that yields annatto

an·neal \ə-ˈnē(ə)l\ *verb* [Middle English *anelen* to set on fire, from Old English *onǣlan,* from *on* + *ǣlan* to set on fire, burn, from *āl* fire; akin to Old English *ǣled* fire, Old Norse *eldr*] (1664)
transitive verb
1 a : to heat and then cool (as steel or glass) usually for softening and making less brittle; *also* **:** to cool slowly usually in a furnace **b :** to heat and then cool (nucleic acid) in order to separate strands and induce combination at lower temperature especially with complementary strands of a different species
2 : STRENGTHEN, TOUGHEN
intransitive verb
: to be capable of combining with complementary nucleic acid by a process of heating and cooling

an·ne·lid \ˈa-nᵊl-əd, ˈa-nə-ˌlid\ *noun* [ultimately from Latin *anellus* little ring — more at ANNULET] (1834)
: any of a phylum (Annelida) of usually elongated segmented coelomate invertebrates (as earthworms, various marine worms, and leeches)
— **annelid** *adjective*
— **an·nel·i·dan** \ə-ˈne-lə-dᵊn, a-\ *adjective or noun*

[1]**an·nex** \ə-ˈneks, ˈa-ˌneks\ *transitive verb* [Middle English, from Middle French *annexer,* from Old French, from *annexe* joined, from Latin *annexus,* past participle of *annectere* to bind to, from *ad-* + *nectere* to bind] (14th century)
1 : to attach as a quality, consequence, or condition
2 *archaic* **:** to join together materially **:** UNITE
3 : to add to something earlier, larger, or more important
4 : to incorporate (a country or other territory) within the domain of a state
5 : to obtain or take for oneself
— **an·nex·ation** \ˌa-ˌnek-ˈsā-shən\ *noun*
— **an·nex·ation·al** \-shnəl, -shə-nᵊl\ *adjective*
— **an·nex·ation·ist** \-sh(ə-)nist\ *noun*

[2]**an·nex** \ˈa-ˌneks, -niks\ *noun* (1501)
: something annexed as an expansion or supplement: as **a :** an added stipulation or statement **:** APPENDIX **b :** a subsidiary or supplementary structure **:** WING

an·nexe \ˈa-ˌneks, -niks\ *chiefly British variant of* [2]ANNEX

An·nie Oak·ley \ˌa-nē-ˈō-klē\ *noun, plural* **Annie Oakleys** [*Annie Oakley* (died 1926) American markswoman; from the resemblance of a punched pass to a playing card with bullet holes through the spots] (circa 1910)
: a free ticket

an·ni·hi·late \ə-ˈnī-ə-ˌlāt\ *verb* **-lat·ed; -lat·ing** [Late Latin *annihilatus,* past participle of *annihilare* to reduce to nothing, from Latin *ad-* + *nihil* nothing — more at NIL] (1525)
transitive verb
1 a : to cause to be of no effect **:** NULLIFY **b :** to destroy the substance or force of
2 : to regard as of no consequence
3 : to cause to cease to exist; *especially* **:** KILL
4 a : to destroy a considerable part of ⟨bombs *annihilated* the city⟩ **b :** to vanquish completely **:** ROUT ⟨*annihilated* the visitors 56–0⟩
intransitive verb
of a particle and its antiparticle **:** to vanish or cease to exist by coming together and changing into other forms of energy (as radiation or particles)
— **an·ni·hi·la·tion** \-ˌnī-ə-ˈlā-shən\ *noun*

\ə\ abut \ᵊ\ kitten \ər\ further \a\ ash \ā\ ace \ä\ mop, mar \au̇\ out \ch\ chin \e\ bet \ē\ easy \g\ go \i\ hit \ī\ ice \j\ job \ŋ\ sing \ō\ go \ȯ\ law \ȯi\ boy \th\ thin \t͟h\ the \ü\ loot \u̇\ foot \y\ yet \zh\ vision *see also* Guide to Pronunciation

— **an·ni·hi·la·tor** \-ˌlā-tər\ *noun*
— **an·ni·hi·la·to·ry** \-ˈnī-ə-lə-ˌtōr-ē, -ˌtȯr-\ *adjective*
an·ni·ver·sa·ry \ˌa-nə-ˈvərs-rē, -ˈvər-sə-\ *noun, plural* **-ries** [Middle English *anniversarie,* from Medieval Latin *anniversarium,* from Latin, neuter of *anniversarius* returning annually, from *annus* year + *versus,* past participle of *vertere* to turn — more at ANNUAL, WORTH] (13th century)
1 : the annual recurrence of a date marking a notable event
2 : the celebration of an anniversary
an·no Do·mi·ni \ˌa-(ˌ)nō-ˈdä-mə-nē, -ˈdō-, -ˌnī\ *adverb, often A capitalized* [Medieval Latin, in the year of the Lord] (1530)
— used to indicate that a time division falls within the Christian era
an·no he·gi·rae \-hi-ˈjī-(ˌ)rē, -ˈhe-jə-ˌrē\ *adverb, often A&H capitalized* [New Latin, in the year of the Hegira] (circa 1889)
— used to indicate that a time division falls within the Islamic era
an·no·tate \ˈa-nə-ˌtāt\ *verb* **-tat·ed; -tat·ing** [Latin *annotatus,* past participle of *annotare,* from *ad-* + *notare* to mark — more at NOTE] (1733)
intransitive verb
: to make or furnish critical or explanatory notes or comment
transitive verb
: to make or furnish annotations for (as a literary work or subject)
— **an·no·ta·tive** \-ˌtā-tiv\ *adjective*
— **an·no·ta·tor** \-ˌtā-tər\ *noun*
an·no·ta·tion \ˌa-nə-ˈtā-shən\ *noun* (15th century)
1 : a note added by way of comment or explanation
2 : the act of annotating
an·nounce \ə-ˈnau̇n(t)s\ *verb* **-nounced; -nounc·ing** [Middle English, from Middle French *annoncer,* from Latin *annuntiare,* from *ad-* + *nuntiare* to report, from *nuntius* messenger] (15th century)
transitive verb
1 : to make known publicly **:** PROCLAIM ⟨*announced* the appointment⟩
2 a : to give notice of the arrival, presence, or readiness of ⟨*announce* dinner⟩ **b :** to indicate beforehand **:** FORETELL
3 : to serve as an announcer of
intransitive verb
1 : to serve as an announcer
2 a : to declare one's candidacy **b :** to declare oneself politically — used with *for* or *against*
synonym see DECLARE
an·nounce·ment \ə-ˈnau̇n(t)-smənt\ *noun* (1798)
1 : the act of announcing or of being announced
2 : a public notification or declaration
3 : a piece of formal stationery designed for a social or business announcement
an·nounc·er \ə-ˈnau̇n(t)-sər\ *noun* (circa 1611)
: one that announces: as **a :** a person who introduces television or radio programs, makes commercial announcements, or gives station identification **b :** a person who describes and comments on the action in a broadcast sports event
an·noy \ə-ˈnȯi\ *verb* [Middle English *anoien,* from Middle French *enuier,* from Late Latin *inodiare* to make loathsome, from Latin *in* + *odium* hatred — more at ODIUM] (13th century)
transitive verb
1 : to disturb or irritate especially by repeated acts
2 : to harass especially by quick brief attacks
intransitive verb
: to cause annoyance ☆
— **an·noy·er** *noun*
an·noy·ance \ə-ˈnȯi-ən(t)s\ *noun* (14th century)
1 : the act of annoying or of being annoyed
2 : the state or feeling of being annoyed **:** VEXATION
3 : a source of vexation or irritation **:** NUISANCE
an·noy·ing *adjective* (14th century)
: causing vexation **:** IRRITATING
— **an·noy·ing·ly** \-iŋ-lē\ *adverb*
¹an·nu·al \ˈan-yə(-wə)l\ *adjective* [Middle English, from Middle French & Late Latin; Middle French *annuel,* from Late Latin *annualis,* blend of Latin *annuus* yearly (from *annus* year) and Latin *annalis* yearly (from *annus* year); probably akin to Gothic *athnam* (dative plural) years, Sanskrit *atati* he walks, goes] (14th century)
1 : covering the period of a year ⟨*annual* rainfall⟩
2 : occurring or happening every year or once a year **:** YEARLY ⟨an *annual* reunion⟩
3 : completing the life cycle in one growing season
— **an·nu·al·ly** *adverb*
²annual *noun* (14th century)
1 : an event that occurs yearly
2 : a publication appearing yearly
3 : something that lasts one year or season; *specifically* **:** an annual plant
an·nu·al·ize \ˈan-yə(-wə)-ˌlīz\ *transitive verb* **-ized; -iz·ing** (1918)
: to calculate or adjust to reflect a rate based on a full year ⟨quarterly returns yielding at an *annualized* rate of 7 percent⟩
annual ring *noun* (circa 1879)
: the layer of wood produced by a single year's growth of a woody plant
an·nu·itant \ə-ˈnü-ət-ənt, -ˈnyü-\ *noun* (1720)
: a beneficiary of an annuity
an·nu·ity \ə-ˈnü-ə-tē, -ˈnyü-\ *noun, plural* **-ities** [Middle English *annuite,* from Middle French *annuité,* from Medieval Latin *annuitat-, annuitas,* from Latin *annuus* yearly] (15th century)
1 : a sum of money payable yearly or at other regular intervals
2 : the right to receive an annuity
3 : a contract or agreement providing for the payment of an annuity
an·nul \ə-ˈnəl\ *transitive verb* **an·nulled; an·nul·ling** [Middle English *annullen,* from Middle French *annuller,* from Late Latin *annullare,* from Latin *ad-* + *nullus* not any — more at NULL] (15th century)
1 : to reduce to nothing **:** OBLITERATE
2 : to make ineffective or inoperative **:** NEUTRALIZE ⟨*annul* the drug's effect⟩
3 : to declare or make legally invalid or void ⟨wants the marriage *annulled*⟩
synonym see NULLIFY
an·nu·lar \ˈan-yə-lər\ *adjective* [Middle French or Medieval Latin; Middle French *annulaire,* from Medieval Latin *anularis,* from Latin *anulus*] (1571)
: of, relating to, or forming a ring ⟨an *annular* skin lesion⟩
annular eclipse *noun* (1771)
: an eclipse in which a thin outer ring of the sun's disk is not covered by the apparently smaller dark disk of the moon
an·nu·late \ˈan-yə-lət, -ˌlāt\ *adjective* (circa 1823)
: furnished with or composed of rings **:** RINGED
an·nu·la·tion \ˌan-yə-ˈlā-shən\ *noun* (1829)
: a ringlike anatomical structure
an·nu·let \ˈan-yə-lət\ *noun* [modification of Middle French *annelet,* diminutive of *anel,* from Latin *anellus,* diminutive of *anulus*] (1598)
1 : a little ring
2 : a small architectural molding or ridge forming a ring
an·nul·ment \ə-ˈnəl-mənt\ *noun* (15th century)
1 : the act of annulling **:** the state of being annulled
2 : a judicial pronouncement declaring a marriage invalid
an·nu·lus \ˈan-yə-ləs\ *noun, plural* **-li** \-ˌlī, -(ˌ)lē\ *also* **-lus·es** [Medieval Latin, from Latin *anulus* finger ring, from *anus* ring — more at ANUS] (1563)
1 : RING
2 : a part, structure, or marking resembling a ring: as **a :** a line of cells around a fern sporangium that ruptures the sporangium by contracting **b :** a growth ring (as on the scale of a fish) that is used in estimating age
an·nun·ci·ate \ə-ˈnən(t)-sē-ˌāt\ *transitive verb* **-at·ed; -at·ing** (circa 1536)
: ANNOUNCE
an·nun·ci·a·tion \ə-ˌnən(t)-sē-ˈā-shən\ *noun* [Middle English *annunciacioun,* from Middle French *anunciation,* from Late Latin *annuntiation-, annuntiatio,* from Latin *annuntiare* — more at ANNOUNCE] (14th century)
1 *capitalized* **:** March 25 observed as a church festival in commemoration of the announcement of the Incarnation to the Virgin Mary
2 : the act of announcing or of being announced **:** ANNOUNCEMENT
an·nun·ci·a·tor \ə-ˈnən(t)-sē-ˌā-tər\ *noun* (circa 1753)
: one that annunciates; *specifically* **:** a usually electrically controlled signal board or indicator
— **an·nun·ci·a·to·ry** \-sē-ə-ˌtōr-ē, -ˌtȯr-\ *adjective*
an·nus mi·ra·bi·lis \ˌa-nəs-mə-ˈrä-bə-ləs, ˌä-\ *noun, plural* **an·ni mi·ra·bi·les** \ˈa-ˌnī-mə-ˈrä-bə-ˌlēz, ˈä-(ˌ)nē-mə-ˈrä-bə-ˌlās\ [New Latin, literally, wonderful year] (1660)
: a remarkable or notable year
an·ode \ˈa-ˌnōd\ *noun* [Greek *anodos* way up, from *ana-* + *hodos* way] (1834)
1 : the electrode of an electrochemical cell at which oxidation occurs: as **a :** the positive terminal of an electrolytic cell **b :** the negative terminal of a galvanic cell
2 : the electron-collecting electrode of an electron tube — compare CATHODE
— **an·od·ic** \a-ˈnä-dik\ *also* **an·od·al** \-ˈnō-dᵊl\ *adjective*
— **an·od·i·cal·ly** \-di-k(ə-)lē\ *also* **an·od·al·ly** \-dᵊl-ē\ *adverb*
an·od·ize \ˈa-nə-ˌdīz\ *transitive verb* **-ized; -iz·ing** (1931)
: to subject (a metal) to electrolytic action as the anode of a cell in order to coat with a protective or decorative film
— **an·od·iza·tion** \ˌa-ˌnō-də-ˈzā-shən, ˌa-nə-\ *noun*
¹an·o·dyne \ˈa-nə-ˌdīn\ *adjective* [Latin *anodynos,* from Greek *anōdynos,* from *a-* + *odynē* pain; probably akin to Old English *etan* to eat] (1543)
1 : serving to assuage pain
2 : not likely to offend or arouse tensions **:** INNOCUOUS
²anodyne *noun* (circa 1550)
1 : something that soothes, calms, or comforts ⟨the *anodyne* of bridge, a comfortable book, or sport —Harrison Smith⟩

☆ SYNONYMS
Annoy, vex, irk, bother mean to upset a person's composure. ANNOY implies a wearing on the nerves by persistent petty unpleasantness ⟨their constant complaining *annoys* us⟩. VEX implies greater provocation and stronger disturbance and usually connotes anger but sometimes perplexity or anxiety ⟨*vexed* by her teenager's failure to pick up his room⟩. IRK stresses difficulty in enduring and the resulting weariness or impatience of spirit ⟨careless waste *irks* the boss⟩. BOTHER suggests interference with comfort or peace of mind ⟨don't *bother* me while I'm reading⟩. See in addition WORRY.

2 : a drug that allays pain

anoint \ə-ˈnȯint\ *transitive verb* [Middle English, from Middle French *enoint,* past participle of *enoindre,* from Latin *inunguere,* from *in-* + *unguere* to smear — more at OINTMENT] (14th century)
1 : to smear or rub with oil or an oily substance
2 a : to apply oil to as a sacred rite especially for consecration **b :** to choose by or as if by divine election; *also* **:** to designate as if by a ritual anointment
— anoint·er *noun*
— anoint·ment \-mənt\ *noun*

anointing of the sick (circa 1884)
: EXTREME UNCTION

ano·le \ə-ˈnō-lē\ *noun* [probably from French *anolis,* from Arawak of the Lesser Antilles] (circa 1753)
: any of a genus (*Anolis* of the family Iguanidae) of arboreal American lizards (as the American chameleon) that have a brightly colored dewlap and the ability to change color

anom·a·lous \ə-ˈnä-mə-ləs\ *adjective* [Late Latin *anomalus,* from Greek *anōmalos,* literally, uneven, from *a-* + *homalos* even, from *homos* same — more at SAME] (1655)
1 : inconsistent with or deviating from what is usual, normal, or expected **:** IRREGULAR, UNUSUAL
2 a : of uncertain nature or classification **b :** marked by incongruity or contradiction **:** PARADOXICAL
synonym see IRREGULAR
— anom·a·lous·ly *adverb*
— anom·a·lous·ness *noun*

anom·a·ly \ə-ˈnä-mə-lē\ *noun, plural* **-lies** (1603)
1 : the angular distance of a planet from its perihelion as seen from the sun
2 : deviation from the common rule **:** IRREGULARITY
3 : something different, abnormal, peculiar, or not easily classified

an·o·mie *also* **an·o·my** \ˈa-nə-mē\ *noun* [French *anomie,* from Middle French, from Greek *anomia* lawlessness, from *anomos* lawless, from *a-* + *nomos* law, from *nemein* to distribute — more at NIMBLE] (1933)
: social instability resulting from a breakdown of standards and values; *also* **:** personal unrest, alienation, and uncertainty that comes from a lack of purpose or ideals
— ano·mic \ə-ˈnä-mik, -ˈnō-\ *adjective*

anon \ə-ˈnän\ *adverb* [Middle English, from Old English *on ān,* from *on* in + *ān* one — more at ON, ONE] (before 12th century)
1 *archaic* **:** at once **:** IMMEDIATELY
2 : SOON, PRESENTLY
3 : after a while **:** LATER

an·o·nym \ˈa-nə-ˌnim\ *noun* (1812)
1 : one who is anonymous
2 : PSEUDONYM

an·o·nym·i·ty \ˌa-nə-ˈni-mə-tē\ *noun, plural* **-ties** (1820)
1 : the quality or state of being anonymous
2 : one that is anonymous

anon·y·mous \ə-ˈnä-nə-məs\ *adjective* [Late Latin *anonymus,* from Greek *anōnymos,* from *a-* + *onyma* name — more at NAME] (1631)
1 : not named or identified ⟨an *anonymous* author⟩ ⟨they wish to remain *anonymous*⟩
2 : of unknown authorship or origin ⟨an *anonymous* tip⟩
3 : lacking individuality, distinction, or recognizability ⟨the *anonymous* faces in the crowd⟩ ⟨the gray *anonymous* streets —William Styron⟩
— anon·y·mous·ly *adverb*
— anon·y·mous·ness *noun*

anoph·e·les \ə-ˈnä-fə-ˌlēz\ *noun* [New Latin, genus name, from Greek *anōphelēs* useless, from *a-* + *ophelos* advantage, help; akin to Greek *ophellein* to increase, Armenian *aweli* more] (1899)
: any of a genus (*Anopheles*) of mosquitoes that includes all mosquitoes which transmit malaria to humans
— anoph·e·line \-ˌlīn\ *adjective or noun*

an·o·rak \ˈa-nə-ˌrak\ *noun* [Danish, from Inuit (Greenland) *annoraaq*] (1922)
: a usually pullover hooded jacket long enough to cover the hips

¹an·o·rec·tic \ˌa-nə-ˈrek-tik\ *also* **an·o·ret·ic** \-ˈre-tik\ *adjective* [Greek *anorektos,* from *a-* + *oregein* to reach after — more at RIGHT] (circa 1894)
1 a : lacking appetite **b :** ANOREXIC 2
2 : causing loss of appetite

²anorectic *also* **anoretic** *noun* (circa 1957)
1 : an anorectic agent
2 : ANOREXIC

an·orex·ia \ˌa-nə-ˈrek-sē-ə, -ˈrek-shə\ *noun* [New Latin, from Greek, from *a-* + *orexis* appetite, from *oregein*] (1598)
1 : loss of appetite especially when prolonged
2 : ANOREXIA NERVOSA

anorexia ner·vo·sa \-(ˌ)nər-ˈvō-sə, -zə\ *noun* [New Latin, nervous anorexia] (1873)
: a serious disorder in eating behavior primarily of young women in their teens and early twenties that is characterized especially by a pathological fear of weight gain leading to faulty eating patterns, malnutrition, and usually excessive weight loss

¹an·orex·ic \ˌa-nə-ˈrek-sik\ *adjective* (circa 1907)
1 : ANORECTIC
2 : affected with or as if with anorexia nervosa

²anorexic *noun* (1912)
: a person affected with anorexia nervosa

an·orex·i·gen·ic \ˌa-nə-ˌrek-sə-ˈje-nik\ *adjective* (1948)
: ANORECTIC 2

an·or·thite \ə-ˈnȯr-ˌthīt\ *noun* [French, from *a-* + Greek *orthos* straight] (1833)
: a white, grayish, or reddish feldspar occurring in many igneous rocks
— an·or·thit·ic \ˌa-nȯr-ˈthi-tik\ *adjective*

an·or·tho·site \ə-ˈnȯr-thə-ˌsīt\ *noun* [French *anorthose,* a feldspar, from *a-* + Greek *orthos* — more at ORTH-] (1863)
: a granular plutonic igneous rock composed almost exclusively of a soda-lime feldspar (as labradorite)
— an·or·tho·si·tic \ə-ˌnȯr-thə-ˈsi-tik\ *adjective*

an·os·mia \a-ˈnäz-mē-ə\ *noun* [New Latin, from *a-* + Greek *osmē* smell — more at ODOR] (circa 1811)
: loss or impairment of the sense of smell
— an·os·mic \-mik\ *adjective*

¹an·oth·er \ə-ˈnə-t͟hər *also* a- *or* ā-\ *adjective* (12th century)
1 : different or distinct from the one first considered ⟨the same scene viewed from *another* angle⟩
2 : some other ⟨do it *another* time⟩
3 : being one more in addition to one or more of the same kind ⟨have *another* piece of pie⟩

²another *pronoun* (13th century)
1 : an additional one of the same kind **:** one more
2 : one that is different from the first or present one
3 : one of a group of unspecified or indefinite things ⟨in one way or *another*⟩

anoth·er–guess \ə-ˈnə-t͟hər-ˌges\ *adjective* [alteration of *anothergates,* from ¹*another* + *gate*] (1644)
archaic **:** of another sort

an·ovu·la·to·ry \(ˌ)an-ˈäv-yə-lə-ˌtōr-ē, -ˈōv-, -ˌtȯr-\ *adjective* (1934)
1 : not involving or accompanied by ovulation ⟨*anovulatory* bleeding⟩
2 : suppressing ovulation

an·ox·emia \ˌa-ˌnäk-ˈsē-mē-ə\ *noun* [New Latin] (circa 1881)
: a condition of subnormal oxygenation of the arterial blood
— an·ox·emic \-mik\ *adjective*

an·ox·ia \a-ˈnäk-sē-ə\ *noun* [New Latin] (1931)
: hypoxia especially of such severity as to result in permanent damage

an·ox·ic \(ˌ)a-ˈnäk-sik\ *adjective* (1920)
1 : of, relating to, or affected with anoxia
2 : greatly deficient in oxygen **:** OXYGENLESS ⟨*anoxic* water⟩

an·ser·ine \ˈan(t)-sə-ˌrīn\ *adjective* [Latin *anserinus,* from *anser* goose — more at GOOSE] (circa 1828)
: of, relating to, or resembling a goose

¹an·swer \ˈan(t)-sər\ *noun* [Middle English, from Old English *andswaru* (akin to Old Norse *andsvar* answer); akin to Old English *and-* against, *swerian* to swear — more at ANTE-] (before 12th century)
1 a : something spoken or written in reply to a question **b :** a correct response
2 : a reply to a legal charge or suit **:** PLEA; *also* **:** DEFENSE
3 : something done in response or reaction ⟨his only *answer* was to walk out⟩
4 : a solution of a problem
5 : one that imitates, matches, or corresponds to another ⟨television's *answer* to the news magazines⟩

²answer *verb* **an·swered; an·swer·ing** \ˈan(t)s-riŋ, ˈan(t)-sə-\ (before 12th century)
intransitive verb
1 : to speak or write in reply
2 a : to be or make oneself responsible or accountable **b :** to make amends **:** ATONE
3 : to be in conformity or correspondence ⟨*answered* to the description⟩
4 : to act in response to an action performed elsewhere or by another
5 : to be adequate **:** SERVE
transitive verb
1 a : to speak or write in reply to **b :** to say or write by way of reply
2 : to reply in rebuttal, justification, or explanation
3 a : to correspond to ⟨*answers* the description⟩ **b :** to be adequate or usable for **:** FULFILL
4 *obsolete* **:** to atone for
5 : to act in response to ⟨*answered* the call to arms⟩
6 : to offer a solution for; *especially* **:** SOLVE ☆
— an·swer·er \ˈan(t)-sər-ər\ *noun*

an·swer·able \ˈan(t)s-rə-bəl, ˈan(t)-sə-\ *adjective* (1536)
1 *archaic* **:** SUITABLE, ADEQUATE
2 : liable to be called to account **:** RESPONSIBLE
3 *archaic* **:** CORRESPONDING, SIMILAR
4 : capable of being refuted
synonym see RESPONSIBLE

answering machine *noun* (1961)
: a machine that receives telephone calls by playing a recorded message and usually by recording messages from callers

☆ SYNONYMS
Answer, respond, reply, rejoin, retort mean to say, write, or do something in return. ANSWER implies the satisfying of a question, demand, call, or need ⟨*answered* all the questions⟩. RESPOND may suggest an immediate or quick reaction ⟨*responded* eagerly to a call for volunteers⟩. REPLY implies making a return commensurate with the original question or demand ⟨an invitation that requires you to *reply*⟩. REJOIN often implies sharpness or quickness in answering ⟨"who asked you?" she *rejoined*⟩. RETORT suggests responding to an explicit charge or criticism by way of retaliation ⟨he *retorted* to the attack with biting sarcasm⟩.

answering service *noun* (1941)
: a commercial service that answers telephone calls for its clients

ant \'ant\ *noun* [Middle English *ante, emete,* from Old English *æmette;* akin to Old High German *āmeiza* ant] (before 12th century)
: any of a family (Formicidae) of colonial hymenopterous insects with a complex social organization and various castes performing special duties
— ants in one's pants : impatience for action or activity **:** RESTLESSNESS

ant- — see ANTI-

¹-ant *noun suffix* [Middle English, from Old French, from *-ant,* present participle suffix, from Latin *-ant-, -ans,* present participle suffix of first conjugation, from *-a-* (stem vowel of first conjugation) + *-nt-, -ns,* present participle suffix; akin to Old English *-nde,* present participle suffix, Greek *-nt-, -n,* participle suffix]
1 a : one that performs (a specified action) **:** personal or impersonal agent ⟨claim*ant*⟩ ⟨cool*ant*⟩ **b :** thing that promotes (a specified action or process) ⟨expector*ant*⟩
2 : one connected with ⟨annuit*ant*⟩
3 : thing acted upon (in a specified manner) ⟨inhal*ant*⟩

²-ant *adjective suffix*
1 : performing (a specified action) or being (in a specified condition) ⟨somnambul*ant*⟩
2 : promoting (a specified action or process) ⟨expector*ant*⟩

an·ta \'an-tə\ *noun, plural* **antas** *or* **an·tae** \-ˌtē, -ˌtī\ [Latin; akin to Old Norse *ǫnd* anteroom] (1598)
: a pier produced by thickening a wall at its termination

A anta

ant·ac·id \(ˌ)ant-'a-səd, 'an-ˌta-\ *noun* (1732)
: an agent that counteracts or neutralizes acidity
— antacid *adjective*

An·tae·an \an-'tē-ən\ *adjective* [*Antaeus,* a giant overcome by Hercules] (1921)
1 : MAMMOTH
2 : having superhuman strength

an·tag·o·nism \an-'ta-gə-ˌni-zəm\ *noun* (circa 1828)
1 a : opposition of a conflicting force, tendency, or principle ⟨the *antagonism* of democracy to dictatorship⟩ **b :** actively expressed opposition or hostility ⟨*antagonism* between factions⟩
2 : opposition in physiological action; *especially* **:** interaction of two or more substances such that the action of any one of them on living cells or tissues is lessened
synonym see ENMITY

an·tag·o·nist \-nist\ *noun* (1594)
1 : one that contends with or opposes another **:** ADVERSARY, OPPONENT
2 : an agent of physiological antagonism: as **a :** a muscle that contracts with and limits the action of an agonist with which it is paired — called also *antagonistic muscle* **b :** a chemical that acts within the body to reduce the physiological activity of another chemical substance (as an opiate); *especially* **:** one that opposes the action on the nervous system of a drug or a substance occurring naturally in the body by combining with and blocking its nervous receptor — compare AGONIST 2b

an·tag·o·nis·tic \(ˌ)an-ˌta-gə-'nis-tik\ *adjective* (1632)
: marked by or resulting from antagonism
— an·tag·o·nis·ti·cal·ly \-ti-k(ə-)lē\ *adverb*

an·tag·o·nize \an-'ta-gə-ˌnīz\ *transitive verb* **-nized; -niz·ing** [Greek *antagōnizesthai,* from *anti-* + *agōnizesthai* to struggle, from *agōn* contest — more at AGONY] (circa 1742)
1 : to act in opposition to **:** COUNTERACT
2 : to incur or provoke the hostility of

ant·arc·tic \(ˌ)ant-'ärk-tik, -'är-tik\ *adjective, often capitalized* [Middle English *antartik,* from Latin *antarcticus,* from Greek *antarktikos,* from *anti-* + *arktikos* arctic] (14th century)
: of or relating to the south pole or to the region near it

antarctic circle *noun, often A&C capitalized* (1556)
: the parallel of latitude that is approximately 66½ degrees south of the equator and that circumscribes the southern frigid zone

An·tar·es \an-'tar-(ˌ)ēz, -'ter-\ *noun* [Greek *Antarēs*]
: a giant red star of very low density that is the brightest star in Scorpio

ant·bear \'ant-ˌbar, -ˌber\ *noun* (circa 1889)
: AARDVARK

ant cow *noun* (1875)
: an aphid from which ants obtain honeydew

¹an·te \'an-tē\ *noun* [*ante-*] (1838)
1 : a poker stake usually put up before the deal to build the pot ⟨the dealer called for a dollar *ante*⟩
2 : COST, PRICE ⟨these improvements would raise the *ante*⟩

²ante *verb* **an·ted; an·te·ing** (1845)
transitive verb
: to put up (an ante); *also* **:** PAY, PRODUCE — often used with *up*
intransitive verb
: PAY UP — often used with *up*

ante- *prefix* [Middle English, from Latin, from *ante* before, in front of; akin to Old English *and-* against, Greek *anti* before, against — more at END]
1 a : prior **:** earlier ⟨*ante*date⟩ **b :** anterior **:** forward ⟨*ante*room⟩
2 a : prior to **:** earlier than ⟨*ante*diluvian⟩ **b :** in front of ⟨*ante*choir⟩

ant·eat·er \'ant-ˌē-tər\ *noun* (1764)
: any of several mammals that feed largely or entirely on ants or termites: as **a :** any of a family (Myrmecophagidae) of New World edentates with a long narrow snout, a long tongue, and large salivary glands that includes the giant anteater and tamandua **b :** PANGOLIN **c :** ECHIDNA **d :** AARDVARK

an·te·bel·lum \ˌan-ti-'be-ləm\ *adjective* [Latin *ante bellum* before the war] (circa 1847)
: existing before a war; *especially* **:** existing before the Civil War

an·te·cede \ˌan-tə-'sēd\ *transitive verb* **-ced·ed; -ced·ing** [Latin *antecedere*] (1624)
: PRECEDE

an·te·ced·ence \-'sē-dᵊn(t)s\ *noun* (1651)
: PRIORITY, PRECEDENCE

¹an·te·ced·ent \ˌan-tə-'sē-dᵊnt\ *noun* [Middle English, from Medieval Latin & Latin; Medieval Latin *antecedent-, antecedens,* from Latin, what precedes, from neuter of *antecedent-, antecedens,* present participle of *antecedere* to go before, from *ante-* + *cedere* to go] (14th century)
1 : a substantive word, phrase, or clause whose denotation is referred to by a pronoun (as *John* in "Mary saw John and called to him"); *broadly* **:** a word or phrase replaced by a substitute
2 : the conditional element in a proposition (as *if A* in "if A, then B")
3 : the first term of a mathematical ratio
4 a : a preceding event, condition, or cause **b** *plural* **:** the significant events, conditions, and traits of one's earlier life
5 a : PREDECESSOR; *especially* **:** a model or stimulus for later developments **b** *plural* **:** ANCESTORS, PARENTS

²antecedent *adjective* (14th century)
: PRIOR
synonym see PRECEDING
— an·te·ced·ent·ly *adverb*

an·te·ces·sor \ˌan-ti-'se-sər\ *noun* [Middle English *antecessour,* from Latin *antecessor* — more at ANCESTOR] (14th century)
: one that goes before **:** PREDECESSOR

an·te·cham·ber \'an-ti-ˌchām-bər\ *noun* [French *antichambre,* from Middle French, from Italian *anti-* (from Latin *ante-*) + Middle French *chambre* room] (circa 1656)
: ANTEROOM

an·te·chap·el \-ˌcha-pəl\ *noun* (1703)
: a vestibule or anteroom to a chapel or church

an·te·choir \'an-ti-ˌkwī(-ə)r\ *noun* (circa 1889)
: a space enclosed or reserved for the clergy and choristers at the entrance to a choir

¹an·te·date \'an-ti-ˌdāt\ *noun* (15th century)
: a date assigned to an event or document earlier than the actual date of the event or document

²an·te·date \'an-ti-ˌdāt, ˌan-ti-'\ *transitive verb* (1572)
1 a : to date as of a time prior to that of execution **b :** to assign to a date prior to that of actual occurrence
2 *archaic* **:** ANTICIPATE
3 : to precede in time

an·te·di·lu·vi·an \ˌan-ti-də-'lü-vē-ən, -(ˌ)dī-\ *adjective* [*ante-* + Latin *diluvium* flood — more at DELUGE] (1646)
1 : of or relating to the period before the flood described in the Bible
2 : made, evolved, or developed a long time ago ⟨an *antediluvian* automobile⟩
— antediluvian *noun*

an·te·fix \'an-ti-ˌfiks\ *noun, plural* **-fix·ae** \-ˌfik-ˌsē\ *or* **-fix·es** \-ˌfik-səz\ [Latin *antefixum,* from neuter of *antefixus,* past participle of *antefigere* to fasten before, from *ante-* + *figere* to fasten — more at FIX] (1832)
: an ornament at the eaves of a classical building concealing the ends of the joint tiles of the roof

an·te·lope \'an-tᵊl-ˌōp\ *noun, plural* **-lope** *or* **-lopes** [Middle English, fabulous heraldic beast, probably from Middle French *antelop* savage animal with sawlike horns, from Medieval Latin *anthalopus,* from Late Greek *antholop-, antholops*] (15th century)
1 a : any of various ruminant mammals (family Bovidae) chiefly of Africa and southwest Asia that differ from the true oxen especially in lighter racier build and horns directed upward and backward **b :** PRONGHORN
2 : leather from antelope hide

an·te me·ri·di·em \ˌan-ti-mə-'ri-dē-əm, -dē-ˌem\ *adjective* [Latin] (1563)
: being before noon — abbreviation *a.m.*

an·te·mor·tem \-'mȯr-təm\ *adjective* [Latin *ante mortem*] (1883)
: preceding death

an·te·na·tal \-'nā-tᵊl\ *adjective* (1817)
: PRENATAL ⟨*antenatal* diagnosis of birth defects⟩
— an·te·na·tal·ly \-ē\ *adverb*

an·ten·na \an-'te-nə\ *noun, plural* **-nae** \-(ˌ)nē\ *or* **-nas** [Medieval Latin, from Latin, sail yard] (1646)
1 *plural* **-nae :** one of a pair of slender movable segmented sensory organs on the head of insects, myriapods, and crustaceans — see INSECT illustration
2 : a usually metallic device (as a rod or wire) for radiating or receiving radio waves
3 *antennae plural* **:** a special sensitivity or receptiveness ⟨his political *antennae* proved to be shrewder than ever —Erich Segal⟩ ◆
— an·ten·nal \-'te-nᵊl\ *adjective*

an·ten·nule \an-'ten-(ˌ)yü(ə)l\ *noun* (1845)
: a small antenna or similar appendage
— an·ten·nu·lar \-'ten-yə-lər\ *adjective*

an·te·nup·tial \ˌan-ti-'nəp-shəl, -chəl; ÷-shə-wəl, ÷-chə-\ *adjective* (1818)
: PRENUPTIAL

an·te·pen·di·um \ˌan-ti-'pen-dē-əm\ *noun, plural* **-di·ums** *or* **-dia** \-dē-ə\ [Medieval Latin, from Latin *ante-* + *pendēre* to hang — more at PENDANT] (circa 1696)
: a hanging for the front of an altar, pulpit, or lectern

an·te·pe·nult \ˌan-ti-'pē-ˌnəlt, -pi-'\ *also* **an·te·pen·ul·ti·ma** \-pi-'nəl-tə-mə\ *noun* [Late

Latin *antepaenultima,* feminine of *antepaenultimus* preceding the next to last, from Latin *ante-* + *paenultimus* penultimate] (1581)
: the third syllable of a word counting from the end (as *cu* in *accumulate*)
— an·te·pen·ul·ti·mate \-pi-ˈnəl-tə-mət\ *adjective or noun*
an·te–post \ˈan-ti-ˌpōst\ *adjective* (1902)
British **:** relating to or being wagers on a horse race made especially before the day of the race
an·te·ri·or \an-ˈtir-ē-ər\ *adjective* [Latin, comparative of *ante* before — more at ANTE-] (1541)
1 a : situated before or toward the front **b :** situated near or toward the head or part most nearly corresponding to a head
2 : coming before in time or development
synonym see PRECEDING
— an·te·ri·or·ly *adverb*
an·te·room \ˈan-ti-ˌrüm, -ˌru̇m\ *noun* (1762)
: an outer room that leads to another room and that is often used as a waiting room
anth- — see ANTI-
an·thel·min·tic \ˌant-ˌhel-ˈmin-tik, ˌan-ˌthel-\ *adjective* [*anti-* + Greek *helminth-, helmis* worm] (1684)
: expelling or destroying parasitic worms especially of the intestine
— anthelmintic *noun*
an·them \ˈan(t)-thəm\ *noun* [Middle English *antem,* from Old English *antefn,* from Late Latin *antiphona,* from Late Greek *antiphōna,* plural of *antiphōnon,* from Greek, neuter of *antiphōnos* responsive, from *anti-* + *phōnē* sound — more at BAN] (before 12th century)
1 a : a psalm or hymn sung antiphonally or responsively **b :** a sacred vocal composition with words usually from the Scriptures
2 : a song or hymn of praise or gladness
an·the·mi·on \an-ˈthē-mē-ən\ *noun, plural* **-mia** \-mē-ə\ [Greek, from diminutive of *anthemon* flower, from *anthos* — more at ANTHOLOGY] (1865)
: a flat ornament of floral form (as in relief sculpture or in painting)
an·ther \ˈan(t)-thər\ *noun* [New Latin *anthera,* from Latin, medicine made from flowers, from Greek *anthēra,* from feminine of *anthēros* flowery, from *anthos*] (circa 1706)
: the part of a stamen that produces and contains pollen and is usually borne on a stalk — see FLOWER illustration
— an·ther·al \-thə-rəl\ *adjective*
an·ther·id·i·um \ˌan(t)-thə-ˈri-dē-əm\ *noun, plural* **-id·ia** \-dē-ə\ [New Latin, from *anthera*] (1854)
: the male reproductive organ of some cryptogamous plants
— an·ther·id·i·al \-dē-əl\ *adjective*
an·the·sis \an-ˈthē-səs\ *noun* [New Latin, from Greek *anthēsis* bloom, from *anthein* to flower, from *anthos*] (circa 1823)
: the action or period of opening of a flower
ant·hill \ˈant-ˌhil\ *noun* (14th century)
: a mound of debris thrown up by ants or termites in digging their nest
an·tho·cy·a·nin \ˌan(t)-thə-ˈsī-ə-nən\ *also* **an·tho·cy·an** \-ˈsī-ən, -ˌan\ *noun* [Greek *anthos* + *kyanos* dark blue] (1839)
: any of various soluble glycoside pigments producing blue to red coloring in flowers and plants
an·thol·o·gist \an-ˈthä-lə-jist\ *noun* (1805)
: a compiler of an anthology
an·thol·o·gize \-ˌjīz\ *transitive verb* **-gized; -giz·ing** (1892)
: to compile, publish, or include in an anthology
— an·thol·o·giz·er \-ˌjī-zər\ *noun*
an·thol·o·gy \an-ˈthä-lə-jē\ *noun, plural* **-gies** [New Latin *anthologia* collection of epigrams, from Middle Greek, from Greek, flower gathering, from *anthos* flower + *logia* collecting, from *legein* to gather; akin to Sanskrit *andha* herb — more at LEGEND] (1640)
1 : a collection of selected literary pieces or passages or works of art or music
2 : ASSORTMENT ⟨an *anthology* of threadbare clichés of . . . bistro cuisine —Jay Jacobs⟩
— an·tho·log·i·cal \ˌan(t)-thə-ˈlä-ji-kəl\ *adjective*
an·thoph·i·lous \an-ˈthä-fə-ləs\ *adjective* [International Scientific Vocabulary, from Greek *anthos* + English *-philous*] (1883)
: feeding upon or living among flowers ⟨*anthophilous* insects⟩
an·tho·phyl·lite \ˌan(t)-thə-ˈfi-ˌlīt, (ˌ)an-ˈthä-fə-\ *noun* [German *Anthophyllit,* from New Latin *anthophyllum,* from Greek *anthos* + *phyllon* leaf — more at BLADE] (circa 1828)
: an orthorhombic mineral of the amphibole group that is essentially a silicate of magnesium and iron and is usually lamellar or fibrous and when fibrous is one of the less common forms of asbestos
an·tho·zo·an \ˌan(t)-thə-ˈzō-ən\ *noun* [ultimately from Greek *anthos* + *zōion* animal; akin to Greek *zōē* life — more at QUICK] (circa 1889)
: any of a class (Anthozoa) of marine coelenterates (as the corals and sea anemones) having polyps with radial partitions
— anthozoan *adjective*
an·thra·cene \ˈan(t)-thrə-ˌsēn\ *noun* (1863)
: a crystalline aromatic hydrocarbon $C_{14}H_{10}$ obtained from coal-tar distillation
an·thra·cite \ˈan(t)-thrə-ˌsīt\ *noun* [Greek *anthrakitis,* from *anthrak-, anthrax* coal] (1812)
: a hard natural coal of high luster differing from bituminous coal in containing little volatile matter and in burning very cleanly — called also *hard coal*
— an·thra·cit·ic \ˌan(t)-thrə-ˈsi-tik\ *adjective*
an·thrac·nose \an-ˈthrak-ˌnōs\ *noun* [French, from Greek *anthrak-, anthrax* + *nosos* disease] (1886)
: any of numerous destructive plant diseases caused by imperfect fungi and characterized especially by necrotic lesions
an·thra·ni·late \an-ˈthra-nᵊl-ˌat, ˌan-thrə-ˈni-ˌlāt\ *noun* (1921)
: a salt or ester of anthranilic acid
an·thra·nil·ic acid \ˌan(t)-thrə-ˈni-lik-\ *noun* [International Scientific Vocabulary *anthracene* + *aniline*] (1881)
: a crystalline acid $NH_2C_6H_4COOH$ used as an intermediate in the manufacture of dyes (as indigo), pharmaceuticals, and perfumes
an·thra·qui·none \ˌan(t)-thrə-kwi-ˈnōn, -ˈkwi-ˌnōn\ *noun* [probably from French, from *anthracene* + *quinone*] (1869)
: a yellow crystalline ketone $C_{14}H_8O_2$ often derived from anthracene and used especially in the manufacture of dyes
an·thrax \ˈan-ˌthraks\ *noun* [Middle English *antrax* carbuncle, from Latin *anthrax,* from Greek, coal, carbuncle] (1876)
: an infectious disease of warm-blooded animals (as cattle and sheep) caused by a spore-forming bacterium (*Bacillus anthracis*), transmissible to humans especially by the handling of infected products (as wool), and characterized by external ulcerating nodules or by lesions in the lungs
anthrop- *or* **anthropo-** *combining form* [Latin *anthropo-,* from Greek *anthrōp-, anthrōpo-,* from *anthrōpos*]
: human being ⟨*anthropo*genic⟩
an·throp·ic \an-ˈthrä-pik\ *or* **an·throp·i·cal** \-pi-kəl\ *adjective* [Greek *anthrōpikos,* from *anthrōpos*] (circa 1806)
: of or relating to human beings or the period of their existence on earth
anthropic principle *noun* (1974)
: either of two principles in cosmology: **a :** conditions that are observed in the universe must allow the observer to exist — called also *weak anthropic principle* **b :** the universe must have properties that make inevitable the existence of intelligent life — called also *strong anthropic principle*
an·thro·po·cen·tric \ˌan(t)-thrə-pə-ˈsen-trik\ *adjective* (1863)
1 : considering human beings as the most significant entity of the universe
2 : interpreting or regarding the world in terms of human values and experiences
— an·thro·po·cen·tri·cal·ly \-tri-k(ə-)lē\ *adverb*
— an·thro·po·cen·tric·i·ty \-pō-(ˌ)sen-ˈtri-sə-tē\ *noun*
— an·thro·po·cen·trism \-ˈsen-ˌtri-zəm\ *noun*
an·thro·po·gen·ic \-pə-ˈje-nik\ *adjective* (1923)
: of, relating to, or resulting from the influence of human beings on nature ⟨*anthropogenic* pollutants⟩
an·thro·poid \ˈan(t)-thrə-ˌpȯid\ *noun* [Greek *anthropoeidēs* resembling a human, from *anthrōpos*] (1832)
1 : APE 1b
2 : a person resembling an ape ⟨the howling *anthropoids* of the Hookworm Belt —H. L. Mencken⟩
— anthropoid *adjective*
anthropoid ape *noun* (circa 1837)
: APE 1b
an·thro·pol·o·gy \ˌan(t)-thrə-ˈpä-lə-jē\ *noun* [New Latin *anthropologia,* from *anthrop-* + *-logia* -logy] (1593)
1 : the science of human beings; *especially* **:** the study of human beings in relation to distribution, origin, classification, and relationship of races, physical character, environmental and social relations, and culture
2 : theology dealing with the origin, nature, and destiny of human beings
— an·thro·po·log·i·cal \-pə-ˈlä-ji-kəl\ *adjective*
— an·thro·po·log·i·cal·ly \-ji-k(ə-)lē\ *adverb*
— an·thro·pol·o·gist \ˌan(t)-thrə-ˈpä-lə-jist\ *noun*
an·thro·pom·e·try \ˌan(t)-thrə-ˈpä-mə-trē\ *noun* [French *anthropométrie,* from *anthrop-* + *-métrie* -metry] (circa 1839)
: the study of human body measurements especially on a comparative basis
— an·thro·po·met·ric \-pə-ˈme-trik\ *adjective*
an·thro·po·morph \ˈan(t)-thrə-pə-ˌmȯrf\ *noun* (1894)
: a stylized human figure (as in prehistoric art)
an·thro·po·mor·phic \ˌan(t)-thrə-pə-ˈmȯr-fik\ *adjective* [Late Latin *anthropomorphus* of

◇ WORD HISTORY

antenna In Classical Latin *antenna* means "sail yard," the long spar tapered toward the ends that supports and spreads the head of a sail on a sailing vessel. The Greek word for a sail yard was *keraia,* a derivative of *keras,* meaning "horn." But *keraia* was also used for a number of other things that thrust outward, such as the projections of the hipbone or the spur of a mountain. Aristotle, in his *Historia Animalium* ("Natural History of Animals") used plural *keraiai* for the "horns" or feelers of insects. In a Latin translation of Aristotle's work made during the Renaissance, the word *antennae* appears for Greek *keraiai.* When scientific writings on insects began appearing in English in the 17th century, entomologists borrowed *antennae* as the word for those feelers.

human form, from Greek *anthrōpomorphos,* from *anthrōp-* + *-morphos* -morphous] (1827)
1 : described or thought of as having a human form or human attributes ⟨*anthropomorphic* deities⟩
2 : ascribing human characteristics to nonhuman things ⟨*anthropomorphic* supernaturalism⟩
— **an·thro·po·mor·phi·cal·ly** \-fi-k(ə-)lē\ *adverb*

an·thro·po·mor·phism \-ˌfi-zəm\ *noun* (1753)
: an interpretation of what is not human or personal in terms of human or personal characteristics **:** HUMANIZATION
— **an·thro·po·mor·phist** \-fist\ *noun*

an·thro·po·mor·phize \-ˌfīz\ *verb* **-phized; -phiz·ing** (1845)
transitive verb
: to attribute human form or personality to
intransitive verb
: to attribute human form or personality to things not human
— **an·thro·po·mor·phi·za·tion** \-ˌmȯr-fə-ˈzā-shən\ *noun*

an·thro·po·pa·thism \ˌan(t)-thrə-ˈpä-pə-ˌthi-zəm, -pō-ˈpa-ˌthi-\ *noun* [Late Greek *anthrōpopatheia* humanity, from Greek *anthrōpopathēs* having human feelings, from *anthrōp-* + *pathos* experience — more at PATHOS] (1847)
: the ascription of human feelings to something not human

an·thro·poph·a·gous \ˌan(t)-thrə-ˈpä-fə-gəs\ *adjective* (circa 1828)
: feeding on human flesh
— **an·thro·poph·a·gy** \-fə-jē\ *noun*

an·thro·poph·a·gus \-fə-gəs\ *noun, plural* **-a·gi** \-fə-ˌgī, -ˌjī, -ˌgē\ [Latin, from Greek *anthrōpophagos,* from *anthrōp-* + *-phagos* -phagous] (1552)
: MAN-EATER, CANNIBAL

an·thro·pos·o·phy \ˌan(t)-thrə-ˈpä-sə-fē\ *noun* (1916)
: a 20th century religious system growing out of theosophy and centering on human development

an·thur·i·um \an-ˈthu̇r-ē-əm, -ˈthyu̇r-\ *noun* [New Latin, from Greek *anthos* flower + *oura* tail — more at ANTHOLOGY, ASS] (circa 1839)
: any of a genus (*Anthurium*) of tropical American plants of the arum family with large often highly colored leaves, a cylindrical spadix, and a colored spathe

[1]an·ti \ˈan-ˌtī, ˈan-tē\ *noun, plural* **antis** [*anti-*] (1788)
: one that is opposed

[2]anti *adjective* (1857)
: OPPOSED

[3]anti *preposition* (1953)
: opposed to **:** AGAINST

anti- *or* **ant-** *or* **anth-** *prefix* [*anti-* from Middle English, from Middle French & Latin; Middle French, from Latin, against, from Greek, from *anti; ant-* from Middle English, from Latin, against, from Greek, from *anti; anth-* from Latin, against, from Greek, from *anti* — more at ANTE-]
1 a : of the same kind but situated opposite, exerting energy in the opposite direction, or pursuing an opposite policy ⟨*anti*clinal⟩ **b :** one that is opposite in kind to ⟨*anti*climax⟩
2 a : opposing or hostile to in opinion, sympathy, or practice ⟨*anti*-Semite⟩ **b :** opposing in effect or activity ⟨*ant*acid⟩
3 : serving to prevent, cure, or alleviate ⟨*anti*anxiety⟩
4 : combating or defending against ⟨*anti*aircraft⟩ ⟨*anti*missile⟩

an·ti·ac·a·dem·ic
an·ti–ac·ne
an·ti·ad·min·is·tra·tion
an·ti·ag·gres·sion
an·ti·ag·ing
an·ti–AIDS
an·ti·al·co·hol
an·ti·al·co·hol·ism
an·ti·alien
an·ti·al·ler·gen·ic
an·ti·ane·mia
an·ti·apart·heid
an·ti·aph·ro·di·si·ac
an·ti·aris·to·crat·ic
an·ti·ar·thrit·ic
an·ti·ar·thri·tis
an·ti·as·sim·i·la·tion
an·ti·asth·ma
an·ti·au·thor·i·tar·i·an
an·ti·au·thor·i·tar·i·an·ism
an·ti·au·thor·i·ty
an·ti·back·lash
an·ti·bi·as
an·ti·bill·board
an·ti–Bol·she·vik
an·ti·boss
an·ti·bour·geois
an·ti·boy·cott
an·ti–Brit·ish
an·ti·bug
an·ti·bu·reau·crat·ic
an·ti·bur·glar
an·ti·bur·glary
an·ti·cak·ing
an·ti·cap·i·tal·ism
an·ti·cap·i·tal·ist
an·ti·car·cin·o·gen
an·ti·car·ci·no·gen·ic
an·ti·car·ies
an·ti–Cath·o·lic
an·ti–Ca·thol·i·cism
an·ti·cel·lu·lite
an·ti·cen·sor·ship
an·ti·cho·les·ter·ol
an·ti–Chris·tian
an·ti–Chris·tian·i·ty
an·ti·church
an·ti·cig·a·rette
an·ti·city
an·ti·clas·si·cal
an·ti·cling
an·ti·clot·ting
an·ti·cold
an·ti·col·li·sion
an·ti·co·lo·nial
an·ti·co·lo·nial·ism
an·ti·co·lo·nial·ist
an·ti·com·mer·cial
an·ti·com·mer·cial·ism
an·ti·com·mu·nism
an·ti·com·mu·nist
an·ti·con·glom·er·ate
an·ti·con·ser·va·tion
an·ti·con·ser·va·tion·ist
an·ti·con·sum·er
an·ti·con·ven·tion·al
an·ti·cor·po·rate
an·ti·cor·ro·sion
an·ti·cor·ro·sive
an·ti·cor·rup·tion
an·ti·coun·ter·feit·ing
an·ti·crack
an·ti·cre·ative
an·ti·crime
an·ti·cru·el·ty
an·ti·cult
an·ti·cul·tur·al
an·ti·dan·druff
an·ti–Dar·win·i·an
an·ti–Dar·win·ism
an·ti·def·a·ma·tion
an·ti·de·pres·sion
an·ti·de·seg·re·ga·tion
an·ti·de·sert·i·fi·ca·tion
an·ti·des·ic·cant
an·ti·de·vel·op·ment
an·ti·di·a·bet·ic
an·ti·di·ar·rhe·al
an·ti·di·lu·tion
an·ti·dis·crim·i·na·tion
an·ti·dog·mat·ic
an·ti·draft
an·ti·eco·nom·ic
an·ti·ed·u·ca·tion·al
an·ti·egal·i·tar·i·an
an·ti·elite
an·ti·elit·ism
an·ti·elit·ist
an·ti·emet·ic
an·ti–En·glish
an·ti·en·tro·pic
an·ti·ep·i·lep·sy
an·ti·ep·i·lep·tic
an·ti·erot·ic
an·ti·es·tab·lish·ment
an·ti·es·tro·gen
an·ti·evo·lu·tion
an·ti·evo·lu·tion·ary
an·ti·evo·lu·tion·ism
an·ti·evo·lu·tion·ist
an·ti·fam·i·ly
an·ti·fas·cism
an·ti·fas·cist
an·ti·fash·ion
an·ti·fash·ion·able
an·ti·fa·tigue
an·ti·fe·male
an·ti·fem·i·nine
an·ti·fem·i·nism
an·ti·fem·i·nist
an·ti·fil·i·bus·ter
an·ti·flu
an·ti·foam
an·ti·foam·ing
an·ti·fog·ging
an·ti·fore·clo·sure
an·ti·for·eign
an·ti·for·eign·er
an·ti·for·mal·ist
an·ti·fraud
an·ti–French
an·ti·fric·tion
an·ti·fur
an·ti·gam·bling
an·ti·gay
an·ti–Ger·man
an·ti·glare
an·ti·gov·ern·ment
an·ti·growth
an·ti·guer·ril·la
an·ti·gun
an·ti·her·pes
an·ti·hi·er·ar·chi·cal
an·ti·hi·jack
an·ti·his·tor·i·cal
an·ti·ho·mo·sex·u·al
an·ti·hu·man·ism
an·ti·hu·man·is·tic
an·ti·hu·man·i·tar·i·an
an·ti·hunt·er
an·ti·hunt·ing
an·ti·hys·ter·ic
an·ti–ic·ing
an·ti–ideo·log·i·cal
an·ti–im·pe·ri·al·ism
an·ti–im·pe·ri·al·ist
an·ti–in·cum·bent
an·ti–in·fec·tive
an·ti–in·fla·tion
an·ti–in·fla·tion·ary
an·ti–in·sti·tu·tion·al
an·ti–in·te·gra·tion
an·ti–in·tru·sion
an·ti–Ital·ian
an·ti·jam
an·ti·jam·ming
an·ti–Jap·a·nese
an·ti–Jew·ish
an·ti·kick·back
an·ti·knock
an·ti·la·bor
an·ti·leak
an·ti·lep·ro·sy
an·ti·lib·er·al
an·ti·lib·er·al·ism
an·ti·lib·er·tar·i·an
an·ti·lit·er·ate
an·ti·lit·ter
an·ti·lit·ter·ing
an·ti·log·i·cal
an·ti·lynch·ing
an·ti·ma·cho
an·ti·ma·lar·ia
an·ti·male
an·ti·man
an·ti·man·age·ment
an·ti·mar·i·jua·na
an·ti·mar·ket
an·ti·ma·te·ri·al·ism
an·ti·ma·te·ri·al·ist
an·ti·mech·a·nist
an·ti·merg·er
an·ti·met·a·bol·ic
an·ti·meta·phys·i·cal
an·ti·mil·i·ta·rism
an·ti·mil·i·ta·rist
an·ti·mil·i·tary
an·ti·mis·ce·ge·na·tion
an·ti·mis·sile
an·ti·mod·ern
an·ti·mod·ern·ist
an·ti·mo·nar·chi·cal
an·ti·mon·ar·chist
an·ti·mo·nop·o·list
an·ti·mo·nop·o·ly
an·ti·mos·qui·to
an·ti·mu·si·cal
an·ti·nar·ra·tive
an·ti·na·tion·al
an·ti·na·tion·al·ist
an·ti·nat·u·ral
an·ti·na·ture
an·ti·nau·sea
an·ti–Na·zi
an·ti–Ne·gro
an·ti·nep·o·tism
an·ti·noise
an·ti·obe·si·ty
an·ti·ob·scen·i·ty
an·ti·or·ga·ni·za·tion
an·ti·pa·pal
an·ti·par·ty
an·ti·pes·ti·cide
an·ti·pi·ra·cy
an·ti·plague
an·ti·plaque
an·ti·plea·sure
an·ti·poach·ing
an·ti·po·lice
an·ti·po·lit·i·cal
an·ti·pol·i·tics
an·ti·pop·u·lar
an·ti·porn
an·ti·por·no·graph·ic
an·ti·por·nog·ra·phy
an·ti·pot
an·ti·pov·er·ty
an·ti·pred·a·tor
an·ti·press
an·ti·prof·i·teer·ing
an·ti·pro·gres·sive
an·ti·pros·ti·tu·tion
an·ti·pru·rit·ic
an·ti·ra·bies
an·ti·rac·ism
an·ti·rac·ist
an·ti·rack·e·teer·ing
an·ti·ra·dar
an·ti·rad·i·cal
an·ti·rad·i·cal·ism
an·ti·rape
an·ti·ra·tio·nal
an·ti·ra·tio·nal·ism
an·ti·ra·tio·nal·ist
an·ti·ra·tio·nal·i·ty
an·ti·re·al·ism
an·ti·re·al·ist
an·ti·re·ces·sion
an·ti·re·ces·sion·ary
an·ti·red
an·ti·re·duc·tion·ism
an·ti·re·duc·tion·ist
an·ti·re·flec·tion
an·ti·re·flec·tive
an·ti·re·form
an·ti·reg·u·la·to·ry
an·ti·re·li·gion
an·ti·re·li·gious
an·ti·rev·o·lu·tion·ary
an·ti·ri·ot
an·ti·rit·u·al·ism
an·ti·rock
an·ti·roll
an·ti·ro·man·tic
an·ti·ro·man·ti·cism
an·ti·roy·al·ist
an·ti–Rus·sian
an·ti·rust
an·ti·sat·el·lite
an·ti·schizo·phre·nia
an·ti·schiz·o·phren·ic
an·ti·sci·ence
an·ti·sci·en·tif·ic
an·ti·se·cre·cy
an·ti·seg·re·ga·tion
an·ti·sei·zure
an·ti·sen·ti·men·tal
an·ti·sep·a·rat·ist
an·ti·sex
an·ti·sex·ist
an·ti·sex·u·al
an·ti·sex·u·al·i·ty
an·ti·shark
an·ti·ship
an·ti·shock
an·ti·shop·lift·ing
an·ti·skid
an·ti·slav·ery
an·ti·sleep
an·ti·slip
an·ti·smog
an·ti·smoke
an·ti·smok·er
an·ti·smok·ing
an·ti·smug·gling
an·ti·smut
an·ti·snob
an·ti·so·cial·ist
an·ti–So·vi·et
an·ti–So·vi·et·ism
an·ti·spec·u·la·tion
an·ti·spec·u·la·tive
an·ti·spend·ing
an·ti–Sta·lin·ist
an·ti·state
an·ti·stick
an·ti·sto·ry
an·ti·stress
an·ti·strike
an·ti·stu·dent
an·ti·sub·ma·rine
an·ti·sub·si·dy
an·ti·sub·ver·sion
an·ti·sub·ver·sive
an·ti·sui·cide
an·ti·syph·i·lit·ic
an·ti·take·over
an·ti·tank
an·ti·tar·nish
an·ti·tax
an·ti·tech·no·log·i·cal
an·ti·tech·nol·o·gy
an·ti·ter·ror·ism
an·ti·ter·ror·ist
an·ti·theft
an·ti·the·o·ret·i·cal
an·ti·to·bac·co
an·ti·to·tal·i·tar·i·an
an·ti·tra·di·tion·al
an·ti·tu·ber·cu·lar
an·ti·tu·ber·cu·lo·sis
an·ti·tu·ber·cu·lous
an·ti·tu·mor

an·ti·tu·mor·al
an·ti·ty·phoid
an·ti·ul·cer
an·ti·un·em·ploy·ment
an·ti·union
an·ti·uni·ver·si·ty
an·ti·ur·ban
an·ti·vi·o·lence
an·ti·vi·ral
an·ti·vi·rus
an·ti·vivi·sec·tion
an·ti·vivi·sec·tion·ist
an·ti·war
an·ti·wear
an·ti·wel·fare
an·ti–West
an·ti–West·ern
an·ti·whal·ing
an·ti·wom·an
an·ti·wrin·kle
an·ti–Zi·on·ist

an·ti·abor·tion \ˌan-tē-ə-ˈbȯr-shən, ˌan-ˌtī-\ *adjective* (1971)
: opposed to abortion ⟨*antiabortion* lobbyists⟩
— **an·ti·abor·tion·ist** \-sh(ə-)nist\ *noun*

an·ti·air \ˌan-tē-ˈar, -ˈer, -ˌtī-\ *adjective* (1925)
: ANTIAIRCRAFT

¹an·ti·air·craft \-ˈar-ˌkraft, -ˈer-\ *adjective* (1914)
: designed for or concerned with defense against air attack

²antiaircraft *noun* (1926)
: antiaircraft guns or their fire

an·ti–Amer·i·can \-ə-ˈmer-ə-kən, -ˈmər-, -ˈmar-, -i-kən\ *adjective* (1773)
: opposed or hostile to the people or the government policies of the U.S.
— **an·ti–Amer·i·can·ism** \-kə-ˌni-zəm\ *noun*

an·ti·anx·i·ety \-(ˌ)aŋ-ˈzī-ə-tē\ *adjective* (1962)
: tending to prevent or relieve anxiety ⟨*antianxiety* drugs⟩

an·ti·ar·rhyth·mic \ˌan-tē-(ˌ)ā-ˈrith-mik, ˌan-ˌtī-\ *adjective* (1954)
: tending to prevent or relieve cardiac arrhythmia ⟨an *antiarrhythmic* agent⟩

an·ti–art \-ˈärt\ *noun* (1937)
: art based on premises antithetical to traditional or popular art forms; *specifically* **:** DADA

an·ti·aux·in \-ˈȯk-sən\ *noun* (1949)
: a plant substance that opposes or suppresses the natural effect of an auxin

an·ti·bac·te·ri·al \ˌan-ti-bak-ˈtir-ē-əl, ˌan-ˌtī-\ *adjective* (circa 1897)
: directed or effective against bacteria
— **antibacterial** *noun*

an·ti·bal·lis·tic missile \ˌan-ti-bə-ˈlis-tik-, ˌan-ˌtī-\ *noun* (circa 1959)
: a missile for intercepting and destroying ballistic missiles

an·ti·bi·o·sis \-bī-ˈō-səs, -bē-\ *noun* [New Latin] (1899)
: antagonistic association between organisms to the detriment of one of them or between one organism and a metabolic product of another

¹an·ti·bi·ot·ic \ˌan-ti-bī-ˈä-tik, -tī-; ˌan-ti-bē-\ *adjective* (1894)
1 : tending to prevent, inhibit, or destroy life
2 : of or relating to antibiotics or to antibiosis
— **an·ti·bi·ot·i·cal·ly** \-ti-k(ə-)lē\ *adverb*

²antibiotic *noun* (1943)
: a substance produced by or a semisynthetic substance derived from a microorganism and able in dilute solution to inhibit or kill another microorganism

an·ti·black \-ˈblak\ *adjective* (1952)
: opposed or hostile to black people
— **an·ti·black·ism** \-ˈbla-ˌki-zəm\ *noun*

an·ti·body \ˈan-ti-ˌbä-dē\ *noun* (1900)
: any of a large number of proteins of high molecular weight that are produced normally by specialized B cells after stimulation by an antigen and act specifically against the antigen in an immune response, that are produced abnormally by some cancer cells, and that typically consist of four subunits including two heavy chains and two light chains — called also *immunoglobulin*

an·ti·busi·ness \ˌan-ti-ˈbiz-nəs, ˌan-ˌtī-, -nəz\ *adjective* (1938)
: antagonistic toward business and especially big business

an·ti·bus·ing \-ˈbə-siŋ\ *adjective* (circa 1969)
: opposed to the busing of schoolchildren ⟨*antibusing* parents⟩ ⟨*antibusing* campaign⟩

¹an·tic \ˈan-tik\ *noun* [Italian *antico* ancient thing or person, from *antico* ancient, from Latin *antiquus* — more at ANTIQUE] (1529)
1 : an attention-drawing often wildly playful or funny act or action **:** CAPER ⟨childish *antics*⟩
2 *archaic* **:** a performer of a grotesque or ludicrous part **:** BUFFOON ◆

²antic *adjective* (1548)
1 *archaic* **:** GROTESQUE, BIZARRE
2 a : characterized by clownish extravagance or absurdity **b :** whimsically gay **:** FROLICSOME
— **an·ti·cal·ly** \-ti-k(ə-)lē\ *adverb*

an·ti·can·cer \ˌan-ˌtī-ˈkan(t)-sər, ˌan-ti-\ *adjective* (1926)
: used against or tending to arrest cancer ⟨*anticancer* drugs⟩ ⟨*anticancer* activity⟩ ⟨*anticancer* effects⟩

an·ti·cho·lin·er·gic \-ˌkō-lə-ˈnər-jik\ *adjective* (1942)
: opposing or blocking the physiologic action of acetylcholine
— **anticholinergic** *noun*

an·ti·cho·lin·es·ter·ase \-ˈnes-tə-ˌrās, -ˌrāz\ *noun* (1942)
: a substance (as neostigmine) that inhibits a cholinesterase by combination with it

An·ti·christ \ˈan-ti-ˌkrīst, -ˌtī-\ *noun* [Middle English *anticrist,* from Old English & Late Latin; Old English *antecrist,* from Late Latin *Antichristus,* from Greek *Antichristos,* from *anti-* + *Christos* Christ] (before 12th century)
1 : one who denies or opposes Christ; *specifically* **:** a great antagonist expected to fill the world with wickedness but to be conquered forever by Christ at his second coming
2 : a false Christ

an·tic·i·pant \an-ˈti-sə-pənt\ *adjective* (1626)
: EXPECTANT, ANTICIPATING — usually used with *of*
— **anticipant** *noun*

an·tic·i·pate \an-ˈti-sə-ˌpāt\ *verb* **-pat·ed; -pat·ing** [Latin *anticipatus,* past participle of *anticipare,* from *ante-* + *-cipare* (from *capere* to take) — more at HEAVE] (1532)
transitive verb
1 : to give advance thought, discussion, or treatment to
2 : to meet (an obligation) before a due date
3 : to foresee and deal with in advance **:** FORESTALL
4 : to use or expend in advance of actual possession
5 : to act before (another) often so as to check or counter
6 : to look forward to as certain **:** EXPECT
intransitive verb
: to speak or write in knowledge or expectation of later matter
synonym see FORESEE, PREVENT
— **an·tic·i·pat·able** \-ˌpā-tə-bəl\ *adjective*
— **an·tic·i·pa·tor** \-ˌpā-tər\ *noun*

an·tic·i·pa·tion \(ˌ)an-ˌti-sə-ˈpā-shən\ *noun* (14th century)
1 a : a prior action that takes into account or forestalls a later action **b :** the act of looking forward; *especially* **:** pleasurable expectation
2 : the use of money before it is available
3 a : visualization of a future event or state **b :** an object or form that anticipates a later type
4 : the early sounding of one or more tones of a succeeding chord to form a temporary dissonance — compare SUSPENSION
synonym see PROSPECT

an·tic·i·pa·to·ry \an-ˈti-sə-pə-ˌtōr-ē, -ˌtȯr-\ *adjective* (1669)
: characterized by anticipation **:** ANTICIPATING

an·ti·cler·i·cal \ˌan-ti-ˈkler-i-kəl, ˌan-ˌtī-\ *adjective* (1845)
: opposed to clericalism or to the interference or influence of the clergy in secular affairs
— **anticlerical** *noun*
— **an·ti·cler·i·cal·ism** \-kə-ˌli-zəm\ *noun*

an·ti·cli·mac·tic \-klī-ˈmak-tik, -klə-\ *also* **an·ti·cli·mac·ti·cal** \-ti-kəl\ *adjective* (1898)
: of, relating to, or marked by anticlimax
— **an·ti·cli·mac·ti·cal·ly** \-ti-k(ə-)lē\ *adverb*

an·ti·cli·max \-ˈklī-ˌmaks\ *noun* (1710)
1 : the usually sudden transition in discourse from a significant idea to a trivial or ludicrous idea; *also* **:** an instance of this transition
2 : an event (as at the end of a series) that is strikingly less important than what has preceded it

an·ti·cli·nal \ˌan-ti-ˈklī-nᵊl\ *adjective* [*anti-* + Greek *klinein* to lean — more at LEAN] (1882)
: occurring at right angles to the surface or circumference of a plant organ ⟨an *anticlinal* pattern of cell walls⟩

an·ti·cline \ˈan-ti-ˌklīn\ *noun* (circa 1861)
: an arch of stratified rock in which the layers bend downward in opposite directions from the crest — compare SYNCLINE

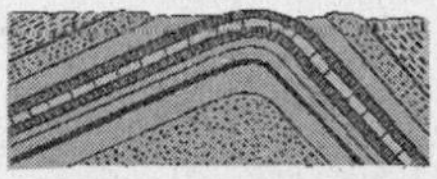
cross section of strata showing anticline

an·ti·clock·wise \ˌan-ti-ˈkläk-ˌwīz, ˌan-ˌtī-\ *adjective or adverb* (1898)
chiefly British **:** COUNTERCLOCKWISE

an·ti·co·ag·u·lant \-kō-ˈa-gyə-lənt\ *noun* (circa 1905)
: a substance that hinders the clotting of blood
— **anticoagulant** *adjective*

an·ti·co·don \-ˈkō-ˌdän\ *noun* (1965)
: a triplet of nucleotide bases in transfer RNA that identifies the amino acid carried and binds to a complementary codon in messenger RNA during protein synthesis at a ribosome

an·ti·com·pet·i·tive \-kəm-ˈpe-tə-tiv\ *adjective* (1952)
: tending to reduce or discourage competition

an·ti·con·vul·sant \-kən-ˈvəl-sənt\ *also* **an·ti·con·vul·sive** \-siv\ *adjective* (1734)
: used or tending to control or to prevent convulsions (as in epilepsy)
— **anticonvulsant** *also* **anticonvulsive** *noun*

an·ti·cy·clone \ˌan-ti-ˈsī-ˌklōn\ *noun* (1877)
1 : a system of winds that rotates about a center of high atmospheric pressure clockwise in the northern hemisphere and counterclockwise in the southern, that usually advances at 20 to 30 miles (about 30 to 50 kilometers) per hour, and that usually has a diameter of 1500 to 2500 miles (2400 to 4000 kilometers)
2 : HIGH 2
— **an·ti·cy·clon·ic** \-sī-ˈklä-nik\ *adjective*

an·ti·dem·o·crat·ic \ˌan-ti-ˌde-mə-ˈkra-tik, ˌan-ˌtī-\ *adjective* (1837)
: opposed or hostile to the theories or policies of democracy

¹an·ti·de·pres·sant \-di-ˈpre-sᵊnt\ *adjective* (1961)

◇ WORD HISTORY

antic The fantastic mural paintings found in the ruins of ancient Roman buildings were, by reason of their age, *antichi*—"ancient things"—to the Italians of the Renaissance. In Renaissance England, on the other hand, any similarly fantastic painting—of whatever vintage—that showed strange combinations of human, animal, and floral forms came to be called *antike* or *anticke,* from the Italian word for "ancient," *antico.* And any odd gesture or strange behavior reminiscent of such artwork became an *antic.*

: used or tending to relieve or prevent psychic depression

[2]antidepressant *noun* (1962)
: an antidepressant drug — compare TRICYCLIC ANTIDEPRESSANT

an·ti·de·riv·a·tive \-di-'ri-və-tiv\ *noun* (circa 1942)
: INDEFINITE INTEGRAL

an·ti·di·uret·ic hormone \ˌan-ti-ˌdī-yu̇-'re-tik-\ *noun* (1942)
: VASOPRESSIN

an·ti·dot·al \ˌan-ti-'dō-tᵊl\ *adjective* (1646)
: of, relating to, or acting as an antidote
— **an·ti·dot·al·ly** \-tᵊl-ē\ *adverb*

an·ti·dote \'an-ti-ˌdōt\ *noun* [Middle English *antidot,* from Latin *antidotum,* from Greek *antidotos,* from feminine of *antidotos* given as an antidote, from *antididonai* to give as an antidote, from *anti-* + *didonai* to give — more at DATE] (15th century)
1 : a remedy to counteract the effects of poison
2 : something that relieves, prevents, or counteracts ⟨an *antidote* to the mechanization of our society⟩
— **antidote** *transitive verb*

an·ti·drom·ic \ˌan-ti-'drä-mik\ *adjective* [*anti-* + Greek *dromos* racecourse, running — more at DROMEDARY] (circa 1927)
: proceeding or conducting in a direction opposite to the usual one — used especially of a nerve impulse or fiber
— **an·ti·drom·i·cal·ly** \-mi-k(ə-)lē\ *adverb*

an·ti·drug \'an-ˌtī-ˌdrəg, ˌan-tī-'\ *adjective* (1970)
: acting against or opposing illicit drugs or their use ⟨*antidrug* activist⟩ ⟨*antidrug* program⟩

an·ti·dump·ing \ˌan-ti-'dəm-piŋ, ˌan-ˌtī-\ *adjective* (1915)
: designed to discourage the importation and sale of foreign goods at prices well below domestic prices ⟨*antidumping* tariffs⟩

an·ti·elec·tron \ˌan-tē-ə-'lek-ˌträn, ˌan-ˌtī-\ *noun* (1931)
: POSITRON

an·ti–fed·er·al·ist \ˌan-ti-'fe-d(ə-)rə-list, ˌan-ˌtī-\ *noun, often A&F capitalized* (1787)
: a member of the group that opposed the adoption of the U.S. Constitution

an·ti·fer·ro·mag·net·ic \-ˌfer-ō-mag-'ne-tik\ *adjective* (1936)
: FERRIMAGNETIC
— **an·ti·fer·ro·mag·net** \-'mag-nət\ *noun*
— **an·ti·fer·ro·mag·net·i·cal·ly** \-mag-'ne-ti-k(ə-)lē\ *adverb*
— **an·ti·fer·ro·mag·net·ism** \-'mag-nə-ˌti-zəm\ *noun*

an·ti·fer·til·i·ty \-fər-'ti-lə-tē\ *adjective* (1953)
: capable of or tending to reduce or destroy fertility **:** CONTRACEPTIVE ⟨*antifertility* agents⟩

an·ti·flu·o·ri·da·tion·ist \-ˌflu̇r-ə-'dā-sh(ə-)nist, -ˌflōr-, -ˌflȯr-\ *noun* (1961)
: a person opposed to the fluoridation of public water supplies

an·ti·foul·ing \-'fau̇-liŋ\ *adjective* (1895)
: intended to prevent fouling of underwater structures (as the bottoms of ships) ⟨*antifouling* paint⟩

an·ti·freeze \'an-ti-ˌfrēz\ *noun* (1924)
: a substance added to a liquid (as the water in an automobile engine) to lower its freezing point

an·ti·fun·gal \ˌan-ti-'fəŋ-gəl, ˌan-ˌtī-\ *adjective* (1945)
: destroying fungi; *also* **:** inhibiting the growth of fungi
— **antifungal** *noun*

an·ti·gen \'an-ti-jən, -ˌjen\ *noun* [German, from French *antigène,* from *anticorps* antibody + *-gène* -gen] (1908)
: a usually protein or carbohydrate substance (as a toxin or enzyme) capable of stimulating an immune response
— **an·ti·gen·ic** \ˌan-ti-'je-nik\ *adjective*
— **an·ti·gen·i·cal·ly** \-ni-k(ə-)lē\ *adverb*
— **an·ti·ge·nic·i·ty** \-jə-'ni-sə-tē\ *noun*

antigenic determinant *noun* (1950)
: EPITOPE

an·ti·glob·u·lin \ˌan-ti-'glä-byə-lən, ˌan-ˌtī-\ *noun* (circa 1909)
: an antibody that combines with and precipitates globulin

An·tig·o·ne \an-'ti-gə-(ˌ)nē\ *noun* [Greek *Antigonē*]
: a daughter of Oedipus and Jocasta who buries her brother Polynices' body against the order of her uncle Creon

[1]an·ti·grav·i·ty \ˌan-ti-'gra-və-tē, ˌan-ˌtī-\ *adjective* (1944)
: reducing, canceling, or protecting against the effect of gravity

[2]antigravity *noun* (1949)
: a hypothetical effect resulting from cancellation or reduction of a gravitational field

an·ti·he·mo·phil·ic factor \-ˌhē-mə-'fi-lik-\ *noun* (1947)
: FACTOR VIII

an·ti·he·ro \'an-ti-ˌhē-(ˌ)rō, 'an-ˌtī-, -ˌhir-(ˌ)ō\ *noun* (1714)
: a protagonist or notable figure who is conspicuously lacking in heroic qualities
— **an·ti·he·ro·ic** \ˌan-ti-hi-'rō-ik, ˌan-ˌtī-\ *adjective*

an·ti·her·o·ine \ˌan-ti-'her-ə-wən, ˌan-ˌtī-\ *noun* (1907)
: a female antihero

an·ti·his·ta·mine \-'his-tə-ˌmēn, -mən\ *noun* (1946)
: any of various compounds that counteract histamine in the body and that are used for treating allergic reactions (as hay fever) and cold symptoms
— **antihistamine** *adjective*
— **an·ti·his·ta·min·ic** \-ˌhis-tə-'mi-nik\ *adjective or noun*

an·ti·hu·man \-'hyü-mən, -'yü-\ *adjective* (1854)
1 : acting or being against humanity
2 : reacting strongly with human antigens

an·ti·hy·per·ten·sive \-ˌhī-pər-'ten(t)-siv\ *noun* (circa 1957)
: a substance that is effective against high blood pressure
— **antihypertensive** *adjective*

an·ti–id·io·type \ˌan-tē-'i-dē-ə-ˌtīp, ˌan-ti-\ *noun* (1975)
: an antibody that treats another antibody as an antigen and suppresses its immunoreactivity
— **an·ti–id·io·typ·ic** \-ˌi-dē-ə-'ti-pik\ *adjective*

an·ti–in·flam·ma·to·ry \-in-'fla-mə-ˌtōr-ē, -ˌtȯr-\ *adjective* (circa 1957)
: counteracting inflammation
— **anti–inflammatory** *noun*

an·ti–in·tel·lec·tu·al \-ˌin-tᵊl-'ek-ch(ə-w)əl, -'ek-shwəl\ *adjective* (1936)
: opposing or hostile to intellectuals or to an intellectual view or approach
— **anti–intellectual** *noun*
— **an·ti–in·tel·lec·tu·al·ism** \-'ek-chə(-wə)-ˌli-zəm, -'ek-shwə-\ *noun*

an·ti·leu·ke·mic \-lü-'kē-mik\ *adjective* (1951)
: counteracting the effects of leukemia

an·ti·life \-'līf\ *adjective* (1929)
: antagonistic or antithetical to life or to normal human values

an·ti·lock \'an-ˌtī-ˌläk, 'an-ti-\ *adjective* (1974)
: being a braking system for a motor vehicle designed to keep the wheels from locking by electronically controlled pulsed application of the brake for each wheel

an·ti·log \'an-ti-ˌlȯg, 'an-ˌtī-, -ˌläg\ *noun* (1910)
: ANTILOGARITHM

an·ti·log·a·rithm \ˌan-ti-'lȯ-gə-ˌri-t͟həm, ˌan-ˌtī-, -'lä-\ *noun* (1675)
: the number corresponding to a given logarithm

an·ti·ma·cas·sar \ˌan-ti-mə-'ka-sər\ *noun* [*anti-* + *Macassar* (*oil*) (a hairdressing)] (1852)
: a cover to protect the back or arms of furniture

an·ti·mag·net·ic \ˌan-ti-mag-'ne-tik, ˌan-ˌtī-\ *adjective* (1946)
of a watch **:** having a balance unit composed of alloys that will not remain magnetized

an·ti·ma·lar·i·al \-mə-'ler-ē-əl\ *adjective* (circa 1893)
: serving to prevent, check, or cure malaria
— **antimalarial** *noun*

an·ti·mat·ter \'an-ti-ˌma-tər, 'an-ˌtī-\ *noun* (1950)
: matter composed of antiparticles

an·ti·me·tab·o·lite \ˌan-ti-mə-'ta-bə-ˌlīt, ˌan-ˌtī-\ *noun* (1945)
: a substance that replaces or inhibits an organism's utilization of a metabolite

an·ti·mi·cro·bi·al \ˌan-ti-mī-'krō-bē-əl\ *adjective* (circa 1910)
: destroying or inhibiting the growth of microorganisms
— **antimicrobial** *noun*

an·ti·mis·sile missile \ˌan-ti-'mi-səl-, ˌan-ˌtī-; *chiefly British* ˌan-ti-'mi-ˌsīl-\ *noun* (circa 1956)
: ANTIBALLISTIC MISSILE

an·ti·mi·tot·ic \ˌan-ti-mī-'tä-tik, ˌan-ˌtī-\ *adjective* (1970)
: inhibiting or disrupting mitosis ⟨*antimitotic* agents⟩ ⟨*antimitotic* activity⟩
— **antimitotic** *noun*

an·ti·mo·ni·al \ˌan-tə-'mō-nē-əl\ *adjective* (1605)
: of, relating to, or containing antimony
— **antimonial** *noun*

an·ti·mo·nide \'an-ti-mə-ˌnīd\ *noun* (1863)
: a binary compound of antimony with a more positive element

an·ti·mo·ny \'an-tə-ˌmō-nē\ *noun* [Middle English *antimonie,* from Medieval Latin *antimonium*] (15th century)
1 : STIBNITE
2 : a trivalent and pentavalent metalloid element that is commonly metallic silvery white, crystalline, and brittle and that is used especially as a constituent of alloys and semiconductors — see ELEMENT table

an·ti·my·cin A \ˌan-ti-'mī-sᵊn-'ā\ *noun* [*anti-* + *-mycin*] (1949)
: a crystalline antibiotic $C_{28}H_{40}N_2O_9$ used especially as a fungicide, insecticide, and miticide — called also *antimycin*

an·ti·neo·plas·tic \ˌan-ti-ˌnē-ə-'plas-tik, ˌan-ˌtī-\ *adjective* (1954)
: inhibiting or preventing the growth and spread of neoplasms or malignant cells

an·ti·neu·tri·no \-nü-'trē-(ˌ)nō, -nyü-\ *noun* (1934)
: the antiparticle of the neutrino

an·ti·neu·tron \-'nü-ˌträn, -'nyü-\ *noun* (1942)
: the antiparticle of the neutron

ant·ing \'an-tiŋ\ *noun* (1936)
: bird behavior in which ants are rubbed on the feathers to obtain chemicals (as formic acid) from the ants

an·ti·node \'an-ti-ˌnōd, 'an-ˌtī-\ *noun* [International Scientific Vocabulary] (1882)
: a region of maximum amplitude situated between adjacent nodes in a vibrating body
— **an·ti·nod·al** \ˌan-ti-'nō-dᵊl, ˌan-ˌtī-\ *adjective*

an·ti·no·mi·an \ˌan-ti-'nō-mē-ən\ *noun* [Medieval Latin *antinomus,* from Latin *anti-* + Greek *nomos* law] (1645)
1 : one who holds that under the gospel dispensation of grace the moral law is of no use or obligation because faith alone is necessary to salvation
2 : one who rejects a socially established morality
— **antinomian** *adjective*
— **an·ti·no·mi·an·ism** \-mē-ə-ˌni-zəm\ *noun*

an·tin·o·my \an-'ti-nə-mē\ *noun, plural* **-mies** [German *Antinomie,* from Latin *antinomia* conflict of laws, from Greek, from *anti-* + *nomos* law — more at NIMBLE] (1592)
1 : a contradiction between two apparently equally valid principles or between inferences correctly drawn from such principles
2 : a fundamental and apparently unresolvable conflict or contradiction ⟨*antinomies* of beauty and evil, freedom and slavery —Stephen Holden⟩
— **an·ti·nom·ic** \,an-ti-'nä-mik\ *adjective*

an·ti·nov·el \'an-ti-,nä-vəl, 'an-,tī-\ *noun* (1958)
: a work of fiction that lacks most or all of the traditional features of the novel
— **an·ti·nov·el·ist** \-,näv-list, -,nä-və-\ *noun*

an·ti·nu·cle·ar \,an-ti-'nü-klē-ər ,an-,tī-, -'nyü-, ÷-kyə-lər\ *adjective* (1958)
1 : tending to react with cell nuclei or their components (as DNA) ⟨*antinuclear* antibodies⟩
2 : opposing the use or production of nuclear power plants

an·ti·nu·cle·on \-'n(y)ü-klē-,än\ *noun* (1946)
: the antiparticle of a nucleon

an·ti·nuke \-'nük, -'nyük\ *adjective* (1975)
: ANTINUCLEAR 2

an·ti·ox·i·dant \,an-tē-'äk-sə-dənt, ,an-,tī-\ *noun* (1926)
: a substance that inhibits oxidation or reactions promoted by oxygen or peroxides
— **antioxidant** *adjective*

an·ti·ozon·ant \-'ō-(,)zō-nənt\ *noun* (1954)
: a substance that opposes ozonization or protects against it

an·ti·par·al·lel \,an-ti-'par-ə-,lel, ,an-,tī-, -ləl\ *adjective* (circa 1660)
: parallel but oppositely directed or oriented ⟨*antiparallel* electron spins⟩ ⟨two *antiparallel* chains of nucleotides comprise DNA⟩

an·ti·par·a·sit·ic \,an-ti-,par-ə-'si-tik, ,an-,tī-\ *adjective* (circa 1860)
: acting against parasites

an·ti·par·ti·cle \'an-ti-,pär-ti-kəl, 'an-,tī-\ *noun* (1934)
: a subatomic particle identical to another subatomic particle in mass but opposite to it in electric and magnetic properties (as sign of charge) that when brought together with its counterpart produces mutual annihilation; *especially* **:** a subatomic particle not found in ordinary matter

an·ti·pas·to \,an-ti-'pas-(,)tō, ,än-ti-, -'päs-\ *noun, plural* **-ti** \-(,)tē\ [Italian, from *anti-* (from Latin *ante-*) + *pasto* food, from Latin *pastus,* from *pascere* to feed — more at FOOD] (1590)
: any of various typically Italian hors d'oeuvres; *also* **:** a plate of these served especially as the first course of a meal

an·ti·pa·thet·ic \,an-ti-pə-'the-tik, (,)an-,ti-pə-\ *adjective* (1640)
1 : having a natural aversion; *also* **:** not sympathetic **:** HOSTILE
2 : arousing antipathy ⟨mountains . . . are *antipathetic* to me —Havelock Ellis⟩
— **an·ti·pa·thet·i·cal·ly** \-ti-k(ə-)lē\ *adverb*

an·tip·a·thy \an-'ti-pə-thē\ *noun, plural* **-thies** [Latin *antipathia,* from Greek *antipatheia,* from *antipathēs* of opposite feelings, from *anti-* + *pathos* experience — more at PATHOS] (1601)
1 *obsolete* **:** opposition in feeling
2 : settled aversion or dislike **:** DISTASTE
3 : an object of aversion
synonym see ENMITY

an·ti·per·son·nel \,an-ti-,pər-sᵊn-'el, ,an-,tī-\ *adjective* (1939)
: designed for use against military personnel ⟨an *antipersonnel* mine⟩

an·ti·per·spi·rant \-'pər-sp(ə-)rənt\ *noun* (1943)
: a preparation used to check perspiration

an·ti·phlo·gis·tic \-flə-'jis-tik\ *adjective* (1769)
: ANTI-INFLAMMATORY

an·ti·phon \'an-tə-fən, -,fän\ *noun* [Middle English *antiphone,* from Middle French, from Late Latin *antiphona* — more at ANTHEM] (15th century)
1 : a psalm, anthem, or verse sung responsively
2 : a verse usually from Scripture said or sung before and after a canticle, psalm, or psalm verse as part of the liturgy

¹an·tiph·o·nal \an-'ti-fə-nᵊl\ *noun* (1537)
: ANTIPHONARY

²antiphonal *adjective* (1719)
: of, relating to, or suggesting an antiphon or antiphony
— **an·tiph·o·nal·ly** \-nᵊl-ē\ *adverb*

an·tiph·o·nary \an-'ti-fə-,ner-ē\ *noun, plural* **-nar·ies** (15th century)
1 : a book containing a collection of antiphons
2 : a book containing the choral parts of the Divine Office

an·tiph·o·ny \an-'ti-fə-nē\ *noun, plural* **-nies** (1592)
: responsive alternation between two groups especially of singers

an·tiph·ra·sis \an-'ti-frə-səs\ *noun, plural* **-ra·ses** \-,sēz\ [Late Latin, from Greek, from *anti-* + *phrasis* diction — more at PHRASE] (1533)
: the usually ironic or humorous use of words in senses opposite to the generally accepted meanings (as in "this giant of 3 feet 4 inches")

¹an·tip·o·dal \an-'ti-pə-dᵊl\ *adjective* (1646)
1 : of or relating to the antipodes; *specifically* **:** situated at the opposite side of the earth or moon ⟨an *antipodal* meridian⟩ ⟨an *antipodal* continent⟩
2 : diametrically opposite ⟨an *antipodal* point on a sphere⟩
3 : entirely opposed ⟨a system *antipodal* to democracy⟩

²antipodal *noun* (1919)
: any of three cells in the female gametophyte of most angiosperms that are grouped at the end of the embryo sac farthest from the micropyle — called also *antipodal cell*

an·ti·pode \'an-tə-,pōd\ *noun, plural* **an·tip·o·des** \an-'ti-pə-,dēz\ [Middle English *antipodes,* plural, persons dwelling at opposite points on the globe, from Latin, from Greek, from plural of *antipod-, antipous* with feet opposite, from *anti-* + *pod-, pous* foot — more at FOOT] (1549)
1 : the parts of the earth diametrically opposite — usually used in plural; often used of Australia and New Zealand
2 : the exact opposite or contrary
— **an·tip·o·de·an** \(,)an-,ti-pə-'dē-ən\ *adjective or noun*

an·ti·po·et·ic \,an-ti-pō-'e-tik, ,an-,tī-\ *adjective* (1847)
: of, relating to, or characterized by opposition to traditional poetic technique or style

an·ti·pol·lu·tion \-pə-'lü-shən\ *adjective* (1924)
: designed to prevent, reduce, or eliminate pollution ⟨*antipollution* laws⟩
— **antipollution** *noun*

an·ti·pope \'an-ti-,pōp\ *noun* [Middle English *antepope,* from Middle French *antipape,* from Medieval Latin *antipapa,* from *anti-* + *papa* pope] (15th century)
: one elected or claiming to be pope in opposition to the pope canonically chosen

an·ti·pro·ton \,an-ti-'prō-,tän, ,an-,tī-\ *noun* (1940)
: the antiparticle of the proton

an·ti·psy·chot·ic \,an-ti-sī-'kä-tik\ *noun* (1966)
: NEUROLEPTIC
— **antipsychotic** *adjective*

an·ti·py·ret·ic \-pī-'re-tik\ *noun* (circa 1681)
: an agent that reduces fever
— **antipyretic** *adjective*

an·ti·py·rine \-'pīr-,ēn\ *noun* [from *Antipyrine,* a trademark] (1884)
: an analgesic and antipyretic $C_{11}H_{12}N_2O$ formerly widely used but now largely replaced in oral use by less toxic substances (as aspirin)

¹an·ti·quar·i·an \,an-tə-'kwer-ē-ən\ *noun* (1610)
: one who collects or studies antiquities

²antiquarian *adjective* (1771)
1 : of or relating to antiquarians or antiquities
2 : dealing in old or rare books
— **an·ti·quar·i·an·ism** \-ē-ə-,ni-zəm\ *noun*

an·ti·quark \'an-ti-,kwärk, 'an-,tī-\ *noun* (1964)
: the antiparticle of the quark

an·ti·quary \'an-tə-,kwer-ē\ *noun, plural* **-quar·ies** (1586)
: ANTIQUARIAN

an·ti·quate \'an-tə-,kwāt\ *transitive verb* **-quat·ed; -quat·ing** [Late Latin *antiquatus,* past participle of *antiquare,* from Latin *antiquus*] (1596)
: to make old or obsolete
— **an·ti·qua·tion** \,an-tə-'kwā-shən\ *noun*

an·ti·quat·ed *adjective* (1623)
1 : OBSOLETE ⟨an *antiquated* calendar⟩
2 : outmoded or discredited by reason of age
: being out of style or fashion ⟨*antiquated* methods of farming⟩
3 : advanced in age
synonym see OLD

¹an·tique \(,)an-'tēk\ *noun* (1530)
1 : a relic or object of ancient times
2 a : a work of art, piece of furniture, or decorative object made at an earlier period and according to various customs laws at least 100 years ago **b :** a manufactured product (as an automobile) from an earlier period

²an·tique \(,)an-'tēk, *in verse often* 'an-tik\ *adjective* [Middle French, from Latin *antiquus,* from *ante* before — more at ANTE-] (1534)
1 : existing since or belonging to earlier times **:** ANCIENT ⟨*antique* trade routes to the Orient⟩
2 a : being in the style or fashion of former times ⟨*antique* manners and graces⟩ **b :** made in or representative of the work of an earlier period ⟨*antique* mirrors⟩; *also* **:** being an antique
3 : selling or exhibiting antiques ⟨an *antique* show⟩
synonym see OLD

³an·tique \(,)an-'tēk\ *verb* **-tiqued; -tiqu·ing** (1923)
transitive verb
: to finish or refinish in antique style **:** give an appearance of age to
intransitive verb
: to shop around for antiques
— **an·tiqu·er** \an-'tē-kər\ *noun*

an·tiq·ui·ty \an-'ti-kwə-tē\ *noun, plural* **-ties** (13th century)
1 : ancient times; *especially* **:** those before the Middle Ages
2 : the quality of being ancient
3 *plural* **a :** relics or monuments (as coins, statues, or buildings) of ancient times **b :** matters relating to the life or culture of ancient times
4 : the people of ancient times

an·ti·ra·chit·ic \,an-ti-rə-'ki-tik, ,an-,tī-\ *adjective* (circa 1860)
: used or tending to prevent the development of rickets ⟨an *antirachitic* vitamin⟩

an·ti·re·jec·tion \,an-ti-ri-'jek-shən, ,an-,tī-\ *adjective* (1968)
: used or tending to prevent organ transplant rejection ⟨*antirejection* drugs⟩ ⟨*antirejection* treatment⟩

an·ti·rheu·mat·ic \-rü-'ma-tik\ *adjective* (1817)

\ə\ **abut** \ᵊ\ **kitten** \ər\ **further** \a\ **ash** \ā\ **ace** \ä\ **mop, mar** \au̇\ **out** \ch\ **chin** \e\ **bet** \ē\ **easy** \g\ **go** \i\ **hit** \ī\ **ice** \j\ **job** \ŋ\ **sing** \ō\ **go** \ȯ\ **law** \ȯi\ **boy** \th\ **thin** \t͟h\ **the** \ü\ **loot** \u̇\ **foot** \y\ **yet** \zh\ **vision** *see also* Guide to Pronunciation

: alleviating or preventing rheumatism ⟨*antirheumatic* therapy⟩ ⟨*antirheumatic* drugs⟩
— **antirheumatic** *noun*

an·ti–roll bar \ˈan-ˌtī-ˌrōl-\ *noun* (1951)
: SWAY BAR

an·tir·rhi·num \ˌan-tə-ˈrī-nəm\ *noun* [New Latin, genus name, from Latin, snapdragon, from Greek *antirrhinon,* from *anti-* like (from *anti* against, equivalent to) + *rhin-, rhis* nose — more at ANTI-] (1548)
: SNAPDRAGON

antis *plural of* ANTI

an·ti·scor·bu·tic \ˌan-ti-skȯr-ˈbyü-tik, ˌan-ˌtī-\ *adjective* (1725)
: counteracting scurvy ⟨the *antiscorbutic* vitamin is vitamin C⟩
— **antiscorbutic** *noun*

an·ti–Sem·i·tism \ˌan-ti-ˈse-mə-ˌti-zəm, ˌan-ˌtī-\ *noun* (1882)
: hostility toward or discrimination against Jews as a religious, ethnic, or racial group
— **an·ti–Se·mit·ic** \-sə-ˈmi-tik\ *adjective*
— **an·ti–Sem·ite** \-ˈse-ˌmīt\ *noun*

an·ti·sense \ˈan-ˌtī-ˌsen(t)s, ˈan-ti-\ *adjective* [*anti-* + non*sense*] (1984)
: having a complementary sequence to a segment of genetic material (as mRNA) and serving to inhibit gene function ⟨*antisense* nucleotides⟩ ⟨*antisense* RNA⟩ — compare MISSENSE, NONSENSE

an·ti·sep·sis \ˌan-tə-ˈsep-səs\ *noun* (1875)
: the inhibiting of the growth and multiplication of microorganisms by antiseptic means

[1]**an·ti·sep·tic** \ˌan-tə-ˈsep-tik\ *adjective* [*anti-* + Greek *sēptikos* putrefying, septic] (1751)
1 a : opposing sepsis, putrefaction, or decay; *especially* **:** preventing or arresting the growth of microorganisms (as on living tissue) **b :** acting or protecting like an antiseptic
2 : relating to or characterized by the use of antiseptics
3 a : scrupulously clean **:** ASEPTIC **b :** extremely neat or orderly; *especially* **:** neat to the point of being bare or uninteresting **c :** free from what is held to be contaminating
4 : coldly impersonal ⟨an *antiseptic* greeting⟩
— **an·ti·sep·ti·cal·ly** \-ti-k(ə-)lē\ *adverb*

[2]**antiseptic** *noun* (1751)
: a substance that checks the growth or action of microorganisms especially in or on living tissue; *also* **:** GERMICIDE

an·ti·se·rum \ˈan-ti-ˌsir-əm, ˈan-ˌtī-, -ˌser-\ *noun* [International Scientific Vocabulary] (1901)
: a serum containing antibodies

an·ti·so·cial \ˌan-ti-ˈsō-shəl, ˌan-ˌtī-\ *adjective* (1797)
1 : averse to the society of others **:** UNSOCIABLE
2 : hostile or harmful to organized society; *especially* **:** being or marked by behavior deviating sharply from the social norm
— **an·ti·so·cial·ly** \-shə-lē\ *adverb*

an·ti·so·lar \-ˈsō-lər\ *noun* (circa 1890)
: being or having a direction away from the sun ⟨the *antisolar* point⟩

an·ti·spas·mod·ic \-spaz-ˈmä-dik\ *adjective* (1763)
: capable of preventing or relieving spasms or convulsions
— **antispasmodic** *noun*

an·ti·stat \-ˈstat\ *or* **an·ti·stat·ic** \-ˈsta-tik\ *adjective* (1952)
: reducing, removing, or preventing the build-up of static electricity

an·tis·tro·phe \an-ˈtis-trə-(ˌ)fē\ *noun* [Late Latin, from Greek *antistrophē,* from *anti-* + *strophē* strophe] (circa 1550)
1 a : the repetition of words in reversed order **b :** the repetition of a word or phrase at the end of successive clauses
2 a : a returning movement in Greek choral dance exactly answering to a previous strophe **b :** the part of a choral song delivered during the antistrophe
— **an·ti·stroph·ic** \ˌan-tə-ˈsträ-fik\ *adjective*
— **an·ti·stroph·i·cal·ly** \-fi-k(ə-)lē\ *adverb*

an·ti·sym·met·ric \-sə-ˈme-trik\ *adjective* (1923)
: relating to or being a relation (as "is a subset of") that implies equality of any two quantities for which it holds in both directions ⟨the relation *R* is *antisymmetric* if *aRb* and *bRa* implies *a* = *b*⟩

an·tith·e·sis \an-ˈti-thə-səs\ *noun, plural* **-e·ses** \-ˌsēz\ [Late Latin, from Greek, literally, opposition, from *antitithenai* to oppose, from *anti-* + *tithenai* to set — more at DO] (1529)
1 a (1) **:** the rhetorical contrast of ideas by means of parallel arrangements of words, clauses, or sentences (as in "action, not words" or "they promised freedom and provided slavery") (2) **:** OPPOSITION, CONTRAST ⟨the *antithesis* of prose and verse⟩ **b** (1) **:** the second of two opposing constituents of an antithesis (2) **:** the direct opposite
2 : the second stage of a dialectic process

an·ti·thet·i·cal \ˌan-tə-ˈthe-ti-kəl\ *also* **an·ti·thet·ic** \-ˈthe-tik\ *adjective* (1583)
1 : constituting or marked by antithesis
2 : being in direct and unequivocal opposition
synonym see OPPOSITE
— **an·ti·thet·i·cal·ly** \-ti-k(ə-)lē\ *adverb*

an·ti·throm·bin \ˌan-ti-ˈthräm-bən, ˌan-ˌtī-\ *noun* (circa 1911)
: any of a group of substances that inhibit blood clotting by inactivating thrombin

an·ti·thy·roid \ˌan-ti-ˈthī-ˌrȯid\ *adjective* (1908)
: able to counteract excessive thyroid activity ⟨*antithyroid* drugs⟩

an·ti·tox·ic \-ˈtäk-sik\ *adjective* (circa 1890)
1 : counteracting toxins
2 : being or containing antitoxins ⟨*antitoxic* serum⟩

an·ti·tox·in \ˌan-ti-ˈtäk-sən\ *noun* [International Scientific Vocabulary] (circa 1890)
: an antibody that is capable of neutralizing the specific toxin (as a specific causative agent of disease) that stimulated its production in the body and is produced in animals for medical purposes by injection of a toxin or toxoid with the resulting serum being used to counteract the toxin in other individuals; *also* **:** an antiserum containing antitoxins

an·ti·trades \ˈan-ti-ˌtrādz, ˈan-ˌtī-\ *noun plural* (1875)
: the westerly winds above the trade winds

an·ti·trust \ˌan-ti-ˈtrəst, ˌan-ˌtī-\ *adjective* (1890)
: of, relating to, or being legislation against or opposition to trusts or combinations; *specifically* **:** consisting of laws to protect trade and commerce from unlawful restraints and monopolies or unfair business practices

an·ti·trust·er \-ˈtrəs-tər\ *noun* (1947)
: one who advocates or enforces antitrust provisions of the law

an·ti·tus·sive \-ˈtə-siv\ *noun* (circa 1909)
: a cough suppressant
— **antitussive** *adjective*

an·ti–uto·pia \ˌan-ti-yu̇-ˈtō-pē-ə, ˌan-ˌtī-\ *noun* (1966)
1 : DYSTOPIA 1
2 : a work describing an anti-utopia
— **an·ti–uto·pi·an** \-pē-ən\ *adjective or noun*

an·ti·ven·in \ˌan-ti-ˈve-nən, ˌan-ˌtī-\ *noun* [International Scientific Vocabulary] (1895)
: an antitoxin to a venom; *also* **:** an antiserum containing such antitoxin

an·ti·vi·ta·min \ˈan-ti-ˌvī-tə-mən, *British usually* -ˌvi-\ *noun* (1927)
: a substance that makes a vitamin metabolically ineffective

an·ti·white \ˌan-ti-ˈhwīt, ˌan-ˌtī-, -ˈwīt\ *adjective* (1906)
: opposed or hostile to people belonging to a light-skinned race

ant·ler \ˈant-lər\ *noun* [Middle English *aunteler,* from Middle French *antoillier,* from (assumed) Vulgar Latin *anteoculare,* from neuter of *anteocularis* located before the eye, from Latin *ante-* + *oculus* eye — more at EYE] (14th century)
: one of the paired deciduous solid bone processes that arise from the frontal bone on the head of an animal of the deer family; *also* **:** a branch of an antler
— **ant·lered** \-lərd\ *adjective*

ant lion *noun* (1815)
: any of various neuropterous insects (as of the genus *Myrmeleon*) having a long-jawed larva that digs a conical pit in which it lies in wait to catch insects (as ants) on which it feeds

ant lion: *1* larva, *2* adult

An·to·ni·an \an-ˈtō-nē-ən\ *noun* [Latin *Antonius* Anthony] (circa 1907)
: a member of one of several monastic communities (as the Armenian Antonians) that follow a rule devised by Saint Anthony

an·ton·o·ma·sia \ˌan-tə-nō-ˈmā-zh(ē-)ə, (ˌ)an-ˌtä-nə-\ *noun* [Latin, use of an epithet for a proper name, from Greek, from *antonomazein* to call by a new name, from *anti-* + *onomazein* to name, from *onoma* name — more at NAME] (circa 1550)
: the use of a proper name to designate a member of a class (as a *Solomon* for a *wise ruler*); *also* **:** the use of an epithet or title in place of a proper name (as *the Bard* for *Shakespeare*)

an·to·nym \ˈan-tə-ˌnim\ *noun* (1870)
: a word of opposite meaning ⟨the usual *antonym* of *good* is *bad*⟩
— **an·to·nym·ic** \ˌan-tə-ˈni-mik\ *adjective*
— **an·ton·y·mous** \an-ˈtä-nə-məs\ *adjective*
— **an·ton·y·my** \-mē\ *noun*

an·tre \ˈan-tər\ *noun* [French, from Latin *antrum*] (1604)
: CAVE 1

an·trum \ˈan-trəm\ *noun, plural* **an·tra** \-trə\ [Late Latin, from Latin, cave, from Greek *antron;* akin to Armenian *ayr* cave] (circa 1751)
: the cavity of a hollow organ or a sinus
— **an·tral** \-trəl\ *adjective*

ant·sy \ˈant-sē\ *adjective* (1838)
1 : impatient of restraint; *especially* **:** FIDGETY
2 : NERVOUS, APPREHENSIVE

Anu·bis \ə-ˈnü-bəs, -ˈnyü-\ *noun* [Latin, from Greek *Anoubis,* from Egyptian *inpw*]
: a jackal-headed god in Egyptian mythology who leads the dead to judgment

an·uran \ə-ˈnyu̇r-ən, a-, -ˈnu̇r-\ *adjective or noun* [ultimately from Greek *a-* + *oura* tail — more at ASS] (1900)
: any of an order (Anura) of amphibians comprising the frogs, toads, and tree toads all of which lack a tail in the adult stage and have long strong hind limbs suited to leaping and swimming
— **anuran** *adjective*

an·uria \ə-ˈnyu̇r-ē-ə, a-, -ˈnu̇r-\ *noun* [New Latin] (1838)
: absence or defective excretion of urine
— **an·uric** \-ˈnyu̇r-ik, -ˈnu̇r-\ *adjective*

anus \ˈā-nəs\ *noun* [Latin, ring, anus; perhaps akin to Old Irish *ánne* ring] (15th century)
: the posterior opening of the alimentary canal

an·vil \ˈan-vəl\ *noun* [Middle English *anfilt,* from Old English; akin to Old High German *anafalz* anvil; akin to Latin *pellere* to beat — more at FELT] (before 12th century)
1 : a heavy usually steel-faced iron block on which metal is shaped (as by hand hammering)
2 : INCUS

anx·i·ety \aŋ-ˈzī-ə-tē\ *noun, plural* **-eties** [Latin *anxietas,* from *anxius*] (circa 1525)

1 a : painful or apprehensive uneasiness of mind usually over an impending or anticipated ill **b :** fearful concern or interest **c :** a cause of anxiety
2 : an abnormal and overwhelming sense of apprehension and fear often marked by physiological signs (as sweating, tension, and increased pulse), by doubt concerning the reality and nature of the threat, and by self-doubt about one's capacity to cope with it
synonym see CARE

an·xi·o·lyt·ic \ˌaŋ-zē-ō-'li-tik, ˌaŋ(k)-sē-\ *noun* [*anxi*ety + *-o-* + *-lytic*] (1965)
: a drug that relieves anxiety
— anxiolytic *adjective*

anx·ious \'aŋ(k)-shəs\ *adjective* [Latin *anxius;* akin to Latin *angere* to strangle, distress — more at ANGER] (circa 1616)
1 : characterized by extreme uneasiness of mind or brooding fear about some contingency **:** WORRIED
2 : characterized by, resulting from, or causing anxiety **:** WORRYING
3 : ardently or earnestly wishing ■
synonym see EAGER
— anx·ious·ly *adverb*
— anx·ious·ness *noun*

[1]any \'e-nē\ *adjective* [Middle English, from Old English *ǣnig;* akin to Old High German *einag* any, Old English *ān* one — more at ONE] (before 12th century)
1 : one or some indiscriminately of whatever kind: **a :** one or another taken at random ⟨ask *any* man you meet⟩ **b :** EVERY — used to indicate one selected without restriction ⟨*any* child would know that⟩
2 : one, some, or all indiscriminately of whatever quantity: **a :** one or more — used to indicate an undetermined number or amount ⟨have you *any* money⟩ **b :** ALL — used to indicate a maximum or whole ⟨needs *any* help he can get⟩ **c :** a or some without reference to quantity or extent ⟨grateful for *any* favor at all⟩
3 a : unmeasured or unlimited in amount, number, or extent ⟨*any* quantity you desire⟩ **b :** appreciably large or extended ⟨could not endure it *any* length of time⟩

[2]any *pronoun, singular or plural in construction* (before 12th century)
1 : any person or persons **:** ANYONE
2 a : any thing or things **b :** any part, quantity, or number

[3]any *adverb* (14th century)
: to any extent or degree **:** AT ALL ⟨was never *any* good⟩

any·body \-ˌbä-dē, -bə-\ *pronoun* (14th century)
: any person **:** ANYONE ■

any·how \-ˌhau̇\ *adverb* (1690)
1 a : in any manner whatever **b :** in a haphazard manner
2 a : at any rate **b :** in any event

any·more \ˌe-nē-'mōr, -'mȯr\ *adverb* (14th century)
1 : any longer ⟨I was not moving *anymore* with my feet —Anaïs Nin⟩
2 : at the present time **:** NOW ⟨hardly a day passes without rain *anymore*⟩ ■

any·one \'e-nē-(ˌ)wən\ *pronoun* (1536)
: any person at all
usage see ANYBODY

any·place \-ˌplās\ *adverb* (1916)
: in any place **:** ANYWHERE

[1]any·thing \-ˌthiŋ\ *pronoun* (before 12th century)
: any thing whatever **:** any such thing

[2]anything *adverb* (before 12th century)
: AT ALL

any·time \'e-nē-ˌtīm\ *adverb* (1926)
: at any time whatever

any·way \-ˌwā\ *adverb* (13th century)
1 : ANYWISE
2 : in any case **:** ANYHOW

any·ways \-ˌwāz\ *adverb* (13th century)
1 a *archaic* **:** ANYWISE **b** *dialect* **:** to any degree at all
2 *chiefly dialect* **:** ANYHOW, ANYWAY

[1]any·where \-ˌ(h)wer, -ˌ(h)war, -(h)wər\ *adverb* (14th century)
1 : at, in, or to any place or point
2 : to any extent **:** AT ALL
3 — used as a function word to indicate limits of variation ⟨*anywhere* from 40 to 60 students⟩

[2]anywhere *noun* (1924)
: any place

any·wheres \-ˌ(h)werz, -ˌ(h)warz, -(h)wərz\ *adverb* (1775)
chiefly dialect **:** ANYWHERE

any·wise \'e-nē-ˌwīz\ *adverb* (13th century)
: in any way whatever **:** AT ALL

An·zac \'an-ˌzak\ *noun* [*A*ustralian and *N*ew *Z*ealand *A*rmy *C*orps] (1915)
: a soldier from Australia or New Zealand

A–OK \ˌā-(ˌ)ō-'kā\ *adverb or adjective* (1959)
: very definitely OK

A1 \'ā-'wən\ *adjective* (1837)
1 : having the highest possible classification — used of a ship
2 : of the finest quality **:** FIRST-RATE

ao·rist \'ā-ə-rəst, 'e-ə-\ *noun* [Late Latin & Greek; Late Latin *aoristos,* from Greek, from *aoristos* undefined, from *a* + *horistos* definable, from *horizein* to define — more at HORIZON] (1581)
: an inflectional form of a verb typically denoting simple occurrence of an action without reference to its completeness, duration, or repetition
— aorist *or* **ao·ris·tic** \ˌā-ə-'ris-tik, ˌe-ə-\ *adjective*
— ao·ris·ti·cal·ly \-ti-k(ə-)lē\ *adverb*

aor·ta \ā-'ȯr-tə\ *noun, plural* **-tas** *or* **-tae** \-tē\ [New Latin, from Greek *aortē,* from *aeirein* to lift] (1543)
: the great arterial trunk that carries blood from the heart to be distributed by branch arteries through the body — see HEART illustration
— aor·tic \-'ȯr-tik\ *adjective*

aortic arch *noun* (1903)
: one of the arterial branches in vertebrate embryos that exist in a series of pairs with one on each side of the embryo, connect the ventral arterial system lying anterior to the heart to the dorsal arterial system above the alimentary tract, and persist in adult fishes but are reduced or much modified in the adult of higher forms

aor·tog·ra·phy \ˌā-ˌȯr-'tä-grə-fē\ *noun* (circa 1935)
: arteriography of the aorta
— aor·to·graph·ic \(ˌ)ā-ˌȯr-tə-'gra-fik\ *adjective*

aou·dad \'au̇-ˌdad, 'ä-u̇-\ *noun* [French, from Berber *audad*] (1861)
: a wild bovine (*Ammotragus lervia*) of North Africa that is closely related to goats and sheep and has been introduced into the southwestern U.S.

à ou·trance \ˌä-ˌü-'träⁿs\ *adverb* [French] (1883)
: to the limit **:** UNSPARINGLY

[1]ap- — see AD-

[2]ap- — see APO-

apace \ə-'pās\ *adverb* [Middle English, probably from Middle French *à pas* on step] (14th century)
1 : at a quick pace **:** SWIFTLY
2 : ABREAST — used with *of* or *with*

Apache \ə-'pa-chē, *in sense 3* ə-'pash\ *noun, plural* **Apache** *or* **Apach·es** \-'pa-chēz; -'pash, -'pa-shəz\ [American Spanish, perhaps from Zuni *ʔa·paču* Navajo, Apachean] (1745)
1 : a member of a group of American Indian peoples of the southwestern U.S.
2 : any of the Athabascan languages of the Apache people
3 *not capitalized* [French, from *Apache* Apache Indian] **a :** a member of a gang of criminals especially in Paris **b :** RUFFIAN
— Apach·e·an \ə-'pa-chē-ən\ *adjective or noun*

ap·a·nage *variant of* APPANAGE

ap·a·re·jo \ˌa-pə-'rā-(ˌ)(h)ō\ *noun, plural* **-jos** [American Spanish] (1844)
: a packsaddle of stuffed leather or canvas

[1]apart \ə-'pärt\ *adverb* [Middle English, from Middle French *a part,* literally, to the side] (14th century)
1 a : at a little distance ⟨tried to keep *apart*

□ USAGE

anxious The opinion that *anxious* must not be used to mean "eager" was formed in the U.S. at the beginning of the 20th century. British commentators have largely ignored the problem—after all, sense 3 of *anxious* had been established in British English for 160 years before American commentators decided it must be wrong ⟨his manner was perhaps the more seductive, because he ne'er seem'd *anxious* to seduce —Lord Byron (1824)⟩. Of course, it was just as firmly established in American English too ⟨at any hour of the day or night, I have been *anxious* to improve the nick of time —Henry David Thoreau (1854)⟩. But the topic is still dear to the hearts of American commentators, few of whom notice that the use is standard.

anybody The indefinite pronouns *anybody* and *anyone* share with other indefinite pronouns the characteristic of taking a singular verb and, more often than not, a plural pronoun in reference. The use of the plural pronoun—*they, their, them*—has traditionally been disapproved by those who insist that agreement in number is more important than avoidance of gender-specific third-person singular pronouns. But use of the plural pronouns has long been established ⟨as *anybody* in their senses would have done —Jane Austen (1814)⟩ ⟨if *anybody* thinks that kings, nobles, or priests are good conservators of the public happiness send them here —Thomas Jefferson (letter from Paris, 1786)⟩ and increasing resistance to generic *he* seems to be easing their acceptance. Usage is still mixed ⟨I had never examined all the papers left by *anybody* after his death —Donald Winch⟩ ⟨isn't interested in serving lunch or having *anyone* get their own —Andy Rooney⟩.

anymore Although both *anymore* and *any more* are found in written use, in the 20th century *anymore* is the more common styling. *Anymore* is regularly used in negative ⟨no one can be natural *anymore* —May Sarton⟩, interrogative ⟨do you read much *anymore*?⟩, and conditional ⟨if you do that *anymore,* I'll leave⟩ contexts and in certain positive constructions ⟨the Washingtonian is too sophisticated to believe *anymore* in solutions —Russell Baker⟩. In many regions of the U.S. the use of *anymore* in sense 2 is quite common in positive constructions, especially in speech ⟨everybody's cool *anymore* —Bill White⟩ ⟨every time we leave the house *anymore,* I play a game called "Stump the Housebreaker" —Erma Bombeck⟩. The positive use appears to have been of Midland origin, but it is now reported to be widespread in all speech areas of the U.S. except New England.

from the family squabbles⟩ **b :** away from one another in space or time ⟨towns 20 miles *apart*⟩
2 a : as a separate unit **:** INDEPENDENTLY ⟨viewed *apart*, his arguments were unsound⟩ **b :** so as to separate one from another ⟨found it hard to tell the twins *apart*⟩
3 : excluded from consideration **:** ASIDE ⟨a few blemishes *apart*, the novel is excellent⟩
4 : in or into two or more parts **:** to pieces ⟨coming *apart* at the seams⟩
[2]apart *adjective* (1786)
1 : SEPARATE, ISOLATED
2 : holding different opinions **:** DIVIDED
— apart·ness *noun*
apart from *preposition* (1833)
: other than **:** BESIDES
apart·heid \ə-ˈpär-ˌtāt, -ˌtīt\ *noun* [Afrikaans, from Dutch, from *apart* apart + *-heid* -hood] (1947)
1 : racial segregation; *specifically* **:** a policy of segregation and political and economic discrimination against non-European groups in the Republic of South Africa
2 : SEPARATION, SEGREGATION ⟨I favor *apartheid* of smokers —L. E. Bellin⟩ ⟨sexual *apartheid*⟩
apart·ment \ə-ˈpärt-mənt\ *noun* [French *appartement*, from Italian *appartamento*] (1641)
1 : a room or set of rooms fitted especially with housekeeping facilities and usually leased as a dwelling
2 : a building containing several individual apartments
— apart·men·tal \ə-ˌpärt-ˈmen-tᵊl\ *adjective*
apartment hotel *noun* (1909)
: a hotel containing apartments as well as accommodations for transients
apartment house *noun* (1874)
: a building containing separate residential apartments — called also *apartment building*
ap·a·thet·ic \ˌa-pə-ˈthe-tik\ *adjective* (1744)
1 : having or showing little or no feeling or emotion **:** SPIRITLESS
2 : having little or no interest or concern **:** INDIFFERENT
synonym see IMPASSIVE
— ap·a·thet·i·cal·ly \-ti-k(ə-)lē\ *adverb*
ap·a·thy \ˈa-pə-thē\ *noun* [Greek *apatheia*, from *apathēs* without feeling, from *a-* + *pathos* emotion — more at PATHOS] (1603)
1 : lack of feeling or emotion **:** IMPASSIVENESS
2 : lack of interest or concern **:** INDIFFERENCE
ap·a·tite \ˈa-pə-ˌtīt\ *noun* [German *Apatit*, from Greek *apatē* deceit] (1803)
: any of a group of calcium phosphate minerals occurring variously as hexagonal crystals, as granular masses, or in fine-grained masses as the chief constituent of phosphate rock and of bones and teeth; *especially* **:** calcium phosphate fluoride
apato·sau·rus \ə-ˌpa-tə-ˈsȯr-əs\ *noun* [New Latin, from Greek *apatē* + *sauros* lizard] (circa 1899)
: BRONTOSAURUS
[1]ape \ˈāp\ *noun* [Middle English, from Old English *apa*; akin to Old High German *affo* ape] (before 12th century)
1 a : MONKEY; *especially* **:** one of the larger tailless or short-tailed Old World forms **b :** any of two families (Pongidae and Hylobatidae) of large tailless semierect primates (as the chimpanzee, gorilla, orangutan, or gibbon) — called also *anthropoid, anthropoid ape*
2 a : MIMIC **b :** a large uncouth person
— ape·like \ˈāp-ˌlīk\ *adjective*
[2]ape *transitive verb* **aped; ap·ing** (1632)
: to copy closely but often clumsily and ineptly
synonym see COPY
— ap·er *noun*
[3]ape *adjective* (circa 1955)
: being beyond restraint **:** CRAZY, WILD — usually used in the phrase *go ape*
apeak \ə-ˈpēk\ *adjective or adverb* [alteration of earlier *apike*, probably from French *à pic* vertically] (1596)
: being in a vertical position ⟨with oars *apeak*⟩
ape–man \ˈāp-ˌman, -ˈman\ *noun* (1879)
: a primate (as an australopithecine) intermediate in character between Homo sapiens and the higher apes
aper·çu \à-per-sǖ, ˌa-pər-ˈsü\ *noun, plural* **aperçus** \-sǖ(z), -ˈsüz\ [French, from *aperçu*, past participle of *apercevoir* to perceive, from Old French *aperceivre*, from *a-* (from Latin *ad-*) + *perceivre* to perceive — more at PERCEIVE] (1828)
1 : a brief survey or sketch **:** OUTLINE
2 : an immediate impression; *especially* **:** INSIGHT 2
ape·ri·ent \ə-ˈpir-ē-ənt\ *adjective* [Latin *aperient-, aperiens*, present participle of *aperire*] (1626)
: gently moving the bowels **:** LAXATIVE
— aperient *noun*
ape·ri·od·ic \ˌā-ˌpir-ē-ˈä-dik\ *adjective* (1879)
1 : of irregular occurrence ⟨*aperiodic* floods⟩
2 : not having periodic vibrations **:** not oscillatory
— ape·ri·od·i·cal·ly \-di-k(ə-)lē\ *adverb*
— ape·ri·o·dic·i·ty \-ē-ə-ˈdi-sə-tē\ *noun*
aper·i·tif \ə-ˌper-ə-ˈtēf, a-; ˌä-pər-(ə-)ˈtēf\ *noun* [French *apéritif* aperient, aperitif, from Middle French *aperitif*, adjective, aperient, from Medieval Latin *aperitivus*, irregular from Latin *aperire*] (1894)
: an alcoholic drink taken before a meal as an appetizer
ap·er·ture \ˈap-ə(r)-ˌchu̇r, -chər, -ˌtyu̇r, -ˌtu̇r\ *noun* [Middle English, from Latin *apertura*, from *apertus*, past participle of *aperire* to open] (15th century)
1 : an opening or open space **:** HOLE
2 a : the opening in a photographic lens that admits the light **b :** the diameter of the stop in an optical system that determines the diameter of the bundle of rays traversing the instrument **c :** the diameter of the objective lens or mirror of a telescope
apet·al·ous \(ˌ)ā-ˈpe-tᵊl-əs\ *adjective* (circa 1706)
: having no petals
apex \ˈā-ˌpeks\ *noun, plural* **apex·es** *or* **api·ces** \ˈā-pə-ˌsēz, ˈa-\ [Latin] (1601)
1 a : the uppermost point **:** VERTEX ⟨the *apex* of a mountain⟩ **b :** the narrowed or pointed end **:** TIP ⟨the *apex* of the tongue⟩
2 : the highest or culminating point ⟨the *apex* of his career⟩
synonym see SUMMIT
Ap·gar score \ˈap-ˌgär-\ *noun* [Virginia *Apgar* (died 1974) American anesthesiologist] (1962)
: an index used to evaluate the condition of a newborn infant based on a rating of 0, 1, or 2 for each of the five characteristics of color, heart rate, response to stimulation of the sole of the foot, muscle tone, and respiration with 10 being a perfect score
aphaer·e·sis *or* **apher·e·sis** \ə-ˈfer-ə-səs\ *noun, plural* **-e·ses** \-ˌsēz\ [Late Latin, from Greek *aphairesis*, literally, taking off, from *aphairein* to take away, from *apo-* + *hairein* to take] (circa 1550)
: the loss of one or more sounds or letters at the beginning of a word (as in *round* for *around* and *coon* for *raccoon*)
— aph·ae·ret·ic \ˌa-fə-ˈre-tik\ *adjective*
aph·a·nite \ˈa-fə-ˌnīt\ *noun* [French, from Greek *aphanēs* invisible, from *a-* + *phainesthai* to appear — more at PHENOMENON] (circa 1828)
: a dark rock of such close texture that its separate grains are invisible to the naked eye
— aph·a·nit·ic \ˌa-fə-ˈni-tik\ *adjective*
apha·sia \ə-ˈfā-zh(ē-)ə\ *noun* [New Latin, from Greek, from *a-* + *-phasia*] (1867)
: loss or impairment of the power to use or comprehend words usually resulting from brain damage
— apha·sic \-zik\ *noun or adjective*
aph·elion \a-ˈfēl-yən\ *noun, plural* **-elia** \-yə\ [New Latin, from *apo-* + Greek *hēlios* sun — more at SOLAR] (1656)
: the point in the path of a celestial body (as a planet) that is farthest from the sun — compare PERIHELION
aph·e·sis \ˈa-fə-səs\ *noun, plural* **-e·ses** \-ˌsēz\ [New Latin, from Greek, release, from *aphienai* to let go, from *apo-* + *hienai* to send — more at JET] (1880)
: aphaeresis consisting of the loss of a short unaccented vowel (as in *lone* for *alone*)
— aphet·ic \ə-ˈfe-tik\ *adjective*
— aphet·i·cal·ly \-ti-k(ə-)lē\ *adverb*
aphid \ˈā-fəd *also* ˈa-fəd\ *noun* (1884)
: any of numerous very small sluggish homopterous insects (superfamily Aphidoidea) that suck the juices of plants
aphid lion *noun* (1949)
: any of several insect larvae (as a lacewing or ladybug larva) that feed on aphids — called also *aphis lion*
aphis \ˈā-fəs *also* ˈa-fəs\ *noun, plural* **aphi·des** \ˈā-fə-ˌdēz, ˈa-fə-\ [New Latin *Aphid-, Aphis*, genus name] (1771)
: any of a genus (*Aphis*) of aphids; *broadly* **:** APHID
apho·nia \(ˌ)ā-ˈfō-nē-ə\ *noun* [New Latin, from Greek *aphōnia*, from *aphōnos* voiceless, from *a-* + *phōnē* sound — more at BAN] (1778)
: loss of voice and of all but whispered speech
— apho·nic \-ˈfä-nik, -ˈfō-\ *adjective*
aph·o·rism \ˈa-fə-ˌri-zəm\ *noun* [Middle French *aphorisme*, from Late Latin *aphorismus*, from Greek *aphorismos* definition, aphorism, from *aphorizein* to define, from *apo-* + *horizein* to bound — more at HORIZON] (1528)
1 : a concise statement of a principle
2 : a terse formulation of a truth or sentiment **:** ADAGE
— aph·o·rist \-rist\ *noun*
— aph·o·ris·tic \ˌa-fə-ˈris-tik\ *adjective*
— aph·o·ris·ti·cal·ly \-ti-k(ə-)lē\ *adverb*
aph·o·rize \ˈa-fə-ˌrīz\ *intransitive verb* **-rized; -riz·ing** (1669)
: to write or speak in or as if in aphorisms
apho·tic \(ˌ)ā-ˈfō-tik\ *adjective* (circa 1900)
: being the deep zone of an ocean or lake receiving too little light to permit photosynthesis
aph·ro·di·si·ac \ˌa-frə-ˈdē-zē-ˌak, -ˈdi-zē-\ *noun* [Greek *aphrodisiakos* sexual, gem with aphrodisiac properties, from *aphrodisia* heterosexual pleasures, from neuter plural of *aphrodisios* of Aphrodite, from *Aphroditē*] (1719)
1 : an agent (as a food or drug) that arouses or is held to arouse sexual desire
2 : something that excites
— aphrodisiac *also* **aph·ro·di·si·a·cal** \ˌa-frə-di-ˈsī-ə-kəl, -ˈzī-\ *adjective*
Aph·ro·di·te \ˌa-frə-ˈdī-tē\ *noun* [Greek *Aphroditē*]
: the Greek goddess of love and beauty — compare VENUS
api·ar·i·an \ˌā-pē-ˈer-ē-ən\ *adjective* (1801)
: of or relating to beekeeping or bees
api·a·rist \ˈā-pē-ə-rist, -pē-ˌer-ist\ *noun* (1816)
: BEEKEEPER
api·ary \ˈā-pē-ˌer-ē\ *noun, plural* **-ar·ies** [Latin *apiarium*, from *apis* bee] (1654)
: a place where bees are kept; *especially* **:** a collection of hives or colonies of bees kept for their honey
api·cal \ˈā-pi-kəl *also* ˈa-pi-\ *adjective* [probably from New Latin *apicalis*, from Latin *apic-, apex*] (1828)
1 : of, relating to, or situated at an apex
2 : of, relating to, or formed with the tip of the tongue ⟨*n, l,* and *r* are *apical* consonants⟩
— api·cal·ly \-k(ə-)lē\ *adverb*
apical dominance *noun* (1947)

: inhibition of the growth of lateral buds by the terminal bud of a shoot
apical meristem *noun* (circa 1934)
: a meristem at the apex of a root or shoot that is responsible for increase in length
apic·u·late \ə-ˈpi-kyə-lət, ā-\ *adjective* [New Latin *apiculus,* diminutive of Latin *apic-, apex*] (1830)
: ending abruptly in a small distinct point ⟨an *apiculate* leaf⟩
api·cul·ture \ˈā-pə-ˌkəl-chər\ *noun* [probably from French, from Latin *apis* bee + French *culture*] (1864)
: the keeping of bees especially on a large scale
— **api·cul·tur·al** \ˌā-pə-ˈkəl-ch(ə-)rəl\ *adjective*
— **api·cul·tur·ist** \-ch(ə-)rist\ *noun*
apiece \ə-ˈpēs\ *adverb* (15th century)
: for each one **:** INDIVIDUALLY
Apis \ˈā-pəs\ *noun* [Latin, from Greek, from Egyptian *hp*]
: a sacred bull worshiped by the ancient Egyptians
ap·ish \ˈā-pish\ *adjective* (1532)
: resembling an ape: as **a :** extremely silly or affected **b :** given to slavish imitation
— **ap·ish·ly** *adverb*
— **ap·ish·ness** *noun*
APL \ˌā-(ˌ)pē-ˈel\ *noun* [*a p*rogramming *l*anguage] (1966)
: a computer programming language designed especially for the concise representation of algorithms
ap·la·nat·ic \ˌa-plə-ˈna-tik\ *adjective* [*a-* + Greek *planasthai* to wander — more at PLANET] (1794)
: free from or corrected for spherical aberration ⟨an *aplanatic* lens⟩
aplas·tic anemia \(ˌ)ā-ˈplas-tik-\ *noun* (1928)
: anemia that is characterized by defective function of the blood-forming organs (as the bone marrow) and is caused by toxic agents (as chemicals or X rays) or is idiopathic in origin
¹aplen·ty \ə-ˈplen-tē\ *adjective* (1830)
: being in plenty or abundance — used postpositively ⟨money *aplenty* for all his needs⟩
²aplenty *adverb* (1846)
1 : in abundance **:** PLENTIFULLY
2 : very much **:** EXTREMELY ⟨scared *aplenty*⟩
ap·lite \ˈa-ˌplīt\ *noun* [probably from German *Aplit,* from Greek *haploos* simple — more at HAPL-] (1879)
: a fine-grained light-colored granite consisting almost entirely of quartz and feldspar
— **ap·lit·ic** \a-ˈpli-tik\ *adjective*
aplomb \ə-ˈpläm, -ˈpləm\ *noun* [French, literally, perpendicularity, from Middle French, from *a plomb,* literally, according to the plummet] (1828)
: complete and confident composure or self-assurance **:** POISE
synonym see CONFIDENCE
ap·nea \ˈap-nē-ə\ *noun* [New Latin, from *a-* + *-pnea*] (circa 1719)
1 : transient cessation of respiration
2 : ASPHYXIA
— **ap·ne·ic** \-nē-ik\ *adjective*
ap·noea *chiefly British variant of* APNEA
apo- *or* **ap-** *prefix* [Middle English, from Middle French & Latin; Middle French, from Latin, from Greek, from *apo* — more at OF]
1 : away from **:** off ⟨*ap*helion⟩
2 : detached **:** separate ⟨*apo*gamy⟩
3 : formed from **:** related to ⟨*apo*morphine⟩
apoc·a·lypse \ə-ˈpä-kə-ˌlips\ *noun* [Middle English, revelation, Revelation, from Late Latin *apocalypsis,* from Greek *apokalypsis,* from *apokalyptein* to uncover, from *apo-* + *kalyptein* to cover — more at HELL] (13th century)
1 a : one of the Jewish and Christian writings of 200 B.C. to A.D. 150 marked by pseudonymity, symbolic imagery, and the expectation of an imminent cosmic cataclysm in which God destroys the ruling powers of evil and raises the righteous to life in a messianic kingdom **b** *capitalized* **:** REVELATION 3
2 a : something viewed as a prophetic revelation **b :** ARMAGEDDON
apoc·a·lyp·tic \ə-ˌpä-kə-ˈlip-tik\ *also* **apoc·a·lyp·ti·cal** \-ti-kəl\ *adjective* (1663)
1 : of, relating to, or resembling an apocalypse
2 : forecasting the ultimate destiny of the world **:** PROPHETIC
3 : foreboding imminent disaster or final doom **:** TERRIBLE
4 : wildly unrestrained **:** GRANDIOSE
5 : ultimately decisive **:** CLIMACTIC
— **apoc·a·lyp·ti·cal·ly** \-ti-k(ə-)lē\ *adverb*
apoc·a·lyp·ti·cism \-tə-ˌsi-zəm\ *or* **apoc·a·lyp·tism** \ə-ˈpä-kə-ˌlip-ˌti-zəm\ *noun* (1884)
: apocalyptic expectation; *especially* **:** a doctrine concerning an imminent end of the world and an ensuing general resurrection and final judgment
apoc·a·lyp·tist \ə-ˈpä-kə-ˌlip-tist\ *noun* (1835)
: the writer of an apocalypse
apo·chro·mat·ic \ˌa-pə-krō-ˈma-tik\ *adjective* [International Scientific Vocabulary] (1887)
: free from chromatic and spherical aberration ⟨an *apochromatic* lens⟩
apoc·o·pe \ə-ˈpä-kə-(ˌ)pē\ *noun* [Late Latin, from Greek *apokopē,* literally, cutting off, from *apokoptein* to cut off, from *apo-* + *koptein* to cut — more at CAPON] (circa 1550)
: the loss of one or more sounds or letters at the end of a word (as in *sing* from Old English *singan*)
apo·crine \ˈa-pə-krən, -ˌkrīn, -ˌkrēn\ *adjective* [International Scientific Vocabulary *apo-* + Greek *krinein* to separate — more at CERTAIN] (1926)
: producing a fluid secretion by pinching off one end of the secretory cell while leaving the rest intact ⟨an *apocrine* gland⟩; *also* **:** produced by an apocrine gland
apoc·ry·pha \ə-ˈpä-krə-fə\ *noun plural but singular or plural in construction* [Medieval Latin, from Late Latin, neuter plural of *apocryphus* secret, not canonical, from Greek *apokryphos* obscure, from *apokryptein* to hide away, from *apo-* + *kryptein* to hide — more at CRYPT] (14th century)
1 : writings or statements of dubious authenticity
2 *capitalized* **a :** books included in the Septuagint and Vulgate but excluded from the Jewish and Protestant canons of the Old Testament — see BIBLE table **b :** early Christian writings not included in the New Testament ◆
apoc·ry·phal \-fəl\ *adjective* (1590)
1 : of doubtful authenticity **:** SPURIOUS
2 *often capitalized* **:** of or resembling the Apocrypha
synonym see FICTITIOUS
— **apoc·ry·phal·ly** \-fə-lē\ *adverb*
— **apoc·ry·phal·ness** *noun*
apo·dic·tic \ˌa-pə-ˈdik-tik\ *also* **apo·deic·tic** \-ˈdīk-tik\ *adjective* [Latin *apodicticus,* from Greek *apodeiktikos,* from *apodeiknynai* to demonstrate, from *apo-* + *deiknynai* to show — more at DICTION] (circa 1645)
: expressing or of the nature of necessary truth or absolute certainty
— **apo·dic·ti·cal·ly** \-ti-k(ə-)lē\ *adverb*
apod·o·sis \ə-ˈpä-də-səs\ *noun, plural* **-o·ses** \-ˌsēz\ [New Latin, from Greek, from *apodidonai* to give back, deliver, from *apo-* + *didonai* to give — more at DATE] (circa 1638)
: the main clause of a conditional sentence — compare PROTASIS
apo·en·zyme \ˌa-pō-ˈen-ˌzīm\ *noun* [International Scientific Vocabulary] (1936)
: a protein that forms an active enzyme system by combination with a coenzyme and determines the specificity of this system for a substrate
apog·a·my \ə-ˈpä-gə-mē\ *noun* [International Scientific Vocabulary] (circa 1878)
: development of a sporophyte from a gametophyte without fertilization
— **apog·a·mous** \ə-ˈpä-gə-məs\ *adjective*
apo·gee \ˈa-pə-(ˌ)jē\ *noun* [French *apogée,* from New Latin *apogaeum,* from Greek *apogaion,* from neuter of *apogeios, apogaios* far from the earth, from *apo-* + *gē, gaia* earth] (1594)
1 : the point in the orbit of an object (as a satellite) orbiting the earth that is at the greatest distance from the center of the earth; *also* **:** the point farthest from a planet or a satellite (as the moon) reached by an object orbiting it — compare PERIGEE

apogee 1

2 : the farthest or highest point **:** CULMINATION ⟨Aegean civilization reached its *apogee* in Crete⟩
— **apo·ge·an** \ˌa-pə-ˈjē-ən\ *adjective*
apo·li·po·pro·tein \ˌa-pō-ˌlī-pō-ˈprō-ˌtēn, -ˌli-, -ˌtē-ən\ *noun* (1970)
: a protein that combines with a lipid to form a lipoprotein
apo·lit·i·cal \ˌā-pə-ˈli-ti-kəl\ *adjective* (1935)
1 : having no interest or involvement in political affairs; *also* **:** having an aversion to politics or political affairs
2 : having no political significance
— **apo·lit·i·cal·ly** \-k(ə-)lē\ *adverb*
Ap·ol·lin·i·an \ˌa-pə-ˈli-nē-ən\ *adjective* (1924)
: APOLLONIAN
Apol·lo \ə-ˈpä-(ˌ)lō\ *noun* [Latin *Apollin-, Apollo,* from Greek *Apollōn*]
: the Greek and Roman god of sunlight, prophecy, music, and poetry
Ap·ol·lo·ni·an \ˌa-pə-ˈlō-nē-ən\ *adjective* (1663)
1 : of, relating to, or resembling the god Apollo

◇ WORD HISTORY
apocrypha *Apocrypha* is the neuter plural form of the Late Latin adjective *apocryphus,* meaning "secret" or "uncanonical." Latin took the word from Greek *apokryphos,* meaning "hidden," which was a derivative of the verb *apokryptein* "to hide." The term *apocrypha* was used in reference to scriptural writings that were "hidden" from the general public. Such books were considered to contain esoteric knowledge accessible to only a small circle of believers. The non-initiates, on the other hand, considered these books to be of doubtful authenticity, or even heretical, so *apocrypha* began to be applied to spurious works. In his Latin translation of the Bible, called the Vulgate, St. Jerome designated as *apocrypha* those books of the Old Testament that were in the Septuagint (a Greek translation of pre-Christian times) but not in the Hebrew-language scriptures to which he had access. Although the Roman Church rejected Jerome's designation and accepted the books as authentic Holy Scripture, Jerome was supported by the German Protestant theologian Karlstadt in 1520. Martin Luther included them in his 1534 German Bible but in a supplement under the heading "Apocrypha."

\ə\ **abut** \ᵊ\ **kitten** \ər\ **further** \a\ **ash** \ā\ **ace** \ä\ **mop, mar** \aů\ **out** \ch\ **chin** \e\ **bet** \ē\ **easy** \g\ **go** \i\ **hit** \ī\ **ice** \j\ **job** \ŋ\ **sing** \ō\ **go** \ȯ\ **law** \ȯi\ **boy** \th\ **thin** \t͟h\ **the** \ü\ **loot** \ů\ **foot** \y\ **yet** \zh\ **vision** *see also* Guide to Pronunciation

2 : harmonious, measured, ordered, or balanced in character — compare DIONYSIAN

Apol·lyon \ə-ˈpäl-yən, -ˈpä-lē-ən\ *noun* [Greek *Apollyōn*]
: the angel of the bottomless pit in the Book of Revelation

¹apol·o·get·ic \ə-ˌpä-lə-ˈje-tik\ *noun* (15th century)
: APOLOGETICS 1

²apologetic *adjective* [Greek *apologētikos,* from *apologeisthai* to defend, from *apo-* + *logos* speech] (1649)
1 a : offered in defense or vindication ⟨the *apologetic* writings of the early Christians⟩ **b :** offered by way of excuse or apology ⟨an *apologetic* smile⟩
2 : regretfully acknowledging fault or failure **:** CONTRITE ⟨was *apologetic* about his mistake⟩
— apol·o·get·i·cal·ly \-ti-k(ə-)lē\ *adverb*

apol·o·get·ics \-tiks\ *noun plural but singular or plural in construction* (circa 1733)
1 : systematic argumentative discourse in defense (as of a doctrine)
2 : a branch of theology devoted to the defense of the divine origin and authority of Christianity

ap·o·lo·gia \ˌa-pə-ˈlō-j(ē-)ə\ *noun* [Late Latin] (1784)
: a defense especially of one's opinions, position, or actions ⟨the finest *apologia* or explanation of what drives a man to devote his life to pure mathematics —*British Book News*⟩
synonym see APOLOGY

apol·o·gise *British variant of* APOLOGIZE

apol·o·gist \ə-ˈpä-lə-jist\ *noun* (1640)
: one who speaks or writes in defense of someone or something

apol·o·gize \-ˌjīz\ *intransitive verb* **-gized; -giz·ing** (1597)
: to make an apology
— apol·o·giz·er *noun*

ap·o·logue \ˈa-pə-ˌlȯg, -ˌläg\ *noun* [French, from Latin *apologus,* from Greek *apologos,* from *apo-* + *logos* speech, narrative] (circa 1555)
: an allegorical narrative usually intended to convey a moral

apol·o·gy \ə-ˈpä-lə-jē\ *noun, plural* **-gies** [Middle French or Late Latin; Middle French *apologie,* from Late Latin *apologia,* from Greek, from *apo-* + *logos* speech — more at LEGEND] (1533)
1 a : a formal justification **:** DEFENSE **b :** EXCUSE 2a
2 : an admission of error or discourtesy accompanied by an expression of regret
3 : a poor substitute **:** MAKESHIFT ☆

apo·lune \ˈa-pə-ˌlün\ *noun* [*apo-* + Latin *luna* moon — more at LUNAR] (circa 1968)
: the point in the path of a body orbiting the moon that is farthest from the center of the moon — compare PERILUNE

apo·mict \ˈa-pə-ˌmikt\ *noun* [probably back-formation from International Scientific Vocabulary *apomictic,* from *apo-* + Greek *mignynai* to mix — more at MIX] (circa 1938)
: one produced or reproducing by apomixis
— apo·mic·tic \ˌa-pə-ˈmik-tik\ *adjective*
— apo·mic·ti·cal·ly \-ti-k(ə-)lē\ *adverb*

apo·mix·is \ˌa-pə-ˈmik-səs\ *noun, plural* **-mix·es** \-ˌsēz\ [New Latin, from *apo-* + Greek *mixis* act of mixing, from *mignynai*] (1913)
: reproduction (as apogamy or parthenogenesis) involving specialized generative tissues but not dependent on fertilization

apo·mor·phine \ˌa-pə-ˈmȯr-ˌfēn\ *noun* [International Scientific Vocabulary] (1888)
: a crystalline morphine derivative $C_{17}H_{17}NO_2$ that is a dopamine agonist and is administered as the hydrochloride for its powerful emetic action

apo·neu·ro·sis \ˌa-pə-nu̇-ˈrō-səs, -nyu̇-\ *noun* [New Latin, from Greek *aponeurōsis,* from *aponeurousthai* to pass into a tendon, from *apo-* + *neuron* sinew — more at NERVE] (1676)
: a broad flat sheet of dense fibrous collagenous connective tissue that covers, invests, and forms the terminations and attachments of various muscles
— apo·neu·rot·ic \-ˈrä-tik\ *adjective*

ap·o·phthegm \ˈa-pə-ˌthem\ *variant of* APOTHEGM

apo·phyl·lite \ˌa-pə-ˈfi-ˌlīt, ə-ˈpä-fə-ˌlīt\ *noun* [French, from *apo-* + Greek *phyllon* leaf — more at BLADE] (1810)
: a mineral composed of a hydrous potassium calcium silicate that is related to the zeolites and is usually found in transparent square prisms or white or grayish masses

apoph·y·sis \ə-ˈpä-fə-səs\ *noun, plural* **-y·ses** \-ˌsēz\ [New Latin, from Greek, from *apo-* + *phyein* to bring forth — more at BE] (1646)
: an expanded or projecting part especially of an organism
— apoph·y·se·al \-ˌpä-fə-ˈsē-əl\ *adjective*

ap·o·plec·tic \ˌa-pə-ˈplek-tik\ *adjective* [French or Late Latin; French *apoplectique,* from Late Latin *apoplecticus,* from Greek *apoplēktikos,* from *apoplēssein*] (1611)
1 : of, relating to, or causing stroke
2 : affected with, inclined to, or showing symptoms of stroke
3 : of a kind to cause or apparently cause stroke ⟨an *apoplectic* rage⟩; *also* **:** greatly excited or angered
— ap·o·plec·ti·cal·ly \-ti-k(ə-)lē\ *adverb*

ap·o·plexy \ˈa-pə-ˌplek-sē\ *noun* [Middle English *apoplexie,* from Middle French & Late Latin; Middle French, from Late Latin *apoplexia,* from Greek *apoplēxia,* from *apoplēssein* to cripple by a stroke, from *apo-* + *plēssein* to strike — more at PLAINT] (15th century)
: STROKE 5

aport \ə-ˈpōrt, -ˈpȯrt\ *adverb* (1627)
: on or toward the left side of a ship ⟨put the helm hard *aport*⟩

apo·se·mat·ic \ˌa-pə-si-ˈma-tik\ *adjective* (1890)
: being conspicuous and serving to warn ⟨*aposematic* coloration⟩
— apo·se·mat·i·cal·ly \-ti-k(ə-)lē\ *adverb*

ap·o·si·o·pe·sis \ˌa-pə-ˌsī-ə-ˈpē-səs\ *noun, plural* **-pe·ses** \-ˌsēz\ [Late Latin, from Greek *aposiōpēsis,* from *aposiōpan* to be fully silent, from *apo-* + *siōpan* to be silent, from *siōpē* silence] (1578)
: the leaving of a thought incomplete usually by a sudden breaking off (as in "his behavior was—but I blush to mention that")
— ap·o·si·o·pet·ic \-ˈpe-tik\ *adjective*

apos·po·ry \ˈa-pə-ˌspōr-ē, -ˌspȯr-; ə-ˈpäs-pə-rē\ *noun* (1884)
: production of gametophytes directly from diploid cells of the sporophytes without spore formation (as in certain ferns and mosses)
— apos·po·rous \ˈa-pə-ˌspōr-əs, -ˌspȯr-; ə-ˈpäs-pə-rəs\ *adjective*

apos·ta·sy \ə-ˈpäs-tə-sē\ *noun, plural* **-sies** [Middle English *apostasie,* from Late Latin *apostasia,* from Greek, literally, revolt, from *aphistasthai* to revolt, from *apo-* + *histasthai* to stand — more at STAND] (14th century)
1 : renunciation of a religious faith
2 : abandonment of a previous loyalty **:** DEFECTION

apos·tate \ə-ˈpäs-ˌtāt, -tət\ *noun* (14th century)
: one who commits apostasy
— apostate *adjective*

apos·ta·tise *British variant of* APOSTATIZE

apos·ta·tize \ə-ˈpäs-tə-ˌtīz\ *intransitive verb* **-tized; -tiz·ing** (1611)
: to commit apostasy

a pos·te·ri·o·ri \ˌä-(ˌ)pō-ˌstir-ē-ˈōr-ē, -ˌster-; ˌā-(ˌ)pä-ˌstir-ē-ˈōr-ˌī, -(ˌ)pō-, -ˈōr-ē; -ˈȯr-\ *adjective* [Latin, literally, from the latter] (1588)
1 : INDUCTIVE
2 : relating to or derived by reasoning from observed facts — compare A PRIORI
— a posteriori *adverb*

apos·tle \ə-ˈpä-səl\ *noun* [Middle English, from Old French & Old English; Old French *apostle* & Old English *apostol,* both from Late Latin *apostolus,* from Greek *apostolos,* from *apostellein* to send away, from *apo-* + *stellein* to send] (before 12th century)
1 : one sent on a mission: as **a :** one of an authoritative New Testament group sent out to preach the gospel and made up especially of Christ's 12 original disciples and Paul **b :** the first prominent Christian missionary to a region or group
2 a : a person who initiates a great moral reform or who first advocates an important belief or system **b :** an ardent supporter **:** ADHERENT ⟨an *apostle* of liberty⟩
3 : the highest ecclesiastical official in some church organizations
4 : one of a Mormon administrative council of 12 men
— apos·tle·ship \-ˌship\ *noun*

Apostles' Creed *noun* (circa 1658)
: a Christian statement of belief ascribed to the Twelve Apostles and used especially in public worship

apos·to·late \ə-ˈpäs-tə-ˌlāt, -lət\ *noun* [Late Latin *apostolatus,* from *apostolus*] (14th century)
1 : the office or mission of an apostle
2 : an association of persons dedicated to the propagation of a religion or a doctrine

ap·os·tol·ic \ˌa-pə-ˈstä-lik\ *adjective* (13th century)
1 a : of or relating to an apostle **b :** of, relating to, or conforming to the teachings of the New Testament apostles
2 a : of or relating to a succession of spiritual authority from the apostles held (as by Roman Catholics, Anglicans, and Eastern Orthodox) to be perpetuated by successive ordinations of bishops and to be necessary for valid sacraments and orders **b :** PAPAL
— apos·to·lic·i·ty \ə-ˌpäs-tə-ˈli-sə-tē\ *noun*

apostolic delegate *noun* (circa 1907)
: an ecclesiastical representative of the Holy See to the Catholic hierarchy of another country

Apostolic Father *noun* (1828)
: a church father of the first or second century A.D.

¹apos·tro·phe \ə-ˈpäs-trə-(ˌ)fē\ *noun* [Latin, from Greek *apostrophē,* literally, active of turning away, from *apostrephein* to turn away, from *apo-* + *strephein* to turn] (1533)

☆ SYNONYMS
Apology, apologia, excuse, plea, pretext, alibi mean matter offered in explanation or defense. APOLOGY usually applies to an expression of regret for a mistake or wrong with implied admission of guilt or fault and with or without reference to palliating circumstances ⟨said by way of *apology* that he would have met them if he could⟩. APOLOGIA implies not admission of guilt or regret but a desire to make clear the grounds for some course, belief, or position ⟨his speech was an *apologia* for his foreign policy⟩. EXCUSE implies an intent to avoid or remove blame or censure ⟨used illness as an *excuse* for missing the meeting⟩. PLEA stresses argument or appeal for understanding or sympathy or mercy ⟨her usual *plea* that she was nearsighted⟩. PRETEXT suggests subterfuge and the offering of false reasons or motives in excuse or explanation ⟨used any *pretext* to get out of work⟩. ALIBI implies a desire to shift blame or evade punishment and imputes plausibility rather than truth to the explanation offered ⟨his *alibi* failed to stand scrutiny⟩.

: the addressing of a usually absent person or a usually personified thing rhetorically ⟨Carlyle's "O Liberty, what things are done in thy name!" is an example of *apostrophe*⟩
— **ap·os·troph·ic** \ˌa-pə-ˈsträ-fik\ *adjective*

²apostrophe *noun* [Middle French & Late Latin; Middle French, from Late Latin *apostrophus,* from Greek *apostrophos,* from *apostrophos* turned away, from *apostrephein*] (1727)
: a mark ' used to indicate the omission of letters or figures, the possessive case, or the plural of letters or figures
— **apostrophic** *adjective*

apos·tro·phise *British variant of* APOSTROPHIZE

apos·tro·phize \ə-ˈpäs-trə-ˌfīz\ *verb* **-phized; -phiz·ing** (1718)
transitive verb
: to address by or in apostrophe
intransitive verb
: to make use of apostrophe

apothecaries' measure *noun* (circa 1900)
: a system of liquid units of measure used chiefly by pharmacists

apothecaries' weight *noun* (1765)
: a system of weights used chiefly by pharmacists — see WEIGHT table

apoth·e·cary \ə-ˈpä-thə-ˌker-ē\ *noun, plural* **-car·ies** [Middle English *apothecarie,* from Medieval Latin *apothecarius,* from Late Latin, shopkeeper, from Latin *apotheca* storehouse, from Greek *apothēkē,* from *apotithenai* to put away, from *apo-* + *tithenai* to put — more at DO] (14th century)
1 : one who prepares and sells drugs or compounds for medicinal purposes
2 : PHARMACY

apo·the·ci·um \ˌa-pə-ˈthē-shē-əm, -sē-\ *noun, plural* **-cia** \-shē-ə, -sē-\ [New Latin, from Latin *apotheca*] (1830)
: a spore-bearing structure in many lichens and fungi consisting of a discoid or cupped body bearing asci on the exposed flat or concave surface
— **apo·the·cial** \-sh(ē-)əl, -sē-əl\ *adjective*

ap·o·thegm \ˈa-pə-ˌthem\ *noun* [Greek *apophthegmat-, apophthegma,* from *apophthengesthai* to speak out, from *apo-* + *phthengesthai* to utter] (circa 1587)
: a short, pithy, and instructive saying or formulation **:** APHORISM
— **ap·o·theg·mat·ic** \ˌa-pə-theg-ˈma-tik\ *adjective*

ap·o·them \ˈa-pə-ˌthem\ *noun* [International Scientific Vocabulary *apo-* + *-them* (from Greek *thema* something laid down, theme)] (circa 1856)
: the perpendicular from the center of a regular polygon to one of the sides

apo·the·o·sis \ə-ˌpä-thē-ˈō-səs, ˌa-pə-ˈthē-ə-səs\ *noun, plural* **-o·ses** \-ˌsēz\ [Late Latin, from Greek *apotheōsis,* from *apotheoun* to deify, from *apo-* + *theos* god] (circa 1580)
1 : elevation to divine status **:** DEIFICATION
2 : the perfect example **:** QUINTESSENCE ⟨this is the literary *apotheosis* of the shaggy dog story —Thomas Sutcliffe⟩
— **apo·the·o·size** \ˌa-pə-ˈthē-ə-ˌsīz, ə-ˈpä-thē-ə-\ *transitive verb*

apo·tro·pa·ic \ˌa-pə-trō-ˈpā-ik\ *adjective* [Greek *apotropaios,* from *apotrepein* to avert, from *apo-* + *trepein* to turn] (1883)
: designed to avert evil ⟨an *apotropaic* ritual⟩
— **apo·tro·pa·i·cal·ly** \-ˈpā-ə-k(ə-)lē\ *adverb*

Ap·pa·la·chian \ˌa-pə-ˈlā-ch(ē-)ən, -ˈla-, -sh(ē-)ən\ *noun* (1888)
: a white native or resident of the Appalachian mountain area

Appalachian dulcimer *noun* (1962)
: DULCIMER 2

ap·pall *also* **ap·pal** \ə-ˈpȯl\ *verb* **ap·palled; ap·pall·ing** [Middle English, from Middle French *apalir,* from Old French, from *a-* (from Latin *ad-*) + *palir* to grow pale, from Latin *pallescere,* inchoative of *pallēre* to be pale — more at FALLOW] (14th century)
intransitive verb
obsolete **:** WEAKEN, FAIL
transitive verb
: to overcome with consternation, shock, or dismay
synonym see DISMAY

ap·pall·ing *adjective* (1817)
: inspiring horror, dismay, or disgust ⟨living under *appalling* conditions⟩
— **ap·pall·ing·ly** *adverb*

Ap·pa·loo·sa \ˌa-pə-ˈlü-sə\ *noun* [origin unknown] (1947)
: any of a breed of rugged saddle horses developed in western North America and usually having a white or solid-colored coat with small spots

Appaloosa

ap·pa·nage \ˈa-pə-nij\ *noun* [French *apanage,* from Old French, from *apaner* to provide for a younger offspring, from Medieval Latin *appanare,* from Latin *ad-* + *panis* bread — more at FOOD] (1602)
1 a : a grant (as of land or revenue) made by a sovereign or a legislative body to a dependent member of the royal family or a principal vassal **b :** a property or privilege appropriated to or by a person as something due
2 : a rightful endowment or adjunct

ap·pa·rat \ˈa-pə-ˌrat, ˌä-pə-ˈrät\ *noun* [Russian] (1941)
: APPARATUS 2

ap·pa·rat·chik \ˌä-pə-ˈrä(t)-chik\ *noun, plural* **-chiks** *also* **-chi·ki** \-chi-kē\ [Russian, from *apparat*] (1941)
1 : a member of a Communist apparat
2 : an official blindly devoted to superiors or to the organization

ap·pa·ra·tus \ˌa-pə-ˈra-təs, -ˈrā-\ *noun, plural* **-tus·es** *or* **-tus** [Latin, from *apparare* to prepare, from *ad-* + *parare* to prepare — more at PARE] (circa 1628)
1 a : a set of materials or equipment designed for a particular use **b :** a group of anatomical or cytological parts functioning together ⟨mitotic *apparatus*⟩ **c :** an instrument or appliance designed for a specific operation
2 : the functional processes by means of which a systematized activity is carried out: as **a :** the machinery of government **b :** the organization of a political party or an underground movement

¹ap·par·el \ə-ˈpar-əl\ *transitive verb* **-eled** *or* **-elled; -el·ing** *or* **-el·ling** [Middle English *appareillen,* from Middle French *apareillier* to prepare, from (assumed) Vulgar Latin *appariculare,* from Latin *apparare*] (14th century)
1 : to put clothes on **:** DRESS
2 : ADORN, EMBELLISH

²apparel *noun* (14th century)
1 : the equipment (as sails and rigging) of a ship
2 : personal attire **:** CLOTHING
3 : something that clothes or adorns ⟨the bright *apparel* of spring⟩

ap·par·ent \ə-ˈpar-ənt, -ˈper-\ *adjective* [Middle English, from Middle French *aparent,* from Latin *apparent-, apparens,* present participle of *apparēre* to appear] (14th century)
1 : open to view **:** VISIBLE
2 : clear or manifest to the understanding
3 : appearing as actual to the eye or mind
4 : having an indefeasible right to succeed to a title or estate
5 : manifest to the senses or mind as real or true on the basis of evidence that may or may not be factually valid ⟨the air of spontaneity is perhaps more *apparent* than real —J. R. Sutherland⟩ ☆
— **ap·par·ent·ness** \-ˈpar-ənt-nəs, -ˈper-\ *noun*

ap·par·ent·ly \-lē\ *adverb* (1566)
: it seems apparent ⟨the window had *apparently* been forced open⟩ ⟨*apparently,* we're supposed to wait here⟩

apparent magnitude *noun* (1875)
: the luminosity of a celestial body (as a star) as observed from the earth — compare ABSOLUTE MAGNITUDE

apparent time *noun* (1694)
: the time of day indicated by the hour angle of the sun or by a sundial

ap·pa·ri·tion \ˌa-pə-ˈri-shən\ *noun* [Middle English *apparicioun,* from Late Latin *apparition-, apparitio* appearance, from Latin *apparēre*] (15th century)
1 a : an unusual or unexpected sight **:** PHENOMENON **b :** a ghostly figure
2 : the act of becoming visible **:** APPEARANCE
— **ap·pa·ri·tion·al** \-ˈrish-nəl, -ˈri-shə-nᵊl\ *adjective*

ap·par·i·tor \ə-ˈpar-ə-tər\ *noun* [Latin, from *apparēre*] (15th century)
: an official formerly sent to carry out the orders of a magistrate, judge, or court

¹ap·peal \ə-ˈpē(ə)l\ *noun* (13th century)
1 : a legal proceeding by which a case is brought before a higher court for review of the decision of a lower court
2 : a criminal accusation
3 a : an application (as to a recognized authority) for corroboration, vindication, or decision **b :** an earnest plea **:** ENTREATY
4 : the power of arousing a sympathetic response **:** ATTRACTION ⟨movies had a great *appeal* for him⟩

²appeal *verb* [Middle English *appelen* to accuse, appeal, from Middle French *apeler,* from Latin *appellare,* from *appellere* to drive to, from *ad-* + *pellere* to drive — more at FELT] (14th century)
transitive verb
1 : to charge with a crime **:** ACCUSE
2 : to take proceedings to have (a lower court's decision) reviewed in a higher court
intransitive verb
1 : to take a lower court's decision to a higher court for review
2 : to call upon another for corroboration, vindication, or decision
3 : to make an earnest request
4 : to arouse a sympathetic response
— **ap·peal·abil·i·ty** \-ˌpē-lə-ˈbi-lə-tē\ *noun*
— **ap·peal·able** \-ˈpē-lə-bəl\ *adjective*
— **ap·peal·er** *noun*

☆ SYNONYMS
Apparent, illusory, seeming, ostensible mean not actually being what appearance indicates. APPARENT suggests appearance to unaided senses that is not or may not be borne out by more rigorous examination or greater knowledge ⟨the *apparent* cause of the accident⟩. ILLUSORY implies a false impression based on deceptive resemblance or faulty observation, or influenced by emotions that prevent a clear view ⟨an *illusory* sense of security⟩. SEEMING implies a character in the thing observed that gives it the appearance, sometimes through intent, of something else ⟨the *seeming* simplicity of the story⟩. OSTENSIBLE suggests a discrepancy between an openly declared or naturally implied aim or reason and the true one ⟨the *ostensible* reason for their visit⟩. See in addition EVIDENT.

ap·peal·ing \ə-ˈpē-liŋ\ *adjective* (1813)
1 : marked by earnest entreaty **:** IMPLORING
2 : having appeal **:** PLEASING
— ap·peal·ing·ly \-liŋ-lē\ *adverb*

ap·pear \ə-ˈpir\ *intransitive verb* [Middle English *apperen,* from Old French *aparoir,* from Latin *apparēre,* from *ad-* + *parēre* to show oneself] (13th century)
1 a : to be or come in sight ⟨the sun *appears* on the horizon⟩ **b :** to show up ⟨*appears* promptly at eight each day⟩
2 : to come formally before an authoritative body ⟨must *appear* in court today⟩
3 : to have an outward aspect **:** SEEM ⟨*appears* happy enough⟩
4 : to become evident or manifest ⟨there *appears* to be evidence to the contrary⟩
5 : to come into public view ⟨first *appeared* on a television variety show⟩ ⟨the book *appeared* in print a few years ago⟩
6 : to come into existence ⟨hominids *appeared* late in the evolutionary chain⟩

ap·pear·ance \ə-ˈpir-ən(t)s\ *noun* (14th century)
1 a : external show **:** SEMBLANCE ⟨although hostile, he preserved an *appearance* of neutrality⟩ **b :** outward aspect **:** LOOK ⟨had a fierce *appearance*⟩ **c** *plural* **:** outward indication ⟨trying to keep up *appearances*⟩
2 a : a sense impression or aspect of a thing ⟨the blue of distant hills is only an *appearance*⟩ **b :** the world of sensible phenomena
3 a : the act, action, or process of appearing **b :** the presentation of oneself in court as a party to an action often through the representation of an attorney
4 a : something that appears **:** PHENOMENON **b :** an instance of appearing **:** OCCURRENCE

ap·pease \ə-ˈpēz\ *transitive verb* **ap·peased; ap·peas·ing** [Middle English *appesen,* from Middle French *apaisier,* from *a-* (from Latin *ad-*) + *pais* peace — more at PEACE] (14th century)
1 : to bring to a state of peace or quiet **:** CALM
2 : to cause to subside **:** ALLAY ⟨*appeased* my hunger⟩
3 : PACIFY, CONCILIATE; *especially* **:** to buy off (an aggressor) by concessions usually at the sacrifice of principles
synonym see PACIFY
— ap·peas·able \-ˈpē-zə-bəl\ *adjective*
— ap·pease·ment \-ˈpēz-mənt\ *noun*
— ap·peas·er *noun*

¹ap·pel·lant \ə-ˈpe-lənt\ *adjective* (14th century)
: of or relating to an appeal **:** APPELLATE

²appellant *noun* (15th century)
: one that appeals; *specifically* **:** one that appeals from a judicial decision or decree

ap·pel·late \ə-ˈpe-lət\ *adjective* [Latin *appellatus,* past participle of *appellare*] (1768)
: of, relating to, or recognizing appeals; *specifically* **:** having the power to review the judgment of another tribunal ⟨an *appellate* court⟩

ap·pel·la·tion \ˌa-pə-ˈlā-shən\ *noun* (15th century)
1 : an identifying name or title **:** DESIGNATION
2 *archaic* **:** the act of calling by a name
3 : a geographical name (as of a region, village or vineyard) under which a winegrower is authorized to identify and market wine

ap·pel·la·tive \ə-ˈpe-lə-tiv\ *adjective* (15th century)
1 : of or relating to a common noun
2 : of, relating to, or inclined to the giving of names
— ap·pel·la·tive *noun*
— ap·pel·la·tive·ly *adverb*

ap·pel·lee \ˌa-pə-ˈlē\ *noun* (1531)
: one against whom an appeal is taken

ap·pend \ə-ˈpend\ *transitive verb* [Middle English, from Middle French *appendre,* from Late Latin *appendere,* from Latin, to weigh, from *ad-* + *pendere* to weigh — more at PENDANT] (14th century)
1 : ATTACH, AFFIX
2 : to add as a supplement or appendix (as in a book)

ap·pend·age \ə-ˈpen-dij\ *noun* (1649)
1 : an adjunct to something larger or more important **:** APPURTENANCE
2 : a subordinate or derivative body part; *especially* **:** a limb or analogous part (as a seta)
3 : a dependent or subordinate person

ap·pen·dant \ə-ˈpen-dənt\ *adjective* (15th century)
1 : belonging as a right by prescription — used of annexed land in English law
2 : associated as an attendant circumstance
3 : attached as an appendage ⟨a seal *appendant* to a document⟩
— appendant *noun*

ap·pen·dec·to·my \ˌa-pən-ˈdek-tə-mē, ˌa-ˌpen-\ *noun, plural* **-mies** [Latin *appendic-, appendix* + English *-ectomy*] (circa 1895)
: surgical removal of the vermiform appendix

ap·pen·di·cec·to·my \ə-ˌpen-də-ˈsek-tə-mē\ *noun, plural* **-mies** (1894)
British **:** APPENDECTOMY

ap·pen·di·ci·tis \ə-ˌpen-də-ˈsī-təs\ *noun* [New Latin] (1886)
: inflammation of the vermiform appendix

ap·pen·dic·u·lar \ˌa-pən-ˈdi-kyə-lər\ *adjective* (1651)
: of or relating to an appendage and especially a limb ⟨the *appendicular* skeleton⟩

ap·pen·dix \ə-ˈpen-diks\ *noun, plural* **-dix·es** *or* **-di·ces** \-də-ˌsēz\ [Latin *appendic-, appendix,* from *appendere*] (1542)
1 a : APPENDAGE **b :** supplementary material usually attached at the end of a piece of writing
2 : a bodily outgrowth or process; *specifically* **:** VERMIFORM APPENDIX

ap·per·ceive \ˌa-pər-ˈsēv\ *transitive verb* **-ceived; -ceiv·ing** [French *apercevoir*] (1876)
: to have apperception of

ap·per·cep·tion \-ˈsep-shən\ *noun* [French *aperception,* from *apercevoir,* from Middle French *aperceivre,* from *a-* (from Latin *ad-*) + *perceivre* to perceive] (1753)
1 : introspective self-consciousness
2 : mental perception; *especially* **:** the process of understanding something perceived in terms of previous experience
— ap·per·cep·tive \-ˈsep-tiv\ *adjective*

ap·per·tain \ˌa-pər-ˈtān\ *intransitive verb* [Middle English *apperteinen,* from Middle French *apartenir,* from Late Latin *appertinēre,* from Latin *ad-* + *pertinēre* to belong — more at PERTAIN] (14th century)
: to belong or be connected as a rightful part or attribute **:** PERTAIN

ap·pe·tence \ˈa-pə-tən(t)s\ *noun* (1610)
: APPETENCY

ap·pe·ten·cy \-tən(t)-sē\ *noun, plural* **-cies** [Latin *appetentia,* from *appetent-, appetens,* present participle of *appetere*] (1631)
: a fixed and strong desire **:** APPETITE
— ap·pe·tent \-tənt\ *adjective*

ap·pe·tis·er, ap·pe·tis·ing *British variant of* APPETIZER, APPETIZING

ap·pe·tite \ˈa-pə-ˌtīt\ *noun* [Middle English *apetit,* from Middle French, from Latin *appetitus,* from *appetere* to strive after, from *ad-* + *petere* to go to — more at FEATHER] (14th century)
1 : any of the instinctive desires necessary to keep up organic life; *especially* **:** the desire to eat
2 a : an inherent craving ⟨an insatiable *appetite* for work⟩ **b :** TASTE, PREFERENCE ⟨the cultural *appetites* of the time —J. D. Hart⟩
— ap·pe·ti·tive \-ˌtī-tiv\ *adjective*

ap·pe·tiz·er \ˈa-pə-ˌtī-zər\ *noun* (1859)
: a food or drink that stimulates the appetite and is usually served before a meal

ap·pe·tiz·ing \-ˌtī-ziŋ\ *adjective* (1653)
: appealing to the appetite especially in appearance or aroma; *also* **:** appealing to one's taste ⟨an *appetizing* display of merchandise⟩
synonym see PALATABLE
— ap·pe·tiz·ing·ly \-ziŋ-lē\ *adverb*

ap·plaud \ə-ˈplȯd\ *verb* [Middle English, from Middle French or Latin; Middle French *applaudir,* from Latin *applaudere,* from *ad-* + *plaudere* to applaud] (15th century)
intransitive verb
: to express approval especially by clapping the hands
transitive verb
1 : to express approval of **:** PRAISE ⟨*applaud* her efforts to lose weight⟩
2 : to show approval of especially by clapping the hands
— ap·plaud·able \-ˈplȯ-də-bəl\ *adjective*
— ap·plaud·ably \-blē\ *adverb*
— ap·plaud·er *noun*

ap·plause \ə-ˈplȯz\ *noun* [Medieval Latin *applausus,* from Latin, beating of wings, from *applaudere*] (15th century)
1 : marked commendation **:** ACCLAIM ⟨the kind of *applause* every really creative writer wants —Robert Tallant⟩
2 : approval publicly expressed (as by clapping the hands)

ap·ple \ˈa-pəl\ *noun, often attributive* [Middle English *appel,* from Old English *æppel;* akin to Old High German *apful* apple, Old Irish *ubull,* Old Church Slavonic *ablŭko*] (before 12th century)
1 : the fleshy usually rounded and red, yellow, or green edible pome fruit of a tree (genus *Malus*) of the rose family; *also* **:** an apple tree
2 : a fruit or other vegetable production suggestive of an apple — compare OAK APPLE
— apple of one's eye : one that is highly cherished ⟨his daughter is the *apple of his eye*⟩

apple butter *noun* (circa 1774)
: a thick brown spread made by cooking apples with sugar and spices usually in cider

ap·ple·cart \-ˌkärt\ *noun* (1788)
: a plan, system, situation, or undertaking that may be disrupted or terminated ⟨upset the *applecart*⟩

ap·ple–cheeked \ˈa-pəl-ˌchēkt\ *adjective* (1864)
: having cheeks the color of red apples

ap·ple·jack \-ˌjak\ *noun* (1816)
: brandy distilled from hard cider; *also* **:** an alcoholic beverage traditionally made by freezing hard cider and siphoning off the concentrated liquor

ap·ple–knock·er \-ˌnä-kər\ *noun* (1919)
: RUSTIC

apple maggot *noun* (1867)
: a dipteran fly (*Rhagoletis pomonella*) whose larva burrows in and feeds especially on apples

ap·ple–pie \ˈa-pəl-ˈpī\ *adjective* (1780)
1 : EXCELLENT, PERFECT ⟨*apple-pie* order⟩
2 : of, relating to, or characterized by traditionally American values (as honesty or simplicity) ⟨is the epitome of *apple-pie* wholesomeness⟩

ap·ple–pol·ish \ˈa-pəl-ˌpä-lish\ *verb* [from the traditional practice of schoolchildren bringing a shiny apple as a gift to their teacher] (1935)
intransitive verb
: to attempt to ingratiate oneself **:** TOADY
transitive verb
: to curry favor with (as by flattery)
— ap·ple–pol·ish·er *noun*

ap·ple·sauce \-ˌsȯs\ *noun* (1739)
1 : a relish or dessert made of apples stewed to a pulp and sweetened
2 *slang* **:** BUNKUM, NONSENSE

apple scab *noun* (circa 1899)
: a disease of apple trees caused by a fungus (*Venturia inaequalis*) producing dark blotches or lesions on the leaves, fruit, and sometimes the young twigs

ap·pli·ance \ə-ˈplī-ən(t)s\ *noun* (1561)
1 : an act of applying
2 a : a piece of equipment for adapting a tool or machine to a special purpose **:** ATTACHMENT

b : an instrument or device designed for a particular use; *specifically* **:** a household or office device (as a stove, fan, or refrigerator) operated by gas or electric current **c** *British* **:** FIRE ENGINE
3 *obsolete* **:** COMPLIANCE
synonym see IMPLEMENT

ap·pli·ca·ble \'a-pli-kə-bəl *also* ə-'pli-kə-\ *adjective* (1660)
: capable of or suitable for being applied **:** APPROPRIATE ⟨statutes *applicable* to the case⟩
synonym see RELEVANT
— ap·pli·ca·bil·i·ty \ˌa-pli-kə-'bi-lə-tē *also* ə-ˌpli-kə-\ *noun*

ap·pli·cant \'a-pli-kənt\ *noun* (circa 1782)
: one who applies ⟨a job *applicant*⟩

ap·pli·ca·tion \ˌa-plə-'kā-shən\ *noun* [Middle English *applicacioun,* from Latin *application-, applicatio* inclination, from *applicare*] (15th century)
1 : an act of applying: **a** (1) **:** an act of putting to use ⟨*application* of new techniques⟩ (2) **:** a use to which something is put ⟨new *applications* for old remedies⟩ (3) **:** a program (as a word processor or a spreadsheet) that performs one of the important tasks for which a computer is used **b :** an act of administering or superposing ⟨*application* of paint to a house⟩ **c :** assiduous attention ⟨succeeds by *application* to her studies⟩
2 a : REQUEST, PETITION ⟨an *application* for financial aid⟩ **b :** a form used in making a request
3 : the practical inference to be derived from a discourse (as a moral tale)
4 : a medicated or protective layer or material ⟨an oily *application* for dry skin⟩
5 : capacity for practical use ⟨words of varied *application*⟩

ap·pli·ca·tive \'a-plə-ˌkā-tiv, ə-'pli-kə-\ *adjective* (1638)
1 : APPLICABLE, PRACTICAL
2 : put to use **:** APPLIED
— ap·pli·ca·tive·ly *adverb*

ap·pli·ca·tor \'a-plə-ˌkā-tər\ *noun* (1659)
: one that applies; *specifically* **:** a device for applying a substance (as medicine or polish)

ap·pli·ca·to·ry \'a-pli-kə-ˌtōr-ē, -ˌtȯr-, ə-'pli-kə-\ *adjective* (1649)
: capable of being applied

ap·plied \ə-'plīd\ *adjective* (1656)
1 : put to practical use ⟨*applied* art⟩; *especially* **:** applying general principles to solve definite problems ⟨*applied* sciences⟩
2 : working in an applied science ⟨an *applied* physicist⟩

[1]ap·pli·qué \ˌa-plə-'kā\ *noun* [French, past participle of *appliquer* to put on, from Latin *applicare*] (1801)
: a cutout decoration fastened to a larger piece of material

[2]appliqué *transitive verb* **-quéd; -qué·ing** (1881)
: to apply (as a decoration or ornament) to a larger surface **:** OVERLAY

ap·ply \ə-'plī\ *verb* **ap·plied; ap·ply·ing** [Middle English *applien,* from Middle French *aplier,* from Latin *applicare,* from *ad-* + *plicare* to fold — more at PLY] (14th century)
transitive verb
1 a : to put to use especially for some practical purpose ⟨*applies* pressure to get what he wants⟩ **b :** to bring into action ⟨*apply* the brakes⟩ **c :** to lay or spread on ⟨*apply* varnish⟩ **d :** to put into operation or effect ⟨*apply* a law⟩
2 : to employ diligently or with close attention ⟨should *apply* yourself to your work⟩
intransitive verb
1 : to have relevance or a valid connection ⟨this rule *applies* to freshmen only⟩
2 : to make an appeal or request especially in the form of a written application ⟨*apply* for a job⟩
— ap·pli·er \-'plī(-ə)r\ *noun*

ap·pog·gia·tu·ra \ə-ˌpä-jə-'tu̇r-ə\ *noun* [Italian, literally, support] (1753)
: an embellishing note or tone preceding an essential melodic note or tone and usually written as a note of smaller size

ap·point \ə-'pȯint\ *verb* [Middle English, from Middle French *apointier* to arrange, from *a-* (from Latin *ad-*) + *point* point] (14th century)
transitive verb
1 a : to fix or set officially ⟨*appoint* a trial date⟩ **b :** to name officially ⟨will *appoint* her director of the program⟩ **c** *archaic* **:** ARRANGE **d :** to determine the disposition of (an estate) to someone by virtue of a power of appointment
2 : to provide with complete and usually appropriate or elegant furnishings or equipment
intransitive verb
: to exercise a power of appointment
synonym see FURNISH

ap·poin·tee \ə-ˌpȯin-'tē, ˌa-\ *noun* (1768)
1 : one who is appointed
2 : one to whom an estate is appointed

ap·point·ive \ə-'pȯin-tiv\ *adjective* (1881)
: of, relating to, or filled by appointment ⟨an *appointive* office⟩

ap·point·ment \ə-'pȯint-mənt\ *noun* (15th century)
1 a : an act of appointing **:** DESIGNATION **b :** the designation by virtue of a vested power of a person to enjoy an estate
2 : an arrangement for a meeting **:** ENGAGEMENT
3 : EQUIPMENT, FURNISHINGS — usually used in plural
4 : a nonelective office or position ⟨holds an academic *appointment*⟩

ap·por·tion \ə-'pōr-shən, -'pȯr-\ *transitive verb* **-tioned; -tion·ing** \-sh(ə-)niŋ\ [Middle French *apportionner,* from *a-* (from Latin *ad-*) + *portionner* to portion] (1574)
: to divide and share out according to a plan; *especially* **:** to make a proportionate division or distribution of
— ap·por·tion·able \-shə-nə-bəl\ *adjective*

ap·por·tion·ment \-shən-mənt\ *noun* (1579)
: an act or result of apportioning; *especially* **:** the apportioning of representatives or taxes among the states according to U.S. law

ap·pose \a-'pōz\ *transitive verb* **ap·posed; ap·pos·ing** [Middle French *aposer,* from Old French, from *a-* + *poser* to put — more at POSE] (1596)
1 *archaic* **:** to put before **:** apply (one thing) to another
2 : to place in juxtaposition or proximity

ap·po·site \'a-pə-zət\ *adjective* [Latin *appositus,* from past participle of *apponere* to place near, from *ad-* + *ponere* to put — more at POSITION] (1621)
: highly pertinent or appropriate **:** APT
synonym see RELEVANT
— ap·po·site·ly *adverb*
— ap·po·site·ness *noun*

ap·po·si·tion \ˌa-pə-'zi-shən\ *noun* (15th century)
1 a : a grammatical construction in which two usually adjacent nouns having the same referent stand in the same syntactical relation to the rest of a sentence (as *the poet* and *Burns* in "a biography of the poet Burns") **b :** the relation of one of such a pair of nouns or noun equivalents to the other
2 a : an act or instance of apposing; *specifically* **:** the deposition of successive layers upon those already present (as in cell walls) **b :** the state of being apposed
— ap·po·si·tion·al \-'zish-nəl, -'zi-shə-nᵊl\ *adjective*

ap·pos·i·tive \ə-'pä-zə-tiv, a-\ *adjective* (1693)
: of, relating to, or standing in grammatical apposition
— appositive *noun*
— ap·pos·i·tive·ly *adverb*

ap·prais·al \ə-'prā-zəl\ *noun* (1817)
: an act or instance of appraising; *especially* **:** a valuation of property by the estimate of an authorized person

ap·praise \ə-'prāz\ *transitive verb* **ap·praised; ap·prais·ing** [Middle English *appreisen,* from Middle French *aprisier* to apprize] (15th century)
1 : to set a value on **:** estimate the amount of
2 : to evaluate the worth, significance, or status of; *especially* **:** to give an expert judgment of the value or merit of
synonym see ESTIMATE
— ap·prais·ee \ə-ˌprā-'zē\ *noun*
— ap·praise·ment \-'prāz-mənt\ *noun*
— ap·prais·er *noun*
— ap·prais·ing·ly \-'prā-ziŋ-lē\ *adverb*
— ap·prais·ive \-'prā-ziv\ *adjective*

ap·pre·cia·ble \ə-'prē-shə-bəl, -'pri-sh(ē-)ə-bəl\ *adjective* (1818)
: capable of being perceived or measured
synonym see PERCEPTIBLE
— ap·pre·cia·bly \-blē\ *adverb*

ap·pre·ci·ate \ə-'prē-shē-ˌāt, -'pri- *also* -'prē-sē-\ *verb* **-at·ed; -at·ing** [Late Latin *appretiatus,* past participle of *appretiare,* from Latin *ad-* + *pretium* price — more at PRICE] (1655)
transitive verb
1 a : to grasp the nature, worth, quality, or significance of ⟨*appreciate* the difference between right and wrong⟩ **b :** to value or admire highly ⟨*appreciates* our work⟩ **c :** to judge with heightened perception or understanding **:** be fully aware of ⟨must see it to *appreciate* it⟩ **d :** to recognize with gratitude ⟨certainly *appreciates* your kindness⟩
2 : to increase the value of
intransitive verb
: to increase in number or value ☆
— ap·pre·ci·a·tor \-ˌā-tər\ *noun*
— ap·pre·cia·to·ry \-'prē-shə-ˌtōr-ē, -'pri-shə-, -ˌtȯr-\ *adjective*

ap·pre·ci·a·tion \ə-ˌprē-shē-'ā-shən, -ˌpri- *also* -ˌprē-sē-\ *noun* (1604)
1 a : JUDGMENT, EVALUATION; *especially* **:** a favorable critical estimate **b :** sensitive awareness; *especially* **:** recognition of aesthetic values **c :** an expression of admiration, approval, or gratitude
2 : increase in value

ap·pre·cia·tive \ə-'prē-shə-tiv, -'pri- *also* -'prē-shē-ˌā-\ *adjective* (circa 1698)
: having or showing appreciation
— ap·pre·cia·tive·ly *adverb*
— ap·pre·cia·tive·ness *noun*

ap·pre·hend \ˌa-pri-'hend\ *verb* [Middle English, from Latin *apprehendere,* literally, to seize, from *ad-* + *prehendere* to seize — more at GET] (15th century)
transitive verb
1 : ARREST, SEIZE ⟨*apprehend* a thief⟩

☆ SYNONYMS
Appreciate, value, prize, treasure, cherish mean to hold in high estimation. APPRECIATE often connotes sufficient understanding to enjoy or admire a thing's excellence ⟨*appreciates* fine wine⟩. VALUE implies rating a thing highly for its intrinsic worth ⟨*values* our friendship⟩. PRIZE implies taking a deep pride in something one possesses ⟨Americans *prize* their freedom⟩. TREASURE emphasizes jealously safeguarding something considered precious ⟨a *treasured* memento⟩. CHERISH implies a special love and care for something ⟨*cherishes* her children above all⟩. See in addition UNDERSTAND.

2 a : to become aware of **:** PERCEIVE **b :** to anticipate especially with anxiety, dread, or fear
3 : to grasp with the understanding **:** recognize the meaning of
intransitive verb
: UNDERSTAND, GRASP

ap·pre·hen·si·ble \ˌa-pri-ˈhen(t)-sə-bəl\ *adjective* (circa 1631)
: capable of being apprehended
— ap·pre·hen·si·bly \-blē\ *adverb*

ap·pre·hen·sion \ˌa-pri-ˈhen(t)-shən\ *noun* [Middle English, from Late Latin *apprehension-, apprehensio,* from Latin *apprehendere*] (14th century)
1 a : the act or power of perceiving or comprehending ⟨a person of dull *apprehension*⟩ **b :** the result of apprehending mentally **:** CONCEPTION ⟨according to popular *apprehension*⟩
2 : seizure by legal process **:** ARREST
3 : suspicion or fear especially of future evil **:** FOREBODING

ap·pre·hen·sive \-ˈhen(t)-siv\ *adjective* (14th century)
1 : capable of apprehending or quick to do so **:** DISCERNING
2 : having apprehension **:** COGNIZANT
3 : viewing the future with anxiety or alarm
synonym see FEARFUL
— ap·pre·hen·sive·ly *adverb*
— ap·pre·hen·sive·ness *noun*

¹ap·pren·tice \ə-ˈpren-təs\ *noun* [Middle English *aprentis,* from Middle French, from Old French, from *aprendre* to learn, from Latin *apprendere, apprehendere*] (14th century)
1 a : one bound by indenture to serve another for a prescribed period with a view to learning an art or trade **b :** one who is learning by practical experience under skilled workers a trade, art, or calling
2 : an inexperienced person **:** NOVICE ⟨an *apprentice* in cooking⟩
— ap·pren·tice·ship \-tə(sh)-ˌship, -təs-ˌship\ *noun*

²apprentice *verb* **-ticed; -tic·ing** (1596)
transitive verb
: to set at work as an apprentice; *especially* **:** to bind to an apprenticeship by contract or indenture
intransitive verb
: to serve as an apprentice

ap·pressed \a-ˈprest\ *adjective* [Latin *appressus,* past participle of *apprimere* to press to, from *ad-* + *premere* to press — more at PRESS] (1791)
: pressed close to or lying flat against something ⟨leaves *appressed* against the stem⟩

ap·pres·so·ri·um \ˌa-pre-ˈsōr-ē-əm, -ˈsȯr-\ *noun, plural* **-ria** \-ē-ə\ [New Latin, from Latin *apprimere*] (1897)
: the flattened thickened tip of a hyphal branch by which some parasitic fungi are attached to their host

ap·prise \ə-ˈprīz\ *transitive verb* **ap·prised; ap·pris·ing** [French *appris,* past participle of *apprendre* to learn, teach, from Old French *aprendre*] (1694)
: to give notice to **:** TELL
synonym see INFORM

ap·prize \ə-ˈprīz\ *transitive verb* **ap·prized; ap·priz·ing** [Middle English *apprisen,* from Middle French *aprisier,* from Old French, from *a-* (from Latin *ad-*) + *prisier* to appraise — more at PRIZE] (14th century)
: VALUE, APPRECIATE

¹ap·proach \ə-ˈprōch\ *verb* [Middle English *approchen,* from Old French *aprochier,* from Late Latin *appropiare,* from Latin *ad-* + *prope* near; akin to Latin *pro* before — more at FOR] (13th century)
transitive verb
1 a : to draw closer to **:** NEAR **b :** to come very near to **:** be almost the same as ⟨its mathematics *approaches* mysticism —Theodore Sturgeon⟩ ⟨as the quantity *x approaches* zero⟩
2 a : to make advances to especially in order to create a desired result ⟨was *approached* by several Broadway producers⟩ **b :** to take preliminary steps toward accomplishment or full knowledge or experience of ⟨*approach* the subject with an open mind⟩
intransitive verb
1 : to draw nearer
2 : to make an approach in golf

²approach *noun* (15th century)
1 a : an act or instance of approaching ⟨the *approach* of summer⟩ **b :** APPROXIMATION ⟨in this book he makes his closest *approach* to greatness⟩
2 a : the taking of preliminary steps toward a particular purpose ⟨experimenting with new lines of *approach*⟩ **b :** a particular manner of taking such steps ⟨a highly individual *approach* to language⟩
3 : a means of access **:** AVENUE
4 a : a golf shot from the fairway toward the green **b :** the steps taken by a bowler before he delivers the ball; *also* **:** the part of the alley behind the foul line from which the bowler delivers the ball
5 : the descent of an aircraft toward a landing place

ap·proach·able \ə-ˈprō-chə-bəl\ *adjective* (1571)
: capable of being approached **:** ACCESSIBLE; *specifically* **:** easy to meet or deal with
— ap·proach·abil·i·ty \-ˌprō-chə-ˈbi-lə-tē\ *noun*

ap·pro·bate \ˈa-prə-ˌbāt\ *transitive verb* **-bat·ed; -bat·ing** [Middle English, from Latin *approbatus,* past participle of *approbare* — more at APPROVE] (15th century)
: APPROVE, SANCTION
— ap·pro·ba·to·ry \ˈa-prə-bə-ˌtōr-ē, ə-ˈprō-bə-, -ˌtȯr-\ *adjective*

ap·pro·ba·tion \ˌa-prə-ˈbā-shən\ *noun* (14th century)
1 *obsolete* **:** PROOF
2 a : an act of approving formally or officially **b :** COMMENDATION, PRAISE

¹ap·pro·pri·ate \ə-ˈprō-prē-ˌāt\ *transitive verb* **-at·ed; -at·ing** [Middle English, from Late Latin *appropriatus,* past participle of *appropriare,* from Latin *ad-* + *proprius* own] (15th century)
1 : to take exclusive possession of **:** ANNEX ⟨no one should *appropriate* a common benefit⟩
2 : to set apart for or assign to a particular purpose or use ⟨*appropriate* money for the research program⟩
3 : to take or make use of without authority or right
— ap·pro·pri·a·ble \-prē-ə-bəl\ *adjective*
— ap·pro·pri·a·tor \-prē-ˌā-tər\ *noun*

²ap·pro·pri·ate \ə-ˈprō-prē-ət\ *adjective* (15th century)
: especially suitable or compatible **:** FITTING
synonym see FIT
— ap·pro·pri·ate·ly *adverb*
— ap·pro·pri·ate·ness *noun*

ap·pro·pri·a·tion \ə-ˌprō-prē-ˈā-shən\ *noun* (14th century)
1 : an act or instance of appropriating
2 : something that has been appropriated; *specifically* **:** money set aside by formal action for a specific use
— ap·pro·pri·a·tive \-ˈprō-prē-ˌā-tiv\ *adjective*

ap·prov·able \ə-ˈprü-və-bəl\ *adjective* (15th century)
: capable or worthy of being approved
— ap·prov·ably \-blē\ *adverb*

ap·prov·al \ə-ˈprü-vəl\ *noun* (1616)
: an act or instance of approving **:** APPROBATION
— on approval : subject to a prospective buyer's acceptance or refusal ⟨stamps sent to collectors *on approval*⟩

ap·prove \ə-ˈprüv\ *verb* **ap·proved; ap·prov·ing** [Middle English, from Middle French *aprover,* from Latin *approbare,* from *ad-* + *probare* to prove — more at PROVE] (14th century)
transitive verb
1 *obsolete* **:** PROVE, ATTEST
2 : to have or express a favorable opinion of ⟨couldn't *approve* such conduct⟩
3 a : to accept as satisfactory ⟨hopes she will *approve* the date of the meeting⟩ **b :** to give formal or official sanction to **:** RATIFY ⟨Congress *approved* the proposed budget⟩
intransitive verb
: to take a favorable view ⟨doesn't *approve* of fighting⟩ ☆
— ap·prov·ing·ly \-ˈprü-viŋ-lē\ *adverb*

approved school *noun* (1932)
British **:** a school for juvenile delinquents

¹ap·prox·i·mate \ə-ˈpräk-sə-mət\ *adjective* [Late Latin *approximatus,* past participle of *approximare* to come near, from Latin *ad-* + *proximare* to come near — more at PROXIMATE] (15th century)
1 : located close together ⟨*approximate* leaves⟩
2 : nearly correct or exact ⟨an *approximate* solution⟩
— ap·prox·i·mate·ly *adverb*

²ap·prox·i·mate \-ˌmāt\ *verb* **-mat·ed; -mat·ing** (15th century)
transitive verb
1 a : to bring near or close **b :** to bring (cut edges of tissue) together
2 : to come near to or be close to in position, value, or characteristics ⟨a child tries to *approximate* his parents' speech⟩
intransitive verb
: to come close — usually used with *to*

ap·prox·i·ma·tion \ə-ˌpräk-sə-ˈmā-shən\ *noun* (15th century)
1 : the act or process of drawing together
2 : the quality or state of being close or near ⟨an *approximation* to the truth⟩ ⟨an *approximation* of justice⟩
3 : something that is approximate; *especially* **:** a mathematical quantity that is close in value to but not the same as a desired quantity
— ap·prox·i·ma·tive \-ˈpräk-sə-ˌmā-tiv\ *adjective*

ap·pur·te·nance \ə-ˈpərt-nən(t)s, -ˈpər-tᵊn-ən(t)s\ *noun* (14th century)
1 : an incidental right (as a right-of-way) attached to a principal property right and passing in possession with it
2 : a subordinate part or adjunct ⟨the *appurtenance* of welcome is fashion and ceremony —Shakespeare⟩
3 *plural* **:** accessory objects **:** APPARATUS

ap·pur·te·nant \ə-ˈpərt-nənt, -ˈpər-tᵊn-ənt\ *adjective* [Middle English *apertenant,* from Middle French, from Old French, present participle of *apartenir* to belong — more at APPERTAIN] (14th century)
1 : constituting a legal accompaniment
2 : AUXILIARY, ACCESSORY
— appurtenant *noun*

aprax·ia \(ˌ)ā-ˈprak-sē-ə\ *noun* [New Latin, from Greek, inaction, from *a-* + *praxis* action, from *prassein* to do — more at PRACTICAL] (circa 1881)
: loss or impairment of the ability to execute complex coordinated movements without impairment of the muscles or senses

☆ SYNONYMS
Approve, endorse, sanction, accredit, certify mean to have or express a favorable opinion of. APPROVE often implies no more than this but may suggest considerable esteem or admiration ⟨the parents *approve* of the marriage⟩. ENDORSE suggests an explicit statement of support ⟨publicly *endorsed* her for Senator⟩. SANCTION implies both approval and authorization ⟨the President *sanctioned* covert operations⟩. ACCREDIT and CERTIFY usually imply official endorsement attesting to conformity to set standards ⟨the board voted to *accredit* the college⟩ ⟨must be *certified* to teach⟩.

— **aprac·tic** \-ˈprak-tik\ *or* **aprax·ic** \-ˈprak-sik\ *adjective*

après \(ˌ)ä-ˈprā, ˈä-ˌprā; ˈa-ˌ\ *preposition* [French *après-,* from *après* after] (1967)
: AFTER ⟨*après* tennis⟩ — usually used in combination ⟨*après*-theater party⟩

après–ski \ˌä-ˌprā-ˈskē, ˌa-\ *noun, often attributive* [French *après* after + *ski* ski, skiing] (1951)
: social activity (as at a ski lodge) after a day's skiing

apri·cot \ˈa-prə-ˌkät, ˈā-\ *noun, often attributive* [alteration of earlier *abrecock,* ultimately from Arabic *al-birqūq* the apricot, ultimately from Latin *(persicum) praecox,* literally, early ripening (peach) — more at PRECOCIOUS] (1551)
1 a : the oval orange-colored fruit of a temperate-zone tree (*Prunus armeniaca*) resembling the related peach and plum in flavor **b :** a tree that bears apricots
2 : a variable color averaging a moderate orange ◆

April \ˈā-prəl\ *noun* [Middle English, from Old French & Latin; Old French *avrill,* from Latin *Aprilis*] (before 12th century)
: the 4th month of the Gregorian calendar

April fool *noun* (1687)
: the butt of a joke or trick played on April Fools' Day; *also* **:** such a joke or trick

April Fools' Day *noun* (1854)
: April 1 characteristically marked by the playing of practical jokes

a pri·o·ri \ˌä-prē-ˈōr-ē, ˌa-; ˌā-(ˌ)prī-ˈōr-ˌī, -ˌprē-ˈōr-ē; -ˈȯr-\ *adjective* [Latin, literally, from the former] (1652)
1 a : DEDUCTIVE **b :** relating to or derived by reasoning from self-evident propositions — compare A POSTERIORI **c :** presupposed by experience
2 a : being without examination or analysis **:** PRESUMPTIVE **b :** formed or conceived beforehand
— **a priori** *adverb*
— **apri·or·i·ty** \-ˈȯr-ə-tē\ *noun*

apron \ˈā-prən, -pərn\ *noun, often attributive* [Middle English, alteration (resulting from false division of *a napron*) of *napron,* from Middle French *naperon,* diminutive of *nape* cloth, modification of Latin *mappa* napkin] (15th century)
1 : a garment usually of cloth, plastic, or leather usually tied around the waist and used to protect clothing or adorn a costume
2 : something that suggests or resembles an apron in shape, position, or use: as **a :** the lower member under the sill of the interior casing of a window **b :** an upward or downward vertical extension of a bathroom fixture (as a sink or tub) **c :** an endless belt for carrying material **d :** an extensive fan-shaped deposit of detritus **e :** the part of the stage in front of the proscenium arch **f :** the area along the waterfront edge of a pier or wharf **g :** a shield (as of concrete or gravel) to protect against erosion (as of a waterway) by water **h :** the extensive paved part of an airport immediately adjacent to the terminal area or hangars ◆
— **aproned** \-prənd, -pərnd\ *adjective*

apron string *noun* (1542)
: the string of an apron — usually used in plural as a symbol of dominance or complete control ⟨though 40 years old he was still tied to his mother's *apron strings*⟩

¹ap·ro·pos \ˌa-prə-ˈpō, ˈa-prə-ˌ\ *adverb* [French *à propos,* literally, to the purpose] (1668)
1 : at an opportune time **:** SEASONABLY
2 : BY THE WAY

²apropos *adjective* (1686)
: being both relevant and opportune
synonym see RELEVANT

³apropos *preposition* (1910)
: APROPOS OF

apropos of *preposition* (1746)
: with regard to **:** CONCERNING

apro·tic \(ˌ)ā-ˈprō-tik\ *adjective* [²*a-* + *proton* + ¹*-ic*] (1931)
of a solvent **:** incapable of acting as a proton donor

apse \ˈaps\ *noun* [Medieval Latin & Latin; Medieval Latin *apsis,* from Latin] (1822)
1 : APSIS 1
2 : a projecting part of a building (as a church) that is usually semicircular in plan and vaulted

ap·si·dal \ˈap-sə-dᵊl\ *adjective* (1846)
: of or relating to an apse or apsis

ap·sis \ˈap-səs\ *noun, plural* **ap·si·des** \-sə-ˌdēz\ [New Latin *apsid-, apsis,* from Latin, arch, orbit, from Greek *hapsid-, hapsis,* from *haptein* to fasten] (1658)
1 : the point in an astronomical orbit at which the distance of the body from the center of attraction is either greatest or least
2 : APSE 2

apt \ˈapt\ *adjective* [Middle English, from Latin *aptus,* literally, fastened, from past participle of *apere* to fasten; akin to Latin *apisci* to grasp, obtain, *apud* near, Hittite *ḫap-* to attach] (14th century)
1 : unusually fitted or qualified **:** READY ⟨proved an *apt* tool in the hands of the conspirators⟩
2 a : having a tendency **:** LIKELY ⟨plants *apt* to suffer from drought⟩ **b :** ordinarily disposed **:** INCLINED ⟨*apt* to accept what is plausible as true⟩
3 : suited to a purpose; *especially* **:** being to the point ⟨an *apt* quotation⟩
4 : keenly intelligent and responsive
synonym see FIT, QUICK
usage see LIABLE
— **apt·ly** \ˈap(t)-lē\ *adverb*
— **apt·ness** \ˈap(t)-nəs\ *noun*

ap·ter·ous \ˈap-tə-rəs\ *adjective* [Greek *apteros,* from *a-* + *pteron* wing — more at FEATHER] (1775)
: lacking wings ⟨*apterous* insects⟩

ap·ter·yx \ˈap-tə-riks\ *noun* [New Latin, from *a-* + Greek *pteryx* wing; akin to Greek *pteron*] (1813)
: KIWI

ap·ti·tude \ˈap-tə-ˌtüd, -ˌtyüd\ *noun* [Middle English, from Medieval Latin *aptitudo,* from Late Latin, fitness, from Latin *aptus*] (15th century)
1 a : INCLINATION, TENDENCY **b :** a natural ability **:** TALENT
2 : capacity for learning **:** APTNESS
3 : general suitability
synonym see GIFT
— **ap·ti·tu·di·nal** \ˌap-tə-ˈtü-dᵊn-əl, -ˈtyü-\ *adjective*
— **ap·ti·tu·di·nal·ly** *adverb*

aptitude test *noun* (1923)
: a standardized test designed to predict an individual's ability to learn certain skills

ap·y·rase \ˈa-pə-ˌrās, -ˌrāz\ *noun* [*a*denosine + *py*rophosphate + *-ase*] (1945)
: any of several enzymes that hydrolyze ATP with the liberation of phosphate and energy

aqua \ˈä-kwə *also* ˈa-\ *noun* [Latin — more at ISLAND] (14th century)
1 *plural* **aquae** \ˈä-ˌkwī *also* ˈa-(ˌ)kwē\ **:** WATER; *especially* **:** WATER 5a(2)
2 *plural* **aquas :** a light greenish blue color

aqua·cade \ˈä-kwə-ˌkād, ˈa-\ *noun* [*Aquacade,* a water spectacle originally at Cleveland, Ohio] (1937)
: a water spectacle that consists usually of exhibitions of swimming and diving with musical accompaniment

aqua·cul·ture *also* **aqui·cul·ture** \ˈä-kwə-ˌkəl-chər, ˈa-\ *noun* [Latin *aqua* + English *-culture* (as in *agriculture*)] (1867)
: the cultivation of the natural produce of water (as fish or shellfish)
— **aqua·cul·tur·al** \ˌä-kwə-ˈkəl-ch(ə-)rəl, ˌa-\ *adjective*
— **aqua·cul·tur·ist** \-ch(ə-)rist\ *noun*

aqua for·tis \ˈä-kwə-ˈfȯr-təs, ˈa-\ *noun* [New Latin *aqua fortis,* literally, strong water] (15th century)
: NITRIC ACID

Aqua–Lung \ˈä-kwə-ˌləŋ, ˈa-\ *trademark*
— used for an underwater breathing apparatus

aqua·ma·rine \ˌä-kwə-mə-ˈrēn, ˌa-\ *noun* [New Latin *aqua marina,* from Latin, sea water] (1677)
1 : a transparent beryl that is blue, blue-green, or green in color
2 : a pale blue to light greenish blue

aqua·naut \ˈä-kwə-ˌnȯt, ˈa-, -ˌnät\ *noun* [Latin *aqua* + English *-naut* (as in *aeronaut*)] (1881)
: a scuba diver who lives and operates both inside and outside an underwater shelter for an extended period

¹aqua·plane \ˈä-kwə-ˌplān, ˈa-\ *noun* (1914)
: a board towed behind a speeding motorboat and ridden by a person standing on it
— **aqua·plan·er** *noun*

²aquaplane *intransitive verb* (circa 1923)
1 : to ride on an aquaplane
2 *British* **:** HYDROPLANE

aqua re·gia \-ˈrē-j(ē-)ə\ *noun* [New Latin, literally, royal water] (1610)
: a mixture of nitric and hydrochloric acids that dissolves gold or platinum

aqua·relle \ˌa-kwə-ˈrel, ˌä-\ *noun* [French,

◇ WORD HISTORY

apricot The apricot is native to China and was probably not introduced to the Mediterranean region before the first century B.C. We assume that the fruit which the Roman poet Martial called *persica praecocia,* literally, "early-ripening peaches," was the apricot. A Greek form of the second word in this phrase, *praikokion,* migrated to the eastern end of the Mediterranean, and when the Arabs entered this area in the early Middle Ages, they turned *praikokion* into *barqūq* or *birqūq.* In areas that they occupied, such as Sicily and the Iberian Peninsula, the Arabs revived apricot cultivation. Arabic *al-barqūq* "the apricot" entered Spanish as *albaricoque,* Portuguese as *albricoque,* and Catalan as *albercoc* or *abercoc.* The earliest English form of the word, *aprecock,* appears to be borrowed from French *abricot* (itself also from Iberian Romance). The change of *b* to *p* (notable also in German *Aprikose*) was perhaps by association with Latin *apricus* "sunny, warmed by sunshine."

apron In late Medieval French a diminutive form of *nape,* meaning "tablecloth," was *naperon,* which was usually used for a small cloth placed over a more elegant tablecloth to protect it from stains. This word appears in English of the 14th century as *napron;* like its French source, *napron* denoted a protective cloth, but one that was placed over clothing rather than on a table. Because in connected speech it is often difficult to tell where word boundaries fall, *a napron* was incorrectly understood as *an apron.* The new form *apron* had effectively replaced *napron* by the 17th century, completely obscuring the etymological relation of *apron* to *napkin,* the name for yet another kind of protective cloth. Curiously, the modern French word for "apron," *tablier,* has undergone a similar sense shift; originally a derivative of *table* "table," in Old French it meant "tablecloth."

from obsolete Italian *acquarella* (now *acquerello*), from *acqua* water, from Latin *aqua*] (1869)
: a drawing usually in transparent watercolor
— **aqua·rell·ist** \-'re-list\ *noun*

Aquar·i·an \ə-'kwar-ē-ən, -'kwer-\ *noun* (1911)
: AQUARIUS 2b
— **Aquarian** *adjective*

aquar·ist \ə-'kwar-ist, -'kwer-\ *noun* (circa 1893)
: a person who keeps or maintains an aquarium

aquar·i·um \ə-'kwar-ē-əm, -'kwer-\ *noun, plural* **-i·ums** *or* **-ia** \-ē-ə\ [probably alteration of *aquatic vivarium*] (circa 1847)
1 : a container (as a glass tank) or an artificial pond in which living aquatic animals or plants are kept
2 : an establishment where aquatic collections of living organisms are kept and exhibited

Aquar·i·us \-ē-əs\ *noun* [Latin (genitive *Aquarii*), literally, water carrier]
1 : a constellation south of Pegasus pictured as a man pouring water
2 a : the 11th sign of the zodiac in astrology — see ZODIAC table **b :** one born under the sign of Aquarius

¹aquat·ic \ə-'kwä-tik, -'kwa-\ *adjective* (1642)
1 : growing or living in or frequenting water ⟨*aquatic* mosquito larvae⟩
2 : taking place in or on water ⟨*aquatic* sports⟩
— **aquat·i·cal·ly** \-ti-k(ə-)lē\ *adverb*

²aquatic *noun* (circa 1600)
1 : an aquatic animal or plant
2 *plural but singular or plural in construction* **:** water sports

aqua·tint \'a-kwə-ˌtint, 'ä-\ *noun* [Italian *acqua tinta* dyed water] (1782)
: a method of etching a printing plate so that tones similar to watercolor washes can be reproduced; *also* **:** a print made from a plate so etched
— **aquatint** *transitive verb*
— **aqua·tint·er** *noun*
— **aqua·tint·ist** \-ˌtin-tist\ *noun*

aqua·vit \'ä-kwə-ˌvēt\ *noun* [Swedish, Danish, & Norwegian *akvavit,* from Medieval Latin *aqua vitae*] (1890)
: a clear Scandinavian liquor flavored with caraway seeds
word history see WHISKEY

aqua vi·tae \ˌa-kwə-'vī-tē, ˌä-\ *noun* [Middle English, from Medieval Latin, literally, water of life] (15th century)
: a strong alcoholic liquor (as brandy)
word history see WHISKEY

aq·ue·duct \'a-kwə-ˌdəkt\ *noun* [Latin *aquaeductus,* from *aquae* (genitive of *aqua*) + *ductus* act of leading — more at DUCT] (1538)
1 a : a conduit for water; *especially* **:** one for carrying a large quantity of flowing water **b :** a structure for conveying a canal over a river or hollow
2 : a canal or passage in a part or organ

aque·ous \'ā-kwē-əs, 'a-\ *adjective* [Medieval Latin *aqueus,* from Latin *aqua*] (1646)
1 a : of, relating to, or resembling water **b :** made from, with, or by water
2 : of or relating to the aqueous humor

aqueous humor *noun* (1643)
: a transparent fluid occupying the space between the crystalline lens and the cornea of the eye

aqui·fer \'a-kwə-fər, 'ä-\ *noun* [New Latin, from Latin *aqua* + *-fer*] (1901)
: a water-bearing stratum of permeable rock, sand, or gravel
— **aquif·er·ous** \a-'kwi-fə-rəs, ä-\ *adjective*

Aq·ui·la \'a-kwə-lə *also* ə-'kwi-lə\ *noun* [Latin (genitive *Aquilae*), literally, eagle]
: a constellation in the northern hemisphere represented by the figure of an eagle

aq·ui·le·gia \ˌa-kwə-'lē-j(ē-)ə\ *noun* [New Latin] (1871)
: COLUMBINE

aq·ui·line \'a-kwə-ˌlīn, -lən\ *adjective* [Latin *aquilinus,* from *aquila* eagle] (1646)
1 : curving like an eagle's beak ⟨an *aquiline* nose⟩
2 : of, relating to, or resembling an eagle
— **aq·ui·lin·i·ty** \ˌa-kwə-'li-nə-tē\ *noun*

aquiv·er \ə-'kwi-vər\ *adjective* (1883)
: marked by trembling or quivering ⟨all *aquiver* with excitement⟩

ar \'är\ *noun* [Middle English] (14th century)
: the letter *r*

-ar *adjective suffix* [Middle English, from Latin *-aris,* alteration of *-alis* -al]
: of or relating to ⟨molecul*ar*⟩ **:** being ⟨spectacul*ar*⟩ **:** resembling ⟨oracul*ar*⟩

Ar·ab \'ar-əb, 'er-; *dialect also* 'ā-ˌrab\ *noun* [Middle English, from Latin *Arabus, Arabs,* from Greek *Arab-, Araps*] (14th century)
1 a : a member of the Semitic people of the Arabian peninsula **b :** a member of an Arabic-speaking people
2 : ARABIAN HORSE
— **Arab** *adjective*

¹ar·a·besque \ˌar-ə-'besk\ *adjective* [French, from Italian *arabesco* Arabian in fashion, from *arabo* Arab, from Latin *Arabus*] (circa 1656)
: of, relating to, or being in the style of arabesque or an arabesque

²arabesque *noun* (1786)
1 : an ornament or style that employs flower, foliage, or fruit and sometimes animal and figural outlines to produce an intricate pattern of interlaced lines
2 : a posture in ballet in which the body is bent forward from the hip on one leg with one arm extended forward and the other arm and leg backward
3 : an elaborate or intricate pattern ⟨richly pierced by an *arabesque* of wormholes —John Chase⟩ ⟨as lawyers . . . perform their procedural *arabesques* —Daniel Okrent⟩

arabesque 1

¹Ara·bi·an \ə-'rā-bē-ən\ *noun* (14th century)
1 : a native or inhabitant of Arabia
2 : ARABIAN HORSE

²Arabian *adjective* (14th century)
: ARABIC 1

Arabian horse *noun* (1737)
: any of a breed of swift compact horses developed in Arabia and usually having gray or chestnut silky hair

¹Ar·a·bic \'ar-ə-bik\ *noun* (14th century)
: a Semitic language originally of the Arabs of the Hejaz and Nejd that is now the prevailing speech of a wide region of southwestern Asia and northern Africa

²Arabic *adjective* (1526)
1 : of, relating to, or characteristic of Arabia or the Arabs
2 : of, relating to, or constituting Arabic
3 : expressed in or utilizing Arabic numerals

arab·i·ca \ə-'ra-bə-kə\ *noun, often attributive* [New Latin, specific epithet of *Coffea arabica,* from Latin, feminine of *Arabicus* Arabian] (1922)
1 : an evergreen shrub or tree (*Coffea arabica*) yielding seeds that produce a high-quality coffee and form a large portion of the coffee of commerce
2 : the seeds of arabica

Arabic alphabet *noun* (1841)
: an alphabet of 28 letters derived from the Aramaic alphabet which is used for writing Arabic and also with adaptations for other languages of the Islamic world

arab·i·cize \ə-'ra-bə-ˌsīz\ *transitive verb* **-cized; -ciz·ing** *often capitalized* (1872)
1 : to adapt (a language or elements of a language) to the phonetic or structural pattern of Arabic
2 : ARABIZE 1
— **arab·i·ci·za·tion** \ə-ˌra-bə-sə-'zā-shən\ *noun*

Arabic numeral *noun* (circa 1847)
: any of the number symbols 0, 1, 2, 3, 4, 5, 6, 7, 8, 9 — see NUMBER table

arab·i·nose \ə-'ra-bə-ˌnōs, -ˌnōz\ *noun* [International Scientific Vocabulary *arabin* (the solid principle in gum arabic, from *gum arabic* + ¹*-in*) + *-ose*] (circa 1884)
: a white crystalline aldose sugar $C_5H_{10}O_5$ occurring especially in vegetable gums

ara·bi·no·side \ˌar-ə-'bi-nə-ˌsīd, ə-'ra-bə-nō-ˌsīd\ *noun* (1927)
: a glycoside that yields arabinose on hydrolysis

Ar·ab·ise *British variant of* ARABIZE

Ar·ab·ism \'ar-ə-ˌbi-zəm\ *noun* (1614)
1 : a characteristic feature of Arabic occurring in another language
2 : devotion to Arab interests, culture, aspirations, or ideals

Ar·ab·ist \'ar-ə-bist\ *noun* (1753)
1 : a specialist in the Arabic language or in Arabic culture
2 : a person who favors Arab interests and positions in international affairs

Ar·ab·ize \'ar-ə-ˌbīz\ *transitive verb* **-ized; -iz·ing** (1883)
1 a : to cause to acquire Arabic customs, manners, speech, or outlook **b :** to modify (a population) by intermarriage with Arabs
2 : ARABICIZE 1
— **Ar·ab·i·za·tion** \ˌar-ə-bə-'zā-shən\ *noun*

¹ar·a·ble \'ar-ə-bəl\ *adjective* [Middle French or Latin; Middle French, from Latin *arabilis,* from *arare* to plow; akin to Old English *erian* to plow, Greek *aroun*] (15th century)
1 : fit for or used for the growing of crops
2 *British* **:** engaged in, produced by, or being the cultivation of arable land
— **ar·a·bil·i·ty** \ˌar-ə-'bi-lə-tē\ *noun*

²arable *noun* (1576)
chiefly British **:** land fit or used for the growing of crops; *also* **:** a plot of such land

ar·a·chi·don·ic acid \ˌar-ə-kə-'dä-nik-\ *noun* [New Latin *Arachid-, Arachis* + English *-onic* (as in *gluconic acid*)] (1913)
: a liquid unsaturated fatty acid $C_{20}H_{32}O_2$ that occurs in most animal fats, is a precursor of prostaglandins, and is considered essential in animal nutrition

ar·a·chis oil \'ar-ə-kəs-\ *noun* [New Latin *Arachis,* genus that includes the peanut, from Greek *arakis,* diminutive of *arakos,* a legume] (circa 1889)
: PEANUT OIL

arach·nid \ə-'rak-nəd, -ˌnid\ *noun* [New Latin *Arachnida,* from Greek *arachnē* spider] (1869)
: any of a class (Arachnida) of arthropods comprising chiefly terrestrial invertebrates, including the spiders, scorpions, mites, and ticks, and having a segmented body divided into two regions of which the anterior bears four pairs of legs but no antennae
— **arachnid** *adjective*

¹arach·noid \ə-'rak-ˌnȯid\ *adjective* (1789)
1 : of or relating to a thin membrane of the brain and spinal cord that lies between the dura mater and the pia mater
2 : covered with or composed of soft loose hairs or fibers

²arachnoid *noun* [New Latin *arachnoides,* from Greek *arachnoeidēs,* like a cobweb, from *arachnē* spider, spider's web] (circa 1847)
: an arachnoid membrane

³arachnoid *adjective* [New Latin *Arachn*ida + English *-oid*] (1852)
: resembling or related to the arachnids

ara·go·nite \ə-'ra-gə-ˌnīt, 'ar-ə-gə-\ *noun* [German *Aragonit,* from *Aragon,* Spain] (1803)
: a mineral similar to calcite in consisting of calcium carbonate but differing from calcite in its orthorhombic crystallization, greater density, and less distinct cleavage

— **ara·go·nit·ic** \ə-ˌra-gə-ˈni-tik, ˌar-ə-gə-\ *adjective*

ar·ak *variant of* ARRACK

Ar·a·mae·an \ˌar-ə-ˈmē-ən\ *noun* [Latin *Aramaeus,* from Greek *Aramaios,* from Hebrew *ʼĂrām* Aram, ancient name for Syria] (1839)
1 : ARAMAIC
2 : a member of a Semitic people of the second millennium B.C. in Syria and Upper Mesopotamia
— **Aramaean** *adjective*

Ar·a·ma·ic \ˌar-ə-ˈmā-ik\ *noun* (1882)
: a Semitic language known since the ninth century B.C. as the speech of the Aramaeans and later used extensively in southwest Asia as a commercial and governmental language and adopted as their customary speech by various non-Aramaean peoples including the Jews after the Babylonian exile

Aramaic alphabet *noun* (1925)
1 : an extinct North Semitic alphabet dating from the ninth century B.C. which was for several centuries the commercial alphabet of southwest Asia and the parent of other alphabets (as Syriac and Arabic)
2 : the square Hebrew alphabet as distinguished from the early Hebrew alphabet

ar·a·mid \ˈar-ə-məd, -ˌmid\ *noun* [*ar*omatic poly*amide*] (1972)
: any of a group of lightweight but very strong heat-resistant synthetic aromatic polyamide materials that are fashioned into fibers, filaments, or sheets and used especially in textiles and plastics

Arap·a·ho *or* **Arap·a·hoe** \ə-ˈra-pə-ˌhō\ *noun, plural* **-ho** *or* **-hos** *or* **-hoe** *or* **-hoes** (1812)
1 : a member of an American Indian people of the plains region ranging from Saskatchewan and Manitoba to New Mexico and Texas
2 : the Algonquian language of the Arapaho people

Arau·ca·ni·an \ə-ˌrau̇-ˈkä-nē-ən, ˌar-ˌȯ-ˈkā\ *also* **Arau·can** \ə-ˈrau̇-kən\ *noun* [Spanish *araucano,* from *Arauco,* former province in Chile] (1809)
1 : a member of a group of Indian peoples of south central Chile and adjacent regions of Argentina
2 : the language of the Araucanian people that constitutes an independent language family
— **Araucanian** *adjective*

ar·au·car·ia \ˌar-ˌȯ-ˈkar-ē-ə\ *noun* [New Latin, from *Arauco*] (1809)
: any of a genus (*Araucaria* of the family Araucariaceae, the araucaria family) of South American or Australian coniferous trees that resemble pines and are often grown as ornamentals
— **ar·au·car·i·an** \-ē-ən\ *adjective*

Ar·a·wak \ˈar-ə-ˌwäk, -ˌwak\ *noun, plural* **Arawak** *or* **Arawaks** (1769)
1 : a member of an Indian people of the Arawakan group now living chiefly along the coast of Guyana
2 : the language of the Arawak people

Ar·a·wak·an \ˌar-ə-ˈwä-kən, -ˈwa-\ *noun, plural* **Arawakan** *or* **Arawakans** (1901)
1 : a member of a group of Indian peoples of South America and the West Indies
2 : the language family of the Arawakan peoples

arb \ˈärb\ *noun* (1979)
: ARBITRAGEUR

ar·ba·lest *or* **ar·ba·list** \ˈär-bə-list\ *noun* [Middle English *arblast,* from Old English, from Old French *arbaleste,* from Late Latin *arcuballista,* from Latin *arcus* bow + *ballista* — more at ARROW] (before 12th century)
: a crossbow especially of medieval times

ar·bi·ter \ˈär-bə-tər\ *noun* [Middle English *arbitre,* from Middle French, from Latin *arbitr-, arbiter*] (14th century)
1 : a person with power to decide a dispute **:** JUDGE
2 : a person or agency having the power of deciding

arbiter el·e·gan·ti·a·rum \-ˌe-lə-ˌgan-shē-ˈar-əm, -ˈer-\ *noun* [Latin, literally, arbiter of refinements] (1818)
: a person who prescribes, rules on, or is a recognized authority on matters of social behavior and taste

ar·bi·tra·ble \ˈär-bə-trə-bəl, är-ˈbi-\ *adjective* (1531)
: subject to decision by arbitration

¹ar·bi·trage \ˈär-bə-ˌträzh\ *noun* [French, from Middle French, arbitration, from Old French, from *arbitrer* to render judgment, from Latin *arbitrari,* from *arbitr-, arbiter*] (1875)
1 : the nearly simultaneous purchase and sale of securities or foreign exchange in different markets in order to profit from price discrepancies
2 : the purchase of the stock of a takeover target especially with a view to selling it profitably to the raider

²arbitrage *intransitive verb* **-traged; -trag·ing** (1900)
: to engage in arbitrage

ar·bi·tra·geur \ˌär-bə-(ˌ)trä-ˈzhər\ *or* **ar·bi·trag·er** \ˈär-bə-ˌträ-zhər\ *noun* [French *arbitrageur,* from *arbitrage*] (1870)
: one that practices arbitrage

ar·bi·tral \ˈär-bə-trəl\ *adjective* (1609)
: of or relating to arbiters or arbitration

ar·bit·ra·ment \är-ˈbi-trə-mənt\ *noun* [Middle English, from Middle French *arbitrement,* from *arbitrer*] (15th century)
1 *archaic* **:** the right or power of deciding
2 : the settling of a dispute by an arbiter
3 : the judgment given by an arbitrator

ar·bi·trary \ˈär-bə-ˌtrer-ē\ *adjective* (15th century)
1 : depending on individual discretion (as of a judge) and not fixed by law ⟨the manner of punishment is *arbitrary*⟩
2 a : not restrained or limited in the exercise of power **:** ruling by absolute authority ⟨an *arbitrary* government⟩ **b :** marked by or resulting from the unrestrained and often tyrannical exercise of power ⟨protection from *arbitrary* arrest and detention⟩
3 a : based on or determined by individual preference or convenience rather than by necessity or the intrinsic nature of something ⟨an *arbitrary* standard⟩ ⟨take any *arbitrary* positive number⟩ ⟨*arbitrary* division of historical studies into watertight compartments —A. J. Toynbee⟩ **b :** existing or coming about seemingly at random or by chance or as a capricious and unreasonable act of will ⟨when a task is not seen in a meaningful context it is experienced as being *arbitrary* —Nehemiah Jordan⟩
— **ar·bi·trari·ly** \ˌär-bə-ˈtrer-ə-lē\ *adverb*
— **ar·bi·trar·i·ness** \ˈär-bə-ˌtrer-ē-nəs\ *noun*

ar·bi·trate \ˈär-bə-ˌtrāt\ *verb* **-trat·ed; -trat·ing** (1592)
transitive verb
1 *archaic* **:** DECIDE, DETERMINE
2 : to act as arbiter upon
3 : to submit or refer for decision to an arbiter ⟨agreed to *arbitrate* their differences⟩
intransitive verb
: to act as arbitrator
— **ar·bi·tra·tive** \-ˌtrā-tiv\ *adjective*

ar·bi·tra·tion \ˌär-bə-ˈtrā-shən\ *noun* (15th century)
: the action of arbitrating; *especially* **:** the hearing and determination of a case in controversy by an arbiter
— **ar·bi·tra·tion·al** \-sh(ə-)nəl\ *adjective*

ar·bi·tra·tor \ˈär-bə-ˌtrā-tər\ *noun* (15th century)
: one that arbitrates **:** ARBITER

¹ar·bor \ˈär-bər\ *noun* [Middle English *erber* plot of grass, arbor, from Middle French *herbier* plot of grass, from *herbe* herb, grass] (14th century)
: a shelter of vines or branches or of latticework covered with climbing shrubs or vines

²arbor *noun* [Latin, tree, shaft] (1659)
1 : a spindle or axle of a wheel
2 : a main shaft or beam
3 : a shaft on which a revolving cutting tool is mounted
4 : a spindle on a cutting machine that holds the work to be cut

arbor- *or* **arbori-** *combining form* [Latin *arbor*]
: tree ⟨*arbori*culture⟩

Arbor Day *noun* [Latin *arbor* tree] (1872)
: a day designated for planting trees

ar·bo·re·al \är-ˈbōr-ē-əl, -ˈbȯr-\ *adjective* [Latin *arboreus* of a tree, from *arbor*] (circa 1667)
1 : of, relating to, or resembling a tree
2 : inhabiting or frequenting trees ⟨*arboreal* monkeys⟩
— **ar·bo·re·al·ly** \-ə-lē\ *adverb*

ar·bo·re·ous \-ē-əs\ *adjective* (1646)
: ARBOREAL ⟨an *arboreous* palm⟩

ar·bo·res·cent \ˌär-bə-ˈre-sᵊnt\ *adjective* (1675)
: resembling a tree in properties, growth, structure, or appearance
— **ar·bo·res·cence** \-sᵊn(t)s\ *noun*

ar·bo·re·tum \ˌär-bə-ˈrē-təm\ *noun, plural* **-retums** *or* **-re·ta** \-ˈrē-tə\ [New Latin, from Latin, plantation of trees, from *arbor*] (1838)
: a place where trees, shrubs, and herbaceous plants are cultivated for scientific and educational purposes

ar·bor·i·cul·ture \ˈär-bər-ə-ˌkəl-chər, är-ˈbȯr-ə-\ *noun* [*arbori-* + *-culture* (as in *agriculture*)] (1828)
: the cultivation of trees and shrubs especially for ornamental purposes
— **ar·bor·i·cul·tur·al** \ˌär-bər-ə-ˈkəl-chər-əl, är-ˌbȯr-ə-\ *adjective*

ar·bor·ist \ˈär-bə-rist\ *noun* (1578)
: a specialist in the care and maintenance of trees

ar·bor·i·za·tion \ˌär-bə-rə-ˈzā-shən\ *noun* (1794)
: formation of or into an arborescent figure or arrangement; *also* **:** such a figure or arrangement (as a dendritic process of a nerve cell)

ar·bor·ize \ˈär-bə-ˌrīz\ *intransitive verb* **-ized; -iz·ing** (1847)
: to branch freely and repeatedly

ar·bor·vi·tae \ˌär-bər-ˈvī-tē\ *noun* [New Latin *arbor vitae,* literally, tree of life] (1646)
: any of various evergreen trees and shrubs (especially genus *Thuja*) of the cypress family that usually have closely overlapping or compressed scale leaves and are often grown for ornament and in hedges

ar·bour *chiefly British variant of* ARBOR

ar·bo·vi·rus \ˌär-bə-ˈvī-rəs\ *noun* [*ar*thropod-*bo*rne *virus*] (1957)
: any of various RNA viruses (as the causative agents of encephalitis, yellow fever, and dengue) transmitted by arthropods

ar·bu·tus \är-ˈbyü-təs\ *noun* [New Latin, from Latin, strawberry tree] (1548)
1 : any of a genus (*Arbutus*) of shrubs and trees of the heath family with white or pink flowers and red or orange berries
2 : a creeping plant (*Epigaea repens*) of the heath family that occurs in eastern North America and bears fragrant pink or white flowers in early spring

¹arc \ˈärk\ *noun* [Middle English *ark,* from Middle French *arc* bow, from Latin *arcus* bow, arch, arc — more at ARROW] (14th century)

1 : the apparent path described above and below the horizon by a celestial body (as the sun)
2 : something arched or curved
3 : a sustained luminous discharge of electricity across a gap in a circuit or between electrodes; *also* **:** ARC LAMP
4 : a continuous portion (as of a circle or ellipse) of a curved line
5 : degree measurement on the circumference of a circle — used especially in the phrase *of arc* ⟨11 minutes 3 seconds of *arc*⟩

[2]arc *intransitive verb* **arced** \ˈärkt\; **arc·ing** \ˈär-kiŋ\ (1893)
1 : to form an electric arc
2 : to follow an arc-shaped course ⟨the missile *arced* across the sky⟩

[3]arc *adjective* [*arc sine* arc or angle (corresponding to the) sine (of so many degrees)] (circa 1949)
: INVERSE 2 — used with the trigonometric functions and hyperbolic functions

ar·cade \är-ˈkād\ *noun* [French, from Italian *arcata,* from *arco* arch, from Latin *arcus*] (1725)
1 : a long arched building or gallery
2 : an arched covered passageway or avenue (as between shops)
3 : a series of arches with their columns or piers
4 : an amusement center having coin-operated games

ar·cad·ed \-ˈkā-dəd\ *adjective* (1805)
: having, formed in, or decorated with arches or arcades ⟨*arcaded* streets⟩ ⟨an *arcaded* bowl⟩

ar·ca·dia \är-ˈkā-dē-ə\ *noun, often capitalized* [*Arcadia,* region of ancient Greece frequently chosen as background for pastoral poetry] (circa 1890)
: a region or scene of simple pleasure and quiet

ar·ca·di·an \är-ˈkā-dē-ən\ *adjective, often capitalized* (1589)
1 : idyllically pastoral; *especially* **:** idyllically innocent, simple, or untroubled
2 a : of or relating to Arcadia or the Arcadians **b :** of or relating to Arcadian

Ar·ca·di·an \är-ˈkā-dē-ən\ *noun* (1590)
1 *often not capitalized* **:** a person who lives a simple quiet life
2 : a native or inhabitant of Arcadia
3 : the dialect of ancient Greek used in Arcadia

ar·cad·ing \är-ˈkā-diŋ\ *noun* (1849)
: a series of arches or arcades used in the construction or decoration especially of a building

Ar·ca·dy \ˈär-kə-dē\ *noun* (14th century)
: ARCADIA

ar·cane \är-ˈkān\ *adjective* [Latin *arcanus*] (1547)
: known or knowable only to the initiate **:** SECRET ⟨the *arcane* rites of a mystery cult⟩; *broadly* **:** MYSTERIOUS, OBSCURE ⟨the technical consultant's *arcane* explanations⟩

ar·ca·num \är-ˈkā-nəm\ *noun, plural* **-na** \-nə\ [Latin, from neuter of *arcanus* secret, from *arca* chest — more at ARK] (15th century)
1 : mysterious knowledge, language, or information accessible only by the initiate — usually used in plural
2 : ELIXIR 1

arc·co·sine \(ˌ)är(k)-ˈkō-ˌsīn\ *noun* (circa 1884)
: the inverse function to the cosine ⟨if *y* is the cosine of *θ,* then *θ* is the *arccosine* of *y*⟩

[1]arch \ˈärch\ *noun* [Middle English *arche,* from Middle French, from (assumed) Vulgar Latin *arca,* from Latin *arcus* — more at ARROW] (14th century)
1 : a typically curved structural member spanning an opening and serving as a support (as for the wall or other weight above the opening)
2 a : something resembling an arch in form or function; *especially* **:** either of two vaulted portions of the bony structure of the foot that impart elasticity to it **b :** a curvature having the form of an arch
3 : ARCHWAY

arch 1: *1* round: *imp* impost, *sp* springer, *v* voussoir, *k* keystone, *ext* extrados, *int* intrados; *2* horseshoe; *3* lancet; *4* ogee; *5* trefoil; *6* basket-handle; *7* Tudor

[2]arch (15th century)
transitive verb
1 : to cover or provide with an arch
2 : to form into an arch
intransitive verb
1 : to form an arch
2 : to take an arch-shaped course

[3]arch *adjective* [ˈ*arch-*] (1547)
1 : PRINCIPAL, CHIEF ⟨your *arch* opponent⟩
2 a : MISCHIEVOUS, SAUCY **b :** marked by a deliberate and often forced irony, brashness, or impudence
— **arch·ly** *adverb*
— **arch·ness** *noun*

[1]arch- *prefix* [Middle English *arche-, arch-,* from Old English & Old French; Old English *arce-,* from Late Latin *arch-* & Latin *archi-;* Old French *arch-,* from Late Latin *arch-* & Latin *archi-,* from Greek *arch-, archi-,* from *archein* to begin, rule; akin to Greek *archē* beginning, rule, *archos* ruler]
1 : chief **:** principal ⟨*arch*fiend⟩
2 : extreme **:** most fully embodying the qualities of the kind ⟨*arch*conservative⟩

[2]arch- — see ARCHI-

[1]-arch *noun combining form* [Middle English *-arche,* from Old French & Late Latin & Latin; Old French *-arche,* from Late Latin *-archa,* from Latin *-arches, -archus,* from Greek *-archēs, -archos,* from *archein*]
: ruler **:** leader ⟨matri*arch*⟩

[2]-arch *adjective combining form* [probably from German, from Greek *archē* beginning]
: having (such) a point or (so many) points of origin ⟨end*arch*⟩

archae- *or* **archaeo-** *also* **archeo-** *combining form* [Greek *archaio-,* from *archaios* ancient, from *archē* beginning]
: ancient **:** primitive ⟨*archaeo*pteryx⟩ ⟨*Archeo*zoic⟩

ar·chae·bac·te·ri·um \ˌär-kē-ˌbak-ˈtir-ē-əm\ *noun* [New Latin] (1977)
: any of a class (Archaeobacteria) or a separate kingdom (Archaebacteria) of primitive bacteria including methane-producing forms, some red halophilic forms, and others of harsh hot acidic environments

ar·chaeo·as·tron·o·my \ˌär-kē-(ˌ)ō-ə-ˈsträ-nə-mē\ *noun* (1971)
: the study of the astronomy of ancient cultures

ar·chae·ol·o·gy *or* **ar·che·ol·o·gy** \ˌär-kē-ˈä-lə-jē\ *noun* [French *archéologie,* from Late Latin *archaeologia* antiquarian lore, from Greek *archaiologia,* from *archaio-* + *-logia* -logy] (1837)
1 : the scientific study of material remains (as fossil relics, artifacts, and monuments) of past human life and activities
2 : remains of the culture of a people **:** ANTIQUITIES
— **ar·chae·o·log·i·cal** \ˌär-kē-ə-ˈlä-ji-kəl\ *adjective*
— **ar·chae·o·log·i·cal·ly** \-k(ə-)lē\ *adverb*
— **ar·chae·ol·o·gist** \ˌär-kē-ˈä-lə-jist\ *noun*

ar·chae·op·ter·yx \ˌär-kē-ˈäp-tə-riks\ *noun* [New Latin, from *archae-* + Greek *pteryx* wing; akin to Greek *pteron* wing — more at FEATHER] (1859)
: a primitive crow-sized bird (genus *Archaeopteryx*) of the Upper Jurassic period of Europe having reptilian characteristics (as teeth and a long bony tail)

ar·cha·ic \är-ˈkā-ik\ *adjective* [French or Greek; French *archaïque,* from Greek *archaïkos,* from *archaios*] (1832)
1 : having the characteristics of the language of the past and surviving chiefly in specialized uses
2 : of, relating to, or characteristic of an earlier or more primitive time **:** ANTIQUATED ⟨*archaic* legal traditions⟩
3 *capitalized* **:** of or belonging to the early or formative phases of a culture or a period of artistic development; *especially* **:** of or belonging to the period leading up to the classical period of Greek culture
4 : surviving from an earlier period; *specifically* **:** typical of a previously dominant evolutionary stage
5 *capitalized* **:** of or relating to the period from about 8000 B.C. to 1000 B.C. and the North American cultures of that time
synonym see OLD
— **ar·cha·i·cal·ly** \-i-k(ə-)lē\ *adverb*

archaic smile *noun* (1889)
: an expression that resembles a smile and is characteristic of early Greek sculpture

ar·cha·ism \ˈär-kē-ˌi-zəm, -(ˌ)kā-ˌi-\ *noun* [New Latin *archaïsmus,* from Greek *archaïsmos,* from *archaios*] (1643)
1 : the use of archaic diction or style
2 : an instance of archaic usage
3 : something archaic; *especially* **:** something (as a practice or custom) that is outmoded or old-fashioned
— **ar·cha·ist** \-ist\ *noun*
— **ar·cha·is·tic** \ˌär-kē-ˈis-tik, -(ˌ)kā-\ *adjective*
— **ar·cha·ize** \ˈär-kē-ˌīz, -(ˌ)kā-\ *verb*

arch·an·gel \ˈär-ˌkān-jəl\ *noun* [Middle English, from Old French or Late Latin; Old French *archangele,* from Late Latin *archangelus,* from Greek *archangelos,* from *archi-* + *angelos* angel] (12th century)
1 : a chief angel
2 *plural* **:** an order of angels — see CELESTIAL HIERARCHY
— **arch·an·gel·ic** \ˌär-(ˌ)kan-ˈje-lik\ *adjective*

arch·bish·op \(ˌ)ärch-ˈbi-shəp\ *noun* [Middle English, from Old English *arcebiscop,* from Late Latin *archiepiscopus,* from Late Greek *archiepiskopos,* from *archi-* + *episkopos* bishop — more at BISHOP] (before 12th century)
: a bishop at the head of an ecclesiastical province or one of equivalent honorary rank

arch·bish·op·ric \-shə-(ˌ)prik\ *noun* (before 12th century)
1 : the see or province over which an archbishop exercises authority
2 : the jurisdiction or office of an archbishop

arch·con·ser·va·tive \(ˌ)ärch-kən-ˈsər-və-tiv\ *noun* (1934)
: an extreme conservative
— **archconservative** *adjective*

arch·dea·con \(ˌ)ärch-ˈdē-kən\ *noun* [Middle English *archedeken,* from Old English *arcediacon,* from Late Latin *archidiaconus,* from Late Greek *archidiakonos,* from Greek *archi-* + *diakonos* deacon] (before 12th century)
: a clergyman having the duty of assisting a diocesan bishop in ceremonial functions or administrative work

arch·dea·con·ry \-kən-rē\ *noun, plural* **-ries** (1779)
: the district or residence of an archdeacon

arch·di·o·cese \(ˌ)ärch-ˈdī-ə-səs, -ˌsēz, -ˌsēs\ *noun, plural* **-ces·es** \-ˈdī-ə-ˌsēz, -ˌsē-zəz, -sə-səz, -sə-ˌsēz\ (1844)
: the diocese of an archbishop

— **arch·di·oc·e·san** \ˌärch-dī-ˈä-sə-sən\ *adjective*

arch·du·cal \(ˌ)ärch-ˈdü-kəl, -ˈdyü-\ *adjective* [French *archiducal,* from *archiduc*] (1665)
: of or relating to an archduke or archduchy

arch·duch·ess \-ˈdə-chəs\ *noun* [French *archiduchesse,* feminine of *archiduc* archduke, from Middle French *archeduc*] (1618)
1 : the wife or widow of an archduke
2 : a woman having in her own right a rank equal to that of an archduke

arch·duchy \-ˈdə-chē\ *noun* [French *archiduché,* from Middle French *archeduché,* from *arche-* arch- + *duché* duchy] (1680)
: the territory of an archduke or archduchess

arch·duke \-ˈdük, -ˈdyük\ *noun* [Middle French *archeduc,* from *arche-* arch- + *duc* duke] (circa 1530)
1 : a sovereign prince
2 : a prince of the imperial family of Austria
— **arch·duke·dom** \-dəm\ *noun*

Ar·che·an *or* **Ar·chae·an** \är-ˈkē-ən\ *adjective* [Greek *archaios*] (1872)
1 : of, relating to, or being the earliest eon of geological history or the corresponding system of rocks — see GEOLOGIC TIME table
2 : PRECAMBRIAN
— **Archean** *noun*

ar·che·go·ni·al \ˌär-ki-ˈgō-nē-əl\ *adjective* (1865)
: of or relating to an archegonium; *also* **:** ARCHEGONIATE

¹ar·che·go·ni·ate \-nē-ət\ *adjective* (circa 1888)
: bearing archegonia

²archegoniate *noun* (circa 1903)
: a plant (as a moss, fern, horsetail, or club moss) that bears archegonia

ar·che·go·ni·um \-nē-əm\ *noun, plural* **-nia** \-nē-ə\ [New Latin, from Greek *archegonos* originator, from *archein* to begin + *gonos* procreation; akin to Greek *gignesthai* to be born — more at ARCH-, KIN] (1854)
: the flask-shaped female sex organ of bryophytes, lower vascular plants (as ferns), and some gymnosperms

arch·en·e·my \(ˌ)ärch-ˈe-nə-mē\ *noun, plural* **-mies** (1550)
: a principal enemy

arch·en·ter·on \är-ˈken-tə-ˌrän, -rən\ *noun* [New Latin] (1877)
: the cavity of a gastrula forming a primitive gut

Ar·cheo·zo·ic *also* **Ar·chaeo·zo·ic** \ˌär-kē-ə-ˈzō-ik\ *adjective* (1872)
: ARCHEAN 1
— **Archeozoic** *noun*

ar·cher \ˈär-chər\ *noun* [Middle English, from Old French, from Late Latin *arcarius,* alteration of *arcuarius,* from *arcuarius* of a bow, from Latin *arcus* bow — more at ARROW] (14th century)
1 : a person who uses a bow and arrow
2 *capitalized* **:** SAGITTARIUS

ar·cher·fish \ˈär-chər-ˌfish\ *noun* (circa 1889)
: a small East Indian fish (*Toxotes jaculator*) that catches insects by stunning them with water ejected from its mouth; *also* **:** any of various related fish of similar habits

ar·chery \ˈär-chə-rē\ *noun* (15th century)
1 : the art, practice, or skill of shooting with bow and arrow
2 : an archer's weapons
3 : a body of archers

ar·che·spo·ri·um \ˌär-ki-ˈspōr-ē-əm, -ˈspȯr-\ *noun, plural* **-spo·ria** \-ē-ə\ [New Latin, from *arche-* (as in *archegonium*) + *-sporium* (from *spora* spore)] (circa 1889)
: the cell or group of cells from which spore mother cells develop
— **ar·che·spo·ri·al** \ˌär-ki-ˈspōr-ē-əl, -ˈspȯr-\ *adjective*

ar·che·type \ˈär-ki-ˌtīp\ *noun* [Latin *archetypum,* from Greek *archetypon,* from neuter of *archetypos* archetypal, from *archein* + *typos* type] (1545)
1 : the original pattern or model of which all things of the same type are representations or copies **:** PROTOTYPE; *also* **:** a perfect example
2 : IDEA 1a
3 : an inherited idea or mode of thought in the psychology of C. G. Jung that is derived from the experience of the race and is present in the unconscious of the individual
— **ar·che·typ·al** \ˌär-ki-ˈtī-pəl\ *also* **ar·che·typ·i·cal** \-ˈti-pi-kəl\ *adjective*
— **ar·che·typ·al·ly** \-pə-lē\ *adverb*

arch·fiend \(ˌ)ärch-ˈfēnd\ *noun* (1667)
: a chief fiend; *especially* **:** SATAN

archi- *or* **arch-** *prefix* [French or Latin; French, from Latin, from Greek — more at ARCH-]
: primitive **:** original **:** primary ⟨*arch*enteron⟩

ar·chi·di·ac·o·nal \ˌär-ki-dī-ˈa-kə-nᵊl\ *adjective* [Late Latin *archidiaconus* archdeacon] (15th century)
: of or relating to an archdeacon

ar·chi·epis·co·pal \ˌär-kē-ə-ˈpis-kə-pəl\ *adjective* [Medieval Latin *archiepiscopalis,* from Late Latin *archiepiscopus* archbishop — more at ARCHBISHOP] (1611)
: of or relating to an archbishop
— **ar·chi·epis·co·pal·ly** \-p(ə-)lē\ *adverb*

ar·chi·epis·co·pate \-pət, -ˌpāt\ *noun* (1792)
: ARCHBISHOPRIC

ar·chil \ˈär-chəl\ *noun* [Middle English *orchell*] (15th century)
1 : a violet dye obtained from lichens (genera *Roccella* and *Lecanora*)
2 : a lichen that yields archil

ar·chi·man·drite \ˌär-kə-ˈman-ˌdrīt\ *noun* [Late Latin *archimandrites,* from Late Greek *archimandritēs,* from Greek *archi-* + Late Greek *mandra* monastery, from Greek, fold, pen] (1591)
: a dignitary in an Eastern church ranking below a bishop; *specifically* **:** the superior of a large monastery or group of monasteries

Ar·chi·me·des' screw \ˌär-kə-ˈmē-dēz-\ *noun* [*Archimedes*] (circa 1859)
: a device made of a tube bent spirally around an axis or of a broad-threaded screw encased by a cylinder and used to raise water

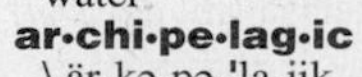
Archimedes' screw

ar·chi·pe·lag·ic \ˌär-kə-pə-ˈla-jik, ˌär-chə-\ *adjective* (1841)
: of, relating to, or located in an archipelago

ar·chi·pel·a·go \ˌär-kə-ˈpe-lə-ˌgō, ˌär-chə-\ *noun, plural* **-goes** *or* **-gos** [*Archipelago* Aegean Sea, from Italian *Arcipelago,* literally, chief sea, from *arci-* (from Latin *archi-*) + Greek *pelagos* sea — more at PLAGAL] (1502)
1 : an expanse of water with many scattered islands
2 : a group of islands

ar·chi·tect \ˈär-kə-ˌtekt\ *noun* [Middle French *architecte,* from Latin *architectus,* from Greek *architektōn* master builder, from *archi-* + *tektōn* builder, carpenter — more at TECHNICAL] (1563)
1 : a person who designs buildings and advises in their construction
2 : a person who designs and guides a plan or undertaking ⟨the *architect* of American foreign policy⟩

ar·chi·tec·ton·ic \ˌär-kə-ˌtek-ˈtä-nik\ *adjective* [Latin *architectonicus,* from Greek *architektonikos,* from *architektōn*] (1645)
1 : of, relating to, or according with the principles of architecture **:** ARCHITECTURAL
2 : having an organized and unified structure that suggests an architectural design
— **ar·chi·tec·ton·i·cal·ly** \-ni-k(ə-)lē\ *adverb*

ar·chi·tec·ton·ics \-ˈtä-niks\ *noun plural but singular or plural in construction, also* **ar·chi·tec·ton·ic** \-nik\ (1660)
1 : the science of architecture
2 a : the unifying structural design of something **b :** the system of structure

ar·chi·tec·tur·al \ˌär-kə-ˈtek-chə-rəl, -ˈtek-shrəl\ *adjective* (circa 1794)
1 : of or relating to architecture **:** conforming to the rules of architecture
2 : having or conceived of as having a single unified overall design, form, or structure
— **ar·chi·tec·tur·al·ly** *adverb*

ar·chi·tec·ture \ˈär-kə-ˌtek-chər\ *noun* (1555)
1 : the art or science of building; *specifically* **:** the art or practice of designing and building structures and especially habitable ones
2 a : formation or construction as or as if as the result of conscious act ⟨the *architecture* of the garden⟩ **b :** a unifying or coherent form or structure ⟨the novel lacks *architecture*⟩
3 : architectural product or work
4 : a method or style of building
5 : the manner in which the components of a computer or computer system are organized and integrated

ar·chi·trave \ˈär-kə-ˌtrāv\ *noun* [Middle French, from Old Italian, from *archi-* + *trave* beam, from Latin *trab-, trabs* — more at THORP] (1563)
1 : the lowest division of an entablature resting in classical architecture immediately on the capital of the column — see ENTABLATURE illustration
2 : the molding around a rectangular opening (as a door)

ar·chi·val \är-ˈkī-vəl\ *adjective* (circa 1828)
: of, relating to, contained in, or constituting archives

¹ar·chive \ˈär-ˌkīv\ *noun* [French & Latin; French, from Latin *archivum,* from Greek *archeion* government house (in plural, official documents), from *archē* rule, government — more at ARCH-] (1603)
: a place in which public records or historical documents are preserved; *also* **:** the material preserved — often used in plural

²archive *transitive verb* **ar·chived; ar·chiv·ing** (1926)
: to file or collect (as records or documents) in or as if in an archive

ar·chi·vist \ˈär-kə-vist, -ˌkī-\ *noun* (1753)
: a person in charge of archives

ar·chi·volt \ˈär-kə-ˌvōlt\ *noun* [Italian *archivolto,* from Medieval Latin *archivoltum*] (circa 1731)
: an ornamental molding around an arch corresponding to an architrave

ar·chon \ˈär-ˌkän, -kən\ *noun* [Latin, from Greek *archōn,* from present participle of *archein*] (1579)
1 : a chief magistrate in ancient Athens
2 : a presiding officer

ar·cho·saur \ˈär-kə-ˌsȯr\ *noun* [New Latin *Archosauria,* from Greek *archōn* + *sauros* lizard] (1933)
: any of a subclass (Archosauria) of reptiles comprising the dinosaurs, pterosaurs, and crocodilians
— **ar·cho·sau·ri·an** \ˌär-kə-ˈsȯr-ē-ən\ *adjective*

arch·priest \(ˌ)ärch-ˈprēst\ *noun* (14th century)
: a priest of preeminent rank

arch·way \ˈärch-ˌwā\ *noun* (1802)
: a way or passage under an arch; *also* **:** an arch over a passage

-archy *noun combining form, plural* **-archies** [Middle English *-archie,* from Middle French, from Latin *-archia,* from Greek, from *archein* to rule — more at ARCH-]

\ə\ **abut** \ᵊ\ kitten \ər\ **further** \a\ **ash** \ā\ **ace** \ä\ **mop, mar** \au̇\ **out** \ch\ **chin** \e\ **bet** \ē\ **easy** \g\ **go** \i\ **hit** \ī\ **ice** \j\ **job** \ŋ\ **sing** \ō\ **go** \ȯ\ **law** \ȯi\ **boy** \th\ **thin** \t͟h\ **the** \ü\ **loot** \u̇\ **foot** \y\ **yet** \zh\ **vision** *see also* Guide to Pronunciation

: rule **:** government ⟨squire*archy*⟩
arc lamp *noun* (1882)
: an electric lamp that produces light by an arc made when a current passes between two incandescent electrodes surrounded by gas — called also *arc light*
ar·co \'är-(ˌ)kō\ *adverb or adjective* [Italian, from *arco* bow, from Latin *arcus* — more at ARROW] (1806)
: with the bow — usually used as a direction in music for players of stringed instruments; compare PIZZICATO
arc·sine \(ˌ)ärk-'sīn\ *noun* (circa 1909)
: the inverse function to the sine ⟨if *y* is the sine of *θ*, then *θ* is the *arcsine* of *y*⟩
arc·tan·gent \(ˌ)ärk-'tan-jənt\ *noun* (circa 1909)
: the inverse function to the tangent ⟨if *y* is the tangent of *θ*, then *θ* is the *arctangent* of *y*⟩
¹arc·tic \'ärk-tik, 'är-tik\ *adjective* [Middle English *artik,* from Latin *arcticus,* from Greek *arktikos,* from *arktos* bear, Ursa Major, north; akin to Latin *ursus* bear, Sanskrit *ṛkṣa*] (14th century)
1 *often capitalized* **:** of or relating to the north pole or the region near it
2 a : a bitter cold **:** FRIGID **b :** cold in temper or mood ⟨an *arctic* smile⟩
— arc·ti·cal·ly \-ti-k(ə-)lē\ *adverb*
²arc·tic \'är-tik, 'ärk-tik\ *noun* (1867)
: a rubber overshoe reaching to the ankle or above
arctic char *noun* (circa 1902)
: a Holarctic char (*Salvelinus alpinus*) of arctic waters occurring in freshwater or anadromous populations
arctic circle *noun, often A&C capitalized* (1834)
: the parallel of latitude that is approximately 66½ degrees north of the equator and that circumscribes the northern frigid zone
arctic fox *noun* (1772)
: a small migratory Holarctic fox (*Alopex lagopus*) especially of coastal arctic and alpine tundra
arctic tern *noun* (1844)
: a Holarctic tern (*Sterna paradisaea*) that breeds in arctic regions and migrates to southern Africa and South America
Arc·tu·rus \ärk-'tu̇r-əs, -'tyu̇r-\ *noun* [Latin, from Greek *Arktouros,* literally, bear watcher]
: a giant fixed star of the first magnitude in Boötes
ar·cu·ate \'är-kyə-wət, -ˌwāt\ *adjective* [Latin *arcuatus,* past participle of *arcuare* to bend like a bow, from *arcus* bow] (1626)
: curved like a bow ⟨an *arcuate* cloud⟩
— ar·cu·ate·ly *adverb*
-ard *also* **-art** *noun suffix* [Middle English, from Old French, of Germanic origin; akin to Old High German *-hart* (in personal names such as *Gērhart* Gerard), Old English *heard* hard]
: one that is characterized by performing some action, possessing some quality, or being associated with some thing especially conspicuously or excessively ⟨bragg*art*⟩ ⟨dull*ard*⟩ ⟨pol*lard*⟩
ar·den·cy \'är-dᵊn(t)-sē\ *noun* (1549)
: the quality or state of being ardent
ar·dent \'är-dᵊnt\ *adjective* [Middle English, from Middle French, from Latin *ardent-, ardens,* present participle of *ardēre* to burn, from *ardor*] (14th century)
1 : characterized by warmth of feeling typically expressed in eager zealous support or activity
2 : FIERY, HOT ⟨an *ardent* sun⟩
3 : SHINING, GLOWING ⟨*ardent* eyes⟩
synonym see IMPASSIONED
— ar·dent·ly *adverb*
ardent spirits *noun plural* (1833)
: strong distilled liquors
ar·dor \'är-dər\ *noun* [Middle English *ardour,* from Middle French & Latin; Middle French, from Latin *ardor* burning, heat, ardor, from *aridus* dry — more at ARID] (14th century)
1 a : an often restless or transitory warmth of feeling ⟨the sudden *ardors* of youth⟩ **b :** extreme vigor or energy **:** INTENSITY **c :** ZEAL **d :** LOYALTY
2 : sexual excitement
synonym see PASSION
ar·dour *chiefly British variant of* ARDOR
ar·du·ous \'är-jə-wəs, -dyu̇-wəs\ *adjective* [Latin *arduus* high, steep, difficult; akin to Old Irish *ard* high] (1538)
1 a : hard to accomplish or achieve **:** DIFFICULT ⟨years of *arduous* training⟩ **b :** marked by great labor or effort **:** STRENUOUS ⟨a life of *arduous* toil —A. C. Cole⟩
2 : hard to climb **:** STEEP
synonym see HARD
— ar·du·ous·ly *adverb*
— ar·du·ous·ness *noun*
¹are [Middle English, from Old English *earun;* akin to Old Norse *eru, erum* are, Old English *is* is] *present 2d singular or present plural of* BE
²are \'ar, 'er, 'är\ *noun* [French, from Latin *area*] (circa 1819)
— see METRIC SYSTEM table
ar·ea \'ar-ē-ə, 'er-; 'ā-rē-ə\ *noun* [Latin, open space, threshing floor; perhaps akin to Latin *arēre* to be dry — more at ARID] (1538)
1 : a level piece of ground
2 : the surface included within a set of lines; *specifically* **:** the number of unit squares equal in measure to the surface — see METRIC SYSTEM table, WEIGHT table
3 : the scope of a concept, operation, or activity **:** FIELD ⟨the whole *area* of foreign policy⟩
4 : AREAWAY ⟨went down the steps into the *area* of a house —James Joyce⟩
5 : a particular extent of space or surface or one serving a special function: as **a :** a part of the surface of the body **b :** a geographic region
6 : a part of the cerebral cortex having a particular function
— ar·e·al \-ē-əl\ *adjective*
— ar·e·al·ly \-ə-lē\ *adverb*
area code *noun* (1961)
: a 3-digit number that identifies each telephone service area in a country (as the U.S. or Canada)
ar·ea·way \'ar-ē-ə-ˌwā, 'er-; 'ā-rē-\ *noun* (1899)
: a sunken space affording access, air, and light to a basement
are·ca \ə-'rē-kə, 'ar-i-kə\ *noun* [New Latin, from Portuguese, from Malayalam *aṭaykka*] (1510)
: any of several tropical Asian palms (*Areca* or related genera); *especially* **:** BETEL PALM
arec·o·line \ə-'re-kə-ˌlēn\ *noun* [International Scientific Vocabulary *areca* + [1]*-ol* + [2]*-ine*] (1899)
: a toxic parasympathomimetic alkaloid $C_8H_{13}NO_2$ that is used as a veterinary anthelmintic and occurs naturally in betel nuts
are·na \ə-'rē-nə\ *noun* [Latin *harena, arena* sand, sandy place] (1600)
1 : an area in a Roman amphitheater for gladiatorial combats
2 a : an enclosed area used for public entertainment **b :** a building containing an arena
3 a : a sphere of interest, activity, or competition ⟨the political *arena*⟩ **b :** a place or situation for controversy ⟨in the public *arena*⟩
ar·e·na·ceous \ˌar-ə-'nā-shəs\ *adjective* [Latin *arenaceus,* from *arena*] (1646)
1 : resembling, made of, or containing sand or sandy particles
2 : growing in sandy places
arena theater *noun* (1943)
: THEATER-IN-THE-ROUND
ar·e·nic·o·lous \ˌar-ə-'ni-kə-ləs\ *adjective* [Latin *arena* + English *-i-* + *-colous*] (circa 1859)
: living, burrowing, or growing in sand
aren't \'ärnt, 'är-ənt\ (1810)
1 : are not
2 : am not — used in questions
ar·eo·cen·tric \ˌar-ē-ō-'sen-trik\ *adjective* [Greek *Areios* of Ares, from *Arēs*] (1877)
: having or relating to the planet Mars as a center
are·o·la \ə-'rē-ə-lə\ *noun, plural* **-lae** \-ˌlē\ *or* **-las** [New Latin, from Latin, small open space, diminutive of *area*] (1664)
: a small area between things or about something; *especially* **:** a colored ring (as about the nipple, a vesicle, or a pustule)
— are·o·lar \-lər\ *adjective*
— are·o·late \-lət\ *adjective*
ar·e·ole \'ar-ē-ˌōl\ *noun* (circa 1934)
: a small pit or cavity
Ar·e·op·a·gite \ˌar-ē-'ä-pə-ˌgīt, -ˌjīt\ *noun* (14th century)
: a member of the Areopagus
— Ar·e·op·a·git·ic \-ˌä-pə-'ji-tik\ *adjective*
Ar·e·op·a·gus \ˌar-ē-'ä-pə-gəs\ *noun* [Latin, from Greek *Areios pagos,* from *Areios pagos* (literally, hill of Ares), a hill in Athens where the tribunal met] (1586)
: the supreme tribunal of Athens
Ar·es \'ar-(ˌ)ēz, 'er-\ *noun* [Greek *Arēs*]
: the Greek god of war — compare MARS
arête \ə-'rāt\ *noun* [French, literally, fish bone, from Late Latin *arista,* from Latin, beard of grain] (1838)
: a sharp-crested ridge in rugged mountains
Ar·e·thu·sa \ˌar-ə-'thü-zə, -'thyü-\ *noun* [Latin, from Greek *Arethousa*]
: a wood nymph who is changed into a spring while fleeing the advances of the river-god Alpheus
ar·ga·li \'är-gə-lē\ *noun* [Mongolian] (circa 1774)
: a large wild sheep (*Ovis ammon*) of Asia that is noted for its large horns
Ar·gand diagram \ˌär-'gän-, -'gan-\ *noun* [Jean Robert *Argand* (died 1825) Swiss mathematician] (1908)
: a system of rectangular coordinates in which the complex number $x + iy$ is represented by the point whose coordinates are x and y
ar·gent \'är-jənt\ *noun* [Middle English, from Middle French & Latin; Middle French, from Latin *argentum;* akin to Greek *argyros* silver, *argos* white, Sanskrit *rajata* whitish, silvery] (15th century)
1 *archaic* **:** the metal silver; *also* **:** WHITENESS
2 : the heraldic color silver or white
— argent *adjective*
ar·gen·tif·er·ous \ˌär-jən-'ti-f(ə-)rəs\ *adjective* (1801)
: containing silver
¹ar·gen·tine \'är-jən-ˌtīn, -ˌtēn\ *adjective* (15th century)
: SILVER, SILVERY
²argentine *noun* (1577)
: SILVER; *also* **:** any of various similar materials
ar·gen·tite \'är-jən-ˌtīt\ *noun* (circa 1837)
: native silver sulfide having a metallic luster and dark lead-gray color and constituting a valuable ore of silver
ar·gil \'är-jəl\ *noun* [Middle English, from Latin *argilla,* from Greek *argillos;* akin to Greek *argos* white] (14th century)
: CLAY; *especially* **:** POTTER'S CLAY
ar·gil·la·ceous \ˌär-jə-'lā-shəs\ *adjective* (circa 1731)
: of, relating to, or containing clay or clay minerals **:** CLAYEY
ar·gil·lite \'är-jə-ˌlīt\ *noun* (1795)
: a compact argillaceous rock cemented by silica and having no slaty cleavage
ar·gi·nase \'är-jə-ˌnās, -ˌnāz\ *noun* [International Scientific Vocabulary] (1904)
: a crystalline enzyme that converts naturally occurring arginine into ornithine and urea
ar·gi·nine \'är-jə-ˌnēn\ *noun* [German *Arginin*] (1886)
: a crystalline basic amino acid $C_6H_{14}N_4O_2$ derived from guanidine

Ar·give \'är-ˌjīv, -ˌgīv\ *adjective* [Latin *Argivus,* from Greek *Argeios,* literally, of Argos, from *Argos* city-state of ancient Greece] (1598)
: of or relating to the Greeks or Greece and especially the Achaean city of Argos or the surrounding territory of Argolis
— Argive *noun*

ar·gle–bar·gle \ˌär-gəl-'bär-gəl\ *noun* [reduplication of Scots & English *argle,* alteration of *argue*] (1872)
chiefly British **:** ARGY-BARGY

Ar·go \'är-(ˌ)gō\ *noun* [Latin (genitive *Argus*), from Greek *Argō*]
: a large constellation in the southern hemisphere lying principally between Canis Major and the Southern Cross

ar·gol \'är-ˌgȯl\ *noun* [Middle English *argoile,* from Anglo-French *argoil*] (14th century)
: crude tartar deposited in wine casks during aging

ar·gon \'är-ˌgän\ *noun* [Greek, neuter of *argos* idle, lazy, from *a-* + *ergon* work; from its relative inertness — more at WORK] (1894)
: a colorless odorless inert gaseous element found in the air and in volcanic gases and used especially in welding, lasers, and electric bulbs — see ELEMENT table

ar·go·naut \'är-gə-ˌnȯt, -ˌnät\ *noun* [Latin *Argonautes,* from Greek *Argonautēs,* from *Argō,* ship in which the Argonauts sailed + *nautēs* sailor — more at NAUTICAL] (14th century)
1 a *capitalized* **:** any of a band of heroes sailing with Jason in quest of the Golden Fleece **b :** an adventurer engaged in a quest
2 : PAPER NAUTILUS

ar·go·sy \'är-gə-sē\ *noun, plural* **-sies** [modification of Italian *ragusea* Ragusan vessel, from *Ragusa,* Dalmatia (now Dubrovnik, Croatia)] (1587)
1 : a large ship; *especially* **:** a large merchant ship ⟨three of your *argosies* are . . . come to harbor —Shakespeare⟩
2 : a fleet of ships
3 : a rich supply ⟨an *argosy* of railway folklore —F. P. Donovan⟩

ar·got \'är-gət, -(ˌ)gō\ *noun* [French] (1860)
: an often more or less secret vocabulary and idiom peculiar to a particular group ⟨shoved into a taxi by a porter whose *argot* I couldn't understand —Allen Tate⟩

ar·gu·able \'är-gyü-ə-bəl\ *adjective* (circa 1611)
1 : open to argument, dispute, or question
2 : that can be plausibly or convincingly argued

ar·gu·ably \'är-gyü-(ə-)blē\ *adverb* (1890)
: it can be argued ⟨the word is *arguably* useful⟩ ⟨*arguably* the busiest airport in the world⟩

ar·gue \'är-(ˌ)gyü\ *verb* **ar·gued; ar·gu·ing** [Middle English, from Middle French *arguer* to accuse, reason & Latin *arguere* to demonstrate, prove; Middle French *arguer,* from Latin *argutare* to prate, frequentative of *arguere;* akin to Hittite *arkuwai-* to plead, respond] (14th century)
intransitive verb
1 : to give reasons for or against something **:** REASON
2 : to contend or disagree in words **:** DISPUTE
transitive verb
1 : to give evidence of **:** INDICATE
2 : to consider the pros and cons of **:** DISCUSS
3 : to prove or try to prove by giving reasons **:** MAINTAIN
4 : to persuade by giving reasons **:** INDUCE
synonym see DISCUSS
— ar·gu·er \-gyə-wər\ *noun*

ar·gu·fy \'är-gyə-ˌfī\ *verb* **-fied; -fy·ing** (1771)
transitive verb
: DISPUTE, DEBATE
intransitive verb
: WRANGLE
— ar·gu·fi·er \-ˌfī(-ə)r\ *noun*

ar·gu·ment \'är-gyə-mənt\ *noun* [Middle English, from Middle French, from Latin *argumentum,* from *arguere*] (14th century)
1 *obsolete* **:** an outward sign **:** INDICATION
2 a : a reason given in proof or rebuttal **b :** discourse intended to persuade
3 a : the act or process of arguing **:** ARGUMENTATION **b :** a coherent series of statements leading from a premise to a conclusion **c :** QUARREL, DISAGREEMENT
4 : an abstract or summary especially of a literary work ⟨a later editor added an *argument* to the poem⟩
5 : the subject matter especially of a literary work
6 a : one of the independent variables upon whose value that of a function depends **b :** a substantive (as the direct object of a transitive verb) that is required by a predicate in grammar **c :** the angle assigned to a complex number when it is plotted in a complex plane using polar coordinates — called also *amplitude;* compare ABSOLUTE VALUE 2

ar·gu·men·ta·tion \ˌär-gyə-mən-'tā-shən, -ˌmen-\ *noun* (15th century)
1 : the act or process of forming reasons and of drawing conclusions and applying them to a case in discussion
2 : DEBATE, DISCUSSION

ar·gu·men·ta·tive \ˌär-gyə-'men-tə-tiv\ *also* **ar·gu·men·tive** \-'men-tiv\ *adjective* (15th century)
1 : characterized by argument **:** CONTROVERSIAL
2 : given to argument **:** DISPUTATIOUS
— ar·gu·men·ta·tive·ly *adverb*

ar·gu·men·tum \ˌär-gyə-'men-təm\ *noun, plural* **-men·ta** \-'men-tə\ [Latin] (1690)
: ARGUMENT 3b

Ar·gus \'är-gəs\ *noun* [Latin, from Greek *Argos*]
1 : a hundred-eyed monster of Greek mythology
2 : a watchful guardian

Ar·gus–eyed \'är-gəs-ˌīd\ *adjective* (1603)
: vigilantly observant

ar·gy–bar·gy \ˌär-jē-'bär-jē, ˌär-gē-'bär-gē\ *noun* [reduplication of Scots & English dialect *argy,* alteration of *argue*] (1887)
chiefly British **:** a lively discussion **:** ARGUMENT, DISPUTE

ar·gyle *also* **ar·gyll** \'är-ˌgī(ə)l, är-'\ *noun, often capitalized* [*Argyle, Argyll,* branch of the Scottish clan of Campbell, from whose tartan the design was adapted] (1899)
: a geometric knitting pattern of varicolored diamonds in solid and outline shapes on a single background color; *also* **:** a sock knit in this pattern

argyle

ar·hat \'är-(ˌ)hət\ *noun* [Sanskrit, from present participle of *arhati* he deserves; akin to Greek *alphein* to gain] (1870)
: a Buddhist who has reached the stage of enlightenment
— ar·hat·ship \-ˌship\ *noun*

aria \'är-ē-ə\ *noun* [Italian, literally, atmospheric air, modification of Latin *aer*] (circa 1724)
1 : AIR, MELODY, TUNE; *specifically* **:** an accompanied elaborate melody sung (as in an opera) by a single voice
2 : a striking solo performance (as in a movie)

Ar·i·ad·ne \ˌar-ē-'ad-nē\ *noun* [Latin, from Greek *Ariadnē*]
: a daughter of Minos who helps Theseus escape from the labyrinth

^{1}**Ar·i·an** \'ar-ē-ən, 'er-\ *adjective* (14th century)
: of or relating to Arius or his doctrines especially that the Son is not of the same substance as the Father but was created as an agent for creating the world
— Ar·i·an·ism \-ə-ˌni-zəm\ *noun*

^{2}**Arian** *noun* (14th century)
: a supporter of Arian doctrines

^{3}**Ar·i·an** \'er-ē-ən, 'ar-\ *noun* (1917)
: ARIES 2b

-ar·i·an *noun suffix* [Latin *-arius* -ary]
1 : believer ⟨necessit*arian*⟩ **:** advocate ⟨latitudin*arian*⟩
2 : producer ⟨disciplin*arian*⟩

ari·bo·fla·vin·osis \ˌā-ˌrī-bə-ˌflā-və-'nō-səs\ *noun* [New Latin] (1939)
: a deficiency disease due to inadequate intake of riboflavin and characterized by sores on the mouth

ar·id \'ar-əd\ *adjective* [French or Latin; French *aride,* from Latin *aridus,* from *arēre* to be dry; akin to Sanskrit *āsa* ash, Old English *asce*] (1652)
1 : excessively dry; *specifically* **:** having insufficient rainfall to support agriculture
2 : lacking in interest and life **:** JEJUNE
— arid·i·ty \ə-'ri-də-tē, a-\ *noun*
— ar·id·ness \'ar-əd-nəs\ *noun*

Ar·i·el \'ar-ē-əl, 'er-\ *noun*
: a prankish spirit in Shakespeare's *The Tempest*

Ar·i·es \'er-ˌēz, 'ar-, -ē-ˌēz\ *noun* [Latin (genitive *Arietis*), literally, ram; perhaps akin to Greek *eriphos* kid, Old Irish *heirp* she-goat]
1 : a constellation between Pisces and Taurus pictured as a ram
2 a : the first sign of the zodiac in astrology — see ZODIAC table **b :** one born under the sign of Aries

ari·et·ta \ˌär-ē-'e-tə, ˌar-\ *noun* [Italian, diminutive of *aria*] (circa 1724)
: a short aria

aright \ə-'rīt\ *adverb* [Middle English, from Old English *ariht,* from ^{1}*a-* + *riht* right] (before 12th century)
: RIGHT, CORRECTLY ⟨if I remember *aright*⟩

Arik·a·ra \ə-'ri-kə-rə\ *noun, plural* **Arikara** (1811)
1 : a member of an American Indian people of the Missouri River Valley in North Dakota
2 : the language of the Arikara

ar·il \'ar-əl\ *noun* [probably from New Latin *arillus,* from Medieval Latin, raisin, grape seed] (1794)
: an exterior covering or appendage of some seeds (as of the yew) that develops after fertilization as an outgrowth from the ovule stalk
— ar·il·late \'ar-ə-ˌlāt\ *adjective*

ari·o·so \ˌär-ē-'ō (ˌ)sō, -(ˌ)zō\ *noun, plural* **-sos** *also* **-si** \-(ˌ)sē, -(ˌ)zē\ [Italian, from *aria*] (circa 1724)
: a musical passage or composition having a mixture of free recitative and metrical song

arise \ə-'rīz\ *intransitive verb* **arose** \-'rōz\; **aris·en** \-'ri-zᵊn\; **aris·ing** \-'rī-ziŋ\ [Middle English, from Old English *ārīsan,* from *ā-,* perfective prefix + *rīsan* to rise — more at ABIDE] (before 12th century)
1 : to get up **:** RISE
2 a : to originate from a source **b :** to come into being or to attention
3 : ASCEND
synonym see SPRING

aris·ta \ə-'ris-tə\ *noun, plural* **-tae** \-(ˌ)tē, -ˌtī\ *or* **-tas** [New Latin, from Latin, beard of grain] (1691)
: a bristlelike structure or appendage
— aris·tate \-ˌtāt\ *adjective*

aris·to \ə-'ris-(ˌ)tō\ *noun, plural* **-tos** [by shortening] (1864)
: ARISTOCRAT

ar·is·toc·ra·cy \ˌar-ə-'stä-krə-sē\ *noun, plural* **-cies** [Middle French & Late Latin; Middle French *aristocratie,* from Late Latin *aristocratia,* from Greek *aristokratia,* from *aristos* best + *-kratia* -cracy] (1561)
1 : government by the best individuals or by a small privileged class

2 a : a government in which power is vested in a minority consisting of those believed to be best qualified **b :** a state with such a government
3 : a governing body or upper class usually made up of an hereditary nobility
4 : the aggregate of those believed to be superior

aris·to·crat \ə-ˈris-tə-ˌkrat, a-; ˈar-ə-stə-\ *noun* (1789)
1 : a member of an aristocracy; *especially* **:** NOBLE
2 a : one who has the bearing and viewpoint typical of the aristocracy **b :** one who favors aristocracy
3 : one believed to be superior of its kind ⟨the *aristocrat* of Southern resorts —*Southern Living*⟩

aris·to·crat·ic \ə-ˌris-tə-ˈkra-tik, (ˌ)a-ˌris-tə-, ˌar-ə-stə-\ *adjective* [Middle French *aristocratique,* from Medieval Latin *aristocraticus,* from Greek *aristokratikos,* from *aristokratia*] (1602)
1 : belonging to, having the qualities of, or favoring aristocracy
2 a : socially exclusive ⟨an *aristocratic* neighborhood⟩ **b :** SNOBBISH
— aris·to·crat·i·cal·ly \-ti-k(ə-)lē\ *adverb*

Ar·is·to·te·lian *also* **Ar·is·to·te·lean** \ˌar-ə-stə-ˈtēl-yən\ *adjective* [Latin *Aristoteles* Aristotle, from Greek *Aristotelēs*] (1607)
: of or relating to the Greek philosopher Aristotle or his philosophy
— Aristotelian *noun*
— Ar·is·to·te·lian·ism \-yə-ˌni-zəm\ *noun*

arith·me·tic \ə-ˈrith-mə-ˌtik\ *noun* [Middle English *arsmetrik,* from Old French *arismetique,* from Latin *arithmetica,* from Greek *arithmētikē,* from feminine of *arithmētikos* arithmetical, from *arithmein* to count, from *arithmos* number; akin to Old English *rīm* number, and perhaps to Greek *arariskein* to fit] (15th century)
1 a : a branch of mathematics that deals usually with the nonnegative real numbers including sometimes the transfinite cardinals and with the application of the operations of addition, subtraction, multiplication, and division to them **b :** a treatise on arithmetic
2 : COMPUTATION, CALCULATION
— ar·ith·met·ic \ˌar-ith-ˈme-tik\ *or* **ar·ith·met·i·cal** \-ti-kəl\ *adjective*
— ar·ith·met·i·cal·ly \-ti-k(ə-)lē\ *adverb*
— arith·me·ti·cian \ə-ˌrith-mə-ˈti-shən\ *noun*

arithmetic mean *noun* (1767)
: a value that is computed by dividing the sum of a set of terms by the number of terms

arithmetic progression *noun* (1594)
: a progression (as 3, 5, 7, 9) in which the difference between any term and its predecessor is constant

-arium *noun suffix, plural* **-ariums** *or* **-aria** [Latin, from neuter of *-arius* -ary]
: thing or place relating to or connected with ⟨planet*arium*⟩

ark \ˈärk\ *noun* [Middle English, from Old English *arc,* from Latin *arca* chest; akin to Latin *arcēre* to hold off, defend, Greek *arkein,* Hittite *ḫark-* to have, hold] (before 12th century)
1 a : a boat or ship held to resemble that in which Noah and his family were preserved from the Flood **b :** something that affords protection and safety
2 a : the sacred chest representing to the Hebrews the presence of God among them **b :** a repository traditionally in or against the wall of a synagogue for the scrolls of the Torah

ar·kose \ˈär-ˌkōs, -ˌkōz\ *noun* [French] (1839)
: a sandstone characterized by feldspar fragments that is derived from granite or gneiss which has disintegrated rapidly
— ar·ko·sic \(ˌ)är-ˈkō-sik, -zik\ *adjective*

¹arm \ˈärm\ *noun* [Middle English, from Old English *earm;* akin to Latin *armus* shoulder, Sanskrit *īrma* arm] (before 12th century)
1 : a human upper limb; *especially* **:** the part between the shoulder and the wrist
2 : something like or corresponding to an arm: as **a :** the forelimb of a vertebrate **b :** a limb of an invertebrate animal **c :** a branch or lateral shoot of a plant **d :** a slender part of a structure, machine, or an instrument projecting from a main part, axis, or fulcrum **e :** the end of a ship's yard; *also* **:** the part of an anchor from the crown to the fluke — see ANCHOR illustration **f :** any of the usually two parts of a chromosome lateral to the centromere
3 : an inlet of water (as from the sea)
4 : a narrow extension of a larger area, mass, or group
5 : POWER, MIGHT ⟨the long *arm* of the law⟩
6 : a support (as on a chair) for the elbow and forearm
7 : SLEEVE
8 : the ability to throw or pitch a baseball well; *also* **:** a player having such ability
9 : a functional division of a group, organization, institution, or activity ⟨the logistical *arm* of the air force⟩
— arm·less \ˈärm-ləs\ *adjective*
— arm·like \-ˌlīk\ *adjective*
— arm in arm : with arms linked together ⟨walked down the street *arm in arm*⟩

²arm *verb* [Middle English *armen,* from Old French *armer,* from Latin *armare,* from *arma* weapons, tools; akin to Latin *ars* skill, Greek *harmos* joint, *arariskein* to fit] (12th century)
transitive verb
1 : to furnish or equip with weapons
2 : to furnish with something that strengthens or protects
3 : to fortify morally
4 : to equip or ready for action or operation ⟨*arm* a bomb⟩
intransitive verb
: to prepare oneself for struggle or resistance

³arm *noun, often attributive* [Middle English *armes* (plural) weapons, from Old French, from Latin *arma*] (13th century)
1 a : a means (as a weapon) of offense or defense; *especially* **:** FIREARM **b :** a combat branch (as of an army) **c :** an organized branch of national defense (as the navy)
2 *plural* **a :** the hereditary heraldic devices of a family **b :** heraldic devices adopted by a government
3 *plural* **a :** active hostilities **:** WARFARE **b :** military service
— up in arms : aroused and ready to undertake hostilities

ar·ma·da \är-ˈmä-də, -ˈmā- *also* -ˈma-\ *noun* [Spanish, from Medieval Latin *armata* army, fleet, from Latin, feminine of *armatus,* past participle of *armare* to arm, from *arma*] (1533)
1 : a fleet of warships
2 : a large force or group usually of moving things ⟨an *armada* of fishing boats⟩

ar·ma·dil·lo \ˌär-mə-ˈdi-(ˌ)lō\ *noun, plural* **-los** [Spanish, from diminutive of *armado* armed one, from Latin *armatus*] (1577)
: any of a family (Dasypodidae) of burrowing edentate mammals found from the southern U.S. to Argentina and having the body and head encased in an armor of small bony plates

armadillo

Ar·ma·ged·don \ˌär-mə-ˈge-dᵊn\ *noun* [Greek *Armageddōn, Harmagedōn,* scene of the battle foretold in Revelation 16:14–16] (1534)
1 a : the site or time of a final and conclusive battle between the forces of good and evil **b :** the battle taking place at Armageddon
2 : a usually vast decisive conflict or confrontation

Ar·ma·gnac \ˈär-mən-ˌyak\ *noun* [French, from *Armagnac,* region in southwest France] (1850)
: a brandy produced in the Gers department of France

ar·ma·ment \ˈär-mə-mənt *also* ˈärm-mənt\ *noun* [French *armement,* from Latin *armamenta* (plural) utensils, military or naval equipment, from *armare*] (1699)
1 : a military or naval force
2 a : the aggregate of a nation's military strength **b :** WEAPONS, ARMS
3 : the process of preparing for war

ar·ma·men·tar·i·um \ˌär-mə-ˌmen-ˈter-ē-əm, -mən-\ *noun, plural* **-tar·ia** \-ē-ə\ [Latin, armory, from *armamenta*] (circa 1860)
1 : the equipment and methods used especially in medicine
2 : matter available or utilized for an undertaking or field of activity ⟨a whole *armamentarium* of devices to create an illusion of real life —Kenneth Rexroth⟩

arm and a leg *noun* (1967)
: an exorbitant price

ar·ma·ture \ˈär-mə-ˌchu̇r, -chər, -ˌtyu̇r, -ˌtu̇r\ *noun* [Latin *armatura* armor, equipment, from *armatus*] (15th century)
1 : an organ or structure (as teeth or thorns) for offense or defense
2 a : a piece of soft iron or steel that connects the poles of a magnet or of adjacent magnets **b :** a usually rotating part of an electric machine (as a generator or motor) which consists essentially of coils of wire around a metal core and in which electric current is induced or in which the input current interacts with a magnetic field to produce torque **c :** the movable part of an electromagnetic device (as a loudspeaker) **d :** a framework used by a sculptor to support a figure being modeled in a plastic material **e :** FRAMEWORK 1b ⟨the *armature* of the book derives from fourteenth century England —Stanley Kauffmann⟩

arm·band \ˈärm-ˌband\ *noun* (1797)
: a band usually worn around the upper part of a sleeve for identification or in mourning

¹arm·chair \ˈärm-ˌcher, -ˌchar\ *noun* (1633)
: a chair with armrests

²armchair *adjective* (1886)
1 : remote from direct dealing with problems **:** theoretical rather than practical ⟨*armchair* strategists⟩
2 : sharing vicariously in another's experiences ⟨an *armchair* traveler⟩

¹armed \ˈärmd\ *adjective* (13th century)
1 a : furnished with weapons **b :** furnished with something that provides security, strength, or efficacy
2 : marked by the maintenance of armed forces in readiness

²armed *adjective* (1606)
: having an arm or arms especially of a specified kind or number — usually used in combination ⟨long-*armed*⟩ ⟨two-*armed*⟩

armed forces *noun plural* (1943)
: the combined military, naval, and air forces of a nation — called also *armed services*

Ar·me·nian \är-ˈmē-nē-ən, -nyən\ *noun* (1537)
1 : a member of a people dwelling chiefly in Armenia and neighboring areas (as Turkey or Azerbaijan)
2 : the Indo-European language of the Armenians — see INDO-EUROPEAN LANGUAGES table
3 : a member of the Armenian church established by Saint Gregory the Illuminator that adheres to the decisions of the first three ecumenical councils
— Armenian *adjective*

arm·ful \ˈärm-ˌfu̇l\ *noun, plural* **arm·fuls** \-ˌfu̇lz\ *or* **arms·ful** \ˈärmz-ˌfu̇l\ (1579)
: as much as the arm or arms can hold

arm·hole \ˈärm-ˌhōl\ *noun* (circa 1775)
: an opening for the arm in a garment

ar·mi·ger \ˈär-mi-jər\ *noun* [Medieval Latin,

from Latin, armor-bearer, from *armiger* bearing arms, from *arma* arms + *gerere* to carry] (1762)
1 : SQUIRE
2 : one entitled to bear heraldic arms
— ar·mig·er·al \är-ˈmi-jə-rəl\ *adjective*

ar·mig·er·ous \är-ˈmi-jər-əs\ *adjective* (circa 1731)
: bearing heraldic arms

ar·mil·la·ry sphere \ˈär-mə-ˌler-ē-, är-ˈmi-lə-rē-\ *noun* [French *sphère armillaire,* from Medieval Latin *armilla,* from Latin, bracelet, iron ring, from *armus* shoulder — more at ARM] (1664)
: an old astronomical instrument composed of rings showing the positions of important circles of the celestial sphere

Ar·min·i·an \är-ˈmi-nē-ən\ *adjective* (1618)
: of or relating to Arminius or his doctrines opposing the absolute predestination of strict Calvinism and maintaining the possibility of salvation for all
— Arminian *noun*
— Ar·min·i·an·ism \-nē-ə-ˌni-zəm\ *noun*

ar·mi·stice \ˈär-mə-stəs\ *noun* [French or New Latin; French, from New Latin *armistitium,* from Latin *arma* + *-stitium* (as in *solstitium* solstice)] (circa 1707)
: temporary suspension of hostilities by agreement between the opponents **:** TRUCE

Armistice Day *noun* [from the armistice terminating World War I on November 11, 1918] (1919)
: VETERANS DAY — used before the official adoption of *Veterans Day* in 1954

arm·let \ˈärm-lət\ *noun* (1535)
1 : a band (as of cloth or metal) worn around the upper arm
2 : a small arm (as of the sea)

arm·load \ˈärm-ˌlōd\ *noun* (1906)
: ARMFUL

arm·lock \-ˌläk\ *noun* (1905)
: HAMMERLOCK

ar·moire \ärm-ˈwär, *Southern also* ˈär-mər\ *noun* [Middle French, from Old French *armaire,* from Latin *armarium,* from *arma*] (1571)
: a usually tall cupboard or wardrobe

ar·mor \ˈär-mər\ *noun* [Middle English *armure,* from Old French, from Latin *armatura* — more at ARMATURE] (13th century)
1 : defensive covering for the body; *especially* **:** covering (as of metal) used in combat
2 : a quality or circumstance that affords protection ⟨the *armor* of prosperity⟩
3 : a protective outer layer (as of a ship, a plant or animal, or a cable)
4 : armored forces and vehicles (as tanks)
— armor *transitive verb*
— ar·mored \-mərd\ *adjective*
— ar·mor·less \-mər-ləs\ *adjective*

armor 1: *1* **helmet,** *2* **gorget,** *3* **shoulder piece,** *4* **pallette,** *5* **breastplate,** *6* **brassard,** *7* **elbow piece,** *8* **skirt of tasses,** *9* **tuille,** *10* **gauntlet,** *11* **cuisse,** *12* **knee piece,** *13* **jambeau,** *14* **solleret**

armored scale *noun* (circa 1903)
: any of a family (Diaspididae) of scale insects having a firm covering of wax best developed in the female

ar·mor·er \ˈär-mər-ər\ *noun* (14th century)
1 : one that makes armor or arms
2 : one that repairs, assembles, and tests firearms

ar·mo·ri·al \är-ˈmōr-ē-əl, -ˈmȯr-\ *adjective* [*armory* (heraldry)] (1576)
: of, relating to, or bearing heraldic arms
— ar·mo·ri·al·ly \-ē-ə-lē\ *adverb*

Ar·mor·i·can \är-ˈmȯr-i-kən, -ˈmär-\ *or* **Ar·mor·ic** \-ik\ *noun* (circa 1645)
: a native or inhabitant of Armorica; *especially* **:** BRETON
— Armorican *or* **Armoric** *adjective*

ar·mory \ˈärm-rē, ˈär-mə-\ *noun, plural* **ar·mor·ies** (14th century)
1 a : a supply of arms for defense or attack **b :** a collection of available resources
2 : a place where arms and military equipment are stored; *especially* **:** one used for training reserve military personnel
3 : a place where arms are manufactured

ar·mour, ar·moury *chiefly British variant of* ARMOR, ARMORY

arm·pit \ˈärm-ˌpit\ *noun* (14th century)
1 : the hollow beneath the junction of the arm and shoulder
2 : the least desirable place **:** PIT ⟨77th Street Station . . . was the *armpit* of detective duty —Joseph Wambaugh⟩

arm·rest \-ˌrest\ *noun* (circa 1889)
: a support for the arm

arm's length *noun* (circa 1909)
1 : a distance discouraging personal contact or familiarity ⟨kept people at *arm's length*⟩
2 : the condition or fact that the parties to a transaction are independent and on an equal footing
— arm's–length *adjective*

arm–twist·ing \-ˌtwis-tiŋ\ *noun* (1948)
: the use of direct personal pressure in order to achieve a desired end ⟨for all the *arm-twisting,* the . . . vote on the measure was unexpectedly tight —*Newsweek*⟩
— arm–twist \-ˌtwist\ *verb*

arm wrestling *noun* (1973)
: a form of wrestling in which two opponents sit face to face gripping usually their right hands, set corresponding elbows firmly on a surface (as a tabletop), and attempt to force each other's arm down — called also *Indian wrestling*

ar·my \ˈär-mē\ *noun, plural* **armies** [Middle English *armee,* from Middle French, from Medieval Latin *armata* — more at ARMADA] (14th century)
1 a : a large organized body of armed personnel trained for war especially on land **b :** a unit capable of independent action and consisting usually of a headquarters, two or more corps, and auxiliary troops **c** *often capitalized* **:** the complete military organization of a nation for land warfare
2 : a great multitude ⟨an *army* of bicycles —Norm Fruchter⟩
3 : a body of persons organized to advance a cause

army ant *noun* (1874)
: any of a subfamily (Dorylinae) of aggressive nomadic tropical ants that prey on insects and spiders

ar·my·worm \ˈär-mē-ˌwərm\ *noun* (1816)
: any of numerous moths whose larvae travel in multitudes from field to field destroying grass, grain, and other crops; *especially* **:** the common armyworm (*Pseudaletia unipuncta*) of the northern U.S.

ar·ni·ca \ˈär-ni-kə\ *noun* [New Latin] (circa 1753)
: any of a genus (*Arnica*) of composite herbs including some with bright yellow ray flowers

ar·oid \ˈar-ˌȯid, ˈer-\ *adjective* [New Latin *Arum*] (circa 1890)
: of or relating to the arum family
— aroid *noun*

aroint \ə-ˈrȯint\ *verb imperative* [origin unknown] (1605)
archaic **:** BEGONE ⟨*aroint* thee, witch —Shakespeare⟩

aro·ma \ə-ˈrō-mə\ *noun* [Middle English *aromat* spice, from Old French, from Latin *aromat-, aroma,* from Greek *arōmat-, arōma*] (1814)
1 a : a distinctive pervasive and usually pleasant or savory smell; *broadly* **:** ODOR **b :** the odor of a wine imparted by the grapes from which it is made
2 : a distinctive quality or atmosphere **:** FLAVOR ⟨the *aroma* of enjoyment —Stella D. Gibbons⟩
synonym see SMELL

aro·ma·ther·a·py \ə-ˌrō-mə-ˈther-ə-pē\ *noun* [French *aromathérapie,* from Latin *aroma* + French *thérapie* therapy] (1969)
: massage of the body and especially of the face with a preparation of fragrant essential oils extracted from herbs, flowers, and fruits
— aro·ma·ther·a·pist \-pist\ *noun*

[1]**ar·o·mat·ic** \ˌar-ə-ˈma-tik\ *adjective* (14th century)
1 : of, relating to, or having aroma: **a :** FRAGRANT **b :** having a strong smell **c :** having a distinctive quality
2 *of an organic compound* **:** characterized by increased chemical stability resulting from the delocalization of electrons in a ring system (as benzene) containing usually multiple conjugated double bonds — compare ALICYCLIC, ALIPHATIC
synonym see ODOROUS
— ar·o·mat·i·cal·ly \-ti-k(ə-)lē\ *adverb*
— ar·o·ma·tic·i·ty \ˌar-ə-mə-ˈti-sə-tē, ə-ˌrō-mə-\ *noun*

[2]**aromatic** *noun* (15th century)
1 : an aromatic plant or plant part; *especially* **:** an aromatic herb or spice
2 : an aromatic organic compound

aro·ma·tize \ə-ˈrō-mə-ˌtīz\ *transitive verb* **-tized; -tiz·ing** (15th century)
1 : to make aromatic **:** FLAVOR
2 : to convert into one or more aromatic compounds
— aro·ma·ti·za·tion \-ˌrō-mə-tə-ˈzā-shən\ *noun*

arose *past of* ARISE

[1]**around** \ə-ˈrau̇nd\ *adverb* [Middle English, from [1]*a-* + [2]*round*] (14th century)
1 a : in a circle or in circumference ⟨the wheel goes *around*⟩ ⟨a tree five feet *around*⟩ **b :** in, along, or through a circuit ⟨the road goes *around* by the lake⟩
2 a : on all or various sides **:** in every or any direction ⟨papers lying *around*⟩ ⟨nothing for miles *around*⟩ **b :** in close from all sides so as to surround ⟨people crowded *around*⟩ **c :** in or near one's present place or situation ⟨wait *around* awhile⟩
3 a : here and there **:** from one place to another ⟨travels *around* on business⟩ **b :** to a particular place ⟨come *around* for dinner⟩
4 a : in rotation or succession ⟨another winter comes *around*⟩ **b :** from beginning to end **:** THROUGH ⟨mild the year *around*⟩ **c :** in order ⟨the other way *around*⟩
5 : in or to an opposite direction or position ⟨turn *around*⟩
6 : with some approach to exactness **:** APPROXIMATELY ⟨cost *around* $5⟩

[2]**around** *preposition* (14th century)
1 a : on all sides of **b :** so as to encircle or enclose ⟨seated *around* the table⟩ **c :** so as to avoid or get past **:** on or to another side of ⟨find a way *around* their objections⟩ ⟨went *around* the lake⟩ ⟨*around* the corner⟩ **d :** NEAR ⟨lives *around* Chicago⟩ ⟨*around* the turn of the century⟩
2 : in all directions outward from ⟨look *around* you⟩
3 : here and there in or throughout ⟨barnstorming *around* the country⟩
4 : so as to have a center or basis in ⟨a society organized *around* kinship ties⟩

[3]**around** *adjective* (1849)
1 : ABOUT 1 ⟨has been up and *around* for two days⟩

2 : being in existence, evidence, or circulation ⟨the most intelligent of the artists *around* today —R. M. Coates⟩
— been around : gone through many varied experiences **:** become worldly-wise
around–the–clock *adjective* (1943)
: being in effect, continuing, or lasting 24 hours a day **:** CONSTANT
arouse \ə-ˈrau̇z\ *verb* **aroused; arous·ing** [*a-* (as in *arise*) + *rouse*] (1593)
transitive verb
1 : to awaken from sleep
2 : to rouse or stimulate to action or to physiological readiness for activity **:** EXCITE ⟨the book *aroused* debate⟩
intransitive verb
: to awake from sleep **:** STIR
— arous·al \ə-ˈrau̇-zəl\ *noun*
ar·peg·gi·ate \är-ˈpe-jē-ˌāt\ *transitive verb* **-at·ed; -at·ing** (1953)
: to play (as a chord or passage) in arpeggio
ar·peg·gio \är-ˈpe-jē-ˌō, -ˈpe-jō\ *noun, plural* **-gios** [Italian, from *arpeggiare* to play on the harp, from *arpa* harp, of Germanic origin; akin to Old High German *harpha* harp] (circa 1724)
1 : production of the tones of a chord in succession and not simultaneously
2 : a chord played in arpeggio
ar·pent \är-ˈpäⁿ\ *noun, plural* **ar·pents** \-ˈpäⁿ(z)\ [Middle French] (1580)
1 : any of various old French units of land area; *especially* **:** one used in French sections of Canada and the U.S. equal to about 0.85 acre (0.34 hectare)
2 : a unit of length equal to one side of a square arpent
ar·que·bus \ˈär-kwi-(ˌ)bəs, -kə-bəs\ *variant of* HARQUEBUS
ar·rack \ˈar-ək, ə-ˈrak\ *noun* [Arabic *ʽaraq* sweet juice, liquor] (1521)
: an Asian alcoholic beverage like rum that is distilled from a fermented mash of malted rice with toddy or molasses
ar·raign \ə-ˈrān\ *transitive verb* [Middle English *arreinen,* from Middle French *araisner,* from Old French, from *a-* (from Latin *ad-*) + *raisnier* to speak, from (assumed) Vulgar Latin *rationare,* from Latin *ration-, ratio* reason — more at REASON] (14th century)
1 : to call (a defendant) before a court to answer to an indictment **:** CHARGE
2 : to accuse of wrong, inadequacy, or imperfection
— ar·raign·ment \-mənt\ *noun*
ar·range \ə-ˈrānj\ *verb* **-ranged; -rang·ing** [Middle English *arangen,* from Middle French *arangier,* from Old French, from *a-* + *rengier* to set in a row, from *reng* row — more at RANK] (1789)
transitive verb
1 : to make preparations for **:** PLAN ⟨*arranged* a reception for the visitor⟩
2 : to put into a proper order or into a correct or suitable sequence, relationship, or adjustment ⟨*arrange* flowers in a vase⟩ ⟨*arrange* cards alphabetically⟩
3 a : to adapt (a musical composition) by scoring for voices or instruments other than those for which originally written **b :** ORCHESTRATE
4 : to bring about an agreement or understanding concerning **:** SETTLE ⟨*arrange* an exchange of war prisoners⟩
intransitive verb
1 : to bring about an agreement or understanding ⟨*arranged* to have a table at the restaurant⟩
2 : to make preparations **:** PLAN ⟨*arranged* for a vacation with his family⟩
synonym see ORDER
— ar·rang·er \ə-ˈrān-jər\ *noun*
ar·range·ment \ə-ˈrānj-mənt\ *noun* (1690)
1 a : the state of being arranged **:** ORDER ⟨everything in neat *arrangement*⟩ **b :** the act of arranging ⟨the *arrangement* of the details was quickly accomplished⟩
2 : something arranged: as **a :** a preliminary measure **:** PREPARATION ⟨travel *arrangements*⟩ **b :** an adaptation of a musical composition by rescoring **c :** an informal agreement or settlement especially on personal, social, or political matters ⟨*arrangements* under the new regime⟩
3 : something made by arranging parts or things together ⟨a floral *arrangement*⟩
ar·rant \ˈar-ənt\ *adjective* [alteration of *errant*] (1553)
: being notoriously without moderation **:** EXTREME ⟨we are *arrant* knaves, all; believe none of us —Shakespeare⟩
— ar·rant·ly *adverb*
ar·ras \ˈar-əs\ *noun, plural* **arras** [Middle English, from *Arras,* France] (15th century)
1 : a tapestry of Flemish origin used especially for wall hangings and curtains
2 : a wall hanging or screen of tapestry
[1]**ar·ray** \ə-ˈrā\ *transitive verb* [Middle English, from Old French *arayer,* from (assumed) Vulgar Latin *arredare,* from Latin *ad-* + a base of Germanic origin; akin to Gothic *garaiths* arranged — more at READY] (14th century)
1 : to dress or decorate especially in splendid or impressive attire **:** ADORN
2 a : to set or place in order **:** DRAW UP, MARSHAL **b :** to set or set forth in order (as a jury) for the trial of a cause
— ar·ray·er *noun*
[2]**array** *noun* (14th century)
1 a : a regular and imposing grouping or arrangement **:** ORDER ⟨lined up . . . in soldierly *array* —Donald Barthelme⟩ **b :** an orderly listing of jurors impaneled
2 a : CLOTHING, ATTIRE **b :** rich or beautiful apparel **:** FINERY
3 : a body of soldiers **:** MILITIA ⟨the baron and his feudal *array*⟩
4 : an imposing group **:** large number ⟨faced a whole *array* of problems⟩
5 a (1) **:** a number of mathematical elements arranged in rows and columns (2) **:** a data structure in which similar elements of data are arranged in a table **b :** a series of statistical data arranged in classes in order of magnitude
6 : a group of elements forming a complete unit ⟨an antenna *array*⟩
ar·rear \ə-ˈrir\ *noun* [Middle English *arrere* behind, backward, from Middle French, from (assumed) Vulgar Latin *ad retro* backward, from Latin *ad* to + *retro* backward, behind — more at AT, RETRO-] (1620)
1 : the state of being behind in the discharge of obligations — usually used in plural ⟨in *arrears* with the rent⟩
2 a : an unfinished duty — usually used in plural ⟨*arrears* of work that have piled up⟩ **b :** an unpaid and overdue debt — usually used in plural ⟨paying off the *arrears* of the past several months⟩
ar·rear·age \-ij\ *noun* (14th century)
1 : the condition of being in arrears
2 : something that is in arrears; *especially* **:** something unpaid and overdue
[1]**ar·rest** \ə-ˈrest\ *transitive verb* [Middle English *aresten,* from Middle French *arester* to rest, arrest, from (assumed) Vulgar Latin *arrestare,* from Latin *ad-* + *restare* to remain — more at REST] (14th century)
1 a : to bring to a stop ⟨sickness *arrested* his activities⟩ **b :** CHECK, SLOW **c :** to make inactive ⟨an *arrested* tumor⟩
2 : SEIZE, CAPTURE; *specifically* **:** to take or keep in custody by authority of law
3 : to catch suddenly and engagingly
— ar·rest·er *or* **ar·res·tor** \-ˈres-tər\ *noun*
— ar·rest·ment \-ˈres(t)-mənt\ *noun*
[2]**arrest** *noun* (14th century)
1 a : the act of stopping **b :** the condition of being stopped or inactive
2 : the taking or detaining in custody by authority of law
— under arrest : in legal custody
ar·res·tant \ə-ˈres-tənt\ *noun* (1962)
: a substance that stimulates an insect to stop locomotion
ar·rest·ee \ə-ˌres-ˈtē\ *noun* (1944)
: one that is under arrest
ar·rest·ing \ə-ˈres-tiŋ\ *adjective* (1792)
: catching the attention **:** STRIKING, IMPRESSIVE
— ar·rest·ing·ly \-tiŋ-lē\ *adverb*
ar·rhyth·mia \ā-ˈrit͟h-mē-ə\ *noun* [New Latin, from Greek, lack of rhythm, from *arrhythmos* unrhythmical, from *a-* + *rhythmos* rhythm] (circa 1860)
: an alteration in rhythm of the heartbeat either in time or force
ar·rhyth·mic \-mik\ *adjective* [Greek *arrhythmos*] (1853)
: lacking rhythm or regularity ⟨*arrhythmic* locomotor activity⟩
ar·ri·ère–ban \ˌar-ē-ˌer-ˈbän, -ˈban\ *noun* [French] (1523)
: a proclamation of a king (as of France) calling his vassals to arms; *also* **:** the body of vassals summoned
ar·ri·ère–pen·sée \-päⁿ-ˈsā\ *noun* [French, from *arrière* in back + *pensée* thought] (1824)
: mental reservation
ar·ris \ˈar-əs\ *noun, plural* **arris** *or* **ar·ris·es** [probably modification of Middle French *areste,* literally, fishbone, from Late Latin *arista* — more at ARÊTE] (1677)
: the sharp edge or salient angle formed by the meeting of two surfaces especially in moldings
ar·riv·al \ə-ˈrī-vəl\ *noun* (14th century)
1 : the act of arriving
2 : the attainment of an end or state
3 : one that has recently reached a destination
ar·rive \ə-ˈrīv\ *intransitive verb* **ar·rived; ar·riv·ing** [Middle English *ariven,* from Old French *ariver,* from (assumed) Vulgar Latin *arripare* to come to shore, from Latin *ad-* + *ripa* shore — more at RIVE] (13th century)
1 a : to reach a destination **b :** to make an appearance ⟨the guests have *arrived*⟩
2 a *archaic* **:** HAPPEN **b :** to be near in time **:** COME ⟨the moment has *arrived*⟩
3 : to achieve success
— ar·riv·er *noun*
— arrive at : to reach by effort or thought ⟨*arrived at* a decision⟩
ar·ri·vé \ˌar-i-ˈvā\ *noun* [French, from past participle of *arriver* to arrive, from Old French *ariver*] (1925)
: one who has risen rapidly to success, power, or fame
ar·ri·viste \-ˈvēst\ *noun* [French, from *arriver*] (1901)
: one that is a new and uncertain arrival (as in social position or artistic endeavor)
ar·ro·ba \ə-ˈrō-bə\ *noun* [Spanish & Portuguese, from Arabic *ar-rubʽ,* literally, the quarter] (1555)
1 : an old Spanish unit of weight equal to about 25 pounds
2 : an old Portuguese unit of weight equal to about 32 pounds
ar·ro·gance \ˈar-ə-gən(t)s\ *noun* (14th century)
: a feeling or an impression of superiority manifested in an overbearing manner or presumptuous claims
ar·ro·gant \-gənt\ *adjective* [Middle English, from Latin *arrogant-, arrogans,* present participle of *arrogare*] (14th century)
1 : exaggerating or disposed to exaggerate one's own worth or importance in an overbearing manner ⟨an *arrogant* official⟩
2 : proceeding from or characterized by arrogance ⟨*arrogant* manners⟩
synonym see PROUD
— ar·ro·gant·ly *adverb*
ar·ro·gate \-ˌgāt\ *transitive verb* **-gat·ed; -gat·ing** [Latin *arrogatus,* past participle of *arrogare,* from *ad-* + *rogare* to ask — more at RIGHT] (1537)
1 a : to claim or seize without justification **b :** to make undue claims to having **:** ASSUME

2 : to claim on behalf of another **:** ASCRIBE
— **ar·ro·ga·tion** \ˌar-ə-ˈgā-shən\ *noun*

ar·ron·disse·ment \ə-ˈrän-də-smənt, ˌar-ˌō^{n}-(ˌ)dē-ˈsmä^{n}\ *noun* [French] (1807)
1 : an administrative district of some large French cities
2 : the largest division of a French department

¹ar·row \ˈar-(ˌ)ō\ *noun* [Middle English *arwe,* from Old English; akin to Gothic *arhwazna* arrow, Latin *arcus* bow, arch, arc] (before 12th century)
1 : a missile weapon shot from a bow and usually having a slender shaft, a pointed head, and feathers at the butt
2 : something shaped like an arrow; *especially* **:** a mark (as on a map or signboard) to indicate direction

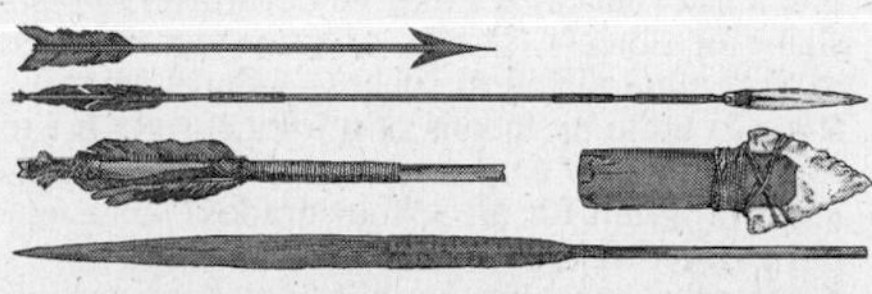

arrow 1

²arrow *intransitive verb* (1827)
: to move fast and straight like an arrow in flight **:** DART

ar·row·head \ˈar-ō-ˌhed, ˈar-ə-\ *noun* (14th century)
1 : a wedge-shaped piercing tip usually fixed to an arrow
2 : something resembling an arrowhead

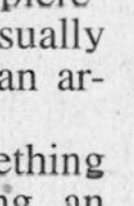

arrowhead 1

3 : any of a genus (*Sagittaria*) of marsh or aquatic plants of the water-plantain family with leaves shaped like arrowheads

ar·row·root \-ˌrüt, -ˌru̇t\ *noun* (1696)
1 a : any of a genus (*Maranta* of the family Marantaceae, the arrowroot family) of tropical American plants with tuberous roots; *especially* **:** one (*M. arundinacea*) whose roots yield an easily digested edible starch **b :** any of several plants (as coontie) that yield starch
2 : starch yielded by an arrowroot

ar·row·wood \-ˌwu̇d\ *noun* (1709)
: any of several common viburnums (especially *Viburnum dentatum*) of the eastern U.S.

ar·row·worm \-ˌwərm\ *noun* (circa 1889)
: any of a phylum (Chaetognatha) of small planktonic wormlike marine organisms having curved bristles on either side of the head for seizing prey

ar·rowy \ˈar-ə-wē\ *adjective* (1637)
1 : resembling or suggesting an arrow; *especially* **:** swiftly moving
2 : consisting of arrows

ar·royo \ə-ˈrȯi-(ˌ)ō, -ə\ *noun, plural* **-royos** [Spanish] (1843)
1 : a watercourse (as a creek) in an arid region
2 : a water-carved gully or channel

ar·roz con po·llo \ä-ˈrōth-(ˌ)kōn-ˈpōl-(ˌ)yō, -ˈpō-(ˌ)yō\ *noun* [Spanish, literally, rice with chicken] (1938)
: chicken cooked with rice and usually flavored with saffron

arse *variant of* ASS

ar·se·nal \ˈärs-nəl, ˈär-s^{ə}n-əl\ *noun* [Italian *arsenale,* ultimately from Arabic *dār ṣināʻah* house of manufacture] (1555)
1 a : an establishment for the manufacture or storage of arms and military equipment **b :** a collection of weapons
2 : STORE, REPERTORY ⟨the team's *arsenal* of veteran players⟩

ar·se·nate \ˈärs-nət, ˈär-s^{ə}n-ət, -s^{ə}n-ˌāt\ *noun* (1800)
: a salt or ester of an arsenic acid

¹ar·se·nic \ˈärs-nik, -s^{ə}n-ik\ *noun* [Middle English, yellow orpiment, from Middle French & Latin; Middle French, from Latin *arsenicum,* from Greek *arsenikon, arrhenikon,* from Syriac *zarnīg,* of Iranian origin; akin to Avestan *zaranya* gold, Sanskrit *hari* yellowish — more at YELLOW] (14th century)
1 : a trivalent and pentavalent solid poisonous element that is commonly metallic steel-gray, crystalline, and brittle — see ELEMENT table
2 : a poisonous trioxide As_2O_3 or As_4O_6 of arsenic used especially as an insecticide or weed killer — called also *arsenic trioxide*

²ar·sen·ic \är-ˈse-nik\ *adjective* (1801)
: of, relating to, or containing arsenic especially with a valence of five

ar·sen·i·cal \är-ˈse-ni-kəl\ *adjective* (1605)
: of, relating to, containing, or caused by arsenic ⟨*arsenical* poisoning⟩
— **arsenical** *noun*

ar·se·nide \ˈär-s^{ə}n-ˌīd\ *noun* (circa 1859)
: a binary compound of arsenic with a more positive element

ar·se·ni·ous \är-ˈsē-nē-əs\ *adjective* (1818)
: of, relating to, or containing arsenic especially when trivalent

ar·se·nite \ˈär-s^{ə}n-ˌīt\ *noun* (1800)
: a salt or ester of an arsenious acid

ar·se·no·py·rite \ˌär-s^{ə}n-ō-ˈpī-ˌrīt\ *noun* (1881)
: a mineral consisting of a combined sulfide and arsenide of iron occurring in prismatic orthorhombic crystals or in masses or grains

ar·sine \är-ˈsēn, ˈär-ˌ\ *noun* [International Scientific Vocabulary, from *ars*enic] (1876)
: a colorless flammable extremely poisonous gas AsH_3 with an odor like garlic; *also* **:** a derivative of arsine

ar·sis \ˈär-səs\ *noun, plural* **ar·ses** \-ˌsēz\ [Late Latin & Greek; Late Latin, raising of the voice, accented part of foot, from Greek, upbeat, less important part of foot; literally, act of lifting, from *aeirein, airein* to lift] (14th century)
1 a : the lighter or shorter part of a poetic foot especially in quantitative verse **b :** the accented or longer part of a poetic foot especially in accentual verse
2 : the unaccented part of a musical measure — compare THESIS

ar·son \ˈär-s^{ə}n\ *noun* [Anglo-French *arsoun,* alteration of Old French *arsin* act of burning, from *ars,* past participle of *ardre* to burn, from Latin *ardēre* — more at ARDOR] (circa 1680)
: the willful or malicious burning of property (as a building) especially with criminal or fraudulent intent
— **ar·son·ist** \-ist\ *noun*
— **ar·son·ous** \-əs\ *adjective*

ars·phen·a·mine \ärs-ˈfe-nə-ˌmēn, -mən\ *noun* [International Scientific Vocabulary *arsenic* + *phenamine*] (1917)
: a light-yellow toxic hygroscopic powder $C_{12}Cl_2H_{14}As_2N_2O_2 \cdot 2H_2O$ formerly used in the treatment especially of syphilis and yaws

¹art \ˈärt, ərt\ [Middle English, from Old English *eart;* akin to Old Norse *est, ert* (thou) art, Old English *is* is] *archaic present 2d singular of* BE

²art \ˈärt\ *noun* [Middle English, from Old French, from Latin *art-, ars* — more at ARM] (13th century)
1 : skill acquired by experience, study, or observation ⟨the *art* of making friends⟩
2 a : a branch of learning: (1) **:** one of the humanities (2) *plural* **:** LIBERAL ARTS **b** *archaic* **:** LEARNING, SCHOLARSHIP
3 : an occupation requiring knowledge or skill ⟨the *art* of organ building⟩
4 a : the conscious use of skill and creative imagination especially in the production of aesthetic objects; *also* **:** works so produced **b** (1) **:** FINE ARTS (2) **:** one of the fine arts (3) **:** a graphic art
5 a *archaic* **:** a skillful plan **b :** the quality or state of being artful
6 : decorative or illustrative elements in printed matter ☆

-art — see -ARD

art de·co \ˌärt-ˈde-ˌkō; ˌär(t)-dā-ˈkō, ˈär(t)-ˈdā-(ˌ)kō\ *noun, often A&D capitalized* [French *Art Déco,* from *Exposition Internationale des Arts Décoratifs et Industriels Modernes,* an exposition of modern decorative and industrial arts held in Paris, France, in 1925] (1966)
: a popular design style of the 1920s and 1930s characterized especially by bold outlines, geometric and zigzag forms, and the use of new materials (as plastic)

ar·te·fact *chiefly British variant of* ARTIFACT

ar·tel \är-ˈtel; -ˈtel^{y}\ *noun* [Russian *artel'*] (1884)
: a workers' or craftsmen's cooperative in Russia and later the U.S.S.R

Ar·te·mis \ˈär-tə-məs\ *noun* [Greek]
: a Greek moon goddess often portrayed as a virgin huntress — compare DIANA

ar·te·mi·sia \ˌär-tə-ˈmi-zh(ē-)ə, -zē-ə\ *noun* [New Latin, from Latin, artemisia, from Greek, wormwood] (14th century)
: any of a genus (*Artemisia*) of aromatic composite herbs and shrubs (as sagebrush) — compare WORMWOOD 1

arteri- *or* **arterio-** *combining form* [Greek *artēri-, artērio-,* from *artēria* artery]
1 : artery ⟨*arterio*gram⟩
2 : arterial and ⟨*arterio*venous⟩

¹ar·te·ri·al \är-ˈtir-ē-əl\ *adjective* (15th century)
1 a : of or relating to an artery **b :** relating to or being the bright red blood present in most arteries that has been oxygenated in lungs or gills
2 : of, relating to, or constituting through traffic
— **ar·te·ri·al·ly** \-ē-ə-lē\ *adverb*

²arterial *noun* (1932)
: a through street or highway

ar·te·rio·gram \är-ˈtir-ē-ə-ˌgram\ *noun* [International Scientific Vocabulary] (1929)
: a roentgenogram of an artery made by arteriography

ar·te·ri·og·ra·phy \är-ˌtir-ē-ˈä-grə-fē\ *noun, plural* **-phies** [International Scientific Vocabulary] (1929)
: the roentgenographic visualization of an artery after injection of a radiopaque substance
— **ar·te·rio·graph·ic** \-ē-ə-ˈgra-fik\ *adjective*

ar·te·ri·ole \är-ˈtir-ē-ˌōl\ *noun* [French or New Latin; French *artériole,* probably from New Latin *arteriola,* diminutive of Latin *arteria*] (circa 1847)
: any of the small terminal twigs of an artery that ends in capillaries
— **ar·te·ri·o·lar** \-ˌtir-ē-ˈō-ˌlär, -lər\ *adjective*

☆ SYNONYMS
Art, skill, cunning, artifice, craft mean the faculty of executing well what one has devised. ART implies a personal, unanalyzable creative power ⟨the *art* of choosing the right word⟩. SKILL stresses technical knowledge and proficiency ⟨the *skill* of a glassblower⟩. CUNNING suggests ingenuity and subtlety in devising, inventing, or executing ⟨a mystery plotted with great *cunning*⟩. ARTIFICE suggests technical skill especially in imitating things in nature ⟨believed realism in film could be achieved only by *artifice*⟩. CRAFT may imply expertness in workmanship ⟨the *craft* of a master goldsmith⟩.

\ə\ **abut** \$^{ə}$\ **kitten** \ər\ **further** \a\ **ash** \ā\ **ace** \ä\ **mop, mar** \au̇\ **out** \ch\ **chin** \e\ **bet** \ē\ **easy** \g\ **go** \i\ **hit** \ī\ **ice** \j\ **job** \ŋ\ **sing** \ō\ **go** \ȯ\ **law** \ȯi\ **boy** \th\ **thin** \t͟h\ **the** \ü\ **loot** \u̇\ **foot** \y\ **yet** \zh\ **vision** *see also* Guide to Pronunciation

ar·te·rio·scle·ro·sis \är-ˌtir-ē-ō-sklə-ˈrō-səs\ *noun* [New Latin] (1881)
: a chronic disease characterized by abnormal thickening and hardening of the arterial walls with resulting loss of elasticity
— **ar·te·rio·scle·rot·ic** \-ˈrä-tik\ *adjective or noun*

ar·te·rio·ve·nous \-ˈvē-nəs\ *adjective* [International Scientific Vocabulary] (circa 1880)
: of, relating to, or connecting the arteries and veins ⟨an *arteriovenous* fistula⟩

ar·ter·i·tis \ˌär-tə-ˈrī-təs\ *noun* [New Latin] (1836)
: arterial inflammation

ar·tery \ˈär-tə-rē, ˈär-trē\ *noun, plural* **-ter·ies** [Middle English *arterie,* from Latin *arteria,* from Greek *artēria;* akin to Greek *aortē* aorta] (14th century)
1 : any of the tubular branching muscular- and elastic-walled vessels that carry blood from the heart through the body
2 : a channel (as a river or highway) of transportation or communication; *especially* **:** the main channel in a branching system

ar·te·sian well \är-ˈtē-zhən-\ *noun* [French *artésien,* literally, of Artois, from Old French, from *Arteis* Artois, France] (1842)
1 : a well in which water is under pressure; *especially* **:** one in which the water flows to the surface naturally
2 : a deep well

art film *noun* (1926)
: a motion picture produced as an artistic effort

art form *noun* (1868)
1 : a recognized form (as a symphony) or medium (as sculpture) of artistic expression
2 : an unconventional form or medium in which impulses regarded as artistic may be expressed ⟨describe pinball as a great American *art form* —Tom Buckley⟩

art·ful \ˈärt-fəl\ *adjective* (1615)
1 : performed with or showing art or skill ⟨an *artful* performance on the violin⟩
2 a : using or characterized by art and skill **:** DEXTEROUS ⟨an *artful* prose stylist⟩ **b :** adroit in attaining an end often by insinuating or indirect means **:** WILY ⟨an *artful* cross-examiner⟩
3 : ARTIFICIAL ⟨trim walks and *artful* bowers —William Wordsworth⟩
synonym see SLY
— **art·ful·ly** \-fə-lē\ *adverb*
— **art·ful·ness** *noun*

art glass *noun* (1926)
: articles of glass designed primarily for decorative purposes; *especially* **:** novelty glassware

art historical *adjective* (1933)
: of or relating to the history of art
— **art historically** *adverb*

art house *noun* (1951)
: ART THEATER

arthr- *or* **arthro-** *combining form* [Latin, from Greek, from *arthron;* akin to Greek *arariskein* to fit — more at ARM]
: joint ⟨*arthro*pathy⟩

ar·thral·gia \är-ˈthral-j(ē-)ə\ *noun* [New Latin] (circa 1848)
: neuralgic pain in one or more joints
— **ar·thral·gic** \-jik\ *adjective*

ar·thrit·ic \är-ˈthri-tik\ *adjective* (14th century)
1 : of, relating to, or affected with arthritis
2 : being or showing effects associated with aging ⟨*arthritic* anxiety⟩
— **arthritic** *noun*
— **ar·thrit·i·cal·ly** \-ti-k(ə-)lē\ *adverb*

ar·thri·tis \är-ˈthrī-təs\ *noun, plural* **-thrit·i·des** \-ˈthri-tə-ˌdēz\ [Latin, from Greek, from *arthron*] (1543)
: inflammation of joints due to infectious, metabolic, or constitutional causes; *also* **:** a specific arthritic condition

ar·throd·e·sis \är-ˈthrä-də-səs\ *noun, plural* **-e·ses** \-ˌsēz\ [New Latin, from *arthr-* + Greek *desis* binding, from *dein* to bind] (circa 1901)
: the surgical immobilization of a joint so that the bones grow solidly together

ar·throp·a·thy \är-ˈthrä-pə-thē\ *noun, plural* **-thies** (circa 1860)
: a disease of a joint

ar·thro·pod \ˈär-thrə-ˌpäd\ *noun* [New Latin *Arthropoda,* from *arthr-* + Greek *pod-, pous* foot — more at FOOT] (1877)
: any of a phylum (Arthropoda) of invertebrate animals (as insects, arachnids, and crustaceans) that have a segmented body and jointed appendages, a usually chitinous exoskeleton molted at intervals, and a dorsal anterior brain connected to a ventral chain of ganglia
— **arthropod** *adjective*
— **ar·throp·o·dan** \är-ˈthrä-pə-dən\ *adjective*

ar·thros·co·py \är-ˈthräs-kə-pē\ *noun* (circa 1935)
: visual examination of the interior of a joint (as the knee) with a special surgical instrument; *also* **:** joint surgery using an arthroscope
— **ar·thro·scope** \ˈär-thrə-ˌskōp\ *noun*
— **ar·thro·scop·ic** \ˌär-thrə-ˈskä-pik\ *adjective*

ar·thro·sis \är-ˈthrō-səs\ *noun, plural* **-thro·ses** \-ˌsēz\ [New Latin, from Greek *arthrōsis* jointing, articulation, from *arthroun* to articulate, from *arthron*] (1634)
1 : an articulation between bones
2 : a degenerative disease of a joint

ar·thro·spore \ˈär-thrə-ˌspōr, -ˌspȯr\ *noun* (1895)
: OIDIUM 1b

Ar·thur \ˈär-thər\ *noun*
: a legendary king of the Britons whose story is based on traditions of a 6th century military leader

Ar·thu·ri·an \är-ˈthu̇r-ē-ən, -ˈthyu̇r-\ *adjective* (1612)
: of or relating to King Arthur and his court

ar·ti·choke \ˈär-tə-ˌchōk\ *noun* [Italian dialect *articiocco,* ultimately from Arabic *al-khurshūf* the artichoke] (1530)
1 : a tall composite herb (*Cynara scolymus*) like a thistle with coarse pinnately incised leaves; *also* **:** its edible immature flower head which is cooked as a vegetable
2 : JERUSALEM ARTICHOKE

[1]**ar·ti·cle** \ˈär-ti-kəl\ *noun* [Middle English, from Old French, from Latin *articulus* joint, division, diminutive of *artus* joint, limb; akin to Greek *arariskein* to fit — more at ARM] (13th century)
1 a : a distinct often numbered section of a writing **b :** a separate clause **c :** a stipulation in a document (as a contract or a creed) **d :** a nonfictional prose composition usually forming an independent part of a publication (as a magazine)
2 : an item of business **:** MATTER
3 : any of a small set of words or affixes (as *a, an,* and *the*) used with nouns to limit or give definiteness to the application
4 : a member of a class of things; *especially* **:** an item of goods ⟨*articles* of value⟩
5 : a thing of a particular and distinctive kind ⟨the genuine *article*⟩

[2]**article** *transitive verb* **ar·ti·cled; ar·ti·cling** \-k(ə-)liŋ\ (1820)
: to bind by articles (as of apprenticeship)

article of faith (15th century)
: a basic belief

ar·tic·u·la·ble \är-ˈti-kyə-lə-bəl\ *adjective* (1833)
: capable of being articulated

ar·tic·u·la·cy \är-ˈti-kyə-lə-sē\ *noun* (1918)
chiefly British **:** the quality or state of being articulate

ar·tic·u·lar \är-ˈti-kyə-lər\ *adjective* [Middle English *articuler,* from Latin *articularis,* from *articulus*] (15th century)
: of or relating to a joint ⟨*articular* cartilage⟩

[1]**ar·tic·u·late** \är-ˈti-kyə-lət\ *adjective* [Latin *articulatus* jointed, past participle of *articulare,* from *articulus*] (1586)
1 a : divided into syllables or words meaningfully arranged **:** INTELLIGIBLE **b :** able to speak **c :** expressing oneself readily, clearly, or effectively; *also* **:** expressed in this manner
2 a : consisting of segments united by joints **:** JOINTED ⟨*articulate* animals⟩ **b :** distinctly marked off
— **ar·tic·u·late·ly** *adverb*
— **ar·tic·u·late·ness** *noun*

[2]**ar·tic·u·late** \-ˌlāt\ *verb* **-lat·ed; -lat·ing** (1594)
transitive verb
1 a : to utter distinctly ⟨*articulating* each note in the musical phrase⟩ **b :** to give clear and effective utterance to **:** put into words ⟨*articulate* one's grievances⟩ **c :** to give definition to (as a shape or object) ⟨shades of gray were chosen to *articulate* different spaces —Carol Vogel⟩
2 a : to unite by means of a joint **:** JOINT **b :** to form or fit into a systematic whole ⟨*articulating* a program for all school grades⟩
intransitive verb
1 : to utter articulate sounds
2 : to become united or connected by or as if by a joint
— **ar·tic·u·la·tive** \-lə-tiv, -ˌlā-\ *adjective*
— **ar·tic·u·la·tor** \-ˌlā-tər\ *noun*

ar·tic·u·lat·ed \-ˌlā-təd\ *adjective* (1899)
of a vehicle **:** having a hinge or pivot connection especially to allow negotiation of sharp turns ⟨*articulated* lorry⟩ ⟨*articulated* bus⟩

ar·tic·u·la·tion \(ˌ)är-ˌti-kyə-ˈlā-shən\ *noun* (15th century)
1 a : a joint or juncture between bones or cartilages in the skeleton of a vertebrate **b :** a movable joint between rigid parts of an animal
2 a : the action or manner of jointing or interrelating **b :** the state of being jointed or interrelated
3 a : the act of giving utterance or expression **b :** the act or manner of articulating sounds **c :** an articulated utterance or sound; *specifically* **:** CONSONANT
4 : OCCLUSION 1b

ar·tic·u·la·to·ry \är-ˈti-kyə-lə-ˌtōr-ē, -ˌtȯr-\ *adjective* (1818)
: of or relating to articulation

ar·ti·fact \ˈär-ti-ˌfakt\ *noun* [Latin *arte* by skill (ablative of *art-, ars* skill) + *factum,* neuter of *factus,* past participle of *facere* to do — more at ARM, DO] (1821)
1 a : something created by humans usually for a practical purpose; *especially* **:** an object remaining from a particular period ⟨caves containing prehistoric *artifacts*⟩ **b :** something characteristic of or resulting from a human institution or activity ⟨self-consciousness . . . turns out to be an *artifact* of our education system —*Times Literary Supplement*⟩
2 : a product of artificial character (as in a scientific test) due usually to extraneous (as human) agency
— **ar·ti·fac·tu·al** \ˌär-ti-ˈfak-chə(-wə)l, -ˈfak-shwəl\ *adjective*

ar·ti·fice \ˈär-tə-fəs\ *noun* [Middle French, from Latin *artificium,* from *artific-, artifex* artificer, from Latin *art-, ars* + *facere*] (circa 1604)
1 a : clever or artful skill **:** INGENUITY ⟨believing that characters had to be created from within rather than with *artifice* —Garson Kanin⟩ **b :** an ingenious device or expedient
2 a : an artful stratagem **:** TRICK **b :** false or insincere behavior ⟨social *artifice*⟩
synonym see TRICK, ART

ar·ti·fi·cer \är-ˈti-fə-sər, ˈär-tə-fə-sər\ *noun* (14th century)
1 : a skilled or artistic worker or craftsman
2 : one that makes or contrives **:** DEVISER ⟨had been the *artificer* of his own fortunes —*Times Literary Supplement*⟩

ar·ti·fi·cial \ˌär-tə-ˈfi-shəl\ *adjective* (14th century)

1 : humanly contrived often on a natural model **:** MAN-MADE ⟨an *artificial* limb⟩ ⟨*artificial* diamonds⟩
2 a : having existence in legal, economic, or political theory **b :** caused or produced by a human and especially social or political agency ⟨an *artificial* price advantage⟩ ⟨*artificial* barriers of discrimination —R. C. Weaver⟩
3 *obsolete* **:** ARTFUL, CUNNING
4 a : lacking in natural or spontaneous quality ⟨an *artificial* smile⟩ ⟨an *artificial* excitement⟩ **b :** IMITATION, SHAM ⟨*artificial* flavor⟩
5 : based on differential morphological characters not necessarily indicative of natural relationships ⟨an *artificial* key for plant identification⟩
— ar·ti·fi·ci·al·i·ty \ˌär-tə-ˌfi-shē-ˈa-lə-tē\ *noun*
— ar·ti·fi·cial·ly \-ˈfi-sh^{ə}l-ē\ *adverb*
— ar·ti·fi·cial·ness \-ˈfi-shəl-nəs\ *noun*

artificial horizon *noun* (1920)
: a gyroscopic flight instrument designed to indicate aircraft attitude with respect to the true horizon

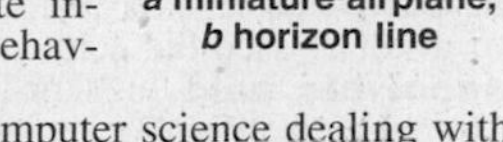
artificial horizon (showing plane banking left): ***a*** **miniature airplane,** ***b*** **horizon line**

artificial insemination *noun* (1897)
: introduction of semen into the uterus or oviduct by other than natural means

artificial intelligence *noun* (1956)
1 : the capability of a machine to imitate intelligent human behavior
2 : a branch of computer science dealing with the simulation of intelligent behavior in computers

artificial respiration *noun* (1852)
: the rhythmic forcing of air into and out of the lungs of a person whose breathing has stopped

ar·til·ler·ist \är-ˈti-lə-rist\ *noun* (1781)
: GUNNER, ARTILLERYMAN

ar·til·lery \är-ˈti-lər-ē, -ˈtil-rē\ *noun, plural* **-ler·ies** [Middle English *artillerie,* from Middle French, from *artillier* to equip, arm, alteration of Old French *atillier,* from (assumed) Vulgar Latin *apticulare,* from Latin *aptare* to don, prepare, fit — more at ADAPT] (15th century)
1 : weapons (as bows, slings, and catapults) for discharging missiles
2 a : large bore crew-served mounted firearms (as guns, howitzers, and rockets) **:** ORDNANCE **b :** a branch of an army armed with artillery
3 : means of impressing, arguing, or persuading

ar·til·lery·man \-mən\ *noun* (1635)
: a soldier in the artillery

ar·tio·dac·tyl \ˌär-tē-ō-ˈdak-t^{ə}l\ *noun* [New Latin *Artiodactyla,* from Greek *artios* fitting, even-numbered + *daktylos* finger, toe; akin to Greek *arariskein* to fit — more at ARM] (circa 1879)
: any of an order (Artiodactyla) of ungulates (as the camel or pig) with an even number of functional toes on each foot
— artiodactyl *adjective*

ar·ti·san \ˈär-tə-zən, -sən, *chiefly British* ˌär-tə-ˈzan\ *noun* [Middle French, ultimately from Old Italian *artigiano,* from *arte* art, from Latin *art-, ars*] (1538)
: CRAFTSMAN
— ar·ti·san·al \-zə-n^{ə}l, -sə-, -ˈza-\ *adjective*
— ar·ti·san·ship \-ˌship\ *noun*

art·ist \ˈär-tist\ *noun* (circa 1507)
1 a *obsolete* **:** one skilled or versed in learned arts **b** *archaic* **:** PHYSICIAN **c** *archaic* **:** ARTISAN
2 a : one who professes and practices an imaginative art **b :** a person skilled in one of the fine arts
3 : a skilled performer; *especially* **:** ARTISTE
4 : one who is adept at something ⟨con *artist*⟩ ⟨strikeout *artist*⟩

ar·tiste \är-ˈtēst\ *noun* [French] (1823)
: a skilled adept public performer; *specifically* **:** a musical or theatrical entertainer

ar·tis·tic \är-ˈtis-tik\ *adjective* (circa 1753)
1 : of, relating to, or characteristic of art or artists ⟨*artistic* subjects⟩ ⟨an *artistic* success⟩
2 : showing imaginative skill in arrangement or execution ⟨*artistic* photography⟩
— ar·tis·ti·cal·ly \-ti-k(ə-)lē\ *adverb*

art·ist·ry \ˈär-tə-strē\ *noun* (1868)
1 : artistic quality of effect or workmanship ⟨the *artistry* of his novel⟩
2 : artistic ability ⟨the *artistry* of the violinist⟩ ⟨a lawyer's *artistry* in persuading juries⟩

art·less \ˈärt-ləs\ *adjective* (1589)
1 : lacking art, knowledge, or skill **:** UNCULTURED
2 a : made without skill **:** CRUDE **b :** free from artificiality **:** NATURAL ⟨*artless* grace⟩
3 : free from guile or craft **:** sincerely simple
synonym see NATURAL
— art·less·ly *adverb*
— art·less·ness *noun*

art mo·derne \ˌär(t)-mō-ˈdern\ *noun, often A&M capitalized* [French — more at ART DECO] (1931)
: ART DECO

art nou·veau \ˌär(t)-nü-ˈvō\ *noun, often A&N capitalized* [French, literally, new art] (1908)
: a design style of late 19th century origin characterized especially by sinuous lines and foliate forms

Arts and Crafts *noun* (1888)
: a movement in European and American design during the late 19th and early 20th centuries promoting handcraftsmanship over industrial mass production

art song *noun* (1890)
: a usually through-composed lyric song with melody and accompaniment

art·sy \ˈärt-sē\ *adjective* (1902)
: ARTY

artsy–craftsy \ˌärt-sē-ˈkraf(t)-sē\ *also* **arty–crafty** \är-tē ˈkraf-tē\ *adjective* [from the phrase *arts and crafts*] (1902)
: ARTY

art theater *noun* (1923)
: a theater that specializes in the presentation of art films

art·work \ˈärt-ˌwərk\ *noun* (1877)
1 a : an artistic production ⟨an 8-foot metal *artwork*⟩ **b :** artistic work ⟨*artwork* being sold on the sidewalk⟩
2 a : ART 6 **b :** material (as a drawing or photograph) prepared for reproduction in printed matter

arty \ˈär-tē\ *adjective* **art·i·er; -est** (1901)
: showily or pretentiously artistic ⟨*arty* lighting and photography⟩
— art·i·ly \ˈär-t^{ə}l-ē\ *adverb*
— art·i·ness \ˈär-tē-nəs\ *noun*

aru·gu·la \ə-ˈrü-gə-lə, -gyə-\ *noun* [probably from Italian dialect; akin to Italian dialect (Lombardy) *arigola* arugula, Italian *ruca* — more at ROCKET] (1967)
: a yellowish-flowered European herb (*Eruca vesicaria*) of the mustard family cultivated for its foliage which is used especially in salads — called also *garden rocket, rocket, rugola*

ar·um \ˈar-əm, ˈer-\ *noun* [New Latin, from Latin *arum,* from Greek *aron*] (14th century)
: any of a genus (*Arum* of the family Araceae, the arum family) of Old World plants having usually arrow-shaped leaves and a showy spathe partially enclosing a spadix; *broadly* **:** a plant of the arum family

¹-ary *US usually* ˌer-ē *when an unstressed syllable precedes,* ə-rē *or* rē *when a stressed syllable precedes; British usually* ə-rē *or* rē *in all cases*\ *noun suffix* [Middle English *-arie,* from Old French & Latin; Old French *-aire, -arie,* from Latin *-arius, -aria, -arium,* from *-arius,* adjective suffix]
1 : thing belonging to or connected with; *especially* **:** place of ⟨ov*ary*⟩
2 : person belonging to, connected with, or engaged in ⟨function*ary*⟩

²-ary *adjective suffix* [Middle English *-arie,* from Middle French & Latin; Middle French *-aire,* from Latin *-arius*]
: of, relating to, or connected with ⟨budget*ary*⟩

¹Ary·an \ˈar-ē-ən, ˈer-; ˈär-yən\ *adjective* [Sanskrit *ārya* noble, belonging to the people speaking an Indo-European dialect who migrated into northern India] (1839)
1 : INDO-EUROPEAN
2 a : of or relating to a hypothetical ethnic type illustrated by or descended from early speakers of Indo-European languages **b :** NORDIC **c** — used in Nazism to designate a supposed master race of non-Jewish Caucasians having especially Nordic features
3 : of or relating to Indo-Iranian or its speakers

²Aryan *noun* (1851)
1 : INDO-EUROPEAN
2 a : NORDIC **b :** GENTILE

ar·yl \ˈar-əl\ *adjective* [International Scientific Vocabulary *ar*omatic + *-yl*] (1906)
: having or being a univalent organic group (as phenyl) derived from an aromatic hydrocarbon by the removal of one hydrogen atom — often used in combination

ar·y·te·noid \ˌar-ə-ˈtē-ˌnȯid, ə-ˈri-t^{ə}n-ˌȯid\ *adjective* [New Latin *arytaenoides,* from Greek *arytainoeidēs,* literally, ladle-shaped, from *arytaina* ladle] (circa 1751)
1 : relating to or being either of two small laryngeal cartilages to which the vocal cords are attached
2 : relating to or being either of a pair of small muscles or an unpaired muscle of the larynx
— arytenoid *noun*

¹as \əz, (ˌ)az\ *adverb* [Middle English, from Old English *eallswā* likewise, just as — more at ALSO] (before 12th century)
1 : to the same degree or amount ⟨*as* deaf as a post⟩ ⟨twice *as* long⟩
2 : for instance ⟨various trees, *as* oak or pine⟩
3 : when considered in a specified form or relation — usually used before a preposition or a participle ⟨my opinion *as* distinguished from his⟩

²as *conjunction* (12th century)
1 : AS IF ⟨looks *as* he had seen a ghost —S. T. Coleridge⟩
2 : in or to the same degree in which ⟨deaf *as* a post⟩ — usually used as a correlative after an adjective or adverb modified by adverbial *as* or *so* ⟨as cool *as* a cucumber⟩
3 : in the way or manner that ⟨do *as* I do⟩
4 : in accordance with what or the way in which ⟨quite good *as* boys go⟩
5 : WHILE, WHEN ⟨spilled the milk *as* she got up⟩
6 : regardless of the degree to which **:** THOUGH ⟨improbable *as* it seems, it's true⟩
7 : for the reason that **:** BECAUSE, SINCE ⟨stayed home *as* she had no car⟩
8 : that the result is ⟨so clearly guilty *as* to leave no doubt⟩
usage see LIKE
— as is : in the presently existing condition without modification ⟨bought the clock at an auction *as is*⟩
— as it were : as if it were so **:** in a manner of speaking

³as *pronoun* (12th century)
1 : THAT, WHO, WHICH — used after *same* or *such* ⟨in the same building *as* my brother⟩ ⟨tears such *as* angels weep —John Milton⟩ and chiefly dialect after a substantive not modified

by *same* or *such* ⟨that kind of fruit *as* maids call medlars —Shakespeare⟩
2 : a fact that ⟨is a foreigner, *as* is evident from his accent⟩

[4]as *preposition* (13th century)
1 a : LIKE 2 ⟨all rose *as* one man⟩ **b :** LIKE 1a ⟨his face was *as* a mask —Max Beerbohm⟩
2 : in the capacity, character, condition, or role of ⟨works *as* an editor⟩

[5]as \ˈas\ *noun, plural* **as·ses** \ˈa-ˌsēz, ˈa-səz\ [Latin] (1540)
1 a : a bronze coin of the ancient Roman republic **b :** a unit of value equivalent to an as coin
2 : LIBRA 2a

as- — see AD-

asa·fet·i·da *or* **asa·foe·ti·da** \ˌa-sə-ˈfe-tə-də, -ˈfē-; *Southern also* -ˈfi-tə-dē\ *noun* [Middle English *asafetida,* from Medieval Latin *asafoetida,* from Persian *azā* mastic + Latin *foetida,* feminine of *foetidus* fetid] (14th century)
: the fetid gum resin of various oriental plants (genus *Ferula*) of the carrot family formerly used in medicine as an antispasmodic and in folk medicine as a general prophylactic against disease

asa·na \ˈä-sə-nə\ *noun* [Sanskrit *āsana* manner of sitting, from *āste* he sits; akin to Greek *hēsthai* to sit, Hittite *es-*] (circa 1934)
: any of various yogic postures

Asan·te \ə-ˈsan-tē, -ˈsän-\ *noun, plural* **Asante** (1721)
: ASHANTI

as·bes·tos \as-ˈbes-təs, az-\ *noun* [Middle English *albestron* mineral supposed to be inextinguishable when set on fire, probably from Middle French, from Medieval Latin *asbeston,* alteration of Latin *asbestos,* from Greek, unslaked lime, from *asbestos* inextinguishable, from *a-* + *sbennynai* to quench] (1607)
: any of several minerals (as chrysotile) that readily separate into long flexible fibers, that have been implicated as causes of certain cancers, and that have been used especially formerly as fireproof insulating materials

as·bes·to·sis \ˌas-ˌbes-ˈtō-səs, ˌaz-\ *noun, plural* **-to·ses** \-ˌsēz\ (1927)
: a pneumoconiosis due to asbestos particles

asc- *or* **asco-** *combining form* [New Latin, from *ascus*]
: ascus ⟨*asco*carp⟩

as·ca·ri·a·sis \ˌas-kə-ˈrī-ə-səs\ *noun, plural* **-a·ses** \-ˌsēz\ [New Latin] (circa 1888)
: infestation with or disease caused by ascarids

as·ca·rid \ˈas-kə-rəd\ *noun* [ultimately from Late Latin *ascarid-, ascaris* intestinal worm, from Greek *askarid-, askaris,* probably by back-formation from *askarizein* to jump, throb, alteration of *skarizein,* from *skairein* to gambol] (circa 1890)
: any of a family (Ascaridae) of nematode worms that includes the common roundworm (*Ascaris lumbricoides*) parasitic in the human intestine

as·ca·ris \ˈas-kə-rəs\ *noun, plural* **as·car·i·des** \a-ˈskar-ə-ˌdēz\ (14th century)
: ASCARID

as·cend \ə-ˈsend\ *verb* [Middle English, from Latin *ascendere,* from *ad-* + *scandere* to climb — more at SCAN] (14th century)
intransitive verb
1 a : to move upward **b :** to slope upward
2 a : to rise from a lower level or degree **b :** to go back in time or in order of genealogical succession
transitive verb
1 : to go or move up ⟨*ascend* a staircase⟩
2 : to succeed to **:** OCCUPY ⟨*ascend* the throne⟩
— **as·cend·able** *or* **as·cend·ible** \-ˈsen-də-bəl\ *adjective*

as·cen·dance *also* **as·cen·dence** \ə-ˈsen-dən(t)s\ *noun* (1742)
: ASCENDANCY

as·cen·dan·cy *also* **as·cen·den·cy** \ə-ˈsen-dən(t)-sē\ *noun* (1712)
: governing or controlling influence **:** DOMINATION

[1]as·cen·dant *also* **as·cen·dent** \ə-ˈsen-dənt\ *noun* [Middle English *ascendent,* from Medieval Latin *ascendent-, ascendens,* from Latin, present participle of *ascendere*] (14th century)
1 : the point of the ecliptic or degree of the zodiac that rises above the eastern horizon at any moment
2 : a state or position of dominant power or importance
3 : a lineal or collateral relative in the ascending line

[2]ascendant *also* **ascendent** *adjective* (1591)
1 a : moving upward **:** RISING **b :** directed upward ⟨an *ascendant* stem⟩
2 a : SUPERIOR **b :** DOMINANT
— **as·cen·dant·ly** *adverb*

as·cend·er \ə-ˈsen-dər, ˈa-\ *noun* (circa 1867)
: the part of a lowercase letter (as b) that rises above the main body of the letter; *also* **:** a letter that has such a part

as·cend·ing \ə-ˈsen-diŋ\ *adjective* (1599)
1 a : mounting or sloping upward **b :** rising or increasing to higher levels, values, or degrees ⟨*ascending* powers of *x*⟩
2 : rising upward usually from a more or less prostrate base or point of attachment

as·cen·sion \ə-ˈsen(t)-shən\ *noun* [Middle English, from Latin *ascension-, ascensio,* from *ascendere*] (14th century)
: the act or process of ascending

as·cen·sion·al \ə-ˈsench-nəl, ə-ˈsen(t)-shə-nᵊl\ *adjective* (1594)
: of or relating to ascension or ascent

Ascension Day *noun* (14th century)
: the Thursday 40 days after Easter observed in commemoration of Christ's ascension into Heaven

as·cen·sive \ə-ˈsen(t)-siv\ *adjective* (1646)
: rising or tending to rise

as·cent \ə-ˈsent, a-\ *noun* [irregular from *ascend*] (circa 1596)
1 a : the act of rising or mounting upward **:** CLIMB **b :** an upward slope or rising grade **:** ACCLIVITY **c :** the degree of elevation **:** INCLINATION, GRADIENT
2 : an advance in social status or reputation **:** PROGRESS
3 : a going back in time or upward in order of genealogical succession

as·cer·tain \ˌa-sər-ˈtān\ *transitive verb* [Middle English *acertainen,* from Middle French *acertainer,* from *a-* (from Latin *ad-*) + *certain*] (15th century)
1 *archaic* **:** to make certain, exact, or precise
2 : to find out or learn with certainty
synonym see DISCOVER
— **as·cer·tain·able** \-ˈtā-nə-bəl\ *adjective*
— **as·cer·tain·ment** \-ˈtān-mənt\ *noun*

as·ce·sis \ə-ˈsē-səs\ *noun, plural* **-ce·ses** \-ˈsē-(ˌ)sēz\ [Late Latin or Greek; Late Latin, from Greek *askēsis,* literally, exercise, from *askein*] (1873)
: SELF-DISCIPLINE, ASCETICISM

as·cet·ic \ə-ˈse-tik, a-\ *also* **as·cet·i·cal** \-ti-kəl\ *adjective* [Greek *askētikos,* literally, laborious, from *askētēs* one that exercises, hermit, from *askein* to work, exercise] (1646)
1 : practicing strict self-denial as a measure of personal and especially spiritual discipline
2 : austere in appearance, manner, or attitude
synonym see SEVERE
— **ascetic** *noun*
— **as·cet·i·cal·ly** \-ti-k(ə-)lē\ *adverb*
— **as·cet·i·cism** \-ˈse-tə-ˌsi-zəm\ *noun*

as·cid·i·an \ə-ˈsi-dē-ən\ *noun* [New Latin *Ascidia,* group comprising tunicates, from *Ascidium,* genus name, from Greek *askidion,* diminutive of *askos* wineskin, bladder] (1856)
: any of a class (Ascidiacea) of solitary or colonial sessile tunicates that have an oral and an atrial siphon — called also *sea squirt*

ASCII \ˈas-(ˌ)kē\ *noun* [American Standard Code for Information Interchange] (1963)
: a code for representing alphanumeric information

as·ci·tes \ə-ˈsī-tēz\ *noun, plural* **ascites** [Middle English *aschytes,* from Late Latin *ascites,* from Greek *askitēs,* from *askos*] (14th century)
: accumulation of serous fluid in the spaces between tissues and organs in the cavity of the abdomen
— **as·cit·ic** \-ˈsi-tik\ *adjective*

as·cle·pi·ad \ə-ˈsklē-pē-əd, a-, -ˌad\ *noun* [ultimately from Greek *asklēpiad-, asklēpias* swallowwort, from *Asklēpios,* Greek god of medicine] (1859)
: MILKWEED

as·co·carp \ˈas-kə-ˌkärp\ *noun* (circa 1887)
: the mature fruiting body of an ascomycetous fungus; *broadly* **:** such a body with its enclosed asci, spores, and paraphyses
— **as·co·car·pic** \ˌas-kə-ˈkär-pik\ *adjective*

as·co·go·ni·um \ˌas-kə-ˈgō-nē-əm\ *noun, plural* **-nia** \-nē-ə\ [New Latin] (1875)
: the female sex organ in ascomycetous fungi

as·co·my·cete \ˌas-kō-ˈmī-ˌsēt, -ˌmī-ˈsēt\ *noun* [ultimately from Greek *askos* + *mykēt-, mykēs* fungus; akin to Greek *myxa* mucus — more at MUCUS] (1875)
: any of a class (Ascomycetes) or subdivision (Ascomycotina) of higher fungi (as yeasts or molds) with septate hyphae and spores formed in asci
— **as·co·my·ce·tous** \-ˌmī-ˈsē-təs\ *adjective*

ascor·bate \ə-ˈskȯr-ˌbāt, -bət\ *noun* (1941)
: a salt of ascorbic acid

ascor·bic acid \ə-ˈskȯr-bik-\ *noun* [International Scientific Vocabulary *a-* + New Latin *scorbutus* scurvy — more at SCORBUTIC] (1933)
: VITAMIN C

as·co·spore \ˈas-kə-ˌspōr, -ˌspȯr\ *noun* (1875)
: any of the spores contained in an ascus
— **as·co·spor·ic** \ˌas-kə-ˈspōr-ik, -ˈspȯr-\ *adjective*

as·cot \ˈas-kət, -ˌkät\ *noun* [*Ascot* Heath, racetrack near Ascot, England] (1898)
: a broad neck scarf that is looped under the chin

as·cribe \ə-ˈskrīb\ *transitive verb* **as·cribed; as·crib·ing** [Middle English, from Latin *ascribere,* from *ad-* + *scribere* to write — more at SCRIBE] (15th century)
: to refer to a supposed cause, source, or author ☆
— **as·crib·able** \-ˈskrī-bə-bəl\ *adjective*

ascribed *adjective* (1972)
: acquired or assigned arbitrarily (as at birth) ⟨*ascribed* social status⟩

as·crip·tion \ə-ˈskrip-shən\ *noun* [Late Latin *ascription-, ascriptio,* from Latin, written addition, from *ascribere*] (1600)
1 : the act of ascribing **:** ATTRIBUTION

☆ SYNONYMS
Ascribe, attribute, assign, impute, credit mean to lay something to the account of a person or thing. ASCRIBE suggests an inferring or conjecturing of cause, quality, authorship ⟨forged paintings formerly *ascribed* to masters⟩. ATTRIBUTE suggests less tentativeness than ASCRIBE, less definiteness than ASSIGN ⟨*attributed* to Rembrandt but possibly done by an associate⟩. ASSIGN implies ascribing with certainty or after deliberation ⟨*assigned* the bones to the Cretaceous Period⟩. IMPUTE suggests ascribing something that brings discredit by way of accusation or blame ⟨tried to *impute* sinister motives to my actions⟩. CREDIT implies ascribing a thing or especially an action to a person or other thing as its agent, source, or explanation ⟨*credited* his teammates for his success⟩.

2 : arbitrary placement (as at birth) in a particular social status

as·crip·tive \ə-ˈskrip-tiv\ *adjective* (1650)
: relating to, marked by, or involving ascription

as·cus \ˈas-kəs\ *noun, plural* **as·ci** \ˈas-ˌkī, -ˌkē, ˈa-ˌsī\ [New Latin, from Greek *askos* wineskin, bladder] (1830)
: the membranous oval or tubular spore case of an ascomycete

as·dic \ˈaz-(ˌ)dik\ *noun* [from *asdics* underwater echo ranging, probably from *Anti-Submarine Division* (British Admiralty department 1916–18) + *-ics*] (1939)
chiefly British **:** SONAR
word history see SONAR

-ase *noun suffix* [French, from *diastase*]
: enzyme ⟨prote*ase*⟩

asep·sis \(ˌ)ā-ˈsep-səs, ə-\ *noun* [New Latin] (1892)
1 : the condition of being aseptic
2 : the methods of making or keeping aseptic

asep·tic \(ˌ)ā-ˈsep-tik, ə-\ *adjective* [International Scientific Vocabulary] (circa 1859)
1 a : preventing infection ⟨*aseptic* techniques⟩ **b :** free or freed from pathogenic microorganisms ⟨an *aseptic* operating room⟩
2 : lacking vitality, emotion, or warmth ⟨the bureaucrat's world is prim and proper and *aseptic* —R. M. Weaver⟩
— asep·ti·cal·ly \-ti-k(ə-)lē\ *adverb*

asex·u·al \(ˌ)ā-ˈsek-sh(ə-)wəl, -ˈsek-shəl\ *adjective* (1830)
1 : lacking sex or functional sex organs ⟨*asexual* plants⟩
2 a : involving or reproducing by reproductive processes (as cell division, spore formation, fission, or budding) that do not involve the union of individuals or germ cells ⟨*asexual* reproduction⟩ **b :** produced by asexual reproduction ⟨*asexual* spores⟩
3 : devoid of sexuality ⟨an *asexual* relationship⟩
— asex·u·al·i·ty \ˌā-ˌsek-shə-ˈwa-lə-tē\ *noun*
— asex·u·al·ly \(ˌ)ā-ˈsek-sh(ə-)wə-lē, -ˌsek-sh(ə-)lē\ *adverb*

[1]as far as *conjunction* (14th century)
: to the extent or degree that ⟨is safe, *as far as* we know⟩ — often used in expressions like "as far as (something) goes" and "as far as (something) is concerned" to mean "with regard to (something)" ⟨we felt pretty safe *as far as* the fire was concerned —Mark Twain⟩ or in expressions like "as far as (someone) is concerned" to mean "in (someone's) opinion" ⟨*as far as* I'm concerned, it's a mistake⟩

[2]as far as *preposition* (1523)
: with regard to **:** CONCERNING ⟨neatly groomed and, *as far as* clothes, casual looking —*N.Y. Times*⟩ ⟨*as far as* being mentioned in the Ten Commandments, I think it is —Billy Graham⟩ — chiefly in oral use

as for *preposition* (1533)
: with regard to **:** CONCERNING ⟨*as for* the others, they'll arrive later⟩

As·gard \ˈas-ˌgärd, ˈaz-\ *noun* [Old Norse *āsgarthr*]
: the home of the Norse gods

[1]ash \ˈash\ *noun* [Middle English *asshe,* from Old English *æsc;* akin to Old High German *ask* ash, Latin *ornus* mountain ash] (before 12th century)
1 : any of a genus (*Fraxinus*) of trees of the olive family with pinnate leaves, thin furrowed bark, and gray branchlets
2 : the tough elastic wood of an ash
3 [Old English *æsc,* name of the corresponding runic letter] **:** the ligature æ used in Old English and some phonetic alphabets to represent a low front vowel \a\

[2]ash *noun, often attributive* [Middle English *asshe,* from Old English *asce* — more at ARID] (before 12th century)
1 : something that symbolizes grief, repentance, or humiliation
2 a : the solid residue left when combustible material is thoroughly burned or is oxidized by chemical means **b :** fine particles of mineral matter from a volcanic vent
3 *plural* **:** the remains of the dead human body after cremation or disintegration
4 *plural* **:** deathly pallor ⟨the lip of *ashes* and the cheek of flame —Lord Byron⟩
5 *plural* **:** RUINS
— ash·less \-ləs\ *adjective*

[3]ash *transitive verb* (1894)
: to convert into ash

ashamed \ə-ˈshāmd\ *adjective* [Middle English, from Old English *āscamod,* past participle of *āscamian* to shame, from *ā-* (perfective prefix) + *scamian* to shame — more at ABIDE, SHAME] (before 12th century)
1 a : feeling shame, guilt, or disgrace **b :** feeling inferior or unworthy
2 : restrained by anticipation of shame ⟨was *ashamed* to beg⟩
— asham·ed·ly \-ˈshā-məd-lē\ *adverb*

Ashan·ti \ə-ˈshan-tē, -ˈshän-\ *noun, plural* **Ashanti** *or* **Ashantis** [Twi *àsàntè*] (1721)
1 : a member of a people of southern Ghana
2 : the dialect of Akan spoken by the Ashanti people

ash–blond *or* **ash–blonde** \ˈash-ˈbländ\ *adjective* (1903)
: pale or grayish blond ⟨*ash-blond* hair⟩

Ash·can \ˈash-ˌkan\ *adjective* (1939)
: of or relating to a group of 20th century American painters who depicted city life realistically ⟨*Ashcan* school⟩

ash can *noun* (1899)
1 : a metal receptacle for refuse
2 *slang* **:** DEPTH CHARGE

[1]ash·en \ˈa-shən\ *adjective* (before 12th century)
: of, relating to, or made from ash wood

[2]ashen *adjective* (14th century)
: resembling ashes (as in color); *especially* **:** deadly pale ⟨a face *ashen* and haggard⟩

Ash·er \ˈa-shər\ *noun* [Hebrew *Āshēr*]
: a son of Jacob and the traditional eponymous ancestor of one of the tribes of Israel

ash·fall \ˈash-ˌfȯl\ *noun* (1923)
: a deposit of volcanic ash

Ash·ke·nazi \ˌäsh-kə-ˈnä-zē, ˌash-kə-ˈna-\ *noun, plural* **-naz·im** \-ˈnä-zəm, -ˈna-\ [Late Hebrew *Ashkĕnāzī,* from *Ashkĕnāz,* medieval rabbinical name for Germany] (1839)
: a member of one of the two great divisions of Jews comprising the eastern European Yiddish-speaking Jews — compare SEPHARDI
— Ash·ke·naz·ic \-ˈnä-zik, -ˈna-\ *adjective*

ash·lar \ˈash-lər\ *noun* [Middle English *asheler,* from Middle French *aisselier* traverse beam, from Old French, from *ais* board, from Latin *axis,* alteration of *assis*] (14th century)
1 : hewn or squared stone; *also* **:** masonry of such stone
2 : a thin squared and dressed stone for facing a wall of rubble or brick

ashore \ə-ˈshōr, -ˈshȯr\ *adverb* (1582)
: on or to the shore

as how *conjunction* (1771)
: THAT ⟨allowed *as how* she was glad to be here⟩

ash·ram \ˈäsh-rəm, -ˌräm; ˈash-ˌram\ *noun* [Sanskrit *āśrama,* from *śrama* religious exercise] (1917)
1 : a secluded dwelling of a Hindu sage; *also* **:** the group of disciples instructed there
2 : a religious retreat

Ash·to·reth \ˈash-tə-ˌreth\ *noun* [Hebrew *ʽAshtōreth*]
: ASTARTE

ash·tray \ˈash-ˌtrā\ *noun* (1887)
: a receptacle for tobacco ashes and for cigar and cigarette butts

Ashur \ˈä-ˌshu̇r\ *noun* [Akkadian *Ashūr*]
: the chief deity of the Assyrians

Ash Wednesday *noun* (14th century)
: the first day of Lent — see EASTER table

ashy \ˈa-shē\ *adjective* **ash·i·er; -est** (14th century)
1 : of or relating to ashes
2 : ASHEN

Asi·a·go \ˌä-zhē-ˈä-(ˌ)gō, ˌä-sē, -shē-\ *noun* [*Asiago,* town in Italy] (1938)
: a pungent hard yellow cheese of Italian origin suitable for grating

[1]Asian \ˈā-zhən *also* -shən\ *adjective* (1550)
: of, relating to, or characteristic of the continent of Asia or its people ◼

[2]Asian *noun* (circa 1890)
: a native or inhabitant of Asia

Asian–Amer·i·can \-ə-ˈmer-ə-kən\ *noun* (1974)
: an American of Asian descent
— Asian–American *adjective*

Asian elephant *noun* (1981)
: ELEPHANT 1b

Asian influenza *noun* (1957)
: influenza caused by a mutant strain of the influenza virus isolated during the 1957 epidemic in Asia — called also *Asian flu*

Asi·at·ic \ˌā-zhē-ˈa-tik, -zē-\ *adjective* (1602)
: ASIAN — sometimes taken to be offensive
— Asiatic *noun*
usage see ASIAN

Asiatic cholera *noun* (1831)
: cholera of Asian origin that is caused by virulent strains of the cholera vibrio (*Vibrio cholerae*)

Asiatic elephant *noun* (1930)
: ELEPHANT 1b

[1]aside \ə-ˈsīd\ *adverb* (14th century)
1 : to or toward the side ⟨stepped *aside*⟩
2 : away from others or into privacy ⟨pulled him *aside*⟩
3 : out of the way especially for future use **:** AWAY ⟨putting *aside* savings⟩
4 : away from one's thought or consideration ⟨jesting *aside*⟩

[2]aside *preposition* (1592)
obsolete **:** BEYOND, PAST

[3]aside *noun* (circa 1751)
1 : an utterance meant to be inaudible to someone; *especially* **:** an actor's speech heard by the audience but supposedly not by other characters
2 : a straying from the theme **:** DIGRESSION

aside from *preposition* (1818)
1 : in addition to **:** BESIDES
2 : EXCEPT FOR

as if *conjunction* (13th century)
1 : as it would be if ⟨it was *as if* he had lost his last friend⟩
2 : as one would do if ⟨he ran *as if* ghosts were chasing him⟩
3 : THAT ⟨it seemed *as if* the day would never end⟩

as·i·nine \ˈa-sᵊn-ˌīn\ *adjective* [Latin *asininus,* from *asinus* ass] (15th century)
1 : marked by inexcusable failure to exercise

☐ USAGE
Asian As we move toward the end of the 20th century, *Asian* is the preferred word especially in ethnic reference, because *Asiatic* is held to be at least mildly offensive. *Asiatic* appears to have been the more common term up until the 1940s, when it began to be considered offensive, especially as associated with the policies of colonialism. While *Asian* prevails today, *Asiatic* still has some use as a geographical designator ⟨the *Asiatic* side of the Bosporus⟩. Such use appears to be more frequent in British English than in American.

intelligence or sound judgment ⟨an *asinine* excuse⟩
2 : of, relating to, or resembling an ass
synonym see SIMPLE
— as·i·nine·ly *adverb*
— as·i·nin·i·ty \ˌa-sə-ˈni-nə-tē\ *noun*

ask \ˈask, ˈȧsk; *dialect* ˈaks\ *verb* **asked** \ˈas(k)t, ˈȧs(k)t, ˈask; *dialect* ˈakst\; **ask·ing** [Middle English, from Old English *āscian;* akin to Old High German *eiscōn* to ask, Lithuanian *eiškoti* to seek, Sanskrit *icchati* he seeks] (before 12th century)
transitive verb
1 a : to call on for an answer **b :** to put a question about **c :** SPEAK, UTTER ⟨*ask* a question⟩
2 a : to make a request of ⟨she *asked* her teacher for help⟩ **b :** to make a request for ⟨she *asked* help from her teacher⟩
3 : to call for **:** REQUIRE
4 : to set as a price ⟨*asked* $3000 for the car⟩
5 : INVITE
intransitive verb
1 : to seek information
2 : to make a request ⟨*asked* for food⟩
3 : LOOK — often used in the phrase *ask for trouble* ☆ ☆ ■
— ask·er *noun*

askance \ə-ˈskan(t)s\ *also* **askant** \-ˈskant\ *adverb* [origin unknown] (circa 1530)
1 : with a side-glance **:** OBLIQUELY
2 : with disapproval or distrust **:** SCORNFULLY

as·ke·sis \ə-ˈskē-səs\ *variant of* ASCESIS

askew \ə-ˈskyü\ *adverb or adjective* [probably from [1]*a-* + *skew*] (1573)
: out of line **:** AWRY ⟨the picture hung *askew*⟩
— askew·ness *noun*

asking price *noun* (1755)
: the price at which something is offered for sale

[1]aslant \ə-ˈslant\ *adverb or adjective* (14th century)
: in a slanting direction **:** OBLIQUELY

[2]aslant *preposition* (1602)
: over or across in a slanting direction

[1]asleep \ə-ˈslēp\ *adjective* [Middle English *aslepe,* from Old English *on slæpe*] (13th century)
1 : being in a state of sleep
2 : DEAD
3 : lacking sensation **:** NUMB
4 a : INACTIVE, DORMANT **b :** not alert **:** INDIFFERENT

[2]asleep *adverb* (13th century)
1 : into a state of sleep
2 : into the sleep of death
3 : into a state of inactivity, sluggishness, or indifference

as long as *conjunction* (15th century)
1 : provided that ⟨can do as they like *as long as* they have a B average⟩
2 : INASMUCH AS, SINCE ⟨*as long as* you're going, I'll go too⟩

aslope \ə-ˈslōp\ *adjective or adverb* (14th century)
: being in a sloping or slanting position or direction

aso·cial \(ˌ)ā-ˈsō-shəl\ *adjective* (1883)
: not social: as **a :** rejecting or lacking the capacity for social interaction **b :** ANTISOCIAL

as of *preposition* (1900)
: ON, AT, FROM — used to indicate a time or date at which something begins or ends ⟨takes effect *as of* July 1⟩

[1]asp \ˈasp\ *noun* [Middle English, from Old English *æspe*] (before 12th century)
: ASPEN

[2]asp *noun* [Middle English *aspis,* from Latin, from Greek] (14th century)
: a small venomous snake of Egypt usually held to be a cobra (*Naja haje*)

as·par·a·gine \ə-ˈspar-ə-ˌjēn\ *noun* [French, from Latin *asparagus*] (1813)
: a nonessential amino acid $C_4H_8N_2O_3$ that is an amide of aspartic acid

as·par·a·gus \ə-ˈspar-ə-gəs\ *noun, plural* **-gus** [New Latin, genus name, from Latin, asparagus plant, from Greek *asparagos;* perhaps akin to Greek *spargan* to swell] (1548)
: any of a genus (*Asparagus*) of Old World perennial plants of the lily family having much-branched stems, minute scalelike leaves, and narrow usually filiform branchlets that function as leaves; *especially* **:** one (*A. officinalis*) widely cultivated for its edible young shoots

asparagus

as·par·tame \ˈas-pər-ˌtām, ə-ˈspär-ˌtām\ *noun* [*aspartic* acid + *phenylalanine* + *methyl* + *ester*] (1972)
: a crystalline compound $C_{14}H_{18}N_2O_5$ that is a diamide synthesized from phenylalanine and aspartic acid and that is used as a low-calorie sweetener

as·par·tate \ə-ˈspär-ˌtāt\ *noun* (1863)
: a salt or ester of aspartic acid

as·par·tic acid \ə-ˈspär-tik-\ *noun* [International Scientific Vocabulary, irregular from Latin *asparagus*] (1863)
: a crystalline amino acid $C_4H_7NO_4$ found especially in plants

as·pect \ˈas-ˌpekt\ *noun* [Middle English, from Latin *aspectus,* from *aspicere* to look at, from *ad-* + *specere* to look — more at SPY] (14th century)
1 a : the position of planets or stars with respect to one another held by astrologers to influence human affairs; *also* **:** the apparent position (as conjunction) of a body in the solar system with respect to the sun **b :** a position facing a particular direction **:** EXPOSURE ⟨the house has a southern *aspect*⟩ **c :** the manner of presentation of a plane to a fluid through which it is moving or to a current
2 a (1) **:** appearance to the eye or mind (2) **:** a particular appearance of countenance **:** MIEN **b :** a particular status or phase in which something appears or may be regarded ⟨studied every *aspect* of the question⟩
3 *archaic* **:** an act of looking **:** GAZE
4 a : the nature of the action of a verb as to its beginning, duration, completion, or repetition and without reference to its position in time **b :** a set of inflected verb forms that indicate aspect
— as·pec·tu·al \a-ˈspek-chə(-wə)l\ *adjective*

aspect ratio *noun* (1907)
: a ratio of one dimension to another: as **a :** the ratio of span to mean chord of an airfoil **b :** the ratio of the width of a television or motion-picture image to its height

as·pen \ˈas-pən\ *noun* [Middle English, of an aspen, from *asp* aspen, from Old English *æspe;* akin to Old High German *aspa* aspen, Russian *osina*] (1596)
: any of several poplars (especially *Populus tremula* of Europe and *P. tremuloides* and *P. grandidentata* of North America) with leaves that flutter in the lightest wind because of their flattened petioles

as per \ˈaz-ˌpər\ *preposition* (1859)
: in accordance with **:** ACCORDING TO
— as per usual : as usual

as·per·ges \ə-ˈspər-(ˌ)jēz\ *noun* [Latin, thou wilt sprinkle, from *aspergere*] (circa 1587)
: a ceremony of sprinkling altar and people with holy water

as·per·gil·lo·sis \ˌas-pər-(ˌ)ji-ˈlō-səs\ *noun, plural* **-lo·ses** \-ˌsēz\ (1898)
: infection with or disease caused (as in poultry) by molds (genus *Aspergillus*)

as·per·gil·lum \ˌas-pər-ˈji-ləm\ *noun, plural* **-la** \-lə\ *or* **-lums** [New Latin, from Latin *aspergere*] (1649)
: a brush or small perforated container with a handle that is used for sprinkling holy water in a liturgical service

aspergillum

as·per·gil·lus \-ˈji-ləs\ *noun, plural* **-gil·li** \-ˈji-ˌlī\ [New Latin, genus name, from *aspergillum*] (circa 1847)
: any of a genus (*Aspergillus*) of ascomycetous fungi with branched radiate sporophores including many common molds

as·per·i·ty \a-ˈsper-ə-tē, ə-\ *noun, plural* **-ties** [Middle English *asprete,* from Old French *aspreté,* from *aspre* rough, from Latin *asper,* from (assumed) Old Latin *absperos,* from *ab-* + *-speros;* akin to Sanskrit *apasphura* repelling, Latin *spernere* to spurn — more at SPURN] (13th century)
1 : RIGOR, SEVERITY
2 a : roughness of surface **:** UNEVENNESS; *also*

☆ SYNONYMS
Ask, question, interrogate, query, inquire mean to address a person in order to gain information. ASK implies no more than the putting of a question ⟨*ask* for directions⟩. QUESTION usually suggests the asking of series of questions ⟨*questioned* them about every detail of the trip⟩. INTERROGATE suggests formal or official systematic questioning ⟨the prosecutor *interrogated* the witness all day⟩. QUERY implies a desire for authoritative information or confirmation ⟨*queried* a librarian about the book⟩. INQUIRE implies a searching for facts or for truth often specifically by asking questions ⟨began to *inquire* of friends and teachers what career she should pursue⟩.

Ask, request, solicit mean to seek to obtain by making one's wants known. ASK implies no more than the statement of the desire ⟨*ask* a favor of a friend⟩. REQUEST implies greater formality and courtesy ⟨*requests* the pleasure of your company⟩. SOLICIT suggests a calling attention to one's wants or desires by public announcement or advertisement ⟨a letter *soliciting* information⟩.

□ USAGE
ask What to some is a usage blunder is to dictionary editors an interesting subject for study called metathesis. A case in point is the positional switch of \s\ and \k\ in *ask.* Though often characterized as a feature of Black English, the pronunciation \ˈaks\ is found also among white speakers, especially in Southern dialects. This form has been competing with the etymologically truer forms \ˈask\ and \ˈȧsk\ for ten centuries or more. In many Old English manuscripts the root is spelled *ax-,* and that the *ask-* form found in other manuscripts is the original is known only from comparisons across several languages. A number of modern English words (as *bird, wasp, horse, task*) are the product of metathesis, as a check of their etymologies shows. Noah Webster defended the \ˈaks\ pronunciation of *ask* in his 1806 dictionary as an acceptable Americanism. Indeed, those who denounce this form should note their own pronunciation of *iron,* in which, unless they speak an uncommon dialect, they will find that they transpose the sounds of the *r* and *o.* Metathesis, then, is not the same as mispronunciation, although in modern English it leads to a noticeable dialectal pronunciation of *ask.*

: a tiny projection from a surface **b :** roughness of sound
3 : roughness of manner or of temper **:** HARSHNESS

as·perse \ə-ˈspərs, a-\ *transitive verb* **aspersed; as·pers·ing** [Latin *aspersus,* past participle of *aspergere,* from *ad-* + *spargere* to scatter — more at SPARK] (15th century)
1 : SPRINKLE; *especially* **:** to sprinkle with holy water
2 : to attack with evil reports or false or injurious charges
synonym see MALIGN

as·per·sion \ə-ˈspər-zhən, -shən\ *noun* (circa 1587)
1 : a sprinkling with water especially in religious ceremonies
2 a : the act of calumniating **b :** a calumnious expression ⟨cast *aspersions* on her integrity⟩

[1]**as·phalt** \ˈas-ˌfȯlt *also* ˈash-, *especially British* -ˌfalt\ *also* **as·phal·tum** \as-ˈfȯl-təm, *especially British* -ˈfal-\ *noun* [Middle English *aspalt,* from Late Latin *aspaltus,* from Greek *asphaltos*] (14th century)
1 : a dark bituminous substance that is found in natural beds and is also obtained as a residue in petroleum refining and that consists chiefly of hydrocarbons
2 : an asphaltic composition used for pavements and as a waterproof cement
— as·phal·tic \as-ˈfȯl-tik, *especially British* -ˈfal-\ *adjective*

[2]**asphalt** *transitive verb* (circa 1859)
: to cover with asphalt **:** PAVE 1

as·phalt·ite \ˈas-ˌfȯl-ˌtīt, *especially British* -ˌfal-\ *noun* (circa 1899)
: a native asphalt occurring in vein deposits below the surface of the ground

asphalt jungle *noun* (1920)
: a big city or a specified part of a big city

aspher·ic \(ˌ)ā-ˈsfir-ik, -ˈsfer-\ *or* **aspher·i·cal** \-i-kəl\ *adjective* (circa 1922)
: departing slightly from the spherical form especially in order to correct for spherical aberration ⟨an *aspheric* lens⟩

as·pho·del \ˈas-fə-ˌdel\ *noun* [Latin *asphodelus,* from Greek *asphodelos*] (1597)
: any of various Old World usually perennial herbs (especially genera *Asphodelus* and *Asphodeline*) of the lily family with flowers in usually long erect racemes

as·phyx·ia \as-ˈfik-sē-ə, əs-\ *noun* [New Latin, from Greek, stopping of the pulse, from *a-* + *sphyzein* to throb] (1778)
: a lack of oxygen or excess of carbon dioxide in the body that is usually caused by interruption of breathing and that causes unconsciousness

as·phyx·i·ate \-sē-ˌāt\ *verb* **-at·ed; -at·ing** (1836)
transitive verb
: to cause asphyxia in; *also* **:** to kill or make unconscious by inadequate oxygen, presence of noxious agents, or other obstruction to normal breathing
intransitive verb
: to become asphyxiated
— as·phyx·i·a·tion \-ˌfik-sē-ˈā-shən\ *noun*

[1]**as·pic** \ˈas-pik\ *noun* [Middle French, alteration of *aspe,* from Latin *aspis*] (1530)
obsolete **:** ASP

[2]**aspic** *noun* [French, literally, asp] (1789)
: a clear savory jelly (as of fish or meat stock) used as a garnish or to make a meat, fish, or vegetable mold

as·pi·dis·tra \ˌas-pə-ˈdis-trə\ *noun* [New Latin, irregular from Greek *aspid-, aspis* shield] (1822)
: an Asian plant (*Aspidistra elatior*) of the lily family that has large basal leaves and is often grown as a foliage plant

[1]**as·pi·rant** \ˈas-p(ə-)rənt, ə-ˈspī-rənt\ *noun* (1738)
: one who aspires ⟨presidential *aspirants*⟩

[2]**aspirant** *adjective* (1814)
: seeking to attain a desired position or status

[1]**as·pi·rate** \ˈas-p(ə-)rət\ *noun* (1617)
1 : an independent sound \h\ or a character (as the letter *h*) representing it
2 : a consonant having aspiration as its final component ⟨in English the \p\ of *pit* is an *aspirate*⟩
3 : material removed by aspiration

[2]**as·pi·rate** \ˈas-pə-ˌrāt\ *transitive verb* **-rat·ed; -rat·ing** [Latin *aspiratus,* past participle of *aspirare*] (circa 1700)
1 : to pronounce (a vowel or a consonant) with aspiration (sense 1a)
2 a : to draw by suction **b :** to remove (as blood) by aspiration **c :** to take into the lungs by aspiration

as·pi·ra·tion \ˌas-pə-ˈrā-shən\ *noun* (14th century)
1 a : audible breath that accompanies or comprises a speech sound **b :** the pronunciation or addition of an aspiration; *also* **:** the symbol of an aspiration
2 : a drawing of something in, out, up, or through by or as if by suction: as **a :** the act of breathing and especially of breathing in **b :** the withdrawal of fluid or tissue from the body **c :** the taking of foreign matter into the lungs with the respiratory current
3 a : a strong desire to achieve something high or great **b :** an object of such desire
synonym see AMBITION
— as·pi·ra·tion·al \-ˈrā-sh(ə-)nəl\ *adjective*

as·pi·ra·tor \ˈas-pə-ˌrā-tər\ *noun* (1804)
: an apparatus for producing suction or moving or collecting materials by suction; *especially* **:** a hollow tubular instrument connected with a partial vacuum and used to remove fluid or tissue or foreign bodies from the body

as·pire \ə-ˈspīr\ *intransitive verb* **as·pired; as·pir·ing** [Middle English, from Middle French or Latin; Middle French *aspirer,* from Latin *aspirare,* literally, to breathe upon, from *ad-* + *spirare* to breathe] (14th century)
1 : to seek to attain or accomplish a particular goal ⟨*aspired* to a career in medicine⟩
2 : ASCEND, SOAR
— as·pir·er *noun*

as·pi·rin \ˈas-p(ə-)rən\ *noun, plural* **aspirin** *or* **aspirins** [International Scientific Vocabulary, from *acetyl* + *spiraeic acid* (former name of salicylic acid), from New Latin *Spiraea,* genus of shrubs — more at SPIREA] (1899)
1 : a white crystalline derivative $C_9H_8O_4$ of salicylic acid used for relief of pain and fever
2 : a tablet of aspirin

as regards *also* **as respects** *preposition* (1840)
: in regard to **:** with respect to

[1]**ass** \ˈas\ *noun* [Middle English, from Old English *assa,* probably from Old Irish *asan,* from Latin *asinus*] (before 12th century)
1 : any of several hardy gregarious African or Asian perissodactyl mammals (genus *Equus*) smaller than the horse and having long ears; *especially* **:** an African mammal (*E. asinus*) that is the ancestor of the donkey
2 : a stupid, obstinate, or perverse person — often compounded with a preceding adjective ⟨don't be a smart-*ass*⟩; often considered vulgar

[2]**ass** \ˈas\ *or* **arse** \ˈas, ˈars\ *noun* [Middle English *ars, ers,* from Old English *ærs, ears;* akin to Old High German & Old Norse *ars* buttocks, Greek *orrhos* buttocks, *oura* tail] (before 12th century)
1 a : BUTTOCKS — often used in emphatic reference to a specific person ⟨get your *ass* over here⟩ ⟨saved my *ass*⟩; often considered vulgar **b :** ANUS — often considered vulgar
2 : SEXUAL INTERCOURSE — usually considered vulgar

[3]**ass** *adverb* [[2]*ass*] (1955)
— used as a postpositive intensive especially with words of derogatory implication ⟨fancy-*ass*⟩; often considered vulgar

as·sai \ä-ˈsī\ *adverb* [Italian, from (assumed) Vulgar Latin *ad satis* enough — more at ASSET] (circa 1724)
: VERY — used with tempo direction in music ⟨allegro *assai*⟩

as·sail \ə-ˈsā(ə)l\ *transitive verb* [Middle English, from Old French *asaillir,* from (assumed) Vulgar Latin *assalire,* alteration of Latin *assilire* to leap upon, from *ad-* + *salire* to leap — more at SALLY] (13th century)
: to attack violently with blows or words
synonym see ATTACK
— as·sail·able \-ˈsā-lə-bəl\ *adjective*
— as·sail·ant \-ˈsā-lənt\ *noun*

As·sam·ese \ˌa-sə-ˈmēz, -ˈmēs\ *noun, plural* **Assamese** (1826)
1 : a native or inhabitant of Assam, India
2 : the Indo-Aryan language of Assam
— Assamese *adjective*

as·sas·sin \ə-ˈsa-sᵊn\ *noun* [Medieval Latin *assassinus,* from Arabic *ḥashshāshīn,* plural of *ḥashshāsh* one who smokes or chews hashish, from *ḥashīsh* hashish] (circa 1520)
1 *capitalized* **:** one of a secret order of Muslims that at the time of the Crusades terrorized Christians and other enemies by secret murder committed under the influence of hashish
2 : a person who commits murder; *especially* **:** one who murders a politically important person either for hire or from fanatical motives

as·sas·si·nate \ə-ˈsa-sᵊn-ˌāt\ *transitive verb* **-nat·ed; -nat·ing** (1607)
1 : to injure or destroy unexpectedly and treacherously
2 : to murder by sudden or secret attack usually for impersonal reasons
synonym see KILL
— as·sas·si·na·tion \-ˌsa-sᵊn-ˈā-shən\ *noun*
— as·sas·si·na·tor \-ˈsa-sᵊn-ˌā-tər\ *noun*

assassin bug *noun* (1895)
: any of a family (Reduviidae) of bugs that are usually predatory on insects though some (as a conenose) suck the blood of mammals — called also *reduviid*

[1]**as·sault** \ə-ˈsȯlt\ *noun* [Middle English *asaut,* from Old French, from (assumed) Vulgar Latin *assaltus,* from *assalire*] (14th century)
1 a : a violent physical or verbal attack **b :** a military attack usually involving direct combat with enemy forces **c :** a concerted effort (as to reach a goal or defeat an adversary)
2 a : a threat or attempt to inflict offensive physical contact or bodily harm on a person (as by lifting a fist in a threatening manner) that puts the person in immediate danger of or in apprehension of such harm or contact — compare BATTERY 1b **b :** RAPE

[2]**assault** (15th century)
transitive verb
1 : to make an assault on
2 : RAPE
intransitive verb
: to make an assault
synonym see ATTACK
— as·sault·er *noun*
— as·saul·tive \-ˈsȯl-tiv\ *adjective*
— as·saul·tive·ly *adverb*
— as·saul·tive·ness *noun*

assault boat *noun* (1941)
: a small portable boat used in an amphibious military attack or in land warfare for crossing rivers or lakes

assault rifle *noun* (1975)
: any of various automatic or semiautomatic rifles designed for military use with large capacity magazines

[1]**as·say** \ˈa-ˌsā, a-ˈsā\ *noun* [Middle English, from Old French *essai, assai* test, effort — more at ESSAY] (14th century)

\ə\ **abut** \ᵊ\ **kitten** \ər\ **further** \a\ **ash** \ā\ **ace** \ä\ **mop, mar** \au̇\ **out** \ch\ **chin** \e\ **bet** \ē\ **easy** \g\ **go** \i\ **hit** \ī\ **ice** \j\ **job** \ŋ\ **sing** \ō\ **go** \ȯ\ **law** \ȯi\ **boy** \th\ **thin** \t͟h\ **the** \ü\ **loot** \u̇\ **foot** \y\ **yet** \zh\ **vision** *see also* Guide to Pronunciation

1 *archaic* **:** TRIAL, ATTEMPT
2 : examination and determination as to characteristics (as weight, measure, or quality)
3 : analysis (as of an ore or drug) to determine the presence, absence, or quantity of one or more components
4 : a substance to be assayed; *also* **:** the tabulated result of assaying

[2]as·say \a-ˈsā, ˈa-ˌsā\ (14th century)
transitive verb
1 : TRY, ATTEMPT
2 a : to analyze (as an ore) for one or more specific components **b :** to judge the worth of **:** ESTIMATE
intransitive verb
: to prove up in an assay
— as·say·er *noun*

assed \ˈast\ *adverb* (1932)
: [3]ASS — used in combination ⟨smart-*assed*⟩; often considered vulgar

as·se·gai *or* **as·sa·gai** \ˈa-si-ˌgī\ *noun* [ultimately from Arabic *az-zaghāya* the assegai, from *al-* the + *zaghāya* assegai] (1600)
: a slender hardwood spear or light javelin usually tipped with iron and used in southern Africa

as·sem·blage \ə-ˈsem-blij, *for 3 also* ˌa-ˌsäm-ˈbläzh\ *noun* (1690)
1 : a collection of persons or things **:** GATHERING
2 : the act of assembling **:** the state of being assembled
3 a : an artistic composition made from scraps, junk, and odds and ends (as of paper, cloth, wood, stone, or metal) **b :** the art of making assemblages

as·sem·blag·ist \-bli-jist, -ˈblä-zhist\ *noun* (1965)
: an artist who specializes in assemblages

as·sem·ble \ə-ˈsem-bəl\ *verb* **as·sem·bled; as·sem·bling** \-b(ə-)liŋ\ [Middle English, from Old French *assembler,* from (assumed) Vulgar Latin *assimulare,* from Latin *ad-* + *simul* together — more at SAME] (13th century)
transitive verb
1 : to bring together (as in a particular place or for a particular purpose)
2 : to fit together the parts of
intransitive verb
: to meet together **:** CONVENE
synonym see GATHER

as·sem·bler \-b(ə-)lər\ *noun* (1635)
1 : one that assembles
2 a : a computer program that automatically converts instructions written in assembly language into machine language **b :** ASSEMBLY LANGUAGE

as·sem·bly \ə-ˈsem-blē\ *noun, plural* **-blies** [Middle English *assemblee,* from Middle French, from Old French, from *assembler*] (14th century)
1 : a company of persons gathered for deliberation and legislation, worship, or entertainment
2 *capitalized* **:** a legislative body; *specifically* **:** the lower house of a legislature
3 : ASSEMBLAGE 1, 2
4 : a signal for troops to assemble or fall in
5 a : the fitting together of manufactured parts into a complete machine, structure, or unit of a machine **b :** a collection of parts so assembled
6 : the translation of assembly language to machine language by an assembler

assembly language *noun* (circa 1964)
: a programming language that consists of instructions that are mnemonic codes for corresponding machine language instructions

assembly line *noun* (1914)
1 : an arrangement of machines, equipment, and workers in which work passes from operation to operation in direct line until the product is assembled
2 : a process for turning out a finished product in a mechanically efficient manner ⟨academic *assembly lines*⟩

as·sem·bly·man \ə-ˈsem-blē-mən\ *noun* (1647)
: a member of an assembly

Assembly of God (1952)
: a congregation belonging to a Pentecostal body founded in the U.S. in 1914

as·sem·bly·wom·an \-ˌwu̇-mən\ *noun* (1969)
: a woman who is a member of an assembly

[1]as·sent \ə-ˈsent, a-\ *intransitive verb* [Middle English, from Old French *assenter,* from Latin *assentari,* from *assentire,* from *ad-* + *sentire* to feel — more at SENSE] (14th century)
: to agree to something especially after thoughtful consideration **:** CONCUR ☆
— as·sen·tor *or* **as·sent·er** \-ˈsen-tər\ *noun*

[2]assent *noun* (14th century)
: an act of assenting **:** ACQUIESCENCE, AGREEMENT

as·sen·ta·tion \ˌa-sᵊn-ˈtā-shən, ˌa-ˌsen-\ *noun* (15th century)
: ready assent especially when insincere or obsequious

as·sert \ə-ˈsərt, a-\ *transitive verb* [Latin *assertus,* past participle of *asserere,* from *ad-* + *serere* to join — more at SERIES] (circa 1604)
1 : to state or declare positively and often forcefully or aggressively
2 a : to demonstrate the existence of ⟨*assert* his manhood —James Joyce⟩ **b :** POSIT, POSTULATE ☆
— assert oneself : to compel recognition especially of one's rights

as·sert·ed·ly \ə-ˈsər-təd-lē, a-\ *adverb* (1937)
: by positive and usually unsubstantiated assertion **:** ALLEGEDLY

as·ser·tion \ə-ˈsər-shən, a-\ *noun* (15th century)
: the act of asserting; *also* **:** DECLARATION, AFFIRMATION

as·ser·tive \ə-ˈsər-tiv, a-\ *adjective* (circa 1619)
1 : disposed to or characterized by bold or confident assertion
2 : having a strong or distinctive flavor or aroma ⟨*assertive* wines⟩
synonym see AGGRESSIVE
— as·ser·tive·ly *adverb*
— as·ser·tive·ness *noun*

assertiveness training *noun* (1975)
: a method of training individuals to act in a bold self-confident manner

asses *plural of* AS *or of* ASS

as·sess \ə-ˈses, a-\ *transitive verb* [Middle English, probably from Medieval Latin *assessus,* past participle of *assidēre,* from Latin, to sit beside, assist in the office of a judge — more at ASSIZE] (15th century)
1 : to determine the rate or amount of (as a tax)
2 a : to impose (as a tax) according to an established rate **b :** to subject to a tax, charge, or levy
3 : to make an official valuation of (property) for the purposes of taxation
4 : to determine the importance, size, or value of
5 : to charge (a player or team) with a foul or penalty
synonym see ESTIMATE
— as·sess·able \-ˈse-sə-bəl\ *adjective*

as·sess·ment \ə-ˈses-mənt, a-\ *noun* (1534)
1 : the action or an instance of assessing **:** APPRAISAL
2 : the amount assessed

as·ses·sor \ə-ˈse-sər\ *noun* (14th century)
1 : an official who assists a judge or magistrate
2 : one that assesses; *especially* **:** an official who assesses property for taxation

as·set \ˈa-ˌset *also* -sət\ *noun* [back-formation from *assets,* singular, sufficient property to pay debts and legacies, from Anglo-French *asetz,* from Old French *assez* enough, from (assumed) Vulgar Latin *ad satis,* from Latin *ad* to + *satis* enough — more at AT, SAD] (1531)
1 *plural* **a :** the property of a deceased person subject by law to the payment of his or her debts and legacies **b :** the entire property of a person, association, corporation, or estate applicable or subject to the payment of debts
2 : ADVANTAGE, RESOURCE ⟨his wit is his chief *asset*⟩
3 a : an item of value owned **b** *plural* **:** the items on a balance sheet showing the book value of property owned

as·sev·er·ate \ə-ˈse-və-ˌrāt\ *transitive verb* **-at·ed; -at·ing** [Latin *asseveratus,* past participle of *asseverare,* from *ad-* + *severus* severe] (1791)
: to affirm or aver positively or earnestly
— as·sev·er·a·tion \-ˌse-və-ˈrā-shən\ *noun*
— as·sev·er·a·tive \-ˈse-və-ˌrā-tiv\ *adjective*

ass·hole \ˈas-ˌ(h)ōl\ *noun* (14th century)
1 : ANUS — usually considered vulgar
2 a : a stupid, incompetent, or detestable person — usually considered vulgar **b :** a despicable place — usually used in the phrase *asshole of the universe;* usually considered vulgar

as·si·du·ity \ˌa-sə-ˈdü-ə-tē, -ˈdyü-\ *noun, plural* **-ities** (1596)
1 : the quality or state of being assiduous **:** DILIGENCE
2 : persistent personal attention — usually used in plural

as·sid·u·ous \ə-ˈsij-wəs, -ˈsi-jə-\ *adjective* [Latin *assiduus,* from *assidēre*] (1660)
: marked by careful unremitting attention or persistent application ⟨an *assiduous* book collector⟩ ⟨tended her garden with *assiduous* attention⟩
synonym see BUSY
— as·sid·u·ous·ly *adverb*
— as·sid·u·ous·ness *noun*

☆ SYNONYMS

Assent, consent, accede, acquiesce, agree, subscribe mean to concur with what has been proposed. ASSENT implies an act involving the understanding or judgment and applies to propositions or opinions ⟨voters *assented* to the proposal⟩. CONSENT involves the will or feelings and indicates compliance with what is requested or desired ⟨*consented* to their daughter's going⟩. ACCEDE implies a yielding, often under pressure, of assent or consent ⟨officials *acceded* to the prisoners' demands⟩. ACQUIESCE implies tacit acceptance or forbearance of opposition ⟨*acquiesced* to his boss's wishes⟩. AGREE sometimes implies previous difference of opinion or attempts at persuasion ⟨finally *agreed* to come along⟩. SUBSCRIBE implies not only consent or assent but hearty approval and active support ⟨*subscribes* wholeheartedly to the idea⟩.

Assert, declare, affirm, protest, avow mean to state positively usually in anticipation of denial or objection. ASSERT implies stating confidently without need for proof or regard for evidence ⟨*asserted* that modern music is just noise⟩. DECLARE stresses open or public statement ⟨*declared* her support for the candidate⟩. AFFIRM implies conviction based on evidence, experience, or faith ⟨*affirmed* the existence of an afterlife⟩. PROTEST emphasizes affirming in the face of denial or doubt ⟨*protested* that he really had been misquoted⟩. AVOW stresses frank declaration and acknowledgment of personal responsibility for what is declared ⟨*avowed* that all investors would be repaid in full⟩. See in addition MAINTAIN.

¹as·sign \ə-ˈsīn\ *transitive verb* [Middle English, from Old French *assigner,* from Latin *assignare,* from *ad-* + *signare* to mark, from *signum* mark, sign] (13th century)
1 : to transfer (property) to another especially in trust or for the benefit of creditors
2 a : to appoint to a post or duty ⟨*assigned* them to light duty⟩ ⟨*assigned* me two clerks⟩ **b :** to appoint as a duty or task ⟨*assigns* 20 pages for homework⟩
3 : to fix or specify in correspondence or relationship ⟨*assign* counsel to the defendant⟩ ⟨*assign* a value to the variable⟩
4 a : to ascribe as a motive, reason, or cause especially after deliberation **b :** to consider to belong to (a specified period of time)
synonym see ASCRIBE
— as·sign·abil·i·ty \-ˌsī-nə-ˈbi-lə-tē\ *noun*
— as·sign·able \-ˈsī-nə-bəl\ *adjective*
— as·sign·er \ə-ˈsī-nər\ *or* **as·sign·or** \ˌa-sə-ˈnȯr, ˌa-ˌsī-, ə-ˌsī-\ *noun*

²assign *noun* (15th century)
: ASSIGNEE 3 ⟨heirs and *assigns*⟩

as·si·gnat \ˌa-(ˌ)sēn-ˈyä, ˈa-sig-ˌnat\ *noun* [French, from Latin *assignatus,* past participle of *assignare*] (1790)
: a bill issued as currency by the French Revolutionary government (1789–96) on the security of expropriated lands

as·sig·na·tion \ˌa-sig-ˈnā-shən\ *noun* (15th century)
1 : the act of assigning or the assignment made
2 : an appointment of time and place for a meeting; *especially* **:** TRYST ⟨returned from an *assignation* with his mistress —W. B. Yeats⟩

assigned risk *noun* (1946)
: a poor risk (as an accident-prone motorist) that insurance companies would normally reject but are forced to insure by state law

as·sign·ee \ˌa-sə-ˈnē, ˌa-ˌsī-, ə-ˌsī-\ *noun* (14th century)
1 : a person to whom an assignment is made
2 : a person appointed to act for another
3 : a person to whom a right or property is legally transferred

as·sign·ment \ə-ˈsīn-mənt\ *noun* (14th century)
1 : the act of assigning
2 a : a position, post, or office to which one is assigned **b :** a specified task or amount of work assigned or undertaken as if assigned by authority
3 : the transfer of property; *especially* **:** the transfer of property to be held in trust or to be used for the benefit of creditors
synonym see TASK

as·sim·i·la·ble \ə-ˈsi-mə-lə-bəl\ *adjective* (1667)
: capable of being assimilated
— as·sim·i·la·bil·i·ty \-ˌsi-mə-lə-ˈbi-lə-tē\ *noun*

¹as·sim·i·late \ə-ˈsi-mə-ˌlāt\ *verb* **-lat·ed; -lat·ing** [Middle English, from Medieval Latin *assimilatus,* past participle of *assimilare,* from Latin *assimulare* to make similar, from *ad-* + *simulare* to make similar, simulate] (15th century)
transitive verb
1 a : to take in and appropriate as nourishment **:** absorb into the system **b :** to take into the mind and thoroughly comprehend
2 a : to make similar **b :** to alter by assimilation **c :** to absorb into the culture or mores of a population or group
3 : COMPARE, LIKEN
intransitive verb
: to become assimilated ■
— as·sim·i·la·tor \-ˌlā-tər\ *noun*

²as·sim·i·late \-lət, -ˌlāt\ *noun* (1935)
: something that is assimilated

as·sim·i·la·tion \ə-ˌsi-mə-ˈlā-shən\ *noun* (15th century)
1 a : an act, process, or instance of assimilating **b :** the state of being assimilated
2 : the incorporation or conversion of nutrients into protoplasm that in animals follows digestion and absorption and in higher plants involves both photosynthesis and root absorption
3 : change of a sound in speech so that it becomes identical with or similar to a neighboring sound ⟨the usual *assimilation* of \z\ to \sh\ in the phrase *his shoe*⟩
4 : the process of receiving new facts or of responding to new situations in conformity with what is already available to consciousness

as·sim·i·la·tion·ist \-sh(ə-)nist\ *noun* (1899)
: a person who advocates a policy of assimilating differing racial or cultural groups
— as·sim·i·la·tion·ism \-shə-ˌni-zəm\ *noun*
— as·sim·i·la·tion·ist *adjective*

as·sim·i·la·tive \ə-ˈsi-mə-ˌlā-tiv, -lə-tiv\ *adjective* (14th century)
: of, relating to, or causing assimilation

as·sim·i·la·to·ry \ə-ˈsi-mə-lə-ˌtōr-ē, -ˌtȯr-\ *adjective* (circa 1847)
: ASSIMILATIVE

As·sin·i·boin *or* **As·sin·i·boine** \ə-ˈsi-nə-ˌbȯin\ *noun, plural* **-boin** *or* **-boins** *or* **boine** *or* **-boines** [Ojibwa dialect *assiniˑpwaˑn,* literally, stone Sioux] (1804)
: a member of an American Indian people originally of the area between the upper Missouri and middle Saskatchewan rivers

¹as·sist \ə-ˈsist\ *verb* [Middle French or Latin; Middle French *assister* to help, stand by, from Latin *assistere,* from *ad-* + *sistere* to cause to stand; akin to Latin *stare* to stand — more at STAND] (15th century)
transitive verb
: to give usually supplementary support or aid to
intransitive verb
1 : to give support or aid ⟨*assisted* at the stove⟩ ⟨another surgeon *assisted* on the operation⟩
2 : to be present as a spectator ⟨the ideal figures *assisting* at Italian holy scenes —Mary McCarthy⟩

²assist *noun* (1597)
1 : an act of assistance **:** AID
2 : the action (as a throw or pass) of a player who enables a teammate to make a putout or score a goal; *also* **:** official credit given for such an action
3 : a mechanical device that provides assistance

as·sis·tance \ə-ˈsis-tən(t)s\ *noun* (14th century)
: the act of assisting or the help supplied **:** AID ⟨financial and technical *assistance*⟩

as·sis·tant \-tənt\ *noun* (15th century)
: a person who assists **:** HELPER; *also* **:** a person holding an assistantship
— assistant *adjective*

assistant professor *noun* (1851)
: a member of a college or university faculty who ranks above an instructor and below an associate professor
— assistant professorship *noun*

as·sis·tant·ship \ə-ˈsis-tən(t)-ˌship\ *noun* (1948)
: a paid appointment awarded annually to a qualified graduate student that requires part-time teaching, research, or residence hall duties

as·size \ə-ˈsīz\ *noun* [Middle English *assise,* from Old French, session, settlement, from *asseoir* to seat, from (assumed) Vulgar Latin *assedēre,* from Latin *assidēre* to sit beside, assist in the office of a judge, from *ad-* + *sedēre* to sit — more at SIT] (14th century)
1 a : a judicial inquest **b :** an action to be decided by such an inquest, the writ for instituting it, or the verdict or finding rendered by the jury
2 a : the former periodical sessions of the superior courts in English counties for trial of civil and criminal cases — usually used in plural **b :** the time or place of holding such a court, the court itself, or a session of it — usually used in plural

¹as·so·ci·ate \ə-ˈsō-shē-ˌāt, -sē-\ *verb* **-at·ed; -at·ing** [Middle English *associat* associated, from Latin *associatus,* past participle of *associare* to unite, from *ad-* + *sociare* to join, from *socius* companion — more at SOCIAL] (14th century)
transitive verb
1 : to join as a partner, friend, or companion
2 *obsolete* **:** to keep company with **:** ATTEND
3 : to join or connect together **:** COMBINE
4 : to bring together or into relationship in any of various intangible ways (as in memory or imagination)
intransitive verb
1 : to come or be together as partners, friends, or companions
2 : to combine or join with other parts **:** UNITE
synonym see JOIN

²as·so·ciate \ə-ˈsō-shē-ət, -sē-, -ˌāt, -shət\ *adjective* (14th century)
1 : closely connected (as in function or office) with another
2 : closely related especially in the mind
3 : having secondary or subordinate status ⟨*associate* membership in a society⟩

³as·so·ciate *same as* ² \ *noun* (1533)
1 : one associated with another: as **a :** PARTNER, COLLEAGUE **b :** COMPANION, COMRADE
2 a : an entry-level member (as of a learned society, professional organization, or profession) **b :** EMPLOYEE, WORKER
3 *often capitalized* **:** a degree conferred especially by a junior college ⟨*associate* in arts⟩
— as·so·ciate·ship \-ˌship\ *noun*

associate professor *noun* (1822)
: a member of a college or university faculty who ranks above an assistant professor and below a professor
— associate professorship *noun*

as·so·ci·a·tion \ə-ˌsō-sē-ˈā-shən, -shē-\ *noun* (1535)
1 a : the act of associating **b :** the state of being associated **:** COMBINATION, RELATIONSHIP
2 : an organization of persons having a common interest **:** SOCIETY
3 : something linked in memory or imagination with a thing or person
4 : the process of forming mental connections or bonds between sensations, ideas, or memories
5 : the aggregation of chemical species to form (as with hydrogen bonds) loosely bound complexes
6 : a major unit in ecological community organization characterized by essential uniformity and usually by two or more dominant species
— as·so·ci·a·tion·al \-sh(ə-)nᵊl\ *adjective*

association area *noun* (circa 1909)
: an area of the cerebral cortex that functions in linking and coordinating the sensory and motor areas

association football *noun* (1873)
: SOCCER

as·so·ci·a·tion·ism \ə-ˌsō-sē-ˈā-shə-ˌni-zəm, -ˌsō-shē-\ *noun* (1875)
: a reductionist school of psychology that

□ USAGE
assimilate When *assimilate* is followed by a preposition, transitive senses 2a and 2c commonly take *to* and *into* and less frequently *with;* 2b regularly takes *to;* sense 3 most often takes *to* and sometimes *with.* The most frequent prepositions used with the intransitive sense are *to* and *into.*

holds that the content of consciousness can be explained by the association and reassociation of irreducible sensory and perceptual elements
— **as·so·ci·a·tion·ist** \-ˈā-sh(ə-)nist\ *noun*
— **as·so·ci·a·tion·is·tic** \-ˌā-shə-ˈnis-tik\ *adjective*

as·so·cia·tive \ə-ˈsō-shē-ˌā-tiv, -sē-, -shə-tiv\ *adjective* (1812)
1 : of or relating to association especially of ideas or images
2 : dependent on or acquired by association or learning
3 : of, having, or being the property of producing the same result no matter which pair of elements next to each other in a mathematical expression is used to perform a given operation first if the elements in the expression are listed in a fixed order ⟨addition is *associative* since $(a + b) + c = a + (b + c)$⟩
— **as·so·cia·tive·ly** *adverb*
— **as·so·cia·tiv·i·ty** \-ˌsō-shē-ə-ˈti-və-tē, -sē-, -shə-ˈti-\ *noun*

associative learning *noun* (1957)
: a learning process in which discrete ideas and percepts become linked to one another

associative neuron *noun* (1935)
: a neuron that conveys impulses from one neuron to another

as·soil \ə-ˈsȯi(ə)l\ *transitive verb* [Middle English, from Old French *assoldre,* from Latin *absolvere* to absolve] (13th century)
1 *archaic* **:** ABSOLVE, PARDON
2 *archaic* **:** ACQUIT, CLEAR
3 *archaic* **:** EXPIATE
— **as·soil·ment** \-mənt\ *noun, archaic*

as·so·nance \ˈa-sə-nən(t)s\ *noun* [French, from Latin *assonare* to answer with the same sound, from *ad-* + *sonare* to sound, from *sonus* sound — more at SOUND] (1727)
1 : resemblance of sound in words or syllables
2 a : relatively close juxtaposition of similar sounds especially of vowels **b :** repetition of vowels without repetition of consonants (as in *stony* and *holy*) used as an alternative to rhyme in verse
— **as·so·nant** \-nənt\ *adjective or noun*
— **as·so·nant·al** \ˌa-sə-ˈnan-tᵊl\ *adjective*

as soon as *conjunction* (14th century)
: immediately at or shortly after the time that

as·sort \ə-ˈsȯrt\ *verb* [Middle French *assortir,* from *a-* (from Latin *ad-*) + *sorte* sort] (15th century)
transitive verb
1 : to distribute into groups of a like kind **:** CLASSIFY
2 : to supply with an assortment (as of goods)
intransitive verb
1 : to agree in kind **:** HARMONIZE
2 : to keep company **:** ASSOCIATE
— **as·sort·er** *noun*

as·sor·ta·tive \ə-ˈsȯr-tə-tiv\ *adjective* (1897)
: being nonrandom mating based on like or unlike characteristics
— **as·sor·ta·tive·ly** \-lē\ *adverb*

as·sort·ed \-ˈsȯr-təd\ *adjective* (circa 1797)
1 : suited especially by nature or character ⟨an ill-*assorted* pair⟩
2 : consisting of various kinds ⟨*assorted* chocolates⟩

as·sort·ment \-ˈsȯrt-mənt\ *noun* (1611)
1 a : the act of assorting **b :** the state of being assorted
2 : a collection of assorted things or persons

as·suage \ə-ˈswāj *also* -ˈswāzh *or* -ˈswäzh\ *transitive verb* **as·suaged; as·suag·ing** [Middle English *aswagen,* from Old French *assouagier,* from (assumed) Vulgar Latin *assuaviare,* from Latin *ad-* + *suavis* sweet — more at SWEET] (14th century)
1 : to lessen the intensity of (something that pains or distresses) **:** EASE
2 : PACIFY, QUIET
3 : to put an end to by satisfying **:** APPEASE, QUENCH
synonym see RELIEVE
— **as·suage·ment** \-mənt\ *noun*

as·sua·sive \ə-ˈswā-siv, -ziv\ *adjective* (1708)
: SOOTHING, CALMING

as·sume \ə-ˈsüm\ *transitive verb* **as·sumed; as·sum·ing** [Middle English, from Latin *assumere,* from *ad-* + *sumere* to take — more at CONSUME] (15th century)
1 a : to take up or in **:** RECEIVE **b :** to take into partnership, employment, or use
2 a : to take to or upon oneself **:** UNDERTAKE **b :** PUT ON, DON
3 : to take control of
4 : to pretend to have or be **:** FEIGN ⟨*assumed* an air of confidence in spite of her dismay⟩
5 : to take as granted or true **:** SUPPOSE
6 : to take over (the debts of another) as one's own ☆
— **as·sum·abil·i·ty** \-ˌsü-mə-ˈbi-lə-tē\ *noun*
— **as·sum·able** \-ˈsü-mə-bəl\ *adjective*
— **as·sum·ably** \-blē\ *adverb*

as·sum·ing *adjective* (1695)
: PRETENTIOUS, PRESUMPTUOUS

as·sump·sit \ə-ˈsəm(p)-sət\ *noun* [New Latin, he undertook, from Latin *assumere* to undertake] (1590)
1 : an express or implied promise or contract not under seal on which an action may be brought
2 a : a former common-law action brought to recover damages alleged from the breach of an assumpsit **b :** an action to recover damages for breach of a contract

as·sump·tion \ə-ˈsəm(p)-shən\ *noun* [Middle English, from Late Latin *assumption-, assumptio* taking up, from Latin *assumere*] (13th century)
1 a : the taking up of a person into heaven **b** *capitalized* **:** August 15 observed in commemoration of the Assumption of the Virgin Mary
2 : a taking to or upon oneself ⟨the *assumption* of a new position⟩
3 : the act of laying claim to or taking possession of something ⟨the *assumption* of power⟩
4 : ARROGANCE, PRETENSION
5 a : an assuming that something is true **b :** a fact or statement (as a proposition, axiom, postulate, or notion) taken for granted
6 : the taking over of another's debts

as·sump·tive \ə-ˈsəm(p)-tiv\ *adjective* (1611)
: of, relating to, or based on assumption

as·sur·ance \ə-ˈshu̇r-ən(t)s\ *noun* (14th century)
1 : the act or action of assuring: as **a :** PLEDGE, GUARANTEE **b :** the act of conveying real property; *also* **:** the instrument by which it is conveyed **c** *chiefly British* **:** INSURANCE
2 : the state of being assured: as **a :** SECURITY **b :** a being certain in the mind ⟨the puritan's *assurance* of salvation⟩ **c :** confidence of mind or manner **:** easy freedom from self-doubt or uncertainty; *also* **:** excessive self-confidence **:** BRASHNESS, PRESUMPTION
3 : something that inspires or tends to inspire confidence ⟨gave repeated *assurances* of goodwill⟩
synonym see CONFIDENCE

as·sure \ə-ˈshu̇r\ *transitive verb* **as·sured; as·sur·ing** [Middle English, from Middle French *assurer,* from Medieval Latin *assecurare,* from Latin *ad-* + *securus* secure] (14th century)
1 : to make safe (as from risks or against overthrow) **:** INSURE
2 : to give confidence to ⟨and hereby we know that we are of the truth, and shall *assure* our hearts —1 John 3:19 (Authorized Version)⟩
3 : to make sure or certain **:** CONVINCE ⟨glancing back to *assure* himself no one was following⟩
4 : to inform positively ⟨I *assure* you that we will do better next time⟩
5 : to make certain the coming or attainment of **:** GUARANTEE ⟨worked hard to *assure* accuracy⟩
synonym see ENSURE

¹as·sured \ə-ˈshu̇rd\ *adjective* (15th century)
1 : characterized by certainty or security **:** GUARANTEED ⟨an *assured* market⟩
2 a : SELF-ASSURED **b :** SELF-SATISFIED
3 : satisfied as to the certainty or truth of a matter ⟨rest *assured* we got what we came for⟩
— **as·sured·ness** \-ˈshu̇r-əd-nəs, -ˈshu̇rd-\ *noun*

²assured *noun, plural* **assured** *or* **assureds** (1755)
: INSURED

as·sured·ly \ə-ˈshu̇r-əd-lē\ *adverb* (14th century)
1 : without a doubt **:** CERTAINLY
2 : in an assured manner **:** CONFIDENTLY

as·sur·er \ə-ˈshu̇r-ər\ *or* **as·sur·or** \ə-ˈshu̇r-ər, ə-ˌshu̇r-ˈȯr\ *noun* (1607)
: one that assures **:** INSURER

as·sur·gent \ə-ˈsər-jənt\ *adjective* [Latin *assurgent-, assurgens,* present participle of *assurgere* to rise, from *ad-* + *surgere* to rise — more at SURGE] (1578)
: moving upward **:** RISING; *especially* **:** ASCENDANT 1b

As·syr·i·an \ə-ˈsir-ē-ən\ *noun* (15th century)
1 : a native or inhabitant of ancient Assyria
2 : the dialect of Akkadian spoken by the Assyrians
— **Assyrian** *adjective*

As·syr·i·ol·o·gy \ə-ˌsir-ē-ˈä-lə-jē\ *noun* (1828)
: the science or study of the history, language, and antiquities of ancient Assyria and Babylonia
— **As·syr·i·o·log·i·cal** \-ˌsir-ē-ə-ˈlä-ji-kəl\ *adjective*
— **As·syr·i·ol·o·gist** \-ˈä-lə-jist\ *noun*

-ast *noun suffix* [Middle English, from Latin *-astes,* from Greek *-astēs,* from verbs in *-azein*]
: one connected with ⟨ecdysi*ast*⟩

astar·board \ə-ˈstär-bərd\ *adverb* (circa 1630)
: toward or on the starboard side of a ship ⟨put the helm hard *astarboard*⟩

As·tar·te \ə-ˈstär-tē\ *noun* [Latin, from Greek *Astartē*]
: the Phoenician goddess of fertility and of sexual love

as·ta·tine \ˈas-tə-ˌtēn\ *noun* [Greek *astatos* unsteady, from *a-* + *statos* standing, from *histanai* to cause to stand — more at STAND] (1947)
: a radioactive halogen element discovered by bombarding bismuth with alpha particles and also formed by radioactive decay — see ELEMENT table

as·ter \ˈas-tər\ *noun* (1664)
1 [New Latin, from Latin, aster, from Greek *aster-, astēr* star, aster — more at STAR] **a :** any of various chiefly fall-blooming leafy-stemmed composite herbs (*Aster* and closely

☆ SYNONYMS
Assume, affect, pretend, simulate, feign, counterfeit, sham mean to put on a false or deceptive appearance. ASSUME often implies a justifiable motive rather than an intent to deceive ⟨*assumed* an air of cheerfulness around the patients⟩. AFFECT implies making a false show of possessing, using, or feeling ⟨*affected* an interest in art⟩. PRETEND implies an overt and sustained false appearance ⟨*pretended* that nothing had happened⟩. SIMULATE suggests a close imitation of the appearance of something ⟨cosmetics that *simulate* a suntan⟩. FEIGN implies more artful invention than PRETEND, less specific mimicry than SIMULATE ⟨*feigned* sickness⟩. COUNTERFEIT implies achieving the highest degree of verisimilitude of any of these words ⟨an actor *counterfeiting* drunkenness⟩. SHAM implies an obvious falseness that fools only the gullible ⟨*shammed* a most unconvincing limp⟩.

related genera) with often showy heads containing disk flowers or both disk and ray flowers **b :** CHINA ASTER
2 [New Latin, from Greek *aster-, astēr*] **:** a system of microtubules arranged radially about a centriole at either end of the mitotic or meiotic spindle

-aster *noun suffix* [Middle English, from Latin, suffix denoting partial resemblance]
: one that is inferior or not genuine ⟨critic*aster*⟩

as•te•ria \a-'stir-ē-ə\ *noun* [Latin, a precious stone, from Greek, feminine of *asterios* starry, from *aster-, astēr*] (1903)
: a gemstone cut to show asterism

as•te•ri•at•ed \-ē-,ā-təd\ *adjective* [Greek *asterios*] (1816)
: exhibiting asterism ⟨*asteriated* sapphire⟩

[1]as•ter•isk \'as-tə-,risk, *especially in plural also* ÷-,rik\ *noun* [Middle English, *astarisc*, from Late Latin *asteriscus*, from Greek *asteriskos*, literally, little star, diminutive of *aster-, astēr*] (14th century)
: the character * used in printing or writing as a reference mark, as an indication of the omission of letters or words, to denote a hypothetical or unattested linguistic form, or for various arbitrary meanings
— as•ter•isk•less \-ləs\ *adjective*

[2]asterisk *transitive verb* (circa 1733)
: to mark with an asterisk **:** STAR

as•ter•ism \'as-tə-,ri-zəm\ *noun* [Greek *asterismos*, from *asterizein* to arrange in constellations, from *aster-, astēr*] (1598)
1 a : CONSTELLATION **b :** a small group of stars
2 : a star-shaped figure exhibited by some crystals by reflected light (as in a star sapphire) or by transmitted light (as in some mica)

astern \ə-'stərn\ *adverb or adjective* (1627)
1 : behind a ship
2 : at or toward the stern of a ship
3 : STERNFOREMOST, BACKWARD

[1]as•ter•oid \'as-tə-,rȯid\ *noun* [Greek *asteroeidēs* starlike, from *aster-, astēr*] (1802)
1 : any of the small celestial bodies found especially between the orbits of Mars and Jupiter
2 : STARFISH
— as•ter•oi•dal \,as-tə-'rȯi-dᵊl\ *adjective*

[2]asteroid *adjective* (1854)
1 : resembling a star ⟨*asteroid* bodies in sporotrichosis⟩
2 : of or resembling a starfish

asteroid belt *noun* (1952)
: the region of interplanetary space between the orbits of Mars and Jupiter in which most asteroids are found

aster yellows *noun plural* (1922)
: a widespread disease that affects more than 40 families of plants, is characterized especially by yellowing and dwarfing, and is caused by a mycoplasma transmitted by leafhoppers

as•the•nia \as-'thē-nē-ə\ *noun* [New Latin, from Greek *astheneia*, from *asthenēs* weak, from *a-* + *sthenos* strength] (1802)
: lack or loss of strength **:** DEBILITY

as•then•ic \as-'the-nik\ *adjective* (1789)
1 : of, relating to, or exhibiting asthenia **:** WEAK
2 : ECTOMORPHIC 2

as•theno•sphere \as-'the-nə-,sfir\ *noun* [Greek *asthenēs* weak + English *-o-* + *sphere*] (1914)
: a hypothetical zone of a celestial body (as the earth) which lies beneath the lithosphere and within which the material is believed to yield readily to persistent stresses
— as•theno•spher•ic \-,the-nə-'sfir-ik, -'sfer-\ *adjective*

asth•ma \'az-mə, *British* 'as-\ *noun* [Middle English *asma*, from Medieval Latin, modification of Greek *asthma*] (14th century)
: a condition often of allergic origin that is marked by continuous or paroxysmal labored breathing accompanied by wheezing, by a sense of constriction in the chest, and often by attacks of coughing or gasping
— asth•mat•ic \az-'ma-tik, *British* as-\ *adjective or noun*
— asth•mat•i•cal•ly \-ti-k(ə-)lē\ *adverb*

as though *conjunction* (13th century)
: AS IF

as•tig•mat•ic \,as-tig-'ma-tik\ *adjective* [*a-* + Greek *stigmat-, stigma* stigma] (1849)
1 : affected with, relating to, or correcting astigmatism
2 : showing incapacity for observation or discrimination ⟨an *astigmatic* fanaticism, a disregard for the facts —*N.Y. Herald Tribune*⟩
— astigmatic *noun*

astig•ma•tism \ə-'stig-mə-,ti-zəm\ *noun* (1849)
1 : a defect of an optical system (as a lens) causing rays from a point to fail to meet in a focal point resulting in a blurred and imperfect image
2 : a defect of vision due to astigmatism of the refractive system of the eye and especially to corneal irregularity
3 : distorted understanding suggestive of the blurred vision of an astigmatic person

astil•be \ə-'stil-(,)bē\ *noun* [New Latin, from [2]*a-* + Greek *stilbē*, feminine of *stilbos* sparkling]
: any of a genus (*Astilbe*) of chiefly Asian perennials of the saxifrage family having simple or usually compound leaves and widely cultivated for their panicles of white or red flowers

astir \ə-'stər\ *adjective* (1823)
1 : exhibiting activity
2 : being out of bed **:** UP

As•ti Spu•man•te \'äs-tē-spü-'män-(,)tē, 'as-tē-, -spü-'man-\ *noun* [Italian, from *Asti*, Italy + Italian *spumante* effervescent, literally, foaming] (1908)
: a sweet sparkling white wine made in and around the village of Asti in Piedmont

as to *preposition* (14th century)
1 : AS FOR, ABOUT ⟨at a loss *as to* how to explain the error⟩
2 : ACCORDING TO, BY ⟨graded *as to* size and color⟩

as•ton•ied \ə-'stä-nēd\ *adjective* [Middle English, from past participle of *astonien*] (14th century)
1 *archaic* **:** deprived briefly of the power to act **:** DAZED
2 *archaic* **:** filled with consternation or dismay

as•ton•ish \ə-'stä-nish\ *transitive verb* [probably from earlier *astony* (from Middle English *astonen, astonien*, from Old French *estoner*, from — assumed — Vulgar Latin *extonare*, from Latin *ex-* + *tonare* to thunder) + *-ish* (as in *abolish*) — more at THUNDER] (1535)
1 *obsolete* **:** to strike with sudden fear
2 : to strike with sudden and usually great wonder or surprise
synonym see SURPRISE

as•ton•ish•ing \-ni-shiŋ\ *adjective* (1543)
: causing astonishment **:** SURPRISING
— as•ton•ish•ing•ly \-shiŋ-lē\ *adverb*

as•ton•ish•ment \ə-'stä-nish-mənt\ *noun* (circa 1586)
1 a : the state of being astonished **b :** CONSTERNATION **c :** AMAZEMENT
2 : a cause of amazement or wonder

[1]as•tound \ə-'stau̇nd\ *adjective* [Middle English *astoned*, from past participle of *astonen*] (14th century)
archaic **:** overwhelmed with astonishment or amazement **:** ASTOUNDED

[2]astound *transitive verb* (1634)
: to fill with bewilderment or wonder
synonym see SURPRISE

as•tound•ing \ə-'stau̇n-diŋ\ *adjective* (1586)
: causing astonishment or amazement
— as•tound•ing•ly \-diŋ-lē\ *adverb*

astr- *or* **astro-** *combining form* [Middle English *astro-*, from Old French, from Latin *astr-, astro-*, from Greek, from *astron* — more at STAR]
: star **:** heavens **:** outer space **:** astronomical ⟨*astro*physics⟩

[1]astrad•dle \ə-'stra-dᵊl\ *adverb* (1703)
: on or above and extending onto both sides **:** ASTRIDE

[2]astraddle *preposition* (1935)
: with one leg on each side of **:** ASTRIDE

as•tra•gal \'as-tri-gəl\ *noun* [Latin *astragalus*, from Greek *astragalos* anklebone, molding; akin to Greek *astakos* lobster, *osteon* bone — more at OSSEOUS] (1563)
1 : a narrow half-round molding
2 : a projecting strip on the edge of a folding door

as•tra•khan \'as-trə-kən, -,kan\ *noun, often capitalized* [*Astrakhan*, Russia] (1766)
1 : karakul of Russian origin
2 : a cloth with a usually wool, curled, and looped pile resembling karakul

as•tral \'as-trəl\ *adjective* [Late Latin *astralis*, from Latin *astrum* star, from Greek *astron*] (1605)
1 : of, relating to, or coming from the stars ⟨*astral* influences⟩ ⟨unusual *astral* occurrences⟩
2 : of or relating to a mitotic or meiotic aster
3 : of or consisting of a supersensible substance held in theosophy to be next above the tangible world in refinement
4 a : VISIONARY **b :** elevated in station or position **:** EXALTED
— as•tral•ly \-trə-lē\ *adverb*

astray \ə-'strā\ *adverb or adjective* [Middle English, from Middle French *estraié* wandering, from *estraier* to stray — more at STRAY] (14th century)
1 : off the right path or route **:** STRAYING
2 : in error **:** away from what is proper or desirable

[1]astride \ə-'strīd\ *adverb* (1664)
1 : with one leg on each side **:** astride a horse ⟨she rode *astride*, not sidesaddle⟩
2 : with the legs stretched wide apart ⟨standing *astride*⟩

[2]astride *preposition* (1713)
1 : on or above and with one leg on each side of
2 : placed or lying on both sides of
3 : extending over or across **:** SPANNING, BRIDGING

as•trin•gen•cy \ə-'strin-jən(t)-sē\ *noun* (1601)
: the quality or state of being astringent

[1]as•trin•gent \-jənt\ *adjective* [probably from Middle French, from Latin *astringent-, astringens*, present participle of *astringere* to bind fast, from *ad-* + *stringere* to bind tight — more at STRAIN] (1541)
1 : able to draw together the soft organic tissues **:** STYPTIC, PUCKERY ⟨*astringent* lotions⟩ ⟨an *astringent* fruit⟩
2 : suggestive of an astringent effect upon tissue **:** rigidly severe **:** AUSTERE ⟨dry *astringent* comments⟩; *also* **:** PUNGENT, CAUSTIC
— as•trin•gent•ly *adverb*

[2]astringent *noun* (1626)
: an astringent agent or substance

as•tro•bi•ol•o•gy \,as-trō-(,)bī-'ä-lə-jē\ *noun* (1955)
: EXOBIOLOGY
— as•tro•bi•ol•o•gist \-(,)bī-'ä-lə-jist\ *noun*

as•tro•cyte \'as-trə-,sīt\ *noun* [International Scientific Vocabulary] (1898)
: a star-shaped cell (as of the neuroglia)
— as•tro•cyt•ic \,as-trə-'si-tik\ *adjective*

as·tro·cy·to·ma \ˌas-trə-sī-ˈtō-mə\ *noun, plural* **-mas** *or* **-ma·ta** \-mə-tə\ [New Latin] (circa 1923)
: a nerve-tissue tumor composed of astrocytes

as·tro·dome \ˈas-trə-ˌdōm\ *noun* [International Scientific Vocabulary] (1941)
: a transparent dome in the upper surface of an airplane from within which the navigator makes celestial observations

as·tro·labe \ˈas-trə-ˌlāb *also* -ˌlab\ *noun* [Middle English, from Middle French & Medieval Latin; Middle French, from Medieval Latin *astrolabium,* from Late Greek *astrolabion,* from Greek *astrolabos,* from *astr-* + *lambanein* to take — more at LATCH] (14th century)
: a compact instrument used to observe and calculate the position of celestial bodies before the invention of the sextant

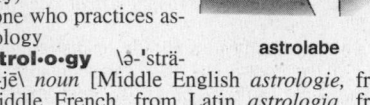

astrolabe

as·trol·o·ger \ə-ˈsträ-lə-jər\ *noun* (14th century)
: one who practices astrology

as·trol·o·gy \ə-ˈsträ-lə-jē\ *noun* [Middle English *astrologie,* from Middle French, from Latin *astrologia,* from Greek, from *astr-* + *-logia* -logy] (14th century)
1 *archaic* **:** ASTRONOMY
2 : the divination of the supposed influences of the stars and planets on human affairs and terrestrial events by their positions and aspects
— **as·tro·log·i·cal** \ˌas-trə-ˈlä-ji-kəl\ *adjective*
— **as·tro·log·i·cal·ly** \-k(ə-)lē\ *adverb*

as·trom·e·try \ə-ˈsträ-mə-trē\ *noun* (circa 1859)
: a branch of astronomy that deals with measurements (as of positions and movements) of celestial bodies
— **as·tro·me·tric** \ˌas-trə-ˈme-trik\ *adjective*

as·tro·naut \ˈas-trə-ˌnȯt, -ˌnät\ *noun* [*astr-* + *-naut* (as in *aeronaut*)] (1929)
: a person who travels beyond the earth's atmosphere; *also* **:** a trainee for spaceflight

as·tro·nau·tics \ˌas-trə-ˈnȯ-tiks, -ˈnä-\ *noun plural but singular or plural in construction* (1928)
: the science of the construction and operation of vehicles for travel in space beyond the earth's atmosphere
— **as·tro·nau·tic** \-tik\ *or* **as·tro·nau·ti·cal** \-ti-kəl\ *adjective*
— **as·tro·nau·ti·cal·ly** \-ti-k(ə-)lē\ *adverb*

as·tron·o·mer \ə-ˈsträ-nə-mər\ *noun* (14th century)
: one who is skilled in astronomy or who makes observations of celestial phenomena

as·tro·nom·i·cal \ˌas-trə-ˈnä-mi-kəl\ *also* **as·tro·nom·ic** \-mik\ *adjective* (1556)
1 : of or relating to astronomy
2 : enormously or inconceivably large or great ⟨*astronomical* numbers⟩
— **as·tro·nom·i·cal·ly** \-mi-k(ə-)lē\ *adverb*

astronomical unit *noun* (1903)
: a unit of length used in astronomy equal to the mean distance of the earth from the sun or about 93 million miles (150 million kilometers)

as·tron·o·my \ə-ˈsträ-nə-mē\ *noun, plural* **-mies** [Middle English *astronomie,* from Old French, from Latin *astronomia,* from Greek, from *astr-* + *-nomia* -nomy] (13th century)
: the study of objects and matter outside the earth's atmosphere and of their physical and chemical properties

as·tro·pho·tog·ra·phy \ˌas-(ˌ)trō-fə-ˈtä-grə-fē\ *noun* [International Scientific Vocabulary] (circa 1858)
: photography as used in astronomical investigations
— **as·tro·pho·to·graph** \-ˈfō-tə-ˌgraf\ *noun*
— **as·tro·pho·tog·ra·pher** \-fə-ˈtä-grə-fər\ *noun*

as·tro·phys·ics \ˌas-trə-ˈfi-ziks\ *noun plural but singular or plural in construction* [International Scientific Vocabulary] (1890)
: a branch of astronomy dealing especially with the behavior, physical properties, and dynamic processes of celestial objects and phenomena
— **as·tro·phys·i·cal** \-zi-kəl\ *adjective*
— **as·tro·phys·i·cal·ly** \-zi-k(ə-)lē\ *adverb*
— **as·tro·phys·i·cist** \-ˈfi-zə-sist, -ˈfiz-sist\ *noun*

as·tute \ə-ˈstüt, a-, -ˈstyüt\ *adjective* [Latin *astutus,* from *astus* craft] (circa 1611)
: having or showing shrewdness and perspicacity ⟨an *astute* observer⟩ ⟨*astute* remarks⟩; *also* **:** CRAFTY, WILY
synonym see SHREWD
— **as·tute·ly** *adverb*
— **as·tute·ness** *noun*

As·ty·a·nax \ə-ˈstī-ə-ˌnaks\ *noun* [Greek]
: a son of Hector and Andromache hurled by the Greeks from the walls of Troy

asun·der \ə-ˈsən-dər\ *adverb or adjective* (14th century)
1 : into parts ⟨torn *asunder*⟩
2 : apart from each other in position ⟨wide *asunder*⟩

aswarm \ə-ˈswȯrm\ *adjective* (circa 1869)
: filled to overflowing **:** SWARMING ⟨streets *aswarm* with people⟩

[1]as well as *conjunction* (15th century)
: and in addition **:** AND ⟨brave *as well as* loyal⟩

[2]as well as *preposition* (1589)
: in addition to **:** BESIDES ⟨the coach, *as well as* the team, is ready⟩

aswirl \ə-ˈswər(-ə)l\ *adjective* (1877)
: being in a swirl **:** SWIRLING

aswoon \ə-ˈswün\ *adjective* (14th century)
: being in a swoon **:** DAZED

asy·lum \ə-ˈsī-ləm\ *noun* [Middle English, from Latin, from Greek *asylon,* neuter of *asylos* inviolable, from *a-* + *sylon* right of seizure] (15th century)
1 : an inviolable place of refuge and protection giving shelter to criminals and debtors **:** SANCTUARY
2 : a place of retreat and security **:** SHELTER
3 a : the protection or inviolability afforded by an asylum **:** REFUGE **b :** protection from arrest and extradition given especially to political refugees by a nation or by an embassy or other agency enjoying diplomatic immunity
4 : an institution for the care of the destitute or sick and especially the insane

asym·met·ri·cal \ˌā-sə-ˈme-tri-kəl\ *or* **asym·met·ric** \-trik\ *adjective* [Greek *asymmetria* lack of proportion, from *asymmetros* ill-proportioned, from *a-* + *symmetros* symmetrical] (1690)
1 : not symmetrical
2 *usually asymmetric, of a carbon atom* **:** bonded to four different atoms or groups
— **asym·met·ri·cal·ly** \-tri-k(ə-)lē\ *adverb*
— **asym·me·try** \(ˌ)ā-ˈsi-mə-trē\ *noun*

asymp·tom·at·ic \ˌā-ˌsim(p)-tə-ˈma-tik\ *adjective* (1927)
: presenting no symptoms of disease
— **asymp·tom·at·i·cal·ly** \-ti-k(ə-)lē\ *adverb*

as·ymp·tote \ˈa-səm(p)-ˌtōt\ *noun* [probably from (assumed) New Latin *asymptotus,* from Greek *asymptōtos* not meeting, from *a-* + *sympiptein* to meet — more at SYMPTOM] (1656)
: a straight line associated with a curve such that as a point moves along an infinite branch of the curve the distance from the point to the line approaches zero and the slope of the curve at the point approaches the slope of the line
— **as·ymp·tot·ic** \ˌa-səm(p)-ˈtä-tik\ *adjective*
— **as·ymp·tot·i·cal·ly** \-ti-k(ə-)lē\ *adverb*

asymptotes to the hyperbola

asyn·ap·sis \ˌā-sə-ˈnap-səs\ *noun, plural* **-ap·ses** \-ˌsēz\ [New Latin [2]*a-* + *synapsis*] (1930)
: failure of pairing of homologous chromosomes in meiosis

asyn·chro·nous \(ˌ)ā-ˈsiŋ-krə-nəs, -ˈsin-\ *adjective* (1748)
1 : not synchronous
2 : of, used in, or being digital communication (as between computers) in which there is no timing requirement for transmission and in which the start of each character is individually signaled by the transmitting device
— **asyn·chro·nous·ly** *adverb*

asyn·chro·ny \-krə-nē\ *or* **asyn·chro·nism** \-krə-ˌni-zəm\ *noun* (1875)
: the quality or state of being asynchronous **:** absence or lack of concurrence in time

as·yn·det·ic \ˌa-sᵊn-ˈde-tik\ *adjective* (circa 1864)
: marked by asyndeton
— **as·yn·det·i·cal·ly** \-ti-k(ə-)lē\ *adverb*

asyn·de·ton \ə-ˈsin-də-ˌtän, (ˌ)ā-ˈsin-\ *noun, plural* **-tons** *or* **-ta** \-də-tə\ [Late Latin, from Greek, from neuter of *asyndetos* unconnected, from *a-* + *syndetos* bound together, from *syndein* to bind together, from *syn-* + *dein* to bind — more at DIADEM] (1555)
: omission of the conjunctions that ordinarily join coordinate words or clauses (as in "I came, I saw, I conquered")

[1]at \ət, ˈat\ *preposition* [Middle English, from Old English *æt;* akin to Old High German *az* at, Latin *ad*] (before 12th century)
1 — used as a function word to indicate presence or occurrence in, on, or near ⟨staying *at* a hotel⟩ ⟨*at* a party⟩ ⟨sick *at* heart⟩
2 — used as a function word to indicate the goal of an indicated or implied action or motion ⟨aim *at* the target⟩ ⟨creditors are *at* him again⟩
3 — used as a function word to indicate that with which one is occupied or employed ⟨*at* work⟩ ⟨*at* the controls⟩ ⟨good *at* chess⟩
4 — used as a function word to indicate situation in an active or passive state or condition ⟨*at* liberty⟩ ⟨*at* rest⟩
5 — used as a function word to indicate the means, cause, or manner ⟨sold *at* auction⟩ ⟨laughed *at* my joke⟩ ⟨act *at* your own discretion⟩
6 a — used as a function word to indicate the rate, degree, or position in a scale or series ⟨the temperature *at* 90⟩ ⟨*at* first⟩ **b —** used as a function word to indicate age or position in time ⟨will retire *at* 65⟩

[2]at \ˈät\ *noun, plural* **at** [Lao] (1955)
— see *kip* at MONEY table

at- — see AD-

At·a·brine \ˈa-tə-brən, -ˌbrēn\ *trademark*
— used for quinacrine

atac·tic \(ˌ)ā-ˈtak-tik\ *adjective* [International Scientific Vocabulary [2]*a-* + *-tactic*] (1957)
: of, relating to, or being a polymer exhibiting no stereochemical regularity of structure ⟨*atactic* polypropylene⟩ — compare ISOTACTIC

At·a·lan·ta \ˌa-tᵊl-ˈan-tə\ *noun* [Latin, from Greek *Atalantē*]
: a fleet-footed huntress in Greek mythology who challenges her suitors to a race and is defeated by Hippomenes when she stops to pick up three golden apples he has dropped

at all *adverb* (14th century)
: in any way or respect **:** to the least extent or degree **:** under any circumstances ⟨doesn't smoke *at all*⟩

at·a·man \ˌa-tə-ˈman\ *noun* [Russian] (1835)
: HETMAN

at·a·mas·co lily \ˌa-tə-ˈmas-(ˌ)kō-\ *noun* [Virginia Algonquian *attamusco*] (1743)
: any of a genus (*Zephyranthes*) of American bulbous herbs of the amaryllis family with pink, white, or yellowish flowers; *especially* **:** one (*Z. atamasco*) of the southeastern U.S. with white flowers usually tinged with purple

at·a·rac·tic \ˌa-tə-ˈrak-tik\ *or* **at·a·rax·ic** \-ˈrak-sik\ *noun* [*ataractic* from Greek *ataraktos* calm, from *a-* + *tarassein* to disturb; *ataraxic* from Greek *ataraxia* calmness, from *a-* + *tarassein*] (1955)
: TRANQUILIZER 2
— ataractic *or* **ataraxic** *adjective*

at·a·vism \ˈa-tə-ˌvi-zəm\ *noun* [French *atavisme,* from Latin *atavus* ancestor, from *at-* (probably akin to *atta* daddy) + *avus* grandfather — more at UNCLE] (1833)
1 a : recurrence in an organism of a trait or character typical of an ancestral form and usually due to genetic recombination **b :** recurrence of or reversion to a past style, manner, outlook, or approach ⟨architectural *atavism*⟩
2 : one that manifests atavism **:** THROWBACK
— at·a·vis·tic \ˌa-tə-ˈvis-tik\ *adjective*
— at·a·vis·ti·cal·ly \-ti-k(ə-)lē\ *adverb*

atax·ia \ə-ˈtak-sē-ə, (ˌ)ā-\ *noun* [Greek, from *a-* + *tassein* to put in order] (1670)
: an inability to coordinate voluntary muscular movements that is symptomatic of some nervous disorders
— atax·ic \-sik\ *adjective*

at bat *noun* (1884)
: an official time at bat charged to a baseball batter except when he walks, sacrifices, is hit by a pitched ball, or is interfered with by the catcher ⟨three hits in five *at bats*⟩

ate *past of* EAT

Ate \ˈä-tē, ˈā-(ˌ)tē; ˈä-ˌtā\ *noun* [Greek *Atē*]
: a Greek goddess personifying foolhardy and ruinous impulse

¹-ate *noun suffix* [Middle English *-at,* from Old French, from Latin *-atus, -atum,* masculine & neuter of *-atus,* past participle ending]
1 : one acted upon (in a specified way) ⟨distil*late*⟩
2 [New Latin *-atum,* from Latin] **:** chemical compound or complex anion derived from a (specified) compound or element ⟨phenol*ate*⟩; *especially* **:** salt or ester of an acid with a name ending in *-ic* and not beginning with *hydro-* ⟨bor*ate*⟩

²-ate *noun suffix* [Middle English *-at,* from Old French, from Latin *-atus,* from *-a-,* stem vowel of 1st conjugation + *-tus,* suffix of verbal nouns]
1 : office **:** function **:** rank **:** group of persons holding a (specified) office or rank or having a (specified) function ⟨vicar*ate*⟩
2 : state **:** dominion **:** jurisdiction ⟨emir*ate*⟩ ⟨khan*ate*⟩

³-ate *adjective suffix* [Middle English *-at,* from Latin *-atus,* from past participle ending of 1st conjugation verbs, from *-a-,* stem vowel of 1st conjugation + *-tus,* past participle suffix — more at -ED]
: marked by having ⟨crani*ate*⟩

⁴-ate *verb suffix* [Middle English *-aten,* from Latin *-atus,* past participle ending]
: act on (in a specified way) ⟨insul*ate*⟩ **:** cause to be modified or affected by ⟨camphor*ate*⟩ **:** cause to become ⟨activ*ate*⟩ **:** furnish with ⟨capacit*ate*⟩

-ated *adjective suffix*
: ³-ATE ⟨pile*ated*⟩

at·el·ec·ta·sis \ˌa-tᵊl-ˈek-tə-səs\ *noun, plural* **-ta·ses** \-ˌsēz\ [New Latin, from Greek *atelēs* incomplete, defective (from *a-* ²a- + *telos* end) + *ektasis* extension, from *ekteinein* to stretch out, from *ex-* + *teinein* to stretch — more at TELOS, THIN] (1859)
: collapse of the expanded lung; *also* **:** defective expansion of the pulmonary alveoli at birth

ate·lier \ˌa-tᵊl-ˈyā\ *noun* [French, from Middle French *astelier* woodpile, from *astele* splinter, from Late Latin *astella,* diminutive of Latin *astula*] (1699)
1 : an artist's or designer's studio or workroom
2 : WORKSHOP

ate·moya \ˌä-tə-ˈmói-ə, ˌa-\ *noun* [*ates* sweetsop (from Tagalog) + cheri*moya*] (1914)
: a white-pulped tropical fruit of a tree that is a hybrid of the sweetsop and the cherimoya

a tem·po \ä-ˈtem-(ˌ)pō\ *adverb or adjective* [Italian] (circa 1740)
: in time — used as a direction in music to return to the original tempo

atem·po·ral \(ˌ)ā-ˈtem-p(ə-)rəl\ *adjective* (1870)
: independent of or unaffected by time **:** TIMELESS

Ate·ri·an \ə-ˈtir-ē-ən\ *adjective* [French *atérien,* from Bir el-*Ater* (Constantine), Algeria] (1928)
: of or relating to a Paleolithic culture of northern Africa characterized by Mousterian features, tanged arrow points, and leaf-shaped spearheads

Ath·a·bas·can \ˌa-thə-ˈbas-kən\ *or* **Ath·a·pas·kan** \-ˈpas-\ *also* **Ath·a·bas·kan** \-ˈbas-\ *or* **Ath·a·pas·can** \-ˈpas-\ *noun* [*Athabasca,* a Cree band, from Cree dialect *aδapaskaˑw,* name for the area east of Lake Athabasca, literally, (where) there are reeds one after another] (1846)
1 : a family of American Indian languages spoken primarily in western Canada, Alaska, and the U.S. Southwest
2 : a member of a people speaking an Athabascan language

Ath·a·na·sian \ˌa-thə-ˈnā-zhən, -shən\ *adjective* (1586)
: of or relating to Athanasius or his advocacy of the homoousian doctrine against Arianism

Athanasian Creed *noun* (1586)
: a Christian creed originating in Europe about A.D. 400 and relating especially to the Trinity and Incarnation

athe·ism \ˈā-thē-ˌi-zəm\ *noun* [Middle French *athéisme,* from *athée* atheist, from Greek *atheos* godless, from *a-* + *theos* god] (1546)
1 *archaic* **:** UNGODLINESS, WICKEDNESS
2 a : a disbelief in the existence of deity **b :** the doctrine that there is no deity

athe·ist \ˈā-thē-ist\ *noun* (1571)
: one who denies the existence of God
— athe·is·tic \ˌā-thē-ˈis-tik\ *or* **athe·is·ti·cal** \ˌā-thē-ˈis-ti-kəl\ *adjective*
— athe·is·ti·cal·ly \-ti-k(ə-)lē\ *adverb*

ath·e·ling \ˈa-thə-liŋ, -t͟hə-\ *noun* [Middle English, from Old English *ætheling,* from *æthelu* nobility, akin to Old High German *adal* nobility] (before 12th century)
: an Anglo-Saxon prince or nobleman; *especially* **:** the heir apparent or a prince of the royal family

Athe·na \ə-ˈthē-nə\ *or* **Athe·ne** \-nē\ *noun* [Latin *Athena,* from Greek *Athēnē*]
: the Greek goddess of wisdom — compare MINERVA

ath·e·nae·um *or* **ath·e·ne·um** \ˌa-thə-ˈnē-əm\ *noun* [Latin *Athenaeum,* a school in ancient Rome for the study of arts, from Greek *Athēnaion,* a temple of Athena, from *Athēnē*] (1799)
1 : a building or room in which books, periodicals, and newspapers are kept for use
2 : a literary or scientific association

athe·o·ret·i·cal \ˌā-ˌthē-ə-ˈre-ti-kəl, -ˌthi(-ə)r-ˈe-\ *adjective* (1969)
: not based on or concerned with theory

athero- *combining form* [Greek *athēra*]
: atheroma ⟨*athero*genic⟩

ath·ero·gen·e·sis \ˌa-thə-rō-ˈje-nə-səs\ *noun* (1953)
: the process of developing atheroma

ath·ero·gen·ic \-ˈje-nik\ *adjective* (1954)
: relating to or producing degenerative changes in arterial walls ⟨*atherogenic* diet⟩

ath·er·o·ma \ˌa-thə-ˈrō-mə\ *noun* [New Latin *atheromat-, atheroma,* from Latin, a tumor containing matter resembling gruel, from Greek *athērōma,* from *athēra* gruel] (1875)
1 : fatty degeneration of the inner coat of the arteries
2 : an abnormal fatty deposit in an artery
— ath·er·o·ma·tous \-ˈrō-mə-təs\ *adjective*

ath·ero·scle·ro·sis \ˌa-thə-ˌrō-sklə-ˈrō-səs\ *noun* [New Latin] (1910)
: an arteriosclerosis characterized by atheromatous deposits in and fibrosis of the inner layer of the arteries
— ath·ero·scle·rot·ic \-sklə-ˈrä-tik\ *adjective*

athirst \ə-ˈthərst\ *adjective* [Middle English, from Old English *ofthyrst,* past participle of *ofthyrstan* to suffer from thirst, from *of* off, from + *thyrstan* to thirst — more at OF] (before 12th century)
1 *archaic* **:** THIRSTY
2 : having a strong eager desire ⟨I that for ever feel *athirst* for glory —John Keats⟩
synonym see EAGER

ath·lete \ˈath-ˌlēt, ÷ˈa-thə-ˌlēt\ *noun* [Middle English, from Latin *athleta,* from Greek *athlētēs,* from *athlein* to contend for a prize, from *athlon* prize, contest] (15th century)
: a person who is trained or skilled in exercises, sports, or games requiring physical strength, agility, or stamina

athlete's foot *noun* (1928)
: ringworm of the feet

ath·let·ic \ath-ˈle-tik, ÷ˌa-thə-ˈle-\ *adjective* (1636)
1 : of or relating to athletes or athletics
2 : characteristic of an athlete; *especially* **:** VIGOROUS, ACTIVE
3 : MESOMORPHIC
4 : used by athletes
— ath·let·i·cal·ly \-ti-k(ə-)lē\ *adverb*
— ath·let·i·cism \-ˈle-tə-ˌsi-zəm\ *noun*

ath·let·ics \ath-ˈle-tiks, ÷ˌa-thə-ˈle-\ *noun plural but singular or plural in construction* (1727)
1 : exercises, sports, or games engaged in by athletes
2 : the practice or principles of athletic activities

athletic supporter *noun* (1927)
: a supporter for the genitals worn by men participating in sports or strenuous activities

at–home \ət-ˈhōm\ *adjective* (1951)
1 : intended or suitable for one's home ⟨an *at-home* dress⟩
2 : being or occurring at one's home ⟨*at-home* entertainment⟩

at home *noun* (1745)
: a reception given at one's home

-athon *noun combining form* [*marathon*]
: event or activity lasting a long time ⟨talk*athon*⟩

ath·ro·cyte \ˈa-thrə-ˌsīt\ *noun* [Greek *athroos* together, collected + International Scientific Vocabulary *-cyte*] (1938)
: a cell that has the ability to store up substances of an excretory nature

¹athwart \ə-ˈthwȯrt, *nautical often* -ˈthȯrt\ *preposition* (15th century)
1 : ACROSS
2 : in opposition to ⟨a procedure directly *athwart* the New England prejudices —R. G. Cole⟩

²athwart *adverb* (circa 1500)
1 : across especially in an oblique direction
2 : in opposition to the right or expected course ⟨and quite *athwart* goes all decorum —Shakespeare⟩

athwart·ship \-ˌship\ *adjective* (1879)

: being across the ship from side to side ⟨*athwartship* and longitudinal framing⟩

athwart·ships \-ˌships\ *adverb* (1718)
: across the ship from side to side

atilt \ə-ˈtilt\ *adverb or adjective* (1562)
1 : in a tilted position
2 : with lance in hand ⟨run *atilt* at death —Shakespeare⟩

atin·gle \ə-ˈtiŋ-gəl\ *adjective* (1855)
: tingling especially with excitement

-ation *noun suffix* [Middle English *-acioun,* from Old French *-ation,* from Latin *-ation-, -atio,* from *-a-,* stem vowel of 1st conjugation + *-tion-, -tio,* noun suffix]
: action or process ⟨flirt*ation*⟩ : something connected with an action or process ⟨discolor*ation*⟩

-ative *adjective suffix* [Middle English, from Middle French *-atif,* from Latin *-ativus,* from *-atus* -ate + *-ivus* -ive]
: of, relating to, or connected with ⟨authorit*ative*⟩ : tending to ⟨talk*ative*⟩

At·ka mackerel \ˈat-kə-, ˈät-\ *noun* [*Atka* Island, Alaska] (1893)
: a greenling (*Pleurogrammus monopterygius*) of Alaska and adjacent regions valued as a food fish

[1]**At·lan·te·an** \ˌat-ˌlan-ˈtē-ən, ət-ˈlan-tē-\ *adjective* (1667)
: of, relating to, or resembling Atlas : STRONG

[2]**Atlantean** *adjective* (circa 1828)
: of or relating to Atlantis

At·lan·tic \ət-ˈlan-tik, at-\ *adjective* (1601)
1 a : of, relating to, or found in, on, or near the Atlantic Ocean **b :** of, relating to, or found on or near the east coast of the U.S.
2 : of or relating to the nations that border the Atlantic Ocean ⟨the *Atlantic* community⟩

Atlantic croaker *noun* (circa 1949)
: a small fish (*Micropogonias undulatus*) of the Gulf coast and the Atlantic coast — called also *hardhead*

At·lan·ti·cism \-ˈlan-tə-ˌsi-zəm\ *noun* (1950)
: a policy of military cooperation between European powers and the U.S.
— **At·lan·ti·cist** \-sist\ *noun*

Atlantic salmon *noun* (1902)
: SALMON 1a

Atlantic time *noun* (circa 1909)
: the time of the fourth time zone west of Greenwich that includes the Canadian Maritime Provinces, Puerto Rico, and the Virgin Islands

At·lan·tis \ət-ˈlan-təs, at-\ *noun* [Latin, from Greek, from *Atlas*]
: a fabled island in the Atlantic that according to legend sank beneath the sea

at·las \ˈat-ləs\ *noun* [Latin *Atlant-, Atlas,* from Greek]
1 *capitalized* **:** a Titan who for his part in the Titans' revolt against the gods is forced by Zeus to support the heavens on his shoulders
2 *capitalized* **:** one who bears a heavy burden
3 a : a bound collection of maps often including illustrations, informative tables, or textual matter **b :** a bound collection of tables, charts, or plates
4 : the first vertebra of the neck
5 *plural usually* **at·lan·tes** \ət-ˈlan-(ˌ)tēz, at-\ **:** a male figure used like a caryatid as a supporting column or pilaster — called also *telamon* ◆

atlas 5

at·latl \ˈät-ˌlä-tᵊl\ *noun* [Nahuatl *ahtlatl*] (1871)
: a device for throwing a spear or dart that consists of a rod or board with a projection (as a hook or thong) at the rear end to hold the weapon in place until released

At·li \ˈät-lē\ *noun* [Old Norse]
: a king of the Huns figuring in Germanic legend and corresponding to the historical Attila

at·man \ˈät-mən, -ˌmän\ *noun, often capitalized* [Sanskrit *ātman,* literally, breath, soul; akin to Old English *ǣthm* breath] (1785)
1 *Hinduism* **:** the innermost essence of each individual
2 *Hinduism* **:** the supreme universal self : BRAHMA 2

at·mom·e·ter \at-ˈmä-mə-tər\ *noun* [Greek *atmos* + English *-meter*] (1815)
: an instrument for measuring the evaporating capacity of the air

at·mo·sphere \ˈat-mə-ˌsfir\ *noun* [New Latin *atmosphaera,* from Greek *atmos* vapor + Latin *sphaera* sphere] (1677)
1 a : the gaseous envelope of a celestial body (as a planet) **b :** the whole mass of air surrounding the earth
2 : the air of a locality
3 : a surrounding influence or environment ⟨an *atmosphere* of hostility⟩
4 : a unit of pressure equal to the pressure of the air at sea level or approximately 14.7 pounds per square inch (101,325 pascals)
5 a : the overall aesthetic effect of a work of art **b :** a dominant aesthetic or emotional effect or appeal
— **at·mo·sphered** \-ˌsfird\ *adjective*

at·mo·spher·ic \ˌat-mə-ˈsfir-ik, -ˈsfer-\ *adjective* (1783)
1 a : of, relating to, or occurring in the atmosphere **b :** resembling the atmosphere : AIRY
2 : having, marked by, or contributing aesthetic or emotional atmosphere
— **at·mo·spher·i·cal·ly** \-i-k(ə-)lē\ *adverb*

at·mo·spher·ics \-iks\ *noun plural* (1915)
1 : audible disturbances produced in radio receiving apparatus by atmospheric electrical phenomena (as lightning); *also* **:** the electrical phenomena causing these disturbances
2 : actions (as official statements) intended to create or suggest a particular atmosphere or mood in politics and especially international relations; *also* **:** the mood so created or suggested
3 : realistic detail added (as to a literary work) to create a mood

atoll \ˈa-ˌtȯl, -ˌtäl, -ˌtōl, ˈā-\ *noun* [Divehi (Indo-Aryan language of the Maldive Islands) *atolu*] (1625)
: a coral island consisting of a reef surrounding a lagoon

atoll

at·om \ˈa-təm\ *noun* [Middle English, from Latin *atomus,* from Greek *atomos,* from *atomos* indivisible, from *a-* + *temnein* to cut] (15th century)
1 : one of the minute indivisible particles of which according to ancient materialism the universe is composed
2 : a tiny particle : BIT
3 : the smallest particle of an element that can exist either alone or in combination
4 : the atom considered as a source of vast potential energy ◆

atom·ic \ə-ˈtä-mik\ *adjective* (1678)
1 a : of, relating to, or concerned with atoms ⟨*atomic* physics⟩ **b :** NUCLEAR 2 ⟨*atomic* energy⟩
2 a : marked by acceptance of the theory of atomism **b :** ATOMISTIC 2
3 : MINUTE
4 *of a chemical element* **:** existing in the state of separate atoms
— **atom·i·cal·ly** \-mi-k(ə-)lē\ *adverb*

atomic bomb *noun* (1914)
1 : a bomb whose violent explosive power is due to the sudden release of energy resulting from the splitting of nuclei of a heavy chemical element (as plutonium or uranium) by neutrons in a very rapid chain reaction — called also *atom bomb*
2 : a nuclear weapon (as a hydrogen bomb)

atomic clock *noun* (1938)
: a precision clock that depends for its operation on an electrical oscillator regulated by the natural vibration frequencies of an atomic system (as a beam of cesium atoms)

atomic mass *noun* (1898)
: the mass of an atom usually expressed in atomic mass units; *also* : ATOMIC WEIGHT

atomic mass unit *noun* (circa 1942)
: a unit of mass for expressing masses of atoms, molecules, or nuclear particles equal to $\frac{1}{12}$ the mass of a single atom of the most abundant carbon isotope ^{12}C — called also *dalton*

atomic number *noun* (1821)
: an experimentally determined number characteristic of a chemical element that represents the number of protons in the nucleus which in a neutral atom equals the number of electrons outside the nucleus and that determines the place of the element in the periodic table — see ELEMENT table

atomic reactor *or* **atomic pile** *noun* (1945)
: REACTOR 3b

atomic theory *noun* (circa 1847)
1 : a theory of the nature of matter: all material substances are composed of minute particles or atoms of a comparatively small number of kinds and all the atoms of the same kind are uniform in size, weight, and other properties

◇ WORD HISTORY

atlas In Greek myth, the Titan Atlas, following the defeat of the Titans in a war with Zeus and the other gods, was punished by being forced to support the heavens on his shoulders. A representation of Atlas as a man bearing on his back the globe of the earth became a conventional title-page illustration in bound volumes of maps during the 16th century. In 1585, the Flemish mathematician and mapmaker Gerhardus Mercator initiated a series of map collections with a flowery Latin title that can be translated as *Atlas, or Cosmographical Reflections on the Edifice of the World and the Representation of its Form.* The word *atlas* eventually became the generic term for a book of maps in most languages of Europe, including English.

atom For ancient Greek thinkers, ideas about the nature of matter and motion, what we think of as physics, were inseparable from philosophical reflections about existence and plurality. On grounds that a modern physicist would consider philosophical rather than experimental, Leucippus and his better-known student Democritus decided that matter was not infinitely divisible, but rather consisted of particles of being separated by empty space. These particles of being that made up all things were called *atoma sōmata* "indivisible bodies." The later Greek word *atomos,* borrowed into Latin as *atomus* and into French as *atome,* was used in Epicurean philosophy. Long after ancient Greece and Rome the concept was taken up again by early experimental science in the 17th century. With the development of modern atomic chemistry in the 19th and 20th centuries, the atoms of Democritus proved real enough, though they were not indivisible after all, and the splitting of their nuclei was discovered to be the source of enormous energy.

2 : any of several theories of the structure of the atom; *especially* **:** one based on experimentation and theoretical considerations holding that the atom is composed essentially of a small positively charged comparatively heavy nucleus surrounded by a comparatively large arrangement of electrons

atomic weight *noun* (1820)
: the mass of one atom of an element; *specifically* **:** the average mass of an atom of an element as it occurs in nature that is expressed in atomic mass units — see ELEMENT table

at·om·ise, at·om·is·er *British variant of* ATOMIZE, ATOMIZER

at·om·ism \'a-tə-ˌmi-zəm\ *noun* (1678)
1 : a doctrine that the physical or physical and mental universe is composed of simple indivisible minute particles
2 : INDIVIDUALISM 1
— **at·om·ist** \-mist\ *noun*

at·om·is·tic \ˌa-tə-'mis-tik\ *adjective* (1809)
1 : of or relating to atoms or atomism
2 : composed of many simple elements; *also* **:** characterized by or resulting from division into unconnected or antagonistic fragments ⟨an *atomistic* society⟩
— **at·om·is·ti·cal·ly** \-ti-k(ə-)lē\ *adverb*

at·om·ize \'a-tə-ˌmīz\ *transitive verb* **-ized; -iz·ing** (1845)
1 : to treat as made up of many discrete units
2 : to reduce to minute particles or to a fine spray
3 : DIVIDE, FRAGMENT
4 : to subject to attack by nuclear weapons
— **at·om·iza·tion** \ˌa-tə-mə-'zā-shən\ *noun*

at·om·iz·er \'a-tə-ˌmī-zər\ *noun* (1865)
: an instrument for atomizing usually a perfume, disinfectant, or medicament

atom smasher *noun* (1937)
: ACCELERATOR d

at·o·my \'a-tə-mē\ *noun, plural* **-mies** [irregular from Latin *atomi,* plural of *atomus* atom] (1591)
: a tiny particle **:** ATOM, MITE

aton·al \(ˌ)ā-'tō-nᵊl, (ˌ)a-\ *adjective* [[2]*a-* + *tonal*] (1922)
: marked by avoidance of traditional musical tonality; *especially* **:** organized without reference to key or tonal center and using the tones of the chromatic scale impartially
— **aton·al·ism** \-nᵊl-ˌi-zəm\ *noun*
— **aton·al·ist** \-ist\ *noun*
— **ato·nal·i·ty** \ˌā-tō-'na-lə-tē, ˌa-\ *noun*
— **aton·al·ly** \(ˌ)ā-'tō-nᵊl-ē, (ˌ)a-\ *adverb*

atone \ə-'tōn\ *verb* **atoned; aton·ing** [Middle English, to become reconciled, from *at on* in harmony, from *at* + *on* one] (1593)
transitive verb
1 *obsolete* **:** RECONCILE
2 : to supply satisfaction for **:** EXPIATE
intransitive verb
: to make amends

atone·ment \ə-'tōn-mənt\ *noun* (1513)
1 *obsolete* **:** RECONCILIATION
2 : the reconciliation of God and man through the sacrificial death of Jesus Christ
3 : reparation for an offense or injury **:** SATISFACTION
4 *Christian Science* **:** the exemplifying of man's oneness with God

aton·ic \(ˌ)ā-'tä-nik, (ˌ)a-\ *adjective* (1792)
1 : characterized by atony
2 : uttered without accent or stress

at·o·ny \'a-tᵊn-ē\ *noun* [Late Latin *atonia,* from Greek, from *atonos* without tone, from *a-* + *tonos* tone] (1693)
: lack of physiological tone especially of a contractile organ

[1]**atop** \ə-'täp\ *preposition* (1655)
: on top of

[2]**atop** *adverb or adjective* (1658)
: on, to, or at the top

at·o·py \'a-tə-pē\ *noun* [Greek *atopia* uncommonness, from *atopos* out of the way, uncommon, from *a-* + *topos* place] (1923)
: a probably hereditary allergy characterized by symptoms (as asthma, hay fever, or hives) produced upon exposure to the exciting antigen without inoculation
— **atop·ic** \(ˌ)ā-'tä-pik, -'tō-\ *adjective*

-ator *noun suffix* [Middle English *-atour,* from Old French & Latin; Old French, from Latin *-ator,* from *-a-,* stem vowel of 1st conjugation + *-tor,* agent suffix]
: one that does ⟨totaliz*ator*⟩

ATP \ˌā-ˌtē-'pē\ *noun* [*a*denosine *t*ri*p*hosphate] (1939)
: a phosphorylated nucleoside $C_{10}H_{16}N_5O_{13}P_3$ of adenine that supplies energy for many biochemical cellular processes by undergoing enzymatic hydrolysis especially to ADP — called also *adenosine triphosphate*

ATPase \ˌā-ˌtē-'pē-ˌās, -ˌāz\ *noun* (1946)
: an enzyme that hydrolyzes ATP; *especially* **:** one that hydrolyzes ATP to ADP and inorganic phosphate

at·ra·bil·ious \ˌa-trə-'bil-yəs\ *adjective* [Latin *atra bilis* black bile] (1651)
1 : given to or marked by melancholy **:** GLOOMY
2 : ILL-NATURED, PEEVISH
— **at·ra·bil·ious·ness** *noun*

at·ra·zine \'a-trə-ˌzēn\ *noun* [perhaps from *a*mino + *triazine*] (1962)
: a photosynthesis-inhibiting persistent herbicide $C_8H_{14}ClN_5$ used especially to kill annual weeds and quack grass

atrem·ble \ə-'trem-bəl\ *adjective* (1862)
: shaking involuntarily **:** TREMBLING ⟨he was white as death and all *atremble* —Robert Coover⟩

atre·sia \ə-'trē-zhə\ *noun* [New Latin, from [2]*a-* + Greek *trēsis* perforation, from *tetrainein* to pierce — more at THROW] (circa 1807)
1 : absence or closure of a natural passage of the body
2 : absence or disappearance of an anatomical part (as an ovarian follicle) by degeneration

Atreus \'ā-ˌtrüs, -trē-əs\ *noun* [Greek]
: a king of Mycenae and father of Agamemnon and Menelaus

atrial natriuretic factor *noun* (1984)
: a peptide hormone secreted by the cardiac atria that stimulates natriuresis and diuresis and helps regulate blood pressure

atrio·ven·tric·u·lar \ˌā-trē-ō-ven-'tri-kyə-lər, -vən-\ *adjective* [New Latin *atrium* + English *ventricular*] (circa 1860)
: of, relating to, or located between an atrium and ventricle of the heart

atrioventricular node *noun* (circa 1934)
: a small mass of tissue in the right atrioventricular region of higher vertebrates through which impulses from the sinoatrial node are passed to the ventricles

atrip \ə-'trip\ *adjective* (1796)
of an anchor **:** AWEIGH

atri·um \'ā-trē-əm\ *noun, plural* **atria** \-trē-ə\ *also* **atri·ums** [Latin] (1577)
1 : the central room of a Roman house
2 *plural usually* *atriums* **:** a rectangularly shaped open patio around which a house is built; *also* **:** a many-storied court in a building (as a hotel) usually with a skylight
3 [New Latin, from Latin] **:** an anatomical cavity or passage; *especially* **:** the chamber or either of the chambers of the heart that receives blood from the veins and forces it into the ventricle or ventricles — see HEART illustration
— **atri·al** \-trē-əl\ *adjective*

atro·cious \ə-'trō-shəs\ *adjective* [Latin *atroc-, atrox* gloomy, atrocious, from *atr-, ater* black + *-oc-, -ox* (akin to Greek *ōps* eye) — more at EYE] (1669)
1 : extremely wicked, brutal, or cruel **:** BARBARIC
2 : APPALLING, HORRIFYING ⟨the *atrocious* weapons of modern war⟩
3 a : utterly revolting **:** ABOMINABLE ⟨*atrocious* working conditions⟩ **b :** of very poor quality ⟨*atrocious* handwriting⟩
— **atro·cious·ly** *adverb*
— **atro·cious·ness** *noun*

atroc·i·ty \ə-'trä-sə-tē\ *noun, plural* **-ties** (1534)
1 : the quality or state of being atrocious
2 : an atrocious act, object, or situation ⟨the . . . sufferings and *atrocities* of trench warfare —Aldous Huxley⟩

at·ro·phy \'a-trə-fē\ *noun, plural* **-phies** [Late Latin *atrophia,* from Greek, from *atrophos* ill fed, from *a-* + *trephein* to nourish] (1601)
1 : decrease in size or wasting away of a body part or tissue; *also* **:** arrested development or loss of a part or organ incidental to the normal development or life of an animal or plant
2 : a wasting away or progressive decline **:** DEGENERATION ⟨the *atrophy* of freedom⟩ ⟨was not a solitude of *atrophy,* of negation, but of perpetual flowering —Willa Cather⟩
— **atro·phic** \(ˌ)ā-'trō-fik\ *adjective*
— **atrophy** \'a-trə-fē, -ˌfī\ *verb*

at·ro·pine \'a-trə-ˌpēn\ *noun* [German *Atropin,* from New Latin *Atropa,* genus name of belladonna, from Greek *Atropos,* one of the three Fates] (1836)
: a racemic mixture of hyoscyamine obtained from any of various solanaceous plants (as belladonna) and used especially in the form of its sulfate for its anticholinergic effects (as pupil dilation or inhibition of smooth muscle spasms)

att *variant of* [2]AT

at·ta·boy \'a-də-ˌbȯi\ *interjection* [probably alteration of *that's the boy*] (1909)
— used to express encouragement, approval, or admiration

at·tach \ə-'tach\ *verb* [Middle English, from Middle French *attacher,* from Old French *estachier,* from *estache* stake, of Germanic origin; akin to Old English *staca* stake] (14th century)
transitive verb
1 : to take by legal authority especially under a writ ⟨*attached* the property⟩
2 a : to bring (oneself) into an association **b :** to assign temporarily
3 : to bind by personal ties (as of affection or sympathy) ⟨was strongly *attached* to his family⟩
4 : to make fast (as by tying or gluing) ⟨*attach* a label to a package⟩
5 : to associate especially as a property **:** ATTRIBUTE ⟨*attached* great importance to public opinion polls⟩
intransitive verb
: to become attached **:** ADHERE
synonym see FASTEN
— **at·tach·able** \-'ta-chə-bəl\ *adjective*

at·ta·ché \ˌa-tə-'shā, ˌa-ˌta-, ə-ˌta-\ *noun* [French, past participle of *attacher*] (1829)
1 : a technical expert on the diplomatic staff of his country at a foreign capital ⟨a military *attaché*⟩
2 : ATTACHÉ CASE

at·ta·ché case \ˌa-ˌta-'shā-, ˌa-tə-; ə-'ta-(ˌ)shā-\ *noun* (1904)
1 : a small thin suitcase used especially for carrying business papers
2 : BRIEFCASE

at·tached \ə-'tacht\ *adjective* (1834)
: permanently fixed when adult ⟨*attached* barnacles⟩

at·tach·ment \ə-'tach-mənt\ *noun* (14th century)
1 : a seizure by legal process; *also* **:** the writ or precept commanding such seizure
2 a : the state of being personally attached **:** FIDELITY ⟨*attachment* to a cause⟩ **b :** affectionate regard ⟨a deep *attachment* to nature⟩

3 : a device attached to a machine or implement
4 : the physical connection by which one thing is attached to another
5 : the process of physically attaching

[1]**at·tack** \ə-ˈtak\ *verb* [Middle French *attaquer,* from (assumed) Old Italian *estaccare* to attach, from *stacca* stake, of Germanic origin; akin to Old English *staca*] (1600)
transitive verb
1 : to set upon or work against forcefully
2 : to assail with unfriendly or bitter words
3 : to begin to affect or to act on injuriously
4 : to set to work on
5 : to threaten (a piece in chess) with immediate capture
intransitive verb
: to make an attack ☆
— at·tack·er *noun*

[2]**attack** *noun* (1661)
1 : the act of attacking with physical force or unfriendly words **:** ASSAULT
2 : a belligerent or antagonistic action
3 a : a fit of sickness; *especially* **:** an active episode of a chronic or recurrent disease **b :** a period of being strongly affected by something (as a desire or mood)
4 a : an offensive or scoring action ⟨won the game with an eight-hit *attack*⟩ **b :** offensive players or the positions taken up by them
5 : the setting to work on some undertaking ⟨made a new *attack* on the problem⟩
6 : the beginning of destructive action (as by a chemical agent)
7 : the act or manner of beginning a musical tone or phrase

at·tack·man \-ˌman\ *noun* (1940)
: a player (as in lacrosse) assigned to an offensive zone or position

at·tain \ə-ˈtān\ *verb* [Middle English *atteynen,* from Old French *ataindre,* from (assumed) Vulgar Latin *attangere,* alteration of Latin *attingere,* from *ad-* + *tangere* to touch — more at TANGENT] (14th century)
transitive verb
1 : to reach as an end **:** GAIN, ACHIEVE ⟨*attain* a goal⟩
2 : to come into possession of **:** OBTAIN ⟨he *attained* preferment over his fellows⟩
3 : to come to as the end of a progression or course of movement ⟨they *attained* the top of the hill⟩ ⟨*attain* a ripe old age⟩
intransitive verb
: to come or arrive by motion, growth, or effort — usually used with *to*
— at·tain·abil·i·ty \-ˌtā-nə-ˈbi-lə-tē\ *noun*
— at·tain·able \-ˈtā-nə-bəl\ *adjective*

at·tain·der \ə-ˈtān-dər\ *noun* [Middle English *attaynder,* from Middle French *ataindre* to accuse, attain] (15th century)
1 : extinction of the civil rights and capacities of a person upon sentence of death or outlawry usually after a conviction of treason
2 *obsolete* **:** DISHONOR

at·tain·ment \ə-ˈtān-mənt\ *noun* (1549)
1 : the act of attaining **:** the condition of being attained
2 : something attained **:** ACCOMPLISHMENT ⟨scientific *attainments*⟩

[1]**at·taint** \ə-ˈtānt\ *transitive verb* [Middle English *attaynten,* from Middle French *ataint,* past participle of *ataindre*] (14th century)
1 : to affect by attainder
2 a : INFECT, CORRUPT **b** *archaic* **:** TAINT, SULLY
3 *archaic* **:** ACCUSE

[2]**attaint** *noun* (1592)
obsolete **:** a stain upon honor or purity **:** DISGRACE

at·tar \ˈa-tər, ˈa-ˌtär\ *noun* [Persian *ʽaṭir* perfumed, from Arabic, from *ʽiṭr* perfume] (1798)
: a fragrant essential oil (as from rose petals); *also* **:** FRAGRANCE

[1]**at·tempt** \ə-ˈtem(p)t\ *transitive verb* [Middle English, from Middle French & Latin; Middle French *attempter,* from Latin *attemptare,* from *ad-* + *temptare* to touch, try — more at TEMPT] (14th century)
1 : to make an effort to do, accomplish, solve, or effect ⟨*attempted* to swim the swollen river⟩
2 *archaic* **:** TEMPT
3 *archaic* **:** to try to subdue or take by force **:** ATTACK ☆
— at·tempt·able \-ˈtem(p)-tə-bəl\ *adjective*

[2]**attempt** *noun* (1534)
1 a : the act or an instance of attempting; *especially* **:** an unsuccessful effort **b :** something resulting from or representing an attempt ⟨surrounded by . . . a few *attempts* at rose bushes —Marian Engel⟩
2 : ATTACK, ASSAULT ⟨an *attempt* on the life of the president⟩

at·tend \ə-ˈtend\ *verb* [Middle English, from Old French *atendre,* from Latin *attendere,* literally, to stretch to, from *ad-* + *tendere* to stretch — more at THIN] (14th century)
transitive verb
1 : to pay attention to
2 : to look after **:** take charge of ⟨campsites . . . *attended* by park rangers —Jackson Rivers⟩
3 a : to go or stay with as a companion, nurse, or servant **b :** to visit professionally as a physician
4 *archaic* **a :** to wait for **b :** to be in store for
5 : to be present with **:** ACCOMPANY
6 : to be present at **:** go to ⟨*attend* law school⟩
intransitive verb
1 : to apply oneself ⟨*attend* to your work⟩
2 : to apply the mind or pay attention **:** HEED
3 a : to be ready for service ⟨ministers who *attend* upon the king⟩ **b :** to be present
4 *obsolete* **:** WAIT, STAY
5 : to direct one's attention **:** SEE ⟨I'll *attend* to that⟩
— at·tend·er *noun*

at·ten·dance \ə-ˈten-dən(t)s\ *noun* (14th century)
1 : the act or fact of attending ⟨a physician in *attendance*⟩
2 a : the persons or number of persons attending ⟨daily *attendance* at the fair⟩ **b :** the number of times a person attends

attendance officer *noun* (1884)
: one employed by a public-school system to investigate the continued absences of pupils

[1]**at·ten·dant** \ə-ˈten-dənt\ *noun* (15th century)
1 : one who attends another to perform a service; *especially* **:** an employee who waits on customers
2 : something that accompanies **:** CONCOMITANT
3 : ATTENDEE

[2]**attendant** *adjective* (15th century)
1 : accompanying, waiting upon, or following in order to perform service ⟨Cherub and Seraph . . . *attendant* on their Lord —John Milton⟩
2 : accompanying or following as a consequence or result ⟨problems *attendant* upon pollution⟩

at·tend·ee \ə-ˌten-ˈdē, ˌa-\ *noun* (1937)
: a person who is present on a given occasion or at a given place ⟨*attendees* at a convention⟩

at·tend·ing \ə-ˈten-diŋ\ *adjective* (circa 1923)
: serving as a physician on the staff of a teaching hospital ⟨*attending* surgeon⟩

at·ten·tion \ə-ˈten(t)-shən; *sense 4 often* (ə-)ˌten(ch)-ˈhət\ *noun* [Middle English *attencioun,* from Latin *attention-, attentio,* from *attendere*] (14th century)
1 a : the act or state of attending especially through applying the mind to an object of sense or thought **b :** a condition of readiness for such attention involving especially a selective narrowing or focusing of consciousness and receptivity
2 : OBSERVATION, NOTICE; *especially* **:** consideration with a view to action ⟨a problem requiring prompt *attention*⟩
3 a : an act of civility or courtesy especially in courtship **b :** sympathetic consideration of the needs and wants of others **:** ATTENTIVENESS
4 : a position assumed by a soldier with heels together, body erect, arms at the sides, and eyes to the front — often used as a command
— at·ten·tion·al \-ˈten(t)-sh(ə-)nəl\ *adjective*

attention deficit disorder *noun* (1980)
: a syndrome of learning and behavioral problems that is not caused by any serious underlying physical or mental disorder and is characterized especially by difficulty in sustaining attention, by impulsive behavior (as in speaking out of turn), and usually by excessive activity

attention line *noun* (1925)
: a line usually placed above the salutation in a business letter directing the letter to one specified

attention span *noun* (1934)
: the length of time during which an individual is able to concentrate or remain interested

at·ten·tive \ə-ˈten-tiv\ *adjective* (14th century)
1 : MINDFUL, OBSERVANT ⟨*attentive* to what he is doing⟩
2 : heedful of the comfort of others **:** SOLICITOUS
3 : offering attentions in or as if in the role of a suitor
— at·ten·tive·ly *adverb*
— at·ten·tive·ness *noun*

[1]**at·ten·u·ate** \ə-ˈten-yə-ˌwāt\ *verb* **-at·ed; -at·ing** [Latin *attenuatus,* past participle of *attenuare* to make thin, from *ad-* + *tenuis* thin — more at THIN] (15th century)
transitive verb
1 : to make thin or slender
2 : to make thin in consistency **:** RAREFY
3 : to lessen the amount, force, magnitude, or value of **:** WEAKEN
4 : to reduce the severity, virulence, or vitality of
intransitive verb
: to become thin, fine, or less
— at·ten·u·a·tion \-ˌten-yə-ˈwā-shən\ *noun*

[2]**at·ten·u·ate** \ə-ˈten-yə-wət\ *adjective* (15th century)

☆ SYNONYMS
Attack, assail, assault, bombard, storm mean to make an onslaught upon. ATTACK implies taking the initiative in a struggle ⟨plan to *attack* the town at dawn⟩. ASSAIL implies attempting to break down resistance by repeated blows or shots ⟨*assailed* the enemy with artillery fire⟩. ASSAULT suggests a direct attempt to overpower by suddenness and violence of onslaught ⟨commandos *assaulted* the building from all sides⟩. BOMBARD applies to attacking with bombs or shells ⟨*bombarded* the city nightly⟩. STORM implies attempting to break into a defended position ⟨preparing to *storm* the fortress⟩.

Attempt, try, endeavor, essay, strive mean to make an effort to accomplish an end. ATTEMPT stresses the initiation or beginning of an effort ⟨will *attempt* to photograph the rare bird⟩. TRY is often close to ATTEMPT but may stress effort or experiment made in the hope of testing or proving something ⟨*tried* to determine which was the better procedure⟩. ENDEAVOR heightens the implications of exertion and difficulty ⟨*endeavored* to find crash survivors in the mountains⟩. ESSAY implies difficulty but also suggests tentative trying or experimenting ⟨will *essay* a dramatic role for the first time⟩. STRIVE implies great exertion against great difficulty and specifically suggests persistent effort ⟨continues to *strive* for peace⟩.

1 : attenuated especially in thickness, density, or force
2 : tapering gradually usually to a long slender point ⟨*attenuate* leaves⟩

at·ten·u·a·tor \-yə-ˌwā-tər\ *noun* (1924)
: a device for attenuating; *especially* **:** one for reducing the amplitude of an electrical signal without appreciable distortion

at·test \ə-ˈtest\ *verb* [Middle French *attester,* from Latin *attestari,* from *ad-* + *testis* witness — more at TESTAMENT] (1596)
transitive verb
1 a : to affirm to be true or genuine; *specifically* **:** to authenticate by signing as a witness **b :** to authenticate officially
2 : to establish or verify the usage of
3 : to be proof of **:** MANIFEST ⟨her record *attests* her integrity⟩
4 : to put on oath
intransitive verb
: to bear witness **:** TESTIFY ⟨*attest* to a belief⟩
synonym see CERTIFY
— at·tes·ta·tion \ˌa-ˌtes-ˈtā-shən, ˌa-tə-ˈstā-\ *noun*
— at·test·er \ə-ˈtes-tər\ *noun*

at·tic \ˈa-tik\ *noun* [French *attique,* from *attique* of Attica, from Latin *Atticus*] (circa 1696)
1 : a low story or wall above the main order of a facade in the classical styles
2 : a room behind an attic
3 : a room or a space immediately below the roof of a building **:** GARRET
4 : something resembling an attic (as in being used for storage) ◆

[1]At·tic \ˈa-tik\ *adjective* [Latin *Atticus* of Attica, from Greek *Attikos,* from *Attikē* Attica, Greece] (1599)
1 : of, relating to, or having the characteristics of Athens or its ancient civilization
2 : marked by simplicity, purity, and refinement ⟨an *Attic* prose style⟩

[2]Attic *noun* (circa 1771)
: a dialect of ancient Greek originally used in Attica and later the literary language of the Greek-speaking world

at·ti·cism \ˈa-tə-ˌsi-zəm\ *noun, often capitalized* (1593)
1 : a witty or well-turned phrase
2 : a characteristic feature of Attic Greek occurring in another language or dialect

[1]at·tire \ə-ˈtīr\ *transitive verb* **at·tired; at·tir·ing** [Middle English, from Old French *atirier,* from *a-* (from Latin *ad-*) + *tire* order, rank, of Germanic origin; akin to Old English *tīr* glory, ornament] (14th century)
: to put garments on **:** DRESS, ARRAY; *especially* **:** to clothe in fancy or rich garments

[2]attire *noun* (14th century)
1 : DRESS, CLOTHES; *especially* **:** splendid or decorative clothing
2 : the antlers or antlers and scalp of a stag or buck

at·ti·tude \ˈa-tə-ˌtüd, -ˌtyüd\ *noun* [French, from Italian *attitudine,* literally, aptitude, from Late Latin *aptitudin-, aptitudo* fitness — more at APTITUDE] (1668)
1 : the arrangement of the parts of a body or figure **:** POSTURE
2 : a position assumed for a specific purpose ⟨a threatening *attitude*⟩
3 : a ballet position similar to the arabesque in which the raised leg is bent at the knee
4 a : a mental position with regard to a fact or state **b :** a feeling or emotion toward a fact or state
5 : the position of an aircraft or spacecraft determined by the relationship between its axes and a reference datum (as the horizon or a particular star)
6 : an organismic state of readiness to respond in a characteristic way to a stimulus (as an object, concept, or situation)
7 a : a negative or hostile state of mind **b :** a cocky or arrogant manner

at·ti·tu·di·nal \ˌa-tə-ˈtü-dᵊn-əl, -ˈtyü-\ *adjective* [*attitude* + *-inal* (as in *aptitudinal,* from Latin *aptitudin-, aptitudo*)] (1831)
: relating to, based on, or expressive of personal attitudes or feelings ⟨*attitudinal* judgment⟩
— at·ti·tu·di·nal·ly \-ə-lē\ *adverb*

at·ti·tu·di·nise *British variant of* ATTITUDINIZE

at·ti·tu·di·nize \ˌa-tə-ˈtü-dᵊn-ˌīz, -ˈtyü-\ *intransitive verb* **-nized; -niz·ing** (1784)
: to assume an affected mental attitude **:** POSE

at·to- \ˈa-(ˌ)tō\ *combining form* [International Scientific Vocabulary, from Danish or Norwegian *atten* eighteen, from Old Norse *āttjān;* akin to Old English *eahtatīene* eighteen]
: one quintillionth (10^{-18}) part of ⟨*atto*gram⟩

at·torn \ə-ˈtərn\ *intransitive verb* [Middle English *attournen,* from Middle French *atorner,* from Old French, from *a-* (from Latin *ad-*) + *torner* to turn] (15th century)
: to agree to be tenant to a new owner or landlord of the same property
— at·torn·ment \-mənt\ *noun*

at·tor·ney \ə-ˈtər-nē\ *noun, plural* **-neys** [Middle English *attourney,* from Middle French *atorné,* past participle of *atorner*] (14th century)
: one who is legally appointed to transact business on another's behalf; *specifically* **:** a legal agent qualified to act for suitors and defendants in legal proceedings
— at·tor·ney·ship \-ˌship\ *noun*

attorney–at–law *noun, plural* **attorneys–at–law** (1768)
: a practitioner in a court of law who is legally qualified to prosecute and defend actions in such court on the retainer of clients

attorney general *noun, plural* **attorneys general** *or* **attorney generals** (1585)
: the chief law officer of a nation or state who represents the government in litigation and serves as its principal legal adviser

at·tract \ə-ˈtrakt\ *verb* [Middle English, from Latin *attractus,* past participle of *attrahere,* from *ad-* + *trahere* to pull, draw] (15th century)
transitive verb
: to cause to approach or adhere: as **a :** to pull to or draw toward oneself or itself ⟨a magnet *attracts* iron⟩ **b :** to draw by appeal to natural or excited interest, emotion, or aesthetic sense **:** ENTICE ⟨*attract* attention⟩
intransitive verb
: to exercise attraction ☆
— at·trac·tor \-ˈtrak-tər\ *noun*

at·trac·tance \-tən(t)s\ *noun* (1948)
: ATTRACTANCY

at·trac·tan·cy \ə-ˈtrak-tən(t)-sē\ *noun* (1948)
: the quality or capacity (as of a pheromone) of attracting

at·trac·tant \ə-ˈtrak-tənt\ *noun* (1920)
: a substance (as of a pheromone) that attracts insects or other animals

at·trac·tion \ə-ˈtrak-shən\ *noun* (14th century)
1 a : the act, process, or power of attracting **b :** personal charm
2 : the action or power of drawing forth a response **:** an attractive quality
3 : a force acting mutually between particles of matter, tending to draw them together, and resisting their separation
4 : something that attracts or is intended to attract people by appealing to their desires and tastes ⟨coming *attractions*⟩ ☆

at·trac·tive \ə-ˈtrak-tiv\ *adjective* (14th century)
1 : having or relating to the power to attract ⟨*attractive* forces between molecules⟩ ⟨an *attractive* offer⟩
2 : arousing interest or pleasure **:** CHARMING ⟨an *attractive* smile⟩
— at·trac·tive·ly *adverb*
— at·trac·tive·ness *noun*

[1]at·tri·bute \ˈa-trə-ˌbyüt\ *noun* [Middle English, from Latin *attributus,* past participle of *attribuere* to attribute, from *ad-* + *tribuere* to bestow — more at TRIBUTE] (14th century)
1 : an inherent characteristic; *also* **:** an accidental quality
2 : an object closely associated with or belonging to a specific person, thing, or office ⟨a scepter is the *attribute* of power⟩; *especially* **:** such an object used for identification in painting or sculpture
3 : a word ascribing a quality; *especially* **:** ADJECTIVE
synonym see QUALITY

[2]at·trib·ute \ə-ˈtri-ˌbyüt, -byət\ *transitive verb* **-ut·ed; -ut·ing** (1530)

☆ SYNONYMS

Attract, allure, charm, captivate, fascinate, enchant mean to draw another by exerting a powerful influence. ATTRACT applies to any degree or kind of ability to exert influence over another ⟨students *attracted* by the school's locale⟩. ALLURE implies an enticing by what is fair, pleasing, or seductive ⟨an *alluring* smile⟩. CHARM implies the power of casting a spell over the person or thing affected and so compelling a response ⟨*charmed* by their hospitality⟩, but it may, like CAPTIVATE, suggest no more than evoking delight or admiration ⟨her performances *captivated* audiences⟩. FASCINATE suggests a magical influence and tends to stress the ineffectiveness of attempts to resist ⟨a story that continues to *fascinate* children⟩. ENCHANT is perhaps the strongest of these terms in stressing the appeal of the agent and the degree of delight evoked in the subject ⟨hopelessly *enchanted* by her beauty⟩.

Attraction, affinity, sympathy mean the relationship existing between things or persons that are naturally or involuntarily drawn together. ATTRACTION implies the possession by one thing of a quality that pulls another to it ⟨felt an *attraction* to danger⟩. AFFINITY implies a susceptibility or predisposition on the part of the one drawn ⟨an *affinity* for mathematics⟩. SYMPATHY implies a reciprocal or natural relation between two things that are both susceptible to the same influence ⟨two minds in *sympathy*⟩.

◇ WORD HISTORY

attic The ancient Greek city-state of Athens included the whole of the Attic peninsula, the region called Attica. Typical of the Attic or Athenian style of architecture is the use of pilasters, rectangular columns projecting from, but attached to, the wall. These take the place of the freestanding and usually rounded pillars common in other architectural styles. Occasionally the large columns and entablature that form the facade of a building are surmounted by a similar but smaller decorative structure, whose columns are usually pilasters, rather than pillars. Because this small upper order is in the Attic style, the French named it *attique.* The English borrowed the word as *attic,* and extended its meaning, originally in the phrases *attic story* and *attic floor,* to designate the top story of any building just under the roof. So a word that was once associated with an elegant architectural style now names what is usually no more than a lowly storage area.

1 : to explain by indicating a cause ⟨*attributed* his success to his coach⟩
2 a : to regard as a characteristic of a person or thing **b :** to reckon as made or originated in an indicated fashion ⟨*attributed* the invention to a Russian⟩ **c :** CLASSIFY, DESIGNATE
synonym see ASCRIBE
— at·trib·ut·able \-byü-tə-bəl, -byə-\ *adjective*

at·tri·bu·tion \ˌa-trə-ˈbyü-shən\ *noun* (1651)
1 : the act of attributing; *especially* **:** the ascribing of a work (as of literature or art) to a particular author or artist
2 : an ascribed quality, character, or right
— at·tri·bu·tion·al \-sh(ə-)nᵊl\ *adjective*

at·trib·u·tive \ə-ˈtri-byə-tiv\ *adjective* (1606)
1 : relating to or of the nature of an attribute **:** ATTRIBUTING
2 : joined directly to a modified noun without a linking verb (as *city* in *city streets*)
— attributive *noun*
— at·trib·u·tive·ly *adverb*

at·trit·ed \ə-ˈtrī-təd\ *adjective* (1760)
: worn by attrition

at·tri·tion \ə-ˈtri-shən, a-\ *noun* [Latin *attrition-, attritio,* from *atterere* to rub against, from *ad-* + *terere* to rub — more at THROW] (14th century)
1 [Middle English *attricioun,* from Medieval Latin *attrition-, attritio,* from Latin] **:** sorrow for one's sins that arises from a motive other than that of the love of God
2 : the act of rubbing together **:** FRICTION; *also* **:** the act of wearing or grinding down by friction
3 : the act of weakening or exhausting by constant harassment, abuse, or attack
4 : a reduction in numbers usually as a result of resignation, retirement, or death
— at·tri·tion·al \-ˈtri-sh(ə-)nᵊl\ *adjective*

at·tune \ə-ˈtün, -ˈtyün\ *transitive verb* (1596)
1 : to bring into harmony **:** TUNE
2 : to make aware or responsive ⟨*attune* businesses to changing trends⟩
— at·tune·ment \-mənt\ *noun*

atwit·ter \ə-ˈtwi-tər\ *adjective* (1833)
: nervously concerned **:** EXCITED ⟨gossips *atwitter* with speculation —*Time*⟩

atyp·i·cal \(ˌ)ā-ˈti-pi-kəl\ *adjective* (1885)
: not typical **:** IRREGULAR, UNUSUAL
— atyp·i·cal·i·ty \ˌā-ˌti-pə-ˈka-lə-tē\ *noun*
— atyp·i·cal·ly \(ˌ)ā-ˈti-pi-k(ə-)lē\ *adverb*

au·bade \ō-ˈbäd\ *noun* [French, from Middle French, from Old Provençal *aubada,* from *alba, auba* dawn, from (assumed) Vulgar Latin *alba,* from Latin, feminine of *albus* white — more at ALB] (circa 1678)
1 : a song or poem greeting the dawn
2 a : a morning love song **b :** a song or poem of lovers parting at dawn
3 : morning music — compare NOCTURNE

au·berge \ō-ˈberzh\ *noun* [French, from Middle French, of Germanic origin; akin to Old High German *heriberga* military quarters — more at HARBOR] (1599)
: INN 1a

au·ber·gine \ˈō-bər-ˌzhēn\ *noun* [French, from Catalan *albergínia,* from Arabic *al-bādhinjān* the eggplant] (1794)
1 *chiefly British* **:** EGGPLANT 1
2 : EGGPLANT 2

[1]au·burn \ˈȯ-bərn\ *adjective* [Middle English *auborne* blond, from Middle French, from Medieval Latin *alburnus* whitish, from Latin *alburnum* sapwood] (15th century)
1 : of the color auburn
2 : of a reddish brown color

[2]auburn *noun* (1613)
: a moderate brown

Au·bus·son \ˌō-bə-ˈsōⁿ\ *noun* [*Aubusson,* town in France] (1851)
: a figured scenic tapestry used for wall hangings and upholstery; *also* **:** a rug woven to resemble Aubusson tapestry

au cou·rant \ˌō-kü-ˈräⁿ\ *adjective* [French, literally, in the current] (1762)
1 a : fully informed **:** UP-TO-DATE **b :** FASHIONABLE, STYLISH
2 : fully familiar **:** CONVERSANT

[1]auc·tion \ˈȯk-shən\ *noun* [Latin *auction-, auctio,* from *augēre* to increase — more at EKE] (1595)
1 : a sale of property to the highest bidder
2 : the act or process of bidding in some card games

[2]auction *transitive verb* **auc·tioned; auc·tion·ing** \-sh(ə-)niŋ\ (circa 1798)
: to sell at auction ⟨*auctioned* off his library⟩

auction bridge *noun* (1908)
: a bridge game differing from contract bridge in that tricks made in excess of the contract are scored toward game

auc·tion·eer \ˌȯk-shə-ˈnir\ *noun* (circa 1708)
: an agent who sells goods at auction

auc·to·ri·al \ȯk-ˈtȯr-ē-əl, -ˈtȯr-\ *adjective* [Latin *auctor* author — more at AUTHOR] (1821)
: of or relating to an author

au·da·cious \ȯ-ˈdā-shəs\ *adjective* [Middle French *audacieux,* from *audace* boldness, from Latin *audacia,* from *audac-, audax* bold, from *audēre* to dare, from *avidus* eager — more at AVID] (1550)
1 a : intrepidly daring **:** ADVENTUROUS ⟨an *audacious* mountain climber⟩ **b :** recklessly bold **:** RASH
2 : contemptuous of law, religion, or decorum **:** INSOLENT
3 : marked by originality and verve
— au·da·cious·ly *adverb*
— au·da·cious·ness *noun*

au·dac·i·ty \ȯ-ˈda-sə-tē\ *noun, plural* **-ties** [Middle English *audacite,* from Latin *audac-, audax*] (15th century)
1 : the quality or state of being audacious: as **a :** intrepid boldness **b :** bold or arrogant disregard of normal restraints
2 : an audacious act — usually used in plural
synonym see TEMERITY

au·di·al \ˈȯ-dē-əl\ *adjective* [audio + [1]-*al*] (1966)
: of, relating to, or affecting the sense of hearing **:** AURAL

[1]au·di·ble \ˈȯ-də-bəl\ *adjective* [Late Latin *audibilis,* from Latin *audire* to hear; akin to Greek *aisthanesthai* to perceive, Sanskrit *āvis* evidently] (1529)
: heard or capable of being heard
— au·di·bil·i·ty \ˌȯ-də-ˈbi-lə-tē\ *noun*
— au·di·bly \ˈȯ-də-blē\ *adverb*

[2]audible *noun* (1962)
: a substitute offensive or defensive play called at the line of scrimmage in football

au·di·ence \ˈȯ-dē-ən(t)s, ˈä-\ *noun* [Middle English, from Middle French, from Latin *audientia,* from *audient-, audiens,* present participle of *audire*] (14th century)
1 : the act or state of hearing
2 a : a formal hearing or interview ⟨an *audience* with the pope⟩ **b :** an opportunity of being heard ⟨I would succeed if I were once given *audience*⟩
3 a : a group of listeners or spectators **b :** a reading, viewing, or listening public
4 : a group of ardent admirers or devotees **:** FOLLOWING

au·dile \ˈȯ-ˌdīl\ *adjective* [*aud*itory + *-ile* (as in *tactile*)] (1897)
: AUDITORY

aud·ing \ˈȯ-diŋ\ *noun* [Latin *aud*ire + English [1]-*ing*] (circa 1949)
: the process of hearing, recognizing, and interpreting spoken language

[1]au·dio \ˈȯ-dē-ˌō\ *adjective* [*audio-*] (1916)
1 : of or relating to acoustic, mechanical, or electrical frequencies corresponding to normally audible sound waves which are of frequencies approximately from 15 to 20,000 hertz
2 a : of or relating to sound or its reproduction and especially high-fidelity reproduction **b :** relating to or used in the transmission or reception of sound — compare VIDEO **c :** of, relating to, or utilizing recorded sound

[2]audio *noun* (1934)
1 : an audio signal; *broadly* **:** SOUND
2 : the section of television or motion-picture equipment that deals with sound
3 : the transmission, reception, or reproduction of sound

audio- *combining form* [Latin *audire* to hear]
1 : hearing ⟨*audio*meter⟩
2 : sound ⟨*audio*phile⟩
3 : auditory and ⟨*audio*visual⟩

au·dio–an·i·ma·tron·ic \ˌȯ-dē-ō-ˌa-nə-mə-ˈträ-nik\ *adjective* [from *Audio-Animatronics,* a trademark] (1964)
: being or consisting of a lifelike electromechanical figure of a person or animal that has synchronized movement and sound

au·dio·cas·sette \ˌȯ-dē-(ˌ)ō-kə-ˈset, -ka-\ *noun* (1971)
: an audiotape recording mounted in a cassette

au·dio·gen·ic \ˌȯ-dē-ō-ˈje-nik\ *adjective* (1941)
: produced by frequencies corresponding to sound waves — used especially of epileptoid responses

au·dio·gram \ˈȯ-dē-ō-ˌgram\ *noun* (1927)
: a graphic representation of the relation of vibration frequency and the minimum sound intensity for hearing

au·dio–lin·gual \ˌȯ-dē-ō-ˈliŋ-gwəl *also* -gyə-wəl\ *adjective* (1960)
: involving a drill routine of listening and speaking in language learning

au·di·ol·o·gy \ˌȯ-dē-ˈä-lə-jē\ *noun* (1946)
: a branch of science dealing with hearing; *specifically* **:** therapy of individuals having impaired hearing
— au·di·o·log·i·cal \-ē-ə-ˈlä-ji-kəl\ *also* **au·di·o·log·ic** \-ē-ə-ˈlä-jik\ *adjective*
— au·di·ol·o·gist \-ē-ˈä-lə-jist\ *noun*

au·di·om·e·ter \ˌȯ-dē-ˈä-mə-tər\ *noun* (1879)
: an instrument used in measuring the acuity of hearing
— au·dio·met·ric \-dē-ō-ˈme-trik\ *adjective*
— au·di·om·e·try \-dē-ˈä-mə-trē\ *noun*

au·dio·phile \ˈȯ-dē-ō-ˌfīl\ *noun* (1951)
: a person who is enthusiastic about high-fidelity sound reproduction

au·dio·tape \ˈȯ-dē-ō-ˌtāp\ *noun* (1958)
: a tape recording of sound

au·dio·vi·su·al \ˌȯ-dē-(ˌ)ō-ˈvi-zhə-wəl, -zhəl\ *adjective* (1937)
1 : designed to aid in learning or teaching by making use of both hearing and sight
2 : of or relating to both hearing and sight

au·dio·vi·su·als \-zhə-wəlz, -zhəlz\ *noun plural* (1955)
: audiovisual teaching materials (as filmstrips accompanied by recordings)

[1]au·dit \ˈȯ-dət\ *noun* [Middle English, from Latin *auditus* act of hearing, from *audire*] (15th century)
1 a : a formal examination of an organization's or individual's accounts or financial situation **b :** the final report of an audit
2 : a methodical examination and review
— au·dit·able \-di-tə-bəl\ *adjective*

[2]audit *transitive verb* (15th century)
1 : to perform an audit of or for ⟨*audit* the books⟩ ⟨*audit* the company⟩
2 : to attend (a course) without working for or expecting to receive formal credit

[1]au·di·tion \ȯ-ˈdi-shən\ *noun* [Middle French or Latin; Middle French, from Latin *audition-, auditio,* from *audire*] (1599)
1 : the power or sense of hearing
2 : the act of hearing; *especially* **:** a critical hearing ⟨an *audition* of new recordings⟩
3 : a trial performance to appraise an entertainer's merits

[2]audition *verb* **au·di·tioned; au·di·tion·ing** \-ˈdi-sh(ə-)niŋ\ (1934)
transitive verb
: to test especially in an audition

intransitive verb
: to give a trial performance

au·di·tor \ˈȯ-də-tər\ *noun* (14th century)
1 : a person authorized to examine and verify accounts
2 : one that hears or listens; *especially* **:** one that is a member of an audience
3 : a person who audits a course of study
4 : a person who hears (as a court case) in the capacity of judge

au·di·to·ri·um \ˌȯ-də-ˈtōr-ē-əm, -ˈtȯr-\ *noun, plural* **-riums** *or* **-ria** \-ē-ə\ [Latin, literally, lecture room] (circa 1751)
1 : the part of a public building where an audience sits
2 : a room, hall, or building used for public gatherings

¹au·di·to·ry \ˈȯ-də-ˌtōr-ē, -ˌtȯr-\ *noun* [Middle English *auditorie,* from Latin *auditorium*] (14th century)
1 *archaic* **:** AUDIENCE
2 *archaic* **:** AUDITORIUM

²au·di·to·ry *adjective* [Late Latin *auditorius*] (1578)
: of, relating to, or experienced through hearing
— au·di·to·ri·ly *adverb*

auditory nerve *noun* (1724)
: either of the 8th pair of cranial nerves connecting the inner ear with the brain and transmitting impulses concerned with hearing and balance — see EAR illustration

auditory tube *noun* (1907)
: EUSTACHIAN TUBE

audit trail *noun* (1954)
: a record of a sequence of events (as actions performed by a computer) from which a history may be reconstructed

Auf·klä·rung \ˈau̇f-ˌkler-əŋ, -u̇ŋ\ *noun* [German] (1842)
: ENLIGHTENMENT 2

auf Wie·der·seh·en \au̇f-ˈvē-dər-ˌzā(-ə)n\ *interjection* [German, literally, till seeing again] (1885)
— used to express farewell

Au·ge·an \ȯ-ˈjē-ən\ *adjective* [Latin *Augeas,* king of Elis, from Greek *Augeias;* from the legend that his stable, left neglected for 30 years, was finally cleaned by Hercules] (1599)
: extremely formidable or difficult and occasionally distasteful ⟨an *Augean* task⟩

Augean stable *noun* (1635)
: a condition or place marked by great accumulation of filth or corruption

au·ger \ˈȯ-gər\ *noun* [Middle English, alteration (resulting from false division of *a nauger*) of *nauger,* from Old English *nafogār;* akin to Old High German *nabugēr* auger, Old English *nafu* nave, *gār* spear — more at NAVE, GORE] (before 12th century)
: any of various tools or devices having a helical shaft or member that are used for boring holes (as in wood, earth, or ice) or moving loose material (as snow) ◆

1 2 3

auger: *1, 2* **screw,** *3* **tapering pod**

Au·ger effect \(ˌ)ō-ˈzhā-\ *noun* [Pierre V. *Auger* (born 1899) French physicist] (1931)
: a process in which an atom that has been ionized through the emission of an electron with energy in the X-ray range undergoes a transition in which a second electron is emitted rather than an X-ray photon — called also *Auger process*

Auger electron *noun* (1939)
: an electron emitted from an atom in the Auger effect

Auger electron spectroscopy *noun* (1970)
: an instrumental method for determining the chemical composition of a material's surface by bombardment with an electron beam to produce Auger electrons whose energy spectra are characteristic of their parent atoms — called also *Auger spectroscopy*

¹aught \ˈȯt, ˈät\ *pronoun* [Middle English, from Old English *āwiht,* from *ā* ever + *wiht* creature, thing — more at AYE, WIGHT] (before 12th century)
1 : ANYTHING
2 : ALL, EVERYTHING ⟨for *aught* I care⟩ ⟨for *aught* we know⟩

²aught *adverb* (13th century)
archaic **:** AT ALL

³aught *noun* [alteration (resulting from false division of *a naught*) of *naught*] (1872)
1 : ZERO, CIPHER
2 *archaic* **:** NONENTITY, NOTHING

au·gite \ˈȯ-ˌjīt\ *noun* [Latin *augites,* a precious stone, from Greek *augitēs*] (1804)
1 : an aluminous usually black or dark green pyroxene that is found in igneous rocks
2 : PYROXENE
— au·git·ic \ȯ-ˈji-tik\ *adjective*

¹aug·ment \ȯg-ˈment\ *verb* [Middle English, from Middle French *augmenter,* from Late Latin *augmentare,* from Latin *augmentum* increase, from *augēre* to increase — more at EKE] (14th century)
transitive verb
1 : to make greater, more numerous, larger, or more intense ⟨the impact of the report was *augmented* by its timing⟩
2 : to add an augment to
3 : SUPPLEMENT ⟨*augmented* her scholarship by working nights⟩
intransitive verb
: to become augmented
synonym see INCREASE
— aug·ment·er *or* **aug·men·tor** \-ˈmen-tər\ *noun*

²aug·ment \ˈȯg-ˌment\ *noun* (circa 1771)
: a vowel prefixed or a lengthening of the initial vowel to mark past time especially in Greek and Sanskrit verbs

aug·men·ta·tion \ˌȯg-mən-ˈtā-shən, -ˌmen-\ *noun* (14th century)
1 a : the act or process of augmenting **b :** the state of being augmented
2 : something that augments **:** ADDITION

¹aug·men·ta·tive \ȯg-ˈmen-tə-tiv\ *adjective* (15th century)
1 : able to augment
2 : indicating large size and sometimes awkwardness or unattractiveness — used of words and affixes; compare DIMINUTIVE

²augmentative *noun* (1804)
: an augmentative word or affix

aug·ment·ed \ȯg-ˈmen-təd\ *adjective* (1825)
of a musical interval **:** made one half step greater than major or perfect ⟨an *augmented* fifth⟩

augmented matrix *noun* (circa 1949)
: a matrix whose elements are the coefficients of a set of simultaneous linear equations with the constant terms of the equations entered in an added column

au gra·tin \ō-ˈgrä-tᵊn, ȯ-, -ˈgra-\ *adjective* [French, literally, with the burnt scrapings from the pan] (1806)
: covered with bread crumbs or grated cheese and browned (as under a broiler)

¹au·gur \ˈȯ-gər\ *noun* [Latin; akin to Latin *augēre*] (14th century)
1 : an official diviner of ancient Rome
2 : one held to foretell events by omens

²augur (1601)
transitive verb
1 : to foretell especially from omens
2 : to give promise of **:** PRESAGE ⟨higher pay *augurs* a better future⟩
intransitive verb
: to predict the future especially from omens

au·gu·ry \ˈȯ-gyə-rē, -gə-\ *noun, plural* **-ries** (14th century)
1 : divination from auspices or omens; *also* **:** an instance of this
2 : OMEN, PORTENT

au·gust \ȯ-ˈgəst, ˈȯ-(ˌ)gəst\ *adjective* [Latin *augustus;* akin to Latin *augur*] (1664)
: marked by majestic dignity or grandeur
— au·gust·ly *adverb*
— au·gust·ness \ȯ-ˈgəs(t)-nəs, ˈȯ-(ˌ)gəs(t)-\ *noun*

Au·gust \ˈȯ-gəst\ *noun* [Middle English, from Old English, from Latin *Augustus,* from *Augustus* Caesar] (before 12th century)
: the 8th month of the Gregorian calendar

Au·gus·tan \ȯ-ˈgəs-tən, ə-\ *adjective* (1704)
1 : of, relating to, or characteristic of Augustus Caesar or his age
2 : of, relating to, or characteristic of the neoclassical period in England
— Augustan *noun*

¹Au·gus·tin·i·an \ˌȯ-gə-ˈsti-nē-ən\ *noun* (1602)
1 : a member of an Augustinian order; *specifically* **:** a friar of the Hermits of Saint Augustine founded in 1256 and devoted to educational, missionary, and parish work
2 : a follower of Saint Augustine

²Augustinian *adjective* (1674)
1 : of or relating to Saint Augustine or his doctrines
2 : of or relating to any of several orders under a rule ascribed to Saint Augustine
— Au·gus·tin·i·an·ism \-nē-ə-ˌni-zəm\ *noun*

au jus \ō-ˈzhü(s), -ˈjüs; ō-zhᵫ\ *adjective* [French, literally, with juice] (circa 1919)
of meat **:** served in the juice obtained from roasting

auk \ˈȯk\ *noun* [Norwegian or Icelandic *alk, alka,* from Old Norse *ālka*] (1674)
: any of several black-and-white short-necked diving seabirds (family Alcidae) that breed in colder parts of the northern hemisphere

auk·let \ˈȯ-klət\ *noun* (1886)
: any of several small auks of the North Pacific coasts

auk

auld \ˈōl(d), ˈäl(d)\ *adjective* (14th century)
chiefly Scottish **:** OLD

auld lang syne \ˌōl(d)-ˌ(l)aŋ-ˈzīn, ˌōl(d)-\ *noun* [Scots, literally, old long ago] (1692)
: the good old times

au na·tu·rel \ˌō-ˌna-tə-ˈrel, -ˌna-chə-\ *adjective* [French] (1817)
1 : cooked or served plainly
2 a : being in natural style or condition **b :** NUDE

aunt \ˈant, ˈänt\ *noun* [Middle English, from Old French *ante,* from Latin *amita;* akin to Old High German *amma* mother, nurse, Greek *amma* nurse] (14th century)

◇ WORD HISTORY

auger *Auger,* like *apron,* arose because it is hard to be sure of word boundaries in connected speech. *Auger* began its life in Old English as *nafogār,* a compound of *nafu* "nave of a wheel" and *gār* "spear," a "nave-spear" being the tool used to bore a hole through the hub of a wheel into which the axle could be inserted. By Middle English, *nafogār* had shrunk to *nauger,* obscuring the original sense of the compound. Already in Middle English, the combination of indefinite article and noun was spelled as both *an auger* and *a nauger,* an indication that younger generations of speakers who heard the word in connected speech were unsure whether it began with a consonant or a vowel. The form *auger* emerged as dominant in the 17th century.

1 : the sister of one's father or mother
2 : the wife of one's uncle
— **aunt·hood** \-ˌhu̇d\ *noun*
— **aunt·like** \-ˌlīk\ *adjective*
— **aunt·ly** *adjective*

Aunt Sal·ly \-ˈsa-lē\ *noun, plural* **Aunt Sal·lies** [*Aunt Sally,* name given to an effigy of a woman smoking a pipe set up as an amusement attraction at English fairs for patrons to throw missiles at] (1898)
British **:** an object of criticism or contention; *especially* **:** one that is set up to invite criticism or be easily refuted

au pair \ˈō-ˈpar, -ˈper\ *noun, plural* **au pairs** \-ˈparz, -ˈperz\ [French, on even terms] (1960)
: a usually young foreign person who does domestic work for a family in return for room and board and the opportunity to learn the family's language

au poivre \ō-ˈpwäv(rᵊ)\ *adjective* [French, with pepper] (1971)
: prepared or served with a generous amount of usually coarsely ground black pepper ⟨steak *au poivre*⟩

au·ra \ˈȯr-ə\ *noun* [Middle English, from Latin, puff of air, breeze, from Greek; probably akin to Greek *aēr* air] (1732)
1 a : a subtle sensory stimulus (as an aroma) **b :** a distinctive atmosphere surrounding a given source ⟨the place had an *aura* of mystery⟩
2 : a luminous radiation **:** NIMBUS
3 : a subjective sensation (as of lights) experienced before an attack of some disorders (as epilepsy or a migraine)
4 : an energy field that is held to emanate from a living being

au·ral \ˈȯr-əl\ *adjective* [Latin *auris* ear — more at EAR] (1847)
: of or relating to the ear or to the sense of hearing
— **au·ral·ly** \-ə-lē\ *adverb*

aurar *plural of* EYRIR

au·re·ate \ˈȯr-ē-ət\ *adjective* [Middle English *aureat,* from Medieval Latin *aureatus* decorated with gold, from Latin *aureus*] (15th century)
1 : of a golden color or brilliance
2 : marked by grandiloquent and rhetorical style

au·re·ole \ˈȯr-ē-ˌōl\ *or* **au·re·o·la** \ȯ-ˈrē-ə-lə, ə-\ *noun* [Middle English *aureole* heavenly crown worn by saints, from Medieval Latin *aureola,* from Latin, feminine of *aureolus* golden, diminutive of *aureus*] (13th century)
1 a : a radiant light around the head or body of a representation of a sacred personage **b :** something resembling an aureole ⟨an *aureole* of hair⟩
2 : RADIANCE, AURA ⟨an *aureole* of youth and health⟩
3 : the luminous area surrounding the sun or other bright light when seen through thin cloud or mist **:** CORONA
4 : a ring-shaped zone around an igneous intrusion
— **aureole** *transitive verb*

Au·reo·my·cin \ˌȯr-ē-ō-ˈmī-sᵊn\ *trademark*
— used for a preparation of the hydrochloride of chlortetracycline

au·re·us \ˈȯr-ē-əs\ *noun, plural* **-rei** \-ē-ˌī\ [Latin, literally, golden, from *aurum* gold; akin to Old Prussian *ausis* gold] (1609)
: a gold coin of ancient Rome varying in weight from 1/30 to 1/70 libra

au re·voir \ˌōr-ə-ˈvwär, ˌȯr-, *French* ōr(-ə)-vwär\ *noun* [French, literally, till seeing again] (1694)
: GOOD-BYE — often used interjectionally

au·ric \ˈȯr-ik\ *adjective* [Latin *aurum*] (circa 1828)
: of, relating to, or derived from gold

au·ri·cle \ˈȯr-i-kəl\ *noun* [Latin *auricula,* from diminutive of *auris* ear — more at EAR] (15th century)
1 a : an atrium of a heart **b :** PINNA 2b **c :** an anterior ear-shaped pouch in each atrium of the human heart
2 : an angular or ear-shaped lobe, process, or appendage

au·ric·u·la \ȯ-ˈri-kyə-lə\ *noun* [New Latin, from Latin, external ear] (1655)
: a yellow-flowered Alpine primrose (*Primula auricula*)

au·ric·u·lar \ȯ-ˈri-kyə-lər\ *adjective* (15th century)
1 : told privately ⟨an *auricular* confession⟩
2 : understood or recognized by the sense of hearing
3 : of, relating to, or using the ear or the sense of hearing
4 : of or relating to an auricle

au·ric·u·late \ȯ-ˈri-kyə-lət\ *adjective* (1713)
: having auricles

au·rif·er·ous \ȯ-ˈri-f(ə-)rəs\ *adjective* [Latin *aurifer,* from *aurum* + *-fer* -ferous] (1727)
: containing gold

Au·ri·ga \ȯ-ˈrī-gə\ *noun* [Latin, literally, charioteer]
: a constellation between Perseus and Gemini

Au·ri·gna·cian \ˌȯr-ēn-ˈyā-shən\ *adjective* [French *aurignacien,* from *Aurignac,* France] (1909)
: of or relating to an Upper Paleolithic culture marked by finely made artifacts of stone and bone, paintings, and engravings

au·rochs \ˈau̇r-ˌäks, ˈȯr-\ *noun, plural* **au·rochs** [German, from Old High German *ūrohso,* from *ūro* aurochs + *ohso* ox; akin to Old English *ūr* aurochs — more at OX] (1766)
1 : an extinct large long-horned wild ox (*Bos primigenius*) of Europe that is the ancestor of domestic cattle
2 : WISENT

au·ro·ra \ə-ˈrōr-ə, ȯ-, -ˈrȯr-\ *noun, plural* **au·roras** *or* **au·ro·rae** \-(ˌ)ē\ [Latin — more at EAST] (14th century)
1 : DAWN
2 *capitalized* **:** the Roman goddess of dawn — compare EOS
3 : a luminous phenomenon that consists of streamers or arches of light appearing in the upper atmosphere of a planet's magnetic polar regions and is caused by the emission of light from atoms excited by electrons accelerated along the planet's magnetic field lines
— **au·ro·ral** \-əl\ *adjective*
— **au·ro·re·an** \-ē-ən\ *adjective*

aurora aus·tra·lis \-ȯ-ˈstrā-ləs, -ä-ˈstrā-\ *noun* [New Latin, literally, southern dawn] (1741)
: an aurora that occurs in earth's southern hemisphere — called also *southern lights*

aurora bo·re·al·is \-ˌbōr-ē-ˈa-ləs, -ˌbȯr-\ *noun* [New Latin, literally, northern dawn] (1717)
: an aurora that occurs in earth's northern hemisphere — called also *northern lights*

aus·cul·tate \ˈȯ-skəl-ˌtāt\ *transitive verb* **-tat·ed; -tat·ing** [back-formation from *auscultation*] (circa 1860)
: to examine by auscultation
— **aus·cul·ta·to·ry** \ȯ-ˈskəl-tə-ˌtōr-ē, -ˌtȯr-\ *adjective*

aus·cul·ta·tion \ˌȯ-skəl-ˈtā-shən\ *noun* [Latin *auscultation-, auscultatio* act of listening, from *auscultare* to listen; akin to Latin *auris* ear — more at EAR] (circa 1828)
: the act of listening to sounds arising within organs (as the lungs) as an aid to diagnosis and treatment

aus·land·er \ˈau̇s-ˌlen-dər, -ˌlan-\ *noun* [German *Ausländer*] (1936)
: OUTSIDER, FOREIGNER

aus·pice \ˈȯs-pəs\ *noun, plural* **aus·pic·es** \-pə-səz, -ˌsēz\ [Latin *auspicium,* from *auspic-, auspex* diviner by birds, from *avis* bird + *specere* to look, look at — more at AVIARY, SPY] (1533)
1 : observation by an augur especially of the flight and feeding of birds to discover omens
2 *plural* **:** kindly patronage and guidance
3 : a prophetic sign; *especially* **:** a favorable sign

aus·pi·cious \ȯ-ˈspi-shəs\ *adjective* (1593)
1 : affording a favorable auspice **:** PROPITIOUS ⟨made an *auspicious* beginning⟩
2 : attended by good auspices **:** PROSPEROUS ⟨an *auspicious* year⟩
synonym see FAVORABLE
— **aus·pi·cious·ly** *adverb*
— **aus·pi·cious·ness** *noun*

Aus·sie \ˈȯ-sē, ˈä-sē, *British & Australian usually* ˈȯ-zē\ *noun* [*Aus*tralian + *-ie*] (1917)
: a native or inhabitant of Australia

aus·ten·ite \ˈȯs-tə-ˌnīt, ˈäs-\ *noun* [French, from Sir W. C. Roberts-*Austen* (died 1902) English metallurgist] (1902)
: a solid solution in iron of carbon and sometimes other solutes that occurs as a constituent of steel under certain conditions
— **aus·ten·it·ic** \ˌȯs-tə-ˈni-tik, ˌäs-\ *adjective*

aus·tere \ȯ-ˈstir *also* -ˈster\ *adjective* [Middle English, from Middle French, from Latin *austerus,* from Greek *austēros* harsh, severe; akin to Greek *hauos* dry — more at SERE] (14th century)
1 a : stern and cold in appearance or manner **b :** SOMBER, GRAVE ⟨an *austere* critic⟩
2 : morally strict **:** ASCETIC
3 : markedly simple or unadorned ⟨an *austere* office⟩ ⟨an *austere* style of writing⟩
4 : giving little or no scope for pleasure ⟨*austere* diets⟩
5 *of a wine* **:** having the flavor of acid or tannin predominant over fruit flavors usually indicating a capacity for aging
synonym see SEVERE
— **aus·tere·ly** *adverb*
— **aus·tere·ness** *noun*

aus·ter·i·ty \ȯ-ˈster-ə-tē *also* -ˈstir-\ *noun, plural* **-ties** (14th century)
1 : the quality or state of being austere
2 a : an austere act, manner, or attitude **b :** an ascetic practice
3 : enforced or extreme economy

¹Austr- *or* **Austro-** *combining form* [Latin, from *Austr-, Auster* south wind]
: south **:** southern ⟨*Austro*asiatic⟩

²Austr- *or* **Austro-** *combining form* [probably from New Latin, from *Austria*]
: Austrian and ⟨*Austro*-Hungarian⟩

¹aus·tral \ˈȯs-trəl, ˈäs-\ *adjective* [Middle English, from Latin *australis,* from *Austr-, Auster*] (14th century)
1 : of or relating to the Southern hemisphere
2 *capitalized* **:** AUSTRALIAN

²aus·tral \au̇-ˈsträl\ *noun, plural* **aus·tral·es** \-ˈsträ-lās\ *also* **australs** [Spanish] (1985)
: the basic monetary unit of Argentina 1985–91

Aus·tra·lia Day \ȯ-ˈstrāl-yə-, ä-, ə-\ *noun* (1911)
: a national holiday in Australia commemorating the landing of the British at Sydney Cove in 1788 and observed on the Monday of or next following Jan. 26

¹Aus·tra·lian \ȯ-ˈstrāl-yən, ä-, ə-\ *adjective* (1814)
1 : of, relating to, or characteristic of the continent or commonwealth of Australia, its inhabitants, or the languages spoken there
2 : of, relating to, or being a biogeographic region that comprises Australia, the islands north of it from Celebes eastward, Tasmania, New Zealand, and Polynesia

²Australian *noun* (1814)
1 : a native or inhabitant of the Australian commonwealth
2 : a group of about 200 languages spoken by the aboriginal inhabitants of Australia

Australian ballot *noun* (1888)
: an official ballot printed at public expense on which the names of all the candidates and proposals appear and which is distributed only at the polling place and marked in secret

Australian cattle dog *noun* (1926)
: any of a breed of medium-sized dogs that were developed in Australia to herd cattle and that have upright ears and a red or blue mottled coat

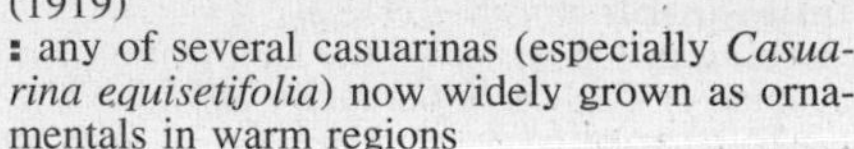
Australian cattle dog

Aus·tra·lian·ism \ˌȯ-ˈstrāl-yə-ˌni-zəm, ˌä-, ə-\ *noun* (1883)
: a characteristic feature of Australian English

Australian pine *noun* (1919)
: any of several casuarinas (especially *Casuarina equisetifolia*) now widely grown as ornamentals in warm regions

Australian Rules football *noun* (1904)
: a game resembling rugby that is played between two teams of 18 players on a field 180–190 yards long that has four goalposts at each end

Australian shepherd *noun* (1964)
: any of a breed of bobtailed medium-sized dogs developed in America and often used in herding

Australian terrier *noun* (1903)
: any of a breed of small rather short-legged wirehaired terriers of Australian origin usually having a tan and blue or sandy coat

Australian terrier

Aus·tra·loid \ˈȯs-trə-ˌlȯid, ˈäs-\ *adjective* [*Australia* + English *-oid*] (1901)
in some classifications **:** of or relating to a racial group including the Australian aborigines and other peoples of southern Asia and Pacific islands
— **Australoid** *noun*

aus·tra·lo·pith·e·cine \ȯ-ˌstrā-lō-ˈpi thə-ˌsīn, ä-; ˌȯs-trə-, ˌäs-\ *noun* [ultimately from Latin *australis* + Greek *pithēkos* ape] (1943)
: any of a genus (*Australopithecus*) of extinct southern and eastern African hominids that include gracile and robust forms with near-human dentition and a relatively small brain
— **australopithecine** *adjective*

Austrian pine *noun* (1858)
: a tall European pine (*Pinus nigra*) widely cultivated for ornament and having needles in clusters of two

Aus·tro·asi·at·ic \ˈȯs-(ˌ)trō-ˌā-zhē-ˈa-tik, -zē-ˈa-, ˈäs- *also* -ˌā-shē-\ *adjective* (1922)
: of, relating to, or constituting a family of languages of south and southeast Asia that includes Mon-Khmer and Munda as subfamilies

Aus·tro·ne·sian \ˌȯs-trə-ˈnē-zhən, ˌäs-, -shən\ *adjective* [*Austronesia,* islands of the southern Pacific] (1925)
: of, relating to, or constituting a family of languages spoken in the area extending from Madagascar eastward through the Malay Peninsula and Archipelago to Hawaii and Easter Island and including practically all the native languages of the Pacific islands with the exception of the Australian and Papuan languages

aut- *or* **auto-** *combining form* [Greek, from *autos* same, -self, self]
1 : self **:** same one ⟨*aut*ism⟩ ⟨*auto*biography⟩
2 : automatic **:** self-acting ⟨*auto*pilot⟩

au·ta·coid \ˈȯ-tə-ˌkȯid\ *noun* [*aut-* + Greek *akos* remedy; probably akin to Old Irish *ícc* cure] (1914)
: a physiologically active substance (as serotonin, bradykinin, or angiotensin) produced by and acting within the body

au·tar·chic \ȯ-ˈtär-kik\ *adjective* (1883)
: AUTARKIC
— **au·tar·chi·cal** \-ki-kəl\ *adjective*

[1]**au·tar·chy** \ˈȯ-ˌtär-kē\ *noun* [by alteration] (1617)
: AUTARKY

[2]**autarchy** *noun, plural* **-chies** [Greek *autarchia,* from *aut-* + *-archia* -archy] (1665)
: absolute sovereignty **:** AUTOCRACY

au·tar·kic \ȯ-ˈtär-kik\ *adjective* (1936)
: of, relating to, or marked by autarky
— **au·tar·ki·cal** \-ki-kəl\ *adjective*

au·tar·ky \ˈȯ-ˌtär-kē\ *noun* [German *Autarkie,* from Greek *autarkeia,* from *autarkēs* self-sufficient, from *aut-* + *arkein* to defend, suffice — more at ARK] (1657)
1 : SELF-SUFFICIENCY, INDEPENDENCE; *specifically* **:** national economic self-sufficiency and independence
2 : a policy of establishing a self-sufficient and independent national economy

aut·ecol·o·gy \ˌȯ-ti-ˈkä-lə-jē, ˌȯt-ē-\ *noun* [International Scientific Vocabulary] (1910)
: ecology dealing with individual organisms or individual species of organisms
— **aut·eco·log·i·cal** \ˌȯt-ˌē-kə-ˈlä-ji-kəl, -ˌe-kə-\ *adjective*

au·teur \ō-ˈtər\ *noun* [French, originator, author, from Old French *autor,* from Latin *auctor* — more at AUTHOR] (1967)
: a film director whose practice accords with the auteur theory
— **au·teur·ist** \-ist\ *adjective or noun*

auteur theory \ō-ˈtər-\ *noun* (1962)
: a view of filmmaking in which the director is considered the primary creative force in a motion picture

au·then·tic \ə-ˈthen-tik, ȯ-\ *adjective* [Middle English *autentik,* from Middle French *autentique,* from Late Latin *authenticus,* from Greek *authentikos,* from *authentēs* perpetrator, master, from *aut-* + *-hentēs* (akin to Greek *anyein* to accomplish, Sanskrit *sanoti* he gains)] (14th century)
1 *obsolete* **:** AUTHORITATIVE
2 a : worthy of acceptance or belief as conforming to or based on fact ⟨paints an *authentic* picture of our society⟩ **b :** conforming to an original so as to reproduce essential features ⟨an *authentic* reproduction of a colonial farmhouse⟩ **c :** made or done the same way as an original ⟨*authentic* Mexican fare⟩
3 : not false or imitation **:** REAL, ACTUAL ⟨based on *authentic* documents⟩ ⟨an *authentic* cockney accent⟩
4 a *of a church mode* **:** ranging upward from the keynote — compare PLAGAL 1 **b** *of a cadence* **:** progressing from the dominant chord to the tonic — compare PLAGAL 2
5 : true to one's own personality, spirit, or character ☆
— **au·then·ti·cal·ly** \-ti-k(ə-)lē\ *adverb*
— **au·then·tic·i·ty** \ˌȯ-ˌthen-ˈti-sə-tē, -thən-\ *noun*

au·then·ti·cate \ə-ˈthen-ti-ˌkāt, ȯ-\ *transitive verb* **-cat·ed; -cat·ing** (1653)
: to prove or serve to prove the authenticity of
synonym see CONFIRM
— **au·then·ti·ca·tion** \-ˌthen-ti-ˈkā-shən\ *noun*
— **au·then·ti·ca·tor** \-ˈthen-ti-ˌkā-tər\ *noun*

[1]**au·thor** \ˈȯ-thər\ *noun* [Middle English *auctour,* from Old North French, from Latin *auctor* promoter, originator, author, from *augēre* to increase — more at EKE] (14th century)
1 a : one that originates or creates **:** SOURCE ⟨software *authors*⟩ ⟨the *author* of this crime⟩ **b** *capitalized* **:** GOD 1
2 : the writer of a literary work (as a book)
— **au·tho·ri·al** \ȯ-ˈthōr-ē-əl, -ˈthȯr-\ *adjective*

[2]**author** *transitive verb* (1596)
: to be the author of

au·thor·ess \ˈȯ-th(ə-)rəs\ *noun* (15th century)
: a woman author

au·tho·rise *British variant of* AUTHORIZE

au·thor·i·tar·i·an \ȯ-ˌthär-ə-ˈter-ē-ən, ə-, -ˌthȯr-\ *adjective* (1879)
1 : of, relating to, or favoring blind submission to authority ⟨had *authoritarian* parents⟩
2 : of, relating to, or favoring a concentration of power in a leader or an elite not constitutionally responsible to the people
— **authoritarian** *noun*
— **au·thor·i·tar·i·an·ism** \-ē-ə-ˌni-zəm\ *noun*

au·thor·i·ta·tive \ə-ˈthär-ə-ˌtā-tiv, ȯ-, -ˈthȯr-\ *adjective* (1605)
1 a : having or proceeding from authority **:** OFFICIAL ⟨*authoritative* church doctrine⟩ **b :** showing evident authority **:** DEFINITIVE ⟨a most *authoritative* literary critique⟩
2 : DICTATORIAL 2
— **au·thor·i·ta·tive·ly** *adverb*
— **au·thor·i·ta·tive·ness** *noun*

au·thor·i·ty \ə-ˈthär-ə-tē, ȯ-, -ˈthȯr-\ *noun, plural* **-ties** [Middle English *auctorite,* from Old French *auctorité,* from Latin *auctoritat-, auctoritas* opinion, decision, power, from *auctor*] (13th century)
1 a (1) **:** a citation (as from a book or file) used in defense or support (2) **:** the source from which the citation is drawn **b** (1) **:** a conclusive statement or set of statements (as an official decision of a court) (2) **:** a decision taken as a precedent (3) **:** TESTIMONY **c :** an individual cited or appealed to as an expert
2 a : power to influence or command thought, opinion, or behavior **b :** freedom granted by one in authority **:** RIGHT
3 a : persons in command; *specifically* **:** GOVERNMENT **b :** a governmental agency or corporation to administer a revenue-producing public enterprise ⟨the transit *authority*⟩
4 a : GROUNDS, WARRANT ⟨had excellent *authority* for believing the claim⟩ **b :** convincing force ⟨lent *authority* to the performance⟩
synonym see INFLUENCE, POWER

au·tho·ri·za·tion \ˌȯ-th(ə-)rə-ˈzā-shən\ *noun* (15th century)
1 : the act of authorizing
2 : an instrument that authorizes **:** SANCTION

au·tho·rize \ˈȯ-thə-ˌrīz\ *transitive verb* **-rized; -riz·ing** (14th century)
1 : to establish by or as if by authority **:** SANCTION ⟨a custom *authorized* by time⟩
2 : to invest especially with legal authority **:** EMPOWER ⟨*authorized* to act for her husband⟩
3 *archaic* **:** JUSTIFY 1a
— **au·tho·riz·er** *noun*

Authorized Version *noun* (1824)
: a revision of the English Bishops' Bible carried out under James I, published in 1611, and widely used by Protestants — called also *King James Version*

au·thor·ship \ˈȯ-thər-ˌship\ *noun* (1710)
1 : the profession of writing

☆ SYNONYMS
Authentic, genuine, bona fide mean being actually and exactly what is claimed. AUTHENTIC implies being fully trustworthy as according with fact ⟨an *authentic* account of the perilous journey⟩; it can also stress painstaking or faithful imitation of an original ⟨an *authentic* reproduction⟩ ⟨*authentic* Vietnamese cuisine⟩. GENUINE implies actual character not counterfeited, imitated, or adulterated ⟨*genuine* piety⟩ ⟨*genuine* maple syrup⟩; it also connotes definite origin from a source ⟨a *genuine* Mark Twain autograph⟩. BONA FIDE implies good faith and sincerity of intention ⟨a *bona fide* offer for the stock⟩.

2 **:** the source (as the author) of a piece of writing, music, or art
3 **:** the state or act of writing, creating, or causing
au·tism \'ȯ-ˌti-zəm\ *noun* (1912)
1 **:** absorption in self-centered subjective mental activity (as daydreams, fantasies, delusions, and hallucinations) usually accompanied by marked withdrawal from reality
2 **:** a mental disorder originating in infancy that is characterized by self-absorption, inability to interact socially, repetitive behavior, and language dysfunction (as echolalia)
— **au·tis·tic** \ȯ-'tis-tik\ *adjective or noun*
— **au·tis·ti·cal·ly** \-ti-k(ə-)lē\ *adverb*
au·to \'ȯ-(ˌ)tō, 'ä-\ *noun, plural* **autos** (1899)
: AUTOMOBILE
auto- — see AUT-
au·to·an·ti·body \ˌȯ-(ˌ)tō-'an-ti-ˌbä-dē\ *noun* (circa 1910)
: an antibody active against a tissue constituent of the individual producing it
au·to·bahn \'ȯ-tō-ˌbän, 'au̇-\ *noun* [German, from *Auto* + *Bahn* road] (1937)
: a German, Swiss, or Austrian expressway
au·to·bi·og·ra·phy \ˌȯ-tə-bī-'ä-grə-fē, -bē-\ *noun* (1771)
: the biography of a person narrated by himself
— **au·to·bi·og·ra·pher** \-fər\ *noun*
— **au·to·bio·graph·i·cal** \-ˌbī-ə-'gra-fi-kəl\ *also* **au·to·bio·graph·ic** \-fik\ *adjective*
— **au·to·bio·graph·i·cal·ly** \-fi-k(ə-)lē\ *adverb*
au·to·bus \'ȯ-tō-ˌbəs\ *noun* [*auto* + *bus*] (1899)
: OMNIBUS 1
au·to·ca·tal·y·sis \ˌȯ-tō-kə-'ta-lə-səs\ *noun, plural* **-y·ses** \-ˌsēz\ [New Latin] (1891)
: catalysis of a reaction by one of its products
— **au·to·cat·a·lyt·ic** \-ˌka-tə-'li-tik\ *adjective*
— **au·to·cat·a·lyt·i·cal·ly** \-ti-k(ə-)lē\ *adverb*
au·to·ceph·a·lous \ˌȯ-tō-'se-f(ə-)ləs\ *adjective* [Late Greek *autokephalos,* from Greek *aut-* + *kephalē* head — more at CEPHALIC] (1863)
: being independent of external and especially patriarchal authority — used especially of Eastern national churches
— **au·to·ceph·a·ly** \-fə-lē\ *noun*
au·toch·thon \ȯ-'täk-thən\ *noun, plural* **-thons** *or* **-tho·nes** \-thə-ˌnēz\ [Greek *autochthōn,* from *aut-* + *chthōn* earth — more at HUMBLE] (1590)
: one (as a person, plant, or animal) that is autochthonous
au·toch·tho·nous \ȯ-'täk-thə-nəs\ *adjective* (1805)
1 **:** INDIGENOUS, NATIVE ⟨an *autochthonous* people⟩
2 **:** formed or originating in the place where found ⟨*autochthonous* rock⟩ ⟨an *autochthonous* infection⟩
— **au·toch·tho·nous·ly** *adverb*
[1]**au·to·clave** \'ȯ-tō-ˌklāv\ *noun* [French, from *aut-* + Latin *clavis* key — more at CLAVICLE] (1876)
: an apparatus in which special conditions (as high or low pressure or temperature) can be established for a variety of applications; *especially* **:** an apparatus (as for sterilizing) using superheated steam under high pressure
[2]**autoclave** *transitive verb* **-claved; -claving** (1911)
: to treat in an autoclave
au·to·cor·re·la·tion \ˌȯ-tō-ˌkȯr-ə-'lā-shən, -ˌkär-\ *noun* (1950)
: the correlation between paired values of a function of a mathematical or statistical variable taken at usually constant intervals that indicates the degree of periodicity of the function
au·toc·ra·cy \ȯ-'tä-krə-sē\ *noun, plural* **-cies** (1655)
1 **:** the authority or rule of an autocrat
2 **:** government in which one person possesses unlimited power
3 **:** a community or state governed by autocracy
au·to·crat \'ȯ-tə-ˌkrat\ *noun* [French *autocrate,* from Greek *autokratēs* ruling by oneself, absolute, from *aut-* + *-kratēs* ruling — more at -CRAT] (1803)
1 **:** a person (as a monarch) ruling with unlimited authority
2 **:** one who has undisputed influence or power
au·to·crat·ic \ˌȯ-tə-'kra-tik\ *also* **au·to·crat·i·cal** \-ti-kəl\ *adjective* (1823)
1 **:** of, relating to, or being an autocracy **:** ABSOLUTE ⟨an *autocratic* government⟩
2 **:** characteristic of or resembling an autocrat **:** DESPOTIC ⟨an *autocratic* ruler⟩
— **au·to·crat·i·cal·ly** \-ti-k(ə-)lē\ *adverb*
au·to·cross \'ȯ-tō-ˌkrȯs, 'ä-tō-\ *noun* [*auto* + moto*cross*] (1963)
: an automobile gymkhana
au·to–da–fé \ˌau̇-tō-də-'fā, ˌȯ-tō-\ *noun, plural* **au·tos–da–fé** \-tōz-də-\ [Portuguese *auto da fé,* literally, act of the faith] (1723)
: the ceremony accompanying the pronouncement of judgment by the Inquisition and followed by the execution of sentence by the secular authorities; *broadly* **:** the burning of a heretic
au·to·di·dact \ˌȯ-tō-'dī-ˌdakt, -dī-', -də-'\ *noun* [Greek *autodidaktos* self-taught, from *aut-* + *didaktos* taught, from *didaskein* to teach] (1748)
: a self-taught person
— **au·to·di·dac·tic** \-dī-'dak-tik, -də-\ *adjective*
au·toe·cious \ȯ-'tē-shəs\ *adjective* [*aut-* + Greek *oikia* house — more at VICINITY] (circa 1882)
: passing through all life stages on the same host ⟨*autoecious* rusts⟩
— **au·toe·cious·ly** *adverb*
— **au·toe·cism** \-'tē-ˌsi-zəm\ *noun*
au·to·er·o·tism \ˌȯ-tō-'er-ə-ˌti-zəm\ *or* **au·to·erot·i·cism** \-i-'rä-tə-ˌsi-zəm\ *noun* (1898)
1 **:** sexual feeling arising without known external stimulation
2 **:** sexual gratification obtained solely through stimulation by oneself of one's own body
— **au·to·erot·ic** \-i-'rä-tik\ *adjective*
au·tog·a·my \ȯ-'tä-gə-mē\ *noun* [International Scientific Vocabulary] (1877)
: SELF-FERTILIZATION: as **a :** pollination of a flower by its own pollen **b :** conjugation of two sister cells or sister nuclei of protozoans or fungi
— **au·tog·a·mous** \-məs\ *adjective*
au·tog·e·nous \ȯ-'tä-jə-nəs\ *also* **au·to·gen·ic** \ˌȯ-tə-'je-nik\ *adjective* [Greek *autogenēs,* from *aut-* + *-genēs* born, produced — more at -GEN] (1846)
1 **:** produced independently of external influence or aid **:** ENDOGENOUS
2 **:** originating or derived from sources within the same individual ⟨an *autogenous* graft⟩ ⟨*autogenous* vaccine⟩
3 **:** not requiring a meal of blood to produce eggs ⟨*autogenous* mosquitoes⟩
— **au·tog·e·nous·ly** *adverb*
— **au·tog·e·ny** \ȯ-'tä-jə-nē\ *noun*
au·to·gi·ro *or* **au·to·gy·ro** \ˌȯ-tō-'jīr-(ˌ)ō\ *noun, plural* **-ros** [from *Autogiro,* a trademark] (1923)
: a rotary-wing aircraft that employs a propeller for forward motion and a freely rotating rotor for lift

autogiro

au·to·graft \'ȯ-tō-ˌgraft\ *noun* (circa 1913)
: a tissue or organ that is transplanted from one part to another of the same body
— **autograft** *transitive verb*
[1]**au·to·graph** \'ȯ-tə-ˌgraf\ *noun* [Late Latin *autographum,* from Latin, neuter of *autographus* written with one's own hand, from Greek *autographos,* from *aut-* + *-graphos* written — more at -GRAPH] (circa 1644)
: something written or made with one's own hand: **a :** an original manuscript or work of art **b :** a person's handwritten signature
— **au·tog·ra·phy** \ȯ-'tä-grə-fē\ *noun*
[2]**autograph** *transitive verb* (1818)
1 **:** to write with one's own hand
2 **:** to write one's signature in or on
[3]**autograph** *adjective* (1832)
: being in the writer's own handwriting **:** not copied or duplicated ⟨an *autograph* letter⟩
au·to·graph·ic \ˌȯ-tə-'gra-fik\ *adjective* (1810)
1 **:** of, relating to, or constituting an autograph
2 **a** *of an instrument* **:** SELF-RECORDING **b** *of a record* **:** recorded by a self-recording instrument
— **au·to·graph·i·cal·ly** \-fi-k(ə-)lē\ *adverb*
Au·to·harp \'ȯ-tō-ˌhärp\ *trademark*
— used for a zither with button-controlled dampers for selected strings
au·to·hyp·no·sis \ˌȯ-tō-hip-'nō-səs\ *noun* [New Latin] (1903)
: self-induced and usually automatic hypnosis
— **au·to·hyp·not·ic** \-'nä-tik\ *adjective*
au·to·im·mune \-im-'yün\ *adjective* (1952)
: of, relating to, or caused by autoantibodies or lymphocytes that attack molecules, cells, or tissues of the organism producing them ⟨*autoimmune* diseases⟩
— **au·to·im·mu·ni·ty** \-'yü-nə-tē\ *noun*
— **au·to·im·mu·ni·za·tion** \-ˌim-yə-nə-'zā-shən *also* -im-ˌyü-\ *noun*
au·to·in·fec·tion \-in-'fek-shən\ *noun* [International Scientific Vocabulary] (1903)
: reinfection with larvae produced by parasitic worms already in the body
au·to·in·tox·i·ca·tion \-in-ˌtäk-sə-'kā-shən\ *noun* [International Scientific Vocabulary] (1887)
: a state of being poisoned by toxic substances produced within the body
au·to·load·ing \ˌȯ-tō-'lō-diŋ\ *adjective* (1923)
: SEMIAUTOMATIC b
au·tol·o·gous \ȯ-'tä-lə-gəs\ *adjective* [*aut-* + *-ologous* (as in *homologous*)] (circa 1921)
: derived from the same individual ⟨*autologous* blood transfusion⟩
au·tol·y·sate \ȯ-'tä-lə-ˌsāt, -ˌzāt\ *also* **au·tol·y·zate** \-ˌzāt\ *noun* (1910)
: a product of autolysis
au·to·lyse *British variant of* AUTOLYZE
au·tol·y·sis \-lə-səs\ *noun* [New Latin] (1902)
: breakdown of all or part of a cell or tissue by self-produced enzymes
— **au·to·lyt·ic** \ˌȯ-tᵊl-'i-tik\ *adjective*
au·to·lyze \'ȯ-tō-ˌlīz\ *verb* **-lyzed; -lyzing** [back-formation from *autolysis*] (1903)
intransitive verb
: to undergo autolysis
transitive verb
: to subject to autolysis
au·to·mak·er \'ȯ-tō-ˌmā-kər, 'ä-tō-\ *noun* (circa 1905)
: a manufacturer of automobiles
au·to·man \'ȯ-tō-ˌman, 'ä-tō-\ *noun* (1952)
: AUTOMAKER
Au·to·mat \'ȯ-tə-ˌmat\ *service mark*
— used for a cafeteria in which food is obtained especially from vending machines
au·to·mate \'ȯ-tə-ˌmāt\ *verb* **-mat·ed; -mat·ing** [back-formation from *automation*] (1952)
transitive verb
1 **:** to operate by automation
2 **:** to convert to largely automatic operation **:** AUTOMATIZE

intransitive verb
: to undergo automation
— **au·to·mat·able** \-ˌmā-tə-bəl\ *adjective*
automated teller machine *noun* (1981)
: a computerized electronic machine that performs basic banking functions (as handling check deposits or issuing cash withdrawals) — called also *automatic teller, automatic teller machine*
[1]**au·to·mat·ic** \ˌȯ-tə-ˈma-tik\ *adjective* [Greek *automatos* self-acting, from *aut-* + *-matos* (akin to Latin *ment-, mens* mind) — more at MIND] (1748)
1 a : largely or wholly involuntary; *especially* **:** REFLEX 5 ⟨*automatic* blinking of the eyelids⟩ **b :** acting or done spontaneously or unconsciously **c :** done or produced as if by machine **:** MECHANICAL ⟨the answers were *automatic*⟩
2 : having a self-acting or self-regulating mechanism
3 *of a firearm* **:** using either gas pressure or force of recoil and mechanical spring action for repeatedly ejecting the empty cartridge shell, introducing a new cartridge, and firing it
synonym see SPONTANEOUS
— **au·to·mat·i·cal·ly** \-ti-k(ə-)lē\ *adverb*
— **au·to·ma·tic·i·ty** \-mə-ˈti-sə-tē, -ma-\ *noun*
[2]**automatic** *noun* (1902)
1 : a machine or apparatus that operates automatically: as **a :** an automatic firearm **b :** an automatic gear-shifting mechanism
2 : a semiautomatic firearm
3 : AUDIBLE
automatic pilot *noun* (1916)
1 : AUTOPILOT 1
2 : a state or condition in which activity or behavior is regulated automatically in a predetermined or instinctive manner ⟨would go on *automatic pilot* and repeat all his standard denunciations —G. J. Church⟩
automatic writing *noun* (1883)
: writing produced without conscious intention as if of telepathic or spiritualistic origin
au·to·ma·tion \ˌȯ-tə-ˈmā-shən\ *noun* [[1]*automatic*] (circa 1948)
1 : the technique of making an apparatus, a process, or a system operate automatically
2 : the state of being operated automatically
3 : automatically controlled operation of an apparatus, process, or system by mechanical or electronic devices that take the place of human organs of observation, effort, and decision
au·tom·a·tism \ȯ-ˈtä-mə-ˌti-zəm\ *noun* [French *automatisme,* from *automate* automaton, from Latin *automaton*] (1838)
1 a : the quality or state of being automatic **b :** an automatic action
2 : the power or fact of moving or functioning without conscious control either independently of external stimuli (as in the beating of the heart) or under the influence of external stimuli (as in pupil dilation)
3 : a theory that views the body as a machine and consciousness as a noncontrolling adjunct of the body
4 : suspension of the conscious mind to release subconscious images ⟨*automatism* —the surrealist trend toward spontaneity and intuition —*Elle*⟩
— **au·tom·a·tist** \-ˈtä-mə-tist\ *noun or adjective*
au·tom·a·tize \ȯ-ˈtä-mə-ˌtīz\ *transitive verb* **-tized; -tiz·ing** [[1]*automatic*] (1952)
: to make automatic
— **au·tom·a·ti·za·tion** \ȯ-ˌtä-mə-tə-ˈzā-shən, -ˌtī-ˈzā-\ *noun*
au·tom·a·ton \ȯ-ˈtä-mə-tən, -mə-ˌtän\ *noun, plural* **-atons** *or* **-a·ta** \-mə-tə, -mə-ˌtä\ [Latin, from Greek, neuter of *automatos*] (1645)
1 : a mechanism that is relatively self-operating; *especially* **:** ROBOT
2 : a machine or control mechanism designed to follow automatically a predetermined sequence of operations or respond to encoded instructions
3 : an individual who acts in a mechanical fashion
[1]**au·to·mo·bile** \ˈȯ-tə-mō-ˌbēl, ˌȯ-tə-mō-ˈbē(ə)l, ˌȯ-tə-ˈmō-ˌbēl\ *adjective* [French, from *aut-* + *mobile*] (1883)
: AUTOMOTIVE
[2]**automobile** *noun* (circa 1889)
: a usually four-wheeled automotive vehicle designed for passenger transportation
— **automobile** *intransitive verb*
— **au·to·mo·bil·ist** \-ˈbē-list, -ˌbē-\ *noun*
au·to·mo·bil·i·ty \ˌȯ-tō-mə-ˈbi-lə-tē, -mō-\ *noun* (1903)
: the use of automobiles as the major means of transportation
au·to·mor·phism \ˌȯ-tə-ˈmȯr-ˌfi-zəm\ *noun* [*aut-* + *isomorphism*] (1903)
: an isomorphism of a set (as a group) with itself
au·to·mo·tive \ˌȯ-tə-ˈmō-tiv\ *adjective* (1898)
1 : of, relating to, or concerned with self-propelled vehicles or machines
2 : SELF-PROPELLED
au·to·nom·ic \ˌȯ-tə-ˈnä-mik\ *adjective* (1898)
1 : acting or occurring involuntarily ⟨*autonomic* reflexes⟩
2 : relating to, affecting, or controlled by the autonomic nervous system or its effects or activity ⟨*autonomic* drugs⟩
— **au·to·nom·i·cal·ly** \-mi-k(ə-)lē\ *adverb*
autonomic nervous system *noun* (1898)
: a part of the vertebrate nervous system that innervates smooth and cardiac muscle and glandular tissues and governs involuntary actions (as secretion and peristalsis) and that consists of the sympathetic nervous system and the parasympathetic nervous system
au·ton·o·mist \ȯ-ˈtä-nə-mist\ *noun* (1865)
: one who advocates autonomy
au·ton·o·mous \ȯ-ˈtä-nə-məs\ *adjective* [Greek *autonomos* independent, from *aut-* + *nomos* law — more at NIMBLE] (1800)
1 : of, relating to, or marked by autonomy
2 a : having the right or power of self-government **b :** undertaken or carried on without outside control **:** SELF-CONTAINED ⟨an *autonomous* school system⟩
3 a : existing or capable of existing independently ⟨an *autonomous* zooid⟩ **b :** responding, reacting, or developing independently of the whole ⟨an *autonomous* growth⟩
4 : controlled by the autonomic nervous system
synonym see FREE
— **au·ton·o·mous·ly** *adverb*
au·ton·o·my \-mē\ *noun, plural* **-mies** (circa 1623)
1 : the quality or state of being self-governing; *especially* **:** the right of self-government
2 : self-directing freedom and especially moral independence
3 : a self-governing state
au·to·pi·lot \ˈȯ-tō-ˌpī-lət\ *noun* (1935)
1 : a device for automatically steering ships, aircraft, and spacecraft
2 : AUTOMATIC PILOT 2
au·to·poly·ploid \ˌȯ-tō-ˈpä-li-ˌplȯid\ *noun* (1930)
: an individual or strain whose chromosome complement consists of more than two complete copies of the genome of a single ancestral species
— **autopolyploid** *adjective*
— **au·to·poly·ploi·dy** \-ˌplȯi-dē\ *noun*
au·top·sy \ˈȯ-ˌtäp-sē, ˈȯ-təp-\ *noun, plural* **-sies** [Greek *autopsia* act of seeing with one's own eyes, from *aut-* + *opsis* sight, appearance — more at OPTIC] (1678)
1 : an examination of a body after death to determine the cause of death or the character and extent of changes produced by disease
2 : a critical examination, evaluation, or assessment of someone or something past
— **autopsy** *transitive verb*
au·to·ra·dio·gram \ˌȯ-tō-ˈrā-dē-ə-ˌgram\ *noun* (1949)
: AUTORADIOGRAPH
au·to·ra·dio·graph \ˌȯ-tō-ˈrā-dē-ə-ˌgraf\ *noun* [International Scientific Vocabulary] (1903)
: an image produced on a photographic film or plate by the radiations from a radioactive substance in an object which is in close contact with the emulsion
— **au·to·ra·dio·graph·ic** \-ˌrā-dē-ə-ˈgra-fik\ *adjective*
— **au·to·ra·di·og·ra·phy** \-ˌrā-dē-ˈä-grə-fē\ *noun*
au·to·ro·ta·tion \-rō-ˈtā-shən\ *noun* (1918)
: the turning of the rotor of an autogiro or a helicopter with the resulting lift caused solely by the aerodynamic forces induced by motion of the rotor along its flight path
— **au·to·ro·tate** \-ˈrō-ˌtāt\ *intransitive verb*
au·to·route \ˈȯ-tō-ˌrüt, -ˌrau̇t\ *noun* [French, from *auto*mobile + *route*] (1951)
: an expressway especially in France
autos–da–fé *plural of* AUTO-DA-FÉ
au·to·sex·ing \ˈȯ-tō-ˌsek-siŋ\ *adjective* (1936)
: exhibiting different characters in the two sexes at birth or hatching
au·to·some \ˈȯ-tə-ˌsōm\ *noun* (circa 1906)
: a chromosome other than a sex chromosome
— **au·to·so·mal** \ˌȯ-tə-ˈsō-məl\ *adjective*
— **au·to·so·mal·ly** \-mə-lē\ *adverb*
au·to·stra·da \ˌau̇-tō-ˈsträ-də, ˌȯ-tō-\ *noun, plural* **-stradas** *or* **-stra·de** \-ˈsträ-(ˌ)dā\ [Italian, from *auto*mobile + *strada* street, from Late Latin *strata* paved road — more at STREET] (1927)
: an expressway especially in Italy
au·to·sug·ges·tion \ˌȯ-tō-sə(g)-ˈjes-chən, -ˈjesh-\ *noun* [International Scientific Vocabulary] (1890)
: an influencing of one's own attitudes, behavior, or physical condition by mental processes other than conscious thought **:** SELF-HYPNOSIS
— **au·to·sug·gest** \-sə(g)-ˈjest\ *transitive verb*
au·to·te·lic \ˌȯ-tō-ˈte-lik, -ˈtē-\ *adjective* [Greek *autotelēs,* from *aut-* + *telos* end — more at TELOS] (circa 1901)
: having a purpose in and not apart from itself
au·to·tet·ra·ploid \ˌȯ-tō-ˈte-trə-ˌplȯid\ *noun* (1930)
: an individual or strain whose chromosome complement consists of four copies of a single genome due to doubling of an ancestral chromosome complement
— **autotetraploid** *adjective*
— **au·to·tet·ra·ploi·dy** \-ˌplȯi-dē\ *noun*
au·tot·o·my \ȯ-ˈtä-tə-mē\ *noun* [International Scientific Vocabulary] (1897)
: reflex separation of a part (as an appendage) from the body **:** division of the body into two or more pieces
— **au·to·to·mous** \-məs\ *adjective*
— **au·tot·o·mize** \-ˌmīz\ *verb*
au·to·trans·form·er \ˌȯ-tō-tran(t)s-ˈfȯr-mər\ *noun* (1895)
: a transformer in which the primary and secondary coils have part or all of their turns in common
au·to·trans·fu·sion \ˌȯ-tō-tran(t)s-ˈfyü-zhən\ *noun* (circa 1923)
: return of autologous blood to the patient's own circulatory system
au·to·troph \ˈȯ-tə-ˌtrōf, -ˌträf\ *noun* [German, from *autotroph,* adjective] (1938)
: an autotrophic organism

au·to·tro·phic \ˌȯ-tə-ˈtrō-fik\ *adjective* [probably from German *autotroph,* from Greek *autotrophos* supplying one's own food, from *aut-* + *trephein* to nourish] (circa 1900)
1 : needing only carbon dioxide or carbonates as a source of carbon and a simple inorganic nitrogen compound for metabolic synthesis
2 : not requiring a specified exogenous factor for normal metabolism
— au·to·tro·phi·cal·ly \-fi-k(ə-)lē\ *adverb*
— au·tot·ro·phy \ȯ-ˈtä-trə-fē\ *noun*

au·to·wor·ker \ˈȯ-tō-ˌwər-kər, ˈä-\ *noun* (1941)
: a person employed in the automobile manufacturing industry

au·tox·i·da·tion \ˌȯ-ˌtäk-sə-ˈdā-shən\ *noun* (1883)
: oxidation by direct combination with oxygen (as in air) at ordinary temperatures

au·tumn \ˈȯ-təm\ *noun* [Middle English *autumpne,* from Latin *autumnus*] (14th century)
1 : the season between summer and winter comprising in the northern hemisphere usually the months of September, October, and November or as reckoned astronomically extending from the September equinox to the December solstice — called also *fall*
2 : a period of maturity or incipient decline ⟨in the *autumn* of life⟩
— au·tum·nal \ȯ-ˈtəm-nəl\ *adjective*
— au·tum·nal·ly \-nə-lē\ *adverb*

autumn crocus *noun* (1885)
: an autumn-blooming colchicum (*Colchicum autumnale*)

au·tun·ite \ō-ˈtə-ˌnīt, ˈō-tᵊn-\ *noun* [*Autun,* town in France] (circa 1852)
: a radioactive lemon-yellow mineral occurring in tabular crystals with basal cleavage and in scales like mica

aux·e·sis \ȯg-ˈzē-səs, ȯk-ˈsē-\ *noun* [New Latin, from Greek *auxēsis* increase, growth, from *auxein* to increase — more at EKE] (circa 1848)
: GROWTH; *specifically* **:** increase of cell size without cell division
— aux·et·ic \-ˈze-tik, -ˈse-\ *adjective or noun*

¹aux·il·ia·ry \ȯg-ˈzil-yə-rē, -ˈzil-rē, -ˈzi-lə-\ *adjective* [Latin *auxiliaris,* from *auxilium* help; akin to Latin *augēre* to increase — more at EKE] (15th century)
1 a : offering or providing help **b :** functioning in a subsidiary capacity ⟨an *auxiliary* branch of the state university⟩
2 *of a verb* **:** accompanying another verb and typically expressing person, number, mood, or tense
3 a : SUPPLEMENTARY **b :** constituting a reserve ⟨an *auxiliary* power plant⟩
4 : equipped with sails and a supplementary inboard engine

²auxiliary *noun, plural* **-ries** (1601)
1 a : an auxiliary person, group, or device; *specifically* **:** a member of a foreign force serving a nation at war **b :** a Roman Catholic titular bishop assisting a diocesan bishop and not having the right of succession — called also *auxiliary bishop*
2 : an auxiliary boat or ship
3 : an auxiliary verb

aux·in \ˈȯk-sən\ *noun* [International Scientific Vocabulary, from Greek *auxein*] (1934)
1 : any of various usually acidic organic substances that promote cell elongation in plant shoots and usually regulate other growth processes (as root initiation): as **a :** INDOLEACETIC ACID **b :** any of various synthetic substances (as 2,4-D) resembling indoleacetic acid in activity and used especially in research and agriculture
2 : PLANT HORMONE
— aux·in·ic \ȯk-ˈsi-nik\ *adjective*

auxo·troph \ˈȯk-sə-ˌtrōf, -ˌträf\ *noun* (1950)
: an auxotrophic strain or individual

auxo·tro·phic \ˌȯk-sə-ˈtrō-fik\ *adjective* [Greek *auxein* to increase + *-o-* + English *-trophic*] (1944)
: requiring a specific growth substance beyond the minimum required for normal metabolism and reproduction by the parental or wild-type strain ⟨*auxotrophic* mutants of bacteria⟩
— aux·ot·ro·phy \ȯk-ˈsä-trə-fē\ *noun*

¹avail \ə-ˈvā(ə)l\ *verb* [Middle English, probably from *a-* (as in *abaten* to abate) + *vailen* to avail, from Old French *valoir* to be of worth, from Latin *valēre* — more at WIELD] (14th century)
intransitive verb
: to be of use or advantage **:** SERVE ⟨our best efforts did not *avail*⟩
transitive verb
1 : to be of use or advantage to **:** PROFIT
2 : to result in **:** bring about ⟨his efforts *availed* him nothing⟩
— avail oneself of *also* **avail of :** to make use of **:** take advantage of

²avail *noun* (15th century)
1 : advantage toward attainment of a goal or purpose **:** USE ⟨effort was of little *avail*⟩
2 *plural* **:** profits or proceeds especially from a business or from the sale of property

avail·abil·i·ty \ə-ˌvā-lə-ˈbi-lə-tē\ *noun, plural* **-ties** (1803)
1 : the quality or state of being available
2 : an available person or thing

avail·able \ə-ˈvā-lə-bəl\ *adjective* (15th century)
1 *archaic* **:** having a beneficial effect
2 : VALID — used of a legal plea or charge
3 : present or ready for immediate use
4 : ACCESSIBLE, OBTAINABLE ⟨articles *available* in any drugstore⟩
5 : qualified or willing to do something or to assume a responsibility ⟨*available* candidates⟩
6 : present in such chemical or physical form as to be usable (as by a plant) ⟨*available* nitrogen⟩ ⟨*available* water⟩
— avail·able·ness *noun*
— avail·ably \-blē\ *adverb*

¹av·a·lanche \ˈa-və-ˌlanch\ *noun* [French, from French dialect *lavantse, avalantse*] (1771)
1 : a large mass of snow, ice, earth, rock, or other material in swift motion down a mountainside or over a precipice
2 : a sudden great or overwhelming rush or accumulation of something ⟨office workers tied down with an *avalanche* of paperwork⟩
3 : a cumulative process in which photons or accelerated charge carriers produce additional photons or charge carriers through collisions (as with gas molecules)

²avalanche *verb* **-lanched; -lanch·ing** (1872)
intransitive verb
: to descend in an avalanche
transitive verb
: OVERWHELM, FLOOD

Av·a·lon \ˈa-və-ˌlän\ *noun*
: a paradise to which Arthur is carried after his death

avant \ˈä-ˌvän(t), ˈa-, -ˌvȯn(t), -ˌvōⁿ\ *adjective* [French *avant-* fore-, front, from *avant* before, from Latin *abante*] (1965)
: culturally or stylistically advanced **:** AVANT-GARDE ⟨*avant* jazz⟩

¹avant–garde \ˌä-ˌvän(t)-ˈgärd, ˌa-, ˌá-; ə-ˈvänt-ˌ; ˌa-ˌvōⁿ-ˈ, ˌa-ˌvȯn(t)-ˈ\ *noun* [French, vanguard] (1910)
: an intelligentsia that develops new or experimental concepts especially in the arts
— avant–gard·ism \-ˈgär-ˌdi-zəm\ *noun*
— avant–gard·ist \-ˈgär-dist\ *noun*

²avant–garde *adjective* (1925)
: of or relating to an avant-garde ⟨*avant-garde* writers⟩

av·a·rice \ˈa-və-rəs, ˈav-rəs\ *noun* [Middle English, from Old French, from Latin *avaritia,* from *avarus* avaricious, from *avēre* to crave — more at AVID] (14th century)
: excessive or insatiable desire for wealth or gain **:** GREEDINESS, CUPIDITY

av·a·ri·cious \ˌa-və-ˈri-shəs\ *adjective* (15th century)
: greedy of gain **:** excessively acquisitive especially in seeking to hoard riches
synonym see COVETOUS
— av·a·ri·cious·ly *adverb*
— av·a·ri·cious·ness *noun*

avas·cu·lar \(ˌ)ā-ˈvas-kyə-lər\ *adjective* (circa 1900)
: having few or no blood vessels ⟨*avascular* tissue⟩
— avas·cu·lar·i·ty \ˌā-ˌvas-kyə-ˈlar-ə-tē\ *noun*

avast \ə-ˈvast\ *verb imperative* [perhaps from Dutch *houd vast* hold fast] (1681)
— a nautical command to stop or cease

av·a·tar \ˈa-və-ˌtär\ *noun* [Sanskrit *avatāra* descent, from *avatarati* he descends, from *ava-* away + *tarati* he crosses over — more at UKASE, THROUGH] (1784)
1 : the incarnation of a Hindu deity (as Vishnu)
2 a : an incarnation in human form **b :** an embodiment (as of a concept or philosophy) often in a person
3 : a variant phase or version of a continuing basic entity

avaunt \ə-ˈvȯnt, -ˈvänt\ *adverb* [Middle English, from Middle French *avant,* from Latin *abante* forward, before, from *ab* from + *ante* before — more at OF, ANTE-] (15th century)
: AWAY, HENCE

ave \ˈä-(ˌ)vā\ *noun* [Middle English, from Latin, hail] (13th century)
1 : an expression of greeting or of leave-taking **:** HAIL, FAREWELL
2 *often capitalized* **:** AVE MARIA

avel·lan \ə-ˈve-lən\ *or* **avel·lane** \ə-ˈve-ˌlān, ˈa-və-ˌlān\ *adjective* [Latin *abellana, avellana* filbert, from feminine of *Abellanus* of Abella, from *Abella,* ancient town in Italy] (1611)
of a heraldic cross **:** having the four arms shaped like conventionalized filberts — see CROSS illustration

Ave Ma·ria \ˌä-(ˌ)vā-mə-ˈrē-ə\ *noun* [Middle English, from Medieval Latin, hail, Mary] (13th century)
: HAIL MARY

avenge \ə-ˈvenj\ *transitive verb* **avenged; aveng·ing** [Middle English *avengen,* probably from *a-* (as in *abaten* to abate) + *vengen* to avenge, from Old French *vengier* — more at VENGEANCE] (14th century)
1 : to take vengeance for or on behalf of
2 : to exact satisfaction for (a wrong) by punishing the wrongdoer
— aveng·er *noun*

av·ens \ˈa-vənz\ *noun, plural* **avens** [Middle English *avence,* from Old French] (13th century)
: any of a genus (*Geum*) of perennial herbs of the rose family with white, purple, or yellow flowers

av·en·tail \ˈa-vən-ˌtāl\ *noun* [Middle English, modification of Old French *ventaille*] (14th century)
: VENTAIL

aven·tu·rine \ə-ˈven-chə-ˌrēn, -rən\ *noun* [French, from *aventure* chance — more at ADVENTURE] (1811)
1 : glass containing opaque sparkling particles of foreign material usually copper or chromic oxide
2 : a translucent quartz spangled throughout with scales of mica or other mineral

av·e·nue \ˈa-və-ˌnü, -ˌnyü\ *noun* [Middle French, from feminine of *avenu,* past participle of *avenir* to come to, from Latin *advenire* — more at ADVENTURE] (1600)
1 : a way of access **:** ROUTE
2 : a channel for pursuing a desired object ⟨*avenues* of communication⟩
3 a *chiefly British* **:** the principal walk or driveway to a house situated off a main road **b**

: a broad passageway bordered by trees
4 : an often broad street or road

aver \ə-ˈvər\ *transitive verb* **averred; aver·ring** [Middle English *averren,* from Middle French *averer,* from Medieval Latin *adverare* to confirm as authentic, from Latin *ad-* + *verus* true — more at VERY] (15th century)
1 a : to verify or prove to be true in pleading a cause **b :** to allege or assert in pleading
2 : to declare positively

[1]av·er·age \ˈa-v(ə-)rij\ *noun* [from earlier *average* proportionally distributed charge for damage at sea, modification of Middle French *avarie* damage to ship or cargo, from Old Italian *avaria,* from Arabic *ʽawārīyah* damaged merchandise] (1735)
1 a : a single value (as a mean, mode, or median) that summarizes or represents the general significance of a set of unequal values **b :** MEAN 1b
2 a : an estimation of or approximation to an arithmetic mean **b :** a level (as of intelligence) typical of a group, class, or series ⟨above the *average*⟩
3 : a ratio expressing the average performance especially of an athletic team or an athlete computed according to the number of opportunities for successful performance ☆
— on average *or* **on the average :** taking the typical example of the group under consideration ⟨prices have increased *on average* by five percent⟩

[2]average *adjective* (1770)
1 : equaling an arithmetic mean
2 a : being about midway between extremes ⟨a man of *average* height⟩ **b :** not out of the ordinary **:** COMMON ⟨the *average* person⟩
— av·er·age·ly *adverb*
— av·er·age·ness *noun*

[3]average *verb* **av·er·aged; av·er·ag·ing** (1769)
intransitive verb
1 a : to be or come to an average ⟨the gain *averaged* out to 20 percent⟩ **b :** to have a medial value of ⟨a color *averaging* a pale purple⟩
2 : to buy on a falling market or sell on a rising market additional shares or commodities so as to obtain a more favorable average price — usually used with *down* or *up*
transitive verb
1 : to do, get, or have on the average or as an average sum or quantity ⟨*averages* 12 hours of work a day⟩
2 : to find the arithmetic mean of (a series of unequal quantities)
3 a : to bring toward the average **b :** to divide among a number proportionately

aver·ment \ə-ˈvər-mənt\ *noun* (15th century)
1 : the act of averring
2 : something that is averred **:** AFFIRMATION

averse \ə-ˈvərs\ *adjective* [Latin *aversus,* past participle of *avertere*] (1597)
: having an active feeling of repugnance or distaste ⟨*averse* to strenuous exercise⟩
synonym see DISINCLINED
usage see ADVERSE
— averse·ly *adverb*
— averse·ness *noun*

aver·sion \ə-ˈvər-zhən, -shən\ *noun* (1596)
1 *obsolete* **:** the act of turning away
2 a : a feeling of repugnance toward something with a desire to avoid or turn from it ⟨regards drunkenness with *aversion*⟩ **b :** a settled dislike **:** ANTIPATHY ⟨expressed an *aversion* to parties⟩ **c :** a tendency to extinguish a behavior or to avoid a thing or situation and especially a usually pleasurable one because it is or has been associated with a noxious stimulus
3 : one that is the object of aversion

aversion therapy *noun* (1946)
: therapy intended to change habits or antisocial behavior by inducing dislike for them through association with a noxious stimulus

aver·sive \ə-ˈvər-siv, -ziv\ *adjective* (1923)
: tending to avoid or causing avoidance of a noxious or punishing stimulus ⟨behavior modification by *aversive* stimulation⟩
— aver·sive·ly *adverb*
— aver·sive·ness *noun*

avert \ə-ˈvərt\ *transitive verb* [Middle English, from Middle French *avertir,* from Latin *avertere,* from *ab-* + *vertere* to turn — more at WORTH] (15th century)
1 : to turn away or aside (as the eyes) in avoidance
2 : to see coming and ward off **:** AVOID

Aves·ta \ə-ˈves-tə\ *noun* [Middle Persian *Avastāk,* literally, original text] (1856)
: the book of the sacred writings of Zoroastrianism

Aves·tan \-tən\ *noun* (1856)
: an ancient Iranian language in which the sacred books of Zoroastrianism were written — see INDO-EUROPEAN LANGUAGES table
— Avestan *adjective*

av·gas \ˈav-ˌgas\ *noun* [*av*iation *gas*oline] (1943)
: gasoline for airplanes

av·go·lem·o·no \ˌäv-gō-ˈle-mə-(ˌ)nō\ *noun* [New Greek *augolemono,* from *augo* egg + *lemoni* lemon] (1961)
: a soup or sauce made of chicken stock, rice, egg yolks, and lemon sauce

avi·an \ˈā-vē-ən\ *adjective* [Latin *avis*] (1870)
: of, relating to, or derived from birds

avi·ary \ˈā-vē-ˌer-ē\ *noun, plural* **-ar·ies** [Latin *aviarium,* from *avis* bird; akin to Greek *aetos* eagle] (1577)
: a place for keeping birds confined

avi·ate \ˈā-vē-ˌāt, ˈa-\ *intransitive verb* **-at·ed; -at·ing** [back-formation from *aviation*] (1887)
: to navigate the air (as in an airplane)

avi·a·tion \ˌā-vē-ˈā-shən, ˌa-\ *noun, often attributive* [French, from Latin *avis*] (1866)
1 : the operation of heavier-than-air aircraft
2 : military airplanes
3 : airplane manufacture, development, and design

avi·a·tor \ˈā-vē-ˌā-tər, ˈa-\ *noun* (1887)
: the operator or pilot of an aircraft and especially an airplane

aviator glasses *noun plural* (1968)
: eyeglasses having a lightweight metal frame and relatively large usually tinted lenses

avi·a·trix \ˌā-vē-ˈā-triks, ˌa-\ *noun, plural* **-trix·es** \-trik-səz\ *or* **-tri·ces** \-trə-ˌsēz\ (1910)
: a woman aviator

avi·cul·ture \ˈā-və-ˌkəl-chər, ˈa-\ *noun* [Latin *avis* + English *culture*] (circa 1879)
: the raising and care of birds and especially of wild birds in captivity
— avi·cul·tur·ist \ˌā-və-ˈkəl-ch(ə-)rist, ˌa-\ *noun*

av·id \ˈa-vəd\ *adjective* [French or Latin; French *avide,* from Latin *avidus,* from *avēre* to desire, crave; akin to Welsh *ewyllys* desire, Old Irish *con-oí* he protects] (1769)
1 : desirous to the point of greed **:** urgently eager **:** GREEDY ⟨*avid* for publicity⟩
2 : characterized by enthusiasm and vigorous pursuit ⟨*avid* readers⟩
synonym see EAGER
— av·id·ly *adverb*
— av·id·ness *noun*

av·i·din \ˈa-və-dən\ *noun* [from its avidity for biotin] (1941)
: a protein found in egg white that inactivates biotin by combining with it

avid·ity \ə-ˈvi-də-tē, a-\ *noun, plural* **-ities** (15th century)
1 : the quality or state of being avid: **a :** keen eagerness **b :** consuming greed
2 : AFFINITY 2b(2)

avi·fau·na \ˌā-və-ˈfȯ-nə, ˌa-, -ˈfä-nə\ *noun* [New Latin, from Latin *avis* + New Latin *fauna*] (1874)
: the birds or the kinds of birds of a region, period, or environment
— avi·fau·nal \-ˈfȯ-nᵊl, -ˈfä-\ *adjective*

avi·on·ics \ˌā-vē-ˈä-niks, ˌa-\ *noun plural* [*avi*ation electr*onics*] (1949)
: electronics designed for use in aerospace vehicles
— avi·on·ic \-nik\ *adjective*

avir·u·lent \(ˌ)ā-ˈvir-(y)ə-lənt\ *adjective* [International Scientific Vocabulary] (circa 1900)
: not virulent — compare NONPATHOGENIC

avi·ta·min·osis \ˌā-ˌvī-tə-mə-ˈnō-səs\ *noun, plural* **-o·ses** \-ˌsēz\ [New Latin] (1919)
: disease (as pellagra) resulting from a deficiency of one or more vitamins
— avi·ta·min·ot·ic \-mə-ˈnä-tik\ *adjective*

A–V node \ˌā-ˈvē-, ˈā-ˌ\ *noun* (1949)
: ATRIOVENTRICULAR NODE

avo \ˈa-(ˌ)vü\ *noun, plural* **avos** [Portuguese, from *avo* fractional part, from *-avo* ordinal suffix (as in *oitavo* eighth, from Latin *octavus*) — more at OCTAVE] (circa 1909)
— see *pataca* at MONEY table

av·o·ca·do \ˌä-və-ˈkä-(ˌ)dō, ˌa-\ *noun, plural* **-dos** *also* **-does** [modification of Spanish *aguacate,* from Nahuatl *āhuacatl* avocado, testicle] (1697)
1 : a pulpy green to purple nutty-flavored edible fruit of any of various tropical American trees (genus *Persea* especially *P. americana*) of the laurel family; *also* **:** a tree bearing avocados
2 : a light yellowish green ◆

☆ SYNONYMS

Average, mean, median, norm mean something that represents a middle point. AVERAGE is exactly or approximately the quotient obtained by dividing the sum total of a set of figures by the number of figures ⟨scored an *average* of 85 on tests⟩. MEAN may be the simple average or it may represent value midway between two extremes ⟨a high of 70° and a low of 50° give a *mean* of 60°⟩. MEDIAN applies to the value that represents the point at which there are as many instances above as there are below ⟨*average* of a group of persons earning 3, 4, 5, 8, and 10 dollars a day is 6 dollars, whereas the *median* is 5 dollars⟩. NORM means the computed or estimated average of performance of a significantly large group, class, or grade ⟨scores about the *norm* for 5th grade arithmetic⟩.

◇ WORD HISTORY

avocado We owe the avocado, along with chocolate and the tomato, to pre-Columbian Middle America, and the words for these tasty items to Nahuatl, the language of several important Indian peoples, including the Aztecs, who lived in the area that is now Mexico City. The Nahuatl word for the avocado was *āhuacatl,* which also means "testicle"—one sense presumably being a metaphor for the other, though we do not know which came first. This word was borrowed into Spanish as *aguacate,* though by a peculiar twist of folk etymology some American Spanish speakers modified it to *abocado,* identical with *abocado* or *abogado* "advocate, lawyer." This modification eventually lost out to *aguacate* in almost all varieties of overseas Spanish, but not before it was borrowed into French (as *avocat*) and English (as *avocato* or *avocado*) in the late 17th century. Nahuatl *āhuacatl* also occurred in the compound *āhuacamōlli,* literally, "avocado sauce," borrowed into Spanish as *guacamole,* which American English borrowed several centuries later.

avocado pear *noun* (1830)
chiefly British **:** AVOCADO 1

av·o·ca·tion \ˌa-və-ˈkā-shən\ *noun* [Latin *avocation-, avocatio,* from *avocare* to call away, from *ab-* + *vocare* to call, from *voc-, vox* voice — more at VOICE] (circa 1617)
1 *archaic* **:** DIVERSION, DISTRACTION
2 : customary employment **:** VOCATION
3 : a subordinate occupation pursued in addition to one's vocation especially for enjoyment **:** HOBBY

av·o·ca·tion·al \ˌa-və-ˈkā-sh(ə-)nəl\ *adjective* (1921)
1 : of or relating to an avocation ⟨an *avocational* interest in sports⟩
2 : being such by avocation ⟨an *avocational* musician⟩
— av·o·ca·tion·al·ly \-sh(ə-)nə-lē\ *adverb*

av·o·cet \ˈa-və-ˌset\ *noun* [French & Italian; French *avocette,* from Italian *avocetta*] (1766)
: any of a genus (*Recurvirostra*) of rather large long-legged shorebirds with webbed feet and slender upward-curving bill

avocet

Avo·ga·dro's num·ber \ˌa-və-ˈgä-(ˌ)drōz-, ˌa-və-, -ˈga-\ *noun* [Count Amedeo *Avogadro*] (1924)
: the number 6.023×10^{23} indicating the number of atoms or molecules in a mole of any substance — called also *Avogadro number*

avoid \ə-ˈvȯid\ *transitive verb* [Middle English, from Middle French *esvuidier,* from *es-* (from Latin *ex-*) + *vuidier* to empty — more at VOID] (14th century)
1 : to make legally void **:** ANNUL ⟨*avoid* a plea⟩
2 *obsolete* **:** VOID, EXPEL
3 a : to keep away from **:** SHUN **b :** to prevent the occurrence or effectiveness of **c :** to refrain from
4 *archaic* **:** to depart or withdraw from **:** LEAVE
synonym see ESCAPE
— avoid·able \-ˈvȯi-də-bəl\ *adjective*
— avoid·ably \-blē\ *adverb*
— avoid·er *noun*

avoid·ance \ə-ˈvȯi-dᵊn(t)s\ *noun* (14th century)
1 *obsolete* **a :** an action of emptying, vacating, or clearing away **b :** OUTLET
2 : ANNULMENT 1
3 : an act or practice of avoiding or withdrawing from something

¹av·oir·du·pois \ˌa-vər-də-ˈpȯiz, ˈa-vər-də-ˌ\ *noun* [Middle English *avoir de pois* goods sold by weight, from Old French, literally, goods of weight] (15th century)
1 : AVOIRDUPOIS WEIGHT
2 : WEIGHT, HEAVINESS; *especially* **:** personal weight

²avoirdupois *adjective* (1755)
: expressed in avoirdupois weight ⟨one ounce *avoirdupois*⟩

avoirdupois weight *noun* (1619)
: the series of units of weight based on the pound of 16 ounces and the ounce of 16 drams — see WEIGHT table

avouch \ə-ˈvau̇ch\ *transitive verb* [Middle English, to cite as authority, from Middle French *avochier* to summon, from Latin *advocare* — more at ADVOCATE] (15th century)
1 : to declare as a matter of fact or as a thing that can be proved **:** AFFIRM
2 : to vouch for **:** CORROBORATE
3 a : to acknowledge (as an act) as one's own **b :** CONFESS, AVOW

avouch·ment \-mənt\ *noun* (1574)
: an act of avouching **:** AVOWAL

avow \ə-ˈvau̇\ *transitive verb* [Middle English, from Middle French *avouer,* from Latin *advocare*] (14th century)
1 : to declare assuredly
2 : to declare openly, bluntly, and without shame ⟨ever ready to *avow* his reactionary outlook⟩
synonym see ACKNOWLEDGE, ASSERT
— avow·er \-ˈvau̇(-ə)r\ *noun*

avow·al \-ˈvau̇(-ə)l\ *noun* (circa 1732)
: an open declaration or acknowledgment

avow·ed·ly \ə-ˈvau̇-əd-lē\ *adverb* (1656)
1 : with open acknowledgment **:** FRANKLY ⟨an *avowedly* hostile review⟩
2 : by unsupported assertion or profession alone **:** ALLEGEDLY ⟨politicians remain skeptical of . . . *avowedly* democratic intentions —Jerry Kirshenbaum⟩

avulse \ə-ˈvəls\ *transitive verb* **avulsed; avuls·ing** [Latin *avulsus,* past participle of *avellere* to tear off, from *ab-* + *vellere* to pluck — more at VULNERABLE] (circa 1765)
: to separate by avulsion

avul·sion \ə-ˈvəl-shən\ *noun* (1622)
: a forcible separation or detachment: as **a :** a tearing away of a body part accidentally or surgically **b :** a sudden cutting off of land by flood, currents, or change in course of a body of water; *especially* **:** one separating land from one person's property and joining it to another's

avun·cu·lar \ə-ˈvəŋ-kyə-lər\ *adjective* [Latin *avunculus* maternal uncle — more at UNCLE] (1831)
1 : of or relating to an uncle
2 : suggestive of an uncle especially in kindliness or geniality ⟨*avuncular* indulgence⟩
— avun·cu·lar·i·ty \ə-ˌvəŋ-kyə-ˈla-rə-tē\ *noun*
— avun·cu·lar·ly \ə-ˈvəŋ-kyə-lər-lē\ *adverb*

aw \ˈȯ\ *interjection* (1852)
— used to express mild disappointment, gentle entreaty, or real or mock sympathy or sentiment

await \ə-ˈwāt\ *verb* [Middle English, from Old North French *awaitier,* from *a-* (from Latin *ad-*) + *waitier* to watch — more at WAIT] (13th century)
transitive verb
1 *obsolete* **:** to lie in wait for
2 a : to wait for **b :** to remain in abeyance until ⟨a treaty *awaiting* ratification⟩
3 : to be in store for ⟨wonders what *awaits* him next⟩
intransitive verb
1 *obsolete* **:** ATTEND
2 : to stay or be in waiting **:** WAIT
3 : to be in store

¹awake \ə-ˈwāk\ *verb* **awoke** \-ˈwōk\ *also* **awaked** \-ˈwākt\; **awo·ken** \-ˈwō-kən\ *or* **awaked** *also* **awoke; awak·ing** [Middle English *awaken* (from Old English *awacan, onwacan,* from ¹*a-, on* + *wacan* to awake) & *awakien,* from Old English *awacian,* from ¹*a-* + *wacian* to be awake — more at WAKE] (before 12th century)
intransitive verb
1 : to cease sleeping
2 : to become aroused or active again
3 : to become conscious or aware of something ⟨*awoke* to the possibilities⟩
transitive verb
1 : to arouse from sleep or a sleeplike state ⟨*awoken* by the storm⟩
2 : to make active **:** stir up ⟨*awoke* old memories⟩ ■

²awake *adjective* (13th century)
: fully conscious, alert, and aware **:** not asleep
synonym see AWARE

awak·en \ə-ˈwā-kən\ *verb* **awak·ened; awak·en·ing** \-ˈwāk-niŋ, -ˈwā-kə-\ [Middle English, from Old English *awæcnian,* from *a-* + *wæcnian* to waken] (before 12th century)
: AWAKE
— awak·en·er \-ˈwāk-nər, -ˈwā-kə-\ *noun*

¹award \ə-ˈwȯrd\ *transitive verb* [Middle English, to decide, from Old North French *eswarder,* from *es-* (from Latin *ex-*) + *warder* to guard, of Germanic origin; akin to Old High German *wartēn* to watch — more at WARD] (14th century)
1 : to give by judicial decree or after careful consideration
2 : to confer or bestow as being deserved or merited or needed ⟨*award* scholarships to disadvantaged students⟩
synonym see GRANT
— award·able \-ˈwȯr-də-bəl\ *adjective*
— award·ee \-ˌwȯr-ˈdē\ *noun*
— award·er \-ˈwȯr-dər\ *noun*

²award *noun* (14th century)
1 a : a judgment or final decision; *especially* **:** the decision of arbitrators in a case submitted to them **b :** the document containing the decision of arbitrators
2 : something that is conferred or bestowed especially on the basis of merit or need

aware \ə-ˈwar, -ˈwer\ *adjective* [Middle English *iwar,* from Old English *gewær,* from *ge-* (associative prefix) + *wær* wary — more at CO-, WARY] (before 12th century)
1 *archaic* **:** WATCHFUL, WARY
2 : having or showing realization, perception, or knowledge ☆
— aware·ness *noun*

awash \ə-ˈwȯsh, -ˈwäsh\ *adjective* (1833)

☆ SYNONYMS
Aware, cognizant, conscious, sensible, alive, awake mean having knowledge of something. AWARE implies vigilance in observing or alertness in drawing inferences from what one experiences ⟨*aware* of changes in climate⟩. COGNIZANT implies having special or certain knowledge as from firsthand sources ⟨not fully *cognizant* of the facts⟩. CONSCIOUS implies that one is focusing one's attention on something or is even preoccupied by it ⟨*conscious* that my heart was pounding⟩. SENSIBLE implies direct or intuitive perceiving especially of intangibles or of emotional states or qualities ⟨*sensible* of a teacher's influence⟩. ALIVE adds to SENSIBLE the implication of acute sensitivity to something ⟨*alive* to the thrill of danger⟩. AWAKE implies that one has become alive to something and is on the alert ⟨a country always *awake* to the threat of invasion⟩.

□ USAGE
awake Our modern verb *awake* represents the amalgamation of two older verbs, one with regular verb parts and one with irregular. As a result, the modern verb has a mixture of regular and irregular parts whose status is still in flux. One old past participle, *awaken,* has become a separate verb. The past participle *awoken* is a curious case. It appears to have been formed from the past *awoke* and to have replaced an older past participle *awake,* now used only as an adjective. By the early 20th century, dictionary editors thought it had become obsolete, but it was staging a comeback under their very noses ⟨I was *awoken* by a very persistent lark —Robert Graves (1915)⟩ ⟨with eyes like sparks and his blood *awoken* —John Masefield (1919)⟩. It is likely that *awoken* is the prevailing past participle in British English today ⟨he had *awoken* early for once, rising before dawn —Salman Rushdie⟩. It appears also to be common in spoken American English—it appears in surveys—but we have little printed American evidence as yet. Here *awaked* prevails in print.

1 a : alternately covered and exposed by waves or tide **b :** washing about **:** AFLOAT **c :** covered with water **:** FLOODED
2 : filled, covered, or completely overrun as if by a flood ⟨a movie *awash* in sentimentality⟩

[1]away \ə-ˈwā\ *adverb* (before 12th century)
1 : on the way **:** ALONG ⟨get *away* early⟩
2 : from this or that place **:** HENCE, THENCE ⟨go *away*⟩
3 a : in a secure place or manner ⟨locked *away*⟩ **b :** in another direction ⟨look *away*⟩
4 : out of existence **:** to an end ⟨echoes dying *away*⟩
5 : from one's possession ⟨gave *away* a fortune⟩
6 : steadily onward **:** UNINTERRUPTEDLY ⟨clocks ticking *away*⟩
7 : by a long distance or interval **:** FAR ⟨*away* back in 1910⟩

[2]away *adjective* (14th century)
1 : absent from a place **:** GONE ⟨*away* for the weekend⟩
2 : distant in space or time ⟨a lake 10 miles *away*⟩ ⟨the season is two months *away*⟩
3 : played on an opponent's grounds ⟨home and *away* games⟩
4 *baseball* **:** OUT ⟨two *away* in the ninth⟩
— away·ness *noun*

[1]awe \ˈȯ\ *noun* [Middle English, from Old Norse *agi;* akin to Old English *ege* awe, Greek *achos* pain] (13th century)
1 : an emotion variously combining dread, veneration, and wonder that is inspired by authority or by the sacred or sublime ⟨stood in *awe* of the king⟩ ⟨regard nature's wonders with *awe*⟩
2 *archaic* **a :** DREAD, TERROR **b :** the power to inspire dread

[2]awe *transitive verb* **awed; aw·ing** (13th century)
: to inspire with awe

aweary \ə-ˈwir-ē\ *adjective* (1537)
archaic **:** being weary

aweath·er \ə-ˈwe-t͟hər\ *adverb* (1599)
: on or toward the weather or windward side — compare ALEE

awed \ˈȯd\ *adjective* (1592)
: showing awe ⟨*awed* respect⟩

aweigh \ə-ˈwā\ *adjective* (1670)
: raised just clear of the bottom — used of an anchor

awe·less *or* **aw·less** \ˈȯ-ləs\ *adjective* (14th century)
1 : feeling no awe
2 *obsolete* **:** inspiring no awe

awe·some \ˈȯ-səm\ *adjective* (1598)
1 : expressive of awe ⟨*awesome* tribute⟩
2 a : inspiring awe ⟨an *awesome* task⟩ **b :** TERRIFIC 2, 3; *especially* **:** EXTRAORDINARY
— awe·some·ly *adverb*
— awe·some·ness *noun*

awe·struck \-ˌstrək\ *also* **awe·strick·en** \-ˌstri-kən\ *adjective* (1634)
: filled with awe

[1]aw·ful \ˈȯ-fəl\ *adjective* (13th century)
1 : inspiring awe
2 : filled with awe: as **a** *obsolete* **:** AFRAID, TERRIFIED **b :** deeply respectful or reverential
3 : extremely disagreeable or objectionable
4 : exceedingly great — used as an intensive ⟨an *awful* lot of money⟩ ■
— aw·ful·ly \ˈȯ-f(ə-)lē, *especially as adverb of adjective senses 3 & 4* -flē\ *adverb*
— aw·ful·ness \-fəl-nəs\ *noun*

[2]awful *adverb* (1818)
: VERY, EXTREMELY ⟨*awful* tired⟩

awhile \ə-ˈhwī(ə)l, ə-ˈwī(ə)l\ *adverb* (before 12th century)
: for a while ■

awhirl \ə-ˈhwər(-ə)l, -ˈwər(-ə)l\ *adjective* (1883)
: being in a whirl

awk·ward \ˈȯ-kwərd\ *adjective* [Middle English *awkeward* in the wrong direction, from *awke* turned the wrong way, from Old Norse *ǫfugr;* akin to Old High German *abuh* turned the wrong way] (1530)
1 *obsolete* **:** PERVERSE
2 *archaic* **:** UNFAVORABLE, ADVERSE
3 a : lacking dexterity or skill (as in the use of hands) ⟨*awkward* with a needle and thread⟩ **b :** showing the result of a lack of expertness ⟨*awkward* pictures⟩
4 a : lacking ease or grace (as of movement or expression) **b :** lacking the right proportions, size, or harmony of parts **:** UNGAINLY
5 a : lacking social grace and assurance **b :** causing embarrassment ⟨an *awkward* moment⟩
6 : not easy to handle or deal with **:** requiring great skill, ingenuity, or care ⟨an *awkward* load⟩ ⟨an *awkward* diplomatic situation⟩ ☆
— awk·ward·ly *adverb*
— awk·ward·ness *noun*

awl \ˈȯl\ *noun* [Middle English *al,* from Old Norse *alr;* akin to Old High German *āla* awl, Sanskrit *ārā*] (before 12th century)
: a pointed tool for marking surfaces or piercing small holes (as in leather or wood)

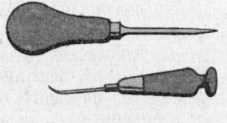

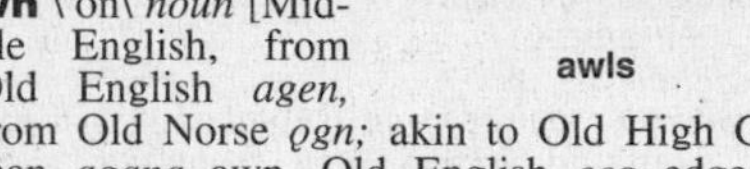
awls

awn \ˈȯn\ *noun* [Middle English, from Old English *agen,* from Old Norse *ǫgn;* akin to Old High German *agana* awn, Old English *ecg* edge — more at EDGE] (12th century)
: one of the slender bristles that terminate the glumes of the spikelet in some cereal and other grasses
— awned \ˈȯnd\ *adjective*
— awn·less \ˈȯn-ləs\ *adjective*

aw·ning \ˈȯ-niŋ, ˈä-niŋ\ *noun* [origin unknown] (1624)
: a rooflike cover extending over or in front of a place (as over the deck or in front of a door or window) as a shelter
— aw·ninged \-niŋd\ *adjective*

awoke *past and past participle of* AWAKE

awoken *past participle of* AWAKE

[1]AWOL \ˈā-ˌwȯl, ˌā ˌdə-bəl-yü-ˌō-ˈel\ *adjective or adverb, sometimes not capitalized* [absent without leave] (1919)
: ABSENT WITHOUT LEAVE; *broadly* **:** absent often without notice or permission ⟨the place looked as if its caretaker had been *AWOL* for some time —Daniel Ford⟩

[2]AWOL *noun, sometimes not capitalized* (1919)
: a person who is AWOL

awry \ə-ˈrī\ *adverb or adjective* (14th century)
1 : in a turned or twisted position or direction **:** ASKEW
2 : off the correct or expected course **:** AMISS

aw–shucks \ˈȯ-ˌshəks\ *adjective* (1951)
: being or marked by an unsophisticated, self-conscious, or self-effacing manner

[1]ax *or* **axe** \ˈaks\ *noun* [Middle English, from Old English *æcs;* akin to Old High German *ackus* ax, Latin *ascia,* Greek *axinē*] (before 12th century)
1 : a cutting tool that consists of a heavy edged head fixed to a handle with the edge parallel to the handle and that is used especially for felling trees and chopping and splitting wood
2 : a hammer with a sharp edge for dressing or spalling stone
3 : abrupt removal (as from employment or from a budget) — sometimes used in the phrase *get the ax*
4 : a musical instrument (as a guitar or a saxophone)
— ax to grind : an ulterior often selfish purpose to further

[2]ax *or* **axe** *transitive verb* **axed; ax·ing** (1679)
1 a : to shape, dress, or trim with an ax **b :** to chop, split, or sever with an ax
2 : to remove abruptly (as from employment or from a budget)

ax·el \ˈak-səl, ˈäk-\ *noun, often capitalized* [*Axel* Paulsen (died 1938) Norwegian figure skater] (1930)
: a jump in figure skating from the outer forward edge of one skate with 1½ turns taken in the air and a return to the outer backward edge of the other skate

axe·nic \(ˌ)ā-ˈze-nik, -ˈzē-\ *adjective* [[2]*a-* + Greek *xenos* strange] (1942)
: free from other living organisms
— axe·ni·cal·ly \-ni-k(ə-)lē\ *adverb*

ax·i·al \ˈak-sē-əl\ *adjective* (circa 1847)
1 : of, relating to, or having the characteristics of an axis
2 a : situated around, in the direction of, on, or along an axis **b :** extending in a direction essentially perpendicular to the plane of a cyclic structure (as of cyclohexane) ⟨*axial* hydrogens⟩ — compare EQUATORIAL
— ax·i·al·i·ty \ˌak-sē-ˈa-lə-tē\ *noun*
— ax·i·al·ly \ˈak-sē-ə-lē\ *adverb*

axial skeleton *noun* (1872)
: the skeleton of the trunk and head

ax·il \ˈak-səl, -ˌsil\ *noun* [New Latin *axilla,* from Latin] (1794)

☆ SYNONYMS
Awkward, clumsy, maladroit, inept, gauche mean not marked by ease (as of performance, movement, or social conduct). AWKWARD is widely applicable and may suggest unhandiness, inconvenience, lack of muscular control, embarrassment, or lack of tact ⟨periods of *awkward* silence⟩. CLUMSY implies stiffness and heaviness and so may connote inflexibility, unwieldiness, or lack of ordinary skill ⟨a *clumsy* mechanic⟩. MALADROIT suggests a tendency to create awkward situations ⟨a *maladroit* politician⟩. INEPT often implies complete failure or inadequacy ⟨a hopelessly *inept* defense attorney⟩. GAUCHE implies the effects of shyness, inexperience, or ill breeding ⟨felt *gauche* and unsophisticated at formal parties⟩.

□ USAGE
awful Many grammarians take issue with the senses of *awful* and *awfully* that do not convey the etymological connection with *awe.* However, senses 3 and 4 of the adjective were used in speech and casual writing by the late 18th century ⟨it is an *awful* while since you have heard from me —John Keats (letter)⟩ ⟨there was an *awful* crowd —Sir Walter Scott (letter)⟩ ⟨this is an *awful* thing to say to oil painters —William Blake⟩. Adverbial use of *awful* as an intensifier began to appear in print in the early 19th century, as did the senses of *awfully* corresponding to senses 3 and 4 of the adjective. Both adverbs remain in widespread use ⟨a sad state of affairs and *awful* tough on art —H. L. Mencken⟩ ⟨the *awfully* rich young American —Henry James⟩ ⟨decided to play it so *awfully* safe —A. M. Schlesinger (born 1917)⟩.

awhile Although considered a solecism by many commentators, *awhile,* like several other adverbs of time and place, is often used as the object of a preposition ⟨for *awhile* there is a silence —Lord Dunsany⟩.

: the angle between a branch or leaf and the axis from which it arises

ax·il·la \ag-ˈzi-lə, ak-ˈsi-\ *noun, plural* **-lae** \-(ˌ)lē, -ˌlī\ *or* **-las** [Latin, diminutive of *ala* wing, upper arm, armpit, axil — more at AISLE] (1616)
: the cavity beneath the junction of a forelimb and the body; *especially* : ARMPIT

ax·il·lar \ag-ˈzi-lər, ak-ˈsi-, ˈag-zə-, ˈak-sə-, -ˌlär\ *noun* (1541)
: an axillary part (as a feather)

¹ax·il·lary \ˈak-sə-ˌler-ē\ *adjective* (1615)
1 : of, relating to, or located near the axilla
2 : situated in or growing from an axil ⟨*axillary* buds⟩

²axillary *noun, plural* **-lar·ies** (circa 1889)
: one of the feathers arising from the axilla and closing the space between the flight feathers and body of a flying bird

ax·i·ol·o·gy \ˌak-sē-ˈä-lə-jē\ *noun* [Greek *axios* + International Scientific Vocabulary *-logy*] (1908)
: the study of the nature, types, and criteria of values and of value judgments especially in ethics
— **ax·i·o·log·i·cal** \ˌak-sē-ə-ˌlä-ji-kəl\ *adjective*
— **ax·i·o·log·i·cal·ly** \-ji-k(ə-)lē\ *adverb*

ax·i·om \ˈak-sē-əm\ *noun* [Latin *axioma,* from Greek *axiōma,* literally, something worthy, from *axioun* to think worthy, from *axios* worth, worthy; akin to Greek *agein* to weigh, drive — more at AGENT] (15th century)
1 : a maxim widely accepted on its intrinsic merit
2 : a statement accepted as true as the basis for argument or inference : POSTULATE 1
3 : an established rule or principle or a self-evident truth

ax·i·om·at·ic \ˌak-sē-ə-ˈma-tik\ *adjective* [Middle Greek *axiōmatikos,* from Greek, honorable, from *axiōmat-, axiōma*] (1797)
1 : taken for granted : SELF-EVIDENT
2 : based on or involving an axiom or system of axioms ⟨*axiomatic* set theory⟩
— **ax·i·om·at·i·cal·ly** \-ti-k(ə-)lē\ *adverb*

ax·i·omat·isa·tion *chiefly British variant of* AXIOMATIZATION

ax·i·omat·iza·tion \ˌak-sē-ə-ˌma-tə-ˈzā-shən, -sē-ˌä-mə-tə-\ *noun* (1931)
: the act or process of reducing to a system of axioms
— **ax·i·om·a·tize** \-sē-ˈä-mə-ˌtīz\ *transitive verb*

axiom of choice (1942)
: an axiom in set theory that is equivalent to Zorn's lemma: for every collection of nonempty sets there is a function which chooses an element from each set

ax·ion \ˈak-sē-(ˌ)än\ *noun* [*axi*al + ²*-on*] (1978)
: a hypothetical subatomic particle of low mass and energy that is postulated to exist because of certain properties of the strong force

ax·is \ˈak-səs\ *noun, plural* **ax·es** \-ˌsēz\ [Latin, axis, axle; akin to Old English *eax* axis, axle, Greek *axōn,* Lithuanian *ašis,* Sanskrit *akṣa*] (14th century)
1 a : a straight line about which a body or a geometric figure rotates or may be supposed to rotate **b :** a straight line with respect to which a body or figure is symmetrical — called also *axis of symmetry* **c :** a straight line that bisects at right angles a system of parallel chords of a curve and divides the curve into two symmetrical parts **d :** one of the reference lines of a coordinate system
2 a : the second vertebra of the neck on which the head and first vertebra turn as on a pivot **b :** any of various central, fundamental, or axial parts
3 : a plant stem
4 : one of several imaginary lines assumed in describing the positions of the planes by which a crystal is bounded and the positions of atoms in the structure of the crystal
5 : a main line of direction, motion, growth, or extension
6 a : an implied line in painting or sculpture through a composition to which elements in the composition are referred **b :** a line actually drawn and used as the basis of measurements in an architectural or other working drawing
7 : any of three fixed lines of reference in an aircraft that run in the longitudinal, lateral, and vertical directions, are mutually perpendicular, and usually pass through the aircraft's center of gravity
8 : PARTNERSHIP, ALLIANCE

Axis *adjective* (1938)
: of or relating to the three powers Germany, Italy, and Japan engaged against the Allied nations in World War II

axi·sym·met·ric \ˌak-si-sə-ˈme-trik\ *also* **axi·sym·met·ri·cal** \-tri-kəl\ *adjective* [*axis* + *symmetric*] (1893)
: symmetric in respect to an axis
— **axi·sym·me·try** \-ˈsi-mə-trē\ *noun*

ax·le \ˈak-səl\ *noun* [Middle English *axel-* (as in *axeltre*)] (14th century)
1 a : a pin or shaft on or with which a wheel or pair of wheels revolves **b** (1) **:** a fixed bar or beam with bearings at its ends on which wheels (as of a cart) revolve (2) **:** the spindle of such a beam
2 *archaic* **:** AXIS

axle·tree \-(ˌ)trē\ *noun* [Middle English *axeltre,* from Old Norse *ǫxultrē,* from *ǫxull* axle + *trē* tree] (14th century)
: AXLE 1b(1)

ax·man \ˈak-smən\ *noun* (1671)
: one who wields an ax

Ax·min·ster \ˈak-ˌsmin(t)-stər\ *noun* [*Axminster,* town in England] (1818)
: a machine-woven carpet with pile tufts inserted mechanically in a variety of textures and patterns

ax·o·lotl \ˈak-sə-ˌlä-tᵊl\ *noun* [Nahuatl *āxōlōtl*] (circa 1768)
: any of several salamanders (genus *Ambystoma* especially *A. mexicanum* and *A. tigrinum*) of mountain lakes of Mexico and the western U.S. that ordinarily live and breed without metamorphosing

ax·on \ˈak-ˌsän\ *also* **ax·one** \-ˌsōn\ *noun* [New Latin *axon,* from Greek *axōn*] (circa 1899)
: a usually long and single nerve-cell process that usually conducts impulses away from the cell body
— **ax·o·nal** \ˈak-sə-nᵊl; ak-ˈsä-, -ˈsō-\ *adjective*

ax·o·neme \ˈak-sə-ˌnēm\ *noun* [Greek *axōn* axis + *nēma* thread, from *nēn* to spin — more at NEEDLE] (1901)
: the fibrillar bundle of a flagellum or cilium that usually consists of nine pairs of microtubules arranged in a ring around a single central pair
— **ax·o·ne·mal** \ˌak-sə-ˈnē-məl\ *adjective*

ax·o·no·met·ric \ˌak-sə-nō-ˈme-trik\ *adjective* [Greek *axōn* axis + English *-metric*] (1908)
: being or prepared by the projection of objects on the drawing surface so that they appear inclined with three sides showing and with horizontal and vertical distances drawn to scale but diagonal and curved lines distorted ⟨an *axonometric* drawing⟩

axo·plasm \ˈak-sə-ˌpla-zəm\ *noun* [*axon* + *-plasm*] (1900)
: the protoplasm of an axon
— **axo·plas·mic** \ˌak-sə-ˈplaz-mik\ *adjective*

ay \ˈī\ *interjection* [Middle French *aymi* ay me] (14th century)
— usually used with following *me* to express sorrow or regret

ayah \ˈī-ə; ˈä-yə, -(ˌ)yä\ *noun* [Hindi *āyā,* from Portuguese *aia,* from Latin *avia* grandmother] (1779)
: a nurse or maid native to India

aya·hua·sca \ˌī-yə-ˈhwäs-kə, -ˈwäs-\ *noun* [American Spanish] (1949)
: an hallucinogenic beverage prepared from the bark of a South American vine (*Banisteriopsis caapi* of the family Malpighiaceae)

aya·tol·lah \ˌī-ə-ˈtō-lə, -ˈtä-, -ˈtə-, ˈī-ə-ˌ\ *noun* [Persian, literally, sign of God, from Arabic *āya* sign, miracle + *allāh* God] (1953)
: a religious leader among Shiite Muslims — used as a title of respect especially for one who is not an imam

¹aye *also* **ay** \ˈā\ *adverb* [Middle English, from Old Norse *ei;* akin to Old English *ā* always, Latin *aevum* age, lifetime, Greek *aiōn* age] (13th century)
: ALWAYS, CONTINUALLY, EVER ⟨love that will *aye* endure —W. S. Gilbert⟩

²aye *also* **ay** \ˈī\ *adverb* [perhaps from Middle English *ye, yie* — more at YEA] (1576)
: YES ⟨*aye, aye,* sir⟩

³aye *also* **ay** \ˈī\ *noun, plural* **ayes** (1589)
: an affirmative vote or voter ⟨the *ayes* have it⟩

aye–aye \ˈī-ˌī\ *noun* [French, from Malagasy *aiay*] (circa 1781)
: a small primitive nocturnal forest-dwelling primate (*Daubentonia madagascariensis*) of northern Madagascar that has a round head, large eyes and ears, and long thin fingers

ayin \ˈī-ən\ *noun* [Hebrew *ʽayin,* literally, eye] (1823)
: the 16th letter of the Hebrew alphabet — see ALPHABET table

Ay·ma·ra \ˌī-mə-ˈrä\ *noun, plural* **Aymara** *or* **Aymaras** [Spanish *aymará*] (1860)
1 : a member of an Indian people of Bolivia and Peru
2 : the language of the Aymara people

Ayr·shire \ˈar-ˌshir, ˈer-, -shər; ˈash-ˌir\ *noun* [*Ayrshire,* Scotland] (1856)
: any of a breed of hardy dairy cattle originated in Ayr and usually marked with blotches of red or brown with white

az- *or* **azo-** *combining form* [International Scientific Vocabulary, from French *azote* nitrogen, from *a-* ²a- + *-zote,* probably from Greek *zōtikos* maintaining life, from *zōē* life — more at QUICK]
: containing nitrogen especially as the bivalent group N=N ⟨*azine*⟩

aza- *or* **az-** *combining form* [International Scientific Vocabulary *az-* + *-a-*]
: containing nitrogen in place of carbon and usually the bivalent group NH for the group CH_2 or a single trivalent nitrogen atom for the group CH ⟨*aza*thioprine⟩

aza·lea \ə-ˈzāl-yə\ *noun* [New Latin, genus name, from Greek, feminine of *azaleos* dry, from *azein* to parch, dry; akin to Hittite *ḫat-* to dry up and probably to Latin *ador* emmer] (1767)
: any of a subgenus (*Azalea*) of rhododendrons with funnel-shaped corollas and usually deciduous leaves including many species and hybrid forms cultivated as ornamentals

aza·thi·o·prine \ˌa-zə-ˈthī-ə-ˌprēn\ *noun* [*aza-* + *thi-* + *purine*] (1962)
: a purine antimetabolite $C_9H_7N_7O_2S$ used especially as an immunosuppressant

Aza·zel \ə-ˈzā-zəl, ˈa-zə-ˌzel\ *noun* [Hebrew *ʽăzāzēl*]
: an evil spirit of the wilderness to which a scapegoat was sent by the ancient Hebrews in a ritual of atonement

azeo·trope \ˈā-zē-ə-ˌtrōp\ *noun* [²*a-* + *zeo-* (from Greek *zein* to boil) + *-trope* something changed, from Greek *tropos* turn — more at YEAST, TROPE] (1938)
: a liquid mixture that is characterized by a constant minimum or maximum boiling point which is lower or higher than that of any of the components

azide \ˈā-ˌzīd, ˈa-\ *noun* (circa 1904)
: a compound containing the group N_3 combined with an element or radical

az·i·do \ˈa-zə-(ˌ)dō\ *adjective* [International Scientific Vocabulary *azide* + *-o-*] (circa 1926)

: relating to or containing the univalent group N_3 — often used in combination

az·i·do·thy·mi·dine \ə-ˌzi-dō-ˈthī-mə-ˌdēn\ *noun* (1974)
: an antiviral drug $C_{10}H_{13}N_5O_4$ that inhibits replication of some retroviruses (as HIV) and is used to treat AIDS — called also *AZT, zidovudine*

az·i·muth \ˈaz-məth, ˈa-zə-\ *noun* [Middle English, from (assumed) Medieval Latin, from Arabic *as-sumūt* the azimuth, plural of *as-samt* the way] (14th century)
1 : an arc of the horizon measured between a fixed point (as true north) and the vertical circle passing through the center of an object usually in astronomy and navigation clockwise from the north point through 360 degrees **2 :** horizontal direction expressed as the angular distance between the direction of a fixed point (as the observer's heading) and the direction of the object
— **az·i·muth·al** \ˌa-zə-ˈmə-thəl\ *adjective*
— **az·i·muth·al·ly** \-ˈmə-thə-lē\ *adverb*

azimuthal equidistant projection *noun* (1942)
: a map projection of the surface of the earth so centered at any given point that a straight line radiating from the center to any other point represents the shortest distance and can be measured to scale

azimuthal equidistant projection, centered on Washington, D.C.: *1* **London,** *2* **Algiers,** *3* **Moscow,** *4* **Rio de Janeiro,** *5* **Tokyo,** *6* **Auckland**

azine \ˈā-ˌzēn, ˈa-\ *noun* (1887)
: a compound of the general formula $RCH{=}NH{=}CHR$ or $R_2C{=}NN{=}CR_2$ formed by the action of hydrazine on aldehydes or ketones

azo \ˈā-(ˌ)zō, ˈa-\ *adjective* [*az-*] (circa 1879)
: relating to or containing the bivalent group $N{=}N$ united at both ends to carbon

azo dye *noun* (1884)
: any of numerous dyes containing azo groups

azo·ic \(ˌ)ā-ˈzō-ik\ *adjective* [²*a-* + Greek *zōē* life — more at QUICK] (circa 1847)
: having no life; *especially* **:** of or relating to the part of geologic time that antedates life — compare ARCHEOZOIC

azole \ˈā-ˌzōl, ˈa-\ *noun* (circa 1899)
: any of numerous compounds characterized by a 5-membered ring containing at least one nitrogen atom

azon·al \(ˌ)ā-ˈzō-nᵊl\ *adjective* (1938)
: of, relating to, or being a soil or a major group of soils lacking well-developed horizons often because of immaturity — compare INTRAZONAL, ZONAL

azo·o·sper·mia \(ˌ)ā-ˌzō-ə-ˈspər-mē-ə\ *noun* [New Latin, from Greek *azoos* lifeless (from *a-* ²a- + *zōē* life) + *sperma* semen, seed — more at SPERM] (circa 1881)
: absence of spermatozoa from the seminal fluid

azo·te·mia \ˌā-zō-ˈtē-mē-ə\ *noun* [International Scientific Vocabulary *azote* nitrogen + New Latin *-emia* — more at AZ-] (circa 1900)
: an excess of nitrogenous bodies in the blood as a result of kidney insufficiency — compare UREMIA
— **azo·te·mic** \-mik\ *adjective*

az·oth \ˈa-ˌzȯth\ *noun* [Arabic *az-zāʼūq* the mercury] (15th century)
1 : mercury regarded by alchemists as the first principle of metals
2 : the universal remedy of Paracelsus

azo·to·bac·ter \ā-ˈzō-tə-ˌbak-tər\ *noun* [New Latin, genus name, from International Scientific Vocabulary *azote* + New Latin *bacterium*] (1910)
: any of a genus (*Azotobacter*) of large rod-shaped or spherical bacteria occurring in soil and sewage and fixing atmospheric nitrogen

azo·tu·ria \ˌā-zō-ˈtu̇r-ē-ə, -ˈtyu̇r-\ *noun* [International Scientific Vocabulary *azote* + New Latin *-uria*] (circa 1838)
: an abnormal condition of horses characterized by an excess of urea or other nitrogenous substances in the urine and muscle damage especially to the hindquarters

AZT \ˌā-(ˌ)zē-ˈtē\ *noun* (1985)
: AZIDOTHYMIDINE

Az·tec \ˈaz-ˌtek\ *noun* [Spanish *azteca,* from Nahuatl *aztēcah,* plural of *aztēcatl*] (1787)
1 a : a member of a Nahuatl-speaking people that founded the Mexican empire conquered by Cortes in 1519 **b :** a member of any people under Aztec influence
2 : NAHUATL
— **Az·tec·an** \ˈaz-ˌte-kən, ˌaz-ˈte-\ *adjective*

azure \ˈa-zhər\ *noun* [Middle English *asur,* from Old French *azur,* probably from Old Spanish, modification of Arabic *lāzaward,* from Persian *lāzhuward*] (14th century)
1 *archaic* **:** LAPIS LAZULI
2 a : the blue color of the clear sky **b :** the heraldic color blue
3 : the unclouded sky
— **azure** *adjective*

azur·ite \ˈa-zhə-ˌrīt\ *noun* [French, from *azur* azure] (circa 1868)
1 : a mineral that consists of blue basic carbonate of copper and is a copper ore
2 : a semiprecious stone derived from azurite

¹azy·gos \ā-ˈzī-gəs\ *noun* [New Latin, from Greek, unyoked, from *a-* + *zygon* yoke — more at YOKE] (1646)
: an azygos anatomical part

²azy·gos *also* **azy·gous** \(ˌ)ā-ˈzī-gəs\ *adjective* (1681)
: not being one of a pair **:** SINGLE ⟨an *azygos* vein⟩

B *is the second letter of the English alphabet and of most alphabets that are closely related to English. It came into these alphabets from the Latin, into the Latin (by way of Etruscan) from the Greek, and into the Greek from the Phoenician, where it was called* beth. B *usually represents the voiced bilabial stop, the consonant sound pronounced at the beginning of the English word* banana; *in a few English words it is silent, as in* debt *and* tomb. *In related languages English* b *corresponds to German* b (Bruder *"*brother*"), Latin* f (frater *"*brother,*"* fero *"I bear"), Greek* ph (phero *"I bear"), and Sanskrit* bh (bharati *"he bears"). The small* b *developed from the capital form through the shrinking and eventual disappearance of the upper lobe, together with a reduction in the overall size of the letter.*

b \'bē\ *noun, plural* **b's** *or* **bs** \'bēz\ *often capitalized, often attributive* (before 12th century)
1 a : the 2d letter of the English alphabet **b :** a graphic representation of this letter **c :** a speech counterpart of orthographic *b*
2 : the 7th tone of a C-major scale
3 : a graphic device for reproducing the letter *b*
4 : one designated *b* especially as the 2d in order or class
5 a : a grade rating a student's work as good but short of excellent **b :** one graded or rated with a B
6 : something shaped like the letter B

baa \'ba, 'bä\ *intransitive verb* **baaed; baa·ing** [imitative] (circa 1586)
: to make the bleat of a sheep
— **baa** *noun*

baal \'bā(-ə)l\ *noun, plural* **baals** *or* **baa·lim** \'bā-(ə-)ləm, 'bā-ə-ˌlim\ *often capitalized* [Hebrew *ba'al* lord] (14th century)
: any of numerous Canaanite and Phoenician local deities
— **baal·ism** \'bā-(ə-)ˌli-zəm\ *noun, often capitalized*

baas \'bäs\ *noun* [Afrikaans, from Dutch] (1785)
South African **:** BOSS, MASTER — used especially by nonwhites when speaking to or about Europeans in positions of authority

ba·ba \'bä-(ˌ)bä, -bə\ *noun* [French, from Polish, literally, old woman] (1827)
: a rich cake soaked in a rum and sugar syrup

ba·bas·su \ˌbä-bə-'sü\ *noun* [Portuguese *babaçú,* from (assumed) Tupi *iwaβasu,* from Tupi *iwa* fruit + *βasu* large] (1917)
: a tall pinnate-leaved palm (*Orbignya barbosiana*) of Brazil with hard-shelled nuts yielding a valuable oil

bab·bitt \'ba-bət\ *noun* [Isaac *Babbitt* (died 1862) American inventor] (1900)
: an alloy used for lining bearings; *especially* **:** one containing tin, copper, and antimony — called also *babbitt metal*
— **bab·bitt·ed** \-bə-təd\ *adjective*

Bab·bitt \'ba-bət\ *noun* [George F. *Babbitt,* character in the novel *Babbitt* (1922) by Sinclair Lewis] (1923)
: a business or professional man who conforms unthinkingly to prevailing middle-class standards ◆
— **Bab·bitt·ry** \-bə-trē\ *noun*
— **Bab·itty** \-tē\ *adjective*

bab·ble \'ba-bəl\ *verb* **bab·bled; bab·bling** \-b(ə-)liŋ\ [Middle English *babelen,* probably of imitative origin] (13th century)
intransitive verb
1 a : to talk enthusiastically or excessively **b :** to utter meaningless or unintelligible sounds
2 : to make sounds as though babbling
transitive verb
1 : to utter in an incoherently or meaninglessly repetitious manner
2 : to reveal by talk that is too free
— **babble** *noun*
— **bab·ble·ment** \-bəl-mənt\ *noun*
— **bab·bler** \-b(ə-)lər\ *noun*

babe \'bāb\ *noun* [Middle English, probably of imitative origin] (14th century)
1 a : INFANT, BABY **b** *slang* **:** GIRL, WOMAN
2 : a naive inexperienced person — used especially in the phrase *babe in the woods*

Ba·bel \'bā-bəl, 'ba-\ *noun* [Hebrew *Bābhel,* from Akkadian *bāb-ilu* gate of god]
1 : a city in Shinar where the building of a tower is held in Genesis to have been halted by the confusion of tongues
2 *often not capitalized* **a :** a confusion of sounds or voices **b :** a scene of noise or confusion

ba·be·sia \bə-'bē-zh(ē-)ə\ *noun* [New Latin, from Victor *Babeş* (died 1926) Romanian bacteriologist] (1911)
: any of a genus (*Babesia*) of sporozoans parasitic in mammalian red blood cells (as in Texas fever) and transmitted by the bite of a tick — called also *piroplasm*

bab·e·si·o·sis \ˌba-bə-'zī-ə-səs\ *noun* [New Latin] (1911)
: an infection with or disease caused by babesias

Ba·bin·ski reflex \bə-'bin-skē-\ *noun* [J.F.F. *Babinski* (died 1932) French neurologist] (1900)
: a reflex movement in which when the sole is tickled the big toe turns upward instead of downward and which is normal in infancy but indicates damage to the central nervous system (as in the pyramidal tracts) later in life — called also *Babinski sign, Ba·bin·ski's reflex* \-skēz-\

bab·i·ru·sa \ˌba-bə-'rü-sə, ˌbä-\ *noun* [Malay, from *babi* pig + *rusa* deer] (1673)
: a large wild swine (*Babyrousa babyrussa*) of Indonesia

ba·boon \ba-'bün, *chiefly British* bə-\ *noun* [Middle English *babewin,* from Middle French *babouin,* from *baboue* grimace] (15th century)
: any of a genus (*Papio*) of large gregarious primates of Africa and southwestern Asia having a long square naked muzzle; *also* **:** any of several closely related primates

ba·bu *also* **ba·boo** \'bä-(ˌ)bü\ *noun, often attributive* [Hindi *bābū,* literally, father] (1776)
1 : a Hindu gentleman — a form of address corresponding to *Mr.*
2 a : an Indian clerk who writes English **b :** an Indian having some education in English — often used disparagingly

ba·bul \bə-'bül\ *noun* [Persian *babūl*] (1780)
: an acacia tree (*Acacia nilotica* synonym *A. arabica*) widespread in India and northern Africa that yields gum arabic and tannins as well as fodder and timber

ba·bush·ka \bə-'büsh-kə, -'bu̇sh-\ *noun* [Russian, grandmother, diminutive of *baba* old woman] (1938)
1 a : a usually triangularly folded kerchief for the head **b :** a head covering resembling a babushka
2 : an elderly Russian woman

babushka 1a

[1]ba·by \'bā-bē\ *noun, plural* **babies** [Middle English, from *babe*] (14th century)
1 a (1) **:** an extremely young child; *especially* **:** INFANT (2) **:** an extremely young animal **b :** the youngest of a group
2 a : one that is like a baby (as in behavior) **b :** something that is one's special responsibility, achievement, or interest
3 *slang* **a :** GIRL, WOMAN — often used in address **b :** BOY, MAN — often used in address
4 : PERSON, THING ⟨is one tough *baby*⟩
— **ba·by·hood** \-bē-ˌhu̇d\ *noun*
— **ba·by·ish** \-ish\ *adjective*

[2]baby *adjective* (1591)
1 : of, relating to, or being a baby
2 : much smaller than the usual ⟨*baby* carrots⟩ ⟨a *baby* flattop⟩ ⟨take two *baby* steps⟩

[3]baby *transitive verb* **ba·bied; ba·by·ing** (1742)
1 : to tend to indulge with often excessive or inappropriate care and solicitude ⟨*babying* their only child⟩
2 : to operate or treat with care ⟨*baby* a new motor⟩ ⟨*babying* a sore knee⟩
synonym see INDULGE

baby blue *noun* (1889)
: a pale blue

baby blue–eyes \-'blü-ˌīz\ *noun plural but singular or plural in construction* (1887)
: a delicate blue-flowered California herb (*Nemophila menziesii*) of the waterleaf family

baby boom *noun* (1941)
: a marked rise in birthrate (as in the U.S. immediately following the end of World War II)
— **baby boomer** *noun*

baby bust *noun* (1971)
: a marked decline in birthrate
— **baby buster** *noun*

baby carriage *noun* (1866)
: a small four-wheeled carriage often with a folding top for pushing a baby around in — called also *baby buggy*

baby grand *noun* (circa 1903)
: a small grand piano

Bab·y·lon \'ba-bə-ˌlän, -lən\ *noun* [*Babylon,* ancient city of Babylonia] (14th century)
: a city devoted to materialism and sensual pleasure

◇ WORD HISTORY
Babbitt *Babbitt* and *Babbittry* are words that sum up all the values, attitudes, and mores associated with the American middle class during the 1920's. Both derive from the protagonist of *Babbitt,* a satirical novel by Sinclair Lewis published in 1922. George F. Babbitt epitomizes the unimaginative and self-important businessmen that Lewis found typical of the provincial cities and towns of Middle America. A real estate agent by profession and a conformist by nature, Babbitt is a compulsive booster and joiner in his native Zenith, "The Zip City." Despite his evident prosperity and status in the community, he remains vaguely dissatisfied with his life. Although he makes tentative attempts at rebellion through brief encounters with artists, bohemians, and socialists and by having an extramarital affair, Babbitt finds his need for social acceptance greater than his desire for escape and he ends as he began.

[1]**Bab·y·lo·nian** \ˌba-bə-ˈlō-nyən, -nē-ən\ *noun* (1564)

1 : a native or inhabitant of ancient Babylonia or Babylon

2 : the form of the Akkadian language used in ancient Babylonia

[2]**Babylonian** *adjective* (1596)

1 : of, relating to, or characteristic of Babylonia or Babylon, the Babylonians, or Babylonian

2 : marked by luxury, extravagance, or the pursuit of sensual pleasure ⟨the *Babylonian* halls of the big hotel —G. K. Chesterton⟩ ⟨the *Babylonian* delights of the city⟩

baby's breath *noun* (circa 1890)

: GYPSOPHILA; *especially* **:** a perennial herb (*Gypsophila paniculata*) or an annual herb (*G. elegans*) commonly used in floral arrangements

ba·by–sit \ˈbā-bē-ˌsit\ *verb* **-sat** \-ˌsat\; **-sit·ting** [back-formation from *baby-sitter*] (1947)

intransitive verb

: to care for children usually during a short absence of the parents; *broadly* **:** to give care ⟨*baby-sit* for a neighbor's pets⟩

transitive verb

: to baby-sit for; *broadly* **:** MIND, TEND ⟨*baby-sit* house plants⟩ ⟨police *baby-sitting* a witness⟩

— ba·by–sit·ter *noun*

baby talk *noun* (1836)

1 a : the syntactically imperfect speech or phonetically modified forms used by small children learning to talk **b :** the consciously imperfect or altered speech used by adults in speaking to small children

2 : oversimplified speech or writing

baby tooth *noun* (1939)

: MILK TOOTH

bac·ca·lau·re·ate \ˌba-kə-ˈlȯr-ē-ət, -ˈlär-\ *noun* [Medieval Latin *baccalaureatus,* from *baccalaureus* bachelor, alteration of *baccalarius*] (circa 1649)

1 : the degree of bachelor conferred by universities and colleges

2 a : a sermon to a graduating class **b :** the service at which this sermon is delivered

bac·ca·rat \ˌbä-kə-ˈrä, ˌba-\ *noun* [French *baccara*] (1865)

: a card game resembling chemin de fer in which three hands are dealt and players may bet either or both hands against the dealer's; *also* **:** a two-handed version in which players may bet on or against the dealer

Bac·chae \ˈba-ˌkē, -ˌkī\ *noun plural* [Latin, from Greek *Bakchai,* from *Bakchos* Bacchus] (circa 1909)

1 : the female attendants or priestesses of Bacchus

2 : the women participating in the Bacchanalia

[1]**bac·cha·nal** \ˈba-kə-nᵊl, ˈbä-; ˌba-kə-ˈnal, ˌbä-kə-ˈnäl\ *noun* [Latin, shrine of Bacchus, probably back-formation from *Bacchanalia*] (1550)

1 : ORGY 2, 3

2 a : a devotee of Bacchus; *especially* **:** one who celebrates the Bacchanalia **b :** REVELER

[2]**bac·cha·nal** \ˈba-kə-nᵊl\ *adjective* (1550)

: of, relating to, or suggestive of the Bacchanalia **:** BACCHANALIAN

bac·cha·na·lia \ˌba-kə-ˈnāl-yə, ˈbä-\ *noun, plural* **bacchanalia** [Latin, from *Bacchus*] (1591)

1 *plural, capitalized* **:** a Roman festival of Bacchus celebrated with dancing, song, and revelry

2 : ORGY 2, 3

— bac·cha·na·lian \-ˈnāl-yən\ *adjective or noun*

bac·chant \bə-ˈkant, -ˈkänt; ˈba-kənt, ˈbä-\ *noun, plural* **bacchants** *or* **bacchantes** \bə-ˈkants, -ˈkänts, -ˈkan-tēz, -ˈkän-tēz\ [Latin *bacchant-, bacchans,* from present participle of *bacchari* to take part in the orgies of Bacchus] (1699)

: BACCHANAL

— bacchant *adjective*

bac·chante \bə-ˈkant, -ˈkänt; -ˈkan-tē, -ˈkän-\ *noun* [French, from Latin *bacchant-, bacchans*] (1579)

: a priestess or female follower of Bacchus **:** MAENAD

bac·chic \ˈba-kik, ˈbä-\ *adjective, often capitalized* (1669)

: of, relating to, or suggestive of Bacchus or the Bacchanalia **:** BACCHANALIAN

Bac·chus \ˈba-kəs, ˈbä-\ *noun* [Latin, from Greek *Bakchos*]

: the Greek god of wine — called also *Dionysus*

[1]**bach** \ˈbach\ *noun* (1855)

: BACHELOR 3a

[2]**bach** *intransitive verb* (1870)

: to live as a bachelor — often used with *it*

[3]**bach** *noun* [[2]*bach*] (1925)

New Zealand **:** a small house or weekend cottage

[1]**bach·e·lor** \ˈbach-lər, ˈba-chə-\ *noun* [Middle English *bacheler,* from Old French] (14th century)

1 : a young knight who follows the banner of another

2 : a person who has received what is usually the lowest degree conferred by a four-year college, university, or professional school ⟨*bachelor* of arts⟩

3 a : an unmarried man **b :** a male animal (as a fur seal) without a mate during breeding time

— bach·e·lor·dom \-dəm\ *noun*

— bach·e·lor·hood \-ˌhu̇d\ *noun*

[2]**bachelor** *adjective* (1857)

1 : suitable for or occupied by a single person ⟨a *bachelor* apartment⟩

2 : UNMARRIED ⟨*bachelor* women⟩ ⟨*bachelor* parents⟩

bach·e·lor·ette \ˌbach-lə-ˈret, ˌba-chə-\ *noun* (1938)

: a bachelor woman

bachelor's button *noun* (1847)

: a European composite (*Centaurea cyanus*) having flower heads with blue, pink, or white rays that is often cultivated in North America — called also *cornflower*

ba·cil·la·ry \ˈba-sə-ˌler-ē, bə-ˈsi-lə-rē\ *or* **ba·cil·lar** \bə-ˈsi lər, ˈba-sə-lər\ *adjective* [Medieval Latin & New Latin *bacillus*] (1865)

1 : shaped like a rod; *also* **:** consisting of small rods

2 : of, relating to, or caused by bacilli

ba·cil·lus \bə-ˈsi-ləs\ *noun, plural* **-li** \-ˌlī *also* -lē\ [New Latin, from Medieval Latin, small staff, rod, diminutive of Latin *baculus* staff, alteration of *baculum*] (circa 1879)

1 : any of a genus (*Bacillus*) of aerobic rod-shaped gram-positive bacteria producing endospores that do not thicken the rod and including many saprophytes and some parasites (as *B. anthracis* of anthrax); *broadly* **:** a straight rod-shaped bacterium

2 : BACTERIUM; *especially* **:** a disease-producing bacterium

bac·i·tra·cin \ˌba-sə-ˈtrā-sᵊn\ *noun* [New Latin *Bacillus subtilis* (species of bacillus producing the toxin) + Margaret *Tracy* (born about 1936) American child in whose tissues it was found] (1945)

: a toxic polypeptide antibiotic isolated from a bacillus (*Bacillus subtilis*) and usually used topically especially against gram-positive bacteria

[1]**back** \ˈbak\ *noun* [Middle English, from Old English *bæc;* akin to Old High German *bah* back] (before 12th century)

1 a (1) **:** the rear part of the human body especially from the neck to the end of the spine (2) **:** the body considered as the wearer of clothes (3) **:** capacity for labor, effort, or endurance (4) **:** the back considered as the seat of one's awareness of duty or failings ⟨get off my *back*⟩ **b :** the part of a lower animal (as a quadruped) corresponding to the human back **c :** SPINAL COLUMN **d :** BACKBONE 4

2 a : the side or surface opposite the front or face **:** the rear part; *also* **:** the farther or reverse side **b :** something at or on the back for support ⟨*back* of a chair⟩ **c :** a place away from the front ⟨sat in *back*⟩

3 : a position in some games (as football or soccer) behind the front line of players; *also* **:** a player in this position

— backed \ˈbakt\ *adjective*

— back·less \ˈbak-ləs\ *adjective*

— back of one's hand *or* **back of the hand :** a show of contempt

— back of one's mind : the remote part of one's mind where thoughts and memories are stored to be drawn on

— behind one's back : without one's knowledge **:** in secret

— in back of : BEHIND

[2]**back** *adverb* (14th century)

1 a : to, toward, or at the rear **b :** in or into the past **:** backward in time; *also* **:** AGO **c :** to or at an angle off the vertical **d** (1) **:** under restraint (2) **:** in a delayed or retarded condition

2 a : to, toward, or in a place from which a person or thing came **b :** to or toward a former state **c :** in return or reply

[3]**back** *adjective* (15th century)

1 a : being at or in the back ⟨*back* door⟩ **b :** distant from a central or main area ⟨*back* roads⟩ **c :** articulated at or toward the back of the oral passage ⟨*back* vowels⟩

2 : having returned or been returned

3 : being in arrears **:** OVERDUE

4 : moving or operating backward **:** REVERSE

5 : not current ⟨*back* numbers of a magazine⟩

6 : constituting the final 9 holes of an 18-hole golf course

[4]**back** (1548)

transitive verb

1 a : to support by material or moral assistance — usually used with *up* **b :** SUBSTANTIATE **c :** to assume financial responsibility for **d :** to provide musical accompaniment for — often used with *up*

2 a : to cause to go back or in reverse **b :** to articulate (a sound) with the tongue farther back

3 a : to furnish with a back **b :** to be at the back of

intransitive verb

1 : to move backward — often used with *up*

2 *of the wind* **:** to shift counterclockwise — compare VEER

3 : to have the back in the direction of something

synonym see SUPPORT, RECEDE

— back·er \ˈba-kər\ *noun*

— back and fill 1 : to manage the sails of a ship so as to keep it clear of obstructions as it floats down with the current of a river or channel **2 :** to take opposite positions alternately **:** SHILLY-SHALLY

— back into : to get into inadvertently ⟨*backed into* the antiques business⟩

back·ache \ˈbak-ˌāk\ *noun* (1601)

: a pain in the lower back

back–and–forth *noun* (1941)

: DISCUSSION 1, GIVE-AND-TAKE; *also* **:** EXCHANGE 1

back and forth *adverb* (1613)

: backward and forward; *also* **:** between two places or persons

back away *intransitive verb* (1919)

: to move away (as from a stand on an issue)

back bacon *noun* (1947)

chiefly British **:** CANADIAN BACON

back·beat \ˈbak-ˌbēt\ *noun* (1928)

: a steady pronounced rhythm stressing the second and fourth beats of a four-beat measure

back·bench *noun* (1874)

\ə\ **abut** \ᵊ\ **kitten** \ər\ **further** \a\ **ash** \ā\ **ace** \ä\ **mop, mar** \au̇\ **out** \ch\ **chin** \e\ **bet** \ē\ **easy** \g\ **go** \i\ **hit** \ī\ **ice** \j\ **job** \ŋ\ **sing** \ō\ **go** \ȯ\ **law** \ȯi\ **boy** \th\ **thin** \t͟h\ **the** \ü\ **loot** \u̇\ **foot** \y\ **yet** \zh\ **vision** *see also* Guide to Pronunciation

: a bench in a British legislature (as the House of Commons) occupied by rank-and-file members — compare FRONT BENCH
— **back•bench•er** \-ˈben-chər\ *noun*

back•bite \-ˌbīt\ *verb* **-bit; -bit•ten; -bit•ing** (12th century)
transitive verb
: to say mean or spiteful things about
intransitive verb
: to backbite a person
— **back•bit•er** *noun*

backbiting \ˈbak-ˌbī-tiŋ\ *noun* (12th century)
: malicious comment about one not present

back•block \-ˌbläk\ *noun* (1870)
Australian & New Zealand : BOONDOCKS 2 — usually used in plural

back•board \-ˌbōrd, -ˌbȯrd\ *noun* (1761)
: a board placed at or serving as the back of something; *especially* : a rounded or rectangular board behind the basket on a basketball court which serves to keep missed shots from going out-of-bounds and from which the ball can be made to rebound into the basket

back•bone \-ˈbōn, -ˌbōn\ *noun* (14th century)
1 : SPINAL COLUMN, SPINE
2 : something that resembles a backbone: as **a** : a chief mountain ridge, range, or system **b** : the foundation or most substantial or sturdiest part of something **c** : the longest chain of atoms or groups of atoms in a usually long molecule (as a polymer or protein)
3 : firm and resolute character
4 : SPINE 1c

back•break•ing \-ˌbrā-kiŋ\ *adjective* (1870)
: extremely tiring or demanding : OPPRESSIVE ⟨*backbreaking* labor⟩ ⟨*backbreaking* rents⟩
— **back•break•er** \-kər\ *noun*

back burner *noun* (1963)
: the condition of being out of active consideration or development — usually used in the phrase *on the back burner*

back channel *noun* (1975)
: a secret, unofficial, or irregular means of communication
— **back–channel** *adjective*

back•chat \ˈbak-ˌchat\ *noun* (1901)
1 : BACK TALK
2 : gossipy or bantering conversation

back–check \-ˌchek\ *intransitive verb* (1937)
: to skate back toward one's own goal while closely defending against the offensive rushes of an opposing player in ice hockey
— **back–check•er** \-ˌche-kər\ *noun*

back•cloth \-ˌklȯth\ *noun* (1886)
chiefly British : BACKDROP

back•coun•try \-ˌkən-trē\ *noun, often attributive* (1746)
: a remote undeveloped rural area

back•court \-ˈkōrt, -ˈkȯrt\ *noun* (1890)
1 : the area near or nearest the back boundary lines or back wall of the playing area in a net or court game
2 : a basketball team's defensive half of the court; *also* : the part of the offensive half of the court farthest from the goal
3 : the basketball players who play the backcourt

back•court•man \-mən\ *noun* (1954)
: a guard on a basketball team

¹back•cross \ˈbak-ˌkrȯs\ *transitive verb* (1904)
: to cross (a first-generation hybrid) with one of the parental types

²backcross *noun* (1918)
: a mating that involves backcrossing; *also* : an individual produced by backcrossing

back•date \ˈbak-ˌdāt\ *transitive verb* (1944)
: to put a date earlier than the actual one on ⟨*backdate* a memo⟩; *also* : to make retroactive ⟨*backdate* pension rights⟩

back dive *noun* (circa 1934)
: a dive from a position facing the diving board

back•door \ˈbak-ˈdōr, -ˈdȯr\ *adjective* (1805)
: INDIRECT, DEVIOUS

back down *intransitive verb* (1849)
: to withdraw from a commitment or position

back•drop \ˈbak-ˌdräp\ *noun* (1913)
1 : a painted cloth hung across the rear of a stage
2 : BACKGROUND
— **backdrop** *transitive verb*

back•field \-ˌfēld\ *noun* (1920)
: the football players whose positions are behind the line of scrimmage; *also* : the positions themselves

back•fill \-ˌfil\ (1908)
transitive verb
: to refill (as an excavation) usually with excavated material
intransitive verb
: to backfill an excavation
— **backfill** *noun*

¹back•fire \-ˌfīr\ *noun* (1839)
1 : a fire started to check an advancing fire by clearing an area
2 : a loud noise caused by the improperly timed explosion of fuel mixture in the cylinder of an internal combustion engine

²backfire *intransitive verb* (1886)
1 : to make or undergo a backfire
2 : to have the reverse of the desired or expected effect

back•fit \ˈbak-ˌfit\ *transitive verb* (1967)
: RETROFIT
— **backfit** *noun*

back•flow \-ˌflō\ *noun* (1884)
: a flowing back or returning toward a source

back–formation *noun* (1889)
1 : a word formed by subtraction of a real or supposed affix from an already existing longer word (as *burgle* from *burglar*)
2 : the formation of back-formations

back•gam•mon \ˈbak-ˌga-mən, bak-ˈ\ *noun* [perhaps from ³*back* + Middle English *gamen, game* game] (circa 1645)
: a board game played with dice and counters in which players try to be the first to gather their pieces into one corner and then systematically remove them from the board

¹back•ground \ˈbak-ˌgrau̇nd\ *noun, often attributive* (1672)
1 a : the scenery or ground behind something **b** : the part of a painting representing what lies behind objects in the foreground
2 : an inconspicuous position
3 a : the conditions that form the setting within which something is experienced **b** (1) : the circumstances or events antecedent to a phenomenon or development (2) : information essential to understanding of a problem or situation **c** : the total of a person's experience, knowledge, and education
4 a : intrusive sound or radiation that interferes with received or recorded electronic signals **b** : a more or less steady level of noise above which the effect (as radioactivity) being measured by an apparatus (as a Geiger counter) is detected; *especially* : a somewhat steady level of radiation in the natural environment (as from cosmic rays) ☆

²background *transitive verb* (1768)
: to provide with background

back•ground•er \ˈbak-ˌgrau̇n-dər\ *noun* (1960)
: an off-the-record briefing for reporters

background music *noun* (1928)
: music to accompany the dialogue or action of a motion picture or radio or television drama

background radiation *noun* (1969)
: the microwave radiation pervading the universe that has the spectral energy distribution of a blackbody having a temperature of 2.7 K and that is the principle evidence supporting the big bang theory — called also *cosmic background radiation*

¹back•hand \ˈbak-ˌhand\ *noun* (1657)
1 a : a stroke (as in tennis) made with the back of the hand turned in the direction of movement; *also* : the side on which such strokes are made **b** : a catch (as in baseball) made to the side of the body opposite the hand being used
2 : handwriting whose strokes slant downward from left to right

backhand 1a

²backhand *adjective* (1695)
: made with a backhand ⟨a *backhand* tennis stroke⟩

³backhand *or* **back•hand•ed** \-ˈhan-dəd\ *adverb* (1889)
: with a backhand

⁴backhand *transitive verb* (circa 1935)
: to do, hit, or catch backhand

back•hand•ed \ˈbak-ˈhan-dəd\ *adjective* (1800)
1 : INDIRECT, DEVIOUS; *especially* : SARCASTIC ⟨a *backhanded* compliment⟩
2 : using or made with a backhand
— **back•hand•ed•ly** *adverb*

back•hand•er \-dər\ *noun* (1960)
British : BRIBE

back•hoe \-ˌhō\ *noun* (1928)
: an excavating machine having a bucket that is attached to a rigid bar hinged to a boom and that is drawn toward the machine in operation

back•house \-ˌhau̇s\ *noun* (circa 1847)
: PRIVY 1a

back•ing \ˈba-kiŋ\ *noun* (1793)
1 : something forming a back
2 a : SUPPORT, AID **b** : endorsement especially of a warrant by a magistrate

back judge *noun* (circa 1966)
: a football official whose duties include keeping the game's official time and identifying eligible pass receivers

back•land \ˈbak-ˌland\ *noun* (1681)
: BACKCOUNTRY, HINTERLAND — usually used in plural

back•lash \ˈbak-ˌlash\ *noun* (1815)
1 a : a sudden violent backward movement or reaction **b** : the play between adjacent movable parts (as in a series of gears); *also* : the jar caused by this when the parts are put into action
2 : a snarl in that part of a fishing line wound on the reel
3 : a strong adverse reaction (as to a recent political or social development)
— **back•lash•er** *noun*

back•light \-ˌlīt\ *noun* (circa 1846)
: illumination from behind; *also* : the source of such illumination
— **backlight** *transitive verb*

back•list \ˈbak-ˌlist\ *noun* (1964)

☆ SYNONYMS
Background, setting, environment, milieu, mise-en-scène mean the place, time, and circumstances in which something occurs. BACKGROUND often refers to the circumstances or events that precede a phenomenon or development ⟨the shocking decision was part of the *background* of the riots⟩. SETTING suggests looking at real-life situations in literary or dramatic terms ⟨a militant reformer who was born into an unlikely social *setting*⟩. ENVIRONMENT applies to all the external factors that have a formative influence on one's physical, mental, or moral development ⟨the kind of *environment* that produces juvenile delinquents⟩. MILIEU applies especially to the physical and social surroundings of a person or group of persons ⟨an intellectual *milieu* conducive to artistic experimentation⟩. MISE-EN-SCENE strongly suggests the use of properties to achieve a particular atmosphere or theatrical effect ⟨a gothic thriller with a carefully crafted *mise-en-scène*⟩.

: a list of books kept in print as distinguished from books newly published

[1]**back·log** \-ˌlȯg, -ˌläg\ *noun* (1684)
1 : a large log at the back of a hearth fire
2 : an accumulation of tasks unperformed or materials not processed

[2]**backlog** *verb* (1963)
: ACCUMULATE

back matter *noun* (1947)
: matter following the main text of a book

back mutation *noun* (1939)
: mutation of a previously mutated gene to its former condition

back of *preposition* (1694)
: BEHIND

back of beyond (1816)
: a remote place

back off *intransitive verb* (1938)
: BACK DOWN

back–office \ˈbak-ˈä-fəs, -ˈȯ-fəs\ *adjective* (1953)
: of or relating to the inner workings of a business or institution **:** INTERNAL ⟨*back-office* operations⟩

back out *intransitive verb* (1807)
: to withdraw especially from a commitment or contest

[1]**back·pack** \ˈbak-ˌpak\ *noun* (1914)
1 a : a load carried on the back **b :** a camping pack (as of canvas or nylon) supported by a usually aluminum frame and carried on the back
2 : a piece of equipment designed for use while being carried on the back

[2]**backpack** (1927)
transitive verb
: to carry (food or equipment) on the back especially in hiking
intransitive verb
: to hike with a backpack
— back·pack·er *noun*

back·ped·al \ˈbak-ˌpe-dᵊl\ *intransitive verb* (1901)
: to retreat or move backward

back·rest \-ˌrest\ *noun* (1859)
: a rest for the back

back·room \ˈbak-ˈrüm, -ˈrum\ *adjective* (1940)
: made or operating in an inconspicuous way **:** BEHIND-THE-SCENES ⟨*backroom* deals⟩ ⟨a *backroom* politician⟩

back room *noun* (1592)
1 : a room situated in the rear
2 : the meeting place of a directing group that exercises its authority in an inconspicuous and indirect way

back·saw \ˈbak-ˌsȯ\ *noun* (circa 1877)
: a saw with a metal rib along its back

back·scat·ter \-ˌska-tər\ *or* **back·scat·ter·ing** \-tə-riŋ\ *noun* (1940)
: the scattering of radiation or particles in a direction opposite to that of the incident radiation due to reflection from particles of the medium traversed; *also* **:** the radiation or particles so reversed in direction
— backscatter *transitive verb*

back–scratch·ing \ˈbak-ˌskra-chiŋ\ *noun* (1924)
: the reciprocal exchange of favors, services, assistance, or praise

back·seat \-ˈsēt\ *noun* (1829)
1 : a seat in the back (as of an automobile)
2 : an inferior position ⟨won't take a *backseat* to anyone⟩

back·set \ˈbak-ˌset\ *noun* (1721)
: SETBACK

back·side \-ˌsīd\ *noun* (circa 1500)
: BUTTOCKS — often used in plural

back·slap \-ˌslap\ (1926)
transitive verb
: to display excessive or effusive goodwill for
intransitive verb
: to display excessive cordiality or goodwill
— backslap *noun*
— back·slap·per *noun*

back·slide \-ˌslīd\ *intransitive verb* **-slid** \-ˌslid\; **-slid** *or* **-slid·den** \-ˌsli-dᵊn\; **-slid·ing** \-ˌslī-diŋ\ (1581)
1 : to lapse morally or in the practice of religion
2 : to revert to a worse condition **:** RETROGRESS
— back·slid·er \-ˌslī-dər\ *noun*

back·space \-ˌspās\ *intransitive verb* (1911)
: to move back one space in a text with each depression of a key
— backspace *noun*

back·spin \-ˌspin\ *noun* (circa 1909)
: a backward rotary motion of a ball

back·splash \ˈbak-ˌsplash\ *noun* (1947)
: a vertical surface (as of tiles) designed to protect the wall behind a stove or countertop

back·stab·bing \-ˌsta-biŋ\ *noun* (1946)
: betrayal (as by a verbal attack against one not present) especially by a false friend
— back·stab \-ˌstab\ *verb*
— back·stab·ber \-ˌsta-bər\ *noun*

[1]**back·stage** \ˈbak-ˌstāj\ *adjective* (1916)
1 : of, relating to, or occurring in the area behind the proscenium and especially in the dressing rooms
2 : of or relating to the private lives of theater people
3 : of or relating to the inner working or operation (as of an organization)

[2]**back·stage** \ˈbak-ˈstāj\ *adverb* (1923)
1 : in or to a backstage area
2 : in private **:** SECRETLY

back·stairs \-ˌstarz, -ˌsterz\ *adjective* (1663)
1 : SECRET, FURTIVE ⟨*backstairs* political deals⟩
2 : SORDID, SCANDALOUS ⟨*backstairs* gossip⟩

back·stay \-ˌstā\ *noun* (1626)
1 : a stay extending from the mastheads to the side of a ship and slanting aft
2 : a strengthening or supporting device at the back (as of a carriage or a shoe)

back·stitch \-ˌstich\ *noun* (1611)
: a stitch sewn one stitch length backward on the front side and two stitch lengths forward on the reverse side to form a solid line of stitching on both sides
— backstitch *verb*

[1]**back·stop** \-ˌstäp\ *noun* (1851)
1 : something at the back serving as a stop: as **a :** a screen or fence for keeping a ball from leaving the field of play **b :** a stop (as a pawl) that prevents a backward movement (as of a wheel)
2 : a player (as the catcher) positioned behind the batter

[2]**backstop** *transitive verb* (1941)
1 : SUPPORT, BOLSTER
2 : to serve as a backstop to

back·street \ˈbak-ˈstrēt\ *noun* (15th century)
: a street away from the main thoroughfares

back·stretch \ˈbak-ˌstrech\ *noun* (1839)
: the side opposite the homestretch on a racecourse

back·stroke \-ˌstrōk\ *noun* (1879)
: a swimming stroke executed on the back and usually consisting of alternating circular arm pulls and a flutter kick

back·swept \-ˌswept\ *adjective* (circa 1918)
: swept or slanting backward

back swimmer *noun* (1862)
: an aquatic bug (family Notonectidae) that swims on its back

back·swing \ˈbak-ˌswiŋ\ *noun* (1899)
: the movement of a club, racket, bat, or arm backward to a position from which the forward or downward swing is made

back·sword \-ˌsōrd, -ˌsȯrd\ *noun* (1609)
: a single-edged sword

back talk *noun* (1858)
: impudent, insolent, or argumentative replies

back–to–back *adjective or adverb* (15th century)
1 : facing in opposite directions and often touching
2 : coming one after the other **:** CONSECUTIVE

back·track \ˈbak-ˌtrak\ *intransitive verb* (1904)
1 a : to retrace one's course **b :** to go back to an earlier point in a sequence
2 : to reverse a position

back·up \-ˌəp\ *noun, often attributive* (1951)
1 a : one that serves as a substitute or support ⟨a *backup* plan⟩ **b :** musical accompaniment
2 : an accumulation caused by a stoppage in the flow ⟨traffic *backup*⟩
3 : the act or an instance of backing up a computer's hard disk

back up (1837)
intransitive verb
: to accumulate in a congested state ⟨traffic *backed up* for miles⟩
transitive verb
1 : to move into a position behind (a teammate) in order to assist on a play
2 : HOLD BACK 1
3 : to make copies of the computer files on (a hard disk) to protect against loss of data

[1]**back·ward** \ˈbak-wərd\ *or* **back·wards** \-wərdz\ *adverb* (14th century)
1 a : toward the back **b :** with the back foremost
2 a : in a reverse or contrary direction or way **b :** toward the past **c :** toward a worse state
— bend over backward *or* **lean over backward :** to make extreme efforts (as at concession)

[2]**backward** *adjective* (14th century)
1 a : directed or turned backward **b :** done or executed backward
2 : DIFFIDENT, SHY
3 : retarded in development
— back·ward·ly *adverb*
— back·ward·ness *noun*

[3]**backward** *noun* (1610)
: the part behind or past

back·wash \ˈbak-ˌwȯsh, -ˌwäsh\ *noun* (1876)
1 : a backward flow or movement (as of water or air) produced especially by a propelling force; *also* **:** the fluid that is moving backward
2 : CONSEQUENCE, AFTERMATH

back·wa·ter \-ˌwȯ-tər, -ˌwä-\ *noun* (1629)
1 a : water backed up in its course by an obstruction, an opposing current, or the tide **b :** a body of water (as an inlet or tributary) that is out of the main current of a larger body
2 : an isolated or backward place or condition

back·woods \-ˈwu̇dz\ *noun plural but singular or plural in construction, often attributive* (1709)
1 : wooded or partly cleared areas far from cities
2 : a remote or culturally backward area
— back·woodsy \-ˈwu̇d-zē\ *adjective*

back·woods·man \ˌbak-ˈwu̇dz-mən, ˈbak-ˌ\ *noun* (1774)
: one who lives in the backwoods

back·wrap \-ˌrap\ *noun* (1951)
: a wraparound garment (as a skirt) that fastens in the back

back·yard \-ˈyärd\ *noun* (1659)
1 : an area at the rear of a house
2 : an area that is one's special domain

ba·con \ˈbā-kən\ *noun* [Middle English, from Middle French, of Germanic origin; akin to Old High German *bahho* side of bacon, *bah* back] (14th century)
: a side of a pig cured and smoked

Ba·co·ni·an \bā-ˈkō-nē-ən\ *adjective* (1812)
1 : of, relating to, or characteristic of Francis Bacon or his doctrines
2 : of or relating to those who believe that Francis Bacon wrote the works usually attributed to Shakespeare
— Baconian *noun*

bac·ter·emia \ˌbak-tə-ˈrē-mē-ə\ *noun* [New Latin, alteration of *bacteriemia,* from *bacteri-* + *-emia*] (circa 1890)

: the usually transient presence of bacteria in the blood
— **bac·ter·emic** \-mik\ *adjective*

bacteri- *or* **bacterio-** *combining form* [New Latin *bacterium*]
: bacteria ⟨*bacterio*lysis⟩

bac·te·ria \bak-'tir-ē-ə\ *noun, plural* **-ri·as** [plural of *bacterium*] (1884)
: a group (as a genus, species, or strain) of bacteria — used chiefly in nontechnical writing and in news broadcasts ■

bac·te·ri·al \bak-'tir-ē-əl\ *adjective* (1871)
: of, relating to, or caused by bacteria ⟨*bacterial* infection⟩
— **bac·te·ri·al·ly** \-ə-lē\ *adverb*

bac·te·ri·cid·al \bak-,tir-ə-'sī-dᵊl\ *adjective* (1878)
: destroying bacteria
— **bac·te·ri·cid·al·ly** \-dᵊl-ē\ *adverb*
— **bac·te·ri·cide** \-'tir-ə-,sīd\ *noun*

bac·ter·in \'bak-tə-rən\ *noun* (circa 1912)
: a suspension of killed or attenuated bacteria for use as a vaccine

bac·te·rio·chlo·ro·phyll \bak-,tir-ē-ō-'klōr-ə-,fil, -'klȯr-, -fəl\ *noun* (1938)
: a pyrrole derivative in photosynthetic bacteria related to the chlorophyll of higher plants

bac·te·rio·cin \bak-'tir-ē-ə-sən\ *noun* [International Scientific Vocabulary *bacteri-* + *-cin* (as in *colicin*)] (1954)
: an antibiotic (as colicin) produced by bacteria

bac·te·ri·ol·o·gy \(,)bak-,tir-ē-'ä-lə-jē\ *noun* [International Scientific Vocabulary] (1884)
1 : a science that deals with bacteria and their relations to medicine, industry, and agriculture
2 : bacterial life and phenomena
— **bac·te·ri·o·log·ic** \bak-,tir-ē-ə-'lä-jik\ *or* **bac·te·ri·o·log·i·cal** \-'lä-ji-kəl\ *adjective*
— **bac·te·ri·o·log·i·cal·ly** \-ji-k(ə-)lē\ *adverb*
— **bac·te·ri·ol·o·gist** \(,)bak-,tir-ē-'ä-lə-jist\ *noun*

bac·te·ri·ol·y·sis \(,)bak-,tir-ē-'ä-lə-səs\ *noun* [New Latin] (1900)
: destruction or lysis of bacterial cells
— **bac·te·ri·o·lyt·ic** \bak-,tir-ē-ə-'li-tik\ *adjective*

bac·te·rio·phage \bak-'tir-ē-ə-,fāj *also* -,fäzh\ *noun* [International Scientific Vocabulary] (1920)
: a virus that infects bacteria
— **bac·te·ri·oph·a·gy** \(,)bak-,tir-ē-'ä-fə-jē\ *noun*

bac·te·rio·rho·dop·sin \bak-,tir-ē-ə-rō-'däp-sin\ *noun* (1974)
: a purple-pigmented protein found in the outer membrane of a bacterium (*Halobacterium halobium*) that converts light energy into chemical energy in the synthesis of ATP

bac·te·rio·sta·sis \bak-,tir-ē-ō-'stā-səs\ *noun* [New Latin] (1920)
: inhibition of the growth of bacteria without destruction

bac·te·rio·stat \-'tir-ē-ō-,stat\ *noun* (1920)
: an agent that causes bacteriostasis
— **bac·te·rio·stat·ic** \-,tir-ē-ō-'sta-tik\ *adjective*

bac·te·ri·um \bak-'tir-ē-əm\ *noun, plural* **-ria** \-ē-ə\ [New Latin, from Greek *baktērion* staff] (circa 1849)
: any of a group (as Kingdom Procaryotae or Kingdom Monera) of prokaryotic unicellular round, spiral, or rod-shaped single-celled microorganisms that are often aggregated into colonies or motile by means of flagella, that live in soil, water, organic matter, or the bodies of plants and animals, and that are autotrophic, saprophytic, or parasitic in nutrition and important because of their biochemical effects and pathogenicity

bac·te·ri·uria \bak-,tir-ē-'yu̇r-ē-ə\ *noun* [New Latin] (1900)
: the presence of bacteria in the urine

bac·te·rize \'bak-tə-,rīz\ *transitive verb* **-rized; -riz·ing** (1914)
: to subject to bacterial action
— **bac·te·ri·za·tion** \,bak-tə-rə-'zā-shən\ *noun*

bac·te·roid \'bak-tə-,ròid\ *noun* (1878)
: an irregularly shaped bacterium (as a rhizobium) found especially in root nodules of legumes

Bac·tri·an camel \'bak-trē-ən-\ *noun* (1609)
: CAMEL 1b

¹bad \'bad\ *adjective* **worse** \'wərs\; **worst** \'wərst\ [Middle English] (14th century)
1 a : failing to reach an acceptable standard **:** POOR **b :** UNFAVORABLE ⟨make a *bad* impression⟩ **c :** not fresh **:** SPOILED ⟨*bad* fish⟩ **d :** not sound **:** DILAPIDATED ⟨the house was in *bad* condition⟩
2 a : morally objectionable **b :** MISCHIEVOUS, DISOBEDIENT
3 : inadequate or unsuited to a purpose ⟨a *bad* plan⟩ ⟨*bad* lighting⟩
4 : DISAGREEABLE, UNPLEASANT ⟨*bad* news⟩
5 a : INJURIOUS, HARMFUL **b :** SERIOUS, SEVERE ⟨in *bad* trouble⟩ ⟨a *bad* cough⟩
6 : INCORRECT, FAULTY ⟨*bad* grammar⟩
7 a : suffering pain or distress ⟨felt generally *bad*⟩ **b :** UNHEALTHY, DISEASED ⟨*bad* teeth⟩
8 : SORROWFUL, SORRY
9 : INVALID, VOID ⟨a *bad* check⟩
10 bad·der; bad·dest *slang* **a :** GOOD, GREAT **b :** TOUGH, MEAN
— **bad·ness** *noun*

²bad *noun* (15th century)
1 : something that is bad
2 : an evil or unhappy state

³bad *adverb* (1681)
: BADLY ■

¹bad·ass \-,as\ *adjective* (1956)
: ready to cause or get into trouble **:** MEAN — often considered vulgar

²badass *noun* (1984)
: a person who is badass — often considered vulgar

bad blood *noun* (1825)
: ill feeling **:** BITTERNESS

bad·die *or* **bad·dy** \'ba-dē\ *noun, plural* **baddies** (1937)
: one that is bad; *especially* **:** an opponent of the hero (as in fiction or motion pictures)

bade *past and past participle of* BID

badge \'baj\ *noun* [Middle English *bage, bagge*] (14th century)
1 : a device or token especially of membership in a society or group
2 : a characteristic mark
3 : an emblem awarded for a particular accomplishment
— **badge** *transitive verb*

¹bad·ger \'ba-jər\ *noun* [probably from *badge;* from the white mark on its forehead] (1523)
1 a : any of various burrowing mammals (especially *Taxidea taxus* and *Meles meles*) that are related to the weasels and are widely distributed in the northern hemisphere **b :** the pelt or fur of a badger
2 *capitalized* **:** a native or resident of Wisconsin — used as a nickname

²badger *transitive verb* [from the sport of baiting badgers] (1794)
: to harass or annoy persistently
synonym see BAIT

ba·di·nage \,ba-dᵊn-'äzh\ *noun* [French] (circa 1658)
: playful repartee **:** BANTER

bad·land \'bad-,land\ *noun* (1851)
: a region marked by intricate erosional sculpturing, scanty vegetation, and fantastically formed hills — usually used in plural

bad·ly \'bad-lē\ *adverb* (14th century)
1 : in a bad manner ⟨played *badly*⟩
2 : to a great or intense degree ⟨want something *badly*⟩
usage see BAD

bad·min·ton \'bad-,min-tᵊn\ *noun* [*Badminton,* residence of the Duke of Beaufort, England] (1874)
: a court game played with light long-handled rackets and a shuttlecock volleyed over a net

bad–mouth \'bad-,mau̇th, -,mau̇th\ *transitive verb* (1942)
: to criticize severely

bad news *noun plural but singular in construction* (1930)
: one that is troublesome, unwelcome, or dangerous ⟨stay away from him, he's *bad news*⟩

Bae·de·ker \'bā-di-kər, 'be-\ *noun* [Karl *Baedeker* (died 1859) German publisher of guidebooks] (1924)
: GUIDEBOOK

¹baf·fle \'ba-fəl\ *transitive verb* **baf·fled; baf·fling** \-f(ə-)liŋ\ [probably alteration of Middle English (Scots) *bawchillen* to denounce, discredit publicly] (circa 1590)
1 : to defeat or check (as a person) by confusing or puzzling **:** DISCONCERT
2 : to check or break the force or flow of by or as if by a baffle
synonym see FRUSTRATE
— **baf·fle·ment** \-fəl-mənt\ *noun*
— **baf·fler** \-f(ə-)lər\ *noun*
— **baf·fling·ly** \'ba-fliŋ-lē\ *adverb*

²baffle *noun* (circa 1900)
: a device (as a plate, wall, or screen) to deflect, check, or regulate flow (as of a fluid, light, or sound)
— **baf·fled** \'ba-fəld\ *adjective*

□ USAGE

bacteria *Bacteria* is regularly a plural in scientific and pedagogical use; in speech and in journalism it is also used as a singular ⟨caused by a *bacteria* borne by certain tiny ticks —*Wall Street Journal*⟩ ⟨more resistant to chlorine and elevated water temperatures than other *bacterias* —Allan Bruckheim, M.D., *Chicago Tribune*⟩. This journalistic use is found in British as well as American sources.

bad *Bad* is a 17th century adverb that is not nearly as old as its relatives *bad,* adjective, and *badly,* adverb. It is interchangeable with *badly* in many constructions, and while these constructions are grammatically correct, they are more common in speech than in edited prose ⟨I didn't do too *bad* —Denny McLain⟩ ⟨I wanted to get a mandolin real *bad* —Bill Holt⟩ ⟨the luxurious . . . irresponsibility that is the nicest thing about being *bad* sick —James Jones⟩. After *feel* both *bad* and *badly* are in common use. There are several strands to this mixed usage. Present-day teaching says that *feel* is a linking verb and must be followed by the adjective *bad,* but 19th century teaching said that *bad* referred to morals or behavior and that *badly* should be used for health. Further, some adverbs, like *differently* or *strongly,* are regularly used after *feel;* other adverbs, like *angrily* or *sadly,* are not. And the choice between *feel bad* and *feel badly* is related to the choice between *feel well* and *feel good.* Many people choose *well* for health and *good* for mental state, and many choose *bad* for health and *badly* for mental state ⟨we feel very *badly* about your only having one turkey —James Thurber⟩. But many others do not make that distinction ⟨I feel *bad* about not having written you —E. B. White⟩. So the controversy over *feel bad* and *feel badly* is not as simple as it is often made out to be. The writer will have to make a choice situation by situation, and, whatever that choice is, some people will agree and some will not.

[1]bag \'bag *also* 'bāg\ *noun* [Middle English *bagge,* from Old Norse *baggi*] (13th century)
1 : a usually flexible container that may be closed for holding, storing, or carrying something: as **a :** PURSE; *especially* **:** HANDBAG **b :** a bag for game **c :** SUITCASE
2 : something resembling a bag: as **a :** a pouched or pendulous bodily part or organ; *especially* **:** UDDER **b :** a puffed-out sag or bulge in cloth **c :** a square white stuffed canvas bag to mark a base in baseball
3 : the amount contained in a bag
4 a : a quantity of game taken; *also* **:** the maximum legal quantity of game **b :** an assortment or collection especially of nonmaterial things ⟨a *bag* of tricks⟩
5 : an unattractive woman
6 : something one likes or does regularly or well; *also* **:** one's characteristic way of doing things
— in the bag : SURE, CERTAIN ⟨her nomination was *in the bag*⟩

[2]bag *verb* **bagged; bag·ging** (15th century)
intransitive verb
1 : to swell out **:** BULGE
2 : to hang loosely
transitive verb
1 : to cause to swell
2 : to put into a bag
3 a : to take (animals) as game **b :** to get possession of especially by strategy or stealth **c :** CAPTURE, SEIZE **d :** to shoot down **:** DESTROY
synonym see CATCH
— bag·ger *noun*

ba·gasse \bə-'gas\ *noun* [French] (circa 1826)
: plant residue (as of sugarcane or grapes) left after a product (as juice) has been extracted

bag·a·telle \ˌba-gə-'tel\ *noun* [French, from Italian *bagattella*] (1633)
1 : TRIFLE 1
2 : any of various games involving the rolling of balls into scoring areas
3 : a short literary or musical piece in light style

ba·gel \'bā-gəl\ *noun* [Yiddish *beygl,* from (assumed) Middle High German *böugel* ring, from Middle High German *bouc* ring, from Old High German; akin to Old English *bēag* ring, *būgan* to bend — more at BOW] (1932)
: a hard glazed doughnut-shaped roll

bag·ful \'bag-ˌfu̇l\ *noun* (15th century)
1 : as much or as many as a bag will hold
2 : a large number or amount ⟨had a *bagful* of tricks⟩

[1]bag·gage \'ba-gij\ *noun* [Middle English *bagage,* from Middle French, from *bague* bundle] (15th century)
1 : suitcases, trunks, and personal belongings of travelers **:** LUGGAGE
2 : transportable equipment especially of a military force
3 : things (as objects, circumstances, or beliefs) that get in the way **:** IMPEDIMENTA

[2]baggage *noun* [probably modification of Middle French *bagasse,* from Old Provençal *bagassa*] (1594)
1 : a contemptible woman; *especially* **:** PROSTITUTE
2 : a young woman

bag·ging \'ba-giŋ\ *noun* (1732)
: material (as cloth) for bags

bag·gy \'ba-gē\ *adjective* **bag·gi·er; -est** (1831)
: loose, puffed out, or hanging like a bag ⟨*baggy* trousers⟩
— bag·gi·ly \'ba-gə-lē\ *adverb*
— bag·gi·ness \'ba-gē-nəs\ *noun*

bag·house \'bag-ˌhau̇s\ *noun* (1914)
: a device or facility in which particulates are removed from a stream of exhaust gases (as from a blast furnace) as the stream passes through a large cloth bag; *also* **:** the bag used to filter the gas stream

bag lady *noun* (1979)
: a homeless woman who roams the streets of a large city carrying her possessions in shopping bags

bag·man \'bag-mən\ *noun* (1765)
1 *chiefly British* **:** TRAVELING SALESMAN
2 : a person who on behalf of another collects or distributes illicitly gained money

ba·gnio \'ban-(ˌ)yō\ *noun, plural* **bagnios** [Italian *bagno,* literally, public baths (from the Turks' use of Roman baths at Constantinople as prisons), from Latin *balneum,* from Greek *balaneion*] (1599)
1 *obsolete* **:** PRISON
2 : BORDELLO

bag of waters (circa 1881)
: the double-walled fluid-filled sac that encloses and protects the fetus in the womb and that breaks releasing its fluid during the birth process

bag·pipe \'bag-ˌpīp\ *noun* (14th century)
: a wind instrument consisting of a reed melody pipe and from one to five drones with air supplied continuously either by a bag with valve-stopped mouth tube or by bellows — often used in plural
— bag·pip·er \-ˌpī-pər\ *noun*

bagpipe

ba·guette \ba-'get\ *noun* [French, literally, rod] (1926)
1 : a gem having the shape of a narrow rectangle; *also* **:** the shape itself
2 : a long thin loaf of French bread

bag·wig \'bag-ˌwig\ *noun* (1717)
: an 18th century wig with the back hair enclosed in a small silk bag

bag·worm \-ˌwərm\ *noun* (1862)
: any of a family (Psychidae) of moths with wingless females and plant-feeding larvae that live in a silk case covered with plant debris; *especially* **:** one (*Thyridopteryx ephemeraeformis*) often destructive to deciduous and evergreen trees of the eastern U.S.

bah \'bä, 'ba\ *interjection* (1600)
— used to express disdain or contempt

Ba·ha'i \bä-'hä-ˌē, -'hī\ *noun, plural* **Baha'is** [Persian *bahā'ī,* literally, of glory, from *bahā* glory] (1889)
: an adherent of a religious movement originating in Iran in the 19th century and emphasizing the spiritual unity of mankind
— Baha'i *adjective*
— Ba·ha·ism \-'hä-ˌi-zəm, -'hī-ˌi-\ *noun*
— Ba·ha·ist \-'hä-(ˌ)ist\ *noun*

Ba·ha·sa In·do·ne·sia \bə-ˌhä-sə-ˌin-də-'nē-zhə, -shə\ *noun* [Indonesian *bahasa indonésia,* literally, Indonesian language] (1952)
: INDONESIAN 2b

Ba·hia grass \bə-'hē-ə-\ *noun* [*Bahia,* state in Brazil] (circa 1927)
: a perennial tropical American grass (*Paspalpum notatum*) used in the southern U.S. as a lawn grass

baht \'bät\ *noun, plural* **baht** *also* **bahts** [Thai *bàad*] (1828)
— see MONEY table

[1]bail \'bā(ə)l\ *noun* [Middle English *baille,* from Middle French, bucket, from Medieval Latin *bajula* water vessel, from feminine of Latin *bajulus*] (14th century)
: a container used to remove water from a boat

[2]bail *transitive verb* (1613)
1 : to clear (water) from a boat by dipping and throwing over the side — usually used with *out*
2 : to clear water from by dipping and throwing — usually used with *out*
— bail·er *noun*

[3]bail *noun* [Middle English, custody, bail, from Middle French, custody, from *baillier* to have in charge, deliver, from Medieval Latin *bajulare* to control, from Latin, to carry a load, from *bajulus* porter] (15th century)
1 : the temporary release of a prisoner in exchange for security given for the due appearance of the prisoner
2 : security given for the release of a prisoner on bail
3 : one who provides bail

[4]bail *transitive verb* (1548)
1 : to release under bail
2 : to procure the release of by giving bail — often used with *out*
3 : to help from a predicament — used with *out* ⟨*bailing* out impoverished countries⟩
— bail·able *adjective*

[5]bail *noun* [Middle English *beil, baile,* probably of Scandinavian origin; akin to Swedish *bygel* bow, hoop; akin to Old English *būgan* to bend — more at BOW] (15th century)
1 a : a supporting half hoop **b :** a hinged bar for holding paper against the platen of a typewriter
2 : a usually arched handle (as of a kettle or pail)

[6]bail *transitive verb* [Anglo-French *baillier,* from French] (1768)
: to deliver (personal property) in trust to another for a special purpose and for a limited period

[7]bail *noun* [perhaps from [5]*bail*] (1844)
chiefly British **:** a device for confining or separating animals

bail·ee \bā-'lē\ *noun* (1528)
: the person to whom personal property is bailed

bai·ley \'bā-lē\ *noun, plural* **baileys** [Middle English *bailli,* palisade, bailey, from Old French *baille, balie*] (13th century)
1 : the outer wall of a castle or any of several walls surrounding the keep
2 : a courtyard within the external wall or between two outer walls of a castle

bai·lie \'bā-lē\ *noun* [Middle English] (14th century)
1 *chiefly dialect* **:** BAILIFF
2 : a Scottish municipal magistrate corresponding to an English alderman

bai·liff \'bā-ləf\ *noun* [Middle English *baillif, bailie,* from Old French *baillif,* from *bail* custody, jurisdiction — more at BAIL] (14th century)
1 a : an official employed by a British sheriff to serve writs and make arrests and executions **b :** a minor officer of some U.S. courts usually serving as a messenger or usher
2 *chiefly British* **:** one who manages an estate or farm
— bai·liff·ship \-ˌship\ *noun*

bai·li·wick \'bā-li-ˌwik\ *noun* [Middle English *baillifwik,* from *baillif* + *wik* dwelling place, village, from Old English *wīc,* from Latin *vīcus* village — more at VICINITY] (15th century)
1 : the office or jurisdiction of a bailiff
2 : a special domain

bail·ment \'bā(ə)l-mənt\ *noun* (1554)
: the act of bailing a person or personal property

bail·or \bā-'lȯr, 'bā-lər\ *or* **bail·er** \'bā-lər\ *noun* (1602)

\ə\ **abut** \ᵊ\ **kitten** \ər\ **further** \a\ **ash** \ā\ **ace** \ä\ **mop, mar** \au̇\ **out** \ch\ **chin** \e\ **bet** \ē\ **easy** \g\ **go** \i\ **hit** \ī\ **ice** \j\ **job** \ŋ\ **sing** \ō\ **go** \ȯ\ **law** \ȯi\ **boy** \th\ **thin** \t͟h\ **the** \ü\ **loot** \u̇\ **foot** \y\ **yet** \zh\ **vision** *see also* Guide to Pronunciation

: a person who delivers personal property to another in trust

bail·out \'bā-ˌlau̇t\ *noun* (1951)
: a rescue (as of a corporation) from financial distress

bail out *intransitive verb* (1930)
1 : to parachute from an aircraft
2 : to abandon a harmful or difficult situation; *also* **:** LEAVE, DEPART

bails·man \'bā(ə)lz-mən\ *noun* (1862)
: one who gives bail for another

bairn \'barn, 'bern\ *noun* [Middle English *bern, barn,* from Old English *bearn* & Old Norse *barn;* akin to Old High German *barn* child] (before 12th century)
chiefly Scottish **:** CHILD

[1]**bait** \'bāt\ *verb* [Middle English, from Old Norse *beita;* akin to Old English *bǣtan* to bait, *bītan* to bite] (13th century)
transitive verb
1 a : to persecute or exasperate with unjust, malicious, or persistent attacks **b :** TEASE
2 a : to harass (as a chained animal) with dogs usually for sport **b :** to attack by biting and tearing
3 a : to furnish with bait **b :** ENTICE, LURE
4 : to give food and drink to (an animal) especially on the road
intransitive verb
archaic **:** to stop for food and rest when traveling ☆
— bait·er *noun*

[2]**bait** *noun* [Middle English, from Old Norse *beit* pasturage & *beita* food; akin to Old English *bītan* to bite] (14th century)
1 a : something used in luring especially to a hook or trap **b :** a poisonous material placed where it will be eaten by usually wild animals considered undesirable or deleterious
2 : LURE, TEMPTATION

bait and switch *noun* (1967)
: a sales tactic in which a customer is attracted by the advertisement of a low-priced item but is then encouraged to buy a higher-priced one

bai·za \'bī-(ˌ)zä\ *noun, plural* **baiza** *or* **bai·zas** [Arabic, from Hindi *paisā*] (1970)
— see *rial* at MONEY table

baize \'bāz\ *noun* [Middle French *baies,* plural of *baie* baize, from feminine of *bai* bay-colored] (1578)
: a coarse woolen or cotton fabric napped to imitate felt

[1]**bake** \'bāk\ *verb* **baked; bak·ing** [Middle English, from Old English *bacan;* akin to Old High German *bahhan* to bake, Greek *phōgein* to roast] (before 12th century)
transitive verb
1 : to cook (as food) by dry heat especially in an oven
2 : to dry or harden by subjecting to heat
intransitive verb
1 : to prepare food by baking it
2 : to become baked
3 : to become extremely hot
— bak·er *noun*

[2]**bake** *noun* (1565)
1 : the act or process of baking
2 : a social gathering at which a baked food is served
3 : baked food ⟨a *bake* sale⟩

Ba·ke·lite \'bā-kə-ˌlīt, -ˌklīt\ *trademark*
— used for any of various synthetic resins and plastics

baker's dozen *noun* (1596)
: THIRTEEN

baker's yeast *noun* (1854)
: a yeast (as *Saccharomyces cerevisiae*) used or suitable for use as leaven

bak·ery \'bā-k(ə-)rē\ *noun, plural* **-er·ies** (circa 1820)
: a place for baking or selling baked goods

bake·shop \'bāk-ˌshäp\ *noun* (1789)
: BAKERY

baking powder *noun* (1850)
: a powder used as a leavening agent in making baked goods (as quick breads) that consists of a carbonate, an acid substance, and starch or flour

baking soda *noun* (1881)
: SODIUM BICARBONATE

bak·la·va \ˌbä-klə-'vä\ *noun* [Turkish] (1653)
: a dessert made of thin pastry, nuts, and honey

bak·sheesh \'bak-ˌshēsh, bak-'\ *noun* [Persian *bakhshīsh,* from *bakhshīdan* to give; akin to Greek *phagein* to eat, Sanskrit *bhajati* he allots] (1775)
: payment (as a tip or bribe) to expedite service

BAL \ˌbē-(ˌ)ā-'el\ *noun* [*B*ritish *A*nti-*L*ewisite] (1942)
: DIMERCAPROL

Ba·laam \'bā-ləm\ *noun* [Greek, from Hebrew *Bil'ām*]
: an Old Testament prophet who is reproached by the ass he is riding and rebuked by God's angel while on the way to meet with an enemy of Israel

bal·a·cla·va \ˌba-lə-'klä-və, -'kla-\ *noun* [*Balaclava,* Crimea, where a battle of the Crimean War was fought] (1881)
: a knit cap for the head and neck — called also *balaclava helmet*

bal·a·lai·ka \ˌba-lə-'lī-kə\ *noun* [Russian] (1788)
: a usually 3-stringed instrument of Russian origin with a triangular body played by plucking or strumming

balalaika

[1]**bal·ance** \'ba-lən(t)s\ *noun* [Middle English, from Old French, from (assumed) Vulgar Latin *bilancia,* from Late Latin *bilanc-, bilanx* having two scalepans, from Latin *bi-* + *lanc-, lanx* plate] (13th century)
1 : an instrument for weighing: as **a :** a beam that is supported freely in the center and has two pans of equal weight suspended from its ends **b :** a device that uses the elasticity of a spiral spring for measuring weight or force **c** *capitalized* **:** LIBRA
2 : a means of judging or deciding
3 : a counterbalancing weight, force, or influence
4 : an oscillating wheel operating with a hairspring to regulate the movement of a timepiece
5 a : stability produced by even distribution of weight on each side of the vertical axis **b :** equipoise between contrasting, opposing, or interacting elements **c :** equality between the totals of the two sides of an account
6 a : an aesthetically pleasing integration of elements **b :** the juxtaposition in writing of syntactically parallel constructions containing similar or contrasting ideas
7 a : physical equilibrium **b :** the ability to retain one's balance
8 a : weight or force of one side in excess of another **b :** something left over **:** REMAINDER **c :** an amount in excess especially on the credit side of an account
9 : mental and emotional steadiness
— bal·anced \-lən(t)st\ *adjective*
— in the balance *or* **in balance :** with the fate or outcome about to be determined
— on balance : with all things considered

[2]**balance** *verb* **bal·anced; bal·anc·ing** (1588)
transitive verb
1 a (1) **:** to compute the difference between the debits and credits of (an account) (2) **:** to pay the amount due on **:** SETTLE **b** (1) **:** to arrange so that one set of elements exactly equals another ⟨*balance* a mathematical equation⟩ (2) **:** to complete (a chemical equation) so that the same number of atoms and electric charges of each kind appears on each side
2 a : COUNTERBALANCE, OFFSET **b :** to equal or equalize in weight, number, or proportion
3 : to weigh in or as if in a balance
4 a : to bring to a state or position of equipoise **b :** to poise in or as if in balance **c :** to bring into harmony or proportion
intransitive verb
1 : to become balanced or established in balance
2 : to be an equal counterpoise
3 : WAVER 1 ⟨*balances* and temporizes on matters that demand action⟩

balance beam *noun* (circa 1949)
1 : a narrow wooden beam supported in a horizontal position approximately four feet above the floor and used for balancing feats in gymnastics
2 : an event in gymnastics competition in which the balance beam is used

balance of payments (1844)
: a summary of the international transactions of a country or region over a period of time including commodity and service transactions, capital transactions, and gold movements

balance of power (1701)
: an equilibrium of power sufficient to discourage or prevent one nation or party from imposing its will on or interfering with the interests of another

balance of trade (1668)
: the difference in value over a period of time between a country's imports and exports

bal·anc·er \'ba-lən(t)-sər\ *noun* (15th century)
: one that balances; *specifically* **:** HALTERE

balance sheet *noun* (circa 1771)
: a statement of financial condition at a given date

balance wheel *noun* (1669)
1 : a wheel that regulates or stabilizes the motion of a mechanism
2 : a balancing or stabilizing force

balancing act *noun* (1954)
: an attempt to cope with several often conflicting factors or situations at the same time

Ba·lante \bə-'länt\ *noun, plural* **Balante** *or* **Balantes** [French, from Balante *Bulanda*] (circa 1895)
1 : a member of a people of Senegal and Guinea-Bissau
2 : the language of the Balante people

bal·as \'ba-ləs\ *noun* [Middle English, from Middle French *balais,* from Arabic *balakhsh,* from *Balakhshān,* ancient region of Afghanistan] (15th century)
: a ruby spinel of a pale rose-red or orange

ba·la·ta \bə-'lä-tə\ *noun* [Spanish, from Carib] (1860)
: a substance like gutta-percha that is the dried juice of tropical American trees (especially *Manilkara bidentata*) of the sapodilla family and is used especially in belting and golf balls; *also* **:** a tree yielding it

☆ SYNONYMS
Bait, badger, heckle, hector, chivy, hound mean to harass by efforts to break down. BAIT implies wanton cruelty or delight in persecuting a helpless victim ⟨*baited* the chained dog⟩. BADGER implies pestering so as to drive a person to confusion or frenzy ⟨*badgered* her father for a car⟩. HECKLE implies persistent annoying or belligerent interruptions of a speaker ⟨drunks *heckled* the stand-up comic⟩. HECTOR carries an implication of bullying and domineering ⟨football players *hectored* by their coach⟩. CHIVY suggests persecution by teasing or nagging ⟨*chivied* the new student mercilessly⟩. HOUND implies unrelenting pursuit and harassing ⟨*hounded* by creditors⟩.

bal·boa \bal-ˈbō-ə\ *noun* [Spanish, from Vasco Núñez de *Balboa*] (circa 1909)
— see MONEY table

bal·brig·gan \bal-ˈbri-gən\ *noun* [*Balbriggan,* town in Ireland] (1885)
: a knitted cotton fabric used especially for underwear or hosiery

bal·co·ny \ˈbal-kə-nē\ *noun, plural* **-nies** [Italian *balcone,* from Old Italian, large window, of Germanic origin; akin to Old High German *balko* beam — more at BALK] (1618)
1 : a platform that projects from the wall of a building and is enclosed by a parapet or railing
2 : an interior projecting gallery in a public building (as a theater)
— bal·co·nied \-nēd\ *adjective*

¹bald \ˈbȯld\ *adjective* [Middle English *balled;* probably akin to Danish *bældet* bald, Latin *fulica* coot, Greek *phalios* having a white spot] (14th century)
1 a : lacking a natural or usual covering (as of hair, vegetation, or nap) **b :** having little or no tread ⟨*bald* tires⟩
2 : UNADORNED
3 : UNDISGUISED, PALPABLE
4 : marked with white
synonym see BARE
— bald·ish \ˈbȯl-dish\ *adjective*
— bald·ly \ˈbȯl(d)-lē\ *adverb*
— bald·ness \ˈbȯl(d)-nəs\ *noun*

²bald (1602)
transitive verb
: to make bald
intransitive verb
: to become bald

bal·da·chin \ˈbȯl-də-kən, ˈbal-\ *or* **bal·da·chi·no** \ˌbal-də-ˈkē-(ˌ)nō, ˌbäl-\ *noun, plural* **baldachins** *or* **baldachinos** [Italian *baldacchino,* from *Baldacco* Baghdad, Iraq] (1537)
1 : a cloth canopy fixed or carried over an important person or a sacred object
2 : a rich embroidered fabric of silk and gold
3 : an ornamental structure resembling a canopy used especially over an altar

bald cypress *noun* (1709)
1 : either of two large swamp trees (*Taxodium distichum* and *T. ascendens* of the family Taxodiaceae, the bald cypress family) of the southern U.S. that are related to the sequoias
2 : the hard red wood of bald cypress that is much used for shingles

bald eagle *noun* (1688)
: an eagle (*Haliaeetus leucocephalus*) of North America that is brown when young with white only on the undersides of the wings but in full adult plumage has white head and neck feathers and a white tail

bald eagle

Bal·der \ˈbȯl-dər\ *noun* [Old Norse *Baldr*]
: the son of Odin and Frigga and Norse god of light and peace slain through the trickery of Loki by a mistletoe sprig

bal·der·dash \ˈbȯl-dər-ˌdash\ *noun* [origin unknown] (1674)
: NONSENSE

bald–faced \ˈbȯl(d)-ˈfāst\ *adjective* (1943)
: BAREFACED

bald·head \ˈbȯld-ˌhed\ *noun* (1535)
: a bald-headed person

balding *adjective* (1938)
: getting bald ⟨bespectacled and *balding*⟩

bald·pate \ˈbȯl(d)-ˌpāt\ *noun* (1592)
1 : BALDHEAD
2 : a North American wigeon (*Anas americana*) with a large white patch on each wing and in the male a white crown

bal·dric \ˈbȯl-drik\ *noun* [Middle English *baudry, baudrik,* from Middle French *baudré,* from Old French *baldrei*] (14th century)
: an often ornamented belt worn over one shoulder to support a sword or bugle

¹bale \ˈbā(ə)l\ *noun* [Middle English, from Old English *bealu;* akin to Old High German *balo* evil, Old Church Slavonic *bolĭ* sick person] (before 12th century)
1 : great evil
2 : WOE, SORROW

²bale *noun* [Middle English, from Middle French, of Germanic origin; akin to Old High German *balla* ball] (14th century)
: a large bundle of goods; *specifically* **:** a large closely pressed package of merchandise bound and usually wrapped ⟨a *bale* of paper⟩ ⟨a *bale* of hay⟩

³bale *transitive verb* **baled; bal·ing** (1760)
: to make up into a bale
— bal·er *noun*

ba·leen \bə-ˈlēn, ˈba-ˌlēn\ *noun* [Middle English *baleine* whale, baleen, from Latin *balaena* whale; akin to Greek *phallaina* whale] (14th century)
: a horny substance found in two rows of plates from 2 to 12 feet long attached along the upper jaws of baleen whales

baleen whale *noun* (1874)
: any of a suborder (Mysticeti) of usually large whales lacking teeth but having baleen which is used to filter chiefly small marine crustaceans (as krill) out of large quantities of seawater

bale·fire \ˈbā(ə)l-ˌfīr\ *noun* [Middle English, from Old English *bǣlfȳr* funeral fire, from *bǣl* pyre + *fȳr* fire] (before 12th century)
: an outdoor fire often used as a signal fire

bale·ful \-fəl\ *adjective* (before 12th century)
1 : deadly or pernicious in influence
2 : foreboding evil **:** OMINOUS
synonym see SINISTER
— bale·ful·ly \-fə-lē\ *adverb*
— bale·ful·ness \-fəl-nəs\ *noun*

Ba·li·nese \ˌbä-li-ˈnēz, ˌba-, -ˈnēs\ *noun* [Dutch *Balinees,* from *Bali* island of Indonesia] (1967)
: any of a breed of slender long-haired cats that originated as a spontaneous mutation of the Siamese

¹balk \ˈbȯk\ *noun* [Middle English *balke,* from Old English *balca;* akin to Old High German *balko* beam, Latin *fulcire* to prop, Greek *phalanx* log, phalanx] (before 12th century)
1 : a ridge of land left unplowed as a dividing line or through carelessness
2 : BEAM, RAFTER
3 : HINDRANCE, CHECK
4 a : the space behind the balkline on a billiard table **b :** any of the outside divisions made by the balklines
5 : failure of a player to complete a motion; *especially* **:** an illegal motion of the pitcher in baseball while in position

²balk (15th century)
transitive verb
1 *archaic* **:** to pass over or by
2 : to check or stop by or as if by an obstacle **:** BLOCK
intransitive verb
1 : to stop short and refuse to proceed
2 : to refuse abruptly — used with *at*
3 : to commit a balk in sports
synonym see FRUSTRATE
— balk·er *noun*

bal·kan·ize \ˈbȯl-kə-ˌnīz\ *transitive verb* **-ized; -iz·ing** *often capitalized* [Balkan Peninsula] (1919)
: to break up (as a region or group) into smaller and often hostile units
— bal·kan·i·za·tion \ˌbȯl-kə-nə-ˈzā-shən\ *noun, often capitalized*

balk·line \ˈbȯ-ˌklīn\ *noun* (1839)
1 : a line across a billiard table near one end behind which the cue balls are placed in making opening shots
2 a : one of four lines parallel to the cushions of a billiard table dividing it into nine compartments **b :** a billiards game that sets restrictions in scoring caroms according to these lines

balky \ˈbȯ-kē\ *adjective* **balk·i·er; -est** (1847)
: refusing or likely to refuse to proceed, act, or function as directed or expected ⟨a *balky* mule⟩
synonym see CONTRARY
— balk·i·ness *noun*

¹ball \ˈbȯl\ *noun, often attributive* [Middle English *bal,* from Old Norse *bǫllr;* akin to Old English *bealluc* testis, Old High German *balla* ball, Old English *blāwan* to blow] (13th century)
1 : a round or roundish body or mass: as **a :** a spherical or ovoid body used in a game or sport **b :** EARTH, GLOBE **c :** a spherical or conical projectile; *also* **:** projectiles used in firearms **d :** a roundish protuberant anatomic structure; *especially* **:** the rounded eminence at the base of the thumb or big toe
2 a : TESTIS — often considered vulgar **b** *plural* (1) **:** NONSENSE — often used interjectionally; often considered vulgar (2) **:** NERVE 3 — often considered vulgar
3 : a game in which a ball is thrown, kicked, or struck; *also* **:** the quality of play in such a game
4 a : a pitch not swung at by the batter that fails to pass through the strike zone **b :** a hit or thrown ball in various games ⟨foul *ball*⟩
— on the ball 1 : COMPETENT, KNOWLEDGEABLE, ALERT ⟨the other introductory essay . . . is much more *on the ball* —*Times Literary Supplement*⟩ ⟨keep *on the ball*⟩ **2 :** of ability or competence ⟨if the teacher has something *on the ball,* the pupils won't squirm much —*New Yorker*⟩

²ball (1658)
transitive verb
1 : to form or gather into a ball ⟨*balled* the paper into a wad⟩
2 : to have sexual intercourse with — usually considered vulgar
intransitive verb
1 : to form or gather into a ball
2 : to engage in sexual intercourse — usually considered vulgar

³ball *noun* [French *bal,* from Old French, from *baller* to dance, from Late Latin *ballare,* from Greek *ballizein*] (circa 1639)
1 : a large formal gathering for social dancing
2 : a very pleasant experience **:** a good time ⟨had a *ball* on their vacation⟩

bal·lad \ˈba-ləd\ *noun* [Middle English *balade* ballade, song, from Middle French, from Old Provençal *balada* dance, song sung while dancing, from *balar* to dance, from Late Latin *ballare*] (14th century)
1 a : a narrative composition in rhythmic verse suitable for singing **b :** an art song accompanying a traditional ballad
2 : a simple song **:** AIR
3 : a popular song; *especially* **:** a slow romantic or sentimental song
— bal·lad·ic \bə-ˈla-dik, ba-\ *adjective*

bal·lade \bə-ˈläd, ba-\ *noun* [Middle English *balade,* from Middle French, ballad, ballade] (14th century)

\ə\ abut \ᵊ\ kitten \ər\ further \a\ ash \ā\ ace \ä\ mop, mar \au̇\ out \ch\ chin \e\ bet \ē\ easy \g\ go \i\ hit \ī\ ice \j\ job \ŋ\ sing \ō\ go \ȯ\ law \ȯi\ boy \th\ thin \t͟h\ the \ü\ loot \u̇\ foot \y\ yet \zh\ vision *see also* Guide to Pronunciation

1 : a fixed verse form consisting usually of three stanzas with recurrent rhymes, an envoi, and an identical refrain for each part
2 : a musical composition usually for piano suggesting the epic ballad
bal·lad·eer \ˌba-lə-ˈdir\ *noun* (1830)
: a singer of ballads
bal·lad·ist \ˈba-lə-dist\ *noun* (1858)
: one who writes or sings ballads
bal·lad·ry \ˈba-lə-drē\ *noun* (1596)
1 : the composing or performing of ballads
2 : BALLADS
ballad stanza *noun* (circa 1934)
: a stanza consisting of four lines with the first and third lines unrhymed iambic tetrameters and the second and fourth lines rhymed iambic trimeters
ball–and–socket joint *noun* (1809)
1 : a joint in which a ball moves within a socket so as to allow rotary motion in every direction within certain limits

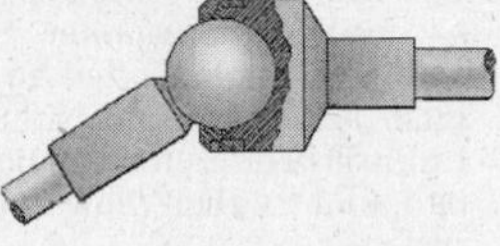
ball-and-socket joint 1

2 : an articulation (as the hip joint) in which the rounded head of one bone fits into a cuplike cavity of the other and admits movement in any direction
¹bal·last \ˈba-ləst\ *noun* [probably from Low German, of Scandinavian origin; akin to Danish & Swedish *barlast* ballast; perhaps akin to Old English *bær* bare & to Old English *hlæst* load, *hladan* to load — more at LADE] (1530)
1 : a heavy substance used to improve the stability and control the draft of a ship or the ascent of a balloon
2 : something that gives stability (as in character or conduct)
3 : gravel or broken stone laid in a railroad bed or used in making concrete
4 : a device used to provide the starting voltage or to stabilize the current in a circuit (as of a fluorescent lamp)
— in ballast *of a ship* **:** having only ballast for a load
²ballast *transitive verb* (1538)
1 : to steady or equip with or as if with ballast
2 : to fill in (as a railroad bed) with ballast
ball bearing *noun* (1883)
: a bearing in which the journal turns upon loose hardened steel balls that roll easily in a race; *also* **:** one of the balls in such a bearing
ball boy *noun* (1903)
: a male tennis court attendant who retrieves balls for the players
ball·car·ri·er \ˈbȯl-ˌkar-ē-ər\ *noun* (1935)
: the football player carrying the ball on an offensive play
ball cock *noun* (1790)
: an automatic valve whose opening and closing are controlled by a spherical float at the end of a lever
ball control *noun* (1928)
: an offensive strategy (as in football) in which a team tries to maintain possession of the ball for extended periods of time
bal·le·ri·na \ˌba-lə-ˈrē-nə\ *noun* [Italian, from *ballare* to dance, from Late Latin] (1815)
: a woman who is a ballet dancer **:** DANSEUSE
bal·let \ˈba-ˌlā, ba-ˈ\ *noun* [French, from Italian *balletto,* diminutive of *ballo* dance, from *ballare*] (1634)
1 a : a theatrical art form using dancing, music, and scenery to convey a story, theme, or atmosphere **b :** dancing in which conventional poses and steps are combined with light flowing figures (as leaps and turns)
2 : music for a ballet
3 : a group that performs ballets
— bal·let·ic \ba-ˈle-tik\ *adjective*
bal·let·o·mane \ba-ˈle-tə-ˌmān\ *noun* [*ballet* + *-o-* + *-mane* (from *mania*)] (1930)
: a devotee of ballet
— bal·let·o·ma·nia \-ˌle-tə-ˈmā-nē-ə, -nyə\ *noun*
ball–flow·er \ˈbȯl-ˌflau̇(-ə)r\ *noun* (1845)
: an architectural ornament consisting of a ball in the flower-shaped hollow of a circular mold
ball game *noun* (1848)
1 : a game played with a ball
2 a : a set of circumstances **:** SITUATION ⟨a whole new *ball game*⟩ **b :** CONTEST 1
ball girl *noun* (1926)
: a female tennis court attendant who retrieves balls for the players
ball handler *noun* (1948)
: a player who controls the ball in any of various games; *especially* **:** a player who is skilled at handling the ball (as in basketball)
— ball·han·dling *noun*
bal·lis·ta \bə-ˈlis-tə\ *noun, plural* **-tae** \-ˌtē\ [Latin, from (assumed) Greek *ballistēs,* from *ballein* to throw — more at DEVIL] (14th century)
: an ancient military engine often in the form of a crossbow for hurling large missiles
bal·lis·tic \bə-ˈlis-tik\ *adjective* [Latin *ballista*] (circa 1775)
1 : of or relating to ballistics or to a body in motion according to the laws of ballistics
2 : suddenly and extremely upset or angry **:** WILD — usually used with *go*
— bal·lis·ti·cal·ly \-ti-k(ə-)lē\ *adverb*
ballistic missile *noun* (1954)
: a missile guided in the ascent of a high-arch trajectory and freely falling in the descent
bal·lis·tics \bə-ˈlis-tiks\ *noun plural but singular or plural in construction* (circa 1753)
1 a : the science of the motion of projectiles in flight **b :** the flight characteristics of a projectile
2 a : the study of the processes within a firearm as it is fired **b :** the firing characteristics of a firearm or cartridge
ball joint *noun* (circa 1884)
: BALL-AND-SOCKET JOINT 1
ball lightning *noun* (1857)
: a rare form of lightning consisting of luminous balls that may move along solid objects or float in the air
ball mill *noun* (1903)
: a pulverizing machine consisting of a rotating drum which contains pebbles or metal balls as the grinding implements
ball of fire (circa 1900)
: a person of unusual energy, vitality, or drive
ball of wax (circa 1953)
: AFFAIR, CONCERN ⟨the whole *ball of wax*⟩
bal·lon \ba-ˈlōⁿ\ *noun* [French, literally, balloon] (1830)
: lightness of movement that exaggerates the duration of a ballet dancer's jump
bal·lo·net \ˌba-lə-ˈnā\ *noun* [French *ballonnet,* diminutive of *ballon*] (1902)
: a compartment of variable volume within the interior of a balloon or airship used to control ascent and descent
¹bal·loon \bə-ˈlün\ *noun* [French *ballon* large football, balloon, from Italian dialect *ballone* large football, augmentative of *balla* ball, of Germanic origin] (1783)
1 : a nonporous bag of tough light material filled with heated air or a gas lighter than air so as to rise and float in the atmosphere
2 : an inflatable bag (as of rubber) usually used as a toy or for decoration
3 : the outline enclosing words spoken or thought by a figure especially in a cartoon
²balloon *adjective* (circa 1786)
1 : relating to, resembling, or suggesting a balloon ⟨a *balloon* sleeve⟩
2 : being or having a final installment that is much larger than preceding ones in a term or installment note
³balloon (1841)
intransitive verb
1 : to swell or puff out **:** EXPAND ⟨*ballooned* to 200 pounds⟩
2 : to ascend or travel in a balloon
3 : to increase rapidly
transitive verb
: INFLATE, INCREASE
balloon angioplasty *noun* (1980)
: dilatation of an atherosclerotically obstructed artery by the passage of a balloon catheter through the vessel to the area of disease where inflation of the catheter compresses the plaque against the vessel wall
balloon catheter *noun* (1961)
: a catheter with an inflatable tip that is used especially to measure blood pressure in a blood vessel or to expand a partly closed or obstructed bodily passage or tube (as a coronary artery)
bal·loon·ing \bə-ˈlü-niŋ\ *noun* (1784)
: the act or sport of riding in a balloon
bal·loon·ist \-nist\ *noun* (1784)
: a person who rides in a balloon
balloon tire *noun* (1923)
: a low-pressure pneumatic tire with a flexible carcass and large cross section designed to provide cushioning
balloon vine *noun* (1836)
: a tropical American vine (*Cardiospermum halicacabum*) of the soapberry family bearing large ornamental pods
¹bal·lot \ˈba-lət\ *noun* [Italian *ballotta,* from Italian dialect, diminutive of *balla* ball] (1549)
1 a : a small ball used in secret voting **b :** a sheet of paper used to cast a secret vote
2 a : the action or system of secret voting **b :** the right to vote **c :** VOTE 1a
3 : the number of votes cast
4 : the drawing of lots ◆
²ballot *intransitive verb* (1580)
: to vote or decide by ballot
— bal·lot·er *noun*
ballot box *noun* (circa 1680)
1 : a box for receiving ballots
2 : BALLOT 2a
¹ball·park \ˈbȯl-ˌpärk\ *noun* (1899)
1 : a park in which ball games (as baseball) are played
2 : a range (as of prices or views) within which comparison or compromise is possible
— in the ballpark : approximately correct
²ballpark *adjective* [from the expression *in the ballpark*] (1967)
: approximately correct ⟨a *ballpark* figure⟩
ball·play·er \ˈbȯl-ˌplā-ər\ *noun* (1619)
: a person who plays ball; *especially* **:** a baseball player
ball·point \-ˌpȯint\ *noun* (1953)
: a pen having as the writing point a small rotating metal ball that inks itself by contact with an inner magazine
ball·room \ˈbȯl-ˌrüm, -ˌru̇m\ *noun* (1736)
: a large room used for dances

◇ WORD HISTORY
ballot Small objects have long been used as a means of tallying votes. In ancient Greece, a word for "pebble," *psēphos,* came to designate any voting token, and then to mean simply "vote," from the practice of dropping a pebble into one of two urns as decisions were made in public assemblies. (Today, the scientific study of elections is sometimes referred to as *psephology.*) In late medieval Venice, small balls were utilized in public lotteries and elections; one lottery involved the placement of gold and silver balls in urns on a pedestal, from which members of the Great Council would draw to determine nominating committees for state officeholders. In the Venetian form of Italian at that time, the word for "small ball" was *ballotta* (corresponding to Tuscan *pallottola*), which was extended to other tokens used in drawing lots and voting, such as scraps of linen or paper. Familiarity with Venetian electoral practices led to the adoption of the word in English.

balls–up \'bȯl-ˌzəp\ *noun* (1939)
British **:** FOUL-UP

ball•sy \'bȯl-zē\ *adjective* **ball•si•er; -est** [[1]*ball*] (1959)
: aggressively tough **:** GUTSY — sometimes considered vulgar

ball up (1885)
transitive verb
: to make a mess of **:** CONFUSE, MUDDLE
intransitive verb
: to become badly muddled or confused

ball valve *noun* (1839)
: a valve in which a ball regulates the aperture by its rise and fall due to fluid pressure, a spring, or its own weight

bally \'ba-lē\ *adjective or adverb* [euphemism for *bloody,* adjective, adverb] (1885)
British — used as an intensive

bal•ly•hoo \'ba-lē-ˌhü\ *noun, plural* **-hoos** [origin unknown] (1901)
1 : a noisy attention-getting demonstration or talk
2 : flamboyant, exaggerated, or sensational advertising or propaganda
— ballyhoo *transitive verb*

bal•ly•rag \-ˌrag\ *variant of* BULLYRAG

balm \'bä(l)m, *New England also* 'bȧm\ *noun* [Middle English *basme, baume,* from Old French, from Latin *balsamum* balsam] (13th century)
1 : a balsamic resin; *especially* **:** one from small tropical evergreen trees (genus *Commiphora* of the family Burseraceae)
2 : an aromatic preparation (as a healing ointment)
3 : any of several aromatic plants of the mint family; *especially* **:** LEMON BALM
4 : a spicy aromatic odor
5 : a soothing restorative agency

bal•ma•caan \ˌbal-mə-'kan, -'kän\ *noun* [*Balmacaan,* estate near Inverness, Scotland] (1919)
: a loose single-breasted overcoat usually having raglan sleeves and a short turnover collar

balm of Gil•e•ad \-'gi-lē-əd\ [*Gilead,* region of ancient Palestine known for its balm] (1703)
1 : a small evergreen African and Asian tree (*Commiphora meccanensis* of the family Burseraceae) with aromatic leaves; *also* **:** a fragrant oleoresin from this tree
2 : an agency that soothes, relieves, or heals
3 : either of two poplars: **a :** a hybrid northern tree (*Populus gileadensis*) with broadly cordate leaves that are pubescent especially on the underside **b :** BALSAM POPLAR

bal•mor•al \bal-'mȯr-əl, -'mär-\ *noun* [*Balmoral* Castle, Scotland] (1859)
1 : a laced boot or shoe
2 *often capitalized* **:** a round flat cap with a top projecting all around

balmy \'bä-mē, 'bäl-mē, *New England also* 'bȧ-mē\ *adjective* **balm•i•er; -est** (15th century)
1 a : having the qualities of balm **:** SOOTHING **b :** MILD
2 : CRAZY, FOOLISH
— balm•i•ly \-mə-lē\ *adverb*
— balm•i•ness \-mē-nəs\ *noun*

bal•ne•ol•o•gy \ˌbal-nē-'ä-lə-jē\ *noun* [International Scientific Vocabulary, from Latin *balneum* bath — more at BAGNIO] (circa 1879)
: the science of the therapeutic use of baths

[1]ba•lo•ney \bə-'lō-nē\ *variant of* BOLOGNA

[2]baloney *noun* [*bologna*] (1923)
: pretentious nonsense **:** BUNKUM — often used as a generalized expression of disagreement

bal•sa \'bȯl-sə\ *noun* [Spanish] (circa 1600)
1 : a small raft or boat; *specifically* **:** one made of tightly bundled reeds and used on Lake Titicaca
2 : a tropical American tree (*Ochroma pyramidale* synonym *O. lagopus*) of the silk-cotton family with extremely light strong wood used especially for floats; *also* **:** its wood

bal•sam \'bȯl-səm\ *noun* [Latin *balsamum,* from Greek *balsamon,* probably of Semitic origin; akin to Hebrew *bāshām* balsam] (before 12th century)
1 a : an aromatic and usually oily and resinous substance flowing from various plants; *especially* **:** any of several resinous substances containing benzoic or cinnamic acid and used especially in medicine **b :** a preparation containing resinous substances and having a balsamic odor
2 a : a balsam-yielding tree; *especially* **:** BALSAM FIR **b :** IMPATIENS; *especially* **:** a common garden ornamental (*Impatiens balsamina*)
3 : BALM 5
— bal•sam•ic \bȯl-'sa-mik\ *adjective*

balsam fir *noun* (1805)
: a resinous American fir (*Abies balsamea*) that is widely used for pulpwood and as a Christmas tree and is the source of Canada balsam

balsamic vinegar *noun* [translation of Italian *aceto balsamico,* literally, curative vinegar] (1982)
: an aged Italian vinegar made from the must of white grapes

balsam poplar *noun* (1819)
: a North American poplar (*Populus balsamifera*) that is often cultivated as a shade tree and has buds thickly coated with an aromatic resin — called also *balm of Gilead, tacamahac*

Bal•ti \'bəl-tē, 'bȯl-\ *noun* (1901)
: a Tibeto-Burman language of northern Kashmir

Bal•tic \'bȯl-tik\ *adjective* [Medieval Latin (*mare*) *balticum* Baltic Sea] (circa 1590)
1 : of or relating to the Baltic Sea or to the states of Lithuania, Latvia, and Estonia
2 : of or relating to a branch of the Indo-European language family containing Latvian, Lithuanian, and Old Prussian — see INDO-EUROPEAN LANGUAGES table

Bal•ti•more oriole \'bȯl-tə-ˌmōr-, -ˌmȯr-, -mər-\ *noun* [George Calvert, Lord *Baltimore*] (1808)
: a northern oriole (*Icterus galbula galbula*) of the eastern and central U.S. and southern Canada in which the male has a solid black head and the female usually has an olive brown back and orange yellow underside

Bal•to–Slav•ic \ˌbȯl-(ˌ)tō-'slä-vik, -'sla-\ *noun* (1896)
: a subgroup of Indo-European languages consisting of the Baltic and the Slavic branches — see INDO-EUROPEAN LANGUAGES table

Ba•lu•chi \bə-'lü-chē\ *noun, plural* **Baluchis** *also* **Baluchi** [Persian *Balūchī* of the Baluchis, from *Balūch, Baloch* Baluchi] (1616)
1 : a member of an Indo-Iranian people of Baluchistan
2 : the Iranian language of the Baluchi people

bal•us•ter \'ba-lə-stər\ *noun* [French *balustre,* from Italian *balaustro,* from *balaustra* wild pomegranate flower, from Latin *balaustium,* from Greek *balaustion;* from its shape] (1602)
1 : an object or vertical member (as the leg of a table, a round in a chair back, or the stem of a glass) having a vaselike or turned outline
2 : an upright often vase-shaped support for a rail

bal•us•trade \'ba-lə-ˌstrād\ *noun* [French, from Italian *balaustrata,* from *balaustro*] (1644)
1 : a row of balusters topped by a rail
2 : a low parapet or barrier
— bal•us•trad•ed \-ˌstrā-dəd\ *adjective*

bam *noun* [imitative] (1930)
: a dull resounding noise (as of a hard blow or impact) — often used interjectionally

Bam•ba•ra \bam-'bär-ə\ *noun, plural* **Bambara** *or* **Bambaras** (1883)
1 : a member of an African people of the upper Niger
2 : a Mande language of the Bambara people

bam•bi•no \bam-'bē-(ˌ)nō, bäm-\ *noun, plural* **-nos** *or* **-ni** \-(ˌ)nē\ [Italian, diminutive of *bambo* child] (1722)
1 *plural usually bambini* **:** a representation of the infant Christ
2 : CHILD, BABY

bam•boo \(ˌ)bam-'bü, 'bam-ˌ\ *noun, plural* **bamboos** *often attributive* [Malay *bambu*] (1586)
: any of various chiefly tropical woody or arborescent grasses (as of the genera *Bambusa, Arundinaria,* and *Dendrocalamus*) including some with hollow stems used for building, furniture, or utensils and young shoots used for food

bamboo

bamboo curtain *noun, often B&C capitalized* (1949)
: a political, military, and ideological barrier isolating an area of the Orient

bam•boo•zle \bam-'bü-zəl\ *transitive verb* **-boo•zled; -boo•zling** \-'büz-liŋ, -'bü-zə-\ [origin unknown] (1703)
: to deceive by underhanded methods **:** DUPE, HOODWINK
— bam•boo•zle•ment \-'bü-zəl-mənt\ *noun*

[1]ban \'ban\ *verb* **banned; ban•ning** [Middle English *bannen* to summon, curse, from Old English *bannan* to summon; akin to Old High German *bannan* to command, Latin *fari* to speak, Greek *phanai* to say, *phōnē* sound, voice] (12th century)
transitive verb
1 *archaic* **:** CURSE
2 : to prohibit especially by legal means ⟨*ban* discrimination⟩; *also* **:** to prohibit the use, performance, or distribution of ⟨*ban* a book⟩ ⟨*ban* a pesticide⟩
intransitive verb
: to utter curses or maledictions

[2]ban *noun* [Middle English, partly from *bannen* & partly from Old French *ban,* of Germanic origin; akin to Old High German *bannan* to command] (14th century)
1 : the summoning in feudal times of the king's vassals for military service
2 : ANATHEMA, EXCOMMUNICATION
3 : MALEDICTION, CURSE
4 : legal or formal prohibition
5 : censure or condemnation especially through social pressure

[3]ban \'bän\ *noun, plural* **ba•ni** \'bä-(ˌ)nē\ [Romanian] (1880)
— see *leu* at MONEY table

Ba•nach space \'bä-ˌnäk-, -nək-\ *noun* [Stefan *Banach* (died 1945) Polish mathematician] (1949)
: a normed vector space for which the field of multipliers comprises the real or complex numbers and in which every Cauchy sequence converges to a point in the space

ba•nal \bə-'nal, ba-, -'näl; bā-'nal; 'bā-nᵊl\ *adjective* [French, from Middle French, of compulsory feudal service, possessed in common, commonplace, from *ban*] (1840)
: lacking originality, freshness, or novelty **:** TRITE
synonym see INSIPID

— **ba•nal•i•ty** \bə-'na-lə-tē *also* bā- *or* ba-\ *noun*
— **ba•nal•ize** \bə-'na-,līz, ba-, -'nā-; bā-'na-; 'bā-nᵊl-,īz\ *transitive verb*
— **ba•nal•ly** \bə-'nal-lē, ba-, -'näl-; bā-'nal-; 'bā-nᵊl-(l)ē\ *adverb*

ba•nana \bə-'na-nə, *especially British* -'nä-\ *noun, often attributive* [Spanish or Portuguese; Spanish, from Portuguese, of African origin; akin to Wolof *banäna* banana] (1597)
1 : an elongated usually tapering tropical fruit with soft pulpy flesh enclosed in a soft usually yellow rind **2 :** any of several widely cultivated perennial herbs (genus *Musa* of the family Musaceae, the banana family) bearing bananas in compact pendent bunches

banana 2

banana oil *noun* (1926)
: a colorless liquid acetate $C_7H_{14}O_2$ of amyl alcohol that has a pleasant fruity odor and is used as a solvent and in the manufacture of artificial fruit essences

banana republic *noun* (1935)
: a small dependent country usually of the tropics; *especially* **:** one run despotically

ba•nan•as \bə-'na-nəz, *especially British* -'nä-\ *adjective* (1968)
: CRAZY ⟨go *bananas*⟩ ⟨drives me *bananas*⟩
word history see NUTS

banana seat *noun* (1965)
: an elongated bicycle saddle

bananas Fos•ter \-'fós-tər, -'fäs-\ *noun* [Richard *Foster,* friend of New Orleans restaurateur Owen E. Brennan, at whose restaurant the dish was first made] (1976)
: a dessert of bananas flamed (as with rum) and served with ice cream

banana split *noun* (1920)
: ice cream served on a banana sliced in half lengthwise and usually garnished with flavored syrups, fruits, nuts, and whipped cream

ba•nau•sic \bə-'nó-sik, -zik\ *adjective* [Greek *banausikos* of an artisan, nonintellectual, vulgar, from *banausos* artisan] (1845)
: relating to or concerned with earning a living — used pejoratively ⟨contempt for the *banausic* occupations —T. S. Eliot⟩; *also* **:** UTILITARIAN, PRACTICAL ⟨such mundane and *banausic* considerations as comfort and durability —G. B. Boyer⟩

¹band \'band\ *noun* [in senses 1 & 2, from Middle English *band, bond* something that constricts, from Old Norse *band;* akin to Old English *bindan* to bind; in other senses, from Middle English *bande* strip, from Middle French, from (assumed) Vulgar Latin *binda,* of Germanic origin; akin to Old High German *binta* fillet; akin to Old English *bindan* to bind, *bend* fetter] (12th century)
1 : something that confines or constricts while allowing a degree of movement
2 : something that binds or restrains legally, morally, or spiritually
3 : a strip serving to join or hold things together: as **a :** BELT 2 **b :** a cord or strip across the back of a book to which the sections are sewn
4 : a thin flat encircling strip especially for binding: as **a :** a close-fitting strip that confines material at the waist, neck, or cuff of clothing **b :** a strip of cloth used to protect a newborn baby's navel — called also *bellyband* **c :** a ring of elastic
5 a : a strip (as of living tissue or rock) or a stripe (as on an animal) differentiable (as by color, texture, or structure) from the adjacent material or area **b :** a more or less well-defined range of wavelengths, frequencies, or energies **c :** RANGE 7a
6 : a narrow strip serving chiefly as decoration: as **a :** a narrow strip of material applied as trimming to an article of dress **b** *plural* **:** a pair of strips hanging at the front of the neck as part of a clerical, legal, or academic dress **c :** a ring without raised portions
7 : TRACK 1e(2)

²band (15th century)
transitive verb
1 : to affix a band to or tie up with a band
2 : to finish or decorate with a band
3 : to gather together **:** UNITE ⟨*banded* themselves together for protection⟩
intransitive verb
: to unite for a common purpose — often used with *together* ⟨have *banded* together in hopes of attacking the blight that is common to them all —J. B. Conant⟩
— **band•er** *noun*

³band *noun* [Middle French *bande* troop, from Old Provençal *banda,* of Germanic origin; akin to Gothic *bandwo* sign, standard — more at BANNER] (15th century)
: a group of persons, animals, or things; *especially* **:** a group of musicians organized for ensemble playing and using chiefly woodwinds, brass, and percussion instruments — compare ORCHESTRA

¹ban•dage \'ban-dij\ *noun* [Middle French, from *bande*] (1599)
1 : a strip of fabric used especially to dress and bind up wounds
2 : a flexible strip or band used to cover, strengthen, or compress something

²bandage *transitive verb* **ban•daged; ban•dag•ing** (1774)
: to bind, dress, or cover with a bandage

Band–Aid \'ban-'dād\ *trademark*
— used for a small adhesive strip with a gauze pad for covering minor wounds

ban•dan•na *or* **ban•dana** \ban-'da-nə\ *noun* [Hindi *bā̃dhnū* tie-dyeing, cloth so dyed, from *bā̃dhnā* to tie, from Sanskrit *badhnāti* he ties; akin to Old English *bindan*] (1741)
: a large often colorfully patterned handkerchief

¹band•box \'ban(d)-,bäks\ *noun* (1631)
1 : a usually cylindrical box of paperboard or thin wood for holding light articles of attire
2 : a structure (as a baseball park) having relatively small interior dimensions

²bandbox *adjective* (1844)
: exquisitely neat, clean, or ordered as if just taken from a bandbox ⟨a *bandbox* perfection of appearance⟩
— **bandbox** *adverb*

ban•deau \ban-'dō\ *noun, plural* **ban•deaux** \-'dōz\ [French, diminutive of *bande*] (1706)
1 : a fillet or band especially for the hair
2 : BRASSIERE; *also* **:** a band-shaped covering for the breasts

band•ed \'ban-dəd\ *adjective* (1787)
: having or marked with bands

ban•de•ri•lla \,ban-də-'rē(l)-yə\ *noun* [Spanish, diminutive of *bandera* banner] (1797)
: a decorated barbed dart that the banderillero thrusts into the neck or shoulders of the bull in a bullfight

ban•de•ri•lle•ro \,ban-də-(,)rē(l)-'yer-(,)ō\ *noun, plural* **-ros** [Spanish, from *banderilla*] (1797)
: one who thrusts in the banderillas in a bullfight

ban•de•role *or* **ban•de•rol** \'ban-də-,rōl\ *noun* [French *banderole,* from Italian *banderuola,* diminutive of *bandiera* banner, of Germanic origin; akin to Gothic *bandwo* sign] (1562)
1 : a long narrow forked flag or streamer
2 : a long scroll bearing an inscription or a device

ban•di•coot \'ban-di-,küt\ *noun* [Telugu *pandikokku*] (1813)
1 : any of several very large rats (genera *Bandicota* and *Nesokia*) of southern Asia destructive to crops
2 : any of a family (Peramelidae) of small insectivorous and herbivorous marsupial mammals of Australia, Tasmania, and New Guinea

ban•dit \'ban-dət\ *noun* [Italian *bandito,* from past participle of *bandire* to banish, of Germanic origin; akin to Old High German *bannan* to command — more at BAN] (1591)
1 *plural also* **ban•dit•ti** \ban-'di-tē\ **:** an outlaw who lives by plunder; *especially* **:** a member of a band of marauders
2 : ROBBER
3 : an enemy plane
— **ban•dit•ry** \'ban-də-trē\ *noun*

band•lead•er \'band-,lē-dər\ *noun* (1894)
: the conductor of a band (as a dance band)

band•mas•ter \'ban(d)-,mas-tər\ *noun* (1858)
: BANDLEADER; *especially* **:** a conductor of a military or concert band

ban•dog \'ban-,dóg\ *noun* [Middle English *bandogge,* from *band* + *dogge* dog] (14th century)
: a dog kept tied to serve as a watchdog or because of its ferocity

ban•do•lier *or* **ban•do•leer** \,ban-də-'lir\ *noun* [Middle French *bandouliere,* ultimately from Old Spanish *bando* band, of Germanic origin; akin to Gothic *bandwo*] (circa 1577)
: a belt worn over the shoulder and across the breast often for the suspending or supporting of some article (as cartridges) or as a part of an official or ceremonial dress

ban•dore \'ban-,dōr, -,dór\ *or* **ban•do•ra** \ban-'dōr-ə, -'dór-\ *noun* [Spanish *bandurria* or Portuguese *bandurra,* from Late Latin *pandura* 3-stringed lute, from Greek *pandoura*] (1566)
: a bass stringed instrument resembling a guitar

band–pass filter \'ban(d)-,pas-\ *noun* (1926)
: a filter that transmits only frequencies within a selected band

band saw *noun* (circa 1864)
: a saw in the form of an endless steel belt running over pulleys; *also* **:** a power sawing machine using this device

band shell *noun* (1926)
: a bandstand having at the rear a sounding board shaped like a huge concave seashell

bands•man \'ban(d)z-mən\ *noun* (circa 1842)
: a member of a musical band

band•stand \'ban(d)-,stand\ *noun* (1859)
1 : a usually roofed platform on which a band or orchestra performs outdoors
2 : a platform in a ballroom or nightclub on which musicians perform

band•wag•on \'band-,wa-gən\ *noun* (1855)
1 : a usually ornate and high wagon for a band of musicians especially in a circus parade
2 : a popular party, faction, or cause that attracts growing support — often used in such phrases as *climb on the bandwagon*
3 : a current or fashionable trend

band•width \'band-,width\ *noun* (circa 1937)
1 : a range within a band of wavelengths, frequencies, or energies; *especially* **:** a range of radio frequencies which is occupied by a modulated carrier wave, which is assigned to a service, or over which a device can operate
2 : the data transfer rate of an electronic communications system

¹ban•dy \'ban-dē\ *verb* **ban•died; ban•dy•ing** [probably from Middle French *bander* to be tight, to bandy, from *bande* strip — more at BAND] (1577)
transitive verb
1 : to bat (as a tennis ball) to and fro
2 a : to toss from side to side or pass about from one to another often in a careless or inappropriate manner **b :** EXCHANGE; *especially* **:** to exchange (words) argumentatively **c :** to discuss lightly or banteringly **d :** to use in a glib or offhand manner — often used with

about ⟨*bandy* these statistics about with considerable bravado —Richard Pollak⟩
3 *archaic* **:** to band together
intransitive verb
1 *obsolete* **:** CONTEND
2 *archaic* **:** UNITE

[2]bandy *noun* [perhaps from Middle French *bandé,* past participle of *bander*] (1693)
: a game similar to hockey and believed to be its prototype

[3]bandy *adjective* [probably from *bandy* (hockey stick)] (1687)
1 *of legs* **:** BOWED
2 : BOWLEGGED
— ban·dy–legged \'ban-dē-ˌlegd; ˌban-dē-'le-gəd, -'lā-\ *adjective*

[1]bane \'bān\ *noun* [Middle English, from Old English *bana;* akin to Old High German *bano* death] (before 12th century)
1 a *obsolete* **:** KILLER, SLAYER **b :** POISON **c :** DEATH, DESTRUCTION ⟨stop the way of those that seek my *bane* —Philip Sidney⟩ **d :** WOE
2 : a source of harm or ruin **:** CURSE ⟨national frontiers have been more of a *bane* than a boon for mankind —D. C. Thomson⟩

[2]bane *transitive verb* **baned; ban·ing** (1578)
obsolete **:** to kill especially with poison

[3]bane *noun* [Middle English (northern dialect) *ban,* from Old English *bān*] (before 12th century)
chiefly Scottish **:** BONE

bane·ber·ry \'bān-ˌber-ē\ *noun* (1755)
: any of several perennial herbs (genus *Actaea*) of the buttercup family having acrid poisonous berries; *also* **:** one of the berries

bane·ful \'bān-fəl\ *adjective* (1579)
1 : productive of destruction or woe **:** seriously harmful ⟨a *baneful* influence⟩
2 *archaic* **:** POISONOUS
synonym see PERNICIOUS
— bane·ful·ly \-fə-lē\ *adverb*

[1]bang \'baŋ\ *verb* [probably of Scandinavian origin; akin to Icelandic *banga* to hammer] (circa 1550)
transitive verb
1 : to strike sharply **:** BUMP
2 : to knock, beat, or thrust vigorously often with a sharp noise
3 : to have sexual intercourse with — often considered vulgar
intransitive verb
1 : to strike with a sharp noise or thump
2 : to produce a sharp often metallic explosive or percussive noise or series of such noises

[2]bang *noun* (circa 1550)
1 : a resounding blow
2 : a sudden loud noise — often used interjectionally
3 a : a sudden striking effect **b :** a quick burst of energy ⟨start off with a *bang*⟩ **c :** THRILL ⟨I get a *bang* out of all this —W. H. Whyte⟩
— bang for the buck *also* **bang for one's buck :** value received from outlay or effort ⟨investment is yielding less *bang for the buck* —*Fortune*⟩

[3]bang *adverb* (1828)
: RIGHT, DIRECTLY ⟨ran *bang* up against more trouble⟩

[4]bang *noun* [probably short for *bangtail* (short tail)] (1878)
: a fringe of banged hair — usually used in plural

[5]bang *transitive verb* (1878)
: to cut (as front hair) short and squarely across

ban·ga·lore torpedo \'baŋ-gə-ˌlōr-, -ˌlȯr-\ *noun* [*Bangalore,* India] (1913)
: a metal tube that contains explosives and a firing mechanism and is used to cut barbed wire and detonate buried mines

bang away *intransitive verb* (circa 1889)
1 : to work with determined effort ⟨students *banging away* at their homework⟩
2 : to attack persistently ⟨police are going to keep *banging away* at you —Erle Stanley Gardner⟩

bang·er \'baŋ-ər\ *noun* (circa 1919)
1 *British* **:** SAUSAGE
2 *British* **:** FIRECRACKER
3 *British* **:** JALOPY

bang·kok \'baŋ-ˌkäk, baŋ-'\ *noun* [earlier *bangkok,* a fine straw, from *Bangkok,* Thailand] (1916)
: a hat woven of fine palm fiber in the Philippines

ban·gle \'baŋ-gəl\ *noun* [Hindi *baṅglī*] (1787)
1 : a stiff usually ornamental bracelet or anklet slipped or clasped on
2 : an ornamental disk that hangs loosely (as on a bracelet)

bang on *adjective* (1936)
chiefly British **:** exactly correct or appropriate

Bang's disease \'baŋz-\ *noun* [Bernhard L. F. *Bang* (died 1932) Danish veterinarian] (circa 1929)
: BRUCELLOSIS; *specifically* **:** contagious abortion of cattle caused by a brucella (*Brucella abortus*)

bang·tail \'baŋ-ˌtāl\ *noun* [*bangtail* (short tail)] (1921)
: RACEHORSE

bang–up \'baŋ-ˌəp\ *adjective* [[3]*bang*] (1810)
: FIRST-RATE, EXCELLENT ⟨a *bang-up* job⟩

bang up *transitive verb* [[1]*bang*] (1920)
: to cause extensive damage to

bani *plural of* [3]BAN

ban·ish \'ba-nish\ *transitive verb* [Middle English, from Middle French *baniss-,* stem of *banir,* of Germanic origin; akin to Old High German *bannan* to command — more at BAN] (14th century)
1 : to require by authority to leave a country
2 : to drive out or remove from a home or place of usual resort or continuance
3 : to clear away **:** DISPEL ⟨his discovery *banishes* anxiety —Stringfellow Barr⟩ ☆
— ban·ish·er *noun*
— ban·ish·ment \-nish-mənt\ *noun*

ban·is·ter \'ba-nəs-tər\ *noun* [alteration of *baluster*] (1667)
1 : BALUSTER 2
2 a : a handrail with its supporting posts **b :** HANDRAIL
— ban·is·tered \-tərd\ *adjective*

ban·jax \'ban-ˌjaks\ *transitive verb* [origin unknown] (1939)
chiefly Irish **:** DAMAGE, RUIN; *also* **:** SMASH

ban·jo \'ban-(ˌ)jō\ *noun, plural* **banjos** *also* **banjoes** [probably of African origin; akin to Kimbundu *mbanza,* a similar instrument] (1739)
: a musical instrument with a drumlike body, a fretted neck, and usually four or five strings which may be plucked or strummed
— ban·jo·ist \-ˌjō-ist\ *noun*

banjo

banjo clock *noun* (1903)
: a pendulum clock whose shape suggests a banjo

[1]bank \'baŋk\ *noun* [Middle English, probably of Scandinavian origin; akin to Old Norse *bakki* bank; akin to Old English *benc* bench — more at BENCH] (13th century)
1 : a mound, pile, or ridge raised above the surrounding level: as **a :** a piled-up mass of cloud or fog **b :** an undersea elevation rising especially from the continental shelf
2 : the rising ground bordering a lake, river, or sea or forming the edge of a cut or hollow
3 a : a steep slope (as of a hill) **b :** the lateral inward tilt of a surface along a curve or of a vehicle (as an airplane) when taking a curve
4 : a protective or cushioning rim or piece ◇

[2]bank (1590)
transitive verb
1 a : to raise a bank about **b :** to cover (as a fire) with fresh fuel and adjust the draft of air so as to keep in an inactive state **c :** to build (a curve) with the roadbed or track inclined laterally upward from the inside edge
2 : to heap or pile in a bank
3 : to drive (a ball in billiards) into a cushion
4 : to form or group in a tier
intransitive verb
1 : to rise in or form a bank — often used with *up* ⟨clouds would *bank up* about midday, and showers fall —William Beebe⟩
2 a : to incline an airplane laterally **b** (1) **:** to incline laterally (2) **:** to follow a curve or incline ⟨skiers *banking* around the turn⟩

[3]bank *noun* [Middle English, from Middle French or Old Italian; Middle French *banque,* from Old Italian *banca,* literally, bench, of Germanic origin; akin to Old English *benc*] (15th century)
1 a : an establishment for the custody, loan, exchange, or issue of money, for the extension of credit, and for facilitating the transmission of funds **b** *obsolete* **:** the table, counter, or place of business of a money changer
2 : a person conducting a gambling house or game; *specifically* **:** DEALER
3 : a supply of something held in reserve: as **a :** the fund of supplies (as money, chips, or pieces) held by the banker or dealer for use in a game **b :** a fund of pieces belonging to a game (as dominoes) from which the players draw
4 : a place where something is held available ⟨memory *bank*⟩; *especially* **:** a depot for the collection and storage of a biological product of human origin for medical use ⟨blood *bank*⟩

[4]bank (circa 1751)

☆ SYNONYMS
Banish, exile, deport, transport mean to remove by authority from a state or country. BANISH implies compulsory removal from a country not necessarily one's own ⟨*banished* for seditious activities⟩. EXILE may imply compulsory removal or an enforced or voluntary absence from one's own country ⟨a writer who *exiled* himself for political reasons⟩. DEPORT implies sending out of the country an alien who has illegally entered or whose presence is judged inimical to the public welfare ⟨illegal aliens will be *deported*⟩. TRANSPORT implies sending a convicted criminal to an overseas penal colony ⟨a convict who was *transported* to Australia⟩.

◇ WORD HISTORY
bank Long before the days of automatic teller machines, money transactions took place between individuals over a table or counter. The medieval Italian word for such a counter was *banco.* As the transactions of financial institutions became more complex than simple exchanges of money across a counter, especially in Italian cities such as Florence and Venice, the word *banco* was extended to the institution itself. Italian banking practices spread to the rest of Europe, and with them went the word *banco,* usually via the feminine variant *banca*—hence late medieval French *banque* and late Middle English *banke.* In Italian, *banca,* possibly influenced by the feminine gender of the French noun, has now largely supplanted *banco* in the sense "bank."

intransitive verb
1 : to manage a bank
2 : to deposit money or have an account in a bank
transitive verb
: to deposit in a bank
— bank on : to depend or rely on
[5]bank *noun* [Middle English *banc* bench, from Old French, of Germanic origin; akin to Old English *benc*] (1614)
1 : a group or series of objects arranged together in a row or a tier: as **a :** a set of elevators **b :** a row or tier of telephones
2 : one of the horizontal and usually secondary or lower divisions of a headline
bank·able \ˈbaŋ-kə-bəl\ *adjective* (1818)
1 : acceptable to or at a bank
2 : sure to bring in a profit ⟨Hollywood's most *bankable* star —Sidney Sheldon⟩
— bank·abil·i·ty \ˌbaŋ-kə-ˈbi-lə-tē\ *noun*
bank·book \ˈbaŋk-ˌbu̇k\ *noun* (1714)
: the depositor's book in which a bank records deposits and withdrawals — called also *passbook*
bank·card \-ˌkärd\ *noun* (1970)
: a credit card issued by a bank
bank discount *noun* (1841)
: the interest discounted in advance on a note and computed on the face value of the note
[1]bank·er \ˈbaŋ-kər\ *noun* (1534)
1 : one that engages in the business of banking
2 : the player who keeps the bank in various games
— bank·er·ly \-lē\ *adjective*
[2]banker *noun* (1666)
: a man or boat employed in the cod fishery on the Newfoundland banks
[3]banker *noun* (1677)
: a sculptor's or mason's workbench
banker's acceptance *noun* (circa 1924)
: a short-term credit instrument issued by an importer's bank that guarantees payment of an exporter's invoice
bank holiday *noun* (1871)
1 *British* **:** LEGAL HOLIDAY
2 : a period when banks in general are closed often by government fiat
bank·ing *noun* (1735)
: the business of a bank or a banker
bank money *noun* (1904)
: a medium of exchange consisting chiefly of checks and drafts
bank·note \ˈbaŋk-ˌnōt\ *noun* (1695)
: a promissory note issued by a bank payable to bearer on demand without interest and acceptable as money
[1]bank·roll \ˈbaŋ-ˌkrōl\ *noun* (1887)
: supply of money **:** FUNDS
[2]bankroll *transitive verb* (1928)
: to supply the capital for or pay the cost of (a business or project)
— bank·roll·er *noun*
[1]bank·rupt \ˈbaŋ-(ˌ)krəpt\ *noun* [modification of Middle French & Old Italian; Middle French *banqueroute* bankruptcy, from Old Italian *bancarotta,* from *banca* bank + *rotta* broken, from Latin *rupta,* feminine of *ruptus,* past participle of *rumpere* to break — more at BANK, REAVE] (1533)
1 a : a person who has done any of the acts that by law entitle his creditors to have his estate administered for their benefit **b :** a person judicially declared subject to having his estate administered under the bankrupt laws for the benefit of his creditors **c :** a person who becomes insolvent
2 : one who is destitute of a particular thing ⟨a moral *bankrupt*⟩
[2]bankrupt *adjective* (1570)
1 a : reduced to a state of financial ruin **:** IMPOVERISHED; *specifically* **:** legally declared a bankrupt ⟨the company went *bankrupt*⟩ **b :** of or relating to bankrupts or bankruptcy ⟨*bankrupt* laws⟩
2 a : BROKEN, RUINED ⟨a *bankrupt* professional career⟩ **b :** exhausted of valuable qualities **:** STERILE ⟨a *bankrupt* old culture⟩ **c :** DESTITUTE — used with *of* or *in* ⟨*bankrupt* of all merciful feelings⟩
[3]bankrupt *transitive verb* (1588)
1 : to reduce to bankruptcy
2 : IMPOVERISH ⟨defections had *bankrupted* the party of its brainpower⟩
synonym see DEPLETE
bank·rupt·cy \ˈbaŋ-(ˌ)krəp(t)-sē\ *noun, plural* **-cies** (1700)
1 : the quality or state of being bankrupt
2 : utter failure or impoverishment
bank shot *noun* (1897)
1 : a shot in billiards and pool in which a player banks the cue ball or the object ball
2 : a shot in basketball played to rebound from the backboard into the basket
bank·sia \ˈbaŋ(k)-sē-ə\ *noun* [New Latin, genus name, from Sir Joseph *Banks*] (1788)
: any of a genus (*Banksia*) of Australian evergreen trees or shrubs of the protea family with alternate leathery leaves and flowers in dense cylindrical heads
bank·side \ˈbaŋk-ˌsīd\ *noun* (15th century)
: the slope of a bank especially of a stream
[1]ban·ner \ˈba-nər\ *noun* [Middle English *banere,* from Old French, of Germanic origin; akin to Gothic *bandwo* sign; probably akin to Greek *phainein* to show — more at FANCY] (13th century)
1 a : a piece of cloth attached by one edge to a staff and used by a leader (as a monarch or feudal lord) as his standard **b :** [2]FLAG 1 **c :** an ensign displaying a distinctive or symbolic device or legend; *especially* **:** one presented as an award of honor or distinction
2 : a headline in large type running across a newspaper page
3 : a strip of cloth on which a sign is painted ⟨welcome *banners* stretched across the street⟩
4 : a name, slogan, or goal associated with a particular group or ideology ⟨the new *banner* is "community control" —F. M. Hechinger⟩ — often used with *under* ⟨every new administration arrives . . . under the *banner* of change —John Cogley⟩
[2]banner *adjective* (1840)
1 : prominent in support of a political party ⟨a *banner* Democratic county⟩
2 : distinguished from all others especially in excellence ⟨a *banner* year for business⟩
[3]banner *transitive verb* (1889)
1 : to furnish with a banner
2 : to print (as a news story) under a banner usually on the front page
[1]ban·ner·et \ˈba-nə-rət, ˌba-nə-ˈret\ *noun, often capitalized* [Middle English *baneret,* from Old French, from *banere*] (14th century)
: a knight leading his vassals into the field under his own banner
[2]banneret *also* **ban·ner·ette** *noun* (14th century)
: a small banner
ban·ne·rol \ˈba-nə-ˌrōl\ *noun* (1548)
: BANDEROLE
bannister *variant of* BANISTER
ban·nock \ˈba-nək\ *noun* [Middle English *bannok,* from Old English *bannuc*] (before 12th century)
1 : a usually unleavened flat bread or biscuit made with oatmeal or barley meal
2 *chiefly New England* **:** CORN BREAD; *especially* **:** a thin cake baked on a griddle
banns \ˈbanz\ *noun plural* [plural of *bann,* from Middle English *bane, ban* proclamation, ban] (14th century)
: public announcement especially in church of a proposed marriage
[1]ban·quet \ˈbaŋ-kwət, ˈban- *also* -ˌkwet\ *noun* [Middle French, from Old Italian *banchetto,* from diminutive of *banca* bench, bank] (15th century)
1 : a sumptuous feast; *especially* **:** an elaborate and often ceremonious meal for numerous people often in honor of a person ⟨a state *banquet*⟩
2 : a meal held in recognition of some occasion or achievement ⟨an awards *banquet*⟩
[2]banquet (circa 1500)
intransitive verb
: to partake of a banquet
transitive verb
: to treat with a banquet **:** FEAST
— ban·quet·er *noun*
banquet room *noun* (1837)
: a large room (as in a restaurant or hotel) suitable for banquets
ban·quette \baŋ-ˈket, ban-, *1b is also* ˈbaŋ-kət\ *noun* [French, from Middle French, from Old Provençal *banqueta,* diminutive of *banc* bench, of Germanic origin; akin to Old English *benc* bench] (1629)
1 a : a raised way along the inside of a parapet or trench for gunners or guns **b** *Southern* **:** SIDEWALK
2 a : a long upholstered bench **b :** a sofa having one roll-over arm **c :** a built-in upholstered bench along a wall
Ban·quo \ˈbaŋ-(ˌ)kwō, ˈban-\ *noun*
: a murdered Scottish thane in Shakespeare's *Macbeth* whose ghost appears to Macbeth
ban·shee \ˈban-(ˌ)shē, ban-ˈ\ *noun* [Irish *bean sídhe* & Scottish Gaelic *bean sìth,* literally, woman of fairyland] (1771)
: a female spirit in Gaelic folklore whose appearance or wailing warns a family that one of them will soon die
[1]ban·tam \ˈban-təm\ *noun* [*Bantam,* former residency in Java] (1749)
1 : any of numerous small domestic fowls that are often miniatures of members of the standard breeds
2 : a person of diminutive stature and often combative disposition
[2]bantam *adjective* (1782)
1 : SMALL, DIMINUTIVE
2 : pertly combative
ban·tam·weight \-ˌwāt\ *noun* (1884)
: a boxer in a weight division having a maximum limit of 118 pounds for professionals and 119 pounds for amateurs — compare FEATHERWEIGHT, FLYWEIGHT
ban·teng \ˈbän-ˌteŋ\ *noun* [Malay of Indonesia, from Javanese *banténg*] (1817)
: a wild ox (*Bos javanicus* synonym *B. banteng*) of southeastern Asia sometimes domesticated for use as a draft animal or for its meat
[1]ban·ter \ˈban-tər\ *verb* [origin unknown] (1676)
transitive verb
1 : to speak to or address in a witty and teasing manner
2 *archaic* **:** DELUDE
3 *chiefly Southern & Midland* **:** CHALLENGE
intransitive verb
: to speak or act playfully or wittily
— ban·ter·er \-tər-ər\ *noun*
— ban·ter·ing·ly \ˈban-tə-riŋ-lē\ *adverb*
[2]banter *noun* (1690)
: good-natured and usually witty and animated joking
bant·ling \ˈbant-liŋ\ *noun* [perhaps modification of German *Bänkling* bastard, from *Bank* bench, from Old High German — more at BENCH] (1593)
: a very young child
Ban·tu \ˈban-(ˌ)tü, ˈbän-\ *noun, plural* **Bantu** *or* **Bantus** (1862)
1 : a family of Niger-Congo languages spoken in central and southern Africa
2 : a member of any of a group of African peoples who speak Bantu languages
Ban·tu·stan \ˌban-tü-ˈstan, ˌbän-tü-ˈstän\ *noun* [*Bantu* + *-stan* land (as in *Hindustan*)] (1949)
: any of several all-black enclaves in the Republic of South Africa that have a limited degree of self-government

ban•yan \'ban-yən\ *noun* [earlier *banyan* Gujarati trader, from Portuguese *banean,* probably from Tamil *vāṇiyan* trader, from Sanskrit *vāṇija;* from a tree of the species in Iran under which such traders conducted business] (1634)
: an East Indian fig tree (*Ficus benghalensis*) of the mulberry family with branches that send out shoots which grow down to the soil and root to form secondary trunks

ban•zai \(ˌ)bän-'zī, 'bän-ˌ\ *noun* [Japanese] (1893)
: a Japanese cheer or war cry

banzai attack *noun* (1944)
: a mass attack by Japanese soldiers in World War II; *also* **:** an all-out usually desperate attack

banzai charge *noun* (1944)
1 : BANZAI ATTACK
2 : a determined often reckless act

bao•bab \'baů-ˌbab, 'bā-ə-\ *noun* [New Latin *bahobab*] (1640)
: a broad-trunked Old World tropical tree (*Adansonia digitata*) of the silk-cotton family with an edible acid fruit resembling a gourd and bark used in making paper, cloth, and rope

bap \'bap\ *noun* [origin unknown] (circa 1575)
British **:** a small bun or roll

bap•ti•sia \bap-'ti-zh(ē-)ə\ *noun* [New Latin, genus name, from Greek *baptisis* a dipping, from *baptein*] (circa 1868)
: any of a genus (*Baptisia*) of North American leguminous plants having showy papilionaceous flowers similar in form to those of the pea plant

bap•tism \'bap-ˌti-zəm, *especially Southern* 'bab-\ *noun* [Middle English *baptisme*] (14th century)
1 a : a Christian sacrament marked by ritual use of water and admitting the recipient to the Christian community **b :** a non-Christian rite using water for ritual purification **c** *Christian Science* **:** purification by or submergence in Spirit
2 : an act, experience, or ordeal by which one is purified, sanctified, initiated, or named
— bap•tis•mal \bap-'tiz-məl, *especially Southern* bab-\ *adjective*
— bap•tis•mal•ly \-mə-lē\ *adverb*

baptismal name *noun* (1869)
: a name given at christening or confirmation

baptism of fire (1857)
1 : an introductory or initial experience that is a severe ordeal; *specifically* **:** a soldier's first exposure to enemy fire
2 : a spiritual baptism by a gift of the Holy Spirit — often used in allusion to Acts 2:3–4; Matthew 3:11 (Revised Standard Version)

bap•tist \'bap-tist, *especially Southern* 'bab-\ *noun* (13th century)
1 : one that baptizes
2 *capitalized* **:** a member or adherent of an evangelical Protestant denomination marked by congregational polity and baptism by immersion of believers only
— Baptist *adjective*

bap•tis•tery *or* **bap•tis•try** \'bap-tə-strē, *especially Southern* 'bab-\ *noun, plural* **-ter•ies** *or* **-tries** (14th century)
: a part of a church or formerly a separate building used for baptism

bap•tize *also* **bap•tise** \bap-'tīz, 'bap-ˌ, *especially Southern* bab- *or* 'bab-\ *verb* **bap•tized** *also* **bap•tised; bap•tiz•ing** *also* **bap•tis•ing** [Middle English, from Old French *baptiser,* from Late Latin *baptizare,* from Greek *baptizein* to dip, baptize, from *baptein* to dip, dye; akin to Old Norse *kvefja* to quench] (13th century)
transitive verb
1 : to administer baptism to
2 a : to purify or cleanse spiritually especially by a purging experience or ordeal **b :** INITIATE
3 : to give a name to (as at baptism) **:** CHRISTEN
intransitive verb
: to administer baptism
— bap•tiz•er *noun*

[1]**bar** \'bär\ *noun, often attributive* [Middle English *barre,* from Middle French] (12th century)
1 a : a straight piece (as of wood or metal) that is longer than it is wide and has any of various uses (as for a lever, support, barrier, or fastening) **b :** a solid piece or block of material that is usually considerably longer than it is wide ⟨a *bar* of gold⟩ ⟨candy *bar*⟩ **c :** a usually rigid piece (as of wood or metal) longer than it is wide that is used as a handle or support; *especially* **:** a handrail used by ballet dancers to maintain balance while exercising
2 : something that obstructs or prevents passage, progress, or action: as **a :** the destruction of an action or claim in law; *also* **:** a plea or objection that effects such destruction **b :** an intangible or nonphysical impediment **c :** a submerged or partly submerged bank (as of sand) along a shore or in a river often obstructing navigation
3 a (1) **:** the railing in a courtroom that encloses the place about the judge where prisoners are stationed or where the business of the court is transacted in civil cases (2) **:** COURT, TRIBUNAL (3) **:** a particular system of courts (4) **:** an authority or tribunal that hands down judgment **b** (1) **:** the barrier in the English Inns of Court that formerly separated the seats of the benchers or readers from the body of the hall occupied by the students (2) **:** the whole body of barristers or lawyers qualified to practice in the courts of any jurisdiction (3) **:** the profession of barrister or lawyer
4 : a straight stripe, band, or line much longer than it is wide: as **a :** one of two or more horizontal stripes on a heraldic shield **b :** a metal or embroidered strip worn on a usually military uniform especially to indicate rank (as of a company officer) or service
5 a : a counter at which food or especially alcoholic beverages are served **b :** BARROOM **c :** SHOP 2b
6 a : a vertical line across the musical staff before the initial measure accent **b :** MEASURE
7 : a lace and embroidery joining covered with buttonhole stitch for connecting various parts of the pattern in needlepoint lace and cutwork
— behind bars : in jail

[2]**bar** *transitive verb* **barred; bar•ring** (13th century)
1 a : to fasten with a bar **b :** to place bars across to prevent ingress or egress
2 : to mark with bars **:** STRIPE
3 a : to confine or shut in by or as if by bars **b :** to set aside **:** RULE OUT **c :** to keep out **:** EXCLUDE
4 a : to interpose legal objection to or to the claim of **b :** PREVENT, FORBID

[3]**bar** *preposition* (1714)
: EXCEPT

[4]**bar** *noun* [German, from Greek *baros*] (1910)
: a unit of pressure equal to one million dynes per square centimeter

bar- *or* **baro-** *combining form* [Greek *baros;* akin to Greek *barys* heavy — more at GRIEVE]
: weight **:** pressure ⟨*baro*meter⟩

Ba•rab•bas \bə-'ra-bəs\ *noun* [Greek, from Aramaic *Bar abbā*]
: a Jewish prisoner according to Matthew, Mark, and John released in preference to Christ at the demand of the multitude

bar•a•thea \ˌbar-ə-'thē-ə\ *noun* [from *Barathea,* a trademark] (1862)
: a fabric that has a broken rib weave and a pebbly texture and that is made of silk, worsted, or synthetic fiber or a combination of these

[1]**barb** \'bärb\ *noun* [Middle English *barbe* barb, beard, from Middle French, from Latin *barba* — more at BEARD] (14th century)
1 : a medieval cloth headdress passing over or under the chin and covering the neck
2 a : a sharp projection extending backward (as from the point of an arrow or fishhook) and preventing easy extraction; *also* **:** a sharp projection with its point similarly oblique to something else **b :** a biting or pointedly critical remark or comment
3 : [2]BARBEL
4 : any of the side branches of the shaft of a feather — see FEATHER illustration
5 : a plant hair or bristle ending in a hook

[2]**barb** *transitive verb* (1759)
: to furnish with a barb

[3]**barb** *noun* [French *barbe,* from Italian *barbero,* from *barbero* of Barbary, from *Barberia* Barbary, coastal region in Africa] (1636)
: any of a northern African breed of horses that are noted for speed and endurance and are related to Arabians

[4]**barb** *noun* (1967)
slang **:** BARBITURATE

bar•bar•i•an \bär-'ber-ē-ən, -'bar-\ *adjective* [Latin *barbarus*] (14th century)
1 : of or relating to a land, culture, or people alien and usually believed to be inferior to another land, culture, or people
2 : lacking refinement, learning, or artistic or literary culture
— barbarian *noun*
— bar•bar•i•an•ism \-ē-ə-ˌni-zəm\ *noun*

bar•bar•ic \bär-'bar-ik\ *adjective* (15th century)
1 a : of, relating to, or characteristic of barbarians **b :** possessing or characteristic of a cultural level more complex than primitive savagery but less sophisticated than advanced civilization
2 a : marked by a lack of restraint **:** WILD **b :** having a bizarre, primitive, or unsophisticated quality
3 : BARBAROUS 3 ◆
— bar•bar•i•cal•ly \-i-k(ə-)lē\ *adverb*

bar•ba•rism \'bär-bə-ˌri-zəm\ *noun* (15th century)
1 a : a barbarian or barbarous social or intellectual condition **:** BACKWARDNESS **b :** the practice or display of barbarian acts, attitudes, or ideas
2 : an idea, act, or expression that in form or use offends against contemporary standards of good taste or acceptability

bar•bar•i•ty \bär-'bar-ə-tē\ *noun, plural* **-ties** (circa 1570)
1 : BARBARISM
2 a : barbarous cruelty **:** INHUMANITY **b :** an act or instance of such cruelty

bar•ba•ri•za•tion \ˌbär-bə-rə-'zā-shən\ *noun* (1822)

◇ WORD HISTORY

barbaric It is unfortunate that the ancient Greeks, in spite of their scientific curiosity and their distinction in the writing of history, left scant records of the languages spoken by peoples with whom they came into contact—languages now mostly lost forever to investigation. On the whole, the Greeks were not very interested in those who were not *Hellēnes,* "Greek-speakers." All non-Greek speakers were collectively *barbaroi,* a word that is onomatopoeic in origin, mimicking the babble-like sound of foreign speech to Greek ears. Especially after the Greeks' wars with the Persians, the *barbaroi* par excellence to the Greeks, the derivative *barbarikos* came to mean "violent" or "savage" in addition to simply "foreign." It is these senses we have inherited in the loanword *barbaric.*

: the act or process of barbarizing : the state of being barbarized

bar·ba·rize \'bär-bə-ˌrīz\ *verb* **-rized; -riz·ing** (1648)
transitive verb
: to make barbarian or barbarous
intransitive verb
: to become barbarous

bar·ba·rous \'bär-b(ə-)rəs\ *adjective* [Latin *barbarus,* from Greek *barbaros* foreign, ignorant] (15th century)
1 a : UNCIVILIZED **b :** lacking culture or refinement : PHILISTINE
2 : characterized by the occurrence of barbarisms
3 : mercilessly harsh or cruel
synonym see FIERCE
— bar·ba·rous·ly *adverb*
— bar·ba·rous·ness *noun*

Bar·ba·ry ape \'bär-b(ə-)rē-\ *noun* [*Barbary,* Africa] (1864)
: a tailless monkey (*Macaca sylvanus*) of northern Africa and Gibraltar

Barbary Coast *noun* (1880)
: a district or section of a city noted as a center of gambling, prostitution, and riotous nightlife

Barbary sheep *noun* (circa 1898)
: AOUDAD

barbe \'bärb\ *noun* [Middle English, from Middle French, literally, beard] (14th century)
: [1]BARB 1

[1]bar·be·cue \'bär-bi-ˌkyü\ *transitive verb* **-cued; -cu·ing** (1690)
1 : to roast or broil on a rack over hot coals or on a revolving spit before or over a source of heat
2 : to cook in a highly seasoned vinegar sauce
— bar·be·cu·er *noun*

[2]barbecue *also* **bar·be·que** *noun* [American Spanish *barbacoa* framework for supporting meat over a fire, probably from Taino] (1709)
1 : a large animal (as a steer) roasted whole or split over an open fire or a fire in a pit; *also* **:** smaller pieces of barbecued meat
2 : a social gathering especially in the open air at which barbecued food is eaten
3 : an often portable fireplace over which meat and fish are roasted

barbed \'bärbd\ *adjective* (1611)
1 : having barbs
2 : characterized by pointed and biting criticism or sarcasm ⟨*barbed* witticisms⟩

barbed wire \'bärb(d)-'wīr, 'bäb(d)-\ *noun* (1881)
: twisted wires armed with barbs or sharp points — called also *barbwire*

[1]bar·bel \'bär-bəl\ *noun* [Middle English, from Middle French, from (assumed) Vulgar Latin *barbellus,* diminutive of Latin *barbus* barbel, from *barba* beard — more at BEARD] (14th century)
: a European freshwater cyprinid fish (*Barbus barbus*) with four barbels on its upper jaw; *also* **:** any of various closely related fishes

[2]barbel *noun* [obsolete French, from Middle French, diminutive of *barbe* barb, beard] (1601)
: a slender tactile process on the lips of certain fishes (as catfishes)

bar·bell \'bär-ˌbel\ *noun* (1887)
: a bar with adjustable weighted disks attached to each end that is used for exercise and in weight lifting

[1]bar·ber \'bär-bər\ *noun* [Middle English, from Middle French *barbeor,* from *barbe* beard — more at BARB] (14th century)
: one whose business is cutting and dressing hair, shaving and trimming beards, and performing related services

[2]barber *verb* **bar·bered; bar·ber·ing** \-b(ə-)riŋ\ (1606)
transitive verb
: to perform the services of a barber for
intransitive verb
: to perform the services of a barber

bar·ber·ry \'bär-ˌber-ē\ *noun* [Middle English *barbere,* from Middle French *barbarin,* from Arabic *barbārīs*] (14th century)
: any of a genus (*Berberis* of the family Berberidaceae, the barberry family) of shrubs usually having spines, yellow flowers, and oblong red berries

[1]bar·ber·shop \'bär-bər-ˌshäp\ *noun* (1579)
: a barber's place of business

[2]barbershop *adjective* [from the old custom of men in barbershops forming quartets for impromptu singing of sentimental songs] (1910)
: of a style of unaccompanied group singing of popular songs usually marked by highly conventionalized close harmony

bar·bet \'bär-bət\ *noun* [probably from [1]*barb*] (1824)
: any of a family (Capitonidae) of nonpasserine tropical birds with a stout bill bearing bristles and usually swollen at the base

bar·bette \bär-'bet\ *noun* [French, diminutive of *barbe* headdress] (1772)
1 : a mound of earth or a protected platform from which guns fire over a parapet
2 : an armored structure protecting a gun turret on a warship

bar·bi·can \'bär-bi-kən\ *noun* [Middle English, from Middle French *barbacane,* from Medieval Latin *barbacana*] (13th century)
: an outer defensive work; *especially* **:** a tower at a gate or bridge

bar·bi·cel \'bär-bə-ˌsel\ *noun* [New Latin *barbicella,* diminutive of Latin *barba*] (1869)
: any of the small hook-bearing processes on a barbule of a feather — see FEATHER illustration

bar·bi·tal \'bär-bə-ˌtȯl\ *noun* [*barbi*turic + *-al* (as in *Veronal,* trademark for barbital)] (1919)
: a white crystalline addictive hypnotic $C_8H_{12}N_2O_3$ often administered in the form of its soluble sodium salt

bar·bi·tone \'bär-bə-ˌtōn\ *noun* [*barbi*turic + *-one*] (1914)
British **:** BARBITAL

bar·bi·tu·rate \bär-'bi-chə-rət, -ˌrāt; ˌbär-bə-'tyu̇r-ət, -'tu̇r-, -ˌāt\ *noun* (1928)
1 : a salt or ester of barbituric acid
2 : any of various derivatives of barbituric acid used especially as sedatives, hypnotics, and antispasmodics

bar·bi·tu·ric acid \ˌbär-bə-'t(y)u̇r-ik-\ *noun* [part translation of German *Barbitursäure,* irregular from the name *Barbara* + International Scientific Vocabulary *uric* + German *Säure* acid] (1866)
: a synthetic crystalline acid $C_4H_4N_2O_3$ derived from pyrimidine

Bar·bi·zon \ˌbär-bə-'zōn, -'zän\ *adjective* [*Barbizon,* France] (1890)
: of, relating to, or being a school of mid-nineteenth century French landscape painters whose naturalistic canvases were based on direct observation of nature

bar·bule \'bär-(ˌ)byü(ə)l\ *noun* (1835)
: a minute barb; *especially* **:** one of the processes that fringe the barbs of a feather — see FEATHER illustration

barb·wire \'bärb-'wīr, 'bäb-\ *noun* (1880)
: BARBED WIRE

bar car *noun* (1945)
: CLUB CAR

bar·ca·role *or* **bar·ca·rolle** \'bär-kə-ˌrōl\ *noun* [French *barcarolle,* from Italian dialect (Venice) *barcarola,* from *barcarolo* gondolier, from *barca* bark, from Late Latin] (circa 1779)
1 : a Venetian boat song usually in 6/8 or 12/8 time characterized by the alternation of a strong and weak beat that suggests a rowing rhythm
2 : music imitating a barcarole

Bar·ce·lo·na chair \ˌbär-sə-'lō-nə-\ *noun* [*Barcelona,* Spain] (1970)
: an armless chair with leather-covered cushions on a stainless steel frame

Barcelona chair

bar·chan \(ˌ)bär-'kän, -'kän\ *noun* [Russian *barkhan,* from Kazakh] (1888)
: a moving crescent-shaped sand dune

bar chart *noun* (1914)
: BAR GRAPH

bar code *noun* (1963)
: a code consisting of a group of printed and variously patterned bars and spaces and sometimes numerals that is designed to be scanned and read into computer memory as identification for the object it labels

[1]bard \'bärd\ *noun* [Middle English, from Scottish Gaelic & Irish] (15th century)
1 a : a tribal poet-singer skilled in composing and reciting verses on heroes and their deeds **b :** a composer, singer, or declaimer of epic or heroic verse
2 : POET
— bard·ic \'bär-dik\ *adjective*

[2]bard *or* **barde** \'bärd\ *noun* [Middle French *barde,* from Old Spanish *barda,* from Arabic *barda'ah*] (15th century)
: a piece of armor or ornament for a horse's neck, breast, or flank

[3]bard *transitive verb* (circa 1521)
1 : to furnish with bards
2 : to dress meat for cooking by covering with strips of fat

bard·ol·a·ter \bär-'dä-lə-tər\ *noun* [*Bard* (*of Avon*), epithet of Shakespeare + *-o-* + *-later*] (1903)
: a person who idolizes Shakespeare
— bard·ol·a·try \-lə-trē\ *noun*

Bar·do·li·no \ˌbär-dᵊl-'ē-(ˌ)nō\ *noun, plural* **-nos** [*Bardolino,* village on Lake Garda, Italy] (1934)
: a light red Italian wine

[1]bare \'bar, 'ber\ *adjective* **bar·er; bar·est** [Middle English, from Old English *bær;* akin to Old High German *bar* naked, Lithuanian *basas* barefoot] (before 12th century)
1 a : lacking a natural, usual, or appropriate covering **b** (1) **:** lacking clothing (2) *obsolete* **:** BAREHEADED **c :** UNARMED
2 : open to view **:** EXPOSED
3 a : unfurnished or scantily supplied **b :** DESTITUTE ⟨*bare* of all safeguards⟩
4 a : having nothing left over or added ⟨the *bare* necessities of life⟩ **b :** MERE ⟨a *bare* two hours away⟩ **c :** devoid of amplification or adornment
5 *obsolete* **:** WORTHLESS ☆
— bare·ness *noun*

[2]bare *transitive verb* **bared; bar·ing** (before 12th century)
: to make or lay bare **:** UNCOVER

[3]bare *archaic past of* BEAR

bare·back \-ˌbak\ *or* **bare·backed** \-'bakt\ *adverb or adjective* (1562)
: on the bare back of a horse **:** without a saddle ⟨likes riding *bareback*⟩ ⟨*bareback* riding⟩

☆ SYNONYMS
Bare, naked, nude, bald, barren mean deprived of naturally or conventionally appropriate covering. BARE implies the removal of what is additional, superfluous, ornamental, or dispensable ⟨an apartment with *bare* walls⟩. NAKED suggests absence of protective or ornamental covering but may imply a state of nature, of destitution, or of defenselessness ⟨poor half-*naked* children⟩. NUDE applies especially to the unclothed human figure ⟨a *nude* model posing for art students⟩. BALD implies actual or seeming absence of natural covering and may suggest a conspicuous bareness ⟨a *bald* mountain peak⟩. BARREN often suggests aridity or impoverishment or sterility ⟨*barren* plains⟩.

bare·boat \-ˌbōt\ *noun* (circa 1949)
: a boat chartered without its crew

bare bones *noun plural* (1915)
: the barest essentials, facts, or elements
— **bare–bones** *adjective*

bare·faced \ˈbar-ˈfāst, ˈber-\ *adjective* (1590)
1 : having the face uncovered: **a :** having no whiskers **:** BEARDLESS **b :** wearing no mask
2 a : OPEN, UNCONCEALED **b :** lacking scruples
— **bare·faced·ly** \-ˈfā-səd-lē, -ˈfāst-lē\ *adverb*
— **bare·faced·ness** \-ˈfā-səd-nəs, -ˈfās(t)-nəs\ *noun*

bare·foot \-ˌfu̇t\ *or* **bare·foot·ed** \-ˈfu̇-təd\ *adverb or adjective* (before 12th century)
: with the feet bare ⟨went *barefoot* most of the summer⟩ ⟨*barefoot* boy, with cheek of tan —J. G. Whittier⟩

barefoot doctor *noun* (1971)
: an auxiliary medical worker trained to provide basic health care in rural areas of China

bare–hand·ed \ˈbar-ˈhan-dəd, ˈber-\ *adverb or adjective* (15th century)
1 : without gloves
2 : without tools or weapons ⟨fight an animal *bare-handed*⟩

bare·head·ed \-ˈhe-dəd\ *adverb or adjective* (14th century)
: without a covering for the head ⟨went *bareheaded* in the hot sun⟩ ⟨a *bareheaded* boy who had lost his cap⟩

bare–knuck·le \-ˈnə-kəl\ *also* **bare–knuck·led** \-kəld\ *adjective or adverb* (1903)
1 : not using boxing gloves ⟨champion *bare-knuckle* prizefighter of England —Dennis Craig⟩ ⟨when men fought *bare-knuckle*⟩
2 : having a fierce unrelenting character ⟨*bare-knuckle* politics⟩

bare·ly *adverb* (before 12th century)
1 : in a meager manner **:** PLAINLY ⟨a *barely* furnished room⟩
2 : SCARCELY, HARDLY ⟨*barely* enough money for lunch⟩

barf \ˈbärf\ *intransitive verb* [origin unknown] (1957)
: VOMIT

bar·fly \ˈbär-ˌflī\ *noun* (1910)
: a person who spends much time in bars

[1]**bar·gain** \ˈbär-gən\ *noun, often attributive* (14th century)
1 : an agreement between parties settling what each gives or receives in a transaction between them or what course of action or policy each pursues in respect to the other
2 : something acquired by or as if by bargaining; *especially* **:** an advantageous purchase
3 : a transaction, situation, or event regarded in the light of its results
— **into the bargain** *also* **in the bargain :** BESIDES ⟨tastes good and is good for you, *into the bargain*⟩

[2]**bargain** *verb* [Middle English, from Middle French *bargaignier,* probably of Germanic origin; akin to Old English *borgian* to borrow — more at BURY] (14th century)
intransitive verb
1 : to negotiate over the terms of a purchase, agreement, or contract **:** HAGGLE
2 : to come to terms **:** AGREE
transitive verb
1 : to bring to a desired level by bargaining ⟨*bargain* a price down⟩
2 : to sell or dispose of by bargaining
— **bar·gain·er** *noun*
— **bargain for :** EXPECT ⟨more work than I *bargained for*⟩

bar·gain–base·ment \ˈbär-gən-ˈbās-mənt\ *adjective* (1948)
1 : of inferior quality or worth
2 : markedly inexpensive ⟨*bargain-basement* rates⟩

bargain basement *noun* (1899)
: a section of a store (as the basement) where merchandise is sold at reduced prices

[1]**barge** \ˈbärj\ *noun* [Middle English, from Middle French, from Late Latin *barca*] (14th century)
: any of various boats: as **a :** a roomy usually flat-bottomed boat used chiefly for the transport of goods on inland waterways and usually propelled by towing **b :** a large motorboat supplied to the flag officer of a flagship **c :** a roomy pleasure boat; *especially* **:** a boat of state elegantly furnished and decorated

[2]**barge** *verb* **barged; barg·ing** (1649)
transitive verb
: to carry by barge
intransitive verb
1 : to move ponderously or clumsily
2 : to thrust oneself heedlessly or unceremoniously ⟨*barged* into the meeting⟩

barge·board \ˈbärj-ˌbōrd, -ˌbȯrd\ *noun* [origin unknown] (1833)
: an often ornamented board that conceals roof timbers projecting over gables

barg·ee \bär-ˈjē\ *noun* (1666)
British **:** BARGEMAN

bar·gel·lo \bär-ˈje-(ˌ)lō\ *noun* [the *Bargello,* museum in Florence, Italy; from the use of this stitch in the upholstery of 17th century chairs at the Bargello] (circa 1924)
: a needlework stitch that produces a zigzag pattern

barge·man \ˈbärj-mən\ *noun* (14th century)
: the master or a deckhand of a barge

bar graph *noun* (1924)
: a graphic means of quantitative comparison by rectangles with lengths proportional to the measure of the data or things being compared — called also *bar chart*

bar·hop \ˈbär-ˌhäp\ *intransitive verb* (1947)
: to visit and drink at a series of bars in the course of an evening

bar·ite \ˈbar-ˌīt, ˈber-\ *noun* [Greek *barytēs* weight, from *barys*] (1868)
: barium sulfate occurring as a mineral

[1]**bari·tone** \ˈbar-ə-ˌtōn\ *noun* [French *baryton* or Italian *baritono,* from Greek *barytonos* deep sounding, from *barys* heavy + *tonos* tone — more at GRIEVE] (1609)
1 : a male singing voice of medium compass between bass and tenor; *also* **:** a person having this voice
2 : a member of a family of instruments having a range between tenor and bass; *especially* **:** the baritone saxhorn or baritone saxophone
— **bari·ton·al** \ˌbar-ə-ˈtō-nᵊl\ *adjective*

[2]**baritone** *adjective* (1729)
: relating to or having the range or part of a baritone

bar·i·um \ˈbar-ē-əm, ˈber-\ *noun* [New Latin, from *bar-*] (1808)
: a silver-white malleable toxic bivalent metallic element of the alkaline-earth group that occurs only in combination — see ELEMENT table

barium sulfate *noun* (1903)
: a colorless crystalline insoluble compound $BaSO_4$ that is used as a pigment and extender, as a filler, and as a substance opaque to X rays in medical photography of the alimentary canal

[1]**bark** \ˈbärk\ *verb* [Middle English *berken,* from Old English *beorcan;* akin to Old Norse *berkja* to bark, Lithuanian *burgėti* to growl] (before 12th century)
intransitive verb
1 a : to make the characteristic short loud cry of a dog **b :** to make a noise resembling a bark
2 : to speak in a curt loud and usually angry tone **:** SNAP
transitive verb
1 : to utter in a curt loud usually angry tone
2 : to advertise by persistent outcry ⟨*barking* their wares⟩
— **bark up the wrong tree :** to proceed under a misapprehension; *also* **:** to misdirect one's efforts

[2]**bark** *noun* (before 12th century)
1 a : the sound made by a barking dog **b :** a similar sound
2 : a short sharp peremptory tone of speech or utterance
— **bark·less** \ˈbär-kləs\ *adjective*

[3]**bark** *noun* [Middle English, from Old Norse *bark-, bǫrkr;* akin to Middle Dutch & Middle Low German *borke* bark] (14th century)
1 : the tough exterior covering of a woody root or stem; *specifically* **:** the tissues outside the cambium that include an inner layer especially of secondary phloem and an outer layer of periderm
2 : CINCHONA 2
— **bark·less** \ˈbär-kləs\ *adjective*

[4]**bark** *transitive verb* (14th century)
1 : to treat with an infusion of tanbark
2 a : to strip the bark from **b :** to rub off or abrade the skin of

[5]**bark** *noun* [Middle English, from Middle French *barque,* from Old Provençal *barca,* from Late Latin] (15th century)
1 a : a small sailing ship **b :** a sailing ship of three or more masts with the aftmost mast fore-and-aft rigged and the others square-rigged
2 : a craft propelled by sails or oars

bark beetle *noun* (1862)
: any of numerous beetles (family Scolytidae) that bore under the bark of trees both as a larva and as an adult

bar·keep \ˈbär-ˌkēp\ *also* **bar·keep·er** \-ˌkē-pər\ *noun* (1712)
: BARTENDER

bar·ken·tine \ˈbär-kən-ˌtēn\ *noun* [[5]*bark* + *-entine,* alteration of *-antine* (as in *brigantine*)] (1693)
: a sailing ship of three or more masts with the foremast square-rigged and the others fore-and-aft rigged

[1]**bark·er** \ˈbär-kər\ *noun* (14th century)
: one that barks; *especially* **:** a person who advertises by hawking at an entrance to a show

[2]**barker** *noun* (1611)
: one that removes or prepares bark

barking deer *noun* (1880)
: MUNTJAC

barky \ˈbär-kē\ *adjective* **bark·i·er; -est** (1590)
: covered with or resembling bark

bar·ley \ˈbär-lē\ *noun* [Middle English *barly,* from Old English *bærlic* of barley; akin to Old English *bere* barley, Latin *far* spelt] (before 12th century)
: a cereal grass (genus *Hordeum,* especially *H. vulgare*) having the flowers in dense spikes with long awns and three spikelets at each joint of the rachis; *also* **:** its seed used especially in malt beverages, breakfast foods, and stock feeds

bar·ley–bree \-ˌbrē\ *also* **bar·ley–broo** \-ˌbrü\ *noun* [*barley* + Scots *bree* or *broo* (bree)] (1724)
chiefly Scottish **:** WHISKY; *also* **:** MALT LIQUOR

bar·ley·corn \-ˌkȯrn\ *noun* (1500)
1 : a grain of barley
2 : an old unit of length equal to a third of an inch

bar·low \ˈbär-ˌlō\ *noun* [*Barlow,* family of 18th century English knife makers] (1884)
: a sturdy inexpensive jackknife

barm \ˈbärm\ *noun* [Middle English *berme,* from Old English *beorma;* akin to Latin *fermentum* yeast, *fervēre* to boil, Old Irish *berbaid* he boils] (before 12th century)
: yeast formed on fermenting malt liquors

bar·maid \ˈbär-ˌmād\ *noun* (circa 1658)
: a woman who serves liquor at a bar

bar·man \-mən\ *noun* (1837)
chiefly British **:** BARTENDER

Bar·me·cid·al \ˌbär-mə-ˈsī-dᵊl\ *or* **Bar·me·cide** \ˈbär-mə-ˌsīd\ *adjective* [*Barmecide,* a wealthy Persian, who, in a tale of *The Arabian Nights' Entertainments,* invited a beggar to a feast of imaginary food] (1842)
: providing only the illusion of abundance ⟨a *Barmecidal* feast⟩

¹bar mitz·vah \bär-ˈmits-və\ *noun, often B&M capitalized* [Hebrew *bar miṣwāh,* literally, son of the (divine) law] (1816)
1 : a Jewish boy who reaches his 13th birthday and attains the age of religious duty and responsibility
2 : the initiatory ceremony recognizing a boy as a bar mitzvah

²bar mitzvah *transitive verb* **bar mitz·vahed; bar mitz·vah·ing** (1947)
: to administer the ceremony of bar mitzvah to

¹barmy \ˈbär-mē\ *adjective* **barm·i·er; -est** (15th century)
: full of froth or ferment

²barmy *adjective* **barm·i·er; -est** [alteration of *balmy*] (1892)
chiefly British **:** BALMY 2

barn \ˈbärn\ *noun* [Middle English *bern,* from Old English *bereærn,* from *bere* barley + *ærn* house, store] (before 12th century)
1 a : a usually large building for the storage of farm products or feed and usually for the housing of farm animals or farm equipment **b :** an unusually large and usually bare building ⟨a great *barn* of a hotel —W. A. White⟩
2 : a large building for the housing of a fleet of vehicles (as trolley cars or trucks) ◆
— barn·like \-ˌlīk\ *adjective*
— barny \ˈbär-nē\ *adjective*

Bar·na·bas \ˈbär-nə-bəs\ *noun* [Greek, from Aramaic *Barnebhū'āh*]
: a companion of the apostle Paul on his first missionary journey

bar·na·cle \ˈbär-ni-kəl\ *noun* [Middle English *barnakille,* alteration of *bernake, bernekke*] (15th century)
1 : BARNACLE GOOSE
2 [from a popular belief that the goose grew from the crustacean] **:** any of numerous marine crustaceans (subclass Cirripedia) with feathery appendages for gathering food that are free-swimming as larvae but permanently fixed (as to rocks, boat hulls, or whales) as adults
— bar·na·cled \-kəld\ *adjective*

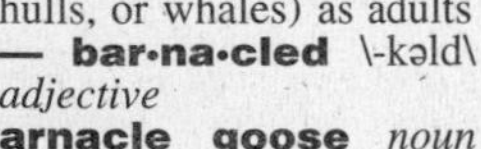

barnacle 2: *1* **peduncle,** *2* **cirri**

barnacle goose *noun* (1768)
: a European goose (*Branta leucopsis*) that breeds in the arctic and is larger than the related brant

barn burner *noun* (circa 1960)
: one that arouses much interest or excitement ⟨the game should be a real *barn burner*⟩

barn dance *noun* (1831)
: an American social dance originally held in a barn and featuring several dance forms (as square dancing)

barn lot *noun* (1724)
chiefly Southern & Midland **:** BARNYARD

barn owl *noun* (1674)
: a widely distributed owl (*Tyto alba*) that has plumage mottled buff brown and gray above and chiefly white below, frequents barns and other buildings, and preys especially on rodents

barn raising *noun* (1856)
: a gathering for the purpose of erecting a barn — compare ³BEE

barn·storm \ˈbärn-ˌstȯrm\ (1883)
intransitive verb
1 : to tour through rural districts staging usually theatrical performances
2 : to travel from place to place making brief stops (as in a political campaign or a promotional tour)
3 : to pilot one's airplane in sight-seeing flights with passengers or in exhibition stunts in an unscheduled course especially in rural districts
transitive verb
: to travel across while barnstorming
— barn·storm·er *noun*

barn swallow *noun* (1851)
: a swallow (*Hirundo rustica*) that is widespread in the northern hemisphere, has a deeply forked tail, and often nests in or near buildings

¹barn·yard \-ˌyärd\ *noun* (14th century)
: a usually fenced area adjoining a barn

²barnyard *adjective* (1927)
: SMUTTY, EARTHY, SCATOLOGICAL ⟨*barnyard* humor⟩

barnyard grass *noun* (1843)
: a coarse annual panicled grass (*Echinochloa crusgalli*) that has flowers borne on only one side of the raceme and is nearly cosmopolitan as a weed in cultivated ground

baro- — see BAR-

baro·gram \ˈbar-ə-ˌgram\ *noun* [International Scientific Vocabulary] (1884)
: a barographic tracing

baro·graph \-ˌgraf\ *noun* [International Scientific Vocabulary] (circa 1864)
: a recording barometer
— baro·graph·ic \ˌbar-ə-ˈgra-fik\ *adjective*

Ba·ro·lo \bä-ˈrō-(ˌ)lō, bə-\ *noun, plural* **-los** [*Barolo,* village in the Piedmont region, Italy] (1875)
: a dry red Italian wine

ba·rom·e·ter \bə-ˈrä-mə-tər\ *noun* (circa 1666)
1 : an instrument for determining the pressure of the atmosphere and hence for assisting in forecasting weather and for determining altitude
2 : one that indicates fluctuations (as in public opinion)
3 : STANDARD 4 ⟨a *barometer* to measure high school talent —Jeff Fellenzer⟩
— baro·met·ric \ˌbar-ə-ˈme-trik\ *adjective*
— baro·met·ri·cal·ly \-tri-k(ə-)lē\ *adverb*
— ba·rom·e·try \bə-ˈrä-mə-trē\ *noun*

barometric pressure *noun* (1827)
: the pressure of the atmosphere usually expressed in terms of the height of a column of mercury

bar·on \ˈbar-ən\ *noun* [Middle English, from Old French, of Germanic origin; akin to Old High German *baro* freeman] (13th century)
1 a : one of a class of tenants holding his rights and title by military or other honorable service directly from a feudal superior (as a king) **b :** a lord of the realm **:** NOBLE, PEER
2 a : a member of the lowest grade of the peerage in Great Britain **b :** a nobleman on the continent of Europe of varying rank **c :** a member of the lowest order of nobility in Japan
3 : a joint of meat consisting of two sirloins or loins and legs not cut apart at the backbone ⟨a *baron* of beef⟩
4 : a man of great power or influence in some field of activity ⟨cattle *baron*⟩

bar·on·age \-ə-nij\ *noun* (13th century)
: the whole body of barons or peers **:** NOBILITY 2

bar·on·ess \ˈbar-ə-nəs, -ˌnes, *US also* ˌbar-ə-ˈnes\ *noun* (15th century)
1 : the wife or widow of a baron
2 : a woman who holds a baronial title in her own right

bar·on·et \ˈbar-ə-nət, *US also* ˌbar-ə-ˈnet\ *noun* (1614)
: the holder of a rank of honor below a baron and above a knight

bar·on·et·age \ˈbar-ə-nə-(ˌ)tij, ˌbar-ə-ˈne-tij\ *noun* (1760)
1 : BARONETCY
2 : the whole body of baronets

bar·on·et·cy \ˈbar-ə-nət-sē, ˌbar-ə-ˈnet-sē\ *noun* (1795)
: the rank of a baronet

ba·rong \bə-ˈrȯŋ, -ˈräŋ\ *noun* [of Austronesian origin; akin to Malay *parang* knife] (1898)
: a thick-backed thin-edged knife or sword used by the Moros

ba·ro·ni·al \bə-ˈrō-nē-əl\ *adjective* (1767)
1 : of or relating to a baron or the baronage
2 : STATELY, AMPLE ⟨a *baronial* room⟩

bar·ony \ˈbar-ə-nē\ *noun, plural* **-on·ies** (14th century)
1 : the domain, rank, or dignity of a baron
2 : a vast private landholding
3 : a field of activity under the sway of an individual or a special group

¹ba·roque \bə-ˈrōk, ba-, -ˈräk, -ˈrȯk\ *adjective, often capitalized* [French, from Middle French *barroque* irregularly shaped (of a pearl), from Portuguese *barroco* irregularly shaped pearl] (1765)
1 : of, relating to, or having the characteristics of a style of artistic expression prevalent especially in the 17th century that is marked generally by use of complex forms, bold ornamentation, and the juxtaposition of contrasting elements often conveying a sense of drama, movement, and tension
2 : characterized by grotesqueness, extravagance, complexity, or flamboyance
3 : irregularly shaped — used of gems ⟨a *baroque* pearl⟩
— ba·roque·ly *adverb*

²baroque *noun, often capitalized* (1877)
: the baroque style or the period in which it flourished

baro·re·cep·tor \ˌbar-ō-ri-ˈsep-tər\ *also* **baro·cep·tor** \ˈbar-ō-ˌsep-tər\ *noun* (1948)
: a neural receptor (as of the arterial walls) sensitive to changes in pressure

ba·rouche \bə-ˈrüsh\ *noun* [German *Barutsche,* from Italian *biroccio,* ultimately from Late Latin *birotus* two-wheeled, from Latin *bi-* + *rota* wheel — more at ROLL] (1801)
: a four-wheeled carriage with a driver's seat high in front, two double seats inside facing each other, and a folding top over the back seat

barque \ˈbärk\, **bar·quen·tine** \ˈbär-kən-ˌtēn\ *variant of* BARK, BARKENTINE

bar·quette \bär-ˈket\ *noun* [French, diminutive of *barque* bark (ship)] (circa 1949)
: a small boat-shaped pastry shell

¹bar·rack \ˈbar-ək, -ik\ *noun* [French *baraque* hut, from Catalan *barraca*] (1686)
1 : a building or set of buildings used especially for lodging soldiers in garrison
2 a : a structure resembling a shed or barn that provides temporary housing **b :** housing characterized by extreme plainness or dreary uniformity — usually used in plural in all senses

◇ WORD HISTORY

barn Some English words that are monosyllabic today were obvious compound words to the Anglo-Saxons. Such a word is *barn,* from Old English *berern,* a compound of *bere, bære* "barley" and *ærn* "house, building." Neither *bere* nor *ærn* have left many direct descendants in Modern English, though in Scotland *bear* survived into the 20th century as a word for barley, and *ærn* "house, building" is akin to Old Norse *rann* "house," which figures in English *ransack.* Already in late Old English the compound *berern* was reduced to *bern,* which by the end of the Middle English period had become *barn.* In England, unlike the United States, a barn is still principally a place for storing fodder, so "barley-house" is not too inappropriate. English cows are typically housed and milked in a separate structure, the *cowhouse* or *cowshed.*

[2]barrack *transitive verb* (1701)
: to lodge in barracks

[3]barrack *verb* [origin unknown] (1887)
transitive verb
chiefly British **:** to shout at derisively or sarcastically
intransitive verb
1 *chiefly British* **:** ROOT, CHEER — usually used with *for*
2 *chiefly British* **:** JEER, SCOFF
— bar·rack·er *noun*

barracks bag *noun* (1938)
: a fabric bag for carrying personal equipment; *especially* **:** DUFFEL BAG

bar·ra·coon \ˌbar-ə-ˈkün\ *noun* [Spanish *barracón,* augmentative of *barraca* hut, from Catalan] (1848)
: an enclosure or barracks formerly used for temporary confinement of slaves or convicts — often used in plural

bar·ra·cou·ta \ˌbar-ə-ˈkü-tə\ *noun* [modification of American Spanish *barracuda*] (1835)
: a large elongate marine fish (*Thyrsites atun* of the family Gempylidae) used for food

bar·ra·cu·da \ˌbar-ə-ˈkü-də\ *noun, plural* **-da** *or* **-das** [American Spanish] (1678)
1 : any of a genus (*Sphyraena* of the family Sphyraenidae) of elongate predaceous often large fishes of warm seas that include fishes used for food as well as some forms regarded as toxic
2 : one that uses aggressive, selfish, and sometimes unethical methods to obtain a goal especially in business

[1]bar·rage \ˈbär-ij\ *noun* [French, from *barrer* to bar, from *barre* bar] (1859)
: a dam placed in a watercourse to increase the depth of water or to divert it into a channel for navigation or irrigation

[2]bar·rage \bə-ˈräzh, -ˈräj\ *noun* [French (*tir de*) *barrage* barrier fire] (1916)
1 : artillery fire laid on a line close to friendly troops to screen and protect them
2 : a vigorous or rapid outpouring or projection of many things at once ⟨a *barrage* of protests⟩

[3]bar·rage \bə-ˈräzh, -ˈräj\ *transitive verb* **bar·raged; bar·rag·ing** (1918)
: to deliver a barrage against

barrage balloon *noun* (circa 1920)
: a small captive balloon used to support wires or nets as protection against air attacks

bar·ra·mun·di \ˌbar-ə-ˈmən-dē\ *also* **bar·ra·mun·da** \-də\ *noun* [probably from an Australian aboriginal language of Queensland] (1864)
: any of several Australian fishes (especially *Lates calcarifer* of the family Centropomidae) used for food

bar·ran·ca \bə-ˈraŋ-kə\ *or* **bar·ran·co** \-(ˌ)kō\ *noun, plural* **-cas** *or* **-cos** [Spanish] (circa 1691)
1 : a deep gully or arroyo with steep sides
2 : a steep bank or bluff

bar·ra·tor *also* **bar·ra·ter** \ˈbar-ə-tər\ *noun* (15th century)
: one who engages in barratry

bar·ra·try \ˈbar-ə-trē\ *noun, plural* **-tries** [Middle English *barratrie,* from Middle French *baraterie* deception, from *barater* to deceive, exchange] (15th century)
1 : the purchase or sale of office or preferment in church or state
2 : an unlawful act or fraudulent breach of duty on the part of a master of a ship or of the mariners to the injury of the owner of the ship or cargo
3 : the persistent incitement of litigation

Barr body \ˈbär-\ *noun* [Murray Llewellyn *Barr* (born 1908) Canadian anatomist] (1963)
: material of the inactivated X chromosome present in each of the female's somatic cells of most mammals used as a test of genetic femaleness (as in a fetus or an athlete) — called also *sex chromatin*

barre \ˈbär\ *noun* [French, from Medieval Latin *barra*] (1936)
: BAR 1c

barred \ˈbärd\ *adjective* (14th century)
: marked by or divided off by bars; *especially* **:** having alternate bands of different color ⟨*barred* feather⟩

barred owl *noun* (1811)
: a large American owl (*Strix varia*) with brown eyes and bars of dark brown on the breast

[1]bar·rel \ˈbar-əl\ *noun* [Middle English *barel,* from Middle French *baril*] (14th century)
1 : a round bulging vessel of greater length than breadth that is usually made of staves bound with hoops and has flat ends of equal diameter
2 a : the amount contained in a barrel; *especially* **:** the amount (as 31 gallons of fermented beverage or 42 gallons of petroleum) fixed for a certain commodity used as a unit of measure **b :** a great quantity
3 : a drum or cylindrical part: as **a :** the discharging tube of a gun **b :** the cylindrical metal box enclosing the mainspring of a timepiece **c :** the part of a fountain pen or of a pencil containing the ink or lead **d :** a cylindrical or tapering housing containing the optical components of a photographic-lens system and the iris diaphragm **e :** TUMBLING BARREL **f :** the fuel outlet from the carburetor on a gasoline engine
4 : the trunk of a quadruped
— bar·reled \-əld\ *adjective*
— on the barrel : asking for or granting no credit
— over a barrel : at a disadvantage **:** in an awkward position

[2]barrel *verb* **-reled** *or* **-relled; -rel·ing** *or* **-rel·ling** (15th century)
transitive verb
: to put or pack in a barrel
intransitive verb
: to move at a high speed or without hesitation

bar·rel·age \ˈbar-ə-lij\ *noun* (1890)
: amount (as of beer) in barrels

barrel cactus *noun* (1881)
: any of a genus (*Ferocactus*) of nearly globular deeply ribbed spiny cacti of Mexico and the adjacent U.S.

barrel cactus

bar·rel–chest·ed \ˈbar-əl-ˌches-təd\ *adjective* (1926)
: having a large rounded chest

barrel cuff *noun* (1926)
: an unfolded cuff (as on a shirt) usually fastened by a button

bar·rel·ful \ˈbar-əl-ˌfu̇l\ *noun, plural* **barrel·fuls** \-ˌfu̇lz\ *or* **bar·rels·ful** \-əlz-ˌfu̇l\ (14th century)
1 : as much or as many as a barrel will hold
2 : a large number or amount

bar·rel·head \-ˌhed\ *noun* (1840)
: the flat end of a barrel
— on the barrelhead : asking for or granting no credit ⟨paid cash *on the barrelhead*⟩

bar·rel·house \-ˌhau̇s\ *noun* (1883)
1 : a cheap drinking and usually dancing establishment
2 : a strident, uninhibited, and forcefully rhythmic style of jazz or blues

barrel organ *noun* (1772)
: an instrument for producing music by the action of a revolving cylinder studded with pegs on a series of valves that admit air from a bellows to a set of pipes

barrel racing *noun* (1972)
: a rodeo event for women in which a mounted rider makes a series of sharp turns around three barrels in a cloverleaf pattern
— barrel race *noun*
— barrel racer *noun*

barrel roll *noun* (circa 1920)
: an airplane maneuver in which a complete revolution about the longitudinal axis is made

barrel vault *noun* (1849)
: a semicylindrical vault
— bar·rel–vault·ed \-ˌvȯl-təd\ *adjective*

[1]bar·ren \ˈbar-ən\ *adjective* [Middle English *bareine,* from Old French *baraine*] (13th century)
1 : not reproducing: as **a :** incapable of producing offspring — used especially of females or matings **b :** not yet or not recently pregnant **c :** habitually failing to fruit
2 : not productive: as **a :** producing little or no vegetation **:** DESOLATE ⟨*barren* deserts⟩ **b :** producing inferior crops ⟨*barren* soil⟩ **c :** unproductive of results or gain **:** FRUITLESS ⟨a *barren* scheme⟩
3 : DEVOID, LACKING — used with *of* ⟨*barren* of excitement⟩
4 : lacking interest, information, or charm
5 : DULL, UNRESPONSIVE
synonym see BARE
— bar·ren·ly *adverb*
— bar·ren·ness \-ə(n)-nəs\ *noun*

[2]barren *noun* (1651)
1 *plural* **:** an extent of usually level land having an inferior growth of trees or little vegetation
2 : a tract of barren land

bar·rette \bä-ˈret, bə-\ *noun* [French, diminutive of *barre* bar] (1901)
: a clip or bar for holding hair in place

[1]bar·ri·cade \ˈbar-ə-ˌkād, ˌbar-ə-ˈ\ *transitive verb* **-cad·ed; -cad·ing** (1592)
1 : to block off or stop up with a barricade
2 : to prevent access to by means of a barricade

[2]barricade *noun* [French, from Middle French, from *barriquer* to barricade, from *barrique* barrel] (1642)
1 : an obstruction or rampart thrown up across a way or passage to check the advance of the enemy
2 : BARRIER 3, OBSTACLE
3 *plural* **:** a field of combat or dispute

bar·ri·ca·do \ˌbar-ə-ˈkā-(ˌ)dō\ *noun, plural* **-does** [modification of Middle French *barricade*] (1590)
archaic **:** BARRICADE
— barricado *transitive verb, archaic*

bar·ri·er \ˈbar-ē-ər\ *noun* [Middle English *barrere,* from Middle French *barriere,* from *barre*] (14th century)
1 a : something material that blocks or is intended to block passage ⟨highway *barriers*⟩ ⟨a *barrier* contraceptive⟩ **b :** a natural formation or structure that prevents or hinders movement or action ⟨geographic *barriers* to species dissemination⟩ ⟨*barrier* beaches⟩ ⟨drugs that cross the placental *barrier*⟩
2 *plural often capitalized* **:** a medieval war game in which combatants fight on foot with a fence or railing between them
3 : something immaterial that impedes or separates **:** OBSTACLE ⟨behavioral *barriers*⟩ ⟨trade *barriers*⟩ ⟨a *barrier* to honest communication⟩

barrier island *noun* (1943)
: a long broad sandy island lying parallel to a shore that is built up by the action of waves, currents, and winds and that protects the shore from the effects of the ocean

barrier reef *noun* (1805)
: a coral reef roughly parallel to a shore and separated from it by a lagoon

bar·ring \ˈbär-iŋ\ *preposition* (15th century)
: excluding by exception **:** EXCEPTING

bar·rio \'bär-ē-ˌō, 'bar-\ *noun, plural* **-ri·os** [Spanish, from Arabic *barrī* of the open country, from *barr* outside, open country] (1841)
1 : a ward, quarter, or district of a city or town in Spanish-speaking countries
2 : a Spanish-speaking quarter or neighborhood in a city or town in the U.S. especially in the Southwest

bar·ris·ter \'bar-ə-stər\ *noun* [Middle English *barrester,* from *barre* bar + *-ster* (as in *legister* lawyer)] (15th century)
: a counsel admitted to plead at the bar and undertake the public trial of causes in an English superior court — compare SOLICITOR

bar·room \'bär-ˌrüm, -ˌru̇m\ *noun* (1797)
: a room or establishment whose main feature is a bar for the sale of liquor

¹bar·row \'bar-(ˌ)ō\ *noun* [Middle English *bergh,* from Old English *beorg;* akin to Old High German *berg* mountain, Sanskrit *bṛhant* high] (before 12th century)
1 : MOUNTAIN, MOUND — used only in the names of hills in England
2 : a large mound of earth or stones over the remains of the dead **:** TUMULUS

²barrow *noun* [Middle English *barow,* from Old English *bearg;* akin to Old High German *barug* barrow] (before 12th century)
: a male hog castrated before sexual maturity

³barrow *noun* [Middle English *barew,* from Old English *bearwe;* akin to Old English *beran* to carry — more at BEAR] (before 12th century)
1 a : HANDBARROW **b :** WHEELBARROW
2 : a cart with a shallow box body, two wheels, and shafts for pushing it

barrow boy *noun* (1939)
British **:** COSTERMONGER

bar sinister *noun* (1823)
1 : a heraldic charge held to be a mark of bastardy
2 : the fact or condition of being of illegitimate birth

bar·tend·er \'bär-ˌten-dər\ *noun* (1836)
: one that serves liquor at a bar

¹bar·ter \'bär-tər\ *verb* [Middle English *bartren,* from Middle French *barater* to deceive, exchange] (15th century)
intransitive verb
: to trade by exchanging one commodity for another
transitive verb
: to trade or exchange by or as if by bartering
— bar·ter·er \-tər-ər\ *noun*

²barter *noun* (15th century)
1 : the act or practice of carrying on trade by bartering
2 : the thing given in exchange in bartering

Bar·tho·lin's gland \'bär-thə-lənz-, 'bär-tᵊl-ənz-\ *noun* [Kaspar *Bartholin* (died 1738) Danish physician] (1901)
: either of two oval racemose glands lying one to each side of the lower part of the vagina and secreting a lubricating mucus — compare COWPER'S GLAND

bar·ti·zan \'bär-tə-zən, ˌbär-tə-'zan\ *noun* [alteration of Middle English *bretasinge,* from *bretais* parapet — more at BRATTICE] (1808)
: a small structure (as a turret) projecting from a building and serving especially for lookout or defense

Ba·ruch \bə-'rük, 'bär-ˌu̇k\ *noun* [Late Latin, from Greek *Barouch,* from Hebrew *Bārūkh*]
: a homiletic book included in the Roman Catholic canon of the Old Testament and in the Protestant Apocrypha — see BIBLE table

bar·ware \'bär-ˌwar, -ˌwer\ *noun* (1941)
: glassware or utensils used in preparing and serving alcoholic beverages

bary·on \'bar-ē-ˌän\ *noun* [International Scientific Vocabulary *bary-* (from Greek *barys* heavy) + ²*-on* — more at GRIEVE] (1953)
: any of a group of elementary particles (as nucleons) that are subject to the strong force and are held to be a combination of three quarks
— bary·on·ic \ˌbar-ē-'ä-nik\ *adjective*

bar·yte \'bar-ˌīt, 'ber-\ *or* **ba·ry·tes** \bə-'rī-tēz\ *variant of* BARITE

bary·tone \'bar-ə-ˌtōn\ *variant of* BARITONE

bas·al \'bā-səl, -zəl\ *adjective* (1645)
1 a : relating to, situated at, or forming the base **b :** arising from the base of a stem ⟨*basal* leaves⟩
2 a : of or relating to the foundation, base, or essence **:** FUNDAMENTAL **b :** of, relating to, or being essential for maintaining the fundamental vital activities of an organism **:** MINIMAL **c :** used for teaching beginners ⟨*basal* readers⟩
— ba·sal·ly *adverb*

basal body *noun* (1902)
: a minute distinctively staining cell organelle found at the base of a flagellum or cilium and identical to a centriole in structure — called also *basal granule, kinetosome*

basal cell *noun* (circa 1903)
: one of the innermost cells of the deeper epidermis of the skin

basal ganglion *noun* (circa 1889)
: any of four deeply placed masses of gray matter (as the amygdala) in each cerebral hemisphere — called also *basal nucleus*

basal metabolic rate *noun* (1922)
: the rate at which heat is given off by an organism at complete rest

basal metabolism *noun* (1913)
: the turnover of energy in a fasting and resting organism using energy solely to maintain vital cellular activity, respiration, and circulation as measured by the basal metabolic rate

ba·salt \bə-'sȯlt, 'bā-ˌ\ *noun* [Latin *basaltes,* manuscript variant of *basanites* touchstone, from Greek *basanitēs* (*lithos*), from *basanos* touchstone, from Egyptian *bḫnw*] (1601)
: a dark gray to black dense to fine-grained igneous rock that consists of basic plagioclase, augite, and usually magnetite
— ba·sal·tic \bə-'sȯl-tik\ *adjective*

bas·cule \'bas-(ˌ)kyü(ə)l\ *noun* [French, seesaw] (1678)
: an apparatus or structure (as a drawbridge) in which one end is counterbalanced by the other on the principle of the seesaw or by weights

¹base \'bās\ *noun, plural* **bas·es** \'bā-səz\ [Middle English, from Middle French, from Latin *basis,* from Greek, step, base, from *bainein* to go — more at COME] (13th century)
1 a (1) **:** the lower part of a wall, pier, or column considered as a separate architectural feature (2) **:** the lower part of a complete architectural design **b :** the bottom of something considered as its support **:** FOUNDATION **c** (1) **:** a side or face of a geometrical figure from which an altitude can be constructed; *especially* **:** one on which the figure stands (2) **:** the length of a base **d :** that part of a bodily organ by which it is attached to another more central structure of the organism

base of a column: *1* **upper torus,** *2* **scotia,** *3* **lower torus,** *4* **plinth,** *5* **shaft,** *6* **fillets**

2 a : a main ingredient ⟨paint having a latex *base*⟩ **b :** a supporting or carrying ingredient (as of a medicine)
3 a : the fundamental part of something **:** GROUNDWORK, BASIS **b :** the economic factors on which in Marxist theory all legal, social, and political relations are formed
4 : the lower part of a heraldic field
5 a : the starting point or line for an action or undertaking **b :** a baseline in surveying **c :** a center or area of operations: as (1) **:** the place from which a military force draws supplies (2) **:** a place where military operations begin (3) **:** a permanent military installation **d** (1) **:** a number (as 5 in $5^{6.44}$ or 5^7) that is raised to a power; *especially* **:** the number that when raised to a power equal to the logarithm of a number yields the number itself ⟨the logarithm of 100 to the *base* 10 is 2 since $10^2 = 100$⟩ (2) **:** a number equal to the number of units in a given digit's place that for a given system of writing numbers is required to give the numeral 1 in the next higher place ⟨the decimal system uses a *base* of 10⟩; *also* **:** such a system of writing numbers using an indicated base ⟨convert from *base* 10 to *base* 2⟩ (3) **:** a number that is multiplied by a rate or of which a percentage or fraction is calculated ⟨to find the interest on $90 at 10% multiply the *base* 90 by .10⟩ **e :** ROOT 6
6 a : the starting place or goal in various games **b :** any one of the four stations at the corners of a baseball infield **c :** a point to be considered ⟨his opening remarks touched every *base*⟩
7 a : any of various typically water-soluble and bitter tasting compounds that in solution have a pH greater than 7, are capable of reacting with an acid to form a salt, and are molecules or ions able to take up a proton from an acid or able to give up an unshared pair of electrons to an acid **b :** any of the five purine or pyrimidine bases of DNA and RNA that include cytosine, guanine, adenine, thymine, and uracil
8 : a price level at which a security previously actively declining in price resists further price decline
9 : the part of a transformational grammar that consists of rules and a lexicon and generates the deep structures of a language
— based \'bāst\ *adjective*
— base·less \'bā-sləs\ *adjective*
— off base 1 : WRONG, MISTAKEN **2 :** UNAWARES

²base *transitive verb* **based; bas·ing** (1587)
1 : to make, form, or serve as a base for
2 : to find a base or basis for — usually used with *on* or *upon*

³base *adjective* [Middle English *bas,* from Middle French, from Medieval Latin *bassus* short, low] (14th century)
1 *archaic* **:** of little height
2 *obsolete* **:** low in place or position
3 *obsolete* **:** BASS
4 *archaic* **:** BASEBORN
5 a : resembling a villein **:** SERVILE ⟨a *base* tenant⟩ **b :** held by villenage ⟨*base* tenure⟩
6 a : being of comparatively low value and having relatively inferior properties (as lack of resistance to corrosion) ⟨a *base* metal such as iron⟩ — compare NOBLE **b :** containing a larger than usual proportion of base metals ⟨*base* silver denarii⟩
7 a : lacking or indicating the lack of higher qualities of mind or spirit **:** IGNOBLE **b :** lacking higher values **:** DEGRADING ⟨a drab *base* way of life⟩ ☆
— base·ly *adverb*
— base·ness *noun*

base angle *noun* (circa 1949)
: either of the angles of a triangle that have one side in common with the base

base·ball \'bās-ˌbȯl\ *noun, often attributive* (circa 1815)
: a game played with a bat and ball between two teams of nine players each on a large field

☆ SYNONYMS
Base, low, vile mean deserving of contempt because of the absence of higher values. BASE stresses the ignoble and may suggest cruelty, treachery, greed, or grossness ⟨*base* motives⟩. LOW may connote crafty cunning, vulgarity, or immorality and regularly implies an outraging of one's sense of decency or propriety ⟨refused to listen to such *low* talk⟩. VILE, the strongest of these words, tends to suggest disgusting depravity or filth ⟨a *vile* remark⟩.

having four bases that mark the course a runner must take to score; *also* **:** the ball used in this game

base·board \-ˌbōrd, -ˌbȯrd\ *noun* (1853)
: a board situated at or forming the base of something; *specifically* **:** a molding covering the joint of a wall and the adjoining floor

base·born \-ˈbȯrn\ *adjective* (1591)
1 : MEAN, IGNOBLE
2 a : of humble birth **b :** of illegitimate birth

base burner *noun* (1874)
: a stove in which the fuel is fed from a hopper as the lower layer is consumed

base exchange *noun* (circa 1956)
: a post exchange at a naval or air force base

base hit *noun* (1874)
: a hit in baseball that enables the batter to reach base safely without benefit of an error or fielder's choice

base·line \ˈbās-ˌlīn\ *noun, often attributive* (1750)
1 : a line serving as a basis; *especially* **:** one of known measure or position used (as in surveying or navigation) to calculate or locate something
2 a : either of the lines leading from home plate to first base and third base that are extended into the outfield as foul lines **b :** BASE PATH
3 : a boundary line at either end of a court (as in tennis or basketball)
4 : a set of critical observations or data used for comparison or a control
5 : a starting point ⟨the *baseline* of this discussion⟩

base·lin·er \ˈbās-ˌlī-nər\ *noun* (circa 1929)
: a tennis player who stays on or near the baseline and seldom moves to the net

base·ment \ˈbā-smənt\ *noun* [probably from [1]*base*] (1730)
1 : the part of a building that is wholly or partly below ground level
2 : the ground floor facade or interior in Renaissance architecture
3 : the lowest or fundamental part of something; *specifically* **:** the rocks underlying stratified rocks
4 *chiefly New England* **:** a toilet or washroom especially in a school
— base·ment·less \-ləs\ *adjective*

basement membrane *noun* (1847)
: a thin membranous layer of connective tissue that separates a layer of epithelial cells from the underlying lamina propia

ba·sen·ji \bə-ˈsen-jē, -ˈzen-\ *noun* [probably from Lingala *basenji,* plural of *mosenji* native] (1933)
: any of a breed of small curly-tailed dogs of African origin that do not bark

base on balls (circa 1891)
: an advance to first base awarded a baseball player who during his turn at bat takes four pitches that are balls

base pair *noun* (1962)
: one of the pairs of chemical bases composed of a purine on one strand of DNA joined by hydrogen bonds to a pyrimidine on the other that hold together the two complementary strands much like the rungs of a ladder and include adenine linked to thymine or sometimes to uracil and guanine linked to cytosine

base path *noun* (1935)
: the area between the bases of a baseball field used by a base runner

base pay *noun* (1920)
: a rate or amount of pay for a standard work period, job, or position exclusive of additional payments or allowances

base runner *noun* (1867)
: a baseball player of the team at bat who is on base or is attempting to reach a base
— base·run·ning *noun*

bases *plural of* BASE *or of* BASIS

[1]**bash** \ˈbash\ *verb* [origin unknown] (1750)
transitive verb
1 : to strike violently **:** HIT; *also* **:** to injure or damage by striking **:** SMASH — often used with *in*
2 : to attack physically or verbally ⟨media *bashing*⟩ ⟨celebrity *bashing*⟩
intransitive verb
: CRASH
— bash·er *noun*

[2]**bash** *noun* (1805)
1 : a forceful blow
2 : a festive social gathering **:** PARTY
3 : TRY, ATTEMPT ⟨have a *bash* at it⟩

ba·shaw \bə-ˈshȯ\ *variant of* PASHA

bash·ful \ˈbash-fəl\ *adjective* [obsolete *bash* (to be abashed)] (1548)
1 : socially shy or timid **:** DIFFIDENT, SELF-CONSCIOUS
2 : resulting from or typical of a bashful nature ⟨a *bashful* smile⟩
synonym see SHY
— bash·ful·ly \-fə-lē\ *adverb*
— bash·ful·ness \-fəl-nəs\ *noun*

[1]**ba·sic** \ˈbā-sik *also* -zik\ *adjective* (1842)
1 : of, relating to, or forming the base or essence **:** FUNDAMENTAL
2 : constituting or serving as the basis or starting point
3 a : of, relating to, containing, or having the character of a chemical base **b :** having an alkaline reaction
4 : containing relatively little silica ⟨*basic* rocks⟩
5 : relating to, made by, used in, or being a process of making steel done in a furnace lined with basic material and under basic slag
— ba·sic·i·ty \bā-ˈsi-sə-tē\ *noun*

[2]**basic** *noun* (1926)
1 : something that is basic **:** FUNDAMENTAL ⟨get back to *basics*⟩
2 : BASIC TRAINING

BA·SIC \ˈbā-sik\ *noun* [*B*eginner's *A*ll-purpose *S*ymbolic *I*nstruction *C*ode] (1964)
: a simplified language for programming a computer

ba·si·cal·ly \ˈbā-si-k(ə-)lē *also* -zi-\ *adverb* (1903)
1 a : at a basic level **:** in fundamental disposition or nature ⟨*basically* correct⟩ ⟨*basically,* they are simple people⟩ **b :** for the most part ⟨they *basically* play zone defense⟩
2 : in a basic manner **:** SIMPLY ⟨live *basically*⟩

basic slag *noun* (1888)
: a slag low in silica and high in base-forming oxides that is used in the basic process of steelmaking and that is subsequently useful as a fertilizer

basic training *noun* (1943)
: the initial period of training of a military recruit

ba·sid·io·my·cete \bə-ˌsi-dē-ō-ˈmī-ˌsēt, -ˌmī-ˈsēt\ *noun* [ultimately from New Latin *basidium* + Greek *mykēt-, mykēs* fungus; akin to Greek *myxa* mucus — more at MUCUS] (1899)
: any of a large class (Basidiomycetes) or subdivision (Basidiomycotina) of higher fungi having septate hyphae, bearing spores on a basidium, and including rusts, smuts, mushrooms, and puffballs
— ba·sid·io·my·ce·tous \-dē-ō-ˌmī-ˈsē-təs\ *adjective*

ba·sid·io·spore \bə-ˈsi-dē-ə-ˌspōr, -ˌspȯr\ *noun* [New Latin *basidium* + English *-o-* + *spore*] (1859)
: a spore produced by a basidium

ba·sid·i·um \bə-ˈsi-dē-əm\ *noun, plural* **-ia** \-dē-ə\ [New Latin, from Latin *basis*] (1859)
: a structure on the fruiting body of a basidiomycete in which karyogamy occurs followed by meiosis to form usually four basidiospores

ba·si·fy \ˈbā-sə-ˌfī\ *transitive verb* **-fied; -fy·ing** (circa 1847)
: to convert into a base or make alkaline
— ba·si·fi·ca·tion \ˌbā-sə-fə-ˈkā-shən\ *noun*

ba·sil \ˈba-zəl, ˈbā-, -səl\ *noun* [Middle French *basile,* from Late Latin *basilicum,* from Greek *basilikon,* from neuter of *basilikos*] (15th century)
1 : any of several aromatic herbs (genus *Ocimum*) of the mint family; *especially* **:** SWEET BASIL
2 : the dried or fresh leaves of a basil used especially as a seasoning

bas·i·lar \ˈba-zᵊl-ər, -sᵊl- *also* ˈbā-\ *adjective* [Middle French *basilaire,* irregular from *base* base] (1541)
: of, relating to, or situated at the base

basilar membrane *noun* (1867)
: a membrane extending from the bony shelf of the cochlea to the outer wall and supporting the organ of Corti

Ba·sil·i·an \bə-ˈzi-lē-ən\ *noun* (1780)
: a member of the monastic order founded by Saint Basil in the 4th century in Cappadocia
— Basilian *adjective*

ba·sil·i·ca \bə-ˈsi-li-kə *also* -ˈzi-\ *noun* [Latin, from Greek *basilikē,* from feminine of *basilikos* royal, from *basileus* king] (1541)
1 : an oblong building ending in a semicircular apse used in ancient Rome especially for a court of justice and place of public assembly
2 : an early Christian church building consisting of nave and aisles with clerestory and a large high transept from which an apse projects
3 : a Roman Catholic church given ceremonial privileges
— ba·sil·i·can \-kən\ *adjective*

bas·i·lisk \ˈba-sə-ˌlisk, ˈba-zə-\ *noun* [Middle English, from Latin *basiliscus,* from Greek *basiliskos,* from diminutive of *basileus*] (14th century)
1 : a legendary reptile with fatal breath and glance
2 : any of several crested tropical American lizards (genus *Basiliscus*) related to the iguanas and noted for their ability to run on their hind legs
— basilisk *adjective*

ba·sin \ˈbā-sᵊn\ *noun* [Middle English, from Old French *bacin,* from Late Latin *bacchinon*] (13th century)
1 a : an open usually circular vessel with sloping or curving sides used typically for holding water for washing **b** *chiefly British* **:** a bowl used especially in cooking **c :** the quantity contained in a basin
2 a : a dock built in a tidal river or harbor **b :** an enclosed or partly enclosed water area
3 a : a large or small depression in the surface of the land or in the ocean floor **b :** the entire tract of country drained by a river and its tributaries **c :** a great depression in the surface of the lithosphere occupied by an ocean
4 : a broad area of the earth beneath which the strata dip usually from the sides toward the center
— ba·sin·al \-sᵊn-əl\ *adjective*
— ba·sined \-sᵊnd\ *adjective*
— ba·sin·ful \-ˌfu̇l\ *noun*

bas·i·net \ˌba-sə-ˈnet\ *noun* [Middle English *bacinet,* from Middle French, diminutive of Old French *bacin*] (14th century)
: a light often pointed steel helmet

ba·sip·e·tal \bā-ˈsi-pə-tᵊl, -ˈzi-\ *adjective* [Latin *basis* + *petere* to go toward — more at FEATHER] (1869)
: proceeding from the apex toward the base or from above downward ⟨*basipetal* maturation of an inflorescence⟩
— ba·sip·e·tal·ly \-tᵊl-ē\ *adverb*

ba·sis \ˈbā-səs\ *noun, plural* **ba·ses** \-ˌsēz\ [Latin — more at BASE] (14th century)

\ə\ **abut** \ᵊ\ **kitten** \ər\ **further** \a\ **ash** \ā\ **ace** \ä\ **mop, mar** \au̇\ **out** \ch\ **chin** \e\ **bet** \ē\ **easy** \g\ **go** \i\ **hit** \ī\ **ice** \j\ **job** \ŋ\ **sing** \ō\ **go** \ȯ\ **law** \ȯi\ **boy** \th\ **thin** \t͟h\ **the** \ü\ **loot** \u̇\ **foot** \y\ **yet** \zh\ **vision** *see also* Guide to Pronunciation

1 : the bottom of something considered as its foundation
2 : the principal component of something
3 a : something on which something else is established or based **b :** an underlying condition or state of affairs ⟨hired on a trial *basis*⟩ ⟨on a first-name *basis*⟩
4 : the basic principle
5 : a set of linearly independent vectors in a vector space such that any vector in the vector space can be expressed as a linear combination of them with appropriately chosen coefficients

basis point *noun* (1967)
: one hundredth of one percent in the yield of an investment

bask \'bask\ *verb* [Middle English, probably from Old Norse *bathask,* reflexive of *batha* to bathe; akin to Old English *bæth* bath] (14th century)
intransitive verb
1 : to lie in or expose oneself to a pleasant warmth or atmosphere
2 : to take pleasure or derive enjoyment
transitive verb
obsolete **:** to warm by continued exposure to heat

bas•ket \'bas-kit, *British also* 'bäs-\ *noun* [Middle English, probably from (assumed) Old North French *baskot;* akin to Old French *baschoue* wooden vessel; both from Latin *bascauda* kind of basin, of Celtic origin; akin to Middle Irish *basc* necklace — more at FASCIA] (14th century)
1 a : a receptacle made of interwoven material (as osiers) **b :** any of various lightweight usually wood containers **c :** the quantity contained in a basket
2 : something that resembles a basket especially in shape or use
3 a : a net open at the bottom and suspended from a metal ring that constitutes the goal in basketball **b :** a field goal in basketball
4 a : an aggregate of values (as of selected currencies) the average of which serves as a monetary standard **b :** a selection of financial instruments (as equities, futures, or options) the values of which reflect market fluctuations
5 : a ring around the lower end of a ski pole that keeps the pole from sinking too deep in snow
— bas•ket•like \-,līk\ *adjective*

bas•ket•ball \-,bȯl\ *noun, often attributive* (1892)
: a usually indoor court game between two teams of usually five players each who score by tossing an inflated ball through a raised goal; *also* **:** the ball used in this game

basket case *noun* (1919)
1 : a person who has all four limbs amputated
2 : one that is completely incapacitated, inoperative, or worn out (as from nervous tension)

bas•ket•ful \'bas-kit-,fu̇l\ *noun, plural* **bas•ket•fuls** \-,fu̇lz\ *also* **bas•kets•ful** \-kits-,fu̇l\ (14th century)
: as much or as many as a basket will hold; *also* **:** a considerable quantity

basket hilt *noun* (circa 1550)
: a hilt with a basket-shaped guard to protect the hand
— bas•ket–hilt•ed \,bas-kit-'hil-təd\ *adjective*

Basket Maker *noun* (1897)
: any of three stages of an ancient culture of the plateau area of southwestern U.S.; *also* **:** a member of the people who produced the Basket Maker culture

basket–of–gold *noun* (1930)
: a European perennial herb (*Aurinia saxatilis* synonym *Alyssum saxatile*) widely cultivated for its grayish foliage and yellow flowers

bas•ket•ry \'bas-ki-trē\ *noun, plural* **-ries** (1851)
1 : BASKETWORK
2 : the art or craft of making baskets or objects woven like baskets

basket star *noun* (circa 1923)
: any of various brittle stars (suborder Euryalina) with slender complexly branched interlacing arms

basket weave *noun* (circa 1915)
: a textile weave resembling the checkered pattern of a plaited basket; *also* **:** something resembling this weave

bas•ket•work \'bas-kit-,wərk\ *noun* (1769)
: objects produced by basketry

bask•ing shark \'bas-kiŋ-\ *noun* (circa 1769)
: a large plankton-feeding shark (*Cetorhinus maximus*) that has an oil-rich liver and sometimes attains a length of 40 feet (12 meters)

bas•ma•ti rice \,bäz-'mä-tē- *also* ,baz-\ *noun* [Hindi *bāsmatī* kind of rice, literally, something fragrant] (1845)
: a cultivated aromatic long-grain rice of South Asian origin — called also *basmati*

bas mitz•vah \bäs-'mits-və\ *noun, often B&M capitalized* [Hebrew *bath miṣwāh,* literally, daughter of the (divine) law] (1952)
1 : a Jewish girl who at about 13 years of age assumes religious responsibilities
2 : the initiatory ceremony recognizing a girl as a bas mitzvah

ba•so•phil \'bā-sə-,fil, -zə-\ *or* **ba•so•phile** \-,fīl\ *noun* (circa 1890)
: a basophilic substance or structure; *especially* **:** a leukocyte containing basophilic granules that is similar in function to a mast cell

ba•so•phil•ia \,bā-sə-'fi-lē-ə, -zə-\ *noun* [New Latin] (1905)
1 : tendency to stain with basic dyes
2 : an abnormal condition in which some tissue element has increased basophilia

ba•so•phil•ic \-'fi-lik\ *adjective* [International Scientific Vocabulary *base* + *-o-* + *-philic*] (circa 1894)
: staining readily with basic stains

Ba•so•tho \bä-'sō-,tō, -'sü-,tü\ *noun plural* (1895)
: a Bantu-speaking people of Lesotho

Basque \'bask\ *noun* [French, from Middle French, from Latin *Vasco*] (1835)
1 : a member of a people inhabiting the western Pyrenees on the Bay of Biscay
2 : the language of the Basques of unknown relationship
3 *not capitalized* **:** a tight-fitting bodice for women
— Basque *adjective*

bas–re•lief \,bä-ri-'lēf, 'bä-ri-,\ *noun* [French, from *bas* low + *relief* raised work] (1667)
: sculptural relief in which the projection from the surrounding surface is slight and no part of the modeled form is undercut; *also* **:** sculpture executed in bas-relief

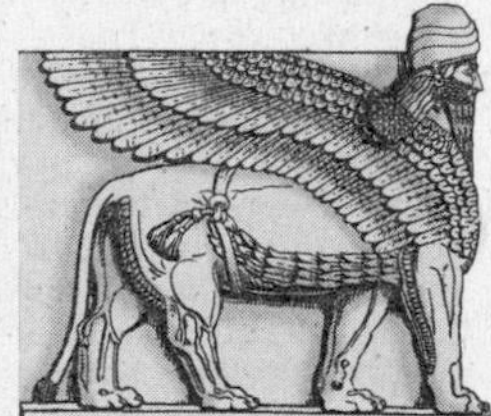
bas-relief

¹bass \'bas\ *noun, plural* **bass** *or* **bass•es** [Middle English *base, bærs,* from Old English *bærs;* akin to Old High German *bersich* perch] (before 12th century)
: any of numerous edible spiny-finned marine or freshwater fishes (especially families Centrarchidae, Serranidae, and Percichthyidae)

²bass \'bās\ *adjective* [Middle English *bas* base] (15th century)
1 : deep or grave in tone
2 a : of low pitch **b :** relating to or having the range or part of a bass

³bass \'bās\ *noun* (15th century)
1 a : the lowest voice part in a 4-part chorus **b :** the lower half of the whole vocal or instrumental tonal range — compare TREBLE **c :** the lowest adult male singing voice; *also* **:** a person having this voice **d :** a member of a family of instruments having the lowest range; *especially* **:** DOUBLE BASS
2 : a deep or grave tone **:** a low-pitched sound

⁴bass \'bas\ *noun* [alteration of *bast*] (1691)
1 : BASSWOOD 1
2 : a coarse tough fiber from palms

bass clef *noun* (circa 1771)
1 : a clef placing the F below middle C on the fourth line of the staff
2 : the bass staff

bass drum *noun* (1804)
: a large drum having two heads and giving a booming sound of low indefinite pitch — see DRUM illustration

bas•set hound \'ba-sət-\ *noun* [French *basset,* from Middle French, from *basset* short, from *bas* low — more at BASE] (1883)
: any of an old breed of short-legged hunting dogs of French origin having very long ears and crooked front legs — called also *basset*

basset hound

bass fiddle *noun* (1951)
: DOUBLE BASS

bass horn *noun* (circa 1846)
: an obsolete wind instrument shaped like a bassoon but with a cup-shaped mouthpiece

bas•si•net \,ba-sə-'net\ *noun* [probably modification of French *barcelonnette,* diminutive of *berceau* cradle] (1854)
1 : a baby's basketlike bed (as of wickerwork or plastic) often with a hood over one end
2 : a perambulator that resembles a bassinet

bass•ist \'bā-sist\ *noun* (circa 1909)
: a person who plays an acoustic or electric bass

bas•so \'ba-(,)sō, 'bä-\ *noun, plural* **bassos** *or* **bas•si** \'bä-,sē\ [Italian, from Medieval Latin *bassus,* from *bassus* short, low] (circa 1724)
1 : a bass singer; *especially* **:** an operatic bass
2 : a low deep voice

bas•soon \bə-'sün, ba-\ *noun* [French *basson,* from Italian *bassone,* from *basso*] (1724)
: a double-reed woodwind instrument having a long U-shaped conical tube connected to the mouthpiece by a thin metal tube and a usual range two octaves lower than that of the oboe
— bas•soon•ist \-'sü-nist\ *noun*

bassoon

bas•so pro•fun•do \,ba-(,)sō-prə-'fən-(,)dō, ,bä-, -'fün-\ *noun, plural* **basso profundos** [Italian, literally, deep bass] (1860)
: a deep heavy bass voice with an exceptionally low range; *also* **:** a person having this voice

bas•so–re•lie•vo *also* **bas•so–ri•lie•vo** \,ba-(,)sō-ri-'lē-(,)vō, ,bä-(,)sō-rēl-'yā-(,)vō\ *noun* [Italian *bassorilievo,* from *basso* low + *rilievo* relief] (1644)
: BAS-RELIEF

bass viol *noun* (1590)
1 : VIOLA DA GAMBA
2 : DOUBLE BASS

bass•wood \'bas-,wu̇d\ *noun* (1670)
1 : any of several New World lindens; *especially* **:** LINDEN 1b
2 : the straight-grained white wood of a basswood

bast \'bast\ *noun* [Middle English, from Old English *bæst;* akin to Old High German & Old Norse *bast* bast] (before 12th century)
1 : PHLOEM
2 : BAST FIBER

[1]**bas·tard** \'bas-tərd\ *noun* [Middle English, from Old French, probably of Germanic origin; akin to Old Frisian *bost* marriage, Old English *bindan* to bind] (14th century)
1 : an illegitimate child
2 : something that is spurious, irregular, inferior, or of questionable origin
3 a : an offensive or disagreeable person — used as a generalized term of abuse **b :** MAN, FELLOW
— bas·tard·ly *adjective*

[2]**bastard** *adjective* (14th century)
1 : ILLEGITIMATE
2 : of mixed or ill-conceived origin ⟨known for coining *bastard* words⟩
3 : of abnormal shape or irregular size
4 : of a kind similar to but inferior to or less typical than some standard ⟨*bastard* measles⟩
5 : lacking genuineness or authority **:** FALSE

bas·tard·ise *British variant of* BASTARDIZE

bas·tard·ize \'bas-tər-ˌdīz\ *transitive verb* **-ized; -iz·ing** (1587)
1 : to reduce from a higher to a lower state or condition **:** DEBASE
2 : to declare or prove to be a bastard
3 : to modify especially by introducing discordant or disparate elements
— bas·tard·i·za·tion \ˌbas-tər-də-'zā-shən\ *noun*

bastard wing *noun* (1772)
: ALULA

bas·tardy \'bas-tər-dē\ *noun, plural* **-tard·ies** (15th century)
1 : the quality or state of being a bastard **:** ILLEGITIMACY
2 : the begetting of an illegitimate child

[1]**baste** \'bāst\ *transitive verb* **bast·ed; bast·ing** [Middle English, from Middle French *bastir,* of Germanic origin; akin to Old High German *besten* to patch, Old English *bæst* bast] (15th century)
: to sew with long loose stitches in order to hold something in place temporarily
— bas·ter *noun*

[2]**baste** *transitive verb* **bast·ed; bast·ing** [Middle English *baisten*] (15th century)
: to moisten (as meat) at intervals with a liquid (as melted butter, fat, or pan drippings) especially during cooking
— bast·er *noun*

[3]**baste** *transitive verb* **bast·ed; bast·ing** [probably from Old Norse *beysta;* akin to Old English *bēatan* to beat] (1533)
1 : to beat severely or soundly **:** THRASH
2 : to scold vigorously **:** BERATE

bast fiber *noun* (circa 1885)
: a strong woody fiber obtained chiefly from the phloem of plants and used especially in cordage, matting, and fabrics

bas·tille *noun* \ba-'stē(ə)l\ [French *bastille,* from the *Bastille,* fortress in Paris, from Middle French *bastille,* modification of Old Provençal *bastida* fortified town, from *bastir* to build, of Germanic origin; akin to Old High German *besten* to patch] (1741)
: PRISON, JAIL

Bastille Day *noun* (1920)
: July 14 observed in France as a national holiday in commemoration of the fall of the Bastille in 1789

[1]**bas·ti·na·do** \ˌbas-tə-'nā-(ˌ)dō, -'nä\ *or* **bas·ti·nade** \ˌbas-tə-'nād, -'näd\ *noun, plural* **-na·does** *or* **-nades** [Spanish *bastonada,* from *bastón* stick, from Late Latin *bastum*] (1572)
1 : a blow with a stick or cudgel
2 a : a beating especially with a stick **b :** a punishment consisting of beating the soles of the feet with a stick
3 : STICK, CUDGEL

[2]**bastinado** *transitive verb* **-doed; -do·ing** (1599)
: to subject to repeated blows

[1]**bast·ing** \'bā-stiŋ\ *noun* (15th century)
1 : the action of a sewer who bastes
2 a : the thread used in basting **b :** the stitching made by basting

[2]**basting** *noun* (1530)
1 : the action of one that bastes food
2 : the liquid used in basting

[3]**basting** *noun* (1590)
: a severe beating

bas·tion \'bas-chən\ *noun* [Middle French, from Old Italian *bastione,* augmentative of *bastia* fortress, derivative from dialect form of *bastire* to build, of Germanic origin; akin to Old High German *besten* to patch] (1562)
1 : a projecting part of a fortification
2 : a fortified area or position
3 : something that is considered a stronghold **:** BULWARK
— bas·tioned \-chənd\ *adjective*

Ba·su·to \bə-'sü-(ˌ)tō\ *noun, plural* **Basuto** *or* **Basutos** (1835)
: a member of the Basotho people

[1]**bat** \'bat\ *noun* [Middle English, from Old English *batt*] (before 12th century)
1 : a stout solid stick **:** CLUB
2 : a sharp blow **:** STROKE
3 a : a usually wooden implement used for hitting the ball in various games **b :** a paddle used in various games (as table tennis) **c :** the short whip used by a jockey
4 a : BATSMAN **b :** a turn at batting — usually used in the phrase *at bat*
5 : BATT
6 *British* **:** rate of speed **:** GAIT
7 : BINGE
— off one's own bat *chiefly British* **:** through one's own efforts
— off the bat : without delay **:** IMMEDIATELY

[2]**bat** *verb* **bat·ted; bat·ting** (15th century)
transitive verb
1 : to strike or hit with or as if with a bat
2 a : to advance (a base runner) by batting **b :** to have a batting average of
3 : to discuss at length **:** consider in detail
intransitive verb
1 a : to strike or hit a ball with a bat **b :** to take one's turn at bat
2 : to wander aimlessly

[3]**bat** *noun* [alteration of Middle English *bakke,* probably of Scandinavian origin; akin to Old Swedish natt*bakka* bat] (1580)
: any of a widely distributed order (Chiroptera) of nocturnal usually frugivorous or insectivorous flying mammals that have wings formed from four elongated digits of the forelimb covered by a cutaneous membrane and that have adequate visual capabilities but often rely on echolocation

[4]**bat** *transitive verb* **bat·ted; bat·ting** [probably alteration of [2]*bate*] (circa 1838)
: to wink especially in surprise or emotion ⟨never *batted* an eye⟩; *also* **:** FLUTTER ⟨*batted* his eyelashes⟩

bat·boy \'bat-ˌbȯi\ *noun* (circa 1925)
: a boy employed to look after the equipment (as bats) of a baseball team

[1]**batch** \'bach\ *noun* [Middle English *bache;* akin to Old English *bacan* to bake] (15th century)
1 : the quantity baked at one time **:** BAKING
2 a : the quantity of material prepared or required for one operation; *specifically* **:** a mixture of raw materials ready for fusion into glass **b :** the quantity produced at one operation **c :** a group of jobs (as programs) which are submitted for processing on a computer and whose results are obtained at a later time ⟨*batch* processing⟩ — compare TIME-SHARING
3 : a quantity (as of persons or things) considered as a group

[2]**batch** *transitive verb* (1876)
: to bring together or process as a batch
— batch·er *noun*

[3]**batch** *variant of* BACH

[1]**bate** \'bāt\ *verb* **bat·ed; bat·ing** [Middle English, short for *abaten* to abate] (14th century)
transitive verb
1 : to reduce the force or intensity of **:** RESTRAIN ⟨with *bated* breath⟩
2 : to take away **:** DEDUCT
3 *archaic* **:** to lower especially in amount or estimation
4 *archaic* **:** BLUNT
intransitive verb
obsolete **:** DIMINISH, DECREASE

[2]**bate** *intransitive verb* **bat·ed; bat·ing** [Middle English, from Middle French *batre* to beat, from Latin *battuere*] (14th century)
of a falcon or hawk **:** to attempt to fly off something (as a gauntlet) in fear

bat–eared fox \'bat-ˌird-\ *noun* (1946)
: a large-eared yellowish gray fox (*Otocyon megalotis*) that inhabits arid unforested areas of eastern and southern Africa

ba·teau \ba-'tō\ *noun, plural* **ba·teaux** \-'tō(z)\ [Canadian French, from French, from Old French *batel,* from Old English *bāt* boat — more at BOAT] (1711)
: any of various small craft; *especially* **:** a flat-bottomed boat with raked bow and stern and flaring sides

Bates·ian \'bāt-sē-ən\ *adjective* [Henry Walter *Bates* (died 1892) English naturalist] (1896)
: characterized by or being mimicry involving resemblance of an innocuous species to another that is protected from predators by repellent qualities (as unpalatability) ⟨*Batesian* mimic⟩

bat·fish \'bat-ˌfish\ *noun* (1873)
: any of several fishes with winglike processes; *especially* **:** any of a family (Ogcocephalidae) of flattened pediculate fishes (as a common West Indian form *Ogcocephalus vespertillo*)

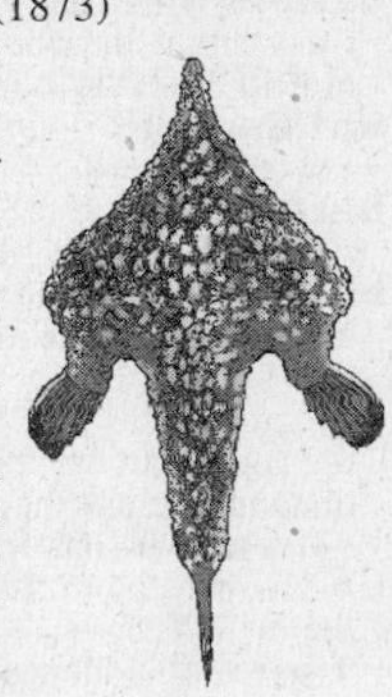
batfish

bat·fowl \-ˌfau̇l\ *intransitive verb* (15th century)
: to catch birds at night by blinding them with a light and knocking them down with a stick or netting them

[1]**bath** \'bath, 'bȧth\ *noun, plural* **baths** \'bat͟hz, 'baths, 'bȧt͟hz, 'bȧths\ [Middle English, from Old English *bæth;* akin to Old High German *bad* bath, Old High German *bāen* to warm] (before 12th century)
1 : a washing or soaking (as in water or steam) of all or part of the body
2 a : water used for bathing **b** (1) **:** a contained liquid for a special purpose (2) **:** a receptacle holding the liquid **c** (1) **:** a medium for regulating the temperature of something placed in or on it (2) **:** a vessel containing this medium
3 a : BATHROOM **b :** a building containing an apartment or a series of rooms designed for bathing **c :** SPA — usually used in plural **d** *British* **:** SWIMMING POOL — often used in plural
4 a : the quality or state of being covered with a liquid **b :** FLOOD 3
5 : BATHTUB
6 : a financial setback **:** LOSS ⟨took a *bath* in the market⟩

[2]**bath** (15th century)
transitive verb
British **:** to give a bath to
intransitive verb
British **:** to take a bath

[3]**bath** *noun* [Hebrew] (14th century)

: an ancient Hebrew liquid measure corresponding to the ephah of dry measure

bath chair \'bath-, 'bȧth-\ *noun, often B capitalized* [*Bath,* England] (1823)
: a hooded and sometimes glassed wheeled chair used especially by invalids; *broadly* : WHEELCHAIR

1bathe \'bāth\ *verb* **bathed; bath·ing** [Middle English, from Old English *bathian;* akin to Old English *bæth* bath] (before 12th century)
transitive verb
1 : MOISTEN, WET
2 : to wash in a liquid (as water)
3 : to apply water or a liquid medicament to
4 : to flow along the edge of **:** LAVE
5 : to suffuse with or as if with light
intransitive verb
1 : to take a bath
2 : to go swimming
3 : to become immersed or absorbed
— bath·er \'bā-thər\ *noun*

2bathe *noun* (1831)
1 *British* **:** 1BATH 1
2 *British* **:** SWIM, DIP

ba·thet·ic \bə-'the-tik\ *adjective* [*bathos* + *-etic* (as in *pathetic*)] (circa 1864)
: characterized by bathos
— ba·thet·i·cal·ly \-ti-k(ə-)lē\ *adverb*

bath·house \'bath-,hau̇s, 'bȧth-\ *noun* (1705)
1 : a building equipped for bathing
2 : a building containing dressing rooms for bathers

bathing beauty *noun* (1920)
: a woman in a bathing suit who is a contestant in a beauty contest

bathing suit *noun* (1873)
: SWIMSUIT

bath mat *noun* (1895)
: a usually washable mat used in a bathroom

bath·o·lith \'ba-thə-,lith\ *noun* [Greek *bathos* depth + International Scientific Vocabulary *-lith*] (circa 1900)
: a great mass of intruded igneous rock that for the most part stopped in its rise a considerable distance below the surface
— bath·o·lith·ic \,ba-thə-'li-thik\ *adjective*

ba·thos \'bā-,thäs\ *noun* [Greek, literally, depth] (1727)
1 a : the sudden appearance of the commonplace in otherwise elevated matter or style **b :** ANTICLIMAX
2 : exceptional commonplaceness **:** TRITENESS
3 : insincere or overdone pathos **:** SENTIMENTALISM

bath·robe \'bath-,rōb, 'bȧth-\ *noun* (1902)
: a loose often absorbent robe worn before and after bathing or as a dressing gown

bath·room \-,rüm, -,ru̇m\ *noun* (1780)
1 : a room containing a bathtub or shower and usually a sink and toilet
2 : LAVATORY 2

bath salts *noun plural* (1907)
: a usually colored crystalline compound for perfuming and softening bathwater

bath·tub \-,təb\ *noun* (1869)
: a usually fixed tub for bathing

bathtub gin *noun* (1930)
: a homemade spirit concocted from raw alcohol, water, essences, and essential oils

bath·wa·ter \'bath-,wȯ-tər, 'bȧth-, -,wä-tər\ *noun* (14th century)
: water for a bath

bathy- *combining form* [International Scientific Vocabulary, from Greek, from *bathys* deep]
1 : deep **:** depth ⟨*bathy*al⟩
2 : deep-sea ⟨*bathy*sphere⟩

bathy·al \'ba-thē-əl\ *adjective* (1921)
: of or relating to the ocean depths or floor usually from 600 to 6000 feet (180 to 1800 meters)

ba·thym·e·try \bə-'thi-mə-trē\ *noun, plural* **-tries** [International Scientific Vocabulary] (circa 1859)
: the measurement of water depth at various places in a body of water; *also* **:** the information derived from such measurements
— bathy·met·ric \,ba-thi-'me-trik\ *also* **bathy·met·ri·cal** \-tri-kəl\ *adjective*
— bathy·met·ri·cal·ly \-tri-k(ə-)lē\ *adverb*

bathy·pe·lag·ic \,ba-thi-pə-'la-jik\ *adjective* (circa 1900)
: of, relating to, or living in the ocean depths especially between 2000 and 12,000 feet (600 and 3600 meters)

bathy·scaphe \'ba-thi-,skaf, -,skāf\ *or* **bathy·scaph** \-,skaf\ *noun* [International Scientific Vocabulary *bathy-* + Greek *skaphē* light boat] (1947)
: a navigable submersible for deep-sea exploration having a spherical watertight cabin attached to its underside

bathy·sphere \-,sfir\ *noun* (1930)
: a strongly built steel diving sphere for deep-sea observation

bathy·ther·mo·graph \-'thər-mə-,graf\ *noun* (1938)
: an instrument designed to record water temperature as a function of depth

ba·tik \bə-'tēk, 'ba-tik\ *noun* [Javanese *baṭik*] (1880)
1 : a fabric printed by an Indonesian method of hand-printing textiles by coating with wax the parts not to be dyed; *also* **:** the method itself
2 : a design executed in batik

bat·ing \'bā-tiŋ\ *preposition* (1647)
: with the exception of **:** EXCEPTING

ba·tiste \bə-'tēst, ba-\ *noun* [French] (1697)
: a fine soft sheer fabric of plain weave made of various fibers

bat·man \'bat-mən\ *noun* [French *bât* pack-saddle] (1755)
: an orderly of a British military officer

bat mitz·vah \bät-'mits-və\ *often B&M capitalized, variant of* BAS MITZVAH

ba·ton \bə-'tän, ba-, -'tō^{n} *also* 'ba-t^{ə}n\ *noun* [French *bâton,* from Old French *baston,* from Late Latin *bastum* stick] (1520)
1 : CUDGEL, TRUNCHEON; *specifically* **:** BILLY CLUB
2 : a staff borne as a symbol of office
3 : a narrow heraldic bend
4 : a slender rod with which a leader directs a band or orchestra
5 : a hollow cylinder carried by each member of a relay team and passed to the succeeding runner
6 : a hollow metal rod with a weighted bulb at one or both ends that is flourished or twirled by a drum major or drum majorette

bat out *transitive verb* (1941)
: to compose especially in a casual, careless, or hurried manner

ba·tra·chi·an \bə-'trā-kē-ən\ *noun* [ultimately from Greek *batrachos* frog] (circa 1828)
: AMPHIBIAN 1; *especially* **:** FROG, TOAD
— batrachian *adjective*

bats \'bats\ *adjective* (1919)
: BATTY 2

bats·man \'bat-smən\ *noun* (1756)
: a batter especially in cricket

batt \'bat\ *noun* (1871)
: BATTING 2; *also* **:** an often square piece of batting

bat·tai·lous \'ba-t^{ə}l-əs\ *adjective* [Middle English *bataillous,* from Middle French *batail-los,* from *bataille* battle] (14th century)
archaic **:** ready for battle **:** WARLIKE

bat·ta·lia \bə-'tāl-yə, -'tal-\ *noun* [Italian *battaglia*] (1569)
1 *archaic* **:** order of battle
2 *obsolete* **:** a large body of men in battle array

bat·tal·ion \bə-'tal-yən\ *noun* [Middle French *bataillon,* from Old Italian *battaglione,* augmentative of *battaglia* company of soldiers, battle, from Late Latin *battalia* combat — more at BATTLE] (1579)
1 : a considerable body of troops organized to act together **:** ARMY
2 : a military unit composed of a headquarters and two or more companies, batteries, or similar units
3 : a large group

bat·teau *variant of* BATEAU

bat·te·ment \,bat-'mä^{n}\ *noun* [French, from *battre* to beat, from Latin *battuere*] (1830)
: a ballet movement in which the foot is extended in any direction usually followed by a beat against the supporting foot

1bat·ten \'ba-t^{ə}n\ *verb* **bat·tened; bat·ten·ing** \'bat-niŋ, 'ba-t^{ə}n-iŋ\ [probably from Old Norse *batna* to improve; akin to Old English *betera* better] (1591)
intransitive verb
1 a : to grow fat **b :** to feed gluttonously
2 : to grow prosperous especially at the expense of another — usually used with *on*
transitive verb
: FATTEN

2batten *noun* [alteration of Middle English *batent, bataunt* finished board, from Middle French *batant,* from present participle of *battre*] (1658)
1 a *British* **:** a piece of lumber used especially for flooring **b :** a thin narrow strip of lumber used especially to seal or reinforce a joint
2 : a strip, bar, or support resembling or used similarly to a batten

3batten *verb* **bat·tened; bat·ten·ing** \'bat-niŋ, 'ba-t^{ə}n-iŋ\ (1663)
transitive verb
1 : to furnish with battens
2 : to fasten with or as if with battens — often used with *down*
intransitive verb
: to make one secure by or as if by battens ⟨*battening* down for the hurricane⟩

1bat·ter \'ba-tər\ *verb* [Middle English *bat-eren,* probably frequentative of *batten* to bat, from *bat*] (14th century)
transitive verb
1 a : to beat with successive blows so as to bruise, shatter, or demolish **b :** BOMBARD
2 : to subject to strong, overwhelming, or repeated attack
3 : to wear or damage by hard usage or blows ⟨a *battered* old hat⟩
intransitive verb
: to strike heavily and repeatedly **:** BEAT
synonym see MAIM

2batter *noun* [Middle English *bater,* probably from *bateren*] (14th century)
1 a : a mixture consisting chiefly of flour, egg, and milk or water and being thin enough to pour or drop from a spoon **b :** a mixture (as of flour and egg) used as a coating for food (as chicken) that is to be fried
2 : an instance of battering

3batter *transitive verb* (1973)
: to coat (food) with batter for frying

4batter *noun* [origin unknown] (1743)
: a receding upward slope of the outer face of a structure

5batter *transitive verb* (circa 1882)
: to give a receding upward slope to (as a wall)

6batter *noun* (1773)
: one that bats; *especially* **:** the player whose turn it is to bat

battered child syndrome *noun* (1962)
: the complex of physical injuries sustained by a grossly abused child

bat·te·rie \,ba-tə-'rē\ *noun* [French, literally, beating — more at BATTERY] (1712)
: a ballet movement consisting of beating together the feet or calves of the legs during a leap

battering ram *noun* (1593)
1 : a military siege engine consisting of a large wooden beam with a head of iron used in ancient times to beat down the walls of a besieged place
2 : a heavy metal bar with handles used (as by firefighters) to batter down doors and walls

bat·tery \ˈba-t(ə-)rē\ *noun, plural* **-ter·ies** [Middle French *batterie,* from Old French, from *battre* to beat, from Latin *battuere*] (1531)
1 a : the act of battering or beating **b :** an offensive touching or use of force on a person without the person's consent — compare ASSAULT 2a
2 a : a grouping of artillery pieces for tactical purposes **b :** the guns of a warship
3 : an artillery unit in the army equivalent to a company
4 a : a combination of apparatus for producing a single electrical effect **b :** a group of two or more cells connected together to furnish electric current; *also* **:** a single cell that furnishes electric current ⟨a flashlight *battery*⟩
5 a : a number of similar articles, items, or devices arranged, connected, or used together **:** SET, SERIES ⟨a *battery* of tests⟩ **b :** a usually impressive or imposing group **:** ARRAY
6 : the position of readiness of a gun for firing
7 : the pitcher and catcher of a baseball team
bat·ting \ˈba-tiŋ\ *noun* (1773)
1 a : the action of one who bats **b :** the use of or ability with a bat
2 : layers or sheets of raw cotton or wool or of synthetic fibrous material used for lining quilts or for stuffing or packaging; *also* **:** a blanket of thermal insulation (as fiberglass)
batting average *noun* (1867)
1 : a ratio (as a rate per thousand) of base hits to official times at bat for a baseball player
2 : a record of achievement or accomplishment
[1]bat·tle \ˈba-tᵊl\ *noun, often attributive* [Middle English *batel,* from Old French *bataille* battle, fortifying tower, battalion, from Late Latin *battalia* combat, alteration of *battualia* fencing exercises, from Latin *battuere* to beat] (13th century)
1 *archaic* **:** BATTALION
2 : a combat between two persons
3 : a general encounter between armies, ships of war, or aircraft
4 : an extended contest, struggle, or controversy
[2]battle *verb* **bat·tled; bat·tling** \ˈbat-liŋ, ˈba-tᵊl-iŋ\ (14th century)
intransitive verb
1 : to engage in battle **:** FIGHT
2 : to contend with full strength, vigor, skill, or resources **:** STRUGGLE
transitive verb
1 : to fight against
2 : to force (as one's way) by battling
— **bat·tler** \-lər, ˈba-tᵊl-ər\ *noun*
[3]battle *transitive verb* **bat·tled; bat·tling** [Middle English *batailen,* from Middle French *bataillier* to fortify, from Old French, from *bataille*] (14th century)
archaic **:** to fortify with battlements
bat·tle–ax \ˈba-tᵊl-ˌaks\ *noun* (14th century)
1 : a broadax formerly used as a weapon of war
2 : a sharp-tongued domineering usually older woman
battle cruiser *noun* (1911)
: a large heavily armed warship that is lighter, faster, and more maneuverable than a battleship
battle cry *noun* (1814)
: WAR CRY
battle fatigue *noun* (1945)
: COMBAT FATIGUE
— **bat·tle–fa·tigued** *adjective*
bat·tle·field \ˈba-tᵊl-ˌfēld\ *noun* (1812)
1 : a place where a battle is fought
2 : an area of conflict
bat·tle·front \-ˌfrənt\ *noun* (1914)
: the military sector in which actual combat takes place
bat·tle·ground \-ˌgrau̇nd\ *noun* (1815)
: BATTLEFIELD
battle line (1814)
1 : a line along which a battle is fought
2 : a line defining the positions of opposing groups in a conflict or controversy — usually used in plural ⟨*battle lines* were drawn over economic policies⟩
bat·tle·ment \ˈba-tᵊl-mənt\ *noun* [Middle English *batelment,* from Middle French *bataille*] (14th century)
: a parapet with open spaces that surmounts a wall and is used for defense or decoration
— **bat·tle·ment·ed** \-ˌmen-təd\ *adjective*

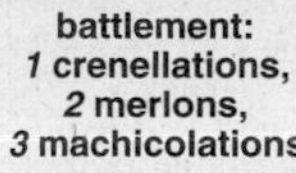

battlement:
1 crenellations,
2 merlons,
3 machicolations

battle royal *noun, plural* **battles royal** *or* **battle royals** (1672)
1 a : a fight participated in by more than two combatants; *especially* **:** one in which the last man in the ring or on his feet is declared the winner **b :** a violent struggle
2 : a heated dispute
bat·tle·ship \ˈba-tᵊl-ˌship\ *noun* [short for *line-of-battle ship*] (1794)
: a warship of the largest and most heavily armed and armored class
bat·tle·wag·on \-ˌwa-gən\ *noun* (circa 1927)
: BATTLESHIP
bat·tu \ba-ˈtü, -ˈtyü\ *adjective* [French, from past participle of *battre* to beat] (1947)
of a ballet movement **:** performed with a striking together of the legs
bat·tue \ba-ˈtü, -ˈtyü\ *noun* [French, from *battre* to beat] (1816)
: the beating of woods and bushes to flush game; *also* **:** a hunt in which this procedure is used
bat·ty \ˈba-tē\ *adjective* **bat·ti·er; -est** (1590)
1 : of, relating to, or resembling a bat
2 : mentally unstable **:** CRAZY
— **bat·ti·ness** *noun*
bau·ble \ˈbȯ-bəl, ˈbä-\ *noun* [Middle English *babel,* from Middle French] (14th century)
1 : TRINKET
2 : a fool's scepter
3 : TRIFLE
Bau·cis \ˈbȯ-səs\ *noun* [Latin, from Greek *Baukis*]
: the wife of Philemon
baud \ˈbȯd, ˈbäd, *British* ˈbōd\ *noun, plural* **baud** *also* **bauds** [*baud* (telegraphic transmission speed unit), from J. M. E. *Baudot* (died 1903) French inventor] (1931)
: a variable unit of data transmission speed (as one bit per second)
Bau·haus \ˈbau̇-ˌhau̇s\ *adjective* [German *Bauhaus,* literally, architecture house, school founded by Walter Gropius] (1923)
: of, relating to, or influenced by a school of design noted especially for a program that synthesized technology, craftsmanship, and design aesthetics
baulk *chiefly British variant of* BALK
Bau·mé \bō-ˈmā\ *adjective* [Antoine *Baumé*] (1877)
: being, calibrated in accordance with, or according to either of two arbitrary hydrometer scales for liquids lighter than water or for liquids heavier than water that indicate specific gravity in degrees
baum marten \ˈbau̇m-ˌmär-tᵊn\ *noun* [part translation of German *Baummarder,* from *Baum* tree + *Marder* marten] (1909)
: the pelt or fur of the European marten (*Martes martes*)
baux·ite \ˈbȯk-ˌsīt, ˈbäk-\ *noun* [French *bauxite,* from Les Baux, near Arles, France] (1861)
: an impure mixture of earthy hydrous aluminum oxides and hydroxides that is the principal source of aluminum
— **baux·it·ic** \bȯk-ˈsi-tik, bäk-\ *adjective*
Ba·var·i·an \bə-ˈver-ē-ən, -ˈvar-\ *noun* (1638)
1 : a native or inhabitant of Bavaria
2 : the High German dialect of southern Bavaria and Austria
— **Bavarian** *adjective*
Bavarian cream *noun* (circa 1879)
: flavored custard or pureed fruit combined with gelatin and whipped cream
baw·bee \ˈbȯ-(ˌ)bē, bȯ-ˈ\ *noun* [probably from Alexander Orrok, laird of Sille*bawbe* (flourished 1538) Scottish master of the mint] (1542)
1 : any of various Scottish coins of small value
2 : an English halfpenny
baw·cock \ˈbȯ-ˌkäk\ *noun* [French *beau coq,* from *beau* fine + *coq* fellow, cock] (1599)
archaic **:** a fine fellow
bawd \ˈbȯd\ *noun* [Middle English *bawde*] (14th century)
1 *obsolete* **:** PANDER
2 a : one who keeps a house of prostitution **:** MADAM **b :** PROSTITUTE
bawd·ry \ˈbȯ-drē\ *noun* [Middle English *bawderie,* from *bawde*] (15th century)
1 *obsolete* **:** UNCHASTITY
2 : suggestive, coarse, or obscene language
[1]bawdy \ˈbȯ-dē\ *adjective* **bawd·i·er; -est** [*bawd*] (1513)
1 : OBSCENE, LEWD
2 : boisterously or humorously indecent
— **bawd·i·ly** \ˈbȯ-dᵊl-ē\ *adverb*
— **bawd·i·ness** \ˈbȯ-dē-nəs\ *noun*
[2]bawdy *noun* [probably from [1]*bawdy*] (1656)
: BAWDRY 2
bawdy house *noun* (1552)
: BORDELLO
[1]bawl \ˈbȯl\ *verb* [Middle English, to bark, probably of Scandinavian origin; akin to Icelandic *baula* to low] (1570)
intransitive verb
1 : to cry out loudly and unrestrainedly **:** YELL, BELLOW
2 : to cry loudly **:** WAIL
transitive verb
: to cry out at the top of one's voice
— **bawl·er** *noun*
[2]bawl *noun* (1792)
: a loud prolonged cry **:** OUTCRY
bawl out *transitive verb* (1905)
: to reprimand loudly or severely
[1]bay \ˈbā\ *adjective* [Middle English, from Middle French *bai,* from Latin *badius;* akin to Old Irish *buide* yellow] (14th century)
: reddish brown ⟨a *bay* mare⟩
[2]bay *noun* (1535)
1 : a bay-colored animal; *specifically* **:** a horse with a bay-colored body and black mane, tail, and points — compare CHESTNUT 4, [1]SORREL 1
2 : a reddish brown
[3]bay *noun* [Middle English, from Middle French *baee* opening, from Old French, from feminine of *baé,* past participle of *baer* to gape, yawn — more at ABEYANCE] (14th century)
1 : a principal compartment of the walls, roof, or other part of a building or of the whole building
2 : a main division of a structure
3 : any of various compartments or sections used for a special purpose (as in an airplane, spacecraft, or service station) ⟨bomb *bay*⟩ ⟨cargo *bay*⟩
4 : BAY WINDOW 1
5 : a support or housing for electronic equipment
[4]bay *verb* [Middle English *baien, abaien,* from Old French *abaiier,* of imitative origin] (14th century)
intransitive verb
1 : to bark with prolonged tones
2 : to cry out **:** SHOUT

transitive verb
1 : to bark at
2 : to bring to bay
3 : to pursue with barking
4 : to utter in deep prolonged tones
[5]**bay** *noun* (14th century)
1 : a baying of dogs
2 : the position of one unable to retreat and forced to face danger ⟨brought his quarry to *bay*⟩
3 : the position of one checked ⟨police kept the rioters at *bay*⟩
[6]**bay** *noun, often attributive* [Middle English *baye,* from Middle French *baie*] (14th century)
1 : an inlet of the sea or other body of water usually smaller than a gulf
2 : a small body of water set off from the main body
3 : any of various terrestrial formations resembling a bay of the sea
[7]**bay** *noun* [Middle English, berry, from Middle French *baie,* from Latin *baca*] (15th century)
1 a : the European laurel (*Laurus nobilis*) **b :** any of several shrubs or trees (as of the genera *Magnolia, Pimenta,* and *Gordonia*) resembling the laurel
2 a : a garland or crown especially of laurel given as a prize for victory or excellence **b :** HONOR, FAME — usually used in plural
ba·ya·dere \'bī-ə-,dir, -,der\ *noun* [French *bayadère* professional female dancer in India] (1856)
: a fabric with horizontal stripes in strongly contrasted colors
bay·ber·ry \'bā-,ber-ē\ *noun* (1687)
1 : any of several wax myrtles; *especially* **:** a hardy shrub (*Myrica pensylvanica*) of coastal eastern North America bearing dense clusters of small berries covered with grayish white wax
2 : the fruit of a bayberry
Bayes·ian \'bā-zē-ən, -zhən\ *adjective* (1961)
: being, relating to, or concerned with a theory (as of decision making or statistical inference) involving the application of Bayes' theorem and the use of probabilities based on prior knowledge and accumulated experience ⟨*Bayesian* probability models⟩
Bayes' theorem \'bāz-\ *noun* [Thomas *Bayes* (died 1761) English mathematician] (1939)
: a theorem about conditional probabilities: the probability that an event A occurs given that another event B has already occurred is equal to the probability that the event B occurs given that A has already occurred multiplied by the probability of occurrence of event A and divided by the probability of occurrence of event B
bay leaf *noun* (15th century)
: the dried leaf of the European laurel (*Laurus nobilis*) used in cooking
bay·man \'bā-mən, -,man\ *noun* (1641)
: a person and especially a fisherman who lives or works on or about a bay
[1]**bay·o·net** \'bā-ə-nət, -,net, ,bā-ə-'net\ *noun* [French *baïonnette,* from *Bayonne,* France] (1704)
: a steel blade attached at the muzzle end of a shoulder arm (as a rifle) and used in hand-to-hand combat
[2]**bayonet** *verb* **-net·ed** *also* **-net·ted; -net·ing** *also* **-net·ting** (1858)
transitive verb
1 : to stab with a bayonet
2 : to compel or drive by or as if by the bayonet
intransitive verb
: to use a bayonet
bay·ou \'bī-(,)ü, -(,)ō\ *noun* [Louisiana French, from Choctaw *bayuk*] (1763)
1 : a creek, secondary watercourse, or minor river that is tributary to another body of water
2 : any of various usually marshy or sluggish bodies of water
bay rum *noun* (1840)
: a fragrant cosmetic and medicinal liquid distilled from the leaves of a West Indian bay tree (*Pimenta racemosa*) or usually prepared from essential oils, alcohol, and water
Bay Stat·er \'bā-,stā-tər\ *noun* (1845)
: a native or resident of Massachusetts — used as a nickname
bay window *noun* (15th century)
1 : a window or series of windows forming a bay in a room and projecting outward from the wall
2 : POTBELLY 1
ba·zaar \bə-'zär\ *noun* [Persian *bāzār*] (1612)
1 : an oriental market consisting of rows of shops or stalls selling miscellaneous goods
2 a : a place for the sale of goods **b :** DEPARTMENT STORE
3 : a fair for the sale of articles especially for charitable purposes
ba·zoo·ka \bə-'zü-kə\ *noun* [*bazooka* (a crude musical instrument made of pipes and a funnel)] (1943)
: a light portable antitank weapon consisting of an open-breech smoothbore firing tube that launches an armor-piercing rocket and is fired from the shoulder
BB \'bē-(,)bē\ *noun* (1874)
1 : a shot pellet 0.18 inch in diameter for use in a shotgun cartridge
2 : a shot pellet 0.175 inch in diameter for use in an air gun
BCD \,bē-(,)sē-'dē\ *noun* [*b*inary *c*oded *d*ecimal] (circa 1962)
: a computer code for representing alphanumeric information
B cell *noun* [*b*one-marrow-derived *cell*] (1968)
: any of the lymphocytes that have antibody molecules on the surface and comprise the antibody-secreting plasma cells when mature — called also *B lymphocyte; compare* T CELL
BCG vaccine \,bē-(,)sē-'jē-\ *noun* [*B*acillus *C*almette-*G*uérin (an attenuated strain of tubercle bacilli), from Albert *Calmette* (died 1933) and Camille *Guérin* (died 1961) French bacteriologists] (1927)
: a vaccine prepared from a living attenuated strain of tubercle bacilli and used to vaccinate human beings against tuberculosis — called also *BCG*
B complex *noun* (1934)
: VITAMIN B COMPLEX
BC soil \'bē-'sē-\ *noun* (circa 1938)
: a soil whose profile has only B horizons and C horizons
bdel·li·um \'de-lē-əm\ *noun* [Middle English, from Latin, from Greek *bdellion*] (14th century)
: a gum resin similar to myrrh obtained from various trees (genus *Commiphora*) of the East Indies and Africa
be \'bē\ *verb, past 1st & 3d singular* **was** \'wəz, 'wäz\; *2d singular* **were** \'wər\; *plural* **were;** *past subjunctive* **were;** *past participle* **been** \'bin, *chiefly British* 'bēn\; *present participle* **be·ing** \'bē(-i)ŋ\; *present 1st singular* **am** \əm, 'am\; *2d singular* **are** \'är, ər\; *3d singular* **is** \'iz, əz\ *plural* **are;** *present subjunctive* **be** [Middle English, from Old English *bēon;* akin to Old High German *bim* am, Latin *fui* I have been, *futurus* about to be, *fieri* to become, be done, Greek *phynai* to be born, be by nature, *phyein* to produce] (before 12th century)
intransitive verb
1 a : to equal in meaning **:** have the same connotation as **:** SYMBOLIZE ⟨God *is* love⟩ ⟨January *is* the first month⟩ ⟨let *x be* 10⟩ **b :** to have identity with ⟨the first person I met *was* my brother⟩ **c :** to constitute the same class as **d :** to have a specified qualification or characterization ⟨the leaves *are* green⟩ **e :** to belong to the class of ⟨the fish *is* a trout⟩ — used regularly in senses 1a through 1e as the copula of simple predication
2 a : to have an objective existence **:** have reality or actuality **:** LIVE ⟨I think, therefore I *am*⟩ ⟨once upon a time there *was* a knight⟩ **b :** to have, maintain, or occupy a place, situation, or position ⟨the book *is* on the table⟩ **c :** to remain unmolested, undisturbed, or uninterrupted — used only in infinitive form ⟨let him *be*⟩ **d :** to take place **:** OCCUR ⟨the concert *was* last night⟩ **e :** to come or go ⟨has already *been* and gone⟩ ⟨has never *been* to the circus⟩ **f** *archaic* **:** BELONG, BEFALL
verbal auxiliary
1 — used with the past participle of transitive verbs as a passive-voice auxiliary ⟨the money *was* found⟩ ⟨the house is *being* built⟩
2 — used as the auxiliary of the present participle in progressive tenses expressing continuous action ⟨he *is* reading⟩ ⟨I have *been* sleeping⟩
3 — used with the past participle of some intransitive verbs as an auxiliary forming archaic perfect tenses ⟨Christ *is* risen from the dead —1 Corinthians 15:20 (Douay Version)⟩
4 — used with the infinitive with *to* to express futurity, arrangement in advance, or obligation ⟨I *am* to interview him today⟩ ⟨she *was* to become famous⟩
be- *prefix* [Middle English, from Old English *bi-, be-;* akin to Old English *bī* by, near — more at BY]
1 : on **:** around **:** over ⟨*be*daub⟩ ⟨*be*smear⟩
2 : to a great or greater degree **:** thoroughly ⟨*be*fuddle⟩ ⟨*be*rate⟩
3 : excessively **:** ostentatiously — in intensive verbs formed from simple verbs ⟨*be*deck⟩ and in adjectives based on adjectives ending in *-ed* ⟨*be*ribboned⟩
4 : about **:** to **:** at **:** upon **:** against **:** across ⟨*be*stride⟩ ⟨*be*speak⟩
5 : make **:** cause to be **:** treat as ⟨*be*little⟩ ⟨*be*fool⟩ ⟨*be*friend⟩
6 : call or dub especially excessively ⟨*be*doctor⟩
7 : affect, afflict, treat, provide, or cover with especially excessively ⟨*be*devil⟩ ⟨*be*fog⟩
[1]**beach** \'bēch\ *noun* [origin unknown] (1535)
1 : shore pebbles **:** SHINGLE
2 a : a shore of a body of water covered by sand, gravel, or larger rock fragments **b :** a seashore area
[2]**beach** *transitive verb* (1840)
1 : to run or drive ashore
2 : to strand on or as if on a beach
beach ball *noun* (1940)
: a large inflated ball for use at the beach
beach·boy \-,bȯi\ *noun* (1941)
: a male beach attendant (as at a club or hotel)
beach buggy *noun* (1943)
: DUNE BUGGY
beach·comb·er \'bēch-,kō-mər\ *noun* (1840)
1 : a white man living as a drifter or loafer especially on the islands of the South Pacific
2 : a person who searches along a shore (as for salable refuse or for seashells)
— beach·comb \-,kōm\ *verb*
beach flea *noun* (1843)
: any of numerous amphipod crustaceans (family Talitridae) living on ocean beaches and leaping like fleas
beach·front \'bēch-,frənt\ *noun* (1921)
: a strip of land that fronts a beach
beach·go·er \-,gō-ər\ *noun* (1954)
: a person who frequently goes to the beach
beach grass *noun* (1681)
: any of several tough strongly rooted grasses that grow on exposed sandy shores; *especially* **:** any of a genus (*Ammophila*) of rhizomatous perennials widely planted to bind sandy slopes
beach·head \'bēch-,hed\ *noun* (1940)
1 : an area on a hostile shore occupied to secure further landing of troops and supplies
2 : FOOTHOLD
beach pea *noun* (1802)
: a wild pea (*Lathyrus japonicus* synonym *L. maritimus*) having tough roots and purple flowers that is found along sandy shores
beach plum *noun* (1784)

: a shrubby plum (*Prunus maritima*) having white flowers and growing along the northeastern coast of North America; *also* **:** its edible usually dark purple fruit that is used especially in preserves

beach·side \'bēch-ˌsīd\ *adjective* (1952)
: located at a beach

beach·wear \-ˌwar, -ˌwer\ *noun* (1928)
: clothing for wear at a beach

beachy \'bē-chē\ *adjective* (1597)
: covered with pebbles or shingle

¹bea·con \'bē-kən\ *noun* [Middle English *beken,* from Old English *bēacen* sign; akin to Old High German *bouhhan* sign] (14th century)
1 : a signal fire commonly on a hill, tower, or pole
2 a : a lighthouse or other signal for guidance **b :** a radio transmitter emitting signals for guidance of aircraft
3 : a source of light or inspiration

²beacon (1821)
transitive verb
: to furnish with a beacon
intransitive verb
: to shine as a beacon

¹bead \'bēd\ *noun* [Middle English *bede* prayer, prayer bead, from Old English *bed, gebed* prayer; akin to Old English *biddan* to entreat, pray — more at BID] (before 12th century)
1 a *obsolete* **:** PRAYER — usually used in plural **b** *plural* **:** a series of prayers and meditations made with a rosary
2 : a small piece of material pierced for threading on a string or wire (as in a rosary)
3 *plural* **a :** ROSARY **b :** a necklace of beads or pearls
4 : a small ball-shaped body: as **a :** a drop of sweat or blood **b :** a bubble formed in or on a beverage **c :** a small metal knob on a firearm used as a front sight **d :** a blob or a line of weld metal
5 : a projecting rim, band, or molding ◆

²bead (1577)
transitive verb
1 : to furnish, adorn, or cover with beads or beading
2 : to string together like beads
intransitive verb
: to form into a bead

bead·ing *noun* (1845)
1 : a beaded molding
2 : material or a part or a piece consisting of a bead
3 : an openwork trimming
4 : BEADWORK

bea·dle \'bē-dᵊl\ *noun* [Middle English *bedel* messenger, from Old English *bydel;* akin to Old High German *butil* bailiff, Old English *bēodan* to command — more at BID] (1581)
: a minor parish official whose duties include ushering and preserving order at services and sometimes civil functions

bead·roll \'bēd-ˌrōl\ *noun* [from the reading in church of a list of names of persons for whom prayers are to be said] (1529)
1 : a list of names **:** CATALOG
2 : ROSARY

beads·man \'bēdz-mən\ *noun* (13th century)
archaic **:** one who prays for another

bead·work \'bēd-ˌwərk\ *noun* (1751)
1 : ornamental work in beads
2 : joinery beading

beady \'bē-dē\ *adjective* **bead·i·er; -est** (1826)
1 a : resembling beads **b :** small, round, and shiny with interest or greed ⟨*beady* eyes⟩
2 : marked by bubbles or beads ⟨a *beady* liquor⟩
— bead·i·ly \'bē-də-lē\ *adverb*

bea·gle \'bē-gəl\ *noun* [Middle English *begle*] (15th century)
: any of a breed of small short-legged smooth-coated often black, white, and tan hounds

beak \'bēk\ *noun* [Middle English *bec,* from Old French, from Latin *beccus,* of Gaulish origin] (13th century)

beagle

1 a : the bill of a bird; *especially* **:** a strong short broad bill **b** (1) **:** the elongated sucking mouth of some insects (as the true bugs) (2) **:** any of various rigid projecting mouth structures (as of a turtle) **c :** the human nose
2 : a pointed structure or formation: **a :** a metal-pointed beam projecting from the bow especially of an ancient galley for piercing an enemy ship **b :** the spout of a vessel **c :** a continuous slight architectural projection ending in an arris — see MOLDING illustration **d :** a process suggesting the beak of a bird
3 *chiefly British* **a :** MAGISTRATE **b :** HEADMASTER
— beaked \'bēkt\ *adjective*
— beaky \'bē-kē\ *adjective*

beaked whale *noun* (1877)
: any of a widely distributed family (Ziphiidae) of medium-sized toothed whales that have an elongated snout and a small dorsal fin

bea·ker \'bē-kər\ *noun* [Middle English *biker,* from Old Norse *bikarr,* probably from Old Saxon *bikeri,* from Medieval Latin *bicarium*] (14th century)
1 : a large drinking cup that has a wide mouth and is sometimes supported on a standard
2 : a deep widemouthed thin-walled vessel usually with a lip for pouring that is used especially in science laboratories

be–all and end–all \'bē-ˌȯ-lən-'(d)en-ˌdȯl\ *noun* (1605)
1 : prime cause **:** essential element
2 : TOTALITY 1

¹beam \'bēm\ *noun* [Middle English *beem,* from Old English *bēam* tree, beam; akin to Old High German *boum* tree] (before 12th century)
1 a : a long piece of heavy often squared timber suitable for use in construction **b :** a wood or metal cylinder in a loom on which the warp is wound **c :** the part of a plow to which handles, standard, and coulter are attached **d :** the bar of a balance from which scales hang **e :** one of the principal horizontal supporting members (as of a building or ship) ⟨a steel *beam* supporting a floor⟩; *also* **:** BOOM, SPAR ⟨the *beam* of a crane⟩ **f :** the extreme width of a ship at the widest part **g :** an oscillating lever on a central axis receiving motion at one end from an engine piston rod and transmitting it at the other
2 a : a ray or shaft of light **b :** a collection of nearly parallel rays (as X rays) or a stream of particles (as electrons) **c :** a constant directional radio signal transmitted for the guidance of pilots; *also* **:** the course indicated by a radio beam
3 : the main stem of a deer's antler
4 : the width of the buttocks
— on the beam 1 : following a guiding beam **2 :** proceeding or operating correctly

²beam (15th century)
transitive verb
1 : to emit in beams or as a beam
2 : to support with beams
3 a : to aim and transmit (a broadcast) by directional antennas or by satellite **b :** to direct to a particular audience
intransitive verb
1 : to send out beams of light
2 : to smile with joy

beam–ends \'bēm-'en(d)z\ *noun plural* (1773)
: the ends of a ship's beams
— on her beam–ends : inclined so much on one side that the beams approach a vertical position

beam·ish \'bē-mish\ *adjective* (1870)
: beaming and bright with optimism, promise, or achievement
— beam·ish·ly *adverb*

beam splitter *noun* (1935)
: a mirror or prism or a combination of the two that is used to divide a beam of radiation into two or more parts

beamy \'bē-mē\ *adjective* (14th century)
1 : emitting beams of light **:** RADIANT
2 : broad in the beam ⟨a *beamy* cargo ship⟩

¹bean \'bēn\ *noun* [Middle English *bene,* from Old English *bēan;* akin to Old High German *bōna* bean] (before 12th century)
1 a : BROAD BEAN **b :** the seed of any of various erect or climbing leguminous plants (especially genera *Phaseolus, Dolichos,* and *Vigna*) other than the broad bean **c :** a plant bearing beans **d :** an immature bean pod used as a vegetable
2 a : a valueless item **b** *plural* **:** the least amount ⟨didn't know *beans* about it⟩
3 : any of various seeds or fruits that resemble beans or bean pods; *also* **:** a plant producing these
4 a *plural* **:** EXUBERANCE — used in the phrase *full of beans* **b** *plural* **:** NONSENSE, BUNKUM — used in the phrase *full of beans*
5 : HEAD, BRAIN
6 : a protuberance on the upper mandible of waterfowl — see DUCK illustration

²bean *transitive verb* (1910)
: to strike (a person) on the head with an object

bean·bag \'bēn-ˌbag\ *noun* (1871)
1 : a cloth bag partially filled typically with dried beans and used as a toy
2 : any of various pellet-filled bags used as furniture (as a chair) or household articles (as an ashtray base)

bean·ball \-ˌbȯl\ *noun* (circa 1905)
: a pitch thrown at a batter's head

bean counter *noun* (1975)
: a person involved in corporate financial decisions and especially one reluctant to spend money

bean curd *noun* (circa 1889)
: a soft vegetable cheese prepared by treating soybean milk with coagulants (as magnesium chloride or dilute acids) — called also *tofu*

bean·ery \'bēn-rē, 'bē-nə-\ *noun, plural* **-er·ies** (1887)
: RESTAURANT

bean·ie \'bē-nē\ *noun* (1940)
: a small round tight-fitting skullcap worn especially by schoolboys and college freshmen

◇ WORD HISTORY
bead Middle English *bede* developed from Old English *gebed.* Originally the word meant "a prayer." The number and order of a series of prayers were kept track of with the aid of what is today called a rosary, a string of variously sized small round balls. Because each of these balls stands for a particular prayer, the name *bede,* Modern English *bead,* was transferred to the balls themselves. Today *bead* is used to refer to any small piece of material pierced for threading on a string or wire. The sense is also extended to refer to any small, round object, such as a drop of sweat.

beano \'bē-(ˌ)nō\ *noun, plural* **beanos** [by alteration] (1935)
: BINGO
bean·pole \'bēn-ˌpō(ə)l\ *noun* (1798)
1 : a pole up which bean vines may climb
2 : a tall thin person
bean sprouts *noun plural* (1921)
: the sprouts of bean seeds especially of the mung bean used as a vegetable
¹bear \'bar, 'ber\ *noun, plural* **bears** *often attributive* [Middle English *bere,* from Old English *bera;* akin to Old English *brūn* brown — more at BROWN] (before 12th century)
1 *or plural* **bear :** any of a family (Ursidae of the order Carnivora) of large heavy mammals of America and Eurasia that have long shaggy hair, rudimentary tails, and plantigrade feet and feed largely on fruit and insects as well as on flesh
2 : a surly, uncouth, or shambling person
3 [probably from the proverb about *selling the bearskin before catching the bear*] **:** one that sells securities or commodities in expectation of a price decline — compare BULL
— **bear·like** \-ˌlīk\ *adjective*
²bear *verb* **bore** \'bōr, 'bȯr\; **borne** \'bōrn, 'bȯrn\ *also* **born** \'bȯrn\; **bear·ing** [Middle English *beren* to carry, bring forth, from Old English *beran;* akin to Old High German *beran* to carry, Latin *ferre,* Greek *pherein*] (before 12th century)
transitive verb
1 a : to move while holding up and supporting **b :** to be equipped or furnished with **c :** BEHAVE, CONDUCT ⟨*bearing* himself well⟩ **d :** to have as a feature or characteristic **e :** to give as testimony ⟨*bear* false witness⟩ **f :** to have as an identification ⟨*bore* the name of John⟩ **g :** to hold in the mind **h :** DISSEMINATE **i :** LEAD, ESCORT **j :** RENDER, GIVE
2 a : to give birth to **b :** to produce as yield **c** (1) **:** to permit growth of (2) **:** CONTAIN ⟨oil-bearing shale⟩
3 a : to support the weight of **:** SUSTAIN **b :** to put up with especially without giving way ⟨couldn't *bear* the pain⟩ **c :** to call for as suitable or essential ⟨it *bears* watching⟩ **d :** to hold above, on top, or aloft **e :** to admit of **:** ALLOW **f :** ASSUME, ACCEPT
4 : THRUST, PRESS
intransitive verb
1 : to produce fruit **:** YIELD
2 a : to force one's way **b :** to extend in a direction indicated or implied **c :** to be situated **:** LIE **d :** to become directed **e :** to go or incline in an indicated direction
3 : to support a weight or strain — often used with *up*
4 a : to exert influence or force **b :** APPLY, PERTAIN ☆
— **bear a hand :** to join in and help out
— **bear arms 1 :** to carry or possess arms **2 :** to serve as a soldier
— **bear fruit :** to come to satisfying fruition, production, or development
— **bear in mind :** to think of especially as a warning **:** REMEMBER
— **bear with :** to be indulgent, patient, or forbearing with
bear·able \'bar-ə-bəl, 'ber-\ *adjective* (circa 1550)
: capable of being borne
— **bear·abil·i·ty** \ˌbar-ə-'bi-lə-tē, ˌber-\ *noun*
— **bear·ably** \-blē\ *adverb*
bear·bait·ing \'bar-ˌbā-tiŋ, 'ber-\ *noun* (14th century)
: the practice of setting dogs on a chained bear
bear·ber·ry \-ˌber-ē\ *noun* (1625)
: a trailing evergreen plant (*Arctostaphylos uva-ursi*) of the heath family with astringent foliage and red berries
¹beard \'bird\ *noun* [Middle English *berd,* from Old English *beard;* akin to Old High German *bart* beard, Latin *barba*] (before 12th century)
1 : the hair that grows on a man's face often excluding the mustache
2 : a hairy or bristly appendage or tuft
3 : FRONT 7a
— **beard·ed** \'bir-dəd\ *adjective*
— **beard·ed·ness** *noun*
— **beard·less** \'bird-ləs\ *adjective*
²beard *transitive verb* (1525)
1 : to confront and oppose with boldness, resolution, and often effrontery **:** DEFY
2 : to furnish with a beard
bearded collie *noun* (1880)
: any of a breed of large working dogs of Scottish origin that have a long rough coat and drooping ears
bearded iris *noun* (1923)
: any of numerous wild or cultivated irises with a growth of short hairs on each fall
bearded seal *noun* (1913)
: a large arctic seal (*Erignathus barbatus*) with a tuft of long whiskers on each side of the muzzle
bear down (14th century)
transitive verb
: OVERCOME, OVERWHELM
intransitive verb
: to exert full strength and concentrated attention
— **bear down on 1 :** EMPHASIZE **2 :** to weigh heavily on **:** BURDEN
beard·tongue \'bird-ˌtəŋ\ *noun* (1821)
: PENSTEMON
bear·er \'bar-ər, 'ber-\ *noun* (13th century)
: one that bears: as **a :** PORTER 1 **b :** a plant yielding fruit **c :** PALLBEARER **d :** one holding a check, draft, or other order for payment especially if marked payable to bearer
bear grass *noun* (1750)
: any of several plants (genera *Yucca, Nolina,* or *Xerophyllum*) of the lily or agave families chiefly of the southern and western U.S. with foliage resembling coarse blades of grass
bear hug *noun* (1921)
: a rough tight embrace
bear·ing *noun* (13th century)
1 : the manner in which one bears or comports oneself
2 a : the act, power, or time of bringing forth offspring or fruit **b :** a product of bearing **:** CROP
3 a : an object, surface, or point that supports **b :** a machine part in which another part (as a journal or pin) turns or slides
4 : a figure borne on a heraldic field
5 : PRESSURE, THRUST
6 a : the situation or horizontal direction of one point with respect to another or to the compass **b :** a determination of position **c** *plural* **:** comprehension of one's position, environment, or situation **d :** RELATION, CONNECTION; *also* **:** PURPORT
7 : the part of a structural member that rests on its supports ☆
bearing rein *noun* (1794)
: CHECKREIN 1
bear·ish \'bar-ish, 'ber-\ *adjective* (1744)
1 : resembling a bear in build or in roughness, gruffness, or surliness
2 a : marked by, tending to cause, or fearful of falling prices (as in a stock market) **b :** PESSIMISTIC
— **bear·ish·ly** *adverb*
— **bear·ish·ness** *noun*
bé·ar·naise sauce \'bā-är-ˌnāz-, -ər-; 'ber-\ *noun* [French *béarnaise,* feminine of *béarnais* of Béarn, France] (1877)
: a sauce of egg yolks and butter flavored with shallots, wine, vinegar, and seasonings
bear out *transitive verb* (15th century)
: CONFIRM, SUBSTANTIATE
bear·skin \'bar-ˌskin, 'ber-\ *noun* (1752)
: an article made of the skin of a bear; *especially* **:** a military hat made of the skin of a bear
bear up (13th century)
transitive verb
: SUPPORT, ENCOURAGE
intransitive verb
: to summon up courage, resolution, or strength ⟨*bearing up* under the strain⟩
beast \'bēst\ *noun* [Middle English *beste,* from Old French, from Latin *bestia*] (13th century)
1 a : a four-footed mammal as distinguished from a human being, a lower vertebrate, and an invertebrate **b :** a lower animal as distinguished from a human being **c :** an animal as distinguished from a plant **d :** an animal under human control
2 : a contemptible person
3 : something formidably difficult to control or deal with
beast epic *noun* (1889)
: a poem with epic conventions in which animals speak and act like human beings
beast fable *noun* (1865)
: a usually didactic prose or verse fable in which animals speak and act like human beings
beast·ie \'bē-stē\ *noun* (circa 1773)
: BEAST, ANIMAL, CRITTER
beas·tings *variant of* BEESTINGS
¹beast·ly \'bēst-lē\ *adjective* **beast·li·er; -est** (13th century)
1 : BESTIAL 1
2 : ABOMINABLE, DISAGREEABLE ⟨*beastly* weather⟩
— **beast·li·ness** *noun*
²beastly *adverb* (1865)
: VERY ⟨a *beastly* cold day⟩
beast of burden (1740)
: an animal employed to carry heavy loads or to perform other heavy work (as pulling a plow)
¹beat \'bēt\ *verb* **beat; beat·en** \'bē-tᵊn\ *or* **beat; beat·ing** [Middle English *beten,* from Old English *bēatan;* akin to Old High German *bōzan* to beat] (before 12th century)
transitive verb

☆ SYNONYMS

Bear, suffer, endure, abide, tolerate, stand mean to put up with something trying or painful. BEAR usually implies the power to sustain without flinching or breaking ⟨forced to *bear* a tragic loss⟩. SUFFER often suggests acceptance or passivity rather than courage or patience in bearing ⟨*suffering* many insults⟩. ENDURE implies continuing firm or resolute through trials and difficulties ⟨*endured* years of rejection⟩. ABIDE suggests acceptance without resistance or protest ⟨cannot *abide* their rudeness⟩. TOLERATE suggests overcoming or successfully controlling an impulse to resist, avoid, or resent something injurious or distasteful ⟨refused to *tolerate* such treatment⟩. STAND emphasizes more strongly the ability to bear without discomposure or flinching ⟨cannot *stand* teasing⟩.

Bearing, deportment, demeanor, mien, manner, carriage mean the outward manifestation of personality or attitude. BEARING is the most general of these words but now usually implies characteristic posture ⟨a woman of regal *bearing*⟩. DEPORTMENT suggests actions or behavior as formed by breeding or training ⟨your *deportment* was atrocious⟩. DEMEANOR suggests one's attitude toward others as expressed in outward behavior ⟨the haughty *demeanor* of the headwaiter⟩. MIEN is a literary term referring both to bearing and demeanor ⟨a *mien* of supreme self-satisfaction⟩. MANNER implies characteristic or customary way of moving and gesturing and addressing others ⟨the imperious *manner* of a man used to giving orders⟩. CARRIAGE applies chiefly to habitual posture in standing or walking ⟨the kind of *carriage* learned at boarding school⟩.

1 : to strike repeatedly: **a :** to hit repeatedly so as to inflict pain — often used with *up* **b :** to walk on **:** TREAD **c :** to strike directly against forcefully and repeatedly **:** dash against **d :** to flap or thrash at vigorously **e :** to strike at in order to rouse game; *also* **:** to range over in or as if in quest of game **f :** to mix by stirring **:** WHIP — often used with *up* **g :** to strike repeatedly in order to produce music or a signal ⟨*beat* a drum⟩
2 a : to drive or force by blows **b :** to pound into a powder, paste, or pulp **c :** to make by repeated treading or driving over **d** (1) **:** to dislodge by repeated hitting (2) **:** to lodge securely by repeated striking **e :** to shape by beating ⟨*beat* swords into plowshares⟩; *especially* **:** to flatten thin by blows **f :** to sound or express especially by drumbeat
3 : to cause to strike or flap repeatedly
4 a : OVERCOME, DEFEAT; *also* **:** SURPASS — often used with *out* **b :** to prevail despite ⟨*beat* the odds⟩ **c :** BEWILDER, BAFFLE **d** (1) **:** FATIGUE, EXHAUST (2) **:** to leave dispirited, irresolute, or hopeless **e :** CHEAT, SWINDLE
5 a (1) **:** to act ahead of usually so as to forestall (2) **:** to report a news item in advance of **b :** to come or arrive before **c :** CIRCUMVENT ⟨*beat* the system⟩ **d :** to outmaneuver (a defender) and get free **e :** to score against (a goalkeeper)
6 : to indicate by beating ⟨*beat* the tempo⟩
intransitive verb
1 a : to become forcefully impelled **:** DASH **b :** to glare or strike with oppressive intensity **c :** to sustain distracting activity **d :** to beat a drum
2 a (1) **:** PULSATE, THROB (2) **:** TICK **b :** to sound upon being struck
3 a : to strike repeated blows **b :** to strike the air **:** FLAP **c :** to strike cover in order to rouse game; *also* **:** to range or scour for or as if for game
4 : to progress with much difficulty
5 : to sail to windward by a series of tacks
— beat·a·ble \ˈbē-tə-bəl\ *adjective*
— beat about the bush *or* **beat around the bush :** to fail or refuse to come to the point in discourse
— beat a retreat : to leave in haste
— beat it 1 : to hurry away **:** SCRAM **2 :** HURRY, RUSH
— beat one's brains out : to try intently to resolve something difficult by thinking
— beat the bushes : to search thoroughly through all possible areas
— beat the drum : to proclaim as meritorious or significant **:** publicize vigorously
— beat the rap : to escape or evade the penalties connected with an accusation or charge
— beat up on : to attack physically or verbally

²beat *noun* (1615)
1 a : a single stroke or blow especially in a series; *also* **:** PULSATION, TICK **b :** a sound produced by or as if by beating **c :** a driving impact or force
2 : one swing of the pendulum or balance of a timepiece
3 a : a regularly traversed round ⟨the cop on the *beat*⟩ **b :** a group of news sources that a reporter covers regularly
4 a : a metrical or rhythmic stress in poetry or music or the rhythmic effect of these stresses **b :** the tempo indicated (as by a conductor) to a musical performer **c :** the pronounced rhythm that is the characteristic driving force in jazz or rock music; *also* **:** ²ROCK 2
5 a : one that excels ⟨I've never seen the *beat* of it⟩ **b :** the reporting of a news story ahead of competitors
6 : DEADBEAT
7 a : an act of beating to windward **b :** one of the reaches so traversed **:** TACK
8 : each of the pulsations of amplitude produced by the union of sound or radio waves or electric currents having different frequencies
9 : an accented stroke (as of one leg or foot against the other) in dancing
— beat·less \-ləs\ *adjective*

³beat *adjective* [Middle English *beten, bete,* from past participle of *beten*] (1832)
1 a : being in a state of exhaustion **:** EXHAUSTED **b :** sapped of resolution or morale
2 *often capitalized* **:** of, relating to, or being beatniks ⟨*beat* poets⟩

⁴beat *noun, often capitalized* (1957)
: BEATNIK

beat·en \ˈbē-tᵊn\ *adjective* (13th century)
1 : hammered into a desired shape
2 : much trodden and worn smooth; *also* **:** FAMILIAR ⟨a *beaten* path⟩
3 : being in a state of exhaustion **:** EXHAUSTED

beaten biscuit *noun* (1876)
: a biscuit whose dough is lightened by beating and folding

beat·er \ˈbē-tər\ *noun* (14th century)
1 : one that beats: as **a :** EGGBEATER **b :** a rotary blade attached to an electric mixer **c :** DRUMSTICK 1
2 : one that strikes cover to rouse game

be·atif·ic \ˌbē-ə-ˈti-fik\ *adjective* [Latin *beatificus* making happy, from *beatus* happy, from past participle of *beare* to bless; perhaps akin to Latin *bonus* good — more at BOUNTY] (1649)
1 : of, possessing, or imparting beatitude
2 : having a blissful appearance ⟨a *beatific* smile⟩
— be·atif·i·cal·ly \-fi-k(ə-)lē\ *adverb*

beatific vision *noun* (1639)
: the direct knowledge of God enjoyed by the blessed in heaven

be·at·i·fy \bē-ˈa-tə-ˌfī\ *transitive verb* **-fied; -fy·ing** [Middle French *beatifier,* from Late Latin *beatificare,* from Latin *beatus* + *facere* to make — more at DO] (1535)
1 : to make supremely happy
2 : to declare to have attained the blessedness of heaven and authorize the title "Blessed" and limited public religious honor
— be·at·i·fi·ca·tion \-ˌa-tə-fə-ˈkā-shən\ *noun*

beat·ing \ˈbē-tiŋ\ *noun* (13th century)
1 : an act of striking with repeated blows so as to injure or damage; *also* **:** the injury or damage thus inflicted
2 : PULSATION
3 : DEFEAT, SETBACK

beating reed *noun* (1879)
: a reed in a musical instrument that vibrates against the edges of an air opening (as in a clarinet or organ pipe) to which it is attached — compare FREE REED

be·at·i·tude \bē-ˈa-tə-ˌtüd, -ˌtyüd\ *noun* [Latin *beatitudo,* from *beatus*] (15th century)
1 a : a state of utmost bliss **b** — used as a title for a primate especially of an Eastern church
2 : any of the declarations made in the Sermon on the Mount (Matthew 5:3–12) beginning in the Authorized Version "Blessed are"

beat·nik \ˈbēt-nik\ *noun* [³*beat* + *-nik*] (1958)
: a person who rejects the mores of established society (as by dressing and behaving unconventionally) and indulges in exotic philosophizing and self-expression

beat off (15th century)
transitive verb
: REPEL
intransitive verb
: MASTURBATE — used of a male; usually considered vulgar

beat out *transitive verb* (1607)
1 : to make or perform by or as if by beating
2 : to mark or accompany by beating
3 : to turn (a routine ground ball) into a hit in baseball by fast running to first base

Be·atrice \ˌbā-ä-ˈtrē-(ˌ)chā, ˈbē-ə-trəs\ *noun* [Italian]
: a Florentine woman idealized in Dante's *Vita Nuova* and *Divina Commedia*

beat–up \ˈbēt-ˌəp, -ˈəp\ *adjective* (1946)
: DILAPIDATED, SHABBY

beau \ˈbō\ *noun, plural* **beaux** \ˈbōz\ *or* **beaus** [French, from *beau* beautiful, from Latin *bellus* pretty] (1684)
1 : DANDY 1
2 : BOYFRIEND 2

Beau Brum·mell \ˈbō-ˈbrə-məl\ *noun* [nickname of G. B. *Brummell*] (1920)
: DANDY 1

beau·coup \ˈbō-(ˌ)kü\ *adjective* [French] (1918)
slang **:** great in quantity or amount **:** MANY, MUCH ⟨spent *beaucoup* dollars⟩

Beau·fort scale \ˈbō-fərt-\ *noun* [Sir Francis *Beaufort*] (1858)
: a scale in which the force of the wind is indicated by numbers from 0 to 12
▶ The Beaufort scale table is on page 156.

beau geste \bō-ˈzhest\ *noun, plural* **beaux gestes** *or* **beau gestes** \bō-ˈzhest\ [French, literally, beautiful gesture] (1914)
1 : a graceful or magnanimous gesture
2 : an ingratiating conciliatory gesture

beau ide·al \ˌbō-ī-ˈdē(-ə)l, ˌbō-ˌē-dā-ˈäl\ *noun, plural* **beau ideals** [French *beau idéal* ideal beauty] (1809)
: the perfect type or model

Beau·jo·lais \ˌbō-zhō-ˈlā, -zhə-\ *noun, plural* **Beaujolais** [French, from *Beaujolais,* region of eastern France] (1863)
: a light fruity red burgundy wine made from the Gamay grape

beau monde \bō-ˈmänd, -mōⁿd\ *noun, plural* **beau mondes** \-ˈmän(d)z\ *or* **beaux mondes** \bō-mōⁿd\ [French, literally, fine world] (1673)
: the world of high society and fashion

¹beaut \ˈbyüt\ *noun* (1896)
: BEAUTY 4

²beaut *adjective* (1918)
Australian & New Zealand **:** EXCELLENT 2

beau·te·ous \ˈbyü-tē-əs\ *adjective* [Middle English, from *beaute*] (15th century)
: BEAUTIFUL
— beau·te·ous·ly *adverb*
— beau·te·ous·ness *noun*

beau·ti·cian \byü-ˈti-shən\ *noun* [*beauty* + *-ician*] (1924)
: COSMETOLOGIST

beau·ti·ful \ˈbyü-ti-fəl\ *adjective* (15th century)
1 : having qualities of beauty **:** exciting aesthetic pleasure
2 : generally pleasing **:** EXCELLENT ☆

☆ SYNONYMS
Beautiful, lovely, handsome, pretty, comely, fair mean exciting sensuous or aesthetic pleasure. BEAUTIFUL applies to whatever excites the keenest of pleasure to the senses and stirs emotion through the senses ⟨*beautiful* mountain scenery⟩. LOVELY is close to BEAUTIFUL but applies to a narrower range of emotional excitation in suggesting the graceful, delicate, or exquisite ⟨a *lovely* melody⟩. HANDSOME suggests aesthetic pleasure due to proportion, symmetry, or elegance ⟨a *handsome* Georgian mansion⟩. PRETTY often applies to superficial or insubstantial attractiveness ⟨a painter of conventionally *pretty* scenes⟩. COMELY is like HANDSOME in suggesting what is coolly approved rather than emotionally responded to ⟨the *comely* grace of a dancer⟩. FAIR suggests beauty because of purity, flawlessness, or freshness ⟨*fair* of face⟩.

BEAUFORT SCALE

BEAUFORT NUMBER	NAME	WIND SPEED		DESCRIPTION
		MPH	KPH	
0	calm	<1	<1	calm; smoke rises vertically
1	light air	1-3	1-5	direction of wind shown by smoke but not by wind vanes
2	light breeze	4-7	6-11	wind felt on face; leaves rustle; wind vane moves
3	gentle breeze	8-12	12-19	leaves and small twigs in constant motion; wind extends light flag
4	moderate breeze	13-18	20-28	wind raises dust and loose paper; small branches move
5	fresh breeze	19-24	29-38	small-leaved trees begin to sway; crested wavelets form on inland waters
6	strong breeze	25-31	39-49	large branches move; overhead wires whistle; umbrellas difficult to control
7	moderate gale *or* near gale	32-38	50-61	whole trees sway; walking against wind is difficult
8	fresh gale *or* gale	39-46	62-74	twigs break off trees; moving cars veer
9	strong gale	47-54	75-88	slight structural damage occurs; shingles may blow away
10	whole gale *or* storm	55-63	89-102	trees uprooted; considerable structural damage occurs
11	storm *or* violent storm	64-72	103-117	widespread damage occurs
12	hurricane*	>72	>117	widespread damage occurs

*The U.S. uses 74 statute mph as the speed criterion for a hurricane.

— **beau·ti·ful·ly** \-f(ə-)lē\ *adverb*
— **beau·ti·ful·ness** \-fəl-nəs\ *noun*

beautiful people *noun plural, often B&P capitalized* (1964)
: wealthy or famous people whose lifestyle is usually expensive and well-publicized

beau·ti·fy \'byü-tə-,fī\ *verb* **-fied; -fy·ing** (1526)
transitive verb
: to make beautiful or add beauty to
intransitive verb
: to grow beautiful
synonym see ADORN
— **beau·ti·fi·ca·tion** \byü-tə-fə-'kā-shən\ *noun*
— **beau·ti·fi·er** \'byü-tə-,fī(-ə)r\ *noun*

beau·ty \'byü-tē\ *noun, plural* **beauties** [Middle English *beaute,* from Old French *biauté,* from *bel, biau* beautiful, from Latin *bellus* pretty; akin to Latin *bonus* good — more at BOUNTY] (14th century)
1 : the quality or aggregate of qualities in a person or thing that gives pleasure to the senses or pleasurably exalts the mind or spirit **:** LOVELINESS
2 : a beautiful person or thing; *especially* **:** a beautiful woman
3 : a particularly graceful, ornamental, or excellent quality
4 : a brilliant, extreme, or egregious example or instance ⟨that mistake was a *beauty*⟩
5 : a quantum characteristic that accounts for the existence and lifetime of the upsilon particle; *also* **:** a particle having this characteristic

beauty bush *noun* (1926)
: a Chinese shrub (*Kolkwitzia amabilis*) of the honeysuckle family with pinkish flowers and bristly fruit

beauty contest *noun* (1899)
1 : an assemblage of girls or women at which judges select the most beautiful
2 : a presidential primary election in which the popular vote does not determine the number of convention delegates a candidate receives

beauty part *noun* (1951)
: the most desirable or beneficial aspect of something

beauty shop *noun* (1901)
: an establishment or department where hairdressing, facials, and manicures are done — called also *beauty parlor, beauty salon*

beauty spot *noun* (1657)
1 : [1]PATCH 2
2 : NEVUS

[1]beaux arts \bō-'zär\ *noun plural* [French *beaux-arts*] (1821)
: FINE ARTS

[2]beaux arts *adjective, often B&A capitalized* [French *École des Beaux-Arts* School of Fine Arts, in Paris] (1924)
: characterized by the use of historic forms, rich decorative detail, and a tendency toward monumental conception in architecture

[1]bea·ver \'bē-vər\ *noun, plural* **beavers** [Middle English *bever,* from Old English *beofor;* akin to Old High German *bibar* beaver, and probably to Old English *brūn* brown — more at BROWN] (before 12th century)
1 *or plural* **beaver a :** either of two large semiaquatic herbivorous rodents (*Castor canadensis* of North America and *C. fiber* of Eurasia) having webbed hind feet and a broad flat scaly tail and constructing dams and partially submerged lodges **b :** the fur or pelt of the beaver
2 a : a hat made of beaver fur or a fabric imitation **b :** SILK HAT
3 : a heavy fabric of felted wool or of cotton napped on both sides
4 : the pudenda of a woman — usually considered vulgar

[2]beaver *noun* [Middle English *baviere,* from Middle French] (15th century)
1 : a piece of armor protecting the lower part of the face
2 : a helmet visor

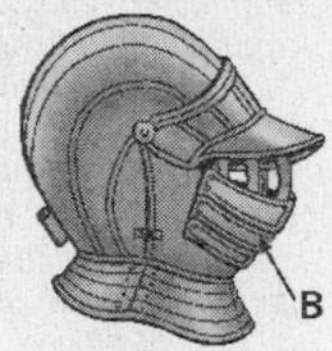

B beaver 1

[3]beaver *intransitive verb* (1946)
: to work energetically ⟨*beavering* away at the problem⟩

bea·ver·board \'bē-vər-,bōrd, -,bȯrd\ *noun* [from *Beaver Board,* a trademark] (1909)
: a fiberboard used for partitions and ceilings

be·bop \'bē-,bäp\ *noun* [imitative] (1944)
: [3]BOP 1
— **be·bop·per** *noun*

be·calm \bi-'kä(l)m, *New England also* -'kȧm\ *transitive verb* (1582)
1 a : to keep motionless by lack of wind **b :** to stop the progress of ⟨an industry *becalmed* by decreasing demand⟩
2 : to make calm **:** SOOTHE

be·cause \bi-'kȯz, -'kəz *also* -'kȯs\ *conjunction* [Middle English *because that, because,* from *by cause that*] (14th century)
1 : for the reason that **:** SINCE ⟨rested *because* he was tired⟩
2 : the fact that **:** THAT ⟨the reason I haven't been fired is *because* my boss hasn't got round to it yet —E. B. White⟩

because of *preposition* (14th century)
: by reason of **:** on account of

bé·cha·mel \,bā-shə-'mel\ *noun* [French *sauce béchamelle,* from Louis de *Béchamel* (died 1703) French courtier] (1796)
: a rich white sauce

be·chance \bi-'chan(t)s\ *verb* (1527)
archaic **:** BEFALL

bêche–de–mer \,besh-də-'mer, ,bāsh-\ *noun* [French, alteration of *biche de mer,* from Portuguese *bicho do mar,* literally, sea worm] (1783)
1 *plural* **bêche–de–mer** *or* **bêches–de–mer** \,besh-də-, ,be-shəz-, ,bāsh-\ **:** TREPANG
2 *B&M capitalized* **:** any of several English-based pidgins spoken on islands of the western Pacific

[1]beck \'bek\ *noun* [Middle English *bek,* from Old Norse *bekkr;* akin to Old English *bæc* brook, Old High German *bah,* Lithuanian *bėgti* to flee — more at PHOBIA] (14th century)
British **:** CREEK 2

[2]beck *transitive verb* [Middle English, alteration of *beknen*] (14th century)
archaic **:** BECKON

[3]beck *noun* (14th century)
1 *chiefly Scottish* **:** BOW, CURTSY
2 a : a beckoning gesture **b :** SUMMONS, BIDDING
— **at one's beck and call :** ready to obey one's command immediately

beck·et \'be-kət\ *noun* [origin unknown] (circa 1769)
: a device for holding something in place: as **a :** a grommet or a loop of rope with a knot at one end to catch in an eye at the other **b :** a ring of rope or metal **c :** a loop of rope (as for a handle)

beck·on \'be-kən\ *verb* **beck·oned; beck·on·ing** [Middle English *beknen,* from Old English *bīecnan,* from *bēacen* sign — more at BEACON] (before 12th century)
intransitive verb

1 : to summon or signal typically with a wave or nod
2 : to appear inviting **:** ATTRACT
transitive verb
: to beckon to
— beckon *noun*

be·cloud \bi-'klau̇d\ *transitive verb* (1598)
1 : to obscure with or as if with a cloud
2 : to prevent clear perception or realization of **:** MUDDLE ⟨prejudices that *becloud* his judgment⟩

be·come \bi-'kəm\ *verb* **-came** \-'kām\; **-come; -com·ing** [Middle English, to come to, become, from Old English *becuman,* from *be-* + *cuman* to come] (before 12th century)
intransitive verb
1 a : to come into existence **b :** to come to be ⟨*become* sick⟩
2 : to undergo change or development
transitive verb
: to suit or be suitable to ⟨seriousness *becoming* the occasion⟩ ⟨her clothes *become* her⟩
— become of : to happen to

be·com·ing \-'kə-miŋ\ *adjective* (15th century)
: SUITABLE, FITTING; *especially* **:** attractively suitable ⟨*becoming* modesty⟩
— be·com·ing·ly \-miŋ-lē\ *adverb*

[1]bed \'bed\ *noun* [Middle English, from Old English *bedd;* akin to Old High German *betti* bed, Latin *fodere* to dig] (before 12th century)
1 a : a piece of furniture on or in which to lie and sleep **b** (1) **:** a place of sex relations (2) **:** marital relationship **c :** a place for sleeping **d :** SLEEP; *also* **:** a time for sleeping ⟨took a walk before *bed*⟩ **e** (1) **:** a mattress filled with soft material (2) **:** BEDSTEAD **f :** the equipment and services needed to care for one hospitalized patient or hotel guest
2 : a flat or level surface: as **a :** a plot of ground prepared for plants; *also* **:** the plants grown in such a plot **b :** the bottom of a body of water; *especially* **:** an area of sea bottom supporting a heavy growth of a particular organism ⟨an oyster *bed*⟩
3 : a supporting surface or structure **:** FOUNDATION
4 : LAYER, STRATUM
5 a : the place or material in which a block or brick is laid **b :** the lower surface of a brick, slate, or tile
6 : a mass or heap resembling a bed ⟨a *bed* of ashes⟩ ⟨served on a *bed* of lettuce⟩
— in bed : in the act of sexual intercourse

[2]bed *verb* **bed·ded; bed·ding** (before 12th century)
intransitive verb
1 a : to find or make sleeping accommodations **b :** to go to bed
2 : to form a layer
3 : to lie flat or flush
transitive verb
1 a : to furnish with a bed or bedding **:** settle in sleeping quarters — often used with *down* **b :** to put, take, or send to bed
2 a : EMBED **b :** to plant or arrange in beds **c :** BASE, ESTABLISH
3 a : to lay flat or in a layer **b :** to make a bed in or of
4 : to have sexual intercourse with — often used with *down*

be·dab·ble \bi-'da-bəl\ *transitive verb* (1590)
archaic **:** to wet or soil by dabbling

bed–and–breakfast *noun* (1910)
: an establishment (as an inn) offering lodging and breakfast

be·daub \bi-'dȯb, -'däb\ *transitive verb* (1558)
1 : to daub over **:** BESMEAR
2 : to ornament with vulgar excess

be·daz·zle \bi-'da-zəl\ *transitive verb* (1596)
1 : to confuse by a strong light **:** DAZZLE
2 : to impress forcefully **:** ENCHANT
— be·daz·zle·ment \-mənt\ *noun*

bed board *noun* (1946)
: a stiff thin wide board inserted usually between a bedspring and mattress especially to give support to one's back or to protect a mattress from sagging springs

bed·bug \'bed-ˌbəg\ *noun* (1808)
: a wingless bloodsucking bug (*Cimex lectularius*) sometimes infesting houses and especially beds and feeding on human blood

bed·cham·ber \-ˌchām-bər\ *noun* (14th century)
: BEDROOM

bed check *noun* (1927)
: a night inspection to check the presence of persons (as soldiers) required by regulations to be in bed or in quarters

bed·clothes \'bed-ˌklō(th)z\ *noun plural* (14th century)
: the covering (as sheets and blankets) used on a bed

bed·cov·er \-ˌkə-vər\ *also* **bed·cov·er·ing** \-ˌkə-v(ə-)riŋ\ *noun* (circa 1656)
1 : BEDSPREAD
2 : BEDCLOTHES — usually used in plural

bed·ded \'be-dəd\ *adjective* (1788)
: having a bed or beds of a specified kind or number — used in combination ⟨a twin-bedded room⟩

bed·der \'be-dər\ *noun* (1803)
1 : one that makes up beds
2 : a bedding plant

[1]bed·ding \'be-diŋ\ *noun* [Middle English, from Old English, from *bedd*] (before 12th century)
1 : BEDCLOTHES
2 : a bottom layer **:** FOUNDATION
3 : material to provide a bed for livestock
4 : STRATIFICATION

[2]bedding *adjective* [from gerund of [2]*bed*] (1856)
: suitable for planting in large groups in flower beds to produce a mass display ⟨*bedding* plants⟩

be·deck \bi-'dek\ *transitive verb* (circa 1566)
1 : to clothe with finery **:** DECK
2 : DECORATE 2

be·dev·il \bi-'de-vəl\ *transitive verb* (1574)
1 : to possess with or as if with a devil
2 : to cause distress **:** TROUBLE
3 : to change for the worse **:** SPOIL
4 : to confuse utterly **:** BEWILDER
— be·dev·il·ment \-mənt\ *noun*

be·dew \bi-'dü, -'dyü\ *transitive verb* (14th century)
: to wet with or as if with dew

bed·fast \'bed-ˌfast\ *adjective* (1560)
: BEDRIDDEN

bed·fel·low \-ˌfe-(ˌ)lō\ *noun* (15th century)
1 : one who shares a bed with another
2 : ASSOCIATE, ALLY ⟨political *bedfellows*⟩

Bed·ford cord \'bed-fərd-\ *noun* [perhaps from New *Bedford,* Massachusetts] (1862)
: a clothing fabric with lengthwise ribs that resembles corduroy; *also* **:** the weave used in making this fabric

be·dight \bi-'dīt\ *transitive verb* **be·dight·ed** *or* **bedight; be·dight·ing** (14th century)
archaic **:** EQUIP, ARRAY

be·dim \bi-'dim\ *transitive verb* (1583)
1 : to make less bright
2 : to make indistinct **:** OBSCURE

Bed·i·vere \'be-də-ˌvir\ *noun*
: a knight of the Round Table

be·di·zen \bi-'dī-zᵊn, -'di-\ *transitive verb* (1661)
: to dress or adorn gaudily
— be·di·zen·ment \-mənt\ *noun*

bed·lam \'bed-ləm\ *noun* [*Bedlam,* popular name for the Hospital of Saint Mary of Bethlehem, London, an insane asylum, from Middle English *Bedlem* Bethlehem] (1522)
1 *obsolete* **:** MADMAN, LUNATIC
2 *often capitalized* **:** a lunatic asylum
3 : a place, scene, or state of uproar and confusion ◆
— bedlam *adjective*

bed·lam·ite \'bed-lə-ˌmīt\ *noun* (1589)
: MADMAN, LUNATIC
— bedlamite *adjective*

Bed·ling·ton terrier \'bed-liŋ-tən-\ *noun* [*Bedlington,* England] (1867)
: a swift lightly built terrier of English origin with a long narrow head, arched back, and usually curly coat — called also *Bedlington*

bed·mate \'bed-ˌmāt\ *noun* (1583)
: one who shares one's bed; *especially* **:** a sexual partner

bed molding *noun* (1703)
: the molding of a cornice below the corona and above the frieze; *also* **:** a molding below a deep projection

bed of roses (1648)
: a place or situation of agreeable ease

bed·ou·in *or* **bed·u·in** \'be-də-wən\ *noun, plural* **bedouin** *or* **bedouins** *or* **beduin** *or* **beduins** *often capitalized* [Middle English *Bedoyne,* from Middle French *bedoïn,* from Arabic *badawī* desert dweller] (15th century)
: a nomadic Arab of the Arabian, Syrian, or North African deserts

bed·pan \'bed-ˌpan\ *noun* (1678)
: a shallow vessel used by a bedridden person for urination or defecation

bed·plate \-ˌplāt\ *noun* (1850)
: a plate or framing used as a support

bed·post \-ˌpōst\ *noun* (1598)
: the usually turned or carved post of a bed

be·drag·gle \bi-'dra-gəl\ *transitive verb* (1727)
: to wet thoroughly

be·drag·gled \bi-'dra-gəld\ *adjective* (1824)
1 : left wet and limp by or as if by rain
2 : soiled and stained by or as if by trailing in mud
3 : DILAPIDATED ⟨*bedraggled* buildings⟩

bed rest *noun* (1944)
: confinement of a sick person to bed

bed·rid·den \'bed-ˌri-dᵊn\ *also* **bed·rid** \-ˌrid\ *adjective* [alteration of Middle English *bedrede, bedreden,* from Old English *bedreda,* from *bedreda* one confined to bed, from *bedd* bed + *-rida, -reda,* from *rīdan* to ride] (before 12th century)
: confined (as by illness) to bed

[1]bed·rock \-'räk, -ˌräk\ *noun* (1850)
1 : the solid rock underlying unconsolidated surface materials (as soil)
2 a : lowest point **:** NADIR **b :** BASIS

[2]bed·rock *adjective* (1896)
: solidly fundamental, basic, or reliable ⟨traditional *bedrock* values⟩ ⟨a *bedrock* constituency⟩

bed·roll \-ˌrōl\ *noun* (1910)
: bedding rolled up for carrying

[1]bed·room \-ˌrüm, -ˌru̇m\ *noun* (1616)

◇ WORD HISTORY

bedlam In 1247 a house was founded in London for the religious order of St. Mary of Bethlehem. By 1330 this priory had become the Hospital of St. Mary of Bethlehem, and it was intended to serve the poor or homeless who where afflicted with any ailment. By 1405 the hospital, now under royal control, was being used, at least in part, as an asylum for the insane—the first such institution in England. Over time everyday speech telescoped *Bethlehem* into *Bedlam*. And over time Bedlam became proverbial for its wretched conditions and tumultuous commotion. By the latter part of the 17th century, the word *bedlam* had begun to be used generically for any lunatic asylum. At about the same time, the term was first applied metaphorically to a scene of wild uproar or confusion, its common meaning today.

: a room furnished with a bed and intended primarily for sleeping
— **bed·roomed** \-ˌrümd, -ˌru̇md\ *adjective*

[2]bed·room *adjective* (1915)
1 : dealing with, suggestive of, or inviting sexual relations ⟨a *bedroom* farce⟩ ⟨*bedroom* eyes⟩
2 : inhabited or used by commuters ⟨a *bedroom* community⟩

bed·sheet \ˈbed-ˌshēt\ *noun* (15th century)
: an oblong piece of usually cotton or linen cloth used as an article of bedding

[1]bed·side \ˈbed-ˌsīd\ *noun* (14th century)
: the side of a bed **:** a place beside a bed

[2]bedside *adjective* (1837)
1 : of, relating to, or conducted at the bedside ⟨a *bedside* diagnosis⟩
2 : suitable for reading in bed ⟨a *bedside* book⟩

bedside manner *noun* (1869)
: the manner that a physician assumes toward patients

bed–sit·ter \ˈbed-ˌsi-tər\ *noun* [*bed-sitting* room + *-er* (as in *fresher* freshman, *rugger* rugby)] (1927)
British **:** a one-room apartment serving as both bedroom and sitting room — called also *bed-sit, bed-sitting-room*

bed·sore \ˈbed-ˌsōr, -ˌsȯr\ *noun* (1861)
: an ulceration of tissue deprived of adequate blood supply by prolonged pressure

bed·spread \-ˌspred\ *noun* (1845)
: a usually ornamental cloth cover for a bed

bed·spring \-ˌspriŋ\ *noun* (1897)
: a spring supporting a mattress

bed·stead \ˈbed-ˌsted\ *noun* [Middle English *bedstede,* from *bed* + *stede* stead, place — more at STEAD] (15th century)
: the framework of a bed

bed·straw \-ˌstrȯ\ *noun* [from its use for mattresses] (1527)
: any of a genus (*Galium*) of herbs of the madder family having squarish stems, whorled leaves, and small flowers

bed table *noun* (1811)
1 : an adjustable table used (as for eating or writing) by a person in bed
2 : a small table beside a bed

bed·time \-ˌtīm\ *noun* (13th century)
: a time for going to bed

bedtime story *noun* (1899)
: a story read or recounted to someone (as a child) at bedtime

be·du \ˈbe-(ˌ)dü\ *noun, plural* **bedu** *often capitalized* [Arabic *badw* Bedouins, literally, desert] (1912)
: BEDOUIN

bed warmer *noun* (1922)
: a covered pan containing hot coals used to warm a bed

bed–wet·ting \-ˌwe-tiŋ\ *noun* (1890)
: enuresis especially when occurring in bed during sleep
— **bed wetter** *noun*

[1]bee \ˈbē\ *noun* [Middle English, from Old English *bēo;* akin to Old High German *bīa* bee, Old Irish *bech,* Lithuanian *bitis*] (before 12th century)
1 : HONEYBEE; *broadly* **:** any of numerous insects (superfamily Apoidea) that differ from the related wasps especially in the heavier hairier body and in having sucking as well as chewing mouthparts, that feed on pollen and nectar, and that store both and often also honey
2 : an eccentric notion **:** FANCY
— **bee·like** \-ˌlīk\ *adjective*
— **bee in one's bonnet :** [1]BEE 2

[2]bee *noun* (14th century)
: the letter *b*

[3]bee *noun* [perhaps from English dialect *been* help given by neighbors, from Middle English *bene* prayer, boon, from Old English *bēn* prayer — more at BOON] (1769)
: a gathering of people for a specific purpose ⟨quilting *bee*⟩

bee balm *noun* (1840)
1 : any of several monardas; *especially* **:** OSWEGO TEA
2 : LEMON BALM

bee·bee *variant of* BB

bee·bread \ˈbē-ˌbred\ *noun* (1657)
: bitter yellowish brown pollen stored up in honeycomb cells and used mixed with honey by bees as food

beech \ˈbēch\ *noun, plural* **beech·es** *or* **beech** [Middle English *beche,* from Old English *bēce;* akin to Old English *bōc* beech, Old High German *buohha,* Latin *fagus,* Greek *phēgos* oak] (before 12th century)
: any of a genus (*Fagus* of the family Fagaceae, the beech family) of hardwood trees with smooth gray bark and small edible nuts; *also* **:** its wood
— **beech·en** \ˈbē-chən\ *adjective*

beech·drops \ˈbēch-ˌdräps\ *noun plural but singular or plural in construction* (1815)
: a plant (*Epifagus virginiana*) of the broomrape family parasitic on the roots of beeches

beech·nut \-ˌnət\ *noun* (1739)
: the nut of the beech

bee–eat·er \ˈbē-ˌē-tər\ *noun* (1668)
: any of a family (Meropidae) of brightly colored slender-billed insectivorous chiefly tropical Old World birds

[1]beef \ˈbēf\ *noun, plural* **beefs** \ˈbēfs\ *or* **beeves** \ˈbēvz\ [Middle English, from Old French *buef* ox, beef, from Latin *bov-, bos* head of cattle — more at COW] (14th century)
1 : the flesh of an adult domestic bovine (as a steer or cow) used as food
2 a : an ox, cow, or bull in a full-grown or nearly full-grown state; *especially* **:** a steer or cow fattened for food ⟨quality Texas *beeves*⟩ ⟨a herd of good *beef*⟩ **b :** a dressed carcass of a beef animal
3 : muscular flesh **:** BRAWN
4 *plural* *beefs* **:** COMPLAINT

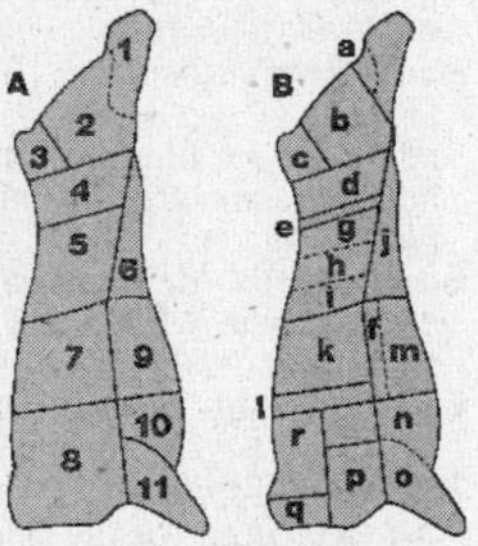

beef 2b: *A* wholesale cuts: *1* shank, *2* round with rump and shank cut off, *3* rump, *4* sirloin, *5* short loin, *6* flank, *7* rib, *8* chuck, *9* plate, *10* brisket, *11* shank; *B* retail cuts: *a* heel pot roast, *b* round steak, *c* rump roast, *d* sirloin steak, *e* pinbone steak, *f* short ribs, *g* porterhouse, *h* T-bone, *i* club steak, *j* flank steak, *k* rib roast, *l* blade rib roast, *m* plate, *n* brisket, *o* crosscut shank, *p* arm pot roast, *q* boneless neck, *r* blade roast

[2]beef (1860)
transitive verb
: to add substance, strength, or power to — usually used with *up* ⟨money to *beef* up its staff of professional economists —John Fischer⟩
intransitive verb
: COMPLAIN

beef·a·lo \ˈbē-fə-ˌlō\ *noun, plural* **-a·los** *or* **-a·loes** [blend of [1]*beef* and *buffalo*] (1973)
: any of a breed of beef cattle developed in the U.S. that is genetically 3/8 North American bison and 5/8 domestic bovine

beef·cake \ˈbēf-ˌkāk\ *noun* (1949)
: a usually photographic display of muscular male physiques; *also* **:** men of the type featured in such a display — compare CHEESECAKE

beef cattle *noun plural* (1758)
: cattle developed primarily for the efficient production of meat and marked by capacity for rapid growth, heavy well-fleshed body, and stocky build

beef·eat·er \ˈbē-ˌfē-tər\ *noun, often capitalized* (1671)
: a yeoman of the guard of an English monarch

bee fly *noun* (1852)
: any of a family (Bombyliidae) of dipteran flies many of which resemble bees

beef·steak \ˈbēf-ˌstāk\ *noun* (1711)
: a steak of beef usually from the hindquarter

beefsteak tomato *noun* (1968)
: a very large globe-shaped red tomato with dense flesh

beef Stro·ga·noff \-ˈstrȯ-gə-ˌnȯf, -ˈstrō-\ *noun* [Count Paul *Stroganoff,* 19th century Russian diplomat] (1932)
: beef sliced thin and cooked in a sour-cream sauce

beef Wel·ling·ton \-ˈwe-liŋ-tən\ *noun* [probably from the name *Wellington*] (1965)
: a fillet of beef covered with pâté de foie gras and baked in a casing of pastry

beef·wood \ˈbēf-ˌwu̇d\ *noun* (1805)
1 : the hard heavy reddish wood of any of various chiefly tropical trees (as the Australian pine)
2 : AUSTRALIAN PINE

beefy \ˈbē-fē\ *adjective* **beef·i·er; -est** (1743)
1 a : heavily and powerfully built **b :** SUBSTANTIAL, STURDY
2 : full of beef

bee·hive \ˈbē-ˌhīv\ *noun* (14th century)
1 : HIVE 1
2 : something resembling a hive for bees: as **a :** a scene of crowded activity **b :** a woman's hairdo that is conical in shape
— **beehive** *adjective*

beehive oven *noun* (circa 1881)
: an arched oven used especially for baking food and formerly for coking coal

bee·keep·er \-ˌkē-pər\ *noun* (1817)
: a person who raises bees
— **bee·keep·ing** *noun*

[1]bee·line \-ˌlīn\ *noun* [from the belief that nectar-laden bees return to their hives in a direct line] (1830)
: a straight direct course

[2]beeline *intransitive verb* (1940)
: to go quickly in a straight direct course

Beel·ze·bub \bē-ˈel-zi-ˌbəb, ˈbēl-zi-, ˈbel-\ *noun* [*Beelzebub,* prince of devils, from Latin, from Greek *Beelzeboub,* from Hebrew *Baʽal zĕbhūbh,* a Philistine god, literally, lord of flies] (before 12th century)
1 : DEVIL
2 : a fallen angel in Milton's *Paradise Lost* ranking next to Satan

been *past participle of* BE

[1]beep \ˈbēp\ *noun* [imitative] (1929)
: a short usually high-pitched sound (as from a horn or an electronic device) that serves as a signal or warning

[2]beep (1936)
intransitive verb
1 : to sound a horn
2 : to make a beep
transitive verb
: to cause (as a horn) to sound

beep·er \ˈbē-pər\ *noun* (1970)
: an electronic device that beeps to make contact with the person carrying it when it receives a special radio signal

beer \ˈbir\ *noun* [Middle English *ber,* from Old English *bēor;* akin to Old High German *bior* beer] (before 12th century)
1 : an alcoholic beverage usually made from malted cereal grain (as barley), flavored with hops, and brewed by slow fermentation
2 : a carbonated nonalcoholic or a fermented slightly alcoholic beverage with flavoring from roots or other plant parts ⟨birch *beer*⟩
3 : fermented mash
4 : a drink of beer

beer and skittles *noun plural but singular or plural in construction* (1857)
: a situation of agreeable ease ⟨won't be all *beer and skittles*⟩

beery \ˈbir-ē\ *adjective* **beer·i·er; -est** (1848)

1 : smelling or tasting of beer ⟨*beery* tavern⟩
2 : affected or caused by beer ⟨*beery* voices⟩

bees·tings \'bē-stiŋz\ *noun plural but singular or plural in construction* [Middle English *bestynge,* from Old English *bȳsting,* from *bēost* beestings; akin to Old High German *biost* beestings] (before 12th century)
: the colostrum especially of a cow

bee–stung \'bē-ˌstəŋ\ *adjective* (1933)
: having a red puffy appearance as if from being stung by a bee ⟨*bee-stung* lips⟩

bees·wax \'bēz-ˌwaks\ *noun* (1676)
: WAX 1

beet \'bēt\ *noun* [Middle English *bete,* from Old English *bēte,* from Latin *beta*] (before 12th century)
: a biennial garden plant (*Beta vulgaris*) of the goosefoot family that has several cultivars (as Swiss chard and sugar beet) and possesses thick long-stalked edible leaves and swollen root used as a vegetable, as a source of sugar, or for forage; *also* **:** its root

beet armyworm *noun* (1902)
: an armyworm (*Spodoptera exigua*) that eats the foliage of beets, alfalfa, and vegetables

[1]bee·tle \'bē-tᵊl\ *noun* [Middle English *betylle,* from Old English *bitula;* akin to *bītan* to bite] (before 12th century)
1 : any of an order (Coleoptera) of insects having four wings of which the outer pair are modified into stiff elytra that protect the inner pair when at rest
2 : any of various insects resembling a beetle

[2]beetle *intransitive verb* **bee·tled; bee·tling** \'bē-tᵊl-iŋ\ (circa 1919)
: to scurry like a beetle ⟨editors *beetled* around the office⟩

[3]beetle *noun* [Middle English *betel,* from Old English *bīetel;* akin to Old English *bēatan* to beat] (before 12th century)
1 : a heavy wooden hammering or ramming instrument
2 : a wooden pestle or bat for domestic tasks

[4]beetle *adjective* [Middle English *bitel-browed* having overhanging brows, probably from *betylle, bitel* beetle] (14th century)
: being prominent and overhanging ⟨*beetle* brows⟩

[5]beetle *intransitive verb* **bee·tled; bee·tling** \'bē-tᵊl-iŋ\ (1602)
: PROJECT, JUT ⟨to scale the *beetling* crags —R. L. Stevenson⟩

beet leafhopper *noun* (1919)
: a leafhopper (*Circulifer tenellus*) that transmits curly top virus to sugar beets and other garden plants

bee tree *noun* (1782)
: a hollow tree in which honeybees nest

beet·root \'bēt-ˌrüt\ *noun* (1579)
chiefly British **:** a beet grown for its edible usually red root; *also* **:** the root

beeves *plural of* BEEF

bee·yard \'bē-ˌyärd\ *noun* (15th century)
: APIARY

be·fall \bi-'fȯl\ *verb* **-fell** \-'fel\; **-fall·en** \-'fȯ-lən\ (13th century)
intransitive verb
: to happen especially as if by fate
transitive verb
: to happen to

be·fit \bi-'fit\ *transitive verb* **be·fit·ted; be·fit·ting** (15th century)
: to be proper or becoming to

be·fit·ting \-'fi-tiŋ\ *adjective* (1564)
1 : SUITABLE, APPROPRIATE
2 : PROPER, DECENT
— be·fit·ting·ly \-tiŋ-lē\ *adverb*

be·fog \bi-'fȯg, -'fäg\ *transitive verb* (1601)
1 : CONFUSE
2 : FOG, OBSCURE

be·fool \bi-'fül\ *transitive verb* (14th century)
1 : to make a fool of
2 : DELUDE, DECEIVE

[1]be·fore \bi-'fōr, -'fȯr\ *adverb or adjective* [Middle English, adverb & preposition, from Old English *beforan,* from *be-* + *foran* before, from *fore*] (before 12th century)
1 : in advance **:** AHEAD ⟨marching on *before*⟩
2 : at an earlier time **:** PREVIOUSLY ⟨the night *before*⟩ ⟨knew her from *before*⟩

[2]before *preposition* (before 12th century)
1 a (1) **:** in front of (2) **:** in the presence of **b :** under the jurisdiction or consideration of ⟨the case *before* the court⟩ **c** (1) **:** at the disposal of (2) **:** in store for
2 : preceding in time **:** earlier than
3 : in a higher or more important position than ⟨put quantity *before* quality⟩

[3]before *conjunction* (13th century)
1 a (1) **:** earlier than the time that ⟨call me *before* you go⟩ (2) **:** sooner or quicker than ⟨I'll be done *before* you know it⟩ (3) **:** so that . . . do not ⟨get out of there *before* you get dirty⟩ **b :** until the time that ⟨miles to go *before* I sleep —Robert Frost⟩ **c** (1) **:** or else . . . not ⟨must be convicted *before* he can be removed from office⟩ (2) **:** or else ⟨get out of here *before* I call a cop⟩
2 : rather or sooner than ⟨would starve *before* he'd steal⟩

be·fore·hand \bi-'fōr-ˌhand, -'fȯr-\ *adverb or adjective* (13th century)
1 a : in anticipation **b :** in advance
2 : ahead of time **:** EARLY

be·fore·time \-ˌtīm\ *adverb* (14th century)
archaic **:** FORMERLY

be·foul \bi-'fau̇(ə)l\ *transitive verb* (before 12th century)
: to make foul with or as if with dirt

be·friend \bi-'frend\ *transitive verb* (1559)
: to act as a friend to

be·fud·dle \bi-'fə-dᵊl\ *transitive verb* (circa 1879)
1 : to muddle or stupefy with or as if with drink
2 : CONFUSE, PERPLEX
— be·fud·dle·ment \-mənt\ *noun*

beg \'beg\ *verb* **begged; beg·ging** [Middle English *beggen*] (13th century)
transitive verb
1 : to ask for as a charity
2 a : to ask earnestly for **:** ENTREAT **b :** to require as necessary or appropriate
3 a : EVADE, SIDESTEP ⟨*begged* the real problems⟩ **b :** to pass over or ignore by assuming to be established or settled ⟨*beg* the question⟩
intransitive verb
1 : to ask for alms
2 : to ask earnestly ⟨*begged* for mercy⟩ ☆

be·get \bi-'get\ *transitive verb* **-got** \-'gät\ *also* **-gat** \-'gat\; **-got·ten** \-'gä-tᵊn\ *or* **-got; -get·ting** [Middle English *begeten,* alteration of *beyeten,* from Old English *bigietan* — more at GET] (13th century)
1 : to procreate as the father **:** SIRE
2 : to produce especially as an effect or outgrowth
— be·get·ter *noun*

[1]beg·gar \'be-gər\ *noun* [Middle English *beggere, beggare,* from *beggen* to beg + *-ere, -are* [2]-er] (13th century)
1 : one that begs; *especially* **:** a person who lives by asking for gifts
2 : PAUPER
3 : FELLOW 4c

[2]beggar *transitive verb* **beg·gared; beg·gar·ing** \'be-gə-riŋ\ (15th century)
1 : to reduce to beggary
2 : to exceed the resources or abilities of **:** DEFY ⟨*beggars* description⟩

beg·gar·ly \'be-gər-lē\ *adjective* (1526)
1 : contemptibly mean, scant, petty, or paltry
2 : befitting or resembling a beggar; *especially* **:** marked by extreme poverty
— beg·gar·li·ness *noun*

beg·gar's–lice \'be-gərz-ˌlīs\ *or* **beg·gar-lice** \-gər-ˌlīs\ *noun plural but singular or plural in construction* (1837)
: any of various plants (as of the genera *Hackelia* and *Cynoglossum*) with prickly or adhesive fruits; *also* **:** one of these fruits

beg·gar–ticks *also* **beg·gar's–ticks** \-ˌtiks\ *noun plural but singular or plural in construction* (circa 1818)
1 : BUR MARIGOLD; *also* **:** its prickly achenes
2 : BEGGAR'S-LICE

beg·gar·weed \'be-gər-ˌwēd\ *noun* (circa 1809)
1 : any of various plants (as a knotgrass, spurrey, or dodder) that grow in waste ground
2 : any of several tick trefoils (genus *Desmodium*); *especially* **:** a West Indian forage plant (*D. tortuosum*) cultivated in the southern U.S.

beg·gary \'be-gə-rē\ *noun, plural* **-gar·ies** (14th century)
1 : POVERTY, PENURY
2 : the class of beggars
3 : the practice of begging

be·gin \bi-'gin\ *verb* **be·gan** \-'gan\; **be·gun** \-'gən\; **be·gin·ning** [Middle English *beginnen,* from Old English *beginnan;* akin to Old High German *biginnan* to begin, Old English *onginnan*] (before 12th century)
intransitive verb
1 : to do the first part of an action **:** go into the first part of a process **:** START
2 a : to come into existence **:** ARISE **b :** to have a starting point
3 : to do or succeed in the least degree ⟨I can't *begin* to tell you how pleased I am⟩
transitive verb
1 : to set about the activity of **:** START
2 a : to bring into being **:** FOUND **b :** ORIGINATE, INVENT ☆

☆ SYNONYMS

Beg, entreat, beseech, implore, supplicate, adjure, importune mean to ask urgently. BEG suggests earnestness or insistence especially in asking for a favor ⟨children *begging* to stay up late⟩. ENTREAT implies an effort to persuade or to overcome resistance ⟨*entreated* him to change his mind⟩. BESEECH implies great eagerness or anxiety ⟨I *beseech* you to have mercy⟩. IMPLORE adds to BESEECH a suggestion of greater urgency or anguished appeal ⟨*implored* her not to leave him⟩. SUPPLICATE suggests a posture of humility ⟨with bowed heads they *supplicated* their Lord⟩. ADJURE implies advising as well as pleading and suggests the invoking of something sacred ⟨in God's name I *adjure* you to cease⟩. IMPORTUNE suggests an annoying persistence in trying to break down resistance to a request ⟨*importuning* viewers for contributions⟩.

Begin, commence, start, initiate, inaugurate, usher in mean to take the first step in a course, process, or operation. BEGIN, START, and COMMENCE are often interchangeable. BEGIN, opposed to *end,* is the most general ⟨*begin* a trip⟩ ⟨*began* dancing⟩. START, opposed to *stop,* applies especially to first actions, steps, or stages ⟨the work *started* slowly⟩. COMMENCE can be more formal or bookish than BEGIN or START ⟨*commence* firing⟩ ⟨*commenced* a conversation⟩. INITIATE implies taking a first step in a process or series that is to continue ⟨*initiated* diplomatic contacts⟩. INAUGURATE suggests a beginning of some formality or notion of significance ⟨the discovery of penicillin *inaugurated* a new era in medicine⟩. USHER IN is somewhat less weighty than INAUGURATE ⟨*ushered in* a period of economic decline⟩.

— to begin with : as the first thing to be considered

be·gin·ner \bi-ˈgi-nər\ *noun* (14th century)
: one that begins something; *especially* **:** an inexperienced person

[1]**be·gin·ning** \bi-ˈgi-niŋ\ *noun* (12th century)
1 : the point at which something begins **:** START
2 : the first part
3 : ORIGIN, SOURCE
4 : a rudimentary stage or early period — usually used in plural

[2]**beginning** *adjective* (1576)
1 : just starting out ⟨a *beginning* writer⟩
2 a : being first or the first part ⟨the *beginning* chapters⟩ **b :** INTRODUCTORY ⟨*beginning* chemistry⟩

beginning rhyme *noun* (1913)
1 : rhyme at the beginning of successive lines of verse
2 : ALLITERATION

be·gird \bi-ˈgərd\ *transitive verb* **-gird·ed** *or* **-girt** \-ˈgərt\; **-girt·ing** (before 12th century)
1 : GIRD 1a
2 : SURROUND, ENCOMPASS

be·glam·our *also* **be·glam·or** \bi-ˈgla-mər\ *transitive verb* (1822)
: to impress or deceive with glamour

beg off *intransitive verb* (1854)
: to ask to be excused from something
— beg–off \ˈbeg-ˌȯf\ *noun*

be·gone \bi-ˈgȯn *also* -ˈgän\ *intransitive verb* [Middle English, from *be gone* (imperative)] (14th century)
: to go away **:** DEPART — used especially in the imperative

be·go·nia \bi-ˈgōn-yə\ *noun* [New Latin, from Michel *Bégon* (died 1710) French governor of Santo Domingo] (1751)
: any of a large genus (*Begonia* of the family Begoniaceae, the begonia family) of tropical herbs and shrubs that have asymmetrical leaves and are widely cultivated as ornamentals

be·gor·ra \bi-ˈgȯr-ə, -ˈgär-\ *interjection* [euphemism for *by God*] (1839)
Irish — used as a mild oath

be·grime \bi-ˈgrīm\ *transitive verb* **be·grimed; be·grim·ing** (circa 1553)
1 : to make dirty with grime
2 : SULLY, CORRUPT

be·grudge \bi-ˈgrəj\ *transitive verb* (14th century)
1 : to give or concede reluctantly or with displeasure
2 : to look upon with disapproval
— be·grudg·ing·ly \-ˈgrə-jiŋ-lē\ *adverb*

be·guile \bi-ˈgī(ə)l\ *verb* **be·guiled; be·guil·ing** (13th century)
transitive verb
1 : to lead by deception
2 : HOODWINK
3 : to while away especially by some agreeable occupation; *also* **:** DIVERT 2
4 : to engage the interest of by or as if by guile
intransitive verb
: to deceive by wiles
synonym see DECEIVE
— be·guile·ment \-ˈgī(ə)l-mənt\ *noun*
— be·guil·er \-ˈgī-lər\ *noun*
— be·guil·ing·ly \-ˈgī-liŋ-lē\ *adverb*

be·guine \bi-ˈgēn\ *noun* [American French *béguine,* from French *béguin* flirtation] (1935)
: a vigorous popular dance of the islands of Saint Lucia and Martinique that somewhat resembles the rumba

Be·guine \ˈbā-ˌgēn, ˌbā-ˈ\ *noun* [Middle French] (15th century)
: a member of one of various ascetic and philanthropic communities of women not under vows founded chiefly in the Netherlands in the 13th century

be·gum \ˈbā-gəm, ˈbē-\ *noun* [Hindi *begam*] (1617)
: a Muslim woman of high rank (as in India or Pakistan)

be·half \bi-ˈhaf, -ˈhȧf\ *noun* [Middle English, from *by* + *half* half, side] (14th century)
: INTEREST, BENEFIT; *also* **:** SUPPORT, DEFENSE ⟨argued in his *behalf*⟩ ■
— on behalf of *or* **in behalf of :** in the interest of; *also* **:** as a representative of

be·have \bi-ˈhāv\ *verb* **be·haved; be·hav·ing** [Middle English *behaven,* from *be-* + *haven* to have, hold] (15th century)
transitive verb
1 : to manage the actions of (oneself) in a particular way
2 : to conduct (oneself) in a proper manner
intransitive verb
1 : to act, function, or react in a particular way
2 : to conduct oneself properly ☆
— be·hav·er *noun*

be·hav·ior \bi-ˈhā-vyər\ *noun* [alteration of Middle English *behavour,* from *behaven*] (15th century)
1 a : the manner of conducting oneself **b :** anything that an organism does involving action and response to stimulation **c :** the response of an individual, group, or species to its environment
2 : the way in which someone behaves; *also* **:** an instance of such behavior
3 : the way in which something functions or operates
— be·hav·ior·al \-vyə-rəl\ *adjective*
— be·hav·ior·al·ly \-rə-lē\ *adverb*

behavioral science *noun* (1951)
: a science (as psychology, sociology, or anthropology) that deals with human action and seeks to generalize about human behavior in society
— behavioral scientist *noun*

be·hav·ior·ism \bi-ˈhā-vyə-ˌri-zəm\ *noun* (1913)
: a school of psychology that takes the objective evidence of behavior (as measured responses to stimuli) as the only concern of its research and the only basis of its theory without reference to conscious experience — compare INTROSPECTIONISM
— be·hav·ior·ist \-vyə-rist\ *adjective or noun*
— be·hav·ior·is·tic \-ˌhā-vyə-ˈris-tik\ *adjective*

behavior therapy *noun* (1961)
: psychotherapy that emphasizes the application of the principles of learning to substitute desirable responses and behavior patterns for undesirable ones — called also *behavior modification*

be·hav·iour *chiefly British variant of* BEHAVIOR

be·head \bi-ˈhed\ *transitive verb* (before 12th century)
: to cut off the head of **:** DECAPITATE

be·he·moth \bi-ˈhē-məth, ˈbē-ə-məth, -ˌmäth, -ˌmȯth\ *noun, often attributive* [Middle English, from Late Latin, from Hebrew *bĕhēmōth*] (14th century)
1 *often capitalized* **:** a mighty animal described in Job 40:15–24 as an example of the power of God
2 : something of monstrous size or power

be·hest \bi-ˈhest\ *noun* [Middle English, promise, command, from Old English *behǣs* promise, from *behātan* to promise, from *be-* + *hātan* to command, promise — more at HIGHT] (12th century)
1 : an authoritative order **:** COMMAND
2 : an urgent prompting ⟨called at the *behest* of my friends⟩

[1]**be·hind** \bi-ˈhīnd\ *adverb or adjective* [Middle English *behinde,* from Old English *behindan,* from *be-* + *hindan* from behind; akin to Old English *hinder* behind — more at HIND] (before 12th century)
1 a : in the place or situation that is being or has been departed from ⟨stay *behind*⟩ **b :** in, to, or toward the back ⟨look *behind*⟩ ⟨came from *behind*⟩ **c :** later in time ⟨can spring be far *behind*⟩
2 a : in a secondary or inferior position **b :** in arrears ⟨*behind* in the rent⟩ **c :** SLOW
3 *archaic* **:** still to come

[2]**behind** *preposition* (before 12th century)
1 a : in or to a place or situation in back of or to the rear of ⟨look *behind* you⟩ ⟨put *behind* bars⟩ **b** — used as a function word to indicate something that screens an observer ⟨the sun went *behind* a cloud⟩ **c :** following in order ⟨marched *behind* the band⟩
2 — used as a function word to indicate backwardness, delay, or deficiency ⟨*behind* the times⟩ ⟨*behind* schedule⟩ ⟨lagged *behind* last year's sales⟩
3 a : in the background of ⟨the conditions *behind* the strike⟩ **b :** out of the mind or consideration of ⟨put our troubles *behind* us⟩ **c :** beyond in depth or time ⟨the story *behind* the story⟩ ⟨go back *behind* Saint Augustine⟩
4 a : in support of **:** on the side of ⟨solidly *behind* the candidate⟩ **b :** with the support of ⟨won 1–0 *behind* brilliant pitching⟩

[3]**behind** *noun* [[1]*behind*] (circa 1830)
: BUTTOCKS

be·hind·hand \bi-ˈhīnd-ˌhand\ *adjective* (1535)
1 : being in arrears
2 a : being in an inferior position **b :** being behind schedule

behind–the–scenes *adjective* (1711)
1 : being or working out of public view or in secret ⟨*behind-the-scenes* wrangling⟩ ⟨a *behind-the-scenes* player⟩
2 : revealing or reporting the hidden workings ⟨a *behind-the-scenes* account⟩ ⟨a *behind-the-scenes* glimpse⟩

be·hold \bi-ˈhōld\ *verb* **-held** \-ˈheld\; **-hold·ing** [Middle English, to keep, behold, from Old English *behealdan,* from *be-* + *healdan* to hold] (before 12th century)

☆ SYNONYMS
Behave, conduct, deport, comport, acquit mean to act or to cause oneself to do something in a certain way. BEHAVE may apply to the meeting of a standard of what is proper or decorous ⟨the children *behaved* in church⟩. CONDUCT implies action or behavior that shows the extent of one's power to control or direct oneself ⟨*conducted* herself with unfailing good humor⟩. DEPORT implies behaving so as to show how far one conforms to conventional rules of discipline or propriety ⟨the hero *deported* himself in accord with the code of chivalry⟩. COMPORT suggests conduct measured by what is expected or required of one in a certain class or position ⟨*comported* themselves as gentlemen⟩. ACQUIT applies to action under stress that deserves praise or meets expectations ⟨*acquitted* herself well in her first assignment⟩.

□ USAGE
behalf A body of opinion favors *in* with the "interest, benefit" sense of *behalf* and *on* with the "support, defense" sense. This distinction has been observed by some writers but overall has never had a sound basis in actual usage. In current British use, *on behalf* (*of*) has replaced *in behalf* (*of*); both are still used in American English, but the distinction is frequently not observed.

transitive verb
1 : to perceive through sight or apprehension **:** SEE
2 : to gaze upon **:** OBSERVE
intransitive verb
— used in the imperative especially to call attention
— be·hold·er *noun*

be·hold·en \bi-ˈhōl-dən\ *adjective* [Middle English, from past participle of *beholden*] (14th century)
: being under obligation for a favor or gift **:** INDEBTED ⟨I'm *beholden* to you⟩

be·hoof \bi-ˈhüf\ *noun* [Middle English *behof,* from Old English *behōf* profit, need; akin to Old English *hebban* to raise — more at HEAVE] (before 12th century)
: ADVANTAGE, PROFIT ⟨for his own *behoof*⟩

be·hoove \bi-ˈhüv\ *verb* **be·hooved; be·hoov·ing** [Middle English *behoven,* from Old English *behōfian,* from *behōf*] (before 12th century)
transitive verb
: to be necessary, proper, or advantageous for ⟨it *behooves* us to go⟩
intransitive verb
: to be necessary, fit, or proper

be·hove \bi-ˈhōv\ *chiefly British variant of* BEHOOVE

¹beige \ˈbāzh\ *noun* [French] (circa 1858)
1 : cloth made of natural undyed wool
2 a : a variable color averaging light grayish yellowish brown **b :** a pale to grayish yellow
— beigy \ˈbā-zhē\ *adjective*

²beige *adjective* (1879)
1 : of the color beige
2 : VANILLA 2, VAPID

bei·gnet \ˌbān-ˈyā, ˌben-\ *noun* [American French & French; American French, from French, from Middle French *bignet,* from *buyne* bump, bruise] (1835)
1 : FRITTER
2 : a light square doughnut

¹be·ing \ˈbē(-i)ŋ\ *noun* (14th century)
1 a : the quality or state of having existence **b** (1) **:** something conceivable as existing (2) **:** something that actually exists (3) **:** the totality of existing things **c :** conscious existence **:** LIFE
2 : the qualities that constitute an existent thing **:** ESSENCE; *especially* **:** PERSONALITY
3 : a living thing; *especially* **:** PERSON

²being *adjective* [present participle of *be*] (14th century)
: PRESENT — used in the phrase *for the time being*

³being *conjunction* (1528)
chiefly dialect **:** SINCE, BECAUSE — usually used with *as, as how,* or *that* ■

Be·ja \ˈbā-jə\ *noun, plural* **Beja** (1819)
1 : a member of a pastoral people living between the Nile and the Red Sea
2 : the Cushitic language of the Beja people

be·je·sus *also* **be·jee·zus** \bi-ˈjē-zəs, -ˈjā-, -zəz\ *interjection* [alteration of *by Jesus*] (circa 1908)
— used as a mild oath; used as a noun for emphasis ⟨scares the *bejesus* out of me⟩

be·jew·eled *or* **be·jew·elled** \bi-ˈjü-əld, -ˈjüld *also* -ˈju̇(ə)ld\ *adjective* (1557)
: ornamented with or as if with jewels

bel \ˈbel\ *noun* [Alexander Graham *Bell*] (1929)
: ten decibels

be·la·bor \bi-ˈlā-bər\ *transitive verb* (1596)
1 a : to attack verbally **b :** to beat soundly
2 : to explain or insist on excessively ⟨*belabor* the obvious⟩

be·la·bour *chiefly British variant of* BELABOR

be·lat·ed \bi-ˈlā-təd\ *adjective* [past participle of *belate* (to make late)] (1670)
1 : delayed beyond the usual time
2 : existing or appearing past the normal or proper time
— be·lat·ed·ly *adverb*
— be·lat·ed·ness *noun*

be·laud \bi-ˈlȯd\ *transitive verb* (circa 1849)
: to praise usually to excess

¹be·lay \bi-ˈlā\ *verb* [Middle English *beleggen* to beset, from Old English *belecgan,* from *be-* + *lecgan* to lay] (1548)
transitive verb
1 a : to secure (as a rope) by turns around a cleat, pin, or bitt **b :** to make fast
2 : STOP
3 a : to secure (a person) at the end of a rope **b :** to secure (a rope) to a person or object
intransitive verb
1 : to be made fast
2 : STOP, QUIT — used in the imperative ⟨*belay* there⟩
3 : to make a line fast by turns around a cleat, pin, or bitt

²belay *noun* (1908)
1 : the securing of a person or a safety rope to an anchor point (as during mountain climbing); *also* **:** a method of so securing a person or rope
2 : something (as a projection of rock) to which a person or rope is anchored

bel can·to \bel-ˈkän-(ˌ)tō, -ˈkan-\ *noun* [Italian, literally, beautiful singing] (1894)
: operatic singing originating in 17th century and 18th century Italy and stressing ease, purity, and evenness of tone production and an agile and precise vocal technique

belch \ˈbelch\ *verb* [Middle English, from Old English *bealcan*] (before 12th century)
intransitive verb
1 : to expel gas suddenly from the stomach through the mouth
2 : to erupt, explode, or detonate violently
3 : to issue forth spasmodically **:** GUSH
transitive verb
1 : to eject or emit violently
2 : to expel (gas) from the stomach suddenly **:** ERUCT
— belch *noun*

bel·dam *or* **bel·dame** \ˈbel-dəm\ *noun* [Middle English *beldam* grandmother, from Middle French *bel* beautiful + Middle English *dam*] (1580)
: an old woman

be·lea·guer \bi-ˈle-gər\ *transitive verb* **-guered; -guer·ing** \-g(ə-)riŋ\ [Dutch *belegeren,* from *be-* (akin to Old English *be-*) + *leger* camp; akin to Old High German *legar* bed — more at LAIR] (1587)
1 : BESIEGE
2 : TROUBLE, HARASS ⟨*beleaguered* parents⟩
— be·lea·guer·ment \-mənt\ *noun*

bel·em·nite \ˈbe-ləm-ˌnīt\ *noun* [New Latin *belemnites,* from Greek *belemnon* dart; akin to Greek *ballein* to throw — more at DEVIL] (1646)
: any of a genus (*Belemnites*) of extinct cephalopods that had bullet-shaped internal shells and were especially abundant in the Mesozoic era

bel·fry \ˈbel-frē\ *noun, plural* **belfries** [Middle English *belfrey,* alteration of *berfrey,* from Middle French *berfrei* bell tower, siege tower, of Germanic origin (akin to Middle High German *bërvrit* siege tower), akin to Old High German *bergan* to shelter and to Old English *frith* peace, refuge — more at BURY] (15th century)
1 : a bell tower; *especially* **:** one surmounting or attached to another structure
2 : a room or framework for enclosing a bell
3 : HEAD 2a ⟨batty in the *belfry*⟩ ◆

Bel·gae \ˈbel-ˌgī, -ˌjē\ *noun plural* [Latin, plural of *Belga*] (circa 1895)
: a people occupying parts of northern Gaul and Britain in Caesar's time
— Bel·gic \-jik\ *adjective*

Bel·gian \ˈbel-jən\ *noun* (circa 1623)
1 : a native or inhabitant of Belgium
2 : any of a breed of heavy muscular usually roan or chestnut draft horses developed in Belgium
— Belgian *adjective*

Belgian endive *noun* (1931)
: ENDIVE 2

Belgian hare *noun* (1900)
: any of a breed of slender chestnut-colored domestic rabbits

Belgian Ma·li·nois \-ˌma-lən-ˈwä\ *noun* (1968)
: any of a breed of squarely built working dogs closely related to the Belgian sheepdog and having relatively short straight hair with a dense undercoat — called also *Malinois*

Belgian sheepdog *noun* (1929)
: any of a breed of hardy black dogs developed in Belgium especially for herding sheep

Belgian Ter·vu·ren \-(ˌ)tər-ˈvyu̇r-ən, -ter-\ *noun* [*Tervuren,* commune in Brabant, Belgium] (1964)
: any of a breed of working dogs closely related to the Belgian sheepdog but having abundant long straight fawn-colored hair with black tips

Be·lial \ˈbē-lē-əl, ˈbēl-yəl\ *noun* [Greek, from Hebrew *bĕlīya'al* worthlessness] (13th century)

□ USAGE

being If your own vocabulary does not include the conjunction *being* or its compound forms *being as, being as how,* and *being that,* you might well wonder at the wide coverage these terms receive in handbooks. The point that almost all of the handbooks miss is that this conjunction is dialectal. Like many words that have dwindled to regional use, *being* was once standard ⟨Sir John, you loiter here too long, *being* you are to take soldiers up in counties as you go —Shakespeare⟩ ⟨I am tired of Lives of Nelson, *being that* I never read any —Jane Austen⟩. *The Dictionary of American Regional English* reports that *being* and its compound forms still survive in the South, South Midland, and New England speech areas.

◇ WORD HISTORY

belfry In our day *belfry* means bell tower; the first syllable appears to be a perfect match with *bell,* whatever *-fry* might mean. But in fact *belfry* does not derive from *bell,* and the original meaning of its medieval French source was not "bell tower," but rather "siege tower." A siege tower was a wheeled wooden structure that was pushed up to the walls of a besieged castle to provide both shelter and a base of attack for the besiegers. A Middle High German word for such a tower was *bercvrit,* which appears to be a compound of the verb *bergen* "to protect, conceal, shelter" and *vride* "peace, refuge," the presumed sense of the compound being "that which affords refuge." The ancient Germanic language that had the greatest influence on early French was Frankish, and the Frankish counterpart to such a word would have been *bergfrithu.* This word would have emerged in Old French as *berfroi,* for which we have evidence from the mid-12th century. In some dialects, *berfroi* underwent alteration to *belfroi,* and it is from an Anglo-French form of this word that Middle English borrowed *belfrey*—hence Modern English *belfry,* with its illusory suggestion of *bell.*

1 — a biblical name of the devil or one of the fiends
2 : one of the fallen angels in Milton's *Paradise Lost*

be·lie \bi-ˈlī\ *transitive verb* **-lied; -ly·ing** (before 12th century)
1 a : to give a false impression of **b :** to present an appearance not in agreement with
2 a : to show (something) to be false or wrong **b :** to run counter to **:** CONTRADICT
3 : DISGUISE 3
— be·li·er \-ˈlī(-ə)r\ *noun*

be·lief \bə-ˈlēf\ *noun* [Middle English *beleave,* probably alteration of Old English *gelēafa,* from *ge-,* associative prefix + *lēafa;* akin to Old English *lȳfan*] (12th century)
1 : a state or habit of mind in which trust or confidence is placed in some person or thing
2 : something believed; *especially* **:** a tenet or body of tenets held by a group
3 : conviction of the truth of some statement or the reality of some being or phenomenon especially when based on examination of evidence ☆

be·liev·able \-ˈlē-və-bəl\ *adjective* (14th century)
: capable of being believed especially as within the range of known possibility or probability
— be·liev·abil·i·ty \-ˌlē-və-ˈbi-lə-tē\ *noun*
— be·liev·ably \-ˈlē-və-blē\ *adverb*

be·lieve \bə-ˈlēv\ *verb* **be·lieved; be·liev·ing** [Middle English *beleven,* from Old English *belēfan,* from *be-* + *lȳfan, lēfan* to allow, believe; akin to Old High German *gilouben* to believe, Old English *lēof* dear — more at LOVE] (before 12th century)
intransitive verb
1 a : to have a firm religious faith **b :** to accept as true, genuine, or real ⟨ideals we *believe* in⟩ ⟨*believes* in ghosts⟩
2 : to have a firm conviction as to the goodness, efficacy, or ability of something ⟨*believe* in exercise⟩
3 : to hold an opinion **:** THINK ⟨I *believe* so⟩
transitive verb
1 a : to consider to be true or honest ⟨*believe* the reports⟩ ⟨you wouldn't *believe* how long it took⟩ **b :** to accept the word or evidence of ⟨I *believe* you⟩ ⟨couldn't *believe* my ears⟩
2 : to hold as an opinion **:** SUPPOSE ⟨I *believe* it will rain soon⟩
— be·liev·er *noun*
— not believe : to be astounded at ⟨I could*n't believe* my luck⟩

be·like \bi-ˈlīk\ *adverb* (circa 1533) *archaic*
: most likely **:** PROBABLY

be·lit·tle \bi-ˈli-tᵊl\ *transitive verb* **-lit·tled; -lit·tling** \-ˈli-tᵊl-iŋ, -ˈlit-liŋ\ (1797)
1 : to speak slightingly of **:** DISPARAGE ⟨*belittles* her efforts⟩
2 : to cause (a person or thing) to seem little or less
synonym see DECRY
— be·lit·tle·ment \-ˈli-tᵊl-mənt\ *noun*
— be·lit·tler \-ˈli-tᵊl-ər, -ˈlit-lər\ *noun*

be·live \bi-ˈlīv\ *adverb* [Middle English *bilive* vigorously, from *by* + *live,* dative of *lif* life] (1594)
Scottish **:** in due time **:** BY AND BY

[1]bell \ˈbel\ *noun* [Middle English *belle,* from Old English; perhaps akin to Old English *bellan* to roar — more at BELLOW] (before 12th century)
1 a : a hollow metallic device that gives off a reverberating sound when struck **b :** DOORBELL
2 a : the sounding of a bell as a signal **b :** a stroke of a bell (as on shipboard) to indicate the time; *also* **:** the time so indicated **c :** a half hour period of a watch on shipboard indicated by the strokes of a bell — see SHIP'S BELLS table below
3 : something having the form of a bell: as **a :** the corolla of a flower **b :** the part of the capital of a column between the abacus and neck molding **c :** the flared end of a wind instrument
4 a : a percussion instrument consisting of metal bars or tubes that when struck give out tones resembling bells — usually used in plural **b :** GLOCKENSPIEL

SHIP'S BELLS

NO. OF BELLS	HOUR (A.M. OR P.M.)		
1	12:30	4:30	8:30
2	1:00	5:00	9:00
3	1:30	5:30	9:30
4	2:00	6:00	10:00
5	2:30	6:30	10:30
6	3:00	7:00	11:00
7	3:30	7:30	11:30
8	4:00	8:00	12:00

[2]bell (14th century)
transitive verb
1 : to provide with a bell
2 : to flare the end of (as a tube) into the shape of a bell
intransitive verb
: to take the form of a bell **:** FLARE
— bell the cat : to do a daring or risky deed

[3]bell *intransitive verb* [Middle English, from Old English *bellan*] (before 12th century)
: to make a resonant bellowing or baying sound ⟨the wild buck *bells* from ferny brake —Sir Walter Scott⟩

[4]bell *noun* (1862)
: BELLOW, ROAR

bel·la·don·na \ˌbe-lə-ˈdä-nə\ *noun* [Italian, literally, beautiful lady] (1597)
1 : an Old World poisonous plant (*Atropa belladonna*) of the nightshade family having purple or green bell-shaped flowers, glossy black berries, and root and leaves that yield atropine — called also *deadly nightshade*
2 : a medicinal extract (as atropine) from the belladonna plant

belladonna lily *noun* (1734)
: AMARYLLIS

bell·bird \ˈbel-ˌbərd\ *noun* (1802)
: any of various birds (as of the genera *Procnias* and *Anthornis*) whose notes suggest the sound of a bell

bell–bot·toms \ˈbel-ˈbä-təmz\ *noun plural* (1898)
: pants with wide flaring bottoms
— bell–bottom *adjective*

bell·boy \ˈbel-ˌbȯi\ *noun* (1861)
: BELLHOP

bell buoy *noun* (1838)
: a buoy with a bell rung by the action of the waves

bell captain *noun* (1926)
: CAPTAIN 2c

bell curve *noun* [from the shape] (circa 1941)
: NORMAL CURVE

belle \ˈbel\ *noun* [French, from feminine of *beau* beautiful — more at BEAU] (1622)
: a popular and attractive girl or woman; *especially* **:** a girl or woman whose charm and beauty make her a favorite ⟨the *belle* of the ball⟩

Bel·leek \bə-ˈlēk\ *noun* [*Belleek,* town in Northern Ireland] (1869)
: a very thin translucent porcelain with a lustrous pearly glaze produced in Ireland — called also *Belleek china, Belleek ware*

belle epoque *or* **belle époque** \ˈbel-ā-ˈpȯk\ *noun, often B&E capitalized* [French, literally, beautiful age] (1954)
: a period of high artistic or cultural development; *especially* **:** such a period in fin de siècle France

Bel·ler·o·phon \bə-ˈler-ə-fən, -ˌfän\ *noun* [Latin, from Greek *Bellerophōn*]
: a legendary Greek hero noted for killing the Chimera

belles let·tres \bel-letrᵊ\ *noun plural but singular in construction* [French, literally, fine letters] (1710)
: literature that is an end in itself and not merely informative; *specifically* **:** light, entertaining, and often sophisticated literature

bel·le·trist \bel-ˈle-trist\ *noun* [*belles lettres*] (1816)
: a writer of belles lettres
— bel·le·tris·tic \ˌbe-lə-ˈtris-tik\ *adjective*

bell·flow·er \ˈbel-ˌflau̇(-ə)r\ *noun* (1578)
: any of a genus (*Campanula* of the family Campanulaceae, the bellflower family) of widely cultivated herbs having alternate leaves and usually showy flowers

bell·hop \-ˌhäp\ *noun* [short for *bell-hopper*] (1910)
: a hotel or club employee who escorts guests to rooms, assists them with luggage, and runs errands

bel·li·cose \ˈbe-li-ˌkōs\ *adjective* [Middle English, from Latin *bellicosus,* from *bellicus* of war, from *bellum* war] (15th century)
: favoring or inclined to start quarrels or wars
synonym see BELLIGERENT
— bel·li·cos·i·ty \ˌbe-li-ˈkä-sə-tē\ *noun*

bel·lied \ˈbe-lēd\ *adjective* (15th century)
: having a belly of a specified kind — used in combination ⟨a big-*bellied* man⟩

bel·lig·er·ence \bə-ˈlij-rən(t)s, -ˈli-jə-\ *noun* (1814)
: an aggressive or truculent attitude, atmosphere, or disposition

bel·lig·er·en·cy \-rən(t)-sē\ *noun* (1863)
1 : the state of being at war or in conflict; *specifically* **:** the status of a legally recognized belligerent state or nation
2 : BELLIGERENCE

bel·lig·er·ent \-rənt\ *adjective* [modification of Latin *belligerant-, belligerans,* present participle of *belligerare* to wage war, from *belliger* waging war, from *bellum* + *gerere* to wage] (1577)
1 : waging war; *specifically* **:** belonging to or recognized as a state at war and protected by and subject to the laws of war
2 : inclined to or exhibiting assertiveness, hostility, or combativeness ☆
— belligerent *noun*
— bel·lig·er·ent·ly *adverb*

bell jar *noun* (circa 1859)

☆ SYNONYMS

Belief, faith, credence, credit mean assent to the truth of something offered for acceptance. BELIEF may or may not imply certitude in the believer ⟨my *belief* that I had caught all the errors⟩. FAITH almost always implies certitude even where there is no evidence or proof ⟨an unshakable *faith* in God⟩. CREDENCE suggests intellectual assent without implying anything about grounds for assent ⟨a theory now given *credence* by scientists⟩. CREDIT may imply assent on grounds other than direct proof ⟨gave full *credit* to the statement of a reputable witness⟩. See in addition OPINION.

Belligerent, bellicose, pugnacious, quarrelsome, contentious mean having an aggressive or fighting attitude. BELLIGERENT implies being actually at war or engaged in hostilities ⟨*belligerent* nations⟩. BELLICOSE suggests a disposition to fight ⟨a drunk in a *bellicose* mood⟩. PUGNACIOUS suggests a disposition that takes pleasure in personal combat ⟨a *pugnacious* thug⟩. QUARRELSOME stresses an ill-natured readiness to fight without good cause ⟨the heat made us all *quarrelsome*⟩. CONTENTIOUS implies perverse and irritating fondness for arguing and quarreling ⟨wearied by his *contentious* disposition⟩.

: a bell-shaped usually glass vessel designed to cover objects or to contain gases or a vacuum

bell jar

bell–ly•ra \ˈbel-ˈlī-rə\ *or* **bell lyre** \-ˌlīr\ *noun* [*lyra* from Latin, lyre] (circa 1943)
: a glockenspiel mounted in a portable lyre-shaped frame and used especially in marching bands

bell•man \ˈbel-mən\ *noun* (14th century)
1 : a man (as a town crier) who rings a bell
2 : BELLHOP

bell metal *noun* (1541)
: bronze that consists usually of three to four parts of copper to one of tin and that is used for making bells

Bel•lo•na \bə-ˈlō-nə\ *noun* [Latin]
: the Roman goddess of war

bel•low \ˈbe-(ˌ)lō\ *verb* [Middle English *belwen,* from Old English *bylgian;* akin to Old English & Old High German *bellan* to roar] (before 12th century)
intransitive verb
1 : to make the loud deep hollow sound characteristic of a bull
2 : to shout in a deep voice
transitive verb
: BAWL ⟨*bellows* the orders⟩
— bellow *noun*

bel•lows \ˈbe-(ˌ)lōz, -ləz\ *noun plural but singular or plural in construction* [Middle English *belu,* from Old English *belg* — more at BELLY] (before 12th century)
1 : an instrument or machine that by alternate expansion and contraction draws in air through a valve or orifice and expels it through a tube; *also* **:** any of various other blowers
2 : LUNGS
3 : the pleated expansible part in a camera; *also* **:** a metallic or plastic flexible and expansible vessel

bell pepper *noun* (1707)
: SWEET PEPPER

bell•pull \ˈbel-ˌpu̇l\ *noun* (1832)
: a handle or knob attached to a cord by which one rings a bell; *also* **:** the cord itself

bell push *noun* (1884)
: a button that is pushed to ring a bell

bells and whistles *noun plural* (1969)
: items or features that are useful or decorative but not essential **:** FRILLS

bell–shaped \ˈbel-ˌshāpt\ *adjective* (circa 1828)
1 : shaped like a bell
2 : relating to or being a normal curve or a normal distribution

Bell's palsy \ˈbelz-\ *noun* [Sir Charles *Bell* (died 1842) Scottish anatomist] (circa 1860)
: paralysis of the facial nerve producing distortion on one side of the face

bell tower *noun* (1614)
: a tower that supports or shelters a bell

bell•weth•er \ˈbel-ˈwe-t͟hər, -ˌwe-\ *noun* [Middle English, leading sheep of a flock, leader, from *belle* bell + *wether;* from the practice of belling the leader of a flock] (13th century)
: one that takes the lead or initiative **:** LEADER; *also* **:** an indicator of trends

bell•wort \ˈbel-ˌwərt, -ˌwȯrt\ *noun* (1784)
: any of a small genus (*Uvularia*) of herbs of the lily family with yellow drooping flowers

[1]**bel•ly** \ˈbe-lē\ *noun, plural* **bellies** [Middle English *bely* bellows, belly, from Old English *belg* bag, skin; akin to Old High German *balg* bag, skin, Old English *blāwan* to blow — more at BLOW] (before 12th century)
1 a : ABDOMEN 1; *also* **:** POTBELLY 1 **b :** the stomach and its adjuncts **c :** the undersurface of an animal's body; *also* **:** hide from this part **d :** WOMB, UTERUS
2 : an internal cavity **:** INTERIOR
3 : appetite for food
4 : a surface or object curved or rounded like a human belly
5 a : the enlarged fleshy body of a muscle **b :** the part of a sail that swells out when filled with wind
6 : GUT 1a(2)

[2]**belly** *verb* **bel•lied; bel•ly•ing** (1606)
transitive verb
: to cause to swell or fill out
intransitive verb
1 : SWELL, FILL
2 a : to slide or crawl on one's belly **b :** BELLY-LAND

[1]**bel•ly•ache** \ˈbe-lē-ˌāk\ *noun* (1552)
: pain in the abdomen and especially in the bowels **:** COLIC

[2]**bellyache** *intransitive verb* (1881)
: to complain whiningly or peevishly **:** find fault
— bel•ly•ach•er *noun*

bel•ly•band \ˈbe-lē-ˌband\ *noun* (15th century)
: a band around or across the belly: as **a :** GIRTH 1 **b :** BAND 4b

belly button *noun* (circa 1877)
: the human navel

belly dance *noun* (1899)
: a usually solo dance emphasizing movements of the belly
— belly dance *intransitive verb*
— belly dancer *noun*

belly flop *noun* (1895)
: a dive (as into water or in coasting prone on a sled) in which the front of the body strikes flat against another surface — called also *belly flopper*
— belly flop *intransitive verb*

bel•ly•ful \ˈbe-lē-ˌfu̇l\ *noun* (1535)
: an excessive amount ⟨a *bellyful* of advice⟩

bel•ly–land \-ˌland\ *intransitive verb* (1943)
: to land an airplane on its undersurface without use of landing gear
— belly landing *noun*

belly laugh *noun* (1921)
: a deep hearty laugh

bel•ly–up \-ˈəp\ *adjective* [from the floating position of a dead fish] (1939)
: hopelessly ruined or defeated; *especially* **:** BANKRUPT ⟨the business went *belly-up*⟩

belly up *intransitive verb* (1948)
: to move close or next to ⟨*bellied up* to the bar⟩

be•long \bi-ˈlȯŋ\ *verb* [Middle English *belongen,* from *be-* + *longen* to be suitable — more at LONG] (14th century)
intransitive verb
1 a : to be suitable, appropriate, or advantageous ⟨a dictionary *belongs* in every home⟩ **b :** to be in a proper situation ⟨a man of his ability *belongs* in teaching⟩
2 a : to be the property of a person or thing — used with *to* **b :** to be attached or bound by birth, allegiance, or dependency **c :** to be a member of a club, organization, or set
3 : to be an attribute, part, adjunct, or function of a person or thing ⟨nuts and bolts *belong* to a car⟩
4 : to be properly classified
verbal auxiliary
chiefly Southern & southern Midland **:** OUGHT, MUST

be•long•ing \-ˈlȯŋ-iŋ\ *noun* (1817)
1 : POSSESSION — usually used in plural
2 : close or intimate relationship ⟨a sense of *belonging*⟩
— be•long•ing•ness *noun*

Belo•rus•sian \ˌbe-lō-ˈrə-shən, ˌbye-\ *noun* (1943)
1 : a native or inhabitant of Belorussia (Belarus)
2 : the Slavic language of the Belorussians
— Belorussian *adjective*

be•loved \bi-ˈləvd, -ˈlə-vəd\ *adjective* [Middle English, from past participle of *beloven* to love, from *be-* + *loven* to love] (14th century)
: dearly loved **:** dear to the heart
— beloved *noun*

[1]**be•low** \bi-ˈlō\ *adverb* [Middle English *biloghe,* from *bi* by + *looghe* [3]low] (14th century)
1 : in or to a lower place
2 a : on earth **b :** in or to Hades or hell
3 : on or to a lower floor or deck
4 a : in, to, at, or by a lower rank or number **b :** below zero ⟨20 degrees *below*⟩
5 : lower on the same page or on a following page
6 : under the surface of the water

[2]**below** *preposition* (1575)
1 a : lower in place, rank, or value than **:** UNDER **b :** down river from **c :** south of
2 : inferior to (as in rank)
3 : not suitable to the rank of **:** BENEATH

[3]**below** *noun* (1697)
: something that is below

[4]**below** *adjective* (1916)
: written or discussed lower on the same page or on a following page

be•low•decks \bi-ˌlō-ˈdeks; -ˈlō-ˌdeks\ *adverb* (1909)
: inside or into the superstructure of a boat **:** down to a lower deck

be•low•ground \-ˈgrau̇nd; -ˌgrau̇nd\ *adjective* (1928)
: being under the ground

Bel•shaz•zar \bel-ˈsha-zər\ *noun* [Hebrew *Bēlshaṣṣar*]
: a son of Nebuchadnezzar and king of Babylon

[1]**belt** \ˈbelt\ *noun* [Middle English, from Old English; akin to Old High German *balz* belt; both from Latin *balteus* belt] (before 12th century)
1 a : a strip of flexible material worn especially around the waist **b :** a similar article worn as a corset or for protection or safety or as a symbol of distinction
2 : a continuous band of tough flexible material for transmitting motion and power or conveying materials
3 : an area characterized by some distinctive feature (as of culture, habitation, geology, or life forms); *especially* **:** one suited to a particular crop ⟨the corn *belt*⟩
4 : BELTWAY
— belt•ed \ˈbel-təd\ *adjective*
— belt•less \ˈbelt-ləs\ *adjective*
— below the belt : UNFAIR, UNFAIRLY
— under one's belt : in one's possession **:** as part of one's experience

[2]**belt** (14th century)
transitive verb
1 a : to encircle or fasten with a belt **b :** to strap on
2 a : to beat with or as if with a belt **:** THRASH **b :** STRIKE, HIT
3 : to mark with a band
4 : to sing in a forceful manner or style ⟨*belting* out popular songs⟩
intransitive verb
1 : to move or act in a speedy, vigorous, or violent manner
2 : to sing loudly
— belt•er \ˈbel-tər\ *noun*

[3]**belt** *noun* (1899)
1 : a jarring blow **:** WHACK
2 : DRINK ⟨a *belt* of gin⟩

Bel•tane \ˈbel-ˌtān, -tin\ *noun* [Middle English (Scots), 1st or 3rd of May, from Scottish Gaelic *bealltain*] (15th century)
: the Celtic May Day festival

belted kingfisher *noun* (1811)
: a North American kingfisher (*Ceryle alcyon* synonym *Megaceryle alcyon*) that is slate blue above and white below with a slate blue breast band and an additional chestnut-colored band in the female

belt·ing \ˈbel-tiŋ\ *noun* (1567)
1 : BELTS
2 : material for belts
belt–tight·en·ing \ˈbelt-ˌtīt-niŋ, -ˌtī-t^{ə}n-iŋ\ *noun* (1937)
: a reduction in spending
belt up *intransitive verb* (1949)
British **:** SHUT UP
belt·way \ˈbelt-ˌwā\ *noun* (circa 1952)
: a highway skirting an urban area — often used to refer specifically to the beltway around Washington D.C. especially as delimiting what is seen as an insular political and social world
be·lu·ga \bə-ˈlü-gə\ *noun* [Russian, from *belyĭ* white; akin to Greek *phalios* having a white spot — more at BALD] (1772)
1 a : a large white sturgeon (*Huso huso* synonym *Acipenser huso*) of the Black Sea, Caspian Sea, and their tributaries **b :** caviar processed from beluga roe
2 [Russian *belukha,* from *belyĭ*] **:** WHITE WHALE
bel·ve·dere \ˈbel-və-ˌdir\ *noun* [Italian, literally, beautiful view] (1593)
: a structure (as a cupola or a summerhouse) designed to command a view
be·ma \ˈbē-mə\ *noun* [Late Latin & Late Greek; Late Latin, from Late Greek *bēma,* from Greek, step, tribunal, from *bainein* to go — more at COME] (1683)
1 : the usually raised part of an Eastern church containing the altar
2 : a raised platform in a synagogue from which the Pentateuch and the Prophets are read
Bem·ba \ˈbem-bə\ *noun, plural* **Bemba** *or* **Bembas** (1940)
1 : a member of a primarily agricultural Bantu-speaking people of northeastern Zambia
2 : a Bantu language of the Bemba people
be·med·aled *or* **be·med·alled** \bi-ˈme-d^{ə}ld\ *adjective* (1880)
: wearing or decorated with medals
be·mire \bi-ˈmīr\ *transitive verb* (circa 1532)
1 : to soil with mud or dirt
2 : to drag through or sink in mire
be·moan \bi-ˈmōn\ *transitive verb* (before 12th century)
1 : to express deep grief or distress over
2 : to regard with displeasure, disapproval, or regret
synonym see DEPLORE
be·mock \bi-ˈmäk, -ˈmȯk\ *transitive verb* (1607)
archaic **:** MOCK
be·muse \bi-ˈmyüz\ *transitive verb* (1735)
1 : to make confused **:** BEWILDER
2 : to occupy the attention of **:** ABSORB
— be·mus·ed·ly \-ˈmyü-zəd-lē\ *adverb*
— be·muse·ment \-ˈmyüz-mənt\ *noun*
[1]**ben** \ˈben\ *adverb* [Middle English, from Old English *binnan,* from *be-* + *innan* within, from within, from *in*] (before 12th century)
Scottish **:** WITHIN
[2]**ben** *preposition* (before 12th century)
Scottish **:** WITHIN
[3]**ben** *noun* (circa 1799)
Scottish **:** the inner room or parlor of a 2-room cottage
Bence–Jones protein \ˈben(t)s-ˈjōnz-\ *noun* [Henry *Bence-Jones* (died 1873) English physician and chemist] (circa 1923)
: an immunoglobulin that is composed only of light chain polypeptides and that is found especially in the urine of persons affected with multiple myeloma
[1]**bench** \ˈbench\ *noun* [Middle English, from Old English *benc;* akin to Old High German *bank* bench] (before 12th century)
1 a : a long seat for two or more persons **b :** a thwart in a boat **c** (1) **:** a seat on which the members of an athletic team await a turn or opportunity to play (2) **:** the reserve players on a team
2 a : the seat where a judge sits in court **b :** the place where justice is administered **:** COURT **c :** the office or dignity of a judge **d :** the persons who sit as judges
3 a : the office or dignity of an official **b :** a seat for an official **c :** the officials occupying a bench
4 a : a long worktable; *also* **:** LABORATORY ⟨*bench* chemist⟩ ⟨*bench* test⟩ **b :** a table forming part of a machine
5 : TERRACE, SHELF: as **a :** a former wave-cut shore of a sea or lake or floodplain of a river **b :** a shelf or ridge formed in working an open excavation on more than one level
6 : a compartmented platform on which dogs or cats are kept at a show when not being judged
[2]**bench** (14th century)
transitive verb
1 : to furnish with benches
2 a : to seat on a bench **b** (1) **:** to remove from or keep out of a game (2) **:** to remove from the starting lineup
3 : to exhibit (dogs or cats) to the public on a bench
intransitive verb
: to form a bench by natural processes
bench·er \ˈben-chər\ *noun* (15th century)
: one who sits on or presides at a bench
bench·land \ˈbench-ˌland\ *noun* (1857)
: BENCH 5a
bench·mark \ˈbench-ˌmärk\ *noun* (circa 1842)
1 *usually* **bench mark :** a mark on a permanent object indicating elevation and serving as a reference in topographical surveys and tidal observations
2 a : a point of reference from which measurements may be made **b :** something that serves as a standard by which others may be measured or judged **c :** a standardized problem or test that serves as a basis for evaluation or comparison (as of computer system performance)
bench press *noun* (1973)
: a press in weight lifting performed by a lifter lying on a bench
— bench–press *transitive verb*
bench seat *noun* (1975)
: a seat in an automotive vehicle that extends the full width of the passenger section
bench·warm·er \ˈbench-ˌwȯr-mər\ *noun* (1892)
: a reserve player on an athletic team
bench warrant *noun* (1696)
: a warrant issued by a presiding judge or by a court against a person guilty of contempt or indicted for a crime
[1]**bend** \ˈbend\ *verb* **bent** \ˈbent\; **bend·ing** [Middle English, from Old English *bendan;* akin to Old English *bend* fetter — more at BAND] (before 12th century)
transitive verb
1 : to constrain or strain to tension by curving ⟨*bend* a bow⟩
2 a : to turn or force from straight or even to curved or angular **b :** to force from a proper shape **c :** to force back to an original straight or even condition
3 : FASTEN ⟨*bend* a sail to its yard⟩
4 a : to cause to turn from a straight course **:** DEFLECT **b :** to guide or turn toward **:** DIRECT **c :** INCLINE, DISPOSE **d :** to adapt to one's purpose **:** DISTORT ⟨*bend* the rules⟩
5 : to direct strenuously or with interest **:** APPLY
6 : to make submissive **:** SUBDUE
intransitive verb
1 : to curve out of a straight line or position; *specifically* **:** to incline the body in token of submission
2 : to apply oneself vigorously ⟨*bending* to their work⟩
3 : INCLINE, TEND
4 : to make concessions **:** COMPROMISE
— bend·a·ble \ˈben-də-bəl\ *adjective*
— bend one's ear : to talk to someone at length
[2]**bend** *noun* (15th century)
1 : the act or process of bending **:** the state of being bent
2 : something that is bent: as **a :** a curved part of a stream or road **b :** [1]WALE 2 — usually used in plural
3 *plural but singular or plural in construction* **:** a sometimes fatal disorder that is marked by neuralgic pains and paralysis, distress in breathing, and often collapse and that is caused by the release of gas bubbles (as of nitrogen) in tissue upon too rapid decrease in air pressure after a stay in a compressed atmosphere — usually used with *the;* called also *caisson disease, decompression sickness;* compare AEROEMBOLISM
— around the bend : MAD, CRAZY
[3]**bend** *noun* [Middle English, from Middle French *bende,* of Germanic origin; akin to Old High German *binta, bant* band — more at BAND] (15th century)
1 : a diagonal band that runs from the dexter chief to the sinister base on a heraldic shield — compare BEND SINISTER
2 [Middle English, band, from Old English *bend* fetter — more at BAND] **:** a knot by which one rope is fastened to another or to some object
ben·day \ˈben-ˈdā\ *adjective, often capitalized* [*Benjamin Day* (died 1916) American printer] (1903)
: involving a process for adding shaded or tinted areas made up of dots for reproduction by line engraving
— benday *transitive verb*
bend·er \ˈben-dər\ *noun* (15th century)
1 : one that bends
2 : SPREE
bend sinister *noun* (1612)
: a diagonal bend that runs from the sinister chief to the dexter base on a heraldic shield
bendy \ˈben-dē\ *adjective* (1928)
: FLEXIBLE, PLIABLE ⟨a *bendy* mast⟩
[1]**be·neath** \bi-ˈnēth\ *adverb* [Middle English *benethe,* from Old English *beneothan,* from *be-* + *neothan* below; akin to Old English *nithera* nether] (before 12th century)
1 : in or to a lower position **:** BELOW
2 : directly under **:** UNDERNEATH
[2]**beneath** *preposition* (before 12th century)
1 a : in or to a lower position than **:** BELOW **b :** directly under **c :** at the foot of
2 : not suitable to the rank of **:** unworthy of
3 : under the control, pressure, or influence of
4 : concealed by **:** under the guise of ⟨a warm heart *beneath* a gruff manner⟩
ben·e·dict \ˈbe-nə-ˌdikt\ *noun* [alteration of *Benedick,* character in Shakespeare's *Much Ado about Nothing*] (1821)
: a newly married man who has long been a bachelor
Ben·e·dic·tine \ˌbe-nə-ˈdik-tən, -ˌtēn\ *noun* (15th century)
: a monk or a nun of one of the congregations following the rule of Saint Benedict and devoted especially to scholarship and liturgical worship
— Benedictine *adjective*
bene·dic·tion \ˌbe-nə-ˈdik-shən\ *noun* [Middle English *benediccioun,* from Late Latin *benediction-, benedictio,* from *benedicere* to bless, from Latin, to speak well of, from *bene* well (akin to Latin *bonus* good) + *dicere* to say — more at BOUNTY, DICTION] (15th century)
1 : the invocation of a blessing; *especially* **:** the short blessing with which public worship is concluded
2 : something that promotes goodness or well-being
3 *often capitalized* **:** a Roman Catholic or Anglo-Catholic devotion including the exposition of the eucharistic Host in the monstrance and the blessing of the people with it
4 : an expression of good wishes
bene·dic·to·ry \-ˈdik-t(ə-)rē\ *adjective* (1710)

: of or expressing benediction

Ben•e•dict's solution \'be-nə-ˌdik(t)s-\ *noun* [Stanley Rossiter *Benedict* (died 1936) American chemist] (1921)
: a blue solution containing a carbonate, citrate, and sulfate which yields a red, yellow, or orange precipitate upon warming with a sugar (as glucose) that is a reducing agent

Ben•e•dic•tus \ˌbe-nə-'dik-təs\ *noun* [Late Latin, blessed, from past participle of *benedicere;* from its first word] (1552)
1 : a canticle from Luke 1:68 beginning "Blessed be the Lord God of Israel"
2 : a canticle from Matthew 21:9 beginning "Blessed is he that cometh in the name of the Lord"

ben•e•fac•tion \ˌbe-nə-'fak-shən\ *noun* [Late Latin *benefaction-, benefactio,* from Latin *bene facere* to do good to, from *bene + facere* to do — more at DO] (circa 1662)
1 : the act of benefiting
2 : a benefit conferred; *especially* **:** a charitable donation

ben•e•fac•tor \'be-nə-ˌfak-tər\ *noun* (15th century)
: one that confers a benefit; *especially* **:** one that makes a gift or bequest

ben•e•fac•tress \'be-nə-ˌfak-tris\ *noun* (1711)
: a woman who is a benefactor

be•nef•ic \bə-'ne-fik\ *adjective* [Latin *beneficus,* from *bene + facere*] (1641)
: BENEFICENT

ben•e•fice \'be-nə-fəs\ *noun* [Middle English, from Middle French, from Medieval Latin *beneficium,* from Latin, favor, promotion, from *beneficus*] (14th century)
1 : an ecclesiastical office to which the revenue from an endowment is attached
2 : a feudal estate in lands **:** FIEF
— benefice *transitive verb*

be•nef•i•cence \bə-'ne-fə-sən(t)s\ *noun* [Latin *beneficentia,* from *beneficus*] (15th century)
1 : the quality or state of being beneficent
2 : BENEFACTION

be•nef•i•cent \-sənt\ *adjective* [back-formation from *beneficence*] (1616)
1 : doing or producing good; *especially* **:** performing acts of kindness and charity
2 : BENEFICIAL
— be•nef•i•cent•ly *adverb*

ben•e•fi•cial \ˌbe-nə-'fi-shəl\ *adjective* [Latin *beneficium* favor, benefit] (15th century)
1 : conferring benefits **:** conducive to personal or social well-being
2 : receiving or entitling one to receive advantage, use, or benefit ⟨a *beneficial* legacy⟩
— ben•e•fi•cial•ly \-'fi-sh(ə-)lē\ *adverb*
— ben•e•fi•cial•ness *noun*

ben•e•fi•cia•ry \ˌbe-nə-'fi-shē-ˌer-ē, -'fi-sh(ə-)rē\ *noun, plural* **-ries** (1662)
1 : one that benefits from something
2 a : the person designated to receive the income of a trust estate **b :** the person named (as in an insurance policy) to receive proceeds or benefits
— beneficiary *adjective*

ben•e•fi•ci•a•tion \ˌbe-nə-ˌfi-shē-'ā-shən\ *noun* (circa 1881)
: the treatment of raw material (as iron ore) to improve physical or chemical properties especially in preparation for smelting
— ben•e•fi•ci•ate \-'fi-shē-ˌāt\ *transitive verb*

[1]ben•e•fit \'be-nə-ˌfit\ *noun* [Middle English, from Anglo-French *benfet,* from Latin *bene factum,* from neuter of *bene factus,* past participle of *bene facere*] (14th century)
1 *archaic* **:** an act of kindness **:** BENEFACTION
2 a : something that promotes well-being **:** ADVANTAGE **b :** useful aid **:** HELP
3 a : financial help in time of sickness, old age, or unemployment **b :** a payment or service provided for under an annuity, pension plan, or insurance policy
4 : an entertainment or social event to raise funds for a person or cause

[2]benefit *verb* **-fit•ed** \-ˌfi-təd\ *also* **-fit•ted; -fit•ing** *also* **-fit•ting** (15th century)
transitive verb
: to be useful or profitable to
intransitive verb
: to receive benefit
— ben•e•fit•er \-ˌfi-tər\ *noun*

benefit of clergy (15th century)
1 : clerical exemption from trial in a civil court
2 : the ministration or sanction of the church

be•nev•o•lence \bə-'nev-lən(t)s, -'ne-və-\ *noun* (14th century)
1 : disposition to do good
2 a : an act of kindness **b :** a generous gift
3 : a compulsory levy by certain English kings with no other authority than the claim of prerogative

be•nev•o•lent \-lənt\ *adjective* [Middle English, from Latin *benevolent-, benevolens,* from *bene + volent-, volens,* present participle of *velle* to wish — more at WILL] (15th century)
1 a : marked by or disposed to doing good ⟨a *benevolent* donor⟩ **b :** organized for the purpose of doing good ⟨a *benevolent* society⟩
2 : marked by or suggestive of goodwill ⟨*benevolent* smiles⟩
— be•nev•o•lent•ly *adverb*
— be•nev•o•lent•ness *noun*

Ben•gali \ben-'gȯ-lē, beŋ-\ *noun* [Hindi *Baṅgālī,* from *Baṅgāl* Bengal] (1848)
1 : a native or resident of Bengal
2 : the modern Indo-Aryan language of Bengal
— Bengali *adjective*

ben•ga•line \'beŋ-gə-ˌlēn\ *noun* [French, from *Bengal*] (1884)
: a fabric with a crosswise rib made from textile fibers (as rayon, nylon, cotton, or wool) often in combination

Ben•gal light \'ben-ˌgȯl-, ˌbeŋ-, -ˌgäl-, -gəl-\ *noun* (1818)
: a usually blue light or flare used formerly especially for signaling and illumination

Bengal tiger *noun* (circa 1864)
: a tiger (*Panthera tigris tigris*) occurring especially in India

be•night•ed \bi-'nī-təd\ *adjective* (15th century)
1 : overtaken by darkness or night
2 : existing in a state of intellectual, moral, or social darkness **:** UNENLIGHTENED
— be•night•ed•ly *adverb*
— be•night•ed•ness *noun*

be•nign \bi-'nīn\ *adjective* [Middle English *benigne,* from Middle French, from Latin *benignus,* from *bene + gignere* to beget — more at KIN] (14th century)
1 : of a gentle disposition **:** GRACIOUS ⟨a *benign* teacher⟩
2 a : showing kindness and gentleness ⟨*benign* faces⟩ **b :** FAVORABLE, WHOLESOME ⟨a *benign* climate⟩
3 a : of a mild type or character that does not threaten health or life ⟨a *benign* tumor⟩ **b :** having no significant effect **:** HARMLESS ⟨environmentally *benign*⟩
— be•nig•ni•ty \-'nig-nə-tē\ *noun*
— be•nign•ly \-'nīn-lē\ *adverb*

be•nig•nan•cy \bi-'nig-nən(t)-sē\ *noun* (1876)
: benignant quality

be•nig•nant \-nənt\ *adjective* [*benign* + *-ant* (as in *malignant*)] (circa 1782)
1 : serenely mild and kindly **:** BENIGN
2 : FAVORABLE, BENEFICIAL ⟨a *benignant* power⟩
— be•nig•nant•ly *adverb*

benign neglect *noun* (1970)
: an attitude or policy of ignoring an often undesirable situation that one is perceived to be responsible for dealing with

ben•i•son \'be-nə-sən, -zən\ *noun* [Middle English *beneson,* from Middle French *beneiçon,* from Late Latin *benediction-, benedictio*] (14th century)
: BLESSING, BENEDICTION

Ben•ja•min \'ben-jə-mən\ *noun* [Hebrew *Binyāmīn*]
: a son of Jacob and the traditional eponymous ancestor of one of the tribes of Israel

ben•ne *also* **bene** \'be-nē\ *noun* [of African origin; akin to Malinke *bĕne* sesame] (1769)
: SESAME 1

ben•ny \'be-nē\ *noun, plural* **bennies** [*Benzedrine* + *-ie*] (1949)
slang **:** a tablet of amphetamine taken as a stimulant

ben•o•myl \'be-nə-ˌmil\ *noun* [*benz-* + *-o-* + *methyl*] (1969)
: a derivative $C_{14}H_{18}N_4O_3$ of carbamate and benzimidazole used especially as a systemic agricultural fungicide

[1]bent \'bent\ *noun* [Middle English, grassy place, bent grass, from Old English *beonot-;* akin to Old High German *binuz* rush] (14th century)
1 : unenclosed grassland
2 a (1) **:** a reedy grass (2) **:** a stalk of stiff coarse grass **b :** BENT GRASS

[2]bent *adjective* [Middle English, from past participle of *benden* to bend] (14th century)
1 : changed by bending out of an originally straight or even condition ⟨*bent* twigs⟩
2 : strongly inclined **:** DETERMINED ⟨was *bent* on going⟩
3 *slang* **a :** different from the normal or usual **b** *chiefly British* **:** DISHONEST, CORRUPT

[3]bent *noun* [irregular from [1]*bend*] (1586)
1 a : a strong inclination or interest **:** BIAS **b :** a special inclination or capacity **:** TALENT
2 : capacity of endurance
3 : a transverse framework (as in a bridge) to carry lateral as well as vertical loads
synonym see GIFT

bent grass *noun* (1778)
: any of a genus (*Agrostis*) of grasses including important chiefly perennial and rhizomatous pasture and lawn grasses with fine velvety or wiry herbage

Ben•tham•ism \'ben(t)-thə-ˌmi-zəm\ *noun* (1829)
: the utilitarian philosophy of Jeremy Bentham and his followers
— Ben•tham•ite \-ˌmīt\ *noun or adjective*

ben•thic \'ben(t)-thik\ *adjective* [*benthos*] (1902)
1 : of, relating to, or occurring at the bottom of a body of water
2 : of, relating to, or occurring in the depths of the ocean

ben•thon•ic \ben-'thä-nik\ *adjective* [irregular from *benthos*] (1897)
: BENTHIC

ben•thos \'ben-ˌthäs\ *noun* [New Latin, from Greek, depth, deep sea; akin to Greek *bathys* deep] (1891)
: organisms that live on or in the bottom of a body of water

ben•ton•ite \'ben-tᵊn-ˌīt\ *noun* [Fort *Benton,* Mont.] (1898)
: an absorptive and colloidal clay used especially as a sealing agent or carrier (as of drugs)
— ben•ton•it•ic \ˌben-tᵊn-'i-tik\ *adjective*

ben tro•va•to \ˌben-trō-'vä-(ˌ)tō\ *adjective* [Italian, literally, well found] (1876)
: characteristic or appropriate even if not true ⟨the story is *ben trovato*⟩

bent•wood \'bent-ˌwu̇d\ *adjective* (1862)
: made of wood that is bent rather than cut into shape ⟨*bentwood* furniture⟩
— bentwood *noun*

be•numb \bi-ˈnəm\ *transitive verb* [Middle English *benomen,* from *benomen,* past participle of *benimen* to deprive, from Old English *beniman,* from *be-* + *niman* to take — more at NIMBLE] (14th century)
1 : to make inactive **:** DEADEN
2 : to make numb especially by cold

benz- *or* **benzo-** *combining form* [International Scientific Vocabulary, from *benzoin*]
: related to benzene or benzoic acid ⟨*benzo*phenone⟩ ⟨*benz*yl⟩

benz•al•de•hyde \ben-ˈzal-də-ˌhīd\ *noun* [International Scientific Vocabulary] (1866)
: a colorless nontoxic aromatic liquid C_6H_5CHO found in essential oils (as in peach kernels) and used in flavoring and perfumery, in pharmaceuticals, and in synthesis of dyes

benz•an•thra•cene \ben-ˈzan(t)-thrə-ˌsēn\ *noun* [International Scientific Vocabulary] (1938)
: a crystalline feebly carcinogenic cyclic hydrocarbon $C_{18}H_{12}$ that is found in small amounts in coal tar

Ben•ze•drine \ˈben-zə-ˌdrēn\ *trademark*
— used for amphetamine

ben•zene \ˈben-ˌzēn, ben-ˈ\ *noun* [International Scientific Vocabulary *benz-* + *-ene*] (circa 1872)
: a colorless volatile flammable toxic liquid aromatic hydrocarbon C_6H_6 used in organic synthesis, as a solvent, and as a motor fuel — called also *benzol*
— ben•ze•noid \ˈben-zə-ˌnȯid\ *adjective*

benzene hexa•chlo•ride \-ˌhek-sə-ˈklōr-ˌīd, -ˈklȯr-\ *noun* (1884)
: BHC

benzene ring *noun* (1877)
: a structural arrangement of atoms held to exist in benzene and other aromatic compounds and marked by six carbon atoms linked in a planar symmetrical hexagon with each carbon attached to hydrogen in benzene itself or to other atoms or groups in substituted benzenes — compare META- 4a, ORTH- 4b, PARA- 2b

ben•zi•dine \ˈben-zə-ˌdēn\ *noun* [International Scientific Vocabulary *benz-* + *-idine*] (1878)
: a crystalline base $C_{12}H_{12}N_2$ prepared from nitrobenzene and used especially in making dyes

benz•imid•azole \ˌben-ˌzi-mə-ˈda-ˌzōl, ˌben-zə-ˈmi-də-ˌzōl\ *noun* [International Scientific Vocabulary] (circa 1929)
: a crystalline base $C_7H_6N_2$ used especially to inhibit the growth of various viruses, parasitic worms, or fungi; *also* **:** one of its derivatives

ben•zine \ˈben-ˌzēn, ben-ˈ\ *noun* [German *Benzin,* from *benz-*] (1835)
: any of various volatile flammable petroleum distillates used especially as solvents or as motor fuels

ben•zo•a•py•rene \ˌben-zō-ˌā-ˈpī-ˌrēn, -zō-ˌal-fə-\ *noun* [*benz-* + *-a-,* denoting attachment of the benzene substituent at carbon atoms 1 and 2 of the pyrene molecule + *pyrene*] (1950)
: a yellow crystalline carcinogenic hydrocarbon $C_{20}H_{12}$ found in coal tar — *a* often italicized and placed in parentheses or brackets or between hyphens

ben•zo•ate \ˈben-zə-ˌwāt\ *noun* (1806)
: a salt or ester of benzoic acid

ben•zo•caine \ˈben-zə-ˌkān\ *noun* [International Scientific Vocabulary *benz-* + *-caine*] (1922)
: a white crystalline ester $C_9H_{11}NO_2$ used as a local anesthetic

ben•zo•di•az•e•pine \ˌben-zō-dī-ˈa-zə-ˌpēn\ *noun* [*benz-* + *di-* + *az-* + *-epine* (from *hepta-* + [2]*-ine*)] (1934)
: any of a group of aromatic lipophilic amines (as diazepam and chlordiazepoxide) used especially as tranquilizers

ben•zo•fu•ran \ˌben-zō-ˈfyu̇r-ˌan, -fyu̇-ˈran\ *noun* (1946)
: a compound C_8H_6O found in coal tar and polymerized with indene to form thermoplastic resins used especially in adhesives and printing inks

ben•zo•ic acid \ben-ˈzō-ik-\ *noun* [International Scientific Vocabulary, from *benzoin*] (1791)
: a white crystalline acid $C_7H_6O_2$ found naturally (as in benzoin or in cranberries) or made synthetically and used especially as a preservative of foods, in medicine, and in organic synthesis

ben•zo•in \ˈben-zə-wən, -ˌwēn; -ˌzȯin\ *noun* [Middle French *benjoin,* from Catalan *benjuí,* from Arabic *lubān jāwi,* literally, frankincense of Java] (1562)
1 : a hard fragrant yellowish balsamic resin from trees (genus *Styrax*) of southeastern Asia used especially in medication, as a fixative in perfumes, and as incense
2 : a white crystalline hydroxy ketone $C_{14}H_{12}O_2$ made from benzaldehyde

ben•zol \ˈben-ˌzȯl, -ˌzōl\ *noun* [German, from *benz-* + *-ol*] (1838)
: BENZENE; *also* **:** a mixture of benzene and other aromatic hydrocarbons

ben•zo•phe•none \ˌben-zō-fi-ˈnōn, -ˈfē-ˌnōn\ *noun* [International Scientific Vocabulary *benz-* + *phen-* + *-one*] (1885)
: a colorless crystalline ketone $C_{13}H_{10}O$ used chiefly in perfumery and sunscreens; *also* **:** a derivative of benzophenone

ben•zo•yl \ˈben-zə-ˌwil, -ˌzȯil\ *noun* [German, from *Benzoësäure* benzoic acid + Greek *hylē* matter, literally, wood] (circa 1855)
: the radical C_6H_5CO of benzoic acid

benzoyl peroxide *noun* (1924)
: a white crystalline flammable compound $C_{14}H_{10}O_4$ used in bleaching and in medicine especially in the treatment of acne

ben•zyl \ˈben-ˌzēl, -zəl\ *noun* [International Scientific Vocabulary *benz-* + *-yl*] (1869)
: a univalent radical $C_6H_5CH_2$ derived from toluene
— ben•zyl•ic \ben-ˈzi-lik\ *adjective*

Be•o•wulf \ˈbā-ə-ˌwu̇lf\ *noun*
: a legendary Geatish warrior and hero of the Old English poem *Beowulf*

be•paint \bi-ˈpānt\ *transitive verb* (circa 1555)
archaic **:** TINGE

be•queath \bi-ˈkwēth, -ˈkwēth\ *transitive verb* [Middle English *bequethen,* from Old English *becwethan,* from *be-* + *cwethan* to say — more at QUOTH] (before 12th century)
1 : to give or leave by will — used especially of personal property
2 : to hand down **:** TRANSMIT
— be•queath•al \-ˈkwē-thəl, -thəl\ *noun*

be•quest \bi-ˈkwest\ *noun* [Middle English, irregular from *bequethen*] (14th century)
1 : the act of bequeathing
2 : something bequeathed **:** LEGACY

be•rate \bi-ˈrāt\ *transitive verb* (1548)
: to scold or condemn vehemently and at length
synonym see SCOLD

Ber•ber \ˈbər-bər\ *noun* [Arabic *Barbar*] (1732)
1 : a member of any of various peoples living in northern Africa west of Tripoli
2 a : a branch of the Afro-Asiatic language family comprising languages spoken by various peoples of northern Africa and the Sahara (as the Tuaregs and the Kabyles) **b :** any one of the Berber languages

ber•ber•ine \ˈbər-bə-ˌrēn\ *noun* [German *Berberin,* from New Latin *berberis*] (circa 1847)
: a bitter crystalline yellow alkaloid $C_{20}H_{19}NO_5$ obtained from the roots of various plants (as barberry) and used as a tonic in medicine

ber•ber•is \ˈbər-bər-əs\ *noun* [New Latin, the genus including barberry, alteration of Medieval Latin *barberis* barberry, from Arabic *barbāris*] (circa 1868)
: BARBERRY

ber•ceuse \ber-ˈsœz, -ˈsüz\ *noun, plural* **berceuses** \-ˈsœz; -ˈsüz, -ˈsü-zəz\ [French] (1876)
1 : LULLABY
2 : a musical composition usually in 6/8 time that resembles a lullaby

be•reave \bi-ˈrēv\ *transitive verb* **-reaved** *or* **-reft** \-ˈreft\; **-reav•ing** [Middle English *bereven,* from Old English *berēafian,* from *be-* + *rēafian* to rob — more at REAVE] (before 12th century)
1 : to deprive of something — usually used with *of* ⟨madam, you have *bereft* me of all words —Shakespeare⟩
2 : to take away (a valued or necessary possession) especially by force

[1]be•reaved \bi-ˈrēvd\ *adjective* (1828)
: suffering the death of a loved one

[2]bereaved *noun, plural* **bereaved** (1943)
: one who is bereaved

be•reave•ment \bi-ˈrēv-mənt\ *noun* (circa 1731)
: the state or fact of being bereaved; *especially* **:** the loss of a loved one by death

be•reft \-ˈreft\ *adjective* (1586)
1 a : deprived or robbed of the possession or use of something — usually used with *of* ⟨both players are instantly *bereft* of their poise —A. E. Wier⟩ **b :** lacking something needed, wanted, or expected — used with *of* ⟨the book is . . . completely *bereft* of an index —*Times Literary Supplement*⟩
2 : BEREAVED ⟨a *bereft* mother⟩

be•ret \bə-ˈrā\ *noun* [French *béret,* from Gascon *berret,* from Old Provençal, cap — more at BIRETTA] (1827)
: a visorless usually woolen cap with a tight headband and a soft full flat top

berg \ˈbərg\ *noun* (1823)
: ICEBERG

ber•ga•mot \ˈbər-gə-ˌmät\ *noun* [French *bergamote,* from Italian *bergamotta,* modification of Turkish *bey armudu,* literally, the bey's pear] (1696)
1 : a pear-shaped orange (*Citrus aurantium bergamia*) having a rind which yields an essential oil used in perfumery; *also* **:** this oil
2 : any of several mints (genus *Monarda*) — compare WILD BERGAMOT

be•rib•boned \bi-ˈri-bənd\ *adjective* (1863)
: adorned with ribbons

beri•beri \ˌber-ē-ˈber-ē\ *noun* [Sinhalese *bæribæri*] (1703)
: a deficiency disease marked by inflammatory or degenerative changes of the nerves, digestive system, and heart and caused by a lack of or inability to assimilate thiamine

Berke•le•ian *or* **Berke•ley•an** \ˈbär-klē-ən, ˈbər-; bär-ˈ, ˌbər-ˈ\ *adjective* (1842)
: of, relating to, or suggestive of Bishop Berkeley or his system of philosophical idealism
— Berkeleian *noun*
— Berke•le•ian•ism \-ə-ˌni-zəm\ *noun*

berke•li•um \ˈbər-klē-əm\ *noun* [New Latin, from *Berkeley,* Calif.] (1950)
: a radioactive metallic element produced by bombarding americium 241 with helium ions — see ELEMENT table

Berk•shire \ˈbərk-ˌshir, -shər\ *noun* [*Berkshire,* England] (1831)
: any of a breed of medium-sized black swine with white markings

berm \ˈbərm\ *noun* [French *berme,* from Dutch *berm* strip of ground along a dike; akin to Middle English *brimme* brim] (1729)
: a narrow shelf, path, or ledge typically at the top or bottom of a slope; *also* **:** a mound or wall of earth ⟨a landscaped *berm*⟩

Ber•mu•da bag \(ˌ)bər-ˈmyü-də-, *especially Southern* -ˈmü-də-\ *noun* [*Bermuda* islands, North Atlantic] (1979)
: a round or oval-shaped handbag with a wooden handle and removable cloth covers

Bermuda grass *noun* (1808)

: a creeping stoloniferous southern European grass (*Cynodon dactylon*) often used as a lawn and pasture grass

Bermuda rig *noun* (1853)
: a fore-and-aft rig marked by a triangular sail and a mast with an extreme rake

Ber·mu·das \(ˌ)bər-ˈmyü-dəz, *especially Southern* -ˈmü-dəz\ *noun plural* (1961)
: BERMUDA SHORTS

Bermuda shorts *noun plural* (1951)
: knee-length walking shorts

Ber·nese mountain dog \ˈbər-ˌnēz-, -ˌnēs-\ *noun* [*Bern,* Switzerland] (1935)
: any of a breed of large powerful long-coated black dogs of Swiss origin that have tan and white markings and were developed as draft animals

Ber·noul·li's principle \bər-ˈnü-lēz-, ˌber-ˌnü-ˈlēz-\ *noun* [Daniel *Bernoulli* (died 1782) Swiss physicist] (1940)
: a principle in hydrodynamics: the pressure in a stream of fluid is reduced as the speed of the flow is increased

Ber·noul·li trial \bər-ˈnü-lē-, ˌber-ˌnü-ˈlē-\ *noun* [Jacques *Bernoulli* (died 1705) Swiss mathematician] (1951)
: one of the repetitions of a statistical experiment having two mutually exclusive outcomes with constant probability of occurrence

ber·ried \ˈber-ēd\ *adjective* (1794)
1 : having or covered with berries ⟨*berried* shrub⟩
2 : bearing eggs ⟨a *berried* lobster⟩

¹ber·ry \ˈber-ē\ *noun, plural* **berries** [Middle English *berye,* from Old English *berie;* akin to Old High German *beri* berry] (before 12th century)
1 a : a pulpy and usually edible fruit (as a strawberry, raspberry, or checkerberry) of small size irrespective of its structure **b :** a simple fruit (as a currant, grape, tomato, or banana) with a pulpy or fleshy pericarp **c :** the dry seed of some plants (as wheat)
2 : an egg of a fish or lobster

²berry *intransitive verb* **ber·ried; ber·ry·ing** (circa 1780)
1 : to bear or produce berries ⟨a *berrying* shrub⟩
2 : to gather or seek berries

ber·ry·like \ˈber-ē-ˌlīk\ *adjective* (1864)
1 : resembling a berry especially in size or structure
2 : being small and rounded **:** COCCOID

ber·seem \(ˌ)bər-ˈsēm\ *noun* [Arabic *barsīm,* from Coptic *bersīm*] (circa 1902)
: a succulent clover (*Trifolium alexandrinum*) cultivated as a forage plant and green-manure crop especially in the alkaline soils of the Nile valley and in the southwestern U.S. — called also *Egyptian clover*

¹ber·serk \bə(r)-ˈsərk, ˌbər-, -ˈzərk; ˈbər-ˌ\ *or* **ber·serk·er** \-ˈsər-kər, -ˈzər-; ˈbər-ˌ\ *noun* [Old Norse *berserkr,* from *bjǫrn* bear + *serkr* shirt] (1818)
1 : an ancient Scandinavian warrior frenzied in battle and held to be invulnerable
2 : one whose actions are recklessly defiant ◆

²berserk *adjective* (1851)
: FRENZIED, CRAZED — usually used in the phrase *go berserk* ⟨sinister ravings of an imagination gone *berserk* —John Gruen⟩
— berserk *adverb*
— ber·serk·ly *adverb*

¹berth \ˈbərth\ *noun* [probably from ²*bear* + *-th*] (15th century)
1 a : sufficient distance for maneuvering a ship **b :** safe distance — used especially with *wide*
2 a : the place where a ship lies when at anchor or at a wharf **b :** a space for an automotive vehicle at rest ⟨a truck-loading *berth*⟩
3 : a place to sit or sleep especially on a ship or vehicle **:** ACCOMMODATION
4 a : a billet on a ship **b :** JOB, POSITION, PLACE ⟨a starting *berth* on the team⟩

²berth (1667)
transitive verb
1 : to bring into a berth
2 : to allot a berth to
intransitive verb
: to come into a berth

ber·tha \ˈbər-thə\ *noun* [French *berthe,* from *Berthe* (Bertha) (died 783) queen of the Franks] (1842)
: a wide round collar covering the shoulders

Ber·til·lon system \ˈbər-tᵊl-ˌän-, ˈber-tē-ˌyōⁿ-\ *noun* [Alphonse *Bertillon* (died 1914) French criminologist] (1896)
: a system of identification of persons by a description based on anthropometric measurements, photographs, and notation of data (as markings, color, and thumb line impressions)

ber·yl \ˈber-əl\ *noun* [Middle English, from Middle French *beril,* from Latin *beryllus,* from Greek *bēryllos,* back-formation from *bēryllion* beryl, of Indo-Aryan origin; akin to Prakrit *verulia, veluriya* beryl] (13th century)
: a mineral consisting of a silicate of beryllium and aluminum of great hardness and occurring in green, bluish green, yellow, pink, or white hexagonal prisms

be·ryl·li·um \bə-ˈri-lē-əm\ *noun* [New Latin, from Greek *bēryllion*] (circa 1847)
: a steel-gray light strong brittle toxic bivalent metallic element used chiefly as a hardening agent in alloys — see ELEMENT table

be·seech \bi-ˈsēch\ *verb* **-seeched** *or* **-sought** \-ˈsȯt\; **-seech·ing** [Middle English *besechen,* from *be-* + *sechen* to seek] (12th century)
transitive verb
1 : to beg for urgently or anxiously
2 : to request earnestly **:** IMPLORE
intransitive verb
: to make supplication
synonym see BEG
— be·seech·ing·ly \-ˈsē-chiŋ-lē\ *adverb*

be·seem \bi-ˈsēm\ (13th century)
intransitive verb
archaic **:** to be fitting or becoming
transitive verb
archaic **:** to be suitable to **:** BEFIT

be·set \bi-ˈset, bē-\ *transitive verb* **-set; -set·ting** [Middle English *besetten,* from Old English *besettan,* from *be-* + *settan* to set] (before 12th century)
1 : to set or stud with or as if with ornaments
2 : TROUBLE, HARASS ⟨inflation *besets* the economy⟩
3 a : to set upon **:** ASSAIL ⟨the settlers were *beset* by savages⟩ **b :** to hem in **:** SURROUND
— be·set·ment \-mənt\ *noun*

be·set·ting *adjective* (1795)
: constantly present or attacking **:** OBSESSIVE

be·shrew \bi-ˈshrü, bē-, *especially Southern* -ˈsrü\ *transitive verb* (14th century)
archaic **:** CURSE

¹be·side \bi-ˈsīd\ *preposition* [Middle English, adverb & preposition, from Old English *be sīdan* at or to the side, from *be* at (from *bī*) + *sīdan,* dative & accusative of *sīde* side — more at BY] (13th century)
1 a : by the side of ⟨walk *beside* me⟩ **b :** in comparison with **c :** on a par with
2 : BESIDES
3 : not relevant to ⟨*beside* the point⟩
— beside oneself : in a state of extreme excitement

²beside *adverb* (14th century)
1 *archaic* **:** NEARBY
2 *archaic* **:** BESIDES

¹be·sides \bi-ˈsīdz\ *preposition* (14th century)
1 : OTHER THAN, EXCEPT
2 : together with

²besides *adverb* (1564)
1 : as well **:** ALSO
2 : MOREOVER, FURTHERMORE

³besides *adjective* (1954)
: ELSE

be·siege \bi-ˈsēj\ *transitive verb* **-sieged; -sieg·ing** (14th century)
1 : to surround with armed forces
2 a : to press with requests **:** IMPORTUNE **b :** to cause worry or distress to **:** BESET ⟨doubts *besieged* him⟩
— be·sieg·er *noun*

be·smear \bi-ˈsmir\ *transitive verb* (before 12th century)
: SMEAR

be·smirch \bi-ˈsmərch\ *transitive verb* (1599)
: SULLY, SOIL

be·som \ˈbē-zəm\ *noun* [Middle English *beseme,* from Old English *besma;* akin to Old High German *besmo* broom] (before 12th century)
: BROOM 2; *especially* **:** one made of twigs

besom pocket *noun* [origin unknown] (1966)
: a pocket with a welted slit opening

be·sot \bi-ˈsät\ *transitive verb* **be·sot·ted; be·sot·ting** [*be-* + *sot* (to stultify)] (1581)
1 : INFATUATE 2
2 : to make dull or stupid; *especially* **:** to muddle with drunkenness

be·spat·ter \bi-ˈspa-tər\ *transitive verb* (1640)
: SPATTER

be·speak \bi-ˈspēk\ *transitive verb* **-spoke** \-ˈspōk\; **-spo·ken** \-ˈspō-kən\; **-speak·ing** (1536)
1 : to hire, engage, or claim beforehand
2 : to speak to especially with formality **:** ADDRESS
3 : REQUEST ⟨*bespeak* a favor⟩
4 a : INDICATE, SIGNIFY ⟨her performance *bespeaks* considerable practice⟩ **b :** to show beforehand **:** FORETELL

be·spec·ta·cled \bi-ˈspek-(ˌ)ti-kəld\ *adjective* (1742)
: wearing spectacles

be·spoke \bi-ˈspōk\ *or* **be·spo·ken** \-ˈspō-kən\ *adjective* [past participle of *bespeak*] (1607)
1 a : CUSTOM-MADE **b :** dealing in or producing custom-made articles
2 *dialect* **:** ENGAGED

be·sprent \bi-ˈsprent\ *adjective* [Middle English *bespreynt,* from past participle of *besprengen* to besprinkle, from Old English *besprengan,* from *be-* + *sprengan* to scatter; akin to Old English *springan* to spring] (14th century)
archaic **:** sprinkled over

◇ WORD HISTORY

berserk During the 19th century, the word *berserk* made a rapid linguistic trip from literary use as a noun—arising out of antiquarian interest in medieval Scandinavia—to colloquial use as an adjective. In Old Norse sagas, a *berserkr* was a man experiencing a sort of ecstatic frenzy called *berserksgangr.* The 13th century Icelandic author Snorri Sturluson describes men seized by *berserksgangr* as going into battle without coats of mail and gnawing the rims of their shields. They supposedly had the strength of bears or oxen, and were invulnerable to iron or fire. The word *berserkr* is a compound of *ber-,* presumed to mean "bear," and *sarkr* "shirt." Originally the "bear-shirt" probably referred to a man dressed in a mask or garment representing a bear who entered a trancelike state in which he magically assumed powers associated with the animal. By the time the legend of the *berserkr* had made its way into Norse saga, it was this ecstatic frenzy—not bear paraphernalia—that characterized him and that is still dimly reflected in English *berserk.*

be·sprin·kle \bi-ˈspriŋ-kəl\ *transitive verb* [Middle English *besprengeln,* frequentative of *besprengen*] (15th century)
: SPRINKLE

Bes·sel function \ˈbe-səl-\ *noun* [Friedrich W. *Bessel* (died 1846) Prussian astronomer] (1872)
: one of a class of transcendental functions expressible as infinite series and occurring in the solution of the differential equation

$$x^2\frac{d^2y}{dx^2} + x\frac{dy}{dx} = (n^2 - x^2)y$$

Bes·se·mer converter \ˈbe-sə-mər-\ *noun* (circa 1887)
: the furnace used in the Bessemer process

Bessemer process *noun* [Sir Henry *Bessemer*] (1875)
: a process of making steel from pig iron by burning out carbon and other impurities by means of a blast of air forced through the molten metal

¹best \ˈbest\ *adjective, superlative of* GOOD [Middle English, from Old English *betst;* akin to Old English *bōt* remedy — more at BETTER] (before 12th century)
1 : excelling all others ⟨the *best* student⟩
2 : most productive of good or of advantage, utility, or satisfaction ⟨what is the *best* thing to do⟩
3 : MOST, LARGEST ⟨it rained for the *best* part of their vacation⟩

²best *adverb, superlative of* WELL (before 12th century)
1 : in the best way **:** to greatest advantage ⟨some things are *best* left unsaid⟩
2 : MOST ⟨those *best* able will provide needed support⟩

³best *noun, plural* **best** (before 12th century)
1 : the best state or part
2 : one that is best ⟨the *best* falls short⟩
3 : the greatest degree of good or excellence
4 : one's maximum effort ⟨do your *best*⟩
5 : best clothes ⟨Sunday *best*⟩
— at best : under the most favorable circumstances

⁴best *transitive verb* (1863)
: to get the better of **:** OUTDO

⁵best *verbal auxiliary* (1941)
: had best ⟨you *best* listen⟩

best–ball \ˈbes(t)-ˈbȯl\ *adjective* (1909)
: relating to or being a golf match in which one player competes against the best individual score of two or more players for each hole — compare FOUR-BALL

best boy *noun* (1937)
: the chief assistant to the gaffer in motion-picture or television production

best–case \ˈbes(t)-ˈkās\ *adjective* (1977)
: being, relating to, or based on a projection of future events that assumes only the best possible circumstances ⟨a *best-case* scenario⟩

¹be·stead *or* **be·sted** \bi-ˈsted\ *adjective* [Middle English *bested,* from *be-* + *sted,* past participle of *steden* to place, from *stede* place — more at STEAD] (14th century)
archaic **:** SITUATED

²bestead *transitive verb* **be·stead·ed; be·stead; be·stead·ing** [*be-* + *stead*] (1581)
1 *archaic* **:** HELP
2 *archaic* **:** to be useful to **:** AVAIL

bes·tial \ˈbes-chəl, ˈbesh-, ˈbēs-, ˈbēsh-\ *adjective* [Middle English, from Middle French, from Latin *bestialis,* from *bestia* beast] (14th century)
1 a : of or relating to beasts **b :** resembling a beast
2 a : lacking intelligence or reason **b :** marked by base or inhuman instincts or desires **:** BRUTAL
synonym see BRUTAL
— bes·tial·ize \-chə-ˌlīz\ *transitive verb*
— bes·tial·ly \-chə-lē\ *adverb*

bes·ti·al·i·ty \ˌbes-chē-ˈa-lə-tē, ˌbesh-, ˌbēs-, ˌbēsh-\ *noun, plural* **-ties** (14th century)
1 : the condition or status of a lower animal
2 : display or gratification of bestial traits or impulses
3 : sexual relations between a human being and a lower animal

bes·ti·ary \ˈbes-chē-ˌer-ē, ˈbesh-, ˈbēs-, ˈbēsh-\ *noun, plural* **-ar·ies** [Medieval Latin *bestiarium,* from Latin, neuter of *bestiarius* of beasts, from *bestia*] (1840)
1 : a medieval allegorical or moralizing work on the appearance and habits of real or imaginary animals
2 : a collection of descriptions of real or imaginary animals

be·stir \bi-ˈstər\ *transitive verb* (14th century)
: to rouse to action **:** get going

best man *noun* (circa 1782)
: the principal groomsman at a wedding

be·stow \bi-ˈstō\ *transitive verb* [Middle English, from *be-* + *stowe* place — more at STOW] (14th century)
1 : to put to use **:** APPLY ⟨*bestowed* his spare time on study⟩
2 : to put in a particular or appropriate place **:** STOW
3 : to provide with quarters **:** PUT UP
4 : to convey as a gift — usually used with *on* or *upon*
synonym see GIVE
— be·stow·al \-ˈstō-əl\ *noun*

be·strew \bi-ˈstrü\ *transitive verb* **-strewed; -strewed** *or* **-strewn** \-ˈstrün\; **-strew·ing** (before 12th century)
1 : STREW
2 : to lie scattered over

be·stride \bi-ˈstrīd\ *transitive verb* **-strode** \-ˈstrōd\; **-strid·den** \-ˈstri-dᵊn\; **-strid·ing** \-ˈstrī-diŋ\ (before 12th century)
1 : to ride, sit, or stand astride **:** STRADDLE
2 : to tower over **:** DOMINATE ⟨the bloated bureaucracy that *bestrides* us all —Edward Ney⟩
3 *archaic* **:** to stride across

best–sell·er \ˈbes(t)-ˈse-lər\ *noun* (1889)
: an article (as a book) whose sales are among the highest of its class
— best·sell·er·dom \-dəm\ *noun*
— best–sell·ing \-ˈse-liŋ\ *adjective*

¹bet \ˈbet\ *noun* [origin unknown] (1592)
1 a : something that is laid, staked, or pledged typically between two parties on the outcome of a contest or a contingent issue **:** WAGER **b :** the act of giving such a pledge
2 : something to wager on
3 : a choice made by consideration of probabilities ⟨your best *bet* is the back road⟩

²bet *verb* **bet** *also* **bet·ted; bet·ting** (1597)
transitive verb
1 a : to stake on the outcome of an issue or the performance of a contestant **b :** to be able to be sure that — usually used in the expression *you bet* ⟨you *bet* I'll be there⟩
2 a : to maintain with or as if with a bet **b :** to make a bet with **c :** to make a bet on
intransitive verb
: to lay a bet

¹be·ta \ˈbā-tə, *chiefly British* ˈbē-\ *noun* [Middle English *betha,* from Latin *beta,* from Greek *bēta,* of Semitic origin; akin to Hebrew *bēth* beth] (14th century)
1 : the 2d letter of the Greek alphabet — see ALPHABET table
2 : BETA PARTICLE
3 : a measure of a stock's or a portfolio's volatility that is expressed numerically as deviation from the market's volatility taken as unity

²beta *adjective* (1899)
: second in position in the structure of an organic molecule from a particular group or atom ⟨*beta* substitution⟩ — often used in combination; symbol β

be·ta–ad·ren·er·gic \-ˌa-drə-ˈnər-jik\ *adjective* (1965)
: of, relating to, or being a beta-receptor ⟨*beta-adrenergic* blocking action⟩

be·ta–block·er \-ˌblä-kər\ *noun* (1968)
: any of a class of drugs (as propranolol) that slow heart action and increase coronary blood flow by blocking the activity of beta-receptors
— be·ta–block·ing \-kiŋ\ *adjective*

be·ta–car·o·tene \-ˈkar-ə-ˌtēn\ *noun* (1938)
: an isomer of carotene found in dark green and dark yellow vegetables and fruits

beta cell *noun* (1926)
: any of the insulin-secreting pancreatic cells in the islets of Langerhans

beta decay *noun* (1934)
: a radioactive nuclear transformation governed by the weak force in which a nucleon (as a neutron) changes into the other type of nucleon (as a proton) with the emission of an electron or positron and a neutrino and which results in a change in the atomic number of the nucleus of +1 or −1

be·ta–en·dor·phin \-en-ˈdȯr-fən\ *noun* (1977)
: an endorphin of the pituitary gland having a much greater analgesic potency than morphine

beta globulin *noun* [International Scientific Vocabulary] (1945)
: any of several globulins of plasma or serum that have at alkaline pH electrophoretic mobilities intermediate between those of the alpha globulins and gamma globulins

be·ta·ine \ˈbē-tə-ˌēn\ *noun* [International Scientific Vocabulary, from Latin *beta* beet] (1879)
: a sweet crystalline quaternary ammonium salt $C_5H_{11}NO_2$ occurring especially in beet juice; *also* **:** its hydrate $C_5H_{13}NO_3$ or its hydrochloride $C_5H_{12}NO_2Cl$

be·take \bi-ˈtāk\ *transitive verb* **-took** \-ˈtu̇k\; **-tak·en** \-ˈtā-kən\; **-tak·ing** (14th century)
1 *archaic* **:** COMMIT
2 : to cause (oneself) to go

be·ta–lac·ta·mase \-ˈlak-tə-ˌmās, -ˌmāz\ *noun* [*lactam,* a cyclic amide (from International Scientific Vocabulary *lact-* + *amide*) + *-ase*] (1965)
: PENICILLINASE

be·ta–ox·i·da·tion \ˈbā-tə-ˌäk-sə-ˈdā-shən\ *noun* (circa 1935)
: stepwise catabolism of fatty acids in which two-carbon fragments are successively removed from the carboxyl end of the chain

beta particle *noun* (1904)
: a high-speed electron; *specifically* **:** one emitted by a radioactive nucleus in beta decay

beta ray *noun* (1902)
1 : BETA PARTICLE
2 : a stream of beta particles — called also *beta radiation*

be·ta–re·cep·tor \ˈbā-tə-ri-ˈsep-tər\ *noun* (1948)
: any of a group of receptors postulated to exist on nerve cell membranes of the sympathetic nervous system to explain the specificity of certain adrenergic agents in affecting only some sympathetic activities (as vasodilation, increase in muscular contraction and beat of the heart, and relaxation of smooth muscle in the bronchi and intestine)

be·ta–thal·as·se·mia \-ˌtha-lə-ˈsē-mē-ə\ *noun* (1962)
: thalassemia in which the longer hemoglobin chain is affected and which comprises Cooley's anemia in the homozygous condition and a less severe thalassemia in the heterozygous condition

be·ta·tron \ˈbā-tə-ˌträn\ *noun* [International Scientific Vocabulary] (1941)
: an accelerator in which electrons are propelled by the inductive action of a rapidly varying magnetic field

beta wave *noun* (1936)
: an electrical rhythm of the brain with a frequency of 13 to 30 cycles per second that is associated with normal conscious waking experience — called also *beta, beta rhythm*

be·tel \ˈbē-tᵊl\ *noun* [Portuguese *bétele,* from Tamil *verṟilai*] (1553)

: a climbing pepper (*Piper betle*) whose leaves are chewed together with betel nut and mineral lime as a stimulant masticatory especially by southeastern Asians

Be·tel·geuse \'bē-tᵊl-ˌjüs, 'be-, -ˌjüz\ *noun* [French *Bételgeuse,* from Arabic *bayt al-jawzā'* Gemini, literally, the house of the twins (confused with Orion & Betelgeuse)]
: a variable red giant star of the first magnitude near one shoulder of Orion

betel nut *noun* [from its being chewed with betel leaves] (1681)
: the astringent seed of the betel palm

betel palm *noun* (1875)
: an Asian pinnate-leaved palm (*Areca catechu*) that has an orange-colored drupe with an outer fibrous husk

bête noire \ˌbet-'nwär, ˌbāt-\ *noun, plural* **bêtes noires** \ˌbet-'nwär(z), ˌbāt-\ [French, literally, black beast] (1844)
: a person or thing strongly detested or avoided : BUGBEAR

beth \'bāth, 'bāt, 'bās\ *noun* [Hebrew *bēth,* from *bayith* house] (circa 1823)
: the 2d letter of the Hebrew alphabet — see ALPHABET table

beth·el \'be-thəl\ *noun* [Hebrew *bēth'ēl* house of God] (circa 1617)
1 : a hallowed spot
2 a : a chapel for Nonconformists **b :** a place of worship for seamen

be·think \bi-'thiŋk\ *transitive verb* **-thought** \-'thȯt\; **-think·ing** (before 12th century)
1 a : REMEMBER, RECALL **b :** to cause (oneself) to be reminded
2 : to cause (oneself) to consider

be·tide \bi-'tīd\ (12th century)
intransitive verb
: to happen especially as if by fate
transitive verb
: to happen to : BEFALL — used chiefly in the phrase *woe betide* ⟨woe *betide* our enemies⟩

be·times \bi-'tīmz\ *adverb* (13th century)
1 : in good time : EARLY
2 *archaic* **:** in a short time : SPEEDILY
3 : at times : OCCASIONALLY

bê·tise \ba-'tēz\ *noun, plural* **bê·tises** \-'tēz\ [French] (1827)
1 : an act of foolishness or stupidity
2 : lack of good sense : STUPIDITY

be·to·ken \bi-'tō-kən\ *transitive verb* **-to·kened; -to·ken·ing** \-'tōk-niŋ, -'tō-kə-\ (15th century)
1 : to typify beforehand : PRESAGE
2 : to give evidence of : SHOW

be·tray \bi-'trā\ *verb* [Middle English, from *be-* + *trayen* to betray, from Old French *trair,* from Latin *tradere* — more at TRAITOR] (13th century)
transitive verb
1 : to lead astray; *especially* **:** SEDUCE
2 : to deliver to an enemy by treachery
3 : to fail or desert especially in time of need
4 a : to reveal unintentionally **b :** SHOW, INDICATE **c :** to disclose in violation of confidence
intransitive verb
: to prove false
synonym see REVEAL
— **be·tray·al** \-'trā(-ə)l\ *noun*
— **be·tray·er** \-'trā-ər\ *noun*

be·troth \bi-'trōth, -'trȯth\ *transitive verb* [Middle English, from *be-* + *trouthe* truth, troth] (14th century)
1 : to promise to marry
2 : to give in marriage

be·troth·al \-'trō-thəl, -'trȯ-, -thəl\ *noun* (1844)
1 : the act of betrothing or fact of being betrothed
2 : a mutual promise or contract for a future marriage

be·trothed \bi-'trōthd, -'trȯtht\ *noun* (1588)
: the person to whom one is betrothed

bet·ta \'be-tə\ *noun* [New Latin] (1927)
: any of a genus (*Betta*) of small brilliantly colored long-finned freshwater fishes of southeastern Asia; *especially* **:** SIAMESE FIGHTING FISH

¹bet·ter \'be-tər\ *adjective, comparative of* GOOD [Middle English *bettre,* from Old English *betera;* akin to Old English *bōt* remedy, Sanskrit *bhadra* fortunate] (before 12th century)
1 : greater than half
2 : improved in health or mental attitude
3 : more attractive, favorable, or commendable
4 : more advantageous or effective
5 : improved in accuracy or performance

²better (before 12th century)
transitive verb
1 : to make better: as **a :** to make more tolerable or acceptable ⟨trying to *better* the lot of slum dwellers⟩ **b :** to make more complete or perfect ⟨looked forward to *bettering* her acquaintance with the new neighbors⟩
2 : to surpass in excellence : EXCEL
intransitive verb
: to become better
synonym see IMPROVE

³better *adverb, comparative of* WELL (12th century)
1 a : in a more excellent manner **b :** to greater advantage : PREFERABLY ⟨some things are *better* left unsaid⟩
2 a : to a higher or greater degree ⟨he knows the story *better* than you do⟩ **b :** MORE ⟨it is *better* than nine miles to the next town⟩

⁴better *noun* (12th century)
1 a : something better **b :** a superior especially in merit or rank
2 : ADVANTAGE, VICTORY ⟨get the *better* of him⟩

⁵better *verbal auxiliary* (1831)
: had better ⟨you *better* hurry⟩

bet·ter·ment \'be-tər-mənt\ *noun* (1598)
1 : a making or becoming better
2 : an improvement that adds to the value of a property or facility

better–off \ˌbe-tə-'rȯf\ *adjective* (circa 1859)
1 : being in comfortable economic circumstances ⟨the *better-off* people live in the older section of town⟩
2 : being in a more advantageous position

betting shop *noun* (1852)
British **:** a shop where bets are taken

bet·tor *or* **bet·ter** \'be-tər\ *noun* (1609)
: one that bets

¹be·tween \bi-'twēn\ *preposition* [Middle English *betwene,* preposition & adverb, from Old English *betwēonum,* from *be-* + *-twēonum* (dative plural) (akin to Gothic *tweihnai* two each); akin to Old English *twā* two] (before 12th century)
1 a : by the common action of : jointly engaging ⟨shared the work *between* the two of them⟩ ⟨talks *between* the three —*Time*⟩ **b :** in common to : shared by ⟨divided *between* his four grandchildren⟩
2 a : in the time, space, or interval that separates **b :** in intermediate relation to
3 a : from one to another of ⟨air service *between* Miami and Chicago⟩ **b :** serving to connect or unite in a relationship (as difference, likeness, or proportion) ⟨a one to one correspondence *between* sets⟩ **c :** setting apart ⟨the line *between* fact and fancy⟩
4 : in point of comparison of ⟨not much to choose *between* the two coats⟩
5 : in confidence restricted to ⟨a secret *between* you and me⟩ ■

²between *adverb* (before 12th century)
: in an intermediate space or interval

be·tween·brain \-ˌbrān\ *noun* (circa 1909)
: DIENCEPHALON

be·tween·ness \bi-'twēn-nəs\ *noun* (1892)
: the quality or state of being between two others in an ordered mathematical set

be·tween·times \bi-'twēn-ˌtīmz\ *adverb* (1907)
: at or during intervals

be·tween·whiles \-ˌhwīlz, -ˌwīlz\ *adverb* (1678)
: BETWEENTIMES

be·twixt \bi-'twikst\ *adverb or preposition* [Middle English, from Old English *betwux,* from *be-* + *-twux* (akin to Gothic *tweihnai*)] (before 12th century)
: BETWEEN

betwixt and between *adverb or adjective* (1832)

□ USAGE

between There is a persistent but unfounded notion that *between* can be used only of two items and that *among* must be used for more than two. *Between* has been used of more than two since Old English; it is especially appropriate to denote a one-to-one relationship, regardless of the number of items. It can be used when the number is unspecified ⟨economic cooperation *between* nations⟩, when more than two are enumerated ⟨*between* you and me and the lamppost⟩ ⟨partitioned *between* Austria, Prussia, and Russia —Nathaniel Benchley⟩, and even when only one item is mentioned (but repetition is implied) ⟨pausing *between* every sentence to rap the floor —George Eliot⟩. *Among* is more appropriate where the emphasis is on distribution rather than individual relationships ⟨discontent *among* the peasants⟩. When *among* is automatically chosen for more than two, English idiom may be strained ⟨a worthy book that nevertheless falls *among* many stools —John Simon⟩ ⟨the author alternates *among* mod slang, clichés and quotes from literary giants —A. H. Johnston⟩.

Grammarians and other commentators have been hard put to explain the occurrence of *I* where *me* would be expected as the proper form after this preposition: *between you and I.* The most common explanation put forward is hypercorrection—the use of an improper form in one expression on analogy with a proper use of the same form in another expression. In this instance it is supposed that many people taught to say "It is I" instead of "It is me" become intimidated by *me* and substitute *I* whenever they have a chance. On the surface of it, this seems to be a far-fetched explanation, and another point against it is that the phrase can be found as early as the end of the 16th century, when there were no schoolteachers harping on "It is I" ⟨an argument too deepe to be discussed *between you and I* —Thomas Deloney (1597)⟩ ⟨all debts are cleared *between you and I* —Shakespeare, *The Merchant of Venice* (1600)⟩. The most plausible explanation is that offered by the linguist Noam Chomsky, who in a 1986 book postulated that compound phrases like *you and I* are barriers to the assignment of case by verb or preposition, which means that pronouns may appear as nominative or objective or may even be reflexive—in our phrase, *I, me,* or *myself.* Chomsky's theory has the advantage of also accounting for other anomalies of pronoun case in compound subjects and objects of verbs. It is to be observed that the expected case—nominative in subject position, objective in object position—predominates, and that the others, while common, are much more likely to be found in speech than in writing.

: in a midway position : neither one thing nor the other

Beu·lah \ˈbyü-lə\ *noun*
: an idyllic land near the end of life's journey in Bunyan's *Pilgrim's Progress*

beurre blanc \ˈbər-ˈbläⁿ\ *noun* [French, literally, white butter] (1931)
: a hot butter sauce (as for fish) flavored with vinegar or lemon juice

beurre ma·nié \-män-ˈyā\ *noun* [French, literally, handled butter] (1939)
: flour and butter kneaded together used as a thickener in sauces

beurre noir \-ˈnwär\ *noun* [French, literally, black butter] (1856)
: butter heated until brown or black and often flavored with vinegar or lemon juice

[1]bev·el \ˈbe-vəl\ *adjective* (circa 1600)
: OBLIQUE, BEVELED

[2]bevel *noun* [(assumed) Middle French, from Old French *baïf* with open mouth, from *baer* to yawn — more at ABEYANCE] (1610)

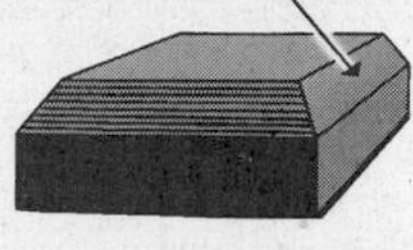
bevel 2

1 : an instrument consisting of two rules or arms jointed together and opening to any angle for drawing angles or adjusting surfaces to be cut at an angle
2 a : the angle that one surface or line makes with another when they are not at right angles **b :** the slant of such a surface or line
3 : the part of printing type extending from face to shoulder

[3]bevel *verb* **-eled** *or* **-elled; -el·ing** *or* **-el·ling** \ˈbev-liŋ, ˈbe-və-\ (1677)
transitive verb
: to cut or shape to a bevel
intransitive verb
: INCLINE, SLANT

bevel gear *noun* (1833)
: either of a pair of toothed wheels whose working surfaces are inclined to nonparallel axes

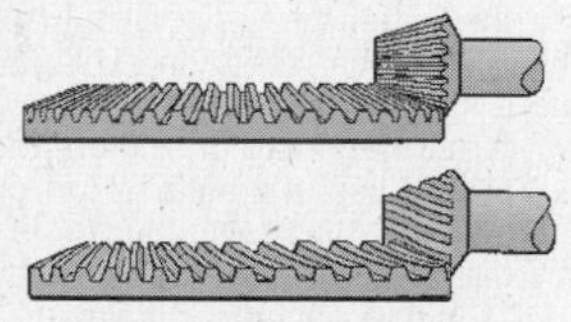
bevel gears

bev·er·age \ˈbev-rij, ˈbe-və-\ *noun* [Middle English, from Middle French *bevrage,* from *beivre* to drink, from Latin *bibere* — more at POTABLE] (14th century)
: a drinkable liquid

bevy \ˈbe-vē\ *noun, plural* **bev·ies** [Middle English *bevey*] (15th century)
1 : a large group or collection ⟨a *bevy* of girls⟩
2 : a group of animals and especially quail together

be·wail \bi-ˈwā(ə)l\ *transitive verb* (14th century)
1 : to wail over
2 : to express deep sorrow for usually by wailing and lamentation ⟨wringing her hands and *bewailing* her fate⟩
synonym see DEPLORE

be·ware \bi-ˈwar, -ˈwer\ *verb* [Middle English *been war,* from *been* to be + *war* careful — more at BE, WARE] (14th century)
intransitive verb
: to be on one's guard ⟨*beware* of the dog⟩
transitive verb
1 : to take care of ⟨*beware* your wallet⟩
2 : to be wary of ⟨we must . . . *beware* the exceedingly tenuous generalization —Matthew Lipman⟩

be·whis·kered \-ˈhwis-kərd, -ˈwis-\ *adjective* (1820)
: wearing whiskers

be·wigged \bi-ˈwigd\ *adjective* (1774)
: wearing a wig

be·wil·der \bi-ˈwil-dər\ *transitive verb* **-wil·dered; -wil·der·ing** \-d(ə-)riŋ\ (1684)
1 : to cause to lose one's bearings
2 : to perplex or confuse especially by a complexity, variety, or multitude of objects or considerations
synonym see PUZZLE
— be·wil·dered·ly *adverb*
— be·wil·dered·ness *noun*
— be·wil·der·ing·ly \-d(ə-)riŋ-lē\ *adverb*

be·wil·der·ment \-dər-mənt\ *noun* (1820)
1 : the quality or state of being bewildered
2 : a bewildering tangle or confusion

be·witch \bi-ˈwich\ (13th century)
transitive verb
1 a : to influence or affect especially injuriously by witchcraft **b :** to cast a spell over
2 : to attract as if by the power of witchcraft : ENCHANT ⟨*bewitched* by her beauty⟩
intransitive verb
: to bewitch someone or something
— be·witch·ery \-ˈwi-ch(ə-)rē\ *noun*
— be·witch·ing·ly \-ˈwi-chiŋ-lē\ *adverb*

be·witch·ment \-ˈwich-mənt\ *noun* (1607)
1 a : the act or power of bewitching **b :** a spell that bewitches
2 : the state of being bewitched

be·wray \bi-ˈrā\ *transitive verb* [Middle English, from *be-* + *wreyen* to accuse, from Old English *wrēgan;* akin to Old High German *ruogen* to accuse] (13th century)
archaic : DIVULGE, BETRAY

bey \ˈbā\ *noun* [Turkish, gentleman, chief] (1595)
1 a : a provincial governor in the Ottoman Empire **b :** the former native ruler of Tunis or Tunisia
2 — used as a courtesy title in Turkey and Egypt

[1]be·yond \bē-ˈänd\ *adverb* [Middle English, preposition & adverb, from Old English *begeondan,* from *be-* + *geondan* beyond, from *geond* yond — more at YOND] (before 12th century)
1 : on or to the farther side : FARTHER
2 : in addition : BESIDES

[2]beyond *preposition* (before 12th century)
1 : on or to the farther side of : at a greater distance than
2 a : out of the reach or sphere of **b :** in a degree or amount surpassing **c :** out of the comprehension of
3 : in addition to : BESIDES

[3]beyond *noun* (14th century)
1 : something that lies beyond
2 : something that lies outside the scope of ordinary experience; *specifically* : HEREAFTER

be·zant \ˈbe-zᵊnt, bə-ˈzant\ *noun* [Middle English *besant,* from Old French, from Medieval Latin *Byzantius* Byzantine, from *Byzantium,* ancient name of Istanbul] (13th century)
1 : SOLIDUS 1
2 : a flat disk used in architectural ornament

be·zel \ˈbe-zəl, ˈbe-\ *noun* [probably from dialect form of French *biseau* bezel, from Middle French] (circa 1616)
1 : a rim that holds a transparent covering (as on a watch, clock, or headlight) or that is rotatable and has special markings (as on a watch)
2 : the oblique side or face of a cut gem; *specifically* : the upper faceted portion of a brilliant projecting from the setting
3 : a usually metal rim of a piece of jewelry in which an ornament (as a gem) is set

be·zique \bə-ˈzēk\ *noun* [French *bésique*] (1861)
: a card game similar to pinochle that is played with a pack of 64 cards

be·zoar \ˈbē-ˌzōr, -ˌzȯr\ *noun* [Middle French, from Medieval Latin, from Arabic dialect *bezuwār,* from Arabic *bāzahr,* from Persian *pād-zahr,* from *pād* protecting (against) + *zahr* poison] (1577)
: any of various calculi found chiefly in the gastrointestinal organs and formerly believed to possess magical properties — called also *bezoar stone*

B–girl *noun* [probably from *bar* + *girl*] (1936)
: a woman who entertains bar patrons and encourages them to spend freely

BHA \ˌbē-(ˌ)ā-ˈchā\ *noun* [*b*utylated *h*ydroxy*a*nisole] (1950)
: a phenolic antioxidant $C_{11}H_{16}O_2$ used especially to preserve fats and oils in food

Bha·ga·vad Gi·ta \ˌbä-gə-ˌväd-ˈgē-tə\ *noun* [Sanskrit *Bhagavadgītā,* literally, song of the blessed one (Krishna)] (circa 1785)
: a Hindu devotional work in poetic form

bhak·ti \ˈbək-tē\ *noun* [Sanskrit, literally, portion] (1832)
: devotion to a deity constituting a way to salvation in Hinduism

bhang \ˈbaŋ, ˈbäŋ\ *noun* [Hindi *bhā̃g*] (1563)
: HEMP 1a, c; *also* : the leaves and flowering tops of uncultivated hemp : CANNABIS — compare MARIJUANA, HASHISH

bhar·al \ˈbər-əl, ˈbə-rəl\ *noun* [Hindi] (1838)
: a goatlike artiodactyl mammal (*Pseudois nayaur*) of the Himalayas and western China having a bluish-gray coat

BHC \ˌbē-ˌāch-ˈsē\ *noun* [*b*enzene *h*exa*c*hloride] (1947)
1 : any of several stereoisomeric chlorine derivatives $C_6H_6Cl_6$ of cyclohexane in which the chlorine atoms are all attached to different carbon atoms
2 : LINDANE

Bhoj·puri \ˈbōj-ˌpu̇r-ē, ˈbäj-, -pə-rē\ *noun* [Hindi *bhojpurī,* from *Bhojpur,* village in Bihar] (1901)
: an Indo-Aryan language spoken in western Bihar and eastern Uttar Pradesh, India

B horizon *noun* (1938)
: a subsurface soil layer that is immediately beneath the A horizon from which it obtains organic matter chiefly by illuviation and is usually distinguished by less weathering

BHT \ˌbē-ˌāch-ˈtē\ *noun* [*b*utylated *h*ydroxy*t*oluene] (1961)
: a phenolic antioxidant $C_{15}H_{24}O$ used especially to preserve fats and oils in food, cosmetics, and pharmaceuticals

bi \ˈbī\ *noun or adjective* (circa 1965)
: BISEXUAL

[1]bi- *prefix* [Middle English, from Latin — more at TWI-]
1 a : two ⟨*bi*lateral⟩ **b :** coming or occurring every two ⟨*bi*centennial⟩ **c :** into two parts ⟨*bi*sect⟩
2 a : twice : doubly : on both sides ⟨*bi*convex⟩ **b :** coming or occurring two times ⟨*bi*annual⟩ — compare SEMI-
3 : between, involving, or affecting two (specified) symmetrical parts ⟨*bi*labial⟩
4 a : containing one (specified) constituent in double the proportion of the other constituent or in double the ordinary proportion ⟨*bi*carbonate⟩ **b :** DI- 2 ⟨*bi*phenyl⟩ ▢

▢ USAGE
bi- Many people are puzzled about *bimonthly* and *biweekly,* which are often ambiguous because they are formed from both senses 1b and 2b of *bi-*. This ambiguity has been in existence for nearly a century and a half and cannot be eliminated by the dictionary. The chief difficulty is that many users of these words assume that others know exactly what they mean, and they do not bother to make their context clear. So if you need *bimonthly* or *biweekly,* you should leave some clues in your context to the sense of *bi-* you mean. And if you need the meaning "twice a," you can substitute *semi-* for *bi-*. *Biannual* and *biennial* are usually differentiated.

BIBLE				
OLD TESTAMENT		NEW TESTAMENT	PROTESTANT APOCRYPHA	JEWISH SCRIPTURE
ROMAN		Matthew	1 & 2 Esdras	*Law*
CATHOLIC	PROTESTANT	Mark	Tobit	Genesis
CANON	CANON	Luke	Judith	Exodus
Genesis	Genesis	John	Additions to	Leviticus
Exodus	Exodus	Acts of the	Esther	Numbers
Leviticus	Leviticus	Apostles	Wisdom of	Deuteronomy
Numbers	Numbers	Romans	Solomon	*Prophets*
Deuteronomy	Deuteronomy	1 & 2	Ecclesiasticus	Joshua
Joshua	Joshua	Corinthians	or the Wisdom	Judges
Judges	Judges	Galatians	of Jesus Son	1 & 2 Samuel
Ruth	Ruth	Ephesians	of Sirach	1 & 2 Kings
1 & 2 Samuel	1 & 2 Samuel	Philippians	Baruch	Isaiah
1 & 2 Kings	1 & 2 Kings	Colossians	Prayer of	Jeremiah
1 & 2 Chronicles	1 & 2 Chronicles	1 & 2	Azariah and	Ezekiel
Ezra	Ezra	Thessalonians	the Song of	Hosea
Nehemiah	Nehemiah	1 & 2 Timothy	the Three Holy	Joel
Tobit		Titus	Children	Amos
Judith		Philemon	Susanna	Obadiah
Esther	Esther	Hebrews	Bel and the	Jonah
Job	Job	James	Dragon	Micah
Psalms	Psalms	1 & 2 Peter	The Prayer of	Nahum
Proverbs	Proverbs	1, 2, 3 John	Manasses	Habakkuk
Ecclesiastes	Ecclesiastes	Jude	1 & 2 Maccabees	Zephaniah
Song of Songs	Song of Solomon	Revelation *or*		Haggai
Wisdom		Apocalypse		Zechariah
Sirach				Malachi
Isaiah	Isaiah			*Hagiographa*
Jeremiah	Jeremiah			Psalms
Lamentations	Lamentations			Proverbs
Baruch				Job
Ezekiel	Ezekiel			Song of Songs
Daniel	Daniel			Ruth
Hosea	Hosea			Lamentations
Joel	Joel			Ecclesiastes
Amos	Amos			Esther
Obadiah	Obadiah			Daniel
Jonah	Jonah			Ezra
Micah	Micah			Nehemiah
Nahum	Nahum			1 & 2 Chronicles
Habakkuk	Habakkuk			
Zephaniah	Zephaniah			
Haggai	Haggai			
Zechariah	Zechariah			
Malachi	Malachi			
1 & 2 Maccabees				

²bi- *or* **bio-** *combining form* [Greek, from *bios* mode of life — more at QUICK]
: life **:** living organisms or tissue ⟨*bio*luminescence⟩

bi·aly \bē-ˈa-lē\ *noun, plural* **bialys** [Yiddish, short for *bialystoker,* from *bialystoker* of Bialystok, city in Poland] (1965)
: a flat breakfast roll that has a depressed center and is usually covered with onion flakes

bi·an·nu·al \(ˌ)bī-ˈan-yə(-wə)l\ *adjective* (1877)
: occurring twice a year; *sometimes* **:** BIENNIAL 1
usage see BI-
— bi·an·nu·al·ly *adverb*

¹bi·as \ˈbī-əs\ *noun* [Middle French *biais*] (1530)
1 : a line diagonal to the grain of a fabric; *especially* **:** a line at a 45° angle to the selvage often utilized in the cutting of garments for smoother fit
2 a : a peculiarity in the shape of a bowl that causes it to swerve when rolled on the green **b :** the tendency of a bowl to swerve; *also* **:** the impulse causing this tendency **c :** the swerve of the bowl
3 a : BENT, TENDENCY **b :** an inclination of temperament or outlook; *especially* **:** a personal and sometimes unreasoned judgment **:** PREJUDICE **c :** an instance of such prejudice **d** (1) **:** deviation of the expected value of a statistical estimate from the quantity it estimates (2) **:** systematic error introduced into sampling or testing by selecting or encouraging one outcome or answer over others
4 a : a voltage applied to a device (as a transistor control electrode) to establish a reference level for operation **b :** a high-frequency voltage combined with an audio signal to reduce distortion in tape recording
synonym see PREDILECTION
— on the bias : ASKEW, OBLIQUELY

²bias *adjective* (1551)
: DIAGONAL, SLANTING — used chiefly of fabrics and their cut
— bi·as·ness *noun*

³bias *adverb* (1575)
1 : DIAGONALLY ⟨cut cloth *bias*⟩
2 *obsolete* **:** AWRY

⁴bias *transitive verb* **bi·ased** *or* **bi·assed; bi·as·ing** *or* **bi·as·sing** (circa 1628)
1 : to give a settled and often prejudiced outlook to ⟨his background *biases* him against foreigners⟩
2 : to apply a slight negative or positive voltage to (as an electron-tube grid)
synonym see INCLINE

bi·as–belt·ed tire \ˈbī-əs-ˌbel-təd-\ *noun* (1968)
: a pneumatic tire with a belt (as of steel or fiberglass) around the tire under the tread and on top of the ply cords set diagonally to the center line of the tread

bi·ased *adjective* (1649)
1 : exhibiting or characterized by bias; *especially* **:** PREJUDICED
2 : tending to yield one outcome more frequently than others in a statistical experiment ⟨a *biased* coin⟩
3 : having an expected value different from the quantity or parameter estimated ⟨a *biased* estimate⟩

bi·as–ply tire \ˈbī-əs-ˌplī-\ *noun* (1968)
: a pneumatic tire having crossed layers of ply-cord set diagonally to the center line of the tread

bias tape *noun* (1926)
: a narrow strip of cloth cut on the bias, folded, and used for finishing or decorating clothing

bi·ath·lete \bī-ˈath-ˌlēt\ *noun* [blend of *athlete* and *biathlon*] (1968)
: a competitor in a biathlon

bi·ath·lon \bī-ˈath-lən, -ˌlän\ *noun* [¹*bi-* + *-athlon* (as in *decathlon*)] (1958)
: a composite athletic contest consisting of cross-country skiing and rifle sharpshooting

bi·ax·i·al \(ˌ)bī-ˈak-sē-əl\ *adjective* (1854)
: having or relating to two axes or optic axes ⟨a *biaxial* crystal⟩
— bi·ax·i·al·ly \-ə-lē\ *adverb*

¹bib \ˈbib\ *verb* **bibbed; bib·bing** [Middle English *bibben*] (14th century)
: DRINK

²bib *noun* (1580)
1 : a cloth or plastic shield tied under the chin to protect the clothes
2 : the part of an apron or of overalls extending above the waist
3 : a patch of differently colored feathers or fur immediately below the bill or chin of a bird or mammal
— bibbed \ˈbibd\ *adjective*
— bib·less \ˈbib-ləs\ *adjective*

bib and tucker *noun* (1747)
: an outfit of clothing — usually used in the phrase *best bib and tucker*

bib·ber \ˈbi-bər\ *noun* (1536)
: a person who regularly drinks alcoholic beverages
— bib·bery \ˈbi-bə-rē\ *noun*

Bibb lettuce \ˈbib-\ *noun* [Major John *Bibb,* 19th century American grower] (1961)
: lettuce of a variety that has a small head and dark green color

bib·cock \ˈbib-ˌkäk\ *also* **bibb cock** *noun* (circa 1853)
: a faucet having a bent-down nozzle

bi·be·lot \ˈbē-bə-ˌlō\ *noun, plural* **bibelots** \-ˌlō(z)\ [French] (1873)
: a small household ornament or decorative object **:** TRINKET

bi·ble \ˈbī-bəl\ *noun* [Middle English, from Old French, from Medieval Latin *biblia,* from Greek, plural of *biblion* book, diminutive of *byblos* papyrus, book, from *Byblos,* ancient Phoenician city from which papyrus was exported] (14th century)
1 *capitalized* **a :** the sacred scriptures of Christians comprising the Old Testament and the New Testament **b :** the sacred scriptures of some other religion (as Judaism)
2 *obsolete* **:** BOOK
3 *capitalized* **:** a copy or an edition of the Bible
4 : a publication that is preeminent especially in authoritativeness or wide readership ⟨the fisherman's *bible*⟩ ⟨the *bible* of the entertainment industry⟩

Bible Belt *noun* (1925)
: an area chiefly in the southern U.S. believed to hold uncritical allegiance to the literal accuracy of the Bible; *broadly* **:** an area characterized by ardent religious fundamentalism

Bible paper *noun* (1903)
: INDIA PAPER 2

Bi·ble–thump·er \ˈbī-bəl-ˌthəm-pər\ *noun* (1937)

: an overzealous proponent of Christian fundamentalism

— **Bi·ble–thump·ing** \-piŋ\ *adjective*

bibli- *or* **biblio-** *combining form* [Middle French, from Latin, from Greek, from *biblion*]
: book ⟨*biblio*logy⟩

bib·li·cal \ˈbi-bli-kəl\ *adjective* [Medieval Latin *biblicus,* from *biblia*] (circa 1775)
1 : of, relating to, or being in accord with the Bible
2 : suggestive of the Bible or Bible times
— **bib·li·cal·ly** \-k(ə-)lē\ *adverb*

bib·li·cism \ˈbi-blə-ˌsi-zəm\ *noun, often capitalized* (1851)
: adherence to the letter of the Bible
— **bib·li·cist** \-lə-sist\ *noun, often capitalized*

bib·li·og·ra·pher \ˌbi-blē-ˈä-grə-fər\ *noun* (1775)
1 : an expert in bibliography
2 : a compiler of bibliographies

bib·li·og·ra·phy \ˌbi-blē-ˈä-grə-fē\ *noun, plural* **-phies** [probably from New Latin *bibliographia,* from Greek, the copying of books, from *bibli-* + *-graphia* -graphy] (1802)
1 : the history, identification, or description of writings or publications
2 a : a list often with descriptive or critical notes of writings relating to a particular subject, period, or author **b :** a list of works written by an author or printed by a publishing house
3 : the works or a list of the works referred to in a text or consulted by the author in its production
— **bib·lio·graph·ic** \ˌbi-blē-ə-ˈgra-fik\ *also* **bib·lio·graph·i·cal** \-fi-kəl\ *adjective*
— **bib·lio·graph·i·cal·ly** \-k(ə-)lē\ *adverb*

bib·li·ol·a·ter \ˌbi-blē-ˈä-lə-tər\ *noun* (1847)
1 : one having excessive reverence for the letter of the Bible
2 : one overly devoted to books
— **bib·li·ol·a·trous** \-ˈä-lə-trəs\ *adjective*
— **bib·li·ol·a·try** \-trē\ *noun*

bib·li·ol·o·gy \ˌbi-blē-ˈä-lə-jē\ *noun* (1806)
1 : the history and science of books as physical objects **:** BIBLIOGRAPHY
2 *often capitalized* **:** the study of the theological doctrine of the Bible

bib·lio·ma·nia \ˌbi-blē-ə-ˈmā-nē-ə, -nyə\ *noun* [French *bibliomanie,* from *bibli-* + *manie* mania, from Late Latin *mania*] (1734)
: extreme preoccupation with collecting books
— **bib·lio·ma·ni·ac** \-nē-ˌak\ *noun or adjective*
— **bib·lio·ma·ni·a·cal** \-lē-ō-mə-ˈnī-ə-kəl\ *adjective*

bib·li·op·e·gy \ˌbi-blē-ˈä-pə-jē\ *noun* [ultimately from Greek *bibli-* + *pēgnynai* to fasten together — more at PACT] (circa 1859)
: the art of binding books
— **bib·li·o·pe·gic** \ˌbi-blē-ə-ˈpe-jik, -ˈpē-\ *adjective*
— **bib·li·op·e·gist** \ˌbi-blē-ˈä-pə-jist\ *noun*

bib·lio·phile \ˈbi-blē-ə-ˌfīl\ *noun* [French, from *bibli-* + *-phile*] (1824)
: a lover of books especially for qualities of format; *also* **:** a book collector
— **bib·lio·phil·ic** \ˌbi-blē-ə-ˈfi-lik\ *adjective*
— **bib·li·oph·i·lism** \-ˈä-fə-ˌli-zəm\ *noun*
— **bib·li·oph·i·ly** \-lē\ *noun*

bib·li·o·pole \ˈbi-blē-ə-ˌpōl\ *or* **bib·li·op·o·list** \ˌbi-blē-ˈä-pə-list\ *noun* [Latin *bibliopola* bookseller, from Greek *bibliopōlēs,* from *bibli-* + *pōlein* to sell] (1775)
: a dealer especially in rare or curious books

bib·lio·the·ca \ˌbi-blē-ə-ˈthē-kə\ *noun, plural* **-cas** *or* **-cae** \-ˌsē, -ˌkē\ [Latin, from Greek *bibliothēkē,* from *bibli-* + *thēkē* case; akin to Greek *tithenai* to put, place — more at DO] (circa 1824)
1 : a collection of books
2 : a list of books
— **bib·lio·the·cal** \-ˈthē-kəl\ *adjective*

bib·lio·ther·a·py \ˌbi-blē-ə-ˈther-ə-pē\ *noun* (1919)
: the use of reading materials for help in solving personal problems or for psychiatric therapy

bib·li·ot·ics \ˌbi-blē-ˈä-tiks\ *noun plural but singular in construction* [*bibli-* + [1]*-otic* + *-ics*] (1901)
: the study of handwriting, documents, and writing materials especially for determining genuineness or authorship
— **bib·li·ot·ic** \-tik\ *adjective*
— **bib·li·o·tist** \ˈbi-blē-ə-tist\ *noun*

bib·u·lous \ˈbi-byə-ləs\ *adjective* [Latin *bibulus,* from *bibere* to drink — more at POTABLE] (1675)
1 : highly absorbent
2 a : fond of alcoholic beverages **b :** of, relating to, or marked by the consumption of alcoholic beverages
— **bib·u·lous·ly** *adverb*
— **bib·u·lous·ness** *noun*

bi·cam·er·al \(ˌ)bī-ˈkam-rəl, -ˈka-mə-\ *adjective* [[1]*bi-* + Late Latin *camera* chamber — more at CHAMBER] (1856)
: having, consisting of, or based on two legislative chambers ⟨a *bicameral* legislature⟩
— **bi·cam·er·al·ism** \-rə-ˌli-zəm\ *noun*

bi·carb \(ˌ)bī-ˈkärb, ˈbī-ˌ\ *noun* (1922)
: SODIUM BICARBONATE

bi·car·bon·ate \(ˌ)bī-ˈkär-bə-ˌnāt, -nət\ *noun* [International Scientific Vocabulary] (1819)
: an acid carbonate

bicarbonate of soda (circa 1887)
: SODIUM BICARBONATE

bi·cen·te·na·ry \ˌbī-(ˌ)sen-ˈte-nə-rē, (ˌ)bī-ˈsen-t^{ə}n-ˌer-ē, ˌbī-(ˌ)sen-ˈtē-nə-rē\ *noun* (1872)
: BICENTENNIAL
— **bicentenary** *adjective*

bi·cen·ten·ni·al \ˌbī-(ˌ)sen-ˈte-nē-əl\ *noun* (1883)
: a 200th anniversary or its celebration
— **bicentennial** *adjective*

bi·ceps \ˈbī-ˌseps\ *noun, plural* **biceps** *also* **bi·ceps·es** [New Latin *bicipit-, biceps,* from Latin, two-headed, from *bi-* + *capit-, caput* head — more at HEAD] (1650)
: a muscle having two heads: as **a :** the large flexor muscle of the front of the upper arm **b :** the large flexor muscle of the back of the upper leg

biceps bra·chii \-ˈbrā-kē-ˌī, -kē-ˌē\ *noun* [New Latin, literally, biceps of the arm] (circa 1860)
: BICEPS a

biceps fe·mo·ris \-ˈfe-mə-rəs\ *noun* [New Latin, literally, biceps of the femur] (circa 1860)
: BICEPS b

bi·chlo·ride of mercury \(ˌ)bī-ˈklōr-ˌīd-, -ˈklȯr-\ [International Scientific Vocabulary] (1810)
: MERCURIC CHLORIDE

bi·chon fri·se \bē-ˌshō^{n}-frē-ˈzā\ *noun, plural* **bi·chons fri·ses** \-ˌshō^{n}-frē-ˈzā(z)\ [modification of French *bichon à poil frisé* curly-haired lapdog] (1966)
: any of a breed of small sturdy dogs of Mediterranean origin having a thick wavy white coat

bi·chro·mate \(ˌ)bī-ˈkrō-ˌmāt, ˈbī-krō-\ *noun* (1836)
: a dichromate especially of sodium or potassium
— **bi·chro·mat·ed** \-ˌmā-təd\ *adjective*

bi·chrome \ˈbī-ˌkrōm\ *adjective* (1924)
: two-colored

bi·cip·i·tal \bī-ˈsi-pə-t^{ə}l\ *adjective* (1646)
: of, relating to, or being a biceps

[1]**bick·er** \ˈbi-kər\ *noun* [Middle English *biker*] (14th century)
1 : petulant quarreling **:** ALTERCATION
2 : a sound of or as if of bickering

[2]**bicker** *intransitive verb* **bick·ered; bick·er·ing** \ˈbi-k(ə-)riŋ\ (15th century)
1 : to engage in a petulant or petty quarrel
2 a : to move with a rapidly repeated noise ⟨a *bickering* stream⟩ **b :** QUIVER, FLICKER
— **bick·er·er** \-kə-rər\ *noun*

bi·coast·al \(ˌ)bī-ˈkōs-təl\ *adjective* (1972)
: of or relating to or living or working on both the East and West coasts of the U.S.

bi·col·ored \ˈbī-ˌkə-lərd\ *or* **bi·col·or** \-lər\ *adjective* [Latin *bicolor,* from *bi-* + *color*] (circa 1843)
: two-colored
— **bicolor** *noun*

bi·com·po·nent \(ˌ)bī-kəm-ˈpō-nənt, -käm-, -ˈkäm-ˌ\ *adjective* (1962)
: being a fiber made of two polymers having slightly different physical properties so that the fiber has a permanent crimp and fabrics made from it have inherent bulk and stretchability

bi·con·cave \(ˌ)bī-(ˌ)kän-ˈkāv, -ˈkän-ˌ\ *adjective* (1833)
: concave on both sides
— **bi·con·cav·i·ty** \ˌbī-(ˌ)kän-ˈka-və-tē\ *noun*

bi·con·di·tion·al \ˌbī-kən-ˈdish-nəl, -ˈdi-shə-\ *noun* (1940)
: a relation between two propositions that is true only when both propositions are simultaneously true or false

bi·con·vex \ˌbī-(ˌ)kän-ˈveks, (ˌ)bī-ˈkän-ˌ\ *adjective* [International Scientific Vocabulary] (circa 1852)
: convex on both sides
— **bi·con·vex·i·ty** \ˌbī-kən-ˈvek-sə-tē, -(ˌ)kän-\ *noun*

bi·corne \ˈbī-ˌkȯrn\ *noun* [French, from Latin *bicornis* two-horned, from *bi-* + *cornu* horn — more at HORN] (1936)
: COCKED HAT 2

bi·cul·tur·al \(ˌ)bī-ˈkəl-chər-əl\ *adjective* (1940)
: of, relating to, or including two distinct cultures ⟨*bicultural* education⟩
— **bi·cul·tur·al·ism** \-ə-ˌli-zəm\ *noun*

[1]**bi·cus·pid** \(ˌ)bī-ˈkəs-pəd\ *adjective* [New Latin *bicuspid-, bicuspis,* from *bi-* + Latin *cuspid-, cuspis* point] (circa 1839)
: having or ending in two points ⟨*bicuspid* teeth⟩

[2]**bicuspid** *noun* (1852)
: a human premolar tooth — see TOOTH illustration

bicuspid valve *noun* (circa 1903)
: a cardiac valve consisting of two triangular flaps which allow only unidirectional blood flow from the left atrium to the ventricle — called also *mitral valve*

[1]**bi·cy·cle** \ˈbī-si-kəl, -ˌsi- *also* -ˌsī-\ *noun* [French, from *bi-* + *-cycle* (as in *tricycle*)] (1868)
: a vehicle with two wheels tandem, a steering handle, a saddle seat, and pedals by which it is propelled; *also* **:** a stationary exercise machine that resembles such a vehicle

[2]**bicycle** *intransitive verb* **bi·cy·cled; bi·cy·cling** \-k(ə-)liŋ\ (1869)
: to ride a bicycle
— **bi·cy·cler** \-klər\ *noun*
— **bi·cy·clist** \-klist\ *noun*

bi·cy·clic \(ˌ)bī-ˈsī-klik, -ˈsi-\ *adjective* [International Scientific Vocabulary] (circa 1909)
1 : consisting of or arranged in two cycles
2 : containing two usually fused rings in the structure of the molecule

[1]**bid** \ˈbid\ *verb* **bade** \ˈbad, ˈbād\ *or* **bid; bid·den** \ˈbi-d^{ə}n\ *or* **bid** *also* **bade; bid·ding** [partly from Middle English *bidden,* from Old English *biddan;* akin to Old High German *bitten* to entreat, and perhaps to Sanskrit *bādhate* he presses; partly from Middle English *beden* to offer, command, from Old English *bēodan;* akin to Old High German *biotan* to offer, Greek *pynthanesthai* to examine, Sanskrit *bodhi* enlightenment] (before 12th century)
transitive verb
1 a *obsolete* **:** BESEECH, ENTREAT **b :** to issue an order to **:** TELL **c :** to request to come **:** INVITE
2 : to give expression to ⟨*bade* a tearful farewell⟩

3 a : OFFER — usually used in the phrase *to bid defiance* **b** *past and past participle bid* (1) **:** to offer (a price) whether for payment or acceptance (2) **:** to make a bid of or in (a suit at cards)
intransitive verb
: to make a bid
synonym see COMMAND
— **bid·der** *noun*
— **bid fair :** to seem likely

²bid *noun* (1788)
1 a : the act of one who bids **b :** a statement of what one will give or take for something; *especially* **:** an offer of a price **c :** something offered as a bid
2 : an opportunity to bid
3 : INVITATION
4 a : an announcement of what a cardplayer proposes to undertake **b :** the amount of such a bid **c :** a biddable bridge hand
5 : an attempt or effort to win, achieve, or attract ⟨a *bid* for reelection⟩

bid·da·ble \'bi-də-bəl\ *adjective* (circa 1768)
1 : easily led, taught, or controlled **:** DOCILE
2 : capable of being bid
— **bid·da·bil·i·ty** \,bi-də-'bi-lə-tē\ *noun*
— **bid·da·bly** \'bi-də-blē\ *adverb*

¹bid·dy \'bi-dē\ *noun, plural* **biddies** [perhaps imitative] (1601)
: HEN 1a; *also* **:** a young chicken

²biddy *noun, plural* **biddies** [diminutive of the name *Bridget*] (circa 1861)
1 : a hired girl or cleaning woman
2 : WOMAN; *especially* **:** an elderly woman — usually used disparagingly

bide \'bīd\ *verb* **bode** \'bōd\ *or* **bid·ed; bid·ed; bid·ing** [Middle English, from Old English *bīdan;* akin to Old High German *bītan* to wait, Latin *fidere* to trust, Greek *peithesthai* to believe] (before 12th century)
transitive verb
1 *past usually* **bided :** to wait for — used chiefly in the phrase *bide one's time*
2 *archaic* **:** WITHSTAND ⟨two men . . . might *bide* the winter storm —W. C. Bryant⟩
3 *chiefly dialect* **:** to put up with **:** TOLERATE
intransitive verb
1 : to continue in a state or condition
2 : to wait awhile **:** TARRY
3 : to continue in a place **:** SOJOURN
— **bid·er** *noun*

bi·det \bi-'dā\ *noun* [French, small horse, bidet, from Middle French, from *bider* to trot] (1766)
: a bathroom fixture used especially for bathing the external genitals and the posterior parts of the body

bi·di·a·lec·tal·ism \,bī-,dī-ə-'lek-tᵊl-,i-zəm\ *noun* (1958)
: facility in using two dialects of the same language; *also* **:** the teaching of Standard English to pupils who normally use a nonstandard dialect
— **bi·di·a·lec·tal** *adjective*

bi·di·rec·tion·al \,bī-də-'rek-shnəl, -dī-, -shə-nᵊl\ *adjective* (1928)
: involving, moving, or taking place in two usually opposite directions ⟨*bidirectional* flow⟩ ⟨*bidirectional* replication of DNA⟩
— **bi·di·rec·tion·al·ly** *adverb*

bi·don·ville \,bē-,dōⁿ-'vē(ə)l\ *noun* [French, from *bidon* metal can or drum + *ville* city] (1952)
: a settlement of jerry-built dwellings on the outskirts of a city (as in France or North Africa)

bid up *transitive verb* (1864)
: to raise the price of (as property at auction) by a succession of offers

Bie·der·mei·er \'bē-dər-,mī(-ə)r\ *adjective* [after Gottlieb *Biedermeier,* satirical name for an uninspired German bourgeois] (1905)
: of a style of unostentatious furniture and interior decoration popular especially with the middle class in early 19th century Germany

bi·en·ni·al \(,)bī-'e-nē-əl\ *adjective* (1562)
1 : occurring every two years
2 : continuing or lasting for two years; *specifically* **:** growing vegetatively during the first year and fruiting and dying during the second
usage see BI-
— **biennial** *noun*
— **bi·en·ni·al·ly** \-ə-lē\ *adverb*

bi·en·ni·um \bī-'e-nē-əm\ *noun, plural* **-ni·ums** *or* **-nia** \-nē-ə\ [Latin, from *bi-* + *annus* year — more at ANNUAL] (1899)
: a period of two years

bier \'bir\ *noun* [Middle English *bere,* from Old English *bǣr;* akin to Old English *beran* to carry — more at BEAR] (before 12th century)
1 *archaic* **:** a framework for carrying
2 : a stand on which a corpse or coffin is placed; *also* **:** a coffin together with its stand

bi·face \'bī-,fās\ *noun* (1934)
: a bifacial stone tool

bi·fa·cial \(,)bī-'fā-shəl\ *adjective* (circa 1847)
: having opposite sides or faces worked on to form an edge for cutting or scraping
— **bi·fa·cial·ly** *adverb*

biff \'bif\ *noun* [probably imitative] (circa 1889)
: WHACK, BLOW
— **biff** *transitive verb*

bi·fid \'bī-,fid, -fəd\ *adjective* [Latin *bifidus,* from *bi-* + *-fidus* -fid] (1661)
: divided into two equal lobes or parts by a median cleft ⟨a *bifid* leaf⟩

bi·fi·lar \(,)bī-'fī-lər\ *adjective* [International Scientific Vocabulary *bi-* + Latin *filum* thread — more at FILE] (1846)
1 : involving two threads or wires ⟨*bifilar* suspension of a pendulum⟩
2 : involving a single thread or wire doubled back upon itself ⟨a *bifilar* resistor⟩
— **bi·fi·lar·ly** *adverb*

bi·fla·gel·late \(,)bī-'fla-jə-lət, -,lāt; -flə-'je-lət\ *adjective* (1856)
: having two flagella ⟨*biflagellate* gametes⟩

¹bi·fo·cal \'bī-,fō-kəl\ *adjective* [International Scientific Vocabulary] (1888)
1 : having two focal lengths
2 : having one part that corrects for near vision and one for distant vision ⟨a *bifocal* eyeglass lens⟩

²bifocal *noun* (1899)
1 *plural* **:** eyeglasses with bifocal lenses
2 : a bifocal glass or lens

bi·func·tion·al \(,)bī-'fəŋ(k)-shnəl, -shə-nᵊl\ *adjective* (1936)
: having two functions; *especially* **:** DIFUNCTIONAL

bi·fur·cate \'bī-(,)fər-,kāt, bī-'fər-\ *verb* **-cat·ed; -cat·ing** [Medieval Latin *bifurcatus,* past participle of *bifurcare,* from Latin *bifurcus* two-pronged, from *bi-* + *furca* fork] (1615)
transitive verb
: to cause to divide into two branches or parts
intransitive verb
: to divide into two branches or parts
— **bi·fur·cate** \(,)bī-'fər-kət, -,kāt; 'bī-(,)fər-,kāt\ *adjective*

bi·fur·ca·tion \,bī-(,)fər-'kā-shən\ *noun* (1615)
1 a : the point at which bifurcating occurs **b :** BRANCH
2 : the act of bifurcating **:** the state of being bifurcated

¹big \'big\ *adjective* **big·ger; big·gest** [Middle English, perhaps of Scandinavian origin; akin to Norwegian dialect *bugge* important man] (14th century)
1 a *obsolete* **:** of great strength **b :** of great force ⟨a *big* storm⟩
2 a : large or great in dimensions, bulk, or extent ⟨a *big* house⟩; *also* **:** large or great in quantity, number, or amount ⟨a *big* fleet⟩ **b :** operating on a large scale ⟨*big* government⟩ **c :** CAPITAL 1
3 a : PREGNANT; *especially* **:** nearly ready to give birth **b :** full to bursting **:** SWELLING ⟨*big* with rage⟩ **c** *of the voice* **:** full and resonant
4 a : CHIEF, PREEMINENT ⟨the *big* issue of the campaign⟩ **b :** outstandingly worthy or able ⟨a truly *big* man⟩ **c :** of great importance or significance ⟨the *big* moment⟩ **d :** IMPOSING, PRETENTIOUS; *also* **:** marked by or given to boasting ⟨*big* talk⟩ **e :** MAGNANIMOUS, GENEROUS ⟨was *big* about it⟩
5 : POPULAR ⟨soft drinks are very *big* in Mexico —Russ Leadabrand⟩
6 : full-bodied and flavorful — used of wine
— **big·ly** *adverb*
— **big·ness** *noun*
— **big on :** strongly favoring or liking; *also* **:** noted for ⟨she is *big on* blushing —Arnold Hano⟩

²big *adverb* (1807)
1 : in a loud or declamatory manner; *also* **:** in a boasting manner ⟨talk *big*⟩
2 a : to a large amount or extent ⟨won *big*⟩ ⟨lost *big*⟩ **b :** on a large scale ⟨think *big*⟩ ⟨worry *big*⟩
3 : HARD ⟨hits her forehand *big*⟩

³big *noun* (1965)
: an individual or organization of outstanding importance or power; *especially* **:** MAJOR LEAGUE — usually used in plural ⟨playing in the *bigs*⟩

big·a·mous \'bi-gə-məs\ *adjective* (1864)
1 : guilty of bigamy
2 : involving bigamy
— **big·a·mous·ly** *adverb*

big·a·my \'bi-gə-mē\ *noun* [Middle English *bigamie,* from Medieval Latin *bigamia,* from Latin *bi-* + Late Latin *-gamia* -gamy] (13th century)
: the act of entering into a marriage with one person while still legally married to another
— **big·a·mist** \-mist\ *noun*

bi·ga·rade \,bē-gä-'räd\ *noun* [French, from Provençal *bigarrado,* from *bigarra* to variegate] (1703)
1 : SOUR ORANGE
2 : a brown sauce flavored with the juice and grated rind of oranges

big band *noun* (1926)
: a band that is larger than a combo and that usually features a mixture of ensemble playing and solo improvisation typical of jazz or swing

big bang *noun* (1950)
: the cosmic explosion that marked the beginning of the universe according to the big bang theory

big bang theory *noun* (1955)
: a theory in astronomy: the universe originated billions of years ago in an explosion from a single point of nearly infinite energy density — compare STEADY STATE THEORY

big beat *noun, often both Bs capitalized* (1958)
: music (as rock) characterized by a heavy persistent beat

Big Ben \-'ben\ *noun* [Sir *Benjamin* Hall (died 1867) English Chief Commissioner of Works] (circa 1895)
1 : a large bell in the clock tower of the Houses of Parliament in London
2 : the tower that houses Big Ben; *also* **:** the clock in the tower

big boy *noun* (1926)
: BIG GUN — usually used in plural

big brother *noun* (1863)
1 : an older brother
2 : a man who befriends a delinquent or friendless boy
3 *both Bs capitalized* [*Big Brother,* personification of the power of the state in *1984* (1949) by George Orwell] **a :** the leader of an authoritarian state or movement **b :** an all-powerful government or organization monitoring and

directing people's actions ⟨data banks that tell *Big Brother* all about us —Herbert Brucker⟩
Big Broth·er·ism \-ˈbrə-thər-ˌi-zəm\ *noun* (1950)
: authoritarian attempts at complete control (as of a person or a nation)
big buck *noun* (1970)
: a large sum of money — usually used in plural; usually hyphenated when attributive
big business *noun* (1905)
1 : an economic group consisting of large profit-making corporations especially with regard to their influence on social or political policy
2 : a very profitable enterprise
big C *noun, often B capitalized* (1968)
: CANCER 2
big daddy *noun, often B&D capitalized* (1958)
: one preeminent especially by reason of power, size, or seniority **:** one representing paternalistic authority
big deal *noun* (1949)
: something of special importance — sometimes used ironically as an interjection
Big Dipper *noun*
: the seven principal stars in the constellation of Ursa Major
bi·gem·i·ny \bī-ˈje-mə-nē\ *noun* [*bigeminal* (double, paired), from Late Latin *bigeminus,* from *bi-* + *geminus* twin] (circa 1923)
: the state of having a pulse characterized by two beats close together with a pause following each pair of beats
— bi·gem·i·nal \-mə-nᵊl\ *adjective*
bi·ge·ner·ic \ˌbī-jə-ˈner-ik\ *adjective* (1885)
: of, relating to, or involving two genera ⟨a *bigeneric* hybrid⟩
big·eye \ˈbig-ˌī\ *noun* (circa 1889)
: either of two small widely distributed reddish to silvery bony fishes (*Priacanthus cruentatus* and *P. arenatus* of the family Priacanthidae) of tropical seas.
big·foot \ˈbig-ˌfu̇t\ *noun, often capitalized* [from the size of the footprints ascribed to it] (1958)
: SASQUATCH
big game *noun* (1864)
1 : large animals sought or taken by hunting or fishing for sport
2 : an important objective especially when involving risk
big·ge·ty *or* **big·gi·ty** \ˈbi-gə-tē\ *adjective* [*big* + *-ety* (as in *persnickety*)] (1880)
1 *Southern & Midland* **:** CONCEITED, VAIN
2 *Southern & Midland* **:** rudely self-important **:** IMPUDENT ⟨never acted *biggety* in court, but she would bow her head only so low —Claude Brown⟩
big·gie \ˈbi-gē\ *noun* (circa 1931)
: one that is big
[1]**big·gin** *or* **big·ging** \ˈbi-gən\ *noun* [Middle English *bigging,* from *biggen* to dwell, from Old Norse *byggja;* akin to Old English *bēon* to be] (14th century)
archaic **:** BUILDING
[2]**biggin** *noun* [Middle French *beguin*] (1530)
archaic **:** CAP: **a :** a child's cap **b :** NIGHTCAP
big·gish \ˈbi-gish\ *adjective* (circa 1626)
: somewhat big
big gun *noun* (1834)
: one having preeminent status or power in a field
big·head \ˈbig-ˌhed\ *noun* (1805)
1 : any of several diseases of animals marked by swelling about the head
2 : an exaggerated opinion of one's importance — usually used with *the*
— big·head·ed \-ˈhe-dəd\ *adjective*
big·heart·ed \-ˈhär-təd\ *adjective* (1868)
: GENEROUS, CHARITABLE
— big·heart·ed·ly *adverb*
— big·heart·ed·ness *noun*
big·horn sheep \ˈbig-ˌhȯrn-\ *noun* (1838)
: a usually grayish brown wild sheep (*Ovis canadensis*) of mountainous and desert regions of western North America — called also *bighorn*

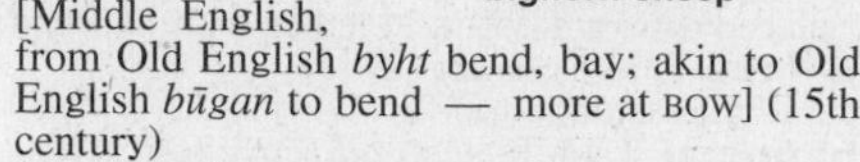
bighorn sheep

big house *noun, often B&H capitalized* (1916)
slang **:** PENITENTIARY
bight \ˈbīt\ *noun* [Middle English, from Old English *byht* bend, bay; akin to Old English *būgan* to bend — more at BOW] (15th century)
1 : a bend in a coast forming an open bay; *also* **:** a bay formed by such a bend
2 : a slack part or loop in a rope
big league *noun* (1899)
1 : MAJOR LEAGUE
2 : BIG TIME 2 — often used in plural
— big–league *adjective*
— big leaguer *noun*
big lie *noun, sometimes B&L capitalized* (1946)
: a deliberate gross distortion of the truth used especially as a propaganda tactic
big·mouthed \ˈbig-ˌmau̇thd, ˈbig-ˌmau̇tht, ˈbig-ˈ\ *adjective* (1642)
1 : having a large mouth
2 : LOUDMOUTHED
big name *noun* (1932)
: a performer or personage of top rank in popular recognition
— big–name *adjective*
big·no·nia \big-ˈnō-nē-ə\ *noun* [New Latin, genus name, from J. P. *Bignon* (died 1743) French royal librarian] (1785)
: any of a genus (*Bignonia*) of American and Japanese woody vines of the trumpet-creeper family with compound leaves and tubular flowers
big·ot \ˈbi-gət\ *noun* [Middle French, hypocrite, bigot] (1661)
: a person obstinately or intolerantly devoted to his or her own opinions and prejudices
— big·ot·ed \-gə-təd\ *adjective*
— big·ot·ed·ly *adverb*
big·ot·ry \ˈbi-gə-trē\ *noun, plural* **-ries** (circa 1674)
1 : the state of mind of a bigot
2 : acts or beliefs characteristic of a bigot
big picture *noun, often B&P capitalized* (circa 1960)
: the entire perspective on a situation or issue — used with *the*
big shot \ˈbig-ˌshät\ *noun* (1929)
: a person of consequence or prominence
big stick *noun* (1904)
: threat especially of military or political intervention
big–tick·et \ˈbig-ˈti-kət\ *adjective* (1945)
: having a high price
big time \-ˌtīm\ *noun* (1910)
1 : a high-paying vaudeville circuit requiring only two performances a day
2 : the top rank of an activity or enterprise
— big–time *adjective*
— big–tim·er \-ˌtī-mər\ *noun*
big toe *noun* (circa 1887)
: the innermost and largest toe of the foot
big top *noun* (1895)
1 : the main tent of a circus
2 : CIRCUS 2a, b, c
big tree *noun* (1853)
: a California evergreen (*Sequoiadendron giganteum*) of the bald cypress family that sometimes exceeds 270 feet (about 82 meters) in height — called also *giant sequoia, sequoia*
big wheel *noun* (1942)
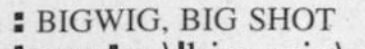
: BIGWIG, BIG SHOT
big·wig \ˈbig-ˌwig\ *noun* (1703)
: an important person
Bi·ha·ri \bi-ˈhär-ē\ *noun* (1882)
1 : a group of Indo-Aryan languages (as Bhojpuri) spoken in Bihar, India, and adjacent areas
2 a : a native or inhabitant of Bihar **b :** a Muslim born in Bihar who emigrated after the partition of India in 1947; *also* **:** a descendant of such a person
bi·jec·tion \(ˌ)bī-ˈjek-shən\ *noun* [[1]*bi-* + *-jection* (as in *injection*)] (1966)
: a mathematical function that is a one-to-one and onto mapping — compare INJECTION, SURJECTION
— bi·jec·tive \-ˈjek-tiv\ *adjective*
bi·jou \ˈbē-ˌzhü\ *noun, plural* **bijous** *or* **bi·joux** \-ˌzhü(z)\ [French, from Breton *bizou* ring, from *biz* finger] (1668)
1 : a small dainty usually ornamental piece of delicate workmanship **:** JEWEL
2 : something delicate, elegant, or highly prized
— bijou *adjective*
bi·jou·te·rie \bi-ˈzhü-tə-(ˌ)rē\ *noun* [French, from *bijou*] (1815)
: a collection of trinkets or ornaments **:** JEWELS; *also* **:** DECORATION
[1]**bike** \ˈbīk\ *noun* [Middle English] (14th century)
1 *chiefly Scottish* **:** a nest of wild bees, wasps, or hornets
2 *chiefly Scottish* **:** a crowd or swarm of people
[2]**bike** *noun* [by shortening & alteration] (1882)
1 : BICYCLE
2 : MOTORCYCLE
3 : MOTORBIKE
[3]**bike** *intransitive verb* **biked; bik·ing** (1895)
: to ride a bike
bik·er \ˈbī-kər\ *noun* (1883)
1 : BICYCLIST
2 : MOTORCYCLIST; *especially* **:** one who is a member of an organized gang
bike·way \ˈbīk-ˌwā\ *noun* (1965)
: a thoroughfare for bicycles
bik·ie \ˈbī-kē\ *noun* [[2]*bike* + *-ie*] (1967)
: BIKER 2
bi·ki·ni \bə-ˈkē-nē\ *noun* [French, from *Bikini,* atoll of the Marshall islands] (1947)
1 a : a woman's scanty two-piece bathing suit **b :** a man's brief swimsuit
2 : a man's or woman's low-cut briefs ◆
— bi·ki·nied \-nēd\ *adjective*
[1]**bi·la·bi·al** \(ˌ)bī-ˈlā-bē-əl\ *adjective* [International Scientific Vocabulary] (1894)
of a consonant **:** produced with both lips
[2]**bilabial** *noun* (1899)
: a bilabial consonant
bi·la·bi·ate \-bē-ət\ *adjective* (1794)
: having two lips ⟨a *bilabiate* corolla⟩
bi·lat·er·al \(ˌ)bī-ˈla-t(ə-)rəl\ *adjective* (1775)
1 : having two sides
2 : affecting reciprocally two nations or parties ⟨a *bilateral* treaty⟩ ⟨a *bilateral* trade agreement⟩
3 a : of, relating to, or affecting the right and left sides of the body or the right and left members of paired organs ⟨*bilateral* nephrectomy⟩ **b :** having bilateral symmetry
— bi·lat·er·al·ism \-t(ə-)rə-ˌli-zəm\ *noun*
— bi·lat·er·al·ly *adverb*
bilateral symmetry *noun* (1860)
: symmetry in which similar anatomical parts are arranged on opposite sides of a median axis so that only one plane can divide the individual into essentially identical halves
bi·lay·er \ˈbī-ˌlā-ər, -ˌle(-ə)r\ *noun* (1963)
: a film or membrane with two molecular layers ⟨a *bilayer* of phospholipid molecules⟩
— bilayer *adjective*
bil·ber·ry \ˈbil-ˌber-ē\ *noun* [*bil-* (probably of Scandinavian origin; akin to Danish *bølle* whortleberry) + *berry*] (1577)

: any of several ericaceous shrubs (genus *Vaccinium*) that resemble blueberries but have flowers which arise solitary or in very small clusters from axillary buds; *also* : its sweet edible fruit

1bil·bo \ˈbil-(ˌ)bō\ *noun, plural* **bilboes** [perhaps from *Bilboa,* Spain] (1557)
: a long bar of iron with sliding shackles used to confine the feet of prisoners especially on shipboard

2bilbo *or* **bil·boa** \ˈbil-(ˌ)bō\ *noun, plural* **bilboes** *or* **bilboas** [*Bilboa, Bilbao,* Spain] (1565)
: SWORD

bil·dungs·ro·man \ˈbil-du̇ŋ(k)s-rō-ˌmän, -du̇ŋz-\ *noun* [German, from *Bildung* education + *Roman* novel] (1910)
: a novel about the moral and psychological growth of the main character

bile \ˈbī(ə)l\ *noun* [Latin *bilis;* akin to Welsh *bustl* bile] (1547)
1 a : either of two humors associated in old physiology with irascibility and melancholy **b :** a yellow or greenish viscid alkaline fluid secreted by the liver and passed into the duodenum where it aids especially in the emulsification and absorption of fats
2 : inclination to anger

bile acid *noun* (circa 1881)
: any of several steroid acids (as cholic acid) of or derived from bile

bile duct *noun* (1774)
: a duct by which bile passes from the liver or gallbladder to the duodenum

bile salt *noun* (1881)
1 : a salt of bile acid
2 *plural* **:** a dry mixture of the principal salts of the gall of the ox used as a liver stimulant and as a laxative

1bi·lev·el \ˈbī-ˈle-vəl\ *adjective* (1960)
1 : having two levels of freight or passenger space
2 : divided vertically into two ground-floor levels

2bi-level \ˈbī-ˌ\ *noun* (1966)
: a bi-level house

1bilge \ˈbilj\ *noun* [probably modification of Middle French *boulge, bouge* leather bag, curved part — more at BUDGET] (1513)
1 : the bulging part of a cask or barrel
2 a : the part of the underwater body of a ship between the flat of the bottom and the vertical topsides **b :** the lowest point of a ship's inner hull
3 : stale or worthless remarks or ideas

2bilge *intransitive verb* **bilged; bilg·ing** (1728)
: to become damaged in the bilge

bilge keel *noun* (1850)
: a projection like a fin extending from the hull near the turn of the bilge on either side to check rolling

bilge·wa·ter \ˈbilj-ˌwȯ-tər, -ˌwä-\ *noun* (1706)
: water that collects in the bilge of a ship

bil·har·zia \bil-ˈhär-zē-ə; -ˈhärt-sē-\ *noun* [New Latin, genus name, from Theodor *Bilharz* (died 1862) German zoologist] (circa 1881)
1 : SCHISTOSOMIASIS
2 : SCHISTOSOME
— bil·har·zi·al \-zē-əl; -sē-\ *adjective*

bil·har·zi·a·sis \ˌbil-ˌhär-ˈzī-ə-səs; -ˌhärt-ˈsī-\ *noun, plural* **-a·ses** \-ˌsēz\ [New Latin] (circa 1900)
: SCHISTOSOMIASIS

bil·i·ary \ˈbi-lē-ˌer-ē\ *adjective* [French *biliare,* from Latin *bilis*] (1731)
: of, relating to, or conveying bile; *also* : affecting the bile-conveying structures

bi·lin·ear \(ˌ)bī-ˈli-nē-ər\ *adjective* (1886)
: linear with respect to each of two mathematical variables; *specifically* : of or relating to an algebraic form each term of which involves one variable to the first degree from each of two sets of variables

bi·lin·gual \(ˌ)bī-ˈliŋ-gwəl *also* -gyə-wəl\ *adjective* [Latin *bilinguis,* from *bi-* + *lingua* tongue — more at TONGUE] (1845)
1 : having or expressed in two languages
2 : using or able to use two languages especially with equal fluency
3 : of or relating to bilingual education
— bilingual *noun*
— bi·lin·gual·ly *adverb*

bilingual education *noun* (1972)
: education in an English-language school system in which minority students with little fluency in English are taught in their native language

bi·lin·gual·ism \-gwə-ˌli-zəm\ *noun* (1873)
: the ability to speak two languages : the frequent oral use of two languages

bil·ious \ˈbil-yəs\ *adjective* [Middle French *bilioux,* from Latin *biliosus,* from *bilis*] (1541)
1 a : of or relating to bile **b :** marked by or suffering from liver dysfunction and especially excessive secretion of bile **c :** appearing as if affected by a bilious disorder
2 : of or indicative of a peevish ill-natured disposition
— bil·ious·ly *adverb*
— bil·ious·ness *noun*

bil·i·ru·bin \ˌbi-li-ˈrü-bən, ˈbi-li-ˌ\ *noun* [Latin *bilis* + *ruber* red — more at RED] (1871)
: a reddish yellow water insoluble pigment occurring especially in bile and blood and causing jaundice if accumulated in excess

bil·i·ver·din \-ˈvər-dᵊn, -ˌvər-\ *noun* [Swedish, from Latin *bilis* + obsolete French *verd* green] (1845)
: a green pigment that occurs in bile and is an intermediate in the degradation of hemoglobin heme groups to bilirubin

1bilk \ˈbilk\ *transitive verb* [perhaps alteration of *2balk*] (1651)
1 : to block the free development of : FRUSTRATE ⟨fate *bilks* their hopes⟩
2 a : to cheat out of something valuable : DEFRAUD **b :** to evade payment of or to ⟨*bilks* his creditors⟩
3 : to slip away from : ELUDE ⟨*bilked* her pursuers⟩
— bilk·er *noun*

2bilk *noun* (1790)
: an untrustworthy tricky individual : CHEAT

1bill \ˈbil\ *noun* [Middle English *bile,* from Old English; akin to Old English *bill*] (before 12th century)
1 : the jaws of a bird together with their horny covering
2 : a mouthpart (as the beak of a turtle) that resembles a bird's bill
3 : the point of an anchor fluke — see ANCHOR illustration
4 : the visor of a cap or hood

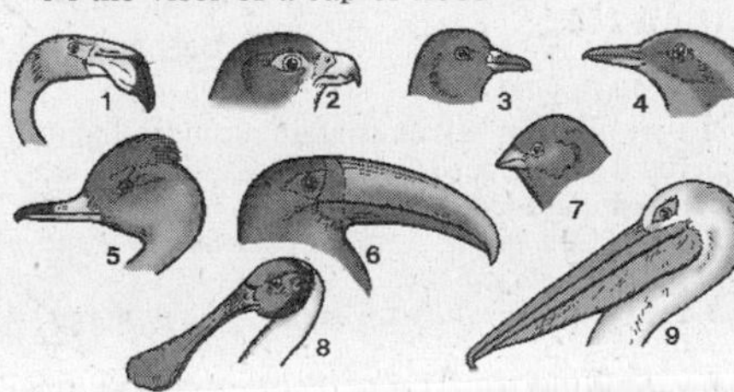

bill 1: ***1*** **flamingo,** ***2*** **falcon,** ***3*** **pigeon,** ***4*** **thrush,** ***5*** **merganser,** ***6*** **toucan,** ***7*** **finch,** ***8*** **spoonbill,** ***9*** **pelican**

2bill *intransitive verb* (1592)
1 : to touch and rub bill to bill
2 : to caress affectionately ⟨*billing* and cooing⟩

3bill *noun* [Middle English *bil,* from Old English *bill* sword; akin to Old High German *bill* pickax] (14th century)
1 : a weapon in use up to the 18th century that consists of a long staff ending in a hook-shaped blade
2 : BILLHOOK

4bill *noun* [Middle English, from Medieval Latin *billa* seal, alteration of *bulla,* from Latin, bubble, boss] (14th century)
1 : an itemized list or a statement of particulars (as a list of materials or of members of a ship's crew)
2 : a written document or note
3 *obsolete* **:** a formal petition
4 a : an itemized account of the separate cost of goods sold, services performed, or work done : INVOICE **b :** an amount expended or owed **c :** a statement of charges for food or drink : CHECK
5 a : a written or printed advertisement posted or otherwise distributed to announce an event of interest to the public; *especially* : an announcement of a theatrical entertainment **b :** a programmed presentation (as a motion picture, play, or concert)
6 : a draft of a law presented to a legislature for enactment; *also* : the law itself ⟨the GI *bill*⟩
7 : a declaration in writing stating a wrong a complainant has suffered from a defendant or stating a breach of law by some person ⟨a *bill* of complaint⟩
8 a : a piece of paper money **b :** an individual or commercial note ⟨*bills* receivable⟩ **c** *slang* **:** one hundred dollars
— fill the bill *or* **fit the bill :** to be exactly what is needed : be suitable

5bill *transitive verb* (14th century)
1 a : to enter in an accounting system : prepare a bill of (charges) **b :** to submit a bill of charges to **c :** to enter (as freight) in a waybill **d :** to issue a bill of lading to or for
2 : to announce (as a performance) especially by posters or placards
3 : ADVERTISE, PROMOTE ⟨the book is *billed* as a "report" —P. G. Altbach⟩
— bill·able *adjective*

bil·la·bong \ˈbi-lə-ˌbȯŋ, -ˌbäŋ\ *noun* [Wiradhuri (Australian aboriginal language of central New South Wales) *bilabaŋ*] (1861)

◇ WORD HISTORY

bikini In July 1946 a joint U.S. Army-Navy task force detonated two thermonuclear bombs at Bikini, an atoll in the Marshall Islands. The second test, which was set off deep within the atoll's lagoon and propelled two million tons of water into an enormous cauliflower-shaped cloud over a mile high, remains one of the most enduring and terrifying images of our century. Nothing so characterizes the innocence of the early Atomic Age as the fact that the Bikini tests, which dominated news around the world from their announcement in June of 1946, became the vehicle for launching a titillating fashion statement. A French couturier, Jacques Heim, introduced an abbreviated two-piece bathing suit billed as the world's smallest and dubbed *l'atome,* "the atom." Another fashion designer, 49-year-old Louis Réard, created an even scantier suit, for which on June 20 he registered *bikini* as a French proprietary mark. The new sensation debuted at the opening of a Paris swimming pool on July 5, worn by a striptease dancer, Micheline Bernardini, whom Réard was forced to hire because Parisian couture models refused to wear the suit. By the time of Réard's death in 1984, this style with its bizarre link to atomic destruction accounted for a fifth of U.S. swimsuit sales, and French *bikini* had become an international word.

1 *Australian* **a :** a blind channel leading out from a river **b :** a usually dry streambed that is filled seasonally
2 *Australian* **:** a backwater forming a stagnant pool

¹bill·board \'bil-ˌbōrd, -ˌbȯrd\ *noun* (1851)
: a flat surface (as of a panel, wall, or fence) on which bills are posted; *specifically* **:** a large panel designed to carry outdoor advertising

²billboard *transitive verb* (1950)
: to promote by a conspicuous display on or as if on a billboard

bill·bug \'bil-ˌbəg\ *noun* [¹*bill* + *bug*] (1861)
: any of various weevils (as of the genus *Sphenophorus*) having larvae that eat the roots of cereal and other grasses

billed \'bild\ *adjective*
: having a bill of a specified kind — used in combination ⟨spoon-*billed*⟩

bill·er \'bi-lər\ *noun* (1920)
: one that bills; *especially* **:** one that makes out bills

¹bil·let \'bi-lət\ *noun* [Middle English *bylet,* from Middle French *billette,* diminutive of *bulle* document, from Medieval Latin *bulla*] (15th century)
1 *archaic* **:** a brief letter **:** NOTE
2 a : an official order directing that a member of a military force be provided with board and lodging (as in a private home) **b :** quarters assigned by or as if by a billet
3 : POSITION, JOB ⟨a lucrative *billet*⟩

²billet *transitive verb* (1594)
1 : to assign lodging to (as soldiers) by or as if by a billet
2 : to serve with a billet ⟨*billet* a householder⟩

³billet *noun* [Middle English *bylet,* from Middle French *billete,* diminutive of *bille* log, of Celtic origin; akin to Old Irish *bile* landmark tree] (15th century)
1 a : a chunky piece of wood (as for firewood) **b** *obsolete* **:** CUDGEL
2 a : a bar of metal **b :** a piece of semifinished iron or steel nearly square in section made by rolling an ingot or bloom **c :** a section of nonferrous metal ingot hot-worked by forging, rolling, or extrusion **d :** a nonferrous casting suitable for rolling or extrusion

bil·let–doux \ˌbi-lē-'dü, ˌbi-(ˌ)lā-\ *noun, plural* **bil·lets–doux** \-'dü(z)\ [French *billet doux,* literally, sweet letter] (1673)
: a love letter

bill·fish \'bil-ˌfish\ *noun* (1782)
: a fish with long slender jaws; *especially* **:** any of a family (Istiophoridae) including marlins, spearfishes, and sailfishes

bill·fold \-ˌfōld\ *noun* [short for earlier *billfolder*] (1895)
: a folding pocketbook for paper money **:** WALLET

bill·hook \-ˌhu̇k\ *noun* (1611)
: a cutting or pruning tool with a hooked blade

bil·liard \'bi(l)-yərd\ *noun* (1580)
— used as an attributive form of *billiards* ⟨*billiard* ball⟩

bil·liards \-yərdz\ *noun plural but singular in construction* [Middle French *billard* billiard cue, billiards, from *bille*] (1580)
: any of several games played on an oblong table by driving small balls against one another or into pockets with a cue; *specifically* **:** a game in which one scores by causing a cue ball to hit in succession two object balls — compare POOL

bil·li–bi *also* **bil·ly–bi** \'bi-lē-ˌbē, ˌbi-lē-'\ *noun* [French, alteration of *Billy B.,* perhaps from William B. Leeds, Jr. (died 1972) American industrialist] (1961)
: a soup of mussel stock, white wine, and cream served hot or cold

bill·ing \'bi-liŋ\ *noun* [⁵*bill*] (1875)
1 : advertising or public promotion (as of a product or personality); *also* **:** relative prominence of a name in such promotion ⟨got top *billing*⟩
2 : total amount of business or investments (as of an advertising agency) within a given period

bil·lings·gate \'bi-liŋz-ˌgāt, *British usually* -git\ *noun* [*Billingsgate,* old gate and fish market, London, England] (1652)
: coarsely abusive language ◆
synonym see ABUSE

bil·lion \'bi(l)-yən\ *noun* [French, from *bi-* + *-illion* (as in *million*)] (1834)
1 — see NUMBER table
2 : a very large number
— billion *adjective*
— bil·lionth \-yən(t)th\ *adjective or noun*

bil·lion·aire \ˌbi(l)-yə-'nar, -'ner, 'bi(l)-yə-ˌ\ *noun* [*billion* + *-aire* (as in *millionaire*)] (1860)
: one whose wealth is estimated at a billion or more (as of dollars or pounds)

bill of exchange (1534)
: an unconditional written order from one person to another to pay a specified sum of money to a designated person

bill of fare (1636)
1 : MENU
2 : PROGRAM

bill of goods (1920)
1 : a consignment of merchandise
2 : something intentionally misrepresented **:** something passed off in a deception or fraud — often used in the phrase *sell a bill of goods*

bill of health (1644)
1 : a certificate given to the ship's master at the time of leaving port that indicates the state of health of a ship's company and of a port with regard to infectious diseases
2 : a usually favorable report following an examination or investigation ⟨gave the criticized textbook a clean *bill of health*⟩

bill of indictment (circa 1530)
: an indictment before it is found or ignored by the grand jury

bill of lading (1532)
: a receipt listing goods shipped that is signed by the agent of the owner of a ship or issued by a common carrier

bill of particulars (circa 1860)
: a detailed listing of charges or claims brought in a legal action or of a defendant's response or counterclaim

bill of rights *often B&R capitalized* (1798)
: a summary of fundamental rights and privileges guaranteed to a people against violation by the state — used especially of the first 10 amendments to the U.S. Constitution

bill of sale (1608)
: a formal instrument for the conveyance or transfer of title to goods and chattels

bil·lon \'bi-lən\ *noun* [French, from Middle French, from *bille* log — more at BILLET] (circa 1727)
1 : gold or silver heavily alloyed with a less valuable metal
2 : an alloy of silver containing more than 50 percent of copper by weight

¹bil·low \'bi-(ˌ)lō\ *noun* [Old Norse *bylgja;* akin to Old High German *balg* bag — more at BELLY] (1552)
1 : WAVE; *especially* **:** a great wave or surge of water
2 : a rolling mass (as of flame or smoke) that resembles a high wave
— bil·lowy \'bi-lə-wē\ *adjective*

²billow (1597)
intransitive verb
1 : to rise or roll in waves or surges
2 : to bulge or swell out (as through action of the wind)
transitive verb
: to cause to billow

¹bil·ly \'bi-lē\ *noun, plural* **billies** [Scots *billy-pot* cooking utensil] (1839)
chiefly Australian & New Zealand **:** a metal or enamelware pail or pot with a lid and wire bail — called also *billycan*

²billy *noun, plural* **billies** [probably from the name *Billy*] (1848)
1 : BILLY CLUB
2 : BILLY GOAT

billy club *noun* [²*billy*] (1949)
: a heavy usually wooden club; *specifically* **:** a police officer's club

bil·ly·cock \'bi-lē-ˌkäk\ *noun* [origin unknown] (1721)
British **:** DERBY 3

billy goat *noun* [from the name *Billy*] (1861)
: a male goat

bi·lobed \'bī-'lōbd\ *adjective* (1756)
: divided into two lobes ⟨a *bilobed* nucleus⟩

bi·lo·ca·tion \'bī-lō-ˌkā-shən\ *noun* (1858)
: the state of being or ability to be in two places at the same time

bil·tong \'bil-ˌtȯŋ, -ˌtäŋ\ *noun* [Afrikaans, from *bil* rump + *tong* tongue] (1815)
chiefly South African **:** jerked meat

bi·man·u·al \(ˌ)bī-'man-yə-wəl, -yəl\ *adjective* (circa 1889)
: done with or requiring the use of both hands
— bi·man·u·al·ly *adverb*

bim·bo \'bim-(ˌ)bō\ *noun, plural* **bimbos** [perhaps from Italian *bimbo* baby] (1919)
1 *slang* **:** MAN, WOMAN — usually used disparagingly ⟨telling a thickheaded pitcher that the *bimbo* at the plate hasn't hit a curve in three seasons —Jay Stuller⟩ and especially of an attractive but empty-headed person ⟨we didn't want a blond *bimbo* in that role . . . we wanted her to be smart —Hugh Wilson⟩
2 *slang* **:** TRAMP 1c ⟨evidence of how her hubby's been cheating on her with various *bimbos* —Dan Greenburg⟩

bi·met·al \'bī-ˌme-tᵊl\ *adjective* (1924)
: BIMETALLIC
— bimetal *noun*

bi·me·tal·lic \ˌbī-mə-'ta-lik\ *adjective* (1876)
1 : relating to, based on, or using bimetallism
2 : composed of two different metals — often used of devices having a part in which two metals that expand differently are bonded together
— bimetallic *noun*

bi·met·al·lism \(ˌ)bī-'me-tᵊl-ˌi-zəm\ *noun* [French *bimétallisme,* from *bi-* + *métal* metal] (1876)
: the use of two metals (as gold and silver) jointly as a monetary standard with both constituting legal tender at a predetermined ratio
— bi·met·al·list \-tᵊl-ist\ *noun*
— bi·met·al·lis·tic \ˌbī-ˌme-tᵊl-'is-tik\ *adjective*

bi·mil·le·na·ry \(ˌ)bī-'mi-lə-ˌner-ē, ˌbī-mə-'le-nə-rē\ *or* **bi·mil·len·ni·al** \ˌbī-mə-'le-nē-əl\ *noun* (1850)
1 : a period of 2000 years

◇ WORD HISTORY

billingsgate Billingsgate was one of two water gates in the wall of the city of London that from Britain's days as a Roman colony gave access to the River Thames. A marketplace developed in the vicinity of the gate, and in 1699, by government decree, it was devoted exclusively to the sale of fish. While Billingsgate was indeed renowned for its fish, it achieved an equal measure of fame for the coarse language that resounded through the stalls. The fishmongers, especially the women, enjoyed a reputation for unrestrained insult that was wittily inventive, vulgarly explicit, and vigorously conveyed. In the 18th century, *billingsgate* came to be used as a simple common noun meaning "abusive language." Today, about all that is left of Billingsgate is its lexical legacy. London's principal fish market was moved in 1982 to a modern warehouse on the Isle of Dogs, overlooking the West India Dock on the Thames.

2 : a 2000th anniversary
— **bimillenary** *adjective*

bi·mod·al \(,)bī-'mō-dᵊl\ *adjective* (1903)
: having or relating to two modes; *especially* **:** having or occurring with two statistical modes
— **bi·mo·dal·i·ty** \,bī-mō-'da-lə-tē\ *noun*

bi·mo·lec·u·lar \,bī-mə-'le-kyə-lər\ *adjective* [International Scientific Vocabulary] (1899)
1 : relating to or formed from two molecules
2 : being two molecules thick ⟨*bimolecular* lipid layers⟩
— **bi·mo·lec·u·lar·ly** *adverb*

[1]bi·month·ly \(,)bī-'mən(t)th-lē\ *adjective* (1846)
1 : occurring every two months
2 : occurring twice a month **:** SEMIMONTHLY
usage see BI-

[2]bimonthly *adverb* (1864)
1 : once every two months
2 : twice a month

[3]bimonthly *noun* (circa 1890)
: a bimonthly publication

bi·mor·phe·mic \,bī-mȯr-'fē-mik\ *adjective* (1942)
: consisting of two morphemes

[1]bin \'bin\ *noun* [Middle English *binn,* from Old English] (before 12th century)
: a box, frame, crib, or enclosed place used for storage

[2]bin *transitive verb* **binned; bin·ning** (1841)
: to put into a bin

bin- *prefix* [Middle English, from Late Latin, from Latin *bini* two by two; akin to Old English *twinn* twofold — more at TWIN]
: [1]BI- ⟨*bin*aural⟩

[1]bi·na·ry \'bī-nə-rē, -,ner-ē\ *noun, plural* **-ries** (15th century)
: something made of or based on two things or parts: as **a :** BINARY STAR **b :** a binary number system

[2]binary *adjective* [Late Latin *binarius,* from Latin *bini*] (1597)
1 : compounded or consisting of or marked by two things or parts
2 a : DUPLE — used of measure or rhythm **b :** having two musical subjects or two complementary sections
3 a : relating to, being, or belonging to a system of numbers having 2 as its base ⟨the *binary* digits 0 and 1⟩ **b :** involving a choice or condition of two alternatives (as on-off or yes-no)
4 a : composed of two chemical elements, an element and a radical that acts as an element, or two such radicals **b :** utilizing two harmless ingredients that upon combining form a lethal substance (as a gas) ⟨*binary* weapon⟩
5 : relating two logical or mathematical elements ⟨*binary* operation⟩

binary fission *noun* (1897)
: reproduction of a cell by division into two approximately equal parts ⟨the *binary fission* of protozoans⟩

binary star *noun* (circa 1847)
: a system of two stars that revolve around each other under their mutual gravitation — called also *binary system*

bi·na·tion·al \(,)bī-'na-sh(ə-)nəl\ *adjective* (1888)
: of or relating to two nations ⟨a *binational* board of directors⟩

bin·au·ral \(,)bī-'nȯr-əl, (,)bi-\ *adjective* [International Scientific Vocabulary] (1861)
1 : of, relating to, or involving two or both ears
2 : of, relating to, or constituting sound reproduction involving the use of two separated microphones and two transmission channels to achieve a stereophonic effect
— **bin·au·ral·ly** \-ə-lē\ *adverb*

[1]bind \'bīnd\ *verb* **bound** \'bau̇nd\; **bind·ing** [Middle English, from Old English *bindan;* akin to Old High German *bintan* to bind, Greek *peisma* cable, Sanskrit *badhnāti* he binds] (before 12th century)
transitive verb
1 a : to make secure by tying **b :** to confine, restrain, or restrict as if with bonds **c :** to put under an obligation ⟨*binds* himself with an oath⟩ **d :** to constrain with legal authority
2 a : to wrap around with something so as to enclose or cover **b :** BANDAGE
3 : to fasten round about
4 : to tie together (as stocks of wheat)
5 a : to cause to stick together **b :** to take up and hold (as by chemical forces) **:** combine with
6 : CONSTIPATE
7 : to make a firm commitment for ⟨a handshake *binds* the deal⟩
8 : to protect, strengthen, or decorate by a band or binding
9 : to apply the parts of the cover to (a book)
10 : to set at work as an apprentice **:** INDENTURE
11 : to cause or bring about an emotional attachment
12 : to fasten together
intransitive verb
1 a : to form a cohesive mass **b :** to combine or be taken up especially by chemical action ⟨antibody *binds* to a specific antigen⟩
2 : to hamper free movement or natural action
3 : to become hindered from free operation
4 : to exert a restraining or compelling effect ⟨a promise that *binds*⟩

[2]bind *noun* (before 12th century)
1 a : something that binds **b :** the act of binding **:** the state of being bound **c :** a place where binding occurs
2 : TIE 3
3 : a position or situation in which one is hampered, constrained, or prevented from free movement or action
— **in a bind :** in trouble

bind·er \'bīn-dər\ *noun* (before 12th century)
1 : a person or machine that binds something (as books)
2 a : something used in binding **b :** a usually detachable cover (as for holding sheets of paper)
3 : something (as tar or cement) that produces or promotes cohesion in loosely assembled substances
4 : a temporary insurance contract that provides coverage until the policy is issued
5 : something (as money) given in earnest; *also* **:** the agreement arrived at

bind·ery \'bīn-d(ə-)rē\ *noun, plural* **-er·ies** (1810)
: a place where books are bound

[1]bind·ing \'bīn-diŋ\ *noun* (13th century)
1 : the action of one that binds
2 : a material or device used to bind: as **a :** the cover and materials that hold a book together **b :** a narrow fabric used to finish raw edges **c :** a set of ski fastenings for holding the boot firm on the ski

[2]binding *adjective* (14th century)
1 : that binds
2 : imposing an obligation
— **bind·ing·ly** \-diŋ-lē\ *adverb*
— **bind·ing·ness** *noun*

binding energy *noun* (1932)
: the energy required to break up a molecule, atom, or atomic nucleus completely into its constituent particles

bin·dle stiff \'bin-dᵊl-,stif\ *noun* [*bindle,* perhaps alteration of *bundle*] (1901)
: HOBO; *especially* **:** one who carries his clothes or bedding in a bundle

bind off *transitive verb* (circa 1939)
: to cast off in knitting

bind over *transitive verb* (1610)
: to put under a bond to do something (as to appear in court)

bind·weed \'bīnd-,wēd\ *noun* (1548)
: any of various twining plants (especially genus *Convolvulus* of the morning-glory family) that mat or interlace with plants among which they grow

bine \'bīn\ *noun* [alteration of [2]*bind*] (1727)
: a twining stem or flexible shoot (as of the hop); *also* **:** a plant (as woodbine) whose shoots are bines

Bi·net–Si·mon scale \bi-,nā-sē-'mōⁿ-\ *noun* [Alfred *Binet* (died 1911) and Théodore *Simon* (died 1961) French psychologists] (1914)
: an intelligence test consisting originally of tasks graded from the level of the average 3-year-old to that of the average 12-year-old but later extended in range

[1]binge \'binj\ *noun* [English dialect *binge* (to drink heavily)] (1854)
1 a : a drunken revel **:** SPREE **b :** an unrestrained and often excessive indulgence ⟨a buying *binge*⟩
2 : a social gathering **:** PARTY

[2]binge *intransitive verb* **binged; binge·ing** *or* **bing·ing** (1910)
: to go on a binge; *especially* **:** to go on an eating binge
— **bing·er** \'bin-jər\ *noun*

[1]bin·go \'biŋ-(,)gō\ *interjection* [alteration of *bing* (interjection suggestive of a ringing sound)] (1925)
1 — used to announce an unexpected event or instantaneous result
2 — used to announce a winning position in bingo

[2]bingo *noun, plural* **bingos** (1932)
: a game of chance played with cards having numbered squares corresponding to numbered balls drawn at random and won by covering five such squares in a row; *also* **:** a social gathering at which bingo is played

bin·na·cle \'bi-ni-kəl\ *noun* [alteration of Middle English *bitakle,* from Old Portuguese or Old Spanish; Old Portuguese *bitácola* & Old Spanish *bitácula,* from Latin *habitaculum* dwelling place, from *habitare* to inhabit — more at HABITATION] (1762)
: a housing for a ship's compass and a lamp

[1]bin·oc·u·lar \bī-'nä-kyə-lər, bə-\ *adjective* (1738)
: of, relating to, using, or adapted to the use of both eyes ⟨*binocular* vision⟩
— **bin·oc·u·lar·i·ty** \(,)bī-,nä kyə 'lar-ə-tē, bə-\ *noun*
— **bin·oc·u·lar·ly** \bī-'nä-kyə-lər-lē, bə-\ *adverb*

[2]bin·oc·u·lar \bə-'nä-kyə-lər, bī-\ *noun* (1871)
1 : a binocular optical instrument
2 : a handheld optical instrument composed of two telescopes and a focusing device and usually having prisms to increase magnifying ability — usually used in plural

bi·no·mi·al \bī-'nō-mē-əl\ *noun* [New Latin *binomium,* from Medieval Latin, neuter of *binomius* having two names, alteration of Latin *binominis,* from *bi-* + *nomin-, nomen* name — more at NAME] (1557)
1 : a mathematical expression consisting of two terms connected by a plus sign or minus sign
2 : a biological species name consisting of two terms
— **binomial** *adjective*
— **bi·no·mi·al·ly** \-mē-ə-lē\ *adverb*

binomial coefficient *noun* (circa 1889)
: a coefficient of a term in the expansion of the binomial $(x + y)^n$ according to the binomial theorem

binomial distribution *noun* (1911)
: a probability function each of whose values gives the probability that an outcome with constant probability of occurrence in a statistical experiment will occur a given number of times in a succession of repetitions of the experiment

binomial nomenclature *noun* (1880)

: a system of nomenclature in which each species of animal or plant receives a name of two terms of which the first identifies the genus to which it belongs and the second the species itself

binomial theorem *noun* (1755)
: a theorem that specifies the expansion of a binomial of the form $(x + y)^n$ as the sum of $n + 1$ terms of which the general term is of the form

$$\frac{n!}{(n-k)!\,k!}x^{(n-k)}y^k$$

where k takes on values from 0 to n

bint \ˈbint\ *noun* [Arabic, girl, daughter] (1855)
British **:** GIRL, WOMAN

bi·nu·cle·ate \(ˌ)bī-ˈnü-klē-ət *also* -ˈnyü-\ *also* **bi·nu·cle·at·ed** \-klē-ˌā-təd\ *adjective* (1881)
: having two cellular nuclei ⟨*binucleate* hepatocytes⟩

bio \ˈbī-(ˌ)ō\ *noun, plural* **bi·os** (1947)
: a biography or biographical sketch

bio- — see BI-

bio·acous·tics \ˌbī-(ˌ)ō-ə-ˈkü-stiks\ *noun plural but singular in construction* (1957)
: a branch of science concerned with the production of sound by and its effects on living systems

bio·ac·tive \ˌbī-ō-ˈak-tiv\ *adjective* (1965)
: having an effect on a living organism ⟨*bioactive* molecules⟩
— **bio·ac·tiv·i·ty** \-ak-ˈti-və-tē\ *noun*

bio·as·say \ˌbī-(ˌ)ō-ˈa-ˌsā, -a-ˈsā\ *noun* (1912)
: determination of the relative strength of a substance (as a drug) by comparing its effect on a test organism with that of a standard preparation
— **bio·as·say** \-a-ˈsā, -ˈa-ˌsā\ *transitive verb*

bio·avail·abil·i·ty \-ə-ˌvā-lə-ˈbi-lə-tē\ *noun* (1971)
: the degree and rate at which a substance (as a drug) is absorbed into a living system or is made available at the site of physiological activity
— **bio·avail·able** \-ˈvā-lə-bəl\ *adjective*

bio·ce·no·sis *or* **bio·coe·no·sis** \ˌbī-ō-sə-ˈnō-səs\ *noun, plural* **-no·ses** \-ˌsēz\ [New Latin, from ²*bi-* + Greek *koinōsis* sharing, from *koinos* common] (1883)
: an ecological community especially when forming a self-regulating unit

bio·chem·i·cal \ˌbī-ō-ˈke-mi-kəl\ *adjective* [International Scientific Vocabulary] (1851)
1 : of or relating to biochemistry
2 : characterized by, produced by, or involving chemical reactions in living organisms
— **biochemical** *noun*
— **bio·chem·i·cal·ly** \-k(ə-)lē\ *adverb*

biochemical oxygen demand *noun* (circa 1927)
: the oxygen used in meeting the metabolic needs of aerobic microorganisms in water rich in organic matter (as water polluted with sewage)

bio·chem·is·try \ˌbī-ō-ˈke-mə-strē\ *noun* [International Scientific Vocabulary] (1881)
1 : chemistry that deals with the chemical compounds and processes occurring in organisms
2 : the chemical characteristics and reactions of a particular living system or biological substance (as chlorophyll) ⟨a change in the patient's *biochemistry*⟩
— **bio·chem·ist** \-mist\ *noun*

bio·chip \ˈbī-ō-ˌchip\ *noun* (1981)
: a hypothetical computer logic circuit or storage device in which the physical or chemical properties of large biological molecules (as proteins) are used to process information

bio·cid·al \ˌbī-ə-ˈsī-dᵊl\ *adjective* (1949)
: destructive to life

bio·cide \ˈbī-ə-ˌsīd\ *noun* (1947)
: a substance (as DDT) that is destructive to many different organisms

bio·cli·mat·ic \ˌbī-ō-klī-ˈma-tik\ *adjective* (1918)
: of or relating to the relations of climate and living matter ⟨*bioclimatic* adaptations⟩

bio·com·pat·i·bil·i·ty \-kəm-ˌpa-tə-ˈbi-lə-tē\ *noun* (1971)
: compatibility with living tissue or a living system by not being toxic or injurious and not causing immunological rejection
— **bio·com·pat·i·ble** \-ˈpa-tə-bəl\ *adjective*

bio·con·trol \ˌbī-ō-kən-ˈtrōl\ *noun* (1967)
: BIOLOGICAL CONTROL

bio·con·ver·sion \ˌbī-(ˌ)ō-kən-ˈvər-zhən, -shən\ *noun* (1960)
: the conversion of organic materials (as wastes) into an energy source (as methane) by processes (as fermentation) involving living organisms

bio·de·grad·able \-di-ˈgrā-də-bəl\ *adjective* (1961)
: capable of being broken down especially into innocuous products by the action of living things (as microorganisms) ⟨*biodegradable* trash bags⟩
— **bio·de·grad·abil·i·ty** \-ˌgrā-də-ˈbi-lə-tē\ *noun*
— **bio·deg·ra·da·tion** \-ˌde-grə-ˈdā-shən\ *noun*
— **bio·de·grade** \-di-ˈgrād\ *verb*

bio·de·te·ri·o·ra·tion \-di-ˌtir-ē-ə-ˈrā-shən\ *noun* (1953)
: the breakdown of materials by microbial action

bio·di·ver·si·ty \-də-ˈvər-sə-tē, -dī-\ *noun* (1986)
: biological diversity in an environment as indicated by numbers of different species of plants and animals

bio·dy·nam·ic \-di-ˈna-mik, -dī-\ *adjective* (1939)
: of or relating to a system of farming that uses only organic materials for fertilizing and soil conditioning

bio·elec·tric \-i-ˈlek-trik\ *also* **bio·elec·tri·cal** \-tri-kəl\ *adjective* (1918)
: of or relating to electric phenomena in animals and plants
— **bio·elec·tric·i·ty** \-ˌlek-ˈtri-sə-tē, -ˈtris-tē\ *noun*

bio·en·er·get·ics \-ˌe-nər-ˈje-tiks\ *noun plural but singular in construction* (1912)
1 : the biology of energy transformations and energy exchanges (as in photosynthesis) within and between living things and their environments
2 : a system of physical and psychological therapy that is held to increase well-being by releasing blocked physical and psychic energy
— **bio·en·er·get·ic** \-ˈje-tik\ *adjective*

bio·en·gi·neer·ing \-ˌen-jə-ˈnir-iŋ\ *noun* (circa 1954)
: biological or medical application of engineering principles or engineering equipment; *broadly* **:** BIOTECHNOLOGY 1
— **bio·en·gi·neer** \-ˈnir\ *noun or transitive verb*

bio·eth·ics \-ˈe-thiks\ *noun plural but singular in construction* (1971)
: a discipline dealing with the ethical implications of biological research and applications especially in medicine
— **bio·eth·i·cal** \-ˈe-thi-kəl\ *adjective*
— **bio·eth·i·cist** \-ˈe-thə-sist\ *noun*

bio·feed·back \-ˈfēd-ˌbak\ *noun* (1970)
: the technique of making unconscious or involuntary bodily processes (as heartbeats or brain waves) perceptible to the senses (as by the use of an oscilloscope) in order to manipulate them by conscious mental control

bio·foul·ing \-ˈfau̇-liŋ\ *noun* (1943)
: the gradual accumulation of waterborne organisms (as bacteria and protozoa) on the surfaces of engineering structures in water that contributes to corrosion of the structures and to a decrease in the efficiency of moving parts

bio·gas \ˈbī-ō-ˌgas\ *noun* (1974)
: a mixture of methane and carbon dioxide produced by the bacterial decomposition of organic wastes and used as a fuel

bio·gen·e·sis \ˌbī-ō-ˈje-nə-səs\ *noun* [New Latin] (1870)
1 : the development of life from preexisting life
2 : a supposed tendency for stages in the evolutionary history of a race to briefly recur during the development and differentiation of an individual of that race
3 : the synthesis of chemical compounds or structures in the living organism — compare BIOSYNTHESIS
— **bio·ge·net·ic** \-jə-ˈne-tik\ *adjective*
— **bio·ge·net·i·cal·ly** \-ti-k(ə-)lē\ *adverb*

biogenetic law *noun* (1882)
: a theory of development much disputed in biology: an organism passes through successive stages resembling the series of ancestral types from which it has descended so that the ontogeny of the individual is a recapitulation of the phylogeny of the group

bio·gen·ic \-ˈje-nik\ *also* **bi·og·e·nous** \bī-ˈä-jə-nəs\ *adjective* (1913)
: produced by living organisms ⟨*biogenic* methane formation⟩

bio·geo·chem·i·cal \-ˌjē-ə-ˈke-mi-kəl\ *noun* (1938)
: of or relating to the partitioning and cycling of chemical elements and compounds between the living and nonliving parts of an ecosystem
— **bio·geo·chem·is·try** \-mə-strē\ *noun*

bio·ge·og·ra·phy \-jē-ˈä-grə-fē\ *noun* [International Scientific Vocabulary] (1895)
: a science that deals with the geographical distribution of animals and plants
— **bio·ge·og·ra·pher** \-grə-fər\ *noun*
— **bio·geo·graph·ic** \-ˌjē-ə-ˈgra-fik\ *or* **bio·geo·graph·i·cal** \-fi-kəl\ *adjective*

bi·og·ra·phee \bī-ˌä-grə-ˈfē *also* bē-\ *noun* (1841)
: a person about whom a biography is written

bi·og·ra·pher \-ˈä-grə-fər\ *noun* (1715)
: a writer of a biography

bio·graph·i·cal \ˌbī-ə-ˈgra-fi-kəl\ *also* **bio·graph·ic** \-fik\ *adjective* (1738)
1 : of, relating to, or constituting biography
2 : consisting of biographies ⟨a *biographical* dictionary⟩
3 : relating to a list briefly identifying persons ⟨*biographical* notes⟩
— **bio·graph·i·cal·ly** \-fi-k(ə-)lē\ *adverb*

bi·og·ra·phy \bī-ˈä-grə-fē *also* bē-\ *noun, plural* **-phies** [Late Greek *biographia,* from Greek *bi-* + *-graphia* -graphy] (1683)
1 : a usually written history of a person's life
2 : biographical writings as a whole
3 : an account of the life of something (as an animal, a coin, or a building)

bio·haz·ard \ˈbī-ō-ˌha-zərd\ *noun* (1967)
: a biological agent or condition that constitutes a hazard to humans or the environment; *also* **:** a hazard posed by such an agent or condition

bi·o·log·ic \ˌbī-ə-ˈlä-jik\ *or* **bi·o·log·i·cal** \-ji-kəl\ *noun* (1921)
: a biological product used in medicine

biological *also* **biologic** *adjective* (1859)
1 : of or relating to biology or to life and living processes
2 : used in or produced by applied biology
3 : related by direct genetic relationship rather than by adoption or marriage ⟨*biological* parents⟩
— **bi·o·log·i·cal·ly** \-ji-k(ə-)lē\ *adverb*

biological clock *noun* (1955)
: an inherent timing mechanism that is inferred to exist in some living systems (as a cell) in order to explain various cyclical behaviors and physiological processes

biological control *noun* (1923)

: reduction in numbers or elimination of pest organisms by interference with their ecology (as by the introduction of parasites or diseases)

biological oxygen demand *noun* (1945)
: BIOCHEMICAL OXYGEN DEMAND

biological warfare *noun* (1946)
: warfare involving the use of living organisms (as disease germs) or their toxic products as weapons; *also* : warfare involving the use of herbicides

bi·ol·o·gism \bī-'ä-lə-,ji-zəm\ *noun* (1924)
: preoccupation with biological explanations in the analysis of social situations
— **bi·ol·o·gis·tic** \-,ä-lə-'jis-tik\ *adjective*

bi·ol·o·gy \bī-'ä-lə-jē\ *noun* [German *Biologie,* from *bi-* + *-logie* -logy] (1819)
1 : a branch of knowledge that deals with living organisms and vital processes
2 a : the plant and animal life of a region or environment **b** : the life processes especially of an organism or group; *broadly* : ECOLOGY
— **bi·ol·o·gist** \-jist\ *noun*

bio·lu·mi·nes·cence \,bī-ō-,lü-mə-'nes-ᵊn(t)s\ *noun* [International Scientific Vocabulary] (1916)
: the emission of light from living organisms; *also* : the light so produced
— **bio·lu·mi·nes·cent** \-ᵊnt\ *adjective*

bio·mass \'bī-ō-,mas\ *noun* (1934)
1 : the amount of living matter (as in a unit area or volume of habitat)
2 : plant materials and animal waste used especially as a source of fuel

bio·ma·te·ri·al \,bī-ō-mə-'tir-ē-əl\ *noun* (1966)
: material used for or suitable for use in prostheses that come in direct contact with living tissues

bio·math·e·mat·ics \-,ma-thə-'ma-tiks, -math-'ma-\ *noun plural but usually singular in construction* (1923)
: mathematics of special use in biology and medicine
— **bio·math·e·mat·i·cal** \-ti-kəl\ *adjective*
— **bio·math·e·ma·ti·cian** \-,math-mə-'ti-shən, -,ma-thə-\ *noun*

bi·ome \'bī-,ōm\ *noun* [²*bi-* + *-ome*] (1916)
: a major ecological community type (as tropical rain forest, grassland, or desert)

bio·me·chan·ics \'bī-ō-mə-'ka-niks\ *noun plural but singular or plural in construction* (circa 1933)
: the mechanics of biological and especially muscular activity (as in locomotion or exercise); *also* : the scientific study of this
— **bio·me·chan·i·cal** \-ni-kəl\ *adjective*
— **bio·me·chan·i·cal·ly** \-ni-k(ə-)lē\ *adverb*

bio·med·i·cal \,bī-ō-'me-di-kəl\ *adjective* (1955)
1 : of or relating to biomedicine
2 : of, relating to, or involving biological, medical, and physical science

bio·med·i·cine \-'me-də-sən, *British usually* -'med-sən\ *noun* (1947)
: medicine based on the application of the principles of the natural sciences and especially biology and biochemistry

bio·me·te·o·rol·o·gy \-,mē-tē-ə-'rä-lə-jē\ *noun* (1946)
: a science that deals with the relationship between living things and atmospheric phenomena
— **bio·me·te·o·ro·log·i·cal** \-rə-'lä-ji-kəl\ *adjective*

bi·om·e·try \bī-'ä-mə-trē\ *noun* [International Scientific Vocabulary] (1831)
: the statistical analysis of biological observations and phenomena
— **bio·met·ric** \,bī-ō-'me-trik\ *or* **bio·met·ri·cal** \-tri-kəl\ *adjective*
— **bio·me·tri·cian** \-me-'tri-shən\ *noun*
— **bio·met·rics** \-'me-triks\ *noun plural but singular or plural in construction*

bio·mol·e·cule \,bī-ō-'mä-li-,kyü(ə)l\ *noun* (1901)
: an organic molecule and especially a macromolecule in living organisms
— **bio·mo·lec·u·lar** \-mə-'le-kyə-lər\ *adjective*

bio·mor·phic \,bī-ō-'mȯr-fik\ *adjective* (1895)
: resembling or suggesting the forms of living organisms ⟨*biomorphic* sculptures⟩ ⟨*biomorphic* images⟩

bi·on·ic \bī-'ä-nik\ *adjective* (1963)
1 : of or relating to bionics
2 : having normal biological capability or performance enhanced by or as if by electronic or electromechanical devices

bi·on·ics \bī-'ä-niks\ *noun plural but singular or plural in construction* [²*bi-* + *-onics* (as in *electronics*)] (1960)
: a science concerned with the application of data about the functioning of biological systems to the solution of engineering problems

bi·o·nom·ics \,bī-ə-'nä-miks\ *noun plural but singular or plural in construction* [*bionomic,* adjective, probably from French *bionomique,* from *bionomie* ecology, from *bi-* + *-nomie* -nomy] (1888)
: ECOLOGY
— **bi·o·nom·ic** \-mik\ *adjective*

bio·phys·ics \,bī-ō-'fi-ziks\ *noun plural but singular or plural in construction* (1892)
: a branch of science concerned with the application of physical principles and methods to biological problems
— **bio·phys·i·cal** \-zi-kəl\ *adjective*
— **bio·phys·i·cist** \-'fi-zə-sist, -'fiz-sist\ *noun*

bio·pic \'bī-(,)ō-,pik\ *noun* [²*bi-* + ¹*pic*] (1951)
: a biographical movie

bio·poly·mer \,bī-ō-'pä-lə-mər\ *noun* (1961)
: a polymeric substance (as a protein or polysaccharide) formed in a biological system

bi·op·sy \'bī-,äp-sē\ *noun, plural* **-sies** [International Scientific Vocabulary ²*bi-* + *-opsy* (as in *autopsy*)] (1895)
: the removal and examination of tissue, cells, or fluids from the living body
— **biopsy** *transitive verb*

bio·re·ac·tor \,bī-ō-rē-'ak-tər\ *noun* (1974)
: a device or apparatus in which living organisms and especially bacteria synthesize useful substances (as interferon) or break down harmful ones (as in sewage)

bio·rhythm \'bī-ō-,ri-t͟həm\ *noun* (1960)
: an inherent rhythm that appears to control or initiate various biological processes
— **bio·rhyth·mic** \-'rit͟h-mik\ *adjective*

bio·safe·ty \,bī-ō-'sāf-tē\ *noun* (1977)
: safety with respect to the effects of biological research on humans and the environment

bio·sci·ence \-'sī-ən(t)s\ *noun* (1941)
: BIOLOGY 1; *also* : LIFE SCIENCE
— **bio·sci·en·tif·ic** \-,sī-ən-'ti-fik\ *adjective*
— **bio·sci·en·tist** \-'sī-ən-tist\ *noun*

bio·sen·sor \-'sen-,sȯr, -'sen(t)-sər\ *noun* (1962)
: a device that is sensitive to a physical or chemical stimulus (as heat or an ion) and transmits information about a life process

-biosis *noun combining form, plural* **-bioses** [New Latin, from Greek *biōsis,* from *bioun* to live, from *bios* life — more at QUICK]
: mode of life ⟨para*biosis*⟩

bio·so·cial \,bī-ō-'sō-shəl\ *adjective* (1897)
: of, relating to, or concerned with the interaction of the biological aspects and social relationships of living organisms ⟨*biosocial* science⟩
— **bio·so·cial·ly** \-'sōsh-lē, -'sō-shə-\ *adverb*

bio·sphere \'bī-ə-,sfir\ *noun* (1899)
1 : the part of the world in which life can exist
2 : living beings together with their environment
— **bio·spher·ic** \,bī-ə-'sfir-ik, -'sfer-\ *adjective*

bio·sta·tis·tics \,bī-ō-stə-'tis-tiks\ *noun plural but singular in construction* (1950)
: statistics applied to the analysis of biological data
— **bio·sta·tis·ti·cal** \-ti-kəl\ *adjective*
— **bio·stat·is·ti·cian** \-,sta-tə-'sti-shən\ *noun*

bio·strati·graph·ic \-,stra-tə-'gra-fik\ *adjective* (1947)
: of or relating to the branch of paleontology dealing with the conditions and order of deposition of sedimentary rocks
— **bio·stra·tig·ra·phy** \-strə-'ti-grə-fē\ *noun*

bio·syn·the·sis \-'sin(t)-thə-səs\ *noun* [New Latin] (1930)
: the production of a chemical compound by a living organism
— **bio·syn·thet·ic** \-sin-'the-tik\ *adjective*
— **bio·syn·thet·i·cal·ly** \-ti-k(ə-)lē\ *adverb*

bio·sys·te·mat·ics \-,sis-tə-'ma-tiks\ *noun plural but singular or plural in construction* (1945)
: experimental taxonomy especially as based on cytogenetics and genetics
— **bio·sys·te·mat·ic** \-tik\ *adjective*
— **bio·sys·tem·atist** \-'sis-tə-mə-tist, -sis-'te-mə-\ *noun*

bi·o·ta \bī-'ō-tə\ *noun* [New Latin, from Greek *biotē* life; akin to Greek *bios*] (1901)
: the flora and fauna of a region

bio·tech \'bī-ō-,tek\ *noun* (1974)
: BIOTECHNOLOGY 1

bio·tech·ni·cal \,bī-ō-'tek-ni-kəl\ *adjective* (1938)
: of or relating to biotechnology

bio·tech·nol·o·gy \,bī-ō-tek-'nä-lə-jē\ *noun* (1941)
1 : applied biological science (as bioengineering or recombinant DNA technology)
2 : ERGONOMICS
— **bio·tech·no·log·i·cal** \-,tek-nə-'lä-ji-kəl\ *adjective*
— **bio·tech·nol·o·gist** \-'nä-lə-jist\ *noun*

bio·te·lem·e·try \-tə-'le-mə-trē\ *noun* (1963)
: the remote detection and measurement of a human or animal function, activity, or condition
— **bio·tel·e·met·ric** \-,te-lə-'me-trik\ *adjective*

bi·ot·ic \bī-'ä-tik\ *adjective* [Greek *biōtikos,* from *bioun*] (1868)
: of or relating to life; *especially* : caused or produced by living beings ⟨*biotic* diversity⟩

-biotic *adjective combining form* [probably from New Latin *-bioticus,* from Greek *biōtikos*]
: having a (specified) mode of life ⟨endo*biotic*⟩

biotic potential *noun* (1935)
: the inherent capacity of an organism or species to reproduce and survive

bi·o·tin \'bī-ə-tən\ *noun* [International Scientific Vocabulary, from Greek *biotos* life, sustenance; akin to Greek *bios*] (1936)
: a colorless crystalline growth vitamin $C_{10}H_{16}N_2O_3S$ of the vitamin B complex found especially in yeast, liver, and egg yolk

bi·o·tite \'bī-ə-,tīt\ *noun* [German *Biotit,* from Jean B. *Biot* (died 1862) French mathematician] (1862)
: a generally black or dark green form of mica that is a constituent of crystalline rocks and consists of a silicate of iron, magnesium, potassium, and aluminum
— **bi·o·tit·ic** \,bī-ə-'ti-tik\ *adjective*

bio·tope \'bī-ə-,tōp\ *noun* [²*bi-* + Greek *topos* place] (1927)

: a region uniform in environmental conditions and in its populations of animals and plants for which it is the habitat

bio·trans·for·ma·tion \ˈbī-ō-ˌtran(t)s-fər-ˈmā-shən, -ˌfȯr-\ *noun* (1955)
: the transformation of chemical compounds within a living system

bio·type \-ˌtīp\ *noun* [International Scientific Vocabulary] (1906)
: the organisms sharing a specified genotype; *also* : the genotype shared or its distinguishing peculiarity
— **bio·typ·ic** \ˌbī-ə-ˈti-pik\ *adjective*

bi·pa·ren·tal \ˌbī-pə-ˈren-tᵊl\ *adjective* (1900)
: of, relating to, involving, or derived from two parents
— **bi·pa·ren·tal·ly** \-tᵊl-ē\ *adverb*

bi·par·ti·san \(ˌ)bī-ˈpär-tə-zən, -sən, -ˌzan, *chiefly British* ˌbī-ˌpär-tə-ˈzan\ *adjective* (circa 1909)
: of, relating to, or involving members of two parties ⟨a *bipartisan* commission⟩
— **bi·par·ti·san·ism** \-zə-ˌni-zəm, -sə-\ *noun*
— **bi·par·ti·san·ship** \-zən-ˌship, -sən-\ *noun*

bi·par·tite \(ˌ)bī-ˈpär-ˌtīt\ *adjective* [Latin *bipartitus*, past participle of *bipartire* to divide in two, from *bi-* + *partire* to divide, from *part-, pars* part] (1574)
1 a : being in two parts **b :** having a correspondent part for each of two parties ⟨a *bipartite* contract⟩ **c :** shared by two ⟨a *bipartite* treaty⟩
2 : divided into two parts almost to the base ⟨a *bipartite* leaf⟩
— **bi·par·tite·ly** *adverb*
— **bi·par·ti·tion** \ˌbī-(ˌ)pär-ˈti-shən\ *noun*

bi·ped \ˈbī-ˌped\ *noun* [Latin *biped-, bipes,* from *bi-* + *ped-, pes* foot — more at FOOT] (1646)
: a two-footed animal
— **bi·ped·al** \(ˌ)bī-ˈpe-dᵊl\ *adjective*
— **bi·ped·al·ly** \-ē\ *adverb*

bi·ped·al·ism \(ˌ)bī-ˈpe-dᵊl-ˌi-zəm\ *noun* (1907)
: the condition of having two feet or of using only two feet for locomotion

bi·pe·dal·i·ty \ˌbī-pə-ˈda-lə-tē\ *noun* (1847)
: BIPEDALISM

bi·phas·ic \(ˌ)bī-ˈfā-zik\ *adjective* (circa 1909)
: having two phases ⟨a *biphasic* immune response⟩

bi·phe·nyl \(ˌ)bī-ˈfe-nᵊl, -ˈfē-\ *noun* [International Scientific Vocabulary] (circa 1923)
: a white crystalline hydrocarbon $C_6H_5 \cdot C_6H_5$ used especially as a heat-transfer medium

bi·pin·nate \-ˈpi-ˌnāt\ *adjective* (1794)
: twice pinnate
— **bi·pin·nate·ly** *adverb*

bi·plane \ˈbī-ˌplān\ *noun* (1874)
: an aircraft with two main supporting surfaces usually placed one above the other

bi·pod \ˈbī-ˌpäd\ *noun* [*bi-* + *-pod* (as in *tripod*)] (1922)
: a two-legged support

bi·po·lar \(ˌ)bī-ˈpō-lər\ *adjective* (1810)
1 : having or marked by two mutually repellent forces or diametrically opposed natures or views
2 a : having or involving the use of two poles or polarities **b :** relating to, being, or using a transistor in which both electrons and holes are utilized as charge carriers
3 : relating to, associated with, or occurring in both polar regions ⟨*bipolar* species of birds⟩
4 : characterized by the alternation of manic and depressive states ⟨a *bipolar* affective disorder⟩
— **bi·po·lar·i·ty** \ˌbī-pō-ˈlar-ə-tē\ *noun*
— **bi·po·lar·iza·tion** \(ˌ)bī-ˌpō-lə-rə-ˈzā-shən\ *noun*
— **bi·po·lar·ize** \(ˌ)bī-ˈpō-lə-ˌrīz\ *transitive verb*

bi·pro·pel·lant \ˌbī-prə-ˈpe-lənt\ *noun* (1947)
: a rocket propellant consisting of separate fuel and oxidizer that come together only in a combustion chamber

bi·pyr·a·mid \(ˌ)bī-ˈpir-ə-mid\ *noun* (1897)
: a crystal consisting of two identical pyramids base to base
— **bi·py·ra·mi·dal** \-pə-ˈra-mə-dᵊl\ *adjective*

bi·qua·drat·ic \ˌbī-kwä-ˈdra-tik\ *adjective or noun* (1668)
: QUARTIC

bi·ra·cial \(ˌ)bī-ˈrā-shəl\ *adjective* (1922)
: of, relating to, or involving members of two races
— **bi·ra·cial·ism** \-shə-ˌli-zəm\ *noun*

bi·ra·di·al \(ˌ)bī-ˈrā-dē-əl\ *adjective* (circa 1909)
: having both bilateral and radial symmetry

bi·ra·mous \(ˌ)bī-ˈrā-məs\ *adjective* (1877)
: having two branches

¹birch \ˈbərch\ *noun* [Middle English, from Old English *beorc;* akin to Old High German *birka* birch, Old English *beorht* bright, and probably to Latin *fraxinus* ash tree — more at BRIGHT] (before 12th century)
1 : any of a genus (*Betula* of the family Betulaceae, the birch family) of monoecious deciduous usually short-lived trees or shrubs having simple petioled leaves and typically a layered membranous outer bark that peels readily
2 : the hard pale close-grained wood of a birch
3 : a birch rod or bundle of twigs for flogging
— **birch** *or* **birch·en** \ˈbər-chən\ *adjective*

²birch *transitive verb* (1830)
: to beat with or as if with a birch : WHIP

Birch·er \ˈbər-chər\ *noun* (1961)
: a member or adherent of the John Birch Society
— **Birch·ism** \ˈbər-ˌchi-zəm\ *noun*
— **Birch·ist** \-chist\ *or* **Birch·ite** \-ˌchīt\ *noun or adjective*

¹bird \ˈbərd\ *noun, often attributive* [Middle English, from Old English *bridd*] (before 12th century)
1 *archaic* **:** the young of a feathered vertebrate
2 : any of a class (Aves) of warm-blooded vertebrates distinguished by having the body more or less completely covered with feathers and the forelimbs modified as wings
3 : a game bird
4 : CLAY PIGEON
5 a : FELLOW **b :** a peculiar person **c** *chiefly British* **:** GIRL
6 : SHUTTLECOCK
7 *chiefly British* **a :** a hissing or jeering expressive of disapproval **b :** dismissal from employment
8 : a thin piece of meat rolled up with stuffing and cooked
9 : something (as an aircraft, rocket, or satellite) resembling a bird especially by flying or being aloft

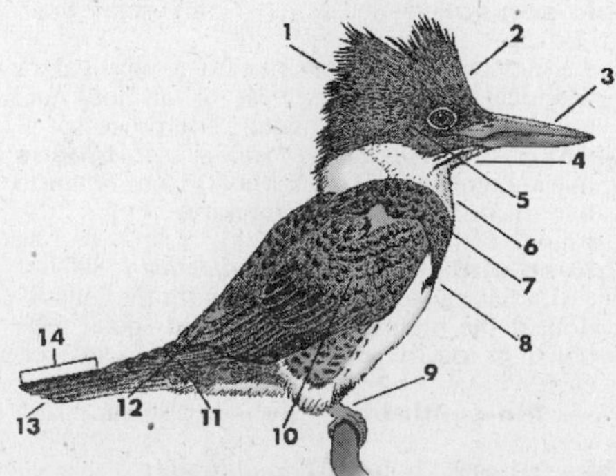

bird 2 (kingfisher): *1* **crest,** *2* **crown,** *3* **bill,** *4* **throat,** *5* **auricular region,** *6* **breast,** *7* **scapulars,** *8* **abdomen,** *9* **tarsus,** *10* **upper wing coverts,** *11* **primaries,** *12* **secondaries,** *13* **rectrix,** *14* **tail**

10 : an obscene gesture of contempt made by pointing the middle finger upward while keeping the other fingers down — usually used with *the* — called also *finger*
— **bird·like** \-ˌlīk\ *adjective*
— **for the birds :** WORTHLESS, RIDICULOUS

²bird *intransitive verb* (1918)
: to observe or identify wild birds in their natural environment

bird·bath \ˈbərd-ˌbath, -ˌbäth\ *noun* (1895)
: a usually ornamental basin set up for birds to bathe in

bird·brain \-ˌbrān\ *noun* (1943)
1 : a stupid person
2 : SCATTERBRAIN
— **bird·brained** \-ˌbrānd\ *adjective*

bird·cage \-ˌkāj\ *noun* (15th century)
: a cage for confining birds

bird·call \-ˌkȯl\ *noun* (circa 1625)
1 : a device for imitating the cry of a bird
2 : the note or cry of a bird; *also* **:** a sound imitative of it

bird colonel *noun* [from the eagle serving as insignia for this rank] (circa 1947)
slang **:** COLONEL 1a

bird–dog \ˈbərd-ˌdȯg\ (1943)
intransitive verb
: to watch closely
transitive verb
: to seek out : FOLLOW, DETECT

bird dog *noun* (circa 1888)
1 : a gundog trained to hunt or retrieve birds
2 a : one (as a canvasser or talent scout) who seeks out something for another **b :** one who steals another's date

bird–dog·ging \-ˌdȯ-giŋ\ *noun* (circa 1941)
1 : the stealing of another's date (as at a party)
2 : the action of one that bird-dogs

bird·er \ˈbər-dər\ *noun* (15th century)
1 : a catcher or hunter of birds especially for market
2 : a person who birds

bird·house \ˈbərd-ˌhau̇s\ *noun* (1870)
: an artificial nesting site for birds; *also* **:** AVIARY

¹bird·ie \ˈbər-dē\ *noun* (1792)
1 : a little bird
2 : a golf score of one stroke less than par on a hole — compare EAGLE

²birdie *transitive verb* **bird·ied; bird·ie·ing** (1948)
: to score a birdie on

bird·lime \ˈbərd-ˌlīm\ *noun* (15th century)
1 : a sticky substance usually made from the bark of a holly (*Ilex aquifolium*) that is smeared on twigs to snare small birds
2 : something that ensnares
— **birdlime** *transitive verb*

bird louse *noun* (1826)
: BITING LOUSE

bird·man \ˈbərd-mən, *especially for 1 also* -ˌman\ *noun* (1697)
1 : a person who deals with birds
2 : a person who flies (as in an aircraft)

bird–of–paradise *noun* (circa 1884)
: an ornamental plant (*Strelitzia reginae* of the family Strelitziaceae) having scapes of orange or yellow and purple flowers with three sepals and three very irregular petals

bird of paradise (1638)
: any of numerous brilliantly colored plumed oscine birds (family Paradiseidae) of the New Guinea area

bird of passage (1728)
1 : a migratory bird
2 : a person who leads a wandering or unsettled life

bird of prey (14th century)
: a carnivorous bird (as a hawk, falcon, or vulture) that feeds wholly or chiefly on meat taken by hunting or on carrion

bird pepper *noun* (1696)
: a capsicum (*Capsicum annuum glabriusculum*) having very small oblong extremely pungent red fruits

bird·seed \ˈbərd-ˌsēd\ *noun* (1840)

: a mixture of seeds (as of hemp, millet, and sunflowers) used for feeding caged and wild birds

[1]**bird's–eye** \'bərd-ˌzī\ *noun* (1597)
1 : any of numerous plants with small bright-colored flowers; *especially* **:** a speedwell (*Veronica chamaedrys*)
2 a : an allover pattern for textiles consisting of a small diamond with a center dot **b :** a fabric woven with this pattern
3 : a small spot in wood surrounded with an ellipse of concentric fibers

[2]**bird's–eye** *adjective* (1665)
1 : marked with spots resembling birds' eyes
2 : having or involving a bird's-eye view ⟨*bird's-eye* perspective⟩ ⟨a *bird's-eye* survey of the topic⟩

bird's–eye view *noun* (1762)
1 : a view from a high angle as if seen by a bird in flight
2 : an overall or cursory look at something

bird's–foot trefoil \'bərdz-ˌfu̇t-\ *noun* (1833)
: a European legume (*Lotus corniculatus*) having claw-shaped pods and widely used especially in the U.S. as a forage and fodder plant or often planted along roadsides

bird's–foot violet *noun* (1839)
: a common violet (*Viola pedata*) of the eastern U.S. with deeply cleft leaves and pale blue to purple flattened flowers

bird's–nest fern \'bərdz-ˌnest-\ *noun* (1858)
: a large epiphytic spleenwort (*Asplenium nidus*) of tropical Asia and Polynesia that has large lance-shaped leaves and is often grown as a houseplant

bird's nest soup *noun* (1871)
: a soup made with the nests of small cave-nesting swifts (genus *Collocalia*) that build them using a glutinous secretion from their salivary glands

bird·song \'bərd-ˌsȯŋ\ *noun* (1896)
: the song of a bird

bird–watch \'bərd-ˌwäch\ *intransitive verb* [back-formation from *bird-watcher*] (1948)
: BIRD

bird–watch·er \-ˌwä-chər\ *noun* (1905)
: BIRDER 2

bi·re·frin·gence \ˌbī-ri-'frin-jən(t)s\ *noun* [International Scientific Vocabulary] (1898)
: the refraction of light in an anisotropic material (as calcite) in two slightly different directions to form two rays
— bi·re·frin·gent \-jənt\ *adjective*

bi·reme \'bī-ˌrēm\ *noun* [Latin *biremis,* from *bi-* + *remus* oar — more at ROW] (1662)
: a galley with two banks of oars used especially by the ancient Greeks and Phoenicians

bi·ret·ta \bə-'re-tə\ *noun* [Italian *berretta,* from Old Provençal *berret* cap, from Medieval Latin *birretum,* from Late Latin *birrus* cloak with a hood, perhaps of Celtic origin; akin to Middle Irish *berr* short] (1598)
: a square cap with three ridges on top worn by clergymen especially of the Roman Catholic Church

biretta

birk \'birk\ *noun* [Middle English *birch, birk*] (14th century)
chiefly Scottish **:** BIRCH

birk·ie \'bir-kē, 'bər-\ *noun* [origin unknown] (1724)
1 *Scottish* **:** a lively smart assertive person
2 *Scottish* **:** FELLOW, BOY

[1]**birl** \'bər(-ə)l, *Scottish also* 'bir(-ə)l\ *intransitive verb* [Middle English from Old English *byrelian;* probably akin to Old English *beran* to carry — more at BEAR] (circa 1585)
chiefly Scottish **:** CAROUSE

[2]**birl** *verb* [perhaps imitative] (1790)
transitive verb
1 : SPIN
2 : to cause (a floating log) to rotate by treading
intransitive verb
: to progress by whirling
— birl·er \'bər-lər, 'bir-\ *noun*

Bi·ro \'bī-(ˌ)rō\ *trademark*
— used for a ballpoint pen

birr \'bər, 'bir\ *noun, plural* **birr** [Arabic] (circa 1978)
— see MONEY table

birse \'birs, 'bərs\ *noun* [(assumed) Middle English *birst,* from Old English *byrst* — more at BRISTLE] (before 12th century)
1 *chiefly Scottish* **:** a bristle or tuft of bristles
2 *chiefly Scottish* **:** ANGER

[1]**birth** \'bərth\ *noun, often attributive* [Middle English, from Old Norse *byrth;* akin to Old English *beran*] (13th century)
1 a : the emergence of a new individual from the body of its parent **b :** the act or process of bringing forth young from the womb
2 : a state resulting from being born especially at a particular time or place ⟨a Southerner by *birth*⟩
3 a : LINEAGE, EXTRACTION ⟨marriage between equals in *birth*⟩ **b :** high or noble birth
4 a *archaic* **:** one that is born **b :** BEGINNING, START ⟨the *birth* of an idea⟩

[2]**birth** (1906)
transitive verb
1 *chiefly dialect* **:** to bring forth
2 : to give rise to **:** ORIGINATE
intransitive verb
: to bring forth a child or young

birth canal *noun* (1927)
: the channel formed by the cervix, vagina, and vulva through which the fetus passes during birth

birth certificate *noun* (1900)
: a copy of an official record of a person's date and place of birth and parentage

birth control *noun* (1914)
: control of the number of children born especially by preventing or lessening the frequency of conception **:** CONTRACEPTION

birth·day \'bərth-ˌdā\ *noun* (14th century)
1 a : the day of a person's birth **b :** a day of origin
2 : an anniversary of a birth ⟨her 21st *birthday*⟩

birthday suit *noun* (1753)
: unclothed skin **:** NAKEDNESS

birth defect *noun* (1971)
: a physical or biochemical defect that is present at birth and may be inherited or environmentally induced

birth·mark \'bərth-ˌmärk\ *noun* (1580)
: an unusual mark or blemish on the skin at birth **:** NEVUS

birth pang *noun* (circa 1887)
1 : one of the regularly recurrent pains that are characteristic of childbirth — usually used in plural
2 *plural* **:** disorder and distress incident especially to a major social change

birth·place \'bərth-ˌplās\ *noun* (1607)
: place of birth or origin

birth·rate \'bərth-ˌrāt\ *noun* (1859)
: the ratio between births and individuals in a specified population and time

birth·right \'bərth-ˌrīt\ *noun* (1535)
: a right, privilege, or possession to which a person is entitled by birth

birth·root \'bərth-ˌrüt, -ˌru̇t\ *noun* (1822)
: any of several trilliums with astringent roots used in folk medicine

birth·stone \'bərth-ˌstōn\ *noun* (1907)
: a gemstone associated symbolically with the month of one's birth

birth·wort \-ˌwərt, -ˌwȯrt\ *noun* (1551)
: any of several herbs or woody vines (genus *Aristolochia* of the family Aristolochiaceae, the birthwort family) with aromatic roots used in folk medicine to aid childbirth

bis \'bis\ *adverb* [Latin, from Old Latin *dvis;* akin to Old High German *zwiro* twice, Latin *duo* two — more at TWO] (1819)
1 : AGAIN — used in music as a direction to repeat
2 : TWICE

Bi·sa·yan \bə-'sī-ən\ *noun* [Bisayan *Bisayâ*] (1951)
1 : a member of any of several peoples in the Visayan islands, Philippines
2 : the group of Austronesian languages of the Bisayans

bis·cuit \'bis-kət\ *noun, plural* **biscuits** *also* **biscuit** [Middle English *bisquite,* from Middle French *bescuit,* from (*pain*) *bescuit* twice-cooked bread] (14th century)
1 a : any of various hard or crisp dry baked products: as (1) *British* **:** CRACKER 4 (2) *British* **:** COOKIE **b :** a small quick bread made from dough that has been rolled out and cut or dropped from a spoon
2 : earthenware or porcelain after the first firing and before glazing
3 a : a light grayish yellowish brown **b :** a grayish yellow ◆

bi·sect \'bī-ˌsekt, bī-'\ *verb* [[1]*bi-* + inter*sect*] (circa 1645)
transitive verb
: to divide into two usually equal parts
intransitive verb
: CROSS, INTERSECT
— bi·sec·tion \'bī-ˌsek-shən, bī-'\ *noun*
— bi·sec·tion·al \-shnəl, -shə-nᵊl\ *adjective*
— bi·sec·tion·al·ly *adverb*

bi·sec·tor \'bī-ˌsek-tər, bī-'\ *noun* (1864)
: one that bisects; *especially* **:** a straight line that bisects an angle or a line segment

bi·sex·u·al \(ˌ)bī-'sek-sh(ə-)wəl, -shəl\ *adjective* (1824)
1 a : possessing characters of both sexes **:** HERMAPHRODITIC **b :** sexually oriented toward both sexes
2 : of, relating to, or involving both sexes
— bisexual *noun*
— bi·sex·u·al·i·ty \ˌbī-ˌsek-shə-'wa-lə-tē\ *noun*
— bi·sex·u·al·ly \(ˌ)bī-'sek-sh(ə-)wə-lē, -sh(ə-)lē\ *adverb*

bish·op \'bi-shəp\ *noun* [Middle English *bisshop,* from Old English *bisceop,* from Late Latin *episcopus,* from Greek *episkopos,* literally, overseer, from *epi-* + *skeptesthai* to look — more at SPY] (before 12th century)
1 : one having spiritual or ecclesiastical supervision: as **a :** an Anglican, Eastern Orthodox, or Roman Catholic clergyman ranking above a

◇ WORD HISTORY

biscuit In earlier ages the preservation of food presented a greater problem than it does today, especially on long journeys. One approach was to preserve flat cakes of bread by baking them a second time in order to dry them out. In medieval French the result of this process was called *pain bescuit,* literally "twice-cooked bread." In the 14th century, the second element of this phrase was borrowed into English. The notion of cooking twice having been lost, *biscuit* came to be used to designate various hard or crisp, dry baked products. (This is the sense that now prevails in England, while Americans would say *cookie* or *cracker.*) Similarities in shape and size led to the use of *biscuit* as the name for a small quick bread made from dough that has been rolled out and cut, or dropped from a spoon.

priest, having authority to ordain and confirm, and typically governing a diocese **b :** any of various Protestant clerical officials who superintend other clergy **c :** a Mormon high priest presiding over a ward or over all other bishops and over the Aaronic priesthood
2 : either of two pieces of each color in a set of chessmen having the power to move diagonally across any number of adjoining unoccupied squares
3 : mulled port wine flavored with roasted oranges and cloves

bish·op·ric \'bi-shə-(,)prik\ *noun* [Middle English *bisshopriche,* from Old English *bisceoprīce,* from *bisceop* + *rīce* kingdom — more at RICH] (before 12th century)
1 : DIOCESE
2 : the office of bishop
3 : the administrative body of a Mormon ward consisting of a bishop and two high priests as counselors

Bishops' Bible *noun* (1835)
: an officially commissioned English translation of the Bible published in 1568

bis·muth \'biz-məth\ *noun* [New Latin *bismuthum,* modification of German *Wismut*] (1668)
: a heavy brittle grayish white chiefly trivalent metallic element that is chemically like arsenic and antimony and that is used in alloys and pharmaceuticals — see ELEMENT table
— bis·mu·thic \biz-'mə-thik, -'myü-\ *adjective*

bi·son \'bī-sᵊn, -zᵊn\ *noun, plural* **bison** [Latin *bisont-, bison,* of Germanic origin; akin to Old High German *wisant* aurochs; akin to Old Prussian *wissambrs* aurochs] (1611)
: any of several large shaggy-maned usually gregarious recent or extinct bovine mammals (genus *Bison*) having a large head with short horns and heavy forequarters surmounted by a large fleshy hump: as **a :** WISENT **b :** BUFFALO 1c(1)
— bi·son·tine \-sᵊn-,tīn, -zᵊn-\ *adjective*

¹bisque \'bisk\ *noun* [French] (1647)
1 a : a thick cream soup made with shellfish or game **b :** a cream soup of pureed vegetables
2 : ice cream containing powdered nuts or macaroons

²bisque *noun* [French] (circa 1656)
: odds allowed an inferior player: as **a :** a point taken when desired in a set of tennis **b :** an extra turn in croquet **c :** one or more strokes off a golf score

³bisque *noun* [by shortening & alteration] (1664)
: BISCUIT 2; *especially* **:** unglazed china that is not to be glazed but is hard-fired and vitreous

bi·state \'bī-,stāt\ *adjective* (1928)
: of or relating to two states

bis·ter *or* **bis·tre** \'bis-tər\ *noun* [French *bistre*] (circa 1751)
1 : a yellowish brown to dark brown pigment used in art
2 : a grayish to yellowish brown
— bis·tered \-tərd\ *adjective*

bis·tort \'bis-,tȯrt, bis-'\ *noun* [Middle French *bistorte,* from (assumed) Medieval Latin *bistorta,* from Latin *bis* + *torta,* feminine of *tortus,* past participle of *torquēre* to twist — more at TORTURE] (1578)
: any of several polygonums; *especially* **:** a European herb (*Polygonum bistorta*) or a related American plant (*P. bistortoides*) with twisted roots used as astringents

bis·tro \'bēs-(,)trō, 'bis-\ *noun, plural* **bistros** [French] (1921)
1 : a small or unpretentious restaurant
2 a : a small bar or tavern **b :** NIGHTCLUB
— bis·tro·ic \bēs-'trō-ik, bis-\ *adjective*

bi·sul·fate \(,)bī-'səl-,fāt\ *noun* (circa 1846)
: an acid sulfate

bi·sul·fide \-,fīd\ *noun* [International Scientific Vocabulary] (1863)
: DISULFIDE

bi·sul·fite \-,fīt\ *noun* [French, from *bi-* + *sulfite*] (circa 1846)
: an acid sulfite

bi–swing \'bī-'swiŋ\ *adjective* [¹*bi-* + *swing;* perhaps from the freedom of movement allowed by this jacket] (1968)
: made with a pleat or gusset at the back of the arms ⟨*bi-swing* jacket⟩

¹bit \'bit\ *noun* [Middle English *bitt,* from Old English *bite* act of biting; akin to Old English *bītan*] (14th century)
1 a (1) **:** the biting or cutting edge or part of a tool (2) **:** a replaceable part of a compound tool that actually performs the function (as drilling or boring) for which the whole tool is designed **b** *plural* **:** the jaws of tongs or pincers
2 : something bitten or held with the teeth: **a :** the usually steel part of a bridle inserted in the mouth of a horse **b :** the rimmed mouth end on the stem of a pipe or cigar holder
3 : something that curbs or restrains
4 : the part of a key that enters the lock and acts on the bolt and tumblers

²bit *transitive verb* **bit·ted; bit·ting** (1583)
1 a : to put a bit in the mouth of (a horse) **b :** to control as if with a bit
2 : to form a bit on (a key)

³bit *noun* [Middle English, from Old English *bita;* akin to Old English *bītan*] (before 12th century)
1 : a small quantity of food; *especially* **:** a small delicacy
2 a : a small piece or quantity of some material thing **b** (1) **:** a small coin (2) **:** a unit of value equal to ⅛ of a dollar ⟨four *bits*⟩
3 : something small or unimportant of its kind: as **a :** a brief period **:** WHILE **b :** an indefinite usually small degree, extent, or amount ⟨a *bit* of a rascal⟩ ⟨every *bit* as powerful⟩ **c** (1) **:** a small part usually with spoken lines in a theatrical performance (2) **:** a usually short theatrical routine ⟨a corny comedy *bit*⟩
4 : the aggregate of items, situations, or activities appropriate to a given style, genre, or role ⟨rejected the whole *bit* about love-marriage-motherhood —Vance Packard⟩
— a bit : SOMEWHAT, RATHER ⟨the play was *a bit* dull⟩
— a bit much : a little more than one wants to endure
— bit by bit : by degrees **:** LITTLE BY LITTLE

⁴bit *noun* [*bi*nary digi*t*] (1948)
1 : a unit of computer information equivalent to the result of a choice between two alternatives (as *yes* or *no, on* or *off*)
2 : the physical representation of a bit by an electrical pulse, a magnetized spot, or a hole whose presence or absence indicates data

bi·tar·trate \(,)bī-'tär-,trāt\ *noun* [International Scientific Vocabulary] (1879)
: an acid tartrate

¹bitch \'bich\ *noun* [Middle English *bicche,* from Old English *bicce*] (before 12th century)
1 : the female of the dog or some other carnivorous mammals
2 a : a lewd or immoral woman **b :** a malicious, spiteful, or domineering woman — sometimes used as a generalized term of abuse
3 : something that is highly objectionable or unpleasant
4 : COMPLAINT

²bitch (1823)
transitive verb
1 : SPOIL, BOTCH ⟨I must have *bitched* up my life —Mavis Gallant⟩
2 : CHEAT, DOUBLECROSS
3 : to complain of or about
intransitive verb
: COMPLAIN

bitch·ery \'bi-chə-rē\ *noun, plural* **-er·ies** (1936)
: malicious, spiteful, or domineering behavior; *also* **:** an instance of such behavior

bitch goddess *noun* (1906)
: SUCCESS; *especially* **:** material or worldly success

bitchy \'bi-chē\ *adjective* **bitch·i·er; -est** (1937)
: characterized by malicious, spiteful, or arrogant behavior
— bitch·i·ly \'bi-chə-lē\ *adverb*
— bitch·i·ness \'bi-chē-nəs\ *noun*

¹bite \'bīt\ *verb* **bit** \'bit\; **bit·ten** \'bi-tᵊn\ *also* **bit; bit·ing** \'bī-tiŋ\ [Middle English, from Old English *bītan;* akin to Old High German *bīzan* to bite, Latin *findere* to split] (before 12th century)
transitive verb
1 a : to seize especially with teeth or jaws so as to enter, grip, or wound **b :** to wound, pierce, or sting especially with a fang or a proboscis
2 : to cut or pierce with or as if with an edged weapon
3 : to cause sharp pain or stinging discomfort to
4 : to take hold of
5 *archaic* **:** to take in **:** CHEAT
intransitive verb
1 : to bite or have the habit of biting something
2 *of a weapon or tool* **:** to cut, pierce, or take hold
3 : to cause irritation or smarting
4 : CORRODE
5 a *of fish* **:** to take a bait **b :** to respond so as to be caught (as by a trick)
6 : to take or maintain a firm hold
— bit·er \'bī-tər\ *noun*
— bite off more than one can chew : to undertake more than one can perform
— bite the bullet : to enter with resignation upon a difficult or distressing course of action
— bite the dust 1 : to fall dead especially in battle **2 :** to suffer humiliation or defeat
— bite the hand that feeds one : to injure a benefactor maliciously

²bite *noun* (15th century)
1 : the act or manner of biting
2 : FOOD: as **a :** the amount of food taken at a bite **:** MORSEL **b :** a small amount of food **:** SNACK
3 *archaic* **a :** CHEAT, TRICK **b :** SHARPER
4 : a wound made by biting
5 : the hold or grip by which friction is created or purchase is obtained
6 : a surface that creates friction or is brought into contact with another for the purpose of obtaining a hold
7 a : a keen incisive quality **b :** a sharp penetrating effect
8 : a single exposure of an etcher's plate to the corrosive action of acid
9 : an amount taken usually in one operation for one purpose **:** CUT

bite·wing \'bīt-,wiŋ\ *noun* (1938)
: a dental X-ray film designed to show the crowns of the upper and lower teeth simultaneously

bit·ing \'bī-tiŋ\ *adjective* (14th century)
: having the power to bite ⟨a *biting* wind⟩; *especially* **:** able to grip and impress deeply ⟨the report is *biting* in its intolerance of deceit⟩
— bit·ing·ly \-tiŋ-lē\ *adverb*

biting louse *noun* (1896)
: any of numerous wingless insects (order Mallophaga) that are parasitic especially on birds — called also *bird louse*

biting midge *noun* (1945)
: any of a family (Ceratopogonidae) of tiny biting dipteran flies of which some are vectors of filarial worms

bit·stock \'bit-,stäk\ *noun* (circa 1887)
: BRACE 3

bit·sy \'bit-sē\ *adjective* [*itsy-bitsy*] (1905)
: TINY

¹bitt \'bit\ *noun* [perhaps from Old Norse *biti* beam; akin to Old English *bōt* boat] (14th century)

1 : a post or pair of posts fixed on the deck of a ship for securing lines
2 : BOLLARD 1

[2]**bitt** *transitive verb* (1769)
: to make (a cable) fast about a bitt

[1]**bit·ter** \ˈbi-tər\ *adjective* [Middle English, from Old English *biter;* akin to Old High German *bittar* bitter, Old English *bītan*] (before 12th century)
1 a : being or inducing the one of the four basic taste sensations that is peculiarly acrid, astringent, or disagreeable and suggestive of an infusion of hops — compare SALT, SOUR, SWEET **b :** distasteful or distressing to the mind **:** GALLING ⟨a *bitter* sense of shame⟩
2 : marked by intensity or severity: **a :** accompanied by severe pain or suffering ⟨a *bitter* death⟩ **b :** being relentlessly determined **:** VEHEMENT ⟨a *bitter* partisan⟩ **c :** exhibiting intense animosity ⟨*bitter* enemies⟩ **d** (1) **:** harshly reproachful ⟨*bitter* complaints⟩ (2) **:** marked by cynicism and rancor ⟨*bitter* contempt⟩ **e :** intensely unpleasant especially in coldness or rawness
3 : expressive of severe pain, grief, or regret ⟨*bitter* tears⟩
— bit·ter·ish \ˈbi-tə-rish\ *adjective*
— bit·ter·ly *adverb*
— bit·ter·ness *noun*

[2]**bitter** *noun* (before 12th century)
1 : bitter quality
2 a *plural* **:** a usually alcoholic solution of bitter and often aromatic plant products used especially in preparing mixed drinks or as a mild tonic **b** *British* **:** a very dry heavily hopped ale

[3]**bitter** *transitive verb* (12th century)
: to make bitter ⟨*bittered* ale⟩

[4]**bitter** *adverb* (1749)
: to a bitter degree ⟨it's *bitter* cold⟩

bitter almond *noun* (1632)
: an almond with a bitter taste that contains amygdalin; *also* **:** a tree (*Prunus dulcis amara*) producing bitter almonds

bit·ter·brush \ˈbi-tər-ˌbrəsh\ *noun* (1910)
: a much-branched silvery shrub (*Purshia tridentata*) of arid western North America that has 3-toothed leaves and yellow flowers and is valuable for forage

bitter cress *noun* (circa 1890)
: any of a genus (*Cardamine*) of cruciferous herbs that produce flat pods and wingless seeds

[1]**bitter end** *noun* [probably from [2]*bitter end*] (1849)
: the last extremity however painful or calamitous
— bit·ter–end·er \ˌbi-tər-ˈen-dər\ *noun*

[2]**bitter end** *noun* [*bitter* (a turn of cable around the bitts)] (circa 1862)
: the inboard end of a ship's anchoring cable

[1]**bit·tern** \ˈbi-tərn\ *noun* [Middle English *bitoure,* from Middle French *butor,* from (assumed) Vulgar Latin *butitaurus,* from Latin *buteo* hawk + *taurus* bull] (1515)
: any of various small or medium-sized usually secretive herons (especially genera *Botaurus* and *Ixobrychus*)

[2]**bittern** *noun* [irregular from [1]*bitter*] (1682)
: the bitter water solution of salts that remains after sodium chloride has crystallized out of a brine

bit·ter·root \ˈbi-tə(r)-ˌrüt, -ˌru̇t\ *noun* (1838)
: a succulent herb (*Lewisia rediviva*) of mountainous slopes of western North America that has starchy roots and pink or white flowers

[1]**bit·ter·sweet** \ˈbi-tər-ˌswēt\ *noun* (14th century)
1 : something that is bittersweet; *especially* **:** pleasure alloyed with pain
2 a : a sprawling poisonous weedy nightshade (*Solanum dulcamara*) with purple flowers and oval reddish orange berries **b :** a North American woody climbing plant (*Celastrus scandens*) of the staff-tree family having clusters of small greenish flowers succeeded by yellow capsules that open when ripe and disclose the scarlet aril

[2]**bittersweet** *adjective* (1611)
1 : being at once bitter and sweet; *especially* **:** pleasant but including or marked by elements of suffering or regret ⟨a *bittersweet* ballad⟩
2 : of or relating to a prepared chocolate containing little sugar
— bit·ter·sweet·ly *adverb*
— bit·ter·sweet·ness *noun*

bit·ter·weed \ˈbi-tər-ˌwēd\ *noun* (1819)
: any of several American plants containing a bitter substance: as **a :** HORSEWEED **b :** a sneezeweed (*Helenium amarum* synonym *H. tenuifolium*) **c :** an erect composite herb (*Hymenoxys odorata*) of the southwestern U.S. that has yellow terminal flowerheads and is poisonous to livestock

bit·tock \ˈbi-tək\ *noun* (circa 1802)
chiefly Scottish **:** a little bit

[1]**bit·ty** \ˈbi-tē\ *adjective* (1892)
chiefly British **:** made up of or containing bits; *especially* **:** not cohesive or flowing

[2]**bitty** *adjective* (1905)
: SMALL, TINY ⟨a little *bitty* room⟩

bi·tu·men \bə-ˈtyü-mən, bī-, -ˈtü-, *especially British also* ˈbit-yə-\ *noun* [Middle English *bithumen* mineral pitch, from Latin *bitumin-, bitumen*] (15th century)
1 : an asphalt of Asia Minor used in ancient times as a cement and mortar
2 : any of various mixtures of hydrocarbons (as tar) often together with their nonmetallic derivatives that occur naturally or are obtained as residues after heat-refining natural substances (as petroleum); *specifically* **:** such a mixture soluble in carbon disulfide
— bi·tu·mi·ni·za·tion \bə-ˌtyü-mə-nə-ˈzā-shən, bī-, -ˌtü-\ *noun*
— bi·tu·mi·nize \-ˈtyü-mə-ˌnīz, -ˈtü-\ *transitive verb*

bi·tu·mi·nous \bə-ˈtyü-mə-nəs, bī-, -ˈtü-\ *adjective* (1620)
1 : containing or impregnated with bitumen
2 : of or relating to bituminous coal

bituminous coal *noun* (1827)
: a coal that when heated yields considerable volatile bituminous matter — called also *soft coal*

bi·unique \ˌbī-yu̇-ˈnēk\ *adjective* (1941)
: being a correspondence between two sets that is one-to-one in both directions
— bi·unique·ness *noun*

[1]**bi·va·lent** \(ˌ)bī-ˈvā-lənt, ˈbī-və-\ *adjective* (1869)
1 : DIVALENT
2 : associated in pairs in synapsis

[2]**bivalent** *noun* (circa 1934)
: a pair of synaptic chromosomes

[1]**bi·valve** \ˈbī-ˌvalv\ *adjective* (1661)
: having a shell composed of two valves

[2]**bivalve** *noun* (1683)
: an animal (as a clam) with a 2-valved shell

bi·var·i·ate \(ˌ)bī-ˈver-ē-ət, -ˈvar-\ *adjective* (1920)
: of, relating to, or involving two variables ⟨a *bivariate* frequency distribution⟩

[1]**biv·ouac** \ˈbi-və-ˌwak, ˈbiv-ˌwak\ *noun* [French, from Low German *biwacht,* from *bi* by + *wacht* guard] (circa 1702)
1 : a usually temporary encampment under little or no shelter
2 a : encampment usually for a night **b :** a temporary shelter or settlement

[2]**bivouac** *intransitive verb* **-ouacked; -ouack·ing** (1809)
: to make a bivouac **:** CAMP

[1]**bi·week·ly** \(ˌ)bī-ˈwē-klē\ *adjective* (1832)
1 : occurring twice a week
2 : occurring every two weeks **:** FORTNIGHTLY
usage see BI-
— biweekly *adverb*

[2]**biweekly** *noun* (circa 1890)
1 : a publication issued every two weeks
2 : SEMIWEEKLY

bi·year·ly \(ˌ)bī-ˈyir-lē\ *adjective* (circa 1909)
1 : BIANNUAL 1
2 : BIENNIAL 1

biz \ˈbiz\ *noun* (1862)
: BUSINESS

[1]**bi·zarre** \bə-ˈzär\ *adjective* [French, from Italian *bizzarro*] (circa 1648)
: strikingly out of the ordinary: as **a :** odd, extravagant, or eccentric in style or mode **b :** involving sensational contrasts or incongruities
synonym see FANTASTIC
— bi·zarre·ly *adverb*
— bi·zarre·ness *noun*

[2]**bizarre** *noun* (circa 1753)
: a flower with atypical striped marking

bi·zar·re·rie \bi-ˌzär-ə-ˈrē\ *noun* [French] (1747)
1 : a bizarre quality
2 : something bizarre

bi·zon·al \(ˌ)bī-ˈzō-n^{ə}l\ *adjective* (1946)
: of or relating to the affairs of a zone governed or administered by two powers acting together
— bi·zone \ˈbī-ˌzōn\ *noun*

[1]**blab** \ˈblab\ *noun* [Middle English *blabbe;* akin to Middle English *blaberen*] (14th century)
1 *archaic* **:** one that blabs **:** TATTLETALE
2 : idle or excessive talk **:** CHATTER
— blab·by \ˈbla-bē\ *adjective*

[2]**blab** *verb* **blabbed; blab·bing** (15th century)
intransitive verb
1 : to talk idly or thoughtlessly **:** PRATTLE
2 : to reveal a secret especially by indiscreet chatter
transitive verb
: to reveal especially without reserve or discretion ⟨*blabbed* the whole affair to the press⟩

[1]**blab·ber** \ˈbla-bər\ *verb* **blab·bered; blab·ber·ing** \-b(ə-)riŋ\ [Middle English *blaberen*] (14th century)
intransitive verb
: to talk foolishly or excessively
transitive verb
: to say indiscreetly

[2]**blabber** *noun* (circa 1913)
: idle talk **:** BABBLE

[3]**blabber** *noun* [[2]*blab*] (1557)
: one that blabs

blab·ber·mouth \ˈbla-bər-ˌmau̇th\ *noun* (1936)
: a person who talks too much; *especially* **:** TATTLETALE

[1]**black** \ˈblak\ *adjective* [Middle English *blak,* from Old English *blæc;* akin to Old High German *blah* black, and probably to Latin *flagrare* to burn, Greek *phlegein*] (before 12th century)
1 a : of the color black **b** (1) **:** very dark in color ⟨his face was *black* with rage⟩ (2) **:** having a very deep or low register ⟨a bass with a *black* voice⟩ (3) **:** HEAVY, SERIOUS ⟨the play was a *black* intrigue⟩
2 a : having dark skin, hair, and eyes **:** SWARTHY ⟨the *black* Irish⟩ **b** (1) *often capitalized* **:** of or relating to any of various population groups having dark pigmentation of the skin ⟨*black* Americans⟩ (2) **:** of or relating to the Afro-American people or their culture ⟨*black* literature⟩ ⟨a *black* college⟩ ⟨*black* pride⟩ ⟨*black* studies⟩ (3) **:** typical or representative of the most readily attended parts of black culture ⟨tried to play *blacker* jazz⟩
3 : dressed in black
4 : DIRTY, SOILED ⟨hands *black* with grime⟩

5 a : characterized by the absence of light ⟨a *black* night⟩ **b :** reflecting or transmitting little or no light ⟨*black* water⟩ **c :** served without milk or cream ⟨*black* coffee⟩
6 a : thoroughly sinister or evil **:** WICKED ⟨a *black* deed⟩ **b :** indicative of condemnation or discredit ⟨got a *black* mark for being late⟩
7 : connected with or invoking the supernatural and especially the devil ⟨*black* magic⟩
8 a : very sad, gloomy, or calamitous ⟨*black* despair⟩ **b :** marked by the occurrence of disaster ⟨*black* Friday⟩
9 : characterized by hostility or angry discontent **:** SULLEN ⟨*black* resentment filled his heart⟩
10 *chiefly British* **:** subject to boycott by trade-union members as employing or favoring nonunion workers or as operating under conditions considered unfair by the trade union
11 a *of propaganda* **:** conducted so as to appear to originate within an enemy country and designed to weaken enemy morale **b :** characterized by or connected with the use of black propaganda ⟨*black* radio⟩
12 : characterized by grim, distorted, or grotesque satire ⟨*black* humor⟩
13 : of or relating to covert intelligence operations ⟨*black* government programs⟩
— black•ish \ˈbla-kish\ *adjective*
— black•ly *adverb*
— black•ness *noun*

[2]black *noun* (before 12th century)
1 : a black pigment or dye; *especially* **:** one consisting largely of carbon
2 : the achromatic color of least lightness characteristically perceived to belong to objects that neither reflect nor transmit light
3 : something that is black: as **a :** black clothing ⟨looks good in *black*⟩ **b :** a black animal (as a horse)
4 a : a person belonging to any of various population groups having dark pigmentation of the skin **b :** AFRO-AMERICAN
5 : the pieces of a dark color in a board game for two players (as chess)
6 : total or nearly total absence of light ⟨the *black* of night⟩
7 : the condition of making a profit — usually used with *the* ⟨operating in the *black*⟩; compare RED

[3]black (13th century)
intransitive verb
: to become black
transitive verb
1 : to make black
2 *chiefly British* **:** to declare (as a business or industry) subject to boycott by trade-union members

black alder *noun* (1805)
: WINTERBERRY 1

black•a•moor \ˈbla-kə-ˌmu̇r\ *noun* [irregular from *black* + *Moor*] (1547)
: a dark-skinned person; *especially* **:** BLACK 4a

black–and–blue \ˌbla-kən-ˈblü\ *adjective* (14th century)
: darkly discolored from blood effused by bruising

black–and–tan \-kən-ˈtan\ *adjective* (1850)
1 : having a predominantly black color pattern with deep red or rusty tan on the feet, breeching, and cheek patches, above the eyes, and inside the ears
2 : favoring or practicing proportional representation of whites and blacks in politics
3 : frequented by both blacks and whites ⟨a *black-and-tan* bar⟩
4 : a member of a black-and-tan political organization (as in the southern U.S.) — compare LILY-WHITE

black and tan *noun* (1870)
1 : a black-and-tan animal (as a dog)
2 *chiefly British* **:** a drink consisting of a dark beer (as stout or porter) and ale or beer
3 *B&T capitalized* [from the color of his uniform] **:** a recruit enlisted in England in 1920–21 for service in the Royal Irish Constabulary against the armed movement for Irish independence

black–and–tan coonhound *noun* (1948)
: any of a breed of strong vigorous coonhounds of American origin with black-and-tan markings

black–and–white \ˌbla-kən-ˈhwīt, -ˈwīt\ *adjective* (1612)
1 : being in writing or print
2 : partly black and partly white in color
3 : executed in dark pigment on a light background or in light pigment on a dark ground ⟨a *black-and-white* drawing⟩
4 : MONOCHROME 2a ⟨*black-and-white* film⟩ ⟨*black-and-white* television⟩
5 a : sharply divided into good and evil **b :** evaluating or viewing things as either all good or all bad ⟨*black-and-white* morality⟩ **c :** sharply defined **:** CLEAR-CUT ⟨the truth is not always *black-and-white*⟩

black and white *noun* (1599)
1 : WRITING, PRINT
2 : a drawing or print done in black and white or in monochrome
3 : monochrome reproduction of visual images (as by photography or television)

Black Angus *noun* (1948)
: ANGUS

black–a–vised \ˈbla-kə-ˌvīst\ *adjective* [*black* + French *à vis* as to face] (circa 1758)
: having a dark complexion

[1]black•ball \ˈblak-ˌbȯl\ *transitive verb* (1770)
1 : to vote against; *especially* **:** to exclude from membership by casting a negative vote
2 a : to exclude socially **:** OSTRACIZE **b :** BOYCOTT

[2]blackball *noun* (circa 1847)
1 : a small black ball for use as a negative vote in a ballot box
2 : an adverse vote especially against admitting someone to membership in an organization

black bass *noun* (1815)
: any of a genus (*Micropterus*) of freshwater sunfishes native to eastern and central North America and including the largemouth bass and smallmouth bass

black bean *noun* (1792)
1 : a black kidney bean commonly used in Latin American cuisine
2 : a black soybean commonly used in oriental cuisine

black bear *noun* (1781)
: the common American bear (*Ursus americanus*) ranging in color from brown or typical black to white

black bear

black beast *noun* (1926)
: BÊTE NOIRE

[1]black belt \ˈblak-ˌbelt\ *noun* (1870)
1 : an area characterized by rich black soil
2 *often both Bs capitalized* **:** an area densely populated by blacks

[2]black belt *noun* (1954)
: one who holds the rating of expert in various arts of self-defense (as judo and karate); *also* **:** the rating itself

black•ber•ry \ˈblak-ˌber-ē\ *noun* (before 12th century)
1 : the usually black or dark purple juicy but seedy edible aggregate fruit of various brambles (genus *Rubus*) of the rose family
2 : a plant that bears blackberries

black bile *noun* (1797)
: a humor of medieval physiology believed to be secreted by the kidneys or spleen and to cause melancholy

black birch *noun* (1674)
: SWEET BIRCH

[1]black•bird \ˈblak-ˌbərd\ *noun* (14th century)
1 : any of various birds of which the males are largely or entirely black: as **a :** a common thrush (*Turdus merula*) of Eurasia and northern Africa having an orange bill and eye rim **b :** any of several American birds (family Icteridae)
2 : a Pacific islander kidnapped for use as a plantation laborer

[2]blackbird *intransitive verb* (1894)
: to engage in the slave trade especially in the South Pacific

black•bird•er *noun* (1880)
1 : a person who blackbirds
2 : a ship used in blackbirding

black•board \ˈblak-ˌbōrd, -ˌbȯrd\ *noun* (1823)
: a hard smooth usually dark surface used especially in a classroom for writing or drawing on with chalk

black•body \ˈblak-ˈbä-dē\ *noun* (1710)
: an ideal body or surface that completely absorbs all radiant energy falling upon it with no reflection and that radiates at all frequencies with a spectral energy distribution dependent on its absolute temperature

black book *noun* (1592)
: a book containing a blacklist

black box *noun* (circa 1945)
1 : a usually complicated electronic device that functions and is packaged as a unit and whose internal mechanism is usually hidden from or mysterious to the user; *broadly* **:** anything that has mysterious or unknown internal functions or mechanisms
2 : a crashworthy device in aircraft for recording cockpit conversations and flight data

black•cap \ˈblak-ˌkap\ *noun* (1678)
1 : any of several birds with black heads or crowns: as **a :** a small European warbler (*Sylvia atricapilla*) with a black crown **b :** CHICKADEE
2 : BLACK RASPBERRY

black–capped \-ˈkapt\ *adjective* (1781)
of a bird **:** having the top of the head black

black•cock \-ˌkäk\ *noun* (15th century)
chiefly British **:** BLACK GROUSE; *specifically* **:** the male black grouse

black cohosh *noun* (1828)
: a bugbane (*Cimicifuga racemosa*) of the eastern U.S.

black crappie *noun* (circa 1827)
: a silvery black-mottled sunfish (*Pomoxis nigromaculatus*) of the Mississippi drainage and eastern U.S. having seven or eight protruding spines on the dorsal fins

black–crowned night heron \ˈblak-ˌkrau̇nd-\ *noun* (1844)
: a widely distributed night heron (*Nycticorax nycticorax*) having a black crown and back

black damp *noun* (1836)
: a carbon dioxide mixture occurring as a mine gas and incapable of supporting life or flame

black death *noun, often B&D capitalized* [from the black patches formed on the skin of its victims] (1780)
1 : PLAGUE 2b
2 : a severe epidemic of plague and especially bubonic plague that occurred in Asia and Europe in the 14th century

black diamond *noun* (1763)
1 : dense black hematite
2 *plural* **:** COAL 3a
3 : [3]CARBONADO

black duck *noun* (1637)
: any of several ducks that are dark in color; *especially* **:** a common brown duck (*Anas rubripes*) of eastern North America

black dwarf *noun, plural* **black dwarfs** (1945)
: a very small star that emits no detectable light

black•en \ˈbla-kən\ *verb* **black•ened; black•en•ing** \ˈblak-niŋ, ˈbla-kə-\ (14th century)
intransitive verb

: to become dark or black ⟨the sky *blackens*⟩
transitive verb
1 : to make black
2 : DEFAME, SULLY
— black·en·er \ˈblak-nər, ˈbla-kə-\ *noun*

black·ened \-kənd\ *adjective* (1984)
: coated with a mixture of cayenne pepper and spices and cooked in a frying pan over extremely high heat ⟨*blackened* redfish⟩

Black English *noun* (1969)
: a nonstandard variety of English spoken by some American blacks — called also *Black English vernacular*

black·en·ing \ˈblak-niŋ, ˈbla-kə-\ *noun* (circa 1909)
: BLACKING

black eye *noun* (1604)
1 : a discoloration of the skin around the eye from bruising
2 : a bad reputation

black–eyed pea \ˈblak-ˌīd-\ *noun* (1728)
: COWPEA

black–eyed Su·san \-ˈsü-zᵊn\ *noun* (1892)
: either of two North American coneflowers (*Rudbeckia hirta* and *R. serotina*) having flower heads with deep yellow to orange rays and dark conical disks

black·face \ˈblak-ˌfās\ *noun* (1869)
: makeup applied to a performer playing a black person especially in a minstrel show; *also* **:** a performer wearing such makeup

black·fish \-ˌfish\ *noun* (1754)
1 : any of numerous dark-colored fishes: as **a :** TAUTOG **b :** a small food fish (*Dallia pectoralis*) of Alaska and Siberia that is noted for its resistance to cold
2 : any of several small toothed whales (genus *Globicephala*) found in warm seas

black–flag \-ˈflag\ *transitive verb* (1963)
: to signal (a race-car driver) to go immediately to the pits

black flag *noun* (1720)
: a pirate's flag usually bearing a skull and crossbones

black·fly \ˈblak-ˌflī\ *noun, plural* **-flies** *or* **-fly** (1608)
: any of various small dark-colored insects; *especially* **:** any of a family (Simuliidae and especially genus *Simulium*) of bloodsucking dipteran flies whose larvae usually live in flowing streams

Black·foot \ˈblak-ˌfu̇t\ *noun, plural* **Black·feet** *or* **Blackfoot** (1794)
1 : a member of an American Indian people of Montana, Alberta, and Saskatchewan
2 : the Algonquian language of the Blackfeet

black–foot·ed albatross \ˈblak-ˌfu̇-təd-\ *noun* (1839)
: an albatross (*Diomedea nigripes*) of the Pacific that is chiefly blackish with dusky bill and black feet and legs — called also *gooney*

black–footed ferret *noun* (1846)
: a rare weasel (*Mustela nigripes*) of western North American prairies having a yellowish coat, dark feet and facial mask, and a dark-tipped tail

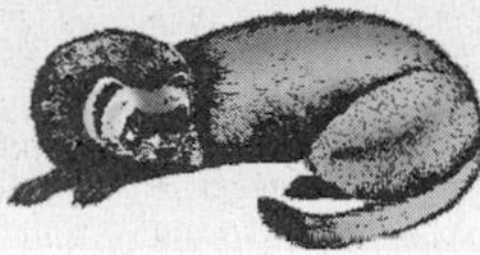
black-footed ferret

black gold *noun* (1910)
: PETROLEUM

black grouse *noun* (1678)
: a large grouse (*Lyrurus tetrix*) of western Asia and Europe of which the male is black with white wing patches and the female is barred and mottled

[1]black·guard \ˈbla-gərd, -ˌgärd; ˈblak-ˌgärd\ *noun* (1535)
1 *obsolete* **:** the kitchen servants of a household
2 a : a rude or unscrupulous person **b :** a person who uses foul or abusive language
— black·guard·ism \-gər-ˌdi-zəm, -ˌgär-\ *noun*
— black·guard·ly \-gərd-lē, -ˌgärd-\ *adjective or adverb*

[2]blackguard *transitive verb* (1823)
: to talk about or address in abusive terms

black gum *noun* (1709)
: a tupelo (*Nyssa sylvatica*) of the eastern U.S. with light and soft but tough wood

black hand *noun, often B&H capitalized* [*Black Hand,* a Sicilian and Italian-American secret society of the late 19th and 20th centuries] (1904)
: a lawless secret society engaged in criminal activities (as terrorism or extortion)
— black·hand·er \ˈblak-ˌhan-dər\ *noun*

black·head \ˈblak-ˌhed\ *noun* (circa 1837)
1 : a small plug of sebum blocking the duct of a sebaceous gland especially on the face
2 : a destructive disease of turkeys and related birds caused by a protozoan (*Histomonas meleagridis*) that invades the intestinal ceca and liver
3 : a larval clam or mussel attached to the skin or gills of a freshwater fish

black·heart \-ˌhärt\ *noun* (1909)
: a plant disease in which the central tissues blacken

black hole *noun* (1968)
1 : a hypothetical celestial object with a gravitational field so strong that light cannot escape from it which is believed to be created in the collapse of a very massive star
2 : something resembling a black hole: as **a :** something that consumes a resource continually ⟨a financial *black hole*⟩ **b :** an empty space **:** VOID

black ice *noun* (1961)
: a thin film of ice on paved surfaces (as roads) that is difficult to see

black·ing \ˈbla-kiŋ\ *noun* (1571)
: a substance (as a paste or polish) that is applied to an object to make it black

[1]black·jack \-ˌjak\ *noun* (1591)
1 [*black* + *jack* (vessel)] **:** a tankard for beer or ale usually of tar-coated leather
2 : SPHALERITE
3 : a hand weapon typically consisting of a piece of leather-enclosed metal with a strap or springy shaft for a handle
4 : an often scrubby oak (*Quercus marilandica*) chiefly of the southeastern U.S. — called also *blackjack oak*
5 : a card game the object of which is to be dealt cards having a higher count than those of the dealer up to but not exceeding 21 — called also *twenty-one, vingt-et-un*

[2]blackjack *transitive verb* (1905)
1 : to strike with a blackjack
2 : to coerce with threats or pressure

black·land \ˈblak-ˌland\ *noun* (1803)
1 : a heavy sticky black soil such as that covering large areas in Texas
2 *plural* **:** a region of blackland

black·lead \ˈblak-ˌled\ *noun* (1583)
chiefly British **:** GRAPHITE 1

black·leg \ˈblak-ˌleg, -ˌlāg\ *noun* (circa 1722)
1 : a usually fatal toxemia especially of young cattle caused by a soil bacterium (*Clostridium chauvoei*)
2 : a cheating gambler **:** SWINDLER
3 *chiefly British* **:** a worker hostile to trade unionism or acting in opposition to union policies **:** SCAB

black letter *noun* (circa 1644)
: a heavy angular condensed typeface used especially by the earliest European printers and based on handwriting used chiefly in the 13th to 15th centuries; *also* **:** this style of handwriting

mnopqrstuv

black letter

black light *noun* (1927)
1 : invisible ultraviolet light
2 : a lamp that radiates black light

black·light trap \ˈblak-ˌlīt-\ *noun* (1961)
: a trap for insects that uses a form of black light perceptible to particular insects as an attractant

[1]black·list \-ˌlist\ *noun* (circa 1619)
: a list of persons who are disapproved of or are to be punished or boycotted

[2]blacklist *transitive verb* (1718)
: to put on a blacklist
— black·list·er *noun*

black locust *noun* (1787)
: a tall tree (*Robinia pseudoacacia*) of eastern North America with pinnately compound leaves, drooping racemes of fragrant white flowers, and strong stiff wood

black lung *noun* (1837)
: pneumoconiosis caused by habitual inhalation of coal dust

black·mail \ˈblak-ˌmāl\ *noun* [*black* + [1]*mail*] (1552)
1 : a tribute anciently exacted on the Scottish border by freebooting chiefs for immunity from pillage
2 a : extortion or coercion by threats especially of public exposure or criminal prosecution **b :** the payment that is extorted
— blackmail *transitive verb*
— black·mail·er *noun*

Black Ma·ria \ˌblak-mə-ˈrī-ə\ *noun* (1847)
: PADDY WAGON

black–mar·ket (1943)
intransitive verb
: to buy or sell goods in the black market
transitive verb
: to sell in the black market
— black marketer *or* **black mar·ke·teer** \-ˌmär-kə-ˈtir\ *noun*

black market *noun* (1931)
: illicit trade in goods or commodities in violation of official regulations; *also* **:** a place where such trade is carried on

Black Mass *noun* (1893)
: a travesty of the Christian mass ascribed to worshipers of Satan

Black Muslim *noun* (1960)
: a member of a chiefly black group that professes Islamic religious belief

black nationalist *noun, often B&N capitalized* (1963)
: a member of a group of militant blacks who advocate separatism from the whites and the formation of self-governing black communities
— black nationalism *noun, often B&N capitalized*

black·out \ˈblak-ˌau̇t\ *noun* (1913)
1 a : a turning off of the stage lighting to separate scenes in a play or end a play or skit; *also* **:** a skit that ends with a blackout **b :** a period of darkness enforced as a precaution against air raids **c :** a period of darkness (as in a city) caused by a failure of electrical power
2 : a transient dulling or loss of vision, consciousness, or memory ⟨an alcoholic *blackout*⟩
3 a : a wiping out **:** OBLITERATION **b :** a blotting out by censorship **:** SUPPRESSION ⟨a news *blackout*⟩
4 : a usually temporary loss of a radio signal
5 : the prohibition or restriction of the telecasting of a sports event

black out (1921)
intransitive verb
1 : to become enveloped in darkness

2 : to undergo a temporary loss of vision, consciousness, or memory
transitive verb
1 : to cause to black out ⟨*black out* the stage⟩
2 : to make inoperative (as by a power failure)
3 a : BLOT OUT, ERASE ⟨*blacked out* the event from his mind⟩ **b :** to suppress by censorship ⟨*black out* the news⟩
4 : to impose a blackout on

Black Panther *noun* (1965)
: a member of an organization of militant American blacks

black pepper *noun* (before 12th century)
: a spice that consists of the fruit of an East Indian plant (*Piper nigrum*) ground with the black husk still on

black•poll \'blak-ˌpōl\ *noun* (1783)
: a North American warbler (*Dendroica striata*) with the male having a black cap when in breeding plumage

black powder *noun* (circa 1909)
: an explosive mixture of potassium nitrate or sodium nitrate, charcoal, and sulfur used especially in fireworks and as a propellant in antique firearms — compare GUNPOWDER

black power *noun, often B&P capitalized* (1966)
: the mobilization of the political and economic power of American blacks especially to compel respect for their rights and improve their condition

black pudding *noun* (1568)
chiefly British **:** BLOOD SAUSAGE

black racer *noun* (1849)
: an American blacksnake (*Coluber constrictor constrictor*) common in the eastern U.S.

black raspberry *noun* (circa 1782)
: a raspberry (*Rubus occidentalis*) of eastern North America that has a purplish black fruit and is the source of several cultivated varieties — called also *blackcap*

black rhinoceros *noun* (1850)
: a rhinoceros (*Diceros bicornis*) of sub-Saharan Africa having a prehensile upper lip that protrudes in the middle — called also *black rhino*

Black Rod *noun* (circa 1607)
: the principal usher of the House of Lords

black rot *noun* (1849)
: a bacterial or fungal rot of plants marked by dark brown discoloration

black sheep *noun* (1792)
: a discreditable member of a respectable group

Black•shirt \'blak-ˌshərt\ *noun* (1922)
: a member of a fascist organization having a black shirt as a distinctive part of its uniform; *especially* **:** a member of the Italian Fascist party

black•smith \'blak-ˌsmith\ *noun* [from a distinction between black metal (iron) and white metal (tin)] (15th century)
: a smith who forges iron
— black•smith•ing \-ˌsmi-thiŋ\ *noun*

black smoker *noun* (1980)
: a vent in a geologically active region of the sea floor from which issues superheated water laden with minerals (as sulfide precipitates)

black•snake \-ˌsnāk\ *noun* (1634)
1 : any of several snakes that are largely black or very dark in color; *especially* **:** either of two harmless snakes (*Coluber constrictor* and *Elaphe obsoleta*) of the U.S.
2 : a long tapering braided whip of rawhide or leather

black spot *noun* (circa 1889)
: any of several bacterial or fungal diseases of plants characterized by black spots or blotches especially on the leaves

black•tail \'blak-ˌtāl\ *noun* (1828)
: BLACK-TAILED DEER

black–tailed deer \'blak-ˌtāl(d)-\ *noun* (1806)
: MULE DEER; *specifically* **:** one of a subspecies (*Odocoileus hemionus columbianus*) especially of British Columbia, Oregon, and Washington

black tea *noun* (1789)
: tea that is dark in color from complete fermentation of the leaf before firing

black•thorn \'blak-ˌthórn\ *noun* (14th century)
: a European spiny plum (*Prunus spinosa*) with hard wood and small white flowers

black–tie *adjective* (1933)
: characterized by or requiring the wearing of semiformal evening clothes consisting of a usually black tie and tuxedo for men and a formal dress for women ⟨a *black-tie* dinner⟩ — compare WHITE-TIE

black•top \'blak-ˌtäp\ *noun* (1931)
: a bituminous material used especially for surfacing roads; *also* **:** a surface paved with blacktop
— blacktop *transitive verb*

Blackwall hitch \'blak-ˌwól-\ *noun* [*Blackwall,* shipyard in London, England] (circa 1862)
: a hitch for securing a rope to a hook — see KNOT illustration

black walnut *noun* (1612)
: a walnut (*Juglans nigra*) of eastern North America with hard strong heavy dark brown wood and oily edible nuts; *also* **:** its wood or nut

black•wa•ter \'blak-ˌwó-tər, -ˌwä-\ *noun* (1800)
: any of several diseases (as blackwater fever) characterized by dark-colored urine

blackwater fever *noun* (1884)
: a febrile complication of repeated malarial attacks that is characterized especially by extensive kidney damage and urine discolored by heme from blood

black widow *noun* (1915)
: a venomous New World spider (*Latrodectus mactans*) the female of which is black with an hourglass-shaped red mark on the underside of the abdomen

black widow

black•wood \'blak-ˌwůd\ *noun* (1631)
: any of several hardwood trees (as the legumes *Acacia melanoxylon* and *Dalbergia latifolia*) or their dark-colored wood

blad•der \'bla-dər\ *noun* [Middle English, from Old English *blǣdre;* akin to Old High German *blātara* bladder, Old English *blāwan* to blow] (before 12th century)
1 a : a membranous sac in animals that serves as the receptacle of a liquid or contains gas; *especially* **:** URINARY BLADDER **b :** CYST 2b
2 : something (as the rubber bag inside a football) resembling a bladder
— blad•der•like \-ˌlīk\ *adjective*

bladder campion *noun* (1817)
: an Old World campion (*Silene vulgaris* synonym *S. cucubalus*) with an inflated calyx introduced into temperate North America

blad•der•nut \'bla-dər-ˌnət\ *noun* (1578)
: any of a genus (*Staphylea* of the family Staphyleaceae, the bladdernut family) of ornamental shrubs or small trees with panicles of small white flowers followed by inflated capsules; *also* **:** one of the capsules

bladder worm *noun* (1858)
: a bladderlike larval tapeworm (as a cysticercus)

blad•der•wort \'bla-dər-ˌwərt, -ˌwórt\ *noun* (circa 1815)
: any of a genus (*Utricularia* of the family Lentibulariaceae, the bladderwort family) of chiefly aquatic plants with leaves usually having insect-trapping bladders

bladder wrack *noun* (1810)
: a common rockweed (*Fucus vesiculosus*) used in preparing kelp and as a manure

blade \'blād\ *noun* [Middle English, from Old English *blæd;* akin to Old High German *blat* leaf, Latin *folium,* Greek *phyllon,* Old English *blōwan* to blossom — more at BLOW] (before 12th century)
1 a : LEAF 1a(1); *especially* **:** the leaf of an herb or a grass **b :** the flat expanded part of a leaf as distinguished from the petiole
2 : something resembling the blade of a leaf: as **a :** the broad flattened part of an oar or paddle **b :** an arm of a screw propeller, electric fan, or steam turbine **c :** the broad flat or concave part of a machine (as a bulldozer or snowplow) that comes into contact with the material to be moved **d :** a broad flat body part; *specifically* **:** SCAPULA — used chiefly in naming cuts of meat **e :** the flat portion of the tongue immediately behind the tip; *also* **:** this portion together with the tip
3 a : the cutting part of an implement **b** (1) **:** SWORD (2) **:** SWORDSMAN (3) **:** a dashing lively man **c :** the runner of an ice skate
— blade•like \-ˌlīk\ *adjective*

blad•ed \'blā-dəd\ *adjective* (1578)
: having blades — often used in combination ⟨five-*bladed* propeller⟩

blae \'blā\ *adjective* [Middle English *bla, blo,* from Old Norse *blār;* akin to Old High German *blāo* blue — more at BLUE] (13th century)
chiefly Scottish **:** dark blue or bluish gray

blae•ber•ry \'blā-ˌber-ē, -ˌb(ə-)rē\ *noun* (15th century)
chiefly Scottish **:** WHORTLEBERRY 1

¹blah \'blä\ *noun* [imitative] (1918)
1 *also* **blah–blah** \-ˌblä\ **:** silly or pretentious chatter or nonsense
2 *plural* [perhaps influenced in meaning by *blasé*] **:** a feeling of boredom, lethargy, or general dissatisfaction

²blah *adjective* (1923)
: lacking interest **:** MEDIOCRE ⟨a *blah* winter day⟩

blain \'blān\ *noun* [Middle English, from Old English *blegen;* akin to Middle Low German *bleine* blain, Old English *blāwan* to blow] (before 12th century)
: an inflammatory swelling or sore

blam•able \'blā-mə-bəl\ *adjective* (14th century)
: deserving blame **:** REPREHENSIBLE
synonym see BLAMEWORTHY
— blam•ably \-blē\ *adverb*

¹blame \'blām\ *transitive verb* **blamed; blam•ing** [Middle English, from Old French *blamer,* from Late Latin *blasphemare* to blaspheme, from Greek *blasphēmein*] (13th century)
1 : to find fault with **:** CENSURE ⟨the right to praise or *blame* a literary work⟩
2 a : to hold responsible ⟨they *blame* me for everything⟩ **b :** to place responsibility for ⟨*blames* it on me⟩ ■
— blam•er *noun*
— to blame : at fault **:** RESPONSIBLE

²blame *noun* (13th century)
1 : an expression of disapproval or reproach **:** CENSURE
2 a : a state of being blameworthy **:** CULPABILITY **b** *archaic* **:** FAULT, SIN

□ USAGE
blame Use of *blame* in sense 2b with *on* has occasionally been disparaged as wrong. Such disparagement is without basis; *blame on* occurs as frequently in carefully edited prose as *blame for.* Both forms are standard.

3 : responsibility for something believed to deserve censure ⟨they must share the *blame*⟩
— **blame·less** \-ləs\ *adjective*
— **blame·less·ly** *adverb*
— **blame·less·ness** *noun*

blame·ful \ˈblām-fəl\ *adjective* (14th century)
: BLAMABLE
— **blame·ful·ly** \-fə-lē\ *adverb*

blame·wor·thy \-ˌwər-t͟hē\ *adjective* (14th century)
: being at fault **:** deserving blame ☆
— **blame·wor·thi·ness** *noun*

blanc fixe \ˈblaŋk-ˈfiks\ *noun* [French, literally, fixed white] (1866)
: barium sulfate prepared as a heavy white powder and used especially as a filler in paper, rubber, and linoleum or as a pigment

blanch \ˈblanch\ *verb* [Middle English *blaunchen,* from Middle French *blanchir,* from Old French *blanche,* feminine of *blanc,* adjective, white — more at BLANK] (15th century)
transitive verb
: to take the color out of: as **a :** to bleach by excluding light ⟨*blanch* celery⟩ **b :** to scald or parboil in water or steam in order to remove the skin from, whiten, or stop enzymatic action in (as food for freezing) **c :** to make ashen or pale ⟨fear *blanches* the cheek⟩
intransitive verb
: to become white or pale
— **blanch·er** *noun*

blanc·mange \blə-ˈmänj, -ˈmä^{n}zh\ *noun* [Middle English *blancmanger,* from Middle French *blanc manger,* literally, white food] (14th century)
: a usually sweetened and flavored dessert made from gelatinous or starchy ingredients and milk

bland \ˈbland\ *adjective* [Latin *blandus*] (1661)
1 a : smooth and soothing in manner or quality ⟨a *bland* smile⟩ **b :** exhibiting no personal concern or embarrassment **:** UNPERTURBED ⟨a *bland* confession of guilt⟩
2 a : not irritating, stimulating, or invigorating **:** SOOTHING **b :** DULL, INSIPID ⟨*bland* stories with little plot or action⟩
synonym see SUAVE
— **bland·ly** \ˈblan(d)-lē\ *adverb*
— **bland·ness** \ˈblan(d)-nəs\ *noun*

blan·dish \ˈblan-dish\ *verb* [Middle English, from Middle French *blandiss-,* stem of *blandir,* from Latin *blandiri,* from *blandus* mild, flattering] (14th century)
transitive verb
: to coax with flattery **:** CAJOLE
intransitive verb
: to act or speak in a flattering or coaxing manner
— **blan·dish·er** *noun*

blan·dish·ment \-dish-mənt\ *noun* (1591)
: something that tends to coax or cajole **:** ALLUREMENT — often used in plural

¹blank \ˈblaŋk\ *adjective* [Middle English, from Middle French *blanc* colorless, white, of Germanic origin; akin to Old High German *blanch* white; probably akin to Latin *flagrare* to burn — more at BLACK] (14th century)
1 *archaic* **:** COLORLESS
2 a : appearing or causing to appear dazed, confounded, or nonplussed ⟨stared in *blank* dismay⟩ **b :** EXPRESSIONLESS ⟨a *blank* stare⟩
3 a : devoid of covering or content; *especially* **:** free from writing or marks ⟨*blank* paper⟩ **b :** having spaces to be filled in **c :** lacking interest, variety, or change ⟨*blank* hours⟩
4 : ABSOLUTE, UNQUALIFIED ⟨a *blank* refusal⟩
5 : UNFINISHED; *especially* **:** having a plain or unbroken surface where an opening is usual ⟨a *blank* key⟩ ⟨a *blank* arch⟩
synonym see EMPTY
— **blank·ly** *adverb*
— **blank·ness** *noun*

²blank *noun* (1554)
1 *obsolete* **:** the bull's-eye of a target
2 a : an empty space (as on a paper) **b :** a paper with spaces for the entry of data ⟨an order *blank*⟩
3 a : a piece of material prepared to be made into something (as a key) by a further operation **b :** a cartridge loaded with propellant and a seal but no projectile
4 a : an empty or featureless place or space ⟨my mind was a *blank*⟩ **b :** a vacant or uneventful period ⟨a long *blank* in history⟩
5 : a dash substituting for an omitted word

³blank (circa 1765)
transitive verb
1 a : OBSCURE, OBLITERATE ⟨*blank* out a line⟩ **b :** to stop access to **:** SEAL ⟨*blank* off a tunnel⟩
2 : to keep (an opposing team) from scoring ⟨were *blanked* for eight innings⟩
intransitive verb
1 : FADE — usually used with *out* ⟨the music *blanked* out⟩
2 : to become confused or abstracted — often used with *out* ⟨his mind *blanked* out momentarily⟩

blank check *noun* (1884)
1 : complete freedom of action or control **:** CARTE BLANCHE
2 : a signed check with the amount unspecified

¹blan·ket \ˈblaŋ-kət\ *noun* [Middle English, from Old French *blankete,* from *blanc*] (14th century)
1 a : a large usually oblong piece of woven fabric used as a bed covering **b :** a similar piece of fabric used as a body covering (as for an animal) ⟨a horse *blanket*⟩
2 : something that resembles a blanket ⟨a *blanket* of fog⟩ ⟨a *blanket* of gloom⟩
3 : a rubber or plastic sheet on the cylinder in an offset press that transfers the image to the surface being printed
— **blan·ket·like** \-ˌlīk\ *adjective*

²blanket *transitive verb* (1605)
1 : to cover with or as if with a blanket ⟨new grass *blankets* the slope⟩
2 a : to cover so as to obscure, interrupt, suppress, or extinguish ⟨*blanket* a fire with foam⟩ **b :** to apply or cause to apply to uniformly despite wide separation or diversity among the elements included ⟨freight rates that *blanket* a region⟩ **c :** to cause to be included ⟨automatically *blanketed* into the insurance program⟩

³blanket *adjective* (1886)
1 : covering all members of a group or class ⟨a *blanket* wage increase⟩
2 : effective or applicable in all instances

blan·ket·flow·er \ˈblaŋ-kət-ˌflau̇(-ə)r\ *noun* (1879)
: GAILLARDIA

blanket stitch *noun* (1880)
: a buttonhole stitch with spaces of variable width used on materials too thick to hem
— **blanket–stitch** *transitive verb*

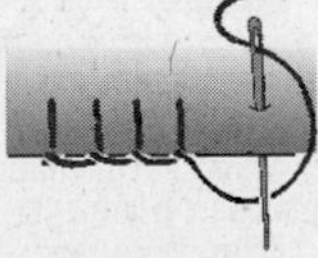
blanket stitch

blank verse *noun* (1588)
: unrhymed verse; *specifically* **:** unrhymed iambic pentameter verse

blan·quette \blä^{n}-ˈket\ *noun* [French, from Provençal *blanqueto,* from *blanc* white, of Germanic origin; akin to Old High German *blanch* white] (1727)
: a stew of light meat or seafood in a white sauce ⟨*blanquette* of veal⟩ ⟨*blanquette* of lobster⟩

¹blare \ˈblar, ˈbler\ *verb* **blared; blar·ing** [Middle English *bleren;* akin to Middle Dutch *blēren* to shout] (15th century)
intransitive verb
: to sound loud and strident ⟨radios *blaring*⟩
transitive verb
1 : to sound or utter raucously ⟨sat *blaring* the car horn⟩
2 : to proclaim flamboyantly ⟨headlines *blared* his defeat⟩

²blare *noun* (1809)
1 : a loud strident noise
2 : dazzling often garish brilliance
3 : FLAMBOYANCE

blar·ney \ˈblär-nē\ *noun* [*Blarney stone,* a stone in Blarney Castle, near Cork, Ireland, held to bestow skill in flattery on those who kiss it] (1796)
1 : skillful flattery **:** BLANDISHMENT
2 : NONSENSE, HUMBUG ◆
— **blarney** *verb*

bla·sé \blä-ˈzā\ *also* **bla·se** *adjective* [French] (1819)
1 : apathetic to pleasure or excitement as a result of excessive indulgence or enjoyment **:** WORLD-WEARY
2 : SOPHISTICATED, WORLDLY-WISE
3 : UNCONCERNED
synonym see SOPHISTICATED

blas·pheme \blas-ˈfēm, ˈblas-ˌ\ *verb* **blas·phemed; blas·phem·ing** [Middle English *blasfemen,* from Late Latin *blasphemare* — more at BLAME] (14th century)
transitive verb
1 : to speak of or address with irreverence
2 : REVILE, ABUSE
intransitive verb
: to utter blasphemy
— **blas·phem·er** \-ˈfē-mər; ˈblas-ˌfē-mər, -fə-mər\ *noun*

blas·phe·mous \ˈblas-fə-məs\ *adjective* (15th century)
: impiously irreverent **:** PROFANE
— **blas·phe·mous·ly** *adverb*
— **blas·phe·mous·ness** *noun*

blas·phe·my \ˈblas-fə-mē\ *noun, plural* **-mies** (13th century)

☆ SYNONYMS
Blameworthy, blamable, guilty, culpable mean deserving reproach or punishment. BLAMEWORTHY and BLAMABLE apply to any degree of reprehensibility ⟨conduct adjudged *blameworthy*⟩ ⟨an accident for which no one is *blamable*⟩. GUILTY implies responsibility for or consciousness of crime, sin, or, at the least, grave error or misdoing ⟨*guilty* of a breach of etiquette⟩. CULPABLE is weaker than *guilty* and is likely to connote malfeasance or errors of ignorance, omission, or negligence ⟨*culpable* neglect⟩.

◇ WORD HISTORY
blarney *Blarney* is the name of a town, a stream, and especially a castle, built in 1446, in County Cork, Ireland. Embedded in the southern wall of the castle's keep is a block of limestone, known as the Blarney Stone. Legend has it that anyone who manages to kiss the Blarney Stone (its position makes it not easily accessible) will be blessed with the gift of flattery or cajolery. The tradition of making a pilgrimage to the castle to kiss the Blarney Stone is supposedly based on a particular event in the castle's history. The story goes that in 1602 the lord of the castle contrived an endless series of plausible excuses and expressions of flattery to forestall a promised surrender of the castle to attacking English forces until help could arrive. The lack of historical evidence for the story has failed to stifle the claim that anyone who kisses the stone will acquire comparable persuasive powers.

1 a : the act of insulting or showing contempt or lack of reverence for God **b :** the act of claiming the attributes of deity
2 : irreverence toward something considered sacred or inviolable

[1]**blast** \'blast\ *noun* [Middle English, from Old English *blǣst;* akin to Old High German *blāst* blast, *blāsan* to blow, Old English *blāwan* — more at BLOW] (before 12th century)
1 a : a violent gust of wind **b :** the effect or accompaniment (as sleet) of such a gust
2 : the sound produced by an impulsion of air through a wind instrument or whistle
3 : something resembling a gust of wind: as **a :** a stream of air or gas forced through a hole **b :** a vehement outburst **c :** the continuous blowing to which a charge of ore or metal is subjected in a blast furnace
4 a : a sudden pernicious influence or effect ⟨the *blast* of a huge epidemic⟩ **b :** a disease of plants that causes the foliage or flowers to wither
5 a : an explosion or violent detonation **b :** the violent effect produced in the vicinity of an explosion that consists of a wave of increased atmospheric pressure followed by a wave of decreased atmospheric pressure
6 : SPEED, CAPACITY, OPERATION ⟨go full *blast*⟩ ⟨in full *blast*⟩
7 : an enjoyably exciting experience, occasion, or event; *especially* **:** PARTY

[2]**blast** (14th century)
intransitive verb
1 : BLARE ⟨music *blasting* from the radio⟩
2 : to make a vigorous attack
3 a : to use an explosive **b :** SHOOT
4 : to hit a golf ball out of a sand trap with explosive force
transitive verb
1 a : to injure by or as if by the action of wind **b :** BLIGHT
2 a : to shatter by or as if by an explosive **b :** to remove, open, or form by or as if by an explosive **c :** SHOOT
3 : to attack vigorously
4 : to cause to blast off ⟨will *blast* themselves from the moon's surface⟩
5 : to hit vigorously and effectively
— blast·er *noun*

blast- *or* **blasto-** *combining form* [German, from Greek, from *blastos*]
: bud **:** budding **:** germ ⟨*blasto*disc⟩ ⟨*blast*ula⟩

-blast \ˌblast\ *noun combining form* [New Latin *-blastus,* from Greek *blastos* bud, shoot, from *blastanein* to bud, sprout]
: formative unit especially of living matter **:** germ **:** cell **:** cell layer ⟨epi*blast*⟩

blast·ed *adjective* (1552)
1 : damaged by or as if by an explosive, lightning, wind, or supernatural force ⟨upon this *blasted* heath —Shakespeare⟩ ⟨a *blasted* apple tree⟩
2 : DAMNED, DETESTABLE ⟨this *blasted* weather⟩

blas·te·ma \bla-'stē-mə\ *noun, plural* **-mas** *or* **-ma·ta** \-mə-tə\ [New Latin, from Greek *blastēma* offshoot, from *blastanein*] (circa 1823)
: a mass of living substance capable of growth and differentiation
— blas·te·mal \-məl\ *or* **blas·te·mat·ic** \ˌblas-tə-'ma-tik\ *adjective*

blast furnace *noun* (1706)
: a furnace in which combustion is forced by a current of air under pressure; *especially* **:** one for the reduction of iron ore

-blastic *adjective combining form* [International Scientific Vocabulary, from *-blast*]
: having (such or so many) buds, germs, cells, or cell layers ⟨diplo*blastic*⟩

blast·ie \'blas-tē\ *noun* [Scots *blast* to wither, from [2]*blast*] (1787)
Scottish **:** an ugly little creature

blast·ment \'blas(t)-mənt\ *noun* (1602)
archaic **:** a blighting influence

blas·to·coel *or* **blas·to·coele** \'blas-tə-ˌsēl\ *noun* [International Scientific Vocabulary] (1877)
: the cavity of a blastula — see BLASTULA illustration
— blas·to·coe·lic \ˌblas-tə-'sē-lik\ *adjective*

blas·to·cyst \'blas-tə-ˌsist\ *noun* (circa 1881)
: the modified blastula of a placental mammal

blas·to·derm \-ˌdərm\ *noun* [German, from *blast-* + *-derm*] (circa 1843)
: a blastodisc after completion of cleavage and formation of the blastocoel

blas·to·der·mic vesicle \'blas-tə-ˌdər-mik-\ *noun* (circa 1860)
: BLASTOCYST

blas·to·disc \'blas-tə-ˌdisk\ *noun* (circa 1881)
: the embryo-forming portion of an egg with discoidal cleavage usually appearing as a small disc on the upper surface of the yolk mass — see EGG illustration

blast–off \'blas-ˌtȯf\ *noun* (1951)
: a blasting off (as of a rocket)

blast off *intransitive verb* (1951)
: TAKE OFF 2d — used especially of rocket-propelled missiles and vehicles

blas·to·mere \'blas-tə-ˌmir\ *noun* [International Scientific Vocabulary] (1877)
: a cell produced during cleavage of a fertilized egg

blas·to·my·co·sis \ˌblas-tə-ˌmī-'kō-səs\ *noun* [New Latin, from *Blastomyces,* fungus genus, from *blast-* + Greek *mykēs* fungus; akin to Greek *myxa* mucus — more at MUCUS] (circa 1900)
: any of several fungal infections; *especially* **:** an infectious disease caused by a fungus (*Blastomyces dermtitides*) that affects especially the lungs and skin

blas·to·pore \'blas-tə-ˌpōr, -ˌpȯr\ *noun* (1880)
: the opening of the archenteron
— blas·to·por·ic \-'pōr-ik, -'pȯr-\ *adjective*

blas·to·spore \'blas-tə-ˌspōr, -ˌspȯr\ *noun* [*blast-* + *spore*] (circa 1923)
: a fungal spore produced by budding

blas·tu·la \'blas-chə-lə\ *noun, plural* **-las** *or* **-lae** \-ˌlē\ [New Latin, from Greek *blastos*] (1887)
: an early metazoan embryo typically having the form of a hollow fluid-filled rounded cavity bounded by a single layer of cells — compare GASTRULA, MORULA
— blas·tu·la·tion \ˌblas-chə-'lā-shən\ *noun*

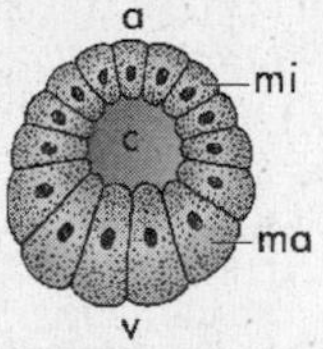

section of blastula: *c* **blastocoel,** *ma* **macromere,** *mi* **micromere,** *a* **animal pole,** *v* **vegetal pole**

blat \'blat\ *verb* **blat·ted; blat·ting** [perhaps alteration of *bleat*] (1846)
intransitive verb
1 : to cry like a calf or sheep **:** BLEAT
2 a : to make a raucous noise **b :** BLAB
transitive verb
: to utter loudly or foolishly **:** BLURT
— blat *noun*

bla·tan·cy \'blā-tᵊn(t)-sē\ *noun, plural* **-cies** (1610)
1 : the quality or state of being blatant
2 : something that is blatant

bla·tant \'blā-tᵊnt\ *adjective* [perhaps from Latin *blatire* to chatter] (circa 1656)
1 : noisy especially in a vulgar or offensive manner **:** CLAMOROUS
2 : completely obvious, conspicuous, or obtrusive especially in a crass or offensive manner **:** BRAZEN
synonym see VOCIFEROUS
— bla·tant·ly *adverb*

blate \'blāt\ *adjective* [Middle English] (1535)
chiefly Scottish **:** TIMID, SHEEPISH

[1]**blath·er** \'bla-t͟hər\ *intransitive verb* **blath·ered; blath·er·ing** \-t͟h(ə-)riŋ\ [Old Norse *blathra;* akin to Middle High German *blōdern* to chatter] (1524)
: to talk foolishly at length — often used with *on*
— blath·er·er \-t͟hər-ər\ *noun*

[2]**blather** *noun* (1719)
1 : voluble nonsensical or inconsequential talk or writing
2 : STIR, COMMOTION

blath·er·skite \'bla-t͟hər-ˌskīt\ *noun* [*blather* + Scots dialect *skate* a contemptible person] (circa 1650)
1 : a person who blathers a lot
2 : NONSENSE, BLATHER

blat·ter \'bla-tər\ *intransitive verb* [perhaps from Latin *blaterare* to chatter] (circa 1555)
dialect **:** to talk noisily and fast

blaw \'blȯ\ *verb* **blawed; blawn** \'blȯn\; **blaw·ing** \'blȯ(-)iŋ\ [Middle English (northern dialect) *blawen,* from Old English *blāwan*] (before 12th century)
chiefly Scottish **:** BLOW

blax·ploi·ta·tion \ˌblak-(ˌ)splȯi-'tā-shən\ *noun* [blend of *blax-* (alteration of *blacks*) and *exploitation*] (1972)
: the exploitation of blacks by producers of black-oriented films

[1]**blaze** \'blāz\ *noun* [Middle English *blase,* from Old English *blæse* torch] (before 12th century)
1 a : an intensely burning fire **b :** intense direct light often accompanied by heat ⟨the *blaze* of TV lights⟩ **c :** an active burning; *especially* **:** a sudden bursting forth of flame
2 : something that resembles the blaze of a fire: as **a :** a dazzling display ⟨a *blaze* of color⟩ **b :** a sudden outburst ⟨a *blaze* of fury⟩ **c** *plural* **:** HELL ⟨go to *blazes*⟩ ⟨as hot as *blazes*⟩

[2]**blaze** *intransitive verb* **blazed; blaz·ing** (13th century)
1 a : to burn brightly ⟨the sun *blazed* overhead⟩ **b :** to flare up **:** FLAME ⟨inflation *blazed* up⟩
2 : to be conspicuously brilliant or resplendent ⟨fields *blazing* with flowers⟩
3 : to shoot rapidly and repeatedly — usually used with *away*

[3]**blaze** *transitive verb* **blazed; blaz·ing** [Middle English *blasen,* from Middle Dutch *blāsen* to blow; akin to Old High German *blāst* blast] (1541)
: to make public or conspicuous

[4]**blaze** *noun* [German *Blas,* from Old High German *plas;* akin to Old English *blæse*] (1639)
1 a : a usually white stripe down the center of the face of an animal **b :** a white or gray streak in the hair of the head
2 : a trail marker; *especially* **:** a mark made on a tree by chipping off a piece of the bark

[5]**blaze** *transitive verb* **blazed; blaz·ing** (1750)
1 : to mark (as a trail) with blazes
2 : to lead in some direction or activity ⟨*blaze* new trails in education⟩

blaz·er \'blā-zər\ *noun* (circa 1635)
1 : one that blazes
2 : a sports jacket often with notched collar and patch pockets

blaz·ing *adjective* (1596)
: of outstanding power, speed, heat, or intensity ⟨*blazing* eyes⟩ ⟨a *blazing* fastball⟩ ⟨*blazing* gunfire⟩
— blaz·ing·ly *adverb*

blazing star *noun* (15th century)
1 *archaic* **:** COMET
2 : any of various plants having conspicuous flower clusters or star-shaped flowers: as **a :** any of a genus (*Liatris*) of North American composite herbs with spikes of rosy-purple rayless flowers — called also *button snakeroot* **b :** any of several North American rough-

leaved herbs (genus *Mentzelia* of the family Losaceae)

1bla·zon \ˈblā-zᵊn\ *noun* [Middle English *blason,* from Middle French] (14th century)
1 a : armorial bearings **:** COAT OF ARMS **b :** the proper description or representation of heraldic or armorial bearings
2 : ostentatious display

2blazon *transitive verb* **bla·zoned; bla·zon·ing** \ˈblāz-niŋ; ˈblā-zᵊn-iŋ\ (1534)
1 : to publish widely **:** PROCLAIM
2 a : to describe (heraldic or armorial bearings) in technical terms **b :** to represent (armorial bearings) in drawing or engraving
3 a : DISPLAY **b :** DECK, ADORN ⟨the town was *blazoned* with flags⟩
— **bla·zon·er** \-nər; -zᵊn-ər\ *noun*
— **blazoning** *noun*

bla·zon·ry \ˈblā-zᵊn-rē\ *noun, plural* **-ries** (1622)
1 a : BLAZON 1b **b :** BLAZON 1a
2 : a dazzling display

1bleach \ˈblēch\ *verb* [Middle English *blechen,* from Old English *blǣcean;* akin to Old English *blāc* pale; probably akin to Latin *flagrare* to burn — more at BLACK] (before 12th century)
transitive verb
1 : to remove color or stains from
2 : to make whiter or lighter especially by physical or chemical removal of color
intransitive verb
: to grow white or lose color
— **bleach·able** \ˈblē-chə-bəl\ *adjective*

2bleach *noun* (1887)
1 : the act or process of bleaching
2 : a preparation used in bleaching
3 : the degree of whiteness obtained by bleaching

bleach·er \ˈblē-chər\ *noun* (1550)
1 : one that bleaches or is used in bleaching
2 : a usually uncovered stand of tiered planks providing seating for spectators — usually used in plural
— **bleach·er·ite** \-chə-ˌrīt\ *noun*

bleaching powder *noun* (circa 1846)
: a white powder consisting chiefly of calcium hydroxide, calcium chloride, and calcium hypochlorite and used as a bleach, disinfectant, or deodorant

bleak \ˈblēk\ *adjective* [Middle English *bleke* pale; probably akin to Old English *blāc*] (1574)
1 : exposed and barren and often windswept
2 : COLD, RAW ⟨a *bleak* November evening⟩
3 a : lacking in warmth, life, or kindliness **:** GRIM **b :** not hopeful or encouraging **:** DEPRESSING ⟨a *bleak* outlook⟩ **c :** severely simple or austere
— **bleak·ish** \ˈblē-kish\ *adjective*
— **bleak·ly** *adverb*
— **bleak·ness** *noun*

1blear \ˈblir\ *transitive verb* [Middle English *bleren*] (14th century)
1 : to make (the eyes) sore or watery
2 : DIM, BLUR

2blear *adjective* (14th century)
1 : dim with water or tears
2 : obscure to the view or imagination

blear–eyed \-ˌīd\ *adjective* (14th century)
: BLEARY-EYED

bleary \ˈblir-ē\ *adjective* (14th century)
1 *of the eyes or vision* **:** dull or dimmed especially from fatigue or sleep
2 : poorly outlined or defined **:** DIM
3 : tired to the point of exhaustion
— **blear·i·ly** \ˈblir-ə-lē\ *adverb*
— **blear·i·ness** \ˈblir-ē-nəs\ *noun*

bleary–eyed \-ˌīd\ *adjective* (circa 1927)
: having the eyes dimmed and watery (as from fatigue, drink, or emotion)

1bleat \ˈblēt, *Northern also* ˈblat, *Southern usually* ˈblāt\ *verb* [Middle English *bleten,* from Old English *blǣtan;* akin to Latin *flēre* to weep, Old English *bellan* to roar — more at BELLOW] (before 12th century)
intransitive verb
1 a : to utter the natural cry of a sheep or goat or a similar sound **b :** WHIMPER
2 a : to talk complainingly or with a whine **b :** BLATHER
transitive verb
: to utter in a bleating manner
— **bleat·er** *noun*

2bleat *noun* (circa 1505)
1 : the cry of a sheep or goat or a similar sound
2 : a feeble outcry, protest, or complaint

bleb \ˈbleb\ *noun* [perhaps alteration of *blob*] (1607)
1 : a small blister
2 : BUBBLE; *also* **:** a small particle
— **bleb·by** \ˈble-bē\ *adjective*

1bleed \ˈblēd\ *verb* **bled** \ˈbled\; **bleed·ing** [Middle English *bleden,* from Old English *blēdan,* from *blōd* blood] (before 12th century)
intransitive verb
1 a : to emit or lose blood **b :** to sacrifice one's blood especially in battle
2 : to feel anguish, pain, or sympathy ⟨a heart that *bleeds* at a friend's misfortune⟩
3 : to escape by oozing or flowing (as from a wound)
4 : to give up some constituent (as sap or dye) by exuding or diffusing it
5 a : to pay out or give money **b :** to have money extorted
6 : to be printed so as to run off one or more edges of the page after trimming
transitive verb
1 : to remove or draw blood from
2 : to get or extort money from especially over a prolonged period
3 : to draw sap from (a tree)
4 a : to extract or let out some or all of a contained substance from ⟨*bleed* a brake line⟩ **b :** to extract or cause to escape from a container **c :** to diminish gradually — usually used with *off* ⟨a pilot *bleeding* off airspeed⟩
5 : to cause (as a printed illustration) to bleed
— **bleed white :** to drain of blood or resources

2bleed *noun* (circa 1937)
: printed matter (as an illustration) that bleeds; *also* **:** the part of a bleed trimmed off

bleed·er *noun* (1803)
1 : one that bleeds; *especially* **:** HEMOPHILIAC
2 *British* **:** ROTTER; *also* **:** BLOKE

bleed·ing \ˈblē-diŋ, -dᵊn\ *adjective or adverb* (1858)
chiefly British **:** BLOODY — used as an intensive

bleeding heart *noun* (1691)
1 : a garden plant (*Dicentra spectabilis*) of the fumitory family with racemes of usually deep pink drooping heart-shaped flowers; *broadly* **:** any of several plants (genus *Dicentra*)
2 : a person who shows extravagant sympathy especially for an object of alleged persecution

1bleep \ˈblēp\ *noun* [imitative] (1953)
: a short high-pitched sound (as from electronic equipment)

2bleep *transitive verb* (1968)
: BLIP

3bleep *interjection* (1970)
— used in place of an expletive

1blem·ish \ˈble-mish\ *transitive verb* [Middle English *blemisshen,* from Middle French *blesmiss-,* stem of *blesmir* to make pale, wound] (14th century)
: to spoil by a flaw

2blemish *noun* (1535)
: a noticeable imperfection; *especially* **:** one that seriously impairs appearance

1blench \ˈblench\ *intransitive verb* [Middle English, to deceive, blench, from Old English *blencan* to deceive; akin to Old Norse *blekkja* to impose on] (13th century)
: to draw back or turn aside from lack of courage **:** FLINCH
synonym see RECOIL

2blench *verb* [alteration of *blanch*] (1813)
: BLEACH, WHITEN

1blend \ˈblend\ *verb* **blend·ed** *also* **blent** \ˈblent\; **blend·ing** [Middle English, modification of Old Norse *blanda;* akin to Old English *blandan* to mix, Lithuanian *blandus* impure, cloudy] (14th century)
transitive verb
1 : MIX; *especially* **:** to combine or associate so that the separate constituents or the line of demarcation cannot be distinguished
2 : to prepare by thoroughly intermingling different varieties or grades
intransitive verb
1 a : to mingle intimately or unobtrusively **b :** to combine into an integrated whole
2 : to produce a harmonious effect
synonym see MIX

2blend *noun* (1883)
1 : something produced by blending: as **a :** a product prepared by blending **b :** a word (as *brunch*) produced by combining other words or parts of words
2 : a group of two or more consecutive consonants that begin a syllable

blende \ˈblend\ *noun* [German, from *blenden* to blind, from Old High German *blenten;* akin to Old English *blind*] (circa 1753)
: SPHALERITE

blended whiskey *noun* (1940)
: whiskey blended from two or more straight whiskeys or from whiskey and neutral spirits

blend·er \ˈblen-dər\ *noun* (circa 1798)
: one that blends; *especially* **:** an electric appliance for grinding or mixing ⟨a food *blender*⟩

blending inheritance *noun* (1922)
: the expression in offspring of phenotypic characters (as pink flower color from red and white parents) intermediate between those of the parents; *also* **:** inheritance in a now discarded theory in which the genetic material of offspring was held to be a uniform blend of that of the parents

blen·ny \ˈble-nē\ *noun, plural* **blennies** [Latin *blennius,* a sea fish, from Greek *blennos*] (1774)
: any of numerous usually small and elongated marine fishes (especially families Blenniidae and Clinidae) including both scaled and scaleless forms

blephar- *or* **blepharo-** *combining form* [New Latin, from Greek, from *blepharon*]
1 : eyelid ⟨*blepharo*spasm⟩
2 : cilium **:** flagellum ⟨*blepharo*plast⟩

bleph·a·ro·plast \ˈble-fə-rō-ˌplast\ *noun* (1897)
: a basal body especially of a flagellated cell

bleph·a·ro·plas·ty \-ˌplas-tē\ *noun* (circa 1881)
: plastic surgery on the eyelid especially to remove fatty or excess tissue

bleph·a·ro·spasm \-ˌspa-zəm\ *noun* (1872)
: spasmodic winking from involuntary contraction of the orbicular muscle of the eyelids

bles·bok \ˈbles-ˌbäk\ *noun* [Afrikaans, from *bles* blaze + *bok* male antelope] (1824)
: a South African antelope (*Damaliscus dorcas*) having a large white patch down the center of the face

bless \ˈbles\ *transitive verb* **blessed** \ˈblest\ *also* **blest** \ˈblest\; **bless·ing** [Middle English, from Old English *blētsian,* from *blōd* blood; from the use of blood in consecration] (before 12th century)
1 : to hallow or consecrate by religious rite or word
2 : to hallow with the sign of the cross

3 : to invoke divine care for ⟨*bless* your heart⟩
4 a : PRAISE, GLORIFY ⟨*bless* his holy name⟩ **b :** to speak well of **:** APPROVE
5 : to confer prosperity or happiness upon
6 *archaic* **:** PROTECT, PRESERVE
7 : ENDOW, FAVOR ⟨*blessed* with athletic ability⟩

bless·ed \'ble-səd\ *also* **blest** \'blest\ *adjective* (before 12th century)
1 a : held in reverence **:** VENERATED ⟨the *blessed* saints⟩ **b :** honored in worship **:** HALLOWED ⟨the *blessed* Trinity⟩ **c :** BEATIFIC ⟨a *blessed* visitation⟩
2 : of or enjoying happiness; *specifically* **:** enjoying the bliss of heaven — used as a title for a beatified person
3 : bringing pleasure, contentment, or good fortune
4 — used as an intensive ⟨no one gave us a *blessed* penny —*Saturday Review*⟩
— bless·ed·ly *adverb*
— bless·ed·ness *noun*

Bless·ed Sacrament \'ble-səd-\ *noun* (15th century)
: the Communion elements; *specifically* **:** the consecrated host

bless·ing *noun* (before 12th century)
1 a : the act or words of one that blesses **b :** APPROVAL, ENCOURAGEMENT
2 : a thing conducive to happiness or welfare
3 : grace said at a meal

bleth·er \'ble-t͟hər\ *variant of* BLATHER

blew *past of* BLOW

[1]blight \'blīt\ *noun* [origin unknown] (1611)
1 a : a disease or injury of plants resulting in withering, cessation of growth, and death of parts without rotting **b :** an organism that causes blight
2 : something that frustrates plans or hopes
3 : something that impairs or destroys
4 : a deteriorated condition ⟨urban *blight*⟩

[2]blight (1695)
transitive verb
1 : to affect (as a plant) with blight
2 : to cause to deteriorate
intransitive verb
: to suffer from or become affected with blight

blight·er \'blī-tər\ *noun* (1822)
1 : one that blights
2 *chiefly British* **a :** a disliked or contemptible person **b :** FELLOW, GUY

blimp \'blimp\ *noun* [imitative; perhaps from the sound made by striking the gas bag with the thumb] (1916)
1 : a nonrigid airship
2 *capitalized* **:** COLONEL BLIMP

blimp·ish \'blim-pish\ *adjective, often capitalized* (1938)
: of, relating to, or suggesting a Colonel Blimp
— blimp·ish·ly *adverb*
— blimp·ish·ness *noun*

blin \'blin\ *noun, plural* **bli·ni** \'blē-nē, 'bli-; blə-'nē\ *or* **bli·nis** \'blē-nēz, 'bli-; blə-'nēz\ [Russian] (1888)
: a thin often buckwheat pancake usually filled (as with sour cream) and folded

[1]blind \'blīnd\ *adjective* [Middle English, from Old English; akin to Old High German *blint* blind, Old English *blandan* to mix — more at BLEND] (before 12th century)
1 a (1) **:** SIGHTLESS (2) **:** having less than 1/10 of normal vision in the more efficient eye when refractive defects are fully corrected by lenses **b :** of or relating to sightless persons
2 a : unable or unwilling to discern or judge ⟨*blind* to a lover's faults⟩ **b :** UNQUESTIONING ⟨*blind* loyalty⟩ ⟨*blind* faith⟩
3 a : having no regard to rational discrimination, guidance, or restriction ⟨*blind* choice⟩ **b :** lacking a directing or controlling consciousness ⟨*blind* chance⟩ **c :** DRUNK
4 : made or done without sight of certain objects or knowledge of certain facts that could serve for guidance ⟨a *blind* taste test⟩; *especially* **:** performed solely by the aid of instruments within an airplane ⟨a *blind* landing⟩
5 : DEFECTIVE: as **a :** lacking a growing point or producing leaves instead of flowers **b :** lacking a complete or legible address ⟨*blind* mail⟩
6 a : difficult to discern, make out, or discover **b :** hidden from sight **:** COVERED ⟨*blind* seam⟩
7 : having but one opening or outlet ⟨*blind* sockets⟩
8 : having no opening for light or passage **:** BLANK ⟨*blind* wall⟩
— blind·ly \'blīn(d)-lē\ *adverb*
— blind·ness \'blīn(d)-nəs\ *noun*

[2]blind *transitive verb* (before 12th century)
1 a : to make blind **b :** DAZZLE
2 a : to withhold light from **b :** HIDE, CONCEAL
— blind·ing·ly \'blīn-diŋ-lē\ *adverb*

[3]blind *noun* (1702)
1 : something to hinder sight or keep out light: as **a :** a window shutter **b :** a roller window shade **c :** VENETIAN BLIND **d :** BLINDER
2 : a place of concealment; *especially* **:** a concealing enclosure from which one may shoot game or observe wildlife
3 a : something put forward for the purpose of misleading **:** SUBTERFUGE **b :** a person who acts as a decoy or distraction

[4]blind *adverb* (circa 1775)
1 : BLINDLY: as **a :** to the point of insensibility ⟨*blind* drunk⟩ **b :** without seeing outside an airplane ⟨fly *blind*⟩
2 — used as an intensive ⟨was robbed *blind*⟩

blind alley *noun* (1583)
: a fruitless or mistaken course or direction

blind date *noun* (1925)
1 : a date between two persons who have not previously met
2 : either participant in a blind date

blind·er \'blīn-dər\ *noun* (1809)
1 : either of two flaps on a horse's bridle to keep it from seeing objects at its sides
2 *plural* **:** a limitation or obstruction to sight or discernment

blind·fish \'blīn(d)-ˌfish\ *noun* (1843)
: any of several small fishes with vestigial functionless eyes found usually in the waters of caves

[1]blind·fold \-ˌfōld\ *transitive verb* [alteration of Middle English *blindfellen, blindfelden* to strike blind, blindfold, from *blind* + *fellen* to fell] (1599)
1 : to cover the eyes of with or as if with a bandage
2 : to hinder from seeing; *especially* **:** to keep from comprehension
— blindfold *adjective*

[2]blindfold *noun* (1880)
1 : a bandage for covering the eyes
2 : something that obscures mental or physical vision

blind gut *noun* (15th century)
: a digestive cavity open at only one end; *especially* **:** the cecum of the large intestine

blind·man's buff \'blīn(d)-'manz-\ *noun* (1600)
: a group game in which a blindfolded player tries to catch and identify another player — called also *blindman's bluff*

blind pig *noun* (1887)
: BLIND TIGER

blind·side \'blīn(d)-ˌsīd\ *transitive verb* (1968)
1 : to hit unexpectedly from or as if from the blind side
2 : to surprise unpleasantly

blind side *noun* (1606)
1 : the side away from which one is looking
2 : the side on which one that is blind in one eye cannot see

blind spot *noun* (1872)
1 a : the nearly circular light-colored area at the back of the retina where the optic nerve enters the eyeball and which is not sensitive to light — called also *optic disk;* see EYE illustration **b :** a portion of a field that cannot be seen or inspected with available equipment
2 : an area in which one fails to exercise judgment or discrimination

blind tiger *noun* (1857)
: a place that sells intoxicants illegally

blind trust *noun* (1969)
: an arrangement in which the financial holdings of a person in an influential position (as a government official) are placed in the control of a fiduciary in order to avoid a possible conflict of interest

blind·worm \'blīnd-ˌwərm\ *noun* (15th century)
: SLOWWORM

[1]blink \'bliŋk\ *verb* [Middle English, to open one's eyes] (14th century)
intransitive verb
1 a *obsolete* **:** to look glancingly **:** PEEP **b :** to look with half-shut eyes **c :** to close and open the eyes involuntarily (as when dazzled)
2 : to shine dimly or intermittently
3 a : to look with too little concern **b :** to look with surprise or dismay
transitive verb
1 a : to cause to blink **b :** to remove (as tears) from the eye by blinking
2 : to deny recognition to

[2]blink *noun* (1594)
1 *chiefly Scottish* **:** GLIMPSE, GLANCE
2 : GLIMMER, SPARKLE
3 : a usually involuntary shutting and opening of the eye
4 : ICEBLINK
— on the blink : in or into a disabled or useless condition

[1]blink·er \'bliŋ-kər\ *noun* (1636)
1 : one that blinks; *especially* **:** a light that flashes off and on (as for the directing of traffic or the coded signaling of messages)
2 a : BLINDER 1 **b :** a cloth hood with shades projecting at the sides of the eye openings used on skittish racehorses — usually used in plural
3 *plural* **:** BLINDER 2

[2]blinker *transitive verb* (1865)
: to put blinders on

blink·ered \'bliŋ-kərd\ *adjective* (1867)
1 : limited in scope or understanding **:** NARROW-MINDED
2 : fitted with blinders

blin·tze \'blin(t)-sə\ *or* **blintz** \'blin(t)s\ *noun* [Yiddish *blintse,* of Slavic origin; akin to Ukrainian *mlynets',* diminutive of *mlyn* pancake] (1903)
: a thin usually wheat-flour pancake folded to form a casing and then sautéed or baked

[1]blip \'blip\ *noun* [imitative] (1945)
1 : a trace on an oscilloscope; *especially* **:** a spot on a radar screen
2 : a short crisp sound
3 : an interruption of the sound received in a radio or television program or occurring in a recording as a result of blipping
4 : a transient sharp movement up or down (as of a quantity commonly shown on a graph)

[2]blip *transitive verb* **blipped; blip·ping** (1968)
: to remove (recorded sound) from a recording so that there is an interruption of the sound in the reproduction ⟨a censor *blipped* the swear-words⟩

bliss \'blis\ *noun* [Middle English *blisse,* from Old English *bliss;* akin to Old English *blīthe* blithe] (before 12th century)
1 : complete happiness
2 : PARADISE, HEAVEN

blissed–out \'blist-'au̇t\ *adjective* (1973)
: being in a state of bliss **:** ECSTATIC

bliss·ful \'blis-fəl\ *adjective* (12th century)
1 : full of, marked by, or causing bliss
2 : happily benighted ⟨*blissful* ignorance⟩
— bliss·ful·ly \-fə-lē\ *adverb*
— bliss·ful·ness *noun*

[1]blis·ter \'blis-tər\ *noun* [Middle English, modification of Old French or Middle Dutch;

Old French *blostre* leprous nodule, from Middle Dutch *bluyster* blister; akin to Old English *blǣst* blast] (14th century)
1 : an elevation of the epidermis containing watery liquid
2 : an enclosed raised spot (as in paint) resembling a blister
3 : an agent that causes blistering
4 : a disease of plants marked by large swollen patches on the leaves
5 : any of various structures that bulge out (as a gunner's compartment on a bomber)
— **blis·tery** \-t(ə-)rē\ *adjective*

[2]**blister** *verb* **blis·tered; blis·ter·ing** \-t(ə-)riŋ\ (15th century)
intransitive verb
: to become affected with a blister
transitive verb
1 : to raise a blister on
2 : LAMBASTE 2

blister beetle *noun* (1816)
: a beetle (as the Spanish fly) used medicinally dried and powdered to raise blisters on the skin; *broadly* **:** any of a family (Meloidae) of soft-bodied beetles

blister copper *noun* (1861)
: metallic copper of a black blistered surface that is the product of converting copper matte and is about 98.5 to 99.5 percent pure

blis·ter·ing *adjective* (1562)
: extremely intense or severe
— **blistering** *adverb*
— **blis·ter·ing·ly** \-t(ə-)riŋ-lē\ *adverb*

blister pack *noun* (1955)
: a package holding and displaying merchandise in a clear plastic case sealed to a sheet of cardboard

blister rust *noun* (1916)
: any of several diseases of pines that are caused by rust fungi (genus *Cronartium*) in the aecial stage and that affect the sapwood and inner bark and produce blisters externally

blithe \ˈblīth, ˈblīt͟h\ *adjective* **blith·er; blith·est** [Middle English, from Old English *blīthe;* akin to Old High German *blīdi* joyous] (before 12th century)
1 : of a happy lighthearted character or disposition
2 : lacking due thought or consideration **:** CASUAL, HEEDLESS ⟨*blithe* unconcern⟩
synonym see MERRY
— **blithe·ly** *adverb*

blith·er \ˈbli-t͟hər\ *intransitive verb or noun* [by alteration] (1868)
: BLATHER

blithe·some \ˈblīth-səm, ˈblīt͟h-\ *adjective* (1724)
: GAY, MERRY
— **blithe·some·ly** *adverb*

blitz \ˈblits\ *noun* (circa 1939)
1 a : BLITZKRIEG 1 **b** (1) **:** an intensive aerial campaign (2) **:** AIR RAID
2 a : an intensive nonmilitary campaign **:** a sudden overwhelming bombardment ⟨an advertising *blitz*⟩ **b :** a rush of the passer by the defensive linebackers in football
— **blitz** *verb*

blitz·krieg \-ˌkrēg\ *noun* [German, from *Blitz* lightning + *Krieg* war] (1939)
1 : war conducted with great speed and force; *specifically* **:** a violent surprise offensive by massed air forces and mechanized ground forces in close coordination
2 : BLITZ 2a

bliz·zard \ˈbli-zərd\ *noun* [origin unknown] (1870)
1 : a long severe snowstorm
2 : an intensely strong cold wind filled with fine snow
3 : an overwhelming rush or deluge ⟨the *blizzard* of mail at Christmas⟩ ◆
— **bliz·zardy** \ˈbli-zər-dē\ *also* **bliz·zard·ly** \-ˌzərd-lē\ *adjective*

[1]**bloat** \ˈblōt\ *adjective* [alteration of Middle English *blout*] (14th century)
: BLOATED, PUFFY

[2]**bloat** (1677)
transitive verb
1 : to make turgid or swollen
2 : to fill to capacity or overflowing
intransitive verb
: SWELL

[3]**bloat** *noun* (1860)
1 : one that is bloated
2 : a flatulent digestive disturbance of domestic animals and especially cattle marked by abdominal bloating

bloat·ed *adjective* (1711)
1 : being much larger than what is warranted ⟨a *bloated* estimate⟩
2 : obnoxiously vain

[1]**bloat·er** \ˈblō-tər\ *noun* [obsolete *bloat* (to cure)] (1832)
: a large fat herring or mackerel lightly salted and briefly smoked

[2]**bloater** *noun* [[2]*bloat*] (1896)
: a small but common cisco (*Coregonus hoyi*) of the Great Lakes

[1]**blob** \ˈbläb\ *noun* [Middle English] (15th century)
1 a : a small drop or lump of something viscid or thick **b :** a daub or spot of color
2 : something ill-defined or amorphous

[2]**blob** *transitive verb* **blobbed; blob·bing** (15th century)
: to mark with blobs **:** SPLOTCH

bloc \ˈbläk\ *noun* [French, literally, block] (1903)
1 a : a temporary combination of parties in a legislative assembly **b :** a group of legislators who act together for some common purpose irrespective of party lines
2 a : a combination of persons, groups, or nations forming a unit with a common interest or purpose **b :** a group of nations united by treaty or agreement for mutual support or joint action

[1]**block** \ˈbläk\ *noun, often attributive* [Middle English *blok,* from Middle French *bloc,* from Middle Dutch *blok;* akin to Old High German *bloh* block] (14th century)
1 : a compact usually solid piece of substantial material especially when worked or altered to serve a particular purpose: as **a :** the piece of wood on which the neck of a person condemned to be beheaded is laid for execution **b :** a mold or form on which articles are shaped or displayed **c :** a hollow rectangular building unit usually of artificial material **d :** a lightweight usually cubical and solid wooden or plastic building toy that is usually provided in sets **e :** the casting that contains the cylinders of an internal-combustion engine
2 a : OBSTACLE **b :** an obstruction of an opponent's play in sports; *especially* **:** a halting or impeding of the progress or movement of an opponent in football by use of the body **c** (1) **:** interruption of normal physiological function (as of a tissue or organ); *especially* **:** HEART BLOCK (2) **:** local anesthesia (as by injection) produced by interruption of the flow of impulses along a nerve **d :** interruption or cessation especially of train of thought by competing thoughts or psychological suppression
3 *slang* **:** HEAD 1 ⟨threatened to knock his *block* off⟩
4 : a wooden or metal case enclosing one or more pulleys and having a hook, eye, or strap by which it may be attached
5 : a piece of material (as wood or linoleum) having on its surface a hand-cut design from which impressions are to be printed
6 a (1) **:** a usually rectangular space (as in a city) enclosed by streets and occupied by or intended for buildings (2) **:** the distance along one of the sides of such a block **b** (1) **:** a large building divided into separate functional units (2) **:** a line of row houses (3) **:** a distinctive part of a building or integrated group of buildings **c :** a short section of railroad track in the block system
7 : a platform from which property is sold at auction
8 a : a quantity, number, or section of things dealt with as a unit **b :** BLOC 2
— **on the block :** for sale

[2]**block** (1580)
transitive verb
1 a : to make unsuitable for passage or progress by obstruction **b** *archaic* **:** BLOCKADE **c :** to hinder the passage, progress, or accomplishment of by or as if by interposing an obstruction **d :** to shut off from view ⟨forest canopy *blocking* the sun⟩ **e :** to interfere usually legitimately with (as an opponent) in various games or sports **f :** to prevent normal functioning of **g :** to restrict the exchange of (as currency or checks)
2 : to mark or indicate the outline or chief lines of ⟨*block* out a design⟩ ⟨*block* in a sketched figure⟩
3 : to shape on, with, or as if with a block ⟨*block* a hat⟩
4 : to secure, support, or provide with a block
5 : to work out or chart the movements of (as stage performers) — often used with *out*
6 : to make (two or more lines of writing or type) flush at the left or at both left and right
intransitive verb
: to block an opponent in sports
synonym see HINDER
— **block·er** *noun*

[1]**block·ade** \blä-ˈkād\ *transitive verb* **block·ad·ed; block·ad·ing** (1680)
1 : to subject to a blockade
2 : BLOCK, OBSTRUCT
— **block·ad·er** *noun*

[2]**blockade** *noun* (1693)
1 : the isolation by a warring nation of an enemy area (as a harbor) by troops or warships to prevent passage of persons or supplies; *broadly* **:** a restrictive measure designed to obstruct the commerce and communications of an unfriendly nation
2 : something that blocks
3 : interruption of normal physiological function (as transmission of nerve impulses) of a tissue or organ

block·ade–run·ner \-ˈkād-ˌrə-nər\ *noun* (1863)
: a ship or person that runs through a blockade

◇ WORD HISTORY

blizzard The word *blizzard,* when used in the meaning "severe snowstorm," first turned up in 19th century northwestern Iowa. Its earliest recorded appearance in printed form was in the April 23, 1870 issue of the *Northern Vindicator,* a newspaper published in the town of Estherville. Spelled *blizard*—the form *blizzard* appeared in the paper a week later—it was cautiously enclosed in quotation marks. *Blizzard* crops up again during the following years in several other newspapers of Iowa and neighboring states, and by 1888, when a March storm paralyzed the Eastern seaboard for days and took 400 lives, the word was well-known nationally. In other senses, however, *blizzard* existed decades earlier, and the ultimate origin of the word is unclear. The frontiersman and politician Davy Crockett used it on two occasions in the 1830s, once to mean a rifle blast, and once to refer figuratively to a blast of words. From a gun blast to a verbal blast to a wintry blast is a reasonable enough development, but evidence for *blizzard* is so scattered in time and place that it is difficult to draw any firm conclusions about the word's history.

— **block·ade–run·ning** \-ˌrə-niŋ\ *noun*
block·age \ˈblä-kij\ *noun* (1874)
: an act or instance of obstructing **:** the state of being blocked ⟨a *blockage* in a coronary artery⟩
block and tackle *noun* (1838)
: pulley blocks with associated rope or cable for hoisting or hauling
block·bust·er \ˈbläk-ˌbəs-tər\ *noun* (1942)
1 : a very large high-explosive bomb
2 : one that is notably effective, successful, large, or violent
3 : one who engages in blockbusting
block·bust·ing \-tiŋ\ *noun* (1954)
: profiteering by inducing property owners to sell hastily and often at a loss by appeals to fears of depressed values because of threatened minority encroachment and then reselling at inflated prices
block diagram *noun* (1944)
: a diagram (as of a system, process, or program) in which labeled figures (as rectangles) and interconnecting lines represent the relationship of parts
blocked \ˈbläkt\ *adjective* (1945)
: exhibiting or affected by a psychological block ⟨a *blocked* writer⟩
block grant *noun* (1900)
: an unrestricted federal grant
block·head \ˈbläk-ˌhed\ *noun* (1549)
: a stupid person
block·house \-ˌhau̇s\ *noun* (1512)
1 a : a structure of heavy timbers formerly used for military defense with sides loopholed and pierced for gunfire and often with a projecting upper story **b :** a small easily defended building for protection from enemy fire

blockhouse 1a

2 : a building usually of reinforced concrete serving as an observation point for an operation likely to be accompanied by heat, blast, or radiation hazard
block·ish \ˈblä-kish\ *adjective* (1565)
: resembling a block
block letter *noun* (1908)
: an often hand-drawn simple capital letter composed of strokes of uniform thickness
block party *noun* (1941)
: an outdoor public party put on by the residents of a city block or neighborhood
block plane *noun* (circa 1884)
: a small plane made with the blade set at a lower pitch than other planes and used chiefly on end grains of wood
block system *noun* (1864)
: a system by which a railroad track is divided into short sections and trains are run by guidance signals
blocky \ˈblä-kē\ *adjective* **block·i·er; -est** (circa 1879)
1 : resembling a block in form **:** CHUNKY
2 : filled with or made up of blocks or patches
bloke \ˈblōk\ *noun* [origin unknown] (1851)
chiefly British **:** MAN, FELLOW
[1]**blond** *or* **blonde** \ˈbländ\ *adjective* [Middle French *blond,* masculine, *blonde,* feminine] (15th century)
1 a : of a flaxen, golden, light auburn, or pale yellowish brown color ⟨*blond* hair⟩ **b :** of a pale white or rosy white color ⟨*blond* skin⟩ **c :** being a blond
2 a : of a light color **b :** of the color blond **c :** made light-colored by bleaching ⟨a table of *blond* walnut⟩
— **blond·ish** \ˈblän-dish\ *adjective*
[2]**blond** *or* **blonde** *noun* (1822)
1 : a person having blond hair and usually a light complexion and blue or gray eyes
2 : a light yellowish brown to dark grayish yellow

[1]**blood** \ˈbləd\ *noun, often attributive* [Middle English, from Old English *blōd;* akin to Old High German *bluot* blood] (before 12th century)
1 a : the fluid that circulates in the heart, arteries, capillaries, and veins of a vertebrate animal carrying nourishment and oxygen to and bringing away waste products from all parts of the body **b :** a comparable fluid of an invertebrate **c :** a fluid resembling blood
2 : the shedding of blood; *also* **:** the taking of life
3 a : LIFEBLOOD; *broadly* **:** LIFE **b :** human stock or lineage; *especially* **:** royal lineage ⟨a prince of the *blood*⟩ **c :** relationship by descent from a common ancestor **:** KINSHIP **d :** persons related through common descent **:** KINDRED **e** (1) **:** honorable or high birth or descent (2) **:** descent from parents of recognized breed or pedigree
4 a : blood regarded as the seat of the emotions **:** TEMPER **b** *obsolete* **:** LUST **c :** a showy foppish man **:** RAKE
5 : PERSONNEL ⟨a company in need of new *blood*⟩
6 : a black American male — used especially among blacks
[2]**blood** *transitive verb* (1633)
1 *archaic* **:** BLEED 1
2 : to stain or wet with blood
3 : to expose (a hunting dog) to sight, scent, or taste of the blood of its prey
4 : to give experience to ⟨troops *blooded* in battle⟩
blood–and–guts \ˈblə-dᵊn-ˈgəts\ *adjective* (1937)
: marked by great vigor, violence, or fierceness ⟨*blood-and-guts* competition⟩
blood·bath \ˈbləd-ˌbath, -ˌbȧth\ *noun* (1867)
: a great slaughter
blood–brain barrier \ˈbləd-ˈbrān-\ *noun* (1944)
: a barrier created by the modification of brain capillaries (as by reduction in fenestration and formation of tight cell-to-cell contacts) that prevents many substances from leaving the blood and crossing the capillary walls into the brain tissues
blood brother *noun* (1890)
1 : a brother by birth
2 : one of two men pledged to mutual loyalty by a ceremonial use of each other's blood
— **blood brotherhood** *noun*
blood cell *noun* (1846)
: a cell normally present in blood
blood count *noun* (circa 1900)
: the determination of the blood cells in a definite volume of blood; *also* **:** the number of cells so determined
blood·cur·dling \ˈbləd-ˌkərd-liŋ, -ˌkər-dᵊl-iŋ\ *adjective* (1904)
: arousing fright or horror ⟨*bloodcurdling* screams⟩
blood doping *noun* (1973)
: a technique for temporarily improving athletic performance in which oxygen-carrying red blood cells from blood previously withdrawn from an athlete are reinjected just before an event — called also *blood packing*
blood·ed \ˈblə-dəd\ *adjective* (1595)
1 : having blood of a specified kind — used in combination ⟨cold-*blooded*⟩
2 : being entirely or largely purebred ⟨a herd of *blooded* stock⟩
blood feud *noun* (1858)
: a feud between different clans or families
blood fluke *noun* (1872)
: SCHISTOSOME
blood group *noun* (1916)
: one of the classes (as those designated A, B, AB, or O) into which individuals or their blood can be separated on the basis of the presence or absence of specific antigens in the blood — called also *blood type*
blood·guilt \ˈbləd-ˌgilt\ *noun* (1593)
: guilt resulting from bloodshed

— **blood·guilt·i·ness** \-ˌgil-tē-nəs\ *noun*
— **blood·guilty** \-tē\ *adjective*
blood·hound \ˈbləd-ˌhau̇nd\ *noun* (14th century)
1 : any of a breed of large powerful hounds of European origin remarkable for acuteness of smell
2 : a person keen in pursuit

bloodhound 1

blood·less \ˈbləd-ləs\ *adjective* (before 12th century)
1 : deficient in or free from blood
2 : not accompanied by loss or shedding of blood ⟨a *bloodless* victory⟩
3 : lacking in spirit or vitality
4 : lacking in human feeling ⟨*bloodless* statistics⟩
— **blood·less·ly** *adverb*
— **blood·less·ness** *noun*
blood·let·ting \-ˌle-tiŋ\ *noun* (13th century)
1 : PHLEBOTOMY
2 : BLOODSHED
3 : elimination of personnel or resources
blood·line \-ˌlīn\ *noun* (circa 1909)
: a sequence of direct ancestors especially in a pedigree; *also* **:** FAMILY, STRAIN
blood·mo·bile \-mō-ˌbēl\ *noun* (1948)
: an automotive vehicle staffed and equipped for collecting blood from donors
blood money *noun* (1535)
1 : money obtained at the cost of another's life
2 : money paid by a manslayer or members of his family, clan, or tribe to the next of kin of a person killed by him
blood orange *noun* (1855)
: a cultivated sweet orange having fruit with usually red pulp; *also* **:** its fruit
blood platelet *noun* (1898)
: one of the minute protoplasmic disks of vertebrate blood that assist in blood clotting
blood poisoning *noun* (1863)
: SEPTICEMIA
blood pressure *noun* (1874)
: pressure that is exerted by the blood upon the walls of the blood vessels and especially arteries and that varies with the muscular efficiency of the heart, the blood volume and viscosity, the age and health of the individual, and the state of the vascular wall
blood–red \ˈbləd-ˈred\ *adjective* (before 12th century)
: having the color of blood
blood·root \-ˌrüt, -ˌru̇t\ *noun* (1722)
: a plant (*Sanguinaria canadensis*) of the poppy family having a red root and sap and bearing a solitary lobed leaf and white flower in early spring — called also *sanguinaria*
blood sausage *noun* (1868)
: very dark sausage containing a large proportion of blood — called also *blood pudding*
blood serum *noun* (circa 1909)
: blood from which the fibrin and suspended material (as cells) have been removed
blood·shed \ˈbləd-ˌshed\ *noun* (15th century)
1 : the shedding of blood
2 : the taking of life **:** SLAUGHTER
blood·shot \-ˌshät\ *adjective* (1552)
of an eye **:** inflamed to redness
blood sport *noun* (1895)
: a sport or contest (as hunting or cockfighting) involving bloodshed
blood·stain \-ˌstān\ *noun* (1820)
: a discoloration caused by blood
blood·stained \-ˌstānd\ *adjective* (1596)
1 : stained with blood
2 : involved with slaughter ⟨a *bloodstained* chronicle of war⟩
blood·stock \-ˌstäk\ *noun* (1830)
: horses of Thoroughbred breeding

blood•stone \-ˌstōn\ *noun* (1551)
: a green chalcedony sprinkled with red spots resembling blood — called also *heliotrope*

blood•stream \-ˌstrēm\ *noun* (1873)
1 : the flowing blood in a circulatory system
2 : a mainstream of power or vitality ⟨introduce into the economic *bloodstream* a large amount of money —*Harper's*⟩

blood•suck•er \-ˌsə-kər\ *noun* (14th century)
1 : an animal that sucks blood; *especially* **:** LEECH
2 : a person who sponges or preys on another
— **blood•suck•ing** \-kiŋ\ *adjective*

blood sugar *noun* (1918)
: the glucose in the blood; *also* **:** its concentration (as in milligrams per 100 milliliters)

blood test *noun* (1912)
: a test of the blood; *especially* **:** a serologic test for the presence of substances indicative of disease (as syphilis) or disease-causing agents (as HIV)

blood•thirsty \ˈbləd-ˌthər-stē\ *adjective* (1535)
: eager for or marked by the shedding of blood, violence, or killing
— **blood•thirst•i•ly** \-stə-lē\ *adverb*
— **blood•thirst•i•ness** \-stē-nəs\ *noun*

blood–typ•ing \-ˌtī-piŋ\ *noun* (1926)
: the action or process of determining an individual's blood group

blood vessel *noun* (1694)
: any of the vessels through which blood circulates in the body

blood•worm \ˈbləd-ˌwərm\ *noun* (1714)
1 : any of various reddish annelid worms (as of genera *Glycera* or *Tubifex*) often used as bait
2 : any of several strongyle worms (genus *Strongylus*) that are parasitic in the large intestine of horses

¹bloody \ˈblə-dē\ *adjective* **blood•i•er; -est** (before 12th century)
1 a : containing or made up of blood **b :** of or contained in the blood
2 : smeared or stained with blood
3 : accompanied by or involving bloodshed; *especially* **:** marked by great slaughter
4 a : MURDEROUS **b :** MERCILESS, CRUEL
5 : BLOODRED
6 — used as an intensive; sometimes considered vulgar ☆
— **blood•i•ly** \ˈblə-dᵊl-ē\ *adverb*
— **blood•i•ness** \ˈblə-dē-nəs\ *noun*

²bloody *transitive verb* **blood•ied; bloody•ing** (1530)
: to make bloody or bloodred

³bloody *adverb* (1676)
— used as an intensive; sometimes considered vulgar

Bloody Mary *noun, plural* **Bloody Marys** [probably from *Bloody Mary,* appellation of Mary I of England] (1947)
: a cocktail consisting essentially of vodka and usually spiced tomato juice

bloody–mind•ed \ˌblə-dē-ˈmīn-dəd\ *adjective* (1584)
1 : inclined towards violence or bloodshed
2 *chiefly British* **:** stubbornly contrary or obstructive **:** CANTANKEROUS
— **bloody–mind•ed•ness** *noun*

bloody murder *adverb* (1925)
: in a loud and violent manner ⟨ran off, screaming *bloody murder*⟩; *also* **:** in vehement protest ⟨screaming *bloody murder* over the pay cut⟩

bloody shirt *noun* (1874)
: something intended to stir up or revive partisan animosity — usually used in the phrase *wave the bloody shirt*

¹bloom \ˈblüm\ *noun* [Middle English *blome* lump of metal, from Old English *blōma*] (before 12th century)
1 : a mass of wrought iron from the forge or puddling furnace
2 : a bar of iron or steel hammered or rolled from an ingot

²bloom *noun* [Middle English *blome,* from Old Norse *blōm;* akin to Old English *blōwan* to blossom — more at BLOW] (13th century)
1 a : FLOWER **b :** the flowering state ⟨the roses in *bloom*⟩ **c :** a period of flowering ⟨the spring *bloom*⟩ **d :** an excessive growth of plankton
2 a : a state or time of beauty, freshness, and vigor **b :** a state or time of high development or achievement ⟨a career in full *bloom*⟩
3 : a surface coating or appearance: as **a :** a delicate powdery coating on some fruits and leaves **b :** a rosy appearance of the cheeks; *broadly* **:** an outward evidence of freshness or healthy vigor **c :** a cloudiness on a film of varnish or lacquer **d :** a grayish discoloration on chocolate **e :** glare caused by an object reflecting too much light into a television camera

³bloom (13th century)
intransitive verb
1 a : to produce or yield flowers **b :** to support abundant plant life ⟨make the desert *bloom*⟩
2 a (1) **:** to mature into achievement of one's potential (2) **:** to flourish in youthful beauty, freshness or excellence **b :** to shine out **:** GLOW
3 : to appear or occur unexpectedly or in remarkable quantity or degree
4 : to become densely populated with microorganisms and especially plankton — used of bodies of water
transitive verb
1 *obsolete* **:** to cause to bloom
2 : to give bloom to

¹bloom•er \ˈblü-mər\ *noun* (circa 1736)
1 : a plant that blooms
2 : a person who reaches full competence or maturity
3 [*blooming*] **:** a stupid blunder

²bloo•mer \ˈblü-mər\ *noun* [Amelia *Bloomer*] (1851)
1 : a costume for women consisting of a skirt over long loose trousers gathered closely about the ankles
2 *plural* **a :** full loose trousers gathered at the knee formerly worn by women for athletics **b :** underpants of similar design worn chiefly by girls and women ◆

bloom•ing \ˈblü-mən, -miŋ\ *adjective or adverb* [probably euphemism for *bloody*] (1882)
chiefly British — used as a generalized intensive ⟨*blooming* fool⟩

bloomy \ˈblü-mē\ *adjective* (1593)
1 : full of bloom
2 : covered with bloom ⟨*bloomy* plums⟩
3 : showing freshness or vitality ⟨all the *bloomy* flush of life is fled —Oliver Goldsmith⟩

bloop \ˈblüp\ *noun* [back-formation from *blooper*] (1967)
: BLOOPER 1a
— **bloop** *transitive verb*

bloop•er \ˈblü-pər\ *noun* [*bloop* (an unpleasant sound)] (1937)
1 a : a fly ball hit barely beyond a baseball infield **b :** a high baseball pitch lobbed to the batter
2 : an embarrassing public blunder

¹blos•som \ˈblä-səm\ *noun* [Middle English *blosme,* from Old English *blōstm;* akin to Old English *blōwan*] (before 12th century)
1 a : the flower of a seed plant; *also* **:** the mass of such flowers on a single plant **b :** the state of bearing flowers
2 : a peak period or stage of development
— **blos•somy** \-sə-mē\ *adjective*

²blossom *intransitive verb* (before 12th century)
1 : BLOOM
2 a : to come into one's own **:** DEVELOP ⟨a *blossoming* talent⟩ **b :** to become evident **:** make an appearance

¹blot \ˈblät\ *noun* [Middle English] (14th century)
1 : a soiling or disfiguring mark **:** SPOT
2 : a mark of reproach **:** moral flaw
3 : a usually nitrocellulose sheet that contains spots of immobilized macromolecules (as of DNA, RNA, or protein) or their fragments and is used to identify specific components of the spots by applying a molecular probe (as a complementary nucleic acid or a radioactively labeled antibody)

²blot *verb* **blot•ted; blot•ting** (15th century)
transitive verb
1 : to spot, stain, or spatter with a discoloring substance
2 *obsolete* **:** MAR; *especially* **:** to stain with infamy
3 a : to dry (as writing) with an absorbing agent **b :** to remove with absorbing material ⟨*blotting* up spilled water⟩
intransitive verb
1 : to make a blot
2 : to become marked with a blot

³blot *noun* [origin unknown] (1595)
1 : a lone backgammon man exposed to capture
2 *archaic* **:** a weak or exposed point

¹blotch \ˈbläch\ *transitive verb* (1604)
: to mark or mar with blotches

²blotch *noun* [perhaps blend of ¹*blot* and ³*botch*] (1669)
1 : IMPERFECTION, BLEMISH
2 : a spot or mark (as of color or ink) especially when large or irregular
— **blotch•i•ly** \ˈblä-chə-lē\ *adverb*
— **blotchy** \ˈblä-chē\ *adjective*

☆ SYNONYMS
Bloody, sanguinary, gory mean affected by or involving the shedding of blood. BLOODY is applied especially to things that are actually covered with blood or are made up of blood ⟨*bloody* hands⟩. SANGUINARY applies especially to something attended by, or someone inclined to, bloodshed ⟨the Civil War was America's most *sanguinary* conflict⟩. GORY suggests a profusion of blood and slaughter ⟨exceptionally *gory,* even for a horror movie⟩.

◇ WORD HISTORY
bloomer *Bloomers* are named after a small-town feminist who unwittingly sparked a fashion revolution. A resident of Seneca Falls, New York—the site of a historic women's rights convention in 1848—Amelia Jenks Bloomer (1818–1894) was a feminist long before the cause was popular or the term even invented. A newspaper journalist already writing about various women's issues, she became involved with the right of women to liberate themselves from traditional clothing in 1849, when she defended the wearing of pantalets by the actress Fanny Kemble. The following year she rushed to the defense of visiting feminists who appeared in Seneca Falls wearing full Turkish pantaloons under a shortened skirt. Her articles attracted the attention of newspapers in New York City and elsewhere, and soon a fad of national proportions developed. Correspondents from all over solicited her for patterns and information about the new attire, which was first known in 1851 as the "Bloomer Costume" and immediately shortened to *bloomer* or *bloomers.* Bloomer herself wore the new attire exclusively for the next several years. Aided by their suitability for women's bicycling, bloomers remained fashionable for the rest of the 19th century.

blot out *transitive verb* (1530)
1 : to make obscure, insignificant, or inconsequential
2 : WIPE OUT, DESTROY
blot·ter \'blä-tər\ *noun* (1591)
1 : a piece of blotting paper
2 : a book in which entries (as of transactions or occurrences) are made temporarily pending their transfer to permanent record books ⟨police *blotter*⟩
blotting paper *noun* (15th century)
: a spongy unsized paper for absorbing ink
blot·to \'blä-(,)tō\ *adjective* [probably irregular from ²*blot*] (1917)
: DRUNK 1a
¹blouse \'blau̇s *also* 'blau̇z\ *noun, plural* **blous·es** \'blau̇-səz, -zəz\ [French] (1828)
1 a : a long loose overgarment that resembles a shirt or smock and is worn especially by workmen, artists, and peasants **b :** the jacket of a uniform
2 : a usually loose-fitting garment especially for women that covers the body from the neck to the waist
— blousy \'blau̇-sē, -zē\ *adjective*
²blouse \'blau̇s, 'blau̇z\ *verb* **bloused; blous·ing** (1904)
intransitive verb
: to fall in a fold ⟨coats that *blouse* above the hip⟩
transitive verb
: to cause to blouse ⟨trousers are *bloused* over the boots⟩
blou·son \'blau̇-,sän, 'blü-,zän\ *noun* [French, from *blouse*] (1904)
: a garment (as a dress) having a close waistband with blousing of material over it
¹blow \'blō\ *verb* **blew** \'blü\; **blown** \'blōn\; **blow·ing** [Middle English, from Old English *blāwan;* akin to Old High German *blāen* to blow, Latin *flare,* Greek *phallos* penis] (before 12th century)
intransitive verb
1 *of air* **:** to move with speed or force
2 : to send forth a current of air or other gas
3 a : to make a sound by or as if by blowing **b** *of a wind instrument* **:** SOUND
4 a : BOAST **b :** to talk windily
5 a : PANT, GASP ⟨the horse *blew* heavily⟩ **b** *of a cetacean* **:** to eject moisture-laden air from the lungs through the blowhole
6 : to move or be carried by or as if by wind ⟨just *blew* into town⟩
7 *of an electric fuse* **:** to melt when overloaded — often used with *out*
8 *of a tire* **:** to release the contained air through a spontaneous rupture — usually used with *out*
transitive verb
1 a : to set (gas or vapor) in motion **b :** to act on with a current of gas or vapor
2 : to play or sound on (a wind instrument)
3 a : to spread by report **b** *past participle* **blowed** \'blōd\ **:** DAMN ⟨*blow* the expense⟩ ⟨I'm *blowed* if I know⟩
4 a : to drive with a current of gas or vapor **b :** to clear of contents by forcible passage of a current of air **c :** to project (a gesture or sound made with the mouth) by blowing ⟨*blew* him a kiss⟩
5 a : to distend with or as if with gas **b :** to produce or shape by the action of blown or injected air ⟨*blowing* bubbles⟩ ⟨*blowing* glass⟩
6 *of insects* **:** to deposit eggs or larvae on or in
7 : to shatter, burst, or destroy by explosion
8 a : to put out of breath with exertion **b :** to let (as a horse) pause to catch the breath
9 a : to expend (as money) extravagantly ⟨*blew* $50 on lunch⟩ **b :** to treat with unusual expenditure ⟨I'll *blow* you to a steak⟩
10 : to cause (a fuse) to blow
11 : to rupture by too much pressure ⟨*blew* a gasket⟩
12 : to foul up hopelessly ⟨*blew* her lines⟩ ⟨*blew* his chance⟩
13 : to leave hurriedly ⟨*blew* town⟩
14 : to propel with great force or speed ⟨*blew* a fastball by the batter⟩
— blow hot and cold : to be favorable at one moment and adverse the next
— blow off steam : to release pent-up emotions
— blow one's cool : to lose one's composure
— blow one's cover : to reveal one's real identity
— blow one's mind : to overwhelm one with wonder or bafflement
— blow one's top *or* **blow one's stack 1 :** to become violently angry **2 :** to go crazy
— blow smoke : to speak idly, misleadingly, or boastfully
— blow the whistle : to call public or official attention to something (as a wrongdoing) kept secret — usually used with *on*
²blow *noun* (1660)
1 : a blowing of wind especially when strong or violent
2 : BRAG, BOASTING
3 : an act or instance of blowing
4 a : the time during which air is forced through molten metal to refine it **b :** the quantity of metal refined during that time
5 *slang* **:** COCAINE
³blow *intransitive verb* **blew** \'blü\; **blown** \'blōn\; **blow·ing** [Middle English, from Old English *blōwan;* akin to Old High German *bluoen* to bloom, Latin *florēre* to bloom, *flor-, flos* flower] (before 12th century)
: FLOWER, BLOOM
⁴blow *noun* (1710)
1 : BLOSSOMS
2 : ²BLOOM 1b ⟨lilacs in full *blow*⟩
⁵blow *noun* [Middle English (northern dialect) *blaw;* probably akin to Old High German *bliuwan* to beat] (15th century)
1 : a forcible stroke delivered with a part of the body or with an instrument
2 : a hostile act or state **:** COMBAT ⟨come to *blows*⟩
3 : a forcible or sudden act or effort **:** ASSAULT
4 : an unfortunate or calamitous happening ⟨failure to land the job came as a *blow*⟩
blow away *transitive verb* (1776)
1 : to kill by gunfire **:** shoot dead
2 : to dissipate or remove as if with a current of air ⟨their doubts were *blown away*⟩
3 : to impress very strongly and usually favorably
4 : to defeat soundly ⟨*blew* their rivals *away* in the first game⟩
blow·by \'blō-,bī\ *noun* (1926)
: leakage of combustion gases between a piston and the cylinder wall into the crankcase in an automobile
blow–by–blow \'blō-bī-'blō, -bə-\ *adjective* (1933)
: minutely detailed ⟨a *blow-by-blow* account⟩
blow·down \'blō-,dau̇n\ *noun* (1895)
: an instance of trees being blown down by the wind; *also* **:** a tree blown down
blow–dried \'blō-,drīd\ *adjective* (1976)
: having blow-dried hair; *also* **:** being or appearing well-groomed but superficial or vacuous
blow–dry \-,drī\ (1966)
transitive verb
: to dry and usually style (hair) with a blow-dryer
intransitive verb
: to dry hair with a blow-dryer
— blow–dry *noun*
blow–dryer \-,drī(-ə)r\ *noun* (1976)
: a handheld hair dryer
blow·er \'blō(-ə)r\ *noun* (before 12th century)
1 : one that blows
2 : a device for producing a current of air or gas
3 : BRAGGART
4 *British* **:** TELEPHONE
blow·fish \'blō-,fish\ *noun* (circa 1893)
: PUFFER 2a
blow·fly \-,flī\ *noun* (1821)
: any of a family (Calliphoridae) of dipteran flies (as the bluebottle or screwworm) that deposit their eggs especially on meat or in wounds
blow·gun \-,gən\ *noun* (1864)
: a tube through which a projectile (as a dart) may be impelled by the force of the breath
blow·hard \-,härd\ *noun* (1857)
: BRAGGART
blow·hole \-,hōl\ *noun* (1691)
1 : a hole in metal caused by a bubble of gas captured during solidification
2 : a nostril in the top of the head of a cetacean and especially a whale
blow in *intransitive verb* (1895)
: to arrive casually or unexpectedly
blow job *noun* (1956)
: an act of fellatio — usually considered vulgar
blown \'blōn\ *adjective* [Middle English *blowen,* from past participle of *blowen* to blow] (14th century)
1 a : SWOLLEN **b :** affected with bloat
2 : being out of breath
blow off *transitive verb* (1987)
1 : to refuse to take notice of, honor, or deal with **:** IGNORE ⟨decided to *blow off* two billion viewers —Harry Homburg⟩
2 : to fail to attend or show up for ⟨*blew off* an official dinner⟩
blow·out \'blō-,au̇t\ *noun* (1824)
1 : a festive social affair
2 : a bursting of a container (as a tire) by pressure of the contents on a weak spot
3 : an uncontrolled eruption of an oil or gas well
4 : an easy or one-sided victory
blow out (14th century)
transitive verb
1 : to extinguish by a gust
2 : to dissipate (itself) by blowing — used of storms
3 : to defeat easily
intransitive verb
1 : to become extinguished by a gust
2 : to erupt out of control — used of an oil or gas well
blow over *intransitive verb* (1617)
: to pass away without effect
blow·pipe \'blō-,pīp\ *noun* (1685)
1 : a small tubular instrument for directing a jet of air or other gas into a flame so as to concentrate and increase the heat
2 : BLOWGUN
3 : a tubular instrument used for revealing or cleaning a bodily cavity by forcing air into it
4 : a long metal tube on the end of which a glassmaker gathers a quantity of molten glass and through which he blows to expand and shape it
blow·sy *also* **blow·zy** \'blau̇-zē\ *adjective* [English dialect *blowse, blowze* wench] (circa 1770)
1 : having a sloppy appearance or aspect **:** FROWSY
2 : being coarse and ruddy of complexion
blow·torch \'blō-,tȯrch\ *noun* (1897)
: a small burner having a device to intensify combustion by means of a blast of air or oxygen, usually including a fuel tank pressurized by a hand pump, and used especially in plumbing
blow·up \'blō-,əp\ *noun* (1807)
: a blowing up: as **a :** EXPLOSION **b :** an outburst of temper **c :** a photographic enlargement
blow up (1536)
transitive verb
1 : to build up or tout to an unreasonable extent ⟨advertisers *blowing up* their products⟩
2 : to rend apart, shatter, or destroy by explosion
3 : to fill up with a gas (as air) ⟨*blow up* a balloon⟩
4 : to make a photographic enlargement of

5 : to bring into existence by blowing of wind ⟨it may *blow up* a storm⟩
intransitive verb
1 a : EXPLODE **b :** to be disrupted or destroyed (as by explosion) **c :** to lose self-control; *especially* **:** to become violently angry
2 : to become or come into being by or as if by blowing of wind
3 a : to become filled with a gas (as air) **b :** to become expanded to unreasonable proportions

blowy \ˈblō-ē\ *adjective* (1830)
1 : WINDY ⟨a *blowy* March day⟩
2 : readily blown about ⟨*blowy* desert sand⟩

BLT *noun* (1952)
: a bacon, lettuce, and tomato sandwich

blub \ˈbləb\ *intransitive verb* **blubbed; blub·bing** (1804)
chiefly British **:** BLUBBER

[1]blub·ber \ˈblə-bər\ *verb* **blub·bered; blub·ber·ing** \ˈblə-b(ə-)riŋ\ [Middle English *bluberen* to make a bubbling sound, from *bluber*] (15th century)
intransitive verb
: to weep noisily
transitive verb
1 : to swell, distort, or wet with weeping
2 : to utter while weeping

[2]blubber *noun* [Middle English *bluber* bubble, foam, probably of imitative origin] (15th century)
1 a : the fat of whales and other large marine mammals **b :** excessive fat on the body
2 : the action of blubbering

[3]blubber *adjective* (1667)
: puffed out **:** THICK ⟨*blubber* lips⟩

blub·bery \ˈblə-b(ə-)rē\ *adjective* (1791)
1 : having or characterized by blubber
2 : puffed out **:** THICK

blu·cher \ˈblü-chər *also* -kər\ *noun* [G. L. von *Blücher*] (1831)
: a shoe having the tongue and vamp cut in one piece and the quarters lapped over the vamp and laced together for closing

[1]blud·geon \ˈblə-jən\ *noun* [origin unknown] (circa 1730)
1 : a short stick that usually has one thick or loaded end and is used as a weapon
2 : something used to attack or bully ⟨the *bludgeon* of satire⟩

[2]bludgeon *transitive verb* (1868)
1 : to hit with heavy impact
2 : to overcome by aggressive argument

blud·ger \ˈblə-jər\ *noun* [probably contraction of *bludgeoner* pimp, from [1]*bludgeon*] (1939)
chiefly Australian & New Zealand **:** LOAFER, SHIRKER

[1]blue \ˈblü\ *adjective* **blu·er; blu·est** [Middle English, from Old French *blou,* of Germanic origin; akin to Old High German *blāo* blue; akin to Latin *flavus* yellow] (13th century)
1 : of the color blue
2 a : BLUISH **b :** discolored by or as if by bruising ⟨*blue* with cold⟩ **c :** bluish gray ⟨*blue* cat⟩
3 a : low in spirits **:** MELANCHOLY **b :** marked by low spirits **:** DEPRESSING ⟨a *blue* funk⟩ ⟨things looked *blue*⟩
4 : wearing blue
5 *of a woman* **:** LEARNED, INTELLECTUAL
6 : PURITANICAL
7 a : PROFANE, INDECENT ⟨*blue* movie⟩ **b :** OFF-COLOR, RISQUÉ ⟨*blue* jokes⟩
8 : of, relating to, or used in blues ⟨a *blue* song⟩
— blue·ly *adverb*
— blue·ness *noun*
— blue in the face : extremely exasperated

[2]blue *noun* (13th century)
1 : a color whose hue is that of the clear sky or that of the portion of the color spectrum lying between green and violet
2 a : a pigment or dye that colors blue **b :** BLUING
3 a : blue clothing or cloth **b** *plural* **:** a blue costume or uniform
4 a : a Union soldier in the Civil War **b :** the Union army
5 a (1) **:** SKY (2) **:** the far distance **b :** SEA
6 : a blue object
7 : BLUESTOCKING
8 : any of numerous small chiefly blue butterflies (family Lycaenidae)
9 : BLUEFISH
10 : BLUE CHEESE
— out of the blue : without advance notice **:** UNEXPECTEDLY ⟨the job offer came *out of the blue*⟩

[3]blue *verb* **blued; blue·ing** *or* **blu·ing** (1606)
transitive verb
: to make blue
intransitive verb
: to turn blue

blue baby *noun* (circa 1899)
: an infant with a bluish tint usually from a congenital heart defect in which mingling of venous and arterial blood occurs

blue·beard \ˈblü-ˌbird\ *noun, often capitalized* [*Bluebeard,* a fairy-tale character] (1822)
: a man who marries and kills one wife after another

blue·bell \-ˌbel\ *noun* (1578)
1 : any of various bellflowers; *especially* **:** HAREBELL
2 : any of various plants bearing blue bell-shaped flowers: as **a :** a European herb (*Endymion nonscriptus* synonym *Scilla nonscripta*) of the lily family having scapose racemes of drooping bell-shaped flowers — called also *wild hyacinth* **b** *plural* **:** a glabrous erect eastern U.S. herb (*Mertensia virginica*) of the borage family with entire leaves and showy blue flowers pink in the bud — called also *Virginia bluebells*

blue·ber·ry \ˈblü-ˌber-ē, -b(ə-)rē\ *noun* (1709)
: the edible blue or blackish berry of any of several plants (genus *Vaccinium*) of the heath family; *also* **:** a low or tall shrub producing these berries

blue·bird \-ˌbərd\ *noun* (1688)
: any of several small North American thrushes (genus *Sialia*) that are blue above and reddish brown or pale blue below

blue blood *noun* (1834)
1 \ˈblü-ˈbləd\ **:** membership in a noble or socially prominent family
2 \-ˌbləd\ **:** a member of a noble or socially prominent family
— blue–blood·ed \-ˈblə-dəd\ *adjective*

blue·bon·net \ˈblü-ˌbä-nət\ *noun* (1682)
1 a : a wide flat round cap of blue wool formerly worn in Scotland **b :** one that wears such a cap; *specifically* **:** SCOT
2 : either of two low-growing annual lupines (*Lupinus subcarnosus* or *Latin texensis*) of Texas with silky foliage and blue flowers

blue book *noun* (1836)
1 : a register especially of socially prominent persons
2 : a book of specialized information often published under government auspices
3 : a blue-covered booklet used for writing examinations
4 : a periodically issued price list (as of used cars)

blue·bot·tle \ˈblü-ˌbä-tᵊl\ *noun* (15th century)
1 : BACHELOR'S BUTTON
2 : any of several blowflies (genus *Calliphora*) that have the abdomen or the whole body iridescent blue in color and that make a loud buzzing noise in flight

blue catfish *noun* (1835)
: a large bluish catfish (*Ictalurus furcatus*) of the Mississippi valley that may weigh over 100 pounds (45 kilograms)

blue cheese *noun* (1925)
: cheese having veins of greenish blue mold

blue chip *noun* (1929)
1 a : a stock issue of high investment quality that usually pertains to a substantial well-established company and enjoys public confidence in its worth and stability **b :** a consistently successful and profitable venture or enterprise
2 a : an outstandingly worthwhile or valuable property or asset **b :** an athlete rated as excellent or as an excellent prospect
— blue–chip *adjective*

blue–chip·per \ˈblü-ˌchi-pər\ *noun* (1968)
: BLUE CHIP 2b

blue·coat \ˈblü-ˌkōt\ *noun* (1593)
: a person who wears a blue coat: as **a :** a Union soldier during the Civil War **b :** POLICE OFFICER

blue cohosh *noun* (1821)
: a perennial herb (*Caulophyllum thalictroides*) of the barberry family that has greenish yellow or purplish flowers and large blue berrylike fruits

blue–col·lar \ˈblü-ˈkä-lər\ *adjective* (1946)
: of, relating to, or constituting the class of wage earners whose duties call for the wearing of work clothes or protective clothing — compare WHITE-COLLAR

blue crab *noun* (1883)
: a large bluish green edible crab (*Callinectes sapidus*) of the Atlantic and Gulf coasts

blue crab

blue curls *noun plural but singular or plural in construction* (1817)
: any of several mints (genus *Trichostema*) with irregular blue flowers

blue devils *noun plural* (1781)
: low spirits **:** DESPONDENCY

blue–eyed \ˈblü-ˌīd\ *adjective* (1610)
1 : having blue eyes
2 : performed by whites ⟨*blue-eyed* soul⟩; *also* **:** WHITE ⟨a *blue-eyed* soul singer⟩

blue–eyed grass *noun* (1784)
: any of several herbs (genus *Sisyrinchium*) of the iris family with grasslike foliage and blue flowers

blue·fin tuna \ˈblü-ˌfin-\ *noun* (1922)
: a very large tuna (*Thunnus thynnus*) that is an important food and game fish — called also *bluefin*

blue·fish \-ˌfish\ *noun* (circa 1622)
1 : an active food and game marine fish (*Pomatomus saltatrix*) that is bluish above with silvery sides
2 : any of various dark or bluish fishes (as the pollack)

blue flag *noun* (1784)
: a blue-flowered iris; *especially* **:** a common iris (*Iris versicolor*) of the eastern U.S. with a root formerly used medicinally

blue flu *noun* [from the color of a police uniform] (1970)
: a sick-out staged by police officers

blue·gill \ˈblü-ˌgil\ *noun* (1881)
: a common sunfish (*Lepomis macrochirus*) of the eastern and central U.S. sought for food and sport

blue·grass \-ˌgras\ *noun* (1751)
1 : any of several grasses (genus *Poa*) of which some have bluish green culms; *especially* **:** KENTUCKY BLUEGRASS
2 [from the *Blue Grass Boys,* performing group, from *Bluegrass state,* nickname of Kentucky] **:** country music played on unamplified stringed instruments (as banjo, fiddle, guitar, and mandolin) and characterized by free improvisation and close usually high-pitched harmony

blue–green alga \ˈblü-ˈgrēn-\ *noun* (1899)

\ə\ **abut** \ᵊ\ **kitten** \ər\ **further** \a\ **ash** \ā\ **ace** \ä\ **mop, mar** \au̇\ **out** \ch\ **chin** \e\ **bet** \ē\ **easy** \g\ **go** \i\ **hit** \ī\ **ice** \j\ **job** \ŋ\ **sing** \ō\ **go** \ȯ\ **law** \ȯi\ **boy** \th\ **thin** \t͟h\ **the** \ü\ **loot** \u̇\ **foot** \y\ **yet** \zh\ **vision** *see also* Guide to Pronunciation

: any of a group of photosynthetic microorganisms now usually classified as bacteria (division Cyanobacteria) or sometimes as plants (division Cyanophyta) — called also *blue-green bacterium, cyanobacterium*

blue gum *noun* (1799)
: any of several Australian trees (genus *Eucalyptus*) that yield valuable timber

blue heron *noun* (circa 1730)
: either of two herons with bluish or slaty plumage; *especially* **:** GREAT BLUE HERON

blue•jack•et \-ˌja-kət\ *noun* (1830)
: an enlisted man in the navy **:** SAILOR

blue jay \-ˌjā\ *noun* (1709)
: a crested bright blue North American jay (*Cyanocitta cristata*)

blue jeans *noun plural* (1901)
: pants usually made of blue denim
— **blue–jeaned** \-ˌjēnd\ *adjective*

blue law *noun* (1781)
1 : one of numerous extremely rigorous laws designed to regulate morals and conduct in colonial New England
2 : a statute regulating work, commerce, and amusements on Sundays

blue line *noun* (1937)
: either of two blue lines that divide an ice-hockey rink into three equal zones and that separate the offensive and defensive zones from the center-ice neutral zone

blue mold *noun* (1664)
: any of various fungi (genus *Penicillium*) that produce blue or blue-green surface growths

blue moon *noun* (1821)
: a very long period of time ⟨such people happen along only once in a *blue moon* — *Saturday Review*⟩

blue•nose \ˈblü-ˌnōz\ *noun* (1903)
: a person who advocates a rigorous moral code

blue note *noun* [from its frequent use in blues music] (1919)
: a variable microtonal lowering of the third, seventh, and occasionally fifth degrees of the major scale

blue–pen•cil \ˈblü-ˈpen(t)-səl\ *transitive verb* (1888)
: to edit especially by shortening or deletion
— **blue penciller** *noun*

blue pencil *noun* (1893)
: a writing instrument used for editing; *also* **:** the act or practice of blue-penciling

blue pe•ter \-ˈpē-tər\ *noun* [probably from the name *Peter*] (1823)
: a blue signal flag with a white square in the center used to indicate that a merchant vessel is ready to sail

blue pike *noun* (1842)
: a grayish-blue walleye categorized as either a morph or a subspecies (*Stizostedion vitreum glaucum*)

blue plate *adjective* (1926)
: being a main course usually offered at a special price in a restaurant ⟨a *blue plate* luncheon⟩

blue•point \ˈblü-ˌpȯint\ *noun* [*Blue Point,* Long Island] (1789)
: a small oyster (*Crassostrea virginica*) typically from the south shore of Long Island

blue point \-ˌpȯint\ *adjective* (1944)
of a domestic cat **:** having a bluish cream body coat with dark gray points
— **blue point** *noun*

blue•print \-ˌprint\ *noun* (1886)
1 : a photographic print in white on a bright blue ground or blue on a white ground used especially for copying maps, mechanical drawings, and architects' plans
2 : something resembling a blueprint; *especially* **:** a detailed plan or program of action ⟨a *blueprint* for victory⟩
— **blueprint** *transitive verb*

blue racer *noun* (1886)
: a blue or greenish blue blacksnake (*Coluber constrictor* subspecies) occurring from Michigan and Wisconsin to Texas

blue–ribbon *adjective* (1926)
: of outstanding quality; *especially* **:** consisting of individuals selected for quality, reputation, or authority ⟨a *blue-ribbon* committee⟩ ⟨a *blue-ribbon* jury⟩

blue ribbon *noun* (1651)
1 : an honor or award gained for preeminence
2 : a blue ribbon awarded as an honor (as to the first-place winner in a competition)

blues \ˈblüz\ *noun plural but singular or plural in construction* [*blue devils*] (1741)
1 : low spirits **:** MELANCHOLY
2 : a song often of lamentation characterized by usually 12-bar phrases, 3-line stanzas in which the words of the second line usually repeat those of the first, and continual occurrence of blue notes in melody and harmony
3 : jazz or popular music using harmonic and phrase structures of blues

blue shark *noun* (circa 1672)
: a chiefly pelagic shark (*Prionace glauca*) found in all tropical and temperate seas that occasionally attacks humans

blue sheep *noun* (1911)
: BHARAL

blue•shift \ˈblü-ˈshift\ *noun* (1951)
: the displacement of the spectrum of an approaching celestial body toward shorter wavelengths
— **blue•shift•ed** *adjective*

blue–sky \ˈblü-ˈskī\ *adjective* (1906)
1 : having little or no value ⟨*blue-sky* stock⟩
2 : not grounded in the realities of the present **:** VISIONARY ⟨*blue-sky* thinking⟩

blue–sky law *noun* (1912)
: a law providing for the regulation of the sale of securities (as stock)

blues•man \ˈblüz-mən\ *noun* (1966)
: a man who plays or sings the blues

blue spruce *noun* (1884)
: a spruce (*Picea pungens*) of the Rocky Mountains having sharp usually bluish gray needles and often planted as an ornamental

blue•stem \ˈblü-ˌstem\ *noun* (circa 1862)
1 : a tall North American grass (*Andropogon gerardii* synonym *A. furcatus*) that has smooth bluish leaf sheaths and slender spikes borne in pairs or clusters and is used for hay and forage
2 : LITTLE BLUESTEM

blue•stock•ing \-ˌstä-kiŋ\ *noun* [*Bluestocking* society, 18th century literary clubs] (1790)
: a woman having intellectual or literary interests ◆

blue•stone \-ˌstōn\ *noun* (1709)
: a building stone of bluish gray color

blue streak *noun* (1830)
1 : something that moves very fast
2 : a constant stream of words ⟨talked a *blue streak*⟩

bluesy \ˈblü-zē\ *adjective* **blues•i•er; -est** (1946)
: resembling, characteristic of, or suited to the blues

blu•et \ˈblü-ət\ *noun* [probably from [1]*blue*] (circa 1821)
: a North American plant (*Hedyotis caerulea* synonym *Houstonia caerulea*) of the madder family having tufted stems and bluish or white flowers with yellow centers

blue•tick \-ˌtik\ *noun* (1945)
: any of a breed of tricolor coonhounds of American origin having the white areas of the coat heavily ticked with black

blue•tongue \ˈblü-ˌtəŋ\ *noun* (1863)
: a serious virus disease especially of sheep characterized by hyperemia, cyanosis, and punctate hemorrhages and by swelling and sloughing of the epithelium especially about the mouth and tongue

blue vitriol *noun* (1728)
: a hydrated copper sulfate $CuSO_4 \cdot 5H_2O$

blue water *noun* (1582)
: the open sea
— **blue–wa•ter** \ˈblü-ˌwȯ-tər, -ˌwä-\ *adjective*

blue•weed \ˈblü-ˌwēd\ *noun* (circa 1837)
1 : VIPER'S BUGLOSS
2 : a small weedy sunflower (*Helianthus ciliaris*) of the southwestern U.S. with blue-green or gray-green foliage

blue whale *noun* (1851)
: a very large baleen whale (*Balaenoptera musculus* synonym *Sibbaldus musculus*) that may reach a weight of 150 tons (135 metric tons) and a length of 100 feet (30 meters) and is generally considered the largest living animal

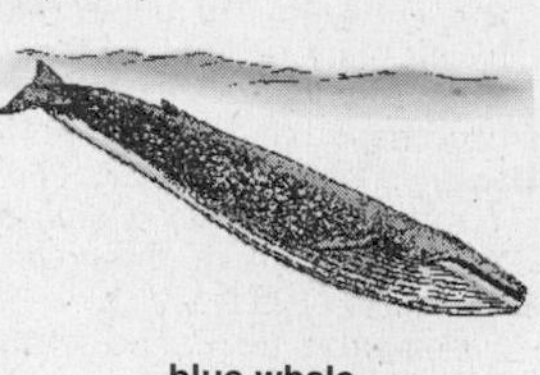
blue whale

bluey \ˈblü-ē\ *noun* [from the blue blanket commonly used to wrap the bundle] (1877)
Australian **:** a swagman's bundle of personal effects; *also* **:** a bag of clothing carried in travel

[1]**bluff** \ˈbləf\ *adjective* [obsolete Dutch *blaf* flat; akin to Middle Low German *blaff* smooth] (1627)
1 a : having a broad flattened front **b :** rising steeply with a broad flat or rounded front
2 : good-naturedly frank and outspoken ☆

☆ SYNONYMS
Bluff, blunt, brusque, curt, crusty, gruff mean abrupt and unceremonious in speech and manner. BLUFF connotes good-natured outspokenness and unconventionality ⟨a *bluff* manner⟩. BLUNT suggests directness of expression in disregard of others' feelings ⟨a *blunt* appraisal⟩. BRUSQUE applies to a sharpness or ungraciousness ⟨a *brusque* response⟩. CURT implies disconcerting shortness or rude conciseness ⟨a *curt* command⟩. CRUSTY suggests a harsh or surly manner sometimes concealing an inner kindliness ⟨a *crusty* exterior⟩. GRUFF suggests a hoarse or husky speech which may imply bad temper but more often implies embarrassment or shyness ⟨puts on a *gruff* pose⟩.

◇ WORD HISTORY
bluestocking The origin of *bluestocking* goes back to the mid-18th century, to a circle of English ladies who decided to replace their social evenings of cardplaying and idle chatter to which tradition had confined them, with more intellectual pursuits. Taking the literary salons of Paris as their model, they decided to hold discussions at which illustrious men of letters would be the honored guests. On one occasion, they invited the English botanist and sometime poet Benjamin Stillingfleet. A poor scholar who lacked fancy clothes, including the black silk stockings customarily worn to such high-society events, he felt obliged to decline the invitation, until he was assured that his everyday clothes, including his blue worsted stockings, would be quite acceptable. Thus attired, Stillingfleet became a popular fixture at the evening conversations and was dubbed "blue stockings." As it was then considered unseemly for women to aspire to learning, the circle presided over by Stillingfleet acquired the derisive nickname the "Blue Stocking Society." Rather defiantly, the members of the group embraced the intended put-down, adopting the name *Bluestocking,* or its French equivalent *Bas Bleu.* By the early 19th century, any woman with intellectual interests was called, often derogatorily still, a *bluestocking.*

— **bluff·ly** *adverb*
— **bluff·ness** *noun*
²bluff *noun* (1666)
: a high steep bank : CLIFF
³bluff *verb* [probably from Dutch *bluffen* to boast, play a kind of card game] (1839)
transitive verb
1 a : to deter or frighten by pretense or a mere show of strength **b :** DECEIVE **c :** FEIGN ⟨the catcher *bluffed* a throw to first⟩
2 : to deceive (an opponent) in cards by a bold bet on an inferior hand
intransitive verb
: to bluff someone : act deceptively
— **bluff·er** *noun*
⁴bluff *noun* (1845)
1 a : an act or instance of bluffing **b :** the practice of bluffing
2 : one who bluffs
blu·ing *or* **blue·ing** \'blü-iŋ\ *noun* (1669)
: a preparation used in laundering to counteract yellowing of white fabrics
blu·ish \'blü-ish\ *adjective* (14th century)
: somewhat blue : having a tinge of blue
— **blu·ish·ness** *noun*
¹blun·der \'blən-dər\ *verb* **blun·dered; blun·der·ing** \-d(ə-)riŋ\ [Middle English *blundren*] (14th century)
intransitive verb
1 : to move unsteadily or confusedly
2 : to make a mistake through stupidity, ignorance, or carelessness
transitive verb
1 : to utter stupidly, confusedly, or thoughtlessly
2 : to make a stupid, careless, or thoughtless mistake in
— **blun·der·er** \-dər-ər\ *noun*
— **blun·der·ing·ly** \-d(ə-)riŋ-lē\ *adverb*
²blunder *noun* (circa 1706)
: a gross error or mistake resulting usually from stupidity, ignorance, or carelessness
synonym see ERROR
blun·der·buss \'blən-dər-ˌbəs\ *noun* [by folk etymology from obsolete Dutch *donderbus,* from Dutch *donder* thunder + obsolete Dutch *bus* gun] (1654)
1 : a muzzle-loading firearm with a short barrel and flaring muzzle to facilitate loading

blunderbuss 1

2 : a blundering person
¹blunt \'blənt\ *adjective* [Middle English] (13th century)
1 a : slow or deficient in feeling : INSENSITIVE **b :** obtuse in understanding or discernment : DULL
2 : having an edge or point that is not sharp
3 a : abrupt in speech or manner **b :** being straight to the point : DIRECT
synonym see DULL, BLUFF
— **blunt·ly** *adverb*
— **blunt·ness** *noun*
²blunt (14th century)
transitive verb
: to make less sharp or definite
intransitive verb
: to become blunt
¹blur \'blər\ *noun* [perhaps akin to Middle English *bleren* to blear] (1548)
1 : a smear or stain that obscures
2 : something vaguely or indistinctly perceived; *especially* **:** something moving or occurring too quickly to be clearly perceived
²blur *verb* **blurred; blur·ring** (1581)
transitive verb
1 : to obscure or blemish by smearing
2 : SULLY
3 : to make dim, indistinct, or vague in outline or character
4 : to make cloudy or confused
intransitive verb
1 : to make blurs
2 : to become vague or indistinct
— **blur·ring·ly** \'blər-iŋ-lē\ *adverb*
¹blurb \'blərb\ *noun* [coined by Gelett Burgess] (1914)
: a short publicity notice (as on a book jacket) ◇
²blurb *transitive verb* (1915)
: to describe or praise in a blurb
blur·ry \'blər-ē\ *adjective* **blur·ri·er; -est** (1884)
: marked by blurring
— **blur·ri·ly** \'blər-ə-lē\ *adverb*
— **blur·ri·ness** \'blər-ē-nəs\ *noun*
blurt \'blərt\ *transitive verb* [probably imitative] (1573)
: to utter abruptly and impulsively — usually used with *out*
— **blurt·er** *noun*
¹blush \'bləsh\ *noun* [Middle English, probably from *blusshen*] (14th century)
1 : outward appearance : VIEW ⟨at first *blush*⟩
2 : a reddening of the face especially from shame, modesty, or confusion
3 : a red or rosy tint
— **blush·ful** \-fəl\ *adjective*
²blush *intransitive verb* [Middle English *blusshen,* from Old English *blyscan* to redden; akin to Old English *blȳsa* flame, Old High German *bluhhen* to burn brightly] (15th century)
1 : to become red in the face especially from shame, modesty, or confusion
2 : to feel shame or embarrassment
3 : to have a rosy or fresh color : BLOOM
— **blush·ing·ly** \'blə-shiŋ-lē\ *adverb*
blush·er \'blə-shər\ *noun* (1665)
1 : one that blushes
2 : a cosmetic applied to the face to give a usually pink color or to accent the cheekbones
blush wine *noun* (1985)
: any of various pinkish table wines
¹blus·ter \'bləs-tər\ *verb* **blus·tered; blus·ter·ing** \-t(ə-)riŋ\ [Middle English *blustren,* probably from Middle Low German *blüsteren*] (15th century)
intransitive verb
1 : to talk or act with noisy swaggering threats
2 a : to blow in stormy noisy gusts **b :** to be windy and boisterous
transitive verb
1 : to utter with noisy self-assertiveness
2 : to drive or force by blustering
— **blus·ter·er** \-tər-ər\ *noun*
— **blus·ter·ing·ly** \-t(ə-)riŋ-lē\ *adverb*
²bluster *noun* (1583)
1 : a violent boisterous blowing
2 : violent commotion
3 : loudly boastful or threatening speech
— **blus·ter·ous** \-t(ə-)rəs\ *adjective*
— **blus·tery** \-t(ə-)rē\ *adjective*
B lymphocyte *noun* (1971)
: B CELL
B movie *noun* (1948)
: a cheaply produced motion picture
boa \'bō-ə\ *noun* [Latin, a water snake] (14th century)
1 : a large snake (as the boa constrictor, anaconda, or python) that kills by constriction
2 : a long fluffy scarf of fur, feathers, or delicate fabric
boa constrictor *noun* (1809)
: a tropical American boa (*Constrictor constrictor* synonym *Boa constrictor*) that is light brown barred or mottled with darker brown and reaches a length of 10 feet (3 meters) or more; *broadly* **:** BOA 1
boar \'bōr, 'bȯr\ *noun* [Middle English *bor,* from Old English *bār;* akin to Old High German & Old Saxon *bēr* boar] (before 12th century)
1 a : an uncastrated male swine **b :** the male of any of several mammals (as a guinea pig or raccoon)
2 : WILD BOAR
— **boar·ish** \-ish\ *adjective*
¹board \'bōrd, 'bȯrd\ *noun* [Middle English *bord* piece of sawed lumber, border, ship's side, from Old English; akin to Old High German *bort* ship's side] (before 12th century)
1 *obsolete* **:** BORDER, EDGE
2 : the side of a ship
3 a : a piece of sawed lumber of little thickness and a length greatly exceeding its width **b** *plural* **:** STAGE 2a(2)
4 a *archaic* **:** TABLE 3a **b :** a table spread with a meal **c :** daily meals especially when furnished for pay **d :** a table at which a council or magistrates sit **e** (1) **:** a group of persons having managerial, supervisory, investigatory, or advisory powers ⟨*board* of directors⟩ ⟨*board* of examiners⟩ (2) **:** an examination given by an examining board — often used in plural ⟨pass the medical *boards*⟩ **f :** LEAGUE, ASSOCIATION **g** (1) **:** the exposed hands of all the players in a stud poker game (2) **:** an exposed dummy hand in bridge
5 a : a flat usually rectangular piece of material (as wood) designed for a special purpose: as (1) **:** SPRINGBOARD 1 (2) **:** SURFBOARD **b :** BACKBOARD; *also* **:** a rebound in basketball **c :** a surface, frame, or device for posting notices **d :** BLACKBOARD **e :** SWITCHBOARD
6 a : PAPERBOARD **b :** the stiff foundation piece for the side of a book cover
7 : a securities or commodities exchange
8 *plural* **:** the low wooden wall enclosing a hockey rink
9 : a sheet of insulating material carrying circuit elements and terminals so that it can be inserted in an electronic apparatus (as a computer)
— **board·like** \-ˌlīk\ *adjective*
— **across the board :** so as to include or affect all classes or categories ⟨cut spending *across the board*⟩; *also* **:** in all areas or respects ⟨considered an average player *across the board*⟩
— **on board :** ABOARD
²board (15th century)
transitive verb
1 *archaic* **:** to come up against or alongside (a ship) usually to attack
2 : ACCOST, ADDRESS

◇ WORD HISTORY
blurb The origin of the word *blurb* can be traced to a single individual and single occasion. At the annual dinner of the American Booksellers' Association in 1907 humorist Gelett Burgess was one of the guests. It was the custom at these dinners for the guest authors to present to the assembled company souvenir copies of their latest books. Seizing the occasion to satirize publishing's practice of plastering book jackets with effusive copy, Burgess prepared a mock-solemn commemorative jacket of his latest, placing on the front cover a doctored picture of a woman that he had lifted from a dental advertisement. He dubbed his creation Miss Belinda Blurb and appended a self-congratulatory text. Seven years later, Burgess defined *blurb* itself as "a flamboyant advertisement; an inspired testimonial . . . . On the 'jacket' of the 'latest' fiction, we find the blurb . . . attesting that this book is the 'sensation' of the year." Burgess's coinage filled a widely felt need, and the very sound of the word was suggestive of the "mush, gush and tosh" (in the words of one writer) that book publicists tend to favor.

3 a : to go aboard (as a ship, train, airplane, or bus) **b :** to put aboard ⟨an airliner *boarding* passengers⟩
4 : to cover with boards ⟨*board* up a window⟩
5 : to provide with regular meals and often also lodging usually for compensation
6 : to check with a board check
intransitive verb
: to take one's meals usually as a paying customer
board check *noun* (circa 1936)
: a body check of an opposing player against the rink boards in ice hockey
board·er \ˈbōr-dər, ˈbȯr-\ *noun* (1530)
: one that boards; *especially* **:** one that is provided with regular meals or regular meals and lodging
board foot *noun* (1896)
: a unit of quantity for lumber equal to the volume of a board 12 × 12 × 1 inches — abbreviation *bd ft*
board game *noun* (1934)
: a game of strategy (as checkers, chess, or backgammon) played by moving pieces on a board
board·ing·house \ˈbōr-diŋ-ˌhau̇s, ˈbȯr-\ *noun* (1728)
: a lodging house at which meals are provided
boarding school *noun* (1677)
: a school at which meals and lodging are provided
board·man \ˈbōrd-ˌman, ˈbȯrd-, *especially for 2* -mən\ *noun* (circa 1923)
1 : a member of a board
2 : one who works at a board
board of trade (1780)
1 *B&T capitalized* **:** a British governmental department concerned with commerce and industry
2 : a commodities exchange
board·room \ˈbōrd-ˌrüm, ˈbȯrd-, -ˌru̇m\ *noun* (1836)
: a room that is designated for meetings of a board
board·sail·ing \-ˌsā-liŋ\ *noun* (1980)
: WINDSURFING
— board·sail·or \-ˌsā-lər\ *noun*
board·walk \-ˌwȯk\ *noun* (1872)
1 : a walk constructed of planking
2 : a walk constructed along a beach
boart \ˈbōrt, ˈbȯrt\ *variant of* BORT
[1]boast \ˈbōst\ *noun* [Middle English *boost*] (14th century)
1 : the act or an instance of boasting **:** BRAG
2 : a cause for pride
— boast·ful \ˈbōst-fəl\ *adjective*
— boast·ful·ly \-fə-lē\ *adverb*
— boast·ful·ness *noun*
[2]boast (14th century)
intransitive verb
1 : to puff oneself up in speech **:** speak vaingloriously
2 *archaic* **:** GLORY, EXULT
transitive verb
1 : to speak of or assert with excessive pride
2 a : to possess and often call attention to (something that is a source of pride) ⟨*boasts* a new stadium⟩ **b :** HAVE, CONTAIN ⟨a room *boasting* no more than a desk and a chair⟩ ☆
— boast·er *noun*
[3]boast *transitive verb* [origin unknown] (1823)
: to shape (stone) roughly in sculpture and stonecutting as a preliminary to finer work
[1]boat \ˈbōt\ *noun* [Middle English *boot,* from Old English *bāt;* akin to Old Norse *beit* boat] (before 12th century)
1 a : a small vessel for travel on water **b :** SHIP
2 : a boat-shaped container, utensil, or device ⟨a gravy *boat*⟩ ⟨a laboratory *boat*⟩
— boat·ful \-ˌfu̇l\ *noun*
— boat·like \-ˌlīk\ *adjective*
— in the same boat : in the same situation or predicament
[2]boat (1613)
transitive verb
: to place in or bring into a boat
intransitive verb
: to go by boat
boat·build·er \-ˌbil-dər\ *noun* (1679)
: one that builds boats
— boat·build·ing \-ˌbil-diŋ\ *noun*
boat·er \ˈbō-tər\ *noun* (1605)
1 : one who travels in a boat
2 : a stiff hat usually made of braided straw with a brim, hatband, and flat crown
boat hook *noun* (circa 1611)
: a pole-handled hook with a point or knob on the back used especially to pull or push a boat, raft, or log into place
boat·house \-ˌhau̇s\ *noun* (1722)
: a building to house and protect boats
boat·load \ˈbōt-ˌlōd\ *noun* (1680)
1 : a load that fills a boat
2 : an indefinitely large number ⟨a *boatload* of money⟩
boat·man \ˈbōt-mən\ *noun* (14th century)
: a man who works on, deals in, or operates boats
boat people *noun plural* (1977)
: refugees fleeing by boat
boat·swain \ˈbō-sᵊn\ *noun* [Middle English *bootswein,* from *boot* boat + *swein* boy, servant — more at SWAIN] (14th century)
1 : a petty officer on a merchant ship having charge of hull maintenance and related work
2 : a naval warrant officer in charge of the hull and all related equipment
boat train *noun* (1884)
: an express train for transporting passengers between a port and a city
boat·yard \ˈbōt-ˌyärd\ *noun* (1847)
: a yard where boats are built, repaired, and stored and often sold or rented
[1]bob \ˈbäb\ *verb* **bobbed; bob·bing** [Middle English *boben*] (13th century)
transitive verb
1 : to strike with a quick light blow **:** RAP
2 : to move up and down in a short quick movement ⟨*bob* the head⟩
3 : to polish with a bob **:** BUFF
intransitive verb
1 a : to move up and down briefly or repeatedly ⟨a cork *bobbed* in the water⟩ **b :** to emerge, arise, or appear suddenly or unexpectedly ⟨the question *bobbed* up again⟩
2 : to nod or curtsy briefly
3 : to try to seize a suspended or floating object with the teeth ⟨*bob* for apples⟩
[2]bob *noun* (1550)
1 a : a short quick down-and-up motion **b** *Scottish* **:** any of several folk dances
2 *obsolete* **:** a blow or tap especially with the fist
3 a : a modification of the order in change ringing **b :** a method of change ringing using a bob
4 : a small polishing wheel of solid felt or leather with rounded edges
[3]bob *transitive verb* **bobbed; bob·bing** [Middle English *bobben,* from Middle French *bober*] (14th century)
1 *obsolete* **:** DECEIVE, CHEAT
2 *obsolete* **:** to take by fraud **:** FILCH
[4]bob *noun* [Middle English *bobbe*] (14th century)
1 a (1) **:** BUNCH, CLUSTER (2) *Scottish* **:** NOSEGAY **b :** a knob, knot, twist, or curl especially of ribbons, yarn, or hair **c :** a short haircut on a woman or child
2 : FLOAT 2a
3 : a hanging ball or weight (as on a plumb line)
4 : TRIFLE 1
[5]bob *transitive verb* **bobbed; bob·bing** (1675)
1 : to cut shorter **:** CROP ⟨*bob* a horse's tail⟩
2 : to cut (hair) in the style of a bob
[6]bob *noun, plural* **bob** [perhaps from the name *Bob*] (1789)
slang British **:** SHILLING
[7]bob *noun* (1856)
: BOBSLED
[1]bob·ber \ˈbä-bər\ *noun* (1593)
: one that bobs
[2]bobber *noun* (1904)
: a person who rides or races on a bobsled
bob·bery \ˈbä-b(ə-)rē\ *noun, plural* **-ber·ies** [Hindi *bāp re,* literally, oh father!] (1803)
: HUBBUB
bob·bin \ˈbä-bən\ *noun* [origin unknown] (1530)
1 a : a cylinder or spindle on which yarn or thread is wound (as in a sewing machine) **b :** any of various small round devices on which threads are wound for working handmade lace **c :** a coil of insulated wire or the reel it is wound on
2 : a cotton cord formerly used by dressmakers for piping
bob·bi·net \ˈbä-bə-ˌnet\ *noun* [blend of *bobbin* and *net*] (1814)
: a machine-made net of cotton, silk, or nylon usually with hexagonal mesh
[1]bob·ble \ˈbä-bəl\ *verb* **bob·bled; bob·bling** \-b(ə-)liŋ\ [frequentative of [1]*bob*] (1812)
1 : [1]BOB
2 : FUMBLE
[2]bobble *noun* (1880)
1 : a repeated bobbing movement
2 : a small ball of fabric; *especially* **:** one in a series used on an edging ⟨curtains with *bobbles*⟩
3 : ERROR, MISTAKE; *especially* **:** a mishandling of the ball in baseball or football
bob·by \ˈbä-bē\ *noun, plural* **bobbies** [*Bobby,* nickname for *Robert,* after Sir *Robert* Peel, who organized the London police force] (1844)
British **:** POLICE OFFICER
bobby pin *noun* [perhaps from [4]*bob*] (1932)
: a flat wire hairpin with prongs that press close together
bobby socks *or* **bobby sox** *noun plural* [perhaps from *bobby* pin] (1943)
: girls' socks reaching above the ankle
bob·by–sox·er \-ˌsäk-sər\ *noun* (1944)
: an adolescent girl
bob·cat \ˈbäb-ˌkat\ *noun* [[4]*bob;* from the stubby tail] (1888)
: a common North American lynx (*Lynx rufus*) variably reddish in base color with dark markings

bobcat

bo·beche \bō-ˈbesh, -ˈbāsh\ *noun* [French *bobèche*] (circa 1897)
: a usually glass collar on a candle socket to catch drippings or on a candlestick or chandelier to hold suspended glass prisms
bob·o·link \ˈbä-bə-ˌliŋk\ *noun* [imitative] (circa 1801)

☆ SYNONYMS
Boast, brag, vaunt, crow mean to express pride in oneself or one's accomplishments. BOAST often suggests ostentation and exaggeration ⟨*boasts* of every trivial success⟩, but it may imply a claiming with proper and justifiable pride ⟨the town *boasts* one of the best museums in the area⟩. BRAG suggests crudity and artlessness in glorifying oneself ⟨*bragging* of their exploits⟩. VAUNT usually connotes more pomp and bombast than BOAST and less crudity or naïveté than BRAG ⟨*vaunted* his country's military might⟩. CROW usually implies exultant boasting or bragging ⟨*crowed* after winning the championship⟩.

: an American migratory songbird (*Dolichonyx oryzivorus*) with the breeding male chiefly black

bob·sled \ˈbäb-ˌsled\ *noun* [perhaps from [4]*bob*] (1839)
1 : a short sled usually used as one of a pair joined by a coupling
2 : a large usually metal sled used in racing and equipped with two pairs of runners in tandem, a long seat for two or more people, a steering wheel, and a hand brake
— bobsled *intransitive verb*
— bob·sled·der *noun*

bob·sled·ding \-ˌsle-diŋ\ *noun* (1883)
: the act, skill, or sport of riding or racing on a bobsled

bob·stay \ˈbäb-ˌstā\ *noun* [probably from [2]*bob*] (1758)
: a stay to hold a ship's bowsprit down

bob·tail \ˈbäb-ˌtāl\ *noun* [[4]*bob*] (1605)
1 a : a bobbed tail **b :** a horse, dog, or cat with a bobbed or very short tail; *especially* **:** OLD ENGLISH SHEEPDOG
2 : something curtailed
— bobtail *or* **bob·tailed** \-ˌtāld\ *adjective*

bob veal \ˈbäb-\ *noun* [English dialect *bob* young calf] (1855)
: the veal of a very young or unborn calf

bob·white \ˈbäb-ˈhwīt, -ˈwīt\ *noun* [imitative] (1819)
: any of a genus (*Colinus*) of quail; *especially* **:** a popular game bird (*C. virginianus*) of the eastern and central U.S.

bo·cac·cio \bə-ˈkä-chē-ˌō, -chō\ *noun* [perhaps modification of Spanish *bocacha,* augmentative of *boca* mouth] (circa 1890)
: a large rockfish (*Sebastes paucispinis*) of the Pacific coast locally important as a market fish

boc·cie *or* **boc·ci** *or* **boc·ce** \ˈbä-chē\ *noun* [Italian *bocce,* plural of *boccia* ball, from (assumed) Vulgar Latin *bottia* boss] (1926)
: a game of Italian origin similar to lawn bowling played on a long narrow usually dirt court

bock \ˈbäk\ *noun* [German, short for *Bockbier,* by shortening & alteration from *Einbecker Bier,* literally, beer from Einbeck, from *Einbeck,* Germany] (1856)
: a heavy dark rich beer usually sold in the early spring

bod \ˈbäd\ *noun* (1933)
1 *British* **:** FELLOW, GUY
2 : BODY

bo·da·cious \bō-ˈdā-shəs\ *adjective* [probably blend of *bold* and *audacious*] (1845)
1 *Southern & Midland* **:** OUTRIGHT, UNMISTAKABLE
2 : REMARKABLE, NOTEWORTHY ⟨the *bodacious* 1974 comedy —*People*⟩
— bo·da·cious·ly *adverb*

[1]**bode** \ˈbōd\ *transitive verb* **bod·ed; bod·ing** [Middle English, from Old English *bodian;* akin to Old English *bēodan* to proclaim — more at BID] (before 12th century)
1 *archaic* **:** to announce beforehand **:** FORETELL
2 : to indicate by signs **:** PRESAGE ⟨this controversy . . . will *bode* ill for both of us —A. H. Lowe⟩

[2]**bode** *past of* BIDE

bo·de·ga \bō-ˈdā-gə\ *noun* [Spanish, from Latin *apotheca* storehouse — more at APOTHECARY] (1846)
1 : a storehouse for maturing wine
2 a : WINESHOP **b** (1) **:** [1]BAR 5a (2) **:** BARROOM
3 : a store specializing in Hispanic groceries

bode·ment \ˈbōd-mənt\ *noun* (1605)
1 : OMEN, FOREBODING
2 : PREDICTION, PROPHECY

bo·dhi·satt·va *or* **bod·dhi·satt·va** \ˌbō-di-ˈsət-və, -ˈsät-\ *noun* [Sanskrit *bodhisattva* one whose essence is enlightenment, from *bodhi* enlightenment + *sattva* being — more at BID] (1828)
: a being that compassionately refrains from entering nirvana in order to save others and is worshiped as a deity in Mahayana Buddhism

bodh·ran \ˈbò-(ˌ)rän, -rən\ *noun* [Irish *bodhrán*] (1972)
: an Irish goatskin drum

bod·ice \ˈbä-dəs\ *noun* [alteration of *bodies,* plural of [1]*body*] (circa 1567)
1 : the upper part of a woman's dress
2 *archaic* **:** CORSET, STAYS ◆

bodice ripper *noun* (1980)
: a historical or Gothic romance typically featuring scenes in which the heroine is subjected to sexual violence

bod·ied \ˈbä-dēd\ *adjective* (circa 1547)
: having a body of a specified kind — used in combination ⟨full-*bodied*⟩ ⟨glass-*bodied*⟩

bodi·less \ˈbä-di-ləs, ˈbä-dᵊl-əs\ *adjective* (14th century)
: having no body

[1]**bodi·ly** \ˈbä-dᵊl-ē\ *adjective* (14th century)
1 : having a body **:** PHYSICAL
2 : of or relating to the body ⟨*bodily* comfort⟩ ⟨*bodily* organs⟩

[2]**bodily** *adverb* (14th century)
1 : in the flesh
2 : as a whole **:** ALTOGETHER

bod·ing \ˈbō-diŋ\ *noun* (13th century)
: FOREBODING

bod·kin \ˈbäd-kən\ *noun* [Middle English *bodekin*] (14th century)
1 a : DAGGER, STILETTO **b :** a sharp slender instrument for making holes in cloth **c :** an ornamental hairpin shaped like a stiletto
2 : a blunt needle with a large eye for drawing tape or ribbon through a loop or hem

[1]**body** \ˈbä-dē\ *noun, plural* **bod·ies** [Middle English, from Old English *bodig;* akin to Old High German *boteh* corpse] (before 12th century)
1 a : the main part of a plant or animal body especially as distinguished from limbs and head **:** TRUNK **b :** the main, central, or principal part: as (1) **:** the nave of a church (2) **:** the bed or box of a vehicle on or in which the load is placed (3) **:** the enclosed or partly enclosed part of an automobile
2 a : the organized physical substance of an animal or plant either living or dead: as (1) **:** the material part or nature of a human being (2) **:** the dead organism **:** CORPSE **b :** a human being **:** PERSON
3 a : a mass of matter distinct from other masses ⟨a *body* of water⟩ ⟨a celestial *body*⟩ **b :** something that embodies or gives concrete reality to a thing; *also* **:** a sensible object in physical space **c :** AGGREGATE, QUANTITY ⟨a *body* of evidence⟩
4 a : the part of a garment covering the body or trunk **b :** the main part of a literary or journalistic work **:** TEXT 2b **c :** the sound box or pipe of a musical instrument
5 : a group of persons or things: as **a :** a fighting unit **:** FORCE **b :** a group of individuals organized for some purpose ⟨a legislative *body*⟩
6 a : fullness and richness of flavor (as of wine) **b :** VISCOSITY, CONSISTENCY — used especially of oils and grease **c :** compactness or firmness of texture **d :** fullness or resonance of a musical tone

[2]**body** *transitive verb* **bod·ied; body·ing** (15th century)
1 : to give form or shape to **:** EMBODY
2 : REPRESENT, SYMBOLIZE — usually used with *forth*

body bag *noun* (1954)
: a large zippered usually rubber bag in which a human corpse is placed especially for transportation

body blow *noun* (1792)
1 : a blow to the body
2 : a damaging or deeply felt blow ⟨an economic *body blow*⟩

body·build·ing \-ˌbil-diŋ\ *noun* (1904)
: the developing of the body through exercise and diet; *specifically* **:** the developing of the physique for competitive exhibition
— body·build·er *noun*

body cavity *noun* (1875)
: a cavity in an animal body; *specifically* **:** COELOM

body–cen·tered \ˈbä-dē-ˌsen-tərd\ *adjective* (1921)
: relating to or being a crystal space lattice in which each cubic unit cell has an atom at its center and at each vertex — compare FACE-CENTERED

body check *noun* (1892)
: a blocking of an opposing player with the body (as in ice hockey or lacrosse)
— body·check \ˈbä-dē-ˌchek\ *transitive verb*

body clock *noun* (1968)
: the internal mechanisms that schedule periodic bodily functions and activities — usually not used technically

body corporate *noun* (15th century)
: CORPORATION

body count *noun* (1967)
1 : a count of the bodies of killed enemy soldiers
2 : the number of persons involved in a particular activity

body English *noun* (1908)
: bodily motions made in a usually unconscious effort to influence the progress of a propelled object (as a ball)

body·guard \ˈbä-dē-ˌgärd\ *noun* (circa 1735)
: a usually armed attendant or group of attendants whose duty is to protect a person

body language *noun* (1926)
: the gestures, movements, and mannerisms by which a person or animal communicates with others

body louse *noun* (1575)
: a louse feeding primarily on the body; *especially* **:** a sucking louse (*Pediculus humanus humanus*) feeding on the body and living in the clothing of humans — called also *cootie*

body mechanics *noun plural but singular or plural in construction* (circa 1970)
: systematic exercises designed especially to develop coordination, endurance, and poise

body politic *noun* (15th century)
1 : a group of persons politically organized under a single governmental authority
2 *archaic* **:** CORPORATION 2
3 : a people considered as a collective unit

body shirt *noun* (1967)
1 : a close-fitting shirt or blouse
2 : a woman's close-fitting top made with a sewn-in or snapped crotch

body shop *noun* (1954)

◇ WORD HISTORY

bodice The word *bodice* was originally identical with *bodies,* the plural of *body.* In the 16th century, *body* was used for one half of a corsetlike undergarment worn originally by both women and men. The front and back halves together were called "a pair of bodies." In the 14th century, a tendency had begun to pronounce plural *-s* after a vowel sound as \z\, but in the 16th century this rule was not absolute. It was with the \s\ sound that *bodies* in "a pair of bodies" froze, as speakers lost the sense that *bodies* in reference to a corset had any connection with the plural of *body.* Like all other words ending in a vowel *body* came to form its plural exclusively with a final \z\ sound. By 1679 "a pair of bodies" was being spelled "a pair of bodice," and even when still spelled *bodies* it began to be used as a singular noun rather than a plural. By that time it frequently had its modern sense "the upper part of a woman's dress."

: a shop where automotive bodies are made or repaired

body snatcher *noun* (1812)
: one that steals corpses from graves

body stocking *noun* (1965)
: a sheer close-fitting one-piece garment for the torso that often has sleeves and legs

body•suit \ˈbä-dē-ˌsüt\ *noun* (1970)
: a close-fitting one-piece garment for the torso

body•surf \ˈbä-dē-ˌsərf\ *intransitive verb* (1943)
: to ride on a wave without a surfboard by planing on the chest and stomach
— **body•surf•er** *noun*

body wall *noun* (1888)
: the external surface of the animal body consisting of ectoderm and mesoderm and enclosing the body cavity

body•work \ˈbä-dē-ˌwərk\ *noun* (1908)
1 : a vehicle body
2 : the act or process of making or repairing vehicle bodies
3 : therapeutic touching or manipulation of the body by using specialized techniques

boehm•ite \ˈbā-ˌmīt, ˈbə(r)-\ *noun* [German *Böhmit,* from J. *Böhm* (*Boehm*), 20th century German scientist] (circa 1929)
: a mineral consisting of an orthorhombic form of aluminum oxide and hydroxide AlO(OH) found in bauxite

Boer \ˈbōr, ˈbȯr, ˈbu̇r\ *noun* [Dutch, literally, farmer — more at BOOR] (1834)
: a South African of Dutch or Huguenot descent

¹boff \ˈbäf\ *transitive verb* [from *boff* blow, punch, perhaps imitative] (1937)
: to have sexual intercourse with — sometimes considered vulgar

²boff *or* **bof•fo** \ˈbä-(ˌ)fō\ *noun, plural* **boffs** *or* **boffos** [perhaps from *box office*] (1946)
1 : a hearty laugh
2 : a gag or line that produces a hearty laugh
3 : something that is conspicuously successful **:** HIT

bof•fin \ˈbä-fən\ *noun* [origin unknown] (1945)
chiefly British **:** a scientific expert; *especially* **:** one involved in technological research

bof•fo \ˈbä-(ˌ)fō\ *adjective* (1949)
: extremely successful **:** SENSATIONAL

bof•fo•la \bä-ˈfō-lə\ *noun* [irregular from *boff*] (1946)
: BOFF

Bo•fors gun \ˈbō-ˌfȯrz-, ˈbü-\ *noun* [*Bofors,* munition works in Sweden] (1939)
: a double-barreled automatic antiaircraft gun

¹bog \ˈbäg, ˈbȯg\ *noun* [Middle English (Scots), from Scottish Gaelic & Irish *bog-* (as in *bogluachair* bulrushes), from *bog* marshy, literally, soft, from Middle Irish *bocc;* probably akin to Old English *būgan* to bend — more at BOW] (14th century)
: wet spongy ground; *especially* **:** a poorly drained usually acid area rich in accumulated plant material, frequently surrounding a body of open water, and having a characteristic flora (as of sedges, heaths, and sphagnum)
— **bog•gy** \ˈbä-gē, ˈbȯ-\ *adjective*

²bog *verb* **bogged; bog•ging** (1599)
transitive verb
: to cause to sink into or as if into a bog **:** IMPEDE, MIRE — usually used with *down*
intransitive verb
: to become impeded or stuck — usually used with *down*

bog asphodel *noun* (1857)
: either of two bog herbs (*Narthecium ossifragum* of Europe and *N. americanum* of the U.S.) of the lily family

¹bo•gey *also* **bo•gie** *or* **bo•gy** *noun, plural* **bogeys** *also* **bogies** [probably alteration of *bogle*] (1857)
1 \ˈbu̇-gē, ˈbō-, ˈbü-\ **:** SPECTER, PHANTOM
2 \ˈbō-gē *also* ˈbu̇- *or* ˈbü-\ **:** a source of fear, perplexity, or harassment
3 \ˈbō-gē\ **a** *chiefly British* **:** an average golfer's score used as a standard for a particular hole or course **b :** one stroke over par on a hole in golf
4 \ˈbō-gē\ **:** a numerical standard of performance set up as a mark to be aimed at in competition
5 \ˈbō-gē\ **:** an unidentified aircraft; *especially* **:** one not positively identified as friendly and so assumed to be hostile

²bo•gey \ˈbō-gē\ *transitive verb* **bo•geyed; bo•gey•ing** (1948)
: to shoot (a hole in golf) in one over par

bo•gey•man *also* **bo•gy•man** \ˈbu̇-gē-ˌman, ˈbō-, ˈbü-, ˈbu̇-gər-\ *noun* (circa 1890)
1 : a monstrous imaginary figure used in threatening children
2 : a terrifying or dreaded person or thing **:** BUGBEAR

bog•gle \ˈbä-gəl\ *verb* **bog•gled; bog•gling** \-g(ə-)liŋ\ [perhaps from *bogle*] (1598)
intransitive verb
1 : to start with fright or amazement **:** be overwhelmed ⟨the mind *boggles* at the research needed⟩
2 : to hesitate because of doubt, fear, or scruples
transitive verb
1 : MISHANDLE, BUNGLE
2 : to overwhelm with wonder or bewilderment
— **boggle** *noun*

bo•gie *also* **bo•gey** \ˈbō-gē\ *noun, plural* **bogies** *also* **bogeys** [origin unknown] (1835)
1 : a low strongly built cart
2 a *chiefly British* **:** a swiveling railway truck **b :** the driving-wheel assembly consisting of the rear four wheels of a 6-wheel automotive truck
3 : a small supporting or aligning wheel (as on the inside perimeter of a tank tread)

bo•gle \ˈbō-gəl\ *also* **bog•gle** \ˈbä-gəl\ *noun* [origin unknown] (circa 1505)
dialect British **:** GOBLIN, SPECTER; *also* **:** an object of fear or loathing

Bo•go•mil *also* **Bo•go•mile** \ˌbə-gə-ˈmē(ə)l\ *noun* [Middle Greek *Bogomilos,* from *Bogomilos* Bogomil, 10th century Bulgarian priest, founder of the sect] (1841)
: a member of a medieval Bulgarian sect holding that God has two sons, the rebellious Satan and the obedient Jesus

bo•gus \ˈbō-gəs\ *adjective* [*bogus* (a machine for making counterfeit money)] (1825)
: not genuine **:** COUNTERFEIT, SHAM

bo•hea \bō-ˈhē\ *noun, often capitalized* [Chinese (Fujian) *Bú-î,* hills in China where it was grown] (1701)
: a black tea

bo•he•mia \bō-ˈhē-mē-ə\ *noun, often capitalized* [translation of French *bohème*] (1861)
: a community of bohemians **:** the world of bohemians

Bo•he•mi•an \-mē-ən\ *noun* (1603)
1 a : a native or inhabitant of Bohemia **b :** the group of Czech dialects used in Bohemia
2 *often not capitalized* **a :** VAGABOND, WANDERER; *especially* **:** GYPSY **b :** a person (as a writer or an artist) living an unconventional life usually in a colony with others
— **bohemian** *adjective, often capitalized*

Bohemian Brethren *noun plural* (circa 1862)
: a Christian body originating in Bohemia in 1467 and forming a parent body of the Moravian Church

bo•he•mi•an•ism \bō-ˈhē-mē-ə-ˌni-zəm\ *noun, often capitalized* (1861)
: the unconventional way of life of bohemians

Bohr effect \ˈbōr-, ˈbȯr-\ *noun* [Christian *Bohr* (died 1911) Danish physiologist] (1939)
: the decrease in the oxygen affinity of a respiratory pigment (as hemoglobin) in response to decreased blood pH resulting from increased carbon dioxide concentration in the blood

Bohr theory *noun* [Niels *Bohr*] (1922)
: a theory in early quantum physics: an atom consists of a positively charged nucleus about which revolves one or more electrons of quantized energy

bo•hunk \ˈbō-ˌhəŋk\ *noun* [*Bo*hemian + *Hunk* person of central European descent, by shortening & alteration of *Hungarian*] (circa 1903)
: a person of central European descent or birth — usually used disparagingly

¹boil \ˈbȯi(ə)l\ *verb* [Middle English, from Old French *boillir,* from Latin *bullire* to bubble, from *bulla* bubble] (13th century)
intransitive verb
1 a : to come to the boiling point **b :** to generate bubbles of vapor when heated — used of a liquid **c :** to cook in boiling water
2 : to become agitated like boiling water **:** SEETHE
3 : to be moved, excited, or stirred up ⟨made his blood *boil*⟩
4 a : to rush headlong **b :** to burst forth **:** ERUPT ⟨water *boiling* from a spring⟩
5 : to undergo the action of a boiling liquid
transitive verb
1 : to subject to the action of a boiling liquid
2 : to heat to the boiling point
3 : to form or separate (as sugar or salt) by boiling
— **boil•able** \ˈbȯi-lə-bəl\ *adjective*

²boil *noun* [Middle English, alteration of *bile,* from Old English *bȳl;* akin to Old High German *pūlla* bladder] (15th century)
: a localized swelling and inflammation of the skin resulting from infection in a skin gland, having a hard central core, and forming pus

³boil *noun* (15th century)
1 : the act or state of boiling
2 : a swirling upheaval (as of water)

boil down (1845)
transitive verb
1 : to reduce in bulk by boiling
2 : CONDENSE, SUMMARIZE ⟨*boil down* a report⟩
intransitive verb
1 : to undergo reduction in bulk by boiling
2 a : to be equivalent in summary **:** AMOUNT ⟨his speech *boiled down* to a plea for more money⟩ **b :** to reduce ultimately ⟨your choices *boil down* to three⟩

boil•er \ˈbȯi-lər\ *noun* (circa 1540)
1 : one that boils
2 a : a vessel used for boiling **b :** the part of a steam generator in which water is converted into steam and which consists usually of metal shells and tubes **c :** a tank in which water is heated or hot water is stored

boil•er•mak•er \ˈbȯi-lər-ˌmā-kər\ *noun* (1865)
1 : a worker who makes, assembles, or repairs boilers
2 : whiskey with a beer chaser

boil•er•plate \-ˌplāt\ *noun* (1897)
1 : syndicated material supplied especially to weekly newspapers in matrix or plate form
2 a : standardized text **b :** formulaic or hackneyed language ⟨bureaucratic *boilerplate*⟩

boiler room *noun* (1903)
1 : a room in which a boiler is located
2 : a room equipped with telephones used for making high-pressure usually fraudulent sales pitches

boil•er•suit \-ˌsüt\ *noun* (1928)
: COVERALL

¹boil•ing \ˈbȯi-liŋ\ *adjective* (14th century)
1 a : heated to the boiling point **b :** TORRID ⟨a *boiling* sun⟩
2 : intensely agitated ⟨a *boiling* sea⟩ ⟨*boiling* with anger⟩

²boiling *adverb* (1607)
: to an extreme degree **:** VERY ⟨*boiling* mad⟩ ⟨*boiling* hot⟩

boiling point *noun* (1773)
1 : the temperature at which a liquid boils
2 a : the point at which a person becomes uncontrollably angry **b :** the point of crisis **:** HEAD 17b ⟨matters had reached the *boiling point*⟩

boil over *intransitive verb* (15th century)
1 : to overflow while boiling or during boiling
2 : to become so incensed as to lose one's temper

bois d'arc \'bō-,dä(r)k, *dialect also* 'bwä-, 'bȯr-\ *noun, plural* **bois d'arcs** *or* **bois d'arc** [American French, literally, bow wood] (1805)
: OSAGE ORANGE; *also* **:** its wood

bois·ter·ous \'bȯi-st(ə-)rəs\ *adjective* [Middle English *boistous* rough] (14th century)
1 *obsolete* **a :** COARSE **b :** DURABLE, STRONG **c :** MASSIVE
2 a : noisily turbulent **:** ROWDY **b :** marked by or expressive of exuberance and high spirits
3 : STORMY, TUMULTUOUS
synonym see VOCIFEROUS
— bois·ter·ous·ly *adverb*
— bois·ter·ous·ness *noun*

boîte \'bwät\ *noun* [French, literally, box] (1922)
: NIGHTCLUB

bok choy \'bäk-'chȯi\ *noun* [Chinese (Guangdong) *baahk-choi,* literally, white vegetable] (1938)
: a Chinese cabbage (*Brassica chinensis*) forming an open head with long white stalks and green leaves

Bok·mål \'bu̇k-,mȯl, 'bōk-\ *noun* [Norwegian, literally, book language] (1931)
: a literary form of Norwegian developed by the gradual reform of written Danish — compare NYNORSK

bo·la \'bō-lə\ *or* **bo·las** \-ləs\ *noun, plural* **bo·las** \-ləz\ *also* **bo·las·es** [American Spanish *bolas,* from Spanish *bola* ball] (1818)
: a cord with weights attached to the ends for throwing at and entangling an animal

bola

¹bold \'bōld\ *adjective* [Middle English, from Old English *beald;* akin to Old High German *bald* bold] (before 12th century)
1 a : fearless before danger **:** INTREPID **b :** showing or requiring a fearless daring spirit ⟨a *bold* plan⟩
2 : IMPUDENT, PRESUMPTUOUS
3 *obsolete* **:** ASSURED, CONFIDENT
4 : SHEER, STEEP ⟨*bold* cliffs⟩
5 : ADVENTUROUS, FREE ⟨a *bold* thinker⟩
6 : standing out prominently
7 : being or set in boldface
— bold·ly \'bōl(d)-lē\ *adverb*
— bold·ness \'bōl(d)-nəs\ *noun*

²bold *noun* (circa 1871)
: BOLDFACE

bold·face \'bōl(d)-,fās\ *noun* (circa 1889)
: a heavy-faced type; *also* **:** printing in boldface

bold–faced \'bōl(d)-'fāst\ *adjective* (1591)
1 : bold in manner or conduct **:** IMPUDENT
2 *usually* **bold·faced :** being or set in boldface

bole \'bōl\ *noun* [Middle English, from Old Norse *bolr*] (14th century)
: TRUNK 1a

bo·le·ro \bə-'ler-(,)ō\ *noun, plural* **-ros** [Spanish] (1787)
1 : a Spanish dance characterized by sharp turns, stamping of the feet, and sudden pauses in a position with one arm arched over the head; *also* **:** music in ¾ time for a bolero
2 : a loose waist-length jacket open at the front

bo·lete \bō-'lēt\ *noun* [New Latin *Boletus*] (1914)
: any of a family (Boletaceae) of fleshy stalked pore fungi that usually grow on the ground in wooded areas; *especially* **:** BOLETUS

bo·le·tus \bō-'lē-təs\ *noun, plural* **-tus** *or* **-ti** \-'lē-,tī\ [New Latin, genus name, from Latin, mushroom] (1601)
: any of a genus (*Boletus*) of boletes (as a cèpe) some of which are poisonous and others edible

bo·lide \'bō-,līd, -lid\ *noun* [French, from Latin *bolid-, bolis,* from Greek, from *bolē* throw, stroke] (1852)
: a large meteor **:** FIREBALL; *especially* **:** one that explodes

bo·li·var \bə-'lē-,vär, 'bä-lə-vər\ *noun, plural* **-va·res** \,bä-lə-'vär-,ās, ,bō-li-\ *or* **-vars** [American Spanish *bolívar,* from Simón *Bolívar*] (circa 1895)
— see MONEY table

bo·li·vi·a·no \bə-,li-vē-'ä-(,)nō\ *noun, plural* **-nos** [American Spanish] (circa 1872)
— see MONEY table

boll \'bōl\ *noun* [Middle English] (15th century)
: the pod or capsule of a plant (as cotton)

bol·lard \'bä-lərd, *British also* -,lärd\ *noun* [perhaps from *bole*] (circa 1775)
1 : a post of metal or wood on a wharf around which to fasten mooring lines
2 : BITT 1
3 *chiefly British* **:** any of a series of short posts set at intervals to delimit an area (as a traffic island) or to exclude vehicles

bol·lix \'bä-liks\ *transitive verb* [alteration of *ballocks,* plural of *ballock* (testis), from Middle English, from Old English *beallucc* — more at BALL] (1937)
: to throw into disorder; *also* **:** BUNGLE — usually used with *up*
— bollix *noun*

boll weevil *noun* (1895)
: a grayish weevil (*Anthonomus grandis*) that infests the cotton plant and feeds on the squares and bolls

boll·worm \'bōl-,wərm\ *noun* (1847)
: CORN EARWORM; *also* **:** any of several other moths or their immature stages that feed on cotton bolls as larvae

bo·lo \'bō-(,)lō\ *noun, plural* **bolos** [Philippine Spanish] (circa 1900)
: a long heavy single-edged knife of Philippine origin used to cut vegetation and as a weapon

bo·lo·gna \bə-'lō-nē *also* -nyə, -nə\ *noun* [short for *Bologna sausage,* from *Bologna,* Italy] (1596)
: a large smoked sausage of beef, veal, and pork; *also* **:** a sausage made (as of turkey) to resemble bologna

bo·lom·e·ter \bō-'lä-mə-tər\ *noun* [Greek *bolē* stroke, beam of light (from *ballein* to throw) + English *-o-* + *-meter* — more at DEVIL] (1881)
: a very sensitive thermometer whose electrical resistance varies with temperature and which is used in the detection and measurement of feeble thermal radiation and is especially adapted to the study of infrared spectra
— bo·lo·met·ric \,bō-lə-'me-trik\ *adjective*
— bo·lo·met·ri·cal·ly \-tri-k(ə-)lē\ *adverb*

bo·lo·ney \bə-'lō-nē\ *variant of* BALONEY

bo·lo tie \'bō-lō-\ *or* **bo·la tie** \-lə-\ *noun* [probably from *bola*] (1964)
: a cord fastened around the neck with an ornamental clasp and worn as a necktie

Bol·she·vik \'bōl-shə-,vik, 'bȯl-, 'bäl-, -,vēk\ *noun, plural* **Bolsheviks** *also* **Bol·she·vi·ki** \,bōl-shə-'vi-kē, ,bȯl-, ,bäl-, -'vē-kē\ [Russian *bol'shevik,* from *bol'shiĭ* greater] (1917)
1 : a member of the extremist wing of the Russian Social Democratic party that seized power in Russia by the Revolution of November 1917
2 : COMMUNIST 3
— Bolshevik *adjective*

bol·she·vism \'bōl-shə-,vi-zəm, 'bȯl-, 'bäl-\ *noun, often capitalized* (1917)
1 : the doctrine or program of the Bolsheviks advocating violent overthrow of capitalism
2 : Russian communism

Bol·she·vist \-vist\ *noun or adjective* (1917)
: BOLSHEVIK

bol·she·vize \-,vīz\ *transitive verb* **-vized; -viz·ing** (1919)
: to make Bolshevist
— Bol·she·vi·za·tion \,bōl-shə-və-'zā-shən, ,bȯl-, ,bäl-\ *noun*

bol·shie *or* **bol·shy** \'bōl-shē, 'bȯl-, 'bäl-\ *noun or adjective, often capitalized* [by shortening & alteration] (1918)
: BOLSHEVIK

¹bol·ster \'bōl-stər\ *noun* [Middle English, from Old English; akin to Old English *belg* bag — more at BELLY] (before 12th century)
1 : a long pillow or cushion
2 : a structural part designed to eliminate friction or provide support or bearing

²bolster *transitive verb* **bol·stered; bol·ster·ing** \-st(ə-)riŋ\ (1508)
1 : to support with or as if with a bolster **:** REINFORCE
2 : to give a boost to ⟨news that *bolstered* his spirits⟩
— bol·ster·er \-stər-ər\ *noun*

¹bolt \'bōlt\ *noun* [Middle English, from Old English; akin to Old High German *bolz* crossbow bolt, and perhaps to Lithuanian *beldėti* to beat] (before 12th century)
1 a : a shaft or missile designed to be shot from a crossbow or catapult; *especially* **:** a short stout usually blunt-headed arrow

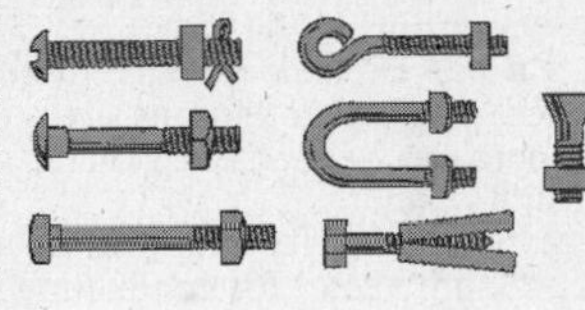
bolt 4

b : a lightning stroke; *also* **:** THUNDERBOLT
2 a : a wood or metal bar or rod used to fasten a door **b :** the part of a lock that is shot or withdrawn by the key
3 : a roll of cloth or wallpaper of specified length
4 : a metal rod or pin for fastening objects together that usually has a head at one end and a screw thread at the other and is secured by a nut
5 a : a block of timber to be sawed or cut **b :** a short round section of a log
6 : a metal cylinder that drives the cartridge into the chamber of a firearm, locks the breech, and usually contains the firing pin and extractor

²bolt (13th century)
intransitive verb
1 : to move suddenly or nervously **:** START
2 : to move rapidly **:** DASH
3 a : to dart off or away **:** FLEE **b :** to break away from control or a set course
4 : to break away from or oppose one's previous affiliation (as with a political party or candidate)
5 : to produce seed prematurely
transitive verb
1 a *archaic* **:** SHOOT, DISCHARGE **b :** FLUSH, START ⟨*bolt* rabbits⟩
2 : to say impulsively **:** BLURT
3 : to secure with a bolt
4 : to attach or fasten with bolts
5 : to swallow hastily or without chewing
6 : to break away from or refuse to support (as a political party)

³bolt *adverb* (14th century)
1 : in an erect or straight-backed position **:** RIGIDLY ⟨sat *bolt* upright⟩
2 *archaic* **:** DIRECTLY, STRAIGHT

⁴bolt *noun* (1550)
: the act or an instance of bolting

⁵bolt *transitive verb* [Middle English *bulten,* from Old French *buleter,* modification of Middle High German *biuteln* to sift, from *biutel* bag, from Old High German *būtil*] (13th century)
1 : to sift (as flour) usually through fine-meshed cloth
2 *archaic* **:** SIFT 2

bolt–ac·tion \'bōlt-'ak-shən\ *adjective* (1896) *of a firearm* **:** loaded by means of a manually operated bolt

bolt·er \'bōl-tər\ *noun* (circa 1699)
: one that bolts: as **a :** a horse given to running away **b :** a voter who bolts from a political party

bolt–hole \'bōlt-,hōl\ *noun* (circa 1851)
chiefly British **:** a place of escape or refuge

bolt·rope \'bōlt-,rōp\ *noun* (circa 1626)
: a strong rope stitched to the edges of a sail to strengthen it

bo·lus \'bō-ləs\ *noun* [Late Latin, from Greek *bōlos* lump] (1562)
: a rounded mass: as **a :** a large pill **b :** a soft mass of chewed food

¹bomb \'bäm\ *noun* [French *bombe,* from Italian *bomba,* probably from Latin *bombus* deep hollow sound, from Greek *bombos,* of imitative origin] (1684)
1 a : an explosive device fused to detonate under specified conditions **b :** ATOMIC BOMB; *also* **:** nuclear weapons in general — usually used with *the*
2 : a vessel for compressed gases: as **a :** a pressure vessel for conducting chemical experiments **b :** a container for an aerosol **:** SPRAY CAN
3 : a rounded mass of lava exploded from a volcano
4 : a lead-lined container for radioactive material
5 : FAILURE, FLOP ⟨the play was a *bomb*⟩
6 *British* **:** a large sum of money
7 *British* **:** a great success **:** HIT
8 : a long pass in football

²bomb (1688)
transitive verb
1 : to attack with or as if with bombs **:** BOMBARD
2 : to defeat decisively
intransitive verb
1 : to fall flat **:** FAIL
2 *slang* **:** to move rapidly ⟨*bombed* down the hill⟩
— bomb·ing *noun*

¹bom·bard \'bäm-,bärd\ *noun* [Middle English *bombarde,* from Middle French, probably from Latin *bombus*] (15th century)
: a late medieval cannon used to hurl large stones

²bom·bard \bäm-'bärd *also* bəm-\ *transitive verb* (1686)
1 : to attack especially with artillery or bombers
2 : to assail vigorously or persistently (as with questions)
3 : to subject to the impact of rapidly moving particles (as electrons)
synonym see ATTACK
— bom·bard·ment \-mənt\ *noun*

bom·bar·dier \,bäm-bə-'dir, -bər-\ *noun* (1560)
1 a *archaic* **:** ARTILLERYMAN **b :** a noncommissioned officer in the British artillery
2 : a bomber-crew member who releases the bombs

bom·bar·don \'bäm-bər-,dōn, bäm-'bär-dᵊn\ *noun* [French, from Italian *bombardone*] (1856)
1 : a bass tuba
2 : the bass member of the shawm family

bom·bast \'bäm-,bast\ *noun* [Middle English *bombast* cotton padding, from Middle French *bombace,* from Medieval Latin *bombac-, bombax* cotton, alteration of Latin *bombyc-, bombyx* silkworm, silk, from Greek *bombyk-, bombyx*] (1589)
: pretentious inflated speech or writing

bom·bas·tic \bäm-'bas-tik\ *adjective* (1704)
: marked by or given to bombast **:** POMPOUS, OVERBLOWN
— bom·bas·ti·cal·ly \-ti-k(ə-)lē\ *adverb*

bom·ba·zine \,bäm-bə-'zēn\ *noun* [Middle French *bombasin,* from Medieval Latin *bombacinum, bombycinum* silken texture, from Latin, neuter of *bombycinus* of silk, from *bombyc-, bombyx*] (1572)
1 : a twilled fabric with silk warp and worsted filling
2 : a silk fabric in twill weave dyed black

bombe \'bäm, 'bōⁿ(m)b\ *noun* [French, literally, bomb] (1892)
: a frozen dessert usually containing ice cream and formed in layers in a mold

bom·bé *or* **bom·be** \(,)bäm-'bā, (,)bōⁿ-\ *adjective* [French, from *bombe*] (1904)
: having outward curving lines — usually used of furniture

bombed \'bämd\ *adjective* (1959)
: affected by alcohol or drugs **:** DRUNK, HIGH

bomb·er \'bä-mər\ *noun* (1915)
1 : one that bombs; *specifically* **:** an airplane designed for bombing
2 : BOMBER JACKET

bomber jacket *noun* (1952)
: a zippered usually leather jacket with front pockets and knitted cuffs and waistband

bom·bi·nate \'bäm-bə-,nāt\ *intransitive verb* **-nat·ed; -nat·ing** [New Latin *bombinatus,* past participle of *bombinare,* alteration of Latin *bombilare,* from *bombus*] (1880)
: BUZZ, DRONE
— bom·bi·na·tion \,bäm-bə-'nā-shən\ *noun*

bomb·proof \'bäm-,prüf\ *adjective* (1702)
: safe from the force of bombs

bomb·shell \'bäm-,shel\ *noun* (1708)
1 : BOMB 1a
2 : one that is stunning, amazing, or devastating ⟨the book was a political *bombshell*⟩

bomb·sight \-,sīt\ *noun* (1917)
: a sighting device for aiming bombs

bo·na fide \'bō-nə-,fīd, 'bä-; ,bō-nə-'fī-dē, -'fī-də\ *adjective* [Latin, literally, in good faith] (1788)
1 : made in good faith without fraud or deceit ⟨a *bona fide* offer to buy a farm⟩
2 : made with earnest intent **:** SINCERE
3 : neither specious nor counterfeit **:** GENUINE
synonym see AUTHENTIC

bo·na fi·des \,bō-nə-'fī-,dēz, ÷'bō-nə-,fīdz\ *noun* [Latin, literally, good faith] (1798)
1 : good faith **:** SINCERITY
2 : the fact of being genuine — often plural in construction
3 : evidence of one's good faith or genuineness — often plural in construction
4 : evidence of one's qualifications or achievements — often plural in construction

bo·nan·za \bə-'nan-zə\ *noun* [Spanish, literally, calm sea, from Medieval Latin *bonacia,* alteration of Latin *malacia,* from Greek *malakia,* literally, softness, from *malakos* soft] (1844)
1 : an exceptionally large and rich mineral deposit (as of an ore, precious metal, or petroleum)
2 a : something that is very valuable, profitable, or rewarding ⟨a box-office *bonanza*⟩ **b :** an extremely large amount ⟨expected a *bonanza* of sympathy⟩

Bo·na·part·ism \'bō-nə-,pär-,ti-zəm\ *noun* (1815)
1 : support of the French emperors Napoleon I, Napoleon III, or their dynasty
2 : a political movement associated chiefly with authoritarian rule usually by a military leader ostensibly supported by a popular mandate
— Bo·na·part·ist \-,pär-tist\ *noun or adjective*

bon·bon \'bän-,bän\ *noun* [French, reduplication of *bon* good, from Latin *bonus* — more at BOUNTY] (1796)
: a candy with chocolate or fondant coating and fondant center that sometimes contains fruits and nuts

¹bond \'bänd\ *noun* [Middle English *band, bond* — more at BAND] (12th century)
1 : something that binds or restrains **:** FETTER
2 : a binding agreement **:** COVENANT
3 a : a band or cord used to tie something **b :** a material or device for binding **c :** an attractive force that holds together the atoms, ions, or groups of atoms in a molecule or crystal **d :** an adhesive, cementing material, or fusible ingredient that combines, unites, or strengthens
4 : a uniting or binding element or force **:** TIE ⟨the *bonds* of friendship⟩
5 a : an obligation made binding by a money forfeit; *also* **:** the amount of the money guarantee **b :** one who acts as bail or surety **c :** an interest-bearing certificate of public or private indebtedness **d :** an insurance agreement pledging surety for financial loss caused to another by the act or default of a third person or by some contingency over which the third person may have no control
6 : the systematic lapping of brick in a wall
7 : the state of goods made, stored, or transported under the care of bonded agencies until the duties or taxes on them are paid
8 : a 100-proof straight whiskey aged at least four years under government supervision before being bottled — called also *bonded whiskey*
9 : BOND PAPER

²bond (1677)
transitive verb
1 : to lap (as brick) for solidity of construction
2 a : to secure payment of duties and taxes on (goods) by giving a bond **b :** to convert into a debt secured by bonds **c :** to provide a bond for or cause to provide such a bond ⟨*bond* an employee⟩
3 a : to cause to adhere firmly **b :** to embed in a matrix **c :** to hold together in a molecule or crystal by chemical bonds
intransitive verb
: to hold together or solidify by or as if by means of a bond or binder **:** COHERE
— bond·able \'bän-də-bəl\ *adjective*
— bond·er *noun*

³bond *adjective* [Middle English *bonde,* from *bonde* peasant, serf, from Old English *bōnda* householder, from Old Norse *bōndi*] (14th century)
archaic **:** bound in slavery

bond·age \'bän-dij\ *noun* (14th century)
1 : the tenure or service of a villein, serf, or slave
2 : a state of being bound usually by compulsion (as of law or mastery): as **a :** CAPTIVITY, SERFDOM **b :** servitude or subjugation to a controlling person or force ⟨young people in *bondage* to drugs⟩

bond·ed \'bän-dəd\ *adjective* (1945)
: composed of two or more layers of the same or different fabrics held together by an adhesive ⟨*bonded* jersey⟩

bond·hold·er \'bänd-,hōl-dər\ *noun* (1823)
: one that holds a government or corporation bond

bond·ing *noun* (1976)
1 : the formation of a close personal relationship (as between a mother and child) especially through frequent or constant association
2 : the attaching of a material (as porcelain) to a tooth surface especially for cosmetic purposes

bond·maid \'bän(d)-,mād\ *noun* (1526)
archaic **:** a female bond servant

bond·man \'bän(d)-mən\ *noun* (13th century)
: SLAVE, SERF

bond paper *noun* (circa 1877)

: a durable paper originally used for documents

bond servant *noun* (15th century)
: one bound to service without wages; *also* **:** SLAVE

¹bonds·man \'bän(d)z-mən\ *noun* (1713)
: one who assumes the responsibility of a bond **:** SURETY

²bondsman *noun* (1735)
: BONDMAN

bond·stone \'bän(d)-ˌstōn\ *noun* (circa 1845)
: a stone long enough to extend through the full thickness of a wall to bind it together

bond·wom·an \'bän-ˌdwu̇-mən\ *noun* (14th century)
: a female slave

¹bone \'bōn\ *noun, often attributive* [Middle English *bon,* from Old English *bān;* akin to Old High German & Old Norse *bein* bone, and perhaps to Old Irish *benaid* he hews] (before 12th century)
1 a : one of the hard parts of the skeleton of a vertebrate **b :** any of various hard animal substances or structures (as baleen or ivory) akin to or resembling bone **c :** the hard largely calcareous connective tissue of which the adult skeleton of most vertebrates is chiefly composed
2 a : ESSENCE, CORE ⟨cut costs to the *bone*⟩ ⟨a liberal to the *bone*⟩ **b :** the most deeply ingrained part **:** HEART — usually used in plural ⟨knew in his *bones* that it was wrong⟩
3 *plural* **a** (1) **:** SKELETON (2) **:** BODY ⟨rested my weary *bones*⟩ (3) **:** CORPSE ⟨inter a person's *bones*⟩ **b :** the basic design or framework (as of a play or novel)
4 : MATTER, SUBJECT ⟨a *bone* of contention⟩
5 a *plural* **:** thin bars of bone, ivory, or wood held in pairs between the fingers and used to produce musical rhythms **b :** a strip of whalebone or steel used to stiffen a corset or dress **c** *plural* **:** DICE
6 : something that is designed to placate **:** SOP
7 : a light beige
— **boned** \'bōnd\ *adjective*
— **bone·less** \'bōn-ləs\ *adjective*
— **bone to pick :** a matter to argue or complain about

²bone *verb* **boned; bon·ing** (15th century)
transitive verb
1 : to remove the bones from ⟨*bone* a fish⟩
2 : to provide (a garment) with stays
intransitive verb
: to study hard **:** GRIND ⟨*bone* through medical school⟩

³bone *adverb* (circa 1825)
: EXTREMELY, VERY ⟨*bone* tired⟩; *also* **:** TOTALLY

bone ash *noun* (1622)
: the white porous residue chiefly of tribasic calcium phosphate from bones calcined in air used especially in making pottery and glass and in cleaning jewelry

bone black *noun* (1815)
: the black residue chiefly of tribasic calcium phosphate and carbon from bones calcined in closed vessels used especially as a pigment or as a decolorizing adsorbent in sugar manufacturing — called also *bone char*

bone china *noun* (circa 1895)
: translucent white china made with bone ash or calcium phosphate and characterized by whiteness

bone–dry \'bōn-'drī\ *adjective* (circa 1825)
1 : very dry
2 : DRY 5

bone·fish \'bōn-ˌfish\ *noun* (1884)
1 : a slender silvery small-scaled fish (*Albula vulpes*) that is a notable sport and food fish of warm seas
2 : LADYFISH 2
— **bone·fish·ing** *noun*

¹bone·head \-ˌhed\ *noun* (1909)
: a stupid person **:** NUMSKULL
— **bone·head·ed** \-'he-dəd\ *adjective*
— **bone·head·ed·ness** *noun*

²bonehead *adjective* (1915)
: being a college course for students lacking fundamental skills ⟨teaches *bonehead* English⟩

bone·meal \'bōn-ˌmē(ə)l\ *noun* (1850)
: crushed or ground bone used especially as fertilizer or feed

bon·er \'bō-nər\ *noun* (circa 1899)
1 : one that bones
2 : HOWLER 2

bone·set \'bōn-ˌset\ *noun* (1764)
: any of several composite herbs (genus *Eupatorium*); *especially* **:** a perennial (*E. perfoliatum*) with opposite perfoliate leaves and white-rayed flower heads used in folk medicine

bone·set·ter \-ˌse-tər\ *noun* (15th century)
: a person who sets broken or dislocated bones usually without being a licensed physician

bone up *intransitive verb* (1887)
1 : to try to master necessary information quickly **:** CRAM ⟨*bone up* for the exam⟩
2 : to renew one's skill or refresh one's memory ⟨*boned up* on the speech just before giving it⟩

bone·yard \-ˌyärd\ *noun* (1866)
1 : CEMETERY
2 : a place where worn-out or damaged objects (as cars) are collected to await disposal

bon·fire \'bän-ˌfīr\ *noun* [Middle English *bonefire* a fire of bones, from *bon* bone + *fire*] (15th century)
: a large fire built in the open air

¹bong \'bäŋ, 'bȯŋ\ *noun* [imitative] (1860)
: the deep resonant sound especially of a bell
— **bong** *verb*

²bong *noun* [Thai *bhaung*³] (1971)
: a simple water pipe consisting of a bottle or vertical tube partially filled with a liquid (as water or liqueur) and a smaller offset tube ending in a bowl

¹bon·go \'bäŋ-(ˌ)gō, 'bȯŋ-\ *noun, plural* **bongos** *also* **bongoes** [American Spanish *bongó*] (1920)
: one of a pair of small connected drums of different sizes and pitches played with the hands
— **bon·go·ist** \-ˌgō-ist\ *noun*

²bongo *noun, plural* **bongo** *or* **bongos** [probably from Kele (Bantu language of Gabon)] (1861)
: an African antelope (*Tragelaphus euryceros* synonym *Boocerus euryceros*) that is chestnut-red with narrow white vertical stripes and is found in forests from Sierra Leone to Kenya

bon·ho·mie \ˌbä-nə-'mē, ˌbō-\ *noun* [French *bonhomie,* from *bonhomme* good-natured man, from *bon* good + *homme* man] (1779)
: good-natured easy friendliness
— **bon·ho·mous** \'bä-nə-məs\ *adjective*

bon·i·face \'bä-nə-fəs, -ˌfās\ *noun* [*Boniface,* innkeeper in *The Beaux' Stratagem* (1707) by George Farquhar] (1803)
: the proprietor of a hotel, nightclub, or restaurant

bo·ni·to \bə-'nē-(ˌ)tō, -'nē-tə\ *noun, plural* **-tos** *or* **-to** [Spanish, from *bonito* pretty, diminutive of *bueno* good, from Latin *bonus*] (circa 1565)
: any of several scombroid fishes (especially genera *Sarda* and *Euthynnus*) intermediate between the smaller mackerels and the larger tunas

bon·kers \'bäŋ-kərz, 'bȯŋ-\ *adjective* [perhaps from *bonk* to hit, bang + *-ers* (as in *crackers*)] (circa 1948)
: CRAZY, MAD ⟨if I don't work, I go *bonkers* —Zoe Caldwell⟩
word history see NUTS

bon mot \bōⁿ-'mō\ *noun, plural* **bons mots** \bōⁿ-'mō(z)\ *or* **bon mots** \-'mō(z)\ [French, literally, good word] (circa 1730)
: a clever remark **:** WITTICISM

bonne \'bȯn\ *noun* [French, from feminine of *bon*] (1771)
: a French nursemaid or maidservant

¹bon·net \'bä-nət\ *noun* [Middle English *bonet,* from Middle French, probably of Germanic origin; akin to Old Saxon *gibund* bundle, Old English *bindan* to bind] (14th century)
1 a (1) *chiefly Scottish* **:** a man's or boy's cap (2) **:** a brimless Scotch cap of seamless woolen fabric — compare TAM-O'-SHANTER 2 **b :** a cloth or straw hat tied under the chin and worn by women and children
2 a *British* **:** an automobile hood **b :** a metal covering or cowl (as for a fireplace, valve chamber, or ventilator)

²bonnet *transitive verb* (1858)
: to provide with or dress in a bonnet

bon·ny *also* **bon·nie** \'bä-nē\ *adjective* **bon·ni·er; -est** [Middle English *bonie,* from Middle French *bon* good, from Latin *bonus*] (15th century)
chiefly British **:** ATTRACTIVE, FAIR; *also* **:** FINE, EXCELLENT
— **bon·ni·ly** \'bä-nə-lē\ *adverb*

bon·ny·clab·ber \'bä-nē-ˌkla-bər\ *noun* [Irish *bainne clabair,* from *bainne* milk + *clabair,* genitive of *clabar* sour thick milk] (1616)
Northern & Midland **:** CLABBER

bon·sai \(ˌ)bōn-'sī, 'bōn-ˌ, 'bän-, *also* 'bän-ˌzī\ *noun, plural* **bonsai** [Japanese, literally, tray planting] (1900)
: a potted plant (as a tree) dwarfed and trained to an artistic shape by special methods of culture; *also* **:** the art of growing such a plant

bonsai

bon·spiel \'bän-ˌspēl\ *noun* [perhaps from Dutch *bond* league + *spel* game] (circa 1772)
: a match or tournament between curling clubs

bon ton \(ˌ)bän-'tän, 'bän-ˌ\ *noun* [French, literally, good tone] (1747)
1 a : fashionable manner or style **b :** the fashionable or proper thing
2 : high society

bo·nus \'bō-nəs\ *noun* [Latin, literally, good — more at BOUNTY] (1773)
: something in addition to what is expected or strictly due: as **a :** money or an equivalent given in addition to an employee's usual compensation **b :** a premium (as of stock) given by a corporation to a purchaser of its securities, to a promoter, or to an employee **c :** a government payment to war veterans **d :** a sum in excess of salary given an athlete for signing with a team

bon vi·vant \ˌbän-vē-'vänt, ˌbōⁿ-vē-'väⁿ\ *noun, plural* **bons vivants** \ˌbän-vē-'vän(t)s, ˌbōⁿ-vē-'väⁿ(z)\ *or* **bon vivants** *same*\ [French, literally, good liver] (circa 1695)
: a person having cultivated, refined, and sociable tastes especially in respect to food and drink

bon voy·age \ˌbōⁿ-ˌvȯi-'äzh, ˌbän-; ˌbōⁿ-ˌvwī-'äzh, -ˌvwä-'yäzh\ *noun* [French, literally, good journey!] (15th century)
: FAREWELL — often used interjectionally

bony *also* **bon·ey** \'bō-nē\ *adjective* **bon·i·er; -est** (14th century)
1 a : consisting of bone **b :** resembling bone
2 a : full of bones ⟨a *bony* piece of fish⟩ **b :** having prominent bones ⟨a rugged *bony* face⟩
3 a : SKINNY, SCRAWNY **b :** BARREN, LEAN

bony fish *noun* (circa 1890)
: any of a major taxon (class Osteichthyes or superclass Teleostomi) comprising fishes (as a sturgeon, salmon, marlin, or ocean sunfish)

with a bony rather than a cartilaginous skeleton — called also *teleost;* compare CARTILAGINOUS FISH, JAWLESS FISH

bonze \'bänz\ *noun* [Middle French, from Portuguese *bonzo,* from Japanese *bonsō*] (1588)
: a Buddhist monk

¹boo \'bü\ *interjection* [Middle English *bo*] (15th century)
— used to express contempt or disapproval or to startle or frighten

²boo *noun, plural* **boos** (1575)
1 : a shout of disapproval or contempt
2 : any sound at all — usually used in negative constructions ⟨never said *boo*⟩

³boo (1884)
intransitive verb
: to deride especially by uttering *boo*
transitive verb
: to express disapproval of by booing ⟨the crowd *booed* the referee⟩

⁴boo *noun* [origin unknown] (circa 1959)
: MARIJUANA

¹boob \'büb\ *noun* [short for ¹*booby*] (1909)
1 : a stupid awkward person **:** SIMPLETON
2 : BOOR, PHILISTINE
— boob·ish \'bü-bish\ *adjective*

²boob *noun* [³*boob*] (1934)
British **:** MISTAKE, BLUNDER

³boob *intransitive verb* [¹*boob*] (1935)
British **:** GOOF 2

⁴boob *noun* [short for ²*booby*] (1945)
: BREAST — sometimes considered vulgar

boob·oi·sie \ˌbüb-ˌwä-'zē\ *noun* [blend of ¹*boob* and *bourgeoisie*] (1922)
: the general public regarded as consisting of boobs

boo–boo \'bü-(ˌ)bü\ *noun, plural* **boo–boos** [probably baby-talk alteration of *boohoo,* imitation of the sound of weeping] (1953)
1 : a usually trivial injury (as a bruise or scratch) — used especially by or of a child
2 : MISTAKE, BLUNDER

boob tube *noun* (1966)
: TELEVISION — used with *the*

¹boo·by \'bü-bē\ *noun, plural* **boobies** [modification of Spanish *bobo,* from Latin *balbus* stammering, probably of imitative origin] (circa 1603)
1 : an awkward foolish person **:** DOPE
2 : any of several tropical seabirds (family Sulidae, especially genus *Sula*)

²boo·by \'bü-bē, 'bu̇-\ *noun, plural* **boobies** [alteration of *bubby*] (1934)
: BREAST — sometimes considered vulgar

booby hatch *noun* (1840)
1 : a raised framework with a sliding cover over a small hatch on a ship
2 : an insane asylum

booby prize *noun* (1889)
1 : an award for the poorest performance in a game or competition
2 : an acknowledgment of notable inferiority

booby trap *noun* (1850)
1 : a trap for the unwary or unsuspecting **:** PITFALL
2 : a concealed explosive device contrived to go off when some harmless-looking object is touched
— boo·by–trap *transitive verb*

boo·dle \'bü-dᵊl\ *noun* [Dutch *boedel* estate, lot, from Middle Dutch; akin to Old Norse *būth* booth] (1833)
1 : a collection or lot of persons **:** CABOODLE
2 a : bribe money **b :** a large amount especially of money

boog·er \'bu̇-gər, 'bü-\ *noun* [alteration of English dialect *buggard, boggart,* from ¹*bug* + *-ard*] (1866)
1 : BOGEYMAN
2 : a piece of dried nasal mucus

boo·gey·man \'bu̇-gē-ˌman, 'bü-\ *also* **boog·er·man** \'bu̇-gər-, 'bü-\ *noun* [by alteration] (circa 1850)
: BOGEYMAN

¹boo·gie \'bu̇-gē, 'bü-\ *noun* (1929)
1 : BOOGIE-WOOGIE
2 : earthy and strongly rhythmic rock music conducive to dancing; *also* **:** a period of or occasion for dancing to this music

²boogie *also* **boo·gy** *or* **boo·gey** \'bu̇-gē, 'bü-\ *intransitive verb* **boo·gied** *also* **boo·geyed; boo·gy·ing** *also* **boo·gey·ing** (1930)
1 : to dance to rock music; *also* **:** REVEL, PARTY
2 a : to move quickly **b :** to get going

boo·gie–woo·gie \ˌbu̇-gē-'wu̇-gē, ˌbü-gē-'wü-gē\ *noun* [origin unknown] (1928)
: a percussive style of playing blues on the piano characterized by a steady rhythmic ground bass of eighth notes in quadruple time and a series of improvised melodic variations

¹book \'bu̇k\ *noun* [Middle English, from Old English *bōc;* akin to Old High German *buoh* book, Gothic *boka* letter] (before 12th century)
1 a : a set of written sheets of skin or paper or tablets of wood or ivory **b :** a set of written, printed, or blank sheets bound together into a volume **c :** a long written or printed literary composition **d :** a major division of a treatise or literary work **e :** a record of a business's financial transactions or financial condition — often used in plural ⟨the *books* show a profit⟩
2 *capitalized* **:** BIBLE 1
3 : something that yields knowledge or understanding ⟨the great *book* of nature⟩ ⟨her face was an open *book*⟩
4 a : the total available knowledge and experience that can be brought to bear on a task or problem ⟨tried every trick in the *book*⟩ ⟨the *book* on him is that he can't hit a curveball⟩ **b :** the standards or authority relevant in a situation ⟨run by the *book*⟩
5 a : all the charges that can be made against an accused person ⟨threw the *book* at him⟩ **b :** a position from which one must answer for certain acts **:** ACCOUNT ⟨bring criminals to *book*⟩
6 a : LIBRETTO **b :** the script of a play **c :** a book of arrangements for a musician or dance orchestra **:** musical repertory
7 : a packet of items bound together like a book ⟨a *book* of stamps⟩ ⟨a *book* of matches⟩
8 a : BOOKMAKER **b :** the bets registered by a bookmaker; *also* **:** the business or activity of giving odds and taking bets
9 : the number of tricks a cardplayer or side must win before any trick can have scoring value
— book·ful \'bu̇k-ˌfu̇l\ *noun*
— in one's book : in one's own opinion
— in one's good books : in favor with one
— one for the book : an act or occurrence worth noting
— on the books : on the records

²book *adjective* (13th century)
1 : derived from books and not from practical experience ⟨*book* learning⟩
2 : shown by books of account ⟨*book* assets⟩

³book (1807)
transitive verb
1 a : to register (as a name) for some future activity or condition (as to engage transportation or reserve lodgings) ⟨*booked* to sail on Monday⟩ **b :** to schedule engagements for ⟨*book* the band for a week⟩ **c :** to set aside time for **d :** to reserve in advance ⟨*book* two seats at the theater⟩ ⟨were all *booked* up⟩
2 a : to enter charges against in a police register **b** *chiefly British* **:** to charge (as a soccer player) with an infraction of the rules
intransitive verb
1 : to make a reservation ⟨*book* through your travel agent⟩
2 *chiefly British* **:** to register in a hotel — usually used with *in*
— book·able \'bu̇-kə-bəl\ *adjective, chiefly British*
— book·er *noun*

book·bind·ing \'bu̇k-ˌbīn-diŋ\ *noun* (1771)
1 : the art or trade of binding books
2 : the binding of a book
— book·bind·er \-ˌbīn-dər\ *noun*
— book·bind·ery \-d(ə-)rē\ *noun*

book·case \-ˌkās\ *noun* (1726)
: a piece of furniture consisting of shelves to hold books

book club *noun* (1905)
: an organization that ships selected books to members usually on a regular schedule and often at discount prices

book·end \-ˌend\ *noun* (1907)
: a support placed at the end of a row of books

book·ie \'bu̇-kē\ *noun* [by shortening & alteration] (1885)
: BOOKMAKER 2

book·ing \'bu̇-kiŋ\ *noun* (1881)
1 : the act of one that books
2 : an engagement or scheduled performance
3 : RESERVATION 1c

booking office *noun* (circa 1837)
chiefly British **:** a ticket office; *especially* **:** one in a railroad station

book·ish \'bu̇-kish\ *adjective* (1567)
1 a : of or relating to books **b :** fond of books and reading
2 a : inclined to rely on book knowledge **b** *of words* **:** literary and formal as opposed to colloquial and informal **c :** given to literary or scholarly pursuits; *also* **:** affectedly learned
— book·ish·ly *adverb*
— book·ish·ness *noun*

book·keep·er \'bu̇k-ˌkē-pər\ *noun* (1555)
: one who records the accounts or transactions of a business
— book·keep·ing \-piŋ\ *noun*

book·let \'bu̇k-lət\ *noun* (1859)
: a little book; *especially* **:** PAMPHLET

book louse *noun* (1867)
: any of various tiny usually wingless insects (order Psocoptera) that feed on organic matter; *especially* **:** an insect (as *Liposcelis divinatorius*) injurious especially to books

book lung *noun* (1879)
: a saccular breathing organ in many arachnids containing thin folds of membrane arranged like the leaves of a book

book·mak·er \'bu̇k-ˌmā-kər\ *noun* (1515)
1 : a printer, binder, or designer of books
2 : one who determines odds and receives and pays off bets
— book·mak·ing \-kiŋ\ *noun*

book·man \-mən\ *noun* (1583)
1 : one who has a love of books and especially of reading
2 : one who is involved in the writing, publishing, or selling of books

book·mark \-ˌmärk\ *or* **book·mark·er** \-ˌmär-kər\ *noun* (1838)
: a marker for finding a place in a book

book–match \-ˌmach\ *transitive verb* (1942)
: to match the grains of (as two sheets of veneer) so that one sheet seems to be the mirrored image of the other

book·mo·bile \'bu̇k-mō-ˌbēl\ *noun* (1926)
: a truck that serves as a traveling library

Book of Common Prayer (1549)
: the service book of the Anglican Communion

book off *intransitive verb* (1971)
Canadian **:** to notify an employer that one is not reporting for work (as because of sickness)

book·plate \'bu̇k-ˌplāt\ *noun* (1791)
: a book owner's identification label that is usually pasted to the inside front cover of a book

book·sell·er \'bu̇k-ˌse-lər\ *noun* (15th century)
: one who sells books; *especially* **:** the proprietor of a bookstore
— book·sell·ing \-ˌse-liŋ\ *noun*

book·shelf \-ˌshelf\ *noun* (1818)
: an open shelf for holding books

book·shop \-ˌshäp\ *noun* (1862)
: BOOKSTORE

book·stall \-ˌstȯl\ *noun* (1800)
1 : a stall where books are sold

2 *chiefly British* **:** NEWSSTAND

book·store \-ˌstōr, -ˌstȯr\ *noun* (circa 1763)
: a place of business where books are the main item offered for sale — called also *bookshop*

book value *noun* (1899)
: the value of something as shown on bookkeeping records as distinguished from market value: **a :** the value of an asset equal to cost less depreciation **b :** the value of a corporation's capital stock equal to its book value less its liabilities

book·worm \ˈbu̇k-ˌwərm\ *noun* (1592)
: a person unusually devoted to reading and study

Bool·ean \ˈbü-lē-ən\ *adjective* [George *Boole* (died 1864) English mathematician] (1851)
: of, relating to, or being a logical combinatorial system (as Boolean algebra) that represents symbolically relationships (as those implied by the logical operators AND, OR, and NOT) between entities (as sets, propositions, or on-off computer circuit elements) ⟨*Boolean* expression⟩ ⟨*Boolean* search strategy for information retrieval⟩

Boolean algebra *noun* (1889)
: a set that is closed under two commutative binary operations and that can be described by any of various systems of postulates all of which can be deduced from the postulates that an identity element exists for each operation, that each operation is distributive over the other, and that for every element in the set there is another element which when combined with the first under one of the operations yields the identity element of the other operation

¹boom \ˈbüm\ *verb* [imitative] (15th century)
intransitive verb
1 : to make a deep hollow sound
2 a : to increase in importance or esteem **b :** to experience a sudden rapid growth and expansion usually with an increase in prices ⟨business was *booming*⟩ **c :** to develop rapidly in population and importance ⟨California *boomed* when gold was discovered there⟩
transitive verb
1 : to cause to resound — often used with *out* ⟨his voice *booms* out the lyrics⟩
2 : to cause a rapid growth or increase of **:** BOOST

²boom *noun* (circa 1500)
1 : a booming sound or cry
2 : a rapid expansion or increase: as **a :** a general movement in support of a candidate for office **b :** rapid settlement and development of a town or district **c :** a rapid widespread expansion of economic activity

³boom *noun* [Dutch, tree, beam; akin to Old High German *boum* tree — more at BEAM] (1627)
1 : a long spar used to extend the foot of a sail
2 a : a chain cable or line of connected floating timbers extended across a river, lake, or harbor (as to obstruct passage or catch floating objects) **b :** a temporary floating barrier used to contain an oil spill
3 a : a long beam projecting from the mast of a derrick to support or guide cargo **b :** a long more or less horizontal supporting arm or brace (as for holding a microphone or for supporting an antenna)
4 : a spar or outrigger connecting the tail surfaces and the main supporting structure of an aircraft

boom box *noun* (1981)
: a large portable radio and often tape player with two attached speakers

boom·er *noun* (1880)
1 : one that booms
2 : one that joins a rush of settlers to a boom area
3 : a transient worker (as a bridge builder)

boo·mer·ang \ˈbü-mə-ˌraŋ\ *noun* [Dharuk (Australian aboriginal language of the Port Jackson area) *bumarinʸ*] (1825)
1 : a bent or angular throwing club typically flat on one side and rounded on the other so that it soars or curves in flight; *especially* **:** one designed to return near the thrower
2 : an act or utterance that backfires on its originator
— **boomerang** *intransitive verb*

boom·let \ˈbüm-lət\ *noun* (1880)
: a small boom; *specifically* **:** a sudden increase in business activity ⟨a stock market *boomlet*⟩

boom·town \ˈbüm-ˌtau̇n\ *noun* (1896)
: a town enjoying a business and population boom

boomy \ˈbü-mē\ *adjective* **boom·i·er; -est** (1888)
1 : of, relating to, or characterized by an economic boom
2 : having an excessive accentuation on the tones of lower pitch in reproduced sound

¹boon \ˈbün\ *noun* [Middle English, from Old Norse *bōn* petition; akin to Old English *bēn* prayer, *bannan* to summon — more at BAN] (12th century)
1 : BENEFIT, FAVOR; *especially* **:** one that is given in answer to a request
2 : a timely benefit **:** BLESSING

²boon *adjective* [Middle English *bon,* from Middle French, good — more at BONNY] (14th century)
1 *archaic* **:** FAVORABLE
2 : CONVIVIAL ⟨a *boon* companion⟩

boon·docks \ˈbün-ˌdäks\ *noun plural* [Tagalog *bundok* mountain] (circa 1909)
1 : rough country filled with dense brush
2 : a rural area **:** STICKS

boon·dog·gle \ˈbün-ˌdä-gəl, -ˌdȯ-\ *noun* [coined by Robert H. Link (died 1957) American scoutmaster] (1929)
1 : a braided cord worn by Boy Scouts as a neckerchief slide, hatband, or ornament
2 : a wasteful or impractical project or activity often involving graft ◆
— **boondoggle** *intransitive verb*
— **boon·dog·gler** \-g(ə-)lər\ *noun*

boon·ies \ˈbü-nēz\ *noun plural* (1956)
slang **:** BOONDOCKS 2

boor \ˈbu̇r\ *noun* [Dutch *boer;* akin to Old English *būan* to dwell — more at BOWER] (1551)
1 : PEASANT
2 : a rude or insensitive person

boor·ish \ˈbu̇r-ish\ *adjective* (1562)
: resembling or befitting a boor (as in crude insensitivity) ☆
— **boor·ish·ly** *adverb*
— **boor·ish·ness** *noun*

¹boost \ˈbüst\ *verb* [origin unknown] (circa 1815)
transitive verb
1 : to push or shove up from below
2 : INCREASE, RAISE ⟨plans to *boost* production⟩ ⟨an extra holiday to *boost* morale⟩
3 : to promote the cause or interests of **:** PLUG ⟨a campaign to *boost* the new fashions⟩
4 : to raise the voltage of or across (an electric circuit)
5 *slang* **:** STEAL, SHOPLIFT
intransitive verb
slang **:** SHOPLIFT
synonym see LIFT

²boost *noun* (1825)
1 : a push upward
2 : an act that brings help or encouragement **:** ASSIST
3 : an increase in amount

boost·er \ˈbü-stər\ *noun* (1890)
: one that boosts: as **a :** an enthusiastic supporter **b :** an auxiliary device for increasing force, power, pressure, or effectiveness **c** *slang* **:** SHOPLIFTER **d :** a radio-frequency amplifier for a radio or television receiving set **e :** the first stage of a multistage rocket providing thrust for the launching and the initial part of the flight **f :** a substance that increases the effectiveness of a medicament; *especially* **:** BOOSTER SHOT

boost·er·ism \-stə-ˌri-zəm\ *noun* (circa 1913)
: the activities and attitudes characteristic of boosters

booster shot *noun* (1944)
: a supplementary dose of an immunizing agent — called also *booster, booster dose*

¹boot \ˈbüt\ *noun* [Middle English, from Old English *bōt* remedy; akin to Old English *betera* better] (before 12th century)
1 *archaic* **:** DELIVERANCE
2 *chiefly dialect* **:** something to equalize a trade
3 *obsolete* **:** AVAIL
— **to boot :** BESIDES

²boot *verb* (15th century)
archaic **:** AVAIL, PROFIT

³boot *noun* [Middle English, from Middle French *bote*] (14th century)
1 : a fitted covering (as of leather or rubber) for the foot and usually reaching above the ankle
2 : an instrument of torture used to crush the leg and foot
3 : something that resembles or is likened to a boot; *especially* **:** an enclosing or protective casing or sheath (as for a rifle or over an electrical or mechanical connection)
4 : a navy or marine corps recruit undergoing basic training

☆ SYNONYMS
Boorish, churlish, loutish, clownish mean uncouth in manners or appearance. BOORISH implies rudeness of manner due to insensitiveness to others' feelings and unwillingness to be agreeable ⟨a drunk's *boorish* behavior⟩. CHURLISH suggests surliness, unresponsiveness, and ungraciousness ⟨*churlish* remarks⟩. LOUTISH implies bodily awkwardness together with stupidity ⟨a *loutish* oaf⟩. CLOWNISH suggests ill-bred awkwardness, ignorance or stupidity, ungainliness, and often a propensity for absurd antics ⟨an adolescent's *clownish* conduct⟩.

◇ WORD HISTORY
boondoggle *Boondoggle* is one of those colorful concoctions that, like *hornswoggle* or *sockdolager,* seem redolent of the 19th century American frontier. It is, in fact, a 20th century coinage, although there is an outdoorsy connection. At the World Jamboree of Scouts in England in 1929, Robert H. Link, usually identified as a scoutmaster from Rochester, N.Y., dubbed a plaited lanyard—a standard item of Boy Scout handicraft—a *boondoggle.* So designated, the lanyard was presented to the Prince of Wales at the Jamboree. Owing to the public nature of the presentation, the term spread through scouting circles. In the fall of 1929 the British magazine *Punch* discussed the word in a report on the presentation of a boondoggle to Lord Baden-Powell, the founder of the Boy Scouts. In 1935, during a municipal hearing in New York City, *boondoggle* was used by a government employee to describe the handicrafts, apparently of dubious value, that he was teaching to people on relief. The appearance of the word in news reports of the hearing did much to popularize it. Soon opponents of President Roosevelt's New Deal were using it to refer to any government project whose major or sole achievement was the expenditure of public funds.

5 *British* **:** an automobile trunk
6 a : a kick with the foot **b :** summary dismissal — used with *the* **c :** momentary pleasure or enjoyment **:** BANG ⟨got a big *boot* out of the joke⟩
7 : a sheath enclosing the inflorescence

[4]**boot** (15th century)
transitive verb
1 : to put boots on
2 a : KICK **b :** to eject or discharge summarily — often used with *out* ⟨was *booted* out of office⟩
3 : to make an error on (a grounder in baseball); *broadly* **:** BOTCH
4 : to ride (a horse) in a race ⟨*booted* home three winners⟩
5 [[2]*bootstrap*] **a :** to load (a program) into a computer from a disk **b :** to start or ready for use especially by booting a program ⟨*boot* a microcomputer⟩ — often used with *up*
intransitive verb
: to become loaded into a computer's memory from a disk ⟨the program *boots* automatically⟩
— boot·able \'bü-tə-bəl\ *adjective*

[5]**boot** *noun* [[1]*boot*] (1593)
archaic **:** BOOTY, PLUNDER

boot·black \'büt-ˌblak\ *noun* (1817)
: one who shines shoes

boot camp *noun* (circa 1942)
: a navy or marine corps camp for basic training

boot·ed \'bü-təd\ *adjective* (1552)
: wearing boots

boo·tee *or* **boo·tie** \bü-'tē, *of infants' footwear* 'bü-tē\ *noun* (1799)
: a usually ankle-length boot, slipper, or sock; *especially* **:** an infant's knitted or crocheted sock

Bo·ö·tes \bō-'ō-tēz\ *noun* [Latin (genitive *Boötis*), from Greek *Boōtēs,* literally, plowman, from *bous* head of cattle — more at cow]
: a northern constellation containing the bright star Arcturus

booth \'büth, *especially British* 'büth\ *noun, plural* **booths** \'büthz, 'büths\ [Middle English *bothe,* of Scandinavian origin; akin to Old Norse *būth* booth; akin to Old English *būan* to dwell — more at BOWER] (13th century)
1 : a temporary shelter for livestock or field workers
2 a : a stall or stand (as at a fair) for the sale or exhibition of goods **b** (1) **:** a small enclosure affording privacy for one person at a time ⟨a telephone *booth*⟩ ⟨polling *booth*⟩ (2) **:** a small enclosure that isolates its occupant especially from patrons or customers ⟨a ticket *booth*⟩ (3) **:** an isolated enclosure used in sound recording or in broadcasting ⟨a radio *booth*⟩ **c :** a restaurant seating arrangement consisting of a table between two high-back benches

boot·jack \'büt-ˌjak\ *noun* (circa 1841)
: a device (as with a V-shaped notch) used for pulling off boots

bootjack

boot·lace \-ˌlās\ *noun* (circa 1887)
British **:** SHOELACE

[1]**boot·leg** \-ˌleg, -ˌlāg\ *noun* (1634)
1 : the upper part of a boot
2 : something bootlegged; *specifically* **:** MOONSHINE
3 : a football play in which the quarterback fakes a handoff, hides the ball against his hip, and rolls out — compare DRAW 8
— bootleg *adjective*

[2]**bootleg** (1900)
transitive verb
1 a : to carry (alcoholic liquor) on one's person illegally **b :** to manufacture, sell, or transport for sale (alcoholic liquor) illegally
2 a : to produce, reproduce, or distribute illicitly or without authorization **b :** SMUGGLE
intransitive verb
1 : to engage in bootlegging
2 : to run a bootleg play in football
— boot·leg·ger *noun*

boot·less \'büt-ləs\ *adjective* (1559)
: USELESS, UNPROFITABLE
— boot·less·ly *adverb*
— boot·less·ness *noun*

boot·lick \-ˌlik\ (1845)
transitive verb
: to try to gain favor with through a servile or obsequious manner
intransitive verb
: to act obsequiously
— boot·lick·er *noun*

boots \'büts\ *noun plural but singular or plural in construction* [from plural of [3]*boot*] (circa 1837)
British **:** a servant who shines shoes especially in a hotel

[1]**boot·strap** \'büt-ˌstrap\ *noun* (1913)
1 *plural* **:** unaided efforts — often used in the phrase *by one's own bootstraps*
2 : a looped strap sewed at the side or the rear top of a boot to help in pulling it on

[2]**bootstrap** *adjective* (1926)
1 : designed to function independently of outside direction **:** capable of using one internal function or process to control another ⟨a *bootstrap* operation to load a computer⟩
2 : carried out with minimum resources or advantages **:** SELF-RELIANT ⟨the city recovered from the flood by the *bootstrap* method⟩

[3]**bootstrap** *transitive verb* (1951)
: to promote or develop by initiative and effort with little or no assistance ⟨*bootstrapped* herself to the top⟩

boo·ty \'bü-tē\ *noun, plural* **booties** [modification of Middle French *butin,* from Middle Low German *būte* exchange] (15th century)
1 : plunder taken (as in war); *especially* **:** plunder taken on land as distinguished from prizes taken at sea
2 : a rich gain or prize
synonym see SPOIL

[1]**booze** \'büz\ *intransitive verb* **boozed; booz·ing** [Middle English *bousen,* from Middle Dutch *būsen*] (14th century)
: to drink intoxicating liquor especially to excess — often used in the phrase *booze it up*

[2]**booze** *noun* (14th century)
: intoxicating drink; *especially* **:** hard liquor
— booz·i·ly \-zə-lē\ *adverb*
— boozy \-zē\ *adjective*

booz·er \'bü-zər\ *noun* (circa 1819)
1 : a person who boozes **:** DRUNK
2 *British* **:** a drinking place **:** PUB

[1]**bop** \'bäp\ *transitive verb* **bopped; bop·ping** [imitative] (1931)
: HIT, SOCK

[2]**bop** *noun* (1947)
: a blow (as with the fist or a club) that strikes a person

[3]**bop** *noun* [short for *bebop*] (1947)
1 : jazz characterized by harmonic complexity, convoluted melodic lines, and constant shifting of accent and often played at very rapid tempos
2 : JIVE 2
— bop·per *noun*

[4]**bop** *intransitive verb* **bopped; bop·ping** (1952)
1 : to go quickly or unceremoniously **:** POP ⟨*bop* into the corner store⟩ — often used with *off*
2 : to dance or shuffle along to or as if to bop music

bo·peep \(ˌ)bō-'pēp\ *noun* [[1]*boo* + [3]*peep*] (1528)
: PEEKABOO

bo·ra \'bōr-ə, 'bȯr-\ *noun* [Italian, from Italian dialect (Trieste), from Latin *boreas*] (1864)
: a violent cold northerly wind of the Adriatic

bo·rac·ic acid \bə-'ras-ik-\ *noun* [Medieval Latin *borac-, borax* borax] (1801)
: BORIC ACID

bor·age \'bȯr-ij, 'bär-\ *noun* [Middle English, from Old French *bourage*] (14th century)
: a coarse hairy blue-flowered European herb (*Borago officinalis* of the family Boraginaceae, the borage family) used medicinally and in salads

bo·rane \'bōr-ˌān, 'bȯr-\ *noun* [International Scientific Vocabulary, from *boron*] (1916)
1 : a compound of boron and hydrogen; *specifically* **:** a compound BH_3 known only in the form of its derivatives
2 : a derivative of borane

bo·rate \-ˌāt\ *noun* (1816)
: a salt or ester of a boric acid

bo·rat·ed \-ˌā-təd\ *adjective* (circa 1901)
: mixed or impregnated with borax or boric acid

[1]**bo·rax** \'bōr-ˌaks, 'bȯr-, -əks\ *noun* [Middle English *boras,* from Middle French, from Medieval Latin *borac-, borax,* from Arabic *būraq,* from Persian *būrah*] (14th century)
: a white crystalline compound that consists of a hydrated sodium borate $Na_2B_4O_7{\cdot}10H_2O$, that occurs as a mineral or is prepared from other minerals, and that is used especially as a flux, cleansing agent, and water softener, as a preservative, and as a fireproofing agent

[2]**borax** *noun* [origin unknown] (1932)
: cheap shoddy merchandise

Bo·ra·zon \'bōr-ə-ˌzän, 'bȯr-\ *trademark*
— used for a boron nitride abrasive

bor·bo·ryg·mus \ˌbȯr-bə-'rig-məs\ *noun, plural* **-mi** \-ˌmī\ [New Latin, from Greek *borborygmos,* from *borboryzein* to rumble] (circa 1796)
: intestinal rumbling caused by moving gas

Bor·deaux \bȯr-'dō\ *noun, plural* **Bordeaux** \-'dōz\ *often not capitalized* (circa 1570)
1 : white or red wine of the Bordeaux region of France
2 : CLARET 2

bor·deaux mixture \bȯr-'dō-, 'bȯr-ˌ\ *noun, often B capitalized* (1892)
: a fungicide made by reaction of copper sulfate, lime, and water

bor·de·laise sauce \'bȯr-dᵊl-'āz-\ *noun, often B capitalized* [French *bordelaise,* feminine of *bordelais* of Bordeaux] (1902)
: a sauce consisting of stock thickened with roux and flavored typically with red wine and shallots

bor·del·lo \bȯr-'de-(ˌ)lō\ *noun, plural* **-los** [Italian, from Old French *bordel,* from *borde* hut, of Germanic origin; akin to Old English *bord* board] (1593)
: a building in which prostitutes are available

[1]**bor·der** \'bȯr-dər\ *noun* [Middle English *bordure,* from Middle French, from Old French, from *border* to border, from *bort* border, of Germanic origin; akin to Old English *bord*] (14th century)
1 : an outer part or edge
2 : an ornamental design at the edge of a fabric or rug
3 : a narrow bed of planted ground along the edge of a garden or walk ⟨a *border* of tulips⟩
4 : BOUNDARY ⟨crossed the *border* into Italy⟩
5 : a plain or decorative margin around printed matter
— bor·dered \-dərd\ *adjective*

[2]**border** *verb* **bor·dered; bor·der·ing** \'bȯr-d(ə-)riŋ\ (14th century)
transitive verb
1 : to put a border on
2 : to touch at the edge or boundary **:** BOUND ⟨*borders* the city on the south⟩
intransitive verb
1 : to lie on the border ⟨the U.S. *borders* on Canada⟩
2 : to approach the nature of a specified thing **:** VERGE ⟨*borders* on the ridiculous⟩
— bor·der·er \-dər-ər\ *noun*

Border collie *noun* (1941)
: any of a breed of medium-sized sheepdogs of British origin noted for their herding abilities

bor·de·reau \ˌbȯr-də-ˈrō\ *noun, plural* **-reaux** \-ˈrō(z)\ [French, from Middle French *bordrel,* probably from *bord* border, from Old French *bort*] (circa 1858)
: a detailed note or memorandum of account; *especially* **:** one containing an enumeration of documents

bor·der·land \ˈbȯr-dər-ˌland\ *noun* (1813)
1 a : territory at or near a border **b :** FRINGE 3a ⟨lives on the *borderland* of society⟩
2 : a vague intermediate state or region ⟨the *borderland* between fantasy and reality⟩

[1]**bor·der·line** \-ˌlīn\ *adjective* (1907)
1 : being in an intermediate position or state **:** not fully classifiable as one thing or its opposite; *especially* **:** not quite up to what is usual, standard, or expected ⟨*borderline* intelligence⟩
2 : situated at or near a border
— borderline *noun*

[2]**borderline** *adverb* (1925)
: ALMOST, NEARLY ⟨*borderline* tacky⟩ ⟨*borderline* suicidal⟩

Border terrier *noun* (1894)
: any of a breed of small terriers of British origin having a harsh dense coat and close undercoat

bor·dure \ˈbȯr-jər\ *noun* [Middle English] (14th century)
: a border on a heraldic shield

[1]**bore** \ˈbōr, ˈbȯr\ *verb* **bored; bor·ing** [Middle English, from Old English *borian;* akin to Old High German *borōn* to bore, Latin *forare* to bore, *ferire* to strike] (before 12th century)
transitive verb
1 : to pierce with a turning or twisting movement of a tool
2 : to make (as a cylindrical hole) by boring or digging away material ⟨*bored* a tunnel⟩
intransitive verb
1 a : to make a hole by or as if by boring **b :** to sink a mine shaft or well
2 : to make one's way steadily especially against resistance ⟨we *bored* through the jostling crowd⟩

[2]**bore** *noun* (14th century)
1 a : a usually cylindrical hole made by or as if by boring **b** *chiefly Australian & New Zealand* **:** a borehole drilled especially to make an artesian well
2 a : the long usually cylindrical hollow part of something (as a tube or gun barrel) **b :** the inner surface of a hollow cylindrical object
3 : the size of a bore: as **a :** the interior diameter of a gun barrel; *specifically, chiefly British* **:** GAUGE 1a(2) **b :** the diameter of an engine cylinder

[3]**bore** *past of* BEAR

[4]**bore** *noun* [(assumed) Middle English *bore* wave, from Old Norse *bāra*] (1601)
: a tidal flood with a high abrupt front

[5]**bore** *noun* [origin unknown] (1766)
: one that causes boredom: as **a :** a tiresome person **b :** something that is devoid of interest

[6]**bore** *transitive verb* **bored; bor·ing** (1768)
: to cause to feel boredom

bo·re·al \ˈbōr-ē-əl, ˈbȯr-\ *adjective* [Middle English *boriall,* from Late Latin *borealis,* from Latin *boreas* north wind, north, from Greek, from *Boreas*] (15th century)
1 : of, relating to, or located in northern regions ⟨*boreal* waters⟩
2 : of, relating to, or comprising the northern biotic area characterized especially by dominance of coniferous forests

Bo·re·as \-ē-əs\ *noun* [Latin, from Greek]
1 : the Greek god of the north wind
2 : the north wind personified

bore·dom \ˈbōr-dəm, ˈbȯr-\ *noun* (1852)
: the state of being weary and restless through lack of interest

bo·reen \bōr-ˈēn, bȯr-\ *noun* [Irish *bóthrín,* diminutive of *bóthar* road] (1836)
Irish **:** a narrow country lane

bore·hole \ˈbōr-ˌhōl, ˈbȯr-\ *noun* (1708)
: a hole bored or drilled in the earth: as **a :** an exploratory well **b** *chiefly British* **:** a small-diameter well drilled especially to obtain water

bor·er \ˈbōr-ər, ˈbȯr-\ *noun* (14th century)
1 : a tool used for boring
2 a : SHIPWORM **b :** an insect that bores in the woody parts of plants

bore·scope \ˈbōr-ˌskōp, ˈbȯr-\ *noun* (1941)
: an optical device (as a prism or optical fiber) used to inspect an inaccessible space (as an engine cylinder)

boric acid *noun* [*boron*] (1869)
: a white crystalline acid H_3BO_3 obtained from its salts and used especially as a weak antiseptic and fire-retardant

bo·ride \ˈbōr-ˌīd, ˈbȯr-\ *noun* (1863)
: a binary compound of boron with a more electropositive element or radical

bor·ing \ˈbōr-iŋ, ˈbȯr-\ *adjective* (1840)
: causing boredom **:** TIRESOME
— bor·ing·ly \-iŋ-lē\ *adverb*
— bor·ing·ness *noun*

born \ˈbȯrn\ *adjective* [Middle English, from Old English *boren,* past participle of *beran* to carry — more at BEAR] (before 12th century)
1 a : brought forth by or as if by birth **b :** NATIVE — usually used in combination ⟨American-*born*⟩ **c :** deriving or resulting from — usually used in combination ⟨poverty-*born* crime⟩
2 a : having from birth specified qualities ⟨a *born* leader⟩ **b :** being in specified circumstances from birth ⟨nobly *born*⟩ ⟨*born* to wealth⟩
3 : destined from or as if from birth ⟨*born* to succeed⟩

born–again *adjective* [from the verse "Except a man be *born again,* he cannot see the Kingdom of God" (John 3:3 Authorized Version)] (1961)
1 : of, relating to, or being a usually Christian person who has made a renewed or confirmed commitment of faith especially after an intense religious experience
2 : having returned to or newly adopted an activity, a conviction, or a persona especially with a proselytizing zeal ⟨a *born-again* conservative⟩

[1]**borne** \ˈbōrn, ˈbȯrn\ *past participle of* BEAR

[2]**borne** *adjective* [*borne*] (circa 1559)
: transported or transmitted by — used in combination ⟨soil*borne*⟩ ⟨air*borne*⟩

bor·ne·ol \ˈbȯr-nē-ˌȯl, -ˌōl\ *noun* [International Scientific Vocabulary, from *Borneo,* island of Indonesia] (1876)
: a crystalline cyclic alcohol $C_{10}H_{17}OH$ that occurs in two enantiomeric forms, is found in essential oils, and is used especially in perfumery

born·ite \ˈbȯr-ˌnīt\ *noun* [German *Bornit,* from Ignaz von *Born* (died 1791) Austrian mineralogist] (circa 1847)
: a brittle metallic-looking mineral that consists of a sulfide of copper and iron and is a valuable copper ore

boro- *combining form*
: boron ⟨*boro*silicate⟩

bo·ro·hy·dride \ˌbōr-ō-ˈhī-ˌdrīd, ˌbȯr-\ *noun* (1940)
: the anion BH_4^- of boron and hydrogen that is used especially as a reducing agent and as a source of hydrogen atoms; *also* **:** any of various compounds (as of metals) containing the borohydride anion

bo·ron \ˈbōr-ˌän, ˈbȯr-\ *noun* [*borax* + *-on* (as in *carbon*)] (1812)
: a trivalent metalloid element found in nature only in combination and used in metallurgy and in composite structural materials — see ELEMENT table
— bo·ron·ic \bȯr-ˈä-nik, bōr-\ *adjective*

boron carbide *noun* (circa 1909)
: a refractory shiny black crystalline compound B_4C that is one of the hardest known materials and is used especially in abrasives and as a structural reinforcing material

bo·ro·sil·i·cate \ˌbōr-ō-ˈsi-lə-ˌkāt, ˌbȯr-, -ˈsi-li-kət\ *noun* [International Scientific Vocabulary] (1817)
1 : a silicate containing boron in the anion and occurring naturally
2 : BOROSILICATE GLASS

borosilicate glass *noun* (1933)
: a silicate glass that is composed of at least five percent oxide of boron and is used especially in heat-resistant glassware

bor·ough \ˈbər-(ˌ)ō, ˈbə-(ˌ)rō\ *noun* [Middle English *burgh,* from Old English *burg* fortified town; akin to Old High German *burg* fortified place, and probably to Old English *beorg* hill — more at BARROW] (before 12th century)
1 a : a medieval fortified group of houses forming a town with special duties and privileges **b :** a town or urban constituency in Great Britain that sends a member to Parliament **c :** an urban area in Great Britain incorporated for purposes of self-government
2 a : a municipal corporation proper in some states (as New Jersey and Minnesota) corresponding to the incorporated town or village of the other states **b :** one of the five constituent political divisions of New York City
3 : a civil division of the state of Alaska corresponding to a county in most other states

borough English *noun* (14th century)
: a custom formerly existing in parts of England by which the lands of an intestate descended to the youngest son

borough hall *noun* (1939)
: the chief administrative building of a borough

bor·row \ˈbär-(ˌ)ō, ˈbȯr-\ *verb* [Middle English *borwen,* from Old English *borgian;* akin to Old English *beorgan* to preserve — more at BURY] (before 12th century)
transitive verb
1 a : to receive with the implied or expressed intention of returning the same or an equivalent ⟨*borrow* a book⟩ ⟨*borrowed* a dollar⟩ **b :** to borrow (money) with the intention of returning the same plus interest
2 a : to appropriate for one's own use ⟨*borrow* a metaphor⟩ **b :** DERIVE, ADOPT
3 : to take (one) from a digit of the minuend in arithmetical subtraction in order to add as 10 to the digit holding the next lower place
4 : to adopt into one language from another
5 *dialect* **:** LEND
intransitive verb
: to borrow something
— bor·row·er \-ə-wər\ *noun*
— borrow trouble : to do something unnecessarily that may result in adverse reaction or repercussions

bor·rowed time *noun* (1898)
: an uncertain and usually uncontrolled postponement of something inevitable — used with *living on*

bor·row·ing \ˈbär-ə-wiŋ, ˈbȯr-\ *noun* (circa 1630)
: something borrowed; *especially* **:** a word or phrase adopted from one language into another

borrow pit *noun* (1893)
: an excavated area where material has been dug for use as fill at another location

Bors \ˈbȯrz\ *noun* [Middle English, from Middle French *Bohort*]
: a knight of the Round Table and nephew of Lancelot

borscht *or* **borsch** \'bȯrsh(t)\ *noun* [Yiddish *borsht* & Ukrainian & Russian *borshch*] (1829)
: a soup made primarily of beets and served hot or cold often with sour cream

borscht belt *also* **borsch belt** *noun* (1938)
: BORSCHT CIRCUIT

borscht circuit *or* **borsch circuit** *noun, often B&C capitalized* [from the popularity of borscht on menus of the resorts] (1938)
: the theaters and nightclubs associated with the Jewish summer resorts in the Catskills

Bor·stal \'bȯr-stᵊl\ *noun* [*Borstal,* English village where the first such institution was set up] (1907)
British **:** REFORMATORY

bort \'bȯrt\ *noun* [probably from Dutch *boort*] (1622)
: imperfectly crystallized diamond or diamond fragments used as an abrasive

bor·zoi \'bȯr-ˌzȯi\ *noun* [Russian *borzoĭ,* from *borzoĭ* swift] (1887)
: any of a breed of large long-haired dogs of greyhound type developed in Russia especially for pursuing wolves — called also *Russian wolfhound*

bos·cage *also* **bosk·age** \'bäs-kij\ *noun* [Middle English *boskage,* from Middle French *boscage,* from Old French, from *bois, bosc* forest, of Germanic origin; akin to Old High German *busk* forest, bush] (14th century)
: a growth of trees or shrubs **:** THICKET

bosh \'bäsh\ *noun* [Turkish *boş* empty] (1834)
: foolish talk or activity **:** NONSENSE — often used interjectionally

bosk *or* **bosque** \'bäsk\ *noun* [probably back-formation from *bosky*] (1814)
: a small wooded area

bosky \'bäs-kē\ *adjective* [English dialect *bosk* bush, from Middle English *bush, bosk*] (1593)
1 : having abundant trees or shrubs
2 : of or relating to a woods

bo·s'n *or* **bo'·s'n** *or* **bo·sun** *or* **bo'·sun** \'bō-sᵊn\ *variant of* BOATSWAIN

[1]**bos·om** \'bu̇-zəm *also* 'bü-\ *noun* [Middle English, from Old English *bōsm;* akin to Old High German *buosam* bosom] (before 12th century)
1 a : the human chest and especially the front part of the chest ⟨hugged the child to his *bosom*⟩ **b :** a woman's breasts regarded especially as a single feature ⟨a woman with an ample *bosom*⟩; *also* **:** BREAST
2 a : the chest conceived of as the seat of the emotions and intimate feelings ⟨a story you will take to your *bosom*⟩ **b :** the security and intimacy of or like that of being hugged to someone's bosom ⟨lived in the *bosom* of her family⟩
3 : the part of a garment that covers the chest or the breasts

[2]**bosom** *adjective* (1590)
: CLOSE, INTIMATE ⟨*bosom* friends⟩

[3]**bosom** *transitive verb* (1605)
1 : EMBRACE
2 : to enclose or carry in the bosom

bos·omed \-zəmd\ *adjective* (1603)
: having a bosom of a specified kind — used in combination ⟨full-*bosomed*⟩

bos·omy \'bu̇-zə-mē *also* 'bü-\ *adjective* (1860)
1 : swelling upward or outward ⟨*bosomy* hills⟩
2 : having prominent breasts

bo·son \'bō-ˌsän, -ˌzän\ *noun* [Satyendranath *Bose* (died 1974) Indian physicist + English [2]*-on*] (1947)
: a particle (as a photon or meson) whose spin is zero or an integral number — compare FERMION

bos·quet \'bäs-kət\ *noun* [French, from Italian *boschetto,* diminutive of *bosco* forest, of Germanic origin; akin to Old High German *busk* forest, bush] (circa 1737)
: THICKET

[1]**boss** \'bäs, 'bȯs\ *noun* [Middle English *boce,* from Middle French, from (assumed) Vulgar Latin *bottia*] (14th century)
1 a : a protuberant part or body ⟨a *boss* of granite⟩ ⟨a *boss* on an animal's horn⟩ **b :** a raised ornamentation **:** STUD **c :** an ornamental projecting block used in architecture
2 : a soft pad used in ceramics and glassmaking
3 : the hub of a propeller

boss 1c

[2]**boss** *transitive verb* (15th century)
1 : to ornament with bosses **:** EMBOSS
2 : to treat (as the surface of porcelain) with a boss

[3]**boss** *noun* [Dutch *baas* master] (1679)
1 : a person who exercises control or authority; *specifically* **:** one who directs or supervises workers
2 : a politician who controls votes in a party organization or dictates appointments or legislative measures ◆
— boss·dom \-dəm\ *noun*
— boss·ism \'bä-ˌsi-zəm, 'bȯ-\ *noun*

[4]**boss** \'bȯs\ *adjective* (1836)
slang **:** EXCELLENT, FIRST-RATE

[5]**boss** \'bȯs\ *transitive verb* (1856)
1 : to act as boss of
2 : to give usually arbitrary orders to — usually used with *around*

[6]**boss** \'bȯs, 'bäs\ *noun* [English dialect, young cow] (1790)
: COW, CALF

bos·sa no·va \ˌbä-sə-'nō-və\ *noun* [Portuguese, literally, new trend] (1962)
1 : popular music of Brazilian origin that is rhythmically related to the samba but with complex harmonies and improvised jazzlike passages
2 : a dance performed to bossa nova music

boss man *noun* (1934)
: [3]BOSS

[1]**bossy** \'bä-sē, 'bȯ-\ *adjective* (1543)
1 : marked by a swelling or roundness
2 : marked by bosses **:** STUDDED

[2]**bossy** \'bȯ-sē, 'bä-\ *noun, plural* **boss·ies** (1843)
: COW, CALF

[3]**bossy** \'bȯ-sē, 'bä-\ *adjective* **boss·i·er; -est** (1882)
: inclined to domineer **:** DICTATORIAL
— boss·i·ness *noun*

Bos·ton \'bȯs-tən\ *noun* [French, from *Boston,* Mass.] (1800)
1 : a variation of whist played with two decks of cards
2 [*Boston,* Mass.] **:** a dance somewhat like a waltz

Boston cream pie *noun* (circa 1933)
: a round cake that is split and filled with a custard or cream filling and usually frosted with chocolate

Boston fern *noun* (circa 1900)
: a luxuriant fern (*Nephrolepis exaltata bostoniensis*) often with drooping much-divided fronds

Boston ivy *noun* (circa 1900)
: a woody Asian vine (*Parthenocissus tricuspidata*) of the grape family typically having 3-lobed leaves

Boston rocker *noun* (1856)
: a wooden rocking chair with a high spindle back, a decorative top panel, and a seat and arms that curve down at the front

Boston rocker

Boston terrier *noun* (1894)
: any of a breed of small smooth-coated terriers originating as a cross of the bulldog and bullterrier and being brindled or black with white markings — called also *Boston bull*

Bos·well \'bäz-ˌwel, -wəl\ *noun* [James *Boswell*] (1858)
: a person who records in detail the life of a usually famous contemporary
— Bos·well·ian \bäz-'we-lē-ən\ *adjective*
— Bos·well·ize \'bäz-wə-ˌlīz, -ˌwe-\ *verb*

bot *also* **bott** \'bät\ *noun* [Middle English; akin to Dutch lever*bot* liver fluke] (15th century)
: the larva of a botfly; *especially* **:** one infesting the horse

bo·ta \'bō-tə\ *noun* [Spanish, from Late Latin *buttis* cask] (1832)
: a leather bottle (as for wine)

bo·tan·i·ca \bə-'ta-ni-kə\ *noun* [American Spanish *botánica,* from feminine of Spanish *botánico* botanical] (1969)
: a shop that deals in herbs and charms used especially by adherents of Santeria

[1]**bo·tan·i·cal** \bə-'ta-ni-kəl\ *adjective* [French *botanique,* from Greek *botanikos* of herbs, from *botanē* pasture, herb, from *boskein* to feed, graze; probably akin to Lithuanian *guotas* flock] (1658)
1 : of or relating to plants or botany
2 : derived from plants
3 : SPECIES ⟨*botanical* tulips⟩
— bo·tan·i·cal·ly \-k(ə-)lē\ *adverb*

[2]**botanical** *noun* (circa 1926)
: a plant part or extract used especially in skin and hair care products

botanical garden *noun* (1785)
: a garden often with greenhouses for the culture, study, and exhibition of special plants — called also *bo·tan·ic garden* \bə-'ta-nik-\

bot·an·ise *British variant of* BOTANIZE

bot·a·nist \'bä-tᵊn-ist, 'bät-nist\ *noun* (circa 1682)
: a specialist in botany or in a branch of botany

bot·a·nize \-tᵊn-ˌīz\ *verb* **-nized; -niz·ing** (1767)
intransitive verb
: to collect plants for botanical investigation; *also* **:** to study plants especially on a field trip
transitive verb
: to explore for botanical purposes

bot·a·ny \'bä-tᵊn-ē, 'bät-nē\ *noun, plural* **-nies** [*botan*ic botanical + [2]-y] (1696)
1 : a branch of biology dealing with plant life
2 a : plant life **b :** the properties and life phenomena exhibited by a plant, plant type, or plant group
3 : a botanical treatise or study; *especially* **:** a particular system of botany

[1]**botch** \'bäch\ *noun* [Middle English *boche,* from Old North French, from (assumed) Vulgar Latin *bottia* boss] (14th century)
: an inflammatory sore

◇ WORD HISTORY

boss The Dutch settlements in North America all came under English rule in the 1660s and 1670s, but Dutch continued to be spoken in the region long afterwards. Noah Webster heard Dutch sermons at Albany, New York, in 1786, and a small number of people of Dutch descent in northern New Jersey retained some knowledge of the language into the 20th century. In American English, the lexical heritage of the Netherlands' colonies includes *dope, stoop* (in the sense "porch"), *Santa Claus,* and a host of distinctive Eastern placenames such as *Schuylkill, Yonkers,* and *Spuyten Duyvil.* Perhaps the most pervasive Dutch loanword is *boss,* from Dutch *baas* "master." The word first appears in the form *boss* in 1806, used by Washington Irving. Part of its success seems to have resulted from an American aversion to *master,* which was common in British use.

[2]botch *transitive verb* [Middle English *bocchen*] (1530)
1 : to foul up hopelessly — often used with *up*
2 : to put together in a makeshift way
— botch·er *noun*

[3]botch *noun* (1605)
1 : something that is botched **:** MESS
2 : PATCHWORK, HODGEPODGE
— botchy \'bä-chē\ *adjective*

bot·fly \'bät-ˌflī\ *noun* (1819)
: any of various stout dipteran flies (family Oestridae) with larvae parasitic in cavities or tissues of various mammals including humans

[1]both \'bōth\ *pronoun, plural in construction* [Middle English *bothe,* from Old Norse *bāthir;* akin to Old High German *beide* both] (12th century)
: the one as well as the other ⟨*both* of us⟩ ⟨we are *both* well⟩ ⟨$1000 fine or 30 days in jail, or *both*⟩

[2]both *conjunction* (12th century)
— used as a function word to indicate and stress the inclusion of each of two or more things specified by coordinated words, phrases, or clauses ⟨prized *both* for its beauty and for its utility⟩ ⟨he prayeth well, who loveth well *both* man and bird and beast —S. T. Coleridge⟩

[3]both *adjective* (13th century)
: being the two **:** affecting or involving the one and the other ⟨*both* feet⟩ ⟨*both* his eyes⟩ ⟨*both* these armies⟩

[1]both·er \'bä-th̲ər\ *verb* **both·ered; both·er·ing** \'bäth̲-riŋ, 'bä-th̲ə-\ [origin unknown] (circa 1745)
transitive verb
1 : to annoy especially by petty provocation **:** IRK
2 : to intrude upon **:** PESTER
3 : to cause to be anxious or concerned — often used interjectionally
intransitive verb
1 : to become concerned
2 : to take pains **:** take the trouble
synonym see ANNOY

[2]bother *noun* (1834)
1 a : a state of petty discomfort, annoyance, or worry **b :** something that causes petty annoyance or worry
2 : FUSS

both·er·ation \ˌbä-th̲ə-'rā-shən\ *noun* (1797)
1 : the act of bothering **:** the state of being bothered
2 : something that bothers — often used interjectionally

both·er·some \'bä-th̲ər-səm\ *adjective* (1834)
: causing bother **:** VEXING

bo·thy \'bä-thē, 'bō-\ *noun* [Scots, probably from obsolete Scots *both* booth] (1771)
chiefly Scottish **:** HUT

bot·o·née *or* **bot·on·née** \ˌbä-tᵊn-'ā\ *adjective* [Middle French *botonné*] (15th century)
of a heraldic cross **:** having a cluster of three balls or knobs at the end of each arm — see CROSS illustration

bo tree \'bō-\ *noun* [Sinhalese *bō,* from Sanskrit *bodhi* enlightenment; from Buddha receiving enlightenment under this tree — more at BID] (1862)
: PIPAL

bot·ry·oi·dal \ˌbä-trē-'ȯi-dᵊl\ *adjective* [Greek *botryoeidēs,* from *botrys* bunch of grapes] (1816)
: having the form of a bunch of grapes ⟨*botryoidal* garnets⟩

bo·try·tis \bō-'trī-təs\ *noun* [New Latin, from Greek *botrys*] (1900)
: any of a genus (*Botrytis*) of imperfect fungi having botryoidal conidia and including several serious plant pathogens

[1]bot·tle \'bä-tᵊl\ *noun, often attributive* [Middle English *botel,* from Middle French *bouteille,* from Medieval Latin *butticula,* diminutive of Late Latin *buttis* cask] (14th century)
1 a : a rigid or semirigid container typically of glass or plastic having a comparatively narrow neck or mouth and usually no handle **b :** a usually bottle-shaped container made of skin for storing a liquid
2 : the quantity held by a bottle
3 a : intoxicating drink **:** the practice of drinking **b :** liquid food (as milk) used in place of mother's milk
— bot·tle·ful \-ˌfu̇l\ *noun*

[2]bottle *transitive verb* **bot·tled; bot·tling** \'bä-tᵊl-iŋ, 'bät-liŋ\ (1600)
1 : to confine as if in a bottle **:** RESTRAIN — usually used with *up* ⟨*bottling* up their anger⟩
2 : to put into a bottle
— bot·tler \'bä-tᵊl-ər, 'bät-lər\ *noun*

bot·tle·brush \'bä-tᵊl-ˌbrəsh\ *noun* [from the shape of the flowers] (circa 1841)
: any of a genus (*Callistemon*) of Australian trees and shrubs of the myrtle family widely cultivated in warm regions especially for their spikes of brightly colored flowers

bottle club *noun* (1943)
: a club serving patrons previously purchased or reserved alcoholic drinks after normal legal closing hours

bottled gas *noun* (1930)
: gas under pressure in portable cylinders

bot·tle–feed \'bä-tᵊl-ˌfēd\ *transitive verb* **-fed; -feed·ing** (circa 1865)
: to feed (as an infant) with a bottle

bottle gourd *noun* (circa 1828)
: a common cultivated gourd (*Lagenaria siceraria*) with a variably shaped fruit that is sometimes used as a container

bottle green *noun* (1804)
: a dark green

[1]bot·tle·neck \'bä-tᵊl-ˌnek\ *adjective* (1896)
: NARROW ⟨*bottleneck* harbors⟩

[2]bottleneck *noun* (1907)
1 a : a narrow route **b :** a point of traffic congestion
2 a : a condition or situation that retards or halts free movement and progress **b :** IMPASSE
3 : a style of guitar playing in which glissando effects are produced by sliding an object (as a knife blade or the neck of a bottle) along the strings

[3]bottleneck *transitive verb* (1933)
: to slow or halt by causing a bottleneck

bot·tle–nosed dolphin \'bä-tᵊl-ˌnōz-\ *noun* (circa 1909)
: a relatively small stout-bodied chiefly gray toothed whale (*Tursiops truncatus*) with a prominent beak and falcate dorsal fin

bottlenose dolphin *noun* (1940)
: BOTTLE-NOSED DOLPHIN

bot·tling \'bä-tᵊl-iŋ, 'bät-liŋ\ *noun* (1954)
: a beverage and especially a wine that is bottled at a particular time

[1]bot·tom \'bä-təm\ *noun* [Middle English *botme,* from Old English *botm;* akin to Old High German *bodam* bottom, Latin *fundus,* Greek *pythmēn*] (before 12th century)
1 a : the underside of something **b :** a surface (as the seat of a chair) designed to support something resting on it **c :** the posterior end of the trunk **:** BUTTOCKS, RUMP
2 : the surface on which a body of water lies
3 a : the part of a ship's hull lying below the water **b :** BOAT, SHIP
4 a : the lowest part or place **b :** the remotest or inmost point **c :** the lowest or last place in point of precedence ⟨started work at the *bottom*⟩ **d :** the pants of pajamas — usually used in plural **e :** the last half of an inning of baseball **f :** the bass or baritone instruments of a band
5 : BOTTOMLAND — usually used in plural
6 : BASIS, SOURCE
7 : capacity (as of a horse) to endure strain
8 : a foundation color applied to textile fibers before dyeing
— bot·tomed \-təmd\ *adjective*
— at bottom : REALLY, BASICALLY

[2]bottom (1544)
transitive verb
1 : to furnish with a bottom
2 : to provide a foundation for
3 : to bring to the bottom
4 : to get to the bottom of **:** FATHOM
intransitive verb
1 : to become based
2 : to reach the bottom
3 : to reach a low point before rebounding — usually used with *out*
— bot·tom·er *noun*

[3]bottom *adjective* (1561)
1 : of, relating to, or situated at the bottom ⟨*bottom* rock⟩
2 : frequenting the bottom ⟨*bottom* fish⟩
3 : having a quantum characteristic that accounts for the existence and lifetime of upsilon particles and that has a value of zero for most known particles ⟨*bottom* quark⟩

bot·tom·land \'bä-təm-ˌland\ *noun* (1728)
: low-lying land along a watercourse

bot·tom·less \-ləs\ *adjective* (14th century)
1 : having no bottom ⟨a *bottomless* chair⟩
2 a : extremely deep **b :** impossible to comprehend **:** UNFATHOMABLE ⟨a *bottomless* mystery⟩ **c :** BOUNDLESS, UNLIMITED
3 a [from the absence of lower as well as upper garments] **:** NUDE ⟨*bottomless* dancers⟩ **b :** featuring nude entertainers
— bot·tom·less·ly *adverb*
— bot·tom·less·ness *noun*

bottom–line \'bä-təm-ˌlīn\ *adjective* (1972)
1 : concerned only with cost or profits
2 : PRAGMATIC, REALISTIC

bottom line *noun* (1967)
1 a : the essential or salient point **:** CRUX **b :** the primary or most important consideration
2 a : the line at the bottom of a financial report that shows the net profit or loss **b :** financial considerations (as cost or profit or loss) **c :** the final result **:** OUTCOME, UPSHOT

bot·tom·most \'bä-tə(m)-ˌmōst\ *adjective* (1861)
1 a : situated at the very bottom **:** LOWEST, DEEPEST **b :** LAST ⟨the *bottommost* part of the day —Alfred Kazin⟩
2 : most basic ⟨the *bottommost* problems facing the world⟩

bottom round *noun* (1923)
: meat (as steak) from the outer part of a round of beef

bot·u·lin \'bä-chə-lən\ *noun* [probably from New Latin *botulinum*] (circa 1900)
: a neurotoxin formed by botulinum and causing botulism

bot·u·li·num \ˌbä-chə-'lī-nəm\ *also* **bot·u·li·nus** \-nəs\ *noun* [New Latin, from Latin *botulus* sausage] (1902)
: a spore-forming bacterium (*Clostridium botulinum*) that secretes botulin
— bot·u·li·nal \-'lī-nᵊl\ *adjective*

bot·u·lism \'bä-chə-ˌli-zəm\ *noun* (1887)
: an acute paralytic disease caused by botulin especially in food

bou·bou \'bü ˌbü\ *noun* [French, from Malinke *bubu*] (1961)
: a long flowing garment worn in parts of Africa

bou·chée \bü-'shā\ *noun* [French, literally, mouthful, from Old French *buchiee,* from (assumed) Vulgar Latin *buccata,* from Latin *bucca* cheek, mouth] (1846)
: a small patty shell usually containing a creamed filling

bou·clé *or* **bou·cle** \bü-ˈklā\ *noun* [French *bouclé* curly, from past participle of *boucler* to curl, from *bocle* buckle, curl] (1895)
1 : an uneven yarn of three plies one of which forms loops at intervals
2 : a fabric of bouclé yarn

bou·doir \ˈbü-ˌdwär, ˈbu̇-, ˌbü-ˈ, ˌbu̇-ˈ\ *noun* [French, from *bouder* to pout] (1781)
: a woman's dressing room, bedroom, or private sitting room

bouf·fant \bü-ˈfänt, ˈbü-ˌ\ *adjective* [French, from Middle French, from present participle of *bouffer* to puff] (1880)
: puffed out ⟨*bouffant* hairdos⟩ ⟨a *bouffant* veil⟩

bou·gain·vil·lea *also* **bou·gain·vil·laea** \ˌbü-gən-ˈvil-yə, ˌbō-, ˌbu̇-, -ˈvē-ə\ *noun* [New Latin, from Louis Antoine de *Bougainville*] (1881)
: any of a genus (*Bougainvillaea*) of the four-o'clock family of ornamental tropical American woody vines with brilliant purple or red floral bracts

bough \ˈbau̇\ *noun* [Middle English, shoulder, bough, from Old English *bōg;* akin to Old High German *buog* shoulder, Greek *pēchys* forearm] (before 12th century)
: a branch of a tree; *especially* **:** a main branch
— boughed \ˈbau̇d\ *adjective*

¹bought \ˈbȯt\ *past and past participle of* BUY

²bought *adjective* [past participle of *buy*] (1599)
: STORE 2 ⟨*bought* clothes⟩

bought·en \ˈbȯ-tᵊn\ *adjective* [*bought* + *-en* (as in *forgotten*)] (1793)
chiefly dialect **:** BOUGHT ⟨the only *boughten* carpet in the region —H. W. Thompson⟩

bou·gie \ˈbü-ˌzhē, -ˌjē\ *noun* [French, from *Bougie,* seaport in Algeria] (1755)
1 : a wax candle
2 a : a tapering cylindrical instrument for introduction into a tubular passage of the body **b :** SUPPOSITORY

bouil·la·baisse \ˌbü-yə-ˈbās, -ˈbāz, ˈbü-yə-ˌ, -ˌbāz\ *noun* [French] (1855)
1 : a highly seasoned fish stew made with at least two kinds of fish
2 : POTPOURRI

bouil·lon \ˈbü(l)-ˌyän, ˈbu̇(l)-; ˈbu̇l-yən; ˈbü-ˌyō^{n}\ *noun* [French, from Old French *boillon,* from *boillir* to boil] (circa 1656)
: a clear seasoned soup made usually from lean beef; *broadly* **:** BROTH

bouillon cube *noun* (circa 1922)
: a cube of evaporated meat extract

boul·der \ˈbōl-dər\ *noun* [short for *boulder stone,* from Middle English *bulder ston,* part translation of a word of Scandinavian origin; akin to Swedish dialect *bullersten* large stone in a stream, from *buller* noise + *sten* stone] (1617)
: a detached and rounded or much-worn mass of rock
— boul·dered \-dərd\ *adjective*
— boul·dery \-d(ə-)rē\ *adjective*

¹bou·le \ˈbü-(ˌ)lē, bü-ˈlā\ *noun* [Greek *boulē,* literally, will, from *boulesthai* to wish] (1846)
: a legislative council of ancient Greece consisting first of an aristocratic advisory body and later of a representative senate

²boule \ˈbül\ *noun* [French, ball — more at BOWL] (1918)
: a pear-shaped synthetically formed mass (as of sapphire) with the atomic structure of a single crystal

bou·le·vard \ˈbu̇-lə-ˌvärd, ˈbü- *also* ˈbə-\ *noun* [French, modification of Middle Dutch *bolwerc* bulwark] (1769)
: a broad often landscaped thoroughfare ◆

bou·le·vard·ier \ˌbu̇-lə-ˌvär-ˈdyā, ˌbü-, -ˈdir\ *noun* [French, from *boulevard*] (1879)
: a frequenter of the Parisian boulevards; *broadly* **:** MAN-ABOUT-TOWN

bou·le·ver·se·ment \bül-ver-sə-mä^{n}\ *noun* [French] (1782)
1 : REVERSAL
2 : a violent disturbance **:** DISORDER

boulle \ˈbül, ˈbyü(ə)l\ *noun* [André Charles *Boulle* (died 1732) French cabinetmaker] (1823)
: inlaid decoration of tortoiseshell, yellow metal, and white metal in cabinetwork

¹bounce \ˈbau̇n(t)s\ *verb* **bounced; bounc·ing** [Middle English *bounsen*] (13th century)
transitive verb
1 *obsolete* **:** BEAT, BUMP
2 : to cause to rebound ⟨*bounce* a ball⟩
3 a : DISMISS, FIRE **b :** to expel precipitately from a place
4 : to issue (a check) drawn on an account with insufficient funds
5 : to present (as an idea) to another person to elicit comments or to gain approval — usually used with *off*
intransitive verb
1 : to rebound after striking a surface (as the ground)
2 : to recover from a blow or a defeat quickly — usually used with *back*
3 : to be returned by a bank because of insufficient funds in a checking account ⟨his checks *bounce*⟩
4 a : to leap suddenly **:** BOUND **b :** to walk with springing steps
5 : to hit a baseball so that it hits the ground before it reaches an infielder

²bounce *noun* (1523)
1 a : a sudden leap or bound **b :** REBOUND
2 : BLUSTER
3 : VERVE, LIVELINESS

bounc·er \ˈbau̇n(t)-sər\ *noun* (1865)
: one that bounces; *especially* **:** a person employed to restrain or eject disorderly persons

bounc·ing \-siŋ\ *adjective* (circa 1563)
1 : LIVELY, ANIMATED
2 : enjoying good health **:** ROBUST
— bounc·ing·ly \-siŋ-lē\ *adverb*

bouncing bet \-ˈbet\ *noun, often 2nd B capitalized* [from *Bet,* nickname for *Elizabeth*] (circa 1818)
: a European perennial herb (*Saponaria officinalis*) of the pink family that is widely naturalized in the U.S. and has pink or white flowers and leaves which yield a detergent when bruised

bouncy \ˈbau̇n(t)-sē\ *adjective* **bounc·i·er; -est** (1921)
1 : BUOYANT, EXUBERANT
2 : RESILIENT
3 : marked by or producing bounces
— bounc·i·ly \-sə-lē\ *adverb*

¹bound \ˈbau̇nd\ *adjective* [Middle English *boun,* from Old Norse *būinn,* past participle of *būa* to dwell, prepare; akin to Old High German *būan* to dwell — more at BOWER] (13th century)
1 *archaic* **:** READY
2 : intending to go **:** GOING ⟨*bound* for home⟩ ⟨college-*bound*⟩

²bound *noun* [Middle English, from Old French *bodne,* from Medieval Latin *bodina*] (13th century)
1 a : a limiting line **:** BOUNDARY — usually used in plural **b :** something that limits or restrains ⟨beyond the *bounds* of decency⟩
2 *usually plural* **a :** BORDERLAND **b :** the land within certain bounds
3 : a number greater than or equal to every number in a set (as the range of a function); *also* **:** a number less than or equal to every number in a set

³bound *past and past participle of* BIND

⁴bound *transitive verb* (14th century)
1 : to set limits or bounds to **:** CONFINE
2 : to form the boundary of **:** ENCLOSE
3 : to name the boundaries of

⁵bound *adjective* [Middle English *bounden,* from past participle of *binden* to bind] (14th century)
1 : placed under legal or moral restraint or obligation **:** OBLIGED ⟨duty-*bound*⟩
2 a : fastened by or as if by a band **:** CONFINED ⟨desk-*bound*⟩ **b :** very likely **:** SURE ⟨*bound* to rain soon⟩
3 : made costive **:** CONSTIPATED
4 *of a book* **:** secured to the covers by cords, tapes, or glue
5 : DETERMINED, RESOLVED
6 : held in chemical or physical combination
7 : always occurring in combination with another linguistic form ⟨*un-* in *unknown* and *-er* in *speaker* are *bound* forms⟩ — compare FREE 11d

⁶bound *noun* [Middle French *bond,* from *bondir* to leap, from (assumed) Vulgar Latin *bombitire* to hum, from Latin *bombus* deep hollow sound — more at BOMB] (circa 1553)
1 : LEAP, JUMP
2 : the action of rebounding **:** BOUNCE

⁷bound *intransitive verb* (1592)
1 : to move by leaping
2 : REBOUND, BOUNCE

bound·ary \ˈbau̇n-d(ə-)rē\ *noun, plural* **-aries** (1626)
: something (as a line, point, or plane) that indicates or fixes a limit or extent

boundary layer *noun* (1921)
: a region of retarded fluid near the surface of a body which moves through a fluid or past which a fluid moves

bound·ed·ness \ˈbau̇n-dəd-nəs\ *noun* (1674)
: the quality or state of being bounded

bound·en \ˈbau̇n-dən\ *adjective* [Middle English] (14th century)
1 *archaic* **:** being under obligation **:** BEHOLDEN
2 : made obligatory **:** BINDING ⟨our *bounden* duty⟩

bound·er \-dər\ *noun* (1505)
1 : one that bounds
2 : a man of objectionable social behavior **:** CAD
— bound·er·ish \-d(ə-)rish\ *adjective*

bound·less \ˈbau̇n(d)-ləs\ *adjective* (1592)
: having no boundaries **:** VAST
— bound·less·ly *adverb*
— bound·less·ness *noun*

bound up *adjective* (1611)
: closely involved or associated — usually used with *with*

boun·te·ous \ˈbau̇n-tē-əs\ *adjective* [Middle English *bountevous,* from Middle French *bontif* kind, from Old French, from *bonté*] (14th century)
1 : giving or disposed to give freely

◇ WORD HISTORY

boulevard The history of the word *boulevard* is actually a snippet of urban history in the West. In French, *boulevard* originally denoted the exterior rampart of a city wall. In the late Middle Ages, the dialects of French adjacent to the Netherlands had acquired the word from Middle Dutch *bolwerc,* and from the north of France it spread to other dialects. (Middle Dutch *bolwerc* is a compound of *bolle* "plank, beam" and *werc* "work, structure," alluding most likely to the use of wooden planks as supporting elements in earthworks; English *bulwark* is a direct borrowing of the same Middle Dutch word.) In 1675, the 14th century rampart of Charles V that had once formed the northern limit of Paris was razed and replaced by a wide tree-lined promenade, which contrasted pleasantly with the city's narrow streets. The same word *boulevard* that denoted the rampart was now applied to this promenade, and hence to similar avenues, which were created in 19th century Paris by demolishing existing buildings rather than old walls. As an English loanword, *boulevard* has been particularly successful in the United States; streets so named are now sometimes busy commercial thoroughfares or even highways, with no hint of the word's earlier applications.

2 : liberally bestowed
— **boun·te·ous·ly** *adverb*
— **boun·te·ous·ness** *noun*

boun·tied \ˈbau̇n-tēd\ *adjective* (1788)
1 : having the benefit of a bounty
2 : rewarded or rewardable by a bounty

boun·ti·ful \ˈbau̇n-ti-fəl\ *adjective* (1508)
1 : liberal in bestowing gifts or favors
2 : given or provided abundantly ⟨a *bountiful* harvest⟩
synonym see LIBERAL
— **boun·ti·ful·ly** \-f(ə-)lē\ *adverb*
— **boun·ti·ful·ness** \-fəl-nəs\ *noun*

boun·ty \ˈbau̇n-tē\ *noun, plural* **bounties** [Middle English *bounte* goodness, from Old French *bonté,* from Latin *bonitat-, bonitas,* from *bonus* good, from Old Latin *duenos;* akin to Sanskrit *duva* reverence, favor] (13th century)
1 : something that is given generously
2 : liberality in giving **:** GENEROSITY
3 : yield especially of a crop
4 : a reward, premium, or subsidy especially when offered or given by a government: as **a :** an extra allowance to induce entry into the armed services **b :** a grant to encourage an industry **c :** a payment to encourage the destruction of noxious animals **d :** a payment for the capture of or assistance in the capture of an outlaw

bounty hunter *noun* (1930)
1 : one that tracks down and captures outlaws for whom a reward is offered
2 : one that hunts predatory animals for the reward offered

bou·quet \bō-ˈkā, bü-\ *noun* [French, from Middle French, thicket, from Old North French *bosquet,* from Old French *bosc* forest — more at BOSCAGE] (circa 1718)
1 a : flowers picked and fastened together in a bunch **:** NOSEGAY **b :** MEDLEY ⟨*bouquet* of songs⟩
2 : COMPLIMENT
3 a : a distinctive and characteristic fragrance (as of wine) **b :** a subtle aroma or quality (as of an artistic performance)

bou·quet gar·ni \-gär-ˈnē\ *noun, plural* **bou·quets gar·nis** \-ˈkā(z)-gär-ˈnē\ [French, literally, garnished bouquet] (circa 1852)
: an herb mixture that is either tied together or enclosed in a porous container and is cooked with a dish but removed before serving

bour·bon \ˈbu̇r-bən, ˈbōr-, ˈbȯr-; *usually* ˈbər- *in sense 3*\ *noun* [*Bourbon,* seigniory in France] (1600)
1 *capitalized* **:** a member of a French family founded in 1272 to which belong the rulers of France from 1589 to 1793 and from 1814 to 1830, of Spain from 1700 to 1808, from 1814 to 1868, from 1875 to 1931, and from 1975, of Naples from 1735 to 1805, and of the Two Sicilies from 1815 to 1860
2 [*Bourbon* (now Réunion), French island in the Indian Ocean] **:** a rose (*Rosa borboniana*) of compact upright growth with shining leaves, prickly branches, and clustered flowers
3 [*Bourbon* county, Kentucky] **:** a whiskey distilled from a mash made up of not less than 51 percent corn plus malt and rye — compare CORN WHISKEY
4 *often capitalized* **:** a person who clings obstinately to old social and political ideas; *specifically* **:** an extremely conservative member of the U.S. Democratic party usually from the South
— **bour·bon·ism** \-bə-ˌni-zəm\ *noun, often capitalized*

bourg \ˈbu̇r(g)\ *noun* [Middle English, from Middle French, from Old French *borc,* from Latin *burgus* fortified place, of Germanic origin; akin to Old High German *burg* fortified place — more at BOROUGH] (12th century)
: TOWN, VILLAGE: as **a :** one neighboring a castle **b :** a market town

¹bour·geois \ˈbu̇rzh-ˌwä *also* ˈbu̇zh- *or* ˈbüzh- *or* bu̇rzh-ˈ\ *adjective* [Middle French, from Old French *borjois,* from *borc*] (circa 1565)
1 : of, relating to, or characteristic of the townsman or of the social middle class
2 : marked by a concern for material interests and respectability and a tendency toward mediocrity
3 : dominated by commercial and industrial interests **:** CAPITALISTIC
— **bour·geois·ifi·ca·tion** \ˌbu̇(r)zh-ˌwä-zə-fə-ˈkā-shən\ *noun*
— **bour·geois·ify** \bu̇(r)zh-ˈwä-zə-ˌfī\ *verb*

²bourgeois *noun, plural* **bourgeois** \-ˌwä(z), -ˈwä(z)\ (circa 1674)
1 a : BURGHER **b :** a middle-class person
2 : a person with social behavior and political views held to be influenced by private-property interest **:** CAPITALIST
3 *plural* **:** BOURGEOISIE

bour·geoise \ˈbu̇rzh-ˌwäz *also* ˈbu̇zh- *or* ˈbüzh- *or* bu̇rzh-ˈ\ *noun* [French, feminine of *bourgeois*] (1794)
: a woman of the middle class

bour·geoi·sie \ˌbu̇(r)zh-ˌwä-ˈzē\ *noun* [French, from *bourgeois*] (1707)
1 : MIDDLE CLASS
2 : a social order dominated by bourgeois

bour·geon \ˈbər-jən\ *variant of* BURGEON

bour·gui·gnonne \ˌbu̇r-gēn-ˈyȯn\ *or* **bour·gui·gnon** \-ˈyōⁿ\ *adjective, often capitalized* [French, from *Bourgogne* Burgundy, region in France] (circa 1919)
: prepared or served in the manner of Burgundy (as with a sauce made with red wine) ⟨beef *bourguignonne*⟩

bourn *or* **bourne** \ˈbōrn, ˈbȯrn, ˈbu̇rn\ *noun* [Middle English *burn, bourne* — more at BURN] (12th century)
: STREAM, BROOK

bourne *also* **bourn** *noun* [Middle French *bourne,* from Old French *bodne* — more at BOUND] (1523)
1 : BOUNDARY, LIMIT
2 : GOAL, DESTINATION

bour·rée \bu̇ ˈrā, ˈbü-ˌ\ *noun* [French] (1706)
1 : a 17th century French dance usually in quick duple time; *also* **:** a musical composition with the rhythm of this dance
2 : PAS DE BOURRÉE

bour·ride \bu̇-ˈrēd, bə-\ *noun* [French, from Provençal *bourrido,* alteration of *boulido* something boiled, from *bouli* to boil, from Latin *bullire* — more at BOIL] (circa 1919)
: a fish stew similar to bouillabaisse that is usually thickened with egg yolks and strongly flavored with garlic

bourse \ˈbu̇rs\ *noun* [Middle French, literally, purse, from Medieval Latin *bursa* — more at PURSE] (1597)
1 : EXCHANGE 5a; *specifically* **:** a European stock exchange
2 : a sale of numismatic or philatelic items on tables (as at a convention)

bouse *variant of* BOWSE

bou·stro·phe·don \ˌbü-strə-ˈfē-ˌdän, -dᵊn\ *noun* [Greek *boustrophēdon,* adverb, literally, turning like oxen in plowing, from *bous* ox, cow + *strephein* to turn — more at COW] (1699)
: the writing of alternate lines in opposite directions (as from left to right and from right to left)
— **boustrophedon** *adjective or adverb*
— **bou·stro·phe·don·ic** \-fē-ˈdä-nik\ *adjective*

bout \ˈbau̇t\ *noun* [English dialect, a trip going and returning in plowing, from Middle English *bought* bend] (1575)
: a spell of activity: as **a :** an athletic match (as of boxing) **b :** OUTBREAK, ATTACK ⟨a *bout* of lumbago⟩ **c :** SESSION

bou·tique \bü-ˈtēk\ *noun, often attributive* [French, shop, probably from Old Provençal *botica,* ultimately from Greek *apothēkē* storehouse — more at APOTHECARY] (1767)
1 a : a small fashionable specialty shop or business **b :** a small shop within a large department store
2 : a small company that offers highly specialized services ⟨an independent investment *boutique*⟩

bou·ton·niere \ˌbü-tᵊn-ˈir, ˌbü-tən-ˈyer\ *noun* [French *boutonnière* buttonhole, from Middle French, from *bouton* button] (circa 1867)
: a flower or bouquet worn in a buttonhole

Bou·vi·er des Flan·dres \ˌbü-vē-ˌā-də-ˈflandərz, -ˈfläⁿdrᵊ\ *noun* [French, literally, cowherd of Flanders] (1929)
: any of a breed of large powerfully built rough-coated dogs of Belgian origin used especially for herding and in guard work — called also *Bouvier*

Bouvier des Flandres

bou·zou·ki \bu̇-ˈzü-kē\ *noun, plural* **-kis** *also* **-kia** \-kē-ə\ [New Greek *mpouzouki*] (1952)
: a long-necked stringed instrument of Greek origin that resembles a mandolin

bo·vid \ˈbō-vid\ *noun* [New Latin *Bovidae,* from *Bov-, Bos,* type genus, from Latin *bov-, bos*] (1939)
: any of a family (Bovidae) of ruminants that have hollow unbranched permanently attached horns present in both sexes and that include antelopes, oxen, sheep, and goats

¹bo·vine \ˈbō-ˌvīn, -ˌvēn\ *adjective* [Late Latin *bovinus,* from Latin *bov-, bos* ox, cow — more at COW] (1817)
1 : of, relating to, or resembling bovines and especially the ox or cow
2 : having qualities (as placidity or dullness) characteristic of oxen or cows
— **bo·vine·ly** *adverb*
— **bo·vin·i·ty** \bō-ˈvi-nə-tē\ *noun*

²bovine *noun* (1865)
: any of a subfamily (Bovinae) of bovids including oxen, bison, buffalo, and their close relatives

¹bow \ˈbau̇\ *verb* [Middle English, from Old English *būgan;* akin to Old High German *biogan* to bend, Sanskrit *bhujati* he bends] (before 12th century)
intransitive verb
1 : to cease from competition or resistance **:** SUBMIT, YIELD ⟨refusing to *bow* to the inevitable —John O'Hara⟩; *also* **:** to suffer defeat ⟨*bowed* to the champion⟩
2 : to bend the head, body, or knee in reverence, submission, or shame
3 : to incline the head or body in salutation or assent or to acknowledge applause
transitive verb
1 : to cause to incline
2 : to incline (as the head) especially in respect or submission
3 : to crush with a heavy burden
4 a : to express by bowing **b :** to usher in or out with a bow

²bow *noun* (circa 1656)
: a bending of the head or body in respect, submission, assent, or salutation; *also* **:** a show of respect or submission

³bow \ˈbō\ *noun* [Middle English *bowe,* from Old English *boga;* akin to Old English *būgan*] (before 12th century)
1 a : something bent into a simple curve **:** BEND, ARCH **b :** RAINBOW

2 : a weapon that is made of a strip of flexible material (as wood) with a cord connecting the two ends and holding the strip bent and that is used to propel an arrow
3 : ARCHER
4 a : a metal ring or loop forming a handle (as of a key) **b :** a knot formed by doubling a ribbon or string into two or more loops **c :** BOW TIE **d :** a frame for the lenses of eyeglasses; *also* **:** the sidepiece of the frame passing over the ear
5 a : a wooden rod with horsehairs stretched from end to end used in playing an instrument of the viol or violin family **b :** a stroke of such a bow

⁴bow \'bō\ (before 12th century)
intransitive verb
1 : to bend into a curve
2 : to play a stringed musical instrument with a bow
transitive verb
1 : to cause to bend into a curve
2 : to play (a stringed instrument) with a bow

⁵bow \'bau̇\ *noun* [Middle English *bowe, bowgh,* probably from Middle Dutch *boech* bow, shoulder; akin to Old English *bōg* bough] (15th century)
1 : the forward part of a ship
2 : ²BOWMAN

Bow bells \'bō-\ *noun plural* (1600)
: the bells of the Church of Saint Mary-le-Bow in London

bowd·ler·ise *British variant of* BOWDLERIZE

bowd·ler·ize \'bōd-lə-ˌrīz, 'bau̇d-\ *transitive verb* **-ized; -iz·ing** [Thomas *Bowdler* (died 1825) English editor] (1836)
1 : to expurgate (as a book) by omitting or modifying parts considered vulgar
2 : to modify by abridging, simplifying, or distorting in style or content ◆
— bowd·ler·iza·tion \ˌbōd-lə-rə-'zā-shən, ˌbau̇d-\ *noun*
— bowd·ler·iz·er *noun*

¹bowed \'bau̇d\ *adjective* [past participle of ¹*bow*] (14th century)
1 : bent downward and forward ⟨listened with *bowed* heads⟩
2 : having the back and head inclined

²bowed \'bōd\ *adjective* [partly from ³*bow* + *-ed;* partly from past participle of ⁴*bow*] (15th century)
: furnished with or shaped like a bow

bow·el \'bau̇(-ə)l\ *noun* [Middle English, from Middle French *boel,* from Medieval Latin *botellus,* from Latin, diminutive of *botulus* sausage] (14th century)
1 : INTESTINE, GUT; *also* **:** one of the divisions of the intestines — usually used in plural except in medical use ⟨the large *bowel*⟩ ⟨move your *bowels*⟩
2 *archaic* **:** the seat of pity, tenderness, or courage — usually used in plural
3 *plural* **:** the interior parts; *especially* **:** the deep or remote parts ⟨*bowels* of the earth⟩
— bow·el·less \'bau̇(-ə)l-ləs\ *adjective*

¹bow·er \'bau̇(-ə)r\ *noun* [Middle English *bour* dwelling, from Old English *būr;* akin to Old English & Old High German *būan* to dwell, Old English *bēon* to be — more at BE] (before 12th century)
1 : an attractive dwelling or retreat
2 : a lady's private apartment in a medieval hall or castle
3 : a shelter (as in a garden) made with tree boughs or vines twined together **:** ARBOR
— bow·ery \-ē\ *adjective*

²bower *transitive verb* (1592)
: EMBOWER, ENCLOSE

³bower *noun* (1652)
: an anchor carried at the bow of a ship

bow·er·bird \'bau̇(-ə)r-ˌbərd\ *noun* (1845)
: any of a family (Ptilonorhynchidae) of passerine birds of the Australian region in which the male builds a chamber or passage arched over with twigs and grasses, often adorned with bright-colored objects, and used especially to attract the female

bow·ery \'bau̇(-ə)r-ē\ *noun, plural* **-er·ies** [Dutch *bouwerij,* from *bouwer* farmer, from *bouwen* to till; akin to Old High German *būan* to dwell] (1650)
1 : a colonial Dutch plantation or farm
2 [*Bowery,* street in New York City] **:** a city district known for cheap bars and derelicts

bow·fin \'bō-ˌfin\ *noun* (1845)
: a predaceous dull-green iridescent North American freshwater fish (*Amia calva*) that is the only surviving member of an order (Amiiformes) dating back to the Jurassic

bow·front \-ˌfrənt\ *adjective* (1925)
1 : having an outward curving front ⟨*bowfront* furniture⟩
2 : having a bow window in front ⟨*bowfront* houses⟩

bow·head whale \-ˌhed-\ *noun* (1887)
: a baleen whale (*Balaena mysticetus*) of arctic and subarctic seas — called also *bowhead*

bow·ie knife \'bü-ē-, 'bō-\ *noun* [James *Bowie*] (1836)
: a stout single-edged hunting knife with part of the back edge curved concavely to a point and sharpened

bow·ing \'bō-iŋ\ *noun* (1838)
: the technique or manner of managing the bow in playing a stringed musical instrument

bow·knot \'bō-ˌnät\ *noun* (1547)
: a knot with decorative loops

¹bowl \'bōl\ *noun* [Middle English *bolle,* from Old English *bolla;* akin to Old High German *bolla* blister] (before 12th century)
1 : a concave usually hemispherical vessel often used for holding liquids; *especially* **:** a drinking vessel (as for wine)
2 : the contents of a bowl
3 : a bowl-shaped or concave part: as **a :** the hollow of a spoon or tobacco pipe **b :** the receptacle of a toilet
4 a : a natural formation or geographical region shaped like a bowl **b :** a bowl-shaped structure; *especially* **:** an athletic stadium
5 : a postseason football game between specially invited teams
— bowled \'bōld\ *adjective*
— bowl·ful \-ˌfu̇l\ *noun*

²bowl *noun* [Middle English *boule,* from Middle French, from Latin *bulla* bubble] (15th century)
1 a : a ball (as of lignum vitae) weighted or shaped to give it a bias when rolled in lawn bowling **b** *plural but singular in construction* **:** LAWN BOWLING
2 : a delivery of the ball in bowling
3 : a cylindrical roller or drum (as for a mechanical device)

³bowl (15th century)
intransitive verb
1 a : to participate in a game of bowling **b :** to roll a ball in bowling
2 : to travel smoothly and rapidly (as in a wheeled vehicle)
transitive verb
1 a : to roll (a ball) in bowling **b** (1) **:** to complete by bowling ⟨*bowl* a string⟩ (2) **:** to score by bowling ⟨*bowls* 150⟩
2 : to strike with a swiftly moving object
3 : to overwhelm with surprise

bowl·der *variant of* BOULDER

bow·leg \'bō-ˌleg, -ˌlāg, 'bō-'\ *noun* (circa 1864)
: a leg bowed outward at or below the knee
— bow·legged \'bō-ˌle-gəd, -ˌlā-; -ˌlegd, -ˌlāgd\ *adjective*

¹bowl·er \'bō-lər\ *noun* (circa 1500)
: a person who bowls; *specifically* **:** the player that delivers the ball to the batsman in cricket

²bow·ler \'bō-lər\ *noun* [*Bowler,* 19th century family of English hatters] (1861)
: DERBY 3

bow·line \'bō-lən, -ˌlīn\ *noun* [Middle English *bouline,* perhaps from *bowe* bow + *line*] (13th century)
1 : a rope used to keep the weather edge of a square sail taut forward
2 : a knot used to form a loop that neither slips nor jams — see KNOT illustration

bowl·ing \'bō-liŋ\ *noun* (1535)
: any of several games in which balls are rolled on a green or down an alley at an object or group of objects

bowl over *transitive verb* (1867)
1 : to take unawares
2 : ¹IMPRESS 2

¹bow·man \'bō-mən\ *noun* (13th century)
: ARCHER 1

²bow·man \'bau̇-mən\ *noun* (1829)
: a boatman, oarsman, or paddler stationed in the front of a boat

Bow·man's capsule \'bō-mənz-\ *noun* [Sir William *Bowman* (died 1892) English surgeon] (circa 1860)
: a thin membranous double-walled capsule surrounding the glomerulus of a vertebrate nephron

bow out *intransitive verb* (1942)
: RETIRE, WITHDRAW

bow saw \'bō-\ *noun* (1677)
: a saw having a narrow blade held under tension by a light bow-shaped frame

bowse \'bau̇z\ *verb* **bowsed; bows·ing** [origin unknown] (1593)
transitive verb
: to haul by means of a tackle
intransitive verb
: to bowse something

bow shock \'bau̇-\ *noun* (1950)
: the shock wave formed by the collision of a stellar wind with another medium (as the magnetosphere of a planet)

bow·sprit \'bau̇-ˌsprit, *British usually* 'bō-\ *noun* [Middle English *bouspret,* probably from Middle Low German *bōchsprēt,* from *bōch* bow + *sprēt* pole] (13th century)
: a large spar projecting forward from the stem of a ship

bow·string \'bō-ˌstriŋ\ *noun* (14th century)
: a waxed or sized cord joining the ends of a shooting bow

bowstring hemp *noun* (circa 1858)
: any of various Asian and African sansevierias; *also* **:** its soft tough leaf fiber used especially in cordage

bow tie \'bō-\ *noun* (1897)
: a short necktie tied in a bowknot

bow window \'bō-\ *noun* (1753)
: a usually curved bay window

◇ WORD HISTORY

bowdlerize Editors of books typically languish in obscurity, while authors bask in fame. Few editors have ever gained the public attention achieved by Thomas Bowdler (1754–1825). Born near Bath, England, and trained as a physician, Bowdler engaged in mostly nonmedical pursuits from the 1780s onward. His travels through Europe led to his first foray into publishing, a report warning English tourists against the uncleanliness of watering places in France. Expanding the scope of his interest in purification, he next tackled works of literature. His first effort was *The Family Shakespeare* (1818), the title page of which promised that "those words and expressions are omitted which cannot with propriety be read aloud in a family." Although Bowdler's edition met with adverse critical reaction, it pleased the 19th century public. Bowdler then went on to expurgate Gibbon's *History of the Decline and Fall of the Roman Empire* of all passages of "an irreligious or immoral tendency." A decade after his death Bowdler's name had become synonymous with censorship on moral grounds, and by 1836 *bowdlerize* was used to describe the expurgation of literary works.

bow·wow \'bau̇-ˌwau̇, bau̇-'\ *noun* [imitative] (1576)
1 : the bark of a dog; *also* **:** DOG
2 : noisy clamor
3 : arrogant dogmatic manner

bow·yer \'bō-yər\ *noun* [Middle English *bowyere*] (14th century)
: a person who makes shooting bows

[1]**box** \'bäks\ *noun, plural* **box** *or* **box·es** [Middle English, from Old English, from Latin *buxus,* from Greek *pyxos*] (before 12th century)
: an evergreen shrub or small tree (genus *Buxus* of the family Buxaceae, the box family) with opposite entire leaves and capsular fruits; *especially* **:** a widely cultivated shrub (*B. sempervirens*) used for hedges, borders, and topiary figures

[2]**box** *noun* [Middle English, from Old English, from Late Latin *buxis,* from Greek *pyxis,* from *pyxos* box tree] (before 12th century)
1 : a rigid typically rectangular container with or without a cover ⟨a cigar *box*⟩: as **a :** an open cargo container of a vehicle **b :** COFFIN
2 : the contents of a box especially as a measure of quantity
3 : a box or boxlike container and its contents: as **a** *British* **:** a gift in a box **b :** an automobile transmission **c :** TELEVISION **d :** a signaling apparatus ⟨alarm *box*⟩ **e :** a usually self-contained piece of electronic equipment **f :** BOOM BOX
4 : an often small space, compartment, or enclosure: as **a :** an enclosed group of seats for spectators (as in a theater or stadium) **b :** a driver's seat on a carriage or coach **c :** a cell for holding mail **d** *British* **:** BOX STALL **e :** PENALTY BOX
5 : a usually rectangular space that is frequently outlined or demarcated on a surface: as **a :** any of six spaces on a baseball diamond where the batter, coaches, pitcher, and catcher stand **b :** a space on a page for printed matter or in which to make a mark
6 : PREDICAMENT, FIX
7 : a cubical building
— box·ful \-ˌfu̇l\ *noun*
— box·like \-ˌlīk\ *adjective*

[3]**box** *transitive verb* (15th century)
1 : to enclose in or as if in a box
2 : to hem in (as an opponent) — usually used with *in, out,* or *up* ⟨*boxed* out the opposing tackle⟩

[4]**box** *noun* [Middle English] (14th century)
: a punch or slap especially on the ear

[5]**box** (1519)
transitive verb
1 : to hit (as the ears) with the hand
2 : to engage in boxing with
intransitive verb
: to fight with the fists **:** engage in boxing

[6]**box** *transitive verb* [probably from Spanish *bojar* to circumnavigate, from Catalan *vogir* to turn, from Latin *volvere* to roll — more at VOLUBLE] (1753)
: to name the 32 points of (the compass) in their order — used figuratively in the phrase *box the compass* to describe making a complete reversal

box·board \'bäks-ˌbōrd, -ˌbȯrd\ *noun* (1841)
: paperboard used for making boxes and cartons

box camera *noun* (1902)
: a box-shaped camera with a simple lens and rotary shutter

[1]**box·car** \'bäks-ˌkär\ *noun* (1856)
: a roofed freight car usually with sliding doors in the sides

[2]**boxcar** *adjective* [from the high numbers stenciled on the sides of boxcars] (1944)
: very large ⟨the judge awarded her a *boxcar* figure⟩

box coat *noun* (1822)
1 : a heavy overcoat formerly worn for driving
2 : a loose coat usually fitted at the shoulders

box elder *noun* (1787)
: a North American maple (*Acer negundo*) with compound leaves

[1]**box·er** \'bäk-sər\ *noun* (1742)
1 : a person who engages in the sport of boxing
2 *plural* **:** SHORT 4b

[2]**boxer** *noun* (1871)
: one that makes boxes or packs things in boxes

[3]**boxer** *noun* [German, from English [1]*boxer*] (circa 1904)
: a compact medium-sized short-haired usually fawn or brindled dog of a breed originating in Germany

Box·er \'bäk-sər\ *noun* [approximate translation of Chinese (Beijing) *yìhé juǎn,* literally, righteous harmonious fist] (1899)
: a member of a Chinese secret society that in 1900 attempted by violence to drive foreigners out of China and to force Chinese converts to renounce Christianity

boxer shorts *noun plural* (1944)
: SHORT 4b

[1]**box·ing** \'bäk-siŋ\ *noun* (1607)
1 : an act of enclosing in a box
2 : a boxlike enclosure **:** CASING
3 : material used for boxes and casings

[2]**boxing** *noun* (1711)
: the art of attack and defense with the fists practiced as a sport

Boxing Day *noun* (1833)
: the first weekday after Christmas observed as a legal holiday in parts of the Commonwealth and marked by the giving of Christmas boxes to service workers (as postal workers)

boxing glove *noun* (1875)
: one of a pair of leather mittens heavily padded on the back and worn in boxing

box kite *noun* (1897)
: a tailless kite consisting of two or more open-ended connected boxes

box kite

box lunch *noun* (1950)
: a lunch packed in a container (as a box)

box office *noun* (1786)
1 a : an office (as in a theater) where tickets of admission are sold **b :** income from ticket sales (as for a film)
2 : the ability or power (as of a show) to attract ticket buyers; *also* **:** something that enhances that ability ⟨any publicity is good *box office*⟩

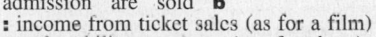

box pleat *noun* (1883)
: a pleat made by forming two folded edges one facing right and the other left

box score *noun* [from its arrangement in a newspaper box] (1913)
: a printed score of a game (as baseball) giving the names and positions of the players and a record of the play arranged in tabular form; *broadly* **:** total count **:** SUMMARY

box seat *noun* (1849)
1 : [2]BOX 4b
2 a : a seat in a box (as in a theater or grandstand) **b :** a position favorable for viewing something

box social *noun* (1928)
: a fund raising affair at which box lunches are auctioned to the highest bidder

box spring *noun* (1895)
: a bedspring that consists of spiral springs attached to a foundation and enclosed in a cloth-covered frame

box stall *noun* (1885)
: an individual enclosure within a barn or stable in which an animal may move about freely without a restraining device

box·thorn \'bäks-ˌthȯrn\ *noun* (1678)
: MATRIMONY VINE

box turtle *noun* (circa 1804)
: any of several North American land turtles (genus *Terrapene*) capable of withdrawing into its shell and closing it by hinged joints in the lower half — called also *box tortoise*

box turtle

box·wood \'bäks-ˌwu̇d\ *noun* (1652)
1 : the very close-grained heavy tough hard wood of the box (*Buxus*); *also* **:** a wood of similar properties
2 : a plant producing boxwood

boxy \'bäk-sē\ *adjective* **box·i·er; -est** (circa 1861)
: resembling a box
— box·i·ness *noun*

boy \'bȯi\ *noun, often attributive* [Middle English] (13th century)
1 : a male servant — sometimes taken to be offensive
2 a : a male child from birth to puberty **b :** SON **c :** an immature male **:** YOUTH **d :** SWEETHEART, BEAU
3 a : one native to a given place ⟨local *boy*⟩ **b :** FELLOW, PERSON ⟨the *boys* at the office⟩ **c** — used interjectionally to express intensity of feeling ⟨*boy,* what a game⟩
— boy·hood \-ˌhu̇d\ *noun*
— boy·ish \-ish\ *adjective*
— boy·ish·ly *adverb*
— boy·ish·ness *noun*

bo·yar *also* **bo·yard** \bō-'yär\ *noun* [Russian *boyarin*] (1591)
: a member of a Russian aristocratic order next in rank below the ruling princes until its abolition by Peter the Great

boy·chick \'bȯi-ˌchik\ *noun* [American Yiddish *boytshik,* from English *boy* + Yiddish *-tshik,* diminutive suffix] (circa 1951)
: a young man **:** BOY

boy·cott \'bȯi-ˌkät\ *transitive verb* [Charles C. *Boycott* (died 1897) English land agent in Ireland who was ostracized for refusing to reduce rents] (1880)
: to engage in a concerted refusal to have dealings with (as a person, store, or organization) usually to express disapproval or to force acceptance of certain conditions ◆

◇ WORD HISTORY

boycott In September 1880 the recently formed Irish National Land League, which grew out of unrest following an agricultural depression in Ireland during the late 1870s, initiated a campaign against the eviction of tenants unable to pay their rents. Charles S. Boycott, a retired English army captain who rented a farm in County Mayo and worked as an agent for an absentee landlord, became the first victim of the Land League's campaign. When the landlord's tenants refused to pay their rents unless they received a 25 percent abatement, Boycott attempted to serve eviction notices. As a result, he was ostracized by the community, his laborers and servants quit, and the crops on his farm began to rot. His plight became well-known through press coverage and Boycott's own letters to *The Times* in London. A force of fifty volunteers from Ulster was raised and sent to Mayo to harvest his crops, guarded by nearly a thousand troops and police. The newly popular tactic of shunning a person to assert a grievance needed a name and Boycott's name was at hand. The Land League was banned by Parliament in 1881, but the word *boycott,* noun and verb, is still part of English.

— **boycott** *noun*
— **boy·cott·er** *noun*
boy·friend \'bȯi-ˌfrend\ *noun* (1896)
1 : a male friend
2 : a frequent or regular male companion of a girl or woman
3 : a male lover
Boyle's law \'bȯilz-\ *noun* [Robert *Boyle*] (circa 1860)
: a statement in physics: the volume of a gas at constant temperature varies inversely with the pressure exerted on it
boyo \'bȯi-(ˌ)ō\ *noun, plural* **boy·os** [*boy* + [1]*-o*] (circa 1870)
Irish **:** BOY, LAD
Boy Scout *noun* (1909)
1 : a member of any of various national scouting programs (as the Boy Scouts of America) for boys usually 11 to 17 years of age
2 : a person whose values or actions are characteristic of a Boy Scout
boy·sen·ber·ry \'bȯi-z^{ə}n-ˌber-ē, -s^{ə}n-, -z^{ə}m-\ *noun* [Rudolph *Boysen* (died 1950) American horticulturist + English *berry*] (1935)
: a large bramble fruit with a raspberry flavor; *also* **:** the trailing hybrid bramble yielding this fruit and developed by crossing several blackberries and raspberries
boy wonder *noun* (1946)
: a young man of noteworthy achievements
bo·zo \'bō-(ˌ)zō\ *noun, plural* **bozos** [origin unknown] (1920)
: a foolish or incompetent person
B picture *noun* (circa 1937)
: B MOVIE
bra \'brä\ *noun* (1936)
: BRASSIERE
— **bra·less** *adjective*
brab·ble \'bra-bəl\ *intransitive verb* **brab·bled; brab·bling** \-b(ə-)liŋ\ [perhaps from Middle Dutch *brabbelen,* of imitative origin] (circa 1530)
: SQUABBLE
— **brabble** *noun*
[1]**brace** \'brās\ *verb* **braced; brac·ing** [Middle English, from Middle French *bracier* to embrace, from *brace*] (14th century)
transitive verb
1 *archaic* **:** to fasten tightly **:** BIND
2 a : to prepare for use by making taut **b :** PREPARE, STEEL ⟨*brace* yourself for the shock⟩ **c :** INVIGORATE, FRESHEN
3 : to turn (a sail yard) by means of a brace
4 a : to furnish or support with a brace ⟨heavily *braced* against the wind⟩ **b :** to make stronger **:** REINFORCE
5 : to put or plant firmly ⟨*braces* his foot in the stirrup⟩
6 : to waylay especially with demands or questions
intransitive verb
1 : to take heart — used with *up*
2 : to get ready (as for an attack)
[2]**brace** *noun, plural* **brac·es** [Middle English, clasp, pair, from Middle French, two arms, from Latin *bracchia,* plural of *bracchium* arm, from Greek *brachiōn,* from comparative of *brachys* short — more at BRIEF] (14th century)
1 : something (as a clasp) that connects or fastens
2 *or plural* **brace :** two of a kind **:** PAIR ⟨several *brace* of quail⟩
3 : a crank-shaped instrument for turning a bit
4 : something that transmits, directs, resists, or supports weight or pressure: as **a :** a diagonal piece of structural material that serves to strengthen something (as a framework) **b :** a rope rove through a block at the end of a ship's yard to swing it horizontally **c** *plural* **:** SUSPENDERS **d :** an appliance for supporting a body part **e** *plural* **:** dental appliances used to exert pressure to straighten misaligned teeth
5 a : one of two marks { } used to connect words or items to be considered together **b :** one of these marks connecting two or more musical staffs carrying parts to be performed simultaneously **c :** BRACKET 3a
6 : a position of rigid attention
7 : something that arouses energy or strength-ens morale
brace·let \'brās-lət\ *noun* [Middle English, from Middle French, diminutive of *bras* arm, from Latin *bracchium*] (15th century)
1 : an ornamental band or chain worn around the wrist
2 : something (as handcuffs) resembling a bracelet
[1]**bra·cer** \'brā-sər\ *noun* [Middle English, from Middle French *braciere,* from Old French, from *braz* arm, from Latin *bracchium*] (14th century)
: an arm or wrist protector especially for use by an archer
[2]**brac·er** \'brā-sər\ *noun* (1579)
1 : one that braces, binds, or makes firm
2 : a drink (as of liquor) taken as a stimulant
bra·ce·ro \brä-'ser-(ˌ)ō\ *noun, plural* **-ros** [Spanish, laborer, from *brazo* arm, from Latin *bracchium*] (1920)
: a Mexican laborer admitted to the U.S. especially for seasonal contract labor in agriculture — compare WETBACK
brace root *noun* (1892)
: PROP ROOT
brachi- *or* **brachio-** *combining form* [Latin *bracchium, brachium*]
1 : arm ⟨*brachi*al⟩
2 : brachial and ⟨*brachio*cephalic⟩
bra·chi·al \'brā-kē-əl\ *adjective* (1578)
: of, relating to, or situated in the arm or an armlike process ⟨the *brachial* artery of the upper arm⟩
brachial plexus *noun* (circa 1860)
: a network of nerves lying mostly in the armpit and supplying nerves to the chest, shoulder, and arm
bra·chi·ate \'brā-kē-ˌāt\ *intransitive verb* **-at·ed; -at·ing** (1932)
: to progress by swinging from one hold to another by the arms ⟨a *brachiating* gibbon⟩
— **bra·chi·a·tion** \ˌbrā-kē-'ā-shən\ *noun*
— **bra·chi·a·tor** \'brā-kē-ˌā-tər\ *noun*
bra·chio·ce·phal·ic artery \ˌbrā-kē-(ˌ)ō-sə-'fa-lik-\ *noun* (circa 1839)
: a short artery that arises from the arch of the aorta and divides into the carotid and subclavian arteries of the right side — called also *innominate artery*
brachiocephalic vein *noun* (circa 1852)
: either of two large veins that occur one on each side of the neck, receive blood from the head and neck, and unite to form the superior vena cava — called also *innominate vein*
bra·chio·pod \'brā-kē-ə-ˌpäd\ *noun* [ultimately from Latin *bracchium* + Greek *pod-, pous* foot — more at FOOT] (1836)
: any of a phylum (Brachiopoda) of marine invertebrates with bivalve shells within which is a pair of arms bearing tentacles by which a current of water is made to bring microscopic food to the mouth
— **brachiopod** *adjective*
brachy- *combining form* [Greek, from *brachys* — more at BRIEF]
: short ⟨*brachy*cephalic⟩
brachy·ce·phal·ic \ˌbra-ki-sə-'fa-lik\ *adjective* [New Latin *brachycephalus,* from Greek *brachy-* + *kephalē* head — more at CEPHALIC] (circa 1852)
: short-headed or broad-headed with a cephalic index of over 80
— **brachy·ceph·a·ly** \-'se-fə-lē\ *noun*
bra·chyp·ter·ous \bra-'kip-tə-rəs\ *adjective* [Greek *brachypteros,* from *brachy-* + *pteron* wing — more at FEATHER] (circa 1847)
: having rudimentary or abnormally small wings ⟨*brachypterous* insects⟩
brac·ing \'brā-siŋ\ *adjective* (1750)
: giving strength, vigor, or freshness ⟨a *bracing* breeze⟩
— **brac·ing·ly** *adverb*
bra·ci·o·la \ˌbrä-chē-'ō-lə, -'chō-lə\ *or* **bra·ci·o·le** \-'ō-ˌlā\ *noun* [Italian, literally, slice of meat roasted over coals, from *brace* live coals, probably of Germanic origin; akin to Swedish *brasa* fire] (circa 1945)
: a thin slice of meat wrapped around a seasoned filling and often cooked in wine
brack·en \'bra-kən\ *noun* [Middle English *braken,* probably of Scandinavian origin; akin to Old Swedish *brække* fern] (14th century)
1 : a large coarse fern; *especially* **:** a nearly cosmopolitan brake (*Pteridium aquilinum*) found in most tropical and temperate regions
2 : a growth of brakes
[1]**brack·et** \'bra-kət\ *noun* [perhaps from Middle French *braguette* codpiece, from diminutive of *brague* breeches, from Old Provençal *braga,* from Latin *braca,* of Celtic origin — more at BREECH] (1580)
1 : an overhanging member that projects from a structure (as a wall) and is usually designed to support a vertical load or to strengthen an angle
2 : a fixture (as for holding a lamp) projecting from a wall or column
3 a : one of a pair of marks [ ] used in writing and printing to enclose matter or in mathematics and logic as signs of aggregation — called also *square bracket* **b :** one of the pair of marks ⟨ ⟩ used to enclose matter — called also *angle bracket* **c :** PARENTHESIS 3 **d :** BRACE 5b
4 : a section of a continuously numbered or graded series (as age ranges or income levels)
[2]**bracket** *transitive verb* (circa 1847)
1 a : to place within or as if within brackets ⟨editorial comments are *bracketed*⟩ ⟨news stories *bracketed* by commercials⟩ **b :** to eliminate from consideration ⟨*bracket* off politics⟩ **c :** to extend around so as to encompass **:** INCLUDE ⟨test pressures . . . which *bracket* virtually the entire range of passenger-car tire pressures —*Consumer Reports*⟩
2 : to furnish or fasten with brackets
3 : to put in the same category or group ⟨*bracketed* in a tie for third⟩
4 a : to get the range on (a target) by firing over and short ⟨there were mortar rounds *bracketing* the area —Ed Bradley⟩ **b :** to establish the limits of ⟨*bracketed* the problem neatly⟩ **c :** to take photographs at more than one exposure in order to ensure that the desired exposure is obtained
bracket creep *noun* (1980)
: movement into a higher tax bracket as a result of income rises intended to offset the effects of inflation
brack·et·ed *adjective* (1885)
of a serif **:** joined to the stroke by a curved line
bracket fungus *noun* (1899)
: a basidiomycete that forms shelflike sporophores
brack·ish \'bra-kish\ *adjective* [Dutch *brac* salty; akin to Middle Low German *brac* salty] (1538)
1 : somewhat salty ⟨*brackish* water⟩
2 a : not appealing to the taste ⟨*brackish* tea⟩ **b :** REPULSIVE
— **brack·ish·ness** *noun*
brac·o·nid \'bra-kə-(ˌ)nid\ *noun* [ultimately from Greek *brachys*] (circa 1893)
: any of a large family (Braconidae) of ichneumon flies
— **braconid** *adjective*
bract \'brakt\ *noun* [New Latin *bractea,* from Latin, thin metal plate] (1770)

1 : a leaf from the axil of which a flower or floral axis arises
2 : a leaf borne on a floral axis; *especially* **:** one subtending a flower or flower cluster
— **brac·te·al** \'brak-tē-əl\ *adjective*
— **brac·te·ate** \-tē-ət, -ˌāt\ *adjective*
— **bract·ed** \-təd\ *adjective*

brac·te·ole \'brak-tē-ˌōl\ *noun* [New Latin *bracteola,* from Latin, diminutive of *bractea*] (circa 1828)
: a small bract especially on a floral axis

¹brad \'brad\ *noun* [Middle English, from Old Norse *broddr* spike; perhaps akin to Old English *byrst* bristle — more at BRISTLE] (13th century)
1 : a thin nail of the same thickness throughout but tapering in width and having a slight projection at the top of one side instead of a head
2 : a slender wire nail with a small barrel-shaped head

²brad *transitive verb* **brad·ded; brad·ding** (1794)
: to fasten with brads

brad·awl \'brad-ˌȯl\ *noun* (1823)
: an awl with chisel edge used to make holes for brads or screws

brady- *combining form* [Greek *bradys*]
: slow ⟨*brady*kinin⟩

bra·dy·car·dia \ˌbrā-di-'kär-dē-ə *also* ˌbra-\ *noun* [New Latin] (circa 1890)
: relatively slow heart action — compare TACHYCARDIA

bra·dy·ki·nin \-'kī-nən\ *noun* (1949)
: a kinin that is formed locally in injured tissue, acts in vasodilation of small arterioles, is considered to play a part in inflammatory processes, and is composed of a chain of nine amino-acid residues

brae \'brā\ *noun* [Middle English *bra,* from Old Norse *brā* eyelid; akin to Old English *brǣw* eyebrow, and probably to Old English *bregdan* to move quickly — more at BRAID] (13th century)
chiefly Scottish **:** a hillside especially along a river

¹brag \'brag\ *noun* [Middle English] (14th century)
1 : a pompous or boastful statement
2 : arrogant talk or manner **:** COCKINESS
3 : BRAGGART

²brag *verb* **bragged; brag·ging** (14th century)
intransitive verb
: to talk boastfully **:** engage in self-glorification
transitive verb
: to assert boastfully
synonym see BOAST
— **brag·ger** \'bra-gər\ *noun*
— **brag·gy** \'bra-gē\ *adjective*

³brag *adjective* **brag·ger; brag·gest** (1836)
: FIRST-RATE

brag·ga·do·cio \ˌbra-gə-'dō-sē-ˌō, -shē-, -chē-, -(ˌ)shō, -(ˌ)chō\ *noun, plural* **-cios** [*Braggadochio,* personification of boasting in *Faerie Queene* by Edmund Spenser] (1594)
1 : BRAGGART
2 a : empty boasting **b :** arrogant pretension **:** COCKINESS

brag·gart \'bra-gərt\ *noun* (circa 1577)
: a loud arrogant boaster
— **braggart** *adjective*

¹Brah·ma \'brä-mə\ *noun* [Sanskrit *brahman*] (1690)
1 : the creator god of the Hindu sacred triad — compare SIVA, VISHNU
2 : the ultimate ground of all being in Hinduism

²Brah·ma \'brā-mə, 'brä-, 'bra-\ *noun* (1938)
: BRAHMAN 2

Brah·man *or* **Brah·min** \'brä-mən; *2 is* 'brā-, 'brä-, 'bra-\ *noun* [Middle English *Bragman* inhabitant of India, from Latin *Bracmanus,* from Greek *Brachman,* from Sanskrit *brāhmaṇa* of the Brahman caste, from *brahman* Brahman] (15th century)
1 a : a Hindu of the highest caste traditionally assigned to the priesthood **b :** ¹BRAHMA 2
2 : any of an Indian breed of humped cattle **:** ZEBU; *especially* **:** a large vigorous heat-resistant and tick-resistant usually silvery gray animal developed in the southern U.S. from the zebu
3 *usually Brahmin* **:** a person of high social standing and cultivated intellect and taste ⟨Boston *Brahmins*⟩
— **Brah·man·ic** \brä-'ma-nik\ *or* **Brah·man·i·cal** \-ni-kəl\ *adjective*

Brah·man·ism \'brä-mə-ˌni-zəm\ *noun* (1816)
: orthodox Hinduism adhering to the pantheism of the Vedas and to the ancient sacrifices and family ceremonies

¹braid \'brād\ *transitive verb* [Middle English *breyden* to move suddenly, snatch, plait, from Old English *bregdan;* akin to Old High German *brettan* to draw (a sword)] (before 12th century)
1 a : to make from braids ⟨*braid* a rug⟩ **b :** to form (three or more strands) into a braid
2 : to do up (the hair) by interweaving three or more strands
3 : MIX, INTERMINGLE ⟨*braid* fact with fiction⟩
4 : to ornament especially with ribbon or braid
— **braid·er** *noun*

²braid *noun* (1530)
1 a : a length of braided hair **b :** a cord or ribbon having usually three or more component strands forming a regular diagonal pattern down its length; *especially* **:** a narrow fabric of intertwined threads used especially for trimming
2 : high-ranking naval officers

braid·ed *adjective* (15th century)
1 a : made by intertwining three or more strands **b :** ornamented with braid
2 : forming an interlacing network of channels ⟨a *braided* river⟩

braid·ing \'brā diŋ\ *noun* (15th century)
: something made of braided material

¹brail \'brā(ə)l\ *noun* [Middle English *brayle,* from Anglo-French *braiel,* from Old French, strap] (15th century)
1 : a rope fastened to the leech of a sail and used for hauling the sail up or in
2 : a dip net with which fish are hauled aboard a boat from a purse seine or trap

²brail *transitive verb* (1625)
1 : to take in (a sail) by the brails
2 : to hoist (fish) by means of a brail

braille \'brā(ə)l\ *noun, often capitalized* [Louis *Braille*] (1853)
: a system of writing for the blind that uses characters made up of raised dots ◆
— **braille** *transitive verb*
— **braill·ist** \'brā-list\ *noun*

a b c d e f g h i j
1 2 3 4 5 6 7 8 9 0
k l m n o p q r s t
u v w x y z
Capital Sign Numeral Sign

braille alphabet

braille·writ·er \-ˌrī-tər\ *noun, often capitalized* (1942)
: a machine for writing braille

¹brain \'brān\ *noun* [Middle English, from Old English *brægen;* akin to Middle Low German *bregen* brain, and perhaps to Greek *brechmos* front part of the head] (before 12th century)
1 a : the portion of the vertebrate central nervous system that constitutes the organ of thought and neural coordination, includes all the higher nervous centers receiving stimuli from the sense organs and interpreting and correlating them to formulate the motor impulses, is made up of neurons and supporting and nutritive structures, is enclosed within the skull, and is continuous with the spinal cord through the foramen magnum **b :** a nervous center in invertebrates comparable in position and function to the vertebrate brain

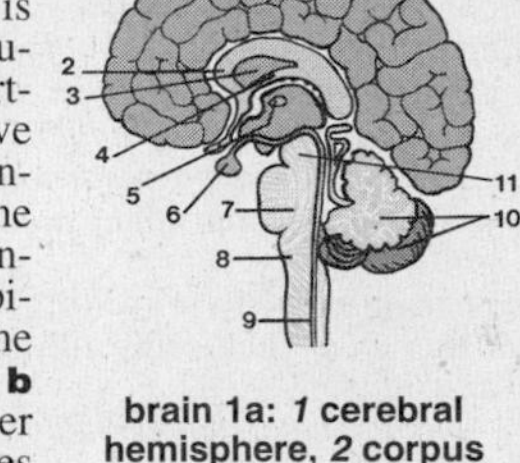

brain 1a: *1* **cerebral hemisphere,** *2* **corpus callosum,** *3* **ventricle,** *4* **fornix,** *5* **thalamus,** *6* **pituitary gland,** *7* **pons,** *8* **medulla oblongata,** *9* **spinal cord,** *10* **cerebellum,** *11* **midbrain**

2 a (1) **:** INTELLECT, MIND ⟨has a clever *brain*⟩ (2) **:** intellectual endowment **:** INTELLIGENCE — often used in plural ⟨plenty of *brains* in that family⟩ **b** (1) **:** a very intelligent or intellectual person (2) **:** the chief planner of an organization or enterprise — usually used in plural
3 : something that performs the functions of a brain; *especially* **:** an automatic device (as a computer) for control or computation

²brain *transitive verb* (14th century)
1 : to kill by smashing the skull
2 : to hit on the head

brain·case \'brān-ˌkās\ *noun* (1741)
: the cranium enclosing the brain

brain·child \-ˌchīld\ *noun* (1631)
: a product of one's creative imagination

brain coral *noun* (circa 1711)
: a massive reef-building coral (as genus *Maeandra*) having the surface covered by ridges and furrows

brain death *noun* (1964)
: final cessation of activity in the central nervous system especially as indicated by a flat electroencephalogram for a predetermined length of time
— **brain–dead** *adjective*

brain drain *noun* (1963)
: the departure of educated or professional people from one country, economic sector, or field for another usually for better pay or living conditions

brained \'brānd\ *adjective* (15th century)
: having a brain of a specified kind — used in combination ⟨feather*brained*⟩

brain hormone *noun* (1957)
: a hormone that is secreted by neurosecretory cells of the insect brain and that stimulates the prothoracic glands to secrete ecdysone

brain·ish \'brā-nish\ *adjective* (circa 1530)

◇ WORD HISTORY
braille The system of writing known as *braille* is named after its principal inventor, the Frenchman Louis Braille (1809–1852). Blinded at the age of three in an accident, Braille entered the Institution Nationale des Jeunes Aveugles (National Institute for Blind Children) in Paris in 1819. While there he became interested in a writing system developed by Charles Barbier in which messages were encoded in dots embossed on cardboard. Braille worked out an adaptation using a six-dot code and a simple instrument for producing the dots. He published treatises on his system in 1829 and 1837.

archaic **:** IMPETUOUS, HOTHEADED ⟨and in this *brainish* apprehension kills the unseen good old man —Shakespeare⟩

brain·less \ˈbrān-ləs\ *adjective* (15th century)
: devoid of intelligence **:** STUPID
— **brain·less·ly** *adverb*
— **brain·less·ness** *noun*

brain·pan \ˈbrān-ˌpan\ *noun* (14th century)
: BRAINCASE

brain–pick·ing \-ˌpi-kiŋ\ *noun* (1954)
: the act of gathering information from another's mind
— **brain–pick·er** \-kər\ *noun*

brain·pow·er \-ˌpau̇(-ə)r\ *noun* (1878)
1 : intellectual ability
2 : people with developed intellectual ability

brain·sick \-ˌsik\ *adjective* (15th century)
1 : mentally disordered
2 : arising from mental disorder ⟨a *brainsick* frenzy⟩
— **brain·sick·ly** *adverb*

brain stem *noun* (1879)
: the part of the brain composed of the mesencephalon, pons, and medulla oblongata and connecting the spinal cord with the forebrain and cerebrum

brain·storm \-ˌstȯrm\ *noun* (circa 1894)
1 : a violent transient fit of insanity
2 a : a sudden bright idea **b :** a harebrained idea

brain·storm·ing \-ˌstȯr-miŋ\ *noun* (circa 1955)
: a group problem-solving technique that involves the spontaneous contribution of ideas from all members of the group
— **brain·storm** *verb*
— **brain·storm·er** *noun*

brains trust *noun* (1934)
chiefly British **:** BRAIN TRUST

brain·teas·er \-ˌtē-zər\ *noun* (1923)
: something (as a puzzle) that demands mental effort and acuity for its solution

brain trust *noun* (1910)
: a group of official or unofficial advisers concerned especially with planning and strategy
— **brain trust·er** \-ˌtrəs-tər\ *noun*

brain·wash·ing \ˈbrān-ˌwȯ-shiŋ, -ˌwä-\ *noun* [translation of Chinese (Beijing) *xǐnǎo*] (1950)
1 : a forcible indoctrination to induce someone to give up basic political, social, or religious beliefs and attitudes and to accept contrasting regimented ideas
2 : persuasion by propaganda or salesmanship
— **brain·wash** *transitive verb*
— **brainwash** *noun*
— **brain·wash·er** *noun*

brain wave *noun* (1890)
1 : BRAINSTORM 2a
2 a : rhythmic fluctuations of voltage between parts of the brain resulting in the flow of an electric current **b :** a current produced by brain waves

brainy \ˈbrā-nē\ *adjective* **brain·i·er; -est** (1874)
: having or showing a well-developed intellect **:** INTELLIGENT
— **brain·i·ness** *noun*

[1]**braise** \ˈbrāz\ *transitive verb* **braised; brais·ing** [French *braiser,* from *braise* live coals, from Old French *breze,* probably of Germanic origin; akin to Swedish *brasa* fire] (1797)
: to cook slowly in fat and little moisture in a closed pot

[2]**braise** *noun* (circa 1885)
: an item of braised food

[1]**brake** \ˈbrāk\ *archaic past of* BREAK

[2]**brake** *noun* [Middle English, fern, probably back-formation from *braken* bracken] (14th century)
: the common bracken fern (*Pteridium aquilinum*)

[3]**brake** *noun* [Middle English, from Middle Low German; akin to Old English *brecan* to break] (15th century)
1 : a toothed instrument or machine for separating out the fiber of flax or hemp by breaking up the woody parts
2 : a machine for bending, flanging, folding, and forming sheet metal

[4]**brake** *noun* [Middle English *-brake*] (1563)
: rough or marshy land overgrown usually with one kind of plant
— **braky** \ˈbrā-kē\ *adjective*

[5]**brake** *noun* [perhaps from obsolete *brake* bridle] (circa 1782)
1 : a device for arresting or preventing the motion of a mechanism usually by means of friction
2 : something used to slow down or stop movement or activity ⟨use interest rates as a *brake* on spending⟩
— **brake·less** \ˈbrā-kləs\ *adjective*

[6]**brake** *verb* **braked; brak·ing** (1868)
transitive verb
: to retard or stop by a brake
intransitive verb
1 : to operate or manage a brake; *especially* **:** to apply the brake on a vehicle
2 : to become checked by a brake

brake·man \ˈbrāk-mən\ *noun* (1833)
1 : a freight or passenger train crew member who inspects the train and assists the conductor
2 : the end man on a bobsled team who operates the brake

bram·ble \ˈbram-bəl\ *noun* [Middle English *brembel,* from Old English *brēmel;* akin to Old English *brōm* broom] (before 12th century)
: any of a genus (*Rubus*) of usually prickly shrubs of the rose family including the raspberries and blackberries; *broadly* **:** a rough prickly shrub or vine
— **bram·bly** \-b(ə-)lē\ *adjective*

bran \ˈbran\ *noun* [Middle English, from Middle French] (14th century)
: the edible broken seed coats of cereal grain separated from the flour or meal by sifting or bolting

[1]**branch** \ˈbranch\ *noun, often attributive* [Middle English, from Middle French *branche,* from Late Latin *branca* paw] (14th century)
1 : a natural subdivision of a plant stem; *especially* **:** a secondary shoot or stem (as a bough) arising from a main axis (as of a tree)
2 : something that extends from or enters into a main body or source: as **a** (1) **:** a stream that flows into another usually larger stream **:** TRIBUTARY (2) *Southern & Midland* **:** CREEK 2 **b :** a side road or way **c :** a slender projection (as the tine of an antler) **d :** a distinctive part of a mathematical curve **e :** a part of a computer program executed as a result of a program decision
3 : a part of a complex body: as **a :** a division of a family descending from a particular ancestor **b :** an area of knowledge that may be considered apart from related areas ⟨pathology is a *branch* of medicine⟩ **c** (1) **:** a division of an organization (2) **:** a separate but dependent part of a central organization ⟨the neighborhood *branch* of the city library⟩ **d :** a language group less inclusive than a family ⟨the Germanic *branch* of the Indo-European language family⟩
— **branched** \ˈbrancht\ *adjective*
— **branch·less** \ˈbranch-ləs\ *adjective*
— **branchy** \ˈbran-chē\ *adjective*

[2]**branch** (14th century)
intransitive verb
1 : to put forth branches **:** RAMIFY
2 : to spring out (as from a main stem) **:** DIVERGE
3 : to be an outgrowth — used with *from* ⟨poetry that *branched* from religious prose⟩
4 : to extend activities — usually used with *out* ⟨the business is *branching* out⟩
5 : to follow one of two or more branches (as in a computer program)
transitive verb
1 : to ornament with designs of branches
2 : to divide up **:** SECTION

bran·chi·al \ˈbraŋ-kē-əl\ *adjective* [Greek *branchia* gills] (1801)
: of, relating to, or supplying the gills or associated structures or their embryonic precursors

bran·chio·pod \ˈbraŋ-kē-ə-ˌpäd\ *noun* [ultimately from Greek *branchia* gills + *pod-, pous* foot — more at FOOT] (1836)
: any of a group (Branchiopoda) of small usually freshwater crustaceans (as fairy shrimp or water fleas) with usually many pairs of setae-bearing appendages
— **branchiopod** *adjective*

branch·let \ˈbranch-lət\ *noun* (circa 1731)
: a small usually terminal branch

branch·line \-ˌlīn\ *noun* (1846)
: a secondary line usually of a railroad

branch water *noun* [[1]*branch* (creek)] (1835)
: plain water ⟨bourbon and *branch water*⟩

[1]**brand** \ˈbrand\ *noun* [Middle English, torch, sword, from Old English; akin to Old English *bærnan* to burn] (before 12th century)
1 a : a charred piece of wood **b :** FIREBRAND 1 **c :** something (as lightning) that resembles a firebrand
2 : SWORD
3 a (1) **:** a mark made by burning with a hot iron to attest manufacture or quality or to designate ownership (2) **:** a printed mark made for similar purposes **:** TRADEMARK **b** (1) **:** a mark put on criminals with a hot iron (2) **:** a mark of disgrace **:** STIGMA ⟨the *brand* of poverty⟩
4 a : a class of goods identified by name as the product of a single firm or manufacturer **:** MAKE **b :** a characteristic or distinctive kind ⟨a lively *brand* of theater⟩
5 : a tool used to produce a brand

[2]**brand** *transitive verb* (15th century)
1 : to mark with a brand
2 : to mark with disapproval **:** STIGMATIZE
3 : to impress indelibly ⟨*brand* the lesson on his mind⟩
— **brand·er** *noun*

brand·ed \ˈbran-dəd\ *adjective* (1897)
: having a brand name ⟨*branded* products⟩

[1]**bran·dish** \ˈbran-dish\ *transitive verb* [Middle English *braundisshen,* from Middle French *brandiss-,* stem of *brandir,* from Old French, from *brand* sword, of Germanic origin; akin to Old English *brand*] (14th century)
1 : to shake or wave (as a weapon) menacingly
2 : to exhibit in an ostentatious or aggressive manner
synonym see SWING

[2]**brandish** *noun* (1599)
: an act or instance of brandishing

brand–name \ˈbran(d)-ˈnām\ *adjective* (1949)
1 : of or relating to a trade name ⟨*brand-name* products⟩
2 : having a reputation and a loyal following ⟨*brand-name* authors⟩

brand name *noun* (1922)
: TRADE NAME 1b

brand–new \-ˈnü, -ˈnyü\ *adjective* [[1]*brand*] (circa 1570)
: conspicuously new and unused; *also* **:** recently introduced ⟨a *brand-new* executive officer⟩

[1]**bran·dy** \ˈbran-dē\ *noun, plural* **brandies** [short for *brandywine,* from Dutch *brandewijn,* from Middle Dutch *brantwijn,* from *brant* distilled + *wijn* wine] (1657)
: an alcoholic beverage distilled from wine or fermented fruit juice

[2]**brandy** *transitive verb* **bran·died; bran·dy·ing** (circa 1848)
: to flavor, blend, or preserve with brandy

Bran·gus \ˈbraŋ-gəs\ *trademark*
— used for polled solid black beef cattle of a breed developed from a Brahman-Angus cross

bran·ni·gan \ˈbra-ni-gən\ *noun* [probably from the name *Brannigan*] (1927)
1 : a drinking spree
2 : SQUABBLE

brant \'brant\ *noun, plural* **brant** *or* **brants** [Middle English *brand gos*] (14th century)
: any of several wild geese (especially genus *Branta*); *especially* **:** a small black-necked goose (*Branta bernicla*) about the size of a mallard

¹brash \'brash\ *adjective* [origin unknown] (1566)
1 : BRITTLE ⟨*brash* wood⟩
2 a : heedless of the consequences **:** AUDACIOUS ⟨the *brashest* bush pilot of them all⟩ **b :** done in haste without regard for consequences **:** RASH ⟨*brash* acts⟩
3 a : full of fresh raw vitality ⟨a *brash* frontier town⟩ **b :** uninhibitedly energetic or demonstrative **:** BUMPTIOUS ⟨a *brash* comedian⟩
4 a : lacking restraint and discernment **:** TACTLESS ⟨*brash* remarks⟩ **b :** aggressively self-assertive **:** IMPUDENT ⟨*brash* to the point of arrogance⟩
5 : piercingly sharp **:** HARSH ⟨a *brash* squeal of brakes⟩
6 : marked by vivid contrast **:** BOLD ⟨*brash* colors⟩
— brash·ly *adverb*
— brash·ness *noun*

²brash *noun* [obsolete English *brash* to breach a wall] (1787)
: a mass of fragments (as of ice)

brass \'bras\ *noun* [Middle English *bras,* from Old English *bræs;* akin to Middle Low German *bras* metal] (before 12th century)
1 : an alloy consisting essentially of copper and zinc in variable proportions
2 a : the brass instruments of an orchestra or band — often used in plural **b :** a usually brass memorial tablet **c :** bright metal fittings, utensils, or ornaments **d :** empty cartridge shells
3 : brazen self-assurance **:** GALL
4 *singular or plural in construction* **a :** high-ranking members of the military **b :** persons in high positions (as in a business or the government)
— brass *adjective*

bras·sard \brə-'sard, 'bra-ˌ\ *noun* [French *brassard,* from Middle French *brassal,* from Old Italian *bracciale,* from *braccio* arm, from Latin *bracchium* — more at BRACE] (1830)
1 : armor for protecting the arm — see ARMOR illustration
2 : a cloth band worn around the upper arm usually bearing an identifying mark

brass band *noun* (1834)
: a band consisting chiefly or solely of brass and percussion instruments

brass·bound \'bras-ˌbau̇nd, -'bau̇nd\ *adjective* (1867)
1 : having trim made of brass or a metal resembling brass
2 a (1) **:** tradition-bound and opinionated (2) **:** making no concessions **:** INFLEXIBLE **b :** BRAZEN, PRESUMPTUOUS

brass collar Democrat *noun* (1951)
: a conservative Democrat especially in the south who votes the straight Democratic ticket

bras·se·rie \ˌbras-'rē, ˌbra-sə-\ *noun* [French, literally, brewery, from Middle French *brasser* to brew, from Old French *bracier,* from (assumed) Vulgar Latin *braciare,* of Celtic origin; akin to Welsh *brag* malt] (1864)
: an informal usually French restaurant serving simple hearty food

brass hat *noun* [from the gold braid worn on the cap] (1893)
: a member of the brass

bras·si·ca \'bra-si-kə\ *noun* [New Latin, genus name, from Latin, cabbage] (1832)
: any of a large genus (*Brassica*) of Old World temperate zone herbs (as cabbages) of the mustard family with beaked cylindrical pods

bras·siere \brə-'zir *also* ˌbra-sē-'er\ *noun* [obsolete French *brassière* bodice, from Old French *braciere* arm protector, from *bras* arm — more at BRACELET] (1911)
: a woman's undergarment to cover and support the breasts

brass instrument *noun* (1854)
: any of a group of wind instruments (as a French horn, trombone, trumpet, or tuba) that is usually characterized by a long cylindrical or conical metal tube commonly curved two or more times and ending in a flared bell, that produces tones by the vibrations of the player's lips against a usually cup-shaped mouthpiece, and that usually has valves or a slide by which the player may produce all the tones within the instrument's range

brass knuckles *noun plural but singular or plural in construction* (1855)
: KNUCKLE 4

brass ring *noun* (1950)
: a rich opportunity **:** PRIZE

brass tacks *noun plural* (1897)
: details of immediate practical importance — usually used in the phrase *get down to brass tacks*

brassy \'bra-sē\ *adjective* **brass·i·er; -est** (1576)
1 a : being shamelessly bold **b :** OBSTREPEROUS
2 : resembling brass especially in color
3 : resembling the sound of a brass instrument
— brass·i·ly \'bra-sə-lē\ *adverb*
— brass·i·ness \'bra-sē-nəs\ *noun*

brat \'brat\ *noun* [perhaps from English dialect *brat* (coarse garment)] (circa 1505)
1 a : CHILD; *specifically* **:** an ill-mannered annoying child **b :** an ill-mannered immature person
2 : the son or daughter of a career military person
— brat·ti·ness \'bra-tē-nəs\ *noun*
— brat·tish \'bra-tish\ *adjective*
— brat·ty \-tē\ *adjective*

brat·tice \'bra-təs, 'bra-tish\ *noun* [Middle English *bretais* parapet, from Old French *bretesche,* from Medieval Latin *breteschia*] (circa 1846)
: an often temporary partition of planks or cloth used especially to control mine ventilation
— brattice *transitive verb*

¹brat·tle \'bra-tᵊl\ *noun* [probably imitative] (circa 1500)
chiefly Scottish **:** CLATTER, SCAMPER

²brattle *intransitive verb* **brat·tled; brat·tling** (1513)
chiefly Scottish **:** to make a clattering or rattling sound

brat·wurst \'brät-(ˌ)wərst *also* 'brat- *also* -ˌvu̇(r)st; *sometimes* -ˌvu̇sht\ *noun* [German, from Old High German *brātwurst,* from *brāt* meat without waste + *wurst* sausage] (circa 1888)
: fresh pork sausage for frying

braun·schweig·er \'brau̇n-ˌshwī-gər, -ˌshvī- *also* 'brän-\ *noun* [German *Braunschweiger* (*Wurst*), literally, Brunswick sausage] (1934)
: smoked liverwurst

bra·va \'brä-(ˌ)vä, brä-'vä\ *noun* [Italian, feminine of *bravo*] (1803)
: BRAVO — used interjectionally in applauding a woman

bra·va·do \brə-'vä-(ˌ)dō\ *noun, plural* **-does** *or* **-dos** [Middle French *bravade* & Old Spanish *bravata,* from Old Italian *bravata,* from *bravare* to challenge, show off, from *bravo*] (circa 1580)
1 a : blustering swaggering conduct **b :** a pretense of bravery
2 : the quality or state of being foolhardy

¹brave \'brāv\ *adjective* **brav·er; brav·est** [Middle French, from Old Italian & Old Spanish *bravo* courageous, wild, probably from Latin *barbarus* barbarous] (15th century)
1 : having courage **:** DAUNTLESS
2 : making a fine show **:** COLORFUL ⟨*brave* banners flying in the wind⟩
3 : EXCELLENT, SPLENDID ⟨the *brave* fire I soon had going —J. F. Dobie⟩
— brave·ly *adverb*

²brave *verb* **braved; brav·ing** (1546)
transitive verb
1 : to face or endure with courage
2 *obsolete* **:** to make showy
intransitive verb
archaic **:** to make a brave show
— brav·er *noun*

³brave *noun* (1590)
1 *archaic* **:** BRAVADO
2 : one who is brave; *specifically* **:** an American Indian warrior
3 *archaic* **:** BULLY, ASSASSIN

brave new world *noun* [from the dystopian novel *Brave New World* (1932) by Aldous Huxley] (1933)
: a future world, situation, or development; *also* **:** a recent development or recently changed situation

brav·ery \'brāv-rē, 'brā-və-\ *noun, plural* **-er·ies** (1548)
1 : the quality or state of being brave **:** COURAGE
2 a : fine clothes **b :** showy display

¹bra·vo \'brä-(ˌ)vō\ *noun, plural* **bravos** *or* **bravoes** [Italian, from *bravo* brave] (1597)
: VILLAIN, DESPERADO; *especially* **:** a hired assassin

²bra·vo \'brä-(ˌ)vō, brä-'vō\ *noun, plural* **bravos** (1761)
: a shout of approval — often used interjectionally in applauding a performance

³bra·vo *see* ²\ *transitive verb* **bra·voed; bra·vo·ing** (1831)
: to applaud by shouts of *bravo*

Bra·vo \'brä-(ˌ)vō\ (1952)
— a communications code word for the letter *b*

¹bra·vu·ra \brə-'vyu̇r-ə, brä-, -'vu̇r-\ *noun* [Italian, literally, bravery, from *bravare*] (1757)
1 : a musical passage requiring exceptional agility and technical skill in execution
2 : a florid brilliant style
3 : a show of daring or brilliance

²bravura *adjective* (1920)
1 : marked by an ostentatious display of skill
2 : ORNATE, SHOWY

braw \'bró, 'brä\ *adjective* [modification of Middle French *brave*] (circa 1565)
1 *chiefly Scottish* **:** GOOD, FINE
2 *chiefly Scottish* **:** well dressed

¹brawl \'bról\ *intransitive verb* [Middle English] (14th century)
1 : to quarrel or fight noisily **:** WRANGLE
2 : to make a loud confused noise
— brawl·er *noun*

²brawl *noun* (15th century)
1 : a noisy quarrel or fight
2 : a loud tumultuous noise

brawly \'bró-lē\ *adjective* **brawl·i·er; -est** (1940)
1 : inclined to brawl
2 : characterized by brawls or brawling

brawn \'brón\ *noun* [Middle English, from Middle French *braon* muscle, of Germanic origin; akin to Old English *brǣd* flesh] (13th century)
1 a *British* **:** the flesh of a boar **b :** HEADCHEESE
2 a : full strong muscles **b :** muscular strength

brawny \'bró-nē\ *adjective* **brawn·i·er; -est** (1599)
1 : MUSCULAR; *also* **:** STRONG, POWERFUL
2 : being swollen and hard ⟨a *brawny* infected foot⟩
— brawn·i·ly \-nə-lē\ *adverb*
— brawn·i·ness \-nē-nəs\ *noun*

¹bray \'brā\ *verb* [Middle English, from Middle French *braire* to cry, from (assumed) Vulgar Latin *bragere,* of Celtic origin; akin to Old Irish *braigid* he breaks wind] (14th century)
intransitive verb
: to utter the characteristic loud harsh cry of a donkey; *also* **:** to utter a sound like a donkey's
transitive verb
: to utter or play loudly, harshly, or discordantly
— bray *noun*

²bray *transitive verb* [Middle English, from Anglo-French *braier,* from Old French *breier,* of Germanic origin; akin to Old High German *brehhan* to break — more at BREAK] (14th century)
1 : to crush or grind fine ⟨*bray* seeds in a mortar⟩
2 : to spread thin ⟨*bray* printing ink⟩

bray·er \'brā-ər\ *noun* (1688)
: a printer's hand inking roller

¹braze \'brāz\ *transitive verb* **brazed; braz·ing** [irregular from *brass*] (1602)
archaic **:** HARDEN

²braze *transitive verb* **brazed; braz·ing** [French *braser,* from Old French, to burn, from *breze* hot coals — more at BRAISE] (1677)
: to solder with a nonferrous alloy that melts at a lower temperature than that of the metals being joined
— braz·er *noun*

¹bra·zen \'brā-zᵊn\ *adjective* [Middle English *brasen,* from Old English *bræsen,* from *bræs* brass] (before 12th century)
1 : made of brass
2 a : sounding harsh and loud like struck brass **b :** of the color of polished brass
3 : marked by contemptuous boldness
— bra·zen·ly *adverb*
— bra·zen·ness \'brā-zᵊn-(n)əs\ *noun*

²brazen *transitive verb* **bra·zened; bra·zen·ing** \'brāz-niŋ, 'brā-zᵊn-iŋ\ (circa 1555)
: to face with defiance or impudence — usually used in the phrase *brazen it out*

bra·zen–faced \'brā-zᵊn-,fāst\ *adjective* (1571)
: marked by insolence and bold disrespect ⟨*brazen-faced* assertions⟩

¹bra·zier \'brā-zhər\ *noun* [Middle English *brasier,* from *bras* brass] (14th century)
: one that works in brass

²brazier *noun* [French *brasier,* from Old French, fire of hot coals, from *breze*] (circa 1690)
1 : a pan for holding burning coals
2 : a utensil in which food is exposed to heat through a wire grill

Bra·zil nut \brə-'zil-\ *noun* [*Brazil,* South America] (1830)
: a tall South American tree (*Bertholletia excelsa* of the family Lecythidaceae) that bears large globular capsules each containing several closely packed roughly triangular oily edible nuts; *also* **:** its nut

bra·zil·wood \brə-'zil-,wu̇d\ *noun* [Spanish *brasil,* from *brasa* live coals (from the wood's color), probably of Germanic origin; akin to Swedish *brasa* fire] (1559)
: the heavy wood of any of various tropical leguminous trees (especially genus *Caesalpinia*) that is used as red and purple dyewood and in cabinetwork

¹breach \'brēch\ *noun* [Middle English *breche,* from Old English *brǣc* act of breaking; akin to Old English *brecan* to break] (before 12th century)
1 : infraction or violation of a law, obligation, tie, or standard
2 a : a broken, ruptured, or torn condition or area **b :** a gap (as in a wall) made by battering
3 a : a break in accustomed friendly relations **b :** a temporary gap in continuity **:** HIATUS
4 : a leap especially of a whale out of water

²breach (1547)
transitive verb
1 : to make a breach in
2 : BREAK, VIOLATE ⟨*breach* an agreement⟩
intransitive verb
: to leap out of water ⟨a whale *breaching*⟩

breach of promise (1590)
: violation of a promise especially to marry

¹bread \'bred\ *noun* [Middle English *breed,* from Old English *brēad;* akin to Old High German *brōt* bread, Old English *brēowan* to brew] (before 12th century)
1 : a usually baked and leavened food made of a mixture whose basic constituent is flour or meal
2 : FOOD, SUSTENANCE ⟨our daily *bread*⟩
3 a : LIVELIHOOD ⟨earns his *bread* as a laborer⟩ **b** *slang* **:** MONEY
— bready \'bre-dē\ *adjective*
— bread upon the waters : resources risked or charitable deeds performed without expectation of return

²bread *transitive verb* (1629)
: to cover with bread crumbs ⟨a *breaded* pork chop⟩

bread–and–butter *adjective* (circa 1837)
1 a : being as basic as the earning of one's livelihood ⟨*bread-and-butter* issues⟩ **b :** that can be depended on ⟨our *bread-and-butter* repertoire⟩
2 : sent or given as thanks for hospitality ⟨a *bread-and-butter* letter⟩

bread and butter *noun* (1732)
: a means of sustenance or livelihood

bread and circuses *noun plural* [translation of Latin *panis et circenses*] (1914)
: a palliative offered especially to avert potential discontent

bread·bas·ket \'bred-,bas-kət\ *noun* (1753)
1 *slang* **:** STOMACH
2 : a major cereal-producing region

¹bread·board \'bred-,bōrd, -,bȯrd\ *noun* (1857)
1 : a board on which dough is kneaded or bread cut
2 : a board on which components are mounted for breadboarding

²breadboard *transitive verb* (1956)
: to make an experimental arrangement of (as an electronic circuit or a mechanical system) to test feasibility

bread·fruit \'bred-,früt\ *noun* (1697)
: a round starchy usually seedless fruit that resembles bread in color and texture when baked; *also* **:** a tall tropical evergreen tree (*Artocarpus altilis*) of the mulberry family that bears this fruit

bread·line \-,līn\ *noun* (1900)
: a line of people waiting to receive free food

bread mold *noun* (1914)
: any of various molds found especially on bread; *especially* **:** a rhizopus (*Rhizopus nigricans* synonym *R. stolonifer*)

bread·stuff \-,stəf\ *noun* (1793)
1 : a cereal product (as grain or flour)
2 : BREAD

breadth \'bretth, 'bredth, ÷'breth\ *noun* [obsolete English *brede* breadth (from Middle English, from Old English *brǣdu,* from *brād* broad) + *-th* (as in *length*)] (15th century)
1 : distance from side to side **:** WIDTH
2 : something of full width
3 a : comprehensive quality **:** SCOPE ⟨*breadth* of his learning⟩ **b :** liberality of views or taste ⟨*breadth* of mind⟩
— breadth·wise \-,wīz\ *adverb or adjective*

bread·win·ner \'bred-,wi-nər\ *noun* (1818)
1 : a means (as a tool or craft) of livelihood
2 : a member of a family whose wages supply its livelihood
— bread·win·ning \-,wi-niŋ\ *noun*

¹break \'brāk\ *verb* **broke** \'brōk\; **bro·ken** \'brō-kən\; **break·ing** [Middle English *breken,* from Old English *brecan;* akin to Old High German *brehhan* to break, Latin *frangere*] (before 12th century)
transitive verb
1 a : to separate into parts with suddenness or violence **b :** FRACTURE ⟨*break* an arm⟩ **c :** RUPTURE ⟨*break* the skin⟩ **d :** to cut into and turn over the surface of ⟨*break* the soil⟩ **e :** to render inoperable ⟨*broke* his watch⟩
2 a : VIOLATE, TRANSGRESS ⟨*break* the law⟩ ⟨*break* a promise⟩ **b :** to invalidate (a will) by action at law
3 a *archaic* **:** to force entry into **b :** to burst and force a way through ⟨*break* the sound barrier⟩ ⟨*break* a racial barrier⟩ **c :** to escape by force from ⟨*break* jail⟩ **d :** to make or effect by cutting, forcing, or pressing through ⟨*break* a trail through the woods⟩
4 : to disrupt the order or compactness of ⟨*break* ranks⟩
5 : to make ineffective as a binding force ⟨*break* the spell⟩
6 a : to defeat utterly and end as an effective force **:** DESTROY **b :** to crush the spirit of **c :** to make tractable or submissive: as (1) *past participle often* **broke :** to train (an animal) to adjust to the service or convenience of humans ⟨a halter-*broke* horse⟩ (2) **:** INURE, ACCUSTOM **d :** to exhaust in health, strength, or capacity
7 a : to stop or bring to an end suddenly **:** HALT ⟨*break* a deadlock⟩ **b :** INTERRUPT, SUSPEND ⟨*break* the silence with a cry⟩ **c :** to open and bring about suspension of operation ⟨*break* an electric circuit⟩ **d :** to destroy unity or completeness of ⟨*break* a dining room set by buying a chair⟩ **e :** to change the appearance of uniformity of ⟨a dormer *breaks* the level roof⟩ **f :** to split the surface of ⟨fish *breaking* water⟩ **g :** to cause to discontinue a habit ⟨tried to *break* him of smoking⟩
8 a : to make known **:** TELL ⟨*break* the bad news gently⟩ **b :** to bring to attention or prominence initially ⟨radio stations *breaking* new musicians⟩ ⟨*break* a news story⟩
9 a : to ruin financially **b :** to reduce in rank
10 a : to split into smaller units, parts, or processes **:** DIVIDE **b** (1) **:** to give or get the equivalent of (a bill) in smaller denominations (2) **:** to use as the denomination in paying a bill ⟨didn't want to *break* a $20 bill⟩ — often used with *into, up,* or *down*
11 a : to check the speed, force, or intensity of ⟨the bushes will *break* his fall⟩ ⟨without *breaking* her stride⟩ **b :** to cause failure and discontinuance of (a strike) by measures outside bargaining processes
12 : to cause a sudden significant decrease in the price, value, or volume of ⟨news likely to *break* the market sharply⟩
13 a : EXCEED, SURPASS ⟨*break* the record⟩ **b :** to score less than (a specified total) ⟨golfer trying to *break* 90⟩ **c :** to win against (an opponent's service) in a racket game
14 : to open the action of (a breechloader)
15 a : to find an explanation or solution for **:** SOLVE ⟨the detective will *break* the case⟩ **b :** to discover the essentials of (a code or cipher system)
16 : to demonstrate the falsity of ⟨*break* an alibi⟩
17 : to ruin the prospects of ⟨could make or *break* her career⟩
18 : to produce visibly ⟨barely *breaks* a sweat⟩
intransitive verb
1 a : to escape with sudden forceful effort — often used with *out* ⟨*break* out of jail⟩ **b :** to come into being by or as if by bursting forth ⟨day was *breaking*⟩ **c :** to effect a penetration ⟨*break* through security lines⟩ **d :** to emerge through the surface of the water **e :** to start abruptly ⟨when the storm *broke*⟩ **f :** to become known or published ⟨when the news *broke*⟩ **g :** to make a sudden dash ⟨*break* for cover⟩ **h :** to separate after a clinch in boxing
2 a : to come apart or split into pieces **:** BURST, SHATTER **b :** to open spontaneously or by pressure from within ⟨his boil finally *broke*⟩ **c** *of a wave* **:** to curl over and fall apart in surf or foam

3 : to interrupt one's activity or occupation for a brief period ⟨*break* for lunch⟩
4 : to alter sharply in tone, pitch, or intensity ⟨a voice *breaking* with emotion⟩
5 : to become fair **:** CLEAR ⟨when the weather *breaks*⟩
6 : to make the opening shot of a game of pool
7 : to end a relationship, connection, or agreement — usually used with *with*
8 : to give way in disorderly retreat
9 a : to swerve suddenly **b :** to curve, drop, or rise sharply ⟨a pitch that *breaks* away from the batter⟩
10 a : to fail in health, strength, vitality, resolve, or control ⟨may *break* under questioning⟩ — often used with *down* **b :** to become inoperative because of damage, wear, or strain
11 : to fail to keep a prescribed gait — used of a horse
12 : to undergo a sudden significant decrease in price, value, or volume ⟨transportation stocks may *break* sharply⟩
13 : HAPPEN, DEVELOP ⟨for the team to succeed, everything has to *break* right⟩
14 : to win against an opponent's service in a racket game
15 a : to divide into classes, categories, or types — usually used with *into* and often with *down* ⟨the topic *breaks* down into three questions⟩ **b :** to fold, bend, lift, or come apart at a seam, groove, or joint **c** *of cream* **:** to separate during churning into liquid and fat
— break a leg — used to wish good luck especially to a performer
— break bread : to dine together
— break camp : to pack up gear and leave a camp or campsite
— break cover *also* **break covert :** to start from a covert or lair
— break even : to achieve a balance; *especially* **:** to operate a business or enterprise without either loss or profit
— break ground 1 : to begin construction — usually used with *for* **2** *or* **break new ground :** to make or show discoveries **:** PIONEER
— break into 1 : to begin with or as if with a sudden throwing off of restraint ⟨*broke into* tears⟩ ⟨face *breaking into* a smile⟩ ⟨the horse *breaks into* a gallop⟩ **2 :** to make entry or entrance ⟨*broke into* the house⟩ ⟨*break into* show business⟩ **3 :** INTERRUPT ⟨*break into* a TV program with a news flash⟩
— break one's heart : to crush emotionally with sorrow
— break one's wrists : to turn the wrists as part of the swing of a club or bat
— break ranks *also* **break rank :** to differ in opinion or action from one's peers; often used with *with*
— break the back of : to subdue the main force of ⟨*break the back of* inflation⟩
— break the ice 1 : to make a beginning **2 :** to get through the first difficulties in starting a conversation or discussion
— break wind : to expel gas from the intestine

²break *noun* (14th century)
1 a : an act or action of breaking **b :** the opening shot in a game of pool or billiards
2 a : a condition produced by or as if by breaking **:** GAP ⟨a *break* in the clouds⟩ **b :** a gap in an otherwise continuous electric circuit
3 : the action or act of breaking in, out, or forth ⟨at *break* of day⟩ ⟨a jail *break*⟩
4 : a place or situation at which a break occurs; *especially* **:** the place at which a word is divided especially at the end of a line of print or writing
5 : an interruption in continuity ⟨a *break* in the weather⟩: as **a :** a notable change of subject matter, attitude, or treatment **b** (1) **:** an abrupt, significant, or noteworthy change or interruption in a continuous process, trend, or surface (2) **:** a respite from work, school, or duty ⟨coffee *break*⟩ ⟨spring *break*⟩ (3) **:** a planned interruption in a radio or television program ⟨a *break* for the commercial⟩ **c :** deviation of a pitched baseball from a straight line **d** *mining* **:** FAULT, DISLOCATION **e :** failure of a horse to maintain the prescribed gait **f :** an abrupt change in musical or vocal pitch or quality **g :** the action or an instance of breaking service in a racket game **h :** a usually solo instrumental passage in jazz, folk, or popular music
6 a : DASH, RUSH ⟨a base runner making a *break* for home⟩ **b :** FAST BREAK
7 : a sudden and abrupt decline of prices or values
8 a : the start of a race **b :** the act of separating after a clinch in boxing
9 a : a stroke of luck and especially of good luck ⟨a bad *break*⟩ ⟨got the *breaks*⟩ **b :** a favorable or opportune situation **:** CHANCE ⟨waiting for a big *break* in show business⟩ **c :** favorable consideration or treatment ⟨a tax *break*⟩ ⟨a *break* on the price⟩
10 a : a rupture in previously agreeable relations ⟨a *break* between the two countries⟩ **b :** an abrupt split or difference with something previously adhered to or followed ⟨a sharp *break* with tradition⟩
11 : BREAKDOWN 1c ⟨suffered a mental *break*⟩

break·able \'brā-kə-bəl\ *adjective* (1570)
: capable of being broken
— breakable *noun*

break·age \'brā-kij\ *noun* (1813)
1 a : the action or an instance of breaking **b :** a quantity broken
2 : loss due to things broken

¹break·away \'brā-kə-,wā\ *noun* (1881)
1 a : one that breaks away **b :** a departure from or rejection of (as a group or tradition)
2 a : a play (as in hockey) in which an offensive player breaks free of the defenders and rushes toward the goal **b :** a sudden acceleration by one or more bicyclists pulling away from the pack in a race
3 : an object made to shatter or collapse under pressure or impact

²breakaway *adjective* (1927)
1 : favoring independence from an affiliation **:** SECEDING ⟨a *breakaway* faction formed a new party⟩
2 : made to break, shatter, or bend easily ⟨*breakaway* road signs for highway safety⟩

break away *intransitive verb* (1535)
1 : to detach oneself especially from a group **:** get away
2 : to depart from former or accustomed ways
3 : to pull away with a burst of speed

break·bone fever \'brāk-'bōn-\ *noun* (circa 1860)
: DENGUE

break dancing *noun* [perhaps from ²*break* (solo passage)] (1983)
: dancing in which solo dancers perform acrobatics that involve touching various parts of the body (as the back or head) to the ground
— break–dance \'brāk-,dan(t)s\ *verb*
— break–danc·er \-,dan(t)-sər\ *noun*

break·down \'brāk-,daun\ *noun* (1832)
1 : the action or result of breaking down: as **a :** a failure to function **b :** failure to progress or have effect **:** DISINTEGRATION ⟨a *breakdown* of negotiations⟩ **c :** a physical, mental, or nervous collapse **d :** the process of decomposing **e :** division into categories **:** CLASSIFICATION; *also* **:** an account analyzed into categories
2 : a fast shuffling dance; *also* **:** music for such a dance

break down (14th century)
transitive verb
1 a : to cause to fall or collapse by breaking or shattering **b :** to make ineffective ⟨*break down* legal barriers⟩
2 a : to divide into parts or categories **b :** to separate (as a chemical compound) into simpler substances **:** DECOMPOSE **c :** to take apart especially for storage or shipment and for later reassembling
intransitive verb
1 a : to stop functioning because of breakage or wear **b :** to become inoperative or ineffective **:** FAIL ⟨negotiations *broke down*⟩
2 *of horses* **:** to tear or strain the tendons of the lower leg
3 a : to be susceptible to or undergo analysis or subdivision ⟨the statistics *break down* like this⟩ **b :** to undergo decomposition
synonym see ANALYZE

¹break·er \'brā-kər\ *noun* (12th century)
1 a : one that breaks **b :** a machine or plant for breaking rocks or coal **c** *chiefly British* **:** one who breaks up ships or cars for salvage
2 : a wave breaking into foam (as against the shore)
3 : a strip of fabric under the tread of a tire for extra protection of the carcass

²brea·ker \'brā-kər\ *noun* [by folk etymology from Spanish *barrica*] (1833)
: a small water cask

break·even \'brāk-'ē-vən\ *noun* (1958)
: the point at which cost and income are equal and there is neither profit nor loss; *also* **:** a financial result reflecting neither profit nor loss

break–even \'brāk-'ē-vən\ *adjective* (1931)
: having equal cost and income

break·fast \'brek-fəst\ *noun* (15th century)
1 : the first meal of the day especially when taken in the morning
2 : the food prepared for a breakfast ⟨eat your *breakfast*⟩
— breakfast *verb*
— break·fast·er *noun*

break·front \'brāk-,frənt\ *noun* (1928)
: a large cabinet or bookcase whose center section projects beyond the flanking end sections

breakfront

break–in \'brā-,kin\ *noun* (1856)
1 : the act or action of breaking in ⟨a rash of *break-ins* at the new apartment house⟩
2 : a performance or a series of performances serving as a trial run
3 : an initial period of operation during which working parts begin to function efficiently

break in (circa 1552)
intransitive verb
1 : to enter a house or building by force
2 a : INTRUDE ⟨*break in* upon his privacy⟩ **b :** to interrupt a conversation
3 : to start in an activity or enterprise ⟨*breaking in* as a cub reporter⟩
transitive verb
1 : to accustom to a certain activity or occurrence ⟨*break in* the new clerk⟩
2 : to overcome the stiffness or newness of

breaking and entering *noun* (1797)
: the act of forcing or otherwise gaining unlawful passage into and entering another's building

breaking point *noun* (1908)
1 : the point at which a person gives way under stress
2 : the point at which a situation becomes critical

break·neck \'brāk-'nek\ *adjective* (1562)
: very fast or dangerous ⟨*breakneck* speed⟩

break off (14th century)

intransitive verb
: to stop abruptly ⟨*break off* in the middle of a sentence⟩
transitive verb
: DISCONTINUE ⟨*break off* diplomatic relations⟩

break•out \'brā-ˌkau̇t\ *noun* (1820)
: a violent or forceful break from a restraining condition or situation; *especially* **:** a military attack to break from encirclement

break out (before 12th century)
intransitive verb
1 : to develop or emerge with suddenness or force ⟨fire *broke out*⟩ ⟨a riot *broke out*⟩
2 a : to become covered ⟨*break out* in a sweat⟩
b : to become affected with a skin eruption
3 : to make a break from a restraining condition or situation ⟨*broke out* of a slump⟩
transitive verb
1 a : to make ready for action or use ⟨*break out* the tents and make camp⟩ **b :** to produce for consumption ⟨*break out* a bottle⟩
2 : to display flying and unfurled
3 : to separate from a mass of data ⟨*break out* newsstand sales⟩

break•through \'brāk-ˌthrü\ *noun, often attributive* (1918)
1 : an offensive thrust that penetrates and carries beyond a defensive line in warfare
2 : an act or instance of breaking through an obstruction
3 a : a sudden advance especially in knowledge or technique ⟨a medical *breakthrough*⟩ **b :** a person's first notable success

break through *intransitive verb* (1955)
: to make a breakthrough

break•up \'brā-ˌkəp\ *noun* (1794)
1 : an act or instance of breaking up
2 : the breaking, melting, and loosening of ice in the spring

break up (15th century)
intransitive verb
1 a : to cease to exist as a unified whole **:** DISPERSE ⟨their partnership *broke up*⟩ **b :** to end a romance
2 : to lose morale, composure, or resolution; *especially* **:** to become abandoned to laughter ⟨*breaks up* completely, laughing himself into a coughing fit —Gene Williams⟩
transitive verb
1 : to break into pieces
2 : to bring to an end ⟨*broke up* the fight⟩
3 : to do away with **:** DESTROY ⟨*break up* a monopoly⟩
4 : to disrupt the continuity or flow of ⟨*break up* a dull routine⟩
5 : DECOMPOSE ⟨*break up* a chemical⟩
6 : to cause to laugh heartily ⟨that joke *breaks* me *up*⟩

breakup value *noun* (1902)
: the value especially of shares of stock of a corporation liquidating its assets

break•wa•ter \'brāk-ˌwȯ-tər, -ˌwä-\ *noun* (circa 1769)
: an offshore structure (as a wall) protecting a harbor or beach from the force of waves

[1]**bream** \'brim, 'brēm\ *noun, plural* **bream** *or* **breams** [Middle English *breme,* from Middle French, of Germanic origin; akin to Old High German *brahsima* bream, Middle High German *brehen* to shine] (14th century)
1 : a European freshwater cyprinid fish (*Abramis brama*); *broadly* **:** any of various related fishes
2 a : a porgy or related fish (family Sparidae) **b :** any of various freshwater sunfishes (*Lepomis* and related genera); *especially* **:** BLUEGILL

[2]**bream** \'brēm\ *transitive verb* [probably from Dutch *brem* furze] (1626)
: to clean (a ship's bottom) by heating and scraping

[1]**breast** \'brest\ *noun* [Middle English *brest,* from Old English *brēost;* akin to Old High German *brust* breast, Old Irish *brú* belly, Russian *bryukho*] (before 12th century)
1 : either of the pair of mammary glands extending from the front of the chest in pubescent and adult human females and some other mammals; *also* **:** either of the analogous but rudimentary organs of the male chest especially when enlarged
2 a : the fore or ventral part of the body between the neck and the abdomen **b :** the part of an article of clothing covering the breast
3 : the seat of emotion and thought **:** BOSOM
4 a : something (as a front, swelling, or curving part) resembling a breast **b :** FACE 6
— breast•ed \'bres-təd\ *adjective*

[2]**breast** *transitive verb* (1599)
1 : to contend with resolutely **:** CONFRONT ⟨*breasting* the waves⟩
2 *chiefly British* **:** CLIMB, ASCEND
3 : to thrust the chest against ⟨the sprinter *breasted* the tape⟩

breast–beat•ing \'bres(t)-ˌbē-tiŋ\ *noun* (1940)
: noisy demonstrative protestation (as of grief, anger, or self-recrimination)

breast•bone \'bres(t)-ˌbōn\ *noun* (before 12th century)
: STERNUM

breast drill *noun* (1865)
: a portable drill with a plate that is pressed by the breast in forcing the drill against the work

breast–feed \'brest-ˌfēd\ *transitive verb* (1903)
: to feed (a baby) from a mother's breast rather than from a bottle

breast•plate \'bres(t)-ˌplāt\ *noun* (14th century)
1 : a usually metal plate worn as defensive armor for the breast — see ARMOR illustration
2 : a vestment worn in ancient times by a Jewish high priest and set with 12 gems bearing the names of the tribes of Israel
3 : a piece against which a worker's breast is pressed in operating a tool (as a breast drill)

breast•stroke \'bres(t)-ˌstrōk\ *noun* (1867)
: a swimming stroke executed in a prone position by coordinating a kick in which the legs are brought forward with the knees together and the feet are turned outward and whipped back with a glide and a backward sweeping movement of the arms
— breast•strok•er \-ˌstrō-kər\ *noun*

breast•work \'brest-ˌwərk\ *noun* (1642)
: a temporary fortification

breath \'breth\ *noun* [Middle English *breth,* from Old English *brǣth;* akin to Old High German *brādam* breath, and perhaps to Old English *beorma* yeast — more at BARM] (before 12th century)
1 a : air filled with a fragrance or odor **b :** a slight indication **:** SUGGESTION ⟨the faintest *breath* of scandal⟩
2 a : the faculty of breathing ⟨recovering his *breath* after the race⟩ **b :** an act of breathing ⟨fought to the last *breath*⟩ **c :** opportunity or time to breathe **:** RESPITE
3 : a slight breeze
4 a : air inhaled and exhaled in breathing ⟨bad *breath*⟩ **b :** something (as moisture on a cold surface) produced by breath or breathing **c :** INHALATION
5 : a spoken sound **:** UTTERANCE
6 : SPIRIT, ANIMATION
— in one breath *or* **in the same breath :** almost simultaneously
— out of breath : breathing very rapidly (as from strenuous exercise)

breath•able \'brē-thə-bəl\ *adjective* (circa 1731)
1 : suitable for breathing ⟨*breathable* air⟩
2 : allowing air to pass through **:** POROUS ⟨a *breathable* synthetic fabric⟩
— breath•abil•i•ty \ˌbrē-thə-'bi-lə-tē\ *noun*

Breath•a•lyz•er \'bre-thə-ˌlī-zər\ *trademark*
— used for a device that is used to determine the alcohol content of a breath sample

breathe \'brēth\ *verb* **breathed; breath•ing** [Middle English *brethen,* from *breth*] (14th century)
intransitive verb
1 a : to draw air into and expel it from the lungs **:** RESPIRE; *broadly* **:** to take in oxygen and give out carbon dioxide through natural processes **b :** to inhale and exhale freely
2 : LIVE
3 a *obsolete* **:** to emit a fragrance or aura **b :** to become perceptible **:** be expressed ⟨a personality that *breathes* and that distinguishes his work —Bennett Schiff⟩
4 : to pause and rest before continuing
5 : to blow softly
6 : to feel free of restraint ⟨needs room to *breathe*⟩
7 *of wine* **:** to develop flavor and bouquet by exposure to air
8 a : to permit passage of air or vapor ⟨a fabric that *breathes*⟩ **b** *of an internal-combustion engine* **:** to use air to support combustion
transitive verb
1 a : to send out by exhaling **b :** to instill by or as if by breathing ⟨*breathe* new life into the movement⟩
2 : to give rest from exertion to
3 : to take in in breathing ⟨*breathe* the scent of pines⟩
4 : to inhale and exhale ⟨*breathe* air⟩
5 a : UTTER, EXPRESS ⟨don't *breathe* a word of it to anyone⟩ **b :** to make manifest **:** EVINCE ⟨the novel *breathes* despair⟩
— breathe down one's neck 1 : to threaten especially in attack or pursuit **2 :** to keep one under close or constant surveillance ⟨parents always *breathing down his neck*⟩
— breathe easily *or* **breathe freely :** to enjoy relief (as from pressure or danger)

breathed \'bretht\ *adjective* (1555)
1 : having breath especially of a specified kind — usually used in combination ⟨sweet-*breathed*⟩
2 : VOICELESS 2

breath•er \'brē-thər\ *noun* (14th century)
1 : one that breathes
2 : a break in activity for rest or relief
3 : a small vent in an otherwise airtight enclosure

breath•ing \'brē-thiŋ\ *noun* (1746)
: either of the marks ʽ and ʼ used in writing Greek to indicate aspiration or its absence

breathing space *noun* (1650)
: some time in which to recover, get organized, or get going — called also *breathing room, breathing spell*

breath•less \'breth-ləs\ *adjective* (14th century)
1 a : not breathing **b :** DEAD
2 a : panting or gasping for breath **b :** leaving one breathless **:** very rapid or strenuous ⟨go at a *breathless* pace⟩ **c :** holding one's breath from emotion ⟨*breathless* in anticipation⟩ **d :** marked by or as if by a holding of one's breath **:** INTENSE, GRIPPING ⟨*breathless* tension⟩
3 : oppressive because of no fresh air or breeze
— breath•less•ly *adverb*
— breath•less•ness *noun*

breath•tak•ing \'breth-ˌtā-kiŋ\ *adjective* (1880)
1 : making one out of breath
2 a : EXCITING, THRILLING ⟨a *breathtaking* stock car race⟩ **b :** very great **:** ASTONISHING ⟨his *breathtaking* ignorance⟩
— breath•tak•ing•ly \-kiŋ-lē\ *adverb*

breathy \'bre-thē\ *adjective* **breath•i•er; -est** (1883)
: characterized or accompanied by or as if by the audible passage of breath
— breath•i•ly \-thə-lē\ *adverb*
— breath•i•ness \-thē-nəs\ *noun*

brec•cia \'bre-ch(ē-)ə\ *noun* [Italian] (1774)
: a rock consisting of sharp fragments embedded in a fine-grained matrix (as sand or clay)

brec•ci•ate \'bre-chē-ˌāt\ *transitive verb* **-at•ed; -at•ing** (1772)
1 : to form (rock) into breccia
2 : to break (rock) into fragments

— **brec·ci·a·tion** \ˌbre-chē-ˈā-shən\ *noun*

brede \ˈbrēd\ *noun* [variant of *braid*] (1640)
archaic **:** EMBROIDERY

bred–in–the–bone \ˈbre-dᵊn-thə-ˌbōn\ *adjective* (15th century)
1 : DEEP-ROOTED ⟨*bred-in-the-bone* honesty⟩
2 : INVETERATE ⟨a *bred-in-the-bone* gambler⟩

breech \ˈbrēch\ *noun* [Middle English, breeches, from Old English *brēc,* plural of *brōc* leg covering; akin to Old High German *bruoh* breeches, Latin *brāca* pants] (before 12th century)
1 *plural* \ˈbri-chəz *also* ˈbrē-\ **a :** short pants covering the hips and thighs and fitting snugly at the lower edges at or just below the knee **b :** PANTS
2 a : the hind end of the body **:** BUTTOCKS **b :** BREECH PRESENTATION; *also* **:** a fetus that is presented breech first
3 : the part of a firearm at the rear of the barrel
word history see SILLY

breech·block \ˈbrēch-ˌbläk\ *noun* (1881)
: the block in breech-loading firearms that closes the rear of the barrel against the force of the charge and prevents gases from escaping

breech·cloth \ˈbrēch-ˌklȯth, ˈbrich-\ *noun* (1793)
: LOINCLOTH

breech·clout \-ˌklau̇t\ *noun* (1757)
: LOINCLOTH

breech·es buoy \ˈbrē-chəz- *also* ˈbri-\ *noun* (1880)
: a canvas seat in the form of breeches hung from a life buoy running on a hawser and used to haul persons from one ship to another or from ship to shore especially in rescue operations

breech·ing \ˈbrē-chiŋ, ˈbri-\ *noun* (circa 1524)
1 : the part of a harness that passes around the rump of a draft animal
2 : the short coarse wool on the rump and hind legs of a sheep or goat; *also* **:** the hair on the corresponding part of a dog

breech·load·er \ˈbrēch-ˈlō-dər\ *noun* (1858)
: a firearm that loads at the breech
— **breech–load·ing** \-ˈlō-diŋ\ *adjective*

breech presentation *noun* (1811)
: presentation of the fetus in which the breech is the first part to appear at the uterine cervix

¹breed \ˈbrēd\ *verb* **bred** \ˈbred\; **breed·ing** [Middle English *breden,* from Old English *brēdan;* akin to Old English *brōd* brood] (before 12th century)
transitive verb
1 : to produce (offspring) by hatching or gestation
2 a : BEGET 1 **b :** PRODUCE, ENGENDER ⟨despair often *breeds* violence⟩
3 : to propagate (plants or animals) sexually and usually under controlled conditions ⟨*bred* several strains of corn together to produce a superior variety⟩
4 a : BRING UP, NURTURE ⟨born and *bred* in the country⟩ **b :** to inculcate by training ⟨*breed* good manners into one's children⟩
5 a : ⁴MATE 3 **b :** to mate with **:** INSEMINATE **c :** IMPREGNATE 2
6 : to produce (a fissionable element) by bombarding a nonfissionable element with neutrons from a radioactive element
intransitive verb
1 a : to produce offspring by sexual union **b :** COPULATE, MATE
2 : to propagate animals or plants

²breed *noun* (1553)
1 : a group of animals or plants presumably related by descent from common ancestors and visibly similar in most characters; *especially* **:** such a group differentiated from the wild type under domestication
2 : a number of persons of the same stock
3 : CLASS, KIND ⟨a new *breed* of athlete⟩

breed·er \ˈbrē-dər\ *noun* (1531)
: one that breeds: as **a :** an animal or plant kept for propagation **b :** one engaged in the breeding of a specified organism **c :** a nuclear reactor designed to produce more fissionable material than it uses as fuel — called also *breeder reactor*

breed·ing *noun* (14th century)
1 : the action or process of bearing or generating
2 : ANCESTRY
3 a *archaic* **:** EDUCATION ⟨she had her *breeding* at my father's charge —Shakespeare⟩ **b :** training in or observance of the proprieties
4 : the sexual propagation of plants or animals

breeding ground *noun* (1856)
1 : the place to which animals go to breed
2 : a place or set of circumstances suitable for or favorable to growth and development ⟨hurricane *breeding grounds*⟩

breeks \ˈbrēks, ˈbriks\ *noun plural* [Middle English (northern dialect) *breke,* from Old English *brēc*] (14th century)
chiefly Scottish **:** BREECHES

¹breeze \ˈbrēz\ *noun* [probably from Spanish *brisa* northeast wind] (1626)
1 a : a light gentle wind **b :** a wind of from 4 to 31 miles (6 to 50 kilometers) an hour
2 : something easily done **:** CINCH
— **breeze·less** \-ləs\ *adjective*
— **in a breeze :** EASILY

²breeze *intransitive verb* **breezed; breez·ing** (1907)
1 : to move swiftly and airily ⟨*breezed* past the protesters⟩
2 : to make progress quickly and easily ⟨*breezed* through the exam⟩ ⟨*breezed* to victory⟩

³breeze *noun* [probably modification of French *braise* cinders — more at BRAISE] (1726)
: residue from the making of coke or charcoal

breeze·way \ˈbrēz-ˌwā\ *noun* (1931)
: a roofed often open passage connecting two buildings (as a house and garage) or halves of a building

breezy \ˈbrē zē\ *adjective* **breez·i·er; -est** (1718)
1 : swept by breezes
2 a : briskly informal **b :** AIRY, NONCHALANT
— **breez·i·ly** \-zə-lē\ *adverb*
— **breez·i·ness** \-zē-nəs\ *noun*

breg·ma \ˈbreg-mə\ *noun, plural* **-ma·ta** \-mə-tə\ [New Latin *bregmat-, bregma,* from Late Latin, front part of the head, from Greek; akin to Greek *brechmos* front part of the head — more at BRAIN] (1578)
: the point of junction of the coronal and sagittal sutures of the skull

brems·strah·lung \ˈbrem(p)-ˌshträ-ləŋ\ *noun* [German, literally, decelerated radiation] (1939)
: the electromagnetic radiation produced by the sudden retardation of a charged particle in an intense electric field (as of an atomic nucleus); *also* **:** the process that produces such radiation

brent goose \ˈbrent-\ *noun* (1570)
chiefly British **:** BRANT

breth·ren \ˈbreth-rən; ˈbre-thə-, -thərn\ *plural of* BROTHER (before 12th century)
— used chiefly in formal or solemn address or in referring to the members of a profession, society, or sect

Brethren *noun plural* (1822)
: members of various sects originating chiefly in 18th century German Pietism; *especially* **:** DUNKERS

Bret·on \ˈbre-tᵊn\ *noun* [French, from Medieval Latin *Briton-, Brito,* from Latin, Briton] (1653)
1 : a native or inhabitant of Brittany
2 : the Celtic language of the Breton people
— **Breton** *adjective*

breve \ˈbrēv, ˈbrev\ *noun* [Latin, neuter of *brevis* brief — more at BRIEF] (15th century)
1 : a note equivalent to two whole notes
2 : a curved mark ˘ used to indicate a short vowel or a short or unstressed syllable

¹bre·vet \bri-ˈvet, *chiefly British* ˈbre-vit\ *noun* [Middle English, an official message, from Middle French, from Old French, diminutive of *brief* letter — more at BRIEF] (1689)
: a commission giving a military officer higher nominal rank than that for which pay is received

²brevet *transitive verb* **bre·vet·ted** *or* **brev·et·ed; bre·vet·ting** *or* **brev·et·ing** (1824)
: to confer rank upon by brevet

bre·via·ry \ˈbrē-və-rē, -vyə-, -vē-ˌer-ē *also* ˈbre-\ *noun, plural* **-ries** [Middle English *breviarie,* from Medieval Latin *breviarium,* from Latin, summary, from *brevis*] (15th century)
1 *often capitalized* **a :** a book containing the prayers, hymns, psalms, and readings for the canonical hours **b :** DIVINE OFFICE
2 [Latin *breviarium*] **:** a brief summary **:** ABRIDGMENT

brev·i·ty \ˈbre-və-tē\ *noun, plural* **-ties** [Latin *brevitas,* from *brevis*] (15th century)
: shortness of duration; *especially* **:** shortness or conciseness of expression

¹brew \ˈbrü\ *verb* [Middle English, from Old English *brēowan;* akin to Latin *fervēre* to boil — more at BARM] (before 12th century)
transitive verb
1 : to prepare (as beer or ale) by steeping, boiling, and fermentation or by infusion and fermentation
2 a : to bring about **:** FOMENT ⟨*brew* trouble⟩ **b :** CONTRIVE, PLOT
3 : to prepare (as tea) by infusion in hot water
intransitive verb
1 : to brew beer or ale
2 : to be in the process of formation ⟨a storm is *brewing* in the east⟩
— **brew·er** \ˈbrü-ər, ˈbru̇(-ə)r\ *noun*

²brew *noun* (circa 1510)
1 a : a brewed beverage (as beer) **b :** a serving of a brewed beverage ⟨quaff a few *brews*⟩ **c :** something produced by or as if by brewing
2 : the process of brewing

brew·age \ˈbrü-ij\ *noun* (1542)
: BREW 1a, 2

brewer's yeast *noun* (1871)
: a yeast used or suitable for use in brewing; *specifically* **:** the dried pulverized cells of such a yeast (*Saccharomyces cerevisiae*) used especially as a source of B-complex vitamins

brew·ery \ˈbrü-ə-rē, ˈbru̇(-ə)r-ē\ *noun, plural* **-er·ies** (1658)
: a plant where malt liquors are produced

¹bri·ar \ˈbrī(-ə)r\ *variant of* BRIER

²briar *noun* (1882)
: a tobacco pipe made from the root or stem of a brier

bri·ard \brē-ˈär(d)\ *noun* [French, from *Brie,* district in France] (circa 1929)
: any of an old French breed of large long-coated sheepdogs

¹bribe \ˈbrīb\ *noun* [Middle English, something stolen, from Middle French, bread given to a beggar] (15th century)
1 : money or favor given or promised in order to influence the judgment or conduct of a person in a position of trust
2 : something that serves to induce or influence

²bribe *verb* **bribed; brib·ing** (1528)
transitive verb
: to induce or influence by or as if by bribery
intransitive verb
: to practice bribery
— **brib·able** \ˈbrī-bə-bəl\ *adjective*
— **brib·ee** \ˌbrī-ˈbē\ *noun*

— **brib·er** *noun*
brib·ery \'brī-b(ə-)rē\ *noun, plural* **-er·ies** (1549)
: the act or practice of giving or taking a bribe
bric–a–brac \'bri-kə-ˌbrak\ *noun, plural* **bric–a–brac** [French *bric-à-brac*] (1840)
1 : a miscellaneous collection of small articles commonly of ornamental or sentimental value **:** CURIOS
2 : something suggesting bric-a-brac especially in extraneous decorative quality
[1]**brick** \'brik\ *noun, often attributive* [Middle English *bryke,* from Middle French *brique,* from Middle Dutch *bricke*] (15th century)
1 *plural* **bricks** *or* **brick :** a handy-sized unit of building or paving material typically being rectangular and about 2¼ × 3¾ × 8 inches (57 × 95 × 203 millimeters) and of moist clay hardened by heat
2 : a good-hearted person
3 : a rectangular compressed mass (as of ice cream)
4 : a semisoft cheese with numerous small holes, smooth texture, and often mild flavor
5 : GAFFE, BLUNDER — used especially in the phrase *drop a brick*
[2]**brick** *transitive verb* (1592)
: to close, face, or pave with bricks — usually used with *up, in,* or *over*
brick·bat \'brik-ˌbat\ *noun* [*brick* + [1]*bat* (lump, fragment)] (circa 1587)
1 : a fragment of a hard material (as a brick); *especially* **:** one used as a missile
2 : an uncomplimentary remark
brick·field \-ˌfēld\ *noun* (1801)
British **:** BRICKYARD
brick·lay·er \'brik-ˌlā-ər, -ˌle(-ə)r\ *noun* (15th century)
: a person who lays brick
— **brick·lay·ing** \-ˌlā-iŋ\ *noun*
brick·le \'bri-kəl\ *adjective* [Middle English *brekyl*] (13th century)
dialect **:** BRITTLE
brick red *noun* (1810)
: a moderate reddish brown
brick·work \'brik-ˌwərk\ *noun* (1580)
: work of or with bricks and mortar
brick·yard \-ˌyärd\ *noun* (1731)
: a place where bricks are made
bri·co·lage \ˌbrē-kō-'läzh, ˌbri-\ *noun* [French, from *bricoler* to putter about] (1966)
: construction or something constructed by using whatever comes to hand
[1]**brid·al** \'brī-dᵊl\ *noun* [Middle English *bridale,* from Old English *brȳdealu,* from *brȳd* + *ealu* ale — more at ALE] (before 12th century)
: a marriage festival or ceremony ◆
[2]**bridal** *adjective* (13th century)
1 : of or relating to a bride or a wedding **:** NUPTIAL
2 : intended for a newly married couple ⟨a *bridal* suite⟩
bridal wreath *noun* (circa 1889)
: a spirea (*Spiraea prunifolia*) widely grown for its umbels of small white flowers borne in spring
bride \'brīd\ *noun* [Middle English, from Old English *brȳd;* akin to Old High German *brūt* bride] (before 12th century)
: a woman just married or about to be married
bride·groom \'brīd-ˌgrüm, -ˌgru̇m\ *noun* [Middle English (Scots) *brydegrome,* by folk etymology from Middle English *bridegome,* from Old English *brȳdguma,* from *brȳd* + *guma* man; akin to Old High German *brūtgomo* bridegroom — more at HOMAGE] (14th century)
: a man just married or about to be married
bride–price \'brīd-ˌprīs\ *noun* (1876)
: a payment given by or in behalf of a prospective husband to the bride's family in many cultures
brides·maid \'brīdz-ˌmād\ *noun* (1552)
1 : a woman attendant of a bride
2 : one that finishes just behind the winner
bride·well \'brīd-ˌwel, -wəl\ *noun* [*Bridewell,* London jail] (circa 1593)
: PRISON
[1]**bridge** \'brij\ *noun* [Middle English *brigge,* from Old English *brycg;* akin to Old High German *brucka* bridge, Old Church Slavonic *brŭvŭno* beam] (before 12th century)
1 a : a structure carrying a pathway or roadway over a depression or obstacle **b :** a time, place, or means of connection or transition
2 : something resembling a bridge in form or function: as **a :** the upper bony part of the nose; *also* **:** the part of a pair of glasses that rests upon it **b :** a piece raising the strings of a musical instrument — see VIOLIN illustration **c :** the forward part of a ship's superstructure from which the ship is navigated **d :** GANTRY 2b **e :** the hand as a rest for a billiards or pool cue; *also* **:** a device used as a cue rest
3 a : a musical passage linking two sections of a composition **b :** a partial denture anchored to adjacent teeth **c :** a connection (as an atom or group of atoms) that joins two different parts of a molecule (as opposite sides of a ring)
4 : an electrical instrument or network for measuring or comparing resistances, inductances, capacitances, or impedances by comparing the ratio of two opposing voltages to a known ratio
— **bridge·less** \-ləs\ *adjective*

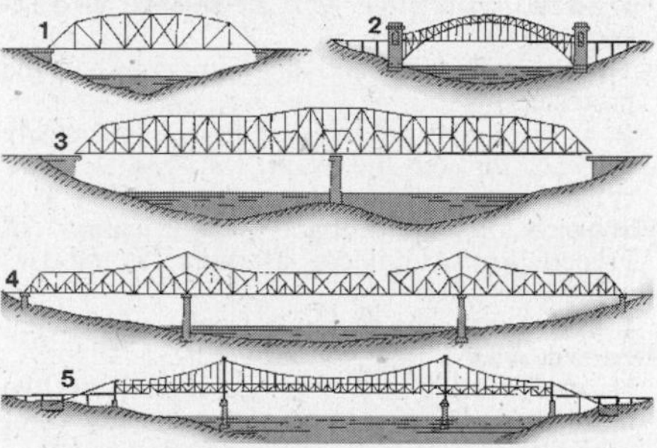

bridge 1a: ***1*** **simple truss,** ***2*** **steel arch,** ***3*** **continuous truss,** ***4*** **cantilever,** ***5*** **suspension**

[2]**bridge** *transitive verb* **bridged; bridg·ing** (before 12th century)
1 : to make a bridge over or across ⟨*bridge* the gap⟩; *also* **:** to join by a bridge
2 : to provide with a bridge
— **bridge·able** \'bri-jə-bəl\ *adjective*
[3]**bridge** *noun* [alteration of earlier *biritch,* of unknown origin] (circa 1897)
: any of various card games for usually four players in two partnerships that bid for the right to declare a trump suit, seek to win tricks equal to the final bid, and play with the hand of declarer's partner exposed and played by declarer; *especially* **:** CONTRACT BRIDGE
bridge·head \-ˌhed\ *noun* (1812)
1 a : a fortification protecting the end of a bridge nearest an enemy **b :** an area around the end of a bridge
2 : an advanced position seized in hostile territory
bridge·work \-ˌwərk\ *noun* (1883)
: dental bridges
[1]**bri·dle** \'brī-dᵊl\ *noun* [Middle English *bridel,* from Old English *brīdel;* akin to Old English *bregdan* to move quickly — more at BRAID] (before 12th century)
1 : the headgear with which a horse is governed and which carries a bit and reins
2 : a length of line or cable attached to two parts of something to spread the force of a pull; *especially* **:** rigging on a kite for attaching line
3 : CURB, RESTRAINT ⟨set a *bridle* on his power⟩
[2]**bridle** *verb* **bri·dled; bri·dling** \'brīd-liŋ, 'brī-dᵊl-iŋ\ (before 12th century)
transitive verb
1 : to put a bridle on
2 : to restrain, check, or control with or as if with a bridle ⟨*bridle* your tongue⟩
intransitive verb
: to show hostility or resentment (as to an affront to one's pride or dignity) especially by drawing back the head and chin
synonym see RESTRAIN
bridle path *noun* (1811)
: a trail suitable for horseback riding
Brie \'brē\ *noun* [French, from *Brie,* district in France] (1876)
: a soft surface-ripened cheese with a whitish rind and a pale yellow interior
[1]**brief** \'brēf\ *adjective* [Middle English *bref, breve,* from Middle French *brief,* from Latin *brevis;* akin to Old High German *murg* short, Greek *brachys*] (14th century)
1 : short in duration, extent, or length
2 a : CONCISE **b :** CURT, ABRUPT
— **brief·ness** *noun*
[2]**brief** *noun* [Middle English *bref,* from Middle French, from Medieval Latin *brevis,* from Late Latin, summary, from Latin *brevis,* adjective] (14th century)
1 : an official letter or mandate; *especially* **:** a papal letter less formal than a bull
2 a : a concise article **b :** SYNOPSIS, SUMMARY **c :** a concise statement of a client's case made out for the instruction of counsel in a trial at law
3 : an outline of an argument; *especially* **:** a formal outline especially in law that sets forth the main contentions with supporting statements or evidence
4 *plural* **:** short snug pants or underpants
— **in brief :** in a few words **:** BRIEFLY
[3]**brief** *transitive verb* (15th century)
1 : to make an abstract or abridgment of
2 a : to give final precise instructions to **b :** to coach thoroughly in advance **c :** to give essential information to
— **brief·er** *noun*
brief·case \'brēf-ˌkās\ *noun* (1917)
: a flat flexible case for carrying papers or books
brief·ing \'brē-fiŋ\ *noun* (1910)
: an act or instance of giving precise instructions or essential information
brief·less \'brē-fləs\ *adjective* (1824)
: having no legal clients
brief·ly \'brē-flē\ *adverb* (14th century)
1 a : in a brief way **b :** in brief
2 : for a short time
[1]**bri·er** \'brī(-ə)r\ *noun* [Middle English *brere,* from Old English *brēr*] (before 12th century)

◇ WORD HISTORY
bridal A secondary sense of Old English *ealu,* the ancestor of Modern English *ale,* was "feast, banquet," at which the drinking of ale was a prominent if not the main activity. The Middle and Modern English descendants of *ealu* preserved this sense, at least into the 19th century, especially in compounds of which the first element named the occasion for the feast. There was the *leet-ale,* held before a session of the leet (a kind of court), the *Whitsunale,* a parish festival held at Whitsuntide, and, incongruous as it may seem, the *church-ale,* a benefit at which ale brewed from malt provided to a parish would be sold to raise money for church upkeep. The oldest and most institutionalized such "ale" was the *bride-ale,* or wedding feast, attested in Old English as *brydealu.* It seems that already in Middle English the *ale* half of the compound had lost its secondary stress, becoming \əl\ or \ələ\. This ending was subsequently associated both with the noun suffix *-al* of *espousal* and *funeral* and with the adjective suffix *-al* of *nuptial, parental,* and *autumnal.* By the 18th century, *bridal* was perceived primarily as an adjective, which is its most common use today.

: a plant (as of the genera *Rosa, Rubus,* and *Smilax*) with a woody and thorny or prickly stem; *also* **:** a mass or twig of these
— **bri•ery** \'brī(-ə)r-ē\ *adjective*

[2]**brier** *noun* [French *bruyère* heath, from Middle French *bruiere,* from (assumed) Vulgar Latin *brucaria,* from Late Latin *brucus* heather, of Celtic origin; akin to Old Irish *froech* heather; akin to Greek *ereikē* heather] (1868)
: a heath (*Erica arborea*) of southern Europe whose roots and knotted stems are used for making briar tobacco pipes

[1]**brig** \'brig\ *noun* [short for *brigantine*] (1712)
: a 2-masted square-rigged ship

brig

[2]**brig** *noun* [probably from [1]*brig*] (1852)
1 : a place (as on a ship) for temporary confinement of offenders in the U.S. Navy
2 : GUARDHOUSE, PRISON

[1]**bri•gade** \bri-'gād\ *noun* [French, from Italian *brigata,* from *brigare*] (1637)
1 a : a large body of troops **b :** a tactical and administrative unit composed of a headquarters, one or more units of infantry or armor, and supporting units
2 : a group of people organized for special activity

[2]**brigade** *transitive verb* **bri•gad•ed; bri•gad•ing** (1781)
: to form or unite into a brigade

brig•a•dier *noun* \,bri-gə-'dir\ [French, from *brigade*] (1678)
1 : an officer in the British army commanding a brigade and ranking immediately below a major general
2 : BRIGADIER GENERAL

brigadier general *noun* (1690)
: a commissioned officer in the army, air force, or marine corps who ranks above a colonel and whose insignia is one star

Brig•a•doon \,bri-gə-'dün\ *noun* [from *Brigadoon,* village in the musical *Brigadoon* (1947) by A. J. Lerner and F. Loewe] (1968)
: a place that is idyllic, unaffected by time, or remote from reality

brig•and \'bri-gənd\ *noun* [Middle English *brigaunt,* from Middle French *brigand,* from Old Italian *brigante,* from *brigare* to fight, from *briga* strife, of Celtic origin; akin to Old Irish *bríg* strength] (14th century)
: one who lives by plunder usually as a member of a band **:** BANDIT
— **brig•and•age** \-gən-dij\ *noun*

brig•an•dine \'bri-gən-,dēn\ *noun* [Middle English, from Middle French, from *brigand*] (15th century)
: medieval body armor of scales or plates

brig•an•tine \'bri-gən-,tēn\ *noun* [Middle French *brigantin,* from Old Italian *brigantino,* from *brigante*] (1525)
: a 2-masted sailing ship that is square rigged except for a fore-and-aft mainsail

[1]**bright** \'brīt\ *adjective* [Middle English, from Old English *beorht;* akin to Old High German *beraht* bright, Sanskrit *bhrājate* it shines] (before 12th century)
1 a : radiating or reflecting light **:** SHINING, SPARKLING ⟨*bright* lights⟩ ⟨*bright* eyes⟩ **b :** SUNNY ⟨a *bright* day⟩; *also* **:** radiant with happiness ⟨*bright* smiling faces⟩ ⟨*bright* moments⟩
2 : ILLUSTRIOUS, GLORIOUS ⟨*brightest* star of the opera⟩
3 : BEAUTIFUL
4 : of high saturation or brilliance ⟨*bright* colors⟩
5 a : LIVELY, CHEERFUL ⟨be *bright* and jovial among your guests —Shakespeare⟩ **b :** INTELLIGENT, CLEVER ⟨a *bright* idea⟩ ⟨*bright* children⟩
6 : AUSPICIOUS, PROMISING ⟨*bright* prospects for the future⟩ ☆
— **bright** *adverb*
— **bright•ly** *adverb*

[2]**bright** *noun* (1969)
: a bright color — usually used in plural ⟨rich earth tones and crisp *brights* —Patricia Peterson⟩

bright•en \'brī-tᵊn\ *verb* **bright•ened; bright•en•ing** \'brīt-niŋ, 'brī-tᵊn-iŋ\ (14th century)
intransitive verb
: to become bright or brighter
transitive verb
: to make bright or brighter
— **bright•en•er** \'brīt-nər, 'brī-tᵊn-ər\ *noun*

bright•ness *noun* (before 12th century)
1 a : the quality or state of being bright; *also* **:** an instance of such a quality or state **b :** LUMINANCE
2 : the attribute of light-source colors by which emitted light is ordered continuously from light to dark in correlation with its intensity — compare HUE 2c, [1]LIGHTNESS 2, SATURATION 4

Bright's disease \'brīts-\ *noun* [Richard *Bright* (died 1858) English physician] (1831)
: any of several kidney diseases marked especially by albumin in the urine

bright•work \'brīt-,wərk\ *noun* (1841)
: polished or plated metalwork

brill \'bril\ *noun, plural* **brill** [Middle English *brell*] (15th century)
: a European flatfish (*Scophthalmus rhombus* synonym *Bothus rhombus* of the family Bothidae); *broadly* **:** TURBOT

bril•liance \'bril-yən(t)s\ *noun* (1755)
: the quality or state of being brilliant

bril•lian•cy \-yən(t)-sē\ *noun, plural* **-cies** (1747)
1 : BRILLIANCE
2 : an instance of brilliance

[1]**bril•liant** \'bril-yənt\ *adjective* [French *brillant,* present participle of *briller* to shine, from Italian *brillare*] (circa 1681)
1 : very bright **:** GLITTERING ⟨a *brilliant* light⟩
2 a : STRIKING, DISTINCTIVE ⟨a *brilliant* example⟩ **b :** distinguished by unusual mental keenness or alertness
synonym see BRIGHT
— **bril•liant•ly** *adverb*

[2]**brilliant** *noun* (1690)
: a gem (as a diamond) cut in a particular form with numerous facets so as to have special brilliance

bril•lian•tine \'bril-yən-,tēn\ *noun* (1873)
1 : a light lustrous fabric that is similar to alpaca and is woven usually with a cotton warp and mohair or worsted filling
2 : a preparation for making hair glossy

[1]**brim** \'brim\ *noun* [Middle English *brimme;* akin to Middle High German *brem* edge] (13th century)
1 a (1) **:** an upper or outer margin **:** VERGE (2) *archaic* **:** the upper surface of a body of water **b :** the edge or rim of a hollow vessel, a natural depression, or a cavity
2 : the projecting rim of a hat
— **brim•less** \-ləs\ *adjective*

[2]**brim** *verb* **brimmed; brim•ming** (1611)
transitive verb
: to fill to the brim
intransitive verb
1 : to be or become full often to overflowing
2 : to reach or overflow a brim

brim•ful \'brim-'fu̇l, -,fu̇l\ *adjective* (circa 1530)
: full to the brim **:** ready to overflow

brimmed \'brimd\ *adjective* (1606)
: having a brim of a specified nature — used in combination ⟨a wide-*brimmed* hat⟩

brim•mer \'bri-mər\ *noun* (1663)
: a brimming cup or glass

brim•stone \'brim-,stōn\ *noun* [Middle English *brinston,* probably from *birnen* to burn + *ston* stone] (12th century)
: SULFUR

brind•ed \'brin-dəd\ *adjective* [Middle English *brended*] (15th century)
archaic **:** BRINDLED

brin•dle \'brin-dᵊl\ *noun* [*brindle,* adjective] (1696)
1 : a brindled color
2 : a brindled animal

brin•dled \-dᵊld\ *or* **brindle** *adjective* [alteration of *brinded*] (1679)
: having obscure dark streaks or flecks on a gray or tawny ground

[1]**brine** \'brīn\ *noun* [Middle English, from Old English *brȳne;* akin to Middle Dutch *brīne* brine] (before 12th century)
1 a : water saturated or strongly impregnated with common salt **b :** a strong saline solution (as of calcium chloride)
2 : the water of a sea or salt lake

[2]**brine** *transitive verb* **brined; brin•ing** (1552)
: to treat (as by steeping) with brine
— **brin•er** *noun*

Bri•nell hardness \brə-'nel-\ *noun* [Johann A. *Brinell* (died 1925) Swedish engineer] (1915)
: the hardness of a metal or alloy measured by hydraulically pressing a hard ball under a standard load into the specimen

Brinell hardness number *noun* (1915)
: a number expressing Brinell hardness and denoting the load applied in testing in kilograms divided by the spherical area of indentation produced in the specimen in square millimeters — called also *Brinell number*

brine shrimp *noun* (1836)
: any of a genus (*Artemia*) of branchiopod crustaceans that can exist in strongly saline environments

bring \'briŋ\ *verb* **brought** \'bro̊t\; **bring•ing** \'briŋ-iŋ\ [Middle English, from Old English *bringan,* akin to Old High German *bringan* to bring, Welsh he*brwng* to accompany] (before 12th century)
transitive verb
1 a : to convey, lead, carry, or cause to come along with one toward the place from which the action is being regarded **b :** to cause to be, act, or move in a special way: as (1) **:** ATTRACT ⟨her screams *brought* the neighbors⟩ (2) **:** PERSUADE, INDUCE (3) **:** FORCE, COMPEL (4) **:** to cause to come into a particular state or condition ⟨*bring* water to a boil⟩ **c** *dialect* **:** ESCORT, ACCOMPANY
2 : to cause to exist or occur: as **a :** to be the occasion of ⟨winter *brings* snow⟩ **b :** to result in **c :** INSTITUTE ⟨*bring* legal action⟩ **d :** ADDUCE ⟨*bring* an argument⟩
3 : PREFER ⟨*bring* charges⟩
4 : to procure in exchange **:** sell for

☆ SYNONYMS
Bright, brilliant, radiant, luminous, lustrous mean shining or glowing with light. BRIGHT implies emitting or reflecting a high degree of light. BRILLIANT implies intense often sparkling brightness. RADIANT stresses the emission or seeming emission of rays of light. LUMINOUS implies emission of steady, suffused, glowing light by reflection or in surrounding darkness. LUSTROUS stresses an even, rich light from a surface that reflects brightly without sparkling or glittering.

intransitive verb
chiefly Midland **:** YIELD, PRODUCE ■
— **bring·er** *noun*
— **bring forth 1 :** BEAR ⟨*brought forth* fruit⟩ **2 :** to give birth to **:** PRODUCE **3 :** ADDUCE ⟨*bring forth* persuasive arguments⟩
— **bring forward 1 :** to produce to view **:** INTRODUCE ⟨*brought* new evidence *forward*⟩ **2 :** to carry (a total) forward
— **bring home :** to make unmistakably clear
— **bring to account 1 :** to bring to book **2 :** REPRIMAND
— **bring to bear :** to use with effect ⟨*bring* pressure *to bear*⟩
— **bring to book :** to compel to give an account
— **bring to light :** DISCLOSE, REVEAL
— **bring to mind :** RECALL
— **bring to terms :** to compel to agree, assent, or submit
— **bring up the rear :** to come last or behind

bring about *transitive verb* (14th century)
: to cause to take place **:** EFFECT

bring around *transitive verb* (1862)
1 : to restore to consciousness **:** REVIVE
2 : PERSUADE

bring·down \'briŋ-ˌdau̇n\ *noun* (circa 1944)
: COMEDOWN, LETDOWN

bring down *transitive verb* (14th century)
1 : to cause to fall by or as if by shooting
2 : to carry (a total) forward
— **bring down the house :** to win the enthusiastic approval of the audience

bring in *transitive verb* (14th century)
1 : INCLUDE, INTRODUCE
2 : to produce as profit or return ⟨each sale *brought in* $5⟩
3 : to enable (a base runner) to reach home plate by hitting the ball
4 : to report to a court ⟨the jury *brought in* a verdict⟩
5 a : to cause (as an oil well) to be productive **b :** to win tricks with the cards of (a long suit) in bridge
6 : EARN ⟨*brings in* a good salary⟩

bring off *transitive verb* (1606)
1 : to cause to escape **:** RESCUE
2 : to carry to a successful conclusion **:** ACHIEVE, ACCOMPLISH

bring on *transitive verb* (1602)
: to cause to appear or occur

bring out *transitive verb* (1605)
1 a : to make apparent **b :** to effectively develop (as a quality)
2 a : to present to the public **b :** to introduce formally to society
3 : UTTER

bring to *transitive verb* (1753)
1 : to cause (a boat) to lie to or come to a standstill
2 : to restore to consciousness **:** REVIVE

bring up (14th century)
transitive verb
1 : to bring (a person) to maturity through nurturing care and education
2 : to cause to stop suddenly
3 : to bring to attention **:** INTRODUCE
4 : VOMIT
intransitive verb
: to stop suddenly

brink \'briŋk\ *noun* [Middle English, of Scandinavian origin; akin to Old Norse *brekka* slope; akin to Middle Dutch *brink* grassland] (13th century)
1 : EDGE; *especially* **:** the edge at the top of a steep place
2 : a bank especially of a river
3 : the point of onset **:** VERGE ⟨on the *brink* of war⟩
4 : the threshold of danger

brink·man·ship \'briŋk-mən-ˌship\ *also* **brinks·man·ship** \'briŋ(k)s-mən-\ *noun* (1956)
: the art or practice of pushing a dangerous situation or confrontation to the limit of safety especially to force a desired outcome
word history see GAMESMANSHIP

briny \'brī-nē\ *adjective* **brin·i·er; -est** (1590)
: of, relating to, or resembling brine or the sea **:** SALTY
— **brin·i·ness** *noun*

brio \'brē-(ˌ)ō\ *noun* [Italian] (1734)
: enthusiastic vigor **:** VIVACITY, VERVE

bri·oche \brē-'ōsh, -'ȯsh\ *noun* [French, from Middle French dialect, from *brier* to knead, of Germanic origin; akin to Old High German *brehhan* to break — more at BREAK] (1826)
: light slightly sweet bread made with a rich yeast dough

bri·o·lette \ˌbrē-ə-'let\ *noun* [French] (1865)
: an oval or pear-shaped gemstone cut in triangular facets

bri·quette *or* **bri·quet** \bri-'ket\ *noun* [French *briquette,* diminutive of *brique* brick] (1883)
: a compacted often brick-shaped mass of usually fine material ⟨a charcoal *briquette*⟩
— **briquette** *transitive verb*

bri·sance \bri-'zän(t)s, -'zäⁿs\ *noun* [French, from *brisant,* present participle of *briser* to break, from Old French *brisier,* of Celtic origin; akin to Old Irish *brisid* he breaks; perhaps akin to Latin *fricare* to rub — more at FRICTION] (1915)
: the shattering or crushing effect of an explosive
— **bri·sant** \-'zänt, -'zäⁿ\ *adjective*

¹brisk \'brisk\ *adjective* [probably modification of Middle French *brusque*] (1560)
1 : keenly alert **:** LIVELY
2 a : pleasingly tangy ⟨*brisk* tea⟩ **b :** FRESH, INVIGORATING ⟨*brisk* weather⟩
3 : sharp in tone or manner
4 : ENERGETIC, QUICK ⟨a *brisk* pace⟩
— **brisk·ly** *adverb*
— **brisk·ness** *noun*

²brisk (1628)
transitive verb
: to make brisk
intransitive verb
: to become brisk — usually used with *up* ⟨business *brisked* up⟩

bris·ket \'bris-kət\ *noun* [Middle English *brusket;* akin to Old English *brēost* breast] (14th century)
: the breast or lower chest of a quadruped animal; *also* **:** a cut of beef from the brisket — see BEEF illustration

bris·ling \'briz-liŋ, 'bris-\ *noun* [Norwegian *brisling,* from Low German *bretling,* from *bret* broad; akin to Old English *brād* broad] (circa 1868)
: SPRAT 1a

¹bris·tle \'bri-səl\ *noun* [Middle English *bristil,* from *brust* bristle, from Old English *byrst;* akin to Old High German *burst* bristle, and perhaps to Latin *fastigium* top] (14th century)
: a short stiff coarse hair or filament
— **bris·tle·like** \'bri-sə(l)-ˌlīk\ *adjective*

²bristle *verb* **bris·tled; bris·tling** \'bris-liŋ, 'bri-sə-\ (15th century)
transitive verb
1 : to furnish with bristles
2 : to make bristly **:** RUFFLE
intransitive verb
1 a : to rise and stand stiffly erect ⟨quills *bristling*⟩ **b :** to raise the bristles (as in anger)
2 : to take on an aggressive attitude or appearance (as in response to a slight)
3 a : to be full of or covered with especially something suggestive of bristles ⟨roofs *bristled* with chimneys⟩ **b :** to be full of something specified ⟨book *bristles* with detail and irony —W. J. Broad⟩

bris·tle·cone pine \'bri-səl-ˌkōn-\ *noun* (1894)
: either of two pines (*Pinus longaeva* and *P. aristata*) of the western U.S. that include the oldest living trees

bristlecone pine

bris·tle·tail \-ˌtāl\ *noun* (1706)
: any of an order (Thysanura) of wingless insects with two or three slender caudal bristles

bris·tly \'bris-lē, 'bri-sə-\ *adjective* **bris·tli·er; -est** (circa 1591)
1 a : consisting of or resembling bristles **b :** thickly set with bristles
2 : inclined to or exhibiting aggressiveness or anger

bris·tol board \'bris-tᵊl-\ *noun* [*Bristol,* England] (1809)
: a paperboard with a smooth surface suitable especially for artwork — called also *bristol*

Bristol fashion *adjective* [*Bristol,* England] (1823)
: being in good order **:** SHIPSHAPE

brit *also* **britt** \'brit\ *noun* [perhaps from Cornish *brȳthel* mackerel] (1851)
: minute marine animals (as crustaceans and pteropods) on which right whales feed

Brit \'brit\ *noun* (1901)
: BRITON 2

Bri·tan·nia metal \bri-'tan-yə-, -'ta-nē-ə-\ *noun* [*Britannia,* poetic name for Great Britain, from Latin] (1817)
: a silver-white alloy largely of tin, antimony, and copper that is similar to pewter

Bri·tan·nic \bri-'ta-nik\ *adjective* (1641)
: BRITISH

britch·es \'bri-chəz\ *noun plural* [alteration of *breeches*] (circa 1803)
: BREECHES, TROUSERS

Brith Mi·lah \'brit-mē-'lä, 'brith-, 'bris-, -'mē-(ˌ)lä\ *noun* [Late Hebrew *bĕrīth mīlāh* covenant of circumcision] (circa 1902)
: the Jewish rite of circumcision

Brit·i·cism \'bri-tə-ˌsi-zəm\ *noun* [*Brit*ish + *-icism* (as in *gallicism*)] (1868)
: a characteristic feature of British English

□ USAGE
bring Although almost every native speaker of English has mastered the directional complexities of *bring* and *take* before they are old enough for school, a surprisingly large number of usage commentators have felt it necessary to explain the distinction to adults. Their basic points are these: *bring* implies movement toward the speaker or writer, and *take* implies movement away. These are points well made, and they hold for all cases to which they apply. Unfortunately, they do not apply to all cases in the real world. There is a third point to be made that few commentators have noticed: either verb can be used when the point of view is irrelevant. This can be easily shown ⟨copies will be given to pupils to *bring* home to their parents —*N. Y. Times*⟩. The readers of the newspaper will not be misled by either *bring* or *take* in this sentence because they know the writer is neither a school administrator expecting the copies to be taken home nor a parent expecting them to be brought home. When the direction of movement does not matter to the reader or hearer, it need not matter to the writer or speaker. Either verb will do. Such cases are common and have been at least since Shakespeare's time. But *bring* and *take* are also used in a great many idiomatic phrases—*bring to bear, take to task*—in which the verbs are never interchanged.

Brit·ish \'bri-tish\ *noun* [Middle English *Bruttische* of Britain, from Old English *Brettisc,* from *Brettas* Britons, of Celtic origin; akin to Welsh *Brython* Briton] (13th century)
1 a : the Celtic language of the ancient Britons **b :** BRITISH ENGLISH
2 *plural in construction* **:** the people of Great Britain or the Commonwealth
— British *adjective*
— Brit·ish·ism \'bri-ti-ˌshi-zəm\ *noun*
— Brit·ish·ness *noun*
British English *noun* (1869)
: the native language of most inhabitants of England; *especially* **:** English characteristic of England and clearly distinguishable from that used elsewhere (as in the U.S. or Australia)
Brit·ish·er \'bri-ti-shər\ *noun* (1829)
: BRITON 2
British thermal unit *noun* (1876)
: the quantity of heat required to raise the temperature of one pound of water one degree Fahrenheit at a specified temperature (as 39°F)
Brit·on \'bri-tᵊn\ *noun* [Middle English *Breton,* from Middle French & Latin; Middle French, from Latin *Britton-, Britto,* of Celtic origin; akin to Welsh *Brython*] (13th century)
1 : a member of one of the peoples inhabiting Britain prior to the Anglo-Saxon invasions
2 : a native or subject of Great Britain; *especially* **:** ENGLISHMAN
Brit·ta·ny \'bri-tᵊn-ē\ *noun, plural* **Brittanys** *also* **Brittanies** [*Brittany,* region in France] (1967)
: any of a breed of medium-sized pointing spaniels of French origin — called also *Brittany spaniel*
¹brit·tle \'bri-tᵊl\ *adjective* **brit·tler** \'brit-lər, 'bri-tᵊl-ər\; **brit·tlest** \-ləst, -tᵊl-əst\ [Middle English *britil;* akin to Old English *brēotan* to break, Old Norse *brjōta*] (14th century)
1 a : easily broken, cracked, or snapped ⟨*brittle* clay⟩ ⟨*brittle* glass⟩ **b :** easily disrupted, overthrown, or damaged **:** FRAIL ⟨a *brittle* friendship⟩
2 a : PERISHABLE, MORTAL **b :** TRANSITORY, EVANESCENT
3 : easily hurt or offended **:** SENSITIVE ⟨a *brittle* personality⟩
4 : SHARP ⟨*brittle* staccato of snare drums⟩
5 : lacking warmth, depth, or generosity of spirit **:** COLD ⟨a *brittle* selfish person⟩
6 : affected with or being a form of diabetes characterized by large and unpredictable fluctuations in blood glucose level
synonym see FRAGILE
— brit·tle·ly \'bri-tᵊl-(l)ē\ *adverb*
— brit·tle·ness \'bri-tᵊl-nəs\ *noun*
²brittle *noun* (1913)
: a candy made with caramelized sugar and nuts spread in thin sheets ⟨peanut *brittle*⟩
brittle star *noun* (1843)
: any of a subclass or class (Ophiuroidea) of echinoderms that have slender flexible arms distinct from the central disk
Brit·ton·ic \bri-'tä-nik\ *adjective* [Latin *Britton-, Britto* Briton] (1923)
: BRYTHONIC
Brix \'briks\ *adjective* (1897)
: of or relating to a Brix scale
Brix scale *noun* [Adolf F. *Brix* (died 1870) Austrian scientist] (1897)
: a hydrometer scale for sugar solutions so graduated that its readings at a specified temperature represent percentages by weight of sugar in the solution — called also *Brix*
bro \'brō\ *noun, plural* **bros** [by alteration] (1838)
1 : BROTHER 1
2 : SOUL BROTHER — often used informally as a term of address
¹broach \'brōch\ *noun* [Middle English *broche,* from Middle French, from (assumed) Vulgar Latin *brocca,* from Latin, feminine of *broccus* projecting] (13th century)
1 : BROOCH
2 : any of various pointed or tapered tools, implements, or parts: as **a :** a spit for roasting meat **b :** a tool for tapping casks **c :** a cutting tool for removing material from metal or plastic to shape an outside surface or a hole
²broach (15th century)
transitive verb
1 a : to pierce (as a cask) in order to draw the contents; *also* **:** to open for the first time **b :** to open up or break into (as a mine or stores)
2 : to shape or enlarge (a hole) with a broach
3 a : to make known for the first time **b :** to open up (a subject) for discussion
intransitive verb
: to break the surface from below
synonym see EXPRESS
— broach·er *noun*
³broach *intransitive verb* [perhaps from ²*broach*] (1705)
: to veer or yaw dangerously so as to lie broadside to the waves — often used with *to*
¹broad \'brȯd\ *adjective* [Middle English *brood,* from Old English *brād;* akin to Old High German *breit* broad] (before 12th century)
1 a : having ample extent from side to side or between limits ⟨*broad* shoulders⟩ **b :** having a specified extension from side to side ⟨made the path 10 feet *broad*⟩
2 : extending far and wide **:** SPACIOUS ⟨the *broad* plains⟩
3 a : OPEN, FULL ⟨*broad* daylight⟩ **b :** PLAIN, OBVIOUS ⟨a *broad* hint⟩
4 : dialectal especially in pronunciation
5 : marked by lack of restraint, delicacy, or subtlety: **a** *obsolete* **:** OUTSPOKEN **b :** COARSE, RISQUÉ ⟨*broad* humor⟩
6 *of a vowel* **:** OPEN — used specifically of *a* pronounced as in *father*
7 a : LIBERAL, TOLERANT ⟨*broad* views⟩ **b :** widely applicable or applied **:** GENERAL
8 : relating to the main or essential points ⟨*broad* outlines⟩ ☆
— broad·ly *adverb*
— broad·ness *noun*
²broad *adverb* (before 12th century)
: in a broad manner **:** FULLY
³broad *noun* (1659)
1 *British* **:** an expansion of a river — often used in plural
2 *slang* **:** WOMAN
broad arrow *noun* (14th century)
1 : an arrow with a flat barbed head
2 *British* **:** a mark shaped like a broad arrow that identifies government property including clothing formerly worn by convicts
broad·ax *or* **broad·axe** \'brȯ-ˌdaks\ *noun* (before 12th century)
: a large ax with a broad blade
broad·band \'brȯd-ˌband\ *adjective* (1956)
1 : operating at, responsive to, or comprising a wide band of frequencies ⟨a *broadband* radio antenna⟩
2 : of, relating to, or being a communications network in which a frequency range is divided into multiple independent channels for simultaneous transmission of signals (as voice, data, or video)
broad bean *noun* (1783)
: the large flat edible seed of an Old World upright vetch (*Vicia faba*); *also* **:** this plant widely grown for its seeds and as fodder
broad–brush \'brȯd-ˌbrəsh\ *adjective* (1967)
: GENERAL, NONSPECIFIC
¹broad·cast \'brȯd-ˌkast\ *adjective* (1767)
1 : cast or scattered in all directions
2 : made public by means of radio or television
3 : of or relating to radio or television broadcasting
²broadcast *verb* **broadcast** *also* **broadcast·ed; broad·cast·ing** (1813)
transitive verb
1 : to scatter or sow (as seed) broadcast
2 : to make widely known
3 : to transmit or make public by means of radio or television
intransitive verb
1 : to transmit a broadcast
2 : to speak or perform on a broadcast program
— broad·cast·er *noun*
³broadcast *adverb* (1814)
: to or over a broad area
⁴broadcast *noun* (1922)
1 : the act of transmitting sound or images by radio or television
2 : a single radio or television program
Broad Church *adjective* (1853)
: of or relating to a liberal party in the Anglican communion especially in the later 19th century
broad·cloth \'brȯd-ˌklȯth\ *noun* (15th century)
1 : a twilled napped woolen or worsted fabric with smooth lustrous face and dense texture
2 : a fabric usually of cotton, silk, or rayon made in plain and rib weaves with soft semigloss finish
broad·en \'brȯ-dᵊn\ *verb* **broad·ened; broad·en·ing** \'brȯd-niŋ, 'brȯ-dᵊn-iŋ\ (1726)
transitive verb
: to make broader
intransitive verb
: to become broad
broad–gauge \'brȯd-'gāj\ *or* **broad–gauged** \-'gājd\ *adjective* (1858)
1 : wide in area or scope ⟨a *broad-gauge* effort⟩
2 : comprehensive in outlook, range, or capability ⟨a *broad-gauge* statesman⟩
broad gauge *noun* (1844)
: a railroad gauge wider than standard gauge
broad jump *noun* (1872)
: LONG JUMP
— broad jumper *noun*
broad–leaved \-'lēvd\ *or* **broad·leaf** \-'lēf\ *also* **broad–leafed** \-'lēft\ *adjective* (1552)
1 : having broad leaves; *specifically* **:** having leaves that are not needles
2 : composed of broad-leaved plants ⟨*broad-leaved* forests⟩
¹broad·loom \-ˌlüm\ *adjective* (1925)
: woven on a wide loom; *also* **:** so woven in solid color
²broadloom *noun* (1926)
: a broadloom carpet
broad–mind·ed \'brȯd-'mīn-dəd, -ˌmīn-\ *adjective* (1882)
1 : tolerant of varied views
2 : inclined to condone minor departures from conventional behavior
— broad–mind·ed·ly *adverb*
— broad–mind·ed·ness *noun*
broad·scale \-ˌskā(ə)l\ *adjective* (1939)
: broad in extent, range, or effect
broad·sheet \-ˌshēt\ *noun* (1705)
1 : BROADSIDE 1

☆ SYNONYMS
Broad, wide, deep mean having horizontal extent. BROAD and WIDE apply to a surface measured or viewed from side to side ⟨a *broad* avenue⟩. WIDE is more common when units of measurement are mentioned ⟨rugs eight feet *wide*⟩ or applied to unfilled space between limits ⟨a *wide* doorway⟩. BROAD is preferred when full horizontal extent is considered ⟨*broad* shoulders⟩. DEEP may indicate horizontal extent away from the observer or from a front or peripheral point ⟨a *deep* cupboard⟩ ⟨*deep* woods⟩.

2 *British* **:** a newspaper with full-size pages as distinguished from a tabloid

[1]**broad·side** \-ˌsīd\ *noun* (1575)

1 a (1) **:** a sizable sheet of paper printed on one side (2) **:** a sheet printed on one or both sides and folded **b :** something (as a ballad) printed on a broadside

2 *archaic* **:** the side of a ship above the waterline

3 a : all the guns on one side of a ship; *also* **:** their simultaneous discharge **b :** a volley of abuse or denunciation

4 : a broad or unbroken surface

[2]**broadside** *adjective* (1646)

: directed or placed broadside ⟨a *broadside* attack⟩

[3]**broadside** *adverb* (1870)

1 a : with the side forward or toward a given point **:** SIDEWAYS ⟨turned *broadside*⟩ **b :** directly from the side ⟨the car was hit *broadside*⟩

2 : in one volley

3 : at random

[4]**broadside** *transitive verb* (1981)

: to hit broadside ⟨the car was *broadsided*⟩

broad–spectrum *adjective* (1952)

: effective against a wide range of organisms (as insects or bacteria) ⟨*broad-spectrum* antibiotic⟩

broad·sword \ˈbrȯd-ˌsȯrd, -ˌsȯrd\ *noun* (before 12th century)

: a large heavy sword with a broad blade for cutting rather than thrusting

broad·tail \-ˌtāl\ *noun* (1892)

1 : KARAKUL 1

2 : the fur or skin of a very young or premature karakul lamb having a flat and wavy appearance resembling moiré silk

Broad·way \ˈbrȯd-ˌwā, -ˈwā\ *noun* [*Broadway,* street in New York City] (1835)

: the New York commercial theater and amusement world; *specifically* **:** playhouses located in the area between the Avenue of the Americas and Ninth Avenue and from W. 41st Street to W. 53d Street

— Broadway *adjective*

— Broad·way·ite \-ˌīt\ *noun*

Brob·ding·nag·ian \ˌbräb-diŋ-ˈna-gē-ən, -dig-ˈna-\ *adjective* [*Brobdingnag,* imaginary land of giants in *Gulliver's Travels,* by Jonathan Swift] (1728)

: marked by tremendous size

— Brobdingnagian *noun*

bro·cade \brō-ˈkād\ *noun* [Spanish *brocado,* from Catalan *brocat,* from Italian *broccato,* from *broccare* to spur, brocade, from *brocco* small nail, from Latin *broccus* projecting] (1588)

1 : a rich oriental silk fabric with raised patterns in gold and silver

2 : a fabric characterized by raised designs

— brocade *transitive verb*

— bro·cad·ed *adjective*

Bro·ca's area \ˈbrō-kəz-\ *noun* [Paul P. *Broca* (died 1880) French surgeon] (circa 1898)

: a brain center associated with the motor control of speech and usually located in the left side of the frontal lobe

broc·a·telle \ˌbrä-kə-ˈtel\ *noun* [French, from Italian *broccatello,* diminutive of *broccato*] (1669)

: a stiff decorating fabric with patterns in high relief

broc·co·li \ˈbrä-kə-lē, ˈbrä-klē\ *noun* [Italian, plural of *broccolo* flowering top of a cabbage, diminutive of *brocco* small nail, sprout] (1699)

1 *chiefly British* **:** a large hardy cauliflower

2 a : either of two garden vegetable plants closely related to the cabbage: (1) **:** one with a thick central stem and a compact head of dense usually green florets that is classified with the cauliflower (2) **:** one (*Brassica oleracea italica*) with slender stems and usually green or purple florets not arranged in a central head **b :** the stems and immature florets of broccoli used as food

broccoli rabe \-ˌräb\ *noun* [perhaps modification of Italian *broccoli di rapa,* literally, flowering tops of the turnip] (1976)

: a garden brassica (*Brassica rapa ruvo*) that is related to the turnip and produces edible leafy branching stalks and compact clusters of yellow florets — called also *broccoli raab* \-ˌräb, -ˌrab\, *rapini*

bro·chette \brō-ˈshet\ *noun* [French, from Old French *brochete,* from *broche* pointed tool — more at BROACH] (15th century)

: SKEWER; *also* **:** food broiled on a skewer

bro·chure \brō-ˈshu̇r, *British especially* ˈbrō-ˌ\ *noun* [French, from *brocher* to sew, from Middle French, to prick, from Old French *brochier,* from *broche*] (1748)

: PAMPHLET, BOOKLET; *especially* **:** one containing descriptive or advertising material

brock \ˈbräk\ *noun* [Middle English, from Old English *broc,* of Celtic origin; akin to Welsh *broch* badger] (before 12th century)

British **:** BADGER

brock·age \ˈbrä-kij\ *noun* [English dialect *brock* rubbish + English *-age*] (1879)

: an imperfectly minted coin

brock·et \ˈbrä-kət\ *noun* [Middle English *broket,* from Old North French *broquard;* akin to Old French *broche* tine of an antler, pointed tool — more at BROACH] (1837)

: any of several small deer (genus *Mazama*) of Central and South America with unbranched antlers

bro·gan \ˈbrō-gən, -ˌgan; brō-ˈgan\ *noun* [Irish *brógán,* diminutive of *bróg*] (1835)

: a heavy shoe; *especially* **:** a coarse work shoe reaching to the ankle

[1]**brogue** \ˈbrōg\ *noun* [Irish *bróg* & Scottish Gaelic *bròg,* from Middle Irish *bróc,* probably from Old Norse *brōk* leg covering; akin to Old English *brōc* leg covering — more at BREECH] (1586)

1 : a stout coarse shoe worn formerly in Ireland and the Scottish Highlands

2 : a heavy shoe often with a hobnailed sole **:** BROGAN

3 : a stout oxford shoe with perforations and usually a wing tip

[2]**brogue** *noun* [Irish *barróg* accent, speech impediment, literally, wrestling hold, tight grip] (1705)

: a dialect or regional pronunciation; *especially* **:** an Irish accent

broi·der \ˈbrȯi-dər\ *transitive verb* [Middle English *broideren,* modification of Middle French *broder* — more at EMBROIDER] (14th century)

: EMBROIDER

— broi·dery \ˈbrȯi-d(ə-)rē\ *noun*

[1]**broil** \ˈbrȯi(ə)l\ *verb* [Middle English, from Middle French *bruler* to burn, modification of Latin *ustulare* to singe, from *urere* to burn] (14th century)

transitive verb

: to cook by direct exposure to radiant heat **:** GRILL

intransitive verb

: to be subjected to great or oppressive heat ⟨*broiling* in the sun⟩

[2]**broil** *noun* (1583)

: the act or state of broiling

[3]**broil** *verb* [Middle English, from Middle French *brouiller* to mix, broil, from Old French *brooilier,* from *breu* broth, of Germanic origin; akin to Old High German *brod* broth — more at BROTH] (15th century)

intransitive verb

: BRAWL

transitive verb

: EMBROIL

[4]**broil** *noun* (1525)

: a noisy disturbance **:** TUMULT; *especially* **:** BRAWL ⟨a tavern row . . . widens into a general *broil* —J. R. Green⟩

broil·er \ˈbrȯi-lər\ *noun* (14th century)

1 : one that broils

2 : a bird fit for broiling; *especially* **:** a young chicken of up to 2½ pounds (1.14 kilograms) dressed weight

[1]**broke** \ˈbrōk\ *past of* BREAK

[2]**broke** *adjective* [Middle English, alteration of *broken*] (1710)

: PENNILESS

bro·ken \ˈbrō-kən\ *adjective* [Middle English, from Old English *brocen,* from past participle of *brecan* to break] (13th century)

1 : violently separated into parts **:** SHATTERED

2 : damaged or altered by breaking: as **a :** having undergone or been subjected to fracture ⟨a *broken* leg⟩ **b** *of land surfaces* **:** being irregular, interrupted, or full of obstacles **c :** violated by transgression ⟨a *broken* promise⟩ **d :** DISCONTINUOUS, INTERRUPTED **e :** disrupted by change **f** *of a flower* **:** having an irregular, streaked, or blotched pattern especially from virus infection

3 a : made weak or infirm **b :** subdued completely **:** CRUSHED ⟨a *broken* spirit⟩ **c :** BANKRUPT **d :** reduced in rank

4 a : cut off **:** DISCONNECTED **b :** imperfectly spoken or written ⟨*broken* English⟩

5 : not complete or full

6 : disunited by divorce, separation, or desertion of one parent ⟨children from *broken* homes⟩ ⟨a *broken* family⟩

— bro·ken·ly *adverb*

— bro·ken·ness \-kə(n)-nəs\ *noun*

broken–down *adjective* (1817)

: WORN-OUT, DEBILITATED

bro·ken–field \-ˌfēld\ *adjective* (1923)

: characterized by quick changes in direction to avoid widely scattered tacklers ⟨a halfback known for *broken-field* running⟩

bro·ken·heart·ed \-ˈhär-təd\ *adjective* (1526)

: overcome by grief or despair

bro·ken–wind·ed \-ˈwin-dəd\ *adjective* (1580)

: affected with or as if with heaves

bro·ker \ˈbrō-kər\ *noun* [Middle English, negotiator, from Anglo-French *brocour*] (14th century)

1 : one who acts as an intermediary: as **a :** an agent who arranges marriages **b :** an agent who negotiates contracts of purchase and sale (as of real estate, commodities, or securities)

2 : POWER BROKER

— broker *verb*

— bro·ker·ing \ˈbrō-k(ə-)riŋ\ *noun*

bro·ker·age \ˈbrō-k(ə-)rij\ *noun* (15th century)

1 : the business or establishment of a broker

2 : a broker's fee or commission

bro·kered \-kərd\ *adjective* (1967)

: arranged or controlled by power brokers ⟨a *brokered* political convention⟩

bro·king \-kiŋ\ *noun* (1569)

chiefly British **:** the business of a broker **:** BROKERAGE

brol·ly \ˈbrä-lē\ *noun, plural* **brollies** [by shortening & alteration] (circa 1874)

chiefly British **:** UMBRELLA

brom- *or* **bromo-** *combining form* [probably from French *brome,* from Greek *brōmos* bad smell]

: bromine ⟨*brom*ide⟩

bro·mate \ˈbrō-ˌmāt\ *noun* (circa 1836)

: a salt of bromic acid

brome·grass \ˈbrōm-ˌgras\ *noun* [New Latin *Bromus,* from Latin *bromos* oats, from Greek] (circa 1791)

: any of a large genus (*Bromus*) of tall grasses often having drooping spikelets

bro·me·lain \ˈbrō-mə-lən, -ˌlān\ *also* **bro·me·lin** \ˈbrō-mə-lən, brō-ˈmē-\ *noun* [*bromelain* by alteration of *bromelin,* from New Latin *Bromelia*] (1894)

: a protease obtained especially from the pineapple

bro·me·li·ad \brō-ˈmē-lē-ˌad\ *noun* [New Latin *Bromelia,* genus of tropical American plants, from Olaf *Bromelius* (died 1705) Swedish botanist] (1866)

: any of a family (Bromeliaceae) of chiefly tropical American usually epiphytic herbaceous plants including the pineapple, Spanish moss, and various ornamentals

bro•mic acid \'brō-mik-\ *noun* (1828)
: an unstable strongly oxidizing acid $HBrO_3$ known only in solution or in the form of its salts

bro•mide \'brō-ˌmīd\ *noun* (1836)
1 : a binary compound of bromine with another element or a radical including some (as potassium bromide) used as sedatives
2 a : a commonplace or tiresome person : BORE **b :** a commonplace or hackneyed statement or notion

bro•mid•ic \brō-'mi-dik\ *adjective* (1906)
: lacking in originality : TRITE

bro•mi•nate \'brō-mə-ˌnāt\ *transitive verb* **-nat•ed; -nat•ing** (1873)
: to treat or cause to combine with bromine or a compound of bromine
— bro•mi•na•tion \ˌbrō-mə-'nā-shən\ *noun*

bro•mine \'brō-ˌmēn\ *noun* [French *brome* bromine + English ²*-ine*] (1827)
: a nonmetallic halogen element that is isolated as a deep red corrosive toxic volatile liquid of disagreeable odor — see ELEMENT table

bro•mo \'brō-(ˌ)mō\ *noun, plural* **bromos** [*brom-*] (1923)
: a dose of a proprietary effervescent headache remedy and antacid

bro•mo•crip•tine \ˌbrō-mō-'krip-ˌtēn, -tən\ *noun* [by shortening & alteration from *bromoergocryptine,* from *brom-* + *ergocryptine,* an ergot derivative, from *ergo-* + Greek *kryptos* hidden — more at CRYPT] (1975)
: a polypeptide ergot derivative $C_{32}H_{40}BrN_5O_5$ that mimics the activity of dopamine in inhibiting prolactin secretion

bro•mo•ura•cil \ˌbrō-mō-'yu̇r-ə-ˌsil, -səl\ *noun* (1960)
: a mutagenic uracil derivative $C_4H_3N_2O_2Br$ that is an analogue of thymine and pairs readily with adenine and sometimes with guanine

brom•thy•mol blue \ˌbrōm-'thī-ˌmȯl-\ *noun* (1920)
: a dye derived from thymol that is an acid-base indicator — called also *bro•mo•thy•mol blue* \ˌbrō-mō-'thī-ˌmȯl-\

bronc \'bräŋk\ *noun* [short for *bronco*] (1893)
: an unbroken or imperfectly broken range horse of western North America; *broadly* : MUSTANG

bronch- *or* **broncho-** *combining form* [Late Latin, from Greek, from *bronchos* windpipe]
: bronchial tube : bronchial ⟨*bronch*itis⟩

bronchi- *or* **bronchio-** *combining form* [New Latin, from *bronchia,* plural, branches of the bronchi, from Late Latin, from Greek, diminutive of *bronchos*]
: bronchial tubes ⟨*bronchi*ectasis⟩

bron•chi•al \'bräŋ-kē-əl\ *adjective* (circa 1735)
: of or relating to the bronchi or their ramifications in the lungs
— bron•chi•al•ly \-kē-ə-lē\ *adverb*

bronchial asthma *noun* (circa 1881)
: asthma resulting from spasmodic contraction of bronchial muscles

bronchial tube *noun* (1847)
: a primary bronchus or any of its branches

bron•chi•ec•ta•sis \ˌbräŋ-kē-'ek-tə-səs\ *noun* [New Latin, from *bronchi-* + Greek *ektasis* extension — more at ATELECTASIS] (circa 1860)
: a chronic dilatation of bronchi or bronchioles

bron•chi•ole \'bräŋ-kē-ˌōl\ *noun* [New Latin *bronchiolum,* diminutive of *bronchia*] (circa 1860)
: a minute thin-walled branch of a bronchus
— bron•chi•o•lar \ˌbräŋ-kē-'ō-lər\ *adjective*

bron•chi•tis \brän-'kī-təs, bräŋ-\ *noun* [New Latin] (1808)
: acute or chronic inflammation of the bronchial tubes; *also* : a disease marked by this
— bron•chit•ic \-'ki-tik\ *adjective*

bron•cho•di•la•tor \ˌbräŋ-(ˌ)kō-dī-'lā-tər, -'dī-ˌ\ *noun* (1903)
: a drug that relaxes bronchial muscle resulting in expansion of the bronchial air passages
— bronchodilator *adjective*

bron•cho•gen•ic \ˌbräŋ-kə-'je-nik\ *adjective* (1927)
: of, relating to, or arising in or by way of the air passages of the lungs ⟨*bronchogenic* carcinoma⟩

bron•cho•pneu•mo•nia \ˌbräŋ-(ˌ)kō-nu̇-'mō-nyə, -nyu̇-\ *noun* [New Latin] (1858)
: pneumonia involving many relatively small areas of lung tissue

bron•cho•scope \'bräŋ-kə-ˌskōp\ *noun* [International Scientific Vocabulary] (1899)
: a tubular illuminated instrument used for inspecting or passing instruments into the bronchi
— bron•cho•scop•ic \ˌbräŋ-kə-'skä-pik\ *adjective*
— bron•chos•co•pist \brän-'käs-kə-pist, bräŋ-\ *noun*
— bron•chos•co•py \-pē\ *noun*

bron•cho•spasm \'bräŋ-kə-ˌspa-zəm\ *noun* (circa 1901)
: constriction of the air passages of the lung (as in asthma) by spasmodic contraction of the bronchial muscles
— bron•cho•spas•tic \ˌbräŋ-kə-'spas-tik\ *adjective*

bron•chus \'bräŋ-kəs\ *noun, plural* **bron•chi** \'bräŋ-ˌkī, -ˌkē\ [New Latin, from Greek *bronchos*] (circa 1706)
: either of the two primary divisions of the trachea that lead respectively into the right and the left lung; *broadly* : BRONCHIAL TUBE

bron•co *also* **bron•cho** \'bräŋ-(ˌ)kō\ *noun, plural* **broncos** *also* **bronchos** [Mexican Spanish, from Spanish, literally, rough, wild] (1850)
: BRONC

bron•co•bust•er \-ˌbəs-tər\ *noun* (1887)
: one who breaks wild horses to the saddle

bron•to•sau•rus \ˌbrän-tə-'sȯr-əs\ *also* **bron•to•saur** \'brän-tə-ˌsȯr\ *noun* [New Latin, from Greek *brontē* thunder + *sauros* lizard; akin to Greek *bremein* to roar] (1905)
: any of a genus (*Apatosaurus* synonym *Brontosaurus*) of large sauropod dinosaurs of the Jurassic — called also *apatosaurus* ◆

Bronx cheer \'bräŋ(k)s-\ *noun* [*Bronx,* borough of New York City] (1929)
: RASPBERRY 2

¹bronze \'bränz\ *transitive verb* **bronzed; bronz•ing** (1645)
: to give the appearance of bronze to ⟨a *bronzed* sculpture⟩; *also* : TAN 2
— bronz•er *noun*

²bronze *noun, often attributive* [French, from Italian *bronzo*] (1739)
1 a : an alloy of copper and tin and sometimes other elements **b :** any of various copper-base alloys with little or no tin
2 : a sculpture or artifact of bronze
3 : a moderate yellowish brown
— bronzy \'brän-zē\ *adjective*

Bronze Age *noun* (1865)
: the period of ancient human culture characterized by the use of bronze that began between 4000 and 3000 B.C. and ended with the advent of the Iron Age

Bronze Star Medal *noun* (1944)
: a U.S. military decoration awarded for heroic or meritorious service not involving aerial flights

bronz•ing *noun* (1868)
: a bronze coloring or discoloration (as of leaves)

brooch \'brōch *also* 'brüch\ *noun* [Middle English *broche* pointed tool, brooch — more at BROACH] (13th century)
: an ornament that is held by a pin or clasp and is worn at or near the neck

¹brood \'brüd\ *noun* [Middle English, from Old English *brōd;* akin to Old English *beorma* yeast — more at BARM] (before 12th century)
1 : the young of an animal or a family of young; *especially* **:** the young (as of a bird or insect) hatched or cared for at one time
2 : a group having a common nature or origin

²brood *adjective* (15th century)
: kept for breeding ⟨a *brood* flock⟩

³brood (15th century)
transitive verb
1 a : to sit on or incubate (eggs) **b :** to produce by or as if by incubation : HATCH
2 *of a bird* **:** to cover (young) with the wings
3 : to think anxiously or gloomily about : PONDER
intransitive verb
1 a *of a bird* **:** to brood eggs or young **b :** to sit quietly and thoughtfully : MEDITATE
2 : HOVER, LOOM
3 a : to dwell gloomily on a subject : WORRY **b :** to be in a state of depression
— brood•ing•ly \'brü-diŋ-lē\ *adverb*

brood•er \'brü-dər\ *noun* (1599)
1 : one that broods
2 : a heated structure used for raising young fowl

brood•mare \'brüd-ˌmar, -ˌmer\ *noun* (1821)
: a mare kept for breeding

broody \'brü-dē\ *adjective* (1523)
1 : being in a state of readiness to brood eggs that is characterized by cessation of laying and by marked changes in behavior and physiology ⟨a *broody* hen⟩
2 : given or conducive to introspection : CONTEMPLATIVE, MOODY
— brood•i•ness *noun*

¹brook \'bru̇k\ *noun* [Middle English, from Old English *brōc;* akin to Old High German *bruoh* marshy ground] (before 12th century)
: CREEK 2

²brook *transitive verb* [Middle English *brouken* to use, enjoy, from Old English *brūcan;* akin to Old High German *brūhhan* to use, Latin *frui* to enjoy] (15th century)
: to stand for : TOLERATE ⟨he would *brook* no interference with his plans⟩

brook•ie \'bru̇-kē\ *noun* (1933)
: BROOK TROUT

brook•ite \'bru̇-ˌkīt\ *noun* [Henry J. *Brooke* (died 1857) English mineralogist] (1825)

◇ WORD HISTORY

brontosaurus Although it has been ousted in scientific nomenclature, the word *brontosaurus* is part of the vocabulary of all dinosaur aficionados. *Brontosaurus*—according to a literal translation of its Greek elements, "thunder lizard"—was the name given by American paleontologist Edward Drinker Cope (1840–97) to a genus of gigantic sauropods whose fossilized bones he had unearthed while affiliated with the U.S. Geological Survey in the 1870s. Unbeknownst to Cope, the same sauropod had already been discovered by his archrival, Othniel Charles Marsh (1831–99), who named the animal *Apatosaurus* "deceptive lizard." Not until the specimens were studied in museums years later was it discovered that *Brontosaurus* and *Apatosaurus* were one and the same. Biologists' rules of classification are strict and clear: the first name given to a new species is the correct one. By the time the mistake was discovered, however, *Brontosaurus* was firmly entrenched not just in scientific literature but in the public imagination as well.

: titanium dioxide TiO_2 occurring as a mineral in orthorhombic crystals commonly translucent brown or opaque brown to black

brook·let \'brủ-klət\ *noun* (1813)
: a small brook

Brook·lyn·ese \,brủ-klə-'nēz, -'nēs\ *noun* [*Brooklyn,* borough of New York City] (1939)
: the vernacular speech of greater New York City and environs

brook trout *noun* (1836)
: the common speckled cold-water char (*Salvelinus fontinalis*) of North America

[1]broom \'brüm, 'brủm\ *noun* [Middle English, from Old English *brōm;* akin to Old High German *brāmo* bramble] (before 12th century)
1 : any of various leguminous shrubs (especially genera *Cytisus* and *Genista*) with long slender branches, small leaves, and usually showy yellow flowers; *especially* **:** SCOTCH BROOM
2 : a bundle of firm stiff twigs or fibers bound together on a long handle for sweeping and brushing

[2]broom *transitive verb* (1838)
1 : to sweep with or as if with a broom
2 : to finish (as a concrete surface) by means of a broom

broom·ball \-,bȯl\ *noun* (1933)
: a variation of ice hockey played on ice without skates and with brooms and a soccer ball used instead of sticks and a puck
— broom·ball·er \-,bȯ-lər\ *noun*

broom·corn \-,kȯrn\ *noun* (circa 1782)
: any of several tall cultivated sorghums having stiff-branched panicles used in brooms and brushes

broom·rape \-,rāp\ *noun* (1578)
: any of a genus (*Orobanche* of the family Orobanchaceae, the broomrape family) of herbs that have leaves modified to scales and that grow as parasites on the roots of other plants

broom·stick \-,stik\ *noun* (1683)
: the long thin handle of a broom

brose \'brōz\ *noun* [perhaps alteration of Scots *bruis* broth, from Middle English *brewes,* from Old French *broez,* nominative singular of *broet,* diminutive of *breu* broth — more at BROIL] (1515)
: a chiefly Scottish dish made with a boiling liquid and meal

broth \'brȯth\ *noun, plural* **broths** \'brȯths, 'brȯthz\ [Middle English, from Old English; akin to Old High German *brod* broth, Old English *brēowan* to brew] (before 12th century)
1 : liquid in which meat, fish, cereal grains, or vegetables have been cooked **:** STOCK
2 : a fluid culture medium

broth·el \'brä-thəl, 'brȯ- *also* -thəl\ *noun* [Middle English, worthless fellow, prostitute, from *brothen,* past participle of *brethen* to waste away, go to ruin, from Old English *brēothan* to waste away; akin to Old English *brēotan* to break — more at BRITTLE] (circa 1593)
: BORDELLO

broth·er \'brə-thər\ *noun, plural* **brothers** *also* **breth·ren** \'breth-rən; 'bre-thə-rən, -thərn\ [Middle English, from Old English *brōthor;* akin to Old High German *bruodor* brother, Latin *frater,* Greek *phratēr* member of the same clan] (before 12th century)
1 : a male who has the same parents as another or one parent in common with another
2 : one related to another by common ties or interests
3 : a fellow member — used as a title for ministers in some evangelical denominations
4 : one of a type similar to another
5 a : KINSMAN **b :** one who shares with another a common national or racial origin; *especially* **:** SOUL BROTHER
6 a *capitalized* **:** a member of a congregation of men not in holy orders and usually in hospital or school work **b :** a member of a men's religious order who is not preparing for or is not ready for holy orders ⟨a lay *brother*⟩

broth·er·hood \'brə-thər-,hủd\ *noun* [Middle English *brotherhede, brotherhod,* alteration of *brotherrede,* from Old English *brōthorrǣden,* from *brōthor* + *rǣden* condition — more at KINDRED] (14th century)
1 : the quality or state of being brothers
2 : FELLOWSHIP, ALLIANCE
3 : an association (as a labor union) for a particular purpose
4 : the whole body of persons engaged in a business or profession

broth·er–in–law \'brəth-rən-,lȯ, 'brə-thə-; 'brə-thərn-\ *noun, plural* **broth·ers–in–law** \'brə-thər-zən-\ (14th century)
1 : the brother of one's spouse
2 a : the husband of one's sister **b :** the husband of one's spouse's sister

broth·er·ly \'brə-thər-lē\ *adjective* (before 12th century)
1 : of or relating to brothers
2 : natural or becoming to brothers **:** AFFECTIONATE ⟨*brotherly* love⟩
— broth·er·li·ness *noun*
— brotherly *adverb*

brougham \'brü(-ə)m, 'brō(-ə)m\ *noun* [Henry Peter *Brougham,* Baron Brougham and Vaux (died 1868) Scottish jurist] (1851)
: a light closed horse-drawn carriage with the driver outside in front

brought *past and past participle of* BRING

brou·ha·ha \'brü-,hä-,hä, ,brü-,hä-'hä, brü-'hä-,hä\ *noun* [French] (1890)
: HUBBUB, UPROAR

[1]brow \'braủ\ *noun* [Middle English, from Old English *brū;* akin to Old Norse *brūn* eyebrow, Greek *ophrys,* Sanskrit *bhrū*] (before 12th century)
1 a : EYEBROW **b :** FOREHEAD
2 : the projecting upper part or margin of a steep place
3 : EXPRESSION, MIEN

[2]brow *noun* [perhaps from Danish or Swedish *brother* bridge; akin to Old English *brycg* bridge] (1867)
: GANGPLANK

brow·beat \'braủ-,bēt\ *transitive verb* **-beat; -beat·en** \-'bē-tᵊn\ *or* **-beat; -beat·ing** (1581)
: to intimidate or disconcert by a stern manner or arrogant speech **:** BULLY
synonym see INTIMIDATE

browed \'braủd\ *adjective* (15th century)
: having brows of a specified nature — used in combination ⟨smooth-*browed*⟩

[1]brown \'braủn\ *adjective* [Middle English *broun,* from Old English *brūn;* akin to Old High German *brūn* brown, Greek *phrynē* toad] (before 12th century)
: of the color brown; *especially* **:** of dark or tanned complexion

[2]brown *noun* (13th century)
1 : any of a group of colors between red and yellow in hue, of medium to low lightness, and of moderate to low saturation
2 : a brown-skinned person
— brown·ish \'braủ-nish\ *adjective*
— browny \-nē\ *adjective*

[3]brown (14th century)
intransitive verb
: to become brown
transitive verb
: to make brown

brown alga *noun* (circa 1899)
: any of a division (Phaeophyta) of variable mostly marine algae with chlorophyll masked by brown pigment

brown bagging *noun* (1959)
1 : the practice of carrying (as to work) one's lunch usually in a brown paper bag
2 : the practice of carrying a bottle of liquor into a restaurant or club where setups are available
— brown–bag \'braủn-,bag\ *verb or adjective*
— brown bagger *noun*

brown bear *noun* (1805)
: any of several bears predominantly brown in color that are sometimes considered a single species (*Ursus arctos*) including the grizzly bear and that formerly inhabited western North America from the barrens of Alaska to northern Mexico and much of Europe and Asia but are now much restricted in range

brown Bet·ty \-'be-tē\ *noun* (1864)
: a baked pudding of apples, bread crumbs, and spices

brown bread \-,bred\ *noun* (14th century)
1 : bread made of whole wheat flour
2 : a dark brown steamed bread made usually of cornmeal, white or whole wheat flours, molasses, soda, and milk or water

brown coal *noun* (circa 1828)
: LIGNITE

brown dwarf *noun* (1978)
: a celestial object that is much smaller than a normal star and has insufficient mass for nuclear fusion to begin but that is hot enough to radiate energy especially at infrared wavelengths

brown earth *noun* (1932)
: any of a group of intrazonal soils developed in temperate humid regions under deciduous forests and characterized by a dark brown mull horizon that grades through lighter colored soil into parent material

brown–eyed Su·san \'braủ-,nīd-'sü-zᵊn\ *noun* [*brown-eyed* + *Susan* (as in *black-eyed Susan*)] (1896)
: a dark-centered coneflower (*Rudbeckia triloba*) of eastern North America often having tripartite lower leaves

brown fat *noun* (1951)
: a mammalian heat-producing tissue occurring especially in human newborns and in hibernators — called also *brown adipose tissue*

Brown·ian motion \'braủ-nē-ən-\ *noun* [Robert *Brown* (died 1858) Scottish botanist] (circa 1889)
: a random movement of microscopic particles suspended in liquids or gases resulting from the impact of molecules of the fluid surrounding the particles — called also *Brownian movement*

brown·ie \'braủ-nē\ *noun* [[1]*brown*] (circa 1500)
1 : a legendary good-natured elf that performs helpful services at night
2 *capitalized* **:** a member of a program of the Girl Scouts for girls in the first through third grades in school
3 : a small square or rectangle of rich usually chocolate cake often containing nuts

brownie point *noun, often B capitalized* (circa 1962)
: a credit regarded as earned especially by currying favor (as with a superior)

Brow·ning automatic rifle \'braủ-niŋ-\ *noun* [John M. *Browning* (died 1926) American designer of firearms] (1920)
: a .30 caliber gas-operated air-cooled magazine-fed automatic rifle often provided with a rest for the barrel

Browning machine gun *noun* (1918)
: a .30 or .50 caliber recoil-operated air- or water-cooled machine gun fed by a cartridge belt

brown lung *noun* (1969)
: BYSSINOSIS

brown·nose \'braủ(n)-,nōz\ *transitive verb* [from the implication that servility is equivalent to kissing the hinder parts of the person from whom advancement is sought] (circa 1939) *slang*
: to ingratiate oneself with **:** curry favor with
— brownnose *noun*
— brown·nos·er *noun*

brown·out \'braủ-,naủt\ *noun* [*brown* + *blackout*] (1942)
: a period of reduced voltage of electricity caused especially by high demand and resulting in subnormal illumination

brown pelican *noun* (1823)

: a pelican (*Pelecanus occidentalis*) of American coasts that has a brown body and a chiefly white or white with yellow head

brown rat *noun* (1826)
: a common domestic rat (*Rattus norvegicus*) that has been introduced worldwide — called also *Norway rat*

brown recluse spider *noun* (1964)
: a venomous spider (*Loxosceles reclusa*) of the U.S. that has a violin-shaped mark on the cephalothorax and produces a dangerous cytotoxin

brown rice *noun* (1916)
: hulled but unpolished rice that retains most of the bran layers, endosperm, and germ

brown rot *noun* (1894)
: a disease of stone fruits caused by fungi (genus *Monilinia,* especially *M. fructicola*)

brown sauce *noun* (1878)
: a sauce consisting typically of stock thickened with flour browned in fat

brown·shirt \'braůn-ˌshərt\ *noun, often capitalized* (1932)
: NAZI; *especially* : STORM TROOPER

brown·stone \-ˌstōn\ *noun* (1858)
1 : a reddish brown sandstone used for building
2 : a dwelling faced with brownstone

brown study *noun* (1532)
: a state of serious absorption or abstraction

brown sugar *noun* (1704)
: soft sugar whose crystals are covered by a film of refined dark syrup

Brown Swiss *noun* (1902)
: any of a breed of large hardy brown dairy cattle originating in Switzerland

brown–tail moth \'braůn-ˌtāl-\ *noun* (1782)
: a tussock moth (*Euproctis chrysorrhoea*) introduced in the U.S. and having larvae which feed on foliage and are irritating to the skin

brown trout *noun* (1886)
: a speckled European trout (*Salmo trutta*) widely introduced as a game fish

brow·ridge \'braů-ˌrij\ *noun* (1898)
: a prominence of the frontal bone above the eye caused by the projection of the frontal air sinuses

[1]**browse** \'braůz\ *verb* **browsed; brows·ing** (15th century)
transitive verb
1 a : to consume as browse **b :** GRAZE
2 : to look over casually **:** SKIM
intransitive verb
1 a : to feed on or as if on browse **b :** GRAZE
2 a : to skim through a book reading at random passages that catch the eye **b :** to look over or through an aggregate of things casually especially in search of something of interest
— **brows·er** *noun*

[2]**browse** *noun* [probably modification of Middle French *brouts,* plural of *brout* sprout, from Old French *brost,* of Germanic origin; akin to Old Saxon *brustian* to sprout, and perhaps to Old English *brēost* breast] (1523)
1 : tender shoots, twigs, and leaves of trees and shrubs used by animals for food
2 : an act or instance of browsing

bru·cel·la \brü-'se-lə\ *noun, plural* **-cel·lae** \-'se-(ˌ)lē\ *or* **-cel·las** [New Latin, from Sir David *Bruce*] (1930)
: any of a genus (*Brucella*) of nonmotile pleomorphic bacteria that cause disease in humans and domestic animals

bru·cel·lo·sis \ˌbrü-sə-'lō-səs\ *noun, plural* **-lo·ses** \-ˌsēz\ [New Latin] (1930)
: infection with or disease caused by brucellae

bru·cine \'brü-ˌsēn\ *noun* [probably from French, from New Latin *Brucea* (genus name of *Brucea antidysenterica,* a shrub)] (1823)
: a poisonous alkaloid $C_{23}H_{26}N_2O_4$ found with strychnine especially in nux vomica

bru·in \'brü-ən\ *noun* [Middle Dutch, name of the bear in *Reynard the Fox*] (15th century)
: BEAR 1

[1]**bruise** \'brüz\ *verb* **bruised; bruis·ing** [Middle English *brusen, brisen,* from Middle French & Old English; Middle French *bruisier* to break, of Celtic origin; akin to Old Irish *bruid* he shatters; Old English *brȳsan* to bruise; akin to Old Irish *bruid,* Latin *frustum* piece] (14th century)
transitive verb
1 a *archaic* **:** DISABLE **b :** BATTER, DENT
2 : to inflict a bruise on **:** CONTUSE
3 : to break down (as leaves or berries) by pounding **:** CRUSH
4 : WOUND, INJURE; *especially* **:** to inflict psychological hurt on
intransitive verb
1 : to inflict a bruise
2 : to undergo bruising

[2]**bruise** *noun* (1541)
1 a : an injury involving rupture of small blood vessels and discoloration without a break in the overlying skin **:** CONTUSION **b :** a similar injury to plant tissue
2 : ABRASION, SCRATCH
3 : an injury especially to the feelings

bruis·er \'brü-zər\ *noun* (1744)
: a big husky man

[1]**bruit** *noun* [Middle English, from Middle French, from Old French, noise] (15th century)
1 \'brüt\ *archaic* **a :** NOISE, DIN **b :** REPORT, RUMOR
2 \'brü-ē\ [French, literally, noise] **:** any of several generally abnormal sounds heard on auscultation

[2]**bruit** \'brüt\ *transitive verb* (15th century)
: to noise abroad **:** REPORT

bru·mal \'brü-məl\ *adjective* [Latin *brumalis,* from *bruma* winter] (1513)
archaic **:** indicative of or occurring in the winter

brum·by \'brəm-bē\ *noun, plural* **brumbies** [origin unknown] (1880)
Australian **:** a wild or unbroken horse

brume \'brüm\ *noun* [French, mist, winter, from Old Provençal *bruma,* from Latin, winter solstice, winter; akin to Latin *brevis* short — more at BRIEF] (1808)
: MIST, FOG
— **bru·mous** \'brü-məs\ *adjective*

brum·ma·gem \'brə-mi jəm\ *adjective* [alteration of *Birmingham,* England, the source in the 17th century of counterfeit groats] (1637)
: SPURIOUS; *also* **:** cheaply showy **:** TAWDRY
— **brummagem** *noun*

brunch \'brənch\ *noun* [*b*reakfast + *l*unch] (1896)
: a meal usually taken late in the morning that combines a late breakfast and an early lunch

[1]**bru·net** *or* **bru·nette** \brü-'net\ *noun* (circa 1539)
: a person having brown or black hair and usually a relatively dark complexion

[2]**brunet** *or* **brunette** *adjective* [French *brunet,* masculine, *brunette,* feminine, brownish, from Old French, from *brun* brown, from Medieval Latin *brunus,* of Germanic origin; akin to Old High German *brūn* brown] (1752)
1 : being a brunet
2 : of a dark-brown or black color ⟨*brunet* hair⟩

Brun·hild \'brün-ˌhilt\ *noun* [German]
: a queen in Germanic legend won by Siegfried for Gunther

bru·ni·zem \'brü-nə-ˌzem\ *noun* [*bruni-* (from Medieval Latin *brunus* brown) + *-zem* earth (as in *chernozem*)] (1953)
: any of a zonal group of deep dark prairie soils developed from loess

Bruns·wick stew \'brənz-(ˌ)wik-\ *noun* [*Brunswick* county, Va.] (1856)
: a stew made of vegetables and usually two meats (as chicken and squirrel)

brunt \'brənt\ *noun* [Middle English] (15th century)
1 : the principal force, shock, or stress (as of an attack)
2 : the greater part **:** BURDEN

[1]**brush** \'brəsh\ *noun* [Middle English *brusch,* from Middle French *broce*] (14th century)
1 : BRUSHWOOD
2 a : scrub vegetation **b :** land covered with scrub vegetation

[2]**brush** *noun* [Middle English *brusshe,* from Middle French *broisse,* from Old French *broce*] (14th century)
1 : a device composed of bristles typically set into a handle and used especially for sweeping, smoothing, scrubbing, or painting
2 : something resembling a brush: as **a :** a bushy tail **b :** a feather tuft worn on a hat
3 : an electrical conductor that makes sliding contact between a stationary and a moving part (as of a generator or a motor)
4 a : an act of brushing **b :** a quick light touch or momentary contact in passing

[3]**brush** *transitive verb* (15th century)
1 a : to apply a brush to **b :** to apply with a brush
2 a : to remove with passing strokes (as of a brush) **b :** to dispose of in an offhand way **:** DISMISS ⟨*brushed* him off⟩
3 : to pass lightly over or across **:** touch gently against in passing
— **brush·er** *noun*

[4]**brush** *noun* [Middle English *brusche* rush, hostile collision, from *bruschen*] (14th century)
: a brief encounter or skirmish ⟨a *brush* with disaster⟩ ⟨a *brush* with the law⟩

[5]**brush** *intransitive verb* [Middle English *bruschen* to rush, from Middle French *brosser* to dash through underbrush, from *broce*] (1674)
: to move lightly or heedlessly ⟨*brushed* past the well-wishers waiting to greet him⟩

brush·abil·i·ty \ˌbrə-shə-'bi-lə-tē\ *noun* (1936)
: ease of application with a brush ⟨*brushability* of a paint⟩

brush·back \'brəsh-ˌbak\ *noun* (1954)
: a fastball intentionally thrown near the batter's head or body in baseball
— **brush back** *transitive verb*

brush border *noun* (1903)
: a stria of microvilli on the plasma membrane of an epithelial cell (as in a kidney tubule) that is specialized for absorption

brush cut *noun* (1945)
: CREW CUT

brush discharge *noun* (1849)
: a faintly luminous relatively slow electrical discharge having no spark

brushed \'brəsht\ *adjective* (1926)
1 : finished with a nap ⟨a *brushed* fabric⟩
2 : polished but not shiny ⟨*brushed* aluminum⟩

brush·fire \'brəsh-ˌfīr\ *adjective* [*brush fire* (a fire involving brush but not full-sized trees)] (1954)
: involving mobilization only on a small and local scale ⟨*brushfire* border wars⟩

brush·land \-ˌland\ *noun* (1853)
: an area covered with brush growth

brush–off \-ˌȯf\ *noun* (1941)
: a quietly curt or disdainful dismissal

brush up (circa 1600)
transitive verb
1 : to improve or polish as if by brushing
2 : to renew one's skill in ⟨*brush up* your Spanish⟩
intransitive verb
: to refresh one's memory **:** renew one's skill ⟨*brush up* on math⟩
— **brush·up** \'brəsh-ˌəp\ *noun*

brush·wood \'brəsh-ˌwůd\ *noun* (circa 1613)
1 : wood of small branches especially when cut or broken
2 : a thicket of shrubs and small trees

brush·work \-ˌwərk\ *noun* (1868)

: work done with a brush (as in painting); *especially* : the characteristic work of an artist using a brush

[1]brushy \'brə-shē\ *adjective* **brush·i·er; -est** (1658)
: covered with or abounding in brush or brushwood ⟨*brushy* hills⟩ ⟨a *brushy* habitat⟩

[2]brushy *adjective* **brush·i·er; -est** (1673)
: SHAGGY, ROUGH ⟨a *brushy* moustache⟩

brusque *also* **brusk** \'brəsk\ *adjective* [French *brusque,* from Italian *brusco,* from Medieval Latin *bruscus* butcher's-broom (plant with bristly twigs)] (1651)
1 : markedly short and abrupt
2 : blunt in manner or speech often to the point of ungracious harshness
synonym see BLUFF
— brusque·ly *adverb*
— brusque·ness *noun*

brus·que·rie \ˌbrəs-kə-'rē\ *noun* [French, from *brusque*] (1752)
: abruptness of manner

Brus·sels carpet \'brə-səlz-\ *noun* [*Brussels,* Belgium] (1799)
: a carpet made of colored worsted yarns first fixed in a foundation web of strong linen thread and then drawn up in loops to form the pattern

Brussels griffon *noun* (1904)
: any of a breed of short-faced compact rough- or smooth-coated toy dogs of Belgian origin — called also *griffon*

Brussels griffon

Brussels lace *noun* (1748)
1 : any of various fine needlepoint or bobbin laces with floral designs first made in or near Brussels
2 : a machine-made net of hexagonal mesh

brus·sels sprout \'brə-səl-\ *noun, often B capitalized* (1796)
1 *plural* **:** a plant (*Brassica oleracea gemmifera*) of the mustard family that bears small edible green heads on its stem
2 : any of the edible green heads borne on brussels sprouts — usually used in plural

brussels sprout 1

brut \'brüt, 'brœt\ *adjective* [French, literally, rough] (1891)
of champagne **:** very dry; *specifically* **:** being the driest made by the producer

bru·tal \'brü-tᵊl\ *adjective* [Middle English, from Middle French or Medieval Latin; Middle French, from Medieval Latin *brutalis,* from Latin *brutus*] (15th century)
1 *archaic* **:** typical of beasts **:** ANIMAL
2 : befitting a brute: as **a :** grossly ruthless or unfeeling ⟨a *brutal* slander⟩ **b :** CRUEL, COLD-BLOODED ⟨a *brutal* attack⟩ **c :** HARSH, SEVERE ⟨*brutal* weather⟩ **d :** unpleasantly accurate and incisive ⟨the *brutal* truth⟩ ☆
— bru·tal·ly \-tᵊl-ē\ *adverb*

bru·tal·ise *British variant of* BRUTALIZE

bru·tal·i·ty \brü-'ta-lə-tē\ *noun, plural* **-ties** (1549)
1 : the quality or state of being brutal
2 : a brutal act or course of action

bru·tal·ize \'brü-tᵊl-ˌīz\ *transitive verb* **-ized; -iz·ing** (circa 1704)
1 : to make brutal, unfeeling, or inhuman ⟨people *brutalized* by poverty and disease⟩
2 : to treat brutally ⟨an accord not to *brutalize* prisoners of war⟩
— bru·tal·iza·tion \ˌbrü-tᵊl-ə-'zā-shən\ *noun*

[1]brute \'brüt\ *adjective* [Middle English, from Middle French *brut* rough, from Latin *brutus* brutish, literally, heavy; akin to Latin *gravis* heavy — more at GRIEVE] (15th century)
1 : of or relating to beasts ⟨the ways of the *brute* world⟩
2 : INANIMATE 1a
3 : characteristic of an animal in quality, action, or instinct: as **a :** CRUEL, SAVAGE ⟨*brute* violence⟩ **b :** not working by reason ⟨*brute* instinct⟩
4 : purely physical ⟨*brute* strength⟩
5 : unrelievedly harsh ⟨*brute* facts⟩ ⟨*brute* necessity⟩

[2]brute *noun* (1611)
1 : BEAST
2 : a brutal person

brut·ish \'brü-tish\ *adjective* (1534)
1 : befitting beasts ⟨lived a short and *brutish* life as a slave⟩
2 a : strongly and grossly sensual ⟨*brutish* gluttony⟩ **b :** showing little intelligence or sensibility ⟨a *brutish* lack of understanding⟩
synonym see BRUTAL
— brut·ish·ly *adverb*
— brut·ish·ness *noun*

brux·ism \'brək-ˌsi-zəm\ *noun* [irregular from Greek *brychein* to gnash the teeth + English *-ism*] (circa 1940)
: the habit of unconsciously gritting or grinding the teeth especially in situations of stress or during sleep

Bryn·hild \'brin-ˌhild\ *noun* [Old Norse *Brynhildr*]
: a Valkyrie who is waked from an enchanted sleep by Sigurd and later has him killed when he forgets her

bry·ol·o·gy \brī-'ä-lə-jē\ *noun* [Greek *bryon* moss (akin to Greek *bryein* to grow luxuriantly) + International Scientific Vocabulary *-logy*] (circa 1859)
1 : moss life or biology
2 : a branch of botany that deals with the bryophytes
— bry·o·log·i·cal \brī-ə-'lä-ji-kəl\ *adjective*
— bry·ol·o·gist \brī-'ä-lə-jist\ *noun*

bry·o·ny \'brī-ə-nē\ *noun, plural* **-nies** [Middle English, from Latin *bryonia,* from Greek *bryōnia;* akin to Greek *bryein*] (14th century)
: any of a genus (*Bryonia*) of tendril-bearing vines of the gourd family with large leaves and red or black fruit

bryo·phyl·lum \ˌbrī-ə-'fi-ləm\ *noun* [New Latin, from Greek *bryon* + *phyllon* leaf — more at BLADE] (circa 1868)
: any of various kalanchoes; *especially* **:** a succulent kalanchoe (*Kalanchoe pinnata*) often grown as a foliage plant that propagates new plants from its leaves

bryo·phyte \'brī-ə-ˌfīt\ *noun* [ultimately from Greek *bryon* + *phyton* plant; akin to Greek *phyein* to bring forth — more at BE] (1878)
: any of a division (Bryophyta) of nonflowering plants comprising the mosses, liverworts, and hornworts
— bryo·phyt·ic \ˌbrī-ə-'fi-tik\ *adjective*

bryo·zo·an \ˌbrī-ə-'zō-ən\ *noun* [New Latin *Bryozoa,* from Greek *bryon* + New Latin *-zoa*] (circa 1864)
: any of a phylum (Bryozoa) of aquatic mostly marine invertebrate animals that reproduce by budding and usually form permanently attached branched or mossy colonies
— bryozoan *adjective*

[1]Bry·thon·ic \bri-'thä-nik\ *adjective* (1884)
: of, relating to, or characteristic of the division of the Celtic languages that includes Welsh, Cornish, and Breton

[2]Brythonic *noun* (1884)
: the Brythonic branch of the Celtic languages — see INDO-EUROPEAN LANGUAGES table

Bt \'bē-'tē\ *noun* [New Latin *Bacillus thuringiensis,* species name, literally, Thuringian bacillus] (1971)
: an insecticide composed of a bacterium (*Bacillus thuringiensis*) pathogenic especially for lepidopteran larvae; *also* **:** the bacterium itself

bub \'bəb\ *noun* [probably short for *bubby* little boy] (1839)
: FELLOW, BUDDY — used in informal address ⟨come on, *bub,* get moving⟩

[1]bub·ble \'bə-bəl\ *noun, often attributive* [Middle English *bobel*] (14th century)
1 : a small globule typically hollow and light: as **a :** a small body of gas within a liquid **b :** a thin film of liquid inflated with air or gas **c :** a globule in a transparent solid **d :** something (as a plastic or inflatable structure) that is hemispherical or semicylindrical
2 a : something that lacks firmness, solidity, or reality **b :** a delusive scheme
3 : a sound like that of bubbling
4 : MAGNETIC BUBBLE

[2]bubble *verb* **bub·bled; bub·bling** \'bə-b(ə-)liŋ\ (15th century)
intransitive verb
1 a : to form or produce bubbles **b :** to rise in or as if in bubbles — usually used with *up*
2 : to flow with a gurgling sound ⟨a brook *bubbling* over rocks⟩
3 a : to become lively or effervescent ⟨*bubbling* with good humor⟩ **b :** to speak in a lively and fluent manner
transitive verb
1 : to utter (as words) effervescently
2 : to cause to bubble

bubble and squeak *noun* (circa 1785)
: a British dish consisting of usually leftover potatoes, greens (as cabbage), and sometimes meat fried together

bubble chamber *noun* (1953)
: a chamber of superheated liquid in which the path of an ionizing particle is made visible by a string of vapor bubbles

bub·ble·gum \'bə-bəl-ˌgəm\ *adjective* (1969)
: appealing to or characteristic of preteens or adolescents ⟨*bubblegum* music⟩

bubble gum *noun* (1937)
1 : a chewing gum that can be blown into large bubbles
2 *usually* **bub·ble·gum** \'bə-bəl-ˌgəm\ **:** rock music characterized by simple repetitive phrasings and intended especially for young teenagers

bub·ble·head \'bə-bəl-ˌhed\ *noun* (1949)
: a foolish or stupid person
— bub·ble·head·ed \-ˌhe-dəd\ *adjective*

bubble memory *noun* (1969)
: a computer memory that uses magnetic bubbles to store information

bub·bler \'bə-b(ə-)lər\ *noun* (1914)
1 : a drinking fountain from which a stream of water bubbles upward
2 : one that bubbles

[1]bub·bly \'bə-b(ə-)lē\ *adjective* **bub·bli·er; -est** (1599)
1 : full of bubbles **:** EFFERVESCENT ⟨a *bubbly* bottle of pop⟩
2 : full of or showing good spirits **:** LIVELY, EFFUSIVE
3 : resembling a bubble ⟨a *bubbly* dome⟩

[2]bubbly *noun* (1920)
: CHAMPAGNE

bub·by \'bə-bē\ *noun, plural* **bubbies** [probably of imitative origin] (1686)

☆ SYNONYMS
Brutal, brutish, bestial, feral mean characteristic of an animal in nature, action, or instinct. BRUTAL applies to people, their acts, or their words and suggests a lack of intelligence, feeling, or humanity ⟨a senseless and *brutal* war⟩. BRUTISH stresses likeness to an animal in low intelligence, in base appetites, and in behavior based on instinct ⟨*brutish* stupidity⟩. BESTIAL suggests a state of degradation unworthy of man and fit only for beasts ⟨*bestial* depravity⟩. FERAL suggests the savagery or ferocity of wild animals ⟨the struggle to survive unleashed their *feral* impulses⟩.

: BREAST 1 — sometimes considered vulgar

bu·bo \ˈbyü-(ˌ)bō, ˈbü-\ *noun, plural* **buboes** [Medieval Latin *bubon-, bubo,* from Greek *boubōn*] (14th century)
: an inflammatory swelling of a lymph gland especially in the groin
— **bu·bon·ic** \byü-ˈbä-nik, bü-\ *adjective*

bubonic plague *noun* (1885)
: plague caused by a bacterium (*Yersinia pestis*) and characterized especially by the formation of buboes

buc·cal \ˈbə-kəl\ *adjective* [Latin *bucca* cheek] (circa 1771)
1 : of, relating to, near, involving, or supplying a cheek ⟨the *buccal* surface of a tooth⟩ ⟨the *buccal* branch of the facial nerve⟩
2 : of, relating to, involving, or lying in the mouth ⟨the *buccal* cavity⟩
— **buc·cal·ly** *adverb*

buc·ca·neer \ˌbə-kə-ˈnir\ *noun* [French *boucanier*] (circa 1690)
1 : any of the freebooters preying on Spanish ships and settlements especially in the West Indies in the 17th century; *broadly* **:** PIRATE
2 : an unscrupulous adventurer especially in politics or business ◆
— **buccaneer** *intransitive verb*
— **buc·ca·neer·ish** \-ish\ *adjective*

buc·ci·na·tor \ˈbək-sə-ˌnā-tər\ *noun* [New Latin, from Latin *bucinator* trumpeter, from *bucinare* to sound on the trumpet, from *bucina* trumpet, from *bov-, bos* cow + *canere* to sing, play — more at COW, CHANT] (1671)
: a thin broad muscle forming the wall of the cheek

[1]buck \ˈbək\ *noun, plural* **bucks** [Middle English, from Old English *bucca* stag, he-goat; akin to Old High German *boc* he-goat, Middle Irish *bocc*] (before 12th century)
1 *or plural* **buck :** a male animal; *especially* **:** a male deer or antelope
2 a : a male human being **:** MAN **b :** a dashing fellow **:** DANDY
3 *or plural* **buck :** ANTELOPE
4 a : BUCKSKIN; *also* **:** an article (as a shoe) made of buckskin **b** (1) **:** DOLLAR 3b (2) **:** a sum of money especially to be gained ⟨make a quick *buck*⟩; *also* **:** MONEY — usually used in plural
5 [short for *sawbuck* sawhorse] **a :** a supporting rack or frame **b :** a short thick leather-covered block for gymnastic vaulting

[2]buck (1750)
transitive verb
1 a *archaic* **:** [1]BUTT **b :** OPPOSE, RESIST ⟨*bucking* the system⟩
2 : to throw (as a rider) by bucking
3 : to charge into (as the opponent's line in football)
4 a : to pass especially from one person to another **b :** to move or load (as heavy objects) especially with mechanical equipment
intransitive verb
1 *of a horse or mule* **:** to spring into the air with the back arched
2 : to charge against something (as an obstruction)
3 a : to move or react jerkily **b :** to refuse assent **:** BALK
4 : to strive for advancement sometimes without regard to ethical behavior ⟨*bucking* for a promotion⟩
— **buck·er** *noun*

[3]buck *noun* (circa 1877)
: an act or instance of bucking

[4]buck *noun* [short for earlier *buckhorn knife*] (1865)
1 : an object formerly used in poker to mark the next player to deal; *broadly* **:** a token used as a mark or reminder
2 : RESPONSIBILITY — used especially in the phrases *pass the buck* and *the buck stops here*

[5]buck *adjective* [probably from [1]*buck*] (1918)
: of the lowest grade within a military category ⟨*buck* private⟩

[6]buck *adverb* [origin unknown] (1928)
: STARK ⟨*buck* naked⟩

buck–and–wing \ˌbə-kᵊn-ˈwiŋ\ *noun* (1895)
: a solo tap dance with sharp foot accents, springs, leg flings, and heel clicks

buck·a·roo *also* **buck·er·oo** \ˌbə-kə-ˈrü, ˈbə-kə-ˌ\ *noun, plural* **-aroos** *also* **-eroos** [by folk etymology from Spanish *vaquero,* from *vaca* cow, from Latin *vacca* — more at VACCINE] (1827)
1 : COWBOY
2 : BRONCOBUSTER

buck·bean \ˈbək-ˌbēn\ *noun* (1578)
: a plant (*Menyanthes trifoliata* of the family Menyanthaceae) growing in bogs and having racemes of white or purplish flowers

buck·board \-ˌbōrd, -ˌbȯrd\ *noun* [obsolete English *buck* body of a wagon + English *board*] (1839)
: a four-wheeled vehicle with a floor made of long springy boards

buckboard

[1]buck·et \ˈbə-kət\ *noun* [Middle English, from Anglo-French *buket,* from Old English *būc* pitcher, belly; akin to Old High German *būh* belly] (13th century)
1 : a typically cylindrical vessel for catching, holding, or carrying liquids or solids
2 : something resembling a bucket: as **a :** the scoop of an excavating machine **b :** one of the receptacles on the rim of a waterwheel **c :** one of the cups of an endless-belt conveyor **d :** one of the vanes of a turbine rotor
3 : BUCKETFUL
4 : BUCKET SEAT
5 : BASKET 3b

[2]bucket (1649)
transitive verb
1 : to draw or lift in buckets
2 *British* **a :** to ride (a horse) hard **b :** to drive hurriedly or roughly
3 : to deal with in a bucket shop
intransitive verb
1 : HUSTLE, HURRY
2 a : to move about haphazardly or irresponsibly **b :** to move roughly or jerkily

bucket brigade *noun* (1911)
: a chain of persons acting to put out a fire by passing buckets of water from hand to hand

buck·et·ful \ˈbə-kət-ˌfu̇l\ *noun, plural* **buck·etfuls** \-ˌfu̇lz\ *or* **buck·ets·ful** \-kəts-ˌfu̇l\ (circa 1563)
: as much as a bucket will hold; *broadly* **:** a large quantity

bucket seat *noun* (1908)
: a low separate seat for one person (as in an automobile or an airplane)

bucket shop *noun* (1875)
1 : a saloon in which liquor was formerly sold from or dispensed in open containers (as buckets or pitchers)
2 a : a gambling establishment that formerly used market fluctuations (as in securities or commodities) as a basis for gaming **b :** a dishonest brokerage firm; *especially* **:** one that formerly failed to execute customers' margin orders in expectation of making a profit from market fluctuations adverse to the customers' interests

buck·eye \ˈbə-ˌkī\ *noun* (1763)
1 : a shrub or tree (genus *Aesculus*) of the horse-chestnut family; *also* **:** its large nutlike seed
2 *capitalized* **:** a native or resident of Ohio — used as a nickname

buck fever *noun* (1841)
: nervous excitement of an inexperienced hunter at the sight of game

[1]buck·le \ˈbə-kəl\ *noun* [Middle English *bocle,* from Middle French, boss of a shield, buckle, from Latin *buccula,* diminutive of *bucca* cheek] (14th century)
1 : a fastening for two loose ends that is attached to one and holds the other by a catch
2 : an ornamental device that suggests a buckle
3 *archaic* **:** a crisp curl

[2]buckle *verb* **buck·led; buck·ling** \ˈbə-k(ə-)liŋ\ (14th century)
transitive verb
1 : to fasten with a buckle
2 : to prepare with vigor
3 : to cause to bend, give way, or crumple
intransitive verb
1 : to become fastened with a buckle
2 : to apply oneself with vigor ⟨*buckles* down to the job⟩
3 : to bend, heave, warp, or kink usually under the influence of some external agency ⟨wheat *buckling* in the wind⟩
4 : COLLAPSE ⟨the props *buckled* under the strain⟩
5 : to give way **:** YIELD ⟨he *buckled* under pressure⟩

[3]buckle *noun* (circa 1876)
1 : a product of buckling
2 : a coffee cake baked with berries and a crumbly topping ⟨blueberry *buckle*⟩

[1]buck·ler \ˈbə-klər\ *noun* [Middle English *bocler,* from Middle French, shield with a boss, from *bocle*] (13th century)
1 a : a small round shield held by a handle at arm's length **b :** a shield worn on the left arm
2 : one that shields and protects

[2]buckler *transitive verb* (1590)
: to shield or defend with a buckler

buckle up *intransitive verb* (1976)
: to fasten one's seat belt

buck·min·ster·ful·ler·ene \ˈbək-(ˌ)min-stər-ˌfu̇-lə-ˈrēn\ *noun* [R. *Buckminster Fuller*] (1985)
: a spherical fullerene C_{60} that is an extremely stable form of pure carbon, consists of interconnected pentagons and hexagons suggestive of the geometry of a geodesic dome, and is believed to be a major constituent of soot

bucko \ˈbə-(ˌ)kō\ *noun, plural* **buck·oes** (1883)
1 : a person who is domineering and bullying **:** SWAGGERER

◇ WORD HISTORY

buccaneer The history of the word *buccaneer* begins not in the Caribbean, but rather on the coast of Brazil. The French Protestant missionary Jean de Léry, a member of a colonizing expedition to Guanabara Bay in 1556, wrote a detailed account of the Tupinambá Indians of that region and their customs. Léry describes their use of a raised wooden grill called by a word in their language that he rendered *boucan,* on which they would slowly cook meat, including limbs of their captured enemies. Most likely both this word and the culinary device were already familiar to French mariners, who during the previous half century had made many voyages to the Brazilian coast to collect brazilwood. In the 17th century, the word *boucan* surfaced again in the Caribbean, where the French had refocused their tropical American enterprises, and a new derivative *boucanier* now appeared, referring to Europeans, probably former mariners, who survived in the interior of the Antillean islands by hunting and smoking game and escaped livestock on a *boucan.* Some among these men returned to the sea as pirates, and the word *boucanier* continued to be applied to them in their new profession.

2 *chiefly Irish* **:** young fellow **:** LAD

buck passer *noun* (1920)
: a person who habitually passes the buck
— **buck–pass·ing** \ˈbək-ˌpa-siŋ\ *noun*

[1]**buck·ram** \ˈbə-krəm\ *noun* [Middle English *bukeram,* from Old French *boquerant,* from Old Provençal *bocaran,* probably from *Bokhara,* city of central Asia] (15th century)
1 : a stiff-finished heavily sized fabric of cotton or linen used for interlinings in garments, for stiffening in millinery, and in bookbinding
2 *archaic* **:** STIFFNESS, RIGIDITY

[2]**buckram** *adjective* (circa 1589)
: suggesting buckram especially in stiffness

[3]**buckram** *transitive verb* (1783)
1 : to give strength or stiffness to (as with buckram)
2 *archaic* **:** to make pretentious

Buck Rogers \ˈbək-ˈrä-jərz\ *adjective* [*Buck Rogers,* hero of a science-fiction comic strip created by Phil Nowlan] (1941)
: marked by futuristic and high-tech qualities
: suggestive of science fiction

buck·saw \ˈbək-ˌsȯ\ *noun* (1856)
: a saw set in a usually H-shaped frame for sawing wood

buck·shee \ˈbək-(ˌ)shē, ˌbək-ˈ\ *noun* [Hindi *bakhśiś,* from Persian *bakhshīsh* — more at BAKSHEESH] (circa 1760)
1 *British* **:** something extra obtained free; *especially* **:** extra rations
2 *British* **:** WINDFALL, GRATUITY

[1]**buck·shot** \ˈbək-ˌshät\ *noun* (1775)
: lead shot that is from .24 to .33 inch (about 6.1 to 8.4 millimeters) in diameter

[2]**buckshot** *adjective* (1941)
: SCATTERSHOT

buck·skin \-ˌskin\ *noun, often attributive* (14th century)
1 a : the skin of a buck **b :** a soft pliable usually suede-finished leather
2 a *plural* **:** buckskin breeches **b** *archaic* **:** a person dressed in buckskin; *especially* **:** an early American backwoodsman
3 : a horse of a light yellowish dun color with black mane and tail

buck·skinned \-ˌskind\ *adjective* (1829)
: dressed in buckskin

buck·tail \-ˌtāl\ *noun* (1911)
: an angler's lure made typically of hairs from the tail of a deer

buck·thorn \-ˌthȯrn\ *noun* (1578)
1 : any of a genus (*Rhamnus* of the family Rhamnaceae, the buckthorn family) of often thorny trees or shrubs some of which yield purgatives or pigments
2 : a tree (*Bumelia lycioides*) of the sapodilla family of the southern U.S.

buck·tooth \-ˈtüth\ *noun* (1753)
: a large projecting front tooth
— **buck·toothed** \-ˌtütht\ *adjective*

buck up *verb* [[2]*buck*] (1844)
intransitive verb
: to become encouraged
transitive verb
1 : IMPROVE, SMARTEN
2 : to raise the morale of

buck·wheat \ˈbək-ˌhwēt, -ˌwēt\ *noun* [Dutch *boekweit,* from Middle Dutch *boecweit,* from *boec-* (akin to Old High German *buohha* beech tree) + *weit* wheat — more at BEECH] (1548)
1 : any of a genus (*Fagopyrum* of the family Polygonaceae, the buckwheat family) of herbs with alternate leaves, clusters of apetalous pinkish white flowers, and triangular seeds; *especially* **:** either of two plants (*F. esculentum* and *F. tartaricum*) cultivated for their edible seeds
2 : the seed of a buckwheat used as a cereal grain

bucky·ball \ˈbə-kē-ˌbȯl\ *noun* [R. *Buck*minster Fuller + [4]-y] (1985)
: a molecule of buckminsterfullerene

bu·col·ic \byü-ˈkä-lik\ *adjective* [Latin *bucolicus,* from Greek *boukolikos,* from *boukolos* cowherd, from *bous* head of cattle + *-kolos* (akin to Latin *colere* to cultivate) — more at COW, WHEEL] (circa 1609)
1 : of or relating to shepherds or herdsmen **:** PASTORAL
2 : relating to or typical of rural life ⟨picnicked in a pleasant *bucolic* setting⟩
— **bu·col·i·cal·ly** \-li-k(ə-)lē\ *adverb*

[1]**bud** \ˈbəd\ *noun* [Middle English *budde*] (14th century)
1 : a small lateral or terminal protuberance on the stem of a plant that may develop into a flower, leaf, or shoot
2 : something not yet mature or at full development: as **a :** an incompletely opened flower **b :** CHILD, YOUTH **c :** an outgrowth of an organism that differentiates into a new individual **:** GEMMA; *also* **:** PRIMORDIUM
— **in the bud :** in an early stage of development ⟨nipped the rebellion *in the bud*⟩

[2]**bud** *verb* **bud·ded; bud·ding** (14th century)
intransitive verb
1 *of a plant* **a :** to set or put forth buds **b :** to commence growth from buds
2 : to grow or develop as if from a bud
3 : to reproduce asexually especially by the pinching off of a small part of the parent
transitive verb
1 : to produce or develop from buds
2 : to cause (as a plant) to bud
3 : to insert a bud from a plant of one kind into an opening in the bark of (a plant of another kind) usually in order to propagate a desired variety
— **bud·der** *noun*

Bud·dha \ˈbü-də, ˈbu̇-\ *noun* [Sanskrit, enlightened; akin to Sanskrit *bodhi* enlightenment — more at BID] (1681)
1 : a person who has attained Buddhahood
2 : a representation of Gautama Buddha

Bud·dha·hood \-ˌhu̇d\ *noun* (1837)
: a state of perfect enlightenment sought in Buddhism

Bud·dhism \ˈbü-ˌdi-zəm, ˈbu̇-\ *noun* (1801)
: a religion of eastern and central Asia growing out of the teaching of Gautama Buddha that suffering is inherent in life and that one can be liberated from it by mental and moral self-purification
— **Bud·dhist** \ˈbü-dist, ˈbu̇-\ *noun or adjective*
— **Bud·dhis·tic** \bü-ˈdis-tik, bu̇-\ *adjective*

bud·ding \ˈbə-diŋ\ *adjective* (1581)
: being in an early stage of development ⟨*budding* novelists⟩

bud·dle \ˈbə-dᵊl\ *noun* [origin unknown] (circa 1532)
: an apparatus on which crushed ore is concentrated by washing

bud·dle·ia \ˈbəd-lē-ə, ˌbəd-ˈlē-\ *noun* [New Latin, genus name, from Adam *Buddle* (died 1715) English botanist] (1791)
: any of a genus (*Buddleia* of the family Loganiaceae) of shrubs or trees of warm regions with showy terminal clusters of usually yellow or violet flowers

[1]**bud·dy** \ˈbə-dē\ *noun, plural* **buddies** [probably baby talk alteration of *brother*] (1850)
1 a : COMPANION, PARTNER **b :** FRIEND 1
2 : FELLOW — used especially in informal address

[2]**buddy** *intransitive verb* **bud·died; bud·dy·ing** (1918)
: to become friendly — usually used with *up* or *with*

[3]**buddy** *adjective* (1976)
: featuring a friendship or partnership between the two main usually male characters ⟨a *buddy* movie⟩

bud·dy–bud·dy \ˌbə-dē-ˈbə-dē\ *adjective* (1951)
: familiarly friendly

buddy system *noun* (1942)
: an arrangement in which two individuals are paired (as for mutual safety in a hazardous situation)

[1]**budge** \ˈbəj\ *noun* [Middle English *bugee,* from Anglo-French *bogee*] (14th century)
: a fur formerly prepared from lambskin dressed with the wool outward

[2]**budge** *verb* **budged; budg·ing** [Middle French *bouger,* from (assumed) Vulgar Latin *bullicare,* from Latin *bullire* to boil — more at BOIL] (1590)
intransitive verb
1 : MOVE, SHIFT ⟨the mule wouldn't *budge*⟩
2 : to give way **:** YIELD ⟨wouldn't *budge* on the issue⟩
transitive verb
: to cause to move or change

[3]**budge** *adjective* [origin unknown] (1634)
archaic **:** POMPOUS, SOLEMN

bud·ger·i·gar \ˈbə-jə-rē-ˌgär\ *noun* [modification of Yuwaalaraay (Australian aboriginal language of northern New South Wales) *gijirrigaa*] (1840)
: a small Australian parrot (*Melopsittacus undulatus*) usually light green with black and yellow markings in the wild but bred under domestication in many colors

[1]**bud·get** \ˈbə-jət\ *noun* [Middle English *bowgette,* from Middle French *bougette,* diminutive of *bouge* leather bag, from Latin *bulga,* of Celtic origin; akin to Middle Irish *bolg* bag; akin to Old English *belg* bag — more at BELLY] (15th century)
1 *chiefly dialect* **:** a usually leather pouch, wallet, or pack; *also* **:** its contents
2 : STOCK, SUPPLY
3 : a quantity (as of energy or water) involved in, available for, or assignable to a particular situation; *also* **:** an account of gains and losses of such a quantity
4 a : a statement of the financial position of an administration for a definite period of time based on estimates of expenditures during the period and proposals for financing them **b :** a plan for the coordination of resources and expenditures **c :** the amount of money that is available for, required for, or assigned to a particular purpose ◆
— **bud·get·ary** \ˈbə-jə-ˌter-ē\ *adjective*

[2]**budget** (1618)
transitive verb
1 a : to put or allow for in a budget **b :** to require to adhere to a budget

◇ WORD HISTORY

budget Some loanwords have had a more successful career in the language that took them than in the one that gave them; on occasion this had led to reborrowing by the donor. Such a word is *budget,* borrowed from medieval French *bougette* "leather bag" and recorded in later Middle English as *bowgett.* The French word was a diminutive of *bouge* "travel bag, chest, sack." In French, both *bouge* and *bougette* became largely obsolete or historical in reference by the 18th century, at least in their original senses. Middle English *bowgett,* on the other hand, emerged in Early Modern English as *budget.* From its first sense "satchel," *budget* came to denote the contents of a satchel, especially a bundle of letters or papers—a semantic shift identical to that of [2]*mail.* In the 18th century British Parliament, the Chancellor of the Exchequer was said "to open the budget" when he presented the government's estimates of revenues and expenditures for the following year. By the 19th century, any public or private estimation of a similar nature could be called a *budget.* The usefulness of this word and the influence of English parliamentary institutions led to the adoption of *budget* in a number of European languages, including French.

2 a : to allocate funds for in a budget ⟨*budget* a new hospital⟩ **b :** to plan or provide for the use of in detail ⟨*budgeting* manpower⟩
intransitive verb
: to put oneself on a budget ⟨*budgeting* for a vacation⟩

[3]**budget** *adjective* (1941)
: suitable for one on a budget **:** INEXPENSIVE

bud·ge·teer \ˌbə-jə-ˈtir\ *also* **bud·get·er** \ˈbə-jə-tər\ *noun* (circa 1845)
1 : one who prepares a budget
2 : one who is restricted to a budget

bud·gie \ˈbə-jē\ *noun* [by shortening and alteration] (1935)
: BUDGERIGAR

bud scale *noun* (circa 1880)
: one of the leaves resembling scales that form the sheath of a plant bud

bud·worm \ˈbəd-ˌwərm\ *noun* (1849)
: a moth larva that feeds on the buds of plants — compare SPRUCE BUDWORM

[1]**buff** \ˈbəf\ *noun* [Middle French *buffle* wild ox, from Old Italian *bufalo*] (1580)
1 : a garment (as a uniform) made of buff leather
2 : the state of being nude ⟨sunbathing in the *buff*⟩
3 a : a moderate orange yellow **b :** a light to moderate yellow
4 : a device having a soft absorbent surface (as of cloth) by which polishing material is applied
5 [earlier *buff* an enthusiast about going to fires; perhaps from the buff overcoats worn by volunteer firefighters in New York City *about* 1820] **:** FAN, ENTHUSIAST

[2]**buff** *adjective* (circa 1771)
: of the color buff

[3]**buff** *transitive verb* (1885)
1 : POLISH, SHINE ⟨waxed and *buffed* the floor⟩
2 : to give a velvety surface to (leather)

[1]**buf·fa·lo** \ˈbə-fə-ˌlō\ *noun, plural* **-lo** *or* **-loes** *also* **-los** *often attributive* [Italian *bufalo* & Spanish *búfalo,* from Late Latin *bufalus,* alteration of Latin *bubalus,* from Greek *boubalos* African gazelle] (1562)
1 : any of several wild bovids: as **a :** WATER BUFFALO **b :** CAPE BUFFALO **c** (1) **:** any of a genus (*Bison*) of bovids; *especially* **:** a large shaggy-maned North American bovid (*B. bison*) that has short horns and heavy forequarters with a large muscular hump and that was formerly abundant on the central and western plains — compare WISENT (2) **:** the flesh of the buffalo used as food

buffalo 1c(1)

2 : any of several suckers (genus *Ictiobus*) found mostly in the Mississippi valley — called also *buffalofish*

[2]**buffalo** *transitive verb* **-loed; -lo·ing** (circa 1896)
: BEWILDER, BAFFLE

buf·fa·lo·ber·ry \-ˌber-ē\ *noun* (1805)
: either of two western U.S. shrubs (*Shepherdia argentea* and *S. canadensis*) of the oleaster family with silvery foliage; *also* **:** their edible scarlet berry

buffalo gourd *noun* (circa 1928)
: a perennial cucurbit (*Cucurbita foetidissima*) of arid lands of the central and southwest U.S. and Mexico with a large starchy taproot and seeds rich in an oil suitable after processing for use in cooking or as fuel

buffalo grass *noun* (1784)
: a low-growing grass (*Buchloë dactyloides*) of former feeding grounds of the American buffalo; *also* **:** GRAMA

buffalo plaid *noun* (1979)
: a broad checkered plaid pattern usually of two colors

[1]**buff·er** \ˈbə-fər\ *noun, often attributive* [*buff,* verb, to react like a soft body when struck] (1835)
1 : any of various devices or pieces of material for reducing shock or damage due to contact
2 : a means or device used as a cushion against the shock of fluctuations in business or financial activity
3 : something that serves as a protective barrier: as **a :** BUFFER STATE **b :** a person who shields another especially from annoying routine matters **c :** MEDIATOR 1
4 : a substance capable in solution of neutralizing both acids and bases and thereby maintaining the original acidity or basicity of the solution; *also* **:** a solution containing such a substance
5 : a temporary storage unit (as in a computer); *especially* **:** one that accepts information at one rate and delivers it at another
— **buff·ered** \-fərd\ *adjective*

[2]**buffer** *transitive verb* **buff·ered; buff·er·ing** \-f(ə-)riŋ\ (1894)
1 : to lessen the shock of **:** CUSHION
2 : to treat (as a solution or its acidity) with a buffer; *also* **:** to prepare (aspirin) with an antacid
3 : to collect (as data) in a buffer

[3]**buffer** *noun* (1854)
: one that buffs

buffer state *noun* (1883)
: a usually neutral state lying between two larger potentially rival powers

buffer zone *noun* (1908)
: a neutral area separating conflicting forces; *broadly* **:** an area designed to separate

[1]**buf·fet** \ˈbə-fət\ *noun* [Middle English, from Old French, diminutive of *buffe* blow] (13th century)
1 : a blow especially with the hand
2 : something that strikes with telling force

[2]**buffet** (13th century)
transitive verb
1 : to strike sharply especially with the hand **:** CUFF
2 : to strike repeatedly **:** BATTER ⟨the waves *buffeted* the shore⟩
3 : to drive, force, move, or attack by or as if by repeated blows
intransitive verb
: to make one's way especially under difficult conditions

[3]**buf·fet** \(ˌ)bə-ˈfā, bü-ˈ, *British especially* ˈbü-ˌ\ *noun* [French] (1718)
1 : SIDEBOARD
2 a : a counter for refreshments **b** *chiefly British* **:** a restaurant operated as a public convenience (as in a railway station) **c :** a meal set out on a buffet or table for ready access and informal service

[4]**buf·fet** *same as* [3]\ *adjective* (1898)
: served informally (as from a buffet)

buffing wheel *noun* (circa 1889)
: a wheel covered with polishing material

buff leather *noun* (1580)
: a strong supple oil-tanned leather produced chiefly from cattle hides

buf·fle·head \ˈbə-fəl-ˌhed\ *noun* [archaic English *buffle* buffalo + English *head*] (1731)
: a small North American diving duck (*Bucephala albeola*)

buf·fo \ˈbü-(ˌ)fō\ *noun, plural* **buf·fi** \-(ˌ)fē\ *or* **buffos** [Italian, from *buffone*] (1764)
: CLOWN, BUFFOON; *specifically* **:** a male singer of comic roles in opera

buf·foon \(ˌ)bə-ˈfün\ *noun* [Middle French *bouffon,* from Old Italian *buffone*] (1585)
1 : a ludicrous figure **:** CLOWN
2 : a gross and usually ill-educated or stupid person
— **buf·foon·ish** \-ˈfü-nish\ *adjective*

buf·foon·ery \-ˈfü-nə-rē, -ˈfün-rē\ *noun, plural* **-er·ies** (1621)
: foolish or playful behavior or practice

[1]**bug** \ˈbəg\ *noun* [Middle English *bugge* scarecrow; akin to Norwegian dialect *bugge* important man] (14th century)
obsolete **:** BOGEY, BUGBEAR

[2]**bug** *noun* [origin unknown] (1622)
1 a : an insect or other creeping or crawling invertebrate **b :** any of several insects commonly considered obnoxious: as (1) **:** BEDBUG (2) **:** COCKROACH (3) **:** HEAD LOUSE **c :** any of an order (Hemiptera and especially its suborder Heteroptera) of insects that have sucking mouthparts, forewings thickened at the base, and incomplete metamorphosis and are often economic pests — called also *true bug*
2 : an unexpected defect, fault, flaw, or imperfection
3 a : a germ or microorganism especially when causing disease **b :** an unspecified or nonspecific sickness usually presumed due to a bug
4 : a sudden enthusiasm
5 : ENTHUSIAST ⟨a camera *bug*⟩
6 : a prominent person
7 : a concealed listening device
8 [from its designation by an asterisk on race programs] **:** a weight allowance given apprentice jockeys

[3]**bug** *transitive verb* **bugged; bug·ging** (1949)
1 : BOTHER, ANNOY ⟨don't *bug* me with petty details⟩
2 : to plant a concealed microphone in

[4]**bug** *verb* **bugged; bug·ging** [probably from [2]*bug*] (1877)
intransitive verb
of the eyes **:** PROTRUDE, BULGE — often used with *out*
transitive verb
: to cause to bug ⟨his eyes were *bugged* with horror⟩

bug·a·boo \ˈbə-gə-ˌbü\ *noun, plural* **-boos** [origin unknown] (1740)
1 : an imaginary object of fear
2 : BUGBEAR 2; *also* **:** something that causes fear or distress out of proportion to its importance

bug·bane \ˈbəg-ˌbān\ *noun* (1804)
: any of several perennial herbs (genus *Cimicifuga*) of the buttercup family; *especially* **:** BLACK COHOSH

bug·bear \-ˌbar, -ˌber\ *noun* (1581)
1 : an imaginary goblin or specter used to excite fear
2 a : an object or source of dread **b :** a continuing source of irritation **:** PROBLEM

bug·eye \-ˌī\ *noun* (1881)
: a small boat with a flat bottom, a centerboard, and two raked masts

bug–eyed \-ˌīd\ *adjective* (1922)
: having the eyes bulging (as with astonishment)

[1]**bug·ger** \ˈbə-gər, ˈbu̇-gər\ *noun* [Middle English *bougre* heretic, from Middle French, from Medieval Latin *Bulgarus,* literally, Bulgarian; from the association of Bulgaria with the Bogomils, who were accused of sodomy] (1555)
1 : SODOMITE
2 a : a worthless person **:** RASCAL **b :** FELLOW, CHAP
3 : a small or annoying thing ⟨put down my keys and now I can't find the *buggers*⟩

[2]**bug·ger** \ˈbə-gər\ *transitive verb* (1598)

\ə\ **abut** \ᵊ\ **kitten** \ər\ **further** \a\ **ash** \ā\ **ace** \ä\ **mop, mar** \au̇\ **out** \ch\ **chin** \e\ **bet** \ē\ **easy** \g\ **go** \i\ **hit** \ī\ **ice** \j\ **job** \ŋ\ **sing** \ō\ **go** \ȯ\ **law** \ȯi\ **boy** \th\ **thin** \t͟h\ **the** \ü\ **loot** \u̇\ **foot** \y\ **yet** \zh\ **vision** *see also* Guide to Pronunciation

1 : to commit sodomy with — usually considered vulgar
2 : DAMN

[3]**bugger** *noun* (1955)
: a person who plants electronic bugs

bugger all *noun* (circa 1937)
slang British **:** NOTHING

bugger off *intransitive verb* (1922)
slang British **:** LEAVE, DEPART

bug·gery \'bə-gə-rē\ *noun* (1514)
: SODOMY

[1]**bug·gy** \'bə-gē\ *adjective* **bug·gi·er; -est** (1714)
: infested with bugs

[2]**buggy** *noun, plural* **buggies** [origin unknown] (1773)
1 : a light one-horse carriage made with two wheels in England and with four wheels in the U.S.
2 : a small cart or truck for short transportations of heavy materials
3 : BABY CARRIAGE

[1]**bug·house** \'bəg-,haùs\ *adjective* (1895)
: mentally deranged **:** CRAZY

[2]**bughouse** *noun* (1902)
: an insane asylum

[1]**bu·gle** \'byü-gəl\ *noun* [Middle English, from Old French, from Late Latin *bugula*] (13th century)
: any of a genus (*Ajuga*) of plants of the mint family; *especially* **:** a European annual (*A. reptans*) that has spikes of blue flowers and is naturalized in the U.S.

[2]**bugle** *noun* [Middle English, buffalo, instrument made of buffalo horn, bugle, from Middle French, from Latin *buculus*, diminutive of *bos* head of cattle — more at COW] (14th century)
: a valveless brass instrument that resembles a trumpet and is used especially for military calls

bugle

[3]**bugle** *intransitive verb* **bu·gled; bu·gling** \-g(ə-)liŋ\ (1884)
1 : to sound a bugle
2 : to utter the characteristic rutting call of the bull elk

[4]**bugle** *noun* [perhaps from [2]*bugle*] (1579)
: a small cylindrical bead of glass or plastic used for trimming especially on women's clothing

bu·gler \'byü-glər\ *noun* (1838)
: a person who sounds a bugle

bu·gle·weed \'byü-gəl-,wēd\ *noun* (circa 1818)
1 : any of a genus (*Lycopus*) of mints; *especially* **:** one (*Latin virginicus*) that is mildly narcotic and astringent
2 : [1]BUGLE

bu·gloss \'byü-,gläs, -,glòs\ *noun* [Middle French *buglosse*, from Latin *buglossa*, irregular from Greek *bouglōssos*, from *bous* head of cattle + *glōssa* tongue — more at COW, GLOSS] (14th century)
: any of several coarse hairy plants (genera *Anchusa, Lycopsis*, and *Echium*) of the borage family — compare VIPER'S BUGLOSS

bug off *intransitive verb* [probably short for *bugger off*] (1971)
: LEAVE, DEPART — usually used as a command

buhl \'bül, 'byü(ə)l\ *variant of* BOULLE

buhr·stone \'bər-,stōn\ *noun* [probably from *burr* + *stone*] (1690)
1 : a siliceous rock used for millstones
2 : a millstone cut from buhrstone

[1]**build** \'bild\ *verb* **built** \'bilt\; **build·ing** [Middle English *bilden*, from Old English *byldan;* akin to Old English *būan* to dwell — more at BOWER] (before 12th century)
transitive verb
1 : to form by ordering and uniting materials by gradual means into a composite whole **:** CONSTRUCT
2 : to cause to be constructed
3 : to develop according to a systematic plan, by a definite process, or on a particular base
4 : INCREASE, ENLARGE
intransitive verb
1 : to engage in building
2 a : to progress toward a peak (as of intensity) ⟨*build* to a climax⟩ **b :** to develop in extent ⟨a crowd *building*⟩
— build a fire under : to stimulate to vigorous action
— build into : to make an integral part of ⟨*build* quality *into* the product⟩
— build on : to use as a foundation ⟨*building on* past experience⟩

[2]**build** *noun* (1667)
: form or mode of structure **:** MAKE; *especially* **:** bodily conformation of a person or lower animal

build·able \'bil-də-bəl\ *adjective* (1927)
: suitable for building ⟨*buildable* land⟩

build·ed *archaic past of* BUILD

build·er \'bil-dər\ *noun* (13th century)
1 : one that builds; *especially* **:** one that contracts to build and supervises building operations
2 : a substance added to or used with detergents to increase their cleansing action

build in *transitive verb* (1933)
: to make an integral part of something

build·ing \'bil-diŋ\ *noun* (14th century)
1 : a usually roofed and walled structure built for permanent use (as for a dwelling)
2 : the art or business of assembling materials into a structure

building block *noun* (1846)
: a unit of construction or composition

build-up \'bil-,dəp\ *noun* (1926)
1 : something produced by building up ⟨fluid *buildup* in the lungs⟩
2 : the act or process of building up

build up (1611)
transitive verb
1 : to develop gradually by increments ⟨*building up* endurance⟩ ⟨*built up* a library⟩
2 : to promote the health, strength, esteem, or reputation of
intransitive verb
: to accumulate or develop appreciably
— build·er–up·per \'bil-dər-'ə-pər\ *noun*

built \'bilt\ *adjective* (1621)
: formed as to physique or bodily contours ⟨slimly *built*⟩; *especially* **:** well or attractively formed

[1]**built–in** \'bilt-'in\ *adjective* (1898)
1 : forming an integral part of a structure; *especially* **:** constructed as or in a recess in a wall
2 : INHERENT

[2]**built–in** \-,in\ *noun* (1930)
: a built-in piece of furniture

built–up \'bilt-'əp\ *adjective* (1829)
1 : made of several sections or layers fastened together
2 : covered with buildings

buird·ly \'bùr(d)-lē\ *adjective* [probably alteration of *burly*] (1773)
Scottish **:** STURDY

bulb \'bəlb\ *noun* [Latin *bulbus*, from Greek *bolbos* bulbous plant] (circa 1601)
1 a : a resting stage of a plant (as the lily, onion, hyacinth, or tulip) that is usually formed underground and consists of a short stem base bearing one or more buds enclosed in overlapping membranous or fleshy leaves **b :** a fleshy structure (as a tuber or corm) resembling a bulb in appearance **c :** a plant having or developing from a bulb
2 : a bulb-shaped part; *specifically* **:** a glass envelope enclosing the light source of an electric lamp or such an envelope together with the light source it encloses
3 : a rounded or swollen anatomical structure
4 : a camera setting that indicates that the shutter can be opened by pressing on the release and closed by ending the pressure
— bulbed \'bəlbd\ *adjective*

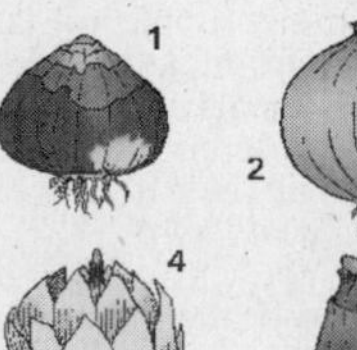

bulb 1a: *1* **hyacinth,** *2* **onion,** *3* **tulip,** *4* **lily**

bul·bar \'bəl-bər, -,bär\ *adjective* (1878)
: of or relating to a bulb; *specifically* **:** involving the medulla oblongata

bul·bil \'bəl-bəl, -,bil\ *noun* [French *bulbille*, diminutive of *bulbe* bulb, from Latin *bulbus*] (1831)
: BULBLET

bulb·let \'bəl-blət\ *noun* (1842)
: a small or secondary bulb; *especially* **:** an aerial deciduous bud produced in a leaf axil or replacing the flowers and capable when separated of producing a new plant

bul·bo·ure·thral gland \,bəl-(,)bō-yù-'rē-thrəl-\ *noun* [Latin *bulbus* bulb + English *urethral*] (circa 1903)
: COWPER'S GLAND

bul·bous \'bəl-bəs\ *adjective* (1578)
1 : having a bulb **:** growing from or bearing bulbs
2 : resembling a bulb especially in roundness ⟨a *bulbous* nose⟩
— bul·bous·ly *adverb*

bul·bul \'bùl-,bùl\ *noun* [Persian, from Arabic] (1665)
1 : a Persian songbird frequently mentioned in poetry that is probably a nightingale (*Erithacus megarhynchos*)
2 : any of a group of gregarious passerine birds (family Pycnonotidae) of Asia and Africa

Bul·gar \'bəl-,gär, 'bùl-\ *noun* [Medieval Latin *Bulgarus*] (1591)
: BULGARIAN

Bul·gar·i·an \,bəl-'gar-ē-ən, bùl-, -'ger-\ *noun* (1555)
1 : a native or inhabitant of Bulgaria
2 : the Slavic language of the Bulgarians
— Bulgarian *adjective*

[1]**bulge** \'bəlj *also* 'bùlj\ *verb* **bulged; bulg·ing** (15th century)
transitive verb
: to cause to bulge
intransitive verb
1 *archaic* **:** BILGE
2 a : to jut out **:** SWELL **b :** to bend outward **c :** to become swollen or protuberant
3 : to be filled to overflowing

[2]**bulge** *noun* [Middle French *boulge, bouge* leather bag, curved part — more at BUDGET] (1622)
1 : BILGE 1, 2
2 : a protuberant or swollen part or place
3 : ADVANTAGE, UPPER HAND
4 : sudden expansion
synonym see PROJECTION
— bulgy \'bəl-jē *also* 'bùl-\ *adjective*

bul·gur \'bəl-gər, 'bùl-\ *noun* [Turkish] (1926)
: parched cracked wheat

bu·lim·ia \bü-'lē-mē-ə, byü-, -'li-\ *noun* [New Latin, from Greek *boulimia* great hunger, from *bou-*, augmentative prefix (from *bous* head of cattle) + *limos* hunger — more at COW] (14th century)
1 : an abnormal and constant craving for food
2 : a serious eating disorder that occurs chiefly in females, is characterized by compulsive overeating usually followed by self-induced vomiting or laxative or diuretic abuse, and is often accompanied by guilt and depression
— bu·lim·ic \-'lē-mik, -'li-\ *adjective or noun*

[1]**bulk** \ˈbəlk *also* ˈbu̇lk\ *noun* [Middle English, heap, bulk, from Old Norse *bulki* cargo] (15th century)
1 a : spatial dimension **:** MAGNITUDE **b :** material that forms a mass in the intestine; *especially* **:** FIBER 1d
2 a : BODY; *especially* **:** a large or corpulent human body **b :** an organized structure especially when viewed primarily as a mass of material **c :** a ponderous shapeless mass
3 : the main or greater part ☆
— in bulk 1 : not divided into parts or packaged in separate units **2 :** in large quantities

[2]**bulk** (1540)
transitive verb
1 : to cause to swell or bulge **:** STUFF
2 : to gather into a mass or aggregate
intransitive verb
1 : SWELL, EXPAND
2 : to appear as a factor **:** LOOM ⟨a consideration that *bulks* large in everyone's thinking⟩

[3]**bulk** *adjective* (1693)
1 : being in bulk ⟨*bulk* cement⟩
2 : of or relating to materials in bulk

bulk·head \ˈbəlk-ˌhed, ˈbəl-ˌked\ *noun* [*bulk* structure projecting from a building + *head*] (15th century)
1 : an upright partition separating compartments
2 : a structure or partition to resist pressure or to shut off water, fire, or gas
3 : a retaining wall along a waterfront
4 : a projecting framework with a sloping door giving access to a cellar stairway or a shaft

bulk up (1981)
intransitive verb
: to gain weight especially by becoming more muscular
transitive verb
: to cause to bulk up

bulky \ˈbəl-kē *also* ˈbu̇l-\ *adjective* **bulk·i·er; -est** (15th century)
1 a : having bulk **b** (1) **:** large of its kind (2) **:** CORPULENT
2 : having great volume in proportion to weight ⟨a *bulky* knit sweater⟩
— bulk·i·ly \-kə-lē\ *adverb*
— bulk·i·ness \-kē-nəs\ *noun*

[1]**bull** \ˈbu̇l, ˈbəl\ *noun* [Middle English *bule,* from Old English *bula;* akin to Old Norse *boli* bull] (before 12th century)
1 a : a male bovine; *especially* **:** an adult uncastrated male ox **b :** a usually adult male of various large animals (as elephants, whales, or seals)
2 : one who buys securities or commodities in expectation of a price rise or who acts to effect such a rise — compare BEAR
3 : one that resembles a bull (as in brawny physique)
4 : BULLDOG
5 *slang* **:** POLICE OFFICER, DETECTIVE
6 *capitalized* **:** TAURUS

[2]**bull** *adjective* (13th century)
1 a : of or relating to a bull **b :** MALE **c :** suggestive of a bull
2 : large of its kind

[3]**bull** (1884)
intransitive verb
: to advance forcefully
transitive verb
1 : to act on with violence
2 : FORCE ⟨*bulled* his way through the crowd⟩

[4]**bull** *noun* [Middle English *bulle,* from Medieval Latin *bulla,* from Latin, bubble, amulet] (14th century)
1 : a solemn papal letter sealed with a bulla or with a red-ink imprint of the device on the bulla
2 : EDICT, DECREE

[5]**bull** (1609)
transitive verb
slang **:** to fool especially by fast boastful talk
intransitive verb
slang **:** to engage in idle and boastful talk

[6]**bull** *noun* [perhaps from obsolete *bull* to mock] (1640)
1 : a grotesque blunder in language
2 *slang* **:** empty boastful talk
3 *slang* **:** NONSENSE 2

bul·la \ˈbu̇-lə\ *noun, plural* **bul·lae** \ˈbu̇-ˌlē, -ˌlī\ (14th century)
1 [Medieval Latin] **:** the round usually lead seal attached to a papal bull
2 [New Latin, from Latin] **:** a hollow thin-walled rounded bony prominence
3 : a large vesicle or blister

bul·lace \ˈbu̇-ləs\ *noun* [Middle English *bolace,* from Middle French *beloce,* from Medieval Latin *bolluca*] (14th century)
: a European plum (*Prunus domestica insititia*) with small ovoid fruit in clusters

bull·bait·ing \ˈbu̇l-ˌbā-tiŋ, ˈbəl-\ *noun* (circa 1580)
: the former practice of baiting bulls with dogs

bull·bat \ˈbu̇l ˌbat, ˈbəl-\ *noun* (1838)
: NIGHTHAWK 1a

[1]**bull·dog** \ˈbu̇l-ˌdȯg *also* ˈbəl-\ *noun* (circa 1500)
1 : any of a breed of compact muscular short-haired dogs having widely separated forelegs and an undershot lower jaw that were developed in England to fight bulls
2 : a handgun with a thick usually short barrel
3 : a proctor's attendant at an English university

[2]**bulldog** *adjective* (1855)
: suggestive of a bulldog ⟨*bulldog* tenacity⟩

[3]**bulldog** *transitive verb* (1907)
: to throw (a steer) by seizing the horns and twisting the neck
— bull·dog·ger *noun*
— bull·dog·ging *noun*

bull·doze \ˈbu̇l-ˌdōz *also* ˈbəl-\ *verb* [perhaps from [1]*bull* + alteration of *dose*] (1876)
transitive verb
1 : to coerce or restrain by threats **:** BULLY
2 : to move, clear, gouge out, or level off by pushing with or as if with a bulldozer
3 : to force insensitively or ruthlessly ⟨*bulldozed* the program through the legislature⟩
intransitive verb
1 : to operate a bulldozer
2 : to force one's way like a bulldozer
synonym see INTIMIDATE

bull·doz·er \-ˌdō-zər\ *noun* (1876)
1 : one that bulldozes
2 : a tractor-driven machine usually having a broad blunt horizontal blade for moving earth (as in road building)

bull dyke *noun* (circa 1942)
: an aggressively masculine lesbian — often used disparagingly

bul·let \ˈbu̇-lət *also* ˈbə-\ *noun, often attributive* [Middle French *boulette* small ball & *boulet* missile, diminutives of *boule* ball — more at BOWL] (1579)
1 : a round or elongated missile (as of lead) to be fired from a firearm; *broadly* **:** CARTRIDGE 1a
2 a : something resembling a bullet (as in curved form) **b :** a large dot placed in printed matter to call attention to a particular passage
3 : a very fast and accurately thrown ball
— bul·let·ed \-lə-təd\ *adjective*
— bul·let·proof \-ˌprüf\ *adjective*

[1]**bul·le·tin** \ˈbu̇-lə-tᵊn *also* ˈbə-\ *noun* [French, from Middle French, from *bullette* seal, notice, diminutive of *bulle* seal, from Medieval Latin *bulla*] (1765)
1 : a brief public notice issuing usually from an authoritative source; *specifically* **:** a brief news item intended for immediate publication or broadcast
2 : PERIODICAL; *especially* **:** the organ of an institution or association

[2]**bulletin** *transitive verb* (1838)
: to make public by bulletin

bulletin board *noun* (1831)
1 : a board for posting notices (as at a school)
2 : a program on a computer system that allows users to read and write public notices and is accessed usually by modem

bullet train *noun* (1966)
: a high-speed passenger train especially of Japan

bull fiddle *noun* (1880)
: DOUBLE BASS
— bull fiddler *noun*

bull·fight \ˈbu̇l-ˌfīt *also* ˈbəl-\ *noun* (1788)
: a spectacle in which men ceremonially fight with and in Hispanic tradition kill bulls in an arena for public entertainment
— bull·fight·er \-ˌfī-tər\ *noun*

bull·fight·ing \-tiŋ\ *noun* (circa 1753)
: the action involved in a bullfight

bull·finch \ˈbu̇l-ˌfinch *also* ˈbəl-\ *noun* (circa 1570)
: a European finch (*Pyrrhula pyrrhula*) having in the male rosy red underparts, blue-gray back, and black cap, chin, tail, and wings; *also* **:** any of several other finches

bull·frog \-ˌfrȯg, -ˌfräg\ *noun* (1698)
: a heavy-bodied deep-voiced frog (*Rana catesbeiana*) of the eastern U.S. and southern Canada that has been introduced elsewhere

bull·head \-ˌhed\ *noun* (15th century)
: any of various large-headed fishes (as a sculpin); *especially* **:** any of several common freshwater catfishes (genus *Ameiurus* sometimes included in the genus *Ictalurus*) of the U.S.

bull·head·ed \-ˈhe-dəd\ *adjective* (1818)
: stupidly stubborn **:** HEADSTRONG
— bull·head·ed·ly *adverb*
— bull·head·ed·ness *noun*

bull·horn \-ˌhȯrn\ *noun* (1942)
1 : a loudspeaker on a naval ship
2 : a handheld combined microphone and loudspeaker

bul·lion \ˈbu̇l-yən, ˌyän\ *noun* [Middle English, from Anglo-French, mint] (14th century)
1 a : gold or silver considered as so much metal; *specifically* **:** uncoined gold or silver in bars or ingots **b :** metal in the mass ⟨lead *bullion*⟩
2 : lace, braid, or fringe of gold or silver threads

bull·ish \ˈbu̇-lish *also* ˈbə-\ *adjective* (1566)
1 : suggestive of a bull (as in brawniness)
2 a : marked by, tending to cause, or hopeful of rising prices (as in a stock market) **b :** optimistic about something's or someone's prospects
— bull·ish·ly *adverb*
— bull·ish·ness *noun*

bull·mas·tiff \ˈbu̇l-ˈmas-təf *also* ˈbəl-\ *noun* (1871)
: any of a breed of large powerful dogs developed by crossing bulldogs with mastiffs

Bull Moose *noun* [*bull moose,* emblem of the Progressive party of 1912] (1912)
: a follower of Theodore Roosevelt in the U.S. presidential campaign of 1912

☆ SYNONYMS
Bulk, mass, volume mean the aggregate that forms a body or unit. BULK implies an aggregate that is impressively large, heavy, or numerous ⟨the darkened *bulk* of the skyscrapers⟩. MASS suggests an aggregate made by piling together things of the same kind ⟨a *mass* of boulders⟩. VOLUME applies to an aggregate without shape or outline and capable of flowing or fluctuating ⟨a tremendous *volume* of water⟩.

Bull Moos·er \-ˈmü-sər\ *noun* (1912)
: BULL MOOSE

bull neck *noun* (1830)
: a thick short powerful neck
— **bull–necked** \ˈbu̇l-ˈnekt *also* ˈbəl-\ *adjective*

bull·ock \ˈbu̇-lək *also* ˈbə-\ *noun* (before 12th century)
1 : a young bull
2 : a castrated bull **:** STEER
— **bull·ocky** \-lə-kē\ *adjective*

bul·lous \ˈbu̇-ləs\ *adjective* (1833)
: resembling or characterized by bullae **:** VESICULAR ⟨*bullous* lesions⟩

bull pen *noun* (1809)
1 : a large cell where prisoners are detained until brought into court
2 a : a place on a baseball field where pitchers warm up before they start pitching **b :** the relief pitchers of a baseball team
3 : an open work area not divided into offices

bull·ring \ˈbu̇l-ˌriŋ\ *noun* (1802)
: an arena for bullfights

bull session *noun* [[6]*bull*] (1920)
: an informal discursive group discussion

bull's–eye \ˈbu̇l-ˌzī *also* ˈbəl-\ *noun, plural* **bull's–eyes** (1825)
1 : a very hard globular candy
2 : a circular piece of glass with a lump in the middle formed at the end of a glassblower's pipe
3 a : the center of a target; *also* **:** something central or crucial **b :** a shot that hits the bull's-eye; *broadly* **:** something that precisely attains a desired end
4 : a simple lens of short focal distance; *also* **:** a lantern with such a lens

bull's–eye window *noun* (1926)
: a circular window; *also* **:** a window made up of glassblower's bull's-eyes

[1]**bull·shit** \ˈbu̇l-ˌshit *also* ˈbəl-\ *noun* [[1]*bull* & [6]*bull*] (circa 1915)
: NONSENSE; *especially* **:** foolish insolent talk — usually considered vulgar

[2]**bullshit** (circa 1942)
intransitive verb
1 : to talk foolishly, boastfully, or idly — usually considered vulgar
2 : to engage in a discursive discussion — usually considered vulgar
transitive verb
: to talk nonsense to especially with the intention of deceiving or misleading — usually considered vulgar

bull·shot \-ˌshät\ *noun* (1964)
: a drink made of vodka and bouillon

bull snake *noun* (1784)
: any of several large harmless North American snakes (genus *Pituophis*) that feed chiefly on rodents — called also *gopher snake, pine snake*

bull·ter·ri·er \ˈbu̇l-ˈter-ē-ər *also* ˈbəl-\ *noun* [*bull*dog + *terrier*] (1848)
: any of a breed of short-haired terriers originated in England by crossing the bulldog with terriers

bull thistle *noun* (1863)
: a European thistle (*Cirsium vulgare*) with rather large heads and prickly leaves that is naturalized as a weed in the U.S.

bull tongue *noun* (1831)
: a wide blade attached to a cultivator or plow to stir the soil, kill weeds, or mark furrows

bull·whip \ˈbu̇l-ˌ(h)wip *also* ˈbəl-\ *noun* (1852)
: a rawhide whip with a very long plaited lash

[1]**bul·ly** \ˈbu̇-lē, ˈbə-\ *noun, plural* **bullies** [probably modification of Dutch *boel* lover, from Middle High German *buole*] (1538)
1 *archaic* **a :** SWEETHEART **b :** a fine chap
2 a : a blustering browbeating person; *especially* **:** one habitually cruel to others who are weaker **b :** PIMP
3 : a hired ruffian

[2]**bully** *adjective* (1681)
1 : EXCELLENT, FIRST-RATE — often used in interjectional expressions ⟨*bully* for you⟩
2 : resembling or characteristic of a bully

[3]**bully** *verb* **bul·lied; bul·ly·ing** (1710)
transitive verb
1 : to treat abusively
2 : to affect by means of force or coercion
intransitive verb
: to use browbeating language or behavior **:** BLUSTER
synonym see INTIMIDATE

[4]**bully** *noun* [probably modification of French (*bœuf*) *boulli* boiled beef] (1753)
: pickled or canned usually corned beef

bul·ly·boy \ˈbu̇-lē-ˌbȯi, ˈbə-\ *noun* (1925)
: a swaggering tough

bully pulpit *noun* (1976)
: a prominent public position (as a political office) that provides an opportunity for expounding one's views; *also* **:** such an opportunity

bul·ly·rag \-ˌrag\ *transitive verb* [origin unknown] (circa 1790)
1 : to intimidate by bullying
2 : to vex by teasing **:** BADGER

bul·rush *also* **bull·rush** \ˈbu̇l-ˌrəsh\ *noun* [Middle English *bulrysche*] (15th century)
: any of several large rushes or sedges growing in wetlands: as **a :** any of a genus of annual or perennial sedges (*Scirpus*, especially *S. lacustris*) that bear solitary or much-clustered spikelets containing perfect flowers with a perianth of six bristles **b** *British* **:** either of two cattails (*Typha latifolia* and *T. angustifolia*) **c :** PAPYRUS

[1]**bul·wark** \ˈbu̇l-(ˌ)wərk, -ˌwȯrk; ˈbəl-(ˌ)wərk; *sense 3 also* ˈbə-ˌläk\ *noun* [Middle English *bulwerke,* from Middle Dutch *bolwerc,* from Middle High German, from *bole* plank + *werc* work] (15th century)
1 a : a solid wall-like structure raised for defense **:** RAMPART **b :** BREAKWATER, SEAWALL
2 : a strong support or protection
3 : the side of a ship above the upper deck — usually used in plural
word history see BOULEVARD

[2]**bulwark** *transitive verb* (15th century)
: to fortify or safeguard with a bulwark

[1]**bum** \ˈbəm\ *noun* [Middle English *bom*] (14th century)
chiefly British **:** BUTTOCKS — sometimes considered vulgar

[2]**bum** *adjective* [perhaps from [4]*bum*] (1859)
1 a : of poor quality or nature **:** INVALID, INFERIOR ⟨*bum* advice⟩ ⟨a *bum* check⟩ **b :** acutely disagreeable ⟨a *bum* trip⟩
2 : not functioning because of damage or injury **:** DISABLED ⟨a *bum* knee⟩

[3]**bum** *verb* **bummed; bum·ming** [probably back-formation from [1]*bummer*] (1863)
intransitive verb
1 : LOAF
2 : to spend time unemployed and often wandering — often used with *around*
transitive verb
: to obtain by begging **:** CADGE

[4]**bum** *noun* [probably short for *bummer*] (1864)
1 a : one who sponges off others and avoids work **b :** one who performs a function poorly ⟨called the umpire a *bum*⟩ **c :** one who devotes his time to a recreational activity ⟨a beach *bum*⟩ ⟨ski *bums*⟩
2 : VAGRANT, TRAMP
— **on the bum :** with no settled residence or means of support

[5]**bum** *noun* [probably from [3]*bum*] (1871)
: a drinking spree **:** BENDER

bum·ber·shoot \ˈbəm-bər-ˌshüt\ *noun* [*bumber-* (alteration of *umbr-* in *umbrella*) + *-shoot* (alteration of *-chute* in *parachute*)] (circa 1896)
: UMBRELLA

[1]**bum·ble** \ˈbəm-bəl\ *intransitive verb* **bumbled; bum·bling** \-b(ə-)liŋ\ [Middle English *bomblen* to boom, of imitative origin] (15th century)
1 : BUZZ
2 : DRONE, RUMBLE

[2]**bumble** *verb* **bumbled; bumbling** [perhaps alteration of *bungle*] (1532)
intransitive verb
1 : BLUNDER; *specifically* **:** to speak ineptly in a stuttering and faltering manner
2 : to proceed unsteadily **:** STUMBLE
transitive verb
: BUNGLE
— **bum·bler** \-b(ə-)lər\ *noun*
— **bum·bling·ly** \-b(ə-)liŋ-lē\ *adverb*

bum·ble·bee \ˈbəm-bəl-ˌbē\ *noun* (1530)
: any of numerous large robust hairy social bees (genus *Bombus*)

bum·boat \ˈbəm-ˌbōt\ *noun* [probably from Low German *bumboot,* from *bum* tree + *boot* boat] (1769)
: a boat that brings provisions and commodities for sale to larger ships in port or offshore

bumf *also* **bumph** \ˈbəm(p)f\ *noun* [from *bumf* toilet paper, short for *bumfodder,* from [1]*bum* + *fodder*] (circa 1889)
chiefly British **:** PAPERWORK

bummed \ˈbəmd\ *adjective* [probably back-formation from [2]*bummer*] (1971)
: DISAPPOINTED, DEPRESSED — usually used with *out*

[1]**bum·mer** \ˈbə-mər\ *noun* [probably modification of German *Bummler* loafer, from *bummeln* to dangle, loaf] (1855)
: one that bums

[2]**bummer** *noun* [[2]*bum* + [2]*-er*] (1967)
1 : an unpleasant experience (as a bad reaction to a hallucinogenic drug)
2 : FAILURE, FLOP

[1]**bump** \ˈbəmp\ *noun* [probably imitative of the sound of a blow] (1592)
1 : a relatively abrupt convexity or protuberance on a surface: as **a :** a swelling of tissue **b :** a cranial protuberance
2 a : a sudden forceful blow, impact, or jolt **b :** DEMOTION
3 : an act of thrusting the hips forward in an erotic manner

[2]**bump** (1611)
transitive verb
1 : to strike or knock with force or violence
2 : to collide with
3 a (1) **:** to dislodge with a jolt (2) **:** to subject to a scalar change ⟨rates being *bumped* up⟩ **b :** to oust usually by virtue of seniority or priority ⟨was *bumped* from the flight⟩
intransitive verb
1 : to knock against something with a forceful jolt
2 : to proceed in or as if in a series of bumps
3 : to encounter something that is an obstacle or hindrance ⟨*bumped* up against a chair⟩
— **bump into :** to encounter especially by chance

[1]**bum·per** \ˈbəm-pər\ *noun* [probably from *bump* to bulge] (1676)
1 : a brimming cup or glass
2 : something unusually large

[2]**bumper** *adjective* (1885)
1 : unusually large ⟨a *bumper* crop⟩
2 : BANNER 2

[3]**bump·er** \ˈbəm-pər\ *noun* (1839)
1 : a device for absorbing shock or preventing damage (as in collision); *specifically* **:** a usually metal bar at either end of an automobile
2 : one that bumps

bumper car *noun* (1959)
: a small electric car made to be driven around in an enclosure and to be bumped into others (as at an amusement park)

bumper sticker *noun* (1967)
: a strip of adhesive paper or plastic bearing a printed message and designed to be stuck on a vehicle's bumper

bumper–to–bumper *adjective* (1951)
: marked by long closed lines of cars

[1]**bump·kin** \ˈbəm(p)-kən\ *noun* [perhaps from Flemish *bommekijn* small cask, from Middle Dutch, from *bomme* cask] (1570)
: an awkward and unsophisticated rustic

— **bump·kin·ish** \-kə-nish\ *adjective*
— **bump·kin·ly** \-kən-lē\ *adjective*

²bump·kin *or* **bum·kin** \'bəm(p)-kən\ *noun* [probably from Flemish *boomken,* diminutive of *boom* tree] (circa 1632)
: a spar projecting from a ship especially at the stern

bump off *transitive verb* (1910)
: to murder casually or cold-bloodedly

bump·tious \'bəm(p)-shəs\ *adjective* [¹*bump* + *-tious* (as in *fractious*)] (1803)
: presumptuously, obtusely, and often noisily self-assertive : OBTRUSIVE
— **bump·tious·ly** *adverb*
— **bump·tious·ness** *noun*

bumpy \'bəm-pē\ *adjective* **bump·i·er; -est** (1865)
1 : having or covered with bumps ⟨a *bumpy* road⟩ ⟨*bumpy* skin⟩
2 a : marked by bumps or jolts ⟨a *bumpy* ride⟩ **b :** marked by or full of difficulties ⟨a *bumpy* relationship⟩
— **bump·i·ly** \-pə-lē\ *adverb*
— **bump·i·ness** \-pē-nəs\ *noun*

bum's rush \'bəmz-\ *noun* (1922)
: forcible eviction or dismissal

¹bun \'bən\ *noun* [Middle English *bunne*] (14th century)
1 : a sweet or plain small bread; *especially* **:** a round roll
2 : a knot of hair shaped like a bun
3 *plural* **:** BUTTOCKS

²bun *noun* [perhaps from ¹*bun*] (1901)
: LOAD 4

Bu·na \'byü-nə, 'bü-\ *trademark*
— used for any of several rubbers made by polymerization or copolymerization of butadiene

¹bunch \'bənch\ *noun* [Middle English *bunche*] (14th century)
1 : PROTUBERANCE, SWELLING
2 a : a number of things of the same kind ⟨a *bunch* of grapes⟩ **b :** GROUP 2a ⟨a *bunch* of friends⟩ **c :** a considerable amount : LOT ⟨a *bunch* of money⟩
— **bunch·i·ly** \'bən-chə-lē\ *adverb*
— **bunchy** \-chē\ *adjective*

²bunch (14th century)
intransitive verb
1 : SWELL, PROTRUDE
2 : to form a group or cluster — often used with *up*
transitive verb
: to form into a bunch

bunch·ber·ry \'bənch-ˌber-ē\ *noun* (1845)
: a creeping perennial herb (*Cornus canadensis*) of the dogwood family that has whorled leaves and white floral bracts and bears red berries in capitate cymes

bunch·grass \-ˌgras\ *noun* (1839)
: any of several grasses (as of the genus *Andropogon*) especially of the western U.S. that grow in tufts

bun·co *or* **bun·ko** \'bəŋ-(ˌ)kō\ *noun, plural* **buncos** *or* **bunkos** [perhaps alteration of Spanish *banca* bench, banking, bank in gambling, from Italian — more at BANK] (1872)
: a swindling game or scheme
— **bunco** *transitive verb*

¹bund \'bənd\ *noun* [Hindi *band,* from Persian] (1810)
1 : an embankment used especially in India to control the flow of water
2 : an embanked thoroughfare along a river or the sea especially in the Far East

²bund \'bu̇nd, 'bənd, 'bu̇nt\ *noun, often capitalized* [Yiddish *bund* & German *Bund,* from Middle High German *bunt;* akin to Old English *byndel*] (1850)
: a political association: as **a :** a Jewish socialist organization founded in czarist Russia in 1897 **b :** a pro-Nazi German-American organization of the 1930s
— **bund·ist** \'bu̇n-dist, 'bən-\ *noun, often capitalized*

¹bun·dle \'bən-dᵊl\ *noun* [Middle English *bundel,* from Middle Dutch; akin to Old English *byndel* bundle, *bindan* to bind] (14th century)
1 a : a group of things fastened together for convenient handling **b :** PACKAGE, PARCEL **c :** a considerable number : LOT **d :** a sizable sum of money **e :** a person embodying a specified quality or characteristic ⟨a *bundle* of energy⟩
2 a : a small band of mostly parallel fibers (as of nerve or muscle) **b :** VASCULAR BUNDLE

²bundle *verb* **bun·dled; bun·dling** \'bən(d)-liŋ, 'bən-dᵊl-iŋ\ (circa 1628)
transitive verb
1 : to make into a bundle or package : WRAP
2 : to hustle or hurry unceremoniously ⟨*bundled* the children off to school⟩
3 : to include (a product or service) with a related product for sale at a single price ⟨software is *bundled* with computer hardware⟩
intransitive verb
1 : HURRY, HUSTLE
2 : to practice bundling
— **bun·dler** \-lər, -dᵊl-ər\ *noun*

bundle up (1918)
transitive verb
: to dress (someone) warmly
intransitive verb
: to dress warmly

bun·dling \'bən(d)-liŋ, 'bən-dᵊl-iŋ\ *noun* (1781)
: a former custom of an unmarried couple's occupying the same bed without undressing especially during courtship

¹bung \'bəŋ\ *noun* [Middle English, from Middle Dutch *bonne, bonghe*] (15th century)
1 : the stopper especially in the bunghole of a cask; *also* **:** BUNGHOLE
2 : the cecum or anus especially of a slaughtered animal

²bung *transitive verb* (1589)
1 : to plug with or as if with a bung
2 *British* **:** THROW, TOSS ⟨*bung* a spanner into the works —P. G. Wodehouse⟩

bun·ga·low \'bəŋ-gə-ˌlō\ *noun* [Hindi *baṅglā,* literally, (house) in the Bengal style] (1676)
: a usually one-storied house with a low-pitched roof

bun·gee cord \'bən-jē-\ *noun* [origin unknown] (1948)
: an elasticized cord used especially as a fastening or shock-absorbing device — called also *bungee*

bung·hole \'bəŋ-ˌhōl\ *noun* (1571)
: a hole for emptying or filling a cask

bun·gle \'bəŋ-gəl\ *verb* **bun·gled; bun·gling** \-g(ə-)liŋ\ [perhaps of Scandinavian origin; akin to Icelandic *banga* to hammer] (1549)
intransitive verb
: to act or work clumsily and awkwardly
transitive verb
: MISHANDLE, BOTCH
— **bungle** *noun*
— **bun·gler** \-g(ə-)lər\ *noun*
— **bun·gling·ly** \-g(ə-)liŋ-lē\ *adverb*

bun·gle·some \-gəl-səm\ *adjective* (circa 1889)
: AWKWARD, CLUMSY

bung up *transitive verb* (1951)
: BATTER

bun·ion \'bən-yən\ *noun* [probably alteration of *bunny* swelling] (circa 1718)
: an inflamed swelling of the small sac on the first joint of the big toe

¹bunk \'bəŋk\ *noun* [probably short for *bunker*] (1758)
1 a : BUNK BED **b :** a built-in bed (as on a ship) that is often one of a tier of berths **c :** a sleeping place
2 : a feeding trough for farm animals and especially cattle

²bunk (1840)
intransitive verb
: to occupy a bunk or bed : stay the night ⟨*bunked* with a friend for the night⟩
transitive verb
: to provide with a bunk or bed

³bunk *noun* (1900)
: BUNKUM, NONSENSE

⁴bunk *noun* [origin unknown] (circa 1870)
British **:** a hurried departure or escape — usually used in the phrase *do a bunk*

bunk bed *noun* (1924)
: one of two single beds usually placed one above the other

¹bun·ker \'bəŋ-kər\ *noun* [Scots *bonker* chest, box] (1839)
1 : a bin or compartment for storage; *especially* **:** one on shipboard for the ship's fuel
2 a : a protective embankment or dugout; *especially* **:** a fortified chamber mostly below ground often built of reinforced concrete and provided with embrasures **b :** a sand trap or embankment constituting a hazard on a golf course
— **bun·kered** \-kərd\ *adjective*

²bunker *verb* **bun·kered; bun·ker·ing** \-k(ə-)riŋ\ (1891)
intransitive verb
: to fill a ship's bunker with coal or oil
transitive verb
1 : to place or store in a bunker
2 : to hit (a golf ball or shot) into a bunker

bunker mentality *noun* (1976)
: a state of mind especially among members of a group that is characterized by chauvinistic defensiveness and self-righteous intolerance of criticism

bunk·house \'bəŋk-ˌhau̇s\ *noun* (1876)
: a rough simple building providing sleeping quarters

bun·kum *or* **bun·combe** \'bəŋ-kəm\ *noun* [*Buncombe* county, N.C.; from a remark made by its congressman, who defended an irrelevant speech by claiming that he was speaking to Buncombe] (1845)
: insincere or foolish talk : NONSENSE ◆

bun·ny \'bə-nē\ *noun, plural* **bunnies** [English dialect *bun* rabbit] (circa 1690)
1 : RABBIT; *especially* **:** a young rabbit
2 : a desirable young woman

Bun·ra·ku \bün-'rä-(ˌ)kü\ *noun* [Japanese] (1920)

◇ WORD HISTORY

bunkum The word *bunkum* or *buncombe* "nonsense" first occurs in 1828 in the phrase *talk to bunkum,* alluding to overly long congressional speeches given to a nearly empty chamber. According to a tale that has become part of American political folklore, this phrase originated around 1820, when Felix Walker (1753–1828), a U.S. Congressman who represented a district in North Carolina that included Buncombe County, persisted in delivering an exceptionally wearisome speech to the Sixteenth Congress, despite the objections of his impatient colleagues. In self-defense Walker explained that he had done what his constituents expected of him, which was "to make a speech for Buncombe." Whatever the veracity of this story, *bunkum* has remained thoroughly entrenched in American English. The noun *bunk,* recorded in 1893 in the sense "something fraudulent" and in 1900 alluding to nonsensical talk, is probably short for *bunkum,* though it has very likely been influenced by *bunco* (or *bunko*), a word that developed in the late 19th century in reference to rigged games of chance.

: Japanese puppet theater featuring large costumed wooden puppets, puppeteers who are onstage, and a chanter who speaks all the lines

Bun·sen burner \'bən(t)-sən-\ *noun* [Robert Welsh *Bunsen*] (circa 1888)
: a gas burner consisting typically of a straight tube with small holes at the bottom where air enters and mixes with the gas to produce an intensely hot blue flame

[1]bunt \'bənt\ *noun* [perhaps from Low German, bundle, from Middle Low German; akin to Old English *byndel* bundle] (circa 1582)
1 a : the middle part of a square sail **b :** the part of a furled sail gathered up in a bunch at the center of the yard
2 : the bagging part of a fishing net

[2]bunt *verb* [alteration of *butt*] (1584)
transitive verb
1 : to strike or push with or as if with the head **:** BUTT
2 : to push or tap (a baseball) lightly without swinging the bat
intransitive verb
: to bunt a baseball
— bunt·er *noun*

[3]bunt *noun* (1767)
1 : an act or instance of bunting
2 : a bunted ball

[4]bunt *noun* [origin unknown] (circa 1790)
: a destructive covered smut of wheat caused by a fungus (*Tilletia foetida* or *T. caries*)

[1]bun·ting \'bən-tiŋ\ *noun* [Middle English] (14th century)
: any of various stout-billed passerine birds (family Emberizidae) included with some of the finches — compare INDIGO BUNTING

[2]bunting *noun* [perhaps from English dialect *bunt* (to sift)] (1711)
1 : a lightweight loosely woven fabric used chiefly for flags and festive decorations
2 a : FLAGS **b :** decorations especially in the colors of the national flag

[3]bunting *noun* [term of endearment in the nursery rhyme "Bye, baby bunting"] (1922)
: an infant's hooded sleeping bag made of napped fabric

bunt·line \'bənt-ˌlīn, -lən\ *noun* (1627)
: one of the lines attached to the foot of a square sail to haul the sail up to the yard for furling

Bun·yan·esque \ˌbən-yə-'nesk\ *adjective* (1888)
1 [John *Bunyan*] **:** of, relating to, or suggestive of the allegorical writings of John Bunyan
2 [Paul *Bunyan*, legendary giant lumberjack of U.S. & Canada] **a :** of, relating to, or suggestive of the tales of Paul Bunyan **b :** of fantastically large size

[1]buoy \'bü-ē, 'bȯi\ *noun* [Middle English *boye*, probably from Middle Dutch *boeye;* akin to Old High German *bouhhan* sign — more at BEACON] (13th century)
1 : FLOAT 2; *especially* **:** a floating object moored to the bottom to mark a channel or something (as a shoal) lying under the water
2 : LIFE BUOY

buoy 1

[2]buoy (1596)
transitive verb
1 : to mark by or as if by a buoy
2 a : to keep afloat **b :** SUPPORT, UPLIFT ⟨an economy *buoyed* by the dramatic postwar growth of industry —*Time*⟩
3 : to raise the spirits of — usually used with *up* ⟨hope *buoys* him up⟩
intransitive verb
: FLOAT — usually used with *up*

buoy·ance \'bȯi-ən(t)s, 'bü-yən(t)s\ *noun* (1821)
: BUOYANCY

buoy·an·cy \'bȯi-ən(t)-sē, 'bü-yən(t)-\ *noun* (1713)
1 a : the tendency of a body to float or to rise when submerged in a fluid **b :** the power of a fluid to exert an upward force on a body placed in it; *also* **:** the upward force exerted
2 : the ability to recover quickly from depression or discouragement **:** RESILIENCE

buoy·ant \'bȯi-ənt, 'bü-yənt\ *adjective* (1578)
: having buoyancy: as **a :** capable of floating **b :** CHEERFUL, GAY **c :** capable of recovering **:** having positive indications ⟨a *buoyant* economy⟩
— buoy·ant·ly *adverb*

bup·pie \'bə-pē\ *noun* [[1]*b*lack + y*uppie*] (1984)
: a college-educated black adult who is employed in a well-paying profession and who lives or works in or near a large city

bur *variant of* BURR

Bur·ber·ry \'bər-b(ə-)rē, 'bər-ˌber-ē\ *trademark*
— used for various fabrics used especially for coats for outdoor wear

[1]bur·ble \'bər-bəl\ *intransitive verb* **bur·bled; bur·bling** \-b(ə-)liŋ\ [Middle English] (14th century)
1 : BUBBLE
2 : BABBLE, PRATTLE
— bur·bler \-b(ə-)lər\ *noun*

[2]burble *noun* (1898)
1 : PRATTLE
2 : the breaking up of the streamline flow of air about a body (as an airplane wing)
— bur·bly \-b(ə-)lē\ *adjective*

bur·bot \'bər-bət\ *noun, plural* **burbot** *also* **burbots** [Middle English *borbot*, from Middle French *bourbotte*, from *bourbeter* to burrow in the mud] (14th century)
: a Holarctic freshwater fish (*Lota lota*) of the cod family having barbels on the nose and chin

burbs \'bərbz\ *noun plural* (1971)
: SUBURBS

[1]bur·den \'bər-dᵊn\ *noun* [Middle English, from Old English *byrthen;* akin to Old English *beran* to carry — more at BEAR] (before 12th century)
1 a : something that is carried **:** LOAD **b :** DUTY, RESPONSIBILITY
2 : something oppressive or worrisome
3 a : the bearing of a load — usually used in the phrase *beast of burden* **b :** capacity for carrying cargo ⟨a ship of a hundred tons *burden*⟩
4 : the amount of a deleterious parasite, growth, or substance present in a human or animal body ⟨worm *burden*⟩ ⟨cancer *burden*⟩

[2]burden *transitive verb* **bur·dened; bur·den·ing** \'bərd-niŋ, 'bər-dᵊn-iŋ\ (1541)
: LOAD, OPPRESS ⟨I will not *burden* you with a lengthy account⟩

[3]burden *noun* [Middle English *burdoun*, from Middle French *bourdon* a drone bass, of imitative origin] (14th century)
1 *archaic* **:** a bass or accompanying part
2 a : CHORUS, REFRAIN **b :** a central topic **:** THEME

burden of proof (1593)
: the duty of proving a disputed assertion or charge

bur·den·some \'bər-dᵊn-səm\ *adjective* (1578)
: imposing or constituting a burden **:** OPPRESSIVE ⟨*burdensome* restrictions⟩
synonym see ONEROUS

bur·dock \'bər-ˌdäk\ *noun* (15th century)
: any of a genus (*Arctium*) of coarse composite herbs bearing globular flower heads with prickly bracts

bu·reau \'byu̇r-(ˌ)ō\ *noun, plural* **bureaus** *also* **bu·reaux** \-(ˌ)ōz\ [French, desk, cloth covering for desks, from Old French *burel* woolen cloth, from (assumed) Old French *bure*, from Late Latin *burra* shaggy cloth] (1699)
1 a *British* **:** WRITING DESK; *especially* **:** one having drawers and a slant top **b :** a low chest of drawers for use in a bedroom
2 a : a specialized administrative unit; *especially* **:** a subdivision of an executive department of a government **b :** a branch of a newspaper, newsmagazine, or wire service in an important news center **c :** a usually commercial agency that serves as an intermediary especially for exchanging information or coordinating activities ⟨credit *bureau*⟩

bu·reau·cra·cy \byu̇-'rä-krə-sē\ *noun, plural* **-cies** [French *bureaucratie*, from *bureau* + *-cratie* -cracy] (1818)
1 a : a body of nonelective government officials **b :** an administrative policy-making group
2 : government characterized by specialization of functions, adherence to fixed rules, and a hierarchy of authority
3 : a system of administration marked by officialism, red tape, and proliferation

bu·reau·crat \'byu̇r-ə-ˌkrat\ *noun* (1842)
: a member of a bureaucracy

bu·reau·crat·ese \ˌbyu̇r-ə-(ˌ)kra-'tēz, -'tēs\ *noun* (1949)
: a style of language held to be characteristic of bureaucrats and marked by the prevalence of abstractions, jargon, euphemisms, and circumlocutions

bu·reau·crat·ic \ˌbyu̇r-ə-'kra-tik\ *adjective* (1836)
: of, relating to, or having the characteristics of a bureaucracy or a bureaucrat ⟨*bureaucratic* government⟩
— bu·reau·crat·i·cal·ly \-ti-k(ə-)lē\ *adverb*

bu·reau·cra·tise *British variant of* BUREAUCRATIZE

bu·reau·crat·ism \'byu̇r-ə-(ˌ)kra-ˌti-zəm\ *noun* (1880)
: a bureaucratic system **:** BUREAUCRACY

bu·reau·cra·tize \byu̇-'rä-krə-ˌtīz\ *transitive verb* **-tized; -tiz·ing** (1892)
: to make bureaucratic
— bu·reau·cra·ti·za·tion \-ˌrä-krə-tə-'zā-shən\ *noun*

bu·rette *or* **bu·ret** \byu̇-'ret\ *noun* [Middle French, cruet, from *buire* pitcher, perhaps of Germanic origin; akin to Old English *būr* storehouse, dwelling — more at BOWER] (1836)
: a graduated glass tube with a small aperture and stopcock for delivering measured quantities of liquid or for measuring the liquid or gas received or discharged

burg \'bərg\ *noun* [Old English — more at BOROUGH] (1753)
1 : an ancient or medieval fortress or walled town
2 [German *Burg*] **:** CITY, TOWN

bur·gage \'bər-gij\ *noun* [Middle English, property held by burgage tenure, from Middle French *bourgage*, from Old French, from *bourg, borc* town — more at BOURG] (15th century)
: a tenure by which real property in England and Scotland was held under the king or a lord for a yearly rent or for watching and warding

bur·gee \ˌbər-'jē, 'bər-ˌ\ *noun* [perhaps from French dialect *bourgeais* shipowner] (1750)
1 : a swallow-tailed flag used especially by ships for signals or identification
2 : the usually triangular identifying flag of a yacht club

bur·geon \'bər-jən\ *intransitive verb* [Middle English *burjonen*, from *burjon* bud, from Old French, from (assumed) Vulgar Latin *burrion-, burrio*, from Late Latin *burra* fluff, shaggy cloth] (14th century)
1 a : to send forth new growth (as buds or branches) **:** SPROUT **b :** BLOOM
2 : to grow and expand rapidly **:** FLOURISH

bur·ger \'bər-gər\ *noun* (1939)
1 : HAMBURGER
2 : a sandwich similar to a hamburger — often used in combination ⟨tofu *burger*⟩

bur·gess \'bər-jəs\ *noun* [Middle English *burgeis,* from Old French *borjois,* from *borc*] (13th century)
1 a : a citizen of a British borough **b :** a representative of a borough, corporate town, or university in the British Parliament
2 : a representative in the popular branch of the legislature of colonial Maryland and Virginia

burgh \'bər-(,)ō, 'bə-(,)rō\ *noun* [Middle English — more at BOROUGH] (12th century)
: BOROUGH; *specifically* **:** an incorporated town in Scotland having local jurisdiction of certain services
— burgh·al \'bər-gəl\ *adjective*

bur·gher \'bər-gər\ *noun* (13th century)
1 : an inhabitant of a borough or a town
2 : a member of the middle class **:** a prosperous solid citizen

bur·glar \'bər-glər *also* -gə-lər\ *noun* [Anglo-French *burgler,* from Medieval Latin *burglator,* probably alteration of *burgator,* from *burgare* to commit burglary] (1541)
: one who commits burglary

bur·glar·ize \'bər-glə-,rīz *also* 'bər-gə-lə-\ *verb* **-ized; -iz·ing** (1871)
transitive verb
1 : to break into and steal from
2 : to commit burglary against
intransitive verb
: to commit burglary

bur·glar·proof \'bər-glər-,prüf\ *adjective* (1856)
: protected against or designed to afford protection against burglary

bur·glary \'bər-glə-re *also* -gə-lə-rē *also* -gəl-rē\ *noun, plural* **-glar·ies** (circa 1523)
: the act of breaking and entering a dwelling at night to commit a felony (as theft); *broadly* **:** the entering of a building with the intent to commit a crime
— bur·glar·i·ous \,bər-'glar-ē-əs, -'gler-\ *adjective*
— bur·glar·i·ous·ly *adverb*

bur·gle \'bər-gəl\ *transitive verb* **bur·gled; bur·gling** \-g(ə-)liŋ\ [back-formation from *burglar*] (1870)
: BURGLARIZE

bur·go·mas·ter \'bər-gə-,mas-tər\ *noun* [part modification, part translation of Dutch *burgemeester,* from *burg* town + *meester* master] (1592)
: the chief magistrate of a town in some European countries **:** MAYOR

bur·go·net \'bər-gə-nət, ,bər-gə-'net\ *noun* [modification of Middle French *bourguignotte*] (circa 1587)
: a helmet of either of two 16th century styles

bur·goo \'bər-,gü, (,)bər-'\ *noun, plural* **burgoos** [origin unknown] (1700)
1 : oatmeal gruel
2 : hardtack and molasses cooked together
3 a : a stew or thick soup of meat and vegetables originally served at outdoor gatherings **b :** a picnic at which burgoo is served

bur·gun·dy \'bər-gən-dē\ *noun, plural* **-dies** [*Burgundy,* region in France] (1668)
1 *often capitalized* **:** a red or white unblended wine from Burgundy; *also* **:** a blended red wine produced elsewhere (as California)
2 : a reddish purple color

buri·al \'ber-ē-əl *also* 'bər-\ *noun, often attributive* [Middle English *beriel, berial,* back-formation from *beriels* (taken as a plural), from Old English *byrgels;* akin to Old Saxon *burgisli* tomb, Old English *byrgan* to bury — more at BURY] (13th century)
1 : GRAVE, TOMB
2 : the act or process of burying

buri·er \'ber-ē-ər *also* 'bər-\ *noun* (before 12th century)
: one that buries

bu·rin \'byůr-ən, 'bər-\ *noun* [French] (1662)
1 : an engraver's steel cutting tool having the blade ground obliquely to a sharp point
2 : a prehistoric flint tool with a beveled point

burke \'bərk\ *transitive verb* **burked; burking** [from *burke* to suffocate, from William *Burke* (died 1829) Irish criminal executed for smothering victims to sell their bodies for dissection] (1840)
1 : to suppress quietly or indirectly ⟨*burke* an inquiry⟩
2 : BYPASS, AVOID ⟨*burke* an issue⟩

Bur·kitt's lymphoma \'bər-kəts-\ *also* **Burkitt lymphoma** \-kət-\ *noun* [Denis Parsons *Burkitt* (died 1993) British surgeon] (1963)
: a malignant lymphoma that occurs especially in children of central Africa and is associated with Epstein-Barr virus

burl \'bər(-ə)l\ *noun* [Middle English *burle,* from (assumed) Middle French *bourle* tuft of wool, from (assumed) Vulgar Latin *burrula,* diminutive of Late Latin *burra* shaggy cloth] (15th century)
1 : a knot or lump in thread or cloth
2 a : a hard woody often flattened hemispherical outgrowth on a tree **b :** veneer made from burls

bur·la·de·ro \,bůr-lə-'der-(,)ō, ,bər-\ *noun, plural* **-ros** [Spanish, from *burlar* to make fun of, elude, from *burla* joke] (1938)
: a wooden shield near the wall in a bullring for bullfighters to take shelter behind if pursued

bur·lap \'bər-,lap\ *noun* [origin unknown] (circa 1696)
1 : a coarse heavy plain-woven fabric usually of jute or hemp used for bagging and wrapping and in furniture and linoleum manufacture
2 : a lightweight material resembling burlap used in interior decoration or for clothing

burled \'bər(-ə)ld\ *adjective* (1924)
: having a distorted grain due to burls

[1]bur·lesque \(,)bər-'lesk\ *noun* [*burlesque,* adjective, comic, droll, from French, from Italian *burlesco,* from *burla* joke, from Spanish] (1667)
1 : a literary or dramatic work that seeks to ridicule by means of grotesque exaggeration or comic imitation
2 : mockery usually by caricature
3 : theatrical entertainment of a broadly humorous often earthy character consisting of short turns, comic skits, and sometimes striptease acts
synonym see CARICATURE
— burlesque *adjective*
— bur·lesque·ly *adverb*

[2]burlesque *verb* **bur·lesqued; bur·lesquing** (1676)
transitive verb
: to imitate in a humorous or derisive manner **:** MOCK
intransitive verb
: to employ burlesque
— bur·lesqu·er *noun*

bur·ley \'bər-lē\ *noun, often capitalized* [probably from the name *Burley*] (1881)
: a thin-bodied air-cured tobacco grown mainly in Kentucky

bur·ly \'bər-lē\ *adjective* **bur·li·er; -est** [Middle English] (13th century)
1 : strongly and heavily built **:** HUSKY
2 : heartily direct and frank **:** BLUFF, FORTHRIGHT ⟨an evocative story less *burly* than the real thing but entertaining —E. A. Weeks⟩
— bur·li·ly \-lə-lē\ *adverb*
— bur·li·ness \-lē-nəs\ *noun*

bur marigold *noun* (circa 1818)
: any of a genus (*Bidens*) of coarse composite herbs with prickly flattened achenes that adhere to clothing

Bur·mese \,bər-'mēz, -'mēs\ *noun, plural* **Burmese** (1824)
1 : a native or inhabitant of Burma (Myanmar)
2 : the Tibeto-Burman language of the Burmese people
3 : any of a U.S.-developed breed of slender, shorthaired cats having gold eyes and a usually dark brown coat
— Burmese *adjective*

[1]burn \'bərn\ *noun* [Middle English, from Old English; akin to Old High German *brunno* spring of water] (before 12th century)
British **:** CREEK 2

[2]burn \'bərn\ *verb* **burned** \'bərnd, 'bərnt\ *or* **burnt** \'bərnt\; **burn·ing** [Middle English *birnen,* from Old English *byrnan,* intransitive verb, *bærnan,* transitive verb; akin to Old High German *brinnan* to burn] (before 12th century)
intransitive verb
1 a : to consume fuel and give off heat, light, and gases ⟨a small fire *burns* on the hearth⟩ **b :** to undergo combustion; *also* **:** to undergo nuclear fission or nuclear fusion **c :** to contain a fire ⟨a little stove *burning* in the corner⟩ **d :** to give off light **:** SHINE, GLOW ⟨a light *burning* in the window⟩
2 a : to be hot ⟨the *burning* sand⟩ **b :** to produce or undergo discomfort or pain ⟨ears *burning* from the cold⟩ **c :** to become emotionally excited or agitated: as (1) **:** to yearn ardently ⟨*burning* to tell the story⟩ (2) **:** to be or become very angry or disgusted ⟨the remark made him *burn*⟩
3 a : to undergo alteration or destruction by the action of fire or heat ⟨the house *burned* down⟩ ⟨the potatoes *burned* to a crisp⟩ **b :** to die in the electric chair
4 : to force or make a way by or as if by burning ⟨her words *burned* into his heart⟩
5 : to receive sunburn ⟨she *burns* easily⟩
transitive verb
1 a : to cause to undergo combustion; *especially* **:** to destroy by fire ⟨*burned* the trash⟩ **b :** to use as fuel ⟨this furnace *burns* gas⟩ **c :** to use up **:** CONSUME ⟨*burn* calories⟩
2 a : to transform by exposure to heat or fire ⟨*burn* clay to bricks⟩ **b :** to produce by burning ⟨*burned* a hole in his sleeve⟩
3 a : to injure or damage by or as if by exposure to fire, heat, or radiation **:** SCORCH ⟨*burned* his hand⟩ **b :** to execute by burning ⟨*burned* heretics at the stake⟩; *also* **:** ELECTROCUTE
4 a : IRRITATE, ANNOY — often used with *up* ⟨really *burns* me up⟩ **b :** to take advantage of: as (1) **:** DECEIVE, CHEAT — often used in passive (2) **:** OUTDO, BEAT ⟨*burned* the defense with a touchdown pass⟩
— burn·able \'bər-nə-bəl\ *adjective*
— burn one's bridges *also* **burn one's boats :** to cut off all means of retreat
— burn one's ears : to rebuke strongly
— burn the candle at both ends : to use one's resources or energies to excess
— burn the midnight oil : to work or study far into the night

[3]burn *noun* (1594)
1 : an act, process, instance, or result of burning: as **a :** injury or damage resulting from exposure to fire, heat, caustics, electricity, or certain radiations **b :** a burned area ⟨a *burn* on the tabletop⟩ **c :** an abrasion (as of the skin) having the appearance of a burn ⟨rope *burns*⟩ **d :** a burning sensation ⟨the *burn* of iodine on a cut⟩
2 : the firing of a spacecraft rocket engine in flight
3 : ANGER; *especially* **:** increasing fury — used chiefly in the phrase *slow burn*

burned–out \'bərn-'daůt, 'bərnt-'aůt\ *or* **burnt–out** \'bərnt-'aůt\ *adjective* (1837)
: WORN-OUT; *also* **:** EXHAUSTED

burn·er \'bər-nər\ *noun* (14th century)
: one that burns; *especially* **:** the part of a fuel-burning or heat-producing device (as a furnace or stove) where the flame or heat is produced

bur·net \(,)bər-'net, 'bər-nət\ *noun* [Middle English, from Middle French *burnete,* from *brun* brown — more at BRUNET] (14th century)

: any of a genus (*Sanguisorba*) of herbs of the rose family with odd-pinnate stipulate leaves and spikes of apetalous flowers

burn–in \ˈbər-ˌnin\ *noun* (1966)
: the continuous operation of a device (as a computer) as a test for defects or failure prior to putting it to use

burn in *transitive verb* (circa 1939)
: to increase the density of (portions of a photographic print) during enlarging by giving extra exposure

burn·ing \ˈbər-niŋ\ *adjective* (before 12th century)
1 a : being on fire **b :** ARDENT, INTENSE ⟨*burning* enthusiasm⟩
2 a : affecting with or as if with heat ⟨a *burning* fever⟩ **b :** resembling that produced by a burn ⟨a *burning* sensation on the tongue⟩
3 : of fundamental importance **:** URGENT ⟨one of the *burning* issues of our time⟩
— burn·ing·ly \-niŋ-lē\ *adverb*

burning bush *noun* (1785)
: any of several plants associated with fire (as by redness): as **a :** [2]WAHOO **b :** SUMMER CYPRESS

burning ghat *noun* (1877)
: a level space at the head of a ghat for cremation

[1]**bur·nish** \ˈbər-nish\ *transitive verb* [Middle English *burnischen,* from Middle French *bruniss-,* stem of *brunir,* literally, to make brown, from *brun*] (14th century)
1 a : to make shiny or lustrous especially by rubbing **b :** POLISH 3
2 : to rub (a material) with a tool for compacting or smoothing or for turning an edge
— bur·nish·er *noun*
— bur·nish·ing *adjective or noun*

[2]**burnish** *noun* (circa 1647)
: LUSTER, GLOSS

burn off (circa 1925)
intransitive verb
: to be dissipated by the sun's warmth ⟨waiting for the fog to *burn off*⟩
transitive verb
: to cause to burn off

bur·noose *or* **bur·nous** \(ˌ)bər-ˈnüs\ *noun* [French *burnous,* from Arabic *burnus*] (1695)
: a one-piece hooded cloak worn by Arabs and Berbers
— bur·noosed \-ˈnüst\ *adjective*

burn·out \ˈbər-ˌnau̇t\ *noun* (1940)
1 : the cessation of operation of a jet or rocket engine; *also* **:** the point at which burnout occurs
2 a : exhaustion of physical or emotional strength or motivation usually as a result of prolonged stress or frustration **b :** a person suffering from burnout
3 : a person showing the results of drug abuse

burn out (1710)
transitive verb
1 : to drive out or destroy the property of by fire
2 : to cause to fail, wear out, or become exhausted especially from overwork or overuse
intransitive verb
: to suffer burnout

burn·sides \ˈbərn-ˌsīdz\ *noun plural* [Ambrose E. *Burnside*] (1875)
: SIDE-WHISKERS; *especially* **:** full muttonchop whiskers
word history see SIDEBURNS

[1]**burp** \ˈbərp\ *noun* [imitative] (circa 1932)
: BELCH

[2]**burp** (circa 1932)
intransitive verb
: BELCH
transitive verb
: to help (a baby) expel gas from the stomach especially by patting or rubbing the back

burp gun *noun* (1944)
: a small submachine gun

[1]**burr** \ˈbər\ *noun* [Middle English *burre;* akin to Old English *byrst* bristle — more at BRISTLE] (14th century)
1 *usually* **bur a :** a rough or prickly envelope of a fruit **b :** a plant that bears burs
2 a : something that sticks or clings ⟨a *burr* in the throat⟩ **b :** HANGER-ON
3 : an irregular rounded mass; *especially* **:** a tree burl
4 : a thin ridge or area of roughness produced in cutting or shaping metal
5 a : a trilled uvular \r\ as used by some speakers of English especially in northern England and in Scotland **b :** a tongue-point trill that is the usual Scottish \r\
6 a : a small rotary cutting tool **b** *usually* **bur :** a bit used on a dental drill
7 : a rough humming sound **:** WHIR
— burred \ˈbərd\ *adjective*

[2]**burr** (1798)
intransitive verb
1 : to speak with a burr
2 : to make a whirring sound
transitive verb
1 : to pronounce with a burr
2 a : to form into a projecting edge **b :** to remove burrs from
— burr·er *noun*

bur reed *noun* (1597)
: any of a genus (*Sparganium* of the family Sparganiaceae) of plants with globose fruits resembling burs

bur·ri·to \bə-ˈrē-(ˌ)tō\ *noun, plural* **-tos** [American Spanish, from Spanish, little donkey, diminutive of *burro*] (1934)
: a flour tortilla rolled or folded around a filling (as of meat, beans, or cheese) and usually baked

bur·ro \ˈbər-(ˌ)ō, ˈbu̇r-; ˈbə-(ˌ)rō\ *noun, plural* **burros** [Spanish, irregular from *borrico,* from Late Latin *burricus* small horse] (1800)
: DONKEY; *especially* **:** a small one used as a pack animal

[1]**bur·row** \ˈbər-(ˌ)ō; ˈbə-(ˌ)rō\ *noun* [Middle English *borow*] (13th century)
: a hole or excavation in the ground made by an animal (as a rabbit) for shelter and habitation

[2]**burrow** (1602)
transitive verb
1 *archaic* **:** to hide in or as if in a burrow
2 a : to construct by tunneling **b :** to penetrate by means of a burrow
3 : to make a motion suggestive of burrowing with **:** NESTLE ⟨*burrows* her hand into mine⟩
intransitive verb
1 : to conceal oneself in or as if in a burrow
2 a : to make a burrow **b :** to progress by or as if by digging
3 : to make a motion suggestive of burrowing **:** SNUGGLE, NESTLE ⟨*burrowed* against his back for warmth⟩
— bur·row·er *noun*

burrstone *variant of* BUHRSTONE

bur·ry \ˈbər-ē\ *adjective* **bur·ri·er; -est** (15th century)
1 : containing burs
2 : PRICKLY
3 *of speech* **:** characterized by a burr

bur·sa \ˈbər-sə\ *noun, plural* **bur·sas** \-səz\ *or* **bur·sae** \-ˌsē, -ˌsī\ [New Latin, from Medieval Latin, bag, purse — more at PURSE] (1803)
: a bodily pouch or sac: as **a :** a small serous sac between a tendon and a bone **b :** BURSA OF FABRICIUS
— bur·sal \-səl\ *adjective*

bursa of Fa·bri·cius \-fə-ˈbrē-sh(ē-)əs, -ˈbri-, -s(ē-)əs\ [Johan C. *Fabricius* (died 1808) Danish entomologist] (1945)
: a lymphoid organ that opens into the cloaca of birds and functions in B cell production

bur·sar \ˈbər-sər, -ˌsär\ *noun* [Medieval Latin *bursarius,* from *bursa*] (13th century)
: an officer (as of a monastery or college) in charge of funds **:** TREASURER

bur·sa·ry \ˈbər-sə-rē, ˈbərs-rē\ *noun, plural* **-ries** [Medieval Latin *bursaria,* from *bursa*] (1695)
1 : the treasury of a college or monastery
2 *British* **:** a monetary grant to a needy student **:** SCHOLARSHIP

burse \ˈbərs\ *noun* [Middle French *bourse,* from Medieval Latin *bursa*] (15th century)
1 a : PURSE **b :** a square cloth case used to carry the corporal in a Communion service
2 *obsolete* **:** EXCHANGE, BOURSE

bur·si·tis \(ˌ)bər-ˈsī-təs\ *noun* [New Latin, from *bursa*] (1857)
: inflammation of a bursa especially of the shoulder or elbow

[1]**burst** \ˈbərst\ *verb* **burst** *also* **burst·ed; burst·ing** [Middle English *bersten,* from Old English *berstan;* akin to Old High German *brestan* to burst] (before 12th century)
intransitive verb
1 : to break open, apart, or into pieces usually from impact or from pressure from within
2 a : to give way from an excess of emotion ⟨my heart will *burst*⟩ **b :** to give vent suddenly to a repressed emotion ⟨*burst* into tears⟩ ⟨*burst* out laughing⟩
3 a : to emerge or spring suddenly ⟨*burst* out of the house⟩ ⟨*burst* onto the scene⟩ **b :** LAUNCH, PLUNGE ⟨*burst* into song⟩
4 : to be filled to the breaking point ⟨*bursting* with excitement⟩ ⟨a crate *bursting* with fruit⟩
transitive verb
1 : to cause to burst
2 : to force open (as a door) by strong or vigorous action
3 : to produce by or as if by bursting
— burst at the seams : to be larger, fuller, or more crowded than could reasonably have been anticipated

[2]**burst** *noun* (1610)
1 a : a sudden outbreak; *especially* **:** a vehement outburst (as of emotion) **b :** EXPLOSION, ERUPTION **c :** a sudden intense effort ⟨a *burst* of speed⟩ **d :** the duration of fire in one engagement of the mechanism of an automatic firearm
2 : an act of bursting
3 : a result of bursting; *especially* **:** a visible puff accompanying the explosion of a shell

burst·er \ˈbərs-tər\ *noun* (1611)
1 : one that bursts
2 : the celestial source of an outburst of radiation (as X rays)

bur·then \ˈbər-t͟hən\ *variant of* BURDEN

bur·weed \ˈbər-ˌwēd\ *noun* (circa 1783)
: any of various plants (as a cocklebur or burdock) having burry fruit

bury \ˈber-ē *also* ˈbər-\ *transitive verb* **buried; bury·ing** [Middle English *burien,* from Old English *byrgan;* akin to Old High German *bergan* to shelter, Russian *berech'* to spare] (before 12th century)
1 : to dispose of by depositing in or as if in the earth; *especially* **:** to inter with funeral ceremonies
2 a : to conceal by or as if by covering with earth **b :** to cover from view ⟨*buried* her face in her hands⟩
3 a : to have done with ⟨*burying* their differences⟩ **b :** to conceal in obscurity ⟨*buried* the retraction among the classified ads⟩ **c :** SUBMERGE, ENGROSS — usually used with *in* ⟨*buried* himself in his books⟩
4 : to put (a playing card) out of play by placing it in or under the dealer's pack
5 : to succeed emphatically or impressively in making (a shot) ⟨*bury* a jumper⟩
6 : to defeat overwhelmingly
synonym see HIDE
— bury the hatchet : to settle a disagreement **:** become reconciled

[1]**bus** \ˈbəs\ *noun, plural* **bus·es** *or* **bus·ses** *often attributive* [short for *omnibus*] (circa 1909)
1 a : a large motor vehicle designed to carry passengers usually along a fixed route according to a schedule **b :** AUTOMOBILE
2 : a small hand truck

3 a : BUS BAR **b :** a set of parallel conductors in a computer system that forms a main transmission path
4 : a spacecraft or missile that carries one or more detachable devices (as warheads) ◊

[2]bus *verb* **bused** *or* **bussed; bus·ing** *or* **bus·sing** (circa 1909)
intransitive verb
1 : to travel by bus
2 : to work as a busboy
transitive verb
1 : to transport by bus
2 a : CLEAR 4d ⟨*bus* dishes⟩ **b :** to remove dirty dishes from ⟨*bus* tables⟩

bus bar *noun* (1893)
: a conductor or an assembly of conductors for collecting electric currents and distributing them to outgoing feeders

bus·boy \ˈbəs-ˌbȯi\ *noun* [*omnibus* busboy] (1913)
: a waiter's assistant; *specifically* **:** one who removes dirty dishes and resets tables in a restaurant

bus·by \ˈbəz-bē\ *noun, plural* **busbies** [probably from the name *Busby*] (1853)
1 : a military full-dress fur hat with a pendent bag on one side usually of the color of regimental facings
2 : the bearskin worn by British guardsmen — not used by the guardsmen themselves

busby 1

[1]bush \ˈbu̇sh\ *noun, often attributive* [Middle English; akin to Old High German *busc* forest] (14th century)
1 a : SHRUB; *especially* **:** a low densely branched shrub **b :** a close thicket of shrubs suggesting a single plant
2 : a large uncleared or sparsely settled area (as in Australia) usually scrub-covered or forested **:** WILDERNESS — usually used with *the*
3 a *archaic* **:** a bunch of ivy formerly hung outside a tavern to indicate wine for sale **b** *obsolete* **:** TAVERN **c :** ADVERTISING ⟨good wine needs no *bush* —Shakespeare⟩
4 : a bushy tuft or mass ⟨a *bush* of hair⟩; *especially* **:** [2]BRUSH 2a
5 : MINOR LEAGUE — usually used in plural

[2]bush (15th century)
transitive verb
: to support, mark, or protect with bushes
intransitive verb
: to extend like a bush **:** resemble a bush

[3]bush *adjective* (1595)
1 : having a low-growing compact bushy habit — used especially of cultivated beans ⟨*bush* snap beans⟩
2 : serving, occurring in, or used in the bush ⟨*bush* planes⟩

[4]bush *noun* [Dutch *bus* bushing, box, from Middle Dutch *busse* box, from Late Latin *buxis* — more at BOX] (1566)
chiefly British **:** BUSHING

[5]bush *adjective* [short for *bush-league*] (1970)
: falling below acceptable standards **:** UNPROFESSIONAL ⟨*bush* behavior⟩

bush baby *noun* (1901)
: GALAGO

bush basil *noun* (1597)
: a sweet basil of a cultivar with small leaves

bush·buck \ˈbu̇sh-ˌbək\ *noun, plural* **bushbuck** *or* **bushbucks** [translation of Afrikaans *bosbok*] (1852)
: a small African striped antelope (*Tragelaphus scriptus*) especially of sub-Saharan forests that has spirally twisted horns

bush clover *noun* (circa 1818)
: any of several usually shrubby lespedezas

[1]bushed \ˈbu̇sht\ *adjective* (14th century)
1 : covered with or as if with a bushy growth
2 *chiefly Australian* **a :** lost especially in the bush **b :** PERPLEXED 1, CONFUSED
3 : TIRED, EXHAUSTED

[2]bushed *adjective* (1907)
: having a bushing

[1]bush·el \ˈbu̇-shəl\ *noun* [Middle English *busshel,* from Old French *boissel,* from (assumed) Old French *boisse* one sixth of a bushel, of Celtic origin; akin to Middle Irish *boss* breadth of the hand] (14th century)
1 : any of various units of dry capacity — see WEIGHT table
2 : a container holding a bushel
3 : a large quantity **:** LOTS ⟨makes *bushels* of money⟩

[2]bushel *verb* **bush·eled; bush·el·ing** \-sh(ə-)liŋ\ [probably from German *bosseln* to do poor work, to patch; akin to Old English *bēatan* to beat] (circa 1877)
: REPAIR, RENOVATE
— bush·el·er \-sh(ə-)lər\ *noun*

bush·fire \ˈbu̇sh-ˌfīr\ *noun* (1832)
Australian **:** an uncontrolled fire in a bush area

Bu·shi·do \ˈbü-shi-ˌdō, ˈbü-\ *noun* [Japanese *bushidō*] (1898)
: a feudal-military Japanese code of chivalry valuing honor above life

bush·ing \ˈbu̇-shiŋ\ *noun* (1839)
1 : a usually removable cylindrical lining for an opening (as of a mechanical part) used to limit the size of the opening, resist abrasion, or serve as a guide
2 : an electrically insulating lining for a hole to protect a through conductor

bush jacket *noun* [from its use in rough country] (circa 1939)
: a long cotton jacket resembling a shirt and having four patch pockets and a belt

bush·land \ˈbu̇sh-ˌland\ *noun* (1827)
: [1]BUSH 2

bush–league *adjective* (1914)
: being of an inferior class or group of its kind

bush league *noun* (1909)
: MINOR LEAGUE
— bush leaguer *noun*

bush·man \ˈbu̇sh mən\ *noun* (1785)
1 *capitalized* [modification of obsolete Afrikaans *boschjesman,* from *boschje* (diminutive of *bosch* forest) + Afrikaans *man*] **a :** a member of a group of short statured peoples of southern Africa who traditionally live by hunting and foraging **b :** the Khoisan languages spoken by these people
2 a : WOODSMAN **b** *chiefly Australian* **:** a person who lives in the bush

bush·mas·ter \-ˌmas-tər\ *noun* (1826)
: a tropical American pit viper (*Lachesis mutus*) that is the largest New World venomous snake

bush·pig \-ˌpig\ *noun* (1840)
: a wild usually reddish to black pig (*Potamochoerus porcus*) of forests and scrubland of sub-Saharan Africa and Madagascar that has much facial hair, long pointed ears, and a light-colored mane along the top of the neck and back

bush pilot *noun* (1936)
: a pilot who flies a small plane into remote areas

bush·rang·er \-ˌrān-jər\ *noun* (1801)
1 *Australian* **:** an outlaw living in the bush
2 : FRONTIERSMAN, WOODSMAN
— bush·rang·ing \-jiŋ\ *noun*

bush shirt *noun* [from its use in rough country] (1909)
: a usually loose-fitting cotton shirt with patch pockets

bush·tit \-ˌtit\ *noun* (circa 1889)
: a small gray titmouse (*Psaltriparus minimus*) of western North America with light underparts that occurs in several geographic forms sometimes placed in separate species

bush·whack \ˈbu̇sh-ˌhwak, -ˌwak\ *verb* [back-formation from *bushwhacker*] (1866)
transitive verb
: AMBUSH; *broadly* **:** to attack suddenly **:** ASSAULT
intransitive verb
: to clear a path through thick woods especially by chopping down bushes and low branches
— bush·whack·er *noun*

bushy \ˈbu̇-shē\ *adjective* **bush·i·er; -est** (14th century)
1 : full of or overgrown with bushes
2 : resembling a bush; *especially* **:** being thick and spreading
— bush·i·ly \ˈbu̇-shə-lē\ *adverb*
— bush·i·ness \ˈbu̇-shē-nəs\ *noun*

busi·ness \ˈbiz-nəs, -nəz, *Southern also* ˈbid-\ *noun, often attributive* (14th century)
1 *archaic* **:** purposeful activity **:** BUSYNESS
2 a : ROLE, FUNCTION ⟨how the human mind went about its *business* of learning —H. A. Overstreet⟩ **b :** an immediate task or objective **:** MISSION ⟨what is your *business* here⟩ **c :** a particular field of endeavor ⟨the best in the *business*⟩
3 a : a usually commercial or mercantile activity engaged in as a means of livelihood **:** TRADE, LINE ⟨in the restaurant *business*⟩ **b :** a commercial or sometimes an industrial enterprise; *also* **:** such enterprises ⟨the *business* district⟩ **c :** usually economic dealings **:** PATRONAGE ⟨took their *business* elsewhere⟩
4 : AFFAIR, MATTER ⟨the whole *business* got out of hand⟩ ⟨*business* as usual⟩
5 : CREATION, CONCOCTION
6 : movement or action (as lighting a cigarette) by an actor intended especially to establish atmosphere, reveal character, or explain a situation — called also *stage business*
7 a : personal concern ⟨none of your *business*⟩ **b :** RIGHT ⟨you have no *business* speaking to me that way⟩
8 a : serious activity requiring time and effort and usually the avoidance of distractions ⟨got down to *business*⟩ ⟨she means *business*⟩ **b :** maximum effort
9 a : a damaging assault **b :** REBUKE, TONGUE-LASHING ⟨gave him the *business* for staying out so late⟩ **c :** DOUBLE CROSS

◇ WORD HISTORY

bus In 1828 a new era in public transportation dawned when horse-drawn boxes on wheels seating fourteen people on facing benches began to run regular routes in Paris. They were christened with the name *omnibus,* taken to be short for *voitures omnibus* "vehicles (open) to all"; *omnibus,* the dative of Latin *omnes* "everybody, all," summed up in one word their general accessibility. An earlier transport line of the same kind had operated in the city of Nantes, and according to one tradition the name *omnibus* occurred to the proprietors because the line terminated at the shop of a certain M. Omnes, who used the slogan *Omnes Omnibus.* In 1829 a horse-drawn line patterned after the French *omnibus* opened in London with the vehicles bearing the same name. By 1832 Londoners had clipped *omnibus* to *bus.* As the vehicles became self-propelled, first by steam and then by internal-combustion engines, the coinage *autobus* appeared, in English by 1899 and in French by 1907. *Bus* and *autobus* have since become international words, current over much of the world and in a number of languages.

10 : a bowel movement — used especially of pets ☆

business administration *noun* (circa 1911)
: a program of studies in a college or university providing general knowledge of business principles and practices

business card *noun* (1840)
: a small card bearing information (as name and address) about a business or business representative

business cycle *noun* (1919)
: a cycle of economic activity usually consisting of recession, recovery, growth, and decline

business end *noun* (1878)
: the end with, from, or through which a thing's function is fulfilled ⟨the *business end* of a revolver⟩

busi·ness·like \'biz-nəs-ˌlīk, -nəz-\ *adjective* (1791)
1 : exhibiting qualities believed to be advantageous in business
2 : SERIOUS, PURPOSEFUL

busi·ness·man \-ˌman, -mən\ *noun* (1826)
: a man who transacts business; *especially* **:** a business executive

busi·ness·peo·ple \-ˌpē-pəl\ *noun plural* (1865)
: persons active in business

busi·ness·per·son \-ˌpər-sᵊn\ *noun* (1974)
: a businessman or businesswoman

business suit *noun* (1870)
: a man's suit consisting of matching coat and trousers and sometimes a vest

busi·ness·wom·an \-ˌwu̇-mən\ *noun* (1844)
: a woman who transacts business; *especially* **:** a female business executive

bus·ing *or* **bus·sing** \'bə-siŋ\ *noun* (1923)
: the act of transporting by bus; *specifically* **:** the transporting of children to a school outside their residential area as a means of establishing racial balance in that school

busk·er \'bəs-kər\ *noun* [*busk,* probably from Italian *buscare* to procure, gain, from Spanish *buscar* to look for] (1857)
chiefly British **:** a person who entertains especially by playing music on the street
— busk \'bəsk\ *intransitive verb*

bus·kin \'bəs-kən\ *noun* [probably modification of Middle French *brozequin*] (1503)
1 : a laced boot reaching halfway or more to the knee
2 a : COTHURNUS 1 **b :** TRAGEDY; *especially* **:** tragedy resembling that of ancient Greek drama

bus·load \'bəs-ˌlōd\ *noun* (1938)
: a load that fills a bus ⟨*busloads* of tourists⟩

bus·man's holiday \'bəs-mənz-\ *noun* (1893)
: a holiday spent in following or observing the practice of one's usual occupation

buss \'bəs\ *noun* [perhaps alteration of Middle English *bassen* to kiss] (1570)
: KISS
— buss *transitive verb*

¹bust \'bəst\ *noun* [French *buste,* from Italian *busto,* from Latin *bustum* tomb] (1645)
1 : a sculptured representation of the upper part of the human figure including the head and neck and usually part of the shoulders and breast
2 : the upper part of the human torso between neck and waist; *especially* **:** the breasts of a woman

²bust *verb* **bust·ed** *also* **bust; bust·ing** [alteration of *burst*] (1806)
transitive verb
1 a : to break or smash especially with force; *also* **:** to make inoperative ⟨*busted* my watch⟩ **b :** to bring an end to **:** BREAK UP ⟨helped *bust* trusts —*Newsweek*⟩ — often used with *up* ⟨better not try to *bust* up his happy marriage —*Forbes*⟩ **c :** to ruin financially **d :** EXHAUST, WEAR OUT — used in phrases like *bust one's butt* to describe making a strenuous effort
2 : TAME ⟨bronco *busting*⟩
3 : DEMOTE
4 *slang* **a :** ARREST ⟨*busted* for carrying guns —Saul Gottlieb⟩ **b :** RAID ⟨*busted* the apartment⟩
5 : HIT, SLUG
intransitive verb
1 : to go broke
2 a : BURST ⟨laughing fit to *bust*⟩ **b :** BREAK DOWN
3 a : to lose at cards by exceeding a limit (as the count of 21 in blackjack) **b :** to fail to complete a straight or flush in poker

³bust *noun* (1840)
1 a : SPREE **b :** a hearty drinking session ⟨a beer *bust*⟩
2 a : a complete failure **:** FLOP **b :** a business depression
3 : PUNCH, SOCK
4 *slang* **a :** a police raid **b :** ARREST 2

⁴bust *or* **bust·ed** \'bəs-təd\ *adjective* (1837)
: BANKRUPT, BROKE ⟨go *bust*⟩

bus·tard \'bəs-tərd\ *noun* [Middle English, modification of Middle French *bistarde,* from Old Italian *bistarda,* from Latin *avis tarda,* literally, slow bird] (15th century)
: any of a family (Otididae) of large chiefly terrestrial Old World and Australian game birds

bust·er \'bəs-tər\ *noun* (1848)
1 *chiefly Australian* **:** a sudden violent wind often coming from the south
2 a : an unusually sturdy child **b** *often capitalized* **:** FELLOW — usually used as a noun of address ⟨hey *buster,* come here⟩
3 : one that breaks, breaks up, or eliminates something ⟨crime *busters*⟩: as **a :** PLOW **b** [short for *broncobuster*] **:** a person who breaks horses
4 : a bad fall

bus·tier \ˌbüs-tē-'ā, ˌbəs-, -'tyā\ *noun* [French, from *buste*] (1979)
: a tight-fitting often strapless top worn as a brassiere or outer garment

¹bus·tle \'bə-səl\ *intransitive verb* **bus·tled; bus·tling** \'bəs-liŋ, 'bə-sə-\ [probably alteration of obsolete *buskle* to prepare, frequentative of *busk,* from Old Norse *būask* to prepare oneself] (1580)
1 : to move briskly and often ostentatiously
2 : to be busily astir **:** TEEM
— bustling *adjective*
— bus·tling·ly \-liŋ-lē\ *adverb*

²bustle *noun* (1634)
: noisy, energetic, and often obtrusive activity ⟨the hustle and *bustle* of the big city⟩

³bustle *noun* [origin unknown] (1786)
: a pad or framework expanding and supporting the fullness and drapery of the back of a woman's skirt or dress; *also* **:** the drapery so supported

bust·line \'bəs(t)-ˌlīn\ *noun* (1926)
1 : an arbitrary line encircling the fullest part of the bust
2 : body circumference at the bust

busty \'bəs-tē\ *adjective* **bust·i·er; -est** (1944)
: having a large bust

bu·sul·fan \byü-'səl-fən\ *noun* [*butane* + *sulfonyl*] (circa 1958)
: an antineoplastic agent $C_6H_{14}O_6S_2$ used in the treatment of chronic myelogenous leukemia

¹busy \'bi-zē\ *adjective* **busi·er; -est** [Middle English *bisy,* from Old English *bisig;* akin to Middle Dutch & Middle Low German *besich* busy] (before 12th century)
1 a : engaged in action **:** OCCUPIED **b :** being in use ⟨found the telephone *busy*⟩
2 : full of activity **:** BUSTLING ⟨a *busy* seaport⟩
3 : foolishly or intrusively active **:** MEDDLING
4 : full of distracting detail ⟨a *busy* design⟩ ☆
— busi·ly \'bi-zə-lē\ *adverb*
— busy·ness \'bi-zē-nəs\ *noun*

²busy *verb* **bus·ied; busy·ing** (before 12th century)
transitive verb
: to make busy **:** OCCUPY
intransitive verb
: BUSTLE ⟨small boats *busied* to and fro —Quentin Crewe⟩

busy·body \'bi-zē-ˌbä-dē\ *noun* (1526)
: an officious or inquisitive person

busy·work \-ˌwərk\ *noun* (1910)
: work that usually appears productive or of intrinsic value but actually only keeps one occupied

¹but \'bət\ *conjunction* [Middle English, from Old English *būtan,* preposition & conjugation, outside, without, except, except that; akin to Old High German *būzan* without, except; akin to Old English *be* by, *ūt* out — more at BY, OUT] (before 12th century)
1 a : except for the fact ⟨would have protested *but* that he was afraid⟩ **b :** THAT — used after a negative ⟨there is no doubt *but* he won⟩ **c :** without the concomitant that ⟨it never rains *but* it pours⟩ **d :** if not **:** UNLESS **e :** THAN ⟨no sooner started *but* it stopped⟩ — not often in formal use
2 a : on the contrary **:** on the other hand **:** NOTWITHSTANDING — used to connect coordinate elements ⟨he was called *but* he did not answer⟩ ⟨not peace *but* a sword⟩ **b :** YET ⟨poor *but* proud⟩ **c :** with the exception of — used before a word often taken to be the subject of a clause ⟨none *but* the brave deserves the fair —John Dryden⟩
— but what : that . . . not — used to indicate possibility or uncertainty ⟨I don't know *but what* I will go⟩

²but *preposition* (before 12th century)
1 a : with the exception of **:** BARRING ⟨no one there *but* me⟩ — compare ¹BUT 2c **b :** other than ⟨this letter is nothing *but* an insult⟩
2 *Scottish* **a :** WITHOUT, LACKING **b :** OUTSIDE

³but *adverb* (12th century)
1 : ONLY, MERELY ⟨he is *but* a child⟩
2 *Scottish* **:** OUTSIDE
3 : to the contrary ⟨who knows *but* that she may succeed⟩
4 — used as an intensive ⟨get there *but* fast⟩

⁴but *pronoun* (1556)
: that not **:** who not ⟨nobody *but* has his fault —Shakespeare⟩

⁵but *noun* [Scots *but,* adjective (outer)] (1724)

☆ SYNONYMS

Business, commerce, trade, industry, traffic mean activity concerned with the supplying and distribution of commodities. BUSINESS may be an inclusive term but specifically designates the activities of those engaged in the purchase or sale of commodities or in related financial transactions. COMMERCE and TRADE imply the exchange and transportation of commodities. INDUSTRY applies to the producing of commodities, especially by manufacturing or processing, usually on a large scale. TRAFFIC applies to the operation and functioning of public carriers of goods and persons. See in addition WORK.

Busy, industrious, diligent, assiduous, sedulous mean actively engaged or occupied. BUSY chiefly stresses activity as opposed to idleness or leisure ⟨too *busy* to spend time with the children⟩. INDUSTRIOUS implies characteristic or habitual devotion to work ⟨*industrious* employees⟩. DILIGENT suggests earnest application to some specific object or pursuit ⟨very *diligent* in her pursuit of a degree⟩. ASSIDUOUS stresses careful and unremitting application ⟨*assiduous* practice⟩. SEDULOUS implies painstaking and persevering application ⟨a *sedulous* investigation of the murder⟩.

Scottish **:** the kitchen or living quarters of a 2-room cottage

bu•ta•di•ene \ˌbyü-tə-ˈdī-ˌēn, -ˌdī-ˈ\ *noun* [International Scientific Vocabulary *butane* + *di-* + *-ene*] (1900)
: a flammable gaseous open chain hydrocarbon C_4H_6 used in making synthetic rubbers

bu•tane \ˈbyü-ˌtān\ *noun* [International Scientific Vocabulary *butyric* + *-ane*] (1875)
: either of two isomeric flammable gaseous alkanes C_4H_{10} obtained usually from petroleum or natural gas and used as fuels

bu•ta•nol \ˈbyü-tᵊn-ˌȯl, -ˌōl\ *noun* (1894)
: either of two flammable four-carbon alcohols C_4H_9OH derived from normal butane

butch \ˈbu̇ch\ *adjective* [probably from *Butch,* male nickname] (1941)
1 : very masculine in appearance or manner
2 : playing the male role in a homosexual relationship

¹butch•er \ˈbu̇-chər\ *noun* [Middle English *bocher,* from Old French *bouchier,* from *bouc* he-goat, probably of Celtic origin; akin to Middle Irish *bocc* he-goat — more at BUCK] (13th century)
1 a : a person who slaughters animals or dresses their flesh **b :** a dealer in meat
2 : one that kills ruthlessly or brutally
3 : one that bungles or botches
4 : a vendor especially on trains or in theaters

²butcher *transitive verb* **butch•ered; butch•er•ing** \ˈbu̇ch-riŋ, ˈbu̇-chə-\ (1562)
1 : to slaughter and dress for market ⟨*butcher* hogs⟩
2 : to kill in a barbarous manner
3 : BOTCH ⟨*butchered* the play⟩

butch•er–bird \ˈbu̇-chər-ˌbərd\ *noun* (1668)
: any of various shrikes

butcher block *noun* (1967)
: a block made with thick strips of usually laminated hardwood
— butcher–block *adjective*

butch•er•ly \ˈbu̇-chər-lē\ *adjective* (1513)
: resembling a butcher **:** SAVAGE

butcher paper *noun* (1944)
: heavy brown or white paper used especially for wrapping meats

butch•ery \ˈbu̇ch-rē, ˈbu̇-chə-\ *noun, plural* **-er•ies** (14th century)
1 *chiefly British* **:** SLAUGHTERHOUSE
2 : the preparation of meat for sale
3 : cruel and ruthless slaughter of human beings
4 : BOTCH

bute \ˈbyüt\ *noun* (1968)
: PHENYLBUTAZONE

bu•tene \ˈbyü-ˌtēn\ *noun* [International Scientific Vocabulary *butyl* + *-ene*] (1885)
: a normal butylene

bu•teo \ˈbyü-tē-ˌō\ *noun, plural* **-te•os** [New Latin, genus name, from Latin, a hawk] (1940)
: any of a genus (*Buteo*) of hawks with broad rounded wings, relatively short tails, and soaring flight

but for *preposition* (12th century)
: EXCEPT FOR

but•ler \ˈbət-lər\ *noun* [Middle English *buteler,* from Old French *bouteillier* bottle bearer, from *bouteille* bottle — more at BOTTLE] (13th century)
1 : a manservant having charge of the wines and liquors
2 : the chief male servant of a household who has charge of other employees, receives guests, directs the serving of meals, and performs various personal services

butler's pantry *noun* (1816)
: a service room between kitchen and dining room

¹butt \ˈbət\ *verb* [Middle English, from Old French *boter,* of Germanic origin; akin to Old High German *bōzan* to beat — more at BEAT] (13th century)
intransitive verb
: to thrust or push head foremost **:** strike with the head or horns
transitive verb
: to strike or shove with the head or horns

²butt *noun* (1647)
: a blow or thrust usually with the head or horns

³butt *noun* [Middle English, from Middle French *botte,* from Old Provençal *bota,* from Late Latin *buttis*] (14th century)
1 : a large cask especially for wine, beer, or water
2 : any of various units of liquid capacity; *especially* **:** a measure equal to 108 imperial gallons (491 liters)

⁴butt *noun* [Middle English, partly from Middle French *but* target, end, of Germanic origin; akin to Old Norse *būtr* log, Low German *butt* blunt; partly from Middle French *bute* backstop, from *but* target] (14th century)
1 a : a backstop (as a mound or bank) for catching missiles shot at a target **b :** TARGET **c** *plural* **:** RANGE 5c **d :** a blind for shooting birds
2 a *obsolete* **:** LIMIT, BOUND **b** *archaic* **:** GOAL ⟨here is my journey's end, here is my *butt* —Shakespeare⟩
3 : an object of abuse or ridicule **:** VICTIM ⟨the *butt* of all their jokes⟩

⁵butt *noun* [Middle English; probably akin to Middle English *buttok* buttock, Low German *butt* blunt] (15th century)
1 : BUTTOCKS — often used as a euphemism for *ass* in idiomatic expressions ⟨get your *butt* over here⟩ ⟨kick *butt*⟩ ⟨saved our *butts*⟩
2 : the large or thicker end part of something: **a :** a lean upper cut of the pork shoulder **b :** the base of a plant from which the roots spring **c :** the thicker or handle end of a tool or weapon
3 : an unused remainder
4 : the part of a hide or skin corresponding to the animal's back and sides

⁶butt *verb* [partly from ⁴*butt,* partly from ⁵*butt*] (1785)
intransitive verb
: ABUT — used with *on* or *against*
transitive verb
1 : to place end to end or side to side without overlapping
2 : to trim or square off (as a log) at the end
3 : to reduce (as a cigarette) to a butt by stubbing or stamping

butte \ˈbyüt\ *noun* [French, knoll, from Middle French *bute*] (1805)
: an isolated hill or mountain with steep or precipitous sides usually having a smaller summit area than a mesa

¹but•ter \ˈbə-tər\ *noun* [Middle English, from Old English *butere,* from Latin *butyrum,* from Greek *boutyron,* from *bous* cow + *tyros* cheese; akin to Avestan *tūiri-* curds — more at COW] (before 12th century)
1 : a solid emulsion of fat globules, air, and water made by churning milk or cream and used as food
2 : a buttery substance: as **a :** any of various fatty oils remaining nearly solid at ordinary temperatures **b :** a creamy food spread; *especially* **:** one made of ground roasted nuts ⟨*peanut butter*⟩
3 : FLATTERY
— but•ter•less \-ləs\ *adjective*

²butter *transitive verb* (15th century)
: to spread with or as if with butter

but•ter–and–eggs \ˌbə-tər-ᵊn-ˈegz, -ˈāgz\ *noun plural but singular or plural in construction* (1776)
: a common Eurasian perennial herb (*Linaria vulgaris*) of the snapdragon family that has showy yellow and orange flowers and is naturalized in much of North America — called also *toadflax*

but•ter•ball \ˈbə-tər-ˌbȯl\ *noun* (1813)
1 : BUFFLEHEAD
2 : a chubby person

butter bean *noun* (circa 1819)
1 : LIMA BEAN: as **a** *chiefly Southern & Midland* **:** a large dried lima bean **b :** SIEVA BEAN
2 : WAX BEAN
3 : a green shell bean especially as opposed to a snap bean

butter clam *noun* (1936)
: either of two clams (*Saxidomus nuttallii* and *S. giganteus*) of the Pacific coast of North America

but•ter•cup \ˈbə-tər-ˌkəp\ *noun* (1777)
: any of a genus (*Ranunculus* of the family Ranunculaceae, the buttercup family) of herbs with yellow or white flowers and alternate leaves

but•ter•fat \-ˌfat\ *noun* (1889)
: the natural fat of milk and chief constituent of butter consisting essentially of a mixture of glycerides (as those derived from butyric, capric, caproic, and caprylic acids)

but•ter•fin•gered \-ˌfiŋ-gərd\ *adjective* (1615)
: apt to let things fall or slip through the fingers **:** CARELESS
— but•ter•fin•gers \-gərz\ *noun plural but singular or plural in construction*

but•ter•fish \-ˌfish\ *noun* (1674)
: any of numerous bony fishes (especially family Stromateidae) with a slippery coating of mucus

¹but•ter•fly \-ˌflī\ *noun, often attributive* (before 12th century)
1 : any of numerous slender-bodied diurnal insects (order Lepidoptera) with broad often brightly colored wings
2 : something that resembles or suggests a butterfly; *especially* **:** a person chiefly occupied with the pursuit of pleasure
3 : a swimming stroke executed in a prone position by moving both arms in a circular motion while kicking both legs up and down
4 *plural* **:** a feeling of hollowness or queasiness caused especially by emotional or nervous tension or anxious anticipation

²butterfly *transitive verb* **-flied; -fly•ing** (1954)
: to split almost entirely and spread apart ⟨a *butterflied* steak⟩ ⟨*butterflied* shrimp⟩

butterfly bush *noun* (1924)
: BUDDLEIA

butterfly chair *noun* (1953)
: a chair for lounging consisting of a cloth sling supported by a frame of metal tubing or bars

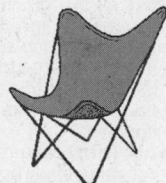

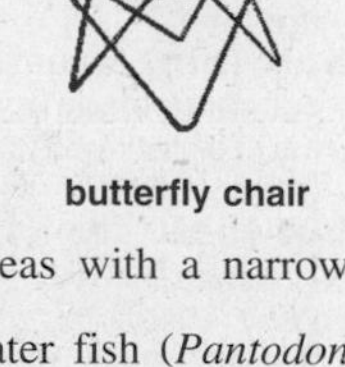
butterfly chair

but•ter•fly•er \ˈbə-tər-ˌflī(-ə)r\ *noun* (1967)
: a swimmer who specializes in the butterfly

butterfly fish *noun* (1740)
1 : any of a family (Chaetodontidae) of small brilliantly colored spiny-finned fishes of tropical seas with a narrow deep body and scaled fins
2 : a small brown freshwater fish (*Pantodon buchholzi* of the family Pantodontidae) of western Africa that has elongate winglike pectoral fins

butterfly valve *noun* (1816)
1 : a valve consisting of two semicircular clappers hinged to a cross rib that permits fluid flow in only one direction
2 : a damper or valve in a pipe consisting of a disk turning on a diametral axis

butterfly weed *noun* (1816)
: an orange-flowered showy milkweed (*Asclepias tuberosa*) of eastern temperate North America

but·ter·milk \'bə-tər-ˌmilk\ *noun* (15th century)
1 : the liquid left after butter has been churned from milk or cream
2 : cultured milk made by the addition of suitable bacteria to sweet milk

but·ter·nut \-ˌnət\ *noun* (1741)
1 : an eastern North American tree of the walnut family with sweet egg-shaped nuts and light brown wood — called also *white walnut*
2 a : a light yellowish brown **b** *plural* **:** homespun overalls dyed brown with a butternut extract **c :** a soldier or partisan of the Confederacy during the Civil War

butternut squash *noun* (1945)
: a smooth somewhat bottle-shaped buff-colored winter squash with usually orange flesh

but·ter·scotch \-ˌskäch\ *noun* (1855)
1 : a candy made from brown sugar, butter, corn syrup, and water; *also* **:** the flavor of such candy
2 : a moderate yellowish brown

butter up *transitive verb* (1819)
: to charm or beguile with lavish flattery or praise

but·ter·weed \'bə-tər-ˌwēd\ *noun* (circa 1845)
: any of several plants having yellow flowers or smooth soft foliage: as **a :** HORSEWEED 1 **b :** a North American groundsel (*Senecio glabellus*)

but·ter·wort \-ˌwərt, -ˌwȯrt\ *noun* (1597)
: any of a genus (*Pinguicula*) of herbs of the bladderwort family with fleshy leaves that produce a viscid secretion serving to capture and digest insects

¹but·tery \'bə-tə-rē, 'bə-trē\ *noun, plural* **-ter·ies** [Middle English *boterie,* from Middle French, from *botte* cask, butt — more at BUTT] (14th century)
1 : a storeroom for liquors
2 a *chiefly dialect* **:** PANTRY **b :** a room (as in an English college) stocking provisions for sale to students

²but·tery \'bə-tə-rē\ *adjective* (14th century)
1 a : having the qualities of butter **b :** containing or spread with butter
2 : marked by flattery

butt hinge *noun* (1815)
: a hinge usually mortised flush into the edge of a door and its jamb

butt in *intransitive verb* (1900)
: to meddle in the affairs of others **:** INTERFERE

butt·in·sky *also* **butt·in·ski** \bə-'din-skē\ *noun, plural* **-skies** [*butt in* + *-sky, -ski* (last element in Slavic surnames)] (1902)
: a person given to butting in **:** a troublesome meddler

butt joint *noun* (1823)
: a joint made by fastening the parts together end-to-end without overlap and often with reinforcement

but·tock \'bə-tək *also* -(ˌ)täk\ *noun* [Middle English *buttok* — more at BUTT] (14th century)
1 : the back of a hip that forms one of the fleshy parts on which a person sits
2 *plural* **a :** the seat of the body **b :** RUMP

¹but·ton \'bə-tᵊn\ *noun, often attributive* [Middle English *boton,* from Middle French, from Old French, from *boter* to thrust — more at BUTT] (14th century)
1 a : a small knob or disk secured to an article (as of clothing) and used as a fastener by passing it through a buttonhole or loop **b :** a usually circular metal or plastic badge bearing a stamped design or printed slogan ⟨campaign *button*⟩
2 : something that resembles a button: as **a :** any of various parts or growths of a plant or of an animal: as (1) **:** an immature whole mushroom (2) **:** the terminal segment of a rattlesnake's rattle **b :** a small globule of metal remaining after fusion (as in assaying) **c :** a guard on the tip of a fencing foil
3 : PUSH BUTTON
4 : the point of the chin especially as a target for a knockout blow
— but·ton·less \-ləs\ *adjective*
— on the button : EXACTLY ⟨arrived at noon *on the button*⟩; *also* **:** exactly on target **:** on the nose ⟨the estimate was right *on the button*⟩

²button *verb* **but·toned; but·ton·ing** \'bət-niŋ, 'bə-tᵊn-iŋ\ (14th century)
transitive verb
1 : to furnish or decorate with buttons
2 a : to pass (a button) through a buttonhole or loop **b :** to close or fasten with buttons — often used with *up* ⟨*button* up your overcoat⟩
3 a : to close (the lips) to prevent speech ⟨*button* your lip⟩ **b :** to close or seal tightly — usually used with *up* ⟨*button* up the house for winter⟩
intransitive verb
: to have buttons for fastening ⟨this dress *buttons* at the back⟩
— but·ton·er \'bət-nər, 'bə-tᵊn-ər\ *noun*

but·ton·ball \'bə-tᵊn-ˌbȯl\ *noun* (1821)
: ²PLANE

but·ton·bush \-ˌbu̇sh\ *noun* (1754)
: a North American shrub (*Cephalanthus occidentalis*) of the madder family with globular flower heads

¹but·ton–down \-ˌdau̇n\ *adjective* (1934)
1 a *of a collar* **:** having the ends fastened to the garment with buttons **b** *of a garment* **:** having a button-down collar
2 *also* **but·toned–down** \-tᵊn-ˌdau̇n\ **:** conservatively traditional or conventional; *especially* **:** adhering to conventional ideals in dress and behavior ⟨*button-down* businessmen⟩

²button–down *noun* (1952)
: a shirt with a button-down collar

but·toned–up \'bə-tənd-'dəp\ *adjective* (1936)
: coldly reserved or standoffish

¹but·ton·hole \'bə-tᵊn-ˌhōl\ *noun* (1561)
1 : a slit or loop through which a button is passed
2 *chiefly British* **:** BOUTONNIERE

²buttonhole *transitive verb* (1828)
1 : to furnish with buttonholes
2 : to work with buttonhole stitch
— but·ton·hol·er \-ˌhō-lər\ *noun*

³buttonhole *transitive verb* [alteration of *buttonhold*] (1862)
: to detain in conversation by or as if by holding on to the outer garments of

buttonhole stitch *noun* (circa 1885)
: a closely worked loop stitch used to make a firm edge (as on a buttonhole)

but·ton·hook \'bə-tᵊn-ˌhu̇k\ *noun* (1870)
1 : a hook for drawing small buttons through buttonholes
2 : an offensive play in football in which the pass receiver runs straight downfield and then abruptly cuts back toward the line of scrimmage
— buttonhook *intransitive verb*

button man *noun* [perhaps from *buttons* bellhop] (1966)
: a low-ranking member of a criminal underworld organization

button mushroom *noun* (1865)
: a small cultivated mushroom (*Agaricus bisporus* synonym *A. brunnescens*)

button quail *noun* (1885)
: any of a family (Turnicidae) of small terrestrial Old World birds that resemble quails and have only three toes on a foot with the hind toe being absent

button snakeroot *noun* (1775)
1 : BLAZING STAR 2a
2 : any of several usually prickly herbs (genus *Eryngium*) of the carrot family

but·ton·wood \'bə-tᵊn-ˌwu̇d\ *noun* (1674)
: ²PLANE

¹but·tress \'bə-trəs\ *noun* [Middle English *butres,* from Middle French *bouterez,* from Old French *boterez,* from *boter* — more at BUTT] (14th century)
1 : a projecting structure of masonry or wood for supporting or giving stability to a wall or building
2 : something that resembles a buttress: as **a :** a projecting part of a mountain or hill **b :** a horny protuberance on a horse's hoof at the heel — see HOOF illustration **c :** the broadened base of a tree trunk or a thickened vertical part of it
3 : something that supports or strengthens ⟨a *buttress* of the cause of peace⟩
— but·tressed \-trəst\ *adjective*

²buttress *transitive verb* (14th century)
: to furnish or shore up with a buttress; *also* **:** SUPPORT, STRENGTHEN ⟨arguments *buttressed* by solid facts⟩

butt shaft *noun* (1588)
: a target arrow without a barb

butt·stock \'bət-ˌstäk\ *noun* (circa 1909)
: the stock of a firearm in the rear of the breech mechanism

butt weld *noun* (circa 1864)
: a butt joint made by welding
— butt–weld *transitive verb*

but·ty \'bə-tē\ *noun, plural* **butties** [origin unknown] (circa 1790)
chiefly British **:** a fellow worker **:** CHUM, PARTNER

bu·tut \bu̇-'tüt\ *noun, plural* **bututs** *or* **butut** [Wolof] (1972)
— see *dalasi* at MONEY table

bu·tyl \'byü-tᵊl\ *noun* [International Scientific Vocabulary *butyric* + *-yl*] (1869)
: any of four isomeric univalent radicals C_4H_9 derived from butanes

butyl alcohol *noun* (circa 1871)
: any of four flammable alcohols C_4H_9OH (as butanol) used in organic synthesis and as solvents

bu·tyl·at·ed \'byü-tᵊl-ˌā-təd\ *adjective* (1942)
: combined with the butyl group
— bu·tyl·a·tion \ˌbyü-tᵊl-'ā-shən\ *noun*

butylated hy·droxy·an·i·sole \-(ˌ)hī-ˌdräk-sē-'a-nə-ˌsōl\ *noun* [*hydroxy* + *anise* + *-ole*] (1950)
: BHA

butylated hy·droxy·tol·u·ene \-'täl-yə-ˌwēn\ *noun* (1961)
: BHT

bu·tyl·ene \'byü-tᵊl-ˌēn\ *noun* (1877)
: any of three isomeric hydrocarbons C_4H_8 of the ethylene series obtained usually by cracking petroleum

butyl rubber *noun* (1940)
: any of a class of synthetic rubbers that are made by copolymerizing isobutylene with a small amount usually of isoprene at low temperature

butyr- *or* **butyro-** *combining form* [International Scientific Vocabulary, from *butyric*]
: butyric ⟨*butyr*aldehyde⟩

bu·tyr·al·de·hyde \ˌbyü-tə-'ral-də-ˌhīd\ *noun* [International Scientific Vocabulary] (circa 1885)
: either of two aldehydes C_4H_8O used especially in making synthetic resins

bu·ty·rate \'byü-tə-ˌrāt\ *noun* (1873)
: a salt or ester of butyric acid

bu·tyr·ic \byü-'tir-ik\ *adjective* [French *butyrique,* from Latin *butyrum* butter — more at BUTTER] (1854)
: relating to or producing butyric acid ⟨*butyric* fermentation⟩

butyric acid *noun* (1826)
: either of two isomeric fatty acids $C_4H_8O_2$; *especially* **:** an acid of unpleasant odor normally found in perspiration and rancid butter

bu·ty·ro·phe·none \ˌbyü-tə-(ˌ)rō-fə-'nōn\ *noun* [*butyr-* + *phen-* + *-one*] (1970)
: any of a class of neuroleptic drugs (as haloperidol) used especially in the treatment of schizophrenia

bux·om \'bək-səm\ *adjective* [Middle English *buxsum,* from (assumed) Old English *būhsum;*

akin to Old English *būgan* to bend — more at BOW] (12th century)
1 *obsolete* **a :** OBEDIENT, TRACTABLE **b :** offering little resistance **:** FLEXIBLE, PLIANT ⟨wing silently the *buxom* air —John Milton⟩
2 *archaic* **:** full of gaiety **:** BLITHE
3 : vigorously or healthily plump; *specifically* **:** full-bosomed
— **bux·om·ly** *adverb*
— **bux·om·ness** *noun*

[1]**buy** \'bī\ *verb* **bought** \'bȯt\; **buy·ing** [Middle English *byen,* from Old English *bycgan;* akin to Gothic *bugjan* to buy] (before 12th century)
transitive verb
1 : to acquire possession, ownership, or rights to the use or services of by payment especially of money **:** PURCHASE
2 a : to obtain in exchange for something often at a sacrifice ⟨they *bought* peace with their freedom⟩ **b :** REDEEM 6
3 : BRIBE, HIRE
4 : to be the purchasing equivalent of ⟨the dollar *buys* less today than it used to⟩
5 : ACCEPT, BELIEVE ⟨I don't *buy* that hooey⟩
intransitive verb
: to make a purchase
— **buy·er** \'bī(-ə)r\ *noun*
— **buy it** *or* **buy the farm :** to get killed **:** DIE
— **buy time :** to delay an imminent action or decision **:** STALL

[2]**buy** *noun* (1879)
1 : something of value at a favorable price; *especially* **:** BARGAIN ⟨it's a real *buy* at that price⟩
2 : an act of buying **:** PURCHASE

buy·back \'bī-ˌbak\ *noun* (1974)
: the act or an instance of buying something back; *especially* **:** the repurchase by a corporation of shares of its own common stock on the open market

buyer's market *noun* (1926)
: a market in which goods are plentiful, buyers have a wide range of choice, and prices tend to be low — compare SELLER'S MARKET

buy off *transitive verb* (1629)
1 : to induce to refrain (as from prosecution) by a payment or other consideration
2 : to free (as from military service) by payment

buy·out \'bī-ˌau̇t\ *noun* (1971)
: an act or instance of buying out

buy out *transitive verb* (1644)
1 : to purchase the share or interest of
2 : to purchase the entire stock-in-trade and the goodwill of (a business)

buy up *transitive verb* (circa 1534)
1 : to buy freely or extensively
2 : to buy the entire available supply of

[1]**buzz** \'bəz\ *verb* [Middle English *bussen,* of imitative origin] (14th century)
intransitive verb
1 : to make a low continuous humming sound like that of a bee
2 a : MURMUR, WHISPER **b :** to be filled with a confused murmur ⟨the room *buzzed* with excitement⟩
3 : to make a signal with a buzzer
4 : to go quickly **:** HURRY ⟨*buzzed* around town in a sports car⟩; *also* **:** SCRAM — usually used with *off*
transitive verb
1 : to utter covertly by or as if by whispering
2 : to cause to buzz
3 : to fly fast and close to ⟨planes *buzz* the crowd⟩
4 : to summon or signal with a buzzer
5 *dialect English* **:** to drink to the last drop ⟨get some more port whilst I *buzz* this bottle —W. M. Thackeray⟩

[2]**buzz** *noun* (1605)
1 : a persistent vibratory sound
2 a : RUMOR, GOSSIP **b :** a confused murmur **c :** a flurry of activity **d :** FAD, CRAZE
3 : a signal conveyed by buzzer; *specifically* **:** a telephone call
4 : HIGH 4

buz·zard \'bə-zərd\ *noun* [Middle English *busard,* from Old French, alteration of *buison,* from Latin *buteon, buteo* hawk] (14th century)
1 *chiefly British* **:** BUTEO
2 : any of various usually large birds of prey (as the turkey vulture)
3 : a contemptible or rapacious person

buzz bomb *noun* (1944)
: an unguided jet-propelled missile used by the Germans against England in World War II

buzz cut *noun* (1983)
: CREW CUT

buzz·er \'bə-zər\ *noun* (1606)
1 : one that buzzes; *specifically* **:** an electric signaling device that makes a buzzing sound
2 : the sound of a buzzer

buzz saw *noun* (1858)
: CIRCULAR SAW

buzz·word \'bəz-ˌwərd\ *noun* (1946)
1 : an important-sounding usually technical word or phrase often of little meaning used chiefly to impress laymen
2 : a voguish word or phrase — called also *buzz phrase*

BVD \ˌbē-(ˌ)vē-'dē\ *trademark*
— used for underwear

B vitamin *noun* (1920)
: any vitamin of the vitamin B complex

bwa·na \'bwä-nə\ *noun* [Swahili, from Arabic *abūna* our father] (1878)
: MASTER, BOSS

[1]**by** \'bī, *before consonants also* bə\ *preposition* [Middle English, preposition & adverb, from Old English, preposition, *be, bī;* akin to Old High German *bī* by, near, Latin *ambi-* on both sides, around, Greek *amphi*] (before 12th century)
1 : in proximity to **:** NEAR ⟨standing *by* the window⟩
2 a : through or through the medium of **:** VIA ⟨enter *by* the door⟩ **b :** in the direction of **:** TOWARD ⟨north *by* east⟩ **c :** into the vicinity of and beyond **:** PAST ⟨went right *by* him⟩
3 a : during the course of ⟨studied *by* night⟩ **b :** not later than ⟨*by* 2 p.m.⟩
4 a : through the agency or instrumentality of ⟨*by* force⟩ **b :** born or begot of **c :** sired or borne by
5 : with the witness or sanction of ⟨swear *by* all that is holy⟩
6 a : in conformity with ⟨acted *by* the rules⟩ **b :** ACCORDING TO ⟨called her *by* name⟩
7 a : on behalf of ⟨did right *by* his children⟩ **b :** with respect to ⟨a lawyer *by* profession⟩
8 a : in or to the amount or extent of ⟨win *by* a nose⟩ **b** *chiefly Scottish* **:** in comparison with **:** BESIDE
9 — used as a function word to indicate successive units or increments ⟨little *by* little⟩ ⟨walk two *by* two⟩
10 — used as a function word in multiplication, in division, and in measurements ⟨divide *a by b*⟩ ⟨multiply 10 *by* 4⟩ ⟨a room 15 feet *by* 20 feet⟩
11 : in the opinion of ⟨okay *by* me⟩
— **by the by** *or* **by the bye :** INCIDENTALLY 2

[2]**by** \'bī\ *adverb* (before 12th century)
1 a : close at hand **:** NEAR **b :** at or to another's home ⟨stop *by*⟩
2 : PAST ⟨saw him go *by*⟩
3 : ASIDE, AWAY

[3]**by** *or* **bye** \'bī\ *adjective* (14th century)
1 : being off the main route **:** SIDE
2 : INCIDENTAL

[4]**by** *or* **bye** \'bī\ *noun, plural* **byes** \'bīz\ (1567)
: something of secondary importance **:** a side issue

[5]**by** *or* **bye** \'bī\ *interjection* [short for *goodbye*] (1709)
— used to express farewell; often used with following *now*

by–and–by \ˌbī-ən-'bī\ *noun* (1591)
: a future time or occasion

by and by *adverb* (1526)
: before long **:** SOON

by and large *adverb* (1706)
: on the whole **:** in general

by–blow \'bī-ˌblō\ *noun* (1594)
1 : an indirect blow
2 : an illegitimate child

bye \'bī\ *noun* [alteration of [2]*by*] (1883)
: the position of a participant in a tournament who advances to the next round without playing

[1]**bye–bye** *or* **by–by** \'bī-ˌbī, bī-'bī\ *interjection* [baby-talk reduplication of *goodbye*] (circa 1736)
— used to express farewell

[2]**bye–bye** *or* **by–by** \'bī-ˌbī\ *adverb* (1917)
: out especially for a walk or ride — used with the verb *go*

[3]**bye–bye** *or* **by–by** \'bī-ˌbī\ *noun* (1867)
: BED, SLEEP ⟨lie down . . . and go to *bye-bye* —Rudyard Kipling⟩

[4]**bye–bye** *or* **by–by** \'bī-ˌbī\ *adverb* (1920)
: to bed or sleep — used with the verb *go*

by–elec·tion *also* **bye–election** \'bī-ə-ˌlek-shən\ *noun* (1880)
: a special election held between regular elections in order to fill a vacancy

by·gone \'bī-ˌgȯn *also* -ˌgän\ *adjective* (15th century)
: gone by **:** PAST; *especially* **:** OUTMODED
— **bygone** *noun*

by·law *or* **bye·law** \'bī-ˌlȯ\ *noun* [Middle English *bilawe,* probably from (assumed) Old Norse *bȳlǫg,* from Old Norse *bȳr* town + *lǫg* law] (14th century)
1 : a rule adopted by an organization chiefly for the government of its members and the regulation of its affairs
2 : a local ordinance

[1]**by·line** \'bī-ˌlīn\ *noun* (1916)
1 : a secondary line **:** SIDELINE
2 : a line at the beginning of a news story, magazine article, or book giving the writer's name

[2]**byline** *transitive verb* (1938)
: to write (an article) under a byline
— **by·lin·er** \-ˌlī-nər\ *noun*

by·name \'bī ˌnām\ *noun* (14th century)
1 : a secondary name
2 : NICKNAME

[1]**by·pass** \'bī-ˌpas\ *noun* (1848)
1 : a passage to one side; *especially* **:** a deflected route usually around a town
2 a : a channel carrying a fluid around a part and back to the main stream **b** (1) **:** SHUNT 1b (2) **:** SHUNT 1c; *also* **:** a surgical procedure for the establishment of a shunt ⟨have a coronary *bypass*⟩

[2]**bypass** *transitive verb* (1886)
1 a : to avoid by means of a bypass **b :** to cause to follow a bypass
2 a : to neglect or ignore usually intentionally **b :** CIRCUMVENT

by·past \'bī-ˌpast\ *adjective* (15th century)
: BYGONE

by·path \-ˌpath, -ˌpäth\ *noun* (14th century)
: BYWAY

by·play \-ˌplā\ *noun* (1812)
: action engaged in on the side while the main action proceeds (as during a dramatic production)

by–prod·uct \-ˌprä-(ˌ)dəkt\ *noun* (1857)
1 : something produced in a usually industrial or biological process in addition to the principal product
2 : a secondary and sometimes unexpected or unintended result ⟨unpleasant *by-products* of civilization⟩

byre \'bīr\ *noun* [Middle English, from Old English *bȳre;* akin to Old English *būr* dwelling

— more at BOWER] (before 12th century) *chiefly British* **:** a cow barn

by·road \'bī-ˌrōd\ *noun* (1673)
: BYWAY

bys·si·no·sis \ˌbi-sə-'nō-səs\ *noun, plural* **-no·ses** \-ˌsēz\ [New Latin, from Latin *byssinus* of fine linen, from Greek *byssinos,* from *byssos*] (1881)
: an occupational respiratory disease associated with inhalation of cotton, flax, or hemp dust and characterized initially by chest tightness, shortness of breath, and cough and eventually by irreversible lung disease

bys·sus \'bi-səs\ *noun, plural* **bys·sus·es** *or* **bys·si** \-ˌsī, -(ˌ)sē\ [Middle English *bissus,* from Latin *byssus,* from Greek *byssos* flax, of Semitic origin; akin to Hebrew *būṣ* linen cloth] (14th century)
1 : a fine probably linen cloth of ancient times
2 [New Latin, from Latin] **:** a tuft of long tough filaments by which some bivalve mollusks (as mussels) adhere to a surface

by·stand·er \'bī-ˌstan-dər\ *noun* (circa 1619)
: one present but not taking part in a situation or event **:** a chance spectator

by·street \-ˌstrēt\ *noun* (1672)
: a street off a main thoroughfare

byte \'bīt\ *noun* [perhaps alteration of ²*bite*] (1959)
: a group of eight binary digits processed as a unit by a computer and used especially to represent an alphanumeric character — compare WORD 2c

by·way \'bī-ˌwā\ *noun* (14th century)
1 : a little traveled side road
2 : a secondary or little known aspect or field ⟨meandering more and more in the fascinating *byways* of learning —*Times Literary Supplement*⟩

by·word \-ˌwərd\ *noun* (before 12th century)
1 : a proverbial saying **:** PROVERB
2 a : one that personifies a type **b :** one that is noteworthy or notorious
3 : EPITHET
4 : a frequently used word or phrase

by–your–leave \ˌbī-yər-'lēv\ *noun* (1914)
: a request for permission ⟨imposed . . . without so much as a *by-your-leave* —J. L. Granatstein⟩

¹Byz·an·tine \'bi-zᵊn-ˌtēn, 'bī-, -ˌtīn; bə-'zan-ˌ, bī-'\ *adjective* (1794)
1 : of, relating to, or characteristic of the ancient city of Byzantium
2 : of, relating to, or having the characteristics of a style of architecture developed in the Byzantine Empire especially in the 5th and 6th centuries featuring the dome carried on pendentives over a square and incrustation with marble veneering and with colored mosaics on grounds of gold
3 : of or relating to the churches using a traditional Greek rite and subject to Eastern canon law
4 *often not capitalized* **a :** of, relating to, or characterized by a devious and usually surreptitious manner of operation ⟨a *Byzantine* power struggle⟩ **b :** intricately involved **:** LABYRINTHINE ⟨rules of *Byzantine* complexity⟩

²Byzantine *noun* (1836)
: a native or inhabitant of Byzantium

By·zan·tin·ist \bi-'zan-ti-nist, bī-\ *noun* (1892)
: a student of Byzantine culture

C is the third letter of the English alphabet and of most alphabets closely related to that of English. It came through the Latin from the Etruscan and there from the Greek, which took it from the Phoenician; it has the alphabetic position, but not the pronunciation, of Greek gamma *and Phoenician* gimel. *In Phoenician and Greek,* C *represented the sound of* g *in* go. *Etruscan assigned* C *the value of* k, *since its speakers did not use the* g *sound. Latin, on the other hand, had both sound values, and so the Romans invented a new letter* G *for the value* g, *while keeping* C *for the value* k. C *retains the* k *pronunciation in modern English before* a, o, u, *or a consonant other than* h, *and has taken on the pronunciation of* s *before* e, i, *or* y. *The small form of* c *emerged through a simple reduction in the size of the capital form.*

c \'sē\ *noun, plural* **c's** *or* **cs** \'sēz\ *often capitalized, often attributive*
1 a : the 3rd letter of the English alphabet **b :** a graphic representation of this letter **c :** a speech counterpart of orthographic *c*
2 a : one hundred — see NUMBER table **b** *slang* **:** a sum of $100
3 : the keynote of a C-major scale
4 : a graphic device for reproducing the letter *c*
5 : one designated *c* especially as the 3rd in order or class
6 a : a grade rating a student's work as fair or mediocre in quality **b :** one graded or rated with a C
7 : something shaped like the letter C
8 : a structured programming language designed to produce a compact and efficient translation of a program into machine language

ca' \'kȯ, 'kä\ *Scottish variant of* CALL

[1]cab \'kab, 'käb\ *noun* [Hebrew *qabh*] (1535)
: an ancient Hebrew unit of capacity equal to about two quarts (2.2 liters)

[2]cab \'kab\ *noun* [short for *cabriolet*] (1827)
1 a (1) **:** CABRIOLET (2) **:** a similar light closed carriage (as a hansom) **b :** a carriage for hire
2 : TAXICAB
3 [short for *cabin*] **a :** the part of a locomotive that houses the engineer and operating controls **b :** a comparable shelter (as on a truck) housing operating controls

[1]ca·bal \kə-'bäl, -'bal\ *noun* [French *cabale* cabala, intrigue, cabal, from Medieval Latin *cabbala* cabala, from Late Hebrew *qabbālāh,* literally, received (lore)] (1614)
: the artifices and intrigues of a group of persons secretly united to bring about an overturn or usurpation especially in public affairs; *also* **:** a group engaged in such artifices and intrigues
synonym see PLOT

[2]cabal *intransitive verb* **ca·balled; ca·bal·ling** (1680)
: to unite in or form a cabal

ca·ba·la *or* **cab·ba·la** *or* **cab·ba·lah** \kə-'bä-lə, 'ka-bə-lə\ *noun, often capitalized* [Medieval Latin *cabbala*] (1521)
1 : a medieval and modern system of Jewish theosophy, mysticism, and thaumaturgy marked by belief in creation through emanation and a cipher method of interpreting Scripture
2 a : a traditional, esoteric, occult, or secret matter **b :** esoteric doctrine or mysterious art
— cab·a·lism \'kä-bə-,li-zəm, 'ka-\ *noun*
— cab·a·lis·tic \,kä-bə-'lis-tik, ,ka-\ *adjective*

ca·ba·let·ta \,ka-bə-'le-tə, ,kä-\ *noun* [Italian] (1842)
1 : an operatic song in simple popular style characterized by a uniform rhythm
2 : the lively bravura concluding section of an extended aria or duet

[1]ca·ba·list \kə-'bä-list, 'ka-bə-\ *noun* (circa 1533)
1 *often capitalized* **:** a student, interpreter, or devotee of the Jewish cabala
2 : one skilled in esoteric doctrine or mysterious art

[2]ca·bal·ist \kə-'ba-list, -'bä-\ *noun* (1569)
: a member of a cabal

ca·bal·le·ro \,ka-bə-'ler-(,)ō, -bə(l)-'yer-\ *noun, plural* **-ros** [Spanish, from Late Latin *caballarius* hostler — more at CAVALIER] (1749)
1 : KNIGHT, CAVALIER
2 *chiefly Southwest* **:** HORSEMAN

ca·bana \kə-'ban-yə, -'ba-nə\ *noun* [Spanish *cabaña,* literally, hut, from Medieval Latin *capanna*] (1890)
1 : a shelter resembling a cabin usually with an open side facing a beach or swimming pool
2 : a lightweight structure with living facilities

cab·a·ret \,ka-bə-'rā, 'ka-bə-,\ *noun* [French, from Old North French] (1655)
1 *archaic* **:** a shop selling wines and liquors
2 a : a restaurant serving liquor and providing entertainment (as by singers or dancers) **:** NIGHTCLUB **b :** the show provided at a cabaret

[1]cab·bage \'ka-bij\ *noun, often attributive* [Middle English *caboche,* from Old North French, head] (15th century)
1 : a leafy garden plant (*Brassica oleracea capitata*) of European origin with a short stem and a dense globular head of usually green leaves that is used as a vegetable
2 *slang* **:** PAPER MONEY, BANKNOTES ◇

[2]cabbage *noun* [perhaps by folk etymology from Middle French *cabas* cheating, theft] (1663)
British **:** pieces of cloth left in cutting out garments and traditionally kept by tailors as perquisites

[3]cabbage *transitive verb* **cab·baged; cab·bag·ing** (1712)
: STEAL, FILCH

cabbage butterfly *noun* (1816)
: any of several largely white butterflies (family Pieridae) whose green larvae are cabbageworms; *especially* **:** a small cosmopolitan butterfly (*Pieris rapae* synonym *Artogeia rapae*) that is a pest on cabbage — called also *cabbage white*

cabbage butterfly: *1* **larva,** *2* **adult**

cabbage looper *noun* (circa 1902)
: a noctuid moth (*Trichoplusia ni*) having pale green white-striped larvae that feed on cruciferous plants (as the cabbage)

cabbage palm *noun* (circa 1784)
: a palm with terminal buds eaten as a vegetable

cabbage palmetto *noun* (1802)
: a cabbage palm (*Sabal palmetto*) with fan-shaped leaves that is native to coastal southeastern U.S. and the Bahamas

cabbage rose *noun* (1795)
: a fragrant garden rose (*Rosa centifolia*) with upright branches and large pink flowers

cab·bage·worm \'ka-bij-,wərm\ *noun* (1688)
: an insect larva (as of a cabbage butterfly) that feeds on cabbages

cab·bie *or* **cab·by** \'ka-bē\ *noun, plural* **cabbies** (1859)
: CABDRIVER

cab·driv·er \'kab-,drī-vər\ *noun* (1830)
: a driver of a cab

ca·ber \'kā-bər, 'kä-\ *noun* [Scottish Gaelic *cabar*] (1505)
: POLE; *especially* **:** a young tree trunk used for tossing as a trial of strength in a Scottish sport

cab·er·net sau·vi·gnon \,ka-bər-'nā-sō-vē-'nyō^{n}\ *noun, often C&S capitalized* [French] (1941)
: a dry red wine made from a single widely cultivated variety of black grape — called also *cabernet*

[1]cab·in \'ka-bən\ *noun* [Middle English *cabane,* from Middle French, from Old Provençal *cabana* hut, from Medieval Latin *capanna*] (14th century)
1 a (1) **:** a private room on a ship or boat (2) **:** a compartment below deck on a boat used for living accommodations **b :** the passenger or cargo compartment of a vehicle (as an airplane or automobile) **c :** the crew compartment of an exploratory vehicle (as a spacecraft)
2 : a small one-story dwelling usually of simple construction
3 *chiefly British* **:** CAB 3

[2]cabin (1586)
intransitive verb
: to live in or as if in a cabin
transitive verb
: CONFINE

cabin boy *noun* (1726)
: a boy working as servant on a ship

cabin car *noun* (1879)
: CABOOSE

cabin class *noun* (1929)
: a class of accommodations on a passenger ship superior to tourist class and inferior to first class

◇ WORD HISTORY
cabbage Old French *caboce* was a jocular word for "head," probably similar in this respect to the English word *noggin.* The form of the word in the dialects of Normandy and Picardy was *caboche,* which spread to the French of the Paris region and retains some currency in Modern French. The origin of *caboce* and *caboche* is not entirely clear, though the second syllable appears to be *boce* "boss, protuberance." While *caboche* has remained limited to very colloquial uses in French, the medieval French word found its way into later Middle English in the sense "head of cabbage." In English it has become the everyday word for the vegetable that the French call *chou.* Middle English *caboche* emerged as *cabage* in Early Modern English, through the same change of a final \ch\ sound to \j\ in an unstressed syllable that is evident in *knowledge* (Middle English *knowleche*) and *sausage* (Middle English *sawsiche*).

cabin cruiser *noun* (1921)
: CRUISER 1b

¹cab·i·net \'kab-nit, 'ka-bə-\ *noun* [Middle French, small room, diminutive of Old North French *cabine* gambling house] (circa 1550)
1 a : a case or cupboard usually having doors and shelves **b :** a collection of specimens especially of biological or numismatic interest **c :** CONSOLE 4a **d :** a chamber having temperature and humidity controls and used especially for incubating biological samples
2 a *archaic* **:** a small room providing seclusion **b :** a small exhibition room in a museum
3 a *archaic* (1) **:** the private room serving as council chamber of the chief councillors or ministers of a sovereign (2) **:** the consultations and actions of these councillors **b** (1) *often capitalized* **:** a body of advisers of a head of state (2) **:** a similar advisory council of a governor of a state or a mayor **c** *British* **:** a meeting of a cabinet

²cabinet *adjective* (1631)
1 : of or relating to a governmental cabinet
2 : suitable by reason of size for a small room or by reason of attractiveness or perfection for preservation and display in a cabinet
3 a : used or adapted for cabinetmaking **b :** done or used by a cabinetmaker

cab·i·net·mak·er \-ˌmā-kər\ *noun* (1681)
: a skilled woodworker who makes fine furniture
— cab·i·net·mak·ing \-ˌmā-kiŋ\ *noun*

cab·i·net·ry \'kab-ni-trē, 'ka-bə-\ *noun* (1926)
: CABINETWORK; *also* **:** CABINETS ⟨kitchen *cabinetry*⟩

cab·i·net·work \-ˌwərk\ *noun* (1732)
: finished woodwork made by a cabinetmaker

cabin fever *noun* (1918)
: extreme irritability and restlessness from living in isolation or a confined indoor area for a prolonged time

¹ca·ble \'kā-bəl\ *noun, often attributive* [Middle English, from Old North French, from Medieval Latin *capulum* lasso, from Latin *capere* to take — more at HEAVE] (13th century)
1 a : a strong rope especially of 10 inches (25 centimeters) or more in circumference **b :** a cable-laid rope **c :** a wire rope or metal chain of great tensile strength **d :** a wire or wire rope by which force is exerted to control or operate a mechanism
2 : CABLE LENGTH
3 a : an assembly of electrical conductors insulated from each other but laid up together usually by being twisted around a central core **b :** CABLEGRAM; *also* **:** a radio message or telegram
4 : something resembling or fashioned like a cable ⟨a fiber-optic *cable*⟩
5 : CABLE TELEVISION

²cable *verb* **ca·bled; ca·bling** \'kā-b(ə-)liŋ\ (circa 1500)
transitive verb
1 : to fasten with or as if with a cable
2 : to provide with cables
3 : to telegraph by submarine cable
4 : to make into a cable or into a form resembling a cable
intransitive verb
: to communicate by a submarine cable

cable car *noun* (1887)
: a vehicle moved by an endless cable: **a :** one suspended from an overhead cable **b :** one that moves along tracks

ca·ble·gram \'kā-bəl-ˌgram\ *noun* (1868)
: a message sent by a submarine telegraph cable

ca·ble–laid \'kā-bəl-'lād\ *adjective* (1723)
: composed of three ropes laid together left-handed with each containing three strands twisted together

cable length *noun* (1555)
: a maritime unit of length variously reckoned as 100 fathoms, 120 fathoms, or 608 feet

cable television *noun* (1965)
: a system of television reception in which signals from distant stations are picked up by a master antenna and sent by cable to the individual receivers of paying subscribers — called also *cable TV*

ca·ble·way \'kā-bəl-ˌwā\ *noun* (1899)
: a suspended cable used as a track along which carriers can be pulled

cab·man \'kab-mən\ *noun* (1834)
: CABDRIVER

cab·o·chon \'ka-bə-ˌshän\ *noun* [Middle French, diminutive of Old North French *caboche* head] (1578)
: a gem or bead cut in convex form and highly polished but not faceted; *also* **:** this style of cutting
— cabochon *adverb*

ca·boo·dle \kə-'bü-dᵊl\ *noun* [probably from *ca-* (intensive prefix) + *boodle*] (circa 1848)
: COLLECTION, LOT ⟨sell the whole *caboodle*⟩

ca·boose \kə-'büs\ *noun* [probably from Dutch *kabuis,* from Middle Low German *kabūse*] (1769)
1 : a ship's galley
2 : a freight-train car attached usually to the rear mainly for the use of the train crew
3 : one that follows or brings up the rear

cab·o·tage \'ka-bə-ˌtäzh\ *noun* [French, from *caboter* to sail along the coast] (1831)
1 : trade or transport in coastal waters or airspace or between two points within a country
2 : the right to engage in cabotage

ca·bret·ta \kə-'bre-tə\ *noun* [modification of Portuguese and Spanish *cabra* goat] (1926)
: a light soft leather from skins of hairy sheep

ca·bril·la \kə-'brē-ə, -'bri-lə\ *noun* [Spanish, diminutive of *cabra* goat, from Latin *capra* she-goat, feminine of *caper* he-goat — more at CAPRIOLE] (1859)
: any of various sea basses (especially of the genera *Epinephelus* and *Paralabrax*) of the Mediterranean, the California coast, and the warmer parts of the western Atlantic

cab·ri·ole \'ka-brē-ˌōl\ *noun* [French, caper] (circa 1797)
1 : a ballet leap in which one leg is extended in midair and the other struck against it
2 : a curved furniture leg ending in an ornamental foot

cab·ri·o·let \ˌka-brē-ə-'lā\ *noun* [French, from diminutive of *cabriole* caper, alteration of Middle French *capriole*] (1763)
1 : a light 2-wheeled one-horse carriage with a folding leather hood, a large apron, and upward-curving shafts
2 : a convertible coupe

cab·stand \'kab-ˌstand\ *noun* (1848)
: a place where cabs await hire

cac- *or* **caco-** *combining form* [New Latin, from Greek *kak-, kako-,* from *kakos* bad]
: bad ⟨*caco*graphy⟩

ca' can·ny \kò-'ka-nē\ *noun* [Scots, verb, to proceed cautiously, from *ca'* (call) + *canny* careful] (1886)
British **:** SLOWDOWN
— ca' canny *intransitive verb, British*

ca·cao \kə-'kaů, kə-'kā-(ˌ)ō\ *noun, plural* **cacaos** [Spanish, from Nahuatl *cacahuatl* cacao beans] (1555)
1 : the dried partly fermented fatty seeds of a South American evergreen tree (*Theobroma cacao* of the family Sterculiaceae) that are used in making cocoa, chocolate, and cocoa butter — called also *cacao bean, cocoa bean*
2 : a tree having small yellowish flowers followed by fleshy pods with many seeds that bears cacao

cacao butter *variant of* COCOA BUTTER

cac·cia·to·re \ˌkä-chə-'tōr-ē, -'tòr-\ *adjective* [Italian, from *cacciatore* hunter] (1942)
: cooked with tomatoes and herbs and sometimes wine ⟨chicken *cacciatore*⟩

cach·a·lot \'ka-shə-ˌlät, -ˌlō\ *noun* [French] (1747)
: SPERM WHALE

¹cache \'kash\ *noun* [French, from *cacher* to press, hide, from (assumed) Vulgar Latin *coacticare* to press together, from Latin *coactare* to compel, frequentative of *cogere* to compel — more at COGENT] (1797)
1 a : a hiding place especially for concealing and preserving provisions or implements **b :** a secure place of storage
2 : something hidden or stored in a cache
3 : a computer memory with very short access time used for storage of frequently used instructions or data — called also *cache memory*

²cache *transitive verb* **cached; cach·ing** (1805)
: to place, hide, or store in a cache

ca·chec·tic \kə-'kek-tik, ka-\ *adjective* [French *cachectique,* from Latin *cachecticus,* from Greek *kachektikos,* from *kak-* + *echein*] (1634)
: affected by cachexia

cache·pot \'kash-ˌpät, 'kash-ˌpō, 'ka-shə-\ *noun* [French, from *cacher* to hide + *pot* pot] (1872)
: an ornamental receptacle to hold and usually to conceal a flowerpot

ca·chet \ka-'shā\ *noun* [French, from *cacher*] (circa 1639)
1 a : a seal used especially as a mark of official approval **b :** an indication of approval carrying great prestige
2 a : a characteristic feature or quality conferring prestige **b :** PRESTIGE ⟨being rich . . . doesn't have the *cachet* it used to —Truman Capote⟩
3 : a medicinal preparation for swallowing consisting of a case usually of rice-flour paste enclosing a medicine
4 a : a design or inscription on an envelope to commemorate a postal or philatelic event **b :** an advertisement forming part of a postage meter impression **c :** a motto or slogan included in a postal cancellation

ca·chex·ia \kə-'kek-sē-ə, ka-\ *noun* [Late Latin *cachexia,* from Greek *kachexia* bad condition, from *kak-* cac- + *hexis* condition, from *echein* to have, be disposed — more at SCHEME] (1541)
: general physical wasting and malnutrition usually associated with chronic disease

cach·in·nate \'ka-kə-ˌnāt\ *intransitive verb* **-nat·ed; -nat·ing** [Latin *cachinnatus,* past participle of *cachinnare,* of imitative origin] (1824)
: to laugh loudly or immoderately
— cach·in·na·tion \ˌka-kə-'nā-shən\ *noun*

ca·chou \ka-'shü, 'ka-ˌ\ *noun* [French, from Portuguese *cachu,* from Malayalam *kāccu*] (circa 1879)
: a pill or pastille used to sweeten the breath

ca·cique \kə-'sēk\ *noun* [Spanish, from Taino, chief] (1555)
1 : a native Indian chief in areas dominated primarily by a Spanish culture
2 : a local political boss in Spain and Latin America
— ca·ciqu·ism \-'sē-ˌki-zəm\ *noun*

cack·le \'ka-kəl\ *intransitive verb* **cack·led; cack·ling** \-k(ə-)liŋ\ [Middle English *cakelen,* of imitative origin] (13th century)
1 : to make the sharp broken noise or cry characteristic of a hen especially after laying
2 : to laugh especially in a harsh or sharp manner
3 : CHATTER
— cackle *noun*
— cack·ler \-k(ə-)lər\ *noun*

caco·de·mon \ˌka-kə-'dē-mən\ *noun* [Greek *kakodaimōn,* from *kak-* cac- + *daimōn* spirit] (1594)
: DEMON
— caco·de·mon·ic \-di-'mä-nik\ *adjective*

cacodylic acid \ˌka-kə-'di-lik-\ *noun* [German *Kakodyl* the radical $As(CH_3)_2$, from Greek *kakōdēs* foul-smelling, from *kak-* +

-ōdēs (akin to Greek *ozein* to smell) — more at ODOR] (1850)
: a toxic crystalline compound of arsenic $C_2H_7AsO_2$ used especially as an herbicide

caco·ë·thes \ˌka-kə-ˈwē-(ˌ)thēz\ *noun* [Latin, from Greek *kakoēthes* wickedness, from neuter of *kakoēthēs* malignant, from *kak-* cac- + *ēthos* character — more at SIB] (circa 1587)
: an insatiable desire **:** MANIA

ca·cog·ra·phy \ka-ˈkä-grə-fē\ *noun* (1580)
1 : bad spelling — compare ORTHOGRAPHY
2 : bad handwriting — compare CALLIGRAPHY
— caco·graph·i·cal \ˌka-kə-ˈgra-fi-kəl\ *adjective*

cac·o·mis·tle \ˈka-kə-ˌmi-səl, ˌka-kə-ˈmis(t)-lē\ *noun* [Mexican Spanish, from Nahuatl *tlahcomiztli,* from *tlahco* half + *miztli* mountain lion] (1869)
: RINGTAIL 2

ca·coph·o·nous \ka-ˈkä-fə-nəs, -ˈkō- *also* -ˈka-\ *adjective* [Greek *kakophōnos,* from *kak-* + *phōnē* voice, sound — more at BAN] (1797)
: marked by cacophony **:** harsh-sounding
— ca·coph·o·nous·ly *adverb*

ca·coph·o·ny \-nē\ *noun, plural* **-nies** (circa 1656)
: harsh or discordant sound **:** DISSONANCE 2; *specifically* **:** harshness in the sound of words or phrases

cac·tus \ˈkak-təs\ *noun, plural* **cac·ti** \-ˌtī, -(ˌ)tē\ *or* **cac·tus·es** *also* **cactus** [New Latin, genus name, from Latin, cardoon, from Greek *kaktos*] (1767)
: any of a family (Cactaceae, the cactus family) of plants that have succulent stems and branches with scales or spines instead of leaves and are found especially in dry areas (as deserts)

ca·cu·mi·nal \ka-ˈkyü-mə-nᵊl, kə-\ *adjective* [International Scientific Vocabulary, from Latin *cacumin-, cacumen* top, point] (1862)
: RETROFLEX 2

cad \ˈkad\ *noun* [English dialect, unskilled assistant, short for Scots *caddie*] (1833)
1 : an omnibus conductor
2 : a man who acts with deliberate disregard for another's feelings or rights

ca·das·tral \kə-ˈdas-trəl\ *adjective* (1858)
1 : of or relating to a cadastre
2 : showing or recording property boundaries, subdivision lines, buildings, and related details
— ca·das·tral·ly \-trə-lē\ *adverb*

ca·das·tre \kə-ˈdas-tər\ *noun* [French, from Italian *catastro,* from Old Italian *catastico,* from Late Greek *katastichon* notebook, from Greek *kata* by + *stichos* row, line — more at CATA-, DISTICH] (1804)
: an official register of the quantity, value, and ownership of real estate used in apportioning taxes

ca·dav·er \kə-ˈda-vər\ *noun* [Latin, from *cadere* to fall] (circa 1500)
: a dead body; *especially* **:** one intended for dissection
— ca·dav·er·ic \-ˈdav-rik, -ˈda-və-\ *adjective*

ca·dav·er·ine \kə-ˈda-və-ˌrēn\ *noun* (1887)
: a syrupy colorless poisonous ptomaine $C_5H_{14}N_2$ formed by decarboxylation of lysine especially in putrefaction of flesh

ca·dav·er·ous \kə-ˈdav-rəs, -ˈda-vər-əs\ *adjective* (1627)
1 a : of or relating to a corpse **b :** suggestive of corpses or tombs
2 a : PALLID, LIVID **b :** GAUNT, EMACIATED
— ca·dav·er·ous·ly *adverb*

cad·die *or* **cad·dy** \ˈka-dē\ *noun, plural* **caddies** [French *cadet* military cadet] (circa 1730)
1 *Scottish* **:** one that waits about for odd jobs
2 a : one who assists a golfer especially by carrying the clubs **b :** a wheeled device for conveying things not readily carried by hand
— caddie *or* **caddy** *intransitive verb*

¹cad·dis *also* **cad·dice** \ˈka-dəs\ *noun* [Middle English *cadas* cotton wool, probably from Middle French *cadaz,* from Old Provençal *cadarz*] (1530)
: worsted yarn; *specifically* **:** a worsted ribbon or binding formerly used for garters and girdles

²caddis *noun* (1651)
: CADDISWORM

caddis fly *noun* (1787)
: any of an order (Trichoptera) of insects with four membranous wings, vestigial mouthparts, slender many-jointed antennae, and aquatic larvae — compare CADDISWORM

cad·dish \ˈka-dish\ *adjective* (1868)
: of, relating to, or resembling a cad
— cad·dish·ly *adverb*
— cad·dish·ness *noun*

cad·dis·worm \ˈka-dəs-ˌwərm\ *noun* [probably alteration of obsolete *codworm;* from the case or tube in which it lives] (1622)
: the larva of a caddis fly that lives in and carries around a silken case covered with bits of debris

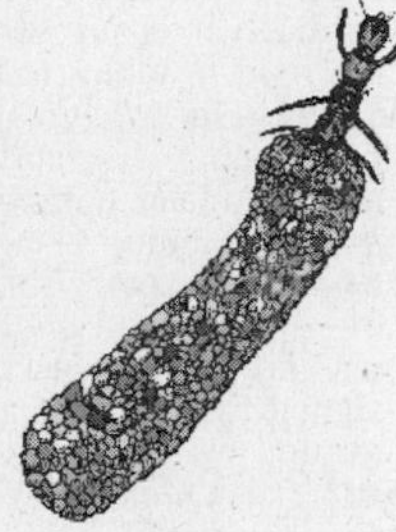
caddisworm

Cad·do \ˈka-(ˌ)dō\ *noun, plural* **Caddo** *or* **Caddos** (1805)
: a member of a group of American Indian peoples of Louisiana, Arkansas, and eastern Texas

cad·dy \ˈka-dē\ *noun, plural* **caddies** [Malay *kati* catty] (1792)
1 : a small box, can, or chest used especially to keep tea in
2 : a container or device for storing or holding objects when they are not in use

cade \ˈkād\ *adjective* [English dialect *cade* pet lamb, from Middle English *cad*] (1551)
: left by its mother and reared by hand **:** PET ⟨a *cade* lamb⟩

-cade *noun combining form* [*cavalcade*]
: procession ⟨motor*cade*⟩

ca·delle \kə-ˈdel\ *noun* [French, from Provençal *cadello,* from Latin *catella,* feminine of *catellus* little dog, diminutive of *catulus* young animal] (circa 1861)
: a small cosmopolitan black beetle (*Tenebroides mauritanicus*) destructive to stored grain

ca·dence \ˈkā-dᵊn(t)s\ *noun* [Middle English, from Old Italian *cadenza,* from *cadere* to fall, from Latin — more at CHANCE] (14th century)
1 a : a rhythmic sequence or flow of sounds in language **b :** the beat, time, or measure of rhythmical motion or activity
2 a : a falling inflection of the voice **b :** a concluding and usually falling strain; *specifically* **:** a musical chord sequence moving to a harmonic close or point of rest and giving the sense of harmonic completion
3 : the modulated and rhythmic recurrence of a sound especially in nature
— ca·denced \-dᵊn(t)st\ *adjective*
— ca·den·tial \kā-ˈden(t)-shəl\ *adjective*

ca·den·cy \ˈkā-dᵊn(t)-sē\ *noun, plural* **-cies** (1627)
: CADENCE

ca·dent \ˈkā-dᵊnt\ *adjective* [Latin *cadent-, cadens,* present participle of *cadere*] (1605)
1 *archaic* **:** being in the process of falling ⟨with *cadent* tears fret channels in her cheeks —Shakespeare⟩
2 : having rhythmic cadence

ca·den·za \kə-ˈden-zə\ *noun* [Italian, cadence, cadenza] (1836)
1 : a parenthetical flourish in an aria or other solo piece commonly just before a final or other important cadence
2 : a technically brilliant sometimes improvised solo passage toward the close of a concerto
3 : an exceptionally brilliant part of an artistic work

cade oil \ˈkād-\ *noun* [*cade* juniper, from Middle French, from Old Provençal, from Medieval Latin *catanus*] (1880)
: JUNIPER TAR

ca·det \kə-ˈdet\ *noun, often attributive* [French, from Gascon *capdet* chief, from Late Latin *capitellum,* diminutive of Latin *capit-, caput* head — more at HEAD] (1610)
1 a : a younger brother or son **b :** youngest son **c :** a younger branch of a family or a member of it
2 : one in training for a military or naval commission; *especially* **:** a student in a service academy
3 *slang* **:** PIMP ◆
— ca·det·ship \-ˌship\ *noun*

Ca·dette \kə-ˈdet\ *noun* [from *cadet*] (1963)
: a member of a program of the Girl Scouts for girls in the sixth through ninth grades in school

cadge \ˈkaj\ *verb* **cadged; cadg·ing** [back-formation from Scots *cadger* carrier, huckster, from Middle English *cadgear*] (circa 1812)
: BEG, SPONGE
— cadg·er *noun*

cad·mi·um \ˈkad-mē-əm\ *noun* [New Latin, from Latin *cadmia* zinc oxide, from Greek *kadmeia,* literally, Theban (earth), from feminine of *kadmeios* Theban, from *Kadmos;* from the occurrence of its ores together with zinc oxide] (1822)
: a bluish white malleable ductile toxic bivalent metallic element used especially in protective platings and in bearing metals — see ELEMENT table

cadmium sulfide *noun* (circa 1893)
: a yellow-brown poisonous salt CdS used especially in electronic parts, in photoelectric cells, and in medicine

Cad·mus \ˈkad-məs\ *noun* [Latin, from Greek *Kadmos*]
: the legendary founder of Thebes

cad·re \ˈka-ˌdrä, ˈkä-, -drē; *especially British* ˈkä-də(r), ˈkā-, -drə\ *noun* [French, from Italian *quadro,* from Latin *quadrum* square — more at QUARREL] (1830)

◇ WORD HISTORY

cadet Until relatively recent times, most inhabitants of France south of an irregular line running from the Gironde to the Alps spoke Occitan (or, to use a more traditional name, Provençal), a Romance language quite different from French. Usually considered a dialect of Occitan is Gascon, spoken in southwestern France, though it is so distinct from the rest of Occitan that it is practically an independent language itself. One Gascon contribution to the lexicon of French, and through French to English, is *cadet,* from 15th century Gascon *capdet* "chief, captain." The younger sons of noble Gascon families served as officers in the 15th century French army and retained the Gascon word *capdet* as a name for their rank, but the socially conscious French took the word as marking genealogical rather than military status, so *capdet* was borrowed in the sense "younger son." Reduced to *cadet,* the word was used additionally in the 16th century for a young gentleman who served as a common soldier to learn the military profession—hence, ultimately, the English use of *cadet* to denote a student in a service academy. *Cadet* entered English in the 17th century.

1 : FRAME, FRAMEWORK
2 : a nucleus or core group especially of trained personnel able to assume control and to train others; *broadly* **:** a group of people having some unifying relationship ⟨a *cadre* of lawyers⟩
3 : a cell of indoctrinated leaders active in promoting the interests of a revolutionary party
4 : a member of a cadre

ca·du·ceus \kə-ˈdü-sē-əs, -ˈdyü-, -shəs\ *noun, plural* **-cei** \-sē-ˌī\ [Latin, modification of Greek *karykeion,* from *karyx, kēryx* herald; akin to Sanskrit *kāru* singer] (1588)
1 : the symbolic staff of a herald; *specifically* **:** a representation of a staff with two entwined snakes and two wings at the top
2 : an insignia bearing a caduceus and symbolizing a physician

ca·du·ci·ty \kə-ˈdü-sə-tē, -ˈdyü-\ *noun* [French *caducité,* from *caduc* transitory, from Latin *caducus*] (1769)
1 : SENILITY
2 : the quality of being transitory or perishable

ca·du·cous \kə-ˈdü-kəs, -ˈdyü-\ *adjective* [Latin *caducus* tending to fall, transitory, from *cadere* to fall — more at CHANCE] (1808)
: falling off easily or before the usual time — used especially of floral organs

cae·cal, cae·cum *variant of* CECAL, CECUM

cae·ci·lian \si-ˈsil-yən, -ˈsēl-, -ˈsi-lē-ən\ *noun* [ultimately from Latin *caecilia* slowworm, from *caecus* blind] (circa 1879)
: any of an order (Gymnophiona) of chiefly tropical burrowing amphibians resembling worms
— caecilian *adjective*

Caer·phil·ly \kär-ˈfi-lē\ *noun* [*Caerphilly,* urban district in Wales] (circa 1893)
: a mild white friable cheese of Welsh origin

Cae·sar \ˈsē-zər\ *noun* [Gaius Julius *Caesar*] (1567)
1 : any of the Roman emperors succeeding Augustus Caesar — used as a title
2 a *often not capitalized* **:** a powerful ruler: (1) **:** EMPEROR (2) **:** AUTOCRAT, DICTATOR **b** [from the reference in Matthew 22:21] **:** the civil power **:** a temporal ruler
— Cae·sar·e·an *or* **Cae·sar·i·an** \si-ˈzar-ē-ən, -zer-\ *adjective*

cae·sar·e·an *also* **cae·sar·i·an** *variant of* CESAREAN

Cae·sar·ism \ˈsē-zə-ˌri-zəm\ *noun* (1857)
: imperial authority or system **:** political absolutism **:** DICTATORSHIP
— Cae·sar·ist \-zə-rist\ *noun*

Caesar salad *noun* [*Caesar*'s, restaurant in Tijuana, Mexico] (1950)
: a tossed salad usually made of romaine, garlic, anchovies, and croutons and dressed with olive oil, coddled egg, lemon juice, and grated cheese

cae·si·um *chiefly British variant of* CESIUM

caes·pi·tose \ˈses-pə-ˌtōs\ *adjective* [New Latin *caespitosus,* from Latin *caespit-, caespes* turf] (1830)
1 : growing in clusters or tufts
2 : forming a dense turf

cae·su·ra \si-ˈzyu̇r-ə, -ˈzhu̇r-\ *noun, plural* **-suras** *or* **-su·rae** \-ˈzyu̇r-(ˌ)ē, -ˈzhu̇r-\ [Late Latin, from Latin, act of cutting, from *caedere* to cut] (1556)
1 *in modern prosody* **:** a usually rhetorical break in the flow of sound in the middle of a line of verse
2 *Greek & Latin prosody* **:** a break in the flow of sound in a verse caused by the ending of a word within a foot
3 : BREAK, INTERRUPTION
4 : a pause marking a rhythmic point of division in a melody
— cae·su·ral \-ˈzyu̇r-əl, -ˈzhu̇r-\ *adjective*

ca·fé *also* **ca·fe** \ka-ˈfā, kə-\ *noun, often attributive* [French *café* coffee, café, from Turkish *kahve* — more at COFFEE] (1802)
1 : a usually small and informal establishment serving various refreshments (as coffee); *broadly* **:** RESTAURANT
2 : BARROOM
3 : CABARET, NIGHTCLUB

ca·fé au lait \(ˌ)ka-ˈfā-ō-ˈlā\ *noun* [French, coffee with milk] (1763)
1 : coffee with usually hot milk in about equal parts
2 : the color of coffee with milk

ca·fé noir \-ˈnwär\ *noun* [French, black coffee] (1841)
: coffee without milk or cream; *also* **:** DEMITASSE

café society *noun* (1937)
: society of persons who are regular patrons of fashionable cafés

¹caf·e·te·ria \ˌka-fə-ˈtir-ē-ə\ *noun* [American Spanish *cafetería* coffeehouse, from *cafetera* coffee maker, from French *cafetière,* from *café*] (1894)
: a restaurant in which the customers serve themselves or are served at a counter and take the food to tables to eat ◆

²cafeteria *adjective* (1951)
: providing a selection from which a choice may be made ⟨*cafeteria* benefit plan⟩ ⟨a *cafeteria* curriculum⟩

caf·e·to·ri·um \-ˈtōr-ē-əm, -ˈtȯr-\ *noun* [blend of *cafeteria* and *auditorium*] (1952)
: a large room (as in a school building) designed for use both as a cafeteria and an auditorium

caff \ˈkaf\ *noun* (1931)
British **:** CAFÉ 1

caf·feine \ka-ˈfēn, ˈka-ˌ\ *noun* [German *Kaffein,* from *Kaffee* coffee, from French *café*] (circa 1823)
: a bitter alkaloid $C_8H_{10}N_4O_2$ found especially in coffee, tea, and kola nuts and used medicinally as a stimulant and diuretic

caf·fe lat·te \ˈkä-(ˌ)fā-ˈlä-(ˌ)tā\ *noun* [Italian *caffelatte,* short for *caffè e latte* coffee and milk] (1927)
: espresso mixed with hot or steamed milk

caf·tan \ˈkaf-(ˌ)tan\ *noun* [Russian *kaftan,* from Turkish, from Persian *qaftān*] (1591)
: a usually cotton or silk ankle-length garment with long sleeves that is common throughout the Levant

¹cage \ˈkāj\ *noun* [Middle English, from Old French, from Latin *cavea* cavity, cage, from *cavus* hollow — more at CAVE] (13th century)
1 : a box or enclosure having some openwork for confining or carrying animals (as birds)
2 a : a barred cell for confining prisoners **b :** a fenced area for prisoners of war
3 : a framework serving as support ⟨the steel *cage* of a skyscraper⟩
4 a : an enclosure resembling a cage in form or purpose ⟨a cashier's *cage*⟩ **b :** an arrangement of atoms or molecules so bonded as to enclose a space in which another atom or ion (as of a metal) can reside
5 a : a screen placed behind home plate to stop baseballs during batting practice **b :** a goal consisting of posts or a frame with a net attached (as in ice hockey)
6 : a large building containing an area for practicing outdoor sports and often adapted for indoor events
— cage·ful \-ˌfu̇l\ *noun*

²cage *transitive verb* **caged; cag·ing** (1577)
1 : to confine or keep in or as if in a cage
2 : to drive (as a puck) into a cage and score a goal

cage·ling \ˈkāj-liŋ\ *noun* (1859)
: a caged bird

ca·gey *also* **ca·gy** \ˈkā-jē\ *adjective* **ca·gi·er; -est** [origin unknown] (circa 1893)
1 : hesitant about committing oneself
2 a : wary of being trapped or deceived **:** SHREWD ⟨a *cagey* consumer⟩ **b :** marked by cleverness ⟨a *cagey* reply⟩
— ca·gi·ly \-jə-lē\ *adverb*
— ca·gi·ness *also* **ca·gey·ness** \-jē-nəs\ *noun*

ca·hier \kä-ˈyā, kī-ˈā\ *noun* [French, from Middle French *quaer, caier* quire — more at QUIRE] (1789)
: a report or memorial concerning policy especially of a parliamentary body

ca·hoot \kə-ˈhüt\ *noun* [perhaps from French *cahute* cabin, hut] (1829)
: PARTNERSHIP, LEAGUE — usually used in plural ⟨they're in *cahoots*⟩

ca·how \kə-ˈhau̇\ *noun* [imitative] (1615)
: a dark-colored petrel (*Pterodroma cahow*) formerly abundant in Bermuda but now nearly extinct

cai·man \ˈkā-mən; kā-ˈman, kī-\ *noun* [Spanish *caimán,* probably from Carib *caymán*] (1577)
: any of several Central and South American crocodilians (genera *Caiman, Melanosuchus,* and *Paleosuchus*) similar to alligators

Cain \ˈkān\ *noun* [Hebrew *Qayin*]
: the brother and murderer of Abel

-caine *noun combining form* [German *-kain,* from *kokain* cocaine]
: synthetic alkaloid anesthetic ⟨pro*caine*⟩

ca·ïque \kä-ˈēk, ˈkīk\ *noun* [French, from Turkish *kayık*] (1625)
1 : a light skiff used on the Bosporus
2 : a Levantine sailing vessel

caird \ˈkerd\ *noun* [Scottish Gaelic *ceard* craftsman; akin to Greek *kerdos* profit] (1663)
Scottish **:** a traveling tinker; *also* **:** TRAMP, GYPSY

cairn \ˈkarn, ˈkern\ *noun* [Middle English (Scots) *carne,* from Scottish Gaelic *carn;* akin to Old Irish & Welsh *carn* cairn] (15th century)
: a heap of stones piled up as a memorial or as a landmark
— cairned \ˈkarnd, ˈkernd\ *adjective*

cairn·gorm \ˈkarn-ˌgȯrm, ˈkern-\ *noun* [*Cairngorm,* mountain in Scotland] (1794)
: a yellow or smoky-brown crystalline quartz

cairn terrier *noun* [from its use in hunting among cairns] (1910)
: any of a breed of small compactly built hard-coated terriers of Scottish origin

cais·son \ˈkā-ˌsän, -sᵊn, *British also* kə-ˈsün\ *noun* [French, from Middle French, from Old

◇ WORD HISTORY

cafeteria Although American English *cafeteria* is in all likelihood borrowed from Spanish, and within Spanish its derivational pathway from *café* "coffee" is fairly straightforward, the details of the word's passage from one language to the other are murky. Early evidence for Spanish *cafetería* is slim; the word was current in Cuba in the 19th century, but it referred to a shop that sold ground coffee, not a restaurant. During the early decades of the century, the heyday of the self-service restaurant, dictionary etymologists believed that both the word and the institution originated in southern California. Evidence for the word first surfaces, however, not on the West Coast, but in Chicago, where a "Cafetiria Catering Company" operated four downtown restaurants—presumably self-service style—in 1895. (There are some indications that self-service restaurants grew out of the World's Columbian Exposition held in Chicago in 1893.) During the 1920s and '30s in the U.S., coinages based on *cafeteria* proliferated: there were marketerias, gaseterias, and washeterias, usually having some self-service feature. These have faded since World War II, while the cafeteria itself has survived mainly in schools, hospitals, and other domains of institutional food service—pushed off America's street corners by the fast-food restaurant.

Provençal, from *caissa* chest, from Latin *capsa* — more at CASE] (circa 1702)
1 a : a chest to hold ammunition **b :** a usually 2-wheeled vehicle for artillery ammunition attachable to a horse-drawn limber; *also* **:** a limber with its attached caisson

caisson 1b
2 a : a watertight chamber used in construction work under water or as a foundation **b :** a hollow floating box or a boat used as a floodgate for a dock or basin
3 : COFFER 3

caisson disease *noun* (1873)
: ²BEND 3

cai·tiff \'kā-təf\ *adjective* [Middle English *caitif,* from Old North French, captive, vile, from Latin *captivus* captive] (14th century)
: being base, cowardly, or despicable
— caitiff *noun*

caj·e·put \'ka-jə-pət, -ˌpu̇t\ *noun* [Malay *kayu putih,* from *kayu* wood, tree + *putih* white] (1832)
: an Australian and southeast Asian tree (*Melaleuca quinquenervia* synonym *M. leucadendron*) of the myrtle family that yields a pungent medicinal oil and has been introduced into Florida

ca·jole \kə-'jōl\ *transitive verb* **ca·joled; ca·jol·ing** [French *cajoler*] (1645)
1 : to persuade with flattery or gentle urging especially in the face of reluctance **:** COAX ⟨had to *cajole* them into going⟩
2 : to deceive with soothing words or false promises
— ca·jole·ment \-'jōl-mənt\ *noun*
— ca·jol·er *noun*
— ca·jol·ery \-'jō-lə-rē\ *noun*

¹Ca·jun *also* **Ca·jan** \'kā-jən\ *noun* [alteration of *Acadian*] (1868)
: a Louisianian descended from French-speaking immigrants from Acadia

²Cajun *adjective* (1885)
1 : of, relating to, or characteristic of the Cajuns
2 : of, relating to, or prepared in a style of cooking originating among the Cajuns and characterized by the use of hot seasonings (as cayenne pepper)

¹cake \'kāk\ *noun* [Middle English, from Old Norse *kaka;* akin to Old High German *kuocho* cake] (13th century)
1 a : a breadlike food made from a dough or batter that is usually fried or baked in small flat shapes and is often unleavened **b :** a sweet baked food made from a dough or thick batter usually containing flour and sugar and often shortening, eggs, and a raising agent (as baking powder) **c :** a flattened usually round mass of food that is baked or fried ⟨a fish *cake*⟩
2 a : a block of compacted or congealed matter ⟨a *cake* of ice⟩ **b :** a hard or brittle layer or deposit
— cak·ey \'kā-kē\ *adjective*

²cake *verb* **caked; cak·ing** (1607)
transitive verb
1 : ENCRUST ⟨*caked* with dust⟩
2 : to fill (a space) with a packed mass
intransitive verb
: to form or harden into a mass

cake·walk \'kāk-ˌwȯk\ *noun* (1879)
1 : a black American entertainment having a cake as prize for the most accomplished steps and figures in walking
2 : a stage dance developed from walking steps and figures typically involving a high prance with backward tilt
3 a : a one-sided contest **b :** an easy task
— cakewalk *intransitive verb*
— cake·walk·er *noun*

Cal·a·bar bean \'ka-lə-ˌbär-\ *noun* [*Calabar,* Nigeria] (circa 1876)
: the dark brown highly poisonous seed of a tropical African woody vine (*Physostigma venenosum*) that is used as a source of physostigmine and as an ordeal poison in native witchcraft trials

cal·a·bash \'ka-lə-ˌbash\ *noun* [French & Spanish; French *calebasse* gourd, from Spanish *calabaza*] (1596)
1 : a tropical American tree (*Crescentia cujete*) of the trumpet-creeper family; *also* **:** its hard globose fruit
2 : GOURD; *especially* **:** one whose hard shell is used for a utensil
3 : a utensil (as a bottle or dipper) made from the shell of a calabash

cal·a·boose \'ka-lə-ˌbüs\ *noun* [Spanish *calabozo* dungeon] (1792)
: JAIL; *especially* **:** a local jail

ca·la·di·um \kə-'lā-dē-əm\ *noun* [New Latin, genus name, from Malay *kĕladi,* an aroid plant] (circa 1845)
: any of a genus (*Caladium,* especially *C. bicolor*) of tropical American plants of the arum family widely cultivated for their showy variably colored leaves

cal·a·man·der \'ka-lə-ˌman-dər, ˌka-lə-'\ *noun* [probably from Dutch *kalamanderhout* calamander wood] (1804)
: the hazel-brown black-striped wood of a southeast Asian tree (genus *Diospyros,* especially *D. quaesita*) that is used in furniture manufacturing

cal·a·mari \ˌkä-lə-'mär-ē, ˈka-lə-ˌmer-ē\ *noun* [Italian, plural of *calamaro, calamaio,* from Medieval Latin *calamarium* ink pot, from Latin *calamus;* from the inky substance the squid secretes] (circa 1961)
: squid used as food

cal·a·mary \'ka-lə-ˌmer-ē\ *noun, plural* **-mar·ies** [Medieval Latin *calamarium*] (1567)
: SQUID

cal·a·mine \'ka-lə-ˌmīn, -mən\ *noun* [Middle English *calamyn* ore of zinc, from Medieval Latin *calamina,* alteration of Latin *cadmia* — more at CADMIUM] (15th century)
: a mixture of zinc oxide with a small amount of ferric oxide used in lotions, liniments, and ointments

cal·a·mint \'ka-lə-ˌmint\ *noun* [Middle English *calament,* from Old French, from Medieval Latin *calamentum,* from Greek *kalaminthē*] (14th century)
: any of a genus (*Satureja,* especially *S. calamintha*) of mints

cal·a·mite \'ka-lə-ˌmīt\ *noun* [New Latin *Calamites,* genus of fossil plants, from Latin *calamus*] (1837)
: a Paleozoic fossil plant (especially genus *Calamites*) resembling a giant horsetail

ca·lam·i·tous \kə-'la-mə-təs\ *adjective* (1545)
: being, causing, or accompanied by calamity
— ca·lam·i·tous·ly *adverb*

ca·lam·i·ty \kə-'la-mə-tē\ *noun, plural* **-ties** [Middle English *calamytey,* from Middle French *calamité,* from Latin *calamitat-, calamitas;* perhaps akin to Latin *clades* destruction] (15th century)
1 : a state of deep distress or misery caused by major misfortune or loss
2 : an extraordinarily grave event marked by great loss and lasting distress and affliction

cal·a·mon·din \ˌka-lə-'män-dən\ *noun* [Tagalog *kalamunding*] (circa 1928)
: a small hybrid citrus tree (*Citrofortunella mitis* synonym *Citrus mitis*); *also* **:** its tart fruit

cal·a·mus \'ka-lə-məs\ *noun, plural* **-mi** \-ˌmī, -ˌmē\ [Latin, reed, reed pen, from Greek *kalamos* — more at HAULM] (14th century)
1 a : SWEET FLAG **b :** the aromatic peeled and dried rhizome of the sweet flag that is the source of a carcinogenic essential oil
2 : the hollow basal portion of a feather below the vane **:** QUILL

ca·lash \kə-'lash\ *noun* [French *calèche,* from German *Kalesche,* from Czech *kolesa* wheels, carriage; akin to Greek *kyklos* wheel — more at WHEEL] (1679)
1 a : a light small-wheeled 4-passenger carriage with a folding top **b :** CALÈCHE 1b
2 : a large hood worn by women in the 18th century

calc- *or* **calci-** *combining form* [Latin *calc-, calx* lime — more at CHALK]
: calcium **:** calcium salt ⟨*calc*ic⟩ ⟨*calci*fy⟩

cal·ca·ne·al \kal-'kā-nē-əl\ *adjective* (circa 1849)
: relating to the heel or calcaneus

cal·ca·ne·um \-nē-əm\ *noun, plural* **-nea** \-nē-ə\ [Latin, heel — more at CALK] (circa 1751)
: CALCANEUS

cal·ca·ne·us \-nē-əs\ *noun, plural* **-nei** \-nē-ˌī\ [Late Latin, heel, alteration of Latin *calcaneum*] (circa 1925)
: a tarsal bone that in humans is the great bone of the heel

cal·car·e·ous \kal-'kar-ē-əs, -'ker-\ *adjective* [Latin *calcarius* of lime, from *calc-, calx*] (1677)
1 a : resembling calcite or calcium carbonate especially in hardness **b :** consisting of or containing calcium carbonate; *also* **:** containing calcium
2 : growing on limestone or in soil impregnated with lime
— cal·car·e·ous·ly *adverb*

calces *plural of* CALX

cal·cic \'kal-sik\ *adjective* (1871)
: derived from or containing calcium or lime **:** rich in calcium

cal·ci·cole \'kal-sə-ˌkōl\ *noun* [French, calcicolous, from *calc-* + *-cole* -colous] (1882)
: a plant normally growing on calcareous soils
— cal·cic·o·lous \kal-'si-kə-ləs\ *adjective*

cal·cif·er·ol \kal-'si-fə-ˌrȯl, -ˌrōl\ *noun* [*calcif*erous + ergost*erol*] (1931)
: an alcohol $C_{28}H_{43}OH$ usually prepared by irradiation of ergosterol and used as a dietary supplement in nutrition and medicinally in the control of rickets and related disorders — called also *vitamin D_2*

cal·cif·er·ous \kal-'si-f(ə-)rəs\ *adjective* (1799)
: producing or containing calcium carbonate

cal·cif·ic \kal-'si-fik\ *adjective* [*calcify*] (1861)
: involving or caused by calcification ⟨*calcific* lesions⟩

cal·ci·fuge \'kal-sə-ˌfyüj\ *noun* [French, calcifugous, from *calc-* + Latin *fugere* to flee — more at FUGITIVE] (1926)
: a plant not normally growing on calcareous soils
— calcifuge *also* **cal·cif·u·gous** \kal-'si-fyə-gəs\ *adjective*

cal·ci·fy \'kal-sə-ˌfī\ *verb* **-fied; -fy·ing** (1854)
transitive verb
1 : to make calcareous by deposit of calcium salts
2 : to make inflexible or unchangeable
intransitive verb
1 : to become calcareous
2 : to become inflexible and changeless **:** HARDEN
— cal·ci·fi·ca·tion \ˌkal-sə-fə-'kā-shən\ *noun*

cal·ci·mine \'kal-sə-ˌmīn\ *noun* [alteration of *kalsomine,* of unknown origin] (circa 1859)
: a white or tinted wash of glue, whiting or zinc white, and water that is used especially on plastered surfaces
— calcimine *transitive verb*

cal·ci·na·tion \ˌkal-sə-ˈnā-shən\ *noun* (14th century)
: the act or process of calcining : the state of being calcined

[1]**cal·cine** \kal-ˈsīn, ˈkal-ˌ\ *verb* **cal·cined; cal·cin·ing** [Middle English *calcenen,* from Medieval Latin *calcinare,* from Late Latin *calcina* lime, from Latin *calc-, calx*] (14th century)
transitive verb
: to heat (as inorganic materials) to a high temperature but without fusing in order to drive off volatile matter or to effect changes (as oxidation or pulverization)
intransitive verb
: to undergo calcination

[2]**cal·cine** \ˈkal-ˌsīn\ *noun* (circa 1909)
: a product (as a metal oxide) of calcination or roasting

cal·ci·no·sis \ˌkal-sə-ˈnō-səs\ *noun, plural* **-no·ses** \-ˌsēz\ [New Latin, irregular (influenced by International Scientific Vocabulary *calcine*) from *calc-* + *-osis*] (circa 1929)
: the abnormal deposition of calcium salts in a part or tissue of the body

cal·cite \ˈkal-ˌsīt\ *noun* (1849)
: a mineral $CaCO_3$ consisting of calcium carbonate crystallized in hexagonal form and including common limestone, chalk, and marble — compare ARAGONITE
— **cal·cit·ic** \kal-ˈsi-tik\ *adjective*

cal·ci·to·nin \ˌkal-sə-ˈtō-nən\ *noun* [*calci-* + *-tonin* (as in *serotonin*)] (1961)
: a polypeptide hormone especially from the thyroid gland that tends to lower the level of calcium in the blood plasma — called also *thyrocalcitonin*

cal·ci·um \ˈkal-sē-əm\ *noun, often attributive* [New Latin, from Latin *calc-, calx* lime] (1808)
: a silver-white bivalent metallic element of the alkaline-earth group occurring only in combination — see ELEMENT table

calcium carbide *noun* (circa 1888)
: a usually dark gray crystalline compound CaC_2 used especially for the generation of acetylene and for making calcium cyanamide

calcium carbonate *noun* (1873)
: a compound $CaCO_3$ found in nature as calcite and aragonite and in plant ashes, bones, and shells and used in making lime and portland cement and as a gastric antacid

calcium channel blocker *noun* (1982)
: any of a class of drugs that prevent or slow the influx of calcium ions into smooth muscle cells and are used to treat some forms of angina pectoris and some cardiac arrhythmias

calcium chloride *noun* (circa 1885)
: a white deliquescent salt $CaCl_2$ used in its anhydrous state as a drying and dehumidifying agent and in a hydrated state for controlling dust and ice on roads

calcium cyanamide *noun* (circa 1893)
: a compound $CaCN_2$ used as a fertilizer and a weed killer and as a source of other nitrogen compounds

calcium gluconate *noun* (1884)
: a white powdery salt $CaC_{12}H_{22}O_{14}$ used especially to supplement bodily calcium stores

calcium hydroxide *noun* (circa 1889)
: a white crystalline strong alkali $Ca(OH)_2$ that is used especially to make mortar and plaster and to soften water

calcium hypochlorite *noun* (circa 1889)
: a white powder $CaCl_2O_2$ used especially as a bleaching agent and disinfectant

calcium oxalate *noun* (1919)
: a crystalline salt CaC_2O_4 normally deposited in many plant cells and in animals sometimes excreted in urine or retained in the form of urinary calculi

calcium oxide *noun* (circa 1885)
: a caustic solid CaO that is white when pure and that is the chief constituent of lime

calcium phosphate *noun* (1869)
: any of various phosphates of calcium: as **a** : the phosphate $CaH_4P_2O_8$ used as a fertilizer and in baking powder **b** : the phosphate $CaHPO_4$ used in pharmaceutical preparations and animal feeds **c** : the phosphate $Ca_3P_2O_8$ used as a fertilizer **d** : the naturally occurring phosphate $Ca_5(F,Cl,OH,½CO_3)(PO_4)_3$ that contains other elements or radicals and is the chief constituent of phosphate rock, bones, and teeth

calcium silicate *noun* (circa 1888)
: any of several silicates of calicum used especially in construction materials (as portland cement)

calcium sulfate *noun* (circa 1885)
: a white salt $CaSO_4$ that occurs especially as anhydrite, gypsum, and plaster of paris and that in hydrated form is used as a building material and in anhydrous form is used as a drying agent

cal·cu·la·ble \ˈkal-kyə-lə-bəl\ *adjective* (circa 1734)
1 : subject to or ascertainable by calculation
2 : that may be counted on : DEPENDABLE

cal·cu·late \ˈkal-kyə-ˌlāt\ *verb* **-lat·ed; -lat·ing** [Latin *calculatus,* past participle of *calculare,* from *calculus* pebble (used in reckoning), perhaps irregular diminutive of *calc-, calx* lime — more at CHALK] (1570)
transitive verb
1 a : to determine by mathematical processes **b** : to reckon by exercise of practical judgment : ESTIMATE **c** : to solve or probe the meaning of : FIGURE OUT ⟨trying to *calculate* his expression —Hugh MacLennan⟩
2 : to design or adapt for a purpose
3 a : to judge to be true or probable **b** : INTEND ⟨I *calculate* to do it or perish in the attempt —Mark Twain⟩
intransitive verb
1 a : to make a calculation **b** : to forecast consequences
2 : COUNT, RELY

cal·cu·lat·ed \-ˌlā-təd\ *adjective* (1722)
1 : APT, LIKELY
2 a : worked out by mathematical calculation **b** : engaged in, undertaken, or displayed after reckoning or estimating the statistical probability of success or failure ⟨a *calculated* risk⟩
3 a : planned or contrived to accomplish a purpose **b** : DELIBERATE, INTENDED
— **cal·cu·lat·ed·ly** *adverb*
— **cal·cu·lat·ed·ness** *noun*

cal·cu·lat·ing \-ˌlā-tiŋ\ *adjective* (1710)
1 : making calculations ⟨*calculating* machine⟩
2 : marked by prudent analysis or by shrewd consideration of self-interest : SCHEMING
— **cal·cu·lat·ing·ly** \-tiŋ-lē\ *adverb*

cal·cu·la·tion \ˌkal-kyə-ˈlā-shən\ *noun* (14th century)
1 a : the process or an act of calculating **b** : the result of an act of calculating
2 a : studied care in analyzing or planning **b** : cold heartless planning to promote self-interest
— **cal·cu·la·tion·al** \-ˈlāsh-nəl, -ˈlā-shə-nᵊl\ *adjective*

cal·cu·la·tor \ˈkal-kyə-ˌlā-tər\ *noun* (14th century)
: one that calculates: as **a** : a usually electronic device for performing mathematical calculations **b** : a person who operates a calculator

cal·cu·lous \ˈkal-kyə-ləs\ *adjective* (1605)
: caused or characterized by a calculus or calculi

cal·cu·lus \-ləs\ *noun, plural* **-li** \-ˌlī, -ˌlē\ *also* **-lus·es** [Latin, stone (used in reckoning)] (1666)
1 a : a method of computation or calculation in a special notation (as of logic or symbolic logic) **b** : the mathematical methods comprising differential and integral calculus
2 : CALCULATION
3 a : a concretion usually of mineral salts around organic material found especially in hollow organs or ducts **b** : TARTAR 2
4 : a system or arrangement of intricate or interrelated parts ◆

calculus of variations (1837)
: a branch of mathematics concerned with applying the methods of calculus to finding the maxima and minima of a function which depends for its values on another function or a curve

cal·de·ra \kal-ˈder-ə, kȯl-, -ˈdir-\ *noun* [Spanish, literally, caldron, from Late Latin *caldaria* — more at CAULDRON] (1691)
: a volcanic crater that has a diameter many times that of the vent and is formed by collapse of the central part of a volcano or by explosions of extraordinary violence

cal·dron *variant of* CAULDRON

ca·lèche *or* **ca·leche** \kə-ˈlesh, -ˈlash\ *noun* [French *calèche* — more at CALASH] (1666)
1 a : CALASH 1a **b** : a 2-wheeled horse-drawn vehicle with a driver's seat on the splashboard used in Quebec
2 : CALASH 2

cal·e·fac·to·ry \ˌka-lə-ˈfak-t(ə-)rē\ *noun, plural* **-ries** [Medieval Latin *calefactorium,* from Latin *calefacere* to warm — more at CHAFE] (circa 1681)
: a monastery room warmed and used as a sitting room

[1]**cal·en·dar** \ˈka-lən-dər\ *noun* [Middle English *calender,* from Anglo-French or Medieval Latin; Anglo-French *calender,* from Medieval Latin *kalendarium,* from Latin, moneylender's account book, from *kalendae* calends] (13th century)
1 : a system for fixing the beginning, length, and divisions of the civil year and arranging days and longer divisions of time (as weeks and months) in a definite order — see MONTH table
2 : a tabular register of days according to a system usually covering one year and referring the days of each month to the days of the week
3 : an orderly list: as **a** : a list of cases to be tried in court **b** : a list of bills or other items reported out of committee for consideration by a legislative assembly **c** : a list or schedule of planned events or activities giving dates and details
4 *British* : a university catalog

[2]**calendar** *transitive verb* **-dared; -dar·ing** \-d(ə-)riŋ\ (15th century)
: to enter in a calendar

calendar year *noun* (circa 1909)
1 : a period of a year beginning and ending with the dates that are conventionally accepted as marking the beginning and end of a numbered year
2 : a period of time equal in length to that of the year in the calendar conventionally in use

[1]**cal·en·der** \ˈka-lən-dər\ *transitive verb* **-dered; -der·ing** \-d(ə-)riŋ\ [Middle French *calandrer,* from *calandre* machine for calendering, from (assumed) Vulgar Latin *colendra*

◇ WORD HISTORY

calculus In Latin *calculus* meant "pebble." Because the Romans used pebbles to do addition and subtraction on a counting board, the word became associated with computation, and the phrase *ponere calculos,* literally, "to place pebbles" meant figuratively "to perform a computation." Latin and Late Latin derivatives from *calculus* include *calculator* "person versed in arithmetic" and *calculare* "to reckon" (hence English *calculate*). *Calculus* itself has been borrowed into English as a medical term to refer to concretions of matter in the body such as kidney and bladder stones—a sense first found in Medieval Latin and a straightforward extension of the meaning "pebble"—and to refer to systems of mathematical computations such as differential and integral calculus.

cylinder, modification of Greek *kylindros* — more at CYLINDER] (1513)
: to press (as cloth, rubber, or paper) between rollers or plates in order to smooth and glaze or to thin into sheets
— cal·en·der·er \-dər-ər\ *noun*

[2]calender *noun* (1688)
: a machine for calendering something

[3]calender *noun* [Persian *qalandar,* from Arabic, from Persian *kalandar* uncouth man] (1621)
: a member of a Sufic order of wandering mendicant dervishes

ca·len·dri·cal \kə-ˈlen-dri-kəl, ka-\ *also* **ca·len·dric** \-drik\ *adjective* (circa 1843)
: of, relating to, characteristic of, or used in a calendar

ca·lends \ˈka-lən(d)z, ˈkā-\ *noun plural but singular or plural in construction* [Middle English *kalendes,* from Latin *kalendae, calendae*] (14th century)
: the first day of the ancient Roman month from which days were counted backward to the ides

ca·len·du·la \kə-ˈlen-jə-lə, -dyu̇-lə\ *noun* [New Latin, genus name, from Medieval Latin, from Latin *calendae* calends] (1789)
: any of a small genus (*Calendula*) of yellow-rayed composite herbs of temperate regions

cal·en·ture \ˈka-lən-ˌchu̇r\ *noun* [Spanish *calentura,* from *calentar* to heat, from Latin *calent-, calens,* present participle of *calēre* to be warm — more at LEE] (1593)
: a fever formerly supposed to affect sailors in the tropics

[1]calf \ˈkaf, ˈkȧf, *dialect also* ˈkāf\ *noun, plural* **calves** \ˈkavz, ˈkȧvz, ˈkāvz\ *also* **calfs** *often attributive* [Middle English, from Old English *cealf;* akin to Old High German *kalb* calf] (before 12th century)
1 a : the young of the domestic cow; *also* **:** that of a closely related mammal (as a bison) **b :** the young of various large animals (as the elephant or whale)
2 *plural* **calfs :** the hide of the domestic calf; *especially* **:** CALFSKIN
3 : an awkward or silly youth
— calf·like \ˈkaf-ˌlīk, ˈkȧf-, *dialect also* ˈkāf-\ *adjective*
— in calf : PREGNANT — used of a cow

[2]calf *noun, plural* **calves** \ˈkavz, ˈkȧvz\ [Middle English, from Old Norse *kālfi*] (14th century)
: the fleshy back part of the leg below the knee

calf–love \-ˌləv\ *noun* (1823)
: PUPPY LOVE

calf's–foot jelly \ˈkavz-ˌfu̇t-, ˈkafs-, ˈkȧvz-, ˈkȧfs- *also* ˈkāvz-\ *noun* (1775)
: jelly made from gelatin obtained by boiling calves' feet

calf·skin \ˈkaf-ˌskin, ˈkȧf- *also* ˈkāf-\ *noun* (15th century)
: leather made of the skin of a calf

Cal·gon \ˈkal-ˌgän\ *trademark*
— used for a water softener

Cal·i·ban \ˈka-lə-ˌban\ *noun*
: a savage and deformed slave in Shakespeare's *The Tempest*

cal·i·ber *or* **cal·i·bre** \ˈka-lə-bər, *British also* kə-ˈlē-\ *noun* [Middle French *calibre,* from Old Italian *calibro,* from Arabic *qālib* shoemaker's last] (1567)
1 a : degree of mental capacity or moral quality **b :** degree of excellence or importance
2 a : the diameter of a bullet or other projectile **b :** the diameter of a bore of a gun usually expressed in hundredths or thousandths of an inch and typically written as a decimal fraction ⟨.32 *caliber*⟩
3 : the diameter of a round body; *especially* **:** the internal diameter of a hollow cylinder

cal·i·brate \ˈka-lə-ˌbrāt\ *transitive verb* **-brat·ed; -brat·ing** (circa 1864)
1 : to ascertain the caliber of (as a thermometer tube)
2 : to determine, rectify, or mark the graduations of (as a thermometer tube)
3 : to standardize (as a measuring instrument) by determining the deviation from a standard so as to ascertain the proper correction factors
4 : to adjust precisely for a particular function
— cal·i·bra·tor \-ˌbrā-tər\ *noun*

cal·i·bra·tion \ˌka-lə-ˈbrā-shən\ *noun* (circa 1859)
1 : the act or process of calibrating **:** the state of being calibrated
2 : a set of graduations to indicate values or positions — usually used in plural ⟨*calibrations* on a gauge⟩

ca·li·che \kə-ˈlē-chē\ *noun* [American Spanish, from Spanish, flake of lime, from *cal* lime, from Latin *calx* — more at CHALK] (circa 1858)
1 : the nitrate-bearing gravel or rock of the sodium nitrate deposits of Chile and Peru
2 : a crust of calcium carbonate that forms on the stony soil of arid regions

cal·i·co \ˈka-li-ˌkō\ *noun, plural* **-coes** *or* **-cos** [*Calicut,* India] (1578)
1 a : cotton cloth imported from India **b** *British* **:** a plain white cotton fabric that is heavier than muslin **c :** any of various cheap cotton fabrics with figured patterns
2 : a blotched or spotted animal; *especially* **:** one that is predominantly white with red and black patches
— calico *adjective*

calico bass *noun* (circa 1882)
: BLACK CRAPPIE

calico bush *noun* (1814)
: MOUNTAIN LAUREL

Cal·i·for·nia condor \ˌka-lə-ˈfȯr-nyə-\ *noun* [*California,* state of U.S.] (circa 1889)
: a large nearly extinct vulture (*Gymnogyps californianus*) found most recently in the mountains of southern California that is related to the condor of South America

California condor

California laurel *noun* (1871)
: an evergreen Pacific coast tree (*Umbellularia californica*) of the laurel family with small umbellate flowers

California poppy *noun* (1891)
: any of a genus (*Eschscholzia*) of herbs of the poppy family; *especially* **:** one (*E. californica*) widely cultivated for its usually yellow or orange flowers

Cal·i·for·nio \ˌka-lə-ˈfȯr-nē-ˌō\ *noun, plural* **-nios** [Spanish, from *California*] (1923)
: one of the original Spanish colonists of California or their descendants

cal·i·for·ni·um \ˌka-lə-ˈfȯr-nē-əm\ *noun* [New Latin, from *California,* U.S.] (1950)
: a radioactive element discovered by bombarding curium 242 with alpha particles — see ELEMENT table

ca·lig·i·nous \kə-ˈli-jə-nəs\ *adjective* [Middle French or Latin; Middle French *caligineux,* from Latin *caliginosus,* from *caligin-, caligo* darkness] (1548)
: MISTY, DARK

[1]cal·i·per \ˈka-lə-pər\ *noun* [alteration of *caliber*] (1588)
1 a : any of various measuring instruments having two usually adjustable arms, legs, or jaws used especially to measure diameter or thickness — usually used in plural ⟨a pair of *calipers*⟩ **b :** a device consisting of two plates lined with a frictional material that press against the sides of a rotating wheel or disc in certain brake systems
2 : thickness especially of paper, paperboard, or a tree

[2]caliper *transitive verb* **-pered; -per·ing** \-p(ə-)riŋ\ (1876)
: to measure by or as if by calipers

ca·liph *or* **ca·lif** \ˈkā-ləf, ˈka-ləf\ *noun* [Middle English *caliphe,* from Middle French *calife,* from Arabic *khalīfah* successor] (14th century)
: a successor of Muhammad as temporal and spiritual head of Islam — used as a title
— ca·liph·al \-lə-fəl\ *adjective*

ca·liph·ate \-lə-ˌfāt, -fət\ *noun* (1614)
: the office or dominion of a caliph

cal·is·then·ic \ˌka-ləs-ˈthe-nik\ *adjective* (1827)
: of or relating to calisthenics

cal·is·then·ics \-niks\ *noun plural but singular or plural in construction* [Greek *kalos* beautiful + *sthenos* strength] (1827)
1 : systematic rhythmic bodily exercises performed usually without apparatus
2 *usually singular in construction* **:** the art or practice of calisthenics

ca·lix \ˈkā-liks, ˈka-\ *noun, plural* **ca·li·ces** \ˈkā-lə-ˌsēz, ˈka-\ [Latin *calic-, calix* — more at CHALICE] (1698)
: CUP

[1]calk \ˈkȯk\, **calk·er** \ˈkȯ-kər\ *variant of* CAULK, CAULKER

[2]calk \ˈkȯk\ *noun* [probably alteration of *calkin,* from Middle English *kakun,* from Middle Dutch or Old North French; Middle Dutch *calcoen* horse's hoof, from Old North French *calcain* heel, from Latin *calcaneum,* from *calc-, calx* heel] (1587)
: a tapered piece projecting downward on the shoe of a horse to prevent slipping; *also* **:** a similar device worn on the sole of a shoe

[3]calk *transitive verb* (1624)
1 : to furnish with calks
2 : to wound with a calk

[1]call \ˈkȯl\ *verb* [Middle English, from Old Norse *kalla;* akin to Old English hilde*calla* battle herald, Old High German *kallōn* to talk loudly, Old Church Slavonic *glasŭ* voice] (before 12th century)
intransitive verb
1 a : to speak in a loud distinct voice so as to be heard at a distance **:** SHOUT ⟨*call* for help⟩ **b :** to make a request or demand ⟨*call* for an investigation⟩ **c** *of an animal* **:** to utter a characteristic note or cry **d :** to get or try to get into communication by telephone — often used with *up* **e :** to make a demand in card games (as for a particular card or for a show of hands) **f :** to give the calls for a square dance
2 : to make a brief visit ⟨*called* to pay his respects⟩ ⟨*called* on a friend⟩
transitive verb
1 a (1) **:** to utter in a loud distinct voice — often used with *out* ⟨*call* out a number⟩ (2) **:** to announce or read loudly or authoritatively ⟨*call* the roll⟩ ⟨*call* off a row of figures⟩ **b** (1) **:** to command or request to come or be present ⟨*called* to testify⟩ (2) **:** to cause to come **:** BRING ⟨*calls* to mind an old saying⟩ **c :** to summon to a particular activity, employment, or office ⟨was *called* to active duty⟩ **d :** to invite or command to meet **:** CONVOKE ⟨*call* a meeting⟩ **e :** to rouse from sleep or summon to get up **f** (1) **:** to give the order for **:** bring into action ⟨*call* a strike against the company⟩ ⟨*call* a pitchout⟩ (2) **:** to manage by giving the signals or orders ⟨that catcher *calls* a good game⟩

\ə\ **abut** \ᵊ\ **kitten** \ər\ **further** \a\ **ash** \ā\ **ace** \ä\ **mop, mar** \au̇\ **out** \ch\ **chin** \e\ **bet** \ē\ **easy** \g\ **go** \i\ **hit** \ī\ **ice** \j\ **job** \ŋ\ **sing** \ō\ **go** \ȯ\ **law** \ȯi\ **boy** \th\ **thin** \t͟h\ **the** \ü\ **loot** \u̇\ **foot** \y\ **yet** \zh\ **vision** *see also* Guide to Pronunciation

g (1) **:** to make a demand in bridge for (a card or suit) (2) **:** to require (a player) to show the hand in poker by making an equal bet (3) **:** to challenge to make good on a statement (4) **:** to charge with or censure for an offense ⟨deserves to be *called* on that⟩ **h :** to attract (as game) by imitating the characteristic cry **i :** to halt (as a baseball game) because of unsuitable conditions **j :** to rule on the status of (as a pitched ball or a player's action) ⟨*call* balls and strikes⟩ ⟨*call* a base runner safe⟩ **k :** to give the calls for (a square dance) — often used with *off* **l** (1) **:** to demand payment of especially by formal notice ⟨*call* a loan⟩ (2) **:** to demand presentation of (as a bond or option) for redemption **m** (1) **:** to get or try to get in communication with by telephone (2) **:** to generate signals for (a telephone number) in order to reach the party to whom the number is assigned ⟨*call* 911⟩ (3) **:** to make a signal to in order to transmit a message ⟨*call* the flagship⟩
2 a : to speak of or address by a specified name **:** give a name to ⟨*call* her Kitty⟩ **b** (1) **:** to regard or characterize as of a certain kind **:** CONSIDER ⟨can hardly be *called* generous⟩ (2) **:** to estimate or consider for purposes of an estimate or for convenience ⟨*call* it an even dollar⟩ **c** (1) **:** to describe correctly in advance of or without knowledge of the event **:** PREDICT (2) **:** to name or specify in advance ⟨*call* the toss of a coin⟩
3 : to temporarily transfer control of computer processing to (as a subroutine or procedure)
synonym see SUMMON
— **call a spade a spade :** to speak frankly
— **call for 1 :** to call (as at one's house) to get ⟨I'll *call for* you after dinner⟩ **2 :** to require as necessary or appropriate ⟨the job *calls for* typing skills⟩ ⟨the design *calls for* three windows⟩
— **call forth :** ELICIT, EVOKE ⟨these events *call forth* great emotions⟩
— **call in question** *or* **call into question :** to cast doubt upon
— **call it a day :** to stop for the remainder of the day or for the present whatever one has been doing
— **call it quits :** to call it a day **:** QUIT
— **call names :** to address or speak of a person or thing contemptuously or offensively
— **call on 1 :** to call upon **2 :** to elicit a response from (as a student) ⟨the teacher *called on* her first⟩
— **call one's bluff :** to challenge in order to expose an empty pretense or threat
— **call the shots :** to be in charge or control **:** determine the policy or procedure
— **call the tune :** to call the shots
— **call time :** to ask for or grant a time-out
— **call to account :** to hold responsible **:** REPRIMAND
— **call upon 1 :** REQUIRE, OBLIGE ⟨may be *called upon* to do several jobs⟩ **2 :** to make a demand on **:** depend on ⟨universities are *called upon* to produce trained professionals⟩

²call *noun* (14th century)
1 a : an act of calling with the voice **:** SHOUT **b :** an imitation of the cry of a bird or other animal made to attract it **c :** an instrument used for calling ⟨a duck *call*⟩ **d :** the cry of an animal (as a bird)
2 a : a request or command to come or assemble **b :** a summons or signal on a drum, bugle, or pipe **c :** admission to the bar as a barrister **d :** an invitation to become the minister of a church or to accept a professional appointment **e :** a divine vocation or strong inner prompting to a particular course of action **f :** a summoning of actors to rehearsal ⟨the *call* is for 11 o'clock⟩ **g :** the attraction or appeal of a particular activity, condition, or place ⟨the *call* of the wild⟩ **h :** an order specifying the number of men to be inducted into the armed services during a specified period **i :** the selection of a play in football
3 a : DEMAND, CLAIM **b :** NEED, JUSTIFICATION **c :** a demand for payment of money **d :** an option to buy a specified amount of a security (as stock) or commodity (as wheat) at a fixed price at or within a specified time — compare PUT 2 **e :** an instance of asking for something **:** REQUEST ⟨many *calls* for Christmas stories⟩
4 : ROLL CALL
5 : a short usually formal visit
6 : the name or thing called ⟨the *call* was heads⟩
7 : the act of calling in a card game
8 : the act of calling on the telephone
9 : a direction or a succession of directions for a square dance rhythmically called to the dancers
10 : a decision or ruling made by an official of a sports contest; *also* **:** DECISION 1 ⟨a tough *call* to make⟩
11 : a temporary transfer of control of computer processing to a particular set of instructions (as a subroutine or procedure)
— **at call** *or* **on call 1 a :** available for use **:** at the service of ⟨thousands of men *at* his *call*⟩ **b :** ready to respond to a summons or command ⟨a doctor *on call*⟩ **2 :** subject to demand for payment or return without previous notice ⟨money lent *at call*⟩
— **within call :** within hearing or reach of a summons **:** subject to summons

call·able \ˈkȯ-lə-bəl\ *adjective* (1826)
: capable of being called; *specifically* **:** subject to a demand for presentation for payment ⟨*callable* bond⟩

cal·la lily \ˈka-lə-\ *noun* [New Latin, genus name, modification of Greek *kallaia* rooster's wattles] (1870)
: any of several herbs (genus *Zantedeschia*) of the arum family; *especially* **:** a house or greenhouse plant (*Z. aethiopica*) with a white showy spathe and yellow spadix — called also *calla*

cal·la·loo \ˌka-lə-ˈlü, ˈka-lə-ˌ\ *noun* [American Spanish *calalú* plant of the genus *Xanthosoma* whose leaves are used as greens] (1892)
: a soup or stew made with greens, onions, and crabmeat

cal·lant \ˈka-lənt, ˈkä-\ *noun* [Dutch or Old North French; Dutch *kalant* customer, fellow, from Old North French *calland* customer, from Latin *calent-, calens,* present participle of *calēre* to be warm — more at LEE] (circa 1592)
chiefly Scottish **:** BOY, LAD

call·back \ˈkȯl-ˌbak\ *noun* (1926)
1 : a return call
2 a : RECALL 5 **b :** a recall of an employee to work after a layoff **c :** a second or additional audition for a theatrical part

call–board \-ˌbōrd, -ˌbȯrd\ *noun* (1886)
: BULLETIN BOARD 1

call box *noun* (1885)
1 *British* **:** a public telephone booth
2 : a telephone usually located on the side of a road for reporting emergencies (as fires or automobile breakdowns)

call·boy \ˈkȯl-ˌbȯi\ *noun* (1794)
: BELLHOP, PAGE

call down *transitive verb* (1810)
1 : to cause or entreat to descend ⟨*call down* a blessing⟩
2 : REPRIMAND ⟨*called* me *down* for being late⟩

called strike *noun* (1887)
: a pitched baseball not struck at by the batter that passes through the strike zone

¹cal·ler \ˈkä-lər\ *adjective* [Middle English *callour*] (14th century)
1 *Scottish* **:** FRESH
2 *Scottish* **:** COOL

²call·er \ˈkȯ-lər\ *noun* (15th century)
: one that calls

cal·let \ˈka-lət\ *noun* [perhaps from Middle French *caillette* frivolous person, from *Caillette* (flourished 1500) French court fool] (15th century)
chiefly Scottish **:** PROSTITUTE

call girl *noun* (circa 1940)
: a prostitute with whom an appointment may be made by telephone

call house *noun* (1929)
: a house or apartment where call girls may be procured

cal·lig·ra·pher \kə-ˈli-grə-fər\ *noun* (1753)
1 : a professional copyist or engrosser
2 : one who writes a beautiful hand
3 : PENMAN ⟨a fair *calligrapher*⟩

cal·lig·ra·phist \-fist\ *noun* (1816)
: CALLIGRAPHER

cal·lig·ra·phy \-fē\ *noun* [French or Greek; French *calligraphie,* from Greek *kalligraphia,* from *kalli-* beautiful (from *kallos* beauty) + *-graphia* -graphy] (1604)
1 a : artistic, stylized, or elegant handwriting or lettering **b :** the art of producing such writing
2 : PENMANSHIP
3 : an ornamental line in drawing or painting
— **cal·li·graph·ic** \ˌka-lə-ˈgra-fik\ *adjective*
— **cal·li·graph·i·cal·ly** \-fi-k(ə-)lē\ *adverb*

call–in \ˈkȯl-ˌin\ *adjective* (1967)
: allowing listeners to engage in broadcast telephone conversations with the host or a guest ⟨a *call-in* show⟩

call in (1597)
transitive verb
1 : to order to return or to be returned: as **a :** to withdraw from an advanced position ⟨*call in* the outposts⟩ **b :** to withdraw from circulation ⟨*call in* bank notes and issue new ones⟩
2 : to summon to one's aid or for consultation ⟨*call in* a mediator⟩
3 : to deliver (a message) by telephone ⟨*call in* an order for pizza⟩
intransitive verb
: to communicate with a person by telephone
— **call in sick :** to report by telephone that one will be absent because of illness

call·ing \ˈkȯ-liŋ\ *noun* (14th century)
1 : a strong inner impulse toward a particular course of action especially when accompanied by conviction of divine influence
2 : the vocation or profession in which one customarily engages
3 : the characteristic cry of a female cat in heat; *also* **:** the period of heat
synonym see WORK

calling card *noun* (1896)
1 : VISITING CARD
2 : a sign or evidence that someone or something is or has been present; *broadly* **:** an identifying mark

cal·li·ope \kə-ˈlī-ə-(ˌ)pē, *in sense 2 also* ˈka-lē-ˌōp\ *noun* [Latin, from Greek *Kalliopē*]
1 *capitalized* **:** the Greek Muse of heroic poetry
2 : a keyboard musical instrument resembling an organ and consisting of a series of whistles sounded by steam or compressed air

cal·li·per *chiefly British variant of* CALIPER

cal·li·pyg·ian \ˌka-lə-ˈpi-j(ē-)ən\ *or* **cal·li·py·gous** \-ˈpī-gəs\ *adjective* [Greek *kallipygos,* from *kalli-* + *pygē* buttocks] (circa 1800)
: having shapely buttocks

Cal·lis·to \kə-ˈlis-(ˌ)tō\ *noun* [Latin, from Greek *Kallistō*]
: a nymph loved by Zeus, changed into a she-bear by Hera, and subsequently changed into the Great Bear constellation

cal·li·thump \ˈka-lə-ˌthəmp\ *noun* [back-formation from *callithumpian,* adjective, alteration of English dialect *gallithumpian* disturber of order at elections in 18th century] (1856)
: a noisy boisterous band or parade
— **cal·li·thump·ian** \ˌka-lə-ˈthəm-pē-ən\ *adjective*

call letters *noun plural* (1913)
: CALL SIGN

call loan *noun* (1852)
: a loan payable at the discretion of the borrower or on demand of the lender

call number *noun* (1876)

: a combination of characters assigned to a library book to indicate its place on a shelf

call off *transitive verb* (1633)
1 : to draw away **:** DIVERT ⟨her attention was *called off* by a new arrival⟩
2 : CANCEL ⟨*call* the trip *off*⟩

call of nature (1763)
: the need to urinate or defecate

cal·los·i·ty \ka-ˈlä-sə-tē, kə-\ *noun, plural* **-ties** (1578)
1 : the quality or state of being callous: as **a :** marked or abnormal hardness and thickness **b :** lack of feeling or capacity for emotion
2 : CALLUS 1

¹cal·lous \ˈka-ləs\ *adjective* [Middle French *calleux,* from Latin *callosus,* from *callum, callus* callous skin] (15th century)
1 a : being hardened and thickened **b :** having calluses
2 a : feeling no emotion **b :** feeling no sympathy for others
— cal·lous·ly *adverb*
— cal·lous·ness *noun*

²callous *transitive verb* (1834)
: to make callous

call out *transitive verb* (15th century)
1 : to summon into action ⟨*call out* troops⟩
2 : to challenge to a duel
3 : to order on strike ⟨*call out* the workers⟩

cal·low \ˈka-(ˌ)lō\ *adjective* [Middle English *calu* bald, from Old English; akin to Old High German *kalo* bald, Old Church Slavonic *golŭ* bare] (1580)
: lacking adult sophistication **:** IMMATURE ⟨*callow* youth⟩
— cal·low·ness \ˈka-lō-nəs, -lə-nəs\ *noun*

call sign *noun* (1919)
: the combination of identifying letters or letters and numbers assigned to an operator, office, activity, or station for use in communication (as in the address of a message sent by radio)

call slip *noun* (1881)
: a form filled out by a library patron for a desired book

call to quarters (circa 1918)
: a bugle call usually shortly before taps that summons soldiers to their quarters

call–up \ˈkȯ-ˌləp\ *noun* (1940)
: an order to report for military service

call up *transitive verb* (1632)
1 : to bring to mind **:** EVOKE
2 : to summon before an authority
3 : to summon together (as for a united effort) ⟨*call up* all his forces for the attack⟩
4 : to summon for active military duty
5 : to bring forward for consideration or action
6 : to retrieve from the memory of a computer especially for display and user interaction

¹cal·lus \ˈka-ləs\ *noun* [Latin] (1563)
1 : a thickening of or a hard thickened area on skin or bark
2 : a mass of exudate and connective tissue that forms around a break in a bone and is converted into bone in healing
3 : soft tissue that forms over a wounded or cut plant surface

²callus (1864)
intransitive verb
: to form callus
transitive verb
: to cause callus to form on

call–waiting *noun* (1986)
: a telephone service that signals (as by a click) to the user when an incoming call is received during a call in progress

¹calm \ˈkäm, ˈkälm, ˈkam, ˈkȯ(l)m, *New England also* ˈkȧm\ *noun* [Middle English *calme,* from Middle French, from Old Italian *calma,* from Late Latin *cauma* heat, from Greek *kauma,* from *kaiein* to burn] (14th century)
1 a : a period or condition of freedom from storms, high winds, or rough activity of water **b :** complete absence of wind or presence of wind having a speed no greater than one mile (1.6 kilometers) per hour — see BEAUFORT SCALE table
2 : a state of tranquillity

²calm (14th century)
intransitive verb
: to become calm — usually used with *down*
transitive verb
: to make calm — often used with *down*

³calm *adjective* (14th century)
1 : marked by calm **:** STILL ⟨a *calm* sea⟩
2 : free from agitation, excitement, or disturbance ⟨a *calm* manner⟩ ☆
— calm·ly *adverb*
— calm·ness *noun*

calm·ative \ˈkä-mə-tiv, ˈkäl-, *also* ˈkȧ-, ˈkȯ(l)-, ˈka-\ *noun or adjective* [²*calm* + *-ative* (as in *sedative*)] (1870)
: SEDATIVE

cal·mod·u·lin \(ˌ)kal-ˈmä-jə-lin, -dyu̇-lin\ *noun* [*cal*cium + *modul*ate + ¹*-in*] (1979)
: a calcium-binding protein that regulates cellular metabolic processes (as the contraction of muscle fibers) by modifying the activity of calcium-sensitive enzymes

ca·ló \kə-ˈlō\ *noun* [Spanish, argot, speech of Spanish Gypsies, from Romany *kalo* Gypsy, literally, black, from Sanskrit *kāla*] (circa 1948)
: any of several Spanish argots; *especially* **:** an argot used by Chicano youths in cities of the U.S. Southwest

cal·o·mel \ˈka-lə-məl, -ˌmel\ *noun* [probably from (assumed) New Latin *calomelas,* from Greek *kalos* beautiful + *melas* black — more at MELAN-] (1676)
: a white tasteless compound Hg_2Cl_2 used especially as a fungicide and insecticide and occasionally in medicine as a purgative — called also *mercurous chloride*

¹ca·lo·ric \kə-ˈlȯr-ik, -ˈlōr-\ *noun* [French *calorique,* from Latin *calor*] (1792)
1 : a supposed form of matter formerly held responsible for the phenomena of heat and combustion
2 *archaic* **:** HEAT

²caloric *adjective* (circa 1828)
1 : of or relating to heat
2 : of or relating to calorics
— ca·lo·ri·cal·ly \kə-ˈlȯr-i-k(ə-)lē, -ˈlōr-\ *adverb*

cal·o·rie *also* **cal·o·ry** \ˈka-lə-rē, ˈkal-rē\ *noun, plural* **-ries** [French *calorie,* from Latin *calor* heat, from *calēre* to be warm — more at LEE] (1866)
1 a : the amount of heat required at a pressure of one atmosphere to raise the temperature of one gram of water one degree Celsius that is equal to about 4.19 joules — called also *gram calorie, small calorie*; abbreviation *cal* **b :** the amount of heat required to raise the temperature of one kilogram of water one degree Celsius **:** 1000 gram calories or 3.968 Btu — called also *large calorie*; abbreviation *Cal*
2 a : a unit equivalent to the large calorie expressing heat-producing or energy-producing value in food when oxidized in the body **b :** an amount of food having an energy-producing value of one large calorie

cal·o·rif·ic \ˌka-lə-ˈri-fik\ *adjective* [French or Latin; French *calorifique,* from Latin *calorificus,* from *calor*] (1812)
1 : CALORIC
2 : of or relating to heat production

cal·o·rim·e·ter \ˌka-lə-ˈri-mə-tər\ *noun* [International Scientific Vocabulary, from Latin *calor*] (1794)
: any of several apparatuses for measuring quantities of absorbed or evolved heat or for determining specific heats
— ca·lo·ri·met·ric \ˌka-lə-rə-ˈme-trik; kə-ˌlȯr-ə-, -ˌlōr-\ *adjective*
— ca·lo·ri·met·ri·cal·ly \-tri-k(ə-)lē\ *adverb*
— cal·o·rim·e·try \ˌka-lə-ˈri-mə-trē\ *noun*

ca·lotte \kə-ˈlät\ *noun* [French] (1632)
: SKULLCAP 1; *especially* **:** ZUCCHETTO

cal·o·type \ˈka-lə-ˌtīp\ *noun* [Greek *kalos* beautiful + *-type* (as in *daguerreotype*)] (1845)
: a photographic process by which a large number of prints could be produced from a paper negative; *also* **:** a positive print so made

cal·pac *or* **cal·pack** \ˈkal-ˌpak, kal-ˈ\ *noun* [Turkish *kalpak*] (1598)
: a high-crowned cap worn in Turkey, Iran, and neighboring countries

calque \ˈkalk\ *noun* [French, literally, copy, from *calquer* to trace, from Italian *calcare* to trample, trace, from Latin, to trample — more at CAULK] (1937)
: LOAN TRANSLATION

cal·trop \ˈkal-trəp, ˈkȯl-\ *also* **cal·throp** \-thrəp\ *noun* [Middle English *caltrappe,* alteration of *calketrappe* star thistle, from Old English *calcatrippe,* from Medieval Latin *calcatrippa*] (15th century)
1 a *plural but singular or plural in construction* **:** STAR THISTLE 1 **b :** PUNCTURE VINE; *also* **:** any of various related herbs (genera *Tribulus* and *Kallstroemia*) of the same family (Zygophyllaceae, the caltrop family)
2 : a device with four metal points so arranged that when any three are on the ground the fourth projects upward as a hazard to the hooves of horses or to pneumatic tires

cal·u·met \ˈkal-yə-ˌmet, -mət\ *noun* [American French, from French dialect, pipe stem, from Late Latin *calamellus,* diminutive of Latin *calamus* reed — more at CALAMUS] (1673)
: a highly ornamented ceremonial pipe of the American Indians

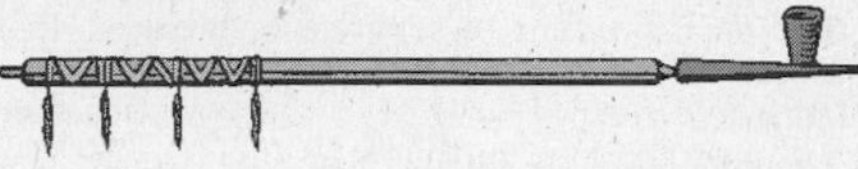
calumet

ca·lum·ni·ate \kə-ˈləm-nē-ˌāt\ *transitive verb* **-at·ed; -at·ing** (1554)
1 : to utter maliciously false statements, charges, or imputations about
2 : to injure the reputation of by calumny
synonym see MALIGN
— ca·lum·ni·a·tion \-ˌləm-nē-ˈā-shən\ *noun*
— ca·lum·ni·a·tor \-ˈləm-nē-ˌā-tər\ *noun*

cal·um·ny \ˈka-ləm-nē *also* ˈkal-yəm-\ *noun, plural* **-nies** [Middle French & Latin; Middle French *calomnie,* from Latin *calumnia,* from *calvi* to deceive; perhaps akin to Old English *hōlian* to slander, Greek *kēlein* to beguile] (15th century)

☆ SYNONYMS
Calm, tranquil, serene, placid, peaceful mean quiet and free from disturbance. CALM often implies a contrast with a foregoing or nearby state of agitation or violence ⟨the protests ended, and the streets were *calm* again⟩. TRANQUIL suggests a very deep quietude or composure ⟨the *tranquil* beauty of a formal garden⟩. SERENE stresses an unclouded and lofty tranquillity ⟨watched the sunset of a *serene* summer's evening⟩. PLACID suggests an undisturbed appearance and often implies a degree of complacency ⟨remained *placid* despite the criticism⟩. PEACEFUL implies a state of repose in contrast with or following strife or turmoil ⟨grown *peaceful* in old age⟩.

1 : a misrepresentation intended to blacken another's reputation
2 : the act of uttering false charges or misrepresentations maliciously calculated to damage another's reputation
— **ca·lum·ni·ous** \kə-ˈləm-nē-əs\ *adjective*
— **ca·lum·ni·ous·ly** *adverb*

cal·u·tron \ˈkal-yə-ˌträn\ *noun* [*Cal*ifornia *U*niversity cyclo*tron*] (1945)
: an electromagnetic apparatus for separating isotopes according to their masses

cal·va·dos \ˌkal-və-ˈdōs, ˌkäl-\ *noun, often capitalized* [French, from *Calvados,* department in Normandy, France] (1906)
: an applejack made in Calvados

cal·var·i·um \kal-ˈvar-ē-əm, -ˈver-\ *noun, plural* **-ia** \-ē-ə\ [New Latin, from Latin *calvaria* skull, from *calvus* bald; probably akin to Sanskrit *kulva* bald] (14th century)
: the portion of a skull including the braincase and excluding the lower jaw or lower jaw and facial portion

cal·va·ry \ˈkal-v(ə-)rē\ *noun, plural* **-ries** [*Calvary,* the hill near Jerusalem where Jesus was crucified] (1738)
1 : an open-air representation of the crucifixion of Jesus
2 : an experience of usually intense mental suffering

Calvary cross *noun* (1826)
: a Latin cross usually mounted on three steps — see CROSS illustration

calve \ˈkav, ˈkäv\ *verb* **calved; calv·ing** [Middle English, from Old English *cealfian,* from *cealf* calf] (before 12th century)
intransitive verb
1 : to give birth to a calf; *also* **:** to produce offspring
2 *of an ice mass* **:** to separate or break so that a part becomes detached
transitive verb
1 : to produce by birth
2 *of an ice mass* **:** to let become detached

calves *plural of* CALF

Cal·vin cycle \ˈkal-vən-\ *noun* [Melvin *Calvin*] (1957)
: the cycle of enzyme-catalyzed dark reactions of photosynthesis that occurs in the chloroplasts of plants and in many bacteria and that involves the fixation of carbon dioxide and the formation of a six-carbon sugar

Cal·vin·ism \ˈkal-və-ˌni-zəm\ *noun* [John *Calvin*] (circa 1570)
: the theological system of Calvin and his followers marked by strong emphasis on the sovereignty of God, the depravity of mankind, and the doctrine of predestination
— **Cal·vin·ist** \-və-nist\ *noun or adjective*
— **Cal·vin·is·tic** \ˌkal-və-ˈnis-tik\ *adjective*
— **Cal·vin·is·ti·cal·ly** \-ti-k(ə-)lē\ *adverb*

calx \ˈkalks\ *noun, plural* **calx·es** *or* **cal·ces** \ˈkal-ˌsēz\ [Middle English *cals,* from Latin *calx* lime — more at CHALK] (15th century)
: the crumbly residue left when a metal or mineral has been subjected to calcination or combustion

¹ca·lyp·so \kə-ˈlip-(ˌ)sō\ *noun* [Latin, from Greek *Kalypsō*]
1 *capitalized* **:** a sea nymph in Homer's *Odyssey* who keeps Odysseus seven years on the island of Ogygia
2 *plural* **calypsos** [New Latin, genus name, probably from Latin] **:** a bulbous bog orchid (*Calypso bulbosa*) of northern regions bearing a single white to purple flower

²calypso *noun, plural* **-sos** *or* **-soes** [origin unknown] (1934)
: a style of music originating in the West Indies, marked by lively duple meter, and having lyrics that are often improvised and usually satirize local personalities and events; *also* **:** a song in this style
— **ca·lyp·so·ni·an** \kə-ˌlip-ˈsō-nē-ən, ˌka-(ˌ)lip-\ *noun or adjective*

ca·lyp·tra \kə-ˈlip-trə\ *noun* [New Latin, from Greek *kalyptra* veil, from *kalyptein* to cover — more at HELL] (circa 1753)
: a hoodlike structure in a plant; *especially* **:** haploid tissue forming a membranous hood over the capsule in a moss

ca·lyx \ˈkā-liks *also* ˈka-\ *noun, plural* **ca·lyx·es** *or* **ca·ly·ces** \ˈkā-lə-ˌsēz *also* ˈka-\ [Latin *calyc-, calyx,* from Greek *kalyx* — more at CHALICE] (1693)
1 : the usually green outer whorl of a flower consisting of sepals
2 : a cuplike animal structure (as the body wall of a crinoid or a division of the kidney pelvis)

cal·zo·ne \kal-ˈzōn, -ˈzō-nē, -ˈzō-nā; käl-ˈzō-nā\ *noun, plural* **calzone** *or* **calzones** [Italian, from *calzone* (singular of *calzoni* pants), augmentative of *calza* stocking, from Medieval Latin *calcea,* from Latin *calceus* shoe, from *calc-, calx* heel] (circa 1950)
: a baked or fried turnover of pizza dough stuffed with various fillings usually including cheese

cam \ˈkam\ *noun* [perhaps from French *came,* from German *Kamm,* literally, comb, from Old High German *kamb*] (1777)
: a rotating or sliding piece in a mechanical linkage used especially in transforming rotary motion into linear motion or vice versa

ca·ma·ra·de·rie \ˌkäm-ˈrä-d(ə-)rē, ˌkam-, ˌkä-mə-, ˌka-, -ˈra-\ *noun* [French, from *camarade* comrade] (1840)
: a spirit of friendly good-fellowship

cam·a·ril·la \ˌka-mə-ˈri-lə, -ˈrē-ə\ *noun* [Spanish, literally, small room] (1839)
: a group of unofficial often secret and scheming advisers; *also* **:** CABAL

cam·as *also* **cam·ass** \ˈka-məs\ *noun* [Chinook Jargon *kamass*] (1805)
: any of a genus (*Camassia* especially *C. quamash*) of plants of the lily family chiefly of the western U.S. with edible bulbs — compare DEATH CAMAS

¹cam·ber \ˈkam-bər\ *verb* **cam·bered; cam·ber·ing** \-b(ə-)riŋ\ [French *cambrer,* from Middle French *cambre* curved, from Latin *camur*] (1627)
intransitive verb
: to curve upward in the middle
transitive verb
1 : to arch slightly
2 : to impart camber to

²camber *noun* (1823)
1 : a slight convexity, arching, or curvature (as of a beam, deck, or road)
2 : the convexity of the curve of an airfoil from the leading edge to the trailing edge
3 : a setting of the wheels of an automotive vehicle closer together at the bottom than at the top

cam·bi·um \ˈkam-bē-əm\ *noun, plural* **-bi·ums** *or* **-bia** \-bē-ə\ [New Latin, from Medieval Latin, exchange, from Latin *cambire* to exchange — more at CHANGE] (1671)
: a thin formative layer between the xylem and phloem of most vascular plants that gives rise to new cells and is responsible for secondary growth
— **cam·bi·al** \-bē-əl\ *adjective*

Cam·bo·di·an \kam-ˈbō-dē-ən\ *noun* (1770)
1 : a native or inhabitant of Cambodia
2 : KHMER 2
— **Cambodian** *adjective*

Cam·bri·an \ˈkam-brē-ən, ˈkām-\ *adjective* [Medieval Latin *Cambria* Wales, from Middle Welsh *Cymry* Wales, Welshmen] (circa 1656)
1 : WELSH
2 : of, relating to, or being the earliest geologic period of the Paleozoic era or the corresponding system of rocks marked by fossils of every great animal type except the vertebrates — see GEOLOGIC TIME table
— **Cambrian** *noun*

cam·bric \ˈkām-brik\ *noun* [obsolete Flemish *Kameryk* Cambrai, city of France] (1530)
1 : a fine thin white linen fabric
2 : a cotton fabric that resembles cambric

cambric tea *noun* (1888)
: a hot drink of water, milk, sugar, and often a small amount of tea

cam·cord·er \ˈkam-ˌkȯr-dər\ *noun* [*cam*era + re*corder*] (1982)
: a small portable combined video camera and videocassette recorder

¹came *past of* COME

²came \ˈkām\ *noun* [origin unknown] (1688)
: a slender grooved lead rod used to hold together panes of glass especially in a stained-glass window

cam·el \ˈka-məl\ *noun* [Middle English, from Old English & Old North French, from Latin *camelus,* from Greek *kamēlos,* of Semitic origin; akin to Hebrew *gāmāl* camel] (before 12th century)
1 : either of two large ruminant mammals (genus *Camelus*) used as draft and saddle animals in desert regions especially of Africa and Asia: **a :** the one-humped camel (*C. dromedarius*) extant only as a domestic or feral animal — called also *dromedary* **b :** the two-humped camel (*C. bactrianus* synonym *C. ferus*) of Chinese Turkestan and Mongolia — called also *Bactrian camel*

camel 1: *1* **dromedary,** *2* **Bactrian camel**

2 : a watertight structure used especially to lift submerged ships
3 : a light yellowish brown

cam·el·back \ˈkam-əl-ˌbak\ *noun* (1860)
: the back of a camel

cam·el·eer \ˌka-mə-ˈlir\ *noun* (1808)
: a camel driver

camel hair *also* **camel's hair** *noun* (14th century)
1 : the hair of the camel or a substitute for it (as hair from squirrels' tails)
2 : cloth made of camel hair or a mixture of camel hair and wool usually light tan and of soft silky texture

ca·mel·lia \kə-ˈmēl-yə\ *noun* [New Latin *Camellia,* from *Camellus* (Georg Josef Kamel (died 1706) Moravian Jesuit missionary)] (circa 1753)
: any of a genus (*Camellia*) of shrubs or trees of the tea family; *especially* **:** an ornamental greenhouse shrub (*C. japonica*) with glossy leaves and roselike flowers

ca·mel·o·pard \kə-ˈme-lə-ˌpärd\ *noun* [Late Latin *camelopardus,* alteration of Latin *camelopardalis,* from Greek *kamēlopardalis,* from *kamēlos* + *pardalis* leopard] (14th century)
1 *archaic* **:** GIRAFFE
2 *capitalized* **:** CAMELOPARDALIS

Ca·mel·o·par·da·lis \kə-ˌme-lə-ˈpär-dᵊl-əs\ *noun* [Latin (genitive *Camelopardalis*), camelopard]
: a northern constellation between Cassiopeia and Ursa Major

Cam·e·lot \ˈka-mə-ˌlät\ *noun*
1 : the site of King Arthur's palace and court
2 : a time, place, or atmosphere of idyllic happiness

Cam·em·bert \ˈka-məm-ˌber\ *noun* [French, from *Camembert,* Normandy, France] (1878)
: a soft surface-ripened cheese with a thin grayish white rind and a yellow interior

cam·eo \ˈka-mē-ˌō\ *noun, plural* **-eos** [Middle English *camew,* from Middle French *camau, kamaheu*] (15th century)
1 a : a gem carved in relief; *especially* **:** a small piece of sculpture on a stone or shell cut

in relief in one layer with another contrasting layer serving as background **b :** a small medallion with a profiled head in relief
2 : a carving or sculpture made in the manner of a cameo
3 : a usually brief literary or filmic piece that brings into delicate or sharp relief the character of a person, place, or event
4 : a small theatrical role usually performed by a well-known actor and often limited to a single scene; *broadly* **:** any brief appearance
— cameo *adjective*
— cameo *transitive verb*

cam•era \'kam-rə, 'ka-mər-ə\ *noun* [Late Latin, room — more at CHAMBER] (1712)
1 : the treasury department of the papal curia
2 a : CAMERA OBSCURA **b :** a device that consists of a lightproof chamber with an aperture fitted with a lens and a shutter through which the image of an object is projected onto a surface for recording (as on film) or for translation into electrical impulses (as for television broadcast)
— on camera : before a live television camera

cam•era lu•ci•da \-'lü-sə-də\ *noun* [New Latin, literally, light chamber] (1831)
: an instrument that by means of a prism or mirrors and often a microscope causes a virtual image of an object to appear as if projected upon a plane surface so that an outline may be traced

cam•era•man \-ˌman, -mən\ *noun* (1908)
: a person who operates a camera (as for motion pictures or television)

cam•era ob•scu•ra \-əb-'skyu̇r-ə\ *noun* [New Latin, literally, dark chamber] (1725)
: a darkened enclosure having an aperture usually provided with a lens through which light from external objects enters to form an image of the objects on the opposite surface

cam•era•per•son \-ˌpər-sᵊn\ *noun* (1976)
: a cameraman or camerawoman

cam•era•wom•an \-ˌwu̇-mən\ *noun* (1971)
: a woman who operates a camera

cam•er•len•go \ˌka-mər-'leŋ-(ˌ)gō\ *noun, plural* **-gos** [Italian *camarlingo*] (1625)
: a cardinal who heads the Apostolic Camera

ca•mion \kȧ-'myōⁿ\ *noun* [French] (1911)
: MOTORTRUCK; *also* **:** BUS

cam•i•sa•do \ˌka-mə-'sā-(ˌ)dō, -'sä-\ *noun, plural* **-does** [probably from obsolete Spanish *camisada*] (1548)
archaic **:** an attack by night

ca•mise \kə-'mēz, -'mēs\ *noun* [Arabic *qamīṣ*, perhaps from Late Latin *camisia*] (1812)
: a light loose long-sleeved shirt, gown, or tunic

cam•i•sole \'ka-mə-ˌsōl\ *noun* [French, from Provençal *camisolla*, diminutive of *camisa* shirt, from Late Latin *camisia*] (1795)
1 : a short negligee jacket for women
2 : a short sleeveless garment for women

cam•let \'kam-lət\ *noun* [Middle English *cameloit*, from Middle French *camelot*, from Arabic *khamlat* woolen plush] (15th century)
1 a : a medieval Asian fabric of camel hair or angora wool **b :** a European fabric of silk and wool **c :** a fine lustrous woolen
2 : a garment made of camlet

camomile *variant of* CHAMOMILE

ca•mor•ra \kə-'mȯr-ə, -'mär-\ *noun* [Italian] (1865)
: a group of persons united for dishonest or dishonorable ends; *especially* **:** a secret organization formed about 1820 at Naples, Italy

ca•mor•ris•ta \ˌkä-mō-'rē-stə\ *noun, plural* **-ti** \-(ˌ)stē\ [Italian, from *camorra*] (1897)
: a member of a camorra

¹cam•ou•flage \'ka-mə-ˌfläzh, -ˌfläj\ *noun* [French, from *camoufler* to disguise] (1917)
1 : the disguising especially of military equipment or installations with paint, nets, or foliage; *also* **:** the disguise so applied
2 a : concealment by means of disguise **b :** behavior or artifice designed to deceive or hide
— cam•ou•flag•ic \ˌka-mə-'flä-zhik, -jik\ *adjective*

²camouflage *verb* **-flaged; -flag•ing** (1917)
transitive verb
: to conceal or disguise by camouflage
intransitive verb
: to practice camouflage
— cam•ou•flage•able \'ka-mə-ˌflä-zhə-bəl, -jə-bəl\ *adjective*

³camouflage *adjective* (1945)
: made in colors or patterns typical of camouflage ⟨a *camouflage* jacket⟩

¹camp \'kamp\ *noun, often attributive* [Middle French, probably from Old North French or Old Provençal, from Latin *campus* plain, field] (1528)
1 a : a place usually away from urban areas where tents or simple buildings (as cabins) are erected for shelter or for temporary residence (as for laborers, prisoners, or vacationers) ⟨migrant labor *camp*⟩ **b :** a group of tents, cabins, or huts ⟨fishing *camps* along the river⟩ **c :** a settlement newly sprung up in a lumbering or mining region **d :** a place usually in the country for recreation or instruction during the summer ⟨goes to *camp* every July⟩ ⟨computer *camp*⟩ ⟨football *camp*⟩
2 a : a body of persons encamped **b** (1) **:** a group of persons; *especially* **:** a group engaged in promoting or defending a theory, doctrine, position, or person (2) **:** an ideological position
3 : military service or life

²camp (1543)
intransitive verb
1 : to make camp or occupy a camp
2 : to live temporarily in a camp or outdoors — often used with *out*
3 : to take up one's quarters **:** LODGE
4 : to take up one's position **:** settle down
transitive verb
: to put into a camp; *also* **:** ACCOMMODATE

³camp *noun* [origin unknown] (circa 1909)
1 : exaggerated effeminate mannerisms exhibited especially by homosexuals
2 : a homosexual displaying camp
3 : something so outrageously artificial, affected, inappropriate, or out-of-date as to be considered amusing
4 : something self-consciously exaggerated or theatrical
— camp•i•ly \'kam-pə-lē\ *adverb*
— camp•i•ness \-pē-nəs\ *noun*
— campy \'kam-pē\ *adjective*

⁴camp *adjective* (1909)
: of, relating to, being, or displaying camp ⟨*camp* send-ups of the songs of the fifties and sixties —John Elsom⟩

⁵camp *intransitive verb* (1925)
: to engage in camp **:** exhibit the qualities of camp ⟨he . . . was *camping*, hands on hips, with a quick eye to notice every man who passed by —R. M. McAlmon⟩

¹cam•paign \(ˌ)kam-'pān\ *noun* [French *campagne*, probably from Italian *campagna* level country, campaign, from Late Latin *campania* level country, from Latin, the level country around Naples] (circa 1656)
1 : a connected series of military operations forming a distinct phase of a war
2 : a connected series of operations designed to bring about a particular result ⟨election *campaign*⟩

²campaign *intransitive verb* (1701)
: to go on, engage in, or conduct a campaign
— cam•paign•er *noun*

cam•pa•ni•le \ˌkam-pə-'nē-lē, ˌkäm-, -(ˌ)lā, *especially of US structures also* -'nē(ə)l\ *noun, plural* **-ni•les** *or* **-ni•li** \-'nē-lē\ [Italian, from *campana* bell, from Late Latin] (1640)
: a usually freestanding bell tower

cam•pa•nol•o•gist \ˌkam-pə-'nä-lə-jist\ *noun* (1857)
: one that practices or is skilled in campanology

cam•pa•nol•o•gy \-jē\ *noun* [New Latin *campanologia*, from Late Latin *campana* + New Latin *-o-* + *-logia* -logy] (circa 1823)
: the art of bell ringing

cam•pan•u•la \kam-'pan-yə-lə\ *noun* [New Latin, diminutive of Late Latin *campana*] (1664)
: any of a genus (*Campanula*) of bellflowers

cam•pan•u•late \-lət, -ˌlāt\ *adjective* [New Latin *campanula* bell-shaped part, diminutive of Late Latin *campana*] (1668)
: shaped like a bell ⟨*campanulate* flower⟩

Camp•bell•ite \'ka-mə-ˌlīt *also* 'kam-bə-\ *noun* [Alexander *Campbell*] (1830)
: DISCIPLE 2 — often taken to be offensive

camp•craft \'kamp-ˌkraft\ *noun* (circa 1893)
: skill and practice in the activities relating to camping

camp•er \'kam-pər\ *noun* (1856)
1 : one that camps
2 : a portable dwelling (as a specially equipped trailer or automotive vehicle) for use during casual travel and camping

cam•pe•si•no \ˌkam-pə-'sē-(ˌ)nō\ *noun, plural* **-nos** [Spanish, from *campo* field, country, from Latin *campus* field] (1898)
: a native of a Latin-American rural area; *especially* **:** a Latin-American Indian farmer or farm laborer

cam•pes•tral \kam-'pes-trəl\ *adjective* [Latin *campestr-*, *campester*, from *campus*] (circa 1750)
: of or relating to fields or open country **:** RURAL

camp•fire \'kamp-ˌfīr\ *noun* (1675)
: a fire built outdoors (as at a camp or a picnic)

Camp Fire girl *noun* [from *Camp Fire Girls, Inc.*, former name of Camp Fire, Inc.] (1912)
: a girl who is a member of a national organization of young people from ages 5 to 18

camp follower *noun* (1810)
1 : a civilian who follows a military unit to attend or exploit military personnel; *specifically* **:** PROSTITUTE
2 : a disciple or follower who is not of the main body of members or adherents; *especially* **:** a politician who joins the party or movement solely for personal gain

camp•ground \'kamp-ˌgrau̇nd\ *noun* (1805)
: the area or place (as a field or grove) used for a camp, for camping, or for a camp meeting

cam•phene \'kam-ˌfēn\ *noun* (circa 1847)
: any of several terpenes related to camphor; *especially* **:** a colorless crystalline terpene $C_{10}H_{16}$ used in insecticides

cam•phor \'kam(p)-fər\ *noun* [Middle English *caumfre*, from Anglo-French, from Medieval Latin *camphora*, from Arabic *kāfūr*, from Malay *kapur*] (14th century)
: a tough gummy volatile aromatic crystalline compound $C_{10}H_{16}O$ obtained especially from the wood and bark of the camphor tree and used as a liniment and mild topical analgesic in medicine, as a plasticizer, and as an insect repellent; *also* **:** any of several similar compounds (as some terpene alcohols and ketones)
— cam•pho•ra•ceous \ˌkam(p)-fə-'rā-shəs\ *adjective*

cam•phor•ate \'kam(p)-fə-ˌrāt\ *transitive verb* **-at•ed; -at•ing** (1641)
: to impregnate or treat with camphor

camphor tree *noun* (1607)
: a large Asian evergreen tree (*Cinnamomum camphora*) of the laurel family grown in warm regions

cam·pi·on \'kam-pē-ən\ *noun* [probably from obsolete *campion* (champion)] (1576)
: any of various plants (genera *Lychnis* and *Silene*) of the pink family

camp meeting *noun* (1803)
: a series of evangelistic meetings usually held outdoors and attended by persons who often camp nearby

cam·po \'kam-(ˌ)pō, 'käm-\ *noun, plural* **campos** [American Spanish, from Spanish, field, from Latin *campus*] (1863)
: a grassland plain in South America with scattered perennial herbs

camp·o·ree \ˌkam-pə-'rē\ *noun* [*camp* + *jamboree*] (1927)
: a gathering of Boy Scouts or Girl Scouts from a given geographic area

camp shirt *noun* (1979)
: a woman's shirt having a notched collar and often patch pockets

camp·site \'kamp-ˌsīt\ *noun* (1910)
: a place suitable for or used as the site of a camp

cam·pus \'kam-pəs\ *noun, often attributive* [Latin, plain] (1774)
1 : the grounds and buildings of a university, college, or school
2 : a university, college, or school viewed as an academic, social, or spiritual entity
3 : grounds that resemble a campus ⟨hospital *campus*⟩ ⟨landscaped corporate *campus*⟩

cam·pylo·bac·ter \ˌkam-pi-lō-'bak-tər, kam-ˌpi-lə-\ *noun* [New Latin, from Greek *kampylos* bent + New Latin *bacterium;* akin to Greek *kampē* bend — more at GAMBIT] (1964)
: any of a genus (*Campylobacter*) of spirally curved motile gram-negative rod-shaped bacteria of which some are pathogenic in domestic animals and humans

cam·py·lot·ro·pous \ˌkam-pi-'lä-trə-pəs\ *adjective* [Greek *kampylos* + International Scientific Vocabulary *-tropous* -tropous] (1835)
: having the ovule curved

cam·shaft \'kam-ˌshaft\ *noun* (circa 1877)
: a shaft to which a cam is fastened or of which a cam forms an integral part

cam wheel *noun* (circa 1853)
: a wheel set or shaped to act as a cam

¹can \kən, 'kan\ *verb, past* **could** \kəd, 'ku̇d\ *pres singular & plural* **can** [Middle English (1st & 3d singular present indicative), from Old English; akin to Old High German *kan* (1st & 3d singular present indicative) know, am able, Old English *cnāwan* to know — more at KNOW] (before 12th century)
transitive verb
1 *obsolete* **:** KNOW, UNDERSTAND
2 *archaic* **:** to be able to do, make, or accomplish
intransitive verb
archaic **:** to have knowledge or skill
verbal auxiliary
1 a : know how to ⟨she *can* read⟩ **b :** be physically or mentally able to ⟨he *can* lift 200 pounds⟩ **c** — used to indicate possibility ⟨do you think he *can* still be alive⟩ ⟨those things *can* happen⟩; sometimes used interchangeably with *may* **d :** be permitted by conscience or feeling to ⟨*can* hardly blame her⟩ **e :** be made possible or probable by circumstances to ⟨he *can* hardly have meant that⟩ **f :** be inherently able or designed to ⟨everything that money *can* buy⟩ **g :** be logically or axiologically able to ⟨2 + 2 *can* also be written 3 + 1⟩ **h :** be enabled by law, agreement, or custom to
2 : have permission to — used interchangeably with *may* ⟨you *can* go now if you like⟩ ■

²can \'kan\ *noun* [Middle English *canne,* from Old English; akin to Old High German *channa*] (before 12th century)
1 : a usually cylindrical receptacle: **a :** a vessel for holding liquids; *specifically* **:** a drinking vessel **b :** a usually metal typically cylindrical receptacle usually with an open top, often with a removable cover, and sometimes with a spout or side handles (as for holding milk or trash) **c :** a container (as of tinplate) in which products (as perishable foods) are hermetically sealed for preservation until use **d :** a jar for packing or preserving fruit or vegetables
2 : JAIL
3 a : TOILET **b :** BATHROOM 1
4 : BUTTOCKS
5 : DESTROYER 2
6 *slang* **:** an ounce of marijuana
— **can·ful** \'kan-ˌfu̇l\ *noun*
— **in the can** *of a film or videotape* **:** completed and ready for release

³can \'kan\ *transitive verb* **canned; canning** (1861)
1 a : to put in a can **:** preserve by sealing in airtight cans or jars **b :** to hit (a golf shot) into the cup **c :** to hit (a shot) in basketball
2 : to discharge from employment
3 *slang* **:** to put a stop or end to
— **can·ner** *noun*

Ca·naan·ite \'kā-nə-ˌnīt\ *noun* [Greek *Kananitēs,* from *Kanaan* Canaan, from Hebrew *Kenaʿan*] (1535)
: a member of a Semitic people inhabiting ancient Palestine and Phoenicia from about 3000 B.C.
— **Canaanite** *adjective*

Can·a·da balsam \'ka-nə-də-\ *noun* [Canada, country in North America] (1811)
: a viscid yellowish to greenish oleoresin exudate of the balsam fir (*Abies balsmea*) that solidifies to a transparent mass and is used as a transparent cement especially in microscopy

Canada Day *noun* (1950)
: July 1 observed as a legal holiday in commemoration of the proclamation of dominion status in 1867

Canada goose *noun* (1731)
: the common wild goose (*Branta canadensis*) of North America that is chiefly gray and brownish with black head and neck and a white patch running from the sides of the head under the throat

Canada goose

Canada thistle *noun* (1799)
: a European thistle (*Cirsium arvense*) with pinkish purple or white flowers naturalized as a weed in North America

Ca·na·di·an \kə-'nā-dē-ən\ *noun* (1568)
: a native or inhabitant of Canada
— **Canadian** *adjective*

Canadian bacon *noun* (circa 1934)
: bacon cut from the loin that has little fat and is cut into round or oblong slices

Canadian football *noun* (1944)
: a game resembling American football that is played on a turfed field between two teams of 12 players each

Canadian French *noun* (1816)
: the language of the French Canadians

Canadian lynx *or* **Canada lynx** *noun* (1840)
: LYNX C

ca·naille \kə-'nī, -'nā(ə)l\ *noun* [French, from Italian *canaglia,* from *cane* dog, from Latin *canis* — more at HOUND] (1661)
1 : RABBLE, RIFFRAFF
2 : PROLETARIAN

¹ca·nal \kə-'nal\ *noun* [Middle English, from Latin *canalis* pipe, channel, from *canna* reed — more at CANE] (15th century)
1 : a tubular anatomical passage or channel **:** DUCT
2 : CHANNEL, WATERCOURSE
3 : an artificial waterway for navigation or for draining or irrigating land
4 : any of various faint narrow lines on the planet Mars seen through telescopes and once thought by some to be canals built by Martians

²canal *transitive verb* **-nalled** *or* **-naled; -nal·ling** *or* **-nal·ing** (1819)
: to construct a canal through or across

can·a·lic·u·lus \ˌka-nə-'li-kyə-ləs\ *noun, plural* **-li** \-ˌlī, -ˌlē\ [Latin, diminutive of *canalis*] (1854)
: a minute canal in a bodily structure
— **can·a·lic·u·lar** \-lər\ *adjective*

can·a·li·za·tion \ˌka-nᵊl-ə-'zā-shən\ *noun* (1844)
1 : an act or instance of canalizing
2 : a system of channels

can·a·lize \'ka-nᵊl-ˌīz\ *verb* **-lized; -liz·ing** (1860)
transitive verb
1 a : to provide with a canal or channel **b :** to make into or similar to a canal
2 : to provide with an outlet; *especially* **:** to direct into preferred channels
intransitive verb
1 : to flow in or into a channel
2 : to establish new channels

can·a·pé \'ka-nə-pē, -ˌpā\ *noun* [French, literally, sofa, from Medieval Latin *canopeum, canapeum* mosquito net — more at CANOPY] (1890)
: an appetizer consisting of a piece of bread or toast or a cracker topped with a savory spread (as caviar or cheese) — compare HORS D'OEUVRE ◆

ca·nard \kə-'närd *also* -'när\ *noun* [French, literally, duck; in sense 1, from Middle French *vendre des canards à moitié* to cheat, literally, to half-sell ducks] (circa 1859)

□ USAGE
can *Can* and *may* are most frequently interchangeable in senses denoting possibility; because the possibility of one's doing something may depend on another's acquiescence, they have also become interchangeable in the sense denoting permission. The use of *can* to ask or grant permission has been common since the 19th century and is well established, although some commentators feel *may* is more appropriate in formal contexts. *May* is relatively rare in negative constructions (*mayn't* is not common); *cannot* and *can't* are usual in such contexts.

◇ WORD HISTORY
canapé A mosquito was called *kōnōps* in ancient Greek, and a bed with curtains for protection against mosquitos was called a *kōnōpion.* Latin *conopeum,* borrowed from Greek, referred to a mosquito net or a bed provided with such nets. In postclassical Latin, the association with insects was gradually lost, and the word, with vowels usually altered to *canopeum* or *canapeum,* named a bed curtain, awning, or veil. It was borrowed into Old French as *conopé* and into Middle English (influenced by the French form) as *canope* or *canape* (hence Modern English *canopy*). After a lapse of four or five centuries, the French word mysteriously reappeared about the mid-17th century in the form *canapé* and now referred not to an awning but rather an article of furniture, a sofa. This sense shift has not been adequately explained and appears to owe nothing to the meaning of the word in the classical languages. In the late 18th century, a piece of toast topped with some savory food was felt to resemble a couch or sofa, and French *canapé* gained a new meaning. English has borrowed the appetizer and its name from the French.

1 : a false or unfounded report or story; *especially* **:** a fabricated report
2 : an airplane with horizontal stabilizing and control surfaces in front of supporting surfaces; *also* **:** a small airfoil in front of the wing of an aircraft that increases the aircraft's stability ◆

ca·nary \kə-ˈner-ē\ *noun, plural* **ca·nar·ies** [Middle French *canarie,* from Old Spanish *canario,* from *Islas Canarias* Canary Islands] (1592)
1 : a lively 16th century court dance
2 : a Canary Islands usually sweet wine similar to Madeira
3 : a small finch (*Serinus canarius*) of the Canary Islands that is usually greenish to yellow and is kept as a cage bird and singer
4 *slang* **:** INFORMER 2

canary seed *noun* (1597)
: seed of a Canary Islands grass (*Phalaris canariensis*) used as food for cage birds

canary yellow *noun* (circa 1865)
: a light to a moderate or vivid yellow

ca·nas·ta \kə-ˈnas-tə\ *noun* [Spanish, literally, basket] (1948)
1 : a form of rummy using two full decks in which players or partnerships try to meld groups of three or more cards of the same rank and score bonuses for 7-card melds
2 : a meld of seven cards of the same rank in canasta

can·can \ˈkan-ˌkan\ *noun* [French] (1848)
: a woman's dance of French origin characterized by high kicking usually while holding up the front of a full ruffled skirt

[1]can·cel \ˈkan(t)-səl\ *verb* **-celed** *or* **-celled; -cel·ing** *or* **-cel·ling** \-s(ə-)liŋ\ [Middle English *cancellen,* from Middle French *canceller,* from Late Latin *cancellare,* from Latin, to make like a lattice, from *cancelli* (plural), diminutive of *cancer* lattice, probably alteration of *carcer* prison] (14th century)
transitive verb
1 a : to destroy the force, effectiveness, or validity of **:** ANNUL ⟨*cancel* a magazine subscription⟩ ⟨a *canceled* check⟩ **b :** to bring to nothingness **:** DESTROY **c :** to match in force or effect **:** OFFSET — often used with *out* ⟨his irritability *canceled* out his natural kindness —Osbert Sitwell⟩ **d :** to call off usually without expectation of conducting or performing at a later time ⟨*cancel* a football game⟩
2 a : to mark or strike out for deletion **b :** OMIT, DELETE
3 a : to remove (a common divisor) from numerator and denominator **b :** to remove (equivalents) on opposite sides of an equation or account
4 : to deface (a postage or revenue stamp) especially with a set of ink lines so as to invalidate for reuse
intransitive verb
: to neutralize each other's strength or effect **:** COUNTERBALANCE
— can·cel·able *or* **can·cel·la·ble** \-s(ə-)lə-bəl\ *adjective*
— can·cel·er *or* **can·cel·ler** \-s(ə-)lər\ *noun*

[2]cancel *noun* (1806)
1 : CANCELLATION
2 a : a deleted part or passage **b** (1) **:** a leaf containing matter to be deleted (2) **:** a new leaf or slip substituted for matter already printed

can·cel·la·tion *also* **can·cel·ation** \ˌkan(t)-sə-ˈlā-shən\ *noun* (1535)
1 : the act or an instance of canceling
2 : a released accommodation
3 : a mark made to cancel something (as a postage stamp)

can·cel·lous \kan-ˈse-ləs, ˈkan(t)-sə-ləs\ *adjective* [New Latin *cancelli* intersecting osseous plates and bars in cancellous bone, from Latin, lattice] (circa 1839)
of bone **:** having a porous structure

can·cer \ˈkan(t)-sər\ *noun* [Middle English, from Latin (genitive *Cancri*), literally, crab; akin to Greek *karkinos* crab, cancer]
1 *capitalized* **a :** a northern zodiacal constellation between Gemini and Leo **b** (1) **:** the 4th sign of the zodiac in astrology — see ZODIAC table (2) **:** one born under the sign of Cancer
2 [Latin, crab, cancer] **a :** a malignant tumor of potentially unlimited growth that expands locally by invasion and systemically by metastasis **b :** an abnormal bodily state marked by such tumors
3 : something evil or malignant that spreads destructively ⟨the *cancer* of hidden resentment —*Irish Digest*⟩
4 a : an enlarged tumorlike plant growth (as that of crown gall) **b :** a plant disease marked by such growths
word history see RANKLE
— can·cer·ous \ˈkan(t)s-rəs, ˈkan(t)-sə-\ *adjective*
— can·cer·ous·ly *adverb*

Can·cer·ian \kan-ˈser-ē-ən, -ˈsir-\ *noun* (1911)
: CANCER 1b(2)

can·de·la \kan-ˈdē-lə, -ˈde-, -ˈdā-; ˈkan-də-lə\ *noun* [Latin, candle] (1949)
: the base unit of luminous intensity in the International System of Units that is equal to the luminous intensity in a given direction of a source which emits monochromatic radiation of frequency 540×10^{12} hertz and has a radiant intensity in that direction of 1/683 watt per unit solid angle — called also *candle*; abbreviation *cd*

can·de·la·bra \ˌkan-də-ˈlä-brə *sometimes* -ˈla-\ *noun* [alteration of Latin *candelabrum,* from *candela*] (1815)
: a branched candlestick or lamp with several lights

can·de·la·brum \-brəm\ *noun, plural* **-bra** \-brə\ *also* **-brums** [Latin] (1811)
: CANDELABRA

candelabra

can·dent \ˈkan-dənt\ *adjective* [Latin *candent-, candens,* present participle of *candēre*] (1577)
: glowing from or as if from great heat

can·des·cence \kan-ˈde-sᵊn(t)s\ *noun* (circa 1864)
: a candescent state **:** glowing whiteness

can·des·cent \-sᵊnt\ *adjective* [Latin *candescent-, candescens, present participle of candescere,* inchoative of *candēre*] (1824)
: glowing or dazzling from or as if from great heat

can·did \ˈkan-dəd\ *adjective* [French & Latin; French *candide,* from Latin *candidus* bright, white, from *candēre* to shine, glow; akin to Welsh *can* white, Sanskrit *candati* it shines] (1630)
1 : WHITE ⟨*candid* flames⟩
2 : free from bias, prejudice, or malice **:** FAIR ⟨a *candid* observer⟩
3 a : marked by honest sincere expression **b :** indicating or suggesting sincere honesty and absence of deception **c :** disposed to criticize severely **:** BLUNT
4 : relating to photography of subjects acting naturally or spontaneously without being posed
synonym see FRANK
— can·did·ly *adverb*
— can·did·ness *noun*

can·di·da \ˈkan-də-də\ *noun* [New Latin, genus name, from Latin, feminine of *candidus*] (1939)
: any of a genus (*Candida*) of parasitic imperfect fungi that resemble yeasts and occur especially in the mouth, vagina, and intestinal tract and that are usually benign but can become pathogenic; *especially* **:** one (*C. albicans*) causing thrush

can·di·da·cy \ˈkan-də-də-sē, ˈka-nə-\ *noun, plural* **-cies** (1864)
: the state of being a candidate

can·di·date \ˈkan-də-ˌdāt, ˈka-nə-, -dət\ *noun* [Latin *candidatus,* from *candidatus* clothed in white, from *candidus* white; from the white toga worn by candidates for office in ancient Rome] (1600)
1 a : one that aspires to or is nominated or qualified for an office, membership, or award **b :** one likely or suited to undergo or be chosen for something specified ⟨a *candidate* for surgery⟩
2 : a student in the process of meeting final requirements for a degree ◆

can·di·da·ture \ˈkan-də-də-ˌchu̇r, ˈka-nə-, -chər\ *noun* (1851)
chiefly British **:** CANDIDACY

candid camera *noun* (1929)
: a camera used to record subjects in a natural, spontaneous, or unposed manner; *also* **:** something likened to a camera used in such a manner
— candid–camera *adjective*

can·di·di·a·sis \ˌkan-də-ˈdī-ə-səs\ *noun, plural* **-a·ses** \-ˌsēz\ [New Latin] (1951)
: infection with or disease caused by a candida — called also *moniliasis*

can·died \ˈkan-dēd\ *adjective* (circa 1606)
1 : encrusted or coated with sugar
2 : baked with sugar or syrup until translucent

[1]can·dle \ˈkan-dᵊl\ *noun* [Middle English *candel,* from Old English, from Latin *candela,* from *candēre*] (before 12th century)
1 : a usually molded or dipped mass of wax or tallow containing a wick that may be burned (as to give light, heat, or scent or for celebration or votive purposes)
2 : something resembling a candle in shape or use ⟨a sulfur *candle* for fumigating⟩

◇ WORD HISTORY

canard In 16th century France, *vendre des canards à moitié,* which means literally "to half-sell ducks," was a colorful way of saying "to cheat." Unfortunately, no one knows just what was meant by this expression; it probably alludes to an incident in some story widely known at the time that has not survived. At any rate, *canard* in the sense of "duck" came to stand for any hoax, especially a made-up report. And French *canard,* in this sense, was borrowed into English.

candidate A person campaigning for public office in ancient Rome traditionally wore a toga that had been whitened with chalk when he greeted voters in the Forum. Whatever the original significance of the whitening might have been, it apparently came to be regarded as a form of electioneering, and according to the historian Livy it was banned by the Senate as early as 432 B.C. As with later curbs on soliciting votes, however, the ban was habitually disregarded, and office seekers wore white togas for some time after the Caesars had replaced the Roman Republic. The Latin word for an office seeker was *candidatus,* from an adjective of the same form meaning "wearing white," itself a derivative of *candidus* "white, bright." In the 16th and 17th centuries, the word *candidatus* was borrowed into French (as *candidat*), German (as *Kandidat*), and English, denoting aspirants either to a political office or to some academic degree.

3 : required effort, expense, or trouble — usually used in the phrase *not worth the candle*
4 : CANDELA

[2]candle *transitive verb* **can·dled; can·dling** \ˈkan(d)-liŋ, ˈkan-dᵊl-iŋ\ (1879)
: to examine by holding between the eye and a light; *especially* **:** to test (eggs) in this way for staleness, blood clots, fertility, and growth
— **can·dler** \ˈkan(d)-lər, ˈkan-dᵊl-ər\ *noun*

can·dle·ber·ry \ˈkan-dᵊl-ˌber-ē\ *noun* (circa 1730)
: a wax myrtle (*Myrica cerifera*); *also* **:** a bayberry (*Myrica pensylvanica*)

can·dle·fish \-ˌfish\ *noun* (1881)
: EULACHON

can·dle·hold·er \-ˌhōl-dər\ *noun* (1932)
: CANDLESTICK

can·dle·light \ˈkan-dᵊl-ˌ(l)īt\ *noun* (before 12th century)
1 a : the light of a candle **b :** a soft artificial light
2 : the time for lighting candles **:** TWILIGHT

can·dle·light·er \-ˌ(l)ī-tər\ *noun* (15th century)
1 : one who lights the candles for a ceremony
2 : a long-handled implement with a taper and a snuffer that is used for the ceremonial lighting and extinguishing of candles

can·dle·lit \-ˌ(l)it\ *or* **can·dle·light·ed** \-ˌ(l)ī-təd\ *adjective* (1917)
: illuminated by candlelight ⟨a *candlelit* dinner⟩

Can·dle·mas \ˈkan-dᵊl-məs\ *noun* [Middle English *candelmasse,* from Old English *candelmæsse,* from *candel* + *mæsse* mass, feast; from the candles blessed and carried in celebration of the feast] (before 12th century)
: February 2 observed as a church festival in commemoration of the presentation of Christ in the temple and the purification of the Virgin Mary

can·dle·nut \-ˌnət\ *noun* (circa 1836)
: the oily seed of a tropical tree (*Aleurites moluccana*) of the spurge family used locally to make candles and commercially as a source of oil; *also* **:** this tree

can·dle·pin \-ˌpin\ *noun* (1901)
1 : a slender bowling pin tapering toward top and bottom
2 *plural but singular in construction* **:** a bowling game using candlepins and a smaller ball than that used in tenpins

can·dle·pow·er \-ˌpau̇(-ə)r\ *noun* (1877)
: luminous intensity expressed in candelas; *also* **:** CANDELA

can·dle·snuff·er \-ˌsnə-fər\ *noun* (1552)
: an implement for snuffing candles that consists of a small hollow cone attached to a handle

can·dle·stick \-ˌstik\ *noun* (before 12th century)
: a holder with a socket for a candle

can·dle·wick \-ˌwik\ *noun* (before 12th century)
1 : the wick of a candle
2 : a soft cotton embroidery yarn; *also* **:** embroidery made with this yarn usually in tufts

can·dle·wood \-ˌwu̇d\ *noun* (1712)
1 : any of several trees or shrubs (as ocotillo) chiefly of resinous character
2 : slivers of resinous wood burned for light

can–do \ˈkan-ˈdü\ *adjective* (1945)
: characterized by eager willingness to accept and meet challenges ⟨a *can-do* attitude⟩

can·dor \ˈkan-dər, -ˌdȯr\ *noun* [French & Latin; French *candeur,* from Latin *candor,* from *candēre* — more at CANDID] (14th century)
1 a : WHITENESS, BRILLIANCE **b** *obsolete* **:** unstained purity
2 : freedom from prejudice or malice **:** FAIRNESS
3 *archaic* **:** KINDLINESS
4 : unreserved, honest, or sincere expression **:** FORTHRIGHTNESS ⟨the *candor* with which he acknowledged a weakness in his own case —Aldous Huxley⟩

can·dour \ˈkan-dər\ *chiefly British variant of* CANDOR

[1]can·dy \ˈkan-dē\ *noun, plural* **candies** [Middle English *sugre candy,* part translation of Middle French *sucre candi,* part translation of Old Italian *zucchero candi,* from *zucchero* sugar + Arabic *qandī* candied, from *qand* cane sugar] (15th century)
1 : crystallized sugar formed by boiling down sugar syrup
2 a : a confection made with sugar and often flavoring and filling **b :** a piece of such confection
— **candy** *adjective*

[2]candy *verb* **can·died; can·dy·ing** (1533)
transitive verb
1 : to encrust in or coat with sugar; *specifically* **:** to cook (as fruit or fruit peel) in a heavy syrup until glazed
2 : to make attractive **:** SWEETEN
3 : to crystallize into sugar
intransitive verb
: to become coated or encrusted with sugar crystals **:** become crystallized into sugar

candy floss *noun* (1951)
1 *British* **:** COTTON CANDY
2 *usually* **candyfloss** *British* **:** something attractive but insubstantial

candy striper *noun* [from the striped uniform worn suggesting the stripes on some sticks of candy] (1963)
: a teenage volunteer worker at a hospital

can·dy·tuft \ˈkan-dē-ˌtəft\ *noun* [*Candy,* alteration of *Candia* Crete, Greek island + English *tuft*] (1664)
: any of a genus (*Iberis*) of plants of the mustard family cultivated for their white, pink, or purple flowers

[1]cane \ˈkān\ *noun* [Middle English, from Middle French, from Old Provençal *cana,* from Latin *canna,* from Greek *kanna,* of Semitic origin; akin to Arabic *qanāh* hollow stick, reed] (14th century)
1 a (1) **:** a hollow or pithy and usually slender and flexible jointed stem (as of a reed) (2) **:** any of various slender woody stems; *especially* **:** an elongated flowering or fruiting stem (as of a rose) usually arising directly from the ground **b :** any of various tall woody grasses or reeds: as (1) **:** any of a genus (*Arundinaria*) of coarse grasses (2) **:** SUGARCANE (3) **:** SORGHUM
2 : cane dressed for use: as **a :** a cane walking stick; *broadly* **:** WALKING STICK **b :** a cane or rod for flogging **c :** RATTAN; *especially* **:** split rattan for wickerwork or basketwork
3 : a tiny glass rod used in decorative glasswork (as in millefiori and paperweights)

[2]cane *transitive verb* **caned; can·ing** (circa 1667)
1 : to beat with a cane ⟨he sat in a professor's chair and *caned* sophomores for blowing spitballs —H. L. Mencken⟩
2 : to weave or furnish with cane ⟨*cane* the seat of a chair⟩

cane·brake \ˈkān-ˌbrāk\ *noun* (1769)
: a thicket of cane

can·er \ˈkā-nər\ *noun* (1868)
: one that canes chairs

ca·nes·cent \kə-ˈne-sᵊnt, ka-\ *adjective* [Latin *canescent-, canescens,* present participle of *canescere,* inchoative of *canēre* to be gray, be white, from *canus* white, hoary — more at HARE] (circa 1828)
: growing white, whitish, or hoary; *especially* **:** having a fine grayish white pubescence ⟨*canescent* leaves⟩

cane sugar *noun* (1841)
: sugar from sugarcane

cane·ware \ˈkān-ˌwar, -ˌwer\ *noun* [from its color] (1878)
: a buff or yellowish stoneware

ca·nic·o·la fever \kə-ˈni-kə-lə-\ *noun* [New Latin *canicola,* from Latin *canis* dog + *-cola* inhabitant — more at HOUND, COLOUS] (1943)
: an acute disease in humans and dogs characterized by gastroenteritis and mild jaundice and caused by a spirochete (*Leptospira canicola*)

ca·nic·u·lar \kə-ˈni-kyə-lər\ *adjective* [Middle English *caniculer* of the star Sirius, from Late Latin *canicularis,* from Latin *Canicula* Sirius, diminutive of *canis*] (12th century)
: of or relating to the dog days

ca·nid \ˈka-nəd, ˈkā-\ *noun* [New Latin *Canidae,* from *Canis,* type genus, from Latin *canis*] (circa 1889)
: any of a family (Canidae) of carnivorous animals that includes the wolves, jackals, foxes, coyote, and the domestic dog

[1]ca·nine \ˈkā-ˌnīn, *British also* ˈka-\ *noun* (15th century)
1 [Middle English, from Latin (*dens*) *caninus* canine tooth] **:** a conical pointed tooth; *especially* **:** one situated between the lateral incisor and the first premolar — see TOOTH illustration
2 : CANID; *also* **:** DOG 1a

[2]canine *adjective* [Latin *caninus,* from *canis* dog — more at HOUND] (1607)
1 : of or resembling that of a dog ⟨*canine* loyalty⟩
2 : of or relating to dogs or to the family (Canidae) including the canids

Ca·nis Ma·jor \ˌkā-nəs-ˈmā-jər, ˌka-\ *noun* [Latin (genitive *Canis Majoris*), literally, greater dog]
: a constellation to the southeast of Orion containing Sirius

Canis Mi·nor \-ˈmī-nər\ *noun* [Latin (genitive *Canis Minoris*), literally, lesser dog]
: a constellation to the east of Orion containing Procyon

can·is·ter *also* **can·nis·ter** \ˈka-nə-stər\ *noun* [Latin *canistrum* basket, from Greek *kanastron* wicker basket, from *kanna* reed — more at CANE] (1711)
1 : an often cylindrical container for holding a usually specified object or substance ⟨a film *canister*⟩
2 : encased shot for close-range artillery fire
3 : a perforated metal box for gas masks with material to adsorb, filter, or detoxify airborne poisons and irritants

[1]can·ker \ˈkaŋ-kər\ *noun* [Middle English, from Old North French *cancre,* from Latin *cancer* crab, cancer] (13th century)
1 a (1) **:** an erosive or spreading sore (2) **:** an area of necrosis in a plant; *also* **:** a plant disease characterized by cankers **b :** any of various disorders of animals marked by chronic inflammatory changes
2 *archaic* **:** a caterpillar destructive to plants
3 *chiefly dialect* **:** RUST 1
4 : a source of corruption or debasement
5 *chiefly dialect* **:** DOG ROSE
— **can·ker·ous** \ˈkaŋ-k(ə-)rəs\ *adjective*

[2]canker *verb* **can·kered; can·ker·ing** \ˈkaŋ-k(ə-)riŋ\ (14th century)
transitive verb
1 *obsolete* **:** to infect with a spreading sore
2 : to corrupt the spirit of
intransitive verb
1 : to become infested with canker
2 : to become corrupted

canker sore *noun* (circa 1909)
: a small painful ulcer especially of the mouth

can·ker·worm \ˈkaŋ-kər-ˌwərm\ *noun* (1530)
: either of two geometrid moths (*Alsophila pometaria* and *Paleacrita vernata*) and especially their larvae which are serious pests of fruit and shade trees

can·na \ˈka-nə\ *noun* [New Latin, genus name, from Latin, reed — more at CANE] (1664)
: any of a genus (*Canna* of the family Cannaceae) of tropical herbs with simple stems, large leaves, and a terminal raceme of irregular flowers

can·na·bi·noid \kə-ˈna-bə-ˌnȯid\ *noun* [Latin *cannabis* + [1]*-in* + [1]*-oid*] (1970)
: any of various chemical constituents (as THC or cannabinol) of cannabis or marijuana

can·na·bi·nol \-ˌnȯl, -ˌnōl\ *noun* [Latin *cannabis* + [1]*-in* + [1]*-ol*] (1896)
: a physiologically inactive crystalline cannabinoid $C_{21}H_{26}O_2$

can·na·bis \ˈka-nə-bəs\ *noun* [Latin, hemp, from Greek *kannabis;* akin to Old English *hænep* hemp] (1783)
1 : HEMP 1a
2 : any of the preparations (as marijuana or hashish) or chemicals (as THC) that are derived from the hemp and are psychoactive

canned \ˈkand\ *adjective* (1904)
1 a : prepared or recorded in advance; *especially* **:** prepared in standardized form for nonspecific use or wide distribution ⟨*canned* laughter⟩ ⟨*canned* music⟩ ⟨*canned* speeches⟩ **b :** lacking originality or individuality as if mass-produced ⟨*canned* sales pitch⟩
2 *slang* **:** DRUNK 1a

can·nel coal \ˈka-nᵊl-\ *noun* [probably from English dialect *cannel* candle, from Middle English *candel*] (1610)
: a bituminous coal containing much volatile matter that burns brightly

can·nel·lo·ni \ˌka-nə-ˈlō-nē\ *noun plural but singular or plural in construction* [Italian, plural of *cannellone,* augmentative of *cannello* segment of cane stalk, from *canna*] (1906)
: boiled tube-shaped or rolled pasta filled with a meat, fish, cheese, or vegetable mixture and baked in a sauce

can·nery \ˈka-nə-rē\ *noun, plural* **-ner·ies** (1870)
: a factory for the canning of foods

can·ni·bal \ˈka-nə-bəl\ *noun* [New Latin *Canibalis* Carib, from Spanish *Caníbal,* from Taino *Caniba,* of Cariban origin; akin to Carib *kariʔna* Carib, person] (1553)
: one that eats the flesh of its own kind ◆

can·ni·bal·ise *British variant of* CANNIBALIZE

can·ni·bal·ism \ˈka-nə-bə-ˌli-zəm\ *noun* (1796)
1 : the usually ritualistic eating of human flesh by a human being
2 : the eating of the flesh of an animal by another animal of the same kind
3 : an act of cannibalizing something
— can·ni·bal·is·tic \ˌka-nə-bə-ˈlis-tik\ *adjective*

can·ni·bal·ize \ˈka-nə-bə-ˌlīz\ *verb* **-ized; -iz·ing** (1943)
transitive verb
1 a : to take salvageable parts from (as a disabled machine) for use in building or repairing another machine **b :** to make use of (a part taken from one thing) in building or repairing something else
2 : to deprive of an essential part or element in creating or sustaining another facility or enterprise ⟨the energy system has begun *cannibalizing* the economic system it is supposed to fuel —Barry Commoner⟩
3 : to use or draw on material of (as another writer or an earlier work) ⟨a volume . . . that not only *cannibalizes* previous publications but is intended itself to be *cannibalized* —R. M. Adams⟩
4 : to take (sales) away from an existing product by selling or being sold as a similar but new product usually from the same manufacturer; *also* **:** to affect (as an existing product) adversely by cannibalizing sales
intransitive verb
1 : to practice cannibalism
2 : to cannibalize one unit for the sake of another of the same kind
— can·ni·bal·i·za·tion \ˌka-nə-bə-lə-ˈzā-shən\ *noun*

can·ni·kin \ˈka-ni-kən\ *noun* [probably from obsolete Dutch *kanneken,* from Middle Dutch *canneken,* diminutive of *canne* can; akin to Old English *canne* can] (1570)
: a small can or drinking vessel

can·no·li \kə-ˈnō-lē, ka-\ *noun plural but singular or plural in construction* [Italian, plural of *cannolo* small tube, diminutive of *canna*] (1943)
: a deep-fried tube of pastry filled with sweetened and flavored ricotta cheese

[1]**can·non** \ˈka-nən\ *noun, plural* **cannons** *or* **cannon** [Middle French *canon,* from Italian *cannone,* literally, large tube, augmentative of *canna* reed, tube, from Latin, cane, reed — more at CANE] (15th century)
1 *plural usually cannon* **a :** a large heavy gun usually mounted on a carriage **b :** a heavy-caliber automatic aircraft gun firing explosive shells
2 *or* **can·on :** the projecting part of a bell by which it is hung **:** EAR
3 : the part of the leg in which the cannon bone is found

[2]**cannon** (1691)
intransitive verb
: to discharge cannon
transitive verb
: CANNONADE

[1]**can·non·ade** \ˌka-nə-ˈnād\ *noun* (1562)
1 : a heavy fire of artillery
2 : an attack (as with words) likened to artillery fire **:** BOMBARDMENT

[2]**cannonade** *verb* **-ad·ed; -ad·ing** (circa 1670)
transitive verb
: to attack with or as if with artillery
intransitive verb
: to deliver artillery fire

[1]**can·non·ball** \ˈka-nən-ˌbȯl\ *noun* (1663)
1 : a usually round solid missile made for firing from a cannon
2 : a jump into water made with the arms holding the knees tight against the chest
3 : a hard flat tennis service
4 : an express train

[2]**cannonball** *intransitive verb* (1951)
: to travel with great speed

cannon bone *noun* [French *canon,* literally, cannon] (1834)
: a bone in hoofed mammals that extends from the knee or hock to the fetlock

can·non·eer \ˌka-nə-ˈnir\ *noun* (1562)
: an artillery gunner

cannon fodder *noun* (circa 1891)
: soldiers regarded or treated as expendable in battle

can·non·ry \ˈka-nən-rē\ *noun, plural* **-ries** (circa 1840)
: a battery of cannons or cannon fire

can·not \ˈka-(ˌ)nät; kə-ˈnät, ka-\ (15th century)
: can not
— cannot but *or* **cannot help but** *also* **cannot help :** to be unable to do otherwise than

can·nu·la \ˈkan-yə-lə\ *noun, plural* **-las** *or* **-lae** \-ˌlē, -ˌlī\ [New Latin, from Latin, diminutive of *canna* reed — more at CANE] (1684)
: a small tube for insertion into a body cavity or into a duct or vessel

can·nu·lar \ˈkan-yə-lər\ *adjective* (1823)
: TUBULAR

[1]**can·ny** \ˈka-nē\ *adjective* **can·ni·er; -est** [[1]*can*] (1596)
1 : CLEVER, SHREWD; *also* **:** PRUDENT
2 *chiefly Scottish* **a :** CAREFUL, STEADY; *also* **:** RESTRAINED **b :** QUIET, SNUG ⟨then *canny,* in some cozy place, they close the day —Robert Burns⟩
— can·ni·ly \ˈka-nᵊl-ē\ *adverb*
— can·ni·ness \ˈka-nē-nəs\ *noun*

[2]**canny** *adverb* (circa 1796)
Scottish **:** in a canny manner **:** CAREFULLY

[1]**ca·noe** \kə-ˈnü\ *noun* [French, from New Latin *canoa,* from Spanish, from Arawakan, of Cariban origin; akin to Carib *kana:wa* canoe] (1555)
: a light narrow boat with both ends sharp that is usually propelled by paddling

[2]**canoe** *verb* **ca·noed; ca·noe·ing** (1794)
transitive verb
: to transport in a canoe; *also* **:** to travel by canoe down (a river)
intransitive verb
: to go or travel in a canoe
— ca·noe·able \-ə-bəl\ *adjective*
— ca·noe·ist *noun*

can of worms (1962)
: PANDORA'S BOX

ca·no·la \kə-ˈnō-lə\ *noun* [*Can*ada *o*il—*l*ow *a*cid] (1979)
1 : a rape plant of an improved variety having seeds that are low in erucic acid and are the source of canola oil
2 : CANOLA OIL

canola oil *noun* (1986)
: an edible vegetable oil obtained from the seeds of canola that is high in monounsaturated fatty acids

[1]**can·on** \ˈka-nən\ *noun* [Middle English, from Old English, from Late Latin, from Latin, ruler, rule, model, standard, from Greek *kanōn*] (before 12th century)
1 a : a regulation or dogma decreed by a church council **b :** a provision of canon law
2 [Middle English, probably from Old French, from Late Latin, from Latin, model] **:** the most solemn and unvarying part of the Mass including the consecration of the bread and wine
3 [Middle English, from Late Latin, from Latin, standard] **a :** an authoritative list of books accepted as Holy Scripture **b :** the authentic works of a writer **c :** a sanctioned or accepted group or body of related works ⟨the *canon* of great literature⟩
4 a : an accepted principle or rule **b :** a criterion or standard of judgment **c :** a body of principles, rules, standards, or norms
5 [Late Greek *kanōn,* from Greek, model] **:** a contrapuntal musical composition in two or more voice parts in which the melody is imitated exactly and completely by the successive voices though not always at the same pitch
synonym see LAW

[2]**canon** *noun* [Middle English *canoun,* from Anglo-French *canunie,* from Late Latin *canonicus* one living under a rule, from Latin, according to rule, from Greek *kanonikos,* from *kanōn*] (13th century)

◇ WORD HISTORY

cannibal On Christopher Columbus's first voyage to the New World, the mainly Arawakan-speaking peoples whom he encountered in Cuba and Hispaniola told him about a people living to their east, who periodically raided them and whom they greatly feared. In his log Columbus recorded a number of phonetically similar names for this people, including *caníbales* and *caribes.* The Spanish court historian Petrus Martyr wrote a Latin account of Columbus's discoveries, first printed in 1516, that used these two words and widely disseminated them throughout Europe. In Petrus Martyr's words, "the inhabitants of these lands assert that the *Canibales* or *Caribes* are eaters of human flesh." Subsequently, the meaning of the two words diverged. *Caribes* was applied to the Carib-speaking peoples of the Lesser Antilles and northern South America who were so feared by the Arawakans; it is also ultimately the base of the word *Caribbean. Canibales,* on the other hand, passed into English as *cannibal* and became a generic word for any creature that eats the flesh of its own kind.

1 : a clergyman belonging to the chapter or the staff of a cathedral or collegiate church
2 : CANON REGULAR
ca·ñon \'kan-yən\ *variant of* CANYON
can·on·ess \'ka-nə-nəs\ *noun* (1682)
1 : a woman living in community under a religious rule but not under a perpetual vow
2 : a member of a Roman Catholic congregation of women corresponding to canons regular
ca·non·ic \kə-'nä-nik\ *adjective* (15th century)
1 : CANONICAL
2 : of or relating to musical canon
ca·non·i·cal \-ni-kəl\ *adjective* (15th century)
1 : of, relating to, or forming a canon
2 : conforming to a general rule or acceptable procedure : ORTHODOX
3 : of or relating to a clergyman who is a canon
4 : reduced to the canonical form ⟨a *canonical* matrix⟩
— **ca·non·i·cal·ly** \-k(ə-)lē\ *adverb*
canonical form *noun* (1851)
: the simplest form of something; *specifically* : the form of a square matrix that has zero elements everywhere except along the principal diagonal
canonical hour *noun* (15th century)
1 : a time of day canonically appointed for an office of devotion
2 : one of the daily offices of devotion that compose the Divine Office and include matins with lauds, prime, terce, sext, none, vespers, and compline
ca·non·i·cals \kə-'nä-ni-kəlz\ *noun plural* (1748)
: the vestments prescribed by canon for an officiating clergyman
can·on·ic·i·ty \,ka-nə-'ni-sə-tē\ *noun* (1797)
: the quality or state of being canonical
can·on·ist \'ka-nə-nist\ *noun* (1542)
: a specialist in canon law
can·on·ize \'ka-nə-,nīz\ *transitive verb* **can·on·ized** \-,nīzd; *in "Hamlet" usually* kə-'nä-,nīzd\; **can·on·iz·ing** [Middle English, from Medieval Latin *canonizare,* from Late Latin *canon* catalog of saints, from Latin, standard] (14th century)
1 : to declare (a deceased person) an officially recognized saint
2 : to make canonical
3 : to sanction by ecclesiastical authority
4 : to attribute authoritative sanction or approval to
5 : to treat as illustrious, preeminent, or sacred
— **can·on·i·za·tion** \,ka-nə-nə-'zā-shən\ *noun*
canon law *noun* (1552)
: the usually codified law governing a church
canon lawyer *noun* (circa 1859)
: CANONIST
canon regular *noun, plural* **canons regular** (14th century)
: a member of one of several Roman Catholic religious institutes of regular priests living in community under a usually Augustinian rule
can·on·ry \'ka-nən-rē\ *noun, plural* **-ries** (15th century)
: the office of a canon; *also* : the endowment that financially supports a canon
ca·no·pic jar \kə-'nō-pik-, -'nä-\ *noun* [*Canopus,* Egypt] (1893)
: a jar in which the ancient Egyptians preserved the viscera of a deceased person usually for burial with the mummy

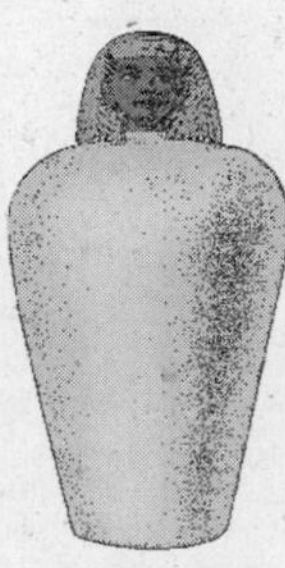
canopic jar

Ca·no·pus \kə-'nō-pəs\ *noun* [Latin, from Greek *Kanōpos*]
: a star of the first magnitude in the constellation Carina not visible north of 37° latitude
¹can·o·py \'ka-nə-pē\ *noun, plural* **-pies** [Middle English *canope,* from Medieval Latin *canopeum* mosquito net, from Latin *conopeum,* from Greek *kōnōpion,* from *kōnōps* mosquito] (14th century)
1 a : a cloth covering suspended over a bed **b** : a cover (as of cloth) fixed or carried above a person of high rank or a sacred object : BALDACHIN **c** : a protective covering: as (1) : the uppermost spreading branchy layer of a forest (2) : AWNING, MARQUEE
2 : an ornamental rooflike structure
3 a : the transparent enclosure over an airplane cockpit **b** : the fabric part of a parachute that catches the air
word history see CANAPÉ
²canopy *transitive verb* **-pied; -py·ing** (1599)
: to cover with or as if with a canopy
ca·no·rous \kə-'nōr-əs, -'nȯr-; 'ka-nə-rəs\ *adjective* [Latin *canorus,* from *canor* melody, from *canere* to sing — more at CHANT] (1646)
: pleasant sounding : MELODIOUS
— **ca·no·rous·ly** *adverb*
— **ca·no·rous·ness** *noun*
canst \kən(t)st, 'kan(t)st\ *archaic present 2d singular of* CAN
¹cant \'kant\ *adjective* [Middle English, probably from (assumed) Middle Low German *kant*] (14th century)
dialect English : LIVELY, LUSTY
²cant (circa 1543)
transitive verb
1 : to give a cant or oblique edge to : BEVEL
2 : to set at an angle : TILT
3 *chiefly British* : to throw with a lurch
intransitive verb
1 : to pitch to one side : LEAN
2 : SLOPE
³cant *noun* [Middle English *cante* edge, probably from Middle Dutch or Old North French; Middle Dutch, edge, corner, from Old North French, from Latin *canthus, cantus* iron tire, perhaps of Celtic origin; akin to Welsh *cant* rim; perhaps akin to Greek *kanthos* corner of the eye] (1603)
1 *obsolete* : CORNER, NICHE
2 : an external angle (as of a building)
3 : a log with one or more squared sides
4 a : an oblique or slanting surface **b** : INCLINATION, SLOPE
⁴cant *adjective* (1663)
1 : having canted corners or sides
2 : INCLINED 2
⁵cant *intransitive verb* [probably from Old North French *canter* to tell, literally, to sing, from Latin *cantare* — more at CHANT] (1567)
1 : to talk or beg in a whining or singsong manner
2 : to speak in cant or jargon
3 : to talk hypocritically
⁶cant *noun* (1640)
1 : affected singsong or whining speech
2 a : the private language of the underworld **b** *obsolete* : the phraseology peculiar to a religious class or sect **c** : JARGON 2
3 : a set or stock phrase
4 : the expression or repetition of conventional or trite opinions or sentiments; *especially* : the insincere use of pious words
can't \'kant, 'kȧnt, 'känt, *especially Southern* 'kānt\ (circa 1652)
: can not
Can·tab \'kan-,tab\ *noun* [by shortening] (1750)
: CANTABRIGIAN
can·ta·bi·le \kän-'tä-bi-,lā, -lē\ *adverb or adjective* [Italian, from Late Latin *cantabilis* worthy to be sung, from Latin *cantare*] (circa 1724)
: in a singing manner — often used as a direction in music
Can·ta·brig·i·an \,kan-tə-'bri-j(ē-)ən\ *noun* [Medieval Latin *Cantabrigia* Cambridge] (circa 1540)
1 : a student or graduate of Cambridge University
2 : a native or resident of Cambridge, Mass.
— **Cantabrigian** *adjective*
can·ta·la \kan-'tä-lə\ *noun* [New Latin, specific epithet of *Agave cantala,* perhaps from Sanskrit *kaṇṭala* babul, from *kaṇṭa* thorn] (1911)
: a hard fiber produced from the leaves of an agave (*Agave cantala*)
can·ta·loupe *also* **can·ta·loup** \'kan-tᵊl-,ōp *also* -,üp\ *noun* [*Cantalupo,* former papal villa near Rome, Italy] (1739)
1 : a muskmelon (*Cucumis melo reticulatus*) having a rind with netted tracery and reddish orange flesh
2 : any of several muskmelons resembling the cantaloupe; *broadly* : MUSKMELON
can·tan·ker·ous \kan-'taŋ-k(ə-)rəs, kən-\ *adjective* [perhaps irregular from obsolete *contack* (contention)] (1772)
: difficult or irritating to deal with
— **can·tan·ker·ous·ly** *adverb*
— **can·tan·ker·ous·ness** *noun*
can·ta·ta \kən-'tä-tə\ *noun* [Italian, from Latin, from feminine of *cantatus,* past participle of *cantare*] (1724)
: a composition for one or more voices usually comprising solos, duets, recitatives, and choruses and sung to an instrumental accompaniment
can·ta·trice \,kän-tə-'trē-(,)chā, ,känⁿ-tə-'trēs\ *noun, plural* **-trices** \-'trē-(,)chāz, -'trēs, -'trē-səz\ *or* **-tri·ci** \,kän-tə-'trē-(,)chē\ [Italian & French, from Italian, from Late Latin *cantatric-, cantatrix,* feminine of Latin *cantator* singer, from *cantare*] (1803)
: a woman who is a singer; *especially* : an opera singer
cant dog *noun* [³*cant*] (1850)
: PEAVEY
can·teen \kan-'tēn\ *noun* [French *cantine* bottle case, sutler's shop, from Italian *cantina* wine cellar, probably from *canto* corner, from Latin *canthus* iron tire — more at CANT] (1737)
1 a : a portable chest with compartments for carrying bottles or for cooking and eating utensils **b** : a flask for carrying liquids (as on a hike) **c** : MESS KIT **d** *British* : a chest for storing flatware
2 a : a bar at a military post or camp **b** : a general store at a military post : EXCHANGE **c** : an establishment that serves as an informal social club (as for soldiers or a community's teenagers) **d** : a small cafeteria or snack bar
¹cant·er \'kan-tər\ *noun* (1609)
: one that uses cant: as **a** : BEGGAR, VAGABOND **b** : a user of professional or religious cant
²can·ter \'kan-tər\ *verb* [short for obsolete *canterbury,* noun (canter), from *Canterbury,* England; from the supposed gait of pilgrims riding to Canterbury] (1706)
intransitive verb
1 : to move at or as if at a canter : LOPE
2 : to ride a horse at a canter ◆

◇ WORD HISTORY
canter In the Middle Ages, the town of Canterbury in southwest England became a popular pilgrimage site after the canonization of Thomas à Becket, who was murdered in Canterbury cathedral in 1170. A classic of medieval literature, Geoffrey Chaucer's *Canterbury Tales,* used a pilgrimage to the town as the frame for a set of stories told by a group of fictional pilgrims in the course of their journey. In the 17th century, the word *canterbury* was applied to a horse's gait that is slower and smoother than a gallop, allegedly because pilgrims traveling to Canterbury on horseback rode at that gait, though by this date the era of such pilgrimages had long passed and was largely remembered through Chaucer. In reference to a horse's gait, *canterbury* came to be used as a verb and was then shortened to *canter.*

transitive verb
: to cause to go at a canter

[3]**can·ter** *noun* (1755)
1 : a 3-beat gait resembling but smoother and slower than the gallop
2 : a ride at a canter

Can·ter·bury bell \'kan-tə(r)-ˌber-ē-, -ˌb(ə-)rē-\ *noun* [*Canterbury,* England] (1578)
: any of several bellflowers (as *Campanula medium*) cultivated for their showy flowers

can·thar·i·din \kan-'thar-ə-dᵊn, -'ther-\ *noun* (1819)
: a bitter crystalline compound $C_{10}H_{12}O_4$ that is the active blister-producing ingredient of cantharides

can·tha·ris \'kan(t)-thə-rəs\ *noun, plural* **can·thar·i·des** \kan-'thar-ə-ˌdēz\ [Middle English & Latin; Middle English *cantharide,* from Latin *cantharid-, cantharis,* from Greek *kantharid-, kantharis*] (14th century)
1 : SPANISH FLY 1
2 *plural but singular or plural in construction* **:** a preparation of dried beetles (as Spanish flies) used in medicine as a counterirritant and formerly as an aphrodisiac

can·tha·xan·thin \ˌkan(t)-thə-'zan-ˌthin\ *noun* [International Scientific Vocabulary *cantha-* (from New Latin *Cantharellus cinnabarinus,* mushroom species from which it was obtained) + *xanth-* + [1]*-in*] (circa 1951)
: a carotenoid $C_{40}H_{52}O_2$ used especially as a color additive for food

cant hook *noun* [[3]*cant*] (circa 1848)
: a lumberman's lever that has a pivoting hooked arm and a blunt often toothed metal cap at one end — compare PEAVEY

can·thus \'kan(t)-thəs\ *noun, plural* **can·thi** \'kan-ˌthī, -ˌthē\ [Late Latin, from Greek *kanthos* — more at CANT] (1646)
: either of the angles formed by the meeting of an eye's upper and lower eyelids

can·ti·cle \'kan-ti-kəl\ *noun* [Middle English, from Latin *canticulum,* diminutive of *canticum* song, from *cantus,* past participle of *canere*] (13th century)
: SONG; *specifically* **:** one of several liturgical songs (as the Magnificat) taken from the Bible

Canticle of Canticles (circa 1934)
: SONG OF SOLOMON

Canticles *noun plural but singular in construction* (15th century)
: SONG OF SOLOMON

can·ti·le·na \ˌkan-tə-'lā-nə, -'lē-\ *noun* [Italian, from Latin, song, from *cantus*] (circa 1740)
: a vocal or instrumental passage of sustained lyricism

[1]**can·ti·le·ver** \'kan-tə-ˌlē-vər, -ˌle-\ *noun* [perhaps from [3]*cant* + *-i-* + *lever*] (1667)
: a projecting beam or member supported at only one end: as **a :** a bracket-shaped member supporting a balcony or a cornice **b :** either of the two beams or trusses that project from piers toward each other and that when joined directly or by a suspended connecting member form a span of a cantilever bridge — see BRIDGE illustration

[2]**cantilever** (1902)
transitive verb
1 : to support by a cantilever ⟨a *cantilevered* shelf⟩
2 : to build as a cantilever
intransitive verb
: to project as a cantilever

can·til·late \'kan-tᵊl-ˌāt\ *transitive verb* **-lat·ed; -lat·ing** [Latin *cantilatus,* past participle of *cantilare* to sing, perhaps from *cantilena*] (circa 1828)
: to recite with musical tones
— **can·til·la·tion** \ˌkan-tᵊl-'ā-shən\ *noun*

can·ti·na \kan-'tē-nə\ *noun* [American Spanish, from Spanish, canteen, from Italian, wine cellar — more at CANTEEN] (1844)
1 *Southwest* **:** a pouch or bag at the pommel of a saddle
2 *Southwest* **:** a small barroom **:** SALOON

cant·ing \'kan-tiŋ\ *adjective* [[5]*cant*] (1663)
: affectedly pious or righteous

can·tle \'kan-tᵊl\ *noun* [Middle English *cantel,* from Old North French, diminutive of *can* edge, corner — more at CANT] (14th century)
1 : a segment cut off or out of something **:** PART, PORTION
2 : the upward projecting rear part of a saddle

can·to \'kan-(ˌ)tō\ *noun, plural* **cantos** [Italian, from Latin *cantus* song, from *canere* to sing — more at CHANT] (1590)
: one of the major divisions of a long poem

[1]**can·ton** \'kan-tᵊn, -ˌtän\ *noun* [Middle French, from Italian *cantone,* from *canto* corner, from Latin *canthus* iron tire — more at CANT] (1522)
: a small territorial division of a country: as **a :** one of the states of the Swiss confederation **b :** a division of a French arrondissement

[2]**canton** *noun* [Middle French, from Old Provençal, from *cant* edge, corner, from Latin *canthus*] (1572)
1 *obsolete* **:** DIVISION, SECTION
2 : the top inner quarter of a flag
3 : the dexter chief region of a heraldic field
— **can·ton·al** \'kan-tᵊn-əl, kan-'tä-nᵊl\ *adjective*

Can·ton·ese \ˌkan-tᵊn-'ēz, -'ēs\ *noun, plural* **Cantonese** (1857)
1 : a native or inhabitant of Canton, China
2 : the dialect of Chinese spoken in and around Canton
3 : a style of Chinese cooking that is typically based on mild spices
— **Cantonese** *adjective*

can·ton flannel \'kan-ˌtän-, -tᵊn-\ *noun, often C capitalized* [*Canton,* China] (circa 1879)
: FLANNEL 1c

can·ton·ment \kan-'tōn-mənt, -'tän- *also* -'tün-\ *noun* (1756)
1 : usually temporary quarters for troops
2 : a permanent military station in India

Can·ton ware \'kan-ˌtän-\ *noun* (circa 1902)
: ceramic ware exported from China especially during the 18th and 19th centuries by way of Canton and including blue-and-white and enameled porcelain and various ornamented stonewares

can·tor \'kan-tər\ *noun* [Latin, singer, from *canere* to sing] (1538)
1 : a choir leader **:** PRECENTOR
2 : a synagogue official who sings or chants liturgical music and leads the congregation in prayer
— **can·to·ri·al** \kan-'tōr-ē-əl, -'tȯr-\ *adjective*

can·trip \'kan-trəp\ *noun* [probably alteration of *caltrop*] (1719)
1 *chiefly Scottish* **:** a witch's trick **:** SPELL
2 *chiefly British* **:** HOCUS-POCUS 2

can·tus \'kan-təs\ *noun, plural* **can·tus** \'kan-təs, 'kan-ˌtüs\ (1590)
1 : CANTUS FIRMUS
2 : the principal melody or voice

cantus fir·mus \-'fir-məs, -'fər-\ *noun* [Medieval Latin, literally, fixed song] (1847)
1 : the plainsong or simple Gregorian melody originally sung in unison and prescribed as to form and use by ecclesiastical tradition
2 : a melodic theme or subject; *especially* **:** one for contrapuntal treatment

canty \'kan-tē\ *adjective* [[1]*cant*] (1720)
dialect British **:** CHEERFUL, SPRIGHTLY

Ca·nuck \kə-'nək *sometimes* -'nu̇k\ *noun* [origin unknown] (1835)
: a Canadian and especially a French Canadian

[1]**can·vas** *also* **can·vass** \'kan-vəs\ *noun* [Middle English *canevas,* from Old North French, from (assumed) Vulgar Latin *cannabaceus* hempen, from Latin *cannabis* hemp — more at CANNABIS] (13th century)
1 : a firm closely woven cloth usually of linen, hemp, or cotton used for clothing and formerly much used for tents and sails
2 : a set of sails **:** SAIL
3 : a piece of canvas used for a particular purpose
4 : TENT; *also* **:** a group of tents
5 a : a piece of cloth backed or framed as a surface for a painting; *also* **:** the painting on such a surface **b :** the background, setting, or scope of an historical or fictional account or narrative
6 : a coarse cloth so woven as to form regular meshes for working with the needle
7 : the canvas-covered floor of a boxing or wrestling ring
— **can·vas·like** \-vəs-ˌlīk\ *adjective*

[2]**canvas** *transitive verb* **-vased** *or* **-vassed; -vas·ing** *or* **-vass·ing** (1556)
: to cover, line, or furnish with canvas

can·vas·back \'kan-vəs-ˌbak\ *noun* (1782)
: a North American wild duck (*Aythya valisineria*) that has a reddish brown head, black breast, and whitish body and is characterized especially by the elongate sloping profile of the bill and head

[1]**can·vass** *also* **can·vas** \'kan-vəs\ (1508)
transitive verb
1 *obsolete* **:** to toss in a canvas sheet in sport or punishment
2 a : to examine in detail; *specifically* **:** to examine (votes) officially for authenticity **b :** DISCUSS, DEBATE
3 : to go through (a district) or go to (persons) in order to solicit orders or political support or to determine opinions or sentiments
intransitive verb
: to seek orders or votes **:** SOLICIT
— **can·vass·er** *also* **can·vas·er** *noun*

[2]**canvass** *also* **canvas** *noun* (circa 1611)
: the act or an instance of canvassing; *especially* **:** a personal solicitation of votes or survey of public opinion

can·yon \'kan-yən\ *noun* [American Spanish *cañón,* probably alteration of obsolete Spanish *callón,* augmentative of *calle* street, from Latin *callis* footpath] (1837)
: a deep narrow valley with steep sides and often with a stream flowing through it

can·zo·ne \kan-'zō-nē, känt-'sō-(ˌ)nā\ *noun, plural* **-nes** \-nēz, -(ˌ)nāz\ *or* **-ni** \-nē\ [Italian, from Latin *cantion-, cantio* song, from *canere* to sing — more at CHANT] (1589)
1 : a medieval Italian or Provençal lyric poem
2 : the musical setting of a canzone

can·zo·net \ˌkan-zə-'net\ *noun* [Italian *canzonetta,* diminutive of *canzone*] (1588)
1 : a light usually strophic song
2 : a part-song resembling but less elaborate than a madrigal

caou·tchouc \'kau̇-ˌchu̇k, -ˌchük, -ˌchü\ *noun* [French, from obsolete Spanish *cauchuc* (now *caucho*), from Quechua *kawchu*] (1775)
: [1]RUBBER 2a

[1]**cap** \'kap\ *noun, often attributive* [Middle English *cappe,* from Old English *cæppe,* from Late Latin *cappa* head covering, cloak] (before 12th century)
1 a : a head covering especially with a visor and no brim **b :** a distinctive head covering emblematic of a position or office: as (1) **:** a cardinal's biretta (2) **:** MORTARBOARD
2 : a natural cover or top: as **a :** an overlying rock layer that is usually hard to penetrate **b** (1) **:** PILEUS (2) **:** CALYPTRA **c :** the top of a bird's head or a patch of distinctively colored feathers in this area
3 a : something that serves as a cover or protection especially for a tip, knob, or end ⟨a bottle *cap*⟩ **b :** a fitting for closing the end of a tube (as a water pipe or electric conduit) **c** *British* **:** CERVICAL CAP
4 : an overlaying or covering structure

5 : a paper or metal container holding an explosive charge (as for a toy pistol)
6 : an upper limit (as on expenditures) **:** CEILING
7 : the symbol ∩ indicating the intersection of two sets — compare CUP 9
8 : a cluster of molecules or chemical groups bound to one end or a region of a cell, virus, or molecule
— **cap in hand :** in a respectful, humble, or sometimes fearful manner

²cap *verb* **capped; cap·ping** (15th century)
transitive verb
1 a : to provide or protect with a cap **b :** to give a cap to as a symbol of honor, rank, or achievement
2 : to form a cap over **:** CROWN ⟨the mountains were *capped* with mist —John Buchan⟩
3 a : to follow with something more noticeable or more significant **:** OUTDO **b :** CLIMAX
4 : to form a chemical cap on
intransitive verb
: to form or produce a chemical cap

ca·pa·bil·i·ty \ˌkā-pə-ˈbi-lə-tē\ *noun, plural* **-ties** (1587)
1 : the quality or state of being capable; *also* **:** ABILITY
2 : a feature or faculty capable of development **:** POTENTIALITY
3 : the facility or potential for an indicated use or deployment ⟨the *capability* of a metal to be fused⟩ ⟨nuclear *capability*⟩

ca·pa·ble \ˈkā-pə-bəl, *in rapid speech* ˈkāp-bəl\ *adjective* [Middle French or Late Latin; Middle French *capable,* from Late Latin *capabilis,* irregular from Latin *capere* to take — more at HEAVE] (1579)
1 : SUSCEPTIBLE ⟨a remark *capable* of being misunderstood⟩
2 *obsolete* **:** COMPREHENSIVE
3 : having attributes (as physical or mental power) required for performance or accomplishment ⟨is *capable* of intense concentration⟩
4 : having traits conducive to or features permitting ⟨this woman is *capable* of murder by violence —Robert Graves⟩ ⟨an outer coat of light color *capable* of reflecting solar heat —*Current Biography*⟩
5 : having legal right to own, enjoy, or perform
6 : having general efficiency and ability
— **ca·pa·ble·ness** \ˈkā-pə-bəl-nəs\ *noun*
— **ca·pa·bly** \-pə-blē\ *adverb*

ca·pa·cious \kə-ˈpā-shəs\ *adjective* [Latin *capac-, capax* capacious, capable, from Latin *capere*] (1614)
: containing or capable of containing a great deal
synonym see SPACIOUS
— **ca·pa·cious·ly** *adverb*
— **ca·pa·cious·ness** *noun*

ca·pac·i·tance \kə-ˈpa-sə-tən(t)s\ *noun* [*capacity*] (1893)
1 a : the property of an electric nonconductor that permits the storage of energy as a result of the separation of charge that occurs when opposite surfaces of the nonconductor are maintained at a difference of potential **b :** the measure of this property that is equal to the ratio of the charge on either surface to the potential difference between the surfaces
2 : a part of a circuit or network that possesses capacitance
— **ca·pac·i·tive** \-ˈpa-sə-tiv\ *adjective*
— **ca·pac·i·tive·ly** *adverb*

ca·pac·i·tate \kə-ˈpa-sə-ˌtāt\ *transitive verb* **-tat·ed; -tat·ing** (1657)
1 *archaic* **:** to make capable
2 : to cause (sperm) to undergo capacitation

ca·pac·i·ta·tion \kə-ˌpa-sə-ˈtā-shən\ *noun* (1951)
: the change undergone by sperm in the female reproductive tract that enables them to penetrate and fertilize an egg

ca·pac·i·tor \kə-ˈpa-sə-tər\ *noun* (1925)
: a device giving capacitance and usually consisting of conducting plates or foils separated by thin layers of dielectric (as air or mica) with the plates on opposite sides of the dielectric layers oppositely charged by a source of voltage and the electrical energy of the charged system stored in the polarized dielectric

¹ca·pac·i·ty \kə-ˈpa-sə-tē, -ˈpas-tē\ *noun, plural* **-ties** [Middle English *capacite,* from Middle French *capacité,* from Latin *capacitat-, capacitas,* from *capac-, capax*] (15th century)
1 : legal competency or fitness
2 a : the potential or suitability for holding, storing, or accommodating **b :** the maximum amount or number that can be contained or accommodated ⟨a jug with a one-gallon *capacity*⟩ ⟨the auditorium was filled to *capacity*⟩ — see METRIC SYSTEM table, WEIGHT table
3 a : an individual's mental or physical ability **:** APTITUDE, SKILL **b :** the faculty or potential for treating, experiencing, or appreciating ⟨*capacity* for love⟩
4 : DUTY, POSITION, ROLE ⟨will be happy to serve in any *capacity*⟩
5 : the facility or power to produce, perform, or deploy **:** CAPABILITY ⟨a plan to double the factory's *capacity*⟩; *also* **:** maximum output ⟨industries running at three-quarter *capacity*⟩
6 a : CAPACITANCE **b :** the quantity of electricity that a battery can deliver under specified conditions

²capacity *adjective* (1897)
: equaling maximum capacity ⟨a *capacity* crowd⟩

cap–a–pie *or* **cap–à–pie** \ˌka-pə-ˈpē, -ˈpā\ *adverb* [Middle French (*de*) *cap a pé* from head to foot] (1523)
: from head to foot ⟨armed *cap-a-pie* for battle⟩

¹ca·par·i·son \kə-ˈpar-ə-sən\ *noun* [Middle French *caparaçon,* from Old Spanish *caparazón*] (1579)
1 a : an ornamental covering for a horse **b :** decorative trappings and harness
2 : rich clothing **:** ADORNMENT

²caparison *transitive verb* (1594)
: to provide with or as if with a rich ornamental covering **:** ADORN

¹cape \ˈkāp\ *noun, often attributive* [Middle English *cap,* from Middle French, from Old Provençal, from Latin *caput* head — more at HEAD] (14th century)
1 : a point or extension of land jutting out into water as a peninsula or as a projecting point
2 *often capitalized* **:** CAPE COD COTTAGE

²cape *noun* [probably from Spanish *capa* cloak, from Late Latin *cappa* head covering, cloak] (circa 1578)
1 : a sleeveless outer garment or part of a garment that fits closely at the neck and hangs loosely over the shoulders
2 : the short feathers covering the shoulders of a fowl — see DUCK illustration

Cape buffalo \ˈkāp-\ *noun* [*Cape* of Good Hope, Africa] (circa 1890)
: the large reddish brown to black wild buffalo (*Syncerus caffer*) of sub-Saharan Africa — called also *African buffalo*

Cape Cod cottage \(ˈ)kāp-ˈkäd-\ *noun* [*Cape Cod,* Mass.] (1916)
: a compact rectangular dwelling of one or one-and-a-half stories usually with a central chimney and steep gable roof

Cape gooseberry *noun* (1833)
: any of several ground-cherries (especially *Physalis peruviana*) bearing edible acid berries; *also* **:** its berry

Cape Horn·er \ˌkāp-ˈhȯr-nər\ *noun* (1840)
: a ship that voyages around Cape Horn

cape·let \ˈkāp-lət\ *noun* (1912)
: a small cape usually covering the shoulders

cap·e·lin \ˈka-p(ə-)lən\ *noun* [Canadian French *capelan,* from French, codfish, from Old Provençal, chaplain, codfish, from Medieval Latin *cappellanus* chaplain — more at CHAPLAIN] (1620)
: a small northern sea fish (*Mallotus villosus*) related to the smelts

Ca·pel·la \kə-ˈpe-lə\ *noun* [Latin, literally, she-goat, from *caper* he-goat — more at CAPRIOLE]
: a star of the first magnitude in Auriga

¹ca·per \ˈkā-pər\ *noun* [back-formation from earlier *capers* (taken as a plural), from Middle English *caperis,* from Latin *capparis,* from Greek *kapparis*] (14th century)
1 : any of a genus (*Capparis* of the family Capparidaceae, the caper family) of low prickly shrubs of the Mediterranean region; *especially* **:** one (*C. spinosa*) cultivated for its buds
2 : one of the greenish flower buds or young berries of the caper pickled and used as a seasoning or garnish

²caper *intransitive verb* **ca·pered; ca·per·ing** \-p(ə-)riŋ\ [probably by shortening & alteration from *capriole*] (1588)
: to leap or prance about in a playful manner

³caper *noun* (1592)
1 : a frolicsome leap
2 : a capricious escapade **:** PRANK
3 : an illegal or questionable act; *especially* **:** THEFT

cap·er·cail·lie \ˌka-pər-ˈkā-lē, -ˈkāl-yē\ *or* **cap·er·cail·zie** \-ˈkāl-zē\ *noun, plural* **-caillie** *or* **-cailzies** *also* **-caillies** *or* **-cailzie** [Scottish Gaelic *capalcoille,* literally, horse of the woods] (1536)
: the largest Old World grouse (*Tetrao urogallus*)

cape·skin \ˈkāp-ˌskin\ *noun* [*Cape* of Good Hope, Africa] (1919)
: a light flexible leather made from sheepskins with the natural grain retained and used especially for gloves and garments

Ca·pe·tian \kə-ˈpē-shən\ *adjective* [Hugh *Capet*] (1836)
: of or relating to the French royal house that ruled from 987 to 1328
— **Capetian** *noun*

cape·work \ˈkāp-ˌwərk\ *noun* (1926)
: the art of the bullfighter in working a bull with the cape

cap·ful \ˈkap-ˌfu̇l\ *noun* (1873)
: as much as a cap will hold ⟨a *capful* of detergent⟩

capful of wind (1719)
: a sudden light breeze

ca·pi·as \ˈkā-pē-əs\ *noun* [Middle English, from Latin, literally, you should seize, from *capere* to take — more at HEAVE] (15th century)
: an arrest warrant

cap·il·lar·i·ty \ˌka-pə-ˈlar-ə-tē\ *noun, plural* **-ties** (1830)
1 : the property or state of being capillary
2 : the action by which the surface of a liquid where it is in contact with a solid (as in a capillary tube) is elevated or depressed depending on the relative attraction of the molecules of the liquid for each other and for those of the solid

¹cap·il·lary \ˈka-pə-ˌler-ē, *British usually* kə-ˈpi-lə-rē\ *adjective* [French or Latin; French *capillaire,* from Latin *capillaris,* from *capillus* hair] (14th century)
1 a : resembling a hair especially in slender elongated form ⟨*capillary* leaves⟩ **b :** having a very small bore ⟨a *capillary* tube⟩
2 : involving, held by, or resulting from surface tension ⟨*capillary* water in the soil⟩
3 : of or relating to capillaries or capillarity

²capillary *noun, plural* **-lar·ies** (1667)
: a capillary tube; *especially* **:** any of the smallest blood vessels connecting arterioles with venules and forming networks throughout the body

capillary attraction *noun* (1813)
: the force of adhesion between a solid and a liquid in capillarity

[1]cap·i·tal \ˈka-pə-t^{ə}l, ˈkap-t^{ə}l\ *noun* [Middle English *capitale,* modification of Old North French *capitel,* from Late Latin *capitellum* small head, top of column, diminutive of Latin *capit-, caput* head — more at HEAD] (13th century)
: the uppermost member of a column or pilaster crowning the shaft and taking the weight of the entablature — see COLUMN illustration

[2]capital *adjective* [Middle English, from Latin *capitalis,* from *capit-, caput*] (14th century)
1 *of a letter* **:** of or conforming to the series A, B, C, etc. rather than a, b, c, etc.
2 a : punishable by death ⟨a *capital* crime⟩ **b :** involving execution ⟨*capital* punishment⟩ **c :** most serious ⟨a *capital* error⟩
3 a : chief in importance or influence ⟨*capital* ships⟩ ⟨the *capital* importance of criticism in the work of creation itself —T. S. Eliot⟩ **b :** being the seat of government
4 : of or relating to capital; *especially* **:** relating to or being assets that add to the long-term net worth of a corporation ⟨*capital* improvements⟩
5 : EXCELLENT ⟨a *capital* book⟩

[3]capital *noun* [French or Italian; French, from Italian *capitale,* from *capitale,* adjective, chief, principal, from Latin *capitalis*] (circa 1639)
1 a (1) **:** a stock of accumulated goods especially at a specified time and in contrast to income received during a specified period; *also* **:** the value of these accumulated goods (2) **:** accumulated goods devoted to the production of other goods (3) **:** accumulated possessions calculated to bring in income **b** (1) **:** net worth (2) **:** CAPITAL STOCK **c :** persons holding capital **d :** ADVANTAGE, GAIN ⟨make *capital* of the situation⟩
2 [[2]*capital*] **a :** a capital letter; *especially* **:** an initial capital letter **b :** a letter belonging to a style of alphabet modeled on the style customarily used in inscriptions
3 [[2]*capital*] **a :** a city serving as a seat of government **b :** a city preeminent in some special activity ⟨the fashion *capital*⟩

capital gain *noun* (1921)
: the increase in value of an asset (as stock or real estate) between the time it is bought and the time it is sold

capital goods *noun plural* (1896)
: [3]CAPITAL 1a(1), 1a(2)

capital–intensive *adjective* (1959)
: having a high capital cost per unit of output; *especially* **:** requiring greater expenditure in the form of capital than of labor

cap·i·tal·ise *British variant of* CAPITALIZE

cap·i·tal·ism \ˈka-pə-t^{ə}l-ˌiz-əm, ˈkap-t^{ə}l-, *British also* kə-ˈpi-t^{ə}l-\ *noun* (1877)
: an economic system characterized by private or corporate ownership of capital goods, by investments that are determined by private decision, and by prices, production, and the distribution of goods that are determined mainly by competition in a free market

[1]cap·i·tal·ist \-ist\ *noun* (1792)
1 : a person who has capital especially invested in business; *broadly* **:** a person of wealth **:** PLUTOCRAT
2 : a person who favors capitalism

[2]capitalist *or* **cap·i·tal·is·tic** \ˌka-pə-t^{ə}l-ˈis-tik, ˌkap-t^{ə}l-, *British also* kə-ˌpi-t^{ə}l-\ *adjective* (1845)
1 : owning capital ⟨the *capitalist* class⟩
2 a : practicing or advocating capitalism ⟨*capitalist* nations⟩ **b :** marked by capitalism ⟨*capitalist* period of history⟩
— cap·i·tal·is·ti·cal·ly \-ti-k(ə-)lē\ *adverb*

cap·i·tal·i·za·tion \ˌka-pə-t^{ə}l-ə-ˈzā-shən, ˌkap-t^{ə}l-, *British also* kə-ˌpi-t^{ə}l-\ *noun* (1860)
1 a : the act or process of capitalizing **b :** a sum resulting from a process of capitalizing **c :** the total liabilities of a business including both ownership capital and borrowed capital **d :** the total par value or the stated value of no-par issues of authorized capital stock
2 : the use of a capital letter in writing or printing

cap·i·tal·ize \ˈka-pə-t^{ə}l-ˌīz, ˈkap-t^{ə}l-, *British also* kə-ˈpi-t^{ə}l-\ *verb* **-ized; -iz·ing** (1764)
transitive verb
1 : to write or print with an initial capital or in capitals
2 a : to convert into capital ⟨*capitalize* the company's reserve fund⟩ **b :** to treat as capital rather than as an expense
3 a : to compute the present value of (an income extended over a period of time) **b :** to convert (a periodic payment) into an equivalent capital sum ⟨*capitalized* annuities⟩
4 : to supply capital for
intransitive verb
: to gain by turning something to advantage **:** PROFIT ⟨*capitalize* on an opponent's mistake⟩

cap·i·tal·ly \ˈka-pə-t^{ə}l-ē, ˈkap-t^{ə}l-\ *adverb* (1619)
1 : in a manner involving capital punishment
2 : in a capital manner **:** EXCELLENTLY

capital stock *noun* (1709)
1 : the outstanding shares of a joint-stock company considered as an aggregate
2 : CAPITALIZATION 1d
3 : the ownership element of a corporation divided into shares and represented by certificates

cap·i·tate \ˈka-pə-ˌtāt\ *adjective* [Latin *capitatus* headed, from *capit-, caput* head] (1661)
1 : forming a head
2 : abruptly enlarged and globose

cap·i·ta·tion \ˌka-pə-ˈtā-shən\ *noun* [Late Latin *capitation-, capitatio* poll tax, from Latin *capit-, caput*] (1641)
1 : a direct uniform tax imposed on each head or person **:** POLL TAX
2 : a uniform per capita payment or fee

cap·i·tol \ˈka-pə-t^{ə}l, ˈkap-t^{ə}l\ *noun* [Latin *Capitolium,* temple of Jupiter at Rome on the Capitoline hill] (1699)
1 a : a building in which a state legislative body meets **b :** a group of buildings in which the functions of state government are carried out
2 *capitalized* **:** the building in which the U.S. Congress meets at Washington

Capitol Hill *noun* [*Capitol Hill,* Washington, site of the U.S. Capitol] (1943)
: the legislative branch of the U.S. government

Cap·i·to·line \ˈka-pə-t^{ə}l-ˌīn, *British usually* kə-ˈpi-tə-ˌlīn\ *adjective* [Latin *capitolinus,* from *Capitolium*] (1667)
: of or relating to the smallest of the seven hills of ancient Rome, the temple on it, or the gods worshiped there

ca·pit·u·lar \kə-ˈpi-chə-lər\ *adjective* [Medieval Latin *capitularis,* from *capitulum*] (circa 1525)
: of or relating to an ecclesiastical chapter

ca·pit·u·lary \-ˌler-ē\ *noun, plural* **-lar·ies** [Medieval Latin *capitulare,* literally, document divided into sections, from Late Latin *capitulum* section, chapter — more at CHAPTER] (1650)
: a civil or ecclesiastical ordinance; *also* **:** a collection of ordinances

ca·pit·u·late \kə-ˈpi-chə-ˌlāt\ *intransitive verb* **-lat·ed; -lat·ing** [Medieval Latin *capitulatus,* past participle of *capitulare* to distinguish by heads or chapters, from Late Latin *capitulum*] (1537)
1 *archaic* **:** PARLEY, NEGOTIATE
2 a : to surrender often after negotiation of terms **b :** to cease resisting **:** ACQUIESCE
synonym see YIELD

ca·pit·u·la·tion \kə-ˌpi-chə-ˈlā-shən\ *noun* (1535)
1 : a set of terms or articles constituting an agreement between governments
2 a : the act of surrendering or yielding **b :** the terms of surrender

ca·pit·u·lum \kə-ˈpi-chə-ləm\ *noun, plural* **-la** \-lə\ [New Latin, from Latin, small head — more at CHAPTER] (circa 1755)
1 : a rounded protuberance of an anatomical part (as a bone)
2 : a racemose inflorescence (as of the buttonbush) with the axis shortened and dilated to form a rounded or flattened cluster of sessile flowers — see INFLORESCENCE illustration

Cap·lets \ˈka-pləts\ *trademark*
— used for capsule-shaped medicinal tablets

[1]ca·po \ˈkā-(ˌ)pō\ *noun, plural* **capos** [short for *capotasto,* from Italian, literally, head of fingerboard] (1926)
: a movable bar attached to the fingerboard of a fretted instrument to uniformly raise the pitch of all the strings

[2]ca·po \ˈkä-(ˌ)pō, ˈka-\ *noun, plural* **capos** [Italian, head, chief, from Latin *caput*] (circa 1959)
: the head of a branch of a crime syndicate

ca·pon \ˈkā-ˌpän, -pən\ *noun* [Middle English, from Old English *capūn,* probably from Old North French *capon,* from Latin *capon-, capo;* akin to Lithuanian *kapoti* to mince, Greek *koptein* to cut] (before 12th century)
: a castrated male chicken

ca·po·na·ta \ˌkä-pə-ˈnä-tə\ *noun* [Italian, from Italian dialect (Sicily) *capunata,* perhaps from *capuni* capon, from Latin *capon-, capo*] (1951)
: a relish of chopped eggplant and assorted vegetables

ca·pote \kə-ˈpōt\ *noun* [French, from *cape* cloak, from Late Latin *cappa*] (1799)
: a usually long and hooded cloak or overcoat

cap·pel·let·ti \ˌka-pə-ˈle-tē\ *noun plural but singular or plural in construction* [Italian, plural of *cappelletto,* diminutive of *cappello* hat, from Medieval Latin *cappellus* cap, diminutive of Late Latin *cappa* head covering — more at CAP] (1945)
: pasta in the form of little peaked hats filled with a savory mixture

cap·per \ˈka-pər\ *noun* (1587)
1 : one that caps: as **a :** a device that fits caps on bottles **b :** FINALE, CLIMAX, CLINCHER
2 : a lure or decoy especially in an illicit or questionable activity **:** SHILL

cap·ping \ˈka-piŋ\ *noun* (14th century)
: something that caps

cap·puc·ci·no \ˌka-pə-ˈchē-(ˌ)nō, ˌkä-pü-\ *noun* [Italian, literally, Capuchin; from the likeness of its color to that of a Capuchin's habit] (1948)
: espresso coffee mixed with frothed hot milk or cream and often flavored with cinnamon

cap·ric acid \ˈka-prik-\ *noun* [International Scientific Vocabulary, from Latin *capr-, caper* goat; from its odor — more at CAPRIOLE] (1836)
: a fatty acid $C_{10}H_{20}O_2$ found in fats and oils and used in flavors and perfumes

ca·pric·cio \kə-ˈprē-ch(ē-ˌ)ō\ *noun, plural* **-cios** [Italian] (1601)
1 : FANCY, WHIMSY
2 : CAPER, PRANK
3 : an instrumental piece in free form usually lively in tempo and brilliant in style

ca·price \kə-ˈprēs\ *noun* [French, from Italian *capriccio* caprice, shudder, perhaps from *capo* head (from Latin *caput*) + *riccio* hedgehog, from Latin *ericius* — more at HEAD, URCHIN] (1667)
1 a : a sudden, impulsive, and seemingly unmotivated notion or action **b :** a sudden usually unpredictable condition, change, or series of changes ⟨the *caprices* of the weather⟩
2 : a disposition to do things impulsively

3 : CAPRICCIO 3 ☆

ca·pri·cious \kə-ˈpri-shəs, -ˈprē-\ *adjective* (1601)
: governed or characterized by caprice **:** IMPULSIVE, UNPREDICTABLE
synonym see INCONSTANT
— ca·pri·cious·ly *adverb*
— ca·pri·cious·ness *noun*

Cap·ri·corn \ˈka-pri-ˌkȯrn\ *noun* [Middle English *Capricorne,* from Latin *Capricornus* (genitive *Capricorni*), from *caper* goat + *cornu* horn — more at HORN]
1 : a southern zodiacal constellation between Sagittarius and Aquarius
2 a : the 10th sign of the zodiac in astrology — see ZODIAC table **b :** one born under the sign of Capricorn

cap·ri·fi·ca·tion \ˌka-prə-fə-ˈkā-shən\ *noun* [Latin *caprification-, caprificatio,* from *caprificare* to pollinate by caprification, from *caprificus*] (1601)
: artificial pollination of figs that usually bear only pistillate flowers by hanging male flowering branches of the caprifig in the trees to facilitate pollen transfer by a wasp

cap·ri·fig \ˈka-prə-ˌfig\ *noun* [Middle English *caprifige,* part translation of Latin *caprificus,* from *capr-, caper* goat + *ficus* fig — more at FIG] (15th century)
: a wild fig (*Ficus carica sylvestris*) of southern Europe and Asia Minor used for caprification of the edible fig; *also* **:** its fruit

cap·rine \ˈka-ˌprīn\ *adjective* [Latin *caprinus,* from *capr-, caper*] (15th century)
: of, relating to, or being a goat ⟨*caprine* serum⟩ ⟨the *caprine* family⟩

cap·ri·ole \ˈka-prē-ˌōl\ *noun* [Middle French or Old Italian; Middle French *capriole,* from Old Italian *capriola,* from *capriolo* roebuck, from Latin *capreolus* goat, roebuck, from *capr-, caper* he-goat; akin to Old English *hæfer* goat, Greek *kapros* wild boar] (1594)
1 : a playful leap **:** CAPER
2 *of a trained horse* **:** a vertical leap with a backward kick of the hind legs at the height of the leap
— capriole *intransitive verb*

ca·pri pants \kə-ˈprē-\ *noun plural, often C capitalized* [*Capri,* Italy] (circa 1956)
: close-fitting women's pants that end above the ankle — called also *capris*

cap·rock \ˈkap-ˌräk\ *noun* (1867)
: CAP 2a

ca·pro·ic acid \kə-ˈprō-ik-\ *noun* [International Scientific Vocabulary, from Latin *capr-, caper*] (circa 1847)
: a liquid fatty acid $C_6H_{12}O_2$ that is found as a glycerol ester in fats and oils or made synthetically and used in pharmaceuticals and flavors

cap·ro·lac·tam \ˌka-prō-ˈlak-ˌtam\ *noun* [*caproic* acid + *lac*tone + *am*ide] (1944)
: a white crystalline cyclic amide $C_6H_{11}NO$ used especially in making one type of nylon

ca·pryl·ic acid \kə-ˈpri-lik-\ *noun* [International Scientific Vocabulary *capryl,* a radical contained in it] (1845)
: a fatty acid $C_8H_{16}O_2$ of rancid odor occurring in fats and oils and used in perfumes

cap·sa·i·cin \kap-ˈsā-ə-sən\ *noun* [irregular from New Latin *Capsicum*] (circa 1890)
: a colorless irritant phenolic amide $C_{18}H_{27}NO_3$ that is found in various capsicums and that gives hot peppers their hotness

Cap·si·an \ˈkap-sē-ən\ *adjective* [French *capsien,* from Latin *Capsa* Gafsa, Tunisia] (1915)
: of or relating to a Paleolithic culture of northern Africa and southern Europe

cap·si·cum \ˈkap-si-kəm\ *noun* [New Latin, perhaps from Latin *capsa*] (1588)
1 : any of a genus (*Capsicum*) of tropical herbs and shrubs of the nightshade family widely cultivated for their many-seeded usually fleshy-walled berries — called also *pepper*
2 : the dried ripe fruit of some capsicums (as *C. frutescens*) used as a gastric and intestinal stimulant

cap·sid \ˈkap-səd\ *noun* [French *capside,* from Latin *capsa* case + French *-ide* [2]-id] (1960)
: the outer protein shell of a virus particle

cap·size \ˈkap-ˌsīz, kap-ˈ\ *verb* **cap·sized; cap·siz·ing** [perhaps from Spanish *capuzar* or Catalan *cabussar* to thrust (the head) underwater] (1788)
transitive verb
: to cause to overturn ⟨*capsize* a canoe⟩
intransitive verb
: to become upset or overturned **:** TURN OVER ⟨the canoe *capsized*⟩

cap sleeve *noun* (1926)
: a very short sleeve (as on a dress) that hangs over the edge of the shoulder without extending along the underside of the arm

cap·stan \ˈkap-stən, -ˌstan\ *noun* [Middle English, probably from Middle French *cabestant*] (14th century)
1 : a machine for moving or raising heavy weights that consists of a vertical drum which can be rotated and around which cable is turned
2 : a rotating shaft that drives tape at a constant speed in a recorder

cap·stone \ˈkap-ˌstōn\ *noun* [[1]*cap*] (14th century)
1 : a coping stone **:** COPING
2 : the high point **:** crowning achievement

cap·su·lar \ˈkap-sə-lər\ *adjective* (circa 1730)
1 : of, relating to, or resembling a capsule
2 : CAPSULATED

cap·su·lat·ed \-ˌlā-təd\ *adjective* (1668)
: enclosed in a capsule

[1]**cap·sule** \ˈkap-səl, -(ˌ)sül *also* -ˌsyü(ə)l\ *noun* [French, from Latin *capsula,* diminutive of *capsa* box — more at CASE] (circa 1693)
1 a : a membrane or sac enclosing a body part **b :** either of two layers of white matter in the cerebrum
2 : a closed receptacle containing spores or seeds: as **a :** a dry dehiscent usually many-seeded fruit composed of two or more carpels **b :** the spore case of a moss
3 : a shell usually of gelatin for packaging something (as a drug or vitamins); *also* **:** a usually medicinal or nutritional preparation for oral use consisting of the shell and its contents
4 : an often polysaccharide envelope surrounding a microorganism
5 : an extremely brief condensation **:** OUTLINE, SURVEY
6 a : a compact often sealed and detachable container or compartment **b :** a small pressurized compartment or vehicle (as for space flight or emergency escape)

[2]**capsule** *transitive verb* **cap·suled; cap·sul·ing** (1859)
1 : to equip with or enclose in a capsule
2 : to condense into or devise in a compact form

[3]**capsule** *adjective* (1938)
1 : extremely brief
2 : small and very compact

cap·sul·ize \ˈkap-sə-ˌlīz\ *transitive verb* **-ized; -iz·ing** (1945)
: CAPSULE

[1]**cap·tain** \ˈkap-tən *also* ˈkap-ᵊm\ *noun* [Middle English *capitane,* from Middle French *capitain,* from Late Latin *capitaneus,* adjective & noun, chief, from Latin *capit-, caput* head — more at HEAD] (14th century)
1 a (1) **:** a military leader **:** the commander of a unit or a body of troops (2) **:** a subordinate officer commanding under a sovereign or general (3) **:** a commissioned officer in the army, air force, or marine corps ranking above a first lieutenant and below a major **b :** a naval officer who is master or commander of a ship; *especially* **:** a commissioned officer in the navy ranking above a commander and below a commodore and in the coast guard ranking above a commander and below a rear admiral **c :** a senior pilot who commands the crew of an airplane **d :** an officer in a police department or fire department in charge of a unit (as a precinct or company) and usually ranking above a lieutenant and below a chief
2 : one who leads or supervises: as **a :** a leader of a sports team or side **b :** HEADWAITER **c :** a person in charge of hotel bellhops — called also *bell captain*
3 : a person of importance or influence in a field ⟨*captains* of industry⟩
— cap·tain·cy \ˈkap-tən-sē\ *noun*
— cap·tain·ship \-ˌship\ *noun*

[2]**captain** *transitive verb* (1598)
: to be captain of **:** LEAD ⟨*captained* the football team⟩

captain's chair *noun* (1946)
: an armchair with a saddle seat and a low curved back with vertical spindles

captain's mast *noun* (1941)
: MAST 3

cap·tan \ˈkap-ˌtan\ *noun* [short for *mercaptan*] (1952)
: a fungicide $C_9H_8Cl_3NO_2S$ used on agricultural crops

[1]**cap·tion** \ˈkap-shən\ *noun* [Middle English *capcioun,* from Latin *caption-, captio* act of taking, from *capere* to take — more at HEAVE] (1670)
1 : the part of a legal document that shows where, when, and by what authority it was taken, found, or executed
2 a : the heading especially of an article or document **:** TITLE **b :** the explanatory comment or designation accompanying a pictorial illustration **c :** a motion-picture subtitle
— cap·tion·less \-ləs\ *adjective*

[2]**caption** *transitive verb* **cap·tioned; cap·tion·ing** \-sh(ə-)niŋ\ (1901)
: to furnish with a caption

cap·tious \ˈkap-shəs\ *adjective* [Middle English *capcious,* from Middle French or Latin; Middle French *captieux,* from Latin *captiosus,* from *captio*] (14th century)
1 : marked by an often ill-natured inclination to stress faults and raise objections
2 : calculated to confuse, entrap, or entangle in argument
synonym see CRITICAL
— cap·tious·ly *adverb*
— cap·tious·ness *noun*

cap·ti·vate \ˈkap-tə-ˌvāt\ *transitive verb* **-vat·ed; -vat·ing** (circa 1555)
1 *archaic* **:** SEIZE, CAPTURE
2 : to influence and dominate by some special charm, art, or trait and with an irresistible appeal
synonym see ATTRACT
— cap·ti·va·tion \ˌkap-tə-ˈvā-shən\ *noun*
— cap·ti·va·tor \ˈkap-tə-ˌvā-tər\ *noun*

cap·tive \ˈkap-tiv\ *adjective* [Middle English, from Latin *captivus,* from *captus,* past participle of *capere*] (14th century)
1 a : taken and held as or as if a prisoner of war **b :** kept within bounds **:** CONFINED
2 : held under control of another but having the appearance of independence; *especially* **:** owned or controlled by another concern and

☆ SYNONYMS

Caprice, whim, vagary, crotchet mean an irrational or unpredictable idea or desire. CAPRICE stresses lack of apparent motivation and suggests willfulness ⟨by sheer *caprice* she quit her job⟩. WHIM implies a fantastic, capricious turn of mind or inclination ⟨an odd antique that was bought on a *whim*⟩. VAGARY stresses the erratic, irresponsible character of the notion or desire ⟨recently he had been prone to strange *vagaries*⟩. CROTCHET implies an eccentric opinion or preference ⟨a serious scientist equally known for his bizarre *crotchets*⟩.

operated for its needs rather than for an open market ⟨a *captive* mine⟩

3 : being such involuntarily because of a situation that makes free choice or departure difficult ⟨the airline passengers were a *captive* audience⟩

— captive *noun*

cap·tiv·i·ty \kap-ˈti-və-tē\ *noun* (14th century)

1 : the state of being captive ⟨some birds thrive in *captivity*⟩

2 *obsolete* **:** a group of captives

cap·to·pril \ˈkap-tə-ˌpril\ *noun* [mer*cap*tan + *-o-* + *pr*ol*i*ne + *-il,* alteration of *-yl*] (1978)

: an antihypertensive drug $C_9H_{15}NO_3S$ that is an ACE inhibitor

cap·tor \ˈkap-tər, -ˌtȯr\ *noun* [Late Latin, from Latin *capere*] (circa 1688)

: one that has captured a person or thing

[1]**cap·ture** \ˈkap-chər, -shər\ *noun* [Middle French, from Latin *captura,* from *captus*] (circa 1542)

1 : an act or instance of capturing: as **a :** an act of catching, winning, or gaining control by force, stratagem, or guile **b :** a move in a board game (as chess or checkers) that gains an opponent's piece **c :** the absorption by an atom, nucleus, or particle of a subatomic particle that often results in subsequent emission of radiation or in fission **d :** the act of recording in a permanent file ⟨data *capture*⟩

2 : one that has been taken (as a prize ship)

[2]**capture** *transitive verb* **cap·tured; cap·tur·ing** \ˈkap-chə-riŋ, ˈkap-shriŋ\ (1795)

1 a : to take captive; *also* **:** to gain control of especially by force ⟨*capture* a city⟩ **b :** to gain or win especially through effort ⟨*captured* 60% of the vote⟩

2 : to emphasize, represent, or preserve (as a scene, mood, or quality) in a more or less permanent form ⟨at any such moment as a photograph might *capture* —C. E. Montague⟩

3 : to captivate and hold the interest of

4 : to take according to the rules of a game

5 : to bring about the capture of (a subatomic particle)

6 : to record in a permanent file (as in a computer)

synonym see CATCH

capture the flag *noun* (circa 1925)

: a game in which players on each of two teams seek to capture the other team's flag and return it to their side without being captured and imprisoned

ca·puche \kə-ˈpüch, -ˈpüsh\ *noun* [Middle French, from Italian *cappuccio,* from *cappa* cloak, from Late Latin] (circa 1600)

: HOOD; *especially* **:** the cowl of a Capuchin friar

ca·pu·chin \ˈka-pyə-shən, -pə-, *especially for 3 also* kə-ˈpyü-, -ˈpü-\ *noun* [Middle French, from Old Italian *cappuccino,* from *cappuccio;* from his cowl] (1589)

1 *capitalized* **:** a member of the Order of Friars Minor Capuchin forming since 1529 an austere branch of the first order of Saint Francis of Assisi engaged in missionary work and preaching

2 : a hooded cloak for women

3 : any of a genus (*Cebus*) of South American monkeys; *especially* **:** one (*C. capucinus*) with the hair on its crown resembling a monk's cowl

Cap·u·let \ˈka-pyə-lət\ *noun*

: the family of Juliet in Shakespeare's *Romeo and Juliet*

cap·y·bara \ˌka-pi-ˈbar-ə, -ˈbär-\ *noun* [Portuguese *capibara,* from Tupi] (1774)

: a tailless largely aquatic South American rodent (*Hydrochaerus hydrochaeris*) often exceeding four feet (1.2 meters) in length

car \ˈkär, *dialect also* ˈkȯr, ˈkyär\ *noun* [Middle English *carre,* from Anglo-French, from Latin *carra,* plural of *carrum,* alteration of *carrus,* of Celtic origin; akin to Old Irish & Middle Welsh *carr* vehicle; akin to Latin *currere* to run] (14th century)

1 : a vehicle moving on wheels: as **a** *archaic* **:** CARRIAGE, CHARIOT **b :** a vehicle designed to move on rails (as of a railroad) **c :** AUTOMOBILE

2 : the passenger compartment of an elevator

3 : the part of an airship or balloon that carries the passengers and cargo

ca·ra·bao \ˌkar-ə-ˈbau̇, ˌkär-\ *noun, plural* **-bao** *or* **-baos** [Philippine Spanish, from Bisayan of Samar and Leyte *karabáw*] (1900)

: WATER BUFFALO

ca·ra·bid \ˈkar-ə-bəd, kə-ˈra-bəd\ *noun* [ultimately from Greek *karabos* horned beetle] (1880)

: GROUND BEETLE

car·a·bi·neer *or* **car·a·bi·nier** \ˌkar-ə-bə-ˈnir\ *noun* [French *carabinier,* from *carabine* carbine] (1672)

: a cavalry soldier armed with a carbine

car·a·bi·ner \ˌkar-ə-ˈbē-nər\ *noun* [German *Karabiner,* short for *Karabinerhaken,* literally, carabineer's hook] (1920)

: an oblong metal ring with one spring-hinged side that is used especially in mountain climbing as a connector and to hold a freely running rope

carabiner

ca·ra·bi·ne·ro \ˌkar-ə-bə-ˈner-(ˌ)ō, ˌkär-\ *noun, plural* **-ros** [Spanish, from *carabina* carbine, from French *carabine*] (1845)

1 : a member of a Spanish national police force serving especially as frontier guards

2 : a customs or coast guard officer in the Philippines

ca·ra·bi·nie·re \ˌkar-ə-bən-ˈyer-(ˌ)ā, ˌkär-\ *noun, plural* **-nie·ri** \-ˈyer-ē\ [Italian, from French *carabinier*] (1847)

: a member of the Italian national police force

car·a·cal \ˈkar-ə-ˌkal\ *noun* [French, from Turkish *karakulak,* from *kara* black + *kulak* ear] (1760)

: a long-legged reddish brown nocturnal cat (*Felis caracal* synonym *Lynx caracal*) of savannas in Africa and parts of Asia that has long pointed ears with a tuft of black hairs at the tip

ca·ra·ca·ra \ˌkar-ə-ˈkar-ə, -ə-kə-ˈrä\ *noun* [Spanish *caracara* & Portuguese *caracará,* from Tupi *caracará*] (1838)

: any of various large long-legged mostly South American hawks classified with the falcons

car·a·cole \ˈkar-ə-ˌkōl\ *noun* [French, from Spanish *caracol* snail, spiral stair, caracole] (1614)

: a half turn to right or left executed by a mounted horse

— caracole *verb*

car·a·cul \ˈkar-ə-kəl\ *noun* [alteration of *karakul*] (1894)

: the pelt of a karakul lamb after the curl begins to loosen

ca·rafe \kə-ˈraf, -ˈräf\ *noun* [French, from Italian *caraffa,* from Arabic *gharrāfah*] (1786)

: a bottle with a flaring lip used to hold beverages and especially wine

car·am·bo·la \ˌkar-əm-ˈbō-lə\ *noun* [Portuguese, from Marathi *karambal,* from Sanskrit *karmaphala*] (1598)

1 : a five-angled green to yellow tropical fruit of star-shaped cross section — called also *starfruit*

2 : a tropical tree (*Averrhoa carambola*) of the wood-sorrel family widely cultivated for carambolas

car·a·mel \ˈkär-məl; ˈkar-ə-məl, -ˌmel\ *noun* [French, from Spanish *caramelo,* from Portuguese, icicle, caramel, from Late Latin *calamellus* small reed — more at SHAWM] (1725)

1 : an amorphous brittle brown and somewhat bitter substance obtained by heating sugar and used as a coloring and flavoring agent

2 : a firm chewy usually caramel-flavored candy

car·a·mel·ise *British variant of* CARAMELIZE

car·a·mel·ize \-mə-ˌlīz\ *verb* **-ized; -iz·ing** (1842)

transitive verb

: to change (as sugar) into caramel

intransitive verb

: to change to caramel

ca·ran·gid \kə-ˈran-jəd, -ˈraŋ-gəd\ *adjective* [ultimately from French *carangue* shad, horse mackerel, from Spanish *caranga*] (1931)

: of or relating to a large family (Carangidae of the order Perciformes) of marine spiny-finned bony fishes including important food fishes

— carangid *noun*

car·a·pace \ˈkar-ə-ˌpās\ *noun* [French, from Spanish *carapacho*] (1836)

1 : a bony or chitinous case or shield covering the back or part of the back of an animal (as a turtle or crab)

2 : a protective, decorative, or disguising shell ⟨the *carapace* of reserve he built around himself —M. M. Mintz⟩

[1]**carat** *variant of* KARAT

[2]**car·at** \ˈkar-ət\ *noun* [Middle English, from Middle French, from Italian *carato,* from Arabic *qīrāṭ* bean pod, a small weight, from Greek *keration* carob bean, a small weight, from diminutive of *kerat-, keras* horn — more at HORN] (1555)

: a unit of weight for precious stones equal to 200 milligrams

[1]**car·a·van** \ˈkar-ə-ˌvan\ *noun* [Italian *caravana,* from Persian *kārwān*] (1588)

1 a : a company of travelers on a journey through desert or hostile regions; *also* **:** a train of pack animals **b :** a group of vehicles traveling together in a file

2 a : a covered wagon or motor vehicle equipped as traveling living quarters **b** *British* **:** TRAILER 3b

[2]**caravan** *intransitive verb* **-vanned** *or* **-vaned; -van·ning** *or* **-van·ing** (1885)

: to travel in a caravan

car·a·van·ner *or* **car·a·van·er** \-ˌva-nər\ *noun* (1909)

1 : one that travels in a caravan

2 *British* **:** one who goes camping with a trailer

car·a·van·sa·ry \ˌkar-ə-ˈvan(t)-sə-rē\ *or* **car·a·van·se·rai** \-sə-ˌrī\ *noun, plural* **-ries** *or* **-rais** *or* **-rai** [Persian *kārwānsarāī,* from *kārwān* caravan + *sarāī* palace, inn] (1599)

1 : an inn surrounding a court in eastern countries where caravans rest at night

2 : HOTEL, INN

car·a·vel \ˈkar-ə-ˌvel, -vəl\ *noun* [Middle French *caravelle,* from Old Portuguese *caravela*] (1527)

: any of several sailing ships; *specifically* **:** a small 15th and 16th century ship that has broad bows, high narrow poop, and usually three masts with lateen or both square and lateen sails

caravel

car·a·way \ˈkar-ə-ˌwā\ *noun* [Middle English, probably from Medieval Latin *carvi,* from Arabic *karawyā,* from Greek *karon*] (13th century)

\ə\ **abut** \ᵊ\ **kitten** \ər\ **further** \a\ **ash** \ā\ **ace** \ä\ **mop, mar** \au̇\ **out** \ch\ **chin** \e\ **bet** \ē\ **easy** \g\ **go** \i\ **hit** \ī\ **ice** \j\ **job** \ŋ\ **sing** \ō\ **go** \ȯ\ **law** \ȯi\ **boy** \th\ **thin** \t͟h\ **the** \ü\ **loot** \u̇\ **foot** \y\ **yet** \zh\ **vision** *see also* Guide to Pronunciation

1 : a biennial usually white-flowered aromatic herb (*Carum carvi*) of the carrot family
2 : the pungent fruit of the caraway used in seasoning and medicine — called also *caraway seed*

[1]carb \'kärb\ *noun* (circa 1942)
slang **:** CARBURETOR

[2]carb \'kärb\ *or* **car•bo** \'kär-(ˌ)bō\ *noun* (1965)
: CARBOHYDRATE; *also* **:** a high-carbohydrate food — usually used in plural

carb- *or* **carbo-** *combining form* [French, from *carbone*]
: carbon **:** carbonic **:** carbonyl **:** carboxyl ⟨*car*bide⟩ ⟨*carbo*hydrate⟩

car•ba•chol \'kär-bə-ˌkȯl, -ˌkōl\ *noun* [*car*bamic acid + *chol*ine] (circa 1940)
: a synthetic parasympathomimetic drug $C_6H_{15}ClN_2O_2$ that is used in veterinary medicine and topically in glaucoma

car•ba•mate \'kär-bə-ˌmāt, kär-'ba-ˌmāt\ *noun* (1888)
: a salt or ester of carbamic acid; *especially* **:** one that is a synthetic organic insecticide

car•bam•ic acid \(ˌ)kär-'ba-mik-\ *noun* [International Scientific Vocabulary *carbam*ide + [1]*-ic*] (1869)
: an acid CH_3NO_2 known in the form of salts and esters that is a half amide of carbonic acid

car•bam•ide \'kär-bə-ˌmīd, kär-'ba-məd\ *noun* [International Scientific Vocabulary *carb-* + *amide*] (1865)
: UREA

carb•ami•no \ˌkär-bə-'mē-(ˌ)nō\ *adjective* (1922)
: relating to any carbamic acid derivative formed by reaction of carbon dioxide with an amino acid or a protein (as hemoglobin)

carb•an•ion \kär-'ba-ˌnī-ən, -ˌnī-ˌän\ *noun* (1933)
: an organic ion carrying a negative charge on a carbon atom — compare CARBONIUM ION

car•barn \'kär-ˌbärn\ *noun* (1880)
: a building that houses the cars of a street railway or the buses of a bus system

car•ba•ryl \'kär-bə-ˌril\ *noun* [*carb*amate + *aryl*] (1963)
: a carbamate insecticide $C_{12}H_{11}NO_2$ effective especially against numerous crop, forage, and forest pests

car•ba•zole \'kär-bə-ˌzōl\ *noun* [International Scientific Vocabulary *carb-* + *az-* + *-ole*] (1887)
: a crystalline slightly basic cyclic compound $C_{12}H_9N$ found in anthracene and used in making dyes

car•bide \'kär-ˌbīd\ *noun* [International Scientific Vocabulary] (circa 1865)
1 : a binary compound of carbon with a more electropositive element; *especially* **:** CALCIUM CARBIDE
2 : a very hard material made of carbon and one or more heavy metals

car•bine \'kär-ˌbēn, -ˌbīn\ *noun* [French *carabine,* from Middle French *carabin* carabineer] (1605)
1 : a short-barreled lightweight firearm originally used by cavalry
2 : a light short-barreled repeating rifle that is used as a supplementary military arm or for hunting in dense brush

car•bi•nol \'kär-bə-ˌnȯl, -ˌnōl\ *noun* [International Scientific Vocabulary, from obsolete German *Karbin* methyl, from German *karb-* carb-] (circa 1885)
: METHANOL; *also* **:** an alcohol derived from it

car•bo•cy•clic \ˌkär-bō-'sī-klik, -'si-\ *adjective* [International Scientific Vocabulary] (1899)
: being or having an organic ring composed of carbon atoms

car•bo•hy•drase \ˌkär-bō-'hī-ˌdrās, -bə-, -ˌdrāz\ *noun* [International Scientific Vocabulary] (1910)
: any of a group of enzymes (as amylase) that promote hydrolysis or synthesis of a carbohydrate (as a disaccharide)

car•bo•hy•drate \-ˌdrāt, -drət\ *noun* (circa 1869)
: any of various neutral compounds of carbon, hydrogen, and oxygen (as sugars, starches, and celluloses) most of which are formed by green plants and which constitute a major class of animal foods

car•bol•ic \kär-'bä-lik\ *noun* (1884)
: PHENOL 1

carbolic acid *noun* [International Scientific Vocabulary *carb-* + Latin *oleum* oil — more at OIL] (circa 1859)
: PHENOL 1

car bomb *noun* (1973)
: an explosive device concealed in an automobile for use as a weapon of terrorism

car•bon \'kär-bən\ *noun, often attributive* [French *carbone,* from Latin *carbon-, carbo* ember, charcoal] (1789)
1 : a nonmetallic chiefly tetravalent element found native (as in the diamond and graphite) or as a constituent of coal, petroleum, and asphalt, of limestone and other carbonates, and of organic compounds or obtained artificially in varying degrees of purity especially as carbon black, lampblack, activated carbon, charcoal, and coke — see ELEMENT table
2 : a carbon rod used in an arc lamp
3 a : a sheet of carbon paper **b :** CARBON COPY

car•bo•na•ceous \ˌkär-bə-'nā-shəs\ *adjective* (1791)
1 : rich in carbon
2 : relating to, containing, or composed of carbon

[1]car•bo•na•do \ˌkär-bə-'nā-(ˌ)dō, -'nä-\ *noun, plural* **-dos** *or* **-does** [Spanish *carbonada*] (1586)
archaic **:** a piece of meat scored before grilling

[2]carbonado *transitive verb* (1599)
1 *archaic* **:** to make a carbonado of
2 *archaic* **:** CUT, SLASH

[3]carbonado *noun, plural* **-dos** [Portuguese, literally, carbonated] (1853)
: an impure opaque dark-colored fine-grained aggregate of diamond particles valuable for its superior toughness

car•bo•nara \ˌkär-bə-'när-ə\ *noun* [Italian dialect (*alla*) *carbonara,* literally, in the manner of a charcoal maker] (1963)
: a dish of hot pasta into which other ingredients (as eggs, bacon or ham, and grated cheese) have been mixed — often used as a postpositive modifier ⟨spaghetti *carbonara*⟩

[1]car•bon•ate \'kär-bə-ˌnāt, -nət\ *noun* (1794)
: a salt or ester of carbonic acid

[2]car•bon•ate \-ˌnāt\ *transitive verb* **-at•ed; -at•ing** (1805)
1 : to convert into a carbonate
2 : to combine or impregnate with carbon dioxide ⟨*carbonated* beverages⟩
— car•bon•ation \ˌkär-bə-'nā-shən\ *noun*

carbon black *noun* (circa 1889)
: any of various colloidal black substances consisting wholly or principally of carbon obtained usually as soot and used especially in tires and as pigments

carbon copy *noun* (1895)
1 : a copy made by carbon paper
2 : DUPLICATE

carbon cycle *noun* (1912)
1 : the cycle of carbon in the earth's ecosystems in which carbon dioxide is fixed by photosynthetic organisms to form organic nutrients and is ultimately restored to the inorganic state by respiration and protoplasmic decay
2 : a cycle of thermonuclear reactions in which four hydrogen atoms synthesize into a helium atom by the catalytic action of carbon with the release of nuclear energy and which is held to be the source of most of the energy radiated by the sun and stars

carbon dating *noun* (1951)
: the determination of the age of old material (as an archaeological or paleontological specimen) by means of the content of carbon 14
— carbon–date \'kär-bən-'dāt\ *transitive verb*

carbon dioxide *noun* (1873)
: a heavy colorless gas CO_2 that does not support combustion, dissolves in water to form carbonic acid, is formed especially in animal respiration and in the decay or combustion of animal and vegetable matter, is absorbed from the air by plants in photosynthesis, and is used in the carbonation of beverages

carbon disulfide *noun* (1869)
: a colorless flammable poisonous liquid CS_2 used as a solvent for rubber and as an insect fumigant — called also *carbon bisulfide*

carbon fiber *noun* (1960)
: a very strong lightweight synthetic fiber made especially by carbonizing acrylic fiber at high temperatures

carbon 14 *see* FOURTEEN\ *noun* (1936)
: a heavy radioactive isotope of carbon of mass number 14 used especially in tracer studies and in dating archaeological and geological materials

car•bon•ic \kär-'bä-nik\ *adjective* (1791)
: of, relating to, or derived from carbon, carbonic acid, or carbon dioxide

carbonic acid *noun* (1791)
: a weak dibasic acid H_2CO_3 known only in solution that reacts with bases to form carbonates

carbonic acid gas *noun* (circa 1880)
: CARBON DIOXIDE

carbonic an•hy•drase \-an-'hī-ˌdrās, -ˌdrāz\ *noun* [*anhydr*ous + *-ase;* from its promotion of dehydration] (1932)
: a zinc-containing enzyme that occurs in living tissues (as red blood cells) and aids carbon-dioxide transport from the tissues and its release from the blood in the lungs by catalyzing the reversible hydration of carbon dioxide to carbonic acid

car•bon•if•er•ous \ˌkär-bə-'ni-f(ə-)rəs\ *adjective* (1799)
1 : producing or containing carbon or coal
2 *capitalized* **:** of, relating to, or being the period of the Paleozoic era between the Devonian and the Permian or the corresponding system of rocks that includes coal beds — see GEOLOGIC TIME table
— Carboniferous *noun*

car•bo•ni•um ion \kär-'bō-nē-əm-\ *noun* [*carb-* + *-onium*] (1942)
: an organic ion carrying a positive charge on a carbon atom — compare CARBANION

car•bon•iza•tion \ˌkär-bə-nə-'zā-shən\ *noun* (1804)
: the process of carbonizing; *especially* **:** DESTRUCTIVE DISTILLATION

car•bon•ize \'kär-bə-ˌnīz\ *verb* **-ized; -iz•ing** (1806)
transitive verb
1 : to convert into carbon or a carbonic residue
2 : CARBURIZE 1
intransitive verb
: to become carbonized **:** CHAR

car•bon•less \'kär-bən-ləs\ *adjective* (1850)
1 : being without carbon
2 : being or composed of paper that makes multiple copies without intervening layers of carbon paper ⟨*carbonless* forms⟩

carbon monoxide *noun* (1873)
: a colorless odorless very toxic gas CO that burns to carbon dioxide with a blue flame and is formed as a product of the incomplete combustion of carbon

car•bon•nade *also* **car•bo•nade** \ˌkär-bə-'näd\ *noun* [French, literally, dish of grilled meat, from Italian *carbonata,* from *carbone* charcoal, coal, from Latin *carbon-, carbo*] (1877)
: a beef stew cooked in beer

carbon paper *noun* (1895)
: a thin paper faced with a waxy pigmented coating so that when placed between two sheets of paper the pressure of writing or typing on the top sheet causes transfer of pigment to the bottom sheet

carbon steel *noun* (1903)
: a strong hard steel that derives its physical properties from the presence of carbon and is used in hand tools and kitchen utensils

carbon tetrachloride *noun* (1866)
: a colorless nonflammable toxic liquid CCl_4 that has an odor resembling that of chloroform and is used as a solvent and a refrigerant

carbon 13 \-thir-'tēn, -'thir-ˌtēn\ *noun* (1939)
: an isotope of carbon of mass number 13 that constitutes about 1/70 of natural carbon and is used as a tracer especially in spectroscopy utilizing nuclear magnetic resonance

carbon 12 \-'twelv\ *noun* (1946)
: an isotope of carbon of mass number 12 that is the most abundant carbon isotope and is used as a standard for measurements of atomic weight

car·bon·yl \'kär-bə-ˌnil, -ˌnēl\ *noun* (1869)
1 : an organic functional group −CO− occurring in aldehydes, ketones, carboxylic acids, esters, and their derivatives
2 : a coordination complex involving the neutral radical CO ⟨chromium *carbonyl*⟩
— **car·bon·yl·ic** \ˌkär-bə-'ni-lik\ *adjective*

car·bon·yl·a·tion \(ˌ)kär-ˌbä-nə-'lā-shən\ *noun* (1946)
: the synthesis of a carbonyl compound especially by a reaction involving carbon monoxide

Car·bo·run·dum \ˌkär-bə-'rən-dəm\ *trademark*
— used for various abrasives

carboxy- *or* **carbox-** *combining form*
: carboxyl ⟨*carboxy*peptidase⟩

car·box·yl \kär-'bäk-səl\ *noun* [International Scientific Vocabulary] (1869)
: a univalent radical COOH typical of organic acids — called also *carboxyl group*
— **car·box·yl·ic** \ˌkär-(ˌ)bäk-'si-lik\ *adjective*

car·box·yl·ase \kär-'bäk-sə-ˌlās, -ˌlāz\ *noun* [International Scientific Vocabulary] (1911)
: an enzyme that catalyzes decarboxylation or carboxylation

¹car·box·yl·ate \-ˌlāt, -lət\ *noun* (1927)
: a salt or ester of a carboxylic acid

²car·box·yl·ate \-ˌlāt\ *transitive verb* **-at·ed; -at·ing** (circa 1928)
: to introduce carboxyl or carbon dioxide into (a compound) with formation of a carboxylic acid
— **car·box·yl·a·tion** \(ˌ)kär-ˌbäk-sə-'lā-shən\ *noun*

carboxylic acid *noun* (1902)
: an organic acid (as acetic acid) containing one or more carboxyl groups

car·boxy·meth·yl·cel·lu·lose \kär-ˌbäk-sē-ˌme-thəl-'sel-yə-ˌlōs, -ˌlōz\ *noun* (1947)
: an acid ether derivative of cellulose that in the form of its sodium salt is used as a thickening, emulsifying, and stabilizing agent and as a bulk laxative in medicine

car·boxy·pep·ti·dase \-'pep-tə-ˌdās, -ˌdāz\ *noun* (1937)
: an enzyme that hydrolyzes peptides and especially polypeptides by splitting off sequentially the amino acids at the end of the peptide chain which contain free carboxyl groups

car·boy \'kär-ˌbȯi\ *noun* [Persian *qarāba,* from Arabic *qarrābah* demijohn] (1753)
: a large container for liquids

car·bun·cle \'kär-ˌbəŋ-kəl\ *noun* [Middle English, from Old French, from Latin *carbunculus* small coal, carbuncle, diminutive of *carbon-, carbo* charcoal, ember] (13th century)
1 a *obsolete* **:** any of several red precious stones **b :** the garnet cut cabochon
2 : a painful local purulent inflammation of the skin and deeper tissues with multiple openings for the discharge of pus and usually necrosis and sloughing of dead tissue
— **car·bun·cled** \-kəld\ *adjective*
— **car·bun·cu·lar** \kär-'bəŋ-kyə-lər\ *adjective*

car·bu·ret·ed \'kär-bə-ˌrā-təd, -byə-, *especially by chemists* -ˌre-təd\ *adjective* [back-formation from *carburetor*] (1972)
: equipped with a carburetor ⟨a *carbureted* engine⟩

car·bu·re·tion \ˌkär-bə-'rā-shən, -byə-\ *noun* [*carburet* to combine chemically with carbon, from obsolete *carburet* carbide] (1896)
: the process of mixing (as in a carburetor) the vapor of a flammable hydrocarbon (as gasoline) with air to form an explosive mixture especially for use in an internal-combustion engine

car·bu·re·tor \'kär-bə-ˌrā-tər, -byə-\ *noun* (1896)
: an apparatus for supplying an internal combustion engine with an explosive mixture of vaporized fuel and air

car·bu·ret·tor *also* **car·bu·ret·ter** \ˌkär-byə-'re-tər, 'kär-byə-ˌ\ *chiefly British variant of* CARBURETOR

car·bu·rise *British variant of* CARBURIZE

car·bu·rize \'kär-bə-ˌrīz, -byə-\ *transitive verb* **-rized; -riz·ing** [obsolete *carburet* carbide] (circa 1889)
: to combine or impregnate (as metal) with carbon
— **car·bu·ri·za·tion** \ˌkär-bə-rə-'zā-shən, -byə-\ *noun*

car·ca·net \'kär-kə-nət\ *noun* [Middle French *carcan*] (circa 1530)
archaic **:** an ornamental necklace, chain, collar, or headband

car·case \'kär-kəs\ *British variant of* CARCASS

car·cass \'kär-kəs\ *noun* [Middle English *carcays,* from Middle French *carcasse,* from Old French *carcois*] (14th century)
1 : a dead body **:** CORPSE; *especially* **:** the dressed body of a meat animal
2 : the living, material, or physical body
3 : the decaying or worthless remains of a structure ⟨the *carcass* of an abandoned automobile⟩
4 : the underlying structure or frame of something (as of a piece of furniture)

carcin- *or* **carcino-** *combining form* [Greek *karkin-, karkino-,* from *karkinos* ulcerous sore, literally, crab — more at CANCER]
: tumor **:** cancer ⟨*carcino*genic⟩

car·ci·no·em·bry·on·ic antigen \ˌkär-sᵊn-ō-ˌem-brē-'ä-nik-\ *noun* (1967)
: a glycoprotein present in fetal gut tissues during the first two trimesters of pregnancy and in peripheral blood of patients with some forms of cancer

car·cin·o·gen \kär-'si-nə-jən, 'kär-sᵊn-ə-ˌjen\ *noun* (1853)
: a substance or agent producing or inciting cancer
— **car·ci·no·gen·ic** \ˌkär-sᵊn-ō-'je-nik\ *adjective*
— **car·ci·no·ge·nic·i·ty** \-jə-'ni-sə-tē\ *noun*

car·ci·no·gen·e·sis \ˌkär-sᵊn-ō-'je-nə-sis\ *noun* (circa 1923)
: the production of cancer

car·ci·noid \'kär-sᵊn-ˌȯid\ *noun* (1925)
: a benign or malignant tumor arising especially from the mucosa of the gastrointestinal tract

car·ci·no·ma \ˌkär-sᵊn-'ō-mə\ *noun, plural* **-mas** *or* **-ma·ta** \-mə-tə\ [Latin, from Greek *karkinōma* cancer, from *karkinos*] (circa 1751)
: a malignant tumor of epithelial origin
— **car·ci·no·ma·tous** \-'ō-mə-təs\ *adjective*

car·ci·no·ma·to·sis \-ˌō-mə-'tō-səs\ *noun* [New Latin, from Latin *carcinomat-, carcinoma*] (1903)
: a condition in which multiple carcinomas develop simultaneously usually after dissemination from a primary source

car·ci·no·sar·co·ma \ˌkär-sᵊn-ō-(ˌ)sär-'kō-mə\ *noun, plural* **-mas** *or* **-ma·ta** \-mə-tə\ (circa 1919)
: a malignant tumor combining elements of carcinoma and sarcoma

car coat *noun* (1958)
: a three-quarter-length overcoat

¹card \'kärd\ *transitive verb* (14th century)
: to cleanse, disentangle, and collect together (as fibers) by the use of cards preparatory to spinning
— **card·er** *noun*

²card *noun* [Middle English *carde,* from Middle French, from Late Latin *cardus* thistle, from Latin *carduus* — more at CHARD] (15th century)
1 : an instrument or machine for carding fibers that consists usually of bent wire teeth set closely in rows in a thick piece of leather fastened to a back
2 : an implement for raising a nap on cloth

³card *noun* [Middle English *carde,* modification of Middle French *carte,* probably from Old Italian *carta,* literally, leaf of paper, from Latin *charta* leaf of papyrus, from Greek *chartēs*] (15th century)
1 : PLAYING CARD
2 *plural but singular or plural in construction* **a :** a game played with cards **b :** card playing
3 : something compared to a valuable playing card in one's hand
4 : a usually clownishly amusing person **:** WAG
5 : COMPASS CARD
6 a : a flat stiff usually small and rectangular piece of material (as paper, paperboard, or plastic) usually bearing information: as (1) **:** POSTCARD (2) **:** VISITING CARD (3) **:** CREDIT CARD (4) **:** one bearing a picture (as of a baseball player) on one side and usually statistical data on the other (5) **:** one on which computer information is stored (as in the form of punched holes or magnetic encoding) (6) **:** one bearing electronic circuit components for insertion into a larger electronic device (as a computer) **b :** PROGRAM; *especially* **:** a sports program **c** (1) **:** a wine list (2) **:** MENU **d :** GREETING CARD ⟨a birthday *card*⟩
— **in the cards** *also* **on the cards :** INEVITABLE

⁴card *transitive verb* (1884)
1 : to place or fasten on or by means of a card
2 : to provide with a card
3 : to list or record on a card
4 : SCORE
5 : to ask for identification (as in a bar)

car·da·mom \'kär-də-məm, -ˌmäm\ *noun* [Latin *cardamomum,* from Greek *kardamōmon,* blend of *kardamon* peppergrass and *amōmon,* an Indian spice plant] (1553)
: the aromatic capsular fruit of an East Indian herb (*Elettaria cardamomum*) of the ginger family with seeds used as a spice or condiment and in medicine; *also* **:** this plant

¹card·board \'kärd-ˌbōrd, -ˌbȯrd\ *noun* (1848)
: PAPERBOARD

²cardboard *adjective* (1893)
1 a : made of or as if of cardboard **b :** FLAT, TWO-DIMENSIONAL
2 : UNREAL, STEREOTYPED ⟨a play with *cardboard* characters⟩

card–car·ry·ing \'kärd-ˌkar-ē-iŋ\ *adjective* [from the assumption that such a person carries an identification card] (1948)
1 : being a full-fledged member especially of a Communist party
2 : being strongly identified with a group (as of people with a common interest) ⟨*card*-

carrying members of the ecology movement —Richard Neuhaus⟩

card catalog *noun* (1854)
: a catalog (as of books) in which the entries are arranged systematically on cards

card•hold•er \'kärd-ˌhōl-dər\ *noun* (1909)
: one who possesses a card and especially a credit card

cardi- *or* **cardio-** *combining form* [Greek *kardi-, kardio-,* from *kardia* — more at HEART]
: heart **:** cardiac **:** cardiac and ⟨*cardio*gram⟩ ⟨*cardio*vascular⟩

car•dia \'kär-dē-ə\ *noun, plural* **-di•ae** \-dē-ˌē\ *or* **-dias** [New Latin, from Greek *kardia* heart, upper orifice of the stomach] (1782)
: the opening of the esophagus into the stomach; *also* **:** the part of the stomach adjoining this opening

-cardia *noun combining form* [New Latin, from Greek *kardia*]
: heart action or location (of a specified type) ⟨tachy*cardia*⟩

¹car•di•ac \'kär-dē-ˌak\ *adjective* [Latin *cardiacus,* from Greek *kardiakos,* from *kardia*] (1601)
1 a : of, relating to, situated near, or acting on the heart **b :** of or relating to the cardia of the stomach
2 : of, relating to, or affected with heart disease

²cardiac *noun* (circa 1929)
: a person with heart disease

cardiac muscle *noun* (circa 1881)
: the principal muscle tissue of the vertebrate heart made up of striated fibers that appear to be separated from each other under the electron microscope but that function in long-term rhythmic contraction as if in protoplasmic continuity

car•di•gan \'kär-di-gən\ *noun* [James Thomas Brudenell, 7th Earl of *Cardigan* (died 1868) English soldier] (1868)
: a usually collarless sweater or jacket that opens the full length of the center front ◆

Cardigan Welsh corgi *noun* [*Cardigan,* former county in Wales] (1935)
: a Welsh corgi with rounded ears, slightly bowed forelegs, and long tail — called also *Cardigan*; see WELSH CORGI illustration

¹car•di•nal \'kärd-nəl, 'kär-dᵊn-əl\ *noun* [Middle English, from Medieval Latin *cardinalis,* from Late Latin *cardinalis,* adjective] (12th century)
1 : a high ecclesiastical official of the Roman Catholic Church who ranks next below the pope and is appointed by him to assist him as a member of the college of cardinals
2 : CARDINAL NUMBER — usually used in plural
3 : a woman's short hooded cloak originally of scarlet cloth
4 [from its color, resembling that of the cardinal's robes] **:** a crested finch (*Cardinalis cardinalis*) of the eastern U.S. and adjacent Canada, the southwestern U.S., and Mexico to Belize which has a black face and heavy red bill in both sexes and is nearly completely red in the male
— car•di•nal•ship \-ˌship\ *noun*

²cardinal *adjective* [Middle English, from Late Latin *cardinalis,* from Latin, serving as a hinge, from *cardin-, cardo* hinge] (14th century)
: of basic importance **:** MAIN, CHIEF, PRIMARY
synonym see ESSENTIAL
— car•di•nal•ly *adverb*

car•di•nal•ate \'kärd-nə-lət, 'kär-dᵊn-ə-let, -ˌlāt\ *noun* (1645)
: the office, rank, or dignity of a cardinal

cardinal flower *noun* (1698)
: a North American lobelia (*Lobelia cardinalis*) that bears a spike of brilliant red flowers

car•di•nal•i•ty \ˌkär-dᵊn-'a-lə-tē\ *noun, plural* **-ties** [¹*cardinal* + *-ity*] (1935)
: the number of elements in a given mathematical set

cardinal number *noun* (1591)
1 : a number (as 1, 5, 15) that is used in simple counting and that indicates how many elements there are in an assemblage — see NUMBER table
2 : the property that a mathematical set has in common with all sets that can be put in one-to-one correspondence with it

cardinal point *noun* (1755)
: one of the four principal compass points north, south, east, and west

cardinal virtue *noun* (14th century)
1 : one of the four classically defined natural virtues prudence, justice, temperance, or fortitude
2 : a quality designated as a major virtue

car•dio•gen•ic \ˌkär-dē-ō-'je-nik\ *adjective* (circa 1923)
: originating in the heart **:** caused by a cardiac condition ⟨*cardiogenic* shock⟩

car•dio•gram \'kär-dē-ə-ˌgram\ *noun* [International Scientific Vocabulary] (1876)
: the curve or tracing made by a cardiograph

car•dio•graph \-ˌgraf\ *noun* [International Scientific Vocabulary] (1870)
: an instrument that registers graphically movements of the heart
— car•dio•graph•ic \ˌkär-dē-ə-'gra-fik\ *adjective*
— car•di•og•ra•phy \ˌkär-dē-'ä-grə-fē\ *noun*

car•di•oid \'kär-dē-ˌȯid\ *noun* (1753)
: a heart-shaped curve that is traced by a point on the circumference of a circle rolling completely around an equal fixed circle and has the general equation $\rho = a(1 + \cos\theta)$ in polar coordinates

cardioid: *ABP* fixed circle; *PCD* first position of rolling circle; *P* tracing point; *PM* diameter through *P*; P_1, P_2, P_3, P_4 various positions of *P*; P_1M_1, P_2M_2, P_3M_3, P_4M_4 various positions of *PM*

car•di•ol•o•gy \ˌkär-dē-'ä-lə-jē\ *noun* [International Scientific Vocabulary] (1847)
: the study of the heart and its action and diseases
— car•di•o•log•i•cal \-dē-ə-'lä-ji-kəl\ *adjective*
— car•di•ol•o•gist \-dē-'ä-lə-jist\ *noun*

car•dio•my•op•a•thy \'kär-dē-ō-(ˌ)mī-'ä-pə-thē\ *noun, plural* **-thies** (1961)
: a typically chronic disorder of heart muscle that may involve hypertrophy and obstructive damage to the heart

car•di•op•a•thy \ˌkär-dē-'ä-pə-thē\ *noun, plural* **-thies** (1885)
: any disease of the heart

car•dio•pul•mo•nary \ˌkär-dē-ō-'pu̇l-mə-ˌner-ē, -'pəl-\ *adjective* (circa 1881)
: of or relating to the heart and lungs

cardiopulmonary resuscitation *noun* (1972)
: a procedure designed to restore normal breathing after cardiac arrest that includes the clearance of air passages to the lungs, mouth-to-mouth method of artificial respiration, and heart massage by the exertion of pressure on the chest

car•dio•re•spi•ra•to•ry \ˌkär-dē-ō-'res-p(ə-)rə-ˌtōr-ē, -ri-'spī-rə-, -ˌtȯr-\ *adjective* (1892)
: of or relating to the heart and the respiratory system

car•dio•tho•rac•ic \-thə-'ra-sik\ *adjective* (1962)
: relating to or involving the heart and chest

car•dio•ton•ic \-'tä-nik\ *adjective* (1927)
: tending to increase the tonus of heart muscle
— cardiotonic *noun*

car•dio•vas•cu•lar \-'vas-kyə-lər\ *adjective* [International Scientific Vocabulary] (1879)
: of, relating to, or involving the heart and blood vessels

-cardium *noun combining form, plural* **-cardia** [New Latin, from Greek *kardia*]
: heart ⟨epi*cardium*⟩

car•doon \kär-'dün\ *noun* [French *cardon,* from Late Latin *cardon-, cardo* thistle, from *cardus,* from Latin *carduus* thistle, cardoon] (1611)
: a large perennial plant (*Cynara cardunculus*) related to the artichoke and cultivated for its edible root and petioles; *also* **:** the root and petioles

card•play•er \'kärd-ˌplā-ər\ *noun* (1589)
: one that plays cards

card•sharp•er \-ˌshär-pər\ *or* **card•sharp** \-ˌshärp\ *noun* (1859)
: one who habitually cheats at cards

¹care \'ker, 'kar\ *noun* [Middle English, from Old English *caru;* akin to Old High German *kara* lament, Latin *garrire* to chatter] (before 12th century)
1 : suffering of mind **:** GRIEF
2 a : a disquieted state of mixed uncertainty, apprehension, and responsibility **b :** a cause for such anxiety
3 a : painstaking or watchful attention **b :** MAINTENANCE ⟨floor-*care* products⟩
4 : regard coming from desire or esteem
5 : CHARGE, SUPERVISION ⟨under a doctor's *care*⟩
6 : a person or thing that is an object of attention, anxiety, or solicitude ☆

²care *verb* **cared; car•ing** (before 12th century)
intransitive verb
1 a : to feel trouble or anxiety **b :** to feel interest or concern ⟨*care* about freedom⟩
2 : to give care ⟨*care* for the sick⟩
3 a : to have a liking, fondness, or taste ⟨don't *care* for your attitude⟩ **b :** to have an inclination ⟨would you *care* for some pie⟩
transitive verb
1 : to be concerned about or to the extent of
2 : WISH ⟨if you *care* to go⟩ ■
— car•er *noun*
— care less : not to care — used positively and negatively with the same meaning ⟨I could *care less* what happens⟩ ⟨I couldn't *care less* what happens⟩

¹ca•reen \kə-'rēn\ *verb* [from *carine* side of a ship, from Middle French, submerged part of a

◇ WORD HISTORY
cardigan The style of sweater known as a cardigan is named after James Thomas Brudenell (1797–1868), the 7th Earl of Cardigan and a major general in the British army during the Crimean War. He is now largely remembered as the leader of the disastrous Charge of the Light Brigade. The garment that bears his name was originally a knitted woolen vest that warded off the chill of a Crimean winter. The Crimean War was an unusually fertile breeding ground for new fashions, which were a response to the harsh field conditions experienced by British troops during the long siege of Sevastopol. The village of Balaclava, the site of a battle on October 25, 1854, is the namesake of the knitted hoodlike cap called a balaclava, and Lord Raglan, commander of Britain's Crimean force, is commemorated by another article of clothing, the raglan.

hull, from Latin *carina* hull, half of a nutshell; perhaps akin to Greek *karyon* nut] (circa 1583)
transitive verb
1 : to put (a ship or boat) on a beach especially in order to clean, caulk, or repair the hull
2 : to cause to heel over
intransitive verb
1 a : to careen a boat **b :** to undergo this process
2 : to heel over
3 : to sway from side to side **:** LURCH ⟨a *careening* carriage being pulled wildly . . . by a team of runaway horses —J. P. Getty⟩
4 : CAREER

[2]careen *noun* (1712)
archaic **:** the act or process of careening **:** the state of being careened

[1]ca·reer \kə-ˈrir\ *noun* [Middle French *carrière,* from Old Provençal *carriera* street, from Medieval Latin *carraria* road for vehicles, from Latin *carrus* car] (1580)
1 a : speed in a course ⟨ran at full *career*⟩ **b :** COURSE, PASSAGE
2 : ENCOUNTER, CHARGE
3 : a field for or pursuit of consecutive progressive achievement especially in public, professional, or business life ⟨Washington's *career* as a soldier⟩
4 : a profession for which one trains and which is undertaken as a permanent calling ⟨a *career* diplomat⟩

[2]career *intransitive verb* (1647)
: to go at top speed especially in a headlong manner ⟨a car *careered* off the road⟩

ca·reer·ism \-ˌi-zəm\ *noun* (1933)
: the policy or practice of advancing one's career often at the cost of one's integrity
— **ca·reer·ist** \-ist\ *noun*

care·free \ˈker-ˌfrē, ˈkar-\ *adjective* (1795)
: free from care: as **a :** having no worries or troubles **b :** IRRESPONSIBLE ⟨is *carefree* with his money⟩

care·ful \-fəl\ *adjective* **care·ful·ler; care·ful·lest** (before 12th century)
1 *archaic* **a :** SOLICITOUS, ANXIOUS **b :** filling with care or solicitude
2 : exercising or taking care
3 a : marked by attentive concern and solicitude **b :** marked by wary caution or prudence ⟨be very *careful* with knives⟩ **c :** marked by painstaking effort to avoid errors or omissions — often used with *of* or an infinitive ⟨*careful* of money⟩ ⟨*careful* to adjust the machine⟩ ☆
— **care·ful·ly** \-f(ə-)lē\ *adverb*
— **care·ful·ness** \-fəl-nəs\ *noun*

care·giv·er \-ˌgi-vər\ *noun* (1975)
: a person who provides direct care (as for children or the chronically ill)
— **care·giv·ing** \-ˌgi-viŋ\ *noun*

care·less \-ləs\ *adjective* (before 12th century)
1 a : free from care **:** UNTROUBLED ⟨*careless* days⟩ **b :** INDIFFERENT, UNCONCERNED ⟨*careless* of the consequences⟩
2 : not taking care
3 : not showing or receiving care: **a :** NEGLIGENT, SLOVENLY ⟨*careless* writing⟩ **b :** UNSTUDIED, SPONTANEOUS ⟨*careless* grace⟩ **c** *obsolete* **:** UNVALUED, DISREGARDED
— **care·less·ly** *adverb*
— **care·less·ness** *noun*

[1]ca·ress \kə-ˈres\ *noun* [French *caresse,* from Italian *carezza,* from *caro* dear, from Latin *carus* — more at CHARITY] (circa 1611)
1 : an act or expression of kindness or affection **:** ENDEARMENT
2 a : a light stroking, rubbing, or patting **b :** KISS
— **ca·res·sive** \-ˈre-siv\ *adjective*
— **ca·res·sive·ly** *adverb*

[2]caress *transitive verb* (1658)
1 : to treat with tokens of fondness, affection, or kindness **:** CHERISH
2 a : to touch or stroke lightly in a loving or endearing manner **b :** to touch or affect as if with a caress ⟨echoes that *caress* the ear⟩
— **ca·ress·er** *noun*
— **ca·ress·ing·ly** \kə-ˈre-siŋ-lē\ *adverb*

car·et \ˈkar-ət\ *noun* [Latin, there is lacking, from *carēre* to lack, be without] (1681)
: a wedge-shaped mark made on written or printed matter to indicate the place where something is to be inserted

care·tak·er \ˈker-ˌtā-kər, ˈkar-\ *noun* (1858)
1 : one that gives physical or emotional care and support ⟨served as *caretaker* to the younger children⟩
2 : one that takes care of the house or land of an owner who may be absent
3 : one temporarily fulfilling the function of office ⟨a *caretaker* government⟩
— **care·take** \-ˌtāk\ *verb*
— **care·tak·ing** *noun*

care·worn \-ˌwōrn, -ˌwȯrn\ *adjective* (1828)
: showing the effect of grief or anxiety ⟨a *careworn* face⟩

car·fare \ˈkär-ˌfar, -ˌfer\ *noun* (1870)
: passenger fare (as on a bus)

car·ful \ˈkär-ˌfu̇l\ *noun* (1832)
: as much or as many as a car will hold

car·go \ˈkär-(ˌ)gō\ *noun, plural* **cargoes** *or* **cargos** [Spanish, load, charge, from *cargar* to load, from Late Latin *carricare* — more at CHARGE] (1657)
: the goods or merchandise conveyed in a ship, airplane, or vehicle **:** FREIGHT

cargo cult *noun* (1949)
: any of various Melanesian religious groups characterized by the belief that material wealth (as money or manufactured goods) can be obtained through ritual worship

cargo pocket *noun* (1974)
: a large pocket usually with a flap and a pleat

car·hop \ˈkär-ˌhäp\ *noun* [*car* + *-hop* (as in *bellhop*)] (1937)
: one who serves customers at a drive-in restaurant

Car·ib \ˈkar-əb\ *noun* [New Latin *Caribes* (plural), from Spanish *Caribe,* of Cariban origin — more at CANNIBAL] (1555)
1 : a member of an Indian people of northern South America and the Lesser Antilles
2 : the language of the Caribs

Ca·ri·ban \ˈkar-ə-bən, kə-ˈrē-bən\ *noun* (1901)
1 : a member of a group of Indian peoples of South America and the Lesser Antilles
2 : the language family comprising the languages of the Cariban peoples

Ca·rib·be·an \ˌkar-ə-ˈbē-ən, kə-ˈri-bē-ən\ *adjective* [New Latin *Caribbaeus,* from *Caribes*] (1777)
: of or relating to the Caribs, the eastern and southern West Indies, or the Caribbean Sea ■

ca·ri·be \kə-ˈrē-bē\ *noun* [American Spanish, from Spanish, Carib, cannibal] (1868)
: PIRANHA

car·i·bou \ˈkar-ə-ˌbü\ *noun, plural* **-bou** *or* **-bous** [Canadian French, from Micmac *ɣalipu*] (circa 1665)
: a large gregarious deer (*Rangifer tarandus*) of Holarctic taiga and tundra that usually has palmate antlers in both sexes — used especially for one of the New World; called also *reindeer*

caribou

[1]car·i·ca·ture \ˈkar-i-kə-ˌchu̇r, -ˌchər, -ˌtyu̇r, -ˌtu̇r\ *noun* [Italian *caricatura,* literally, act of loading, from *caricare* to load, from Late Latin *carricare*] (1712)
1 : exaggeration by means of often ludicrous distortion of parts or characteristics
2 : a representation especially in literature or art that has the qualities of caricature

☆ SYNONYMS

Care, concern, solicitude, anxiety, worry mean a troubled or engrossed state of mind or the thing that causes this. CARE implies oppression of the mind weighed down by responsibility or disquieted by apprehension ⟨a face worn by years of *care*⟩. CONCERN implies a troubled state of mind because of personal interest, relation, or affection ⟨crimes caused *concern* in the neighborhood⟩. SOLICITUDE implies great concern and connotes either thoughtful or hovering attentiveness toward another ⟨acted with typical maternal *solicitude*⟩. ANXIETY stresses anguished uncertainty or fear of misfortune or failure ⟨plagued by *anxiety* and self-doubt⟩. WORRY suggests fretting over matters that may or may not be real cause for anxiety ⟨financial *worries*⟩.

Careful, meticulous, scrupulous, punctilious mean showing close attention to detail. CAREFUL implies attentiveness and cautiousness in avoiding mistakes ⟨a *careful* worker⟩. METICULOUS may imply either commendable extreme carefulness or a hampering finicky caution over small points ⟨*meticulous* scholarship⟩. SCRUPULOUS applies to what is proper or fitting or ethical ⟨*scrupulous* honesty⟩. PUNCTILIOUS implies minute, even excessive attention to fine points ⟨*punctilious* observance of ritual⟩.

□ USAGE

care The idiom *care less* is used with *could* in both positive and negative constructions. The positive form of this idiom, *could care less,* is apparently about 20 years younger than the negative, and it has been disparaged by some critics as illogical and worse. It is, as several commentators have pointed out, intended to be sarcastic. No one knows just how it started. It would be nice to account for it as a development from a form of the negative phrase in which the negative element is separated from the rest of the phrase ⟨none of these writers *could care less* about the "tradition of the novel" —Alfred Kazin⟩. But we do not have enough evidence to make a persuasive case. While *could care less* may be superior in speech for purposes of sarcasm, it is hard to be obviously sarcastic in print. Consequently most writers, faced with putting the words on paper, choose the clearer *couldn't care less.*

Caribbean The two pronunciations of *Caribbean* are equally popular, and one can sometimes hear the same speaker switch between the two variants in the course of a conversation. The variant \kə-ˈri-bē-ən\ is the newer one and seems to have taken hold in the first half of the twentieth century. The fact that it is a newer variant in no way means that it is less acceptable than the variant \ˌkar-ə-ˈbē-ən\.

3 : a distortion so gross as to seem like caricature ☆
— **car·i·ca·tur·al** \ˌkar-i-kə-ˈchu̇r-əl, -ˈchər-, -ˈtyu̇r-, -ˈtu̇r-\ *adjective*
— **car·i·ca·tur·ist** \ˈkar-i-kə-ˌchu̇r-ist, -ˌchər-, -ˌtyu̇r-, -ˌtu̇r-\ *noun*

²caricature *transitive verb* **-tured; -tur·ing** (circa 1771)
: to make or draw a caricature of **:** represent in caricature ⟨the portrait *caricatured* its subject⟩

car·ies \ˈkar-ēz, ˈker-\ *noun, plural* **caries** [Latin, decay; akin to Old Irish ara-*chrinn* it decays] (1634)
: a progressive destruction of bone or tooth; *especially* **:** tooth decay

car·il·lon \ˈkar-ə-ˌlän, -lən; ˈkar-ē-ˌän, -ˌōn; kə-ˈril-yən\ *noun* [French, alteration of Old French *quarregnon,* modification of Late Latin *quaternion-, quaternio* set of four — more at QUATERNION] (1775)
1 a : a set of fixed chromatically tuned bells sounded by hammers controlled from a keyboard **b :** an electronic instrument imitating a carillon
2 : a composition for the carillon

car·il·lon·neur \ˌkar-ə-lə-ˈnər, ˌkar-ē-ə-ˈnər, kə-ˌril-yə-ˈnər\ *noun* [French, from *carillon*] (1772)
: a carillon player

ca·ri·na \kə-ˈrī-nə, -ˈrē-\ *noun, plural* **-ri·nas** *or* **-ri·nae** \-ˈrī-ˌnē, -ˈrē-ˌnī\ [New Latin, from Latin, hull, keel — more at CAREEN] (circa 1704)
1 : a keel-shaped anatomical part, ridge, or process
2 *capitalized* **:** a constellation in the southern hemisphere lying near the Southern Cross

car·i·nate \ˈkar-ə-ˌnāt, -nət\ *or* **car·i·nat·ed** \-ˌnā-təd\ *adjective* (1781)
: having or shaped like a keel or carina

ca·ri·o·ca \ˌkar-ē-ˈō-kə\ *noun* [Brazilian Portuguese] (1830)
1 *capitalized* **:** a native or resident of Rio de Janeiro
2 a : a variation of the samba **b :** the music for this dance

car·i·o·gen·ic \ˌkar-ē-ə-ˈje-nik, ˌker-, -ˈjē-\ *adjective* [*caries* + *-o-* + *-genic*] (1930)
: producing or promoting the development of tooth decay ⟨*cariogenic* foods⟩

car·i·ous \ˈkar-ē-əs, ˈker-\ *adjective* [Latin *cariosus,* from *caries*] (1676)
: affected with caries

car·jack·ing \ˈkär-ˌja-kiŋ\ *noun* [*car* + hi*jack* + *-ing*] (1991)
: the theft of an automobile from its driver by force or intimidation
— **car·jack·er** *noun*

cark·ing \ˈkär-kiŋ\ *adjective* [Middle English, from *carken,* literally, to load, burden, from Old North French *carquier,* from Late Latin *carricare*] (circa 1565)
: BURDENSOME, ANNOYING

carl *or* **carle** \ˈkar(-ə)l\ *noun* [Middle English, from Old English *-carl,* from Old Norse *karl* man, carl — more at CHURL] (before 12th century)
1 : a man of the common people
2 *chiefly dialect* **:** CHURL, BOOR

car·line *or* **car·lin** \ˈkär-lən\ *noun* [Middle English *kerling,* from Old Norse, from *karl* man] (14th century)
chiefly Scottish **:** WOMAN; *especially* **:** an old woman

Car·list \ˈkär-list\ *noun* [Spanish *carlista,* from Don *Carlos*] (1830)
: a supporter of Don Carlos or his successors as having rightful title to the Spanish throne
— **Carlist** *adjective*

car·load \ˈkär-ˈlōd, -ˌlōd\ *noun* (1854)
1 : a load that fills a car
2 : the minimum number of tons required for shipping at carload rates

carload rate *noun* (1906)
: a rate for large shipments lower than that quoted for less-than-carload lots of the same class

Car·lo·vin·gian \ˌkär-lə-ˈvin-j(ē-)ən\ *adjective* [French *carlovingien,* from Medieval Latin *Car*lus Charles + French *-ovingien* (as in *mérovingien* Merovingian)] (1781)
: CAROLINGIAN

car·ma·gnole \ˈkär-mən-ˌyōl\ *noun* [French] (1793)
1 : a lively song popular at the time of the first French Revolution
2 : a street dance in a meandering course to the tune of the carmagnole

car·mak·er \ˈkär-ˌmā-kər\ *noun* (1954)
: an automobile manufacturer

Car·mel·ite \ˈkär-mə-ˌlīt\ *noun* [Middle English, from Medieval Latin *carmelita,* from *Carmel* Mount Carmel, Palestine] (15th century)
: a member of the Roman Catholic mendicant Order of Our Lady of Mount Carmel founded in the 12th century
— **Carmelite** *adjective*

car·mi·na·tive \kär-ˈmi-nə-tiv, ˈkär-mə-ˌnā-\ *adjective* [French *carminatif,* from Latin *carminatus,* past participle of *carminare* to card, from (assumed) Latin *carmin-, carmen* card, from Latin *carrere* to card; akin to Lithuanian *karšti* to card] (15th century)
: expelling gas from the alimentary canal so as to relieve colic or griping
— **carminative** *noun*

car·mine \ˈkär-mən, -ˌmīn\ *noun* [French *carmin,* from Medieval Latin *carminium,* irregular from Arabic *qirmiz* kermes + Latin *minium* cinnabar] (1712)
1 : a rich crimson or scarlet lake made from cochineal
2 : a vivid red

car·nage \ˈkär-nij\ *noun* [French, from Medieval Latin *carnaticum* tribute consisting of animals or meat, from Latin *carn-, caro*] (circa 1656)
1 : the flesh of slain animals or men
2 : great and bloody slaughter (as in battle)

car·nal \ˈkär-nᵊl\ *adjective* [Middle English, from Old North French or Late Latin; Old North French, from Late Latin *carnalis,* from Latin *carn-, caro* flesh; akin to Greek *keirein* to cut — more at SHEAR] (14th century)
1 a : relating to or given to crude bodily pleasures and appetites **b :** marked by sexuality
2 : BODILY, CORPOREAL
3 a : TEMPORAL **b :** WORLDLY ☆
— **car·nal·i·ty** \kär-ˈna-lə-tē\ *noun*
— **car·nal·ly** \ˈkär-nᵊl-ē\ *adverb*

car·nall·ite \ˈkär-nᵊl-ˌīt\ *noun* [German *Carnallit,* from Rudolf von *Carnall* (died 1874) German mining engineer] (1876)
: a mineral consisting of hydrous potassium-magnesium chloride that is an important source of potassium

car·nas·si·al \kär-ˈna-sē-əl\ *adjective* [French *carnassier* carnivorous, ultimately from Latin *carn-, caro*] (circa 1852)
: of, relating to, or being teeth of a carnivore often larger and longer than adjacent teeth and adapted for cutting rather than tearing
— **carnassial** *noun*

car·na·tion \kär-ˈnā-shən\ *noun* [Middle French, from Old Italian *carnagione,* from *carne* flesh, from Latin *carn-, caro*] (circa 1535)
1 a (1) **:** the variable color of human flesh (2) **:** a pale to grayish yellow **b :** a moderate red
2 : a plant of any of numerous often cultivated and usually double-flowered varieties or subspecies of an Old World pink (*Dianthus caryophyllus*) originally flesh-colored but now found in many color variations

car·nau·ba \kär-ˈnȯ-bə, -ˈnau̇-; ˌkär-nə-ˈü-bə\ *noun* [Portuguese] (1866)
: a fan-leaved palm (*Copernicia prunifera* synonym *C. cerifera*) of Brazil that has an edible root and yields a useful leaf fiber and carnauba wax

carnauba wax *noun* (1854)
: a hard brittle high-melting wax obtained from the leaves of the carnauba palm and used chiefly in polishes

car·ne·lian \kär-ˈnēl-yən\ *noun* [alteration of *cornelian,* from Middle English *corneline,* from Middle French, perhaps from *cornelle* cornel] (1695)
: a hard tough chalcedony that has a reddish color and is used in jewelry

car·ni·tine \ˈkär-nə-ˌtēn\ *noun* [International Scientific Vocabulary, from Latin *carn-, caro*] (circa 1922)
: a quaternary ammonium compound $C_7H_{15}NO_3$ that is present especially in vertebrate muscle and is involved in the transfer of fatty acids across mitochondrial membranes

car·ni·val \ˈkär-nə-vəl\ *noun* [Italian *carnevale,* alteration of earlier *carnelevare,* literally, removal of meat, from *carne* flesh (from Latin *carn-, caro*) + *levare* to remove, from Latin, to raise] (1549)
1 : a season or festival of merrymaking before Lent
2 a : an instance of merrymaking, feasting, or masquerading **b :** an instance of riotous excess ⟨a *carnival* of violence⟩
3 a : a traveling enterprise offering amusements **b :** an organized program of entertainment or exhibition **:** FESTIVAL ⟨a winter *carnival*⟩ ◆

car·niv·o·ra \kär-ˈni-və-rə\ *noun plural* [Latin, neuter plural of *carnivorus*] (1830)
: carnivorous mammals

car·ni·vore \ˈkär-nə-ˌvōr, -ˌvȯr\ *noun* [ultimately from Latin *carnivorus*] (1840)
1 : a carnivorous animal; *especially* **:** any of an order (Carnivora) of flesh-eating mammals
2 : a carnivorous plant

car·niv·o·rous \kär-ˈni-v(ə-)rəs\ *adjective* [Latin *carnivorus,* from *carn-, caro* + *-vorus* -vorous] (1592)
1 : subsisting or feeding on animal tissues
2 *of a plant* **:** subsisting on nutrients obtained from the breakdown of animal protoplasm
3 : of or relating to the carnivores
4 : RAPACIOUS
— **car·niv·o·rous·ly** *adverb*
— **car·niv·o·rous·ness** *noun*

☆ SYNONYMS

Caricature, burlesque, parody, travesty mean a comic or grotesque imitation. CARICATURE implies ludicrous exaggeration of the characteristic features of a subject ⟨*caricatures* of politicians in cartoons⟩. BURLESQUE implies mockery especially through giving a serious or lofty subject a frivolous treatment ⟨a nightclub *burlesque* of a trial in court⟩. PARODY applies especially to treatment of a trivial or ludicrous subject in the exactly imitated style of a well-known author or work ⟨a witty *parody* of a popular novel⟩. TRAVESTY implies that the subject remains unchanged but that the style is extravagant or absurd ⟨this production is a *travesty* of the opera⟩.

Carnal, fleshly, sensual, animal mean having a relation to the body. CARNAL may mean only this but more often connotes derogatorily an action or manifestation of a person's lower nature ⟨a slave to *carnal* desires⟩. FLESHLY is less derogatory than CARNAL ⟨a saint who had experienced *fleshly* temptations⟩. SENSUAL may apply to any gratification of a bodily desire or pleasure but commonly implies sexual appetite with absence of the spiritual or intellectual ⟨fleshpots providing *sensual* delights⟩. ANIMAL stresses the physical as distinguished from the rational nature of a person ⟨led a mindless *animal* existence⟩.

car·no·tite \'kär-nə-ˌtīt\ *noun* [French, from M. A. *Carnot* (died 1920) French inspector general of mines] (1899)
: a mineral consisting of a hydrous radioactive vanadate of uranium and potassium that is a source of radium and uranium

car·ny *or* **car·ney** *or* **car·nie** \'kär-nē\ *noun, plural* **carnies** *or* **carneys** *often attributive* (circa 1933)
1 : CARNIVAL 3a
2 : a person who works with a carnival

car·ob \'kar-əb\ *noun* [Middle French *carobe,* from Medieval Latin *carrubium,* from Arabic *kharrūbah*] (1548)
1 : a Mediterranean evergreen leguminous tree (*Ceratonia siliqua*) with racemose red flowers
2 : a carob pod or its sweet pulp having a flavor similar to that of chocolate

ca·roche \kə-'rōch, -'rōsh\ *noun* [Middle French *carroche,* from Old Italian *carroccio,* augmentative of *carro* car, from Latin *carrus*] (1591)
: a luxurious horse-drawn carriage

[1]**car·ol** \'kar-əl\ *noun* [Middle English *carole,* from Middle French, modification of Late Latin *choraula* choral song, from Latin, choral accompanist, from Greek *choraulēs,* from *choros* chorus + *aulein* to play a reed instrument, from *aulos,* a reed instrument — more at ALVEOLUS] (14th century)
1 : an old round dance with singing
2 : a song of joy or mirth ⟨the *carol* of a bird —Lord Byron⟩
3 : a popular song or ballad of religious joy ◆

[2]**carol** *verb* **-oled** *or* **-olled; -ol·ing** *or* **-ol·ling** (14th century)
intransitive verb
1 : to sing especially in a joyful manner
2 : to sing carols; *specifically* **:** to go about outdoors in a group singing Christmas carols
transitive verb
1 : to praise in or as if in song
2 : to sing especially in a cheerful manner **:** WARBLE
— car·ol·er *or* **car·ol·ler** \-ə-lər\ *noun*

Car·o·line \'kar-ə-ˌlīn, -lən\ *or* **Car·o·le·an** \ˌkar-ə-'lē-ən\ *adjective* [New Latin *carolinus,* from Medieval Latin *Carolus* Charles] (1652)
: of or relating to Charles — used especially with reference to Charles I and Charles II of England

Car·o·lin·gi·an \ˌkar-ə-'lin-j(ē-)ən\ *adjective* [French *carolingien,* from Medieval Latin *Karolingi* Carolingians, from *Karolus* Charlemagne + *-ingi* (as in *Merovingi* Merovingians)] (1881)
: of or relating to a Frankish dynasty dating from about A.D. 613 and including among its members the rulers of France from 751 to 987, of Germany from 752 to 911, and of Italy from 774 to 961
— Carolingian *noun*

[1]**car·om** \'kar-əm\ *noun* [by shortening & alteration from obsolete *carambole,* from Spanish *carambola*] (1779)
1 a : a shot in billiards in which the cue ball strikes each of two object balls **b :** a shot in pool in which an object ball strikes another ball before falling into a pocket — compare COMBINATION SHOT
2 : a rebounding especially at an angle

[2]**carom** *intransitive verb* (1860)
1 : to strike and rebound **:** GLANCE ⟨the car *caromed* off a tree⟩
2 : to make a carom
3 : to proceed by or as if by caroms

car·o·tene \'kar-ə-ˌtēn\ *noun* [International Scientific Vocabulary, from Late Latin *carota* carrot] (1861)
: any of several orange or red crystalline hydrocarbon pigments $C_{40}H_{56}$ that occur in the chromoplasts of plants and in the fatty tissues of plant-eating animals and are convertible to vitamin A

ca·rot·en·oid *also* **ca·rot·in·oid** \kə-'rä-tᵊn-ˌȯid\ *noun* (1911)
: any of various usually yellow to red pigments (as carotenes) found widely in plants and animals and characterized chemically by a long aliphatic polyene chain composed of eight isoprene units
— carotenoid *adjective*

ca·rot·id \kə-'rä-təd\ *adjective* [French or Greek; French *carotide,* from Greek *karōtides* carotid arteries, from *karoun* to stupefy; akin to Greek *kara* head — more at CEREBRAL] (1543)
: of, relating to, or being the chief artery or pair of arteries that pass up the neck and supply the head
— carotid *noun*

carotid body *noun* (1940)
: a small body of vascular tissue that adjoins the carotid sinus, functions as a chemoreceptor sensitive to change in the oxygen content of blood, and mediates reflex changes in respiratory activity

carotid sinus *noun* (circa 1923)
: a small but richly innervated arterial enlargement that is located near the point in the neck where either carotid artery divides to form its main branches and that functions in the regulation of heart rate and blood pressure

ca·rous·al \kə-'rau̇-zəl\ *noun* (1765)
: CAROUSE 2

[1]**ca·rouse** \kə-'rau̇z\ *noun* [Middle French *carrousse,* from *carous,* adverb, all out (in *boire carous* to empty the cup), from German *gar aus*] (1559)
1 *archaic* **:** a large draft of liquor **:** TOAST
2 : a drunken revel

[2]**carouse** *verb* **ca·roused; ca·rous·ing** (1567)
intransitive verb
1 : to drink liquor deeply or freely
2 : to take part in a carouse **:** engage in dissolute behavior
transitive verb
obsolete **:** to drink up **:** QUAFF
— ca·rous·er *noun*

car·ou·sel \ˌkar-ə-'sel *also* -'zel; 'kar-ə-ˌ\ *noun* [French *carrousel,* from Italian *carosello*] (1650)
1 : a tournament or exhibition in which horsemen execute evolutions
2 a : MERRY-GO-ROUND **b :** a circular conveyer ⟨the luggage *carousel* at the airport⟩ **c :** a revolving case used for storage or display

[1]**carp** \'kärp\ *intransitive verb* [Middle English, of Scandinavian origin; akin to Icelandic *karpa* to dispute] (14th century)
: to find fault or complain querulously
— carp·er *noun*

[2]**carp** *noun* (1904)
: COMPLAINT

[3]**carp** *noun, plural* **carp** *or* **carps** [Middle English *carpe,* from Middle French, from Late Latin *carpa,* probably of Germanic origin; akin to Old High German *karpfo* carp] (15th century)
1 : a large variable Asian soft-finned freshwater cyprinid fish (*Cyprinus carpio*) of sluggish waters that is often raised for food and has been widely introduced into U.S. waters; *also* **:** any of various related fishes
2 : a fish (as the European sea bream) resembling a carp

carp- *or* **carpo-** *combining form* [French & New Latin, from Greek *karp-, karpo-,* from *karpos* — more at HARVEST]
: fruit ⟨*carpo*gonium⟩

-carp *noun combining form* [New Latin *-carpium,* from Greek *-karpion,* from *karpos*]
: part of a fruit ⟨meso*carp*⟩ **:** fruit ⟨schizo*carp*⟩

car·pac·cio \kär-'pä-ch(ē-)ō\ *noun* [Vittore *Carpaccio;* from the prominent use of red in his painting] (1969)
: thinly sliced raw meat or fish served with a sauce — often used as a postpositive modifier ⟨beef *carpaccio*⟩

[1]**car·pal** \'kär-pəl\ *adjective* [New Latin *carpalis,* from *carpus*] (1743)
: relating to the carpus

[2]**carpal** *noun* (1855)
: a carpal element or bone

carpal tunnel syndrome *noun* (1954)
: a condition caused by compression of a nerve where it passes through the wrist into the hand and characterized especially by weakness, pain, and disturbances of sensation in the hand

car park *noun* (1926)
chiefly British **:** PARKING LOT

car·pe di·em \'kär-pe-'dē-ˌem, -'dī-, -əm\ *noun* [Latin, literally, pluck the day] (1817)
: the enjoyment of the pleasures of the moment without concern for the future

car·pel \'kär-pəl\ *noun* [New Latin *carpellum,* from Greek *karpos* fruit] (1835)
: one of the ovule-bearing structures in an angiosperm that comprises the innermost whorl of a flower — compare PISTIL
— car·pel·lary \-pə-ˌler-ē\ *adjective*
— car·pel·late \-ˌlāt, -lət\ *adjective*

[1]**car·pen·ter** \'kär-pən-tər, 'kärp-ᵊm-tər\ *noun* [Middle English, from Old North French *carpentier,* from Latin *carpentarius* carriage maker, from *carpentum* carriage, of Celtic origin; akin to Old Irish *carpat* chariot, *carr* vehicle — more at CAR] (14th century)
: a worker who builds or repairs wooden structures or their structural parts

◇ WORD HISTORY

carnival In Roman Catholic countries, Lent has traditionally been a period of abstinence from meat, and the days immediately before Lent a time of unrestrained indulgence and merrymaking that ended on the eve of Ash Wednesday. A pattern common to a number of European languages was to apply the name for the very beginning of Lent, the end of meat-eating, to the preceding festive period. In Italian dialects, the word for this period was a compound formed from descendants of Latin *carn-* "meat" and any of several verbs marking removal or cessation. Examples of such compounds are Tuscan *carnasciale,* ultimately from Latin *carn-* plus *laxare* "to let go," and, in the dialect of the area around Rome, *carnevale,* ultimately from Latin *carn-* plus *levare* "to raise, remove." The splendor of pre-Lenten festivities held in Rome during the Renaissance led to the adoption of *carnevale* into other languages of Europe in the 16th century, including French and English.

carol In medieval England and France, the word *carole* referred to a courtly dance that could be either processional or circular, one in which the musical accompaniment was provided by the dancers' own singing. The original choreography of such dances has not survived, nor are there any extant musical or literary texts from the 14th century or earlier that are explicitly given this name. It appears, though, that the leader of the dance sang verses to which the other dancers responded with a refrain. In any event, the Middle English word *carole* is applied in the 15th century to a kind of popular song with a refrain that both begins the song and is repeated after each stanza. Because these songs were often about events of the Nativity, the word *carol* has become closely associated with the Christmas season. Now it may refer not only to traditional compositions of medieval date but also to comparatively modern stanzaic hymns such as "Joy to the World" or "Hark the Herald Angels Sing."

²carpenter *verb* **-tered; -ter·ing** \-t(ə-)riŋ\ (circa 1815)
intransitive verb
: to follow the trade of a carpenter ⟨*carpentered* when he was young⟩
transitive verb
1 : to make by or as if by carpentry
2 : to put together often in a mechanical manner ⟨*carpentered* many television scripts⟩

carpenter ant *noun* (1883)
: an ant (especially genus *Camponotus*) that gnaws galleries especially in dead or decayed wood

carpenter bee *noun* (1838)
: any of various solitary bees (genera *Xylocopa* and *Ceratina*) that nest in wood

car·pen·try \-trē\ *noun* (14th century)
1 : the art or trade of a carpenter; *specifically* **:** the art of shaping and assembling structural woodwork
2 : timberwork constructed by a carpenter
3 : the form or manner of putting together the parts (as of a literary or musical composition)

car·pet \'kär-pət\ *noun* [Middle English, from Middle French *carpite,* from Old Italian *carpita,* from *carpire* to pluck, modification of Latin *carpere* to pluck — more at HARVEST] (15th century)
1 : a heavy often tufted fabric used as a floor covering; *also* **:** a floor covering made of this fabric
2 : a surface resembling or suggesting a carpet
— carpet *transitive verb*
— on the carpet : before an authority for censure or reproof

¹car·pet·bag \-ˌbag\ *noun* (1830)
: a traveler's bag made of carpet and widely used in the U.S. in the 19th century

²carpetbag *or* **car·pet·bag·ging** \-ˌba-giŋ\ *adjective* (1870)
: of, relating to, or characteristic of carpetbaggers ⟨a *carpetbag* government⟩

car·pet·bag·ger \-ˌba-gər\ *noun* [from their carrying all their belongings in carpetbags] (1868)
1 : a Northerner in the South after the American Civil War usually seeking private gain under the reconstruction governments
2 : OUTSIDER; *especially* **:** a nonresident or new resident who meddles in politics
— car·pet·bag·gery \-ˌba-g(ə-)rē\ *noun*

carpetbag steak *noun* (1958)
: a thick piece of steak in which a pocket is cut and stuffed (as with oysters)

carpet beetle *noun* (1889)
: any of several small dermestid beetles (genera *Anthrenus* and *Attagenus*) whose larvae are destructive especially to woolen goods

carpet bombing *noun* (1944)
: the dropping of large numbers of bombs so as to cause uniform devastation over a given area
— carpet bomb *verb*

car·pet·ing \'kär-pə-tiŋ\ *noun* (1758)
: material for carpets; *also* **:** CARPETS

car·pet·weed \'kär-pət-ˌwēd\ *noun* (1784)
: a North American mat-forming weed (*Mollugo verticillata* of the family Aizoaceae, the carpetweed family)

-carpic *adjective combining form* [probably from New Latin *-carpicus,* from Greek *karpos* fruit]
: -CARPOUS ⟨mono*carpic*⟩

carp·ing \'kär-piŋ\ *adjective* (1581)
: marked by or inclined to querulous and often perverse criticism
synonym see CRITICAL
— carp·ing·ly \-piŋ-lē\ *adverb*

car·po·go·ni·um \ˌkär-pə-'gō-nē-əm\ *noun, plural* **-nia** \-nē-ə\ [New Latin] (1882)
: the flask-shaped egg-bearing portion of the female reproductive branch in some thallophytes (as red algae)
— car·po·go·ni·al \-nē-əl\ *adjective*

car·pool \'kär-ˌpü(ə)l\ *intransitive verb* (1962)
: to participate in a car pool
— car·pool·er *noun*

car pool *noun* (1942)
: an arrangement in which a group of people commute together by car; *also* **:** the group entering into such an arrangement

car·po·phore \'kär-pə-ˌfōr, -ˌfȯr\ *noun* [probably from New Latin *carpophorum,* from *carp-* + *-phorum* -phore] (circa 1859)
1 : the stalk of a fungal fruiting body; *also* **:** the entire fruiting body
2 : a slender prolongation of a floral axis from which the carpels are suspended

car·port \'kär-ˌpōrt, -ˌpȯrt\ *noun* (1939)
: an open-sided automobile shelter by the side of a building

car·po·spore \'kär-pə-ˌspōr, -ˌspȯr\ *noun* (1881)
: a diploid spore of a red alga

-carpous *adjective combining form* [New Latin *-carpus,* from Greek *-karpos,* from *karpos* fruit — more at HARVEST]
: having (such) fruit or (so many) fruits ⟨syn*carpous*⟩
— -carpy *noun combining form*

car·pus \'kär-pəs\ *noun, plural* **car·pi** \-ˌpī, -(ˌ)pē\ [New Latin, from Greek *karpos* — more at WHARF] (1679)
1 : WRIST 1
2 : the bones of the wrist

carr \'kä(r)\ *noun* [Middle English *ker,* of Scandinavian origin; akin to Old Norse *kjarr* underbrush] (14th century)
chiefly British **:** ¹FEN

car·rack \'kar-ək, -ik\ *noun* [Middle English *carrake,* from Middle French *caraque,* from Old Spanish *carraca,* from Arabic *qarāqīr,* plural of *qurqūr* merchant ship] (14th century)
: a beamy sailing ship especially of the 15th and 16th centuries

car·ra·geen *also* **car·ra·gheen** \'kar-ə-ˌgēn\ *noun* [*Carragheen,* near Waterford, Ireland] (1829)
1 : IRISH MOSS 2
2 : CARRAGEENAN

car·ra·geen·an *or* **car·ra·geen·in** \ˌkar-ə-'gē-nən\ *noun* [*carrageen* + ³*-an* or ¹*-in*] (circa 1889)
: a colloid extracted from various red algae and especially Irish moss and used especially as a stabilizing or thickening agent

car·re·four \ˌkar-ə-'fu̇r\ *noun* [Middle French, from Late Latin *quadrifurcum,* neuter of *quadrifurcus* having four forks, from Latin *quadri-* + *furca* fork] (15th century)
1 : CROSSROADS
2 : SQUARE, PLAZA ⟨the farmers . . . preferred the open *carrefour* for their transactions —Thomas Hardy⟩

car·rel \'kar-əl\ *noun* [alteration of Middle English *caroll,* from Medieval Latin *carola,* perhaps from *carola* round dance, something circular, from Late Latin *choraula* choral song — more at CAROL] (1593)
: a table that is often partitioned or enclosed and is used for individual study especially in a library

car·riage \'kar-ij\ *noun* [Middle English *cariage,* from Old North French, from *carier* to transport in a vehicle — more at CARRY] (14th century)
1 : the act of carrying
2 a *archaic* **:** DEPORTMENT **b :** manner of bearing the body **:** POSTURE
3 *archaic* **:** MANAGEMENT
4 *chiefly British* **:** the price or expense of carrying
5 *obsolete* **:** BURDEN, LOAD
6 a : a wheeled vehicle; *especially* **:** a horse-drawn vehicle designed for private use and comfort **b** *British* **:** a railway passenger coach
7 : a wheeled support carrying a burden
8 *obsolete* **:** IMPORT, SENSE
9 *obsolete* **:** a hanger for a sword
10 : a movable part of a machine for supporting some other movable object or part ⟨a typewriter *carriage*⟩
synonym see BEARING

carriage trade *noun* (circa 1909)
: trade from well-to-do or upper-class people; *also* **:** well-to-do people

car·riage·way \'kar-ij-ˌwā\ *noun* (1800)
British **:** a road used by vehicular traffic **:** HIGHWAY

car·rick bend \'kar-ik-\ *noun* [probably from obsolete English *carrick* carrack, from Middle English *carrake, carryk*] (1819)
: a knot used to join the ends of two large ropes — see KNOT illustration

car·ri·er \'kar-ē-ər\ *noun* (14th century)
1 : one that carries **:** BEARER, MESSENGER
2 a : an individual or organization engaged in transporting passengers or goods for hire **b :** a transportation line carrying mail between post offices **c :** a postal employee who delivers or collects mail **d :** one that delivers newspapers **e :** an entity (as a hole or an electron) capable of carrying an electric charge
3 a : a container for carrying **b :** a device or machine that carries **:** CONVEYER
4 : AIRCRAFT CARRIER
5 a : a bearer and transmitter of a causative agent of an infectious disease; *especially* **:** one who carries the causative agent of a disease (as typhoid fever) systemically but is immune to it **b :** an individual (as one heterozygous for a recessive) having a specified gene that is not expressed or only weakly expressed in its phenotype
6 a : a usually inactive accessory substance **:** VEHICLE ⟨a *carrier* for a drug or an insecticide⟩ **b :** a substance (as a catalyst) by whose agency some element or group is transferred from one compound to another
7 a : an electromagnetic wave or alternating current whose modulations are used as signals in radio, telephonic, or telegraphic transmission **b :** a telecommunication company
8 : an organization acting as an insurer

carrier pigeon *noun* (1647)
1 : a pigeon used to carry messages; *especially* **:** HOMING PIGEON
2 : any of a breed of large long-bodied show pigeons

car·ri·on \'kar-ē-ən\ *noun* [Middle English *caroine,* from Anglo-French, from (assumed) Vulgar Latin *caronia,* irregular from Latin *carn-, caro* flesh — more at CARNAL] (14th century)
: dead and putrefying flesh; *also* **:** flesh unfit for food

carrion crow *noun* (1528)
: a uniformly black crow (*Corvus corone corone*) occurring in much of western Europe

car·ron·ade \ˌkar-ə-'nād\ *noun* [*Carron,* Scotland] (1779)
: a short-barreled gun of the late 18th and 19th centuries that fired large shot at short range and was used especially on warships

car·rot \'kar-ət\ *noun* [Middle French *carotte,* from Late Latin *carota,* from Greek *karōton*] (1533)
1 : a biennial herb (*Daucus carota* of the family Umbelliferae, the carrot family) with a usually orange spindle-shaped edible root; *also* **:** its root
2 : a reward or advantage offered especially as an inducement

car·rot–and–stick \ˌkar-ət-ᵊn-'stik\ *adjective* [from the traditional alternatives of driving a donkey on by either holding out a carrot or whipping it with a stick] (1951)
: characterized by the use of both reward and punishment to induce cooperation ⟨*carrot-and-stick* foreign policy⟩

car·rot·top \'kar-ə(t)-ˌtäp\ *noun* (circa 1902)
: REDHEAD 1
— car·rot·topped \-ˌtäpt\ *adjective*

car·roty \'kar-ə-tē\ *adjective* (1696)
: resembling carrots in color ⟨*carroty* hair⟩

car·rou·sel *variant of* CAROUSEL

¹car·ry \'kar-ē, 'ker-\ *verb* **car·ried; car·ry·ing** [Middle English *carien,* from Old North French *carier* to transport in a vehicle, from

car vehicle, from Latin *carrus* — more at CAR] (14th century)
transitive verb
1 : to move while supporting **:** TRANSPORT ⟨her legs refused to *carry* her further —Ellen Glasgow⟩
2 : to convey by direct communication ⟨*carry* tales about a friend⟩
3 *chiefly dialect* **:** CONDUCT, ESCORT
4 : to influence by mental or emotional appeal **:** SWAY
5 : to get possession or control of **:** CAPTURE ⟨*carried* off the prize⟩
6 : to transfer from one place (as a column) to another ⟨*carry* a number in adding⟩
7 : to contain and direct the course of ⟨the drain *carries* sewage⟩
8 a : to wear or have on one's person **b :** to bear upon or within one ⟨is *carrying* an unborn child⟩
9 a : to have or bear especially as a mark, attribute, or property ⟨*carry* a scar⟩ **b :** IMPLY, INVOLVE ⟨the crime *carried* a heavy penalty⟩
10 : to hold or comport (as one's person) in a specified manner
11 : to sustain the weight or burden of ⟨pillars *carry* an arch⟩
12 : to bear as a crop
13 : to sing with reasonable correctness of pitch ⟨*carry* a tune⟩
14 a : to keep in stock for sale **b :** to provide sustenance for ⟨land *carrying* 10 head of cattle⟩ **c :** to have or maintain on a list or record ⟨*carry* a person on a payroll⟩
15 : to be chiefly or solely responsible for the success, effectiveness, or continuation of ⟨a player capable of *carrying* a team⟩
16 : to prolong in space, time, or degree ⟨*carry* a principle too far⟩
17 a : to gain victory for; *especially* **:** to secure the adoption or passage of **b :** to win a majority of votes in (as a legislative body or a state)
18 : to present for public consumption ⟨newspapers *carry* weather reports⟩ ⟨channel 9 will *carry* the game⟩
19 a : to bear the charges of holding or having (as stocks or merchandise) from one time to another **b :** to keep on one's books as a debtor ⟨a merchant *carries* a customer⟩
20 : to hold to and follow after (as a scent)
21 : to hoist and maintain (a sail) in use
22 : to pass over (as a hazard) at a single stroke in golf ⟨*carry* a bunker⟩
23 : to propel and control (a puck or ball) along a playing surface
intransitive verb
1 : to act as a bearer
2 a : to reach or penetrate to a distance ⟨voices *carry* well⟩ ⟨fly balls don't *carry* well in cold air⟩ **b :** to convey itself to a reader or audience
3 : to undergo or admit of carriage in a specified way
4 *of a hunting dog* **:** to keep and follow the scent
5 : to win adoption ⟨the motion *carried* by a vote of 71–25⟩
— carry a torch *or* **carry the torch 1 :** CRUSADE **2 :** to be in love especially without reciprocation **:** cherish a longing or devotion ⟨still *carrying a torch* for a former lover⟩
— carry the ball : to perform or assume the chief role **:** bear the major portion of work or responsibility
— carry the day : WIN, PREVAIL

[2]**carry** *noun, plural* **car·ries** (1858)
1 : carrying power; *especially* **:** the range of a gun or projectile or of a struck or thrown ball
2 a : PORTAGE **b :** the act or method of carrying ⟨fireman's *carry*⟩ **c :** the act of rushing with the ball in football ⟨averaged four yards per *carry*⟩
3 : the position assumed by a color-bearer with the flag or guidon held in position for marching
4 : a quantity that is transferred in addition from one number place to the adjacent one of higher place value

car·ry·all \ˈkar-ē-ˌȯl, ˈker-\ *noun* (1714)
1 [by folk etymology from French *carriole,* from Old Provençal *carriola,* ultimately from Latin *carrus* car] **a :** a light covered carriage for four or more persons **b :** a passenger automobile used as a small bus
2 [[1]*carry* + [3]*all*] **:** a capacious bag or carrying case

carry away *transitive verb* (1570)
1 : to arouse to a high and often excessive degree of emotion or enthusiasm
2 : CARRY OFF 1

car·ry·back \ˈkar-ē-ˌbak, ˈker-\ *noun* (1942)
: a loss sustained or a portion of a credit not used in a given period that may be deducted from taxable income of a prior period

car·ry–cot \-ˌkät\ *noun* (1943)
British **:** a portable bed for an infant

car·ry·for·ward \ˈkar-ē-ˈfȯr-wərd, ˈker-, -ˌfȯr-, *Southern also* -ˈfär-, -ˌfär-\ *noun* (1898)
: CARRYOVER

carrying capacity *noun* (1921)
: the population (as of deer) that an area will support without undergoing deterioration

carrying charge *noun* (1914)
1 : expense incident to ownership or use of property
2 : a charge added to the price of merchandise sold on the installment plan

car·ry·ing–on \ˌkar-ē-iŋ-ˈȯn, ˌker-, -ˈän\ *noun, plural* **carryings–on** (1663)
: foolish, excited, or improper behavior; *also* **:** an instance of such behavior

carry off *transitive verb* (circa 1680)
1 : to cause the death of ⟨the plague *carried off* thousands⟩
2 : to perform or manage successfully **:** BRING OFF ⟨tried to look suave but couldn't *carry* it *off*⟩

[1]**car·ry–on** \ˈkar-ē-ˌȯn, ˈker-, -ˌän\ *noun* (1890)
1 *British* **:** CARRYING-ON
2 : a piece of luggage suitable for being carried aboard an airplane by a passenger

[2]**carry–on** *adjective* (1967)
: carried or suitable for being carried aboard ⟨*carry-on* baggage⟩

carry on (1644)
transitive verb
: to continue doing, pursuing, or operating ⟨*carry on* research⟩ ⟨*carried on* the business⟩
intransitive verb
1 : to behave or speak in a foolish, excited, or improper manner ⟨embarrassed by the way he *carries on*⟩
2 : to continue especially in spite of hindrance or discouragement

car·ry·out \ˈkar-ē-ˌau̇t, ˈker-\ *noun* (1964)
1 : prepared food packaged to be consumed away from its place of sale
2 : an establishment selling carryout
— carryout *adjective*

carry out *transitive verb* (1605)
1 : to put into execution ⟨*carry out* a plan⟩
2 : to bring to a successful issue **:** COMPLETE, ACCOMPLISH ⟨you will be paid when you have *carried out* the assignment⟩
3 : to continue to an end or stopping point

car·ry·over \ˈkar-ē-ˌō-vər, ˈker-\ *noun* (1804)
1 : the act or process of carrying over
2 : something retained or carried over

carry over (1745)
transitive verb
1 a : to transfer (an amount) to the next column, page, or book relating to the same account **b :** to hold over (as goods) for another time or season
2 : to deduct (a loss or an unused credit) from taxable income of a later period
intransitive verb
: to persist from one stage or sphere of activity to another

carry through (1605)
transitive verb
: CARRY OUT
intransitive verb
: PERSIST, SURVIVE ⟨feelings that *carry through* to the present⟩

car·sick \ˈkär-ˌsik\ *adjective* (1908)
: affected with motion sickness especially in an automobile
— car sickness *noun*

[1]**cart** \ˈkärt\ *noun* [Middle English, probably from Old Norse *kartr;* akin to Old English *cræt* cart] (13th century)
1 : a heavy usually horse-drawn 2-wheeled vehicle used for farming or transporting freight
2 : a lightweight 2-wheeled vehicle drawn by a horse, pony, or dog
3 : a small wheeled vehicle

[2]**cart** *transitive verb* (14th century)
1 : to carry or convey in or as if in a cart ⟨buses to *cart* the kids to and from school —L. S. Gannett⟩
2 : to take or drag away without ceremony or by force — usually used with *off* ⟨they *carted* him off to jail⟩
— cart·er *noun*

cart·age \ˈkär-tij\ *noun* (15th century)
: the action of or rate charged for carting

carte blanche \ˈkärt-ˈbläⁿsh, -ˈblänch\ *noun, plural* **cartes blanches** \ˈkärt(s)-\ [French, literally, blank document] (1754)
: full discretionary power ⟨was given *carte blanche* to furnish the house⟩

carte du jour \ˌkärt-də-ˈzhu̇r\ *noun, plural* **cartes du jour** \ˌkärt(s)-\ [French, literally, card of the day] (1936)
: MENU

car·tel \kär-ˈtel\ *noun* [French, letter of defiance, from Old Italian *cartello,* literally, placard, from *carta* leaf of paper — more at CARD] (1692)
1 : a written agreement between belligerent nations
2 : a combination of independent commercial or industrial enterprises designed to limit competition or fix prices
3 : a combination of political groups for common action ◆

◇ WORD HISTORY

cartel The Italian word *carta* "paper," descended from Latin *charta* "sheet of papyrus," is related to English *card, chart,* and *charter,* all likewise from Romance descendants of the Latin word. From *carta,* Italian derived a masculine noun *cartello,* which meant originally "public notice, poster," and then "written challenge to a duel, letter of defiance." In the latter senses the word was borrowed into French as *cartel* and from French into English. During the 17th century, *cartel* became current in several European languages as a designation for a written agreement between warring nations regulating the exchange of prisoners. In addition to this sense, German *Kartell* in the late 19th century came to stand first for an agreement between political parties in the Reichstag (the national legislature), and then for a compact between commercial enterprises to control market conditions and fix prices. Italian *cartello* and English and French *cartel* all borrowed this sense from German. Government antitrust laws in the U.S. and other countries ban such commercial compacts, though *cartel* is now often applied to international organizations that fix prices, such as the Organization of Petroleum Exporting Countries (OPEC).

car·tel·ise *British variant of* CARTELIZE

car·tel·ize \ˈkär-tə-ˌlīz\ *transitive verb* **-ized; -iz·ing** (1915)
: to bring under the control of a cartel
— **car·tel·i·za·tion** \ˌkär-tə-lə-ˈzā-shən\ *noun*

Car·te·sian \kär-ˈtē-zhən\ *adjective* [New Latin *cartesianus,* from *Cartesius* Descartes] (1656)
: of or relating to René Descartes or his philosophy ◆
— **Cartesian** *noun*
— **Car·te·sian·ism** \-zhə-ˌni-zəm\ *noun*

Cartesian coordinate *noun* (circa 1888)
1 : either of two coordinates that locate a point on a plane and measure its distance from either of two intersecting straight-line axes along a line parallel to the other axis
2 : any of three coordinates that locate a point in space and measure its distance from any of three intersecting coordinate planes measured parallel to that one of three straight-line axes that is the intersection of the other two planes

Cartesian plane *noun* (1960)
: a plane whose points are labeled with Cartesian coordinates

Cartesian product *noun* (1958)
: a set that is constructed from two given sets and comprises all pairs of elements such that the first element of the pair is from the first set and the second is from the second set

Car·thu·sian \kär-ˈthü-zhən, -ˈthyü-\ *noun* [Medieval Latin *Cartusiensis,* from *Cartusia* Chartreuse, motherhouse of the Carthusian order, near Grenoble, France] (1526)
: a member of an ascetic contemplative religious order founded by Saint Bruno in 1084
— **Carthusian** *adjective*

car·ti·lage \ˈkär-tᵊl-ij, ˈkärt-lij\ *noun* [Latin *cartilagin-, cartilago*] (15th century)
1 : a usually translucent somewhat elastic tissue that composes most of the skeleton of vertebrate embryos and except for a small number of structures (as some joints, respiratory passages, and the external ear) is replaced by bone during ossification in the higher vertebrates
2 : a part or structure composed of cartilage

car·ti·lag·i·nous \ˌkär-tə-ˈla-jə-nəs\ *adjective* (14th century)
: composed of, relating to, or resembling cartilage

cartilaginous fish *noun* (1695)
: any of a class (Chondrichthyes) of fishes (as a shark, ray, or chimaera) having the skeleton wholly or largely composed of cartilage — compare BONY FISH, JAWLESS FISH

cart·load \ˈkärt-ˈlōd, -ˌlōd\ *noun* (14th century)
: as much as a cart will hold

car·tog·ra·pher \kär-ˈtä-grə-fər\ *noun* (circa 1847)
: one that makes maps

car·tog·ra·phy \-fē\ *noun* [French *cartographie,* from *carte* card, map + *-graphie* -graphy — more at CARD] (circa 1847)
: the science or art of making maps
— **car·to·graph·ic** \ˌkär-tə-ˈgra-fik\ *also* **car·to·graph·i·cal** \-fi-kəl\ *adjective*
— **car·to·graph·i·cal·ly** \-fi-k(ə-)lē\ *adverb*

[1]car·ton \ˈkär-tᵊn\ *noun* [French, from Italian *cartone* pasteboard] (circa 1859)
: a box or container usually made of paperboard and often of corrugated paperboard

[2]carton (1921)
transitive verb
: to pack or enclose in a carton
intransitive verb
: to shape cartons from paperboard sheets

car·toon \kär-ˈtün\ *noun, often attributive* [Italian *cartone* pasteboard, cartoon, augmentative of *carta* leaf of paper — more at CARD] (1671)
1 : a preparatory design, drawing, or painting (as for a fresco)
2 a : a drawing intended as satire, caricature, or humor ⟨a political *cartoon*⟩ **b :** COMIC STRIP
3 : ANIMATED CARTOON
4 : a ludicrously simplistic, unrealistic, or one-dimensional portrayal or version ⟨the film's villain is an entertaining *cartoon*⟩
— **cartoon** *verb*
— **car·toon·ing** *noun*
— **car·toon·ish** \-ˈtü-nish\ *adjective*
— **car·toon·ish·ly** *adverb*
— **car·toon·ist** \-ˈtü-nist\ *noun*
— **car·toon·like** \-ˈtün-ˌlīk\ *adjective*
— **car·toony** \-ˈtü-nē\ *adjective*

car·top \ˈkär-ˌtäp\ *adjective* (1946)
: suitable in size and weight for carrying on top of an automobile ⟨a *cartop* fishing boat⟩
— **car·top·per** \-ˌtä-pər\ *noun*

car·touche *also* **car·touch** \kär-ˈtüsh\ *noun* [French *cartouche,* from Italian *cartoccio,* from *carta*] (1611)
1 : a gun cartridge with a paper case
2 : an ornate or ornamental frame
3 : an oval or oblong figure (as on ancient Egyptian monuments) enclosing a sovereign's name

car·tridge \ˈkär-trij, *dialect* ˈka-trij\ *noun* [alteration of earlier *cartage,* modification of Middle French *cartouche*] (1579)
: a case or container that holds a substance, device, or material which is difficult, troublesome, or awkward to handle and that usually can be easily changed: as **a :** a tube (as of metal) containing a complete charge for a firearm and usually an initiating device (as a primer) **b :** a case containing an explosive charge for blasting **c :** an often cylindrical container for insertion into a larger mechanism or apparatus **d :** CASSETTE 2 **e :** a small case that contains a phonograph needle and transducer and is attached to a tonearm **f :** a case containing a reel of magnetic tape arranged for insertion into a recorder or player **g :** a removable case containing a magnetic tape or one or more disks and used as a computer storage medium **h :** a case for holding printed circuit chips containing a computer program ⟨a video-game *cartridge*⟩

cartridge belt *noun* (1874)
1 : a belt having a series of loops for holding cartridges
2 : a belt worn around the waist and designed for carrying various attachable equipment (as a cartridge case, canteen, or compass)

car·tu·lary \ˈkär-chə-ˌler-ē\ *noun, plural* **-lar·ies** [Medieval Latin *chartularium,* from *chartula* charter — more at CHARTER] (1541)
: a collection of charters; *especially* **:** a book holding copies of the charters and title deeds of an estate

[1]cart·wheel \ˈkärt-ˌhwēl, -ˌwēl\ *noun* (1855)
1 : a large coin (as a silver dollar)
2 : a lateral handspring with arms and legs extended

[2]cartwheel *intransitive verb* (1917)
: to move like a turning wheel; *specifically* **:** to perform cartwheels
— **cart·wheel·er** *noun*

car·un·cle \ˈkar-ˌəŋ-kəl, kə-ˈrəŋ-\ *noun* [obsolete French *caruncule,* from Latin *caruncula* little piece of flesh, diminutive of *caro* flesh — more at CARNAL] (1615)
1 : a naked fleshy outgrowth (as a bird's wattle)
2 : an outgrowth on a seed adjacent to the micropyle

car·va·crol \ˈkär-və-ˌkrȯl, -ˌkrōl\ *noun* [International Scientific Vocabulary, from New Latin *carvi* (specific epithet of *Carum carvi* caraway) + Latin *acr-, acer* sharp — more at CARAWAY, EDGE] (1854)
: a liquid phenol $C_{10}H_{14}O$ found in essential oils of various mints (as thyme) and used as a fungicide and disinfectant

carve \ˈkärv\ *verb* **carved; carv·ing** [Middle English *kerven,* from Old English *ceorfan;* akin to Old High German *kerban* to notch, Greek *graphein* to scratch, write] (before 12th century)
transitive verb
1 : to cut with care or precision ⟨*carved* fretwork⟩
2 : to make or get by or as if by cutting — often used with *out* ⟨*carve* out a career⟩
3 : to cut into pieces or slices ⟨*carved* the turkey⟩
intransitive verb
1 : to cut up and serve meat
2 : to work as a sculptor or engraver
— **carv·er** *noun*

car·vel–built \ˈkär-vəl-ˌbilt, -ˌvel-\ *adjective* [probably from Dutch *karveel-,* from *karveel* caravel, from Middle French *carvelle*] (1798)
: built with the planks meeting flush at the seams

carv·en \ˈkär-vən\ *adjective* (14th century)
: wrought or ornamented by carving

carv·ing \ˈkär-viŋ\ *noun* (13th century)
1 : the act or art of one who carves
2 : a carved object, design, or figure

car wash *noun* (1956)
: an area or structure equipped with facilities for washing automobiles

cary- *or* **caryo-** — see KARY-

cary·at·id \ˌkar-ē-ˈa-təd, ˈkar-ē-ə-ˌtid\ *noun, plural* **-ids** *or* **-i·des** \ˌkar-ē-ˈa-tə-ˌdēz\ [Latin *caryatides,* plural, from Greek *karyatides* priestesses of Artemis at Caryae, caryatids, from *Karyai* Caryae in Laconia] (1563)
: a draped female figure supporting an entablature

cary·op·sis \ˌkar-ē-ˈäp-səs\ *noun, plural* **-op·ses** \-ˌsēz\ *also* **-si·des** \-sə-ˌdēz\ [New Latin] (1830)
: a small one-seeded dry indehiscent fruit (as of Indian corn or wheat) in which the fruit and seed fuse in a single grain

ca·sa \ˈkä-sə\ *noun* [Spanish & Italian, from Latin, cottage] (1844)
chiefly Southwest **:** DWELLING

ca·sa·ba \kə-ˈsä-bə\ *noun* [*Kasaba* (now Turgutlu), Turkey] (1889)
: any of several winter melons with yellow rind and sweet flesh

caryatid

Ca·sa·no·va \ˌka-zə-ˈnō-və, ˌka-sə-\ *noun* [Giacomo Girolamo *Casanova*] (1888)
: LOVER; *especially* **:** a man who is a promiscuous and unscrupulous lover

Cas·bah \ˈkaz-ˌbä, ˈkäz-\ *noun* [French, from Arabic dialect *qaṣbah*] (1944)
1 : a North African castle or fortress
2 : the native section of a North African city

cas·ca·bel \ˈkas-kə-ˌbel\ *noun* [Spanish, literally, small bell] (1639)
1 : a projection behind the breech of a muzzle-loading cannon

◇ WORD HISTORY

Cartesian That the derivative adjective for the French philosopher and mathematician René Descartes should be *Cartesian* might seem puzzling at first. The explanation lies in the fact that Descartes lived and wrote in the 17th century, when Latin was the international language of science and letters. Descartes' name was Latinized as Renatus Cartesius, and so *Cartesian* became the word for things relating to him and his work.

2 : a small hollow perforated spherical bell enclosing a loose pellet

[1]**cas·cade** \(ˌ)kas-'kād\ *noun* [French, from Italian *cascata,* from *cascare* to fall, from (assumed) Vulgar Latin *casicare,* from Latin *casus* fall] (1641)
1 : a steep usually small fall of water; *especially* **:** one of a series
2 a : something arranged or occurring in a series or in a succession of stages so that each stage derives from or acts upon the product of the preceding ⟨blood clotting involves a biochemical *cascade*⟩ **b :** a fall of material (as lace) that hangs in a zigzag line
3 : something falling or rushing forth in quantity ⟨a *cascade* of sound⟩ ⟨a *cascade* of events⟩

[2]**cascade** *verb* **cas·cad·ed; cas·cad·ing** (1702)
intransitive verb
: to fall, pour, or rush in or as if in a cascade
transitive verb
1 : to cause to fall like a cascade
2 : to connect in a cascade arrangement

cas·cara \ka-'skar-ə\ *noun* [Spanish *cáscara* husk, bark, probably from *cascar* to crack, break, from (assumed) Vulgar Latin *quassicare* to shake, from Latin *quassare* — more at QUASH] (1879)
1 : CASCARA BUCKTHORN
2 : CASCARA SAGRADA

cascara buckthorn *noun* (circa 1900)
: a buckthorn (*Rhamnus purshiana*) of the Pacific coast of the U.S. yielding cascara sagrada

cascara sa·gra·da \-sə-'grä-də\ *noun* [American Spanish *cáscara sagrada,* literally, sacred bark] (1885)
: the dried bark of cascara buckthorn used as a laxative

cas·ca·ril·la \ˌkas-kə-'ri-lə, -'rē-ə\ *noun* [Spanish, diminutive of *cáscara*] (1686)
: the aromatic bark of a West Indian shrub (*Croton eluteria*) of the spurge family used for making incense and as a tonic; *also* **:** this shrub

[1]**case** \'kās\ *noun* [Middle English *cas,* from Old French, from Latin *casus* fall, chance, from *cadere* to fall — more at CHANCE] (13th century)
1 a : a set of circumstances or conditions ⟨is the statement true in all three *cases*⟩ **b** (1) **:** a situation requiring investigation or action (as by the police) (2) **:** the object of investigation or consideration
2 : CONDITION; *specifically* **:** condition of body or mind
3 [Middle English *cas,* from Middle French, from Latin *casus,* translation of Greek *ptōsis,* literally, fall] **a :** an inflectional form of a noun, pronoun, or adjective indicating its grammatical relation to other words **b :** such a relation whether indicated by inflection or not
4 : what actually exists or happens **:** FACT
5 a : a suit or action in law or equity **b** (1) **:** the evidence supporting a conclusion or judgment (2) **:** ARGUMENT; *especially* **:** a convincing argument
6 a : an instance of disease or injury; *also* **:** PATIENT **b :** an instance that directs attention to a situation or exhibits it in action **:** EXAMPLE **c :** a peculiar person **:** CHARACTER
7 : oneself considered as an object of harassment ⟨get off my *case*⟩
synonym see INSTANCE
— in any case : without regard to or in spite of other considerations **:** whatever else is done or is the case ⟨war is inevitable *in any case*⟩ ⟨*in any case* the report will be made public next month⟩
— in case : as a precaution ⟨took an umbrella, just *in case*⟩
— in case of : in the event of ⟨*in case of* trouble, yell⟩

[2]**case** *noun* [Middle English *cas,* from Old North French *casse,* from Latin *capsa* chest, case, probably from *capere* to take — more at HEAVE] (14th century)
1 a : a box or receptacle for holding something **b :** a box together with its contents **c :** SET; *specifically* **:** PAIR
2 a : an outer covering or housing ⟨a pastry *case*⟩ **b :** a tube into which the components of a round of ammunition are loaded
3 : a divided tray for holding printing type
4 : the frame of a door or window **:** CASING

[3]**case** *transitive verb* **cased; cas·ing** (1575)
1 : to enclose in or cover with or as if with a case **:** ENCASE
2 : to line (as a well) with supporting material (as metal pipe)
3 : to inspect or study especially with intent to rob

ca·se·ation \ˌkā-sē-'ā-shən\ *noun* [Latin *caseus* cheese] (1866)
: necrosis with conversion of damaged tissue into a soft cheesy substance
— ca·se·ate \'kā-sē-ˌāt\ *intransitive verb*

case·bear·er \'kās-ˌbar-ər, -ˌber-\ *noun* (circa 1889)
: an insect larva that forms a protective case (as of silk)

case·book \-ˌbu̇k\ *noun* (1762)
1 : a book containing records of illustrative cases that is used for reference and instruction (as in law or medicine)
2 : a compilation of primary and secondary documents relating to a central topic together with scholarly comment, exercises, and study aids that is designed to serve as a sourcebook for short papers (as in a writing course) or as a point of departure for a research paper

cased glass \'kāst-\ *noun* (1849)
: glass consisting of two or more fused layers of different colors often decorated by cutting so that the inner layers show through — called also *case glass*

case goods *noun plural* (1922)
1 : furniture (as bureaus or bookcases) that provides interior storage space; *also* **:** dining-room and bedroom furniture sold as sets
2 : products often sold by the case

case–hard·en \'kās-ˌhär-dᵊn\ *transitive verb* (1677)
1 : to harden (a ferrous alloy) so that the surface layer is harder than the interior
2 : to make callous or insensible
— case–hard·ened *adjective*

case history *noun* (1894)
: a record of history, environment, and relevant details of a case especially for use in analysis or illustration

ca·sein \'kā-ˌsēn, kā-'\ *noun* [probably from French *caséine,* from Latin *caseus*] (1841)
: a phosphoprotein of milk: as **a :** one that is precipitated from milk by heating with an acid or by the action of lactic acid in souring and is used in making paints and adhesives **b :** one that is produced when milk is curdled by rennet, is the chief constituent of cheese, and is used in making plastics

ca·sein·ate \kā-'sē-ˌnāt, 'kā-si-ˌnāt\ *noun* (1904)
: a compound of casein with a metal (as calcium or sodium)

case in point (1965)
: an illustrative, relevant, or pertinent case

case knife *noun* (1704)
1 : SHEATH KNIFE
2 : a table knife

case law *noun* (1861)
: law established by judicial decision in cases

case·load \'kās-ˌlōd\ *noun* (1938)
: the number of cases handled (as by a court or clinic) usually in a particular period

case·mate \'kās-ˌmāt\ *noun* [Middle French, from Old Italian *casamatta*] (1575)
: a fortified position or chamber or an armored enclosure on a warship from which guns are fired through embrasures

case·ment \'kās-mənt\ *noun* [Middle English, hollow molding, probably from Old North French *encassement* frame, from *encasser* to enchase, frame, from *en-* + *casse*] (15th century)
: a window sash that opens on hinges at the side; *also* **:** a window with such a sash

ca·se·ous \'kā-sē-əs\ *adjective* [Latin *caseus* cheese] (1661)
: marked by caseation; *also* **:** CHEESY

ca·sern *or* **ca·serne** \kə-'zərn\ *noun* [French *caserne*] (1696)
: a military barracks in a garrison town

case study *noun* (1875)
1 : an intensive analysis of an individual unit (as a person or community) stressing developmental factors in relation to environment
2 : CASE HISTORY

case system *noun* (circa 1889)
: a system of teaching law in which instruction is chiefly on the basis of leading or selected cases as primary authorities instead of from textbooks

case·work \'kās-ˌwərk\ *noun* (1886)
: social work involving direct consideration of the problems, needs, and adjustments of the individual case (as a person or family)
— case·work·er \-ˌwər-kər\ *noun*

[1]**cash** \'kash\ *noun* [modification of Middle French or Old Italian; Middle French *casse* money box, from Old Italian *cassa,* from Latin *capsa* chest — more at CASE] (1596)
1 : ready money
2 : money or its equivalent (as a check) paid for goods or services at the time of purchase or delivery
— cash·less \-ləs\ *adjective*

[2]**cash** *adjective* (1622)
: being a method of accounting that includes as income only what has been received in cash and as expenses only those paid in cash

[3]**cash** *transitive verb* (1811)
1 : to pay or obtain cash for ⟨*cash* a check⟩
2 : to lead and win a bridge trick with (a card that is the highest remaining card of its suit)
— cash·able \'ka-shə-bəl\ *adjective*

[4]**cash** *noun, plural* **cash** [Portuguese *caixa,* from Tamil *kācu,* a small copper coin, from Sanskrit *karṣa,* a weight of gold or silver] (1598)
1 : any of various coins of small value in China and southern India; *especially* **:** a Chinese coin with a square hole in the center
2 : a unit of value equivalent to one cash

[1]**cash–and–car·ry** \ˌka-shᵊn-'kar-ē, -'ker-\ *adjective* (1917)
: sold or provided for cash and usually without delivery service

[2]**cash–and–carry** *noun* (1921)
: the policy of selling on a cash-and-carry basis

cash bar *noun* (1968)
: a bar (as at a reception) at which drinks are sold — compare OPEN BAR

cash·book \'kash-ˌbu̇k\ *noun* (1622)
: a book in which record is kept of all cash receipts and disbursements

cash cow *noun* (1979)
: a consistently profitable business, property, or product whose profits are used to finance a company's investments in other areas

cash crop *noun* (1868)
: a readily salable crop (as cotton or tobacco) produced or gathered primarily for market

cash discount *noun* (1917)
: a discount granted in consideration of immediate payment or payment within a prescribed time

cash·ew \'ka-(ˌ)shü, kə-'shü\ *noun* [Portuguese *acajú, cajú,* from Tupi *acajú*] (1598)

\ə\ abut \ᵊ\ kitten \ər\ further \a\ ash \ā\ ace \ä\ mop, mar \au̇\ out \ch\ chin \e\ bet \ē\ easy \g\ go \i\ hit \ī\ ice \j\ job \ŋ\ sing \ō\ go \ȯ\ law \ȯi\ boy \th\ thin \t͟h\ the \ü\ loot \u̇\ foot \y\ yet \zh\ vision *see also* Guide to Pronunciation

: a tropical American tree (*Anacardium occidentale* of the family Anacardiaceae, the cashew family) grown for a phenolic oil and the edible kernel of its nut and for a gum from its stem; *also* : CASHEW NUT

cashew nut *noun* (1796)
: the kidney-shaped kernel of the fruit of the cashew that is edible when roasted

cash flow *noun* (1954)
1 : a measure of an organization's liquidity that usually consists of net income after taxes plus noncash charges (as depreciation) against income
2 : a flow of cash; *especially* : one that provides solvency

[1]ca·shier \ka-'shir, kə-\ *transitive verb* [Dutch *casseren,* from Middle French *casser* to discharge, annul — more at QUASH] (1592)
1 : to dismiss from service; *especially* : to dismiss dishonorably
2 : REJECT, DISCARD

[2]cash·ier \(,)ka-'shir\ *noun* [Dutch or Middle French; Dutch *kassier,* from Middle French *cassier,* from *casse* money box] (1596)
: one that has charge of money: as **a** : a high officer in a bank or trust company responsible for moneys received and expended **b** : one who collects and records payments

cashier's check *noun* (1867)
: a check drawn by a bank on its own funds and signed by the cashier

cash in (1888)
transitive verb
: to obtain cash for ⟨*cashed in* the bonds⟩
intransitive verb
1 a : to retire from a gambling game **b** : to settle accounts and withdraw from an involvement (as a business deal)
2 : to obtain advantage or financial profit — often used with *on* ⟨*cash in* on a best-seller⟩

cash·mere \'kazh-,mir, 'kash-\ *noun* [*Cashmere* (Kashmir)] (1684)
1 : fine wool from the undercoat of the cashmere goat; *also* : a yarn of this wool
2 : a soft twilled fabric made originally from cashmere

cashmere goat *noun* (circa 1890)
: an Indian goat raised especially for its undercoat of fine soft wool that constitutes the cashmere wool of commerce

cash register *noun* (1879)
: a business machine that usually has a money drawer, indicates the amount of each sale, and records the amount of money received and often automatically makes change

cas·ing \'kā-siŋ\ *noun* (1791)
1 : something that encases : material for encasing: as **a** : an enclosing frame especially around a door or window opening **b** : a metal pipe used to case a well **c** : [4]TIRE 2 **d** : a membranous case for processed meat
2 : a space formed between two parallel lines of stitching through at least two layers of cloth into which something (as a rod or string) may be inserted

ca·si·no \kə-'sē-(,)nō\ *noun, plural* **-nos** [Italian, from *casa* house, from Latin, cottage] (1744)
1 : a building or room used for social amusements; *specifically* : one used for gambling
2 *also* **cas·si·no** : a card game in which cards are won by matching or combining cards in a hand with those exposed on the table
3 : SUMMERHOUSE 2

ca·si·ta \kə-'sē-tə\ *noun* [Spanish, diminutive of *casa*] (1923)
: a small house

cask \'kask\ *noun* [Middle English *caske,* perhaps from Middle French *casque* helmet, from Spanish *casco* potsherd, skull, helmet] (15th century)
1 : a barrel-shaped vessel of staves, headings, and hoops usually for liquids
2 : a cask and its contents; *also* : the quantity contained in a cask

cas·ket \'kas-kət\ *noun* [Middle English, perhaps modification of Middle French *cassette*] (15th century)
1 : a small chest or box (as for jewels)
2 : a usually fancy coffin
— **casket** *transitive verb*

casque \'kask\ *noun* [Middle French — more at CASK] (1580)
1 : a piece of armor for the head : HELMET
2 : an anatomic structure suggestive of a helmet

Cas·san·dra \kə-'san-drə, -'sän-\ *noun* [Latin, from Greek *Kassandra*]
1 : a daughter of Priam endowed with the gift of prophecy but fated never to be believed
2 : one that predicts misfortune or disaster

cas·sa·va \kə-'sä-və\ *noun* [Spanish *cazabe* cassava bread, from Taino *caçábi*] (1555)
: any of several plants (genus *Manihot* and especially *M. esculenta*) of the spurge family grown in the tropics for their fleshy edible rootstocks which yield a nutritious starch; *also* : the rootstock

cas·se·role \'ka-sə-,rōl *also* 'ka-zə-\ *noun* [French, saucepan, from Middle French, diminutive of *casse* ladle, dripping pan, from Old Provençal *cassa,* probably ultimately from Greek *kyathos* ladle] (1708)
1 : a dish in which food may be baked and served
2 : food cooked and served in a casserole
3 : a deep round usually porcelain dish with a handle used for heating substances in the laboratory

cas·sette *also* **ca·sette** \kə-'set, ka-\ *noun* [French, from Middle French, diminutive of Old North French *casse* case] (1793)
1 : CASKET 1
2 : a usually flat case or container that is used especially in a device or machine, that holds something which is difficult, troublesome, or awkward to handle, and that can be easily loaded or unloaded: as **a** : a lightproof magazine for holding film or plates for use in a camera **b** : a plastic cartridge containing magnetic tape with the tape on one reel passing to the other

cas·sia \'ka-shə\ *noun* [Middle English, from Old English, from Latin, from Greek *kassia,* of Semitic origin; akin to Hebrew *qěṣīʽāh* cassia] (before 12th century)
1 : a coarse cinnamon bark (as from *Cinnamomum cassia*)
2 : any of a genus (*Cassia*) of leguminous herbs, shrubs, and trees of warm regions

cas·si·mere \'ka-zə-,mir, 'ka-sə-\ *noun* [obsolete *Cassimere* (Kashmir)] (1774)
: a closely woven smooth twilled usually wool fabric (as for suits)

Cas·si·ni division \kə-'sē-nē-\ *noun* [Gian Domenico *Cassini* (died 1712) Italian astronomer] (circa 1909)
: the dark region between the two brightest rings of Saturn — called also *Cas·si·ni's division*

Cas·si·o·pe·ia \,ka-sē-ə-'pē-ə\ *noun* [Latin, from Greek *Kassiopeia*]
1 : the wife of King Cepheus who gives birth to Andromeda and is later changed into a constellation
2 [Latin (genitive *Cassiopeiae*), from Greek *Kassiopeia*] : a northern constellation between Andromeda and Cepheus

cas·sis \kə-'sēs\ *noun* [French, literally, black currants, perhaps from Latin *cassia*] (1899)
: a liqueur made from black currants

cas·sit·er·ite \kə-'si-tə-,rīt\ *noun* [French *cassitérite,* from Greek *kassiteros* tin] (1858)
: a brown or black mineral that consists of tin dioxide and is the chief source of metallic tin

cas·sock \'ka-sək\ *noun* [Middle French *casaque*] (1631)
: a close-fitting ankle-length garment worn especially in Roman Catholic and Anglican churches by the clergy and by laymen assisting in services

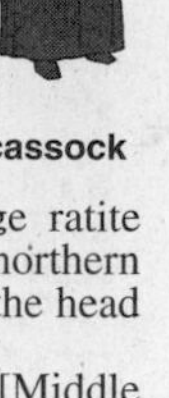
cassock

cas·sou·let \,ka-sə-'lā\ *noun* [French, from Provençal, literally, earthenware dish, diminutive of *cassolo* dish, diminutive of *casso* ladle, from Old Provençal *cassa*] (circa 1929)
: a casserole of white beans baked with herbs and meat (as pork, lamb, and goose or duck)

cas·so·wary \'ka-sə-,wer-ē\ *noun, plural* **-war·ies** [Malay *kĕsuari,* from an Austronesian language of the Moluccas] (1611)
: any of a genus (*Casuarius*) of large ratite birds chiefly of New Guinea and northern Australia that have a horny casque on the head and are closely related to the emu

[1]cast \'kast\ *verb* **cast; cast·ing** [Middle English, from Old Norse *kasta;* akin to Old Norse *kǫs* heap] (13th century)
transitive verb
1 a : to cause to move or send forth by throwing ⟨*cast* a fishing lure⟩ ⟨*cast* dice⟩ **b** : DIRECT ⟨*cast* a glance⟩ **c** (1) : to put forth ⟨the fire *casts* a warm glow⟩ (2) : to place as if by throwing ⟨*cast* doubt on their reliability⟩ **d** : to deposit (a ballot) formally **e** (1) : to throw off or away ⟨the horse *cast* a shoe⟩ (2) : to get rid of : DISCARD ⟨*cast* off all restraint⟩ (3) : SHED, MOLT (4) : to bring forth; *especially* : to give birth to prematurely **f** : to throw to the ground especially in wrestling **g** : to build by throwing up earth
2 a (1) : to perform arithmetical operations on : ADD (2) : to calculate by means of astrology **b** *archaic* : DECIDE, INTEND
3 a : to dispose or arrange into parts or into a suitable form or order **b** (1) : to assign the parts of (a dramatic production) to actors (2) : to assign (as an actor) to a role or part
4 a : to give a shape to (a substance) by pouring in liquid or plastic form into a mold and letting harden without pressure ⟨*cast* steel⟩ **b** : to form by this process
5 : TURN ⟨*cast* the scale slightly⟩
6 : to make (a knot or stitch) by looping or catching up
7 : TWIST, WARP ⟨a beam *cast* by age⟩
intransitive verb
1 : to throw something; *specifically* : to throw out a lure with a fishing rod
2 *dialect British* : VOMIT
3 *dialect English* : to bear fruit : YIELD
4 a : to perform addition **b** *obsolete* : ESTIMATE, CONJECTURE
5 : WARP
6 : to range over land in search of a trail — used of hunting dogs or trackers
7 : VEER
synonym see DISCARD, THROW
— **cast·abil·i·ty** \,kas-tə-'bi-lə-tē\ *noun*
— **cast·able** \'kas-tə-bəl\ *adjective*
— **cast lots** : to draw lots to determine a matter by chance

[2]cast *noun* (14th century)
1 a : an act of casting **b** : something that happens as a result of chance **c** : a throw of dice **d** : a throw of a line (as a fishing line) or net
2 a : the form in which a thing is constructed **b** (1) : the set of actors in a dramatic production (2) : a set (as in a narrative) of characters or persons **c** : the arrangement of draperies in a painting
3 : the distance to which a thing can be thrown; *specifically* : the distance a bow can shoot
4 a : a turning of the eye in a particular direction; *also* : EXPRESSION ⟨this freakish, elfish *cast* came into the child's eye —Nathaniel Hawthorne⟩ **b** : a slight strabismus
5 : something that is thrown or the quantity thrown; *especially, British* : the leader of a fishing line

6 a : something that is formed by casting in a mold or form: as (1) **:** a reproduction (as of a statue) in metal or plaster **:** CASTING (2) **:** a fossil reproduction of the details of a natural object by mineral infiltration **b :** an impression taken from an object with a liquid or plastic substance **:** MOLD **c :** a rigid dressing of gauze impregnated with plaster of paris for immobilizing a diseased or broken part
7 : FORECAST, CONJECTURE
8 a : an overspread of a color or modification of the appearance of a substance by a trace of some added hue **:** SHADE ⟨gray with a greenish *cast*⟩ **b :** TINGE, SUGGESTION
9 a : a ride on one's way in a vehicle **:** LIFT **b** *Scottish* **:** HELP, ASSISTANCE
10 a : SHAPE, APPEARANCE ⟨the delicate *cast* of her features⟩ **b :** characteristic quality ⟨his father's conservative *cast* of mind⟩
11 : something that is shed, ejected, or thrown out or off: as **a :** the excrement of an earthworm **b :** a mass of soft matter formed in cavities of diseased organs and discharged from the body **c :** the skin of an insect
12 : the ranging in search of a trail by a dog, hunting pack, or tracker

cast about (1575)
intransitive verb
: to look around **:** SEEK ⟨*cast about* for a seat⟩
transitive verb
: to lay plans concerning **:** CONTRIVE ⟨*cast about* how he was to go⟩

cas·ta·net \ˌkas-tə-ˈnet\ *noun* [Spanish *castañeta,* from *castaña* chestnut, from Latin *castanea* — more at CHESTNUT] (circa 1647)
: a percussion instrument used especially by dancers that consists of two small shells of hard wood, ivory, or plastic usually fastened to the thumb and clicked together by the other fingers — usually used in plural

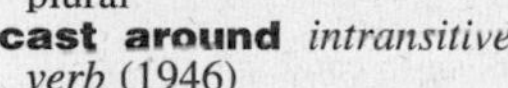

castanets

cast around *intransitive verb* (1946)
: CAST ABOUT

cast·away \ˈkas-tə-ˌwā\ *adjective* (1534)
1 : thrown away **:** REJECTED
2 a : cast adrift or ashore as a survivor of a shipwreck **b :** thrown out or left without friends or resources
— castaway *noun*

cast down *adjective* (14th century)
: DOWNCAST

caste \ˈkast *also* ˈkäst\ *noun* [Portuguese *casta,* literally, race, lineage, from feminine of *casto* pure, chaste, from Latin *castus*] (1613)
1 : one of the hereditary social classes in Hinduism that restrict the occupation of their members and their association with the members of other castes
2 a : a division of society based on differences of wealth, inherited rank or privilege, profession, or occupation **b :** the position conferred by caste standing **:** PRESTIGE
3 : a system of rigid social stratification characterized by hereditary status, endogamy, and social barriers sanctioned by custom, law, or religion
4 : a specialized form (as the worker of an ant or bee) of a polymorphic social insect that carries out a particular function in the colony
— caste·ism \ˈkas-ˌti-zəm\ *noun*

cas·tel·lan \ˈkas-tə-lən\ *noun* [Middle English *castelleyn,* from Old North French *castelain,* from Latin *castellanus* occupant of a castle, from *castellanus* of a castle, from *castellum* castle] (14th century)
: a governor or warden of a castle or fort

cas·tel·lat·ed \ˈkas-tə-ˌlā-təd\ *adjective* [Medieval Latin *castellatus,* past participle of *castellare* to fortify, from Latin *castellum*] (1679)
1 : having battlements like a castle
2 : having or supporting a castle

cast·er \ˈkas-tər\ *noun* (14th century)
1 : one that casts; *especially* **:** a machine that casts type
2 *or* **cas·tor** \-tər\ **a :** a usually silver table vessel with a perforated top for sprinkling a seasoning (as sugar or spice) **b :** a usually revolving metal stand bearing condiment containers (as cruets, mustard pot, and often shakers) for table use **:** a cruet stand
3 : any of a set of wheels or rotating balls mounted in a swivel frame and used for the support and movement of furniture, trucks, and portable equipment
4 : the slight usually backward tilt of the upper end of the kingbolt of an automobile for giving directional stability to the front wheels

cas·ti·gate \ˈkas-tə-ˌgāt\ *transitive verb* **-gat·ed; -gat·ing** [Latin *castigatus,* past participle of *castigare* — more at CHASTEN] (1607)
: to subject to severe punishment, reproof, or criticism
synonym see PUNISH
— cas·ti·ga·tion \ˌkas-tə-ˈgā-shən\ *noun*
— cas·ti·ga·tor \ˈkas-tə-ˌgā-tər\ *noun*

cas·tile soap \(ˌ)kas-ˈtēl-\ *noun, often C capitalized* [Middle English *castell sope,* from *Castell* Castile] (15th century)
: a fine hard bland soap made from olive oil and sodium hydroxide; *also* **:** any of various similar soaps

Cas·til·ian \ka-ˈstil-yən\ *noun* (1796)
1 : a native or inhabitant of Castile; *broadly* **:** SPANIARD
2 a : the dialect of Castile **b :** the official and literary language of Spain based on this dialect
— Castilian *adjective*

cast·ing *noun* (14th century)
1 : something (as excrement) that is cast out or off
2 : the act of one that casts: as **a :** the throwing of a fishing line by means of a rod and reel **b :** the assignment of parts and duties to actors or performers
3 : something cast in a mold

casting director *noun* (1922)
: a person who supervises the casting of dramatic productions (as films and plays)

casting vote *noun* (1678)
: a deciding vote cast by a presiding officer to break a tie

cast–iron *adjective* (1692)
1 : made of cast iron
2 : resembling cast iron: as **a :** capable of withstanding great strain ⟨a *cast-iron* stomach⟩ **b :** not admitting change, adaptation, or exception **:** RIGID ⟨a *cast-iron* will⟩

cast iron *noun* (1664)
: a commercial alloy of iron, carbon, and silicon that is cast in a mold and is hard, brittle, nonmalleable, and incapable of being hammer-welded but more easily fusible than steel

¹cas·tle \ˈka-səl\ *noun* [Middle English *castel,* from Old English, from Old North French, from Latin *castellum* fortress, castle, diminutive of *castrum* fortified place; perhaps akin to Latin *castrare* to castrate] (before 12th century)
1 a : a large fortified building or set of buildings **b :** a massive or imposing house
2 : a retreat safe against intrusion or invasion
3 : ³ROOK

²castle *verb* **cas·tled; cas·tling** \ˈka-s(ə-)liŋ\ (1587)
transitive verb
1 : to establish in a castle
2 : to move (the chess king) in castling
intransitive verb
: to move a chess king two squares toward a rook and in the same move the rook to the square next past the king

cas·tled \ˈka-səld\ *adjective* (1789)
: CASTELLATED

castle in the air (1580)
: an impracticable project **:** DAYDREAM — called also *castle in Spain*

cast–off \ˈkas-ˌtȯf\ *adjective* (1746)
: thrown away or aside
— cast·off *noun*

cast off (1602)
transitive verb
1 : LOOSE ⟨*cast off* a hunting dog⟩
2 : UNFASTEN ⟨*cast off* a boat⟩
3 : to remove (a stitch) from a knitting needle in such a way as to prevent unraveling
intransitive verb
1 : to unfasten or untie a boat or a line
2 : to turn one's partner in a square dance and pass around the outside of the set and back
3 : to finish a knitted fabric by casting off all stitches

cast on *transitive verb* (1840)
: to place (stitches) on a knitting needle for beginning or enlarging knitted work

cas·tor \ˈkas-tər\ *noun* [Middle English, from Latin, from Greek *kastōr,* from *Kastōr* Castor] (14th century)
1 : BEAVER 1a
2 : CASTOREUM
3 : a beaver hat

Cas·tor \ˈkas-tər\ *noun* [Latin, from Greek *Kastōr*]
1 : one of the Dioscuri
2 : the more northern of the two bright stars in Gemini

castor bean *noun* (1819)
: the very poisonous seed of the castor-oil plant; *also* **:** CASTOR-OIL PLANT

cas·to·re·um \ka-ˈstōr-ē-əm, -ˈstȯr-\ *noun* [Middle English *castorium,* from Latin *castoreum,* from *castor*] (14th century)
: a bitter strong-smelling creamy orange-brown substance that consists of the dried perineal glands of the beaver and their secretion and is used especially by perfumers — called also *castor*

castor oil *noun* [probably from its former use as a substitute for castoreum in medicine] (1746)
: a pale viscous fatty oil from castor beans used especially as a cathartic and as a lubricant and plasticizer

castor–oil plant *noun* (1836)
: a tropical Old World herb (*Ricinus communis*) widely grown as an ornamental or for its oil-rich castor beans

castor sugar *or* **caster sugar** *noun* [*caster*] (1855)
chiefly British **:** finely granulated white sugar

cast out *transitive verb* (13th century)
: to drive out **:** EXPEL

cas·trate \ˈkas-ˌtrāt\ *transitive verb* **cas·trat·ed; cas·trat·ing** [Latin *castratus,* past participle of *castrare;* akin to Greek *keazein* to split, Sanskrit *śasati* he slaughters] (1609)
1 a : to deprive of the testes **:** GELD **b :** to deprive of the ovaries **:** SPAY
2 : to render impotent or deprive of vitality especially by psychological means
— castrate *noun*
— cas·tra·tion \kas-ˈtrā-shən\ *noun*
— cas·tra·tor \-ˌtrā-tər\ *noun*
— cas·tra·to·ry \ˈkas-trə-ˌtōr-ē, -ˌtȯr-\ *adjective*

cas·tra·to \ka-ˈsträ-(ˌ)tō, kə-\ *noun, plural* **-ti** \-tē\ [Italian, from past participle of *castrare* to castrate, from Latin] (1763)
: a singer castrated before puberty to preserve the soprano or contralto range of his voice

Cas·tro·ism \ˈkas-(ˌ)trō-ˌi-zəm\ *noun* (1960)
: the political, economic, and social principles and policies of Fidel Castro
— Cas·tro·ite \-ˌīt\ *noun*

¹ca·su·al \ˈkazh-wəl, ˈka-zhə-wəl, ˈka-zhəl\ *adjective* [Middle English, from Middle French & Late Latin; Middle French *casuel,*

\ə\ **abut** \ᵊ\ **kitten** \ər\ **further** \a\ **ash** \ā\ **ace** \ä\ **mop, mar** \au̇\ **out** \ch\ **chin** \e\ **bet** \ē\ **easy** \g\ **go** \i\ **hit** \ī\ **ice** \j\ **job** \ŋ\ **sing** \ō\ **go** \ȯ\ **law** \ȯi\ **boy** \th\ **thin** \t͟h\ **the** \ü\ **loot** \u̇\ **foot** \y\ **yet** \zh\ **vision** *see also* Guide to Pronunciation

from Late Latin *casualis,* from Latin *casus* fall, chance — more at CASE] (14th century)
1 : subject to, resulting from, or occurring by chance
2 a : occurring without regularity **:** OCCASIONAL **b :** employed for irregular periods
3 a : feeling or showing little concern **:** NONCHALANT **b** (1) **:** INFORMAL, NATURAL (2) **:** designed for informal use
synonym see ACCIDENTAL, RANDOM
— ca·su·al·ly *adverb*
— ca·su·al·ness *noun*

²casual *noun* (circa 1852)
1 : a casual or migratory worker
2 : an officer or enlisted man awaiting assignment or transportation to a unit

ca·su·al·ty \'ka-zhəl-tē, 'kazh-wəl-, 'ka-zhə-wəl-\ *noun, plural* **-ties** (15th century)
1 *archaic* **:** CHANCE, FORTUNE ⟨losses that befall them by mere *casualty* —Sir Walter Raleigh⟩
2 : serious or fatal accident **:** DISASTER
3 a : a military person lost through death, wounds, injury, sickness, internment, or capture or through being missing in action **b :** a person or thing injured, lost, or destroyed **:** VICTIM ⟨the ex-senator was a *casualty* of the last election⟩

casual water *noun* (1899)
: a temporary accumulation of water not forming a regular hazard of a golf course

ca·su·a·ri·na \,ka-zhə-(wə-)'rē-nə\ *noun* [New Latin, genus name, from Malay (*pohon*) *kĕsuari,* literally, cassowary tree; from the resemblance of its twigs to cassowary feathers] (1806)
: any of a genus (*Casuarina* of the family Casuarinaceae) of dicotyledonous chiefly Australian trees which have whorls of scalelike leaves and jointed stems resembling horsetails and some of which yield a heavy hard wood

ca·su·ist \'ka-zhü-ist, 'kazh-wist\ *noun* [probably from Spanish *casuista,* from Latin *casus* fall, chance — more at CASE] (1609)
: one skilled in or given to casuistry
— ca·su·is·tic \,ka-zhə-'wis-tik\ *or* **ca·su·is·ti·cal** \-ti-kəl\ *adjective*

ca·su·ist·ry \'kazh-wə-strē, 'ka-zhə-\ *noun, plural* **-ries** (1725)
1 : a resolving of specific cases of conscience, duty, or conduct through interpretation of ethical principles or religious doctrine
2 : specious argument **:** RATIONALIZATION

ca·sus bel·li \'kä-səs-'be-,lē, 'kā-səs-'be-,lī\ *noun, plural* **ca·sus belli** \'kä-,süs-, 'kā-,süs-\ [New Latin, occasion of war] (1849)
: an event or action that justifies or allegedly justifies a war or conflict

¹cat \'kat\ *noun, often attributive* [Middle English, from Old English *catt,* probably from Late Latin *cattus, catta* cat] (before 12th century)
1 a : a carnivorous mammal (*Felis catus*) long domesticated as a pet and for catching rats and mice **b :** any of a family (Felidae) of carnivorous usually solitary and nocturnal mammals (as the domestic cat, lion, tiger, leopard, jaguar, cougar, wildcat, lynx, and cheetah)
2 : a malicious woman
3 : a strong tackle used to hoist an anchor to the cathead of a ship
4 a : CATBOAT **b :** CATAMARAN
5 : CAT-O'-NINE-TAILS
6 : CATFISH
7 a : a player or devotee of jazz **b :** GUY

²cat *verb* **cat·ted; cat·ting** (1769)
transitive verb
: to bring (an anchor) up to the cathead
intransitive verb
: to search for a sexual mate — often used with *around*

Cat \'kat\ *trademark*
— used for a Caterpillar tractor

cata- *or* **cat-** *or* **cath-** *prefix* [Greek *kata-, kat-, kath-,* from *kata* down, in accordance with, by; akin to Old Welsh *cant* with, Hittite *katta*]
: down ⟨*cat*ion⟩ ⟨*cath*ode⟩

ca·tab·o·lism \kə-'ta-bə-,li-zəm\ *noun* [Greek *katabolē* throwing down, from *kataballein* to throw down, from *kata-* + *ballein* to throw — more at DEVIL] (1876)
: destructive metabolism involving the release of energy and resulting in the breakdown of complex materials within the organism — compare ANABOLISM
— cat·a·bol·ic \,ka-tə-'bä-lik\ *adjective*
— cat·a·bol·i·cal·ly \-li-k(ə-)lē\ *adverb*

ca·tab·o·lite \-,līt\ *noun* (circa 1909)
: a product of catabolism

ca·tab·o·lize \-,līz\ *verb* **-lized; -liz·ing** (circa 1926)
transitive verb
: to subject to catabolism
intransitive verb
: to undergo catabolism

cat·a·chre·sis \,ka-tə-'krē-səs\ *noun, plural* **-chre·ses** \-,sēz\ [Latin, from Greek *katachrēsis* misuse, from *katachrēsthai* to use up, misuse, from *kata-* + *chrēsthai* to use] (1553)
1 : use of the wrong word for the context
2 : use of a forced and especially paradoxical figure of speech (as *blind mouths*)
— cat·a·chres·tic \-'kres-tik\ *or* **cat·a·chres·ti·cal** \-ti-kəl\ *adjective*
— cat·a·chres·ti·cal·ly \-ti-k(ə-)lē\ *adverb*

cat·a·clysm \'ka-tə-,kli-zəm\ *noun* [French *cataclysme,* from Latin *cataclysmos,* from Greek *kataklysmos,* from *kataklyzein* to inundate, from *kata-* + *klyzein* to wash — more at CLYSTER] (1637)
1 : FLOOD, DELUGE
2 : CATASTROPHE 3a
3 : a momentous and violent event marked by overwhelming upheaval and demolition; *broadly* **:** an event that brings great changes
— cat·a·clys·mal \,ka-tə-'kliz-məl\ *or* **cat·a·clys·mic** \-mik\ *adjective*
— cat·a·clys·mi·cal·ly \-mi-k(ə-)lē\ *adverb*

cat·a·comb \'ka-tə-,kōm\ *noun* [Middle French *catacombe,* probably from Old Italian *catacomba,* from Late Latin *catacumbae,* plural] (15th century)
1 : a subterranean cemetery of galleries with recesses for tombs — usually used in plural
2 : something resembling a catacomb: as **a :** an underground passageway or group of passageways **b :** a complex set of interrelated things ⟨the endless *catacombs* of formal education —Kingman Brewster, Jr.⟩

cata·di·op·tric \,ka-tə-dī-'äp-trik\ *adjective* (1723)
: belonging to, produced by, or involving both the reflection and the refraction of light

ca·tad·ro·mous \kə-'ta-drə-məs\ *adjective* [probably from New Latin *catadromus,* from *cata-* + *-dromus* -dromous] (1880)
: living in fresh water and going to the sea to spawn ⟨*catadromous* eels⟩ — compare ANADROMOUS

cat·a·falque \'ka-tə-,fȯ(l)k, -,falk\ *noun* [Italian *catafalco,* from (assumed) Vulgar Latin *catafalicum* scaffold, from *cata-* + Latin *fala* siege tower] (1641)
1 : an ornamental structure sometimes used in funerals for the lying in state of the body
2 : a pall-covered coffin-shaped structure used at requiem masses celebrated after burial

Cat·a·lan \'ka-tə-lən, -,lan\ *noun* [Spanish *catalán*] (15th century)
1 : a native or inhabitant of Catalonia
2 : the Romance language of Catalonia, Valencia, Andorra, and the Balearic islands
— Catalan *adjective*

cat·a·lase \'ka-tə-,lās, -,lāz\ *noun* [*catalysis*] (1901)
: a red crystalline enzyme that consists of a protein complex with hematin groups and catalyzes the decomposition of hydrogen peroxide into water and oxygen
— cat·a·lat·ic \,ka-tə-'la-tik\ *adjective*

cat·a·lec·tic \,ka-tə-'lek-tik\ *adjective* [Late Latin *catalecticus,* from Greek *katalēktikos,* from *katalēgein* to leave off, from *kata-* + *lēgein* to stop — more at SLACK] (1589)
: lacking a syllable at the end or ending in an incomplete foot
— catalectic *noun*

cat·a·lep·sy \'ka-tə-,lep-sē\ *noun, plural* **-sies** [Middle English *catalempsi,* from Medieval Latin *catalepsia,* from Late Latin *catalepsis,* from Greek *katalēpsis,* literally, act of seizing, from *katalambanein* to seize, from *kata-* + *lambanein* to take — more at LATCH] (14th century)
: a condition of suspended animation and loss of voluntary motion in which the limbs remain in whatever position they are placed
— cat·a·lep·tic \,ka-tə-'lep-tik\ *adjective or noun*
— cat·a·lep·ti·cal·ly \-ti-k(ə-)lē\ *adverb*

cat·a·lex·is \,ka-tə-'lek-səs\ *noun, plural* **-lex·es** \-,sēz\ [New Latin, from Greek *katalēxis* close, cadence, from *katalēgein*] (1830)
: omission or incompleteness usually in the last foot of a line in metrical verse

¹cat·a·log *or* **cat·a·logue** \'ka-tə-,lȯg, -,läg\ *noun* [Middle English *cateloge,* from Middle French *catalogue,* from Late Latin *catalogus,* from Greek *katalogos,* from *katalegein* to list, enumerate, from *kata-* + *legein* to gather, speak — more at LEGEND] (15th century)
1 : LIST, REGISTER
2 a : a complete enumeration of items arranged systematically with descriptive details **b :** a pamphlet or book that contains such a list **c :** material in such a list

²catalog *or* **catalogue** *verb* **-loged** *or* **-logued; -log·ing** *or* **-logu·ing** (1598)
transitive verb
1 : to make a catalog of
2 a : to enter in a catalog **b :** to classify (as books or information) descriptively
intransitive verb
1 : to make or work on a catalog
2 : to become listed in a catalog at a specified price ⟨this stamp *catalogs* at $2⟩
— cat·a·log·er *or* **cat·a·logu·er** *noun*

catalogue rai·son·né \-,rā-zᵊn-'ā\ *noun, plural* **cat·a·logues rai·son·nés** \-,lȯg(z)-,rā-zᵊn-'ā\ [French, literally, reasoned catalog] (1784)
: a systematic annotated catalog; *especially* **:** a critical bibliography

ca·tal·pa \kə-'tal-pə, -'tȯl-\ *noun* [Creek *katałpa,* from *iká* head + *tałpa* wing] (circa 1730)
: any of a genus (*Catalpa*) of American and Asian trees of the trumpet-creeper family with pale showy flowers in terminal clusters

ca·tal·y·sis \kə-'ta-lə-səs\ *noun, plural* **-y·ses** \-,sēz\ [Greek *katalysis* dissolution, from *katalyein* to dissolve, from *kata-* + *lyein* to dissolve, release — more at LOSE] (1836)
1 : a modification and especially increase in the rate of a chemical reaction induced by material unchanged chemically at the end of the reaction
2 : a reaction between two or more persons or forces precipitated by a separate agent ⟨a representative list of questions . . . valuable for the *catalysis* of class discussions —B. S. Meyer & D. B. Anderson⟩

cat·a·lyst \'ka-tᵊl-əst\ *noun* (1902)
1 : a substance (as an enzyme) that enables a chemical reaction to proceed at a usually faster rate or under different conditions (as at a lower temperature) than otherwise possible
2 : an agent that provokes or speeds significant change or action

cat·a·lyt·ic \,ka-tə-'li-tik\ *adjective* (1836)
: causing, involving, or relating to catalysis
— cat·a·lyt·i·cal·ly \-'li-ti-k(ə-)lē\ *adverb*

catalytic converter *noun* (1964)

: an automobile exhaust-system component containing a catalyst that causes decomposition of harmful gases into mostly harmless products (as water and carbon dioxide)

catalytic cracker *noun* (1947)
: the unit in a petroleum refinery in which cracking is carried out in the presence of a catalyst

cat·a·lyze \'ka-tə-ˌlīz\ *transitive verb* **-lyzed; -lyz·ing** (1902)
1 : to bring about the catalysis of (a chemical reaction)
2 : BRING ABOUT, INSPIRE
3 : to alter significantly by catalysis ⟨innovations in basic chemical theory that have *catalyzed* the field —*Newsweek*⟩
— **cat·a·lyz·er** *noun*

cat·a·ma·ran \ˌka-tə-mə-'ran, 'ka-tə-mə-ˌran\ *noun* [Tamil *kaṭṭumaram,* from *kaṭṭu* to tie + *maram* tree, wood] (1673)
: a vessel (as a sailboat) with twin hulls and usually a deck or superstructure connecting the hulls

cat·a·me·nia \ˌka-tə-'mē-nē-ə\ *noun plural* [New Latin, from Greek *katamēnia,* from neuter plural of *katamēnios* monthly, from *kata* by + *mēn* month — more at CATA-, MOON] (1750)
: MENSES
— **cat·a·me·ni·al** \-nē-əl\ *adjective*

cat·a·mite \'ka-tə-ˌmīt\ *noun* [Latin *catamitus,* from *Catamitus* Ganymede, from Etruscan *Catmite,* from Greek *Ganymēdēs*] (1593)
: a boy kept by a pederast

cat·a·mount \'ka-tə-ˌmau̇nt\ *noun* [short for *cat-a-mountain*] (1664)
: any of various wild cats: as **a :** COUGAR **b :** LYNX

cat–a–moun·tain \ˌka-tə-'mau̇n-tᵊn\ *noun* [Middle English *cat of the mountaine*] (15th century)
: any of various wild cats

cat and mouse *noun* (1887)
: behavior like that of a cat with a mouse: as **a :** the act of toying with or tormenting something before destroying it **b :** a contrived action involving constant pursuit, near captures, and repeated escapes ⟨played a game of *cat and mouse* with the police⟩; *broadly* **:** an evasive action

ca·taph·o·ra \kə-'ta-fə-rə\ *noun* [*cata-* + *anaphora*] (1976)
: the use of a grammatical substitute (as a pronoun) that has the same reference as a following word or phrase

cat·a·pho·re·sis \ˌka-tə-fə-'rē-səs\ *noun, plural* **-re·ses** \-ˌsēz\ [New Latin] (1889)
: ELECTROPHORESIS
— **cat·a·pho·ret·ic** \-'re-tik\ *adjective*
— **cat·a·pho·ret·i·cal·ly** \-ti-k(ə-)lē\ *adverb*

cat·a·phor·ic \ˌka-tə-'fȯr-ik\ *adjective* (1968)
: of or relating to cataphora; *especially* **:** being a word or phrase (as a pronoun) that takes its reference from a following word or phrase (as *her* in *before her Jane saw nothing but desert*) — compare ANAPHORIC

cat·a·plasm \'ka-tə-ˌpla-zəm\ *noun* [Middle French *cataplasme,* from Latin *cataplasma,* from Greek *kataplasma,* from *kataplassein* to plaster over, from *kata-* + *plassein* to mold — more at PLASTER] (1541)
: POULTICE

cat·a·plexy \'ka-tə-ˌplek-sē\ *noun, plural* **-plex·ies** \-sēz\ [German *Kataplexie,* from Greek *kataplēxis,* from *kataplēssein* to strike down, terrify, from *kata-* + *plēssein* to strike — more at PLAINT] (1883)
: sudden loss of muscle power following a strong emotional stimulus

¹cat·a·pult \'ka-tə-ˌpəlt, -ˌpu̇lt\ *noun* [Middle French or Latin; Middle French *catapulte,* from Latin *catapulta,* from Greek *katapaltēs,* from *kata-* + *pallein* to hurl] (1577)
1 : an ancient military device for hurling missiles
2 : a device for launching an airplane at flying speed (as from an aircraft carrier)

catapult 1

²catapult (1848)
transitive verb
: to throw or launch by or as if by a catapult
intransitive verb
: to become catapulted

cat·a·ract \'ka-tə-ˌrakt\ *noun* [Latin *cataracta* waterfall, portcullis, from Greek *kataraktēs,* from *katarassein* to dash down, from *kata-* + *arassein* to strike, dash] (14th century)
1 [Middle English, from Middle French or Medieval Latin; Middle French *cataracte,* from Medieval Latin *cataracta,* from Latin, portcullis] **:** a clouding of the lens of the eye or of its surrounding transparent membrane that obstructs the passage of light
2 a *obsolete* **:** WATERSPOUT **b :** WATERFALL; *especially* **:** a large one over a precipice **c :** steep rapids in a river **d :** DOWNPOUR, FLOOD ⟨*cataracts* of rain⟩ ⟨*cataracts* of information⟩
— **cat·a·rac·tous** \ˌka-tə-'rak-təs\ *adjective*

ca·tarrh \kə-'tär\ *noun* [Middle French or Late Latin; Middle French *catarrhe,* from Late Latin *catarrhus,* from Greek *katarrhous,* from *katarrhein* to flow down, from *kata-* + *rhein* to flow — more at STREAM] (15th century)
: inflammation of a mucous membrane; *especially* **:** one chronically affecting the human nose and air passages
— **ca·tarrh·al** \-əl\ *adjective*
— **ca·tarrh·al·ly** \-ə-lē\ *adverb*

cat·ar·rhine \'ka-tə-ˌrīn\ *adjective* [New Latin *Catarrhina,* from Greek *katarrhina,* neuter plural of *katarrhin* hook-nosed, from *kata-* + *rhin-, rhis* nose] (1863)
: of, relating to, or being any of a division (Catarrhina) of primates comprising the Old World monkeys, higher apes, and hominids that have the nostrils close together and directed downward, 32 teeth, and the tail when present never prehensile
— **catarrhine** *noun*

ca·tas·tro·phe \kə-'tas-trə-(ˌ)fē\ *noun* [Greek *katastrophē,* from *katastrephein* to overturn, from *kata-* + *strephein* to turn] (1540)
1 : the final event of the dramatic action especially of a tragedy
2 : a momentous tragic event ranging from extreme misfortune to utter overthrow or ruin
3 a : a violent and sudden change in a feature of the earth **b :** a violent usually destructive natural event (as a supernova)
4 : utter failure **:** FIASCO ◆
— **cat·a·stroph·ic** \ˌka-tə-'strä-fik\ *adjective*
— **cat·a·stroph·i·cal·ly** \-fi-k(ə-)lē\ *adverb*

catastrophe theory *noun* (1976)
: mathematical theory and conjecture that uses topology to explain events (as an earthquake or a stock market crash) characterized by major abrupt changes

ca·tas·tro·phism \kə-'tas-trə-ˌfi-zəm\ *noun* (1869)
: a geological doctrine that changes in the earth's crust have in the past been brought about suddenly by physical forces operating in ways that cannot be observed today — compare UNIFORMITARIANISM
— **ca·tas·tro·phist** \-fist\ *noun*

cat·a·to·nia \ˌka-tə-'tō-nē-ə\ *noun* [New Latin, from German *katatonie,* from *kata-* cata- + New Latin *-tonia*] (circa 1891)
: catatonic schizophrenia

cat·a·ton·ic \-'tä-nik\ *adjective* (1904)
1 : of, relating to, being, resembling, or affected by schizophrenia characterized especially by a marked psychomotor disturbance that may involve stupor or mutism, negativism, rigidity, purposeless excitement, and inappropriate or bizarre posturing
2 : characterized by a marked lack of movement, activity, or expression
— **catatonic** *noun*
— **cat·a·ton·i·cal·ly** \-ni-k(ə-)lē\ *adverb*

Ca·taw·ba \kə-'tȯ-bə\ *noun* (1715)
1 *plural* **Catawba** *or* **Catawbas :** a member of an American Indian people of North Carolina and South Carolina
2 : the language of the Catawba people
3 : any of various wines produced from a pale red native American grape

cat·bird \'kat-ˌbərd\ *noun* (1709)
: an American songbird (*Dumetella carolinensis*) that is dark gray in color with a black cap and reddish coverts under the tail and is related to the mockingbird

catbird seat *noun* (1942)
: a position of great prominence or advantage

cat·boat \'kat-ˌbōt\ *noun* (1878)
: a sailboat having a cat rig and usually a centerboard and being of light draft and broad beam

catboat

cat·bri·er \-ˌbrī(-ə)r\ *noun* (1839)
: any of a genus (*Smilax*) of dioecious often prickly climbing plants of the lily family

cat burglar *noun* (1907)
: a burglar who is adept at entering and leaving the burglarized place without attracting notice

cat·call \-ˌkȯl\ *noun* (1749)
: a loud or raucous cry made especially to express disapproval (as at a sports event)
— **catcall** *verb*

¹catch \'kach, 'kech\ *verb* **caught** \'kȯt *also* 'kät\; **catch·ing** [Middle English *cacchen,* from Old North French *cachier* to hunt, from (assumed) Vulgar Latin *captiare,* alteration of Latin *captare* to chase, frequentative of *capere* to take — more at HEAVE] (13th century)
transitive verb

◇ WORD HISTORY
catastrophe The Greek noun *katastrophē* meant "end, conclusion," especially one not desired; as a derivative of the verb *katastrephein* "to overturn, ruin, undo," it might more literally be translated "undoing." The word aptly fit the finale of a classical tragedy, which normally ended with the complete ruin of the central character. English *catastrophe,* borrowed from Greek in the 16th century, first denoted the dénouement of a play, comic or tragic. Later its scope was extended beyond drama to any kind of final event in a series, though usually an event of an unhappy nature, better matching the original sense of the Greek word. Since the 18th century, the meaning "unfortunate event" has prevailed over "conclusion," so that *catastrophe* is now nearly synonymous with *disaster.*

1 a : to capture or seize especially after pursuit **b :** to take or entangle in or as if in a snare **c :** DECEIVE **d :** to discover unexpectedly **:** FIND ⟨*caught* in the act⟩ **e :** to check (oneself) suddenly or momentarily **f :** to become suddenly aware of ⟨*caught* me looking at him⟩
2 a : to take hold of **:** SEIZE **b :** to affect suddenly **c :** INTERCEPT **d :** to avail oneself of **:** TAKE **e :** to obtain through effort **:** GET **f :** to get entangled ⟨*catch* a sleeve on a nail⟩
3 : to become affected by: as **a :** CONTRACT ⟨*catch* a cold⟩ **b :** to respond sympathetically to the point of being imbued with ⟨*catch* the spirit of an occasion⟩ **c :** to be struck by **d :** to be subjected to **:** RECEIVE ⟨*catch* hell⟩
4 a : to seize and hold firmly **b :** FASTEN
5 : to take or get usually momentarily or quickly ⟨*catch* a glimpse of a friend⟩ ⟨*catch* a nap⟩
6 a : OVERTAKE **b :** to get aboard in time ⟨*catch* the bus⟩
7 : to attract and hold **:** ARREST, ENGAGE ⟨*caught* my attention⟩ ⟨*caught* her eye⟩
8 : to make contact with **:** STRIKE
9 a : to grasp by the senses or the mind ⟨you *catch* what I mean?⟩ ⟨didn't *catch* the name⟩ **b :** to apprehend and fix by artistic means
10 a : SEE, WATCH ⟨*catch* a game on TV⟩ **b :** to listen to
11 : to serve as a catcher for in baseball
12 : to meet with socially ⟨*catch* you later⟩
intransitive verb
1 : to grasp hastily or try to grasp
2 : to become caught
3 : to catch fire
4 *of a crop* **:** to come up and become established
5 : to play the position of catcher on a baseball team
6 : KICK OVER ☆
— catch·able \ˈka-chə-bəl, ˈke-\ *adjective*
— catch a crab : to fail to raise an oar clear of the water on recovery of a stroke
— catch dead : willing to be publicly exposed — used in negative constructions ⟨wouldn't be *caught dead* in that shirt⟩
— catch fire 1 : to become ignited **2 :** to become fired with enthusiasm **3 :** to increase greatly in scope, popularity, interest, or effectiveness ⟨this stock has not *caught fire*—yet —*Forbes*⟩
— catch it : to incur blame, reprimand, or punishment
— catch one's breath : to rest long enough to restore normal breathing; *broadly* **:** to rest after a period of intense activity

²catch *noun* (15th century)
1 : something caught; *especially* **:** the total quantity caught at one time ⟨a large *catch* of fish⟩
2 a : the act, action, or fact of catching **b :** a game in which a ball is thrown and caught
3 : something that checks or holds immovable ⟨a safety *catch*⟩
4 : one worth catching especially as a spouse
5 : a round for three or more unaccompanied usually male voices often with suggestive or obscene lyrics
6 : FRAGMENT, SNATCH
7 : a concealed difficulty or complication ⟨there must be a *catch*⟩
8 : the germination of a field crop to such an extent that replanting is unnecessary

catch·all \ˈka-ˌchȯl, ˈke-\ *noun, often attributive* (1838)
: something that holds or includes odds and ends or a wide variety of things

catch–as–catch–can *adjective* (1764)
: using any available means or method **:** UNPLANNED ⟨a *catch-as-catch-can* existence begging and running errands —*Time*⟩

catch·er \ˈka-chər, ˈke-\ *noun* (15th century)
: one that catches; *specifically* **:** a baseball player positioned behind home plate

catch·fly \ˈkach-ˌflī, ˈkech-\ *noun* (1597)
: any of various plants (as of the genera *Lychnis* and *Silene*) of the pink family often with viscid stems

catch·ing *adjective* (1590)
1 : INFECTIOUS, CONTAGIOUS ⟨the flu is *catching*⟩ ⟨his spirit is *catching*⟩
2 : CATCHY, ALLURING

catch·ment \ˈkach-mənt, ˈkech-\ *noun* (1847)
1 : something that catches water; *also* **:** the amount of water caught
2 : the action of catching water

catchment area *noun* (1940)
: the geographical area served by an institution

catch on *intransitive verb* (1883)
1 : to become aware **:** LEARN; *also* **:** UNDERSTAND ⟨didn't *catch on* to what was going on⟩
2 : to become popular ⟨this idea has already *caught on*⟩

catch out *transitive verb* (1816)
: to detect in error or wrongdoing ⟨delighting to *catch out* his literary victims in error —John Clive⟩

catch·pen·ny \ˈkach-ˌpe-nē, ˈkech-\ *adjective* (1748)
: using sensationalism or cheapness for appeal ⟨a *catchpenny* newspaper⟩

catch·phrase \-ˌfrāz\ *noun* (circa 1850)
: an expression that has caught on and is used repeatedly

catch·pole *or* **catch·poll** \-ˌpōl\ *noun* [Middle English *cacchepol,* from Old English *cæcepol,* from (assumed) Old North French *cachepol,* literally, chicken chaser, from Old North French *cachier* + *pol* chicken, from Latin *pullus* — more at CATCH, PULLET] (before 12th century)
: a sheriff's deputy; *especially* **:** one who makes arrests for failure to pay a debt

catch–22 \-ˌtwen-tē-ˈtü\ *noun, plural* **catch–22's** *or* **catch–22s** *often capitalized* [from *Catch-22,* paradoxical rule in the novel *Catch-22* (1961) by Joseph Heller] (1961)
1 : a problematic situation for which the only solution is denied by a circumstance inherent in the problem or by a rule ⟨the show-business *catch-22*—no work unless you have an agent, no agent unless you've worked —Mary Murphy⟩; *also* **:** the circumstance or rule that denies a solution
2 a : an illogical, unreasonable, or senseless situation **b :** a measure or policy whose effect is the opposite of what was intended **c :** a situation presenting two equally undesirable alternatives
3 : a hidden difficulty or means of entrapment **:** CATCH

catch·up \ˈke-chəp, ˈka-; ˈkat-səp\ *variant of* KETCHUP

¹catch–up \ˈkach-ˌəp, ˈkech-\ *adjective* (1945)
: intended to catch up to a theoretical norm or a competitor's accomplishments

²catch–up *noun* (1948)
: the act or fact of catching up or trying to catch up (as with a norm or competitor) ⟨had to play *catch-up*⟩; *also* **:** an increase intended to achieve catch-up

catch up (14th century)
transitive verb
1 a : to pick up often abruptly ⟨the thief *caught* the purse *up* and ran⟩ **b :** ENSNARE, ENTANGLE ⟨education has been *caught up* in a stultifying mythology —N. M. Pusey⟩ **c :** ENTHRALL ⟨the . . . public was *caught up* in the car's magic —D. A. Jedlicka⟩
2 : to provide with the latest information ⟨*catch* me *up* on the news⟩
intransitive verb
1 a : to travel fast enough to overtake an advance party **b :** to reach a state of parity or of being able to cope ⟨kids left behind in preschool may never *catch up*⟩
2 : to bring about arrest for illicit activities ⟨the police *caught up* with the thieves⟩
3 a : to complete or compensate for something belatedly ⟨*catch up* on lost sleep⟩ **b :** to acquire belated information ⟨*catch up* on the news⟩

catch·word \ˈkach-ˌwərd, ˈkech-\ *noun* (circa 1736)
1 a : a word under the right-hand side of the last line on a book page that repeats the first word on the following page **b :** GUIDE WORD
2 : a word or expression repeated until it becomes representative of a party, school, or point of view

catchy \ˈka-chē, ˈke-chē\ *adjective* **catch·i·er; -est** (1831)
1 : tending to catch the interest or attention ⟨a *catchy* title⟩
2 : FITFUL, IRREGULAR ⟨*catchy* breathing⟩
3 : TRICKY ⟨a *catchy* question⟩

cat·claw \ˈkat-ˌklȯ\ *noun* (1898)
: a yellow-flowered spiny acacia (*Acacia greggi*) of the southwestern U.S. and Mexico

cat distemper *noun* (circa 1950)
: PANLEUKOPENIA

cate \ˈkāt\ *noun* [Middle English, article of purchased food, short for *acate,* from Old North French *acat* purchase, from *acater* to buy, from (assumed) Vulgar Latin *accaptare,* from Latin *acceptare* to accept] (15th century)
archaic **:** a dainty or choice food

cat·e·che·sis \ˌka-tə-ˈkē-səs\ *noun, plural* **-che·ses** \-ˌsēz\ [Late Latin, from Greek *katēchēsis,* from *katēchein* to teach] (1753)
: oral instruction of catechumens
— cat·e·chet·i·cal \-ˈke-ti-kəl\ *adjective*

cat·e·chin \ˈka-tə-ˌkin\ *noun* [International Scientific Vocabulary *catechu* + ¹*-in*] (1853)
: a crystalline compound $C_{15}H_{14}O_6$ or its derivatives related chemically to the flavones and used in dyeing and tanning

cat·e·chism \ˈka-tə-ˌki-zəm\ *noun* (1502)
1 : oral instruction
2 : a manual for catechizing; *specifically* **:** a summary of religious doctrine often in the form of questions and answers
3 a : a set of formal questions put as a test **b :** something resembling a catechism especially in being a rote response or formulaic statement
— cat·e·chis·mal \ˌka-tə-ˈkiz-məl\ *adjective*
— cat·e·chis·tic \-ˈkis-tik\ *adjective*

cat·e·chist \ˈka-tə-kist\ *noun* (circa 1563)
: one that catechizes: as **a :** a teacher of catechumens **b :** a native in a missionary district who does Christian teaching

cat·e·chize \ˈka-tə-ˌkīz\ *transitive verb* **-chized; -chiz·ing** [Late Latin *catechizare,* from Greek *katēchein* to teach, literally, to din into, from *kata-* cata- + *ēchein* to resound, from *ēchē* sound — more at ECHO] (15th century)
1 : to instruct systematically especially by questions, answers, and explanations and cor-

☆ SYNONYMS
Catch, capture, trap, snare, entrap, ensnare, bag mean to come to possess or control by or as if by seizing. CATCH implies the seizing of something in motion or in flight or in hiding ⟨*caught* the dog as it ran by⟩. CAPTURE suggests taking by overcoming resistance or difficulty ⟨*capture* an enemy stronghold⟩. TRAP, SNARE, ENTRAP, ENSNARE imply seizing by some device that holds the one caught at the mercy of the captor. TRAP and SNARE apply more commonly to physical seizing ⟨*trap* animals⟩ ⟨*snared* butterflies with a net⟩. ENTRAP and ENSNARE more often are figurative ⟨*entrapped* the witness with a trick question⟩ ⟨a sting operation that *ensnared* burglars⟩. BAG implies shooting down a fleeing or distant prey ⟨*bagged* a brace of pheasants⟩.

rections; *specifically* **:** to give religious instruction in such a manner

2 : to question systematically or searchingly

— **cat·e·chi·za·tion** \ˌka-ti-kə-ˈzā-shən\ *noun*

— **cat·e·chiz·er** \ˈka-tə-ˌkī-zər\ *noun*

cat·e·chol \ˈka-tə-ˌkȯl, -ˌkōl\ *noun* (1880)

1 : CATECHIN

2 : PYROCATECHOL

cat·e·chol·amine \ˌka-tə-ˈkō-lə-ˌmēn, -ˈkȯ-\ *noun* (1954)

: any of various amines (as epinephrine, norepinephrine, and dopamine) that function as hormones or neurotransmitters or both

cat·e·chol·amin·er·gic \-ˌkō-lə-mē-ˈnər-jik, -ˌkȯ-, -mi-\ *adjective* (1970)

: involving, liberating, or mediated by catecholamine

cat·e·chu \ˈka-tə-ˌchü, -ˌshü\ *noun* [probably modification of Malay *kachu,* of Dravidian origin; akin to Tamil & Kannada *kācu* catechu] (1683)

: any of several dry, earthy, or resinous astringent substances obtained from tropical plants of Asia: as **a :** an extract of the heartwood of an East Indian acacia (*Acacia catechu*) **b :** GAMBIER

cat·e·chu·men \ˌka-tə-ˈkyü-mən\ *noun* [Middle English *cathecumyn,* from Middle French *cathecumine,* from Late Latin *catechumenus,* from Greek *katēchoumenos,* present passive participle of *katēchein*] (15th century)

1 : a convert to Christianity receiving training in doctrine and discipline before baptism

2 : one receiving instruction in the basic doctrines of Christianity before admission to communicant membership in a church

cat·e·gor·i·cal \ˌka-tə-ˈgȯr-i-kəl, -ˈgär-\ *also* **cat·e·gor·ic** \-ik\ *adjective* [Late Latin *categoricus,* from Greek *katēgorikos,* from *katēgoria*] (1588)

1 : ABSOLUTE, UNQUALIFIED ⟨a *categorical* denial⟩

2 a : of, relating to, or constituting a category **b :** involving, according with, or considered with respect to specific categories

— **cat·e·gor·i·cal·ly** \-i-k(ə-)lē\ *adverb*

categorical imperative *noun* (1827)

: a moral obligation or command that is unconditionally and universally binding

cat·e·go·rise *British variant of* CATEGORIZE

cat·e·go·rize \ˈka-ti-gə-ˌrīz\ *transitive verb* **-rized; -riz·ing** (1705)

: to put into a category **:** CLASSIFY

— **cat·e·go·ri·za·tion** \ˌka-ti-gə-rə-ˈzā-shən\ *noun*

cat·e·go·ry \ˈka-tə-ˌgōr-ē, -ˌgȯr-\ *noun, plural* **-ries** [Late Latin *categoria,* from Greek *katēgoria* predication, category, from *katēgorein* to accuse, affirm, predicate, from *kata-* + *agora* public assembly, from *ageirein* to gather] (1588)

1 : any of several fundamental and distinct classes to which entities or concepts belong

2 : a division within a system of classification

ca·te·na \kə-ˈtē-nə\ *noun, plural* **-nae** \-(ˌ)nē\ *or* **-nas** [Medieval Latin, from Latin, chain] (1641)

: a connected series of related things

cat·e·nary \ˈka-tə-ˌner-ē, *especially British* kə-ˈtē-nə-rē\ *noun, plural* **-nar·ies** [New Latin *catenaria,* from Latin, feminine of *catenarius* of a chain, from *catena*] (1788)

1 : the curve assumed by a cord of uniform density and cross section that is perfectly flexible but not capable of being stretched and that hangs freely from two fixed points

2 : something in the form of a catenary

— **catenary** *adjective*

cat·e·nate \ˈka-tə-ˌnāt\ *transitive verb* **-nat·ed; -nat·ing** [Latin *catenatus,* past participle of *catenare,* from *catena*] (circa 1623)

: to connect in a series **:** LINK

— **cat·e·na·tion** \ˌka-tə-ˈnā-shən\ *noun*

ca·ter \ˈkā-tər\ *verb* [obsolete *cater* buyer of provisions, from Middle English *catour,* short for *acatour,* from Anglo-French, from Old North French *acater* to buy — more at CATE] (1600)

intransitive verb

1 : to provide a supply of food

2 : to supply what is required or desired ⟨*catering* to middle-class tastes⟩

transitive verb

: to provide food and service for ⟨*catered* the banquet⟩

— **ca·ter·er** \-tər-ər\ *noun*

cat·er·an \ˈka-tə-rən\ *noun* [Middle English *ketharan,* probably from Scottish Gaelic *ceathairneach* freebooter, robber] (14th century)

: a former military irregular or brigand of the Scottish Highlands

cat·er·cor·ner \ˈka-tē-ˌkȯr-nər, ˈka-tə-, ˈki-tē-\ *also* **cat·er·cor·nered** \-nərd\ *adverb or adjective* [obsolete *cater* four + *corner*] (1838)

: in a diagonal or oblique position **:** on a diagonal or oblique line ⟨the house stood *catercorner* across the square⟩

ca·ter–cous·in \ˈkā-tər-ˌkə-z^{ə}n\ *noun* [perhaps from obsolete *cater* buyer of provisions] (1519)

: an intimate friend

cat·er·pil·lar \ˈka-tə(r)-ˌpi-lər\ *noun, often attributive* [Middle English *catyrpel,* from Old North French *catepelose,* literally, hairy cat] (15th century)

: the elongated wormlike larva of a butterfly or moth; *also* **:** any of various similar larvae ◇

Caterpillar *trademark*

— used for a tractor made for use on rough or soft ground and moved on two endless metal belts

cat·er·waul \ˈka-tər-ˌwȯl\ *intransitive verb* [Middle English *caterwawen*] (14th century)

1 : to make a harsh cry

2 : to quarrel noisily

— **caterwaul** *noun*

cat·fac·ing \ˈkat-ˌfā-siŋ\ *noun* (1940)

: a disfigurement or malformation of fruit suggesting a cat's face in appearance

cat·fight \-ˌfīt\ *noun* (1919)

: an intense fight or argument between two women

cat·fish \-ˌfish\ *noun* (1612)

: any of an order (Siluriformes) of chiefly freshwater stout-bodied scaleless bony fishes having long tactile barbels

catfish

cat·gut \-ˌgət\ *noun* (1599)

: a tough cord made usually from sheep intestines

cath- — see CATA-

Cath·ar \ˈka-ˌthär\ *noun, plural* **Cath·a·ri** \ˈka-thə-ˌrī, -ˌrē\ *or* **Cathars** [Late Latin *cathari* (plural), from Late Greek *katharoi,* from Greek, plural of *katharos* pure] (1637)

: a member of one of various ascetic and dualistic Christian sects especially of the later Middle Ages teaching that matter is evil and professing faith in an angelic Christ who did not really undergo human birth or death

— **Cath·a·rism** \ˈka-thə-ˌri-zəm\ *noun*

— **Cath·a·rist** \-rist\ *or* **Cath·a·ris·tic** \ˌka-thə-ˈris-tik\ *adjective*

ca·thar·sis \kə-ˈthär-səs\ *noun, plural* **ca·thar·ses** \-ˌsēz\ [New Latin, from Greek *katharsis,* from *kathairein* to cleanse, purge, from *katharos*] (circa 1775)

1 : PURGATION

2 a : purification or purgation of the emotions (as pity and fear) primarily through art **b :** a purification or purgation that brings about spiritual renewal or release from tension

3 : elimination of a complex by bringing it to consciousness and affording it expression

^{1}**ca·thar·tic** \kə-ˈthär-tik\ *adjective* [Late Latin or Greek; Late Latin *catharticus,* from Greek *kathartikos,* from *kathairein*] (1612)

: of, relating to, or producing catharsis

^{2}**cathartic** *noun* (1651)

: a cathartic medicine **:** PURGATIVE

cat·head \ˈkat-ˌhed\ *noun* (1626)

: a projecting piece of timber or iron near the bow of a ship to which the anchor is hoisted and secured

ca·thect \kə-ˈthekt, ka-\ *transitive verb* [back-formation from *cathectic*] (1925)

: to invest with mental or emotional energy

ca·thec·tic \kə-ˈthek-tik, ka-\ *adjective* [New Latin *cathexis*] (1927)

: of, relating to, or invested with mental or emotional energy

ca·the·dra \kə-ˈthē-drə\ *noun* [Latin, chair — more at CHAIR] (circa 1797)

: a bishop's official throne

^{1}**ca·the·dral** \kə-ˈthē-drəl\ *adjective* (14th century)

1 : of, relating to, or containing a cathedra

2 : emanating from a chair of authority

3 : suggestive of a cathedral

^{2}**cathedral** *noun* (1587)

1 : a church that is the official seat of a diocesan bishop

2 : something that resembles or suggests a cathedral

ca·thep·sin \kə-ˈthep-sən\ *noun* [Greek *kathepsein* to digest (from *kata-* cata- + *hepsein* to boil) + English ^{2}*-in*] (1929)

: any of several intracellular proteases of animal tissue that aid in autolysis

Cath·er·ine wheel \ˈka-th(ə-)rin-\ *noun* [Saint *Catherine* of Alexandria (died about 307) Christian martyr] (1584)

1 : a wheel with spikes projecting from the rim

2 : PINWHEEL 1

3 : CARTWHEEL 2

cath·e·ter \ˈka-thə-tər, ˈkath-tər\ *noun* [Late Latin, from Greek *kathetēr,* from *kathienai* to send down, from *kata-* cata- + *hienai* to send — more at JET] (1601)

: a tubular medical device for insertion into canals, vessels, passageways, or body cavities usually to permit injection or withdrawal of fluids or to keep a passage open

cath·e·ter·i·za·tion \ˌka-thə-tə-rə-ˈzā-shən, ˌkath-tə-rə-\ *noun* (circa 1852)

: the use of or introduction of a catheter (as in or into the bladder, trachea, or heart)

◇ WORD HISTORY

caterpillar *Caterpillar,* our common word for a butterfly or moth larva, first appeared in a 15th century English-Latin dictionary as *catirpel.* It was presumably altered from the medieval form of a word which emerges in the modern French dialect of Normandy as *catepeleuse* or *carpeleuse* and which then meant literally "hairy cat." (Similar applications of a name for a furry animal to fuzzy larvae are English *woolly bear* and French *chenille* "caterpillar," descended from Latin *canicula* "little dog.") In the 16th century, *caterpil* is replaced by *catirpiller,* probably influenced by the now obsolete word *piller* "pillager, looter," a not inappropriate association in view of the destructive habits of some insects. The spelling *caterpillar,* which much less appropriately suggests *pillar,* was adopted by the 18th century lexicographer Samuel Johnson and has since remained standard.

\ə\ **abut** \ᵊ\ **kitten** \ər\ **further** \a\ **ash** \ā\ **ace** \ä\ **mop, mar** \au̇\ **out** \ch\ **chin** \e\ **bet** \ē\ **easy** \g\ **go** \i\ **hit** \ī\ **ice** \j\ **job** \ŋ\ **sing** \ō\ **go** \ȯ\ **law** \ȯi\ **boy** \th\ **thin** \t͟h\ **the** \ü\ **loot** \u̇\ **foot** \y\ **yet** \zh\ **vision** *see also* Guide to Pronunciation

— **cath·e·ter·ize** \'ka-thə-tə-ˌrīz, 'kath-tə-\ *transitive verb*

ca·thex·is \kə-'thek-səs, ka-\ *noun, plural* **ca·thex·es** \-ˌsēz\ [New Latin (intended as translation of German *Besetzung*), from Greek *kathexis* holding, from *katechein* to hold fast, occupy, from *kata-* + *echein* to have, hold — more at SCHEME] (1922)
: investment of mental or emotional energy in a person, object, or idea

cath·ode \'ka-ˌthōd\ *noun* [Greek *kathodos* way down, from *kata-* + *hodos* way] (1834)
1 : the electrode of an electrochemical cell at which reduction occurs: **a :** the negative terminal of an electrolytic cell **b :** the positive terminal of a galvanic cell
2 : the electron-emitting electrode of an electron tube — compare ANODE
— **cath·od·al** \'ka-ˌthō-dᵊl\ *adjective*
— **cath·od·al·ly** *adverb*
— **ca·thod·ic** \ka-'thä-dik, -'thō-\ *adjective*
— **ca·thod·i·cal·ly** \-di-k(ə-)lē\ *adverb*

cathode ray *noun* (1880)
1 *plural* **:** the high-speed electrons emitted in a stream from the heated cathode of a vacuum tube
2 : a stream of electrons emitted from the cathode of a vacuum tube — usually used in plural

cathode–ray tube *noun* (1905)
: a vacuum tube in which a beam of electrons is projected on a fluorescent screen to produce a luminous spot at a point on the screen determined by the effect on the electron beam of a variable magnetic field within the tube

cathodic protection *noun* (1930)
: the prevention of electrolytic corrosion of a usually metallic structure (as a pipeline) by causing it to act as the cathode rather than as the anode of an electrochemical cell

cath·o·lic \'kath-lik, 'ka-thə-\ *adjective* [Middle French & Late Latin; Middle French *catholique,* from Late Latin *catholicus,* from Greek *katholikos* universal, general, from *katholou* in general, from *kata* by + *holos* whole — more at CATA-, SAFE] (14th century)
1 a *often capitalized* **:** of, relating to, or forming the church universal **b** *often capitalized* **:** of, relating to, or forming the ancient undivided Christian church or a church claiming historical continuity from it **c** *capitalized* **:** ROMAN CATHOLIC
2 : COMPREHENSIVE, UNIVERSAL; *especially* **:** broad in sympathies, tastes, or interests
— **ca·thol·i·cal·ly** \kə-'thä-li-k(ə-)lē\ *adverb*
— **ca·thol·i·cize** \-'thä-lə-ˌsīz\ *verb*

Cath·o·lic \'kath-lik, 'ka-thə-\ *noun* (15th century)
1 : a person who belongs to the universal Christian church
2 : a member of a Catholic church; *especially* **:** ROMAN CATHOLIC

Catholic Apostolic *adjective* (1888)
: of or relating to a Christian sect founded in 19th century England in anticipation of Christ's second coming

ca·thol·i·cate \kə-'thä-lə-ˌkāt, -kət\ *noun* (1850)
: the jurisdiction of a catholicos

Catholic Epistles *noun plural* (1582)
: the five New Testament letters including James, I and II Peter, I John, and Jude addressed to the early Christian churches at large

Ca·thol·i·cism \kə-'thä-lə-ˌsi-zəm\ *noun* (circa 1617)
1 : ROMAN CATHOLICISM
2 : the faith, practice, or system of Catholic Christianity

cath·o·lic·i·ty \ˌka-thə-'li-sə-tē, ˌkath-'li-\ *noun, plural* **-ties** (1704)
1 *capitalized* **:** the character of being in conformity with a Catholic church
2 a : liberality of sentiments or views ⟨*catholicity* of viewpoint —W. V. O'Connor⟩ **b :** UNIVERSALITY **c :** comprehensive range ⟨*catholicity* of topics⟩

ca·thol·i·con \kə-'thä-lə-ˌkän\ *noun* [Middle English, from Medieval Latin, from Greek *katholikon,* neuter of *katholikos*] (15th century)
: CURE-ALL, PANACEA

ca·thol·i·cos \kə-'thä-li-kəs\ *noun, plural* **-i·cos·es** \-kə-səz\ *or* **-i·coi** \-'thä-lə-ˌkȯi\ *often capitalized* [Late Greek *katholikos,* from Greek, general] (1878)
: a primate of certain Eastern churches and especially of the Armenian or of the Nestorian church

cat·house \'kat-ˌhau̇s\ *noun* (1931)
: BORDELLO

cat·ion \'kat-ˌī-ən, 'ka-(ˌ)tī-ən\ *noun* [Greek *kation,* neuter of *katiōn,* present participle of *katienai* to go down, from *kata-* cata- + *ienai* to go — more at ISSUE] (1834)
: the ion in an electrolyzed solution that migrates to the cathode; *broadly* **:** a positively charged ion

cat·ion·ic \ˌkat-(ˌ)ī-'ä-nik, ˌka-(ˌ)tī-\ *adjective* (circa 1920)
1 : of, relating to, or being a cation
2 : characterized by an active and especially surface-active cation ⟨a *cationic* dye⟩
— **cat·ion·i·cal·ly** \-ni-k(ə-)lē\ *adverb*

cat·kin \'kat-kən\ *noun* [from its resemblance to a cat's tail] (1578)
: a spicate inflorescence (as of the willow, birch, or oak) bearing scaly bracts and unisexual usually apetalous flowers — called also *ament*

cat·like \'kat-ˌlīk\ *adjective* (1600)
: resembling a cat; *especially* **:** STEALTHY ⟨with *catlike* tread, upon our prey we steal —W. S. Gilbert⟩

cat·mint \-ˌmint\ *noun* (13th century)
: CATNIP

cat·nap \-ˌnap\ *noun* (1823)
: a very short light nap
— **catnap** *intransitive verb*

cat·nap·per *also* **cat·nap·er** \'kat-ˌna-pər\ *noun* [[1]*cat* + *-napper* (as in *kidnapper*)] (1942)
: one that steals cats usually to sell them for research

cat·nip \-ˌnip\ *noun* [[1]*cat* + obsolete *nep* catnip, from Middle English, from Old English *nepte,* from Latin *nepeta*] (1712)
1 : a strong-scented mint (*Nepeta cataria*) that has whorls of small pale flowers in terminal spikes and contains a substance attractive to cats
2 : something very attractive

cat–o'–nine–tails \ˌka-tə-'nīn-ˌtālz\ *noun, plural* **cat–o'–nine–tails** [from the resemblance of its scars to the scratches of a cat] (1665)
: a whip made of usually nine knotted lines or cords fastened to a handle

ca·top·tric \kə-'täp-trik\ *adjective* [Greek *katoptrikos,* from *katoptron* mirror, from *katopsesthai* to be going to observe, from *kata-* + *opsesthai* to be going to see — more at OPTIC] (circa 1774)
: being or using a mirror to focus light

cat rig *noun* (1867)
: a rig consisting of a single mast far forward carrying a single large sail extended by a boom
— **cat–rigged** \'kat-'rigd\ *adjective*

cats and dogs *adverb* (1738)
: in great quantities **:** very hard ⟨it was raining *cats and dogs*⟩

CAT scan \'kat-\ *noun* [computerized *a*xial *t*omography] (1975)
: an image made by computed tomography
— **CAT scanning** *noun*

CAT scanner *noun* (1975)
: a medical instrument consisting of integrated X-ray and computing equipment and used for computed tomography

cat's cradle *noun* (1768)
1 : a game in which a string looped in a pattern like a cradle on the fingers of one person's hands is transferred to the hands of another so as to form a different figure
2 : something that is intricate, complicated, or elaborate ⟨a *cat's cradle* of red tape⟩

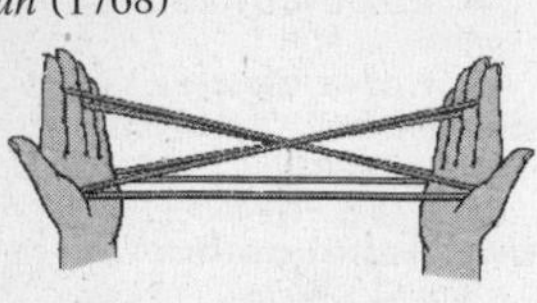
cat's cradle 1

cat scratch disease *noun* (1952)
: an illness characterized by swelling of the lymph glands, fever, and chills and assumed to be caused by a bacterium transmitted especially by a cat scratch — called also *cat scratch fever*

cat's–eye \'kat-ˌsī\ *noun, plural* **cat's–eyes** (circa 1599)
1 : any of various gems (as a chrysoberyl or a chalcedony) exhibiting opalescent reflections from within
2 : a marble with eyelike concentric circles

cat's meow *noun* (1926)
: a highly admired person or thing

cat's–paw \'kats-ˌpȯ\ *noun, plural* **cat's–paws** (circa 1769)
1 : a light air that ruffles the surface of the water in irregular patches during a calm
2 [from the fable of the monkey that used a cat's paw to draw chestnuts from the fire] **:** one used by another as a tool **:** DUPE
3 : a hitch formed with two eyes for attaching a line to a hook — see KNOT illustration

cat·sup *variant of* KETCHUP

cat·tail \'kat-ˌtāl\ *noun* (1548)
: any of a genus (*Typha* of the family Typhaceae, the cattail family) of tall reedy marsh plants with brown furry fruiting spikes; *especially* **:** a plant (*Typha latifolia*) with long flat leaves used especially for making mats and chair seats

cat·tery \'ka-tə-rē\ *noun, plural* **-ter·ies** (circa 1843)
: an establishment for the breeding and boarding of cats

cat·tle \'ka-tᵊl\ *noun plural* [Middle English *catel,* from Old North French, personal property, from Medieval Latin *capitale,* from Latin, neuter of *capitalis* of the head — more at CAPITAL] (14th century)
1 : domesticated quadrupeds held as property or raised for use; *specifically* **:** bovine animals on a farm or ranch
2 : human beings especially en masse

cattle call *noun* (1952)
: a mass audition (as of actors)

cattle egret *noun* (circa 1899)
: a small Old World white egret (*Bubulcus ibis*) introduced into the New World and having a yellow bill and in the breeding season buff on the crown, breast, and back

cattle grub *noun* (1926)
: either of two warble flies (genus *Hypoderma*) especially in the larval stage: **a :** COMMON CATTLE GRUB **b :** a related warble fly (*H. bovis*)

cat·tle·man \-mən, -ˌman\ *noun* (1864)
: one who tends or raises cattle

cattle tick *noun* (1869)
: either of two ixodid ticks (*Boophilus annulatus* and *B. microplus*) that infest cattle and transmit the protozoan which causes Texas fever

cat·tle·ya \'kat-lē-ə; kat-'lā-ə, -'lē-\ *noun* [New Latin, from William *Cattley* (died 1832) English patron of botany] (1828)
: any of a genus (*Cattleya*) of tropical American epiphytic orchids with showy hooded flowers

[1]**cat·ty** \'ka-tē\ *noun, plural* **catties** [Malay *kati*] (1598)

: any of various units of weight of China and southeast Asia varying around 1⅓ pounds (about 600 grams); *also* **:** a standard Chinese unit equal to 1.1023 pounds (500 grams)

[2]catty *adjective* **cat·ti·er; -est** (1886)
1 : resembling a cat; *especially* **:** slyly spiteful **:** MALICIOUS
2 : of or relating to a cat
— cat·ti·ly \ˈka-tᵊl-ē\ *adverb*
— cat·ti·ness \ˈka-tē-nəs\ *noun*

cat·ty–cor·ner *or* **cat·ty–cor·nered** *variant of* CATERCORNER

cat·walk \ˈkat-ˌwȯk\ *noun* (1885)
: a narrow walkway (as along a bridge)

Cau·ca·sian \kȯ-ˈkā-zhən, kä- *also* -ˈka-zhən\ *adjective* (1807)
1 : of or relating to the Caucasus or its inhabitants
2 a : of or relating to the white race of humankind as classified according to physical features **b :** of or relating to the white race as defined by law specifically as composed of persons of European, North African, or southwest Asian ancestry
— Caucasian *noun*
— Cau·ca·soid \ˈkȯ-kə-ˌsȯid\ *adjective or noun*

Cau·chy sequence \kō-ˈshē-, ˈkō-shē-\ *noun* [Augustin-Louis *Cauchy* (died 1857) French mathematician] (circa 1949)
: a sequence of elements in a metric space such that for any positive number no matter how small there exists a term in the sequence for which the distance between any two terms beyond this term is less than the arbitrarily small number

[1]cau·cus \ˈkȯ-kəs\ *noun* [origin unknown] (1763)
: a closed meeting of a group of persons belonging to the same political party or faction usually to select candidates or to decide on policy; *also* **:** a group of people united to promote an agreed-upon cause

[2]caucus *intransitive verb* (1788)
: to meet in or hold a caucus

cau·dad \ˈkȯ-ˌdad\ *adverb* [Latin *cauda*] (1889)
: toward the tail or posterior end

cau·dal \ˈkȯ-dᵊl\ *adjective* [New Latin *caudalis,* from Latin *cauda* tail] (1661)
1 : of, relating to, or being a tail
2 : situated in or directed toward the hind part of the body
— cau·dal·ly \-dᵊl-ē\ *adverb*

cau·date \ˈkȯ-ˌdāt\ *adjective* (1600)
: having a tail or a taillike appendage

caudate nucleus *noun* (circa 1903)
: the most medial of the four basal ganglia in each cerebral hemisphere — called also *caudate*

cau·dex \ˈkȯ-ˌdeks\ *noun, plural* **cau·di·ces** \ˈkȯ-də-ˌsēz\ *or* **cau·dex·es** [Latin, tree trunk or stem] (circa 1797)
1 : the stem of a palm or tree fern
2 : the woody base of a perennial plant

cau·di·llis·mo \ˌkau̇-t͟hē-ˈyēz-(ˌ)mō, -t͟hēl-ˈyēz-\ *noun* [Spanish, from *caudillo*] (1927)
: the doctrine or practice of a caudillo

cau·di·llo \kau̇-ˈt͟hē-(ˌ)yō, -ˈt͟hēl-(ˌ)yō\ *noun, plural* **-llos** [Spanish, from Late Latin *capitellum* small head — more at CADET] (1852)
: a Spanish or Latin-American military dictator

cau·dle \ˈkȯ-dᵊl\ *noun* [Middle English *caudel,* from Old North French, from *caut* warm, from Latin *calidus* — more at CAULDRON] (14th century)
: a drink (as for invalids) usually of warm ale or wine mixed with bread or gruel, eggs, sugar, and spices

[1]caught \ˈkȯt\ *past and past participle of* CATCH

[2]caught *adjective* (1858)
: PREGNANT — often used in the phrase *get caught*

caul \ˈkȯl\ *noun* [Middle English *calle,* from Middle French *cale*] (14th century)
1 : the large fatty omentum covering the intestines (as of a cow, sheep, or pig)
2 : the inner fetal membrane of higher vertebrates especially when covering the head at birth

caul·dron \ˈkȯl-drən\ *noun* [Middle English, alteration of *cauderon,* from Old North French, diminutive of *caudiere,* from Late Latin *caldaria,* from feminine of Latin *caldarius* used for hot water, from *calidus* warm, from *calēre* to be warm — more at LEE] (14th century)
1 : a large kettle or boiler
2 : something resembling a boiling cauldron ⟨a *cauldron* of intense emotions⟩

cau·li·flow·er \ˈkȯ-li-ˌflau̇(-ə)r, ˈkä-\ *noun, often attributive* [Italian *cavolfiore,* from *cavolo* cabbage (from Late Latin *caulus,* from Latin *caulis* stem, cabbage) + *fiore* flower, from Latin *flor-, flos* — more at COLE, BLOW] (1597)
: a garden plant (*Brassica oleracea botrytis*) related to the cabbage and grown for its compact edible head of usually white undeveloped flowers; *also* **:** its flower cluster used as a vegetable

cauliflower ear *noun* (1909)
: an ear deformed from injury and excessive growth of reparative tissue

cau·li·flow·er·et \ˌkȯ-li-ˌflau̇(-ə)-ˈret, ˌkä-\ *noun* (1946)
: a bite-size piece of cauliflower

cau·line \ˈkȯ-ˌlīn\ *adjective* [probably from New Latin *caulinus,* from Latin *caulis*] (1756)
: of, relating to, or growing on a stem and especially on the upper part

[1]caulk \ˈkȯk\ *transitive verb* [Middle English *caulken,* from Old North French *cauquer* to trample, from Latin *calcare,* from *calc-, calx* heel] (15th century)
: to stop up and make tight against leakage (as a boat or its seams, the cracks in a window frame, or the joints of a pipe)
— caulk·er *noun*

[2]caulk *also* **caulk·ing** \ˈkȯ-kiŋ\ *noun* (1954)
: material used to caulk

[3]caulk *variant of* [2]CALK

caus·al \ˈkȯ-zəl\ *adjective* (circa 1530)
1 : expressing or indicating cause **:** CAUSATIVE ⟨a *causal* clause introduced by *since*⟩
2 : of, relating to, or constituting a cause ⟨the *causal* agent of a disease⟩
3 : involving causation or a cause ⟨the relationship . . . was not one of *causal* antecedence so much as one of analogous growth —H. O. Taylor⟩
4 : arising from a cause ⟨a *causal* development⟩
— caus·al·ly \-zə-lē\ *adverb*

cau·sal·gia \kȯ-ˈzal-j(ē-)ə, -ˈsal-\ *noun* [New Latin, from Greek *kausos* fever (from *kaiein* to burn) + New Latin *-algia*] (1872)
: a constant usually burning pain resulting from injury to a peripheral nerve
— cau·sal·gic \-jik\ *adjective*

cau·sal·i·ty \kȯ-ˈza-lə-tē\ *noun, plural* **-ties** (1603)
1 : a causal quality or agency
2 : the relation between a cause and its effect or between regularly correlated events or phenomena

cau·sa·tion \kȯ-ˈzā-shən\ *noun* (1615)
1 a : the act or process of causing **b :** the act or agency which produces an effect
2 : CAUSALITY

caus·a·tive \ˈkȯ-zə-tiv\ *adjective* (15th century)
1 : effective or operating as a cause or agent ⟨*causative* bacteria of cholera⟩
2 : expressing causation; *specifically* **:** being a linguistic form that indicates that the subject causes an act to be performed or a condition to come into being
— causative *noun*
— caus·a·tive·ly *adverb*

[1]cause \ˈkȯz\ *noun* [Middle English, from Old French, from Latin *causa*] (13th century)
1 a : a reason for an action or condition **:** MOTIVE **b :** something that brings about an effect or a result **c :** a person or thing that is the occasion of an action or state; *especially* **:** an agent that brings something about **d :** sufficient reason ⟨discharged for *cause*⟩
2 a : a ground of legal action **b :** CASE
3 : a matter or question to be decided
4 a : a principle or movement militantly defended or supported **b :** a charitable undertaking ⟨for a good *cause*⟩
— cause·less \-ləs\ *adjective*

[2]cause *transitive verb* **caused; caus·ing** (14th century)
1 : to serve as a cause or occasion of **:** MAKE
2 : to effect by command, authority, or force
— caus·er *noun*

'cause \ˈkȯz, ˈkəz\ *conjunction* (15th century)
: BECAUSE

cause cé·lè·bre *also* **cause ce·le·bre** \ˌkȯz-sə-ˈleb, -ˈle-brə, ˌkōz-, -ˈlebrᵊ\ *noun, plural* **causes cé·lè·bres** *also* **causes celebres** *same*\ [French, literally, celebrated case] (1763)
1 : a legal case that excites widespread interest
2 : a notorious person, thing, incident, or episode

cau·se·rie \ˌkōz-ˈrē, ˌkō-zə-\ *noun* [French, from *causer* to chat, from Latin *causari* to plead, discuss, from *causa*] (1827)
1 : an informal conversation **:** CHAT
2 : a short informal essay

cause·way \ˈkȯz-ˌwā\ *noun* [Middle English *cauciwey,* from *cauci* + *wey* way] (15th century)
1 : a raised way across wet ground or water
2 : HIGHWAY; *especially* **:** one of ancient Roman construction in Britain
— causeway *transitive verb*

cau·sey \ˈkȯ-zē\ *noun, plural* **causeys** [Middle English *cauci,* from Old North French *caucie,* from Medieval Latin *calciata* paved highway, probably from Latin *calc-, calx* limestone — more at CHALK] (14th century)
1 : CAUSEWAY 1
2 *obsolete* **:** CAUSEWAY 2

[1]caus·tic \ˈkȯs-tik\ *adjective* [Latin *causticus,* from Greek *kaustikos,* from *kaiein* to burn] (14th century)
1 : capable of destroying or eating away by chemical action **:** CORROSIVE
2 : marked by incisive sarcasm
3 : relating to or being the surface or curve of a caustic ☆
— caus·ti·cal·ly \-ti-k(ə-)lē\ *adverb*
— caus·tic·i·ty \kȯ-ˈsti-sə-tē\ *noun*

[2]caustic *noun* (15th century)
1 : a caustic agent: as **a :** a substance that burns or destroys organic tissue by chemical action **b :** SODIUM HYDROXIDE
2 : the envelope of rays emanating from a point and reflected or refracted by a curved surface

caustic potash *noun* (1869)
: POTASSIUM HYDROXIDE

☆ SYNONYMS
Caustic, mordant, acrid, scathing mean stingingly incisive. CAUSTIC suggests a biting wit ⟨*caustic* comments⟩. MORDANT suggests a wit that is used with deadly effectiveness ⟨*mordant* reviews of the play⟩. ACRID implies bitterness and often malevolence ⟨*acrid* invective⟩. SCATHING implies indignant attacks delivered with fierce severity ⟨a *scathing* satire⟩.

caustic soda *noun* (1876)
: SODIUM HYDROXIDE

cau·ter·ize \'kȯ-tə-ˌrīz\ *transitive verb* **-ized; -iz·ing** (14th century)
1 : to sear with a cautery or caustic
2 : to make insensible **:** DEADEN ⟨must oust the feeling, or *cauterize* it —Robert Craft⟩
— cau·ter·iza·tion \ˌkȯ-tə-rə-'zā-shən\ *noun*

cau·tery \'kȯ-tə-rē\ *noun, plural* **-ter·ies** [Latin *cauterium,* from Greek *kautērion* branding iron, from *kaiein*] (14th century)
1 : the act or effect of cauterizing **:** CAUTERIZATION
2 : an agent (as a hot iron or caustic) used to burn, sear, or destroy tissue

[1]cau·tion \'kȯ-shən\ *noun* [Latin *caution-, cautio* precaution, from *cavēre* to be on one's guard — more at HEAR] (1596)
1 : WARNING, ADMONISHMENT
2 : PRECAUTION
3 : prudent forethought to minimize risk
4 : one that astonishes or commands attention ⟨some shoes you see . . . these days are a *caution* —*Esquire*⟩
— cau·tion·ary \-shə-ˌner-ē\ *adjective*

[2]caution *transitive verb* **cau·tioned; cau·tion·ing** \'kȯ-sh(ə-)niŋ\ (1683)
: to advise caution to

cau·tious \'kȯ-shəs\ *adjective* (circa 1640)
: marked by or given to caution ☆
— cau·tious·ly *adverb*
— cau·tious·ness *noun*

cav·al·cade \ˌka-vəl-'kād, 'ka-vəl-ˌ\ *noun* [French, ride on horseback, from Old Italian *cavalcata,* from *cavalcare* to go on horseback, from Late Latin *caballicare,* from Latin *caballus* horse; akin to Greek *kaballeion* horse, Middle Irish *capall* workhorse] (1644)
1 a : a procession of riders or carriages **b :** a procession of vehicles or ships
2 : a dramatic sequence or procession **:** SERIES

[1]cav·a·lier \ˌka-və-'lir\ *noun* [Middle French, from Old Italian *cavaliere,* from Old Provençal *cavalier,* from Late Latin *caballarius* horseman, from Latin *caballus*] (1589)
1 : a gentleman trained in arms and horsemanship
2 : a mounted soldier **:** KNIGHT
3 *capitalized* **:** an adherent of Charles I of England
4 : GALLANT

[2]cavalier *adjective* (circa 1641)
1 : DEBONAIR
2 : marked by or given to offhand and often disdainful dismissal of important matters
3 a *capitalized* **:** of or relating to the party of Charles I of England in his struggles with the Puritans and Parliament **b :** ARISTOCRATIC **c** *capitalized* **:** of or relating to the English Cavalier poets of the mid-17th century
— ca·va·lier·ism \-ˌi-zəm\ *noun*
— cav·a·lier·ly *adverb*

cavalier King Charles spaniel \-'chär(-ə)lz-\ *noun* [[1]*cavalier* + *King Charles spaniel,* a breed of toy spaniel, from *Charles* II of England] (1969)
: any of a breed of toy spaniels with a tapered muzzle and a long silky coat

ca·val·la \kə-'va-lə\ *noun, plural* **-la** *or* **-las** [Spanish *caballa,* a fish, from Late Latin, mare, feminine of Latin *caballus*] (1624)
1 *also* **ca·val·ly** \-'va-lē\ **:** any of various carangid fishes (especially genus *Caranx*)
2 : CERO

cav·al·let·ti *also* **cav·a·let·ti** \ˌka-və-'le-tē\ *noun plural but singular or plural in construction* [Italian, plural of *cavalletto* trestle, diminutive of *cavallo* horse, from Latin *caballus*] (1950)
: a series of timber jumps that are adjustable in height for schooling horses

cav·al·ry \'ka-vəl-rē, ÷'kal-və-rē\ *noun, plural* **-ries** [Italian *cavalleria* cavalry, chivalry, from *cavaliere*] (1546)
1 a : an army component mounted on horseback **b :** an army component moving in motor vehicles or helicopters and assigned to combat missions that require great mobility
2 : HORSEMEN ⟨a thousand *cavalry* in flight⟩

cav·al·ry·man \-rē-mən, -ˌman\ *noun* (1860)
: a cavalry soldier

cavalry twill *noun* (1942)
: TRICOTINE

cav·a·ti·na \ˌka-və-'tē-nə, ˌkä-\ *noun* [Italian, from *cavata* production of sound from an instrument, extraction, from *cavare* to dig out, from Latin, to make hollow, from *cavus*] (1813)
1 : an operatic solo simpler and briefer than an aria
2 : a songlike instrumental piece or movement

[1]cave \'kāv\ *noun* [Middle English, from Old French, from Latin *cava,* from *cavus* hollow; akin to Greek *koilos* hollow, and probably to Greek *kyein* to be pregnant — more at CYME] (13th century)
1 : a natural underground chamber or series of chambers open to the surface
2 : a usually underground chamber for storage ⟨a wine *cave*⟩; *also* **:** the articles stored there

[2]cave *verb* **caved; cav·ing** (15th century)
transitive verb
: to form a cave in or under **:** HOLLOW, UNDERMINE
intransitive verb
: to explore caves especially as a sport or hobby
— cav·er \'kā-vər\ *noun*

[3]cave \'kāv\ *verb* **caved; cav·ing** [probably alteration of *calve*] (1513)
intransitive verb
1 : to fall in or down especially from being undermined — usually used with *in*
2 : to cease to resist **:** SUBMIT — usually used with *in*
transitive verb
: to cause to fall or collapse — usually used with *in*

ca·ve·at \'ka-vē-ˌät, -ˌat; 'kä-vē-ˌät; 'kā-vē-ˌat\ *noun* [Latin, let him beware, from *cavēre* — more at HEAR] (1533)
1 a : a warning enjoining one from certain acts or practices **b :** an explanation to prevent misinterpretation
2 : a legal warning to a judicial officer to suspend a proceeding until the opposition has a hearing

caveat emp·tor \-'em(p)-tər, -ˌtȯr\ *noun* [New Latin, let the buyer beware] (1523)
: a principle in commerce: without a warranty the buyer takes the risk

cave dweller *noun* (1865)
1 : one (as a prehistoric man) that dwells in a cave
2 : one that lives in a city apartment building

cave–in \'kā-ˌvin\ *noun* (1860)
1 : the action of caving in
2 : a place where earth has caved in

cave·man \'kāv-ˌman\ *noun* (1865)
1 : a cave dweller especially of the Stone Age
2 : a man who acts in a rough or crude manner

[1]cav·ern \'ka-vərn *also* -vrən\ *noun* [Middle English *caverne,* from Middle French, from Latin *caverna,* from *cavus*] (14th century)
: CAVE; *especially* **:** one of large or indefinite extent

[2]cavern *transitive verb* (circa 1630)
1 : to place in or as if in a cavern
2 : to form a cavern of **:** HOLLOW — used with *out*

cav·er·nic·o·lous \ˌka-vər-'ni-kə-ləs\ *adjective* (circa 1889)
: inhabiting caves

cav·ern·ous \'ka-vər-nəs\ *adjective* (15th century)
1 a : having caverns or cavities **b** *of animal tissue* **:** composed largely of vascular sinuses and capable of dilating with blood to bring about the erection of a body part
2 : constituting or suggesting a cavern
— cav·ern·ous·ly *adverb*

ca·vet·to \kə-'ve-(ˌ)tō, kä-\ *noun, plural* **-ti** \-tē\ [Italian, from *cavo* hollow, from Latin *cavus*] (1664)
: a concave molding having a curve that approximates a quarter circle — see MOLDING illustration

cav·i·ar *or* **cav·i·are** \'ka-vē-ˌär *also* 'kä-\ *noun* [earlier *cavery, caviarie,* from obsolete Italian *caviari,* plural of *caviaro,* from Turkish *havyar*] (circa 1560)
1 : processed salted roe of large fish (as sturgeon)
2 : something considered too delicate or lofty for mass appreciation — usually used in the phrase *caviar to the general*
3 : something considered the best of its kind ⟨the *caviar* of porcelain⟩

cav·il \'ka-vəl\ *verb* **-iled** *or* **-illed; -il·ing** *or* **-il·ling** \'ka-və-liŋ, 'kav-liŋ\ [Latin *cavillari* to jest, cavil, from *cavilla* raillery; akin to Latin *calvi* to deceive — more at CALUMNY] (1542)
intransitive verb
: to raise trivial and frivolous objection
transitive verb
: to raise trivial objections to
— cavil *noun*
— cav·il·er *or* **cav·il·ler** \'ka-və-lər, 'kav-lər\ *noun*

cav·ing \'kā-viŋ\ *noun* (1932)
: the sport of exploring caves **:** SPELUNKING

cav·i·tary \'ka-və-ˌter-ē\ *adjective* (1835)
: of, relating to, or characterized by bodily cavitation ⟨*cavitary* tuberculosis⟩ ⟨*cavitary* lesions⟩

cav·i·tate \'ka-və-ˌtāt\ *verb* **-tat·ed; -tat·ing** (1909)
intransitive verb
: to form cavities or bubbles
transitive verb
: to cavitate in

cav·i·ta·tion \ˌka-və-'tā-shən\ *noun* [*cavity* + *-ation*] (1895)
: the process of cavitating: as **a :** the formation of partial vacuums in a liquid by a swiftly moving solid body (as a propeller) or by high-intensity sound waves; *also* **:** the pitting and wearing away of solid surfaces (as of metal or concrete) as a result of the collapse of these vacuums in surrounding liquid **b :** the formation of cavities in an organ or tissue especially in disease

cav·i·ty \'ka-və-tē\ *noun, plural* **-ties** [Middle French *cavité,* from Late Latin *cavitas,* from Latin *cavus*] (1541)
1 : an unfilled space within a mass; *especially* **:** a hollowed-out space
2 : an area of decay in a tooth **:** CARIES

ca·vort \kə-'vȯrt\ *intransitive verb* [perhaps alteration of *curvet*] (1794)
1 : PRANCE
2 : to engage in extravagant behavior

ca·vy \'kā-vē\ *noun, plural* **cavies** [New Latin *Cavia,* genus name, from obsolete Portuguese *çavía* (now *savía*), from Tupi *sawiya* rat] (1796)

> ☆ SYNONYMS
> **Cautious, circumspect, wary, chary** mean prudently watchful and discreet in the face of danger or risk. CAUTIOUS implies the exercise of forethought usually prompted by fear of danger ⟨a *cautious* driver⟩. CIRCUMSPECT suggests less fear and stresses the surveying of all possible consequences before acting or deciding ⟨*circumspect* in his business dealings⟩. WARY emphasizes suspiciousness and alertness in watching for danger and cunning in escaping it ⟨keeps a *wary* eye on the competition⟩. CHARY implies a cautious reluctance to give, act, or speak freely ⟨*chary* of signing papers without having read them first⟩.

: any of several short-tailed rough-haired South American rodents (family Caviidae); *especially* **:** GUINEA PIG

caw \'kȯ\ *intransitive verb* [imitative] (1589)
: to utter the harsh raucous natural call of the crow or a similar cry
— **caw** *noun*

cay \'kē, 'kā\ *noun* [Spanish *cayo* — more at KEY] (1707)
: a low island or reef of sand or coral

cay·enne pepper \(,)kī-'en-, (,)kā-; 'kī-,, 'kā-,\ *noun* [by folk etymology from earlier *cayan,* modification of Tupi *kyinha*] (1756)
1 : a pungent condiment consisting of the ground dried fruits or seeds of hot peppers
2 : HOT PEPPER; *especially* **:** a cultivated pepper with very long twisted pungent red fruits
3 : the fruit of a cayenne pepper

cay·man *variant of* CAIMAN

Ca·yu·ga \kā-'yü-gə, kī-\ *noun, plural* **Cayuga** *or* **Cayugas** (1744)
1 : a member of an American Indian people of New York
2 : the Iroquoian language of the Cayuga people

Cay·use \'kī-,yüs, kī-'\ *noun, plural* **Cayuse** *or* **Cayuses** (1825)
1 : a member of an American Indian people of Oregon and Washington
2 *plural cayuses, not capitalized, West* **:** a native range horse

CB \,sē-'bē\ *noun* (1959)
: CITIZENS BAND; *also* **:** the radio transmitting and receiving set used for citizens-band communications

CBer \(,)sē-'bē-ər, -'bir\ *noun* (1959)
: one that operates a CB radio

CCD \,sē-(,)sē-'dē\ *noun* (1973)
: CHARGE-COUPLED DEVICE

C–clamp \'sē-,klamp\ *noun* (1926)
: a C-shaped general-purpose clamp

C clef *noun* (1596)
: a movable clef indicating middle C by its placement on one of the lines of the staff

CD \,sē-'dē\ *noun* (1979)
: COMPACT DISC

CD4 \,sē-(,)dē-'fōr, -'fȯr\ *noun* [cluster of differentiation] (1985)
: a large glycoprotein that is found on the surface of T4 cells and is the receptor for HIV

cDNA \'sē ,dē-(,)en-'ā, ,sē-\ *noun* [complementary] (1973)
: a strand of DNA that is complementary to a given messenger RNA and that serves as a template for production of the messenger RNA in the presence of reverse transcriptase

CD–ROM \,sē-,dē-'räm\ *noun* [compact *d*isc *r*ead-*o*nly *m*emory] (1983)
: a compact disc containing data that can be read by a computer

ce·a·no·thus \,sē-ə-'nō-thəs\ *noun* [New Latin, from Greek *keanōthos,* a thistle] (1785)
: any of a genus (*Ceanothus*) of American vines, shrubs, and small trees of the buckthorn family having the calyx disk adherent to the ovary

[1]**cease** \'sēs\ *verb* **ceased; ceas·ing** [Middle English *cesen,* from Middle French *cesser,* from Latin *cessare* to hold back, be remiss, frequentative of *cedere*] (14th century)
transitive verb
: to cause to come to an end especially gradually **:** no longer continue
intransitive verb
1 a : to come to an end **b :** to bring an activity or action to an end **:** DISCONTINUE
2 *obsolete* **:** to become extinct **:** DIE OUT
synonym see STOP

[2]**cease** *noun* (14th century)
: CESSATION — usually used with *without*

cease and desist order *noun* (1926)
: an order from an administrative agency to refrain from a method of competition or a labor practice found by the agency to be unfair

cease–fire \'sēs-'fīr\ *noun* (1859)
1 : a military order to cease firing
2 : a suspension of active hostilities

cease·less \'sēs-ləs\ *adjective* (1586)
: continuing without cease **:** CONSTANT
— **cease·less·ly** *adverb*
— **cease·less·ness** *noun*

ce·cro·pia moth \si-'krō-pē-ə-\ *noun* [New Latin *cecropia,* from Latin, feminine of *Cecropius* Athenian, from Greek *Kekropios,* from *Kekrops* Cecrops, legendary king of Athens] (1885)
: a large North American saturniid moth (*Hyalophora cecropia*) that is brown with red, white, and black markings

ce·cum \'sē-kəm\ *noun, plural* **ce·ca** \-kə\ [New Latin, from Latin *intestinum caecum,* literally, blind intestine] (circa 1721)
: a cavity open at one end (as the blind end of a duct); *especially* **:** the blind pouch at the beginning of the large intestine into which the ileum opens from one side and which is continuous with the colon
— **ce·cal** \-kəl\ *adjective*
— **ce·cal·ly** \-kə-lē\ *adverb*

ce·dar \'sē-dər\ *noun* [Middle English *cedre,* from Old French, from Latin *cedrus,* from Greek *kedros*] (14th century)
1 a : any of a genus (*Cedrus*) of usually tall coniferous trees (as the cedar of Lebanon or the deodar) of the pine family noted for their fragrant durable wood **b :** any of numerous coniferous trees (as of the genera *Juniperus, Chamaecyparis,* or *Thuja*) that resemble the true cedars especially in the fragrance and durability of their wood
2 : the wood of a cedar

ce·dar–ap·ple rust \'sē-dər-'a-pəl-\ *noun* (1946)
: a gall-producing disease especially of the apple caused by a rust fungus (*Gymnosporangium juniperi-virginianae*) that completes the first part of its life cycle on the common red cedar (*Juniperus virginiana*) and the second on the leaves and fruit of the apple

ce·dar·bird \-,bərd\ *noun* (1883)
: CEDAR WAXWING

ce·darn \'sē-dərn\ *adjective* (1634)
archaic **:** made or suggestive of cedar

cedar of Leb·a·non \-'leb-nän, -'le-bə-, -nən\ (14th century)
: a long-lived cedar (*Cedrus libani*) native to Asia Minor with short fascicled leaves and erect cones

cedar waxwing *noun* (circa 1844)
: a brown gregarious American waxwing (*Bombycilla cedrorum*) with a yellow band on the tip of the tail and a pale yellow belly

ce·dar·wood \'sē-dər-,wu̇d\ *noun* (14th century)
: the wood of a cedar that is especially repellent to insects

cede \'sēd\ *transitive verb* **ced·ed; ced·ing** [French or Latin; French *céder,* from Latin *cedere* to go, withdraw, yield] (1754)
1 : to yield or grant typically by treaty
2 : ASSIGN, TRANSFER
— **ced·er** *noun*

ce·di \'sā-dē\ *noun* [Twi *sedi* cowry] (1965)
— see MONEY table

ce·dil·la \si-'di-lə\ *noun* [Spanish, the obsolete letter *ç* (actually a medieval form of the letter *z*), cedilla, from diminutive of *ceda, zeda* the letter *z,* from Late Latin *zeta* — more at ZED] (1599)
: the diacritical mark ¸ placed under a letter (as *ç* in French) to indicate an alteration or modification of its usual phonetic value (as in the French word *façade*)

cee \'sē\ *noun* (1542)
: the letter *c*

cei·ba \'sā-bə\ *noun* [Spanish] (1790)
1 : a massive tropical tree (*Ceiba pentandra*) of the silk-cotton family with large pods filled with seeds invested with a silky floss that yields the fiber kapok
2 : KAPOK

ceil \'sē(ə)l\ *transitive verb* [Middle English *celen,* from Medieval Latin *celare, caelare,* perhaps from Latin *caelare* to carve, from *caelum* chisel; akin to Latin *caedere* to cut] (15th century)
1 : to furnish (as a wooden ship) with a lining
2 : to furnish with a ceiling

ceil·ing \'sē-liŋ\ *noun* [Middle English *celing,* from *celen*] (1535)
1 a : the overhead inside lining of a room **b :** material used to ceil a wall or roof of a room
2 : something thought of as an overhanging shelter or a lofty canopy ⟨a *ceiling* of stars⟩
3 a : the height above the ground from which prominent objects on the ground can be seen and identified **b :** the height above the ground of the base of the lowest layer of clouds when over half of the sky is obscured
4 a : ABSOLUTE CEILING **b :** SERVICE CEILING
5 : an upper usually prescribed limit ⟨a *ceiling* on prices, rents, and wages⟩
— **ceil·inged** \-liŋd\ *adjective*

ceil·om·e·ter \sē-'lä-mə-tər\ *noun* [*ceil*ing + *-o-* + *-meter*] (1943)
: a photoelectric instrument for determining by triangulation the height of the cloud ceiling above the earth

cein·ture \sa^n(n)-'tyu̇r, -'tu̇r, 'san-chər\ *noun* [Middle English *seynture,* from Middle French *ceinture,* from Latin *cinctura* — more at CINCTURE] (15th century)
: a belt or sash for the waist

cel *also* **cell** \'sel\ *noun* [short for *celluloid*] (1938)
: a transparent sheet of celluloid on which objects are drawn or painted in the making of animated cartoons

cel·a·don \'se-lə-,dän, -lə-d^{ə}n\ *noun* [French *céladon*] (circa 1768)
1 : a grayish yellow green
2 : a ceramic glaze originated in China that is greenish in color; *also* **:** an article with a celadon glaze

cel·an·dine \'se-lən-,dīn, -,dēn\ *noun* [Middle English *celidoine,* from Middle French, from Latin *chelidonia,* from feminine of *chelidonius* of the swallow, from Greek *chelidonios,* from *chelidon-, chelidōn* swallow] (12th century)
1 : a yellow-flowered Eurasian biennial herb (*Chelidonium majus*) of the poppy family naturalized in the eastern U.S.
2 : LESSER CELANDINE

-cele *noun combining form* [Middle French, from Latin, from Greek *kēlē;* akin to Old English *hēala* hernia]
: tumor **:** hernia ⟨varico*cele*⟩

ce·leb \sə-'leb\ *noun* (circa 1912)
: CELEBRITY 2

cel·e·brant \'se-lə-brənt\ *noun* (1839)
: one who celebrates; *specifically* **:** the priest officiating at the Eucharist

cel·e·brate \'se-lə-,brāt\ *verb* **-brat·ed; -brat·ing** [Latin *celebratus,* past participle of *celebrare* to frequent, celebrate, from *celebr-, celeber* much frequented, famous; perhaps akin to Latin *celer*] (15th century)
transitive verb
1 : to perform (a sacrament or solemn ceremony) publicly and with appropriate rites ⟨*celebrate* the mass⟩
2 a : to honor (as a holiday) by solemn ceremonies or by refraining from ordinary business **b :** to mark (as an anniversary) by festivities or other deviation from routine
3 : to hold up or play up for public notice ⟨her poetry *celebrates* the glory of nature⟩
intransitive verb
1 : to observe a holiday, perform a religious ceremony, or take part in a festival

2 **:** to observe a notable occasion with festivities
synonym see KEEP
— **cel·e·bra·tion** \ˌse-lə-ˈbrā-shən\ *noun*
— **cel·e·bra·tor** \ˈse-lə-ˌbrā-tər\ *noun*
— **cel·e·bra·to·ry** \-brə-ˌtōr-ē, -ˌtȯr-; ˌse-lə-ˈbrā-\ *adjective*
cel·e·brat·ed *adjective* (circa 1669)
: widely known and often referred to
synonym see FAMOUS
— **cel·e·brat·ed·ness** *noun*
ce·leb·ri·ty \sə-ˈle-brə-tē\ *noun, plural* **-ties** (14th century)
1 : the state of being celebrated
2 : a celebrated person
ce·le·ri·ac \sə-ˈler-ē-ˌak, -ˈlir-\ *noun* [irregular from *celery*] (1743)
: a celery (*Apium graveolens rapaceum*) grown for its knobby edible root — called also *celery root*
ce·ler·i·ty \sə-ˈler-ə-tē\ *noun* [Middle English *celerite,* from Middle French, from Latin *celeritat-, celeritas,* from *celer* swift — more at HOLD] (15th century)
: rapidity of motion or action
cel·ery \ˈse-lə-rē, ˈsel-rē\ *noun, plural* **-er·ies** [obsolete French *celeris,* from Italian dialect *seleri,* plural of *selero,* modification of Late Latin *selinon,* from Greek] (1664)
: a European herb (*Apium graveolens*) of the carrot family; *specifically* **:** one of a cultivated variety (*A. graveolens dulce*) with leafstalks eaten raw or cooked
celery cabbage *noun* (1930)
: CHINESE CABBAGE b
ce·les·ta \sə-ˈles-tə, chə-\ *or* **ce·leste** \sə-ˈlest, chə-\ *noun* [French *célesta,* alteration of *céleste,* literally, heavenly, from Latin *caelestis*] (1899)
: a keyboard instrument with hammers that strike steel plates producing a tone similar to that of a glockenspiel
[1]**ce·les·tial** \sə-ˈles-chəl, -ˈlesh-, -ˈles-tē-əl\ *adjective* [Middle English, from Middle French, from Latin *caelestis* celestial, from *caelum* sky] (14th century)
1 : of, relating to, or suggesting heaven or divinity
2 : of or relating to the sky or visible heavens ⟨the sun, moon, and stars are *celestial* bodies⟩
3 a : ETHEREAL, OTHERWORLDLY **b :** OLYMPIAN, SUPREME
4 *capitalized* [*Celestial* Empire, old name for China] **:** of or relating to China or the Chinese
— **ce·les·tial·ly** \-chə-lē, -tē-ə-lē\ *adverb*
[2]**celestial** *noun* (1573)
1 : a heavenly or mythical being
2 *capitalized* **:** CHINESE 1a
celestial equator *noun* (1875)
: the great circle on the celestial sphere midway between the celestial poles
celestial globe *noun* (circa 1771)
: a globe depicting the celestial bodies
celestial hierarchy *noun* (1883)
: a traditional hierarchy of angels ranked from lowest to highest into the following nine orders: angels, archangels, principalities, powers, virtues, dominions, thrones, cherubim, and seraphim
celestial marriage *noun* (1919)
: a special order of Mormon marriage solemnized in a Mormon temple and held to be binding for a future life as well as the present one
celestial navigation *noun* (1939)
: navigation by observation of the positions of celestial bodies
celestial pole *noun* (1868)
: either of the two points on the celestial sphere around which the diurnal rotation of the stars appears to take place
celestial sphere *noun* (1879)
: an imaginary sphere of infinite radius against which the celestial bodies appear to be projected and of which the apparent dome of the visible sky forms half

ce·les·tite \ˈse-ləs-ˌtīt, sə-ˈles-ˌtīt\ *noun* [German *Zölestin,* from Latin *caelestis*] (1854)
: a usually white mineral consisting of the sulfate of strontium
ce·li·ac \ˈsē-lē-ˌak\ *adjective* [Latin *coeliacus,* from Greek *koiliakos,* from *koilia* cavity, from *koilos* hollow — more at CAVE] (1662)
: of or relating to the abdominal cavity
celiac disease *noun* (1911)
: a chronic nutritional disorder especially in young children that is characterized by defective digestion and utilization of fats and often by abdominal distention, diarrhea, and fatty stools
cel·i·ba·cy \ˈse-lə-bə-sē\ *noun* (1663)
1 : the state of not being married
2 a : abstention from sexual intercourse **b :** abstention by vow from marriage
cel·i·bate \ˈse-lə-bət\ *noun* [Latin *caelibatus,* from *caelib-, caelebs* unmarried] (circa 1847)
: a person who lives in celibacy
— **celibate** *adjective*
cell \ˈsel\ *noun* [Middle English, from Old English, religious house and Old French *celle* hermit's cell, from Latin *cella* small room; akin to Latin *celare* to conceal — more at HELL] (12th century)
1 : a small religious house dependent on a monastery or convent
2 a : a one-room dwelling occupied by a solitary person (as a hermit) **b :** a single room (as in a convent or prison) usually for one person
3 : a small compartment, cavity, or bounded space: as **a :** one of the compartments of a honeycomb **b :** a membranous area bounded by veins in the wing of an insect
4 : a small usually microscopic mass of protoplasm bounded externally by a semipermeable membrane, usually including one or more nuclei and various other organelles with their products, capable alone or interacting with other cells of performing all the fundamental functions of life, and forming the smallest structural unit of living matter capable of functioning independently
5 a (1) **:** a receptacle (as a cup or jar) containing electrodes and an electrolyte either for generating electricity by chemical action or for use in electrolysis (2) **:** FUEL CELL **b :** a single unit in a device for converting radiant energy into electrical energy or for varying the intensity of an electrical current in accordance with radiation
6 : a unit in a statistical array (as a spreadsheet) formed by the intersection of a column and a row
7 : the basic and usually smallest unit of an organization or movement ⟨a communist *cell*⟩
8 : a portion of the atmosphere that behaves as a unit
9 : any of the small sections of a geographic area of a cellular telephone system

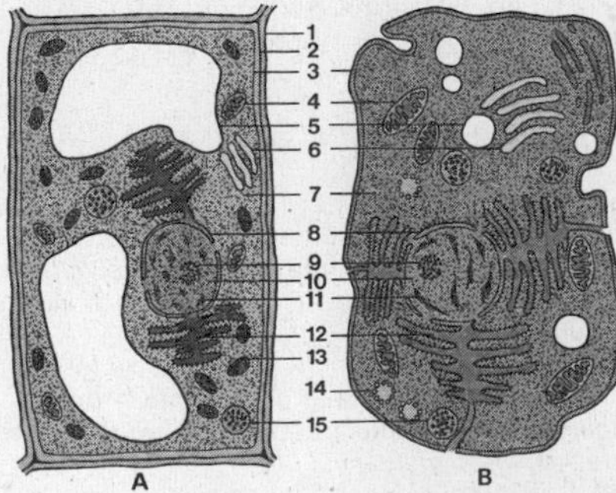

cell 4 (schematic): ***A*** **plant,** ***B*** **animal;** ***1*** **cell wall,** ***2*** **middle lamella,** ***3*** **plasma membrane,** ***4*** **mitochondrion,** ***5*** **vacuole,** ***6*** **Golgi apparatus,** ***7*** **cytoplasm,** ***8*** **nuclear membrane,** ***9*** **nucleolus,** ***10*** **nucleus,** ***11*** **chromatin,** ***12*** **endoplasmic reticulum with associated ribosomes,** ***13*** **chloroplast,** ***14*** **centriole,** ***15*** **lysosome**

[1]**cel·lar** \ˈse-lər\ *noun* [Middle English *celer,* from Anglo-French, from Latin *cellarium* storeroom, from *cella*] (13th century)
1 a : BASEMENT; *also* **:** a covered excavation **b :** the lowest grade or rank; *especially* **:** the lowest place in the standings (as of an athletic league)
2 : a stock of wines
[2]**cellar** *transitive verb* (1677)
: to put into a cellar (as for storage)
cel·lar·age \ˈse-lə-rij\ *noun* (1602)
: cellar space especially for storage
cel·lar·er \ˈse-lər-ər\ *noun* [Middle English *celerer,* from Old French, from Late Latin *cellariarius,* from Latin *cellarium*] (13th century)
: an official (as in a monastery) in charge of provisions
cel·lar·ette *or* **cel·lar·et** \ˌse-lə-ˈret\ *noun* (circa 1807)
: a case or sideboard for holding bottles of wine or liquor
cell body *noun* (1878)
: the nucleus-containing central part of a neuron exclusive of its axons and dendrites
cell cycle *noun* (1961)
: the complete series of events from one cell division to the next — compare G_1 PHASE, G_2 PHASE, M PHASE, S PHASE
cell division *noun* (1882)
: the process by which cells multiply involving both nuclear and cytoplasmic division — compare MEIOSIS, MITOSIS
celled \ˈseld\ *adjective*
: having (such or so many) cells — used in combination ⟨single-*celled* organisms⟩
cell–me·di·at·ed \ˈsel-ˈmē-dē-ˌā-təd\ *adjective* (1967)
: relating to or being the part of immunity or the immune response that is mediated primarily by T cells
cell membrane *noun* (1870)
: PLASMA MEMBRANE
cel·lo \ˈche-(ˌ)lō\ *noun, plural* **cello** *also* **cel·li** \-lē\ [short for *violoncello*] (circa 1876)
: the bass member of the violin family tuned an octave below the viola
— **cel·list** \ˈche-list\ *noun*
cel·lo·bi·ose \ˌse-lə-ˈbī-ˌōs, -ˌōz\ *noun* [International Scientific Vocabulary *cellulose* + *-o-* + *biose* disaccharide, from [1]*bi-* + [2]*-ose*] (1902)
: a faintly sweet disaccharide $C_{12}H_{22}O_{11}$ obtained by partial hydrolysis of cellulose
cel·loi·din \se-ˈlȯi-dᵊn\ *noun* [*cellulose* + [1]*-oid* + [1]*-in*] (1883)
: a purified pyroxylin used chiefly in microscopy
cel·lo·phane \ˈse-lə-ˌfān\ *noun* [French, from *cellulose* + *-phane* (as in *diaphane* diaphanous, from Medieval Latin *diaphanus*)] (1912)
: regenerated cellulose in thin transparent sheets used especially for packaging
cell plate *noun* (1882)
: a disk formed in the phragmoplast of a dividing plant cell that eventually forms the middle lamella of the wall between the daughter cells
cell sap *noun* (circa 1889)
1 : the liquid contents of a plant cell vacuole
2 : CYTOSOL
cell theory *noun* (circa 1890)
: a theory in biology that includes one or both of the statements that the cell is the fundamental structural and functional unit of living matter and that the organism is composed of autonomous cells with its properties being the sum of those of its cells
cel·lu·lar \ˈsel-yə-lər\ *adjective* [New Latin *cellularis,* from *cellula* living cell, from Latin, diminutive of *cella* small room] (circa 1739)
1 : of, relating to, or consisting of cells
2 : containing cavities **:** having a porous texture ⟨*cellular* rocks⟩
3 : of, relating to, or being a radiotelephone

system in which a geographical area (as a city) is divided into small sections each served by a transmitter of limited range so that any available radio channel can be used in different parts of the area simultaneously
— **cel·lu·lar·i·ty** \ˌsel-yə-ˈlar-ə-tē\ *noun*
cel·lu·lase \ˈsel-yə-ˌlās, -ˌlāz\ *noun* [International Scientific Vocabulary] (1903)
: an enzyme that hydrolyzes cellulose
cel·lule \ˈsel-(ˌ)yü(ə)l\ *noun* [Latin *cellula*] (circa 1693)
: a small cell
cel·lu·lite \ˈsel-yə-ˌlīt, -ˌlēt\ *noun* [French, literally, accumulation of subcutaneous fat, cellulitis, from *cellule* cell + *-ite* -itis] (1968)
: lumpy fat found in the thighs, hips, and buttocks of some women
cel·lu·li·tis \ˌsel-yə-ˈlī-təs\ *noun* [New Latin, from *cellula*] (1861)
: diffuse and especially subcutaneous inflammation of connective tissue
cel·lu·loid \ˈsel-yə-ˌlȯid\ *noun* [from *Celluloid*, a trademark] (1871)
1 : a tough flammable thermoplastic composed essentially of cellulose nitrate and camphor
2 : a motion-picture film ⟨a work . . . making its third appearance on *celluloid* —John McCarten⟩
— **celluloid** *adjective*
cel·lu·lo·lyt·ic \ˌsel-yə-lō-ˈli-tik\ *adjective* [*cellulose* + *-o-* + *-lytic*] (1943)
: hydrolyzing or having the capacity to hydrolyze cellulose
cel·lu·lose \ˈsel-yə-ˌlōs, -ˌlōz\ *noun* [French, from *cellule* living cell, from New Latin *cellula*] (1848)
: a polysaccharide $(C_6H_{10}O_5)_x$ of glucose units that constitutes the chief part of the cell walls of plants, occurs naturally in such fibrous products as cotton and kapok, and is the raw material of many manufactured goods (as paper, rayon, and cellophane)
cellulose acetate *noun* (1895)
: any of several compounds insoluble in water that are formed especially by the action of acetic acid, anhydride of acetic acid, and sulfuric acid on cellulose and are used for making textile fibers, packaging sheets, photographic films, and varnishes
cellulose nitrate *noun* (1880)
: NITROCELLULOSE
cel·lu·los·ic \ˌsel-yə-ˈlō-sik, -zik\ *adjective* (1881)
: of, relating to, or made from cellulose
— **cellulosic** *noun*
cell wall *noun* (circa 1849)
: the usually rigid nonliving permeable wall that surrounds the plasma membrane and encloses and supports the cells of most plants, bacteria, fungi, and algae — see CELL illustration
Cel·si·us \ˈsel-sē-əs, -shəs\ *adjective* [Anders *Celsius*] (circa 1850)
: relating to, conforming to, or having the international thermometric scale on which the interval between the triple point of water and the boiling point of water is divided into 99.99 degrees with 0.01° representing the triple point and 100° the boiling point ⟨10° *Celsius*⟩ — abbreviation *C*; compare CENTIGRADE ◆
celt \ˈselt\ *noun* [Late Latin *celtis* chisel] (1715)
: a prehistoric stone or metal implement shaped like a chisel or ax head
Celt \ˈkelt, ˈselt\ *noun* [Latin *Celtae*, plural, from Greek *Keltoi*] (1550)
1 : a member of a division of the early Indo-European peoples distributed from the British Isles and Spain to Asia Minor
2 : a modern Gael, Highland Scot, Irishman, Welshman, Cornishman, or Breton
[1]**Celt·ic** \ˈkel-tik, ˈsel-\ *adjective* (1590)
: of, relating to, or characteristic of the Celts or their languages
[2]**Celtic** *noun* (1739)
: a group of Indo-European languages usually subdivided into Brythonic and Goidelic and now largely confined to Brittany, Wales, western Ireland, and the Scottish Highlands — see INDO-EUROPEAN LANGUAGES table
Celtic cross *noun* (1873)
: a cross having essentially the form of a Latin cross with a ring about the intersection of the crossbar and upright shaft — see CROSS illustration
Celt·i·cist \ˈkel-tə-sist, ˈsel-\ *noun* (1912)
: a specialist in Celtic languages or cultures
cem·ba·lo \ˈchem-bə-ˌlō\ *noun, plural* **-ba·li** \-(ˌ)lē\ *or* **-balos** [Italian] (circa 1801)
: HARPSICHORD
[1]**ce·ment** \si-ˈment *also* ˈsē-ment\ *noun* [Middle English *sement*, from Middle French *ciment*, from Latin *caementum* stone chips used in making mortar, from *caedere* to cut] (14th century)
1 a : a powder of alumina, silica, lime, iron oxide, and magnesium oxide burned together in a kiln and finely pulverized and used as an ingredient of mortar and concrete; *also* **:** any mixture used for a similar purpose **b :** CONCRETE
2 : a binding element or agency: as **a :** a substance to make objects adhere to each other **b :** something serving to unite firmly ⟨justice is the *cement* that holds a political community together —R. M. Hutchins⟩
3 : CEMENTUM
4 : a plastic composition made especially of zinc or silica for filling dental cavities
5 : the fine-grained groundmass or glass of a porphyry
[2]**cement** (14th century)
transitive verb
1 : to unite or make firm by or as if by cement
2 : to overlay with concrete
intransitive verb
: to become cemented
— **ce·ment·er** *noun*
ce·men·ta·tion \ˌsē-ˌmen-ˈtā-shən\ *noun* (1594)
1 : a process of surrounding a solid with a powder and heating the whole so that the solid is changed by chemical combination with the powder
2 : the act or process of cementing **:** the state of being cemented
ce·ment·ite \si-ˈmen-ˌtīt\ *noun* [[1]*cement*] (1888)
: a hard brittle iron carbide Fe_3C that occurs in steel, cast iron, and iron-carbon alloys
ce·men·ti·tious \ˌsē-ˌmen-ˈti-shəs\ *adjective* (circa 1828)
: having the properties of cement
ce·men·tum \si-ˈmen-təm\ *noun* [New Latin, from Latin *caementum*] (1842)
: a specialized external bony layer covering the dentin of the part of a tooth normally within the gum — see TOOTH illustration
cem·e·tery \ˈse-mə-ˌter-ē\ *noun, plural* **-ter·ies** [Middle English *cimitery*, from Middle French *cimitere*, from Late Latin *coemeterium*, from Greek *koimētērion* sleeping chamber, burial place, from *koiman* to put to sleep; akin to Greek *keisthai* to lie, Sanskrit *śete* he lies] (15th century)
: a burial ground
cen·a·cle \ˈse-ni-kəl\ *noun* [Late Latin *cenaculum* the room where Christ and his disciples had the Last Supper, from Latin, top story, probably from *cena* dinner] (1889)
: a retreat house; *especially* **:** one for Roman Catholic women directed by nuns of the Society of Our Lady of the Cenacle
-cene *adjective combining form* [Greek *kainos* new, recent — more at RECENT]
: recent — in names of geologic periods ⟨*Eocene*⟩
cen·o·bite \ˈse-nə-ˌbīt, *especially British* ˈsē-\ *noun* [Late Latin *coenobita*, from *coenobium* monastery, from Late Greek *koinobion*, ultimately from Greek *koin-* coen- + *bios* life — more at QUICK] (circa 1500)
: a member of a religious group living together in a monastic community
— **cen·o·bit·ic** \ˌse-nə-ˈbi-tik, ˌsē-\ *adjective*
ce·no·spe·cies \ˈsē-nə-ˌspē-(ˌ)shēz, ˈse-, -(ˌ)sēz\ *noun* [*coen-* + *species*] (1922)
: a group of related biological taxonomic units capable by reason of closely related genotypes of essentially free gene interchange
ceno·taph \ˈse-nə-ˌtaf, -ˌtäf\ *noun* [French *cénotaphe*, from Latin *cenotaphium*, from Greek *kenotaphion*, from *kenos* empty + *taphos* tomb] (1603)
: a tomb or a monument erected in honor of a person or group of persons whose remains are elsewhere
ce·no·te \si-ˈnō-tē\ *noun* [Mexican Spanish, from Yucatec *ts'onot*] (1841)
: a deep sinkhole in limestone with a pool at the bottom that is found especially in Yucatán
Ce·no·zo·ic \ˌsē-nə-ˈzō-ik, ˌse-\ *adjective* [Greek *kainos* + English *-zoic*] (1854)
: of, relating to, or being an era of geological history that extends from the beginning of the Tertiary period to the present time and is marked by a rapid evolution of mammals and birds and of angiosperms and especially grasses and by little change in the invertebrates; *also* **:** relating to the corresponding system of rocks — see GEOLOGIC TIME table
— **Cenozoic** *noun*
cense \ˈsen(t)s\ *transitive verb* **censed; cens·ing** [Middle English, probably short for *encensen* to incense, from Middle French *encenser*, from Late Latin *incensare*, from *incensum* incense] (14th century)
: to perfume especially with a censer
cen·ser \ˈsen(t)-sər\ *noun* (13th century)
: a vessel for burning incense; *especially* **:** a covered incense burner swung on chains in a religious ritual

censer

[1]**cen·sor** \ˈsen(t)-sər\ *noun* [Latin, from *censēre* to give as one's opinion, assess; perhaps akin to Sanskrit *śaṁsati* he praises] (1531)
1 : one of two magistrates of early Rome acting as census takers, assessors, and inspectors of morals and conduct
2 : one who supervises conduct and morals: as **a :** an official who examines materials (as publications or films) for objectionable matter **b :** an official (as in time of war) who reads communications (as letters) and deletes material considered sensitive or harmful

◇ WORD HISTORY
Celsius The Celsius scale is named after the Swedish astronomer Anders Celsius (1701–44), whose astronomical work is now largely forgotten. Although he was not the first to propose a temperature metric of 100 degrees, Celsius' observations on the usefulness of the freezing and boiling points of water as thermometric constants, published in 1742, were influential. Celsius chose 0° as the boiling point and 100° as the freezing point, but in 1747 the Swedes reversed the order. The Swedish 100-degree system gained widespread scientific acceptance in the 19th century.

3 : a hypothetical psychic agency that represses unacceptable notions before they reach consciousness
— **cen·so·ri·al** \sen-ˈsōr-ē-əl, -ˈsȯr-\ *adjective*

[2]**censor** *transitive verb* **cen·sored; cen·sor·ing** \ˈsen(t)-sə-riŋ, ˈsen(t)s-riŋ\ (1882)
: to examine in order to suppress or delete anything considered objectionable

cen·so·ri·ous \sen-ˈsōr-ē-əs, -ˈsȯr-\ *adjective* [Latin *censorius* of a censor, from *censor*] (1536)
: marked by or given to censure
synonym see CRITICAL
— **cen·so·ri·ous·ly** *adverb*
— **cen·so·ri·ous·ness** *noun*

cen·sor·ship \ˈsen(t)-sər-ˌship\ *noun* (circa 1591)
1 a : the institution, system, or practice of censoring **b :** the actions or practices of censors; *especially* **:** censorial control exercised repressively
2 : the office, power, or term of a Roman censor
3 : exclusion from consciousness by the psychic censor

cen·sur·able \ˈsen(t)-sh(ə-)rə-bəl\ *adjective* (1634)
: deserving or open to censure

[1]**cen·sure** \ˈsen(t)-shər\ *noun* [Latin *censura,* from *censēre*] (14th century)
1 : a judgment involving condemnation
2 *archaic* **:** OPINION, JUDGMENT
3 : the act of blaming or condemning sternly
4 : an official reprimand

[2]**censure** *transitive verb* **cen·sured; cen·sur·ing** \ˈsen(t)-sh(ə-)riŋ\ (1587)
1 *obsolete* **:** ESTIMATE, JUDGE
2 : to find fault with and criticize as blameworthy
synonym see CRITICIZE
— **cen·sur·er** \ˈsen(t)-shər-ər\ *noun*

cen·sus \ˈsen(t)-səs\ *noun* [Latin, from *censēre*] (1634)
1 : a count of the population and a property evaluation in early Rome
2 : a usually complete enumeration of a population; *specifically* **:** a periodic governmental enumeration of population
3 : COUNT, TALLY
— **census** *transitive verb*

cent \ˈsent\ *noun* [Middle French, hundred, from Latin *centum* — more at HUNDRED] (1782)
1 : a monetary unit equal to 1/100 of a basic unit of value — see *birr, dollar, gulden, leone, lilangeni, lira, pound, rand, rupee, shilling* at MONEY table
2 : a coin, token, or note representing one cent
3 : the fen of the People's Republic of China

cen·tal \ˈsen-tᵊl\ *noun* [Latin *centum* + English *-al* (as in *quintal*)] (1870)
chiefly British **:** HUNDREDWEIGHT 1

cen·taur \ˈsen-ˌtȯr\ *noun* [Middle English, from Latin *Centaurus,* from Greek *Kentauros*] (14th century)
: any of a race of creatures fabled to be half man and half horse and to live in the mountains of Thessaly

cen·tau·rea \sen-ˈtȯr-ē-ə\ *noun* [New Latin, genus name, from Medieval Latin] (circa 1829)
: any of a large genus (*Centaurea*) of composite herbs (as knapweed) including several cultivated for their showy heads of tubular florets

Cen·tau·rus \-ˈtȯr-əs\ *noun* [Latin (genitive *Centauri*)]
: a southern constellation between the Southern Cross and Hydra

cen·tau·ry \ˈsen-ˌtȯr-ē\ *noun, plural* **-ries** [Middle English *centaure,* from Middle French *centaurée,* from Medieval Latin *centaurea,* from Latin *centaureum,* from Greek *kentaureion,* from *Kentauros*] (14th century)
1 : any of a genus (*Centaurium*) of low herbs of the gentian family; *especially* **:** an Old World herb (*C. umbellatum*) formerly used as a tonic
2 : an herb (*Sabatia angularis*) of the eastern U.S. closely related to centaury

[1]**cen·ta·vo** \sen-ˈtä-(ˌ)vō\ *noun, plural* **-vos** [Spanish, literally, hundredth, from *ciento* hundred, from Latin *centum*] (1883)
— see *boliviano, colón, cordoba, lempira, peso, quetzal, sol, sucre* at MONEY table

[2]**cen·ta·vo** \-ˈtä-(ˌ)vü, -(ˌ)vō\ *noun, plural* **-vos** [Portuguese, from Spanish] (1920)
— see *escudo, metical* at MONEY table

cen·te·nar·i·an \ˌsen-tə-ˈner-ē-ən\ *noun* (circa 1841)
: one that is 100 years old or older
— **centenarian** *adjective*

cen·te·na·ry \sen-ˈte-nə-rē, ˈsen-tə-ˌner-ē, *especially British* sen-ˈtē-nə-rē\ *noun, plural* **-ries** [Late Latin *centenarium,* from Latin *centenarius* of a hundred, from *centeni* one hundred each, from *centum* hundred — more at HUNDRED] (1788)
: CENTENNIAL
— **centenary** *adjective*

cen·ten·ni·al \sen-ˈte-nē-əl\ *noun* [Latin *centum* + English *-ennial* (as in *biennial*)] (1876)
: a 100th anniversary or its celebration
— **centennial** *adjective*
— **cen·ten·ni·al·ly** \-ə-lē\ *adverb*

[1]**cen·ter** \ˈsen-tər, ˈse-nər\ *noun* [Middle English *centre,* from Middle French, from Latin *centrum,* from Greek *kentron* sharp point, center of a circle, from *kentein* to prick; probably akin to Old High German *hantag* pointed] (14th century)
1 a : the point around which a circle or sphere is described; *broadly* **:** a point that is related to a geometrical figure in such a way that for any point on the figure there is another point on the figure such that a straight line joining the two points is bisected by the original point — called also *center of symmetry* **b :** the center of the circle inscribed in a regular polygon
2 a : a point, area, person, or thing that is most important or pivotal in relation to an indicated activity, interest, or condition ⟨a railroad *center*⟩ ⟨the *center* of the controversy⟩ **b :** a source from which something originates ⟨a propaganda *center*⟩ **c :** a group of nerve cells having a common function ⟨respiratory *center*⟩ **d :** a region of concentrated population ⟨an urban *center*⟩ **e :** a facility providing a place for a particular activity or service ⟨a day-care *center*⟩
3 a : the middle part (as of the forehead or a stage) **b** *often capitalized* (1) **:** a grouping of political figures holding moderate views especially between those of conservatives and liberals (2) **:** the views of such politicians (3) **:** the adherents of such views
4 a : a player occupying a middle position on a team: as (1) **:** the football player in the middle of a line who passes the ball between his legs to a back to start a down (2) **:** the usually tallest player on a basketball team who usually plays near the basket **b :** CENTER FIELD
5 a : either of two tapered rods which support work in a lathe or grinding machine and about or with which the work revolves **b :** a conical recess in the end of work (as a shaft) for receiving such a center
— **cen·ter·less** \-ləs\ *adjective*

[2]**center** *verb* **cen·tered; cen·ter·ing** \ˈsen-t(ə-)riŋ, ˈse-nər-iŋ\ (1610)
transitive verb
1 : to place or fix at or around a center or central area or position ⟨*center* the picture on the wall⟩
2 : to give a central focus or basis ⟨*centers* her hopes on her son⟩ ⟨the plot was *centered* on espionage⟩
3 : to adjust (as lenses) so that the axes coincide
4 a : to pass (a ball or puck) from either side toward the middle of the playing area **b :** to hand or pass (a football) backward between one's legs to a back to start a down
intransitive verb
: to have a specified center **:** FOCUS ■

cen·ter·board \ˈsen-tər-ˌbōrd, ˈse-nər-, -ˌbȯrd\ *noun* (1867)
: a retractable keel used especially in sailboats

cen·tered \ˈsen-tərd, ˈse-nərd\ *adjective* (circa 1893)
1 : having a center — often used in combination ⟨a dark-*centered* coneflower⟩
2 : having a center of curvature — often used in combination ⟨a 3-*centered* arch⟩
3 : emotionally stable and secure
— **cen·tered·ness** \-nəs\ *noun*

center field *noun* (1857)
1 : the position of the player for defending center field
2 : the part of the baseball outfield between right and left field
— **center fielder** *noun*

cen·ter·fold \ˈsen-tər-ˌfōld, ˈse-nər-\ *noun* (1952)
1 : a foldout that is the center spread of a magazine
2 : a picture (as of a nude) on a centerfold; *also* **:** a model featured in such a picture

cen·ter·line \-ˈlīn, -ˌlīn\ *noun* (1807)
: a real or imaginary line that is equidistant from the surface or sides of something

center of curvature (circa 1856)
: the center of the circle whose center lies on the concave side of a curve on the normal to a given point of the curve and whose radius is equal to the radius of curvature at that point

center of gravity (1659)
1 : CENTER OF MASS
2 : the point at which the entire weight of a body may be considered as concentrated so that if supported at this point the body would remain in equilibrium in any position
3 : CENTER 2a

center of mass (1879)
: the point in a body or system of bodies at which the whole mass may be considered as concentrated

cen·ter·piece \-ˌpēs\ *noun* (1803)
1 : an object occupying a central position; *especially* **:** an adornment in the center of a table
2 : one that is of central importance or interest in a larger whole

center punch *noun* (1879)
: a hand punch consisting of a short steel bar with a hardened conical point at one end used for marking the centers of holes to be drilled

center stage *noun* (1954)
1 : the central part of a theatrical stage
2 : a central or highly prominent position
— **center stage** *adjective or adverb*

cen·tes·i·mal \sen-ˈte-sə-məl\ *adjective* [Latin *centesimus* hundredth, from *centum*] (1809)
: marked by or relating to division into hundredths

[1]**cen·tes·i·mo** \chen-ˈte-zə-ˌmō\ *noun, plural* **-mi** \-(ˌ)mē\ [Italian] (1851)

☐ USAGE
center The intransitive verb *center* is most commonly used with the prepositions *in, on, at,* and *around. At* appears to be favored in mathematical contexts; the others are found in a broad range of contexts. *Center around,* a standard idiom, has often been objected to as illogical. The logic on which the objections are based is irrelevant, since *center around* is an idiom and idioms have their own logic. *Center on* is currently more common in edited prose, and *revolve around* and similar verbs are available if you want to avoid *center around.*

— see *lira* at MONEY table

[2]**cen·tes·i·mo** \sen-ˈte-sə-ˌmō\ *noun; plural* **-mos** [Spanish *centésimo*] (circa 1883)
— see *balboa, peso* at MONEY table

centi- *combining form* [French & Latin; French, hundredth, from Latin, hundred, from *centum* — more at HUNDRED]
1 : hundred ⟨*centi*pede⟩
2 : hundredth part ⟨*centi*meter⟩

cen·ti·grade \ˈsen-tə-ˌgrād, ˈsän-\ *adjective* [French, from Latin *centi-* hundred + French *grade*] (1801)
: relating to, conforming to, or having a thermometric scale on which the interval between the freezing point of water and the boiling point of water is divided into 100 degrees with 0° representing the freezing point and 100° the boiling point ⟨10° *centigrade*⟩ — abbreviation *C;* compare CELSIUS

cen·ti·gram \-ˌgram\ *noun* (1801)
— see METRIC SYSTEM table

cen·ti·li·ter \ˈsen-ti-ˌlē-tər, ˈsän-\ *noun* (1801)
— see METRIC SYSTEM table

cen·til·lion \sen-ˈtil-yən\ *noun, often attributive* [Latin *cen*tum + English *-illion* (as in *million*)] (circa 1889)
— see NUMBER table

cen·time \ˈsän-ˌtēm, ˈsen-\ *noun* [French, from *cent* hundred, from Latin *centum*] (1801)
— see *dinar, dirham, franc, gourde* at MONEY table

cen·ti·me·ter \ˈsen-tə-ˌmē-tər, ˈsän-\ *noun* (1801)
— see METRIC SYSTEM table

centimeter–gram–second *adjective* (1875)
: of, relating to, or being a system of units based on the centimeter as the unit of length, the gram as the unit of mass, and the second as the unit of time — abbreviation *cgs*

cen·ti·mo \ˈsen-tə-ˌmō\ *noun, plural* **-mos** [Spanish *céntimo*] (1899)
— see *bolívar, colon, dobra, guarani, peseta* at MONEY table

cen·ti·mor·gan \ˈsen-tə-ˌmȯr-gən, ˈsän-\ *noun* (1919)
: a genetic unit equivalent to 1/100 of a morgan

cen·ti·pede \ˈsen-tə-ˌpēd\ *noun* [Latin *centipeda,* from *centi-* + *ped-, pes* foot — more at FOOT] (1601)
: any of a class (Chilopoda) of long flattened many-segmented predaceous arthropods with each segment bearing one pair of legs of which the foremost pair is modified into poison fangs

cent·ner \ˈsent-nər\ *noun* [probably from Low German] (1683)
: any of various units of weight used especially in Europe and usually equal to about 110 pounds (about 50 kilograms)

cen·to \ˈsen-(ˌ)tō\ *noun, plural* **cen·to·nes** \sen-ˈtō-(ˌ)nēz\ [Late Latin, from Latin, patchwork garment; perhaps akin to Sanskrit *kanthā* patched garment] (1605)
: a literary work made up of parts from other works

centr- *or* **centri-** *or* **centro-** *combining form* [Greek *kentr-, kentro-,* from *kentron* center — more at CENTER]
: center ⟨*centri*fugal⟩ ⟨*centr*oid⟩

centra *plural of* CENTRUM

[1]**cen·tral** \ˈsen-trəl\ *adjective* [Latin *centralis,* from *centrum* center] (1647)
1 : containing or constituting a center
2 : of cardinal importance **:** ESSENTIAL, PRINCIPAL ⟨the *central* character of the novel⟩
3 a : situated at, in, or near the center **b :** easily accessible from outlying districts ⟨a *central* location for the new theater⟩
4 a : centrally placed and superseding separate scattered units ⟨*central* heating⟩ **b :** controlling or directing local or branch activities ⟨the *central* committee⟩
5 : holding to a middle between extremes **:** MODERATE
6 : of, relating to, or comprising the brain and spinal cord; *also* **:** originating within the central nervous system ⟨*central* deafness⟩
— **cen·tral·ly** \-trə-lē\ *adverb*

[2]**central** *noun* (1889)
1 : a telephone exchange or operator
2 : a central office or bureau usually controlling others ⟨weather *central*⟩

central angle *noun* (1904)
: an angle formed by two radii of a circle

central bank *noun* (1922)
: a national bank that operates to establish monetary and fiscal policy and to control the money supply and interest rate
— **central banker** *noun*

central casting *noun* (1957)
: the department of a movie studio responsible for casting actors — usually used figuratively ⟨a politician right out of *central casting*⟩

central city *noun* (1950)
: a city that constitutes the densely populated center of a metropolitan area

cen·tral·ise *British variant of* CENTRALIZE

cen·tral·ism \ˈsen-trə-ˌli-zəm\ *noun* (1831)
: the concentration of power and control in the central authority of an organization (as a political or educational system) — compare FEDERALISM
— **cen·tral·ist** \-list\ *noun or adjective*
— **cen·tral·is·tic** \ˌsen-trə-ˈlis-tik\ *adjective*

cen·tral·i·ty \sen-ˈtra-lə-tē\ *noun, plural* **-ties** (1647)
1 : the quality or state of being central
2 : central situation
3 : tendency to remain in or at the center

cen·tral·ize \ˈsen-trə-ˌlīz\ *verb* **-ized; -iz·ing** (1800)
intransitive verb
: to form a center **:** cluster around a center
transitive verb
1 : to bring to a center **:** CONSOLIDATE ⟨*centralize* all the data in one file⟩
2 : to concentrate by placing power and authority in a center or central organization
— **cen·tral·i·za·tion** \ˌsen-trə-lə-ˈzā-shən\ *noun*
— **cen·tral·iz·er** \ˈsen-trə-ˌlī-zər\ *noun*

central limit theorem *noun* (1951)
: any of several fundamental theorems of probability and statistics that state the conditions under which the distribution of a sum of independent random variables is approximated by the normal distribution; *especially* **:** one which is much applied in sampling and which states that the distribution of a mean of a sample from a population with finite variance is approximated by the normal distribution as the number in the sample becomes large

central nervous system *noun* (circa 1907)
: the part of the nervous system which in vertebrates consists of the brain and spinal cord, to which sensory impulses are transmitted and from which motor impulses pass out, and which coordinates the activity of the entire nervous system — compare AUTONOMIC NERVOUS SYSTEM

central processing unit *noun* (1961)
: PROCESSOR 2a(2) — abbreviation *cpu*

central tendency *noun* (circa 1928)
: the degree of clustering of the values of a statistical distribution that is usually measured by the arithmetic mean, mode, or median

central time *noun, often C capitalized* (1883)
: the time of the 6th time zone west of Greenwich that includes the central U.S. — see TIME ZONE illustration

cen·tre *chiefly British variant of* CENTER

cen·tric \ˈsen-trik\ *adjective* [Greek *kentrikos* of the center, from *kentron*] (circa 1590)
1 : located in or at a center **:** CENTRAL ⟨a *centric* point⟩
2 : concentrated about or directed to a center ⟨a *centric* activity⟩
3 : of, relating to, or having a centromere
4 : of, relating to, or resembling an order (Centrales) of radially symmetrical diatoms
— **cen·tri·cal·ly** \-tri-k(ə-)lē\ *adverb*
— **cen·tric·i·ty** \sen-ˈtri-sə-tē\ *noun*

-centric *adjective combining form* [Medieval Latin *-centricus,* from Latin *centrum* center]
1 : having (such) a center or (such or so many) centers ⟨poly*centric*⟩
2 : having (something specified) as its center ⟨helio*centric*⟩

[1]**cen·trif·u·gal** \sen-ˈtri-fyə-gəl, -ˈtri-fi-, *especially British* ˌsen-tri-ˈfyü-gəl\ *adjective* [New Latin *centrifugus,* from *centr-* + Latin *fugere* to flee — more at FUGITIVE] (circa 1721)
1 : proceeding or acting in a direction away from a center or axis
2 : using or acting by centrifugal force ⟨a *centrifugal* pump⟩
3 : EFFERENT
4 : tending away from centralization **:** SEPARATIST ⟨*centrifugal* tendencies in modern society⟩
— **cen·trif·u·gal·ly** \-gə-lē\ *adverb*

[2]**centrifugal** *noun* (1866)
: a centrifugal machine or a drum in such a machine

centrifugal force *noun* (circa 1721)
1 : the force that tends to impel a thing or parts of a thing outward from a center of rotation
2 : the force that an object moving along a circular path exerts on the body constraining the object and that acts outwardly away from the center of rotation ⟨a stone whirled on a string exerts *centrifugal force* on the string⟩

cen·tri·fu·ga·tion \ˌsen-trə-fyu̇-ˈgā-shən, -f(y)ə-\ *noun* (1903)
: the process of centrifuging

[1]**cen·tri·fuge** \ˈsen-trə-ˌfyüj\ *noun* [French, from *centrifuge* centrifugal, from New Latin *centrifugus*] (1887)
: a machine using centrifugal force for separating substances of different densities, for removing moisture, or for simulating gravitational effects

[2]**centrifuge** *transitive verb* **-fuged; -fug·ing** (circa 1895)
: to subject to centrifugal action especially in a centrifuge

cen·tri·ole \ˈsen-trē-ˌōl\ *noun* [German *Zentriol,* from *Zentrum* center] (circa 1896)
: one of a pair of cellular organelles that occur especially in animals, are adjacent to the nucleus, function in the formation of the spindle apparatus during cell division, and consist of a cylinder with nine microtubules arranged peripherally in a circle — see CELL illustration

cen·trip·e·tal \sen-ˈtri-pə-tᵊl\ *adjective* [New Latin *centripetus,* from *centr-* + Latin *petere* to go to, seek — more at FEATHER] (1709)
1 : proceeding or acting in a direction toward a center or axis
2 : AFFERENT
3 : tending toward centralization **:** UNIFYING
— **cen·trip·e·tal·ly** \-tᵊl-ē\ *adverb*

centripetal force *noun* (1709)
: the force that is necessary to keep an object moving in a circular path and that is directed inward toward the center of rotation ⟨a string on the end of which a stone is whirled about exerts *centripetal force* on the stone⟩

cen·trist \ˈsen-trist\ *noun* (1872)
1 *often capitalized* **:** a member of a center party
2 : one who holds moderate views
— **cen·trism** \-ˌtri-zəm\ *noun*
— **centrist** *adjective*

cen·troid \ˈsen-ˌtrȯid\ *noun* (1882)
1 : CENTER OF MASS

2 : a point whose coordinates are the averages of the corresponding coordinates of a given set of points and which for a given planar or three-dimensional figure (as a triangle or sphere) corresponds to the center of mass of a thin plate of uniform thickness and consistency or a body of uniform consistency having the same boundary

cen·tro·mere \'sen-trə-ˌmir\ *noun* [International Scientific Vocabulary] (circa 1925)
: the point or region on a chromosome to which the spindle attaches during mitosis and meiosis
— cen·tro·mer·ic \ˌsen-trə-'mir-ik, -'mer-\ *adjective*

cen·tro·some \'sen-trə-ˌsōm\ *noun* [International Scientific Vocabulary] (1889)
1 : CENTRIOLE
2 : the centriole-containing region of clear cytoplasm adjacent to the cell nucleus

cen·tro·sym·met·ric \ˌsen-trə-sə-'me-trik\ *adjective* (circa 1909)
: symmetric with respect to a central point ⟨*centrosymmetric* molecules⟩ ⟨a *centrosymmetric* curve⟩

cen·trum \'sen-trəm\ *noun, plural* **cen·trums** *or* **cen·tra** \-trə\ [Latin — more at CENTER] (1854)
1 : CENTER
2 : the body of a vertebra ventral to the neural arch

cen·tum \'ken-təm, -ˌtu̇m\ *adjective* [Latin, hundred; from the fact that its initial sound (a velar stop) is the representative of an Indo-European palatal stop — more at HUNDRED] (1901)
: of, relating to, or constituting an Indo-European language group in which the palatal stops did not in prehistoric times become palatal or alveolar fricatives — compare SATEM

cen·tu·ri·on \sen-'chu̇r-ē-ən, -'tyu̇r-, -'tu̇r-\ *noun* [Middle English, from Middle French & Latin; Middle French, from Latin *centurion-, centurio,* from *centuria*] (13th century)
: an officer commanding a Roman century

cen·tu·ry \'sen(t)-sh(ə-)rē\ *noun, plural* **-ries** [Latin *centuria,* irregular from *centum* hundred] (1533)
1 : a subdivision of the Roman legion
2 : a group, sequence, or series of 100 like things
3 : a period of 100 years especially of the Christian era or of the preceding period of human history
4 : a race over a hundred units (as yards or miles)

century plant *noun* (1764)
: a Mexican agave (*Agave americana*) that takes many years to mature, flowers only once, and then dies

CEO \ˌsē-(ˌ)ē-'ō\ *noun* [chief *e*xecutive *o*fficer] (1975)
: the executive with the chief decision-making authority in an organization or business

ceorl \'chā-ˌȯrl\ *noun* [Old English — more at CHURL] (before 12th century)
: a freeman of the lowest rank in Anglo-Saxon England

cèpe *or* **cepe** \'sēp, sep\ *also* **cep** \'sep\ *noun* [French, from Gascon *cep* tree trunk, mushroom, from Latin *cippus* stake, post] (1865)
: a wild edible boletus mushroom (especially *Boletus edulis*)

cephal- *or* **cephalo-** *combining form* [Latin, from Greek *kephal-, kephalo-,* from *kephalē*]
: head ⟨*cephal*ad⟩ ⟨*cephalo*pod⟩

ceph·a·lad \'se-fə-ˌlad\ *adverb* (1887)
: toward the head or anterior end of the body

ceph·a·lex·in \ˌse-fə-'lek-sən\ *noun* [*cephalo*sporin + *-ex-* (of unknown origin) + [1]*-in*] (1967)
: a semisynthetic cephalosporin $C_{16}H_{17}N_3O_4S$ with a spectrum of antibiotic activity similar to the penicillins

ce·phal·ic \sə-'fa-lik\ *adjective* [Middle French *céphalique,* from Latin *cephalicus,* from Greek *kephalikos,* from *kephalē* head; akin to Old High German *gebal* skull, Old Norse *gafl* gable, Tocharian A *śpāl* head] (1599)
1 : of or relating to the head
2 : directed toward or situated on or in or near the head
— ce·phal·i·cal·ly \-li-k(ə-)lē\ *adverb*

cephalic index *noun* (1866)
: the ratio multiplied by 100 of the maximum breadth from side to side of the head to its maximum length from front to back in living individuals — compare CRANIAL INDEX

ceph·a·lin \'ke-fə-lən, 'se-\ *noun* [International Scientific Vocabulary] (circa 1899)
: PHOSPHATIDYLETHANOLAMINE

ceph·a·li·za·tion \ˌse-fə-lə-'zā-shən\ *noun* (1864)
: a tendency in the evolution of organisms to concentrate the sensory and neural organs in an anterior head

ceph·a·lom·e·try \ˌse-fə-'lä-mə-trē\ *noun* [International Scientific Vocabulary] (circa 1889)
: the science of measuring the head in living individuals — compare CRANIOMETRY
— ceph·a·lo·met·ric \-lō-'me-trik\ *adjective*

ceph·a·lo·pod \'se-fə-lə-ˌpäd\ *noun* [ultimately from *cephal-* + Greek *pod-, pous* foot — more at FOOT] (1826)
: any of a class (Cephalopoda) of marine mollusks including the squids, cuttlefishes, and octopuses that move by expelling water from a tubular siphon under the head and that have a group of muscular usually sucker-bearing arms around the front of the head, highly developed eyes, and usually a sac containing ink which is ejected for defense or concealment
— cephalopod *adjective*

ceph·a·lor·i·dine \ˌse-fə-'lȯr-ə-ˌdēn, -'lär-\ *noun* [probably from *cephalo*sporin + *-idine*] (1965)
: a semisynthetic broad-spectrum antibiotic $C_{19}H_{17}N_3O_4S_2$ derived from cephalosporin

ceph·a·lo·spo·rin \ˌse-fə-lə-'spōr-ən, -'spȯr-\ *noun* [*Cephalosporium,* genus of fungi + [1]*-in*] (1951)
: any of several antibiotics produced by an imperfect fungus (genus *Cephalosporium*)

ceph·a·lo·thin \'se-fə-lə-thən, -ˌthin\ *noun* [*cephalo*sporin + *thi-* + [1]*-in*] (1962)
: a semisynthetic broad-spectrum antibiotic derived from cephalosporin and used as the sodium salt $C_{16}H_{15}N_2NaO_6S_2$

ceph·a·lo·tho·rax \ˌse-fə-lə-'thōr-ˌaks, -'thȯr-\ *noun* [International Scientific Vocabulary] (1835)
: the united head and thorax of an arachnid or higher crustacean

Ce·phe·id \'se-f(ē-)id, 'sē-\ *noun* [International Scientific Vocabulary, from *Cepheus*] (circa 1903)
: any of a class of pulsating stars whose very regular light variations are related directly to their intrinsic luminosities and whose apparent luminosities are used to estimate distances in astronomy

Ce·pheus \'sē-ˌfyüs; 'sē-fē-əs, 'se-\ *noun* [Latin (genitive *Cephei*), from Greek *Kēpheus*]
: a constellation between Cygnus and the north pole

[1]**ce·ram·ic** \sə-'ra-mik, *especially British* kə-\ *adjective* [Greek *keramikos,* from *keramos* potter's clay, pottery] (1850)
: of or relating to the manufacture of any product (as earthenware, porcelain, or brick) made essentially from a nonmetallic mineral (as clay) by firing at a high temperature; *also* **:** of or relating to such a product

[2]**ceramic** *noun* (1859)
1 *plural but singular in construction* **:** the art or process of making ceramic articles
2 : a product of ceramic manufacture

ce·ra·mist \sə-'ra-mist, 'ser-ə-\ *or* **ce·ram·i·cist** \sə-'ra-mə-sist\ *noun* (1855)
: one who engages in ceramics

ce·ras·tes \sə-'ras-(ˌ)tēz\ *noun* [Middle English, from Latin, from Greek *kerastēs,* literally, horned, from *keras* horn] (14th century)
: a venomous viper (*Cerastes cornutus*) of the Near East having a horny process over each eye — called also *horned viper*

ce·rate \'sir-ˌāt\ *noun* [Middle English, from Latin *ceratum* wax salve, from *cera* wax — more at CERUMEN] (15th century)
: an unctuous preparation for external use consisting of wax or resin or spermaceti mixed with oil, lard, and medicinal ingredients

cer·a·top·sian \ˌsər-ə-'täp-sē-ən\ *noun* [New Latin *Ceratopsia,* from *Ceratops,* a genus, from Greek *kerat-, keras* horn + *ōps* facehorn, eye] (1909)
: any of a suborder (Ceratopsia) of ornithischian dinosaurs of the Late Cretaceous having horns, a sharp horny beak, and a bony frill projecting backward from the skull
— ceratopsian *adjective*

Cer·ber·us \'sər-b(ə-)rəs\ *noun* [Latin, from Greek *Kerberos*]
: a 3-headed dog that in Greek mythology guards the entrance to Hades
— Cer·ber·e·an \ˌsər-bə-'rē-ən\ *adjective*

-cercal *adjective combining form* [French *-cerque,* from Greek *kerkos* tail]
: -tailed ⟨homo*cercal*⟩

cer·car·ia \(ˌ)sər-'kar-ē-ə, -'ker-\ *noun, plural* **-i·ae** \-ē-ˌē\ [New Latin, from Greek *kerkos*] (circa 1871)
: a usually tadpole-shaped larval trematode worm that develops in a molluscan host from a redia
— cer·car·i·al \-ē-əl\ *adjective*

cer·cus \'sər-kəs\ *noun, plural* **cer·ci** \'sər-ˌsī, -ˌkī\ [New Latin, from Greek *kerkos*] (1826)
: either of a pair of simple or segmented appendages at the posterior end of various arthropods that usually act as sensory organs

[1]**cere** \'sir\ *transitive verb* **cered; cer·ing** [Middle English, to wax, from Middle French *cirer,* from Latin *cerare,* from *cera*] (15th century)
: to wrap in or as if in a cerecloth

[2]**cere** *noun* [Middle English *sere,* from Middle French *cere,* from Medieval Latin *cera,* from Latin, wax] (15th century)
: a usually waxy protuberance or enlarged area at the base of the bill of a bird

[1]**ce·re·al** \'sir-ē-əl\ *adjective* [French or Latin; French *céréale,* from Latin *cerealis* of Ceres, of grain, from *Ceres*] (1818)
: relating to grain or to the plants that produce it; *also* **:** made of grain

[2]**cereal** *noun* (1832)
1 : a plant (as a grass) yielding starchy grain suitable for food; *also* **:** its grain
2 : a prepared foodstuff of grain

cereal leaf beetle *noun* (1962)
: a small reddish brown black-headed Old World chrysomelid beetle (*Oulema melanopus*) that feeds on cereal grasses and is a serious threat to U.S. grain crops

cer·e·bel·lum \ˌser-ə-'be-ləm\ *noun, plural* **-bellums** *or* **-bel·la** \-'be-lə\ [Medieval Latin, from Latin, diminutive of *cerebrum*] (1543)
: a large dorsally projecting part of the brain concerned especially with the coordination of muscles and the maintenance of bodily equilibrium, situated between the brain stem and the back of the cerebrum, and formed in humans of two lateral lobes and a median lobe — see BRAIN illustration
— cer·e·bel·lar \-'be-lər\ *adjective*

cerebr- *or* **cerebro-** *combining form* [*cerebrum*]
1 : brain **:** cerebrum ⟨*cerebr*ation⟩
2 : cerebral and ⟨*cerebro*spinal⟩

ce·re·bral \sə-'rē-brəl, 'ser-ə-\ *adjective* [French *cérébral,* from Latin *cerebrum* brain; akin to Old High German *hirni* brain, Greek

kara head, *keras* horn, Sanskrit *śiras* head — more at HORN] (1816)
1 a : of or relating to the brain or the intellect **b :** of, relating to, or being the cerebrum
2 a : appealing to intellectual appreciation ⟨*cerebral* drama⟩ **b :** primarily intellectual in nature ⟨a *cerebral* society⟩
— ce•re•bral•ly \-brə-lē\ *adverb*
cerebral cortex *noun* (1926)
: the surface layer of gray matter of the cerebrum that functions chiefly in coordination of sensory and motor information
cerebral hemisphere *noun* (1816)
: either of the two hollow convoluted lateral halves of the cerebrum — see BRAIN illustration
cerebral palsy *noun* (1889)
: a disability resulting from damage to the brain before, during, or shortly after birth and outwardly manifested by muscular incoordination and speech disturbances — compare SPASTIC PARALYSIS
— cerebral–palsied *adjective*
cer•e•brate \'ser-ə-ˌbrāt\ *intransitive verb* **-brat•ed; -brat•ing** [back-formation from *cerebration,* from *cerebrum*] (1915)
: to use the mind **:** THINK
— cer•e•bra•tion \ˌser-ə-'brā-shən\ *noun*
ce•re•bro•side \'ser-ə-brə-ˌsīd, sə-'rē-\ *noun* [*cerebrose* galactose] (1883)
: any of various lipids found especially in nerve tissue
ce•re•bro•spi•nal \sə-ˌrē-brō-'spī-nᵊl, ˌser-ə-brō-\ *adjective* (1826)
: of or relating to the brain and spinal cord or to these together with the cranial and spinal nerves that innervate voluntary muscles
cerebrospinal fluid *noun* (circa 1889)
: a liquid that is comparable to serum, is secreted from the blood into the lateral ventricles of the brain, and serves chiefly to maintain uniform pressure within the brain and spinal cord
cerebrospinal meningitis *noun* (1889)
: inflammation of the meninges of both brain and spinal cord; *specifically* **:** an infectious epidemic and often fatal meningitis caused by the meningococcus
ce•re•bro•vas•cu•lar \sə-ˌrē-brō-'vas-kyə-lər, ˌser-ə-brō-\ *adjective* (1935)
: of or involving the cerebrum and the blood vessels supplying it
ce•re•brum \sə-'rē-brəm, 'ser-ə-brəm\ *noun, plural* **-brums** *or* **-bra** \-brə\ [Latin] (1615)
1 : BRAIN 1a
2 : an enlarged anterior or upper part of the brain; *especially* **:** the expanded anterior portion of the brain that in higher mammals overlies the rest of the brain, consists of cerebral hemispheres and connecting structures, and is considered to be the seat of conscious mental processes **:** TELENCEPHALON
cere•cloth \'sir-ˌklōth\ *noun* [alteration of earlier *cered cloth* (waxed cloth)] (1553)
: cloth treated with melted wax or gummy matter and formerly used especially for wrapping a dead body
cere•ment \'ser-ə-mənt, 'sir-mənt\ *noun* (1602)
: a shroud for the dead; *especially* **:** CERECLOTH — usually used in plural
¹cer•e•mo•ni•al \ˌser-ə-'mō-nē-əl\ *adjective* (14th century)
: marked by, involved in, or belonging to ceremony **:** stressing careful attention to form and detail ☆
— cer•e•mo•ni•al•ism \-ə-ˌli-zəm\ *noun*
— cer•e•mo•ni•al•ist \-ə-list\ *noun*
— cer•e•mo•ni•al•ly \-ə-lē\ *adverb*
²ceremonial *noun* (14th century)
: a ceremonial act, action, or system
cer•e•mo•ni•ous \ˌser-ə-'mō-nē-əs\ *adjective* (1553)
1 : devoted to forms and ceremony **:** PUNCTILIOUS
2 : of, relating to, or constituting a ceremony
3 : according to formal usage or prescribed procedures
4 : marked by ceremony
synonym see CEREMONIAL
— cer•e•mo•ni•ous•ly *adverb*
— cer•e•mo•ni•ous•ness *noun*
cer•e•mo•ny \'ser-ə-ˌmō-nē\ *noun, plural* **-nies** [Middle English *ceremonie,* from Middle French *cérémonie,* from Latin *caerimonia*] (14th century)
1 : a formal act or series of acts prescribed by ritual, protocol, or convention ⟨the marriage *ceremony*⟩
2 a : a conventional act of politeness or etiquette ⟨the *ceremony* of introduction⟩ **b :** an action performed only formally with no deep significance **c :** a routine action performed with elaborate pomp
3 a : prescribed procedures **:** USAGES ⟨the *ceremony* attending an inauguration⟩ **b :** observance of an established code of civility or politeness ⟨opened the door without *ceremony* and strode in⟩
Ce•ren•kov radiation \chə-'reŋ-kòf-, chər-'yeŋ-\ *noun* [P. A. *Cherenkov*] (1939)
: light produced by charged particles (as electrons) traversing a transparent medium at a speed greater than that of light in the same medium — called also *Cerenkov light*
Ce•res \'sir-(ˌ)ēz\ *noun* [Latin]
: the Roman goddess of agriculture — compare DEMETER
ce•re•us \'sir-ē-əs\ *noun* [New Latin, genus name, from Latin, wax candle, from *cera* wax — more at CERUMEN] (1730)
: any of various cacti (as of the genus *Cereus*) of the western U.S. and tropical America
ce•ric \'sir-ik, 'ser-\ *adjective* (circa 1879)
: of, relating to, or containing cerium especially with a valence of four
ce•rise \sə-'rēs, -'rēz\ *noun* [French, literally, cherry, from Late Latin *ceresia* — more at CHERRY] (1844)
: a moderate red
ce•ri•um \'sir-ē-əm\ *noun* [New Latin, from *Ceres*] (1804)
: a malleable ductile metallic element that is the most abundant of the rare-earth group — see ELEMENT table
cer•met \'sər-ˌmet\ *noun* [*cer*amic + *met*al] (1948)
: a composite structural material of a heat-resistant compound (as titanium carbide) and a metal (as nickel) used especially for turbine blades
cero \'ser-(ˌ)ō\ *noun, plural* **cero** *or* **ceros** [modification of Spanish *sierra* saw, cero] (1884)
: either of two large food and sport fishes (*Scomberomorus regalis* and *S. cavalla* of the warmer parts of the western Atlantic
ce•rous \'sir-əs\ *adjective* (circa 1872)
: of, relating to, or containing cerium especially with a valence of three
¹cer•tain \'sər-tᵊn\ *adjective* [Middle English, from Old French, from (assumed) Vulgar Latin *certanus,* from Latin *certus,* from past participle of *cernere* to sift, discern, decide; akin to Greek *krinein* to separate, decide, judge, Old Irish *criathar* sieve] (13th century)
1 : FIXED, SETTLED ⟨a *certain* percentage of the profit⟩
2 : of a specific but unspecified character, quantity, or degree ⟨the house has a *certain* charm⟩
3 a : DEPENDABLE, RELIABLE ⟨a *certain* remedy for the disease⟩ **b :** known or proved to be true **:** INDISPUTABLE ⟨it is *certain* that we exist⟩
4 a : INEVITABLE ⟨the *certain* advance of age⟩ **b :** incapable of failing **:** DESTINED — used with a following infinitive ⟨she is *certain* to do well⟩
5 : assured in mind or action
synonym see SURE
— for certain : as a certainty **:** ASSUREDLY
²certain *pronoun, plural in construction* (15th century)
: certain ones
cer•tain•ly \-lē\ *adverb* (14th century)
1 : in a manner that is certain **:** with certainty
2 : it is certain that **:** ASSUREDLY
cer•tain•ty \'sər-tᵊn-tē\ *noun, plural* **-ties** (14th century)
1 : something that is certain
2 : the quality or state of being certain especially on the basis of evidence ☆
cer•tes \'sər-tēz, 'sərts\ *adverb* [Middle English, from Old French, from *cert* certain, from Latin *certus*] (13th century)
archaic **:** in truth **:** CERTAINLY
¹cer•tif•i•cate \(ˌ)sər-'ti-fi-kət\ *noun* [Middle English *certificat,* from Middle French, from Medieval Latin *certificatum,* from Late Latin, neuter of *certificatus,* past participle of *certificare* to certify] (15th century)
1 : a document containing a certified statement especially as to the truth of something; *specifically* **:** a document certifying that one has fulfilled the requirements of and may practice in a field
2 : something serving the same end as a certificate
3 : a document evidencing ownership or debt ⟨a *certificate* of deposit⟩
²cer•tif•i•cate \-'ti-fə-ˌkāt\ *transitive verb* **-cat•ed; -cat•ing** (1883)
: to testify to or authorize by a certificate; *especially* **:** to recognize as having met special qualifications (as of a governmental agency or professional board) within a field
— cer•tif•i•ca•to•ry \-'ti-fi-kə-ˌtōr-ē, -ˌtòr-\ *adjective*
cer•ti•fi•ca•tion \ˌsər-tə-fə-'kā-shən\ *noun* (15th century)
1 : the act of certifying **:** the state of being certified
2 : a certified statement
certification mark *noun* (1947)

> ☆ SYNONYMS
> **Ceremonial, ceremonious, formal, conventional** mean marked by attention to or adhering strictly to prescribed forms. CEREMONIAL and CEREMONIOUS both imply strict attention to what is prescribed by custom or by ritual, but CEREMONIAL applies to things that are associated with ceremonies ⟨a *ceremonial* offering⟩, CEREMONIOUS to persons given to ceremony or to acts attended by ceremony ⟨made his *ceremonious* entrance⟩. FORMAL applies both to things prescribed by and to persons obedient to custom and may suggest stiff, restrained, or old-fashioned behavior ⟨a *formal* report⟩ ⟨the headmaster's *formal* manner⟩. CONVENTIONAL implies accord with general custom and usage ⟨*conventional* courtesy⟩ and may suggest a stodgy lack of originality or independence ⟨*conventional* fiction⟩.
>
> **Certainty, certitude, conviction** mean a state of being free from doubt. CERTAINTY and CERTITUDE are very close; CERTAINTY may stress the existence of objective proof ⟨claims that cannot be confirmed with scientific *certainty*⟩, while CERTITUDE may emphasize a faith in something not needing or not capable of proof ⟨believes with *certitude* in an afterlife⟩. CONVICTION applies especially to belief strongly held by an individual ⟨holds firm *convictions* on every issue⟩.

: a mark or device used to identify a product or service that has been certified to conform to a particular set of standards

certified mail *noun* (1955)
: first class mail for which proof of delivery is secured but no indemnity value is claimed

certified milk *noun* (1899)
: milk produced in dairies that operate under the rules and regulations of an authorized medical milk commission

certified public accountant *noun* (1896)
: an accountant who has met the requirements of a state law and has been granted a certificate

cer·ti·fy \'sər-tə-ˌfī\ *transitive verb* **-fied; -fy·ing** [Middle English *certifien,* from Middle French *certifier,* from Late Latin *certificare,* from Latin *certus* certain — more at CERTAIN] (14th century)
1 : to attest authoritatively: as **a :** CONFIRM **b :** to present in formal communication **c :** to attest as being true or as represented or as meeting a standard **d :** to attest officially to the insanity of
2 : to inform with certainty **:** ASSURE
3 : to guarantee (a personal check) as to signature and amount by so indicating on the face
4 : CERTIFICATE, LICENSE ☆
— **cer·ti·fi·able** \-ˌfī-ə-bəl\ *adjective*
— **cer·ti·fi·ably** \-blē\ *adverb*
— **cer·ti·fi·er** \-ˌfī(-ə)r\ *noun* See in addition APPROVE.

cer·tio·ra·ri \ˌsər-sh(ē-)ə-'rar-ē, -'rär-ē\ *noun* [Middle English, from Latin, literally, to be informed; from the use of the word in the writ] (15th century)
: a writ of superior court to call up the records of an inferior court or a body acting in a quasi-judicial capacity

cer·ti·tude \'sər-tə-ˌtüd *also* -ˌtyüd\ *noun* [Middle English, from Late Latin *certitudo,* from Latin *certus*] (15th century)
1 : the state of being or feeling certain
2 : certainty of act or event
synonym see CERTAINTY

ce·ru·le·an \sə-'rü-lē-ən\ *adjective* [Latin *caeruleus* dark blue] (1667)
: resembling the blue of the sky

ce·ru·lo·plas·min \sə-ˌrü-lō-'plaz-mən\ *noun* [International Scientific Vocabulary *cerulo-* (from Latin *caeruleus*) + *plasma* + [1]*-in*] (circa 1952)
: a blue copper-binding serum alpha globulin with enzymatic activity

ce·ru·men \sə-'rü-mən\ *noun* [New Latin, irregular from Latin *cera* wax; akin to Greek *kēros* wax] (1741)
: the yellow waxy secretion from the glands of the external ear — called also *earwax*
— **ce·ru·mi·nous** \-mə-nəs\ *adjective*

ce·ruse \sə-'rüs, 'sir-ˌüs\ *noun* [Middle English, from Middle French *céruse,* from Latin *cerussa*] (14th century)
1 : white lead as a pigment
2 : a cosmetic containing white lead

ce·rus·site \sə-'rə-ˌsīt\ *noun* [German *Zerussit,* from Latin *cerussa*] (1850)
: a mineral that consists of lead carbonate occurring in colorless transparent crystals and also in massive form and is a source of white lead

cer·ve·lat \'sər-və-ˌlat, -ˌlä\ *noun* [obsolete French (now *cervelas*)] (1613)
: smoked sausage made from a combination of pork and beef

cer·vi·cal \'sər-vi-kəl\ *adjective* (1681)
: of or relating to a neck or cervix

cervical cap *noun* (1923)
: a contraceptive device in the form of a thimble-shaped molded cap that fits over the uterine cervix and blocks sperm from entering the uterus

cer·vi·ci·tis \ˌsər-və-'sī-təs\ *noun* [New Latin] (1889)
: inflammation of the uterine cervix

cer·vine \'sər-ˌvīn\ *adjective* [Latin *cervinus* of a deer, from *cervus* stag, deer — more at HART] (circa 1828)
: of, relating to, or resembling deer

cer·vix \'sər-viks\ *noun, plural* **cer·vi·ces** \'sər-və-ˌsēz, (ˌ)sər-'vī-(ˌ)sēz\ *or* **cer·vix·es** [Latin *cervic-, cervix*] (15th century)
1 : NECK; *especially* **:** the back part of the neck
2 : a constricted portion of an organ or part; *especially* **:** the narrow outer end of the uterus

ce·sar·e·an *also* **ce·sar·i·an** \si-'zar-ē-ən, -'zer-\ *noun* (circa 1903)
: CESAREAN SECTION
— **cesarean** *also* **cesarian** *adjective*

cesarean section *also* **cesarian section** *noun* [from the belief that Julius Caesar was born this way] (1615)
: surgical incision of the walls of the abdomen and uterus for delivery of offspring ◆

ce·si·um \'sē-zē-əm *also* -zhē-\ *noun* [New Latin, from Latin *caesius* bluish gray] (1861)
: a silver-white soft ductile element of the alkali metal group that is the most electropositive element known and that is used especially in photoelectric cells — see ELEMENT table

cesium 133 *noun* (1966)
: an isotope of cesium used especially in atomic clocks and one of whose atomic transitions is used as a scientific time standard

cess \'ses\ *noun* [probably short for *success*] (1830)
chiefly Irish **:** LUCK — usually used in the phrase *bad cess to you*

ces·sa·tion \se-'sā-shən\ *noun* [Middle English *cessacioun,* from Middle French *cessation,* from Latin *cessation-, cessatio* delay, idleness, from *cessare* to delay, be idle — more at CEASE] (14th century)
: a temporary or final ceasing (as of action) **:** STOP

ces·sion \'se-shən\ *noun* [Middle English, from Middle French, from Latin *cession-, cessio,* from *cedere* to withdraw — more at CEDE] (15th century)
: a yielding to another **:** CONCESSION

cess·pit \'ses-ˌpit\ *noun* [*cess*pool + *pit*] (1777)
: a pit for the disposal of refuse (as sewage)

cess·pool \-ˌpü(ə)l\ *noun* [by folk etymology from Middle English *suspiral* vent, cesspool, from Middle French *souspirail* ventilator, from *souspirer* to sigh, from Latin *suspirare,* literally, to draw a long breath — more at SUSPIRE] (1782)
: an underground reservoir for liquid waste (as household sewage)

ces·ta \'ses-tə\ *noun* [Spanish, literally, basket, from Latin *cista* box, basket] (circa 1902)
: a narrow curved wicker basket used to catch and propel the ball in jai alai

cesta

ces·tode \'ses-ˌtōd\ *noun* [New Latin *Cestoda,* taxonomic group comprising tapeworms, ultimately from Greek *kestos* girdle] (circa 1890)
: TAPEWORM
— **cestode** *adjective*

[1]**ces·tus** \'ses-təs\ *noun, plural* **ces·ti** \-ˌtī\ [Latin, girdle, belt, from Greek *kestos,* from *kestos* stitched, from *kentein* to prick — more at CENTER] (1557)
: a woman's belt; *especially* **:** a symbolic one worn by a bride

[2]**cestus** *noun* [Latin *cestus, caestus*] (circa 1720)
: a hand covering of leather bands often loaded with lead or iron and used by boxers in ancient Rome

ce·su·ra *variant of* CAESURA

ce·ta·cean \si-'tā-shən\ *noun* [ultimately from Latin *cetus* whale, from Greek *kētos*] (1836)
: any of an order (Cetacea) of aquatic mostly marine mammals that include the whales, dolphins, porpoises, and related forms and that have a torpedo-shaped nearly hairless body, paddle-shaped forelimbs but no hind limbs, one or two nares opening externally at the top of the head, and a horizontally flattened tail used for locomotion
— **cetacean** *adjective*
— **ce·ta·ceous** \-shəs\ *adjective*

ce·tane \'sē-ˌtān\ *noun* [from *cetyl,* the radical $C_{16}H_{33}$] (1871)
: a colorless oily hydrocarbon $C_{16}H_{34}$ found in petroleum

cetane number *noun* (1935)
: a measure of the ignition value of a diesel fuel that represents the percentage by volume of cetane in a mixture of liquid methylnaphthalene that gives the same ignition lag as the oil being tested — called also *cetane rating;* compare OCTANE NUMBER

ce·te·ris pa·ri·bus \'kā-tər-əs-'par-ə-bəs, 'ke-, 'se-\ *adverb* [New Latin, other things being equal] (1601)
: if all other relevant things, factors, or elements remain unaltered

ce·tol·o·gy \sē-'tä-lə-jē\ *noun* [Latin *cetus* whale] (circa 1828)
: a branch of zoology dealing with the cetaceans
— **ce·tol·o·gist** \-jist\ *noun*

Ce·tus \'sē-təs\ *noun* [Latin (genitive *Ceti*), literally, whale]
: an equatorial constellation south of Pisces and Aries

☆ SYNONYMS
Certify, attest, witness, vouch mean to testify to the truth or genuineness of something. CERTIFY usually applies to a written statement, especially one carrying a signature or seal ⟨*certified* that the candidate had met all requirements⟩. ATTEST applies to oral or written testimony usually from experts or witnesses ⟨*attested* to the authenticity of the document⟩. WITNESS applies to the subscribing of one's own name to a document as evidence of its genuineness ⟨*witnessed* the signing of the will⟩. VOUCH applies to one who testifies as a competent authority or a reliable person ⟨willing to *vouch* for her integrity⟩.

◇ WORD HISTORY
cesarean section The name *Caesar* is a cognomen, a nickname given to one member of a Roman clan and borne by his descendants as a kind of surname. No one knows who the original Caesar was, but his descendants within his clan, the Julii, continued to use his cognomen and formed a major branch of the clan. According to a legend related by the Roman naturalist Pliny, the first Caesar was so called because he was cut from the womb of his mother (*a caeso matris utero*), *Caesar* supposedly being a derivative of the verb *caedere* "to cut." This etymology is dubious, but the name *Caesar* has continued to be associated with surgery to remove a child that cannot be delivered naturally. In recent centuries, it has been widely believed that the most famous bearer of the cognomen, the politician and general Gaius Julius Caesar (100–44 B.C.), was delivered by cesarean section, but there is no authority for this notion in ancient sources. Moreover, Julius Caesar's mother lived long after his birth, an unlikely outcome if she had undergone such an operation, which few women could have survived in those days.

ce·tyl alcohol \'sē-tᵊl-\ *noun* [International Scientific Vocabulary *cet-* (from Latin *cetus* whale) + *-yl;* from its occurrence in spermaceti] (1873)
: a waxy crystalline alcohol $C_{16}H_{34}O$ obtained by the saponification of spermaceti or the hydrogenation of palmitic acid and used especially in pharmaceutical and cosmetic preparations and in making detergents

ce·vi·che \sə-'vē-(ˌ)chā, -chē\ *variant of* SEVICHE

Cha·blis \sha-'blē, shə-, shä-; 'sha-(ˌ)blē\ *noun, plural* **Cha·blis** \-'blēz, -(ˌ)blēz\ [French, from *Chablis,* town in France] (1668)
1 : a dry sharp white burgundy wine
2 : a semidry soft white California wine

cha–cha \'chä-(ˌ)chä\ *noun* [American Spanish *cha-cha-cha*] (1954)
: a fast rhythmic ballroom dance of Latin-American origin with a basic pattern of three steps and a shuffle

chac·ma baboon \'chäk-mə-\ *noun* [Khoikhoi] (circa 1909)
: a large dusky baboon (*Papio ursinus*) of southern African savannas — called also *chacma*

cha·conne \shä-'kȯn, sha-, -'kän, -'kən\ *noun* [French & Spanish; French *chaconne,* from Spanish *chacona*] (1685)
1 : an old Spanish dance tune of Latin American origin
2 : a musical composition in moderate triple time typically consisting of variations on a repeated succession of chords

chad \'chad\ *noun* [origin unknown] (1947)
: small pieces of paper or cardboard produced in punching paper tape or data cards; *also* **:** a piece of chad
— chad·less \-ləs\ *adjective*

Chad·ic \'cha-dik\ *or* **Chad** \'chad\ *noun* (circa 1950)
: a subfamily of the Afro-Asiatic language family comprising numerous languages of northern Nigeria, northern Cameroon, and Chad
— Chadic *adjective*

cha·dor \'chə-dər, 'chä-\ *noun* [Hindi & Persian; Hindi *caddar,* from Persian *chaddar, chādar*] (1614)
: a large cloth worn as a combination head covering, veil, and shawl usually by Muslim women especially in Iran

chae·ta \'kē-tə\ *noun, plural* **chae·tae** \'kē-ˌtē\ [New Latin, from Greek *chaitē* long flowing hair] (circa 1866)
: BRISTLE, SETA
— chae·tal \'kē-tᵊl\ *adjective*

chae·to·gnath \'kē-ˌtäg-ˌnath, -tə(g)-\ *noun* [New Latin *Chaetognatha,* class or phylum name, ultimately from Greek *chaitē* + *gnathos* jaw — more at -GNATHOUS] (circa 1889)
: ARROWWORM
— chaetognath *adjective*

¹chafe \'chāf\ *verb* **chafed; chaf·ing** [Middle English *chaufen* to warm, from Middle French *chaufer,* from (assumed) Vulgar Latin *calfare,* alteration of Latin *calefacere,* from *calēre* to be warm + *facere* to make — more at LEE, DO] (14th century)
transitive verb
1 : IRRITATE, VEX
2 : to warm by rubbing especially with the hands
3 a : to rub so as to wear away **:** ABRADE ⟨the boat *chafed* its sides against the dock⟩ **b :** to make sore by or as if by rubbing
intransitive verb
1 : to feel irritation or discontent **:** FRET ⟨*chafes* at his restrictive desk job⟩
2 : to rub and thereby cause wear or irritation

²chafe *noun* (1551)
1 : a state of vexation **:** RAGE
2 : injury or wear caused by friction; *also* **:** FRICTION, RUBBING

cha·fer \'chā-fər\ *noun* [Middle English *cheaffer,* from Old English *ceafor;* probably akin to Old English *ceafl* jowl — more at JOWL] (before 12th century)
: any of various scarab beetles (as a cockchafer) that feed on leaves and flowers and whose larvae feed on plant roots

¹chaff \'chaf\ *noun* [Middle English *chaf,* from Old English *ceaf;* akin to Old High German *cheva* husk] (before 12th century)
1 : the seed coverings and other debris separated from the seed in threshing grain
2 : something comparatively worthless
3 : the scales borne on the receptacle among the florets in the heads of many composite plants
4 : material (as strips of foil or clusters of fine wires) ejected into the air for reflecting radar waves (as for confusing an enemy's radar detection or for tracking a descending spacecraft)
— chaffy \'cha-fē\ *adjective*

²chaff *noun* [probably from ¹*chaff*] (1821)
: light jesting talk **:** BANTER

³chaff (1827)
transitive verb
: to tease good-naturedly
intransitive verb
: JEST, BANTER

¹chaf·fer \'cha-fər\ *noun* [Middle English *chaffare,* from *chep* trade + *fare* journey — more at CHEAP, FARE] (13th century)
archaic **:** a haggling about price

²chaffer *verb* **chaf·fered; chaf·fer·ing** \'cha-f(ə-)riŋ\ (14th century)
intransitive verb
1 : HAGGLE
2 *British* **:** to exchange small talk **:** CHATTER
transitive verb
1 : EXCHANGE, BARTER
2 : to bargain for
— chaf·fer·er \-fər-ər\ *noun*

chaf·finch \'cha-(ˌ)finch\ *noun* [Middle English, from Old English *ceaffinc,* from *ceaf* + *finc* finch] (before 12th century)
: a common European finch (*Fringilla coelebs*) of which the male has a pinkish brown breast

chaf·ing dish \'chā-fiŋ-\ *noun* [Middle English *chafing,* present participle of *chaufen, chafen* to warm] (15th century)
: a utensil for cooking or keeping food warm especially at the table

Cha·gas' disease \'shä-gəs-, -gə-səz-\ *noun* [Carlos *Chagas* (died 1934) Brazilian physician] (1912)
: a tropical American disease that is caused by a trypanosome (*Trypanosoma cruzi*) and is marked by prolonged high fever, edema, and enlargement of the spleen, liver, and lymph nodes

¹cha·grin \shə-'grin\ *noun* [French, from *chagrin* sad] (circa 1681)
: disquietude or distress of mind caused by humiliation, disappointment, or failure

²chagrin *transitive verb* **cha·grined** \-'grind\; **cha·grin·ing** \-'gri-niŋ\ (1733)
: to vex or unsettle by disappointing or humiliating

¹chain \'chān\ *noun, often attributive* [Middle English *cheyne,* from Middle French *chaeine,* from Latin *catena*] (14th century)
1 a : a series of usually metal links or rings connected to or fitted into one another and used for various purposes (as support, restraint, transmission of mechanical power, or measurement) **b :** a series of links used or worn as an ornament or insignia **c** (1) **:** a measuring instrument of 100 links used in surveying (2) **:** a unit of length equal to 66 feet (20 meters)
2 : something that confines, restrains, or secures
3 a : a series of things linked, connected, or associated together ⟨a *chain* of events⟩ **b :** a group of enterprises or institutions of the same kind or function usually under a single ownership, management, or control ⟨fast-food *chains*⟩ **c :** a number of atoms or chemical groups united like links in a chain

²chain *transitive verb* (14th century)
1 : to obstruct or protect by a chain
2 : to fasten, bind, or connect with or as if with a chain; *also* **:** FETTER

chaî·né \shā-'nā\ *noun* [French, from past participle of *chaîner* to chain] (1897)
: a series of short usually fast turns by which a ballet dancer moves across the stage

chain gang *noun* (1834)
: a gang of convicts chained together especially as an outside working party

chain letter *noun* (1905)
: a letter sent to several persons with a request that each send copies of the letter to an equal number of persons

chain–link fence *noun* (circa 1927)
: a fence of heavy steel wire woven to form a diamond-shaped mesh

chain mail *noun* (1822)
: flexible armor of interlinked metal rings

chain of command (1898)
: a series of executive positions in order of authority ⟨a military *chain of command*⟩

chain pickerel *noun* [from the markings resembling chains on the sides] (1887)
: a large greenish black pickerel (*Esox niger*) with dark markings along the sides that is common in quiet waters of eastern North America

chain reaction *noun* (circa 1902)
1 a : a series of events so related to each other that each one initiates the next **b :** a number of events triggered by the same initial event
2 : a self-sustaining chemical or nuclear reaction yielding energy or products that cause further reactions of the same kind
— chain–re·act \'chān-rē-'akt\ *intransitive verb*

chain rule *noun* (circa 1847)
: a mathematical rule concerning the differentiation of a function of a function (as $f[u(x)]$) by which under suitable conditions of continuity and differentiability one function is differentiated with respect to the second function considered as an independent variable and then the second function is differentiated with respect to its independent variable

chain saw *noun* (1944)
: a portable power saw that has teeth linked together to form an endless chain
— chain·saw \'chān-ˌsȯ\ *transitive verb*

chain–smoke \'chān-'smōk\ (1930)
intransitive verb
: to smoke especially cigarettes continually
transitive verb
: to smoke (as cigarettes) almost without interruption
— chain–smok·er \-ˌsmō-kər\ *noun*

chain stitch *noun* (1820)
1 : an ornamental stitch like chain links
2 : a machine stitch forming a chain on the underside of the work

chain store *noun* (1910)
: one of numerous usually retail stores having the same ownership and selling the same lines of goods

chain·wheel \'chān-ˌhwēl, -ˌwēl\ *noun* (1845)
: SPROCKET 1

¹chair \'cher, 'char\ *noun* [Middle English *chaiere,* from Old French, from Latin *cathedra,* from Greek *kathedra,* from *kata-* cata- + *hedra* seat — more at SIT] (13th century)
1 a : a seat typically having four legs and a back for one person **b :** ELECTRIC CHAIR — used with *the*

2 a : an official seat or a seat of authority, state, or dignity **b :** an office or position of authority or dignity **c :** PROFESSORSHIP ⟨holds a university *chair*⟩ **d :** CHAIRMAN 1
3 : a sedan chair
4 : a position of employment usually of one occupying a chair or desk; *specifically* **:** the position of a player in an orchestra or band
5 : any of various devices that hold up or support
usage see CHAIRMAN

²chair *transitive verb* (1552)
1 : to install in office
2 *chiefly British* **:** to carry on the shoulders in acclaim ⟨we *chaired* you through the market place —A. E. Housman⟩
3 : to preside as chairman of

chair car *noun* (1880)
1 : a railroad car having pairs of chairs with individually adjustable backs on each side of the aisle
2 : PARLOR CAR

chair·lift \'cher-ˌlift, 'char-\ *noun* (1940)
: a motor-driven conveyor consisting of a series of seats suspended from a cable and used for transporting skiers or sightseers up or down a long slope or mountainside

¹chair·man \-mən\ *noun* (1650)
1 a : the presiding officer of a meeting or an organization or committee **b :** the administrative officer of a department of instruction (as in a college)
2 : a carrier of a sedan chair ■
— chair·man·ship \-ˌship\ *noun*

²chairman *transitive verb* **-maned** *or* **-manned; -man·ing** *or* **-man·ning** (1888)
: CHAIR 3

chair·per·son \-ˌpər-sᵊn\ *noun* (1971)
: CHAIRMAN 1
usage see CHAIRMAN

chair·wom·an \-ˌwu̇-mən\ *noun* (1685)
: a woman who acts as chairman
usage see CHAIRMAN

chaise \'shāz\ *noun* [French, chair, chaise, alteration of Old French *chaiere*] (1701)
1 : any of various light horse-drawn vehicles: as **a :** a two-wheeled carriage for one or two persons with a folding top **b :** POST CHAISE
2 : CHAISE LONGUE

chaise longue \'shāz-'lȯŋ\ *noun, plural* **chaise longues** *also* **chaises longues** \'shāz-'lȯŋ(z)\ [French, literally, long chair] (1800)
: a long reclining chair

chaise lounge \'shāz-'lau̇nj, 'chās-\ *noun* [by folk etymology from French *chaise longue*] (circa 1906)
: CHAISE LONGUE

chak·ra \'chä-krə, 'shä-, 'chə-\ *noun* [Sanskrit *cakra,* literally, wheel — more at WHEEL] (1888)
: any of several points of physical or spiritual energy in the human body according to yoga philosophy

cha·la·za \kə-'lā-zə, -'la-\ *noun, plural* **-zae** \-ˌzē\ *or* **-zas** [New Latin, from Greek, hailstone] (circa 1704)
1 : either of two spiral bands in the white of a bird's egg that extend from the yolk and attach to opposite ends of the lining membrane — see EGG illustration
2 : the basal part of a plant ovule where the nucellus is fused to the surrounding integument and to which the funiculus is usually attached
— cha·la·zal \-'lā-zəl, -'la-\ *adjective*

Chal·ce·do·ni·an \ˌkal-sə-'dō-nē-ən\ *adjective* (1788)
: of or relating to Chalcedon or the ecumenical council held there in A.D. 451 declaring Monophysitism heretical
— Chalcedonian *noun*

chal·ce·do·ny \kal-'se-dᵊn-ē, chal-; 'kal-sə-ˌdō-nē, 'chal-, -ˌdä-\ *noun, plural* **-nies** [Middle English *calcedonie,* a precious stone, from Late Latin *chalcedonius,* from Greek *Chalkēdōn* Chalcedon] (13th century)
: a translucent quartz that is commonly pale blue or gray with nearly waxlike luster
— chal·ce·don·ic \ˌkal-sə-'dä-nik\ *adjective*

chal·cid \'kal-səd\ *noun* [ultimately from Greek *chalkos* copper, bronze] (1893)
: any of a large superfamily (Chalcidoidea) of mostly minute hymenopterous insects parasitic in the larval state on the larvae or pupae of other insects
— chalcid *adjective*

chal·co·cite \'kal-kə-ˌsīt\ *noun* [alteration of *chalcosine,* from French, irregular from Greek *chalkos*] (1868)
: a black or gray lustrous metallic mineral that consists of a sulfide of copper and is an important copper ore

chal·co·gen \'kal-kə-jən\ *noun* [International Scientific Vocabulary *chalk-* bronze, ore (from Greek *chalkos*) + *-gen;* from the occurrence of oxygen and sulfur in many ores] (circa 1961)
: any of the elements oxygen, sulfur, selenium, and tellurium

chal·co·gen·ide \-jə-ˌnīd\ *noun* (1945)
: a binary compound of a chalcogen with a more electropositive element or radical

chal·co·py·rite \ˌkal-kə-'pī-ˌrīt\ *noun* [New Latin *chalcopyrites,* from Greek *chalkos* + Latin *pyrites*] (1835)
: a yellow mineral consisting of copper-iron sulfide and constituting an important ore of copper

Chal·da·ic \kal-'dā-ik, kȯl-, käl-\ *adjective or noun* (1662)
: CHALDEAN

Chal·de·an \kal-'dē-ən, kȯl-, käl-\ *noun* [Latin *Chaldaeus* Chaldean, astrologer, from Greek *Chaldaios,* from *Chaldaia* Chaldea, region of ancient Babylonia] (1581)
1 a : a member of an ancient Semitic people that became dominant in Babylonia **b :** the Semitic language of the Chaldeans
2 : a person versed in the occult arts
— Chaldean *adjective*

Chal·dee \'kal-ˌdē, 'kȯl-, 'käl-\ *noun* [Middle English *Caldey,* probably from Middle French *chaldée,* from Latin *Chaldaeus*] (14th century)
1 : CHALDEAN 1a
2 : the Aramaic vernacular that was the original language of some parts of the Bible

chal·dron \'chȯl-drən, 'chäl-\ *noun* [Middle French *chauderon,* from *chaudere* pot, from Late Latin *caldaria* — more at CAULDRON] (1615)
: any of various old units of measure varying from 32 to 72 imperial bushels

cha·let \sha-'lā, 'sha-(ˌ)lā\ *noun* [French] (1782)
1 : a remote herdsman's hut in the Alps
2 a : a Swiss dwelling with unconcealed structural members and a wide overhang at the front and sides **b :** a cottage or house in chalet style

chalet 2a

chal·ice \'cha-ləs\ *noun* [Middle English, from Anglo-French, from Latin *calic-, calix;* akin to Greek *kalyx* calyx] (14th century)
1 : a drinking cup **:** GOBLET; *especially* **:** the eucharistic cup
2 : the cup-shaped interior of a flower

¹chalk \'chȯk\ *noun* [Middle English, from Old English *cealc,* from Latin *calc-, calx* lime; akin to Greek *chalix* pebble] (before 12th century)
1 a : a soft white, gray, or buff limestone composed chiefly of the shells of foraminifers **b :** chalk or a material resembling chalk especially when used in the form of a crayon
2 a : a mark made with chalk **b** *British* **:** a point scored in a game
— chalky \'chȯ-kē\ *adjective*

²chalk (1580)
transitive verb
1 : to write or draw with chalk
2 : to rub or mark with chalk
3 a : to delineate roughly **:** SKETCH ⟨*chalk* out a plan of attack⟩ **b :** to set down or add up with or as if with chalk **:** RECORD — usually used with *up* ⟨*chalk* up a victory⟩
intransitive verb
: to become chalky

chalk·board \'chȯk-ˌbōrd, -ˌbȯrd\ *noun* (1936)
: BLACKBOARD

chalk talk \'chȯk-ˌtȯk\ *noun* (1881)
: a talk or lecture illustrated at a blackboard

chalk up *transitive verb* (1826)
1 : ASCRIBE, CREDIT
2 : ATTAIN, ACHIEVE ⟨*chalk up* a record score for the season⟩

chal·lah *also* **chal·la** \'k̲ä-lə, 'hä-\ *noun* [Yiddish *khale,* from Hebrew *ḥallāh*] (1927)
: egg-rich yeast-leavened bread that is usually braided or twisted before baking and is traditionally eaten by Jews on the Sabbath and holidays

¹chal·lenge \'cha-lənj\ *verb* **chal·lenged; chal·leng·ing** [Middle English *chalengen* to accuse, from Old French *chalengier,* from Latin *calumniari* to accuse falsely, from *calumnia* calumny] (13th century)
transitive verb
1 : to demand as due or deserved **:** REQUIRE ⟨an event that *challenges* explanation⟩
2 : to order to halt and prove identity ⟨the sentry *challenged* the stranger⟩
3 : to dispute especially as being unjust, invalid, or outmoded **:** IMPUGN ⟨new data that *challenges* old assumptions⟩
4 : to question formally the legality or legal qualifications of
5 a : to confront or defy boldly **:** DARE **b :** to call out to duel or combat **c :** to invite into competition
6 : to arouse or stimulate especially by presenting with difficulties
7 : to administer a physiological and especially an immunologic challenge to (an organism or cell)
intransitive verb
1 : to make or present a challenge

□ USAGE
chairman *Chairman, chair, chairperson,* and *chairwoman* are available alternative terms for a presiding or administrative officer. *Chair* has been disparaged by some as an ugly creation of the women's liberation movement. On the contrary, it is a standard parliamentary term and has been in use since the second half of the 17th century. *Chairperson,* a much more recent coinage, gained rapid acceptance in spite of strong initial opposition especially from usage commentators and newspaper editors; it is now nearly as frequent as *chairman.* Our most recent evidence shows some current variation in use. *Chairman* is preferred in reference to a corporate office, regardless of the sex of the officer. *Chairman, chairperson,* and *chairwoman* are all commonly used in reference to political and social functions. *Chairman* and *chairperson* are common with reference to academic office, but *chairwoman* is not. *Chairman* and *chairperson* are applied to both men and women; when a name is mentioned with *chairperson,* it is slightly more likely to be a woman than a man. *Chair* differs from the rest in that, while it is commonly used as a title and in parliamentary rules, it is much less frequent in running context.

2 : to take legal exception
— **chal·leng·er** *noun*

[2]**challenge** *noun* (14th century)
1 a : a summons that is often threatening, provocative, stimulating, or inciting; *specifically* **:** a summons to a duel to answer an affront **b :** an invitation to compete in a sport
2 a : a calling to account or into question **:** PROTEST **b :** an exception taken to a juror before he is sworn **c :** a sentry's command to halt and prove identity **d :** a questioning of the right or validity of a vote or voter
3 : a stimulating task or problem ⟨looking for new *challenges*⟩
4 : the process of provoking or testing physiological activity by exposure to a specific substance; *especially* **:** a test of immunity by exposure to an antigen

chal·leng·ing \-lən-jiŋ\ *adjective* (1842)
1 : arousing competitive interest, thought, or action ⟨a *challenging* course of study⟩
2 : invitingly provocative **:** FASCINATING ⟨a *challenging* personality⟩
— **chal·leng·ing·ly** \-jiŋ-lē\ *adverb*

chal·lis \'sha-lē\ *noun, plural* **chal·lises** \'sha-lēz\ [probably from the name *Challis*] (circa 1837)
: a lightweight soft clothing fabric made of cotton, wool, or synthetic yarns

cha·lone \'kā-ˌlōn, 'ka-\ *noun* [Greek *chalōn,* present participle of *chalan* to slacken] (1914)
: an internal secretion that is held to inhibit mitosis in a specific tissue — compare HORMONE

[1]**cha·ly·be·ate** \kə-'lē-bē-ət, -'li-\ *adjective* [probably from New Latin *chalybeatus,* irregular from Latin *chalybs* steel, from Greek *chalyb-, chalyps,* from *Chalybes,* ancient people in Asia Minor] (1634)
: impregnated with salts of iron; *also* **:** having a taste due to iron ⟨*chalybeate* springs⟩

[2]**chalybeate** *noun* (1667)
: a chalybeate liquid or medicine

cham \'kam\ *variant of* KHAN

cham·ae·phyte \'ka-mi-ˌfīt\ *noun* [Greek *chamai* on the ground + English *-phyte* — more at HUMBLE] (1913)
: a perennial plant that bears its perennating buds just above the surface of the soil

[1]**cham·ber** \'chām-bər\ *noun* [Middle English *chambre,* from Old French, from Late Latin *camera,* from Latin, arched roof, from Greek *kamara* vault] (13th century)
1 : ROOM; *especially* **:** BEDROOM
2 : a natural or artificial enclosed space or cavity
3 a : a hall for the meetings of a deliberative, legislative, or judicial body ⟨the senate *chamber*⟩ **b :** a room where a judge transacts business — usually used in plural **c :** the reception room of a person of rank or authority
4 a : a legislative or judicial body; *especially* **:** either of the houses of a bicameral legislature **b :** a voluntary board or council
5 a : the part of the bore of a gun that holds the charge **b :** a compartment in the cartridge cylinder of a revolver
— **cham·bered** \-bərd\ *adjective*

[2]**chamber** *transitive verb* **cham·bered; cham·ber·ing** \-b(ə-)riŋ\ (1575)
1 : to place in or as if in a chamber **:** HOUSE
2 : to serve as a chamber for; *especially* **:** to accommodate in the chamber of a firearm

[3]**chamber** *adjective* (1706)
: being, relating to, or performing chamber music

chambered nautilus *noun* (1858)
: NAUTILUS 1

cham·ber·lain \'chām-bər-lən\ *noun* [Middle English, from Old French *chamberlayn,* of Germanic origin; akin to Old High German *chamarling* chamberlain, from *chamara* chamber, from Late Latin *camera*] (13th century)
1 : an attendant on a sovereign or lord in his bedchamber
2 a : a chief officer in the household of a king or nobleman **b :** TREASURER
3 : an often honorary papal attendant; *specifically* **:** a priest having a rank of honor below domestic prelate

cham·ber·maid \-ˌmād\ *noun* (1587)
: a maid who makes beds and does general cleaning of bedrooms (as in a hotel)

chamber music *noun* (circa 1789)
: music and especially instrumental ensemble music intended for performance in a private room or small auditorium and usually having one performer for each part

chamber of commerce (1797)
: an association of businessmen to promote commercial and industrial interests in the community

chamber of horrors (1849)
: a place in which macabre or horrible objects are exhibited; *also* **:** a collection of such exhibits

chamber orchestra *noun* (circa 1927)
: a small orchestra usually with one player for each part

chamber pot *noun* (1540)
: a bedroom vessel for urination and defecation

cham·bray \'sham-ˌbrā, -brē\ *noun* [irregular from *Cambrai,* France] (1814)
: a lightweight clothing fabric with colored warp and white filling yarns

cha·me·leon \kə-'mēl-yən\ *noun, often attributive* [Middle English *camelion,* from Middle French, from Latin *chamaeleon,* from Greek *chamaileōn,* from *chamai* on the ground + *leōn* lion — more at HUMBLE] (14th century)
1 : any of a family (Chamaeleontidae) of chiefly arboreal Old World lizards with prehensile tail, independently movable eyeballs, and unusual ability to change the color of the skin
2 a : a person given to often expedient or facile change in ideas or character **b :** one that is subject to quick or frequent change especially in appearance
3 : AMERICAN CHAMELEON
— **cha·me·le·on·ic** \-ˌmē-lē-'ä-nik\ *adjective*
— **cha·me·leon·like** \-ˌlīk\ *adjective*

[1]**cham·fer** \'cham(p)-fər, 'cham-pər\ *transitive verb* **cham·fered; cham·fer·ing** \-f(ə-)riŋ, -p(ə-)riŋ\ [back-formation from *chamfering,* alteration of Middle French *chanfreint,* from past participle of *chanfraindre* to bevel, from *chant* edge (from Latin *canthus* iron tire) + *fraindre* to break, from Latin *frangere* — more at CANT, BREAK] (circa 1573)
1 : to cut a furrow in (as a column) **:** GROOVE
2 : to make a chamfer on **:** BEVEL

[2]**chamfer** *noun* (circa 1847)
: a beveled edge

cham·fron \'sham-frən, 'cham-\ *noun* [Middle English *shamfron,* from Middle French *chanfrein*] (15th century)
: the headpiece of a horse's bard

cham·ois \'sha-mē, *sense 1 also* sham-'wä\ *noun, plural* **cham·ois** *also* **cham·oix** *sense 1* 'sha-mē(z) *or* sham-'wä(z), *senses 2 & 3* 'sha-mēz\ [Middle French, from Late Latin *camox*] (1560)
1 : a small goatlike antelope (*Rupicapra rupicapra*) of mountainous regions from southern Europe to the Caucasus
2 *also* **cham·my** *or* **sham·my** \'sha-mē\ **:** a soft pliant leather prepared from the skin of the chamois or from sheepskin
3 : a cotton fabric made in imitation of chamois leather

cham·o·mile \'ka-mə-ˌmīl, -ˌmēl\ *noun* [Middle English *camemille,* from Medieval Latin *camomilla,* modification of Latin *chamaemelon,* from Greek *chamaimēlon,* from *chamai* + *mēlon* apple] (13th century)
1 : a composite herb (*Chamaemelum nobile* synonym *Anthemis nobilis*) of Europe and North Africa with strong-scented foliage and flower heads that contain a bitter medicinal substance
2 : any of several composite plants (genera *Matricaria* and *Anthemis*) related to chamomile; *especially* **:** a Eurasian herb (*M. recutita* synonym *M. chamomilla*) naturalized in North America

chamomile

[1]**champ** \'champ, 'chämp, 'chȯmp\ *verb* [perhaps imitative] (14th century)
transitive verb
1 : CHOMP
2 : MASH, TRAMPLE
intransitive verb
1 : to make biting or gnashing movements
2 : to show impatience of delay or restraint — usually used in the phrase *champing at the bit* ⟨he was *champing* at the bit to begin⟩

[2]**champ** \'champ\ *noun* (1868)
: CHAMPION

cham·pac *or* **cham·pak** \'cham-ˌpak, 'chəm-(ˌ)pək\ *noun* [Hindi & Sanskrit; Hindi *campak,* from Sanskrit *campaka*] (circa 1770)
: an Asian tree (*Michelia champaca*) of the magnolia family with yellow flowers

cham·pagne \sham-'pān\ *noun* [French, from *Champagne,* France] (1664)
1 : a white sparkling wine made in the old province of Champagne, France; *also* **:** a similar wine made elsewhere
2 : a pale orange yellow to light grayish yellowish brown

cham·paign \sham-'pān\ *noun* [Middle English *champaine,* from Middle French *champagne,* from Late Latin *campania* — more at CAMPAIGN] (15th century)
1 : an expanse of level open country **:** PLAIN
2 *archaic* **:** BATTLEFIELD
— **champaign** *adjective*

cham·pers \'sham-pərz\ *noun plural but singular in construction* [by alteration] (1955)
British **:** CHAMPAGNE 1

cham·per·ty \'cham-pər-tē\ *noun* [Middle English *champartie,* from Middle French *champart* field rent, from *champ* field (from Latin *campus*) + *part* portion — more at PART] (15th century)
: a proceeding by which a person not a party in a suit bargains to aid in or carry on its prosecution or defense in consideration of a share of the matter in suit
— **cham·per·tous** \-pər-təs\ *adjective*

cham·pi·gnon \(ˌ)sham-pē-'nyōn, (ˌ)shäm-, -'nyō^{n}; sham-'pin-yən, cham-\ *noun* [French, from Middle French, alteration of *champigneul,* ultimately from Late Latin *campania*] (1670)
: an edible fungus; *especially* **:** MEADOW MUSHROOM

[1]**cham·pi·on** \'cham-pē-ən\ *noun* [Middle English, from Old French, from Medieval Latin *campion-, campio,* of West Germanic origin; akin to Old English *cempa* warrior] (13th century)
1 : WARRIOR, FIGHTER
2 : a militant advocate or defender ⟨a *champion* of civil rights⟩
3 : one that does battle for another's rights or honor ⟨God will raise me up a *champion* —Sir Walter Scott⟩
4 : a winner of first prize or first place in competition; *also* **:** one who shows marked superiority ⟨a *champion* at selling⟩

[2]**champion** *transitive verb* (1605)
1 *archaic* **:** CHALLENGE, DEFY

2 : to protect or fight for as a champion
3 : to act as militant supporter of **:** UPHOLD, ADVOCATE ⟨always *champions* the cause of the underdog⟩
synonym see SUPPORT

cham·pi·on·ship \-ˌship\ *noun* (1825)
1 : designation as champion
2 : the act of championing **:** DEFENSE ⟨his *championship* of freedom of speech⟩
3 : a contest held to determine a champion

champ·le·vé \ˌshä^{n}l(-ə)-ˈvā\ *adjective* [French] (1856)
: of, relating to, or being a style of enamel decoration in which the enamel is applied and fired in cells depressed (as by incising) into a metal background — compare CLOISONNÉ
— champlevé *noun*

[1]**chance** \ˈchan(t)s\ *noun* [Middle English, from Old French, from (assumed) Vulgar Latin *cadentia* fall, from Latin *cadent-, cadens,* present participle of *cadere* to fall; perhaps akin to Sanskrit *śad-* to fall off] (14th century)
1 a : something that happens unpredictably without discernible human intention or observable cause **b :** the assumed impersonal purposeless determiner of unaccountable happenings **:** LUCK **c :** the fortuitous or incalculable element in existence **:** CONTINGENCY
2 : a situation favoring some purpose **:** OPPORTUNITY ⟨needed a *chance* to relax⟩
3 : a fielding opportunity in baseball
4 a : the possibility of a particular outcome in an uncertain situation; *also* **:** the degree of likelihood of such an outcome ⟨a small *chance* of success⟩ **b** *plural* **:** the more likely indications ⟨*chances* are he's already gone⟩
5 a : RISK ⟨not taking any *chances*⟩ **b :** a raffle ticket
— chance *adjective*
— by chance : in the haphazard course of events ⟨they met *by chance*⟩

[2]**chance** *verb* **chanced; chanc·ing** (14th century)
intransitive verb
1 a : to take place or come about by chance **:** HAPPEN **b :** to be found by chance **c :** to have the good or bad luck ⟨we *chanced* to meet⟩
2 : to come or light by chance
transitive verb
1 : to leave the outcome of to chance
2 : to accept the hazard of **:** RISK
— chance one's arm *British* **:** to take a risk

chance·ful \ˈchan(t)s-fəl\ *adjective* (1594)
1 *archaic* **:** CASUAL
2 : EVENTFUL

chan·cel \ˈchan(t)-səl\ *noun* [Middle English, from Middle French, from Late Latin *cancellus* lattice, from Latin *cancelli;* from the latticework enclosing it — more at CANCEL] (14th century)
: the part of a church containing the altar and seats for the clergy and choir

chan·cel·lery *or* **chan·cel·lory** \ˈchan(t)-s(ə-)lə-rē, -səl-rē\ *noun, plural* **-ler·ies** *or* **-lor·ies** (14th century)
1 a : the position, court, or department of a chancellor **b :** the building or room where a chancellor has his office
2 : the office of secretary of the court of a person high in authority
3 : the office or staff of an embassy or consulate

chan·cel·lor \ˈchan(t)-s(ə-)lər\ *noun* [Middle English *chanceler,* from Old French *chancelier,* from Late Latin *cancellarius* doorkeeper, secretary, from *cancellus*] (14th century)
1 a : the secretary of a nobleman, prince, or king **b :** the lord chancellor of Great Britain **c** *British* **:** the chief secretary of an embassy **d :** a Roman Catholic priest heading the office in which diocesan business is transacted and recorded
2 a : the titular head of a British university **b** (1) **:** a university president (2) **:** the chief executive officer in some state systems of higher education
3 a : a lay legal officer or adviser of an Anglican diocese **b :** a judge in a court of chancery or equity in various states of the U.S.
4 : the chief minister of state in some European countries
— chan·cel·lor·ship \-ˌship\ *noun*

chancellor of the exchequer *often C&E capitalized* (1672)
: a member of the British cabinet in charge of the public income and expenditure

chance–med·ley \ˈchan(t)s-ˈmed-lē\ *noun* [Anglo-French *chance medlée* mingled chance] (15th century)
1 : accidental homicide not entirely without fault of the killer but without evil intent
2 : haphazard action **:** CONFUSION

chan·cery \ˈchan(t)-sə-rē, ˈchan(t)s-rē\ *noun, plural* **-cer·ies** [Middle English *chancerie,* alteration of *chancellerie* chancellery, from Old French, from *chancelier*] (14th century)
1 a *capitalized* **:** a high court of equity in England and Wales with common-law functions and jurisdiction over causes in equity **b :** a court of equity in the American judicial system **c :** the principles and practice of judicial equity
2 : a record office for public archives or those of ecclesiastical, legal, or diplomatic proceedings
3 a : a chancellor's court or office or the building in which it is located **b :** the office in which the business of a Roman Catholic diocese is transacted and recorded **c :** the office of an embassy **:** CHANCELLERY 3
— in chancery 1 : in litigation in a court of chancery; *also* **:** under the superintendence of the lord chancellor ⟨a ward *in chancery*⟩ **2 :** in a hopeless predicament

chan·cre \ˈshaŋ-kər\ *noun* [French, from Latin *cancer*] (circa 1605)
: a primary sore or ulcer at the site of entry of a pathogen (as in tularemia); *especially* **:** the initial lesion of syphilis
— chan·crous \-k(ə-)rəs\ *adjective*

chan·croid \ˈshaŋ-ˌkrȯid\ *noun* (1861)
: a venereal disease caused by a hemophilic bacterium (*Hemophilus ducreyi*) and characterized by chancres unlike those of syphilis in lacking firm indurated margins — called also *soft chancre*
— chan·croi·dal \shaŋ-ˈkrȯi-d^{ə}l\ *adjective*

chancy \ˈchan(t)-sē\ *adjective* **chanc·i·er; -est** (1513)
1 *Scottish* **:** bringing good luck **:** AUSPICIOUS
2 : uncertain in outcome or prospect **:** RISKY
3 : occurring by chance **:** HAPHAZARD
— chanc·i·ness *noun*

chan·de·lier \ˌshan-də-ˈlir\ *noun* [French, literally, candlestick, modification of Latin *candelabrum*] (1736)
: a branched often ornate lighting fixture suspended from a ceiling
— chan·de·liered \-ˈlird\ *adjective*

chandelier

chan·delle \shan-ˈdel, shä^{n}-\ *noun* [French, literally, candle] (1918)
: an abrupt climbing turn of an airplane in which the momentum of the plane is used to attain a higher rate of climb
— chandelle *intransitive verb*

chan·dler \ˈchan(d)-lər\ *noun* [Middle English *chandeler,* from Middle French *chandelier,* from Old French, from *chandelle* candle, from Latin *candela*] (14th century)
1 : a maker or seller of tallow or wax candles and usually soap
2 : a retail dealer in provisions and supplies or equipment of a specified kind ⟨a yacht *chandler*⟩

chan·dlery \-lə-rē\ *noun, plural* **-dler·ies** (15th century)
1 : a place where candles are kept
2 : the business of a chandler
3 : the commodities sold by a chandler

[1]**change** \ˈchānj\ *verb* **changed; chang·ing** [Middle English, from Old French *changier,* from Latin *cambiare* to exchange, probably of Celtic origin; akin to Old Irish *camm* crooked] (13th century)
transitive verb
1 a : to make different in some particular **:** ALTER ⟨never bothered to *change* the will⟩ **b :** to make radically different **:** TRANSFORM ⟨can't *change* human nature⟩ **c :** to give a different position, course, or direction to
2 a : to replace with another ⟨let's *change* the subject⟩ **b :** to make a shift from one to another **:** SWITCH ⟨always *changes* sides in an argument⟩ **c :** to exchange for an equivalent sum or comparable item **d :** to undergo a modification of ⟨foliage *changing* color⟩ **e :** to put fresh clothes or covering on ⟨*change* a bed⟩
intransitive verb
1 : to become different ⟨her mood *changes* every hour⟩
2 *of the moon* **:** to pass from one phase to another
3 : to shift one's means of conveyance **:** TRANSFER ⟨on the bus trip he had to *change* twice⟩
4 *of the voice* **:** to shift to lower register **:** BREAK
5 : to undergo transformation, transition, or substitution ⟨winter *changed* to spring⟩
6 : to put on different clothes
7 : EXCHANGE, SWITCH ⟨neither liked his seat so they *changed* with each other⟩ ☆
— chang·er *noun*
— change hands : to pass from the possession of one owner to that of another ⟨money *changes hands* many times⟩

[2]**change** *noun* (13th century)
1 : the act, process, or result of changing: as **a :** ALTERATION ⟨a *change* in the weather⟩ **b :** TRANSFORMATION ⟨a time of vast social *change*⟩ ⟨going through *changes*⟩ **c :** SUBSTITUTION ⟨a *change* of scenery⟩ **d :** the passage of the moon from one monthly revolution to another; *also* **:** the passage of the moon from one phase to another
2 : a fresh set of clothes
3 *British* **:** EXCHANGE 5a
4 a : money in small denominations received in exchange for an equivalent sum in larger denominations **b :** money returned when a payment exceeds the amount due **c :** coins especially of low denominations ⟨a pocketful of *change*⟩ **d :** a negligible additional amount ⟨only six minutes and *change* left in the game⟩
5 : an order in which a set of bells is struck in change ringing
6 : CHANGE-UP

☆ SYNONYMS
Change, alter, vary, modify mean to make or become different. CHANGE implies making either an essential difference often amounting to a loss of original identity or a substitution of one thing for another ⟨*changed* the shirt for a larger size⟩. ALTER implies a difference in some particular respect without suggesting loss of identity ⟨slightly *altered* the original design⟩. VARY stresses a breaking away from sameness, duplication, or exact repetition ⟨*vary* your daily routine⟩. MODIFY suggests a difference that limits, restricts, or adapts to a new purpose ⟨*modified* the building for use by the handicapped⟩.

change·able \'chān-jə-bəl\ *adjective* (13th century)
: capable of change: as **a :** able or apt to vary ⟨*changeable* weather⟩ **b :** subject to change **:** ALTERABLE **c :** FICKLE **d :** IRIDESCENT
— change·abil·i·ty \,chān-jə-'bi-lə-tē\ *noun*
— change·able·ness \'chān-jə-bəl-nəs\ *noun*
— change·ably \-blē\ *adverb*

change·ful \'chānj-fəl\ *adjective* (1591)
: notably variable **:** UNCERTAIN
— change·ful·ly \-fə-lē\ *adverb*
— change·ful·ness *noun*

change·less \'chānj-ləs\ *adjective* (1580)
: marked by the absence of change **:** CONSTANT
— change·less·ly *adverb*
— change·less·ness *noun*

change·ling \'chānj-liŋ\ *noun* (1537)
1 *archaic* **:** TURNCOAT
2 : a child secretly exchanged for another in infancy
3 *archaic* **:** IMBECILE
— changeling *adjective*

change off *intransitive verb* (1873)
1 : to alternate with another at doing an act
2 : to alternate between two different acts or instruments or between an action and a rest period

change of heart (circa 1828)
: a reversal in position or attitude

change of life (1834)
: [2]CLIMACTERIC 2

change of pace (1912)
1 : CHANGE-UP
2 : an interruption of continuity by a shift to a different activity

change·over \'chān-,jō-vər\ *noun* (1907)
1 : CONVERSION, TRANSITION
2 : a pause in a tennis match during which the players change sides of the court

change ringing *noun* (1872)
: the art or practice of ringing a set of tuned bells (as in the bell tower of a church) in continually varying order

change–up \'chān-,jəp\ *noun* (1949)
: a slow pitch in baseball thrown with the same motion as a fastball in order to deceive the batter

[1]**chan·nel** \'cha-n^{ə}l\ *noun* [Middle English *chanel,* from Middle French, from Latin *canalis* channel — more at CANAL] (14th century)
1 a : the bed where a natural stream of water runs **b :** the deeper part of a river, harbor, or strait **c :** a strait or narrow sea between two close landmasses **d :** a means of communication or expression: as (1) **:** a path along which information (as data or music) in the form of an electrical signal passes (2) *plural* **:** a fixed or official course of communication ⟨went through established military *channels* with his grievances⟩ **e :** a way, course, or direction of thought or action ⟨new *channels* of exploration⟩ **f :** a band of frequencies of sufficient width for a single radio or television communication **g :** CHANNELER
2 a : a usually tubular enclosed passage **:** CONDUIT **b :** a passage created in a selectively permeable membrane by a conformational change in membrane proteins
3 : a long gutter, groove, or furrow
4 : a metal bar of flattened U-shaped section

[2]**channel** *transitive verb* **-neled** *or* **-nelled; -nel·ing** *or* **-nel·ling** (15th century)
1 a : to form, cut, or wear a channel in **b :** to make a groove in ⟨*channel* a chair leg⟩
2 : to convey or direct into or through a channel ⟨*channel* his energy into constructive activities⟩
3 : to serve as a channeler or intermediary for

[3]**channel** *noun* [alteration of *chainwale,* from [1]*chain* + [1]*wale*] (1769)
: one of the flat ledges of heavy plank or metal bolted edgewise to the outside of a ship to increase the spread of the shrouds

channel bass *noun* (1887)
: a large coppery drum (*Sciaenops ocellatus*) chiefly of the Atlantic coast of North America that has a black spot at the base of the tail and is an important game and food fish — called also *red drum, redfish*

channel catfish *noun* (1820)
: a large black-spotted catfish (*Ictalurus punctatus*) that is an important freshwater food fish of the U.S. and Canada — called also *channel cat*

chan·nel·er \'cha-n^{ə}l-ər\ *noun* (1987)
: a person who conveys thoughts or energy from a source believed to be outside the person's body or conscious mind; *specifically* **:** one who speaks for nonphysical beings or spirits

chan·nel·ize \'cha-n^{ə}l-,īz\ *transitive verb* **-ized; -iz·ing** (1609)
1 : CHANNEL 1, 2
2 : to straighten by means of a channel ⟨*channelize* a stream⟩
— chan·nel·i·za·tion \,cha-n^{ə}l-ə-'zā-shən\ *noun*

channel surfing *noun* (1990)
: the act or practice of scanning through television programs usually by use of a remote control to find something of interest
— channel surf *intransitive verb*
— channel surfer *noun*

chan·son \shä^{n}-'sō^{n}\ *noun, plural* **chansons** \-'sō^{n}(z)\ [French, from Latin *cantion-, cantio,* from *canere*] (1602)
: SONG; *specifically* **:** a music-hall or cabaret song

chanson de geste \-də-'zhest\ *noun, plural* **chansons de geste** *same*\ [French, literally, song of heroic deeds] (1868)
: any of several Old French epic poems of the 11th to the 13th centuries

chan·son·nier \,shä^{n}-sō-'nyā\ *noun* [French, from *chanson*] (1887)
: a writer or singer of chansons; *especially* **:** a cabaret singer

[1]**chant** \'chant\ *verb* [Middle English *chaunten,* from Middle French *chanter,* from Latin *cantare,* frequentative of *canere* to sing; akin to Old English *hana* rooster, Old Irish *canid* he sings] (14th century)
intransitive verb
1 : to make melodic sounds with the voice; *especially* **:** to sing a chant
2 : to recite in a monotonous repetitive tone
transitive verb
1 : to utter as in chanting
2 : to celebrate or praise in song or chant

[2]**chant** *noun* (1671)
1 : SONG
2 a : PLAINSONG **b :** a rhythmic monotonous utterance or song **c :** a composition for chanting

chant·er \'chan-tər\ *noun* (14th century)
1 : one that chants: **a :** CHORISTER **b :** CANTOR
2 : the chief singer in a chantry
3 : the reed pipe of a bagpipe with finger holes on which the melody is played

chan·te·relle \,shan-tə-'rel, ,shän-\ *noun* [French] (1775)
: a fragrant edible mushroom (*Cantharellus cibarius*) of rich yellow color

chan·teuse \shan-'tüz, shä^{n}-'tə(r)z\ *noun, plural* **chan·teuses** \-'tüz, -'tü-zəz, -'tə(r)z, -'tə(r)-zəz\ [French, feminine of *chanteur* singer, from *chanter*] (1888)
: a woman who is a concert or nightclub singer

chan·tey *or* **chan·ty** \'shan-tē, 'chan-\ *noun, plural* **chanteys** *or* **chanties** [modification of French *chanter*] (1856)
: a song sung by sailors in rhythm with their work

chan·ti·cleer \,chan-tə-'klir, ,shan-\ *noun* [Middle English *Chantecleer,* rooster in verse narratives, from Old French *Chantecler,* rooster in the *Roman de Renart*] (14th century)
: ROOSTER

Chan·til·ly lace \shan-'ti-lē-\ *noun* [*Chantilly,* France] (1848)
: a delicate silk, linen, or synthetic lace having a six-sided mesh ground and a floral or scrolled design — called also *Chantilly*

chan·try \'chan-trē\ *noun, plural* **chantries** [Middle English *chanterie,* from Middle French, singing, from *chanter*] (14th century)
1 : an endowment for the chanting of masses commonly for the founder
2 : a chapel endowed by a chantry

Cha·nu·kah \'kä-nə-kə, 'hä-\ *variant of* HANUKKAH

cha·os \'kā-,äs\ *noun* [Latin, from Greek — more at GUM] (15th century)
1 *obsolete* **:** CHASM, ABYSS
2 a *often capitalized* **:** a state of things in which chance is supreme; *especially* **:** the confused unorganized state of primordial matter before the creation of distinct forms — compare COSMOS **b :** the inherent unpredictability in the behavior of a natural system (as the atmosphere, boiling water, or the beating heart)
3 a : a state of utter confusion **b :** a confused mass or mixture ⟨a *chaos* of television antennas⟩
— cha·ot·ic \kā-'ä-tik\ *adjective*
— cha·ot·i·cal·ly \-ti-k(ə-)lē\ *adverb*

[1]**chap** \'chap\ *noun* (14th century)
: a crack in or a sore roughening of the skin caused by exposure to wind or cold

[2]**chap** *verb* **chapped; chap·ping** [Middle English *chappen;* akin to Middle Dutch *cappen* to cut down] (15th century)
intransitive verb
: to open in cracks, slits, or chinks; *also* **:** to become cracked, roughened, or reddened especially by the action of wind or cold ⟨hands often *chap* in winter⟩
transitive verb
: to cause to chap ⟨wind-*chapped* lips⟩

[3]**chap** \'chäp, 'chap\ *noun* [origin unknown] (1555)
1 a : the fleshy covering of a jaw; *also* **:** JAW — usually used in plural ⟨a wolf's *chaps*⟩
2 : the forepart of the face — usually used in plural

[4]**chap** *noun* [short for *chapman*] (1716)
1 : FELLOW 4c
2 *Southern & Midland* **:** BABY, CHILD

chap·a·ra·jos *or* **chap·a·re·jos** \,sha-pə-'rā-(,)ōs, -'rä-\ *noun plural* [modification of Mexican Spanish *chaparreras,* from *chaparro*] (1887)
: CHAPS

chap·ar·ral \,sha-pə-'ral, -'rel\ *noun* [Spanish, from *chaparro* dwarf evergreen oak, from Basque *txapar*] (1845)
1 : a thicket of dwarf evergreen oaks; *broadly* **:** a dense impenetrable thicket of shrubs or dwarf trees
2 : an ecological community composed of shrubby plants adapted to dry summers and moist winters that occurs especially in southern California

chaparral cock *noun* (1853)
: ROADRUNNER — called also *chaparral bird*

cha·pa·ti *also* **chap·pa·ti** \chə-'pä-tē\ *noun, plural* **chapatis** *also* **chappatis** [Hindi *capātī*] (1810)
: a round flat unleavened bread of India that is usually made of whole wheat flour and cooked on a griddle

chap·book \'chap-,bu̇k\ *noun* [*chap*man + *book*] (1798)
: a small book containing ballads, poems, tales, or tracts

\ə\ **abut** \$^{ə}$\ **kitten** \ər\ **further** \a\ **ash** \ā\ **ace** \ä\ **mop, mar** \au̇\ **out** \ch\ **chin** \e\ **bet** \ē\ **easy** \g\ **go** \i\ **hit** \ī\ **ice** \j\ **job** \ŋ\ **sing** \ō\ **go** \ȯ\ **law** \ȯi\ **boy** \th\ **thin** \t͟h\ **the** \ü\ **loot** \u̇\ **foot** \y\ **yet** \zh\ **vision** *see also* Guide to Pronunciation

chape \'chāp, 'chap\ *noun* [Middle English, scabbard, from Middle French, cape, from Late Latin *cappa*] (14th century)
: the metal mounting or trimming of a scabbard or sheath

cha·peau \sha-'pō, shə-\ *noun, plural* **cha·peaus** \-'pōz\ *or* **cha·peaux** \-'pō(z)\ [Middle French, from Old French *chapel* — more at CHAPLET] (1523)
: HAT

chap·el \'cha-pəl\ *noun* [Middle English, from Old French *chapele,* from Medieval Latin *cappella,* from diminutive of Late Latin *cappa* cloak; from the cloak of Saint Martin of Tours preserved as a sacred relic in a chapel built for that purpose] (13th century)
1 : a subordinate or private place of worship: as **a :** a place of worship serving a residence or institution **b :** a small house of worship usually associated with a main church **c :** a room or recess in a church for meditation and prayer or small religious services
2 : a place of worship used by a Christian group other than an established church
3 : a choir of singers belonging to a chapel
4 : a chapel service or assembly at a school or college
5 : an association of the employees in a printing office
6 a : FUNERAL HOME **b :** a room for funeral services in a funeral home ◆

chapel of ease (1538)
: a chapel or dependent church built to accommodate an expanding parish

¹chap·er·on *or* **chap·er·one** \'sha-pə-ˌrōn\ *noun* [French *chaperon,* literally, hood, from Middle French, head covering, from *chape*] (1720)
1 : a person (as a matron) who for propriety accompanies one or more young unmarried women in public or in mixed company
2 : an older person who accompanies young people at a social gathering to ensure proper behavior; *broadly* **:** one delegated to ensure proper behavior

²chaperon *or* **chaperone** *verb* **-oned; -on·ing** (1796)
transitive verb
1 : ESCORT
2 : to act as chaperon to or for
intransitive verb
: to act as a chaperon
— chap·er·on·age \-ˌrō-nij\ *noun*

chap·fall·en \'chap-ˌfȯ-lən, 'chäp-\ *adjective* (1598)
1 : having the lower jaw hanging loosely
2 : cast down in spirit **:** DEPRESSED

chap·i·ter \'cha-pə-tər\ *noun* [Middle English *chapitre,* from Middle French, alteration of Old French *chapitle,* from Latin *capitulum,* literally, little head] (15th century)
: the capital of a column

chap·lain \'cha-plən\ *noun* [Middle English *chapelain,* from Old French, from Medieval Latin *cappellanus,* from *cappella*] (14th century)
1 : a clergyman in charge of a chapel
2 : a clergyman officially attached to a branch of the military, to an institution, or to a family or court
3 : a person chosen to conduct religious exercises (as at a meeting of a club or society)
4 : a clergyman appointed to assist a bishop (as at a liturgical function)
— chap·lain·cy \-sē\ *noun*

chap·let \'chap-lət\ *noun* [Middle English *chapelet,* from Middle French, from Old French, diminutive of *chapel* hat, garland, from Medieval Latin *cappellus* head covering, from Late Latin *cappa*] (14th century)
1 : a wreath to be worn on the head
2 a : a string of beads **b :** a part of a rosary comprising five decades
3 : a small molding carved with small decorative forms
— chap·let·ed \-lə-təd\ *adjective*

chap·man \'chap-mən\ *noun* [Middle English, from Old English *cēapman,* from *cēap* trade + *man* — more at CHEAP] (before 12th century)
1 *archaic* **:** MERCHANT, TRADER
2 *British* **:** PEDDLER

chaps \'shaps, 'chaps\ *noun plural* [modification of Mexican Spanish *chaparreras*] (1844)
: leather leggings joined by a belt or lacing, often having flared outer flaps, and worn over the trousers (as by western ranch hands)

chaps

chap·ter \'chap-tər\ *noun* [Middle English *chapitre,* from Old French, from Late Latin *capitulum* division of a book & Medieval Latin, meeting place of canons, from Latin, diminutive of *capit-, caput* head — more at HEAD] (13th century)
1 a : a main division of a book **b :** something resembling a chapter in being a significant specified unit ⟨a new *chapter* in my life⟩
2 a : a regular meeting of the canons of a cathedral or collegiate church or of the members of a religious house **b :** the body of canons of a cathedral or collegiate church
3 : a local branch of an organization

chapter and verse *noun* (1628)
1 : the exact reference or source of information or justification for an assertion ⟨clinched their arguments by citing *chapter and verse* —J. M. Burns⟩
2 : full precise information or detail ⟨can give *chapter and verse* on the effects of diverting defense spending —Horace Sutton⟩
— chapter and verse *adverb*

¹char \'chär\ *noun, plural* **char** *or* **chars** [origin unknown] (1662)
: any of a genus (*Salvelinus*) of small-scaled trouts with light-colored spots

²char *verb* **charred; char·ring** [*charcoal*] (1679)
transitive verb
1 : to convert to charcoal or carbon usually by heat **:** BURN
2 : to burn slightly or partly **:** SCORCH ⟨the fire *charred* the beams⟩
intransitive verb
: to become charred

³char *noun* (1879)
: a charred substance **:** CHARCOAL; *specifically* **:** a combustible residue remaining after the destructive distillation of coal

⁴char *intransitive verb* **charred; char·ring** [*charwoman*] (1732)
: to work as a cleaning woman

⁵char *noun* [by shortening] (1906)
British **:** CHARWOMAN

char·a·banc \'shar-ə-ˌbaŋ, -ˌbäŋ\ *noun* [French *char à bancs,* literally, wagon with benches] (1914)
British **:** a sight-seeing motor coach

char·a·cin \'kar-ə-sən\ *noun* [ultimately from Greek *charak-, charax* pointed stake, a fish] (1882)
: any of a family (Characidae) of usually small brightly colored tropical freshwater fishes that includes many aquarium fishes
— characin *adjective*

¹char·ac·ter \'kar-ik-tər\ *noun* [Middle English *caracter,* from Middle French *caractère,* from Latin *character* mark, distinctive quality, from Greek *charaktēr,* from *charassein* to scratch, engrave; perhaps akin to Lithuanian *žeyti* to scratch] (14th century)
1 a : a conventionalized graphic device placed on an object as an indication of ownership, origin, or relationship **b :** a graphic symbol (as a hieroglyph or alphabet letter) used in writing or printing **c :** a magical or astrological emblem **d :** ALPHABET **e** (1) **:** WRITING, PRINTING (2) **:** style of writing or printing (3) **:** CIPHER **f :** a symbol (as a letter or number) that represents information; *also* **:** a representation of such a character that may be accepted by a computer
2 a : one of the attributes or features that make up and distinguish an individual **b** (1) **:** a feature used to separate distinguishable things into categories; *also* **:** a group or kind so separated ⟨advertising of a very primitive *character*⟩ (2) **:** the detectable expression of the action of a gene or group of genes (3) **:** the aggregate of distinctive qualities characteristic of a breed, strain, or type ⟨a wine of great *character*⟩ **c :** the complex of mental and ethical traits marking and often individualizing a person, group, or nation **d :** main or essential nature especially as strongly marked and serving to distinguish ⟨excess sewage gradually changed the *character* of the lake⟩
3 : POSITION, CAPACITY ⟨his *character* as a town official⟩
4 : REFERENCE 4b
5 : REPUTATION
6 : moral excellence and firmness ⟨a man of sound *character*⟩
7 a : a person marked by notable or conspicuous traits ⟨quite a *character*⟩ **b :** one of the persons of a drama or novel **c :** the personality or part which an actor recreates **d :** characterization especially in drama or fiction **e :** PERSON, INDIVIDUAL ⟨some *character* just stole her purse⟩
8 : a short literary sketch of the qualities of a social type
synonym see DISPOSITION, QUALITY, TYPE
— char·ac·ter·less \-ləs\ *adjective*
— in character : in accord with a person's usual qualities or traits
— out of character : not in accord with a person's usual qualities or traits

²character *transitive verb* (1591)
1 *archaic* **:** ENGRAVE, INSCRIBE
2 a *archaic* **:** REPRESENT, PORTRAY **b :** CHARACTERIZE

³character *adjective* (1893)
1 : capable of portraying an unusual or eccentric personality often markedly different from the player ⟨a *character* actor⟩
2 : requiring the qualities of a character actor ⟨a *character* role⟩

character assassination *noun* (1949)
: the slandering of a person usually with the intention of destroying public confidence in that person

char·ac·ter·ful \'kar-ik-tər-fəl\ *adjective* (1901)
1 : markedly expressive of character ⟨a *characterful* face⟩

◇ WORD HISTORY
chapel One harsh winter while he was a youthful soldier in the Roman army, St. Martin of Tours (ca. 316–397) is said to have divided his military cloak in half with his sword, giving one part to a naked pauper at the gate of Amiens and keeping the other part for himself. After his death, a cloak supposedly belonging to the saint, to which his demonstration of charity gave special significance, was preserved as a relic by Frankish kings. In early Medieval Latin documents, St. Martin's cloak was referred to as *capella* or *cappella,* a diminutive of Late Latin *cappa,* also meaning "cloak." The meaning of *capella* was eventually expanded to include not only the cloak, but the entire royal treasury of saints' relics, and then the rooms in the royal palace set aside for private worship, where the relics were kept. By the early 9th century, *capella* could refer to any private place of worship. *Chapelle,* the vernacular form of *capella* in the medieval French of the Paris region, is first attested about 1100, and it had been borrowed into English by the early 13th century.

2 : marked by character ⟨a *characterful* decision⟩

¹char·ac·ter·is·tic \ˌkar-ik-tə-ˈris-tik\ *noun* (1664)
1 : a distinguishing trait, quality, or property
2 : the integral part of a common logarithm
3 : the smallest positive integer *n* which for an operation in a ring or field yields 0 when any element is used *n* times with the operation

²characteristic *adjective* (1665)
: revealing, distinguishing, or typical of an individual character ☆
— char·ac·ter·is·ti·cal·ly \-ti-k(ə-)lē\ *adverb*

characteristic equation *noun* (circa 1925)
: an equation in which the characteristic polynomial of a matrix is set equal to 0

characteristic polynomial *noun* (circa 1957)
: the determinant of a square matrix in which an arbitrary variable (as *x*) is subtracted from each of the elements along the principal diagonal

characteristic root *noun* (circa 1957)
: EIGENVALUE

characteristic value *noun* (1956)
: EIGENVALUE

characteristic vector *noun* (1957)
: EIGENVECTOR

char·ac·ter·iza·tion \ˌkar-ik-t(ə-)rə-ˈzā-shən\ *noun* (1814)
: the act of characterizing; *especially* **:** the artistic representation (as in fiction or drama) of human character or motives

char·ac·ter·ize \ˈkar-ik-tə-ˌrīz\ *transitive verb* **-ized; -iz·ing** (1633)
1 : to describe the character or quality of ⟨*characterizes* him as ambitious⟩
2 : to be a characteristic of **:** DISTINGUISH ⟨an era *characterized* by greed⟩

char·ac·ter·olog·i·cal \ˌkar-ik-t(ə-)rə-ˈlä-ji-kəl\ *adjective* [*characterology* study of character] (1916)
: of, relating to, or based on character or the study of character including its development and its differences in different individuals
— char·ac·ter·olog·i·cal·ly \-ˈlä-ji-k(ə-)lē\ *adverb*

character witness *noun* (1952)
: a person who gives evidence in a legal action concerning the reputation, conduct, and moral nature of a party

char·ac·tery \ˈkar-ik-t(ə-)rē, kə-ˈrak-\ *noun, plural* **-ter·ies** (1598)
: a system of written letters or symbols used in the expression of thought

cha·rade \shə-ˈrād, -ˈräd\ *noun* [French, from Provençal *charrado* chat, from *charrá* to chat, chatter] (1776)
1 : a word represented in riddling verse or by picture, tableau, or dramatic action
2 *plural* **:** a game in which some of the players try to guess a word or phrase from the actions of another player who may not speak
3 : an empty or deceptive act or pretense ⟨his concern was a *charade*⟩

cha·ras \ˈchär-əs\ *noun* [Hindi *caras*] (circa 1860)
: HASHISH

char·broil \ˈchär-ˌbrȯi(ə)l\ *transitive verb* (1968)
: to broil on a rack over hot charcoal
— char·broil·er \-ˌbrȯi-lər\ *noun*

¹char·coal \ˈchär-ˌkōl\ *noun* [Middle English *charcole*] (14th century)
1 : a dark or black porous carbon prepared from vegetable or animal substances (as from wood by charring in a kiln from which air is excluded)
2 a : a piece or pencil of fine charcoal used in drawing **b :** a charcoal drawing
3 : a dark gray

²charcoal *transitive verb* (1965)
: CHARBROIL

char·cu·te·rie \(ˌ)shär-ˌkü-tə-ˈrē\ *noun* [French, literally, pork-butcher's shop, from Middle French *chaircuiterie,* from *chaircutier* pork butcher, from *chair cuite* cooked meat] (circa 1858)
: a delicatessen specializing in dressed meats and meat dishes; *also* **:** the products sold in such a shop

chard \ˈchärd\ *noun* [modification of French *carde,* from Provençal *cardo,* from Latin *carduus* thistle, cardoon] (1664)
: SWISS CHARD

char·don·nay \ˌshär-dᵊn-ˈā\ *noun, often capitalized* [French] (circa 1941)
: a dry white table wine of Chablis type

chare \ˈchar, ˈcher\ *or* **char** \ˈchär\ *noun* [Middle English *char* turn, piece of work, from Old English *cierr;* akin to Old English *cierran* to turn] (before 12th century)
: CHORE

¹charge \ˈchärj\ *noun* [Middle English, from Old French, from *chargier*] (13th century)
1 a *obsolete* **:** a material load or weight **b :** a figure borne on a heraldic field
2 a : the quantity that an apparatus is intended to receive and fitted to hold **b :** a store or accumulation of impelling force ⟨the deeply emotional *charge* of the drama⟩ **c :** a definite quantity of electricity; *especially* **:** an excess or deficiency of electrons in a body **d :** THRILL, KICK ⟨got a *charge* out of the game⟩
3 a : OBLIGATION, REQUIREMENT **b :** MANAGEMENT, SUPERVISION ⟨has *charge* of the home office⟩ **c :** the ecclesiastical jurisdiction (as a parish) committed to a clergyman **d :** a person or thing committed to the care of another
4 a : INSTRUCTION, COMMAND **b :** instruction in points of law given by a court to a jury
5 a : EXPENSE, COST ⟨gave the banquet at his own *charge*⟩ **b :** the price demanded for something ⟨no admission *charge*⟩ **c :** a debit to an account ⟨the purchase was a *charge*⟩ **d :** the record of a loan (as of a book from a library) **e** *British* **:** an interest in property granted as security for a loan
6 a : a formal assertion of illegality ⟨a *charge* of murder⟩ **b :** a statement of complaint or hostile criticism ⟨denied the *charges* of nepotism that were leveled against him⟩
7 : a violent rush forward (as to attack)
— in charge : having control or custody of something ⟨he is *in charge* of the training program⟩

²charge *verb* **charged; charg·ing** [Middle English, from Old French *chargier,* from Late Latin *carricare,* from Latin *carrus* wheeled vehicle — more at CAR] (14th century)
transitive verb
1 a *archaic* **:** to lay or put a load on or in **:** LOAD **b** (1) **:** to place a charge (as of powder) in (2) **:** to load or fill to capacity **c** (1) **:** to restore the active materials in (a storage battery) by the passage of a direct current through in the opposite direction to that of discharge (2) **:** to give an electric charge to ⟨*charge* a capacitor⟩ **d** (1) **:** to assume as a heraldic bearing (2) **:** to place a heraldic bearing on **e :** to fill or furnish fully ⟨the music is *charged* with excitement⟩
2 a : to impose a task or responsibility on ⟨*charge* him with the job of finding a new meeting place⟩ **b :** to command, instruct, or exhort with authority ⟨I *charge* you not to go⟩ **c** *of a judge* **:** to give a charge to (a jury)
3 a : to make an assertion against especially by ascribing guilt or blame ⟨*charges* him with armed robbery⟩ ⟨they were *charged* as being instigators⟩ **b :** to place the guilt or blame for ⟨*charge* her failure to negligence⟩ **c :** to assert as an accusation ⟨*charges* that he distorted the data⟩
4 a : to bring (a weapon) into position for attack **:** LEVEL ⟨*charge* a lance⟩ **b :** to rush against **:** ATTACK; *also* **:** to rush into (an opponent) usually illegally in various sports
5 a (1) **:** to impose a financial burden on ⟨*charge* his estate with debts incurred⟩ (2) **:** to impose or record as financial obligation ⟨*charge* debts to an estate⟩ **b** (1) **:** to fix or ask as fee or payment ⟨*charges* $50 for an office visit⟩ (2) **:** to ask payment of (a person) ⟨*charge* a client for expenses⟩ **c :** to record (an item) as an expense, debt, obligation, or liability ⟨*charged* a new sofa⟩
intransitive verb
1 : to rush forward in or as if in assault **:** ATTACK; *also* **:** to charge an opponent in sports
2 : to ask or set a price ⟨do you *charge* for this service?⟩
3 : to charge an item to an account ⟨*charge* now, pay later⟩
synonym see COMMAND

charge·able \ˈchär-jə-bəl\ *adjective* (14th century)
1 *archaic* **:** financially burdensome **:** EXPENSIVE
2 : liable to be charged: as **a :** liable to be accused or held responsible **b :** suitable to be charged to a particular account **c :** qualified to be made a charge on the county or parish

charge account *noun* (1903)
: a customer's account with a creditor (as a merchant) to which the purchase of goods is charged

charge card *noun* (1950)
: CREDIT CARD

charge–coupled device *noun* (1971)
: a semiconductor device that is used especially as an optical sensor and that stores charge and transfers it sequentially to an amplifier and detector — called also *CCD*

charged \ˈchärjd\ *adjective* (1934)
1 : possessing or showing strong emotion ⟨attacked the author in a highly *charged* review⟩
2 : capable of arousing strong emotion ⟨a politically *charged* subject⟩

char·gé d'af·faires \(ˌ)shär-ˌzhā-də-ˈfar, -ˈfer\ *noun, plural* **chargés d'affaires** \-ˌzhā-də-, -ˌzhāz-də-\ [French, literally, one charged with affairs] (1767)
1 : a subordinate diplomat who substitutes for an absent ambassador or minister
2 : a diplomat inferior in rank to an ambassador or minister who heads a mission when no ambassador or minister is assigned

charge·hand \ˈchärj-ˌhand\ *noun* (1916)
British **:** FOREMAN

charge off *transitive verb* (circa 1923)
: to treat as a loss or expense
— charge–off \ˈchärj-ˌȯf\ *noun*

charge of quarters (circa 1918)
: an enlisted man designated to handle administrative matters in his unit especially after duty hours — abbreviation *CQ*

☆ SYNONYMS

Characteristic, individual, peculiar, distinctive mean indicating a special quality or identity. CHARACTERISTIC applies to something that distinguishes or identifies a person or thing or class ⟨responded with her *characteristic* wit⟩. INDIVIDUAL stresses qualities that distinguish one from all other members of the same kind or class ⟨a highly *individual* writing style⟩. PECULIAR applies to qualities possessed only by a particular individual or class or kind and stresses rarity or uniqueness ⟨an eccentricity that is *peculiar* to the British⟩. DISTINCTIVE indicates qualities distinguishing and uncommon and often superior or praiseworthy ⟨a *distinctive* aura of grace and elegance⟩.

¹char•ger \'chär-jər\ *noun* [Middle English *chargeour;* akin to Middle English *chargen* to charge] (14th century)
: a large flat dish or platter

²charg•er *noun* (1539)
1 : one that charges: as **a :** an appliance for holding or inserting a charge of powder or shot in a gun **b :** a cartridge clip **c :** a device for charging storage batteries
2 : a horse for battle or parade

char•i•ness \'char-ē-nəs, 'cher-\ *noun* (1571)
1 : the quality or state of being chary **:** CAUTION
2 : carefully preserved state **:** INTEGRITY

¹char•i•ot \'char-ē-ət\ *noun* [Middle English, from Middle French, from Old French, from *char* wheeled vehicle, from Latin *carrus* — more at CAR] (14th century)
1 : a light four-wheeled pleasure or state carriage
2 : a two-wheeled horse-drawn battle car of ancient times used also in processions and races

²chariot (1550)
intransitive verb
: to drive or ride in or as if in a chariot
transitive verb
: to carry in or as if in a chariot

char•i•ot•eer \,char-ē-ə-'tir\ *noun* (14th century)
1 : one who drives a chariot
2 *capitalized* **:** AURIGA

cha•ris•ma \kə-'riz-mə\ *also* **char•ism** \'kar-,i-zəm\ *noun, plural* **cha•ris•ma•ta** \kə-'riz-mə-tə, ,kər-iz-'mä-tə\ *also* **charisms** [Greek *charisma* favor, gift, from *charizesthai* to favor, from *charis* grace; akin to Greek *chairein* to rejoice — more at YEARN] (circa 1641)
1 : an extraordinary power (as of healing) given a Christian by the Holy Spirit for the good of the church
2 a : a personal magic of leadership arousing special popular loyalty or enthusiasm for a public figure (as a political leader) **b :** a special magnetic charm or appeal ⟨the *charisma* of a popular actor⟩

¹char•is•mat•ic \,kar-əz-'ma-tik\ *adjective* [*charisma*] (circa 1868)
1 : of, relating to, or constituting charisma
2 : having, exhibiting, or based on charisma ⟨*charismatic* sects⟩ ⟨*charismatic* leader⟩

²charismatic *noun* (1951)
: a member of a charismatic religious group or movement

char•i•ta•ble \'char-ə-tə-bəl\ *adjective* (14th century)
1 : full of love for and goodwill toward others **:** BENEVOLENT
2 a : liberal in benefactions to the needy **:** GENEROUS **b :** of or relating to charity ⟨*charitable* institutions⟩
3 : merciful or kind in judging others **:** LENIENT
— char•i•ta•ble•ness *noun*
— char•i•ta•bly \-blē\ *adverb*

char•i•ty \'char-ə-tē\ *noun, plural* **-ties** [Middle English *charite,* from Old French *charité,* from Late Latin *caritat-, caritas* Christian love, from Latin, dearness, from *carus* dear; akin to Old Irish *carae* friend, Sanskrit *kāma* love] (13th century)
1 : benevolent goodwill toward or love of humanity
2 a : generosity and helpfulness especially toward the needy or suffering; *also* **:** aid given to those in need **b :** an institution engaged in relief of the poor **c :** public provision for the relief of the needy
3 a : a gift for public benevolent purposes **b :** an institution (as a hospital) founded by such a gift
4 : lenient judgment of others
synonym see MERCY

cha•ri•va•ri \,shi-və-'rē, 'shi-və-,\ *noun* [French, perhaps from Late Latin *caribaria* headache, from Greek *karēbaria,* from *kara, karē* head + *barys* heavy — more at CEREBRAL, GRIEVE] (circa 1681)
: SHIVAREE

char•la•tan \'shär-lə-tən\ *noun* [Italian *ciarlatano,* alteration of *cerretano,* literally, inhabitant of Cerreto, from *Cerreto,* Italy] (1618)
1 : QUACK 2 ⟨*charlatans* killing their patients with empirical procedures⟩
2 : one making usually showy pretenses to knowledge or ability **:** FRAUD, FAKER ◆
— char•la•tan•ism \-tə-,ni-zəm\ *noun*
— char•la•tan•ry \-rē\ *noun*

Charles's Wain \'chärlz-'wān, ,chärl-zəz-\ *noun* [*Charlemagne*]
: BIG DIPPER

Charles•ton \'chärl-stən\ *noun* [*Charleston,* S. C.] (1925)
: a lively ballroom dance in which the knees are twisted in and out and the heels are swung sharply outward on each step

char•ley horse \'chär-lē-,hȯrs\ *noun* [from *Charley,* nickname for *Charles*] (1888)
: a muscular pain, cramping, or stiffness especially of the quadriceps that results from a strain or bruise

char•lie *also* **char•ley** \'chär-lē\ *noun, often capitalized* [from the name *Charlie*] (circa 1946)
British **:** FOOL

Char•lie \'chär-lē\ [from the name *Charlie*] (1946)
— a communications code word for the letter *c*

char•lock \'chär-,läk, -lək\ *noun* [Middle English *cherlok,* from Old English *cerlic*] (before 12th century)
: a mustard (*Brassica kaber*) that is a common weed in grainfields — called also *wild mustard*

char•lotte \'shär-lət\ *noun* [French] (1796)
: a dessert consisting of a filling (as of fruit, whipped cream, or custard) layered with or placed in a mold lined with strips of bread, ladyfingers, or biscuits

char•lotte russe \'shär-lət-'rüs\ *noun* [French, literally, Russian charlotte] (circa 1845)
: a charlotte made with sponge cake or ladyfingers and a whipped-cream or custard-gelatin filling

¹charm \'chärm\ *noun* [Middle English *charme,* from Middle French, from Latin *carmen* song, from *canere* to sing — more at CHANT] (14th century)
1 a : the chanting or reciting of a magic spell **:** INCANTATION **b :** a practice or expression believed to have magic power
2 : something worn about the person to ward off evil or ensure good fortune **:** AMULET
3 a : a trait that fascinates, allures, or delights **b :** a physical grace or attraction — used in plural **c :** compelling attractiveness ⟨the island possessed great *charm*⟩
4 : a small ornament worn on a bracelet or chain
5 : a quantum characteristic of subatomic particles that accounts for the unexpectedly long lifetime of the J/psi particle, explains difficulties in the theory of the weak force, is conserved in interactions involving electromagnetism or the strong force, and has a value of zero for most known particles
— charm•less \-ləs\ *adjective*

²charm (14th century)
transitive verb
1 a : to affect by or as if by magic **:** COMPEL **b :** to please, soothe, or delight by compelling attraction ⟨*charms* customers with his suave manner⟩
2 : to endow with or as if with supernatural powers by means of charms; *also* **:** to protect by or as if by spells, charms, or supernatural influences
3 : to control (an animal) typically by charms (as the playing of music) ⟨*charm* a snake⟩
intransitive verb
1 : to practice magic and enchantment
2 : to have the effect of a charm **:** FASCINATE
synonym see ATTRACT
— charm•er \'chär-mər\ *noun*

charmed \'chärmd\ *adjective* (1964)
: having the quantum characteristic of charm ⟨a *charmed* antiquark⟩

charmed circle *noun* (1898)
: a group marked by exclusiveness

char•meuse \(,)shär-'müz, -'müs, -'myüz, -'mə(r)z\ *noun* [French, feminine of *charmeur* charmer, from *charmer* to charm] (1907)
: a fine semilustrous crepe in satin weave

charm•ing \'chär-miŋ\ *adjective* (1663)
: extremely pleasing or delightful **:** ENTRANCING
— charm•ing•ly \-miŋ-lē\ *adverb*

char•nel \'chär-nᵊl\ *noun* [Middle English, from Middle French, from Medieval Latin *carnale,* from Late Latin, neuter of *carnalis* of the flesh — more at CARNAL] (14th century)
: a building or chamber in which bodies or bones are deposited — called also *charnel house*
— charnel *adjective*

Cha•ro•lais \,shar-ə-'lā\ *noun* [*Charolais,* district in eastern France] (1893)
: any of a breed of large white cattle developed in France and used primarily for beef and crossbreeding

Char•on \'kar-ən, 'ker-, -än\ *noun* [Latin, from Greek *Charōn*]
: a son of Erebus who in Greek mythology ferries the souls of the dead over the Styx

char•poy \'chär-,pȯi\ *noun, plural* **charpoys** [Hindi *cārpāī*] (1845)
: a bed used especially in India consisting of a frame strung with tapes or light rope

charr \'chär\ *variant of* ¹CHAR

¹chart \'chärt\ *noun* [Middle French *charte,* from Latin *charta* piece of papyrus, document — more at CARD] (1571)
1 : MAP: as **a :** an outline map exhibiting something (as climatic or magnetic variations) in its geographical aspects **b :** a map for the use of navigators
2 a : a sheet giving information in tabular form **b :** GRAPH **c :** DIAGRAM **d :** a sheet of paper ruled and graduated for use in a recording instrument **e :** a listing by rank (as of sales) — usually used in plural ⟨number one on the *charts* —Tim Cahill⟩
3 : a musical arrangement; *also* **:** a part in such an arrangement

²chart (1842)
transitive verb
1 : to lay out a plan for
2 : to make a map or chart of

◇ WORD HISTORY

charlatan There was a wide gap in medieval Italy between the handful of learned physicians who had studied medicine at the great schools of Salerno or Bologna and the average peasant who knew only simple folk remedies. This gap was filled by a large group of pretenders to medical skills, who would perform surgery on demand and traveled from one market town to another advertising cures of dubious efficacy. For reasons now osbcure, the region of Umbria in central Italy was particularly noted as the home of itinerant medicine-peddlers, and the Italian word *cerretano,* which literally meant an inhabitant of Cerreto, a town in eastern Umbria, became an epithet for a quack physician. *Cerretano* was blended with the verb *ciarlare* "to chatter," alluding to the patter of the medicine seller, to form *ciarlatano,* in use by the end of the 15th century. *Ciarlatano* was borrowed into French as *charlatan* and hence into English, where it was in use by the early 17th century.

intransitive verb
: to be ranked on a chart ⟨the song *charted* for three months⟩

¹char·ter \ˈchär-tər\ *noun* [Middle English *chartre,* from Old French, from Medieval Latin *chartula,* from Latin, diminutive of *charta*] (13th century)
1 : a written instrument or contract (as a deed) executed in due form
2 a : a grant or guarantee of rights, franchises, or privileges from the sovereign power of a state or country **b :** a written instrument that creates and defines the franchises of a city, educational institution, or corporation **c :** CONSTITUTION
3 : a written instrument from the authorities of a society creating a lodge or branch
4 : a special privilege, immunity, or exemption
5 : a mercantile lease of a ship or some principal part of it
6 : a charter travel arrangement

²charter *transitive verb* (15th century)
1 a : to establish, enable, or convey by charter **b** *British* **:** CERTIFY ⟨a *chartered* mechanical engineer⟩
2 : to hire, rent, or lease for usually exclusive and temporary use ⟨*chartered* a boat for deep-sea fishing⟩
synonym see HIRE
— char·ter·er \-tər-ər\ *noun*

³charter *adjective* (1922)
: of, relating to, or being a travel arrangement in which transportation (as a bus or plane) is hired by and for one specific group of people ⟨a *charter* flight⟩

chartered accountant *noun* (1855)
British **:** a member of a chartered institute of accountants

charter member *noun* (circa 1909)
: an original member of a group (as a society or corporation)
— charter membership *noun*

Char·tism \ˈchär-ˌti-zəm\ *noun* [Medieval Latin *charta* charter, from Latin, document] (1839)
: the principles and practices of a body of 19th century English political reformers advocating better social and industrial conditions for the working classes
— Char·tist \ˈchär-tist\ *noun or adjective*

chart·ist \ˈchär-tist\ *noun* (1919)
1 : an analyst of market action whose predictions of market courses are based on study of graphic presentations of past market performance
2 : CARTOGRAPHER

char·treuse \shär-ˈtrüz, -ˈtrüs\ *noun* [*Chartreuse*] (1884)
: a variable color averaging a brilliant yellow green

Chartreuse *trademark*
— used for a usually green or yellow liqueur

char·tu·lary \ˈkär-chə-ˌler-ē\ *noun, plural* **-lar·ies** [Medieval Latin *chartularium*] (1571)
: CARTULARY

char·wom·an \ˈchär-ˌwu̇-mən\ *noun* [*chare* + *woman*] (1596)
: a cleaning woman especially in a large building

chary \ˈchar-ē, ˈcher-\ *adjective* **chari·er; -est** [Middle English, sorrowful, dear, from Old English *cearig* sorrowful, from *caru* sorrow — more at CARE] (15th century)
1 *archaic* **:** DEAR, TREASURED
2 : discreetly cautious: as **a :** hesitant and vigilant about dangers and risks **b :** slow to grant, accept, or expend ⟨a person very *chary* of compliments⟩
synonym see CAUTIOUS
— chari·ly \ˈchar-ə-lē, ˈcher-\ *adverb*

Cha·ryb·dis \kə-ˈrib-dəs *also* shə- *or* chə-\ *noun* [Latin, from Greek]
: a whirlpool off the coast of Sicily personified in Greek mythology as a female monster — compare SCYLLA

¹chase \ˈchās\ *noun* (13th century)
1 a : the hunting of wild animals — used with *the* **b :** the act of chasing **:** PURSUIT **c :** an earnest or frenzied seeking after something desired
2 : something pursued **:** QUARRY
3 : a tract of unenclosed land used as a game preserve
4 : STEEPLECHASE 1
5 : a sequence (as in a movie) in which the characters pursue one another

²chase *verb* **chased; chas·ing** [Middle English, from Middle French *chasser,* from (assumed) Vulgar Latin *captiare* — more at CATCH] (14th century)
transitive verb
1 a : to follow rapidly **:** PURSUE **b :** HUNT **c :** to follow regularly or persistently with the intention of attracting or alluring
2 *obsolete* **:** HARASS
3 : to seek out — often used with *down* ⟨detectives *chasing* down clues⟩
4 : to cause to depart or flee **:** DRIVE ⟨*chase* the dog out of the garden⟩
5 : to cause the removal of (a baseball pitcher) by a batting rally
intransitive verb
1 : to chase an animal, person, or thing ⟨*chase* after material possessions⟩
2 : RUSH, HASTEN ⟨*chased* all over town looking for a place to stay⟩ ☆

³chase *transitive verb* **chased; chas·ing** [Middle English *chassen,* modification of Middle French *enchasser* to set] (15th century)
1 a : to ornament (metal) by indenting with a hammer and tools without a cutting edge **b :** to make by such indentation **c :** to set with gems
2 a : GROOVE, INDENT **b :** to cut (a thread) with a chaser

⁴chase *noun* [French *chas* eye of a needle, from Late Latin *capsus* enclosed space, alteration of Latin *capsa* box — more at CASE] (1611)
1 : GROOVE, FURROW
2 : the bore of a cannon
3 a : TRENCH **b :** a channel (as in a wall) for something to lie in or pass through

⁵chase *noun* [probably from French *châsse* frame, reliquary, from Middle French *chasse,* from Latin *capsa*] (1612)
: a rectangular steel or iron frame in which letterpress matter is locked (as for printing)

¹chas·er \ˈchā-sər\ *noun* (13th century)
1 : one that chases
2 : a mild drink (as beer) taken after hard liquor

²chaser *noun* (1707)
1 : a skilled worker who produces ornamental chasing
2 : a tool for cutting screw threads

Cha·sid *or* **Chas·sid** \ˈha-səd, ˈkä-\ *noun, plural* **Cha·si·dim** *or* **Chas·si·dim** \ˈha-sə-dəm, kä-ˈsē-\ *variant of* HASID

chasm \ˈka-zᵊm\ *noun* [Latin *chasma,* from Greek; akin to Latin *hiare* to yawn — more at YAWN] (1596)
1 : a deep cleft in the surface of a planet (as the earth) **:** GORGE
2 : a marked division, separation, or difference ⟨a political *chasm* between the two countries⟩

¹chas·sé \sha-ˈsā\ *intransitive verb* **chas·séd; chas·sé·ing** (1803)
1 : to make a chassé
2 : SASHAY

²chassé *noun* [French, from past participle of *chasser* to chase] (1828)
: a sliding dance step resembling the galop

chasse·pot \ˈshas-ˌpō, ˈsha-sə-\ *noun* [French, from Antoine A. *Chassepot* (died 1905) French inventor] (1869)
: a bolt-action rifle firing a paper cartridge

chas·seur \sha-ˈsər\ *noun* [French, from Middle French *chasser*] (1795)
1 : HUNTER, HUNTSMAN
2 : one of a body of light cavalry or infantry trained for rapid maneuvering
3 : a liveried attendant **:** FOOTMAN

chas·sis \ˈcha-sē, ˈsha-sē *also* ˈcha-səs\ *noun, plural* **chas·sis** \-sēz\ [French *châssis,* from Middle French *chaciz,* from *chasse*] (circa 1864)
: the supporting frame of a structure (as an automobile or television); *also* **:** the frame and working parts (as of an automobile or electronic device) exclusive of the body or housing

chaste \ˈchāst\ *adjective* **chast·er; chast·est** [Middle English, from Old French, from Latin *castus* pure] (13th century)
1 : innocent of unlawful sexual intercourse
2 : CELIBATE
3 : pure in thought and act **:** MODEST
4 a : severely simple in design or execution **:** AUSTERE ⟨*chaste* classicism⟩ **b :** CLEAN, SPOTLESS ☆
— chaste·ly *adverb*
— chaste·ness \ˈchās(t)-nəs\ *noun*

chas·ten \ˈchā-sᵊn\ *transitive verb* **chas·tened; chas·ten·ing** \ˈchās-niŋ, ˈchā-sᵊn-iŋ\ [alteration of obsolete English *chaste* to chasten, from Middle English, from Old French *chastier,* from Latin *castigare,* from *castus* + *-igare* (from *agere* to drive) — more at ACTIVE] (13th century)
1 : to correct by punishment or suffering **:** DISCIPLINE; *also* **:** PURIFY
2 a : to prune (as a work or style of art) of excess, pretense, or falsity **:** REFINE **b :** to cause to be more humble or restrained **:** SUBDUE
synonym see PUNISH
— chas·ten·er \ˈchās-nər, ˈchā-sᵊn-ər\ *noun*

chas·tise \(ˌ)chas-ˈtīz\ *transitive verb* **chas·tised; chas·tis·ing** [Middle English *chastisen,* alteration of *chasten*] (14th century)
1 : to inflict punishment on (as by whipping)
2 : to censure severely **:** CASTIGATE

☆ SYNONYMS

Chase, pursue, follow, trail mean to go after or on the track of something or someone. CHASE implies going swiftly after and trying to overtake something fleeing or running ⟨a dog *chasing* a cat⟩. PURSUE suggests a continuing effort to overtake, reach, or attain ⟨*pursued* the criminal through narrow streets⟩. FOLLOW puts less emphasis upon speed or intent to overtake ⟨friends *followed* me home in their car⟩. TRAIL may stress a following of tracks or traces rather than a visible object ⟨*trail* deer⟩ ⟨*trailed* a suspect across the country⟩.

Chaste, pure, modest, decent mean free from all taint of what is lewd or salacious. CHASTE primarily implies a refraining from acts or even thoughts or desires that are not virginal or not sanctioned by marriage vows ⟨they maintained *chaste* relations⟩. PURE differs from CHASTE in implying innocence and absence of temptation rather than control of one's impulses and actions ⟨the *pure* of heart⟩. MODEST and DECENT apply especially to deportment and dress as outward signs of inward chastity or purity ⟨preferred more *modest* swimsuits⟩ ⟨*decent* people didn't go to such movies⟩.

3 *archaic* **:** CHASTEN 2
synonym see PUNISH
— **chas·tise·ment** \(ˌ)chas-ˈtīz-mənt *also* ˈchas-təz-\ *noun*
— **chas·tis·er** \(ˌ)chas-ˈtī-zər\ *noun*

chas·ti·ty \ˈchas-tə-tē\ *noun* (13th century)
1 : the quality or state of being chaste: as **a :** abstention from unlawful sexual intercourse **b :** abstention from all sexual intercourse **c :** purity in conduct and intention **d :** restraint and simplicity in design or expression
2 : personal integrity

chastity belt *noun* (1931)
: a belt device (as of medieval times) designed to prevent sexual intercourse on the part of the woman wearing it

cha·su·ble \ˈcha-zə-bəl, -zhə-, -sə-\ *noun* [Middle French, from Late Latin *casubla* hooded garment] (14th century)
: a sleeveless outer vestment worn by the officiating priest at mass

chasuble: *1* Gothic, *2* fiddleback

¹chat \ˈchat\ *verb* **chat·ted; chat·ting** [Middle English *chatten,* short for *chatteren*] (15th century)
intransitive verb
1 : CHATTER, PRATTLE
2 : to talk in an informal or familiar manner
transitive verb
chiefly British **:** to talk to; *especially* **:** to talk lightly, glibly, or flirtatiously with — often used with *up*

²chat *noun* (1530)
1 : idle small talk **:** CHATTER
2 : light familiar talk; *especially* **:** CONVERSATION
3 [imitative] **:** any of several songbirds (as of the genera *Saxicola, Granatellus,* or *Icteria*)

châ·teau \sha-ˈtō\ *noun, plural* **châ·teaus** \-ˈtōz\ *or* **châ·teaux** \-ˈtō(z)\ [French, from Latin *castellum* castle] (1739)
1 : a feudal castle or fortress in France
2 : a large country house **:** MANSION
3 : a French vineyard estate

cha·teau·bri·and \(ˌ)sha-ˌtō-brē-ˈäⁿ\ *noun, often capitalized* [François René de *Chateaubriand*] (1877)
: a large tenderloin steak usually grilled or broiled and served with a sauce (as béarnaise)

chat·e·lain \ˈsha-tᵊl-ˌān\ *noun* [Middle French *châtelain,* from Latin *castellanus* occupant of a castle] (15th century)
: CASTELLAN

chat·e·laine \ˈsha-tᵊl-ˌān\ *noun* [French *châtelaine,* feminine of *châtelain*] (1845)
1 a : the wife of a castellan **:** the mistress of a château **b :** the mistress of a household or of a large establishment
2 : a clasp or hook for a watch, purse, or bunch of keys

cha·toy·ance \shə-ˈtȯi-ən(t)s\ *noun* (1910)
: CHATOYANCY

cha·toy·an·cy \-ən(t)-sē\ *noun* (1894)
: the quality or state of being chatoyant

¹cha·toy·ant \shə-ˈtȯi-ənt\ *adjective* [French, from present participle of *chatoyer* to shine like a cat's eyes] (1816)
: having a changeable luster or color with an undulating narrow band of white light ⟨a *chatoyant* gem⟩

²chatoyant *noun* (circa 1828)
: a chatoyant gem

chat show *noun* (1969)
British **:** TALK SHOW

chat·tel \ˈcha-tᵊl\ *noun* [Middle English *chatel* property, from Middle French, from Medieval Latin *capitale* — more at CATTLE] (14th century)
1 : an item of tangible movable or immovable property except real estate, freehold, and things (as buildings) connected with real property
2 : SLAVE, BONDSMAN

¹chat·ter \ˈcha-tər\ *verb* [Middle English *chatteren,* of imitative origin] (13th century)
intransitive verb
1 : to utter rapid short sounds suggestive of language but inarticulate and indistinct ⟨squirrels *chattered* angrily⟩ ⟨a *chattering* stream⟩
2 : to talk idly, incessantly, or fast **:** JABBER
3 a : to click repeatedly or uncontrollably ⟨teeth *chattering* with cold⟩ ⟨machine guns *chattering*⟩ **b :** to vibrate rapidly especially as a consequence of repeated sticking and slipping ⟨*chattering* brakes⟩
transitive verb
: to utter rapidly, idly, or indistinctly
— **chat·ter·er** *noun*

²chatter *noun* (13th century)
1 : the action or sound of chattering
2 : idle talk **:** PRATTLE

chat·ter·box \ˈcha-tər-ˌbäks\ *noun* (1774)
: one who engages in much idle talk

chatter mark *noun* (1888)
1 : a fine undulation formed on the surface of work by a chattering tool
2 : one of a series of short curved cracks on a glaciated rock surface transverse to the glacial striae

chat·ty \ˈcha-tē\ *adjective* **chat·ti·er; -est** (circa 1762)
1 : fond of chatting **:** TALKATIVE ⟨a *chatty* neighbor⟩
2 : having the style and manner of light familiar conversation ⟨a *chatty* letter⟩
— **chat·ti·ly** \ˈcha-tᵊl-ē\ *adverb*
— **chat·ti·ness** \ˈcha-tē-nəs\ *noun*

¹chauf·feur \ˈshō-fər, shō-ˈ\ *noun* [French, literally, stoker, from *chauffer* to heat, from Middle French *chaufer* — more at CHAFE] (1899)
1 : a person employed to drive a motor vehicle
2 : one that transports others by operating a motor vehicle ◆

²chauffeur *verb* **chauf·feured; chauf·feur·ing** \ˈshō-f(ə-)riŋ, shō-ˈfər-iŋ\ (1917)
intransitive verb
: to do the work of a chauffeur
transitive verb
1 : to transport in the manner of a chauffeur ⟨*chauffeurs* the children to school⟩
2 : to operate (as an automobile) as chauffeur

chaul·moo·gra \chȯl-ˈmü-grə\ *noun* [Bengali *cālmugrā*] (circa 1815)
: any of several East Indian trees (family Flacourtiaceae) that yield an acrid oil used especially formerly in treating leprosy and skin diseases

chaunt \ˈchȯnt, ˈchänt\, **chaunt·er** *variant of* CHANT, CHANTER

chaus·sure \shō-ˈsu̇r\ *noun, plural* **chaus·sures** *same*\ [Middle English *chaucer,* from Middle French *chaussure,* from *chausser* to put on footwear, from Latin *calceare,* from *calceus* shoe — more at CALZONE] (14th century)
1 : FOOTGEAR
2 *plural* **:** SHOES

chau·tau·qua \shə-ˈtȯ-kwə\ *noun, often capitalized* [*Chautauqua* Lake] (1873)
: an institution that flourished in the late 19th and early 20th centuries providing popular education combined with entertainment in the form of lectures, concerts, and plays often presented outdoors or in a tent

chau·vin·ism \ˈshō-və-ˌni-zəm\ *noun* [French *chauvinisme,* from Nicolas *Chauvin,* character noted for his excessive patriotism and devotion to Napoleon in Théodore and Hippolyte Cogniard's play *La Cocarde tricolore* (1831)] (1870)
1 : excessive or blind patriotism — compare JINGOISM
2 : undue partiality or attachment to a group or place to which one belongs or has belonged
3 : an attitude of superiority toward members of the opposite sex; *also* **:** behavior expressive of such an attitude
— **chau·vin·ist** \-və-nist\ *noun or adjective*
— **chau·vin·is·tic** \ˌshō-və-ˈnis-tik\ *adjective*
— **chau·vin·is·ti·cal·ly** \-ti-k(ə-)lē\ *adverb*

¹chaw \ˈchȯ\ *verb* [by alteration] (1506)
: CHEW

²chaw *noun* (1709)
: a chew especially of tobacco

chaw·ba·con \ˈchȯ-ˌbā-kən\ *noun* [¹*chaw* + *bacon*] (1537)
: BUMPKIN, HICK

cha·yo·te \chī-ˈyō-tē, chē-, -(ˌ)tā\ *noun* [Spanish, from Nahuatl *chayohtli*] (1887)
: the pear-shaped fruit of a West Indian annual vine (*Sechium edule*) of the gourd family that is widely cultivated as a vegetable; *also* **:** the plant — called also *mirliton*

¹cheap \ˈchēp\ *noun* [Middle English *chep,* from Old English *cēap* trade; akin to Old High German *kouf* trade; both from Latin *caupo* tradesman] (before 12th century)
obsolete **:** BARGAIN
— **on the cheap :** at minimum expense **:** CHEAPLY

²cheap *adjective* (1509)
1 a : purchasable below the going price or the real value **b :** charging or obtainable at a low price **c :** depreciated in value (as by currency inflation) ⟨*cheap* dollars⟩
2 : gained with little effort ⟨a *cheap* victory⟩
3 a : of inferior quality or worth **:** TAWDRY, SLEAZY **b :** contemptible because of lack of any fine, lofty, or redeeming qualities **c :** STINGY
4 a : yielding small satisfaction **b :** paying or able to pay less than going prices
5 *of money* **:** obtainable at a low rate of interest
— **cheap** *adverb*
— **cheap·ish** \ˈchē-pish\ *adjective*
— **cheap·ish·ly** *adverb*
— **cheap·ly** \ˈchē-plē\ *adverb*
— **cheap·ness** *noun*

cheap·en \ˈchē-pən\ *verb* **cheap·ened; cheap·en·ing** \ˈchēp-niŋ, ˈchē-pə-\ (1574)
transitive verb
1 [obsolete English *cheap* to price, bid for] *archaic* **a :** to ask the price of **b :** to bid or bargain for
2 a : to make cheap in price or value **b :** to lower in general esteem **c :** to make tawdry, vulgar, or inferior
intransitive verb
: to become cheap

cheap·ie \ˈchē-pē\ *noun* (circa 1898)
: one that is cheap; *especially* **:** an inexpensively produced motion picture

◇ WORD HISTORY
chauffeur *Chauffeur* in French means literally "stoker," a person who maintains the fire that powers the steam engine of a locomotive or ship. In the 1890s, *chauffeur* also began to be used for the driver of an automobile, who in effect stoked the engine mechanically with the accelerator. Earlier, the compound *chauffeur-mécanicien* "stoker-engineer" was applied to someone who both stoked an engine and ran it, and it is possible that *chauffeur* in the sense "driver" is actually short for *chauffeur-mécanicien.* In any case, the French word has had a successful career in a number of other European languages, from Spanish *chófer* to Russian *shofyor,* usually denoting a professional driver of a motor vehicle. In English, *chauffeur* often refers to someone who is hired to drive a passenger car.

— **cheapie** *adjective*
[1]**cheap·jack** \'chēp-ˌjak\ *noun* [*cheap* + the name *Jack*] (1851)
1 : a haggling huckster
2 : a dealer in cheap merchandise
[2]**cheapjack** *adjective* (1865)
1 : being inferior, cheap, or worthless ⟨*cheapjack* movie companies⟩
2 : unscrupulously opportunistic ⟨*cheapjack* speculators⟩
cheapo \'chē-(ˌ)pō\ *adjective* (1972)
: CHEAP
cheap shot *noun* (1971)
1 : an act of deliberate roughness against a defenseless opponent especially in a contact sport
2 : a critical statement that takes unfair advantage of a known weakness of the target
cheap·skate \'chēp-ˌskāt\ *noun* (1896)
: a miserly or stingy person; *especially* **:** one who tries to avoid paying a fair share of costs or expenses
[1]**cheat** \'chēt\ (1590)
transitive verb
1 : to deprive of something valuable by the use of deceit or fraud
2 : to influence or lead by deceit, trick, or artifice
3 : to elude or thwart by or as if by outwitting ⟨*cheat* death⟩
intransitive verb
1 a : to practice fraud or trickery **b :** to violate rules dishonestly (as at cards or on an examination)
2 : to be sexually unfaithful — usually used with *on* ☆
— **cheat·er** *noun*
[2]**cheat** *noun* [earlier *cheat* forfeited property, from Middle English *chet* escheat, short for *eschete* — more at ESCHEAT] (1631)
1 : the act or an instance of fraudulently deceiving **:** DECEPTION, FRAUD
2 : one that cheats **:** PRETENDER, DECEIVER
3 : any of several grasses; *especially* **:** the common chess (*Bromus secalinus*)
4 : the obtaining of property from another by an intentional active distortion of the truth
[1]**check** \'chek\ *noun* [Middle English *chek,* from Middle French *eschec,* from Arabic *shāh,* from Persian, literally, king; akin to Greek *ktasthai* to acquire, Sanskrit *kṣatra* dominion] (15th century)
1 : exposure of a chess king to an attack from which he must be protected or moved to safety
2 a : a sudden stoppage of a forward course or progress **:** ARREST **b :** a checking of an opposing player (as in ice hockey)
3 : a sudden pause or break in a progression
4 *archaic* **:** REPRIMAND, REBUKE
5 : one that arrests, limits, or restrains **:** RESTRAINT ⟨against all *checks,* rebukes, and manners, I must advance —Shakespeare⟩
6 a : a standard for testing and evaluation **:** CRITERION **b :** EXAMINATION **c :** INSPECTION, INVESTIGATION ⟨a loyalty *check* on government employees⟩ **d :** the act of testing or verifying; *also* **:** the sample or unit used for testing or verifying
7 : a written order directing a bank to pay money as instructed **:** DRAFT
8 a : a ticket or token showing ownership or identity or indicating payment made ⟨a baggage *check*⟩ **b :** a counter in various games **c :** a slip indicating the amount due **:** BILL
9 [Middle English *chek,* short for *cheker* checker] **a :** a pattern in squares that resembles a checkerboard **b :** a fabric woven or printed with such a design
10 : a mark typically √ placed beside an item to show it has been noted, examined, or verified
11 : CRACK, BREAK
— **check·less** \'che-kləs\ *adjective*
— **in check :** under restraint or control
[2]**check** (14th century)
transitive verb
1 : to put (a chess king) in check
2 *chiefly dialect* **:** REBUKE, REPRIMAND
3 a : to slow or bring to a stop **:** BRAKE ⟨hastily *checked* the impulse⟩ **b :** to block the progress of (as a hockey player)
4 a : to restrain or diminish the action or force of **:** CONTROL **b :** to slack or ease off and then belay again (as a rope)
5 a : to compare with a source, original, or authority **:** VERIFY **b :** to inspect, examine, or look at appraisingly — usually used with *out* or *over* ⟨*checking* out new cars⟩ **c :** to mark with a check as examined, verified, or satisfactory — often used with *off* ⟨*checked* off each item⟩
6 a : to consign (as luggage) to a common carrier from which one has purchased a passenger ticket ⟨*checked* our bags before boarding⟩ **b :** to ship or accept for shipment under such a consignment
7 : to mark into squares **:** CHECKER
8 : to leave or accept for safekeeping in a checkroom
9 : to make checks or chinks **:** cause to crack ⟨the sun *checks* timber⟩
intransitive verb
1 a *of a dog* **:** to stop in a chase especially when scent is lost **b :** to halt through caution, uncertainty, or fear **:** STOP
2 a : to investigate conditions ⟨*checked* on the passengers' safety⟩ **b :** to prove to be consistent or truthful ⟨the description *checks* with the photograph⟩ — often used with *out* ⟨the story *checked* out⟩
3 : to draw a check on a bank
4 : to waive the right to initiate the betting in a round of poker
5 : CRACK, SPLIT
synonym see RESTRAIN
— **check into 1 :** to check in at ⟨*check into* a hotel⟩ **2 :** INVESTIGATE
— **check up on :** INVESTIGATE
check·able *adjective* (1877)
1 : capable of being checked ⟨a *checkable* story⟩
2 : held in or being a bank account on which checks can be drawn ⟨*checkable* deposits⟩
check·book \'chek-ˌbu̇k\ *noun* (circa 1846)
: a book containing blank checks to be drawn on a bank
checkbook journalism *noun* (1975)
: the practice of paying someone for a news story and especially for granting an interview
[1]**check·er** \'che-kər\ *noun* [Middle English *cheker,* from Middle French *eschequier,* from *eschec*] (14th century)
1 *archaic* **:** CHESSBOARD
2 : a square or spot resembling the markings of a checkerboard
3 [singular of *checkers*] **:** a piece in checkers
[2]**checker** *transitive verb* **check·ered; check·er·ing** \'che-k(ə-)riŋ\ (15th century)
1 a : to variegate with different colors or shades **b :** to vary with contrasting elements or situations ⟨had a *checkered* career⟩
2 : to mark into squares
[3]**checker** *noun* (1535)
1 : one that checks ⟨a fact *checker*⟩ ⟨spelling *checker*⟩
2 : an employee who checks out purchases in a self-service store
check·er·ber·ry \'che-kər-ˌber-ē\ *noun* [*checker* wild service tree + *berry*] (1776)
1 : the spicy red berrylike fruit of an American wintergreen (*Gaultheria procumbens*)
2 : a plant producing checkerberries
check·er·board \-ˌbōrd, -ˌbȯrd\ *noun* (1775)
1 : a board used in various games (as checkers) with usually 64 squares in 2 alternating colors
2 : something that has a pattern or arrangement like a checkerboard
check·ers \'che-kərz\ *noun plural but singular in construction* (1712)
: a checkerboard game for 2 players each with 12 pieces
check–in \'che-ˌkin\ *noun* (1927)
: an act or instance of checking in
check in (1918)
intransitive verb
1 : to register at a hotel
2 : to report one's presence or arrival ⟨*check in* at a convention⟩
transitive verb
: to satisfy all requirements in returning ⟨*check in* the equipment after using⟩
checking account *noun* (circa 1909)
: a bank account against which the depositor can draw checks
check·list \'chek-ˌlist\ *noun* (1853)
: a list of things to be checked or done ⟨a pilot's *checklist* before takeoff⟩; *also* **:** a comprehensive list ⟨a *checklist* of bird species⟩
check mark *noun* (1917)
: CHECK 10
— **check·mark** *transitive verb*
[1]**check·mate** \'chek-ˌmāt\ *transitive verb* [Middle English *chekmaten,* from *chekmate,* interjection used to announce checkmate, from Middle French *eschec mat,* from Arabic *shāh māt,* from Persian, literally, the king is left unable to escape] (14th century)
1 : to arrest, thwart, or counter completely
2 : to check (a chess opponent's king) so that escape is impossible
[2]**checkmate** *noun* (15th century)
1 a : the act of checkmating **b :** the situation of a checkmated king
2 : a complete check
check·off \'che-ˌkȯf\ *noun* (1911)
1 a : the deduction of union dues from a worker's paycheck by the employer **b :** designation on an income tax return of a small amount of money to be applied to a special fund (as for financing political campaigns)
2 : AUTOMATIC 2
check off (1839)
transitive verb
1 : to eliminate from further consideration
2 : to deduct (union dues) from a worker's paycheck
intransitive verb
: to change a play at the line of scrimmage in football by calling an automatic
check·out \'che-ˌkau̇t\ *noun* (1933)
1 : the action or an instance of checking out
2 : the time at which a lodger must vacate a room (as in a hotel) or be charged for retaining it
3 : a counter or area in a store where goods are checked out
4 a : the action of examining and testing something for performance, suitability, or readiness **b :** the action of familiarizing oneself with the operation of a mechanical thing (as an airplane)
check out (1921)
intransitive verb

☆ SYNONYMS
Cheat, cozen, defraud, swindle mean to get something by dishonesty or deception. CHEAT suggests using trickery that escapes observation ⟨*cheated* me out of a dollar⟩. COZEN implies artful persuading or flattering to attain a thing or a purpose ⟨always able to *cozen* her grandfather out of a few dollars⟩. DEFRAUD stresses depriving one of his or her rights and usually connotes deliberate perversion of the truth ⟨*defrauded* of her inheritance by an unscrupulous lawyer⟩. SWINDLE implies large-scale cheating by misrepresentation or abuse of confidence ⟨*swindled* of their savings by con artists⟩.

: to vacate and pay for one's lodging (as at a hotel)
transitive verb
1 : to satisfy all requirements in taking away ⟨*checked out* a library book⟩
2 a : to itemize and total the cost of and receive payment for (outgoing merchandise) especially in a self-service store **b :** to have the cost totaled and pay for (purchases) at a checkout
check·point \'chek-ˌpȯint\ *noun* (1926)
: a point at which a check is performed ⟨vehicles were inspected at various *checkpoints*⟩
check·rein \-ˌrān\ *noun* (circa 1809)
1 : a short rein looped over a hook on the saddle of a harness to prevent a horse from lowering its head
2 : a branch rein connecting the driving rein of one horse of a pair with the bit of the other
check·room \-ˌrüm, -ˌru̇m\ *noun* (1900)
: a room at which baggage, parcels, or clothing is checked
check·up \'che-ˌkəp\ *noun* (1921)
: EXAMINATION; *especially* **:** a general physical examination
check valve *noun* (circa 1877)
: a valve that permits flow in one direction only
ched·dar \'che-dər\ *noun, often capitalized* [*Cheddar,* England] (circa 1661)
: a hard white, yellow, or orange smooth-textured cheese with a flavor that ranges from mild to strong as the cheese matures — called also *cheddar cheese*
che·der \'kā-dər, 'ke-\ *variant of* HEDER
chee·cha·ko \chē-'chä-(ˌ)kō, -'chȯ- *also* -'cha-\ *noun, plural* **-kos** [Chinook Jargon, from *chee* new (from Lower Chinook *čxi* right away) + *chako* come, from Nootka *čokʷa·* come, imperative] (1897)
: TENDERFOOT 1 — used chiefly in Alaska
[1]cheek \'chēk\ *noun* [Middle English *cheke,* from Old English *cēace;* akin to Middle Low German *kāke* jawbone] (before 12th century)
1 : the fleshy side of the face below the eye and above and to the side of the mouth; *broadly* **:** the lateral aspect of the head
2 : something suggestive of the human cheek in position or form; *especially* **:** one of two laterally paired parts
3 : insolent boldness and self-assurance
4 : BUTTOCK 1
synonym see TEMERITY
— cheek·ful \-ˌfu̇l\ *noun*
[2]cheek *transitive verb* (1840)
chiefly British **:** to speak rudely or impudently to
cheek·bone \'chēk-ˌbōn\ *noun* (circa 1775)
: the prominence below the eye that is formed by the zygomatic bone; *also* **:** ZYGOMATIC BONE
cheek by jowl *adverb* (1577)
: SIDE BY SIDE
cheeked \'chēkt\ *adjective* (1592)
: having cheeks of a specified nature — used in combination ⟨rosy-*cheeked*⟩
cheek tooth *noun* (14th century)
: any of the molar or premolar teeth
cheeky \'chē-kē\ *adjective* **cheek·i·er; -est** (1859)
: having or showing cheek **:** IMPUDENT
— cheek·i·ly \-kə-lē\ *adverb*
— cheek·i·ness \-kē-nəs\ *noun*
cheep \'chēp\ *intransitive verb* [imitative] (1513)
1 : to utter faint shrill sounds **:** PEEP
2 : to utter a single word or sound
— cheep *noun*
[1]cheer \'chir\ *noun* [Middle English *chere* face, cheer, from Old French, face, from Medieval Latin *cara,* probably from Greek *kara* head, face — more at CEREBRAL] (13th century)
1 a *obsolete* **:** FACE **b** *archaic* **:** facial expression
2 : state of mind or heart **:** SPIRIT ⟨be of good *cheer* —Matthew 9:2 (Authorized Version)⟩
3 : lightness of mind and feeling **:** ANIMATION, GAIETY
4 : hospitable entertainment **:** WELCOME
5 : food and drink for a feast **:** FARE
6 : something that gladdens ⟨words of *cheer*⟩
7 : a shout of applause or encouragement
[2]cheer (14th century)
transitive verb
1 a : to instill with hope or courage **:** COMFORT — usually used with *up* **b :** to make glad or happy — usually used with *up*
2 : to urge on or encourage especially by shouts ⟨*cheered* the team on⟩
3 : to applaud with shouts
intransitive verb
1 *obsolete* **:** to be mentally or emotionally disposed
2 : to grow or be cheerful **:** REJOICE — usually used with *up*
3 : to utter a shout of applause or triumph
— cheer·er *noun*
cheer·ful \'chir-fəl\ *adjective* (15th century)
1 a : full of good spirits **:** MERRY **b :** UNGRUDGING ⟨*cheerful* obedience⟩
2 : conducive to cheer **:** likely to dispel gloom or worry ⟨sunny *cheerful* room⟩
— cheer·ful·ly \-f(ə-)lē\ *adverb*
— cheer·ful·ness \-fəl-nəs\ *noun*
cheer·io \ˌchir-ē-'ō\ *interjection* [*cheery* + *-o*] (1910)
chiefly British — usually used as a farewell and sometimes as a greeting or toast
cheer·lead·er \'chir-ˌlē-dər\ *noun* (1903)
: one that calls for and directs organized cheering (as at a football game)
— cheer·lead \-ˌlēd\ *transitive verb*
cheer·less \-ləs\ *adjective* (1579)
: lacking qualities that cheer **:** BLEAK, JOYLESS ⟨a *cheerless* room⟩
— cheer·less·ly *adverb*
— cheer·less·ness *noun*
cheers \'chirz\ *interjection* (1919)
— used as a toast
cheery \'chir-ē\ *adjective* **cheer·i·er; -est** (15th century)
1 : marked by cheerfulness or good spirits
2 : causing or suggesting cheerfulness
— cheer·i·ly \'chir-ə-lē\ *adverb*
— cheer·i·ness \'chir-ē-nəs\ *noun*
[1]cheese \'chēz\ *noun, often attributive* [Middle English *chese,* from Old English *cēse,* from Latin *caseus* cheese] (before 12th century)
1 a : a food consisting of the coagulated, compressed, and usually ripened curd of milk separated from the whey **b :** an often cylindrical cake of this food
2 : something resembling cheese in shape or consistency
[2]cheese *transitive verb* **cheesed; chees·ing** [origin unknown] (circa 1811)
: to put an end to **:** STOP
— cheese it — used in the imperative as a warning of danger ⟨*cheese it,* the cops⟩
[3]cheese *noun* [perhaps from Urdu *chīz* thing] (1920)
slang **:** someone important
cheese·burg·er \'chēz-ˌbər-gər\ *noun* [*cheese* + ham*burger*] (circa 1938)
: a hamburger topped with a slice of cheese
cheese·cake \-ˌkāk\ *noun* (15th century)
1 : a dessert consisting of a creamy filling usually containing cheese baked in a pastry or pressed-crumb shell
2 : a photographic display of shapely and scantily clothed female figures — compare BEEFCAKE
cheese·cloth \-ˌklȯth\ *noun* [from its use in the making of cheese] (14th century)
: a very lightweight unsized cotton gauze
cheese·par·ing \-ˌpar-iŋ, -ˌper-\ *noun* (1597)
1 : something worthless or insignificant
2 : miserly economizing
— cheeseparing *adjective*
cheesy \'chē-zē\ *adjective* **chees·i·er; -est** (14th century)
1 a : resembling or suggesting cheese especially in consistency or odor **b :** containing cheese
2 : SHABBY 3c, CHEAP
— chees·i·ness *noun*
chee·tah \'chē-tə\ *noun, plural* **cheetahs** *also* **cheetah** [Hindi *cītā* leopard, from Sanskrit *citraka,* from *citra* bright, variegated; akin to Old High German *heitar* bright — more at -HOOD] (1610)
: a long-legged spotted swift-moving African and formerly Asian cat (*Acinonyx jubatus*) about the size of a small leopard that has blunt nonretractile claws and is often trained to run down game

cheetah

chef \'shef\ *noun* [French, short for *chef de cuisine* head of the kitchen] (1826)
1 : a skilled cook who manages the kitchen (as of a restaurant)
2 : COOK
— chef *intransitive verb*
— chef·dom \-dəm\ *noun*
chef d'oeu·vre \shā-dœvrᵊ, (ˌ)shā-'də(r)v\ *noun, plural* **chefs d'oeuvre** \-dœvrᵊ, -'də(r)v(z)\ [French *chef-d'oeuvre,* literally, leading work] (1619)
: a masterpiece especially in art or literature
che·la \'kē-lə\ *noun, plural* **che·lae** \-(ˌ)lē\ [New Latin, from Greek *chēlē* claw] (1646)
: a pincerlike organ or claw borne by a limb of a crustacean or arachnid
[1]che·late \'kē-ˌlāt *also* 'chē-\ *adjective* (1826)
1 : resembling or having chelae
2 : of, relating to, or being a chelate
[2]chelate *verb* **che·lat·ed; che·lat·ing** (1922)
transitive verb
: to combine with (a metal) so as to form a chelate ring
intransitive verb
: to react so as to form a chelate ring
— che·lat·able \-ˌlā-tə-bəl\ *adjective*
— che·la·tion \kē-'lā-shən *also* chē-\ *noun*
— che·la·tor \-ˌlā-tər\ *noun*
[3]chelate *noun* (1943)
: a compound having a ring structure that usually contains a metal ion held by coordinate bonds
che·lic·era \ki-'li-sə-rə\ *noun, plural* **-er·ae** \-ˌrē\ [New Latin, from French *chélicère,* from Greek *chēlē* + *keras* horn — more at HORN] (1835)
: one of the anterior pair of appendages of an arachnid often specialized as fangs
— che·lic·er·al \-sə-rəl\ *adjective*
che·li·ped \'kē-lə-ˌped\ *noun* [Greek *chēlē* claw + English *-i-* + *-ped*] (1869)
: one of the pair of legs that bears the large chelae in decapod crustaceans
Chel·le·an *or* **Chel·li·an** \'she-lē-ən\ *adjective* [French *chelléen,* from *Chelles,* France] (1893)
: ABBEVILLIAN
che·lo·ni·an \ki-'lō-nē-ən\ *noun* [Greek *chelōnē* tortoise] (1828)
: TURTLE
— chelonian *adjective*
chem- *or* **chemo-** *also* **chemi-** *combining form* [New Latin, from Late Greek *chēmeia* alchemy — more at ALCHEMY]
1 : chemical **:** chemistry ⟨*chemo*taxis⟩
2 : chemically ⟨*chemi*sorb⟩
chem·ic \'ke-mik\ *adjective* [New Latin *chimicus* alchemist, from Medieval Latin *alchimicus,* from *alchymia* alchemy] (1576)
1 *archaic* **:** ALCHEMIC
2 : CHEMICAL
[1]chem·i·cal \'ke-mi-kəl\ *adjective* (1576)

1 : of, relating to, used in, or produced by chemistry
2 a : acting or operated or produced by chemicals **b :** detectable by chemical means
— **chem·i·cal·ly** \-mi-k(ə-)lē\ *adverb*

[2]chemical *noun* (1747)
: a substance obtained by a chemical process or used for producing a chemical effect

chemical engineering *noun* (1888)
: engineering dealing with the industrial application of chemistry

chemical warfare *noun* (1917)
: tactical warfare using incendiary mixtures, smokes, or irritant, burning, poisonous, or asphyxiating gases

chemical weapon *noun* (1980)
: a weapon used in chemical warfare

chemi·lu·mi·nes·cence \ˌke-mē-ˌlü-mə-ˈne-sᵊn(t)s, ˌkē-\ *noun* [International Scientific Vocabulary] (1889)
: luminescence (as bioluminescence) due to chemical reaction
— **chemi·lu·mi·nes·cent** \-ˈne-sᵊnt\ *adjective*

che·min de fer \shə-ˌman-də-ˈfer\ *noun, plural* **che·mins de fer** *same*\ [French, literally, railroad] (1891)
: a card game in which two hands are dealt, any number of players may bet against the dealer, and the winning hand is the one that comes closer to but does not exceed a count of nine on two or three cards

chemi·os·mot·ic \ˌke-mē-äz-ˈmä-tik, ˌkē-\ *adjective* (1966)
: relating to or being a hypothesis that seeks to explain the mechanism of ATP formation in oxidative phosphorylation by mitochondria and chloroplasts without recourse to the formation of high-energy intermediates by postulating the formation of an energy gradient of hydrogen ions across the organelle membranes that results in the reversible movement of hydrogen ions to the outside and is generated by electron transport or the activity of electron carriers

che·mise \shə-ˈmēz, *sometimes* -ˈmēs\ *noun* [Middle English, from Old French, shirt, from Late Latin *camisia*] (13th century)
1 : a woman's one-piece undergarment
2 : a loose straight hanging dress

chem·i·sette \ˌshe-mi-ˈzet\ *noun* [French, diminutive of *chemise*] (1807)
: a woman's garment; *especially* **:** one (as of lace) to fill the open front of a dress

chem·i·sorb \ˈke-mi-ˌsȯrb, ˈkē-, -ˌzȯrb\ *transitive verb* [*chem-* + *-sorb* (as in *adsorb*)] (1935)
: to take up and hold usually irreversibly by chemical forces
— **chem·i·sorp·tion** \ˌke-mi-ˈsȯrp-shən, ˌkē-, -ˈzȯrp-\ *noun*

chem·ist \ˈke-mist\ *noun* [New Latin *chimista,* short for Medieval Latin *alchimista*] (1562)
1 a *obsolete* **:** ALCHEMIST **b :** one trained in chemistry
2 *British* **:** PHARMACIST

chem·is·try \ˈke-mə-strē\ *noun, plural* **-tries** (1646)
1 : a science that deals with the composition, structure, and properties of substances and with the transformations that they undergo
2 a : the composition and chemical properties of a substance ⟨the *chemistry* of iron⟩ **b :** chemical processes and phenomena (as of an organism) ⟨blood *chemistry*⟩
3 : a strong mutual attraction, attachment, or sympathy ⟨they have a special *chemistry*⟩

che·mo \ˈkē-(ˌ)mō\ *noun* (1982)
: CHEMOTHERAPY

che·mo·au·to·tro·phic \ˌkē-mō-ˌȯ-tə-ˈtrō-fik *also* ˌke-\ *adjective* (1945)
: being autotrophic and oxidizing an inorganic compound as a source of energy ⟨*chemoautotrophic* bacteria⟩
— **che·mo·au·tot·ro·phy** \-ȯ-ˈtä-trə-fē\ *noun*

che·mo·pro·phy·lax·is \-ˌprō-fə-ˈlak-səs *also* -ˌprä-fə-\ *noun* (1936)
: the prevention of infectious disease by the use of chemical agents
— **che·mo·pro·phy·lac·tic** \-ˈlak-tik\ *adjective*

che·mo·re·cep·tion \-ri-ˈsep-shən\ *noun* [International Scientific Vocabulary] (1919)
: the physiological reception of chemical stimuli
— **che·mo·re·cep·tive** \-ˈsep-tiv\ *adjective*

che·mo·re·cep·tor \-ri-ˈsep-tər\ *noun* [International Scientific Vocabulary] (1906)
: a sense organ (as a taste bud) responding to chemical stimuli

che·mo·sur·gery \ˌkē-mō-ˈsərj-rē, -ˈsər-jə-rē\ *noun* (circa 1944)
: chemical removal of diseased or unwanted tissue
— **che·mo·sur·gi·cal** \-ˈsər-ji-kəl\ *adjective*

che·mo·syn·the·sis \-ˈsin(t)-thə-səs\ *noun* [New Latin] (1901)
: synthesis of organic compounds (as in living cells) by energy derived from chemical reactions
— **che·mo·syn·thet·ic** \-sin-ˈthe-tik\ *adjective*

che·mo·tac·tic \-ˈtak-tik\ *adjective* (1893)
: involving, inducing, or exhibiting chemotaxis
— **che·mo·tac·ti·cal·ly** \-ti-k(ə-)lē\ *adverb*

che·mo·tax·is \-ˈtak-səs\ *noun* [New Latin] (circa 1887)
: orientation or movement of an organism or cell in relation to chemical agents

che·mo·tax·on·o·my \-(ˌ)tak-ˈsä-nə-mē\ *noun* (1963)
: the classification of plants and animals based on similarities and differences in biochemical composition
— **che·mo·tax·o·nom·ic** \-ˌtak-sə-ˈnä-mik\ *adjective*
— **che·mo·tax·on·o·mist** \-(ˌ)tak-ˈsä-nə-mist\ *noun*

che·mo·ther·a·peu·tic \-ˌther-ə-ˈpyü-tik\ *adjective* (1907)
: of, relating to, or used in chemotherapy
— **chemotherapeutic** *noun*
— **che·mo·ther·a·peu·ti·cal·ly** \-ti-k(ə-)lē\ *adverb*

che·mo·ther·a·py \-ˈther-ə-pē\ *noun* [International Scientific Vocabulary] (1910)
: the use of chemical agents in the treatment or control of disease or mental illness
— **che·mo·ther·a·pist** \-pist\ *noun*

che·mot·ro·pism \ki-ˈmä-trə-ˌpi-zəm, ke-\ *noun* [International Scientific Vocabulary] (1897)
: orientation of cells or organisms in relation to chemical stimuli

che·nille \shə-ˈnē(ə)l\ *noun* [French, literally, caterpillar, from Latin *canicula,* diminutive of *canis* dog; from its hairy appearance — more at HOUND] (circa 1739)
1 : a wool, cotton, silk, or rayon yarn with protruding pile; *also* **:** a pile-face fabric with a filling of this yarn
2 : an imitation of chenille yarn or fabric

che·no·pod \ˈkē-nə-ˌpäd, ˈke-\ *noun* [ultimately from Greek *chēn* goose + *podion,* diminutive of *pod-, pous* foot — more at GOOSE, FOOT] (1555)
: any plant of the goosefoot family

cheong·sam \ˈchȯŋ-ˌsäm\ *noun* [Chinese (Guangdong) *chèuhng-sāam,* literally, long gown] (1952)
: an oriental dress with a slit skirt and a mandarin collar

cheque \ˈchek\ *chiefly British variant of* [1]CHECK 7

che·quer \ˈche-kər\ *chiefly British variant of* CHECKER

cher·i·moya \ˌcher-ə-ˈmȯi-ə, ˌchir-\ *noun* [Spanish *chirimoya*] (1736)
: a round, oblong, or heart-shaped fruit with a pitted rind that is borne by a widely cultivated tropical American tree (*Annona cherimola*) of the custard-apple family; *also* **:** this tree

cher·ish \ˈcher-ish\ *transitive verb* [Middle English *cherisshen,* from Middle French *cheriss-,* stem of *cherir* to cherish, from Old French, from *chier* dear, from Latin *carus* — more at CHARITY] (14th century)
1 a : to hold dear **:** feel or show affection for **b :** to keep or cultivate with care and affection
2 : to entertain or harbor in the mind deeply and resolutely ⟨still *cherishes* that memory⟩
synonym see APPRECIATE
— **cher·ish·able** \-i-shə-bəl\ *adjective*
— **cher·ish·er** \-i-shər\ *noun*

cher·no·zem \ˌcher-nə-ˈzyȯm, -ˈzem\ *noun* [Russian, from *chërnyĭ* black + *zemlya* earth] (1841)
: any of a group of dark-colored zonal soils with a deep rich humus horizon found in regions (as the grasslands of central North America) of temperate to cool climate
— **cher·no·zem·ic** \-ˈzyȯ-mik, -ˈze-\ *adjective*

Cher·o·kee \ˈcher-ə-(ˌ)kē\ *noun, plural* **Cherokee** *or* **Cherokees** [probably from Creek *tciloki* people of a different speech] (1674)
1 : a member of an American Indian people originally of Tennessee and North Carolina
2 : the language of the Cherokee people

Cherokee rose *noun* (1823)
: a Chinese climbing rose (*Rosa laevigata*) with a fragrant white blossom

che·root \shə-ˈrüt, chə-\ *noun* [Tamil *curuṭṭu,* literally, roll] (circa 1679)
: a cigar cut square at both ends

cher·ry \ˈcher ē\ *noun, plural* **cherries** [Middle English *chery,* from Old North French *cherise* (taken as a plural), from Late Latin *ceresia,* from Latin *cerasus* cherry tree, from Greek *kerasos*] (14th century)
1 a : any of numerous trees and shrubs (genus *Prunus*) of the rose family that bear pale yellow to deep red or blackish smooth-skinned drupes enclosing a smooth seed and that include some cultivated for their fruits or ornamental flowers — compare SWEET CHERRY **b :** the fruit of a cherry **c :** the wood of a cherry
2 : a variable color averaging a moderate red
3 a : HYMEN **b :** VIRGINITY
word history see SHERRY
— **cher·ry·like** \-ˌlīk\ *adjective*

cherry bomb *noun* (1953)
: a powerful globular red firecracker

cherry picker *noun* (circa 1944)
: a traveling crane equipped for holding a passenger at the end of the boom

cher·ry·stone \ˈcher-ē-ˌstōn\ *noun* (1880)
: a small quahog

cherry tomato *noun* (1847)
: a small globose red or orange tomato borne in long dense clusters; *also* **:** a plant (*Lycopersicon lypersicum cerasiforme*) bearing cherry tomatoes

chert \ˈchərt, ˈchat\ *noun* [origin unknown] (1679)
: a rock resembling flint and consisting essentially of a large amount of fibrous chalcedony with smaller amounts of cryptocrystalline quartz and amorphous silica
— **cherty** \ˈchər-tē, ˈcha-\ *adjective*

cher·ub \ˈcher-əb\ *noun, plural usually* **cher·u·bim** \ˈcher-ə-ˌbim, ˈker- *also* ˈcher-yə-\ [Latin, from Greek *cheroub,* from Hebrew *kĕrūbh*] (13th century)

1 *plural* **:** an order of angels — see CELESTIAL HIERARCHY
2 *plural usually* **cherubs a :** a beautiful usually winged child in painting and sculpture **b :** an innocent-looking usually chubby and rosy person
— **che·ru·bic** \chə-ˈrü-bik *also* ˈcher-ə-\ *adjective*
— **che·ru·bi·cal·ly** \-bi-k(ə-)lē\ *adverb*
— **cher·ub·like** \ˈcher-əb-ˌlīk\ *adjective*

cher·vil \ˈchər-vəl\ *noun* [Middle English *cherville,* from Old English *cerfille,* from Latin *caerefolium,* modification of (assumed) Greek *chairephyllon,* from *chairein* to rejoice + *phyllon* leaf — more at YEARN, BLADE] (before 12th century)
: an aromatic herb (*Anthriscus cerefolium*) of the carrot family with divided leaves that are often used in soups and salads; *also* **:** any of several related plants

Ches·a·peake Bay retriever \ˈche-sə-ˌpēk-ˈbā-\ *noun* (1891)
: any of a breed of large brown sporting dogs developed in Maryland and having a dense oily water-shedding coat

Chesh·ire cat \ˈche-shər-\ *noun* [*Cheshire,* England] (1866)
: a broadly grinning cat in Lewis Carroll's *Alice's Adventures in Wonderland*

Cheshire cheese *noun* (1597)
: a cheese similar to cheddar made chiefly in Cheshire, England

¹chess \ˈches\ *noun* [Middle English *ches,* from Middle French *esches,* accusative plural of *eschec* check at chess — more at CHECK] (14th century)
: a game for 2 players each of whom moves 16 pieces according to fixed rules across a checkerboard and tries to checkmate the opponent's king

chessboard with chess pieces arranged as at the beginning of a game

²chess *noun* [origin unknown] (1736)
: a weedy annual bromegrass (*Bromus secalinus*) widely distributed as a weed especially in grain; *broadly* **:** any of several weedy bromegrasses

chess·board \ˈches-ˌbōrd, -ˌbȯrd\ *noun* (15th century)
: a checkerboard used in the game of chess

chess·man \-ˌman, -mən\ *noun* (15th century)
: any of the pieces used in chess

chess pie *noun* [perhaps alteration of *chest*] (1932)
: a pie or tart with a filling made especially of eggs, butter, and sugar

chest \ˈchest\ *noun* [Middle English, from Old English *cest, cist* chest, box, from Latin *cista,* from Greek *kistē* basket, hamper] (before 12th century)
1 a : a container for storage or shipping; *especially* **:** a box with a lid used especially for the safekeeping of belongings **b :** a cupboard used especially for the storing of medicines or first-aid supplies
2 : the place where money of a public institution is kept **:** TREASURY; *also* **:** the fund so kept
3 a : THORAX; *especially* **:** the human thorax ⟨a pain in the *chest*⟩ **b :** BREAST 2a ⟨a hairy *chest*⟩
— **chest·ful** \-ˌfu̇l\ *noun*

chest·ed \ˈches-təd\ *adjective* (1662)
: having a chest of a specified kind — used in combination ⟨flat-*chested*⟩

ches·ter·field \ˈches-tər-ˌfēld\ *noun* [from a 19th century Earl of *Chesterfield*] (1852)
1 : a single-breasted or double-breasted semifitted overcoat with velvet collar
2 : a davenport usually with upright armrests

Ches·ter White \ˈches-tər-\ *noun* [*Chester* County, Pa.] (1856)
: any of a breed of large white swine

¹chest·nut \ˈches(t)-(ˌ)nət\ *noun* [Middle English *chasteine, chesten* chestnut tree, from Middle French *chastaigne,* from Latin *castanea,* from Greek *kastanea*] (14th century)
1 a : any of a genus (*Castanea*) of trees or shrubs of the beech family; *especially* **:** an American tree (*C. dentata*) that was formerly a dominant or codominant member of many deciduous forests of the eastern U.S. but has now been largely eliminated by the chestnut blight and seldom grows beyond the shrub or sapling stage **b :** the edible nut of a chestnut **c :** the wood of a chestnut
2 : a grayish to reddish brown
3 : HORSE CHESTNUT
4 : a chestnut-colored animal; *specifically* **:** a horse having a body color of any shade of pure or reddish brown with mane, tail, and points of the same or a lighter shade — compare ²BAY 1, ¹SORREL 1
5 : a callosity on the inner side of the leg of the horse — see HORSE illustration
6 a : an old joke or story **b :** something (as a musical piece or a saying) repeated to the point of staleness

²chestnut *adjective* (1555)
1 : of the color chestnut
2 : of, relating to, or resembling a chestnut

chestnut blight *noun* (circa 1909)
: a destructive disease of the American chestnut marked by cankers of the bark and cambium and caused by an imported fungus (*Endothia parasitica* synonym *Cryphonectria parasitica*)

chestnut oak *noun* (1703)
: any of several oaks having oblong to lanceolate leaves with crenate or serrate edges: as **a :** CHINQUAPIN OAK **b :** a medium-sized oak (*Quercus prinus*) of eastern North America with large acorns and leaves that are shiny yellow-green above and paler below

chest of drawers (1649)
: a piece of furniture designed to contain a set of drawers (as for holding clothing)

chesty \ˈches-tē\ *adjective* **chest·i·er; -est** (1899)
1 : proudly or arrogantly self-assertive
2 : marked by a large or well-developed chest

chet·rum \ˈchē-trəm, ˈche-\ *noun, plural* **chetrums** *or* **chetrum** [Tibetan] (1973)
— see *ngultrum* at MONEY table

che·val-de-frise \shə-ˌval-də-ˈfrēz\ *noun, plural* **che·vaux-de-frise** \shə-ˌvō-\ [French, literally, horse from Friesland] (1668)
1 : a defense consisting of a timber or an iron barrel covered with projecting spikes and often strung with barbed wire
2 : a protecting line (as of spikes) on top of a wall — usually used in plural

che·val glass \shə-ˈval-\ *noun* [French *cheval* horse, support] (1828)
: a full-length mirror in a frame in which it may be tilted

che·va·lier \ˌshe-və-ˈlir, *especially for 1b & 2 also* shə-ˈval-ˌyā\ *noun* [Middle English, from Middle French, from Late Latin *caballarius* horseman — more at CAVALIER] (14th century)
1 a : CAVALIER 2 **b :** a member of any of various orders of knighthood or of merit (as the Legion of Honor)
2 a : a member of the lowest rank of French nobility **b :** a cadet of the French nobility
3 : a chivalrous man

che·ve·lure \shəv-lœr\ *noun* [French, from Late Latin *capillatura,* from Latin *capillatus* having long hair, from *capillus* hair] (15th century)
: a head of hair

chev·i·ot \ˈshe-vē-ət, *especially British* ˈche-\ *noun, often capitalized* (1815)
1 : any of a breed of hardy hornless relatively small sheep that are a source of quality mutton and have their origin in the Cheviot hills
2 a : a fabric of cheviot wool **b :** a heavy rough napped plain or twill fabric of coarse wool or worsted **c :** a sturdy soft-finished plain or twill cotton shirting

chèvre \ˈshev(rᵊ), ˈshev-rə\ *noun* [French, literally, goat, from Old French *chievre,* from Latin *capra* she-goat, from *caper* he-goat — more at CAPRIOLE] (1950)
: GOAT CHEESE

chev·ron \ˈshev-rən\ *noun* [Middle English, from Middle French, rafter, chevron, from (assumed) Vulgar Latin *caprion-, caprio* rafter; akin to Latin *caper* goat] (14th century)
: a figure, pattern, or object having the shape of a V or an inverted V: as **a** *or* **chev·er·on** \ˈshev-rən, ˈshe-və-\ **:** a heraldic charge consisting of two diagonal stripes meeting at an angle usually with the point up **b :** a sleeve badge that usually consists of one or more chevron-shaped stripes that indicates the wearer's rank and service (as in the armed forces)

chevrons b: *1* **marine staff sergeant,** *2* **air force staff sergeant,** *3* **army staff sergeant**

¹chew \ˈchü\ *verb* [Middle English *chewen,* from Old English *cēowan;* akin to Old High German *kiuwan* to chew, Russian *zhevat'*] (before 12th century)
transitive verb
: to crush, grind, or gnaw (as food) with or as if with the teeth **:** MASTICATE
intransitive verb
: to chew something; *specifically* **:** to chew tobacco
— **chew·able** \-ə-bəl\ *adjective*
— **chew·er** *noun*
— **chewy** \ˈchü-ē\ *adjective*
— **chew the rag** *or* **chew the fat** *slang*
: to make friendly familiar conversation **:** CHAT

²chew *noun* (13th century)
1 : the act of chewing
2 : something for chewing

chewing gum *noun* (1850)
: a sweetened and flavored insoluble plastic material (as a preparation of chicle) used for chewing

che·wink \chi-ˈwiŋk\ *noun* [imitative] (1793)
: TOWHEE 1

chew out *transitive verb* (1943)
: REPRIMAND, BAWL OUT

chew over *transitive verb* (1939)
: to meditate on **:** think about reflectively

Chey·enne \shī-ˈan, -ˈen\ *noun, plural* **Cheyenne** *or* **Cheyennes** [Canadian French, from Dakota *šahíyena*] (1778)
1 : a member of an American Indian people of the western plains of the U.S.
2 : the Algonquian language of the Cheyenne people

chez \ˈshā\ *preposition* [French, from Latin *casae* at home, locative of *casa* cottage] (1740)
: at or in the home or business place of

chi \ˈkī, ˈkē\ *noun* [Greek *chei, chi*] (15th century)
: the 22nd letter of the Greek alphabet — see ALPHABET table

Chi·a·ni·na \ˌkē-ə-ˈnē-nə\ *noun, plural* **Chianina** *or* **Chianinas** [Italian, from feminine of *chianino* of the *Chiani* River valley, Italy] (1914)
: any of a breed of tall white cattle of Italian origin noted especially for producing lean meat

Chi·an·ti \kē-ˈän-tē, -ˈan-\ *noun* [Italian, from the *Chianti* region, Italy] (1833)
: a dry usually red wine from the Tuscany region of Italy; *also* **:** a similar wine made elsewhere

chiar·oscu·rist \kē-ˌär-ə-ˈskyu̇r-ist, kē-ˌar-, -ˈsku̇r-\ *noun* (circa 1798)
: an artist who specializes in chiaroscuro

chiar·oscu·ro \-ˈskyu̇r-(ˌ)ō, -ˈsku̇r-\ *noun, plural* **-ros** [Italian, from *chiaro* clear, light + *oscuro* obscure, dark] (1686)
1 : pictorial representation in terms of light and shade without regard to color
2 a : the arrangement or treatment of light and dark parts in a pictorial work of art **b :** the interplay or contrast of dissimilar qualities (as of mood or character)
3 : a 16th century woodcut technique involving the use of several blocks to print different tones of the same color; *also* **:** a print made by this technique
4 : the interplay of light and shadow on or as if on a surface
5 : the quality of being veiled or partly in shadow

chi·asm \ˈkī-ˌa-zᵊm, ˈkē-\ *noun* [New Latin *chiasma*] (1870)
: CHIASMA 1

chi·as·ma \kī-ˈaz-mə, kē-\ *noun, plural* **-ma·ta** \-mə-tə\ [New Latin, X-shaped configuration, from Greek, crosspiece, from *chiazein* to mark with a chi, from *chi* (x)] (1839)
1 : an anatomical intersection or decussation — compare OPTIC CHIASMA
2 : a cross-shaped configuration of paired chromatids visible in the diplotene of meiotic prophase and considered the cytological equivalent of genetic crossing-over
— chi·as·mat·ic \ˌkī-əz-ˈma-tik, ˌkē-\ *adjective*

chi·as·mus \kī-ˈaz-məs, kē-\ *noun* [New Latin, from Greek *chiasmos,* from *chiazein* to mark with a chi] (1871)
: an inverted relationship between the syntactic elements of parallel phrases (as in Goldsmith's *to stop too fearful, and too faint to go*)

chiaus \ˈchau̇s, ˈchau̇sh\ *noun* [Turkish *çavuş*] (1595)
: a Turkish messenger or sergeant

Chib·cha \ˈchib-(ˌ)chä\ *noun, plural* **Chibcha** *or* **Chibchas** [Spanish] (1814)
1 : a member of an Indian people of central Colombia
2 : the extinct language of the Chibcha people

Chib·chan \-chən\ *adjective* (1902)
: of, relating to, or constituting a language family of Colombia and Central America

chi·bouk *or* **chi·bouque** \chə-ˈbük, shə-\ *noun* [French *chibouque,* from Turkish *çubuk*] (1813)
: a long-stemmed Turkish tobacco pipe with a clay bowl

¹chic \ˈshēk\ *noun* [French] (1856)
1 : smart elegance and sophistication especially of dress or manner **:** STYLE ⟨wears her clothes with superb *chic*⟩
2 : a distinctive mode of dress or manner associated with a fashionable lifestyle, ideology, or pursuit

²chic *adjective* (1865)
1 : cleverly stylish **:** SMART ⟨the woman who is *chic* adapts fashion to her own personality —Elizabeth L. Post⟩
2 : currently fashionable ⟨a *chic* restaurant⟩
— chic·ly *adverb*
— chic·ness *noun*

Chi·ca·na \chi-ˈkä-nə *also* shi-\ *noun* [Mexican Spanish, feminine of *chicano*] (1967)
: an American woman or girl of Mexican descent
— Chicana *adjective*

¹chi·cane \shi-ˈkān, chi-\ *verb* **chi·caned; chi·can·ing** [French *chicaner,* from Middle French, to quibble, prevent justice] (circa 1672)
intransitive verb
: to use chicanery ⟨a wretch he had taught to lie and *chicane* —George Meredith⟩
transitive verb
: TRICK, CHEAT

²chicane *noun* (1686)
1 : CHICANERY
2 a : an obstacle on a racecourse **b :** a series of tight turns in opposite directions in an otherwise straight stretch of a road-racing course
3 : the absence of trumps in a hand of cards

chi·ca·nery \-ˈkān-rē, -ˈkā-nə-\ *noun, plural* **-ner·ies** (1609)
1 : deception by artful subterfuge or sophistry **:** TRICKERY
2 : a piece of sharp practice (as at law) **:** TRICK

Chi·ca·no \chi-ˈkä-(ˌ)nō *also* shi-\ *noun, plural* **-nos** [Mexican Spanish, alteration of Spanish *mexicano* Mexican] (1947)
: an American of Mexican descent
— Chicano *adjective*

¹chi·chi \ˈshē-(ˌ)shē *also* ˈchē-(ˌ)chē\ *noun* [French] (1908)
1 : frilly or elaborate ornamentation
2 : AFFECTATION, PRECIOSITY
3 : CHIC

²chichi *adjective* (1926)
1 : elaborately ornamented **:** SHOWY, FRILLY ⟨a *chichi* dress⟩
2 : ARTY, PRECIOUS ⟨*chichi* poetry⟩
3 : CHIC, FASHIONABLE ⟨a *chichi* nightclub⟩

chick \ˈchik\ *noun* [Middle English *chyke,* alteration of *chiken*] (15th century)
1 a : a domestic chicken; *especially* **:** one newly hatched **b :** the young of any bird
2 : CHILD
3 *slang* **:** a young woman

chick·a·dee \ˈchi-kə-(ˌ)dē\ *noun* [imitative] (1838)
: any of several crestless American titmice (genus *Parus*) usually with the crown of the head sharply demarked and darker than the body

chick·a·ree \ˈchi-kə-ˌrē\ *noun* [imitative] (1829)
: RED SQUIRREL; *also* **:** a related squirrel (*Tamiasciurus douglasii*) of forests from southwestern British Columbia to central California

Chick·a·saw \ˈchi-kə-ˌsȯ\ *noun, plural* **Chickasaw** *or* **Chickasaws** (1674)
1 : a member of an American Indian people of Mississippi and Alabama
2 : a dialect of Choctaw spoken by the Chickasaw

¹chick·en \ˈchi-kᵊn, *sometimes* -kᵊŋ\ *noun* [Middle English *chiken,* from Old English *cicen* young chicken; akin to Old English *cocc* cock] (14th century)
1 a : the common domestic fowl (*Gallus gallus*) especially when young; *also* **:** its flesh used as food — compare JUNGLE FOWL **b :** any of various birds or their young
2 : a young woman
3 a : COWARD **b :** any of various contests in which the participants risk personal safety in order to see which one will give up first
4 [short for *chickenshit*] *slang* **:** petty details
5 : a young male homosexual

²chicken *adjective* (1941)
1 a : SCARED **b :** TIMID, COWARDLY
2 *slang* **a :** insistent on petty details of duty or discipline **b :** PETTY, UNIMPORTANT

³chicken *intransitive verb* **chick·ened; chick·en·ing** \ˈchi-kᵊn-iŋ, ˈchik-niŋ\ (1943)
: to lose one's nerve — usually used with *out* ⟨seemed to exhibit courage, manliness, and conviction when others *chickened* out —J. R. Seeley⟩

chicken–and–egg *adjective* [from the proverbial question "which came first, the chicken or the egg?"] (1959)
: of, relating to, or being a cause-and-effect dilemma

chicken colonel *noun* [from the eagle serving as insignia of the rank] (1947)
slang **:** COLONEL 1a

chicken feed *noun* (1836)
slang **:** a paltry sum (as in profits or wages)

chick·en–fried steak \ˈchi-kᵊn-ˈfrīd-\ *noun* (1952)
: steak coated with batter, fried, and served with gravy

chicken hawk *noun* (1827)
: a hawk that preys or is believed to prey on chickens

chick·en·heart·ed \ˈchi-kən-ˌhär-təd\ *adjective* (1681)
: TIMID, COWARDLY

chick·en–liv·ered \-ˌli-vərd\ *adjective* (1872)
: FAINTHEARTED, COWARDLY

chicken pox *noun* (circa 1738)
: an acute contagious disease especially of children marked by low-grade fever and formation of vesicles and caused by a herpesvirus — compare HERPES ZOSTER

¹chick·en·shit \ˈchi-kᵊn-ˌshit\ *noun* (1947)
1 : the petty details of a duty or discipline — usually considered vulgar
2 : COWARD, CHICKEN — usually considered vulgar

²chickenshit *adjective* (1951)
1 : PETTY, INSIGNIFICANT — usually considered vulgar
2 : lacking courage, manliness, or effectiveness — usually considered vulgar

chicken snake *noun* (1709)
: RAT SNAKE

chicken wire *noun* [from its use for making enclosures for chickens] (circa 1904)
: a light galvanized wire netting of hexagonal mesh

chick·pea \ˈchik-ˌpē\ *noun* [alteration of *chich pea,* from Middle English *chiche,* from Middle French, from Latin *cicer*] (circa 1722)
: an Asian leguminous herb (*Cicer arietinum*) cultivated for its short pods with one or two seeds; *also* **:** its seed

chick·weed \-ˌwēd\ *noun* (14th century)
: any of various low-growing small-leaved weedy plants of the pink family (especially genera *Cerastium* and *Stellaria*); *especially* **:** a cosmopolitan weed (*Stelleria media*) naturalized in the U.S. from Eurasia

chi·cle \ˈchi-kəl, ˈchi-klē\ *noun* [American Spanish, from Nahuatl *tzictli*] (circa 1889)
: a gum from the latex of the sapodilla used as the chief ingredient of chewing gum

chic·o·ry *also* **chick·o·ry** \ˈchi-k(ə-)rē\ *noun, plural* **-ries** [Middle English *cicoree,* from Middle French *cichorée, chicorée,* from Latin *cichoreum,* from Greek *kichoreia*] (15th century)
1 : a thick-rooted usually blue-flowered European perennial composite herb (*Cichorium intybus*) widely grown for its roots and as a salad plant — compare ENDIVE 2, RADICCHIO
2 : the dried ground roasted root of chicory used to flavor or adulterate coffee

chide \ˈchīd\ *verb* **chid** \ˈchid\ *or* **chid·ed** \ˈchī-dəd\; **chid** *or* **chid·den** \ˈchi-dᵊn\ *or* **chided; chid·ing** \ˈchī-diŋ\ [Middle English, from Old English *cīdan* to quarrel, chide, from *cīd* strife] (before 12th century)
intransitive verb
: to speak out in angry or displeased rebuke
transitive verb
: to voice disapproval to **:** reproach in a usually mild and constructive manner **:** SCOLD
synonym see REPROVE

¹chief \ˈchēf\ *adjective* (14th century)
1 : accorded highest rank or office ⟨*chief* librarian⟩
2 : of greatest importance or influence ⟨the *chief* reasons⟩

²chief *adverb* (14th century)
archaic **:** CHIEFLY

³chief *noun* [Middle English, from Middle French, head, chief, from Latin *caput* head — more at HEAD] (15th century)

1 : the upper part of a heraldic field
2 : the head of a body of persons or an organization **:** LEADER ⟨*chief* of police⟩
3 : the principal or most valuable part
— **chief·dom** \-dəm\ *noun*
— **chief·ship** \-ˌship\ *noun*
— **in chief :** in the chief position or place — often used in titles ⟨commander *in chief*⟩

chief executive *noun* (1833)
: a principal executive officer: as **a :** the president of a republic **b :** the governor of a state

chief justice *noun* (1534)
: the presiding or principal judge of a court of justice

[1]chief·ly \ˈchē-flē\ *adverb* (14th century)
1 : most importantly **:** PRINCIPALLY, ESPECIALLY
2 : for the most part **:** MOSTLY, MAINLY

[2]chiefly *adjective* (1870)
: of or relating to a chief ⟨*chiefly* duties⟩

chief master sergeant *noun* (1959)
: a noncommissioned officer in the air force ranking above a senior master sergeant

chief master sergeant of the air force (circa 1961)
: the ranking noncommissioned officer in the air force serving as adviser to the chief of staff

chief of naval operations (1915)
: the commanding officer of the navy and a member of the Joint Chiefs of Staff

chief of staff (circa 1881)
1 : the ranking officer of a staff in the armed forces serving as principal adviser to a commander
2 : the commanding officer of the army or air force and a member of the Joint Chiefs of Staff

chief of state (1950)
: the formal head of a national state as distinguished from the head of the government

chief petty officer *noun* (circa 1887)
: an enlisted man in the navy or coast guard ranking above a petty officer first class and below a senior chief petty officer

chief·tain \ˈchēf-tən\ *noun* [Middle English *chieftaine,* from Middle French *chevetain,* from Late Latin *capitaneus* chief — more at CAPTAIN] (14th century)
: a chief especially of a band, tribe, or clan
— **chief·tain·ship** \-ˌship\ *noun*

chief·tain·cy \-sē\ *noun, plural* **-cies** (1788)
1 : the rank, dignity, office, or rule of a chieftain
2 : a region or a people ruled by a chief **:** CHIEFDOM

chief warrant officer *noun* (1956)
: a warrant officer of senior rank in the armed forces; *also* **:** a commissioned officer in the navy or coast guard ranking below an ensign

chiel \ˈchē(ə)l\ *or* **chield** \ˈchē(ə)ld\ *noun* [Middle English (Scots) *cheld,* alteration of Middle English *child* child] (1728)
chiefly Scottish **:** FELLOW, LAD

chiff·chaff \ˈchif-ˌchaf\ *noun* [imitative] (1780)
: a small grayish European warbler (*Phylloscopus collybita*)

[1]chif·fon \shi-ˈfän, ˈshi-ˌ\ *noun* [French, literally, rag, from *chiffe* old rag, alteration of Middle French *chipe,* from Middle English *chip* chip] (1765)
1 : an ornamental addition (as a knot of ribbons) to a woman's dress
2 : a sheer fabric especially of silk

[2]chiffon *adjective* (1903)
1 : resembling chiffon in sheerness or softness
2 : having a light delicate texture achieved usually by adding whipped egg whites or whipped gelatin ⟨lemon *chiffon* pie⟩

chif·fo·nade \ˌshi-fə-ˈnäd, -ˈnäd\ *noun* [French *chiffonnade,* from *chiffonner* to crumple, from *chiffon*] (1877)
: shredded or finely cut vegetables used especially as a garnish

chif·fo·nier \ˌshi-fə-ˈnir\ *noun* [French *chiffonnier,* from *chiffon*] (1765)
: a high narrow chest of drawers

chif·fo·robe \ˈshi-fə-ˌrōb\ *noun* [*chiffonier* + ward*robe*] (1908)
: a combination of wardrobe and chest of drawers

chifforobe

chig·ger \ˈchi-gər, ˈji-\ *noun* (1756)
1 : CHIGOE 1
2 [alteration of *[2]jigger*] **:** a 6-legged mite larva (family Trombiculidae) that sucks the blood of vertebrates and causes intense irritation

chi·gnon \ˈshēn-ˌyän *also* -ˌyȯn\ *noun* [French, from Middle French *chaignon* chain, collar, nape] (1783)
: a knot of hair that is worn at the back of the head and especially at the nape of the neck

chi·goe \ˈchi-(ˌ)gō, ˈchē-\ *noun* [Carib *chico*] (1691)
1 : a tropical flea (*Tunga penetrans*) of which the fertile female causes great discomfort by burrowing under the skin
2 : CHIGGER 2

Chi·hua·hua \chə-ˈwä-(ˌ)wä, shə-, -wə\ *noun* [Mexican Spanish, from *Chihuahua,* Mexico] (1858)
: any of a breed of very small round-headed dogs that occur in short-coated and long-coated varieties

Chihuahua

chil·blain \ˈchil-ˌblān\ *noun* [[1]*chill*] (1547)
: an inflammatory swelling or sore caused by exposure (as of the feet or hands) to cold

child \ˈchī(ə)ld\ *noun, plural* **chil·dren** \ˈchil-drən, -dərn\ *often attributive* [Middle English, from Old English *cild;* akin to Gothic *kilthei* womb, and perhaps to Sanskrit *jaṭhara* belly] (before 12th century)
1 a : an unborn or recently born person **b** *dialect* **:** a female infant
2 a : a young person especially between infancy and youth **b :** a childlike or childish person **c :** a person not yet of age
3 *usually* **childe** \ˈchī(ə)ld\ *archaic* **:** a youth of noble birth
4 a : a son or daughter of human parents **b :** DESCENDANT
5 : one strongly influenced by another or by a place or state of affairs
6 : PRODUCT, RESULT ⟨barbed wire . . . is truly a *child* of the plains —W. P. Webb⟩
— **child·less** \ˈchī(ə)l(d)-ləs\ *adjective*
— **child·less·ness** *noun*
— **with child :** PREGNANT

child·bear·ing \ˈchīl(d)-ˌbar-iŋ, -ˌber-\ *adjective* (14th century)
: of or relating to the process of conceiving, being pregnant with, and giving birth to children ⟨women of *childbearing* age⟩
— **childbearing** *noun*

child·bed \-ˌbed\ *noun* (13th century)
: the condition of a woman in childbirth

childbed fever *noun* (1928)
: PUERPERAL FEVER

child·birth \ˈchīl(d)-ˌbərth\ *noun* (15th century)
: PARTURITION

child·hood \ˈchīld-ˌhu̇d\ *noun* (before 12th century)
1 : the state or period of being a child
2 : the early period in the development of something

child·ish \ˈchīl-dish\ *adjective* (before 12th century)
1 : of, relating to, or befitting a child or childhood
2 a : marked by or suggestive of immaturity and lack of poise ⟨a *childish* spiteful remark⟩ **b :** lacking complexity **:** SIMPLE ⟨it's a *childish* device, but it works⟩ **c :** deteriorated with age especially in mind **:** SENILE
— **child·ish·ly** *adverb*
— **child·ish·ness** *noun*

child·like \ˈchī(ə)l(d)-ˌlīk\ *adjective* (1586)
: of, relating to, or resembling a child or childhood; *especially* **:** marked by innocence, trust, and ingenuousness
— **child·like·ness** *noun*

child·ly \ˈchī(ə)l(d)-lē\ *adjective* (before 12th century)
: CHILDLIKE

child·proof \ˈchīl(d)-ˌprüf\ *adjective* (1956)
: designed to prevent tampering or opening by children ⟨*childproof* pill bottles⟩

child's play *noun* (14th century)
1 : an extremely simple task or act
2 : something that is insignificant ⟨his injury was *child's play* compared with the damage he inflicted⟩

Chile saltpeter \ˈchi-lē-, ˈchē-(ˌ)lā-\ *noun* [*Chile,* South America] (circa 1909)
: sodium nitrate especially occurring naturally (as in caliche)

chili *or* **chile** *or* **chil·li** \ˈchi-lē\ *noun, plural* **chil·ies** *or* **chil·es** *or* **chil·lies** [Spanish *chile,* from Nahuatl *chīlli*] (1604)
1 a : HOT PEPPER **b** *usually chilli, chiefly British* **:** a pepper whether hot or sweet
2 a : a thick sauce of meat and chilies **b :** CHILI CON CARNE

chil·i·ad \ˈki-lē-ˌad, -əd\ *noun* [Late Latin *chiliad-, chilias,* from Greek, from *chilioi* thousand] (1598)
1 : a group of 1000
2 : MILLENNIUM 2a

chil·i·asm \ˈki-lē-ˌa-zᵊm\ *noun* [New Latin *chiliasmus,* from Late Latin *chiliastes* one that believes in chiliasm, from *chilias*] (1610)
: MILLENARIANISM
— **chil·i·ast** \-lē-ˌast, -lē-əst\ *noun*
— **chil·i·as·tic** \ˌki-lē-ˈas-tik\ *adjective*

chili con car·ne \ˌchi-lē-ˌkän-ˈkär-nē, -kən-\ *noun* [American Spanish *chile con carne* chili with meat] (1857)
: a spiced stew of ground beef and minced chilies or chili powder usually with beans

chili dog *noun* (1969)
: a hot dog topped with chili

chili powder *noun* (1938)
: a condiment made with chilies ground to a powder

chili sauce *noun* (1880)
: a spiced tomato sauce usually made with red and green peppers

[1]chill \ˈchil\ *noun* [Middle English *chile* chill, frost, from Old English *ciele;* akin to Old English *ceald* cold] (before 12th century)
1 a : a sensation of cold accompanied by shivering **b :** a disagreeable sensation of coldness
2 : a moderate but disagreeable degree of cold
3 : a check to enthusiasm or warmth of feeling ⟨felt the *chill* of his opponent's stare⟩

[2]chill *adjective* (14th century)
1 a : moderately cold **b :** COLD, RAW
2 : affected by cold ⟨*chill* travelers⟩
3 : DISTANT, FORMAL ⟨a *chill* reception⟩
4 : DEPRESSING, DISPIRITING ⟨*chill* penury —Thomas Gray⟩
— **chill·ness** *noun*

[3]chill (14th century)
intransitive verb
1 a : to become cold **b :** to shiver or quake with or as if with cold
2 : to become taken with a chill
transitive verb
1 a : to make cold or chilly **b :** to make cool especially without freezing
2 : to affect as if with cold **:** DISPIRIT, DISCOURAGE ⟨were *chilled* by the drab austerity —William Attwood⟩

— **chill·ing·ly** \'chi-liŋ-lē\ *adverb*

chill·er \'chi-lər\ *noun* (1798)
1 : one that chills
2 : an eerie or frightening story of murder, violence, or the supernatural

chill factor *noun* (1965)
: WINDCHILL

chill out *intransitive verb* (1983) *slang*
: to calm down **:** go easy **:** RELAX — often used in the imperative

chil·lum \'chi-ləm\ *noun* [Hindi *cilam,* from Persian *chilam*] (1781)
1 : the part of a water pipe that contains the substance (as tobacco or hashish) which is smoked; *also* **:** a quantity of a substance thus smoked
2 : a funnel-shaped clay pipe for smoking

chilly \'chi-lē\ *adjective* **chill·i·er; -est** (1570)
1 : noticeably cold **:** CHILLING
2 : unpleasantly affected by cold
3 : lacking warmth of feeling
4 : tending to arouse fear or apprehension ⟨*chilly* suspicions⟩
— **chill·i·ly** \'chi-lə-lē\ *adverb*
— **chill·i·ness** \'chi-lē-nəs\ *noun*

[1]**chi·mae·ra** \kī-'mir-ə, kə-\ *variant of* CHIMERA

[2]**chimaera** *noun* [New Latin, genus name, from Latin, chimera] (1804)
: any of a family (Chimaeridae) of marine cartilaginous fishes with a tapering or threadlike tail and usually no anal fin

chi·mae·ric, chi·mae·rism *British variant of* CHIMERIC, CHIMERISM

[1]**chime** \'chīm\ *noun* [Middle English *chimbe,* from Old English *cimb-;* akin to Middle Dutch *kimme* edge of a cask] (14th century)
: the edge or rim of a cask or drum

[2]**chime** *verb* **chimed; chim·ing** (14th century)
intransitive verb
1 a : to make a musical and especially a harmonious sound **b :** to make the sounds of a chime
2 : to be or act in accord ⟨the music and the mood *chimed* well together⟩
transitive verb
1 : to cause to sound musically by striking
2 : to produce by chiming
3 : to call or indicate by chiming ⟨the clock *chimed* midnight⟩
4 : to utter repetitively **:** DIN 2
— **chim·er** *noun*

[3]**chime** *noun* [Middle English, cymbal, from Middle French *chimbe,* from Old French, from Latin *cymbalum* cymbal] (15th century)
1 : an apparatus for chiming a bell or set of bells
2 a : a musically tuned set of bells **b :** one of a set of objects giving a bell-like sound when struck
3 a : the sound of a set of bells — usually used in plural **b :** a musical sound suggesting that of bells
4 : ACCORD, HARMONY ⟨such happy *chime* of fact and theory —Henry Maudsley⟩

chime in (1681)
intransitive verb
1 : to combine harmoniously ⟨the artist's illustrations *chime in* perfectly with the text —*Book Production*⟩
2 : to break into a conversation or discussion especially to express an opinion
transitive verb
: to remark while chiming in

chi·me·ra \kī-'mir-ə, kə-\ *noun* [Latin *chimaera,* from Greek *chimaira* she-goat, chimera; akin to Old Norse *gymbr* yearling ewe, Greek *cheimōn* winter — more at HIBERNATE]
1 a *cap* **:** a fire-breathing she-monster in Greek mythology having a lion's head, a goat's body, and a serpent's tail **b :** an imaginary monster compounded of incongruous parts
2 : an illusion or fabrication of the mind; *especially* **:** an unrealizable dream ⟨a fancy, a *chimera* in my brain, troubles me in my prayer —John Donne⟩
3 : an individual, organ, or part consisting of tissues of diverse genetic constitution

chi·mere \shə-'mir, chə-\ *noun* [Middle English *chimmer, chemeyr*] (14th century)
: a loose sleeveless robe worn by Anglican bishops over the rochet

chi·me·ric \kī-'mir-ik, kə-, -'mer-\ *adjective* (1973)
: relating to, derived from, or being a genetic chimera or its genetic material ⟨a *chimeric* cat⟩

chi·me·ri·cal \kī-'mer-i-kəl, kə-, -'mir-\ *also* **chi·me·ric** \-ik\ *adjective* [*chimera*] (1638)
1 : existing only as the product of unchecked imagination **:** fantastically visionary or improbable
2 : given to fantastic schemes
synonym see IMAGINARY
— **chi·me·ri·cal·ly** \-i-k(ə-)lē\ *adverb*

chi·me·rism \kī-'mir-ˌi-zəm, kə-; 'kī-mə-ˌri-\ *noun* (1961)
: the state of being a genetic chimera

chi·mi·chan·ga \ˌchi-mē-'chäŋ-gə\ *noun* [Mexican Spanish, trinket] (1982)
: a tortilla wrapped around a filling (as of meat) and deep-fried

chim·ney \'chim-nē\ *noun, plural* **chimneys** [Middle English, from Middle French *cheminée,* from Late Latin *caminata,* from Latin *caminus* furnace, fireplace, from Greek *kaminos;* perhaps akin to Greek *kamara* vault] (14th century)
1 *dialect* **:** FIREPLACE, HEARTH
2 : a vertical structure incorporated into a building and enclosing a flue or flues that carry off smoke; *especially* **:** the part of such a structure extending above a roof
3 : SMOKESTACK
4 : a tube usually of glass placed around a flame (as of a lamp)
5 : something resembling a chimney: as **a :** a narrow cleft in rock **b :** a tall column of rock on the ocean floor that is formed by the precipitation of minerals from superheated water issuing from a vent in the earth's crust and rising through the column of rock
— **chim·ney·like** \-ˌlīk\ *adjective*

chim·ney·piece \'chim-nē-ˌpēs\ *noun* (1680)
: an ornamental construction over and around a fireplace that includes the mantel

chimney pot *noun* (circa 1806)
: a usually earthenware pipe placed at the top of a chimney

chimney sweep *noun* (1727)
: a person whose occupation is cleaning soot from chimney flues — called also *chimney sweeper*

chimney swift *noun* (1849)
: a small sooty-gray swift (*Chaetura pelagica*) with long narrow wings that often builds its nest inside an unused chimney — called also *chimney swallow*

chimp \'chimp *sometimes* 'shimp\ *noun* (1877)
: CHIMPANZEE

chim·pan·zee \(ˌ)chim-ˌpan-'zē, -pən-'zē, -'pan-zē *sometimes* (ˌ)shim-\ *noun* [Kongo dialect *chimpenzi*] (1738)
: an anthropoid ape (*Pan troglodytes*) of equatorial Africa that is smaller and more arboreal than the gorilla — compare PYGMY CHIMPANZEE

chimpanzee

[1]**chin** \'chin\ *noun* [Middle English, from Old English *cinn;* akin to Old High German *kinni* chin, Latin *gena* cheek, Greek *genys* jaw, cheek] (before 12th century)
1 : the lower portion of the face lying below the lower lip and including the prominence of the lower jaw
2 : the surface beneath or between the branches of the lower jaw
— **chin·less** \-ləs\ *adjective*

[2]**chin** *verb* **chinned; chin·ning** (1869)
transitive verb
1 : to bring to or hold with the chin ⟨*chin* a violin⟩
2 : to raise (oneself) while hanging by the hands until the chin is level with the support
intransitive verb
slang **:** to talk idly

chi·na \'chī-nə\ *noun* [Persian *chīnī* Chinese porcelain] (1579)
1 : PORCELAIN; *also* **:** vitreous porcelain wares (as dishes, vases, or ornaments) for domestic use
2 : earthenware or porcelain tableware

China aster *noun* (1794)
: a common annual garden aster (*Callistephus chinensis*) native to northern China that occurs in many showy forms

chi·na·ber·ry \'chī-nə-ˌber-ē, *Southern also* 'chā-nē-ˌber-ē\ *noun* (1890)
: a small Asian tree (*Melia azedarach*) of the mahogany family naturalized in the southern U.S. where it is widely planted for shade or ornament

china clay *noun* (1840)
: KAOLIN

china closet *noun* (1771)
: a cabinet or cupboard for the storage or display of household china

Chi·na·man \'chī-nə-mən\ *noun* (1849)
: a native of China **:** CHINESE — often taken to be offensive

China rose *noun* (circa 1731)
: any of numerous garden roses derived from a shrubby Chinese rose (*Rosa chinensis*)

China Syndrome *noun* [from the notion that the molten reactor contents could theoretically sink through the earth to reach China] (1970)
: MELTDOWN 1

Chi·na·town \'chī-nə-ˌtau̇n\ *noun* (1857)
: the Chinese quarter of a city

China tree *noun* (1819)
: CHINABERRY

chi·na·ware \'chī-nə-ˌwar, -ˌwer\ *noun* (1634)
: tableware made of china

chin·bone \'chin-ˌbōn\ *noun* (before 12th century)
: MANDIBLE; *especially* **:** the median anterior part of the human mandible

chinch \'chinch\ *noun* [Spanish *chinche,* from Latin *cimic-, cimex*] (1616)
: BEDBUG

chinch bug *noun* (1785)
: a small black-and-white bug (*Blissus leucopterus*) very destructive to cereal grasses

chin·che·rin·chee \ˌchin-chə-ri(n)-'chē, ˌchiŋ-kə-\ *noun, plural* **chincherinchee** *or* **chincherinchees** [Afrikaans *tjienkerientjee*] (1904)
: a southern African perennial bulbous herb (*Ornithogalum thyrsoides*) of the lily family with spikes of white to golden-yellow blossoms

chin·chil·la \chin-'chi-lə\ *noun* [Spanish] (1604)
1 : either of two small South American rodents (*Chinchilla laniger* and *C. brevicaudata*) of the high Andes that are the size of large squirrels, have very soft pearly gray fur, and are extensively bred in captivity; *also* **:** the fur of a chinchilla

chinchilla

2 : a heavy twilled woolen coating

¹chine \'chīn\ *noun* [Middle English, from Middle French *eschine,* of Germanic origin; akin to Old High German *scina* shinbone, needle — more at SHIN] (14th century)
1 : BACKBONE, SPINE; *also* **:** a cut of meat including all or part of the backbone
2 : the intersection of the bottom and the sides of a flat or V-bottomed boat

²chine *transitive verb* **chined; chin·ing** (circa 1611)
: to cut through the backbone of (as in butchering)

Chi·nese \chī-'nēz, -'nēs\ *noun, plural* **Chinese** (1606)
1 a : a native or inhabitant of China **b :** a person of Chinese descent
2 : a group of related languages used by the people of China that are often mutually unintelligible in their spoken form but share a single system of writing and that constitute a branch of the Sino-Tibetan language family; *especially* **:** MANDARIN
— Chinese *adjective*

Chinese boxes *noun plural* (1829)
1 : a set of boxes graduated in size so that each fits into the next larger one
2 : something that resembles a set of Chinese boxes especially in complexity

Chinese cabbage *noun* (1842)
: either of two Asian brassicas now grown in the U.S. and widely used as greens: **a :** BOK CHOY **b :** one (*Brassica rapa pekinensis*) that forms elongate more or less solid cylindrical heads and has pale green or cream-colored leaves

Chinese checkers *noun plural but singular or plural in construction* (1938)
: a game in which each player seeks to be the first to transfer a set of marbles from a home point to the opposite point of a pitted 6-pointed star by single moves or jumps

Chinese chestnut *noun* (circa 1909)
: an Asian chestnut (*Castanea mollissima*) that is resistant to chestnut blight

Chinese copy *noun* (1920)
: an exact imitation or duplicate that includes defects as well as desired qualities

Chinese crested *noun* (1976)
: any of a breed of hairless or coated dogs with a plumed tail and a crest of hair on the head

Chinese gooseberry *noun* (1925)
: a subtropical vine (*Actinidia chinensis* of the family Actinidiaceae) that bears kiwifruit; *also* **:** KIWIFRUIT

Chinese lantern *noun* (1825)
: a collapsible translucent covering for a light

Chinese parsley *noun* (circa 1953)
: CILANTRO

Chinese puzzle *noun* (circa 1815)
1 : an intricate or ingenious puzzle
2 : something intricate and obscure

Chinese restaurant syndrome *noun* (1968)
: a group of symptoms (as numbness of the neck, arms, and back with headache, dizziness, and palpitations) that is held to affect susceptible persons eating food and especially Chinese food heavily seasoned with monosodium glutamate

chinese shar–pei *noun* (1975)
: SHAR-PEI

Chinese wall *noun* [*Chinese Wall,* a defensive wall built in the 3d century B.C. between China and Mongolia] (1900)
: a strong barrier; *especially* **:** a serious obstacle to understanding

Chinese white *noun* (circa 1884)
: ZINC WHITE

Ching *or* **Ch'ing** \'chiŋ\ *noun* [Chinese (Beijing) *Qīng*] (1795)
: a Manchu dynasty in China dated 1644–1912 and the last imperial dynasty

¹chink \'chiŋk\ *noun* [probably alteration of Middle English *chine* crack, fissure] (1535)
1 : a small cleft, slit, or fissure ⟨a *chink* in the curtain⟩
2 : a weak spot that may leave one vulnerable
3 : a narrow beam of light shining through a chink

²chink *transitive verb* (1748)
: to fill the chinks of (as by caulking) ⟨*chink* a log cabin⟩

³chink *noun* [imitative] (1573)
1 *archaic* **:** COIN, MONEY
2 : a short sharp sound

⁴chink (1589)
intransitive verb
: to make a slight sharp metallic sound
transitive verb
: to cause to make a chink

Chink \'chiŋk\ *noun or adjective* [perhaps alteration of *Chinese*] (1887)
: CHINESE — usually taken to be offensive

chi·no \'chē-(,)nō, 'shē-\ *noun, plural* **chinos** [origin unknown] (1943)
1 : a usually khaki cotton or synthetic-fiber twill of the type used for military uniforms
2 *plural* **:** an article of clothing made of chino

chi·noi·se·rie \shēn-'wäz-rē, -'wä-zə-; ,shēn-,wäz-'rē, -,wä-zə-\ *noun* [French, from *chinois* Chinese, from *Chine* China] (1883)
: a style in art (as in decoration) reflecting Chinese qualities or motifs; *also* **:** an object or decoration in this style

Chi·nook \shə-'nu̇k, chə-, -'nük\ *noun, plural* **Chinook** *or* **Chinooks** [Lower Chehalis (Salishan language of western Washington) *činúk,* name of a Chinook village] (1805)
1 : a member of an American Indian people of the north shore of the Columbia River at its mouth
2 : a Chinookan language of the Chinook and other nearby peoples
3 *often not capitalized* **a :** a warm moist southwest wind of the coast from Oregon northward **b :** a warm dry wind that descends the eastern slopes of the Rocky Mountains
4 *not capitalized* **:** CHINOOK SALMON

Chi·nook·an \-'nu̇-kən, -'nü-\ *noun* (circa 1890)
: an American Indian language family of Washington and Oregon
— Chinookan *adjective*

Chinook jargon *noun, often J capitalized* (1840)
: a pidgin language based on Chinook and other Indian languages, French, and English and formerly used as a lingua franca in the northwestern U.S. and on the Pacific coast of Canada and Alaska

chinook salmon *noun* (1851)
: a large commercially important Pacific salmon (*Oncorhynchus tshawytscha*) with red flesh that occurs in the northern Pacific Ocean — called also *king salmon*

chin·qua·pin *or* **chin·ka·pin** \'chiŋ-ki-,pin\ *noun* [probably modification of Virginia Algonquian *chechinquamin* chinquapin nut] (1612)
1 : the edible nut of a chinquapin
2 : any of several trees (genera *Castanea* and *Castanopsis*); *especially* **:** a dwarf chestnut (*Castanea pumila*) of the U.S.

chinquapin oak *noun* (1785)
: either of two chestnut oaks (*Quercus muhlenbergii* and *Q. prinoides*) of the eastern U.S.

chintz \'chin(t)s\ *noun* [earlier *chints,* plural of *chint,* from Hindi *chī̃ṭ*] (1614)
1 : a printed calico from India
2 : a usually glazed printed cotton fabric

chintzy \'chin(t)-sē\ *adjective* **chintz·i·er; -est** (1851)
1 : decorated with or as if with chintz
2 a : GAUDY, CHEAP ⟨*chintzy* toys⟩ **b :** STINGY

chin–up \'chi-(,)nəp\ *noun* (1954)
: the act or an instance of chinning oneself especially as a conditioning exercise

chin–wag \'chin-,wag\ *noun* (1879)
slang **:** CONVERSATION, CHAT

chi·o·no·doxa \,kī-ə-nō-'däk-sə, kī-,ä-nə-\ *noun* [New Latin, genus name, from Greek *chion-, chiōn* snow (akin to Greek *cheimōn* winter) + *doxa* glory — more at HIBERNATE, DOXOLOGY] (1879)
: GLORY-OF-THE-SNOW

¹chip \'chip\ *noun* [Middle English; akin to Old English *-cippian*] (14th century)
1 a : a small usually thin and flat piece (as of wood or stone) cut, struck, or flaked off **b** (1) **:** a small thin slice of food; *especially* **:** POTATO CHIP (2) **:** FRENCH FRY
2 : something small, worthless, or trivial
3 a : one of the counters used as a token for money in poker and other games **b** *plural* **:** MONEY — used especially in the phrase *in the chips* **c :** something valuable that can be used for advantage in negotiation or trade ⟨a bargaining *chip*⟩
4 : a piece of dried dung — usually used in combination ⟨cow *chip*⟩
5 : a flaw left after a chip is removed
6 a : INTEGRATED CIRCUIT **b :** a small wafer of semiconductor material that forms the base for an integrated circuit
7 : CHIP SHOT
— chip off the old block : a child that resembles his or her parent
— chip on one's shoulder : a challenging or belligerent attitude

²chip *verb* **chipped; chip·ping** [Middle English *chippen,* from Old English *-cippian* (as in *forcippian* to cut off); akin to Old English *cipp* beam, Old High German *chipfa* stave] (15th century)
transitive verb
1 a : to cut or hew with an edged tool **b** (1) **:** to cut or break (a small piece) from something (2) **:** to cut or break a fragment from (3) **:** to cut into chips ⟨*chip* a tree stump⟩
2 *British* **:** CHAFF, BANTER
3 : to hit (a return in tennis) with backspin
intransitive verb
1 : to break off in small pieces
2 : to play a chip shot

chip·board \'chip-,bōrd, -,bȯrd\ *noun* (1919)
: a paperboard usually made entirely from wastepaper

chip in (1861)
intransitive verb
1 : CONTRIBUTE ⟨everyone *chipped* in for the gift⟩
2 *chiefly British* **:** CHIME IN 2
transitive verb
: CONTRIBUTE

chip·munk \'chip-,məŋk\ *noun* [alteration of earlier *chitmunk,* probably from Ojibwa *ačitamo·nʔ* red squirrel] (1832)
: any of a genus (*Tamias*) of small striped North American and Asian rodents of the squirrel family

chipped beef \'chip(t)-\ *noun* (1859)
: smoked dried beef sliced thin

Chip·pen·dale \'chi-pən-,dāl\ *adjective* [Thomas *Chippendale*] (1876)
: of or relating to an 18th century English furniture style characterized by graceful outline and often ornate rococo ornamentation

¹chip·per \'chi-pər\ *noun* (1513)
: one that chips

²chipper *adjective* [perhaps alteration of English dialect *kipper* (lively)] (1838)
: SPRIGHTLY

Chip·pe·wa \'chi-pə-,wȯ, -,wä, -,wā, -wə\ *noun, plural* **Chippewa** *or* **Chippewas** (1671)
: OJIBWA

chip·ping sparrow \'chi-piŋ-\ *noun* [*chip* to cheep] (1791)
: a small gray-breasted North American sparrow (*Spizella passerina*) with a black line through the eye, a white line above it, and in breeding plumage a reddish patch on the top of the head

chip·py \'chi-pē\ *adjective* **chip·pi·er; -est** [*chip* on one's shoulder] (1898)

: aggressively belligerent ⟨a *chippy* hockey player⟩; *also* **:** marked by much fighting ⟨a *chippy* game⟩

chip shot *noun* (1909)
: a short usually low approach shot in golf that lofts the ball to the green and allows it to roll

chir- *or* **chiro-** *combining form* [Latin, from Greek *cheir-, cheiro-,* from *cheir;* akin to Hittite *keššar* hand]
: hand ⟨*chiro*practic⟩

chi·ral \ˈkī-rəl\ *adjective* [*chir-* + [1]*-al*] (1894)
: of or relating to a molecule that is not superimposable on its mirror image
— **chi·ral·i·ty** \kī-ˈra-lə-tē, kə-\ *noun*

chiral center *noun* (1970)
: an atom especially in an organic molecule that has four different atoms or groups attached to it

Chi–Rho \ˈkī-ˈrō, ˈkē-\ *noun, plural* **Chi–Rhos** [*chi* + *rho*] (1868)
: a Christian monogram and symbol formed from the first two letters X and P of the Greek word for *Christ* — called also *Christogram*

Chir·i·ca·hua \ˌchir-ə-ˈkä-wə\ *noun, plural* **Chiricahua** *or* **Chiricahuas** (1885)
: a member of an Apache people of Arizona

chir·i·moya *variant of* CHERIMOYA

chirk \ˈchərk\ *verb* [Middle English *charken, chirken* to creak, chirp, from Old English *cearcian* to creak; akin to Old English *cracian* to crack] (1843)
: CHEER ⟨play with her and *chirk* her up a little —Harriet B. Stowe⟩

chi·rog·ra·phy \kī-ˈrä-grə-fē\ *noun* (1654)
1 : HANDWRITING, PENMANSHIP
2 : CALLIGRAPHY 1
— **chi·rog·ra·pher** \-fər\ *noun*
— **chi·ro·graph·ic** \ˌkī-rə-ˈgra-fik\ *or* **chi·ro·graph·i·cal** \-fi-kəl\ *adjective*

chi·ro·man·cy \ˈkī-rə-ˌman(t)-sē\ *noun* [probably from Middle French *chiromancie,* from Medieval Latin *chiromantia,* from Greek *cheir-* chir- + *-manteia* -mancy] (circa 1528)
: PALMISTRY
— **chi·ro·man·cer** \-ˌman(t)-sər\ *noun*

chi·ron·o·mid \kī-ˈrä-nə-məd\ *noun* [ultimately from Greek *cheironomos* one who gestures with his hands] (1915)
: any of a family (Chironomidae) of midges that lack piercing mouthparts
— **chironomid** *adjective*

chi·rop·o·dy \kə-ˈrä-pə-dē, shə- *also* kī-\ *noun* [*chir-* + *pod-;* from its original concern with both hands and feet] (1886)
: PODIATRY
— **chi·rop·o·dist** \-dist\ *noun*

chi·ro·prac·tic \ˈkī-rə-ˌprak-tik, ˌkī-rə-ˈ\ *noun* [*chir-* + Greek *praktikos* practical, operative — more at PRACTICAL] (1898)
: a system of therapy which holds that disease results from a lack of normal nerve function and which employs manipulation and specific adjustment of body structures (as the spinal column)
— **chi·ro·prac·tor** \-tər\ *noun*

chi·rop·ter·an \kī-ˈräp-tə-rən\ *noun* [ultimately from Greek *cheir* hand + *pteron* wing — more at FEATHER] (1835)
: [3]BAT

chirp \ˈchərp\ *noun* [imitative] (circa 1755)
: the characteristic short sharp sound especially of a small bird or insect
— **chirp** *intransitive verb*

chirpy \ˈchər-pē\ *adjective* **chirp·i·er; -est** (1837)
1 : cheerfully lively ⟨a *chirpy* manner⟩
2 a : making chirps **b :** suggestive of chirping ⟨a *chirpy* voice⟩
— **chirp·i·ly** \-pə-lē\ *adverb*

chirr \ˈchər\ *noun* [imitative] (circa 1600)
: the short vibrant or trilled sound characteristic of an insect (as a grasshopper or cicada)
— **chirr** *intransitive verb*

chir·rup \ˈchər-əp, ˈchir-\ *noun* [imitative] (1788)
: CHIRP
— **chirrup** *intransitive verb*

chir·rupy \ˈchər-ə-pē, ˈchir-\ *adjective* (1874)
: CHIRPY

chi·rur·geon \kī-ˈrər-jən\ *noun* [Middle English *cirurgian,* from Old French *cirurgien,* from *cirurgie* surgery] (13th century)
archaic **:** SURGEON

[1]**chis·el** \ˈchi-zᵊl\ *noun* [Middle English, from Old North French, from (assumed) Vulgar Latin *cisellum,* alteration of *caesellum,* from Latin *caesus,* past participle of *caedere* to cut] (14th century)
: a metal tool with a sharpened edge at one end used to chip, carve, or cut into a solid material (as wood, stone, or metal)

[2]**chisel** *verb* **-eled** *or* **-elled; -el·ing** *or* **-el·ling** \ˈchi-zᵊl-iŋ, ˈchiz-liŋ\ (1509)
transitive verb
1 : to cut or work with or as if with a chisel
2 : to employ shrewd or unfair practices on in order to obtain one's end; *also* **:** to obtain by such practices ⟨*chisel* a job⟩
intransitive verb
1 : to work with or as if with a chisel
2 a : to employ shrewd or unfair practices **b :** to thrust oneself **:** INTRUDE ⟨*chisel* in on a racket⟩
— **chis·el·er** *or* **chis·el·ler** \ˈchi-zᵊl-ər, ˈchiz-lər\ *noun*

chis·eled *or* **chis·elled** \ˈchi-zᵊld\ *adjective* (1821)
: formed or crafted as if with a chisel ⟨*chiseled* good looks⟩ ⟨a *chiseled* essay⟩

chi–square \ˈkī-ˈskwar, -ˈskwer\ *noun, often attributive* (circa 1934)
: a statistic that is a sum of terms each of which is a quotient obtained by dividing the square of the difference between the observed and theoretical values of a quantity by the theoretical value

chi–square distribution *noun* (circa 1956)
: a probability density function that gives the distribution of the sum of the squares of a number of independent random variables each with a normal distribution with zero mean and unit variance, that has the property that the sum of two or more random variables with such a distribution also has one, and that is widely used in testing statistical hypotheses especially about the theoretical and observed values of a quantity and about population variances and standard deviations

[1]**chit** \ˈchit\ *noun* [Middle English *chitte* kitten, cub] (circa 1624)
1 : CHILD
2 : a pert young woman

[2]**chit** *noun* [Hindi *ciṭṭhī*] (1757)
1 : a short letter or note; *especially* **:** a signed voucher of a small debt (as for food)
2 : a small slip of paper with writing on it

chit·chat \ˈchit-ˌchat\ *noun* [reduplication of *chat*] (1710)
: SMALL TALK, GOSSIP
— **chitchat** *intransitive verb*

chi·tin \ˈkī-tᵊn\ *noun* [French *chitine,* from Greek *chitōn*] (circa 1839)
: a horny polysaccharide that forms part of the hard outer integument especially of insects, arachnids, and crustaceans
— **chi·tin·ous** \ˈkī-tᵊn-əs, ˈkīt-nəs\ *adjective*

chi·ton \ˈkī-tᵊn, ˈkī-ˌtän\ *noun* [New Latin, genus name, from Greek *chitōn* tunic, of Semitic origin; akin to Hebrew *kuttōneth* tunic] (1816)
1 : any of a class (Polyplacophora) of elongated bilaterally symmetrical marine mollusks with a dorsal shell of calcareous plates
2 [Greek *chitōn*] **:** the basic garment of ancient Greece worn usually knee-length by men and full-length by women

chit·ter \ˈchi-tər\ *intransitive verb* [Middle English *chiteren,* probably of imitative origin] (13th century)
: TWITTER, CHIRP; *also* **:** CHATTER

chit·ter·lings *or* **chit·lins** \ˈchit-lənz\ *noun plural* [Middle English *chiterling*] (13th century)
: the intestines of hogs especially when prepared as food

chi·val·ric \shə-ˈval-rik\ *adjective* (1797)
: relating to chivalry **:** CHIVALROUS

chiv·al·rous \ˈshi-vəl-rəs\ *adjective* (14th century)
1 : VALIANT
2 : of, relating to, or characteristic of chivalry and knight-errantry
3 a : marked by honor, generosity, and courtesy **b :** marked by gracious courtesy and high-minded consideration especially to women
synonym see CIVIL
— **chiv·al·rous·ly** *adverb*
— **chiv·al·rous·ness** *noun*

chiv·al·ry \ˈshi-vəl-rē\ *noun, plural* **-ries** [Middle English *chivalrie,* from Middle French *chevalerie,* from *chevalier* knight — more at CHEVALIER] (14th century)
1 : mounted men-at-arms
2 *archaic* **a :** martial valor **b :** knightly skill
3 : gallant or distinguished gentlemen
4 : the system, spirit, or customs of medieval knighthood
5 : the qualities of the ideal knight **:** chivalrous conduct

chive \ˈchīv\ *noun* [Middle English, from Old North French, from Latin *cepa* onion] (14th century)
: a perennial plant (*Allium schoenoprasum*) related to the onion and having slender leaves used as a seasoning — usually used in plural

chivy *or* **chiv·vy** \ˈchi-vē\ *transitive verb* **chiv·ied** *or* **chiv·vied; chivy·ing** *or* **chiv·vy·ing** [*chivy,* noun (chase, hunt), probably from English dialect *Chevy Chase* chase, confusion, from the name of a ballad describing the battle of Otterburn (1388)] (1918)
1 : to tease or annoy with persistent petty attacks
2 : to move or obtain by small maneuvers
synonym see BAIT

chla·myd·ia \klə-ˈmi-dē-ə\ *noun, plural* **-i·ae** \-dē-ˌē\ [New Latin, from Greek *chlamyd-, chlamys*] (1966)
1 : any of a genus (*Chlamydia,* family Chlamydiaceae) of spherical gram-negative intracellular bacteria; *especially* **:** one (*C. trachomatis*) that causes or is associated with various diseases of the eye and urogenital tract including trachoma, lymphogranuloma venereum, cervicitis, and some forms of urethritis
2 : a disease or infection caused by chlamydiae
— **chla·myd·i·al** \-dē-əl\ *adjective*

chla·mydo·spore \klə-ˈmi-də-ˌspōr, -ˌspȯr\ *noun* [Latin *chlamyd-, chlamys* + International Scientific Vocabulary *spore*] (1884)
: a thick-walled usually resting fungal spore

chla·mys \ˈkla-məs, ˈklā-\ *noun, plural* **chla·mys·es** *or* **chla·my·des** \-mə-ˌdēz\ [Latin *chlamyd-, chlamys,* from Greek] (1699)
: a short oblong mantle worn by young men of ancient Greece

chlamys

Chloe \ˈklō-ē\ *noun* [Latin, from Greek *Chloē*]
: a lover of Daphnis in a Greek pastoral romance

chlor- *or* **chloro-** *combining form* [New Latin, from Greek, from *chlōros* greenish yellow — more at YELLOW]
1 : green ⟨*chlor*ine⟩ ⟨*chlor*osis⟩
2 : chlorine **:** containing chlorine ⟨*chloro*prene⟩

chlor·ac·ne \klō-'rak-nē, klȯ-\ *noun* (circa 1928)
: a skin eruption resembling acne and resulting from exposure to chlorine or its compounds

chlo·ral \'klōr-əl, 'klȯr-\ *noun* [French, from *chlor-* + *al*cool alcohol] (1838)
1 : a pungent colorless oily aldehyde CCl_3CHO used in making DDT and chloral hydrate
2 : CHLORAL HYDRATE

chloral hydrate *noun* (1874)
: a bitter white crystalline drug $C_2H_3Cl_3O_2$ used as a hypnotic and sedative or in knockout drops

chlo·ral·ose \'klōr-ə-,lōs, 'klȯr-, -,lōz\ *noun* (1893)
: a bitter crystalline compound $C_8H_{11}Cl_3O_6$ used especially to anesthetize animals
— **chlo·ral·osed** \-,lōst, -,lōzd\ *adjective*

chlo·ra·mine \'klōr-ə-,mēn, 'klȯr-\ *noun* [International Scientific Vocabulary] (1893)
: any of various compounds containing nitrogen and chlorine

chlor·am·phen·i·col \,klōr-,am-'fe-ni-,kȯl, ,klȯr-, -,kōl\ *noun* [*chlor-* + *am*ide + *phen-* + *ni*tr- + gly*col*] (1949)
: a broad-spectrum antibiotic $C_{11}H_{12}Cl_2N_2O_5$ isolated from cultures of a soil actinomycete (*Streptomyces venezuelae*) or prepared synthetically

chlo·rate \'klōr-,āt, 'klȯr-\ *noun* (1823)
: a salt containing the anion ClO_3^- ⟨*chlorate* of potassium⟩

chlor·dane \'klȯr-,dān\ *noun* [*chlor-* + *indane* (C_9H_{10})] (1947)
: a highly chlorinated viscous volatile liquid insecticide $C_{10}H_6Cl_8$

chlor·di·az·epox·ide \,klōr-dī-,a-zə-'päk-,sīd, ,klȯr-\ *noun* [*chlor-* + *di-* + *az-* + *epoxide*] (1961)
: a benzodiazepine $C_{16}H_{14}ClN_3O$ related to diazepam and used in the form of its hydrochloride especially as a tranquilizer and in the treatment of alcoholism

chlo·rel·la \klə-'re-lə\ *noun* [New Latin, genus name, from Greek *chlōros*] (1904)
: any of a genus (*Chlorella*) of unicellular green algae

chlor·en·chy·ma \klōr-'eŋ-kə-mə, klȯr-\ *noun* (1894)
: chlorophyll-containing parenchyma of plants

chlo·ride \'klōr-,īd, 'klȯr-\ *noun* [German *Chlorid,* from *chlor-* + *-id* -ide] (1812)
1 : a compound of chlorine with another element or group; *especially* **:** a salt or ester of hydrochloric acid
2 : a univalent anion consisting of one atom of chlorine

chloride of lime (1826)
: BLEACHING POWDER

chlo·ri·nate \'klōr-ə-,nāt, 'klȯr-\ *transitive verb* **-nat·ed; -nat·ing** (1856)
: to treat or combine with chlorine or a chlorine compound
— **chlo·ri·na·tion** \,klōr-ə-'nā-shən, ,klȯr-\ *noun*
— **chlo·ri·na·tor** \'klōr-ə-,nā-tər, 'klȯr-\ *noun*

chlorinated lime *noun* (1876)
: BLEACHING POWDER

chlo·rine \'klōr-,ēn, 'klȯr-, -ən\ *noun* (1810)
: a halogen element that is isolated as a heavy greenish yellow gas of pungent odor and is used especially as a bleach, oxidizing agent, and disinfectant in water purification — see ELEMENT table

chlo·rin·i·ty \klō-'ri-nə-tē, klȯ-\ *noun* (circa 1931)
: a measure of the concentration of halides in one kilogram of seawater

[1]chlo·rite \'klōr-,īt, 'klȯr-\ *noun* [German *Chlorit,* from Latin *chloritis,* a green stone, from Greek *chlōritis,* from *chlōros*] (1794)
: any of a group of monoclinic usually green minerals associated with and resembling the micas
— **chlo·rit·ic** \klō-'ri-tik, klȯ-\ *adjective*

[2]chlorite *noun* [probably from French, from *chlor-*] (1853)
: a salt containing the anion ClO_2^- ⟨*chlorite* of sodium⟩

chloro- — see CHLOR-

chlo·ro·ben·zene \,klōr-ō-'ben-,zēn, ,klȯr-, -ben-'\ *noun* [International Scientific Vocabulary] (circa 1889)
: a colorless flammable volatile toxic liquid C_6H_5Cl used in organic synthesis (as of DDT) and as a solvent

chlo·ro·fluo·ro·car·bon \,klōr-ō-,flōr-ō-'kär-bən, ,klȯr-, -,flu̇r-, -,flȯr-\ *noun* (1949)
: any of several simple gaseous compounds that contain carbon, chlorine, fluorine, and sometimes hydrogen, that are used as refrigerants, cleaning solvents, and aerosol propellants and in the manufacture of plastic foams, and that are suspected to be a major cause of stratospheric ozone depletion — abbreviation *CFC*

chlo·ro·fluo·ro·meth·ane \-'me-,thān, *British usually* -'mē-\ *noun* (1965)
: a chlorofluorocarbon derived from methane

[1]chlo·ro·form \'klōr-ə-,fȯrm, 'klȯr-\ *noun* [French *chloroforme,* from *chlor-* + *formyle* formyl; from its having been regarded as a trichloride of this group] (1838)
: a colorless volatile heavy toxic liquid $CHCl_3$ with an ether odor used especially as a solvent or as a veterinary anesthetic

[2]chloroform *transitive verb* (1848)
: to treat with or as if with chloroform especially so as to produce anesthesia, insensibility, or death

chlo·ro·gen·ic acid \,klōr-ə-'je-nik-, ,klȯr-\ *noun* (circa 1889)
: a crystalline acid $C_{16}H_{18}O_9$ occurring in various plant parts (as coffee beans)

chlo·ro·hy·drin \,klōr-ə-'hī-drən, ,klȯr-\ *noun* [International Scientific Vocabulary *chlor-* + *hydr-* + [1]*-in*] (circa 1890)
: any of various organic compounds derived from diols or polyhydroxy alcohols by substitution of chlorine for part of the hydroxyl groups

Chlo·ro·my·ce·tin \,klōr-ō-mī-'sē-tᵊn, ,klȯr-\ *trademark*
— used for chloramphenicol

chlo·ro·phyll \'klōr-ə-,fil, 'klȯr-, -fəl\ *noun* [French *chlorophylle,* from *chlor-* + Greek *phyllon* leaf — more at BLADE] (1819)
1 : the green photosynthetic pigment found chiefly in the chloroplasts of plants and occurring especially as a blue-black ester $C_{55}H_{72}MgN_4O_5$ or a dark green ester $C_{55}H_{70}MgN_4O_6$ — called also respectively *chlorophyll a, chlorophyll b*
2 : a waxy green chlorophyll-containing substance extracted from green plants and used as a coloring agent or deodorant
— **chlo·ro·phyl·lous** \,klōr-ə-'fi-ləs, ,klȯr-\ *adjective*

chlo·ro·pic·rin \,klōr-ə-'pi-krən, ,klȯr-\ *noun* [German *Chlorpikrin,* from *chlor-* + Greek *pikros* sharp — more at PAINT] (circa 1889)
: a colorless liquid CCl_3NO_2 that causes tears and vomiting and is used especially as a soil fumigant

chlo·ro·plast \'klōr-ə-,plast, 'klȯr-\ *noun* [International Scientific Vocabulary] (1887)
: a plastid that contains chlorophyll and is the site of photosynthesis — see CELL illustration
— **chlo·ro·plas·tic** \,klōr-ə-'plas-tik, ,klȯr-\ *adjective*

chlo·ro·prene \-,prēn\ *noun* [*chlor-* + *isoprene*] (1931)
: a colorless liquid C_4H_5Cl used especially in making neoprene by polymerization

chlo·ro·quine \'klōr-ə-,kwēn, 'klȯr-\ *noun* [*chlor-* + *quinoline*] (1946)
: an antimalarial drug $C_{18}H_{26}ClN_3$ administered as the bitter crystalline diphosphate

chlo·ro·sis \klə-'rō-səs\ *noun* (1678)
1 : an iron-deficiency anemia especially of adolescent girls that may impart a greenish tint to the skin — called also *greensickness*
2 : a diseased condition in green plants marked by yellowing or blanching
— **chlo·rot·ic** \-'rä-tik\ *adjective*

chlo·ro·thi·a·zide \,klōr-ə-'thī-ə-,zīd, ,klȯr-, -zəd\ *noun* (1957)
: a thiazide diuretic $C_7H_6ClN_3O_4S_2$ used especially to treat edema and to increase the effectiveness of antihypertensive drugs

chlor·prom·a·zine \klōr-'prä-mə-,zēn, klȯr-\ *noun* [*chlor-* + *pro*pyl + *m*ethyl + *azine*] (1952)
: a phenothiazine $C_{17}H_{19}ClN_2S$ used as a tranquilizer especially in the form of its hydrochloride to suppress the more flagrant symptoms of psychotic disorders (as in schizophrenia)

chlor·prop·amide \-'prä-pə-,mīd, -'prō-\ *noun* [*chlor-* + *propane* + *amide*] (1960)
: a sulfonylurea drug $C_{10}H_{13}ClN_2O_3S$ used orally to reduce blood sugar in the treatment of mild diabetes

chlor·tet·ra·cy·cline \,klōr-,te-trə-'sī-,klēn, ,klȯr-\ *noun* (1953)
: a yellow crystalline broad-spectrum antibiotic $C_{22}H_{23}ClN_2O_8$ that is produced by a soil actinomycete (*Streptomyces aureofaciens*) and is sometimes used in animal feeds to stimulate growth

cho·ano·cyte \kō-'a-nə-,sīt\ *noun* [International Scientific Vocabulary *choan-* funnel-shaped (from Greek *choanē* funnel) + *-cyte*] (1888)
: COLLAR CELL

[1]chock \'chäk\ *noun* [origin unknown] (1769)
1 : a wedge or block for steadying a body (as a cask) and holding it motionless, for filling in an unwanted space, or for blocking the movement of a wheel
2 : a heavy metal casting (as on the bow or stern of a ship) with two short horn-shaped arms curving inward between which ropes or hawsers may pass for mooring or towing

[2]chock *adverb* (1834)
: as close or as completely as possible

[3]chock *transitive verb* (1854)
: to stop or make fast with or as if with chocks

[1]chock·a·block \'chä-kə-,bläk\ *adverb* (1850)
: CHOCK ⟨*chockablock* full⟩

[2]chockablock *adjective* (circa 1890)
1 : brought close together
2 : very full

chock–full *or* **chock·ful** \'chək-'fu̇l, 'chäk-, -,fu̇l\ *adjective* [Middle English *chokkefull,* probably from *choken* to choke + *full*] (15th century)
: full to the limit

choc·o·hol·ic \,chä-kə-'hȯ-lik, ,chȯ-, -'hä-\ *noun* [*choco*late + *-oholic* (as in *alcoholic*)] (1968)
: a person who craves or compulsively consumes chocolate

choc·o·late \'chä-k(ə-)lət, 'chȯ-\ *noun* [Spanish, from Nahuatl *chocolātl*] (1604)
1 : a beverage made by mixing chocolate with water or milk
2 : a food prepared from ground roasted cacao beans
3 : a small candy with a center (as a fondant) and a chocolate coating
4 : a brownish gray
— **chocolate** *adjective*

chocolate–box *adjective* [from the pictures formerly commonly seen on boxes of chocolates] (1901)
: superficially pretty or sentimental

choc·o·la·tier \,chä-k(ə-)lə-'tir, ,chȯ-\ *noun* [French, from *chocolat* chocolate] (1888)
: a maker or seller of chocolate candy

choc·o·laty *or* **choc·o·lat·ey** \'chä-k(ə-)lə-tē, 'chȯ-\ *adjective* (1926)
: made of or like chocolate; *also* **:** having a rich chocolate flavor

Choc·taw \'chäk-(,)tȯ\ *noun, plural* **Choctaw** *or* **Choctaws** [Choctaw *čahta*] (1722)
1 : a member of an American Indian people of Mississippi, Alabama, and Louisiana
2 : the language of the Choctaw and Chickasaw people

¹choice \'chȯis\ *noun* [Middle English *chois,* from Old French, from *choisir* to choose, of Germanic origin; akin to Old High German *kiosan* to choose — more at CHOOSE] (13th century)
1 : the act of choosing **:** SELECTION
2 : power of choosing **:** OPTION
3 a : the best part **:** CREAM **b :** a person or thing chosen
4 : a sufficient number and variety to choose among
5 : care in selecting
6 : a grade of meat between prime and good
— of choice : to be preferred ☆

²choice *adjective* **choic·er; choic·est** (14th century)
1 : worthy of being chosen
2 : selected with care
3 a : of high quality **b :** of a grade between prime and good ⟨*choice* meat⟩ ☆
— choice·ly *adverb*
— choice·ness *noun*

¹choir \'kwī(-ə)r\ *noun* [Middle English *quer,* from Middle French *cuer,* from Medieval Latin *chorus,* from Latin, chorus — more at CHORUS] (14th century)
1 : an organized company of singers (as in a church service)
2 : a group of instruments of the same class ⟨a brass *choir*⟩
3 : an organized group of persons or things
4 : a division of angels
5 : the part of a church occupied by the singers or by the clergy; *also* **:** the part of a church where the services are performed
6 : a group organized for ensemble speaking

²choir *intransitive verb* (1596)
: to sing or sound in chorus or concert

choir·boy \'kwī(-ə)r-,bȯi\ *noun* (1837)
: a boy member of a choir

choir loft *noun* (1929)
: a gallery occupied by a church choir

choir·mas·ter \-,mas-tər\ *noun* (1860)
: the director of a choir (as in a church)

¹choke \'chōk\ *verb* **choked; chok·ing** [Middle English, alteration of *achoken,* from Old English *āceōcian,* from *ā-,* perfective prefix + *cēoce, cēace* jaw, cheek — more at ABIDE, CHEEK] (14th century)
transitive verb
1 : to check normal breathing of by compressing or obstructing the trachea or by poisoning or adulterating available air
2 a : to check the growth, development, or activity of ⟨the flowers were *choked* by the weeds⟩ **b :** to obstruct by filling up or clogging ⟨leaves *choked* the drain⟩ **c :** to fill completely **:** JAM ⟨roads *choked* with traffic⟩
3 : to enrich the fuel mixture of (a motor) by partially shutting off the air intake of the carburetor
4 : to grip (as a baseball bat) some distance from the end of the handle
intransitive verb
1 : to become choked in breathing
2 a : to become obstructed or checked **b :** to become or feel constricted in the throat (as from strong emotion) — usually used with *up* ⟨*choked* up and couldn't finish the speech⟩
3 : to shorten one's grip especially on the handle of a bat — usually used with *up*
4 : to lose one's composure and fail to perform effectively in a critical situation

²choke *noun* (1736)
1 : the filamentous inedible center of an artichoke head; *broadly* **:** an artichoke head
2 : something that obstructs passage or flow: as **a :** a valve for choking a gasoline engine **b :** a constriction in an outlet (as of an oil well) that restricts flow **c :** REACTOR 2 **d :** a constriction (as a narrowing of the barrel or an attachment) at the muzzle of a shotgun that serves to limit the spread of shot
3 : the act of choking

choke·ber·ry \-,ber-ē\ *noun* (1778)
: a small berrylike astringent fruit; *also* **:** any of a genus (*Aronia*) of shrubs of the rose family bearing chokeberries

choke chain *noun* (1955)
: a collar that may be tightened as a noose and that is used especially in training and controlling powerful or stubborn dogs — called also *choke collar*

choke·cher·ry \'chōk-,cher-ē\ *noun* (1784)
: an American wild cherry (*Prunus virginiana*) with bitter or astringent red to black edible fruit; *also* **:** this fruit

choke coil *noun* (circa 1896)
: REACTOR 2

choke off *transitive verb* (1818)
: to bring to a stop or an end as if by choking

choke point *noun* (1944)
: a strategic narrow route providing passage through or to another region

chok·er \'chō-kər\ *noun* (circa 1552)
1 : one that chokes
2 : something (as a collar or necklace) worn closely about the throat or neck

chok·ing \'chō-kiŋ\ *adjective* (1562)
1 : producing the feeling of strangulation ⟨a *choking* cloud of smog⟩
2 : indistinct in utterance — used especially of a person's voice ⟨a low *choking* laugh⟩
— chok·ing·ly \-kiŋ-lē\ *adverb*

choky \'chō-kē\ *adjective* (1579)
: tending to cause choking or to become choked

chol- *or* **chole-** *or* **cholo-** *combining form* [Greek *chol-, cholē-, cholo-,* from *cholē, cholos* — more at GALL]
: bile **:** gall ⟨*chol*ate⟩

chol·an·gi·og·ra·phy \kə-,lan-jē-'ä-grə-fē, (,)kō-\ *noun* (1936)
: radiographic visualization of the bile ducts after injection of a radiopaque substance
— chol·an·gio·graph·ic \-jē-ə-'gra-fik\ *adjective*
— chol·an·gio·gram \-'lan-jē-ə-,gram\ *noun*

cho·late \'kō-,lāt\ *noun* (circa 1846)
: a salt or ester of cholic acid

cho·le·cal·cif·er·ol \,kō-lə-(,)kal-'si-fə-,rȯl, -,rōl\ *noun* [International Scientific Vocabulary] (1955)
: a sterol $C_{27}H_{43}OH$ that is a natural form of vitamin D found especially in fish, egg yolks, and fish-liver oils and is formed in the skin on exposure to sunlight or ultraviolet rays — called also *vitamin D_3*

cho·le·cys·tec·to·my \-(,)sis-'tek-tə-mē\ *noun, plural* **-mies** [New Latin *cholecystis* gallbladder (from *chol-* + Greek *kystis* bladder) + International Scientific Vocabulary *-ectomy* — more at CYST] (1885)
: surgical excision of the gallbladder
— cho·le·cys·tec·to·mized \-,mīzd\ *adjective*

cho·le·cys·ti·tis \-(,)sis-'tī-təs\ *noun* [New Latin, from *cholecystis*] (1866)
: inflammation of the gallbladder

cho·le·cys·to·ki·nin \-,sis-tə-'kī-nən\ *noun* [New Latin *cholecystis* + English *-o-* + *kinin*] (circa 1929)
: a hormone secreted especially by the duodenal mucosa that regulates the emptying of the gallbladder and secretion of enzymes by the pancreas and that has been found in the brain — called also *cholecystokinin-pancreozymin, pancreozymin*

cho·le·li·thi·a·sis \,kō-lə-li-'thī-ə-səs\ *noun* [New Latin] (circa 1860)
: production of gallstones; *also* **:** the resulting abnormal condition

cho·ler \'kä-lər, 'kō-\ *noun* [Middle English *coler,* from Middle French *colere,* from Latin *cholera* cholera, from Greek] (14th century)
1 a *archaic* **:** YELLOW BILE **b** *obsolete* **:** BILE 1a
2 *obsolete* **:** the quality or state of being bilious
3 : ready disposition to irritation **:** IRASCIBILITY; *also* **:** ANGER

chol·era \'kä-lə-rə\ *noun* [Latin] (1601)
: any of several diseases of humans and domestic animals usually marked by severe gastrointestinal symptoms; *especially* **:** an acute diarrheal disease caused by an enterotoxin produced by a comma-shaped gram-negative bacillus (*Vibrio cholerae* synonym *V. comma*) when it is present in large numbers in the proximal part of the human small intestine ◆

☆ SYNONYMS

Choice, option, alternative, preference, selection, election mean the act or opportunity of choosing or the thing chosen. CHOICE suggests the opportunity or privilege of choosing freely ⟨freedom of *choice*⟩. OPTION implies a power to choose that is specifically granted or guaranteed ⟨the *option* of paying now or later⟩. ALTERNATIVE implies a necessity to choose one and reject another possibility ⟨equally attractive *alternatives*⟩. PREFERENCE suggests the guidance of choice by one's judgment or predilections ⟨a *preference* for cool weather⟩. SELECTION implies a wide range of choice ⟨a varied *selection* of furniture⟩. ELECTION implies an end or purpose which requires exercise of judgment ⟨a tax return forces certain *elections* on you⟩.

Choice, exquisite, elegant, rare, delicate, dainty mean having qualities that appeal to a cultivated taste. CHOICE stresses preeminence in quality or kind ⟨*choice* fabric⟩. EXQUISITE implies a perfection in workmanship or design that appeals only to very sensitive taste ⟨an *exquisite* gold bracelet⟩. ELEGANT applies to what is rich and luxurious but restrained by good taste ⟨a sumptuous but *elegant* dining room⟩. RARE suggests an uncommon excellence ⟨*rare* beauty⟩. DELICATE implies exquisiteness, subtlety, and fragility ⟨*delicate* craftsmanship⟩. DAINTY sometimes carries an additional suggestion of smallness and of appeal to the eye or palate ⟨*dainty* sandwiches⟩.

◇ WORD HISTORY

cholera The word *cholera* now usually denotes Asiatic cholera, a still dreaded disease that typically assumes epidemic proportions when a concentrated population, such as that of a city or refugee camp, draws its drinking water from a source contaminated by fecal matter. Nineteeth-century medicine also recognized other forms of cholera ("summer cholera" or "European cholera") that manifested the same symptoms—vomiting and diarrhea—but in milder form. The antiquity of severe gastrointestinal disorders is evidenced by the word *cholera* itself, which is part of the medical vocabulary inherited by us from the ancient Greeks. Classical authors were divided on whether the Greek word was a derivative of *cholē* "bile" or *cholades* "intestines." In Late Latin, *cholera,* by association with Greek *cholē,* was used in the sense "bile." Through the medieval theory of humors, one of which was yellow bile (the fiery humor), it gave us English *choler* and *choleric.* See in addition HUMOR.

chol·era mor·bus \-ˈmȯr-bəs, *dialect* -ˈmä(r)-; *dialect* ˈkä-lē-\ *noun* [New Latin, literally, the disease cholera] (1673)
: gastrointestinal illness characterized by griping, diarrhea, and sometimes vomiting — not used technically

chol·er·ic \ˈkä-lə-rik, kə-ˈler-ik\ *adjective* (1583)
1 : easily moved to often unreasonable or excessive anger **:** hot-tempered
2 : ANGRY, IRATE
— chol·er·i·cal·ly \-ri-k(ə-)lē, -i-k(ə-)lē\ *adverb*

cho·le·sta·sis \ˌkō-lə-ˈstā-səs\ *noun, plural* **-sta·ses** \-ˌsēz\ [New Latin] (circa 1935)
: a checking or failure of bile flow
— cho·le·stat·ic \-ˈsta-tik\ *adjective*

cho·le·ster·ic \ˌkō-lə-ˈster-ik, kə-ˈles-tə-rik\ *adjective* [*cholesteric* relating to cholesterol, from French *cholesterique*] (1942)
: of, relating to, or being the phase of a liquid crystal characterized by arrangement of molecules in layers with the long molecular axes parallel to one another in the plane of each layer and incrementally displaced in successive layers to give helical stacking — compare NEMATIC, SMECTIC

cho·les·ter·ol \kə-ˈles-tə-ˌrōl, -ˌrȯl\ *noun* [International Scientific Vocabulary, from *chol-* + Greek *stereos* solid] (1894)
: a steroid alcohol $C_{27}H_{45}OH$ that is present in animal cells and body fluids, regulates membrane fluidity, and functions as a precursor molecule in various metabolic pathways and as a constituent of LDL may cause arteriosclerosis

cho·le·styr·amine \(ˌ)kō-ˌles-tə-ˈra-ˌmēn, kə-; ˌkō-lə-ˈstir-ə-ˌmēn\ *noun* [*chol-* + *styr*ene + *amine*] (circa 1962)
: a strongly basic synthetic resin that forms insoluble complexes with bile acids and has been used to lower cholesterol levels in hypercholesterolemic patients

cho·lic acid \ˈkō-lik-\ *noun* [Greek *cholikos* bilious, from *cholē*] (1846)
: a crystalline bile acid $C_{24}H_{40}O_5$

cho·line \ˈkō-ˌlēn\ *noun* [International Scientific Vocabulary] (circa 1871)
: a base $C_5H_{15}NO_2$ that occurs as a component of phospholipids especially in animals, is a precursor of acetylcholine, and is essential to liver function

cho·lin·er·gic \ˌkō-lə-ˈnər-jik\ *adjective* [International Scientific Vocabulary] (1934)
1 : liberating, activated by, or involving acetylcholine ⟨*cholinergic* nerve fiber⟩ ⟨*cholinergic* functions⟩
2 : resembling acetylcholine especially in physiologic action ⟨a *cholinergic* drug⟩
— cho·lin·er·gi·cal·ly \-ji-k(ə-)lē\ *adverb*

cho·lin·es·ter·ase \ˌkō-lə-ˈnes-tə-ˌrās, -ˌrāz\ *noun* (1932)
1 : ACETYLCHOLINESTERASE
2 : an enzyme that hydrolyzes choline esters and that is found especially in blood plasma — called also *pseudocholinesterase*

chol·la \ˈchȯi-yə\ *noun* [Mexican Spanish, from Spanish, head] (1846)
: any of numerous shrubby opuntias chiefly of the southwestern U.S. and Mexico that have needlelike spines partly enclosed in a papery sheath and cylindrical joints

cholla

chomp \ˈchämp, ˈchȯmp\ *verb* [alteration of *champ*] (circa 1847)
intransitive verb
: to chew or bite on something
transitive verb
: to chew or bite on

chon \ˈchän\ *noun, plural* **chon** [Korean *chŏn*]
— see *won* at MONEY table

chon·drio·some \ˈkän-drē-ə-ˌsōm\ *noun* [Greek *chondrion,* diminutive of *chondros* grain + International Scientific Vocabulary *-some*] (1910)
: MITOCHONDRION

chon·drite \ˈkän-ˌdrīt\ *noun* [International Scientific Vocabulary, from Greek *chondros*] (1883)
: a meteoric stone characterized by the presence of chondrules
— chon·drit·ic \kän-ˈdri-tik\ *adjective*

chon·dro·cra·ni·um \ˌkän-drō-ˈkrā-nē-əm\ *noun* [Greek *chondros* grain, cartilage] (1875)
: the cartilaginous parts of an embryonic cranium; *also* **:** the part of the adult skull derived therefrom

chon·droi·tin \kän-ˈdrȯi-tᵊn, -ˈdrō-ə-tᵊn\ *noun* [International Scientific Vocabulary *chondroit*ic acid, an acid found in cartilage + [1]*-in*] (1895)
: a mucopolysaccharide occurring in sulfated form in various animal tissues (as cartilage)

chon·drule \ˈkän-(ˌ)drül\ *noun* [Greek *chondros* grain] (circa 1889)
: a rounded granule of cosmic origin often found embedded in meteoric stones and sometimes free in marine sediments

chook \ˈchu̇k\ *noun* [imitative] (1880)
Australian & New Zealand **:** CHICKEN 1

choose \ˈchüz\ *verb* **chose** \ˈchōz\; **cho·sen** \ˈchō-zᵊn\; **choos·ing** \ˈchü-ziŋ\ [Middle English *chosen,* from Old English *cēosan;* akin to Old High German *kiosan* to choose, Latin *gustare* to taste] (before 12th century)
transitive verb
1 a : to select freely and after consideration ⟨*choose* a career⟩ **b :** to decide on especially by vote **:** ELECT ⟨*chose* her as captain⟩
2 a : to have a preference for ⟨*choose* one car over another⟩ **b :** DECIDE ⟨*chose* to go by train⟩
intransitive verb
1 : to make a selection
2 : to take an alternative — used after *cannot* and usually followed by *but* ⟨when earth is so kind, men cannot *choose* but be happy —J. A. Froude⟩
— choos·er \ˈchü-zər\ *noun*

choose up (1910)
transitive verb
: to form (sides) especially for a game by having opposing captains choose their players
intransitive verb
: to form sides for a game ⟨let's *choose up* and play ball⟩

choosy *or* **choos·ey** \ˈchü-zē\ *adjective* **choos·i·er; -est** (1862)
: fastidiously selective **:** PARTICULAR

[1]chop \ˈchäp\ *verb* **chopped; chop·ping** [Middle English *chappen, choppen* — more at CHAP] (14th century)
transitive verb
1 a : to cut into or sever usually by repeated blows of a sharp instrument **b :** to cut into pieces — often used with *up* **c :** to weed and thin out (young cotton) **d :** to cut as if by chopping ⟨*chop* prices⟩ ⟨a bridge *chops* the lake in two⟩
2 : to strike (as a ball) with a short quick downward stroke
3 : to subject to the action of a chopper ⟨*chop* a beam of light⟩
intransitive verb
1 : to make a quick stroke or repeated strokes with or as if with a sharp instrument (as an ax)
2 *archaic* **:** to move or act suddenly or violently

[2]chop *noun* (14th century)
1 a : a forceful usually slanting blow with or as if with an ax or cleaver **b :** a sharp downward blow or stroke
2 : a small cut of meat often including part of a rib — see LAMB illustration
3 : a mark made by or as if by chopping
4 : material that has been chopped up
5 a : a short abrupt motion (as of a wave) **b :** a stretch of choppy sea
6 : CHOPPER 6
7 *chiefly British* **:** AX 3

[3]chop *intransitive verb* **chopped; chop·ping** [Middle English *chappen, choppen* to barter, from Old English *cēapian*] (1540)
1 : to change direction
2 : to veer with or as if with wind
— chop logic : to argue with sophistical reasoning and minute distinctions

[4]chop *noun* [Hindi *chāp* stamp] (1614)
1 a : a seal or official stamp or its impression **b :** a license validated by a seal
2 a : a mark on goods or coins to indicate nature or quality **b :** a kind, brand, or lot of goods bearing the same chop **c :** QUALITY, GRADE ⟨of the first *chop*⟩

chop–chop \ˌchäp-ˈchäp, ˈchäp-ˈ\ *adverb* [Chinese Pidgin English, reduplication of *chop* fast] (1834)
: without delay **:** QUICKLY

chop·fall·en \ˈchäp-ˌfȯ-lən\ *variant of* CHAPFALLEN

chop·house \ˈchäp-ˌhau̇s\ *noun* (circa 1690)
: RESTAURANT

cho·pine \shä-ˈpēn, chä-\ *noun* [Middle French *chapin,* from Old Spanish] (1577)
: a woman's shoe of the 16th and 17th centuries with a very high sole designed to increase stature and protect the feet from mud and dirt

chop·log·ic \ˈchäp-ˌlä-jik\ *noun* [obsolete *chop* to exchange, trade, from Middle English *choppen* to barter] (1533)
: involved and often specious argumentation
— choplogic *adjective*

chop mark *noun* (1949)
: an indentation made on a coin to attest weight, silver content, or legality
— chop–marked \ˈchäp-ˌmärkt\ *adjective*

chopped liver *noun* (1980)
slang **:** one that is insignificant or not worth considering

[1]chop·per \ˈchä-pər\ *noun* (1552)
1 : one that chops
2 *plural, slang* **:** TEETH
3 : a device that interrupts an electric current or a beam of radiation (as light) at short regular intervals
4 : MACHINE GUN
5 : HELICOPTER
6 : a high-bouncing batted baseball
7 : a customized motorcycle

[2]chopper *verb* (1955)
: HELICOPTER

chop·pi·ness \ˈchä-pē-nəs\ *noun* (1881)
: the quality or state of being choppy

chopping block *noun* (1703)
: a wooden block on which material (as meat, wood, or vegetables) is cut, split, or diced

[1]chop·py \ˈchä-pē\ *adjective* **chop·pi·er; -est** [[2]*chop*] (1605)
1 : being roughened **:** CHAPPED
2 : rough with small waves
3 a : interrupted by ups and downs ⟨*choppy* country⟩ **b :** JERKY ⟨short *choppy* strides⟩ **c :** DISCONNECTED ⟨*choppy* writing⟩
— chop·pi·ly \ˈchä-pə-lē\ *adverb*

[2]choppy *adjective* **chop·pi·er; -est** [[3]*chop*] (1865)
: CHANGEABLE, VARIABLE ⟨a *choppy* wind⟩

chops \ˈchäps\ *noun plural* [alteration of [3]*chap*] (1589)
1 : JAW
2 a : MOUTH **b :** the fleshy covering of the jaws ⟨a dog licking its *chops*⟩
3 *slang* **:** EMBOUCHURE; *broadly* **:** the technical facility of a musical performer

chop shop *noun* (1977)
: a place where stolen automobiles are stripped of salable parts

chop·stick \ˈchäp-ˌstik\ *noun* [Chinese Pidgin English *chop* fast + English *stick*] (1699)

chopsticks

: one of a pair of slender sticks held between thumb and fingers and used chiefly in oriental countries to lift food to the mouth

chop su·ey \ˌchäp-ˈsü-ē\ *noun, plural* **chop sueys** [Chinese (Guangdong) *jaahp-seui* odds and ends, from *jaahp* miscellaneous + *seui* bits] (1888)
: a dish prepared chiefly from bean sprouts, bamboo shoots, water chestnuts, onions, mushrooms, and meat or fish and served with rice and soy sauce

cho·ra·gus \kə-ˈrā-gəs\ *or* **cho·re·gus** \-ˈrē-, -ˈrā-\ *noun* [Latin & Greek; Latin *choragus,* from Greek *choragos, chorēgos,* from *choros* chorus + *agein* to lead — more at AGENT] (1625)
1 : the leader of a chorus or choir; *broadly* **:** the leader of any group or movement
2 : a leader of a dramatic chorus in ancient Greece
— **cho·rag·ic** \-ˈra-jik\ *adjective*

cho·ral \ˈkōr-əl, ˈkȯr-\ *adjective* [French or Medieval Latin; French *choral,* from Medieval Latin *choralis,* from Latin *chorus*] (1587)
1 : of or relating to a chorus or choir ⟨a *choral* group⟩
2 : sung or designed for singing by a choir ⟨a *choral* arrangement⟩
— **cho·ral·ly** \-ə-lē\ *adverb*

cho·rale \kə-ˈral, -ˈräl\ *noun* [German *Choral,* short for *Choralgesang* choral song] (1841)
1 : a hymn or psalm sung to a traditional or composed melody in church; *also* **:** a harmonization of a chorale melody ⟨a Bach *chorale*⟩
2 : CHORUS, CHOIR

chorale prelude *noun* (circa 1924)
: a composition usually for organ based on a chorale

¹chord \ˈkȯrd\ *noun* [alteration of Middle English *cord,* short for *accord*] (1608)
: three or more musical tones sounded simultaneously

²chord (14th century)
intransitive verb
1 : ACCORD
2 : to play chords especially on a stringed instrument
transitive verb
1 : to make chords on
2 : HARMONIZE

³chord *noun* [alteration of ¹*cord*] (1543)
1 : CORD 3a
2 : a straight line segment joining and included between two points on a circle; *broadly* **:** a straight line joining two points on a curve
3 : an individual emotion or disposition ⟨struck a responsive *chord*⟩
4 : either of the two outside members of a truss connected and braced by the web members
5 : the straight line distance joining the leading and trailing edges of an airfoil

chord·al \ˈkȯr-dᵊl\ *adjective* (1848)
1 : of, relating to, or suggesting a chord
2 : relating to music characterized more by harmony than by counterpoint

chor·da·meso·derm \ˌkȯr-də-ˈme-zə-ˌdərm *also* -ˈme-sə-\ *noun* [New Latin *chorda* cord + English *mesoderm*] (1939)
: the portion of the embryonic mesoderm that forms notochord and related structures and induces the formation of neural structures
— **chor·da·meso·der·mal** \-ˌme-zə-ˈdər-məl, -ˌme-sə-\ *adjective*

chor·date \ˈkȯr-ˌdāt, -dət\ *noun* [ultimately from Latin *chorda* cord] (1897)
: any of a phylum (Chordata) of animals having at least at some stage of development a notochord, dorsally situated central nervous system, and gill clefts and including the vertebrates, lancelets, and tunicates
— **chordate** *adjective*

chore \ˈchōr, ˈchȯr\ *noun* [alteration of *chare*] (1746)
1 *plural* **:** the regular or daily light work of a household or farm
2 : a routine task or job
3 : a difficult or disagreeable task
synonym see TASK

cho·rea \kə-ˈrē-ə\ *noun* [New Latin, from Latin, dance, from Greek *choreia,* from *choros* chorus] (1804)
: a nervous disorder (as of humans or dogs) marked by spasmodic movements of limbs and facial muscles and by incoordination
— **cho·re·ic** \-ˈrē-ik\ *adjective*

cho·re·i·form \kə-ˈrē-ə-ˌfȯrm\ *adjective* [International Scientific Vocabulary] (circa 1899)
: resembling chorea ⟨*choreiform* convulsions⟩

cho·reo·graph \ˈkōr-ē-ə-ˌgraf, ˈkȯr-\ (1943)
transitive verb
1 : to compose the choreography of
2 : to arrange or direct the movements, progress, or details of
intransitive verb
: to engage in choreography
— **cho·re·og·ra·pher** \ˌkōr-ē-ˈä-grə-fər, ˌkȯr-\ *noun*

cho·re·og·ra·phy \ˌkōr-ē-ˈä-grə-fē, ˌkȯr-\ *noun, plural* **-phies** [French *chorégraphie,* from Greek *choreia* + French *-graphie* -graphy] (circa 1789)
1 : the art of symbolically representing dancing
2 a : the composition and arrangement of dances especially for ballet **b :** a composition created by this art
3 : something resembling choreography ⟨a snail-paced *choreography* of delicate high diplomacy —Wolfgang Saxon⟩
— **cho·reo·graph·ic** \ˌkōr-ē-ə-ˈgra-fik, ˌkȯr-\ *adjective*
— **cho·reo·graph·i·cal·ly** \-fi-k(ə-)lē\ *adverb*

cho·ric \ˈkōr-ik, ˈkȯr-, ˈkär-\ *adjective* (1830)
: of, relating to, or being in the style of a chorus and especially a Greek chorus

cho·rine \ˈkōr-ˌēn, ˈkȯr-\ *noun* [*chorus* + *-ine,* feminine noun suffix (as in *Pauline*)] (1922)
: CHORUS GIRL

cho·rio·al·lan·to·is \ˌkōr-ē-(ˌ)ō-ə-ˈlan-tə-wəs, ˌkȯr-\ *noun* [New Latin, from Greek *chorion* + New Latin *allantois*] (1933)
: a vascular fetal membrane composed of the fused chorion and adjacent wall of the allantois that in the hen's egg is used as a living culture medium for viruses and for tissues — called also *chorioallantoic membrane*
— **cho·rio·al·lan·to·ic** \-ˌa-lən-ˈtō-ik\ *adjective*

cho·rio·car·ci·no·ma \-ˌkär-sᵊn-ˈō-mə\ *noun* [New Latin, from *chorion* + *carcinoma*] (1901)
: a malignant tumor developing in the uterus from the trophoblast and rarely in the testes from a neoplasm

cho·ri·on \ˈkōr-ē-ˌän, ˈkȯr-\ *noun* [New Latin, from Greek] (1545)
: the highly vascular outer embryonic membrane of reptiles, birds, and mammals that in placental mammals is associated with the allantois in the formation of the placenta

cho·ri·on·ic \ˌkōr-ē-ˈä-nik, ˌkȯr-\ *adjective* (1892)
1 : of, relating to, or being part of the chorion ⟨*chorionic* villi⟩
2 : secreted or produced by chorionic or related tissue (as in the placenta or a choriocarcinoma) ⟨human *chorionic* gonadotropin⟩

cho·ris·ter \ˈkōr-ə-stər, ˈkȯr-, ˈkär-\ *noun* [Middle English *querister,* from Anglo-French *cueristre,* from Medieval Latin *chorista,* from Latin *chorus*] (14th century)
1 : a singer in a choir; *specifically* **:** CHOIRBOY
2 : the leader of a church choir

cho·ri·zo \chə-ˈrē-(ˌ)zō, -(ˌ)sō\ *noun, plural* **-zos** [Spanish] (1846)
: a pork sausage highly seasoned especially with chili powder and garlic

C horizon *noun* (1935)
: the soil layer lying beneath the B horizon and consisting essentially of more or less weathered parent rock

cho·rog·ra·phy \kə-ˈrä-grə-fē\ *noun* [Latin *chorographia,* from Greek *chōrographia,* from *chōros* place + *-graphia* -graphy] (1559)
1 : the art of describing or mapping a region or district
2 : a description or map of a region; *also* **:** the physical conformation and features of such a region
— **cho·rog·ra·pher** \-grə-fər\ *noun*
— **cho·ro·graph·ic** \ˌkȯr-ə-ˈgra-fik, ˌkär-\ *adjective*

cho·roid \ˈkōr-ˌȯid, ˈkȯr-\ *also* **cho·ri·oid** \ˈkōr-ē-ˌȯid, ˈkȯr-\ *noun* [New Latin *choroides* resembling the chorion, from Greek *chorioeidēs,* from *chorion* chorion] (1683)
: a vascular membrane containing large branched pigmented cells that lies between the retina and the sclera of the vertebrate eye — called also *choroid coat;* see EYE illustration
— **choroid** *or* **cho·roi·dal** \kə-ˈrȯi-dᵊl\ *adjective*

chor·tle \ˈchȯr-tᵊl\ *verb* **chor·tled; chor·tling** \ˈchȯrt-liŋ, ˈchȯr-tᵊl-iŋ\ [probably blend of *chuckle* and *snort*] (1872)
intransitive verb
1 : to sing or chant exultantly ⟨he *chortled* in his joy —Lewis Carroll⟩
2 : to laugh or chuckle especially in satisfaction or exultation
transitive verb
: to say or sing with a chortling intonation
— **chortle** *noun*
— **chor·tler** \ˈchȯrt-lər, ˈchȯr-tᵊl-ər\ *noun*

¹cho·rus \ˈkōr-əs, ˈkȯr-\ *noun* [Latin, ring dance, chorus, from Greek *choros*] (1586)
1 a : a company of singers and dancers in Athenian drama participating in or commenting on the action; *also* **:** a similar company in later plays **b :** a character in Elizabethan drama who speaks the prologue and epilogue and comments on the action **c :** an organized company of singers who sing in concert **:** CHOIR; *especially* **:** a body of singers who sing the choral parts of a work (as in opera) **d :** a group of dancers and singers supporting the featured players in a musical comedy or revue
2 a : a part of a song or hymn recurring at intervals **b :** the part of a drama sung or spoken by the chorus **c :** a composition to be sung by a number of voices in concert **d :** the main part of a popular song; *also* **:** a jazz variation on a melodic theme
3 a : something performed, sung, or uttered simultaneously or unanimously by a number of persons or animals ⟨a *chorus* of boos⟩ ⟨that eternal *chorus* of: "Are we there yet?" from the back seat —Sheila More⟩ **b :** sounds so uttered ⟨visitors are taken to the woods by car to hear the mournful *choruses* of howling wolves —Bob Gaines⟩
— **in chorus :** in unison

²chorus *transitive verb* (1826)
: to sing or utter in chorus

chorus boy *noun* (1943)
: a young man who sings or dances in the chorus of a theatrical production (as a musical or revue)

chorus girl *noun* (1894)

: a young woman who sings or dances in the chorus of a theatrical production (as a musical or revue)
¹chose *past of* CHOOSE
²chose \'shōz\ *noun* [French, from Latin *causa* cause, reason] (1670)
: a piece of personal property : THING
¹cho·sen \'chō-zᵊn\ *noun, plural* **chosen** (13th century)
: one who is the object of choice or of divine favor : an elect person
²chosen *adjective* [Middle English, from past participle of *chosen* to choose] (14th century)
1 : ELECT
2 : selected or marked for favor or special privilege ⟨a *chosen* few⟩
Chou \'jō\ *noun* [Chinese (Beijing) *Zhōu*] (1771)
: a Chinese dynasty traditionally dated 1122 to about 256 B.C. and marked by the development of the philosophical schools of Confucius, Mencius, Lao-tzu, and Mo Ti
chough \'chəf\ *noun* [Middle English] (13th century)
: either of two Old World birds (*Pyrrhocorax pyrrhocorax* and *P. graculus*) that are related to the crows and have red legs and glossy blue-black plumage
¹chouse \'chaủs\ *transitive verb* **choused; chous·ing** [perhaps from Turkish *çavuş* doorkeeper, messenger] (1708)
: CHEAT, TRICK
²chouse *transitive verb* **choused; chous·ing** [origin unknown] (1904)
West : to drive or herd roughly
¹chow \'chaủ\ *noun* [short for *chowchow*] (1856)
: FOOD, VICTUALS
²chow *intransitive verb* (1917)
: EAT — often used with *down*
³chow *noun* [by shortening] (1889)
: CHOW CHOW
chow·chow \'chaủ-ˌchaủ\ *noun* [Chinese Pidgin English *chowchow* food] (1850)
1 : a Chinese preserve of ginger, fruits, and peels in heavy syrup
2 : a relish of chopped mixed pickles in mustard sauce
chow chow \'chaủ-ˌchaủ\ *noun, often both Cs capitalized* [perhaps from *chow-chow* Chinese person, from Chinese Pidgin English *chowchow* food] (1886)
: any of a breed of heavy-coated blocky dogs of Chinese origin having a broad head and muzzle, a distinctive blue-black tongue and black-lined mouth, and either a long dense coat with a full ruff or a short smooth coat — called also *chow*

chow chow

¹chow·der \'chaủ-dər\ *transitive verb* (1732)
: to make chowder of
²chowder *noun* [French *chaudière* kettle, contents of a kettle, from Late Latin *caldaria* — more at CAULDRON] (1751)
: a soup or stew of seafood (as clams or fish) usually made with milk or tomatoes, salt pork, onions, and other vegetables (as potatoes); *also* : a soup resembling chowder ⟨corn *chowder*⟩
chow·der·head \-ˌhed\ *noun* [alteration of dialect *jolterhead* blockhead] (1833)
: DOLT, BLOCKHEAD
— **chow·der·head·ed** \'chaủ-dər-ˌhe-dəd\ *adjective*
chow·hound \'chaủ-ˌhaủnd\ *noun* (1942)
: one fond of eating
chow line *noun* (1919)
: a line of people waiting to be served food
chow mein \'chaủ-'mān\ *noun* [Chinese (Guangdong) *cháau-mihn* fried noodles] (1903)
: a seasoned stew of shredded or diced meat, mushrooms, and vegetables that is usually served with fried noodles
chres·tom·a·thy \kre-'stä-mə-thē\ *noun, plural* **-thies** [New Latin *chrestomathia,* from Greek *chrēstomatheia,* from *chrēstos* useful + *manthanein* to learn — more at MATHEMATICAL] (1832)
1 : a selection of passages compiled as an aid to learning a language
2 : a volume of selected passages or stories of an author
chrism \'kri-zəm\ *noun* [Middle English *crisme,* from Old English *crisma,* from Late Latin *chrisma,* from Greek, ointment, from *chriein* to anoint] (before 12th century)
: consecrated oil used in Greek and Latin churches especially in baptism, chrismation, confirmation, and ordination
chris·ma·tion \kriz-'mā-shən\ *noun* [Medieval Latin *chrismation-, chrismatio* anointment with chrism, from Late Latin *chrismare* to anoint with chrism, from *chrisma*] (1642)
: a confirmatory sacrament of the Eastern Orthodox Church in which a baptized member is anointed with chrism
chris·mon \'kriz-ˌmän\ *noun, plural* **chrisma** \-mə\ *or* **chrismons** [Medieval Latin, from Latin *Chris*tus Christ + Late Latin *mono*gramma monogram] (1872)
: CHI-RHO
chris·om \'kri-zəm\ *noun* [Middle English *crisom,* short for *crisom cloth,* from *crisom* chrism + *cloth*] (13th century)
: a white cloth or robe put on a person at baptism as a symbol of innocence
chrisom child *noun* (1593)
: a child that dies in its first month
Christ \'krīst\ *noun* [Middle English *Crist,* from Old English, from Latin *Christus,* from Greek *Christos,* literally, anointed, from *chriein*] (before 12th century)
1 : MESSIAH
2 : JESUS
3 : an ideal type of humanity
4 *Christian Science* **:** the ideal truth that comes as a divine manifestation of God to destroy incarnate error
— **Christ·like** \-ˌlīk\ *adjective*
— **Christ·ly** \-lē\ *adjective*
chris·ten \'kri-sᵊn\ *transitive verb* **christened; chris·ten·ing** \'kris-niŋ, 'kri-sᵊn-iŋ\ [Middle English *cristnen,* from Old English *cristnian,* from *cristen* Christian, from Latin *christianus*] (before 12th century)
1 a : BAPTIZE **b :** to name at baptism
2 : to name or dedicate (as a ship) by a ceremony suggestive of baptism
3 : NAME
4 : to use for the first time
Chris·ten·dom \'kri-sᵊn-dəm\ *noun* [Middle English *cristendom,* from Old English *cristendōm,* from *cristen*] (before 12th century)
1 : CHRISTIANITY 1
2 : the part of the world in which Christianity prevails
chris·ten·ing *noun* (14th century)
: the ceremony of baptizing and naming a child
¹Chris·tian \'kris-chən, 'krish-\ *noun* [Latin *christianus,* adjective & noun, from Greek *christianos,* from *Christos*] (1526)
1 a : one who professes belief in the teachings of Jesus Christ **b** (1) **:** DISCIPLE 2 (2) **:** a member of one of the Churches of Christ separating from the Disciples of Christ in 1906 (3) **:** a member of the Christian denomination having part in the union of the United Church of Christ concluded in 1961
2 : the hero in Bunyan's *Pilgrim's Progress*
²Christian *adjective* (1553)
1 a : of or relating to Christianity ⟨*Christian* scriptures⟩ **b :** based on or conforming with Christianity ⟨*Christian* ethics⟩
2 a : of or relating to a Christian ⟨*Christian* responsibilities⟩ **b :** professing Christianity ⟨a *Christian* affirmation⟩
3 : commendably decent or generous ⟨has a very *Christian* concern for others⟩
— **Chris·tian·ly** *adjective or adverb*
Christian Brother *noun* (1883)
: a member of the Roman Catholic institute of Brothers of the Christian Schools founded by Saint John Baptist de la Salle in France in 1684 and dedicated to education
Christian era *noun* (1657)
: the period dating from the birth of Christ
chris·ti·ania \ˌkris-chē-'a-nē-ə, ˌkrish-, ˌkris-tē-, -'ä-nē-ə\ *noun* [*Christiania,* former name of Oslo, Norway] (1905)
: CHRISTIE
Chris·tian·i·ty \ˌkris-chē-'a-nə-tē, ˌkrish-, -'cha-nə-, ˌkris-tē-'a-\ *noun* (14th century)
1 : the religion derived from Jesus Christ, based on the Bible as sacred scripture, and professed by Eastern, Roman Catholic, and Protestant bodies
2 : conformity to the Christian religion
3 : CHRISTENDOM 2
Chris·tian·ize \'kris-chə-ˌnīz, 'krish-\ *transitive verb* **-ized; -iz·ing** (1593)
: to make Christian
— **Chris·tian·i·za·tion** \ˌkris-chə-nə-'zā-shən, ˌkrish-\ *noun*
— **Chris·tian·iz·er** \'kris-chə-ˌnī-zər, 'krish-\ *noun*
Christian name *noun* (1549)
: GIVEN NAME
Christian Science *noun* (circa 1867)
: a religion founded by Mary Baker Eddy in 1866 that was organized under the official name of the Church of Christ, Scientist, that derives its teachings from the Scriptures as understood by its adherents, and that includes a practice of spiritual healing based on the teaching that cause and effect are mental and that sin, sickness, and death will be destroyed by a full understanding of the divine principle of Jesus's teaching and healing
— **Christian Scientist** *noun*
chris·tie *or* **chris·ty** \'kris-tē\ *noun, plural* **christies** [by shortening & alteration from *christiania*] (1925)
: a skiing turn used for altering the direction of hill descent or for stopping and executed usually at high speed by shifting the body weight forward and skidding into a turn with parallel skis
Christ·mas \'kris-məs\ *noun, often attributive* [Middle English *Christemasse,* from Old English *Cristes mæsse,* literally, Christ's mass] (before 12th century)
1 : a Christian feast on December 25 or among some Eastern Orthodox on January 7 that commemorates the birth of Christ and is usually observed as a legal holiday
2 : CHRISTMASTIDE
— **Christ·mas·sy** *or* **Christ·masy** \-mə-sē\ *adjective*
Christmas cactus *noun* [from its annual blooming around Christmastime] (circa 1900)
: a branching Brazilian cactus (*Schlumbergera truncata* synonym *Zygocactus truncatus*) with flat stems, joints with margins having one to several blunt teeth, and showy usually red, pink, white, or violet zygomorphic flowers — called also *crab cactus*
Christmas card *noun* (1883)
: a greeting card sent at Christmas
Christmas club *noun* (circa 1925)
: a savings account in which regular deposits are made throughout the year to provide money for Christmas shopping
Christmas fern *noun* (1878)
: a North American evergreen fern (*Polystichum acrostichoides*)
Christmas rose *noun* (1688)

: a European evergreen herb (*Helleborus niger*) of the buttercup family that has usually white flowers produced in winter

Christ·mas·tide \'kris-məs-,tīd\ *noun* (1626)
: the festival season from Christmas Eve till after New Year's Day or especially in England till Epiphany

Christ·mas·time \-,tīm\ *noun* (1837)
: the Christmas season

Christmas tree *noun* (1835)
1 : a usually evergreen tree decorated at Christmas
2 : an oil-well control device consisting of an assembly of fittings placed at the top of the well

Chris·to·cen·tric \,kris-tə-'sen-trik, ,krīs-\ *adjective* [Greek *Christos* Christ + English *-centric*] (1873)
: centering theologically on Christ

Chris·to·gram \'kris-tə-,gram, 'krīs-\ *noun* [Greek *Christos* + English *-gram*] (1900)
: a graphic symbol of Christ; *especially* **:** CHI-RHO

Chris·tol·o·gy \kris-'tä-lə-jē, krīs-\ *noun, plural* **-gies** [Greek *Christos* + English *-logy*] (1673)
: theological interpretation of the person and work of Christ
— **Chris·to·log·i·cal** \,kris-tə-'lä-ji-kəl, 'krīs-\ *adjective*

Christ's-thorn \'krīs(ts)-'thȯrn\ *or* **Christ-thorn** \'krīs(t)-\ *noun* (1562)
: any of several prickly or thorny shrubs (as the shrub *Paliurus spina-christi* or the jujube *Ziziphus jujuba*)

chrom- *or* **chromo-** *combining form* [International Scientific Vocabulary, from Greek *chrōma* color]
1 : chromium ⟨*chrom*ize⟩
2 a : color **:** colored ⟨*chromo*lithograph⟩ **b :** pigment ⟨*chromo*gen⟩

chro·ma \'krō-mə\ *noun* [Greek *chrōma*] (circa 1889)
1 : SATURATION 4a
2 : a quality of color combining hue and saturation

chro·maf·fin \'krō-mə-fən\ *adjective* [International Scientific Vocabulary *chrom-* + Latin *affinis* bordering on, related — more at AFFINITY] (1903)
: staining deeply with chromium salts ⟨*chromaffin* cells of the adrenal medulla⟩

chromat- *or* **chromato-** *combining form* [Greek *chrōmat-, chrōma*]
1 : color ⟨*chromat*id⟩
2 : chromatin ⟨*chromato*lysis⟩

chro·mate \'krō-,māt\ *noun* [French, from Greek *chrōma*] (1819)
: a salt of chromic acid

[1]**chro·mat·ic** \krō-'ma-tik\ *noun* (1708)
: ACCIDENTAL 2

[2]**chromatic** *adjective* [Greek *chrōmatikos,* from *chrōmat-, chrōma* skin, color, modified tone; akin to Greek *chrōs* color] (circa 1798)
1 a : of or relating to color or color phenomena or sensations **b :** highly colored
2 : of or relating to chroma
3 a : of, relating to, or giving all the tones of the chromatic scale **b :** characterized by frequent use of accidentals
— **chro·mat·i·cal·ly** \-ti-k(ə-)lē\ *adverb*
— **chro·mat·i·cism** \-'ma-tə-,si-zəm\ *noun*

chromatic aberration *noun* (1831)
: aberration caused by the differences in refraction of the colored rays of the spectrum

chro·ma·tic·i·ty \,krō-mə-'ti-sə-tē\ *noun* (1922)
: the quality of color characterized by its dominant or complementary wavelength and purity taken together

chro·mat·ics \krō-'ma-tiks\ *noun plural but singular in construction* (circa 1790)
: the branch of colorimetry that deals with hue and saturation

chromatic scale *noun* (circa 1789)
: a musical scale consisting entirely of half steps

chro·ma·tid \'krō-mə-təd\ *noun* (1900)
: one of the usually paired and parallel strands of a duplicated chromosome joined by a single centromere

chro·ma·tin \'krō-mə-tən\ *noun* (1882)
: a complex of nucleic acid and basic proteins (as histone) in eukaryotic cells that is usually dispersed in the interphase nucleus and condensed into chromosomes in mitosis and meiosis — compare EUCHROMATIN, HETEROCHROMATIN; see CELL illustration
— **chro·ma·tin·ic** \,krō-mə-'ti-nik\ *adjective*

chro·mato·gram \krō-'ma-tə-,gram, krə-\ *noun* (1922)
1 : the pattern formed on the adsorbent medium by the layers of components separated by chromatography
2 : a time-based graphic record (as of concentration of eluted materials) of a chromatographic separation

chro·mato·graph \krō-'ma-tə-,graf, krə-\ *noun* (1946)
: an instrument for performing chromatographic separations and producing chromatograms
— **chromatograph** *verb*
— **chro·ma·tog·ra·pher** \,krō-mə-'tä-grə-fər\ *noun*

chro·ma·tog·ra·phy \,krō-mə-'tä-grə-fē\ *noun* (1937)
: a process in which a chemical mixture carried by a liquid or gas is separated into components as a result of differential distribution of the solutes as they flow around or over a stationary liquid or solid phase
— **chro·mato·graph·ic** \krō-,ma-tə-'gra-fik, krə-\ *adjective*
— **chro·mato·graph·i·cal·ly** \-fi-k(ə-)lē\ *adverb*

chro·ma·tol·y·sis \,krō-mə-'tä-lə-səs\ *noun* [New Latin] (1901)
: the dissolution and breaking up of chromophil material (as chromatin) of a cell and especially a nerve cell
— **chro·mato·lyt·ic** \krō-,ma-tə-'li-tik, krə-\ *adjective*

chro·mato·phore \krō-'ma-tə-,fōr, krə-, -,fȯr\ *noun* [International Scientific Vocabulary] (circa 1859)
1 : a pigment-bearing cell; *especially* **:** one of the cells of an animal integument capable of causing integumentary color changes by expanding or contracting
2 : the organelle of photosynthesis in blue-green algae and photosynthetic bacteria; *broadly* **:** CHROMOPLAST, CHLOROPLAST

[1]**chrome** \'krōm\ *noun* [French, from Greek *chrōma*] (1800)
1 a : CHROMIUM **b :** a chromium pigment
2 : something plated with an alloy of chromium

[2]**chrome** *transitive verb* **chromed; chroming** (1876)
1 : to treat with a compound of chromium (as in dyeing)
2 : CHROMIZE

-chrome \,krōm\ *noun combining form or adjective combining form* [Medieval Latin *-chromat-, -chroma* colored thing, from Greek *chrōmat-, chrōma*]
1 : colored thing ⟨helio*chrome*⟩
2 : coloring matter ⟨uro*chrome*⟩

chrome green *noun* (circa 1859)
: any of various brilliant green pigments containing or consisting of chromium compounds

chrome yellow *noun* (1819)
: a yellow pigment consisting essentially of neutral lead chromate $PbCrO_4$

chro·mic \'krō-mik\ *adjective* (circa 1828)
: of, relating to, or derived from chromium especially with a valence of three

chromic acid *noun* (1800)
: an acid H_2CrO_4 analogous to sulfuric acid but known only in solution and especially in the form of its salts

chro·mi·nance \'krō-mə-nən(t)s\ *noun* [*chrom-* + lum*inance*] (1952)
: the difference between a color and a chosen reference color of the same luminous intensity in color television

chro·mite \'krō-,mīt\ *noun* [German *Chromit,* from *chrom-*] (1840)
1 : a mineral that consists of an oxide of iron and chromium
2 : an oxide of bivalent chromium

chro·mi·um \'krō-mē-əm\ *noun* [New Latin, from French *chrome*] (1807)
: a blue-white metallic element found naturally only in combination and used especially in alloys and in electroplating — see ELEMENT table

chro·mize \'krō-,mīz\ *transitive verb* **chromized; chro·miz·ing** (1939)
: to treat (metal) with chromium in order to form a protective surface alloy

chro·mo \'krō-(,)mō\ *noun, plural* **chromos** (1869)
: CHROMOLITHOGRAPH

chro·mo·cen·ter \'krō-mə-,sen-tər\ *noun* (1926)
: a densely staining aggregation of heterochromatic regions in the nucleus of some cells

chro·mo·dy·nam·ics \,krō-mə-dī-'na-miks\ *noun plural but singular in construction* (1976)
: QUANTUM CHROMODYNAMICS

chro·mo·gen \'krō-mə-jən\ *noun* [International Scientific Vocabulary] (1858)
1 : a precursor of a biochemical pigment
2 : a pigment-producing microorganism

chro·mo·gen·ic \,krō-mə-'je-nik\ *adjective* (1884)
1 : of or relating to a chromogen
2 : being a process of photographic film development in which silver halides activate precursors of chemical dyes that form the final image while the silver is removed; *also* **:** being a film developed by this process

chro·mo·litho·graph \,krō-mə-'li-thə-,graf\ *noun* (1860)
: a picture printed in colors from a series of lithographic stones or plates
— **chromolithograph** *transitive verb*
— **chro·mo·litho·graph·ic** \-,li-thə-'gra-fik\ *adjective*
— **chro·mo·li·thog·raph·er** \-li-'thä-grə-fər\ *noun*
— **chro·mo·li·thog·ra·phy** \-grə-fē\ *noun*

chro·mo·mere \'krō-mə-,mir\ *noun* [International Scientific Vocabulary] (1896)
: one of the small bead-shaped and heavily staining masses of coiled chromatin that are linearly arranged along the chromosome
— **chro·mo·mer·ic** \,krō-mə-'mer-ik, -'mir-\ *adjective*

chro·mo·ne·ma \,krō-mə-'nē-mə\ *noun, plural* **-ne·ma·ta** \-'nē-mə-tə\ [New Latin, from *chrom-* + Greek *nēmat-, nēma* thread — more at NEMAT-] (circa 1925)
: the coiled filamentous core of a chromatid
— **chro·mo·ne·mat·ic** \-ni-'ma-tik\ *adjective*

chro·mo·phil \'krō-mə-,fil\ *adjective* [International Scientific Vocabulary] (1899)
: staining readily with dyes

chro·mo·phobe \-,fōb\ *adjective* (circa 1909)
: resisting staining with dyes

chro·mo·phore \'krō-mə-,fōr, -,fȯr\ *noun* [International Scientific Vocabulary] (1879)
: a chemical group (as an azo group) that absorbs light at a specific frequency and so im-

parts color to a molecule; *also* **:** a colored chemical compound
— **chro·mo·phor·ic** \ˌkrō-mə-ˈfȯr-ik, -ˈfär-\ *adjective*

chro·mo·plast \ˈkrō-mə-ˌplast\ *noun* [International Scientific Vocabulary] (1900)
: a colored plastid usually containing red or yellow pigment (as carotene)

chro·mo·pro·tein \ˌkro-mə-ˈprō-ˌtēn, -ˈprō-tē-ən\ *noun* (1924)
: any of various proteins (as hemoglobins, carotenoids, or flavoproteins) having a pigment as a prosthetic group

chro·mo·some \ˈkrō-mə-ˌsōm, -ˌzōm\ *noun* [International Scientific Vocabulary] (1889)
: one of the linear or sometimes circular DNA-containing bodies of viruses, prokaryotic organisms, and the cell nucleus of eukaryotic organisms that contain most or all of the genes of the individual — compare CHROMATIN
— **chro·mo·som·al** \ˌkrō-mə-ˈsō-məl, -ˈzō-\ *adjective*
— **chro·mo·som·al·ly** \-mə-lē\ *adverb*

chromosome number *noun* (1910)
: the usually constant number of chromosomes characteristic of a particular kind of animal or plant

chro·mo·sphere \ˈkrō-mə-ˌsfir\ *noun* (1868)
: the region of the atmosphere of a star (as the sun) between the star's photosphere and its corona
— **chro·mo·spher·ic** \ˌkrō-mə-ˈsfir-ik, -ˈsfer-\ *adjective*

chro·mous \ˈkrō-məs\ *adjective* (1840)
: of, relating to, or derived from chromium especially with a valence of two

chron- *or* **chrono-** *combining form* [Greek, from *chronos*]
: time ⟨*chrono*gram⟩

chron·ax·ie *also* **chron·axy** \ˈkrō-ˌnak-sē, ˈkrä-\ *noun* [French *chronaxie,* from *chron-* + Greek *axia* value, from *axios* worthy — more at AXIOM] (1917)
: the minimum time required for excitation of a structure (as a nerve cell) by a constant electric current of twice the threshold voltage

chron·ic \ˈkrä-nik\ *adjective* [French *chronique,* from Greek *chronikos* of time, from *chronos*] (1601)
1 a : marked by long duration or frequent recurrence **:** not acute ⟨*chronic* indigestion⟩ ⟨*chronic* experiments⟩ **b :** suffering from a chronic disease ⟨the special needs of *chronic* patients⟩
2 a : always present or encountered; *especially* **:** constantly vexing, weakening, or troubling ⟨*chronic* petty warfare⟩ **b :** being such habitually ⟨a *chronic* grumbler⟩
synonym see INVETERATE
— **chronic** *noun*
— **chron·i·cal·ly** \-ni-k(ə-)lē\ *adverb*
— **chro·nic·i·ty** \krä-ˈni-sə-tē, krō-\ *noun*

[1]**chron·i·cle** \ˈkrä-ni-kəl\ *noun* [Middle English *cronicle,* from Anglo-French, alteration of Old French *chronique,* from Latin *chronica,* from Greek *chronika,* from neuter plural of *chronikos*] (14th century)
1 : a usually continuous historical account of events arranged in order of time without analysis or interpretation
2 : NARRATIVE

[2]**chronicle** *transitive verb* **-cled; -cling** \-k(ə-)liŋ\ (15th century)
1 : to record in or as if in a chronicle
2 : LIST, DESCRIBE
— **chron·i·cler** \-k(ə-)lər\ *noun*

chronicle play *noun* (1902)
: a play with a theme from history consisting usually of rather loosely connected episodes chronologically arranged

Chron·i·cles \ˈkrä-ni-kəlz\ *noun plural but singular in construction*
: either of two historical books of canonical Jewish and Christian Scripture — called also *Paralipomenon;* see BIBLE table

chro·no·bi·ol·o·gy \ˌkrä-nō-bī-ˈä-lə-jē, ˌkrō-\ *noun* (1972)
: the study of biological rhythms
— **chro·no·bi·o·log·ic** \-bī-ə-ˈlä-jik\ *or* **chro·no·bi·o·log·i·cal** \-ji-kəl\ *adjective*
— **chro·no·bi·o·lo·gist** \-bī-ˈä-lə-jist\ *noun*

chro·no·gram \ˈkrä-nə-ˌgram, ˈkrō-\ *noun* (1621)
: an inscription, sentence, or phrase in which certain letters express a date or epoch

chro·no·graph \ˈkrä-nə-ˌgraf, ˈkrō-\ *noun* (1868)
: an instrument for measuring and recording time intervals: as **a :** an instrument having a revolving drum on which a stylus makes marks **b :** STOPWATCH; *also* **:** a watch incorporating the functions of a stopwatch **c :** an instrument for measuring the time of flight of projectiles
— **chro·no·graph·ic** \ˌkrä-nə-ˈgra-fik, ˌkrō-\ *adjective*
— **chro·nog·ra·phy** \krə-ˈnä-grə-fē\ *noun*

chro·nol·o·ger \krə-ˈnä-lə-jər\ *noun* (circa 1572)
: CHRONOLOGIST

chro·no·log·i·cal \ˌkrä-nə-ˈlä-ji-kəl, ˌkrō-\ *also* **chro·no·log·ic** \-jik\ *adjective* (1614)
: of, relating to, or arranged in or according to the order of time ⟨*chronological* tables of American history⟩; *also* **:** reckoned in units of time ⟨*chronological* age⟩
— **chro·no·log·i·cal·ly** \-ji-k(ə-)lē\ *adverb*

chro·nol·o·gist \krə-ˈnä-lə-jist\ *noun* (1611)
: an expert in chronology

chro·nol·o·gy \-jē\ *noun, plural* **-gies** [New Latin *chronologia,* from *chron-* + *-logia* -logy] (1593)
1 : the science that deals with measuring time by regular divisions and that assigns to events their proper dates
2 : a chronological table, list, or account
3 : an arrangement (as of events) in order of occurrence ⟨reconstruct the *chronology* of the trip⟩

chro·nom·e·ter \krə-ˈnä-mə-tər\ *noun* (circa 1735)
: TIMEPIECE; *especially* **:** one designed to keep time with great accuracy

chro·no·met·ric \ˌkrä-nə-ˈme-trik, ˌkrō-\ *also* **chro·no·met·ri·cal** \-tri-kəl\ *adjective* (1830)
: of or relating to a chronometer or chronometry
— **chro·no·met·ri·cal·ly** \-tri-k(ə-)lē\ *adverb*

chro·nom·e·try \krə-ˈnä-mə-trē\ *noun* (1833)
: the measuring of time

chro·no·ther·a·py \ˌkrä-nō-ˈther-ə-pē, ˌkrō-\ *noun* (1973)
: treatment of a sleep disorder (as insomnia) by changing sleeping and waking times in an attempt to reset the patient's biological clock

chrys- *or* **chryso-** *combining form* [Greek, from *chrysos*]
: gold **:** yellow ⟨*chrys*arobin⟩

chrys·a·lid \ˈkri-sə-ləd\ *noun* (1777)
: CHRYSALIS
— **chrysalid** *adjective*

chrys·a·lis \ˈkri-sə-ləs\ *noun, plural* **chry·sal·i·des** \kri-ˈsa-lə-ˌdēz\ *or* **chrys·a·lis·es** [Latin *chrysallid-, chrysallis* gold-colored pupa of butterflies, from Greek, from *chrysos* gold, of Semitic origin; akin to Hebrew *ḥārūṣ* gold] (1601)
1 : a pupa of a butterfly; *broadly* **:** an insect pupa
2 : a protecting covering **:** a sheltered state or stage of being or growth ⟨a budding writer could not emerge from his *chrysalis* too soon —William Du Bois⟩

chry·san·the·mum \kri-ˈsan(t)-thə-məm *also* -ˈzan(t)-\ *noun* [Latin, from Greek *chrysanthemon,* from *chrys-* + *anthemon* flower; akin to Greek *anthos* flower] (1551)
1 : any of various composite plants (genus *Chrysanthemum*) including weeds, ornamentals grown for their brightly colored often double flower heads, and others important as sources of medicinals and insecticides
2 : a flower head of an ornamental chrysanthemum

chrysanthemum 2

chrys·a·ro·bin \ˌkri-sə-ˈrō-bən\ *noun* [*chrys-* + *araroba,* powder found in the wood of a Brazilian tree (*Andira araroba*) + [1]*-in*] (1887)
: a powder derived from the wood of a tropical tree used to treat skin diseases

Chry·se·is \krī-ˈsē-əs\ *noun* [Latin, from Greek *Chrysēis*]
: a daughter of a priest of Apollo in the *Iliad* narrative taken at Troy by Agamemnon but later restored to her father

chryso·ber·yl \ˈkri-sə-ˌber-əl\ *noun* [Latin *chrysoberyllus,* from Greek *chrysobēryllos,* from *chrys-* + *bēryllos* beryl] (1661)
1 *obsolete* **:** a yellowish beryl
2 : a usually yellow or pale green mineral consisting of beryllium aluminum oxide with a little iron and sometimes used as a gem

chrys·o·lite \ˈkri-sə-ˌlīt\ *noun* [Middle English *crisolite,* from Middle French, from Latin *chrysolithos,* from Greek, from *chrys-* + *-lithos* -lite] (13th century)
: OLIVINE

chrys·o·me·lid \ˌkri-sə-ˈme-ləd, -ˈmē-\ *noun* [ultimately from Greek *chrysomēlolonthē* golden cockchafer] (circa 1904)
: any of a large family (Chrysomelidae) of small, usually oval and smooth, shining, and brightly colored beetles (as the Colorado potato beetle)
— **chrysomelid** *adjective*

chryso·phyte \ˈkri-sə-ˌfīt\ *noun* [ultimately from Greek *chrysos* + *phyton* plant — more at PHYT-] (1959)
: GOLDEN-BROWN ALGA

chrys·o·prase \ˈkri-sə-ˌprāz\ *noun* [alteration of Middle English *crisopace,* from Old French, from Latin *chrysoprasus,* from Greek *chrysoprasos,* from *chrys-* + *prason* leek; akin to Latin *porrum* leek] (1646)
: an apple-green chalcedony valued as a gem

chrys·o·tile \-ˌtīl\ *noun* [German *Chrysotil,* from *chrys-* + *-til* fiber, from Greek *tillein* to pluck] (1850)
: a mineral consisting of a fibrous silky serpentine and constituting a kind of asbestos

chthon·ic \ˈthä-nik\ *also* **chtho·ni·an** \ˈthō-nē-ən\ *adjective* [Greek *chthon-, chthōn* earth — more at HUMBLE] (1882)
: of or relating to the underworld **:** INFERNAL ⟨*chthonic* deities⟩

chub \ˈchəb\ *noun, plural* **chub** *or* **chubs** [Middle English *chubbe*] (15th century)
1 : any of numerous freshwater cyprinid fishes (as of the genera *Gila* and *Nocomis*)
2 : any of several marine or freshwater fishes that are not cyprinids

chub·bi·ly \ˈchə-bə-lē\ *adverb* (1909)
: in the manner of one that is chubby

chub·by \ˈchə-bē\ *adjective* **chub·bi·er; -est** [*chub*] (1722)
: PLUMP ⟨a *chubby* boy⟩
— **chub·bi·ness** \ˈchə-bē-nəs\ *noun*

[1]**chuck** \ˈchək\ *verb* [Middle English *chukken,* of imitative origin] (14th century)
: CLUCK

[2]**chuck** *noun* [perhaps from *chuck* chicken] (1595)
— used as an endearment

[3]**chuck** *transitive verb* [origin unknown] (1583)
1 : PAT, TAP
2 a : TOSS **b :** DISCARD ⟨*chucked* his old shirt⟩ **c :** DISMISS, OUST — used especially with *out* ⟨was *chucked* out of office⟩

3 : to have done with ⟨*chucked* his job⟩ ⟨*chuck* it all and retire⟩

[4]chuck *noun* (1611)
1 : a pat or nudge under the chin
2 : an abrupt movement or toss

[5]chuck *noun* [English dialect *chuck* lump] (1723)
1 : a cut of beef that includes most of the neck, the parts about the shoulder blade, and those about the first three ribs — see BEEF illustration
2 *chiefly West* **:** FOOD
3 : an attachment for holding a workpiece or tool in a machine (as a drill or lathe)

chuck•hole \'chək-ˌhōl\ *noun* [perhaps from [3]*chuck* + *hole*] (1836)
: a hole or rut in a road **:** POTHOLE

chuck•le \'chə-kᵊl\ *intransitive verb* **chuck•led; chuck•ling** [probably frequentative of [1]*chuck*] (1803)
1 : to laugh inwardly or quietly
2 : to make a continuous gentle sound resembling suppressed mirth ⟨the clear bright water *chuckled* over gravel —B. A. Williams⟩
— chuckle *noun*
— chuck•le•some \-səm\ *adjective*
— chuck•ling•ly \-iŋ-lē\ *adverb*

chuck•le•head \'chə-kᵊl-ˌhed\ *noun* [*chuckle* lumpish + *head*] (1748)
: BLOCKHEAD
— chuck•le•head•ed \-ˌhe-dəd\ *adjective*

chuck wagon *noun* [[5]*chuck*] (1890)
: a wagon carrying supplies and provisions for cooking (as on a ranch)

chuck•wal•la \'chək-ˌwä-lə\ *or* **chuck•a•wal•la** \'chə-kə-ˌwä-lə\ *noun* [American Spanish *chacahuala,* from Cahuilla (Uto-Aztecan language of southeast California) *čáxwal*] (1893)
: a large herbivorous lizard (*Sauromalus obesus* of the family Iguanidae) of desert regions of the southwestern U.S.

chuckwalla

chuck–will's–wid•ow \ˌchək-ˌwilz-'wi-(ˌ)dō\ *noun* [imitative] (1791)
: a nightjar (*Caprimulgus carolinensis*) of the southern U.S.

[1]chuff \'chəf\ *noun* [Middle English *chuffe*] (15th century)
: BOOR, CHURL

[2]chuff *intransitive verb* [imitative] (1914)
: to produce noisy exhaust or exhalations **:** proceed or operate with chuffs ⟨the *chuffing* and snorting of switch engines —Paul Gallico⟩

[3]chuff *noun* (1915)
: the sound of noisy exhaust or exhalations

chuffed \'chəft\ *adjective* [English dialect *chuff* pleased, puffed with fat] (1957)
British **:** PROUD, SATISFIED

chuf•fy \'chə-fē\ *adjective* **chuf•fi•er; -est** [English dialect *chuff* puffed with fat] (1611)
: FAT, CHUBBY

[1]chug \'chəg\ *noun* [imitative] (1866)
: a dull explosive sound made by or as if by a laboring engine

[2]chug *intransitive verb* **chugged; chug•ging** (1896)
: to move or go with or as if with chugs ⟨a locomotive *chugging* along⟩
— chug•ger *noun*

[3]chug *transitive verb* **chugged; chug•ging** (circa 1968)
: CHUGALUG

chug•a•lug \'chə-gə-ˌləg\ *verb* **-lugged; -lug•ging** [imitative] (1956)
transitive verb
: to drink a container of (as beer) without pause; *also* **:** GUZZLE
intransitive verb
: to drink a container (as of beer) without pause

chu•kar \chə-'kär *also* chə-'kär\ *noun* [Hindi *cakor*] (1814)
: a grayish brown Eurasian partridge (*Alectoris chukar*) introduced as a game bird into arid mountainous regions of the western U.S. — called also *chukar partridge*

chuk•ka \'chə-kə\ *noun* [*chukka,* alteration of *chukker;* from a similar polo player's boot] (1948)
: a usually ankle-length leather boot with two or three pairs of eyelets or a buckle and strap

chuk•ker \'chə-kər\ *also* **chuk•ka** \'chə-kə\ *noun* [Hindi *cakkar* circular course, from Sanskrit *cakra* wheel, circle — more at WHEEL] (1898)
: a playing period of a polo game

[1]chum \'chəm\ *noun* [perhaps by shortening & alteration from *chamber fellow* roommate] (1684)
: a close friend **:** PAL
— chum•ship \-ˌship\ *noun*

[2]chum *intransitive verb* **chummed; chum•ming** (1730)
1 : to room together
2 a : to be a close friend **b :** to show affable friendliness

[3]chum *noun* [origin unknown] (1857)
: animal or vegetable matter (as chopped fish or corn) thrown overboard to attract fish

[4]chum *verb* **chummed; chum•ming** (1857)
transitive verb
: to attract with chum
intransitive verb
: to throw chum overboard to attract fish

[5]chum *noun* [Chinook Jargon *cəm* spotted, striped, from Lower Chinook *č́ə́m* variegated] (1902)
: CHUM SALMON

Chu•mash \'chü-ˌmash\ *noun* (1891)
: a member of an American Indian people of southwestern California

chum•my \'chə-mē\ *adjective* **chum•mi•er; -est** (1884)
: INTIMATE, SOCIABLE
— chum•mi•ly \'chə-mə-lē\ *adverb*
— chum•mi•ness \'chə-mē-nəs\ *noun*

chump \'chəmp\ *noun* [perhaps blend of *chunk* and *lump*] (1883)
: FOOL, DUPE

chump change *noun* (circa 1968)
: a relatively small or insignificant amount of money

chum salmon *noun* [[5]*chum*] (1907)
: a salmon (*Oncorhynchus keta*) of the northern Pacific — called also *chum*

[1]chunk \'chəŋk\ *noun* [perhaps alteration of *chuck* short piece of wood] (1691)
1 : a short thick piece or lump (as of wood or coal)
2 : a large noteworthy quantity ⟨bet a sizable *chunk* of money on the race⟩
3 : a strong thickset horse usually smaller than a draft horse

[2]chunk *intransitive verb* [imitative] (1890)
: to make a dull plunging or explosive sound ⟨the rhythmic *chunking* of thrown quoits —John Updike⟩

chunky \'chəŋ-kē\ *adjective* **chunk•i•er; -est** (1751)
1 : short and thick or broad; *especially* **:** STOCKY
2 : filled with chunks ⟨*chunky* peanut butter⟩
— chunk•i•ly \-kə-lē\ *adverb*

chun•ter \'chən-tər\ *intransitive verb* [probably of imitative origin] (1599)
British **:** to talk in a low inarticulate way **:** MUTTER

[1]church \'chərch\ *noun* [Middle English *chirche,* from Old English *cirice,* ultimately from Late Greek *kyriakon,* from Greek, neuter of *kyriakos* of the lord, from *kyrios* lord, master; akin to Sanskrit *śūra* hero, warrior] (before 12th century)
1 : a building for public and especially Christian worship
2 : the clergy or officialdom of a religious body
3 : a body or organization of religious believers: as **a :** the whole body of Christians **b :** DENOMINATION **c :** CONGREGATION
4 : a public divine worship ⟨goes to *church* every Sunday⟩
5 : the clerical profession ⟨considered the *church* as a possible career⟩

[2]church *adjective* (before 12th century)
1 : of or relating to a church ⟨*church* government⟩
2 *chiefly British* **:** of or relating to the established church

[3]church *transitive verb* (14th century)
: to bring to church to receive one of its rites

churched \'chərcht\ *adjective* (14th century)
: affiliated with a church

church father *noun* (1856)
: FATHER 4

church•go•er \'chərch-ˌgō(-ə)r\ *noun* (1687)
: one who habitually attends church
— church•go•ing \-ˌgō-iŋ, -ˌgȯ(-)iŋ\ *adjective or noun*

church•i•an•i•ty \ˌchər-chē-'a-nə-tē\ *noun* [*church* + *-ianity* (as in *Christianity*)] (1837)
: the usually excessive or sectarian attachment to the practices and interests of a particular church

church•ing *noun* (15th century)
: the administration or reception of a rite of the church; *specifically* **:** a ceremony in some churches by which women after childbirth are received in the church with prayers, blessings, and thanksgiving

church key *noun* (circa 1953)
: an implement with a triangular pointed head at one end for piercing the tops of cans and often with a rounded head at the other end for opening bottles

church•less \'chərch-ləs\ *adjective* (1641)
: not affiliated with a church

church•ly \'chərch-lē\ *adjective* (before 12th century)
1 : of or relating to a church
2 : suitable to or suggestive of a church
3 : adhering to a church
4 : CHURCHY 1
— church•li•ness *noun*

church•man \'chərch-mən\ *noun* (14th century)
1 : CLERGYMAN
2 : a member of a church

church•man•ship \-mən-ˌship\ *noun* (circa 1680)
: the attitude, belief, or practice of a churchman

church mode *noun* (circa 1864)
: one of eight scales prevalent in medieval music each utilizing a different pattern of intervals and beginning on a different tone

Church of England (1534)
: the established episcopal church of England

church register *noun* (1846)
: a parish register of baptisms, marriages, and deaths

church school *noun* (1862)
1 : a school providing a general education but supported by a particular church in contrast to a public school or a nondenominational private school
2 : an organization of officers, teachers, and pupils for purposes of moral and religious education under the supervision of a local church

Church Slavonic *noun* (1850)
: any of several Slavic literary and liturgical languages that continue Old Church Slavonic

but vary regionally under influence of vernacular languages

church·war·den \'chərch-ˌwȯr-dᵊn\ *noun* (15th century)
1 : one of two lay parish officers in Anglican churches with responsibility especially for parish property and alms
2 : a long-stemmed clay pipe

church·wom·an \-ˌwu̇-mən\ *noun* (1722)
: a woman who is a member of a church

churchy \'chər-chē\ *adjective* (1843)
1 : marked by strict conformity or zealous adherence to the forms or beliefs of a church
2 : of or suggestive of a church or church services

church·yard \'chərch-ˌyärd\ *noun* (12th century)
: a yard that belongs to a church and is often used as a burial ground

churl \'chər(-ə)l\ *noun* [Middle English, from Old English *ceorl* man, ceorl; akin to Old Norse *karl* man, husband] (before 12th century)
1 : CEORL
2 : a medieval peasant
3 : RUSTIC, COUNTRYMAN
4 a : a rude ill-bred person **b :** a stingy morose person

churl·ish \'chər-lish\ *adjective* (before 12th century)
1 : of, resembling, or characteristic of a churl **:** VULGAR
2 : marked by a lack of civility or graciousness **:** SURLY
3 : difficult to work with or deal with **:** INTRACTABLE ⟨*churlish* soil⟩
synonym see BOORISH
— churl·ish·ly *adverb*
— churl·ish·ness *noun*

¹churn \'chərn\ *noun* [Middle English *chyrne,* from Old English *cyrin;* akin to Old Norse *kjarni* churn] (before 12th century)
: a vessel for making butter in which milk or cream is agitated in order to separate the oily globules from the watery medium

²churn (15th century)
transitive verb
1 : to agitate (milk or cream) in a churn in order to make butter
2 a : to stir or agitate violently ⟨an old sternwheeler *churning* the muddy river⟩ **b :** to make (as foam) by so doing
3 : to make (the account of a client) excessively active by frequent purchases and sales primarily in order to generate commissions
intransitive verb
1 : to work a churn
2 a : to produce or be in violent motion **b :** to proceed by means of rotating members (as wheels)

churn out *transitive verb* (1912)
: to produce mechanically or copiously **:** GRIND OUT ⟨the usual pap which has been *churned out* about this superstar —W. S. Murphy⟩

churr \'chər\ *intransitive verb* [imitative] (1555)
: to make a vibrant or whirring noise like that made by some insects (as the cockchafer) or some birds (as the partridge)
— churr *noun*

chur·ri·gue·resque \ˌchu̇r-i-gə-'resk\ *adjective, often capitalized* [Spanish *churrigueresco,* from José *Churriguera* (died 1725) Spanish architect] (1845)
: of or relating to a Spanish baroque architectural style characterized by elaborate surface decoration or its Latin-American adaptation

¹chute \'shüt\ *noun* [French, from Old French, from *cheoir* to fall, from Latin *cadere* — more at CHANCE] (1805)
1 a : FALL 6b **b :** a quick descent (as in a river) **:** RAPID
2 : an inclined plane, sloping channel, or passage down or through which things may pass **:** SLIDE
3 : PARACHUTE

²chute *verb* **chut·ed; chut·ing** (1884)
transitive verb
: to convey by a chute
intransitive verb
1 : to go in or as if in a chute
2 : to utilize a chute (as by passing ore down it)

chut·ist \'shü-tist\ *noun* (1920)
: PARACHUTIST

chut·ney \'chət-nē\ *noun, plural* **chutneys** [Hindi *caṭnī*] (1813)
: a thick sauce of Indian origin that contains fruits, vinegar, sugar, and spices and is used as a condiment

chutz·pah *also* **chutz·pa** \'hu̇t-spə, 'ku̇t-, -(ˌ)spä\ *noun* [Yiddish *khutspe,* from Late Hebrew *ḥuṣpāh*] (1892)
: supreme self-confidence **:** NERVE, GALL
synonym see TEMERITY

chyle \'kī(ə)l\ *noun* [Late Latin *chylus,* from Greek *chylos* juice, chyle; akin to Greek *chein* to pour — more at FOUND] (1541)
: lymph that is milky from emulsified fats, characteristically present in the lacteals, and most apparent during intestinal absorption of fats
— chy·lous \'kī-ləs\ *adjective*

chy·lo·mi·cron \ˌkī-lō-'mī-ˌkrän\ *noun* [Greek *chylos* + *mikron,* neuter of *mikros* small — more at MICRO-] (1921)
: a microscopic lipid particle common in the blood during fat digestion and assimilation

chyme \'kīm\ *noun* [New Latin *chymus,* from Late Latin, chyle, from Greek *chymos* juice; akin to Greek *chein*] (1607)
: the semifluid mass of partly digested food expelled by the stomach into the duodenum

chy·mo·tryp·sin \ˌkī-mō-'trip-sən\ *noun* [*chyme* + *-o-* + *trypsin*] (1933)
: a protease that hydrolyzes peptide bonds and is formed in the intestine from chymotrypsinogen
— chy·mo·tryp·tic \-tik\ *adjective*

chy·mo·tryp·sin·o·gen \-ˌtrip-'si-nə-jən\ *noun* (1933)
: a zymogen that is secreted by the pancreas and is converted by trypsin to chymotrypsin

ciao \'chau̇\ *interjection* [Italian, from Italian dialect, literally, (I am your) slave, from Medieval Latin *sclavus* — more at SLAVE] (1929)
— used conventionally as an utterance at meeting or parting

ci·bo·ri·um \sə-'bōr-ē-əm, -'bȯr-\ *noun, plural* **-ria** \-ē-ə\ [Medieval Latin, from Latin, cup, from Greek *kibōrion*] (1651)
1 : a goblet-shaped vessel for holding eucharistic bread
2 : BALDACHIN; *specifically* **:** a freestanding vaulted canopy supported by four columns over a high altar

ci·ca·da \sə-'kā-də, -'kä-; sī-'kā-\ *noun, plural* **-das** *also* **-dae** \-'kā-(ˌ)dē, -'kä-\ [New Latin, genus name, from Latin, cicada] (14th century)
: any of a family (Cicadidae) of homopterous insects which have a stout body, wide blunt head, and large transparent wings and the males of which produce a loud buzzing noise usually by stridulation

ci·ca·la \sə-'kä-lə\ *noun* [Italian, from Medieval Latin, alteration of Latin *cicada*] (1820)
: CICADA

cic·a·tri·cial \ˌsi-kə-'tri-shəl\ *adjective* (1881)
: of or relating to a cicatrix

ci·ca·trix \'si-kə-ˌtriks, sə-'kā-triks\ *noun, plural* **ci·ca·tri·ces** \ˌsi-kə-'trī-(ˌ)sēz, sə-'kā-trə-ˌsēz\ [Latin *cicatric-, cicatrix*] (1641)
1 : a scar resulting from formation and contraction of fibrous tissue in a wound
2 : a mark resembling a scar especially when caused by the previous attachment of an organ or part (as a leaf)

cic·a·tri·za·tion \ˌsi-kə-trə-'zā-shən\ *noun* (15th century)
: scar formation at the site of a healing wound
— cic·a·trize \'si-kə-ˌtrīz\ *transitive verb*

ci·ce·ro·ne \ˌsi-sə-'rō-nē, ˌchē-chə-\ *noun, plural* **-ni** \-(ˌ)nē\ [Italian, from *Cicerone* Cicero] (1726)
: a guide who conducts sightseers

cich·lid \'si-kləd\ *noun* [ultimately from Greek *kichlē* thrush, a kind of wrasse; perhaps akin to Greek *chelidōn* swallow] (1884)
: any of a family (Cichlidae) of mostly tropical spiny-finned usually freshwater fishes including several kept in tropical aquariums
— cichlid *adjective*

ci·cis·beo \ˌchē-chəz-'bā-(ˌ)ō\ *noun, plural* **-bei** \-'bā-ˌē\ [Italian] (1718)
: LOVER, GALLANT
— ci·cis·be·ism \-'bā-ˌi-zəm\ *noun*

-cidal *adjective combining form* [Late Latin *-cidalis,* from Latin *-cida*]
: killing **:** having power to kill ⟨insecti*cidal*⟩

-cide *noun combining form* [Middle French, from Latin *-cida,* from *caedere* to cut, kill]
1 : killer ⟨insecti*cide*⟩
2 [Middle French, from Latin *-cidium,* from *caedere*] **:** killing ⟨sui*cide*⟩

ci·der \'sī-dər\ *noun* [Middle English *sidre,* from Middle French, from Late Latin *sicera* strong drink, from Greek *sikera,* from Hebrew *shēkhār*] (13th century)
1 : fermented apple juice often made sparkling by carbonation or fermentation in a sealed container
2 : the expressed juice of fruit (as apples) used as a beverage or for making other products (as applejack)

cider vinegar *noun* (1851)
: vinegar made from fermented cider

ci–de·vant \ˌsē-də-'vän\ *adjective* [French, literally, formerly] (1790)
: FORMER

ci·gar \si-'gär\ *noun* [Spanish *cigarro*] (1730)
: a small roll of tobacco leaf for smoking

cig·a·rette *also* **cig·a·ret** \ˌsi-gə-'ret, 'si-gə-ˌ\ *noun* [French *cigarette,* diminutive of *cigare* cigar, from Spanish *cigarro*] (1835)
: a slender roll of cut tobacco enclosed in paper and meant to be smoked; *also* **:** a similar roll of another substance (as marijuana)

cig·a·ril·lo \ˌsi-gə-'ri-(ˌ)lō, -'rē-(ˌ)yō\ *noun, plural* **-los** [Spanish *cigarrillo* cigarette, diminutive of *cigarro* cigar] (1832)
1 : a very small cigar
2 : a cigarette wrapped in tobacco rather than paper

ci·gua·tera \ˌsē-gwə-'ter-ə, ˌsi-\ *noun* [American Spanish, from *ciguato* person ill with ciguatera, perhaps from *cigua* sea snail] (1862)
: poisoning caused by the ingestion of various normally edible tropical fish in whose flesh a toxic substance has accumulated

ci·lan·tro \si-'län-(ˌ)trō, -'lan-\ *noun* [Spanish, coriander, from Medieval Latin *celiandrum,* alteration of Latin *coriandrum* — more at CORIANDER] (1903)
: leaves of coriander used as a flavoring or garnish; *also* **:** CORIANDER 1

cil·i·ary \'si-lē-ˌer-ē\ *adjective* (1691)
1 : of or relating to cilia
2 : of, relating to, or being the annular suspension of the lens of the eye

cil·i·ate \'si-lē-ət, -lē-ˌāt\ *noun* (1916)
: any of a phylum or subphylum (Ciliophora) of ciliated protozoans (as paramecia)

cil·i·at·ed \'si-lē-ˌā-təd\ *or* **cil·i·ate** \-lē-ət, -lē-ˌāt\ *adjective* (1753)
: possessing cilia ⟨*ciliated* epithelial cells⟩ ⟨leaves with *ciliate* petioles⟩
— cil·i·a·tion \ˌsi-lē-'ā-shən\ *noun*

cil·i·um \'si-lē-əm\ *noun, plural* **cil·ia** \-lē-ə\ [New Latin, from Latin, eyelid; akin to Latin *celare* to conceal — more at HELL] (1794)
1 : a minute short hairlike process often forming part of a fringe; *especially* **:** one on a cell that is capable of lashing movement and serves especially in free unicellular organisms to produce locomotion or in higher forms a current of fluid

2 : EYELASH

ci·met·i·dine \sī-ˈme-tə-ˌdēn\ *noun* [*ci-* (alteration of *cyan-*) + *meth*yl + *-idine*] (1975)
: a histamine analogue $C_{10}H_{16}N_6S$ used especially in the treatment of duodenal ulcers

¹Cim·me·ri·an \sə-ˈmir-ē-ən\ *adjective* (1580)
: very dark or gloomy ⟨under ebon shades . . . in dark *Cimmerian* desert ever dwell —John Milton⟩

²Cimmerian *noun* [Latin *Cimmerii,* a mythical people, from Greek *Kimmerioi*] (1584)
: any of a mythical people described by Homer as dwelling in a remote realm of mist and gloom

¹cinch \ˈsinch\ *noun* [Spanish *cincha,* from Latin *cingula* girdle, girth, from *cingere*] (1859)
1 : a girth for a pack or saddle
2 : a tight grip
3 a : a thing done with ease **b :** a certainty to happen

²cinch (1866)
transitive verb
1 : to put a cinch on
2 : to make certain **:** ASSURE
intransitive verb
: to tighten the cinch — often used with *up*

cin·cho·na \siŋ-ˈkō-nə, sin-ˈchō-\ *noun* [New Latin, genus name, from the countess of *Chinchón* (died 1641) wife of the Peruvian viceroy] (1786)
1 : any of a genus (*Cinchona*) of South American trees and shrubs of the madder family
2 : the dried bark of a cinchona (as *C. ledgeriana*) containing alkaloids (as quinine) and formerly used as a specific in malaria

cin·cho·nine \ˈsiŋ-kə-ˌnēn, ˈsin-chə-\ *noun* (1825)
: a bitter white crystalline alkaloid $C_{19}H_{22}N_2O$ found especially in cinchona bark and used like quinine

cin·cho·nism \ˈsiŋ-kə-ˌni-zəm, ˈsin-chə-\ *noun* (1857)
: a disorder due to excessive or prolonged use of cinchona or its alkaloids and marked by temporary deafness, ringing in the ears, headache, dizziness, and rash

cinc·ture \ˈsiŋ(k)-chər\ *noun* [Latin *cinctura* girdle, from *cinctus,* past participle of *cingere* to gird; probably akin to Sanskrit *kāñcī* girdle] (1600)
1 : the act of encircling
2 a : an encircling area **b :** GIRDLE, BELT; *especially* **:** a cord or sash of cloth worn around an ecclesiastical vestment (as an alb) or the habit of a religious

cin·der \ˈsin-dər\ *noun* [Middle English *sinder,* from Old English; akin to Old High German *sintar* dross, slag, Serbo-Croatian *sedra* calcium carbonate] (before 12th century)
1 : the slag from a metal furnace **:** DROSS
2 a *plural* **:** ASHES **b :** a fragment of ash
3 a : a partly burned combustible in which fire is extinct **b :** a hot coal without flame **c :** a partly burned coal capable of further burning without flame
4 : a fragment of lava from an erupting volcano
— **cin·dery** \-d(ə-)rē\ *adjective*

cinder block *noun* (1926)
: a hollow rectangular building block made of cement and coal cinders

cinder cone *noun* (1849)
: a conical hill formed by the accumulation of volcanic debris around a vent

Cin·der·el·la \ˌsin-də-ˈre-lə\ *noun* [after *Cinderella,* fairy-tale heroine who is used as a drudge by her stepmother but ends up married to a prince] (1840)
: one resembling the fairy-tale Cinderella: as **a :** one suffering undeserved neglect **b :** one suddenly lifted from obscurity to honor or significance

cine \ˈsi-nē\ *noun* [probably from French *ciné,* short for *cinéma* cinema] (1920)
: MOTION PICTURE

cine·ast \ˈsi-nē-ˌast, -nē-əst\ *or* **cine·aste** \-ˌast\ *or* **ciné·aste** \-ˌast\ *noun* [French *cinéaste,* from *ciné* + *-aste* (as in *enthousiaste* enthusiast)] (1926)
: a devotee of motion pictures; *also* **:** MOVIEMAKER

cin·e·ma \ˈsi-nə-mə, *British also* -ˌmä\ *noun* [short for *cinematograph*] (1909)
1 a : MOTION PICTURE — usually used attributively **b :** a motion-picture theater
2 a : MOVIES; *especially* **:** the film industry **b :** the art or technique of making motion pictures

cin·e·ma·go·er \-ˌgō(-ə)r\ *noun* (1920)
: MOVIEGOER

cin·e·ma·theque \ˌsi-nə-mə-ˈtek\ *noun* [French *cinémathèque* film library, from *cinéma* + *-thèque* (as in *bibliothèque* library)] (1966)
: a small movie house specializing in avant-garde films

cin·e·mat·ic \ˌsi-nə-ˈma-tik\ *adjective* (1916)
1 : of, relating to, suggestive of, or suitable for motion pictures or the filming of motion pictures ⟨*cinematic* principles and techniques⟩
2 : filmed and presented as a motion picture ⟨*cinematic* fantasies⟩
— **cin·e·mat·i·cal·ly** \-ˈma-ti-k(ə-)lē\ *adverb*

cin·e·ma·tize \ˈsi-nə-mə-ˌtīz\ *transitive verb* **-tized; -tiz·ing** (1916)
: to make a motion picture of (as a novel) **:** adapt for motion pictures

cin·e·mat·o·graph \ˌsi-nə-ˈma-tə-ˌgraf\ *noun* [French *cinématographe,* from Greek *kinēmat-, kinēma* movement (from *kinein* to move) + French *-o-* + *-graphe* -graph — more at -KINESIS] (1896)
chiefly British **:** a motion-picture camera, projector, theater, or show

cin·e·ma·tog·ra·pher \ˌsi-nə-mə-ˈtä-grə-fər\ *noun* (1897)
1 : a motion-picture cameraman
2 : a motion-picture projectionist

cin·e·ma·tog·ra·phy \ˌsi-nə-mə-ˈtä-grə-fē\ *noun* (1897)
: the art or science of motion-picture photography
— **cin·e·mat·o·graph·ic** \-ˌma-tə-ˈgra-fik\ *adjective*
— **cin·e·mat·o·graph·i·cal·ly** \-fi-k(ə-)lē\ *adverb*

ci·né·ma vé·ri·té \ˈsi-nə-mə-ˌver-i-ˈtā; si-nā-ˈmä-ve-rē-ˈtā\ *noun* [French *cinéma-vérité,* literally, cinema-truth, translation of Russian *kinopravda*] (1963)
: the art or technique of filming a motion picture so as to convey candid realism

cin·e·ole \ˈsi-nē-ˌōl\ *noun* [International Scientific Vocabulary, by transposition from New Latin *oleum cinae* wormseed oil] (1885)
: a liquid $C_{10}H_{18}O$ with a camphor odor contained in many essential oils (as of eucalyptus) and used especially as an expectorant and flavoring agent — called also *eucalyptol*

cin·er·ar·ia \ˌsi-nə-ˈrer-ē-ə, -ˈrar-\ *noun* [New Latin, from Latin, feminine of *cinerarius* of ashes, from *ciner-, cinis* ashes — more at INCINERATE] (1597)
: any of several garden or potted plants derived from a perennial composite herb (*Senecio cruentus*) of the Canary Islands and having heart-shaped leaves and clusters of bright flower heads

cin·er·ar·i·um \-ē-əm\ *noun, plural* **-ia** \-ē-ə\ [Latin, from *ciner-, cinis*] (1880)
: a place to receive the ashes of the cremated dead
— **cin·er·ary** \ˈsi-nə-ˌrer-ē\ *adjective*

ci·ne·re·ous \sə-ˈnir-ē-əs\ *adjective* [Latin *cinereus,* from *ciner-, cinis*] (1661)
1 : gray tinged with black
2 : resembling or consisting of ashes

cin·er·in \ˈsi-nə-rən\ *noun* [Latin *ciner-, cinis*] (1948)
: either of two oily liquid esters $C_{20}H_{28}O_3$ and $C_{21}H_{28}O_5$ occurring especially in the flowers of pyrethrum and possessing insecticidal properties

cin·gu·lum \ˈsiŋ-gyə-ləm\ *noun, plural* **-la** \-lə\ [New Latin, from Latin, girdle, from *cingere* to gird — more at CINCTURE] (1845)
: an anatomical band or encircling ridge
— **cin·gu·late** \-lət\ *adjective*

cin·na·bar \ˈsi-nə-ˌbär\ *noun* [Middle English *cynabare,* from Middle French & Latin; Middle French *cenobre,* from Latin *cinnabaris,* from Greek *kinnabari,* of non-Indo-European origin; akin to Arabic *zinjafr* cinnabar] (14th century)
1 : artificial red mercuric sulfide used especially as a pigment
2 : native red mercuric sulfide HgS that is the only important ore of mercury
— **cin·na·bar·ine** \-bə-ˌrīn, -ˌbär-in\ *adjective*

cinnabar moth *noun* (circa 1893)
: a European moth (*Tyria jacobeae*) that has been introduced into the western U.S. in attempts to control the tansy ragwort on which its larvae feed — called also *cinnabar*

cin·nam·ic acid \sə-ˈna-mik-\ *noun* [French *cinnamique* of cinnamon, from *cinname* cinnamon, from Latin *cinnamomum*] (circa 1864)
: a white crystalline odorless acid $C_9H_8O_2$ found especially in cinnamon oil and storax

cin·na·mon \ˈsi-nə-mən\ *noun, often attributive* [Middle English *cynamone,* from Latin *cinnamomum, cinnamon,* from Greek *kinnamōmon, kinnamon,* of non-Indo-European origin; akin to Hebrew *qinnāmōn* cinnamon] (14th century)
1 a : any of several Asian trees (genus *Cinnamomum*) of the laurel family **b :** an aromatic spice prepared from the dried inner bark of a cinnamon (especially *C. zeylanicum*); *also* **:** the bark
2 : a light yellowish brown

cinnamon fern *noun* (1818)
: a large fern (*Osmunda cinnamomea*) with cinnamon-colored spore-bearing fronds shorter than and separate from the green foliage fronds

cinnamon stone *noun* (1805)
: ESSONITE

cin·quain \ˈsiŋ-ˌkān, ˈsaŋ-\ *noun* [French, from *cinq* five, from Latin *quinque* — more at FIVE] (1882)
: a 5-line stanza

cin·que·cen·tist \ˌchiŋ-kwi-ˈchen-tist\ *noun* (1871)
: an Italian of the cinquecento; *especially* **:** a poet or artist of this period

cin·que·cen·to \ˌchiŋ-kwi-ˈchen-(ˌ)tō\ *noun* [Italian, literally, five hundred, from *cinque* five (from Latin *quinque*) + *cento* hundred, from Latin *centum* — more at HUNDRED] (1760)
: the 16th century especially in Italian art and literature

cinque·foil \ˈsiŋk-ˌfȯil, ˈsaŋk-\ *noun* [Middle English *sink foil,* from Middle French *cincfoille,* from Latin *quinquefolium,* from *quinque* five + *folium* leaf — more at BLADE] (14th century)
1 : any of a genus (*Potentilla*) of herbs and shrubs of the rose family with 5 lobed leaves and 5-petaled flowers — called also *potentilla*
2 : a design enclosed by five joined foils

cinquefoil 2

ciop·pi·no \chə-ˈpē-(ˌ)nō\ *noun* [perhaps from Italian dialect] (1936)
: a dish of fish and shellfish cooked usually with tomatoes, wine, spices, and herbs

^{1}ci·pher \'sī-fər\ *noun, often attributive* [Middle English, from Middle French *cifre,* from Medieval Latin *cifra,* from Arabic *ṣifr* empty, cipher, zero] (14th century)
1 a : ZERO 1a **b :** one that has no weight, worth, or influence **:** NONENTITY
2 a : a method of transforming a text in order to conceal its meaning — compare CODE 3b **b :** a message in code
3 : ARABIC NUMERAL
4 : a combination of symbolic letters; *especially* **:** the interwoven initials of a name

^{2}cipher *verb* **ci·phered; ci·pher·ing** \-f(ə-)riŋ\ (circa 1530)
intransitive verb
: to use figures in a mathematical process
transitive verb
1 : ENCIPHER
2 : to compute arithmetically

ci·pher·text \'sī-fər-ˌtekst\ *noun* (1939)
: the enciphered form of a text or of its elements — compare PLAINTEXT

cir·ca \'sər-kə\ *preposition* [Latin, from *circum* around — more at CIRCUM-] (1861)
: at, in, or of approximately — used especially with dates ⟨born *circa* 1600⟩

cir·ca·di·an \sər-'kā-dē-ən\ *adjective* [Latin *circa* about + *dies* day + English ^{2}*-an* — more at DEITY] (1959)
: being, having, characterized by, or occurring in approximately 24-hour periods or cycles (as of biological activity or function) ⟨*circadian* rhythms in activity⟩

Cir·cas·sian \(ˌ)sər-'ka-sh(ē-)ən\ *noun* [*Circassia,* region of the Caucasus] (1555)
1 : a member of a group of peoples of the northwestern Caucasus
2 : the language of the Circassian peoples
— Circassian *adjective*

Circassian walnut *noun* (1914)
: the light brown irregularly black-veined wood of the English walnut much used for veneer and cabinetwork

Cir·ce \'sər-(ˌ)sē\ *noun* [Latin, from Greek *Kirkē*]
: a sorceress who changes Odysseus' men into swine but is forced by Odysseus to change them back

cir·ci·nate \'sər-s^{ə}n-ˌāt\ *adjective* [Latin *circinatus,* past participle of *circinare* to round, from *circinus* pair of compasses, from *circus*] (1830)
: ROUNDED, COILED; *especially* **:** rolled in the form of a flat coil with the apex as a center ⟨*circinate* fern fronds unfolding⟩
— cir·ci·nate·ly *adverb*

^{1}cir·cle \'sər-kəl\ *noun, often attributive* [Middle English *cercle,* from Old French, from Latin *circulus,* diminutive of *circus* circle, circus, from or akin to Greek *krikos, kirkos* ring; akin to Old English *hring* ring — more at RING] (14th century)
1 a : RING, HALO **b :** a closed plane curve every point of which is equidistant from a fixed point within the curve **c :** the plane surface bounded by such a curve
2 : the orbit of a celestial body
3 : something in the form of a circle or section of a circle: as **a :** DIADEM **b :** an instrument of astronomical observation the graduated limb of which consists of an entire circle **c :** a balcony or tier of seats in a theater **d :** a circle formed on the surface of a sphere by the intersection of a plane that passes through it ⟨*circle* of latitude⟩ **e :** ROTARY 2

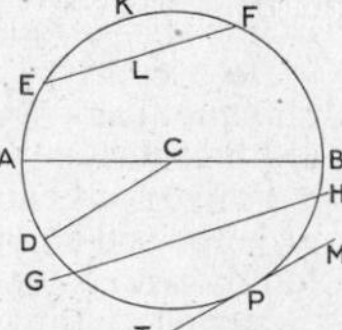

circle 1b: *AB* **diameter;** *C* **center;** *CD, CA, CB* **radii;** *EKF* **arc on chord** *EF; EFKL* **(area) segment on chord** *EF; ACD* **(area) sector;** *GH* **secant;** *TPM* **tangent at point** *P; EKFBPDA* **circumference**

4 : an area of action or influence **:** REALM
5 a : CYCLE, ROUND ⟨the wheel has come full *circle*⟩ **b :** fallacious reasoning in which something to be demonstrated is covertly assumed
6 : a group of persons sharing a common interest or revolving about a common center ⟨the sewing *circle* of her church⟩ ⟨family *circle*⟩ ⟨the gossip of court *circles*⟩
7 : a territorial or administrative division or district
8 : a curving side street

^{2}circle *verb* **cir·cled; cir·cling** \-k(ə-)liŋ\ (14th century)
transitive verb
1 : to enclose in or as if in a circle
2 : to move or revolve around
intransitive verb
1 a : to move in or as if in a circle **b :** CIRCULATE
2 : to describe or extend in a circle
— cir·cler \-k(ə-)lər\ *noun*

circle graph *noun* (1928)
: PIE CHART

cir·clet \'sər-klət\ *noun* (15th century)
: a little circle; *especially* **:** a circular ornament

^{1}cir·cuit \'sər-kət\ *noun, often attributive* [Middle English, from Middle French *circuite,* from Latin *circuitus,* from *circumire, circuire* to go around, from *circum-* + *ire* to go — more at ISSUE] (14th century)
1 a : a usually circular line encompassing an area **b :** the space enclosed within such a line
2 a : a course around a periphery **b :** a circuitous or indirect route
3 a : a regular tour (as by a traveling judge or preacher) around an assigned district or territory **b :** the route traveled **c :** a group of church congregations ministered to by one pastor
4 a : the complete path of an electric current including usually the source of electric energy **b :** an assemblage of electronic elements **:** HOOKUP **c :** a two-way communication path between points (as in a computer)
5 a : an association of similar groups **:** LEAGUE **b :** a number or series of public outlets (as theaters, radio shows, or arenas) offering the same kind of presentation **c :** a number of similar social gatherings ⟨cocktail *circuit*⟩
— cir·cuit·al \-kə-t^{ə}l\ *adjective*

^{2}circuit (15th century)
transitive verb
: to make a circuit about
intransitive verb
: to make a circuit

circuit board *noun* (1948)
: BOARD 9

circuit breaker *noun* (1872)
: a switch that automatically interrupts an electric circuit under an infrequent abnormal condition

circuit court *noun* (1708)
: a court that sits at two or more places within one judicial district

circuit judge *noun* (1801)
: a judge who holds a circuit court

cir·cu·i·tous \(ˌ)sər-'kyü-ə-təs\ *adjective* [perhaps from Medieval Latin *circuitosus,* from Latin *circuitus*] (1664)
1 : having a circular or winding course ⟨a *circuitous* route⟩
2 : not being forthright or direct in language or action
— cir·cu·i·tous·ly *adverb*
— cir·cu·i·tous·ness *noun*

circuit rider *noun* (1837)
: a clergyman assigned to a circuit especially in a rural area

cir·cuit·ry \'sər-kə-trē\ *noun, plural* **-ries** (1946)
1 : the detailed plan or arrangement of an electric circuit
2 : the components of an electric circuit

cir·cu·ity \(ˌ)sər-'kyü-ə-tē\ *noun, plural* **-ities** [*circuitous*] (circa 1626)
: lack of straightforwardness **:** INDIRECTION

^{1}cir·cu·lar \'sər-kyə-lər\ *adjective* [Middle English *circuler,* from Middle French, from Late Latin *circularis,* from Latin *circulus* circle] (15th century)
1 a : having the form of a circle **:** ROUND **b :** moving in or describing a circle or spiral
2 a : of or relating to a circle or its mathematical properties ⟨a *circular* arc⟩ **b :** having a circular base or bases ⟨a *circular* cylinder⟩
3 : CIRCUITOUS, INDIRECT
4 : marked by or moving in a cycle
5 : being or involving reasoning that uses in the argument or proof a conclusion to be proved or one of its unproved consequences
6 : intended for circulation
— cir·cu·lar·i·ty \ˌsər-kyə-'lar-ə-tē\ *noun*
— cir·cu·lar·ly \'sər-kyə-lər-lē\ *adverb*
— cir·cu·lar·ness *noun*

^{2}circular *noun* (1789)
: a paper (as a leaflet) intended for wide distribution

circular dichroism *noun* (circa 1961)
1 : the property (as of an optically active medium) of unequal absorption of right and left plane-polarized light so that the emergent light is elliptically polarized
2 : a spectroscopic technique that makes use of circular dichroism

circular file *noun* (1967)
: WASTEBASKET

circular function *noun* (1884)
: TRIGONOMETRIC FUNCTION

cir·cu·lar·ise *British variant of* CIRCULARIZE

cir·cu·lar·ize \'sər-kyə-lə-ˌrīz\ *transitive verb* **-ized; -iz·ing** (1848)
1 a : to send circulars to **b :** to poll by questionnaire
2 : PUBLICIZE
— cir·cu·lar·i·za·tion \ˌsər-kyə-lər-ə-'zā-shən\ *noun*

circular saw *noun* (1817)
: a power saw with a circular cutting blade; *also* **:** the blade itself

cir·cu·late \'sər-kyə-ˌlāt\ *verb* **-lat·ed; -lat·ing** [Latin *circulatus,* past participle of *circulare,* from *circulus*] (circa 1650)
intransitive verb
1 : to move in a circle, circuit, or orbit; *especially* **:** to follow a course that returns to the starting point ⟨blood *circulates* through the body⟩
2 : to pass from person to person or place to place: as **a :** to flow without obstruction **b :** to become well-known or widespread ⟨rumors *circulated* through the town⟩ **c :** to go from group to group at a social gathering **d :** to come into the hands of readers; *specifically* **:** to become sold or distributed
transitive verb
: to cause to circulate
— cir·cu·lat·able \-ˌlā-tə-bəl\ *adjective*
— cir·cu·la·tive \-ˌlā-tiv\ *adjective*
— cir·cu·la·tor \-ˌlā-tər\ *noun*

circulating decimal *noun* (1768)
: REPEATING DECIMAL

cir·cu·la·tion \ˌsər-kyə-'lā-shən\ *noun* (1654)
1 : orderly movement through a circuit; *especially* **:** the movement of blood through the vessels of the body induced by the pumping action of the heart
2 : FLOW
3 a : passage or transmission from person to person or place to place; *especially* **:** the interchange of currency ⟨coins in *circulation*⟩ **b :** the extent of dissemination: as (1) **:** the average number of copies of a publication sold over a given period (2) **:** the total number of items borrowed from a library

cir·cu·la·to·ry \'sər-kyə-lə-ˌtōr-ē, -ˌtȯr-\ *adjective* (1605)
: of or relating to circulation or the circulatory system ⟨*circulatory* failure⟩

circulatory system *noun* (1862)

: the system of blood, blood vessels, lymphatics, and heart concerned with the circulation of the blood and lymph

circum- *prefix* [Old French or Latin; Old French, from Latin, from *circum,* from *circus* circle — more at CIRCLE]
: around : about ⟨*circum*polar⟩

cir·cum·am·bi·ent \ˌsər-kᵊm-ˈam-bē-ənt\ *adjective* [Late Latin *circumambient-, circumambiens,* present participle of *circumambire* to surround in a circle, from Latin *circum-* + *ambire* to go around — more at AMBIENT] (1633)
: being on all sides : ENCOMPASSING
— **cir·cum·am·bi·ent·ly** *adverb*

cir·cum·am·bu·late \-ˈam-byə-ˌlāt\ *transitive verb* **-lat·ed; -lat·ing** [Late Latin *circumambulatus,* past participle of *circumambulare,* from Latin *circum-* + *ambulare* to walk] (circa 1656)
: to circle on foot especially ritualistically
— **cir·cum·am·bu·la·tion** \-ˌam-byə-ˈlā-shən\ *noun*

cir·cum·cen·ter \ˈsər-kᵊm-ˌsen-tər\ *noun* (circa 1889)
: the point at which the perpendicular bisectors of the sides of a triangle intersect and which is equidistant from the three vertices

cir·cum·cir·cle \-ˌsər-kəl\ *noun* (1885)
: a circle which passes through all the vertices of a polygon (as a triangle)

cir·cum·cise \ˈsər-kᵊm-ˌsīz\ *transitive verb* **-cised; -cis·ing** [Middle English, from Latin *circumcisus,* past participle of *circumcidere,* from *circum-* + *caedere* to cut] (13th century)
: to cut off the prepuce of (a male) or the clitoris of (a female)
— **cir·cum·cis·er** *noun*

cir·cum·ci·sion \ˌsər-kᵊm-ˈsi-zhən, ˈsər-kᵊm-ˌ\ *noun* (12th century)
1 a : the act of circumcising; *especially* **:** a Jewish rite performed on male infants as a sign of inclusion in the Jewish religious community **b :** the condition of being circumcised
2 *capitalized* **:** January 1 observed as a church festival in commemoration of the circumcision of Jesus

cir·cum·fer·ence \sə(r)-ˈkəm(p) fərn(t)s, -f(ə-)rən(t)s\ *noun* [Middle English, from Middle French, from Latin *circumferentia,* from *circumferre* to carry around, from *circum-* + *ferre* to carry — more at BEAR] (14th century)
1 : the perimeter of a circle
2 : the external boundary or surface of a figure or object **:** PERIPHERY
— **cir·cum·fer·en·tial** \-ˌkəm(p)-fə-ˈren(t)-shəl\ *adjective*

¹cir·cum·flex \ˈsər-kᵊm-ˌfleks\ *adjective* [Latin *circumflexus,* past participle of *circumflectere* to bend around, mark with a circumflex, from *circum-* + *flectere* to bend] (circa 1577)
1 : characterized by the pitch, quantity, or quality indicated by a circumflex
2 : marked with a circumflex

²circumflex *noun* (1609)
: a mark ˆ, ˜, or ˜ originally used in Greek over long vowels to indicate a rising-falling tone and in other languages to mark length, contraction, or a particular vowel quality

cir·cum·flu·ent \(ˌ)sər-ˈkəm-flü-ənt\ *adjective* [from Latin *circumfluent-, circumfluens,* present participle of *circumfluere* to flow around, from *circum-* + *fluere* to flow — more at FLUID] (1577)
: flowing round or surrounding in the manner of a fluid
— **cir·cum·flu·ous** \(ˌ)sər-ˈkəm-flü-əs\ *adjective*

cir·cum·fuse \ˌsər-kᵊm-ˈfyüz\ *transitive verb* **-fused; -fus·ing** [Latin *circumfusus,* past participle of *circumfundere* to pour around, from *circum-* + *fundere* to pour — more at FOUND] (1605)
: SURROUND, ENVELOP
— **cir·cum·fu·sion** \-ˈfyü-zhən\ *noun*

cir·cum·ja·cent \ˌsər-kᵊm-ˈjā-sᵊnt\ *adjective* [Latin *circumjacent-, circumjacens,* present participle of *circumjacēre* to lie around, from *circum-* + *jacēre* to lie — more at ADJACENT] (15th century)
: lying adjacent on all sides : SURROUNDING

cir·cum·lo·cu·tion \ˌsər-kᵊm-lō-ˈkyü-shən\ *noun* [Latin *circumlocution-, circumlocutio,* from *circum-* + *locutio* speech, from *loqui* to speak] (15th century)
1 : the use of an unnecessarily large number of words to express an idea
2 : evasion in speech
— **cir·cum·loc·u·to·ry** \-ˈlä-kyə-ˌtōr-ē, -ˌtȯr-\ *adjective*

cir·cum·lu·nar \ˌsər-kəm-ˈlü-nər\ *adjective* (circa 1909)
: revolving about or surrounding the moon

cir·cum·nav·i·gate \-ˈna-və-ˌgāt\ *transitive verb* [Latin *circumnavigatus,* past participle of *circumnavigare* to sail around, from *circum-* + *navigare* to navigate] (1634)
: to go completely around (as the earth) especially by water; *also* **:** to go around instead of through **:** BYPASS ⟨*circumnavigate* a congested area⟩
— **cir·cum·nav·i·ga·tion** \-ˌna-və-ˈgā-shən\ *noun*
— **cir·cum·nav·i·ga·tor** \-ˈna-və-ˌgā-tər\ *noun*

cir·cum·po·lar \ˌsər-kᵊm-ˈpō-lər\ *adjective* (1686)
1 : continually visible above the horizon ⟨a *circumpolar* star⟩
2 : surrounding or found in the vicinity of a terrestrial pole

cir·cum·scis·sile \-ˈsi-səl, -ˌsīl\ *adjective* [Latin *circumscissus,* past participle of *circumscindere* to tear around, from *circum-* + *scindere* to cut, split — more at SHED] (1835)
: dehiscing by fissure around the capsule of the fruit

cir·cum·scribe \ˈsər-kᵊm-ˌskrīb\ *transitive verb* [Latin *circumscribere,* from *circum-* + *scribere* to write, draw — more at SCRIBE] (1835)
1 a : to constrict the range or activity of definitely and clearly **b :** to define or mark off carefully
2 a : to draw a line around **b :** to surround by or as if by a boundary
3 : to construct or be constructed around (a geometrical figure) so as to touch as many points as possible
synonym see LIMIT

cir·cum·scrip·tion \ˌsər-kᵊm-ˈskrip-shən\ *noun* [Latin *circumscription-, circumscriptio,* from *circumscribere*] (1531)
1 : the act of circumscribing **:** the state of being circumscribed: as **a :** DEFINITION, DELIMITATION **b :** LIMITATION
2 : something that circumscribes: as **a :** LIMIT, BOUNDARY **b :** RESTRICTION
3 : a circumscribed area or district

cir·cum·spect \ˈsər-kᵊm-ˌspekt\ *adjective* [Middle English, from Middle French or Latin; Middle French *circonspect,* from Latin *circumspectus,* from past participle of *circumspicere* to look around, be cautious, from *circum-* + *specere* to look — more at SPY] (15th century)
: careful to consider all circumstances and possible consequences **:** PRUDENT
synonym see CAUTIOUS
— **cir·cum·spec·tion** \ˌsər-kᵊm-ˈspek-shən\ *noun*
— **cir·cum·spect·ly** \ˈsər-kᵊm-ˌspek(t)-lē\ *adverb*

cir·cum·stance \ˈsər-kᵊm-ˌstan(t)s, -stən(t)s\ *noun* [Middle English, from Old French, from Latin *circumstantia,* from *circumstant-, circumstans,* present participle of *circumstare* to stand around, from *circum-* + *stare* to stand — more at STAND] (13th century)
1 a : a condition, fact, or event accompanying, conditioning, or determining another **:** an essential or inevitable concomitant ⟨the weather is a *circumstance* to be taken into consideration⟩ **b :** a subordinate or accessory fact or detail ⟨cost is a minor *circumstance* in this case⟩ **c :** a piece of evidence that indicates the probability or improbability of an event (as a crime) ⟨the *circumstance* of the missing weapon told against him⟩ ⟨the *circumstances* suggest murder⟩
2 a : the sum of essential and environmental factors (as of an event or situation) ⟨constant and rapid change in economic *circumstance* —G. M. Trevelyan⟩ **b :** state of affairs **:** EVENTUALITY ⟨open rebellion was a rare *circumstance*⟩ — often used in plural ⟨a victim of *circumstances*⟩ **c** *plural* **:** situation with regard to wealth ⟨he was in easy *circumstances*⟩
3 : attendant formalities and ceremonial ⟨pride, pomp, and *circumstance* of glorious war —Shakespeare⟩
4 : an event that constitutes a detail (as of a narrative or course of events) ⟨considering each *circumstance* in turn⟩
synonym see OCCURRENCE

cir·cum·stanced \-ˌstan(t)st, -stən(t)st\ *adjective* (circa 1611)
: placed in particular circumstances especially in regard to property or income

cir·cum·stan·tial \ˌsər-kᵊm-ˈstan(t)-shəl\ *adjective* (1600)
1 : belonging to, consisting in, or dependent on circumstances
2 : pertinent but not essential **:** INCIDENTAL
3 : marked by careful attention to detail **:** abounding in factual details ⟨a *circumstantial* account of the fight⟩
4 : CEREMONIAL ☆
— **cir·cum·stan·ti·al·i·ty** \-ˌstan(t)-shē-ˈa-lə-tē\ *noun*
— **cir·cum·stan·tial·ly** \-ˈstan(t)-sh(ə-)lē\ *adverb*

circumstantial evidence *noun* (1736)
: evidence that tends to prove a fact by proving other events or circumstances which afford a basis for a reasonable inference of the occurrence of the fact at issue

cir·cum·stan·ti·ate \ˌsər-kᵊm-ˈstan(t)-shē-ˌāt\ *transitive verb* **-at·ed; -at·ing** (circa 1652)
: to supply with circumstantial evidence or support

cir·cum·stel·lar \ˌsər-kᵊm-ˈste-lər\ *adjective* (1951)
: surrounding or occurring in the vicinity of a star

¹cir·cum·val·late \-ˈva-ˌlāt\ *transitive verb* **-lat·ed; -lat·ing** [Latin *circumvallatus,* past participle of *circumvallare* to surround with siege works, from *circum-* + *vallum* rampart — more at WALL] (circa 1798)
: to surround by or as if by a rampart
— **cir·cum·val·la·tion** \-ˌva-ˈlā-shən\ *noun*

²cir·cum·val·late \-ˈva-ˌlāt, -ˈva-lət\ *adjective* (circa 1852)

☆ SYNONYMS
Circumstantial, minute, particular, detailed mean dealing with a matter fully and usually point by point. CIRCUMSTANTIAL implies fullness of detail that fixes something described in time and space ⟨a *circumstantial* account of our visit⟩. MINUTE implies close and searching attention to the smallest details ⟨a *minute* examination of a fossil⟩. PARTICULAR implies a precise attention to every detail ⟨a *particular* description of the scene of the crime⟩. DETAILED stresses abundance or completeness of detail ⟨a *detailed* analysis of the event⟩.

: being any of approximately 12 large papillae near the back of the tongue each of which is surrounded with a marginal sulcus and supplied with taste buds responsive especially to bitter flavors

cir·cum·vent \ˌsər-kᵊm-ˈvent\ *transitive verb* [Latin *circumventus,* past participle of *circumvenire,* from *circum-* + *venire* to come — more at COME] (1539)
1 a : to hem in **b :** to make a circuit around
2 : to manage to get around especially by ingenuity or stratagem ⟨the setup *circumvented* the red tape —Lynne McTaggart⟩
— cir·cum·ven·tion \-ˈven(t)-shən\ *noun*

cir·cum·vo·lu·tion \sər-ˌkəm-və-ˈlü-shən, ˌsər-kᵊm-vō-\ *noun* [Middle English *circumvolucioun,* from Medieval Latin *circumvolution-, circumvolutio,* from Latin *circumvolvere* to revolve, from *circum-* + *volvere* to roll — more at VOLUBLE] (15th century)
: an act or instance of turning around an axis

cir·cus \ˈsər-kəs\ *noun, often attributive* [Latin, circle, circus — more at CIRCLE] (14th century)
1 a : a large arena enclosed by tiers of seats on three or all four sides and used especially for sports or spectacles (as athletic contests, exhibitions of horsemanship, or in ancient times chariot racing) **b :** a public spectacle
2 a : an arena often covered by a tent and used for variety shows usually including feats of physical skill, wild animal acts, and performances by clowns **b :** a circus performance **c :** the physical plant, livestock, and personnel of such a circus **d :** something suggestive of a circus (as in frenzied activity, sensationalism, theatricality, or razzle-dazzle)
3 a *obsolete* **:** CIRCLE, RING **b** *British* **:** a usually circular area at an intersection of streets
— cir·cusy \-kə-sē\ *adjective*

circus catch *noun* (1893)
: a catch (as in baseball or football) requiring an extraordinary or spectacular effort

ci·ré *also* **ci·re** \sə-ˈrā, sē-\ *noun* [French, from past participle of *cirer* to wax, from *cire* wax, from Latin *cera* — more at CERUMEN] (1921)
1 : a highly glazed finish for fabrics usually achieved by applying wax to the fabric
2 : a fabric or garment with a ciré finish

cirque \ˈsərk\ *noun* [French, from Latin *circus*] (1601)
1 *archaic* **:** CIRCUS
2 : CIRCLE, CIRCLET
3 : a deep steep-walled basin on a mountain usually forming the blunt end of a valley

cirque 3

cir·rho·sis \sə-ˈrō-səs\ *noun, plural* **-rho·ses** \-ˌsēz\ [New Latin, from Greek *kirrhos* orange-colored; akin to Old English *hār* gray — more at HOAR] (circa 1847)
: widespread disruption of normal liver structure by fibrosis and the formation of regenerative nodules that is caused by any of various chronic progressive conditions affecting the liver (as long term alcohol abuse or hepatitis)
— cir·rhot·ic \-ˈrä-tik\ *adjective or noun*

cirro- *combining form* [New Latin *cirrus*]
: cirrus ⟨*cirro*stratus⟩

cir·ro·cu·mu·lus \ˌsir-ō-ˈkyü-myə-ləs\ *noun* [New Latin] (circa 1803)
: a cloud form of small white rounded masses at a high altitude usually in regular groupings forming a mackerel sky — see CLOUD illustration

cir·ro·stra·tus \ˌsir-ō-ˈstrā-təs, -ˈstra-\ *noun* [New Latin] (circa 1803)
: a fairly uniform layer of high stratus darker than cirrus — see CLOUD illustration

cir·rus \ˈsir-əs\ *noun, plural* **cir·ri** \ˈsir-ˌī\ [New Latin, from Latin, curl] (1708)
1 : TENDRIL
2 : a slender usually flexible animal appendage: as **a :** an arm of a barnacle **b :** a filament of a crinoid **c :** a fused group of cilia functioning like a limb on some protozoans **d :** the male copulatory organ of various invertebrate animals
3 : a wispy white cloud usually of minute ice crystals formed at altitudes of 20,000 to 40,000 feet (6,000 to 12,000 meters) — see CLOUD illustration

cis \ˈsis\ *adjective* [Latin, literally, on this side] (1888)
: characterized by having certain atoms or groups of atoms on the same side of the longitudinal axis of a double bond or of the plane of a ring in a molecule

cis- *prefix* [Latin, from *cis* — more at HE]
1 : on this side ⟨*cis*lunar⟩
2 *usually italic* **:** cis ⟨*cis*-dichloroethylene⟩ — compare TRANS- 2b

cis·al·pine \(ˌ)sis-ˈal-ˌpīn\ *adjective* (1542)
: situated on the south side of the Alps ⟨*Cisalpine* Gaul⟩ — compare TRANSALPINE

cis·co \ˈsis-(ˌ)kō\ *noun, plural* **ciscoes** [short for Canadian French *ciscoette*] (1848)
: any of various whitefishes (genus *Coregonus*); *especially* **:** LAKE HERRING

cis·lu·nar \(ˌ)sis-ˈlü-nər\ *adjective* (circa 1877)
: lying between the earth and the moon or the moon's orbit ⟨*cislunar* space⟩

cis·plat·in \(ˌ)sis-ˈpla-tᵊn\ *noun* [*cis-* + *platinum*] (1977)
: a platinum-containing antineoplastic drug $Cl_2H_6N_2Pt$ used especially in the treatment of testicular and ovarian tumors and advanced bladder cancer

cis–plat·i·num \-ˈplat-nəm, -ˈpla-tᵊn-əm\ *noun* (1977)
: CISPLATIN

cissy *British variant of* SISSY

cist \ˈsist, ˈkist\ *noun* [Welsh, chest, from Latin *cista*] (1804)
: a neolithic or Bronze Age burial chamber typically lined with stone

Cis·ter·cian \sis-ˈtər-shən\ *noun* [Medieval Latin *Cistercium* Cîteaux] (1611)
: a member of a monastic order founded by Saint Robert of Molesme in 1098 at Cîteaux, France, under Benedictine rule
— Cistercian *adjective*

cis·tern \ˈsis-tərn\ *noun* [Middle English, from Old French *cisterne,* from Latin *cisterna,* from *cista* box, chest — more at CHEST] (13th century)
1 : an artificial reservoir (as an underground tank) for storing liquids and especially water (as rainwater)
2 : a large usually silver vessel formerly used (as in cooling wine) at the dining table
3 : a fluid-containing sac or cavity in an organism

cis·ter·na \sis-ˈtər-nə\ *noun, plural* **-nae** \-ˌnē\ [New Latin, from Latin, reservoir] (circa 1860)
: CISTERN 3: as **a :** one of the large spaces under the arachnoid membrane **b :** one of the interconnected flattened vesicles comprising the part of the endoplasmic reticulum that is studded with ribosomes
— cis·ter·nal \-nəl\ *adjective*

cis·tron \ˈsis-ˌträn\ *noun* [*cis-* + trans- + ²*-on*] (1957)
: a segment of DNA that is equivalent to a gene and that specifies a single functional unit (as a protein or enzyme)
— cis·tron·ic \sis-ˈträ-nik\ *adjective*

cit·a·del \ˈsi-tə-dᵊl, -ˌdel\ *noun* [Middle French *citadelle,* from Old Italian *cittadella,* diminutive of *cittade* city, from Medieval Latin *civitat-, civitas* — more at CITY] (1562)
1 : a fortress that commands a city
2 : STRONGHOLD

ci·ta·tion \sī-ˈtā-shən\ *noun* (13th century)
1 : an official summons to appear (as before a court)
2 a : an act of quoting; *especially* **:** the citing of a previously settled case at law **b :** EXCERPT, QUOTE
3 : MENTION: as **a :** a formal statement of the achievements of a person receiving an academic honor **b :** specific reference in a military dispatch to meritorious performance of duty
synonym see ENCOMIUM
— ci·ta·tion·al \-shnəl, -shə-nᵊl\ *adjective*

cite \ˈsīt\ *transitive verb* **cit·ed; cit·ing** [Middle French *citer* to cite, summon, from Latin *citare* to put in motion, rouse, summon, from frequentative of *ciēre* to stir, move — more at -KINESIS] (15th century)
1 : to call upon officially or authoritatively to appear (as before a court)
2 : to quote by way of example, authority, or proof
3 a : to refer to; *especially* **:** to mention formally in commendation or praise **b :** to name in a citation
4 : to bring forward or call to another's attention especially as an example, proof, or precedent
synonym see SUMMON
— cit·able \ˈsī-tə-bəl\ *adjective*

cith·a·ra \ˈsi-thə-rə, ˈki-\ *variant of* KITHARA

cith·er \ˈsi-t͟hər, -thər\ *noun* [French *cithare,* from Latin *cithara* kithara, from Greek *kithara*] (1606)
: CITTERN

cit·ied \ˈsi-tēd\ *adjective* (1612)
: occupied by cities

cit·i·fied \ˈsi-ti-ˌfīd\ *adjective* (1828)
: of, relating to, or characteristic of a sophisticated urban style of living — often used disparagingly

cit·i·fy \-ˌfī\ *transitive verb* **-fied; -fy·ing** (1828)
: URBANIZE
— cit·i·fi·ca·tion \ˌsi-tə-fi-ˈkā-shən\ *noun*

cit·i·zen \ˈsi-tə-zən *also* -sən\ *noun* [Middle English *citizein,* from Anglo-French *citezein,* alteration of Old French *citeien,* from *cité* city] (14th century)
1 : an inhabitant of a city or town; *especially* **:** one entitled to the rights and privileges of a freeman
2 a : a member of a state **b :** a native or naturalized person who owes allegiance to a government and is entitled to protection from it
3 : a civilian as distinguished from a specialized servant of the state ☆
— cit·i·zen·ly \-zən-lē *also* -sən-\ *adjective*

cit·i·zen·ess \-zə-nəs *also* -sə-\ *noun* (1796)
: a female citizen

cit·i·zen·ry \-zən-rē *also* -sən-\ *noun, plural* **-ries** (1819)
: a whole body of citizens

citizen's arrest *noun* (1941)
: an arrest made not by a law officer but by a citizen who derives authority from the fact of being a citizen

citizens band *noun* (1948)
: a range of radio-wave frequencies that in the U.S. is allocated officially for private radio communications

cit·i·zen·ship \ˈsi-tə-zən-ˌship\ *noun* (1611)

☆ SYNONYMS
Citizen, subject, national mean a person owing allegiance to and entitled to the protection of a sovereign state. CITIZEN is preferred for one owing allegiance to a state in which sovereign power is retained by the people and sharing in the political rights of those people ⟨the rights of a free *citizen*⟩. SUBJECT implies allegiance to a personal sovereign such as a monarch ⟨the king's *subjects*⟩. NATIONAL designates one who may claim the protection of a state and applies especially to one living or traveling outside that state ⟨American *nationals* working in the Middle East⟩.

1 : the status of being a citizen
2 a : membership in a community (as a college) **b :** the quality of an individual's response to membership in a community

citr- *or* **citri-** *or* **citro-** *combining form* [New Latin, from *Citrus,* genus name]
1 : citrus ⟨*citri*culture⟩
2 : citric acid ⟨*citr*ate⟩

cit·ral \'si-ˌtral\ *noun* [International Scientific Vocabulary] (1891)
: an unsaturated liquid isomeric aldehyde $C_{10}H_{16}O$ of many essential oils that has a strong lemon odor and is used especially in perfumery and as a flavoring

cit·rate \'si-ˌtrāt\ *noun* [International Scientific Vocabulary] (1794)
: a salt or ester of citric acid

cit·ric acid \'si-trik-\ *noun* [International Scientific Vocabulary] (1813)
: a tricarboxylic acid $C_6H_8O_7$ occurring in cellular metabolism, obtained especially from lemon and lime juices or by fermentation of sugars, and used as a flavoring

citric acid cycle *noun* (1942)
: KREBS CYCLE

cit·ri·cul·ture \'si-trə-ˌkəl-chər\ *noun* (1916)
: the cultivation of citrus fruits
— cit·ri·cul·tur·ist \ˌsi-trə-'kəl-ch(ə-)rist\ *noun*

¹cit·rine \'si-ˌtrīn\ *adjective* [Middle English, from Middle French *citrin,* from Medieval Latin *citrinus,* from Latin *citrus* citron tree] (14th century)
: resembling a citron or lemon especially in color

²ci·trine \si-'trēn\ *noun* (1748)
: a semiprecious yellow stone resembling topaz formed by heating a black quartz in order to change its color

ci·tri·nin \si-'trī-nən\ *noun* [New Latin *citrinum,* specific epithet of *Penicillium citrinum*] (1931)
: a toxic antibiotic $C_{13}H_{14}O_5$ that is produced especially by a penicillium (*Penicillium citrinum*) and an aspergillus (*Aspergillus niveus*) and is effective against some gram-positive bacteria

cit·ron \'si-trən\ *noun* [Middle English, from Middle French, modification of Latin *citrus*] (15th century)
1 a : a citrus fruit resembling a lemon but larger with little pulp and a very thick rind **b :** a small shrubby tree (*Citrus medica*) that produces citrons and is cultivated in tropical regions **c :** the preserved rind of the citron used especially in cakes and puddings
2 : a small hard-fleshed watermelon used especially in pickles and preserves

cit·ro·nel·la \ˌsi-trə-'ne-lə\ *noun* [New Latin, from French *citronnelle* lemon balm, from *citron*] (circa 1858)
: a fragrant grass (*Cymbopogon nardus*) of southern Asia that yields an oil used in perfumery and as an insect repellent; *also* **:** its oil

cit·ro·nel·lal \-'ne-ˌlal\ *noun* [International Scientific Vocabulary, from New Latin *citronella*] (1893)
: a lemon-odored aldehyde $C_{10}H_{18}O$ that is derived especially from citronella oil and is used in perfumery and as an insect repellent

cit·ro·nel·lol \-'ne-ˌlȯl, -ˌlōl\ *noun* [International Scientific Vocabulary, from New Latin *citronella*] (1872)
: an unsaturated liquid alcohol $C_{10}H_{20}O$ with a roselike odor that is found in two optically active forms in many essential oils (as rose oil) and is used in perfumery and soaps

ci·trov·o·rum factor \sə-'trä-və-rəm-\ *noun* [New Latin *citrovorum,* specific epithet of *Leuconostoc citrovorum,* bacterium that requires this form of folic acid] (1948)
: a metabolically active form of folic acid that has been used in cancer therapy to protect normal cells against methotrexate

cit·rul·line \'si-trə-ˌlēn\ *noun* [International Scientific Vocabulary, from New Latin *Citrullus,* genus name of the watermelon] (1930)
: a crystalline amino acid $C_6H_{13}N_3O_3$ formed especially as an intermediate in the conversion of ornithine to arginine in the living system

cit·rus \'si-trəs\ *noun, plural* **citrus** *or* **cit·rus·es** *often attributive* [New Latin, genus name, from Latin] (1825)
: any of a group of often thorny trees and shrubs (*Citrus* and related genera) of the rue family grown in warm regions for their edible fruit (as the orange or lemon) with firm usually thick rind and pulpy flesh; *also* **:** the fruit
— cit·rusy \'si-trə-sē\ *adjective*

citrus canker *noun* (1916)
: a destructive disease of citrus caused by a bacterium (*Xanthomonas campestris citri*) that produces lesions on the leaves, twigs, and fruits

citrus red mite *noun* (1935)
: a relatively large mite (*Panonychus citri*) that is a destructive pest on the foliage of citrus

cit·tern \'si-tərn\ *also* **cith·ern** \'si-t͟hərn, -thərn\ *or* **cith·ren** \'si-thrən\ *noun* [blend of *cither* and *gittern*] (1566)
: a Renaissance stringed instrument like a guitar with a flat pear-shaped body

city \'si-tē\ *noun, plural* **cit·ies** *often attributive* [Middle English *citie* large or small town, from Old French *cité,* from Medieval Latin *civitat-, civitas,* from Latin, citizenship, state, city of Rome, from *civis* citizen — more at HIND] (13th century)
1 a : an inhabited place of greater size, population, or importance than a town or village **b :** an incorporated British town usually of major size or importance having the status of an episcopal see **c** *capitalized* (1) **:** the financial district of London (2) **:** the influential financial interests of the British economy **d :** a usually large or important municipality in the U.S. governed under a charter granted by the state **e :** an incorporated municipal unit of the highest class in Canada
2 : CITY-STATE
3 : the people of a city
4 *slang* — used with a preceding adjective or noun naming an abundant or quintessential feature or quality ⟨the movie was shoot-out *city*⟩

city clerk *noun* (1919)
: a public officer charged with recording the official proceedings and vital statistics of a city

city council *noun* (1789)
: the legislative body of a city

city editor *noun* (1834)
: a newspaper editor usually in charge of local news and staff assignments

city father *noun* (1845)
: a member (as an alderman or councilman) of the governing body of a city

city hall *noun* (1675)
1 : the chief administrative building of a city
2 a : a municipal government **b :** city officialdom or bureaucracy ⟨you can't fight *city hall*⟩

city manager *noun* (1913)
: an official employed by an elected council to direct the administration of a city government

city planning *noun* (1912)
: the drawing up of an organized arrangement (as of streets, parks, and business and residential areas) of a city
— city planner *noun*

city room *noun* (1919)
: the department where local news is handled in a newspaper editorial office

city·scape \'si-tē-ˌskāp\ *noun* (1856)
1 : a city viewed as a scene
2 : an artistic representation of a city

city slicker *noun* (1924)
: SLICKER 2b

city–state \'si-tē-ˌstāt *also* -'stāt\ *noun* (1893)
: an autonomous state consisting of a city and surrounding territory

city·wide \'si-tē-ˌwīd\ *adjective* (1961)
: including all parts of a city

civ·et \'si-vət\ *noun* [Middle French *civette,* from Old Italian *zibetto,* from Arabic *zabād* civet perfume] (1532)
1 : any of various Old World carnivorous mammals (family Viverridae) with long bodies, short legs, and a usually long tail
2 : a thick yellowish musky-odored substance found in a pouch near the sexual organs of the civet (especially genera *Civettictis, Viverra,* and *Viverricula*) and used in perfume

civet

civet cat *noun* (1607)
1 : CIVET 1
2 : RINGTAIL 2
3 : any of the small spotted skunks (genus *Spilogale*) of western North America

civ·ic \'si-vik\ *adjective* [Latin *civicus,* from *civis* citizen] (circa 1656)
: of or relating to a citizen, a city, citizenship, or civil affairs
— civ·i·cal·ly \'si-vi-k(ə-)lē\ *adverb*

civ·ic–mind·ed \ˌsi-vik-'mīn-dəd\ *adjective* (1947)
: disposed to look after civic needs and interests
— civ·ic–mind·ed·ness *noun*

civ·ics \'si-viks\ *noun plural but singular or plural in construction* (1886)
: a social science dealing with the rights and duties of citizens

civ·il \'si-vəl\ *adjective* [Middle English, from Middle French, from Latin *civilis,* from *civis*] (14th century)
1 a : of or relating to citizens **b :** of or relating to the state or its citizenry
2 a : CIVILIZED ⟨*civil* society⟩ **b :** adequate in courtesy and politeness **:** MANNERLY
3 a : of, relating to, or based on civil law **b :** relating to private rights and to remedies sought by action or suit distinct from criminal proceedings **c :** established by law
4 : of, relating to, or involving the general public, their activities, needs, or ways, or civic affairs as distinguished from special (as military or religious) affairs
5 *of time* **:** based on the mean sun and legally recognized for use in ordinary affairs ☆

☆ SYNONYMS
Civil, polite, courteous, gallant, chivalrous mean observant of the forms required by good breeding. CIVIL often suggests little more than the avoidance of overt rudeness ⟨owed the questioner a *civil* reply⟩. POLITE commonly implies polish of speech and manners and sometimes suggests an absence of cordiality ⟨if you can't be pleasant, at least be *polite*⟩. COURTEOUS implies more actively considerate or dignified politeness ⟨clerks who were unfailingly *courteous* to customers⟩. GALLANT and CHIVALROUS imply courteous attentiveness especially to women. GALLANT suggests spirited and dashing behavior and ornate expressions of courtesy ⟨a *gallant* suitor of the old school⟩. CHIVALROUS suggests high-minded and self-sacrificing behavior ⟨a *chivalrous* display of duty⟩.

civil death *noun* (1767)
: the status of a living person equivalent in its legal consequences to natural death; *specifically* **:** deprivation of civil rights
civil defense *noun* (1939)
: the system of protective measures and emergency relief activities conducted by civilians in case of hostile attack, sabotage, or natural disaster
civil disobedience *noun* (1866)
: refusal to obey governmental demands or commands especially as a nonviolent and usually collective means of forcing concessions from the government
civil engineer *noun* (circa 1792)
: an engineer whose training or occupation is in the design and construction especially of public works (as roads or harbors)
— **civil engineering** *noun*
ci·vil·ian \sə-ˈvil-yən *also* -ˈvi-yən\ *noun* (14th century)
1 : a specialist in Roman or modern civil law
2 a : one not on active duty in a military, police, or fire-fighting force **b :** OUTSIDER 1
— **civilian** *adjective*
ci·vil·ian·ize \-yə-ˌnīz\ *transitive verb* **-ized; -iz·ing** (1870)
: to convert from military to civilian status or control
— **ci·vil·ian·i·za·tion** \-ˌvil-yə-nə-ˈzā-shən\ *noun*
civ·i·li·sa·tion, civ·i·lise *chiefly British variant of* CIVILIZATION, CIVILIZE
ci·vil·i·ty \sə-ˈvi-lə-tē\ *noun, plural* **-ties** (1533)
1 *archaic* **:** training in the humanities
2 a : COURTESY, POLITENESS **b :** a polite act or expression
civ·i·li·za·tion \ˌsi-və-lə-ˈzā-shən\ *noun* (1772)
1 a : a relatively high level of cultural and technological development; *specifically* **:** the stage of cultural development at which writing and the keeping of written records is attained **b :** the culture characteristic of a particular time or place
2 : the process of becoming civilized
3 a : refinement of thought, manners, or taste **b :** a situation of urban comfort
— **civ·i·li·za·tion·al** \-shnəl, -shə-nᵊl\ *adjective*
civ·i·lize \ˈsi-və-ˌlīz\ *verb* **-lized; -liz·ing** (1601)
transitive verb
1 : to cause to develop out of a primitive state; *especially* **:** to bring to a technically advanced and rationally ordered stage of cultural development
2 a : EDUCATE, REFINE **b :** SOCIALIZE 1
intransitive verb
: to acquire the customs and amenities of a civil community
— **civ·i·liz·er** *noun*
civilized *adjective* (1611)
: characteristic of a state of civilization ⟨*civilized* society⟩; *especially* **:** characterized by taste, refinement, or restraint ⟨a *civilized* way to spend the evening⟩
civil law *noun, often C&L capitalized* (14th century)
1 : Roman law especially as set forth in the Justinian code
2 : the body of private law developed from Roman law and used in Louisiana and in many countries outside the English-speaking world
3 : the law established by a nation or state for its own jurisdiction
4 : the law of civil or private rights
civil liberty *noun* (1644)
: freedom from arbitrary governmental interference (as with the right of free speech) specifically by denial of governmental power and in the U.S. especially as guaranteed by the Bill of Rights — usually used in plural
— **civil libertarian** *noun or adjective*
civ·il·ly \ˈsi-və(l)-lē\ *adverb* (15th century)
1 : in terms of civil rights, law, or matters ⟨*civilly* dead⟩
2 : in a civil manner **:** POLITELY
civil marriage *noun* (circa 1889)
: a marriage performed by a magistrate
civil rights *noun plural* (1721)
: the nonpolitical rights of a citizen; *especially* **:** the rights of personal liberty guaranteed to U.S. citizens by the 13th and 14th amendments to the Constitution and by acts of Congress
civil servant *noun* (1800)
1 : a member of a civil service
2 : a member of the administrative staff of an international agency (as the United Nations)
civil service *noun* (circa 1785)
: the administrative service of a government or international agency exclusive of the armed forces; *especially* **:** one in which appointments are determined by competitive examination
civil war *noun* (15th century)
: a war between opposing groups of citizens of the same country
Civ·i·tan \ˈsi-və-ˌtan\ *noun* [*Civitan (Club)*] (1926)
: a member of a major national and international service club
civ·vy *also* **civ·ie** \ˈsi-vē\ *noun, plural* **civvies** *also* **civies** (circa 1889)
1 *plural* **:** civilian clothes as distinguished from a particular uniform (as of the military)
2 : CIVILIAN
civvy street *noun, often C&S capitalized* (1943)
British **:** civilian life
[1]**clab·ber** \ˈkla-bər\ *noun* [short for *bonnyclabber*] (1634)
chiefly dialect **:** sour milk that has thickened or curdled
[2]**clabber** *intransitive verb* (circa 1879)
chiefly dialect **:** CURDLE
clach·an \ˈkla-k̲ən\ *noun* [Middle English, from Scottish Gaelic] (15th century)
Scottish & Irish **:** HAMLET
[1]**clack** \ˈklak\ *verb* [Middle English, of imitative origin] (13th century)
intransitive verb
1 : CHATTER, PRATTLE
2 : to make an abrupt striking sound or series of sounds
3 *of fowl* **:** CACKLE, CLUCK
transitive verb
1 : to cause to make a clatter
2 : to produce with a chattering sound; *specifically* **:** BLAB
— **clack·er** *noun*
[2]**clack** *noun* (15th century)
1 a : rapid continuous talk **:** CHATTER **b :** TONGUE
2 *archaic* **:** an object (as a valve) that produces clapping or rattling noises usually in regular rapid sequence
3 : a sound of clacking ⟨the *clack* of a typewriter⟩
Clac·to·ni·an \klak-ˈtō-nē-ən\ *adjective* [*Clacton*-on-Sea, England] (1932)
: of or relating to a Lower Paleolithic culture usually characterized by stone flakes with a half cone at the point of striking
[1]**clad** \ˈklad\ *past and past participle of* CLOTHE
[2]**clad** *adjective* [Middle English, past participle of *clothen* to clothe] (14th century)
1 : being covered or clothed ⟨ivy-*clad* buildings⟩
2 *of a coin* **:** consisting of outer layers of one metal bonded to a core of a different metal
[3]**clad** *transitive verb* **clad; clad·ding** (1939)
: SHEATHE, FACE; *specifically* **:** to cover (a metal) with another metal by bonding
[4]**clad** *noun* (1941)
1 a : a composite material formed by cladding **b :** a clad coin
2 : CLADDING; *specifically* **:** the outer layer of a clad coin
clad- *or* **clado-** *combining form* [New Latin, from Greek *klad-, klado-,* from *klados* branch, shoot of a tree; akin to Old English *holt* woods — more at HOLT]
: slip **:** sprout ⟨*clado*phyll⟩
clad·ding \ˈkla-diŋ\ *noun* (1936)
: something that covers or overlays; *specifically* **:** metal coating bonded to a metal core
clade \ˈklād\ *noun* [Greek *klados*] (1911)
: a group of biological taxa (as species) that includes all descendants of one common ancestor
cla·dis·tics \klə-ˈdis-tiks, kla-\ *noun plural but singular in construction* (1965)
: a system of biological taxonomy that defines taxa uniquely by shared characteristics not found in ancestral groups and uses inferred evolutionary relationships to arrange taxa in a branching hierarchy such that all members of a given taxon have the same ancestors
— **cla·dist** \ˈkla-dist\ *noun*
— **cla·dis·tic** \-ˈdis-tik\ *adjective*
— **cla·dis·ti·cal·ly** \-ti-k(ə-)lē\ *adverb*
cla·doc·er·an \klə-ˈdä-sə-rən\ *noun* [New Latin *Cladocera,* from *clad-* + Greek *keras* horn — more at HORN] (1909)
: any of an order (Cladocera) of minute chiefly freshwater branchiopod crustaceans that includes the water fleas
clad·ode \ˈkla-ˌdōd\ *noun* [New Latin *cladodium,* from Greek *klados*] (1870)
: CLADOPHYLL
— **cla·do·di·al** \kla-ˈdō-dē-əl\ *adjective*
clad·o·gen·e·sis \ˌkla-də-ˈje-nə-səs\ *noun* (1953)
: evolutionary change characterized by treelike branching of taxa — compare ANAGENESIS
— **clad·o·ge·net·ic** \ˌkla-dō-jə-ˈne-tik\ *adjective*
— **clad·o·ge·net·i·cal·ly** \-ti-k(ə-)lē\ *adverb*
clad·o·gram \ˈkla-də-ˌgram\ *noun* (1966)
: a branching diagrammatic tree used in cladistic classification to illustrate phylogenetic relationships
clad·o·phyll \ˈkla-də-ˌfil\ *noun* (1879)
: a flattened photosynthetic branch assuming the form of and closely resembling an ordinary foliage leaf
[1]**claim** \ˈklām\ *transitive verb* [Middle English, from Middle French *clamer,* from Latin *clamare* to cry out, shout; akin to Latin *calare* to call — more at LOW] (14th century)
1 a : to ask for especially as a right ⟨*claimed* the inheritance⟩ **b :** to call for **:** REQUIRE ⟨this matter *claims* our attention⟩ **c :** TAKE 16b ⟨the accident *claimed* her life⟩
2 : to take as the rightful owner ⟨went to *claim* their bags at the station⟩
3 a : to assert in the face of possible contradiction **:** MAINTAIN ⟨*claimed* that he'd been cheated⟩ **b :** to claim to have ⟨organization . . . which *claims* 11,000 . . . members —*Rolling Stone*⟩ **c :** to assert to be rightfully one's own ⟨*claimed* responsibility for the attack⟩
synonym see DEMAND
— **claim·able** \ˈklā-mə-bəl\ *adjective*
[2]**claim** *noun* (14th century)
1 : a demand for something due or believed to be due ⟨insurance *claim*⟩
2 a : a right to something; *specifically* **:** a title to a debt, privilege, or other thing in the possession of another **b :** an assertion open to challenge ⟨a *claim* of authenticity⟩
3 : something that is claimed; *especially* **:** a tract of land staked out
claim·ant \ˈklā-mənt\ *noun* (15th century)
: one that asserts a right or title ⟨a *claimant* to an estate⟩
claim·er \ˈklā-mər\ *noun* (15th century)
1 : one that claims

2 : a horse running in a claiming race

claiming race *noun* (1935)
: a horse race in which each entry is offered for sale for a specified price that must be deposited before the race

clair·au·di·ence \klar-ˈȯ-dē-ən(t)s, kler-, -ˈä-\ *noun* [*clair-* (as in *clairvoyance*) + *audience* (act of hearing)] (1864)
: the power or faculty of hearing something not present to the ear but regarded as having objective reality
— **clair·au·di·ent** \-ənt\ *adjective*
— **clair·au·di·ent·ly** *adverb*

clair·voy·ance \klar-ˈvȯi-ən(t)s, kler-\ *noun* (1840)
1 : the power or faculty of discerning objects not present to the senses
2 : ability to perceive matters beyond the range of ordinary perception **:** PENETRATION

[1]clair·voy·ant \-ənt\ *adjective* [French, from *clair* clear (from Latin *clarus*) + *voyant,* present participle of *voir* to see, from Latin *vidēre* — more at WIT] (1671)
1 : unusually perceptive **:** DISCERNING
2 : of or relating to clairvoyance
— **clair·voy·ant·ly** *adverb*

[2]clairvoyant *noun* (1851)
: one having the power of clairvoyance

[1]clam \ˈklam\ *noun* [Middle English, from Old English *clamm* bond, fetter; akin to Old High German *klamma* constriction, Latin *glomus* ball] (before 12th century)
: CLAMP, CLASP

[2]clam *noun, often attributive* [[1]*clam;* from the clamping action of the shells] (circa 1520)
1 a : any of numerous edible marine bivalve mollusks living in sand or mud **b :** a freshwater mussel
2 : a stolid or closemouthed person
3 : CLAMSHELL
4 : DOLLAR 3

clam 1a: *a* incurrent orifice, *b* siphon, *c* excurrent orifice, *d* mantle, *e* shell, *f* foot

[3]clam *intransitive verb* **clammed; clam·ming** (1636)
: to gather clams especially by digging
— **clam·mer** \ˈkla-mər\ *noun*

cla·mant \ˈklā-mənt, ˈkla-\ *adjective* [Latin *clamant-, clamans,* present participle of *clamare* to cry out] (1639)
1 : CLAMOROUS, BLATANT
2 : demanding attention **:** URGENT
— **cla·mant·ly** *adverb*

clam·bake \ˈklam-ˌbāk\ *noun* (1835)
1 a : an outdoor party; *especially* **:** a seashore outing where food is usually cooked on heated rocks covered by seaweed **b :** the food served at a clambake
2 : a gathering characterized by noisy sociability; *especially* **:** a political rally

clam·ber \ˈklam-bər, ˈkla-mər\ *intransitive verb* **clam·bered; clam·ber·ing** \ˈklam-b(ə-)riŋ, ˈklam-riŋ, ˈkla-mər-iŋ\ [Middle English *clambren;* akin to Old English *climban* to climb] (14th century)
: to climb awkwardly (as by scrambling) ⟨*clambered* over the rocks⟩
— **clam·ber·er** \-bər-ər, -mər-ər\ *noun*

clam·my \ˈkla-mē\ *adjective* **clam·mi·er; -est** [Middle English, probably from *clammen* to smear, stick, from Old English *clǣman;* akin to Old English *clǣg* clay] (14th century)
1 : being damp, soft, sticky, and usually cool ⟨a *clammy* and intensely cold mist —Charles Dickens⟩
2 : lacking normal human warmth ⟨the *clammy* atmosphere of an institution⟩
— **clam·mi·ly** \ˈkla-mə-lē\ *adverb*
— **clam·mi·ness** \ˈkla-mē-nəs\ *noun*

[1]clam·or \ˈkla-mər\ *noun* [Middle English, from Middle French *clamour,* from Latin *clamor,* from *clamare* to cry out — more at CLAIM] (14th century)
1 a : noisy shouting **b :** a loud continuous noise
2 : insistent public expression (as of support or protest) ⟨a *clamor* against increased taxes⟩

[2]clamor *verb* **clam·ored; clam·or·ing** \ˈklam-riŋ, ˈkla-mər-iŋ\ (14th century)
intransitive verb
1 : to make a din
2 : to become loudly insistent ⟨*clamored* for his impeachment⟩
transitive verb
1 : to utter or proclaim insistently and noisily
2 : to influence by means of clamor

[3]clamor *transitive verb* [origin unknown] (1611)
obsolete **:** SILENCE

clam·or·ous \ˈklam-rəs, ˈkla-mər-əs\ *adjective* (15th century)
1 : marked by confused din or outcry **:** TUMULTUOUS ⟨the busy *clamorous* market⟩
2 : noisily insistent
synonym see VOCIFEROUS
— **clam·or·ous·ly** *adverb*
— **clam·or·ous·ness** *noun*

clam·our \ˈkla-mər\ *chiefly British variant of* CLAMOR

[1]clamp \ˈklamp\ *noun* [Middle English, probably from (assumed) Middle Dutch *klampe;* akin to Old English *clamm* bond, fetter — more at CLAM] (14th century)
1 : a device designed to bind or constrict or to press two or more parts together so as to hold them firmly
2 : any of various instruments or appliances having parts brought together for holding or compressing something

[2]clamp *transitive verb* (circa 1696)
1 : to fasten with or as if with a clamp
2 a : to place by decree **:** IMPOSE — often used with *on* ⟨*clamped* on a curfew after the riots⟩ **b :** to hold tightly

clamp·down \ˈklamp-ˌdau̇n\ *noun* (1940)
: the act or action of making regulations and restrictions more stringent ⟨a *clampdown* on charge accounts, bank loans, and other inflationary influences —*Time*⟩

clamp down *intransitive verb* (1940)
: to impose restrictions **:** CRACK DOWN ⟨the police are *clamping down* on speeders⟩

clams casino *noun plural but singular or plural in construction, often 2d C capitalized* (1952)
: clams on the half shell usually topped with green pepper and baked or broiled

clam·shell \ˈklam-ˌshel\ *noun* (circa 1520)
1 : the shell of a clam
2 a : a bucket or grapple (as on a dredge) having two hinged jaws **b :** an excavating machine having a clamshell **c :** either of a pair of doors (as in an airplane tail) that open out and away from each other

clam up *intransitive verb* (1916)
: to become silent

clam worm *noun* (1885)
: any of several large burrowing polychaete worms (as a nereid) often used as bait

clan \ˈklan\ *noun* [Middle English, from Scottish Gaelic *clann* offspring, clan, from Old Irish *cland* plant, offspring, from Latin *planta* plant] (15th century)
1 a : a Celtic group especially in the Scottish Highlands comprising a number of households whose heads claim descent from a common ancestor **b :** a group of people tracing descent from a common ancestor
2 : a group united by a common interest or common characteristics

clan·des·tine \klan-ˈdes-tən *also* -ˌtīn *or* -ˌtēn *or* ˈklan-dəs-\ *adjective* [Middle French or Latin; Middle French *clandestin,* from Latin *clandestinus,* from *clam* secretly; akin to Latin *celare* to hide — more at HELL] (1566)
: marked by, held in, or conducted with secrecy **:** SURREPTITIOUS
synonym see SECRET
— **clan·des·tine·ly** *adverb*
— **clan·des·tine·ness** *noun*
— **clan·des·tin·i·ty** \ˌklan-də-ˈsti-nə-tē, -des-ˈti-\ *noun*

[1]clang \ˈklaŋ\ *verb* [Latin *clangere;* akin to Greek *klazein* to scream, bark, Old English *hliehhan* to laugh] (1576)
intransitive verb
1 a : to make a loud metallic ringing sound ⟨anvils *clanged*⟩ **b :** to go with a clang
2 : to utter the characteristic harsh cry of a bird
transitive verb
: to cause to clang ⟨*clang* a bell⟩

[2]clang *noun* (1596)
1 : a loud ringing metallic sound ⟨the *clang* of a fire alarm⟩
2 : a harsh cry of a bird (as a crane or goose)

clang·er \ˈklaŋ-ər\ *noun* (1948)
British **:** a conspicuous blunder — often used in the phrase *drop a clanger*

[1]clan·gor \ˈklaŋ-ər *also* -gər\ *noun* [Latin *clangor,* from *clangere*] (1593)
: a resounding clang or medley of clangs ⟨the *clangor* of hammers⟩
— **clan·gor·ous** \-(g)ə-rəs\ *adjective*
— **clan·gor·ous·ly** *adverb*

[2]clangor *intransitive verb* (1837)
: to make a clangor

clan·gour \ˈklaŋ-ər, -gər\ *chiefly British variant of* CLANGOR

[1]clank \ˈklaŋk\ *verb* [probably imitative] (1656)
intransitive verb
1 : to make a clank or series of clanks ⟨the radiator hissed and *clanked*⟩
2 : to go with or as if with a clank ⟨tanks *clanking* through the streets⟩
transitive verb
: to cause to clank
— **clank·ing·ly** \ˈklaŋ-kiŋ-lē\ *adverb*

[2]clank *noun* (1656)
: a sharp brief metallic ringing sound

clan·nish \ˈkla-nish\ *adjective* (1776)
1 : of or relating to a clan
2 : tending to associate only with a select group of similar background or status
— **clan·nish·ly** *adverb*
— **clan·nish·ness** *noun*

clans·man \ˈklanz-mən\ *noun* (1810)
: a member of a clan

[1]clap \ˈklap\ *verb* **clapped** *also* **clapt; clap·ping** [Middle English *clappen,* from Old English *clæppan* to throb; akin to Old High German *klaphōn* to beat] (14th century)
transitive verb
1 : to strike (as two flat hard surfaces) together so as to produce a sharp percussive noise
2 a : to strike (the hands) together repeatedly usually in applause **b :** APPLAUD
3 : to strike with the flat of the hand in a friendly way ⟨*clapped* his friend on the shoulder⟩
4 : to place, put, or set especially energetically ⟨*clap* him into jail⟩ ⟨since I first *clapped* eyes on it⟩
5 : to improvise hastily
intransitive verb
1 : to produce a percussive sound; *especially* **:** SLAM
2 : to go abruptly or briskly
3 : APPLAUD

[2]clap *noun* (13th century)
1 : a device that makes a clapping noise

2 *obsolete* **:** a sudden stroke of fortune and especially ill fortune
3 : a loud percussive noise; *specifically* **:** a sudden crash of thunder
4 a : a sudden blow **b :** a friendly slap
5 : the sound of clapping hands; *especially* **:** APPLAUSE

[3]**clap** *noun* [Middle French *clapoir* bubo] (1587)
: GONORRHEA — often used with *the*

clap•board \'kla-bərd; 'kla(p)-ˌbōrd, -ˌbȯrd\ *noun* [part translation of Dutch *klaphout* stave wood] (circa 1520)
1 *archaic* **:** a size of board for making staves and wainscoting
2 : a narrow board usually thicker at one edge than the other used for siding
3 \'klap-ˌbōrd, -ˌbȯrd\ **:** a pair of hinged boards one of which has a slate with data identifying a piece of film and which are banged together in front of a motion-picture camera at the start of a take to facilitate editing — called also *clapper board*
— clapboard *transitive verb*

clapped–out \ˌklapt-'au̇t\ *adjective* (1946)
British **:** WORN-OUT; *also* **:** TIRED

clap•per \'kla-pər\ *noun* (14th century)
: one that claps: as **a :** the tongue of a bell **b :** a mechanical device that makes noise especially by the banging of one part against another **c :** a person who applauds

clap•per•claw \'kla-pər-ˌklȯ\ *transitive verb* [perhaps from *clapper* + *claw* (verb)] (1590)
1 *dialect English* **:** to claw with the nails
2 *dialect English* **:** SCOLD, REVILE

[1]**clap•trap** \'klap-ˌtrap\ *noun* [[2]*clap;* from its attempt to win applause] (1799)
: pretentious nonsense **:** TRASH

[2]**claptrap** *adjective* (1815)
: characterized by or suggestive of claptrap; *especially* **:** of a cheap showy nature ⟨*claptrap* sentiment⟩

claque \'klak\ *noun* [French, from *claquer* to clap, of imitative origin] (1864)
1 : a group hired to applaud at a performance
2 : a group of sycophants

cla•queur \kla-'kər\ *noun* [French, from *claquer* to clap] (1837)
: a member of a claque

clar•et \'klar-ət\ *noun* [Middle English, from Middle French (*vin*) *claret* clear wine, from *claret* clear, from *cler* clear] (1707)
1 : a red Bordeaux wine; *also* **:** a similar wine produced elsewhere
2 : a dark purplish red
— claret *adjective*

clar•i•fy \'klar-ə-ˌfī\ *verb* **-fied; -fy•ing** [Middle English *clarifien,* from Middle French *clarifier,* from Late Latin *clarificare,* from Latin *clarus* clear — more at CLEAR] (14th century)
transitive verb
1 : to make (as a liquid) clear or pure usually by freeing from suspended matter
2 : to free of confusion
3 : to make understandable
intransitive verb
: to become clear
— clar•i•fi•ca•tion \ˌklar-ə-fə-'kā-shən\ *noun*
— clar•i•fi•er \'klar-ə-ˌfī(-ə)r\ *noun*

clar•i•net \ˌklar-ə-'net, 'klar-ə-nət\ *noun* [French *clarinette,* probably ultimately from Medieval Latin *clarion-, clario*] (1796)
: a single-reed woodwind instrument having a cylindrical tube with a moderately flared bell and a usual range from D below middle C upward for 3½ octaves
— clar•i•net•ist *or* **clar•i•net•tist** \ˌklar-ə-'ne-tist\ *noun*

[1]**clar•i•on** \'klar-ē-ən\ *noun* [Middle English, from Middle French & Medieval Latin; Middle French *clairon,* from Medieval Latin *clarion-, clario,* from Latin *clarus*] (14th century)
1 : a medieval trumpet with clear shrill tones
2 : the sound of or as if of a clarion

[2]**clarion** *adjective* (1841)
: brilliantly clear ⟨her *clarion* top notes⟩; *also* **:** loud and clear ⟨a *clarion* call to action⟩

clar•i•ty \'klar-ə-tē\ *noun* [Middle English *clarite,* from Latin *claritat-, claritas,* from *clarus*] (1616)
: the quality or state of being clear **:** LUCIDITY

clark•ia \'klär-kē-ə\ *noun* [New Latin, from William *Clark*] (1827)
: any of a genus (*Clarkia*) of showy annual herbs of the evening-primrose family that are native to the Pacific slope of North America and southwestern South America

cla•ro \'klär-(ˌ)ō\ *noun, plural* **claros** [Spanish, from *claro* light, from Latin *clarus*] (1891)
: a light-colored usually mild cigar

clary \'klar-ē, 'kler-\ *noun, plural* **clar•ies** [Middle English *clarie,* from Middle French *sclaree,* from Medieval Latin *sclareia*] (14th century)
: an aromatic mint (*Salvia sclarea*) of southern Europe that is widely cultivated especially as an ornamental

[1]**clash** \'klash\ *verb* [imitative] (circa 1500)
intransitive verb
1 : to make a clash ⟨cymbals *clashed*⟩
2 : to come into conflict ⟨where ignorant armies *clash* by night —Matthew Arnold⟩; *also* **:** to be incompatible ⟨the colors *clashed*⟩
transitive verb
: to cause to clash
— clash•er *noun*

[2]**clash** *noun* (1513)
1 : a noisy usually metallic sound of collision
2 a : a hostile encounter **:** SKIRMISH **b :** a sharp conflict ⟨a *clash* of opinions⟩

[1]**clasp** \'klasp\ *noun* [Middle English *claspe*] (14th century)
1 a : a device (as a hook) for holding objects or parts together **b :** a device (as a bar) attached to a military medal to indicate an additional award of the medal or the action or service for which it was awarded
2 : a holding or enveloping with or as if with the hands or arms

[2]**clasp** *transitive verb* (14th century)
1 : to fasten with or as if with a clasp ⟨a robe *clasped* with a brooch⟩
2 : to enclose and hold with the arms; *specifically* **:** EMBRACE
3 : to seize with or as if with the hand **:** GRASP

clasp•er \'klas-pər\ *noun* (circa 1847)
: a male copulatory structure: **a :** one of a pair of external anal processes of an insect that are used to grasp a female **b :** one of a pair of organs that are extensions of the pelvic fins of cartilaginous fishes

clasp knife *noun* (circa 1755)
: POCKETKNIFE; *especially* **:** a large one-bladed folding knife having a catch to hold the blade open

[1]**class** \'klas\ *noun, often attributive* [French *classe,* from Latin *classis* group called to military service, fleet, class; perhaps akin to Latin *calare* to call — more at LOW] (1602)
1 a : a body of students meeting regularly to study the same subject **b :** the period during which such a body meets **c :** a course of instruction **d :** a body of students or alumni whose year of graduation is the same
2 a : a group sharing the same economic or social status ⟨the working *class*⟩ **b :** social rank; *especially* **:** high social rank **c :** high quality **:** ELEGANCE
3 : a group, set, or kind sharing common attributes: as **a :** a major category in biological taxonomy ranking above the order and below the phylum or division **b :** a collection of adjacent and discrete or continuous values of a random variable **c :** SET 21
4 : a division or rating based on grade or quality
5 : the best of its kind ⟨the *class* of the league⟩

[2]**class** *transitive verb* (1705)
: CLASSIFY

class act *noun* (1976)
: an example of outstanding quality or prestige

class action *noun* (1952)
: a legal action undertaken by one or more plaintiffs on behalf of themselves and all other persons having an identical interest in the alleged wrong

class–conscious *adjective* (1903)
1 : actively aware of one's common status with others in a particular economic or social level of society
2 : believing in class struggle
— class consciousness *noun*

[1]**clas•sic** \'kla-sik\ *adjective* [French or Latin; French *classique,* from Latin *classicus* of the highest class of Roman citizens, of the first rank, from *classis*] (circa 1604)
1 a : serving as a standard of excellence **:** of recognized value **b :** TRADITIONAL, ENDURING **c :** characterized by simple tailored lines in fashion year after year ⟨a *classic* suit⟩
2 : of or relating to the ancient Greeks and Romans or their culture **:** CLASSICAL
3 a : historically memorable **b :** noted because of special literary or historical associations ⟨Paris is the *classic* refuge of expatriates⟩
4 a : AUTHENTIC, AUTHORITATIVE **b :** TYPICAL ⟨a *classic* example of chicanery⟩
5 *capitalized* **:** of or relating to the period of highest development of Mesoamerican and especially Mayan culture about A.D. 300–900

[2]**classic** *noun* (1711)
1 : a literary work of ancient Greece or Rome
2 a : a work of enduring excellence; *also* **:** its author **b :** an authoritative source
3 : a typical or perfect example
4 : a traditional event ⟨a football *classic*⟩

clas•si•cal \'kla-si-kəl\ *adjective* [Latin *classicus*] (1599)
1 : STANDARD, CLASSIC
2 a : of or relating to the ancient Greek and Roman world and especially to its literature, art, architecture, or ideals **b :** versed in the classics
3 a : of or relating to music of the late 18th and early 19th centuries characterized by an emphasis on balance, clarity, and moderation **b :** of, relating to, or being music in the educated European tradition that includes such forms as art song, chamber music, opera, and symphony as distinguished from folk or popular music or jazz
4 a : AUTHORITATIVE, TRADITIONAL **b** (1) **:** of or relating to a form or system considered of first significance in earlier times ⟨*classical* Mendelian genetics⟩ (2) **:** not involving relativity, wave mechanics, or quantum theory ⟨*classical* physics⟩ **c :** conforming to a pattern of usage sanctioned by a body of literature rather than by everyday speech
5 : concerned with or giving instruction in the humanities, the fine arts, and the broad aspects of science ⟨a *classical* curriculum⟩

classical conditioning *noun* (1949)
: conditioning in which the conditioned stimulus (as the sound of a bell) is paired with and precedes the unconditioned stimulus (as the sight of food) until the conditioned stimulus alone is sufficient to elicit the response (as salivation in a dog) — compare OPERANT CONDITIONING

clas•si•cal•i•ty \ˌkla-sə-'ka-lə-tē\ *noun* (1819)
1 : the quality or state of being classic
2 : classical scholarship

clas•si•cal•ly \'kla-si-k(ə-)lē\ *adverb* (1772)
1 : in a classic or classical manner ⟨*classically* exact forms of the dance⟩ ⟨*classically* trained⟩
2 a : in classic or traditional circumstances **:** TYPICALLY ⟨*classically,* the whole fish is stuffed⟩ **b :** as a classic example ⟨*classically* bad writing⟩

clas·si·cism \ˈkla-sə-ˌsi-zəm\ *noun* (1830)
1 a : the principles or style embodied in the literature, art, or architecture of ancient Greece and Rome **b :** classical scholarship **c :** a classical idiom or expression
2 : adherence to traditional standards (as of simplicity, restraint, and proportion) that are universally and enduringly valid

clas·si·cist \-sist\ *noun* (1830)
1 : an advocate or follower of classicism
2 : a classical scholar
— **clas·si·cis·tic** \ˌkla-sə-ˈsis-tik\ *adjective*

clas·si·cize \ˈkla-sə-ˌsīz\ *verb* **-cized; -ciz·ing** (1854)
transitive verb
: to make classic or classical
intransitive verb
: to follow classic style

clas·si·co \ˈkla-si-(ˌ)kō\ *adjective* [Italian, from Latin *classicus*] (1968)
: produced in a delimited area of Italy known for its standards of quality ⟨Chianti *classico*⟩

clas·si·fi·ca·tion \ˌkla-sə-fə-ˈkā-shən\ *noun* (1790)
1 : the act or process of classifying
2 a : systematic arrangement in groups or categories according to established criteria; *specifically* **:** TAXONOMY **b :** CLASS, CATEGORY
— **clas·si·fi·ca·to·ry** \ˈkla-sə-fi-kə-ˌtōr-ē, kla-ˈsi-fə-, -ˌtȯr-; ˈkla-sə-fə-ˌkā-tə-rē\ *adjective*

clas·si·fied \ˈkla-sə-ˌfīd\ *adjective* (1889)
1 : divided into classes or placed in a class ⟨*classified* ads⟩
2 : withheld from general circulation for reasons of national security ⟨*classified* information⟩

clas·si·fi·er \ˈkla-sə-ˌfī(-ə)r\ *noun* (1819)
1 : one that classifies; *specifically* **:** a machine for sorting out the constituents of a substance (as ore)
2 : a word or morpheme used with numerals or with nouns designating countable or measurable objects

clas·si·fy \ˈkla-sə-ˌfī\ *transitive verb* **-fied; -fy·ing** (1799)
1 : to arrange in classes ⟨*classifying* books according to subject matter⟩
2 : to assign (as a document) to a category
— **clas·si·fi·able** \ˌkla-sə-ˈfī-ə-bəl\ *adjective*

class interval *noun* (1929)
: CLASS 3b; *also* **:** its numerical width

clas·sis \ˈkla-sis\ *noun, plural* **clas·ses** \ˈkla-ˌsēz\ [New Latin, from Latin, class] (1593)
1 : a governing body in some Reformed churches (as in the former Reformed Church in the U.S.) corresponding to a presbytery
2 : the district governed by a classis

class·ism \ˈkla-ˌsi-zəm\ *noun* (1842)
: prejudice or discrimination based on class
— **class·ist** \ˈkla-sist\ *adjective*

class·less \ˈklas-ləs\ *adjective* (1878)
1 : belonging to no particular social class
2 : free from distinctions of social class ⟨a *classless* society⟩
— **class·less·ness** *noun*

class·mate \-ˌmāt\ *noun* (1713)
: a member of the same class in a school or college

class·room \-ˌrüm, -ˌru̇m\ *noun* (1870)
: a place where classes meet

classy \ˈkla-sē\ *adjective* **class·i·er; -est** (1891)
: ELEGANT, STYLISH ⟨a *classy* clientele⟩; *also* **:** notably superior ⟨a *classy* outfielder⟩
— **class·i·ness** *noun*

clast \ˈklast\ *noun* [Greek *klastos* broken, from *klan* to break; perhaps akin to Latin *clades* disaster] (1952)
: a fragment of rock

clas·tic \ˈklas-tik\ *adjective* [International Scientific Vocabulary] (1877)
: made up of fragments of preexisting rocks ⟨a *clastic* sediment⟩
— **clastic** *noun*

clath·rate \ˈkla-ˌthrāt\ *adjective* [Latin *clathratus,* furnished with a lattice, from *clathri* (plural) lattice, from Greek *klēithron* bar, from *kleiein* to close — more at CLAVICLE] (1906)
: relating to or being a compound formed by the inclusion of molecules of one kind in cavities of the crystal lattice of another
— **clathrate** *noun*

[1]**clat·ter** \ˈkla-tər\ *verb* [Middle English *claterén,* from (assumed) Old English *clatrian;* of imitative origin] (13th century)
intransitive verb
1 : to make a rattling sound ⟨the dishes *clattered* on the shelf⟩
2 : to talk noisily or rapidly
3 : to move or go with a clatter ⟨*clattered* down the stairs⟩
transitive verb
: to cause to clatter
— **clat·ter·er** \-tər-ər\ *noun*
— **clat·ter·ing·ly** \ˈkla-tə-riŋ-lē\ *adverb*

[2]**clatter** *noun* (14th century)
1 : a rattling sound (as of hard bodies striking together) ⟨the *clatter* of pots and pans⟩
2 : COMMOTION ⟨the midday *clatter* of the business district⟩
3 : noisy chatter
— **clat·tery** \ˈkla-tə-rē\ *adjective*

clau·di·ca·tion \ˌklȯ-də-ˈkā-shən\ *noun* [Latin *claudication-, claudicatio,* from *claudicare* to limp, from *claudus* lame] (15th century)
: the quality or state of being lame **:** LIMPING

claus·al \ˈklȯ-zəl\ *adjective* (1904)
: relating to or of the nature of a clause

clause \ˈklȯz\ *noun* [Middle English, from Old French, from Medieval Latin *clausa* close of a rhetorical period, from Latin, feminine of *clausus,* past participle of *claudere* to close — more at CLOSE] (13th century)
1 : a group of words containing a subject and predicate and functioning as a member of a complex or compound sentence
2 : a separate section of a discourse or writing; *specifically* **:** a distinct article in a formal document

claus·tral \ˈklȯs-trəl\ *adjective* [Middle English, from Medieval Latin *claustralis,* from *claustrum* cloister, from Latin, bar, bolt, confining space, from *claudere*] (15th century)
: CLOISTRAL

claus·tro·pho·bia \ˌklȯs-trə-ˈfō-bē-ə\ *noun* [New Latin, from Latin *claustrum* + New Latin *-phobia*] (1879)
: abnormal dread of being in closed or narrow spaces
— **claus·tro·phobe** \ˈklȯs-trə-ˌfōb\ *noun*

claus·tro·pho·bic \ˌklȯs-trə-ˈfō-bik\ *adjective* (circa 1889)
1 : affected with or inclined to claustrophobia
2 : inducing or suggesting claustrophobia
— **claus·tro·pho·bi·cal·ly** \-bi-k(ə-)lē\ *adverb*

claus·trum \ˈklȯs-trəm, ˈklau̇s-\ *noun, plural* **claus·tra** \-trə\ [New Latin, from Latin, bar] (1848)
: the one of the four basal ganglia in each cerebral hemisphere that consists of a thin lamina of gray matter separated from the lenticular nucleus by a layer of white matter

cla·vate \ˈklā-ˌvāt\ *adjective* [New Latin *clavatus,* from Latin *clava* club, from *clavus* nail, knot in wood] (1813)
: gradually thickening near the distal end **:** shaped like a club

[1]**clave** *past of* CLEAVE

[2]**clave** \ˈklä-(ˌ)vā, ˈklāv\ *noun* [American Spanish, from Spanish, keystone, clef, from Latin *clavis*] (1928)
: one of a pair of cylindrical hardwood sticks that are used as a percussion instrument

cla·ver \ˈklā-vər\ *intransitive verb* [probably of Celtic origin; akin to Scottish Gaelic *clabaire* babbler] (circa 1605)
chiefly Scottish **:** PRATE, GOSSIP
— **claver** *noun, chiefly Scottish*

clav·i·chord \ˈkla-və-ˌkȯrd\ *noun* [Medieval Latin *clavichordium,* from Latin *clavis* key + *chorda* string — more at CORD] (15th century)
: an early keyboard instrument having strings struck by tangents attached directly to the key ends
— **clav·i·chord·ist** \-ˌkȯr-dist\ *noun*

clav·i·cle \ˈkla-vi-kəl\ *noun* [French *clavicule,* from New Latin *clavicula,* from Latin, diminutive of Latin *clavis;* akin to Greek *kleid-, kleis* key, *kleiein* to close] (1615)
: a bone of the vertebrate pectoral girdle typically serving to link the scapula and sternum — called also *collarbone*
— **cla·vic·u·lar** \kla-ˈvi-kyə-lər, klə-\ *adjective*

cla·vier \klə-ˈvir, ˈklā-vē-ər, ˈklä-\ *noun* [French, from Old French, key bearer, from Latin *clavis*] (1708)
1 : the keyboard of a musical instrument
2 [German *Klavier,* from French *clavier*] **:** an early keyboard instrument
— **cla·vier·ist** \klə-ˈvir-ist; ˈklā-vē-ə-rist, ˈkla-\ *noun*
— **cla·vier·is·tic** \klə-ˌvi-ˈris-tik; ˌklā-vē-ə-ˈris-tik, ˌkla-\ *adjective*

[1]**claw** \ˈklȯ\ *noun, often attributive* [Middle English *clawe,* from Old English *clawu* hoof, claw; akin to Old Norse *klō* claw, and probably to Old English *cliewen* ball — more at CLEW] (before 12th century)
1 : a sharp usually slender and curved nail on the toe of an animal
2 : any of various sharp curved processes especially at the end of a limb (as of an insect); *also* **:** a limb ending in such a process
3 : one of the pincerlike organs terminating some limbs of various arthropods (as a lobster or scorpion)
4 : something that resembles a claw; *specifically* **:** the forked end of a tool (as a hammer)
— **clawed** \ˈklȯd\ *adjective*
— **claw·like** \-ˌlīk\ *adjective*

[2]**claw** (before 12th century)
transitive verb
: to rake, seize, dig, or progress with or as if with claws
intransitive verb
: to scrape, scratch, dig, or pull with or as if with claws

claw back *transitive verb* (1953)
chiefly British **:** to get back (as money) by strenuous or forceful means (as taxation)
— **claw–back** \ˈklȯ-ˌbak\ *noun*

claw·ham·mer \ˈklȯ-ˌha-mər\ *adjective* (1964)
: of or relating to a style of banjo playing using the thumb and one or more fingers picking or strumming in a downward direction

claw hammer *noun* (circa 1769)
1 : a hammer with one end of the head forked for pulling out nails
2 : TAILCOAT

clay \ˈklā\ *noun, often attributive* [Middle English, from Old English *clǣg;* akin to Old High German *klīwa* bran, Latin *gluten* glue, Middle Greek *glia*] (before 12th century)
1 a : an earthy material that is plastic when moist but hard when fired, that is composed mainly of fine particles of hydrous aluminum silicates and other minerals, and that is used for brick, tile, and pottery; *specifically* **:** soil composed chiefly of this material having particles less than a specified size **b :** EARTH, MUD
2 a : a substance that resembles clay in plasticity and is used for modeling **b :** the human body as distinguished from the spirit **c :** fundamental nature or character ⟨the common *clay*⟩

\ə\ **abut** \ᵊ\ kitten \ər\ **further** \a\ **ash** \ā\ **ace**
\ä\ **mop, mar** \au̇\ **out** \ch\ **chin** \e\ **bet** \ē\ **easy**
\g\ **go** \i\ **hit** \ī\ **ice** \j\ **job** \ŋ\ **sing** \ō\ **go**
\ȯ\ **law** \ȯi\ **boy** \th\ **thin** \t͟h\ **the** \ü\ **loot** \u̇\ **foot**
\y\ **yet** \zh\ **vision** *see also* Guide to Pronunciation

3 : CLAY COURT
— **clay·ey** \'klā-ē\ *adjective*
— **clay·ish** \'klā-ish\ *adjective*
— **clay·like** \-ˌlīk\ *adjective*

clay·bank \'klā-ˌbaŋk\ *noun* (1851)
: a horse of yellowish color

clay court *noun* (1885)
: a tennis court with a clay surface or a synthetic surface that resembles clay

clay loam *noun* (circa 1889)
: a loam containing from 20 to 30 percent clay

clay mineral *noun* (1947)
: any of a group of hydrous silicates of aluminum and sometimes other metals formed chiefly in weathering processes and occurring especially in clay and shale

clay·more \'klā-ˌmōr, -ˌmȯr\ *noun* [Scottish Gaelic *claidheamh mór,* literally, great sword] (1772)
: a large 2-edged sword formerly used by Scottish Highlanders; *also* **:** their basket-hilted broadsword

claymore mine *noun* [perhaps from *claymore*] (1961)
: a usually electrically fired land mine that contains steel fragments which are discharged in a predetermined direction

clay·pan \-ˌpan\ *noun* (1837)
1 : hardpan consisting mainly of clay
2 *Australian* **:** a shallow depression in which water collects after rain

clay pigeon *noun* (1888)
: a saucer-shaped target usually made of baked clay or limestone and pitch and thrown from a trap in skeet and trapshooting

clay·ware \'klā-ˌwar, -ˌwer\ *noun* (1896)
: articles made of fired clay

¹clean \'klēn\ *adjective* [Middle English *clene,* from Old English *clǣne;* akin to Old High German *kleini* delicate, dainty] (before 12th century)
1 a : free from dirt or pollution ⟨changed to *clean* clothes⟩ ⟨*clean* solar energy⟩ **b :** free from contamination or disease **c :** free or relatively free from radioactivity ⟨a *clean* atomic explosion⟩
2 a : UNADULTERATED, PURE ⟨the *clean* thrill of one's first flight⟩ **b** *of a precious stone* **:** having no interior flaws visible **c :** free from growth that hinders tillage
3 a : free from moral corruption or sinister connections of any kind ⟨a candidate with a *clean* record⟩ **b :** free from offensive treatment of sexual subjects and from the use of obscenity ⟨a *clean* joke⟩ **c :** observing the rules **:** FAIR ⟨a *clean* fight⟩
4 : ceremonially or spiritually pure ⟨and all who are *clean* may eat flesh —Leviticus 7:19 (Revised Standard Version)⟩
5 a : THOROUGH, COMPLETE ⟨a *clean* break with the past⟩ **b :** deftly executed **:** SKILLFUL ⟨*clean* ballet technique⟩ **c :** hit beyond the reach of an opponent ⟨a *clean* single to center⟩
6 a : relatively free from error or blemish **:** CLEAR; *specifically* **:** LEGIBLE ⟨*clean* copy⟩ **b :** UNENCUMBERED ⟨*clean* bill of sale⟩
7 a : characterized by clarity and precision **:** TRIM ⟨a *clean* prose style⟩ ⟨architecture with *clean* almost austere lines⟩ **b :** EVEN, SMOOTH ⟨a *clean* edge⟩ ⟨a sharp blow causing a *clean* break⟩ **c :** free from impedances to smooth flow (as of water or air) ⟨a *clean* airplane⟩ ⟨a ship with a *clean* bottom⟩
8 a : EMPTY ⟨the ship returned with a *clean* hold⟩ **b :** free from drug addiction **c** *slang* **:** having no contraband (as weapons or drugs) in one's possession
9 : habitually neat
— **clean·ness** \'klēn-nəs\ *noun*

²clean *adverb* (before 12th century)
1 a : so as to clean ⟨a new broom sweeps *clean*⟩ **b :** in a clean manner ⟨play the game *clean*⟩
2 : all the way **:** COMPLETELY ⟨the bullet went *clean* through his arm⟩

³clean (15th century)
transitive verb
1 a : to make clean: as (1) **:** to rid of dirt, impurities, or extraneous matter (2) **:** to rid of corruption ⟨vowing to *clean* up city hall⟩ **b :** REMOVE, ERADICATE — usually used with *up* or *off* ⟨*clean* up that mess⟩
2 a : STRIP, EMPTY ⟨a tree *cleaned* of fruit⟩ **b :** to remove the entrails from ⟨*clean* fish⟩ **c :** to deprive of money or possessions — often used with *out* ⟨they *cleaned* him out completely⟩
intransitive verb
: to undergo or perform a process of cleaning ⟨*clean* up before dinner⟩
— **clean·abil·i·ty** \ˌklē-nə-'bi-lə-tē\ *noun*
— **clean·able** \'klē-nə-bəl\ *adjective*
— **clean house 1 :** to clean a house and its furniture **2 :** to make sweeping reforms or changes (as of personnel)
— **clean one's clock :** to beat one badly in a fight or competition
— **clean up one's act :** to behave in a more acceptable manner

⁴clean *noun* (circa 1889)
: an act of cleaning dirt especially from the surface of something

clean and jerk *noun* (1939)
: a lift in weight lifting in which the weight is raised to shoulder height, held momentarily, and then quickly thrust overhead usually with a lunge or a spring from the legs — compare PRESS, SNATCH

clean–cut \'klēn-'kət\ *adjective* (1843)
1 : cut so that the surface or edge is smooth and even
2 : sharply defined
3 : of wholesome appearance

clean·er \'klē-nər\ *noun* (circa 1792)
1 a : one whose work is cleaning **b :** DRY CLEANER
2 : a preparation for cleaning
3 : an implement or machine for cleaning

clean·hand·ed \'klēn-'han-dəd\ *adjective* (1728)
: innocent of wrongdoing

clean–limbed \'klēn-'limd\ *adjective* (15th century)
: well proportioned **:** TRIM ⟨*clean-limbed* youths⟩

¹clean·ly \'klēn-lē\ *adverb* (13th century)
: in a clean manner

²clean·ly \'klen-lē\ *adjective* **clean·li·er; -est** (circa 1500)
1 : careful to keep clean **:** FASTIDIOUS
2 : habitually kept clean
— **clean·li·ness** *noun*

clean room \'klēn-ˌrüm, -ˌru̇m\ *noun* (1963)
: a room for the manufacture or assembly of objects (as precision parts) that is maintained at a high level of cleanliness by special means

cleanse \'klenz\ *verb* **cleansed; cleans·ing** [Middle English *clensen,* from Old English *clǣnsian* to purify, from *clǣne* clean] (before 12th century)
: CLEAN; *especially* **:** to rid of impurities by or as if by washing

cleans·er \'klen-zər\ *noun* (before 12th century)
1 : one that cleanses
2 : a preparation (as a scouring powder or a skin cream) used for cleaning

¹clean·up \'klē-ˌnəp\ *noun* (1872)
1 : an act or instance of cleaning
2 : an exceptionally large profit **:** KILLING

²cleanup *adjective* (1937)
: being in the fourth position in the batting order of a baseball team ⟨a *cleanup* hitter⟩
— **cleanup** *adverb*

clean up *intransitive verb* (1920)
: to make a spectacular profit in a business enterprise or a killing in speculation or gambling

¹clear \'klir\ *adjective* [Middle English *clere,* from Old French *cler,* from Latin *clarus* clear, bright; akin to Latin *calare* to call — more at LOW] (13th century)
1 a : BRIGHT, LUMINOUS **b :** CLOUDLESS; *specifically* **:** less than one-tenth covered ⟨a *clear* sky⟩ **c :** free from mist, haze, or dust ⟨a *clear* day⟩ **d :** UNTROUBLED, SERENE ⟨a *clear* gaze⟩
2 : CLEAN, PURE: as **a :** free from blemishes **b :** easily seen through **:** TRANSPARENT **c :** free from abnormal sounds on auscultation
3 a : easily heard **b :** easily visible **:** PLAIN **c :** free from obscurity or ambiguity **:** easily understood **:** UNMISTAKABLE
4 a : capable of sharp discernment **:** KEEN **b :** free from doubt **:** SURE
5 : free from guile or guilt **:** INNOCENT
6 : unhampered by restriction or limitation: as **a :** unencumbered by debts or charges **b :** NET ⟨a *clear* profit⟩ **c :** UNQUALIFIED, ABSOLUTE **d :** free from obstruction **e :** emptied of contents or cargo **f :** free from entanglement **g :** BARE, DENUDED ☆ ☆
— **clear·ness** *noun*

²clear *adverb* (14th century)
1 : in a clear manner ⟨to cry loud and *clear*⟩
2 : all the way ⟨drove *clear* across the state⟩

³clear (14th century)
transitive verb
1 a : to make clear or translucent **b :** to free from pollution or cloudiness
2 : to free from accusation or blame **:** EXONERATE, VINDICATE ⟨the opportunity to *clear* himself⟩
3 a : to give insight to **:** ENLIGHTEN **b :** to make intelligible **:** EXPLAIN ⟨*clear* up the mystery⟩
4 a : to free from what obstructs or is unneeded: as (1) **:** OPEN 1b ⟨*clear* a path⟩ (2) **:** to remove unwanted growth or items from ⟨*clear* the land of timber⟩ (3) **:** to rid or make a rasping noise as if ridding (the throat) of phlegm (4) **:** to erase stored or displayed data from (as a computer or calculator) **b :** to empty of occupants ⟨*clear* the room⟩ **c :** DISENTANGLE ⟨*clear* a fishing line⟩ **d :** to remove from an area or place ⟨*clear* the dishes from the table⟩ **e :** TRANSMIT, DISPATCH
5 a : to submit for approval ⟨*clear* it with me first⟩ **b :** AUTHORIZE, APPROVE ⟨*cleared* the article for publication⟩: as (1) **:** to certify as trustworthy ⟨*clear* a person for classified information⟩ (2) **:** to permit (an aircraft) to proceed usually with a specified action
6 a : to free from obligation or encumbrance **b :** SETTLE, DISCHARGE ⟨*clear* an account⟩ **c** (1) **:** to free (a ship or shipment) by payment of duties or harbor fees (2) **:** to pass through (customs) **d :** to gain without deduction **:** NET ⟨*clear* a profit⟩ **e :** to put through a clearing-house

☆ SYNONYMS
Clear, transparent, translucent, limpid mean capable of being seen through. CLEAR implies absence of cloudiness, haziness, or muddiness ⟨*clear* water⟩. TRANSPARENT implies being so clear that objects can be seen distinctly ⟨a *transparent* sheet of film⟩. TRANSLUCENT implies the passage of light but not a clear view of what lies beyond ⟨*translucent* frosted glass⟩. LIMPID suggests the soft clearness of pure water ⟨her eyes were *limpid* pools of blue⟩.

Clear, perspicuous, lucid mean quickly and easily understood. CLEAR implies freedom from obscurity, ambiguity, or undue complexity ⟨*clear* instructions⟩. PERSPICUOUS applies to a style that is simple and elegant as well as clear ⟨a *perspicuous* style⟩. LUCID suggests a clear logical coherence and evident order of arrangement ⟨a *lucid* explanation⟩. See in addition EVIDENT.

7 a : to go over, under, or by without touching **b :** PASS ⟨the bill *cleared* the legislature⟩
intransitive verb
1 a : to become clear ⟨it *cleared* up quickly after the rain⟩ **b :** to go away **:** VANISH ⟨the symptoms *cleared* gradually⟩ **c :** SELL
2 a : to obtain permission to discharge cargo **b :** to conform to regulations or pay requisite fees prior to leaving port
3 : to pass through a clearinghouse
4 : to go to an authority (as for approval) before becoming effective
— **clear·able** \ˈklir-ə-bəl\ *adjective*
— **clear·er** \ˈklir-ər\ *noun*
— **clear the air** *also* **clear the atmosphere :** to remove elements of hostility, tension, confusion, or uncertainty ⟨had a long meeting to *clear the air*⟩
— **clear the decks :** to make sweeping preparations for action

[4]clear *noun* (1674)
1 : a clear space or part
2 : a high arcing shot over an opponent's head in badminton
— **in the clear 1 :** in inside measurement **2 :** free from guilt or suspicion **3 :** in plaintext **:** not in code or cipher ⟨a message sent *in the clear*⟩

clear–air turbulence *noun* (1955)
: sudden severe turbulence occurring in cloudless regions that causes violent jarring or buffeting of aircraft

clear·ance \ˈklir-ən(t)s\ *noun* (1540)
1 : an act or process of clearing: as **a :** the removal of buildings from an area (as a city slum) **b :** the act of clearing a ship at the customhouse; *also* **:** the papers showing that a ship has cleared **c :** the offsetting of checks and other claims among banks through a clearinghouse **d :** certification as clear of objection **:** AUTHORIZATION ⟨security *clearance*⟩ **e :** a sale to clear out stock **f :** authorization for an aircraft to proceed especially with a specified action ⟨*clearance* to land⟩
2 : the distance by which one object clears another or the clear space between them
3 : the volume of blood or plasma that could be freed of a specified constituent in a specified time (usually one minute) by its excretion into the urine through the kidneys — called also *renal clearance*

[1]clear–cut \ˈklir-ˈkət\ *adjective* (1855)
1 : sharply outlined **:** DISTINCT
2 : free from ambiguity or uncertainty **:** UNAMBIGUOUS

[2]clear–cut \-ˌkət\ *noun* (circa 1958)
: an area of forest which has been clear-cut; *also* **:** CLEAR-CUTTING

clear–cut·ting \-ˌkə-tiŋ\ *noun* (1922)
: removal of all the trees in a stand of timber
— **clear–cut** *verb*

clear–eyed \ˈklir-ˌīd\ *adjective* (1530)
: CLEAR-SIGHTED

clear·head·ed \-ˌhe-dəd\ *adjective* (1709)
1 : having a clear understanding **:** PERCEPTIVE
2 : able to think clearly
— **clear·head·ed·ly** *adverb*
— **clear·head·ed·ness** *noun*

clear·ing \ˈklir-iŋ\ *noun* (14th century)
1 : the act or process of making or becoming clear
2 : a tract of land cleared of wood and brush
3 a : a method of exchanging and offsetting commercial papers or accounts with cash settlement only of the balances due after the clearing **b** *plural* **:** the gross amount of balances so adjusted

clear·ing·house \-ˌhau̇s\ *noun* (1832)
1 : an establishment maintained by banks for settling mutual claims and accounts
2 : a central agency for the collection, classification, and distribution especially of information; *broadly* **:** an informal channel for distributing information or assistance

clear·ly \ˈklir-lē\ *adverb* (14th century)
1 : in a clear manner ⟨speaking *clearly*⟩
2 : it is clear ⟨*clearly*, a new approach is needed⟩
usage see HOPEFULLY

clear out (1792)
intransitive verb
: DEPART
transitive verb
: to drive out or away usually forcibly

clear–sight·ed \ˈklir-ˌsī-təd\ *adjective* (1586)
1 : having clear vision
2 : DISCERNING
— **clear–sight·ed·ly** *adverb*
— **clear–sight·ed·ness** *noun*

clear·sto·ry *variant of* CLERESTORY

clear·wing \-ˌwiŋ\ *noun* (1868)
: a moth (as of the families Aegeriidae or Sphingidae) having the wings largely transparent and devoid of scales

[1]cleat \ˈklēt\ *noun* [Middle English *clete* wedge, from (assumed) Old English *clēat;* akin to Middle High German *klōz* lump — more at CLOUT] (14th century)
1 a : a wedge-shaped piece fastened to or projecting from something and serving as a support or check **b :** a wooden or metal fitting usually with two projecting horns around which a rope may be made fast
2 a : a strip fastened across something to give strength or hold in position **b** (1) **:** a projecting piece (as on the bottom of a shoe) that furnishes a grip (2) *plural* **:** shoes equipped with cleats

[2]cleat *transitive verb* (1794)
1 : to secure to or by a cleat
2 : to provide with a cleat

cleav·able \ˈklē-və-bəl\ *adjective* (circa 1864)
: capable of being split

cleav·age \ˈklē-vij\ *noun* (1816)
1 a : the quality of a crystallized substance or rock of splitting along definite planes; *also* **:** the occurrence of such splitting **b :** a fragment (as of a diamond) obtained by splitting
2 : the action of cleaving **:** the state of being cleft
3 : the series of synchronized mitotic cell divisions of the fertilized egg that results in the formation of the blastomeres and changes the single-celled zygote into a multicellular embryo; *also* **:** one of these cell divisions
4 : the splitting of a molecule into simpler molecules
5 : the depression between a woman's breasts especially when made visible by the wearing of a low-cut dress

[1]cleave \ˈklēv\ *intransitive verb* **cleaved** \ˈklēvd\ *or* **clove** \ˈklōv\ *also* **clave** \ˈklāv\; **cleaved; cleav·ing** [Middle English *clevien,* from Old English *clifian;* akin to Old High German *kleben* to stick] (before 12th century)
: to adhere firmly and closely or loyally and unwaveringly
synonym see STICK

[2]cleave *verb* **cleaved** \ˈklēvd\ *also* **cleft** \ˈkleft\ *or* **clove** \ˈklōv\; **cleaved** *also* **cleft** *or* **clo·ven** \ˈklō-vən\; **cleav·ing** [Middle English *cleven,* from Old English *clēofan;* akin to Old Norse *kljūfa* to split, Latin *glubere* to peel, Greek *glyphein* to carve] (before 12th century)
transitive verb
1 : to divide by or as if by a cutting blow **:** SPLIT
2 : to separate into distinct parts and especially into groups having divergent views
3 : to subject to chemical cleavage ⟨a protein *cleaved* by an enzyme⟩
intransitive verb
1 : to split especially along the grain
2 : to penetrate or pass through something by or as if by cutting
synonym see TEAR

cleav·er \ˈklē-vər\ *noun* (15th century)
1 : one that cleaves; *especially* **:** a butcher's implement for cutting animal carcasses into joints or pieces
2 : a prehistoric stone tool having a sharp edge at one end

cleav·ers \ˈklē-vərz\ *noun plural but singular or plural in construction* [Middle English *clivre,* alteration of Old English *clife* burdock, cleavers; akin to Old English *clifian*] (14th century)
: an annual bedstraw (*Galium aparine*) that has numerous stalked white flowers, stems covered with curved prickles, and whorls of bristle-tipped leaves; *also* **:** any of several related plants

cleek \ˈklēk\ *noun* [Middle English (northern) *cleke,* from *cleken* to clutch] (15th century)
chiefly Scottish **:** a large hook (as for a pot over a fire)

clef \ˈklef\ *noun* [French, literally, key, from Latin *clavis — more at* CLAVICLE] (circa 1577)
: a sign placed at the beginning of a musical staff to determine the pitch of the notes

1

2

clef: *1* treble clef, *2* bass clef

[1]cleft \ˈkleft\ *noun* [Middle English *clift,* from Old English *geclyft;* akin to Old English *clēofan* to cleave] (14th century)
1 : a space or opening made by or as if by splitting **:** FISSURE
2 : a usually V-shaped indented formation **:** a hollow between ridges or protuberances ⟨the anal *cleft* of the human body⟩

[2]cleft *adjective* [Middle English, from past participle of *cleven*] (14th century)
: partially split or divided; *specifically* **:** divided about halfway to the midrib ⟨a *cleft* leaf⟩

cleft lip *noun* (circa 1946)
: HARELIP

cleft palate *noun* (1847)
: congenital fissure of the roof of the mouth

clei·do·ic \klī-ˈdō-ik\ *adjective* [Greek *kleidoun* to fasten, lock in, from *kleid-, kleis* key — more at CLAVICLE] (1931)
of an egg **:** enclosed in a relatively impervious shell which reduces free exchange with the environment

cleis·tog·a·mous \klī-ˈstä-gə-məs\ *also* **cleis·to·gam·ic** \ˌklī-stə-ˈga-mik\ *adjective* [Greek *kleistos* closed (from *kleiein* to close) + International Scientific Vocabulary *-gamous* — more at CLAVICLE] (1874)
: characterized by or being small inconspicuous closed self-pollinating flowers additional to and often more fruitful than showier ones on the same plant ⟨violets are *cleistogamous*⟩
— **cleis·tog·a·mous·ly** \klī-ˈstä-gə-məs-lē\ *adverb*
— **cleis·tog·a·my** \-ˈstä-gə-mē\ *noun*

cle·ma·tis \ˈkle-mə-təs; kli-ˈma-təs, -ˈmā-, -ˈmä-\ *noun* [New Latin, genus name, from Latin, from Greek *klēmatis* brushwood, clematis, from *klēmat-, klēma* twig, from *klan* to break — more at CLAST] (1578)
: any of a genus (*Clematis*) of vines or herbs of the buttercup family having three leaflets on each leaf and usually white, red, pink, or purple flowers

clem·en·cy \ˈkle-mən(t)-sē\ *noun, plural* **-cies** (15th century)
1 a : disposition to be merciful and especially to moderate the severity of punishment due **b :** an act or instance of leniency
2 : pleasant mildness of weather
synonym see MERCY

clem·ent \ˈkle-mənt\ *adjective* [Middle English, from Latin *clement-, clemens*] (15th century)

\ə\ **abut** \ᵊ\ **kitten** \ər\ **further** \a\ **ash** \ā\ **ace** \ä\ **mop, mar** \au̇\ **out** \ch\ **chin** \e\ **bet** \ē\ **easy** \g\ **go** \i\ **hit** \ī\ **ice** \j\ **job** \ŋ\ **sing** \ō\ **go** \ȯ\ **law** \ȯi\ **boy** \th\ **thin** \t͟h\ **the** \ü\ **loot** \u̇\ **foot** \y\ **yet** \zh\ **vision** *see also* Guide to Pronunciation

1 : inclined to be merciful **:** LENIENT ⟨a *clement* judge⟩
2 : MILD ⟨*clement* weather for November⟩
— clem·ent·ly *adverb*

clench \'klench\ *transitive verb* [Middle English, from Old English *-clencan;* akin to Old English *clingan* to cling] (13th century)
1 : CLINCH 2
2 : to hold fast **:** CLUTCH ⟨*clenched* the arms of the chair⟩
3 : to set or close tightly ⟨*clench* one's teeth⟩ ⟨*clench* one's fists⟩
— clench *noun*

clepe \'klep\ *transitive verb* **cleped** \'klept\; **yclept** \i-'klept\ *also* **cleped** *or* **ycleped** \i-'klept\; **clep·ing** \'kle-piŋ\ [Middle English, from Old English *clipian* to speak, call; akin to Old Frisian *kleppa* to ring] (before 12th century)
archaic **:** NAME, CALL

clep·sy·dra \'klep-sə-drə\ *noun, plural* **-dras** *or* **-drae** \-ˌdrē, -ˌdrī\ [Latin, from Greek *klepsydra,* from *kleptein* to steal + *hydōr* water — more at KLEPT-, WATER] (1646)
: WATER CLOCK

clere·sto·ry \'klir-ˌstōr-ē, -ˌstȯr-, -st(ə-)rē\ *noun* [Middle English, from *clere* clear + *story*] (15th century)
1 : an outside wall of a room or building that rises above an adjoining roof and contains windows
2 : GALLERY

clerestory 1

cler·gy \'klər-jē\ *noun, plural* **clergies** [Middle English *clergie,* from Old French, from *clerc* clergyman] (13th century)
1 : a group ordained to perform pastoral or sacerdotal functions in a Christian church
2 : the official or sacerdotal class of a non-Christian religion

cler·gy·man \-mən\ *noun* (1577)
: a member of the clergy

cler·gy·wom·an \-ˌwu̇-mən\ *noun* (1673)
: a woman who is a member of the clergy

cler·ic \'kler-ik\ *noun* [Late Latin *clericus*] (1621)
: a member of the clergy

¹cler·i·cal \'kler-i-kəl\ *adjective* (1592)
1 : of, relating to, or characteristic of the clergy
2 : of or relating to a clerk
— cler·i·cal·ly \-i-k(ə-)lē\ *adverb*

²clerical *noun* (1605)
1 : a member of the clergy
2 : CLERICALIST
3 : CLERK
4 *plural* **:** clerical garments

clerical collar *noun* (1948)
: a narrow stiffly upright white collar worn buttoned at the back of the neck by members of the clergy

cler·i·cal·ism \'kler-i-kə-ˌli-zəm\ *noun* (1864)
: a policy of maintaining or increasing the power of a religious hierarchy

cler·i·cal·ist \-list\ *noun* (1881)
: one that favors maintained or increased ecclesiastical power and influence

cler·i·hew \'kler-i-ˌhyü\ *noun* [Edmund *Clerihew* Bentley (died 1956) English writer] (1928)
: a light verse quatrain rhyming *aabb* and usually dealing with a person named in the initial rhyme

cler·i·sy \'kler-ə-sē\ *noun* [German *Klerisei* clergy, from Medieval Latin *clericia,* from Late Latin *clericus* cleric] (1818)
: INTELLIGENTSIA

¹clerk \'klərk, *British usually* 'klärk\ *noun* [Middle English, from Old French *clerc* & Old English *cleric, clerc,* both from Late Latin *clericus,* from Late Greek *klērikos,* from Greek *klēros* lot, inheritance (in allusion to Deuteronomy 18:2), stick of wood; akin to Greek *klan* to break — more at CLAST] (before 12th century)
1 : CLERIC
2 *archaic* **:** SCHOLAR
3 a : an official responsible (as to a government agency) for correspondence, records, and accounts and vested with specified powers or authority (as to issue writs or other processes as ordered by a court) ⟨city *clerk*⟩ **b :** one employed to keep records or accounts or to perform general office work **c :** one who works at a sales or service counter
— clerk·ship \-ˌship\ *noun*

²clerk *intransitive verb* (1551)
: to act or work as a clerk

clerk·ly \'klər-klē, *British usually* 'klär-\ *adjective* (15th century)
1 : of, relating to, or characteristic of a clerk
2 *archaic* **:** SCHOLARLY
— clerkly *adverb*

Cleve·land bay \'klēv-lənd-\ *noun* [*Cleveland,* district in Yorkshire, England] (1796)
: a large strong horse of a breed of English origin that is uniformly bay with black legs, mane, and tail

clev·er \'kle-vər\ *adjective* [Middle English *cliver,* perhaps of Scandinavian origin; akin to Danish dialect *kløver* alert, skillful] (circa 1595)
1 a : skillful or adroit in using the hands or body **:** NIMBLE **b :** mentally quick and resourceful but often lacking in depth and soundness
2 : marked by wit or ingenuity
3 *dialect* **a :** GOOD **b :** easy to use or handle ☆
— clev·er·ish \-v(ə-)rish\ *adjective*
— clev·er·ly \-vər-lē\ *adverb*
— clev·er·ness \-vər-nəs\ *noun*

clev·is \'kle-vəs\ *noun* [earlier *clevi,* perhaps of Scandinavian origin; akin to Old Norse *kljufa* to split — more at CLEAVE] (1592)
: SHACKLE 3

¹clew \'klü\ *noun* [Middle English *clewe,* from Old English *cliewen;* akin to Old High German *kliuwa* ball, Sanskrit *glau* lump] (before 12th century)
1 : a ball of thread, yarn, or cord
2 : CLUE
3 a : a lower corner or only the after corner of a sail **b :** a metal loop attached to the lower corner of a sail **c** *plural* **:** a combination of lines by which a hammock is suspended

²clew *transitive verb* (15th century)
1 : to roll into a ball
2 : CLUE
3 : to haul (a sail) up or down by ropes through the clews

cli·ché *also* **cli·che** \klē-'shā, 'klē-ˌ, kli-'\ *noun* [French, literally, printer's stereotype, from past participle of *clicher* to stereotype, of imitative origin] (1892)
1 : a trite phrase or expression; *also* **:** the idea expressed by it
2 : a hackneyed theme, characterization, or situation
3 : something (as a menu item) that has become overly familiar or commonplace ◆
— cliché *adjective*

cli·chéd \-'shād\ *adjective* (1928)
1 : marked by or abounding in clichés
2 : HACKNEYED

¹click \'klik\ *verb* [probably imitative] (1581)
transitive verb
: to strike, move, or produce with a click ⟨*clicked* his heels together⟩
intransitive verb
1 : to make a click
2 a : to fit or agree exactly **b :** to fit together **:** hit it off ⟨they did not *click* as friends⟩ **c :** to function smoothly **d :** SUCCEED ⟨a movie that *clicks*⟩

²click *noun* (1611)
1 a : a slight sharp noise **b :** a speech sound in some languages made by enclosing air between two stop articulations of the tongue, enlarging the enclosure to rarefy the air, and suddenly opening the enclosure
2 : DETENT

click beetle *noun* (circa 1864)
: any of a family (Elateridae) of beetles able to right themselves with a click when inverted by flexing the articulation between the prothorax and mesothorax

click stop *noun* (1950)
: a turnable control device (as for a camera diaphragm opening) that engages with a definite click at specific settings

cli·ent \'klī-ənt\ *noun* [Middle English, from Middle French & Latin; Middle French *client,* from Latin *client-, cliens;* perhaps akin to Latin *clinare* to lean — more at LEAN] (14th century)
1 : one that is under the protection of another **:** DEPENDENT
2 a : a person who engages the professional advice or services of another ⟨a lawyer's *clients*⟩ **b :** CUSTOMER ⟨hotel *clients*⟩ **c :** a person served by or utilizing the services of a social agency ⟨a welfare *client*⟩
3 : CLIENT STATE
— cli·ent·age \-ən-tij\ *noun*
— cli·en·tal \klī-'en-tᵊl, 'klī-ən-\ *adjective*
— cli·ent·less \'klī-ənt-ləs\ *adjective*

cli·en·tele \ˌklī-ən-'tel, ˌklē-ən- *also* ˌklē-ˌän-\ *noun* [French *clientèle,* from Latin *clientela,* from *client-, cliens*] (circa 1587)
: a body of clients ⟨a shop that caters to an exclusive *clientele*⟩

client state *noun* (1918)
: a country that is economically, politically, or militarily dependent on another country

cliff \'klif\ *noun* [Middle English *clif,* from Old English; akin to Old High German *klep*] (before 12th century)
: a very steep, vertical, or overhanging face of rock, earth, or ice **:** PRECIPICE

☆ SYNONYMS
Clever, adroit, cunning, ingenious mean having or showing practical wit or skill in contriving. CLEVER stresses physical or mental quickness, deftness, or great aptitude ⟨a person *clever* with horses⟩. ADROIT often implies a skillful use of expedients to achieve one's purpose in spite of difficulties ⟨an *adroit* negotiator⟩. CUNNING implies great skill in constructing or creating ⟨a filmmaker *cunning* in his use of special effects⟩. INGENIOUS suggests the power of inventing or discovering a new way of accomplishing something ⟨an *ingenious* software engineer⟩. See in addition INTELLIGENT.

◇ WORD HISTORY
cliché *Cliché* is the French word for a stereotype, a duplicate printing plate cast from a mold or a matrix of composed type. The production of such plates became standard in 19th century printing to save wear on type and permit longer press runs. Passing from printers' jargon to the world of letters, *cliché* was applied to a turn of phrase or metaphor that through repeated use loses its effectiveness and becomes the mark of unimaginative writing. The word *stereotype* itself, from French *stéréotype,* has undergone a similar shift in sense, and now usually refers to a conventional mental image of a person or thing that is passed about without critical reappraisal.

— **cliffy** \'kli-fē\ *adjective*

cliff dweller *noun* (1881)
1 *often C&D capitalized* **a :** a member of a prehistoric American Indian people of the southwestern U.S. who built their homes on rock ledges or in the natural recesses of canyon walls and cliffs **b :** a member of any cliff-dwelling people
2 : a resident of a large usually metropolitan apartment building
— **cliff dwelling** *noun*

cliff–hang \'klif-ˌhaŋ\ *intransitive verb* [back-formation from *cliff-hanger*] (1946)
: to await the outcome of a suspenseful situation

cliff–hang·er \-ˌhaŋ-ər\ *noun* (circa 1937)
1 : an adventure serial or melodrama; *especially* **:** one presented in installments each ending in suspense
2 : a contest whose outcome is in doubt up to the very end; *broadly* **:** a suspenseful situation
— **cliff–hang·ing** \-iŋ\ *adjective*

cliff swallow *noun* (1825)
: a colonial swallow (*Petrochelidon pyrrhonota*) of the New World with a pale buff rump and dark throat patch that builds mud nests like jugs especially under eaves and on cliffs

[1]cli·mac·ter·ic \klī-'mak-t(ə-)rik; ˌklī-ˌmak-'ter-ik, -'tir-\ *adjective* [Latin *climactericus,* from Greek *klimaktērikos,* from *klimaktēr* critical point, literally, rung of a ladder, from *klimak-, klimax* ladder] (1582)
1 : constituting or relating to a climacteric
2 : CRITICAL, CRUCIAL

[2]climacteric *noun* (circa 1630)
1 : a major turning point or critical stage
2 : MENOPAUSE; *also* **:** a corresponding period in the male usually occurring with less well-defined physiological and psychological changes
3 : the marked and sudden rise in the respiratory rate of fruit just prior to full ripening

cli·mac·tic \klī-'mak-tik, klə-\ *adjective* (1872)
: of, relating to, or constituting a climax
— **cli·mac·ti·cal·ly** \-ti-k(ə-)lē\ *adverb*

cli·mate \'klī-mət\ *noun* [Middle English *climat,* from Middle French, from Late Latin *climat-, clima,* from Greek *klimat-, klima* inclination, latitude, climate, from *klinein* to lean — more at LEAN] (14th century)
1 : a region of the earth having specified climatic conditions
2 a : the average course or condition of the weather at a place usually over a period of years as exhibited by temperature, wind velocity, and precipitation **b :** the prevailing set of conditions (as of temperature and humidity) indoors ⟨a *climate*-controlled office⟩
3 : the prevailing influence or environmental conditions characterizing a group or period **:** ATMOSPHERE ⟨a *climate* of fear⟩

cli·mat·ic \klī-'ma-tik, klə-\ *adjective* (circa 1828)
1 : of or relating to climate
2 : resulting from or influenced by the climate rather than the soil ⟨forests that had reverted to the *climatic* type⟩ — compare EDAPHIC 2
— **cli·mat·i·cal·ly** \-ti-k(ə-)lē\ *adverb*

climatic climax *noun* (circa 1928)
: the one of the ecological climaxes possible in a particular climatic area whose stability is directly due to the influence of climate — compare EDAPHIC CLIMAX

cli·ma·tol·o·gy \ˌklī-mə-'tä-lə-jē\ *noun* (1843)
: the science that deals with climates and their phenomena
— **cli·ma·to·log·i·cal** \ˌklī-mə-tə-'lä-ji-kəl\ *adjective*
— **cli·ma·to·log·i·cal·ly** \-k(ə-)lē\ *adverb*
— **cli·ma·tol·o·gist** \-mə-'tä-lə-jist\ *noun*

[1]cli·max \'klī-ˌmaks\ *noun* [Latin, from Greek *klimax* ladder, from *klinein* to lean] (circa 1550)
1 : a figure of speech in which a series of phrases or sentences is arranged in ascending order of rhetorical forcefulness
2 a : the highest point **:** CULMINATION **b :** the point of highest dramatic tension or a major turning point in the action (as of a play) **c :** ORGASM **d :** MENOPAUSE
3 : a relatively stable ecological stage or community especially of plants that is achieved through successful adjustment to an environment; *especially* **:** the final stage in ecological succession
synonym see SUMMIT
— **cli·max·less** *adjective*

[2]climax (1835)
transitive verb
: to bring to a climax ⟨*climaxed* his boxing career with a knockout⟩
intransitive verb
: to come to a climax ⟨a riot *climaxing* in the destruction of several houses⟩

[1]climb \'klīm\ *verb* [Middle English, from Old English *climban;* probably akin to Old English *clifian* to adhere — more at CLEAVE] (before 12th century)
intransitive verb
1 a : to go upward with gradual or continuous progress **:** RISE, ASCEND **b :** to increase gradually **c :** to slope upward
2 a : to go upward or raise oneself especially by grasping or clutching with the hands ⟨*climbed* upon her father's knee⟩ **b** *of a plant* **:** to ascend in growth (as by twining)
3 : to go about or down usually by grasping or holding with the hands ⟨*climb* down the ladder⟩
4 : to get into or out of clothing usually with some haste or effort ⟨the firemen *climbed* into their clothes⟩
transitive verb
1 : to go upward on or along, to the top of, or over ⟨*climb* a hill⟩
2 : to draw or pull oneself up, over, or to the top of by using hands and feet ⟨children *climbing* the tree⟩
3 : to grow up or over
— **climb·able** \'klī-mə-bəl\ *adjective*

[2]climb *noun* (circa 1587)
1 : a place where climbing is necessary to progress
2 : the act or an instance of climbing **:** RISE, ASCENT

climb·er \'klī-mər\ *noun* (15th century)
1 a : one that climbs or helps in climbing **b :** a vine or twining plant (as a rose or sweet pea) that readily grows up a support or over other plants
2 : one who attempts to gain a superior social or business position

climbing iron *noun* (1857)
: a steel framework with spikes attached that may be affixed to one's boots for climbing (as a pole or tree)

clime \'klīm\ *noun* [Late Latin *clima*] (14th century)
: CLIMATE ⟨traveled to warmer *climes*⟩

-clinal *adjective combining form* [International Scientific Vocabulary, from Greek *klinein* to lean — more at LEAN]
: sloping ⟨iso*clinal*⟩

[1]clinch \'klinch\ *verb* [probably alteration of *clench*] (1542)
transitive verb
1 : CLENCH 3
2 a : to turn over or flatten the protruding pointed end of (a driven nail); *also* **:** to treat (as a screw, bolt, or rivet) in a similar way **b :** to fasten in this way
3 a : to make final or irrefutable **:** SETTLE ⟨that *clinched* the argument⟩ **b :** to secure conclusively **:** WIN
intransitive verb
1 : to hold an opponent (as in boxing) at close quarters with one or both arms
2 : to hold fast or firmly
— **clinch·ing·ly** \'klin-chiŋ-lē\ *adverb*

[2]clinch *noun* (1659)
1 : a fastening by means of a clinched nail, rivet, or bolt; *also* **:** the clinched part of a nail, rivet, or bolt
2 *archaic* **:** PUN
3 : an act or instance of clinching in boxing
4 : EMBRACE

clinch·er \'klin-chər\ *noun* (1737)
: one that clinches: as **a :** a decisive fact, argument, act, or remark ⟨the expense was the *clincher* that persuaded us to give up the enterprise⟩ **b :** a tire with flanged beads fitting into the wheel rim

cline \'klīn\ *noun* [Greek *klinein*] (1938)
: a gradient of morphological or physiological change in a group of related organisms usually along a line of environmental or geographic transition
— **clin·al** \'klī-nᵊl\ *adjective*
— **clin·al·ly** \-nᵊl-ē\ *adverb*

-cline *noun combining form* [back-formation from *-clinal*]
: slope ⟨mono*cline*⟩

[1]cling \'kliŋ\ *intransitive verb* **clung** \'kləŋ\; **cling·ing** [Middle English, from Old English *clingan;* akin to Old High German *klunga* tangled ball of thread] (before 12th century)
1 a : to hold together **b :** to adhere as if glued firmly **c :** to hold or hold on tightly or tenaciously
2 a : to have a strong emotional attachment or dependence **b :** to remain or linger as if resisting complete dissipation or dispersal ⟨the odor *clung* to the room for hours⟩
synonym see STICK
— **cling·er** \'kliŋ-ər\ *noun*
— **clingy** \'kliŋ-ē\ *adjective*

[2]cling *noun* (circa 1625)
: an act or instance of clinging **:** ADHERENCE

cling·stone \'kliŋ-ˌstōn\ *noun* (1705)
: any of various stone fruits (as some peaches or plums) with flesh that adheres strongly to the pit

clin·ic \'kli-nik\ *noun* [French *clinique,* from Greek *klinikē* medical practice at the sickbed, from feminine of *klinikos* of a bed, from *klinē* bed, from *klinein* to lean, recline — more at LEAN] (circa 1843)
1 : a class of medical instruction in which patients are examined and discussed
2 : a group meeting devoted to the analysis and solution of concrete problems or to the acquiring of specific skills or knowledge ⟨writing *clinics*⟩ ⟨golf *clinics*⟩
3 a : a facility (as of a hospital) for diagnosis and treatment of outpatients **b :** a group practice in which several physicians work cooperatively

-clinic *adjective combining form* [International Scientific Vocabulary, from Greek *klinein*]
1 : inclining **:** dipping
2 : having (so many) oblique intersections of the axes ⟨mono*clinic*⟩ ⟨tri*clinic*⟩

clin·i·cal \'kli-ni-kəl\ *adjective* (circa 1780)
1 : of, relating to, or conducted in or as if in a clinic: as **a :** involving direct observation of the patient **b :** diagnosable by or based on clinical observation
2 : analytical or coolly dispassionate ⟨a *clinical* attitude⟩
— **clin·i·cal·ly** \-k(ə-)lē\ *adverb*

clinical thermometer *noun* (1878)
: a thermometer for measuring body temperature that has a constriction in the tube where the column of liquid breaks when the temperature drops from its maximum and that continues to indicate the maximum temperature by

the part of the column above the constriction until reset by shaking

cli·ni·cian \kli-ˈni-shən\ *noun* (1875)
1 : a person qualified in the clinical practice of medicine, psychiatry, or psychology as distinguished from one specializing in laboratory or research techniques or in theory
2 : a person who conducts a clinic

clin·i·co·path·o·log·ic \ˌkli-ni-(ˌ)kō-ˌpa-thə-ˈlä-jik\ *or* **clin·i·co·path·o·log·i·cal** \-ˈlä-ji-kəl\ *adjective* [*clinical*] (1898)
: relating to or concerned both with the signs and symptoms directly observable by the physician and with the results of laboratory examination
— **clin·i·co·path·o·log·i·cal·ly** \-ji-k(ə-)lē\ *adverb*

[1]**clink** \ˈkliŋk\ *verb* [Middle English, of imitative origin] (14th century)
intransitive verb
: to give out a slight sharp short metallic sound
transitive verb
: to cause to clink

[2]**clink** *noun* (15th century)
: a clinking sound

[3]**clink** *noun* [probably from *Clink,* a prison in Southwark, London, England] (1515)
1 *slang* **:** a prison cell
2 *slang* **:** JAIL, PRISON

[1]**clin·ker** \ˈkliŋ-kər\ *noun* [alteration of earlier *klincard* a hard yellowish Dutch brick] (1641)
1 : a brick that has been burned too much in the kiln
2 : stony matter fused together **:** SLAG

[2]**clink·er** \ˈkliŋ-kər\ *noun* [[1]*clink*] (1733)
1 *British* **:** something first-rate
2 a : a wrong note **b :** a serious mistake or error **:** BONER **c :** an utter failure **:** FLOP **d :** something of poor quality

clink·er–built \-ˌbilt\ *adjective* [*clinker,* noun, clinch] (1769)
: having the external planks or plates overlapping like the clapboards on a house ⟨a *clinker-built* boat⟩

clink·ety–clank \ˈkliŋ-kə-tē-ˈklaŋk\ *noun* [imitative] (1901)
: a repeated usually rhythmic clanking sound ⟨the *clinkety-clank* of a loose tire chain⟩

cli·nom·e·ter \klī-ˈnä-mə-tər\ *noun* [Greek *klinein* to lean] (1811)
: any of various instruments for measuring angles of elevation or inclination

-clinous *adjective combining form* [probably from New Latin *-clinus,* from Greek *klinē* bed — more at CLINIC]
: having the androecium and gynoecium in a (single or different) flower or (two separate) flowers ⟨di*clinous*⟩

[1]**clin·quant** \ˈkliŋ-kənt, klaⁿ-ˈkäⁿ\ *adjective* [Middle French, from present participle of *clinquer* to glitter, literally, to clink, of imitative origin] (1591)
: glittering with gold or tinsel

[2]**clinquant** *noun* [French, from *clinquant,* adjective] (1682)
: TINSEL

clin·to·nia \klin-ˈtō-nē-ə\ *noun* [New Latin, genus name, from DeWitt *Clinton*] (1843)
: any of a genus (*Clintonia*) of herbs of the lily family with yellow or white flowers

Clio \ˈklī-(ˌ)ō, ˈklē-\ *noun* [Latin, from Greek *Kleiō*]
1 : the Greek Muse of history
2 *plural* **Cli·os :** a statuette awarded annually by a professional organization for notable achievement in radio and television commercials

clio·met·rics \ˌklī-ə-ˈme-triks\ *noun plural but singular in construction* [*Clio* + *-metrics* (as in *econometrics*)] (1966)
: the application of methods developed in other fields (as economics, statistics, and data processing) to the study of history
— **clio·met·ric** \-trik\ *adjective*
— **clio·met·ri·cian** \-me-ˈtri-shən\ *noun*

[1]**clip** \ˈklip\ *transitive verb* **clipped; clip·ping** [Middle English *clippen,* from Old English *clyppan;* akin to Old High German *klāftra* fathom, Lithuanian *globti* to embrace] (before 12th century)
1 : ENCOMPASS
2 a : to hold in a tight grip **:** CLUTCH **b :** to clasp or fasten with a clip

[2]**clip** *noun* (15th century)
1 : any of various devices that grip, clasp, or hook
2 : a device to hold cartridges for charging the magazines of some rifles; *also* **:** a magazine from which ammunition is fed into the chamber of a firearm
3 : a piece of jewelry held in position by a clip

[3]**clip** *verb* **clipped; clip·ping** [Middle English *clippen,* from Old Norse *klippa*] (13th century)
transitive verb
1 a : to cut or cut off with or as if with shears ⟨*clip* a dog's hair⟩ ⟨*clip* an hour off traveling time⟩ **b :** to cut off the distal or outer part of **c** (1) **:** [3]EXCISE (2) **:** to cut items out of (as a newspaper)
2 a : CURTAIL, DIMINISH **b :** to abbreviate in speech or writing
3 : HIT, PUNCH
4 : to illegally block (an opposing player) in football
5 : to take money from unfairly or dishonestly especially by overcharging ⟨the nightclub *clipped* the tourist for $200⟩
intransitive verb
1 : to clip something
2 : to travel or pass rapidly
3 : to clip an opposing player in football

[4]**clip** *noun* (15th century)
1 a *plural, Scottish* **:** SHEARS **b :** a 2-bladed instrument for cutting especially the nails
2 : something that is clipped: as **a :** the product of a single shearing (as of sheep) **b :** a crop of wool of a sheep, a flock, or a region **c :** a section of filmed or videotaped material **d :** a clipping especially from a newspaper
3 : an act of clipping
4 : a sharp blow
5 : RATE 4a ⟨continues at a brisk *clip*⟩
6 : a single instance or occasion **:** TIME ⟨he charged $10 a *clip*⟩ — often used in the phrase *at a clip* ⟨trained 1000 workers at a *clip*⟩

clip·board \ˈklip-ˌbōrd, -ˌbȯrd\ *noun* (1896)
: a small writing board with a clip at the top for holding papers

clip–clop \ˈklip-ˌkläp\ *noun* [imitative] (1884)
: the sound made by or as if by a horse walking on a hard surface
— **clip–clop** *intransitive verb*

clip joint *noun* (1933)
1 *slang* **:** a place of public entertainment (as a nightclub) that makes a practice of defrauding patrons (as by overcharging)
2 *slang* **:** a business that makes a practice of overcharging

clip–on \ˈkli-ˌpȯn, -ˌpän\ *adjective* (1909)
: of or relating to something that attaches with a clip ⟨a *clip-on* tie⟩ ⟨*clip-on* earrings⟩
— **clip–on** *noun*

clip·per \ˈkli-pər\ *noun* (14th century)
1 : one that clips something
2 : an implement for clipping especially hair, fingernails, or toenails — usually used in plural
3 a : one that moves swiftly **b :** a fast sailing ship; *especially* **:** one with long slender lines, an overhanging bow, tall masts, and a large sail area

clip·ping \ˈkli-piŋ\ *noun* (15th century)
: something that is clipped off or out of something; *especially* **:** an item clipped from a publication

clip·sheet \ˈklip-ˌshēt\ *noun* (1926)
: a sheet of newspaper material issued by an organization for clipping and reprinting

clique \ˈklēk, ˈklik\ *noun* [French] (1711)
: a narrow exclusive circle or group of persons; *especially* **:** one held together by common interests, views, or purposes
— **cliqu·ey** *also* **cliquy** \ˈklē-kē, ˈkli-\ *adjective*
— **cliqu·ish** \ˈkli-kish\ *adjective*
— **cliqu·ish·ly** *adverb*
— **cliqu·ish·ness** *noun*

cli·tel·lum \klī-ˈte-ləm\ *noun, plural* **-la** \-lə\ [New Latin, modification of Latin *clitellae* packsaddle] (1839)
: a thickened glandular section of the body wall of some annelids that secretes a viscid sac in which the eggs are deposited

clit·ic \ˈkli-tik\ *noun* [en*clitic* or pro*clitic*] (1946)
: a word that is treated in pronunciation as forming a part of a neighboring word and that is often unaccented or contracted

clit·o·ri·dec·to·my \ˌkli-tə-ri-ˈdek-tə-mē\ *also* **clit·o·rec·to·my** \ˌkli-tə-ˈrek-tə-mē\ *noun, plural* **-mies** (1866)
: excision of all or part of the clitoris

clit·o·ris \ˈkli-tə-rəs, kli-ˈtȯr-əs\ *noun, plural* **clit·o·ri·des** \kli-ˈtȯr-ə-ˌdēz\ [New Latin, from Greek *kleitoris*] (1615)
: a small erectile organ at the anterior or ventral part of the vulva homologous to the penis
— **clit·o·ral** \ˈkli-tə-rəl\ *or* **clit·or·ic** \kli-ˈtȯr-ik, -ˈtär-\ *adjective*

clo·a·ca \klō-ˈā-kə\ *noun, plural* **-a·cae** \-ˌkē, -ˌsē\ [Latin; akin to Greek *klyzein* to wash — more at CLYSTER] (1599)
1 : [3]SEWER
2 [New Latin, from Latin] **:** the common chamber into which the intestinal, urinary, and generative canals discharge in birds, reptiles, amphibians, and many fishes; *also* **:** a comparable chamber of an invertebrate
— **clo·a·cal** \-ˈā-kəl\ *adjective*

[1]**cloak** \ˈklōk\ *noun* [Middle English *cloke,* from Old North French *cloque* bell, cloak, from Medieval Latin *clocca* bell; from its shape] (13th century)
1 : a loose outer garment
2 : something that conceals **:** PRETENSE, DISGUISE

[2]**cloak** *transitive verb* (1509)
: to cover or hide with or as if with a cloak
synonym see DISGUISE

cloak–and–dagger *adjective* (1860)
: dealing in or suggestive of melodramatic intrigue and action usually involving secret agents and espionage
— **cloak–and–dagger** *noun*

cloak·room \ˈklōk-ˌrüm, -ˌrům\ *noun* (circa 1852)
1 a : a room in which outdoor clothing may be placed during one's stay **b :** a room or cubicle where garments, parcels, and luggage may be checked for temporary safekeeping
2 : an anteroom of a legislative chamber where members may relax and confer with colleagues
3 *British* **:** LAVATORY 2

[1]**clob·ber** \ˈklä-bər\ *noun* [origin unknown] (1879)
slang British **:** CLOTHES 1

[2]**clobber** *transitive verb* **clob·bered; clob·ber·ing** \-b(ə-)riŋ\ [origin unknown] (circa 1943)
1 : to pound mercilessly; *also* **:** to hit with force **:** SMASH
2 : to defeat overwhelmingly

clo·chard \klō-ˈshär\ *noun* [French, from *clocher* to limp, from (assumed) Vulgar Latin *cloppicare,* from Late Latin *cloppus* lame] (1937)
: TRAMP, VAGRANT

cloche \ˈklōsh\ *noun* [French, literally, bell, from Medieval Latin *clocca*] (1882)
1 : a transparent plant cover used outdoors especially for protection against cold
2 : a woman's small close-fitting hat usually with deep rounded crown and very narrow brim

[1]clock \'kläk\ *noun, often attributive* [Middle English *clok,* from Middle Dutch *clocke* bell, clock, from Old North French or Medieval Latin; Old North French *cloque* bell, from Medieval Latin *clocca,* of Celtic origin; akin to Middle Irish *clocc* bell] (14th century)
1 : a device other than a watch for indicating or measuring time commonly by means of hands moving on a dial; *broadly* **:** any periodic system by which time is measured
2 : a registering device usually with a dial; *specifically* **:** ODOMETER
3 : TIME CLOCK
4 : a synchronizing device (as in a computer) that produces pulses at regular intervals
5 : BIOLOGICAL CLOCK
— around the clock *also* **round the clock 1 :** continuously for 24 hours **:** day and night without cessation **2 :** without relaxation and heedless of time
— kill the clock *or* **run out the clock :** to use up as much as possible of the playing time remaining in a game (as football) while retaining possession of the ball or puck especially to protect a lead
[2]clock (1883)
transitive verb
1 : to time with a stopwatch or by an electric timing device
2 : to register on a mechanical recording device ⟨wind velocities were *clocked* at 80 miles per hour⟩
3 : to hit hard
4 *chiefly British* **:** ATTAIN, REALIZE — usually used with *up* ⟨just *clocked* up a million . . . paperback sales —*Punch*⟩
intransitive verb
: to register on a time sheet or time clock **:** PUNCH — used with *in, out, on, off* ⟨he *clocked* in late⟩
— clock·er *noun*
[3]clock *noun* [perhaps from [1]*clock*] (1530)
: an ornamental figure on the ankle or side of a stocking or sock
clock·like \'kläk-ˌlīk\ *adjective* (circa 1770)
: unusually regular, undeviating, and precise ⟨does his job with *clocklike* efficiency⟩
clock radio *noun* (1949)
: a combination clock and radio device in which the clock can be set to turn on the radio at a designated time
clock–watch·er \-ˌwä-chər\ *noun* (1911)
: a person (as a worker or student) who keeps close watch on the passage of time
— clock–watch·ing \-chiŋ\ *noun*
clock·wise \'kläk-ˌwīz\ *adverb* (1888)
: in the direction in which the hands of a clock rotate as viewed from in front or as if standing on a clock face
— clockwise *adjective*
clock·work \-ˌwərk\ *noun* (1628)
1 : the inner workings of something
2 : the machinery (as springs and a train of gears) that run a clock; *also* **:** a similar mechanism running a mechanical device (as a toy)
3 : the precision, regularity, or absence of variation associated with a clock or clockwork ⟨a *clockwork* operation⟩ ⟨the planning went like *clockwork*⟩
clod \'kläd\ *noun* [Middle English *clodde,* from Old English *clod* (in *clodhamer* fieldfare)] (13th century)
1 a : a lump or mass especially of earth or clay **b :** SOIL, EARTH
2 : OAF, DOLT
— clod·dish \'klä-dish\ *adjective*
— clod·dish·ness *noun*
— clod·dy \'klä-dē\ *adjective*
clod·hop·per \'kläd-ˌhä-pər\ *noun* (1721)
1 : a clumsy and uncouth rustic
2 : a large heavy work shoe or boot
clod·hop·ping \-ˌhä-piŋ\ *adjective* (1843)
: BOORISH, RUDE
clod·poll *or* **clod·pole** \'kläd-ˌpōl\ *noun* (1601)
: BLOCKHEAD
clo·fi·brate \klō-'fī-ˌbrāt, -'fi-\ *noun* [perhaps from *chlor-* + *fibr-* + [1]*-ate*] (1964)
: a compound $C_{12}H_{15}ClO_3$ used especially to lower abnormally high concentrations of fats and cholesterol in the blood
[1]clog \'kläg, 'klȯg\ *noun* [Middle English *clogge* short thick piece of wood] (14th century)
1 a : a weight attached especially to an animal to hinder motion **b :** something that shackles or impedes **:** ENCUMBRANCE 1
2 : a shoe, sandal, or overshoe having a thick typically wooden sole
[2]clog *verb* **clogged; clog·ging** (1526)
transitive verb
1 a : to impede with a clog **:** HINDER **b :** to halt or retard the progress, operation, or growth of **:** ENCUMBER ⟨restraints that have been *clogging* the market —T. W. Arnold⟩
2 : to fill beyond capacity **:** OVERLOAD ⟨cars *clogged* the main street⟩ ⟨a speech *clogged* with clichés⟩
intransitive verb
1 : to become filled with extraneous matter
2 : to unite in a mass **:** CLOT
3 : to dance a clog dance
synonym see HAMPER
— clog·ger \'klä-gər, 'klȯ-\ *noun*
clog dance *noun* (1869)
: a dance in which the performer wears clogs and beats out a clattering rhythm on the floor
— clog dancer *noun*
— clog dancing *noun*
cloi·son·né *also* **cloi·son·ne** \ˌklȯi-zə-'nā, ˌklwä-\ *adjective* [French, from past participle of *cloisonner* to partition] (1863)
: of, relating to, or being a style of enamel decoration in which the enamel is applied and fired in raised cells (as of soldered wires) on a usually metal background — compare CHAMPLEVÉ
— cloisonné *noun*
[1]clois·ter \'klȯi-stər\ *noun* [Middle English *cloistre,* from Old French, from Medieval Latin *claustrum,* from Latin, bar, bolt, from *claudere* to close — more at CLOSE] (13th century)
1 a : a monastic establishment **b :** an area within a monastery or convent to which the religious are normally restricted **c :** monastic life **d :** a place or state of seclusion

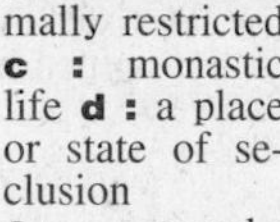

cloister 2

2 : a covered passage on the side of a court usually having one side walled and the other an open arcade or colonnade
[2]cloister *transitive verb* **clois·tered; clois·ter·ing** \-st(ə-)riŋ\ (1581)
1 : to seclude from the world in or as if in a cloister ⟨a scientist who *cloisters* herself in a laboratory⟩
2 : to surround with a cloister ⟨*cloistered* gardens⟩
clois·tered \'klȯi-stərd\ *adjective* (1581)
1 : being or living in or as if in a cloister
2 : providing shelter from contact with the outside world ⟨the *cloistered* atmosphere of a small college⟩
clois·tral \'klȯi-strəl\ *adjective* (1605)
: of, relating to, or suggestive of a cloister
clois·tress \'klȯi-strəs\ *noun* (1601)
obsolete **:** NUN
clo·mi·phene \'klō-mə-ˌfēn\ *noun* [*chlor-* + *amine* + *-phene* (from *phenyl*)] (1963)
: a synthetic drug $C_{26}H_{28}ClNO$ used in the form of its citrate to induce ovulation
clomp \'klämp, 'klȯmp, 'kləmp\ *intransitive verb* [by alteration] (1829)
: CLUMP 1
[1]clone \'klōn\ *noun* [Greek *klōn* twig, slip; akin to Greek *klan* to break — more at CLAST] (1903)
1 a : the aggregate of the asexually produced progeny of an individual; *also* **:** a group of replicas of all or part of a macromolecule (as DNA or an antibody) **b :** an individual grown from a single somatic cell of its parent and genetically identical to it
2 : one that appears to be a copy of an original form ⟨a *clone* of a personal computer⟩
— clon·al \'klō-nᵊl\ *adjective*
— clon·al·ly \-nᵊl-ē\ *adverb*
[2]clone *verb* **cloned; clon·ing** (circa 1948)
transitive verb
1 : to propagate a clone from
2 : to make a copy of
intransitive verb
: to produce a clone
clo·ni·dine \'klä-nə-ˌdēn, 'klō-, -ˌdīn\ *noun* [*chlor-* + *aniline* + *imide* + [2]*-ine*] (1970)
: an antihypertensive drug $C_9H_9Cl_2N_3$ used to treat essential hypertension, to prevent migraine headache, and to diminish withdrawal symptoms
clonk \'kläŋk, 'klȯŋk\ *verb* [imitative] (1930)
intransitive verb
: to make a dull hollow thumping sound
transitive verb
: to cause to clonk
— clonk *noun*
clo·nus \'klō-nəs\ *noun* [New Latin, from Greek *klonos* agitation — more at HOLD] (1817)
: a rapid succession of alternating contractions and partial relaxations of a muscle occurring in some nervous diseases
— clon·ic \'klä-nik\ *adjective*
— clo·nic·i·ty \klō-'ni-sə-tē, klä-\ *noun*
cloot \'klüt\ *noun* [probably of Scandinavian origin; akin to Old Norse *klō* claw — more at CLAW] (1725)
1 *Scottish* **:** a cloven hoof
2 *plural, capitalized, Scottish* **:** CLOOTIE
Cloot·ie \'klü-tē\ *noun* [diminutive of *cloot*]
chiefly Scottish — used as a name of the devil
clop \'kläp\ *noun* [imitative] (1899)
: a sound made by or as if by a hoof or wooden shoe against the pavement
— clop *intransitive verb*
clop–clop \'kläp-ˌkläp\ *noun* (1901)
: a sound of rhythmically repeated clops
— clop–clop *intransitive verb*
clo·que *also* **clo·qué** \klō-'kā, 'klō-ˌ\ *noun* [French *cloqué,* from past participle of *cloquer* to become blistered, from French dialect (Picard) *cloque* bell, bubble, from Medieval Latin *clocca* bell — more at CLOCK] (1936)
1 : a fabric with an embossed design
2 : a fabric especially of piqué with small woven figures
[1]close \'klōz\ *verb* **closed; clos·ing** [Middle English, from Old French *clos-,* stem of *clore,* from Latin *claudere* to shut, close; perhaps akin to Greek *kleiein* to close — more at CLAVICLE] (13th century)
transitive verb
1 a : to move so as to bar passage through something ⟨*close* the gate⟩ **b :** to block against entry or passage ⟨*close* a street⟩ **c :** to deny access to ⟨the city *closed* the beach⟩ **d :** SCREEN, EXCLUDE ⟨*close* a view⟩ **e :** to suspend or stop the operations of ⟨*close* school⟩ — often used with *down*
2 *archaic* **:** ENCLOSE, CONTAIN
3 a : to bring to an end or period ⟨*close* an account⟩ **b :** to conclude discussion or negotiation about ⟨the question is *closed*⟩; *also* **:** to

\ə\ **abut** \ᵊ\ **kitten** \ər\ **further** \a\ **ash** \ā\ **ace** \ä\ **mop, mar** \au̇\ **out** \ch\ **chin** \e\ **bet** \ē\ **easy** \g\ **go** \i\ **hit** \ī\ **ice** \j\ **job** \ŋ\ **sing** \ō\ **go** \ȯ\ **law** \ȯi\ **boy** \th\ **thin** \t͟h\ **the** \ü\ **loot** \u̇\ **foot** \y\ **yet** \zh\ **vision** *see also* Guide to Pronunciation

consummate by performing something previously agreed ⟨*close* a transfer of real estate title⟩ **c :** to terminate access to (a computer file)
4 a : to bring or bind together the parts or edges of ⟨a *closed* book⟩ **b :** to fill up (as an opening) **c :** to make complete by circling or enveloping or by making continuous ⟨*close* a circuit⟩ **d :** to reduce to nil ⟨*closed* the distance to the lead racer⟩
intransitive verb
1 a : to contract, fold, swing, or slide so as to leave no opening ⟨the door *closed* quietly⟩ **b :** to cease operation ⟨the factory *closed* down⟩ ⟨the stores *close* at 9 p.m.⟩
2 a : to draw near ⟨the ship was *closing* with the island⟩ **b :** to engage in a struggle at close quarters **:** GRAPPLE ⟨*close* with the enemy⟩
3 a : to come together **:** MEET **b :** to draw the free foot up to the supporting foot in dancing
4 : to enter into or complete an agreement
5 : to come to an end or period
6 : to reduce a gap ⟨*closed* to within two points⟩ ☆
— clos·able *or* **close·able** \ˈklō-zə-bəl\ *adjective*
— clos·er *noun*
— close one's doors 1 : to refuse admission ⟨the nation *closed its doors* to immigrants⟩ **2 :** to go out of business
— close one's eyes to : to ignore deliberately
— close ranks : to unite in a concerted stand especially to meet a challenge
— close the door : to be uncompromisingly obstructive ⟨*closed the door* to further negotiation⟩

²close \ˈklōz\ *noun* (14th century)
1 a : a coming or bringing to a conclusion ⟨at the *close* of the party⟩ **b :** a conclusion or end in time or existence **:** CESSATION ⟨the decade drew to a *close*⟩ **c :** the concluding passage (as of a speech or play)
2 : the conclusion of a musical strain or period **:** CADENCE
3 *archaic* **:** a hostile encounter
4 : the movement of the free foot in dancing toward or into contact with the supporting foot

³close \ˈklōs, *U.S. also* ˈklōz\ *noun* [Middle English *clos,* literally, enclosure, from Old French *clos,* from Latin *clausum,* from neuter of *clausus,* past participle] (13th century)
1 a : an enclosed area **b** *chiefly British* **:** the precinct of a cathedral
2 *chiefly British* **a :** a narrow passage leading from a street to a court and the houses within or to the common stairway of tenements **b :** a road closed at one end

⁴close \ˈklōs\ *adjective* **clos·er; clos·est** [Middle English *clos,* from Middle French, from Latin *clausus,* past participle of *claudere*] (14th century)
1 : having no openings **:** CLOSED
2 a : confined or carefully guarded ⟨*close* arrest⟩ **b** (1) *of a vowel* **:** HIGH 13 (2) **:** formed with the tongue in a higher position than for the other vowel of a pair
3 : restricted to a privileged class
4 a : SECLUDED, SECRET **b :** SECRETIVE ⟨she could tell us something if she would . . . but she was as *close* as wax —A. Conan Doyle⟩
5 : STRICT, RIGOROUS ⟨keep *close* watch⟩
6 : hot and stuffy
7 : not generous in giving or spending **:** TIGHT
8 : having little space between items or units ⟨a *close* weave⟩
9 a : fitting tightly or exactly **b :** very short or near to the surface ⟨a *close* haircut⟩
10 : being near in time, space, effect, or degree
11 : INTIMATE, FAMILIAR
12 a : ACCURATE, PRECISE ⟨a *close* study⟩ **b :** marked by fidelity to an original ⟨a *close* copy of an old master⟩ **c :** TERSE, COMPACT
13 : decided or won by a narrow margin ⟨a *close* baseball game⟩
14 : difficult to obtain ⟨money is *close*⟩
15 *of punctuation* **:** characterized by liberal use especially of commas
synonym see STINGY
— close·ly *adverb*
— close·ness *noun*
— close to home : within one's personal interests so that one is strongly affected ⟨the speaker's remarks hit *close to home*⟩
— close to the vest : in a reserved or cautious manner

⁵close \ˈklōs\ *adverb* (15th century)
: in a close position or manner

close call \ˈklōs-\ *noun* (1881)
: a narrow escape

close corporation \ˈklōs-\ *noun* (circa 1902)
: a corporation whose stock is not publicly traded but held by a few persons (as those in management)

close–cropped \ˈklōs-ˌkräpt\ *adjective* (circa 1893)
1 : clipped short
2 : having the hair clipped short

closed \ˈklōzd\ *adjective* (13th century)
1 a : not open **b :** ENCLOSED **c :** composed entirely of closed tubes or vessels ⟨a *closed* circulatory system⟩
2 a : forming a self-contained unit allowing no additions ⟨*closed* association⟩ **b** (1) **:** traced by a moving point that returns to an arbitrary starting point ⟨*closed* curve⟩; *also* **:** so formed that every plane section is a closed curve ⟨*closed* surface⟩ (2) **:** characterized by mathematical elements that when subjected to an operation produce only elements of the same set ⟨the set of whole numbers is *closed* under addition and multiplication⟩ (3) **:** containing all the limit points of every possible subset ⟨a *closed* interval contains its endpoints⟩ **c :** characterized by continuous return and reuse of the working substance ⟨a *closed* cooling system⟩ **d** *of a racecourse* **:** having the same starting and finishing point
3 a : confined to a few ⟨*closed* membership⟩ **b :** excluding participation of outsiders or witnesses **:** conducted in strict secrecy **c :** rigidly excluding outside influence ⟨a *closed* mind⟩
4 : ending in a consonant ⟨*closed* syllable⟩

closed book *noun* (1913)
: something beyond comprehension **:** ENIGMA

closed–cap·tioned \ˈklōz(d)-ˈkap-shənd\ *adjective* (1980)
of a television program **:** broadcast with captions that appear only on the screen of a receiver equipped with a decoder

closed–cir·cuit \-ˈsər-kət\ *adjective* (1949)
: used in, shown on, or being a television installation in which the signal is transmitted by wire to a limited number of receivers

closed corporation *noun* (1924)
: CLOSE CORPORATION

closed couplet *noun* (1910)
: a rhymed couplet in which the sense is complete

closed–door \ˈklōz(d)-ˈdōr, -ˈdȯr\ *adjective* (1950)
: barring public and press ⟨a *closed-door* session of the investigating committee⟩

closed–end \ˈklōzd-ˈend\ *adjective* (circa 1938)
: having a fixed capitalization of shares that are traded on the market at prices determined by the operation of the law of supply and demand ⟨a *closed-end* investment company⟩ — compare OPEN-END

closed loop *noun* (1951)
: an automatic control system for an operation or process in which feedback in a closed path or group of paths acts to maintain output at a desired level

close·down \ˈklōz-ˌdau̇n\ *noun* (1889)
: an instance of suspending or stopping operations

closed shop *noun* (1904)
: an establishment in which the employer by agreement hires only union members in good standing

closed stance *noun* (circa 1934)
: a stance (as in golf) in which the forward foot is closer to the line of play than the back foot — compare OPEN STANCE

close·fist·ed \ˈklōs-ˌfis-təd\ *adjective* (1608)
: STINGY, TIGHTFISTED

close–grained \-ˌgrānd\ *adjective* (1754)
: having a compacted smooth texture; *especially* **:** having narrow annual rings or small wood elements

close–hauled \-ˈhȯld\ *adjective* (1769)
: having the sails set for sailing as nearly against the wind as the vessel will go

close–in \ˈklōs-ˈin\ *adjective* (1945)
: near a center of activity and especially a city ⟨*close-in* suburbs⟩

close in \ˈklōz-\ (14th century)
transitive verb
1 : to encircle closely and isolate
2 : to enshroud to such an extent as to preclude entrance or exit ⟨the airport was *closed in* by the storm⟩
intransitive verb
1 : to gather in close all around with an oppressing or isolating effect ⟨despair *closed in* on her⟩
2 : to approach to close quarters especially for an attack, raid, or arrest ⟨the police *closed in*⟩
3 : to grow dark ⟨the short November day was already *closing in* —Ellen Glasgow⟩

close–knit \ˈklōs-ˈnit\ *adjective* (1926)
: bound together by intimate social or cultural ties or by close economic or political ties ⟨*close-knit* families⟩

closely held *adjective* (1946)
: having most stock shares and voting rights in the hands of a few ⟨a *closely held* business⟩

close·mouthed \ˈklōs-ˈmau̇t͟hd, -ˈmau̇tht\ *adjective* (1881)
: cautious in speaking **:** UNCOMMUNICATIVE; *also* **:** SECRETIVE

close order *noun* (circa 1797)
: an arrangement of troops in a typical marching formation

close·out \ˈklō-ˌzau̇t\ *noun* (1925)
1 : a clearing out by a sale usually at reduced prices of the whole remaining stock (as of a business)
2 : an article offered or bought at a closeout

close out \ˈklōz-\ (14th century)
transitive verb
1 a : EXCLUDE **b :** PRECLUDE ⟨*close out* his chances⟩
2 a : to dispose of a whole stock of by sale **b :** to dispose of (a business) **c :** SELL ⟨*closed out* his share of the business⟩ **d :** to put (an account) in order for disposal or transfer
3 a : to bring to an often rapid or abrupt conclusion **b :** to discontinue operation of
intransitive verb
1 : to sell out a business
2 : to buy or sell securities or commodities in order to terminate an account

☆ SYNONYMS
Close, end, conclude, finish, complete, terminate mean to bring or come to a stopping point or limit. CLOSE usually implies that something has been in some way open as well as unfinished ⟨*close* a debate⟩. END conveys a strong sense of finality ⟨*ended* his life⟩. CONCLUDE may imply a formal closing (as of a meeting) ⟨the service *concluded* with a blessing⟩. FINISH may stress completion of a final step in a process ⟨after it is painted, the house will be *finished*⟩. COMPLETE implies the removal of all deficiencies or a successful finishing of what has been undertaken ⟨the resolving of this last issue *completes* the agreement⟩. TERMINATE implies the setting of a limit in time or space ⟨your employment *terminates* after three months⟩.

close quarters \ˈklōs-\ *noun plural* (1809)
: immediate contact or close range ⟨fought at *close quarters*⟩

close shave \ˈklōs-, ˈklōsh-\ *noun* (1834)
: a narrow escape

close·stool \ˈklōs-ˌstül\ *noun* (15th century)
: a stool holding a chamber pot

1clos·et \ˈklä-zət, ˈklȯ-\ *noun* [Middle English, from Middle French, diminutive of *clos* enclosure — more at CLOSE] (14th century)
1 a : an apartment or small room for privacy **b :** a monarch's or official's private chamber
2 : a cabinet or recess for especially china, household utensils, or clothing
3 : a place of retreat or privacy
4 : WATER CLOSET
5 : a state or condition of secrecy, privacy, or obscurity ⟨came out of the *closet*⟩
— clos·et·ful \-ˌfu̇l\ *noun*

2closet *transitive verb* (1595)
1 : to shut up in or as if in a closet
2 : to take into a closet for a secret interview

3closet *adjective* (circa 1615)
1 : closely private
2 : working in or suited to the closet as the place of seclusion or study **:** THEORETICAL
3 : being so in private ⟨a *closet* racist⟩

closet drama *noun* (1922)
: drama suited primarily for reading rather than production

closet queen *noun* (1967)
: one who secretly engages in homosexual activities while leading an ostensibly heterosexual life — often used disparagingly

1close–up \ˈklō-ˌsəp\ *noun* (1913)
1 : a photograph or movie shot taken at close range
2 : an intimate view or examination of something

2close–up \ˌklō-ˈsəp\ *adverb or adjective* (1926)
: at close range

clos·ing \ˈklō-ziŋ\ *noun* (1596)
1 : a concluding part (as of a speech)
2 : a closable gap (as in an article of clothing)
3 : a meeting of parties to a real-estate deal for formally transferring title

clos·trid·i·um \klä-ˈstri-dē-əm\ *noun, plural* **-ia** \-dē-ə\ [New Latin, genus name, from Greek *klōstēr* spindle, from *klōthein* to spin] (1884)
: any of a genus (*Clostridium*) of spore-forming mostly anaerobic soil or intestinal bacteria — compare BOTULINUM
— clos·trid·i·al \-dē-əl\ *adjective*

clo·sure \ˈklō-zhər\ *noun* [Middle English, from Middle French, from Latin *clausura,* from *clausus,* past participle of *claudere* to close — more at CLOSE] (14th century)
1 *archaic* **:** means of enclosing **:** ENCLOSURE
2 : an act of closing **:** the condition of being closed ⟨*closure* of the eyelids⟩
3 : something that closes ⟨pocket with zipper *closure*⟩
4 [translation of French *clôture*] **:** CLOTURE
5 : the property that a number system or a set has when it is mathematically closed under an operation
6 : a set that contains a given set together with all the limit points of the given set

1clot \ˈklät\ *noun* [Middle English, from Old English *clott,* akin to Middle High German *klōz* lump, ball — more at CLOUT] (before 12th century)
1 : a portion of a substance cleaving together in a thick nondescript mass (as of clay or gum)
2 a : a roundish viscous lump formed by coagulation of a portion of liquid or by melting **b :** a coagulated mass produced by clotting of blood
3 *British* **:** BLOCKHEAD
4 : CLUSTER, GROUP ⟨a *clot* of spectators⟩

2clot *verb* **clot·ted; clot·ting** (15th century)
intransitive verb
1 : to become a clot **:** form clots
2 : to undergo a sequence of complex chemical and physical reactions that results in conversion of fluid blood into a coagulated mass **:** COAGULATE
transitive verb
1 : to cause to form into or as if into a clot
2 : to fill with clots

cloth \ˈklȯth\ *noun, plural* **cloths** \ˈklȯthz, ˈklȯths\ *often attributive* [Middle English, from Old English *clāth* cloth, garment; akin to Middle High German *kleit* garment] (before 12th century)
1 a : a pliable material made usually by weaving, felting, or knitting natural or synthetic fibers and filaments **b :** a similar material (as of glass)
2 : a piece of cloth adapted for a particular purpose; *especially* **:** TABLECLOTH
3 a : a distinctive dress of a profession or calling **b :** the dress of the clergy; *also* **:** CLERGY

cloth·bound \-ˌbau̇nd\ *adjective* (1860)
of a book **:** bound in stiff boards covered with cloth

clothe \ˈklōth\ *transitive verb* **clothed** *or* **clad** \ˈklad\; **cloth·ing** [Middle English, from Old English *clāthian,* from *clāth*] (before 12th century)
1 a : to cover with or as if with cloth or clothing **:** DRESS **b :** to provide with clothes
2 : to express or enhance by suitably significant language **:** COUCH ⟨treaties *clothed* in stately phraseology⟩
3 : to endow especially with power or a quality

clothes \ˈklōz *also* ˈklōthz\ *noun plural, often attributive* [Middle English, from Old English *clāthas,* plural of *clāth*] (before 12th century)
1 : CLOTHING
2 : BEDCLOTHES
3 : all the cloth articles of personal and household use that can be washed

clothes·horse \-ˌhȯrs\ *noun* (1775)
1 : a frame on which to hang clothes
2 : a conspicuously dressy person

1clothes·line \-ˌlīn\ *noun* (1830)
: a line (as of cord) on which clothes may be hung to dry

2clothesline *transitive verb* (1964)
: to knock down (a football player) by catching by the neck with an outstretched arm

clothes moth *noun* (1753)
: any of several small yellowish or buff-colored moths (especially *Tinea pellionella* and *Tineola bisselliella* of the family Tineidae) whose larvae eat wool, fur, or feathers

clothes–peg \ˈklōz-ˌpeg *also* ˈklōthz-\ *noun* (1825)
British **:** CLOTHESPIN

clothes·pin \-ˌpin\ *noun* (1846)
: a forked piece of wood or plastic or a small spring clamp used for fastening clothes on a clothesline

clothes·press \-ˌpres\ *noun* (1713)
: a receptacle for clothes

cloth·ier \ˈklōth-yər, ˈklō-thē-ər\ *noun* [Middle English, alteration of *clother,* from *cloth*] (14th century)
: one who makes or sells clothing

cloth·ing \ˈklō-thiŋ\ *noun* (13th century)
: garments in general; *also* **:** COVERING

cloth yard *noun* (15th century)
: a yard especially for measuring cloth; *specifically* **:** a unit of 37 inches equal to the Scottish ell and used also as a length for arrows

clotted cream *noun* (1878)
: a thick cream made chiefly in England by slowly heating whole milk on which the cream has been allowed to rise and then skimming the cooled cream from the top — called also *Devonshire cream*

clo·ture \ˈklō-chər\ *noun* [French *clôture,* literally, closure, alteration of Middle French *closure*] (1871)
: the closing or limitation of debate in a legislative body especially by calling for a vote
— cloture *transitive verb*

1cloud \ˈklau̇d\ *noun, often attributive* [Middle English, rock, cloud, from Old English *clūd;* perhaps akin to Greek *gloutos* buttock] (14th century)
1 : a visible mass of particles of condensed vapor (as water or ice) suspended in the atmosphere of a planet (as the earth) or moon
2 : something resembling or suggesting a cloud: as **a :** a light filmy, puffy, or billowy mass seeming to float in the air ⟨a *cloud* of blond hair⟩ ⟨a ship under a *cloud* of sail⟩ **b** (1) **:** a usually visible mass of minute particles suspended in the air or a gas (2) **:** an aggregation of usually obscuring matter especially in interstellar space (3) **:** an aggregate of charged particles (as electrons) **c :** a great crowd or multitude **:** SWARM ⟨*clouds* of mosquitoes⟩
3 : something that has a dark, lowering, or threatening aspect ⟨*clouds* of war⟩ ⟨a *cloud* of suspicion⟩
4 : something that obscures or blemishes ⟨a *cloud* of ambiguity⟩
5 : a dark or opaque vein or spot (as in marble or a precious stone)

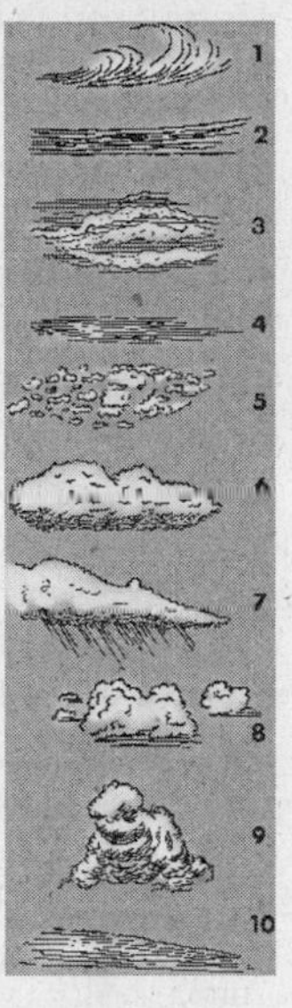

cloud 1: *1* **cirrus,** *2* **cirrostratus,** *3* **cirrocumulus,** *4* **altostratus,** *5* **altocumulus,** *6* **stratocumulus,** *7* **nimbostratus,** *8* **cumulus,** *9* **cumulonimbus,** *10* **stratus**

2cloud (1562)
intransitive verb
1 : to grow cloudy — usually used with *over* or *up* ⟨*clouded* over before the storm⟩
2 a *of facial features* **:** to become troubled, apprehensive, or distressed in appearance **b :** to become blurry, dubious, or ominous — often used with *over*
3 : to billow up in the form of a cloud
transitive verb
1 a : to envelop or hide with or as if with a cloud ⟨smog *clouded* our view⟩ **b :** to make opaque especially by condensation of moisture **c :** to make murky especially with smoke or mist
2 : to make unclear or confused ⟨*cloud* the issue⟩
3 : TAINT, SULLY ⟨a *clouded* reputation⟩
4 : to cast gloom over

cloud·ber·ry \ˈklau̇d-ˌber-ē\ *noun* (1597)
: a creeping herbaceous raspberry (*Rubus chamaemorus*) of north temperate regions; *also* **:** its pale amber-colored edible fruit

cloud·burst \-ˌbərst\ *noun* (1869)
1 : a sudden copious rainfall
2 : DELUGE 2

cloud chamber *noun* (1897)
: a vessel containing air saturated with water vapor whose sudden expansion reveals the passage of an ionizing particle by a trail of visible droplets

cloud–cuck·oo–land \ˌklau̇d-ˈkü-(ˌ)kü-ˌland, -ˈku̇-\ *noun* [translation of Greek *nephelokokkygia*] (1899)
: a realm of fantasy or of whimsical or foolish behavior

cloud·ed leopard \ˈklau̇-dəd-\ *noun* (1910)
: a medium-sized cat (*Neofelis nebulosa*) that occurs from Nepal to southeast China and Indonesia and is grayish to yellowish with darker irregular rings, ovals, and rosettes

cloud forest *noun* (1922)
: a wet tropical mountain forest at an altitude

usually between 3000 and 8000 feet (1000 and 2500 meters) that is characterized by a profusion of epiphytes and the presence of clouds even in the dry season

cloud·land \'klau̇d-ˌland\ *noun* (1817)
1 : the region of the clouds
2 : the realm of visionary speculation or poetic imagination

cloud·less \-ləs\ *adjective* (14th century)
: free from clouds **:** CLEAR
— cloud·less·ly *adverb*
— cloud·less·ness *noun*

cloud·let \'klau̇d-lət\ *noun* (1788)
: a small cloud

cloud nine *noun* (1959)
: a feeling of well-being or elation — usually used with *on*

cloud·scape \'klau̇d-ˌskāp\ *noun* (1880)
: a view or pictorial representation of a cloud formation

cloud street *noun* (1943)
: a row of cumulus clouds

cloudy \'klau̇-dē\ *adjective* **cloud·i·er; -est** (14th century)
1 : of, relating to, or resembling cloud
2 : darkened by gloom or anxiety
3 a : overcast with clouds **b :** having a cloudy sky
4 : obscure in meaning ⟨*cloudy* issues⟩; *also* **:** uncertain as to fact or outcome ⟨a *cloudy* future⟩
5 : dimmed or dulled as if by clouds ⟨a *cloudy* mirror⟩
6 : uneven in color or texture
7 : having visible material in suspension **:** MURKY
— cloud·i·ly \'klau̇-dᵊl-ē\ *adverb*
— cloud·i·ness \'klau̇-dē-nəs\ *noun*

¹clout \'klau̇t\ *noun* [Middle English, from Old English *clūt;* akin to Middle High German *klōz* lump, Russian *gluda*] (before 12th century)
1 *dialect chiefly British* **:** a piece of cloth or leather **:** RAG
2 : a blow especially with the hand; *also* **:** a hard hit in baseball
3 : a white cloth on a stake or frame used as a target in archery
4 : PULL, INFLUENCE ⟨political *clout*⟩

²clout *transitive verb* (14th century)
1 : to cover or patch with a clout
2 : to hit forcefully

¹clove \'klōv\ *noun* [Middle English, from Old English *clufu;* akin to Old English *clēofan* to cleave] (before 12th century)
: one of the small bulbs (as in garlic) developed in the axils of the scales of a large bulb

²clove *past of* CLEAVE

³clove \'klōv\ *noun* [alteration of Middle English *clowe,* from Old French *clou* (*de girofle*), literally, nail of clove, from Latin *clavus* nail] (13th century)
: the dried flower bud of a tropical tree (*Syzygium aromaticum* synonym *Eugenia aromatica*) of the myrtle family that is used as a spice and is the source of an oil; *also* **:** this tree

clove hitch \'klōv-\ *noun* [Middle English *cloven, clove* divided, from past participle of *clevien* to cleave] (circa 1769)
: a knot securing a rope temporarily to an object (as a post or spar) and consisting of a turn around the object, over the standing part, around the object again, and under the last turn — see KNOT illustration

clo·ven \'klō-vən\ *past participle of* CLEAVE

cloven foot *noun* (13th century)
: CLOVEN HOOF
— clo·ven–foot·ed \'klō-vən-ˌfu̇-təd\ *adjective*

cloven hoof *noun* (1870)
1 : a foot (as of a sheep) divided into two parts at its distal extremity
2 [from the traditional representation of Satan as cloven-hoofed] **:** the sign of devilish character

— clo·ven–hoofed \'klō-vən-ˌhu̇ft, -ˌhüft, -ˌhu̇vd, -ˌhüvd\ *adjective*

clove pink *noun* (circa 1855)
: CARNATION 2

clo·ver \'klō-vər\ *noun* [Middle English, from Old English *clāfre;* akin to Old High German *klēo* clover] (before 12th century)
1 : any of a genus (*Trifolium*) of low leguminous herbs having trifoliolate leaves and flowers in dense heads and including many that are valuable for forage and attractive to bees
2 : any of various leguminous plants (as of the genera *Melilotus, Lespedeza,* or *Medicago*) other than the clovers (genus *Trifolium*)
— in clover *or* **in the clover :** in prosperity or in pleasant circumstances

¹clo·ver·leaf \-ˌlēf\ *adjective* (1917)
: resembling a clover leaf in shape

²cloverleaf *noun, plural* **-leafs** \-ˌlēfs\ *or* **-leaves** \-ˌlēvz\ (1931)
: an interchange between two major highways that allows traffic to change from one to the other without requiring any left turns or crossings and that from above resembles a four-leaf clover

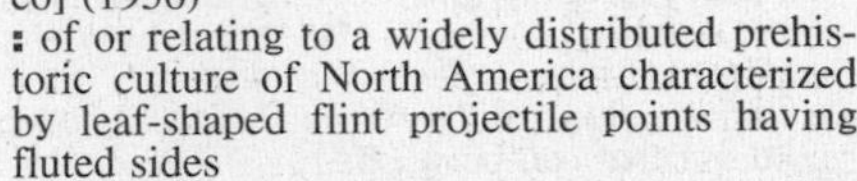

cloverleaf

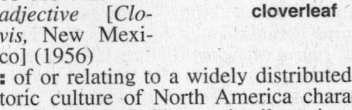

Clo·vis \'klō-vəs\ *adjective* [*Clovis,* New Mexico] (1956)
: of or relating to a widely distributed prehistoric culture of North America characterized by leaf-shaped flint projectile points having fluted sides

¹clown \'klau̇n\ *noun* [probably of Low German origin; akin to Frisian *klönne* clumsy fellow, Old English *clyne* lump of metal] (1563)
1 : FARMER, COUNTRYMAN
2 : a rude ill-bred person **:** BOOR
3 a : a fool, jester, or comedian in an entertainment (as a play); *specifically* **:** a grotesquely dressed comedy performer in a circus **b :** a person who habitually plays the buffoon **c :** JOKER 1b

²clown *intransitive verb* (1599)
: to act as or like a clown

clown·ery \'klau̇-nə-rē\ *noun, plural* **-er·ies** (1589)
: clownish behavior or an instance of clownishness **:** BUFFOONERY

clown·ish \'klau̇-nish\ *adjective* (circa 1570)
: resembling or befitting a clown (as in ignorance and lack of sophistication)
synonym see BOORISH
— clown·ish·ly *adverb*
— clown·ish·ness *noun*

clox·a·cil·lin \ˌkläk-sə-'si-lən\ *noun* [*chlor-* + *oxacillin*] (1963)
: a semisynthetic oral penicillin $C_{19}H_{17}ClN_3NaO_5S$ effective especially against staphylococci which secrete penicillinase

cloy \'klȯi\ *verb* [alteration of Middle English *acloien* to lame, from Middle French *encloer* to drive in a nail, from Medieval Latin *inclavare,* from Latin *in* + *clavus* nail] (1528)
transitive verb
: to surfeit with an excess usually of something originally pleasing
intransitive verb
: to cause surfeit
synonym see SATIATE

cloy·ing \'klȯi-iŋ, 'klȯiŋ\ *adjective* (1594)
: disgusting or distasteful by reason of excess ⟨*cloying* sweetness⟩; *also* **:** excessively sweet or sentimental ⟨a *cloying* romantic comedy⟩
— cloy·ing·ly *adverb*

cloze \'klōz\ *adjective* [by shortening & alteration from *closure*] (1953)
: of, relating to, or being a test of reading comprehension that involves having the person being tested supply words which have been systematically deleted from a text

¹club \'kləb\ *noun, often attributive* [Middle English *clubbe,* from Old Norse *klubba;* akin to Old High German *kolbo* club] (13th century)
1 a : a heavy usually tapering staff especially of wood wielded as a weapon **b :** a stick or bat used to hit a ball in any of various games **c :** something resembling a club
2 a : a playing card marked with a stylized figure of a black clover **b** *plural but singular or plural in construction* **:** the suit comprising cards marked with clubs
3 a : an association of persons for some common object usually jointly supported and meeting periodically; *also* **:** a group identified by some common characteristic ⟨nations in the nuclear *club*⟩ **b :** the meeting place of a club **c :** an association of persons participating in a plan by which they agree to make regular payments or purchases in order to secure some advantage **d :** NIGHTCLUB **e :** an athletic association or team
— club·bish \'klə-bish\ *adjective*

²club *verb* **clubbed; club·bing** (1593)
transitive verb
1 a : to beat or strike with or as if with a club **b :** to gather into a club-shaped mass ⟨*clubbed* her hair⟩
2 a : to unite or combine for a common cause **b :** to contribute to a common fund
intransitive verb
1 : to form a club **:** COMBINE
2 : to pay a share of a common expense

club·ba·ble *also* **club·able** \'klə-bə-bəl\ *adjective* (1783)
: SOCIABLE

clubbed \'kləbd\ *adjective* (14th century)
: shaped like a club ⟨*clubbed* antennae⟩

club·ber \'klə-bər\ *noun* (1633)
: a member of a club

club·by \'klə-bē\ *adjective* **club·bi·er; -est** (1859)
: characteristic of a club or club members: as **a :** displaying friendliness especially to other members of the same social group **:** SOCIABLE **b :** open only to qualified or approved persons **:** SELECT, ELITE
— club·bi·ness *noun*

club car *noun* (1886)
: a railroad passenger car with facilities for serving refreshments and drinks — called also *bar car, lounge car*

club chair *noun* (1919)
: a deep low thickly upholstered easy chair often with rather low back and heavy sides and arms

club cheese *noun* (1916)
: a process cheese made by grinding cheddar and other cheeses usually with added condiments and seasoning

club·foot \'kləb-ˌfu̇t\ *noun* (1538)
: a misshapen foot twisted out of position from birth; *also* **:** this deformity
— club·foot·ed \-ˌfu̇-təd\ *adjective*

club fungus *noun* (1899)
: any of a family (Clavariaceae) of basidiomycetes with a simple or branched often club-shaped sporophore

club·house \'kləb-ˌhau̇s\ *noun* (1818)
1 : a house occupied by a club or used for club activities
2 : locker rooms used by an athletic team

club·man \'kləb-mən, -ˌman\ *noun* (1851)
: a usually wealthy man given to club life

club moss *noun* (1597)
: any of an order (Lycopodiales) of primitive vascular plants (as ground pine) often with the sporangia borne in club-shaped strobili

club·root \'kləb-ˌrüt, -ˌru̇t\ *noun* (1846)
: a disease of cruciferous plants and especially of cabbage caused by a slime mold (*Plasmodiophora brassicae*) producing swellings or distortions of the root

club sandwich *noun* (1903)
: a sandwich of three slices of bread with two layers of meat (as turkey) and lettuce, tomato, and mayonnaise

club soda *noun* (1942)
: SODA WATER 2a

club steak *noun* (1915)
: a small steak cut from the end of the short loin — see BEEF illustration

¹cluck \'klək\ *verb* [imitative] (15th century)
intransitive verb
1 : to make a cluck
2 : to make a clicking sound with the tongue
3 : to express interest or concern ⟨critics *clucked* over the new developments⟩
transitive verb
1 : to call with a cluck
2 : to express with interest or concern

²cluck *noun* (1703)
1 : the characteristic sound made by a hen especially in calling her chicks
2 : a stupid or naive person

¹clue \'klü\ *noun* [alteration of *clew*] (1596)
: something that guides through an intricate procedure or maze of difficulties; *specifically* **:** a piece of evidence that leads one toward the solution of a problem ◆

²clue *transitive verb* **clued; clue·ing** *or* **clu·ing** (1663)
1 : to provide with a clue
2 : to give reliable information to ⟨*clue* me in on how it happened⟩

clue·less \'klü-ləs\ *adjective* (1862)
1 : providing no clue
2 : completely or hopelessly bewildered, unaware, ignorant, or foolish

clum·ber spaniel \'kləm-bər-\ *noun, often C capitalized* [*Clumber,* estate in Nottinghamshire, England] (1883)
: any of a breed of large massive heavyset spaniels with a dense silky largely white coat

¹clump \'kləmp\ *noun* [probably from Low German *klump*] (circa 1586)
1 : a group of things clustered together ⟨a *clump* of bushes⟩
2 : a compact mass
3 : a heavy tramping sound
— clumpy \'kləm-pē\ *adjective*

²clump (1665)
intransitive verb
1 : to walk or move clumsily and noisily
2 : to form clumps
transitive verb
: to arrange in or cause to form clumps ⟨the serum *clumps* the bacteria⟩

clum·sy \'kləm-zē\ *adjective* **clum·si·er; -est** [probably from obsolete English *clumse* benumbed with cold] (circa 1598)
1 a : lacking dexterity, nimbleness, or grace ⟨*clumsy* fingers⟩ **b :** lacking tact or subtlety ⟨a *clumsy* joke⟩
2 : awkward or inefficient in use or construction **:** UNWIELDY
synonym see AWKWARD
— clum·si·ly \-zə-lē\ *adverb*
— clum·si·ness \-zē-nəs\ *noun*

clung *past and past participle of* CLING

¹clunk \'kləŋk\ *verb* [imitative] (circa 1796)
intransitive verb
1 : to make a clunk
2 : to hit something with a clunk
transitive verb
: to strike or hit with a clunk

²clunk *noun* (1823)
1 : a blow or the sound of a blow **:** THUMP
2 : a dull or stupid person

clunk·er \'kləŋ-kər\ *noun* (1943)
1 : an old or badly working piece of machinery; *especially* **:** a dilapidated automobile
2 : someone or something notably unsuccessful

clunky \'kləŋ-kē\ *adjective* **clunk·i·er; -est** (1968)
: clumsy in style, form, or execution ⟨a *clunky* thriller⟩ ⟨*clunky* earrings⟩

clu·pe·id \'klü-pē-əd\ *noun* [ultimately from Latin *clupea,* a small river fish] (1880)
: any of a large family (Clupeidae) of soft-finned bony fishes (as herrings) that have a laterally compressed body and a forked tail and usually occur in schools
— clupeid *adjective*

¹clus·ter \'kləs-tər\ *noun* [Middle English, from Old English *clyster;* akin to Old English *clott* clot] (before 12th century)
: a number of similar individuals that occur together: as **a :** two or more consecutive consonants or vowels in a segment of speech **b :** a group of buildings and especially houses built close together on a sizable tract in order to preserve open spaces larger than the individual yard for common recreation **c :** an aggregation of stars or galaxies that appear close together in the sky and are gravitationally associated
— clus·tery \-t(ə-)rē\ *adjective*

²cluster *verb* **clus·tered; clus·ter·ing** \-t(ə-)riŋ\ (14th century)
transitive verb
1 : to collect into a cluster ⟨*cluster* the tents together⟩
2 : to furnish with clusters
intransitive verb
: to grow, assemble, or occur in a cluster

cluster analysis *noun* (1948)
: a statistical classification technique for discovering whether the individuals of a population fall into different groups by making quantitative comparisons of multiple characteristics

cluster bomb *noun* (1967)
: an aircraft-dropped canister of small fragmentation bombs

cluster headache *noun* (1953)
: a headache that is characterized by severe pain in the eye or temple and tends to recur in a series of attacks

¹clutch \'kləch\ *verb* [Middle English *clucchen,* from Old English *clyccan*] (before 12th century)
transitive verb
1 : to grasp or hold with or as if with the hand or claws usually strongly, tightly, or suddenly
2 *obsolete* **:** CLENCH
intransitive verb
1 : to seek to grasp and hold
2 : to operate an automobile clutch
synonym see TAKE

²clutch *noun* (13th century)
1 a : the claws or a hand in the act of grasping or seizing firmly **b :** an often cruel or unrelenting control, power, or possession ⟨the fell *clutch* of circumstance —W. E. Henley⟩ **c :** the act of grasping, holding, or restraining
2 a : a coupling used to connect and disconnect a driving and a driven part of a mechanism **b :** a lever (as a pedal) operating such a clutch
3 : a tight or critical situation **:** PINCH ⟨come through in the *clutch*⟩
4 : CLUTCH BAG

³clutch *adjective* (1944)
1 : made or done in a crucial situation ⟨a *clutch* hit⟩
2 : successful in a crucial situation ⟨a *clutch* pitcher⟩

⁴clutch *noun* [alteration of dialect English *cletch* hatching, brood] (1721)
1 : a nest of eggs or a brood of chicks
2 : GROUP, BUNCH

clutch bag *noun* (1949)
: a woman's small usually strapless handbag

¹clut·ter \'klə-tər\ *verb* [Middle English *clotteren* to clot, from *clot*] (1556)
intransitive verb
chiefly dialect **:** to run in disorder
transitive verb
: to fill or cover with scattered or disordered things that impede movement or reduce effectiveness — often used with *up*

²clutter *noun* (1649)
1 a : a crowded or confused mass or collection **b :** things that clutter a place
2 : interfering radar echoes caused by reflection from objects (as on the ground) other than the target
3 *chiefly dialect* **:** DISTURBANCE, HUBBUB

Clyde \'klīd\ *noun* (1894)
: CLYDESDALE

Clydes·dale \'klīdz-ˌdāl\ *noun* (1786)
: a heavy draft horse with feathering on the legs of a breed originally from Clydesdale, Scotland

clyp·e·us \'kli-pē-əs\ *noun, plural* **clyp·ei** \-pē-ˌī, -pē-ˌē\ [New Latin, from Latin, round shield] (1834)
: a plate on the anterior median aspect of an insect's head

clys·ter \'klis-tər\ *noun* [Middle English, from Middle French or Latin; Middle French *clistere,* from Latin *clyster,* from Greek *klystēr,* from *klyzein* to wash out; akin to Welsh *clir* pure, Old English *hlūtor* clean] (14th century)
: ENEMA

Cly·tem·nes·tra \ˌklī-təm-'nes-trə\ *noun* [Latin, from Greek *Klytaimnēstra*]
: the wife and murderess of Agamemnon

cni·dar·i·an \nī-'dar-ē-ən, -'der-\ *noun* [ultimately from New Latin *cnida* nematocyst, from Greek *knidē* nettle] (circa 1909)
: COELENTERATE
— cnidarian *adjective*

co- *prefix* [Middle English, from Latin, from *com-;* akin to Old English *ge-,* perfective and collective prefix, Old Irish *com-* with]
1 : with **:** together **:** joint **:** jointly ⟨*co*exist⟩ ⟨*co*heir⟩
2 : in or to the same degree ⟨*co*extensive⟩
3 a : one that is associated in an action with another **:** fellow **:** partner ⟨*co*author⟩ ⟨*co*worker⟩ **b :** having a usually lesser share in duty or responsibility **:** alternate **:** deputy ⟨*co*pilot⟩
4 : of, relating to, or constituting the complement of an angle ⟨*co*sine⟩ ⟨*co*declination⟩

co·act
co·ac·tive
co·ac·tor
co·ad·min·is·tra·tion
co·an·chor
co·au·thor
co·au·thor·ship
co·cap·tain
co·cat·a·lyst
co·chair
co·chair·man
co·chair·per·son
co·chair·wom·an
co·cham·pi·on
co·com·pos·er
co·con·spir·a·tor
co·coun·sel
co·cre·ate
co·cre·ator
co·cul·ti·vate
co·cul·ti·va·tion
co·cul·ture

◇ WORD HISTORY

clue *Clue* was originally a variant spelling of *clew,* meaning "ball of thread or yarn." Our usual modern sense of *clue* and *clew,* "guide to the solution of a mystery," is the metaphorical application of a motif in myth and folklore, the ball of thread that helps in finding one's way out of a maze. Of stories containing such a motif, the best-known is the Greek myth of Theseus and Ariadne. Ariadne, daughter of King Minos of Crete, gives the Athenian hero Theseus a ball of thread and instructs him to unravel it as he searches the Labyrinth of King Minos for the monstrous Minotaur. Theseus follows Ariadne's instructions, and after killing the Minotaur, he retraces his steps out of the maze by rewinding the thread. The old literal "ball of thread" sense of *clew* and *clue* is now little used, though *clew* in the nautical sense "corner of a sail" (if it is indeed the same word and not of independent origin) is still current in the language of yachtsmen.

co•cu•ra•tor
co•de•fen•dant
co•de•sign
co•de•vel•op
co•de•vel•op•er
co•di•rect
co•di•rec•tion
co•di•rec•tor
co•dis•cov•er
co•dis•cov•er•er
co•drive
co•driv•er
co•ed•it
co•ed•i•tor
co•eter•nal
co•ex•ec•u•tor
co•fa•vor•ite
co•fea•ture
co•fi•nance
co•found
co•found•er
co•head
co•heir
co•heir•ess
co•hold•er
co•host
co•host•ess
co•in•vent
co•in•ven•tor
co•in•ves•ti•ga•tor
co•in•ves•tor
co•join
co•lead
co•lead•er
co•man•age
co•man•age•ment
co•man•ag•er
co•mem•ber
co•nom•i•nee
co–oc•cur
co–oc•cur•rence
co–oc•cur•rent
co–of•fi•cial
co–or•ga•niz•er
co–own
co–own•er
co–own•er•ship
co•part•ner
co•part•ner•ship
co•pre•sent
co•pres•i•dent
co•prince
co•prin•ci•pal
co•pris•on•er
co•pro•cess•ing
co•pro•duce
co•pro•duc•er
co•pro•duc•tion
co•pro•mot•er
co•pro•pri•etor
co•pros•per•i•ty
co•pub•lish
co•pub•lish•er
co•re•cip•i•ent
co•re•search•er
co•res•i•dent
co•res•i•dent•tial
co•rul•er
co•script
co•spon•sor
co•spon•sor•ship
co•star
co•sur•fac•tant
co•ten•ant
co•trans•duce
co•trans•duc•tion
co•trans•fer
co•trans•port
co•trust•ee
co•win•ner
co•work•er
co•write

co•ac•er•vate \kō-'a-sər-ˌvāt\ *noun* [Latin *coacervatus,* past participle of *coacervare* to heap up, from *co-* + *acervus* heap] (1929)
: an aggregate of colloidal droplets held together by electrostatic attractive forces
— **co•acer•vate** \ˌkō-ə-'sər-vət\ *adjective*
— **co•ac•er•va•tion** \(ˌ)kō-ˌa-sər-'vā-shən\ *noun*

[1]**coach** \'kōch\ *noun, often attributive* [Middle English *coche,* from Middle French, from German *Kutsche,* from Hungarian *kocsi* (*szekér*), literally, wagon from *Kocs,* Hungary] (1556)
1 a : a large usually closed four-wheeled horse-drawn carriage having doors in the sides and an elevated seat in front for the driver **b :** a railroad passenger car intended primarily for day travel **c :** BUS 1a **d :** TRAILER 3b **e :** a two-door enclosed automobile **f :** a class of passenger air transportation at a lower fare than first class
2 [from the concept that the tutor conveys the student through his examinations] **a :** a private tutor **b :** one who instructs or trains a performer or a team of performers; *specifically* **:** one who instructs players in the fundamentals of a competitive sport and directs team strategy ⟨football *coach*⟩

[2]**coach** (1630)
intransitive verb
1 : to go in a coach
2 : to instruct, direct, or prompt as a coach
transitive verb
1 : to train intensively (as by instruction and demonstration) ⟨*coach* pupils⟩
2 : to act as coach of ⟨*coach* tennis⟩ ⟨*coach* a team⟩
— **coach•able** \'kō-chə-bəl\ *adjective*
— **coach•er** *noun*

coach dog *noun* (1840)
: DALMATIAN

coach•man \'kōch-mən\ *noun* (1579)
: a man who drives a coach or carriage

coach•work \-ˌwərk\ *noun* (1906)
: an automobile body **:** BODYWORK

co•ac•tion \-'ak-shən\ *noun* (1625)
1 : joint action
2 : the interaction between individuals or kinds (as species) in an ecological community

co•adapt•ed \ˌkō-ə-'dap-təd\ *adjective* (1836)
: mutually adapted especially by natural selection ⟨*coadapted* gene complexes⟩
— **co•ad•ap•ta•tion** \ˌkō-ˌa-ˌdap-'tā-shən, -dəp-'\ *noun*

co•ad•ju•tor \ˌkō-ə-'jü-tər, kō-'a-jə-tər\ *noun* [Middle English *coadjutour,* from Middle French *coadjuteur,* from Late Latin *coadjutor,* from Latin *co-* + *adjutor* helper, from *adjuvare* to help — more at AID] (15th century)
1 : one who works together with another **:** ASSISTANT
2 : a bishop assisting a diocesan bishop and often having the right of succession
— **coadjutor** *adjective*

co•ad•ju•trix \ˌkō-ə-'jü-triks, kō-'a-jə-(ˌ)triks\ *noun, plural* **co•ad•ju•tri•ces** \ˌkō-ə-'jü-trə-ˌsēz, (ˌ)kō-ˌa-jə-'trī-(ˌ)sēz\ [New Latin, feminine of *coadjutor*] (1646)
: a female coadjutor

co•ag•u•lant \kō-'a-gyə-lənt\ *noun* (1770)
: something that produces coagulation

co•ag•u•lase \kō-'a-gyə-ˌlās, -ˌlāz\ *noun* (1914)
: any of several enzymes that cause coagulation (as of blood)

[1]**co•ag•u•late** \-lət, -ˌlāt\ *adjective* (14th century)
archaic **:** being clotted or congealed

[2]**co•ag•u•late** \kō-'a-gyə-ˌlāt\ *verb* **-lat•ed; -lat•ing** [Latin *coagulatus,* past participle of *coagulare* to curdle, from *coagulum* curdling agent, from *cogere* to drive together — more at COGENT] (15th century)
transitive verb
1 : to cause to become viscous or thickened into a coherent mass **:** CURDLE, CLOT
2 : to gather together or form into a mass or group
intransitive verb
: to become coagulated
— **co•ag•u•la•bil•i•ty** \kō-ˌa-gyə-lə-'bi-lə-tē\ *noun*
— **co•ag•u•la•ble** \-'a-gyə-lə-bəl\ *adjective*
— **co•ag•u•la•tion** \-'lā-shən\ *noun*

co•ag•u•lum \kō-'a-gyə-ləm\ *noun, plural* **-u•la** \-lə\ *or* **-ulums** [Latin] (1658)
: a coagulated mass or substance **:** CLOT

[1]**coal** \'kōl\ *noun, often attributive* [Middle English *col,* from Old English; akin to Old High German & Old Norse *kol* burning ember, Middle Irish *gúal* coal] (before 12th century)
1 : a piece of glowing carbon or charred wood **:** EMBER
2 : CHARCOAL 1
3 a : a black or brownish black solid combustible substance formed by the partial decomposition of vegetable matter without free access of air and under the influence of moisture and often increased pressure and temperature that is widely used as a natural fuel **b** *plural, British* **:** pieces or a quantity of the fuel broken up for burning

[2]**coal** (1602)
transitive verb
1 : to burn to charcoal **:** CHAR
2 : to supply with coal
intransitive verb
: to take in coal

co•a•lesce \ˌkō-ə-'les\ *verb* **co•a•lesced; co•a•lesc•ing** [Latin *coalescere,* from *co-* + *alescere* to grow — more at OLD] (circa 1656)
intransitive verb
1 : to grow together
2 a : to unite into a whole **:** FUSE ⟨separate townships have *coalesced* into a single, sprawling colony —Donald Gould⟩ **b :** to unite for a common end **:** join forces ⟨people with different points of view *coalesce* into opposing factions —I. L. Horowitz⟩
3 : to arise from the combination of distinct elements ⟨an organized and a popular resistance immediately *coalesced* —C. C. Menges⟩
transitive verb
: to cause to unite ⟨sometimes a book *coalesces* a public into a mass market —Walter Meade⟩
synonym see MIX
— **co•a•les•cence** \-'le-sᵊn(t)s\ *noun*
— **co•a•les•cent** \-sᵊnt\ *adjective*

coal•field \'kōl-ˌfēld\ *noun* (1813)
: a region rich in coal deposits

coal•fish \-ˌfish\ *noun* (1603)
: any of several blackish or dark-backed fishes (as a pollack, cobia, or sablefish)

coal gas *noun* (1809)
: gas made from coal: as **a :** the mixture of gases thrown off by burning coal **b :** gas made by carbonizing bituminous coal in retorts and used for heating and lighting

coal•hole \'kōl-ˌhōl\ *noun* (circa 1662)
1 *British* **:** a compartment for storing coal
2 : a hole for coal (as an opening in a sidewalk leading to a coal bin)

coal•i•fi•ca•tion \ˌkō-lə-fə-'kā-shən\ *noun* (1911)
: a process in which vegetable matter becomes converted into coal of increasingly higher rank with anthracite as the final product
— **coal•i•fy** \'kō-lə-ˌfī\ *transitive verb*

co•a•li•tion \ˌkō-ə-'li-shən\ *noun* [French, from Latin *coalescere*] (1612)
1 a : the act of coalescing **:** UNION **b :** a body formed by the coalescing of originally distinct elements **:** COMBINATION
2 : a temporary alliance of distinct parties, persons, or states for joint action
— **co•a•li•tion•ist** \-'li-sh(ə-)nist\ *noun*

coal measures *noun plural* (1832)
: beds of coal with the associated rocks

coal oil *noun* (1858)
1 : petroleum or a refined oil prepared from it
2 : KEROSENE

coal seam *noun* (1849)
: a bed of coal usually thick enough to be profitably mined

coal tar *noun* (1785)
: tar obtained by distillation of bituminous coal and used especially in making dyes and drugs

coam•ing \'kō-miŋ\ *noun* [probably irregular from *comb*] (1611)
: a raised frame (as around a hatchway in the deck of a ship) to keep out water

co•apt \kō-'apt\ *transitive verb* [Late Latin *coaptare,* from Latin *co-* + *aptus* fastened, fit — more at APT] (1570)
: to fit together and make fast
— **co•ap•ta•tion** \(ˌ)kō-ˌap-'tā-shən\ *noun*

co•arc•ta•tion \(ˌ)kō-ˌärk-'tā-shən\ *noun* (1545)
: a stricture or narrowing especially of a canal or vessel (as the aorta)

coarse \'kōrs, 'kȯrs\ *adjective* **coars•er; coars•est** [Middle English *cors,* perhaps from *course,* noun] (14th century)
1 : of ordinary or inferior quality or value **:** COMMON
2 a (1) **:** composed of relatively large parts or particles ⟨*coarse* sand⟩ (2) **:** loose or rough in texture ⟨*coarse* cloth⟩ **b :** adjusted or designed for heavy, fast, or less delicate work ⟨a *coarse* saw with large teeth⟩ **c :** not precise or detailed with respect to adjustment or discrimination
3 : crude or unrefined in taste, manners, or language
4 : harsh, raucous, or rough in tone
5 *chiefly British* **:** of or relating to coarse fish ⟨*coarse* fishing⟩ ☆
— **coarse•ly** *adverb*
— **coarse•ness** *noun*

coarse fish *noun* (1886)
1 *chiefly British* **:** a freshwater fish other than a salmonid
2 : ROUGH FISH

coarse–grained \'kōrs-ˌgrānd, 'kȯrs-\ *adjective* (circa 1774)

1 : having a coarse grain
2 : CRUDE

coars·en \'kōr-sᵊn, 'kȯr-\ *verb* **coars·ened; coars·en·ing** (1805)
transitive verb
: to make coarse
intransitive verb
: to become coarse

¹coast \'kōst\ *noun* [Middle English *cost,* from Middle French *coste,* from Latin *costa* rib, side; akin to Old Church Slavonic *kostĭ* bone] (14th century)
1 : the land near a shore **:** SEASHORE
2 *obsolete* **:** BORDER, FRONTIER
3 a : a hill or slope suited to coasting **b :** a slide down a slope (as on a sled)
4 *often capitalized* **:** the Pacific coast of the U.S.
5 : the immediate area of view — used in the phrase *the coast is clear*
— **coast·al** \'kōs-tᵊl\ *adjective*
— **coast·wise** \'kōst-,wīz\ *adverb or adjective*
— **from coast to coast :** across an entire nation or continent

²coast (14th century)
transitive verb
1 *obsolete* **:** to move along or past the side of **:** SKIRT
2 : to sail along the shore of
intransitive verb
1 a *archaic* **:** to travel on land along a coast or along or past the side of something **b :** to sail along the shore
2 a : to slide, run, or glide downhill by the force of gravity **b :** to move along without or as if without further application of propulsive power (as by momentum or gravity) **c :** to proceed easily without special application of effort or concern ⟨*coasted* through school⟩

coast·er \'kōs-tər\ *noun* (1574)
1 : one that coasts: as **a :** a person engaged in coastal traffic or commerce **b :** a ship sailing along a coast or engaged in trade between ports of the same country
2 : a resident of a seacoast
3 a : a tray or decanter stand usually of silver and sometimes on wheels **b :** a shallow container or a plate or mat to protect a surface
4 a : a small vehicle (as a sled or wagon) used in coasting **b :** ROLLER COASTER

coaster brake *noun* (1899)
: a brake in the hub of the rear wheel of a bicycle operated by reverse pressure on the pedals

coaster wagon *noun* (1911)
: a child's toy wagon often used for coasting

coast guard *noun* (1833)
1 : a military or naval force employed in guarding a coast or responsible for the safety, order, and operation of maritime traffic in neighboring waters
2 *usually* **coast·guard** *chiefly British* **:** COASTGUARDSMAN

coast·guards·man \'kōs(t)-,gärdz-mən\ *or* **coast·guard·man** \-,gärd-mən\ *noun* (circa 1891)
: a member of a coast guard

coast·land \-,land\ *noun* (1852)
: land bordering the sea

coast·line \-,līn\ *noun* (circa 1859)
1 : a line that forms the boundary between the land and the ocean or a lake
2 : the outline of a coast

coast redwood *noun* (circa 1897)
: REDWOOD 3a

coast–to–coast \'kōs-tə-'kōst\ *adjective* (1911)
: extending or airing across an entire nation or continent

coast·ward \'kōs-twərd\ *or* **coast·wards** \-twərdz\ *adverb* (1853)
: toward the coast
— **coastward** *adjective*

¹coat \'kōt\ *noun, often attributive* [Middle English *cote,* from Middle French, of Germanic origin; akin to Old High German *kozza* coarse wool mantle] (14th century)
1 a : an outer garment worn on the upper body and varying in length and style according to fashion and use **b :** something resembling a coat
2 : the external growth on an animal
3 : a layer of one substance covering another ⟨a *coat* of paint⟩
— **coat·ed** \'kō-təd\ *adjective*
— **coat·less** *adjective*

²coat *transitive verb* (14th century)
1 : to cover with a coat
2 : to cover or spread with a finishing, protecting, or enclosing layer
— **coat·er** *noun*

coat·dress \'kōt-,dres\ *noun* (1854)
: a dress styled like a coat usually with a front buttoning from neckline to hemline

coat hanger *noun* (1895)
: a device which is shaped like the outline of a person's shoulders and over which garments may be hung

co·a·ti \kə-'wä-tē, kwä-'tē\ *noun* [Portuguese *coatí,* from Tupi] (1676)
: either of two tropical American mammals (*Nasua nasua* and *N. nelsoni*) related to the raccoon but with a longer body and tail and a long flexible snout

coati

co·a·ti·mun·di \kə-,wä-ti-'mən-dē, ,kwä-, -'mu̇n-\ *noun* [Tupi] (1676)
: COATI

coat·ing \'kō-tiŋ\ *noun* (1768)
1 : cloth for coats
2 : COAT, COVERING

coat of arms [Middle English *cote of armes,* translation of Middle French *cote d'armes*] (14th century)
1 : a tabard or surcoat embroidered with armorial bearings
2 a : the particular heraldic bearings (as of a person) usually depicted on an escutcheon often with accompanying adjuncts (as a crest, motto, and supporters) **b :** a similar symbolic emblem

coat of mail (15th century)
: a garment of metal scales or chain mail worn as armor

coat·rack \'kōt-,rak\ *noun* (1915)
: a stand or rack fitted with pegs, hooks, or hangers and used for the temporary storage of garments

coat·room \-,rüm, -,ru̇m\ *noun* (1870)
: CLOAKROOM

coat·tail \'kōt-,tāl\ *noun* (circa 1600)
1 : the rear flap of a man's coat
2 *plural* **:** the skirts of a dress coat, cutaway, or frock coat
3 *plural* **:** the influence or pulling power of a popular movement or candidate ⟨congressmen riding into office on the president's *coattails*⟩

¹coax \'kōks\ *transitive verb* [earlier *cokes,* from *cokes* simpleton] (1589)
1 *obsolete* **:** FONDLE, PET
2 : to influence or gently urge by caressing or flattering **:** WHEEDLE
3 : to draw, gain, or persuade by means of gentle urging or flattery
4 : to manipulate with great perseverance and usually with considerable effort toward a desired state or activity ⟨*coax* a fire to burn⟩

²co·ax \'kō-,aks\ *noun* (1949)
: COAXIAL CABLE

co·ax·i·al \(,)kō-'ak-sē-əl\ *adjective* (1881)
1 : having coincident axes
2 : mounted on concentric shafts
— **co·ax·i·al·ly** \-sē-ə-lē\ *adverb*

coaxial cable *noun* (1936)
: a transmission line that consists of a tube of electrically conducting material surrounding a central conductor held in place by insulators and that is used to transmit telegraph, telephone, and television signals of high frequency — called also *coax cable*

cob \'käb\ *noun* [Middle English *cobbe* leader] (15th century)
1 : a male swan
2 *dialect English* **:** a rounded mass, lump, or heap
3 : a crudely struck old Spanish coin of irregular shape
4 : CORNCOB 1
5 : a stocky short-legged riding horse
— **cob·by** \'kä-bē\ *adjective*

co·bal·a·min \kō-'ba-lə-mən\ *noun* [*cobal*t + vit*amin*] (1956)
: VITAMIN B_{12}

co·balt \'kō-,bȯlt\ *noun* [German *Kobalt,* alteration of *Kobold,* literally, goblin, from Middle High German *kobolt;* from its occurrence in silver ore, believed to be due to goblins] (1683)
1 : a tough lustrous silver-white magnetic metallic element that is related to and occurs with iron and nickel and is used especially in alloys — see ELEMENT table
2 : COBALT BLUE 2 ◆

☆ SYNONYMS
Coarse, vulgar, gross, obscene, ribald mean offensive to good taste or morals. COARSE implies roughness, rudeness, or crudeness of spirit, behavior, or language ⟨found the *coarse* humor of coworkers offensive⟩. VULGAR often implies boorishness or ill-breeding ⟨a loud *vulgar* belch⟩. GROSS implies extreme coarseness and insensitiveness ⟨*gross* eating habits⟩. OBSCENE applies to anything strongly repulsive to the sense of decency and propriety especially in sexual matters ⟨*obscene* language not allowed on the air⟩. RIBALD applies to what is amusingly or picturesquely vulgar or irreverent or mildly indecent ⟨entertained the campers with *ribald* folk songs⟩.

◇ WORD HISTORY
cobalt The metallic element cobalt ultimately draws its name from folklore. In Middle High German, *kobolt* denoted a usually helpful household elf that engaged in nasty pranks only when it was offended. Later the more malicious side of the spirit came to the forefront, and early Modern German *Kobold* was extended in sense to cover a variety of goblins inhabiting fields and mountains. The variant *Kobolt* in the 16th century was applied by German miners to ores containing the metal cobalt, which they considered to be worthless. This application was based on the superstition that mountain goblins had spoiled adjacent silver ores, or had stolen the silver within the ore. The metal itself in relatively pure form was not produced and described until the 18th century, when German *Kobalt,* with its first letter influenced by New Latin *cobaltum,* became part of the international language of science.

cobalt blue *noun* (1835)
1 : a greenish blue pigment consisting essentially of cobalt oxide and alumina
2 : a strong greenish blue
cobalt chloride *noun* (1885)
: a chloride of cobalt; *especially* **:** the dichloride $CoCl_2$ that is blue when dehydrated, turns red in the presence of moisture, and is used to indicate humidity
co·bal·tic \kō-ˈbȯl-tik\ *adjective* (1782)
: of, relating to, or containing cobalt especially with a valence of three
co·balt·ite \ˈkō-ˌbȯl-ˌtīt, kō-ˈ\ *or* **co·balt·ine** \-ˌtēn\ *noun* [*cobaltite,* alteration of *cobaltine,* from French, from *cobalt*] (1868)
: a mineral consisting of a grayish to silver-white cobalt sulfur arsenide used in making smalt
co·bal·tous \kō-ˈbȯl-təs\ *adjective* (1863)
: of, relating to, or containing cobalt especially with a valence of two
cobalt 60 *noun* (1946)
: a heavy radioactive isotope of cobalt of the mass number 60 produced in nuclear reactors and used as a source of gamma rays (as for radiotherapy)
cob·ber \ˈkä-bər\ *noun* [origin unknown] (1893)
Australian & New Zealand **:** BUDDY
[1]**cob·ble** \ˈkä-bəl\ *transitive verb* **cob·bled; cob·bling** \-b(ə-)liŋ\ [Middle English *coblen,* perhaps back-formation from *cobelere* cobbler] (15th century)
1 *chiefly British* **:** to mend or patch coarsely
2 : REPAIR, MAKE ⟨*cobble* shoes⟩
3 : to make or put together roughly or hastily — often used with *together* or *up*
[2]**cobble** *noun* (1600)
1 : COBBLESTONE
2 *plural, chiefly British* **:** lump coal about the size of small cobblestones
cobbled *adjective* (1853)
: paved with cobblestones
cob·bler \ˈkä-blər\ *noun* [Middle English *cobelere*] (13th century)
1 : a mender or maker of shoes and often of other leather goods
2 *archaic* **:** a clumsy workman
3 : a tall iced drink consisting usually of wine, rum, or whiskey and sugar garnished with mint or a slice of lemon or orange
4 : a deep-dish fruit dessert with a thick top crust
cob·ble·stone \ˈkä-bəl-ˌstōn\ *noun* [Middle English, from *cobble-* (probably from *cob*) + *stone*] (15th century)
: a naturally rounded stone larger than a pebble and smaller than a boulder; *especially* **:** such a stone used in paving a street or in construction
— cob·ble·stoned \-ˌstōnd\ *adjective*
cob·by \ˈkä-bē\ *adjective* **cob·bi·er; -est** (1883)
: STOCKY ⟨a *cobby* horse⟩
co·bel·lig·er·ent \ˌkō-bə-ˈlij-rənt, -ˈli-jə-\ *noun* (1813)
: a country fighting with another power against a common enemy
— cobelligerent *adjective*
co·bia \ˈkō-bē-ə\ *noun* [origin unknown] (circa 1873)
: a large bony fish (*Rachycentron canadum* of the family Rachycentridae) of warm seas that is a food and sport fish
co·ble \ˈkō-bəl\ *noun* [Middle English] (14th century)
: a flat-bottomed boat propelled chiefly by oars and used in Scotland and northern England especially for fishing
cob·nut \ˈkäb-ˌnət\ *noun* (1580)
: the fruit of a European hazel (*Corylus avellana grandis*); *also* **:** the plant bearing this fruit
CO·BOL *or* **Co·bol** \ˈkō-ˌbȯl\ *noun* [*co*mmon *b*usiness *o*riented *l*anguage] (1960)
: a computer programming language designed for business applications
co·bra \ˈkō-brə\ *noun* [Portuguese *cobra* (*de capello*), literally, hooded snake, from Latin *colubra* snake] (1802)
: any of several venomous Asian and African elapid snakes (genera *Naja* and *Ophiophagus*) that when excited expand the skin of the neck into a hood by movement of the anterior ribs; *also* **:** any of several related African snakes
cob·web \ˈkäb-ˌweb\ *noun* [Middle English *coppeweb,* from *coppe* spider (from Old English ātor*coppe*) + *web;* akin to Middle Dutch *coppe* spider] (14th century)
1 a : the network spread by a spider **:** SPIDERWEB **b :** tangles of the silken threads of a cobweb usually covered with accumulated dirt and dust
2 : something that entangles, obscures, or confuses ⟨a *cobweb* of law and politics⟩
— cob·webbed \-ˌwebd\ *adjective*
— cob·web·by \-ˌwe-bē\ *adjective*
co·ca \ˈkō-kə\ *noun* [Spanish, from Quechua *kuka*] (1577)
1 : any of several South American shrubs (genus *Erythroxylon,* family Erythroxylaceae); *especially* **:** one (*E. coca*) that is the primary source of cocaine
2 : dried leaves of a coca (especially *Erythroxylon coca*) containing alkaloids including cocaine
co·caine \kō-ˈkān, ˈkō-ˌ\ *noun* (1874)
: a bitter crystalline alkaloid $C_{17}H_{21}NO_4$ obtained from coca leaves that is used medically as a topical anesthetic and illicitly for its euphoric effects and that may result in a compulsive psychological need
co·cain·ize \kō-ˈkā-ˌnīz\ *transitive verb* **-ized; -iz·ing** (1887)
: to treat or anesthetize with cocaine
— co·cain·i·za·tion \-ˌkā-nə-ˈzā-shən\ *noun*
co·car·box·yl·ase \ˌkō-kär-ˈbäk-sə-ˌlās, -ˌlāz\ *noun* (1932)
: a coenzyme $C_{12}H_{19}ClN_4O_7P_2S{\cdot}H_2O$ that is a pyrophosphate of thiamine and is important in metabolic reactions (as decarboxylation in the Krebs cycle)
co·car·cin·o·gen \ˌkō-kär-ˈsi-nə-jən, kō-ˈkär-sᵊn-ə-ˌjen\ *noun* (1938)
: an agent that aggravates the carcinogenic effects of another substance
— co·car·cin·o·gen·ic \ˌkō-ˌkär-sᵊn-ō-ˈje-nik\ *adjective*
coc·cid \ˈkäk-səd\ *noun* [New Latin *Coccus,* genus of scales, from Greek *kokkos* grain, kermes] (circa 1889)
: SCALE INSECT
coc·cid·i·oi·do·my·co·sis \(ˌ)käk-ˌsi-dē-ˌȯi-dō-(ˌ)mī-ˈkō-səs\ *noun* [New Latin, from *Coccidioides,* genus of fungi (from *coccidium*) + *mycosis*] (1937)
: a disease especially of humans and domestic animals caused by a fungus (*Coccidioides immitis*) and marked especially by fever and pulmonary symptoms
coc·cid·i·o·sis \(ˌ)käk-ˌsi-dē-ˈō-səs\ *noun, plural* **-o·ses** \-ˌsēz\ [New Latin] (1892)
: infestation with or disease caused by coccidia
coc·cid·i·um \käk-ˈsi-dē-əm\ *noun, plural* **-ia** \-dē-ə\ [New Latin, diminutive of *coccus*] (circa 1879)
: any of an order (Coccidia) of protozoans usually parasitic in the digestive epithelium of vertebrates
coc·coid \ˈkä-ˌkȯid\ *adjective* (1893)
: of, relating to, or resembling a coccus **:** GLOBULAR 1a(1)
— coccoid *noun*
coc·cus \ˈkä-kəs\ *noun, plural* **coc·ci** \ˈkä-ˌkī, -ˌkē; ˈkäk-ˌsī, -ˌsē\ [New Latin, from Greek *kokkos*] (1888)
: a spherical bacterium
— coc·cal \ˈkä-kəl\ *adjective*
-coccus *noun combining form, plural* **-cocci** [New Latin]
: berry-shaped microorganism ⟨micro*coccus*⟩
coc·cy·geal \käk-ˈsi-j(ē-)əl\ *adjective* [Medieval Latin *coccygeus* of the coccyx, from Greek *kokkyk-, kokkyx*] (1836)
: of or relating to the coccyx
coc·cyx \ˈkäk-siks\ *noun, plural* **coc·cy·ges** \ˈkäk-sə-ˌjēz\ *also* **coc·cyx·es** \ˈkäk-sik-səz\ [New Latin, from Greek *kokkyx* cuckoo, coccyx; from its resemblance to a cuckoo's beak] (1615)
: a small bone that articulates with the sacrum and that usually consists of four fused vertebrae which form the terminus of the spinal column in humans and tailless apes
Co·chin \ˈkō-chin\ *noun* [*Cochin* China, part of French Indochina] (1853)
: any of an Asian breed of large domestic fowl with thick plumage, small wings and tail, and densely feathered legs and feet — called also *Cochin China*
co·chi·neal \ˈkä-chə-ˌnēl, ˈkō-\ *noun* [Middle French & Spanish; Middle French *cochenille,* from Old Spanish *cochinilla* cochineal insect] (1582)
1 : a red dye consisting of the dried bodies of female cochineal insects
2 : COCHINEAL INSECT
cochineal insect *noun* (1801)
: a small red cactus-feeding scale insect (*Dactylopius coccus*) the females of which are the source of cochineal
co·chlea \ˈkō-klē-ə, ˈkä-klē-\ *noun, plural* **co·chle·as** *or* **co·chle·ae** \-klē-ˌē, -ˌī\ [New Latin, from Latin, snail, snail shell, from Greek *kochlias,* from *kochlos* snail; probably akin to Greek *konchē* mussel] (1688)
: a division of the bony labyrinth of the inner ear of higher vertebrates that is usually coiled like a snail shell and is the seat of the hearing organ — see EAR illustration
— co·chle·ar \ˈkō-klē-ər, ˈkä-\ *adjective*
[1]**cock** \ˈkäk\ *noun* [Middle English *cok,* from Old English *cocc,* of imitative origin] (before 12th century)
1 a : the adult male of the domestic fowl (*Gallus gallus*) **b :** the male of birds other than the domestic fowl **c :** WOODCOCK **d** *archaic* **:** the crowing of a cock; *also* **:** COCKCROW **e :** WEATHERCOCK
2 : a device (as a faucet or valve) for regulating the flow of a liquid
3 a : a chief person **:** LEADER **b :** a person of spirit and often of a certain swagger or arrogance
4 a : the hammer in the lock of a firearm **b :** the cocked position of the hammer
5 : PENIS — usually considered vulgar
— cock of the walk : one that dominates a group or situation especially overbearingly
[2]**cock** (1575)
intransitive verb
1 : STRUT, SWAGGER
2 : to turn, tip, or stick up
3 : to position the hammer of a firearm for firing
transitive verb
1 a : to draw the hammer of (a firearm) back and set for firing; *also* **:** to set (the trigger) for firing **b :** to draw or bend back in preparation for throwing or hitting **c :** to set a mechanism (as a camera shutter) for tripping
2 a : to set erect **b :** to turn, tip, or tilt usually to one side
3 : to turn up (as a hat brim)
— cock a snook *also* **cock snooks :** to thumb the nose
[3]**cock** *noun* (1717)
: TILT, SLANT ⟨*cock* of the head⟩
[4]**cock** *noun* [Middle English *cok,* of Scandinavian origin] (14th century)
: a small pile (as of hay)
[5]**cock** *transitive verb* (14th century)
: to put (as hay) into cocks
cock·ade \kä-ˈkād\ *noun* [modification of

French *cocarde,* from feminine of *cocard* vain, from *coq* cock, from Old French *coc,* of imitative origin] (1709)
: an ornament (as a rosette) usually worn on a hat as a badge
— **cock·ad·ed** \-ˈkā-dəd\ *adjective*

cock–a–hoop \ˌkä-kə-ˈhüp, -ˈhu̇p\ *adjective* [from the phrase *to set cock a hoop* to be festive] (1663)
1 : triumphantly boastful **:** EXULTING
2 : AWRY

Cock·aigne \kä-ˈkān\ *noun* [Middle English *cokaygne,* from Middle French (*pais de*) *cocaigne* land of plenty] (13th century)
: an imaginary land of great luxury and ease

cock–a–leek·ie \ˌkä-ki-ˈlē-kē\ *noun* [alteration of *cockie* (diminutive of [1]*cock*) + *leekie,* diminutive of *leek*] (1737)
: a soup made of chicken and leeks

cock·a·lo·rum \ˌkä-kə-ˈlōr-əm, -ˈlȯr-\ *noun, plural* **-rums** [probably modification of obsolete Flemish *kockeloeren* to crow, of imitative origin] (circa 1715)
1 : a boastful and self-important person
2 : LEAPFROG
3 : boastful talk

cock·a·ma·my *or* **cock·a·ma·mie** \ˌkä-kə-ˈmā-mē\ *adjective* [perhaps alteration of *decalcomania*] (1960)
: RIDICULOUS, INCREDIBLE ⟨of all the *cockamamy* excuses I ever heard —Leo Rosten⟩

cock–and–bull story \ˈkä-kən-ˈbu̇l-\ *noun* (1795)
: an incredible story told as true

cock·a·poo \ˈkä-kə-ˌpü\ *noun* [alteration of *cocker (spaniel)* + *poo*dle] (1970)
: a dog that is a cross between a cocker spaniel and a poodle

cock·a·tiel \ˈkäk-ə-ˌtē(ə)l\ *noun* [Dutch *kaketielje,* from Portuguese *cacatilha,* from *cacatua* cockatoo] (1877)
: a crested small gray Australian parrot (*Nymphicus hollandicus*) with a yellow head

cock·a·too \ˈkä-kə-ˌtü\ *noun, plural* **-toos** [Dutch *kaketoe,* from Malay *kakatua*] (1634)
: any of various large noisy chiefly Australasian crested parrots (family Cacatuidae and especially genus *Cacatua*)

cockatoo

cock·a·trice \ˈkä-kə-trəs, -ˌtrīs\ *noun* [Middle English *cocatrice,* from Middle French *cocatris* ichneumon, cockatrice, from Medieval Latin *cocatric-, cocatrix* ichneumon] (14th century)
: a legendary serpent that is hatched by a reptile from a cock's egg and that has a deadly glance

cock·boat \ˈkäk-ˌbōt\ *noun* (15th century)
: a small boat; *especially* **:** one used as a tender to a larger boat

cock·cha·fer \ˈkäk-ˌchā-fər\ *noun* [[1]*cock* + *chafer*] (1712)
: a large European beetle (*Melolontha melolontha*) destructive to vegetation as an adult and to roots as a larva; *also* **:** any of various related beetles

cock·crow \ˈkäk-ˌkrō\ *noun* (13th century)
: DAWN

cocked hat \ˈkäkt-\ *noun* (1673)
1 : a hat with brim turned up to give a 3-cornered appearance
2 : a hat with brim turned up on two sides and worn either front to back or sideways

[1]**cock·er** \ˈkä-kər\ *transitive verb* [Middle English *cokeren*] (15th century)
: INDULGE, PAMPER

[2]**cocker** *noun* (1689)
: a keeper or handler of fighting cocks

[3]**cocker** *noun* (circa 1811)
: COCKER SPANIEL

cock·er·el \ˈkä-k(ə-)rəl\ *noun* [Middle English *cokerelle,* from Middle French dialect *kokerel,* diminutive of Old French *coc*] (15th century)
: a young male of the domestic fowl (*Gallus gallus*)

cock·er spaniel \ˈkä-kər-\ *noun* [*cocking* woodcock hunting] (1840)
1 : ENGLISH COCKER SPANIEL
2 : any of a breed of spaniels developed in the U.S. from the English cocker spaniel that are smaller in size and have a shorter muzzle and longer thicker coat

cock·eye \ˈkäk-ˈī, -ˌī\ *noun* (circa 1825)
: a squinting eye

cock·eyed \ˈkäk-ˌīd\ *adjective* (1821)
1 : having a cockeye
2 a : ASKEW, AWRY **b :** slightly crazy **:** TOPSY-TURVY ⟨a *cockeyed* scheme⟩ **c :** DRUNK 1a
— **cock·eyed·ly** \ˈkäk-ˌī(-ə)d-lē\ *adverb*
— **cock·eyed·ness** \-ˌīd-nəs\ *noun*

cock·fight \ˈkäk-ˌfīt\ *noun* (circa 1566)
: a contest in which gamecocks usually fitted with metal spurs are pitted against each other
— **cock·fight·ing** \-ˌfī-tiŋ\ *adjective or noun*

cock·horse \ˈkäk-ˌhȯrs\ *noun* [perhaps from *cock,* adjective, (male) + *horse*] (circa 1541)
: ROCKING HORSE

[1]**cock·le** \ˈkä-kəl\ *noun* [Middle English, from Old English *coccel*] (before 12th century)
: any of several weedy plants of the pink family; *especially* **:** CORN COCKLE

[2]**cockle** *noun* [Middle English *cokille,* from Middle French *coquille* shell, modification of Latin *conchylia,* plural of *conchylium,* from Greek *konchylion,* from *konchē* conch] (14th century)
1 : any of various chiefly marine bivalve mollusks (family Cardiidae) having a shell with convex radially ribbed valves; *especially* **:** a common edible European bivalve (*Cerastoderma edule* synonym *Cardium edule*)
2 : COCKLESHELL

[3]**cockle** *noun* [Middle English *kokell,* ultimately from Middle French *coquillé* wavy or rounded like a shell, from *coquille*] (15th century)
: PUCKER, WRINKLE
— **cockle** *verb*

cock·le·bur \ˈkä-kəl-ˌbər, ˈkə-\ *noun* (1804)
: any of a genus (*Xanthium*) of prickly-fruited composite plants; *also* **:** one of its stiff-spined fruits

cock·le·shell \ˈkä-kəl-ˌshel\ *noun* (15th century)
1 a : the shell or one of the shell valves of a cockle **b :** a shell (as a scallop shell) suggesting a cockleshell
2 : a light flimsy boat

cockleshell 1a

cockles of the heart [perhaps from [2]*cockle*] (1671)
: the core of one's being — usually used in the phrase *warm the cockles of the heart*

cock·loft \ˈkäk-ˌlȯft\ *noun* [probably from [1]*cock*] (1589)
: a small garret

cock·ney \ˈkäk-nē\ *noun, plural* **cockneys** [Middle English *cokeney,* literally, cocks' egg, from *coken* (genitive plural of *cok* cock) + *ey* egg, from Old English *ǣg*] (14th century)
1 *obsolete* **a :** a spoiled child **b :** a squeamish woman
2 *often capitalized* **a :** a native of London and especially of the East End of London **b :** the dialect of London or of the East End of London
— **cockney** *adjective*
— **cock·ney·fy** \ˈkäk-ni-ˌfī\ *transitive verb*
— **cock·ney·ish** \-nē-ish\ *adjective*
— **cock·ney·ism** \-ˌi-zəm\ *noun*

cock·pit \ˈkäk-ˌpit\ *noun* (1587)
1 a : a pit or enclosure for cockfights **b :** a place noted for especially bloody, violent, or long-continued conflict
2 *obsolete* **:** the pit of a theater
3 : a compartment in a sailing warship used as quarters for junior officers and for treatment of the wounded in an engagement
4 : a space or compartment in a usually small vehicle (as a boat, airplane, or automobile) from which it is steered, piloted, or driven — see AIRPLANE illustration

cock·roach \ˈkäk-ˌrōch\ *noun* [by folk etymology from Spanish *cucaracha* cockroach, from *cuca* caterpillar] (1623)
: any of an order or suborder (Blattodea synonym Blattaria) of chiefly nocturnal insects including some that are domestic pests

cocks·comb \ˈkäks-ˌkōm\ *noun* (1542)
1 : COXCOMB
2 : a garden plant (*Celosia cristata*) of the amaranth family grown for its flowers

cocks·foot \-ˌfu̇t\ *noun* (1697)
: ORCHARD GRASS

cock·shut \ˈkäk-ˌshət\ *noun* [from the time poultry are shut in to rest] (1598)
dialect English **:** evening twilight

cock·shy \-ˌshī\ *noun, plural* **cockshies** [[1]*cock* + *shy,* noun] (1836)
1 a : a throw at an object set up as a mark **b :** a mark or target so set up
2 : an object or person taken as a butt (as of criticism)

cock·suck·er \-ˌsə-kər\ *noun* (circa 1891)
: one who performs fellatio — usually considered obscene; often used as a generalized term of abuse

cock·sure \ˈkäk-ˈshu̇r\ *adjective* [probably from [1]*cock* + *sure*] (1672)
1 : feeling perfect assurance sometimes on inadequate grounds
2 : marked by overconfidence or presumptuousness **:** COCKY
synonym see SURE
— **cock·sure·ly** *adverb*
— **cock·sure·ness** *noun*

[1]**cock·tail** \ˈkäk-ˌtal\ *noun* [probably from [1]*cock* + *tail*] (1806)
1 a : an iced drink of wine or distilled liquor mixed with flavoring ingredients **b :** something resembling or suggesting such a drink **c :** a solution of agents taken or used together especially for medical treatment or diagnosis
2 : an appetizer served as a first course at a meal

[2]**cocktail** *adjective* (1865)
1 : of, relating to, or set aside for cocktails ⟨a *cocktail* hour⟩
2 : designed for semiformal wear ⟨*cocktail* dress⟩

[3]**cocktail** *noun* [[1]*cock* + *tail*] (1808)
: a horse with its tail docked

cocktail glass *noun* (1907)
: a bell-shaped drinking glass usually having a foot and stem and holding about three ounces (90 milliliters)

cocktail lounge *noun* (1939)
: a public room (as in a hotel, club, or restaurant) where cocktails and other drinks are served

cocktail party *noun* (1928)
: an informal or semiformal party or gathering at which cocktails are served

cocktail table *noun* (1946)
: COFFEE TABLE

cock–up \ˈkä-ˌkəp\ *noun* (circa 1948)
British **:** MESS 3b

cocky \ˈkä-kē\ *adjective* **cock·i·er; -est** (1768)

1 **:** boldly or brashly self-confident
2 **:** JAUNTY
— **cock·i·ly** \'kä-kə-lē\ *adverb*
— **cock·i·ness** \'kä-kē-nəs\ *noun*

co·co \'kō-(,)kō\ *noun, plural* **cocos** [Spanish *coco* & Portuguese *côco* bogeyman, grimace, coconut] (1555)
: the coconut palm; *also* **:** its fruit

co·coa \'kō-(,)kō\ *noun* [modification of Spanish *cacao*] (1788)
1 a : powdered ground roasted cacao beans from which a portion of the fat has been removed **b :** a beverage prepared by heating cocoa with water or milk
2 : CACAO 2
3 : a medium brown color

cocoa bean *noun* (1855)
: CACAO 1

cocoa butter *noun* (1872)
: a pale vegetable fat with a low melting point obtained from cacao beans

co·co·nut \'kō-kə-(,)nət\ *noun* (1613)
1 : the drupaceous fruit of the coconut palm whose outer fibrous husk yields coir and whose nut contains thick edible meat and coconut milk
2 : the edible meat of the coconut

coconut crab *noun* (circa 1889)
: a large edible coconut-eating burrowing land crab (*Birgus latro*) widely distributed about islands of the tropical Indian and Pacific Oceans

coconut oil *noun* (1838)
: a fatty oil or semisolid fat extracted from fresh coconuts and used especially in making soaps and food products

coconut palm *noun* (1852)
: a tall pinnate-leaved coconut-bearing palm (*Cocos nucifera*) that grows along tropical coasts

¹co·coon \kə-'kün\ *noun* [French *cocon,* from Provençal *coucoun,* from *coco* shell, probably from Latin *coccum* kermes (thought to be a gall or berry), from Greek *kokkos* berry, kermes] (1699)
1 a : an envelope often largely of silk which an insect larva forms about itself and in which it passes the pupa stage **b :** any of various other protective coverings produced by animals
2 a : a covering suggesting a cocoon **b :** a protective covering placed or sprayed over military or naval equipment in storage

²cocoon *transitive verb* (1881)
: to wrap or envelop in or as if in a cocoon

co·coon·ing \-'kü-niŋ\ *noun* (1987)
: the practice of spending leisure time at home in preference to going out

co·cotte \kò-'kòt\ *noun, plural* **cocottes** \-'kòt(s)\ [French] (1867)
1 : PROSTITUTE
2 : a shallow individual baking dish usually with one or two handles

co·co·yam \'kō-(,)kō-yam\ *noun* [probably *coco* + *yam*] (1922)
1 : TARO
2 : YAUTIA

co·cur·ric·u·lar \,kō-kə-'ri-kyə-lər\ *adjective* (1949)
: being outside of but usually complementing the regular curriculum

cod \'käd\ *noun, plural* **cod** *also* **cods** [Middle English] (14th century)
1 : any of various bottom-dwelling fishes (family Gadidae, the cod family) that usually occur in cold marine waters and often have barbels and three dorsal fins: as **a :** one (*Gadus morhua*) of the North Atlantic that is an important food fish **b :** one (*Gadus macrocephalus*) of the Pacific Ocean
2 : any of various bony fishes resembling the true cods

co·da \'kō-də\ *noun* [Italian, literally, tail, from Latin *cauda*] (circa 1753)
1 a : a concluding musical section that is formally distinct from the main structure **b :** a concluding part of a literary or dramatic work
2 : something that serves to round out, conclude, or summarize yet has its own interest

cod·dle \'kä-dᵊl\ *transitive verb* **cod·dled; cod·dling** \'käd-liŋ, 'kä-dᵊl-iŋ\ [perhaps from *caudle*] (1598)
1 : to cook (as eggs) in liquid slowly and gently just below the boiling point
2 : to treat with extreme care **:** PAMPER
— **cod·dler** \'käd-lər, 'kä-dᵊl-ər\ *noun*

¹code \'kōd\ *noun* [Middle English, from Middle French, from Latin *caudex, codex* trunk of a tree, document formed originally from wooden tablets] (14th century)
1 : a systematic statement of a body of law; *especially* **:** one given statutory force
2 : a system of principles or rules ⟨moral *code*⟩
3 a : a system of signals or symbols for communication **b :** a system of symbols (as letters or numbers) used to represent assigned and often secret meanings
4 : GENETIC CODE
5 : a set of instructions for a computer
— **code·less** \-ləs\ *adjective*

²code *verb* **cod·ed; cod·ing** (1815)
transitive verb
: to put in or into the form or symbols of a code
intransitive verb
: to specify the genetic code ⟨a gene that *codes* for a protein⟩
— **cod·able** \'kō-də-bəl\ *adjective*
— **cod·er** *noun*

code·book *noun* (1884)
: a book containing an alphabetical list of words or expressions with their code equivalents

co·deine \'kō-,dēn\ *noun* [French *codéine,* from Greek *kōdeia* poppyhead] (1881)
: a morphine derivative $C_{18}H_{21}NO_3 \cdot H_2O$ that is found in opium, is weaker in action than morphine, and is used especially in cough remedies

code name *noun* (1919)
: a designation having a secret meaning
— **code–name** \'kōd-,nām\ *transitive verb*

co·de·pen·dence \,kō-di-'pen-dən(t)s\ *noun* (1989)
: CODEPENDENCY

co·de·pen·den·cy \-dən(t)-sē\ *noun* (1987)
: a psychological condition or a relationship in which a person is controlled or manipulated by another who is affected with a pathological condition (as an addiction to alcohol or heroin)
— **co·de·pen·dent** \-dənt\ *noun or adjective*

co·de·ter·mi·na·tion \,kō-di-,tər-mə-'nā-shən\ *noun* (1949)
: the participation of labor with management in determining business policy

code word *noun* (1884)
1 : CODE NAME
2 : EUPHEMISM

co·dex \'kō-,deks\ *noun, plural* **co·di·ces** \'kō-də-,sēz, 'kä-\ [Latin] (1670)
: a manuscript book especially of Scripture, classics, or ancient annals

cod·fish \'käd-,fish\ *noun* (14th century)
: COD; *also* **:** its flesh used as food

cod·ger \'kä-jər\ *noun* [probably alteration of *cadger*] (1756)
: an often mildly eccentric and usually elderly fellow ⟨old *codger*⟩

cod·i·cil \'kä-də-səl, -,sil\ *noun* [Middle French *codicille,* from Latin *codicillus,* diminutive of *codic-, codex*] (15th century)
1 : a legal instrument made to modify an earlier will
2 : APPENDIX, SUPPLEMENT
— **cod·i·cil·la·ry** \,kä-də-'si-lə-rē\ *adjective*

co·di·col·o·gy \,kō-də-'kä-lə-jē, ,kä-\ *noun* [Latin *codic-, codex* + *-o-* + English *-logy*] (1953)
: the study of manuscripts as cultural artifacts for historical purposes
— **co·di·co·log·i·cal** \-kə-'lä-ji-kəl\ *adjective*

cod·i·fy \'kä-də-,fī, 'kō-\ *transitive verb* **-fied; -fy·ing** (circa 1800)
1 : to reduce to a code
2 a : SYSTEMATIZE **b :** CLASSIFY
— **cod·i·fi·abil·i·ty** \,kä-də-,fī-ə-'bi-lə-tē, ,kō-\ *noun*
— **cod·i·fi·ca·tion** \-fə-'kā-shən\ *noun*

¹cod·ling \'käd-liŋ\ *noun* (13th century)
1 : a young cod
2 : any of several hakes (especially genus *Urophycis*)

²cod·ling \'käd-liŋ\ *or* **cod·lin** \-lən\ *noun* [alteration of Middle English *querdlyng*] (15th century)
: a small immature apple; *also* **:** any of several elongated greenish English cooking apples

codling moth *noun* (1747)
: a small tortricid moth (*Cydia pomonella*) having larvae that live in apples, pears, quinces, and English walnuts

cod–liver oil *noun* (1783)
: an oil obtained from the liver of the cod and closely related fishes and used as a source of vitamins A and D

co·dom·i·nant \,kō-'dä-mə-nənt, -'däm-nənt\ *adjective* (circa 1900)
1 a : forming part of the main canopy of a forest ⟨*codominant* trees⟩ **b :** sharing in the controlling influence of a biotic community
2 : being fully expressed in the heterozygous condition ⟨*codominant* alleles⟩
— **codominant** *noun*

co·don \'kō-,dän\ *noun* [¹*code* + ²*-on*] (1963)
: a specific sequence of three consecutive nucleotides that is part of the genetic code and that specifies a particular amino acid in a protein or starts or stops protein synthesis

cod·piece \'käd-,pēs\ *noun* [Middle English *codpese,* from *cod* bag, scrotum (from Old English *codd*) + *pese* piece] (15th century)
: a flap or bag concealing an opening in the front of men's breeches especially in the 15th and 16th centuries

cods·wal·lop \'kòdz-,wä-ləp, 'kädz-\ *noun* [origin unknown] (1963)
British **:** NONSENSE

¹co·ed \'kō-(,)ed\ *adjective* (1889)
1 : of or relating to a coed
2 : of or relating to coeducation
3 : open to or used by both men and women

²coed *noun* [short for *coeducational student*] (circa 1893)
: a student and especially a female student in a coeducational institution

co–edi·tion \,kō-ə-'di-shən\ *noun* (1964)
: an edition of a book published simultaneously by more than one publisher usually in different countries and in different languages

co·ed·u·ca·tion \(,)kō-,e-jə-'kā-shən\ *noun* (1852)
: the education of students of both sexes at the same institution
— **co·ed·u·ca·tion·al** \-shnəl, -shə-nᵊl\ *adjective*
— **co·ed·u·ca·tion·al·ly** *adverb*

co·ef·fi·cient \,kō-ə-'fi-shənt\ *noun* [New Latin *coefficient-, coefficiens,* from Latin *co-* + *efficient-, efficiens* efficient] (circa 1715)
1 : any of the factors of a product considered in relation to a specific factor; *especially* **:** a constant factor of a term as distinguished from a variable
2 a : a number that serves as a measure of some property or characteristic (as of a substance, device, or process) ⟨*coefficient* of expansion of a metal⟩ **b :** MEASURE

coefficient of correlation (1896)
: CORRELATION COEFFICIENT

coefficient of viscosity (1866)
: VISCOSITY 3

coe·la·canth \'sē-lə-,kan(t)th\ *noun* [ultimately from Greek *koilos* hollow + *akantha* spine — more at CAVE] (1857)

: any of an order (Coelacanthiformes) of lobe-finned fishes known chiefly from Paleozoic and Mesozoic fossils

coelacanth

-coele *or* **-coel** *noun combining form* [probably from New Latin *-coela,* from neuter plural of *-coelus* hollow, concave, from Greek *-koilos,* from *koilos*]
: cavity **:** chamber **:** ventricle ⟨blasto*coel*⟩ ⟨entero*coele*⟩

coe·len·ter·ate \si-ˈlen-tə-ˌrāt, -rət\ *noun* [ultimately from Greek *koilos* + *enteron* intestine — more at INTER-] (1872)
: any of a phylum (Cnidaria synonym Coelenterata) of radially symmetrical invertebrate animals including the corals, sea anemones, jellyfishes, and hydroids — called also *cnidarian*
— **coelenterate** *adjective*

coe·len·ter·on \-ˌrän, -rən\ *noun, plural* **-tera** \-rə\ [New Latin, from Greek *koilos* + *enteron*] (1893)
: the internal cavity of a coelenterate

coe·li·ac \ˈsē-lē-ˌak\ *variant of* CELIAC

coe·lom \ˈsē-ləm\ *noun, plural* **coeloms** *or* **coe·lo·ma·ta** \si-ˈlō-mə-tə\ [German, from Greek *koilōma* cavity, from *koilos*] (1878)
: the usually epithelium-lined space between the body wall and the digestive tract of metazoans above the lower worms
— **coe·lo·mate** \ˈsē-lə-ˌmāt\ *adjective or noun*
— **coe·lo·mic** \si-ˈlä-mik, -ˈlō-\ *adjective*

coen- *or* **coeno-** *combining form* [New Latin, from Greek *koin-, koino-,* from *koinos*]
: common **:** general ⟨*coeno*cyte⟩

coe·no·bite \ˈsē-nə-ˌbīt\ *variant of* CENOBITE

coe·no·cyte \ˈsē-nə-ˌsīt\ *noun* [International Scientific Vocabulary] (1897)
1 a : a multinucleate mass of protoplasm resulting from repeated nuclear division unaccompanied by cell fission **b :** an organism consisting of such a structure
2 : SYNCYTIUM 1
— **coe·no·cyt·ic** \ˌsē-nə-ˈsi-tik\ *adjective*

coe·nu·rus \si-ˈnu̇r-əs, -ˈnyu̇r-\ *noun, plural* **-nu·ri** \-ˌī\ [New Latin, from *coen-* + Greek *oura* tail] (1876)
: a complex tapeworm larva consisting of a sac from the inner wall of which numerous scolices develop

co·en·zyme \(ˌ)kō-ˈen-ˌzīm\ *noun* (1908)
: a thermostable nonprotein compound that forms the active portion of an enzyme system after combination with an apoenzyme
— **co·en·zy·mat·ic** \(ˌ)kō-ˌen-zə-ˈma-tik, -(ˌ)zī-\ *adjective*
— **co·en·zy·mat·i·cal·ly** \-ti-k(ə-)lē\ *adverb*

coenzyme A *noun* (1949)
: a coenzyme $C_{21}H_{36}N_7O_{16}P_3S$ that occurs in all living cells and is essential to the metabolism of carbohydrates, fats, and some amino acids — compare ACETYL COENZYME A

coenzyme Q *noun* (1958)
: UBIQUINONE

co·equal \(ˌ)kō-ˈē-kwəl\ *adjective* (14th century)
: equal with one another
— **coequal** *noun*
— **co·equal·i·ty** \ˌkō-ē-ˈkwä-lə-tē\ *noun*
— **co·equal·ly** \(ˌ)kō-ˈē-kwə-lē\ *adverb*

co·erce \kō-ˈərs\ *transitive verb* **co·erced; co·erc·ing** [Latin *coercēre,* from *co-* + *arcēre* to shut up, enclose — more at ARK] (15th century)
1 : to restrain or dominate by force ⟨religion in the past has tried to *coerce* the irreligious —W. R. Inge⟩
2 : to compel to an act or choice
3 : to bring about by force or threat ⟨*coerce* the compliance of the rest of the community —Scott Buchanan⟩
synonym see FORCE
— **co·erc·ible** \-ˈər-sə-bəl\ *adjective*

co·er·cion \-ˈər-zhən, -shən\ *noun* (15th century)
: the act, process, or power of coercing

co·er·cive \-ˈər-siv\ *adjective* (circa 1600)
: serving or intended to coerce
— **co·er·cive·ly** *adverb*
— **co·er·cive·ness** *noun*

coercive force *noun* (1827)
: the opposing magnetic intensity that must be applied to a magnetized material to remove the residual magnetism

co·er·civ·i·ty \ˌkō-ˌər-ˈsi-və-tē\ *noun* (1898)
: the property of a material determined by the value of the coercive force when the material has been magnetized to saturation

co·e·ta·ne·ous \ˌkō-ə-ˈtā-nē-əs\ *adjective* [Latin *coaetaneus,* from *co-* + *aetas* age — more at AGE] (1608)
: COEVAL

co·e·val \kō-ˈē-vəl\ *adjective* [Latin *coaevus,* from *co-* + *aevum* age, lifetime — more at AYE] (circa 1662)
: of the same or equal age, antiquity, or duration
synonym see CONTEMPORARY
— **coeval** *noun*
— **co·e·val·i·ty** \ˌkō-(ˌ)ē-ˈva-lə-tē\ *noun*

co·evo·lu·tion \ˌkō-ˌe-və-ˈlü-shən *also* -ˌē-və-\ *noun* (1965)
: evolution involving successive changes in two or more ecologically interdependent species (as of a plant and its pollinators) that affect their interactions
— **co·evo·lu·tion·ary** \-shə-ˌner-ē\ *adjective*
— **co·evolve** \ˌkō-i-ˈvälv, -ˈvȯlv\ *intransitive verb*

co·ex·ist \ˌkō-ig-ˈzist\ *intransitive verb* (1667)
1 : to exist together or at the same time
2 : to live in peace with each other especially as a matter of policy
— **co·ex·is·tence** \-ˈzis-tən(t)s\ *noun*
— **co·ex·is·tent** \-tənt\ *adjective*

co·ex·ten·sive \ˌkō-ik-ˈsten(t)-siv\ *adjective* (1771)
: having the same spatial or temporal scope or boundaries
— **co·ex·ten·sive·ly** *adverb*

co·fac·tor \ˈkō-ˌfak-tər\ *noun* (circa 1909)
1 : the signed minor of an element of a square matrix or of a determinant with the sign positive if the sum of the column number and row number of the element is even and with the sign negative if it is odd
2 : a substance that acts with another substance to bring about certain effects; *especially* **:** COENZYME

cof·fee \ˈkȯ-fē, ˈkä-\ *noun, often attributive* [Italian & Turkish; Italian *caffè,* from Turkish *kahve,* from Arabic *qahwa*] (1598)
1 a : a beverage made by percolation, infusion, or decoction from the roasted and ground seeds of a coffee plant **b :** any of several Old World tropical plants (genus *Coffea* and especially *C. arabica* and *C. canephora*) of the madder family that are widely cultivated in warm regions for their seeds from which coffee is prepared **c :** coffee seeds especially roasted and often ground — compare ARABICA, ROBUSTA **d :** a dehydrated product made from brewed coffee ⟨instant *coffee*⟩; *also* **:** a beverage made from this
2 : a cup of coffee ⟨two *coffees*⟩
3 : COFFEE HOUR

coffee 1b

coffee break *noun* (1951)
: a short period for rest and refreshments

coffee cake *noun* (1879)
: a sweet rich bread often with added fruit, nuts, and spices that is sometimes glazed after baking

coffee hour *noun* (1952)
1 : a usually fixed occasion of informal meeting and chatting at which refreshments are served
2 : COFFEE BREAK

cof·fee·house \-ˌhau̇s\ *noun* (1612)
: an establishment that sells coffee and usually other refreshments and that commonly serves as an informal club for its regular customers

coffee klatch *also* **cof·fee–klatsch** \-ˌklach, -ˌkläch, -ˌkləch\ *noun* [part translation of German *Kaffeeklatsch*] (1895)
: KAFFEEKLATSCH

cof·fee·mak·er \-ˌmā-kər\ *noun* (1930)
: a utensil or appliance in which coffee is brewed

coffee mill *noun* (1691)
: a mill for grinding coffee beans

cof·fee·pot \-ˌpät\ *noun* (1704)
: a pot for brewing and serving coffee

coffee ring *noun* (1924)
: coffee cake in the shape of a ring

coffee roll *noun* (1945)
: a sweet roll

coffee room *noun* (1712)
: a room where refreshments are served

coffee royal *noun* (1921)
: a drink of black coffee and a liquor

coffee shop *noun* (1836)
: a small restaurant

coffee–table *adjective* (1962)
: of, relating to, or being an article (as a book or magazine) intended for display (as on a coffee table)

coffee table *noun* (1877)
: a low table customarily placed in front of a sofa — called also *cocktail table*

coffee tree *noun* (1741)
1 : a tree (as arabica) that produces coffee
2 : KENTUCKY COFFEE TREE

¹cof·fer \ˈkȯ-fər, ˈkä-\ *noun* [Middle English *coffre,* from Old French, from Latin *cophinus* basket, from Greek *kophinos*] (13th century)
1 : CHEST, BOX; *especially* **:** STRONGBOX
2 : TREASURY — usually used in plural
3 : a recessed panel in a vault, ceiling, or soffit

²coffer *transitive verb* (14th century)
1 : to store or hoard up in a coffer
2 : to form (as a ceiling) with recessed panels

cof·fer·dam \-ˌdam\ *noun* (1736)
1 : a watertight enclosure from which water is pumped to expose the bottom of a body of water and permit construction (as of a pier)
2 : a watertight structure for making repairs below the waterline of a ship

¹cof·fin \ˈkȯ-fən\ *noun* [Middle English, basket, receptacle, from Middle French *cofin,* from Latin *cophinus*] (1525)
: a box or chest for burying a corpse

²coffin *transitive verb* (1564)
: to enclose in or as if in a coffin

coffin bone *noun* (circa 1720)
: the principal bone enclosed within the hoof of a horse — called also *pedal bone*

coffin corner *noun* (1940)
: one of the corners formed by a goal line and a sideline on a football field into which a punt is often aimed so that it may go out of bounds close to the defender's goal line

coffin nail *noun* (1888)
slang **:** CIGARETTE

cof·fle \ˈkȯ-fəl, ˈkä-\ *noun* [Arabic *qāfila* caravan] (1799)

\ə\ **abut** \ᵊ\ **kitten** \ər\ **further** \a\ **ash** \ā\ **ace** \ä\ **mop, mar** \au̇\ **out** \ch\ **chin** \e\ **bet** \ē\ **easy** \g\ **go** \i\ **hit** \ī\ **ice** \j\ **job** \ŋ\ **sing** \ō\ **go** \ȯ\ **law** \ȯi\ **boy** \th\ **thin** \t͟h\ **the** \ü\ **loot** \u̇\ **foot** \y\ **yet** \zh\ **vision** *see also* Guide to Pronunciation

: a train of slaves or animals fastened together

co·func·tion \'kō-ˌfəŋ(k)-shən\ *noun* (1909)
: a trigonometric function whose value for the complement of an angle is equal to the value of a given trigonometric function of the angle itself ⟨the sine is the *cofunction* of the cosine⟩

¹cog \'käg\ *noun* [Middle English *cogge,* of Scandinavian origin; akin to Norwegian *kug* cog] (13th century)
1 : a tooth on the rim of a wheel or gear
2 : a subordinate but vital person or part
— cogged \'kägd\ *adjective*

²cog *verb* **cogged; cog·ging** [obsolete *cog* a trick] (1532)
intransitive verb
1 *obsolete* **:** to cheat in throwing dice
2 *obsolete* **:** DECEIVE
3 *obsolete* **:** to use venal flattery
transitive verb
1 : to direct the fall of (dice) fraudulently
2 *obsolete* **:** WHEEDLE

³cog *transitive verb* **cogged; cog·ging** [probably alteration of *cock* to cog] (1823)
: to connect (as timbers or joists) by means of mortises and tenons

⁴cog *noun* (circa 1858)
: a tenon on a beam or timber

co·gen·cy \'kō-jən(t)-sē\ *noun* (1667)
: the quality or state of being cogent

co·gen·er·a·tion \ˌkō-ˌje-nə-'rā-shən\ *noun* (1978)
: the production of electricity using waste heat (as in steam) from an industrial process or the use of steam from electric power generation as a source of heat
— co·gen·er·a·tor \-'je-nə-ˌrā-tər\ *noun*

co·gent \'kō-jənt\ *adjective* [Latin *cogent-, cogens,* present participle of *cogere* to drive together, collect, from *co-* + *agere* to drive — more at AGENT] (1659)
1 : having power to compel or constrain ⟨*cogent* forces⟩
2 a : appealing forcibly to the mind or reason **:** CONVINCING ⟨*cogent* evidence⟩ **b :** PERTINENT, RELEVANT ⟨a *cogent* analysis of a problem⟩
synonym see VALID
— co·gent·ly *adverb*

cog·i·ta·ble \'kä-jə-tə-bəl\ *adjective* (15th century)
: CONCEIVABLE, THINKABLE

cog·i·tate \'kä-jə-ˌtāt\ *verb* **-tat·ed; -tat·ing** [Latin *cogitatus,* past participle of *cogitare* to think, think about, from *co-* + *agitare* to drive, agitate] (circa 1587)
transitive verb
: to ponder or meditate on usually intently
intransitive verb
: to meditate deeply or intently
synonym see THINK

cog·i·ta·tion \ˌkä-jə-'tā-shən\ *noun* (13th century)
1 a : the act of cogitating **:** MEDITATION **b :** the capacity to think or reflect
2 : a single thought

cog·i·ta·tive \'kä-jə-ˌtā-tiv\ *adjective* (15th century)
1 : of or relating to cogitation
2 : capable of or given to cogitation

co·gi·to \'kä-gi-ˌtō, 'kō-, 'kä-ji-\ *noun* [New Latin *cogito, ergo sum,* literally, I think, therefore I am, principle stated by René Descartes] (1838)
1 : the philosophic principle that one's existence is demonstrated by the fact that one thinks
2 : the intellectual processes of the self or ego

co·gnac \'kōn-ˌyak *also* 'kȯn- *or* 'kän-\ *noun* [French, from *Cognac,* France] (1755)
: a brandy from the departments of Charente and Charente-Maritime distilled from white wine

¹cog·nate \'käg-ˌnāt\ *adjective* [Latin *cognatus,* from *co-* + *gnatus, natus,* past participle of *nasci* to be born; akin to Latin *gignere* to beget — more at KIN] (circa 1645)
1 : of the same or similar nature **:** generically alike
2 : related by blood; *also* **:** related on the mother's side
3 a : related by descent from the same ancestral language **b** *of a word or morpheme* **:** related by derivation, borrowing, or descent **c** *of a substantive* **:** related to a verb usually by derivation and serving as its object to reinforce the meaning
— cog·nate·ly *adverb*

²cognate *noun* (1754)
: one that is cognate with another

cog·na·tion \käg-'nā-shən\ *noun* (14th century)
: cognate relationship

cog·ni·tion \käg-'ni-shən\ *noun* [Middle English *cognicioun,* from Latin *cognition-, cognitio,* from *cognoscere* to become acquainted with, know, from *co-* + *gnoscere* to come to know — more at KNOW] (15th century)
: the act or process of knowing including both awareness and judgment; *also* **:** a product of this act
— cog·ni·tion·al \-'nish-nəl, -'ni-shə-nᵊl\ *adjective*

cog·ni·tive \'käg-nə-tiv\ *adjective* (1586)
1 : of, relating to, or involving cognition ⟨the *cognitive* elements of perception —C. H. Hamburg⟩
2 : based on or capable of being reduced to empirical factual knowledge
— cog·ni·tive·ly *adverb*

cognitive dissonance *noun* (1957)
: psychological conflict resulting from incongruous beliefs and attitudes held simultaneously

cognitive science *noun* (1976)
: an interdisciplinary science that draws on many fields (as psychology, artificial intelligence, linguistics, and philosophy) in developing theories about human perception, thinking, and learning

cog·ni·za·ble \'käg-nə-zə-bəl, käg-'nī-\ *adjective* (1678)
1 : capable of being known
2 : capable of being judicially heard and determined
— cog·ni·za·bly \-blē\ *adverb*

cog·ni·zance \'käg-nə-zən(t)s\ *noun* [Middle English *conisaunce,* from Middle French *conoissance,* from *conoistre* to know, from Latin *cognoscere*] (14th century)
1 : a distinguishing mark or emblem (as a heraldic bearing)
2 a : KNOWLEDGE, AWARENESS ⟨had no *cognizance* of the situation⟩ **b :** NOTICE, ACKNOWLEDGMENT ⟨take *cognizance* of their achievement⟩
3 : JURISDICTION, RESPONSIBILITY

cog·ni·zant \-zənt\ *adjective* (1820)
: knowledgeable of something especially through personal experience; *also* **:** MINDFUL
synonym see AWARE

cog·nize \käg-'nīz, 'käg-ˌ\ *transitive verb* **cog·nized; cog·niz·ing** [back-formation from *cognizance*] (circa 1837)
: KNOW, UNDERSTAND
— cog·niz·er *noun*

cog·no·men \käg-'nō-mən, 'käg-nə-\ *noun, plural* **cognomens** *or* **cog·no·mi·na** \käg-'nä-mə-nə, -'nō-\ [Latin, irregular from *co-* + *nomen* name — more at NAME] (1809)
1 : SURNAME; *especially* **:** the third of usually three names borne by a male citizen of ancient Rome
2 : NAME; *especially* **:** a distinguishing nickname or epithet
— cog·nom·i·nal \käg-'nä-mə-nᵊl\ *adjective*

co·gno·scen·te \ˌkän-yə-'shen-tē, ˌkäg-nə-, -'sen-\ *noun, plural* **-scen·ti** \-tē\ [obsolete Italian (now *conoscente*), from *cognoscente,* adjective, wise, from Latin *cognoscent-, cognoscens,* present participle of *cognoscere*] (1776)
: a person who is especially knowledgeable in a subject **:** CONNOISSEUR

cog·nos·ci·ble \käg-'nä-sə-bəl\ *adjective* [Late Latin *cognoscibilis,* from Latin *cognoscere*] (circa 1644)
: COGNIZABLE, KNOWABLE

co·gon \kō-'gōn\ *noun* [Spanish *cogón,* from Tagalog & Bisayan *kugon*] (1898)
: any of several coarse tall grasses (genus *Imperata*) used especially in the Philippines for thatching

cog railway *noun* (1896)
: a steep mountain railroad that has a rail with cogs engaged by a cogwheel on the locomotive to ensure traction

cog·wheel \'käg-ˌhwēl, -ˌwēl\ *noun* (14th century)
: a wheel with cogs **:** GEAR 6a(2)

co·hab·it \(ˌ)kō-'ha-bət\ *intransitive verb* [Late Latin *cohabitare,* from Latin *co-* + *habitare* to inhabit, from frequentative of *habēre* to have — more at GIVE] (circa 1530)
1 : to live together as or as if a married couple
2 a : to live together or in company ⟨buffaloes *cohabiting* with crossbred cows —*Biological Abstracts*⟩ **b :** to exist together ⟨two strains in his philosophy . . . *cohabit* in each of his major works —Justus Buchler⟩
— co·hab·i·tant \-bə-tənt\ *noun*
— co·hab·i·ta·tion \(ˌ)kō-ˌha-bə-'tā-shən\ *noun*

co·here \kō-'hir\ *verb* **co·hered; co·her·ing** [Latin *cohaerēre,* from *co-* + *haerēre* to stick] (1598)
intransitive verb
1 a : to hold together firmly as parts of the same mass; *broadly* **:** STICK, ADHERE **b :** to display cohesion of plant parts
2 : to hold together as a mass of parts that cohere
3 a : to become united in principles, relationships, or interests **b :** to be logically or aesthetically consistent
transitive verb
: to cause (parts or components) to cohere
synonym see STICK

co·her·ence \kō-'hir-ən(t)s, -'her-\ *noun* (1580)
1 : the quality or state of cohering: as **a :** systematic or logical connection or consistency **b :** integration of diverse elements, relationships, or values
2 : the property of being coherent

co·her·en·cy \kō-'hir-ən(t)-sē, -'her-\ *noun, plural* **-cies** (1630)
: COHERENCE

co·her·ent \-ənt\ *adjective* [Middle French or Latin; Middle French *cohérent,* from Latin *cohaerent-, cohaerens,* present participle of *cohaerēre*] (circa 1555)
1 a : logically or aesthetically ordered or integrated **:** CONSISTENT ⟨*coherent* style⟩ ⟨a *coherent* argument⟩ **b :** having clarity or intelligibility **:** UNDERSTANDABLE ⟨a *coherent* person⟩ ⟨a *coherent* passage⟩
2 : having the quality of cohering; *especially* **:** COHESIVE, COORDINATED ⟨a *coherent* plan for action⟩
3 a : relating to or composed of waves having a constant difference in phase ⟨*coherent* light⟩ **b :** producing coherent light ⟨a *coherent* source⟩
— co·her·ent·ly *adverb*

co·her·er \kō-'hir-ər\ *noun* (1894)
: a radio detector in which an imperfectly conducting contact between pieces of conductive material loosely resting against each other is materially improved in conductance by the passage of high-frequency current

co·he·sion \kō-'hē-zhən\ *noun* [Latin *cohaesus,* past participle of *cohaerēre*] (1660)
1 : the act or state of sticking together tightly; *especially* **:** UNITY ⟨the lack of *cohesion* in the Party —*Times Literary Supplement*⟩
2 : union between similar plant parts or organs

3 : molecular attraction by which the particles of a body are united throughout the mass
— **co·he·sion·less** \-ləs\ *adjective*

co·he·sive \kō-ˈhē-siv, -ziv\ *adjective* (circa 1731)
: exhibiting or producing cohesion or coherence ⟨a *cohesive* social unit⟩ ⟨*cohesive* soils⟩
— **co·he·sive·ly** *adverb*
— **co·he·sive·ness** *noun*

co·ho \ˈkō-(ˌ)hō\ *noun, plural* **cohos** *or* **coho** [Halkomelem (Salishan language of southwest British Columbia) *kʷə́xʷəθ*] (1869)
: a rather small Pacific salmon (*Oncorhynchus kisutch*) that has light-colored flesh and is native to both coasts of the North Pacific and is stocked in the Great Lakes

co·ho·mol·o·gy \(ˌ)kō-hō-ˈmä-lə-jē\ *noun* (circa 1959)
: a part of the theory of topology in which groups are used to study the properties of topological spaces and which is related in a complementary way to homology theory — called also *cohomology theory*
— **co·ho·mo·log·i·cal** \-ˌhō-mə-ˈlä-ji-kəl, -ˌhä-\ *adjective*

co·hort \ˈkō-ˌhȯrt\ *noun* [Middle French & Latin; Middle French *cohorte,* from Latin *cohort-, cohors* — more at COURT] (15th century)
1 a : one of 10 divisions of an ancient Roman legion **b :** a group of warriors or soldiers **c :** BAND, GROUP **d :** a group of individuals having a statistical factor (as age or class membership) in common in a demographic study ⟨a *cohort* of premedical students⟩
2 : COMPANION, COLLEAGUE ⟨a few of their . . . *cohorts* decided to form a company —Burt Hochberg⟩ ▢

co·hosh \ˈkō-ˌhäsh\ *noun* [Eastern Abenaki *kkʷὰhas*] (1789)
: any of several American medicinal or poisonous plants: as **a :** BLACK COHOSH **b :** BLUE COHOSH **c :** BANEBERRY

¹coif \ˈkȯif, *in sense 2 also* ˈkwäf\ *noun* [Middle English *coife,* from Middle French, from Late Latin *cofea*] (14th century)
1 : a close-fitting cap: as **a :** a hoodlike cap worn under a veil by nuns **b :** a protective usually metal skullcap formerly worn under a hood of mail **c :** a white cap formerly worn by English lawyers and especially by serjeants-at-law; *also* **:** the order or rank of a serjeant-at-law
2 : COIFFURE

²coif \ˈkȯif, ˈkwäf\ *transitive verb* **coiffed** *or* **coifed; coif·fing** *or* **coif·ing** (15th century)
1 : to cover or dress with or as if with a coif
2 : to arrange (hair) by brushing, combing, or curling

coif·feur \kwä-ˈfər\ *noun* [French, from *coiffer*] (1847)
: a man who is a hairdresser

coif·feuse \kwä-ˈfə(r)z, -ˈfyüz, -ˈfüz\ *noun* [French, feminine of *coiffeur*] (1870)
: a woman who is a hairdresser

coif·fure \kwä-ˈfyu̇r\ *noun* [French, from *coiffer* to cover with a coif, arrange (hair), from *coife*] (circa 1631)
: a style or manner of arranging the hair

coif·fured \-ˈfyu̇rd\ *adjective* (1907)
1 : being dressed ⟨beautifully *coiffured* hair⟩
2 : having the hair brushed, combed, and curled ⟨stylishly *coiffured* women⟩

coign of van·tage \ˌkȯi-nə-ˈvan-tij\ [*coign,* earlier spelling of ¹*coin* (corner)] (1605)
: an advantageous position

¹coil \ˈkȯi(ə)l\ *noun* [origin unknown] (1567)
1 : TURMOIL
2 : TROUBLE; *also* **:** everyday cares and worries ⟨when we have shuffled off this mortal *coil* —Shakespeare⟩

²coil *verb* [French *coillir, cuillir* to gather — more at CULL] (1611)
transitive verb
1 : to wind into rings or spirals
2 : to roll or twist into a shape resembling a coil
intransitive verb
1 : to move in a circular or spiral course
2 : to form or lie in a coil
— **coil·abil·i·ty** \ˌkȯi-lə-ˈbi-lə-tē\ *noun*

³coil *noun* (1627)
1 a (1) **:** a series of loops (2) **:** SPIRAL **b :** a single loop of such a coil
2 a : a number of turns of wire wound around a core (as of iron) to create a magnetic field for an electromagnet or an induction coil **b :** INDUCTION COIL
3 : a series of connected pipes in rows, layers, or windings
4 : a roll of postage stamps; *also* **:** a stamp from such a roll

¹coin \ˈkȯin\ *noun* [Middle English, from Middle French, wedge, corner, from Latin *cuneus* wedge] (14th century)
1 *archaic* **a :** CORNER, CORNERSTONE, QUOIN **b :** WEDGE
2 a : a usually flat piece of metal issued by governmental authority as money **b :** metal money **c :** something resembling a coin especially in shape
3 : something used as if it were money (as in verbal or intellectual exchange) ⟨perhaps wisecracks . . . are respectable literary *coin* in the U.S. —*Times Literary Supplement*⟩ ⟨would repay him with the full *coin* of his mind —Ian Fleming⟩
4 : something having two different and usually opposing sides — usually used in the phrase *the other side of the coin*
5 : MONEY ⟨I'm in it for the *coin* —Sinclair Lewis⟩

²coin *transitive verb* (14th century)
1 a : to make (a coin) especially by stamping **:** MINT **b :** to convert (metal) into coins
2 : CREATE, INVENT ⟨*coin* a phrase⟩
— **coin·er** \ˈkȯi-nər\ *noun*
— **coin money :** to get rich quickly

³coin *adjective* (circa 1566)
1 : of or relating to coins
2 : operated by coins

coin·age \ˈkȯi-nij\ *noun* (14th century)
1 : the act or process of coining
2 a : COINS **b :** something (as a word) made up or invented

co·in·cide \ˌkō-ən-ˈsīd, ˈkō-ən-ˌ\ *intransitive verb* **-cid·ed; -cid·ing** [Medieval Latin *coincidere,* from Latin *co-* + *incidere* to fall on, from *in-* + *cadere* to fall — more at CHANCE] (1719)
1 a : to occupy the same place in space or time **b :** to occupy exactly corresponding or equivalent positions on a scale or in a series
2 : to correspond in nature, character, or function
3 : to be in accord or agreement **:** CONCUR
synonym see AGREE

co·in·ci·dence \kō-ˈin(t)-sə-dən(t)s, -səˌden(t)s\ *noun* (1605)
1 : the act or condition of coinciding **:** CORRESPONDENCE
2 : the occurrence of events that happen at the same time by accident but seem to have some connection; *also* **:** any of these occurrences

co·in·ci·dent \-sə-dənt, -ˌdent\ *adjective* [French *coincident,* from Medieval Latin *coincident-, coincidens,* present participle of *coincidere*] (circa 1587)
1 : of similar nature **:** HARMONIOUS ⟨a theory *coincident* with the facts⟩
2 : occupying the same space or time ⟨*coincident* events⟩
synonym see CONTEMPORARY
— **co·in·ci·dent·ly** \-sə-dənt-lē, -ˌdent-; (ˌ)kō-ˌin(t)-sə-ˈdent-lē\ *adverb*

co·in·ci·den·tal \(ˌ)kō-ˌin(t)-sə-ˈden-tᵊl\ *adjective* (circa 1800)
1 : resulting from a coincidence
2 : occurring or existing at the same time
— **co·in·ci·den·tal·ly** \-ˈdent-lē, -ˈden-tᵊl-ē\ *adverb*

coin–op \ˈkȯi-ˌnäp\ *noun* (1961)
: a self-service laundry where the machines are operated by coins

co·in·sur·ance \ˌkō-ən-ˈshu̇r-ən(t)s *also* -ˈin-ˌ\ *noun* (circa 1889)
1 : joint assumption of risk (as by two underwriters) with another
2 : insurance (as fire insurance) in which the insured is obligated to maintain coverage on a risk at a stipulated percentage of its total value or in the event of loss suffer a penalty in proportion to the deficiency

co·in·sure \ˌkō-ən-ˈshu̇r\ *transitive verb* (circa 1899)
: to insure jointly
— **co·in·sur·er** *noun*

coir \ˈkȯi(-ə)r\ *noun* [Tamil *kayiru* rope] (1582)
: a stiff coarse fiber from the outer husk of a coconut

cois·trel \ˈkȯi-strəl\ *noun* [Middle French *coustillier* soldier carrying a short sword, from *coustille* short sword, from Latin *cultellus* knife — more at CUTLASS] (1581)
archaic **:** a mean fellow **:** VARLET

co·i·tion \kō-ˈi-shən\ *noun* [Late Latin, from Latin *coition-, coitio* a coming together, from *coire* to come together, from *co-* + *ire* to go — more at ISSUE] (1615)
: COITUS
— **co·i·tion·al** \-ˈish-nəl, -ˈi-shə-nᵊl\ *adjective*

co·i·tus \ˈkō-ə-təs, kō-ˈē-, ˈkȯi-təs\ *noun* [Latin, from *coire*] (1855)
: physical union of male and female genitalia accompanied by rhythmic movements usually leading to the ejaculation of semen from the

▢ USAGE

cohort Sense 2 is an American development with a curious history. The first extension of the meaning of *cohort* was from the Roman army to other armies ⟨the Assyrian came down like the wolf on the fold, and his *cohorts* were gleaming in purple and gold —Lord Byron⟩. Then the word was extended to any group or band ⟨Harry, with a loyal *cohort* of adherents, forcibly barred his entrance —S. N. Behrman⟩. This sense could also be used in the plural ⟨Dr. Mossadegh and his *cohorts* nationalized the British-owned oil concession in southern Iran —*N. Y. Times*⟩. On reflection, it would seem clear that these cohorts are large and probably organized bands, but the word can easily be read as meaning "followers." Some uses could seem even more ambiguous ⟨as Carrie Nation and her *cohorts* swarmed forward and pounded the panels —Herbert Asbury⟩. Here evidently a mob scene is being described, a far cry from the disciplined Roman army units. And again *cohorts* can readily be interpreted as "friends, followers." It is presumably from uses like these that our sense 2 developed ⟨formulated a second constitution which virtually restricted the suffrage to the freedmen and their white *cohorts* —*Dictionary of American History*⟩ ⟨had left, accompanied by two of his *cohorts* —Mary McCarthy⟩. Sense 2 was becoming established in the 1940s, and ten years or so later some critics began to notice it disapprovingly. Disapproval came too late, though, for by then the sense was established as standard—at least in American English—and it is standard now.

penis into the female reproductive tract; *also* : INTERCOURSE 3 — compare ORGASM
— **co·i·tal** \-tᵊl\ *adjective*
— **co·i·tal·ly** \-tᵊl-ē\ *adverb*

coitus in·ter·rup·tus \-ˌin-tə-ˈrəp-təs\ *noun* [New Latin, interrupted coitus] (1900)
: coitus in which the penis is withdrawn prior to ejaculation to prevent the deposit of sperm into the vagina

¹coke \ˈkōk\ *noun* [perhaps from dialect *coke, colk* core, from Middle English; akin to Swedish *kälk* pith] (1669)
: the residue of coal left after destructive distillation and used as fuel; *also* : a similar residue left by other materials (as petroleum) distilled to dryness

²coke *verb* **coked; cok·ing** (1763)
transitive verb
: to change into coke
intransitive verb
: to become coke or like coke

³coke *noun* [by shortening & alteration] (circa 1903)
: COCAINE

Coke \ˈkōk\ *trademark*
— used for a cola drink

coke·head \ˈkō-ked, ˈkōk-ˌhed\ *noun* (circa 1927)
: a person who uses cocaine compulsively

col \ˈkäl\ *noun* [French, from Middle French, neck, from Latin *collum*] (1853)
: SADDLE 3

¹col- — see COM-

²col- *or* **coli-** *or* **colo-** *combining form* [New Latin, from Latin *colon*]
1 : colon ⟨*co*litis⟩ ⟨*colo*stomy⟩
2 [New Latin *Escherichia coli,* species of colon bacillus] : colon bacillus ⟨*coli*form⟩

¹cola *plural of* COLON

²co·la \ˈkō-lə\ *noun* [from *Coca-Cola,* a trademark] (1920)
: a carbonated soft drink colored usually with caramel and flavored usually with extracts from kola nuts

col·an·der \ˈkäl-ən-dər, ˈkəl-\ *noun* [Middle English *colyndore,* probably modification of Old Provençal *colador,* from Medieval Latin *colatorium,* from Latin *colare* to sieve, from *colum* sieve] (14th century)
: a perforated utensil for washing or draining food

co·lat·i·tude \(ˌ)kō-ˈla-tə-ˌtüd, -ˌtyüd\ *noun* (1790)
: the complement of the latitude

Col·by \ˈkōl-bē\ *noun* [probably from the name *Colby*] (1942)
: a moist mild cheese similar to cheddar

col·can·non \käl-ˈka-nən\ *noun* [Irish *cál ceannan,* literally, white-headed cabbage] (circa 1785)
: potatoes and cabbage boiled and mashed together with butter and seasoning

col·chi·cine \ˈkäl-chə-ˌsēn, ˈkäl-kə-\ *noun* (circa 1847)
: a poisonous alkaloid $C_{22}H_{25}NO_6$ that inhibits mitosis, is extracted from the corms or seeds of the autumn crocus (*Colchicum autumnale*), and is used especially in the treatment of gout and to produce polyploidy in plants

col·chi·cum \ˈkäl-chi-kəm, ˈkäl-ki-\ *noun* [New Latin, genus name, from Latin, a kind of plant with a poisonous root, from Greek *kolchikon,* literally, product of Colchis] (1597)
1 : any of a genus (*Colchicum*) of Old World corm-producing herbs of the lily family with flowers that resemble crocuses
2 : the dried corm or dried seeds of autumn crocus containing colchicine, possessing emetic, diuretic, and cathartic action, and used to treat gout

¹cold \ˈkōld\ *adjective* [Middle English, from Old English *ceald, cald;* akin to Old High German *kalt* cold, Latin *gelu* frost, *gelare* to freeze] (before 12th century)
1 a : having or being a temperature that is uncomfortably low for humans ⟨it is *cold* outside today⟩ **b** : having a relatively low temperature or one lower than normal or expected ⟨the bath water has gotten *cold*⟩ **c** : not heated: as (1) *of food* : served without heating especially after initial cooking or processing ⟨*cold* cereal⟩ ⟨*cold* roast beef⟩ (2) : served chilled or with ice ⟨a *cold* drink⟩ (3) : involving processing without the use of heat ⟨*cold* working of steel⟩
2 a : marked by a lack of the warmth of normal human emotion, friendliness, or compassion ⟨got a *cold* reception⟩; *also* : not moved to enthusiasm ⟨the movie leaves me *cold*⟩ **b** : not colored or affected by personal feeling or bias : DETACHED, INDIFFERENT ⟨*cold* chronicles recorded by an outsider —Andrew Sarris⟩; *also* : IMPERSONAL, OBJECTIVE ⟨*cold* facts⟩ ⟨*cold* reality⟩ **c** : marked by sure familiarity : PAT ⟨had her lines *cold* weeks before opening night⟩
3 : conveying the impression of being cold: as **a** : DEPRESSING, GLOOMY ⟨*cold* gray skies⟩ **b** : COOL 6a
4 a : marked by the loss of normal body heat ⟨*cold* hands⟩; *especially* : DEAD **b** : giving the appearance of being dead : UNCONSCIOUS ⟨passed out *cold*⟩
5 a : having lost freshness or vividness : STALE ⟨dogs trying to pick up a *cold* scent⟩ **b** : far off the mark : not close to finding or solving — used especially in children's games **c** : marked by poor or unlucky performance ⟨the team's shooting turned *cold* in the second half⟩ **d** : not prepared or suitably warmed up
— **cold·ish** \ˈkōl-dish\ *adjective*
— **cold·ly** \ˈkōl(d)-lē\ *adverb*
— **cold·ness** \ˈkōl(d)-nəs\ *noun*
— **in cold blood** : with premeditation : DELIBERATELY

²cold *noun* (13th century)
1 : bodily sensation produced by loss or lack of heat
2 : a condition of low temperature; *especially* : cold weather
3 : a bodily disorder popularly associated with chilling; *specifically* : COMMON COLD
— **out in the cold** : deprived of benefits given others ⟨the plan benefits management but leaves labor *out in the cold*⟩

³cold *adverb* (1889)
1 : with utter finality : ABSOLUTELY, COMPLETELY ⟨turned down *cold*⟩; *also* : ABRUPTLY ⟨stopped them *cold*⟩
2 a : without introduction or advance notice **b** : without preparation or warm-up

cold–blood·ed \ˈkōl(d)-ˈblə-dəd\ *adjective* (1595)
1 a : done or acting without consideration, compunction, or clemency ⟨*cold-blooded* murder⟩ **b** : MATTER-OF-FACT, EMOTIONLESS ⟨a *cold-blooded* assessment⟩
2 : having cold blood; *specifically* : having a body temperature not internally regulated but approximating that of the environment
3 *or* **cold·blood** \-ˈbləd\ : of mixed or inferior breeding
4 : noticeably sensitive to cold
— **cold–blood·ed·ly** *adverb*
— **cold–blood·ed·ness** *noun*

cold call *noun* (1966)
: a telephone call soliciting business made directly to a potential customer without prior contact or without a lead
— **cold–call** *verb*

cold cash *noun* (1925)
: money in hand

cold chisel *noun* (1699)
: a chisel made of tool steel of a strength, shape, and temper suitable for chipping or cutting cold metal

cold·cock \ˈkōl(d)-ˈkäk, ˌkōl(d)-\ *transitive verb* [perhaps ²*cock*] (circa 1927)
: to knock unconscious

cold comfort *noun* (14th century)
: quite limited sympathy, consolation, or encouragement

cold cream *noun* (1709)
: a soothing and cleansing cosmetic

cold cuts *noun plural* (1945)
: sliced assorted cold cooked meats

cold duck *noun* [translation of German *Kalte Ente*] (1969)
: a beverage that consists of a blend of sparkling burgundy and champagne

cold–eyed \ˈkōl-ˈdīd, -ˌdīd\ *adjective* (1950)
: cold in manner or appearance; *especially* : coolly dispassionate ⟨*cold-eyed* analysis⟩

cold feet *noun plural* (1893)
: apprehension or doubt strong enough to prevent a planned course of action

cold fish *noun* (1924)
: a cold aloof person

cold frame *noun* (1851)
: a usually glass- or plastic-covered frame without artificial heat used to protect plants and seedlings outdoors

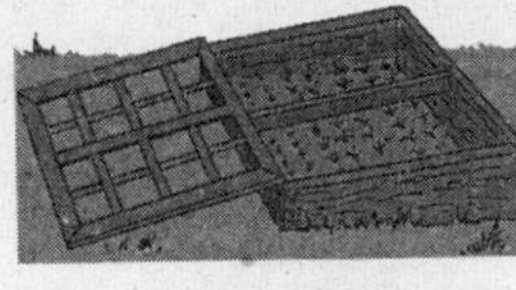
cold frame

cold front *noun* (1921)
: an advancing edge of a cold air mass

cold·heart·ed \ˈkōld-ˈhär-təd\ *adjective* (1606)
: marked by lack of sympathy, interest, or sensitivity
— **cold·heart·ed·ly** *adverb*
— **cold·heart·ed·ness** *noun*

cold shoulder *noun* (1816)
: intentionally cold or unsympathetic treatment
— **cold–shoulder** *transitive verb*

cold sore *noun* (circa 1889)
: a group of blisters appearing about or within the mouth and caused by a herpes simplex virus

cold storage *noun* (1877)
1 : storage (as of food) in a cold place for preservation
2 : a condition of being held or continued without being acted on : ABEYANCE

cold store *noun* (1895)
: a building for cold storage

cold sweat *noun* (1706)
: concurrent perspiration and chill usually associated with fear, pain, or shock

¹cold turkey *noun* (1921)
1 : abrupt complete cessation of the use of an addictive drug; *also* : the symptoms experienced by a person undergoing withdrawal from a drug
2 : unrelieved blunt language or procedure
3 : a cold aloof person
— **cold turkey** *transitive verb*

²cold turkey *adverb* (1941)
: without a period of gradual adjustment, adaptation, or withdrawal : all at once ⟨quit smoking *cold turkey*⟩

cold type *noun* (1949)
: composition or typesetting (as photocomposition) done without the casting of metal; *specifically* : such composition produced directly on paper by a typewriter mechanism

cold war *noun* (1945)
1 : a conflict over ideological differences carried on by methods short of sustained overt military action and usually without breaking off diplomatic relations; *specifically, often C&W capitalized* : the ideological conflict between the U.S. and the U.S.S.R — compare HOT WAR
2 : a condition of rivalry, mistrust, and often open hostility short of violence especially between power groups (as labor and management)

cold warrior *noun* (1949)
: one that supports or is engaged in a cold war

cold–water *adjective* (1942)
: having only running water without heat or utility services provided ⟨a *cold-water* flat⟩

cold water *noun* (1808)
: depreciation of something as being ill-advised, unwarranted, or worthless ⟨threw *cold water* on our hopes⟩

cold wave *noun* (1876)
1 : an unusually large and rapid drop in temperature over a short period of time (as 24 hours)
2 : a permanent wave set by a chemical preparation without the use of heat

cole \'kōl\ *noun* [Middle English, from Old English *cāl,* from Latin *caulis* stem, cabbage; akin to Greek *kaulos* stem, Latvian *kauls*] (before 12th century)
: any of several brassicas; *especially* **:** any of various crop plants (as broccoli, kale, brussels sprouts, cabbage, cauliflower, and kohlrabi) derived from the same wild ancestor (*Brassica oleracea*)

cole·man·ite \'kōl-mə-,nīt\ *noun* [William T. *Coleman* (died 1893) American businessman and mine owner] (1884)
: a mineral consisting of a hydrous calcium borate occurring in brilliant colorless or white massive monoclinic crystals

co·le·op·tera \,kō-lē-'äp-tə-rə\ *noun plural* [New Latin, ultimately from Greek *koleon* sheath + *pteron* wing — more at FEATHER] (1873)
: insects that are beetles
— co·le·op·ter·ist \-tə-rist\ *noun*
— co·le·op·ter·ous \-tə-rəs\ *adjective*

co·le·op·ter·an \-tə-rən\ *noun* (circa 1847)
: [1]BEETLE 1
— coleopteran *adjective*

co·le·op·tile \-'äp-tᵊl\ *noun* [New Latin *coleoptilum,* from Greek *koleon* + *ptilon* down; akin to Greek *pteron*] (circa 1866)
: the first leaf of a monocotyledon forming a protective sheath about the plumule

co·le·o·rhi·za \,kō-lē-ə-'rī-zə\ *noun, plural* **-zae** \-(,)zē\ [New Latin, from Greek *koleon* + New Latin *-rhiza*] (circa 1866)
: the sheath investing the hypocotyl in some monocotyledonous plants through which the roots burst

cole·slaw \'kōl-,slȯ\ *noun* [Dutch *koolsla,* from *kool* cabbage + *sla* salad] (1794)
: a salad made of raw sliced or chopped cabbage

co·le·us \'kō-lē-əs\ *noun* [New Latin, genus name, from Greek *koleos, koleon* sheath] (1885)
: any of a large genus (*Coleus*) of herbs of the mint family

cole·wort \'kōl-,wərt, -,wȯrt\ *noun* (14th century)
: COLE; *especially* **:** one (as kale) that forms no head

coli- — see COL-

[1]**col·ic** \'kä-lik\ *noun* [Middle English, from Middle French *colique,* from Late Latin *colicus* colicky, from Greek *kōlikos,* from *kōlon,* alteration of *kolon* colon] (15th century)
: a paroxysm of acute abdominal pain localized in a hollow organ and often caused by spasm, obstruction, or twisting

[2]**colic** *adjective* (15th century)
: of or relating to colic **:** COLICKY ⟨*colic* crying⟩

[3]**co·lic** \'kō-lik, 'kä-\ *adjective* (1615)
: of or relating to the colon ⟨*colic* lymph nodes⟩

co·li·cin \'kō-lə-sən\ *also* **co·li·cine** \-,sēn\ *noun* [French *colicine,* from *col-* + *-cine* (as in *streptomycine* streptomycin)] (1946)
: any of various antibacterial substances that are produced by strains of intestinal bacteria (as of *Escherichia coli*) having a specific plasmid and that often act to inhibit macromolecular synthesis in related strains

col·icky \'kä-li-kē\ *adjective* (1742)
1 : relating to or associated with colic ⟨*colicky* pain⟩
2 : suffering from colic ⟨*colicky* babies⟩

col·ic·root \'kä-lik-,rüt, -,ru̇t\ *noun* (1833)
: any of several plants having roots used in folk medicine to treat colic; *especially* **:** either of two bitter herbs (*Aletris farinosa* and *A. aurea*) of the lily family

co·li·form \'kō-lə-,fȯrm, 'kä-\ *adjective* (1906)
: relating to, resembling, or being the colon bacillus
— coliform *noun*

co·lin·ear \(,)kō-'li-nē-ər\ *adjective* (1927)
1 : COLLINEAR
2 : having corresponding parts arranged in the same linear order ⟨a gene and the protein it determines are *colinear*⟩
— co·lin·ear·i·ty \(,)kō-,li-nē-'ar-ə-tē\ *noun*

co·li·phage \'kō-lə-,fāj, -,fäzh\ *noun* (1944)
: any bacteriophage active against the colon bacillus

col·i·se·um \,kä-lə-'sē-əm\ *noun* [Medieval Latin *Colosseum, Colisseum*] (circa 1715)
1 *capitalized* **:** COLOSSEUM 1
2 : a large sports stadium or building designed like the Colosseum for public entertainments

co·lis·tin \kə-'lis-tən, kō-\ *noun* [New Latin *colistinus,* specific epithet of the bacterium producing it] (1951)
: a polymyxin produced by a bacterium (*Bacillus polymyxa* variety *colistinus*) from Japanese soil

co·li·tis \kō-'lī-təs, kə-\ *noun* (circa 1860)
: inflammation of the colon

coll- *or* **collo-** *combining form* [New Latin, from Greek *koll-, kollo-,* from *kolla* — more at PROTOCOL]
1 : glue ⟨*coll*enchyma⟩
2 : colloid ⟨*collo*type⟩

col·lab·o·rate \kə-'la-bə-,rāt\ *intransitive verb* **-rat·ed; -rat·ing** [Late Latin *collaboratus,* past participle of *collaborare* to labor together, from Latin *com-* + *laborare* to labor] (1871)
1 : to work jointly with others or together especially in an intellectual endeavor
2 : to cooperate with or willingly assist an enemy of one's country and especially an occupying force
3 : to cooperate with an agency or instrumentality with which one is not immediately connected
— col·lab·o·ra·tion \-,la-bə-'rā-shən\ *noun*
— col·lab·o·ra·tive \-'la-bə-,rā-tiv, -b(ə-)rə-\ *adjective or noun*
— col·lab·o·ra·tive·ly \-lē\ *adverb*
— col·lab·o·ra·tor \-'la-bə-,rā-tər\ *noun*

col·lab·o·ra·tion·ism \kə-,la-bə-'rā-shə-,ni-zəm\ *noun* (1923)
: the advocacy or practice of collaboration with an enemy
— col·lab·o·ra·tion·ist \-sh(ə-)nist\ *adjective or noun*

col·lage \kə-'läzh, kō-, kō-\ *noun* [French, literally, gluing, from *coller* to glue, from *colle* glue, from (assumed) Vulgar Latin *colla,* from Greek *kolla*] (1919)
1 : an artistic composition made of various materials (as paper, cloth, or wood) glued on a surface
2 : the art of making collages
3 : an assembly of diverse fragments ⟨a *collage* of ideas⟩
4 : a work (as a film) having disparate scenes in rapid succession without transitions
— collage *transitive verb*
— col·lag·ist \-'lä-zhist\ *noun*

col·la·gen \'kä-lə-jən\ *noun* [Greek *kolla* + International Scientific Vocabulary *-gen*] (circa 1865)
: an insoluble fibrous protein that occurs in vertebrates as the chief constituent of connective tissue fibrils and in bones and yields gelatin and glue on prolonged heating with water
— col·lag·e·nous \kə-'la-jə-nəs\ *adjective*

col·la·ge·nase \kə-'la-jə-,nās, 'kä-lə-jə-, -,nāz\ *noun* (1926)
: any of a group of proteolytic enzymes that decompose collagen and gelatin

[1]**col·lapse** \kə-'laps\ *verb* **col·lapsed; col·laps·ing** [Latin *collapsus,* past participle of *collabi,* from *com-* + *labi* to fall, slide — more at SLEEP] (1732)
intransitive verb
1 : to fall or shrink together abruptly and completely **:** fall into a jumbled or flattened mass through the force of external pressure ⟨a blood vessel that *collapsed*⟩
2 : to break down completely **:** DISINTEGRATE ⟨his case had *collapsed* in a mass of legal wreckage —Erle Stanley Gardner⟩
3 : to cave or fall in or give way
4 : to suddenly lose force, significance, effectiveness, or worth
5 : to break down in vital energy, stamina, or self-control through exhaustion or disease; *especially* **:** to fall helpless or unconscious
6 : to fold down into a more compact shape ⟨a chair that *collapses*⟩
transitive verb
: to cause to collapse
— col·laps·ibil·i·ty \-,lap-sə-'bi-lə-tē\ *noun*
— col·laps·ible \-'lap-sə-bəl\ *adjective*

[2]**collapse** *noun* (1801)
1 a : a breakdown in vital energy, strength, or stamina **b :** a state of extreme prostration and physical depression (as from circulatory failure or great loss of body fluids) **c :** an airless state of all or part of a lung originating spontaneously or induced surgically
2 : the act or action of collapsing ⟨the cutting of many tent ropes, the *collapse* of the canvas —Rudyard Kipling⟩
3 : a sudden failure **:** BREAKDOWN, RUIN
4 : a sudden loss of force, value, or effect ⟨the *collapse* of respect for ancient law and custom —L. S. B. Leakey⟩

[1]**col·lar** \'kä-lər\ *noun* [Middle English *coler,* from Old French, from Latin *collare,* from *collum* neck; akin to Old English *heals* neck, and probably to Old English *hwēol* wheel — more at WHEEL] (14th century)
1 : a band, strip, or chain worn around the neck: as **a :** a band that serves to finish or decorate the neckline of a garment **b :** a short necklace **c :** a band placed about the neck of an animal **d :** a part of the harness of draft animals fitted over the shoulders and taking strain when a load is drawn **e :** an indication of control **:** a token of subservience **f :** a protective or supportive device (as a brace or cast) worn around the neck **g :** CLERICAL COLLAR
2 : something resembling a collar in shape or use (as a ring or round flange to restrain motion or hold something in place)
3 : any of various animal structures or markings similar to a collar
4 : an act of collaring **:** ARREST, CAPTURE
— col·lared \-lərd\ *adjective*
— col·lar·less \-lər-ləs\ *adjective*

[2]**collar** *transitive verb* (1613)
1 a : to seize by the collar or neck **b :** ARREST, GRAB **c :** to get control of **:** PREEMPT ⟨we can *collar* nearly the whole of this market —Roald Dahl⟩ **d :** to stop and detain in unwilling conversation
2 : to put a collar on

col·lar·bone \'kä-lər-,bōn\ *noun* (15th century)
: CLAVICLE

collar cell *noun* (circa 1889)
: a flagellated endodermal cell that lines the cavity of a sponge and has a contractile protoplasmic cup surrounding the flagellum — called also *choanocyte*

col·lard \'kä-lərd\ *noun* [alteration of *colewort*] (1755)

: a stalked smooth-leaved kale — usually used in plural

col·late \kə-ˈlāt, kä-, kō-; ˈkä-ˌ, ˈkō-ˌ\ *transitive verb* **col·lat·ed; col·lat·ing** [back-formation from *collation*] (1612)
1 a : to compare critically **b :** to collect, compare carefully in order to verify, and often to integrate or arrange in order
2 [Latin *collatus,* past participle] **:** to institute (a cleric) to a benefice
3 a : to verify the order of (printed sheets) **b :** to assemble in proper order; *especially* **:** to assemble (as printed sheets) in order for binding
synonym see COMPARE
— col·la·tor \-ˈlā-tər, -ˌlā-\ *noun*

¹col·lat·er·al \kə-ˈla-t(ə-)rəl\ *adjective* [Middle English, probably from Middle French, from Medieval Latin *collateralis,* from Latin *com-* + *lateralis* lateral] (14th century)
1 a : accompanying as secondary or subordinate **:** CONCOMITANT ⟨digress into *collateral* matters⟩ **b :** INDIRECT **c :** serving to support or reinforce **:** ANCILLARY
2 : belonging to the same ancestral stock but not in a direct line of descent — compare LINEAL 3a
3 : parallel, coordinate, or corresponding in position, order, time, or significance ⟨*collateral* states like Athens and Sparta⟩
4 a : of, relating to, or being collateral used as security (as for payment of a debt or performance of a contract) **b :** secured by collateral
— col·lat·er·al·i·ty \-ˌla-tə-ˈra-lə-tē\ *noun*
— col·lat·er·al·ly \-ˈla-t(ə-)rə-lē\ *adverb*

²collateral *noun* (1691)
1 : a collateral relative
2 : property (as securities) pledged by a borrower to protect the interests of the lender
3 : a branch of a bodily part (as a vein)

col·lat·er·al·ize \kə-ˈla-t(ə-)rə-ˌlīz\ *transitive verb* **-ized; -iz·ing** (1941)
1 : to make (a loan) secure with collateral
2 : to use (as securities) for collateral

col·la·tion \kə-ˈlā-shən, kä-, kō-\ *noun* (14th century)
1 [Middle English, from Medieval Latin *collation-, collatio,* from Late Latin, conference, from Latin, bringing together, comparison, from *conferre* (past participle *collatus*) to bring together — more at CONFER, TOLERATE] **a :** a light meal allowed on fast days in place of lunch or supper **b :** a light meal
2 [Middle English, from Latin *collation-, collatio*] **:** the act, process, or result of collating

col·league \ˈkä-(ˌ)lēg\ *noun* [Middle French *collegue,* from Latin *collega,* from *com-* + *legare* to depute — more at LEGATE] (circa 1533)
: an associate in a profession or in a civil or ecclesiastical office
— col·league·ship \-ˌship\ *noun*

¹col·lect \ˈkä-likt *also* -ˌlekt\ *noun* [Middle English *collecte,* from Old French, from Medieval Latin *collecta* (short for *oratio ad collectam* prayer upon assembly), from Late Latin, assembly, from Latin, assemblage, from feminine of *collectus*] (13th century)
1 : a short prayer comprising an invocation, petition, and conclusion; *specifically, often capitalized* **:** one preceding the eucharistic Epistle and varying with the day
2 : COLLECTION

²col·lect \kə-ˈlekt\ *verb* [Latin *collectus,* past participle of *colligere* to collect, from *com-* + *legere* to gather — more at LEGEND] (1563)
transitive verb
1 a : to bring together into one body or place **b :** to gather or exact from a number of persons or sources ⟨*collect* taxes⟩
2 : INFER, DEDUCE
3 : to gain or regain control of ⟨*collect* his thoughts⟩
4 : to claim as due and receive payment for
5 : to get and bring with one; *specifically* **:** PICK UP ⟨went to *collect* her at the train station⟩
intransitive verb
1 : to come together in a band, group, or mass **:** GATHER
2 a : to collect objects **b :** to receive payment ⟨*collecting* on the insurance⟩
synonym see GATHER

³col·lect \kə-ˈlekt\ *adverb or adjective* (1893)
: to be paid for by the receiver

col·lec·ta·nea \ˌkä-ˌlek-ˈtā-nē-ə\ *noun plural* [Latin, neuter plural of *collectaneus* collected, from *collectus,* past participle] (1791)
: collected writings; *also* **:** literary items forming a collection

col·lect·ed \kə-ˈlek-təd\ *adjective* (1610)
1 : possessed of calmness and composure often through concentrated effort
2 : gathered together ⟨the *collected* works of Scott⟩
3 *of a horse's gait* **:** performed slowly and restrainedly with the animal's center of gravity toward the hindquarters — compare EXTENDED
synonym see COOL
— col·lect·ed·ly *adverb*
— col·lect·ed·ness *noun*

¹col·lect·ible *or* **col·lect·able** \kə-ˈlek-tə-bəl\ *adjective* (1660)
1 : suitable for being collected
2 : due for present payment **:** PAYABLE

²collectible *or* **collectable** *noun* (1953)
: an object that is collected by fanciers; *especially* **:** one other than such traditionally collectible items as art, stamps, coins, and antiques

col·lec·tion \kə-ˈlek-shən\ *noun* (14th century)
1 : the act or process of collecting
2 a : something collected; *especially* **:** an accumulation of objects gathered for study, comparison, or exhibition or as a hobby **b :** GROUP, AGGREGATE **c :** a set of apparel designed for sale usually in a particular season

¹col·lec·tive \kə-ˈlek-tiv\ *adjective* (15th century)
1 : denoting a number of persons or things considered as one group or whole ⟨*flock* is a *collective* word⟩
2 a : formed by collecting **:** AGGREGATED **b** *of a fruit* **:** MULTIPLE
3 a : of, relating to, or being a group of individuals **b :** involving all members of a group as distinct from its individuals
4 : marked by similarity among or with the members of a group
5 : collectivized or characterized by collectivism
6 : shared or assumed by all members of the group ⟨*collective* responsibility⟩
— col·lec·tive·ly *adverb*

²collective *noun* (1655)
1 : a collective body **:** GROUP
2 : a cooperative unit or organization; *specifically* **:** COLLECTIVE FARM

collective bargaining *noun* (1891)
: negotiation between an employer and a labor union usually on wages, hours, and working conditions

collective farm *noun* (1925)
: a farm especially in a communist country formed from many small holdings collected into a single unit for joint operation under governmental supervision

collective mark *noun* (1938)
: a trademark or a service mark of a group (as a cooperative association)

collective security *noun* (1934)
: the maintenance by common action of the security of all members of an association of nations

collective unconscious *noun* (1917)
: the genetically determined part of the unconscious that especially in the psychoanalytic theory of C. G. Jung occurs in all the members of a people or race

col·lec·tiv·ise *chiefly British variant of* COLLECTIVIZE

col·lec·tiv·ism \kə-ˈlek-ti-ˌvi-zəm\ *noun* (1857)
: a political or economic theory advocating collective control especially over production and distribution; *also* **:** a system marked by such control
— col·lec·tiv·ist \-vist\ *adjective or noun*
— col·lec·tiv·is·tic \-ˌlek-ti-ˈvis-tik\ *adjective*
— col·lec·tiv·is·ti·cal·ly \-ti-k(ə-)lē\ *adverb*

col·lec·tiv·i·ty \kə-ˌlek-ˈti-və-tē, ˌkä-\ *noun, plural* **-ties** (1862)
1 : the quality or state of being collective
2 : a collective whole; *especially* **:** the people as a body

col·lec·tiv·ize \kə-ˈlek-ti-ˌvīz\ *transitive verb* **-ized; -iz·ing** (1893)
: to organize by collectivism
— col·lec·tiv·i·za·tion \kə-ˌlek-ti-və-ˈzā-shən\ *noun*

col·lec·tor \kə-ˈlek-tər\ *noun* (14th century)
: one that collects: as **a :** an official who collects funds or moneys **b :** a person who makes a collection ⟨stamp *collector*⟩ **c :** an object or device that collects ⟨the statuette was a dust *collector*⟩ **d :** SOLAR COLLECTOR
— col·lec·tor·ship \-ˌship\ *noun*

collector's item *noun* (1932)
: an item whose rarity or excellence makes it especially worth collecting; *broadly* **:** COLLECTIBLE

col·leen \kä-ˈlēn, ˈkä-ˌ\ *noun* [Irish *cailín*] (1828)
: an Irish girl

col·lege \ˈkä-lij\ *noun, often attributive* [Middle English, from Middle French, from Latin *collegium* society, from *collega* colleague — more at COLLEAGUE] (14th century)
1 : a body of clergy living together and supported by a foundation
2 : a building used for an educational or religious purpose
3 a : a self-governing constituent body of a university offering living quarters and instruction but not granting degrees ⟨Balliol and Magdalen *Colleges* at Oxford⟩ **b :** a preparatory or high school **c :** an independent institution of higher learning offering a course of general studies leading to a bachelor's degree **d :** a part of a university offering a specialized group of courses **e :** an institution offering instruction usually in a professional, vocational, or technical field ⟨business *college*⟩
4 : COMPANY, GROUP; *specifically* **:** an organized body of persons engaged in a common pursuit or having common interests or duties
5 a : a group of persons considered by law to be a unit **b :** a body of electors — compare ELECTORAL COLLEGE
6 : the faculty, students, or administration of a college

College Board *service mark*
— used for administration of tests of aptitude and achievement considered by some colleges in determining admission and placement of students

college try *noun* [from the phrase "give it the old *college try*"] (1947)
: a zealous all-out effort

col·le·gial \kə-ˈlē-j(ē-)əl, *especially for 2a also* -ˈlē-gē-əl\ *adjective* (14th century)
1 : COLLEGIATE 2
2 a : marked by power or authority vested equally in each of a number of colleagues **b :** characterized by equal sharing of authority especially by Roman Catholic bishops
3 : marked by camaraderie among colleagues
— col·le·gial·ly *adverb*

col·le·gi·al·i·ty \-ˌlē-jē-ˈa-lə-tē, -ˌlē-gē-\ *noun* (1887)
: the relationship of colleagues; *specifically* **:** the participation of bishops in the govern-

ment of the Roman Catholic Church in collaboration with the pope

col·le·gian \kə-ˈlē-j(ē-)ən\ *noun* (15th century)
: a student or recent graduate of a college

col·le·giate \kə-ˈlē-jət, -jē-ət\ *adjective* [Medieval Latin *collegiatus,* from Latin *collegium*] (15th century)
1 : of or relating to a collegiate church
2 : of, relating to, or comprising a college
3 : COLLEGIAL 2
4 : designed for or characteristic of college students
— col·le·giate·ly *adverb*

collegiate church *noun* (15th century)
1 : a church other than a cathedral that has a chapter of canons
2 : a church or corporate group of churches under the joint pastorate of two or more ministers

col·le·gi·um \kə-ˈle-gē-əm, -ˈlā-\ *noun, plural* **-gia** \-gē-ə\ *or* **-giums** [modification of Russian *kollegiya,* from Latin *collegium*] (1917)
: a group in which each member has approximately equal power and authority

col·lem·bo·lan \kə-ˈlem-bə-lən\ *noun* [ultimately from *coll-* + Greek *embolos* wedge, stopper — more at EMBOLUS] (1873)
: SPRINGTAIL
— collembolan *or* **col·lem·bo·lous** \-ləs\ *adjective*

col·len·chy·ma \kə-ˈleŋ-kə-mə, kä-\ *noun* [New Latin] (1857)
: a plant tissue that consists of living usually elongated cells with unevenly thickened walls and acts as support especially in areas of primary growth — compare SCLERENCHYMA
— col·len·chy·ma·tous \ˌkä-lən-ˈki-mə-təs, -ˈkī-\ *adjective*

col·let \ˈkä-lət\ *noun* [Middle French, diminutive of *col* collar, from Latin *collum* neck — more at COLLAR] (1528)
: a metal band, collar, ferrule, or flange: as **a :** a casing or socket for holding a tool (as a drill bit) **b :** a circle or flange in which a gem is set

col·le·te·ri·al gland \ˌkä-lə-ˈtir-ē-əl-, ˈter-\ *noun* [New Latin *colleterium* colleterial gland, irregular from Greek *kollan* to glue — more at PROTOCOL] (1870)
: a gland in female insects that secretes a cement by which the eggs are glued together or attached to an external object

col·lide \kə-ˈlīd\ *intransitive verb* **col·lid·ed; col·lid·ing** [Latin *collidere,* from *com-* + *laedere* to injure by striking] (1700)
1 : to come together with solid or direct impact
2 : CLASH

col·lid·er \kə-ˈlī-dər\ *noun* (1980)
: a particle accelerator in which two beams of particles moving in opposite directions are made to collide

col·lie \ˈkä-lē\ *noun* [probably from English dialect *colly* black] (circa 1651)
: any of a breed of large dogs developed in Scotland that occur in rough-coated and smooth-coated varieties

collie

col·lier \ˈkäl-yər\ *noun* [Middle English *colier,* from *col* coal] (13th century)
1 : one that produces charcoal
2 : a coal miner
3 : a ship for transporting coal

col·liery \ˈkäl-yə-rē\ *noun, plural* **-lier·ies** (1635)
: a coal mine and its connected buildings

col·lie·shang·ie \ˈkä-lē-ˌshaŋ-ē, ˈkə-\ *noun* [perhaps from *collie* + *shang* kind of meal] (1737)
Scottish **:** SQUABBLE, BRAWL

col·li·gate \ˈkä-lə-ˌgāt\ *verb* **-gat·ed; -gat·ing** [Latin *colligatus,* past participle of *colligare,* from *com-* + *ligare* to tie — more at LIGATURE] (1545)
transitive verb
1 : to bind, unite, or group together
2 : to subsume (isolated facts) under a general concept
intransitive verb
: to be or become a member of a group or unit
— col·li·ga·tion \ˌkä-lə-ˈgā-shən\ *noun*

col·li·ga·tive \ˈkä-lə-ˌgā-tiv, kə-ˈli-gə-\ *adjective* (1901)
: depending on the number of particles (as molecules) and not on the nature of the particles ⟨pressure is a *colligative* property⟩

col·li·mate \ˈkä-lə-ˌmāt\ *transitive verb* **-mat·ed; -mat·ing** [Latin *collimatus,* past participle of *collimare,* manuscript variant of *collineare* to make straight, from *com-* + *linea* line] (1878)
: to make (as light rays) parallel
— col·li·ma·tion \ˌkä-lə-ˈmā-shən\ *noun*

col·li·ma·tor \ˈkä-lə-ˌmā-tər\ *noun* (1865)
1 : a device for producing a beam of parallel rays (as of light) or for forming an infinitely distant virtual image that can be viewed without parallax
2 : a device for obtaining a beam (as of particles) of limited cross section

col·lin·ear \kə-ˈli-nē-ər, kä-\ *adjective* [International Scientific Vocabulary] (1863)
1 : lying on or passing through the same straight line
2 : having axes lying end to end in a straight line ⟨*collinear* antenna elements⟩
— col·lin·ear·i·ty \-ˌli-nē-ˈar-ə-tē\ *noun*

col·lins \ˈkä-lənz\ *noun* [probably from the name *Collins*] (circa 1887)
: a tall iced drink of soda water, sugar, lemon or lime juice, and liquor (as gin)

col·li·sion \kə-ˈli-zhən\ *noun* [Middle English, from Latin *collision-, collisio,* from *collidere*] (15th century)
1 : an act or instance of colliding **:** CLASH
2 : an encounter between particles (as atoms or molecules) resulting in exchange or transformation of energy
— col·li·sion·al \-ˈlizh-nəl, -ˈli-zhə-nᵊl\ *adjective*
— col·li·sion·al·ly *adverb*

collision course *noun* (1944)
: a course (as of moving bodies or antithetical philosophies) that will result in collision or conflict if continued unaltered

collo- — see COLL-

col·lo·cate \ˈkä-lə-ˌkāt\ *verb* **-cat·ed; -cat·ing** [Latin *collocatus,* past participle of *collocare,* from *com-* + *locare* to place, from *locus* place — more at STALL] (1513)
transitive verb
: to set or arrange in a place or position; *especially* **:** to set side by side
intransitive verb
: to occur in conjunction with something

col·lo·ca·tion \ˌkä-lə-ˈkā-shən\ *noun* (1605)
: the act or result of placing or arranging together; *specifically* **:** a noticeable arrangement or conjoining of linguistic elements (as words)
— col·lo·ca·tion·al \-shnəl, -shə-nᵊl\ *adjective*

col·lo·di·on \kə-ˈlō-dē-ən\ *noun* [modification of New Latin *collodium,* from Greek *kollōdēs* glutinous, from *kolla* glue — more at PROTOCOL] (1851)
: a viscous solution of pyroxylin used especially as a coating for wounds or for photographic films

col·logue \kə-ˈlōg\ *intransitive verb* **col·logued; col·logu·ing** [origin unknown] (1646)
1 *dialect* **:** INTRIGUE, CONSPIRE
2 : to talk privately **:** CONFER

col·loid \ˈkä-ˌlȯid\ *noun* [International Scientific Vocabulary *coll-* + *-oid*] (circa 1852)
1 : a gelatinous or mucinous substance found in tissues in disease (as in the thyroid) or normally
2 a : a substance that consists of particles dispersed throughout another substance which are too small for resolution with an ordinary light microscope but are incapable of passing through a semipermeable membrane **b :** a mixture consisting of a colloid together with the medium in which it is dispersed ⟨smoke is a *colloid*⟩
— col·loi·dal \kə-ˈlȯi-dᵊl, kä-\ *adjective*
— col·loi·dal·ly *adverb*

col·lop \ˈkä-ləp\ *noun* [Middle English] (14th century)
1 : a small piece or slice especially of meat
2 : a fold of fat flesh

col·lo·qui·al \kə-ˈlō-kwē-əl\ *adjective* (1751)
1 : of or relating to conversation **:** CONVERSATIONAL
2 a : used in or characteristic of familiar and informal conversation; *also* **:** unacceptably informal **b :** using conversational style
— colloquial *noun*
— col·lo·qui·al·i·ty \-ˌlō-kwē-ˈa-lə-tē\ *noun*
— col·lo·qui·al·ly \-ˈlō-kwē-ə-lē\ *adverb*

col·lo·qui·al·ism \-ˈlō-kwē-ə-ˌli-zəm\ *noun* (1810)
1 a : a colloquial expression **b :** a local or regional dialect expression
2 : colloquial style

col·lo·quist \ˈkä-lə-kwist\ *noun* (1792)
: SPEAKER

col·lo·qui·um \kə-ˈlō-kwē-əm\ *noun, plural* **-qui·ums** *or* **-quia** \-kwē-ə\ [Latin, colloquy] (1844)
: a usually academic meeting at which specialists deliver addresses on a topic or on related topics and then answer questions relating to them

col·lo·quy \ˈkä-lə-kwē\ *noun, plural* **-quies** [Latin *colloquium,* from *colloqui* to converse, from *com-* + *loqui* to speak] (15th century)
1 : CONVERSATION, DIALOGUE
2 : a high-level serious discussion **:** CONFERENCE

col·lo·type \ˈkä-lə-ˌtīp\ *noun* [International Scientific Vocabulary] (1883)
1 : a photomechanical process for making prints directly from a hardened film of gelatin or other colloid that has ink-receptive and ink-repellent parts
2 : a print made by collotype

col·lude \kə-ˈlüd\ *intransitive verb* **col·lud·ed; col·lud·ing** [Latin *colludere,* from *com-* + *ludere* to play, from *ludus* game — more at LUDICROUS] (1535)
: CONSPIRE, PLOT

col·lu·sion \kə-ˈlü-zhən\ *noun* [Middle English, from Middle French, from Latin *collusion-, collusio,* from *colludere*] (14th century)
: secret agreement or cooperation especially for an illegal or deceitful purpose
— col·lu·sive \-ˈlü-siv, -ziv\ *adjective*
— col·lu·sive·ly *adverb*

col·lu·vi·um \kə-ˈlü-vē-əm\ *noun, plural* **-via** \-vē-ə\ *or* **-vi·ums** [New Latin, from Medieval Latin, offscourings, alteration of Latin

colluvies, from *colluere* to wash, from *com-* + *lavere* to wash — more at LYE] (circa 1936)
: rock detritus and soil accumulated at the foot of a slope
— **col·lu·vi·al** \-vē-əl\ *adjective*

col·ly \'kä-lē\ *transitive verb* **col·lied; col·ly·ing** [alteration of Middle English *colwen,* from (assumed) Old English *colgian,* from Old English *col* coal] (1590)
dialect chiefly British : to blacken with or as if with soot

col·lyr·i·um \kə-'lir-ē-əm\ *noun, plural* **-ia** \-ē-ə\ *or* **-i·ums** [Middle English *collirium,* from Latin *collyrium,* from Greek *kollyrion* pessary, eye salve, from diminutive of *kollyra* roll of bread] (14th century)
: EYEWASH 1

col·ly·wob·bles \'kä-lē-,wä-bəlz\ *noun plural but singular or plural in construction* [perhaps by folk etymology from New Latin *cholera morbus,* literally, the disease cholera] (circa 1823)
: BELLYACHE

colo- — see COL-

col·o·bus monkey \'kä-lə-bəs-\ *noun* [New Latin *colobus,* from Greek *kolobos* docked, mutilated, from *kolos* docked; probably akin to Greek *klan* to break — more at CLAST] (1889)
: any of various long-tailed African monkeys (genus *Colobus* and related genera) — called also *colobus*

co·lo·cate \(,)kō-'lō-,kāt, -lō-'kāt\ *transitive verb* (1965)
: to locate together; *especially* : to place two or more units close together so as to share common facilities

col·o·cynth \'kä-lə-,sin(t)th\ *noun* [Latin *colocynthis,* from Greek *kolokynthis*] (1543)
: a Mediterranean and African herbaceous vine (*Citrullus colocynthis*) related to the watermelon; *also* : its spongy fruit from which a powerful cathartic is prepared

co·log·a·rithm \(,)kō-'lò-gə-,ri-thəm, -'lä-\ *noun* (1881)
: the logarithm of the reciprocal of a number

co·logne \kə-'lōn\ *noun* [*Cologne,* Germany] (1814)
1 : a perfumed liquid composed of alcohol and fragrant oils
2 : a cream or paste of cologne sometimes formed into a semisolid stick
— **co·logned** \-'lōnd\ *adjective*

¹co·lon \'kō-lən\ *noun, plural* **colons** *or* **co·la** \-lə\ [Middle English, from Latin, from Greek *kolon*] (14th century)
: the part of the large intestine that extends from the cecum to the rectum
— **co·lon·ic** \kō-'lä-nik, kə-\ *adjective*

²colon *noun, plural* **colons** *or* **co·la** \-lə\ [Latin, part of a poem, from Greek *kōlon* limb, part of a strophe] (circa 1550)
1 *plural cola* : a rhythmical unit of an utterance; *specifically, in Greek or Latin verse* : a system or series of from two to not more than six feet having a principal accent and forming part of a line
2 *plural colons* **a** : a punctuation mark : used chiefly to direct attention to matter (as a list, explanation, or quotation) that follows **b** : the sign : used between the parts of a numerical expression of time in hours and minutes (as in 1:15) or in hours, minutes, and seconds (as in 8:25:30), in a bibliographical reference (as in *Nation* 130:20), in a ratio where it is usually read as "to" (as in 4:1 read "four to one"), or in a proportion where it is usually read as "is to" or when doubled as "as" (as in 2:1::8:4 read "two is to one as eight is to four")

³co·lon \kò-'lōⁿ, kə-'lōn\ *noun* [French, from Latin *colonus*] (1888)
: a colonial farmer or plantation owner

co·lón *also* **co·lone** \kə-'lōn\ *noun, plural* **co·lo·nes** \-'lō-,nās\ [Spanish *colón,* from Cristóbal *Colón* Christopher Columbus] (1916)
— see MONEY table

colon bacillus *noun* (circa 1909)
: any of various bacilli (especially genera *Escherichia* and *Aerobacter*) that are normally commensal in vertebrate intestines; *especially* : one (*E. coli*) used extensively in genetic research

col·o·nel \'kər-nᵊl\ *noun* [alteration of *coronel,* from Middle French, modification of Old Italian *colonnello* column of soldiers, colonel, diminutive of *colonna* column, from Latin *columna*] (1548)
1 a : a commissioned officer in the army, air force, or marine corps ranking above a lieutenant colonel and below a brigadier general **b** : LIEUTENANT COLONEL
2 : a minor titular official of a state especially in southern or midland U.S. — used as an honorific title ◆
— **col·o·nel·cy** \-nᵊl-sē\ *noun*

Colonel Blimp \-'blimp\ *noun* [*Colonel Blimp,* cartoon character created by David Low] (1937)
: a pompous person with out-of-date or ultraconservative views; *broadly* : REACTIONARY
— **Colonel Blimp·ism** \-'blim-,pi-zəm\ *noun*

¹co·lo·nial \kə-'lō-nē-əl, -nyəl\ *adjective* (1776)
1 : of, relating to, or characteristic of a colony
2 *often capitalized* : of or relating to the original 13 colonies forming the United States: as **a** : made or prevailing in America during the colonial period ⟨*colonial* architecture⟩ **b** : adapted from or reminiscent of an American colonial mode of design ⟨*colonial* furniture⟩
3 : forming or existing in a colony ⟨*colonial* organisms⟩
4 : possessing or composed of colonies ⟨a *colonial* empire⟩
— **co·lo·nial·ize** \-nē-ə-,līz, -nyə-,līz\ *transitive verb*
— **co·lo·nial·ly** *adverb*
— **co·lo·nial·ness** *noun*

²colonial *noun* (1798)
1 : a member or inhabitant of a colony
2 a : a product made for use in a colony **b** : a product (as a house) exhibiting colonial style

co·lo·nial·ism \kə-'lō-nē-ə-,li-zəm, -nyə-,li-\ *noun* (1853)
1 : the quality or state of being colonial
2 : something characteristic of a colony
3 a : control by one power over a dependent area or people **b** : a policy advocating or based on such control
— **co·lo·nial·ist** \-list\ *noun or adjective*
— **co·lo·nial·is·tic** \-,lō-nē-ə-'lis-tik, -nyə-'lis-\ *adjective*

¹co·lon·ic \kō-'lä-nik, kə-\ *adjective* (circa 1885)
: of or relating to the colon of the intestine

²colonic *noun* (1939)
: irrigation of the colon : ENEMA

col·o·ni·sa·tion, col·o·nise *British variant of* COLONIZATION, COLONIZE

col·o·nist \'kä-lə-nist\ *noun* (1701)
1 : a member or inhabitant of a colony
2 : one that colonizes or settles in a new country

col·o·ni·za·tion \,kä-lə-nə-'zā-shən\ *noun* (1770)
: an act or instance of colonizing
— **col·o·ni·za·tion·ist** \-sh(ə-)nist\ *noun*

col·o·nize \'kä-lə-,nīz\ *verb* **-nized; -niz·ing** (1622)
transitive verb
1 a : to establish a colony in or on or of **b** : to establish in a colony
2 : to send illegal or irregularly qualified voters into ⟨*colonizing* doubtful districts⟩
3 : to infiltrate with usually subversive militants for propaganda and strategy reasons ⟨*colonize* industries⟩
intransitive verb
: to make or establish a colony : SETTLE
— **col·o·niz·er** *noun*

col·on·nade \,kä-lə-'nād\ *noun* [French, from Italian *colonnato,* from *colonna* column] (1718)
: a series of columns set at regular intervals and usually supporting the base of a roof structure
— **col·on·nad·ed** \-'nā-dəd\ *adjective*

co·lo·nus \kə-'lō-nəs\ *noun, plural* **-ni** \-,nī, -(,)nē\ [Latin, literally, farmer] (1888)
: a free-born serf in the later Roman Empire who could sometimes own property but who was bound to the land and obliged to pay a rent usually in produce

col·o·ny \'kä-lə-nē\ *noun, plural* **-nies** [Middle English *colonie,* from Middle French & Latin; Middle French, from Latin *colonia,* from *colonus* farmer, colonist, from *colere* to cultivate — more at WHEEL] (14th century)
1 a : a body of people living in a new territory but retaining ties with the parent state **b** : the territory inhabited by such a body
2 : a distinguishable localized population within a species ⟨*colony* of termites⟩
3 a : a circumscribed mass of microorganisms usually growing in or on a solid medium **b** : the aggregation of zooids of a compound animal
4 a : a group of individuals or things with common characteristics or interests situated in close association ⟨an artist *colony*⟩ **b** : the section occupied by such a group
5 : a group of persons institutionalized away from others ⟨a leper *colony*⟩ ⟨a penal *colony*⟩; *also* : the land or buildings occupied by such a group

colony–stimulating factor *noun* (1969)
: any of several glycoproteins that promote the differentiation of stem cells especially into blood granulocytes and macrophages and that stimulate their proliferation into colonies in culture

col·o·phon \'kä-lə-fən, -,fän\ *noun* [Latin, from Greek *kolophōn* summit, finishing touch; perhaps akin to Latin *culmen* top — more at HILL] (1774)
1 : an inscription placed at the end of a book or manuscript usually with facts relative to its production
2 : an identifying device used by a printer or a publisher

co·lo·pho·ny \kə-'lä-fə-nē, 'kä-lə-,fō-\ *noun, plural* **-nies** [Middle English *colophonie,* ultimately from Greek *Kolophōn* Colophon, an Ionian city] (14th century)
: ROSIN

¹col·or \'kə-lər\ *noun, often attributive* [Middle English *colour,* from Old French, from Latin

◇ WORD HISTORY

colonel The pronunciation of *colonel* as if it were spelled *kernel* is one of the more striking oddities of the English vocabulary. The source of this discrepancy was not sheer contrariness, for at one time a spelling *coronel* existed to match the pronunciation with \r\ and a pronunciation \'kəl-nᵊl\ existed to match the spelling *colonel.* For a time both forms competed, with *colonel* holding the ground long enough for this spelling to become standard. Later the pronunciation with \r\ made a resurgence and finally won the day, but a corresponding spelling change to *r* never took place. Both *coronel* and *colonel* were borrowed from 16th century French; the French form *coronel* resulted from a phonetic change called dissimilation, by which one of two identical or similar sounds in the same word changes to a contrasting sound. In this case the first \l\ became \r\.

color; akin to Latin *celare* to conceal — more at HELL] (13th century)
1 a : a phenomenon of light (as red, brown, pink, or gray) or visual perception that enables one to differentiate otherwise identical objects **b :** the aspect of objects and light sources that may be described in terms of hue, lightness, and saturation for objects and hue, brightness, and saturation for light sources **c :** a hue as contrasted with black, white, or gray
2 a : an outward often deceptive show **:** APPEARANCE ⟨his story has the *color* of truth⟩ **b :** a legal claim to or appearance of a right, authority, or office **c :** a pretense offered as justification **:** PRETEXT ⟨she could have drawn from the Versailles treaty the *color* of legality for any action she chose —*Yale Review*⟩ **d :** an appearance of authenticity **:** PLAUSIBILITY ⟨lending *color* to this notion⟩
3 : complexion tint: **a :** the tint characteristic of good health **b :** BLUSH
4 a : vividness or variety of effects of language **b :** LOCAL COLOR
5 a : an identifying badge, pennant, or flag — usually used in plural ⟨a ship sailing under Swedish *colors*⟩ **b :** colored clothing distinguishing one as a member of a particular group or representative of a particular person or thing — usually used in plural ⟨a jockey wearing the *colors* of the stable⟩
6 a *plural* **:** position as to a question or course of action **:** STAND ⟨the USSR changed neither its *colors* nor its stripes during all of this —Norman Mailer⟩ **b :** CHARACTER, NATURE — usually used in plural ⟨showed himself in his true *colors*⟩
7 a : the use or combination of colors **b :** two or more hues employed in a medium of presentation ⟨movies in *color*⟩ ⟨*color* television⟩
8 *plural* **a :** a naval or nautical salute to a flag being hoisted or lowered **b :** ARMED FORCES
9 : VITALITY, INTEREST ⟨the play had a good deal of *color* to it⟩
10 : something used to give color **:** PIGMENT
11 : tonal quality in music ⟨the *color* and richness of the cello⟩
12 : skin pigmentation especially other than white characteristic of race ⟨a person of *color*⟩
13 : a small particle of gold in a gold miner's pan after washing
14 : analysis of game action or strategy, statistics and background information on participants, and often anecdotes provided by a sportscaster to give variety and interest to the broadcast of a game or contest
15 : a hypothetical property of quarks that differentiates each type into three forms having a distinct role in binding quarks together
— **col·or·ism** \-lə-ˌri-zəm\ *noun*

²color (14th century)
transitive verb
1 a : to give color to **b :** to change the color of (as by dyeing, staining, or painting)
2 : to change as if by dyeing or painting: as **a :** MISREPRESENT, DISTORT **b :** GLOSS, EXCUSE ⟨*color* a lie⟩ **c :** INFLUENCE ⟨the lives of most of us have been *colored* by politics —Christine Weston⟩
3 : CHARACTERIZE, LABEL ⟨call it progress, *color* it inevitable with shades of job security —C. E. Price⟩
intransitive verb
: to take on color; *specifically* **:** BLUSH
— **col·or·er** \ˈkə-lər-ər\ *noun*

col·or·able \ˈkə-lə-rə-bəl, ˈkəl-rə-\ *adjective* (14th century)
1 : seemingly valid or genuine
2 : intended to deceive **:** COUNTERFEIT
— **col·or·ably** \-blē\ *adverb*

Col·o·ra·do blue spruce \ˌkä-lə-ˈra-dō-, -ˈrä-\ *noun* [*Colorado,* state of the U.S.] (1897)
: BLUE SPRUCE

Colorado potato beetle *noun* (1868)
: a black-and-yellow striped beetle (*Leptinotarsa decemlineata*) that feeds on the leaves of the potato — called also *potato beetle, potato bug*

Colorado potato beetle

col·or·ant \ˈkə-lə-rənt\ *noun* (1884)
: a substance used for coloring a material **:** DYE, PIGMENT

col·or·a·tion \ˌkə-lə-ˈrā-shən\ *noun* (1617)
1 a : the state of being colored ⟨the dark *coloration* of his skin⟩ **b :** use or choice of colors (as by an artist) **c :** arrangement of colors ⟨the *coloration* of a butterfly's wing⟩
2 a : characteristic quality ⟨the newspapers . . . took on the former *coloration* of the magazine —L. B. Seltzer⟩ **b :** aspect suggesting an attitude **:** PERSUASION ⟨the chameleon talent for taking on the intellectual *coloration* of whatever idea he happened to fasten onto —Budd Schulberg⟩
3 : subtle variation of intensity or quality of tone ⟨a wide range of *coloration* from the orchestra⟩

col·or·a·tu·ra \ˌkə-lə-rə-ˈtu̇r-ə, -ˈtyu̇r-\ *noun* [obsolete Italian, literally, coloring, from Late Latin, from Latin *coloratus,* past participle of *colorare* to color, from *color*] (circa 1740)
1 : elaborate embellishment in vocal music; *broadly* **:** music with ornate figuration
2 : a soprano with a light agile voice specializing in coloratura

color bar *noun* (1913)
: a barrier preventing colored persons from participating with whites in various activities — called also *color line*

col·or–bear·er \ˈkə-lər-ˌbar-ər, -ˌber-\ *noun* (circa 1891)
: one that carries a color or standard especially in a military parade or drill

col·or–blind \-ˌblīnd\ *adjective* (1854)
1 : affected with partial or total inability to distinguish one or more chromatic colors
2 : INSENSITIVE, OBLIVIOUS
3 : not recognizing differences of race ⟨tried to get the welfare establishment . . . to abandon its *color blind* policy —D. P. Moynihan⟩; *especially* **:** free from racial prejudice ⟨a white man with an invisible black skin in a *color-blind* community —James Farmer⟩
— **color blindness** *noun*

col·or·bred \-ˌbred\ *adjective* (1948)
: selectively bred for the development of particular colors ⟨pure *colorbred* dogs⟩

co·lo·rec·tal \ˌkō-lō-ˈrek-tᵊl\ *adjective* (1962)
: relating to or affecting the colon and rectum ⟨*colorectal* cancer⟩

¹col·ored \ˈkə-lərd\ *adjective* (14th century)
1 : having color
2 a : COLORFUL **b :** marked by exaggeration or bias
3 a : of a race other than the white; *especially* **:** BLACK 2b — sometimes taken to be offensive **b :** of mixed race — sometimes taken to be offensive
4 : of or relating to colored persons — sometimes taken to be offensive

²colored *noun, plural* **colored** *or* **coloreds** *often capitalized* (1916)
: a colored person — sometimes taken to be offensive

col·or·fast \ˈkə-lər-ˌfast\ *adjective* (1926)
: having color that retains its original hue without fading or running
— **col·or·fast·ness** \-ˌfas(t)-nəs\ *noun*

col·or–field \-ˌfē(ə)ld\ *noun, often attributive* (1964)
: abstract painting in which color is emphasized and form and surface are correspondingly de-emphasized

color filter *noun* (1900)
: FILTER 3b

col·or·ful \ˈkə-lər-fəl\ *adjective* (1889)
1 : having striking colors
2 : full of variety or interest
— **col·or·ful·ly** \-f(ə-)lē\ *adverb*
— **col·or·ful·ness** \-fəl-nəs\ *noun*

color guard *noun* (circa 1823)
: an honor guard for the colors of an organization

col·or·if·ic \ˌkə-lə-ˈri-fik\ *adjective* (1676)
: capable of communicating color

col·or·im·e·ter \ˌkə-lə-ˈri-mə-tər\ *noun* [International Scientific Vocabulary] (circa 1872)
: an instrument or device for determining and specifying colors; *specifically* **:** one used for chemical analysis by comparison of a liquid's color with standard colors
— **col·or·i·met·ric** \ˌkə-lə-rə-ˈme-trik\ *adjective*
— **col·or·i·met·ri·cal·ly** \-tri-k(ə-)lē\ *adverb*
— **col·or·im·e·try** \ˌkə-lə-ˈri-mə-trē\ *noun*

col·or·ing \ˈkə-lə-riŋ\ *noun* (14th century)
1 a : the act of applying colors **b :** something that produces color or color effects **c** (1) **:** the effect produced by applying or combining colors (2) **:** natural color (3) **:** COMPLEXION, COLORATION **d :** change of appearance (as by adding color)
2 : INFLUENCE, BIAS
3 : COLOR 4
4 : TIMBRE, QUALITY

col·or·ist \ˈkə-lə-rist\ *noun* (1686)
: one that colors or deals with color
— **col·or·is·tic** \ˌkə-lə-ˈris-tik\ *adjective*
— **col·or·is·ti·cal·ly** \-ti-k(ə-)lē\ *adverb*

col·or·ize \ˈkə-lə-ˌrīz\ *transitive verb* **-ized; -iz·ing** (1979)
: to add color to (a black-and-white film) by means of a computer
— **col·or·i·za·tion** \ˌkə-lə-rə-ˈzā-shən\ *noun*

col·or·less \ˈkə-lər-ləs\ *adjective* (14th century)
: lacking color: as **a :** PALLID, BLANCHED **b :** DULL, UNINTERESTING
— **col·or·less·ly** *adverb*
— **col·or·less·ness** *noun*

col·or·man \-ˌman\ *noun* (1973)
: a sportscaster who provides color

color phase *noun* (1927)
1 : a seasonally variant pelage color
2 a : a genetic variant manifested by the occurrence of a skin or pelage color unlike the wild type of the animal group in which it appears **b :** an individual marked by such a variant

col·or·point shorthair \ˈkə-lər-ˌpȯint-\ *noun* (1974)
: any of a breed of domestic cats of Siamese type and coat pattern but occurring in different colors — called also *colorpoint*

color temperature *noun* (1916)
: the temperature at which a blackbody emits radiant energy competent to evoke a color the same as that evoked by radiant energy from a given source (as a lamp)

color wheel *noun* (circa 1893)
: a circular diagram of the spectrum used to show the relationships between the colors

co·los·sal \kə-ˈlä-səl\ *adjective* (1712)
1 : of, relating to, or resembling a colossus
2 : of a bulk, extent, power, or effect approaching or suggesting the stupendous or incredible
3 : of an exceptional or astonishing degree
synonym see ENORMOUS
— **co·los·sal·ly** \-sə-lē\ *adverb*

col·os·se·um \ˌkä-lə-ˈsē-əm\ *noun* [Medieval Latin, from Latin, neuter of *colosseus* colossal, from *colossus*] (circa 1715)

1 *capitalized* **:** an amphitheater built in Rome in the first century A.D.
2 : COLISEUM 2

Co·los·sians \kə-ˈlä-shənz *also* -shē-ənz, -sē-\ *noun plural but singular in construction*
: a letter written by Saint Paul to the Christians of Colossae and included as a book in the New Testament — see BIBLE table

co·los·sus \kə-ˈlä-səs\ *noun, plural* **co·los·si** \-ˈlä-ˌsī\ [Latin, from Greek *kolossos*] (14th century)
1 : a statue of gigantic size and proportions
2 : one that resembles a colossus in size or scope

co·los·to·my \kə-ˈläs-tə-mē\ *noun, plural* **-mies** [International Scientific Vocabulary [2]*col-* + *-stomy*] (1888)
: surgical formation of an artificial anus

co·los·trum \kə-ˈläs-trəm\ *noun* [Latin, beestings] (1577)
: milk secreted for a few days after parturition and characterized by high protein and antibody content
— co·los·tral \-trəl\ *adjective*

col·our \ˈkə-lər\ *chiefly British variant of* COLOR

-colous *adjective combining form* [Latin *-cola* inhabitant; akin to Latin *colere* to inhabit — more at WHEEL]
: living or growing in or on ⟨areni*colous*⟩

col·por·tage \ˈkäl-ˌpōr-tij, -ˌpȯr-; ˌkäl-pōr-ˈtäzh, -pȯr-\ *noun* (circa 1846)
: a colporteur's work

col·por·teur \ˈkäl-ˌpōr-tər, -ˌpȯr-; ˌkäl-pōr-ˈtər, -pȯr-\ *noun* [French, alteration of Middle French *comporteur,* from *comporter* to bear, peddle] (1796)
: a peddler of religious books

colt \ˈkōlt\ *noun* [Middle English, from Old English; akin to Swedish dialect *kult* half-grown pig] (before 12th century)
1 a : FOAL **b :** a young male horse that is either sexually immature or has not attained an arbitrarily designated age
2 : a young untried person

colt·ish \ˈkōl-tish\ *adjective* (14th century)
1 a : not subjected to discipline **b :** FRISKY, PLAYFUL
2 : of, relating to, or resembling a colt
— colt·ish·ly *adverb*
— colt·ish·ness *noun*

colts·foot \ˈkōlts-ˌfu̇t\ *noun, plural* **colts·foots** (14th century)
: any of various plants with large rounded leaves resembling the foot of a colt; *especially* **:** a perennial composite herb (*Tussilago farfara*) with yellow flower heads appearing before the leaves

col·u·brid \ˈkäl-yə-brəd, ˈkä-lə-\ *noun* [ultimately from Latin *colubra* snake] (1887)
: any of a large cosmopolitan family (Colubridae) of chiefly nonvenomous snakes
— colubrid *adjective*

col·u·brine \-ˌbrīn\ *adjective* (circa 1528)
1 : of, relating to, or resembling a snake
2 : of or relating to colubrids **:** COLUBRID

co·lu·go \kə-ˈlü-(ˌ)gō\ *noun, plural* **-gos** [perhaps from a language of the Philippines] (circa 1890)
: FLYING LEMUR

col·um·bar·i·um \ˌkä-ləm-ˈbar-ē-əm, -ˈber-\ *noun, plural* **-ia** \-ē-ə\ [Latin, literally, dovecote, from *columba* dove] (1846)
1 : a structure of vaults lined with recesses for cinerary urns
2 : a recess in a columbarium

Co·lum·bia \kə-ˈləm-bē-ə\ *noun* [New Latin, from Christopher *Columbus*] (1775)
: the United States

Co·lum·bi·an \-bē-ən\ *adjective* (1757)
: of or relating to the United States or to Christopher Columbus

col·um·bine \ˈkä-ləm-ˌbīn\ *noun* [Middle English, from Medieval Latin *columbina,* from Latin, feminine of *columbinus* like a dove, from *columba* dove; akin to Old High German *holuntar* elder tree, Greek *kolymbos* a small grebe, *kelainos* black] (14th century)
: any of a genus (*Aquilegia*) of plants of the buttercup family with irregular showy spurred flowers: as **a :** a red-flowered plant (*A. canadensis*) of eastern North America **b :** a blue-flowered plant (*A. caerulea*) of the Rocky Mountains

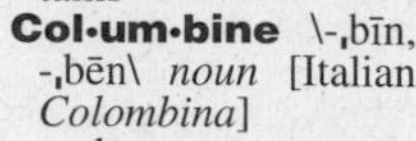

columbine a

Col·um·bine \-ˌbīn, -ˌbēn\ *noun* [Italian *Colombina*]
: the saucy sweetheart of Harlequin in comedy and pantomime

co·lum·bite \kə-ˈləm-ˌbīt, ˈkä-ləm-\ *noun* [New Latin *columbium*] (1805)
: a black mineral consisting mostly of iron and niobium

co·lum·bi·um \kə-ˈləm-bē-əm\ *noun* [New Latin, from *Columbia*] (1801)
: NIOBIUM

Co·lum·bus Day \kə-ˈləm-bəs-\ *noun* (1893)
1 : October 12 formerly observed as a legal holiday in many states of the U.S. in commemoration of the landing of Columbus in the Bahamas in 1492
2 : the second Monday in October observed as a legal holiday in many states of the U.S.

col·u·mel·la \ˌkäl-yə-ˈme-lə\ *noun, plural* **-mel·lae** \-ˈme-(ˌ)lē, -ˌlī\ [New Latin, from Latin, diminutive of *columna*] (circa 1755)
1 : the central column or axis of a spiral univalve shell
2 : the axis of the capsule in mosses and in some liverworts
3 a : the bony or partly cartilaginous rod connecting the tympanic membrane with the internal ear in birds and in many reptiles and amphibians **b :** the bony central axis of the cochlea
4 : the central sterile portion of the sporangium in various fungi (*Mucor* and related genera)
— col·u·mel·lar \-ˈme-lər\ *adjective*

col·umn \ˈkä-ləm *also* ˈkäl-yəm\ *noun* [Middle English *columne,* from Middle French *colomne,* from Latin *columna,* from *columen* top; akin to Latin *collis* hill — more at HILL] (15th century)
1 a : a vertical arrangement of items printed or written on a page **b :** one of two or more vertical sections of a printed page separated by a rule or blank space **c :** an accumulation arranged vertically **:** STACK **d :** one in a usually regular series of newspaper or magazine articles ⟨gossip *column*⟩
2 : a supporting pillar; *especially* **:** one consisting of a usually round shaft, a capital, and a base
3 a : something resembling a column in form, position, or function ⟨a *column* of water⟩ **b :** a tube or cylinder in which a chromatographic separation takes place
4 : a long row (as of soldiers)
5 : one of the vertical lines of elements of a determinant or matrix
— col·umned \-ləmd, -yəmd\ *adjective*

CORNICE
FRIEZE
ARCHITRAVE
CAPITAL
SHAFT
BASE
ENTABLATURE
COLUMN
PEDESTAL

column 2

co·lum·nar \kə-ˈləm-nər\ *adjective* (1728)
1 : of, relating to, resembling, or characterized by columns
2 : of, relating to, being, or composed of tall narrow somewhat cylindrical or prismatic epithelial cells

co·lum·ni·a·tion \kə-ˌləm-nē-ˈā-shən\ *noun* [probably from *intercolumniation*] (1664)
: the employment or the arrangement of columns in a structure

col·um·nist \ˈkä-ləm-nist, -lə-mist *also* ˈkäl-yəm-nist, -yə-mist\ *noun* (1920)
: one who writes a newspaper or magazine column
— col·um·nis·tic \ˌkä-ləm-ˈnis-tik *also* ˌkäl-yəm-\ *adjective*

col·za \ˈkäl-zə, ˈkōl-\ *noun* [French, from Dutch *koolzaad,* from Middle Dutch *coolsaet,* from *coole* cabbage + *saet* seed] (1712)
1 : any of several coles; *especially* **:** one (as rape) producing seed used as a source of oil
2 : RAPESEED

com- *or* **col-** *or* **con-** *prefix* [Middle English, from Old French, from Latin, with, together, thoroughly — more at CO-]
: with **:** together **:** jointly — usually *com-* before *b, p,* or *m* ⟨*com*mingle⟩, *col-* before *l* ⟨*col*linear⟩, and *con-* before other sounds ⟨*con*centrate⟩

[1]**co·ma** \ˈkō-mə\ *noun* [New Latin, from Greek *kōma* deep sleep] (1646)
1 : a state of profound unconsciousness caused by disease, injury, or poison
2 : a state of mental or physical sluggishness **:** TORPOR

[2]**coma** *noun, plural* **co·mae** \-ˌmē, -ˌmī\ [Latin, hair, from Greek *komē*] (1669)
1 : a tufted bunch (as of branches, bracts, or seed hairs)
2 : the head of a comet usually containing a nucleus
3 : an optical aberration in which the image of a point source is a comet-shaped blur
— co·mat·ic \kō-ˈma-tik\ *adjective*

Co·ma Ber·e·ni·ces \ˈkō-mə-ˌber-ə-ˈnī-(ˌ)sēz\ *noun* [Latin (genitive *Comae Berenices*), literally, Berenice's hair]
: a constellation north of Virgo and between Boötes and Leo

co·mak·er \(ˌ)kō-ˈmā-kər\ *noun* (circa 1934)
: one that participates in an agreement; *specifically* **:** one who stands to meet a financial obligation in the event of the maker's default

Co·man·che \kə-ˈman-chē\ *noun, plural* **Comanche** *or* **Comanches** [American Spanish, from Southern Paiute *kimmancinʷi* Shoshones, strangers] (1806)
1 : a member of an American Indian people ranging from Wyoming and Nebraska south into New Mexico and northwestern Texas
2 : the Uto-Aztecan language of the Comanche people

co·mate \(ˌ)kō-ˈmāt, ˈkō-ˌ\ *noun* (1576)
: COMPANION

co·ma·tose \ˈkō-mə-ˌtōs, ˈkä-\ *adjective* [French *comateux,* from Greek *kōmat-, kōma*] (1755)
1 : of, resembling, or affected with coma
2 : characterized by lethargic inertness **:** TORPID ⟨a *comatose* economy⟩

[1]**comb** \ˈkōm\ *noun* [Middle English, from Old English *camb;* akin to Old High German *kamb* comb, Greek *gomphos* tooth] (before 12th century)
1 a : a toothed instrument used especially for adjusting, cleaning, or confining hair **b :** a structure resembling such a comb; *especially* **:** any of several toothed devices used in handling or ordering textile fibers **c :** CURRYCOMB
2 a : a fleshy crest on the head of the domestic fowl and other gallinaceous birds **b :** something (as the ridge of a roof) resembling the comb of a cock
3 : HONEYCOMB
— combed \ˈkōmd\ *adjective*
— comb·like \ˈkōm-ˌlīk\ *adjective*

[2]**comb** (14th century)
transitive verb

1 : to draw a comb through for the purpose of arranging or cleaning
2 : to pass across with a scraping or raking action
3 a : to eliminate (as with a comb) by a thorough going-over **b :** to search or examine systematically
4 : to use in a combing action
intransitive verb
1 *of a wave or its crest* **:** to roll over or break into foam
2 : to make a thorough search ⟨*comb* through the classified ads⟩

¹com·bat \'käm-ˌbat\ *noun* (1546)
1 : a fight or contest between individuals or groups
2 : CONFLICT, CONTROVERSY
3 : active fighting in a war **:** ACTION ⟨casualties suffered in *combat*⟩

²com·bat \kəm-'bat, 'käm-ˌ\ *verb* **-bat·ed** *or* **-bat·ted; -bat·ing** *or* **-bat·ting** [Middle French *combattre,* from (assumed) Vulgar Latin *combattere,* from Latin *com-* + *battuere* to beat] (1564)
intransitive verb
: to engage in combat **:** FIGHT
transitive verb
1 : to fight with **:** BATTLE
2 : to struggle against; *especially* **:** to strive to reduce or eliminate ⟨*combat* pollution⟩
synonym see OPPOSE

³com·bat \'käm-ˌbat\ *adjective* (1825)
1 : relating to combat ⟨*combat* missions⟩
2 : designed or destined for combat ⟨*combat* boots⟩ ⟨*combat* troops⟩

com·bat·ant \kəm-'ba-tᵊnt *also* 'käm bə-tənt\ *noun* (15th century)
: one that is engaged in or ready to engage in combat
— combatant *adjective*

combat fatigue *noun* (1943)
: a traumatic psychoneurotic reaction or an acute psychotic reaction occurring under conditions (as wartime combat) that cause intense stress — called also *battle fatigue*

com·bat·ive \kəm-'ba-tiv\ *adjective* (circa 1834)
: marked by eagerness to fight or contend
— com·bat·ive·ly *adverb*
— com·bat·ive·ness *noun*

combe \'küm, 'kōm\ *noun* [Middle English *coumbe, cumbe,* from Old English *cumb,* of Celtic origin; akin to Welsh *cwm* valley] (before 12th century)
1 *British* **:** a deep narrow valley
2 *British* **:** a valley or basin on the flank of a hill

comb·er \'kō-mər\ *noun* (circa 1682)
1 : one that combs
2 : a long curling wave of the sea

com·bi·na·tion \ˌkäm-bə-'nā-shən\ *noun, often attributive* (14th century)
1 a : a result or product of combining; *especially* **:** an alliance of individuals, corporations, or states united to achieve a social, political, or economic end **b :** two or more persons working as a team
2 : an ordered sequence: as **a :** a sequence of letters or numbers chosen in setting a lock; *also* **:** the mechanism operating or moved by the sequence **b :** a rapid sequence of punches in boxing **c :** any subset of a set considered without regard to order within the subset
3 : any of various one-piece undergarments for the upper and lower parts of the body — usually used in plural
4 : an instrument designed to perform two or more tasks
5 a : the act or process of combining; *especially* **:** that of uniting to form a chemical compound **b :** the quality or state of being combined
— com·bi·na·tion·al \-shnəl, -shə-nᵊl\ *adjective*

combination shot *noun* (circa 1909)
: a shot in pool in which a ball is pocketed by an object ball

com·bi·na·tive \'käm-bə-ˌnā-tiv, kəm-'bī-nə-\ *adjective* (1855)
1 : tending or able to combine
2 : resulting from combination

com·bi·na·to·ri·al \ˌkäm-bə-nə-'tōr-ē-əl, kəm-ˌbī-nə-, -(ˌ)bi-, -'tȯr-\ *adjective* (1818)
1 : of, relating to, or involving combinations
2 : of or relating to the arrangement of, operation on, and selection of discrete mathematical elements belonging to finite sets or making up geometric configurations
— com·bi·na·to·ri·al·ly \-ē-ə-lē\ *adverb*

com·bi·na·tor·ics \-'tȯr-iks, -'tär-\ *noun plural but singular in construction* (1951)
: combinatorial mathematics

com·bi·na·to·ry \kəm-'bī-nə-ˌtōr-ē, -ˌtȯr-\ *adjective* (1647)
: COMBINATIVE

¹com·bine \kəm-'bīn\ *verb* **com·bined; com·bin·ing** [Middle English, from Middle French *combiner,* from Late Latin *combinare,* from Latin *com-* + *bini* two by two — more at BIN-] (15th century)
transitive verb
1 a : to bring into such close relationship as to obscure individual characters **:** MERGE **b :** to cause to unite into a chemical compound **c :** to unite into a single number or expression ⟨*combine* fractions and simplify⟩
2 : INTERMIX, BLEND
3 : to possess in combination
intransitive verb
1 a : to become one **b :** to unite to form a chemical compound
2 : to act together
synonym see JOIN
— com·bin·able \-'bī-nə-bəl\ *adjective*
— com·bin·er *noun*

²com·bine \'käm-ˌbīn\ *noun* (1886)
1 : a combination especially of business or political interests
2 : a harvesting machine that heads, threshes, and cleans grain while moving over a field

³com·bine \'käm-ˌbīn\ *verb* **com·bined; com·bin·ing** (1926)
transitive verb
: to harvest with a combine
intransitive verb
: to combine a crop

comb·ing \'kō-miŋ\ *variant of* COAMING

comb·ings \'kō-miŋz\ *noun plural* (1656)
: loose hair removed by a comb

combing wool *noun* (1757)
: long-staple strong-fibered wool found suitable for combing and used especially in the manufacture of worsteds

combining form *noun* (1884)
: a linguistic form that occurs only in compounds or derivatives and can be distinguished descriptively from an affix by its ability to occur as one immediate constituent of a form whose only other immediate constituent is an affix (as *cephal-* in *cephalic*) or by its being an allomorph of a morpheme having another allomorph that may occur alone or can be distinguished historically from an affix by the fact that it is borrowed from another language in which it is descriptively a word or a combining form

comb jelly *noun* (circa 1889)
: CTENOPHORE

com·bo \'käm-(ˌ)bō\ *noun, plural* **combos** [*combination* + ¹*-o*] (1924)
1 : a usually small jazz or dance band
2 : COMBINATION

com·bust \kəm-'bəst\ *verb* [Latin *combustus,* past participle of *comburere* to burn up, irregular from *com-* + *urere* to burn — more at EMBER] (15th century)
: BURN

com·bus·ti·ble \kəm-'bəs-tə-bəl\ *adjective* (1529)
1 : capable of combustion
2 : easily excited
— com·bus·ti·bil·i·ty \-ˌbəs-tə-'bi-lə-tē\ *noun*
— combustible *noun*
— com·bus·ti·bly \-'bəs-tə-blē\ *adverb*

com·bus·tion \kəm-'bəs-chən\ *noun* (15th century)
1 : an act or instance of burning
2 : a usually rapid chemical process (as oxidation) that produces heat and usually light; *also* **:** a slower oxidation (as in the body)
3 : violent agitation **:** TUMULT
— com·bus·tive \-'bəs-tiv\ *adjective*

com·bus·tor \-'bəs-tər\ *noun* (1945)
: a chamber (as in a gas turbine or a jet engine) in which combustion occurs — called also *combustion chamber*

¹come \'kəm\ *verb* **came** \'kām\; **come; com·ing** \'kə-miŋ\ [Middle English, from Old English *cuman;* akin to Old High German *queman* to come, Latin *venire,* Greek *bainein* to walk, go] (before 12th century)
intransitive verb
1 a : to move toward something **:** APPROACH ⟨*come* here⟩ **b :** to move or journey to a vicinity with a specified purpose ⟨*come* see us⟩ ⟨*come* and see what's going on⟩ **c** (1) **:** to reach a particular station in a series ⟨now we *come* to the section on health⟩ (2) **:** to arrive in due course ⟨the time has *come*⟩ **d** (1) **:** to approach in kind or quality ⟨this *comes* near perfection⟩ (2) **:** to reach a condition or conclusion ⟨*came* to regard him as a friend⟩ ⟨*come* to think of it, you may be right⟩ **e** (1) **:** to advance toward accomplishment ⟨learning new ways doesn't *come* easy⟩ ⟨the job is *coming* nicely⟩ (2) **:** to advance in a particular manner ⟨*come* running when I call⟩ (3) **:** to advance, rise, or improve in rank or condition ⟨has *come* a long way⟩ **f :** to get along **:** FARE — often used with *along* **g :** EXTEND ⟨her dress *came* to her ankles⟩
2 a (1) **:** to arrive at a particular place, end, result, or conclusion ⟨*came* to his senses⟩ ⟨*come* untied⟩ (2) **:** AMOUNT ⟨the taxes on it *come* to more than it's worth⟩ **b** (1) **:** to appear to the mind ⟨the answer *came* to them⟩ (2) **:** to appear on a scene **:** make an appearance ⟨children *come* equipped to learn any language⟩ **c** (1) **:** HAPPEN, OCCUR ⟨no harm will *come* to you⟩ (2) **:** to come to pass **:** take place — used in the subjunctive with inverted subject and verb to express the particular time or occasion ⟨*come* spring the days will be longer⟩ **d :** ORIGINATE, ARISE ⟨wine *comes* from grapes⟩ ⟨*come* of sturdy stock⟩ **e :** to enter or assume a condition ⟨artillery *came* into action⟩ **f :** to fall within a field of view or a range of application ⟨this *comes* within the terms of the treaty⟩ **g :** to issue forth ⟨a sob *came* from her throat⟩ **h :** to take form ⟨churn till the butter *comes*⟩ **i :** to be available ⟨this model *comes* in several sizes⟩ ⟨as good as they *come*⟩ **j :** to experience orgasm — often considered vulgar
3 : to fall to a person in a division or inheritance of property
4 *obsolete* **:** to become moved favorably **:** RELENT
5 : to turn out to be ⟨good clothes *come* high⟩
6 : BECOME ⟨a dream that *came* true⟩
transitive verb
1 : to approach or be near (an age) ⟨a child *coming* eight years old⟩
2 : to take on the aspect of ⟨*come* the stern parent⟩
— come a cropper : to fail completely
— come clean : to tell the whole story **:** CONFESS
— come into : to acquire as a possession or achievement ⟨*come into* a fortune⟩
— come into one's own : to achieve one's potential; *also* **:** to gain recognition

— **come of age :** to reach maturity
— **come off it :** to cease foolish or pretentious talk or behavior
— **come over :** to seize suddenly and strangely ⟨what's *come over* you⟩
— **come to :** to be a question of ⟨when it *comes to* pitching horseshoes, he's the champ⟩
— **come to grips with :** to meet or deal with firmly, frankly, or straightforwardly ⟨*come to grips with* the unemployment problem⟩
— **come to oneself :** to get hold of oneself **:** regain self-control
— **come to pass :** HAPPEN
— **come to terms 1 :** to reach an agreement — often used with *with* **2 :** to become adjusted especially emotionally or intellectually — usually used with *with* ⟨*come to terms* with modern life⟩

²come *noun* (1923)
1 : SEMEN — often considered vulgar
2 : ORGASM — often considered vulgar

come about *intransitive verb* (14th century)
1 : HAPPEN
2 : to change direction ⟨the wind has *come about* into the north⟩
3 : to shift to a new tack

come across *intransitive verb* (1910)
1 : to give over or furnish something demanded; *especially* **:** to pay over money
2 : to produce an impression ⟨*comes across* as a good speaker⟩
3 : COME THROUGH 2

come along *intransitive verb* (1694)
1 : to accompany someone who leads the way ⟨asked me to *come along* on the trip⟩
2 : to make progress ⟨work is *coming along* well⟩
3 : to make an appearance ⟨won't just marry the first man that *comes along*⟩

come around *intransitive verb* (1934)
1 : COME ROUND
2 : MENSTRUATE

come·back \'kəm-ˌbak\ *noun* (1889)
1 a : a sharp or witty reply **:** RETORT **b :** a cause for complaint
2 : a return to a former position or condition (as of success or prosperity) **:** RECOVERY, REVIVAL ⟨staging his ultimate *comeback* from self-imposed exile —Howard Mandel⟩

come back *intransitive verb* (1850)
1 : to return to life or vitality
2 : to return to memory ⟨it's all *coming back* to me now⟩
3 : REPLY, RETORT
4 : to regain a former favorable condition or position

come by (1863)
intransitive verb
: to make a visit
transitive verb
: to get possession of **:** ACQUIRE ⟨a good job is hard to *come by*⟩

co·me·di·an \kə-'mē-dē-ən\ *noun* (1581)
1 *archaic* **a :** a writer of comedies **b :** an actor who plays comic roles
2 : a comical individual; *specifically* **:** a professional entertainer who uses any of various physical or verbal means to be amusing

co·me·dic \-'mē-dik\ *adjective* (1639)
1 : of or relating to comedy
2 : COMICAL 2
— **co·me·dic·al·ly** \-di-k(ə-)lē\ *adverb*

co·me·di·enne \-ˌmē-dē-'en\ *noun* [French *comédienne,* feminine of *comédien* comedian, from *comédie*] (circa 1859)
: a woman who is a comedian

com·e·do \'kä-mə-ˌdō\ *noun, plural* **com·e·do·nes** \ˌkä-mə-'dō-(ˌ)nēz\ [New Latin, from Latin, glutton, from *comedere* to eat — more at COMESTIBLE] (1866)
: BLACKHEAD 1

come·down \'kəm-ˌdau̇n\ *noun* (1840)
: a descent in rank or dignity

come down *intransitive verb* (1711)
1 a : to pass by tradition ⟨a story that has *come down* from medieval times⟩ **b :** to pass from a usually high source ⟨word *came down* that the strike was over⟩
2 a : to reduce itself **:** AMOUNT ⟨it *comes down* to this⟩ **b :** to deal with a subject directly ⟨when you *come down* to it, we all depend on others⟩
3 : to lose or fall in estate or condition ⟨has *come down* in the world⟩
4 : to place oneself in opposition ⟨*came down* hard on gambling⟩
5 : to become ill ⟨*came down* with measles⟩
6 : COME OUT 2

com·e·dy \'kä-mə-dē\ *noun, plural* **-dies** [Middle English, from Middle French *comedie,* from Latin *comoedia,* from Greek *kōmōidia,* from *kōmos* revel + *aeidein* to sing — more at ODE] (14th century)
1 a : a medieval narrative that ends happily ⟨Dante's Divine *Comedy*⟩ **b :** a literary work written in a comic style or treating a comic theme
2 a : a drama of light and amusing character and typically with a happy ending **b :** the genre of dramatic literature dealing with the comic or with the serious in a light or satirical manner — compare TRAGEDY
3 : a ludicrous or farcical event or series of events
4 a : the comic element ⟨the *comedy* of many life situations⟩ **b :** humorous entertainment ⟨nightclub *comedy*⟩

comedy drama *noun* (1885)
: serious drama that is interspersed with comedy

comedy of manners (1822)
: comedy that satirically portrays the manners and fashions of a particular class or set

come–hith·er \(ˌ)kəm-'hi-t͟hər, (ˌ)kə-'mi-\ *adjective* (1925)
: sexually provocative ⟨that *come-hither* look in your eyes⟩

come in *intransitive verb* (14th century)
1 a : to arrive on a scene ⟨new models *coming in*⟩ **b :** to become available ⟨data began *coming in*⟩
2 : to place among those finishing ⟨*came in* last⟩
3 a : to function in an indicated manner ⟨*come in* handy⟩ **b** *of a telecommunications signal* **:** to be received ⟨*came in* loud and clear⟩
4 : to assume a role or function ⟨that's where you *come in*⟩
5 : to attain maturity, fruitfulness, or production
— **come in for :** to become subject to ⟨*coming in for* increasing criticism⟩

come·ly \'kəm-lē *also* 'kōm- *or* 'käm-\ *adjective* **come·li·er; -est** [Middle English *comly,* alteration of Old English *cȳmlic* glorious, from *cȳme* lively, fine; akin to Old High German *kūmig* weak] (13th century)
1 : pleasurably conforming to notions of good appearance, suitability, or proportion
2 : having a pleasing appearance **:** not homely or plain
synonym see BEAUTIFUL
— **come·li·ness** *noun*

come off (1596)
intransitive verb
1 a : to acquit oneself **:** FARE ⟨*came off* well in the contest⟩ **b :** APPEAR, SEEM
2 : SUCCEED ⟨a television series that never *came off* —*TV Guide*⟩
3 : HAPPEN, OCCUR
transitive verb
: to have recently completed or recovered from ⟨*coming off* a good year⟩ ⟨*coming off* an injury⟩

come–on \'kə-ˌmȯn, -ˌmän\ *noun* (1902)
1 : something (as an advertising promotion) intended to entice or allure
2 : a usually sexual advance

come on *intransitive verb* (15th century)
1 a : to advance by degrees ⟨darkness *came on*⟩ **b :** to begin by degrees ⟨rain *came on* toward noon⟩
2 a : PLEASE — used in cajoling or pleading **b** — used interjectionally to express astonishment, incredulity, or recognition of a put-on
3 : to project an indicated personal image ⟨*comes on* as a conservative⟩
4 : to show sexual interest in someone; *also* **:** to make sexual advances ⟨tried to *come on* to her⟩

come out *intransitive verb* (13th century)
1 a : to come into public view **:** make a public appearance ⟨a new magazine has *come out*⟩ **b :** to become evident ⟨his pride *came out* in his refusal to accept help⟩
2 : to declare oneself especially in public utterance ⟨*came out* in favor of the proposal⟩
3 : to turn out in an outcome **:** end up ⟨everything *came out* all right⟩
4 : to make a debut
5 : to openly declare one's homosexuality
— **com·ing–out** \ˌkə-miŋ-'au̇t\ *adjective*
— **come out with 1 :** to give expression to ⟨*came out with* a new proposal⟩ **2 :** PUBLISH

come–out·er \(ˌ)kə-'mau̇-tər\ *noun* (1840)
1 : a person who withdraws from something established (as a religious body)
2 : a person who advocates political reform

come over *intransitive verb* (1576)
1 a : to change from one side (as of a controversy) to the other **b :** to visit casually **:** DROP IN ⟨*come over* whenever you like⟩
2 *British* **:** BECOME

com·er \'kə-mər\ *noun* (14th century)
1 : one that comes or arrives ⟨all *comers*⟩
2 : one making rapid progress or showing promise

come round *intransitive verb* (1818)
1 : to change direction
2 : to return to a former condition; *especially* **:** COME TO 1
3 : to accede to a particular opinion or course of action ⟨knew you'd *come round* to our side eventually⟩

¹co·mes·ti·ble \kə-'mes-tə-bəl\ *adjective* [Middle French, from Medieval Latin *comestibilis,* from Latin *comestus,* past participle of *comedere* to eat, from *com-* + *edere* to eat — more at EAT] (15th century)
: EDIBLE

²comestible *noun* (1837)
: FOOD — usually used in plural

com·et \'kä-mət\ *noun* [Middle English *comete,* from Old English *cometa,* from Latin, from Greek *komētēs,* literally, long-haired, from *koman* to wear long hair, from *komē* hair] (before 12th century)
: a celestial body that consists of a fuzzy head usually surrounding a bright nucleus, that has a usually highly eccentric orbit, and that often when in the part of its orbit near the sun develops a long tail which points away from the sun
— **com·e·tary** \-mə-ˌter-ē\ *adjective*
— **co·met·ic** \kə-'me-tik, kä-\ *adjective*

come through *intransitive verb* (1914)
1 : to do what is needed or expected ⟨*came through* in the clutch⟩
2 : to become communicated

come to *intransitive verb* (circa 1572)
1 : to recover consciousness
2 a : to bring a ship's head nearer the wind **:** LUFF **b :** to come to anchor or to a stop

come up *intransitive verb* (14th century)
1 : RISE 6
2 : to come near **:** make an approach ⟨*came up* and introduced himself⟩
3 : to rise in rank or status ⟨an officer who *came up* from the ranks⟩
4 a : to come to attention or consideration ⟨the question never *came up*⟩ **b :** to occur in the course of time ⟨any problem that may *come up*⟩
5 : to turn out to be ⟨the coin *came up* tails⟩
— **come up with :** to produce especially in

dealing with a problem or challenge ⟨*came up with* a solution⟩

come·up·pance \(ˌ)kə-ˈmə-pən(t)s\ *noun* [*come up* + *-ance*] (1859)
: a deserved rebuke or penalty **:** DESERTS

com·fit \ˈkəm(p)-fət, ˈkäm(p)-\ *noun* [Middle English *confit,* from Middle French, from past participle of *confire* to prepare, from Latin *conficere,* from *com-* + *facere* to make — more at DO] (14th century)
: a candy consisting of a piece of fruit, a root (as licorice), a nut, or a seed coated and preserved with sugar

¹com·fort \ˈkəm(p)-fərt\ *transitive verb* [Middle English, from Old French *conforter,* from Late Latin *confortare* to strengthen greatly, from Latin *com-* + *fortis* strong] (13th century)
1 : to give strength and hope to **:** CHEER
2 : to ease the grief or trouble of **:** CONSOLE
— com·fort·ing·ly \-fər-tiŋ-lē\ *adverb*

²comfort *noun* (13th century)
1 : strengthening aid: **a :** ASSISTANCE, SUPPORT ⟨accused of giving aid and *comfort* to the enemy⟩ **b :** consolation in time of trouble or worry **:** SOLACE
2 a : a feeling of relief or encouragement **b :** contented well-being
3 : a satisfying or enjoyable experience
4 : one that gives or brings comfort ⟨all the *comforts* of home⟩
— com·fort·less \-ləs\ *adjective*

com·fort·able \ˈkəm(p)(f)-tə(r)-bəl, ˈkəm(p)-fə(r)-tə-bəl, ˈkəm-fə(r)-bəl\ *adjective* (1769)
1 a : affording or enjoying contentment and security ⟨a *comfortable* income⟩ **b :** affording or enjoying physical comfort ⟨a *comfortable* chair⟩ ⟨was too *comfortable* to move⟩
2 a : free from vexation or doubt ⟨*comfortable* assumptions⟩ **b :** free from stress or tension ⟨a *comfortable* routine⟩ ☆
— com·fort·able·ness *noun*
— com·fort·ably \-blē\ *adverb*

com·fort·er \ˈkəm(p)-fə(r)-tər\ *noun* (14th century)
1 a *capitalized* **:** HOLY SPIRIT **b :** one that gives comfort
2 a : a long narrow usually knitted neck scarf **b :** a thick bed covering made of two layers of cloth containing a filling (as down)

comfort station *noun* (circa 1913)
: REST ROOM

com·frey \ˈkəm(p)-frē\ *noun, plural* **com·freys** [Middle English *cumfirie,* from Old French, from Latin *conferva* a water plant, from *confervēre* to grow together (of bones), from *com-* + *fervēre* to boil — more at BARM] (13th century)
: any of a genus (*Symphytum*) of perennial herbs of the borage family with coarse hairy entire leaves and flowers in one-sided racemes

com·fy \ˈkəm(p)-fē\ *adjective* [by shortening & alteration] (1829)
: COMFORTABLE

¹com·ic \ˈkä-mik\ *adjective* [Latin *comicus,* from Greek *kōmikos,* from *kōmos* revel] (1576)
1 : of, relating to, or marked by comedy
2 : causing laughter or amusement **:** FUNNY
3 : of or relating to comic strips
synonym see LAUGHABLE

²comic *noun* (1581)
1 : COMEDIAN
2 : the comic element
3 a : COMIC STRIP **b :** COMIC BOOK **c** *plural* **:** the part of a newspaper devoted to comic strips

com·i·cal \ˈkä-mi-kəl\ *adjective* (15th century)
1 *obsolete* **:** of or relating to comedy
2 : causing laughter especially because of a startlingly or unexpectedly humorous impact
synonym see LAUGHABLE
— com·i·cal·i·ty \ˌkä-mi-ˈka-lə-tē\ *noun*
— com·i·cal·ly \ˈkä-mi-k(ə-)lē\ *adverb*

comic book *noun* (1941)
: a magazine containing sequences of comic strips

comic–opera *adjective* (1906)
: not to be taken seriously ⟨a *comic-opera* regime⟩

comic opera *noun* (1762)
: opera of a humorous character with a happy ending and usually some spoken dialogue

comic relief *noun* (1875)
: a relief from the emotional tension of a drama that is provided by the interposition of a comic episode or element

comic strip *noun* (1920)
: a group of cartoons in narrative sequence

¹com·ing \ˈkə-miŋ\ *noun* (13th century)
: an act or instance of arriving

²coming *adjective* (15th century)
1 : immediately due in sequence or development ⟨in the *coming* year⟩
2 : gaining importance

co·min·gle \kə-ˈmiŋ-gəl\ *transitive verb* (1602)
: COMMINGLE

coming–of–age *noun* (1916)
: the attainment of prominence, respectability, recognition, or maturity

Com·in·tern \ˈkä-mən-ˌtərn\ *noun* [Russian *Komintern,* from *Kom*munisticheskiĭ *Intern*atsional Communist International] (1923)
: the Communist International established in 1919 and dissolved in 1943

co·mi·tia \kə-ˈmi-sh(ē-)ə\ *noun, plural* **comitia** [Latin, plural of *comitium,* from *com-* + *-it-* (akin to *ire* to go) — more at ISSUE] (1600)
: any of several public assemblies of the people in ancient Rome for legislative, judicial, and electoral purposes
— co·mi·tial \-ˈmi-shəl\ *adjective*

co·mi·ty \ˈkä-mə-tē, ˈkō-\ *noun, plural* **-ties** [Latin *comitat-, comitas,* from *comis* courteous, probably from Old Latin *cosmis,* from *com-* + *-smis* (akin to Sanskrit *smayate* he smiles) — more at SMILE] (1543)
1 a : friendly social atmosphere **:** social harmony ⟨group activities promoting *comity*⟩ **b :** a loose widespread community based on common social institutions ⟨the *comity* of civilization⟩ **c :** COMITY OF NATIONS **d :** the informal and voluntary recognition by courts of one jurisdiction of the laws and judicial decisions of another
2 : avoidance of proselytizing members of another religious denomination

comity of nations (1862)
1 : the courtesy and friendship of nations marked especially by mutual recognition of executive, legislative, and judicial acts
2 : the group of nations practicing international comity

com·ma \ˈkä-mə\ *noun* [Late Latin, from Latin, part of a sentence, from Greek *komma* segment, clause, from *koptein* to cut — more at CAPON] (1554)
1 : a punctuation mark , used especially as a mark of separation within the sentence
2 : PAUSE, INTERVAL
3 : any of several nymphalid butterflies (genus *Polygonia*) with a silvery comma-shaped mark on the underside of the hind wing

comma fault *noun* (circa 1934)
: COMMA SPLICE

¹com·mand \kə-ˈmand\ *verb* [Middle English *comanden,* from Middle French *comander,* from (assumed) Vulgar Latin *commandare,* alteration of Latin *commendare* to commit to one's charge — more at COMMEND] (14th century)
transitive verb
1 : to direct authoritatively **:** ORDER
2 : to exercise a dominating influence over **:** have command of: as **a :** to have at one's immediate disposal **b :** to demand or receive as one's due ⟨*commands* a high fee⟩ **c :** to overlook or dominate from or as if from a strategic position **d :** to have military command of as senior officer
3 *obsolete* **:** to order or request to be given
intransitive verb
1 : to have or exercise direct authority **:** GOVERN
2 : to give orders
3 : to be commander
4 : to dominate as if from an elevated place ☆
— com·mand·able \-ˈman-də-bəl\ *adjective*

²command *noun* (15th century)
1 a : an order given **b :** a signal that actuates a device (as a control mechanism in a spacecraft or one step in a computer); *also* **:** the activation of a device by means of such a signal
2 a : the ability to control **:** MASTERY **b :** the authority or right to command ⟨the officer in *command*⟩ **c** (1) **:** the power to dominate (2) **:** scope of vision **d :** facility in use ⟨a good *command* of French⟩
3 : the act of commanding
4 : the personnel, area, or organization under a commander; *specifically* **:** a unit of the U.S. Air Force higher than an air force
5 : a position of highest usually military authority
synonym see POWER

³command *adjective* (1826)
: done on command or request ⟨a *command* performance⟩

com·man·dant \ˈkä-mən-ˌdänt, -ˌdant\ *noun* (1687)
: COMMANDING OFFICER

☆ SYNONYMS

Comfortable, cozy, snug, easy, restful mean enjoying or providing a position of contentment and security. COMFORTABLE applies to anything that encourages serenity, well-being, or complacency as well as physical ease ⟨started feeling *comfortable* in our new surroundings⟩. COZY suggests warmth, shelter, assured ease, and friendliness ⟨a *cozy* neighborhood coffee shop⟩. SNUG suggests having just enough space for comfort and safety but no more ⟨a *snug* little cottage⟩. EASY implies relief from or absence of anything likely to cause physical or mental discomfort or constraint ⟨living in *easy* circumstances⟩. RESTFUL applies to whatever induces or contributes to rest or relaxation ⟨a quiet *restful* vacation⟩.

Command, order, bid, enjoin, direct, instruct, charge mean to issue orders. COMMAND and ORDER imply authority and usually some degree of formality and impersonality. COMMAND stresses official exercise of authority ⟨a general *commanding* troops⟩. ORDER may suggest peremptory or arbitrary exercise ⟨*ordered* his employees about like slaves⟩. BID suggests giving orders peremptorily (as to children or servants) ⟨she *bade* him be seated⟩. ENJOIN implies giving an order or direction authoritatively and urgently and often with admonition or solicitude ⟨a sign *enjoining* patrons to be quiet⟩. DIRECT and INSTRUCT both connote expectation of obedience and usually concern specific points of procedure or method, INSTRUCT sometimes implying greater explicitness or formality ⟨*directed* her assistant to hold all calls⟩ ⟨the judge *instructed* the jury to ignore the remark⟩. CHARGE adds to ENJOIN an implication of imposing as a duty or responsibility ⟨*charged* by the President with a secret mission⟩.

command car *noun* (1941)
: an open armored car designed especially for military reconnaissance and capable of traveling over rough terrain

com·man·deer \ˌkä-mən-ˈdir\ *transitive verb* [Afrikaans *kommandeer,* from French *commander* to command, from Old French *comander*] (1881)
1 a : to compel to perform military service **b :** to seize for military purposes
2 : to take arbitrary or forcible possession of

com·mand·er \kə-ˈman-dər\ *noun* (14th century)
1 : one in an official position of command or control: as **a :** COMMANDING OFFICER **b :** the presiding officer of a society or organization
2 : a commissioned officer in the navy or coast guard ranking above a lieutenant commander and below a captain
— com·mand·er·ship \-ˌship\ *noun*

commander in chief (1654)
: one who holds the supreme command of an armed force

com·mand·ery \kə-ˈman-d(ə-)rē\ *noun, plural* **-er·ies** (15th century)
1 : a district under the control of a commander of an order of knights
2 : an assembly or lodge in a secret order

com·mand·ing \kə-ˈman-diŋ\ *adjective* (1591)
1 : drawing attention or priority ⟨a *commanding* presence⟩
2 : difficult to overcome ⟨a *commanding* lead⟩
— com·mand·ing·ly \-diŋ-lē\ *adverb*

commanding officer *noun* (1720)
: an officer in command; *especially* **:** an officer in the armed forces in command of an organization or installation

com·mand·ment \kə-ˈman(d)-mənt\ *noun* (13th century)
1 : the act or power of commanding
2 : something that is commanded; *especially* **:** one of the biblical Ten Commandments

command module *noun* (1962)
: a space vehicle module designed to carry the crew, the chief communication equipment, and the equipment for reentry

com·man·do \kə-ˈman-(ˌ)dō\ *noun, plural* **-dos** *or* **-does** [Afrikaans *kommando,* from Dutch *commando* command, from Spanish *comando,* from *comandar* to command, from Late Latin *commandare*] (1884)
1 *South African* **a :** a military unit or command of the Boers **b :** a raiding expedition
2 a : a military unit trained and organized as shock troops especially for hit-and-run raids into enemy territory **b :** a member of such a unit

command post *noun* (circa 1918)
: a post at which the commander of a unit in the field receives orders and exercises command

command sergeant major *noun* (1967)
: a noncommissioned officer in the army ranking above a first sergeant

comma splice *noun* (1924)
: the careless or unjustified use of a comma between coordinate main clauses not connected by a conjunction (as in "nobody goes there anymore, it's boring")

com·me·dia del·l'ar·te \kə-ˌmā-dē-ə-(ˌ)del-ˈär-tē, kə-ˌme-\ *noun* [Italian, literally, comedy of art] (1877)
: Italian comedy of the 16th to 18th centuries improvised from standardized situations and stock characters

comme il faut \ˌkə-mē(l)-ˈfō\ *adjective* [French, literally, as it should be] (1756)
: conforming to accepted standards **:** PROPER

com·mem·o·rate \kə-ˈme-mə-ˌrāt\ *transitive verb* **-rat·ed; -rat·ing** [Latin *commemoratus,* past participle of *commemorare,* from *com-* + *memorare* to remind of, from *memor* mindful — more at MEMORY] (1599)
1 : to call to remembrance
2 : to mark by some ceremony or observation **:** OBSERVE
3 : to serve as a memorial of ⟨a plaque that *commemorates* the battle⟩
synonym see KEEP
— com·mem·o·ra·tor \-ˌrā-tər\ *noun*

com·mem·o·ra·tion \kə-ˌme-mə-ˈrā-shən\ *noun* (14th century)
1 : the act of commemorating
2 : something that commemorates

com·mem·o·ra·tive \kə-ˈmem-rə-tiv, -ˈme-mə-; -ˈme-mə-ˌrā-tiv\ *adjective* (1612)
: intended as a commemoration; *especially* **:** issued in limited quantities for a relatively short period in commemoration of a person, place, or event ⟨a *commemorative* stamp⟩
— commemorative *noun*
— com·mem·o·ra·tive·ly *adverb*

com·mence \kə-ˈmen(t)s\ *verb* **com·menced; com·menc·ing** [Middle English *comencen,* from Middle French *comencer,* from (assumed) Vulgar Latin *cominitiare,* from Latin *com-* + Late Latin *initiare* to begin, from Latin, to initiate] (14th century)
transitive verb
: to enter upon **:** BEGIN ⟨*commence* proceedings⟩
intransitive verb
1 : to have or make a beginning **:** START
2 *chiefly British* **:** to begin to be or to act as
3 *chiefly British* **:** to take a degree at a university
synonym see BEGIN
— com·menc·er *noun*

com·mence·ment \kəm-ˈmen(t)s-mənt\ *noun* (13th century)
1 : an act, instance, or time of commencing
2 a : the ceremonies or the day for conferring degrees or diplomas **b :** the period of activities at this time

com·mend \kə-ˈmend\ *verb* [Middle English, from Latin *commendare,* from *com-* + *mandare* to entrust — more at MANDATE] (14th century)
transitive verb
1 : to entrust for care or preservation
2 : to recommend as worthy of confidence or notice
3 : to mention with approbation **:** PRAISE
intransitive verb
: to commend or serve as a commendation of something
— com·mend·able \-ˈmen-də-bəl\ *adjective*
— com·mend·ably \-blē\ *adverb*
— com·mend·er *noun*

com·men·da·tion \ˌkä-mən-ˈdā-shən, -ˌmen-\ *noun* (14th century)
1 a : an act of commending **b :** something (as a formal citation) that commends
2 *archaic* **:** COMPLIMENT

com·men·da·to·ry \kə-ˈmen-də-ˌtōr-ē, -ˌtȯr-\ *adjective* (1555)
: serving to commend

com·men·sal \kə-ˈmen(t)-səl\ *adjective* [Middle English, from Medieval Latin *commensalis,* from Latin *com-* + Late Latin *mensalis* of the table, from Latin *mensa* table] (1877)
: of, relating to, or living in a state of commensalism
— commensal *noun*
— com·men·sal·ly \-sə-lē\ *adverb*

com·men·sal·ism \-sə-ˌli-zəm\ *noun* (1870)
: a relation between two kinds of organisms in which one obtains food or other benefits from the other without damaging or benefiting it

com·men·su·ra·ble \kə-ˈmen(t)s-rə-bəl, -ˈmen(t)sh-; -ˈmen(t)-sə-, -shə-\ *adjective* (1557)
1 : having a common measure; *specifically* **:** divisible by a common unit an integral number of times
2 : COMMENSURATE 2
— com·men·su·ra·bil·i·ty \-ˌmen(t)s-rə-ˈbi-lə-tē, -ˌmen(t)sh-; -ˌmen(t)-sə-, -shə-\ *noun*
— com·men·su·ra·bly \-ˈmen(t)s-rə-blē, -ˈmen(t)sh-; -ˈmen(t)-sə-, -shə-\ *adverb*

com·men·su·rate \kə-ˈmen(t)s-rət, -ˈmen(t)sh-; -ˈmen(t)-sə-, -shə-\ *adjective* [Late Latin *commensuratus,* from Latin *com-* + Late Latin *mensuratus,* past participle of *mensurare* to measure, from Latin *mensura* measure — more at MEASURE] (1641)
1 : equal in measure or extent **:** COEXTENSIVE ⟨lived a life *commensurate* with the early years of the republic⟩
2 : corresponding in size, extent, amount, or degree **:** PROPORTIONATE ⟨was given a job *commensurate* with her abilities⟩
3 : COMMENSURABLE 1
— com·men·su·rate·ly *adverb*
— com·men·su·ra·tion \-ˌmen(t)-sə-ˈrā-shən, -shə-\ *noun*

¹com·ment \ˈkä-ˌment\ *noun* [Middle English, from Late Latin *commentum,* from Latin, invention, from neuter of *commentus,* past participle of *comminisci* to invent, from *com-* + *-minisci* (akin to *ment-, mens* mind) — more at MIND] (14th century)
1 : COMMENTARY
2 : a note explaining, illustrating, or criticizing the meaning of a writing
3 a : an observation or remark expressing an opinion or attitude **b :** a judgment expressed indirectly

²comment (15th century)
transitive verb
: to make a comment on ⟨the discovery . . . is hardly *commented* by the press —*Nation*⟩
intransitive verb
: to explain or interpret something by comment ⟨*commenting* on recent developments⟩

com·men·tary \ˈkä-mən-ˌter-ē\ *noun, plural* **-tar·ies** (15th century)
1 a : an explanatory treatise — usually used in plural **b :** a record of events usually written by a participant — usually used in plural
2 a : a systematic series of explanations or interpretations (as of a writing) **b :** COMMENT 2
3 a : something that serves for illustration or explanation ⟨the dark, airless apartments and sunless factories . . . are a sad *commentary* upon our civilization —H. A. Overstreet⟩ **b :** an expression of opinion

com·men·tate \ˈkä-mən-ˌtāt\ *verb* **-tat·ed; -tat·ing** [back-formation from *commentator*] (1794)
transitive verb
: to give a commentary on
intransitive verb
: to comment in a usually expository or interpretive manner; *also* **:** to act as a commentator

com·men·ta·tor \-ˌtā-tər\ *noun* (14th century)
: one who gives a commentary; *especially* **:** one who reports and discusses news on radio or television

¹com·merce \ˈkä-(ˌ)mərs\ *noun* [Middle French, from Latin *commercium,* from *com-* + *merc-, merx* merchandise] (1537)
1 : social intercourse **:** interchange of ideas, opinions, or sentiments
2 : the exchange or buying and selling of commodities on a large scale involving transportation from place to place
3 : SEXUAL INTERCOURSE
synonym see BUSINESS

²com·merce \ˈkä-(ˌ)mərs, kə-ˈmərs\ *intransitive verb* **com·merced; com·merc·ing** (1596)
archaic **:** COMMUNE

¹com·mer·cial \kə-ˈmər-shəl\ *adjective* (1598)
1 a (1) **:** occupied with or engaged in commerce or work intended for commerce ⟨a *commercial* artist⟩ (2) **:** of or relating to commerce ⟨*commercial* regulations⟩ (3) **:** characteristic of commerce ⟨*commercial* weights⟩ (4) **:** suitable, adequate, or prepared for commerce ⟨found oil in *commercial* quantities⟩ **b** (1) **:** being of an average or inferior quality ⟨*commercial* oxalic

acid) ⟨show-quality versus *commercial* cattle⟩ (2) **:** producing artistic work of low standards for quick market success
2 a : viewed with regard to profit ⟨a *commercial* success⟩ **b :** designed for a large market
3 : emphasizing skills and subjects useful in business
4 : supported by advertisers ⟨*commercial* TV⟩
— com·mer·ci·al·i·ty \kə-ˌmər-shē-ˈa-lə-tē\ *noun*
— com·mer·cial·ly \-ˈmər-sh(ə-)lē\ *adverb*

²commercial *noun* (1935)
: an advertisement broadcast on radio or television

commercial bank *noun* (1910)
: a bank organized chiefly to handle the everyday financial transactions of businesses (as through demand deposit accounts and short-term commercial loans)

com·mer·cial·ise *British variant of* COMMERCIALIZE

com·mer·cial·ism \kə-ˈmər-shə-ˌli-zəm\ *noun* (1849)
1 : commercial spirit, institutions, or methods
2 : excessive emphasis on profit
— com·mer·cial·ist \-ˈmər-sh(ə-)list\ *noun*
— com·mer·cial·is·tic \-ˌmər-shə-ˈlis-tik\ *adjective*

com·mer·cial·ize \kə-ˈmər-shə-ˌlīz\ *transitive verb* **-ized; -iz·ing** (1830)
1 a : to manage on a business basis for profit **b :** to develop commerce in
2 : to exploit for profit ⟨*commercialize* Christmas⟩
3 : to debase in quality for more profit
— com·mer·cial·i·za·tion \-ˌmər-sh(ə-)lə-ˈzā-shən\ *noun*

commercial paper *noun* (1836)
: short-term unsecured discounted paper usually sold by one company to another for immediate cash needs

commercial traveler *noun* (1807)
: TRAVELING SALESMAN

com·mie \ˈkä-mē\ *noun, often capitalized* [by shortening & alteration] (1940)
: COMMUNIST

com·mi·na·tion \ˌkä-mə-ˈnā-shən\ *noun* [Middle English, from Middle French or Latin; Middle French, from Latin *commination-, comminatio,* from *comminari* to threaten, from *com-* + *minari* to threaten — more at MOUNT] (15th century)
: DENUNCIATION
— com·mi·na·to·ry \ˈkä-mə-nə-ˌtōr-ē, -ˌtȯr-; kə-ˈmi-nə-, -ˈmī-\ *adjective*

com·min·gle \kə-ˈmiŋ-gəl, kä-\ (circa 1626)
transitive verb
1 : to blend thoroughly into a harmonious whole
2 : to combine (funds or properties) into a common fund or stock
intransitive verb
: to become commingled
synonym see MIX

com·mi·nute \ˈkä-mə-ˌnüt, -ˌnyüt\ *transitive verb* **-nut·ed; -nut·ing** [Latin *comminutus,* past participle of *comminuere,* from *com-* + *minuere* to lessen — more at MINOR] (1626)
: to reduce to minute particles **:** PULVERIZE
— com·mi·nu·tion \ˌkä-mə-ˈnü-shən, -ˈnyü-\ *noun*

com·mis·er·ate \kə-ˈmi-zə-ˌrāt\ *verb* **-at·ed; -at·ing** [Latin *commiseratus,* past participle of *commiserari,* from *com-* + *miserari* to pity, from *miser* wretched] (1599)
intransitive verb
: to feel or express sympathy **:** CONDOLE ⟨*commiserates* with them on their loss⟩
transitive verb
: to feel or express sorrow or compassion for
— com·mis·er·at·ing·ly *adverb*
— com·mis·er·a·tion \-ˌmi-zə-ˈrā-shən\ *noun*
— com·mis·er·a·tive \-ˈmi-zə-ˌrā-tiv\ *adjective*

com·mis·sar \ˈkä-mə-ˌsär\ *noun* [Russian *komissar,* from German *Kommissar,* from Medieval Latin *commissarius*] (1918)
1 a : a Communist party official assigned to a military unit to teach party principles and policies and to ensure party loyalty **b :** one that attempts to control public opinion or its expression
2 : the head of a government department in the U.S.S.R. until 1946
— com·mis·sar·i·al \ˌkä-mə-ˈsär-ē-əl, -ˈser-\ *adjective*

com·mis·sar·i·at \ˌkä-mə-ˈser-ē-ət, -ˈsar-, *especially for 3* -ˈsär-\ *noun* [New Latin *commissariatus,* from Medieval Latin *commissarius*] (1779)
1 : a system for supplying an army with food
2 : food supplies
3 [Russian *komissariat,* from German *Kommissariat,* from New Latin *commissariatus*] **:** a government department in the U.S.S.R. until 1946
4 : a board of commissioners

com·mis·sary \ˈkä-mə-ˌser-ē\ *noun, plural* **-sar·ies** [Middle English *commissarie,* from Medieval Latin *commissarius,* from Latin *commissus,* past participle of *committere*] (14th century)
1 : one delegated by a superior to execute a duty or an office
2 a : a store for equipment and provisions; *especially* **:** a supermarket for military personnel **b :** food supplies **c :** a lunchroom especially in a motion-picture studio

¹com·mis·sion \kə-ˈmi-shən\ *noun* [Middle English, from Middle French, from Latin *commission-, commissio* act of bringing together, from *committere*] (14th century)
1 a : a formal written warrant granting the power to perform various acts or duties **b :** a certificate conferring military rank and authority; *also* **:** the rank and authority so conferred
2 : an authorization or command to act in a prescribed manner or to perform prescribed acts **:** CHARGE
3 a : authority to act for, in behalf of, or in place of another **b :** a task or matter entrusted to one as an agent for another
4 a : a group of persons directed to perform some duty **b :** a government agency having administrative, legislative, or judicial powers **c :** a city council having legislative and executive functions
5 : an act of committing something
6 : a fee paid to an agent or employee for transacting a piece of business or performing a service; *especially* **:** a percentage of the money received from a total paid to the agent responsible for the business
7 : an act of entrusting or giving authority
— in commission *or* **into commission 1 :** under the authority of commissioners **2** *of a ship* **:** ready for active service **3 :** in use or in condition for use
— on commission : with commission serving as partial or full pay for work done
— out of commission 1 : out of active service or use **2 :** out of working order

²commission *transitive verb* **-mis·sioned; -mis·sion·ing** \-ˈmi-sh(ə-)niŋ\ (circa 1661)
1 : to furnish with a commission: as **a :** to confer a formal commission on ⟨was *commissioned* lieutenant⟩ **b :** to appoint or assign to a task or function ⟨was *commissioned* to do the biography⟩
2 : to order to be made ⟨*commissioned* a portrait⟩
3 : to put (a ship) in commission

com·mis·sion·aire \kə-ˌmi-shə-ˈnar, -ˈner\ *noun* [French *commissionnaire,* from *commission*] (1641)
chiefly British **:** a uniformed attendant

commissioned officer *noun* (15th century)
: an officer of the armed forces holding by a commission a rank of second lieutenant or ensign or above

com·mis·sion·er \kə-ˈmi-sh(ə-)nər\ *noun* (15th century)
: a person with a commission: as **a :** a member of a commission **b :** the representative of the governmental authority in a district, province, or other unit often having both judicial and administrative powers **c :** the officer in charge of a department or bureau of the public service **d :** the administrative head of a professional sport
— com·mis·sion·er·ship \-ˌship\ *noun*

commission merchant *noun* (1796)
: BROKER 1b

commission plan *noun* (1919)
: a method of municipal government under which a small elective commission exercises both executive and legislative powers and each commissioner directly administers one or more municipal departments

com·mis·sure \ˈkä-mə-ˌshu̇r\ *noun* [Middle English, from Middle French or Latin; Middle French, from Latin *commissura* a joining, from *commissus,* past participle] (15th century)
1 : a point or line of union or junction especially between two anatomical parts (as adjacent heart valves)
2 : a connecting band of nerve tissue in the brain or spinal cord
— com·mis·sur·al \ˌkä-mə-ˈshu̇r-əl\ *adjective*

com·mit \kə-ˈmit\ *verb* **com·mit·ted; com·mit·ting** [Middle English *committen,* from Latin *committere* to connect, entrust, from *com-* + *mittere* to send] (14th century)
transitive verb
1 a : to put into charge or trust **:** ENTRUST **b :** to place in a prison or mental institution **c :** to consign or record for preservation ⟨*commit* it to memory⟩ **d :** to put into a place for disposal or safekeeping **e :** to refer (as a legislative bill) to a committee for consideration and report
2 : to carry into action deliberately **:** PERPETRATE ⟨*commit* a crime⟩
3 a : OBLIGATE, BIND **b :** to pledge or assign to some particular course or use ⟨*commit* all troops to the attack⟩ **c :** to reveal the views of ⟨refused to *commit* himself on the issue⟩
intransitive verb
1 *obsolete* **:** to perpetrate an offense
2 : to obligate or pledge oneself ☆
— com·mit·ta·ble \-ˈmi-tə-bəl\ *adjective*

☆ SYNONYMS
Commit, entrust, confide, consign, relegate mean to assign to a person or place for a definite purpose. COMMIT may express the general idea of delivering into another's charge or the special sense of transferring to a superior power or to a special place of custody ⟨*committed* the felon to prison⟩. ENTRUST implies committing with trust and confidence ⟨the president is *entrusted* with broad powers⟩. CONFIDE implies entrusting with great assurance or reliance ⟨*confided* complete control of my affairs to my attorney⟩. CONSIGN suggests removing from one's control with formality or finality ⟨*consigned* the damaging notes to the fire⟩. RELEGATE implies a consigning to a particular class or sphere often with a suggestion of getting rid of ⟨*relegated* to an obscure position in the company⟩.

com·mit·ment \kə-ˈmit-mənt\ *noun* (1621)
1 a : an act of committing to a charge or trust: as (1) **:** a consignment to a penal or mental institution (2) **:** an act of referring a matter to a legislative committee **b :** MITTIMUS
2 a : an agreement or pledge to do something in the future; *especially* **:** an engagement to assume a financial obligation at a future date **b :** something pledged **c :** the state or an instance of being obligated or emotionally impelled ⟨a *commitment* to a cause⟩

com·mit·tal \kə-ˈmi-tᵊl\ *noun* (1818)
: COMMITMENT, CONSIGNMENT

com·mit·tee \kə-ˈmi-tē, *sense 1 also* ˌkä-mi-ˈtē\ *noun* (15th century)
1 *archaic* **:** a person to whom a charge or trust is committed
2 a : a body of persons delegated to consider, investigate, take action on, or report on some matter; *especially* **:** a group of fellow legislators chosen by a legislative body to give consideration to legislative matters **b :** a self-constituted organization for the promotion of a common object

com·mit·tee·man \kə-ˈmi-tē-mən, -ˌman\ *noun* (1654)
1 : a member of a committee
2 : a party leader of a ward or precinct

committee of the whole (1775)
: the whole membership of a legislative house sitting as a committee and operating under informal rules

com·mit·tee·wom·an \-ˌwu̇-mən\ *noun* (1853)
1 : a woman who is a member of a committee
2 : a woman who is a party leader of a ward or precinct

com·mix \kə-ˈmiks, kä-\ *verb* [back-formation from Middle English *comixt* blended, from Latin *commixtus,* past participle of *commiscēre* to mix together, from *com-* + *miscēre* to mix — more at MIX] (15th century)
transitive verb
: MINGLE, BLEND
intransitive verb
: to become mingled or blended

com·mix·ture \-chər\ *noun* [Latin *commixtura,* from *commixtus*] (1592)
1 : the act or process of mixing **:** the state of being mixed
2 : COMPOUND, MIXTURE

com·mode \kə-ˈmōd\ *noun* [French, from *commode,* adjective, suitable, convenient, from Latin *commodus,* from *com-* + *modus* measure — more at METE] (1688)
1 : a woman's ornate cap popular in the late 17th and early 18th centuries
2 a : a low chest of drawers **b :** a movable washstand with a cupboard underneath **c :** a boxlike structure holding a chamber pot under an open seat; *also* **:** CHAMBER POT **d :** TOILET 3b

commode 1

com·mod·i·fy \kə-ˈmä-də-ˌfī\ *transitive verb* **-fied; -fy·ing** (1982)
: to turn (as an intrinsic value or a work of art) into a commodity
— com·mod·i·fi·ca·tion \-ˌmä-də-fə-ˈkā-shən\ *noun*

com·mo·di·ous \kə-ˈmō-dē-əs\ *adjective* [Middle English, useful, from Middle French *commodieux,* from Medieval Latin *commodiosus,* irregular from Latin *commodum* convenience, from neuter of *commodus*] (15th century)
1 : comfortably or conveniently spacious **:** ROOMY ⟨a *commodious* closet⟩
2 *archaic* **:** HANDY, SERVICEABLE
synonym see SPACIOUS
— com·mo·di·ous·ly *adverb*
— com·mo·di·ous·ness *noun*

com·mod·i·ty \kə-ˈmä-də-tē\ *noun, plural* **-ties** [Middle English *commoditee,* from Middle French *commodité,* from Latin *commoditat-, commoditas,* from *commodus*] (15th century)
1 : an economic good: as **a :** a product of agriculture or mining **b :** an article of commerce especially when delivered for shipment ⟨*commodities* futures⟩ **c :** a mass-produced unspecialized product ⟨*commodity* chemicals⟩ ⟨*commodity* memory chips⟩
2 a : something useful or valued ⟨that valuable *commodity* patience⟩ **b :** CONVENIENCE, ADVANTAGE
3 *obsolete* **:** QUANTITY, LOT

com·mo·dore \ˈkä-mə-ˌdōr, -ˌdȯr\ *noun* [probably modification of Dutch *commandeur* commander, from French, from Old French *comandeor,* from *comander* to command] (1695)
1 a : a captain in the navy in command of a squadron **b :** a commissioned officer in the navy formerly ranking above captain and below rear admiral and having an insignia of one star
2 : the ranking officer commanding a body of merchant ships
3 : the chief officer of a yacht club or boating association

¹com·mon \ˈkä-mən\ *adjective* [Middle English *commun,* from Old French, from Latin *communis* — more at MEAN] (13th century)
1 a : of or relating to a community at large **:** PUBLIC ⟨work for the *common* good⟩ **b :** known to the community ⟨*common* nuisances⟩
2 a : belonging to or shared by two or more individuals or things or by all members of a group ⟨a *common* friend⟩ ⟨buried in a *common* grave⟩ **b :** belonging equally to two or more mathematical entities ⟨triangles with a *common* base⟩ **c :** having two or more branches ⟨*common* carotid artery⟩
3 a : occurring or appearing frequently **:** FAMILIAR ⟨a *common* sight⟩ **b :** of the best known kind **c :** VERNACULAR 2 ⟨*common* names⟩
4 a : WIDESPREAD, GENERAL ⟨*common* knowledge⟩ **b :** characterized by a lack of privilege or special status ⟨*common* people⟩ **c :** just satisfying accustomed criteria **:** ELEMENTARY ⟨*common* decency⟩
5 a : falling below ordinary standards **:** SECOND-RATE **b :** lacking refinement **:** COARSE
6 : denoting nominal relations by a single linguistic form that in a more highly inflected language might be denoted by two or more different forms ⟨*common* gender⟩ ⟨*common* case⟩
7 : of, relating to, or being common stock ☆
— com·mon·ly *adverb*
— com·mon·ness \-mən-nəs\ *noun*

²common *noun* (14th century)
1 *plural* **:** the common people
2 *plural but singular in construction* **:** a dining hall
3 *plural but singular or plural in construction, often capitalized* **a :** the political group or estate comprising the commoners **b :** the parliamentary representatives of the commoners **c :** HOUSE OF COMMONS
4 : the legal right of taking a profit in another's land in common with the owner or others
5 : a piece of land subject to common use: as **a :** undivided land used especially for pasture **b :** a public open area in a municipality
6 a : a religious service suitable for any of various festivals **b :** ORDINARY 2
7 : COMMON STOCK
— in common : shared together

com·mon·age \ˈkä-mə-nij\ *noun* (1649)
1 : community land
2 : COMMONALTY 1a(2)

com·mon·al·i·ty \ˌkä-mə-ˈna-lə-tē\ *noun, plural* **-ties** [Middle English *communalite* commonwealth, alteration of *communalte*] (1582)
1 : the common people
2 a : possession of common features or attributes **:** COMMONNESS **b :** a common feature or attribute

com·mon·al·ty \ˈkä-mə-nᵊl-tē\ *noun, plural* **-ties** [Middle English *communalte,* from Middle French *comunalté,* from *comunal* communal] (14th century)
1 a (1) **:** the common people (2) **:** the political estate formed by the common people **b :** a usage or practice common to members of a group
2 : a general group or body

common carrier *noun* (15th century)
: a business or agency that is available to the public for transportation of persons, goods, or messages

common cattle grub *noun* (1947)
: a cattle grub (*Hypoderma lineatum*) which is found throughout the U.S. and whose larva is particularly destructive to cattle

common cold *noun* (1786)
: an acute virus disease of the upper respiratory tract marked by inflammation of the mucous membranes of the nose, throat, eyes, and eustachian tubes and by a watery then purulent discharge

common denominator *noun* (1594)
1 : a common multiple of the denominators of a number of fractions
2 : a common trait or theme

common difference *noun* (circa 1891)
: the difference between two consecutive terms of an arithmetic progression

common divisor *noun* (circa 1847)
: a number or expression that divides two or more numbers or expressions without remainder — called also *common factor*

com·mon·er \ˈkä-mə-nər\ *noun* (14th century)
1 a : one of the common people **b :** one who is not of noble rank
2 : a student (as at Oxford) who pays for his own board

Common Era *noun* (circa 1889)
: CHRISTIAN ERA

common fraction *noun* (circa 1897)
: a fraction in which the numerator and denominator are both integers and are separated by a horizontal or slanted line — compare DECIMAL FRACTION

common ground *noun* (1874)
: a basis of mutual interest or agreement

common–law *adjective* (1848)
1 : of, relating to, or based on the common law
2 : relating to or based on a common-law marriage

common law *noun* (14th century)
: the body of law developed in England prima-

☆ SYNONYMS
Common, ordinary, plain, familiar, popular, vulgar mean generally met with and not in any way special, strange, or unusual. COMMON implies usual everyday quality or frequency of occurrence ⟨a *common* error⟩ ⟨lacked *common* honesty⟩ and may additionally suggest inferiority or coarseness ⟨*common* manners⟩. ORDINARY stresses conformance in quality or kind with the regular order of things ⟨an *ordinary* pleasant summer day⟩ ⟨a very *ordinary* sort of man⟩. PLAIN is likely to suggest homely simplicity ⟨*plain* hard-working people⟩. FAMILIAR stresses the fact of being generally known and easily recognized ⟨a *familiar* melody⟩. POPULAR applies to what is accepted by or prevalent among people in general sometimes in contrast to upper classes or special groups ⟨a writer of *popular* romances⟩. VULGAR, otherwise similar to POPULAR, is likely to carry derogatory connotations (as of inferiority or coarseness) ⟨souvenirs designed to appeal to the *vulgar* taste⟩.

rily from judicial decisions based on custom and precedent, unwritten in statute or code, and constituting the basis of the English legal system and of the system in all of the U.S. except Louisiana

common–law marriage *noun* (1900)

1 : a marriage recognized in some jurisdictions and based on the parties' agreement to consider themselves married and sometimes also on their cohabitation

2 : the cohabitation of a couple even when it does not constitute a legal marriage

common logarithm *noun* (circa 1903)

: a logarithm whose base is 10

common market *noun* (1952)

: an economic association (as of nations) formed to remove trade barriers among its members

common measure *noun* (1922)

: a meter consisting chiefly of iambic lines of 7 accents each arranged in rhymed pairs usually printed in 4-line stanzas — called also *common meter*

common multiple *noun* (circa 1890)

: a multiple of each of two or more numbers or expressions ⟨90 is a *common multiple* of 6 and 10⟩

common noun *noun* (circa 1864)

: a noun that may occur with limiting modifiers (as *a* or *an, some, every,* and *my*) and that designates any one of a class of beings or things

common or garden *adjective* (1892)

chiefly British **:** ORDINARY

[1]com·mon·place \'kä-mən-ˌplās\ *noun* [translation of Latin *locus communis* widely applicable argument, translation of Greek *koinos topos*] (1561)

1 *archaic* **:** a striking passage entered in a commonplace book

2 a : an obvious or trite comment **:** TRUISM **b :** something commonly found

[2]commonplace *adjective* (1609)

: commonly found **:** ORDINARY, UNREMARKABLE

— com·mon·place·ness *noun*

commonplace book *noun* (1578)

: a book of memorabilia

common pleas *noun plural* (1531)

1 *singular in construction* **:** COURT OF COMMON PLEAS

2 a : actions over which the English crown did not exercise exclusive jurisdiction **b :** civil actions between English subjects

common ratio *noun* (1875)

: the ratio of each term of a geometric progression to the term preceding it

common room *noun* (1683)

1 : a lounge available to all members of a residential community

2 : a room in a college for faculty use

common salt *noun* (1676)

: SALT 1a

common school *noun* (circa 1657)

: a free public school

common sense *noun* (1535)

1 : the unreflective opinions of ordinary people

2 : sound and prudent but often unsophisticated judgment

synonym see SENSE

— com·mon·sense \'käm-ən-'sen(t)s\ *adjective*

— com·mon·sen·si·ble \-'sen(t)-sə-bəl\ *adjective*

— com·mon·sen·si·cal \-'sen(t)-si-kəl\ *adjective*

— com·mon·sen·si·cal·ly \-si-k(ə-)lē\ *adverb*

common situs picketing *noun* (1965)

: the picketing of an entire construction site by a trade union having a grievance with only a single subcontractor working there

common stock *noun* (1888)

: capital stock other than preferred stock

common time *noun* (1674)

: a musical meter marked by four beats per measure with the quarter note receiving a single beat

common touch *noun* (1944)

: the gift of appealing to or arousing the sympathetic interest of the common people

com·mon·weal \'kä-mən-ˌwēl\ *noun* (14th century)

1 *archaic* **:** COMMONWEALTH

2 : the general welfare

com·mon·wealth \-ˌwelth *also* -ˌweltth\ *noun* (15th century)

1 *archaic* **:** COMMONWEAL 2

2 : a nation, state, or other political unit: as **a :** one founded on law and united by compact or tacit agreement of the people for the common good **b :** one in which supreme authority is vested in the people **c :** REPUBLIC

3 *capitalized* **a :** the English state from the death of Charles I in 1649 to the Restoration in 1660 **b :** PROTECTORATE 1b

4 : a state of the U.S. — used officially of Kentucky, Massachusetts, Pennsylvania, and Virginia

5 *capitalized* **:** a federal union of constituent states — used officially of Australia

6 *often capitalized* **:** an association of self-governing autonomous states more or less loosely associated in a common allegiance (as to the British crown)

7 *often capitalized* **:** a political unit having local autonomy but voluntarily united with the U.S. — used officially of Puerto Rico and of the Northern Mariana Islands

Commonwealth Day *noun* (1959)

: May 24 observed in parts of the Commonwealth of Nations as the anniversary of Queen Victoria's birthday

common year *noun* (circa 1909)

: a calendar year containing no intercalary period

com·mo·tion \kə-'mō-shən\ *noun* [Middle English, from Middle French, from Latin *commotion-, commotio,* from *commovēre*] (15th century)

1 : a condition of civil unrest or insurrection

2 : steady or recurrent motion

3 : mental excitement or confusion

4 a : an agitated disturbance **:** TO-DO **b :** noisy confusion **:** AGITATION

com·move \kə-'müv, kä-\ *transitive verb* **com·moved; com·mov·ing** [Middle English *commoeven,* from Middle French *commuev-*, present stem of *commovoir,* from Latin *commovēre,* from *com-* + *movēre* to move] (14th century)

1 : to move violently **:** AGITATE

2 : to rouse intense feeling in **:** excite to passion

com·mu·nal \kə-'myü-nᵊl, 'käm-yə-nᵊl\ *adjective* [French, from Late Latin *communalis,* from Latin *communis*] (1811)

1 : of or relating to one or more communes

2 : of or relating to a community

3 a : characterized by collective ownership and use of property **b :** participated in, shared, or used in common by members of a group or community

4 : of, relating to, or based on racial or cultural groups

— com·mu·nal·ize \kə-'myü-nᵊl-ˌīz, 'käm-yə-\ *transitive verb*

— com·mu·nal·ly *adverb*

com·mu·nal·ism \-nᵊl-ˌi-zəm\ *noun* (1871)

1 : social organization on a communal basis

2 : loyalty to a sociopolitical grouping based on religious or ethnic affiliation

— com·mu·nal·ist \-nᵊl-ist\ *noun or adjective*

com·mu·nal·i·ty \ˌkäm-yu̇-'na-lə-tē\ *noun, plural* **-ties** (1901)

1 : communal state or character

2 : a feeling of group solidarity

com·mu·nard \ˌkäm-yu̇-'när(d)\ *noun* [French] (1874)

1 *capitalized* **:** one who supported or participated in the Commune of Paris in 1871

2 : a person who lives in a commune

[1]com·mune \kə-'myün\ *verb* **com·muned; com·mun·ing** [Middle English, to converse, administer Communion, from Middle French *comunier* to converse, administer or receive Communion, from Late Latin *communicare,* from Latin] (15th century)

transitive verb

obsolete **:** TALK OVER, DISCUSS ⟨have more to *commune* —Shakespeare⟩

intransitive verb

1 : to receive Communion

2 : to communicate intimately ⟨*commune* with nature⟩

[2]com·mune \'käm-ˌyün; kə-'myün, kä-\ *noun* [French, alteration of Middle French *comugne,* from Medieval Latin *communia,* from Latin, neuter plural of *communis*] (1673)

1 : the smallest administrative district of many countries especially in Europe

2 : COMMONALTY 1a

3 : COMMUNITY: as **a :** a medieval usually municipal corporation **b** (1) **:** MIR (2) **:** an often rural community organized on a communal basis

com·mu·ni·ca·ble \kə-'myü-ni-kə-bəl\ *adjective* (1534)

1 : capable of being communicated **:** TRANSMITTABLE ⟨*communicable* disease⟩

2 : COMMUNICATIVE

— com·mu·ni·ca·bil·i·ty \-ˌmyü-ni-kə-'bi-lə-tē\ *noun*

— com·mu·ni·ca·ble·ness \-'myü-ni-kə-bəl-nəs\ *noun*

— com·mu·ni·ca·bly \-blē\ *adverb*

com·mu·ni·cant \-'myü-ni-kənt\ *noun* (1552)

1 : a church member entitled to receive Communion; *broadly* **:** a member of a fellowship

2 : one that communicates; *specifically* **:** INFORMANT

— communicant *adjective*

com·mu·ni·cate \kə-'myü-nə-ˌkāt\ *verb* **-cat·ed; -cat·ing** [Latin *communicatus,* past participle of *communicare* to impart, participate, from *communis* common — more at MEAN] (1526)

transitive verb

1 *archaic* **:** SHARE

2 a : to convey knowledge of or information about **:** make known ⟨*communicate* a story⟩ **b :** to reveal by clear signs ⟨his fear *communicated* itself to his friends⟩

3 : to cause to pass from one to another ⟨some diseases are easily *communicated*⟩

intransitive verb

1 : to receive Communion

2 : to transmit information, thought, or feeling so that it is satisfactorily received or understood

3 : to open into each other **:** CONNECT ⟨the rooms *communicate*⟩

— com·mu·ni·ca·tee \-ˌmyü-ni-kə-'tē\ *noun*

— com·mu·ni·ca·tor \-'myü-nə-ˌkā-tər\ *noun*

com·mu·ni·ca·tion \kə-ˌmyü-nə-'kā-shən\ *noun* (14th century)

1 : an act or instance of transmitting

2 a : information communicated **b :** a verbal or written message

3 a : a process by which information is exchanged between individuals through a common system of symbols, signs, or behavior ⟨the function of pheromones in insect *communication*⟩; *also* **:** exchange of information **b :** personal rapport ⟨a lack of *communication* between old and young persons⟩

4 *plural* **a :** a system (as of telephones) for

\ə\ **abut** \ᵊ\ **kitten** \ər\ **further** \a\ **ash** \ā\ **ace** \ä\ **mop, mar** \au̇\ **out** \ch\ **chin** \e\ **bet** \ē\ **easy** \g\ **go** \i\ **hit** \ī\ **ice** \j\ **job** \ŋ\ **sing** \ō\ **go** \ȯ\ **law** \ȯi\ **boy** \th\ **thin** \t͟h\ **the** \ü\ **loot** \u̇\ **foot** \y\ **yet** \zh\ **vision** *see also* Guide to Pronunciation

communicating **b :** a system of routes for moving troops, supplies, and vehicles **c :** personnel engaged in communicating
5 *plural but singular or plural in construction* **a :** a technique for expressing ideas effectively (as in speech) **b :** the technology of the transmission of information (as by print or telecommunication)
— **com·mu·ni·ca·tion·al** \-shnəl, -shə-nᵊl\ *adjective*

com·mu·ni·ca·tive \kə-'myü-nə-ˌkā-tiv, -ni-kə-tiv\ *adjective* (14th century)
1 : tending to communicate **:** TALKATIVE
2 : of or relating to communication
— **com·mu·ni·ca·tive·ly** *adverb*
— **com·mu·ni·ca·tive·ness** *noun*

com·mu·ni·ca·to·ry \kə-'myü-ni-kə-ˌtōr-ē, -ˌtȯr-\ *adjective* (1646)
1 : designed to communicate information ⟨*communicatory* letters⟩
2 : COMMUNICATIVE 2

com·mu·nion \kə-'myü-nyən\ *noun* [Middle English, from Latin *communion-, communio* mutual participation, from *communis*] (14th century)
1 : an act or instance of sharing
2 a *capitalized* **:** a Christian sacrament in which consecrated bread and wine are consumed as memorials of Christ's death or as symbols for the realization of a spiritual union between Christ and communicant or as the body and blood of Christ **b :** the act of receiving Communion **c** *capitalized* **:** the part of a Communion service in which the sacrament is received
3 : intimate fellowship or rapport **:** COMMUNICATION
4 : a body of Christians having a common faith and discipline ⟨the Anglican *communion*⟩

com·mu·ni·qué \kə-'myü-nə-ˌkā, -ˌmyü-nə-'\ *noun* [French, from past participle of *communiquer* to communicate, from Latin *communicare*] (1852)
: BULLETIN 1

com·mu·nise *British variant of* COMMUNIZE

com·mu·nism \'käm-yə-ˌni-zəm\ *noun* [French *communisme,* from *commun* common] (1840)
1 a : a theory advocating elimination of private property **b :** a system in which goods are owned in common and are available to all as needed
2 *capitalized* **a :** a doctrine based on revolutionary Marxian socialism and Marxism-Leninism that was the official ideology of the U.S.S.R. **b :** a totalitarian system of government in which a single authoritarian party controls state-owned means of production **c :** a final stage of society in Marxist theory in which the state has withered away and economic goods are distributed equitably **d :** communist systems collectively

com·mu·nist \'käm-yə-nist\ *noun* (1840)
1 : an adherent or advocate of communism
2 *capitalized* **:** COMMUNARD
3 a *capitalized* **:** a member of a Communist party or movement **b** *often capitalized* **:** an adherent or advocate of a Communist government, party, or movement
4 *often capitalized* **:** one held to engage in left-wing, subversive, or revolutionary activities
— **communist** *adjective, often capitalized*
— **com·mu·nis·tic** \ˌkäm-yə-'nis-tik\ *adjective, often capitalized*
— **com·mu·nis·ti·cal·ly** \-ti-k(ə-)lē\ *adverb*

com·mu·ni·tar·i·an \kə-ˌmyü-nə-'ter-ē-ən\ *adjective* (circa 1909)
: of or relating to social organization in small cooperative partially collectivist communities
— **communitarian** *noun*
— **com·mu·ni·tar·i·an·ism** \-ē-ə-ˌni-zəm\ *noun*

com·mu·ni·ty \kə-'myü-nə-tē\ *noun, plural* **-ties** *often attributive* [Middle English *comunete,* from Middle French *comuneté,* from Latin *communitat-, communitas,* from *communis*] (14th century)
1 : a unified body of individuals: as **a :** STATE, COMMONWEALTH **b :** the people with common interests living in a particular area; *broadly* **:** the area itself ⟨the problems of a large *community*⟩ **c :** an interacting population of various kinds of individuals (as species) in a common location **d :** a group of people with a common characteristic or interest living together within a larger society ⟨a *community* of retired persons⟩ **e :** a group linked by a common policy **f :** a body of persons or nations having a common history or common social, economic, and political interests ⟨the international *community*⟩ **g :** a body of persons of common and especially professional interests scattered through a larger society ⟨the academic *community*⟩
2 : society at large
3 a : joint ownership or participation ⟨*community* of goods⟩ **b :** common character **:** LIKENESS ⟨*community* of interests⟩ **c :** social activity **:** FELLOWSHIP **d :** a social state or condition

community antenna television *noun* (1953)
: CABLE TELEVISION

community center *noun* (1915)
: a building or group of buildings for a community's educational and recreational activities

community chest *noun* (1919)
: a general fund accumulated from individual subscriptions to defray demands on a community for charity and social welfare

community college *noun* (1948)
: a 2-year government-supported college that offers an associate degree

community property *noun* (circa 1925)
: property held jointly by husband and wife

com·mu·nize \'käm-yə-ˌnīz\ *transitive verb* **-nized; -niz·ing** [back-formation from *communization*] (1888)
1 a : to make common **b :** to make into state-owned property
2 : to subject to Communist principles of organization
— **com·mu·ni·za·tion** \ˌkäm-yə-nə-'zā-shən\ *noun*

com·mu·tate \'käm-yə-ˌtāt\ *transitive verb* **-tat·ed; -tat·ing** [back-formation from *commutation*] (1893)
: to reverse every other half cycle of (an alternating current) so as to form a unidirectional current

com·mu·ta·tion \ˌkäm-yə-'tā-shən\ *noun* [Middle English, from Middle French, from Latin *commutation-, commutatio,* from *commutare*] (15th century)
1 : EXCHANGE, TRADE
2 : REPLACEMENT; *specifically* **:** a substitution of one form of payment or charge for another
3 : a change of a legal penalty or punishment to a lesser one
4 : an act or process of commuting
5 : the action of commutating

commutation ticket *noun* (1848)
: a transportation ticket sold for a fixed number of trips over the same route during a limited period

com·mu·ta·tive \kə-'myü-tə-tiv, 'käm-yə-ˌtā-tiv\ *adjective* (1612)
1 : of, relating to, or showing commutation
2 : of, relating to, having, or being the property that a given mathematical operation and set have when the result obtained using any two elements of the set with the operation does not differ with the order in which the elements are used ⟨a *commutative* group⟩ ⟨addition of the positive integers is *commutative*⟩

com·mu·ta·tiv·i·ty \kə-ˌmyü-tə-'ti-və-tē, ˌkäm-yə-tə-\ *noun* (1929)
: the property of being commutative ⟨the *commutativity* of a mathematical operation⟩

com·mu·ta·tor \'käm-yə-ˌtā-tər\ *noun* (1880)
1 : a series of bars or segments so connected to armature coils of a generator or motor that rotation of the armature will in conjunction with fixed brushes result in unidirectional current output in the case of a generator and in the reversal of the current into the coils in the case of a motor
2 : an element of a mathematical group that when used to multiply the product of two given elements either on the right side or on the left side but not necessarily on both sides yields the product of the two given elements in reverse order

¹com·mute \kə-'myüt\ *verb* **com·mut·ed; com·mut·ing** [Latin *commutare* to change, exchange, from *com-* + *mutare* to change — more at MUTABLE] (15th century)
transitive verb
1 a : CHANGE, ALTER **b :** to give in exchange for another **:** EXCHANGE
2 : to convert (as a payment) into another form
3 : to exchange (a penalty) for another less severe
4 : COMMUTATE
intransitive verb
1 : MAKE UP, COMPENSATE
2 : to pay in gross
3 : to travel back and forth regularly (as between a suburb and a city)
4 : to yield the same mathematical result regardless of order — used of two elements undergoing an operation or of two operations on elements
— **com·mut·able** \-'myü-tə-bəl\ *adjective*

²commute *noun* (1954)
1 : an act or an instance of commuting
2 : the distance covered in commuting

com·mut·er \kə-'myü-tər\ *noun* (circa 1859)
1 : a person who commutes (as between a suburb and a city)
2 : a small airline that carries passengers relatively short distances on a regular schedule

co·mo·no·mer \(ˌ)kō-'mä-nə-mər, -'mō-\ *noun* [*co-* + *monomer*] (1945)
: one of the constituents of a copolymer

¹comp \'kämp\ *noun* [short for *complimentary*] (1887)
: a complimentary ticket; *broadly* **:** something provided free of charge

²comp \'kəmp, 'kämp\ *intransitive verb* [short for *accompany*] (1949)
: to punctuate and support a jazz solo with irregularly spaced chords

¹com·pact \kəm-'pakt, käm-', 'käm-ˌ\ *adjective* [Middle English, firmly put together, from Latin *compactus,* from past participle of *compingere* to put together, from *com-* + *pangere* to fasten — more at PACT] (14th century)
1 : predominantly formed or filled **:** COMPOSED, MADE
2 a : having a dense structure or parts or units closely packed or joined ⟨a *compact* woolen⟩ ⟨*compact* bone⟩ **b :** not diffuse or verbose ⟨a *compact* statement⟩ **c :** occupying a small volume by reason of efficient use of space ⟨a *compact* camera⟩ ⟨a *compact* formation of troops⟩ **d :** short-bodied, solid, and without excess flesh
3 : being a topological space and especially a metric space with the property that for any collection of open sets which contains it there is a subset of the collection with a finite number of elements which also contains it
— **com·pact·ly** \-'pak(t)-lē, -ˌpak(t)-\ *adverb*
— **com·pact·ness** \-'pak(t)-nəs, -ˌpak(t)-\ *noun*

²compact (15th century)
transitive verb
1 : to make up by connecting or combining **:** COMPOSE
2 a : to knit or draw together **:** COMBINE **b :** to press together **:** COMPRESS
intransitive verb
: to become compacted

— **com·pact·ible** \-'pak-tə-bəl, -ˌpak-\ *adjective*
— **com·pac·tor** *also* **com·pact·er** \-'pak-tər, -ˌpak-\ *noun*
[3]**com·pact** \'käm-ˌpakt\ *noun* (1601)
: something that is compact or compacted: **a :** a small cosmetic case (as for compressed powder) **b :** an automobile smaller than an intermediate but larger than a subcompact
[4]**com·pact** \'käm-ˌpakt\ *noun* [Latin *compactum,* from neuter of *compactus,* past participle of *compacisci* to make an agreement, from *com-* + *pacisci* to contract — more at PACT] (1591)
: an agreement or covenant between two or more parties
com·pact disc \'käm-ˌpakt-\ *noun* (1979)
: a small plastic optical disc usually containing recorded music or computer data
com·pac·tion \kəm-'pak-shən, käm-\ *noun* (14th century)
: the act or process of compacting **:** the state of being compacted
[1]**com·pan·ion** \kəm-'pan-yən\ *noun* [Middle English *compainoun,* from Old French *compagnon,* from Late Latin *companion-, companio,* from Latin *com-* + *panis* bread, food — more at FOOD] (13th century)
1 : one that accompanies another **:** COMRADE, ASSOCIATE; *also* **:** one that keeps company with another
2 *obsolete* **:** RASCAL
3 a : one that is closely connected with something similar **b :** one employed to live with and serve another ◇
[2]**companion** (1622)
transitive verb
: ACCOMPANY
intransitive verb
: to keep company
[3]**companion** *noun* [by folk etymology from Dutch *kampanje* poop deck] (1762)
1 : a hood covering at the top of a companionway
2 : COMPANIONWAY
com·pan·ion·able \kəm-'pan-yə-nə-bəl\ *adjective* (14th century)
: marked by, conducive to, or suggestive of companionship **:** SOCIABLE
— **com·pan·ion·abil·i·ty** \-ˌpan-yə-nə-'bi-lə-tē\ *noun*
— **com·pan·ion·able·ness** *noun*
— **com·pan·ion·ably** \-'pan-yə-nə-blē\ *adverb*
com·pan·ion·ate \kəm-'pan-yə-nət\ *adjective* (1926)
: relating to or having the manner of companions; *specifically* **:** harmoniously or suitably accompanying
companion cell *noun* (1887)
: a living nucleated cell that is closely associated in origin, position, and probably function with a cell making up part of a sieve tube of a vascular plant
companion piece *noun* (1844)
: a work (as of literature) that is associated with and complements another
com·pan·ion·ship \kəm-'pan-yən-ˌship\ *noun* (1548)
: the fellowship existing among companions **:** COMPANY
com·pan·ion·way \-yən-ˌwā\ *noun* [[3]*companion*] (1840)
: a ship's stairway from one deck to another
[1]**com·pa·ny** \'kəmp-nē, 'kəm-pə-\ *noun, plural* **-nies** *often attributive* [Middle English *companie,* from Old French *compagnie,* from *compain* companion, from Late Latin *companio*] (13th century)
1 a : association with another **:** FELLOWSHIP ⟨enjoy a person's *company*⟩ **b :** COMPANIONS, ASSOCIATES ⟨know a person by the *company* she keeps⟩ **c :** VISITORS, GUESTS ⟨having *company* for dinner⟩
2 a : a group of persons or things ⟨a *company* of horsemen⟩ **b :** a body of soldiers; *especially* **:** a unit (as of infantry) consisting usually of a headquarters and two or more platoons **c :** an organization of performing artists **d :** the officers and crew of a ship **e :** a fire-fighting unit
3 a : a chartered commercial organization or medieval trade guild **b :** an association of persons for carrying on a commercial or industrial enterprise **c :** those members of a partnership firm whose names do not appear in the firm name ⟨John Doe and *Company*⟩
[2]**company** *verb* **-nied; -ny·ing** (14th century)
transitive verb
: ACCOMPANY ⟨may . . . fair winds *company* your safe return —John Masefield⟩
intransitive verb
: ASSOCIATE
company man *noun* (circa 1921)
: a worker who acquiesces in company policy without complaint
company officer *noun* (1844)
: a commissioned officer in the army, air force, or marine corps of the rank of captain, first lieutenant, or second lieutenant — called also *company grade officer;* compare FIELD OFFICER, GENERAL OFFICER
company town *noun* (1927)
: a community that is dependent on one firm for all or most of the necessary services or functions of town life (as employment, housing, and stores)
company union *noun* (1917)
: an unaffiliated labor union of the employees of a single firm; *especially* **:** one dominated by the employer
com·pa·ra·bil·i·ty \ˌkäm-p(ə-)rə-'bi-lə-tē, ÷kəm-ˌpar-ə-\ *noun* (1843)
: the quality or state of being comparable
com·pa·ra·ble \'käm-p(ə-)rə-bəl, ÷kəm-'par-ə-bəl\ *adjective* (15th century)
1 : capable of or suitable for comparison
2 : SIMILAR, LIKE ⟨fabrics of *comparable* quality⟩
— **com·pa·ra·ble·ness** *noun*
— **com·pa·ra·bly** \-blē\ *adverb*
comparable worth *noun* (1983)
: the concept that women and men should receive equal pay for jobs calling for comparable skill and responsibility
com·par·a·tist \kəm-'par-ə-tist\ *noun* [*comparative* + *-ist*] (1933)
: one that uses a comparative method (as in the study of literature)
[1]**com·par·a·tive** \kəm-'par-ə-tiv\ *adjective* (15th century)
1 : of, relating to, or constituting the degree of comparison in a language that denotes increase in the quality, quantity, or relation expressed by an adjective or adverb
2 : considered as if in comparison to something else as a standard not quite attained **:** RELATIVE ⟨a *comparative* stranger⟩
3 : characterized by systematic comparison especially of likenesses and dissimilarities ⟨*comparative* anatomy⟩
— **com·par·a·tive·ly** *adverb*
— **com·par·a·tive·ness** *noun*
[2]**comparative** *noun* (15th century)
1 a : one that compares with another especially on equal footing **:** RIVAL **b :** one that makes witty or mocking comparisons
2 : the comparative degree or form in a language
com·par·a·tiv·ist \kəm-'par-ə-ti-vist\ *noun* (1887)
: COMPARATIST
com·par·a·tor \kəm-'par-ə-tər\ *noun* (1883)
: a device for comparing something with a similar thing or with a standard measure
[1]**com·pare** \kəm-'par, -'per\ *verb* **com·pared; com·par·ing** [Middle English, from Middle French *comparer,* from Latin *comparare* to couple, compare, from *compara* like, from *com-* + *par* equal] (14th century)
transitive verb
1 : to represent as similar **:** LIKEN ⟨shall I *compare* thee to a summer's day? —Shakespeare⟩
2 a : to examine the character or qualities of especially in order to discover resemblances or differences ⟨*compare* your responses with the answers⟩ **b :** to view in relation to ⟨tall *compared* to me⟩
3 : to inflect or modify (an adjective or adverb) according to the degrees of comparison
intransitive verb
1 : to bear being compared
2 : to make comparisons
3 : to be equal or alike ⟨nothing *compares* to you⟩ ☆ ■
[2]**compare** *noun* (1589)
: the possibility of comparing ⟨beauty beyond *compare*⟩; *also* **:** something with which to be compared ⟨a city without *compare*⟩

☆ SYNONYMS
Compare, contrast, collate mean to set side by side in order to show differences and likenesses. COMPARE implies an aim of showing relative values or excellences by bringing out characteristic qualities whether similar or divergent ⟨*compared* the convention facilities of the two cities⟩. CONTRAST implies an emphasis on differences ⟨*contrasted* the computerized system with the old filing cards⟩. COLLATE implies minute and critical inspection in order to note points of agreement or divergence ⟨data from districts around the country will be *collated*⟩.

■ USAGE
compare A great many authors of handbooks and other commentators are eager to point out the difference between *compare to* and *compare with.* Their basic rule is easy enough: use *to* with *transitive verb* sense 1 and use *with* with *transitive verb* sense 2a. But in the less tidy world of actual usage these rules state only broad tendencies: more writers use *to* than *with* for sense 1; somewhat more writers use *with* than *to* for sense 2a. When *compared* is used as a detached participle, *to* and *with* are used with about equal frequency ⟨you always seemed so firm to me *compared* to myself —E. B. White⟩ ⟨*compared* with the fables, my own work is insignificant —Marianne Moore⟩. Sense 2b is unnoticed by the commentators, who probably include it within 2a; it takes either *to* or *with.* Intransitive senses use both prepositions.

◇ WORD HISTORY
companion *Companion* descends ultimately from Late Latin *companio,* formed from Latin *com-* "with" and *panis* "bread," and so literally means "a person with whom one shares bread, a messmate." The word in all likelihood arose in the soldier's Latin of the declining Roman Empire and is a loan translation from a word in an early Germanic language such as Gothic *gahlaiba,* formed from *ga-* "with" and *hlaifs* "bread." The Roman army of that time recruited many troops from the Goths and other Germanic peoples who lived on the borders of the Empire. In the 4th century A.D. translation of the Bible into Gothic, *gahlaifa* renders Greek *systratiōtēs,* literally, "fellow-soldier."

com·par·i·son \kəm-ˈpar-ə-sən\ *noun* [Middle English, from Middle French *comparaison,* from Latin *comparation-, comparatio,* from *comparare*] (14th century)
1 : the act or process of comparing: as **a :** the representing of one thing or person as similar to or like another **b :** an examination of two or more items to establish similarities and dissimilarities
2 : identity of features **:** SIMILARITY ⟨several points of *comparison* between the two⟩
3 : the modification of an adjective or adverb to denote different levels of quality, quantity, or relation

comparison shop *intransitive verb* (1970)
: to compare prices (as of competing brands) in order to find the best value

¹com·part·ment \kəm-ˈpärt-mənt\ *noun* [Middle French *compartiment,* from Italian *compartimento,* from *compartire* to mark out in parts, from Late Latin *compartiri* to share out, from Latin *com-* + *partiri* to share, from *part-, pars* part, share] (circa 1578)
1 : a separate division or section
2 : one of the parts into which an enclosed space is divided
— com·part·men·tal \kəm-ˌpärt-ˈmen-tᵊl, ˌkäm-\ *adjective*

²com·part·ment \-ˌment, -mənt\ *transitive verb* (1918)
: COMPARTMENTALIZE

com·part·men·tal·ise *British variant of* COMPARTMENTALIZE

com·part·men·tal·ize \kəm-ˌpärt-ˈmen-tᵊl-ˌīz, ˌkäm-\ *transitive verb* **-ized; -iz·ing** (1925)
: to separate into isolated compartments or categories
— com·part·men·tal·i·za·tion \-ˌmen-tᵊl-ə-ˈzā-shən\ *noun*

com·part·men·ta·tion \kəm-ˌpärt-mən-ˈtā-shən, -ˌmen-\ *noun* (1926)
: division into separate sections or units

¹com·pass \ˈkəm-pəs *also* ˈkäm-\ *transitive verb* [Middle English, from Old French *compasser* to measure, from (assumed) Vulgar Latin *compassare* to pace off, from Latin *com-* + *passus* pace] (14th century)
1 : to devise or contrive often with craft or skill
2 : ENCOMPASS
3 a : BRING ABOUT, ACHIEVE **b :** to get into one's possession or power **:** OBTAIN
4 : COMPREHEND
— com·pass·able \-pə-sə-bəl\ *adjective*

²compass *noun* (14th century)
1 a : BOUNDARY, CIRCUMFERENCE ⟨within the *compass* of the city walls⟩ **b :** a circumscribed space ⟨within the narrow *compass* of 21 pages —V. L. Parrington⟩ **c :** RANGE, SCOPE ⟨the *compass* of a voice⟩
2 : a curved or roundabout course ⟨a *compass* of seven days' journey —2 Kings 3:9 (Authorized Version)⟩
3 a : a device for determining directions by means of a magnetic needle or group of needles turning freely on a pivot and pointing to the magnetic north **b :** any of various nonmagnetic devices that indicate direction **c :** an instrument for describing circles or transferring measurements that consists of two pointed branches joined at the top by a pivot — usually used in plural; called also *pair of compasses*
4 : DIRECTION 6c
synonym see RANGE

compass 3a

³compass *adjective* (1523)
: forming a curve **:** CURVED ⟨a *compass* timber⟩

compass card *noun* (circa 1859)
: the circular card attached to the needles of a mariner's compass on which are marked the 32 points of the compass and the 360° of the circle

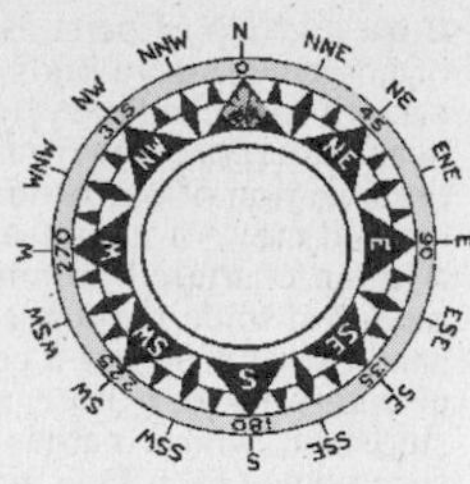

compass card

com·pas·sion \kəm-ˈpa-shən\ *noun* [Middle English, from Middle French or Late Latin; Middle French, from Late Latin *compassion-, compassio,* from *compati* to sympathize, from Latin *com-* + *pati* to bear, suffer — more at PATIENT] (14th century)
: sympathetic consciousness of others' distress together with a desire to alleviate it
synonym see PITY
— com·pas·sion·less \-ləs\ *adjective*

¹com·pas·sion·ate \kəm-ˈpa-sh(ə-)nət\ *adjective* (1587)
1 : having or showing compassion **:** SYMPATHETIC
2 : granted because of unusual distressing circumstances affecting an individual — used of some military privileges (as leave)
— com·pas·sion·ate·ly *adverb*
— com·pas·sion·ate·ness *noun*

²com·pas·sion·ate \-ˈpa-shə-ˌnāt\ *transitive verb* **-at·ed; -at·ing** (1592)
: PITY

compass plant *noun* (1848)
: a coarse yellow-flowered composite plant (*Silphium laciniatum*) with large pinnatifid leaves that occurs in prairies and along roadsides and railroad tracks of the central U.S.

compass rose *noun* (circa 1891)
: a circle graduated to degrees or quarters and printed on a chart to show direction

com·pat·i·ble \kəm-ˈpa-tə-bəl\ *adjective* [Middle English, from Middle French, from Medieval Latin *compatibilis,* literally, sympathetic, from Late Latin *compati*] (15th century)
1 : capable of existing together in harmony ⟨*compatible* theories⟩
2 : capable of cross-fertilizing freely or uniting vegetatively
3 : capable of forming a homogeneous mixture that neither separates nor is altered by chemical interaction
4 : capable of being used in transfusion or grafting without immunological reaction (as agglutination or tissue rejection)
5 : designed to work with another device or system without modification; *especially* **:** being a computer designed to operate in the same manner and use the same software as another computer
— com·pat·i·bil·i·ty \-ˌpa-tə-ˈbi-lə-tē\ *noun*
— compatible *noun*
— com·pat·i·ble·ness \-ˈpa-tə-bəl-nəs\ *noun*
— com·pat·i·bly \-blē\ *adverb*

com·pa·tri·ot \kəm-ˈpā-trē-ət, käm-, -trē-ˌät, *chiefly British* -ˈpa-\ *noun* [French *compatriote,* from Late Latin *compatriota,* from Latin *com-* + Late Latin *patriota* fellow countryman — more at PATRIOT] (1611)
1 : a person born, residing, or holding citizenship in the same country as another
2 : COMPEER, COLLEAGUE
— com·pa·tri·ot·ic \kəm-ˌpā-trē-ˈä-tik, ˌkäm-, *chiefly British* -ˌpa-\ *adjective*

¹com·peer \ˈkäm-ˌpir, käm-ˈ, kəm-ˈ\ *noun* [Middle English, from Old French *compere,* literally, godfather, from Medieval Latin *compater,* from Latin *com-* + *pater* father — more at FATHER] (13th century)
: COMPANION

²compeer *noun* [Middle English *compere,* from Old French, from Latin *compar,* from *compar,* adjective, like — more at COMPARE] (15th century)
: EQUAL, PEER
— compeer *transitive verb, obsolete*

com·pel \kəm-ˈpel\ *transitive verb* **com·pelled; com·pel·ling** [Middle English *compellen,* from Middle French *compellir,* from Latin *compellere,* from *com-* + *pellere* to drive — more at FELT] (14th century)
1 : to drive or urge forcefully or irresistibly
2 : to cause to do or occur by overwhelming pressure
3 *archaic* **:** to drive together
synonym see FORCE
— com·pel·la·ble \-ˈpe-lə-bəl\ *adjective*

com·pel·la·tion \ˌkäm-pə-ˈlā-shən, -ˌpe-\ *noun* [Latin *compellation-, compellatio,* from *compellare* to address, from *com-* + *-pellare* (as in *appellare* to accost, appeal to)] (1603)
1 : an act or action of addressing someone
2 : APPELLATION 1

com·pel·ling \kəm-ˈpe-liŋ\ *adjective* (1606)
: that compels: as **a :** FORCEFUL ⟨a *compelling* personality⟩ **b :** demanding attention ⟨for *compelling* reasons⟩ **c :** CONVINCING ⟨no *compelling* evidence⟩
— com·pel·ling·ly *adverb*

com·pend \ˈkäm-ˌpend\ *noun* [Medieval Latin *compendium*] (1596)
: COMPENDIUM

com·pen·di·ous \kəm-ˈpen-dē-əs\ *adjective* (14th century)
: marked by brief expression of a comprehensive matter **:** concise and comprehensive; *also* **:** COMPREHENSIVE ■
synonym see CONCISE
— com·pen·di·ous·ly *adverb*
— com·pen·di·ous·ness *noun*

com·pen·di·um \kəm-ˈpen-dē-əm\ *noun, plural* **-di·ums** *or* **-dia** \-dē-ə\ [Medieval Latin, from Latin, saving, shortcut, from *compendere* to weigh together, from *com-* + *pendere* to weigh — more at PENDANT] (1589)
1 : a brief summary of a larger work or of a field of knowledge **:** ABSTRACT
2 a : a list of a number of items **b :** COLLECTION, COMPILATION

com·pen·sa·ble \kəm-ˈpen(t)-sə-bəl\ *adjective* (1661)

□ USAGE
compendious Noah Webster's *A Compendious Dictionary of the English Language* defines 37,000 words in 355 pages; he intended its vocabulary to be larger than that of Samuel Johnson's large two-volume dictionary. Webster's *compendious* meant "comprehensive and concise." And "comprehensive and concise" is how most dictionaries have since defined the word. This sense is still in use ⟨gives a *compendious* summary in an appendix —S. Schoenbaum⟩ ⟨newest effort to reap from the *compendious* "news magazine" format —George V. Higgins⟩. But users of the language have never been slaves to dictionaries. Some writers have stressed the notion of conciseness ⟨shall I teach you a *compendious* method of courtship, that will give you no trouble whatever? —Thomas Love Peacock⟩. But often the reader cannot tell if conciseness is intended along with comprehensiveness ⟨a *compendious* volume of thousands of articles about meter —Dennis Taylor⟩ and so a use stressing comprehensiveness has developed ⟨it is doubtful that even the most *compendious* traditional or teaching grammar notes such simple facts —Noam Chomsky⟩ ⟨what must be one of the most *compendious* collections of jackknives in the known universe —Jay Jacobs⟩. Both the original and the merely "comprehensive" senses are in current reputable use.

: that is to be or can be compensated ⟨a *compensable* job-related injury⟩
— **com·pen·sa·bil·i·ty** \kəm-ˌpen(t)-sə-ˈbi-lə-tē, ˌkäm-\ *noun*

com·pen·sate \ˈkäm-pən-ˌsāt, -ˌpen-\ *verb* **-sat·ed; -sat·ing** [Latin *compensatus,* past participle of *compensare,* frequentative of *compendere*] (1646)
transitive verb
1 : to be equivalent to **:** COUNTERBALANCE
2 : to make an appropriate and usually counterbalancing payment to
3 a : to provide with means of counteracting variation **b :** to neutralize the effect of (variations)
intransitive verb
1 : to supply an equivalent — used with *for*
2 : to offset an error, defect, or undesired effect
3 : to undergo or engage in psychological or physiological compensation
synonym see PAY
— **com·pen·sa·tive** \ˈkäm-pən-ˌsā-tiv, -ˌpen-; kəm-ˈpen(t)-sə-\ *adjective*
— **com·pen·sa·tor** \ˈkäm-pən-ˌsā-tər, -ˌpen-\ *noun*
— **com·pen·sa·to·ry** \kəm-ˈpen(t)-sə-ˌtōr-ē, -ˌtȯr-\ *adjective*

com·pen·sa·tion \ˌkäm-pən-ˈsā-shən, -ˌpen-\ *noun* (14th century)
1 a : the act of compensating **:** the state of being compensated **b :** correction of an organic defect or loss by hypertrophy or by increased functioning of another organ or unimpaired parts of the same organ **c :** a psychological mechanism by which feelings of inferiority, frustration, or failure in one field are counterbalanced by achievement in another
2 a (1) **:** something that constitutes an equivalent or recompense ⟨age has its *compensations*⟩ (2) **:** payment to unemployed or injured workers or their dependents **b :** PAYMENT, REMUNERATION
— **com·pen·sa·tion·al** \-shnəl, -shə-nᵊl\ *adjective*

compensatory education *noun* (1965)
: educational programs intended to make up for experiences (as cultural) lacked by disadvantaged children

¹com·pere *or* **com·père** \ˈkäm-ˌper\ *noun* [French *compère,* literally, godfather — more at COMPEER] (1914)
chiefly British **:** the master of ceremonies of an entertainment (as a television program)

²compere *or* **compère** *verb* **com·pered** *or* **com·pèred; com·per·ing** *or* **com·pèr·ing** (1933)
transitive verb
chiefly British **:** to act as compere for
intransitive verb
chiefly British **:** to act as a compere

com·pete \kəm-ˈpēt\ *intransitive verb* **com·pet·ed; com·pet·ing** [Late Latin *competere* to seek together, from Latin, to come together, agree, be suitable, from *com-* + *petere* to go to, seek — more at FEATHER] (1620)
: to strive consciously or unconsciously for an objective (as position, profit, or a prize) **:** be in a state of rivalry

com·pe·tence \ˈkäm-pə-tən(t)s\ *noun* (1632)
1 : a sufficiency of means for the necessities and conveniences of life ⟨provided his family with a comfortable *competence* —Rex Ingamells⟩
2 : the quality or state of being competent: as **a :** the properties of an embryonic field that enable it to respond in a characteristic manner to an organizer **b :** readiness of bacteria to undergo genetic transformation
3 : the knowledge that enables a person to speak and understand a language — compare PERFORMANCE

com·pe·ten·cy \-pə-tən(t)-sē\ *noun, plural* **-cies** (1596)
: COMPETENCE

com·pe·tent \ˈkäm-pə-tənt\ *adjective* [Middle English, suitable, from Middle French & Latin; Middle French, from Latin *competent-, competens,* from present participle of *competere*] (15th century)
1 : proper or rightly pertinent
2 : having requisite or adequate ability or qualities **:** FIT ⟨a *competent* teacher⟩ ⟨a *competent* piece of work⟩
3 : legally qualified or adequate ⟨a *competent* witness⟩
4 : having the capacity to function or develop in a particular way; *specifically* **:** having the capacity to respond (as by producing an antibody) to an antigenic determinant ⟨immunologically *competent* cells⟩
synonym see SUFFICIENT
— **com·pe·tent·ly** *adverb*

com·pe·ti·tion \ˌkäm-pə-ˈti-shən\ *noun* [Late Latin *competition-, competitio,* from Latin *competere*] (1605)
1 : the act or process of competing **:** RIVALRY: as **a :** the effort of two or more parties acting independently to secure the business of a third party by offering the most favorable terms **b :** active demand by two or more organisms or kinds of organisms for some environmental resource in short supply
2 : a contest between rivals; *also* **:** one's competitors ⟨faced tough *competition*⟩

com·pet·i·tive \kəm-ˈpe-tə-tiv\ *adjective* (1829)
1 : relating to, characterized by, or based on competition ⟨*competitive* sports⟩
2 : inclined, desiring, or suited to compete ⟨a *competitive* personality⟩ ⟨salary benefits must be *competitive* —M. S. Eisenhower⟩
3 : depending for effectiveness on the relative concentration of two or more substances ⟨*competitive* inhibition of an enzyme⟩
— **com·pet·i·tive·ly** *adverb*
— **com·pet·i·tive·ness** *noun*

com·pet·i·tor \kəm-ˈpe-tə-tər\ *noun* (1534)
: one that competes: as **a :** RIVAL **b :** one selling or buying goods or services in the same market as another **c :** an organism that lives in competition with another

com·pi·la·tion \ˌkäm-pə-ˈlā-shən *also* -ˌpī-\ *noun* (15th century)
1 : the act or process of compiling
2 : something compiled

com·pile \kəm-ˈpī(ə)l\ *transitive verb* **com·piled; com·pil·ing** [Middle English, from Middle French *compiler,* from Latin *compilare* to plunder] (14th century)
1 : to compose out of materials from other documents
2 : to collect and edit into a volume
3 : to run (as a program) through a compiler
4 : to build up gradually ⟨*compiled* a record of four wins and two losses⟩

com·pil·er \kəm-ˈpī-lər\ *noun* (14th century)
1 : one that compiles
2 : a computer program that translates an entire set of instructions written in a higher-level symbolic language (as Pascal) into machine language before the instructions can be executed

com·pla·cence \kəm-ˈplā-sᵊn(t)s\ *noun* (15th century)
1 : calm or secure satisfaction with oneself or one's lot **:** SELF-SATISFACTION
2 *obsolete* **:** COMPLAISANCE
3 : UNCONCERN

com·pla·cen·cy \-sᵊn(t)-sē\ *noun, plural* **-cies** (1650)
1 : COMPLACENCE; *especially* **:** self-satisfaction accompanied by unawareness of actual dangers or deficiencies
2 : an instance of complacency

com·pla·cent \kəm-ˈplā-sᵊnt\ *adjective* [Latin *complacent-, complacens,* present participle of *complacēre* to please greatly, from *com-* + *placēre* to please — more at PLEASE] (1767)
1 : SELF-SATISFIED ⟨a *complacent* smile⟩
2 : COMPLAISANT 1
3 : UNCONCERNED
— **com·pla·cent·ly** *adverb*

com·plain \kəm-ˈplān\ *intransitive verb* [Middle English *compleynen,* from Middle French *complaindre,* from (assumed) Vulgar Latin *complangere,* from Latin *com-* + *plangere* to lament — more at PLAINT] (14th century)
1 : to express grief, pain, or discontent
2 : to make a formal accusation or charge
— **com·plain·er** *noun*
— **com·plain·ing·ly** \-ˈplā-niŋ-lē\ *adverb*

com·plain·ant \kəm-ˈplā-nənt\ *noun* (15th century)
: the party who makes the complaint in a legal action or proceeding

com·plaint \kəm-ˈplānt\ *noun* [Middle English *compleynte,* from Middle French *complainte,* from Old French, from *complaindre*] (14th century)
1 : expression of grief, pain, or dissatisfaction
2 a : something that is the cause or subject of protest or outcry **b :** a bodily ailment or disease
3 : a formal allegation against a party

com·plai·sance \kəm-ˈplā-sᵊn(t)s, -zᵊn(t)s; ˌkäm-plā-ˈzan(t)s, -plə-, -ˈzän(t)s\ *noun* (1651)
: disposition to please or comply **:** AFFABILITY

com·plai·sant \-sᵊnt, -zᵊnt, -ˈzant, -ˈzänt\ *adjective* [French, from Middle French, from present participle of *complaire* to gratify, acquiesce, from Latin *complacēre*] (1647)
1 : marked by an inclination to please or oblige
2 : tending to consent to others' wishes
synonym see AMIABLE
— **com·plai·sant·ly** *adverb*

com·pleat \kəm-ˈplēt\ *adjective* [archaic variant of *complete* in *The Compleat Angler* (1653) by Izaak Walton] (1526)
: COMPLETE 3

com·plect·ed \kəm-ˈplek-təd\ *adjective* [irregular from *complexion*] (1806)
: having a specified facial complexion ⟨a tall, thin man, fairly dark *complected* —E. J. Kahn⟩ ■

¹com·ple·ment \ˈkäm-plə-mənt\ *noun* [Middle English, from Latin *complementum,* from *complēre* to fill up, complete, from *com-* + *plēre* to fill — more at FULL] (14th century)
1 a : something that fills up, completes, or makes perfect **b :** the quantity or number required to make a thing complete ⟨the usual *complement* of eyes and ears —Francis Parkman⟩; *especially* **:** the whole force or personnel of a ship **c :** one of two mutually completing parts **:** COUNTERPART
2 a : the angle or arc that when added to a given angle or arc equals a right angle in measure **b :** the set of all elements that do not belong to a given set and are contained in a particular mathematical set containing the given

□ USAGE
complected Not an error, nor a dialectal term, nor nonstandard—all of which it has been labeled—*complected* still manages to raise hackles. It is an Americanism, apparently nonexistent in British English. Its currency in American English is attested as early as 1806 (by Meriwether Lewis) and it appears in the works of such notable American writers as Mark Twain, O. Henry, James Whitcomb Riley, and William Faulkner. *Complexioned,* recommended by handbooks, has less use than *complected.* Literary use, old and new, slightly favors *complected.*

set **c :** a number that when added to another number of the same sign yields zero if the significant digit farthest to the left is discarded — used especially in assembly language programming
3 : the musical interval required with a given interval to complete the octave
4 : an added word or expression by which a predication is made complete (as *president* in "they elected him president" and *beautiful* in "he thought her beautiful")
5 : the thermolabile group of proteins in normal blood serum and plasma that in combination with antibodies causes the destruction especially of particulate antigens (as bacteria and foreign blood corpuscles)

A D C B

complement 2a: ***ACB*** **right angle,** ***ACD*** **complement of** ***DCB*** **(and vice versa),** ***AD*** **complement of** ***DB*** **(and vice versa)**

[2]com·ple·ment \-ˌment\ (1602)
intransitive verb
obsolete **:** to exchange formal courtesies
transitive verb
1 : to be complementary to
2 *obsolete* **:** COMPLIMENT

com·ple·men·tal \ˌkäm-plə-ˈmen-tᵊl\ *adjective* (1597)
1 : relating to or being a complement
2 *obsolete* **:** CEREMONIOUS, COMPLIMENTARY

com·ple·men·tar·i·ty \ˌkäm-plə-(ˌ)men-ˈtar-ə-tē, -mən-\ *noun* (1911)
: the quality or state of being complementary

com·ple·men·ta·ry \ˌkäm-plə-ˈmen-t(ə-)rē\ *adjective* (1829)
1 : relating to or constituting one of a pair of contrasting colors that produce a neutral color when combined in suitable proportions
2 : serving to fill out or complete
3 : mutually supplying each other's lack
4 : being complements of each other ⟨*complementary* acute angles⟩
5 : characterized by the capacity for precise pairing of purine and pyrimidine bases between strands of DNA and sometimes RNA such that the structure of one strand determines the other
— **com·ple·men·ta·ri·ly** \-ˈmen-t(ə-)rə-lē, -(ˌ)men-ˈter-ə-lē\ *adverb*
— **com·ple·men·ta·ri·ness** \-ˈmen-t(ə-)rē-nəs\ *noun*
— **complementary** *noun*

com·ple·men·ta·tion \ˌkäm-plə-(ˌ)men-ˈtā-shən, -mən-\ *noun* (1942)
1 : the operation of determining the complement of a mathematical set
2 : production of normal phenotype in an individual heterozygous for two closely related mutations with one on each homologous chromosome and at a slightly different position

complement fixation *noun* (1906)
: the process of binding serum complement to the product formed by the union of an antibody and the antigen for which it is specific that occurs when complement is added to a suitable mixture of such an antibody and antigen

complement–fixation test *noun* (1911)
: a diagnostic test for the presence of a particular antibody in the serum of a patient that involves inactivation of the complement in the serum, addition of measured amounts of the antigen for which the antibody is specific and of foreign complement, and detection of the presence or absence of complement fixation by the addition of a suitable indicator system — compare WASSERMANN TEST

com·ple·men·ti·zer \ˈkäm-plə-mən-ˌtī-zər, -(ˌ)men-\ *noun* (1965)
: a function word or morpheme that combines with a clause or verbal phrase to form a subordinate clause

[1]com·plete \kəm-ˈplēt\ *adjective* **com·plet·er; -est** [Middle English *complet,* from Middle French, from Latin *completus,* from past participle of *complēre*] (14th century)
1 a : having all necessary parts, elements, or steps ⟨*complete* diet⟩ **b :** having all four sets of floral organs **c** *of a subject or predicate* **:** including modifiers, complements, or objects
2 : brought to an end **:** CONCLUDED ⟨a *complete* period of time⟩
3 : highly proficient ⟨a *complete* artist⟩
4 a : fully carried out **:** THOROUGH ⟨a *complete* renovation⟩ **b :** TOTAL, ABSOLUTE ⟨*complete* silence⟩
5 *of insect metamorphosis* **:** characterized by the occurrence of a pupal stage between the motile immature stages and the adult — compare INCOMPLETE 1b
6 *of a metric space* **:** having the property that every Cauchy sequence of elements converges to a limit in the space
synonym see FULL
— **com·plete·ly** *adverb*
— **com·plete·ness** *noun*
— **com·ple·tive** \-ˈplē-tiv\ *adjective*
— **complete with :** made complete by the inclusion of ⟨a birthday cake *complete with* candles⟩

[2]complete *transitive verb* **com·plet·ed; com·plet·ing** (15th century)
1 : to bring to an end and especially into a perfected state ⟨*complete* a painting⟩
2 a : to make whole or perfect ⟨its song *completes* the charm of this bird⟩ **b :** to mark the end of ⟨a rousing chorus *completes* the show⟩ **c :** EXECUTE, FULFILL ⟨*complete* a contract⟩
3 : to carry out (a forward pass) successfully
synonym see CLOSE

complete fertilizer *noun* (1900)
: a fertilizer that contains the three chief plant nutrients nitrogen, phosphoric acid, and potash

com·ple·tion \kəm-ˈplē-shən\ *noun* (1657)
1 : the act or process of completing
2 : the quality or state of being complete
3 : a completed forward pass in football

[1]com·plex \ˈkäm-ˌpleks\ *noun* [Late Latin *complexus* totality, from Latin, embrace, from *complecti*] (1643)
1 : a whole made up of complicated or interrelated parts ⟨a *complex* of university buildings⟩ ⟨a *complex* of welfare programs⟩ ⟨the military-industrial *complex*⟩
2 a : a group of culture traits relating to a single activity (as hunting), process (as use of flint), or culture unit **b** (1) **:** a group of repressed desires and memories that exerts a dominating influence upon the personality (2) **:** an exaggerated reaction to a subject or situation **c :** a group of obviously related units of which the degree and nature of the relationship is imperfectly known
3 : a chemical association of two or more species (as ions or molecules) joined usually by weak electrostatic bonds rather than covalent bonds

[2]com·plex \käm-ˈpleks, kəm-ˈ, ˈkäm-ˌ\ *adjective* [Latin *complexus,* past participle of *complecti* to embrace, comprise (a multitude of objects), from *com-* + *plectere* to braid — more at PLY] (circa 1652)
1 a : composed of two or more parts **:** COMPOSITE **b** (1) *of a word* **:** having a bound form as one or more of its immediate constituents ⟨*unmanly* is a *complex* word⟩ (2) *of a sentence* **:** consisting of a main clause and one or more subordinate clauses
2 : hard to separate, analyze, or solve
3 : of, concerned with, being, or containing complex numbers ⟨a *complex* root⟩ ⟨*complex* analysis⟩ ☆
— **com·plex·ly** *adverb*
— **com·plex·ness** *noun*

[3]com·plex *same as* [2]\ *transitive verb* (1658)
1 : to make complex or into a complex
2 : CHELATE
— **com·plex·a·tion** \ˌkäm-ˌplek-ˈsā-shən, kəm-\ *noun*

complex fraction *noun* (1827)
: a fraction with a fraction or mixed number in the numerator or denominator or both — compare SIMPLE FRACTION

com·plex·i·fy \käm-ˈplek-sə-ˌfī, kəm-\ *verb* **-fied; -fy·ing** (1830)
transitive verb
: to make complex
intransitive verb
: to become complex

com·plex·ion \kəm-ˈplek-shən\ *noun* [Middle English, from Middle French, from Medieval Latin *complexion-, complexio,* from Latin, combination, from *complecti*] (14th century)
1 : the combination of the hot, cold, moist, and dry qualities held in medieval physiology to determine the quality of a body
2 a : an individual complex of ways of thinking or feeling **b :** a complex of attitudes and inclinations
3 : the hue or appearance of the skin and especially of the face ⟨a dark *complexion*⟩
4 : overall aspect or character ⟨by changing the *complexion* of the legislative branch —Trevor Armbrister⟩
— **com·plex·ion·al** \-shnəl, -shə-nᵊl\ *adjective*
— **com·plex·ioned** \-shənd\ *adjective*

com·plex·i·ty \kəm-ˈplek-sə-tē, käm-\ *noun, plural* **-ties** (1685)
1 : the quality or state of being complex
2 : something complex

complex number *noun* (1860)
: a number of the form $a + b\sqrt{-1}$ where *a* and *b* are real numbers

complex plane *noun* (circa 1909)
: a plane whose points are identified by means of complex numbers; *especially* **:** ARGAND DIAGRAM

com·pli·ance \kəm-ˈplī-ən(t)s\ *noun* (1647)
1 a : the act or process of complying to a desire, demand, or proposal or to coercion **b :** conformity in fulfilling official requirements
2 : a disposition to yield to others
3 : the ability of an object to yield elastically when a force is applied **:** FLEXIBILITY

com·pli·an·cy \-ən(t)-sē\ *noun* (1643)
: COMPLIANCE

com·pli·ant \-ənt\ *adjective* (1642)
: ready or disposed to comply **:** SUBMISSIVE
— **com·pli·ant·ly** *adverb*

com·pli·ca·cy \ˈkäm-pli-kə-sē\ *noun, plural* **-cies** [[2]*complicate*] (circa 1828)
1 : the quality or state of being complicated
2 : something that is complicated

[1]com·pli·cate \ˈkäm-plə-ˌkāt\ *transitive verb* **-cat·ed; -cat·ing** (1621)
1 : to combine especially in an involved or inextricable manner
2 : to make complex or difficult
3 : INVOLVE; *especially* **:** to cause to be more

☆ SYNONYMS
Complex, complicated, intricate, involved, knotty mean having confusingly interrelated parts. COMPLEX suggests the unavoidable result of a necessary combining and does not imply a fault or failure ⟨a *complex* recipe⟩. COMPLICATED applies to what offers great difficulty in understanding, solving, or explaining ⟨*complicated* legal procedures⟩. INTRICATE suggests such interlacing of parts as to make it nearly impossible to follow or grasp them separately ⟨an *intricate* web of deceit⟩. INVOLVED implies extreme complication and often disorder ⟨a rambling, *involved* explanation⟩. KNOTTY suggests complication and entanglement that make solution or understanding improbable ⟨*knotty* ethical questions⟩.

complex or severe ⟨a virus disease *complicated* by bacterial infection⟩
²com·pli·cate \-pli-kət\ *adjective* [Latin *complicatus,* past participle of *complicare* to fold together, from *com-* + *plicare* to fold — more at PLY] (1638)
1 : COMPLEX, INTRICATE
2 : CONDUPLICATE
com·pli·cat·ed \ˈkäm-plə-ˌkā-təd\ *adjective* (1656)
1 : consisting of parts intricately combined
2 : difficult to analyze, understand, or explain
synonym see COMPLEX
— com·pli·cat·ed·ly *adverb*
— com·pli·cat·ed·ness *noun*
com·pli·ca·tion \ˌkäm-plə-ˈkā-shən\ *noun* (15th century)
1 a : COMPLEXITY, INTRICACY; *especially* **:** a situation or a detail of character complicating the main thread of a plot **b :** a making difficult, involved, or intricate **c :** a complex or intricate feature or element **d :** a difficult factor or issue often appearing unexpectedly and changing existing plans, methods, or attitudes
2 : a secondary disease or condition developing in the course of a primary disease or condition
com·plice \ˈkäm-pləs, ˈkəm-\ *noun* [Middle English, from Middle French, from Late Latin *complic-, complex,* from Latin, closely connected, from *complicare*] (15th century)
archaic **:** ASSOCIATE
com·plic·it \kəm-ˈpli-sət\ *adjective* (1973)
: having complicity
com·plic·it·ous \-ˈpli-sə-təs\ *adjective* (1860)
: COMPLICIT
com·plic·i·ty \kəm-ˈpli-s(ə-)tē\ *noun, plural* **-ties** (circa 1656)
1 : association or participation in or as if in a wrongful act
2 : an instance of complicity
com·pli·er \-ˈplī(-ə)r\ *noun* (1660)
: one that complies
¹com·pli·ment \ˈkäm-plə-mənt\ *noun* [French, from Italian *complimento,* from Spanish *cumplimiento,* from *cumplir* to be courteous — more at COMPLY] (1654)
1 a : an expression of esteem, respect, affection, or admiration; *especially* **:** an admiring remark **b :** formal and respectful recognition **:** HONOR
2 *plural* **:** best wishes **:** REGARDS ⟨accept my *compliments*⟩ ⟨*compliments* of the season⟩
²com·pli·ment \-ˌment\ *transitive verb* (1735)
1 : to pay a compliment to
2 : to present with a token of esteem
com·pli·men·ta·ry \ˌkäm-plə-ˈmen-t(ə-)rē\ *adjective* (1716)
1 a : expressing or containing a compliment **b :** FAVORABLE ⟨the novel received *complimentary* reviews⟩
2 : given free as a courtesy or favor ⟨*complimentary* tickets⟩
— com·pli·men·ta·ri·ly \-ˈmen-t(ə-)rə-lē, -(ˌ)men-ˈter-ə-lē\ *adverb*
complimentary close *noun* (1919)
: the words (as *sincerely yours*) that conventionally come immediately before the signature of a letter and express the sender's regard for the receiver — called also *complimentary closing*
com·pline \ˈkäm-plən, -ˌplīn\ *noun, often capitalized* [Middle English *compline, complie,* from Old French *complie,* modification of Late Latin *completa,* from Latin, feminine of *completus* complete] (13th century)
: the seventh and last of the canonical hours
¹com·plot \ˈkäm-ˌplät\ *noun* [Middle French *complot* crowd, plot] (1577)
archaic **:** PLOT, CONSPIRACY
²com·plot \kəm-ˈplät, käm-\ *verb* (1579)
archaic **:** PLOT
com·ply \kəm-ˈplī\ *intransitive verb* **complied; com·ply·ing** [Italian *complire,* from Spanish *cumplir* to complete, perform what is due, be courteous, modification of Latin *complēre* to complete] (1602)
1 *obsolete* **:** to be ceremoniously courteous
2 : to conform or adapt one's actions to another's wishes, to a rule, or to necessity
com·po \ˈkäm-(ˌ)pō\ *noun, plural* **compos** [short for *composition*] (1823)
: any of various composition materials
¹com·po·nent \kəm-ˈpō-nənt, ˈkäm-ˌ, käm-ˈ\ *noun* [Latin *component-, componens,* present participle of *componere* to put together — more at COMPOUND] (1645)
1 : a constituent part **:** INGREDIENT
2 a : any one of the vector terms added to form a vector sum or resultant **b :** a coordinate of a vector; *also* **:** either member of an ordered pair of numbers
synonym see ELEMENT
— com·po·nen·tial \ˌkäm-pə-ˈnen(t)-shəl\ *adjective*
²component *adjective* (1664)
: serving or helping to constitute **:** CONSTITUENT
¹com·port \kəm-ˈpōrt, -ˈpȯrt\ *verb* [Middle French *comporter* to bear, conduct, from Latin *comportare* to bring together, from *com-* + *portare* to carry — more at FARE] (1589)
intransitive verb
: to be fitting **:** ACCORD ⟨actions that *comport* with policy⟩
transitive verb
: BEHAVE; *especially* **:** to behave in a manner conformable to what is right, proper, or expected ⟨*comported* himself well in the crisis⟩
synonym see BEHAVE
— com·port·ment \-mənt\ *noun*
²com·port \ˈkäm-ˌpōrt, -ˌpȯrt\ *noun* (1771)
: COMPOTE 2
com·pose \kəm-ˈpōz\ *verb* **com·posed; com·pos·ing** [Middle English, from Middle French *composer,* from Latin *componere* (perfect indicative *composui*) — more at COMPOUND] (15th century)
transitive verb
1 a : to form by putting together **:** FASHION ⟨a committee *composed* of three representatives —*Current Biography*⟩ **b :** to form the substance of **:** CONSTITUTE ⟨*composed* of many ingredients⟩ **c :** to produce (as columns or pages of type) by composition
2 a : to create by mental or artistic labor **:** PRODUCE ⟨*compose* a sonnet⟩ **b** (1) **:** to formulate and write (a piece of music) (2) **:** to compose music for
3 : to deal with or act on so as to reduce to a minimum ⟨*compose* their differences⟩
4 : to arrange in proper or orderly form
5 : to free from agitation **:** CALM, SETTLE ⟨*composed* himself⟩
intransitive verb
: to practice composition
com·posed \-ˈpōzd\ *adjective* (1607)
: free from agitation **:** CALM; *especially* **:** SELF-POSSESSED
synonym see COOL
— com·pos·ed·ly \-ˈpō-zəd-lē\ *adverb*
— com·pos·ed·ness \-ˈpō-zəd-nəs\ *noun*
com·pos·er \kəm-ˈpō-zər\ *noun* (1597)
: one that composes; *especially* **:** a person who writes music
composing room *noun* (1737)
: the department in a printing office where typesetting and related operations are performed
composing stick *noun* (1679)
: a tray with an adjustable slide that a compositor holds in one hand and sets type into with the other

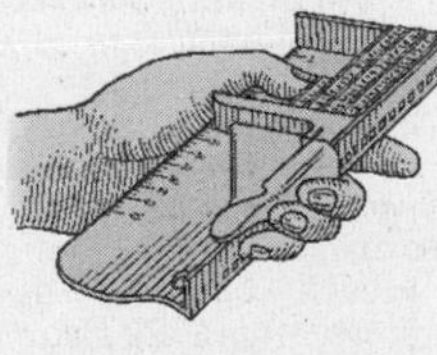
composing stick

¹com·pos·ite \käm-ˈpä-zət, kəm-ˈ, *especially British* ˈkäm-pə-zit\ *adjective* [Latin *compositus,* past participle of *componere*] (1563)
1 : made up of distinct parts: as **a** *capitalized* **:** relating to or being a modification of the Corinthian order combining angular Ionic volutes with the acanthus-circled bell of the Corinthian **b :** of or relating to a very large family (Compositae) of dicotyledonous herbs, shrubs, and trees often considered to be the most highly evolved plants and characterized by florets arranged in dense heads that resemble single flowers **c :** factorable into two or more prime factors other than 1 and itself ⟨8 is a positive *composite* integer⟩
2 : combining the typical or essential characteristics of individuals making up a group ⟨the *composite* man called the Poet —Richard Poirier⟩
3 *of a statistical hypothesis* **:** specifying a range of values for one or more statistical parameters — compare SIMPLE 10
— com·pos·ite·ly *adverb*
²composite *noun* (1656)
1 : something composite **:** COMPOUND
2 : a composite plant
3 : COMPOSITE FUNCTION
4 : a solid material which is composed of two or more substances having different physical characteristics and in which each substance retains its identity while contributing desirable properties to the whole; *especially* **:** a structural material made of plastic within which a fibrous material (as silicon carbide) is embedded
³composite *transitive verb* **-it·ed; -it·ing** (1923)
: to make composite or into something composite ⟨*composited* four soil samples⟩
composite function *noun* (1965)
: a function whose values are found from two given functions by applying one function to an independent variable and then applying the second function to the result and whose domain consists of those values of the independent variable for which the result yielded by the first function lies in the domain of the second
com·po·si·tion \ˌkäm-pə-ˈzi-shən\ *noun* [Middle English *composicioun,* from Middle French *composition,* from Latin *composition-, compositio,* from *componere*] (14th century)
1 a : the act or process of composing; *specifically* **:** arrangement into specific proportion or relation and especially into artistic form **b** (1) **:** the arrangement of type for printing ⟨hand *composition*⟩ (2) **:** the production of type or typographic characters (as in photocomposition) arranged for printing
2 a : the manner in which something is composed **b :** general makeup ⟨the changing ethnic *composition* of the city —Leonard Buder⟩ **c :** the qualitative and quantitative makeup of a chemical compound
3 : mutual settlement or agreement
4 : a product of mixing or combining various elements or ingredients
5 : an intellectual creation: as **a :** a piece of writing; *especially* **:** a school exercise in the form of a brief essay **b :** a written piece of music especially of considerable size and complexity
6 : the quality or state of being compound
7 : the operation of forming a composite function; *also* **:** COMPOSITE FUNCTION
— com·po·si·tion·al \-ˈzish-nəl, -ˈzi-shə-nᵊl\ *adjective*
— com·po·si·tion·al·ly *adverb*
com·pos·i·tor \kəm-ˈpä-zə-tər\ *noun* (1569)
: one who sets type

com·pos men·tis \'käm-pəs-'men-təs\ *adjective* [Latin, literally, having mastery of one's mind] (1616)
: of sound mind, memory, and understanding

¹com·post \'käm-ˌpōst, *especially British* -ˌpäst\ *noun* [Middle French, from Medieval Latin *compostum,* from Latin, neuter of *compositus, compostus,* past participle of *componere*] (1587)
1 : a mixture that consists largely of decayed organic matter and is used for fertilizing and conditioning land
2 : MIXTURE, COMPOUND

²compost *transitive verb* (1829)
: to convert (as plant debris) to compost

com·po·sure \kəm-'pō-zhər\ *noun* (1647)
: a calmness or repose especially of mind, bearing, or appearance **:** SELF-POSSESSION
synonym see EQUANIMITY

com·pote \'käm-ˌpōt\ *noun* [French, from Old French *composte,* from Latin *composta,* feminine of *compostus,* past participle] (1693)
1 : a dessert of fruit cooked in syrup
2 : a bowl of glass, porcelain, or metal usually with a base and stem from which compotes, fruits, nuts, or sweets are served

¹com·pound \käm-'paund, kəm-', 'käm-ˌ\ *verb* [Middle English *compounen,* from Middle French *compondre,* from Latin *componere,* from *com-* + *ponere* to put — more at POSITION] (14th century)
transitive verb
1 : to put together (parts) so as to form a whole **:** COMBINE ⟨*compound* ingredients⟩
2 : to form by combining parts ⟨*compound* a medicine⟩
3 a : to settle amicably **:** adjust by agreement **b :** to agree for a consideration not to prosecute (an offense) ⟨*compound* a felony⟩
4 a : to pay (interest) on both the accrued interest and the principal **b :** to add to **:** AUGMENT ⟨we *compounded* our error in later policy —Robert Lekachman⟩
intransitive verb
1 : to become joined in a compound
2 : to come to terms of agreement
— com·pound·able \-'paun-də-bəl, -ˌpaun-\ *adjective*
— com·pound·er *noun*

²com·pound \'käm-ˌpaund, käm-', kəm-'\ *adjective* [Middle English *compouned,* past participle of *compounen*] (14th century)
1 : composed of or resulting from union of separate elements, ingredients, or parts: as **a :** composed of united similar elements especially of a kind usually independent ⟨a *compound* plant ovary⟩ **b :** having the blade divided to the midrib and forming two or more leaflets on a common axis ⟨a *compound* leaf⟩
2 : involving or used in a combination
3 a *of a word* **:** constituting a compound **b** *of a sentence* **:** having two or more main clauses

³com·pound \'käm-ˌpaund\ *noun* (1530)
1 a : a word consisting of components that are words (as *rowboat, high school, devil-may-care*) **b :** a word (as *anthropology, kilocycle, builder*) consisting of any of various combinations of words, combining forms, or affixes
2 : something formed by a union of elements or parts; *especially* **:** a distinct substance formed by chemical union of two or more ingredients in definite proportion by weight

⁴com·pound \'käm-ˌpaund\ *noun* [by folk etymology from Malay *kampung* group of buildings, village] (1679)
: a fenced or walled-in area containing a group of buildings and especially residences

compound–complex *adjective* (1923)
of a sentence **:** having two or more main clauses and one or more subordinate clauses

compound eye *noun* (1836)
: an eye (as of an insect) made up of many separate visual units

compound fracture *noun* (1543)
: a bone fracture resulting in an open wound through which bone fragments usually protrude

compound interest *noun* (1660)
: interest computed on the sum of an original principal and accrued interest

compound microscope *noun* (circa 1859)
: a microscope consisting of an objective and an eyepiece mounted in a drawtube

compound number *noun* (15th century)
: a number (as 2 feet 5 in.) involving different denominations or more than one unit

com·pra·dor *or* **com·pra·dore** \ˌkäm-prə-'dōr, -'dȯr\ *noun* [Portuguese *comprador,* literally, buyer] (1840)
1 : a Chinese agent engaged by a foreign establishment in China to have charge of its Chinese employees and to act as an intermediary in business affairs
2 : INTERMEDIARY

com·pre·hend \ˌkäm-pri-'hend\ *transitive verb* [Middle English, from Latin *comprehendere,* from *com-* + *prehendere* to grasp — more at GET] (14th century)
1 : to grasp the nature, significance, or meaning of
2 : to contain or hold within a total scope, significance, or amount ⟨philosophy's scope *comprehends* the truth of everything which man may understand —H. O. Taylor⟩
3 : to include by construction or implication ⟨does not prudence *comprehend* all the virtues? —Thomas B. Silver⟩
synonym see UNDERSTAND, INCLUDE
— com·pre·hend·ible \-'hen-də-bəl\ *adjective*

com·pre·hen·si·ble \-'hen(t)-sə-bəl\ *adjective* (1598)
: capable of being comprehended **:** INTELLIGIBLE
— com·pre·hen·si·bil·i·ty \-ˌhen(t)-sə-'bi-lə-tē\ *noun*
— com·pre·hen·si·ble·ness \-'hen(t)-sə-bəl-nəs\ *noun*
— com·pre·hen·si·bly \-blē\ *adverb*

com·pre·hen·sion \ˌkäm-pri-'hen(t)-shən\ *noun* [Middle French & Latin; Middle French, from Latin *comprehension-, comprehensio,* from *comprehendere* to understand, comprise] (15th century)
1 a : the act or action of grasping with the intellect **:** UNDERSTANDING **b :** knowledge gained by comprehending **c :** the capacity for understanding fully
2 a : the act or process of comprising **b :** the faculty or capability of including **:** COMPREHENSIVENESS
3 : CONNOTATION 3

com·pre·hen·sive \-'hen(t)-siv\ *adjective* (1614)
1 : covering completely or broadly **:** INCLUSIVE ⟨*comprehensive* examinations⟩ ⟨*comprehensive* insurance⟩
2 : having or exhibiting wide mental grasp ⟨*comprehensive* knowledge⟩
— com·pre·hen·sive·ly *adverb*
— com·pre·hen·sive·ness *noun*

¹com·press \kəm-'pres\ *verb* [Middle English, from Late Latin *compressare* to press hard, frequentative of Latin *comprimere* to compress, from *com-* + *premere* to press — more at PRESS] (14th century)
transitive verb
1 : to press or squeeze together
2 : to reduce in size or volume as if by squeezing
intransitive verb
: to undergo compression
synonym see CONTRACT

²com·press \'käm-ˌpres\ *noun* [Middle French *compresse,* from *compresser* to compress, from Late Latin *compressare*] (1599)
1 : a folded cloth or pad applied so as to press upon a body part
2 : a machine for compressing

com·pressed \kəm-'prest *also* 'käm-ˌ\ *adjective* (14th century)
1 : pressed together **:** reduced in size or volume (as by pressure)
2 : flattened as though subjected to compression: **a :** flattened laterally ⟨petioles *compressed*⟩ **b :** narrow from side to side and deep in a dorsoventral direction
— com·pressed·ly \kəm-'prest-lē, -'pre-səd-lē\ *adverb*

compressed air *noun* (1669)
: air under pressure greater than that of the atmosphere

com·press·ible \kəm-'pre-sə-bəl\ *adjective* (circa 1691)
: capable of being compressed
— com·press·ibil·i·ty \-ˌpre-sə-'bi-lə-tē\ *noun*

com·pres·sion \kəm-'pre-shən\ *noun* (15th century)
1 a : the act, process, or result of compressing **b :** the state of being compressed
2 : the process of compressing the fuel mixture in a cylinder of an internal combustion engine (as in an automobile)
3 : a much compressed fossil plant
— com·pres·sion·al \-'presh-nəl, -'pre-shə-nᵊl\ *adjective*

compressional wave *noun* (1887)
: a longitudinal wave (as a sound wave) propagated by the elastic compression of the medium — called also *compression wave*

com·pres·sive \kəm-'pre-siv\ *adjective* (1572)
1 : of or relating to compression
2 : tending to compress
— com·pres·sive·ly *adverb*

com·pres·sor \-'pre-sər\ *noun* (1839)
: one that compresses: as **a :** a muscle that compresses a part **b :** a machine that compresses gases

com·prise \kəm-'prīz\ *transitive verb* **com·prised; com·pris·ing** [Middle English, from Middle French *compris,* past participle of *comprendre,* from Latin *comprehendere*] (15th century)
1 : to include especially within a particular scope ⟨civilization as Lenin used the term would then certainly have *comprised* the changes that are now associated in our minds with "developed" rather than "developing" states —*Times Literary Supplement*⟩
2 : to be made up of ⟨a vast installation, *comprising* fifty buildings —Jane Jacobs⟩
3 : COMPOSE, CONSTITUTE ⟨a misconception as to what *comprises* a literary generation —William Styron⟩ ⟨about 8 percent of our military forces are *comprised* of women —Jimmy Carter⟩ ▫

¹com·pro·mise \'käm-prə-ˌmīz\ *noun* [Middle English, mutual promise to abide by an arbiter's decision, from Middle French *compromis,* from Latin *compromissum,* from neuter of *compromissus,* past participle of *compromittere* to promise mutually, from *com-* + *promittere* to promise — more at PROMISE] (15th century)
1 a : settlement of differences by arbitration or by consent reached by mutual concessions **b :** something intermediate between or blending qualities of two different things
2 : a concession to something derogatory or prejudicial ⟨a *compromise* of principles⟩

²compromise *verb* **-mised; -mis·ing** (1598)
transitive verb
1 *obsolete* **:** to bind by mutual agreement
2 : to adjust or settle by mutual concessions
3 a : to expose to suspicion, discredit, or mischief **b :** to reveal or expose to an unauthorized person and especially to an enemy ⟨confidential information was *compromised*⟩ **c :** to cause the impairment of ⟨a *compromised* immune system⟩ ⟨a seriously *compromised* patient⟩

intransitive verb
1 a : to come to agreement by mutual concession **b :** to find or follow a way between extremes
2 : to make a shameful or disreputable concession ⟨wouldn't *compromise* with their principles⟩
— com·pro·mis·er *noun*

compt \ˈkau̇nt, ˈkäm(p)t\ *archaic variant of* COUNT

comp·trol·ler \kən-ˈtrō-lər, ˈkäm(p)-ˌ, ˌkäm(p)-ˈ\ *noun* [Middle English, alteration of *conterroller* controller] (15th century)
1 : a royal-household official who examines and supervises expenditures
2 : a public official who audits government accounts and sometimes certifies expenditures
3 : CONTROLLER 1c
— comp·trol·ler·ship \-ˌship\ *noun*

com·pul·sion \kəm-ˈpəl-shən\ *noun* [Middle English, from Middle French or Late Latin; Middle French, from Late Latin *compulsion-, compulsio,* from Latin *compellere* to compel] (15th century)
1 a : an act of compelling **:** the state of being compelled **b :** a force that compels
2 : an irresistible impulse to perform an irrational act

com·pul·sive \-ˈpəl-siv\ *adjective* (1588)
1 : having power to compel
2 : of, relating to, caused by, or suggestive of psychological compulsion or obsession ⟨*compulsive* actions⟩
— com·pul·sive·ly *adverb*
— com·pul·sive·ness *noun*
— com·pul·siv·i·ty \kəm-ˌpəl-ˈsi-və-tē, ˌkäm-\ *noun*

com·pul·so·ry \kəm-ˈpəls-rē, -ˈpəl-sə-\ *adjective* (1581)
1 : MANDATORY, ENFORCED
2 : COERCIVE, COMPELLING
— com·pul·so·ri·ly \-rə-lē\ *adverb*

com·punc·tion \kəm-ˈpəŋ(k)-shən\ *noun* [Middle English *compunccioun,* from Middle French *componction,* from Late Latin *compunction-, compunctio,* from Latin *compungere* to prick hard, sting, from *com-* + *pungere* to prick — more at PUNGENT] (14th century)
1 a : anxiety arising from awareness of guilt ⟨*compunctions* of conscience⟩ **b :** distress of mind over an anticipated action or result ⟨showed no *compunction* in planning devilish engines of . . . destruction —Havelock Ellis⟩
2 : a twinge of misgiving **:** SCRUPLE ⟨cheated without *compunction*⟩ ◆
synonym see PENITENCE, QUALM
— com·punc·tious \-shəs\ *adjective*

com·pur·ga·tion \ˌkäm-(ˌ)pər-ˈgā-shən\ *noun* [Late Latin *compurgation-, compurgatio,* from Latin *compurgare* to clear completely, from *com-* + *purgare* to purge] (circa 1658)
: the clearing of an accused person by oaths of others who swear to the veracity or innocence of the accused

com·pur·ga·tor \ˈkäm-(ˌ)pər-ˌgā-tər\ *noun* (1533)
: one that under oath vouches for the character or conduct of an accused person

com·put·able \kəm-ˈpyü-tə-bəl\ *adjective* (1646)
: capable of being computed
— com·put·abil·i·ty \-ˌpyü-tə-ˈbi-lə-tē\ *noun*

com·pu·ta·tion \ˌkäm-pyu̇-ˈtā-shən\ *noun* (15th century)
1 a : the act or action of computing **:** CALCULATION **b :** the use or operation of a computer
2 : a system of reckoning
3 : an amount computed
— com·pu·ta·tion·al \-shnəl, -shə-nᵊl\ *adjective*
— com·pu·ta·tion·al·ly *adverb*

com·pute \kəm-ˈpyüt\ *verb* **com·put·ed; com·put·ing** [Latin *computare* — more at COUNT] (1616)
transitive verb
: to determine especially by mathematical means ⟨*compute* your income tax⟩; *also* **:** to determine or calculate by means of a computer
intransitive verb
1 : to make calculation **:** RECKON
2 : to use a computer

computed tomography *noun* (1975)
: radiography in which a three-dimensional image of a body structure is constructed by computer from a series of plane cross-sectional images made along an axis — called also *computed axial tomography, computerized axial tomography, computerized tomography*

com·put·er \kəm-ˈpyü-tər\ *noun, often attributive* (1646)
: one that computes; *specifically* **:** a programmable electronic device that can store, retrieve, and process data
— com·put·er·dom \-dəm\ *noun*
— com·put·er·less \-ləs\ *adjective*
— com·put·er·like \-ˌlīk\ *adjective*

com·put·er·ese \-ˌpyü-tə-ˈrēz, -ˈrēs\ *noun* (circa 1960)
: jargon used by computer technologists

com·put·er·ise *chiefly British variant of* COMPUTERIZE

com·put·er·ist \kəm-ˈpyü-tə-rist\ *noun* (1973)
: a person who uses or operates a computer

com·put·er·ize \kəm-ˈpyü-tə-ˌrīz\ *transitive verb* **-ized; -iz·ing** (1957)
1 : to carry out, control, or produce by means of a computer
2 : to equip with computers
3 a : to store in a computer **b :** to put in a form that a computer can use
— com·put·er·iz·able \-ˌpyü-tə-ˈrī-zə-bəl\ *adjective*
— com·put·er·i·za·tion \-ˌpyü-tə-rə-ˈzā-shən\ *noun*

com·put·er·nik \kəm-ˈpyü-tər-ˌnik\ *noun* (1968)
: a computer enthusiast or expert

com·put·er·phobe \-ˌfōb\ *noun* (1976)
: a person who experiences anxiety about computers and especially about their use
— com·put·er·pho·bia \-ˌpyü-tər-ˈfō-bē-ə\ *noun*
— com·put·er·pho·bic \-ˈfō-bik\ *adjective*

com·rade \ˈkäm-ˌrad, -rəd, *especially British* -ˌrād\ *noun* [Middle French *camarade* group sleeping in one room, roommate, companion, from Old Spanish *camarada,* from *cámara* room, from Late Latin *camera, camara* — more at CHAMBER] (1544)
1 a : an intimate friend or associate **:** COMPANION **b :** a fellow soldier
2 [from its use as a form of address by communists] **:** COMMUNIST ◆
— com·rade·li·ness \-lē-nəs\ *noun*
— com·rade·ly *adjective*
— com·rade·ship \-ˌship\ *noun*

com·rad·ery \ˈkäm-ˌra-d(ə-)rē, -rə-drē, -ˌrä-d(ə-)rē\ *noun* (1879)
: CAMARADERIE

Comsat \ˈkäm-ˌsat\ *service mark*
— used for communications services involving an artificial satellite

Com·stock·ery \ˈkäm-ˌstä-kə-rē *also* ˈkəm-\ *noun* [Anthony *Comstock* + English *-ery*] (1905)
1 : strict censorship of materials considered obscene
2 : censorious opposition to alleged immorality (as in literature)

Com·stock·ian \käm-ˈstä-kē-ən *also* ˌkəm-\ *adjective* (1921)
: of or relating to Comstockery

com·symp \ˈkäm-ˌsimp\ *noun* [*com*munist + *symp*athizer] (circa 1961)
: a person sympathetic to communist causes — usually used disparagingly

Comt·ian *or* **Comt·ean** \ˈkäm(p)-tē-ən, ˈkōⁿ(n)-tē-\ *adjective* (1846)
: of or relating to Auguste Comte or his doctrines
— Comt·ism \ˈkäm(p)-ˌti-zəm, ˈkōⁿ(n)-\ *noun*
— Comt·ist \ˈkäm(p)-tist, ˈkōⁿ(n)-\ *adjective or noun*

¹con \ˈkän\ *transitive verb* **conned; con·ning** [Middle English *connen* to know, learn, study, alteration of *cunnen* to know, infinitive of *can* — more at CAN] (13th century)
1 : to commit to memory
2 : to study or examine closely

²con *variant of* CONN

³con *adverb* [Middle English, short for *contra*] (15th century)
: on the negative side **:** in opposition ⟨so much has been written pro and *con*⟩

⁴con *noun* (1589)
1 : an argument or evidence in opposition
2 : the negative position or one holding it ⟨an appraisal of the pros and *cons*⟩

□ USAGE

comprise Although it has been in use since the late 18th century, sense 3 is still attacked as wrong. Why it has been singled out is not clear, but until comparatively recent times it was found chiefly in scientific or technical writing rather than belles lettres. Our current evidence shows a slight shift in usage: sense 3 is somewhat more frequent in recent literary use than the earlier senses. You should be aware, however, that if you use sense 3 you may be subject to criticism for doing so, and you may want to choose a safer synonym such as *compose* or *make up*.

◇ WORD HISTORY

compunction When we express the hope that someone doing, or about to do, wrong will feel some compunction—or will experience a sting of conscience—we may not realize that both expressions embody the same metaphor. *Compunction* derives ultimately from the Latin verb *compungere* "to prick hard" or "to sting." Because some people seem not to experience even a momentary twinge of guilt, *compunction* is often used in negative constructions, as "acted without compunction."

comrade In classical Latin, *camara* or *camera* denoted a vaulted ceiling or roof. By the time of the Latin Church Fathers, the word meant simply "room, chamber," and was inherited by most of the Romance languages, ultimately yielding English *chamber* (through French *chambre*). Through scientific writing in New Latin, English has also acquired *camera lucida, camera obscura,* and the word *camera* itself. In Spanish, the outcome of the Latin word was *cámara*; a 16th century derivative of *cámara* was *camarada* "group of soldiers quartered in a room" and hence "fellow-soldier, companion." Spanish *camarada* was borrowed into French as *camarade,* and from French into Elizabethan English as both *camerade* and *comerade.* The form with *-o-* probably arose by association with the Latin prefix *com-* "together."

[5]con *adjective* [by shortening] (1889)
: CONFIDENCE

[6]con *transitive verb* **conned; con·ning** (1896)
1 : SWINDLE
2 : MANIPULATE 2b
3 : PERSUADE, CAJOLE

[7]con *noun* (1901)
: something (as a ruse) used deceptively to gain another's confidence; *also* **:** a confidence game **:** SWINDLE

[8]con *noun* [by shortening] (1893)
: CONVICT

[9]con *noun* [short for *consumption*] (1915)
slang **:** a destructive disease of the lungs; *especially* **:** TUBERCULOSIS

con- — see COM-

con amo·re \ˌkän-ə-ˈmōr-ē, ˌkōn-ə-ˈmōr-(ˌ)ā, -ˈmȯr-\ *adverb* [Italian] (1739)
1 : with love, devotion, or zest
2 : in a tender manner — used as a direction in music

con ani·ma \kän-ˈa-nə-ˌmä, kōn-ˈä-ni-\ *adverb* [Italian, literally, with spirit] (circa 1906)
: in a spirited manner — used as a direction in music

co·na·tion \kō-ˈnā-shən\ *noun* [Latin *conation-, conatio* act of attempting, from *conari* to attempt — more at DEACON] (circa 1837)
: an inclination (as an instinct, a drive, a wish, or a craving) to act purposefully **:** IMPULSE 3
— co·na·tive \ˈkō-nə-tiv, ˈkä-, ˈkō-ˌnā-\ *adjective*

con brio \kän-ˈbrē-(ˌ)ō, kōn-\ *adverb* [Italian, literally, with vigor] (circa 1889)
: in a vigorous or brisk manner — used as a direction in music

con·ca·nav·a·lin \ˌkän-kə-ˈna-və-lən\ *noun* [*com-* + *canavalin,* a noncrystalline globulin found in the jack bean, from New Latin *Canavalia,* genus name of the jack bean] (1917)
: either of two crystalline globulins occurring in the jack bean; *especially* **:** one that is a potent hemagglutinin

[1]con·cat·e·nate \kän-ˈka-tə-nət, kən-\ *adjective* [Middle English, from Late Latin *concatenatus,* past participle of *concatenare* to link together, from Latin *com-* + *catena* chain] (15th century)
: linked together

[2]concatenate \-ˌnāt\ *transitive verb* **-nat·ed; -nat·ing** (1598)
: to link together in a series or chain
— con·cat·e·na·tion \(ˌ)kän-ˌka-tə-ˈnā-shən, kən-\ *noun*

[1]con·cave \kän-ˈkāv, ˈkän-ˌ\ *adjective* [Middle English, from Middle French, from Latin *concavus,* from *com-* + *cavus* hollow — more at CAVE] (15th century)
1 : hollowed or rounded inward like the inside of a bowl
2 : arched in **:** curving in — used of the side of a curve or surface on which neighboring normals to the curve or surface converge and on which lies the chord joining two neighboring points of the curve or surface

[2]con·cave \ˈkän-ˌkāv\ *noun* (1552)
: a concave line or surface

con·cav·i·ty \kän-ˈka-və-tē\ *noun, plural* **-ties** (15th century)
1 : a concave line, surface, or space **:** HOLLOW
2 : the quality or state of being concave

con·ca·vo–con·vex \kän-ˈkā-vō-kän-ˈveks; ˈkän-, kən-ˈ\ *adjective* (1676)
1 : concave on one side and convex on the other
2 : having the concave side curved more than the convex ⟨a *concavo-convex* lens⟩

con·ceal \kən-ˈsē(ə)l\ *transitive verb* [Middle English *concelen,* from Middle French *conceler,* from Latin *concelare,* from *com-* + *celare* to hide — more at HELL] (14th century)
1 : to prevent disclosure or recognition of
2 : to place out of sight
synonym see HIDE
— con·ceal·able \-ˈsē-lə-bəl\ *adjective*
— con·ceal·er \-ˈsē-lər\ *noun*
— con·ceal·ing·ly \-ˈsē-liŋ-lē\ *adverb*
— con·ceal·ment \-ˈsē(ə)l-mənt\ *noun*

con·cede \kən-ˈsēd\ *verb* **con·ced·ed; con·ced·ing** [French or Latin; French *concéder,* from Latin *concedere,* from *com-* + *cedere* to yield] (1632)
transitive verb
1 : to grant as a right or privilege
2 a : to accept as true, valid, or accurate ⟨the right of the state to tax is generally *conceded*⟩ **b :** to acknowledge grudgingly or hesitantly
intransitive verb
: to make concession **:** YIELD
synonym see GRANT
— con·ced·ed·ly \-ˈsē-dəd-lē\ *adverb*
— con·ced·er *noun*

[1]con·ceit \kən-ˈsēt\ *noun* [Middle English, from *conceiven*] (14th century)
1 a (1) **:** a result of mental activity **:** THOUGHT (2) **:** individual opinion **b :** favorable opinion; *especially* **:** excessive appreciation of one's own worth or virtue
2 : a fancy article
3 a : a fanciful idea **b :** an elaborate or strained metaphor **c :** use or presence of such conceits in poetry **d :** an organizing theme or concept ⟨found his *conceit* for the film early —Peter Wilkinson⟩

[2]conceit *transitive verb* (1557)
1 *obsolete* **:** CONCEIVE, UNDERSTAND
2 *chiefly dialect* **:** IMAGINE
3 *dialect British* **:** to take a fancy to

con·ceit·ed \-ˈsē-təd\ *adjective* [[1]*conceit*] (1593)
1 : ingeniously contrived **:** FANCIFUL
2 : having an excessively high opinion of oneself
— con·ceit·ed·ly *adverb*
— con·ceit·ed·ness *noun*

con·ceiv·able \kən-ˈsē-və-bəl\ *adjective* (15th century)
: capable of being conceived **:** IMAGINABLE
— con·ceiv·abil·i·ty \kən-ˌsē-və-ˈbi-lə-tē\ *noun*
— con·ceiv·able·ness \-ˈsē-və-bəl-nəs\ *noun*

con·ceiv·ably \-blē\ *adverb* (1625)
1 : in a conceivable manner
2 : it may be conceived **:** POSSIBLY

con·ceive \kən-ˈsēv\ *verb* **con·ceived; con·ceiv·ing** [Middle English, from Middle French *conceivre,* from Latin *concipere* to take in, conceive, from *com-* + *capere* to take — more at HEAVE] (14th century)
transitive verb
1 a : to become pregnant with (young) **b :** to cause to begin **:** ORIGINATE
2 a : to take into one's mind ⟨*conceive* a prejudice⟩ **b :** to form a conception of **:** IMAGINE
3 : to apprehend by reason or imagination **:** UNDERSTAND
4 : to be of the opinion
intransitive verb
1 : to become pregnant
2 : to have a conception — usually used with *of* ⟨*conceives* of death as emptiness⟩
synonym see THINK
— con·ceiv·er *noun*

con·cel·e·brant \kən-ˈse-lə-brənt, kän-\ *noun* (circa 1931)
: one that concelebrates a Eucharist or Mass

con·cel·e·brate \kən-ˈse-lə-ˌbrāt, kän-\ *verb* [Medieval Latin *concelebratus,* past participle of *concelebrare,* from Latin, to frequent, celebrate, from *com-* + *celebrare* to celebrate] (1879)
transitive verb
: to participate in (a Eucharist) as a joint celebrant who recites the canon in unison with other celebrants
intransitive verb
: to participate as a celebrant in a concelebrated Eucharist
— con·cel·e·bra·tion \(ˌ)kän-ˌse-lə-ˈbrā-shən, kən-\ *noun*

con·cent \kən-ˈsent\ *noun* [Latin *concentus,* from *concinere* to sing together, from *com-* + *canere* to sing — more at CHANT] (1585)
archaic **:** HARMONY

con·cen·ter \kən-ˈsen-tər, kän-\ *verb* [French *concentrer,* from *com-* + *centre* center] (1598)
transitive verb
: to draw or direct to a common center **:** CONCENTRATE
intransitive verb
: to come to a common center

[1]con·cen·trate \ˈkän(t)-sən-ˌtrāt, -ˌsen-\ *verb* **-trat·ed; -trat·ing** [*com-* + Latin *centrum* center] (1646)
transitive verb
1 a : to bring or direct toward a common center or objective **:** FOCUS **b :** to gather into one body, mass, or force ⟨power was *concentrated* in a few able hands⟩ **c :** to accumulate (a toxic substance) in bodily tissues ⟨fish *concentrate* mercury⟩
2 a : to make less dilute ⟨*concentrate* syrup⟩ **b :** to express or exhibit in condensed form
intransitive verb
1 : to draw toward or meet in a common center
2 : GATHER, COLLECT
3 : to concentrate one's powers, efforts, or attention ⟨*concentrate* on a problem⟩
— con·cen·trat·ed·ly \-ˌtrā-təd-lē; ˌkän(t)-sən-ˈtrā-\ *adverb*
— con·cen·tra·tive \-ˌtrā-tiv\ *adjective*

[2]concentrate *noun* (1883)
1 : something concentrated: as **a :** a mineral-rich product obtained after an initial processing of ore **b :** a food reduced in bulk by elimination of fluid ⟨orange juice *concentrate*⟩
2 : a feedstuff (as grains) relatively rich in digestible nutrients — compare FIBER

con·cen·tra·tion \ˌkän(t)-sən-ˈtrā-shən, -ˌsen-\ *noun* (1634)
1 a : the act or process of concentrating **:** the state of being concentrated; *especially* **:** direction of attention to a single object **b :** MAJOR 4a
2 : a concentrated mass or thing
3 : the amount of a component in a given area or volume

concentration camp *noun* (1901)
: a camp where persons (as prisoners of war, political prisoners, or refugees) are detained or confined

con·cen·tra·tor \ˈkän(t)-sən-ˌtrā-tər\ *noun* (1833)
: one that concentrates: as **a :** an industrial plant that produces concentrates from ores **b :** a mirror or group of mirrors that focus sunlight for use as an energy source **c :** a device in a computer network that collects data from separate low-volume transmission channels and retransmits it over a single high-volume channel

con·cen·tric \kən-ˈsen-trik, ˌkän-\ *adjective* [Medieval Latin *concentricus,* from Latin *com-* + *centrum* center] (14th century)
1 : having a common center ⟨*concentric* circles⟩
2 : having a common axis **:** COAXIAL
— con·cen·tri·cal·ly \-tri-k(ə-)lē\ *adverb*
— con·cen·tric·i·ty \ˌkän-ˌsen-ˈtri-sə-tē\ *noun*

con·cept \ˈkän-ˌsept\ *noun* [Latin *conceptum,* neuter of *conceptus,* past participle of *concipere* to conceive — more at CONCEIVE] (1556)
1 : something conceived in the mind **:** THOUGHT, NOTION
2 : an abstract or generic idea generalized from particular instances
synonym see IDEA

con·cep·ta·cle \kən-ˈsep-ti-kəl\ *noun* [New Latin *conceptaculum,* from Latin, receptacle,

from *conceptus,* past participle of *concipere* to take in] (1835)
: an external cavity containing reproductive cells in algae (as of the genus *Fucus*)

con·cep·tion \kən-ˈsep-shən\ *noun* [Middle English *concepcioun,* from Old French *conception,* from Latin *conception-, conceptio,* from *concipere*] (14th century)
1 a (1) **:** the act of becoming pregnant **:** the state of being conceived (2) **:** EMBRYO, FETUS **b :** BEGINNING ⟨joy had the like *conception* in our eyes —Shakespeare⟩
2 a : the capacity, function, or process of forming or understanding ideas or abstractions or their symbols **b :** a general idea **:** CONCEPT **c :** a complex product of abstract or reflective thinking **d :** the sum of a person's ideas and beliefs concerning something
3 : the originating of something in the mind
synonym see IDEA
— **con·cep·tion·al** \-shnəl, -shə-nᵊl\ *adjective*
— **con·cep·tive** \-ˈsep-tiv\ *adjective*

con·cep·tu·al \kən-ˈsep-chə-wəl, kän-, -chəl, -shwəl\ *adjective* [Medieval Latin *conceptualis* of thought, from Late Latin *conceptus* act of conceiving, thought, from Latin *concipere*] (circa 1834)
: of, relating to, or consisting of concepts
— **con·cep·tu·al·i·ty** \-ˌsep-chə-ˈwa-lə-tē, -shə-\ *noun*
— **con·cep·tu·al·ly** *adverb*

conceptual art *noun* (circa 1969)
: an art form in which the artist's intent is to convey a concept rather than to create an art object
— **conceptual artist** *noun*

con·cep·tu·al·ise *British variant of* CONCEPTUALIZE

con·cep·tu·al·ism \-ˈsep-chə-wə-ˌli-zəm, -chə-ˌli-, -shwə-ˌli-\ *noun* (circa 1838)
: a theory in philosophy intermediate between realism and nominalism that universals exist in the mind as concepts of discourse or as predicates which may be properly affirmed of reality
— **con·cep·tu·al·is·tic** \-ˌsep-chə-wə-ˈlis-tik, -chə-ˈlis-, -shwə-ˈlis-\ *adjective*
— **con·cep·tu·al·is·ti·cal·ly** \-ti-k(ə-)lē\ *adverb*

con·cep·tu·al·ist \-ˈsep-chə-wə-list, -chə-list, -shwə-list\ *noun* (1785)
: an adherent to the tenets of conceptualism or of conceptual art

con·cep·tu·al·ize \-ˈsep-chə-wə-ˌlīz, -chə-ˌlīz, -shwə-ˌlīz\ *transitive verb* **-ized; -iz·ing** (1878)
: to form a concept of; *especially* **:** to interpret conceptually
— **con·cep·tu·al·i·za·tion** \-ˌsep-chə-wə-lə-ˈzā-shən, -chə-lə-ˈ, -shwə-lə-ˈ\ *noun*
— **con·cep·tu·al·iz·er** \-ˈsep-chə-wə-ˌlī-zər, chə-ˌlī-, -shwə-ˌlī-\ *noun*

con·cep·tus \kən-ˈsep-təs\ *noun* [Latin, one conceived, from past participle of *concipere* to conceive] (circa 1860)
: a fertilized egg, embryo, or fetus

¹con·cern \kən-ˈsərn\ *verb* [Middle English, from Middle French & Medieval Latin; Middle French *concerner,* from Medieval Latin *concernere,* from Late Latin, to sift together, mingle, from Latin *com-* + *cernere* to sift — more at CERTAIN] (15th century)
transitive verb
1 a : to relate to **:** be about ⟨the novel *concerns* three soldiers⟩ **b :** to bear on
2 : to have an influence on **:** INVOLVE; *also* **:** to be the business or affair of ⟨the problem *concerns* us all⟩
3 : to be a care, trouble, or distress to ⟨her ill health *concerns* me⟩
4 : ENGAGE, OCCUPY ⟨he *concerns* himself with trivia⟩
intransitive verb
obsolete **:** to be of importance **:** MATTER

²concern *noun* (1655)
1 a : marked interest or regard usually arising through a personal tie or relationship **b :** an uneasy state of blended interest, uncertainty, and apprehension
2 : something that relates or belongs to one **:** AFFAIR
3 : matter for consideration
4 : an organization or establishment for business or manufacture
5 : CONTRIVANCE, GADGET
synonym see CARE

concerned *adjective* (1656)
1 a : ANXIOUS, WORRIED ⟨*concerned* for their safety⟩ **b :** INTERESTED ⟨*concerned* to prove the point⟩
2 a : interestedly engaged ⟨*concerned* with books and music⟩ **b :** culpably involved **:** IMPLICATED ⟨arrested all *concerned*⟩

concerning *preposition* (15th century)
: relating to **:** REGARDING

con·cern·ment \kən-ˈsərn-mənt\ *noun* (1610)
1 : something in which one is concerned
2 : IMPORTANCE, CONSEQUENCE
3 *archaic* **:** INVOLVEMENT, PARTICIPATION
4 : SOLICITUDE, ANXIETY

¹con·cert \ˈkän(t)-sərt, ˈkän-ˌsərt\ *noun* [French, from Italian *concerto,* from *concertare*] (1674)
1 *obsolete* **:** musical harmony **:** CONCORD
2 a : agreement in design or plan **:** union formed by mutual communication of opinion and views **b :** a concerted action ⟨the sacrifice was hailed with a *concert* of praise⟩
3 : a public performance (as of music or dancing)
— **concert** *adjective*
— **in concert :** TOGETHER

²con·cert \kən-ˈsərt\ *verb* [Middle French *concerter,* from Old Italian *concertare,* perhaps from *com-* + *certo* certain, decided, from Latin *certus* — more at CERTAIN] (1694)
transitive verb
1 : to settle or adjust by conferring and reaching an agreement ⟨*concerted* their differences⟩
2 : to make a plan for ⟨*concert* measures for aiding the poor⟩
intransitive verb
: to act in harmony or conjunction

con·cert·ed \kən-ˈsər-təd\ *adjective* (1716)
1 a : mutually contrived or agreed on ⟨a *concerted* effort⟩ **b :** performed in unison ⟨*concerted* artillery fire⟩
2 : arranged in parts for several voices or instruments
— **con·cert·ed·ly** *adverb*
— **con·cert·ed·ness** *noun*

con·cert·go·er \ˈkän(t)-sərt-ˌgō(-ə)r, ˈkän-ˌsərt-\ *noun* (1855)
: one who frequently attends concerts
— **con·cert·go·ing** \-ˌgō-iŋ, -ˌgȯ(-)iŋ\ *noun or adjective*

concert grand *noun* (circa 1891)
: a grand piano of the largest size adapted in volume, timbre, and brilliance of tone to concert use

con·cer·ti·na \ˌkän(t)-sər-ˈtē-nə\ *noun* [probably from ¹*concert* + Italian *-ina,* diminutive suffix] (1837)
1 : a musical instrument of the accordion family
2 : a coiled barbed wire for use as an obstacle — called also *concertina wire*

concertina 1

con·cer·ti·no \ˌkän-chər-ˈtē-(ˌ)nō\ *noun, plural* **-nos** [Italian, diminutive of *concerto*] (circa 1801)
1 : the solo instruments in a concerto grosso
2 : a short concerto

con·cert·ize \ˈkän(t)-sər-ˌtīz\ *intransitive verb* **-ized; -iz·ing** (1883)
: to perform professionally in concerts

con·cert·mas·ter \ˈkän(t)-sərt-ˌmas-tər\ *or* **con·cert·meis·ter** \-ˌmī-stər\ *noun* [German *Konzertmeister,* from *Konzert* concert + *Meister* master] (circa 1889)
: the leader of the first violins of an orchestra and by custom usually the assistant to the conductor

con·cer·to \kən-ˈcher-(ˌ)tō *also* -ˈchər-\ *noun, plural* **-ti** \-(ˌ)tē\ *or* **-tos** [Italian, from *concerto* concert] (1730)
: a piece for one or more soloists and orchestra with three contrasting movements

concerto gros·so \-ˈgrō-(ˌ)sō, -ˈgrȯ-\ *noun, plural* **concerti gros·si** \-(ˌ)sē\ [Italian, literally, big concerto] (1724)
: a baroque orchestral composition featuring a small group of solo instruments contrasting with the full orchestra

concert pitch *noun* (1767)
1 : INTERNATIONAL PITCH
2 : a high state of fitness, tension, or readiness

con·ces·sion \kən-ˈse-shən\ *noun* [Middle English, from Middle French or Latin; Middle French, from Latin *concession-, concessio,* from *concedere* to concede] (15th century)
1 a : the act or an instance of conceding **b :** the admitting of a point claimed in argument
2 : something conceded: **a :** ACKNOWLEDGMENT, ADMISSION **b :** GRANT **c** (1) **:** a grant of land or property especially by a government in return for services or for a particular use (2) **:** a right to undertake and profit by a specified activity (3) **:** a lease of a portion of premises for a particular purpose; *also* **:** the portion leased or the activities carried on
— **con·ces·sion·al** \-ˈsesh-nəl, -ˈse-shə-nᵊl\ *adjective*
— **con·ces·sion·ary** \-ˈse-shə-ˌner-ē\ *adjective*

con·ces·sion·aire \kən-ˌse-shə-ˈnar, -ˈner\ *noun* [French *concessionnaire,* from *concession*] (1862)
: the owner or operator of a concession; *especially* **:** one that operates a refreshment stand at a recreational center

con·ces·sion·er \kən-ˈse-sh(ə-)nər\ *noun* (circa 1891)
: CONCESSIONAIRE

con·ces·sive \kən-ˈse-siv\ *adjective* (1711)
1 : denoting concession ⟨a *concessive* clause⟩
2 : making for or being a concession
— **con·ces·sive·ly** *adverb*

conch \ˈkäŋk, ˈkänch, ˈkȯŋk\ *noun, plural* **conchs** \ˈkäŋks, ˈkȯŋks\ *or* **conches** \ˈkän-chəz\ [Middle English, from Latin *concha* mussel, mussel shell, from Greek *konchē;* akin to Sanskrit *śaṅkha* conch shell] (15th century)
1 : any of various large spiral-shelled marine gastropod mollusks (as of the genus *Strombus*); *also* **:** its shell used especially for cameos
2 *often capitalized* **:** a native or resident of the Florida Keys
3 : CONCHA 2

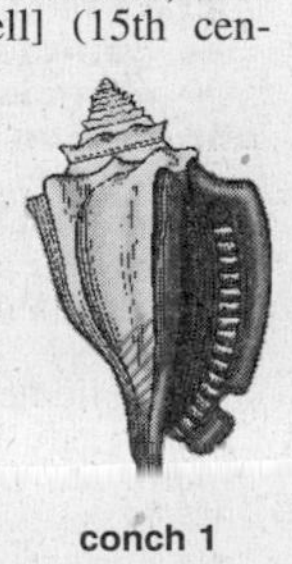
conch 1

con·cha \ˈkäŋ-kə\ *noun, plural* **con·chae** \-ˌkē, -ˌkī\ [Italian & Latin; Italian *conca* semidome, apse, from Late Latin *concha,* from Latin, shell] (circa 1639)
1 a : the plain semidome of an apse **b :** APSE
2 : something shaped like a shell; *especially* **:** the largest and deepest concavity of the external ear
— **con·chal** \-kəl\ *adjective*

\ə\ **abut** \ᵊ\ **kitten** \ər\ **further** \a\ **ash** \ā\ **ace**
\ä\ **mop, mar** \au̇\ **out** \ch\ **chin** \e\ **bet** \ē\ **easy**
\g\ **go** \i\ **hit** \ī\ **ice** \j\ **job** \ŋ\ **sing** \ō\ **go**
\ȯ\ **law** \ȯi\ **boy** \th\ **thin** \t͟h\ **the** \ü\ **loot** \u̇\ **foot**
\y\ **yet** \zh\ **vision** *see also* Guide to Pronunciation

con·choi·dal \käŋ-ˈkȯi-dᵊl, kän-\ *adjective* [Greek *konchoeidēs* like a mussel, from *konchē*] (1666)
: having elevations or depressions shaped like the inside surface of a bivalve shell
— **con·choi·dal·ly** \-dᵊl-ē\ *adverb*

con·chol·o·gy \käŋ-ˈkä-lə-jē\ *noun* [Greek *konchē*] (1776)
1 : a branch of zoology that deals with shells
2 : a treatise on shells
— **con·chol·o·gist** \-jist\ *noun*

con·cierge \kōⁿ-ˈsyerzh\ *noun, plural* **con·cierges** \-ˈsyerzh, -ˈsyer-zhəz\ [French, from Old French, probably from (assumed) Vulgar Latin *conservius,* alteration of Latin *conservus* fellow slave, from *com-* + *servus* slave] (circa 1697)
1 : a resident in an apartment building especially in France who serves as doorkeeper, landlord's representative, and janitor
2 : a usually multilingual hotel staff member who handles luggage and mail, makes reservations, and arranges tours

con·cil·i·ar \kən-ˈsi-lē-ər\ *adjective* [Latin *concilium* council] (circa 1677)
: of, relating to, or issued by a council
— **con·cil·i·ar·ly** *adverb*

con·cil·i·ate \kən-ˈsi-lē-ˌāt\ *verb* **-at·ed; -at·ing** [Latin *conciliatus,* past participle of *conciliare* to assemble, unite, win over, from *concilium* assembly, council — more at COUNCIL] (1545)
transitive verb
1 : to gain (as goodwill) by pleasing acts
2 : to make compatible **:** RECONCILE
3 : APPEASE
intransitive verb
: to become friendly or agreeable
synonym see PACIFY
— **con·cil·i·a·tion** \-ˌsi-lē-ˈā-shən\ *noun*
— **con·cil·i·a·tive** \-ˈsi-lē-ˌā-tiv\ *adjective*
— **con·cil·i·a·tor** \-ˌā-tər\ *noun*
— **con·cil·ia·to·ry** \-ˈsil-yə-ˌtȯr-ē, -ˈsi-lē-ə-, -ˌtȯr-\ *adjective*

con·cin·ni·ty \kən-ˈsi-nə-tē\ *noun, plural* **-ties** [Latin *concinnitas,* from *concinnus* skillfully put together] (1531)
: harmony or elegance of design especially of literary style in adaptation of parts to a whole or to each other

con·cise \kən-ˈsīs\ *adjective* [Latin *concisus,* from past participle of *concidere* to cut up, from *com-* + *caedere* to cut, strike] (circa 1590)
1 : marked by brevity of expression or statement **:** free from all elaboration and superfluous detail
2 : cut short **:** BRIEF ☆
— **con·cise·ly** *adverb*
— **con·cise·ness** *noun*

con·ci·sion \kən-ˈsi-zhən\ *noun* [Middle English, from Latin *concision-, concisio,* from *concidere*] (14th century)
1 *archaic* **:** a cutting up or off
2 : the quality or state of being concise

con·clave \ˈkän-ˌklāv\ *noun* [Middle English, from Middle French or Medieval Latin; Middle French, from Medieval Latin, from Latin, room that can be locked up, from *com-* + *clavis* key — more at CLAVICLE] (1524)
1 : a private meeting or secret assembly; *especially* **:** a meeting of Roman Catholic cardinals secluded continuously while choosing a pope
2 : a gathering of a group or association

con·clude \kən-ˈklüd\ *verb* **con·clud·ed; con·clud·ing** [Middle English, from Latin *concludere* to shut up, end, infer, from *com-* + *claudere* to shut — more at CLOSE] (14th century)
transitive verb
1 *obsolete* **:** to shut up **:** ENCLOSE
2 : to bring to an end especially in a particular way or with a particular action ⟨*conclude* a meeting with a prayer⟩
3 a : to reach as a logically necessary end by reasoning **:** infer on the basis of evidence ⟨*concluded* that her argument was sound⟩ **b :** to make a decision about **:** DECIDE ⟨*concluded* he would wait a little longer⟩ **c :** to come to an agreement on **:** EFFECT ⟨*conclude* a sale⟩
4 : to bring about as a result **:** COMPLETE
intransitive verb
1 : END
2 a : to form a final judgment **b :** to reach a decision or agreement
synonym see CLOSE, INFER
— **con·clud·er** *noun*

con·clu·sion \kən-ˈklü-zhən\ *noun* [Middle English, from Middle French, from Latin *conclusion-, conclusio,* from *concludere*] (14th century)
1 a : a reasoned judgment **:** INFERENCE **b :** the necessary consequence of two or more propositions taken as premises; *especially* **:** the inferred proposition of a syllogism
2 : the last part of something: as **a :** RESULT, OUTCOME **b** *plural* **:** trial of strength or skill — used in the phrase *try conclusions* **c :** a final summation **d :** the final decision in a law case **e :** the final part of a pleading in law
3 : an act or instance of concluding

con·clu·sion·ary \kən-ˈklü-zhə-ˌner-ē\ *adjective* (1948)
: CONCLUSORY

con·clu·sive \-ˈklü-siv, -ziv\ *adjective* (1536)
1 : of, relating to, or being a conclusion
2 : putting an end to debate or question especially by reason of irrefutability ☆
— **con·clu·sive·ly** *adverb*
— **con·clu·sive·ness** *noun*

con·clu·so·ry \kən-ˈklüs-rē, -ˈklü-sə-\ *adjective* (1923)
: consisting of or relating to a conclusion or assertion for which no supporting evidence is offered ⟨*conclusory* allegations⟩

con·coct \kən-ˈkäkt, kän-\ *transitive verb* [Latin *concoctus,* past participle of *concoquere* to cook together, from *com-* + *coquere* to cook — more at COOK] (1675)
1 : to prepare by combining raw materials
2 : DEVISE, FABRICATE
— **con·coct·er** *noun*
— **con·coc·tion** \-ˈkäk-shən\ *noun*
— **con·coc·tive** \-ˈkäk-tiv\ *adjective*

con·com·i·tance \kən-ˈkä-mə-tən(t)s, kän-\ *noun* (circa 1535)
: ACCOMPANIMENT; *especially* **:** a conjunction that is regular and is marked by correlative variation of accompanying elements

¹con·com·i·tant \-mə-tənt\ *adjective* [Latin *concomitant-, concomitans,* present participle of *concomitari* to accompany, from *com-* + *comitari* to accompany, from *comit-, comes* companion — more at COUNT] (1607)
: accompanying especially in a subordinate or incidental way
— **con·com·i·tant·ly** *adverb*

²concomitant *noun* (1621)
: something that accompanies or is collaterally connected with something else **:** ACCOMPANIMENT

con·cord \ˈkän-ˌkȯrd, ˈkäŋ-\ *noun* [Middle English, from Middle French *concorde,* from Latin *concordia,* from *concord-, concors* agreeing, from *com-* + *cord-, cor* heart — more at HEART] (14th century)
1 a : a state of agreement **:** HARMONY **b :** a simultaneous occurrence of two or more musical tones that produces an impression of agreeableness or resolution on a listener — compare DISCORD
2 : agreement by stipulation, compact, or covenant
3 : grammatical agreement

con·cor·dance \kən-ˈkȯr-dᵊn(t)s, kän-\ *noun* [Middle English, from Middle French, from Medieval Latin *concordantia,* from Latin *concordant-, concordans,* present participle of *concordare* to agree, from *concord-, concors*] (14th century)
1 : an alphabetical index of the principal words in a book or the works of an author with their immediate contexts
2 : CONCORD, AGREEMENT

con·cor·dant \-dᵊnt\ *adjective* [Middle English, from Middle French, from Latin *concordant-, concordans*] (15th century)
: CONSONANT, AGREEING
— **con·cor·dant·ly** *adverb*

con·cor·dat \kən-ˈkȯr-ˌdat\ *noun* [French, from Medieval Latin *concordatum,* from Latin, neuter of *concordatus,* past participle of *concordare*] (1616)
: COMPACT, COVENANT; *specifically* **:** an agreement between a pope and a sovereign or government for the regulation of ecclesiastical matters

con·cours d'e·le·gance \(ˌ)kōⁿ-ˌkůr-ˌdā-lā-ˈgäⁿs\ *noun* [French *concours d'élégance,* literally, competition of elegance] (1950)
: a show or contest of vehicles and accessories in which the entries are judged chiefly on excellence of appearance and turnout

con·course \ˈkän-ˌkōrs, ˈkäŋ-, -ˌkȯrs\ *noun* [Middle English, from Middle French & Latin; Middle French *concours,* from Latin *concursus,* from *concurrere* to run together — more at CONCUR] (14th century)
1 : an act or process of coming together and merging
2 : a meeting produced by voluntary or spontaneous coming together
3 a : an open space where roads or paths meet **b :** an open space or hall (as in a railroad terminal) where crowds gather

con·cres·cence \kən-ˈkre-sᵊn(t)s, kän-\ *noun* [Latin *concrescentia,* from *concrescent-, concrescens,* present participle of *concrescere* to grow together, from *com-* + *crescere* to grow — more at CRESCENT] (1614)
1 : increase by the addition of particles
2 : a growing together **:** COALESCENCE; *especially* **:** convergence and fusion of the lateral lips of the blastopore to form the primordium of an embryo

☆ SYNONYMS

Concise, terse, succinct, laconic, summary, pithy, compendious mean very brief in statement or expression. CONCISE suggests the removal of all that is superfluous or elaborative ⟨a *concise* description⟩. TERSE implies pointed conciseness ⟨a *terse* reply⟩. SUCCINCT implies the greatest possible compression ⟨a *succinct* letter of resignation⟩. LACONIC implies brevity to the point of seeming rude, indifferent, or mysterious ⟨an aloof and *laconic* stranger⟩. SUMMARY suggests the statement of main points with no elaboration or explanation ⟨a *summary* listing of the year's main events⟩. PITHY adds to SUCCINCT or TERSE the implication of richness of meaning or substance ⟨a comedy sharpened by *pithy* one-liners⟩. COMPENDIOUS applies to what is at once full in scope and brief and concise in treatment ⟨a *compendious* dictionary⟩.

Conclusive, decisive, determinative, definitive mean bringing to an end. CONCLUSIVE applies to reasoning or logical proof that puts an end to debate or questioning ⟨*conclusive* evidence⟩. DECISIVE may apply to something that ends a controversy, a contest, or any uncertainty ⟨a *decisive* battle⟩. DETERMINATIVE adds an implication of giving a fixed character or direction ⟨the *determinative* factor in the court's decision⟩. DEFINITIVE applies to what is put forth as final and permanent ⟨the *definitive* biography⟩.

— **con·cres·cent** \-sᵊnt\ *adjective*

^{1}**con·crete** \(ˌ)kän-ˈkrēt, ˈkän-ˌ, kən-ˈ\ *adjective* [Middle English, from Latin *concretus,* from past participle of *concrescere*] (14th century)
1 : naming a real thing or class of things ⟨the word *poem* is *concrete, poetry* is abstract⟩
2 : formed by coalition of particles into one solid mass
3 a : characterized by or belonging to immediate experience of actual things or events **b :** SPECIFIC, PARTICULAR **c :** REAL, TANGIBLE
4 : relating to or made of concrete
— **con·crete·ly** *adverb*
— **con·crete·ness** *noun*

^{2}**con·crete** \ˈkän-ˌkrēt, kän-ˈ\ *verb* **con·cret·ed; con·cret·ing** (1590)
transitive verb
1 a : to form into a solid mass **:** SOLIDIFY **b :** COMBINE, BLEND
2 : to make actual or real **:** cause to take on the qualities of reality
3 : to cover with, form of, or set in concrete
intransitive verb
: to become concreted

^{3}**con·crete** \ˈkän-ˌkrēt, (ˌ)kän-ˈ\ *noun* (1656)
1 : a mass formed by concretion or coalescence of separate particles of matter in one body
2 : a hard strong building material made by mixing a cementing material (as portland cement) and a mineral aggregate (as sand and gravel) with sufficient water to cause the cement to set and bind the entire mass
3 : a waxy essence of flowers prepared by extraction and evaporation and used in perfumery

concrete music *noun* (1953)
: MUSIQUE CONCRÈTE

concrete poetry *noun* (1958)
: poetry in which the poet's intent is conveyed by the graphic patterns of letters, words, or symbols rather than by the conventional arrangement of words

con·cre·tion \kän-ˈkrē-shən, kən-\ *noun* (1541)
1 : something concreted: as **a :** a hard usually inorganic mass (as a bezoar or tophus) formed in a living body **b :** a mass of mineral matter found generally in rock of a composition different from its own and produced by deposition from aqueous solution in the rock
2 : the act or process of concreting **:** the state of being concreted ⟨*concretion* of ideas in an hypothesis⟩
— **con·cre·tion·ary** \-shə-ˌner-ē\ *adjective*

con·cret·ism \kän-ˈkrē-ˌti-zəm, ˈkän-ˌ\ *noun* (1865)
: representation of abstract things as concrete; *especially* **:** the theory or practice of concrete poetry
— **con·cret·ist** \-ˈkrē-tist, -ˌkrē-\ *noun*

con·cret·ize \-ˌtīz\ *verb* **-ized; -iz·ing** (1884)
transitive verb
: to make concrete, specific, or definite ⟨tried to *concretize* his ideas⟩
intransitive verb
: to become concrete
— **con·cret·i·za·tion** \(ˌ)kän-ˌkrē-tə-ˈzā-shən\ *noun*

con·cu·bi·nage \kän-ˈkyü-bə-nij, kən-\ *noun* (14th century)
1 : cohabitation of persons not legally married
2 : the state of being a concubine

con·cu·bine \ˈkäŋ-kyü-ˌbīn, ˈkän-\ *noun* [Middle English, from Middle French, from Latin *concubina,* from *com-* + *cubare* to lie] (14th century)
: a woman with whom a man cohabits without being married: as **a :** one having a recognized social status in a household below that of a wife **b :** MISTRESS 4a

con·cu·pis·cence \kän-ˈkyü-pə-sən(t)s, kən-\ *noun* [Middle English, from Middle French, from Late Latin *concupiscentia,* from Latin *concupiscent-, concupiscens,* present participle of *concupiscere* to desire ardently, from *com-* + *cupere* to desire] (14th century)
: strong desire; *especially* **:** sexual desire
— **con·cu·pis·cent** \-sənt\ *adjective*

con·cu·pis·ci·ble \-ˈkyü-pə-sə-bəl\ *adjective* [Middle English, from Middle French or Late Latin; Middle French, from Late Latin *concupiscibilis,* from Latin *concupiscere*] (14th century)
: LUSTFUL, DESIROUS

con·cur \kən-ˈkər, kän-\ *intransitive verb* **con·curred; con·cur·ring** [Middle English *concurren,* from Latin *concurrere,* from *com-* + *currere* to run — more at CAR] (15th century)
1 : to act together to a common end or single effect
2 a : APPROVE ⟨*concur* in a statement⟩ **b :** to express agreement ⟨*concur* with an opinion⟩
3 *obsolete* **:** to come together **:** MEET
4 : to happen together **:** COINCIDE
synonym see AGREE

con·cur·rence \-ˈkər-ən(t)s, -ˈkə-rən(t)s\ *noun* (15th century)
1 a : the simultaneous occurrence of events or circumstances **b :** the meeting of concurrent lines in a point
2 a : agreement or union in action **:** COOPERATION **b** (1) **:** agreement in opinion or design (2) **:** CONSENT
3 : a coincidence of equal powers in law

con·cur·ren·cy \-ən(t)-sē, -rən(t)-sē\ *noun* (1597)
: CONCURRENCE

con·cur·rent \-ˈkər-ənt, -ˈkə-rənt\ *adjective* [Middle English, from Middle French & Latin; Middle French, from Latin *concurrent-, concurrens,* present participle of *concurrere*] (14th century)
1 : operating or occurring at the same time
2 a : running parallel **b :** CONVERGENT; *specifically* **:** meeting or intersecting in a point
3 : acting in conjunction
4 : exercised over the same matter or area by two different authorities ⟨*concurrent* jurisdiction⟩
— **concurrent** *noun*
— **con·cur·rent·ly** *adverb*

concurrent resolution *noun* (1802)
: a resolution passed by both houses of a legislative body that lacks the force of law

con·cuss \kən-ˈkəs\ *transitive verb* [Latin *concussus,* past participle] (1597)
: to affect with or as if with concussion

con·cus·sion \kən-ˈkə-shən\ *noun* [Middle English *concussioun,* from Middle French or Latin; Middle French *concussion,* from Latin *concussion-, concussio,* from *concutere* to shake violently, from *com-* + *quatere* to shake] (14th century)
1 a : a stunning, damaging, or shattering effect from a hard blow; *especially* **:** a jarring injury of the brain resulting in disturbance of cerebral function **b :** a hard blow or collision
2 : AGITATION, SHAKING
— **con·cus·sive** \-ˈkə-siv\ *adjective*

con·demn \kən-ˈdem\ *transitive verb* [Middle English, from Middle French *condemner,* from Latin *condemnare,* from *com-* + *damnare* to condemn — more at DAMN] (14th century)
1 : to declare to be reprehensible, wrong, or evil usually after weighing evidence and without reservation
2 a : to pronounce guilty **:** CONVICT **b :** SENTENCE, DOOM
3 : to adjudge unfit for use or consumption
4 : to declare convertible to public use under the right of eminent domain
synonym see CRITICIZE
— **con·dem·nable** \-ˈdem-nə-bəl, -ˈde-mə-\ *adjective*
— **con·dem·na·to·ry** \-ˈdem-nə-ˌtōr-ē, -ˈde-mə-, -ˌtȯr-\ *adjective*
— **con·demn·er** \-ˈde-mər\ *or* **con·dem·nor** \kən-ˈde-mər; kən-ˌdem-ˈnȯr, ˌkän-\ *noun*

con·dem·na·tion \ˌkän-ˌdem-ˈnā-shən, -dəm-\ *noun* (14th century)
1 : CENSURE, BLAME
2 : the act of judicially condemning
3 : the state of being condemned
4 : a reason for condemning

con·den·sate \ˈkän-dən-ˌsāt, -ˌden-; kən-ˈden-\ *noun* (1889)
: a product of condensation; *especially* **:** a liquid obtained by condensation of a gas or vapor ⟨steam *condensate*⟩

con·den·sa·tion \ˌkän-ˌden-ˈsā-shən, -dən-\ *noun* (1603)
1 : the act or process of condensing: as **a :** a chemical reaction involving union between molecules often with elimination of a simple molecule (as water) to form a new more complex compound of often greater molecular weight **b :** the conversion of a substance (as water) from the vapor state to a denser liquid or solid state usually initiated by a reduction in temperature of the vapor **c :** compression of a written or spoken work into more concise form
2 : the quality or state of being condensed
3 : a product of condensing
— **con·den·sa·tion·al** \-shnəl, -shə-nᵊl\ *adjective*

con·dense \kən-ˈden(t)s\ *verb* **con·densed; con·dens·ing** [Middle English, from Middle French *condenser,* from Latin *condensare,* from *com-* + *densare* to make dense, from *densus* dense] (15th century)
transitive verb
: to make denser or more compact; *especially* **:** to subject to condensation
intransitive verb
: to undergo condensation
synonym see CONTRACT
— **con·dens·able** *also* **con·dens·ible** \-ˈden(t)-sə-bəl\ *adjective*

condensed *adjective* (15th century)
: reduced to a more compact or dense form; *also* **:** having a face narrower than that of a standard typeface

condensed milk *noun* (1863)
: evaporated milk with sugar added

con·dens·er \kən-ˈden(t)-sər\ *noun* (1686)
1 : one that condenses: as **a :** a lens or mirror used to concentrate light on an object **b :** an apparatus in which gas or vapor is condensed
2 : CAPACITOR

con·de·scend \ˌkän-di-ˈsend\ *intransitive verb* [Middle English, from Middle French *condescendre,* from Late Latin *condescendere,* from Latin *com-* + *descendere* to descend] (14th century)
1 a : to descend to a less formal or dignified level **:** UNBEND **b :** to waive the privileges of rank
2 : to assume an air of superiority

con·de·scen·dence \-ˈsen-dən(t)s\ *noun* (1638)
: CONDESCENSION

condescending *adjective* (1707)
: showing or characterized by condescension **:** PATRONIZING
— **con·de·scend·ing·ly** \-ˈsen-diŋ-lē\ *adverb*

con·de·scen·sion \ˌkän-di-ˈsen(t)-shən\ *noun* [Late Latin *condescension-, condescensio,* from *condescendere*] (1647)
1 : voluntary descent from one's rank or dignity in relations with an inferior

\ə\ **abut** \ᵊ\ **kitten** \ər\ **further** \a\ **ash** \ā\ **ace** \ä\ **mop, mar** \au̇\ **out** \ch\ **chin** \e\ **bet** \ē\ **easy** \g\ **go** \i\ **hit** \ī\ **ice** \j\ **job** \ŋ\ **sing** \ō\ **go** \ȯ\ **law** \ȯi\ **boy** \th\ **thin** \t͟h\ **the** \ü\ **loot** \u̇\ **foot** \y\ **yet** \zh\ **vision** *see also* Guide to Pronunciation

2 : patronizing attitude or behavior

con·dign \kən-ˈdīn, ˈkän-ˌ\ *adjective* [Middle English *condigne,* from Middle French, from Latin *condignus* very worthy, from *com-* + *dignus* worthy — more at DECENT] (15th century)
: DESERVED, APPROPRIATE ⟨*condign* punishment⟩
— **con·dign·ly** *adverb*

con·di·ment \ˈkän-də-mənt\ *noun* [Middle English, from Middle French, from Latin *condimentum,* from *condire* to season] (15th century)
: something used to enhance the flavor of food; *especially* **:** a pungent seasoning
— **con·di·men·tal** \ˌkän-də-ˈmen-tᵊl\ *adjective*

¹con·di·tion \kən-ˈdi-shən\ *noun* [Middle English *condicion,* from Middle French, from Latin *condicion-, condicio* terms of agreement, condition, from *condicere* to agree, from *com-* + *dicere* to say, determine — more at DICTION] (14th century)
1 a : a premise upon which the fulfillment of an agreement depends **:** STIPULATION **b** *obsolete* **:** COVENANT **c :** a provision making the effect of a legal instrument contingent upon an uncertain event; *also* **:** the event itself
2 : something essential to the appearance or occurrence of something else **:** PREREQUISITE: as **a :** an environmental requirement ⟨available oxygen is an essential *condition* for animal life⟩ **b :** the subordinate clause of a conditional sentence
3 a : a restricting or modifying factor **:** QUALIFICATION **b :** an unsatisfactory academic grade that may be raised by doing additional work
4 a : a state of being **b :** social status **:** RANK **c :** a usually defective state of health ⟨a serious heart *condition*⟩ **d :** a state of physical fitness or readiness for use ⟨the car was in good *condition*⟩ ⟨exercising to get into *condition*⟩ **e** *plural* **:** attendant circumstances
5 a *obsolete* **:** temper of mind **b** *obsolete* **:** TRAIT **c** *plural, archaic* **:** MANNERS, WAYS

²condition *verb* **con·di·tioned; con·di·tion·ing** \-ˈdi-sh(ə-)niŋ\ (15th century)
intransitive verb
archaic **:** to make stipulations
transitive verb
1 : to agree by stipulating
2 : to make conditional
3 a : to put into a proper state for work or use **b :** AIR-CONDITION
4 : to give a grade of condition to
5 a : to adapt, modify, or mold so as to conform to an environing culture **b :** to modify so that an act or response previously associated with one stimulus becomes associated with another
— **con·di·tion·able** \-sh(ə-)nə-bəl\ *adjective*
— **con·di·tion·er** \-sh(ə-)nər\ *noun*

¹con·di·tion·al \kən-ˈdish-nəl, -ˈdi-shə-nᵊl\ *adjective* (14th century)
1 : subject to, implying, or dependent upon a condition ⟨a *conditional* promise⟩
2 : expressing, containing, or implying a supposition ⟨the *conditional* clause *if he speaks*⟩
3 a : true only for certain values of the variables or symbols involved ⟨*conditional* equations⟩ **b :** stating the case when one or more random variables are fixed or one or more events are known ⟨*conditional* frequency distribution⟩
4 a : CONDITIONED 2 ⟨*conditional* reflex⟩ ⟨*conditional* response⟩ **b :** established by conditioning as the stimulus eliciting a conditional response
— **con·di·tion·al·i·ty** \-ˌdi-shə-ˈna-lə-tē\ *noun*
— **con·di·tion·al·ly** \-ˈdish-nə-lē, -ˈdi-shə-nᵊl-ē\ *adverb*

²conditional *noun* (1828)
1 : a conditional word, clause, verb form, or morpheme
2 : IMPLICATION 2b

conditional probability *noun* (1961)
: the probability that a given event will occur if it is certain that another event has taken place or will take place

conditioned *adjective* (1537)
1 : brought or put into a specified state
2 : determined or established by conditioning

con·do \ˈkän-(ˌ)dō\ *noun* (1964)
: CONDOMINIUM 3

con·dole \kən-ˈdōl\ *verb* **con·doled; con·dol·ing** [Late Latin *condolēre,* from Latin *com-* + *dolēre* to feel pain] (1590)
intransitive verb
1 *obsolete* **:** GRIEVE
2 : to express sympathetic sorrow
transitive verb
archaic **:** LAMENT, GRIEVE
— **con·do·la·to·ry** \-ˈdō-lə-ˌtōr-ē, -ˌtȯr-\ *adjective*

con·do·lence \kən-ˈdō-lən(t)s *also* ˈkän-də-\ *noun* (1603)
1 : sympathy with another in sorrow
2 : an expression of sympathy
synonym see PITY

con·dom \ˈkän-dəm, ˈkən-, *dialect* -drəm\ *noun* [origin unknown] (circa 1706)
: a sheath commonly of rubber worn over the penis (as to prevent conception or venereal infection during coitus); *also* **:** a similar device inserted into the vagina

con·do·min·i·um \ˌkän-də-ˈmi-nē-əm\ *noun, plural* **-ums** [New Latin, from Latin *com-* + *dominium* domain] (circa 1714)
1 a : joint dominion; *especially* **:** joint sovereignty by two or more nations **b :** a government operating under joint rule
2 : a politically dependent territory under condominium
3 a : individual ownership of a unit in a multiunit structure (as an apartment building) or on land owned in common (as a town house complex); *also* **:** a unit so owned **b :** a building containing condominiums ◆

con·do·na·tion \ˌkän-də-ˈnā-shən, -dō-\ *noun* (1625)
: implied pardon of an offense by treating the offender as if it had not been committed

con·done \kən-ˈdōn\ *transitive verb* **condoned; con·don·ing** [Latin *condonare* to absolve, from *com-* + *donare* to give — more at DONATION] (1857)
: to pardon or overlook voluntarily; *especially* **:** to treat as if trivial, harmless, or of no importance ⟨*condone* corruption in politics⟩
synonym see EXCUSE
— **con·don·able** \-ˈdō-nə-bəl\ *adjective*
— **con·don·er** *noun*

con·dor \ˈkän-dər, -ˌdȯr\ *noun* [Spanish *cóndor,* from Quechua *kuntur*] (1604)
1 a : a very large American vulture (*Vultur gryphus*) of the high Andes having the head and neck bare and the plumage dull black with a downy white neck ruff and white patches on the wings **b :** CALIFORNIA CONDOR
2 *plural* **condors** *or* **con·do·res** \kən-ˈdȯr-ˌās, -ˈdȯr-\ **:** a coin (as the centesimo of Chile) bearing the picture of a condor

con·dot·tie·re \ˌkän-də-ˈtyer-ē, ˌkän-ˌdä-tē-ˈer-\ *noun, plural* **-tie·ri** \-ē\ [Italian, from *condotta* troop of mercenaries, from feminine of *condotto,* past participle of *condurre* to conduct, hire, from Latin *conducere*] (1794)
1 : a leader of a band of mercenaries common in Europe between the 14th and 16th centuries; *also* **:** a member of such a band
2 : a mercenary soldier

con·duce \kən-ˈdüs, -ˈdyüs\ *intransitive verb* **con·duced; con·duc·ing** [Middle English, to conduct, from Latin *conducere* to conduct, conduce, from *com-* + *ducere* to lead — more at TOW] (1586)
: to lead or tend to a particular and usually desirable result **:** CONTRIBUTE

con·du·cive \-ˈdü-siv, -ˈdyü-\ *adjective* (1646)
: tending to promote or assist ⟨an atmosphere *conducive* to education⟩
— **con·du·cive·ness** *noun*

¹con·duct \ˈkän-(ˌ)dəkt\ *noun* [Middle English, alteration of *conduit,* from Middle French, act of leading, escort, from Medieval Latin *conductus,* from Latin *conducere*] (15th century)
1 *obsolete* **:** ESCORT, GUIDE
2 : the act, manner, or process of carrying on **:** MANAGEMENT
3 : a mode or standard of personal behavior especially as based on moral principles

²con·duct \kən-ˈdəkt *also* ˈkän-ˌdəkt\ (15th century)
transitive verb
1 : to bring by or as if by leading **:** GUIDE ⟨*conduct* tourists through a museum⟩
2 a : to lead from a position of command ⟨*conduct* a siege⟩ ⟨*conduct* a class⟩ **b :** to direct or take part in the operation or management of ⟨*conduct* an experiment⟩ ⟨*conduct* a business⟩ ⟨*conduct* an investigation⟩ **c :** to direct the performance of ⟨*conduct* an orchestra⟩ ⟨*conduct* an opera⟩
3 a : to convey in a channel **b :** to act as a medium for conveying or transmitting
4 : to cause (oneself) to act or behave in a particular and especially in a controlled manner
intransitive verb
1 *of a road or passage* **:** to show the way **:** LEAD
2 a : to act as leader or director **b :** to have the quality of transmitting light, heat, sound, or electricity ☆
— **con·duct·ibil·i·ty** \kən-ˌdək-tə-ˈbi-lə-tē\ *noun*
— **con·duct·ible** \-ˈdək-tə-bəl\ *adjective*

con·duc·tance \kən-ˈdək-tən(t)s\ *noun* (1885)
1 : conducting power
2 : the readiness with which a conductor transmits an electric current expressed as the reciprocal of electrical resistance

con·duc·tion \kən-ˈdək-shən\ *noun* (1534)
1 : the act of conducting or conveying
2 a : transmission through or by means of a conductor; *also* **:** the transfer of heat through matter by communication of kinetic energy from particle to particle with no net displacement of the particles — compare CONVECTION, RADIATION **b :** CONDUCTIVITY
3 : the transmission of excitation through living tissue and especially nervous tissue

◇ WORD HISTORY

condominium The word *condominium* was coined by European jurists in the 17th century and used in Latin treatises on international law to refer to dual sovereignty exercised by two countries over the same territory; in this sense it was borrowed into English by the early 18th century. Probably the best-known such condominium historically was the joint British and Egyptian rule of the Sudan in Africa, which began in 1899 and ended in 1956 with Sudanese independence. *Condominium* was probably introduced into U.S. real estate terminology in 1961, when the federal government first insured mortgage loans for individually owned units in apartment buildings. Three years earlier condominium mortgages had received U.S. government sanction in Puerto Rico, where the concept of joint ownership of land combined with individual ownership of living space had most likely spread from South America. In the 1960s and 1970s, *condominium* became a common word across the U.S. as state after state passed enabling legislation setting up guidelines for this type of ownership.

conduction band *noun* (1953)
: the range of permissible energy values that allow an electron of an atom to dissociate from the atom and become a free charge carrier — compare VALENCE BAND

con·duc·tive \kən-ˈdək-tiv\ *adjective* (1528)
: having conductivity **:** relating to conduction (as of electricity)

con·duc·tiv·i·ty \ˌkän-ˌdək-ˈti-və-tē, kən-\ *noun, plural* **-ties** (1837)
: the quality or power of conducting or transmitting: as **a :** the reciprocal of electrical resistivity **b :** the quality of living matter responsible for the transmission of and progressive reaction to stimuli

con·duc·to·met·ric *also* **con·duc·ti·met·ric** \kən-ˌdək-tə-ˈme-trik\ *adjective* (circa 1926)
1 : of or relating to the measurement of conductivity
2 : being or relating to titration based on determination of changes in the electrical conductivity of the solution

con·duc·tor \kən-ˈdək-tər\ *noun* (15th century)
: one that conducts: as **a :** GUIDE **b :** a collector of fares in a public conveyance **c :** the leader of a musical ensemble **d** (1) **:** a material or object that permits an electric current to flow easily — compare INSULATOR, SEMICONDUCTOR (2) **:** a material capable of transmitting another form of energy (as heat or sound)
— **con·duc·to·ri·al** \ˌkän-ˌdək-ˈtōr-ē-əl, kən-, -ˈtȯr-\ *adjective*

con·duc·tress \kən-ˈdək-trəs\ *noun* (1624)
: a woman who is a conductor

con·duit \ˈkän-ˌdü-ət, -ˌdyü- *also* -dwət, -dət\ *noun* [Middle English — more at CONDUCT] (14th century)
1 : a natural or artificial channel through which something (as a fluid) is conveyed
2 *archaic* **:** FOUNTAIN
3 : a pipe, tube, or tile for protecting electric wires or cables
4 : a means of transmitting or distributing ⟨a *conduit* for illicit payments⟩ ⟨a *conduit* of information⟩

con·du·pli·cate \(ˌ)kän-ˈdü-pli-kət, -ˈdyü-\ *adjective* [Latin *conduplicatus,* past participle of *conduplicare* to double, from *com-* + *duplic-, duplex* double — more at DUPLEX] (1777)
: folded lengthwise ⟨*conduplicate* petals in the bud⟩

con·dy·lar \ˈkän-də-lər\ *adjective* (1876)
: of or relating to a condyle

con·dyle \ˈkän-ˌdīl *also* -dᵊl\ *noun* [French & Latin; French, from Latin *condylus* knuckle, from Greek *kondylos*] (1634)
: an articular prominence of a bone; *especially* **:** one resembling a pair of knuckles
— **con·dy·loid** \-də-ˌlȯid\ *adjective*

con·dy·lo·ma \ˌkän-də-ˈlō-mə\ *noun, plural* **-ma·ta** \-mə-tə\ *also* **-mas** [New Latin, from Greek *kondylōma,* from *kondylos*] (circa 1526)
: a warty growth on the skin or adjoining mucous membrane usually near the anus and genital organs
— **con·dy·lo·ma·tous** \-mə-təs\ *adjective*

¹cone \ˈkōn\ *noun* [Middle French or Latin; Middle French, from Latin *conus,* from Greek *kōnos*] (1545)
1 a : a solid generated by rotating a right triangle about one of its legs — called also *right circular cone* **b :** a solid bounded by a circular or other closed plane base and the surface formed by line segments joining every point of the boundary of the base to a common vertex — see VOLUME table **c :** a surface traced by a moving straight line passing through a fixed vertex
2 a : a mass of ovule-bearing or pollen-bearing scales or bracts in trees of the pine family or in cycads that are arranged usually on a somewhat elongated axis **b :** any of several flower or fruit clusters suggesting a cone
3 : something that resembles a cone in shape: as **a :** any of the conical photosensitive receptor cells of the vertebrate retina that function in color vision — compare ROD 3 **b :** any of a family (Conidae) of tropical marine gastropod mollusks that inject their prey with a potent toxin **c :** the apex of a volcano **d :** a crisp usually cone-shaped wafer for holding ice cream

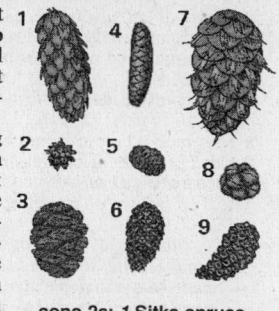

cone 2a: *1* Sitka spruce, *2* cryptomeria, *3* big tree, *4* white spruce, *5* redwood, *6* lodgepole pine, *7* Douglas fir, *8* bald cypress, *9* jack pine

²cone *transitive verb* **coned; con·ing** (circa 1859)
1 : to make cone-shaped
2 : to bevel like the slanting surface of a cone ⟨*cone* a tire⟩

cone·flow·er \ˈkōn-ˌflau̇(-ə)r\ *noun* (circa 1818)
: any of several composite plants having cone-shaped flower disks; *especially* **:** RUDBECKIA

cone·nose \ˈkōn-ˌnōz\ *noun* (circa 1891)
: any of various large bloodsucking bugs and especially some assassin bugs (genus *Triatoma*) including some capable of inflicting painful bites — called also *kissing bug*

con es·pres·sio·ne \ˌkän-ˌes-(ˌ)pre-sē-ˈō-nē, ˌkōn-, -ˈō-(ˌ)nā\ *adverb* [Italian, literally, with expression] (circa 1891)
: with feeling — used as a direction in music

Con·es·to·ga wagon \ˌkä-nə-ˈstō-gə-\ *noun* [*Conestoga,* Pa.] (1717)
: a broad-wheeled covered wagon drawn usually by six horses and used especially for transporting freight across the prairies — called also *Conestoga*

co·ney \ˈkō-nē, *1 also* ˈkə-nē\ *noun, plural* **coneys** [Middle English *conies,* plural, from Old French *conis,* plural of *conil,* from Latin *cuniculus*] (12th century)
1 a : rabbit fur **b** (1) **:** RABBIT; *especially* **:** the European rabbit (*Oryctolagus cuniculus*) (2) **:** PIKA **c :** HYRAX
2 *archaic* **:** DUPE
3 : any of several fishes; *especially* **:** a dusky black-spotted reddish-finned grouper (*Epinephelus fulvus*) of the tropical Atlantic

con·fab \kən-ˈfab, ˈkän-ˌ\ *intransitive verb* **con·fabbed; con·fab·bing** (1741)
: CONFABULATE
— **con·fab** \ˈkän-ˌfab, kən-ˈ\ *noun*

con·fab·u·late \kən-ˈfa-byə-ˌlāt\ *intransitive verb* **-lat·ed; -lat·ing** [Latin *confabulatus,* past participle of *confabulari,* from *com-* + *fabulari* to talk, from *fabula* story — more at FABLE] (circa 1604)
1 : CHAT
2 : to hold a discussion **:** CONFER
3 : to fill in gaps in memory by fabrication
— **con·fab·u·la·tion** \kən-ˌfa-byə-ˈlā-shən\ *noun*
— **con·fab·u·la·tor** \kən-ˈfa-byə-ˌlā-tər\ *noun*
— **con·fab·u·la·to·ry** \-lə-ˌtōr-ē, -ˌtȯr-\ *adjective*

con·fect \kən-ˈfekt\ *transitive verb* [Middle English, from Latin *confectus,* past participle of *conficere* to prepare — more at COMFIT] (14th century)
1 : to put together from varied material
2 a : PREPARE **b :** PRESERVE
— **con·fect** \ˈkän-ˌ\ *noun*

con·fec·tion \kən-ˈfek-shən\ *noun* (15th century)
1 : the act or process of confecting
2 : something confected: as **a :** a fancy dish or sweetmeat; *also* **:** a sweet food **b :** a medicinal preparation usually made with sugar, syrup, or honey **c :** a piece of fine craftsmanship

con·fec·tion·ary \-shə-ˌner-ē\ *noun, plural* **-ar·ies** (1605)
1 *archaic* **:** CONFECTIONER
2 : CONFECTIONERY 3
3 : SWEETS
— **confectionary** *adjective*

con·fec·tion·er \-sh(ə-)nər\ *noun* (1591)
: a manufacturer of or dealer in confections

confectioners' sugar *noun* (circa 1889)
: a refined finely powdered sugar

con·fec·tion·ery \-shə-ˌner-ē\ *noun, plural* **-er·ies** (1769)
1 : sweet foods (as candy or pastry)
2 : the confectioner's art or business
3 : a confectioner's shop

con·fed·er·a·cy \kən-ˈfe-d(ə-)rə-sē\ *noun, plural* **-cies** (14th century)
1 : a league or compact for mutual support or common action **:** ALLIANCE
2 : a combination of persons for unlawful purposes **:** CONSPIRACY
3 : the body formed by persons, states, or nations united by a league; *specifically, capitalized* **:** the 11 southern states seceding from the U.S. in 1860 and 1861

con·fed·er·al \-d(ə-)rəl\ *adjective* (1782)
: of or relating to a confederation

¹con·fed·er·ate \kən-ˈfe-d(ə-)rət\ *adjective* [Middle English *confederat,* from Late Latin *confoederatus,* past participle of *confoederare* to unite by a league, from Latin *com-* + *foeder-, foedus* compact — more at FEDERAL] (14th century)
1 : united in a league **:** ALLIED
2 *capitalized* **:** of or relating to the Confederate States of America

²confederate *noun* (15th century)
1 : ALLY, ACCOMPLICE
2 *capitalized* **:** an adherent of the Confederate States of America or their cause

³con·fed·er·ate \-ˈfe-də-ˌrāt\ *verb* **-at·ed; -at·ing** (1531)
transitive verb
: to unite in a confederacy
intransitive verb
: to band together
— **con·fed·er·a·tive** \-ˈfe-d(ə-)rə-tiv, -də-ˌrā-\ *adjective*

Confederate Memorial Day *noun* (1899)
: any of several days appointed for the commemoration of servicemen of the Confederacy

☆ SYNONYMS
Conduct, manage, control, direct mean to use one's powers to lead, guide, or dominate. CONDUCT implies taking responsibility for the acts and achievements of a group ⟨*conducted* negotiations⟩. MANAGE implies direct handling and manipulating or maneuvering toward a desired result ⟨*manages* a meat market⟩. CONTROL implies a regulating or restraining in order to keep within bounds or on a course ⟨*controlling* his appetite⟩. DIRECT implies constant guiding and regulating so as to achieve smooth operation ⟨*directs* the store's day-to-day business⟩. See in addition BEHAVE.

\ə\ abut \ᵊ\ kitten \ər\ further \a\ ash \ā\ ace \ä\ mop, mar \au̇\ out \ch\ chin \e\ bet \ē\ easy \g\ go \i\ hit \ī\ ice \j\ job \ŋ\ sing \ō\ go \ȯ\ law \ȯi\ boy \th\ thin \t͟h\ the \ü\ loot \u̇\ foot \y\ yet \zh\ vision *see also* Guide to Pronunciation

con·fed·er·a·tion \kən-ˌfe-də-ˈrā-shən\ *noun* (15th century)
1 : an act of confederating **:** a state of being confederated **:** ALLIANCE
2 : LEAGUE

con·fer \kən-ˈfər\ *verb* **con·ferred; con·fer·ring** [Latin *conferre* to bring together, from *com-* + *ferre* to carry — more at BEAR] (1570)
transitive verb
1 : to bestow from or as if from a position of superiority ⟨*conferred* an honorary degree on her⟩ ⟨knowing how to read was a gift *conferred* with manhood —Murray Kempton⟩
2 : to give (as a property or characteristic) to someone or something ⟨a reputation for power will *confer* power —John Spanier⟩
intransitive verb
: to compare views or take counsel **:** CONSULT
synonym see GIVE
— **con·fer·ment** \-ˈfər-mənt\ *noun*
— **con·fer·ra·ble** \-ˈfər-ə-bəl\ *adjective*
— **con·fer·ral** \-ˈfər-əl\ *noun*
— **con·fer·rer** \-ˈfər-ər\ *noun*

con·fer·ee \ˌkän-fə-ˈrē\ *noun* (1771)
: one taking part in a conference

con·fer·ence \ˈkän-f(ə-)rən(t)s, -fərn(t)s, *for 2 usually* kən-ˈfər-ən(t)s\ *noun* (1527)
1 a : a meeting of two or more persons for discussing matters of common concern **b :** a usually formal interchange of views **:** CONSULTATION **c :** a meeting of members of the two branches of a legislature to adjust differences **d :** CAUCUS
2 *also* **con·fer·rence** \kən-ˈfər-ən(t)s\ **:** BESTOWAL, CONFERMENT
3 a : a representative assembly or administrative organization of a religious denomination **b :** a territorial division of a religious denomination
4 : an association of athletic teams
— **con·fer·en·tial** \ˌkän-fə-ˈren(t)-shəl\ *adjective*

conference call *noun* (1941)
: a telephone call by which a caller can speak with several people at the same time

con·fer·enc·ing \ˈkän-f(ə-)rən(t)-siŋ, -fərn(t)-\ *noun* (1865)
: the holding of conferences especially by means of an electronic communications system ⟨computer *conferencing*⟩

con·fess \kən-ˈfes\ *verb* [Middle English, from Middle French *confesser,* from Old French, from *confes* having confessed, from Latin *confessus,* past participle of *confitēri* to confess, from *com-* + *fatēri* to confess; akin to Latin *fari* to speak — more at BAN] (14th century)
transitive verb
1 : to tell or make known (as something wrong or damaging to oneself) **:** ADMIT
2 a : to acknowledge (sin) to God or to a priest **b :** to receive the confession of (a penitent)
3 : to declare faith in or adherence to **:** PROFESS
4 : to give evidence of
intransitive verb
1 a : to disclose one's faults; *specifically* **:** to unburden one's sins or the state of one's conscience to God or to a priest **b :** to hear a confession
2 : ADMIT, OWN
synonym see ACKNOWLEDGE
— **con·fess·able** \-ˈfe-sə-bəl\ *adjective*

con·fessed·ly \-ˈfe-səd-lē, -ˈfest-lē\ *adverb* (1640)
: by confession

con·fes·sion \kən-ˈfe-shən\ *noun* (14th century)
1 : an act of confessing; *especially* **:** a disclosure of one's sins in the sacrament of reconciliation
2 : a statement of what is confessed: as **a :** a written acknowledgment of guilt by a party accused of an offense **b :** a formal statement of religious beliefs **:** CREED
3 : an organized religious body having a common creed

[1]**con·fes·sion·al** \-ˈfesh-nəl, -ˈfe-shə-nᵊl\ *noun* (1727)
1 : a place where a priest hears confessions
2 : the practice of confessing to a priest

[2]**confessional** *adjective* (1817)
1 : of, relating to, or being a confession especially of faith
2 : of, relating to, or being intimately autobiographical ⟨*confessional* fiction⟩
— **con·fes·sion·al·ism** \-nə-ˌli-zəm\ *noun*
— **con·fes·sion·al·ist** \-list\ *noun*
— **con·fes·sion·al·ly** \-nə-lē, -nᵊl-ē\ *adverb*

con·fes·sor \kən-ˈfe-sər, *1 & 3 also* ˈkän-ˌfe-sər, *3 also* ˈkän-fə-ˌsȯr\ *noun* (12th century)
1 : one who gives heroic evidence of faith but does not suffer martyrdom
2 : one that confesses
3 a : a priest who hears confessions **b :** a priest who is one's regular spiritual guide

con·fet·ti \kən-ˈfe-tē\ *noun* [Italian, plural of *confetto* sweetmeat, from Medieval Latin *confectum,* from Latin, neuter of *confectus,* past participle of *conficere* to prepare — more at COMFIT] (1815)
: small bits or streamers of brightly colored paper made for throwing (as at weddings)

con·fi·dant \ˈkän-fə-ˌdänt *also* -ˌdant, -dənt\ *noun* [French *confident,* from Italian *confidente,* from *confidente* confident, trustworthy, from Latin *confident-, confidens*] (1646)
: one to whom secrets are entrusted; *especially* **:** INTIMATE

con·fi·dante *same as* CONFIDANT\ *noun* [French *confidente,* feminine of *confident*] (1696)
: CONFIDANT; *especially* **:** one who is a woman

con·fide \kən-ˈfīd\ *verb* **con·fid·ed; con·fid·ing** [Middle English, from Middle French or Latin; Middle French *confider,* from Latin *confidere,* from *com-* + *fidere* to trust — more at BIDE] (15th century)
intransitive verb
1 : to have confidence **:** TRUST
2 : to show confidence by imparting secrets
transitive verb
1 : to tell confidentially
2 : to give to the care or protection of another **:** ENTRUST
synonym see COMMIT
— **con·fid·er** *noun*

[1]**con·fi·dence** \ˈkän-fə-dən(t)s, -ˌden(t)s\ *noun* (14th century)
1 a : a feeling or consciousness of one's powers or of reliance on one's circumstances ⟨had perfect *confidence* in her ability to succeed⟩ ⟨met the risk with brash *confidence*⟩ **b :** faith or belief that one will act in a right, proper, or effective way ⟨have *confidence* in a leader⟩
2 : the quality or state of being certain **:** CERTITUDE ⟨they had every *confidence* of success⟩
3 a : a relation of trust or intimacy ⟨took his friend into his *confidence*⟩ **b :** reliance on another's discretion ⟨their story was told in strictest *confidence*⟩ **c :** support especially in a legislative body ⟨vote of *confidence*⟩
4 : a communication made in confidence **:** SECRET ☆

[2]**confidence** *adjective* (1849)
: of, relating to, or adept at swindling by false promises ⟨a *confidence* game⟩ ⟨a *confidence* man⟩

confidence interval *noun* (1934)
: a group of continuous or discrete adjacent values that is used to estimate a statistical parameter (as a mean or variance) and that tends to include the true value of the parameter a predetermined proportion of the time if the process of finding the group of values is repeated a number of times

confidence limits *noun plural* (1939)
: the end points of a confidence interval

con·fi·dent \ˈkän-fə-dənt, -ˌdent\ *adjective* [Latin *confident-, confidens,* from present participle of *confidere*] (1576)
1 : characterized by assurance; *especially* **:** SELF-RELIANT
2 *obsolete* **:** TRUSTFUL, CONFIDING
3 a : full of conviction **:** CERTAIN **b :** COCKSURE
— **con·fi·dent·ly** *adverb*

con·fi·den·tial \ˌkän-fə-ˈden(t)-shəl\ *adjective* (1759)
1 : marked by intimacy or willingness to confide ⟨a *confidential* tone⟩
2 : PRIVATE, SECRET ⟨*confidential* information⟩
3 : entrusted with confidences ⟨*confidential* clerk⟩
4 : containing information whose unauthorized disclosure could be prejudicial to the national interest — compare SECRET, TOP SECRET
— **con·fi·den·ti·al·i·ty** \-ˌden(t)-shē-ˈa-lə-tē\ *noun*
— **con·fi·den·tial·ly** \-ˈden(t)-sh(ə-)lē\ *adverb*

con·fid·ing \kən-ˈfī-diŋ\ *adjective* (1829)
: tending to confide **:** TRUSTFUL
— **con·fid·ing·ly** \-diŋ-lē\ *adverb*
— **con·fid·ing·ness** *noun*

con·fig·u·ra·tion \kən-ˌfi-gyə-ˈrā-shən, ˌkän-, -gə-ˈ\ *noun* [Late Latin *configuration-, configuratio* similar formation, from Latin *configurare* to form from or after, from *com-* + *figurare* to form, from *figura* figure] (1646)
1 a : relative arrangement of parts or elements: as (1) **:** SHAPE (2) **:** contour of land ⟨*configuration* of the mountains⟩ (3) **:** functional arrangement ⟨a small business computer system in its simplest *configuration*⟩ **b :** something (as a figure, contour, pattern, or apparatus) that results from a particular arrangement of parts or components **c :** the stable structural makeup of a chemical compound especially with reference to the space relations of the constituent atoms
2 : GESTALT ⟨personality *configuration*⟩
— **con·fig·u·ra·tion·al** \-shnəl, -shə-nᵊl\ *adjective*
— **con·fig·u·ra·tion·al·ly** *adverb*
— **con·fig·u·ra·tive** \-ˈfi-gyə-rə-tiv, -ˈfi-gə-\ *adjective*

con·fig·ure \kən-ˈfi-gyər, *especially British* -ˈfi-gər\ *transitive verb* **-ured; -ur·ing** (1677)
: to set up for operation especially in a particular way ⟨a fighter plane *configured* for the Malaysian air force⟩

[1]**con·fine** \ˈkän-ˌfīn *also* kən-ˈ\ *noun* [Middle English, from Middle French or Latin; Middle French *confines,* plural, from Latin *confine* border, from neuter of *confinis* adjacent, from *com-* + *finis* end] (15th century)
1 *plural* **a :** something (as borders or walls) that encloses ⟨outside the *confines* of the office or hospital —W. A. Nolen⟩; *also* **:** something that restrains ⟨escape from the *confines*

☆ SYNONYMS
Confidence, assurance, self-possession, aplomb mean a state of mind or a manner marked by easy coolness and freedom from uncertainty, diffidence, or embarrassment. CONFIDENCE stresses faith in oneself and one's powers without any suggestion of conceit or arrogance ⟨the *confidence* that comes from long experience⟩. ASSURANCE carries a stronger implication of certainty and may suggest arrogance or lack of objectivity in assessing one's own powers ⟨handled the cross-examination with complete *assurance*⟩. SELF-POSSESSION implies an ease or coolness under stress that reflects perfect self-control and command of one's powers ⟨answered the insolent question with complete *self-possession*⟩. APLOMB implies a manifest self-possession in trying or challenging situations ⟨handled the reporters with great *aplomb*⟩.

of soot and clutter —E. S. Muskie⟩ **b :** SCOPE 3 ⟨work within the *confines* of a small group —Frank Newman⟩
2 a *archaic* **:** RESTRICTION **b** *obsolete* **:** PRISON

[2]**con·fine** \kən-'fīn\ *verb* **con·fined; con·fin·ing** (1523)
intransitive verb
archaic **:** BORDER
transitive verb
1 a : to hold within a location **b :** IMPRISON
2 : to keep within limits ⟨will *confine* my remarks to one subject⟩
synonym see LIMIT
— **con·fin·er** *noun*

con·fined \kən-'fīnd\ *adjective* (1772)
: undergoing childbirth

con·fine·ment \kən-'fīn-mənt\ *noun* (1646)
: an act of confining **:** the state of being confined ⟨solitary *confinement*⟩; *especially* **:** LYING-IN

con·firm \kən-'fərm\ *transitive verb* [Middle English, from Old French *confirmer,* from Latin *confirmare,* from *com-* + *firmare* to make firm, from *firmus* firm] (13th century)
1 : to give approval to **:** RATIFY
2 : to make firm or firmer **:** STRENGTHEN
3 : to administer the rite of confirmation to
4 : to give new assurance of the validity of **:** remove doubt about by authoritative act or indisputable fact ☆
— **con·firm·abil·i·ty** \-,fər-mə-'bi-lə-tē\ *noun*
— **con·firm·able** \-'fər-mə-bəl\ *adjective*

con·fir·mand \,kän-fər-'mand\ *noun* [Latin *confirmandus,* gerundive of *confirmare*] (1884)
: a candidate for religious confirmation

con·fir·ma·tion \,kän-fər-'mā-shən\ *noun* (14th century)
1 : an act or process of confirming: as **a** (1) **:** a Christian rite conferring the gift of the Holy Spirit and among Protestants full church membership (2) **:** a ceremony especially of Reform Judaism confirming youths in their faith **b :** the ratification of an executive act by a legislative body
2 a : confirming proof **:** CORROBORATION **b :** the process of supporting a statement by evidence
— **con·fir·ma·tion·al** \-shnəl, -shə-nᵊl\ *adjective*

con·fir·ma·to·ry \kən-'fər-mə-,tōr-ē, -,tȯr-\ *adjective* (1636)
: serving to confirm **:** CORROBORATIVE

con·firmed \kən-'fərmd\ *adjective* (14th century)
1 a : marked by long continuance and likely to persist ⟨a *confirmed* habit⟩ **b :** fixed in habit and unlikely to change ⟨a *confirmed* do-gooder⟩
2 : having received the rite of confirmation
synonym see INVETERATE
— **con·firm·ed·ly** \-'fər-məd-lē\ *adverb*
— **con·firmed·ness** \-'fər-məd-nəs, -'fərm(d)-nəs\ *noun*

con·fis·ca·ble \kən-'fis-kə-bəl\ *adjective* (circa 1736)
: liable to confiscation

con·fis·cat·able \'kän-fə-,skā-tə-bəl\ *adjective* (1863)
: CONFISCABLE

[1]**con·fis·cate** \'kän-fə-,skāt, kən-'fis-kət\ *adjective* [Latin *confiscatus,* past participle of *confiscare* to confiscate, from *com-* + *fiscus* treasury] (circa 1533)
1 : appropriated by the government **:** FORFEITED
2 : deprived of property by confiscation

[2]**con·fis·cate** \'kän-fə-,skāt\ *transitive verb* **-cat·ed; -cat·ing** (1552)
1 : to seize as forfeited to the public treasury
2 : to seize by or as if by authority
— **con·fis·ca·tion** \,kän-fə-'skā-shən\ *noun*
— **con·fis·ca·tor** \'kän-fə-,skā-tər\ *noun*
— **con·fis·ca·to·ry** \kən-'fis-kə-,tōr-ē, -,tȯr-\ *adjective*

con·fit \kō̃n-'fē, kȯn-, kän-\ *noun* [French, from Old French, preparation, preserves, from past participle of *confire* to prepare — more at COMFIT] (1951)
: meat (as goose, duck, or pork) that has been cooked and preserved in its own fat

con·fi·te·or \kən-'fē-tē-,ȯr, -ər\ *noun* [Middle English, from Latin, literally, I confess, from the opening words — more at CONFESS] (13th century)
: a liturgical form in which sinfulness is acknowledged and intercession for God's mercy requested

con·fi·ture \'kän-fə-,chu̇r, -,tyu̇r, -,tu̇r\ *noun* [French, from Middle French, from *confit*] (1802)
: preserved or candied fruit **:** JAM

con·fla·grant \kən-'flā-grənt\ *adjective* [Latin *conflagrant-, conflagrans,* present participle of *conflagrare* to burn, from *com-* + *flagrare* to burn — more at BLACK] (circa 1656)
: BURNING, BLAZING

con·fla·gra·tion \,kän-flə-'grā-shən\ *noun* [Latin *conflagration-, conflagratio,* from *conflagrare*] (circa 1656)
1 : FIRE; *especially* **:** a large disastrous fire
2 : CONFLICT, WAR

con·flate \kən-'flāt\ *transitive verb* **con·flat·ed; con·flat·ing** [Latin *conflatus,* past participle of *conflare* to blow together, fuse, from *com-* + *flare* to blow — more at BLOW] (1610)
1 a : to bring together **:** FUSE **b :** CONFUSE
2 : to combine (as two readings of a text) into a composite whole

con·fla·tion \-'flā-shən\ *noun* (15th century)
: BLEND, FUSION; *especially* **:** a composite reading or text

[1]**con·flict** \'kän-,flikt\ *noun* [Middle English, from Latin *conflictus* act of striking together, from *confligere* to strike together, from *com-* + *fligere* to strike — more at PROFLIGATE] (15th century)
1 : FIGHT, BATTLE, WAR
2 a : competitive or opposing action of incompatibles **:** antagonistic state or action (as of divergent ideas, interests, or persons) **b :** mental struggle resulting from incompatible or opposing needs, drives, wishes, or external or internal demands
3 : the opposition of persons or forces that gives rise to the dramatic action in a drama or fiction
synonym see DISCORD
— **con·flict·ful** \'kän-,flikt-fəl\ *adjective*
— **con·flic·tu·al** \kän-'flik-chə-wəl, kən-, -chəl, -shwəl\ *adjective*

[2]**con·flict** \kən-'flikt, 'kän-,\ *intransitive verb* (15th century)
1 *archaic* **:** to contend in warfare
2 : to show antagonism or irreconcilability
— **con·flic·tion** \kən-'flik-shən, kän-\ *noun*
— **con·flic·tive** \kən-'flik-tiv, 'kän-,\ *adjective*

con·flict·ed \kən-'flik-təd\ *adjective* (1967)
: having or experiencing an emotional conflict ⟨this unhappy and *conflicted* modern woman —John Updike⟩

conflicting *adjective* (1607)
: being in conflict, collision, or opposition **:** INCOMPATIBLE
— **con·flict·ing·ly** \-'flik-tiŋ-lē, -,flik-\ *adverb*

conflict of interest (1951)
: a conflict between the private interests and the official responsibilities of a person in a position of trust

con·flu·ence \'kän-,flü-ən(t)s, kən-'\ *noun* (15th century)
1 : a coming or flowing together, meeting, or gathering at one point ⟨a happy *confluence* of weather and scenery⟩
2 a : the flowing together of two or more streams **b :** the place of meeting of two streams **c :** the combined stream formed by conjunction

[1]**con·flu·ent** \-ənt\ *adjective* [Latin *confluent-, confluens,* present participle of *confluere* to flow together, from *com-* + *fluere* to flow — more at FLUID] (15th century)
1 : flowing or coming together; *also* **:** run together ⟨*confluent* pustules⟩
2 : characterized by confluent lesions ⟨*confluent* smallpox⟩

[2]**confluent** *noun* (1850)
: a confluent stream; *broadly* **:** TRIBUTARY

con·flux \'kän-,fləks\ *noun* [Medieval Latin *confluxus,* from Latin *confluere*] (1606)
: CONFLUENCE

con·fo·cal \(,)kän-'fō-kəl\ *adjective* (1867)
: having the same foci ⟨*confocal* ellipses⟩ ⟨*confocal* lenses⟩
— **con·fo·cal·ly** \-kə-lē\ *adverb*

con·form \kən-'fȯrm\ *verb* [Middle English, from Middle French *conformer,* from Latin *conformare,* from *com-* + *formare* to form, from *forma* form] (14th century)
transitive verb
: to give the same shape, outline, or contour to **:** bring into harmony or accord ⟨*conform* furrows to the slope of the land⟩
intransitive verb
1 : to be similar or identical; *also* **:** to be in agreement or harmony — used with *to* or *with*
2 a : to be obedient or compliant — usually used with *to* **b :** to act in accordance with prevailing standards or customs
synonym see ADAPT
— **con·form·er** *noun*
— **con·form·ism** \-'fȯr-,mi-zəm\ *noun*
— **con·form·ist** \-mist\ *noun or adjective*

con·form·able \kən-'fȯr-mə-bəl\ *adjective* (15th century)
1 : corresponding in form or character **:** SIMILAR — usually used with *to*
2 : SUBMISSIVE, COMPLIANT
3 : following in unbroken sequence — used of geologic strata formed under uniform conditions
— **con·form·ably** \-blē\ *adverb*

con·for·mal \kən-'fȯr-məl, (,)kän-\ *adjective* [Late Latin *conformalis* having the same shape, from Latin *com-* + *formalis* formal, from *forma*] (1893)
1 : leaving the size of the angle between corresponding curves unchanged ⟨*conformal* transformation⟩
2 *of a map* **:** representing small areas in their true shape

con·for·mance \kən-'fȯr-mən(t)s\ *noun* (1606)

☆ SYNONYMS
Confirm, corroborate, substantiate, verify, authenticate, validate mean to attest to the truth or validity of something. CONFIRM implies the removing of doubts by an authoritative statement or indisputable fact ⟨*confirmed* the reports⟩. CORROBORATE suggests the strengthening of what is already partly established ⟨witnesses *corroborated* his story⟩. SUBSTANTIATE implies the offering of evidence that sustains the contention ⟨the claims have yet to be *substantiated*⟩. VERIFY implies the establishing of correspondence of actual facts or details with those proposed or guessed at ⟨all statements of fact in the article have been *verified*⟩. AUTHENTICATE implies establishing genuineness by adducing legal or official documents or expert opinion ⟨handwriting experts *authenticated* the diaries⟩. VALIDATE implies establishing validity by authoritative affirmation or by factual proof ⟨*validated* the hypothesis by experiments⟩.

: CONFORMITY

con·for·ma·tion \ˌkän-(ˌ)fȯr-ˈmā-shən, -fər-\ *noun* (1511)
1 : the act of conforming or producing conformity **:** ADAPTATION
2 : formation of something by appropriate arrangement of parts or elements **:** an assembling into a whole ⟨the gradual *conformation* of the embryo⟩
3 a : correspondence especially to a model or plan **b :** STRUCTURE **c :** the shape or proportionate dimensions especially of an animal **d :** any of the spatial arrangements of a molecule that can be obtained by rotation of the atoms about a single bond
— **con·for·ma·tion·al** \-shnəl, -shə-n^{ə}l\ *adjective*

con·for·mi·ty \kən-ˈfȯr-mə-tē\ *noun, plural* **-ties** (15th century)
1 : correspondence in form, manner, or character **:** AGREEMENT ⟨behaved in *conformity* with her beliefs⟩
2 : an act or instance of conforming
3 : action in accordance with some specified standard or authority ⟨*conformity* to social custom⟩

con·found \kən-ˈfau̇nd, kän-\ *transitive verb* [Middle English, from Middle French *confondre,* from Latin *confundere* to pour together, confuse, from *com-* + *fundere* to pour — more at FOUND] (14th century)
1 a *archaic* **:** to bring to ruin **:** DESTROY **b :** BAFFLE, FRUSTRATE ⟨conferences . . . are not for accomplishment but to *confound* knavish tricks —J. K. Galbraith⟩
2 *obsolete* **:** CONSUME, WASTE
3 a : to put to shame **:** DISCOMFIT ⟨a performance that *confounded* the critics⟩ **b :** REFUTE ⟨sought to *confound* his arguments⟩
4 : DAMN
5 : to throw (a person) into confusion or perplexity
6 a : to fail to discern differences between **:** mix up **b :** to increase the confusion of
synonym see PUZZLE
— **con·found·er** \-ˈfau̇n-dər\ *noun*
— **con·found·ing·ly** \-diŋ-lē\ *adverb*

con·found·ed \kən-ˈfau̇n-dəd, (ˌ)kän-ˈ, ˈkän-ˌ\ *adjective* (14th century)
1 : CONFUSED, PERPLEXED
2 : DAMNED
— **con·found·ed·ly** *adverb*

con·fra·ter·ni·ty \ˌkän-frə-ˈtər-nə-tē\ *noun* [Middle English *confraternite,* from Middle French *confraternité,* from Medieval Latin *confraternitat-, confraternitas,* from *confrater* fellow, brother, from Latin *com-* + *frater* brother — more at BROTHER] (15th century)
1 : a society devoted especially to a religious or charitable cause
2 : fraternal union

con·frere *also* **con·frère** \ˈkän-ˌfrer, kō^{n}-ˌ, kän-ˈ, kō^{n}-ˈ, kən-ˈ\ *noun* [Middle English, from Middle French, translation of Medieval Latin *confrater*] (15th century)
: COLLEAGUE, COMRADE

con·front \kən-ˈfrənt\ *transitive verb* [Middle French *confronter* to border on, confront, from Medieval Latin *confrontare* to bound, from Latin *com-* + *front-, frons* forehead, front] (circa 1568)
1 : to face especially in challenge **:** OPPOSE
2 a : to cause to meet **:** bring face-to-face ⟨*confront* a reader with statistics⟩ **b :** to meet face-to-face **:** ENCOUNTER ⟨*confronted* the possibility of failure⟩
— **con·front·al** \-ˈfrən-t^{ə}l\ *noun*
— **con·front·er** *noun*

con·fron·ta·tion \ˌkän-(ˌ)frən-ˈtā-shən\ *noun* (1632)
: the act of confronting **:** the state of being confronted: as **a :** a face-to-face meeting **b :** the clashing of forces or ideas **:** CONFLICT **c :** COMPARISON ⟨the flashbacks bring into meaningful *confrontation* present and past, near and far —R. J. Clements⟩
— **con·fron·ta·tion·al** \-shnəl, -shə-n^{ə}l\ *adjective*
— **con·fron·ta·tion·ist** \-sh(ə-)nist\ *noun or adjective*

Con·fu·cian \kən-ˈfyü-shən\ *adjective* (1837)
: of or relating to the Chinese philosopher Confucius or his teachings or followers
— **Confucian** *noun*
— **Con·fu·cian·ism** \-shə-ˌni-zəm\ *noun*
— **Con·fu·cian·ist** \-nist\ *noun or adjective*

con·fuse \kən-ˈfyüz\ *transitive verb* **con·fused; con·fus·ing** [back-formation from Middle English *confused* perplexed, from Middle French *confus,* from Latin *confusus,* past participle of *confundere*] (14th century)
1 *archaic* **:** to bring to ruin
2 a : to make embarrassed **:** ABASH **b :** to disturb in mind or purpose **:** THROW OFF
3 a : to make indistinct **:** BLUR ⟨stop *confusing* the issue⟩ **b :** to mix indiscriminately **:** JUMBLE **c :** to fail to differentiate from an often similar or related other ⟨*confuse* money with comfort⟩
— **con·fus·ing·ly** \-ˈfyü-ziŋ-lē\ *adverb*

con·fused \-ˈfyüzd\ *adjective* (14th century)
1 : being perplexed or disconcerted ⟨the *confused* students⟩
2 : INDISTINGUISHABLE ⟨a zigzag, crisscross, *confused* trail —Harry Hervey⟩
3 : being disordered or mixed up ⟨a contradictory and often *confused* story⟩
— **con·fused·ly** \-ˈfyü-zəd-lē, -ˈfyüzd-\ *adverb*
— **con·fused·ness** \-ˈfyü-zəd-nəs, -ˈfyüz(d)-\ *noun*

con·fu·sion \kən-ˈfyü-zhən\ *noun* (14th century)
1 : an act or instance of confusing
2 a : the quality or state of being confused **b :** a confused mass or mixture ⟨a *confusion* of voices⟩
— **con·fu·sion·al** \-ˈfyüzh-nəl, -ˈfyü-zhə-n^{ə}l\ *adjective*

con·fu·ta·tion \ˌkän-fyu̇-ˈtā-shən\ *noun* (15th century)
1 : the act or process of confuting **:** REFUTATION
2 : something (as an argument or statement) that confutes
— **con·fu·ta·tive** \kən-ˈfyü-tə-tiv\ *adjective*

con·fute \kən-ˈfyüt\ *transitive verb* **con·fut·ed; con·fut·ing** [Latin *confutare* to check, silence] (1529)
1 : to overwhelm in argument **:** refute conclusively ⟨Elijah . . . *confuted* the prophets of Baal —G. B. Shaw⟩
2 *obsolete* **:** CONFOUND
— **con·fut·er** *noun*

con·ga \ˈkäŋ-gə\ *noun* [American Spanish, probably from feminine of *congo* Negro, from *Congo,* region in Africa] (1935)
1 : a Cuban dance of African origin involving three steps followed by a kick and performed by a group usually in single file
2 : a tall barrel-shaped or tapering drum of Afro-Cuban origin that is played with the hands

conga 2

conga line *noun* (1947)
: SNAKE DANCE 2

con·gé \kō^{n}-ˈzhā, ˈkän-ˌzhā\ *noun* [French, from Latin *commeatus* going back and forth, leave, from *commeare* to go back and forth, from *com-* + *meare* to go — more at PERMEATE] (1702)
1 a : a formal permission to depart **b :** DISMISSAL
2 : a ceremonious bow
3 : FAREWELL
4 : an architectural molding of concave profile — see MOLDING illustration

con·geal \kən-ˈjē(ə)l\ *verb* [Middle English *congelen,* from Middle French *congeler,* from Latin *congelare,* from *com-* + *gelare* to freeze — more at COLD] (14th century)
transitive verb
1 : to change from a fluid to a solid state by or as if by cold
2 : to make viscid or curdled **:** COAGULATE
3 : to make rigid, fixed, or immobile
intransitive verb
: to become congealed **:** SOLIDIFY
— **con·geal·ment** \-mənt\ *noun*

con·gee \ˈkän-(ˌ)jē, -(ˌ)zhē\ *noun* (14th century)
: CONGÉ

con·ge·la·tion \ˌkän-jə-ˈlā-shən\ *noun* (15th century)
: the process or result of congealing

con·ge·ner \ˈkän-jə-nər, kən-ˈjē-\ *noun* [Latin, of the same kind, from *com-* + *gener-, genus* kind — more at KIN] (circa 1736)
1 : a member of the same taxonomic genus as another plant or animal
2 : a person or thing resembling another in nature or action ⟨the New England private schools and their *congeners* west of the Alleghenies —Oliver La Farge⟩
— **con·ge·ner·ic** \ˌkän-jə-ˈner-ik\ *adjective*
— **con·ge·ner·ous** \kən-ˈjē-nə-rəs, -ˈjenə-, (ˌ)kän-\ *adjective*

con·ge·nial \kən-ˈjē-nē-əl, -ˈjēn-yəl\ *adjective* [*com-* + *genius*] (circa 1625)
1 : having the same nature, disposition, or tastes **:** KINDRED
2 a : existing or associated together harmoniously **b :** PLEASANT; *especially* **:** agreeably suited to one's nature, tastes, or outlook **c :** SOCIABLE, GENIAL
— **con·ge·nial·i·ty** \-ˌjē-nē-ˈa-lə-tē, -ˌjēn-ˈya-\ *noun*
— **con·ge·nial·ly** \-ˈjē-nē-ə-lē, -ˈjēn-yə-\ *adverb*

con·gen·i·tal \kən-ˈje-nə-t^{ə}l, kän-\ *adjective* [Latin *congenitus,* from *com-* + *genitus,* past participle of *gignere* to bring forth — more at KIN] (1796)
1 a : existing at or dating from birth ⟨*congenital* deafness⟩ **b :** constituting an essential characteristic **:** INHERENT ⟨*congenital* fear of snakes⟩ **c :** acquired during development in the uterus and not through heredity ⟨*congenital* syphilis⟩
2 : being such by nature ⟨*congenital* liar⟩
synonym see INNATE
— **con·gen·i·tal·ly** \-t^{ə}l-ē\ *adverb*

con·ger eel \ˈkäŋ-gər-\ *noun* [Middle English *congre,* from Old French, from Latin *congr-, conger,* probably from Greek *gongros*] (1602)
: a large strictly marine scaleless eel (*Conger oceanicus*) of the Atlantic; *broadly* **:** any of various related eels (family Congridae)

con·ge·ries \ˈkän-jə-(ˌ)rēz\ *noun, plural* **congeries** *same*\ [Latin, from *congerere*] (circa 1619)
: AGGREGATION, COLLECTION

con·gest \kən-ˈjest\ *verb* [Latin *congestus,* past participle of *congerere* to bring together, from *com-* + *gerere* to bear] (1758)
transitive verb
1 : to cause an excessive fullness of the blood vessels of (as an organ)
2 : CLOG ⟨traffic *congested* the highways⟩
3 : to concentrate in a small or narrow space
intransitive verb
: to become congested
— **con·ges·tion** \-ˈjes-chən, -ˈjesh-\ *noun*
— **con·ges·tive** \-ˈjes-tiv\ *adjective*

congestive heart failure *noun* (1930)
: heart failure in which the heart is unable to maintain an adequate circulation of blood in the bodily tissues or to pump out the venous blood returned to it by the veins

con·glo·bate \kän-ˈglō-ˌbāt, kən-\ *transitive verb* **-bat·ed; -bat·ing** [Latin *conglobatus,* past participle of *conglobare,* from *com-* + *globus* globe] (1635)

: to form into a round compact mass
— **con·glo·bate** \-bət, -ˌbāt\ *adjective*
— **con·glo·ba·tion** \ˌkän-(ˌ)glō-'bā-shən\ *noun*

con·globe \kän-'glōb, kən-\ *transitive verb* **con·globed; con·glob·ing** (1535)
: CONGLOBATE

[1]con·glom·er·ate \kən-'gläm-rət, -'glä-mə-\ *adjective* [Latin *conglomeratus,* past participle of *conglomerare* to roll together, from *com-* + *glomerare* to wind into a ball, from *glomer-, glomus* ball — more at CLAM] (1572)
: made up of parts from various sources or of various kinds

[2]con·glom·er·ate \-'glä-mə-ˌrāt\ *verb* **-at·ed; -at·ing** (1642)
intransitive verb
: to gather into a mass or coherent whole ⟨numbers of dull people *conglomerated* round her —Virginia Woolf⟩
transitive verb
: ACCUMULATE
— **con·glom·er·a·tive** \-'gläm-rə-tiv, -'glä-mə-; -mə-ˌrā-\ *adjective*
— **con·glom·er·a·tor** \-'glä-mə-ˌrā-tər\ *noun*

[3]con·glom·er·ate \-'gläm-rət, -'glä-mə-\ *noun* (1818)
1 : a composite mass or mixture; *especially* **:** rock composed of rounded fragments varying from small pebbles to large boulders in a cement (as of hardened clay)
2 : a widely diversified corporation
— **con·glom·er·at·ic** \kən-ˌglä-mə-'ra-tik, ˌkän-\ *adjective*

con·glom·er·a·teur \kən-ˌglä-mə-rə-'tər, -'tyu̇r, -'tu̇r\ *noun* [[3]*conglomerate* + *-eur* (as in *entrepreneur*)] (1969)
: a person who forms or heads a conglomerate
: CONGLOMERATOR

con·glom·er·a·tion \kən-ˌglä-mə-'rā-shən, ˌkän-\ *noun* (1626)
1 : the act of conglomerating **:** the state of being conglomerated
2 : something conglomerated **:** a mixed mass or collection

con·glu·ti·nate \kən-'glü-tᵊn-ˌāt, kän-\ *verb* **-nat·ed; -nat·ing** [Latin *conglutinatus,* past participle of *conglutinare* to glue together, from *com-* + *glutin-, gluten* glue — more at GLUTEN] (1546)
transitive verb
: to unite by or as if by a glutinous substance
intransitive verb
: to become conglutinated ⟨blood platelets *conglutinate* in blood clotting⟩
— **con·glu·ti·na·tion** \kən-ˌglü-tᵊn-'ā-shən, kän-\ *noun*

Congo red *noun* [*Congo,* territory in Africa] (1885)
: an azo dye $C_{32}H_{22}N_6Na_2O_6S_2$ that is red in alkaline and blue in acid solution and that is used especially as an indicator and as a biological stain

con·gou \'käŋ-(ˌ)gō, -(ˌ)gü\ *noun* [probably from Chinese (Xiamen) *kong-hu* pains taken] (1725)
: a black tea from China

con·grat·u·late \kən-'gra-chə-ˌlāt, -'gra-jə-\ *transitive verb* **-lat·ed; -lat·ing** [Latin *congratulatus,* past participle of *congratulari* to wish joy, from *com-* + *gratulari* to wish joy, from *gratus* pleasing — more at GRACE] (1539)
1 *archaic* **:** to express sympathetic pleasure at (an event)
2 : to express vicarious pleasure to (a person) on the occasion of success or good fortune; *also* **:** to feel pleased with ⟨*congratulating* herself for a job well done⟩
3 *obsolete* **:** SALUTE, GREET
— **con·grat·u·la·tor** \-ˌlā-tər\ *noun*
— **con·grat·u·la·to·ry** \-lə-ˌtōr-ē, -ˌtȯr-\ *adjective*

con·grat·u·la·tion \kən-ˌgra-chə-'lā-shən, -ˌgra-jə-\ *noun* (15th century)
1 : the act of congratulating
2 : a congratulatory expression — usually used in plural

con·gre·gant \'käŋ-gri-gənt\ *noun* (1886)
: one that congregates; *specifically* **:** a member of a congregation

[1]con·gre·gate \-ˌgāt\ *verb* **-gat·ed; -gat·ing** [Middle English, from Latin *congregatus,* past participle of *congregare,* from *com-* + *greg-, grex* flock] (15th century)
transitive verb
: to collect into a group or crowd **:** ASSEMBLE
intransitive verb
: to come together into a group, crowd, or assembly
synonym see GATHER
— **con·gre·ga·tor** \-ˌgā-tər\ *noun*

[2]con·gre·gate \-gət\ *adjective* (1900)
: providing or being group services or facilities designed especially for elderly persons requiring supportive services ⟨*congregate* housing⟩

con·gre·ga·tion \ˌkäŋ-gri-'gā-shən\ *noun* (14th century)
1 a : an assembly of persons **:** GATHERING; *especially* **:** an assembly of persons met for worship and religious instruction **b :** a religious community: as (1) **:** an organized body of believers in a particular locality (2) **:** a Roman Catholic religious institute with only simple vows (3) **:** a group of monasteries forming an independent subdivision of an order
2 : the act or an instance of congregating or bringing together **:** the state of being congregated
3 : a body of cardinals and officials forming an administrative division of the papal curia

con·gre·ga·tion·al \-shnəl, -shə-nᵊl\ *adjective* (1639)
1 : of or relating to a congregation
2 *capitalized* **:** of or relating to a body of Protestant churches deriving from the English Independents of the 17th century and affirming the essential importance and the autonomy of the local congregation
3 : of or relating to church government placing final authority in the assembly of the local congregation
— **con·gre·ga·tion·al·ism** \-shnə-ˌli-zəm, -shə nᵊl-ˌi-\ *noun, often capitalized*
— **con·gre·ga·tion·al·ist** \-shnə-list, -shə-nᵊl ist\ *noun or adjective, often capitalized*

con·gress \'käŋ-grəs *also* -rəs, *British usually* 'käŋ-ˌgres\ *noun* [Latin *congressus,* from *congredi* to come together, from *com-* + *gradi* to go — more at GRADE] (1528)
1 a : the act or action of coming together and meeting **b :** COITUS
2 : a formal meeting of delegates for discussion and usually action on some question
3 : the supreme legislative body of a nation and especially of a republic
4 : an association usually made up of delegates from constituent organizations
5 : a single meeting or session of a group
— **con·gres·sio·nal** \kən-'gresh-nəl, kän-, -'gre-shə-nᵊl\ *adjective*
— **con·gres·sio·nal·ly** *adverb*

congressional district *noun* (1812)
: a territorial division of a state from which a member of the U.S. House of Representatives is elected

Congressional Medal *noun* (1910)
: MEDAL OF HONOR

con·gress·man \'käŋ-(g)rəs-mən\ *noun* (1780)
: a member of a congress; *especially* **:** a member of the U.S. House of Representatives

con·gress·peo·ple \-(g)rəs-ˌpē-pəl\ *noun plural* (1973)
: congressmen or congresswomen

con·gress·per·son \-(g)rəs-ˌpər-sən\ *noun* (1972)
: a congressman or congresswoman

con·gress·wom·an \-(g)rəs-ˌwu̇-mən\ *noun* (1917)
: a female member of a congress; *especially* **:** a female member of the U.S. House of Representatives

con·gru·ence \kən-'grü-ən(t)s, 'käŋ-grü-ən(t)s\ *noun* (1533)
1 : the quality or state of agreeing, coinciding, or being congruent
2 : a statement that two numbers or geometric figures are congruent

con·gru·en·cy \-ən(t)-sē\ *noun, plural* **-cies** (15th century)
: CONGRUENCE

con·gru·ent \kən-'grü-ənt, 'käŋ-grü-ənt\ *adjective* [Middle English, from Latin *congruent-, congruens,* present participle of *congruere*] (15th century)
1 : CONGRUOUS
2 : superposable so as to be coincident throughout
3 : having the difference divisible by a given modulus ⟨12 is *congruent* to 2 (modulo 5) since $12-2=2\cdot5$⟩
— **con·gru·ent·ly** *adverb*

con·gru·i·ty \kən-'grü-ə-tē, kän-\ *noun, plural* **-ties** (14th century)
1 : the quality or state of being congruent or congruous
2 : a point of agreement

con·gru·ous \'käŋ-grü-əs\ *adjective* [Latin *congruus,* from *congruere* to come together, agree] (1599)
1 a : being in agreement, harmony, or correspondence **b :** conforming to the circumstances or requirements of a situation **:** APPROPRIATE ⟨a *congruous* room to work in —G. B. Shaw⟩
2 : marked or enhanced by harmonious agreement among constituent elements ⟨a *congruous* theme⟩
— **con·gru·ous·ly** *adverb*
— **con·gru·ous·ness** *noun*

[1]con·ic \'kä-nik\ *adjective* (1570)
1 : of or relating to a cone
2 : CONICAL
— **co·nic·i·ty** \kō-'ni-sə-tē\ *noun*

[2]conic *noun* (1879)
: CONIC SECTION

con·i·cal \'kä-ni-kəl\ *adjective* (1570)
: resembling a cone especially in shape
— **con·i·cal·ly** \-k(ə-)lē\ *adverb*

conic section *noun* (1664)
1 : a plane curve, line, pair of intersecting lines, or point that is the intersection of or bounds the intersection of a plane and a cone with two nappes
2 : a curve generated by a point which always moves so that the ratio of its distance from a fixed point to its distance from a fixed line is constant

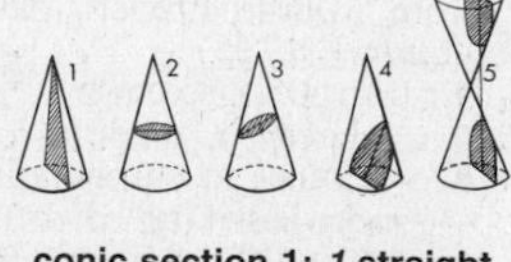

conic section 1: *1* **straight lines,** *2* **circle,** *3* **ellipse,** *4* **parabola,** *5* **hyperbola**

co·nid·i·o·phore \kə-'ni-dē-ə-ˌfōr, -ˌfȯr\ *noun* [New Latin *conidium* + International Scientific Vocabulary *-phore*] (1874)
: a structure that bears conidia; *specifically* **:** a specialized hyphal branch of some fungi that produces conidia usually by the successive cutting off of parts of the sporophore through the growth of septa

co·nid·i·um \kə-'ni-dē-əm\ *noun, plural* **-ia** \-dē-ə\ [New Latin, from Greek *konis* dust — more at INCINERATE] (1856)
: an asexual spore produced on a conidiophore
— **co·nid·i·al** \-dē-əl\ *adjective*

co·ni·fer \'kä-nə-fər *also* 'kō-\ *noun* [ultimately from Latin *conifer* cone-bearing, from *conus* cone + *-fer*] (circa 1841)
: any of an order (Coniferales) of mostly evergreen trees and shrubs including forms (as pines) with true cones and others (as yews) with an arillate fruit
— **co·nif·er·ous** \kō-'ni-f(ə-)rəs, kə-\ *adjective*

co·ni·ine \'kō-nē-,ēn\ *noun* [German *Koniin,* from Late Latin *conium* hemlock, from Greek *kōneion*] (1831)
: a poisonous alkaloid $C_8H_{17}N$ found in poison hemlock (*Conium maculatum*)

con·jec·tur·al \kən-'jek-chə-rəl, -'jek-shrəl\ *adjective* (1553)
1 : of the nature of or involving or based on conjecture
2 : given to conjectures
— **con·jec·tur·al·ly** *adverb*

¹con·jec·ture \kən-'jek-chər\ *noun* [Middle English, from Middle French or Latin; Middle French, from Latin *conjectura,* from *conjectus,* past participle of *conicere,* literally, to throw together, from *com-* + *jacere* to throw — more at JET] (14th century)
1 *obsolete* **a :** interpretation of omens **b :** SUPPOSITION
2 a : inference from defective or presumptive evidence **b :** a conclusion deduced by surmise or guesswork **c :** a proposition (as in mathematics) before it has been proved or disproved

²conjecture *verb* **-tured; -tur·ing** \-'jek-chə-riŋ, -'jek-shriŋ\ (15th century)
transitive verb
1 : to arrive at by conjecture
2 : to make conjectures as to
intransitive verb
: to form conjectures
— **con·jec·tur·er** \-'jek-chər-ər\ *noun*

con·join \kən-'jȯin, kän-\ *verb* [Middle English, from Middle French *conjoindre,* from Latin *conjungere,* from *com-* + *jungere* to join — more at YOKE] (14th century)
transitive verb
: to join together (as separate entities) for a common purpose
intransitive verb
: to join together for a common purpose

con·joined \-'jȯind\ *adjective* (1570)
: being, coming, or brought together so as to meet, touch, or overlap ⟨*conjoined* heads on a coin⟩

con·joint \-'jȯint\ *adjective* [Middle English, from Middle French, past participle of *conjoindre*] (1725)
1 : UNITED, CONJOINED
2 : related to, made up of, or carried on by two or more in combination **:** JOINT
— **con·joint·ly** *adverb*

con·ju·gal \'kän-ji-gəl *also* kən-'jü-\ *adjective* [Middle French or Latin; Middle French, from Latin *conjugalis,* from *conjug-, conjux* husband, wife, from *conjungere* to join, unite in marriage] (1545)
: of or relating to the married state or to married persons and their relations **:** CONNUBIAL
— **con·ju·gal·i·ty** \,kän-ji-'ga-lə-tē, -jü-\ *noun*
— **con·ju·gal·ly** \'kän-ji-gə-lē *also* kən-'jü-\ *adverb*

conjugal rights *noun plural* (circa 1891)
: the sexual rights or privileges implied by and involved in the marriage relationship **:** the right of sexual intercourse between husband and wife

con·ju·gant \'kän-ji-gənt\ *noun* (1910)
: either of a pair of conjugating gametes or organisms

¹con·ju·gate \'kän-ji-gət, -jə-,gāt\ *adjective* [Middle English *conjugat,* from Latin *conjugatus,* past participle of *conjugare* to unite, from *com-* + *jugare* to join, from *jugum* yoke — more at YOKE] (15th century)
1 a : joined together especially in pairs **:** COUPLED **b :** acting or operating as if joined
2 a : having features in common but opposite or inverse in some particular **b :** relating to or being conjugate complex numbers ⟨complex roots occurring in *conjugate* pairs⟩
3 *of an acid or base* **:** related by the difference of a proton ⟨the acid NH_4 and the base NH_3 are *conjugate* to each other⟩
4 : having the same derivation and therefore usually some likeness in meaning ⟨*conjugate* words⟩
5 *of two leaves of a book* **:** forming a single piece
— **con·ju·gate·ly** *adverb*
— **con·ju·gate·ness** *noun*

²con·ju·gate \-jə-,gāt\ *verb* **-gat·ed; -gat·ing** (1530)
transitive verb
1 : to give in prescribed order the various inflectional forms of — used especially of a verb
2 : to join together
intransitive verb
1 : to become joined together
2 a : to pair and fuse in conjugation **b :** to pair in synapsis

³conjugate *same as* ¹\ *noun* (circa 1586)
1 : something conjugate **:** a product of conjugating
2 : CONJUGATE COMPLEX NUMBER
3 : an element of a mathematical group that is equal to a given element of the group multiplied on the right by another element and on the left by the inverse of the latter element

conjugate complex number *noun* (circa 1909)
: one of two complex numbers differing only in the sign of the imaginary part

con·ju·gat·ed \-,gā-təd\ *adjective* (1882)
1 : formed by the union of two compounds or united with another compound ⟨*conjugated* bile acids⟩
2 : relating to or containing a system of two double bonds separated by a single bond ⟨*conjugated* fatty acids⟩

conjugated protein *noun* (circa 1909)
: a compound of a protein with a nonprotein ⟨hemoglobin is a *conjugated protein*⟩ — compare SIMPLE PROTEIN

con·ju·ga·tion \,kän-jə-'gā-shən\ *noun* (15th century)
1 a : a schematic arrangement of the inflectional forms of a verb **b :** verb inflection **c :** a class of verbs having the same type of inflectional forms ⟨the weak *conjugation*⟩ **d :** a set of the simple or derivative inflectional forms of a verb especially in Sanskrit or the Semitic languages ⟨the causative *conjugation*⟩
2 : the act of conjugating **:** the state of being conjugated
3 a : fusion of usually similar gametes with ultimate union of their nuclei that among lower thallophytes replaces the typical fertilization of higher forms **b :** temporary cytoplasmic union with exchange of nuclear material that is the usual sexual process in ciliated protozoans **c :** the one-way transfer of DNA between bacteria in cellular contact
— **con·ju·ga·tion·al** \-shnəl, -shə-nᵊl\ *adjective*
— **con·ju·ga·tion·al·ly** *adverb*

¹con·junct \kən-'jəŋ(k)t, kän-\ *adjective* [Middle English, from Latin *conjunctus,* past participle of *conjungere*] (15th century)
1 : UNITED, JOINED
2 : JOINT
3 : relating to melodic progression by intervals of no more than a major second — compare DISJUNCT

²con·junct \'kän-,jəŋ(k)t\ *noun* (1667)
1 : something joined or associated with another; *specifically* **:** one of the components of a conjunction
2 : an adverb or adverbial (as *so, in addition, however, secondly*) that indicates the speaker's or writer's assessment of the connection between linguistic units (as clauses)

con·junc·tion \kən-'jəŋ(k)-shən\ *noun* (14th century)
1 : the act or an instance of conjoining **:** the state of being conjoined **:** COMBINATION
2 : occurrence together in time or space **:** CONCURRENCE
3 a : the apparent meeting or passing of two or more celestial bodies in the same degree of the zodiac **b :** a configuration in which two celestial bodies have their least apparent separation
4 : an uninflected linguistic form that joins together sentences, clauses, phrases, or words
5 : a complex sentence in logic true if and only if each of its components is true
— **con·junc·tion·al** \-shnəl, -shə-nᵊl\ *adjective*
— **con·junc·tion·al·ly** *adverb*

con·junc·ti·va \,kän-,jəŋ(k)-'tī-və, kən-\ *noun, plural* **-vas** *or* **-vae** \-(,)vē\ [New Latin, from Late Latin, feminine of *conjunctivus* conjoining, from Latin *conjunctus*] (14th century)
: the mucous membrane that lines the inner surface of the eyelids and is continued over the forepart of the eyeball — see EYE illustration
— **con·junc·ti·val** \-vəl\ *adjective*

con·junc·tive \kən-'jəŋ(k)-tiv\ *adjective* (1581)
1 : CONNECTIVE
2 : CONJUNCT, CONJOINED
3 : being or functioning like a conjunction
4 : COPULATIVE 1a
— **conjunctive** *noun*
— **con·junc·tive·ly** *adverb*

con·junc·ti·vi·tis \kən-,jəŋ(k)-ti-'vī-təs\ *noun* (1835)
: inflammation of the conjunctiva

con·junc·ture \kən-'jəŋ(k)-chər\ *noun* (1605)
1 : CONJUNCTION, UNION
2 : a combination of circumstances or events usually producing a crisis **:** JUNCTURE

con·ju·ra·tion \,kän-jü-'rā-shən, ,kən-\ *noun* (14th century)
1 : the act or process of conjuring **:** INCANTATION
2 : an expression or trick used in conjuring
3 : a solemn appeal **:** ADJURATION

con·jure *verb transitive 2 & verb intransitive senses* 'kän-jər *also* 'kən-; *verb transitive 1* kən-'jůr\ *verb* **con·jured; con·jur·ing** \'känj-riŋ, 'kän-jə-, 'kənj-, 'kən-jə-; kən-'jůr-iŋ\ [Middle English, from Old French *conjurer,* from Latin *conjurare* to swear together, from *com-* + *jurare* to swear — more at JURY] (13th century)
transitive verb
1 : to charge or entreat earnestly or solemnly
2 a : to summon by or as if by invocation or incantation **b** (1) **:** to affect or effect by or as if by magic (2) **:** IMAGINE, CONTRIVE — often used with *up* ⟨we *conjure* up our own metaphors for our own needs —R. J. Kaufmann⟩ (3) **:** to bring to mind ⟨words that *conjure* pleasant images⟩
intransitive verb
1 a : to summon a devil or spirit by invocation or incantation **b :** to practice magical arts
2 : to use a conjurer's tricks **:** JUGGLE

con·jur·er *or* **con·ju·ror** \'kän-jər-ər, 'kən-\ *noun* (14th century)
1 : one that practices magic arts **:** WIZARD
2 : one that performs feats of sleight of hand and illusion **:** MAGICIAN, JUGGLER

¹conk \'käŋk, 'kȯŋk\ *transitive verb* [English slang *conk* head] (1821)
: to hit especially on the head **:** KNOCK OUT

²conk *noun* [probably alteration of *conch*] (1851)
: the visible fruiting body of a bracket fungus; *also* **:** decay caused by such a fungus
— **conky** \'käŋ-kē, 'kȯŋ-\ *adjective*

³conk *intransitive verb* [probably imitative] (1918)

1 : BREAK DOWN; *especially* **:** STALL — usually used with *out* ⟨the motor suddenly *conked* out⟩
2 a : FAINT **b :** to go to sleep — usually used with *off* or *out* ⟨*conked* out for a while after lunch⟩ **c :** DIE ⟨I caught pneumonia. I almost *conked* —Truman Capote⟩

⁴conk *transitive verb* [probably by shortening & alteration from *congolene* preparation used for straightening hair] (1950)
: to straighten out (hair) usually by the use of chemicals

⁵conk *noun* (1965)
: a hairstyle in which the hair is straightened out and flattened down or lightly waved — called also *process*

conk·er \'käŋ-kər\ *noun* [*conch* + ²*-er,* from the original use of a snail shell on a string in the game] (circa 1886)
1 : a horse chestnut especially when used in conkers
2 *plural* **:** a game in which each player swings a horse chestnut on a string to try to break one held by the opponent

con mo·to \kän-'mō-(ˌ)tō, kōn-\ *adverb* [Italian] (circa 1854)
: with movement **:** in a spirited manner — used as a direction in music

¹conn \'kän\ *transitive verb* [alteration of Middle English *condien* to conduct, from Middle French *conduire,* from Latin *conducere* — more at CONDUCE] (1626)
: to conduct or direct the steering of (as a ship)

²conn *noun* (1825)
: the control exercised by one who conns a ship

con·nate \kä-'nāt, 'kä-ˌ\ *adjective* [Late Latin *connatus,* past participle of *connasci* to be born together, from Latin *com-* + *nasci* to be born — more at NATION] (1641)
1 : AKIN, CONGENIAL
2 : INNATE, INBORN
3 : congenitally or firmly united ⟨*connate* leaves⟩
4 : born or originated together
5 : entrapped in sediments at the time of their deposition ⟨*connate* water⟩
— con·nate·ly *adverb*

con·nat·u·ral \kä-'nach-rəl, kə-, -'na-chə-\ *adjective* [Medieval Latin *connaturalis,* from Latin *com-* + *naturalis* natural] (1592)
1 : connected by nature **:** INBORN
2 : of the same nature
— con·nat·u·ral·i·ty \-ˌna-chə-'ra-lə-tē\ *noun*
— con·nat·u·ral·ly \-'nach-rə-lē, -'na-chə-\ *adverb*

con·nect \kə-'nekt\ *verb* [Middle English, from Latin *conectere, connectere,* from *com-* + *nectere* to bind] (15th century)
intransitive verb
1 : to become joined ⟨the two rooms *connect* by a hallway⟩ ⟨ideas that *connect* easily to form a theory⟩
2 : to make a successful hit, shot, or throw ⟨*connected* for a home run⟩ ⟨*connected* on 60 percent of his shots —*N.Y. Times*⟩
3 : to have or establish a rapport ⟨tried to *connect* with the younger generation⟩
transitive verb
1 : to join or fasten together usually by something intervening
2 : to place or establish in relationship
synonym see JOIN
— con·nect·able *also* **con·nect·ible** \-'nek-tə-bəl\ *adjective*
— con·nec·tor *also* **con·nect·er** \-'nek-tər\ *noun*

connected *adjective* (1712)
1 : joined or linked together
2 : having the parts or elements logically linked together ⟨presented a thoroughly *connected* view of the problem⟩
3 : related by blood or marriage
4 : having social, professional, or commercial relationships ⟨a well-*connected* lawyer⟩
5 *of a set* **:** having the property that any two of its points can be joined by a line completely contained in the set; *also* **:** incapable of being separated into two or more closed disjoint subsets
— con·nect·ed·ly *adverb*
— con·nect·ed·ness *noun*

connecting rod *noun* (1839)
: a rod that transmits motion from a reciprocating part of a machine (as a piston) to a rotating part or vice versa

con·nec·tion \kə-'nek-shən\ *noun* [Latin *connexion-, connexio,* from *conectere*] (14th century)
1 : the act of connecting **:** the state of being connected: as **a :** causal or logical relation or sequence ⟨the *connection* between two ideas⟩ **b** (1) **:** contextual relation or association ⟨in this *connection* the word has a different meaning⟩ (2) **:** relationship in fact ⟨wanted in *connection* with a robbery⟩ **c :** a relation of personal intimacy (as of family ties) **d :** COHERENCE, CONTINUITY
2 a : something that connects **:** LINK ⟨a loose *connection* in the wiring⟩ **b :** a means of communication or transport
3 : a person connected with another especially by marriage, kinship, or common interest ⟨has powerful *connections*⟩
4 : a political, social, professional, or commercial relationship: as **a :** POSITION, JOB **b :** an arrangement to execute orders or advance interests of another ⟨a firm's foreign *connections*⟩ **c :** a source of contraband (as illegal drugs)
5 : a set of persons associated together: as **a :** DENOMINATION **b :** CLAN
— con·nec·tion·al \-shnəl, -shə-nᵊl\ *adjective*

¹con·nec·tive \kə-'nek-tiv\ *adjective* (circa 1660)
: serving to connect
— con·nec·tive·ly *adverb*

²connective *noun* (1751)
: something that connects: as **a :** a linguistic form that connects words or word groups **b :** the tissue connecting the pollen sacs of an anther **c :** a logical term (as *or, if then, and, not*) or a symbol for it that relates propositions in such a way that the truth or falsity of the resulting statement is determined by the truth or falsity of the components

connective tissue *noun* (1846)
: a tissue of mesodermal origin rich in intercellular substance or interlacing processes with little tendency for the cells to come together in sheets or masses; *specifically* **:** connective tissue of stellate or spindle-shaped cells with interlacing processes that pervades, supports, and binds together other tissues and forms ligaments, tendons, and aponeuroses

con·nec·tiv·i·ty \(ˌ)kä-ˌnek-'ti-və-tē, kə-\ *noun, plural* **-ties** (1893)
: the quality or state of being connective or connected ⟨*connectivity* of a surface⟩; *especially* **:** the ability to connect to or communicate with another computer or computer system

Con·ne·ma·ra \ˌkä-nə-'mär-ə, ˌkó-\ *noun* [*Connemara,* Ireland] (circa 1952)
: any of a breed of rugged ponies developed in Ireland

con·nex·ion \kə-'nek-shən\ *chiefly British variant of* CONNECTION

con·ning tower \'kä-niŋ-\ *noun* (1886)
: a raised structure on the deck of a submarine used especially formerly for navigation and attack direction

con·nip·tion \kə-'nip-shən\ *noun* [origin unknown] (1833)
: a fit of rage, hysteria, or alarm ⟨went into *conniptions*⟩

con·niv·ance \kə-'nī-vən(t)s\ *noun* (1593)
: the act of conniving; *especially* **:** knowledge of and active or passive consent to wrongdoing

conning tower

con·nive \kə-'nīv\ *intransitive verb* **con·nived; con·niv·ing** [French or Latin; French *conniver,* from Latin *connivēre, connivēre* to close the eyes, connive, from *com-* + *-nivēre* (akin to *nictare* to wink); akin to Old English & Old High German *hnīgan* to bow] (1601)
1 : to pretend ignorance of or fail to take action against something one ought to oppose
2 a : to be indulgent or in secret sympathy **:** WINK **b :** to cooperate secretly or have a secret understanding
3 : CONSPIRE, INTRIGUE
— con·niv·er *noun*

con·ni·vent \-'nī-vənt\ *adjective* [Latin *convivent-, conivens,* present participle of *conivēre*] (1757)
: converging but not fused ⟨*connivent* stamens⟩

con·nois·seur \ˌkä-nə-'sər *also* -'su̇r\ *noun* [obsolete French (now *connaisseur*), from Old French *connoisseor,* from *connoistre* to know, from Latin *cognoscere* — more at COGNITION] (1714)
1 : EXPERT; *especially* **:** one who understands the details, technique, or principles of an art and is competent to act as a critical judge
2 : one who enjoys with discrimination and appreciation of subtleties ⟨a *connoisseur* of fine wines⟩
— con·nois·seur·ship \-ˌship\ *noun*

con·no·ta·tion \ˌkä-nə-'tā-shən\ *noun* (1532)
1 a : the suggesting of a meaning by a word apart from the thing it explicitly names or describes **b :** something suggested by a word or thing **:** IMPLICATION ⟨the *connotations* of comfort that surrounded that old chair⟩
2 : the signification of something ⟨that abuse of logic which consists in moving counters about as if they were known entities with a fixed *connotation* —W. R. Inge⟩
3 : an essential property or group of properties of a thing named by a term in logic — compare DENOTATION
— con·no·ta·tion·al \-shnəl, -shə-nᵊl\ *adjective*

con·no·ta·tive \'kä-nə-ˌtā-tiv, kə-'nō-tə-tiv\ *adjective* (1614)
1 : connoting or tending to connote
2 : relating to connotation
— con·no·ta·tive·ly *adverb*

con·note \kə-'nōt, kä-\ *transitive verb* **con·not·ed; con·not·ing** [Medieval Latin *connotare,* from Latin *com-* + *notare* to note] (1665)
1 : to be associated with or inseparable from as a consequence or concomitant ⟨the remorse so often *connoted* by guilt⟩
2 a : to convey in addition to exact explicit meaning ⟨all the misery that poverty *connotes*⟩ **b :** to imply as a logical connotation

con·nu·bi·al \kə-'nü-bē-əl, -'nyü-\ *adjective* [Latin *conubialis,* from *conubium, connubium* marriage, from *com-* + *nubere* to marry — more at NUPTIAL] (circa 1656)
: of or relating to the married state **:** CONJUGAL
— con·nu·bi·al·ism \-bē-ə-ˌli-zəm\ *noun*
— con·nu·bi·al·i·ty \-ˌnü-bē-'a-lə-tē, -ˌnyü-\ *noun*

— **con·nu·bi·al·ly** \-ˈnü-bē-ə-lē, -ˈnyü-\ *adverb*

co·no·dont \ˈkō-nə-ˌdänt, ˈkä-\ *noun* [International Scientific Vocabulary *con-* (from Greek *kōnos* cone) + *-odont*] (1859)
: a Paleozoic toothlike fossil that is probably the remains of an extinct marine invertebrate

co·noid \ˈkō-ˌnȯid\ *or* **co·noi·dal** \kō-ˈnȯi-dᵊl\ *adjective* (1668)
: shaped like or nearly like a cone ⟨*conoid* shells⟩ ⟨*conoid* pottery⟩
— **conoid** *noun*

con·quer \ˈkäŋ-kər\ *verb* **con·quered; con·quer·ing** \-k(ə-)riŋ\ [Middle English, to acquire, conquer, from Old French *conquerre,* from (assumed) Vulgar Latin *conquaerere,* alteration of Latin *conquirere* to search for, collect, from *com-* + *quaerere* to ask, search] (14th century)
transitive verb
1 : to gain or acquire by force of arms **:** SUBJUGATE
2 : to overcome by force of arms **:** VANQUISH
3 : to gain mastery over or win by overcoming obstacles or opposition ⟨*conquered* the mountain⟩
4 : to overcome by mental or moral power **:** SURMOUNT ⟨*conquered* her fear⟩
intransitive verb
: to be victorious ☆
— **con·quer·or** \-kər-ər\ *noun*

con·quest \ˈkän-ˌkwest, ˈkäŋ-; ˈkäŋ-kwəst\ *noun* [Middle English, from Middle French, from (assumed) Vulgar Latin *conquaesitus,* alteration of Latin *conquisitus,* past participle of *conquirere*] (14th century)
1 : the act or process of conquering
2 a : something conquered; *especially* **:** territory appropriated in war **b :** a person whose favor or hand has been won

con·qui·an \ˈkäŋ-kē-ən\ *noun* [Mexican Spanish *conquián* — more at COONCAN] (circa 1911)
: a card game for two played with 40 cards from which all games of rummy developed

con·quis·ta·dor \kän-ˈkēs-tə-ˌdȯr, kən- *also* -ˈkwis-, -ˈkis-\ *noun, plural* **con·quis·ta·do·res** \(ˌ)kän-ˌkēs-tə-ˈdȯr-ēz, -ˈdȯr-ˌās, kən-; (ˌ)kän-ˌkwis-, -ˌkis-\ *or* **con·quis·ta·dors** [Spanish, ultimately from Latin *conquirere*] (1830)
: one that conquers; *specifically* **:** a leader in the Spanish conquest of America and especially of Mexico and Peru in the 16th century

con·san·guine \kän-ˈsaŋ-gwən, kən-\ *adjective* (1610)
: CONSANGUINEOUS

con·san·guin·e·ous \ˌkän-ˌsan-ˈgwi-nē-əs, -ˌsaŋ-\ *adjective* [Latin *consanguineus,* from *com-* + *sanguin-, sanguis* blood] (1601)
: of the same blood or origin; *specifically* **:** descended from the same ancestor
— **con·san·guin·e·ous·ly** *adverb*

con·san·guin·i·ty \-ˈgwi-nə-tē\ *noun, plural* **-ties** (14th century)
1 : the quality or state of being consanguineous
2 : a close relation or connection

con·science \ˈkän(t)-shən(t)s\ *noun* [Middle English, from Old French, from Latin *conscientia,* from *conscient-, consciens,* present participle of *conscire* to be conscious, be conscious of guilt, from *com-* + *scire* to know — more at SCIENCE] (13th century)
1 a : the sense or consciousness of the moral goodness or blameworthiness of one's own conduct, intentions, or character together with a feeling of obligation to do right or be good **b :** a faculty, power, or principle enjoining good acts **c :** the part of the superego in psychoanalysis that transmits commands and admonitions to the ego
2 *archaic* **:** CONSCIOUSNESS
3 : conformity to the dictates of conscience **:** CONSCIENTIOUSNESS
4 : sensitive regard for fairness or justice **:** SCRUPLE
— **con·science·less** \-ləs\ *adjective*
— **in all conscience** *or* **in conscience** **:** in all fairness

conscience money *noun* (1848)
: money paid usually anonymously to relieve the conscience by restoring what has been wrongfully acquired

con·sci·en·tious \ˌkän(t)-shē-ˈen(t)-shəs\ *adjective* (1611)
1 : governed by or conforming to the dictates of conscience **:** SCRUPULOUS ⟨a *conscientious* public servant⟩
2 : METICULOUS, CAREFUL ⟨a *conscientious* listener⟩
synonym see UPRIGHT
— **con·sci·en·tious·ly** *adverb*
— **con·sci·en·tious·ness** *noun*

conscientious objection *noun* (1916)
: objection on moral or religious grounds (as to service in the armed forces or to bearing arms)

conscientious objector *noun* (1899)
: a person who refuses to serve in the armed forces or bear arms on moral or religious grounds

con·scio·na·ble \ˈkän(t)-sh(ə-)nə-bəl\ *adjective* [irregular from *conscience*] (1549)
: CONSCIENTIOUS

¹con·scious \ˈkän(t)-shəs\ *adjective* [Latin *conscius,* from *com-* + *scire* to know] (1592)
1 : perceiving, apprehending, or noticing with a degree of controlled thought or observation
2 *archaic* **:** sharing another's knowledge or awareness of an inward state or outward fact
3 : personally felt ⟨*conscious* guilt⟩
4 : capable of or marked by thought, will, design, or perception
5 : SELF-CONSCIOUS
6 : having mental faculties undulled by sleep, faintness, or stupor **:** AWAKE ⟨became *conscious* after the anesthesia wore off⟩
7 : done or acting with critical awareness ⟨a *conscious* effort to do better⟩
8 a : likely to notice, consider, or appraise ⟨a bargain-*conscious* shopper⟩ **b :** being concerned or interested ⟨a budget-*conscious* businessman⟩ **c :** marked by strong feelings or notions ⟨a race-*conscious* society⟩
synonym see AWARE
— **con·scious·ly** *adverb*

²conscious *noun* (1919)
: CONSCIOUSNESS 5

con·scious·ness \-nəs\ *noun* (1632)
1 a : the quality or state of being aware especially of something within oneself **b :** the state or fact of being conscious of an external object, state, or fact **c :** AWARENESS; *especially* **:** concern for some social or political cause
2 : the state of being characterized by sensation, emotion, volition, and thought **:** MIND
3 : the totality of conscious states of an individual
4 : the normal state of conscious life ⟨regained *consciousness*⟩
5 : the upper level of mental life of which the person is aware as contrasted with unconscious processes

consciousness–raising *noun* (1971)
: an increasing of concerned awareness especially of some social or political issue

con·scribe \kən-ˈskrīb\ *transitive verb* **con·scribed; con·scrib·ing** [Latin *conscribere*] (1613)
1 : LIMIT, CIRCUMSCRIBE ⟨ill-health . . . *conscribed* the force of his intentions —*Times Literary Supplement*⟩
2 : to enlist forcibly **:** CONSCRIPT

¹con·script \ˈkän-ˌskript\ *noun* (1800)
: a conscripted person (as a military recruit)

²con·script \kən-ˈskript\ *transitive verb* (1813)
: to enroll into service by compulsion **:** DRAFT ⟨was *conscripted* into the army⟩

³con·script \ˈkän-ˌskript\ *adjective* [alteration of French *conscrit,* from Latin *conscriptus,* past participle of *conscribere* to enroll, enlist, from *com-* + *scribere* to write — more at SCRIBE] (1823)
1 : enrolled into service by compulsion **:** DRAFTED
2 : made up of conscripted persons

con·scrip·tion \kən-ˈskrip-shən\ *noun* (1800)
: compulsory enrollment of persons especially for military service **:** DRAFT

¹con·se·crate \ˈkän(t)-sə-ˌkrāt\ *adjective* (14th century)
: dedicated to a sacred purpose

²consecrate *transitive verb* **-crat·ed; -crat·ing** [Middle English, from Latin *consecratus,* past participle of *consecrare,* from *com-* + *sacrare* to consecrate — more at SACRED] (14th century)
1 : to induct (a person) into a permanent office with a religious rite; *especially* **:** to ordain to the office of bishop
2 a : to make or declare sacred; *especially* **:** to devote irrevocably to the worship of God by a solemn ceremony **b :** to effect the liturgical transubstantiation of (eucharistic bread and wine) **c :** to devote to a purpose with or as if with deep solemnity or dedication
3 : to make inviolable or venerable ⟨principles *consecrated* by the weight of history⟩
synonym see DEVOTE
— **con·se·cra·tive** \-ˌkrā-tiv\ *adjective*
— **con·se·cra·tor** \-ˌkrā-tər\ *noun*
— **con·se·cra·to·ry** \ˈkän(t)-si-krə-ˌtōr-ē, -ˌtȯr-, -ˌkrā-tə-rē\ *adjective*

con·se·cra·tion \ˌkän(t)-sə-ˈkrā-shən\ *noun* (14th century)
1 : the act or ceremony of consecrating
2 : the state of being consecrated
3 *capitalized* **:** the part of a Communion rite in which the bread and wine are consecrated

con·se·cu·tion \ˌkän(t)-si-ˈkyü-shən\ *noun* [Latin *consecution-, consecutio,* from *consequi* to follow along — more at CONSEQUENT] (1651)
: SEQUENCE

con·sec·u·tive \kən-ˈse-kyə-tiv, -kə-tiv\ *adjective* (1611)
: following one after the other in order **:** SUCCESSIVE
— **con·sec·u·tive·ly** *adverb*
— **con·sec·u·tive·ness** *noun*

con·sen·su·al \kən-ˈsen(t)-sh(ə-)wəl, -shəl\ *adjective* [Latin *consensus* + English *-al*] (1754)
1 a : existing or made by mutual consent without an act of writing ⟨a *consensual* contract⟩ **b :** involving or based on mutual consent ⟨*consensual* acts⟩
2 : relating to or being the constrictive pupillary response of an eye that is covered when the other eye is exposed to light

☆ SYNONYMS
Conquer, vanquish, defeat, subdue, reduce, overcome, overthrow mean to get the better of by force or strategy. CONQUER implies gaining mastery of ⟨Caesar *conquered* Gaul⟩. VANQUISH implies a complete overpowering ⟨*vanquished* the enemy and ended the war⟩. DEFEAT does not imply the finality or completeness of VANQUISH which it otherwise equals ⟨the Confederates *defeated* the Union forces at Manassas⟩. SUBDUE implies a defeating and suppression ⟨*subdued* the native tribes after years of fighting⟩. REDUCE implies a forcing to capitulate or surrender ⟨the city was *reduced* after a month-long siege⟩. OVERCOME suggests getting the better of with difficulty or after hard struggle ⟨*overcame* a host of bureaucratic roadblocks⟩. OVERTHROW stresses the bringing down or destruction of existing power ⟨violently *overthrew* the old regime⟩.

— **con·sen·su·al·ly** *adverb*
con·sen·sus \kən-ˈsen(t)-səs\ *noun, often attributive* [Latin, from *consentire*] (1858)
1 a : general agreement **:** UNANIMITY ⟨the *consensus* of their opinion, based on reports . . . from the border —John Hersey⟩ **b :** the judgment arrived at by most of those concerned ⟨the *consensus* was to go ahead⟩
2 : group solidarity in sentiment and belief ■
[1]**con·sent** \kən-ˈsent\ *intransitive verb* [Middle English, from Latin *consentire,* from *com-* + *sentire* to feel — more at SENSE] (13th century)
1 : to give assent or approval **:** AGREE
2 *archaic* **:** to be in concord in opinion or sentiment
synonym see ASSENT
— **con·sent·ing·ly** \-ˈsen-tiŋ-lē\ *adverb*
[2]**consent** *noun* (14th century)
1 : compliance in or approval of what is done or proposed by another **:** ACQUIESCENCE ⟨he shall have power, by and with the advice and *consent* of the Senate, to make treaties —*U.S. Constitution*⟩
2 : agreement as to action or opinion; *specifically* **:** voluntary agreement by a people to organize a civil society and give authority to the government
— **con·sent·er** *noun*
con·sen·ta·ne·ous \ˌkän(t)-sən-ˈtā-nē-əs, ˌkän-ˌsen-\ *adjective* [Latin *consentaneus,* from *consentire* to agree] (1625)
1 : expressing agreement **:** SUITED
2 : done or made by the consent of all
— **con·sen·ta·ne·ous·ly** *adverb*
consent decree *noun* (1904)
: a judicial decree that sanctions a voluntary agreement between parties in dispute
con·se·quence \ˈkän(t)-sə-ˌkwen(t)s, -si-kwən(t)s\ *noun* (14th century)
1 : a conclusion derived through logic **:** INFERENCE
2 : something produced by a cause or necessarily following from a set of conditions
3 a : importance with respect to power to produce an effect **:** MOMENT **b :** social importance
4 : the appearance of importance; *especially* **:** SELF-IMPORTANCE
synonym see IMPORTANCE
— **in consequence :** as a result **:** CONSEQUENTLY
[1]**con·se·quent** \-kwənt, -ˌkwent\ *noun* (14th century)
1 a : DEDUCTION 2b **b :** the conclusion of a conditional sentence
2 : the second term of a ratio
[2]**consequent** *adjective* [Middle English, from Middle French, from Latin *consequent-, consequens,* present participle of *consequi* to follow along, from *com-* + *sequi* to follow — more at SUE] (15th century)
1 : following as a result or effect ⟨her new job and *consequent* relocation⟩
2 : observing logical sequence **:** RATIONAL
con·se·quen·tial \ˌkän(t)-sə-ˈkwen(t)-shəl\ *adjective* (1626)
1 : of the nature of a secondary result **:** INDIRECT
2 : CONSEQUENT
3 : having significant consequences **:** IMPORTANT ⟨a grave and *consequential* event⟩
4 : SELF-IMPORTANT
— **con·se·quen·ti·al·i·ty** \-ˌkwen(t)-shē-ˈa-lə-tē\ *noun*
— **con·se·quen·tial·ly** \-ˈkwen(t)-sh(ə-)lē\ *adverb*
— **con·se·quen·tial·ness** \-ˈkwen(t)-shəl-nəs\ *noun*
con·se·quent·ly \ˈkän(t)-sə-ˌkwent-lē, -si-kwənt-\ *adverb* (15th century)
: as a result **:** in view of the foregoing **:** ACCORDINGLY
con·ser·van·cy \kən-ˈsər-vən(t)-sē\ *noun, plural* **-cies** [alteration of obsolete *conservacy* conservation, from Anglo-French *conservacie,* from Medieval Latin *conservatia,* from Latin *conservare*] (1755)
1 *British* **:** a board regulating fisheries and navigation in a river or port
2 a : CONSERVATION **b :** an organization or area designated to conserve and protect natural resources
con·ser·va·tion \ˌkän(t)-sər-ˈvā-shən\ *noun* [Middle English, from Middle French, from Latin *conservation-, conservatio,* from *conservare*] (14th century)
1 : a careful preservation and protection of something; *especially* **:** planned management of a natural resource to prevent exploitation, destruction, or neglect
2 : the preservation of a physical quantity during transformations or reactions
— **con·ser·va·tion·al** \-shnəl, -shə-nᵊl\ *adjective*
con·ser·va·tion·ist \-sh(ə-)nist\ *noun* (1870)
: a person who advocates conservation especially of natural resources
conservation of charge (1949)
: a principle in physics: the total electric charge of an isolated system remains constant irrespective of whatever internal changes may take place
conservation of energy (1853)
: a principle in physics: the total energy of an isolated system remains constant irrespective of whatever internal changes may take place with energy disappearing in one form reappearing in another
conservation of mass (1884)
: a principle in classical physics: the total mass of any material system is neither increased nor diminished by reactions between the parts — called also *conservation of matter*
con·ser·va·tism \kən-ˈsər-və-ˌti-zəm\ *noun* (1835)
1 *capitalized* **a :** the principles and policies of a Conservative party **b :** the Conservative party
2 a : disposition in politics to preserve what is established **b :** a political philosophy based on tradition and social stability, stressing established institutions, and preferring gradual development to abrupt change
3 : the tendency to prefer an existing or traditional situation to change
[1]**con·ser·va·tive** \kən-ˈsər-və-tiv\ *adjective* (14th century)
1 : PRESERVATIVE
2 a : of or relating to a philosophy of conservatism **b** *capitalized* **:** of or constituting a political party professing the principles of conservatism: as (1) **:** of or constituting a party of the United Kingdom advocating support of established institutions (2) **:** PROGRESSIVE CONSERVATIVE
3 a : tending or disposed to maintain existing views, conditions, or institutions **:** TRADITIONAL **b :** marked by moderation or caution ⟨a *conservative* estimate⟩ **c :** marked by or relating to traditional norms of taste, elegance, style, or manners ⟨a *conservative* suit⟩
4 : of or relating to Conservative Judaism
— **con·ser·va·tive·ly** *adverb*
— **con·ser·va·tive·ness** *noun*
[2]**conservative** *noun* (1831)
1 a : an adherent or advocate of political conservatism **b** *capitalized* **:** a member or supporter of a conservative political party
2 a : one who adheres to traditional methods or views **b :** a cautious or discreet person
Conservative Judaism *noun* (1946)
: Judaism as practiced especially among some U.S. Jews with adherence to the Torah and Talmud but with allowance for some departures in keeping with differing times and circumstances — compare ORTHODOX JUDAISM, REFORM JUDAISM
con·ser·va·tize \-ˌtīz\ *verb* **-tized; -tiz·ing** (1849)
intransitive verb
: to grow conservative
transitive verb
: to make conservative
con·ser·va·toire \kən-ˈsər-və-ˌtwär\ *noun* [French, from Italian *conservatorio*] (1845)
: CONSERVATORY 2
con·ser·va·tor \kən-ˈsər-və-tər, -və-ˌtȯr; ˈkän(t)-sər-ˌvā-tər\ *noun* (15th century)
1 a : one that preserves from injury or violation **:** PROTECTOR **b :** one that is responsible for the care, restoration, and repair of archival or museum articles
2 : a person, official, or institution designated to take over and protect the interests of an incompetent
3 : an official charged with the protection of something affecting public welfare and interests
— **con·ser·va·to·ri·al** \kən-ˌsər-və-ˈtōr-ē-əl, (ˌ)kän-, -ˈtȯr-\ *adjective*
— **con·ser·va·tor·ship** \kən-ˈsər-və-tər-ˌship, -və-ˌtȯr-; ˈkän(t)-sər-ˌvā-tər-\ *noun*
con·ser·va·to·ry \kən-ˈsər-və-ˌtōr-ē, -ˌtȯr-\ *noun, plural* **-ries** (1664)
1 : a greenhouse for growing or displaying plants
2 [Italian *conservatorio* home for foundlings, music school, from Latin *conservare*] **:** a school specializing in one of the fine arts ⟨a music *conservatory*⟩
[1]**con·serve** \kən-ˈsərv\ *transitive verb* **conserved; con·serv·ing** [Middle English, from Middle French *conserver,* from Latin *conservare,* from *com-* + *servare* to keep, guard, observe; akin to Avestan *haurvaiti* he guards] (14th century)
1 : to keep in a safe or sound state ⟨he *conserved* his inheritance⟩; *especially* **:** to avoid wasteful or destructive use of ⟨*conserve* natural resources⟩
2 : to preserve with sugar
3 : to maintain (a quantity) constant during a process of chemical, physical, or evolutionary change
— **con·serv·er** *noun*
[2]**con·serve** \ˈkän-ˌsərv\ *noun* (15th century)
1 : SWEETMEAT; *especially* **:** a candied fruit
2 : PRESERVE; *specifically* **:** one prepared from a mixture of fruits
con·sid·er \kən-ˈsi-dər\ *verb* **con·sid·ered; con·sid·er·ing** \-d(ə-)riŋ\ [Middle English, from Middle French *considerer,* from Latin *considerare* to observe, think about, from *com-* + *sider-, sidus* star] (14th century)
transitive verb
1 : to think about carefully: as **a :** to think of especially with regard to taking some action ⟨is *considering* you for the job⟩ ⟨*considered* moving to the city⟩ **b :** to take into account ⟨defendant's age must be *considered*⟩
2 : to regard or treat in an attentive or kindly way ⟨he *considered* her every wish⟩
3 : to gaze on steadily or reflectively
4 : to come to judge or classify ⟨*consider* thrift essential⟩
5 : REGARD ⟨his works are well *considered* abroad⟩
6 : SUPPOSE

☐ **USAGE**
consensus The phrase *consensus of opinion,* which is not actually redundant (see sense 1a; the sense that takes the phrase is slightly older), has been so often claimed to be a redundancy that many writers avoid it. You are safe in using *consensus* alone when it is clear you mean consensus of opinion, and most writers in fact do so.

intransitive verb
: REFLECT, DELIBERATE ⟨paused a moment to *consider*⟩ ☆

1con·sid·er·able \-ˈsi-dər(-ə)-bəl, -ˈsi-drə-bəl\ *adjective* (circa 1619)
1 : worth consideration **:** SIGNIFICANT
2 : large in extent or degree ⟨a *considerable* number⟩
— con·sid·er·ably \-blē\ *adverb*

2considerable *noun* (1685)
: a considerable amount, degree, or extent

con·sid·er·ate \kən-ˈsi-d(ə-)rət\ *adjective* (1572)
1 : marked by or given to careful consideration **:** CIRCUMSPECT
2 : thoughtful of the rights and feelings of others
— con·sid·er·ate·ly *adverb*
— con·sid·er·ate·ness *noun*

con·sid·er·ation \kən-ˌsi-də-ˈrā-shən\ *noun* (14th century)
1 : continuous and careful thought ⟨after long *consideration* he agreed to their requests⟩
2 a : a matter weighed or taken into account when formulating an opinion or plan ⟨economic *considerations* forced her to leave college⟩ **b :** a taking into account
3 : thoughtful and sympathetic regard
4 : an opinion obtained by reflection
5 : ESTEEM, REGARD ⟨the family built themselves a large, ugly villa . . . and became people of *consideration* —V. S. Pritchett⟩
6 a : RECOMPENSE, PAYMENT **b :** the inducement to a contract or other legal transaction; *specifically* **:** an act or forbearance or the promise thereof done or given by one party in return for the act or promise of another
— in consideration of : as payment or recompense for ⟨a small fee *in consideration of* many kind services⟩

con·sid·ered \kən-ˈsi-dərd\ *adjective* (circa 1677)
1 : matured by extended deliberative thought ⟨a *considered* opinion⟩
2 : viewed with respect or esteem

1con·sid·er·ing \-d(ə-)riŋ\ *preposition* (14th century)
: in view of **:** taking into account ⟨he did well *considering* his limitations⟩

2considering *conjunction* (15th century)
: INASMUCH AS ⟨*considering* he was new at the job, he did quite well⟩

con·sig·li·e·re \kōn-(ˌ)sil-ˈye-re, -ˈyer-ē; kän-(ˌ)si-glē-ˈye-rā, -rē, -ˈyer\ *noun, plural* **-ri** \-rē\ [Italian, from *consiglio* advice, counsel, from Latin *consilium* — more at COUNSEL] (1615)
: COUNSELOR, ADVISER ⟨*consigliere* of a Mafia family⟩

con·sign \kən-ˈsīn\ *verb* [Middle French *consigner,* from Latin *consignare,* from *com-* + *signum* sign, mark, seal — more at SIGN] (1528)
transitive verb
1 : to give over to another's care
2 : to give, transfer, or deliver into the hands or control of another; *also* **:** to commit especially to a final destination or fate ⟨a writer *consigned* to oblivion⟩
3 : to send or address to an agent to be cared for or sold
intransitive verb
obsolete **:** AGREE, SUBMIT
synonym see COMMIT
— con·sign·able \-ˈsī-nə-bəl\ *adjective*
— con·sig·na·tion \ˌkän-ˌsī-ˈnā-shən, ˌkän(t)-sig-\ *noun*
— con·sign·or \ˌkän-ˌsī-ˈnȯr, kən-\ *noun*

con·sign·ee \ˌkän(t)-sə-ˈnē, ˌkän-ˌsī-, kən-ˌsī-\ *noun* (1789)
: one to whom something is consigned or shipped

1con·sign·ment \kən-ˈsīn-mənt\ *noun* (circa 1668)
1 : the act or process of consigning
2 : something consigned especially in a single shipment
— on consignment : shipped to a dealer who pays only for what he sells and who may return what is unsold ⟨goods shipped *on consignment*⟩

2consignment *adjective* (1913)
: of, relating to, or received as goods on consignment ⟨a *consignment* sale⟩

1con·sist \kən-ˈsist\ *intransitive verb* [Middle French & Latin; Middle French *consister,* from Latin *consistere,* literally, to stop, stand still, from *com-* + *sistere* to take a stand; akin to Latin *stare* to stand — more at STAND] (1526)
1 : LIE, RESIDE — usually used with *in* ⟨liberty *consists* in the absence of obstructions —A. E. Housman⟩
2 *archaic* **a :** EXIST, BE **b :** to be capable of existing
3 : to be composed or made up — usually used with *of* ⟨breakfast *consisted* of cereal, milk, and fruit⟩
4 : to be consistent ⟨it *consists* with the facts⟩

2con·sist \ˈkän-ˌsist\ *noun* (1898)
: makeup or composition (as of coal sizes or a railroad train) by classes, types, or grades and arrangement

con·sis·tence \kən-ˈsis-tən(t)s\ *noun* (1601)
: CONSISTENCY

con·sis·ten·cy \kən-ˈsis-tən(t)-sē\ *noun, plural* **-cies** (1594)
1 a *archaic* **:** condition of adhering together **:** firmness of material substance **b :** firmness of constitution or character **:** PERSISTENCY
2 : degree of firmness, density, viscosity, or resistance to movement or separation of constituent particles ⟨boil the juice to the *consistency* of a thick syrup⟩
3 a : agreement or harmony of parts or features to one another or a whole **:** CORRESPONDENCE; *specifically* **:** ability to be asserted together without contradiction **b :** harmony of conduct or practice with profession ⟨followed her own advice with *consistency*⟩

con·sis·tent \kən-ˈsis-tənt\ *adjective* [Latin *consistent-, consistens,* present participle of *consistere*] (1647)
1 *archaic* **:** possessing firmness or coherence
2 a : marked by harmony, regularity, or steady continuity **:** free from variation or contradiction ⟨a *consistent* style in painting⟩ **b :** COMPATIBLE — usually used with *with* **c :** showing steady conformity to character, profession, belief, or custom ⟨a *consistent* patriot⟩
3 : tending to be arbitrarily close to the true value of the parameter estimated as the sample becomes large ⟨a *consistent* statistical estimator⟩
— con·sis·tent·ly \-lē\ *adverb*

con·sis·to·ry \kən-ˈsis-t(ə-)rē\ *noun, plural* **-ries** [Middle English *consistorie,* from Middle French, from Medieval Latin & Late Latin; Medieval Latin *consistorium* church tribunal, from Late Latin, imperial council, from Latin *consistere*] (14th century)
1 : a solemn assembly **:** COUNCIL
2 : a church tribunal or governing body: as **a :** a solemn meeting of Roman Catholic cardinals convoked and presided over by the pope **b :** a church session in some Reformed churches
3 : the organization that confers the degrees of the Ancient and Accepted Scottish Rite of Freemasonry usually from the 19th to the 32d inclusive; *also* **:** a meeting of such an organization
— con·sis·tor·i·al \ˌkän-si-ˈstōr-ē-əl, -ˈstȯr-\ *adjective*

con·so·ci·ate \kən-ˈsō-sē-ˌāt, -shē-ˌāt\ *verb* **-at·ed; -at·ing** [Latin *consociatus,* past participle of *consociare,* from *com-* + *socius* companion — more at SOCIAL] (1566)
transitive verb
: to bring into association
intransitive verb
: to associate especially in fellowship or partnership

con·so·ci·a·tion \-ˌsō-sē-ˈā-shən, -shē-\ *noun* (1593)
1 : association in fellowship or alliance
2 : an association of churches or religious societies
3 : an ecological community with a single dominant species
— con·so·ci·a·tion·al \-shnəl, -shə-nᵊl\ *adjective*

con·so·la·tion \ˌkän(t)-sə-ˈlā-shən\ *noun* (14th century)
1 : the act or an instance of consoling **:** the state of being consoled **:** COMFORT
2 : something that consoles; *specifically* **:** a contest held for those who have lost early in a tournament ⟨the losers met in a *consolation* game⟩
— con·so·la·to·ry \kən-ˈsō-lə-ˌtōr-ē, -ˈsä-, -ˌtȯr-\ *adjective*

consolation prize *noun* (1886)
: a prize given to a runner-up or a loser in a contest

1con·sole \ˈkän-ˌsōl\ *noun* [French] (1664)
1 : an architectural member projecting from a wall to form a bracket or from a keystone for ornament
2 : CONSOLE TABLE
3 a : an upright case which houses the keyboards and controlling mechanisms of an organ and from which the organ is played **b :** a combination of readouts or displays and an input device (as a keyboard or switches) by which an operator can monitor and interact with a system (as a computer or dubber)
4 a : a cabinet (as for a radio or television set) designed to rest directly on the floor **b :** a small storage cabinet between bucket seats in an automobile

2con·sole \kən-ˈsōl\ *transitive verb* **con·soled; con·sol·ing** [French *consoler,* from Latin *consolari,* from *com-* + *solari* to console] (1693)
: to alleviate the grief, sense of loss, or trouble of **:** COMFORT ⟨*console* a widow⟩
— con·sol·ing·ly \-ˈsō-liŋ-lē\ *adverb*

console table *noun* (1813)
: a table fixed to a wall with its top supported by consoles or front legs; *broadly* **:** a table designed to fit against a wall

console table

con·sol·i·date \kən-ˈsä-lə-ˌdāt\ *verb* **-dat·ed; -dat·ing** [Latin *consolidatus,* past participle of *consolidare* to make solid, from *com-* + *solidus* solid] (circa 1512)
transitive verb
1 : to join together into one whole **:** UNITE ⟨*consolidate* several small school districts⟩
2 : to make firm or secure **:** STRENGTHEN ⟨*consolidate* their hold on first place⟩

☆ SYNONYMS
Consider, study, contemplate, weigh mean to think about in order to arrive at a judgment or decision. CONSIDER may suggest giving thought to in order to reach a suitable conclusion, opinion, or decision ⟨refused even to *consider* my proposal⟩. STUDY implies sustained purposeful concentration and attention to details and minutiae ⟨*study* the plan closely⟩. CONTEMPLATE stresses focusing one's thoughts on something but does not imply coming to a conclusion or decision ⟨*contemplate* the consequences of refusing⟩. WEIGH implies attempting to reach the truth or arrive at a decision by balancing conflicting claims or evidence ⟨*weigh* the pros and cons of the case⟩.

3 : to form into a compact mass
intransitive verb
: to become consolidated; *specifically* **:** MERGE ⟨the two companies *consolidated*⟩
— con·sol·i·da·tor \-ˌdā-tər\ *noun*

consolidated school *noun* (1911)
: a public school formed by merging other schools

con·sol·i·da·tion \kən-ˌsä-lə-ˈdā-shən\ *noun* (15th century)
1 : the act or process of consolidating **:** the state of being consolidated
2 : the process of uniting **:** the quality or state of being united; *specifically* **:** the unification of two or more corporations by dissolution of existing ones and creation of a single new corporation
3 : pathological alteration of lung tissue from an aerated condition to one of solid consistency

con·som·mé \ˌkän(t)-sə-ˈmā\ *noun* [French, from past participle of *consommer* to complete, boil down, from Latin *consummare* to complete — more at CONSUMMATE] (1815)
: clear soup made from well-seasoned stock

con·so·nance \ˈkän(t)-s(ə-)nən(t)s\ *noun* (15th century)
1 : harmony or agreement among components
2 a : correspondence or recurrence of sounds especially in words; *specifically* **:** recurrence or repetition of consonants especially at the end of stressed syllables without the similar correspondence of vowels (as in the final sounds of "stroke" and "luck") **b :** CONCORD 1b **c :** SYMPATHETIC VIBRATION, RESONANCE

con·so·nan·cy \-s(ə-)nən(t)-sē\ *noun, plural* **-cies** (14th century)
: CONSONANCE 1

¹con·so·nant \ˈkän(t)-s(ə-)nənt\ *adjective* [Middle French, from Latin *consonant-, consonans,* present participle of *consonare* to sound together, agree, from *com-* + *sonare* to sound — more at SOUND] (15th century)
1 : being in agreement or harmony **:** free from elements making for discord
2 : marked by musical consonances
3 : having similar sounds ⟨*consonant* words⟩
4 : relating to or exhibiting consonance **:** RESONANT
— con·so·nant·ly *adverb*

²consonant *noun* [Middle English, from Latin *consonant-, consonans,* from present participle of *consonare*] (14th century)
: one of a class of speech sounds (as \p\, \g\, \n\, \l\, \s\, \r\) characterized by constriction or closure at one or more points in the breath channel; *also* **:** a letter representing a consonant — usually used in English of any letter except *a, e, i, o,* and *u*

con·so·nan·tal \ˌkän(t)-sə-ˈnan-tᵊl\ *adjective* (1795)
: relating to, being, or marked by a consonant or group of consonants

consonant shift *noun* (1888)
: a set of regular changes in consonant articulation in the history of a language or dialect: **a :** such a set affecting the Indo-European stops and distinguishing the Germanic languages from the other Indo-European languages — compare GRIMM'S LAW **b :** such a set affecting the Germanic stops and distinguishing High German from the other Germanic languages

¹con·sort \ˈkän-ˌsȯrt\ *noun* [Middle English, from Middle French, from Latin *consort-, consors* partner, sharer, from *com-* + *sort-, sors* lot, share — more at SERIES] (15th century)
1 : ASSOCIATE
2 : a ship accompanying another
3 : SPOUSE — compare PRINCE CONSORT

²consort *noun* [Middle French *consorte,* from *consort*] (1584)
1 : GROUP, ASSEMBLY ⟨a *consort* of specialists⟩
2 : CONJUNCTION, ASSOCIATION ⟨he ruled in *consort* with his father⟩
3 a : a group of singers or instrumentalists performing together **b :** a set of musical instruments of the same family

³con·sort \kən-ˈsȯrt, kän-ˈ, ˈkän-ˌ\ (1588)
transitive verb
1 : UNITE, ASSOCIATE
2 *obsolete* **:** ESCORT
intransitive verb
1 : to keep company ⟨*consorting* with criminals⟩
2 *obsolete* **:** to make harmony **:** PLAY
3 : ACCORD, HARMONIZE ⟨the illustrations *consort* admirably with the text —*Times Literary Supplement*⟩

con·sor·tium \kən-ˈsȯr-sh(ē-)əm, -ˈsȯr-tē-əm\ *noun, plural* **-sor·tia** \-ˈsȯr-sh(ē-)ə, -ˈsȯr-tē-ə\ *also* **-sortiums** [Latin, fellowship, from *consort-, consors*] (1829)
1 : an agreement, combination, or group (as of companies) formed to undertake an enterprise beyond the resources of any one member
2 : ASSOCIATION, SOCIETY
3 : the legal right of one spouse to the company, affection, and assistance of and to sexual relations with the other

con·spe·cif·ic \ˌkän(t)-spi-ˈsi-fik\ *adjective* (1859)
: of the same species
— conspecific *noun*

con·spec·tus \kən-ˈspek-təs\ *noun* [Latin, from *conspectus,* from *conspicere*] (circa 1837)
1 : a usually brief survey or summary (as of an extensive subject) often providing an overall view
2 : OUTLINE, SYNOPSIS

con·spi·cu·i·ty \ˌkän(t)-spə-ˈkyü-ə-tē\ *noun* (1601)
: the quality or state of being conspicuous **:** CONSPICUOUSNESS

con·spic·u·ous \kən-ˈspi-kyə-wəs\ *adjective* [Latin *conspicuus,* from *conspicere* to get sight of, from *com-* + *specere* to look — more at SPY] (1545)
1 : obvious to the eye or mind
2 : attracting attention **:** STRIKING
3 : marked by a noticeable violation of good taste
synonym see NOTICEABLE
— con·spic·u·ous·ly *adverb*
— con·spic·u·ous·ness \-nəs\ *noun*

conspicuous consumption *noun* (1899)
: lavish or wasteful spending thought to enhance social prestige

con·spir·a·cy \kən-ˈspir-ə-sē\ *noun, plural* **-cies** [Middle English *conspiracie,* from Latin *conspirare*] (14th century)
1 : the act of conspiring together
2 a : an agreement among conspirators **b :** a group of conspirators
synonym see PLOT

conspiracy of silence (1865)
: a secret agreement to keep silent about an occurrence, situation, or subject especially in order to promote or protect selfish interests

con·spi·ra·tion \ˌkän(t)-spə-ˈrā-shən, ˌkän-(ˌ)spi-ˈrā-\ *noun* (14th century)
1 : the act or action of plotting or secretly combining
2 : a joint effort toward a particular end
— con·spi·ra·tion·al \-shnəl, -shə-nᵊl\ *adjective*

con·spir·a·tor \kən-ˈspir-ə-tər\ *noun* (15th century)
: one that conspires **:** PLOTTER

con·spir·a·to·ri·al \kən-ˌspir-ə-ˈtōr-ē-əl, -ˈtȯr-\ *adjective* (1855)
: of, relating to, or suggestive of a conspiracy
— con·spir·a·to·ri·al·ly \-ē-ə-lē\ *adverb*

con·spire \kən-ˈspīr\ *verb* **con·spired; con·spir·ing** [Middle English, from Middle French *conspirer,* from Latin *conspirare* to be in harmony, conspire, from *com-* + *spirare* to breathe] (14th century)
transitive verb
: PLOT, CONTRIVE
intransitive verb
1 a : to join in a secret agreement to do an unlawful or wrongful act or an act which becomes unlawful as a result of the secret agreement **b :** SCHEME
2 : to act in harmony toward a common end ⟨circumstances *conspired* to defeat his efforts⟩

con spi·ri·to \kän-ˈspir-ə-ˌtō, kōn-\ *adverb* [Italian] (circa 1891)
: with spirit or animation — used as a direction in music

con·sta·ble \ˈkän(t)-stə-bəl, ˈkən(t)-\ *noun* [Middle English *conestable,* from Old French, from Late Latin *comes stabuli,* literally, officer of the stable] (13th century)
1 : a high officer of a medieval royal or noble household
2 : the warden or governor of a royal castle or a fortified town
3 a : a public officer usually of a town or township responsible for keeping the peace and for minor judicial duties **b** *British* **:** POLICE OFFICER; *especially* **:** one ranking below sergeant ◆

¹con·stab·u·lary \kən-ˈsta-byə-ˌler-ē\ *adjective* (1824)
: of or relating to a constable or constabulary

²constabulary *noun, plural* **-lar·ies** (1837)
1 : the organized body of constables of a particular district or country
2 : an armed police force organized on military lines but distinct from the regular army

con·stan·cy \ˈkän(t)-stən(t)-sē\ *noun, plural* **-cies** (15th century)
1 a : steadfastness of mind under duress **:** FORTITUDE **b :** FIDELITY, LOYALTY
2 : a state of being constant or unchanging

¹con·stant \ˈkän(t)-stənt\ *adjective* [Middle English, from Middle French, from Latin *constant-, constans,* from present participle of *constare* to stand firm, be consistent, from *com-* + *stare* to stand — more at STAND] (14th century)
1 : marked by firm steadfast resolution or faithfulness **:** exhibiting constancy of mind or attachment
2 : INVARIABLE, UNIFORM
3 : continually occurring or recurring **:** REGULAR
synonym see FAITHFUL, CONTINUAL
— con·stant·ly *adverb*

²constant *noun* (1832)

◇ WORD HISTORY
constable In the late Roman Empire, *comes stabuli,* literally, "count of the stable," became the title of various high-ranking civil and military officers whose duties extended far beyond the care of the imperial horses. (A word whose history shows a similar broadening of meaning is *marshal.*) Shrunk in early Medieval Latin to *constabulum,* it appears in Old French as *conestable,* which under the Capetian rulers of France could denote the sovereign's chief executive officer, the commander of an army, or the governor of a royal castle or domain. *Conestable* became the designation for the same functionaries in England after the Norman Conquest. In the 13th century, it was also applied to lower-ranking officers of the king's peace in English shires and towns. With the gradual reshaping of the constabulary in recent centuries into a police force, *constable* in Britain has now become virtually synonymous with "police officer." See in addition MARSHAL.

: something invariable or unchanging: as **a :** a number that has a fixed value in a given situation or universally or that is characteristic of some substance or instrument **b :** a number that is assumed not to change value in a given mathematical discussion **c :** a term in logic with a fixed designation

con·stan·tan \'kän(t)-stən-ˌtan\ *noun* [from the fact that its resistance remains constant under change of temperature] (1903)
: an alloy of copper and nickel used for electrical resistors and in thermocouples

con·sta·tive \kən-'stā-tiv, 'kän-stə-tiv\ *adjective* [*constate* to assert positively, from French *constater,* from Latin *constat* it is certain, 3d singular present indicative of *constare*] (1901)
1 : of, relating to, or being a verbal form that expresses past completed action
2 : being or relating to an utterance (as an assertion, question, or command) that is capable of being judged true or false
— constative *noun*

con·stel·late \'kän(t)-stə-ˌlāt\ *verb* **-lat·ed; -lat·ing** (1643)
transitive verb
1 : to unite in a cluster
2 : to set or adorn with or as if with constellations
intransitive verb
: CLUSTER

con·stel·la·tion \ˌkän(t)-stə-'lā-shən\ *noun* [Middle English *constellacioun,* from Middle French *constellation,* from Late Latin *constellation-, constellatio,* from Latin *com-* + *stella* star — more at STAR] (14th century)
1 : the configuration of stars especially at one's birth
2 : any of 88 arbitrary configurations of stars or an area of the celestial sphere covering one of these configurations
3 : an assemblage, collection, or gathering of usually related persons, qualities, or things ⟨a *constellation* of . . . relatives, friends, and hangers-on —Brendan Gill⟩
4 : PATTERN, ARRANGEMENT ⟨taking advantage of the shifting *constellation* of power throughout the known world —H. D. Lasswell⟩
— con·stel·la·to·ry \kən-'ste-lə-ˌtōr-ē, -ˌtȯr-\ *adjective*

con·ster·nate \'kän(t)-stər-ˌnāt\ *transitive verb* **-nat·ed; -nat·ing** (1651)
: to fill with consternation

con·ster·na·tion \ˌkän(t)-stər-'nā-shən\ *noun* [French or Latin; French, from Latin *consternation-, consternatio,* from *consternare* to throw into confusion, from *com-* + *-sternare,* probably from *sternere* to spread, strike down — more at STREW] (circa 1611)
: amazement or dismay that hinders or throws into confusion ⟨the two . . . stared at each other in *consternation,* and neither knew what to do —Pearl Buck⟩

con·sti·pate \'kän(t)-stə-ˌpāt\ *transitive verb* **-pat·ed; -pat·ing** [Medieval Latin *constipatus,* past participle of *constipare,* from Latin, to crowd together, from *com-* + *stipare* to pack tight — more at STIFF] (1533)
1 : to cause constipation in
2 : to make immobile, inactive, or dull **:** STULTIFY ⟨so much clutter . . . will tend to *constipate* the novel's working order —*Times Literary Supplement*⟩

con·sti·pat·ed \-ˌpā-təd\ *adjective* (1547)
1 : affected with constipation
2 : stilted or stodgy in appearance, expression, or action

con·sti·pa·tion \ˌkän(t)-stə-'pā-shən\ *noun* (15th century)
1 : abnormally delayed or infrequent passage of usually dry hardened feces
2 : STULTIFICATION

con·stit·u·en·cy \kən-'stich-wən(t)-sē, -'sti-chə-\ *noun, plural* **-cies** (1831)
1 a : a body of citizens entitled to elect a representative (as to a legislative or executive position) **b :** the residents in an electoral district **c :** an electoral district
2 a : a group or body that patronizes, supports, or offers representation ⟨creating . . . a grass-roots *constituency* for continuing the project —Fred Reed⟩ **b :** the people involved in or served by an organization (as a business or institution) ⟨regards its corporate customers as its prime *constituency* —Andrew Hacker⟩

¹con·stit·u·ent \-wənt\ *noun* [French *constituant,* from Middle French, from present participle of *constituer* to constitute, from Latin *constituere*] (1622)
1 : one who authorizes another to act as agent **:** PRINCIPAL
2 : a member of a constituency
3 : an essential part **:** COMPONENT, ELEMENT
4 : a structural unit of a definable syntactic, semantic, or phonological category that consists of one or more linguistic elements (as words, morphemes, or features) and that can occur as a component of a larger construction
synonym see ELEMENT

²constituent *adjective* [Latin *constituent-, constituens,* present participle of *constituere*] (1660)
1 : serving to form, compose, or make up a unit or whole **:** COMPONENT
2 : having the power to create a government or frame or amend a constitution ⟨a *constituent* assembly⟩
— con·stit·u·ent·ly *adverb*

con·sti·tute \'kän(t)-stə-ˌtüt, -ˌtyüt\ *transitive verb* **-tut·ed; -tut·ing** [Middle English, from Latin *constitutus,* past participle of *constituere* to set up, constitute, from *com-* + *statuere* to set — more at STATUTE] (15th century)
1 : to appoint to an office, function, or dignity
2 : SET UP, ESTABLISH: as **a :** ENACT **b :** FOUND **c** (1) **:** to give due or lawful form to (2) **:** to legally process
3 : MAKE UP, FORM, COMPOSE ⟨12 months *constitute* a year⟩ ⟨high school dropouts who *constitute* a major problem in large city slums —J. B. Conant⟩

con·sti·tu·tion \ˌkän(t)-stə-'tü-shən, -'tyü-\ *noun* (14th century)
1 : an established law or custom **:** ORDINANCE
2 a : the physical makeup of the individual comprising inherited qualities modified by environment **b :** the structure, composition, physical makeup, or nature of something
3 : the act of establishing, making, or setting up
4 : the mode in which a state or society is organized; *especially* **:** the manner in which sovereign power is distributed
5 a : the basic principles and laws of a nation, state, or social group that determine the powers and duties of the government and guarantee certain rights to the people in it **b :** a written instrument embodying the rules of a political or social organization
— con·sti·tu·tion·less \-ləs\ *adjective*

¹con·sti·tu·tion·al \-shnəl, -shə-nᵊl\ *adjective* (1682)
1 : relating to, inherent in, or affecting the constitution of body or mind
2 : of, relating to, or entering into the fundamental makeup of something **:** ESSENTIAL
3 : being in accordance with or authorized by the constitution of a state or society ⟨a *constitutional* government⟩
4 : regulated by or ruling according to a constitution ⟨a *constitutional* monarchy⟩
5 : of or relating to a constitution
6 : loyal to or supporting an established constitution or form of government

²constitutional *noun* (1829)
: a walk taken for one's health

con·sti·tu·tion·al·ism \-sh(ə-)nə-ˌli-zəm\ *noun* (1832)
: adherence to or government according to constitutional principles; *also* **:** a constitutional system of government
— con·sti·tu·tion·al·ist \-list\ *noun*

con·sti·tu·tion·al·i·ty \-ˌtü-shə-'na-lə-tē, -ˌtyü-\ *noun* (1787)
: the quality or state of being constitutional; *especially* **:** accordance with the provisions of a constitution ⟨questioned the *constitutionality* of the law⟩

con·sti·tu·tion·al·ize \-'tü-shnə-ˌlīz, -shə-nᵊl-ˌīz\ *transitive verb* **-ized; -iz·ing** (1831)
: to provide with a constitution **:** organize along constitutional principles
— con·sti·tu·tion·al·i·za·tion \-ˌtü-shnə-lə-'zā-shən, -shə-nᵊl-ə-\ *noun*

con·sti·tu·tion·al·ly \-'tü-shnə-lē, -'tyü-, -shə-nᵊl-ē\ *adverb* (1742)
1 a : in accordance with one's constitution ⟨*constitutionally* unable to grasp subtleties⟩ **b :** in structure, composition, or constitution ⟨despite repeated heatings the material remained *constitutionally* the same⟩
2 : in accordance with a political constitution ⟨was not *constitutionally* eligible to fill the office⟩

con·sti·tu·tive \'kän(t)-stə-ˌtü-tiv, -ˌtyü-; kən-'sti-chə-tiv\ *adjective* (1592)
1 : having the power to enact or establish **:** CONSTRUCTIVE
2 : CONSTITUENT, ESSENTIAL
3 : relating to or dependent on constitution ⟨a *constitutive* property of all electrolytes⟩
— con·sti·tu·tive·ly *adverb*

con·strain \kən-'strān\ *transitive verb* [Middle English, from Middle French *constraindre,* from Latin *constringere* to constrict, constrain, from *com-* + *stringere* to draw tight — more at STRAIN] (14th century)
1 a : to force by imposed stricture, restriction, or limitation **b :** to restrict the motion of (a mechanical body) to a particular mode
2 : COMPRESS; *also* **:** to clasp tightly
3 : to secure by or as if by bonds **:** CONFINE; *broadly* **:** LIMIT
4 : to force or produce in an unnatural or strained manner ⟨a *constrained* smile⟩
5 : to hold back by or as if by force ⟨*constraining* my mind not to wander from the task —Charles Dickens⟩
synonym see FORCE
— con·strained·ly \-'strā-nəd-lē, -'strānd-lē\ *adverb*

con·straint \kən-'strānt\ *noun* [Middle English, from Middle French *constrainte,* from *constraindre*] (15th century)
1 a : the act of constraining **b :** the state of being checked, restricted, or compelled to avoid or perform some action ⟨the *constraint* and monotony of a monastic life —Matthew Arnold⟩ **c :** a constraining condition, agency, or force **:** CHECK ⟨put legal *constraints* on the board's activities⟩
2 a : repression of one's own feelings, behavior, or actions **b :** a sense of being constrained **:** EMBARRASSMENT

con·strict \kən-'strikt\ *verb* [Latin *constrictus,* past participle of *constringere*] (1732)
transitive verb
1 a : to make narrow or draw together **b :** COMPRESS, SQUEEZE ⟨*constrict* a nerve⟩
2 : to stultify, stop, or cause to falter **:** INHIBIT
intransitive verb
: to become constricted
synonym see CONTRACT
— con·stric·tive \-'strik-tiv\ *adjective*

con·stric·tion \-'strik-shən\ *noun* (15th century)
1 : an act or product of constricting
2 : the quality or state of being constricted
3 : something that constricts

con·stric·tor \-'strik-tər\ *noun* (1735)
1 : a muscle that contracts a cavity or orifice or compresses an organ
2 : a snake (as a boa constrictor) that kills prey by compression in its coils
3 : one that constricts

con·stringe \kən-'strinj\ *transitive verb* **con·stringed; con·string·ing** [Latin *constringere*] (1604)

1 : to cause to shrink ⟨cold *constringes* the pores⟩
2 : CONSTRICT
— **con·strin·gent** \-'strin-jənt\ *adjective*

[1]**con·struct** \kən-'strəkt\ *transitive verb* [Latin *constructus,* past participle of *construere,* from *com-* + *struere* to build — more at STRUCTURE] (1663)
1 : to make or form by combining or arranging parts or elements **:** BUILD; *also* **:** CONTRIVE, DEVISE
2 : to draw (a geometrical figure) with suitable instruments and under specified conditions
3 : to set in logical order
— **con·struct·ible** \-'strək-tə-bəl\ *adjective*
— **con·struc·tor** \-tər\ *noun*

[2]**con·struct** \'kän-ˌstrəkt\ *noun* (1933)
1 : something constructed by the mind: as **a :** a theoretical entity ⟨the deductive study of abstract *constructs* —Daniel J. Boorstin⟩ **b :** a working hypothesis or concept ⟨a point of view which made "abroad" singularly containable as a literary *construct* —Jonathan Raban⟩ **c :** a product of mental invention ⟨the novel . . . a verbal *construct* in which invented human characters appear —Anthony Burgess⟩
2 : something produced by human effort ⟨the East bloc was always an unnatural *construct* —Walter Isaacson⟩

con·struc·tion \kən-'strək-shən\ *noun* (14th century)
1 : the act or result of construing, interpreting, or explaining
2 a : the process, art, or manner of constructing something; *also* **:** a thing constructed **b :** the construction industry ⟨working in *construction*⟩
3 : the arrangement and connection of words or groups of words in a sentence **:** syntactical arrangement
4 : a sculpture that is put together out of separate pieces of often disparate materials
— **con·struc·tion·al** \-shnəl, -shə-nᵊl\ *adjective*
— **con·struc·tion·al·ly** *adverb*

con·struc·tion·ist \-sh(ə-)nist\ *noun* (1838)
: one who construes a legal document (as the U.S. Constitution) in a specific way ⟨a strict *constructionist*⟩

construction paper *noun* (circa 1924)
: a thick groundwood paper available in many colors and used especially for school artwork

con·struc·tive \kən-'strək-tiv\ *adjective* (circa 1680)
1 : declared such by judicial construction or interpretation ⟨*constructive* fraud⟩
2 : of or relating to construction or creation
3 : promoting improvement or development ⟨*constructive* criticism⟩
— **con·struc·tive·ly** *adverb*
— **con·struc·tive·ness** *noun*

con·struc·tiv·ism \kən-'strək-ti-ˌvi-zəm\ *noun, often capitalized* (1925)
: a nonobjective art movement originating in Russia and concerned with formal organization of planes and expression of volume in terms of modern industrial materials (as glass and plastic)
— **con·struc·tiv·ist** \-vist\ *adjective or noun, often capitalized*

[1]**con·strue** \kən-'strü\ *verb* **con·strued; con·stru·ing** [Middle English, from Late Latin *construere,* from Latin, to construct] (14th century)
transitive verb
1 : to analyze the arrangement and connection of words in (a sentence or sentence part)
2 : to understand or explain the sense or intention of usually in a particular way or with respect to a given set of circumstances ⟨*construed* my actions as hostile⟩
intransitive verb
: to construe a sentence or sentence part especially in connection with translating
— **con·stru·able** \-'strü-ə-bəl\ *adjective*

[2]**con·strue** \'kän-ˌstrü\ *noun* (1844)
: an act or the result of construing especially by piecemeal translation

con·sub·stan·tial \ˌkän(t)-səb-'stan(t)-shəl\ *adjective* [Late Latin *consubstantialis,* from Latin *com-* + *substantia* substance] (14th century)
: of the same substance

con·sub·stan·ti·a·tion \ˌkän(t)-səb-ˌstan(t)-shē-'ā-shən\ *noun* (1597)
: the actual substantial presence and combination of the body and blood of Christ with the eucharistic bread and wine according to a teaching associated with Martin Luther — compare TRANSUBSTANTIATION

con·sue·tude \'kän(t)-swi-ˌtüd, kən-'sü-ə-, -ˌtyüd\ *noun* [Middle English, from Latin *consuetudo* — more at CUSTOM] (14th century)
: social usage **:** CUSTOM
— **con·sue·tu·di·nary** \ˌkän(t)-swi-'tü-də-ˌner-ē, kən-ˌsü-ə-, -'tyü-\ *adjective*

con·sul \'kän(t)-səl\ *noun* [Middle English, from Latin; perhaps akin to Latin *consulere* to consult] (14th century)
1 a : either of two annually elected chief magistrates of the Roman republic **b :** one of three chief magistrates of the French republic from 1799 to 1804
2 : an official appointed by a government to reside in a foreign country to represent the commercial interests of citizens of the appointing country
— **con·sul·ar** \-s(ə-)lər\ *adjective*
— **con·sul·ship** \-səl-ˌship\ *noun*

con·sul·ate \-s(ə-)lət\ *noun* (14th century)
1 : a government by consuls
2 : the office, term of office, or jurisdiction of a consul
3 : the residence or official premises of a consul

consulate general *noun, plural* **consulates general** (1883)
: the residence, office, or jurisdiction of a consul general

consul general *noun, plural* **consuls general** (1753)
: a consul of the first rank stationed in an important place or having jurisdiction in several places or over several consuls

[1]**con·sult** \kən-'səlt\ *verb* [Middle French or Latin; Middle French *consulter,* from Latin *consultare,* frequentative of *consulere* to deliberate, counsel, consult] (1527)
transitive verb
1 : to have regard to **:** CONSIDER
2 a : to ask the advice or opinion of ⟨*consult* a doctor⟩ **b :** to refer to ⟨*consult* a dictionary⟩
intransitive verb
1 : to consult an individual
2 : to deliberate together **:** CONFER
3 : to serve as a consultant
— **con·sult·er** *noun*

[2]**con·sult** \kən-'səlt, 'kän-ˌ\ *noun* (1560)
: CONSULTATION

con·sul·tan·cy \kən-'səl-tᵊn(t)-sē\ *noun, plural* **-cies** (1955)
1 : CONSULTATION
2 : an agency that provides consulting services
3 : the position of a consultant

con·sul·tant \kən-'səl-tᵊnt\ *noun* (1697)
1 : one who consults another
2 : one who gives professional advice or services **:** EXPERT
— **con·sul·tant·ship** \-ˌship\ *noun*

con·sul·ta·tion \ˌkän(t)-səl-'tā-shən\ *noun* (15th century)
1 : COUNCIL, CONFERENCE; *specifically* **:** a deliberation between physicians on a case or its treatment
2 : the act of consulting or conferring

con·sul·ta·tive \kən-'səl-tə-tiv, 'kän(t)-səl-ˌtā-tiv\ *adjective* (1583)
: of, relating to, or intended for consultation **:** ADVISORY ⟨*consultative* committee⟩

con·sult·ing \kən-'səl-tiŋ\ *adjective* (1801)
1 : providing professional or expert advice ⟨a *consulting* architect⟩
2 : of or relating to consultation or a consultant ⟨the *consulting* room of a psychiatrist⟩

con·sul·tive \kən-'səl-tiv\ *adjective* (1616)
: CONSULTATIVE

con·sul·tor \kən-'səl-tər\ *noun* (1611)
: one that consults or advises; *especially* **:** an adviser to a Roman Catholic bishop, provincial, or sacred congregation

[1]**con·sum·able** \kən-'sü-mə-bəl\ *adjective* (1641)
: capable of being consumed

[2]**consumable** *noun* (1802)
: something (as food or fuel) that is consumable — usually used in plural

con·sume \kən-'süm\ *verb* **con·sumed; con·sum·ing** [Middle English, from Middle French or Latin; Middle French *consumer,* from Latin *consumere,* from *com-* + *sumere* to take up, take, from *sub-* up + *emere* to take — more at SUB-, REDEEM] (14th century)
transitive verb
1 : to do away with completely **:** DESTROY ⟨fire *consumed* several buildings⟩
2 a : to spend wastefully **:** SQUANDER **b :** USE UP ⟨writing *consumed* much of his time⟩
3 : to eat or drink especially in great quantity ⟨*consumed* several kegs of beer⟩
4 : to engage fully **:** ENGROSS ⟨*consumed* with curiosity⟩
intransitive verb
1 : to waste or burn away **:** PERISH
2 : to utilize economic goods

con·sum·ed·ly \-'sü-məd-lē\ *adverb* (1707)
: as if consumed **:** EXCESSIVELY

con·sum·er \kən-'sü-mər\ *noun, often attributive* (15th century)
: one that consumes: as **a :** one that utilizes economic goods **b :** an organism requiring complex organic compounds for food which it obtains by preying on other organisms or by eating particles of organic matter — compare PRODUCER 4
— **con·sum·er·ship** \-ˌship\ *noun*

consumer credit *noun* (1927)
: credit granted to an individual especially to finance the purchase of consumer goods or to defray personal expenses

consumer goods *noun plural* (1890)
: goods that directly satisfy human wants

con·sum·er·ism \kən-'sü-mə-ˌri-zəm\ *noun* (1944)
1 : the promotion of the consumer's interests
2 : the theory that an increasing consumption of goods is economically desirable; *also* **:** a preoccupation with and an inclination toward the buying of consumer goods
— **con·sum·er·ist** \-rist\ *noun*
— **con·sum·er·is·tic** \kən-ˌsü-mə-'ris-tik\ *adjective*

consumer price index *noun* (1948)
: an index measuring the change in the cost of typical wage-earner purchases of goods and services expressed as a percentage of the cost of these same goods and services in some base period — called also *cost-of-living index*

con·sum·ing \kən-'sü-miŋ\ *adjective* (1920)
: deeply felt **:** ARDENT ⟨a *consuming* interest⟩; *also* **:** ENGROSSING

[1]**con·sum·mate** \'kän(t)-sə-mət, kən-'sə-mət\ *adjective* [Middle English *consummat* fulfilled, from Latin *consummatus,* past participle of *consummare* to sum up, finish, from *com-* + *summa* sum] (1527)
1 : complete in every detail **:** PERFECT
2 : extremely skilled and accomplished ⟨a *consummate* liar⟩
3 : of the highest degree ⟨*consummate* skill⟩ ⟨*consummate* cruelty⟩

— **con·sum·mate·ly** *adverb*

[2]con·sum·mate \'kän(t)-sə-ˌmāt\ *verb* **-mat·ed; -mat·ing** (1530)
transitive verb
1 a : FINISH, COMPLETE ⟨*consummate* a business deal⟩ **b :** to make perfect **c :** ACHIEVE
2 : to make (marital union) complete by sexual intercourse ⟨*consummate* a marriage⟩
intransitive verb
: to become perfected
— **con·sum·ma·tive** \'kän(t)-sə-ˌmā-tiv, kən-'sə-mə-tiv\ *adjective*
— **con·sum·ma·tor** \'kän(t)-sə-ˌmāt-ər\ *noun*

con·sum·ma·tion \ˌkän-sə-'mā-shən\ *noun* (14th century)
1 : the act of consummating ⟨the *consummation* of a contract by mutual signature⟩; *specifically* **:** the consummating of a marriage
2 : the ultimate end **:** FINISH

con·sum·ma·to·ry \kən-'sə-mə-ˌtōr-ē, -ˌtȯr-\ *adjective* (1648)
1 : of or relating to consummation **:** CONCLUDING
2 : of, relating to, or being a response or act (as eating or copulating) that terminates a period of usually goal-directed behavior

con·sump·tion \kən-'səm(p)-shən\ *noun* [Middle English *consumpcioun,* from Latin *consumption-, consumptio,* from *consumere*] (14th century)
1 a : a progressive wasting away of the body especially from pulmonary tuberculosis **b :** TUBERCULOSIS
2 : the act or process of consuming
3 : the utilization of economic goods in the satisfaction of wants or in the process of production resulting chiefly in their destruction, deterioration, or transformation

[1]con·sump·tive \-'səm(p)-tiv\ *adjective* (1664)
1 : tending to consume
2 : of, relating to, or affected with consumption
— **con·sump·tive·ly** *adverb*

[2]consumptive *noun* (1666)
: a person affected with consumption

[1]con·tact \'kän-ˌtakt\ *noun* [French or Latin; French, from Latin *contactus,* from *contingere* to have contact with — more at CONTINGENT] (1626)
1 a : union or junction of surfaces **b :** the apparent touching or mutual tangency of the limbs of two celestial bodies or of the disk of one body with the shadow of another during an eclipse, transit, or occultation **c** (1) **:** the junction of two electrical conductors through which a current passes (2) **:** a special part made for such a junction
2 a : ASSOCIATION, RELATIONSHIP **b :** CONNECTION, COMMUNICATION **c :** an establishing of communication with someone or an observing or receiving of a significant signal from a person or object ⟨radar *contact* with Mars⟩
3 : a person serving as a go-between, messenger, connection, or source of special information ⟨business *contacts*⟩
4 : CONTACT LENS

[2]con·tact \'kän-ˌtakt, kən-'\ (1834)
intransitive verb
: to make contact
transitive verb
1 : to bring into contact
2 a : to enter or be in contact with **:** JOIN **b :** to get in communication with ⟨*contact* your local dealer⟩ ■

[3]con·tact \'kän-ˌtakt\ *adjective* (1859)
: maintaining, involving, or activated or caused by contact ⟨*contact* poisons⟩ ⟨*contact* sports⟩

contact binary *noun* (1952)
: a binary star system in which the two stars are close enough together for material to pass between them

contact hitter *noun* (1982)
: a hitter in baseball who seldom strikes out

contact inhibition *noun* (1965)
: cessation of cellular undulating movements upon contact with other cells with accompanying cessation of cell growth and division

contact language *noun* (1950)
: PIDGIN

contact lens *noun* (1888)
: a thin lens designed to fit over the cornea and usually worn to correct defects in vision

contact print *noun* (1890)
: a photographic print made with the negative in contact with the sensitized paper, plate, or film

con·ta·gion \kən-'tā-jən\ *noun* [Middle English, from Middle French & Latin; Middle French, from Latin *contagion-, contagio,* from *contingere* to have contact with, pollute] (14th century)
1 a : a contagious disease **b :** the transmission of a disease by direct or indirect contact **c :** a disease-producing agent (as a virus)
2 a : POISON **b :** contagious influence, quality, or nature **c :** corrupting influence or contact
3 a : rapid communication of an influence (as a doctrine or emotional state) **b :** an influence that spreads rapidly

con·ta·gious \-jəs\ *adjective* (14th century)
1 : communicable by contact **:** CATCHING ⟨*contagious* diseases⟩
2 : bearing contagion
3 : used for contagious diseases ⟨a *contagious* ward⟩
4 : exciting similar emotions or conduct in others ⟨*contagious* enthusiasm⟩
— **con·ta·gious·ly** *adverb*
— **con·ta·gious·ness** *noun*

contagious abortion *noun* (1910)
: a contagious or infectious disease (as a brucellosis) of domestic animals characterized by abortion

con·ta·gium \kən-'tā-j(ē-)əm\ *noun, plural* **-gia** \-j(ē-)ə\ [Latin, contagion, from *contingere*] (1870)
: a virus or living organism capable of causing a communicable disease

con·tain \kən-'tān\ *verb* [Middle English *conteinen,* from Old French *contenir,* from Latin *continēre* to hold together, hold in, contain, from *com-* + *tenēre* to hold — more at THIN] (14th century)
transitive verb
1 : to keep within limits: as **a :** RESTRAIN, CONTROL **b :** CHECK, HALT **c :** to follow successfully a policy of containment toward **d :** to prevent (as an enemy or opponent) from advancing or from making a successful attack
2 a : to have within **:** HOLD **b :** COMPRISE, INCLUDE
3 a : to be divisible by usually without a remainder **b :** ENCLOSE, BOUND
intransitive verb
: to restrain oneself ☆
— **con·tain·able** \-'tā-nə-bəl\ *adjective*

contained *adjective* (1653)
: RESTRAINED; *also* **:** CALM

con·tain·er \kən-'tā-nər\ *noun* (15th century)
: one that contains; *especially* **:** a receptacle (as a box or jar) for holding goods
— **con·tain·er·less** \-ləs\ *adjective*

con·tain·er·board \-ˌbōrd, -ˌbȯrd\ *noun* (circa 1924)
: corrugated or solid paperboard used for making containers

con·tain·er·i·sa·tion, con·tain·er·ise *British variant of* CONTAINERIZATION, CONTAINERIZE

con·tain·er·i·za·tion \kən-ˌtā-nə-rə-'zā-shən\ *noun* (1956)
: a shipping method in which a large amount of material (as merchandise) is packaged into large standardized containers

con·tain·er·ize \kən-'tā-nə-ˌrīz\ *transitive verb* **-ized; -iz·ing** (1956)
1 : to ship by containerization
2 : to pack in containers

con·tain·er·port \-nər-ˌpōrt, -ˌpȯrt\ *noun* (1970)
: a shipping port specially equipped to handle containerized cargo

con·tain·er·ship \-nər-ˌship\ *noun* (1966)
: a ship specially designed or equipped for carrying containerized cargo

con·tain·ment \kən-'tān-mənt\ *noun* (1655)
1 : the act, process, or means of containing
2 : the policy, process, or result of preventing the expansion of a hostile power or ideology

con·tam·i·nant \kən-'ta-mə-nənt\ *noun* (1922)
: something that contaminates

con·tam·i·nate \kən-'ta-mə-ˌnāt\ *transitive verb* **-nat·ed; -nat·ing** [Middle English, from Latin *contaminatus,* past participle of *contaminare;* akin to Latin *contagio* contagion] (15th century)
1 a : to soil, stain, corrupt, or infect by contact or association ⟨bacteria *contaminated* the wound⟩ **b :** to make inferior or impure by admixture ⟨iron *contaminated* with phosphorus⟩
2 : to make unfit for use by the introduction of unwholesome or undesirable elements ☆
— **con·tam·i·na·tive** \-ˌnā-tiv\ *adjective*
— **con·tam·i·na·tor** \-ˌnā-tər\ *noun*

con·tam·i·na·tion \kən-ˌta-mə-'nā-shən\ *noun* (15th century)
1 : a process of contaminating **:** a state of being contaminated
2 : CONTAMINANT

conte \'kō^{n}(n)t\ *noun* [French] (1891)
: a usually short tale of adventure

con·temn \kən-'tem\ *transitive verb* [Middle English *contempnen,* from Middle French *contempner,* from Latin *contemnere,* from *com-* + *temnere* to despise] (15th century)
: to view or treat with contempt **:** SCORN
synonym see DESPISE
— **con·tem·ner** *also* **con·tem·nor** \-'tem-nər, -'te-mər\ *noun*

☆ SYNONYMS

Contain, hold, accommodate mean to have or be capable of having within. CONTAIN implies the actual presence of a specified substance or quantity within something ⟨the can *contains* a quart of oil⟩. HOLD implies the capacity of containing or the usual or permanent function of containing or keeping ⟨the bookcase will *hold* all my textbooks⟩. ACCOMMODATE stresses holding without crowding or inconvenience ⟨the hall can *accommodate* 500 people⟩.

Contaminate, taint, pollute, defile mean to make impure or unclean. CONTAMINATE implies intrusion of or contact with dirt or foulness from an outside source ⟨water *contaminated* by industrial wastes⟩. TAINT stresses the loss of purity or cleanliness that follows contamination ⟨*tainted* meat⟩ ⟨a politician's *tainted* reputation⟩. POLLUTE, sometimes interchangeable with *contaminate,* distinctively may imply that the process which begins with contamination is complete and that what was pure or clean has been made foul, poisoned, or filthy ⟨the *polluted* waters of the river⟩. DEFILE implies befouling of what could or should have been kept clean and pure or held sacred and commonly suggests violation or desecration ⟨*defile* a hero's memory with slanderous innuendo⟩.

□ USAGE

contact The use of *contact* as a verb, especially in sense 2b, is accepted as standard by almost all commentators except those who write college handbooks.

con·tem·plate \'kän-təm-ˌplāt, -ˌtem-\ *verb* **-plat·ed; -plat·ing** [Latin *contemplatus,* past participle of *contemplari,* from *com-* + *templum* space marked out for observation of auguries — more at TEMPLE] (1537)
transitive verb
1 : to view or consider with continued attention **:** meditate on
2 : to view as contingent or probable or as an end or intention
intransitive verb
: PONDER, MEDITATE
synonym see CONSIDER
— **con·tem·pla·tor** \-ˌplā-tər\ *noun*

con·tem·pla·tion \ˌkän-təm-'plā-shən, -ˌtem-\ *noun* (13th century)
1 a : concentration on spiritual things as a form of private devotion **b :** a state of mystical awareness of God's being
2 : an act of considering with attention **:** STUDY
3 : the act of regarding steadily
4 : INTENTION, EXPECTATION

[1]con·tem·pla·tive \kən-'tem-plə-tiv; 'kän-təm-ˌplā-, -ˌtem-\ *adjective* (14th century)
: marked by or given to contemplation; *specifically* **:** of or relating to a religious order devoted to prayer and penance
— **con·tem·pla·tive·ly** *adverb*
— **con·tem·pla·tive·ness** *noun*

[2]contemplative *noun* (14th century)
: a person who practices contemplation

con·tem·po·ra·ne·i·ty \kən-ˌtem-p(ə-)rə-'nē-ə-tē, -'nā-\ *noun* (1772)
: the quality or state of being contemporaneous or contemporary

con·tem·po·ra·ne·ous \kən-ˌtem-pə-'rā-nē-əs\ *adjective* [Latin *contemporaneus,* from *com-* + *tempor-, tempus* time] (circa 1656)
: existing, occurring, or originating during the same time
synonym see CONTEMPORARY
— **con·tem·po·ra·ne·ous·ly** *adverb*
— **con·tem·po·ra·ne·ous·ness** *noun*

[1]con·tem·po·rary \kən-'tem-pə-ˌrer-ē\ *adjective* [*com-* + Latin *tempor-, tempus*] (1631)
1 : happening, existing, living, or coming into being during the same period of time
2 a : SIMULTANEOUS **b :** marked by characteristics of the present period **:** MODERN, CURRENT ☆
— **con·tem·po·rar·i·ly** \-ˌtem-pə-'rer-ə-lē\ *adverb*

[2]contemporary *noun, plural* **-rar·ies** (1646)
1 : one that is contemporary with another
2 : one of the same or nearly the same age as another

con·tem·po·rize \kən-'tem-pə-ˌrīz\ *transitive verb* **-rized; -riz·ing** (1646)
: to make contemporary

con·tempt \kən-'tem(p)t\ *noun* [Middle English, from Latin *contemptus,* from *contemnere*] (14th century)
1 a : the act of despising **:** the state of mind of one who despises **:** DISDAIN **b :** lack of respect or reverence for something
2 : the state of being despised
3 : willful disobedience to or open disrespect of a court, judge, or legislative body ⟨*contempt* of court⟩

con·tempt·ible \kən-'tem(p)-tə-bəl\ *adjective* (14th century)
1 : worthy of contempt
2 *obsolete* **:** SCORNFUL, CONTEMPTUOUS ☆
— **con·tempt·i·bil·i·ty** \-ˌtem(p)-tə-'bi-lə-tē\ *noun*
— **con·tempt·ible·ness** *noun*
— **con·tempt·ibly** \-'tem(p)-tə-blē\ *adverb*

con·temp·tu·ous \-'tem(p)-chə-wəs, -chəs, -shwəs\ *adjective* [Latin *contemptus*] (1595)
: manifesting, feeling, or expressing contempt
— **con·temp·tu·ous·ly** *adverb*
— **con·temp·tu·ous·ness** *noun*

con·tend \kən-'tend\ *verb* [Middle English, from Middle French or Latin; Middle French *contendre,* from Latin *contendere,* from *com-* + *tendere* to stretch — more at THIN] (15th century)
intransitive verb
1 : to strive or vie in contest or rivalry or against difficulties **:** STRUGGLE
2 : to strive in debate **:** ARGUE
transitive verb
1 : MAINTAIN, ASSERT ⟨*contended* that he was right⟩
2 : to struggle for **:** CONTEST

con·tend·er \-'ten-dər\ *noun* (1547)
: one that contends; *especially* **:** a competitor for a championship or high honor ⟨a heavyweight title *contender*⟩

[1]con·tent \kən-'tent\ *adjective* [Middle English, from Middle French, from Latin *contentus,* from past participle of *continēre* to hold in, contain — more at CONTAIN] (15th century)
: CONTENTED, SATISFIED

[2]content *transitive verb* (15th century)
1 : to appease the desires of
2 : to limit (oneself) in requirements, desires, or actions

[3]content *noun* (1579)
: CONTENTMENT ⟨ate to his heart's *content*⟩

[4]con·tent \'kän-ˌtent\ *noun* [Middle English, from Latin *contentus,* past participle of *continēre* to contain] (15th century)
1 a : something contained — usually used in plural ⟨the jar's *contents*⟩ ⟨the drawer's *contents*⟩ **b :** the topics or matter treated in a written work ⟨table of *contents*⟩
2 a : SUBSTANCE, GIST **b :** MEANING, SIGNIFICANCE **c :** the events, physical detail, and information in a work of art — compare FORM 10b
3 a : the matter dealt with in a field of study **b :** a part, element, or complex of parts
4 : the amount of specified material contained **:** PROPORTION

content analysis *noun* (1945)
: analysis of the manifest and latent content of a body of communicated material (as a book or film) through a classification, tabulation, and evaluation of its key symbols and themes in order to ascertain its meaning and probable effect

con·tent·ed \kən-'ten-təd\ *adjective* (1526)
: feeling or manifesting satisfaction with one's possessions, status, or situation ⟨a *contented* smile⟩
— **con·tent·ed·ly** *adverb*
— **con·tent·ed·ness** *noun*

con·ten·tion \kən-'ten(t)-shən\ *noun* [Middle English *contencioun,* from Middle French, from Latin *contention-, contentio,* from *contendere*] (14th century)
1 : an act or instance of contending
2 : a point advanced or maintained in a debate or argument
3 : RIVALRY, COMPETITION
synonym see DISCORD

con·ten·tious \kən-'ten(t)-shəs\ *adjective* (15th century)
1 : likely to cause contention ⟨a *contentious* argument⟩
2 : exhibiting an often perverse and wearisome tendency to quarrels and disputes ⟨a man of a most *contentious* nature⟩
synonym see BELLIGERENT
— **con·ten·tious·ly** *adverb*
— **con·ten·tious·ness** *noun*

con·tent·ment \kən-'tent-mənt\ *noun* (15th century)
1 : the quality or state of being contented
2 : something that contents

con·tent word \'kän-ˌtent-\ *noun* (1940)
: a word that primarily expresses lexical meaning — compare FUNCTION WORD

con·ter·mi·nous \kən-'tər-mə-nəs, kän-\ *adjective* [Latin *conterminus,* from *com-* + *terminus* boundary — more at TERM] (1631)
1 : having a common boundary
2 : COTERMINOUS
3 : enclosed within one common boundary ⟨the 48 *conterminous* states⟩
— **con·ter·mi·nous·ly** *adverb*

[1]con·test \kən-'test, 'kän-ˌ\ *verb* [Middle French *contester,* from Latin *contestari* (*litem*) to bring an action at law, from *contestari* to call to witness, from *com-* + *testis* witness — more at TESTAMENT] (1603)
intransitive verb
: STRIVE, VIE
transitive verb
: to make the subject of dispute, contention, or litigation; *especially* **:** DISPUTE, CHALLENGE
— **con·test·able** \-'tes-tə-bəl\ *adjective*
— **con·test·er** *noun*

[2]con·test \'kän-ˌtest\ *noun* (1647)
1 : a struggle for superiority or victory **:** COMPETITION
2 : a competition in which each contestant performs without direct contact with or interference from his competitors

con·tes·tant \kən-'tes-tənt *also* 'kän-ˌ\ *noun* (1665)
1 : one that participates in a contest
2 : one that contests an award or decision

con·tes·ta·tion \ˌkän-ˌtes-'tā-shən\ *noun* (1580)
: CONTROVERSY, DEBATE

con·text \'kän-ˌtekst\ *noun* [Middle English, weaving together of words, from Latin *contextus* connection of words, coherence, from *contexere* to weave together, from *com-* + *texere* to weave — more at TECHNICAL] (circa 1568)
1 : the parts of a discourse that surround a word or passage and can throw light on its meaning
2 : the interrelated conditions in which something exists or occurs **:** ENVIRONMENT, SETTING
— **con·text·less** \-ˌtekst-ləs\ *adjective*

☆ SYNONYMS

Contemporary, contemporaneous, coeval, synchronous, simultaneous, coincident mean existing or occurring at the same time. CONTEMPORARY is likely to apply to people and what relates to them ⟨Abraham Lincoln was *contemporary* with Charles Darwin⟩. CONTEMPORANEOUS applies to events ⟨*contemporaneous* accounts of the kidnapping⟩. COEVAL refers usually to periods, ages, eras, eons ⟨two stars thought to be *coeval*⟩. SYNCHRONOUS implies exact correspondence in time and especially in periodic intervals ⟨*synchronous* timepieces⟩. SIMULTANEOUS implies correspondence in a moment of time ⟨the two shots were *simultaneous*⟩. COINCIDENT is applied to events and may be used in order to avoid implication of causal relationship ⟨the end of World War II was *coincident* with a great vintage year⟩.

Contemptible, despicable, pitiable, sorry, scurvy mean arousing or deserving scorn. CONTEMPTIBLE may imply any quality provoking scorn or a low standing in any scale of values ⟨a *contemptible* liar⟩. DESPICABLE may imply utter worthlessness and usually suggests arousing an attitude of moral indignation ⟨a *despicable* crime⟩. PITIABLE applies to what inspires mixed contempt and pity ⟨a *pitiable* attempt at tragedy⟩. SORRY may stress pitiable inadequacy or may suggest wretchedness or sordidness ⟨this rattletrap is a *sorry* excuse for a car⟩. SCURVY adds to DESPICABLE an implication of arousing disgust ⟨a *scurvy* crew of hangers-on⟩.

— **con·tex·tu·al** \kän-ˈteks-chə-wəl, kən-, -chəl\ *adjective*
— **con·tex·tu·al·ly** *adverb*
con·text–free \ˈkän-ˌtekst-ˈfrē\ *adjective* (1964)
: of, relating to, or being a grammar or language based on rules that describe a change in a string without reference to elements outside of the string; *also* **:** being such a rule
con·tex·tu·al·ize \kən-ˈteks-chə-wə-ˌlīz, -chə-ˌlīz\ *transitive verb* **-ized; -iz·ing** (1934)
: to place (as a word or activity) in a context
con·tex·ture \kən-ˈteks-chər, ˈkän-ˌ, kän-ˈ\ *noun* [French, from Latin *contextus,* past participle of *contexere*] (1603)
1 : the act, process, or manner of weaving parts into a whole; *also* **:** a structure so formed ⟨a *contexture* of lies⟩
2 : CONTEXT
con·ti·gu·i·ty \ˌkän-tə-ˈgyü-ə-tē\ *noun, plural* **-ties** (1612)
: the quality or state of being contiguous **:** PROXIMITY
con·tig·u·ous \kən-ˈti-gyə-wəs\ *adjective* [Latin *contiguus,* from *contingere* to have contact with — more at CONTINGENT] (circa 1609)
1 : being in actual contact **:** touching along a boundary or at a point
2 *of angles* **:** ADJACENT 2
3 : next or near in time or sequence
4 : touching or connected throughout in an unbroken sequence ⟨*contiguous* row houses⟩
synonym see ADJACENT
— **con·tig·u·ous·ly** *adverb*
— **con·tig·u·ous·ness** *noun*
con·ti·nence \ˈkän-tᵊn-ən(t)s\ *noun* (14th century)
1 : SELF-RESTRAINT; *especially* **:** a refraining from sexual intercourse
2 : the ability to retain a bodily discharge voluntarily ⟨fecal *continence*⟩
¹con·ti·nent \ˈkän-tᵊn-ənt\ *adjective* [Middle English, from Middle French, from Latin *continent-, continens,* from present participle of *continēre* to hold in — more at CONTAIN] (14th century)
1 : exercising continence
2 *obsolete* **:** RESTRICTIVE
— **con·ti·nent·ly** *adverb*
²con·ti·nent \ˈkän-tᵊn-ənt, ˈkänt-nənt\ *noun* [in senses 1 & 2, from Latin *continent-, continens,* present participle of *continēre,* to hold together, contain; in senses 3 & 4, from Latin *continent-, continens* continuous mass of land, mainland, from *continent-, continens,* present participle] (1541)
1 *archaic* **:** CONTAINER, CONFINES
2 *archaic* **:** EPITOME
3 : MAINLAND
4 a : one of the six or seven great divisions of land on the globe **b** *capitalized* **:** the continent of Europe — used with *the*
¹con·ti·nen·tal \ˌkän-tᵊn-ˈen-tᵊl\ *adjective* (1760)
1 a : of, relating to, or characteristic of a continent ⟨*continental* waters⟩; *specifically, often capitalized* **:** of or relating to the continent of Europe excluding the British Isles **b** *often capitalized* **:** of, relating to, or being a cuisine derived from the classic dishes of Europe and especially France
2 a *often capitalized* **:** of or relating to the colonies later forming the U.S. ⟨*Continental* Congress⟩ **b :** being the part of the U.S. on the North American continent; *also* **:** being the part of the U.S. comprising the lower 48 states
— **con·ti·nen·tal·ly** \-tᵊl-ē\ *adverb*
²continental *noun* (1777)
1 a *often capitalized* **:** an American soldier of the Revolution in the Continental army **b** (1) **:** a piece of Continental paper currency (2) **:** the least bit ⟨not worth a *continental*⟩
2 : an inhabitant of a continent and especially the continent of Europe
continental breakfast *noun, often C capitalized* (1911)
: a light breakfast (as of rolls or toast and coffee)
continental drift *noun* (1926)
: a hypothetical slow movement of the continents on a deep-seated viscous zone within the earth — compare PLATE TECTONICS
continental shelf *noun* (1892)
: a shallow submarine plain of varying width forming a border to a continent and typically ending in a steep slope to the oceanic abyss
continental slope *noun* (1900)
: the usually steep slope from a continental shelf to the ocean floor
con·tin·gence \kən-ˈtin-jən(t)s\ *noun* (circa 1530)
1 : CONTINGENCY
2 : TANGENCY
con·tin·gen·cy \kən-ˈtin-jən(t)-sē\ *noun, plural* **-cies** (1561)
1 : the quality or state of being contingent
2 : a contingent event or condition: as **a :** an event (as an emergency) that may but is not certain to occur ⟨trying to provide for every *contingency*⟩ **b :** something liable to happen as an adjunct to or result of something else
synonym see JUNCTURE
contingency fee *noun* (1945)
: a fee for services (as of a lawyer) paid upon successful completion of the services and usually calculated as a percentage of the gain realized for the client — called also *contingent fee*
contingency table *noun* (circa 1947)
: a table of data in which the row entries tabulate the data according to one variable and the column entries tabulate it according to another variable and which is used especially in the study of the correlation between variables
¹con·tin·gent \kən-ˈtin-jənt\ *adjective* [Middle English, from Middle French, from Latin *contingent-, contingens,* present participle of *contingere* to have contact with, befall, from *com-* + *tangere* to touch — more at TANGENT] (14th century)
1 : likely but not certain to happen **:** POSSIBLE
2 : not logically necessary; *especially* **:** EMPIRICAL
3 a : happening by chance or unforeseen causes **b :** subject to chance or unseen effects **:** UNPREDICTABLE **c :** intended for use in circumstances not completely foreseen
4 : dependent on or conditioned by something else
5 : not necessitated **:** determined by free choice
synonym see ACCIDENTAL
— **con·tin·gent·ly** *adverb*
²contingent *noun* (1548)
1 : something contingent **:** CONTINGENCY
2 : a representative group **:** DELEGATION, DETACHMENT
con·tin·u·al \kən-ˈtin-yü-əl, -yəl\ *adjective* [Middle English, from Middle French, from Latin *continuus* continuous] (14th century)
1 : continuing indefinitely in time without interruption ⟨*continual* fear⟩
2 : recurring in steady usually rapid succession ⟨a history of *continual* invasions⟩ ☆
— **con·tin·u·al·ly** *adverb*
con·tin·u·ance \kən-ˈtin-yü-ən(t)s\ *noun* (14th century)
1 : CONTINUATION
2 : the extent of continuing **:** DURATION
3 : the quality of enduring **:** PERMANENCE
4 : an adjournment of a court case to a future day
con·tin·u·ant \-yü-ənt\ *noun* (1861)
1 : something that continues or serves as a continuation
2 : a speech sound (as a fricative or vowel) that is produced without a complete closure of the breath passage — compare STOP
— **continuant** *adjective*
con·tin·u·ate *adjective* (1555)
obsolete **:** CONTINUOUS, UNINTERRUPTED
con·tin·u·a·tion \kən-ˌtin-yə-ˈwā-shən\ *noun* (14th century)
1 : the act or fact of continuing in or the prolongation of a state or activity
2 : resumption after an interruption
3 : something that continues, increases, or adds
con·tin·u·a·tive \kən-ˈtin-yə-ˌwā-tiv, -wə-tiv\ *adjective* (1684)
: expressing continuity or continuation (as of an idea or action)
con·tin·u·a·tor \-ˌwā-tər\ *noun* (1646)
: one that continues
con·tin·ue \kən-ˈtin-(ˌ)yü\ *verb* **-tin·ued; -tinu·ing** [Middle English, from Middle French *continuer,* from Latin *continuare,* from *continuus*] (14th century)
intransitive verb
1 : to maintain without interruption a condition, course, or action
2 : to remain in existence **:** ENDURE
3 : to remain in a place or condition **:** STAY
4 : to resume an activity after interruption
transitive verb
1 a : KEEP UP, MAINTAIN ⟨*continues* walking⟩ **b :** to keep going or add to **:** PROLONG; *also* **:** to resume after intermission
2 : to cause to continue
3 : to allow to remain in a place or condition **:** RETAIN
4 : to postpone (a legal proceeding) by a continuance ☆
— **con·tinu·er** \-yü-ər\ *noun*
continued *adjective* (15th century)
1 : lasting or extending without interruption ⟨*continued* success⟩
2 : resumed after interruption ⟨a *continued* story⟩
continued fraction *noun* (circa 1856)
: a fraction whose numerator is an integer and whose denominator is an integer plus a fraction whose numerator is an integer and whose denominator is an integer plus a fraction and so on
continuing *adjective* (14th century)

☆ SYNONYMS
Continual, continuous, constant, incessant, perpetual, perennial mean characterized by continued occurrence or recurrence. CONTINUAL often implies a close prolonged succession or recurrence ⟨*continual* showers the whole weekend⟩. CONTINUOUS usually implies an uninterrupted flow or spatial extension ⟨football's oldest *continuous* rivalry⟩. CONSTANT implies uniform or persistent occurrence or recurrence ⟨lived in *constant* pain⟩. INCESSANT implies ceaseless or uninterrupted activity ⟨annoyed by the *incessant* quarreling⟩. PERPETUAL suggests unfailing repetition or lasting duration ⟨a land of *perpetual* snowfall⟩. PERENNIAL implies enduring existence often through constant renewal ⟨a *perennial* source of controversy⟩.

Continue, last, endure, abide, persist mean to exist over a period of time or indefinitely. CONTINUE applies to a process going on without ending ⟨the search for peace will *continue*⟩. LAST, especially when unqualified, may stress existing beyond what is normal or expected ⟨buy shoes that will *last*⟩. ENDURE adds an implication of resisting destructive forces or agencies ⟨in spite of everything, her faith *endured*⟩. ABIDE implies stable and constant existing especially as opposed to mutability ⟨a love that *abides* through 40 years of marriage⟩. PERSIST suggests outlasting the normal or appointed time and often connotes obstinacy or doggedness ⟨the sense of guilt *persisted*⟩.

1 : CONTINUOUS, CONSTANT ⟨*continuing* poverty⟩
2 : needing no renewal **:** ENDURING ⟨*continuing* fame⟩
— **con·tin·u·ing·ly** *adverb*

continuing education *noun* (1954)
: formal courses of study for adult part-time students

con·ti·nu·i·ty \ˌkän-tᵊn-ˈü-ə-tē, -ˈyü-\ *noun, plural* **-ties** (15th century)
1 a : uninterrupted connection, succession, or union **b :** uninterrupted duration or continuation especially without essential change
2 : something that has, exhibits, or provides continuity: as **a :** a script or scenario in the performing arts **b :** transitional spoken or musical matter especially for a radio or television program **c :** the story and dialogue of a comic strip
3 : the property of being mathematically continuous

con·tin·uo \kən-ˈtin-yə-ˌwō, -ˈti-nə-\ *noun, plural* **-u·os** [Italian, from *continuo* continuous, from Latin *continuus*] (1724)
: a bass part (as for a keyboard or stringed instrument) used especially in baroque ensemble music and consisting of a succession of bass notes with figures that indicate the required chords — called also *figured bass, thorough-bass*

con·tin·u·ous \kən-ˈtin-yü-əs\ *adjective* [Latin *continuus,* from *continēre* to hold together — more at CONTAIN] (1673)
1 : marked by uninterrupted extension in space, time, or sequence
2 *of a function* **:** having the property that the absolute value of the numerical difference between the value at a given point and the value at any point in a neighborhood of the given point can be made as close to zero as desired by choosing the neighborhood small enough
synonym see CONTINUAL
— **con·tin·u·ous·ly** *adverb*
— **con·tin·u·ous·ness** *noun*

con·tin·u·um \kən-ˈtin-yü-əm\ *noun, plural* **-ua** \-yü-ə\ *also* **-u·ums** [Latin, neuter of *continuus*] (1646)
1 : a coherent whole characterized as a collection, sequence, or progression of values or elements varying by minute degrees ⟨"good" and "bad" . . . stand at opposite ends of a *continuum* —Wayne Shumaker⟩
2 : the set of real numbers including both the rationals and the irrationals; *broadly* **:** a compact set which cannot be separated into two sets neither of which contains a limit point of the other

con·tort \kən-ˈtȯrt\ *verb* [Middle English, from Latin *contortus,* past participle of *contorquēre,* from *com-* + *torquēre* to twist — more at TORTURE] (15th century)
transitive verb
: to twist in a violent manner ⟨features *contorted* with fury⟩
intransitive verb
: to twist into or as if into a strained shape or expression
synonym see DEFORM
— **con·tor·tion** \-ˈtȯr-shən\ *noun*
— **con·tor·tive** \-ˈtȯr-tiv\ *adjective*

con·tor·tion·ist \kən-ˈtȯr-sh(ə-)nist\ *noun* (1859)
: one who contorts; *specifically* **:** an acrobat able to twist the body into unusual postures
— **con·tor·tion·is·tic** \-ˌtȯr-shə-ˈnis-tik\ *adjective*

[1]**con·tour** \ˈkän-ˌtu̇r\ *noun* [French, from Italian *contorno,* from *contornare* to round off, from Medieval Latin, to turn around, from Latin *com-* + *tornare* to turn on a lathe — more at TURN] (1662)
1 : an outline especially of a curving or irregular figure **:** SHAPE; *also* **:** the line representing this outline
2 : the general form or structure of something **:** CHARACTERISTIC — often used in plural ⟨*contours* of a melody⟩ ⟨to delineate the tortured psychological *contours* of the tribal past —B. J. Phillips⟩
3 : a usually meaningful change in intonation in speech
synonym see OUTLINE

[2]**contour** *adjective* (1844)
1 : following contour lines or forming furrows or ridges along them ⟨*contour* flooding⟩ ⟨*contour* farming⟩
2 : made to fit the contour of something ⟨a *contour* couch⟩ ⟨*contour* sheets⟩

[3]**contour** *transitive verb* (1871)
1 a : to shape the contour of **b :** to shape so as to fit contours
2 : to construct (as a road) in conformity to a contour

contour feather *noun* (1867)
: one of the medium-sized feathers that form the general covering of a bird and determine the external contour

contour line *noun* (1844)
: a line (as on a map) connecting the points on a land surface that have the same elevation

contour map *noun* (1862)
: a map having contour lines

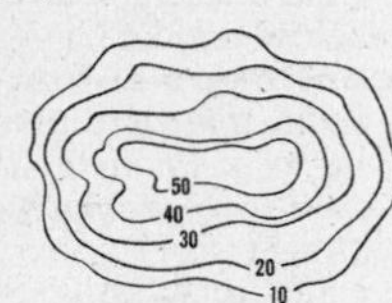

contour map

[1]**con·tra** \ˈkän-trə\ *preposition* [Latin] (15th century)
1 : AGAINST — used chiefly in the phrase *pro and contra*
2 : in opposition or contrast to — used before a proper name

[2]**con·tra** \ˈkän-trə, ˈkōn-, -ˌträ\ *noun* [American Spanish, short for *contra-revolucionario* counterrevolutionary] (1981)
: a member of a guerrilla group opposed to the Sandinista government in Nicaragua

contra- *prefix* [Middle English, from Latin, from *contra* against, opposite — more at COUNTER]
1 : against **:** contrary **:** contrasting ⟨*contra*distinction⟩
2 : pitched below normal bass ⟨*contra*octave⟩

con·tra·band \ˈkän-trə-ˌband\ *noun* [Italian *contrabbando,* from Medieval Latin *contrabannum,* from *contra-* + *bannus, bannum* decree, of Germanic origin; akin to Old High German *bannan* to command — more at BAN] (circa 1529)
1 : illegal or prohibited traffic in goods **:** SMUGGLING
2 : goods or merchandise whose importation, exportation, or possession is forbidden; *also* **:** smuggled goods
3 : a slave who during the Civil War escaped to or was brought within the Union lines
— **contraband** *adjective*

con·tra·band·ist \-ˌban-dist\ *noun* (circa 1818)
: SMUGGLER

con·tra·bass \ˈkän-trə-ˌbās\ *noun* [Italian *contrabbasso,* from *contra-* + *basso* bass] (circa 1611)
: DOUBLE BASS
— **con·tra·bass·ist** \-ˌbā-sist\ *noun*

con·tra·bas·soon \ˌkän-trə-bə-ˈsün, -ba-\ *noun* (1891)
: a double-reed woodwind instrument having a range an octave lower than that of the bassoon — called also *double bassoon*

con·tra·cep·tion \ˌkän-trə-ˈsep-shən\ *noun* [*contra-* + con*ception*] (1886)
: deliberate prevention of conception or impregnation
— **con·tra·cep·tive** \-ˈsep-tiv\ *adjective or noun*

[1]**con·tract** \ˈkän-ˌtrakt\ *noun* [Middle English, from Latin *contractus,* from *contrahere* to draw together, make a contract, reduce in size, from *com-* + *trahere* to draw] (14th century)
1 a : a binding agreement between two or more persons or parties; *especially* **:** one legally enforceable **b :** a business arrangement for the supply of goods or services at a fixed price ⟨make parts on *contract*⟩ **c :** the act of marriage or an agreement to marry
2 : a document describing the terms of a contract
3 : the final bid to win a specified number of tricks in bridge
4 : an order or arrangement for a hired assassin to kill someone

[2]**con·tract** *verb transitive 2a & verb intransitive 1 usually* ˈkän-ˌtrakt, *others usually* kən-ˈ\ *verb* [Middle English, from Middle French or Latin; Middle French *contracter* to agree upon, from Latin *contractus*] (14th century)
transitive verb
1 a : to bring on oneself especially inadvertently **:** INCUR ⟨*contracting* debts⟩ **b :** to become affected with ⟨*contract* pneumonia⟩
2 a : to establish or undertake by contract **b :** BETROTH; *also* **:** to establish (a marriage) formally **c** (1) **:** to hire by contract (2) **:** to purchase (as goods or services) on a contract basis — often used with *out*
3 a : LIMIT, RESTRICT **b :** KNIT, WRINKLE ⟨frown *contracted* his brow⟩ **c :** to draw together **:** CONCENTRATE
4 : to reduce to smaller size by or as if by squeezing or forcing together
5 : to shorten (as a word) by omitting one or more sounds or letters
intransitive verb
1 : to make a contract
2 : to draw together so as to become diminished in size ⟨metal *contracts* on cooling⟩; *also* **:** to become less in compass, duration, or length ⟨muscle *contracts* in tetanus⟩ ☆
— **con·tract·ibil·i·ty** \kən-ˌtrak-tə-ˈbi-lə-tē, ˌkän-\ *noun*
— **con·tract·ible** \kən-ˈtrak-tə-bəl, ˈkän-ˌ\ *adjective*

contract bridge \ˈkän-ˌtrakt-\ *noun* (1924)
: a bridge game distinguished by the fact that overtricks do not count toward game or slam bonuses

con·trac·tile \kən-ˈtrak-tᵊl, -ˌtīl\ *adjective* (circa 1706)
: having or concerned with the power or property of contracting ⟨*contractile* proteins of muscle fibrils⟩
— **con·trac·til·i·ty** \ˌkän-ˌtrak-ˈti-lə-tē\ *noun*

contractile vacuole *noun* (1877)

☆ SYNONYMS
Contract, shrink, condense, compress, constrict, deflate mean to decrease in bulk or volume. CONTRACT applies to a drawing together of surfaces or particles or a reduction of area or length ⟨caused her muscles to *contract*⟩. SHRINK implies a contracting or a loss of material and stresses a falling short of original dimensions ⟨the sweater will *shrink* when washed⟩. CONDENSE implies a reducing of something homogeneous to greater compactness without significant loss of content ⟨*condense* the essay into a paragraph⟩. COMPRESS implies a pressing into a small compass and definite shape usually against resistance ⟨*compressed* cotton into bales⟩. CONSTRICT implies a tightening that reduces diameter ⟨the throat is *constricted* by a tight collar⟩. DEFLATE implies a contracting by reducing the internal pressure of contained air or gas ⟨*deflate* the balloon⟩.

: a vacuole in a unicellular organism that contracts regularly to discharge fluid from the body

con·trac·tion \kən-ˈtrak-shən\ *noun* (15th century)
1 a : the action or process of contracting **:** the state of being contracted **b :** the shortening and thickening of a functioning muscle or muscle fiber **c :** a reduction in business activity or growth
2 : a shortening of a word, syllable, or word group by omission of a sound or letter; *also* **:** a form produced by such shortening
— con·trac·tion·al \-shnəl, -shə-nᵊl\ *adjective*
— con·trac·tive \kən-ˈtrak-tiv, ˈkän-ˌ\ *adjective*
— con·trac·tion·ary \kən-ˈtrak-shə-ˌner-ē\ *adjective*

con·trac·tor *1 usually* ˈkän-ˌtrak-tər, *2 usually* kən-ˈ\ *noun* (1548)
1 : one that contracts or is party to a contract: as **a :** one that contracts to perform work or provide supplies **b :** one that contracts to erect buildings
2 : something (as a muscle) that contracts or shortens

con·trac·tu·al \kən-ˈtrak-chə-wəl, -chəl, -shwəl\ *adjective* (1861)
: of, relating to, or constituting a contract
— con·trac·tu·al·ly *adverb*

con·trac·ture \kən-ˈtrak-chər\ *noun* (1601)
: a permanent shortening (as of muscle, tendon, or scar tissue) producing deformity or distortion

con·tra·dict \ˌkän-trə-ˈdikt\ *transitive verb* [Latin *contradictus,* past participle of *contradicere,* from *contra-* + *dicere* to say, speak — more at DICTION] (circa 1576)
1 : to assert the contrary of **:** take issue with
2 : to imply the opposite or a denial of ⟨your actions *contradict* your words⟩
synonym see DENY
— con·tra·dict·able \-ˈdik-tə-bəl\ *adjective*
— con·tra·dic·tor \-ˈdik-tər\ *noun*

con·tra·dic·tion \ˌkän-trə-ˈdik-shən\ *noun* (14th century)
1 : act or an instance of contradicting
2 a : a proposition, statement, or phrase that asserts or implies both the truth and falsity of something **b :** a statement or phrase whose parts contradict each other ⟨a round square is a *contradiction* in terms⟩
3 a : logical incongruity **b :** a situation in which inherent factors, actions, or propositions are inconsistent or contrary to one another

con·tra·dic·tious \-shəs\ *adjective* (1604)
1 : CONTRADICTORY, OPPOSITE
2 : given to or marked by contradiction **:** CONTRARY

¹con·tra·dic·to·ry \ˌkän-trə-ˈdik-t(ə-)rē\ *noun, plural* **-ries** (14th century)
: a proposition so related to another that if either of the two is true the other is false and if either is false the other must be true

²contradictory *adjective* (1534)
: involving, causing, or constituting a contradiction
synonym see OPPOSITE
— con·tra·dic·to·ri·ly \-t(ə-)rə-lē\ *adverb*
— con·tra·dic·to·ri·ness \-t(ə-)rē-nəs\ *noun*

con·tra·dis·tinc·tion \ˌkän-trə-dis-ˈtiŋ(k)-shən\ *noun* (1647)
: distinction by means of contrast ⟨painting in *contradistinction* to sculpture⟩
— con·tra·dis·tinc·tive \-ˈtiŋ(k)-tiv\ *adjective*
— con·tra·dis·tinc·tive·ly *adverb*

con·tra·dis·tin·guish \-ˈtiŋ-gwish\ *transitive verb* (1622)
: to distinguish by contrasting qualities

con·trail \ˈkän-ˌtrāl\ *noun* [*con*densation *trail*] (1943)
: streaks of condensed water vapor created in the air by an airplane or rocket at high altitudes

con·tra·in·di·cate \ˌkän-trə-ˈin-də-ˌkāt\ *transitive verb* (1666)
: to make (a treatment or procedure) inadvisable

con·tra·in·di·ca·tion \-ˌin-də-ˈkā-shən\ *noun* (1623)
: something (as a symptom or condition) that makes a particular treatment or procedure inadvisable

con·tra·lat·er·al \-ˈla-t(ə-)rəl\ *adjective* [International Scientific Vocabulary] (1882)
: occurring on or acting in conjunction with a part on the opposite side of the body

con·tral·to \kən-ˈtral-(ˌ)tō\ *noun, plural* **-tos** [Italian, from *contra-* + *alto*] (1730)
1 a : a singing voice having a range between tenor and mezzo-soprano **b :** a person having this voice
2 : the part sung by a contralto

con·tra·oc·tave \ˌkän-trə-ˈäk-tiv, -təv, -ˌtāv\ *noun* (circa 1891)
: the musical octave that begins on the third C below middle C — see PITCH illustration

con·tra·po·si·tion \-pə-ˈzi-shən\ *noun* [Late Latin *contraposition-, contrapositio,* from Latin *contraponere* to place opposite, from *contra-* + *ponere* to place — more at POSITION] (1551)
: the relationship between two propositions when the subject and predicate of one are respectively the negation of the predicate and the negation of the subject of the other

con·tra·pos·i·tive \-ˈpä-zə-tiv, -ˈpäz-tiv\ *noun* (1870)
: a proposition resulting from an operation of immediate inference in which the terms of a given proposition are permuted and negated ⟨"all not-*P* is not-*S*" is the *contrapositive* of "all *S* is *P*"⟩

con·trap·tion \kən-ˈtrap-shən\ *noun* [perhaps blend of *contrivance, trap,* and *invention*] (circa 1825)
: DEVICE, GADGET

con·tra·pun·tal \ˌkän-trə-ˈpən-tᵊl\ *adjective* [Italian *contrappunto* counterpoint, from Medieval Latin *contrapunctus* — more at COUNTERPOINT] (1845)
1 : POLYPHONIC
2 : of, relating to, or marked by counterpoint
— con·tra·pun·tal·ly \-tᵊl-ē\ *adverb*

con·tra·pun·tist \-ˈpən-tist\ *noun* (1776)
: one who writes counterpoint

con·trar·i·an \kən-ˈtrer-ē-ən, kän-\ *noun* (1657)
: a person who takes a contrary position or attitude; *specifically* **:** an investor who buys shares of stock when most others are selling and sells when others are buying
— contrarian *adjective*

con·tra·ri·ety \ˌkän-trə-ˈrī-ə-tē\ *noun, plural* **-eties** [Middle English *contrariete,* from Middle French *contrarieté,* from Late Latin *contrarietat-, contrarietas,* from Latin *contrarius* contrary] (14th century)
1 : the quality or state of being contrary
2 : something contrary

con·trar·i·ous \kən-ˈtrer-ē-əs, kän-\ *adjective* (13th century)
: PERVERSE, ANTAGONISTIC

con·trari·wise \ˈkän-ˌtrer-ē-ˌwīz, kən-ˈ\ *adverb* (14th century)
1 : on the contrary
2 : VICE VERSA
3 : in a contrary manner

¹con·trary \ˈkän-ˌtrer-ē\ *noun, plural* **-traries** (13th century)
1 : a fact or condition incompatible with another **:** OPPOSITE — usually used with *the*
2 : one of a pair of opposites
3 a : a proposition so related to another that though both may be false they cannot both be true — compare SUBCONTRARY **b :** either of two terms (as *good* and *evil*) that cannot both be affirmed of the same subject
— by contraries *obsolete* **:** in a manner opposite to what is logical or expected
— on the contrary : just the opposite
— to the contrary 1 : on the contrary **2 :** NOTWITHSTANDING

²con·trary \ˈkän-ˌtrer-ē, *4 often* kən-ˈtrer-ē\ *adjective* [Middle English *contrarie,* from Middle French *contraire,* from Latin *contrarius,* from *contra* opposite] (14th century)
1 : being so different as to be at opposite extremes **:** OPPOSITE ⟨come to the *contrary* conclusion⟩ ⟨went off in *contrary* directions⟩; *also* **:** being opposite to or in conflict with each other ⟨*contrary* viewpoints⟩
2 : being not in conformity with what is usual or expected ⟨actions *contrary* to company policy⟩ ⟨*contrary* evidence⟩
3 : UNFAVORABLE — used of wind or weather
4 : temperamentally unwilling to accept control or advice ☆
— con·trari·ly \-ˌtrer-ə-lē, -ˈtrer-\ *adverb*
— con·trari·ness \-ˌtrer-ē-nəs, -ˈtrer-\ *noun*

³contrary *same as* ²\ *adverb* (15th century)
: CONTRARIWISE, CONTRARILY

contrary to *preposition* (14th century)
: in conflict with **:** DESPITE ⟨*contrary to* orders, he set out alone⟩

¹con·trast \kən-ˈtrast, ˈkän-ˌ\ *verb* [French *contraster,* from Middle French, to oppose, resist, alteration of *contrester,* from (assumed) Vulgar Latin *contrastare,* from Latin *contra-* + *stare* to stand — more at STAND] (1695)
transitive verb
: to set off in contrast **:** compare or appraise in respect to differences ⟨*contrast* European and American manners⟩ — often used with *to* or *with* ⟨*contrasting* her with other women —Victoria Sackville-West⟩
intransitive verb
: to form a contrast
synonym see COMPARE
— con·trast·able \-ˈtras-tə-bəl, -ˌtras-\ *adjective*

²con·trast \ˈkän-ˌtrast\ *noun* (1711)
1 a : juxtaposition of dissimilar elements (as color, tone, or emotion) in a work of art **b :** degree of difference between the lightest and darkest parts of a picture
2 a : the difference or degree of difference between things having similar or comparable natures ⟨the *contrast* between the two forms of government⟩ **b :** comparison of similar objects to set off their dissimilar qualities **:** the state of being so compared ⟨the enforced simplicity in this diary . . . is in *contrast* to the intensity of his former life —*Times Literary Supplement*⟩
3 : a person or thing that exhibits differences when compared with another

con·tras·tive \kən-ˈtras-tiv, ˈkän-ˌ\ *adjective* (1841)
: forming or consisting of a contrast
— con·tras·tive·ly *adverb*

con·trasty \ˈkän-ˌtras-tē\ *adjective* (1891)

☆ SYNONYMS
Contrary, perverse, restive, balky, wayward mean inclined to resist authority or control. CONTRARY implies a temperamental unwillingness to accept orders or advice ⟨a *contrary* child⟩. PERVERSE may imply wrongheaded, determined, or cranky opposition to what is reasonable or normal ⟨a *perverse,* intractable critic⟩. RESTIVE suggests unwillingness or inability to submit to discipline or follow orders ⟨tired soldiers growing *restive*⟩. BALKY suggests a refusing to proceed in a desired direction or course of action ⟨a *balky* witness⟩. WAYWARD suggests strong-willed capriciousness and irregularity in behavior ⟨a school for *wayward* youths⟩. See in addition OPPOSITE.

: having or producing in photography great contrast between highlights and shadows

con·tra·vene \ˌkän-trə-ˈvēn\ *transitive verb* **-vened; -ven·ing** [Middle French or Late Latin; Middle French *contrevenir,* from Late Latin *contravenire,* from Latin *contra-* + *venire* to come — more at COME] (1567)
1 : to go or act contrary to **:** VIOLATE ⟨*contravene* a law⟩
2 : to oppose in argument **:** CONTRADICT ⟨*contravene* a proposition⟩
synonym see DENY
— **con·tra·ven·er** *noun*

con·tra·ven·tion \ˌkän-trə-ˈven(t)-shən\ *noun* [Middle French, from Late Latin *contravenire*] (1579)
: the act of contravening **:** VIOLATION

con·tre·danse \ˈkän-trə-ˌdan(t)s, kō^{n}-trə-dä^{n}s\ *or* **con·tra dance** \ˈkän-trə-ˌdan(t)s\ *noun* [French *contredanse,* by folk etymology from English *country-dance*] (1803)
1 : a folk dance in which couples face each other in two lines or in a square
2 : a piece of music for a contredanse

con·tre·temps \ˈkän-trə-ˌtä^{n}, kō^{n}-trə-tä^{n}\ *noun, plural* **con·tre·temps** \-(ˌ)tä^{n}(z)\ [French, from *contre-* counter- + *temps* time, from Latin *tempus*] (1769)
: an inopportune or embarrassing occurrence or situation

con·trib·ute \kən-ˈtri-byət, -(ˌ)byüt *also & especially before -ed or -ing* -ˈtri-bət; *chiefly British also* ˈkän-trə-ˌbyüt\ *verb* **-ut·ed; -ut·ing** [Latin *contributus,* past participle of *contribuere,* from *com-* + *tribuere* to grant — more at TRIBUTE] (1530)
transitive verb
1 : to give or supply in common with others
2 : to supply (as an article) for a publication
intransitive verb
1 a : to give a part to a common fund or store **b :** to play a significant part in bringing about an end or result
2 : to submit articles to a publication
— **con·trib·u·tor** \-byə-tər, -bə-, -ˌbyü-\ *noun*

con·tri·bu·tion \ˌkän-trə-ˈbyü-shən\ *noun* (14th century)
1 : a payment (as a levy or tax) imposed by military, civil, or ecclesiastical authorities usually for a special or extraordinary purpose
2 : the act of contributing; *also* **:** the thing contributed
— **con·trib·u·tive** \kən-ˈtri-byə-tiv\ *adjective*
— **con·trib·u·tive·ly** *adverb*

con·trib·u·to·ry \kən-ˈtri-byə-ˌtōr-ē, -ˌtȯr-\ *adjective* (15th century)
1 a : subject to a levy of supplies, money, or men **b :** contributing to a common fund or enterprise
2 : of, relating to, or forming a contribution

con·trite \ˈkän-ˌtrīt, kən-ˈ\ *adjective* [Middle English *contrit,* from Middle French, from Medieval Latin *contritus,* from Latin, past participle of *conterere* to grind, bruise, from *com-* + *terere* to rub — more at THROW] (14th century)
1 : grieving and penitent for sin or shortcoming
2 : proceeding from contrition ⟨*contrite* sighs⟩
— **con·trite·ly** *adverb*
— **con·trite·ness** *noun*

con·tri·tion \kən-ˈtri-shən\ *noun* (14th century)
: the state of being contrite **:** REPENTANCE
synonym see PENITENCE

con·triv·ance \kən-ˈtrī-vən(t)s\ *noun* (circa 1628)
1 a : a thing contrived; *especially* **:** a mechanical device **b :** an artificial arrangement or development
2 : the act or faculty of contriving **:** the state of being contrived

con·trive \kən-ˈtrīv\ *verb* **con·trived; con·triv·ing** [Middle English *controven, contreven,* from Middle French *controver,* from Medieval Latin *contropare* to compare, from Latin *com-* + (assumed) Vulgar Latin *tropare* to compose, find — more at TROUBADOR] (14th century)
transitive verb
1 a : DEVISE, PLAN ⟨*contrive* ways of handling the situation⟩ **b :** to form or create in an artistic or ingenious manner ⟨*contrived* household utensils from stone⟩
2 : to bring about by stratagem or with difficulty **:** MANAGE
intransitive verb
: to make schemes
— **con·triv·er** *noun*

contrived *adjective* (15th century)
: ARTIFICIAL, LABORED

¹con·trol \kən-ˈtrōl\ *transitive verb* **con·trolled; con·trol·ling** [Middle English *controllen,* from Middle French *contreroller,* from *contrerolle* copy of an account, audit, from Medieval Latin *contrarotulus,* from Latin *contra-* + Medieval Latin *rotulus* roll — more at ROLL] (15th century)
1 a *archaic* **:** to check, test, or verify by evidence or experiments **b :** to incorporate suitable controls in ⟨a *controlled* experiment⟩
2 a : to exercise restraining or directing influence over **:** REGULATE **b :** to have power over **:** RULE **c :** to reduce the incidence or severity of especially to innocuous levels ⟨*control* an insect population⟩ ⟨*control* a disease⟩
synonym see CONDUCT
— **con·trol·la·bil·i·ty** \-ˌtrō-lə-ˈbi-lə-tē\ *noun*
— **con·trol·la·ble** \-ˈtrō-lə-bəl\ *adjective*
— **con·trol·ment** \-ˈtrōl-mənt\ *noun*

²control *noun, often attributive* (1590)
1 a : an act or instance of controlling; *also* **:** power or authority to guide or manage **b :** skill in the use of a tool, instrument, technique, or artistic medium **c :** the regulation of economic activity especially by government directive — usually used in plural ⟨price *controls*⟩
2 : RESTRAINT, RESERVE
3 : one that controls: as **a** (1) **:** an experiment in which the subjects are treated as in a parallel experiment except for omission of the procedure or agent under test and which is used as a standard of comparison in judging experimental effects — called also *control experiment* (2) **:** one (as an organism, culture, or group) that is part of a control **b :** a device or mechanism used to regulate or guide the operation of a machine, apparatus, or system **c :** an organization that directs a spaceflight ⟨mission *control*⟩ **d :** a personality or spirit believed to actuate the utterances or performances of a spiritualist medium
synonym see POWER

con·trolled \kən-ˈtrōld\ *adjective* (1586)
1 : RESTRAINED
2 : regulated by law with regard to possession and use ⟨*controlled* drugs⟩

con·trol·ler \kən-ˈtrō-lər, ˈkän-ˌ\ *noun* [Middle English *contreroller,* from Middle French *contrerolleur,* from *contrerolle*] (15th century)
1 a : COMPTROLLER 1 **b :** COMPTROLLER 2 **c :** the chief accounting officer of a business enterprise or an institution (as a college)
2 a : one that controls or has power or authority to control ⟨air traffic *controller*⟩ **b :** CONTROL 3b
— **con·trol·ler·ship** \-ˌship\ *noun*

controlling interest *noun* (circa 1924)
: sufficient stock ownership in a corporation to exert control over policy

control surface *noun* (1917)
: a movable airfoil designed to change the attitude of an aircraft

con·tro·ver·sial \ˌkän-trə-ˈvər-shəl, -ˈvər-sē-əl\ *adjective* (1583)
1 : of, relating to, or arousing controversy
2 : given to controversy **:** DISPUTATIOUS
— **con·tro·ver·sial·ism** \-shə-ˌli-zəm, -sē-ə-ˌ\ *noun*
— **con·tro·ver·sial·ist** \-list\ *noun*
— **con·tro·ver·sial·ly** *adverb*

con·tro·ver·sy \ˈkän-trə-ˌvər-sē, *British also* kən-ˈträ-vər-sē\ *noun, plural* **-sies** [Middle English *controversie,* from Latin *controversia,* from *controversus* disputable, literally, turned against, from *contro-* (akin to *contra-*) + *versus,* past participle of *vertere* to turn — more at WORTH] (14th century)
1 : a discussion marked especially by the expression of opposing views **:** DISPUTE
2 : QUARREL, STRIFE

con·tro·vert \ˈkän-trə-ˌvərt, ˌkän-trə-ˈ\ *verb* [*controversy*] (1609)
transitive verb
: to dispute or oppose by reasoning ⟨*controvert* a point in a discussion⟩
intransitive verb
: to engage in controversy
— **con·tro·vert·er** \-ˌvər-tər, -ˈvər-\ *noun*
— **con·tro·vert·ible** \-tə-bəl\ *adjective*

con·tu·ma·cious \ˌkän-tü-ˈmā-shəs, -tyü-ˈ, -chə-ˈ\ *adjective* (circa 1600)
: stubbornly disobedient **:** REBELLIOUS
— **con·tu·ma·cious·ly** *adverb*

con·tu·ma·cy \kən-ˈtü-mə-sē, -ˈtyü-; ˈkän-tü-, -tyü-, -chə-\ *noun* [Middle English *contumacie,* from Latin *contumacia,* from *contumac-, contumax* rebellious] (13th century)
: stubborn resistance to authority; *specifically* **:** willful contempt of court

con·tu·me·li·ous \ˌkän-tü-ˈmē-lē-əs, -tyü-ˈ, -chə-ˈ\ *adjective* (15th century)
: insolently abusive and humiliating
— **con·tu·me·li·ous·ly** *adverb*

con·tume·ly \kän-ˈtü-mə-lē, kən-, -ˈtyü-; ˈkän-tü-ˌmē-lē, -tyü-ˌ, -chə-ˌ; *in "Hamlet"* ˈkän-(ˌ)tyüm-lē *or* ˈkän-chəm-\ *noun, plural* **-lies** [Middle English *contumelie,* from Middle French, from Latin *contumelia*] (14th century)
: harsh language or treatment arising from haughtiness and contempt; *also* **:** an instance of such language or treatment

con·tu·sion \kən-ˈtü-zhən, -ˈtyü-\ *noun* [Middle English *conteschown,* from Middle French *contusion,* from Latin *contusion-, contusio,* from *contundere* to pound, bruise, from *com-* + *tundere* to beat; akin to Gothic *stautan* to strike, Sanskrit *tudati* he pushes] (15th century)
: injury to tissue usually without laceration **:** BRUISE 1a
— **con·tuse** \-ˈtüz, -ˈtyüz\ *transitive verb*

co·nun·drum \kə-ˈnən-drəm\ *noun* [origin unknown] (1645)
1 : a riddle whose answer is or involves a pun
2 a : a question or problem having only a conjectural answer **b :** an intricate and difficult problem

con·ur·ba·tion \ˌkä-(ˌ)nər-ˈbā-shən\ *noun* [*com-* + Latin *urb-, urbs* city] (1915)
: an aggregation or continuous network of urban communities

co·nus ar·te·ri·o·sus \ˈkō-nəs-är-ˌtir-ē-ˈō-səs\ *noun, plural* **co·ni ar·te·ri·o·si** \-ˌnī-är-ˌtir-ē-ˈō-ˌsī\ [New Latin, literally, arterial cone] (circa 1860)
1 : a conical prolongation of the right ventricle in mammals from which the pulmonary arteries emerge — called also *conus*
2 : a prolongation of the ventricle of amphibians and some fishes that has a spiral valve separating venous blood going to the respiratory arteries from blood going to the aorta and systemic arteries

con·va·lesce \ˌkän-və-ˈles\ *intransitive verb* **-lesced; -lesc·ing** [Latin *convalescere,* from *com-* + *valescere* to grow strong, from

valēre to be strong, be well — more at WIELD] (15th century)
: to recover health and strength gradually after sickness or weakness
— **con·va·les·cence** \-'le-sᵊn(t)s\ *noun*
— **con·va·les·cent** \-sᵊnt\ *adjective or noun*

con·vect \kən-'vekt\ *verb* [back-formation from *convection*] (1881)
intransitive verb
: to transfer heat by convection
transitive verb
: to circulate (as air) by convection
— **con·vec·tive** \-'vek-tiv\ *adjective*

con·vec·tion \kən-'vek-shən\ *noun* [Late Latin *convection-, convectio,* from Latin *convehere* to bring together, from *com-* + *vehere* to carry — more at WAY] (circa 1623)
1 : the action or process of conveying
2 a : the circulatory motion that occurs in a fluid at a nonuniform temperature owing to the variation of its density and the action of gravity **b :** the transfer of heat by convection — compare CONDUCTION, RADIATION
— **con·vec·tion·al** \-shnəl, -shə-nᵊl\ *adjective*

convection oven *noun* (1973)
: an oven having a fan that circulates hot air uniformly and continuously around food

con·vec·tor \-'vek-tər\ *noun* (1907)
: a heating unit in which air heated by contact with a heating device (as a radiator or a tube with fins) in a casing circulates by convection

con·vene \kən-'vēn\ *verb* **con·vened; con·ven·ing** [Middle English, from Middle French *convenir* to come together, from Latin *convenire*] (15th century)
intransitive verb
: to come together in a body
transitive verb
1 : to summon before a tribunal
2 : to cause to assemble
synonym see SUMMON
— **con·ven·er** *or* **con·ve·nor** \-'vē-nər\ *noun*

¹con·ve·nience \kən-'vēn-yən(t)s\ *noun* (14th century)
1 : fitness or suitability for performing an action or fulfilling a requirement
2 a : something (as an appliance, device, or service) conducive to comfort or ease **b** *chiefly British* **:** TOILET 3
3 : a suitable or convenient time ⟨at your *convenience*⟩
4 : freedom from discomfort **:** EASE

²convenience *adjective* (1917)
: designed for quick and easy preparation or use ⟨*convenience* foods⟩

convenience store *noun* (1965)
: a small often franchised market that is open long hours

con·ve·nien·cy \-yən(t)-sē\ *noun* (1601)
archaic **:** CONVENIENCE

con·ve·nient \kən-'vēn-yənt\ *adjective* [Middle English, from Latin *convenient-, conveniens,* from present participle of *convenire* to come together, be suitable, from *com-* + *venire* to come — more at COME] (14th century)
1 *obsolete* **:** SUITABLE, PROPER
2 a : suited to personal comfort or to easy performance **b :** suited to a particular situation **c :** affording accommodation or advantage
3 : being near at hand **:** HANDY
— **con·ve·nient·ly** *adverb*

¹con·vent \'kän-vənt, -ˌvent\ *noun* [Middle English *covent,* from Old French, from Medieval Latin *conventus,* from Latin, assembly, from *convenire*] (13th century)
: a local community or house of a religious order or congregation; *especially* **:** an establishment of nuns

²con·vent \kən-'vent\ *verb* [Latin *conventus,* past participle of *convenire*] (1514)
obsolete **:** CONVENE

con·ven·ti·cle \kən-'ven-ti-kəl\ *noun* [Middle English, from Latin *conventiculum,* diminutive of *conventus* assembly] (14th century)
1 : ASSEMBLY, MEETING
2 : an assembly of an irregular or unlawful character
3 : an assembly for religious worship; *especially* **:** a secret meeting for worship not sanctioned by law
4 : MEETINGHOUSE
— **con·ven·ti·cler** \-k(ə-)lər\ *noun*

con·ven·tion \kən-'ven(t)-shən\ *noun* [Middle English, from Middle French or Latin; Middle French, from Latin *convention-, conventio,* from *convenire*] (15th century)
1 a : AGREEMENT, CONTRACT **b :** an agreement between states for regulation of matters affecting all of them **c :** a compact between opposing commanders especially concerning prisoner exchange or armistice **d :** a general agreement about basic principles or procedures; *also* **:** a principle or procedure accepted as true or correct by convention
2 a : the summoning or convening of an assembly **b :** an assembly of persons met for a common purpose; *especially* **:** a meeting of the delegates of a political party for the purpose of formulating a platform and selecting candidates for office **c :** the usually state or national organization of a religious denomination
3 a : usage or custom especially in social matters **b :** a rule of conduct or behavior **c :** a practice in bidding or playing that conveys information between partners in a card game (as bridge) **d :** an established technique, practice, or device (as in literature or the theater)

con·ven·tion·al \kən-'vench-nəl, -'ven(t)-shə-nᵊl\ *adjective* (15th century)
1 : formed by agreement or compact
2 a : according with, sanctioned by, or based on convention **b :** lacking originality or individuality **:** TRITE **c** (1) **:** ORDINARY, COMMONPLACE (2) **:** NONNUCLEAR 1 ⟨*conventional* warfare⟩
3 a : according with a mode of artistic representation that simplifies or provides symbols or substitutes for natural forms **b :** of traditional design
4 : of, resembling, or relating to a convention, assembly, or public meeting
synonym see CEREMONIAL
— **con·ven·tion·al·ism** \-nə-ˌli-zəm, -nᵊl-ˌi-zəm\ *noun*
— **con·ven·tion·al·ist** \-list\ *noun or adjective*
— **con·ven·tion·al·i·za·tion** \-ˌvench-nə-lə-'zā-shən, -ˌven(t)-shə-nᵊl-ə-'zā-\ *noun*
— **con·ven·tion·al·ize** \-'vench-nə-ˌlīz, -'ven(t)-shə-nᵊl-ˌīz\ *transitive verb*
— **con·ven·tion·al·ly** *adverb*

con·ven·tion·al·i·ty \-ˌven(t)-shə-'na-lə-tē\ *noun, plural* **-ties** (circa 1834)
1 : a conventional usage, practice, or thing
2 : the quality or state of being conventional; *especially* **:** adherence to conventions

con·ven·tion·eer \kən-ˌven(t)-shə-'nir\ *noun* (1926)
: a person attending a convention

¹con·ven·tu·al \kən-'ven-chə-wəl, kän-, -'vench-wəl\ *adjective* [Middle English, from Middle French or Medieval Latin; Middle French, from Medieval Latin *conventualis,* from *conventus* convent] (15th century)
1 : of, relating to, or befitting a convent or monastic life **:** MONASTIC
2 *capitalized* **:** of or relating to the Conventuals
— **con·ven·tu·al·ly** *adverb*

²conventual *noun* (1533)
1 *capitalized* **:** a member of the Order of Friars Minor Conventual forming a branch of the first order of Saint Francis of Assisi under a mitigated rule
2 : a member of a conventual community

con·verge \kən-'vərj\ *verb* **con·verged; con·verg·ing** [Late Latin *convergere,* from Latin *com-* + *vergere* to bend, incline — more at WRENCH] (1691)
intransitive verb
1 : to tend or move toward one point or one another **:** come together **:** MEET
2 : to come together and unite in a common interest or focus
3 : to approach a limit as the number of terms increases without limit
transitive verb
: to cause to converge

con·ver·gence \kən-'vər-jən(t)s\ *noun* (1713)
1 : the act of converging and especially moving toward union or uniformity; *especially* **:** coordinated movement of the two eyes so that the image of a single point is formed on corresponding retinal areas
2 : the state or property of being convergent
3 : independent development of similar characters (as of bodily structure or cultural traits) often associated with similarity of habits or environment

con·ver·gen·cy \-jən(t)-sē\ *noun* (1709)
: CONVERGENCE

con·ver·gent \-jənt\ *adjective* (circa 1751)
1 : tending to move toward one point or to approach each other **:** CONVERGING ⟨*convergent* lines⟩
2 : exhibiting convergence in form, function, or development ⟨*convergent* evolution⟩
3 a *of an improper integral* **:** having a value that is a real number **b :** characterized by having the *n*th term or the sum of the first *n* terms approach a finite limit ⟨a *convergent* sequence⟩ ⟨a *convergent* series⟩

converging lens *noun* (1860)
: a lens that causes parallel rays (as of light) to come to a focus

con·vers·able \kən-'vər-sə-bəl\ *adjective* (circa 1631)
1 *archaic* **:** relating to or suitable for social interaction
2 : pleasant and easy to converse with

con·ver·sance \kən-'vər-sᵊn(t)s *also* 'kän-vər-sən(t)s\ *noun* (1609)
: the quality or state of being conversant

con·ver·san·cy \-sᵊn(t)-sē, -sən(t)-\ *noun* (1798)
: CONVERSANCE

con·ver·sant \kən-'vər-sᵊnt *also* 'kän-vər-sənt\ *adjective* (14th century)
1 *archaic* **:** having frequent or familiar association
2 *archaic* **:** CONCERNED, OCCUPIED
3 : having knowledge or experience — used with *with*

con·ver·sa·tion \ˌkän-vər-'sā-shən\ *noun* [Middle English *conversacioun,* from Middle French *conversation,* from Latin *conversation-, conversatio,* from *conversari* to associate with, frequentative of *convertere* to turn around] (14th century)
1 *obsolete* **:** CONDUCT, BEHAVIOR
2 a (1) **:** oral exchange of sentiments, observations, opinions, or ideas (2) **:** an instance of such exchange **:** TALK **b :** an informal discussion of an issue by representatives of governments, institutions, or groups **c :** an exchange similar to conversation
— **con·ver·sa·tion·al** \-shnəl, -shə-nᵊl\ *adjective*
— **con·ver·sa·tion·al·ly** *adverb*

con·ver·sa·tion·al·ist \-shnə-list, -shə-nᵊl-ist\ *noun* (1836)
: one who converses a great deal or who excels in conversation

conversation piece *noun* (1712)
1 : a painting of a group of persons in their customary surroundings
2 : something (as a novel or unusual object) that stimulates conversation

con·ver·sa·zi·o·ne \ˌkän-vər-ˌsät-sē-'ō-nē, ˌkōn-\ *noun, plural* **-o·nes** *or* **-o·ni** \-'ō-(ˌ)nē\ [Italian, literally, conversation, from Latin *conversation-, conversatio*] (1739)

: a meeting for conversation especially about art, literature, or science

[1]**con·verse** \'kän-ˌvərs\ *noun* (15th century)
1 *obsolete* **:** social interaction
2 : CONVERSATION

[2]**con·verse** \kən-'vərs\ *intransitive verb* **con·versed; con·vers·ing** [Middle English, to live (with), from Middle French *converser,* from Latin *conversari*] (1586)
1 *archaic* **a :** to become occupied or engaged **b :** to have acquaintance or familiarity
2 a : to exchange thoughts and opinions in speech **:** TALK **b :** to carry on an exchange similar to a conversation (as with a computer)
— **con·vers·er** \-'vər-sər\ *noun*

[3]**con·verse** \'kän-ˌvərs\ *noun* [Latin *conversus,* past participle of *convertere*] (1570)
: something reversed in order, relation, or action: as **a :** a theorem formed by interchanging the hypothesis and conclusion of a given theorem **b :** a proposition obtained by interchange of the subject and predicate of a given proposition ⟨"no *P* is *S*" is the *converse* of "no *S* is *P*"⟩

[4]**con·verse** \kən-'vərs, 'kän-ˌ\ *adjective* (1794)
1 : reversed in order, relation, or action
2 : being a logical or mathematical converse ⟨the *converse* theorem⟩
— **con·verse·ly** *adverb*

con·ver·sion \kən-'vər-zhən, -shən\ *noun* [Middle English, from Middle French, from Latin *conversion-, conversio,* from *convertere*] (14th century)
1 : the act of converting **:** the process of being converted — compare GENE CONVERSION
2 : an experience associated with a definite and decisive adoption of religion
3 a : the operation of finding a converse in logic or mathematics **b :** reduction of a mathematical expression by clearing of fractions
4 : a successful try for point or free throw
5 : something converted from one use to another
— **con·ver·sion·al** \-'vərzh-nəl; -'vər-zhə-, -'vər-shnəl, -shə-nᵊl\ *adjective*

conversion reaction *noun* (1945)
: a psychoneurosis in which bodily symptoms (as paralysis of the limbs) appear without physical basis — called also *conversion hysteria*

[1]**con·vert** \kən-'vərt\ *verb* [Middle English, from Middle French *convertir,* from Latin *convertere* to turn around, transform, convert, from *com-* + *vertere* to turn — more at WORTH] (14th century)
transitive verb
1 a : to bring over from one belief, view, or party to another **b :** to bring about a religious conversion in
2 a : to alter the physical or chemical nature or properties of especially in manufacturing **b** (1) **:** to change from one form or function to another (2) **:** to alter for more effective utilization (3) **:** to appropriate without right **c :** to exchange for an equivalent
3 *obsolete* **:** TURN
4 : to subject to logical conversion
5 a : to make a goal after receiving (a pass) from a teammate **b :** to score on (as a try for point or free throw) **c :** to make (a spare) in bowling
intransitive verb
1 : to undergo conversion
2 : to make good on a try for point, field goal, or free throw
synonym see TRANSFORM

[2]**con·vert** \'kän-ˌvərt\ *noun* (1561)
: one that is converted

con·vert·er \kən-'vər-tər\ *noun* (1533)
: one that converts: as **a :** the furnace used in the Bessemer process **b** *or* **con·ver·tor** \-'vər-tər\ **:** a device employing mechanical rotation for changing electrical energy from one form to another (as from direct current to alternating current or vice versa); *also* **:** a radio device for converting one frequency to another **c :** a device for adapting a television or radio receiver to receive channels or frequencies for which it was not originally designed ⟨a cable *converter*⟩ ⟨FM *converter*⟩ **d :** a device that accepts data in one form and converts it to another ⟨analog-digital *converter*⟩ **e :** CATALYTIC CONVERTER

[1]**con·vert·ible** \kən-'vər-tə-bəl\ *adjective* (14th century)
1 : capable of being converted
2 : having a top that may be lowered or removed ⟨*convertible* coupe⟩
3 : capable of being exchanged for a specified equivalent (as another currency or security) ⟨a bond *convertible* to 12 shares of common stock⟩
— **con·vert·ibil·i·ty** \-ˌvər-tə-'bi-lə-tē\ *noun*
— **con·vert·ible·ness** \-'vər-tə-bəl-nəs\ *noun*
— **con·vert·ibly** \-blē\ *adverb*

[2]**convertible** *noun* (1615)
: something convertible; *especially* **:** a convertible automobile

con·verti·plane *also* **con·verta·plane** \kən-'vər-tə-ˌplān\ *noun* (1949)
: an aircraft that takes off and lands like a helicopter and is convertible to a fixed-wing configuration for forward flight

con·vex \kän-'veks; 'kän-ˌ, kən-'\ *adjective* [Middle French or Latin; Middle French *convexe,* from Latin *convexus* vaulted, concave, convex, from *com-* + *-vexus;* perhaps akin to Latin *vehere* to carry — more at WAY] (1571)
1 a : curved or rounded like the exterior of a sphere or circle **b :** being a continuous function or part of a continuous function with the property that a line joining any two points on its graph lies on or above the graph
2 a *of a set of points* **:** containing all points in a line joining any two constituent points **b** *of a geometric figure* **:** comprising a convex set when combined with its interior ⟨a *convex* polygon⟩

con·vex·i·ty \kən-'vek-sət-ē, kän-\ *noun, plural* **-ties** (1599)
1 : the quality or state of being convex
2 : a convex surface or part

con·vexo–con·cave \kən-ˌvek-(ˌ)sō-kän-'kāv, kän-ˌvek-, -'kän-ˌkāv\ *adjective* (1693)
1 : CONCAVO-CONVEX
2 : having the convex side of greater curvature than the concave

con·vey \kən-'vā\ *transitive verb* **conveyed; con·vey·ing** [Middle English, from Middle French *conveier* to accompany, escort, from (assumed) Vulgar Latin *conviare,* from Latin *com-* + *via* way — more at WAY] (14th century)
1 *obsolete* **:** LEAD, CONDUCT
2 a : to bear from one place to another; *especially* **:** to move in a continuous stream or mass **b :** to impart or communicate by statement, suggestion, gesture, or appearance **c** (1) *archaic* **:** STEAL (2) *obsolete* **:** to carry away secretly **d :** to transfer or deliver to another especially by a sealed writing **e :** to cause to pass from one place or person to another

con·vey·ance \kən-'vā-ən(t)s\ *noun* (15th century)
1 : the action of conveying
2 : a means or way of conveying: as **a :** an instrument by which title to property is conveyed **b :** a means of transport **:** VEHICLE

con·vey·anc·er \-ən(t)-sər\ *noun* (1650)
: one whose business is conveyancing

con·vey·anc·ing \-ən(t)-siŋ\ *noun* (1714)
: the act or business of drawing deeds, leases, or other writings for transferring the title to property

con·vey·or *also* **con·vey·er** \kən-'vā-ər\ *noun* (circa 1514)
: one that conveys: as **a :** a person who transfers property **b** *usually conveyor* **:** a mechanical apparatus for moving articles or bulk material from place to place (as by an endless moving belt or a chain of receptacles)

con·vey·or·ise *British variant of* CONVEYORIZE

con·vey·or·ize \-ə-ˌrīz\ *transitive verb* **-ized; -iz·ing** (1941)
: to equip with a conveyor
— **con·vey·or·i·za·tion** \-ˌvā-ə-rə-'zā-shən\ *noun*

[1]**con·vict** \kən-'vikt\ *adjective* (14th century) *archaic*
: having been convicted

[2]**con·vict** \kən-'vikt\ *transitive verb* [Middle English, from Latin *convictus,* past participle of *convincere* to refute, convict] (14th century)
1 : to find or prove to be guilty
2 : to convince of error or sinfulness

[3]**con·vict** \'kän-ˌvikt\ *noun* (15th century)
1 : a person convicted of and under sentence for a crime
2 : a person serving a usually long prison sentence

con·vic·tion \kən-'vik-shən\ *noun* (15th century)
1 : the act or process of convicting of a crime especially in a court of law
2 a : the act of convincing a person of error or of compelling the admission of a truth **b :** the state of being convinced of error or compelled to admit the truth
3 a : a strong persuasion or belief **b :** the state of being convinced
synonym see CERTAINTY, OPINION

con·vince \kən-'vin(t)s\ *transitive verb* **convinced; con·vinc·ing** [Latin *convincere* to refute, convict, prove, from *com-* + *vincere* to conquer — more at VICTOR] (1530)
1 *obsolete* **a :** to overcome by argument **b :** OVERPOWER, OVERCOME
2 *obsolete* **:** DEMONSTRATE, PROVE
3 : to bring (as by argument) to belief, consent, or a course of action **:** PERSUADE ⟨*convinced* himself that she was all right —William Faulkner⟩ ⟨something I could never *convince* him to read —John Lahr⟩
— **con·vinc·er** *noun*

con·vinc·ing \kən-'vin(t)-siŋ\ *adjective* (1624)
1 : satisfying or assuring by argument or proof ⟨a *convincing* test of a new product⟩
2 : having power to convince of the truth, rightness, or reality of something **:** PLAUSIBLE ⟨told a *convincing* story⟩
synonym see VALID
— **con·vinc·ing·ly** \-siŋ-lē\ *adverb*
— **con·vinc·ing·ness** *noun*

con·viv·ial \kən-'viv-yəl, -'vi-vē-əl\ *adjective* [Late Latin *convivialis,* from Latin *convivium* banquet, from *com-* + *vivere* to live — more at QUICK] (circa 1668)
: relating to, occupied with, or fond of feasting, drinking, and good company
— **con·viv·i·al·i·ty** \-ˌvi-vē-'a-lə-tē\ *noun*
— **con·viv·ial·ly** \-'viv-yə-lē, -'vi-vē-ə-lē\ *adverb*

con·vo·ca·tion \ˌkän-və-'kā-shən\ *noun* [Middle English, from Middle French, from Latin *convocation-, convocatio,* from *convocare*] (14th century)
1 a : an assembly of persons convoked **b** (1) **:** an assembly of bishops and representative clergy of the Church of England (2) **:** a consultative assembly of clergy and lay delegates from one part of an Episcopal diocese; *also* **:** a territorial division of an Episcopal diocese **c :** a ceremonial assembly of members of a college or university
2 : the act or process of convoking
— **con·vo·ca·tion·al** \-shnəl, -shə-nᵊl\ *adjective*

con·voke \kən-ˈvōk\ *transitive verb* **con·voked; con·vok·ing** [Middle French *convoquer,* from Latin *convocare,* from *com-* + *vocare* to call, from *voc-, vox* voice — more at VOICE] (1598)
: to call together to a meeting
synonym see SUMMON

con·vo·lute \ˈkän-və-ˌlüt\ *verb* **-lut·ed; -lut·ing** [Latin *convolutus,* past participle of *convolvere*] (1698)
: TWIST, COIL

con·vo·lut·ed \-ˌlü-təd\ *adjective* (1766)
1 : having convolutions
2 : INVOLVED, INTRICATE

convoluted tubule *noun* (1923)
: all or part of the coiled sections of a nephron: **a :** PROXIMAL CONVOLUTED TUBULE **b :** DISTAL CONVOLUTED TUBULE

con·vo·lu·tion \ˌkän-və-ˈlü-shən\ *noun* (1545)
1 : a form or shape that is folded in curved or tortuous windings
2 : one of the irregular ridges on the surface of the brain and especially of the cerebrum of higher mammals
3 : a complication or intricacy of form, design, or structure

con·volve \kən-ˈvälv, -ˈvȯlv *also* -ˈväv *or* -ˈvȯv\ *verb* **con·volved; con·volv·ing** [Latin *convolvere,* from *com-* + *volvere* to roll — more at VOLUBLE] (1650)
transitive verb
: to roll together **:** WRITHE
intransitive verb
: to roll together or circulate involvedly

con·vol·vu·lus \kən-ˈväl-vyə-ləs, -ˈvȯl- *also* -ˈväv-yə- *or* -ˈvȯv-yə-\ *noun, plural* **-lus·es** *or* **-li** \-ˌlī, -ˌlē\ [New Latin, from Latin *convolvere*] (1548)
: any of a genus (*Convolvulus*) of erect, trailing, or twining herbs and shrubs of the morning-glory family

¹con·voy \ˈkän-ˌvȯi, kən-ˈ\ *transitive verb* [Middle English, from Middle French *conveier, convoier* — more at CONVEY] (14th century)
: ACCOMPANY; *especially* **:** to escort for protection

²con·voy \ˈkän-ˌvȯi\ *noun* (1523)
1 : one that convoys; *especially* **:** a protective escort (as for ships)
2 : the act of convoying
3 : a group convoyed or organized for convenience or protection in moving

con·vul·sant \kən-ˈvəl-sənt\ *adjective* (1875)
: causing convulsions **:** CONVULSIVE 1a
— convulsant *noun*

con·vulse \kən-ˈvəls\ *verb* **con·vulsed; con·vuls·ing** [Latin *convulsus,* past participle of *convellere* to pluck up, convulse, from *com-* + *vellere* to pluck — more at VULNERABLE] (1643)
transitive verb
: to shake or agitate violently; *especially* **:** to shake with or as if with irregular spasms
intransitive verb
: to become affected with convulsions
synonym see SHAKE

con·vul·sion \kən-ˈvəl-shən\ *noun* (1547)
1 : an abnormal violent and involuntary contraction or series of contractions of the muscles
2 a : a violent disturbance **b :** an uncontrolled fit **:** PAROXYSM
— con·vul·sion·ary \-shə-ˌner-ē\ *adjective*

con·vul·sive \kən-ˈvəl-siv\ *adjective* (1615)
1 a : constituting or producing a convulsion **b :** attended or affected with convulsions
2 : resembling a convulsion in being violent, sudden, frantic, or spasmodic
synonym see FITFUL
— con·vul·sive·ly *adverb*
— con·vul·sive·ness *noun*

cony *variant of* CONEY

coo \ˈkü\ *intransitive verb* [imitative] (1670)
1 : to make the low soft cry of a dove or pigeon or a similar sound
2 : to talk fondly, amorously, or appreciatively ⟨an album that will be *cooed* over by condescending classical music critics —Ellen Sander⟩
— coo *noun*

¹cook \ˈku̇k\ *noun* [Middle English, from Old English *cōc,* from Latin *coquus,* from *coquere* to cook; akin to Old English *āfigen* fried, Greek *pessein* to cook] (before 12th century)
1 : a person who prepares food for eating
2 : a technical or industrial process comparable to cooking food; *also* **:** a substance so processed

²cook (14th century)
intransitive verb
1 : to prepare food for eating by means of heat
2 : to undergo the action of being cooked ⟨the rice is *cooking* now⟩
3 : OCCUR, HAPPEN ⟨find out what was *cooking* in the committee⟩
4 : to perform, do, or proceed well ⟨the jazz quartet was *cooking* along⟩ ⟨the party *cooked* right through the night⟩
transitive verb
1 : CONCOCT, FABRICATE — usually used with *up* ⟨*cooked* up a scheme⟩
2 : to prepare for eating by a heating process
3 : FALSIFY, DOCTOR ⟨*cooked* the books with phony spending cuts and accounting gimmickry —Colleen O'Connor⟩
4 : to subject to the action of heat or fire
— cook·able \ˈku̇-kə-bəl\ *adjective*
— cook one's goose : to ruin one irretrievably

¹cook·book \-ˌbu̇k\ *noun* (1809)
: a book of cooking directions and recipes; *broadly* **:** a book of detailed instructions

²cookbook *adjective* (1944)
: involving or using step-by-step procedures whose rationale is usually not explained ⟨a *cookbook* approach⟩

cook cheese *noun* (1941)
: an unripened cheese made from curd that has been cooked to a soft consistency — called also *cooked cheese*

cook·er \ˈku̇-kər\ *noun* (1869)
: one that cooks: as **a :** a utensil, device, or apparatus for cooking **b :** a person who tends a cooking process **:** COOK **c** *British* **:** STOVE

cook·ery \ˈku̇-k(ə-)rē\ *noun, plural* **-er·ies** (14th century)
1 : the art or practice of cooking
2 : an establishment for cooking

cookery book *noun* (1639)
chiefly British **:** COOKBOOK

cook·house \ˈku̇k-ˌhau̇s\ *noun* (1795)
: a building for cooking

cook·ie *or* **cooky** \ˈku̇-kē\ *noun, plural* **cook·ies** [Dutch *koekje,* diminutive of *koek* cake] (1703)
1 : a small flat or slightly raised cake
2 a : an attractive woman ⟨a buxom French *cookie* who haunts the . . . colony's one night spot —*Newsweek*⟩ **b :** PERSON, GUY ⟨a tough *cookie*⟩

cookie–cutter *adjective* (1963)
: marked by lack of originality or distinction ⟨*cookie-cutter* shopping malls⟩

cookie cutter *noun* (1903)
: a device used to cut rolled cookie dough into shapes before baking

cookie sheet *noun* (1926)
: a flat rectangle of metal with at least one rolled edge used especially for the baking of cookies or biscuits

cooking *adjective* (circa 1813)
: suitable for or used in cooking ⟨*cooking* apples⟩

cook–off \ˈku̇k-ˌȯf, -ˌäf\ *noun* (1937)
: a cooking competition

cook off *intransitive verb* (1945)
of a cartridge **:** to fire as a result of overheating

cook·out \ˈku̇k-ˌau̇t\ *noun* (1947)
: an outing at which a meal is cooked and served in the open; *also* **:** the meal cooked

cook·shack \-ˌshak\ *noun* (1909)
: a shack used for cooking

cook·shop \-ˌshäp\ *noun* (circa 1552)
: a shop supplying or serving cooked food

Cook's tour \ˈku̇ks-\ *noun* [Thomas Cook & Son, English travel agency] (circa 1909)
: a rapid or cursory survey or review

cook·stove \ˈku̇k-ˌstōv\ *noun* (1824)
: a stove for cooking

cook·top \ˈku̇k-ˌtäp\ *noun* (1948)
1 : the flat top of a range
2 : a built-in cabinet-top cooking apparatus containing usually four heating units

cook·ware \ˈku̇k-ˌwar, -ˌwer\ *noun* (1953)
: utensils used in cooking

¹cool \ˈkül\ *adjective* [Middle English *col,* from Old English *cōl;* akin to Old High German *kuoli* cool, Old English *ceald* cold — more at COLD] (before 12th century)
1 : moderately cold **:** lacking in warmth
2 a : marked by steady dispassionate calmness and self-control ⟨a *cool* and calculating administrator —*Current Biography*⟩ **b :** lacking ardor or friendliness ⟨a *cool* impersonal manner⟩ **c** *of jazz* **:** marked by restrained emotion and the frequent use of counterpoint **d :** free from tensions or violence ⟨meeting with minority groups in an attempt to keep the city *cool*⟩
3 — used as an intensive ⟨a *cool* million dollars⟩
4 : marked by deliberate effrontery or lack of due respect or discretion ⟨a *cool* reply⟩
5 : facilitating or suggesting relief from heat ⟨a *cool* dress⟩
6 a *of a color* **:** producing an impression of being cool; *specifically* **:** of a hue in the range violet through blue to green **b** *of a musical tone* **:** relatively lacking in timbre or resonance
7 *slang* **a :** very good **:** EXCELLENT; *also* **:** ALL RIGHT **b :** FASHIONABLE 1 ☆
— cool·ish \ˈkü-lish\ *adjective*
— cool·ly *also* **cooly** \ˈkü(l)-lē\ *adverb*
— cool·ness \ˈkül-nəs\ *noun*

²cool (before 12th century)
intransitive verb
1 : to become cool **:** lose heat or warmth ⟨placed the pie in the window to *cool*⟩ — sometimes used with *off* or *down*
2 : to lose ardor or passion ⟨his anger *cooled*⟩
transitive verb
1 : to make cool **:** impart a feeling of coolness to ⟨*cooled* the room with a fan⟩ — often used with *off* or *down* ⟨a swim *cooled* us off a little⟩
2 a : to moderate the heat, excitement, or force of **:** CALM ⟨*cooled* her growing anger⟩ **b :** to slow or lessen the growth or activity of — usually used with *off* or *down* ⟨wants to *cool* off the economy without freezing it —*Newsweek*⟩

☆ SYNONYMS
Cool, composed, collected, unruffled, imperturbable, nonchalant mean free from agitation or excitement. COOL may imply calmness, deliberateness, or dispassionateness ⟨kept a *cool* head⟩. COMPOSED implies freedom from agitation as a result of self-discipline or a sedate disposition ⟨the *composed* pianist gave a flawless concert⟩. COLLECTED implies a concentration of mind that eliminates distractions especially in moments of crisis ⟨the nurse stayed calm and *collected*⟩. UNRUFFLED suggests apparent serenity and poise in the face of setbacks or in the midst of excitement ⟨harried but *unruffled*⟩. IMPERTURBABLE implies coolness or assurance even under severe provocation ⟨the speaker remained *imperturbable* despite the heckling⟩. NONCHALANT stresses an easy coolness of manner or casualness that suggests indifference or unconcern ⟨a *nonchalant* driver⟩.

— **cool it :** to calm down **:** go easy ⟨the word went out to the young to *cool it* —W. M. Young⟩
— **cool one's heels :** to wait or be kept waiting for a long time especially from or as if from disdain or discourtesy

³cool *noun* (15th century)
1 : a cool time, place, or situation ⟨the *cool* of the evening⟩
2 a : absence of excitement or emotional involvement **:** DETACHMENT ⟨must surrender his fine *cool* and enter the closed crazy world of suicide —Wilfrid Sheed⟩ **b :** POISE, COMPOSURE ⟨press questions . . . seemed to rattle him and he lost his *cool* —*New Republic*⟩

⁴cool *adverb* (1951)
: in a casual and nonchalant manner ⟨play it *cool*⟩

cool•ant \ˈkü-lənt\ *noun* (1926)
: a usually fluid cooling agent

cool•down \ˈkül-ˌdau̇n\ *noun* (1980)
: the act or an instance of allowing physiological activity to return to normal gradually after strenuous exercise by engaging in less strenuous exercise

cool•er \ˈkü-lər\ *noun* (1575)
1 : one that cools: as **a :** a container for cooling liquids **b :** REFRIGERATOR
2 : LOCKUP, JAIL; *especially* **:** a cell for violent or unmanageable prisoners
3 : an iced drink usually with an alcoholic beverage as a base

Coo•ley's anemia \ˈkü-lēz-\ *noun* [Thomas B. *Cooley* (died 1945) American pediatrician] (circa 1935)
: a severe thalassemic anemia that is associated with the presence of microcytes, enlargement of the liver and spleen, increase in the erythroid bone marrow, and jaundice and that occurs especially in children of Mediterranean parents

cool•head•ed \ˈkül-ˌhe-dəd\ *adjective* (1777)
: not easily excited

coo•lie \ˈkü-lē\ *noun* [Hindi *kulī*] (1638)
: an unskilled laborer or porter usually in or from the Far East hired for low or subsistence wages

coolie hat *noun* (1924)
: a conical-shaped usually straw hat worn especially to protect the head from the heat of the sun

cool•ing–off \ˈkü-liŋ-ˈȯf\ *adjective* (1926)
: designed to allow passions to cool or to permit negotiation between parties ⟨a *cooling-off* period⟩

cooling tower *noun* (1901)
: a structure over which circulated water is trickled to reduce its temperature by partial evaporation

coombe *or* **coomb** \ˈküm\ *variant of* COMBE

coon \ˈkün\ *noun* (1742)
1 : RACCOON
2 : BLACK 4 — usually taken to be offensive

coon•can \ˈkün-ˌkan\ *noun* [by folk etymology from Mexican Spanish *conquián* conquian, from Spanish *¿con quién?* with whom?] (1889)
: a game of rummy played with two packs including two jokers

coon cat *noun* (1901)
: MAINE COON

coon cheese \ˈkün-\ *noun* [probably from *coon* (black person); from the color of the coating] (1953)
: a sharp cheddar cheese that has been cured at higher than usual temperature and humidity and that is usually coated with black wax

coon•hound \ˈkün-ˌhau̇nd\ *noun* (1920)
: a sporting dog trained to hunt raccoons; *especially* **:** BLACK-AND-TAN COONHOUND

coon's age *noun* (1844)
: a long while ⟨best fried chicken I've tasted for a *coon's age* —Sinclair Lewis⟩

coon•skin \ˈkün-ˌskin\ *noun* (1818)
1 : the skin or pelt of the raccoon
2 : an article (as a cap or coat) made of coonskin

coon•tie \ˈkün-tē\ *noun* [Creek (Florida dialect) *kuntí*] (1791)
: any of several tropical American woody plants (genus *Zamia*) of the cycad family whose roots and stems yield a starchy foodstuff — called also *arrowroot*

¹coop \ˈküp, ˈku̇p\ *noun* [Middle English *cupe;* akin to Old English *cȳpe* basket] (14th century)
1 : a cage or small enclosure (as for poultry); *also* **:** a small building for housing poultry
2 a : a confined area **b :** JAIL

²coop *transitive verb* (1583)
1 : to confine in a restricted and often crowded area — usually used with *up*
2 : to place or keep in a coop **:** PEN — often used with *up*

co–op \ˈkō-ˌäp, kō-ˈ\ *noun* (1869)
: COOPERATIVE

¹coo•per \ˈkü-pər, ˈku̇-\ *noun* [Middle English *couper, cowper,* from Middle Dutch *cūper* (from *cūpe* cask) or Middle Low German *kūper,* from *kūpe* cask; Middle Dutch *cūpe* & Middle Low German *kūpe,* from Latin *cupa;* akin to Greek *kypellon* cup — more at HIVE] (14th century)
: one that makes or repairs wooden casks or tubs

²cooper *verb* **coo•pered; coo•per•ing** \ˈkü-p(ə-)riŋ, ˈku̇-\ (1742)
transitive verb
: to work as a cooper on
intransitive verb
: to work at or do coopering

coo•per•age \ˈkü-p(ə-)rij, ˈku̇-\ *noun* (1705)
1 : a cooper's work or products
2 : a cooper's place of business

co•op•er•ate \kō-ˈä-pə-ˌrāt\ *intransitive verb* [Late Latin *cooperatus,* past participle of *cooperari,* from Latin *co-* + *operari* to work — more at OPERATE] (1582)
1 : to act or work with another or others **:** act together
2 : to associate with another or others for mutual benefit
— **co•op•er•a•tor** \-ˌrā-tər\ *noun*

co•op•er•a•tion \(ˌ)kō-ˌä-pə-ˈrā-shən\ *noun* (14th century)
1 : the action of cooperating **:** common effort
2 : association of persons for common benefit
— **co•op•er•a•tion•ist** \-sh(ə-)nist\ *noun*

¹co•op•er•a•tive \kō-ˈä-p(ə-)rə-tiv, -ˈä-pə-ˌrā-\ *adjective* (1603)
1 a : marked by cooperation ⟨*cooperative* efforts⟩ **b :** marked by a willingness and ability to work with others ⟨*cooperative* neighbors⟩
2 : of, relating to, or organized as a cooperative
3 : relating to or comprising a program of combined liberal arts and technical studies at different schools
— **co•op•er•a•tive•ly** *adverb*
— **co•op•er•a•tive•ness** *noun*

²cooperative *noun* (1883)
: an enterprise or organization owned by and operated for the benefit of those using its services

Coo•per's hawk \ˈkü-pərz-, ˈku̇-\ *noun* [William *Cooper* (died 1864) American naturalist] (1828)
: an American hawk (*Accipiter cooperii*) that is larger than the similarly colored sharp-shinned hawk and has a more rounded tail

co–opt \kō-ˈäpt\ *transitive verb* [Latin *cooptare,* from *co-* + *optare* to choose] (1651)
1 a : to choose or elect as a member **b :** to appoint as a colleague or assistant
2 a : to take into a group (as a faction, movement, or culture) **:** ABSORB, ASSIMILATE ⟨the students are *co-opted* by a system they serve even in their struggle against it —A. C. Danto⟩ **b :** TAKE OVER, APPROPRIATE
— **co–op•ta•tion** \ˌkō-ˌäp-ˈtā-shən\ *noun*
— **co–op•ta•tive** \kō-ˈäp-tə-tiv\ *adjective*
— **co–op•tion** \-ˈäp-shən\ *noun*
— **co–op•tive** \-ˈäp-tiv\ *adjective*

¹co•or•di•nate \kō-ˈȯrd-nət; -ˈȯr-dᵊn-ət, -də-ˌnāt\ *adjective* [probably back-formation from *coordination*] (1641)
1 a : equal in rank, quality, or significance **b :** being of equal rank in a sentence ⟨*coordinate* clauses⟩
2 : relating to or marked by coordination
3 a : being a university that awards degrees to men and women taught usually by the same faculty but attending separate classes often on separate campuses **b :** being one of the colleges and especially the women's branch of a coordinate university
4 : of, relating to, or being a system of indexing by two or more terms so that documents may be retrieved through the intersection of index terms
— **co•or•di•nate•ly** *adverb*
— **co•or•di•nate•ness** *noun*

²co•or•di•nate \kō-ˈȯr-dᵊn-ˌāt\ *verb* **-nat•ed; -nat•ing** (1665)
transitive verb
1 : to put in the same order or rank
2 : to bring into a common action, movement, or condition **:** HARMONIZE
3 : to attach so as to form a coordination complex
intransitive verb
1 : to be or become coordinate especially so as to act together in a smooth concerted way
2 : to combine by means of a coordinate bond
— **co•or•di•na•tive** \kō-ˈȯrd-nə-tiv; -ˈȯr-dᵊn-ə-tiv, -də-ˌnā-\ *adjective*
— **co•or•di•na•tor** \-ˈȯr-də-ˌnā-tər\ *noun*

³co•or•di•nate *same as* ¹\ *noun* (circa 1823)
1 a : any of a set of numbers used in specifying the location of a point on a line, on a surface, or in space **b :** any one of a set of variables used in specifying the state of a substance or the motion of a particle or momentum
2 : one who is of equal rank, authority, or importance with another
3 *plural* **:** articles (as of clothing) designed to be used together and to attain their effect through pleasing contrast (as of color, material, or texture)

coordinate bond *noun* (1947)
: a covalent bond that consists of a pair of electrons supplied by only one of the two atoms it joins

co•or•di•nat•ed \-ˈȯr-də-ˌnā-təd\ *adjective* (1939)
: able to use more than one set of muscle movements to a single end ⟨a well-*coordinated* athlete⟩

Coordinated Universal Time *noun* (1969)
: the international standard of time that is kept by atomic clocks around the world — abbreviation *UTC*

coordinate geometry *noun* (1855)
: ANALYTIC GEOMETRY

coordinating conjunction *noun* (1916)
: a conjunction that joins together words or word groups of equal grammatical rank

co•or•di•na•tion \(ˌ)kō-ˌȯr-də-ˈnā-shən\ *noun* [Late Latin *coordination-, coordinatio,* from Latin *co-* + *ordination-, ordinatio* arrangement, from *ordinare* to arrange — more at ORDAIN] (circa 1643)
1 : the act or action of coordinating
2 : the harmonious functioning of parts for effective results

coordination complex *noun* (1951)
: a compound or ion with a central usually metallic atom or ion combined by coordinate bonds with a definite number of surrounding

ions, groups, or molecules — called also *coordination compound*

coordination number *noun* (1908)
1 : the number of attachments to the central atom in a coordination complex
2 : a number used in specifying the spatial arrangement of the constituent groups of crystals

coot \'küt\ *noun* [Middle English *coote;* akin to Dutch *koet* coot] (15th century)
1 : any of various sluggish slow-flying slaty-black birds (genus *Fulica*) of the rail family that somewhat resemble ducks and have lobed toes and the upper mandible prolonged on the forehead as a horny frontal shield
2 : any of several North American scoters
3 : a harmless simple person; *broadly* **:** FELLOW 4c

coo·ter \'kü-tər *also* 'kü-tə\ *noun* [of African origin; akin to Bambara & Malinke *kuta* turtle] (1832)
chiefly Southern **:** any of several turtles (genus *Pseudemys* and especially *P. concinna*) especially of the southern and eastern U.S.

coo·tie \'kü-tē\ *noun* [perhaps modification of Malay *kutu*] (1917)
: BODY LOUSE

[1]cop \'käp\ *noun* [Middle English, from Old English *copp*] (before 12th century)
1 *dialect chiefly English* **:** TOP, CREST
2 : a cylindrical or conical mass of thread, yarn, or roving wound on a quill or tube; *also* **:** a quill or tube upon which it is wound

[2]cop *transitive verb* **copped; cop·ping** [perhaps from Dutch *kapen* to steal, from Frisian *kāpia* to take away; akin to Old High German *kouf* trade — more at CHEAP] (1704)
1 *slang* **:** to get hold of **:** CATCH, CAPTURE; *also* **:** PURCHASE
2 *slang* **:** STEAL, SWIPE
— cop a plea : to plead guilty to a lesser charge in order to avoid standing trial for a more serious one; *broadly* **:** to admit fault and plead for mercy

[3]cop *noun* [short for [3]*copper*] (1859)
: POLICE OFFICER

co·pa·cet·ic *or* **co·pe·set·ic** *also* **co·pa·set·ic** \,kō-pə-'se-tik\ *adjective* [origin unknown] (1919)
: very satisfactory

co·pai·ba \kō-'pī-bə, -'pā-\ *noun* [Spanish & Portuguese; Spanish, from Portuguese *copaíba,* from Tupi] (1712)
: a stimulant oleoresin obtained from several pinnate-leaved South American leguminous trees (genus *Copaifera*); *also* **:** one of these trees

co·pal \'kō-pəl, -,pal; kō-'pal\ *noun* [Spanish, from Nahuatl *copalli* resin] (1577)
: a recent or fossil resin from various tropical trees

co·par·ce·nary \kō-'pär-sᵊn-,er-ē\ *noun, plural* **-nar·ies** (circa 1504)
1 : joint heirship
2 : joint ownership

co·par·ce·ner \-'pärs-nər, -'pär-sᵊn-ər\ *noun* (15th century)
: a joint heir

co–pay·ment \'kō-,pā-mənt, ,kō-'\ *noun* (1975)
: a relatively small fixed fee required by a health insurer (as an HMO) to be paid by the patient at the time of each office visit, outpatient service, or filling of a prescription

[1]cope \'kōp\ *noun* [Middle English, from Old English *-cāp,* from Late Latin *cappa* head covering] (13th century)
1 : a long enveloping ecclesiastical vestment
2 a : something resembling a cope (as by concealing or covering) ⟨the dark sky's starry *cope* —P. B. Shelley⟩ **b :** COPING

[2]cope *transitive verb* **coped; cop·ing** (14th century)
: to cover or furnish with a cope

[3]cope *verb* **coped; cop·ing** [Middle English *copen, coupen,* from Middle French *couper* to strike, cut, from Old French, from *coup* blow, from Late Latin *colpus,* alteration of Latin *colaphus,* from Greek *kolaphos* buffet] (14th century)
intransitive verb
1 *obsolete* **:** STRIKE, FIGHT
2 a : to maintain a contest or combat usually on even terms or with success — used with *with* **b :** to deal with and attempt to overcome problems and difficulties — often used with *with*
3 *archaic* **:** MEET, ENCOUNTER
transitive verb
1 *obsolete* **:** to meet in combat
2 *obsolete* **:** to come in contact with
3 *obsolete* **:** MATCH

[4]cope *transitive verb* **coped; cop·ing** [probably from French *couper* to cut] (circa 1901)
1 : to shape (a structural member) to fit a coping or fit with the shape of another member
2 : NOTCH

co·peck *variant of* KOPECK

co·pe·pod \'kō-pə-,päd\ *noun* [ultimately from Greek *kōpē* oar, handle + *pod-, pous* foot; probably akin to Latin *capere* to take — more at HEAVE, FOOT] (1836)
: any of a large subclass (Copepoda) of usually minute freshwater and marine crustaceans
— copepod *adjective*

cop·er \'kō-pər\ *noun* [English dialect *cope* to trade] (1825)
British **:** a horse dealer; *especially* **:** a dishonest one

Co·per·ni·can \kə-'pər-ni-kən, kō-\ *adjective* [Nicolaus *Copernicus*] (1667)
1 : of or relating to Copernicus or the belief that the earth rotates daily on its axis and the planets revolve in orbits around the sun
2 : of radical or major importance or degree ⟨effected a *Copernican* revolution in philosophy —*Times Literary Supplement*⟩
— Copernican *noun*
— Co·per·ni·can·ism \-kə-,ni-zəm\ *noun*

cope·stone \'kōp-,stōn\ *noun* (1567)
1 : a stone forming a coping
2 : a finishing touch **:** CROWN

copi·er \'kä-pē-ər\ *noun* (1597)
: one that copies; *specifically* **:** a machine for making copies of graphic matter (as printing, drawings, or pictures)

co·pi·lot \'kō-,pī-lət\ *noun* (1927)
: a qualified pilot who assists or relieves the pilot but is not in command

cop·ing \'kō-piŋ\ *noun* (1601)
: the covering course of a wall usually with a sloping top

cop·ing saw \'kō-piŋ-\ *noun* [from present participle of [4]*cope*] (1925)
: a handsaw with a very narrow blade held under tension in a U-shaped frame and used especially for cutting curves in wood

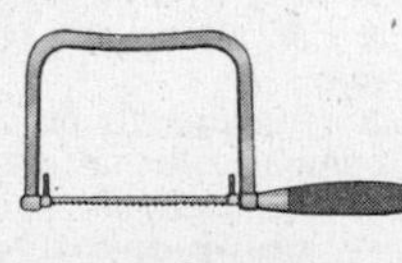
coping saw

cop·ing·stone \'kō-piŋ-,stōn\ *noun* (1778)
chiefly British **:** COPESTONE

co·pi·ous \'kō-pē-əs\ *adjective* [Middle English, from Latin *copiosus,* from *copia* abundance, from *co-* + *ops* wealth — more at OPULENT] (14th century)
1 a : yielding something abundantly ⟨a *copious* harvest⟩ ⟨*copious* springs⟩ **b :** plentiful in number ⟨*copious* references to other writers⟩
2 a : full of thought, information, or matter **b :** profuse or exuberant in words, expression, or style ⟨a *copious* talker⟩
3 : present in large quantity **:** taking place on a large scale ⟨*copious* weeping⟩ ⟨*copious* food and drink⟩
synonym see PLENTIFUL
— co·pi·ous·ly *adverb*
— co·pi·ous·ness *noun*

co·pla·nar \(,)kō-'plā-nər, -,när\ *adjective* (1862)
: lying or acting in the same plane
— co·pla·nar·i·ty \,kō-plā-'nar-ə-tē\ *noun*

co·pol·y·mer \(,)kō-'pä-lə-mər\ *noun* (1936)
: a product of copolymerization
— co·pol·y·mer·ic \,kō-,pä-lə-'mer-ik\ *adjective*

co·po·ly·mer·i·za·tion \,kō-pə-,li-mə-rə-'zā-shən, ,kō-,pä-lə-mə-\ *noun* (1936)
: the polymerization of two substances (as different monomers) together
— co·po·ly·mer·ize \,kō-pə-'li-mə-,rīz, ,kō-'pä-lə-mə-\ *verb*

cop–out \'käp-,au̇t\ *noun* (circa 1942)
1 : the act or an instance of copping out
2 : an excuse or means for copping out **:** PRETEXT
3 : a person who cops out

cop out *intransitive verb* (circa 1961)
1 : to back out (as of an unwanted responsibility) ⟨*cop out* on jury duty⟩
2 : to avoid or neglect problems, responsibilities, or commitments ⟨accused the mayor of *copping out* on the issue of homelessness⟩

[1]cop·per \'kä-pər\ *noun, often attributive* [Middle English *coper,* from Old English, from Late Latin *cuprum* copper, from Latin *(aes) Cyprium,* literally, Cyprian metal] (before 12th century)
1 : a common reddish metallic element that is ductile and malleable and is one of the best conductors of heat and electricity — see ELEMENT table
2 : a coin or token made of copper or bronze
3 *chiefly British* **:** a large boiler (as for cooking)
4 : any of a subfamily (Lycaeninae of the family Lycaenidae) of small butterflies with usually copper-colored wings

[2]copper *transitive verb* **cop·pered; cop·per·ing** \'kä-p(ə-)riŋ\ (1530)
: to coat or sheathe with or as if with copper

[3]copper *noun* [[2]*cop*] (1846)
: POLICE OFFICER

cop·per·as \'kä-p(ə-)rəs\ *noun* [Middle English *coperas,* from Old French *couperose,* from Medieval Latin *cuprosa,* probably from *aqua cuprosa,* literally, copper water, from Late Latin *cuprum*] (14th century)
: a green hydrated ferrous sulfate $FeSO_4 \cdot 7H_2O$ used especially in making inks and pigments

cop·per·head \'kä-pər-,hed\ *noun* (1775)
1 : a common pit viper (*Agkistrodon contortrix*) of the eastern and central U.S. usually having a copper-colored head and often a reddish brown hourglass pattern on the body

copperhead 1

2 : a person in the northern states who sympathized with the South during the Civil War

cop·per·plate \'kä-pər-,plāt\ *noun* (1663)
1 : an engraved or etched copper printing plate; *also* **:** a print made from such a plate
2 : a neat script handwriting based on engraved models

copper pyrites *noun* (1776)
: CHALCOPYRITE

cop·per·smith \'kä-pər-,smith\ *noun* (14th century)
: a worker in copper

copper sulfate *noun* (circa 1893)
: a sulfate of copper; *especially* **:** the normal sulfate that is white in the anhydrous form but blue in the crystalline hydrous form $CuSO_4 \cdot 5H_2O$ and that is often used as an algicide and fungicide

cop·pery \'kä-p(ə-)rē\ *adjective* (circa 1775)
: resembling or suggesting copper; *especially* **:** having the reddish to brownish orange color of copper ⟨*coppery* leaves⟩

1cop·pice \'kä-pəs\ *noun* [Middle French *copeiz,* from *couper* to cut — more at COPE] (1534)
1 : a thicket, grove, or growth of small trees
2 : forest originating mainly from shoots or root suckers rather than seed

2coppice *verb* **cop·piced; cop·pic·ing** (1538)
transitive verb
: to cut back so as to regrow in the form of a coppice
intransitive verb
: to form a coppice; *specifically, of a tree* **:** to sprout freely from the base

copr- *or* **copro-** *combining form* [New Latin, from Greek *kopr-, kopro-,* from *kopros* akin to Sanskrit *śakṛt* dung]
: dung **:** feces ⟨*copro*lite⟩

co·pra \'kō-prə *also* 'kä-\ *noun* [Portuguese, from Malayalam *koppara*] (1584)
: dried coconut meat yielding coconut oil

co·pro·ces·sor \(,)kō-'prä-se-sər, -'prō-\ *noun* (1980)
: an extra processor in a computer that is designed to perform specialized tasks (as mathematical calculations)

co·prod·uct \(,)kō-'prä-(,)dəkt\ *noun* (1942)
: BY-PRODUCT 1

cop·ro·lite \'kä-prə-,līt\ *noun* (1829)
: fossilized excrement
— cop·ro·lit·ic \,kä-prə-'li-tik\ *adjective*

co·proph·a·gous \kə-'prä-fə-gəs\ *adjective* [Greek *koprophagos,* from *kopr-* + *-phagos* -phagous] (1826)
: feeding on dung
— co·proph·a·gy \-fə-jē\ *noun*

cop·ro·phil·ia \,kä-prə-'fi-lē-ə\ *noun* [New Latin] (1923)
: marked interest in excrement; *especially* **:** the use of feces or filth for sexual excitement
— cop·ro·phil·i·ac \-lē-,ak\ *noun*

cop·roph·i·lous \kə-'prä-fə-ləs\ *adjective* (circa 1900)
: growing or living on dung ⟨*coprophilous* fungi⟩

copse \'käps\ *noun* [by alteration] (1578)
: COPPICE 1

Copt \'käpt\ *noun* [Arabic *qubṭ* Copts, from Coptic *gyptios* Egyptian, from Greek *Aigyptios*] (1615)
1 : a member of the traditional Monophysite Christian church originating and centering in Egypt
2 : a member of a people descended from the ancient Egyptians

cop·ter \'käp-tər\ *noun* (1943)
: HELICOPTER

1Cop·tic \'käp-tik\ *adjective* (1677)
: of or relating to the Copts, their liturgical language, or their church

2Coptic *noun* (1711)
: an Afro-Asiatic language descended from ancient Egyptian and used as the liturgical language of the Coptic church

cop·u·la \'kä-pyə-lə\ *noun* [Latin, bond — more at COUPLE] (1619)
: something that connects: as **a :** the connecting link between subject and predicate of a proposition **b :** LINKING VERB

cop·u·late \'kä-pyə-,lāt\ *intransitive verb* **-lat·ed; -lat·ing** [Latin *copulatus,* past participle of *copulare* to join, from *copula*] (1632)
: to engage in sexual intercourse
— cop·u·la·tion \,kä-pyə-'lā-shən\ *noun*
— cop·u·la·to·ry \'kä-pyə-lə-,tōr-ē, -,tòr-\ *adjective*

1cop·u·la·tive \'kä-pyə-lə-tiv, -,lā-\ *adjective* (14th century)
1 a : joining together coordinate words or word groups and expressing addition of their meanings ⟨a *copulative* conjunction⟩ **b :** functioning as a copula
2 : relating to or serving for copulation

2copulative *noun* (1530)
: a copulative word

1copy \'kä-pē\ *noun, plural* **cop·ies** [Middle English *copie,* from Middle French, from Medieval Latin *copia,* from Latin, abundance — more at COPIOUS] (14th century)
1 : an imitation, transcript, or reproduction of an original work (as a letter, a painting, a table, or a dress)
2 : one of a series of especially mechanical reproductions of an original impression; *also* **:** an individual example of such a reproduction
3 *archaic* **:** something to be imitated **:** MODEL
4 a : matter to be set especially for printing **b :** something considered printable or newsworthy — used without an article ⟨remarks that make good *copy* —Norman Cousins⟩ **c :** text especially of an advertisement
synonym see REPRODUCTION

2copy *verb* **cop·ied; copy·ing** (14th century)
transitive verb
1 : to make a copy of
2 : to model oneself on
intransitive verb
1 : to make a copy
2 : to undergo copying ⟨the document did not *copy* well⟩ ☆

copy·book \'kä-pē-,bu̇k\ *noun* (1588)
: a book formerly used in teaching penmanship and containing models for imitation

copy·boy \-,bȯi\ *noun* (1888)
: one who carries copy and runs errands

1copy·cat \-,kat\ *noun, often attributive* (1896)
1 : one who imitates or adopts the behavior or practices of another
2 : an imitative act or product ⟨*copycat* board games⟩

2copycat *verb* **copy·cat·ted; copy·cat·ting** (1926)
intransitive verb
: to act as a copycat
transitive verb
: IMITATE

copy·desk \-,desk\ *noun* (1921)
: the desk at which newspaper copy is edited

copy editor *noun* (1899)
: an editor who prepares copy for the printer; *also* **:** one who edits and headlines newspaper copy
— copy·ed·it \'kä-pē-,e-dət\ *transitive verb*

copy·hold \'kä-pē-,hōld\ *noun* (15th century)
1 : a former tenure of land in England and Ireland by right of being recorded in the court of the manor
2 : an estate held by copyhold

copy·hold·er \-,hōl-dər\ *noun* (1874)
1 : a device for holding copy especially for a typesetter
2 : one who reads copy for a proofreader

copy·ist \'kä-pē-ist\ *noun* (1699)
1 : one who makes copies
2 : IMITATOR

copy·read·er \-,rē-dər\ *noun* (1892)
: COPY EDITOR
— copy·read \-,rēd\ *transitive verb*

1copy·right \-,rīt\ *noun* (1735)
: the exclusive legal right to reproduce, publish, and sell the matter and form (as of a literary, musical, or artistic work)

2copyright *transitive verb* (circa 1806)
: to secure a copyright on
— copy·right·able \-,rī-tə-bəl\ *adjective*

3copyright *adjective* (1870)
: secured by copyright

copy·writ·er \'kä-pē-,rī-tər\ *noun* (1911)
: a writer of advertising or publicity copy

coq au vin \,kōk-ō-'vaⁿ, ,käk-ō-\ *noun* [French, cock with wine] (circa 1938)
: chicken cooked in usually red wine

1co·quet *noun* [French, diminutive of *coq* cock] (1691)
1 \kō-'ket, -'kā\ **:** a man who indulges in coquetry
2 \-'ket\ **:** COQUETTE

2co·quet \kō-'ket\ *adjective* (1697)
: characteristic of a coquette **:** COQUETTISH

3co·quet *or* **co·quette** \-'ket\ *intransitive verb* **co·quet·ted; co·quet·ting** (1701)
1 : to play the coquette **:** FLIRT
2 : to deal with something playfully rather than seriously
synonym see TRIFLE

co·que·try \'kō-kə-trē, kō-'ke-trē\ *noun, plural* **-tries** (circa 1656)
: a flirtatious act or attitude

co·quette \kō-'ket\ *noun* [French, feminine of *coquet*] (circa 1611)
: a woman who endeavors without sincere affection to gain the attention and admiration of men
— co·quett·ish \-'ke-tish\ *adjective*
— co·quett·ish·ly *adverb*
— co·quett·ish·ness *noun*

co·qui·na \kō-'kē-nə\ *noun* [Spanish, probably diminutive of *coca* head, alteration of *coco* bogeyman, coconut] (1837)
1 : a soft whitish limestone formed of broken shells and corals cemented together and used for building
2 : a small clam (*Donax variabilis*) used for broth or chowder and occurring in the intertidal zone of sandy Atlantic beaches from Delaware to the Gulf of Mexico

cor \'kȯr\ *noun* [Middle English, from Late Latin *corus,* from Hebrew *kōr*] (14th century)
: an ancient Hebrew and Phoenician unit of measure of capacity

cor·a·cle \'kȯr-ə-kəl, 'kär-\ *noun* [Welsh *corwgl*] (circa 1547)
: a small boat used in Britain from ancient times and made of a frame (as of wicker) covered usually with hide or tarpaulin

cor·a·coid \'kȯr-ə-,kȯid, 'kär-\ *adjective* [New Latin *coracoides,* from Greek *korakoeidēs,* literally, like a raven, from *korak-, korax* raven — more at RAVEN] (1741)
: of, relating to, or being a process of the scapula in most mammals or a well-developed cartilage bone of many lower vertebrates that extends from the scapula to or toward the sternum
— coracoid *noun*

cor·al \'kȯr-əl, 'kär-\ *noun* [Middle English, from Middle French, from Latin *corallium,* from Greek *korallion*] (14th century)
1 a : the calcareous or horny skeletal deposit produced by anthozoan or rarely hydrozoan polyps; *especially* **:** a richly red precious coral secreted by a gorgonian (genus *Corallium*) **b :** a polyp or polyp colony together with its membranes and skeleton
2 : a piece of coral and especially of red coral
3 a : a bright reddish ovary (as of a lobster or scallop) **b :** a deep pink
— coral *adjective*
— cor·al·loid \-ə-,lȯid\ *adjective*

☆ SYNONYMS

Copy, imitate, mimic, ape, mock mean to make something so that it resembles an existing thing. COPY suggests duplicating an original as nearly as possible ⟨*copied* the painting and sold the fake as an original⟩. IMITATE suggests following a model or a pattern but may allow for some variation ⟨*imitate* a poet's style⟩. MIMIC implies a close copying (as of voice or mannerism) often for fun, ridicule, or lifelike imitation ⟨pupils *mimicking* their teacher⟩. APE may suggest presumptuous, slavish, or inept imitating of a superior original ⟨American fashion designers *aped* their European colleagues⟩. MOCK usually implies imitation with derision ⟨*mocking* a vain man's pompous manner⟩.

cor·al·bells \ˈkȯr-əl-ˌbelz, ˈkär-\ *noun plural but singular or plural in construction* (circa 1900)
: a perennial alumroot (*Heuchera sanguinea*) widely cultivated for its feathery spikes of tiny bright red campanulate flowers

cor·al·ber·ry \-ˌber-ē\ *noun* (circa 1859)
: an American dwarf shrub (*Symphoricarpos orbiculatus*) that bears clusters of small flowers succeeded by red berries

[1]cor·al·line \ˈkȯr-ə-ˌlīn, ˈkär-\ *noun* (1543)
1 : a coralline red alga
2 : a bryozoan or hydroid that resembles a coral

[2]coralline *adjective* [Middle French, from feminine of *corallin* coral-like, from Late Latin *corallinus,* from Latin *corallium*] (circa 1633)
1 : of, relating to, or resembling coral
2 : of, relating to, or being any of a family (Corallinaceae) of calcareous red algae

coral snake *noun* (circa 1772)
1 : any of several venomous chiefly tropical New World elapid snakes (genus *Micrurus*) brilliantly banded in red, black, and yellow or white that include two (*M. fulvius* and *M. euryxanthus*) ranging northward into the southern U.S.
2 : any of several harmless snakes resembling the coral snakes

co·ran·to \kə-ˈran-(ˌ)tō\ *noun, plural* **-tos** *or* **-toes** [modification of French *courante*] (1564)
: COURANTE

cor·ban \ˈkȯr-ˌban\ *noun* [Hebrew *qorbān* offering] (14th century)
: a sacrifice or offering to God among the ancient Hebrews

cor·beil *or* **cor·beille** \ˈkȯr-bəl, kȯr-ˈbā\ *noun* [French *corbeille,* literally, basket, from Late Latin *corbicula*] (circa 1734)
: a sculptured basket of flowers or fruit as an architectural decoration

[1]cor·bel \ˈkȯr-bəl\ *noun* [Middle English, from Middle French, from diminutive of *corp* raven, from Latin *corvus* — more at RAVEN] (15th century)
: an architectural member that projects from within a wall and supports a weight; *especially* **:** one that is stepped upward and outward from a vertical surface

[2]corbel *transitive verb* **-beled** *or* **-belled; -bel·ing** *or* **-bel·ling** (1843)
: to furnish with or make into a corbel

corbeling *noun* (1548)
1 : corbel work
2 : the construction of a corbel

cor·bic·u·la \kȯr-ˈbi-kyə-lə\ *noun, plural* **-lae** \-(ˌ)lē, -ˌlī\ [Late Latin, diminutive of Latin *corbis* basket] (1816)
: POLLEN BASKET

cor·bie \ˈkȯr-bē\ *noun* [Middle English, modification of Middle French *corbin,* from Latin *corvinus* of a raven — more at CORVINE] (15th century)
chiefly Scottish **:** CARRION CROW; *also* **:** RAVEN

cor·bi·na \kȯr-ˈbē-nə\ *noun* [Mexican Spanish, from Spanish *corvina,* a marine fish (*Argyrosomus regius*), from feminine of *corvino* of a raven, from Latin *corvinus*] (1901)
: a coastal marine croaker (*Menticirrhus undulatus*) favored by surf casters along the California coast

[1]cord \ˈkȯrd\ *noun* [Middle English, from Middle French *corde,* from Latin *chorda* string, from Greek *chordē* — more at YARN] (14th century)
1 a : a long slender flexible material usually consisting of several strands (as of thread or yarn) woven or twisted together **b :** the hangman's rope
2 : a moral, spiritual, or emotional bond
3 a : an anatomical structure (as a nerve or the umbilical cord) resembling a cord **b :** a small flexible insulated electrical cable having a plug at one or both ends used to connect a lamp or other appliance with a receptacle
4 : a unit of wood cut for fuel equal to a stack 4 x 4 x 8 feet or 128 cubic feet
5 a : a rib like a cord on a textile **b** (1) **:** a fabric made with such ribs or a garment made of such a fabric (2) *plural* **:** trousers made of such a fabric

[2]cord *transitive verb* (15th century)
1 : to furnish, bind, or connect with a cord
2 : to pile up (wood) in cords
— cord·er *noun*

cord·age \ˈkȯr-dij\ *noun* (1598)
1 : ropes or cords; *especially* **:** the ropes in the rigging of a ship
2 : the number of cords (as of wood) on a given area

cor·date \ˈkȯr-ˌdāt\ *adjective* [New Latin *cordatus,* from Latin *cord-, cor*] (1769)
: shaped like a heart ⟨a *cordate* leaf⟩ — see LEAF illustration
— cor·date·ly *adverb*

cord·ed \ˈkȯr-dəd\ *adjective* (14th century)
1 a : made of or provided with cords or ridges; *specifically* **:** muscled in ridges **b** *of a muscle* **:** TENSE, TAUT
2 : bound, fastened, or wound about with cords
3 : striped or ribbed with or as if with cord **:** TWILLED

cord·grass \ˈkȯrd-ˌgras\ *noun* (1857)
: any of a genus (*Spartina*) of chiefly salt-marsh grasses of coastal regions of Europe, northern Africa, and the New World that have stiff culms and panicled spikelets

[1]cor·dial \ˈkȯr-jəl\ *adjective* [Middle English, from Medieval Latin *cordialis,* from Latin *cord-, cor* heart — more at HEART] (14th century)
1 *obsolete* **:** of or relating to the heart **:** VITAL
2 : tending to revive, cheer, or invigorate
3 a : sincerely or deeply felt ⟨a *cordial* dislike for each other⟩ **b :** warmly and genially affable ⟨*cordial* relations⟩
synonym see GRACIOUS
— cor·dial·ly \ˈkȯrj-lē, ˈkȯr-jə-\ *adverb*
— cor·dial·ness \ˈkȯr-jəl-nəs\ *noun*

[2]cordial *noun* (14th century)
1 : a stimulating medicine or drink
2 : LIQUEUR

cor·dial·i·ty \ˌkȯr-jē-ˈa-lə-tē, kȯr-ˈja- *also* kȯr-ˈdya-\ *noun* (1611)
: sincere affection and kindness **:** cordial regard

cordia pulmonalia *plural of* COR PULMONALE

cor·di·er·ite \ˈkȯr-dē-ə-ˌrīt\ *noun* [French, from Pierre L. A. *Cordier* (died 1861) French geologist] (circa 1814)
: a blue mineral of vitreous luster and strong dichroism that consists of a silicate of aluminum, iron, and magnesium

cor·di·form \ˈkȯr-də-ˌfȯrm\ *adjective* [French *cordiforme,* from Latin *cord-, cor* + French *-iforme* -iform] (1828)
: shaped like a heart ⟨a *cordiform* sea-urchin shell⟩

cor·dil·le·ra \ˌkȯr-d^{ə}l-ˈyer-ə, -də-ˈler- *also* kȯr-ˈdi-lə-rə\ *noun* [Spanish] (1704)
: a system of mountain ranges often consisting of a number of more or less parallel chains
— cor·dil·le·ran \-ˈyer-ən, -ˈler-, -ˈdi-lə-rən\ *adjective*

cord·ite \ˈkȯr-ˌdīt\ *noun* (1889)
: a smokeless powder composed of nitroglycerin, guncotton, and a petroleum substance usually gelatinized by addition of acetone and pressed into cords resembling brown twine

cord·less \ˈkȯr-dləs\ *adjective* (1906)
: having no cord; *especially* **:** powered by a battery ⟨*cordless* tools⟩

cor·do·ba \ˈkȯr-də-bə, -və\ *noun* [Spanish *córdoba,* from Francisco Fernández de *Córdoba* (died 1526) Spanish explorer] (1913)
— see MONEY table

[1]cor·don \ˈkȯr-d^{ə}n, -ˌdän\ *noun* [French, diminutive of *corde* cord] (15th century)
1 a : an ornamental cord or ribbon **b :** STRINGCOURSE
2 a : a line of troops or of military posts enclosing an area to prevent passage **b :** a line of persons or objects around a person or place ⟨a *cordon* of police⟩
3 : an espalier especially of a fruit tree trained as a single horizontal shoot or two diverging horizontal shoots in a single line

[2]cordon *transitive verb* (1561)
: to form a protective or restrictive cordon around — usually used with *off*

cor·don sa·ni·taire \kȯr-ˌdō^{n}-sȧ-nē-ˈter\ *noun* [French, literally, sanitary cordon (quarantine line)] (1920)
: a protective barrier (as of buffer states) against a potentially aggressive nation or a dangerous influence (as an ideology)

[1]cor·do·van \ˈkȯr-də-vən\ *adjective* [Old Spanish *cordovano,* from *Córdova* (now *Córdoba*), Spain] (1591)
1 *capitalized* **:** of or relating to Córdoba and especially Córdoba, Spain
2 : made of cordovan leather

[2]cordovan *noun* (circa 1625)
1 : a soft fine-grained colored leather
2 : dense nonporous leather tanned from the inner layer of horsehide

[1]cor·du·roy \ˈkȯr-də-ˌrȯi\ *noun, plural* **-roys** [origin unknown] (circa 1791)
1 a *plural* **:** trousers of corduroy fabric **b :** a durable usually cotton pile fabric with vertical ribs or wales
2 : logs laid side by side transversely to make a road surface

[2]corduroy *transitive verb* **-royed; -roy·ing** (1854)
: to build (a road) of logs laid side by side transversely; *also* **:** to build a corduroy road across

cord·wain \ˈkȯrd-ˌwān\ *noun* [Middle English *cordwane,* from Middle French *cordoan,* from Old Spanish *cordovano, cordován*] (14th century) *archaic*
: cordovan leather

cord·wain·er \-ˌwā-nər\ *noun* (14th century)
1 *archaic* **:** a worker in cordovan leather
2 : SHOEMAKER
— cord·wain·ery \-ˌwā-nə-rē\ *noun*

cord·wood \ˈkȯrd-ˌwu̇d\ *noun* (circa 1639)
: wood piled or sold in cords

[1]core \ˈkōr, ˈkȯr\ *noun, often attributive* [Middle English] (14th century)
1 : a central and often foundational part usually distinct from the enveloping part by a difference in nature ⟨*core* of the city⟩: as **a :** the usually inedible central part of some fruits (as a pineapple); *especially* **:** the papery or leathery carpels composing the ripened ovary in a pome fruit **b :** the portion of a foundry mold that shapes the interior of a hollow casting **c :** a vertical space (as for elevator shafts, stairways, or plumbing apparatus) in a multistory building **d** (1) **:** a mass of iron serving to concentrate and intensify the magnetic field resulting from a current in a surrounding coil (2) **:** a tiny doughnut-shaped piece of magnetic material (as ferrite) used in computer memories (3) **:** a computer memory consisting of an array of cores strung on fine wires; *broadly* **:** the internal memory of a computer **e :** the central part of a celestial body (as the earth or sun) usually having different physical properties from the surrounding parts **f :** a nodule of stone (as flint or obsidian) from which flakes have been struck for making implements **g :** the conducting wire with its insulation in an electric cable **h :** an arrangement of a course of studies that combines under basic topics material from subjects conventionally separated and aims to provide a common background for all students ⟨*core* curriculum⟩ **i :** the place in a nuclear reactor where fission occurs
2 a : a basic, essential, or enduring part (as of an individual, a class, or an entity) ⟨the staff had a *core* of experts⟩ ⟨the *core* of her beliefs⟩

b : the essential meaning **:** GIST ⟨the *core* of the argument⟩ **c :** the inmost or most intimate part ⟨honest to the *core*⟩
3 : a part (as a thin cylinder of material) removed from the interior of a mass especially to determine composition

²core *transitive verb* **cored; cor·ing** (15th century)
: to remove a core from
— cor·er *noun*

³core *noun* [perhaps alteration of Middle English *chore* chorus, company, perhaps from Latin *chorus*] (1622)
chiefly Scottish **:** a group of people

core city *noun* (1965)
: INNER CITY

cored \'kōrd, 'kȯrd\ *adjective* (1945)
: having a core of a specified kind — usually used in combination

co·re·li·gion·ist \ˌkō-ri-'lij-nist, -'li-jə-\ *noun* (1826)
: a person of the same religion

co·re·mi·um \kə-'rē-mē-əm\ *noun, plural* **-mia** \-mē-ə\ [New Latin, from Greek *korēma* broom, from *korein* to sweep] (1929)
: a fruiting body characteristic of some fungi (as the one causing Dutch elm disease) that consists of a sterile stalk of parallel or fascicled hyphae and a terminal fertile or spore-bearing head

co·re·op·sis \ˌkōr-ē-'äp-səs, ˌkȯr-\ *noun, plural* **coreopsis** [New Latin, genus name, from Greek *koris* bedbug + New Latin *-opsis;* akin to Greek *keirein* to cut — more at SHEAR] (circa 1753)
: any of a genus (*Coreopsis*) of widely cultivated composite herbs with showy flower heads and pinnately lobed or dissected leaves

co·re·pres·sor \ˌkō-ri-'pre-sər\ *noun* (1963)
: a substance that activates a particular genetic repressor by combining with it

co·req·ui·site \kō-'re-kwə-zət\ *noun* (circa 1948)
: a formal course of study required to be taken simultaneously with another

co·re·spon·dent \ˌkō-ri-'spän-dənt\ *noun* (1857)
: a person named as guilty of adultery with the defendant in a divorce suit

corf \'kȯrf\ *noun, plural* **corves** \'kȯrvz\ [Middle English, basket, from Middle Dutch *corf* or Middle Low German *korf,* from Latin *corbis* basket] (1653)
British **:** a basket, tub, or truck used in a mine

cor·gi \'kȯr-gē\ *noun, plural* **corgis** [Welsh, from *cor* dwarf + *ci* dog; akin to Old Irish *cú* dog, Old English *hund* — more at HOUND] (1926)
1 : CARDIGAN WELSH CORGI
2 : PEMBROKE WELSH CORGI

co·ri·a·ceous \ˌkōr-ē-'ā-shəs, ˌkȯr-\ *adjective* [Late Latin *coriaceus* — more at CUIRASS] (1674)
: resembling leather ⟨*coriaceous* foliage⟩

co·ri·an·der \'kōr-ē-ˌan-dər, ˌkōr-ē-', 'kȯr-, ˌkȯr-\ *noun* [Middle English *coriandre,* from Old French, from Latin *coriandrum,* from Greek *koriandron, koriannon*] (14th century)
1 : an Old World herb (*Coriandrum sativum*) of the carrot family with aromatic fruits
2 : the ripened dried fruit of coriander used as a flavoring — called also *coriander seed*

¹Co·rin·thi·an \kə-'rin(t)-thē-ən\ *noun* (1526)
1 : a native or resident of Corinth, Greece
2 : a merry profligate man

²Corinthian *adjective* (1594)
1 : of, relating to, or characteristic of Corinth or Corinthians
2 : of or relating to the lightest and most ornate of the three ancient Greek architectural orders distinguished especially by its large capitals decorated with carved acanthus leaves — see ORDER illustration

Co·rin·thi·ans \-thē-ənz\ *noun plural but singular in construction*
: either of two letters written by Saint Paul to the Christians of Corinth and included as books in the New Testament — see BIBLE table

Co·ri·o·lis effect \ˌkōr-ē-'ō-ləs-, ˌkȯr-\ *noun* (circa 1946)
: the apparent deflection of a moving object that is the result of the Coriolis force

Coriolis force *noun* [Gaspard G. *Coriolis* (died 1843) French civil engineer] (1923)
: an apparent force that as a result of the earth's rotation deflects moving objects (as projectiles or air currents) to the right in the northern hemisphere and to the left in the southern hemisphere

co·ri·um \'kōr-ē-əm, 'kȯr-\ *noun, plural* **co·ria** \-ē-ə\ [New Latin, from Latin, leather — more at CUIRASS] (1836)
: DERMIS

¹cork \'kȯrk\ *noun* [Middle English, cork, bark, probably from Arabic *qurq,* from Latin *cortic-, cortex* bark — more at CUIRASS] (14th century)
1 a : the elastic tough outer tissue of the cork oak that is used especially for stoppers and insulation **b :** PHELLEM
2 : a usually cork stopper for a bottle or jug
3 : a fishing float

²cork *transitive verb* (1535)
1 : to furnish or fit with cork or a cork
2 : to stop up with a cork
3 : to blacken with burnt cork

cork·age \'kȯr-kij\ *noun* (1838)
: a charge (as by a restaurant) for opening a bottle of wine bought elsewhere

cork·board \'kȯrk-ˌbōrd, -ˌbȯrd\ *noun* (circa 1893)
: a heat-insulating material made of compressed granulated cork; *also* **:** a bulletin board made with this material

cork cambium *noun* (1878)
: PHELLOGEN

corked \'kȯrkt\ *adjective* (1830)
: CORKY 2

cork·er \'kȯr-kər\ *noun* (1881)
1 : one that corks containers (as bottles)
2 : one that is excellent or remarkable

cork·ing \'kȯr-kiŋ\ *adjective or adverb* (1895)
: extremely fine — often used as an intensive especially before *good* ⟨had a *corking* good time⟩

cork oak *noun* (1873)
: an oak (*Quercus suber*) of southern Europe and northern Africa that is the source of the cork of commerce

¹cork·screw \'kȯrk-ˌskrü\ *noun* (1720)
: a device for drawing corks from bottles that has a pointed spiral piece of metal turned by a handle

²corkscrew *adjective* (1815)
: resembling a corkscrew **:** SPIRAL

³corkscrew (1837)
transitive verb
1 : WIND
2 : to draw out with difficulty
3 : to twist into a spiral
intransitive verb
: to move in a winding course

cork·wood \'kȯrk-ˌwu̇d\ *noun* (1756)
: any of several trees having light or corky wood, *especially* **:** a small or shrubby tree (*Leitneria floridana*) of the southeastern U.S. that has extremely light soft wood

corky \'kȯr-kē\ *adjective* **cork·i·er; -est** (1756)
1 : resembling cork
2 : having an unpleasant odor and taste (as from a tainted cork) ⟨*corky* wine⟩
— cork·i·ness \-kē-nəs\ *noun*

corm \'kȯrm\ *noun* [New Latin *cormus,* from Greek *kormos* tree trunk, from *keirein* to cut — more at SHEAR] (1830)
: a rounded thick modified underground stem base bearing membranous or scaly leaves and buds and acting as a vegetative reproductive structure — compare BULB, TUBER

corm·el \'kȯr-məl, kȯr-'mel\ *noun* [diminutive of *corm*] (circa 1900)
: a small or secondary corm produced by a larger corm

cor·mo·rant \'kȯrm-rənt, 'kȯr-mə-, 'kȯr-mə-ˌrant\ *noun* [Middle English *cormeraunt,* from Middle French *cormorant,* from Old French *cormareng,* from *corp* raven + *marenc* of the sea, from Latin *marinus* — more at CORBEL, MARINE] (14th century)
1 : any of a family (Phalacrocoracidae) of dark-colored web-footed water birds that have a long neck, hooked bill, and distensible throat pouch
2 : a gluttonous, greedy, or rapacious person

cormorant

¹corn \'kȯrn\ *noun, often attributive* [Middle English, from Old English; akin to Old High German & Old Norse *korn* grain, Latin *granum*] (before 12th century)
1 *chiefly dialect* **:** a small hard particle **:** GRAIN
2 : a small hard seed
3 a : the seeds of a cereal grass and especially of the important cereal crop of a particular region (as wheat in Britain, oats in Scotland and Ireland, and Indian corn in the New World and Australia) **b :** the kernels of sweet corn served as a vegetable while still soft and milky
4 : a plant that produces corn
5 : CORN WHISKEY
6 a : something (as writing, music, or acting) that is corny **b :** the quality or state of being corny **:** CORNINESS

²corn *transitive verb* (1560)
1 : to form into grains **:** GRANULATE
2 a : to preserve or season with salt in grains **b :** to cure or preserve in brine containing preservatives and often seasonings ⟨*corned* beef⟩
3 : to feed with corn ⟨*corn* the horses⟩

³corn *noun* [Middle English *corne,* from Middle French, horn, corner, from Latin *cornu* horn, point — more at HORN] (15th century)
: a local hardening and thickening of epidermis (as on a toe)

¹corn·ball \'kȯrn-ˌbȯl\ *noun* (circa 1949)
: an unsophisticated person; *also* **:** something corny

²cornball *adjective* (1951)
: CORNY

corn borer *noun* (1919)
: any of several insects that bore in maize: as **a :** EUROPEAN CORN BORER **b :** SOUTHWESTERN CORN BORER

corn bread *noun* (1750)
: bread made with cornmeal

corn chip *noun* (1950)
: a piece of a dry crisp snack food prepared from a seasoned cornmeal batter

corn·cob \'kȯrn-ˌkäb\ *noun* (1793)
1 : the axis on which the kernels of Indian corn are arranged
2 : an ear of Indian corn

corncob pipe *noun* (1832)
: a tobacco pipe with a bowl made of a corncob

corn cockle *noun* (1713)
: an annual hairy weed (*Agrostemma githago*) of the pink family with purplish red flowers that is found in grainfields

corn·crake \'kȯrn-ˌkrāk\ *noun* (15th century)
: a common Eurasian short-billed rail (*Crex crex*) that frequents grainfields

corn·crib \-ˌkrib\ *noun* (1681)
: a crib for storing ears of Indian corn

corn dodger *noun* (1834)
chiefly Southern & Midland **:** a cake of corn bread that is fried, baked, or boiled as a dumpling

corn dog *noun* (1967)
: a frankfurter dipped in cornmeal batter, fried, and served on a stick

cor·nea \ˈkȯr-nē-ə\ *noun* [Middle English, from Medieval Latin, from Latin, feminine of *corneus* horny, from *cornu*] (14th century)
: the transparent part of the coat of the eyeball that covers the iris and pupil and admits light to the interior — see EYE illustration
— cor·ne·al \-əl\ *adjective*

corn ear·worm \-ˈir-ˌwərm\ *noun* (1802)
: a noctuid moth (*Heliothis zea*) whose large striped yellow-headed larva is especially destructive to Indian corn, tomatoes, tobacco, and cotton bolls

cor·nel \ˈkȯr-nᵊl, -ˌnel\ *noun* [ultimately from Latin *cornus* cornel cherry tree; akin to Greek *kranon* cornel cherry tree] (1551)
: any of various shrubs or trees (genus *Cornus*) of the dogwood family with very hard wood and perfect flowers; *specifically* **:** DOGWOOD

cor·ne·lian \kȯr-ˈnēl-yən\ *noun* (15th century)
: CARNELIAN

cor·ne·ous \ˈkȯr-nē-əs\ *adjective* [Latin *corneus*] (1646)
: of a horny texture

¹cor·ner \ˈkȯr-nər\ *noun* [Middle English, from Old French *cornere,* from *corne* horn, corner] (13th century)
1 a : the point where converging lines, edges, or sides meet **:** ANGLE **b :** the place of intersection of two streets or roads **c :** a piece designed to form, mark, or protect a corner
2 : the angular part or space between meeting lines, edges, or borders near the vertex of the angle ⟨the southwest *corner* of the state⟩ ⟨the *corners* of the tablecloth⟩: as **a :** the area of a playing field or court near the intersection of the sideline and the goal line or baseline **b** (1) **:** either of the four angles of a boxing ring; *especially* **:** the area in which a boxer rests or is worked on by his seconds during periods between rounds (2) **:** a group of supporters, well-wishers, or adherents associated especially with a contestant **c :** the side of home plate nearest to or farthest from a batter ⟨a fast ball over the outside *corner*⟩ **d :** CORNER KICK **e** (1) **:** the outside of a football formation (2) **:** CORNERBACK
3 a : a private, secret, or remote place ⟨a quiet *corner* of New England⟩ ⟨to every *corner* of the earth⟩ ⟨dark *corners* of the mind⟩ **b :** a difficult or embarrassing situation **:** a position from which escape or retreat is difficult or impossible ⟨was backed into a *corner*⟩
4 : control or ownership of enough of the available supply of a commodity or security especially to permit manipulation of the price
5 : a point at which significant change occurs — often used in the phrase *turn the corner*
— cor·nered \-nərd\ *adjective*
— around the corner : at hand **:** IMMINENT ⟨good times are just *around the corner*⟩

²corner *adjective* (13th century)
1 : situated at a corner ⟨the *corner* drugstore⟩
2 : used or fitted for use in or on a corner ⟨a *corner* table⟩

³corner (1824)
transitive verb
1 a : to drive into a corner ⟨the animal is dangerous when *cornered*⟩ **b :** to catch and hold the attention of especially to force an interview
2 : to get a corner on ⟨*corner* the market⟩
intransitive verb
1 : to meet or converge at a corner or angle
2 : to turn a corner ⟨the car *corners* well⟩

cor·ner·back \ˈkȯr-nər-ˌbak\ *noun* (1955)
: a defensive halfback in football who defends the flank

corner kick *noun* (1882)
: a free kick from a corner of a soccer field awarded to an attacker when a defender plays the ball out-of-bounds over the end line

cor·ner·man \ˈkȯr-nər-ˌman\ *noun* (1957)
1 : one who plays in or near the corner: as **a :** CORNERBACK **b :** a basketball forward
2 : a boxer's second

cor·ner·stone \-ˌstōn\ *noun* (13th century)
1 : a stone forming a part of a corner or angle in a wall; *specifically* **:** such a stone laid at a formal ceremony
2 : a basic element **:** FOUNDATION ⟨a *cornerstone* of foreign policy⟩

cor·ner·ways \-ˌwāz\ *adverb* (1922)
: DIAGONALLY

cor·ner·wise \-ˌwīz\ *adverb* (15th century)
: DIAGONALLY

cor·net \kȯr-ˈnet, *British usually* ˈkȯr-nit\ *noun* [Middle English, from Middle French, from diminutive of *corn* horn, from Latin *cornu*] (14th century)
1 : a valved brass instrument resembling a trumpet in design and range but having a shorter partly conical tube and less brilliant tone
2 : something shaped like a cone: as **a :** a piece of paper twisted for use as a container **b :** a cone-shaped pastry shell that is often filled with whipped cream **c** *British* **:** an ice-cream cone
— cor·net·ist *or* **cor·net·tist** \kȯr-ˈne-tist, ˈkȯr-ni-\ *noun*

corn–fed \ˈkȯrn-ˌfed\ *adjective* (14th century)
1 : fed or fattened on grain (as corn) ⟨*corn-fed* hogs⟩
2 : looking well-fed **:** PLUMP
3 : rustically wholesome or corny

corn·field \-ˌfēld\ *noun* (14th century)
: a field in which corn is grown

corn·flakes \-ˌflāks\ *noun plural* (1907)
: toasted flakes made from the coarse meal of hulled corn for use as a breakfast cereal

corn flour *noun* (1791)
British **:** CORNSTARCH

corn·flow·er \ˈkȯrn-ˌflau̇(-ə)r\ *noun* (1527)
1 : CORN COCKLE
2 : BACHELOR'S BUTTON

cornflower blue *noun* (1907)
: a moderate purplish blue

Corn·husk·er \ˈkȯrn-ˌhəs-kər\ *noun* (circa 1948)
: a native or resident of Nebraska — used as a nickname

corn·husk·ing \-ˌhəs-kiŋ\ *noun* (1692)
: a social gathering especially of farm families to husk corn

¹cor·nice \ˈkȯr-nəs, -nish\ *noun* [Middle French, from Italian, frame, cornice, from Latin *cornic-, cornix* crow; akin to Greek *korax* raven — more at RAVEN] (1563)
1 a : the molded and projecting horizontal member that crowns an architectural composition **b :** a top course that crowns a wall
2 : a decorative band of metal or wood used to conceal curtain fixtures
3 : an overhanging mass of snow, ice, or rock usually on a ridge

c **cornice 1a**

²cornice *transitive verb* **cor·niced; cor·nic·ing** (1744)
: to furnish or crown with a cornice

cor·niche \kȯr-ˈnēsh\ *noun* [French *cornice, corniche,* literally, cornice, from Italian *cornice*] (1837)
: a road built along a coast and especially along the face of a cliff

cor·nic·u·late cartilage \kȯr-ˈni-kyə-lət-\ *noun* [Latin *corniculatus* horned, from *corniculum,* diminutive of *cornu* horn] (circa 1909)
: a small nodule of yellow elastic cartilage articulating with the apex of the arytenoid

cor·ni·fi·ca·tion \ˌkȯr-nə-fə-ˈkā-shən\ *noun* [Latin *cornu* horn] (circa 1843)
1 : conversion into horn or a horny substance or tissue
2 : the conversion of the vaginal epithelium from the columnar to the squamous type

¹Cor·nish \ˈkȯr-nish\ *adjective* [Middle English *Cornysshe,* from *Cornwaile* Cornwall, England] (14th century)
: of, relating to, or characteristic of Cornwall, Cornishmen, or Cornish

²Cornish *noun* (1547)
1 : a Celtic language of Cornwall extinct since the late 18th century
2 : any of an English breed of domestic fowls much used in crossbreeding for meat production

Cor·nish·man \-mən\ *noun* (15th century)
: a native or resident of Cornwall, England

Corn Law *noun* (1766)
: one of a series of laws in force in Great Britain before 1846 prohibiting or discouraging the importation of grain

corn leaf aphid *noun* (circa 1939)
: a dusky greenish or brownish aphid (*Rhopalosiphum maidis*) that feeds on the flowers and foliage of various commercially important grasses (as Indian corn)

corn·meal \ˈkȯrn-ˌmēl\ *noun* (1749)
: meal ground from corn

corn oil *noun* (1900)
: a yellow fatty oil obtained from the germ of Indian corn kernels and used chiefly as salad oil, in soft soap, and in margarine

corn·pone \ˈkȯrn-ˌpōn\ *adjective* (1972)
: DOWN-HOME, COUNTRIFIED ⟨*cornpone* humor⟩

corn pone *noun* (1859)
Southern & Midland **:** corn bread often made without milk or eggs and baked or fried

corn poppy *noun* (circa 1859)
: an annual red-flowered Eurasian poppy (*Papaver rhoeas*) common in fields and cultivated in several varieties

corn rootworm *noun* (1892)
: any of several chrysomelid beetles (genus *Diabrotica*) whose root-eating larvae are pests especially of corn

corn·row \ˈkȯrn-ˌrō\ *noun* (1946)
1 : a section of hair which is braided usually flat to the scalp
2 : a hairstyle in which the hair is divided into cornrow sections arranged in rows
— cornrow *verb*

cornrow 2

corn salad *noun* [from its occurrence as a weed in fields of grain] (1597)
: any of several herbs (genus *Valerianella*) of the valerian family; *especially* **:** a low European herb (*V. locusta* synonym *V. olitoria*) that is widely cultivated as a salad plant and potherb

corn silk *noun* (1861)
: the silky styles on an ear of Indian corn

corn snow *noun* (1935)
: granular snow formed by alternate thawing and freezing

corn·stalk \ˈkȯrn-ˌstȯk\ *noun* (1645)
: a stalk of Indian corn

corn·starch \-ˌstärch\ *noun* (1853)
: starch made from corn and used in foods as a thickening agent, in making corn syrup and sugars, and in the manufacture of adhesives and sizes for paper and textiles

corn sugar *noun* (1850)
: DEXTROSE

corn syrup *noun* (1903)
: a syrup containing dextrins, maltose, and dextrose that is obtained by partial hydrolysis of cornstarch

cor·nu \'kȯr-(,)nü, -(,)nyü\ *noun, plural* **cor·nua** \-nü-ə, -nyü-\ [Latin] (1691)
: a horn-shaped anatomical part (as of the uterus)
— **cor·nu·al** \-nü-əl, -nyü-\ *adjective*

cor·nu·co·pia \,kȯr-nə-'kō-pē-ə, -nyə-'\ *noun* [Late Latin, from Latin *cornu copiae* horn of plenty] (1508)
1 : a curved goat's horn overflowing with fruit and ears of grain that is used as a decorative motif emblematic of abundance
2 : an inexhaustible store **:** ABUNDANCE
3 : a receptacle shaped like a horn or cone
— **cor·nu·co·pi·an** \-pē-ən\ *adjective*

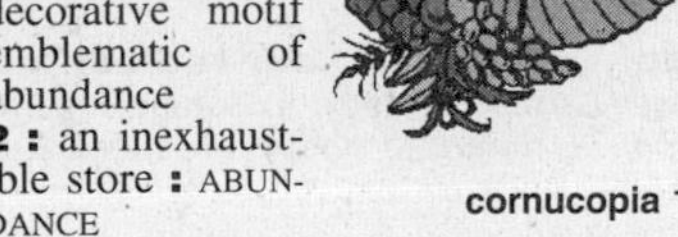
cornucopia 1

cor·nu·to \kȯr-'nü-(,)tō, -'nyü-\ *noun, plural* **-tos** [Italian, from Latin *cornutus* having horns, from *cornu*] (1598)
: CUCKOLD

corn whiskey *noun* (1780)
: whiskey distilled from a mash made up of not less than 80 percent corn — compare BOURBON

[1]**corny** \'kȯr-nē\ *adjective* **corn·i·er; -est** (14th century)
1 *archaic* **:** tasting strongly of malt
2 : of or relating to corn
3 : mawkishly old-fashioned **:** tiresomely simple and sentimental
— **corn·i·ly** \'kȯr-nᵊl-ē\ *adverb*
— **corn·i·ness** \'kȯr-nē-nəs\ *noun*

[2]**corny** *adjective* **corn·i·er; -est** (1707)
: relating to or having corns on the feet

cor·o·dy \'kȯr-ə-dē, 'kär-\ *noun, plural* **-dies** [Middle English *corrodie,* from Medieval Latin *corrodium*] (15th century)
: an allowance of provisions for maintenance dispensed as a charity

co·rol·la \kə-'rä-lə, -'rō-\ *noun* [New Latin, from Latin, diminutive of *corona*] (circa 1753)
: the part of a flower that consists of the separate or fused petals and constitutes the inner whorl of the perianth
— **co·rol·late** \kə-'rä-lət; 'kȯr-ə-,lāt, 'kär-\ *adjective*

cor·ol·lary \'kȯr-ə-,ler-ē, 'kär-, *British* kə-'rä-lə-rē\ *noun, plural* **-lar·ies** [Middle English *corolarie,* from Late Latin *corollarium,* from Latin, money paid for a garland, gratuity, from *corolla*] (14th century)
1 : a proposition inferred immediately from a proved proposition with little or no additional proof
2 a : something that naturally follows **:** RESULT **b :** something that incidentally or naturally accompanies or parallels
— **corollary** *adjective*

cor·o·man·del \,kȯr-ə-'man-dᵊl, ,kär-\ *noun* [*Coromandel* coast region, India] (1845)
: CALAMANDER

coromandel screen *noun, often C capitalized* (1926)
: a Chinese lacquered folding screen

co·ro·na \kə-'rō-nə\ *noun* [Latin, garland, crown, cornice — more at CROWN] (1563)
1 : the projecting part of a classic cornice
2 a : a usually colored circle often seen around and close to a luminous body (as the sun or moon) caused by diffraction produced by suspended droplets or occasionally particles of dust **b :** the tenuous outermost part of the atmosphere of a star (as the sun) **c :** a circle of light made by the apparent convergence of the streamers of the aurora borealis **d :** the upper portion of a bodily part (as a tooth or the skull) **e :** an appendage or series of united appendages on the inner side of the corolla in some flowers (as the daffodil, jonquil, or milkweed) **f :** a faint glow adjacent to the surface of an electrical conductor at high voltage
3 [from *La Corona,* a trademark] **:** a long cigar having the sides straight to the end to be lit and being roundly blunt at the other end

a corona 2e

Corona Aus·tra·lis \-ō-'strā-ləs, -ä-\ *noun* [New Latin (genitive *Coronae Australis*), literally, southern crown]
: a southern constellation adjoining Sagittarius on the south

Corona Bo·re·al·is \-,bōr-ē-'a-ləs, -,bȯr-\ *noun* [New Latin (genitive *Coronae Borealis*), literally, northern crown]
: a northern constellation between Hercules and Boötes

cor·o·nach \'kȯr-ə-nək, 'kär-\ *noun* [Scottish Gaelic *corranach* & Irish *coránach*] (1530)
: a funeral dirge sung or played on the bagpipes in Scotland and Ireland

co·ro·na·graph *also* **co·ro·no·graph** \kə-'rō-nə-,graf\ *noun* (1885)
: a telescope for observation of the sun's corona

[1]**cor·o·nal** *also* **cor·o·nel** \'kȯr-ə-nᵊl, 'kär-\ *noun* [Middle English *coronal,* from Anglo-French, from Latin *coronalis* of a crown, from *corona*] (14th century)
: a circlet for the head usually implying rank or dignity

[2]**co·ro·nal** \'kȯr-ə-nᵊl, 'kär-; kə-'rō-\ *adjective* (14th century)
1 a : lying in the direction of the coronal suture **b :** of or relating to the frontal plane that passes through the long axis of the body
2 : of or relating to a corona or crown

coronal suture *noun* (1615)
: a suture extending across the skull between the parietal and frontal bones

co·ro·na ra·di·a·ta \kə-'rō-nə-,rā-dē-'ā tə, 'ä tə\ *noun, plural* **co·ro·nae ra·di·a·tae** \-(,)nē-,rā-dē-'ā-(,)tē, -'ä-(,)tē\ [New Latin, literally, crown with rays] (1892)
: the zone of small follicular cells immediately surrounding the ovum in the graafian follicle and accompanying the ovum on its discharge from the follicle

[1]**cor·o·nary** \'kȯr-ə-,ner-ē, 'kär-\ *adjective* (1610)
1 : of, relating to, resembling, or being a crown or coronal
2 : of, relating to, or being the coronary arteries or veins of the heart; *broadly* **:** of or relating to the heart

[2]**coronary** *noun, plural* **-nar·ies** (1893)
1 a : CORONARY ARTERY **b :** CORONARY VEIN
2 : CORONARY THROMBOSIS; *broadly* **:** HEART ATTACK

coronary artery *noun* (1741)
: either of two arteries that arise one from the left and one from the right side of the aorta immediately above the semilunar valves and supply the tissues of the heart itself

coronary occlusion *noun* (1946)
: the partial or complete blocking (as by a thrombus, by spasm, or by sclerosis) of a coronary artery

coronary sinus *noun* (1831)
: a venous channel that is derived from the sinus venosus, is continuous with the largest of the cardiac veins, receives most of the blood from the walls of the heart, and empties into the right atrium

coronary thrombosis *noun* (1926)
: the blocking of a coronary artery of the heart by a thrombus

coronary vein *noun* (circa 1828)
: any of several veins that drain the tissues of the heart and empty into the coronary sinus

cor·o·nate \'kȯr-ə-,nāt, 'kär-\ *transitive verb* **-nat·ed; -nat·ing** [Latin *coronatus,* past participle of *coronare* to crown, from *corona*] (circa 1623)
: CROWN 1a

cor·o·na·tion \,kȯr-ə-'nā-shən, ,kär-\ *noun* [Middle English *coronacion,* from Middle French *coronation,* from *coroner* to crown] (14th century)
: the act or occasion of crowning; *also* **:** accession to the highest office

cor·o·ner \'kȯr-ə-nər, 'kär-\ *noun* [Middle English, an officer of the crown, from Anglo-French, from Old French *corone* crown, from Latin *corona*] (circa 1630)
: a public officer whose principal duty is to inquire by an inquest into the cause of any death which there is reason to suppose is not due to natural causes

cor·o·net \,kȯr-ə-'net, ,kär-\ *noun* [Middle French *coronette,* from Old French *coronete,* from *corone*] (15th century)
1 : a small or lesser crown usually signifying a rank below that of a sovereign
2 : an ornamental wreath or band for the head usually for wear by women on formal occasions
3 : the lower part of a horse's pastern where the horn terminates in skin — see HORSE illustration

co·ro·tate \(,)kō-'rō-,tāt\ *intransitive verb* (1962)
: to rotate in conjunction with or at the same rate as another rotating body
— **co·ro·ta·tion** \,kō-rō-'tā-shən\ *noun*

corpora *plural of* CORPUS

[1]**cor·po·ral** \'kȯr-p(ə-)rəl\ *noun* [Middle English, from Middle French, from Medieval Latin *corporale,* from Latin, neuter of *corporalis;* from the doctrine that the bread of the Eucharist becomes or represents the body of Christ] (14th century)
: a linen cloth on which the eucharistic elements are placed

[2]**corporal** *adjective* [Middle English, from Middle French, from Latin *corporalis,* from *corpor-, corpus* body — more at MIDRIFF] (14th century)
1 *obsolete* **:** CORPOREAL, PHYSICAL
2 : of, relating to, or affecting the body ⟨*corporal* punishment⟩
— **cor·po·ral·ly** \-p(ə-)rə-lē\ *adverb*

[3]**corporal** *noun* [Middle French, lowest noncommissioned officer, alteration of *caporal,* from Old Italian *caporale,* from *capo* head, from Latin *caput* — more at HEAD] (1579)
: a noncommissioned officer ranking in the army above a private first class and below a sergeant and in the marine corps above a lance corporal and below a sergeant

cor·po·ral·i·ty \,kȯr-pə-'ra-lə-tē\ *noun, plural* **-ties** (14th century)
: the quality or state of being or having a body or a material or physical existence

corporal's guard *noun* (1844)
1 : the small detachment commanded by a corporal
2 : a small group

cor·po·rate \'kȯr-p(ə-)rət\ *adjective* [Latin *corporatus,* past participle of *corporare* to make into a body, from *corpor-, corpus*] (1512)
1 a : formed into an association and endowed by law with the rights and liabilities of an individual **:** INCORPORATED **b :** of or relating to a corporation ⟨a plan to reorganize the *corporate* structure⟩
2 : of, relating to, or formed into a unified body of individuals ⟨human law arises by the *corporate* action of a people —G. H. Sabine⟩

3 : CORPORATIVE 2
— **cor·po·rate·ly** *adverb*

cor·po·ra·tion \ˌkȯr-pə-ˈrā-shən\ *noun* (15th century)
1 a *obsolete* **:** a group of merchants or traders united in a trade guild **b :** the municipal authorities of a town or city
2 : a body formed and authorized by law to act as a single person although constituted by one or more persons and legally endowed with various rights and duties including the capacity of succession
3 : an association of employers and employees in a basic industry or of members of a profession organized as an organ of political representation in a corporative state
4 : POTBELLY 1

cor·po·rat·ism \ˈkȯr-p(ə-)rə-ˌti-zəm\ *noun* (1890)
: the organization of a society into industrial and professional corporations serving as organs of political representation and exercising some control over persons and activities within their jurisdiction
— **cor·po·rat·ist** \-p(ə-)rə-tist\ *adjective*

cor·po·ra·tive \ˈkȯr-pə-ˌrā-tiv, -p(ə-)rə-\ *adjective* (1833)
1 : of or relating to a corporation
2 : of or relating to corporatism ⟨a *corporative* state⟩

cor·po·ra·tiv·ism \ˈkȯr-pə-ˌrā-ti-ˌvi-zəm, -p(ə-)rə-\ *noun* (1930)
: CORPORATISM

cor·po·ra·tor \ˈkȯr-pə-ˌrā-tər\ *noun* (1784)
: a corporation organizer, member, or stockholder

cor·po·re·al \kȯr-ˈpōr-ē-əl, -ˈpȯr-\ *adjective* [Latin *corporeus* of the body, from *corpor-, corpus*] (15th century)
1 : having, consisting of, or relating to a physical material body: as **a :** not spiritual **b :** not immaterial or intangible **:** SUBSTANTIAL
2 *archaic* **:** CORPORAL
synonym see MATERIAL
— **cor·po·re·al·ly** \-ē-ə-lē\ *adverb*
— **cor·po·re·al·ness** *noun*

cor·po·re·al·i·ty \(ˌ)kȯr-ˌpōr-ē-ˈa-lə-tē, -ˌpȯr-\ *noun, plural* **-ties** (1651)
: corporeal existence

cor·po·re·i·ty \ˌkȯr-pə-ˈrē-ə-tē, -ˈrā-\ *noun, plural* **-ties** (1621)
: the quality or state of having or being a body **:** MATERIALITY

cor·po·sant \ˈkȯr-pə-ˌsant, -ˌzant\ *noun* [Portuguese *corpo-santo,* literally, holy body] (1655)
: SAINT ELMO'S FIRE

corps \ˈkōr, ˈkȯr\ *noun, plural* **corps** \ˈkōrz, ˈkȯrz\ [French, from Latin *corpus* body] (1711)
1 a : an organized subdivision of the military establishment ⟨Marine *Corps*⟩ ⟨Signal *Corps*⟩ **b :** a tactical unit usually consisting of two or more divisions and auxiliary arms and services
2 : a group of persons associated together or acting under common direction; *especially* **:** a body of persons having a common activity or occupation ⟨the press *corps*⟩
3 : CORPS DE BALLET

corps de bal·let \ˌkōr-də-(ˌ)ba-ˈlā, ˌkȯr-\ *noun, plural* **corps de ballet** *same or* ˌkōrz-də-, ˌkȯrz-\ [French] (1826)
: the ensemble of a ballet company

corps d'elite \ˌkōr-dā-ˈlēt, ˌkȯr-\ *noun, plural* **corps d'elite** *same or* ˌkōrz-dā-, ˌkȯrz-\ [French *corps d'élite*] (1884)
1 : a body of picked troops
2 : a group of the best people in a category

corpse \ˈkȯrps\ *noun* [Middle English *corps,* from Middle French, from Latin *corpus*] (13th century)
1 *obsolete* **:** a human or animal body whether living or dead
2 a : a dead body especially of a human being **b :** the remains of something discarded or defunct ⟨the *corpses* of rusting cars⟩

corps·man \ˈkōr(z)-mən, ˈkȯr(z)-\ *noun* (1901)
1 : an enlisted man trained to give first aid and minor medical treatment
2 : a member of a government-sponsored service corps

cor·pu·lence \ˈkȯr-pyə-lən(t)s\ *noun* (1581)
: the state of being corpulent

cor·pu·len·cy \-lən(t)-sē\ *noun, plural* **-cies** (1577)
: CORPULENCE

cor·pu·lent \-lənt\ *adjective* [Middle English, from Latin *corpulentus,* from *corpus*] (14th century)
: having a large bulky body **:** OBESE
— **cor·pu·lent·ly** *adverb*

cor pul·mo·na·le \ˌkȯr-ˌpu̇l-mə-ˈnä-lē, -ˌpəl-, -ˈna-\ *noun, plural* **cor·dia pul·mo·na·lia** \ˈkȯr-dē-ə-ˌpu̇l-mə-ˈnä-lē-ə, -ˌpəl-, -ˈna-\ [New Latin, literally, pulmonary heart] (1857)
: disease of the heart characterized by hypertrophy and dilatation of the right ventricle and secondary to disease of the lungs or their blood vessels

cor·pus \ˈkȯr-pəs\ *noun, plural* **cor·po·ra** \-p(ə-)rə\ [Middle English, from Latin] (15th century)
1 : the body of a human or animal especially when dead
2 a : the main part or body of a bodily structure or organ ⟨the *corpus* of the uterus⟩ **b :** the main body or corporeal substance of a thing; *specifically* **:** the principal of a fund or estate as distinct from income or interest
3 a : all the writings or works of a particular kind or on a particular subject; *especially* **:** the complete works of an author **b :** a collection or body of knowledge or evidence; *especially* **:** a collection of recorded utterances used as a basis for the descriptive analysis of a language

corpus al·la·tum \-ə-ˈlā-təm, -ˈlä-\ *noun, plural* **corpora al·la·ta** \-ˈlā-tə, -ˈlä-tə\ [New Latin, literally, applied body] (1947)
: one of a pair of separate or fused bodies in many insects that are sometimes closely associated with the corpora cardiaca and that secrete hormones (as juvenile hormone)

corpus cal·lo·sum \-ka-ˈlō-səm\ *noun, plural* **corpora cal·lo·sa** \-sə\ [New Latin, literally, callous body] (1706)
: the great band of commissural fibers uniting the cerebral hemispheres of higher mammals including humans — see BRAIN illustration

corpus car·di·a·cum \-kär-ˈdī-ə-kəm\ *noun, plural* **corpora car·di·a·ca** \-ə-kə\ [New Latin, literally, cardiac body] (1960)
: one of a pair of separate or fused bodies of nervous tissue in many insects that lie posterior to the brain and dorsal to the esophagus and that function in the storage and secretion of brain hormone

Cor·pus Chris·ti \ˈkȯr-pəs-ˈkris-tē\ *noun* [Middle English, from Medieval Latin, literally, body of Christ] (14th century)
: the Thursday after Trinity observed as a Roman Catholic festival in honor of the Eucharist

cor·pus·cle \ˈkȯr-(ˌ)pə-səl\ *noun* [Latin *corpusculum,* diminutive of *corpus*] (1660)
1 : a minute particle
2 a : a living cell; *especially* **:** one (as a red or white blood cell or a cell in cartilage or bone) not aggregated into continuous tissues **b :** any of various small circumscribed multicellular bodies
— **cor·pus·cu·lar** \kȯr-ˈpəs-kyə-lər\ *adjective*

cor·pus de·lic·ti \ˈkȯr-pəs-di-ˈlik-ˌtī, -(ˌ)tē\ *noun, plural* **corpora delicti** [New Latin, literally, body of the crime] (1832)
1 : the substantial and fundamental fact necessary to prove the commission of a crime
2 : the material substance (as the body of the victim of a murder) upon which a crime has been committed

corpus lu·te·um \-ˈlü-tē-əm\ *noun, plural* **corpora lu·tea** \-tē-ə\ [New Latin, literally, yellowish body] (1788)
: a yellowish mass of progesterone-secreting endocrine tissue that forms immediately after ovulation from the ruptured graafian follicle in the mammalian ovary

corpus stri·a·tum \-(ˌ)strī-ˈā-təm\ *noun, plural* **corpora stri·a·ta** \-ˈā-tə\ [New Latin, literally, striated body] (1851)
: either of a pair of masses of nervous tissue within the brain that contain two large nuclei of gray matter separated by sheets of white matter

cor·rade \kə-ˈrād\ *verb* **cor·rad·ed; cor·rad·ing** [Latin *corradere* to scrape together, from *com-* + *radere* to scrape — more at RODENT] (1646)
transitive verb
: to wear away by abrasion
intransitive verb
: to crumble away through abrasion
— **cor·ra·sion** \-ˈrā-zhən\ *noun*
— **cor·ra·sive** \-ˈrā-siv, -ziv\ *adjective*

[1]**cor·ral** \kə-ˈral, -ˈrel\ *noun* [Spanish, from (assumed) Vulgar Latin *currale* enclosure for vehicles, from Latin *currus* cart, from *currere* to run — more at CAR] (1582)
1 : a pen or enclosure for confining or capturing livestock
2 : an enclosure made with wagons for defense of an encampment

[2]**corral** *transitive verb* **cor·ralled; cor·ral·ling** (1847)
1 : to enclose in a corral
2 : to arrange (wagons) so as to form a corral
3 : COLLECT, GATHER ⟨*corralling* votes for the upcoming election⟩

[1]**cor·rect** \kə-ˈrekt\ *transitive verb* [Middle English, from Latin *correctus,* past participle of *corrigere,* from *com-* + *regere* to lead straight — more at RIGHT] (14th century)
1 a : to make or set right **:** AMEND **b :** COUNTERACT, NEUTRALIZE **c :** to alter or adjust so as to bring to some standard or required condition ⟨*correct* a lens for spherical aberration⟩
2 a : to punish (as a child) with a view to reforming or improving **b :** to point out usually for amendment the errors or faults of ⟨spent the day *correcting* tests⟩ ☆
— **cor·rect·able** \-ˈrek-tə-bəl\ *adjective*
— **cor·rec·tor** \-ˈrek-tər\ *noun*

[2]**correct** *adjective* [Middle English, corrected, from Latin *correctus,* from past participle of *corrigere*] (1676)
1 : conforming to an approved or conventional standard

☆ SYNONYMS
Correct, rectify, emend, remedy, redress, amend, reform, revise mean to make right what is wrong. CORRECT implies taking action to remove errors, faults, deviations, defects ⟨*correct* your spelling⟩. RECTIFY implies a more essential changing to make something right, just, or properly controlled or directed ⟨*rectify* a misguided policy⟩. EMEND specifically implies correction of a text or manuscript ⟨*emend* a text⟩. REMEDY implies removing or making harmless a cause of trouble, harm, or evil ⟨set out to *remedy* the evils of the world⟩. REDRESS implies making compensation or reparation for an unfairness, injustice, or imbalance ⟨*redress* past social injustices⟩. AMEND, REFORM, REVISE imply an improving by making corrective changes, AMEND usually suggesting slight changes ⟨*amend* a law⟩, REFORM implying drastic change ⟨plans to *reform* the court system⟩, and REVISE suggesting a careful examination of something and the making of necessary changes ⟨*revise* the schedule⟩. See in addition PUNISH.

2 : conforming to or agreeing with fact, logic, or known truth
3 : conforming to a set figure ⟨enclosed the *correct* return postage⟩ ☆
— **cor·rect·ly** \kə-'rek(t)-lē\ *adverb*
— **cor·rect·ness** \-'rek(t)-nəs\ *noun*

corrected time *noun* (circa 1891)
: a boat's elapsed time less its time allowance in yacht racing

cor·rec·tion \kə-'rek-shən\ *noun* (14th century)
1 : the action or an instance of correcting: as **a :** AMENDMENT, RECTIFICATION **b :** REBUKE, PUNISHMENT **c :** a bringing into conformity with a standard **d :** NEUTRALIZATION, COUNTERACTION ⟨*correction* of acidity⟩
2 : a decline in market price or business activity following and counteracting a rise
3 a : something substituted in place of what is wrong ⟨marking *corrections* on the students' papers⟩ **b :** a quantity applied by way of correcting (as for adjustment of an instrument)
4 : the treatment and rehabilitation of offenders through a program involving penal custody, parole, and probation; *also* **:** the administration of such treatment as a matter of public policy — usually used in plural
— **cor·rec·tion·al** \-shnəl, -shə-nᵊl\ *adjective*

cor·rec·ti·tude \kə-'rek-tə-ˌtüd, -ˌtyüd\ *noun* [blend of *correct* and *rectitude*] (1893)
: correctness or propriety of conduct

cor·rec·tive \kə-'rek-tiv\ *adjective* (1531)
: intended to correct ⟨*corrective* lenses⟩ ⟨*corrective* punishment⟩
— **corrective** *noun*
— **cor·rec·tive·ly** *adverb*

¹cor·re·late \'kȯr-ə-lət, 'kär-, -ˌlāt\ *noun* [back-formation from *correlation*] (1643)
1 : either of two things so related that one directly implies or is complementary to the other (as husband and wife)
2 : a phenomenon (as brain activity) that accompanies another phenomenon (as behavior), is usually parallel to it (as in form, type, development, or distribution), and is related in some way to it
— **correlate** *adjective*

²cor·re·late \-ˌlāt\ *verb* **-lat·ed; -lat·ing** (circa 1742)
intransitive verb
: to bear reciprocal or mutual relations **:** CORRESPOND
transitive verb
1 a : to establish a mutual or reciprocal relation between ⟨*correlate* activities in the lab and the field⟩ **b :** to show correlation or a causal relationship between
2 : to present or set forth so as to show relationship ⟨he *correlates* the findings of the scientists, the psychologists, and the mystics —Eugene Exman⟩
— **cor·re·lat·able** \-ˌlā-tə-bəl\ *adjective*
— **cor·re·lat·or** \-ˌlā-tər\ *noun*

cor·re·la·tion \ˌkȯr-ə-'lā-shən, ˌkär-\ *noun* [Medieval Latin *correlation-, correlatio,* from Latin *com-* + *relation-, relatio* relation] (1561)
1 : the state or relation of being correlated; *specifically* **:** a relation existing between phenomena or things or between mathematical or statistical variables which tend to vary, be associated, or occur together in a way not expected on the basis of chance alone ⟨the obviously high positive *correlation* between scholastic aptitude and college entrance —J. B. Conant⟩
2 : the act of correlating
— **cor·re·la·tion·al** \-shnəl, -shə-nᵊl\ *adjective*

correlation coefficient *noun* (circa 1909)
: a number or function that indicates the degree of correlation between two sets of data or between two random variables and that is equal to their covariance divided by the product of their standard deviations

cor·rel·a·tive \kə-'re-lə-tiv\ *adjective* (1530)
1 : naturally related **:** CORRESPONDING
2 : reciprocally related
3 : regularly used together but typically not adjacent ⟨the *correlative* conjunctions *either . . . or*⟩
— **correlative** *noun*
— **cor·rel·a·tive·ly** *adverb*

cor·re·spond \ˌkȯr-ə-'spänd, ˌkär-\ *intransitive verb* [Middle French or Medieval Latin; Middle French *correspondre,* from Medieval Latin *correspondēre,* from Latin *com-* + *respondēre* to respond] (1529)
1 a : to be in conformity or agreement ⟨the ideal failed . . . to *correspond* with the reality —James Sutherland⟩ **b :** to compare closely **:** MATCH — usually used with *to* or *with* **c :** to be equivalent or parallel
2 : to communicate with a person by exchange of letters

cor·re·spon·dence \-'spän-dən(t)s\ *noun* (15th century)
1 a : the agreement of things with one another **b :** a particular similarity **c :** a relation between sets in which each member of one set is associated with one or more members of the other — compare FUNCTION 5a
2 a : communication by letters; *also* **:** the letters exchanged **b :** the news, information, or opinion contributed by a correspondent to a newspaper or periodical

correspondence course *noun* (1902)
: a course offered by a correspondence school

correspondence school *noun* (1889)
: a school that teaches nonresident students by mailing them lessons and exercises which upon completion are returned to the school for grading

cor·re·spon·den·cy \ˌkȯr-ə-'spän-dən(t)-sē, ˌkär-\ *noun, plural* **-cies** (1589)
: CORRESPONDENCE

¹cor·re·spon·dent \ˌkȯr-ə-'spän-dənt, ˌkär-\ *adjective* [Middle English, from Middle French or Medieval Latin; Middle French, from Medieval Latin *correspondent-, correspondens,* present participle of *correspondēre*] (15th century)
1 : CORRESPONDING
2 : FITTING, CONFORMING — used with *with* or *to* ⟨the outcome was entirely *correspondent* with my wishes⟩

²correspondent *noun* (circa 1630)
1 a : one who communicates with another by letter **b :** one who has regular commercial relations with another **c :** one who contributes news or commentary to a publication (as a newspaper) or a radio or television network often from a distant place ⟨a war *correspondent*⟩
2 : something that corresponds

corresponding *adjective* (1579)
1 a : having or participating in the same relationship (as kind, degree, position, correspondence, or function) especially with regard to the same or like wholes (as geometric figures or sets) ⟨*corresponding* parts of similar triangles⟩ **b :** RELATED, ACCOMPANYING ⟨all rights carry with them *corresponding* responsibilities —W. P. Paepcke⟩
2 a : charged with the duty of writing letters ⟨*corresponding* secretary⟩ **b :** participating or serving at a distance and by mail ⟨a *corresponding* member of the society⟩
— **cor·re·spond·ing·ly** \-'spän-diŋ-lē\ *adverb*

corresponding angles *noun plural* (circa 1804)
: any pair of angles each of which is on the same side of one of two lines cut by a transversal and on the same side of the transversal

cor·re·spon·sive \ˌkȯr-ə-'spän(t)-siv, ˌkär-\ *adjective* (1606)
: mutually responsive

cor·ri·da \kō-'rē-t͟hə\ *noun* [Spanish, literally, act of running] (1898)
: BULLFIGHT

cor·ri·dor \'kȯr-ə-dər, 'kär-, -ˌdȯr\ *noun* [Middle French, from Old Italian *corridore,* from *correre* to run, from Latin *currere* — more at CAR] (1814)
1 : a passageway (as in a hotel) into which compartments or rooms open
2 : a usually narrow passageway or route: as **a :** a narrow strip of land through foreign-held territory **b :** a restricted lane for air traffic
3 : a densely populated strip of land including two or more major cities ⟨the Northeast *corridor* stretching from Washington into New England —S. D. Browne⟩

cor·rie \'kȯr-ē, 'kär-ē\ *noun* [Scottish Gaelic *coire,* literally, kettle] (1795)
: CIRQUE 3

Cor·rie·dale \-ˌdāl\ *noun* [*Corriedale,* ranch in New Zealand] (1902)
: any of a dual-purpose breed of rather large usually hornless sheep developed in New Zealand

cor·ri·gen·dum \ˌkȯr-ə-'jen-dəm, ˌkär-\ *noun, plural* **-da** \-də\ [Latin, neuter of *corrigendus,* gerundive of *corrigere* to correct] (circa 1850)
: an error in a printed work discovered after printing and shown with its correction on a separate sheet

cor·ri·gi·ble \'kȯr-ə-jə-bəl, 'kär-\ *adjective* [Middle English, from Middle French, from Medieval Latin *corrigibilis,* from Latin *corrigere*] (15th century)
: capable of being set right **:** REPARABLE ⟨a *corrigible* defect⟩
— **cor·ri·gi·bil·i·ty** \ˌkȯr-ə-jə-'bi-lə-tē, ˌkär-\ *noun*

cor·ri·val \kə-'rī-vəl, kȯ-, kō-\ *noun* [Middle French, from Latin *corrivalis,* from *com-* + *rivalis* rival] (1579)
: RIVAL, COMPETITOR
— **corrival** *adjective*

cor·rob·o·rant \kə-'rä-bə-rənt\ *adjective* (1626) *archaic*
: having an invigorating effect — used of a medicine

cor·rob·o·rate \kə-'rä-bə-ˌrāt\ *transitive verb* **-rat·ed; -rat·ing** [Latin *corroboratus,* past participle of *corroborare,* from *com-* + *robor-, robur* strength] (1530)
: to support with evidence or authority **:** make more certain
synonym see CONFIRM
— **cor·rob·o·ra·tion** \-ˌrä-bə-'rā-shən\ *noun*
— **cor·rob·o·ra·tive** \-'rä-bə-ˌrā-tiv, -'rä-b(ə-)rə-\ *adjective*
— **cor·rob·o·ra·tor** \-'rä-bə-ˌrā-tər\ *noun*
— **cor·rob·o·ra·to·ry** \-'rä-b(ə-)rə-ˌtōr-ē, -ˌtȯr-\ *adjective*

☆ SYNONYMS
Correct, accurate, exact, precise, nice, right mean conforming to fact, standard, or truth. CORRECT usually implies freedom from fault or error ⟨*correct* answers⟩ ⟨socially *correct* dress⟩. ACCURATE implies fidelity to fact or truth attained by exercise of care ⟨an *accurate* description⟩. EXACT stresses a very strict agreement with fact, standard, or truth ⟨*exact* measurements⟩. PRECISE adds to EXACT an emphasis on sharpness of definition or delimitation ⟨*precise* calibration⟩. NICE stresses great precision and delicacy of adjustment or discrimination ⟨makes *nice* distinctions⟩. RIGHT is close to CORRECT but has a stronger positive emphasis on conformity to fact or truth rather than mere absence of error or fault ⟨the *right* thing to do⟩.

cor·rob·o·ree \kə-'rä-bə-rē, -'rä-\ *noun* [Dharuk (Australian aboriginal language of the Port Jackson area) *garaabara*] (1811)
1 : a nocturnal festivity with songs and symbolic dances by which the Australian aborigines celebrate events of importance
2 *Australian* **a :** a noisy festivity **b :** TUMULT

cor·rode \kə-'rōd\ *verb* **cor·rod·ed; cor·rod·ing** [Middle English, from Latin *corrodere* to gnaw to pieces, from *com-* + *rodere* to gnaw — more at RODENT] (14th century)
transitive verb
1 : to eat away by degrees as if by gnawing; *especially* **:** to wear away gradually usually by chemical action ⟨the metal was *corroded* beyond repair⟩
2 : to weaken or destroy gradually **:** UNDERMINE ⟨manners and miserliness that *corrode* the human spirit —Bernard DeVoto⟩
intransitive verb
: to undergo corrosion
— cor·rod·ible \-'rō-də-bəl\ *adjective*

cor·ro·dy *variant of* CORODY

cor·ro·sion \kə-'rō-zhən\ *noun* [Middle English, from Late Latin *corrosion-, corrosio* act of gnawing, from Latin *corrodere*] (14th century)
1 : the action, process, or effect of corroding
2 : a product of corroding

cor·ro·sive \-'rō-siv, -ziv\ *adjective* (14th century)
1 : tending or having the power to corrode ⟨*corrosive* acids⟩ ⟨*corrosive* action⟩ ⟨the *corrosive* effects of alcoholism⟩
2 : bitingly sarcastic ⟨*corrosive* satire⟩
— corrosive *noun*
— cor·ro·sive·ly *adverb*
— cor·ro·sive·ness *noun*

corrosive sublimate *noun* (circa 1751)
: MERCURIC CHLORIDE

cor·ru·gate \'kȯr-ə-,gāt, 'kär-\ *verb* **-gat·ed; -gat·ing** [Latin *corrugatus,* past participle of *corrugare,* from *com-* + *ruga* wrinkle; probably akin to Lithuanian *raukas* wrinkle — more at ROUGH] (1620)
transitive verb
: to form or shape into wrinkles or folds or into alternating ridges and grooves **:** FURROW
intransitive verb
: to become corrugated

corrugated *adjective* (1590)
: having corrugations ⟨*corrugated* paper⟩; *also* **:** made of corrugated material (as paperboard) ⟨*corrugated* boxes⟩

cor·ru·ga·tion \,kȯr-ə-'gā-shən, ,kär-\ *noun* (1528)
1 : the act of corrugating
2 : a ridge or groove of a surface that has been corrugated

[1]cor·rupt \kə-'rəpt\ *verb* [Middle English, from Latin *corruptus,* past participle of *corrumpere,* from *com-* + *rumpere* to break — more at REAVE] (14th century)
transitive verb
1 a : to change from good to bad in morals, manners, or actions; *also* **:** BRIBE **b :** to degrade with unsound principles or moral values
2 : ROT, SPOIL
3 : to subject (a person) to corruption of blood
4 : to alter from the original or correct form or version
intransitive verb
1 a : to become tainted or rotten **b :** to become morally debased
2 : to cause disintegration or ruin
synonym see DEBASE
— cor·rupt·er *also* **cor·rup·tor** \-'rəp-tər\ *noun*
— cor·rupt·ibil·i·ty \-,rəp-tə-'bi-lə-tē\ *noun*
— cor·rupt·ible \-'rəp-tə-bəl\ *adjective*
— cor·rupt·ibly \-blē\ *adverb*

[2]corrupt *adjective* [Middle English, from Middle French or Latin; Middle French, from Latin *corruptus*] (14th century)
1 a : morally degenerate and perverted **:** DEPRAVED **b :** characterized by improper conduct (as bribery or the selling of favors) ⟨*corrupt* judges⟩
2 : PUTRID, TAINTED
3 : adulterated or debased by change from an original or correct condition
synonym see VICIOUS
— cor·rupt·ly \-'rəp(t)-lē\ *adverb*
— cor·rupt·ness \-'rəp(t)-nəs\ *noun*

cor·rup·tion \kə-'rəp-shən\ *noun* (14th century)
1 a : impairment of integrity, virtue, or moral principle **:** DEPRAVITY **b :** DECAY, DECOMPOSITION **c :** inducement to wrong by improper or unlawful means (as bribery) **d :** a departure from the original or from what is pure or correct
2 *archaic* **:** an agency or influence that corrupts
3 *chiefly dialect* **:** PUS

cor·rup·tion·ist \-sh(ə-)nist\ *noun* (1810)
: one who practices or defends corruption especially in politics

corruption of blood (1563)
: the effect of an attainder which bars a person from inheriting, retaining, or transmitting any estate, rank, or title

cor·rup·tive \kə-'rəp-tiv\ *adjective* (15th century)
: producing or tending to produce corruption
— cor·rup·tive·ly *adverb*

cor·sage \kȯr-'säzh, -'säj, 'kȯr-,\ *noun* [French, bust, bodice, from Old French, bust, from *cors* body, from Latin *corpus*] (1843)
1 : the waist or bodice of a dress
2 : an arrangement of flowers worn as a fashion accessory

cor·sair \'kȯr-,sar, -,ser; kȯr-'\ *noun* [Middle French & Old Italian; Middle French *corsaire* pirate, from Old Provençal *corsari,* from Old Italian *corsaro,* from Medieval Latin *cursarius,* from Latin *cursus* course — more at COURSE] (1549)
: PIRATE; *especially* **:** a privateer of the Barbary Coast

corse \'kȯrs\ *noun* [Middle English *cors,* from Middle French] (13th century)
archaic **:** CORPSE

corse·let *for 1* 'kȯr-slət, *for 2* ,kȯr-sə-'let\ *noun* (1563)
1 *or* **cors·let** [Middle French, diminutive of *cors* body, bodice] **:** a piece of armor covering the trunk
2 *or* **cor·se·lette** [from *Corselette,* a trademark] **:** an undergarment combining girdle and brassiere

[1]cor·set \'kȯr-sət\ *noun* [Middle English, from Old French, diminutive of *cors*] (13th century)
1 : a usually close-fitting and often laced medieval jacket
2 : a woman's close-fitting boned supporting undergarment that is often hooked and laced and that extends from above or beneath the bust or from the waist to below the hips and has garters attached

[2]corset *transitive verb* (circa 1847)
1 : to dress in or fit with a corset
2 : to restrict closely **:** control rigidly

cor·se·tiere \,kȯr-sə-'tir, -'tyer\ *noun* [French *corsetière,* feminine of *corsetier,* from *corset*] (1848)
: one who makes, fits, or sells corsets, girdles, or brassieres

cor·tege *also* **cor·tège** \kȯr-'tezh, 'kȯr-,\ *noun* [French *cortège,* from Italian *corteggio,* from *corteggiare* to court, from *corte* court, from Latin *cohort-, cohors* enclosure — more at COURT] (1648)
1 : a train of attendants **:** RETINUE
2 : PROCESSION; *especially* **:** a funeral procession

cor·tex \'kȯr-,teks\ *noun, plural* **cor·ti·ces** \'kȯr-tə-,sēz\ *or* **cor·tex·es** [Latin *cortic-, cortex* bark — more at CUIRASS] (1677)
1 a : the outer or superficial part of an organ or bodily structure (as the kidney, adrenal gland, or a hair); *especially* **:** the outer layer of gray matter of the cerebrum and cerebellum **b :** the outer part of some organisms (as paramecia)
2 : a plant bark or rind (as cinchona) used medicinally
3 a : the typically parenchymatous layer of tissue external to the vascular tissue and internal to the corky or epidermal tissues of a green plant; *broadly* **:** all tissues external to the xylem **b :** an outer or investing layer of various algae, lichens, or fungi

cor·ti·cal \'kȯr-ti-kəl\ *adjective* (1671)
1 : of, relating to, or consisting of cortex
2 : involving or resulting from the action or condition of the cerebral cortex
— cor·ti·cal·ly \-k(ə-)lē\ *adverb*

cortico- *combining form*
: cortex ⟨*cortico*tropin⟩

cor·ti·coid \'kȯr-ti-,kȯid\ *noun* (1941)
: CORTICOSTEROID
— corticoid *adjective*

cor·ti·co·ste·roid \,kȯr-ti-kō-'stir-,ȯid *also* -'ster-\ *noun* (1944)
: any of various adrenal-cortex steroids (as corticosterone, cortisone, and aldosterone) used medically especially as anti-inflammatory agents

cor·ti·co·ste·rone \,kȯr-tə-'käs-tə-,rōn, -ti-kō-stə-'; ,kȯr-ti-kō-'stir-,ōn, -'ster-\ *noun* (1937)
: a colorless crystalline corticosteroid $C_{21}H_{30}O_4$ that is important in protein and carbohydrate metabolism

cor·ti·co·tro·pin \,kȯr-ti-kō-'trō-pən\ *also* **cor·ti·co·tro·phin** \-fən\ *noun* (1946)
: ACTH; *also* **:** a preparation of ACTH that is used especially in the treatment of rheumatoid arthritis and rheumatic fever

cor·tin \'kȯr-tᵊn\ *noun* (1928)
: the active principle of the adrenal cortex

cor·ti·sol \'kȯr-tə-,sȯl, -,zȯl, -,sōl, -,zōl\ *noun* [*cortisone* + [1]*-ol*] (1951)
: HYDROCORTISONE

cor·ti·sone \-,sōn, -,zōn\ *noun* [alteration of *corticosterone*] (1949)
: a glucocorticoid $C_{21}H_{28}O_5$ of the adrenal cortex used especially in the treatment of rheumatoid arthritis

co·run·dum \kə-'rən-dəm\ *noun* [Tamil *kuruntam;* akin to Sanskrit *kuruvinda* ruby] (1804)
: a very hard mineral that consists of aluminum oxide and occurs in massive form and as variously colored crystals (as ruby and sapphire), that can be synthesized, and that is used as an abrasive

co·rus·cant \kə-'rəs-kənt\ *adjective* (15th century)
: SHINING, GLITTERING

cor·us·cate \'kȯr-ə-,skāt, 'kär-\ *intransitive verb* **-cat·ed; -cat·ing** [Latin *coruscatus,* past participle of *coruscare* to flash] (1705)
1 : to give off or reflect light in bright beams or flashes **:** SPARKLE
2 : to be brilliant or showy in technique or style

cor·us·ca·tion \,kȯr-ə-'skā-shən, ,kär-\ *noun* (15th century)
1 : GLITTER, SPARKLE
2 : a flash of wit

cor·vée \'kȯr-,vā, kȯr-'\ *noun* [Middle English *corvee,* from Middle French, from Medieval Latin *corrogata,* from Latin, feminine of *corrogatus,* past participle of *corrogare* to collect, requisition, from *com-* + *rogare* to ask — more at RIGHT] (14th century)
1 : unpaid labor (as toward constructing roads) due from a feudal vassal to his lord
2 : labor exacted in lieu of taxes by public authorities especially for highway construction or repair

corves *plural of* CORF

cor·vette \kȯr-'vet\ *noun* [French, from Middle French, probably from Middle Dutch *corf,* a kind of ship, literally, basket — more at CORF] (1636)
1 : a warship ranking in the old sailing navies next below a frigate
2 : a highly maneuverable armed escort ship that is smaller than a destroyer

cor·vi·na \kȯr-'vē-nə\ *variant of* CORBINA

cor·vine \'kȯr-ˌvīn\ *adjective* [Latin *corvinus,* from *corvus* raven — more at RAVEN] (circa 1656)
: of or relating to the crows **:** resembling a crow

Cor·vus \'kȯr-vəs\ *noun* [Latin (genitive *Corvi*), literally, raven]
: a small constellation adjoining Virgo on the south

Cor·y·bant \'kȯr-ə-ˌbant, 'kär-\ *noun, plural* **Cor·y·bants** \-ˌban(t)s\ *or* **Cor·y·ban·tes** \ˌkȯr-ə-'ban-tēz, ˌkär-\ [Middle French *Corybante,* from Latin *Corybas,* from Greek *Korybas*] (14th century)
: one of the attendants or priests of Cybele noted for wildly emotional processions and rites

cor·y·ban·tic \ˌkȯr-ē-'ban-tik, ˌkär-\ *adjective* (1642)
: like or in the spirit of a Corybant; *especially* **:** WILD, FRENZIED

co·ryd·a·lis \kə-'ri-d^{ə}l-əs\ *noun* [New Latin, genus name, from Greek *korydallis* crested lark (*Galerida cristata*), from *korydos* crested lark] (1818)
: any of a large genus (*Corydalis*) of chiefly temperate herbs of the fumitory family with racemose irregular flowers and dissected leaves

cor·ymb \'kȯr-ˌim(b), 'kär-, -əm(b)\ *noun, plural* **corymbs** \-ˌimz, -əmz\ [French *corymbe,* from Latin *corymbus* cluster of fruit or flowers, from Greek *korymbos*] (1776)
: a flat topped inflorescence; *specifically* **:** one in which the flower stalks arise at different levels on the main axis and reach about the same height and in which the outer flowers open first and the inflorescence is indeterminate
— **cor·ymbed** \-ˌimd, -əmd\ *adjective*
— **cor·ym·bose** \-əm-ˌbōs\ *adjective*
— **cor·ym·bose·ly** *adverb*

corymb

co·ry·ne·bac·te·ri·um \ˌkȯr-ə-(ˌ)nē-bak-'tir-ē-əm, kə-ˌri-nə-\ *noun* [New Latin, from Greek *korynē* club] (1909)
: any of a large genus (*Corynebacterium*) of usually gram-positive nonmotile bacteria that occur as irregular or branching rods and include numerous important animal and plant parasites
— **co·ry·ne·bac·te·ri·al** \-ē-əl\ *adjective*

co·ryn·e·form \kə-'ri-nə-ˌfȯrm\ *adjective* (1952)
: being or resembling corynebacteria

cor·y·phae·us \ˌkȯr-ə-'fē-əs, ˌkär-\ *noun, plural* **-phaei** \-'fē-ˌī\ [Latin, leader, from Greek *koryphaios,* from *koryphē* summit] (1610)
1 : the leader of a party or school of thought
2 : the leader of a chorus

co·ry·phée \ˌkȯr-i-'fā\ *noun* [French, from Latin *coryphaeus*] (1828)
: a ballet dancer who dances in a small group instead of in the corps de ballet or as a soloist

co·ry·za \kə-'rī-zə\ *noun* [Late Latin, from Greek *koryza* nasal mucus; akin to Old English *hrot* nasal mucus, Sanskrit *kardama* mud] (1634)
: an acute inflammatory contagious disease involving the upper respiratory tract; *especially* **:** COMMON COLD
— **co·ry·zal** \-zəl\ *adjective*

co·se·cant \(ˌ)kō-'sē-ˌkant, -kənt\ *noun* [New Latin *cosecant-, cosecans,* from *co-* + *secant-, secans* secant] (circa 1706)
1 : a trigonometric function that for an acute angle is the ratio between the hypotenuse of a right triangle of which the angle is considered part and the leg opposite the angle
2 : a trigonometric function csc θ that is the reciprocal of the sine for all real numbers θ for which the sine is not zero and that is exactly equal to the cosecant of an angle of measure θ in radians

co·set \'kō-ˌset\ *noun* (1910)
: a subset of a mathematical group that consists of all the products obtained by multiplying either on the right or the left a fixed element of the group by each of the elements of a given subgroup

[1]cosh \'käsh\ *noun* [perhaps from Romany *kaš, kašt* stick, piece of wood] (1869)
chiefly British **:** a weighted weapon similar to a blackjack

[2]cosh *transitive verb* (1896)
chiefly British **:** to strike or assault with or as if with a cosh

co·sig·na·to·ry \(ˌ)kō-'sig-nə-ˌtōr-ē, -ˌtȯr-\ *noun* (1865)
: a joint signer

co·sign·er \'kō-ˌsī-nər\ *noun* (circa 1903)
: COSIGNATORY; *especially* **:** a joint signer of a promissory note
— **co-sign** \'kō-ˌsīn\ *verb*

co·sine \'kō-ˌsīn\ *noun* [New Latin *cosinus,* from *co-* + Medieval Latin *sinus* sine] (1635)
1 : a trigonometric function that for an acute angle is the ratio between the leg adjacent to the angle when it is considered part of a right triangle and the hypotenuse
2 : a trigonometric function cos θ that for all real numbers θ is given by the sum of the alternating series

$$\cos\theta = 1 - \frac{\theta^2}{2!} + \frac{\theta^4}{4!} - \frac{\theta^6}{6!} + \frac{\theta^8}{8!} - \cdots$$

and that is exactly equal to the cosine of an angle of measure θ in radians

cos lettuce \'käs-, 'kōs-\ *noun* [*Kos, Cos,* Greek island] (1699)
: ROMAINE

[1]cos·met·ic \käz-'me-tik\ *adjective* [Greek *kosmētikos* skilled in adornment, from *kosmein* to arrange, adorn, from *kosmos* order] (1650)
1 : of, relating to, or making for beauty especially of the complexion **:** BEAUTIFYING ⟨*cosmetic* salves⟩
2 : done or made for the sake of appearance: as **a :** correcting defects especially of the face ⟨*cosmetic* surgery⟩ **b :** DECORATIVE, ORNAMENTAL **c :** not substantive **:** SUPERFICIAL ⟨*cosmetic* changes⟩
3 : visually appealing
— **cos·met·i·cal·ly** \-ti-k(ə-)lē\ *adverb*

[2]cosmetic *noun* (1650)
: something that is cosmetic: as **a :** a cosmetic preparation for external use **b** *plural* **:** superficial features ⟨a poem without rhetorical *cosmetics* —Guy Davenport⟩

cosmetic case *noun* (1948)
: a small piece of luggage especially for cosmetics

cos·me·ti·cian \ˌkäz-mə-'ti-shən\ *noun* (1924)
: a person who is professionally trained in the use of cosmetics

cos·met·i·cize \käz-'me-tə-ˌsīz\ *transitive verb* **-cized; -ciz·ing** (1824)
: to make (something unpleasant or ugly) superficially attractive

cos·me·tol·o·gist \ˌkäz-mə-'tä-lə-jist\ *noun* (1926)
: a person who gives beauty treatments (as to skin and hair) — called also *beautician*

cos·me·tol·o·gy \-jē\ *noun* [French *cosmétologie,* from *cosmétique* cosmetic (from English *cosmetic*) + *-logie* -logy] (1926)
: the cosmetic treatment of the skin, hair, and nails

cos·mic \'käz-mik\ *also* **cos·mi·cal** \-mi-kəl\ *adjective* [Greek *kosmikos,* from *kosmos* order, universe] (1685)
1 : of or relating to the cosmos, the extraterrestrial vastness, or the universe in contrast to the earth alone
2 : characterized by greatness especially in extent, intensity, or comprehensiveness ⟨a *cosmic* thinker⟩ ⟨*cosmic* boredom⟩
— **cos·mi·cal·ly** \-mi-k(ə-)lē\ *adverb*

cosmic background radiation *noun* (circa 1976)
: BACKGROUND RADIATION

cosmic dust *noun* (1881)
: very fine particles of solid matter found in any part of the universe

cosmic noise *noun* (1947)
: unidentified celestial radio-frequency radiation; *especially* **:** such radiation originating from outside the Milky Way

cosmic ray *noun* (1925)
: a stream of atomic nuclei of extremely penetrating character that enter the earth's atmosphere from outer space at speeds approaching that of light

cosmic string *noun* (1984)
: any of a class of hypothetical supermassive astronomical objects that are extremely thin but are millions of light years long and that are postulated to have formed very early in the history of the universe

cos·mo·chem·is·try \ˌkäz-mō-'ke-mə-strē\ *noun* [Greek *kosmos* universe] (1940)
: a branch of chemistry that deals with the chemical composition of and changes in the universe
— **cos·mo·chem·i·cal** \-'ke-mi-kəl\ *adjective*
— **cos·mo·chem·ist** \-mist\ *noun*

cos·mo·gen·ic \ˌkäz-mə-'je-nik\ *adjective* [*cosmic* ray + *-o-* + *-genic*] (1962)
: produced by the action of cosmic rays ⟨*cosmogenic* carbon 14⟩

cos·mog·o·ny \käz-'mä-gə-nē\ *noun, plural* **-nies** [New Latin *cosmogonia,* from Greek *kosmogonia,* from *kosmos* + *gonos* offspring; akin to Greek *genos* race — more at KIN] (1766)
1 : the creation or origin of the world or universe
2 : a theory of the origin of the universe
— **cos·mo·gon·ic** \ˌkäz-mə-'gä-nik\ *or* **cos·mo·gon·i·cal** \-ni-kəl\ *adjective*
— **cos·mog·o·nist** \käz-'mä-gə-nist\ *noun*

cos·mog·ra·phy \käz-'mä-grə-fē\ *noun, plural* **-phies** [Middle English *cosmographie,* from Late Latin *cosmographia,* from Greek *kosmographia,* from *kosmos* + *-graphia* -graphy] (14th century)
1 : a general description of the world or of the universe
2 : the science that deals with the constitution of the whole order of nature
— **cos·mog·ra·pher** \-fər\ *noun*
— **cos·mo·graph·ic** \ˌkäz-mə-'gra-fik\ *or* **cos·mo·graph·i·cal** \-fi-kəl\ *adjective*

Cos·mo·line \'käz-mə-ˌlēn\ *trademark*
— used for petroleum jelly

cos·mol·o·gy \käz-'mä-lə-jē\ *noun, plural* **-gies** [New Latin *cosmologia,* from Greek *kosmos* + New Latin *-logia* -logy] (circa 1656)
1 a : a branch of metaphysics that deals with the nature of the universe **b :** a theory or doctrine describing the natural order of the universe

\ə\ **abut** \$^{ə}$\ **kitten** \ər\ **further** \a\ **ash** \ā\ **ace** \ä\ **mop, mar** \au̇\ **out** \ch\ **chin** \e\ **bet** \ē\ **easy** \g\ **go** \i\ **hit** \ī\ **ice** \j\ **job** \ŋ\ **sing** \ō\ **go** \ȯ\ **law** \ȯi\ **boy** \th\ **thin** \th\ **the** \ü\ **loot** \u̇\ **foot** \y\ **yet** \zh\ **vision** *see also* Guide to Pronunciation

2 : a branch of astronomy that deals with the origin, structure, and space-time relationships of the universe; *also* **:** a theory dealing with these matters

— **cos·mo·log·i·cal** \ˌkäz-mə-ˈlä-ji-kəl\ *adjective*

— **cos·mo·log·i·cal·ly** \-ji-k(ə-)lē\ *adverb*

— **cos·mol·o·gist** \käz-ˈmä-lə-jist\ *noun*

cos·mo·naut \ˈkäz-mə-ˌnȯt, -ˌnät\ *noun* [Russian *kosmonavt,* from Greek *kosmos* + Russian *-navt* (as in *aeronavt* aeronaut)] (1959)
: an astronaut of the Soviet or Russian space program

cos·mop·o·lis \käz-ˈmä-pə-ləs\ *noun* [New Latin, back-formation from *cosmopolites*] (1849)
: a cosmopolitan city

1cos·mo·pol·i·tan \ˌkäz-mə-ˈpä-lə-tᵊn\ *noun* (circa 1645)
: COSMOPOLITE

2cosmopolitan *adjective* (1844)
1 : having worldwide rather than limited or provincial scope or bearing
2 : having wide international sophistication **:** WORLDLY
3 : composed of persons, constituents, or elements from all or many parts of the world
4 : found in most parts of the world and under varied ecological conditions ⟨a *cosmopolitan* herb⟩

— **cos·mo·pol·i·tan·ism** \-tə-ˌni-zəm\ *noun*

cos·mop·o·lite \käz-ˈmä-pə-ˌlīt\ *noun* [New Latin *cosmopolites,* from Greek *kosmopolitēs,* from *kosmos* + *politēs* citizen] (circa 1618)
: a cosmopolitan person or organism

— **cos·mo·po·li·tism** \käz-ˈmä-pə-ˌlī-ˌti-zəm, -lə-ˌti-; ˌkäz-mə-ˈpä-lə-ˌti-\ *noun*

cos·mos \ˈkäz-məs, *1 & 2 also* -ˌmōs, -ˌmäs\ *noun* [Greek *kosmos*] (1650)
1 a (1) **:** an orderly harmonious systematic universe — compare CHAOS (2) **:** ORDER, HARMONY **b :** UNIVERSE 1
2 : a complex orderly self-inclusive system
3 *plural* **cosmos** \-məs, -məz\ *also* **cos·mos·es** \-mə-səz\ [New Latin, genus name, from Greek *kosmos*] **:** any of a genus (*Cosmos*) of tropical American composite herbs; *especially* **:** a widely cultivated tall annual (*C. bipinnatus*) with yellow or red disks and showy ray flowers

cos·sack \ˈkä-ˌsak, -sək\ *noun, often capitalized* [Polish & Ukrainian *kozak,* of Turkic origin; akin to Volga Tatar *kazak* free person] (1589)
: a member of a group of frontiersmen of southern Russia organized as cavalry in the czarist army

1cos·set \ˈkä-sət\ *noun* [origin unknown] (1579)
: a pet lamb; *broadly* **:** PET

2cosset *transitive verb* (1659)
: to treat as a pet **:** PAMPER

1cost \ˈkȯst\ *noun* (13th century)
1 a : the amount or equivalent paid or charged for something **:** PRICE **b :** the outlay or expenditure (as of effort or sacrifice) made to achieve an object
2 : loss or penalty incurred especially in gaining something
3 *plural* **:** expenses incurred in litigation; *especially* **:** those given by the law or the court to the prevailing party against the losing party

— **cost·less** \-ləs\ *adjective*

— **cost·less·ly** *adverb*

2cost *verb* **cost; cost·ing** [Middle English, from Middle French *coster,* from Latin *constare* to stand firm, cost — more at CONSTANT] (14th century)
intransitive verb
1 : to require expenditure or payment ⟨the best goods *cost* more⟩
2 : to require effort, suffering, or loss
transitive verb
1 : to have a price of
2 : to cause to pay, suffer, or lose something ⟨frequent absences *cost* him his job⟩
3 *past* **cost·ed :** to estimate or set the cost of — often used with *out*

cos·ta \ˈkäs-tə\ *noun, plural* **cos·tae** \-(ˌ)tē, -ˌtī\ [Latin — more at COAST] (circa 1864)
1 : 1RIB 1a
2 : a part (as the midrib of a leaf or the anterior vein of an insect wing) that resembles a rib

— **cos·tal** \-tᵊl\ *adjective*

cost accountant *noun* (1918)
: a specialist in cost accounting

cost accounting *noun* (1913)
: the systematic recording and analysis of the costs of material, labor, and overhead incident to production

cos·tard \ˈkäs-tərd\ *noun* [Middle English] (13th century)
1 : any of several large English cooking apples
2 *archaic* **:** NODDLE, PATE

cost–ben·e·fit \ˈkȯs(t)-ˈbe-nə-ˌfit\ *adjective* (1928)
: of, relating to, or being economic analysis that assigns a numerical value to the cost-effectiveness of an operation, procedure, or program

cost–ef·fec·tive \ˈkȯst-ə-ˈfek-tiv, -ˌfek-\ *adjective* (1967)
: economical in terms of tangible benefits produced by money spent ⟨*cost-effective* measures to combat poverty⟩

— **cost–ef·fec·tive·ness** *noun*

cost–ef·fi·cient \-i-ˈfi-shənt, -ˌfi-\ *adjective* (1979)
: COST-EFFECTIVE

cos·ter \ˈkäs-tər\ *noun* (1851)
British **:** COSTERMONGER

cos·ter·mon·ger \-ˌməŋ-gər, -ˌmäŋ-\ *noun* [*costard* + *monger*] (1514)
British **:** a hawker of fruit or vegetables

cos·tive \ˈkäs-tiv, ˈkȯs-\ *adjective* [Middle English, from Middle French *costivé,* past participle of *costiver* to constipate, from Latin *constipare*] (14th century)
1 a : affected with constipation **b :** causing constipation
2 : slow in action or expression
3 : NIGGARDLY, STINGY

— **cos·tive·ly** *adverb*

— **cos·tive·ness** *noun*

cost·ly \ˈkȯs(t)-lē\ *adjective* **cost·li·er; -est** (14th century)
1 a : commanding a high price especially because of intrinsic worth ⟨*costly* gems⟩ **b :** RICH, SPLENDID
2 : made or done at heavy expense or sacrifice ⟨a *costly* mistake⟩

— **cost·li·ness** *noun*

cost·mary \ˈkȯst-ˌmer-ē, ˈkäst-\ *noun, plural* **-mar·ies** [Middle English *costmarie,* from *coste* costmary (from Old English *cost,* from Latin *costum,* from Greek *kostos,* a fragrant root) + *Marie* the Virgin Mary] (14th century)
: an aromatic composite herb (*Chrysanthemum balsamita*) used as a potherb and in flavoring

cost of living (1896)
: the cost of purchasing those goods and services which are included in an accepted standard level of consumption

cost–of–living index *noun* (1913)
: CONSUMER PRICE INDEX

cost–plus \ˈkȯs(t)-ˈpləs\ *adjective* (1918)
1 : paid on the basis of a fixed fee or a percentage added to actual cost ⟨a *cost-plus* contract⟩
2 : of or relating to a cost-plus contract

cost–push \ˈkȯs(t)-ˌpu̇sh\ *noun* (1951)
: an increase or upward trend in production costs (as wages) that tends to result in increased consumer prices irrespective of the level of demand — compare DEMAND-PULL

— **cost–push** *adjective*

cos·trel \ˈkäs-trəl\ *noun* [Middle English, from Middle French *costerel,* from *costier* at the side, from *coste* rib, side — more at COAST] (14th century)
: a flat usually earthenware container for liquids with loops through which a belt or cord may be passed for easy carrying — called also *pilgrim bottle*

1cos·tume \ˈkäs-ˌtüm, -ˌtyüm *also* -təm *or* -ˌchüm\ *noun* [French, from Italian, custom, dress, from Latin *consuetudin-, consuetudo* custom — more at CUSTOM] (1715)
1 : the prevailing fashion in coiffure, jewelry, and apparel of a period, country, or class
2 : an outfit worn to create the appearance characteristic of a particular period, person, place, or thing ⟨Halloween *costumes*⟩
3 : a person's ensemble of outer garments; *especially* **:** a woman's ensemble of dress with coat or jacket

— **cos·tum·ey** *adjective*

2cos·tume \käs-ˈtüm, -ˈtyüm *also* -ˈchüm; *or* ˈkäs-ˌ *or* -təm\ *transitive verb* **cos·tumed; cos·tum·ing** (1823)
1 : to provide with a costume
2 : to design costumes for ⟨*costume* a play⟩

3costume *same as* 1\ *adjective* (1884)
1 : characterized by the use of costumes ⟨a *costume* ball⟩ ⟨a *costume* drama⟩
2 : suitable for or enhancing the effect of a particular costume ⟨a *costume* handbag⟩

costume jewelry *noun* (1933)
: jewelry designed for wear with current fashions and usually made of inexpensive materials

cos·tum·er \ˈkäs-ˌtü-mər, -ˌtyü- *also* -ˌchü-; käs-ˈ\ *noun* (circa 1859)
1 : one that deals in or makes costumes
2 : an upright stand with hooks or pegs on which to hang clothes

cos·tum·ery \-mə-rē\ *noun* (1838)
1 : articles of costume
2 : the art of costuming

cos·tu·mi·er \käs-ˈtü-mē-ˌā, -ˈtyü-, -mē-ər\ *noun* [French] (1831)
chiefly British **:** COSTUMER 1

co·sy \ˈkō-zē\ *variant of* COZY

1cot \ˈkät\ *noun* [Middle English, from Old English; akin to Old Norse *kot* small hut] (before 12th century)
1 : a small house
2 : COVER, SHEATH; *especially* **:** STALL 4

2cot *noun* [Hindi *khāṭ* bedstead, from Sanskrit *khaṭvā,* perhaps of Dravidian origin; akin to Tamil *kaṭṭil* bedstead] (1634)
1 : a small usually collapsible bed often of fabric stretched on a frame
2 *British* **:** CRIB 2b

co·tan·gent \(ˌ)kō-ˈtan-jənt, ˈkō-ˌ\ *noun* [New Latin *cotangent-, cotangens,* from *co-* + *tangent-, tangens* tangent] (1635)
1 : a trigonometric function that for an acute angle is the ratio between the leg adjacent to the angle when it is considered part of a right triangle and the leg opposite
2 : a trigonometric function cot θ that is equal to the cosine divided by the sine for all real numbers θ for which the sine is not equal to zero and is exactly equal to the cotangent of an angle of measure θ in radians

1cote \ˈkōt, ˈkät\ *noun* [Middle English, from Old English] (before 12th century)
1 *dialect English* **:** 1COT 1
2 : a shed or coop for small domestic animals and especially pigeons

2cote \ˈkōt\ *transitive verb* [probably from Middle French *cotoyer*] (1555)
obsolete **:** to pass by

co·te·rie \ˈkō-tə-(ˌ)rē, ˌkō-tə-ˈ\ *noun* [French, from Middle French, tenants, from Old French *cotier* cotter, of Germanic origin; akin to Old English *cot* hut] (1738)
: an intimate and often exclusive group of persons with a unifying common interest or purpose

co·ter·mi·nous \(ˌ)kō-ˈtər-mə-nəs\ *adjective* [alteration of *conterminous*] (1799)

1 : having the same or coincident boundaries ⟨*coterminous* states⟩
2 : coextensive in scope or duration ⟨an experience of life *coterminous* with the years of his father —Elizabeth Hardwick⟩
— co·ter·mi·nous·ly *adverb*

co·thur·nus \kō-ˈthər-nəs\ *noun, plural* **-ni** \-ˌnī, -(ˌ)nē\ [Latin, from Greek *kothornos*] (1606)
1 : a high thick-soled laced boot worn by actors in Greek and Roman tragic drama — called also *co·thurn* \ˈkō-ˌthərn, kō-ˈ\
2 : the dignified somewhat stilted style of ancient tragedy

co·tid·al \(ˌ)kō-ˈtī-dᵊl\ *adjective* (1833)
: indicating equality in the tides or a coincidence in the time of high or low tide

co·til·lion \kō-ˈtil-yən, kə-\ *also* **co·til·lon** \kō-ˈtil-yən, kə-, kȯ-tē-(y)ō^{n}\ *noun* [French *cotillon,* literally, petticoat, from Old French, from *cote* coat] (1766)
1 : a ballroom dance for couples that resembles the quadrille
2 : an elaborate dance with frequent changing of partners carried out under the leadership of one couple at formal balls
3 : a formal ball

co·to·ne·as·ter \kə-ˈtō-nē-ˌas-tər, ˈkä-tᵊn-ˌēs-\ *noun* [New Latin, genus name, from Latin *cotoneum* quince + New Latin *-aster*] (1796)
: any of a genus (*Cotoneaster*) of Old World flowering shrubs of the rose family

cot·quean \ˈkät-ˌkwēn\ *noun* [[1]*cot* + *quean*] (1547)
1 *archaic* **:** a coarse masculine woman
2 *archaic* **:** a man who busies himself with women's work or affairs

Cots·wold \ˈkät-ˌswōld, -swəld\ *noun* [*Cotswold* Hills, England] (circa 1658)
: any of an English breed of large long-wooled sheep

cot·ta \ˈka-tə\ *noun* [Medieval Latin, of Germanic origin; akin to Old High German *kozza* coarse mantle — more at COAT] (1848)
: a waist-length surplice

cot·tage \ˈkä-tij\ *noun* [Middle English *cotage,* from (assumed) Anglo-French, from Middle English *cot* — more at COT] (14th century)
1 : the dwelling of a farm laborer or small farmer
2 : a usually small frame one-family house
3 : a small detached dwelling unit at an institution
4 : a usually small house for vacation use
— cot·tag·ey \-ti-jē\ *adjective*

cottage cheese *noun* (1848)
: a bland soft white cheese made from the curds of skim milk — called also *Dutch cheese, pot cheese, smearcase*

cottage curtains *noun plural* (1943)
: a double set of upper and lower straight-hanging window curtains

cottage industry *noun* (1921)
1 : an industry whose labor force consists of family units or individuals working at home with their own equipment
2 : a small and often informally organized industry

cottage pie *noun* (1791)
: a shepherd's pie made especially with beef

cottage pudding *noun* (circa 1854)
: plain cake covered with a hot sweet sauce

cot·tag·er \ˈkä-ti-jər\ *noun* (1550)
: a person who lives in a cottage

cottage tulip *noun* (1928)
: any of various tall late-flowering tulips

cot·tar *or* **cot·ter** \ˈkä-tər\ *noun* [Middle English *cottar,* from Medieval Latin *cotarius,* from Middle English *cot*] (14th century)
: a peasant or farm laborer who occupies a cottage and sometimes a small holding of land usually in return for services

cot·ter \ˈkä-tər\ *noun* [origin unknown] (14th century)
1 : a wedge-shaped or tapered piece used to fasten together parts of a structure
2 : COTTER PIN
— cot·tered \-tərd\ *adjective*
— cot·ter·less \-tər-ləs\ *adjective*

cotter pin *noun* (1881)
: a half-round metal strip bent into a pin whose ends can be flared after insertion through a slot or hole

[1]cot·ton \ˈkä-tᵊn\ *noun, often attributive* [Middle English *coton,* from Middle French, from Arabic *quṭun*] (14th century)
1 a : a soft usually white fibrous substance composed of the hairs surrounding the seeds of various erect freely branching tropical plants (genus *Gossypium*) of the mallow family **b :** a plant producing cotton; *especially* **:** one grown for its cotton **c :** a crop of cotton
2 a : fabric made of cotton **b :** yarn spun from cotton
3 : a downy cottony substance produced by various plants (as the cottonwood)

cotton: *1* flowering branch; *2* fruit, unopened; *3* fruit, partly opened

[2]cotton *intransitive verb* **cot·toned; cot·ton·ing** \ˈkät-niŋ, ˈka-tᵊn-iŋ\ (1605)
1 : to take a liking — used with *to* ⟨*cottons* to people easily⟩
2 : to come to understand — used with *to* or *on to* ⟨*cottoned* on to the fact that our children work furiously —H. M. McLuhan⟩

cotton candy *noun* (1926)
1 : a candy made of spun sugar
2 : something attractive but insubstantial

cotton gin *noun* (1796)
: a machine that separates the seeds, hulls, and foreign material from cotton

cotton grass *noun* (1597)
: any of a genus (*Eriophorum*) of sedges with tufted spikes

cot·ton·mouth \ˈkä-tᵊn-ˌmau̇th\ *noun* [from the white interior of its mouth] (1832)
: WATER MOCCASIN

cottonmouth moccasin *noun* (1879)
: WATER MOCCASIN

cot·ton–pick·ing \ˈkä-tᵊn-ˌpi-kiŋ, -ˌpi-kən\ *adjective* (circa 1952)
1 : DAMNED — used as a generalized expression of disapproval ⟨a *cotton-picking* hypocrite⟩
2 : DAMNED — used as an intensive ⟨out of his *cotton-picking* mind —Irving Kristol⟩

cot·ton·seed \ˈkä-tᵊn-ˌsēd\ *noun* (1774)
: the seed of the cotton plant

cottonseed oil *noun* (1833)
: a pale yellow semidrying fatty oil that is obtained from the cottonseed and is used chiefly in salad and cooking oils and after hydrogenation in shortenings and margarine

cotton stainer *noun* (1856)
: any of several bugs (genus *Dysdercus*) that damage and stain the lint of developing cotton; *especially* **:** a red and brown bug (*D. suturellus*) that attacks cotton in the southern U.S.

cot·ton·tail \ˈkä-tᵊn-ˌtāl\ *noun* (1869)
: any of several rather small North American rabbits (genus *Sylvilagus*) sandy to grayish brown in color with a white-tufted underside of the tail

cot·ton·weed \-ˌwēd\ *noun* (1562)
: any of various weedy plants (as cudweed) with hoary pubescence or cottony seeds

cot·ton·wood \-ˌwu̇d\ *noun* (1802)
: any of several poplars having seeds with cottony hairs; *especially* **:** one (*Populus deltoides*) of the eastern and central U.S. often cultivated for its rapid growth and luxuriant foliage

cotton wool *noun* (14th century)
: raw cotton; *especially* **:** cotton batting

cot·tony \ˈkät-nē, ˈkä-tᵊn-ē\ *adjective* (1578)
: resembling cotton in appearance or character: as **a :** covered with hairs or pubescence **b :** SOFT

cot·tony–cush·ion scale \-ˈku̇-shən-\ *noun* (1886)
: a scale insect (*Icerya purchasi*) introduced into the U.S. from Australia that infests citrus and other plants

-cotyl *noun combining form* [*cotyledon*]
: cotyledon ⟨hypo*cotyl*⟩

cot·y·le·don \ˌkä-tə-ˈlē-dᵊn\ *noun* [New Latin, from Greek *kotylēdōn* cup-shaped hollow, from *kotylē* cup, anything hollow] (1540)
1 : a lobule of the mammalian placenta
2 : the first leaf or one of the first pair or whorl of leaves developed by the embryo of a seed plant or of some lower plants (as ferns) — see PLUMULE illustration
— cot·y·le·don·ary \-ˈlē-də-ˌner-ē\ *adjective*

co·ty·lo·saur \ˈkä-tᵊl-ō-ˌsȯr, kə-ˈti-lə-\ *noun* [ultimately from Greek *kotylē* + *sauros* lizard] (circa 1909)
: any of an order (Cotylosauria) of extinct primitive reptiles with short legs and massive bodies that were probably the earliest truly terrestrial vertebrate animals

[1]couch \ˈkau̇ch\ *verb* [Middle English, from Middle French *coucher,* from Latin *collocare* to set in place — more at COLLOCATE] (14th century)
transitive verb
1 : to lay (oneself) down for rest or sleep
2 : to embroider (a design) by laying down a thread and fastening it with small stitches at regular intervals
3 : to place or hold level and pointed forward ready for use
4 : to phrase or express in a specified manner ⟨the comments were *couched* in strong terms⟩
5 : to treat (a cataract) by displacing the lens of the eye into the vitreous humor
intransitive verb
1 : to lie down or recline for sleep or rest
2 : to lie in ambush

[2]couch *noun* [Middle English *couche* bed, from Middle French, from *coucher*] (14th century)
1 a : an article of furniture for sitting or reclining **b :** a couch on which a patient reclines when undergoing psychoanalysis
2 : the den of an animal (as an otter)
— on the couch : receiving psychiatric treatment

couch·ant \ˈkau̇-chənt\ *adjective* [Middle English, from Middle French, from present participle of *coucher*] (15th century)
: lying down especially with the head up ⟨a heraldic lion *couchant*⟩

couch grass \ˈkau̇ch-, ˈku̇ch-\ *noun* [alteration of *quitch*] (1578)
1 : QUACK GRASS
2 : any of several grasses that resemble quack grass in spreading by creeping rhizomes

couch potato *noun* (1982)
: a lazy and inactive person; *especially* **:** one who spends a great deal of time watching television

cou·dé \kü-ˈdā\ *adjective* [French *coudé* bent like an elbow, from *coude* elbow, from Latin *cubitum*] (circa 1889)
: of, relating to, or being a telescope constructed so that the light is reflected along the polar axis to come to a focus at a point where the

holder for a photographic plate or a spectrograph may be mounted

cou·gar \'kü-gər *also* -,gär\ *noun, plural* **cougars** *also* **cougar** [French *couguar,* modification of New Latin *cuguacuarana,* from Tupi *siwasuarána,* from *siwasú* deer + *-rana* resembling] (1774)
: a large powerful tawny brown cat (*Felis concolor*) formerly widespread in the Americas but now reduced in number or extinct in many areas — called also *catamount, mountain lion, panther, puma*

cougar

[1]cough \'kôf\ *verb* [Middle English, from (assumed) Old English *cohhian;* akin to Middle High German *kūchen* to breathe heavily] (14th century)
intransitive verb
1 : to expel air from the lungs suddenly with an explosive noise
2 : to make a noise like that of coughing
transitive verb
: to expel by coughing — often used with *up* ⟨*cough* up mucus⟩

[2]cough *noun* (14th century)
1 : a condition marked by repeated or frequent coughing
2 : an act or sound of coughing

cough drop *noun* (1831)
: a lozenge or troche used to relieve coughing

cough syrup *noun* (1877)
: any of various sweet usually medicated liquids used to relieve coughing

cough up *transitive verb* (1894)
: HAND OVER, DELIVER ⟨*cough up* the money⟩

could \kəd, 'kùd\ [Middle English *couthe, coude,* from Old English *cūthe;* akin to Old High German *konda* could] *past of* CAN
— used in auxiliary function in the past ⟨we found we *could* go⟩, in the past conditional ⟨we said we would go if we *could*⟩, and as an alternative to *can* suggesting less force or certainty or as a polite form in the present ⟨if you *could* come we would be pleased⟩

could·est \'kù-dəst\ *archaic past 2d singular of* CAN

couldn't \'kù-dᵊnt, -dᵊn, *dialect also* 'kùt-ᵊn(t) *or* 'kùnt\ (1646)
: could not

couldst \kədst, 'kùdst, kətst, 'kùtst\ *archaic past 2d singular of* CAN

cou·lee \'kü-lē\ *noun* [Canadian French *coulée,* from French, flowing, flow of lava, from *couler* to flow, from Old French, from Latin *colare* to strain, from *colum* sieve] (1807)
1 a : a small stream **b :** a dry streambed **c :** a usually small or shallow ravine **:** GULLY
2 : a thick sheet or stream of lava

cou·lis \kü-'lē\ *noun* [French, from Old French *coleïs,* from *coleïs, coleïz* flowing] (1978)
: a thick sauce made with pureed vegetable or fruit and often used as a garnish

cou·lisse \kü-'lēs, -'lis\ *noun* [French, from Old French *coulice* portcullis, from feminine of *coleïz* flowing, sliding, from *couler*] (1819)
1 a : a side scene of a stage; *also* **:** the space between the side scenes **b :** a backstage area **c :** HALLWAY
2 : a piece of timber having a groove in which something glides

cou·loir \kül-'wär\ *noun* [French, literally, passage, from *couler*] (1822)
: a steep mountainside gorge

[1]cou·lomb \'kü-,läm, -,lōm\ *noun* [Charles A. de *Coulomb*] (1881)
: the practical meter-kilogram-second unit of electric charge equal to the quantity of electricity transferred by a current of one ampere in one second

[2]coulomb *or* **cou·lom·bic** \kü-'läm-bik, -'lōm-, -'lä-mik, -'lō-\ *adjective* (1930)
: of, relating to, or being the electrostatic force of attraction or repulsion between charged particles

Cou·lomb's law \'kü-,lämz-, -,lōmz-\ *noun* (1854)
: a statement in physics: the force of attraction or repulsion acting along a straight line between two electric charges is directly proportional to the product of the charges and inversely to the square of the distance between them

cou·lo·me·ter \kü-'lä-mə-tər, 'kü-lə-,mē-tər\ *noun* [alteration of *coulombmeter,* from *coulomb* + *-meter*] (circa 1889)
: an instrument of chemical analysis that determines the amount of a substance released in electrolysis by measurement of the quantity of electricity used
— **cou·lo·met·ric** \,kü-lə-'me-trik\ *adjective*
— **cou·lo·met·ri·cal·ly** \-tri-k(ə-)lē\ *adverb*
— **cou·lom·e·try** \kü-'lä-mə-trē\ *noun*

coul·ter \'kōl-tər\ *noun* [Middle English *colter,* from Old English *culter* & Old French *coltre,* both from Latin *culter* plowshare; akin to Greek *skallein* to hoe — more at SHELL] (before 12th century)
: a cutting tool (as a knife or sharp disc) that is attached to the beam of a plow, makes a vertical cut in the surface, and permits clean separation and effective covering of the soil and materials being turned under

cou·ma·rin \'kü-mə-rən\ *noun* [French *coumarine,* from *coumarou* tonka bean tree, from Spanish or Portuguese; Spanish *cumarú,* from Portuguese, from Tupi *kumarú*] (1830)
: a toxic white crystalline lactone $C_9H_6O_2$ with an odor of new-mown hay found in plants or made synthetically and used especially in perfumery; *also* **:** a derivative of this compound

[1]coun·cil \'kaùn(t)-səl\ *noun* [Middle English *counceil,* from Old French *concile,* from Latin *concilium,* from *com-* + *calare* to call — more at LOW] (12th century)
1 : an assembly or meeting for consultation, advice, or discussion
2 : a group elected or appointed as an advisory or legislative body
3 a : a usually administrative body **b :** an executive body whose members are equal in power and authority **c :** a governing body of delegates from local units of a federation
4 : deliberation in a council
5 a : a federation of or a central body uniting a group of organizations **b :** a local chapter of an organization **c :** CLUB, SOCIETY

[2]council *adjective* (14th century)
1 : used for councils especially by or with North American Indians ⟨a *council* ground⟩
2 *British* **:** built, maintained, or operated by a local governing agency ⟨*council* housing⟩ ⟨*council* flats⟩

coun·cil·lor *or* **coun·cil·or** \'kaùn(t)-s(ə-)lər\ *noun* (15th century)
: a member of a council
— **coun·cil·lor·ship** \-,ship\ *noun*

coun·cil·man \'kaùn(t)-səl-mən\ *noun* (1659)
: a member of a council (as of a town or city)
— **coun·cil·man·ic** \,kaùn(t)-səl-'ma-nik\ *adjective*

council of ministers *often C&M capitalized* (circa 1909)
: CABINET 3b

coun·cil·wom·an \'kaùn(t)-səl-,wù-mən\ *noun* (circa 1928)
: a woman who is a member of a council

[1]coun·sel \'kaùn(t)-səl\ *noun* [Middle English *conseil,* from Old French, from Latin *consilium,* from *consulere* to consult] (13th century)
1 a : advice given especially as a result of consultation **b :** a policy or plan of action or behavior
2 : DELIBERATION, CONSULTATION
3 a *archaic* **:** PURPOSE **b :** guarded thoughts or intentions
4 a *plural* **counsel** (1) **:** a lawyer engaged in the trial or management of a case in court (2) **:** a lawyer appointed to advise and represent in legal matters an individual client or a corporate and especially a public body **b :** CONSULTANT 2

[2]counsel *verb* **-seled** *or* **-selled; -sel·ing** *or* **-sel·ling** \-s(ə-)liŋ\ (14th century)
transitive verb
: ADVISE ⟨*counseled* them to avoid rash actions —George Orwell⟩
intransitive verb
: CONSULT ⟨*counseled* with her husband⟩

coun·sel·ee \,kaùn(t)-sə-'lē\ *noun* (1923)
: one who is being counseled

counseling *or* **counselling** *noun* (1927)
: professional guidance of the individual by utilizing psychological methods especially in collecting case history data, using various techniques of the personal interview, and testing interests and aptitudes

coun·sel·or *or* **coun·sel·lor** \'kaùn(t)-s(ə-)lər\ *noun* (13th century)
1 : a person who gives advice or counseling ⟨marriage *counselor*⟩
2 : LAWYER; *specifically* **:** one that gives advice in law and manages cases for clients in court
3 : one who has supervisory duties at a summer camp
— **coun·sel·or·ship** \-,ship\ *noun*

counselor–at–law *noun, plural* **counselors–at–law** (1617)
: COUNSELOR 2

[1]count \'kaùnt, *dialect* 'kyaùnt\ *verb* [Middle English, from Middle French *conter, compter,* from Latin *computare,* from *com-* + *putare* to consider] (14th century)
transitive verb
1 a : to indicate or name by units or groups so as to find the total number of units involved **:** NUMBER **b :** to name the numbers in order up to and including ⟨*count* ten⟩ **c :** to include in a tallying and reckoning ⟨about 100 present, *counting* children⟩ **d :** to call aloud (beats or time units) ⟨*count* cadence⟩ ⟨*count* eighth notes⟩
2 a : CONSIDER, ACCOUNT ⟨*count* oneself lucky⟩ **b :** to record as of an opinion or persuasion ⟨*count* me as uncommitted⟩
3 : to include or exclude by or as if by counting ⟨*count* me in⟩
intransitive verb
1 a : to recite or indicate the numbers in order by units or groups ⟨*count* by fives⟩ **b :** to count the units in a group
2 : to rely or depend on someone or something ⟨*counted* on his parents to help with the expenses⟩
3 : ADD, TOTAL ⟨it *counts* up to a sizable amount⟩
4 : to have value or significance ⟨these are the people who really *count*⟩
— **count heads** *or* **count noses :** to count the number present
— **count on :** to look forward to as certain **:** ANTICIPATE ⟨*counted on* winning⟩

[2]count *noun* (14th century)
1 a : the action or process of counting **b :** a total obtained by counting **:** TALLY
2 *archaic* **a :** RECKONING, ACCOUNT **b :** CONSIDERATION, ESTIMATION
3 a : ALLEGATION, CHARGE; *specifically* **:** one separately stating the cause of action or prosecution in a legal declaration or indictment ⟨guilty on all *counts*⟩ **b :** a specific point under consideration **:** ISSUE
4 : the total number of individual things in a given unit or sample obtained by counting all or a subsample of them ⟨bacteria *count*⟩
5 a : the calling off of the seconds from one to ten when a boxer has been knocked down **b**

: the number of balls and strikes charged to a baseball batter during one turn ⟨the *count* stood at 3 and 2⟩ **c :** SCORE ⟨tied the *count* with a minute to play⟩
6 : a measurement of the thickness or fineness of yarn by determining the number of hanks or yards per pound it produces

³count *noun* [Middle English, from Middle French *comte,* from Late Latin *comit-, comes,* from Latin, companion, one of the imperial court, from *com-* + *ire* to go — more at ISSUE] (15th century)
: a European nobleman whose rank corresponds to that of a British earl

count·able \ˈkau̇n-tə-bəl\ *adjective* (1581)
: capable of being counted; *especially* **:** DENUMERABLE ⟨a *countable* set⟩
— **count·abil·i·ty** \ˌkau̇n-tə-ˈbi-lə-tē\ *noun*
— **count·ably** \ˈkau̇n-tə-blē\ *adverb*

count·down \ˈkau̇nt-ˌdau̇n\ *noun* (circa 1952)
: an audible backward counting in fixed units (as seconds) from an arbitrary starting number to mark the time remaining before an event; *also* **:** preparations carried on during such a count
— **count down** \-ˈdau̇n\ *intransitive verb*

¹coun·te·nance \ˈkau̇n-tᵊn-ən(t)s, ˈkau̇nt-nən(t)s\ *noun* [Middle English *contenance,* from Middle French, from Medieval Latin *continentia,* from Latin, restraint, from *continent-, continens,* present participle of *continēre* to hold together — more at CONTAIN] (13th century)
1 *obsolete* **:** BEARING, DEMEANOR
2 a : calm expression **b :** mental composure **c :** LOOK, EXPRESSION
3 *archaic* **a :** ASPECT, SEMBLANCE **b :** PRETENSE
4 : FACE, VISAGE; *especially* **:** the face as an indication of mood, emotion, or character
5 : bearing or expression that offers approval or sanction **:** moral support

²countenance *transitive verb* **-nanced; -nanc·ing** (1568)
: to extend approval or toleration to **:** SANCTION
— **coun·te·nanc·er** *noun*

¹count·er \ˈkau̇n-tər\ *noun* [Middle English *countour,* from Middle French *comptouer,* from Medieval Latin *computatorium* computing place, from Latin *computare*] (14th century)
1 : a piece (as of metal or plastic) used in reckoning or in games
2 : something of value in bargaining **:** ASSET
3 : a level surface (as a table, shelf or display case) over which transactions are conducted or food is served or on which goods are displayed or work is conducted ⟨jewelry *counter*⟩ ⟨a lunch *counter*⟩
— **over the counter 1 :** in or through a broker's office rather than through a stock exchange ⟨stock bought *over the counter*⟩ **2 :** without a prescription ⟨drugs available *over the counter*⟩
— **under the counter :** by surreptitious means **:** in an illicit and private manner

²count·er \ˈkau̇n-tər\ *noun* [Middle English, from Middle French *conteor,* from *compter* to count] (14th century)
: one that counts; *especially* **:** a device for indicating a number or amount

³coun·ter *verb* **coun·tered; coun·ter·ing** \ˈkau̇n-t(ə-)riŋ\ [Middle English *countren,* from Middle French *contre* against, opposite, from Latin *contra;* akin to Latin *com-* with, together — more at CO-] (14th century)
transitive verb
1 a : to act in opposition to **:** OPPOSE **b :** OFFSET, NULLIFY ⟨tried to *counter* the trend toward depersonalization⟩
2 : to adduce in answer ⟨we *countered* that our warnings had been ignored⟩
intransitive verb
: to meet attacks or arguments with defensive or retaliatory steps

⁴coun·ter *adverb* [Middle English *contre,* from Middle French] (15th century)
1 : in an opposite or wrong direction
2 : to or toward a different or opposite direction, result, or effect ⟨values that run *counter* to those of society⟩

⁵coun·ter *noun* (15th century)
1 : CONTRARY, OPPOSITE
2 : the after portion of a boat from the waterline to the extreme outward swell or stern overhang
3 a : the act of making an attack while parrying one (as in boxing or fencing); *also* **:** a blow thus given in boxing **b :** an agency or force that offsets **:** CHECK
4 : a stiffener to give permanent form to a boot or shoe upper around the heel
5 : an area within the face of a letter wholly or partly enclosed by strokes
6 : a football play in which the ballcarrier goes in a direction opposite to the movement of the play

⁶coun·ter *adjective* (1596)
1 : marked by or tending toward or in an opposite direction or effect
2 : given to or marked by opposition, hostility, or antipathy
3 : situated or lying opposite ⟨the *counter* side⟩
4 : recalling or ordering back by a superseding contrary order **:** COUNTERMANDING ⟨*counter* orders from the colonel⟩

counter- *prefix* [Middle English *contre-,* from Middle French, from *contre*]
1 a : contrary **:** opposite ⟨*counter*clockwise⟩ ⟨*counter*march⟩ **b :** opposing **:** retaliatory ⟨*counter*force⟩ ⟨*counter*offensive⟩
2 : complementary **:** corresponding ⟨*counter*weight⟩ ⟨*counter*part⟩
3 : duplicate **:** substitute ⟨*counter*foil⟩

coun·ter·ac·cu·sa·tion
coun·ter·ad·ap·ta·tion
coun·ter·ad·ver·tis·ing
coun·ter·agent
coun·ter·ag·gres·sion
coun·ter·ar·gue
coun·ter·ar·gu·ment
coun·ter·as·sault
coun·ter·at·tack
coun·ter·at·tack·er
coun·ter·bid
coun·ter·blast
coun·ter·block·ade
coun·ter·blow
coun·ter·cam·paign
coun·ter·charge
coun·ter·com·mer·cial
coun·ter·com·plaint
coun·ter·con·spir·a·cy
coun·ter·con·ven·tion
coun·ter·coun·ter·mea·sure
coun·ter·coup
coun·ter·crit·i·cism
coun·ter·cry
coun·ter·de·mand
coun·ter·dem·on·strate
coun·ter·dem·on·stra·tion
coun·ter·dem·on·stra·tor
coun·ter·de·ploy·ment
coun·ter·ed·u·ca·tion·al
coun·ter·ef·fort
coun·ter·ev·i·dence
coun·ter·fire
coun·ter·force
coun·ter·gov·ern·ment
coun·ter·hy·poth·e·sis
coun·ter·im·age
coun·ter·in·cen·tive
coun·ter·in·fla·tion
coun·ter·in·fla·tion·ary
coun·ter·in·flu·ence
coun·ter·in·stance
coun·ter·in·sti·tu·tion
coun·ter·in·ter·pre·ta·tion
coun·ter·memo
coun·ter·mo·bi·li·za·tion
coun·ter·move
coun·ter·move·ment
coun·ter·myth
coun·ter·or·der
coun·ter·pe·ti·tion
coun·ter·pick·et
coun·ter·play
coun·ter·play·er
coun·ter·ploy
coun·ter·pow·er
coun·ter·pres·sure
coun·ter·pro·ject
coun·ter·pro·pa·gan·da
coun·ter·pro·test
coun·ter·ques·tion
coun·ter·raid
coun·ter·ral·ly
coun·ter·re·ac·tion
coun·ter·re·form
coun·ter·re·form·er
coun·ter·re·sponse
coun·ter·re·tal·i·a·tion
coun·ter·sci·en·tif·ic
coun·ter·shot
coun·ter·snip·er
coun·ter·spell
coun·ter·state
coun·ter·state·ment
coun·ter·step
coun·ter·strat·e·gist
coun·ter·strat·e·gy
coun·ter·stream
coun·ter·strike
coun·ter·stroke
coun·ter·style
coun·ter·sue
coun·ter·sug·ges·tion
coun·ter·suit
coun·ter·sur·veil·lance
coun·ter·tac·tics
coun·ter·ten·den·cy
coun·ter·ter·ror
coun·ter·ter·ror·ism
coun·ter·ter·ror·ist
coun·ter·threat
coun·ter·thrust
coun·ter·tra·di·tion
coun·ter·trend
coun·ter·vi·o·lence
coun·ter·world

coun·ter·act \ˌkau̇n-tə-ˈrakt\ *transitive verb* (1678)
: to make ineffective or restrain or neutralize the usually ill effects of by an opposite force
— **coun·ter·ac·tion** \-ˈrak-shən\ *noun*
— **coun·ter·ac·tive** \-ˈrak-tiv\ *adjective*

¹coun·ter·bal·ance \ˈkau̇n-tər-ˌba-lən(t)s, ˌkau̇n-tər-ˈ\ *noun* (1611)
1 : a weight that balances another
2 : a force or influence that offsets or checks an opposing force

²counterbalance \ˌkau̇n-tər-ˈ, ˈkau̇n-tər-ˌ\ *transitive verb* (1611)
1 : to oppose or balance with an equal weight or force
2 : to equip with counterbalances

coun·ter·change \ˈkau̇n-tər-ˌchānj\ *transitive verb* (circa 1604)
1 : INTERCHANGE, TRANSPOSE
2 : CHECKER 1a

¹coun·ter·check \-ˌchek\ *noun* (1559)
: a check or restraint often operating against something that is itself a check

²countercheck *transitive verb* (1587)
1 : CHECK, COUNTERACT
2 : to check a second time for verification

counter check *noun* (1856)
: a check obtainable at a bank usually to be cashed only at the bank by the drawer

¹coun·ter·claim \ˈkau̇n-tər-ˌklām\ *noun* (1784)
: an opposing claim; *especially* **:** a claim brought by a defendant against a plaintiff in a legal action

²counterclaim (1881)
intransitive verb
: to enter or plead a counterclaim
transitive verb
: to ask in a counterclaim

coun·ter·clock·wise \ˌkau̇n-tər-ˈkläk-ˌwīz\ *adverb* (1888)
: in a direction opposite to that in which the hands of a clock rotate as viewed from in front
— **counterclockwise** *adjective*

coun·ter·con·di·tion·ing \-kən-ˈdi-sh(ə-)niŋ\ *noun* (1962)
: conditioning in order to replace an undesirable response (as fear) to a stimulus (as an engagement in public speaking) by a favorable one

coun·ter·cul·ture \ˈkau̇n-tər-ˌkəl-chər\ *noun* (1968)
: a culture with values and mores that run counter to those of established society
— **coun·ter·cul·tur·al** \ˌkau̇n-tər-ˈkəlch-rəl, -ˈkəl-chə-\ *adjective*
— **coun·ter·cul·tur·al·ism** \-rə-ˌli-zəm\ *noun*
— **coun·ter·cul·tur·ist** \-rist\ *noun*

¹coun·ter·cur·rent \ˈkau̇n-tər-ˌkər-ənt, -ˌkə-rənt\ *noun* (1684)
: a current flowing in a direction opposite that of another current

²countercurrent \ˌkau̇n-tər-ˈ\ *adjective* (1799)
1 : flowing in an opposite direction

2 : involving flow of materials in opposite directions ⟨*countercurrent* dialysis⟩
— **coun·ter·cur·rent·ly** *adverb*
coun·ter·cy·cli·cal \-ˈsī-kli-kəl, -ˈsi-\ *adjective* (1944)
: calculated to check excessive developments in a business cycle **:** COMPENSATORY ⟨*countercyclical* budget policies⟩
— **coun·ter·cy·cli·cal·ly** \-k(ə-)lē\ *adverb*
coun·ter·es·pi·o·nage \ˌkau̇n-tər-ˈes-pē-ə-ˌnäzh, -ˌnäj, -nij, *Canadian also* -ˌnazh; -ˌes-pē-ə-ˈnäzh; -ə-ˈspē-ə-nij\ *noun* (1899)
: the activity concerned with detecting and thwarting enemy espionage
coun·ter·ex·am·ple \ˈkau̇n-tər-ig-ˌzam-pəl\ *noun* (1957)
: an example that disproves a proposition or theory
coun·ter·fac·tu·al \ˌkau̇n-tər-ˈfak-chə-wəl, -chəl, -shwəl\ *adjective* (1946)
: contrary to fact ⟨*counterfactual* assumptions⟩
[1]**coun·ter·feit** \ˈkau̇nt-ər-ˌfit\ *adjective* [Middle English *countrefet,* from Middle French *contrefait,* from past participle of *contrefaire* to imitate, from *contre-* + *faire* to make, from Latin *facere* — more at DO] (14th century)
1 : made in imitation of something else with intent to deceive **:** FORGED ⟨*counterfeit* money⟩
2 a : INSINCERE, FEIGNED ⟨*counterfeit* sympathy⟩ **b :** IMITATION ⟨*counterfeit* Georgian houses⟩
[2]**counterfeit** (14th century)
transitive verb
: to imitate or feign especially with intent to deceive; *also* **:** to make a fraudulent replica of ⟨*counterfeiting* $20 bills⟩
intransitive verb
1 : to try to deceive by pretense or dissembling
2 : to engage in counterfeiting something of value
synonym see ASSUME
— **coun·ter·feit·er** *noun*
[3]**counterfeit** *noun* (15th century)
1 : something counterfeit **:** FORGERY
2 : something likely to be mistaken for something of higher value ⟨pity was a *counterfeit* of love —Harry Hervey⟩
synonym see IMPOSTURE
coun·ter·flow \ˈkau̇n-tər-ˌflō\ *noun* (1870)
: the flow of a fluid in opposite directions (as in an apparatus)
coun·ter·foil \-ˌfȯil\ *noun* (1706)
: a detachable stub (as on a check or ticket) usually serving as a record or receipt
coun·ter·guer·ril·la *also* **coun·ter·gue·ril·la** \ˌkau̇n-tər-gə-ˈri-lə, -gi-, -gyi-, -ge-\ *noun* (1901)
: a guerrilla who is trained to thwart enemy guerrilla operations
coun·ter·in·sur·gen·cy \ˌkau̇n-tər-in-ˈsər-jən(t)-sē\ *noun* (1962)
: organized military activity designed to combat insurgency
— **coun·ter·in·sur·gent** \-jənt\ *noun*
coun·ter·in·tel·li·gence \ˌkau̇n-tər-in-ˈte-lə-jən(t)s\ *noun* (1940)
: organized activity of an intelligence service designed to block an enemy's sources of information, to deceive the enemy, to prevent sabotage, and to gather political and military information
coun·ter·in·tu·i·tive \-in-ˈtü-ə-tiv, -ˈtyü-\ *adjective* (1955)
: contrary to what one would intuitively expect
— **coun·ter·in·tu·i·tive·ly** \-lē\ *adverb*
coun·ter·ion \ˈkau̇n-tər-ˌī-ən, -ˌän\ *noun* (1940)
: an ion having a charge opposite to that of the substance with which it is associated
coun·ter·ir·ri·tant \ˌkau̇n-tər-ˈir-ə-tənt\ *noun* (1854)
1 : an agent applied locally to produce superficial inflammation with the object of reducing inflammation in deeper adjacent structures
2 : an irritation or discomfort that diverts attention from another
— **counterirritant** *adjective*
count·er·man \ˈkau̇n-tər-ˌman, -mən\ *noun* (1853)
: one who tends a counter
[1]**coun·ter·mand** \ˈkau̇n-tər-ˌmand, ˌkau̇n-tər-ˈ\ *transitive verb* [Middle English *countermaunden,* from Middle French *contremander,* from *contre-* counter- + *mander* to command, from Latin *mandare* — more at MANDATE] (15th century)
1 : to revoke (a command) by a contrary order
2 : to recall or order back by a superseding contrary order ⟨*countermand* reinforcements⟩
[2]**coun·ter·mand** \ˈkau̇n-tər-ˌmand\ *noun* (1548)
1 : a contrary order
2 : the revocation of an order or command
coun·ter·march \ˈkau̇n-tər-ˌmärch\ *noun* (1598)
1 : a marching back; *specifically* **:** a movement in marching by which a unit of troops reverses direction while marching but keeps the same order
2 : a march (as of political demonstrators) designed to counter the effect of another march
— **countermarch** *intransitive verb*
coun·ter·mea·sure \-ˌme-zhər, -ˌmā-\ *noun* (1923)
: an action or device designed to negate or offset another; *especially* **:** a military system or device intended to thwart a sensing mechanism (as radar) ⟨electronic *countermeasures*⟩
coun·ter·mel·o·dy \-ˌme-lə-dē\ *noun* (1926)
: a secondary melody that is sounded simultaneously with the principal one
coun·ter·mine \-ˌmīn\ *noun* (1548)
: a tunnel for intercepting an enemy mine
coun·ter·of·fen·sive \ˈkau̇n-tər-ə-ˌfen(t)-siv\ *noun* (1909)
: a large-scale military offensive undertaken by a force previously on the defensive
coun·ter·of·fer \-ˌȯ-fər, -ˌä-fər\ *noun* (1788)
: a return offer made by one who has rejected an offer
coun·ter·pane \ˈkau̇n-tər-ˌpān\ *noun* [alteration of Middle English *countrepointe,* modification of Middle French *coute pointe,* literally, embroidered quilt] (15th century)
: BEDSPREAD
coun·ter·part \-ˌpärt\ *noun* (15th century)
1 : one of two corresponding copies of a legal instrument **:** DUPLICATE
2 a : a thing that fits another perfectly **b :** something that completes **:** COMPLEMENT
3 a : one remarkably similar to another **b :** one having the same function or characteristics as another ⟨college presidents and their *counterparts* in business⟩
coun·ter·plan \ˈkau̇n-tər-ˌplan\ *noun* (1788)
1 : a plan designed to counter another plan
2 : an alternate or substitute plan
coun·ter·plea \-ˌplē\ *noun* (1523)
: a replication to a legal plea
[1]**coun·ter·plot** \-ˌplät\ *noun* (circa 1611)
: a plot designed to thwart an opponent's plot
[2]**counterplot** *transitive verb* (1662)
: to intrigue against **:** foil with a plot
[1]**coun·ter·point** \ˈkau̇n-tər-ˌpȯint\ *noun* [Middle French *contrepoint,* from Medieval Latin *contrapunctus,* from Latin *contra-* counter- + Medieval Latin *punctus* musical note, melody, from Latin, act of pricking, from *pungere* to prick — more at PUNGENT] (15th century)
1 a : one or more independent melodies added above or below a given melody **b :** the combination of two or more independent melodies into a single harmonic texture in which each retains its linear character **:** POLYPHONY
2 a : a complementing or contrasting item **:** OPPOSITE **b :** use of contrast or interplay of elements in a work of art (as a drama)
[2]**counterpoint** *transitive verb* (1875)
1 : to compose or arrange in counterpoint
2 : to set off or emphasize by juxtaposition **:** set in contrast ⟨*counterpoints* the public and the private man —Tom Bishop⟩
[1]**coun·ter·poise** \-ˌpȯiz\ *transitive verb* [Middle English *countrepesen,* from Middle French *contrepeser,* from *contre-* + *peser* to weigh — more at POISE] (14th century)
: COUNTERBALANCE
[2]**counterpoise** *noun* (15th century)
1 : COUNTERBALANCE
2 : an equivalent power or force acting in opposition
3 : a state of balance
coun·ter·pose \ˌkau̇n-tər-ˈpōz\ *transitive verb* [*counter-* + *-pose* (as in *compose*)] (1594)
: to place in opposition, contrast, or equilibrium ⟨*counterpose* a positive view to the negative assessment⟩
coun·ter·pro·duc·tive \-prə-ˈdək-tiv\ *adjective* (1959)
: tending to hinder the attainment of a desired goal ⟨violence as a means to achieve an end is *counterproductive* —W. E. Brock (born 1930)⟩
coun·ter·pro·gram·ming \ˌkau̇n-tər-ˈprō-ˌgra-miŋ, -grə-\ *noun* (circa 1966)
: the scheduling of programs by television networks so as to attract audiences away from simultaneously telecast programs of competitors
coun·ter·pro·pos·al \ˈkau̇n-tər-prə-ˌpō-zəl\ *noun* (1885)
: a return proposal made by one who has rejected a proposal
coun·ter·punch \ˈkau̇n-tər-ˌpənch\ *noun* (1942)
: a counter in boxing; *also* **:** a countering blow or attack
— **counterpunch** *intransitive verb*
— **coun·ter·punch·er** \-ˌpən-chər\ *noun*
coun·ter·ref·or·ma·tion \ˌkau̇n-tə(r)-ˌre-fər-ˈmā-shən\ *noun* (1840)
1 *usually* **Counter-Reformation :** the reform movement in the Roman Catholic Church following the Reformation
2 : a reformation designed to counter the effects of a previous reformation
coun·ter·rev·o·lu·tion \-ˌre-və-ˈlü-shən\ *noun* (1793)
1 : a revolution directed toward overthrowing a government or social system established by a previous revolution
2 : a movement to counteract revolutionary trends
— **coun·ter·rev·o·lu·tion·ary** \-shə-ˌner-ē\ *adjective or noun*
coun·ter·shad·ing \ˈkau̇n-tər-ˌshā-diŋ\ *noun* (1896)
: cryptic coloration of an animal with parts normally in shadow being light and parts normally illuminated being dark thereby reducing shadows and contours
coun·ter·sign \-ˌsīn\ *noun* (1591)
1 : a signature attesting the authenticity of a document already signed by another
2 : a sign given in reply to another; *specifically* **:** a military secret signal that must be given by one wishing to pass a guard
— **countersign** *transitive verb*
— **coun·ter·sig·na·ture** \ˌkau̇n-tər-ˈsig-nə-ˌchu̇r, -chər, -ˌtyu̇r, -ˌtu̇r\ *noun*
[1]**coun·ter·sink** \ˈkau̇n-tər-ˌsiŋk\ *transitive verb* **-sunk** \-ˌsəŋk\; **-sink·ing** (1816)
1 : to make a countersink on
2 : to set the head of (as a screw) at or below the surface
[2]**countersink** *noun* (1816)
1 : a bit or drill for making a funnel-shaped enlargement at the outer end of a drilled hole
2 : the enlargement made by a countersink
coun·ter·spy \ˈkau̇n-tər-ˌspī\ *noun* (1939)
: a spy engaged in counterespionage

coun·ter·stain \-ˌstān\ *transitive verb* (1895)
: to stain (as a microscopy specimen) so as to color parts (as the cytoplasm of cells) not colored by another stain (as a nuclear stain)
— counterstain *noun*

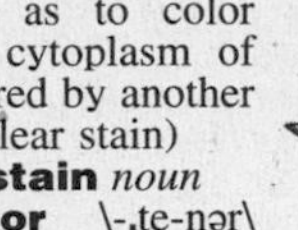
countersink

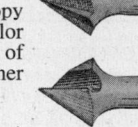

coun·ter·ten·or \-ˌte-nər\ *noun* [Middle English *countretenour,* from Middle French *contreteneur,* from *contre-* + *teneur* tenor] (15th century)
: a tenor with an unusually high range (as an alto range)

coun·ter·top \-ˌtäp\ *noun* [[1]*counter* + [1]*top*] (1897)
: the flat working surface on top of waist-level kitchen cabinets

coun·ter·trade \-ˌtrād\ *noun* (1978)
: a form of international trade in which purchases made by an importing nation are linked to offsetting purchases made by the exporting nation

coun·ter·trans·fer·ence \ˌkau̇n-tər-(ˌ)tran(t)s-ˈfər-ən(t)s, -ˈtran(t)s-(ˌ)\ *noun* (1920)
1 : psychological transference especially by a psychotherapist during the course of treatment; *especially* **:** the psychotherapist's reactions to the patient's transference
2 : the complex of feelings of a psychotherapist toward the patient

coun·ter·vail \ˌkau̇n-tər-ˈvā(ə)l\ *verb* [Middle English *countrevailen,* from Middle French *contrevaloir,* from *contre-* counter- + *valoir* to be worth, from Latin *valēre* — more at WIELD] (14th century)
transitive verb
1 : to compensate for
2 *archaic* **:** EQUAL, MATCH
3 : to exert force against **:** COUNTERACT
intransitive verb
: to exert force against an opposing and often bad or harmful force or influence

coun·ter·view \ˈkau̇n-tər-ˌvyü\ *noun* (1590)
1 *archaic* **:** CONFRONTATION
2 : an opposite point of view

coun·ter·weight \-ˌwāt\ *noun* (1693)
: an equivalent weight or force **:** COUNTERBALANCE
— counterweight *transitive verb*

count·ess \ˈkau̇n-təs\ *noun* (12th century)
1 : the wife or widow of an earl or count
2 : a woman who holds in her own right the rank of earl or count

coun·ti·an \ˈkau̇n-tē-ən\ *noun* (15th century)
: a native or resident of a usually specified county

count·ing·house \ˈkau̇n-tiŋ-ˌhau̇s\ *noun* (15th century)
: a building, room, or office used for keeping books and transacting business

counting number *noun* (circa 1965)
: NATURAL NUMBER

counting room *noun* (1712)
: COUNTINGHOUSE

count·less \ˈkau̇nt-ləs\ *adjective* (1588)
: too numerous to be counted **:** MYRIAD, MANY
— count·less·ly *adverb*

count noun *noun* (1952)
: a noun (as *bean* or *sheet*) that forms a plural and is used with a numeral, with words such as *many* or *few,* or with the indefinite article *a* or *an* — compare MASS NOUN

count palatine *noun* (1539)
1 a : a count of the Holy Roman Empire having imperial powers in his own domain **b :** a high judicial official in the Holy Roman Empire
2 : the proprietor of a county palatine in England or Ireland

coun·tri·fied *also* **coun·try·fied** \ˈkən-tri-ˌfīd\ *adjective* [*country* + *-fied* (as in *glorified*)] (1653)
1 : RURAL, RUSTIC
2 : UNSOPHISTICATED
3 : played or sung in the manner of country music ⟨*countrified* rock⟩

[1]**coun·try** \ˈkən-trē\ *noun, plural* **countries** [Middle English *contree,* from Old French *contrée,* from Medieval Latin *contrata,* from Latin *contra* against, on the opposite side] (13th century)
1 : an indefinite usually extended expanse of land **:** REGION
2 a : the land of a person's birth, residence, or citizenship **b :** a political state or nation or its territory
3 a : the people of a state or district **:** POPULACE **b :** JURY **c :** ELECTORATE 2
4 : rural as distinguished from urban areas
5 : COUNTRY MUSIC ◆
— coun·try·ish \-trē-ish\ *adjective*

[2]**country** *adjective* (14th century)
1 : of, relating to, or characteristic of the country
2 a : of or relating to a decorative style associated with life in the country ⟨an English *country* look⟩; *also* **:** possessing a style of rustic simplicity ⟨*country* furniture⟩ **b :** prepared or processed with farm supplies and procedures ⟨*country* ham⟩
3 : of, relating to, suitable for, or featuring country music ⟨*country* singers⟩

country and western *noun* (1960)
: COUNTRY MUSIC — usually hyphenated in attributive use

country club *noun* (1867)
: a suburban club for social life and recreation; *especially* **:** one having a golf course — usually hyphenated in attributive use

coun·try–dance \ˈkən-trē-ˌdan(t)s\ *noun* (1579)
: any of various native English dances in which partners face each other especially in rows

country gentleman *noun* (1632)
1 : a well-to-do country resident **:** an owner of a country estate
2 : one of the English landed gentry

country house *noun* (14th century)
: a house and especially a mansion in the country

coun·try·man \ˈkən-trē-mən, *3 often* -ˌman\ *noun* (14th century)
1 : an inhabitant or native of a specified country
2 : COMPATRIOT
3 : one living in the country or marked by country ways **:** RUSTIC

country mile *noun* (1950)
: a long distance

country music *noun* (1952)
: music derived from or imitating the folk style of the Southern U.S. or of the Western cowboy

country rock *noun* (1968)
: ROCKABILLY

coun·try·seat \ˌkən-trē-ˈsēt\ *noun* (1583)
: a house or estate in the country

coun·try·side \ˈkən-trē-ˌsīd\ *noun* (1727)
1 : a rural area
2 : the inhabitants of a countryside

coun·try·wide \ˈkən-trē-ˈwīd\ *adjective* (1915)
: extending throughout a country

coun·try·wom·an \ˈkən-trē-ˌwu̇-mən\ *noun* (15th century)
1 : a woman compatriot
2 : a woman resident of the country

[1]**coun·ty** \ˈkau̇n-tē\ *noun, plural* **counties** [Middle English *counte,* from Middle French *conté,* from Medieval Latin *comitatus,* from Late Latin, office of a count, from *comit-, comes* count — more at COUNT] (14th century)
1 : the domain of a count
2 a : one of the territorial divisions of England and Wales and formerly also of Scotland and Northern Ireland constituting the chief units for administrative, judicial, and political purposes **b** (1) **:** the people of a county (2) *British* **:** the gentry of a county
3 : the largest territorial division for local government within a state of the U.S.
4 : the largest local administrative unit in various countries
— county *adjective*

[2]**county** *noun, plural* **counties** [modification of Middle French *comte*] (1550)
archaic **:** [3]COUNT

county agent *noun* (1705)
: a consultant employed jointly by federal and state governments to provide information about agriculture and home economics

county court *noun* (1639)
: a court in some states that has a designated jurisdiction usually both civil and criminal within the limits of a county

county fair *noun* (1856)
: a fair usually held annually at a set location in a county especially to exhibit local agricultural products and livestock

county palatine *noun* (15th century)
: the territory of a count palatine

county seat *noun* (1803)
: a town that is the seat of county administration

county town *noun* (1670)
chiefly British **:** COUNTY SEAT

[1]**coup** \ˈkōp\ *verb* [Middle English, to strike, from Middle French *couper* — more at COPE] (circa 1572)
chiefly Scottish **:** OVERTURN, UPSET

[2]**coup** \ˈkü\ *noun, plural* **coups** \ˈküz\ [French, blow, stroke — more at COPE] (1791)
1 : a brilliant, sudden, and usually highly successful stroke or act
2 : COUP D'ÉTAT

coup de grâce *or* **coup de grace** \ˌkü-də-ˈgräs\ *noun, plural* **coups de grâce** *or* **coups de grace** \ˌkü-də-\ [French *coup de grâce,* literally, stroke of mercy] (1699)
1 : a death blow or shot administered to end the suffering of one mortally wounded
2 : a decisive finishing blow, act, or event

coup de main \-ˈmaⁿ\ *noun, plural* **coups de main** \ˌkü-də-\ [French, literally, hand stroke] (1758)
: a sudden attack in force

coup d'état *or* **coup d'etat** \ˌkü-(ˌ)dā-ˈtä, ˈkü-(ˌ)dā-ˌ, -də-\ *noun, plural* **coups d'état** *or* **coups d'etat** \-ˈtä(z), -ˌtä(z)\ [French, literally, stroke of state] (1646)
: a sudden decisive exercise of force in politics; *especially* **:** the violent overthrow or alteration of an existing government by a small group

coup de thé·âtre *or* **coup de the·atre** \ˌkü-də-tā-ˈätrᵊ\ *noun, plural* **coups de théâtre** *or* **coups de theatre** \ˌkü-də-\

◇ WORD HISTORY

country In Latin the adverb and preposition *contra* means "against" or "on the opposite side." The spoken Latin from which the Romance languages descend formed the derivative *contrata,* which literally meant "that which is situated on the opposite side." But in the Romance words, such as Old Italian *contrata* and Old French *contree,* the attested sense is invariably "environs," "area of land," or "region." Borrowed from French as *contree* in the 13th century, English *country* has had a rich development in its adopted home, acquiring significant new senses, such as "populace" and "rural as opposed to urban areas," that are not shared by Modern French *contrée.*

[French *coup de théâtre,* literally, stroke of theater] (1747)
1 : a sudden sensational turn in a play; *also* **:** a sudden dramatic effect or turn of events
2 : a theatrical success

coup d'oeil \kü-'də(r), -'dəi\ *noun, plural* **coups d'oeil** *same*\ [French, literally, stroke of the eye] (1739)
: a brief survey **:** GLANCE

cou·pé *or* **coupe** \kü-'pā, *2 often* 'küp\ *noun* [French *coupé,* from past participle of *couper* to cut, strike] (1834)
1 : a four-wheeled closed horse-drawn carriage for two persons inside with an outside seat for the driver in front
2 *usually* **coupe :** a 2-door automobile often seating only two persons; *also* **:** one with a tight-spaced rear seat — compare SEDAN

[1]cou·ple \'kə-pəl; *"couple of" is often* ˌkə-plə(v)\ *noun* [Middle English, pair, bond, from Old French *cople,* from Latin *copula* bond, from *co-* + *apere* to fasten — more at APT] (13th century)
1 a : a man and woman married, engaged, or otherwise paired **b :** two persons paired together
2 : PAIR, BRACE
3 : something that joins or links two things together: as **a :** two equal and opposite forces that act along parallel lines **b :** a pair of substances that in contact with an electrolyte participate in a transfer of electrons which causes an electric current to flow
4 : an indefinite small number **:** FEW ⟨a *couple* of days ago⟩ ▫

[2]cou·ple \'kə-pəl\ *verb* **cou·pled; cou·pling** \-p(ə-)liŋ\ (13th century)
transitive verb
1 a : to connect for consideration together **b :** to join for combined effect
2 a : to fasten together **:** LINK **b :** to bring (two electric circuits) into such close proximity as to permit mutual influence
3 : to join in marriage or sexual union
intransitive verb
1 : to unite in sexual union
2 : JOIN
3 : to unite chemically

[3]couple *adjective* (1924)
: TWO; *also* **:** FEW — used with *a* ⟨a *couple* drinks⟩ ▫

cou·ple·ment \'kə-pəl-mənt\ *noun* [Middle French, from *coupler* to join, from Latin *copulare,* from *copula*] (1548)
archaic **:** the act or result of coupling

cou·pler \'kə-p(ə-)lər\ *noun* (1552)
1 : one that couples
2 : a contrivance on a keyboard instrument by which keyboards or keys are connected to play together

cou·plet \'kə-plət\ *noun* [Middle French, diminutive of *cople*] (1580)
1 : two successive lines of verse forming a unit marked usually by rhythmic correspondence, rhyme, or the inclusion of a self-contained utterance **:** DISTICH
2 : COUPLE
3 : one of the musical episodes alternating with the main theme (as in a rondo)

cou·pling \'kə-pliŋ (*usual for 2*), -pə-liŋ\ *noun* (14th century)
1 : the act of bringing or coming together **:** PAIRING; *specifically* **:** sexual union
2 : a device that serves to connect the ends of adjacent parts or objects
3 : the joining of or the part of the body that joins the hindquarters to the forequarters of a quadruped
4 : a means of electric connection of two electric circuits by having a part common to both

cou·pon \'kü-ˌpän, 'kyü-\ *noun* [French, from Old French, piece, from *couper* to cut — more at COPE] (1822)
1 : a statement of due interest to be cut from a bearer bond when payable and presented for payment; *also* **:** the interest rate of a coupon
2 : a form surrendered in order to obtain an article, service, or accommodation: as **a :** one of a series of attached tickets or certificates often to be detached and presented as needed **b :** a ticket or form authorizing purchases of rationed commodities **c :** a certificate or similar evidence of a purchase redeemable in premiums **d :** a part of a printed advertisement to be cut off to use as an order blank or inquiry form or to obtain a discount on merchandise ▫

cou·pon·ing \'kü-ˌpä-niŋ, 'kyü-\ *noun* (1954)
: the distribution or redemption of coupons

cour·age \'kər-ij, 'kə-rij\ *noun* [Middle English *corage,* from Old French, from *cuer* heart, from Latin *cor* — more at HEART] (14th century)
: mental or moral strength to venture, persevere, and withstand danger, fear, or difficulty ☆

cou·ra·geous \kə-'rā-jəs\ *adjective* (14th century)
: having or characterized by courage **:** BRAVE
— cou·ra·geous·ly *adverb*
— cou·ra·geous·ness *noun*

cou·rante \kü-'ränt, -'rant\ *noun* [Middle French, from *courir* to run, from Latin *currere*] (1586)
1 : a dance of Italian origin marked by quick running steps
2 : music in quick triple time or in a mixture of 3/2 and 6/4 time

cou·reur de bois \kü-ˌrər-də-'bwä\ *noun, plural* **coureurs de bois** *same*\ [Canadian French, literally, woods runner] (1700)
: a French or métis trapper of North America and especially of Canada

cour·gette \ku̇r-'zhet\ *noun* [French, diminutive of *courge* gourd, from Latin *cucurbita*] (1931)
chiefly British **:** ZUCCHINI

cou·ri·er \'ku̇r-ē-ər, 'kər-ē-, 'kə-rē-\ *noun* [Middle French *courrier,* from Old Italian *corriere,* from *correre* to run, from Latin *currere*] (1579)
1 : MESSENGER: as **a :** a member of a diplomatic service entrusted with bearing messages **b** (1) **:** an espionage agent transferring secret information (2) **:** a runner of contraband **c :** a member of the armed services whose duties include carrying mail, information, or supplies
2 : a traveler's paid attendant; *especially* **:** a tourists' guide employed by a travel agency

[1]course \'kōrs, 'kȯrs\ *noun* [Middle English, from Old French, from Latin *cursus,* from *currere* to run — more at CAR] (14th century)
1 : the act or action of moving in a path from point to point
2 : the path over which something moves or extends: as **a :** RACECOURSE **b** (1) **:** the direction of travel of a vehicle (as a ship or airplane) usually measured as a clockwise angle from north; *also* **:** the projected path of travel (2) **:** a point of the compass **c :** WATERCOURSE **d :** GOLF COURSE
3 a : accustomed procedure or normal action ⟨the law taking its *course*⟩ **b :** a chosen manner of conducting oneself **:** way of acting ⟨our wisest *course* is to retreat⟩ **c** (1) **:** progression through a development or period or a series of acts or events (2) **:** LIFE HISTORY, CAREER
4 : an ordered process or succession: as **a :** a number of lectures or other matter dealing with a subject; *also* **:** a series of such courses constituting a curriculum ⟨a premed *course*⟩ **b :** a series of doses or medications administered over a designated period
5 a : a part of a meal served at one time **b :** LAYER; *especially* **:** a continuous level range of brick or masonry throughout a wall **c :** the lowest sail on a square-rigged mast
— in due course : after a normal passage of time **:** in the expected or allotted time
— of course 1 : following the ordinary way or procedure **2 :** as might be expected

[2]course *verb* **coursed; cours·ing** (15th century)
transitive verb
1 : to follow close upon **:** PURSUE

☆ SYNONYMS
Courage, mettle, spirit, resolution, tenacity mean mental or moral strength to resist opposition, danger, or hardship. COURAGE implies firmness of mind and will in the face of danger or extreme difficulty ⟨the *courage* to support unpopular causes⟩. METTLE suggests an ingrained capacity for meeting strain or difficulty with fortitude and resilience ⟨a challenge that will test your *mettle*⟩. SPIRIT also suggests a quality of temperament enabling one to hold one's own or keep up one's morale when opposed or threatened ⟨her *spirit* was unbroken by failure⟩. RESOLUTION stresses firm determination to achieve one's ends ⟨the *resolution* of pioneer women⟩. TENACITY adds to RESOLUTION implications of stubborn persistence and unwillingness to admit defeat ⟨held to their beliefs with great *tenacity*⟩.

▫ USAGE
couple When used of two people, the noun *couple* can take either a singular or a plural verb. The plural is used when the writer or speaker thinks of the couple as two persons ⟨the *couple* are on their honeymoon —E. L. Doctorow⟩ and the singular when the writer or speaker thinks of the couple as a unit ⟨a *couple* that has loud parties all the time —Andy Rooney⟩. In either case, the pronoun of reference is *they* ⟨one *couple* mentioned a loan they were trying to get —Andy Rooney⟩.

The phrase *a couple of* has been in use for centuries ⟨finding myself beginning to be troubled with wind as I used to be . . . I took a *couple* of pills that I had by me —Samuel Pepys (diary, 1663)⟩. In the 19th century an etymological objection was raised against it on the basis of the fact that its Latin ancestor meant not just any pair but a pair closely united or bonded. When the commentators lost interest in etymology, they objected to its use for an indefinite small number ⟨a *couple* of days later we arrived in Northville —*New Yorker*⟩. A few handbooks still cling to the strictures of the past, but this phrase has long been established as standard.

couple The adjective use of *a couple,* without *of,* has been called nonstandard, but it is not. In both British and American English it is standard before a word (as *more* or *less*) indicating degree ⟨a *couple* more examples of Middle English writing —Charles Barber⟩. Its use before an ordinary plural noun is an Americanism, common in speech and in writing that is not meant to be formal or elevated ⟨the first *couple* chapters are pretty good —E. B. White (letter)⟩ ⟨still operated a *couple* wagons for hire —Garrison Keillor⟩. It is most frequently used with periods of time ⟨a *couple* weeks⟩ and numbers ⟨a *couple* hundred⟩ ⟨a *couple* dozen⟩.

coupon The pronunciation \'kyü-ˌpän\ has drawn some attention for its unusual rendering of the letters *ou* as \yü\. The \'kyü-\ variant may possibly have developed by analogy to other words (as *culotte, cupric, kudos, Ku Klux Klan*) whose \'kü-\ syllable has a second pronunciation \'kyü-\ in some dialects. The \'kyü-\ variant of *coupon* is heard almost as much as the \'kü-\ variant in American speech. Despite the earlier misgivings of some commentators, \'kyü-ˌpän\ should be recognized as a standard form.

2 a : to hunt or pursue (game) with hounds **b :** to cause (dogs) to run (as after game)
3 : to run or move swiftly through or over **:** TRAVERSE ⟨jets *coursed* the area daily⟩
intransitive verb
: to run or pass rapidly along or as if along an indicated path ⟨blood *coursing* through the veins⟩

course of study (1781)
1 : CURRICULUM
2 : COURSE 4a

¹cours·er \'kōr-sər, 'kȯr-\ *noun* [Middle English, from Middle French *coursier,* from Old French *course* course, run] (14th century)
: a swift or spirited horse **:** CHARGER

²courser *noun* (1600)
1 : a dog for coursing
2 : one that courses **:** HUNTSMAN
3 : any of various Old World birds (subfamily Cursoriinae of the family Glareolidae) noted for their speed in running

course·ware \'kōrs-ˌwar, 'kȯrs-, -ˌwer\ *noun* (1972)
: educational software

coursing *noun* (1538)
1 : the pursuit of running game with dogs that follow by sight instead of by scent
2 : the act of one that courses

¹court \'kōrt, 'kȯrt\ *noun, often attributive* [Middle English, from Old French, from Latin *cohort-, cohors* enclosure, group, retinue, cohort, from *co-* + *-hort-, -hors* (akin to *hortus* garden) — more at YARD] (12th century)
1 a : the residence or establishment of a sovereign or similar dignitary **b :** a sovereign's formal assembly of councillors and officers **c :** the sovereign and officers and advisers who are the governing power **d :** the family and retinue of a sovereign **e :** a reception held by a sovereign
2 a (1) **:** a manor house or large building surrounded by usually enclosed grounds (2) **:** MOTEL **b :** an open space enclosed wholly or partly by buildings or circumscribed by a single building **c :** a quadrangular space walled or marked off for playing one of various games with a ball (as lawn tennis, handball, or basketball); *also* **:** a division of such a court **d :** a wide alley with only one opening onto a street
3 a : an official assembly for the transaction of judicial business **b :** a session of such a court ⟨*court* is now adjourned⟩ **c :** a place (as a chamber) for the administration of justice **d :** a judge or judges in session; *also* **:** a faculty or agency of judgment or evaluation ⟨rest our case in the *court* of world opinion —L. H. Marks⟩
4 a : an assembly or board with legislative or administrative powers **b :** PARLIAMENT, LEGISLATURE
5 : conduct or attention intended to win favor or dispel hostility **:** HOMAGE ⟨pay *court* to the king⟩

²court (1571)
transitive verb
1 a : to seek to gain or achieve **b** (1) **:** ALLURE, TEMPT (2) **:** to act so as to invite or provoke ⟨*courts* disaster⟩
2 a : to seek the affections of; *especially* **:** to seek to win a pledge of marriage from **b** *of an animal* **:** to perform actions in order to attract for mating
3 a : to seek to attract (as by solicitous attention or offers of advantages) ⟨college teams *courting* high school basketball stars⟩ **b :** to seek an alliance with
intransitive verb
1 : to engage in social activities leading to engagement and marriage
2 *of an animal* **:** to engage in activity leading to mating

court bouil·lon \'kür-(ˌ)bü-'yōⁿ, -(ˌ)bwē-, 'kōr-, 'kȯr-\ *noun* [French *court-bouillon,* literally, short bouillon] (1723)
: a liquid made usually with water, white wine, vegetables, and seasonings and used to poach fish

cour·te·ous \'kər-tē-əs, *British also* 'kȯr-\ *adjective* [Middle English *corteis,* from Old French, from *court*] (13th century)
1 : marked by polished manners, gallantry, or ceremonial usage of a court
2 : marked by respect for and consideration of others
synonym see CIVIL
— cour·te·ous·ly *adverb*
— cour·te·ous·ness *noun*

cour·te·san \'kōr-tə-zən, 'kȯr-, -ˌzan *also* 'kər-, -ˌzän; *especially British* ˌkȯr-tə-'zan\ *noun* [Middle French *courtisane,* from northern Italian dialect form of Italian *cortigiana* woman courtier, feminine of *cortigiano* courtier, from *corte* court, from Latin *cohort-, cohors*] (1533)
: a prostitute with a courtly, wealthy, or upper-class clientele

¹cour·te·sy \'kər-tə-sē, *British also* 'kȯr-\ *noun, plural* **-sies** [Middle English *corteisie,* from Old French, from *corteis*] (13th century)
1 a : courteous behavior **b :** a courteous act or expression
2 a : general allowance despite facts **:** INDULGENCE ⟨hills called mountains by *courtesy* only⟩ **b :** consideration, cooperation, and generosity in providing (as a gift or privilege); *also* **:** AGENCY, MEANS — used chiefly in the phrases *through the courtesy of* or *by courtesy of* or sometimes simply *courtesy of*

²courtesy *adjective* (1613)
: granted, provided, or performed as a courtesy or by way of courtesy ⟨made a *courtesy* call on the ambassador⟩

courtesy card *noun* (1934)
: a card entitling its holder to some special privilege

courtesy title *noun* (1865)
1 : a title (as "Lord" added to the Christian name of a peer's younger son) used in addressing certain lineal relatives of British peers
2 : a title (as "Professor" for any teacher) taken by the user and commonly accepted without consideration of official right

court·house \'kōrt-ˌhau̇s, 'kȯrt-\ *noun* (15th century)
1 a : a building in which courts of law are regularly held **b :** the principal building in which county offices are housed
2 : COUNTY SEAT

cour·tier \'kōr-tē-ər, 'kȯr-; 'kōrt-yər, 'kȯrt-; 'kōr-chər, 'kȯr-\ *noun* (14th century)
1 : one in attendance at a royal court
2 : one who practices flattery

¹court·ly \'kōrt-lē, 'kȯrt-\ *adjective* **court·li·er; -est** (15th century)
1 a : of a quality befitting the court **:** ELEGANT **b :** insincerely flattering
2 : favoring the policy or party of the court
— court·li·ness *noun*

²courtly *adverb* (circa 1592)
: in a courtly manner **:** POLITELY

courtly love *noun* (1896)
: a late medieval conventionalized code prescribing conduct and emotions of ladies and their lovers

¹court–mar·tial \'kōrt-ˌmär-shəl, 'kȯrt-, -'mär-\ *noun, plural* **courts–martial** *also* **court–martials** (1651)
1 : a court consisting of commissioned officers and in some instances enlisted personnel for the trial of members of the armed forces or others within its jurisdiction
2 : a trial by court-martial

²court–martial *transitive verb* **-mar·tialed** *also* **-mar·tialled; -mar·tial·ing** *also* **-mar·tial·ling** \-ˌmär-sh(ə-)liŋ, -'mär-\ (1859)
: to subject to trial by court-martial

court of appeals *often C&A capitalized* (1777)
: a court hearing appeals from the decisions of lower courts — called also *court of appeal*

court of claims (1691)
: a court that has jurisdiction over claims (as against a government)

court of common pleas (1687)
1 : a former English superior court having civil jurisdiction
2 : an intermediate court in some American states that usually has civil and criminal jurisdiction

court of domestic relations (1926)
: a court that has jurisdiction and often special advisory powers over family disputes involving the rights and duties of husband, wife, parent, or child especially in matters affecting the support, custody, and welfare of children

court of honor (1687)
: a tribunal (as a military court) for investigating questions of personal honor

court of inquiry (1757)
: a military court that inquires into and reports on some military matter (as an officer's questionable conduct)

court of law (14th century)
: a court that hears cases and decides them on the basis of statutes or the common law

court of record (15th century)
: a court whose acts and proceedings are kept on permanent record

Court of St. James's \-sənt-'jāmz, -'jām-zəz, -ˌsānt-\ [from *Saint James's* Palace, London, former seat of the British court] (1848)
: the British court

court of sessions (circa 1889)
: any of various state criminal courts of record

court order *noun* (1650)
: an order issuing from a competent court that requires a party to do or abstain from doing a specified act

court plaster *noun* [from its use for beauty spots by ladies at royal courts] (1772)
: an adhesive plaster especially of silk coated with isinglass and glycerin

court reporter *noun* (1894)
: a stenographer who records and transcribes a verbatim report of all proceedings in a court of law

court·room \'kōrt-ˌrüm, 'kȯrt-, -ˌru̇m\ *noun* (1677)
: a room in which a court of law is held

court·ship \-ˌship\ *noun* (1596)
: the act, process, or period of courting

court·side \-ˌsīd\ *noun* (1969)
: the area at the edge of a court (as for tennis or basketball)

court tennis *noun* (circa 1890)
: a game played with a ball and racket in an enclosed court divided by a net

court·yard \'kōrt-ˌyärd, 'kȯrt-\ *noun* (1552)
: a court or enclosure adjacent to a building (as a house or palace)

cous·cous \'küs-ˌküs\ *noun* [French *couscous, couscoussou,* from Arabic *kuskus, kuskusū*] (1759)
: a North African dish of steamed semolina usually served with meat or vegetables; *also* **:** the semolina itself

cous·in \'kə-zən\ *noun* [Middle English *cosin,* from Old French, from Latin *consobrinus,* from *com-* + *sobrinus* second cousin, from *soror* sister — more at SISTER] (13th century)
1 a : a child of one's uncle or aunt **b :** a relative descended from one's grandparent or more remote ancestor by two or more steps and in a different line **c :** KINSMAN, RELATIVE ⟨a distant *cousin*⟩
2 : one associated with or related to another **:** COUNTERPART

3 — used as a title by a sovereign in addressing a nobleman
4 : a person of a race or people ethnically or culturally related ⟨our English *cousins*⟩
— **cous·in·hood** \-ˌhu̇d\ *noun*
— **cous·in·ly** *adjective*
— **cous·in·ship** \-ˌship\ *noun*

cous·in·age \ˈkə-zə-nij\ *noun* (14th century)
1 : relationship of cousins **:** KINSHIP
2 : a collection of cousins **:** KINFOLK

cous·in–ger·man \ˌkə-zən-ˈjər-mən\ *noun, plural* **cous·ins–ger·man** \-zənz-\ [Middle English *cosin germain,* from Middle French, from Old French, from *cosin* + *germain* german] (14th century)
: COUSIN 1a

Cousin Jack \ˌkə-zən-ˈjak\ *noun* (1890)
: CORNISHMAN; *especially* **:** a Cornish miner

¹couth \ˈküth\ *adjective* [back-formation from *uncouth*] (1896)
: SOPHISTICATED, POLISHED

²couth *noun* (1956)
: POLISH, REFINEMENT ⟨I expected kindness and gentility . . . but there is such a thing as too much *couth* —S. J. Perelman⟩

couth·ie \ˈkü-thē\ *adjective* [Middle English *couth* familiar, from Old English *cūth* — more at UNCOUTH] (1719)
chiefly Scottish **:** PLEASANT, KINDLY

cou·ture \kü-ˈtu̇r, -ˈtᵫr\ *noun* [French, from Old French *cousture* sewing, from (assumed) Vulgar Latin *consutura,* from Latin *consutus,* past participle of *consuere* to sew together, from *com-* + *suere* to sew — more at SEW] (1908)
1 : the business of designing, making, and selling fashionable custom-made women's clothing
2 : the designers and establishments engaged in couture
3 : the clothes created by couture

cou·tu·ri·er \kü-ˈtu̇r-ē-ər, -ē-ˌā\ *noun* [French, dressmaker, from Old French *cousturier* tailor's assistant, from *cousture*] (1899)
: an establishment engaged in couture; *also* **:** the proprietor of or designer for such an establishment

cou·tu·ri·ere \kü-ˈtu̇r-ē-ər, -ē-ˌer\ *noun* [French *couturière,* from Old French *cousturiere,* feminine of *cousturier*] (1818)
: a woman who is a couturier

cou·vade \kü-ˈväd\ *noun* [French, from Middle French, cowardly inactivity, from *cover* to sit on, brood over — more at COVEY] (1865)
: a custom in some cultures in which when a child is born the father takes to bed as if bearing the child and submits himself to fasting, purification, or taboos

co·va·lence \ˌkō-ˈvā-lən(t)s, ˈkō-ˌ\ *noun* (1919)
: valence characterized by the sharing of electrons

co·va·len·cy \-lən(t)-sē\ *noun* (1919)
: COVALENCE

co·va·lent \ˌkō-ˈvā-lənt, ˈkō-ˌ\ *adjective* (circa 1926)
: of, relating to, or characterized by covalent bonds
— **co·va·lent·ly** \-lē\ *adverb*

covalent bond *noun* (1939)
: a chemical bond formed between atoms by the sharing of electrons

co·vari·ance \ˌkō-ˈver-ē-ən(t)s, -ˈvar-; ˈkō-ˌ\ *noun* (1931)
1 : the expected value of the product of the deviations of two random variables from their respective means
2 : the arithmetic mean of the products of the deviations of corresponding values of two quantitative variables from their respective means

co·var·i·ant \-ənt\ *adjective* [International Scientific Vocabulary] (1893)
: varying with something else so as to preserve certain mathematical interrelations

co·vari·a·tion \ˌkō-ˌver-ē-ˈā-shən, -ˌvar-\ *noun* (1906)
: correlated variation of two or more variables

¹cove \ˈkōv\ *noun* [Middle English, den, from Old English *cofa;* akin to Old High German *chubisi* hut] (before 12th century)
1 : a recessed place **:** CONCAVITY: as **a :** an architectural member with a concave cross section **b :** a trough for concealed lighting at the upper part of a wall
2 : a small sheltered inlet or bay
3 a : a deep recess or small valley in the side of a mountain **b :** a level area sheltered by hills or mountains

²cove *transitive verb* **coved; cov·ing** (1756)
: to make in a hollow concave form

³cove *noun* [Romany *kova* thing, person] (1567)
British **:** MAN, FELLOW

co·vel·lite \kō-ˈve-ˌlīt, ˈkō-və-ˌ\ *also* **co·vel·line** \-ˌlēn\ *noun* [French *covelline,* from Niccolò *Covelli* (died 1829) Italian chemist] (1850)
: a native copper sulfide

co·ven \ˈkə-vən *also* ˈkō-\ *noun* [Middle English *covin* band, from Middle French, from Medieval Latin *convenium* agreement, from Latin *convenire* to agree — more at CONVENIENT] (circa 1520)
1 : a collection of individuals with similar interests or activities ⟨a *coven* of intellectuals⟩
2 : an assembly or band of usually 13 witches

¹cov·e·nant \ˈkəv-nənt, ˈkə-və-\ *noun* [Middle English, from Middle French, from present participle of *covenir* to agree, from Latin *convenire*] (14th century)
1 : a usually formal, solemn, and binding agreement **:** COMPACT
2 a : a written agreement or promise usually under seal between two or more parties especially for the performance of some action **b :** the common-law action to recover damages for breach of such a contract
— **cov·e·nan·tal** \ˌkə-və-ˈnan-tᵊl\ *adjective*

²cov·e·nant \-nənt, -ˌnant\ (14th century)
transitive verb
: to promise by a covenant **:** PLEDGE
intransitive verb
: to enter into a covenant **:** CONTRACT

cov·e·nan·tee \ˌkə-və-ˌnan-ˈtē, -nən-\ *noun* (1649)
: the person to whom a promise in the form of a covenant is made

cov·e·nant·er \ˈkə-və-ˌnan-tər, *1 also* ˌkə-və-ˈ\ *noun* (1638)
1 *capitalized* **:** a signer or adherent of the Scottish National Covenant of 1638
2 : one that makes a covenant

cov·e·nan·tor \ˈkə-və-ˌnan-tər; ˌkə-və-ˌnan-ˈtȯr, -nən-\ *noun* (1649)
: a party bound by a covenant

Cov·en·try \ˈkə-vən-trē *also* ˈkä-\ *noun* [*Coventry,* England] (1765)
: a state of ostracism or exclusion ⟨sent to *Coventry*⟩

¹cov·er \ˈkə-vər\ *verb* **cov·ered; cov·er·ing** \ˈkəv-riŋ, ˈkə-və-\ [Middle English, from Old French *covrir,* from Latin *cooperire,* from *co-* + *operire* to close, cover] (13th century)
transitive verb
1 a : to guard from attack **b** (1) **:** to have within the range of one's guns **:** COMMAND (2) **:** to hold within range of an aimed firearm **c** (1) **:** to afford protection or security to **:** INSURE (2) **:** to afford protection against or compensation for **d** (1) **:** to guard (an opponent) in order to obstruct a play (2) **:** to be in position to receive a throw to (a base in baseball) **e** (1) **:** to make provision for (a demand or charge) by means of a reserve or deposit ⟨your balance is insufficient to *cover* the check⟩ (2) **:** to maintain a check on especially by patrolling (3) **:** to protect by contrivance or expedient
2 a : to hide from sight or knowledge **:** CONCEAL ⟨*cover* up a scandal⟩ **b :** to lie over **:** ENVELOP
3 : to lay or spread something over **:** OVERLAY
4 a : to spread over **b :** to appear here and there on the surface of
5 : to place or set a cover or covering over
6 a : to copulate with (a female animal) ⟨a horse *covers* a mare⟩ **b :** to sit on and incubate (eggs)
7 : to invest with a large or excessive amount of something ⟨*covered* herself with glory⟩
8 : to play a higher-ranking card on (a previously played card)
9 : to have sufficient scope to include or take into account **:** COMPRISE
10 : to deal with **:** TREAT
11 a : to have as one's territory or field of activity ⟨one sales rep *covers* the whole state⟩ **b :** to report news about
12 : to pass over **:** TRAVERSE
13 : to place one's stake in equal jeopardy with in a bet
14 : to buy securities or commodities for delivery against (an earlier short sale)
15 : to record a cover of (a song)
intransitive verb
1 : to conceal something illicit, blameworthy, or embarrassing from notice — usually used with *up*
2 : to act as a substitute or replacement during an absence
— **cov·er·able** \ˈkəv-rə-bəl, ˈkə-və-\ *adjective*
— **co·ver·er** \ˈkə-vər-ər\ *noun*
— **cover one's tracks :** to conceal traces in order to elude pursuers
— **cover the ground** *or* **cover ground :** to deal with a subject or assignment in a particular manner ⟨the new book *covers* a lot of *ground*⟩

²cover *noun, often attributive* (14th century)
1 : something that protects, shelters, or guards: as **a :** natural shelter for an animal; *also* **:** the factors that provide such shelter **b** (1) **:** a position or situation affording protection from enemy fire (2) **:** the protection offered by airplanes in tactical support of a military operation **c** *British* **:** COVERAGE 1a, b, 2a
2 : something that is placed over or about another thing: **a :** LID, TOP **b :** a binding or case for a book or the analogous part of a magazine; *also* **:** the front or back of such a binding **c :** an overlay or outer layer especially for protection ⟨a mattress *cover*⟩ **d :** a tablecloth and the other table accessories **e :** COVER CHARGE **f :** ROOF **g :** a cloth used on a bed **h :** something (as vegetation or snow) that covers the ground **i :** the extent to which clouds obscure the sky
3 a : something that conceals or obscures ⟨under *cover* of darkness⟩ **b :** a masking device **:** PRETEXT ⟨the project was a *cover* for intelligence operations⟩
4 : an envelope or wrapper for mail
5 : one who substitutes for another during an absence
6 : a recording of a song previously recorded usually by another performer
— **cov·er·less** \ˈkə-vər-ləs\ *adjective*
— **under cover 1 :** in an envelope or wrapper **2 :** under concealment **:** in secret

cov·er·age \ˈkəv-rij, ˈkə-və-\ *noun* (1912)
1 : something that covers: as **a :** inclusion within the scope of an insurance policy or protective plan **:** INSURANCE **b :** the amount available to meet liabilities **c :** inclusion within the scope of discussion or reporting ⟨the news *coverage* of the trial⟩
2 : the total group covered **:** SCOPE: as **a :** all the risks covered by the terms of an insurance contract **b :** the number or percentage of persons reached by a communications medium
3 : the act or fact of covering

cov·er·all \ˈkə-və-ˌrȯl\ *noun* (1824)
: a one-piece outer garment worn to protect other garments — usually used in plural
— **cov·er·alled** \-ˌrȯld\ *adjective*

cov·er–all \ˈkə-və-ˌrȯl\ *adjective* (1895)
: COMPREHENSIVE ⟨*cover-all* provisions⟩

cover charge *noun* (1921)
: a charge made by a restaurant or nightclub in addition to the charge for food and drink

cover crop *noun* (1899)
: a crop planted to prevent soil erosion and to provide humus

covered bridge *noun* (1809)
: a bridge that has its roadway protected by a roof and enclosing sides

covered smut *noun* (1900)
: a smut disease of grains in which the spore masses are held together by the persistent grain membrane and glumes

covered wagon *noun* (1745)
: a wagon with a canvas top supported by bowed strips of wood or metal

cover girl *noun* (1915)
: an attractive young woman whose picture appears on a magazine cover

cover glass *noun* (1881)
: a piece of very thin glass used to cover material on a glass microscope slide

¹cov·er·ing \ˈkəv-riŋ, ˈkə-və-\ *noun* (14th century)
: something that covers or conceals

²covering *adjective* (1887)
: containing explanation of or additional information about an accompanying communication ⟨a *covering* letter⟩

cov·er·let \ˈkə-vər-lət, -(ˌ)lid\ *noun* [Middle English, alteration of *coverlite,* from Anglo-French *coverelyth,* from Old French *covrir* + *literally* bed, from Latin *lectus* — more at LIE] (14th century)
: BEDSPREAD
word history see CURFEW

cov·er·slip \ˈkə-vər-ˌslip\ *noun* (1875)
: COVER GLASS

cover story *noun* (1948)
: a story accompanying a magazine-cover illustration

¹co·vert \ˈkō-(ˌ)vərt, kō-ˈ; ˈkə-vərt\ *adjective* [Middle English, from Middle French, past participle of *covrir* to cover] (14th century)
1 : not openly shown, engaged in, or avowed **:** VEILED ⟨a *covert* alliance⟩
2 : covered over **:** SHELTERED
synonym see SECRET
— co·vert·ly *adverb*
— co·vert·ness *noun*

²co·vert \ˈkə-vərt, ˈkō-vərt *also* ˈkə-vər\ *noun* (14th century)
1 a : hiding place **:** SHELTER **b :** a thicket affording cover for game **c :** a masking or concealing device
2 : a feather covering the bases of the quills of the wings and tail of a bird — see BIRD illustration
3 : a firm durable twilled sometimes waterproofed cloth usually of mixed-color yarns

cov·er·ture \ˈkə-vər-ˌchu̇r, -chər, -ˌtyu̇r, -ˌtu̇r\ *noun* (13th century)
1 a : COVERING **b :** SHELTER
2 : the status a woman acquires upon marriage under common law

cov·er–up \ˈkə-və-ˌrəp\ *noun* (1927)
1 : a device or stratagem for masking or concealing ⟨his garrulousness is a *cover-up* for insecurity⟩; *also* **:** a usually concerted effort to keep an illegal or unethical act or situation from being made public
2 : a loose outer garment

cov·et \ˈkə-vət\ *verb* [Middle English *coveiten,* from Old French *coveitier,* from *coveitié* desire, modification of Latin *cupiditat-, cupiditas,* from *cupidus* desirous, from *cupere* to desire] (14th century)
transitive verb
1 : to wish for enviously
2 : to desire (what belongs to another) inordinately or culpably
intransitive verb
: to feel inordinate desire for what belongs to another
synonym see DESIRE
— cov·et·able \-və-tə-bəl\ *adjective*
— cov·et·er \-tər\ *noun*
— cov·et·ing·ly \-tiŋ-lē\ *adverb*

cov·et·ous \ˈkə-və-təs\ *adjective* (13th century)
1 : marked by inordinate desire for wealth or possessions or for another's possessions
2 : having a craving for possession ⟨*covetous* of power⟩ ☆
— cov·et·ous·ly *adverb*
— cov·et·ous·ness *noun*

cov·ey \ˈkə-vē\ *noun, plural* **coveys** [Middle English, from Middle French *covee,* from Old French, from *cover* to sit on, brood over, from Latin *cubare* to lie] (14th century)
1 : a mature bird or pair of birds with a brood of young; *also* **:** a small flock
2 : COMPANY, GROUP

¹cow \ˈkau̇\ *noun* [Middle English *cou,* from Old English *cū;* akin to Old High German *kuo* cow, Latin *bos* head of cattle, Greek *bous,* Sanskrit *go*] (before 12th century)
1 a : the mature female of cattle (genus *Bos*) **b :** the mature female of various usually large animals (as an elephant, whale, or moose)
2 : a domestic bovine animal regardless of sex or age
— cowy \-ē\ *adjective*

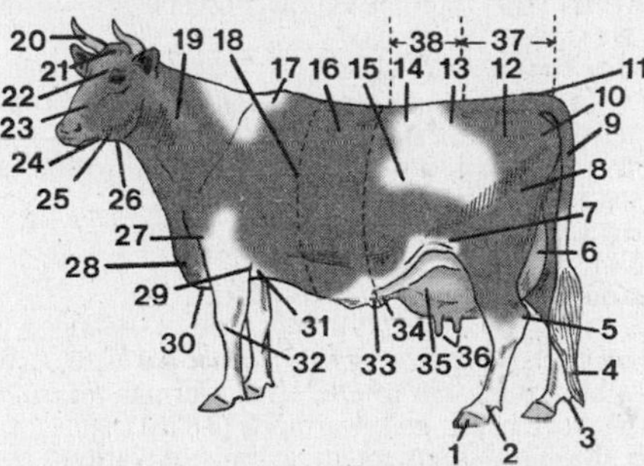

cow 1: *1* **hoof,** *2* **pastern,** *3* **dewclaw,** *4* **switch,** *5* **hock,** *6* **rear udder,** *7* **flank,** *8* **thigh,** *9* **tail,** *10* **pinbone,** *11* **tail head,** *12* **thurl,** *13* **hip,** *14* **barrel,** *15* **ribs,** *16* **crops,** *17* **withers,** *18* **heart girth,** *19* **neck,** *20* **horn,** *21* **poll,** *22* **forehead,** *23* **bridge of nose,** *24* **muzzle,** *25* **jaw,** *26* **throat,** *27* **point of shoulder,** *28* **dewlap,** *29* **point of elbow,** *30* **brisket,** *31* **chest floor,** *32* **knee,** *33* **milk well,** *34* **milk vein,** *35* **fore udder,** *36* **teats,** *37* **rump,** *38* **loin**

²cow *transitive verb* [probably of Scandinavian origin; akin to Danish *kue* to subdue] (1605)
: to destroy the resolve or courage of ⟨the party that Stalin had *cowed* —*World Press Review*⟩; *also* **:** to bring to a state or an action by intimidation — used with *into* ⟨like too many Asian armies, adept at *cowing* a population into feeding them —Edward Lansdale⟩
synonym see INTIMIDATE
— cowed·ly \ˈkau̇(-ə)d-lē\ *adverb*

cow·ard \ˈkau̇(-ə)rd\ *noun* [Middle English, from Old French *coart,* from *coe* tail, from Latin *cauda*] (13th century)
: one who shows disgraceful fear or timidity ◆
— coward *adjective*

cow·ard·ice \ˈkau̇(-ə)r-dəs, *dialect* -(ˌ)dīs\ *noun* [Middle English *cowardise,* from Middle French *coardise,* from *coart*] (14th century)
: lack of courage or resolution

¹cow·ard·ly \-lē\ *adverb* (14th century)
: in a cowardly manner

²cowardly *adjective* (1551)
: being, resembling, or befitting a coward ⟨a *cowardly* retreat⟩ ☆
— cow·ard·li·ness *noun*

cow·bane \ˈkau̇-ˌbān\ *noun* (1776)
: any of several poisonous plants (as a water hemlock) of the carrot family

cow·bell \-ˌbel\ *noun* (1652)
: a bell hung around the neck of a cow to make a sound by which the cow can be located

cow·ber·ry \-ˌber-ē\ *noun* (1800)
: MOUNTAIN CRANBERRY; *also* **:** its fruit

cow·bird \-ˌbərd\ *noun* (1810)
: a small North American blackbird (*Molothrus ater*) that lays its eggs in the nests of other birds

cow·boy \-ˌbȯi\ *noun* (1623)
1 : one who tends cattle or horses; *especially* **:** a usually mounted cattle-ranch hand
2 : a rodeo performer

cowboy boot *noun* (1895)
: a boot made with a high arch, a high Cuban heel, and usually fancy stitching

cowboy hat *noun* (1895)
: a wide-brimmed hat with a large soft crown — called also *ten-gallon hat*

cow·catch·er \ˈkau̇-ˌka-chər, -ˌke-\ *noun* (1838)
: an inclined frame on the front of a railroad locomotive for throwing obstacles off the track

cow college *noun* (circa 1915)
1 : a college that specializes in agriculture

☆ SYNONYMS

Covetous, greedy, acquisitive, grasping, avaricious mean having or showing a strong desire for especially material possessions. COVETOUS implies inordinate desire often for another's possessions ⟨*covetous* of his brother's country estate⟩. GREEDY stresses lack of restraint and often of discrimination in desire ⟨*greedy* for status symbols⟩. ACQUISITIVE implies both eagerness to possess and ability to acquire and keep ⟨an eagerly *acquisitive* mind⟩. GRASPING adds to COVETOUS and GREEDY an implication of selfishness and often suggests unfair or ruthless means ⟨a hard *grasping* trader who cheated the natives⟩. AVARICIOUS implies obsessive acquisitiveness especially of money and strongly suggests stinginess ⟨an *avaricious* miser⟩.

Cowardly, pusillanimous, craven, dastardly mean having or showing a lack of courage. COWARDLY implies a weak or ignoble lack of courage ⟨a *cowardly* failure to stand up for principle⟩. PUSILLANIMOUS suggests a contemptible lack of courage ⟨the *pusillanimous* fear of a future full of possibility⟩. CRAVEN suggests extreme defeatism and complete lack of resistance ⟨secretly despised her own *craven* yes-men⟩. DASTARDLY often implies behavior that is both cowardly and treacherous or skulking or outrageous ⟨a *dastardly* attack on unarmed civilians⟩.

◇ WORD HISTORY

coward A frightened animal may draw its tail between its hind legs—or it may simply turn tail and run. If the fleeing animal has a white tail, as does the hare, the flash of white can leave a keen impression. But fear can be found in humans as well as animals, and armies have their tail ends also. A traditional belief holds that it is in the tail end of an advancing army that one is most likely to find the cowards lurking. Although it is not known whether the original reference was to the tail of an army or an animal, it is certain that the Old French word *cuart* or *coart,* the source of our word *coward,* comes from *coe* or *coue,* meaning "tail."

2 : a provincial college or university that lacks culture, sophistication, and tradition

cow·er \'kau̇(-ə)r\ *intransitive verb* [Middle English *couren,* of Scandinavian origin; akin to Old Norse *kura* to cower] (14th century)
: to shrink away or crouch especially for shelter from something that menaces, domineers, or dismays
synonym see FAWN

cow·fish \'kau̇-ˌfish\ *noun* (1870)
: any of various small bright-colored fishes (family Ostraciidae) with hornlike projections over the eyes

cow·girl \-ˌgər(-ə)l\ *noun* (1884)
1 : a girl or woman who tends cattle or horses
2 : a girl or woman who is a rodeo performer

cow·hage *also* **cow·age** \'kau̇-ij\ *noun* [Hindi *kavā̃c*] (1640)
: a tropical leguminous woody vine (*Mucuna pruriens*) with crooked pods covered with barbed hairs that cause severe itching; *also* **:** these hairs formerly used as a vermifuge

cow·hand \'kau̇-ˌhand\ *noun* (1852)
: COWBOY 1

cow·herd \-ˌhərd\ *noun* (before 12th century)
: one who tends cows

[1]cow·hide \-ˌhīd\ *noun* (14th century)
1 : the hide of a cow; *also* **:** leather made from this hide
2 : a coarse whip of rawhide or braided leather

[2]cowhide *transitive verb* **cow·hid·ed; cow·hid·ing** (1794)
: to flog with a cowhide whip

cow horse *noun* (1853)
: COW PONY

[1]cowl \'kau̇(ə)l\ *noun* [Middle English *cowle,* from Old English *cugele,* from Late Latin *cuculla* monk's hood, from Latin *cucullus* hood] (before 12th century)
1 a : a hood or long hooded cloak especially of a monk **b :** a draped neckline on a woman's garment
2 a : a chimney covering designed to improve the draft **b :** the top portion of the front part of an automobile body forward of the two front doors to which are attached the windshield and instrument board **c :** COWLING

[2]cowl *transitive verb* (1536)
: to cover with or as if with a cowl

cow·lick \'kau̇-ˌlik\ *noun* (1598)
: a lock or tuft of hair growing in a different direction from the rest of the hair

cowl·ing \'kau̇-liŋ\ *noun* (1917)
: a removable metal covering that houses the engine and sometimes a part of the fuselage or nacelle of an airplane; *also* **:** a metal cover for an engine

cowl·staff \'kōl-ˌstaf, 'kau̇(ə)l-, 'kül-\ *noun* [Middle English *cuvelstaff,* from *cuvel* vessel (from Old English *cȳfel,* from Old North French *cuvele* small vat) + [1]*staff*] (13th century)
archaic **:** a staff from which a vessel is suspended and carried between two persons

cow·man \'kau̇-mən, -ˌman\ *noun* (1677)
1 : COWHERD, COWBOY
2 : a cattle owner or rancher

cow parsnip *noun* (1548)
: a tall perennial North American plant (*Heracleum lanatum*) of the carrot family with large compound leaves and broad umbels of white or purplish flowers; *also* **:** a related Eurasian plant (*H. sphondylium*) naturalized in the U.S.

cow·pat \'kau̇-ˌpat\ *noun* (1937)
: a dropping of cow dung

cow·pea \'kau̇-ˌpē\ *noun* (1776)
: a sprawling leguminous herb (*Vigna unguiculata* synonym *V. sinensis*) related to the bean and widely cultivated in the southern U.S. especially for forage and green manure; *also* **:** its edible seed — called also *black-eyed pea*

Cow·per's gland \'kau̇-pərz-, 'kü-, 'ku̇-\ *noun* [William *Cowper* (died 1709) English surgeon] (1738)
: either of two small glands lying on either side of the male urethra below the prostate gland and discharging a secretion into the semen — called also *bulbourethral gland;* compare BARTHOLIN'S GLAND

cow·poke \'kau̇-ˌpōk\ *noun* (circa 1881)
: COWBOY 1

cow pony *noun* (1874)
: a strong and agile saddle horse trained for herding cattle

cow·pox \'kau̇-ˌpäks\ *noun* (1798)
: a mild eruptive disease of the cow that when communicated to humans protects against smallpox

cow·punch·er \-ˌpən-chər\ *noun* (1878)
: COWBOY 1

cow·rie *or* **cow·ry** \'kau̇-rē\ *noun, plural* **cowries** [Hindi *kaurī*] (1662)
: any of various marine gastropods (family Cypraeidae) that are widely distributed in warm seas and have glossy and often brightly colored shells; *also* **:** the shell of a cowrie

cow·shed \'kau̇-ˌshed\ *noun* (1835)
: a shed for the housing of cows

cow·slip \'kau̇-ˌslip\ *noun* [Middle English *cowslyppe,* from Old English *cūslyppe,* literally, cow dung, from *cū* cow + *slypa, slyppe* paste] (before 12th century)
1 : a common European primrose (*Primula veris*) with fragrant yellow flowers
2 : MARSH MARIGOLD

cow town *noun* (1885)
1 : a town or city that serves as a market center or shipping point for cattle
2 : a small unsophisticated town within a cattle-raising area

[1]cox \'käks\ *noun* (1869)
: COXSWAIN

[2]cox *verb* (1881)
: COXSWAIN

coxa \'käk-sə\ *noun, plural* **cox·ae** \-ˌsē, -ˌsī\ [Latin, hip; akin to Old High German *hāhsina* hock, Sanskrit *kakṣa* armpit] (1826)
: the basal segment of a limb of various arthropods (as an insect)
— cox·al \-səl\ *adjective*

cox·comb \'käk-ˌskōm\ *noun* [Middle English *cokkes comb,* literally, cock's comb] (1573)
1 a *obsolete* **:** a jester's cap adorned with a strip of red **b** *archaic* **:** PATE, HEAD
2 a *obsolete* **:** FOOL **b :** a conceited foolish person **:** FOP
— cox·comb·i·cal \käk-'skō-mi-kəl, -'skä-\ *adjective*

cox·comb·ry \'käk-skəm-rē, -ˌskōm-\ *noun, plural* **-ries** (1608)
: behavior that is characteristic of a coxcomb **:** FOPPERY

Cox·sack·ie virus \(ˌ)käk-'sa-kē-\ *noun* [*Coxsackie,* N.Y.] (1949)
: any of several enteroviruses associated with human diseases (as meningitis)

[1]cox·swain \'käk-sən, -ˌswān\ *noun* [Middle English *cokswayne,* from *cok* cockboat + *swain* servant] (15th century)
1 : a sailor who has charge of a ship's boat and its crew and who usually steers
2 : a steersman of a racing shell who usually directs the rowers

[2]coxswain (1928)
transitive verb
: to direct as coxswain
intransitive verb
: to act as coxswain

[1]coy \'kȯi\ *adjective* [Middle English, quiet, shy, from Middle French *coi* calm, from Latin *quietus* quiet] (14th century)
1 a : shrinking from contact or familiarity **b :** marked by cute, coquettish, or artful playfulness
2 : showing reluctance to make a definite commitment
synonym see SHY
— coy·ly *adverb*
— coy·ness *noun*

[2]coy (1583)
transitive verb
obsolete **:** CARESS
intransitive verb
archaic **:** to act coyly

coy·dog \'kȯi-ˌdȯg\ *noun* [*coy*ote + *dog*] (1950)
: a hybrid between a coyote and a feral dog

coy·ote \'kī-ˌōt, kī-'ō-tē\ *noun, plural* **coy·otes** *or* **coyote** [Mexican Spanish, from Nahuatl *coyōtl*] (1759)
1 : a buff-gray to reddish gray North American canid (*Canis latrans*) closely related to but smaller than the wolf
2 : one who smuggles immigrants into the U.S.

coyote 1

coy·o·til·lo \ˌkī-ə-'ti-(ˌ)lō, ˌkȯi-ə-, -'tē-(ˌ)yō\ *noun* [Mexican Spanish, diminutive of *coyote*] (circa 1892)
: a shrub (*Karwinskia humboldtiana*) of the buckthorn family of the southwestern U.S. and Mexico having poisonous berries

coy·pu \'kȯi-(ˌ)pü, kȯi-'\ *noun* [American Spanish *coipú,* from Araucanian *coipu*] (1793)
: NUTRIA

coz \'kəz\ *noun* [by shortening & alteration] (1559)
: COUSIN

coz·en \'kə-zən\ *transitive verb* **coz·ened; coz·en·ing** \'kəz-niŋ, 'kə-zə-\ [perhaps from obsolete Italian *cozzonare,* from Italian *cozzone* horse trader, from Latin *cocion-, cocio* trader] (1573)
1 : to deceive, win over, or induce to do something by artful coaxing and wheedling or shrewd trickery
2 : to gain by cozening someone ⟨*cozened* his supper out of the old couple⟩
synonym see CHEAT
— coz·en·er \'kəz-nər, 'kə-zə-\ *noun*

coz·en·age \'kəz-nij, 'kə-zə-\ *noun* (1583)
1 : the art or practice of cozening **:** FRAUD
2 : an act or an instance of cozening

[1]co·zy \'kō-zē\ *adjective* **co·zi·er; -est** [probably of Scandinavian origin; akin to Norwegian *kose*lig cozy] (1709)
1 : enjoying or affording warmth and ease **:** SNUG
2 a : marked by the intimacy of the family or a close group **b :** marked by or suggesting close association or connivance ⟨a *cozy* agreement⟩
3 : marked by a discreet and cautious attitude or procedure
synonym see COMFORTABLE
— co·zi·ly \-zə-lē\ *adverb*
— co·zi·ness \-zē-nəs\ *noun*

[2]cozy *noun, plural* **co·zies** (1863)
: a padded covering especially for a teapot to keep the contents hot

[3]cozy *adverb* (1946)
: in a cautious manner ⟨play it *cozy* and wait for the other team to make a mistake —Bobby Dodd⟩

cozy up *intransitive verb* (1937)
: to attain or try to attain familiarity, friendship, or intimacy **:** ingratiate oneself ⟨*cozying up* to the boss⟩

[1]crab \'krab\ *noun, often attributive* [Middle English *crabbe,* from Old English *crabba;* akin to Old High German *krebiz* crab, Old English *ceorfan* to carve — more at CARVE] (before 12th century)
1 : any of numerous chiefly marine broadly built decapod crustaceans: **a :** any of a group (Brachyura) with a short broad usually flattened carapace, a small abdomen that curls forward beneath the body, short antennae, and the anterior pair of limbs modified as grasping pincers **b :** any of various crustaceans of a

group (Anomura) resembling true crabs in the more or less reduced condition of the abdomen
2 *capitalized* **:** CANCER 1
3 *plural* **:** infestation with crab lice
4 : the angular difference between an aircraft's course and the heading necessary to make that course in the presence of a crosswind

[2]crab *verb* **crabbed; crab•bing** (1918)
transitive verb
1 : to cause to move sideways or in an indirect or diagonal manner; *specifically* **:** to head (an airplane) into a crosswind to counteract drift
2 : to subject to crabbing
intransitive verb
1 a (1) **:** to move sideways indirectly or diagonally (2) **:** to crab an airplane **b :** to scuttle or scurry sideways
2 : to fish for crabs
— crab•ber *noun*

[3]crab *noun* [Middle English *crabbe,* perhaps from *crabbe* [1]crab] (14th century)
: CRAB APPLE

[4]crab *noun* (1580)
: an ill-tempered person **:** CROSSPATCH

[5]crab *verb* **crabbed; crab•bing** [Middle English *crabben,* probably back-formation from *crabbed*] (1662)
transitive verb
1 : to make sullen **:** SOUR ⟨old age has *crabbed* his nature⟩
2 : to complain about peevishly
3 : SPOIL, RUIN
intransitive verb
: CARP, GROUSE ⟨always *crabs* about the weather⟩
— crab•ber *noun*

crab apple *noun* [[3]*crab*] (1712)
: any of various wild or cultivated trees (genus *Malus*) that are cultivars or relatives of the cultivated apple and that produce small sour fruit; *also* **:** the fruit

crab•bed \'kra-bəd\ *adjective* [Middle English, partly from *crabbe* [1]crab, partly from *crabbe* crab apple] (14th century)
1 : marked by a forbidding moroseness
2 : difficult to read or understand ⟨*crabbed* handwriting⟩
synonym see SULLEN
— crab•bed•ness *noun*

crab•by \'kra-bē\ *adjective* **crab•bi•er; -est** [[4]*crab*] (1776)
: CROSS, ILL-NATURED

crab cactus *noun* (1900)
: CHRISTMAS CACTUS

crab•grass \'krab-,gras\ *noun* (1743)
: a grass (especially *Digitaria sanguinalis*) that has creeping or decumbent stems which root freely at the nodes and that is often a pest in turf or cultivated lands

crab louse *noun* (1547)
: a sucking louse (*Phthirus pubis*) infesting the pubic region of the human body

crab•meat \'krab-,mēt\ *noun* (1876)
: the edible part of a crab

crab•stick \-,stik\ *noun* (1703)
1 : a stick, cane, or cudgel of crab apple tree wood
2 : a crabbed ill-natured person

crab•wise \-,wīz\ *adverb* (1898)
1 : SIDEWAYS
2 : in a sidling or cautiously indirect manner

[1]crack \'krak\ *verb* [Middle English *crakken,* from Old English *cracian;* akin to Old High German *chrāhhōn* to resound] (before 12th century)
intransitive verb
1 : to make a very sharp explosive sound ⟨the whip *cracks* through the air⟩
2 : to break, split, or snap apart
3 : FAIL: as **a :** to lose control or effectiveness under pressure — often used with *up* **b :** to fail in tone ⟨his voice *cracked*⟩
4 : to go or travel at good speed — usually used with *on* ⟨the steamboat *cracked* on⟩
transitive verb
1 a : to break so that fissures appear on the surface ⟨*crack* a mirror⟩ **b :** to break with a sudden sharp sound ⟨*crack* nuts⟩
2 : to tell especially suddenly or strikingly ⟨*crack* a joke⟩
3 : to strike with a sharp noise **:** RAP ⟨then *cracks* him over the head⟩ ⟨*cracked* a two-run homer in the fifth —*N.Y. Times*⟩
4 a (1) **:** to open (as a bottle) for drinking (2) **:** to open (a book) for studying **b :** to puzzle out and expose, solve, or reveal the mystery of ⟨*crack* a code⟩ **c :** to break into ⟨*crack* a safe⟩ **d :** to open slightly ⟨*crack* the throttle⟩ **e :** to break through (as a barrier) so as to gain acceptance or recognition
5 a : to impair seriously or irreparably **:** WRECK ⟨*crack* a car up⟩ **b :** to destroy the tone of (a voice) **c :** DISORDER, CRAZE **d :** to interrupt sharply or abruptly ⟨the criticism *cracked* our complacency⟩
6 : to cause to make a sharp noise ⟨*cracks* his knuckles⟩
7 a (1) **:** to subject (hydrocarbons) to cracking (2) **:** to produce by cracking ⟨*cracked* gasoline⟩ **b :** to break up (chemical compounds) into simpler compounds by means of heat

[2]crack *noun* (14th century)
1 a : a loud roll or peal ⟨a *crack* of thunder⟩ **b :** a sudden sharp noise ⟨the *crack* of rifle fire⟩
2 : a sharp witty remark **:** QUIP
3 a : a narrow break **:** FISSURE ⟨a *crack* in the ice⟩ **b :** a narrow opening ⟨leave the door open a *crack*⟩
4 a : a weakness or flaw caused by decay, age, or deficiency **:** UNSOUNDNESS **b :** a broken tone of the voice **c :** CRACKPOT
5 : MOMENT, INSTANT ⟨the *crack* of dawn⟩
6 : HOUSEBREAKING, BURGLARY
7 : a sharp resounding blow ⟨gave him a *crack* on the head⟩
8 : ATTEMPT, TRY ⟨her first *crack* at writing a novel⟩
9 : highly purified cocaine in small chips used illicitly usually for smoking

[3]crack *adjective* (1793)
: of superior excellence or ability ⟨a *crack* marksman⟩

crack•back \'krak-,bak\ *noun* (1967)
: a blind-side block on a defensive back in football by a pass receiver who starts downfield and then cuts back to the middle of the line

crack•brain \-,brān\ *noun* (circa 1570)
: an erratic person **:** CRACKPOT
— crack•brained \-'brānd\ *adjective*

crack•down \-,daůn\ *noun* (1935)
: an act or instance of cracking down

crack down *intransitive verb* (1939)
: to take positive regulatory or disciplinary action

cracked \'krakt\ *adjective* (1503)
1 a : broken (as by a sharp blow) so that the surface is fissured ⟨*cracked* china⟩ **b :** broken into coarse particles ⟨*cracked* wheat⟩ **c :** marked by harshness, dissonance, or failure to sustain a tone ⟨a *cracked* voice⟩
2 : mentally disturbed **:** CRAZY

crack•er \'kra-kər\ *noun* (15th century)
1 *chiefly dialect* **:** a bragging liar **:** BOASTER
2 : something that makes a cracking or snapping noise: as **a :** FIRECRACKER **b :** the snapping end of a whiplash **:** SNAPPER **c :** a paper holder for a party favor that pops when the ends are pulled sharply
3 *plural* **:** NUTCRACKER
4 : a dry thin crispy baked bread product that may be leavened or unleavened
5 a : a poor usually Southern white — usually used disparagingly **b** *capitalized* **:** a native or resident of Florida or Georgia — used as a nickname
6 : the equipment in which cracking (as of petroleum) is carried out

crack•er–bar•rel \-,bar-əl\ *adjective* [from the cracker barrel in country stores around which customers lounged for informal conversation] (1916)
: suggestive of the friendly homespun character of a country store ⟨a *cracker-barrel* philosopher⟩

crack•er•jack \'kra-kər-,jak\ *also* **crack•a•jack** \-kə-,jak\ *noun* [probably alteration of [3]*crack* + *jack* (man)] (1895)
: a person or thing of marked excellence
— crackerjack *adjective*

Cracker Jack *trademark*
— used for a candied popcorn confection

crack•ers \'kra-kərz\ *adjective* [probably *cracked* + *-ers* (as in *starkers*)] (1928)
: CRAZY
word history see NUTS

[1]crack•ing \'kra-kiŋ\ *adjective* (1830)
: very impressive or effective **:** GREAT

[2]cracking *adverb* (1903)
: VERY, EXTREMELY ⟨a *cracking* good book⟩

[3]cracking *noun* (1868)
: a process in which relatively heavy hydrocarbons are broken up by heat into lighter products (as gasoline)

[1]crack•le \'kra-kəl\ *verb* **crack•led; crack•ling** \-k(ə-)liŋ\ [frequentative of [1]*crack*] (circa 1560)
intransitive verb
1 a : to make small sharp sudden repeated noises ⟨the fire *crackles* on the hearth⟩ **b :** to show animation **:** SPARKLE ⟨the essays *crackle* with wit⟩
2 : CRAZE 3
transitive verb
: to crush or crack with snapping noises

[2]crackle *noun* (1833)
1 a : the noise of repeated small cracks or reports **b :** SPARKLE, EFFERVESCENCE
2 : a network of fine cracks on an otherwise smooth surface

crack•le•ware \'kra-kəl-,war, -,wer\ *noun* (1881)
: ceramic ware with a decorative crazed glaze

crack•ling *noun* (1599)
1 \'kra-k(ə-)liŋ\ **:** a series of small sharp noises ⟨the *crackling* of frozen snow as we walk⟩
2 \'kra-klən, -kliŋ\ **:** the crisp residue left after the rendering of lard from fat or the frying or roasting of the skin (as of pork) — usually used in plural

crack•ly \'kra-k(ə-)lē\ *adjective* (1859)
: inclined to crackle **:** CRISP

crack•nel \'krak-nᵊl\ *noun* [Middle English *krakenelle*] (14th century)
1 : a hard brittle biscuit
2 : CRACKLING 2 — usually used in plural

crack•pot \'krak-,pät\ *noun* (1883)
: one given to eccentric or lunatic notions
— crackpot *adjective*

cracks•man \'krak-smən\ *noun* (circa 1812)
: BURGLAR; *also* **:** SAFECRACKER

crack–up \'krak-,əp\ *noun* (circa 1926)
1 : CRASH, WRECK ⟨an automobile *crack-up*⟩
2 a : a mental collapse **:** NERVOUS BREAKDOWN ⟨his wife's death brought on his *crack-up*⟩ **b :** COLLAPSE, BREAKDOWN

crack up (1829)
transitive verb
1 : PRAISE, TOUT 4 ⟨wasn't all that it was *cracked up* to be⟩
2 : to cause to laugh out loud ⟨that joke really *cracks* him *up*⟩
intransitive verb
1 : to smash up a vehicle (as by losing control) ⟨*cracked up* on a curve⟩
2 : to laugh out loud

-cracy *noun combining form* [Middle French & Late Latin; Middle French *-cratie,* from Late Latin *-cratia,* from Greek *-kratia,* from *kratos* strength, power — more at HARD]

1 : form of government; *also* **:** state having such a form ⟨mono*cracy*⟩
2 : social or political class (as of powerful persons) ⟨mobo*cracy*⟩
3 : theory of social organization ⟨techno*cracy*⟩

¹cra·dle \'krā-dᵊl\ *noun* [Middle English *cradel,* from Old English *cradol;* perhaps akin to Old High German *kratto* basket, Sanskrit *grantha* knot] (before 12th century)
1 a : a bed or cot for a baby usually on rockers or pivots **b :** a framework or support suggestive of a baby's cradle: as (1) **:** a framework of bars and rods (2) **:** the support for a telephone receiver or handset **c :** an implement with rods like fingers attached to a scythe and used formerly for harvesting grain **d :** a frame to keep the bedclothes from contact with an injured part of the body
2 a : the earliest period of life **:** INFANCY ⟨from the *cradle* to the grave⟩ **b :** a place of origin ⟨the *cradle* of civilization⟩
3 : a rocking device used in panning for gold

²cradle *verb* **cra·dled; cra·dling** \'krād-liŋ, 'krā-dᵊl-iŋ\ (15th century)
transitive verb
1 a : to place or keep in or as if in a cradle **b :** SHELTER, REAR **c :** to support protectively or intimately ⟨*cradling* the injured man's head in her arms⟩
2 : to cut (grain) with a cradle scythe
3 : to place, raise, support, or transport on a cradle
intransitive verb
obsolete **:** to rest in or as if in a cradle

cradle cap *noun* (circa 1890)
: a seborrheic condition in infants that usually affects the scalp and is characterized by greasy gray or dark brown adherent scaly crusts

cra·dle·song \'krā-dᵊl-ˌsȯŋ\ *noun* (14th century)
: LULLABY

¹craft \'kraft\ *noun* [Middle English, strength, skill, from Old English *cræft;* akin to Old High German *kraft* strength] (before 12th century)
1 : skill in planning, making, or executing **:** DEXTERITY
2 : an occupation or trade requiring manual dexterity or artistic skill ⟨the carpenter's *craft*⟩ ⟨the *craft* of writing plays⟩ ⟨*crafts* such as pottery, carpentry, and sewing⟩
3 : skill in deceiving to gain an end ⟨used *craft* and guile to close the deal⟩
4 : the members of a trade or trade association
5 *plural usually* **craft a :** a boat especially of small size **b :** AIRCRAFT **c :** SPACECRAFT
synonym see ART

²craft *transitive verb* (15th century)
: to make or produce with care, skill, or ingenuity ⟨is *crafting* a new sculpture⟩ ⟨a carefully *crafted* story⟩

crafts·man \'kraf(t)s-mən\ *noun* (13th century)
1 : a worker who practices a trade or handicraft
2 : one who creates or performs with skill or dexterity especially in the manual arts ⟨jewelry made by European *craftsmen*⟩
— crafts·man·like \-ˌlīk\ *adjective*
— crafts·man·ly \-lē\ *adjective*
— crafts·man·ship \-ˌship\ *noun*

crafts·peo·ple \-ˌpē-pəl\ *noun plural* (1953)
: workers who practice a trade or craft

crafts·per·son \-ˌpər-sᵊn\ *noun* (1920)
: a craftsman or craftswoman

crafts·wom·an \'kraf(t)s-ˌwu̇-mən\ *noun* (1886)
1 : a woman artisan
2 : a woman who is skilled in a craft

craft union *noun* (1922)
: a labor union with membership limited to workers of the same craft — compare INDUSTRIAL UNION

crafty \'kraf-tē\ *adjective* **craft·i·er; -est** (before 12th century)
1 *dialect chiefly British* **:** SKILLFUL, CLEVER
2 a : adept in the use of subtlety and cunning **b :** marked by subtlety and guile ⟨a *crafty* scheme⟩
synonym see SLY
— craft·i·ly \'kraf-tə-lē\ *adverb*
— craft·i·ness \-tē-nəs\ *noun*

¹crag \'krag\ *noun* [Middle English, of Celtic origin; akin to Welsh *craig* rock] (14th century)
1 : a steep rugged rock or cliff
2 *archaic* **:** a sharp detached fragment of rock
— crag·ged \'kra-gəd\ *adjective*

²crag *noun* [Middle English, from Middle Dutch *crāghe;* akin to Old English *cræga* throat — more at CRAW] (14th century)
chiefly Scottish **:** NECK, THROAT

crag·gy \'kra-gē\ *adjective* **crag·gi·er; -est** (15th century)
1 : full of crags ⟨*craggy* slopes⟩
2 : ROUGH, RUGGED ⟨a *craggy* face⟩
— crag·gi·ly \'kra-gə-lē\ *adverb*
— crag·gi·ness \'kra-gē-nəs\ *noun*

crags·man \'kragz-mən\ *noun* (1816)
: one that is expert in climbing crags or cliffs

crake \'krāk\ *noun* [Middle English, probably from Old Norse *krāka* crow or *krākr* raven; akin to Old English *crāwan* to crow] (14th century)
1 : any of various rails; *especially* **:** a short-billed rail (as the corncrake)
2 : the corncrake's cry

¹cram \'kram\ *verb* **crammed; cram·ming** [Middle English *crammen,* from Old English *crammian;* akin to Old Norse *kremja* to squeeze] (before 12th century)
transitive verb
1 : to pack tight **:** JAM ⟨*cram* a suitcase with clothes⟩
2 a : to fill with food to satiety **:** STUFF **b :** to eat voraciously **:** BOLT ⟨the child *crams* her food⟩
3 : to thrust in or as if in a rough or forceful manner ⟨*crammed* the letters into his pocket⟩
4 : to prepare hastily for an examination ⟨*cram* the students for the test⟩
intransitive verb
1 : to eat greedily or to satiety **:** STUFF
2 : to study hastily for an imminent examination
— cram·mer *noun*

²cram *noun* (1810)
1 : a compressed multitude or crowd **:** CRUSH
2 : last-minute study for an examination

cram·bo \'kram-(ˌ)bō\ *noun, plural* **cram·boes** [alteration of earlier *crambe,* from Latin, cabbage, from Greek *krambē*] (1660)
: a game in which one player gives a word or line of verse to be matched in rhyme by other players

cram·oi·sie *or* **cram·oi·sy** \'kra-ˌmȯi-zē, 'kra-mə-zē\ *noun, plural* **-sies** [Middle English *crammassy,* from Middle French *cramoisi,* from *cramoisi* crimson] (15th century)
: crimson cloth

¹cramp \'kramp\ *noun* [Middle English *crampe,* from Middle French, of Germanic origin; akin to Middle Dutch *crampe;* akin to Old High German *krampf* bent] (14th century)
1 : a painful involuntary spasmodic contraction of a muscle
2 : a temporary paralysis of muscles from overuse — compare WRITER'S CRAMP
3 a : sharp abdominal pain — usually used in plural **b :** persistent and often intense though dull lower abdominal pain associated with dysmenorrhea — usually used in plural

²cramp *noun* [Middle English *crampe,* from Middle Dutch] (15th century)
1 a : a usually iron device bent at the ends and used to hold timbers or blocks of stone together **b :** CLAMP
2 a : something that confines **:** SHACKLE **b :** the state of being confined

³cramp (15th century)
transitive verb
1 : to affect with or as if with a cramp or cramps
2 a : CONFINE, RESTRAIN ⟨was *cramped* in the tiny apartment⟩ **b :** to restrain from free expression — used especially in the phrase *cramp one's style*
3 : to turn (the front wheels of a vehicle) to right or left
4 : to fasten or hold with a cramp
intransitive verb
: to suffer from cramps

⁴cramp *adjective* (1674)
1 : hard to understand or figure out ⟨*cramp* law terms⟩ ⟨*cramp* handwriting⟩
2 : being cramped ⟨a *cramp* corner⟩

cram·pon \'kram-ˌpän\ *noun* [Middle English, from Middle French *crampon,* of Germanic origin; akin to Middle Dutch *crampe*] (15th century)
1 : a hooked clutch or dog for raising heavy objects — usually used in plural
2 : a climbing iron used especially on ice and snow in mountaineering — usually used in plural

crampon 2

cran·ber·ry \'kran-ˌber-ē, -b(ə-)rē\ *noun* [part translation of Low German *kraanbere,* from *kraan* crane + *bere* berry] (1647)
1 : the red acid berry produced by some plants (as *Vaccinium oxycoccos* and *V. macrocarpon*) of the heath family; *also* **:** a plant producing these
2 : any of various plants with a fruit that resembles a cranberry

cranberry bush *noun* (1778)
: a shrubby or arborescent viburnum (*Viburnum opulus*) of North America and Europe with prominently 3-lobed leaves and red fruit

cranch \'kränch\ *variant of* CRAUNCH

¹crane \'krān\ *noun* [Middle English *cran,* from Old English; akin to Old High German *krano* crane, Greek *geranos,* Latin *grus*] (before 12th century)
1 : any of a family (Gruidae of the order Gruiformes) of tall wading birds superficially resembling the herons but structurally more nearly related to the rails
2 : any of several herons
3 : an often horizontal projection swinging about a vertical axis: as **a :** a machine for raising, shifting, and lowering heavy weights by means of a projecting swinging arm or with the hoisting apparatus supported on an overhead track **b :** an iron arm in a fireplace for supporting kettles **c :** a boom for holding a motion-picture or television camera

crane 1

²crane *verb* **craned; cran·ing** (1570)
transitive verb
1 : to raise or lift by or as if by a crane
2 : to stretch (as the neck) toward an object of attention ⟨*craning* her neck to get a better view⟩
intransitive verb
1 : to stretch one's neck toward an object of attention ⟨I *craned* out of the window of my compartment —Webb Waldron⟩
2 : HESITATE

crane fly *noun* (1658)
: any of a family (Tipulidae) of long-legged slender dipteran flies that resemble large mosquitoes but do not bite

cranes·bill \ˈkrānz-ˌbil\ *noun* (1548)
: GERANIUM 1

crani- *or* **cranio-** *combining form* [Medieval Latin *cranium*]
: cranium ⟨*crani*ate⟩ **:** cranial and ⟨*cranio*sacral⟩

cra·ni·al \ˈkrā-nē-əl\ *adjective* (1800)
1 : of or relating to the skull or cranium
2 : CEPHALIC
— cra·ni·al·ly \-ə-lē\ *adverb*

cranial index *noun* (1868)
: the ratio multiplied by 100 of the maximum breadth of the bare skull to its maximum length from front to back

cranial nerve *noun* (1840)
: any of the nerves that arise in pairs from the lower surface of the brain one on each side and pass through openings in the skull to the periphery of the body and that comprise 12 pairs in reptiles, birds, and mammals and usually 10 in fishes and amphibians

cra·ni·ate \ˈkrā-nē-ət, -ˌāt\ *adjective* (1879)
: having a cranium
— craniate *noun*

cra·nio·ce·re·bral \ˌkrā-nē-ō-sə-ˈrē-brəl, -ˈser-ə-\ *adjective* (circa 1903)
: involving both cranium and brain ⟨*craniocerebral* injury⟩

cra·nio·fa·cial \-ˈfā-shəl\ *adjective* (circa 1859)
: of, relating to, or involving both the cranium and the face

cra·ni·ol·o·gy \ˌkrā-nē-ˈä-lə-jē\ *noun* [probably from German *Kraniologie,* from *kranio-* crani- + *-logie* -logy] (1851)
: a science dealing with variations in size, shape, and proportions of skulls among human races

cra·ni·om·e·try \-ˈä-mə-trē\ *noun* [International Scientific Vocabulary] (circa 1828)
: a science dealing with cranial measurement — compare CEPHALOMETRY

cra·nio·sa·cral \ˌkrā-nē-ō-ˈsa-krəl, -ˈsā-\ *adjective* (circa 1923)
: PARASYMPATHETIC

cra·ni·ot·o·my \ˌkrā-nē-ˈä-tə-mē\ *noun* [International Scientific Vocabulary] (1855)
: surgical opening of the skull

cra·ni·um \ˈkrā-nē-əm\ *noun, plural* **-ni·ums** *or* **-nia** \-nē-ə\ [Medieval Latin, from Greek *kranion;* akin to Greek *kara* head — more at CEREBRAL] (15th century)
: SKULL; *specifically* **:** the part that encloses the brain **:** BRAINCASE

[1]**crank** \ˈkraŋk\ *noun* [Middle English *cranke,* from Old English *cranc-* (as in *crancstæf,* a weaving instrument); probably akin to Middle High German *krank* weak, sick — more at CRINGE] (13th century)
1 : a bent part of an axle or shaft or an arm keyed at right angles to the end of a shaft by which circular motion is imparted to or received from the shaft or by which reciprocating motion is changed into circular motion or vice versa
2 a *archaic* **:** BEND **b :** a twist or turn of speech **:** CONCEIT — used especially in the phrase *quips and cranks* **c** (1) **:** CAPRICE, CROTCHET (2) **:** an annoyingly eccentric person; *also* **:** one that is overly enthusiastic about a particular subject or activity **d :** a bad-tempered person **:** GROUCH
3 : CRYSTAL 7
— crank·ish \ˈkraŋ-kish\ *adjective*

[2]**crank** (1592)
intransitive verb
1 : to move with a winding course **:** ZIGZAG
2 a : to turn a crank **b :** to get started by or as if by the turning of a crank **c :** to gain speed, momentum, or intensity — usually used with *up* ⟨the campaign is *cranking* up⟩
transitive verb
1 : to move or operate by or as if by a crank ⟨*crank* the window down⟩
2 a : to cause to start (as an automobile) **b :** to rotate the shaft (as a crankshaft) of especially with a starter ⟨*crank* over an engine⟩ **c :** to use in trying to start an engine ⟨*crank* the starter⟩
3 a : to start as if by use of a crank — usually used with *up* ⟨*crank* up an atom smasher⟩ **b :** TURN UP 2 — usually used with *up* ⟨*crank* up the volume⟩

[3]**crank** *adjective* (1924)
: of, relating to, or being a cranky or eccentric person; *also* **:** made or sent by such a person ⟨*crank* calls⟩ ⟨a *crank* letter⟩

[4]**crank** *adjective* [Middle English *cranke,* of unknown origin] (15th century)
1 *chiefly dialect* **:** MERRY, HIGH-SPIRITED
2 *chiefly dialect* **:** COCKY, CONFIDENT

[5]**crank** *adjective* [short for *crank-sided* (easily tipped)] (circa 1649)
of a boat **:** easily tipped by an external force

crank·case \ˈkraŋk-ˌkās\ *noun* (circa 1878)
: the housing of a crankshaft

crank out *transitive verb* (1956)
: to produce especially in a mechanical manner

crank·pin \ˈkraŋk-ˌpin\ *noun* (1839)
: the cylindrical piece which forms the handle of a crank or to which a connecting rod is attached

crank·shaft \ˈkraŋk-ˌshaft\ *noun* (1854)
1 : a shaft driven by or driving a crank
2 : a shaft consisting of a series of cranks and crankpins to which the connecting rods of an engine are attached

[1]**cranky** \ˈkraŋ-kē\ *adjective* **crank·i·er; -est** [[1]*crank*] (1821)
1 a : given to fretful fussiness **:** readily angered when opposed **:** CROTCHETY **b :** marked by eccentricity
2 : full of twists and turns **:** TORTUOUS ⟨a *cranky* road⟩
3 : working erratically **:** UNPREDICTABLE ⟨a *cranky* old tractor⟩
4 *dialect* **:** CRAZY, IMBECILE
— crank·i·ly \-kə-lē\ *adverb*
— crank·i·ness \-kē-nəs\ *noun*

[2]**cranky** *adjective* (1841) *of a boat*
: [5]CRANK

cran·nog \ˈkra ˌnòg, kra-ˈnōg\ *noun* [Scottish Gaelic *crannag* & Irish *crannóg*] (1608)
: an artificial fortified island constructed in a lake or marsh originally in prehistoric Ireland and Scotland

cran·ny \ˈkra-nē\ *noun, plural* **crannies** [Middle English *crany,* from Middle French *cren, cran* notch] (15th century)
1 : a small break or slit **:** CREVICE
2 : an obscure nook or corner
— cran·nied \-nēd\ *adjective*

cran·reuch \ˈkran-ˌrük\ *noun* [probably modification of Scottish Gaelic *crannreotha*] (1682)
Scottish **:** HOARFROST, RIME

[1]**crap** \ˈkrap\ *intransitive verb* **crapped; crap·ping** (1846)
: DEFECATE — usually considered vulgar

[2]**crap** *noun* [Middle English *crappe* chaff, residue from rendered fat, perhaps from Old French *crappe* chaff, residue, from Medieval Latin *crappa*] (circa 1897)
1 a : EXCREMENT — usually considered vulgar **b :** the act of defecating — usually considered vulgar
2 : NONSENSE, RUBBISH; *also* **:** STUFF 4b — sometimes considered vulgar

[3]**crap** *noun* [singular of *craps*] (1885)
1 — used as an attributive form of *craps* ⟨*crap* game⟩ ⟨*crap* table⟩
2 : a throw of 2, 3, or 12 in the game of craps losing the shooter his bet unless he has a point — called also *craps*; compare NATURAL

[4]**crap** *intransitive verb* **crapped; crap·ping** (circa 1930)
1 : to throw a crap
2 : to throw a seven while trying to make a point — usually used with *out*

[1]**crape** \ˈkrāp\ *noun* [alteration of French *crêpe,* from Middle French *crespe,* from *crespe* curly, from Latin *crispus* — more at CRISP] (1633)
1 : CREPE
2 : a band of crepe worn on a hat or sleeve as a sign of mourning

[2]**crape** *transitive verb* **craped; crap·ing** (1815)
: to cover or shroud with or as if with crape

[3]**crape** *transitive verb* **craped; crap·ing** [French *crêper,* from Latin *crispare,* from *crispus*] (1774)
: to make (the hair) curly

crape myrtle *noun* (1850)
: an Asian shrub (*Lagerstroemia indica*) of the loosestrife family widely grown in warm regions for its flowers

crap·per \ˈkra-pər\ *noun* [[1]*crap*] (circa 1932)
: TOILET — usually considered vulgar

crap·pie \ˈkrä-pē\ *noun* [Canadian French *crapet*] (circa 1827)
1 : BLACK CRAPPIE
2 : WHITE CRAPPIE

crap·py \ˈkra-pē\ *adjective* **crap·pi·er; -est** (1846)
slang **:** markedly inferior in quality **:** LOUSY

craps \ˈkraps\ *noun plural but singular or plural in construction* [Louisiana French, from French *crabs, craps,* from English *crabs* lowest throw at hazard, from plural of [1]*crab*] (1843)
1 : a gambling game played with two dice
2 : [3]CRAP 2

crap·shoot \ˈkrap-ˌshüt\ *noun* (1971)
: something (as a business venture) that has an unpredictable outcome

crap·shoot·er \ˈkrap-ˌshü-tər\ *noun* (1895)
: one who plays craps

crap·u·lous \ˈkra-pyə-ləs\ *adjective* [Late Latin *crapulosus,* from Latin *crapula* intoxication, from Greek *kraipalē*] (1536)
1 : marked by intemperance especially in eating or drinking
2 : sick from excessive indulgence in liquor

[1]**crash** \ˈkrash\ *verb* [Middle English *crasschen*] (15th century)
transitive verb
1 a : to break violently and noisily **:** SMASH **b :** to damage (an airplane) in landing
2 a : to cause to make a loud noise ⟨*crash* the cymbals together⟩ **b :** to force (as one's way) through with loud crashing noises
3 : to enter or attend without invitation or without paying ⟨*crash* the party⟩
intransitive verb
1 a : to break or go to pieces with or as if with violence and noise **b :** to fall, land, or hit with destructive force **c :** to decline suddenly and steeply **d** *of a computer system or program* **:** to suffer a sudden major failure usually with attendant loss of data
2 : to make a smashing noise ⟨thunder *crashing* overhead⟩
3 : to move or force one's way with or as if with a crash ⟨*crashes* into the room⟩
4 *slang* **:** to experience the aftereffects (as dysphoria or depression) of drug intoxication
5 *slang* **:** to go to bed or fall asleep; *also* **:** to reside temporarily **:** STAY ⟨*crashing* with friends for a few days⟩
— crash·er *noun*

[2]**crash** *noun* (circa 1580)
1 : a loud sound (as of things smashing) ⟨a *crash* of thunder⟩
2 a : a breaking to pieces by or as if by collision **b :** an instance of crashing ⟨a plane *crash*⟩ ⟨a system *crash*⟩

3 : a sudden decline (as of a population) or failure (as of a business) ⟨a stockmarket *crash*⟩
4 *slang* **:** the process of crashing after drug intoxication
³crash *adjective* (1945)
: marked by a concerted effort and effected in the shortest possible time ⟨a *crash* program⟩
⁴crash *noun* [probably from Russian *krashenina* colored linen] (1812)
: a coarse fabric used for draperies, toweling, and clothing and for strengthening joints of cased-in books
crash dive *noun* (1918)
: a dive made by a submarine in the least possible time
— crash–dive *intransitive verb*
crash helmet *noun* (circa 1918)
: a helmet that is worn (as by motorcyclists) as protection for the head in the event of an accident
crash·ing \'kra-shiŋ\ *adjective* (1924)
1 : UTTER, ABSOLUTE ⟨a *crashing* bore⟩
2 : SUPERLATIVE ⟨a *crashing* effect⟩
— crash·ing·ly \-lē\ *adverb*
crash–land \'krash-'land\ (1941)
transitive verb
: to land (an airplane or spacecraft) under emergency conditions usually with damage to the craft
intransitive verb
: to crash-land an airplane or spacecraft
— crash landing *noun*
crash pad *noun* (1939)
1 : protective padding (as on the inside of an automobile or a tank)
2 : a place to stay temporarily
crash·wor·thy \'krash-,wər-t͟hē\ *adjective* (1966)
: resistant to the effects of collision ⟨*crashworthy* cars⟩
— crash·wor·thi·ness *noun*
crass \'kras\ *adjective* [Latin *crassus* thick, gross] (1653)
1 a : GROSS 6a; *especially* **:** having or indicating such grossness of mind as precludes delicacy and discrimination **b :** being beneath one's dignity ⟨*crass* concerns of daily life⟩ **c** — used as a pejorative intensifier ⟨*crass* flattery⟩ ⟨*crass* propaganda⟩
2 : guided by or indicative of base or materialistic values ⟨*crass* commercialism⟩ ⟨*crass* measures of success⟩
synonym see STUPID
— crass·ly *adverb*
— crass·ness *noun*
cras·si·tude \'kra-sə-,tüd, -,tyüd\ *noun* (1679)
: the quality or state of being crass **:** GROSSNESS; *also* **:** an instance of grossness
-crat *noun combining form* [French *-crate,* back-formation from *-cratie* -cracy]
1 : advocate or partisan of a (specified) theory of government ⟨theo*crat*⟩
2 : member of a (specified) dominant class ⟨pluto*crat*⟩
— -cratic *adjective combining form*
cratch \'krach\ *noun* [Middle English *cracche,* from Old French *creche* manger — more at CRÈCHE] (13th century)
1 *archaic* **:** MANGER
2 : a crib or rack especially for fodder; *also* **:** FRAME
¹crate \'krāt\ *noun* [Middle English, from Latin *cratis*] (15th century)
1 : an open box of wooden slats or a usually wooden protective case or framework for shipping
2 : JALOPY
²crate *transitive verb* **crat·ed; crat·ing** (1871)
: to pack in a crate
¹cra·ter \'krā-tər\ *noun* [Latin, mixing bowl, crater, from Greek *kratēr,* from *kerannynai* to mix; akin to Sanskrit *śrīṇāti* he mixes] (1613)
1 \'krā-tər\ **a :** the bowl-shaped depression around the orifice of a volcano **b :** a depression formed by the impact of a meteorite **c :** a hole in the ground made by the explosion of a bomb or shell **d :** an eroded lesion **e :** a dimple in a painted surface
2 \'krā-tər, krä-'ter\ **:** KRATER
— cra·ter·like \'krā-tər-,līk\ *adjective*
²crater (1884)
intransitive verb
: to exhibit or form craters
transitive verb
: to form craters in
cra·ter·let \'krā-tər-lət\ *noun* (1881)
: a small crater
C ration *noun* (1942)
: a canned field ration of the U.S. Army
cra·ton \'krā-,tän, 'kra-\ *noun* [German *Kraton,* modification of Greek *kratos* strength — more at HARD] (1944)
: a stable relatively immobile area of the earth's crust that forms the nuclear mass of a continent or the central basin of an ocean
— cra·ton·ic \krə-'tä-nik, krā-, kra-\ *adjective*
craunch \'krönch, 'kränch\ *verb* [probably imitative] (1631)
: CRUNCH
— craunch *noun*
cra·vat \krə-'vat\ *noun* [French *cravate,* from *Crabate, Cravate* Croatian] (circa 1656)
1 : a band or scarf worn around the neck
2 : NECKTIE
crave \'krāv\ *verb* **craved; crav·ing** [Middle English, from Old English *crafian;* akin to Old Norse *krefja* to crave, demand] (before 12th century)
transitive verb
1 : to ask for earnestly **:** BEG, DEMAND ⟨*crave* a pardon for neglect⟩
2 a : to want greatly **:** NEED ⟨*craves* drugs⟩ **b :** to yearn for ⟨*crave* a vanished youth⟩
intransitive verb
: to have a strong or inward desire ⟨*craves* after affection⟩
synonym see DESIRE
— crav·er *noun*
cra·ven \'krā-vən\ *adjective* [Middle English *cravant*] (13th century)
1 *archaic* **:** DEFEATED, VANQUISHED
2 : lacking the least bit of courage **:** contemptibly fainthearted
synonym see COWARDLY
— craven *noun*
— cra·ven·ly *adverb*
— cra·ven·ness \-vən-nəs\ *noun*
crav·ing \'krā-viŋ\ *noun* (1633)
: an intense, urgent, or abnormal desire or longing
craw \'krö\ *noun* [Middle English *crawe,* from (assumed) Old English *cræga;* perhaps akin to Latin *vorare* to devour — more at VORACIOUS] (14th century)
1 : the crop of a bird or insect
2 : the stomach especially of a lower animal
craw·dad \'krö-,dad\ *noun* [alteration of *crawfish*] (circa 1905)
: CRAYFISH 1 — used chiefly west of the Appalachians
¹craw·fish \'krö-,fish\ *noun* [by folk etymology from Middle English *crevis, kraveys*] (1624)
1 : CRAYFISH 1
2 : SPINY LOBSTER
²crawfish *intransitive verb* (1842)
: to retreat from a position **:** back out
¹crawl \'kröl\ *verb* [Middle English, from Old Norse *krafla*] (14th century)
intransitive verb
1 : to move slowly in a prone position without or as if without the use of limbs ⟨the snake *crawled* into its hole⟩
2 : to move or progress slowly or laboriously ⟨traffic *crawls* along at 10 miles an hour⟩
3 : to advance by guile or servility ⟨*crawling* into favor by toadying to his boss⟩
4 : to spread by extending stems or tendrils
5 a : to be alive or swarming with or as if with creeping things ⟨a kitchen *crawling* with ants⟩ **b :** to have the sensation of insects creeping over one ⟨the story made her flesh *crawl*⟩
6 : to fail to stay evenly spread — used of paint, varnish, or glaze
transitive verb
1 : to move upon in or as if in a creeping manner
2 : to reprove harshly ⟨they got no good right to *crawl* me for what I wrote —Marjorie K. Rawlings⟩
²crawl *noun* (1818)
1 a : the act or action of crawling **b :** slow or laborious progress **c** *chiefly British* **:** a going from one pub to another
2 : a prone speed swimming stroke consisting of alternating overarm strokes and a flutter kick
3 : lettering that moves vertically or horizontally across a television or motion-picture screen to give information (as performer credits or news bulletins)
crawl·er \'krö-lər\ *noun* (circa 1629)
1 : one that crawls
2 : a vehicle (as a crane) that travels on endless chain belts
crawl space *noun* (1946)
: a shallow unfinished space beneath the first floor or under the roof of a building especially for access to plumbing or wiring
crawl·way \'kröl-,wā\ *noun* (1909)
: a low passageway (as in a cave) that can be traversed only by crawling
crawly \'krö-lē\ *adjective* (1857)
1 : CREEPY 1
2 : marked by crawling or slow motion ⟨*crawly* creatures⟩
cray·fish \'krā-,fish\ *noun* [by folk etymology from Middle English *crevis,* from Middle French *crevice,* of Germanic origin; akin to Old High German *krebiz* crab — more at CRAB] (14th century)
1 : any of numerous freshwater decapod crustaceans (superfamily Astacoidea) resembling the lobster but usually much smaller
2 : SPINY LOBSTER

crayfish 1

¹cray·on \'krā-,än, -ən *also* 'kran\ *noun* [French, crayon, pencil, from diminutive of *craie* chalk, from Latin *creta*] (1644)
1 : a stick of white or colored chalk or of colored wax used for writing or drawing
2 : a crayon drawing
²crayon *transitive verb* (1662)
: to draw with a crayon
— cray·on·ist \'krā-ə-nist\ *noun*
¹craze \'krāz\ *verb* **crazed; craz·ing** [Middle English *crasen* to crush, craze, of Scandinavian origin; akin to Old Swedish *krasa* to crush] (14th century)
transitive verb
1 *obsolete* **:** BREAK, SHATTER
2 : to produce minute cracks on the surface or glaze of
3 : to make insane or as if insane ⟨*crazed* by pain and fear⟩
intransitive verb
1 *archaic* **:** SHATTER, BREAK
2 : to become insane
3 : to develop a mesh of fine cracks
²craze *noun* (1813)
1 : an exaggerated and often transient enthusiasm **:** MANIA
2 : a crack in a surface or coating (as of glaze or enamel)
synonym see FASHION
¹cra·zy \'krā-zē\ *adjective* **cra·zi·er; -est** (1583)

1 a : full of cracks or flaws **:** UNSOUND **b :** CROOKED, ASKEW
2 a : MAD, INSANE **b** (1) **:** IMPRACTICAL (2) **:** ERRATIC **c :** being out of the ordinary **:** UNUSUAL ⟨a taste for *crazy* hats⟩
3 a : distracted with desire or excitement ⟨a thrill-*crazy* mob⟩ **b :** absurdly fond **:** INFATUATED ⟨he's *crazy* about the girl⟩ **c :** passionately preoccupied **:** OBSESSED ⟨*crazy* about boats⟩
— cra·zi·ly \-zə-lē\ *adverb*
— cra·zi·ness \-zē-nəs\ *noun*
— like crazy : to an extreme degree ⟨everyone dancing *like crazy*⟩

[2]crazy *noun, plural* **cra·zies** (1867)
: one who is or acts crazy; *especially* **:** such a one associated with a radical or extremist political cause

crazy bone *noun* (1876)
: FUNNY BONE

crazy quilt *noun* (1886)
1 : a patchwork quilt without a design
2 : JUMBLE, HODGEPODGE ⟨a *crazy quilt* of regulations⟩

cra·zy·weed \'krā-zē-,wēd\ *noun* (circa 1889)
: LOCOWEED

C–re·ac·tive protein \'sē-rē-'ak-tiv-\ *noun* [*C-polysaccharide,* a polysaccharide found in the cell wall of pneumococci and precipitated by this protein, from *carbohydrate*] (1955)
: a protein present in blood serum in various abnormal states (as inflammation or neoplasia)

[1]creak \'krēk\ *intransitive verb* [Middle English *creken* to croak, of imitative origin] (1583)
: to make a prolonged grating or squeaking sound often as a result of being worn-out; *also* **:** to proceed slowly with or as if with creaking wheels ⟨the story *creaks* along to a dull conclusion⟩

[2]creak *noun* (1605)
: a rasping or grating noise

creaky \'krē-kē\ *adjective* **creak·i·er; -est** (1834)
1 : marked by creaking **:** SQUEAKY ⟨*creaky* shoes⟩
2 : showing signs of deterioration or decrepitude ⟨a *creaky* old house⟩ ⟨a *creaky* economy⟩
— creak·i·ly \-kə-lē\ *adverb*
— creak·i·ness \-nəs\ *noun*

[1]cream \'krēm\ *noun, often attributive* [Middle English *creime, creme,* from Middle French *craime, cresme,* from Late Latin *cramum,* of Celtic origin; akin to Welsh *cramen* scab, crust] (14th century)
1 : the yellowish part of milk containing from 18 to about 40 percent butterfat
2 a : a food prepared with cream **b :** something having the consistency of cream; *especially* **:** a usually emulsified medicinal or cosmetic preparation
3 : the choicest part **:** BEST ⟨the *cream* of the crop⟩
4 : CREAMER 1
5 a : a pale yellow **b :** a cream-colored animal
— cream·i·ly \'krē-mə-lē\ *adverb*
— cream·i·ness \-mē-nəs\ *noun*
— creamy \-mē\ *adjective*

[2]cream (1596)
intransitive verb
1 : to form cream or a surface layer like the cream on standing milk
2 : to break into or cause something to break into a creamy froth; *also* **:** to move like froth
transitive verb
1 a : SKIM 1c **b :** to remove (something choice) from an aggregate ⟨she has *creamed* off her favorite stories from her earlier books —*Times Literary Supplement*⟩
2 : to furnish, prepare, or treat with cream; *also* **:** to dress with a cream sauce
3 a : to beat into a creamy froth **b :** to work or blend to the consistency of cream ⟨*cream* butter and sugar together⟩ **c :** to cause to form a surface layer of or like cream
4 a : to defeat decisively ⟨was *creamed* in the first round⟩ **b :** WRECK ⟨*creamed* the car on the turnpike⟩

cream cheese *noun* (1583)
: a mild soft unripened cheese made from whole sweet milk enriched with cream

cream·cups \'krēm-,kəps\ *noun plural but singular or plural in construction* (1888)
: a California annual (*Platystemon californicus*) of the poppy family

cream·er \'krē-mər\ *noun* (1858)
1 : a small vessel for serving cream
2 : a nondairy product used as a substitute for cream (as in coffee)

cream·ery \'krēm-rē, 'kre-mə-\ *noun, plural* **-er·ies** (1872)
: an establishment where butter and cheese are made or where milk and cream are prepared or sold

cream of tartar (1662)
: a white crystalline salt $C_4H_5KO_6$ used especially in baking powder and in certain treatments of metals

cream puff *noun* (1880)
1 : a round shell of light pastry filled with whipped cream or a cream filling
2 : an ineffectual person
3 : something trifling, inconsiderable, or easily dealt with
4 : a usually used motor vehicle that is in especially good condition

cream soda *noun* (1854)
: a carbonated soft drink flavored with vanilla

cream·ware \'krēm-,war, -,wer\ *noun* (1780)
: earthenware having a cream-colored glaze

[1]crease \'krēs\ *noun* [probably alteration of earlier *creaste,* from Middle English *creste* crest] (1578)
1 : a line, mark, or ridge made by or as if by folding a pliable substance
2 : a specially marked area in various sports; *especially* **:** an area surrounding or in front of a goal (as in lacrosse or hockey)
— crease·less \-ləs\ *adjective*

[2]crease *verb* **creased; creas·ing** (1588)
transitive verb
1 : to make a crease in or on **:** WRINKLE ⟨a smile *creased* her face⟩
2 : to wound slightly especially by grazing
intransitive verb
: to become creased
— creas·er *noun*

[1]cre·ate \krē-'āt, 'krē-,\ *verb* **cre·at·ed; cre·at·ing** [Middle English, from Latin *creatus,* past participle of *creare;* akin to Latin *crescere* to grow — more at CRESCENT] (14th century)
transitive verb
1 : to bring into existence ⟨God *created* the heaven and the earth —Genesis 1:1 (Authorized Version)⟩
2 a : to invest with a new form, office, or rank ⟨was *created* a lieutenant⟩ **b :** to produce or bring about by a course of action or behavior ⟨her arrival *created* a terrible fuss⟩ ⟨*create* new jobs⟩
3 : CAUSE, OCCASION ⟨famine *creates* high food prices⟩
4 a : to produce through imaginative skill ⟨*create* a painting⟩ **b :** DESIGN ⟨*creates* dresses⟩
intransitive verb
: to make or bring into existence something new

[2]create *adjective* (15th century)
archaic **:** CREATED

cre·a·tine \'krē-ə-,tēn, -tᵊn\ *noun* [French *créatine,* from Greek *kreat-, kreas* flesh — more at RAW] (1840)
: a white crystalline nitrogenous substance $C_4H_9N_3O_2$ found especially in the muscles of vertebrates either free or as phosphocreatine

creatine phosphate *noun* (1945)
: PHOSPHOCREATINE

cre·at·i·nine \krē-'a-tᵊn-,ēn, -ən\ *noun* [German *Kreatinin,* from *Kreatin* creatine] (1851)
: a white crystalline strongly basic compound $C_4H_7N_3O$ formed from creatine and found especially in muscle, blood, and urine

cre·a·tion \krē-'ā-shən\ *noun* (14th century)
1 : the act of creating; *especially* **:** the act of bringing the world into ordered existence
2 : the act of making, inventing, or producing: as **a :** the act of investing with a new rank or office **b :** the first representation of a dramatic role
3 : something that is created: as **a :** WORLD **b :** creatures singly or in aggregate **c :** an original work of art **d :** a new usually striking article of clothing

cre·a·tion·ism \-shə-,ni-zəm\ *noun* (1880)
: a doctrine or theory holding that matter, the various forms of life, and the world were created by God out of nothing and usually in the way described in Genesis — compare EVOLUTION 4b
— cre·a·tion·ist \-shə-nist\ *noun or adjective*

creation science *noun* (1979)
: CREATIONISM; *also* **:** scientific evidence or arguments put forth in support of creationism

cre·a·tive \krē-'ā-tiv, 'krē-,\ *adjective* (1678)
1 : marked by the ability or power to create **:** given to creating ⟨the *creative* impulse⟩
2 : having the quality of something created rather than imitated **:** IMAGINATIVE ⟨the *creative* arts⟩
3 : managed so as to get around legal or conventional limits ⟨*creative* financing⟩; *also* **:** deceptively arranged so as to conceal or defraud ⟨*creative* accounting⟩
— cre·a·tive·ly *adverb*
— cre·a·tive·ness *noun*

creative evolution *noun* (1909)
: evolution that is a creative product of a vital force rather than a spontaneous process explicable in terms of scientific laws — compare EMERGENT EVOLUTION

cre·a·tiv·i·ty \,krē-(,)ā-'ti-və-tē, ,krē-ə-\ *noun* (1875)
1 : the quality of being creative
2 : the ability to create

cre·a·tor \krē-'ā-tər\ *noun* (13th century)
: one that creates usually by bringing something new or original into being; *especially, capitalized* **:** GOD 1

crea·ture \'krē-chər\ *noun* [Middle English, from Old French, from Late Latin *creatura,* from Latin *creatus,* past participle of *creare*] (14th century)
1 : something created either animate or inanimate: as **a :** a lower animal; *especially* **:** a farm animal **b :** a human being **c :** a being of anomalous or uncertain aspect or nature ⟨*creatures* of fantasy⟩
2 : one that is the servile dependent or tool of another **:** INSTRUMENT
— crea·tur·al \'krē-chə-rəl\ *adjective*
— crea·ture·hood \'krē-chər-,hu̇d\ *noun*
— crea·ture·li·ness \-lē-nəs\ *noun*
— crea·ture·ly \-lē\ *adjective*

creature comfort *noun* (1659)
: something (as food, warmth, or special accommodations) that gives bodily comfort

crèche \'kresh, 'krāsh\ *noun* [French, from Old French *creche* manger, crib, of Germanic origin; akin to Old High German *krippa* manger — more at CRIB] (1792)
1 : a representation of the Nativity scene
2 : DAY NURSERY
3 : a foundling hospital

cre·dence \'krē-dᵊn(t)s\ *noun* [Middle English, from Middle French or Medieval Latin;

Middle French, from Medieval Latin *credentia,* from Latin *credent-, credens,* present participle of *credere* to believe, trust — more at CREED] (14th century)
1 a : mental acceptance as true or real ⟨give *credence* to gossip⟩ **b :** CREDIBILITY 1 ⟨lends *credence* to the theory⟩
2 : CREDENTIALS — used in the phrase *letters of credence*
3 [Middle French, from Old Italian *credenza*] **:** a Renaissance sideboard used chiefly for valuable plate
4 : a small table where the bread and wine rest before consecration
synonym see BELIEF

cre·dent \'krē-dᵊnt\ *adjective* [Latin *credent-, credens,* present participle] (1602)
1 *archaic* **:** giving credence **:** CONFIDING
2 *obsolete* **:** CREDIBLE

¹cre·den·tial \kri-'den(t)-shəl\ *adjective* (15th century)
: warranting credit or confidence — used chiefly in the phrase *credential letters*

²credential *noun* (circa 1674)
1 : something that gives a title to credit or confidence
2 *plural* **:** testimonials showing that a person is entitled to credit or has a right to exercise official power
3 : CERTIFICATE, DIPLOMA

³credential *transitive verb* **-tialed** *also* **-tialled; -tial·ing** *also* **-tial·ling** (1888)
: to furnish with credentials ⟨to *credential* adequate academic performance —K. Patricia Cross⟩

cre·den·tial·ism \-chə-ˌli-zəm\ *noun* (1967)
: undue emphasis on credentials (as college degrees) as prerequisites to employment

cre·den·za \kri-'den-zə\ *noun* [Italian, literally, belief, confidence, from Medieval Latin *credentia*] (1880)
1 : CREDENCE 3
2 : a sideboard, buffet, or bookcase patterned after a Renaissance credence; *especially* **:** one without legs ◆

cred·i·bil·i·ty \ˌkre-də-'bi-lə-tē\ *noun* (1594)
1 : the quality or power of inspiring belief ⟨an account lacking in *credibility*⟩
2 : capacity for belief ⟨strains her reader's *credibility* —*Times Literary Supplement*⟩

credibility gap *noun* (1966)
1 a : lack of trust ⟨a *credibility gap* between generations⟩ **b :** lack of believability ⟨a *credibility gap* created by contradictory official statements —Samuel Ellenport⟩
2 : DISCREPANCY ⟨the *credibility gap* between the professed ideals . . . and their actual practices —Jeanne L. Noble⟩

cred·i·ble \'kre-də-bəl\ *adjective* [Middle English, from Latin *credibilis,* from *credere*] (14th century)
1 : offering reasonable grounds for being believed ⟨a *credible* account of an accident⟩ ⟨*credible* witnesses⟩
2 : of sufficient capability to be militarily effective ⟨a *credible* deterrent⟩ ⟨*credible* forces⟩
— cred·i·bly \'kre-də-blē\ *adverb*

¹cred·it \'kre-dit\ *noun* [Middle French, from Old Italian *credito,* from Latin *creditum* something entrusted to another, loan, from neuter of *creditus,* past participle of *credere* to believe, entrust — more at CREED] (1537)
1 : reliance on the truth or reality of something ⟨gave *credit* to everything he said⟩
2 a : the balance in a person's favor in an account **b :** an amount or sum placed at a person's disposal by a bank **c :** time given for payment for goods or services sold on trust ⟨long-term *credit*⟩ **d** (1) **:** an entry on the right-hand side of an account constituting an addition to a revenue, net worth, or liability account (2) **:** a deduction from an expense or asset account **e :** any one of or the sum of the items entered on the right-hand side of an account **f :** a deduction from an amount otherwise due
3 a : influence or power derived from enjoying the confidence of another or others **b :** good name **:** ESTEEM; *also* **:** financial or commercial trustworthiness
4 *archaic* **:** CREDIBILITY
5 : a source of honor ⟨a *credit* to the school⟩
6 a : something that gains or adds to reputation or esteem **:** HONOR ⟨took no *credit* for his kindly act⟩ **b :** RECOGNITION, ACKNOWLEDGMENT ⟨quite willing to accept undeserved *credit*⟩
7 : recognition by name of a person contributing to a performance (as a film or telecast) ⟨the opening *credits*⟩
8 a : recognition by a school or college that a student has fulfilled a requirement leading to a degree **b :** CREDIT HOUR
synonym see BELIEF, INFLUENCE

²credit *transitive verb* [partly from ¹*credit;* partly from Latin *creditus,* past participle] (1541)
1 : to supply goods on credit to
2 : to trust in the truth of **:** BELIEVE
3 *archaic* **:** to bring credit or honor upon
4 a : to enter upon the credit side of an account **b :** to place an amount to the credit of ⟨*credit* his account with ten dollars⟩
5 a : to consider usually favorably as the source, agent, or performer of an action or the possessor of a trait ⟨*credits* him with an excellent sense of humor⟩ **b :** to attribute to some person ⟨they *credit* the invention to him⟩
synonym see ASCRIBE

cred·it·able \'kre-di-tə-bəl\ *adjective* (1526)
1 : worthy of belief
2 : sufficiently good to bring esteem or praise
3 : worthy of commercial credit
4 : capable of being assigned
— cred·it·abil·i·ty \ˌkre-di-tə-'bi-lə-tē\ *noun*
— cred·it·able·ness \'kre-di-tə-bəl-nəs\ *noun*
— cred·it·ably \-blē\ *adverb*

credit card *noun* (1888)
: a card authorizing purchases on credit

credit hour *noun* (circa 1927)
: the unit of measuring educational credit usually based on the number of classroom hours per week throughout a term

credit line *noun* (1914)
1 : a line, note, or name that acknowledges the source of an item (as a news dispatch or television program)
2 : LINE OF CREDIT

cred·i·tor \'kre-di-tər, -ˌtȯr\ *noun* (15th century)
: one to whom a debt is owed; *especially* **:** a person to whom money or goods are due

credit union *noun* (1881)
: a cooperative association that makes small loans to its members at low interest rates

cred·it·wor·thy \'kre-dit-ˌwər-t͟hē\ *adjective* (1924)
: financially sound enough to justify the extension of credit
— cred·it·wor·thi·ness *noun*

cre·do \'krē-(ˌ)dō, 'krā-\ *noun, plural* **credos** [Middle English, from Latin, I believe] (12th century)
: CREED

cre·du·li·ty \kri-'dü-lə-tē, -'dyü-\ *noun* (15th century)
: readiness or willingness to believe especially on slight or uncertain evidence

cred·u·lous \'kre-jə-ləs\ *adjective* [Latin *credulus,* from *credere*] (1576)
1 : ready to believe especially on slight or uncertain evidence
2 : proceeding from credulity
— cred·u·lous·ly *adverb*
— cred·u·lous·ness *noun*

Cree \'krē\ *noun, plural* **Cree** *or* **Crees** [Canadian French *Cris,* plural, short for *Cristinaux,* from Ojibwa dialect *kirištino·,* singular, member of a band living south of James Bay] (1744)
1 : a member of an American Indian people of Quebec, Ontario, Manitoba, and Saskatchewan
2 : the Algonquian language of the Cree people

creed \'krēd\ *noun* [Middle English *crede,* from Old English *crēda,* from Latin *credo* (first word of the Apostles' and Nicene Creeds), from *credere* to believe, trust, entrust; akin to Old Irish *cretid* he believes, Sanskrit *śrad-dadhāti*] (before 12th century)
1 : a brief authoritative formula of religious belief
2 : a set of fundamental beliefs; *also* **:** a guiding principle
— creed·al *or* **cre·dal** \'krē-dᵊl\ *adjective*

creek \'krēk, 'krik\ *noun* [Middle English *crike, creke,* from Old Norse *-kriki* bend] (13th century)
1 *chiefly British* **:** a small inlet or bay narrower and extending farther inland than a cove
2 : a natural stream of water normally smaller than and often tributary to a river
3 *archaic* **:** a narrow or winding passage
— up the creek : in a difficult or perplexing situation

Creek \'krēk\ *noun* (1725)
1 : an American Indian confederacy of peoples chiefly of Muskogean stock of Alabama, Georgia, and Florida
2 : a member of any of the Creek peoples
3 : the Muskogean language of the Creek Indians

¹creel \'krē(ə)l\ *noun* [Middle English *creille, crele*] (14th century)
1 : a wicker basket (as for carrying newly caught fish)
2 : a bar with skewers for holding bobbins in a spinning machine

²creel *transitive verb* (1844)
: to put (caught fish) in a creel

¹creep \'krēp\ *intransitive verb* **crept** \'krept\; **creep·ing** [Middle English *crepen,* from Old English *crēopan;* akin to Old Norse *krjūpa* to creep] (before 12th century)
1 a : to move along with the body prone and close to the ground **b :** to move slowly on hands and knees
2 a : to go very slowly ⟨the hours *crept* by⟩ **b :** to go timidly or cautiously so as to escape notice ⟨she *crept* away from the festive scene⟩ **c :** to enter or advance gradually so as to be almost unnoticed ⟨age *creeps* up on us⟩ ⟨a note of irritation *crept* into her voice⟩
3 : to have the sensation of being covered with creeping things ⟨the thought made his flesh *creep*⟩
4 *of a plant* **:** to spread or grow over a surface rooting at intervals or clinging with tendrils, stems, or aerial roots
5 a : to slip or gradually shift position **b :** to change shape permanently from prolonged stress or exposure to high temperatures

²creep *noun* (1818)

◇ WORD HISTORY

credenza In Italian *credenza* literally means "belief" or "confidence," the latter being just what a member of a royal or noble household needed to have in his food and drink during the Middle Ages and Renaissance. Being poisoned by one's enemies, employees, or even one's own family was an ever-present danger during those times. As a result, it became customary for royalty and noblemen to have a servant taste their food and drink after it had left the kitchen and just before it was served. The tasting was accomplished at a dining room sideboard reserved for the purpose. If the servant survived the tasting, the master had every reason to believe that he would also. The name for the sideboard became *credenza* in Italian. Sideboards and buffets of Italian Renaissance style are still called credenzas.

1 : a movement of or like creeping ⟨traffic moving at a *creep*⟩
2 : a distressing sensation like that caused by the creeping of insects over one's flesh; *especially* **:** a feeling of apprehension or horror — usually used in plural with *the* ⟨that gives me the *creeps*⟩
3 : a feed trough or enclosure that young animals can enter while adults are excluded
4 : the slow change of dimensions of an object from prolonged exposure to high temperature or stress
5 : an unpleasant or obnoxious person

creep·age \ˈkrē-pij\ *noun* (1903)
: gradual movement **:** CREEP

creep·er \ˈkrē-pər\ *noun* (before 12th century)
1 : one that creeps: as **a :** a creeping plant **b :** a bird (as of the family Certhiidae) that creeps about on trees or bushes searching for insects **c :** a creeping insect or reptile
2 : any of various devices used for creeping: as **a :** a fixture with iron points worn on the shoe to prevent slipping **b :** a low wheeled platform for supporting the body when working under an automobile
3 : a usually one-piece garment for a child at the crawling age

creep·ing \ˈkrē-piŋ\ *adjective* (14th century)
: developing or advancing by slow imperceptible degrees ⟨a period of *creeping* inflation⟩

creeping eruption *noun* (1926)
: a skin disorder marked by a spreading red line of eruption and caused especially by larvae (as of hookworms not normally parasitic in humans) burrowing beneath the human skin

creepy \ˈkrē-pē\ *adjective* **creep·i·er; -est** (1883)
1 : producing a nervous shivery apprehension ⟨*creepy* things were crawling over us⟩ ⟨a *creepy* horror story⟩
2 : of, relating to, or being a creep **:** annoyingly unpleasant
— creep·i·ly \-pə-lē\ *adverb*
— creep·i·ness \-pē-nəs\ *noun*

creese *variant of* KRIS

cre·mains \kri-ˈmānz\ *noun plural* [blend of *cremated* and *remains*] (1947)
: the ashes of a cremated human body

cre·mate \ˈkrē-ˌmāt, kri-ˈ\ *transitive verb* **cre·mat·ed; cre·mat·ing** [Latin *crematus,* past participle of *cremare* to burn up, cremate] (1874)
: to reduce (as a dead body) to ashes by burning
— cre·ma·tion \kri-ˈmā-shən\ *noun*

cre·ma·to·ri·um \ˌkrē-mə-ˈtōr-ē-əm, ˌkre-, -ˈtȯr-\ *noun, plural* **-ri·ums** *or* **-ria** \-ē-ə\ (1880)
: CREMATORY

cre·ma·to·ry \ˈkrē-mə-ˌtōr-ē, ˈkre-, -ˌtȯr-\ *noun, plural* **-ries** (1876)
: a furnace for cremating; *also* **:** an establishment containing such a furnace
— crematory *adjective*

crème *or* **creme** \ˈkrem *also* ˈkrēm *or* ˈkrām\ *noun, plural* **crèmes** *or* **cremes** \ˈkrem(z), ˈkrēmz\ [French, from Old French *cresme* — more at CREAM] (circa 1821)
1 : a sweet liqueur
2 : cream or a preparation made with or resembling cream used in cooking
3 : CREAM 2b

crème an·glaise \-äŋ-ˈglāz, -aŋ-, -äⁿ-ˈglez\ *noun* [French, literally, English cream] (1975)
: a vanilla-flavored custard sauce usually served with desserts

crème brû·lée \-brue-ˈlā, -brü-\ *noun* [French, literally, scorched cream] (1886)
: a rich custard topped with caramelized sugar

crème car·a·mel \-kär-ä-ˈmel, *or see* CARAMEL\ *noun* [French, literally, caramel cream] (1906)
: a custard that has been baked with caramel sauce

crème de ca·cao \ˈkrēm-də-ˈkō-(ˌ)kō; ˌkrem-də-kə-ˈkau̇, -kə-ˈkā-(ˌ)ō\ *noun* [French, literally, cream of cacao] (1904)
: a sweet brown or colorless liqueur flavored with cacao beans and vanilla

crème de la crème \ˈkrem-də-lä-ˈkrem, -lə-\ *noun* [French, literally, cream of the cream] (1848)
: the very best

crème de menthe \ˈkrem-də-ˈmen(t)th, ˈkrēm-, -ˈmint, ˈkrem-də-ˈmänt\ *noun* [French, literally, cream of mint] (1901)
: a sweet green or colorless mint-flavored liqueur

crème fraîche *or* **crème fraiche** \ˈkrem-ˈfresh *also* ˈkrēm- *or* ˈkrām-\ *noun* [French, literally, fresh cream] (1950)
: heavy cream thickened and slightly soured with buttermilk and often served on fruit

cre·nate \ˈkrē-ˌnāt\ *or* **cre·nat·ed** \-ˌnā-təd\ *adjective* [New Latin *crenatus,* from Medieval Latin *crena* notch] (1794)
: having the margin or surface cut into rounded scallops ⟨a *crenate* leaf⟩

cre·na·tion \kri-ˈnā-shən\ *noun* (1846)
1 a : a crenate formation; *especially* **:** one of the rounded projections on an edge (as of a coin) **b :** the quality or state of being crenate
2 : shrinkage of red blood cells resulting in crenate margins

cren·el·lat·ed *or* **cren·el·at·ed** \ˈkre-nᵊl-ˌā-təd\ *adjective* [French *créneler* to furnish with embrasures, from Old French *querneler,* from *kernel, crenel* embrasure, diminutive of *cren* notch, from *crener* to notch; akin to Medieval Latin *crena*] (circa 1823)
: having crenellations

cren·el·la·tion *also* **cren·el·a·tion** \ˌkre-nᵊl-ˈā-shən\ *noun* (1849)
1 : BATTLEMENT
2 : any of the embrasures alternating with merlons in a battlement — see BATTLEMENT illustration

cren·u·lat·ed \ˈkren-yə-ˌlā-təd\ *or* **cren·u·late** \-lət, -ˌlāt\ *adjective* [New Latin *crenulatus,* from *crenula,* diminutive of Medieval Latin *crena*] (1794)
: having an irregularly wavy or serrate outline ⟨a *crenulated* shoreline⟩
— cren·u·la·tion \ˌkren-yə-ˈlā-shən\ *noun*

cre·ole \ˈkrē-ˌōl\ *adjective* (1748)
1 *often capitalized* **:** of or relating to Creoles or their language
2 *often capitalized* **:** relating to or being highly seasoned food typically prepared with rice, okra, tomatoes, and peppers ⟨shrimp *creole*⟩

Cre·ole \ˈkrē-ˌōl\ *noun* [French *créole,* from Spanish *criollo,* from Portuguese *crioulo* white person born in the colonies] (1737)
1 : a person of European descent born especially in the West Indies or Spanish America
2 : a white person descended from early French or Spanish settlers of the U.S. Gulf states and preserving their speech and culture
3 : a person of mixed French or Spanish and black descent speaking a dialect of French or Spanish
4 a : a language evolved from pidginized French that is spoken by blacks in southern Louisiana **b :** HAITIAN **c** *not capitalized* **:** a language that has evolved from a pidgin but serves as the native language of a speech community

cre·ol·ise *British variant of* CREOLIZE

cre·ol·ize \ˈkrē-ə-ˌlīz, ˈkrē-ˌō-\ *transitive verb* **-ized; -iz·ing** (circa 1932)
: to cause (a pidginized language) to become a creole in a speech community
— cre·ol·i·za·tion \ˌkrē-ə-lə-ˈzā-shən, ˌkrē-(ˌ)ō-\ *noun*

¹cre·o·sote \ˈkrē-ə-ˌsōt\ *noun* [German *Kreosot,* from Greek *kreas* flesh + *sōtēr* preserver, from *sōzein* to preserve, from *sōs* safe (probably akin to Sanskrit *tavīti* he is strong); from its antiseptic properties — more at RAW] (1835)
1 : a clear or yellowish flammable oily liquid mixture of phenolic compounds obtained by the distillation of wood tar especially from beech wood
2 : a brownish oily liquid consisting chiefly of aromatic hydrocarbons obtained by distillation of coal tar and used especially as a wood preservative
3 : a dark brown or black flammable tar deposited from especially wood smoke on the walls of a chimney

²creosote *transitive verb* **-sot·ed; -sot·ing** (1846)
: to treat with creosote

creosote bush *noun* (1846)
: a resinous desert shrub (*Larrea tridentata*) of the caltrop family found in the southwestern U.S. and Mexico

crepe *or* **crêpe** \ˈkrāp\ *noun* [French *crêpe* — more at CRAPE] (1797)
1 : a light crinkled fabric woven of any of various fibers
2 : CRAPE 2
3 : crude rubber in the form of nearly white to brown crinkled sheets used especially for shoe soles ⟨*crepe*-soled shoes⟩
4 : a small very thin pancake
— crepe *adjective*
— crep·ey *or* **crepy** \ˈkrā-pē\ *adjective*

crepe de chine *or* **crêpe de chine** \ˌkrāp-də-ˈshēn\ *noun, often 2d C capitalized* [French *crêpe de Chine,* literally, China crepe] (1872)
: a soft fine or sheer clothing crepe especially of silk

crepe myrtle *or* **crêpe myrtle** *noun* (1916)
: CRAPE MYRTLE

crepe paper *noun* (1897)
: paper with a crinkled or puckered texture

crêpe su·zette \ˌkrāp-sü-ˈzet\ *noun, plural* **crêpes suzette** \ˌkrāp(s)-sü-ˈzet\ *or* **crêpe suzettes** \ˌkrap-su-ˈzets\ *often S capitalized* [French *crêpe Suzette,* probably from *Suzette,* nickname of Suzanne Reichenberg (died 1924) French actress] (1922)
: a thin folded or rolled pancake in a hot orange-butter sauce that is sprinkled with a liqueur (as cognac or curaçao) and set ablaze for serving

crep·i·tant \ˈkre-pə-tənt\ *adjective* (1855)
: having or making a crackling sound ⟨*crepitant* sounds in breathing⟩

crep·i·tate \ˈkre-pə-ˌtāt\ *intransitive verb* **-tat·ed; -tat·ing** [Latin *crepitatus,* past participle of *crepitare* to crackle, frequentative of *crepare* to rattle, crack] (circa 1828)
: to make a crackling sound **:** CRACKLE
— crep·i·ta·tion \ˌkre-pə-ˈtā-shən\ *noun*

cre·pon \ˈkrā-ˌpän\ *noun* [French, from *crêpe*] (circa 1864)
: a heavy crepe fabric with lengthwise crinkles

crept *past and past participle of* CREEP

cre·pus·cu·lar \kri-ˈpəs-kyə-lər\ *adjective* (1668)
1 : of, relating to, or resembling twilight **:** DIM
2 : active in the twilight ⟨*crepuscular* insects⟩

cre·pus·cule \kri-ˈpəs-(ˌ)kyü(ə)l\ *or* **cre·pus·cle** \-ˈpə-səl\ *noun* [Latin *crepusculum,* from *creper* dusky] (14th century)
: TWILIGHT

¹cre·scen·do \krə-ˈshen-(ˌ)dō\ *noun, plural* **-dos** *or* **-does** *also* **-di** \-dē\ [Italian, from *crescendo,* adjective, increasing, gerund of *crescere* to grow, increase, from Latin] (1776)
1 a : a gradual increase; *specifically* **:** a gradual increase in volume of a musical passage **b :** the peak of a gradual increase **:** CLIMAX ⟨complaints about stifling smog conditions reach a *crescendo* —*Down Beat*⟩

2 : a crescendo musical passage ■
— **crescendo** *intransitive verb*

mark indicating crescendo 2

²crescendo *adverb or adjective* (1859)
: with an increase in volume — used as a direction in music

¹cres·cent \ˈkre-sᵊnt\ *noun* [Middle English *cressant,* from Middle French *creissant,* from present participle of *creistre* to grow, increase, from Latin *crescere;* akin to Old High German *hirsi* millet, Lithuanian *šerti* to feed, Greek *koros* boy] (15th century)
1 a : the moon at any stage between new moon and first quarter and between last quarter and the succeeding new moon when less than half of the illuminated hemisphere is visible **b :** the figure of the moon at such a stage defined by a convex and a concave edge
2 : something shaped like a crescent
— **cres·cen·tic** \kre-ˈsen-tik, krə-\ *adjective*

²crescent *adjective* [Latin *crescent-, crescens,* present participle of *crescere*] (1574)
: marked by an increase

Cres·cent \ˈkre-sᵊnt\ *trademark*
— used for an adjustable open-end wrench

cres·cive \ˈkre-siv\ *adjective* [Latin *crescere* to grow] (1566)
: marked by gradual spontaneous development
— **cres·cive·ly** *adverb*

cre·sol \ˈkrē-ˌsȯl, -ˌsōl\ *noun* [International Scientific Vocabulary, irregular from *creosote*] (circa 1869)
: any of three poisonous colorless crystalline or liquid isomeric phenols C_7H_8O

cress \ˈkres\ *noun* [Middle English *cresse,* from Old English *cærse, cressa;* akin to Old High German *kressa* cress] (before 12th century)
: any of numerous crucifers (especially genera *Lepidium* and *Nasturtium*) with moderately pungent leaves used especially in salads

cres·set \ˈkre-sət\ *noun* [Middle English, from Middle French, from Old French *craisset,* from *craisse* grease — more at GREASE] (14th century)
: an iron vessel or basket used for holding an illuminant (as oil) and mounted as a torch or suspended as a lantern

Cres·si·da \ˈkre-sə-də\ *noun*
: a Trojan woman of medieval legend who pledges herself to Troilus but while a captive of the Greeks gives herself to Diomedes

¹crest \ˈkrest\ *noun* [Middle English *creste,* from Middle French, from Latin *crista;* probably akin to Latin *crinis* hair] (14th century)
1 a : a showy tuft or process on the head of an animal and especially a bird — see BIRD illustration **b :** the plume or identifying emblem worn on a knight's helmet; *also* **:** the top of a helmet **c** (1) **:** a heraldic representation of the crest (2) **:** a heraldic device depicted above the escutcheon but not upon a helmet (3) **:** COAT OF ARMS 2a **d :** a ridge or prominence on a part of an animal body
2 : something suggesting a crest especially in being an upper prominence, edge, or limit: as **a :** PEAK; *especially* **:** the top line of a mountain or hill **b :** the ridge of a roof **c :** the top of a wave
3 a : a high point of an action or process and especially of one that is rhythmic **b :** CLIMAX, CULMINATION ⟨at the *crest* of his fame⟩
— **crest·al** \ˈkres-tᵊl\ *adjective*
— **crest·less** \-ləs\ *adjective*

²crest (15th century)
transitive verb
1 : to furnish with a crest; *also* **:** CROWN
2 : to reach the crest of ⟨*crested* the hill and looked around⟩
intransitive verb
: to rise to a crest ⟨waves *cresting* in the storm⟩

crest·ed \ˈkres-təd\ *adjective* (14th century)
: having a crest ⟨a *crested* bird⟩

crested wheatgrass *noun* (1923)
: either of two Eurasian grasses (*Agropyron cristatum* or *A. sibericum* synonym *A. desertorum*) that are grown in the U.S. for forage and for erosion control

crest·fall·en \ˈkrest-ˌfȯ-lən\ *adjective* (1589)
1 : having a drooping crest or hanging head
2 : feeling shame or humiliation **:** DEJECTED
— **crest·fall·en·ly** *adverb*
— **crest·fall·en·ness** \-lən-nəs\ *noun*

crest·ing \ˈkres-tiŋ\ *noun* (1908)
: a decorative edging or railing (as on pottery or furniture)

cre·syl \ˈkre-səl, ˈkrē-ˌsil\ *noun* [International Scientific Vocabulary *cres*ol + *-yl*] (circa 1872)
: TOLYL

Cre·ta·ceous \kri-ˈtā-shəs\ *adjective* [Latin *cretaceus* resembling chalk, from *creta* chalk] (1832)
: of, relating to, or being the last period of the Mesozoic era characterized by continued dominance of reptiles, emergent dominance of angiosperms, diversification of mammals, and the extinction of many types of organisms at the close of the period; *also* **:** of, relating to, or being the corresponding system of rocks — see GEOLOGIC TIME table
— **Cretaceous** *noun*

cre·tin \ˈkrē-tᵊn\ *noun* [French *crétin,* from French dialect *cretin,* literally, wretch, innocent victim, from Latin *christianus* Christian] (1779)
1 : one afflicted with cretinism
2 : a stupid, vulgar, or insensitive person **:** CLOD, LOUT ◆
— **cre·tin·ous** \-tᵊn-əs\ *adjective*

cre·tin·ism \-tᵊn-ˌi-zəm\ *noun* (1801)
: a usually congenital abnormal condition marked by physical stunting and mental retardation and caused by severe hypothyroidism

cre·tonne \ˈkrē-ˌtän, kri-ˈ\ *noun* [French, from *Creton,* Normandy] (1870)
: a strong cotton or linen cloth used especially for curtains and upholstery

Creutz·feldt–Ja·kob disease *also* **Creutz·feld–Ja·cob disease** \ˈkrȯits-ˌfelt-ˈyä-(ˌ)kōb-\ *noun* [Hans G. *Creutzfeldt* (died 1964) German psychiatrist and Alfons M. *Jakob* (died 1931) German psychiatrist] (1963)
: a rare progressive fatal encephalopathy caused by a slow virus and marked by development of porous brain tissue, premature dementia in middle age, and gradual loss of muscular coordination

cre·val·le \kri-ˈva-lē\ *noun* [by alteration] (1879)
: CAVALLA 1; *especially* **:** JACK CREVALLE

cre·vasse \kri-ˈvas\ *noun* [French, from Old French *crevace*] (1813)
1 : a breach in a levee
2 : a deep crevice or fissure (as in a glacier or the earth)

crev·ice \ˈkre-vəs\ *noun* [Middle English, from Middle French *crevace,* from Old French, from *crever* to break, from Latin *crepare* to crack] (14th century)
: a narrow opening resulting from a split or crack (as in a cliff) **:** FISSURE

¹crew \ˈkrü\ *chiefly British past of* CROW

²crew \ˈkrü\ *noun* [Middle English *crue,* literally, reinforcement, from Middle French *creue* increase, from *creistre* to grow — more at CRESCENT] (15th century)
1 *archaic* **:** a band or force of armed men
2 : a group of people associated together in a common activity or by common traits or interests
3 a : a company of people working on one job or under one foreman or operating a machine **b :** the whole company belonging to a ship sometimes including the officers and master; *also* **:** one who assists the skipper of a sailboat **c :** the persons who have duties on an aircraft in flight **d :** the rowers and coxswain of a racing shell; *also* **:** ROWING
— **crew·less** \-ləs\ *adjective*

³crew (1935)
intransitive verb
: to act as a member of a crew
transitive verb
: to serve as a crew member on (as a ship or aircraft)

crew cut *noun* (1942)
: a very short haircut usually for men or boys in which the hair resembles the bristle surface of a brush

crew·el \ˈkrü-əl\ *noun* [Middle English *crule*] (15th century)
1 : slackly twisted worsted yarn used for embroidery
2 : CREWELWORK

crew·el·work \-ˌwərk\ *noun* (1863)
: embroidery work with crewel

crew·man \ˈkrü-mən\ *noun* (circa 1927)
: a member of a crew

crew·mate \-ˌmāt\ *noun* (1938)
: a fellow crewman

crew neck *noun* [from the sweaters with this neckline worn by oarsmen] (1940)
1 : a round collarless neckline
2 *usually* **crew·neck** \ˈkrü-ˌnek\ **:** a sweater with a crew neck

crew sock *noun* (1948)
: a short bulky usually ribbed sock

¹crib \ˈkrib\ *noun* [Middle English, from Old English *cribb;* akin to Old High German *krippa* manger, and perhaps to Greek *griphos* reed basket] (before 12th century)
1 : a manger for feeding animals
2 : an enclosure especially of framework: as **a :** a stall for a stabled animal **b :** a small child's bedstead with high enclosing usually slatted sides **c :** any of various devices resembling a crate or framework in structure **d :** a building for storage **:** BIN

□ USAGE
crescendo Sense 1b is apparently an American extension of the meaning of *crescendo* ⟨he was ordered abroad, and their tenderness and desire reached a *crescendo* —F. Scott Fitzgerald⟩ ⟨the infrequent lamps mounted to *crescendo* beneath the arcade of a filling-station —William Faulkner⟩. A handful of critics—most recently British—have found fault with this use for not being sense 1a. It is still a minority use and can easily be avoided, but it is nonetheless a fully established meaning.

◇ WORD HISTORY
cretin Chronic iodine deficiencies in diet can result in malfunctions of the thyroid gland, which produces hormones necessary for normal human development. Some mountainous regions of the world, such as parts of the Alps, cannot naturally provide their inhabitants with a diet rich enough in iodine, and the resultant hypothyroidism causes dwarfism and severe mental retardation to become prevalent. In Franco-Provençal (the Romance speech of French Switzerland and adjacent parts of France), an individual malformed due to hypothyroidism was called a *cretin,* literally, "wretch, innocent victim," a word that etymologically meant simply "Christian" and emphasized their humanity. Borrowed into French and English, this word and its derivative *cretinism* (French *crétinisme*) have become standard medical labels for the condition. With the original benevolent sense now long lost, *cretin* has also become a general term of opprobrium.

3 a : a small narrow room or dwelling **:** HUT, SHACK **b :** a room or shack used for prostitution
4 : the cards discarded in cribbage for the dealer to use in scoring
5 a : a small theft **b :** PLAGIARISM **c :** a literal translation; *especially* **:** PONY 3 **d :** a summary and key to understanding a literary work **e :** something used for cheating in an examination
6 : CRÈCHE 1

²crib *verb* **cribbed; crib·bing** (1605)
transitive verb
1 : CONFINE, CRAMP
2 : to provide with or put into a crib; *especially* **:** to line or support with a framework of timber
3 : PILFER, STEAL; *especially* **:** PLAGIARIZE
intransitive verb
1 a : STEAL, PLAGIARIZE **b :** to use a crib **:** CHEAT
2 : to have the vice of cribbing
— crib·ber *noun*

crib·bage \'kri-bij\ *noun* [¹*crib*] (1630)
: a card game for two players in which each player tries to form various counting combinations of cards

crib·bing \'kri-biŋ\ *noun* (1841)
1 : material for use in making a crib
2 : a vice of horses in which they gnaw (as at the manger) while slobbering and salivating

crib biting *noun* (1831)
: CRIBBING 2

crib death *noun* (1965)
: SUDDEN INFANT DEATH SYNDROME

crib·ri·form \'kri-brə-,fȯrm\ *adjective* [Latin *cribrum* sieve; akin to Latin *cernere* to sift — more at CERTAIN] (1741)
: pierced with small holes

cri·ce·tid \krī-'sē-təd, -'se-\ *noun* [ultimately from New Latin *Cricetus,* genus name, of Slavic origin; akin to Czech *křeček* hamster] (1960)
: any of a family (Cricetidae) of small rodents including the hamsters
— cricetid *adjective*

¹crick \'krik\ *noun* [Middle English *cryk*] (15th century)
: a painful spasmodic condition of muscles (as of the neck or back)

²crick *transitive verb* (1884)
1 : to cause a crick in (as the neck)
2 : to turn or twist (as the head) especially into a strained position

¹crick·et \'kri-kət\ *noun* [Middle English *criket,* from Middle French *criquet,* of imitative origin] (14th century)
1 : any of a family (Gryllidae) of leaping orthopteran insects noted for the chirping notes produced by the male by rubbing together specially modified parts of the forewings

cricket 1

2 : a low wooden footstool
3 : a small metal toy or signaling device that makes a sharp click or snap when pressed

²cricket *noun* [Middle French *criquet* goal stake in a bowling game] (1598)
1 : a game played with a ball and bat by two sides of usually 11 players each on a large field centering upon two wickets each defended by a batsman
2 : fair and honorable behavior ⟨it wasn't *cricket* for her to break her contract —Gerry Nadel⟩

³cricket *intransitive verb* (circa 1809)
: to play the game of cricket
— crick·et·er *noun*

cri·coid \'krī-,kȯid\ *adjective* [New Latin *cricoides,* from Greek *krikoeidēs* ring-shaped, from *krikos* ring — more at CIRCLE] (1746)
: of, relating to, or being a cartilage of the larynx with which arytenoid cartilages articulate

cri de coeur \,krē-də-'kər\ *noun, plural* **cris de coeur** *same*\ [French, literally, cry from the heart] (1904)
: a passionate outcry (as of appeal or protest)

cri·er \'krī(-ə)r\ *noun* (14th century)
: one that cries: **a :** an officer who proclaims the orders of a court **b :** TOWN CRIER

cri·key \'krī-kē\ *or* **crick·ey** \'kri-kē\ *interjection* [euphemism for *Christ*] (1838)
— used as a mild oath

crime \'krīm\ *noun* [Middle English, from Middle French, from Latin *crimen* accusation, reproach, crime; probably akin to Latin *cernere* to sift, determine] (14th century)
1 : an act or the commission of an act that is forbidden or the omission of a duty that is commanded by a public law and that makes the offender liable to punishment by that law; *especially* **:** a gross violation of law
2 : a grave offense especially against morality
3 : criminal activity ⟨efforts to fight *crime*⟩
4 : something reprehensible, foolish, or disgraceful ⟨it's a *crime* to waste good food⟩
synonym see OFFENSE
— crime·less \-ləs\ *adjective*

crime against humanity (1945)
: atrocity (as extermination or enslavement) that is directed especially against an entire population or part of a population on specious grounds and without regard to individual guilt or responsibility even on such grounds

crime against nature (1828)
: SODOMY

¹crim·i·nal \'kri-mə-nᵊl, 'krim-nəl\ *adjective* [Middle English, from Middle French or Late Latin; Middle French *criminel,* from Late Latin *criminalis,* from Latin *crimin-* *crimen* crime] (15th century)
1 : relating to, involving, or being a crime ⟨*criminal* neglect⟩
2 : relating to crime or to the prosecution of suspects in a crime ⟨*criminal* statistics⟩ ⟨brought *criminal* action⟩
3 : guilty of crime; *also* **:** of or befitting a criminal ⟨a *criminal* mind⟩
4 : DISGRACEFUL
— crim·i·nal·ly *adverb*

²criminal *noun* (circa 1626)
1 : one who has committed a crime
2 : a person who has been convicted of a crime

criminal conversation *noun* (1768)
: adultery considered as a tort

criminal court *noun* (1678)
: a court that has jurisdiction to try and punish offenders against criminal law

crim·i·nal·is·tics \,kri-mə-nᵊl-'is-tiks, ,krim-nə-'lis-\ *noun plural but singular in construction* (1943)
: application of scientific techniques in collecting and analyzing physical evidence in criminal cases

crim·i·nal·i·ty \,kri-mə-'na-lə-tē\ *noun* (1611)
1 : the quality or state of being criminal
2 : criminal activity ⟨urban *criminality*⟩

crim·i·nal·ize \'kri-mə-nᵊl-,īz, 'krim-nə-,līz\ *transitive verb* **-ized; -iz·ing** (circa 1956)
: to make illegal **:** OUTLAW; *also* **:** to turn into or treat as a criminal
— crim·i·nal·i·za·tion \,kri-mə-nᵊl-ə-'zā-shən, ,krim-nə-lə-'zā-\ *noun*

criminal law *noun* (1864)
: the law of crimes and their punishments

criminal lawyer *noun* (1869)
: a lawyer who specializes in criminal law; *especially* **:** a lawyer who represents defendants in criminal cases

crim·i·nate \'kri-mə-,nāt\ *transitive verb* **-nat·ed; -nat·ing** [Latin *criminatus,* past participle of *criminari,* from *crimin-, crimen* accusation] (1645)
: INCRIMINATE
— crim·i·na·tion \,kri-mə-'nā-shən\ *noun*

crim·i·nol·o·gy \,kri-mə-'nä-lə-jē\ *noun* [Italian *criminologia,* from Latin *crimin-, crimen* + Italian *-o-* + *-logia* -logy] (1890)
: the scientific study of crime as a social phenomenon, of criminals, and of penal treatment
— crim·i·no·log·i·cal \-mə-nᵊl-'ä-ji-kəl\ *adjective*
— crim·i·no·log·i·cal·ly \-k(ə-)lē\ *adverb*
— crim·i·nol·o·gist \,kri-mə-'nä-lə-jist\ *noun*

crim·i·nous \'kri-mə-nəs\ *adjective* (15th century)
: CRIMINAL

¹crimp \'krimp\ *transitive verb* [Dutch or Low German *krimpen* to shrivel; akin to Middle Dutch *crampe* hook, cramp] (1712)
1 : to cause to become wavy, bent, or pinched: as **a :** to form (leather) into a desired shape **b :** to give (synthetic fibers) a curl or wave like that of natural fibers **c :** to pinch or press together (as the margins of a pie crust) in order to seal
2 : to be an inhibiting or restraining influence on **:** CRAMP ⟨sales had been *crimped* by credit controls —*Time*⟩
— crimp·er \'krim-pər\ *noun*

²crimp *noun* (1863)
1 : something produced by or as if by crimping: as **a :** a section of hair artificially waved or curled **b :** a succession of waves (as in wool fiber) **c :** a bend or crease formed in something
2 : something that cramps or inhibits **:** RESTRAINT, CURB

³crimp *noun* [perhaps from ¹*crimp*] (1758)
: a person who entraps or forces men into shipping as sailors or into enlisting in an army or navy

⁴crimp *transitive verb* (1812)
: to trap into military or sea service **:** IMPRESS

crimpy \'krim-pē\ *adjective* **crimp·i·er; -est** (1888)
: FRIZZY

¹crim·son \'krim-zən\ *noun* [Middle English *crimisin,* from Old Spanish *cremesín,* from Arabic *qirmizī,* from *qirmiz* kermes] (15th century)
: any of several deep purplish reds

²crimson *adjective* (15th century)
: of the color crimson

³crimson (1601)
transitive verb
: to make crimson
intransitive verb
: to become crimson; *especially* **:** BLUSH

¹cringe \'krinj\ *intransitive verb* **cringed; cring·ing** [Middle English *crengen;* akin to Old English *cringan* to yield, Middle High German *krank* weak] (13th century)
1 : to draw in or contract one's muscles involuntarily (as from cold or pain)
2 : to shrink in fear or servility
3 : to behave in an excessively humble or servile way
synonym see FAWN
— cring·er *noun*

²cringe *noun* (1597)
: a cringing act; *specifically* **:** a servile bow

crin·gle \'kriŋ-gəl\ *noun* [Low German *kringel,* diminutive of *kring* ring] (1627)
: a loop or grommet at the corner of a sail to which a line is attached

¹crin·kle \'kriŋ-kəl\ *verb* **crin·kled; crin·kling** \-k(ə-)liŋ\ [Middle English *crynkelen;* akin to Old English *cringan* to yield] (14th century)
intransitive verb
1 a : to form many short bends or ripples **b :** WRINKLE

\ə\ **abut** \ᵊ\ **kitten** \ər\ **further** \a\ **ash** \ā\ **ace** \ä\ **mop, mar** \au̇\ **out** \ch\ **chin** \e\ **bet** \ē\ **easy** \g\ **go** \i\ **hit** \ī\ **ice** \j\ **job** \ŋ\ **sing** \ō\ **go** \ȯ\ **law** \ȯi\ **boy** \th\ **thin** \t͟h\ **the** \ü\ **loot** \u̇\ **foot** \y\ **yet** \zh\ **vision** *see also* Guide to Pronunciation

2 : to give forth a thin crackling sound : RUSTLE ⟨*crinkling* silks⟩
transitive verb
: to cause to crinkle : make crinkles in

²crinkle *noun* (1596)
1 : WRINKLE, CORRUGATION, PUCKER
2 : any of several plant diseases marked by crinkling of leaves
— **crin·kly** \-k(ə-)lē\ *adjective*

cri·noid \'krī-ˌnȯid\ *noun* [ultimately from Greek *krinon* lily] (1847)
: any of a large class (Crinoidea) of echinoderms usually having a somewhat cup-shaped body with five or more feathery arms
— **crinoid** *adjective*

crin·o·line \'kri-nᵊl-ən\ *noun* [French, from Italian *crinolino,* from *crino* horsehair (from Latin *crinis* hair) + *lino* flax, linen, from Latin *linum* — more at CREST] (1830)
1 : an open-weave fabric of horsehair or cotton that is usually stiffened and used especially for interlinings and millinery
2 : a full stiff skirt or underskirt made of crinoline; *also* : HOOPSKIRT
— **crinoline** *or* **crin·o·lined** \-ənd\ *adjective*

cri·o·llo \krē-'ōl-(ˌ)yō, -'ō-(ˌ)yō\ *noun, plural* **-llos** [Spanish — more at CREOLE] (1604)
1 a : a person of pure Spanish descent born in Spanish America **b** : a person born and usually raised in a Spanish-American country
2 : a domestic animal of a breed or strain developed in Latin America; *especially, often capitalized* : a hardy muscular pony of a breed originally developed in Argentina
— **criollo** *adjective*

cripes \'krīps\ *interjection* [euphemism for *Christ*] (1910)
— used as a mild oath

¹crip·ple \'kri-pəl\ *noun* [Middle English *cripel,* from Old English *crypel;* akin to Old English *crēopan* to creep — more at CREEP] (before 12th century)
1 a : a lame or partly disabled person or animal — sometimes taken to be offensive **b** : one that is disabled or deficient in a specified manner ⟨a social *cripple*⟩
2 : something flawed or imperfect

²cripple *adjective* (13th century)
: being a cripple : LAME

³cripple *transitive verb* **crip·pled; crip·pling** \-p(ə-)liŋ\ (14th century)
1 : to deprive of the use of a limb and especially a leg
2 : to deprive of capability for service or of strength, efficiency, or wholeness
synonym see MAIM, WEAKEN
— **crip·pler** \-p(ə-)lər\ *noun*
— **crip·pling·ly** \-p(ə-)liŋ-lē\ *adverb*

cri·sis \'krī-səs\ *noun, plural* **cri·ses** \'krī-ˌsēz\ [Middle English, from Latin, from Greek *krisis,* literally, decision, from *krinein* to decide — more at CERTAIN] (15th century)
1 a : the turning point for better or worse in an acute disease or fever **b** : a paroxysmal attack of pain, distress, or disordered function **c** : an emotionally significant event or radical change of status in a person's life ⟨a midlife *crisis*⟩
2 : the decisive moment (as in a literary plot)
3 a : an unstable or crucial time or state of affairs in which a decisive change is impending; *especially* : one with the distinct possibility of a highly undesirable outcome ⟨a financial *crisis*⟩ **b** : a situation that has reached a critical phase ⟨the environmental *crisis*⟩
synonym see JUNCTURE

¹crisp \'krisp\ *adjective* [Middle English, from Old English, from Latin *crispus;* akin to Welsh *crych* curly] (before 12th century)
1 : CURLY, WAVY; *also* : having close stiff or wiry curls or waves
2 a : easily crumbled : BRITTLE **b** : being desirably firm and fresh ⟨*crisp* lettuce⟩
3 a : being sharp, clean-cut, and clear ⟨a *crisp* illustration⟩; *also* : concise and to the point ⟨a *crisp* reply⟩ **b** : noticeably neat ⟨*crisp* new clothes⟩ **c** : BRISK, LIVELY ⟨a *crisp* tale of intrigue⟩ ⟨*crisp* musical tempi⟩ **d** : FROSTY, SNAPPY ⟨*crisp* winter weather⟩; *also* : FRESH, INVIGORATING ⟨*crisp* autumn air⟩ ⟨a *crisp* white wine⟩
synonym see FRAGILE
— **crisp·ly** *adverb*
— **crisp·ness** *noun*

²crisp (14th century)
transitive verb
1 : CURL, CRIMP
2 : to cause to ripple : WRINKLE
3 : to make or keep crisp
intransitive verb
1 : CURL
2 : RIPPLE
3 : to become crisp

³crisp *noun* (14th century)
1 a : something crisp or brittle ⟨burned to a *crisp*⟩ **b** *chiefly British* : POTATO CHIP — usually used in plural
2 : a baked dessert of fruit with crumb topping ⟨apple *crisp*⟩

crisp·bread \'krisp-ˌbred\ *noun* (circa 1927)
: a plain dry unsweetened cracker made from crushed grain (as wheat or rye)

crisp·en \'kris-pən\ (1943)
transitive verb
: to make crisp
intransitive verb
: to become crisp

crisp·er \'kris-pər\ *noun* (1835)
: one that crisps; *specifically* : a closed container in a refrigerator intended to prevent loss of moisture from fresh produce

crispy \'kris-pē\ *adjective* **crisp·i·er; -est** (14th century)
: CRISP
— **crisp·i·ness** *noun*

¹criss·cross \'kris-ˌkrȯs\ *verb* [obsolete *christcross, crisscross* mark of a cross] (1818)
transitive verb
1 : to mark with intersecting lines
2 : to pass back and forth through or over
intransitive verb
: to go or pass back and forth

²crisscross *adjective* (1846)
: marked or characterized by crisscrossing
— **crisscross** *adverb*

³crisscross *noun* (1876)
1 : a crisscross pattern : NETWORK
2 : the state of being at cross-purposes; *also* : a confused state

cris·ta \'kris-tə\ *noun, plural* **cris·tae** \-ˌtē, -ˌtī\ [New Latin, from Latin, crest — more at CREST] (1959)
: any of the inwardly projecting folds of the inner membrane of a mitochondrion

cri·te·ri·on \krī-'tir-ē-ən *also* krə-\ *noun, plural* **-ria** \-ē-ə\ *also* **-ri·ons** [Greek *kritērion,* from *krinein* to judge, decide — more at CERTAIN] (1622)
1 : a standard on which a judgment or decision may be based
2 : a characterizing mark or trait
synonym see STANDARD ■

cri·te·ri·um \krī-'tir-ē-əm, krē-ter-'yȯm\ *noun* [French *critérium* competition, literally, criterion, from Late Latin *criterium,* from Greek *kritērion*] (1970)
: a bicycle race of a specified number of laps on a closed course over public roads closed to normal traffic

¹crit·ic \'kri-tik\ *noun* [Latin *criticus,* from Greek *kritikos,* from *kritikos* able to discern or judge, from *krinein*] (1588)
1 a : one who expresses a reasoned opinion on any matter especially involving a judgment of its value, truth, righteousness, beauty, or technique **b** : one who engages often professionally in the analysis, evaluation, or appreciation of works of art or artistic performances
2 : one given to harsh or captious judgment

²critic *noun* [Greek *kritikē* art of the critic, from feminine of *kritikos*] (1656)
1 *archaic* : CRITICISM
2 *archaic* : CRITIQUE

crit·i·cal \'kri-ti-kəl\ *adjective* (1547)
1 a : of, relating to, or being a turning point or specially important juncture ⟨*critical* phase⟩: as (1) : relating to or being the stage of a disease at which an abrupt change for better or worse may be expected; *also* : being or relating to an illness or condition involving danger of death ⟨*critical* care⟩ (2) : relating to or being a state in which or a measurement or point at which some quality, property, or phenomenon suffers a definite change ⟨*critical* temperature⟩ **b** : CRUCIAL, DECISIVE ⟨*critical* test⟩ **c** : INDISPENSABLE, VITAL ⟨a *critical* waterfowl habitat⟩ ⟨a component *critical* to the operation of a machine⟩ **d** : being in or approaching a state of crisis ⟨a *critical* shortage⟩ ⟨a *critical* situation⟩
2 a : inclined to criticize severely and unfavorably **b** : consisting of or involving criticism ⟨*critical* writings⟩; *also* : of or relating to the judgment of critics ⟨the play was a *critical* success⟩ **c** : exercising or involving careful judgment or judicious evaluation **d** : including variant readings and scholarly emendations ⟨a *critical* edition⟩
3 : characterized by risk or uncertainty
4 a : of sufficient size to sustain a chain reaction — used of a mass of fissionable material **b** : sustaining a chain reaction — used of a nuclear reactor ☆
— **crit·i·cal·i·ty** \ˌkri-tə-'ka-lə-tē\ *noun*
— **crit·i·cal·ly** \'kri-ti-k(ə-)lē\ *adverb*
— **crit·i·cal·ness** \-kəl-nəs\ *noun*

critical angle *noun* (1873)
: the least angle of incidence at which total reflection takes place

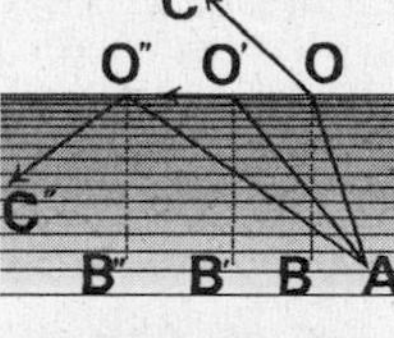

critical angle: *AOB, AO′B′, AO″B″* angles of incidence; *AOC* refracted ray; *AO″C″* totally reflected ray; *AO′B′* critical angle

critical mass *noun* (1964)
: a size, number, or amount large enough to produce a particular result ⟨the *critical mass* of activity needed for a retail store⟩

☆ SYNONYMS
Critical, hypercritical, faultfinding, captious, carping, censorious mean inclined to look for and point out faults and defects. CRITICAL may also imply an effort to see a thing clearly and truly in order to judge it fairly ⟨a *critical* essay⟩. HYPERCRITICAL suggests a tendency to judge by unreasonably strict standards ⟨*hypercritical* disparagement of other people's work⟩. FAULTFINDING implies a querulous or exacting temperament ⟨a *faultfinding* reviewer⟩. CAPTIOUS suggests a readiness to detect trivial faults or raise objections on trivial grounds ⟨a *captious* critic⟩. CARPING implies an ill-natured or perverse picking of flaws ⟨a *carping* editorial⟩. CENSORIOUS implies a disposition to be severely critical and condemnatory ⟨the *censorious* tone of the review⟩. See in addition ACUTE.

□ USAGE
criterion The plural *criteria* has been used as a singular for nearly half a century ⟨let me now return to the third *criteria* —R. M. Nixon⟩ ⟨that really is the *criteria* —Burt Lance⟩. Many of our examples, like the two foregoing, are taken from speech. But singular *criteria* is not uncommon in edited prose, and its use both in speech and writing seems to be increasing. Only time will tell whether it will reach the unquestioned acceptability of *agenda*.

critical point *noun* (circa 1889)
: a point on the graph of a function where the derivative is zero or infinite

critical region *noun* (1951)
: the set of outcomes of a statistical test for which the null hypothesis is to be rejected

critical value *noun* (circa 1909)
: the value of an independent variable corresponding to a critical point of a function

crit·ic·as·ter \ˈkri-ti-ˌkas-tər\ *noun* (1684)
: an inferior or petty critic

crit·i·cise *British variant of* CRITICIZE

crit·i·cism \ˈkri-tə-ˌsi-zəm\ *noun* (1607)
1 a : the act of criticizing usually unfavorably **b :** a critical observation or remark **c :** CRITIQUE
2 : the art of evaluating or analyzing works of art or literature
3 : the scientific investigation of literary documents (as the Bible) in regard to such matters as origin, text, composition, or history

crit·i·cize \ˈkri-tə-ˌsīz\ *verb* **-cized; -ciz·ing** (1649)
intransitive verb
: to act as a critic
transitive verb
1 : to consider the merits and demerits of and judge accordingly **:** EVALUATE
2 : to find fault with **:** point out the faults of ☆
— crit·i·ciz·able \-ˌsī-zə-bəl\ *adjective*
— crit·i·ciz·er *noun*

[1]cri·tique \krə-ˈtēk, kri-\ *noun* [alteration of [2]*critic*] (1702)
: an act of criticizing; *especially* **:** a critical estimate or discussion

[2]critique *transitive verb* **cri·tiqued; cri·tiqu·ing** (1751)
: REVIEW, CRITICIZE

crit·ter \ˈkri-tər\ *noun* [by alteration] (1815)
: CREATURE

[1]croak \ˈkrōk\ *verb* [Middle English *croken,* of imitative origin] (15th century)
intransitive verb
1 a : to make a deep harsh sound **b :** to speak in a hoarse throaty voice
2 : GRUMBLE 1
3 *slang* **:** DIE
transitive verb
1 : to utter in a hoarse raucous voice
2 *slang* **:** KILL

[2]croak *noun* (1561)
: a hoarse harsh cry or sound
— croaky \ˈkrō-kē\ *adjective*

croak·er \ˈkrō-kər\ *noun* (1651)
1 : an animal that croaks
2 : any of various fishes (especially family Sciaenidae) that produce croaking or grunting noises
3 *slang* **:** DOCTOR

Croat \ˈkrō-ˌat *also* ˈkrō(-ə)t\ *noun* [New Latin *Croata,* from Serbo-Croatian *Hrvat*] (circa 1702)
: CROATIAN

Cro·atian \krō-ˈā-shən\ *noun* (1555)
1 : a native or inhabitant of Croatia
2 : a south Slavic language spoken by the Croatian people and distinct from Serbian chiefly in its use of the Latin alphabet
— Croatian *adjective*

croc \ˈkräk\ *noun* (1884)
: CROCODILE

[1]cro·chet \krō-ˈshā\ *noun* [French, hook, crochet, from Middle French, diminutive of *croche* hook, of Scandinavian origin; akin to Old Norse *krōkr* hook] (1846)
: needlework consisting of the interlocking of looped stitches formed with a single thread and a hooked needle

[2]crochet (1858)
transitive verb
: to make of crochet ⟨*crocheted* a doily⟩
intransitive verb
: to work with crochet
— cro·chet·er \-ˈshā-ər\ *noun*

cro·cid·o·lite \krō-ˈsi-d^{ə}l-ˌīt\ *noun* [German *Krokydolith,* from Greek *krokyd-, krokys* nap on cloth (akin to Greek *krekein* to weave) + German *-lith* -lite — more at REEL] (1835)
: a lavender-blue or light green mineral of the amphibole group that occurs in silky fibers and in massive form and is a type of asbestos — compare TIGEREYE

[1]crock \ˈkräk\ *noun* [Middle English, from Old English *crocc;* akin to Middle High German *krūche* crock] (before 12th century)
1 : a thick earthenware pot or jar
2 [from its formation on cooking pots] *dialect* **:** SOOT, SMUT
3 : coloring matter that rubs off from cloth or dyed leather
4 : BUNKUM — usually used with *a*

[2]crock (1594)
transitive verb
1 : to put or preserve in a crock
2 *dialect* **:** to soil with crock **:** SMUDGE
intransitive verb
: to transfer color (as when rubbed or washed) ⟨a suede that will not *crock*⟩

[3]crock *noun* [Middle English *crok,* probably of Scandinavian origin; akin to Norwegian dialect *krokje* crock] (1528)
1 : one that is broken-down, disabled, or impaired
2 : a complaining medical patient whose illness is largely imaginary or psychosomatic

[4]crock (1896)
transitive verb
: to cause to become disabled
intransitive verb
: BREAK DOWN

crocked \ˈkräkt\ *adjective* (circa 1927)
: DRUNK

crock·ery \ˈkrä-k(ə-)rē\ *noun* (circa 1755)
: EARTHENWARE

crock·et \ˈkrä-kət\ *noun* [Middle English *croket,* from Old North French *croquet* crook, diminutive of *croc* hook, of Scandinavian origin; akin to Old Norse *krōkr* hook] (1673)
: an ornament usually in the form of curved and bent foliage used on the edge of a gable or spire
— crock·et·ed \-kə-təd\ *adjective*

Crock–Pot \ˈkräk-ˌpät\ *trademark*
— used for an electric cooking pot

croc·o·dile \ˈkrä-kə-ˌdīl\ *noun* [Middle English & Latin; Middle English *cocodrille,* from Middle French, from Medieval Latin *cocodrillus,* alteration of Latin *crocodilus,* from Greek *krokodilos* lizard, crocodile, from *krokē* shingle, pebble + *drilos* worm; akin to Sanskrit *śarkara* pebble] (1684)
1 a : any of several large carnivorous thick-skinned long-bodied aquatic reptiles (family Crocodilidae) of tropical and subtropical waters; *broadly* **:** CROCODILIAN **b :** the skin or hide of a crocodile

crocodile 1a

2 *chiefly British* **:** a line of people (as schoolchildren) usually walking in pairs

crocodile bird *noun* (1868)
: an African bird (*Pluvianus aegyptius*) that is related to the pratincoles and lights on the crocodile and eats its insect parasites

crocodile tears *noun plural* (1563)
: false or affected tears; *also* **:** hypocritical sorrow

croc·o·dil·ian \ˌkrä-kə-ˈdi-lē-ən, -ˈdil-yən\ *noun* (1837)
: any of an order (Crocodylia) of reptiles including the crocodiles, alligators, caimans, gavials, and related extinct forms
— crocodilian *adjective*

cro·cus \ˈkrō-kəs\ *noun, plural* **cro·cus·es** [Middle English, the saffron plant, from Latin, from Greek *krokos,* of Semitic origin; akin to Akkadian *kurkānū* saffron] (14th century)
1 a *plural also* **crocus** *or* **cro·ci** \-ˌkē, -ˌkī, -ˌsī\ **:** any of a large genus (*Crocus*) of herbs of the iris family developing from corms and having solitary long-tubed flowers and slender linear leaves **b :** SAFFRON 1a
2 : a dark red ferric oxide used for polishing metals

Croe·sus \ˈkrē-səs\ *noun* [*Croesus,* king of Lydia, famed for his wealth] (1650)
: a very rich man

croft \ˈkrȯft\ *noun* [Middle English, from Old English; akin to Middle Dutch *krocht* hill] (before 12th century)
1 *chiefly British* **:** a small enclosed field usually adjoining a house
2 *chiefly British* **:** a small farm worked by a tenant
— croft·er \ˈkrȯf-tər\ *noun, chiefly British*

crois·sant \krȯ-ˈsänt, krə-; krwä-ˈsä^{n}\ *noun, plural* **croissants** \-ˈsänt(s), -ˈsä^{n}(z)\ [French, literally, crescent, from Middle French *creissant*] (1899)
: a flaky rich crescent-shaped roll

Croix de Guerre \ˌkrwä-di-ˈger\ *noun* [French, literally, war cross] (1915)
: a French military decoration awarded for gallant action in war

cro·ker sack \ˈkrō-kər-\ *noun* [alteration of *crocus sack, crocus bag,* of unknown origin] (1895)
chiefly Southern **:** a sack of a coarse material (as burlap)

Cro–Ma·gnon \krō-ˈmag-nən, -ˈman-yən\ *noun* [*Cro-Magnon,* a cave near Les Eyzies, France] (1869)
: a tall erect race of hominids of the Upper Paleolithic known from skeletal remains found chiefly in southern France and classified as the same species (*Homo sapiens*) as present day humans

crom·lech \ˈkräm-ˌlek\ *noun* [Welsh, literally, bent stone] (1695)
1 : DOLMEN
2 : a circle of monoliths usually enclosing a dolmen or mound

crone \ˈkrōn\ *noun* [Middle English, from Old North French *carogne,* literally, carrion, from (assumed) Vulgar Latin *caronia* — more at CARRION] (14th century)
: a withered old woman

Cro·nus \ˈkrō-nəs, ˈkrä-\ *noun* [Latin, from Greek *Kronos*]
: a Titan dethroned by his son Zeus

cro·ny \ˈkrō-nē\ *noun, plural* **cronies** [perhaps from Greek *chronios* long-lasting, from *chronos* time] (1663)
: a close friend especially of long standing **:** PAL

cro·ny·ism \-nē-ˌi-zəm\ *noun* (1840)

☆ SYNONYMS
Criticize, reprehend, censure, reprobate, condemn, denounce mean to find fault with openly. CRITICIZE implies finding fault especially with methods or policies or intentions ⟨*criticized* the police for using violence⟩. REPREHEND implies both criticism and severe rebuking ⟨*reprehends* the self-centeredness of today's students⟩. CENSURE carries a strong suggestion of authority and of reprimanding ⟨a Senator formally *censured* by his peers⟩. REPROBATE implies strong disapproval or firm refusal to sanction ⟨*reprobated* his son's unconventional lifestyle⟩. CONDEMN usually suggests an unqualified and final unfavorable judgment ⟨*condemned* the government's racial policies⟩. DENOUNCE adds to CONDEMN the implication of a public declaration ⟨a pastoral letter *denouncing* abortion⟩.

: partiality to cronies especially as evidenced in the appointment of political hangers-on to office without regard to their qualifications

[1]**crook** \ˈkru̇k\ (12th century)
transitive verb
: BEND
intransitive verb
: CURVE, WIND

[2]**crook** *noun* [Middle English *crok,* from Old Norse *krōkr* hook] (13th century)
1 : an implement having a bent or hooked form: as **a :** POTHOOK **b** (1) **:** a shepherd's staff (2) **:** CROSIER 1
2 : a part of something that is hook-shaped, curved, or bent ⟨the *crook* of an umbrella handle⟩
3 : BEND, CURVE
4 : a person who engages in fraudulent or criminal practices

[3]**crook** *adjective* [probably short for *crooked*] (1898)
Australian & New Zealand **:** not right: **a :** UNSATISFACTORY **b :** DISHONEST, CROOKED **c :** IRRITABLE, ANGRY — used especially in the phrase *go crook* **d :** ILL, UNWELL

crook·back \ˈkru̇k-ˌbak\ *noun* (1508)
1 *obsolete* **:** a crooked back
2 *obsolete* **:** HUNCHBACK
— **crook·backed** \-ˈbakt\ *adjective*

crook·ed \ˈkru̇-kəd\ *adjective* (13th century)
1 : not straight ⟨a *crooked* road⟩ ⟨your tie is *crooked*⟩
2 : DISHONEST ⟨a *crooked* election⟩ ⟨*crooked* politicians⟩
— **crook·ed·ly** *adverb*
— **crook·ed·ness** *noun*

crook·ery \ˈkru̇-kə-rē\ *noun* (1927)
: crooked dealings or practices

crook·neck \ˈkru̇k-ˌnek\ *noun* (1784)
: a squash with a long recurved neck

croon \ˈkrün\ *verb* [Middle English *croynen,* from Middle Dutch *cronen;* akin to Old High German *crōnen* to chatter] (15th century)
intransitive verb
1 *chiefly Scottish* **:** BELLOW, BOOM
2 : to sing or speak in a gentle murmuring manner; *especially* **:** to sing in a soft intimate manner adapted to amplifying systems
transitive verb
: to sing (as a popular song or a lullaby) in a crooning manner
— **croon** *noun*

croon·er \ˈkrü-nər\ *noun* (1930)
: one that croons; *especially* **:** a singer of popular songs

[1]**crop** \ˈkräp\ *noun* [Middle English, craw, head of a plant, yield of a field, from Old English *cropp* craw, head of a plant; akin to Old High German *kropf* goiter, craw] (before 12th century)
1 : a pouched enlargement of the gullet of many birds that serves as a receptacle for food and for its preliminary maceration; *also* **:** an enlargement of the gullet of another animal (as an insect)
2 a (1) **:** a plant or animal or plant or animal product that can be grown and harvested extensively for profit or subsistence ⟨an apple *crop*⟩ ⟨a *crop* of wool⟩ (2) **:** the total yearly production from a specified area **b :** the product or yield of something formed together ⟨the ice *crop*⟩ **c :** a batch or lot of something produced during a particular cycle ⟨the current *crop* of films⟩ **d :** COLLECTION ⟨a *crop* of lies⟩
3 : the stock or handle of a whip; *also* **:** a riding whip with a short straight stock and a loop
4 [2*crop*] **a :** the part of the chine of a quadruped (as a domestic cow) lying immediately behind the withers — usually used in plural; see COW illustration **b :** an earmark on an animal; *especially* **:** one made by a straight cut squarely removing the upper part of the ear **c :** a close cut of the hair

[2]**crop** *verb* **cropped; crop·ping** (13th century)
transitive verb
1 a : to remove the upper or outer parts of ⟨*crop* a hedge⟩ ⟨*crop* a dog's ears⟩ **b :** HARVEST ⟨*crop* trout⟩ **c :** to cut off short **:** TRIM ⟨*crop* a photograph⟩
2 : to cause (land) to bear a crop ⟨will *crop* another 40 acres⟩; *also* **:** to grow as a crop
intransitive verb
1 : to feed by cropping something
2 : to yield or make a crop
3 : to appear unexpectedly or casually ⟨problems *crop* up daily⟩

crop duster *noun* (1939)
: a person who sprays crops with fungicidal or insecticidal dusts from an airplane; *also* **:** the airplane used for such spraying

crop–eared \ˈkräp-ˌird\ *adjective* (1530)
: having the ears cropped

crop·land \-ˌland\ *noun* (1846)
: land that is suited to or used for crops

[1]**crop·per** \ˈkrä-pər\ *noun* (15th century)
1 : one that crops
2 : one that raises crops; *specifically* **:** SHARECROPPER

[2]**cropper** *noun* [probably from English dialect *crop* neck, from [1]*crop*] (1858)
1 : a severe fall
2 : a sudden or violent failure or collapse

crop rotation *noun* (1909)
: the practice of growing different crops in succession on the same land chiefly to preserve the productive capacity of the soil

cro·quet \krō-ˈkā\ *noun* [French dialect, hockey stick, from Old North French, crook — more at CROCKET] (1858)
1 : a game in which players using mallets drive wooden balls through a series of wickets set out on a lawn
2 : the act of driving away an opponent's croquet ball by striking one's own ball placed against it
— **croquet** *transitive verb*

cro·quette \krō-ˈket\ *noun* [French, from *croquer* to crunch, of imitative origin] (1706)
: a small often rounded mass consisting usually of minced meat, fish, or vegetable coated with egg and bread crumbs and deep-fried

cro·qui·gnole \ˈkrō-kə-ˌnōl *also* -kən-ˌyōl\ *noun* [French, light blow, fillip] (1932)
: a method used in waving the hair by winding it on curlers from the ends of the hair toward the scalp

cro·quis \krō-ˈkē\ *noun, plural* **cro·quis** \-ˈkē(z)\ [French, from *croquer* to sketch, rough out, literally, to crunch] (1805)
: a rough draft **:** SKETCH

crore \ˈkrōr, ˈkrȯr\ *noun, plural* **crores** *also* **crore** [Hindi *karoṛ*] (1609)
: a unit of value equal to ten million rupees or 100 lakhs

cro·sier \ˈkrō-zhər\ *noun* [Middle English *crocer* crosier bearer, from Middle French *crossier,* from *crosse* crosier, of Germanic origin; akin to Old English *crycc* crutch — more at CRUTCH] (1500)
1 : a staff resembling a shepherd's crook carried by bishops and abbots as a symbol of office
2 : a plant structure with a coiled end

[1]**cross** \ˈkrȯs\ *noun* [Middle English, from Old English, from Old Norse or Old Irish; Old Norse *kross,* from Old Irish *cros,* from Latin *cruc-, crux*] (before 12th century)
1 a : a structure consisting of an upright with a transverse beam used especially by the ancient Romans for execution **b** *often capitalized* **:** the cross on which Jesus was crucified
2 a : CRUCIFIXION **b :** an affliction that tries one's virtue, steadfastness, or patience
3 : a cruciform sign made to invoke the blessing of Christ especially by touching the forehead, breast, and shoulders
4 a : a device composed of an upright bar traversed by a horizontal one; *specifically* **:** one used as a Christian symbol **b** *capitalized* **:** the Christian religion
5 : a structure (as a monument) shaped like or surmounted by a cross
6 : a figure or mark formed by two intersecting lines crossing at their midpoints; *specifically* **:** such a mark used as a signature
7 : a cruciform badge, emblem, or decoration
8 : the intersection of two ways or lines **:** CROSSING
9 : ANNOYANCE, THWARTING ⟨a *cross* in love⟩
10 a : an act of crossing dissimilar individuals **b :** a crossbred individual or kind **c :** one that combines characteristics of two different types or individuals
11 a : a fraudulent or dishonest contest **b :** dishonest or illegal practices — used especially in the phrase *on the cross*
12 : a movement from one part of a theater stage to another
13 : a hook thrown over the opponent's lead in boxing
14 : a security transaction in which a broker acts for both buyer and seller (as in the placing of a large lot of common stock) — called also *cross-trade*

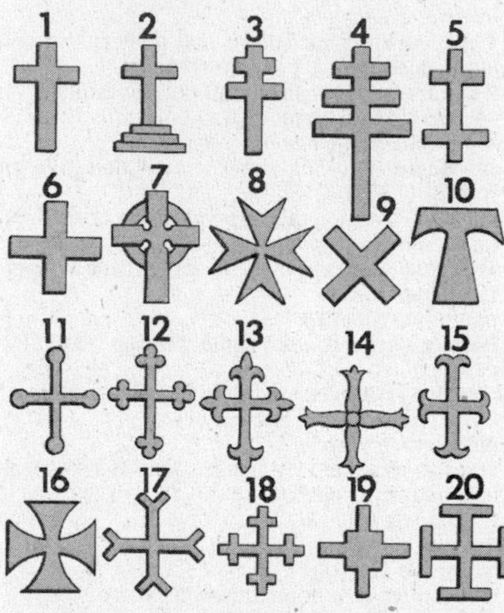

cross 4a: *1* **Latin,** *2* **Calvary,** *3* **patriarchal,** *4* **papal,** *5* **Lorraine,** *6* **Greek,** *7* **Celtic,** *8* **Maltese,** *9* **Saint Andrew's,** *10* **tau,** *11* **pommée,** *12* **botonée,** *13* **fleury,** *14* **avellan,** *15* **moline,** *16* **formée,** *17* **fourchée,** *18* **crosslet,** *19* **quadrate,** *20* **potent**

[2]**cross** (14th century)
transitive verb
1 a : to lie or be situated across **b :** INTERSECT
2 : to make the sign of the cross upon or over
3 : to cancel by marking a cross on or drawing a line through **:** strike out ⟨*cross* names off a list⟩
4 : to place or fold crosswise one over the other ⟨*cross* the arms⟩
5 a (1) **:** to run counter to **:** OPPOSE (2) **:** to deny the validity of **:** CONTRADICT **b :** to confront in a troublesome manner **:** OBSTRUCT **c** (1) **:** to spoil completely **:** DISRUPT — used with *up* ⟨his failure to appear *crossed* up the whole program⟩ (2) **:** to turn against **:** BETRAY ⟨*crossed* me up on the deal⟩
6 a : to extend across or over **:** TRAVERSE ⟨a highway *crossing* the entire state⟩ **b :** REACH, ATTAIN ⟨only two *crossed* the finish line⟩ **c :** to go from one side of to the other ⟨*cross* a street⟩ ⟨*crosses* racial barriers⟩
7 a : to draw a line across **b :** to mark or figure with lines **:** STREAK
8 : to cause (an animal or plant) to interbreed with one of a different kind **:** HYBRIDIZE
9 : to meet and pass on the way ⟨our letters must have *crossed* each other⟩
10 : to occur to ⟨it never *crossed* my mind⟩
11 : to carry or take across something ⟨*crossed* the children at the intersection⟩
intransitive verb

1 a : to move, pass, or extend across something ⟨*crossed* through France⟩ ⟨*crossed* over to the other side of the river⟩ **b :** to move or pass from one character, condition, or allegiance to another — used with *over*
2 : to lie or be athwart each other
3 : to meet in passing especially from opposite directions
4 : INTERBREED, HYBRIDIZE
— cross•er *noun*
— cross swords : to engage in a dispute

[3]cross *adjective* (14th century)
1 a : lying across or athwart **b :** moving across ⟨*cross* traffic⟩
2 a : running counter **:** OPPOSITE **b :** mutually opposed ⟨*cross* purposes⟩
3 : involving mutual interchange **:** RECIPROCAL
4 : marked by typically transitory bad temper
5 : extending over or treating several groups or classes ⟨a *cross* sample from 25 colleges⟩
6 : CROSSBRED, HYBRID
— cross•ly *adverb*
— cross•ness *noun*

[4]cross *preposition* (1551)
: ACROSS

[5]cross *adverb* (1577)
: not parallel **:** CRISSCROSS, CROSSWISE

cross•abil•i•ty \ˌkrȯ-sə-ˈbi-lə-tē\ *noun* (1916)
: the ability of different species or varieties to cross with each other

cross•able \ˈkrȯ-sə-bəl\ *adjective* (1865)
: capable of being crossed

cross action *noun* (circa 1859)
: a legal action in which the defendant in an existing action files a suit against the plaintiff on the same subject matter **:** COUNTERSUIT

cross•band•ing \ˈkrȯs-ˌban-diŋ\ *noun* (1904)
: a veneer border (as on furniture) with its grain at right angles to the grain of the adjacent wood
— cross•band•ed \-ˌban-dəd\ *adjective*

cross•bar \ˈkrȯs-ˌbär\ *noun* (1562)
: a transverse bar or stripe

cross•bear•er \ˈkrȯs-ˌbar-ər, -ˌber-\ *noun* (1568)
: CRUCIFER 1

cross•bill \-ˌbil\ *noun* (circa 1672)
: any of a small genus (*Loxia*) of finches with strongly curved mandibles that cross each other

cross•bones \-ˌbōnz\ *noun plural* (1798)
: two leg or arm bones placed or depicted crosswise — compare SKULL AND CROSSBONES

cross•bow \-ˌbō\ *noun* (15th century)
: a weapon for discharging quarrels and stones that consists chiefly of a short bow mounted crosswise near the end of a stock

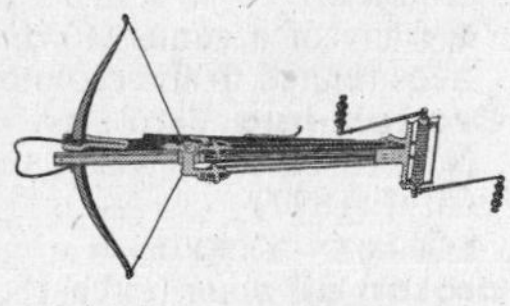
crossbow

cross•bow•man \-mən\ *noun* (circa 1500)
: a person (as a soldier or a hunter) whose weapon is a crossbow

cross•bred \ˈkrȯs-ˈbred\ *adjective* (1856)
: produced by crossbreeding **:** HYBRID
— crossbred \-ˌbred\ *noun*

[1]cross•breed \ˈkrȯs-ˌbrēd, -ˈbrēd\ *verb* **-bred** \-ˌbred, -ˈbred\; **-breed•ing** (1675)
transitive verb
: HYBRIDIZE, CROSS; *especially* **:** to cross (two varieties or breeds) within the same species
intransitive verb
: to engage in or undergo hybridization

[2]cross•breed \-ˌbrēd\ *noun* (1774)
: HYBRID

[1]cross–check \-ˌchek\ *transitive verb* (circa 1930)
1 : to obstruct in ice hockey or lacrosse by thrusting one's stick held in both hands across an opponent's face or body
2 : to check (as data or reports) from various angles or sources to determine validity or accuracy

[2]cross–check *noun* (1937)
: an act or instance of cross-checking

cross–claim \-ˌklām\ *noun* (1952)
: a claim against a party on the same side of a legal action

[1]cross–coun•try \ˈkrȯs-ˈkən-trē, -ˌkən-\ *adjective* (1767)
1 : extending or moving across a country ⟨a *cross-country* concert tour⟩
2 : proceeding over countryside (as across fields and through woods) and not by roads
3 : of or relating to racing or skiing over the countryside instead of over a track or run
— cross–country *adverb*

[2]cross–country *noun* (1925)
: cross-country racing or skiing

cross•court \ˈkrȯs-ˈkōrt, -ˈkȯrt\ *adverb or adjective* (1915)
: to or toward the opposite side of a court (as in tennis or basketball)

cross–cul•tur•al \ˈkrȯs-ˈkəlch-rəl, -ˈkəl-chə-\ *adjective* (circa 1942)
: dealing with or offering comparison between two or more different cultures or cultural areas
— cross–cul•tur•al•ly \-rə-lē\ *adverb*

cross•cur•rent \ˈkrȯs-ˌkər-ənt, -ˌkə-rənt\ *noun* (1598)
1 : a current running counter to the general forward direction
2 : a conflicting tendency — usually used in plural ⟨political *crosscurrents*⟩

[1]cross•cut \ˈkrȯs-ˌkət, -ˈkət\ *transitive verb* (1590)
1 : to cut, go, or move across or through
2 : to cut with a crosscut saw
3 : to subject (as movie scenes) to crosscutting

[2]crosscut *adjective* (1645)
1 : made or used for cutting transversely ⟨a saw with *crosscut* teeth⟩
2 : cut across or transversely ⟨a *crosscut* incision⟩

[3]cross•cut \ˈkrȯs ˌkət\ *noun* (1789)
1 : something that cuts across or through; *specifically* **:** a mine working driven horizontally and at right angles to an adit, drift, or level
2 : CROSS SECTION
3 : CROSSCUT SAW
4 : an instance of crosscutting (as in a movie)

crosscut saw *noun* (1645)
: a saw designed chiefly to cut across the grain of wood — compare RIPSAW

cross•cut•ting \ˈkrȯs-ˌkə-tiŋ\ *noun* (1930)
: a technique especially in filmmaking of interweaving bits of two or more separate scenes

cross–dress•ing \ˈkrȯs-ˌdre-siŋ\ *noun* (1911)
: the wearing of clothes designed for the opposite sex
— cross–dress *intransitive verb*
— cross–dress•er *noun*

crosse \ˈkrȯs\ *noun* [French, literally, crosier — more at CROSIER] (1867)
: a stick with a small net at one end that is used in lacrosse

cross–ex•am•i•na•tion \ˈkrȯs-ig-ˌza-mə-ˈnā-shən\ *noun* (1824)
: the examination of a witness who has already testified in order to check or discredit the witness's testimony, knowledge, or credibility — compare DIRECT EXAMINATION
— cross–ex•am•ine \-ig-ˈza-mən\ *transitive verb*
— cross–ex•am•in•er \-ˈzam-nər, -ˈza-mə-\ *noun*

cross–eye \ˈkrȯ-ˌsī\ *noun* (1826)
1 : strabismus in which the eye turns inward toward the nose
2 *plural* **:** eyes affected with cross-eye
— cross–eyed \-ˌsīd\ *adjective*

cross–fer•tile \ˈkrȯs-ˈfər-tᵊl, -ˌfər-\ *adjective* (1929)
: fertile in a cross or capable of cross-fertilization

cross–fer•til•i•za•tion \-ˌfər-tᵊl-ə-ˈzā-shən\ *noun* (1876)
1 a : fertilization in which the gametes are produced by separate individuals or sometimes by individuals of different kinds **b :** CROSS-POLLINATION
2 : interchange or interaction (as between different ideas, cultures, or categories) especially of a broadening or productive nature

cross–fer•til•ize \-ˈfər-tᵊl-ˌīz\ (1876)
transitive verb
: to accomplish cross-fertilization of
intransitive verb
: to undergo cross-fertilization

cross–file \-ˈfī(ə)l\ (1949)
intransitive verb
: to register as a candidate in the primary elections of more than one political party
transitive verb
: to register (a person) as a candidate for more than one party

cross fire *noun* (circa 1859)
1 a : firing (as in combat) from two or more points so that the lines of fire cross **b :** a situation wherein the forces of opposing factions meet, cross, or clash ⟨caught in a political *cross fire*⟩
2 : rapid or heated exchange of words

cross–grained \ˈkrȯs-ˌgrānd\ *adjective* (1647)
1 : difficult to deal with
2 : having the grain or fibers running diagonally, transversely, or irregularly

cross•hair \-ˌhar, -ˌher\ *noun* (circa 1884)
: a fine wire or thread in the focus of the eyepiece of an optical instrument used as a reference line in the field or for marking the instrumental axis

cross•hatch \ˈkrȯs-ˌhach\ *transitive verb* (1822)
: to mark with two series of parallel lines that intersect
— crosshatch *noun*
— cross–hatch•ing *noun*

cross•head \-ˌhed\ *noun* (1827)
1 : a metal block to which one end of a piston rod is secured
2 : a heading centered usually between portions of text

cross–in•dex \ˈkrȯs-ˈin-ˌdeks\ *transitive verb* (1892)
1 : to index (an item) under a second or under more than one heading
2 : to supply (as a book) with a cross-referenced index
— cross–index *noun*

cross•ing \ˈkrȯ-siŋ\ *noun* (1575)
1 : the act or action of crossing: as **a :** a traversing or traveling across **b :** an opposing, blocking, or thwarting especially in an unfair or dishonest manner
2 a : a place or structure (as on a street or over a river) where pedestrians or vehicles cross; *especially* **:** CROSSWALK **b :** the place in a cruciform church where the transept crosses the nave **c :** a place where a railroad track crosses a street

cross•ing–over \ˌkrȯ-siŋ-ˈō-vər\ *noun* (1912)
: an interchange of genes or segments between homologous chromosomes

cross–legged \ˈkrȯs-ˌle-gəd, -ˌlā-; -ˌlegd, -ˌlāgd\ *adverb or adjective* (circa 1530)
1 : with legs crossed and knees spread wide apart
2 : with one leg placed over and across the other

cross•let \ˈkrȯs-lət\ *noun* (15th century)
: a small cross usually with crossed arms; *especially* **:** one used as a heraldic bearing — see CROSS illustration

cross·lin·guis·tic \-liŋ-ˈgwis-tik\ *adjective* (1954)
: of or relating to languages of different families and types; *especially* **:** relating to the comparison of different languages
— cross·lin·guis·ti·cal·ly \-ti-k(ə-)lē\ *adverb*

cross–link \ˈkrȯs-ˌliŋk\ *noun* (1936)
: a crosswise connecting part (as an atom or group) that connects parallel chains in a complex chemical molecule (as a polymer)
— cross–link *verb*

cross–link·age \ˈkrȯs-ˈliŋ-kij\ *noun* (1937)
: the process of forming cross-links; *also* **:** CROSS-LINK

cross multiply *intransitive verb* (1951)
: to find the two products obtained by multiplying the numerator of each of two fractions by the denominator of the other
— cross multiplication *noun*

cross–na·tion·al \ˈkrȯs-ˈnash-nəl, -ˈna-shə-nᵊl\ *adjective* (1965)
: of or relating to two or more nations

cross of Lor·raine \-lə-ˈrān, -lȯ-\ [*Lorraine,* France] (circa 1889)
: a cross with two crossbars the lower one of which is longer than the upper one — see CROSS illustration

cros·sop·ter·yg·ian \ˌkrȯ-ˌsäp-tə-ˈri-j(ē-)ən\ *noun* [New Latin *Crossopterygii,* subclass name, from Greek *krossoi* fringe + *pterygion,* diminutive of *pteryg-, pteryx* wing, fin; akin to Greek *pteron* wing — more at FEATHER] (1861)
: any of a subclass (Crossopterygii) of fishes (as a coelacanth) that have paired fins suggesting limbs, that may be ancestral to the terrestrial vertebrates, and that are mostly extinct — called also *lobe-fin*
— crossopterygian *adjective*

¹cross·over \ˈkrȯs-ˌō-vər\ *noun* (1884)
1 : CROSSING 2a
2 : an instance or product of genetic crossing-over
3 : a voter registered as a member of one political party who votes in the primary of the other party
4 : a broadening of the popular appeal of an artist (as a musician) or an artist's work that is often the result of a change of the artist's medium or style; *also* **:** an artist or artistic work that has achieved a crossover
5 : an instance of breaking into another category ⟨trying to make the *crossover* to serious acting —Michael Neill⟩

²crossover *adjective* (1893)
1 : having two pieces that cross especially one over the other ⟨a *crossover* vest⟩
2 : CRITICAL 1 ⟨the *crossover* point⟩

cross–own·er·ship \-ˈō-nər-ˌship\ *noun* (1969)
: single ownership of two or more related businesses (as a newspaper and a television station) that allows the owner to control competition

cross·patch \ˈkrȯs-ˌpach\ *noun* [³*cross* + ³*patch*] (circa 1700)
: GROUCH 2

cross·piece \ˈkrȯs-ˌpēs\ *noun* (1607)
: a horizontal member (as of a structure)

cross–pol·li·nate \ˈkrȯs-ˈpä-lə-ˌnāt\ *transitive verb* (circa 1900)
: to subject to cross-pollination

cross–pol·li·na·tion \ˌkrȯs-ˌpä-lə-ˈnā-shən\ *noun* (1882)
1 : the transfer of pollen from one flower to the stigma of another
2 : CROSS-FERTILIZATION 2

cross product *noun* (1929)
1 : VECTOR PRODUCT
2 : either of the two products obtained by multiplying the two means or the two extremes of a proportion

cross–pur·pose \ˈkrȯs-ˈpər-pəs\ *noun* (1681)
: a purpose usually unintentionally contrary to another purpose of oneself or of someone or something else — usually used in plural ⟨the two were always working at *cross-purposes*⟩

cross–ques·tion \ˈkrȯs-ˈkwes-chən, -ˈkwesh-\ *noun* (circa 1694)
: a question asked in cross-examination
— cross–question *transitive verb*

cross–re·ac·tion \ˌkrȯs-rē-ˈak-shən\ *noun* (1946)
: reaction of one antigen with antibodies developed against another antigen
— cross–re·act \-rē-ˈakt\ *intransitive verb*
— cross–re·ac·tive \-ˈak-tiv\ *adjective*
— cross–re·ac·tiv·i·ty \-(ˌ)rē-ˌak-ˈti-və-tē\ *noun*

cross–re·fer \ˌkrȯs-ri-ˈfər\ (1879)
transitive verb
: to refer (a reader) by a notation or direction from one place to another (as in a book, list, or catalog)
intransitive verb
: to make a cross-reference

¹cross–ref·er·ence \ˈkrȯs-ˈre-fərn(t)s, -ˈre-f(ə-)rən(t)s\ *noun* (1834)
: a notation or direction at one place (as in a book or filing system) to pertinent information at another place

²cross–reference *transitive verb* (1902)
: to supply with cross-references

cross–re·sis·tance \ˌkrȯs-ri-ˈzis-tən(t)s\ *noun* (1946)
: tolerance (as of an insect population) to a normally toxic substance (as an insecticide) that is acquired not as a result of direct exposure but by exposure to a related substance

cross·road \ˈkrȯs-ˌrōd *also* -ˈrōd\ *noun* (1719)
1 : a road that crosses a main road or runs cross-country between main roads
2 *usually plural but singular or plural in construction* **a :** the place of intersection of two or more roads **b** (1) **:** a small community located at such a crossroads (2) **:** a central meeting place **c :** a crucial point especially where a decision must be made

cross·ruff \ˈkrȯs-ˌrəf, -ˈrəf\ *noun* (1862)
: a series of plays in a card game (as bridge) in which partners alternately trump different suits and lead to each other for that purpose
— crossruff *verb*

cross section *noun* (1835)
1 a : a cutting or piece of something cut off at right angles to an axis; *also* **:** a representation of such a cutting **b :** SECTION 3b
2 : a measure of the probability of an encounter between particles such as will result in a specified effect (as scattering or capture)
3 : a composite representation typifying the constituents of a thing in their relations
— cross–section *transitive verb*
— cross–sec·tion·al \ˈkrȯs-ˈsek-shnəl, -shə-nᵊl\ *adjective*

cross–stitch \ˈkrȯ(s)-ˌstich\ *noun* (circa 1710)
1 : a needlework stitch that forms an X
2 : work having cross-stitch
— cross–stitch *verb*

cross talk *noun* (1887)
: unwanted signals in a communication channel (as in a telephone, radio, or computer) caused by transference of energy from another circuit (as by leakage or coupling)

cross–tol·er·ance \ˈkrȯs-ˈtäl-rən(t)s, -ˈtä-lə-\ *noun* (circa 1923)
: tolerance or resistance to a drug that develops through continued use of another drug with similar pharmacological action

cross·town \ˈkrȯs-ˈtau̇n\ *adjective* (1886)
1 : situated at opposite points of a town
2 : extending or running across a town ⟨a *crosstown* street⟩ ⟨a *crosstown* bus⟩

cross–trade \ˈkrȯs-ˌtrād\ *noun* (circa 1923)
: CROSS 14

cross·trees \-(ˌ)trēz\ *noun plural* (1626)
: two horizontal crosspieces of timber or metal supported by trestletrees at a masthead that spread the upper shrouds in order to support the mast

cross vault *noun* (1850)
: a vault formed by the intersection of two or more simple vaults — called also *cross vaulting*

cross·walk \ˈkrȯs-ˌwȯk\ *noun* (1744)
: a specially paved or marked path for pedestrians crossing a street or road

cross·way \-ˌwā\ *noun* (14th century)
: CROSSROAD — often used in plural

cross·ways \-ˌwāz\ *adverb* (1564)
: CROSSWISE, DIAGONALLY

cross·wind \-ˌwind\ *noun* (1677)
: a wind blowing in a direction not parallel to a course (as of an airplane)

¹cross·wise \-ˌwīz\ *adverb* (14th century)
1 *archaic* **:** in the form of a cross
2 : so as to cross something **:** ACROSS ⟨logs laid *crosswise*⟩

²crosswise *adjective* (1903)
: TRANSVERSE, CROSSING

cross·word \ˈkrȯs-ˌwərd\ *noun* (1914)
: a puzzle in which words are filled into a pattern of numbered squares in answer to correspondingly numbered clues and in such a way that the words read across and down

crotch \ˈkräch\ *noun* [probably alteration of ¹*crutch*] (1573)
1 : a pole with a forked end used especially as a prop
2 : an angle formed by the parting of two legs, branches, or members
— crotched \ˈkrächt\ *adjective*

crotch·et \ˈkrä-chət\ *noun* [Middle English *crochet,* from Middle French — more at CROCHET] (14th century)
1 *obsolete* **a :** a small hook or hooked instrument **b :** BROOCH
2 a : a highly individual and usually eccentric opinion or preference **b :** a peculiar trick or device
3 : QUARTER NOTE
synonym see CAPRICE

crotch·ety \ˈkrä-chə-tē\ *adjective* (1825)
1 : given to crotchets **:** subject to whims, crankiness, or ill temper ⟨a *crotchety* old man⟩
2 : full of or arising from crotchets
— crotch·et·i·ness \-tē-nəs\ *noun*

cro·ton \ˈkrō-tᵊn\ *noun* [New Latin, genus name, from Greek *krotōn* castor-oil plant] (1751)
1 : any of a genus (*Croton*) of herbs, shrubs, and trees of the spurge family: as **a :** one (*C. eluteria*) of the Bahamas yielding cascarilla bark **b :** an Asian plant (*C. tiglium*) yielding croton oil
2 : any of a genus (*Codiaeum*) of shrubs and trees related to the crotons

Cro·ton bug \ˈkrō-tᵊn-\ *noun* [*Croton* River, N.Y., used as a water supply for New York City] (1877)
: GERMAN COCKROACH

croton oil *noun* (circa 1831)
: a viscid acrid fixed oil from an Asian croton (*Croton tiglium*) formerly used as a drastic cathartic but now used especially in pharmacological experiments as an irritant

crouch \ˈkrau̇ch\ *verb* [Middle English] (14th century)
intransitive verb
1 a : to lower the body stance especially by bending the legs ⟨a sprinter *crouched* ready to go⟩ **b :** to lie close to the ground with the legs bent ⟨a pair of cats, *crouching* on the brink of a fight —Aldous Huxley⟩
2 : to bend or bow servilely **:** CRINGE
transitive verb
: to bow especially in humility or fear **:** BEND
— crouch *noun*

¹croup \ˈkrüp\ *noun* [Middle English *croupe,* from Old French, of Germanic origin; akin to Old High German *kropf* craw — more at CROP] (14th century)
: the rump of a quadruped

²croup *noun* [English dialect *croup* to cry hoarsely, cough, probably of imitative origin] (1765)

: a spasmodic laryngitis especially of infants marked by episodes of difficult breathing and hoarse metallic cough
— **croup·ous** \'krü-pəs\ *adjective*
— **croupy** \-pē\ *adjective*

crou·pi·er \'krü-pē-ər, -pē-ˌā\ *noun* [French, literally, rider on the croup of a horse, from *croupe* croup] (1709)
: an employee of a gambling casino who collects and pays bets and assists at the gaming tables

crouse \'krüs\ *adjective* [Middle English] (15th century)
chiefly Scottish **:** BRISK, LIVELY

crous·tade \krü-'städ\ *noun* [French, probably from Italian *crostata* tart, from *crosta* pastry shell, crust, from Latin *crusta* — more at CRUST] (circa 1845)
: a crisp shell (as of toasted or puff pastry) in which to serve food

crou·ton \'krü-ˌtän, krü-'\ *noun* [French *croûton,* diminutive of *croûte* crust, from Middle French *crouste,* from Old French, from Latin *crusta*] (1806)
: a small cube of toasted or crisply fried bread

¹crow \'krō\ *noun* [Middle English *crowe,* from Old English *crāwe;* akin to Old High German *krāwa* crow, Old English *crāwan* to crow] (before 12th century)
1 : any of various large usually entirely glossy black passerine birds (family Corvidae and especially genus *Corvus*)
2 *capitalized* **a :** a member of an American Indian people of the region between the Platte and Yellowstone rivers **b :** the language of the Crow people
3 *capitalized* **:** CORVUS
— **as the crow flies :** in a straight line

²crow *verb* **crowed** \'krōd\ *also in sense 1 chiefly British* **crew** \'krü\; **crow·ing** [Middle English, from Old English *crāwan*] (before 12th century)
intransitive verb
1 : to make the loud shrill sound characteristic of a cock
2 : to utter a sound expressive of pleasure
3 a : to exult gloatingly especially over the distress of another **b :** to brag exultantly or blatantly
transitive verb
: to say with self-satisfaction
synonym see BOAST

³crow *noun* (13th century)
1 : the cry of the cock
2 : a triumphant cry

crow·bar \'krō-ˌbär\ *noun* (1748)
: an iron or steel bar that is usually wedge-shaped at the working end for use as a pry or lever
— **crowbar** *transitive verb*

crow·ber·ry \'krō-ˌber-ē\ *noun* (1597)
1 : an evergreen subshrub (*Empetrum nigrum* of the family Empetraceae) of arctic and alpine regions with an insipid black berry
2 : the fruit of a crowberry

¹crowd \'kraùd\ *verb* [Middle English *crouden,* from Old English *crūdan;* akin to Middle High German *kroten* to crowd, Old English *crod* multitude, Middle Irish *gruth* curds] (before 12th century)
intransitive verb
1 a : to press on **:** HURRY **b :** to press close
2 : to collect in numbers
transitive verb
1 a : to fill by pressing or thronging together **b :** to press, force, or thrust into a small space
2 : PUSH, FORCE — often used with *off* or *out*
3 a : to urge on **b :** to put on (sail) in excess of the usual for greater speed
4 : to put pressure on ⟨don't *crowd* me, I'll pay⟩
5 : THRONG, JOSTLE
6 : to press or stand close to

²crowd *noun* (1567)
1 : a large number of persons especially when collected together **:** THRONG
2 a : the great body of the people **:** POPULACE **b :** most of one's peers ⟨follow the *crowd*⟩
3 : a large number of things close together
4 : a group of people having something (as a habit, interest, or occupation) in common ⟨in with the wrong *crowd*⟩ ☆

³crowd \'kraùd, 'krüd\ *noun* [Middle English *crowde,* from Middle Welsh *crwth*] (14th century)

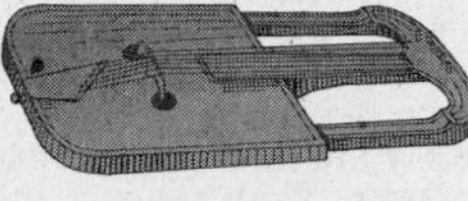
crowd 1

1 : an ancient Celtic stringed instrument that is plucked or bowed — called also *crwth*
2 *dialect English* **:** VIOLIN

crowd·ed·ness \'kraù-dəd-nəs\ *noun* (1823)
: the quality or state of being crowded

crow·die \'kraù-dē\ *noun* [alteration of *crud* curd] (1820)
: a Scottish cottage cheese that is partially cooked

crow·foot \'krō-ˌfùt\ *noun, plural* **crow·feet** \-ˌfēt\ (14th century)
1 *plural usually* **crow·foots :** any of numerous plants having leaves with cleft lobes; *especially* **:** BUTTERCUP
2 : CROW'S-FOOT 1 — usually used in plural

crow·keep·er \'krō-ˌkē-pər\ *noun* (1562)
dialect English **:** a person employed to scare off crows

¹crown \'kraùn\ *noun, often attributive* [Middle English *coroune, crowne,* from Old French *corone,* from Latin *corona* wreath, crown, from Greek *korōnē* culmination, something curved like a crow's beak, literally, crow; akin to Latin *cornix* crow, Greek *korax* raven — more at RAVEN] (12th century)
1 : a reward of victory or mark of honor; *especially* **:** the title representing the championship in a sport
2 : a royal or imperial headdress or cap of sovereignty **:** DIADEM
3 : the highest part: as **a :** the topmost part of the skull or head **b :** the summit of a mountain **c :** the head of foliage of a tree or shrub **d :** the part of a hat or other headgear covering the crown of the head **e :** the part of a tooth external to the gum or an artificial substitute for this — see TOOTH illustration
4 : a wreath, band, or circular ornament for the head
5 a : something resembling a wreath or crown **b :** the knurled cap on top of a watch stem
6 *often capitalized* **a** (1) **:** imperial or regal power **:** SOVEREIGNTY (2) **:** the government under a constitutional monarchy **b :** MONARCH
7 : something that imparts splendor, honor, or finish **:** CULMINATION
8 a : any of several old gold coins with a crown as part of the device **b :** an old usually silver British coin worth five shillings
9 a : KORUNA **b :** KRONA **c :** KRONE
10 a : the region of a seed plant at which stem and root merge **b :** the thick arching end of the shank of an anchor where the arms join it — see ANCHOR illustration
— **crowned** \'kraùnd\ *adjective*
— **crown·less** \-ləs\ *adjective*

²crown *verb* [Middle English *corounen,* from Old French *coroner,* from Latin *coronare,* from *corona*] (12th century)
transitive verb
1 a : to place a crown or wreath on the head of; *specifically* **:** to invest with regal dignity and power **b :** to recognize officially as ⟨they *crowned* her athlete of the year⟩
2 : to bestow something on as a mark of honor or recompense **:** ADORN
3 : SURMOUNT, TOP; *especially* **:** to top (a checker) with a checker to make a king
4 : to bring to a successful conclusion **:** CLIMAX
5 : to provide with something like a crown: as **a :** to fill so that the surface forms a crown **b :** to put an artificial crown on (a tooth)
6 : to hit on the head
intransitive verb
of a forest fire **:** to burn rapidly through the tops of trees

crown colony *noun, often both Cs capitalized* (1845)
: a colony of the Commonwealth over which the Crown retains some control

crown court *noun* (1827)
: a court in England and Wales that exercises jurisdiction over matters formerly heard by the quarter sessions and criminal matters formerly heard by the courts of assize

crow·ner \'krü-nər, 'kraù-\ *noun* [Middle English, alteration of *coroner*] (14th century)
dialect chiefly English **:** CORONER

crown·et \'kraù-nət\ *noun* (15th century)
archaic **:** CORONET

crown gall *noun* (1900)
: a disease that affects many species of plants and is caused by a bacterium (*Agrobacterium tumefaciens*) which forms tumorous enlargements just below the ground on the stem

crown glass *noun* (1706)
1 : a glass blown and whirled into the form of a disk with a center lump left by the worker's rod
2 : alkali-lime silicate optical glass having relatively low index of refraction and low dispersion value

crown jewel *noun* (1649)
1 *plural* **:** the jewels (as crown and scepter) belonging to a sovereign's regalia
2 : the most attractive or valuable one of a collection or group

crown land *noun* (circa 1625)
1 : land belonging to the crown and yielding revenues that the reigning sovereign is entitled to
2 : public land in some British dominions or colonies

crown of thorns (1964)
: a starfish (*Acanthaster planci*) of the Pacific region that is covered with long spines and is destructive to the coral of coral reefs — called also *crown-of-thorns starfish*

crown prince *noun* (1791)
: a male heir apparent to a crown or throne

crown princess *noun* (1863)
1 : the wife of a crown prince
2 : a female heir apparent or heir presumptive to a crown or throne

crown roast *noun* (circa 1909)
: a fancy roast of lamb, veal, or pork made from the rib portions of two loins skewered together at the ends to form a circle

crown rust *noun* (circa 1899)

☆ SYNONYMS
Crowd, throng, crush, mob, horde mean an assembled multitude usually of people. CROWD implies a close gathering and pressing together ⟨a small *crowd* gathered⟩. THRONG strongly suggests movement and pushing ⟨a *throng* of reporters⟩. CRUSH emphasizes the compactness of the group, the difficulty of individual movement, and the attendant discomfort ⟨a *crush* of fans⟩. MOB implies a disorderly crowd with the potential for violence ⟨an angry *mob* outside the jail⟩. HORDE suggests a rushing or tumultuous crowd ⟨a *horde* of shoppers⟩.

: a leaf rust of oats and other grasses that is caused by a fungus (*Puccinia coronata*) and is characterized by rounded light-orange uredinia and buried telia

crown vetch *noun* (circa 1900)
: a European leguminous herb (*Coronilla varia*) that is naturalized in the eastern U.S. and has umbels of pink-and-white flowers and sharp-angled pods

crow's–foot \'krōz-ˌfu̇t\ *noun, plural* **crow's–feet** \-ˌfēt\ (14th century)
1 : a wrinkle extending from the outer corner of the eye — usually used in plural
2 : CROWFOOT 1

crow's nest *noun* (1818)
: a partly enclosed platform high on a ship's mast for use as a lookout; *also* **:** a similar lookout (as for traffic control)

crow•step \'krō-ˌstep\ *noun* (1822)
: any of a series of steps at the top of a gable wall
— crow•stepped \-ˌstept\ *adjective*

cro•zier *variant of* CROSIER

CRT \ˌsē-(ˌ)är-'tē\ *noun, plural* **CRTs** *or* **CRT's** (circa 1945)
: CATHODE-RAY TUBE; *also* **:** a display device incorporating a cathode-ray tube

cruces *plural of* CRUX

cru•cial \'krü-shəl\ *adjective* [French, from Latin *cruc-, crux* cross] (1706)
1 *archaic* **:** CRUCIFORM
2 a : important or essential as resolving a crisis **:** DECISIVE ⟨a *crucial* step⟩ **b :** marked by final determination of a doubtful issue ⟨the *crucial* game of a series⟩ **c :** IMPORTANT, SIGNIFICANT ⟨what use we make of them will be the *crucial* question —Stanley Kubrick⟩
synonym see ACUTE

cru•cial•ly \'krü-sh(ə-)lē\ *adverb* (1879)
1 : in a crucial manner
2 : very importantly

cru•cian carp \'krü-shən-\ *noun* [modification of Low German *karuse,* from Middle Low German *karusse,* perhaps of Baltic origin; akin to Lithuanian *karušis* carp] (1836)
: a European carp (*Carassius carassius*) — called also *crucian*

cru•ci•ate \'krü-shē-ˌāt\ *adjective* [New Latin *cruciatus,* from Latin *cruc-, crux*] (1826)
: cross-shaped **:** CRUCIFORM

cru•ci•ble \'krü-sə-bəl\ *noun* [Middle English *corusible,* from Medieval Latin *crucibulum* earthen pot for melting metals] (15th century)
1 : a vessel of a very refractory material (as porcelain) used for melting and calcining a substance that requires a high degree of heat
2 : a severe test
3 : a place or situation in which concentrated forces interact to cause or influence change or development ⟨conditioned by having grown up within the *crucible* of Chinatown —Tom Wolfe⟩

crucible steel *noun* (circa 1879)
: hard cast steel made in pots that are lifted from the furnace before the metal is poured into molds

cru•ci•fer \'krü-sə-fər\ *noun* [Late Latin, from Latin *cruc-, crux* + *-fer*] (1574)
1 : one who carries a cross especially at the head of an ecclesiastical procession
2 : any of a family (Cruciferae) of plants including the cabbage, turnip, and mustard
— cru•cif•er•ous \krü-'si-f(ə-)rəs\ *adjective*

cru•ci•fix \'krü-sə-ˌfiks\ *noun* [Middle English, from Late Latin *crucifixus* the crucified Christ, from *crucifixus,* past participle of *crucifigere* to crucify, from Latin *cruc-, crux* + *figere* to fasten — more at FIX] (13th century)
: a representation of Christ on the cross

cru•ci•fix•ion \ˌkrü-sə-'fik-shən\ *noun* (15th century)
1 a *capitalized* **:** the crucifying of Christ **b :** the act of crucifying
2 : extreme and painful punishment, affliction, or suffering

cru•ci•form \'krü-sə-ˌfȯrm\ *adjective* [Latin *cruc-, crux* + English *-form*] (1661)
: forming or arranged in a cross
— cruciform *noun*

cru•ci•fy \'krü-sə-ˌfī\ *transitive verb* **-fied; -fy•ing** [Middle English *crucifien,* from Old French *crucifier,* from Late Latin *crucifigere*] (14th century)
1 : to put to death by nailing or binding the wrists or hands and feet to a cross
2 : to destroy the power of **:** MORTIFY ⟨*crucify* the flesh⟩
3 a : to treat cruelly **:** TORMENT **b :** PILLORY 2

cruck \'krək\ *noun* [probably from dialect form of ²*crook* (curved timber)] (1888)
: one of a pair of curved timbers forming a principal support of a roof in primitive English house construction

¹crud \'krəd\ *noun* [Middle English *curd, crudd*] (14th century)
1 *dialect* **:** CURD
2 a : a deposit or incrustation of filth, grease, or refuse **b :** something disagreeable or disgusting **:** RUBBISH **c** *slang* **:** a despicable or contemptible person
3 : a usually ill-defined or imperfectly identified bodily disorder
— crud•dy \'krə-dē\ *adjective*

²crud *verb* **crud•ded; crud•ding** (14th century)
dialect **:** ²CURD

¹crude \'krüd\ *adjective* **crud•er; crud•est** [Middle English, from Latin *crudus* raw, crude, undigested — more at RAW] (14th century)
1 : existing in a natural state and unaltered by cooking or processing ⟨*crude* oil⟩
2 *archaic* **:** UNRIPE, IMMATURE
3 : marked by the primitive, gross, or elemental or by uncultivated simplicity or vulgarity ⟨a *crude* stereotype⟩
4 : rough or inexpert in plan or execution ⟨a *crude* shelter⟩
5 : lacking a covering, glossing, or concealing element **:** OBVIOUS ⟨*crude* facts⟩
6 : tabulated without being broken down into classes ⟨the *crude* death rate⟩
synonym see RUDE
— crude•ly *adverb*
— crude•ness *noun*

²crude *noun* (1904)
: a substance in its natural unprocessed state; *especially* **:** unrefined petroleum

cru•di•tés \krœ-dē-tā, ˌkrü-di-'tā\ *noun plural* [French, from plural of *crudité* rawness, from Latin *cruditas* indigestion, from *crudus*] (1960)
: pieces of raw vegetables (as celery or carrot sticks) served as an hors d'oeuvre often with a dip

cru•di•ty \'krü-də-tē\ *noun, plural* **-ties** (1547)
1 : the quality or state of being crude
2 : something that is crude

cru•el \'krü(-ə)l\ *adjective* **cru•el•er** *or* **cru•el•ler; cru•el•est** *or* **cru•el•lest** [Middle English, from Old French, from Latin *crudelis,* from *crudus*] (14th century)
1 : disposed to inflict pain or suffering **:** devoid of humane feelings
2 a : causing or conducive to injury, grief, or pain ⟨a *cruel* joke⟩ **b :** unrelieved by leniency
synonym see FIERCE
— cru•el•ly \'krü-(ə-)lē\ *adverb*
— cru•el•ness *noun*

cru•el•ty \'krü(-ə)l-tē\ *noun, plural* **-ties** [Middle English *cruelte,* from Old French *cruelté,* from Latin *crudelitat-, crudelitas,* from *crudelis*] (13th century)
1 : the quality or state of being cruel
2 a : a cruel action **b :** inhuman treatment
3 : marital conduct held (as in a divorce action) to endanger life or health or to cause mental suffering or fear

cru•et \'krü-ət\ *noun* [Middle English, from Anglo-French, diminutive of Old French *crue,* of Germanic origin; akin to Old English *crūce*] (14th century)
1 : a vessel to hold wine or water for the Eucharist
2 : a usually glass bottle used to hold a condiment (as oil or vinegar) for use at the table

¹cruise \'krüz\ *verb* **cruised; cruis•ing** [Dutch *kruisen* to make a cross, cruise, from Middle Dutch *crucen,* from *crūce* cross, from Latin *cruc-, crux*] (1651)
intransitive verb
1 : to sail about touching at a series of ports
2 : to move or proceed speedily, smoothly, or effortlessly ⟨I'll *cruise* over to her house to see if she's home⟩
3 : to travel without destination or purpose
4 a : to go about the streets at random but on the lookout for possible developments ⟨the cabdriver *cruised* for an hour before being hailed⟩ **b :** to search (as in public places) for a sexual partner
5 a *of an airplane* **:** to fly at the most efficient operating speed **b** *of an automobile* **:** to travel at a speed suitable for being maintained for a long distance
transitive verb
1 : to cruise over or about
2 : to inspect (as land) with reference to possible lumber yield
3 a : to search in (a public place) for a sexual partner **b :** to approach and suggest sexual relations to

²cruise *noun* (circa 1706)
: an act or an instance of cruising; *especially* **:** a tour by ship

cruise control *noun* (1968)
1 : an electronic device in an automobile that controls the throttle so as to maintain a constant speed
2 : a relaxed and seemingly automatic pace that is easily maintained

cruise missile *noun* (1959)
: a guided missile that has a terrain-following radar system and that flies at moderate speed and low altitude

cruis•er \'krü-zər\ *noun* (1695)
1 : a vehicle that cruises: as **a :** SQUAD CAR **b :** a powerboat with facilities (as a cabin and plumbing) necessary for living aboard — called also *cabin cruiser*
2 : a large fast moderately armored and gunned warship usually of 6000 to 15,000 tons displacement
3 : a person who cruises

crul•ler \'krə-lər\ *noun* [Dutch *krulle,* a twisted cake, from *krul* curly, from Middle Dutch *crul* — more at CURL] (1818)
1 : a small sweet cake in the form of a twisted strip fried in deep fat
2 *Northern & Midland* **:** an unraised doughnut

¹crumb \'krəm\ *noun* [Middle English *crumme,* from Old English *cruma;* akin to Middle High German *krume* crumb] (before 12th century)
1 a : a small fragment especially of something baked (as bread) **b :** a porous aggregate of soil particles
2 : BIT ⟨a *crumb* of good news⟩
3 : the soft part of bread
4 *slang* **:** a worthless person

²crumb *transitive verb* (14th century)
1 : to break into crumbs
2 : to cover or thicken with crumbs
3 : to remove crumbs from ⟨*crumb* a table⟩

¹crum•ble \'krəm-bəl\ *verb* **crum•bled; crum•bling** \-b(ə-)liŋ\ [alteration of Middle English *kremelen,* frequentative of Old English *gecrymian* to crumble, from *cruma*] (1570)
transitive verb
: to break into small pieces
intransitive verb
1 : to fall into small pieces **:** DISINTEGRATE
2 : to break down completely **:** COLLAPSE ⟨marriages *crumble*⟩

²crumble *noun* (1820)

1 : something crumbled **:** fine debris
2 : CRISP 2

crum·blings \ˈkrəm-b(ə-)liŋz\ *noun plural* (1660)
: crumbled particles **:** CRUMBS

crum·bly \-b(ə-)lē\ *adjective* **crum·bli·er; -est** (1523)
: easily crumbled **:** FRIABLE ⟨*crumbly* soil⟩
— crum·bli·ness *noun*

crumb structure *noun* (circa 1906)
: a soil condition suitable for farming in which the soil particles are aggregated into crumbs

crumhorn *variant of* KRUMMHORN

crum·mie *or* **crum·my** \ˈkrə-mē\ *noun, plural* **crummies** [Scots *crumb* crooked, from Middle English, from Old English] (1724)
chiefly Scottish **:** cow; *especially* **:** one with crumpled horns

crum·my *also* **crumby** \ˈkrə-mē\ *adjective* **crum·mi·er** *also* **crumb·i·er; -est** [Middle English *crumme*] (1567)
1 *obsolete* **:** CRUMBLY
2 a : MISERABLE, FILTHY **b :** CHEAP, WORTHLESS
— crum·mi·ness *noun*

[1]crump \ˈkrəmp\ *intransitive verb* [imitative] (1646)
1 : CRUNCH
2 : to explode heavily

[2]crump *noun* (1914)
1 : a crunching sound
2 : SHELL, BOMB

[3]crump *adjective* [perhaps alteration of *crimp* friable] (1787)
chiefly Scottish **:** BRITTLE

crum·pet \ˈkrəm-pət\ *noun* [perhaps from Middle English *crompid* (*cake*) wafer, literally, curled-up cake, from *crumped,* past participle of *crumpen* to curl up, from *crump, crumb* crooked, from Old English *crumb;* akin to Old High German *krump* crooked] (1769)
: a small round unsweetened bread cooked on a griddle and usually split and toasted before serving

[1]crum·ple \ˈkrəm-pəl\ *verb* **crum·pled; crum·pling** \-p(ə-)liŋ\ [Middle English *crumplen,* frequentative of Middle English *crumpen*] (14th century)
transitive verb
1 : to press, bend, or crush out of shape **:** RUMPLE
2 : to cause to collapse
intransitive verb
1 : to become crumpled
2 : COLLAPSE

[2]crumple *noun* (15th century)
: a wrinkle or crease made by crumpling
— crum·ply \ˈkrəm-p(ə-)lē\ *adjective*

[1]crunch \ˈkrənch\ *verb* [alteration of *craunch*] (1814)
intransitive verb
1 : to chew or press with a crushing noise
2 : to make one's way with a crushing noise
transitive verb
1 : to chew, press, or grind with a crunching sound
2 : PROCESS; *especially* **:** to perform mathematical computations on ⟨*crunch* numbers⟩
— crunch·able \ˈkrən-chə-bəl\ *adjective*

[2]crunch *noun* (1836)
1 : an act of crunching
2 : a sound made by crunching
3 : a tight or critical situation: as **a :** a critical point in the buildup of pressure between opposing elements **:** SHOWDOWN **b :** a severe economic squeeze (as on credit) **c :** SHORTAGE ⟨an energy *crunch*⟩

crunch·er \ˈkrən-chər\ *noun* (1946)
1 : one that crunches
2 : a finishing blow

crunchy \ˈkrən-chē\ *adjective* **crunch·i·er; -est** (1913)
: making a crunching sound when chewed or pressed
— crunch·i·ly \-chə-lē\ *adverb*
— crunch·i·ness \-chē-nəs\ *noun*

crup·per \ˈkrə-pər, ˈkrů-\ *noun* [Middle English *cruper,* from Old French *crupiere,* from *croupe* hindquarters — more at CROUP] (14th century)
1 : a leather loop passing under a horse's tail and buckled to the saddle
2 : [1]CROUP; *broadly* **:** BUTTOCKS

cru·ral \ˈkrůr-əl\ *adjective* [Latin *crur-, crus* leg] (1599)
: of or relating to the thigh or leg; *specifically* **:** FEMORAL ⟨*crural* artery⟩

crus \ˈkrüs, ˈkrəs\ *noun, plural* **cru·ra** \ˈkrůr-ə\ [Latin *crur-, crus*] (circa 1751)
: any of various anatomical parts that resemble a leg or a pair of legs

[1]cru·sade \krü-ˈsād\ *noun* [blend of Middle French *croisade* & Spanish *cruzada;* both ultimately from Latin *cruc-, crux* cross] (circa 1708)
1 *capitalized* **:** any of the military expeditions undertaken by Christian powers in the 11th, 12th, and 13th centuries to win the Holy Land from the Muslims
2 : a remedial enterprise undertaken with zeal and enthusiasm ◆

[2]crusade *intransitive verb* **cru·sad·ed; cru·sad·ing** (1732)
: to engage in a crusade
— cru·sad·er *noun*

cru·sa·do \krü-ˈsā-(ˌ)dō\ *also* **cru·za·do** \-ˈzä-(ˌ)dō, -(ˌ)dü\ *noun, plural* **-does** *or* **-dos** [Portuguese *cruzado,* literally, marked with a cross] (1544)
: an old gold or silver coin of Portugal having a cross on the reverse

cruse \ˈkrüz, ˈkrüs\ *noun* [Middle English; akin to Old English *crūse* pitcher] (13th century)
: a small vessel (as a jar or pot) for holding a liquid (as water or oil)

[1]crush \ˈkrəsh\ *verb* [Middle English *crusshen,* from Middle French *cruisir,* of Germanic origin; akin to Middle Low German *krossen* to crush] (15th century)
transitive verb
1 a : to squeeze or force by pressure so as to alter or destroy structure **b :** to squeeze together into a mass
2 : HUG, EMBRACE
3 : to reduce to particles by pounding or grinding
4 a : to suppress or overwhelm as if by pressure or weight **b :** to oppress or burden grievously **c :** to subdue completely
5 : CROWD, PUSH
6 *archaic* **:** DRINK
intransitive verb
1 *obsolete* **:** CRASH
2 : to become crushed
3 : to advance with or as if with crushing
— crush·able \ˈkrə-shə-bəl\ *adjective*
— crush·er *noun*
— crush·ing·ly *adverb*

[2]crush *noun* (1599)
1 : an act of crushing
2 : the quantity of material crushed
3 a : a crowding together (as of people) **b :** CROWD, MOB; *especially* **:** a crowd of people pressing against one another
4 : an intense and usually passing infatuation ⟨have a *crush* on someone⟩; *also* **:** the object of infatuation
synonym see CROWD
— crush·proof \-ˌprüf\ *adjective*

crust \ˈkrəst\ *noun* [Middle English, from Latin *crusta;* akin to Old English *hrūse* earth, Greek *kryos* icy cold, *krystallos* ice, crystal] (14th century)
1 a : the hardened exterior or surface part of bread **b :** a piece of this or of bread grown dry or hard
2 : the pastry cover of a pie
3 : a hard or brittle external coat or covering: as **a :** a hard surface layer (as of soil or snow) **b :** the outer part of a planet, moon, or asteroid composed essentially of crystalline rocks **c :** a deposit built up on the interior surface of a wine bottle during long aging **d :** an encrusting deposit of dried secretions or exudate; *especially* **:** SCAB
4 : GALL, NERVE
— crust *verb*
— crust·al \ˈkrəs-tᵊl\ *adjective*
— crust·less \ˈkrəs(t)-ləs\ *adjective*

crus·ta·cea \ˌkrəs-ˈtā-sh(ē-)ə\ *noun plural* [New Latin, group name, from neuter plural of *crustaceus*] (1814)
: arthropods that are crustaceans

crus·ta·cean \ˌkrəs-ˈtā-shən\ *noun* (1835)
: any of a large class (Crustacea) of mostly aquatic mandibulate arthropods that have a chitinous or calcareous and chitinous exoskeleton, a pair of often much modified appendages on each segment, and two pairs of antennae and that include the lobsters, shrimps, crabs, wood lice, water fleas, and barnacles
— crustacean *adjective*

crus·ta·ceous \-shəs\ *adjective* [New Latin *crustaceus,* from Latin *crusta* crust, shell] (circa 1656)
: of, relating to, having, or forming a crust or shell; *especially* **:** CRUSTOSE

crus·tose \ˈkrəs-ˌtōs\ *adjective* [Latin *crustosus* crusted, from *crusta*] (circa 1879)
: having a thin thallus adhering closely to a substrate (as of rock, bark, or soil) ⟨*crustose* lichens⟩ — compare FOLIOSE, FRUTICOSE

crusty \ˈkrəs-tē\ *adjective* **crust·i·er; -est** (14th century)
1 : having or being a crust
2 : giving an effect of surly incivility in address or disposition
synonym see BLUFF
— crust·i·ly \-tə-lē\ *adverb*
— crust·i·ness \-tē-nəs\ *noun*

[1]crutch \ˈkrəch\ *noun* [Middle English *crucche,* from Old English *crycc;* akin to Old High German *krucka* crutch] (before 12th century)
1 a : a support typically fitting under the armpit for use by the disabled in walking **b :** PROP, STAY
2 : a forked leg rest constituting the pommel of a sidesaddle
3 : the crotch of a human being or an animal

◇ WORD HISTORY

crusade In 1095 when Palestinian lands sacred to Christians were held by Muslims, Pope Urban II convoked a council at Clermont-Ferrand in France and exhorted the assembly to go forth and reclaim the Holy Sepulcher and other venerated sites from the infidels. Those who responded to the call wore on their breasts crosses of cloth to identify themselves as soldiers in Christ's army. These crosses would provide the name for their enterprise and succeeding invasions of the Holy Land in following centuries. Medieval French words for such a holy war against infidels were *croisement, croiserie, croisée,* and *croisade,* all derivatives of *crois* "cross." The last word, *croisade,* was formed under the influence of Old Provençal *crozata* and Spanish *cruzada,* likewise formed from words meaning "cross." English borrowed both French *croisade* and Spanish *cruzada,* blending the two to produce *crusade* in the 18th century, long after the crusades themselves had ceased.

\ə\ **abut** \ᵊ\ **kitten** \ər\ **further** \a\ **ash** \ā\ **ace** \ä\ **mop, mar** \aů\ **out** \ch\ **chin** \e\ **bet** \ē\ **easy** \g\ **go** \i\ **hit** \ī\ **ice** \j\ **job** \ŋ\ **sing** \ō\ **go** \ȯ\ **law** \ȯi\ **boy** \th\ **thin** \th̲\ **the** \ü\ **loot** \ů\ **foot** \y\ **yet** \zh\ **vision** *see also* Guide to Pronunciation

4 : a forked support

²crutch *transitive verb* (1681)
: to support on crutches **:** prop up

crux \'krəks, 'krúks\ *noun, plural* **crux·es** *also* **cru·ces** \'krü-,sēz\ [Latin *cruc-, crux* cross, torture] (1718)
1 : a puzzling or difficult problem **:** an unsolved question
2 : an essential point requiring resolution or resolving an outcome ⟨the *crux* of the problem⟩
3 : a main or central feature (as of an argument)

cru·za·do \krü-'zä-(,)dō, -(,)dü\ *noun, plural* **-dos** [Portuguese] (1986)
: the basic monetary unit of Brazil 1986–90

Cru·zan \krü-'zan\ *noun* [(assumed) American Spanish *cruzano,* from *Santa Cruz* Saint Croix] (1958)
: a native or inhabitant of Saint Croix
— Cruzan *adjective*

cru·zei·ro \krü-'zer-(,)ō, -(,)ü\ *noun, plural* **-ros** [Portuguese] (1927)
: the basic monetary unit of Brazil 1942–85 and 1991–93

crwth \'krüth\ *noun* [Welsh] (circa 1837)
: ³CROWD 1

¹cry \'krī\ *verb* **cried; cry·ing** [Middle English *crien,* from Old French *crier,* from Latin *quiritare* to make a public outcry, perhaps from *Quirit-, Quiris,* name for the Roman citizen] (13th century)
transitive verb
1 : to utter loudly **:** SHOUT
2 *archaic* **:** BEG, BESEECH
3 : to proclaim publicly **:** ADVERTISE ⟨*cry* their wares⟩
intransitive verb
1 : to call loudly **:** SHOUT
2 : to shed tears often noisily **:** WEEP, SOB
3 : to utter a characteristic sound or call
4 : to require or suggest strongly a remedy or disposition ⟨a hundred things which *cry* out for planning —Roger Burlingame⟩
— cry havoc : to sound an alarm
— cry over spilled milk : to express vain regrets for what cannot be recovered or undone
— cry wolf : to give alarm unnecessarily

²cry *noun, plural* **cries** (13th century)
1 : an instance of crying: as **a :** an inarticulate utterance of distress, rage, or pain **b** *obsolete* **:** OUTCRY, CLAMOR
2 a *obsolete* **:** PROCLAMATION **b** *plural, Scottish* **:** BANNS
3 : ENTREATY, APPEAL
4 : a loud shout
5 : WATCHWORD, SLOGAN
6 a : common report **b :** a general opinion
7 : the public voice raised in protest or approval
8 : a fit of weeping
9 : the characteristic sound or call of an animal
10 a : a pack of hounds **b :** PURSUIT — used in the phrase *in full cry*
11 : DISTANCE — usually used in the phrase *a far cry*

cry- *or* **cryo-** *combining form* [German *kryo-,* from Greek, from *kryos* — more at CRUST]
: cold **:** freezing ⟨*cry*onics⟩ ⟨*cryo*gen⟩

cry·ba·by \'krī-,bā-bē\ *noun* (1851)
: one who cries or complains easily or often

cry down *transitive verb* (1598)
: DISPARAGE, DEPRECIATE

cry·ing \'krī-iŋ\ *adjective* (1607)
1 : calling for notice ⟨a *crying* need⟩
2 : NOTORIOUS, HEINOUS ⟨a *crying* shame⟩

cryo·bi·ol·o·gy \,krī-ō-bī-'ä-lə-jē\ *noun* (1960)
: the study of the effects of extremely low temperature on biological systems
— cryo·bi·o·log·i·cal \-,bī-ə-'lä-ji-kəl\ *adjective*
— cryo·bi·ol·o·gist \-bī-'ä-lə-jist\ *noun*

cry off (1928)
transitive verb
: to call off (as a bargain)
intransitive verb
chiefly British **:** to beg off

cryo·gen \'krī-ə-jən\ *noun* (1875)
: a substance for obtaining low temperatures **:** REFRIGERANT — called also *cryogenic*

cryo·gen·ic \,krī-ə-'je-nik\ *adjective* (1896)
1 a : of or relating to the production of very low temperatures **b :** being or relating to very low temperatures
2 a : requiring or involving the use of a cryogenic temperature **b :** requiring cryogenic storage **c :** suitable for storage of a cryogenic substance
— cryo·gen·i·cal·ly \-ni-k(ə-)lē\ *adverb*

cryo·gen·ics \-niks\ *noun plural but singular in construction* (circa 1934)
: a branch of physics that deals with the production and effects of very low temperatures

cryo·lite \'krī-ə-,līt\ *noun* (1801)
: a mineral consisting of a fluoride of sodium and aluminum found in Greenland usually in white cleavable masses and used as a source of aluminum

cry·on·ics \krī-'ä-niks\ *noun plural but usually singular in construction* [*cry-* + *-onics* (as in *electronics*)] (1967)
: the practice of freezing a dead diseased human in hopes of restoring life at some future time when a cure for the disease has been developed
— cry·on·ic \-nik\ *adjective*

cryo·phil·ic \,krī-ə-'fi-lik\ *adjective* (1942)
: thriving at low temperatures

cryo·pres·er·va·tion \,krī-ō-,pre-zər-'vā-shən\ *noun* (1968)
: preservation (as of cells) by subjection to extremely low temperatures
— cryo·pre·serve \,krī-ō-pri-'zərv, 'krī-ō-pri-,\ *transitive verb*

cryo·probe \'krī-ə-,prōb\ *noun* (1965)
: a blunt chilled instrument used to freeze tissues in cryosurgery

cryo·pro·tec·tive \,krī-ō-prə-'tek-tiv\ *adjective* (1967)
: serving to protect from freezing ⟨an extracellular *cryoprotective* agent⟩
— cryo·pro·tec·tant \-'tek-tənt\ *noun or adjective*

cryo·scope \'krī-ə-,skōp\ *noun* (1920)
: an instrument for determining freezing points

cry·os·co·py \krī-'äs-kə-pē\ *noun* [International Scientific Vocabulary] (circa 1900)
: the determination of the lowered freezing points produced in liquid by dissolved substances in order to determine molecular weights of solutes and various properties of solutions
— cryo·scop·ic \,krī-ə-'skä-pik\ *adjective*

cryo·stat \'krī-ə-,stat\ *noun* [International Scientific Vocabulary] (1913)
: an apparatus for maintaining a constant low temperature especially below 0°C
— cryo·stat·ic \,krī-ə-'sta-tik\ *adjective*

cryo·sur·gery \,krī-ō-'sərj-rē, -'sər-jə-\ *noun* (1962)
: surgery in which the tissue to be treated or operated on is frozen (as by liquid nitrogen)
— cryo·sur·geon \-'sər-jən\ *noun*
— cryo·sur·gi·cal \-ji-kəl\ *adjective*

cryo·ther·a·py \-'ther-ə-pē\ *noun* (1926)
: the therapeutic use of cold

crypt \'kript\ *noun* [Latin *crypta,* from Greek *kryptē,* from feminine of *kryptos* hidden, from *kryptein* to hide; perhaps akin to Lithuanian *krauti* to pile up] (1789)
1 a : a chamber (as a vault) wholly or partly underground; *especially* **:** a vault under the main floor of a church **b :** a chamber in a mausoleum
2 a : an anatomical pit or depression **b :** a simple tubular gland

crypt- *or* **crypto-** *combining form* [New Latin, from Greek *kryptos*]
1 : hidden **:** covered ⟨*crypto*genic⟩
2 : CRYPTOGRAPHIC ⟨*crypt*analysis⟩

crypt·anal·y·sis \,krip-tə-'na-lə-səs\ *noun* (1923)
1 : the solving of cryptograms or cryptographic systems
2 : the theory of solving cryptograms or cryptographic systems **:** the art of devising methods for this
— crypt·an·a·lyt·ic \,krip-,ta-nᵊl-'i-tik\ *also* **crypt·an·a·lyt·i·cal** \-ti-kəl\ *adjective*

crypt·an·a·lyst \krip-'ta-nᵊl-ist\ *noun* (1921)
: a specialist in cryptanalysis

cryp·ta·rithm \'krip-tə-,ri-thəm\ *noun* [*crypt-* + *-arithm* (as in *logarithm*)] (1943)
: an arithmetic problem in which letters have been substituted for numbers and which is solved by finding all possible pairings of digits with letters that produce a numerically correct answer

cryp·tic \'krip-tik\ *adjective* [Late Latin *crypticus,* from Greek *kryptikos,* from *kryptos*] (circa 1638)
1 : SECRET, OCCULT
2 a : having or seeming to have a hidden or ambiguous meaning **:** MYSTERIOUS ⟨*cryptic* messages⟩ **b :** marked by an often perplexing brevity ⟨*cryptic* marginal notes⟩
3 : serving to conceal ⟨*cryptic* coloration in animals⟩; *also* **:** exhibiting cryptic coloration ⟨*cryptic* animals⟩
4 : UNRECOGNIZED ⟨a *cryptic* infection⟩
5 : employing cipher or code
synonym see OBSCURE
— cryp·ti·cal·ly \-ti-k(ə-)lē\ *adverb*

¹cryp·to \'krip-(,)tō\ *noun, plural* **cryptos** [*crypt-*] (1946)
: a person who adheres or belongs secretly to a party, sect, or other group

²crypto *adjective* (1952)
1 : CRYPTOGRAPHIC
2 : not openly avowed or declared — often used in combination ⟨*crypto*-fascist⟩

cryp·to·coc·co·sis \,krip-tə-(,)kä-'kō-səs\ *noun, plural* **-co·ses** \-(,)sēz\ [New Latin] (1938)
: an infectious disease that is caused by a fungus (*Cryptococcus neoformans*) and is characterized by the production of nodular lesions or abscesses in the lungs, subcutaneous tissues, joints, and especially the brain and meninges

cryp·to·coc·cus \-'kä-kəs\ *noun, plural* **-coc·ci** \-'käk-,sī, -,sē, -'kä-,kī, -,kē\ [New Latin] (circa 1902)
: any of a genus (*Cryptococcus*) of budding imperfect fungi that resemble yeasts and include a number of saprophytes and a few serious pathogens
— cryp·to·coc·cal \-'kä-kəl\ *adjective*

cryp·to·crys·tal·line \,krip-tō-'kris-tə-lən\ *adjective* [International Scientific Vocabulary] (1862)
: having a crystalline structure so fine that no distinct particles are recognizable under the microscope ⟨*cryptocrystalline* quartz⟩

cryp·to·gam \'krip-tə-,gam\ *noun* [ultimately from Greek *kryptos* + *-gamia* -gamy] (1847)
: a plant (as a fern, moss, alga, or fungus) reproducing by spores and not producing flowers or seed
— cryp·to·gam·ic \,krip-tə-'ga-mik\ *or* **cryp·tog·a·mous** \krip-'tä-gə-məs\ *adjective*

cryp·to·gen·ic \,krip-tə-'je-nik\ *adjective* (1908)
: of obscure or unknown origin ⟨a *cryptogenic* disease⟩

cryp·to·gram \'krip-tə-,gram\ *noun* [French *cryptogramme,* from *crypt-* + *-gramme* -gram] (1880)
1 : a communication in cipher or code
2 : a figure or representation having a hidden significance

cryp·to·graph \'krip-tə-,graf\ *noun* (1849)
: CRYPTOGRAM

cryp·tog·ra·pher \krip-'tä-grə-fər\ *noun* (1641)

: a specialist in cryptography: as **a :** a clerk who enciphers and deciphers messages **b :** one who devises cryptographic methods or systems **c :** CRYPTANALYST

cryp·to·graph·ic \ˌkrip-tə-ˈgra-fik\ *adjective* (1824)
: of, relating to, or using cryptography
— **cryp·to·graph·i·cal·ly** \-fi-k(ə-)lē\ *adverb*

cryp·tog·ra·phy \krip-ˈtä-grə-fē\ *noun* [New Latin *cryptographia,* from *crypt-* + *-graphia* -graphy] (1658)
1 : secret writing
2 : the enciphering and deciphering of messages in secret code or cipher
3 : CRYPTANALYSIS

cryp·tol·o·gy \krip-ˈtä-lə-jē\ *noun* (1935)
: the scientific study of cryptography and cryptanalysis
— **cryp·to·log·i·cal** \ˌkrip-tə-ˈlä-ji-kəl\ *or* **cryp·to·log·ic** \-jik\ *adjective*
— **cryp·tol·o·gist** \krip-ˈtä-lə-jist\ *noun*

cryp·to·me·ria \ˌkrip-tə-ˈmir-ē-ə\ *noun* [New Latin, genus name, from *crypt-* + Greek *meros* part — more at MERIT] (1841)
: an evergreen tree (*Cryptomeria japonica*) of the pine family that is a valuable timber tree of Japan

cryp·to·nym \ˈkrip-tə-ˌnim\ *noun* (1876)
: a secret name

crypt·or·chid \krip-ˈtȯr-kəd\ *noun* [New Latin *cryptorchid-, cryptorchis,* from *crypt-* + *orchid-, orchis* testicle, from Greek *orchis* — more at ORCHIS] (1874)
: one affected with cryptorchidism
— **cryptorchid** *adjective*

crypt·or·chi·dism \-kə-ˌdi-zəm\ *or* **crypt·or·chism** \-ˌki-zəm\ *noun* (circa 1882)
: a condition in which one or both testes fail to descend normally

cryp·to·zo·ol·o·gy \ˌkrip-tə-zō-ˈä-lə-jē, -zə-ˈwä-\ *noun* (1969)
: the study of the lore concerning legendary animals (as Sasquatch) especially in order to evaluate the possibility of their existence
— **cryp·to·zo·ol·o·gist** \-ˈä-lə-jist, ˈwä-\ *noun*

[1]**crys·tal** \ˈkris-t^{ə}l\ *noun* [Middle English *cristal,* from Old French, from Latin *crystallum,* from Greek *krystallos* — more at CRUST] (13th century)
1 : quartz that is transparent or nearly so and that is either colorless or only slightly tinged
2 : something resembling crystal in transparency and colorlessness
3 : a body that is formed by the solidification of a chemical element, a compound, or a mixture and has a regularly repeating internal arrangement of its atoms and often external plane faces
4 : a clear colorless glass of superior quality; *also* **:** objects or ware of such glass
5 : the glass or transparent plastic cover over a watch or clock dial
6 : a crystalline material used in electronics as a frequency-determining element or for rectification
7 : powdered methamphetamine

[2]**crystal** *adjective* (14th century)
1 : consisting of or resembling crystal **:** CLEAR, LUCID
2 : relating to or using a crystal ⟨a *crystal* radio receiver⟩

crystal ball *noun* (1855)
1 : a sphere especially of quartz crystal traditionally used by fortune-tellers
2 : a means or method of predicting future events

crystal gazing *noun* (1889)
1 : the art or practice of concentrating on a glass or crystal globe with the aim of inducing a psychic state in which divination can be performed
2 : the attempt to predict future events or make difficult judgments especially without adequate data
— **crystal gazer** *noun*

crystall- *or* **crystallo-** *combining form* [Greek *krystallos*]
: crystal ⟨*crystallo*graphy⟩

crys·tal·line \ˈkris-tə-lən *also* -ˌlīn, -ˌlēn\ *adjective* [Middle English *cristallin,* from Middle French & Latin; Middle French, from Latin *crystallinus,* from Greek *krystallinos,* from *krystallos*] (15th century)
1 : resembling crystal: as **a :** strikingly clear or sparkling ⟨*crystalline* air⟩ ⟨a *crystalline* lake⟩ **b :** CLEAR-CUT
2 : made of crystal **:** composed of crystals
3 : constituting or relating to a crystal
— **crys·tal·lin·i·ty** \ˌkris-tə-ˈli-nə-tē\ *noun*

crystalline lens *noun* (1794)
: the lens of the eye in vertebrates

crys·tal·lise *British variant of* CRYSTALLIZE

crys·tal·lite \ˈkris-tə-ˌlīt\ *noun* [German *Kristallit,* from Greek *krystallos*] (1805)
1 a : a minute mineral form (as in glassy volcanic rocks) that marks the beginning of crystallization **b :** a single grain in a polycrystalline substance
2 : MICELLE

crys·tal·lize *also* **crys·tal·ize** \ˈkris-tə-ˌlīz\ *verb* **-lized; -liz·ing** (1598)
transitive verb
1 : to cause to form crystals or assume crystalline form
2 : to cause to take a definite form ⟨tried to *crystallize* his thoughts⟩
3 : to coat with crystals especially of sugar ⟨*crystallize* grapes⟩
intransitive verb
: to become crystallized
— **crys·tal·liz·able** \-ˌlī-zə-bəl, ˌkris-tə-ˈ\ *adjective*
— **crys·tal·li·za·tion** \ˌkris-tə-li-ˈzā-shən\ *noun*
— **crys·tal·liz·er** *noun*

crys·tal·lo·graph·ic \ˌkris-tə-lə-ˈgra-fik\ *adjective* (1804)
: of or relating to crystals or crystallography
— **crys·tal·lo·graph·i·cal·ly** \-fi-k(ə-)lē\ *adverb*

crys·tal·log·ra·phy \ˌkris-tə-ˈlä-grə-fē\ *noun* (1802)
: a science that deals with the forms and structures of crystals
— **crys·tal·log·ra·pher** \-fər\ *noun*

crys·tal·loid \ˈkris-tə-ˌlȯid\ *noun* (1861)
: a substance that forms a true solution and is capable of being crystallized
— **crystalloid** *or* **crys·tal·loi·dal** \ˌkris-tə-ˈlȯi-d^{ə}l\ *adjective*

crystal pleat *noun* (1976)
: any of a series of narrow sharply pressed pleats all turned in one direction
— **crystal pleated** *adjective*

crystal violet *noun* (circa 1893)
: a triphenylmethane dye found in gentian violet

cry up *transitive verb* (1593)
: to praise publicly in order to enhance in value or repute

CS \ˌsē-ˈes\ *noun* [Ben B. *C*orson (born 1896) and Roger W. *S*taughton (died 1957) American chemists] (1960)
: a potent tear gas used especially for riot control

C–sec·tion \ˈsē-ˌsek-shən\ *noun* (1973)
: CESAREAN SECTION

cte·noid \ˈte-ˌnȯid, ˈtē-\ *adjective* [International Scientific Vocabulary, from Greek *ktenoeidēs,* from *kten-, kteis* comb — more at PECTINATE] (1872)
: having the margin toothed ⟨*ctenoid* scale⟩; *also* **:** having or consisting of ctenoid scales ⟨*ctenoid* fishes⟩

cteno·phore \ˈte-nə-ˌfōr, -ˌfȯr\ *noun* [ultimately from Greek *kten-, kteis* + *pherein* to carry — more at BEAR] (circa 1882)
: any of a phylum (Ctenophora) of marine animals superficially resembling jellyfishes but having biradial symmetry and swimming by means of eight meridional bands of transverse ciliated plates — called also *comb jelly*
— **cte·noph·o·ran** \tə-ˈnä-fə-rən\ *noun or adjective*

CT scan \ˌsē-ˈtē-\ *noun* [computerized *t*omography] (1974)
: CAT SCAN
— **CT scanner** *noun*

C–type \ˈsē-ˌtīp\ *adjective* (1967)
: relating to or being any of the oncornaviruses in which the structure containing the nucleic acid is spherical and centrally located ⟨*C-type* virus particles⟩

cua·dri·lla \kwä-ˈdrē-yə, -ˈdrēl-\ *noun* [Spanish, diminutive of *cuadra* square, from Latin *quadra*] (1893)
: the team assisting the matador in the bullring

cub \ˈkəb\ *noun* [origin unknown] (1530)
1 a : a young carnivorous mammal (as a bear, fox, or lion) **b :** a young shark
2 : a young person
3 : APPRENTICE; *especially* **:** an inexperienced newspaper reporter

cub·age \ˈkyü-bij\ *noun* (1840)
: cubic content, volume, or displacement

Cu·ban heel \ˈkyü-bən-\ *noun* [*Cuba,* West Indies] (1908)
: a broad medium-high heel with a moderately curved back

cu·ba·ture \ˈkyü-bə-ˌchu̇r, -chər, -ˌtyu̇r, -ˌtu̇r\ *noun* [*cube* + *-ature* (as in *quadrature*)] (1679)
: the process of determining the volume of a solid

cub·by \ˈkə-bē\ *noun, plural* **cubbies** [obsolete English *cub* pen, from Dutch *kub* fish basket; akin to Old English *cofa* den] (circa 1859)
: CUBBYHOLE

cub·by·hole \ˈkə-bē-ˌhōl\ *noun* (circa 1842)
: a small snug place (as for hiding or storage); *also* **:** a cramped space

[1]**cube** \ˈkyüb\ *noun* [Latin *cubus,* from Greek *kybos* die, cube] (1551)
1 a : the regular solid of six equal square sides — see VOLUME table **b :** something shaped like a cube ⟨ice *cube*⟩
2 : the product got by taking a number three times as a factor

[2]**cube** *adjective* (1570)
: raised to the third power

[3]**cube** *transitive verb* **cubed; cub·ing** (1588)
1 : to raise to the third power
2 : to form into a cube
3 : to cut partly through (a steak) in a checkered pattern to increase tenderness by breaking the fibers
— **cub·er** *noun*

[4]**cu·be** *or* **cu·bé** \ˈkyü-ˌbā, kyü-ˈ\ *noun* [American Spanish *cubé*] (1924)
: any of several tropical American plants (genus *Lonchocarpus*) furnishing rotenone

cu·beb \ˈkyü-ˌbeb\ *noun* [Middle French *cubebe,* from Old French, from Medieval Latin *cubeba,* from Arabic *kubābah*] (14th century)
: the dried unripe berry of a tropical shrub (*Piper cubeba*) of the pepper family that is crushed and smoked in cigarettes for catarrh

cube root *noun* (circa 1696)
: a number whose cube is a given number

cube steak *noun* (1930)
: a thin slice of beef that has been cubed

[1]**cu·bic** \ˈkyü-bik\ *adjective* (15th century)
1 : having the form of a cube **:** CUBICAL
2 a : relating to the cube considered as a crystal form **b :** ISOMETRIC 1
3 a : THREE-DIMENSIONAL **b :** being the volume of a cube whose edge is a specified unit ⟨*cubic* inch⟩

4 : of third degree, order, or power ⟨a *cubic* polynomial⟩

[2]**cubic** *noun* (1799)
: a cubic curve, equation, or polynomial

cu·bi·cal \'kyü-bi-kəl\ *adjective* (15th century)
1 : CUBIC; *especially* **:** shaped like a cube
2 : relating to volume
— cu·bi·cal·ly \-k(ə-)lē\ *adverb*

cubic equation *noun* (circa 1751)
: a polynomial equation in which the highest sum of exponents of variables in any term is three

cu·bi·cle \'kyü-bi-kəl\ *noun* [Latin *cubiculum,* from *cubare* to lie, recline] (15th century)
1 : a sleeping compartment partitioned off from a large room
2 : a small partitioned space; *especially* **:** CARREL

cubic measure *noun* (1660)
: a unit (as cubic inch or cubic centimeter) for measuring volume — see METRIC SYSTEM table, WEIGHT table

cub·ism \'kyü-ˌbi-zəm\ *noun, often capitalized* (1911)
: a style of art that stresses abstract structure at the expense of other pictorial elements especially by displaying several aspects of the same object simultaneously and by fragmenting the form of depicted objects
— cub·ist \-bist\ *noun or adjective, often capitalized*
— cu·bis·tic \kyü-'bis-tik\ *adjective*

cu·bit \'kyü-bət\ *noun* [Middle English, from Latin *cubitum* elbow, cubit] (14th century)
: any of various ancient units of length based on the length of the forearm from the elbow to the tip of the middle finger and usually equal to about 18 inches (45.7 centimeters)

[1]**cu·boid** \'kyü-ˌbȯid\ *adjective* (circa 1828)
1 : approximately cubical in shape
2 : relating to or being the cuboid

[2]**cuboid** *noun* (1839)
: the outermost bone in the distal row of tarsal bones of many higher vertebrates

cu·boi·dal \kyü-'bȯi-dᵊl\ *adjective* (1803)
1 : somewhat cubical
2 : composed of nearly cubical elements ⟨*cuboidal* epithelium⟩

Cub Scout *noun* (circa 1935)
: a member of the scouting program of the Boy Scouts for boys in the first through fifth grades in school

cuck·ing stool \'kə-kiŋ-\ *noun* [Middle English *cucking stol,* literally, defecating chair] (12th century)
: a chair formerly used for punishing offenders (as dishonest tradesmen) by public exposure or ducking in water

cuck·old \'kə-kəld, -(ˌ)kōld\ *noun* [Middle English *cokewold*] (13th century)
: a man whose wife is unfaithful
— cuckold *transitive verb*

cuck·old·ry \-kəl-drē\ *noun* (1529)
1 : the practice of making cuckolds
2 : the state of being a cuckold

[1]**cuck·oo** \'kü-(ˌ)kü, 'ku̇-\ *noun, plural* **cuck·oos** [Middle English *cuccu,* of imitative origin] (13th century)
1 : a largely grayish brown European bird (*Cuculus canorus*) that is a parasite given to laying its eggs in the nests of other birds which hatch them and rear the offspring; *broadly* **:** any of a large family (Cuculidae of the order Cuculiformes) to which this bird belongs
2 : the call of the cuckoo
3 : a silly or slightly crackbrained person

cuckoo 1

[2]**cuckoo** *transitive verb* (1648)
: to repeat monotonously as a cuckoo does its call

[3]**cuckoo** *adjective* (1627)
1 : of, relating to, or resembling the cuckoo
2 : deficient in sense or intelligence **:** SILLY

cuckoo clock *noun* (1789)
: a wall or shelf clock that announces the hours by sounds resembling a cuckoo's call

cuck·oo·flow·er \'kü-(ˌ)kü-ˌflau̇(-ə)r, 'ku̇-\ *noun* (1578)
1 : a bitter cress (*Cardamine pratensis*) of Eurasia and North America
2 : RAGGED ROBIN

cuck·oo·pint \-ˌpint\ *noun* [Middle English *cuccupintel,* from *cuccu* + *pintel* pintle] (1551)
: a European arum (*Arum maculatum*) with erect spathe and short purple spadix

cuckoo spit *noun* (1592)
1 : a frothy secretion exuded on plants by the nymphs of spittle insects
2 : SPITTLEBUG

cu·cul·late \'kyü-kə-ˌlāt, kyü-'kə-lət\ *adjective* [Medieval Latin *cucullatus,* from Latin *cucullus* hood] (1794)
: having the shape of a hood ⟨a *cucullate* leaf⟩

cu·cum·ber \'kyü-(ˌ)kəm-bər\ *noun* [Middle English, from Middle French *cocombre,* from Latin *cucumer-, cucumis*] (14th century)
: the fruit of a vine (*Cucumis sativus*) of the gourd family cultivated as a garden vegetable; *also* **:** this vine

cucumber mosaic *noun* (1916)
: a virus disease especially of cucumbers that is transmitted chiefly by an aphid and produces mottled foliage and often pale warty fruits

cucumber tree *noun* (circa 1782)
: a magnolia (*Magnolia acuminata*) of the eastern U.S. having fruit resembling a small cucumber

cu·cur·bit \kyü-'kər-bət\ *noun* [Middle English *cucurbite,* from Middle French, from Latin *cucurbita* gourd] (14th century)
1 : a vessel or flask for distillation used with or forming part of an alembic
2 : a plant of the gourd family

cud \'kəd, *chiefly Southern* 'ku̇d *or* 'küd\ *noun* [Middle English *cudde,* from Old English *cwudu;* akin to Old High German *kuti* glue, Sanskrit *jatu* gum] (before 12th century)
1 : food brought up into the mouth by a ruminating animal from its first stomach to be chewed again
2 : [2]QUID

cud·bear \'kəd-ˌbar, -ˌber\ *noun* [irregular from Dr. *Cuthbert* Gordon, 18th century Scottish chemist] (1764)
: a reddish coloring matter from lichens

[1]**cud·dle** \'kə-dᵊl\ *verb* **cud·dled; cud·dling** \'kəd-liŋ, 'kə-dᵊl-iŋ\ [origin unknown] (1520)
transitive verb
: to hold close for warmth or comfort or in affection
intransitive verb
: to lie close or snug **:** NESTLE, SNUGGLE
— cud·dler \'kəd-lər, 'kə-dᵊl-ər\ *noun*

[2]**cuddle** *noun* (1825)
: a close embrace

cud·dle·some \'kə-dᵊl-səm\ *adjective* (1876)
: CUDDLY

cud·dly \'kəd-lē, 'kə-dᵊl-ē\ *adjective* **cud·dli·er; -est** (1863)
: fit for or inviting cuddling

[1]**cud·dy** \'kə-dē\ *noun, plural* **cuddies** [origin unknown] (1660)
1 : a usually small cabin or shelter (as on a sailboat)
2 : a small room or cupboard

[2]**cud·dy** *or* **cud·die** \'ku̇-dē, 'kə-\ *noun, plural* **cuddies** [perhaps from *Cuddy,* nickname for *Cuthbert*] (circa 1715)
1 *dialect British* **:** DONKEY
2 *dialect British* **:** BLOCKHEAD

[1]**cud·gel** \'kə-jəl\ *noun* [Middle English *kuggel,* from Old English *cycgel;* perhaps akin to Middle High German *kugele* ball] (before 12th century)
: a short heavy club

[2]**cudgel** *transitive verb* **-geled** *or* **-gelled; -gel·ing** *or* **-gel·ling** \'kəj-liŋ, 'kə-jə-\ (1596)
: to beat with or as if with a cudgel
— cudgel one's brains : to think hard (as for a solution to a problem)

cud·weed \'kəd-ˌwēd, *chiefly Southern* 'ku̇d- *or* 'küd-\ *noun* (1548)
: any of several composite plants (as of the genus *Gnaphalium*) with silky or woolly foliage

[1]**cue** \'kyü\ *noun* [Middle English *cu* half a farthing (spelled form of *q,* abbreviation for Latin *quadrans* quarter of an as)] (circa 1755)
: the letter *q*

[2]**cue** *noun* [probably from *qu,* abbreviation (used as a direction in actors' copies of plays) of Latin *quando* when] (1553)
1 a : a signal (as a word, phrase, or bit of stage business) to a performer to begin a specific speech or action **b :** something serving a comparable purpose **:** HINT
2 : a feature indicating the nature of something perceived
3 *archaic* **:** the part one has to perform in or as if in a play
4 *archaic* **:** MOOD, HUMOR

[3]**cue** *transitive verb* **cued; cu·ing** *or* **cue·ing** (1922)
1 : to give a cue to **:** PROMPT
2 : to insert into a continuous performance ⟨*cue* in sound effects⟩

[4]**cue** *noun* [French *queue,* literally, tail, from Latin *cauda*] (circa 1749)
1 a : a leather-tipped tapering rod for striking the cue ball (as in billiards and pool) **b :** a long-handled instrument with a concave head for shoving disks in shuffleboard
2 : QUEUE 2

[5]**cue** *verb* **cued; cu·ing** *or* **cue·ing** (circa 1784)
transitive verb
1 : QUEUE
2 : to strike with a cue
intransitive verb
1 : QUEUE
2 : to use a cue

cue ball *noun* (1881)
: the ball a player strikes with the cue in billiards and pool

cues·ta \'kwes-tə\ *noun* [Spanish, from Latin *costa* side, rib — more at COAST] (1818)
: a hill or ridge with a steep face on one side and a gentle slope on the other

[1]**cuff** \'kəf\ *noun* [Middle English *coffe, cuffe* mitten] (1522)
1 : something (as a part of a sleeve or glove) encircling the wrist
2 : the turned-back hem of a trouser leg
3 a : HANDCUFF — usually used in plural **b :** a usually wide metal band worn as a bracelet
4 : an inflatable band that is wrapped around an extremity to control the flow of blood through the part when recording blood pressure with a sphygmomanometer
— cuff·less \'kə-fləs\ *adjective*
— off the cuff : without preparation **:** AD LIB
— on the cuff : on credit

[2]**cuff** *transitive verb* (1693)
1 : to furnish with a cuff
2 : HANDCUFF

[3]**cuff** *verb* [perhaps from obsolete English, glove, from Middle English] (1530)
transitive verb
: to strike especially with or as if with the palm of the hand **:** BUFFET
intransitive verb
: FIGHT, SCUFFLE

[4]**cuff** *noun* (1570)
: a blow with the hand especially when open **:** SLAP

cuff link *noun* (1897)
: a usually ornamental device consisting of two parts joined by a shank, chain, or bar for

passing through buttonholes to fasten shirt cuffs — usually used in plural

cui bo·no \'kwē-'bō-(,)nō\ *noun* [Latin, to whose advantage?] (1604)
1 : a principle that probable responsibility for an act or event lies with one having something to gain
2 : usefulness or utility as a principle in estimating the value of an act or policy

cui·rass \kwi-'ras, kyu̇-\ *noun* [Middle English *curas,* from Middle French *curasse,* from Late Latin *coreacea,* feminine of *coreaceus* leathern, from Latin *corium* skin, leather; akin to Old English *heortha* deerskin, Latin *cortex* bark, Greek *keirein* to cut — more at SHEAR] (15th century)
1 : a piece of armor covering the body from neck to waist; *also* **:** the breastplate of such a piece
2 : something (as bony plates covering an animal) resembling a cuirass
— cui·rassed *adjective*

cui·ras·sier \,kwir-ə-'sir, ,kyu̇r-\ *noun* (1625)
: a mounted soldier wearing a cuirass

Cui·se·naire rod \,kwē-zə-'nar-, -'ner-\ *noun* [from *Cuisenaire,* a trademark] (1954)
: any of a set of colored rods usually of 1 centimeter cross section and of 10 lengths from 1 to 10 centimeters that are used for teaching number concepts and the basic operations of arithmetic

cui·sine \kwi-'zēn, kwē-\ *noun* [French, literally, kitchen, from Late Latin *coquina* — more at KITCHEN] (1786)
: manner of preparing food **:** style of cooking; *also* **:** the food prepared

cuisse \'kwis\ *also* **cuish** \'kwish\ *noun* [Middle English *cusseis,* plural, from Middle French *cuissaux,* plural of *cuissel,* from *cuisse* thigh, from Latin *coxa* hip — more at COXA] (15th century)
: a piece of plate armor for the front of the thigh — see ARMOR illustration

cuke \'kyük\ *noun* (1903)
: CUCUMBER

cul–de–sac \'kəl-di-,sak, 'ku̇l-; ,kəl-di-', ,ku̇l-\ *noun, plural* **culs–de–sac** \'kəl(z)-, 'ku̇l(z)-, ,kəl(z)-, ,ku̇l(z)-\ *also* **cul–de–sacs** \-,saks, -'saks\ [French, literally, bottom of the bag] (1738)
1 : a blind diverticulum or pouch
2 : a street or passage closed at one end
3 : BLIND ALLEY

cu·let \'kyü-lət, 'kə-lət\ *noun* [French, from diminutive of *cul* backside, from Latin *culus;* akin to Old Irish *cúl* back] (1678)
1 : the small flat facet at the bottom of a brilliant parallel to the table
2 : plate armor covering the buttocks

cu·lex \'kyü-,leks\ *noun* [New Latin, from Latin, gnat; akin to Old Irish *cuil* fly] (15th century)
: any of a large cosmopolitan genus (*Culex*) of mosquitoes that includes the common house mosquito (*C. pipiens*) of Europe and North America
— cu·li·cine \'kyü-lə-,sīn\ *adjective or noun*

cu·li·nar·i·an \,kə-lə-'ner-ē-ən, ,kyü-\ *noun* (1949)
: COOK, CHEF

cu·li·nary \'kə-lə-,ner-ē, 'kyü-\ *adjective* [Latin *culinarius,* from *culina* kitchen — more at KILN] (1638)
: of or relating to the kitchen or cookery
— cu·li·nar·i·ly \,kə-lə-'ner-ə-lē, ,kyü-\ *adverb*

[1]**cull** \'kəl\ *transitive verb* [Middle English, from Middle French *cuillir,* from Latin *colligere* to bind together — more at COLLECT] (13th century)
1 : to select from a group **:** CHOOSE ⟨*culled* the best passages from the poet's work⟩
2 : to identify and remove the culls from
— cull·er *noun*

[2]**cull** *noun* (1809)
: something rejected especially as being inferior or worthless ⟨how to separate good-looking pecans from *culls* —*Washington Post*⟩

cul·len·der \'kə-lən-dər\ *variant of* COLANDER

cul·let \'kə-lət\ *noun* [perhaps from French *cueillette* act of gathering, from Latin *collecta,* from feminine of *collectus,* past participle of *colligere*] (1817)
: broken or refuse glass usually added to new material to facilitate melting in making glass

cul·lion \'kəl-yən\ *noun* [Middle English *coillon* testicle, from Middle French, from (assumed) Vulgar Latin *coleon-, coleo,* from Latin *coleus* scrotum] (1575) *archaic*
: a mean or base fellow

[1]**cul·ly** \'kə-lē\ *noun, plural* **cullies** [perhaps alteration of *cullion*] (1664)
: one easily tricked or imposed on **:** DUPE

[2]**cully** *transitive verb* **cul·lied; cul·ly·ing** (1676)
archaic **:** CHEAT, DECEIVE

[1]**culm** \'kəlm\ *noun* [Middle English] (14th century)
: refuse coal screenings **:** SLACK

[2]**culm** *noun* [Latin *culmus* stalk — more at HAULM] (circa 1657)
: a monocotyledonous stem (as of a grass or sedge)

cul·mi·nant \'kəl-mə-nənt\ *adjective* (1605)
1 : being at greatest altitude or on the meridian
2 : fully developed

cul·mi·nate \'kəl-mə-,nāt\ *verb* **-nat·ed; -nat·ing** [Medieval Latin *culminatus,* past participle of *culminare,* from Late Latin, to crown, from Latin *culmin-, culmen* top — more at HILL] (1647)
intransitive verb
1 *of a celestial body* **:** to reach its highest altitude; *also* **:** to be directly overhead
2 a : to rise to or form a summit **b :** to reach the highest or a climactic or decisive point
transitive verb
: to bring to a head or to the highest point

cul·mi·na·tion \,kəl-mə-'nā-shən\ *noun* (1633)
1 : the action of culminating
2 : culminating position **:** CLIMAX
synonym see SUMMIT

cu·lotte \'kü-,lät, 'kyü-; kü-'lät, kyü-'\ *noun* [French, breeches, from diminutive of *cul* backside — more at CULET] (1911)
: a divided skirt; *also* **:** a garment having a divided skirt — often used in plural

cul·pa·ble \'kəl-pə-bəl\ *adjective* [Middle English *coupable,* from Middle French, from Latin *culpabilis,* from *culpare* to blame, from *culpa* guilt] (14th century)
1 *archaic* **:** GUILTY, CRIMINAL
2 : meriting condemnation or blame especially as wrong or harmful ⟨*culpable* negligence⟩
synonym see BLAMEWORTHY
— cul·pa·bil·i·ty \,kəl-pə-'bi-lə-tē\ *noun*
— cul·pa·ble·ness \'kəl-pə-bəl-nəs\ *noun*
— cul·pa·bly \-blē\ *adverb*

cul·prit \'kəl-prət, -,prit\ *noun* [Anglo-French *cul.* (abbreviation of *culpable* guilty) + *prest, prit* ready (i.e., to prove it), from Latin *praestus* — more at PRESTO] (1670)
1 : one accused of or charged with a crime
2 : one guilty of a crime or a fault
3 : the source or cause of a problem

cult \'kəlt\ *noun, often attributive* [French & Latin; French *culte,* from Latin *cultus* care, adoration, from *colere* to cultivate — more at WHEEL] (1617)
1 : formal religious veneration **:** WORSHIP
2 : a system of religious beliefs and ritual; *also* **:** its body of adherents
3 : a religion regarded as unorthodox or spurious; *also* **:** its body of adherents
4 : a system for the cure of disease based on dogma set forth by its promulgator ⟨health *cults*⟩
5 a : great devotion to a person, idea, object, movement, or work (as a film or book); *especially* **:** such devotion regarded as a literary or intellectual fad **b :** a usually small group of people characterized by such devotion
— cul·tic \'kəl-tik\ *adjective*
— cult·ish \-tish\ *adjective*
— cult·ish·ly \-lē\ *adverb*
— cult·ish·ness \-nəs\ *noun*
— cult·ism \'kəl-,ti-zəm\ *noun*
— cult·ist \'kəl-tist\ *noun*
— cult·like \-,līk\ *adjective*

cultch *also* **culch** \'kəlch\ *noun* [perhaps from a French dialect form of French *couche* couch] (1667)
1 : material (as oyster shells) laid down on oyster grounds to furnish points of attachment for the spat
2 *chiefly New England* **:** CLUTTER, TRASH

cul·ti·gen \'kəl-tə-jən\ *noun* [*culti*vated + *-gen*] (1924)
1 : a cultivated or domestic organism of a variety or species for which a wild ancestor is unknown
2 : CULTIVAR

cul·ti·va·ble \'kəl-tə-və-bəl\ *adjective* (1682)
: capable of being cultivated
— cul·ti·va·bil·i·ty \,kəl-tə-və-'bi-lə-tē\ *noun*

cul·ti·var \'kəl-tə-,vär, -,ver, -,var\ *noun* [*culti*vated + *var*iety] (1923)
: an organism of a kind originating and persistent under cultivation

cul·ti·vate \'kəl-tə-,vāt\ *transitive verb* **-vat·ed; -vat·ing** [Medieval Latin *cultivatus,* past participle of *cultivare,* from *cultivus* cultivable, from Latin *cultus,* past participle of *colere*] (circa 1655)
1 : to prepare or prepare and use for the raising of crops; *also* **:** to loosen or break up the soil about (growing plants)
2 a : to foster the growth of ⟨*cultivate* vegetables⟩ **b :** CULTURE 2a **c :** to improve by labor, care, or study **:** REFINE ⟨*cultivate* the mind⟩
3 : FURTHER, ENCOURAGE ⟨*cultivate* the arts⟩
4 : to seek the society of **:** make friends with
— cul·ti·vat·able \-,vā-tə-bəl\ *adjective*

cultivated *adjective* (1665)
: REFINED, EDUCATED ⟨*cultivated* speech⟩ ⟨*cultivated* tastes⟩

cul·ti·va·tion \,kəl-tə-'vā-shən\ *noun* (circa 1716)
1 : CULTURE, REFINEMENT
2 : the act or art of cultivating or tilling

cul·ti·va·tor \'kəl-tə-,vā-tər\ *noun* (1665)
: one that cultivates; *especially* **:** an implement for loosening the soil while crops are growing

cul·tur·al \'kəlch-rəl, 'kəl-chə-\ *adjective* (circa 1864)
1 : of or relating to culture or culturing
2 : concerned with the fostering of plant or animal growth
— cul·tur·al·ly \-rə-lē\ *adverb*

cultural anthropology *noun* (1933)
: anthropology that deals with human culture especially with respect to social structure, language, law, politics, religion, magic, art, and technology — compare PHYSICAL ANTHROPOLOGY
— cultural anthropologist *noun*

cul·tur·a·ti \,kəl-chə-'rä-(,)tē\ *noun plural* [[1]*culture* + *-ati* (as in *literati*)] (1964)
: people intensely interested in cultural affairs

[1]**cul·ture** \'kəl-chər\ *noun* [Middle English, from Middle French, from Latin *cultura,* from *cultus,* past participle] (15th century)
1 : CULTIVATION, TILLAGE
2 : the act of developing the intellectual and moral faculties especially by education
3 : expert care and training ⟨beauty *culture*⟩

4 a : enlightenment and excellence of taste acquired by intellectual and aesthetic training **b :** acquaintance with and taste in fine arts, humanities, and broad aspects of science as distinguished from vocational and technical skills **5 a :** the integrated pattern of human knowledge, belief, and behavior that depends upon man's capacity for learning and transmitting knowledge to succeeding generations **b :** the customary beliefs, social forms, and material traits of a racial, religious, or social group **c :** the set of shared attitudes, values, goals, and practices that characterizes a company or corporation
6 : cultivation of living material in prepared nutrient media; *also* **:** a product of such cultivation

²culture *transitive verb* **cul·tured; cul·tur·ing** \ˈkəlch-riŋ, ˈkəl-chə-\ (1510)
1 : CULTIVATE
2 a : to grow in a prepared medium **b :** to start a culture from

cultured *adjective* (circa 1746)
1 : CULTIVATED
2 : produced under artificial conditions ⟨*cultured* viruses⟩ ⟨*cultured* pearls⟩

culture shock *noun* (circa 1940)
: a sense of confusion and uncertainty sometimes with feelings of anxiety that may affect people exposed to an alien culture or environment without adequate preparation

cul·tus \ˈkəl-təs\ *noun* [Latin, adoration] (1640)
: CULT

cul·ver \ˈkəl-vər, ˈku̇l-\ *noun* [Middle English, from Old English *culfer,* from (assumed) Vulgar Latin *columbra,* from Latin *columbula,* diminutive of Latin *columba* dove — more at COLUMBINE] (before 12th century)
: PIGEON

cul·ver·in \ˈkəl-və-rən\ *noun* [Middle English, from Middle French *couleuvrine,* from *couleuvre* snake, from Latin *colubra*] (15th century)
: an early firearm: **a :** a rude musket **b :** a long cannon (as an 18-pounder) of the 16th and 17th centuries

cul·vert \ˈkəl-vərt\ *noun* [origin unknown] (1773)
1 : a transverse drain
2 : a conduit for a culvert
3 : a bridge over a culvert

cum \ˈku̇m, ˈkəm\ *conjunction* [Latin, with; akin to Latin *com-* — more at CO-] (circa 1873)
: along with being **:** AND — used to form usually hyphenated phrases ⟨a credible mining camp elder-*cum*-publican —G. B. Shaw⟩ ⟨Christian and Christian-*cum*-voodoo churches —David Binder⟩

¹cum·ber \ˈkəm-bər\ *transitive verb* **cum·bered; cum·ber·ing** \-b(ə-)riŋ\ [Middle English *combren,* perhaps from Old French *combrer* to prevent, from (assumed) Old French *combre* abatis] (14th century)
1 *archaic* **:** TROUBLE, HARASS
2 a : to hinder or encumber by being in the way ⟨*cumbered* with heavy clothing⟩ **b :** to clutter up ⟨rocks *cumbering* the yard⟩

²cumber *noun* (14th century)
: something that cumbers; *especially* **:** HINDRANCE

cum·ber·some \ˈkəm-bər-səm\ *adjective* (1535)
1 *dialect* **:** BURDENSOME, TROUBLESOME
2 : unwieldy because of heaviness and bulk
3 : slow-moving **:** PONDEROUS
synonym see HEAVY
— cum·ber·some·ly *adverb*
— cum·ber·some·ness *noun*

cum·brous \ˈkəm-b(ə-)rəs\ *adjective* (15th century)
: CUMBERSOME
synonym see HEAVY
— cum·brous·ly *adverb*
— cum·brous·ness *noun*

cum·in \ˈkə-mən, ˈkyü-\ *noun* [Middle English, from Old English *cymen,* from Latin *cuminum,* from Greek *kyminon,* of Semitic origin; akin to Akkadian *kamūnu* cumin] (before 12th century)
: a low plant (*Cuminum cyminum*) of the carrot family long cultivated for its aromatic seeds; *also* **:** the fruit or seed of cumin used as a spice

cum lau·de \ku̇m-ˈlau̇-də, -dē; ˌkəm-ˈlȯ-dē\ *adverb or adjective* [New Latin, with praise] (1893)
: with distinction ⟨graduated *cum laude*⟩ — compare MAGNA CUM LAUDE, SUMMA CUM LAUDE

cum·mer·bund \ˈkə-mər-ˌbənd, ˈkəm-bər-\ *also* **cum·ber·bund** \ˈkəm-bər-\ *noun* [Hindi *kamarband,* from Persian, from *kamar* waist + *band* band] (1616)
: a broad waistband usually worn in place of a vest with men's dress clothes and adapted in various styles of women's clothes

cum·shaw \ˈkəm-ˌshȯ\ *noun* [Chinese (Xiamen) *kam siā* grateful thanks] (1839)
: PRESENT, GRATUITY

cumul- *or* **cumuli-** *or* **cumulo-** *combining form* [New Latin, from Latin *cumulus*]
: cumulus and ⟨*cumulo*nimbus⟩

cu·mu·late \ˈkyü-myə-ˌlāt\ *verb* **-lat·ed; -lat·ing** [Latin *cumulatus,* past participle of *cumulare,* from *cumulus* mass] (1534)
transitive verb
1 : to gather or pile in a heap
2 : to combine into one
3 : to build up by addition of new material
intransitive verb
: to become massed
— cu·mu·late \-lət, -ˌlāt\ *adjective*
— cu·mu·la·tion \ˌkyü-myə-ˈlā-shən\ *noun*

cu·mu·la·tive \ˈkyü-myə-lə-tiv, -ˌlā-\ *adjective* (1605)
1 a : made up of accumulated parts **b :** increasing by successive additions
2 a : tending to prove the same point ⟨*cumulative* evidence⟩ **b :** additional rather than repeated ⟨*cumulative* legacy⟩
3 a : taking effect upon completion of another penal sentence ⟨*cumulative* sentence⟩ **b :** increasing in severity with repetition of the offense ⟨*cumulative* penalty⟩
4 : formed by the addition of new material of the same kind ⟨*cumulative* book index⟩
5 : summing or integrating overall data or values of a random variable less than or less than or equal to a specified value ⟨*cumulative* normal distribution⟩ ⟨*cumulative* frequency distribution⟩
— cu·mu·la·tive·ly *adverb*
— cu·mu·la·tive·ness *noun*

cumulative distribution function *noun* (circa 1965)
: DISTRIBUTION FUNCTION

cu·mu·li·form \ˈkyü-myə-lə-ˌfȯrm\ *adjective* (1885)
: of the form of a cumulus

cu·mu·lo·nim·bus \-ˈnim-bəs\ *noun* [New Latin] (1887)
: cumulus cloud often spread out in the shape of an anvil extending to great heights — see CLOUD illustration

cu·mu·lous \ˈkyü-myə-ləs\ *adjective* (1815)
: resembling cumulus

cu·mu·lus \-ləs\ *noun, plural* **-li** \-ˌlī, -ˌlē\ [Latin] (1659)
1 : HEAP, ACCUMULATION
2 [New Latin, from Latin] **:** a massy cloud form having a flat base and rounded outlines often piled up like a mountain — see CLOUD illustration

cunc·ta·tion \ˌkəŋ(k)-ˈtā-shən\ *noun* [Latin *cunctation-, cunctatio,* from *cunctari* to hesitate; akin to Sanskrit *śaṅkate* he wavers, Old English *hangian* to hang] (1585)
: DELAY
— cunc·ta·tive \ˈkeŋ(k)-ˌtā-tiv, -tə-tiv\ *adjective*

cu·ne·ate \ˈkyü-nē-ˌāt, -ət\ *adjective* [Latin *cuneatus,* from *cuneus* wedge] (1810)
: narrowly triangular with the acute angle toward the base — see LEAF illustration

¹cu·ne·i·form \kyu̇-ˈnē-ə-ˌfȯrm, ˈkyü-n(ē-)ə-\ *adjective* [probably from French *cunéiforme,* from Middle French, from Latin *cuneus* + Middle French *-iforme* -iform] (1677)
1 : having the shape of a wedge
2 : composed of or written in wedge-shaped characters ⟨*cuneiform* syllabary⟩

²cuneiform *noun* (1808)
1 : a cuneiform part; *specifically* **:** a cuneiform bone or cartilage
2 : cuneiform writing

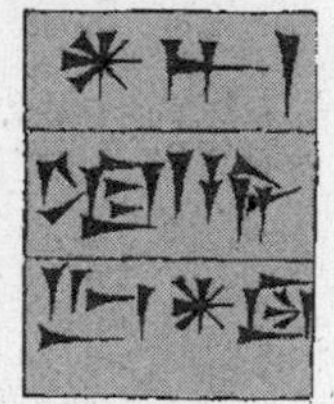
cuneiform 2

cun·ner \ˈkə-nər\ *noun* [origin unknown] (1602)
: either of two wrasses: **a :** a European wrasse (*Crenilabrus melops*) **b :** a wrasse (*Tautogolabrus adspersus*) common along the northeastern U.S. and adjacent Canadian coast

cun·ni·lin·gus \ˌkə-ni-ˈliŋ-gəs\ *also* **cun·ni·linc·tus** \-ˈliŋ(k)-təs\ *noun* [*cunnilingus,* New Latin, from Latin, one who licks the vulva, from *cunnus* vulva + *lingere* to lick; *cunnilinctus,* New Latin, from Latin *cunnus* + *linctus,* act of licking, from *lingere* — more at LICK] (1887)
: oral stimulation of the vulva or clitoris

¹cun·ning \ˈkə-niŋ\ *adjective* [Middle English, from present participle of *can* know] (14th century)
1 : dexterous or crafty in the use of special resources (as skill or knowledge) or in attaining an end ⟨a *cunning* plotter⟩
2 : displaying keen insight ⟨a *cunning* observation⟩
3 : characterized by wiliness and trickery ⟨*cunning* schemes⟩
4 : prettily appealing **:** CUTE
synonym see CLEVER, SLY
— cun·ning·ly \-niŋ-lē\ *adverb*
— cun·ning·ness *noun*

²cunning *noun* (14th century)
1 *obsolete* **a :** KNOWLEDGE, LEARNING **b :** magic art
2 : dexterous skill and subtlety (as in inventing, devising, or executing) ⟨high-ribbed vault . . . with perfect *cunning* framed —William Wordsworth⟩
3 : CRAFT, SLYNESS
synonym see ART

cunt \ˈkənt\ *noun* [Middle English *cunte;* akin to Middle Low German *kunte* female pudenda] (14th century)
1 : the female pudenda; *also* **:** coitus with a woman — usually considered obscene
2 : WOMAN 1a — usually used disparagingly; usually considered obscene

¹cup \ˈkəp\ *noun* [Middle English *cuppe,* from Old English, from Late Latin *cuppa* cup, alteration of Latin *cupa* tub — more at HIVE] (before 12th century)
1 : an open usually bowl-shaped drinking vessel
2 a : a drinking vessel and its contents **b :** the consecrated wine of the Communion
3 : something that falls to one's lot
4 : an ornamental cup offered as a prize (as in a championship)
5 a : something resembling a cup **b :** a cup-shaped plant organ **c :** an athletic supporter reinforced usually with plastic to provide extra protection to the wearer **d :** either of two parts of a brassiere that are shaped like and fit over the breasts **e :** the metal case inside a hole in golf; *also* **:** the hole itself
6 : a usually iced beverage resembling punch but served from a pitcher rather than a bowl
7 : a half pint **:** eight fluid ounces
8 : a food served in a cup-shaped usually footed vessel ⟨fruit *cup*⟩

9 : the symbol ∪ indicating the union of two sets — compare CAP 7
— **cup·like** \-ˌlīk\ *adjective*
— **in one's cups :** DRUNK

[2]**cup** *transitive verb* **cupped; cup·ping** (14th century)
1 : to treat by cupping
2 a : to curve into the shape of a cup ⟨*cupped* his hands around his mouth⟩ **b :** to place in or as if in a cup

cup·bear·er \ˈkəp-ˌbar-ər, -ˌber-\ *noun* (15th century)
: one who has the duty of filling and handing around the cups in which wine is served

cup·board \ˈkə-bərd\ *noun* (1530)
: a closet with shelves where dishes, utensils, or food is kept; *also* **:** a small closet

cup·cake \ˈkəp-ˌkāk\ *noun* (1828)
: a small cake baked in a cuplike mold

[1]**cu·pel** \kyü-ˈpel, ˈkyü-pəl\ *noun* [French *coupelle,* diminutive of *coupe* cup, from Late Latin *cuppa*] (1605)
: a small shallow porous cup especially of bone ash used in assaying to separate precious metals from lead

[2]**cupel** *transitive verb* **-pelled** *or* **-peled; -pel·ling** *or* **-pel·ing** (1644)
: to refine by means of a cupel
— **cu·pel·ler** *noun*

cu·pel·la·tion \ˌkyü-pə-ˈlā-shən, -ˌpe-\ *noun* (circa 1691)
: refinement (as of gold or silver) in a cupel by exposure to high temperature in a blast of air by which the lead, copper, tin, and other unwanted metals are oxidized and partly sink into the porous cupel

cup·ful \ˈkəp-ˌfu̇l\ *noun, plural* **cup·fuls** \-ˌfu̇lz\ *also* **cups·ful** \ˈkəps-ˌfu̇l\ (12th century)
1 : as much as a cup will hold
2 : CUP 7
usage see [2]-FUL

cup fungus *noun* (circa 1905)
: any of an order (Pezizales) of mostly saprophytic fungi with a fleshy or horny apothecium that is often colored and is typically shaped like a cup, saucer, or disk

Cu·pid \ˈkyü-pəd\ *noun* [Latin *Cupido*]
1 : the Roman god of erotic love — compare EROS
2 *not capitalized* **:** a figure that represents Cupid as a naked usually winged boy often holding a bow and arrow
word history see EROTIC

cu·pid·i·ty \kyü-ˈpi-də-tē\ *noun, plural* **-ties** [Middle English *cupidite,* from Middle French *cupidité,* from Latin *cupiditat-, cupiditas* — more at COVET] (15th century)
1 : inordinate desire for wealth **:** AVARICE, GREED
2 : strong desire **:** LUST

Cupid's bow *noun* (1858)
: a bow that consists of two convex curves usually with recurved ends

cup of tea (1932)
1 : something one likes or excels in ⟨I see already that storytelling isn't my *cup of tea* —John Barth⟩; *also* **:** a person suited to one's taste
2 : a thing to be reckoned with **:** MATTER ⟨poltergeists are a different *cup of tea* —D. B. W. Lewis⟩

cu·po·la \ˈkyü-pə-lə, ÷-ˌlō\ *noun* [Italian, from Latin *cupula,* diminutive of *cupa* tub] (1549)
1 a : a rounded vault resting on a usually circular base and forming a roof or a ceiling **b :** a small structure built on top of a roof
2 : a vertical cylindrical furnace for melting iron in the foundry that has tuyeres and tapping spouts near the bottom
3 : a raised observation post in the roof of a railroad caboose
— **cu·po·laed** \-ləd, ÷-ˌlōd\ *adjective*

cup·pa \ˈkə-pə\ *noun* [short for *cuppa tea,* pronunciation spelling of *cup of tea*] (1934)
chiefly British **:** a cup of tea

cup·ping *noun* (14th century)
: an operation of drawing blood to the surface of the body by use of a glass vessel evacuated by heat

cup·py \ˈkə-pē\ *adjective* **cup·pi·er; -est** (1882)
1 : resembling a cup
2 : full of small depressions ⟨a *cuppy* racetrack⟩

cupr- *or* **cupri-** *or* **cupro-** *combining form* [Late Latin *cuprum* — more at COPPER]
1 : copper ⟨*cupri*ferous⟩
2 : copper and ⟨*cupro*nickel⟩

cu·pric \ˈkyü-prik, ˈkü-\ *adjective* (1799)
: of, relating to, or containing copper with a valence of two

cu·prif·er·ous \kyü-ˈpri-f(ə-)rəs, kü-\ *adjective* (1784)
: containing copper

cu·prite \ˈkyü-ˌprīt, ˈkü-\ *noun* [German *Kuprit,* from Late Latin *cuprum*] (circa 1850)
: a mineral consisting of copper oxide and constituting an ore of copper

cu·pro·nick·el \ˌkyü-prō-ˈni-kəl, ˌkü-\ *noun* (1900)
: an alloy of copper and nickel; *especially* **:** one containing about 70 percent copper and 30 percent nickel

cu·prous \ˈkyü-prəs, ˈkü-\ *adjective* (1669)
: of, relating to, or containing copper with a valence of one

cu·pu·late \ˈkyü-pyə-ˌlāt, -lət\ *adjective* (1835)
: shaped like, having, or bearing a cupule

cu·pule \ˈkyü-(ˌ)pyü(ə)l\ *noun* [New Latin *cupula,* from Late Latin, diminutive of Latin *cupa* tub — more at HIVE] (1826)
: a cup-shaped anatomical structure: as **a :** an involucre characteristic of the oak in which the bracts are indurated and coherent **b :** an outer integument partially enclosing the seed of some seed ferns

cur \ˈkər\ *noun* [Middle English, short for *curdogge,* from (assumed) Middle English *curren* to growl (perhaps from Old Norse *kurra* to grumble) + Middle English *dogge* dog] (13th century)
1 : a mongrel or inferior dog
2 : a surly or cowardly fellow

cur·able \ˈkyu̇r-ə-bəl\ *adjective* (14th century)
: capable of being cured
— **cur·abil·i·ty** \ˌkyu̇r-ə-ˈbi-lə-tē\ *noun*
— **cur·able·ness** \ˈkyu̇r-ə-bəl-nəs\ *noun*
— **cur·ably** \-blē\ *adverb*

cu·ra·çao \ˈkyu̇r-ə-ˌsō, -ˌsau̇, ˈku̇r-; ˌk(y)u̇r-ə-ˈ\ *also* **cu·ra·çoa** *same, or* ˌkyu̇r-ə-ˈsō-ə\ *noun* [Dutch *curaçao,* from *Curaçao,* Netherlands Antilles] (1813)
: a liqueur flavored with the dried peel of the sour orange

cu·ra·cy \ˈkyu̇r-ə-sē\ *noun, plural* **-cies** (1682)
: the office or term of office of a curate

cu·ra·re *also* **cu·ra·ri** \kyu̇-ˈrär-ē, ku̇-\ *noun* [Portuguese & Spanish *curare,* from Carib *kurari*] (1777)
: a dried aqueous extract especially of a vine (as *Strychnos toxifera* of the family Loganiaceae or *Chondodendron tomentosum* of the family Menispermaceae) used by South American Indians to poison arrow tips and in medicine to produce muscular relaxation

cu·ra·rine \-ˈrär-ən, -ˌēn\ *noun* (circa 1872)
: any of several alkaloids from curare

cu·ra·rize \-ˈrär-ˌīz\ *transitive verb* **-rized; -riz·ing** (1875)
: to treat with curare
— **cu·ra·ri·za·tion** \-ˌrär-ə-ˈzā-shən\ *noun*

cu·ras·sow \ˈkyu̇r-ə-ˌsō, ˈku̇r-\ *noun* [alteration of *Curaçao*] (1685)
: any of several large arboreal game birds (family Cracidae, especially genus *Crax*) of South and Central America related to the domestic fowls

[1]**cu·rate** \ˈkyu̇r-ət *also* ˈkyu̇r-ˌāt\ *noun* [Middle English, from Medieval Latin *curatus,* from *cura* cure of souls, from Latin, care] (14th century)
1 : a clergyman in charge of a parish
2 : a clergyman serving as assistant (as to a rector) in a parish

[2]**cu·rate** \ˈkyu̇r-ˌāt, kyu̇-ˈrāt\ *transitive verb* **cu·rat·ed; cu·rat·ing** (1909)
: to act as curator of

curate's egg *noun* [from the story of a curate who was given a stale egg by his bishop and declared that parts of it were excellent] (1905)
British **:** something with both good and bad parts or qualities

cu·ra·tive \ˈkyu̇r-ə-tiv\ *adjective* (15th century)
: relating to or used in the cure of diseases **:** tending to cure
— **curative** *noun*
— **cu·ra·tive·ly** *adverb*

cu·ra·tor \ˈkyu̇r-ˌā-tər, kyu̇-ˈrā-, ˈkyu̇r-ə-\ *noun* [Latin, from *curatus,* past participle of *curare* to care, from *cura* care] (1632)
: one that has the care and superintendence of something; *especially* **:** one in charge of a museum, zoo, or other place of exhibit
— **cu·ra·to·ri·al** \ˌkyu̇r-ə-ˈtōr-ē-əl, -ˈtȯr-\ *adjective*
— **cu·ra·tor·ship** \ˈkyu̇r-ˌā-tər-ˌship, kyu̇-ˈrā-, ˈkyu̇r-ə-\ *noun*

[1]**curb** \ˈkərb\ *noun* [Middle French *courbe* curve, curved piece of wood or iron, from *courbe* curved, from Latin *curvus*] (15th century)
1 : a bit that exerts severe pressure on a horse's jaws; *also* **:** the chain or strap attached to it
2 : an enclosing frame, border, or edging
3 : CHECK, RESTRAINT ⟨a price *curb*⟩
4 : a raised edge or margin to strengthen or confine
5 : an edging (as of concrete) built along a street to form part of a gutter
6 [from the fact that it originally transacted its business on the street] **:** a market for trading in securities not listed on a stock exchange

[2]**curb** *transitive verb* (1530)
1 : to furnish with a curb
2 : to check or control with or as if with a curb ⟨trying to *curb* her curiosity⟩
3 : to lead (a dog) to a suitable place (as a gutter) for defecation
synonym see RESTRAIN

curb·ing \ˈkər-biŋ\ *noun* (1838)
1 : the material for a curb
2 : CURB

curb service *noun* (1931)
: service extended (as by a restaurant) to persons sitting in parked automobiles

curb·side \ˈkərb-ˌsīd\ *noun* (1946)
1 : the side of a pavement bordered by a curb
2 : SIDEWALK

[1]**curb·stone** \-ˌstōn\ *noun* (1791)
: a stone or edging of concrete forming a curb

[2]**curbstone** *adjective* (1848)
1 : operating on the street without maintaining an office ⟨a *curbstone* broker⟩
2 : not having the benefit of training or experience ⟨a *curbstone* critic⟩

curb weight *noun* (1949)
: the weight of an automobile with standard equipment and fuel, oil, and coolant

curch \ˈkərch\ *noun* [Middle English] (14th century)
Scottish **:** KERCHIEF 1

cur·cu·lio \(ˌ)kər-ˈkyü-lē-ˌō\ *noun, plural* **-li·os** [Latin, grain weevil] (1756)

: any of various weevils; *especially* : one that injures fruit

[1]**curd** \'kərd\ *noun* [Middle English *crud;* probably akin to Old English *crūdan* to press — more at CROWD] (15th century)
1 : the thick casein-rich part of coagulated milk
2 : something suggesting the curd of milk
— **curdy** \'kər-dē\ *adjective*

[2]**curd** *verb* (15th century)
: COAGULATE, CURDLE

cur·dle \'kər-dᵊl\ *verb* **cur·dled; cur·dling** \'kərd-liŋ, 'kər-dᵊl-iŋ\ [frequentative of [2]*curd*] (1590)
intransitive verb
1 : to form curds; *also* : to congeal as if by forming curds ⟨a scream *curdled* in her throat⟩
2 : to go bad or wrong : SPOIL
transitive verb
1 : to cause curds to form in
2 : SPOIL, SOUR

[1]**cure** \'kyůr\ *noun* [Middle English, from Old French, from Medieval Latin & Latin; Medieval Latin *cura,* cure of souls, from Latin, care] (14th century)
1 a : spiritual charge : CARE **b** : pastoral charge of a parish
2 a : recovery or relief from a disease **b** : something (as a drug or treatment) that cures a disease **c** : a course or period of treatment ⟨take the *cure* for alcoholism⟩ **d** : SPA
3 : a complete or permanent solution or remedy ⟨seeking a *cure* for unemployment⟩
4 : a process or method of curing
— **cure·less** \-ləs\ *adjective*

[2]**cure** *verb* **cured; cur·ing** (14th century)
transitive verb
1 a : to restore to health, soundness, or normality **b** : to bring about recovery from
2 a : to deal with in a way that eliminates or rectifies ⟨his small size, which time would *cure* for him —William Faulkner⟩ **b** : to free from something objectionable or harmful
3 : to prepare or alter especially by chemical or physical processing for keeping or use
intransitive verb
1 a : to undergo a curing process **b** : SET 11
2 : to effect a cure
— **cur·er** *noun*

cu·ré \kyů-'rā, 'kyůr-,ā\ *noun* [French, from Old French, from Medieval Latin *curatus* — more at CURATE] (1655)
: a parish priest

cure–all \'kyůr-,ȯl\ *noun* (1821)
: a remedy for all ills : PANACEA

cu·ret·tage \,kyůr-ə-'täzh\ *noun* (1897)
: a surgical scraping or cleaning by means of a curette

[1]**cu·rette** *also* **cu·ret** \kyů-'ret\ *noun* [French *curette,* from *curer* to cure, from Latin *curare,* from *cura*] (1753)
: a surgical instrument that has a scoop, ring, or loop at the tip and is used in performing curettage

[2]**curette** *also* **curet** *transitive verb* **cu·rett·ed; cu·rett·ing** (1888)
: to perform curettage on
— **cu·rette·ment** \kyů-'ret-mənt\ *noun*

cur·few \'kər-(,)fyü\ *noun* [Middle English, from Middle French *covrefeu,* signal given to bank the hearth fire, curfew, from *covrir* to cover + *feu* fire, from Latin *focus* hearth] (14th century)
1 : the sounding of a bell at evening ⟨the *Curfew* tolls the knell of parting day —Thomas Gray⟩
2 a : a regulation enjoining the withdrawal of usually specified persons (as juveniles or military personnel) from the streets or the closing of business establishments or places of assembly at a stated hour **b** : a signal to announce the beginning of a curfew **c** : the hour at which a curfew becomes effective **d** : the period during which a curfew is in effect ◆

cu·ria \'kyůr-ē-ə, 'kůr-\ *noun, plural* **cu·ri·ae** \'kyůr-ē-,ē, 'kůr-ē-,ī\ [Latin, perhaps from *co-* + *vir* man — more at VIRILE] (1600)
1 a : a division of the ancient Roman people comprising several gentes of a tribe **b** : the place of assembly of one of these divisions
2 a : the court of a medieval king **b** : a court of justice
3 *often capitalized* : the body of congregations, tribunals, and offices through which the pope governs the Roman Catholic Church
— **cu·ri·al** \'kyůr-ē-əl, 'kůr-\ *adjective*

cu·rie \'kyůr-(,)ē, kyů-'rē\ *noun* [Marie & Pierre *Curie*] (1910)
1 : a unit quantity of any radioactive nuclide in which 3.7×10^{10} disintegrations occur per second
2 : a unit of radioactivity equal to 3.7×10^{10} disintegrations per second ◆

Curie point *noun* (1911)
1 : the temperature at which there is a transition between the ferromagnetic and paramagnetic phases
2 : a temperature at which the anomalies that characterize a ferroelectric substance disappear — called also *Curie temperature*

cu·rio \'kyůr-ē-,ō\ *noun, plural* **cu·ri·os** [short for *curiosity*] (1851)
: something (as a decorative object) considered novel, rare, or bizarre : CURIOSITY; *also* : an unusual or bizarre person

cu·ri·o·sa \,kyůr-ē-'ō-sə, -'ō-zə\ *noun plural* [New Latin, from Latin, neuter plural of *curiosus*] (1883)
: CURIOSITIES, RARITIES; *especially* : unusual or erotic books

cu·ri·os·i·ty \,kyůr-ē-'ä-s(ə-)tē\ *noun, plural* **-ties** (14th century)
1 : desire to know: **a** : inquisitive interest in others' concerns : NOSINESS **b** : interest leading to inquiry ⟨intellectual *curiosity*⟩
2 *archaic* : undue nicety or fastidiousness
3 a : one that arouses interest especially for uncommon or exotic characteristics **b** : an unusual knickknack : CURIO **c** : a curious trait or aspect

cu·ri·ous \'kyůr-ē-əs\ *adjective* [Middle English, from Middle French *curios,* from Latin *curiosus* careful, inquisitive, from *cura* cure] (14th century)
1 a *archaic* : made carefully **b** *obsolete* : ABSTRUSE **c** *archaic* : precisely accurate
2 a : marked by desire to investigate and learn **b** : marked by inquisitive interest in others' concerns : NOSY
3 : exciting attention as strange, novel, or unexpected : ODD ☆
— **cu·ri·ous·ness** *noun*

cu·ri·ous·ly \-lē\ *adverb* (1665)
1 : in a curious manner ⟨seemed *curiously* calm⟩
2 : as is curious ⟨*curiously,* he continues to win reelection⟩

cu·ri·um \'kyůr-ē-əm\ *noun* [New Latin, from Marie & Pierre *Curie*] (1946)
: a metallic radioactive element produced artificially — see ELEMENT table

[1]**curl** \'kər(-ə)l\ *verb* [Middle English, from *crul* curly, probably from Middle Dutch; akin to Old High German *krol* curly] (14th century)
transitive verb
1 : to form (as the hair) into coils or ringlets
2 : to form into a curved shape : TWIST ⟨*curled* his lip in a sneer⟩
3 : to furnish with curls
intransitive verb
1 a : to grow in coils or spirals **b** : to form ripples or crinkles ⟨bacon *curling* in a pan⟩
2 : to move or progress in curves or spirals : WIND ⟨the path *curled* along the mountainside⟩
3 : TWIST, CONTORT
4 : to play the game of curling

[2]**curl** *noun* (1602)
1 : a lock of hair that coils : RINGLET
2 : something having a spiral or winding form : COIL
3 : the action of curling : the state of being curled
4 : a curved or spiral marking in the grain of wood
5 : a hollow arch of water formed when the crest of a breaking wave spills forward

curl·er \'kər-lər\ *noun* (1638)
1 : a player of curling
2 : one that curls; *especially* : a device on which hair is wound for curling

cur·lew \'kər-(,)lü, 'kərl-(,)yü\ *noun, plural* **curlews** *or* **curlew** [Middle English, from Middle French *corlieu,* of imitative origin] (14th century)
: any of various largely brownish chiefly migratory birds (especially genus *Numenius*) having long legs and a long slender down-curved bill and related to the sandpipers and snipes

[1]**cur·li·cue** *also* **cur·ly·cue** \'kər-lē-,kyü, -li-\ *noun* [*curly* + *cue* a braid of hair] (1843)
: a fancifully curved or spiral figure : FLOURISH

[2]**curlicue** *verb* **-cued; -cu·ing** (1844)
intransitive verb
: to form curlicues

☆ SYNONYMS

Curious, inquisitive, prying mean interested in what is not one's personal or proper concern. CURIOUS, a neutral term, basically connotes an active desire to learn or to know ⟨children are *curious* about everything⟩. INQUISITIVE suggests impertinent and habitual curiosity and persistent quizzing ⟨dreaded the visits of their *inquisitive* relatives⟩. PRYING implies busy meddling and officiousness ⟨*prying* neighbors who refuse to mind their own business⟩.

◇ WORD HISTORY

curfew In Europe in the Middle Ages people were required to put out or cover their hearth fires by a certain time in the evening. Given the fact that houses were usually of wood and closely adjoining, a hearth fire left unattended overnight might spread and destroy neighboring houses or even the entire town. A bell was rung to let people know that the time had come for covering their fires. In Middle French this signal was called *covrefeu,* a compound of *covrir* "to cover," and *feu* "fire." The meaning of *covrefeu* was extended to any public signal calling on people to retire in the evening, and also to the time of evening when the signal was given. It was in these senses that English borrowed the word as *corfu* or *curfew* in the 14th century. Another English word based on French *covre* and a following noun object is *coverlet.*

curie The unit of measurement called the *curie* is named after Pierre (1859–1906) and Marie (1867–1934) Curie, the French chemists and physicists. The Curies were two of the most important and influential figures in modern physics. Their major joint contributions include the discovery (with Henri Becquerel) of radioactivity and the discovery and isolation of radium and polonium in 1898. In 1910 the first International Congress of Radiology honored the husband and wife team by establishing *curie* as a standard unit of measurement for radioactive material. Later it was used as a unit of radioactivity itself. The Curies were awarded the Nobel Prize for Physics in 1903, and Marie Curie was awarded the Nobel Prize for Chemistry in 1911.

transitive verb
: to embellish with curlicues

curl·ing \'kər-liŋ\ *noun* (1620)
: a game in which two teams of four players each slide curling stones over a stretch of ice toward a target circle

curling iron *noun* (1616)
: a rod-shaped usually metal instrument which is heated and around which a lock of hair to be curled or waved is wound

curling stone *noun* (1620)
: an ellipsoid stone or occasionally piece of iron with a gooseneck handle used in the game of curling

curl·pa·per \'kər(-ə)l-,pā-pər\ *noun* (circa 1817)
: a strip or piece of paper around which a lock of hair is wound for curling

curl up *intransitive verb* (1861)
: to arrange oneself in or as if in a ball or curl ⟨*curl up* by the fire⟩ ⟨*curl up* with a good book⟩

curly \'kər-lē\ *adjective* **curl·i·er; -est** (circa 1784)
1 : tending to curl; *also* **:** having curls
2 : having the grain composed of fibers that undulate without crossing and that often form alternating light and dark lines ⟨*curly* maple⟩
— curl·i·ness *noun*

curly–coat·ed retriever \'kər-lē-'kō-təd-\ *noun* (1885)
: any of a breed of sporting dogs with a short curly black or liver-colored coat

curly top *noun* (1901)
: a destructive virus disease especially of beets that kills young plants and causes curling and puckering of the leaves in older plants

cur·mud·geon \(,)kər-'mə-jən\ *noun* [origin unknown] (1577)
1 *archaic* **:** MISER
2 : a crusty, ill-tempered, and usually old man
— cur·mud·geon·li·ness \-lē-nəs\ *noun*
— cur·mud·geon·ly \-lē\ *adjective*

curn \'kərn\ *or* **cur·ran** \'kə-rən\ *noun* [Middle English *curn;* akin to Middle English *corn*] (14th century)
1 *Scottish* **:** GRAIN
2 *Scottish* **:** a small number **:** FEW

curr \'kər\ *intransitive verb* [imitative] (1677)
: to make a murmuring sound (as of doves)

cur·ragh *or* **cur·rach** \'kə-rə, -rək\ *noun* [Middle English *currok,* from Irish *currach, curach*] (15th century)
: a usually large coracle used especially on the west coast of Ireland

cur·rant \'kər-ənt, 'kə-rənt\ *noun* [Middle English *raison of Coraunte,* literally, raisin of Corinth] (14th century)
1 : a small seedless raisin grown chiefly in the Levant
2 : the acid edible fruit of several shrubs (genus *Ribes*) of the saxifrage family; *also* **:** a plant bearing currants

cur·ren·cy \'kər-ən(t)-sē, 'kə-rən(t)-\ *noun, plural* **-cies** (1699)
1 a : circulation as a medium of exchange **b :** general use, acceptance, or prevalence **c :** the quality or state of being current **:** CURRENTNESS
2 a : something (as coins, treasury notes, and banknotes) that is in circulation as a medium of exchange **b :** paper money in circulation **c :** a common article for bartering **d :** a medium of verbal or intellectual expression

1cur·rent \'kər-ənt, 'kə-rənt\ *adjective* [Middle English *curraunt,* from Old French *curant,* present participle of *courre* to run, from Latin *currere* — more at CAR] (14th century)
1 a *archaic* **:** RUNNING, FLOWING **b** (1) **:** presently elapsing (2) **:** occurring in or existing at the present time (3) **:** most recent ⟨the *current* issue⟩
2 : used as a medium of exchange
3 : generally accepted, used, practiced, or prevalent at the moment
— cur·rent·ly *adverb*
— cur·rent·ness *noun*

2current *noun* (14th century)
1 a : the part of a fluid body (as air or water) moving continuously in a certain direction **b :** the swiftest part of a stream **c :** a tidal or nontidal movement of lake or ocean water **d :** flow marked by force or strength
2 a : a tendency or course of events that is usually the result of an interplay of forces ⟨*currents* of public opinion⟩ **b :** a prevailing mood **:** STRAIN
3 : a flow of electric charge; *also* **:** the rate of such flow
synonym see TENDENCY

current assets *noun plural* (circa 1909)
: assets of a short-term nature that are readily convertible to cash

cur·ri·cle \'kər-i-kəl, 'kə-ri-\ *noun* [Latin *curriculum* running, chariot, from *currere*] (1752)
: a 2-wheeled chaise usually drawn by two horses

cur·ric·u·lar \kə-'ri-kyə-lər\ *adjective* (circa 1909)
: of or relating to a curriculum

cur·ric·u·lum \-ləm\ *noun, plural* **-la** \-lə\ *also* **-lums** [New Latin, from Latin, running, course] (1633)
1 : the courses offered by an educational institution
2 : a set of courses constituting an area of specialization

cur·ric·u·lum vi·tae \kə-'ri-kyə-ləm-'vē-,tī, -kə-ləm-, -'wē-,tī, -'vī-,tē\ *noun, plural* **cur·ric·u·la vitae** \-lə-\ [Latin, course of (one's) life] (1902)
: a short account of one's career and qualifications prepared typically by an applicant for a position

cur·ri·ery \'kər-ē-ə-rē, 'kə-rē-\ *noun, plural* **-er·ies** (circa 1889)
1 : the trade of a currier of leather
2 : a place where currying is done

cur·rish \'kər-ish\ *adjective* (15th century)
1 : IGNOBLE
2 : resembling a cur **:** MONGREL
— cur·rish·ly *adverb*

1cur·ry \'kər-ē, 'kə-rē\ *transitive verb* **cur·ried; cur·ry·ing** [Middle English *curraýen,* from Old French *correer* to prepare, curry, from (assumed) Vulgar Latin *conredare,* from Latin *com-* + a base of Germanic origin; akin to Gothic *garaiths* arrayed — more at READY] (13th century)
1 : to clean the coat of (as a horse) with a currycomb
2 : to treat (tanned leather) especially by incorporating oil or grease
3 : BEAT, THRASH
— cur·ri·er *noun*
— curry favor [Middle English *currayen favel* to curry a chestnut horse] **:** to seek to gain favor by flattery or attention

2cur·ry *also* **cur·rie** \'kər-ē, 'kə-rē\ *noun, plural* **curries** [Tamil *kari*] (1681)
1 : a food, dish, or sauce in Indian cuisine seasoned with a mixture of pungent spices; *also* **:** a food or dish seasoned with curry powder
2 : CURRY POWDER

3curry *transitive verb* **cur·ried; cur·ry·ing** (1839)
: to flavor or cook with curry powder or a curry sauce

cur·ry·comb \-,kōm\ *noun* (1573)
: a comb made of rows of metallic teeth or serrated ridges and used especially to curry horses
— currycomb *transitive verb*

curry powder *noun* (1810)
: a condiment consisting of several pungent ground spices (as cayenne pepper, fenugreek, and turmeric)

1curse \'kərs\ *noun* [Middle English *curs,* from Old English] (before 12th century)
1 : a prayer or invocation for harm or injury to come upon one **:** IMPRECATION
2 : something that is cursed or accursed
3 : evil or misfortune that comes as if in response to imprecation or as retribution
4 : a cause of great harm or misfortune **:** TORMENT
5 : MENSTRUATION — used with *the*

2curse *verb* **cursed; curs·ing** (before 12th century)
transitive verb
1 : to use profanely insolent language against **:** BLASPHEME
2 a : to call upon divine or supernatural power to send injury upon **b :** to execrate in fervent and often profane terms
3 : to bring great evil upon **:** AFFLICT
intransitive verb
: to utter imprecations **:** SWEAR
synonym see EXECRATE

cursed \'kər-səd, 'kərst\ *also* **curst** \'kərst\ *adjective* (13th century)
: being under or deserving a curse
— cursed·ly *adverb*
— cursed·ness *noun*

1cur·sive \'kər-siv\ *adjective* [French or Medieval Latin; French *cursif,* from Medieval Latin *cursivus,* literally, running, from Latin *cursus,* past participle of *currere* to run] (1784)
: RUNNING, COURSING: as **a** *of writing* **:** flowing often with the strokes of successive characters joined and the angles rounded **b :** having a flowing, easy, impromptu character
— cur·sive·ly *adverb*
— cur·sive·ness *noun*

2cursive *noun* (1861)
1 : a manuscript written in cursive writing; *also* **:** cursive writing
2 : a style of printed letter resembling handwriting

cur·sor \'kər-sər, -,sȯr\ *noun* [Latin, runner, from *currere*] (1594)
: a movable item used to mark a position: as **a :** a transparent slide with a line attached to a slide rule **b :** a visual cue (as a flashing rectangle) on a video display that indicates position (as for data entry)

cur·so·ri·al \,kər-'sōr-ē-əl, -'sȯr-\ *adjective* (1836)
: adapted to or involving running

cur·so·ry \'kərs-rē, 'kər-sə-\ *adjective* [Late Latin *cursorius* of running, from Latin *currere*] (1601)
: rapidly and often superficially performed or produced **:** HASTY ⟨a *cursory* glance⟩
synonym see SUPERFICIAL
— cur·so·ri·ly \-rə-lē\ *adverb*
— cur·so·ri·ness \-rē-nəs\ *noun*

curt \'kərt\ *adjective* [Latin *curtus* shortened — more at SHEAR] (1630)
1 a : sparing of words **:** TERSE **b :** marked by rude or peremptory shortness **:** BRUSQUE
2 : shortened in linear dimension
synonym see BLUFF
— curt·ly *adverb*
— curt·ness *noun*

cur·tail \(,)kər-'tā(ə)l\ *transitive verb* [by folk etymology from earlier *curtal* to dock an animal's tail, from *curtal,* noun, animal with a docked tail, from Middle French *courtault* — more at CURTAL] (1580)
: to make less by or as if by cutting off or away some part ⟨*curtail* the power of the executive branch⟩ ⟨*curtail* inflation⟩
synonym see SHORTEN
— cur·tail·er \-'tā-lər\ *noun*

cur·tail·ment \-'tā(ə)l-mənt\ *noun* (1794)
: the act of curtailing **:** the state of being curtailed

1cur·tain \'kər-tᵊn\ *noun* [Middle English *curtine,* from Old French, from Late Latin *cortina* (translation of Greek *aulaia,* from *aulē* court),

from Latin *cohort-, cohors* enclosure, court — more at COURT] (14th century)
1 : a hanging screen usually capable of being drawn back or up; *especially* **:** window drapery
2 : a device or agency that conceals or acts as a barrier — compare IRON CURTAIN
3 a : the part of a bastioned front that connects two neighboring bastions **b** (1) **:** a similar stretch of plain wall (2) **:** a nonbearing exterior wall
4 a : the movable screen separating the stage from the auditorium of a theater **b :** the ascent or opening (as at the beginning of a play) of a stage curtain; *also* **:** its descent or closing (as at the end of an act) **c :** the final situation, line, or scene of an act or play **d :** the time at which a theatrical performance begins **e** *plural* **:** END; *especially* **:** DEATH ⟨it will be *curtains* for us if we're caught⟩
— cur·tain·less \-ləs\ *adjective*

²curtain *transitive verb* **cur·tained; cur·tain·ing** \ˈkər-tᵊn-iŋ, -ˈkər-tᵊn-iŋ\ (14th century)
1 : to furnish with or as if with curtains
2 : to veil or shut off with or as if with a curtain

curtain call *noun* (1884)
: an appearance by a performer (as after the final curtain of a play) in response to the applause of the audience

curtain lecture *noun* [from its originally being given behind the curtains of a bed] (1633)
: a private lecture by a wife to her husband

cur·tain–rais·er \ˈkər-tᵊn-ˌrā-zər\ *noun* (1886)
1 : a short play usually of one scene that is presented before the main full-length drama
2 : a usually short preliminary to a main event

curtain wall *noun* (1853)
: a nonbearing exterior wall between columns or piers

cur·tal \ˈkər-tᵊl\ *adjective* [Middle French *courtault,* from *court* short, from Latin *curtus*] (1576)
1 *obsolete* **:** having a docked tail
2 *obsolete* **:** BRIEF, CURTAILED
3 *archaic* **:** wearing a short frock

cur·tal ax *or* **cur·tle ax** \ˈkər-tᵊl-\ *noun* [modification of Middle French *coutelas*] (circa 1580)
archaic **:** CUTLASS

cur·te·sy \ˈkər-tə-sē\ *noun, plural* **-sies** [Middle English *corteisie* courtesy] (1523)
: a husband's interest upon the death of his wife in the real property of an estate that she either solely owned or inherited provided they bore a child capable of inheriting the estate — compare DOWER

cur·ti·lage \ˈkər-tᵊl-ij\ *noun* [Middle English, from Middle French *cortillage,* from *cortil* courtyard, from *cort* court] (14th century)
: a piece of ground (as a yard or courtyard) within the fence surrounding a house

¹curt·sy *also* **curt·sey** \ˈkərt-sē\ *intransitive verb* **curt·sied** *also* **curt·seyed; curt·sy·ing** *also* **curt·sey·ing** [alteration of *courtesy*] (circa 1553)
: to make a curtsy

²curtsy *also* **curtsey** *noun, plural* **curtsies** *also* **curtseys** (1575)
: an act of civility, respect, or reverence made mainly by women and consisting of a slight lowering of the body with bending of the knees

cu·rule \ˈkyu̇r-ˌül\ *adjective* [Latin *curulis,* perhaps alteration of *currulis* of a chariot, from *currus* chariot, from *currere* to run] (1600)
: privileged to sit in a seat reserved in ancient Rome for the use of the highest dignitaries; *also* **:** of or relating to a curule chair

cur·va·ceous *also* **cur·va·cious** \ˌkər-ˈvā-shəs\ *adjective* (circa 1935)
: having curves suggestive of a well-proportioned feminine figure

cur·va·ture \ˈkər-və-ˌchu̇r, -chər, -ˌtyu̇r, -ˌtu̇r\ *noun* (1603)
1 : the act of curving **:** the state of being curved
2 : a measure or amount of curving; *specifically* **:** the rate of change of the angle through which the tangent to a curve turns in moving along the curve and which for a circle is equal to the reciprocal of the radius
3 a : an abnormal curving (as of the spine) **b :** a curved surface of an organ

¹curve \ˈkərv\ *adjective* [Middle English, from Latin *curvus;* akin to Greek *kyrtos* convex, Middle Irish *cruinn* round] (15th century)
archaic **:** bent or formed into a curve

²curve *verb* **curved; curv·ing** [Latin *curvare,* from *curvus*] (1594)
intransitive verb
: to have or take a turn, change, or deviation from a straight line or plane surface without sharp breaks or angularity
transitive verb
1 : to cause to curve
2 : to throw a curveball to (a batter)
3 : to grade (as an examination) on a curve

³curve *noun* (1696)
1 a : a line especially when curved: as (1) **:** the path of a moving point (2) **:** a line defined by an equation so that the coordinates of its points are functions of a single independent variable or parameter **b :** the graph of a variable
2 : something curved: as **a :** a curving line of the human body **b** *plural* **:** PARENTHESIS
3 a : CURVEBALL **b :** TRICK, DECEPTION
4 : a distribution indicating the relative performance of individuals measured against each other that is used especially in assigning good, medium, or poor grades to usually predetermined proportions of students rather than in assigning grades based on predetermined standards of achievement
— curvy \ˈkər-vē\ *adjective*

curve·ball \ˈkərv-ˌbȯl\ *noun* (1936)
: a slow or moderately fast baseball pitch thrown with spin to make it swerve downward and usually to the left when thrown from the right hand or to the right when thrown from the left hand
— curveball *verb*

curve fitting *noun* (circa 1924)
: the empirical determination of a curve or function that approximates a set of data

¹cur·vet \(ˌ)kər-ˈvet\ *noun* [Italian *corvetta,* from Middle French *courbette,* from *courber* to curve, from Latin *curvare*] (1575)
: a prancing leap of a horse in which the hind legs are raised just before the forelegs touch the ground

²curvet *intransitive verb* **-vet·ted** *or* **-vet·ed; -vet·ting** *or* **-vet·ing** (1592)
: to make a curvet; *also* **:** PRANCE, CAPER

cur·vi·lin·ear \ˌkər-və-ˈli-nē-ər\ *adjective* [Latin *curvus* + *linea* line] (1710)
1 : consisting of or bounded by curved lines **:** represented by a curved line
2 : marked by flowing tracery ⟨*curvilinear* Gothic⟩
— cur·vi·lin·ear·i·ty \-ˌli-nē-ˈar-ə-tē\ *noun*

cu·sec \ˈkyü-ˌsek\ *noun* [*cu*bic foot per *sec*ond] (circa 1903)
: a volumetric unit of flow equal to a cubic foot per second

cush·at \ˈkə-shət\ *noun* [Middle English *cowschote,* from Old English *cūscote*] (before 12th century)
chiefly Scottish **:** RINGDOVE 1

cu·shaw \ku̇-ˈshȯ, ˈkü-ˌ\ *noun* [origin unknown] (1698)
: a squash of any of several cultivars of a winter squash (*Cucurbita mixta*)

Cush·ing's disease \ˈku̇-shiŋz-\ *noun* [Harvey *Cushing*] (circa 1935)
: Cushing's syndrome especially when caused by excessive production of ACTH by the pituitary gland

Cushing's syndrome *noun* (1937)
: an abnormal bodily condition characterized by obesity and muscular weakness due to excess corticosteroids and especially hydrocortisone from adrenal or pituitary hyperfunction

¹cush·ion \ˈku̇-shən\ *noun* [Middle English *cusshin,* from Middle French *coissin,* from (assumed) Vulgar Latin *coxinus,* from Latin *coxa* hip — more at COXA] (14th century)
1 : a soft pillow or pad usually used for sitting, reclining, or kneeling
2 : a bodily part resembling a pad
3 : something resembling a cushion: as **a :** PILLOW 2 **b :** RAT 3 **c :** a pad of springy rubber along the inside of the rim of a billiard table **d :** a padded insert in a shoe **e :** an elastic body for reducing shock **f :** a mat laid under a large rug to ease the effect of wear
4 a : something (as an economic factor or a medical procedure) serving to mitigate the effects of disturbances or disorders **b :** a reserve supply (as of money) **c :** a comfortable lead ⟨a 4–0 *cushion* in the ninth inning⟩
— cush·ion·less \-ləs\ *adjective*
— cush·iony \-shə-nē\ *adjective*

²cushion *transitive verb* **cush·ioned; cush·ion·ing** \ˈku̇-sh(ə-)niŋ\ (circa 1738)
1 : to seat or place on a cushion
2 : to suppress by ignoring
3 : to furnish with a cushion
4 a : to mitigate the effects of **b :** to protect against force or shock
5 : to check gradually so as to minimize shock of moving parts

Cush·it·ic \kə-ˈshi-tik, ku̇-\ *noun* [*Cush* (Kush), Africa] (circa 1903)
: a subfamily of the Afro-Asiatic language family comprising various languages spoken in eastern Africa and especially in Ethiopia, Djibouti, and Somalia
— Cushitic *adjective*

cushy \ˈku̇-shē\ *adjective* **cush·i·er; cush·i·est** [Hindi *khush* pleasant, from Persian *khūsh*] (1915)
: entailing little hardship or difficulty ⟨a *cushy* job with a high salary⟩
— cush·i·ly \ˈku̇-shə-lē\ *adverb*

cusk \ˈkəsk\ *noun, plural* **cusk** *or* **cusks** [probably alteration of *tusk,* a kind of codfish] (1616)
1 : a large edible North Atlantic fish (*Brosme brosme*) of the cod family
2 : BURBOT

cusp \ˈkəsp\ *noun* [Latin *cuspis* point] (1585)
: POINT, APEX: as **a :** either horn of a crescent moon **b :** a fixed point on a mathematical curve at which a point tracing the curve would exactly reverse its direction of motion **c :** an ornamental pointed projection formed by or arising from the intersection of two arcs or foils **d** (1) **:** a point on the grinding surface of a tooth (2) **:** a fold or flap of a cardiac valve **e :** a point of transition (as from one astrological sign to another or one historical period to the next) **:** TURNING POINT; *also* **:** EDGE, VERGE ⟨on the *cusp* of stardom⟩

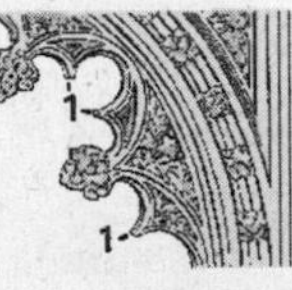

1 **cusp c**

— cus·pate \ˈkəs-ˌpāt, -pət\ *adjective*
— cusped \ˈkəspt\ *adjective*

cus·pid \ˈkəs-pəd\ *noun* [bi*cuspid*] (1878)
: CANINE 1

cus·pi·date \ˈkəs-pə-ˌdāt\ *adjective* [Latin *cuspidatus,* past participle of *cuspidare* to make pointed, from *cuspid-, cuspis* point] (1692)
: having a cusp **:** terminating in a point ⟨a *cuspidate* leaf⟩ ⟨*cuspidate* molars⟩

cus·pi·da·tion \ˌkəs-pə-ˈdā-shən\ *noun* (1848)
: decoration with cusps

cus·pi·dor \ˈkəs-pə-ˌdȯr, -ˌdōr\ *noun* [Portuguese *cuspidouro* place for spitting, from *cuspir* to spit, from Latin *conspuere,* from *com-* + *spuere* to spit — more at SPEW] (1735)
: SPITTOON

[1]**cuss** \ˈkəs\ *noun* [alteration of *curse*] (1775)
1 : FELLOW 4c
2 : CURSE 1, 2

[2]**cuss** *verb* (1815)
: CURSE
— **cuss·er** *noun*

cuss·ed \ˈkə-səd\ *adjective* (1840)
1 : CURSED
2 : OBSTINATE, CANTANKEROUS
— **cuss·ed·ly** *adverb*
— **cuss·ed·ness** \-nəs\ *noun*

cuss·word \ˈkəs-ˌwərd\ *noun* (1872)
1 : SWEARWORD
2 : a term of abuse : a derogatory term

cus·tard \ˈkəs-tərd\ *noun* [Middle English, a kind of pie, alteration of *crustarde, crustade,* probably from (assumed) Anglo-French *crustade,* from Old French *crouste* crust, from Latin *crusta* — more at CRUST] (1740)
: a pudding-like usually sweetened mixture made with eggs and milk
— **cus·tardy** \-tər-dē\ *adjective*

custard apple *noun* (1657)
1 a : any of several chiefly tropical American soft-fleshed edible fruits **b** : any of a genus (*Annona* of the family Annonaceae, the custard-apple family) of trees or shrubs bearing this fruit; *especially* : a small West Indian tree (*A. reticulata*)
2 : PAPAW 2

cus·to·di·al \ˌkəs-ˈtō-dē-əl\ *adjective* (1772)
: relating to guardianship; *specifically* : marked by or given to watching and protecting rather than seeking to cure ⟨*custodial* care⟩

cus·to·di·an \ˌkəs-ˈtō-dē-ən\ *noun* (1781)
: one that guards and protects or maintains; *especially* : one entrusted with guarding and keeping property or records or with custody or guardianship of prisoners or inmates
— **cus·to·di·an·ship** \-ˌship\ *noun*

cus·to·dy \ˈkəs-tə-dē\ *noun, plural* **-dies** [Middle English *custodie,* from Latin *custodia* guarding, from *custod-, custos* guardian] (15th century)
: immediate charge and control (as over a ward or a suspect) exercised by a person or an authority; *also* : SAFEKEEPING

[1]**cus·tom** \ˈkəs-təm\ *noun* [Middle English *custume,* from Old French, from Latin *consuetudin-, consuetudo,* from *consuescere* to accustom, from *com-* + *suescere* to accustom; akin to *suus* one's own — more at SUICIDE] (13th century)
1 a : a usage or practice common to many or to a particular place or class or habitual with an individual **b** : long-established practice considered as unwritten law **c** : repeated practice **d** : the whole body of usages, practices, or conventions that regulate social life
2 *plural* **a** : duties, tolls, or imposts imposed by the sovereign law of a country on imports or exports **b** *usually singular in construction* : the agency, establishment, or procedure for collecting such customs
3 a : business patronage **b** : usually habitual patrons : CUSTOMERS
synonym see HABIT

[2]**custom** *adjective* (1830)
1 : made or performed according to personal order
2 : specializing in custom work or operation ⟨a *custom* tailor⟩

cus·tom·ary \ˈkəs-tə-ˌmer-ē\ *adjective* (1535)
1 : based on or established by custom
2 : commonly practiced, used, or observed
synonym see USUAL
— **cus·tom·ar·i·ly** \ˌkəs-tə-ˈmer-ə-lē\ *adverb*
— **cus·tom·ar·i·ness** \ˈkəs-tə-ˌmer-ē-nəs\ *noun*

cus·tom–built \ˈkəs-təm-ˈbilt\ *adjective* (1925)
: built to individual specifications

cus·tom·er \ˈkəs-tə-mər\ *noun* [Middle English *custumer,* from *custume*] (15th century)
1 : one that purchases a commodity or service
2 : an individual usually having some specified distinctive trait ⟨a real tough *customer*⟩

cus·tom·house \ˈkəs-təm-ˌhau̇s\ *also* **cus·toms·house** \-təmz-\ *noun* (15th century)
: a building where customs and duties are paid or collected and where vessels are entered and cleared

cus·tom·ise *British variant of* CUSTOMIZE

cus·tom·ize \ˈkəs-tə-ˌmīz\ *transitive verb* **-ized; -iz·ing** (1926)
: to build, fit, or alter according to individual specifications
— **cus·tom·iz·er** *noun*

cus·tom–made \ˈkəs-tə(m)-ˈmād\ *adjective* (1855)
: made to individual specifications

cus·tom–tai·lor \-ˈtā-lər\ *transitive verb* (1895)
: to alter, plan, or build according to individual specifications or needs

[1]**cut** \ˈkət\ *verb* **cut; cut·ting** [Middle English *cutten*] (13th century)
transitive verb
1 a : to penetrate with or as if with an edged instrument **b** : to hurt the feelings of **c** : to strike sharply with a cutting effect **d** : to strike (a ball) with a glancing blow that imparts a reverse spin **e** : to experience the growth of (a tooth) through the gum
2 a : TRIM, PARE ⟨*cut* one's nails⟩ **b** : to shorten by omissions **c** : DISSOLVE, DILUTE, ADULTERATE **d** : to reduce in amount ⟨*cut* costs⟩
3 a : MOW, REAP **b** (1) : to divide into parts with an edged tool ⟨*cut* bread⟩ (2) : FELL, HEW **c** (1) : to separate or discharge from an organization : DETACH (2) : to single out and isolate ⟨*cut* a calf out from the herd⟩ **d** : to change the direction of sharply **e** : to go or pass around or about
4 a : to divide into segments **b** : INTERSECT, CROSS **c** : BREAK, INTERRUPT ⟨*cut* our supply lines⟩ **d** (1) : to divide (a deck of cards) into two portions (2) : to draw (a card) from the deck **e** : to divide into shares : SPLIT
5 a : to make by or as if by cutting: as (1) : CARVE ⟨*cut* stone⟩ (2) : to shape by grinding ⟨*cut* a diamond⟩ (3) : ENGRAVE (4) : to shear or hollow out **b** : to record sounds (as speech or music) on ⟨*cut* a record⟩ **c** : to type on a stencil
6 a : STOP, CEASE ⟨*cut* the nonsense⟩ **b** : to refuse to recognize (an acquaintance) : OSTRACIZE **c** : to absent oneself from (as a class) **d** : to stop (a motor) by opening a switch **e** : to stop the filming of (a motion-picture scene)
7 a : to engage in (a frolicsome or mischievous action) ⟨on summer nights strange capers are *cut* under the thin guise of a Christian festival —D. C. Peattie⟩ **b** : to give the appearance or impression of ⟨*cut* a fine figure⟩
8 : to be able to manage or handle — usually used in negative constructions ⟨can't *cut* that kind of work anymore⟩
intransitive verb
1 a : to function as or as if as an edged tool **b** : to undergo incision or severance ⟨cheese *cuts* easily⟩ **c** : to perform the operation of dividing, severing, incising, or intersecting **d** : to make a stroke with a whip, sword, or other weapon **e** : to wound feelings or sensibilities **f** : to cause constriction or chafing **g** : to be of effect, influence, or significance ⟨an analysis that *cuts* deep⟩
2 a (1) : to divide a pack of cards especially in order to decide the deal or settle a bet (2) : to draw a card from the pack **b** : to divide spoils : SPLIT
3 a : to proceed obliquely from a straight course ⟨*cut* across the yard⟩ **b** : to move swiftly ⟨a yacht *cutting* through the water⟩ **c** : to describe an oblique or diagonal line **d** : to change sharply in direction : SWERVE **e** : to make an abrupt transition from one sound or image to another in motion pictures, radio, or television
4 : to stop photographing motion pictures
— **cut a deal** : to negotiate an agreement
— **cut both ways** : to have both favorable and unfavorable results or implications
— **cut corners** : to perform some action in the quickest, easiest, or cheapest way
— **cut ice** : to be of importance — usually used in negative constructions
— **cut it** : to cut the mustard
— **cut the mustard** : to achieve the standard of performance necessary for success

[2]**cut** *noun* (1548)
1 : a product of cutting: as **a** : a creek, channel, or inlet made by excavation or worn by natural action **b** (1) : an opening made with an edged instrument (2) : a wound made by something sharp : GASH **c** : a surface or outline left by cutting **d** : a passage cut as a roadway **e** : a grade or step especially in a social scale ⟨a *cut* above the ordinary⟩ **f** : a subset of a set such that when it is subtracted from the set the remainder is not connected **g** : a pictorial illustration **h** : TRACK 1e(2)
2 : the act or an instance of cutting: as **a** : a gesture or expression that hurts the feelings ⟨made an unkind *cut*⟩ **b** : a straight passage or course **c** : a stroke or blow with the edge of a knife or other edged tool **d** : a lash with or as if with a whip **e** : the act of reducing or removing a part ⟨a *cut* in pay⟩ **f** : an act or turn of cutting cards; *also* : the result of cutting **g** : the elimination of part of a large field from further competition (as in a golf tournament) — often used with *miss* or *make* to denote respectively being or not being among those eliminated ⟨played poorly and missed the *cut*⟩
3 : something that is cut or cut off: as **a** : a length of cloth varying from 40 to 100 yards (36.6 to 91.4 meters) **b** : the yield of products cut especially during one harvest **c** : a segment or section of a meat carcass or a part of one **d** : a group of animals selected from a herd **e** : SHARE ⟨took his *cut* of the profits⟩
4 : a voluntary absence from a class
5 a : a stroke that cuts a ball; *also* : the spin imparted by such a stroke **b** : a swing by a batter at a pitched baseball **c** : an exchange of captures in checkers
6 : an abrupt transition from one sound or image to another in motion pictures, radio, or television
7 a : the shape and style in which a thing is cut, formed, or made ⟨clothes of the latest *cut*⟩ **b** : PATTERN, TYPE **c** : HAIRCUT
— **cut of one's jib** : APPEARANCE, STYLE

cut·abil·i·ty \ˌkə-tə-ˈbi-lə-tē\ *noun* (1965)
: the proportion of lean salable meat yielded by a carcass

cut–and–dried \ˌkət-ᵊn-ˈdrīd\ *also* **cut–and–dry** \-ˈdrī\ *adjective* (1710)
: being or done according to a plan, set procedure, or formula : ROUTINE

cut–and–paste \-ˈpāst\ *adjective* (1953)
: pieced together by excerpting and combining passages from many sources ⟨the book was a *cut-and-paste* job⟩

cut–and–try \-ˈtrī\ *adjective* (1903)
: marked by experimental procedure : EMPIRICAL

cu·ta·ne·ous \kyu̇-ˈtā-nē-əs\ *adjective* [New Latin *cutaneus,* from Latin *cutis* skin — more at HIDE] (1578)

\ə\ **abut** \ᵊ\ **kitten** \ər\ **further** \a\ **ash** \ā\ **ace** \ä\ **mop, mar** \au̇\ **out** \ch\ **chin** \e\ **bet** \ē\ **easy** \g\ **go** \i\ **hit** \ī\ **ice** \j\ **job** \ŋ\ **sing** \ō\ **go** \ȯ\ **law** \ȯi\ **boy** \th\ **thin** \t͟h\ **the** \ü\ **loot** \u̇\ **foot** \y\ **yet** \zh\ **vision** *see also* Guide to Pronunciation

: of, relating to, or affecting the skin
— **cu·ta·ne·ous·ly** *adverb*

¹cut·away \'kə-tə-,wā\ *adjective* (1841)
: having or showing parts cut away

²cutaway *noun* (1849)
1 : a coat with skirts tapering from the front waistline to form tails at the back
2 a : a cutaway picture or representation **b :** a shot that interrupts the main action of a film or television program to take up a related subject or to depict action supposed to be going on at the same time as the main action
3 : a back dive in which the head is lowered toward the board after the takeoff

cut·back \'kət-,bak\ *noun* (1897)
1 : something cut back
2 : REDUCTION

cut back (1871)
transitive verb
: to shorten by cutting **:** PRUNE
intransitive verb
1 : to interrupt the sequence of a plot (as of a movie) by introducing events prior to those last presented
2 : CUT DOWN ⟨*cut back* on sugar⟩

cutch \'kəch\ *noun* [modification of Malay *kachu*] (1759)
: CATECHU a

cut down (1821)
transitive verb
1 a : to strike down and kill or incapacitate **b :** KNOCK DOWN
2 a : to remodel by removing extras or unwanted furnishings and fittings **b :** to remake in a smaller size
intransitive verb
: to reduce or curtail volume or activity ⟨*cut down* on smoking⟩
— **cut down to size :** to reduce from an inflated or exaggerated importance to true or suitable stature

cute \'kyüt\ *adjective* **cut·er; cut·est** [short for *acute*] (circa 1731)
1 a : clever or shrewd often in an underhanded manner **b :** IMPERTINENT, SMART-ALECKY ⟨don't get *cute* with me⟩
2 : attractive or pretty especially in a dainty or delicate way
3 : obviously straining for effect
— **cute·ly** *adverb*
— **cute·ness** *noun*

cute·sy \'kyüt-sē\ *adjective* **cute·si·er; -est** [*cute* + *-sy* (as in *folksy*)] (1914)
: self-consciously or excessively cute

cut glass *noun* (1800)
: glass ornamented with patterns cut into its surface by an abrasive wheel and polished

cut–grass \'kət-,gras\ *noun* (circa 1818)
: a grass (especially genus *Leersia*) with minute hooked bristles along the edges of the leaf blade

cu·ti·cle \'kyü-ti-kəl\ *noun* [Latin *cuticula,* diminutive of *cutis* skin — more at HIDE] (1615)
1 : SKIN, PELLICLE: as **a :** an external investment (as of an insect) secreted usually by epidermal cells **b :** the outermost layer of animal integument (as in humans) when composed of epidermis **c :** a thin continuous fatty or waxy film on the external surface of many higher plants that consists chiefly of cutin
2 : dead or horny epidermis
— **cu·tic·u·lar** \kyü-'ti-kyə-lər\ *adjective*

cut·ie *or* **cut·ey** \'kyü-tē\ *noun, plural* **cuties** *or* **cuteys** [*cute* + *-ie*] (1908)
: an attractive person; *especially* **:** a pretty girl

cu·tin \'kyü-tᵊn\ *noun* [International Scientific Vocabulary, from Latin *cutis*] (circa 1872)
: an insoluble mixture containing waxes, fatty acids, soaps, and resinous material that forms a continuous layer on the outer epidermal wall of a plant

cut–in \'kət-,in\ *noun* (1883)
: something cut in
— **cut–in** *adjective*

cut in (1612)
intransitive verb
1 : to thrust oneself into a position between others or belonging to another
2 : to join in something suddenly ⟨*cut in* on the conversation⟩
3 : to interrupt a dancing couple and take one as one's partner
4 : to become automatically connected or started in operation
transitive verb
1 : to mix with cutting motions ⟨after sifting the flour into a mixing bowl, *cut* the lard *in*⟩
2 : to introduce into a number, group, or sequence
3 : to connect into an electrical circuit to a mechanical apparatus so as to permit operation
4 : to include especially among those benefiting or favored ⟨*cut* them *in* on the profits⟩

cu·tin·ized \'kyü-tᵊn-,īzd\ *adjective* (1901)
: infiltrated with cutin ⟨*cutinized* epidermal cells⟩

cu·tis \'kyü-təs\ *noun, plural* **cu·tes** \'kyü-,tēz\ *or* **cu·tis·es** [Latin] (1603)
: DERMIS

cut·lass \'kət-ləs\ *noun* [Middle French *coutelas,* augmentative of *coutel* knife, from Latin *cultellus,* diminutive of *culter* knife, plowshare] (1594)
1 : a short curving sword formerly used by sailors on warships
2 : MACHETE

cut·ler \'kət-lər\ *noun* [Middle English, from Middle French *coutelier,* from Late Latin *cultellarius,* from Latin *cultellus*] (14th century)
: one who makes, deals in, or repairs cutlery

cut·lery \'kət-lə-rē\ *noun* (15th century)
1 : the business of a cutler
2 : edged or cutting tools; *specifically* **:** implements for cutting and eating food

cut·let \'kət-lət\ *noun* [French *côtelette,* from Old French *costelette,* diminutive of *coste* rib, side, from Latin *costa* — more at COAST] (circa 1706)
1 : a small slice of meat ⟨a veal *cutlet*⟩
2 : a flat croquette of chopped meat or fish

cut·line \'kət-,līn\ *noun* (1943)
: CAPTION, LEGEND

cut·off \'kət-,ȯf\ *noun* (1741)
1 : the act or action of cutting off
2 a : the new and relatively short channel formed when a stream cuts through the neck of an oxbow **b :** SHORTCUT 1 **c :** a channel made to straighten a stream
3 : a device for cutting off
4 a : something cut off **b** *plural* **:** shorts originally made from jeans with the legs cut off at the knees or higher
5 : the point, date, or period for a cutoff
— **cutoff** *adjective*

cut off (14th century)
transitive verb
1 : to bring to an untimely end
2 : to stop the passage of
3 : SHUT OFF, BAR
4 : DISCONTINUE, TERMINATE
5 : SEPARATE, ISOLATE
6 : DISINHERIT
7 a : to stop the operation of **:** TURN OFF **b :** to stop or interrupt while in communication ⟨the operator *cut* me *off*⟩
intransitive verb
: to cease operating

cutoff man *noun* (1967)
: a player in baseball who relays a ball from an outfielder to the infield

cut·out \'kət-,au̇t\ *noun* (1851)
1 : something cut out or off from something else; *also* **:** the space or hole left after cutting
2 : one that cuts out
3 : one serving as an intermediary for clandestine operations
— **cutout** *adjective*

¹cut out (15th century)
transitive verb
1 : to form by erosion
2 : to determine or assign through necessity ⟨you've got your work *cut out* for you⟩
3 : to take the place of **:** SUPPLANT
4 : to put an end to **:** desist from
5 : DEPRIVE, DEFRAUD
6 a : to remove from a series or circuit **:** DISCONNECT **b :** to make inoperative
intransitive verb
1 : to depart in haste
2 : to cease operating
3 : to swerve out of a traffic line

²cut out *adjective* (1645)
: naturally fitted or suited ⟨not *cut out* to be a vet⟩

cut·over \'kət-'ō-vər\ *adjective* (1899)
: having most of the salable timber cut down

cut–price \-'prīs\ *adjective* (1897)
chiefly British **:** CUT-RATE

cut·purse \'kət-,pərs\ *noun* (14th century)
: PICKPOCKET

cut–rate \'kət-'rāt\ *adjective* (1881)
1 : marked by, offering, or making use of a reduced rate or price ⟨*cut-rate* stores⟩
2 : SECOND-RATE, CHEAP

cut·ta·ble \'kə-tə-bəl\ *adjective* (15th century)
: capable of being cut **:** ready for cutting

cut·ter \'kə-tər\ *noun* (15th century)
1 : one that cuts: as **a :** one whose work is cutting or involves cutting **b** (1) **:** an instrument, machine, machine part, or tool that cuts (2) **:** a device for vibrating a cutting stylus in disc recording; *also* **:** the stylus or its point
2 a : a ship's boat for carrying stores or passengers **b :** a single-masted fore-and-aft rigged sailing vessel **c :** a small armed vessel in government service
3 : a light sleigh

¹cut·throat \'kət-,thrōt\ *noun* (1535)
1 : KILLER, MURDERER
2 : a cruel unprincipled person

²cutthroat *adjective* (1567)
1 : MURDEROUS, CRUEL
2 : marked by unprincipled practices **:** RUTHLESS ⟨*cutthroat* competition⟩
3 : characterized by each player playing independently rather than having a permanent partner — used especially of partnership games adapted for three players ⟨*cutthroat* bridge⟩

cutthroat contract *noun* (circa 1944)
: contract bridge in which partnerships are determined by the bidding

cutthroat trout *noun* (circa 1891)
: a large spotted trout (*Oncorhynchus clarki* synonym *Salmo clarki*) chiefly of northwestern North America that has reddish streaks on the integument of the lower jaw — called also *cutthroat*

cutthroat trout

cut time *noun* (1951)
: duple or quadruple time with the beat represented by a half note

¹cut·ting \'kə-tiŋ\ *noun* (14th century)
1 : something cut or cut off or out: as **a :** a plant section originating from stem, leaf, or root and capable of developing into a new plant **b :** HARVEST
2 : something made by cutting; *especially* **:** RECORD 4

²cutting *adjective* (15th century)
1 : given to or designed for cutting; *especially* **:** SHARP, EDGED
2 : marked by sharp piercing cold
3 : inclined or likely to wound the feelings of others especially because of a ruthless incisiveness ⟨a *cutting* remark⟩
4 : INTENSE, PIERCING ⟨a *cutting* pain⟩
— **cut·ting·ly** \-tiŋ-lē\ *adverb*

cutting board *noun* (1825)
: a board on which something (as food or cloth) is placed for cutting

cutting edge *noun* (1951)
1 : the foremost part or place **:** VANGUARD

2 : a sharp effect or quality

cutting horse *noun* (1881)
: an agile saddle horse trained to separate individual animals from a cattle herd

cut·tle·bone \ˈkə-tᵊl-ˌbōn\ *noun* [Middle English *cotul* cuttlefish (from Old English *cudele*) + English *bone*] (1547)
: the shell of cuttlefishes that is sometimes used for polishing powder or for supplying cage birds with lime and salts

cut·tle·fish \-ˌfish\ *noun* [Middle English *cotul* + English *fish*] (circa 1828)
: any of various 10-armed marine cephalopod mollusks (order Sepioidea, especially genus *Sepia*) differing from the related squid in having a calcified internal shell

cut·ty sark \ˈkə-tē-ˈsärk\ *noun* [English dialect *cutty* short + *sark*] (1779)
chiefly Scottish **:** a short garment; *especially* **:** a woman's short undergarment

cutty stool *noun* (1820)
1 *chiefly Scottish* **:** a low stool
2 : a seat in a Scottish church where offenders formerly sat for public rebuke

cut·up \ˈkət-ˌəp\ *noun* (1843)
: a person who clowns or acts boisterously

cut up (1580)
transitive verb
1 a : to cut into parts or pieces **b :** to injure or damage by or as if by cutting **:** GASH, SLASH
2 : to subject to hostile criticism **:** CENSURE
intransitive verb
1 : to undergo being cut up
2 : to behave in a comic, boisterous, or unruly manner **:** CLOWN

cut·wa·ter \ˈkət-ˌwȯ-tər, -ˌwä-\ *noun* (1644)
: the forepart of a ship's stem

cut·work \-ˌwərk\ *noun* (15th century)
: embroidery usually on linen in which a design is outlined in buttonhole stitch and the enclosed material cut away

cut·worm \-ˌwərm\ *noun* (1816)
: any of various smooth-bodied chiefly nocturnal noctuid moth caterpillars which often feed on young plant stems near ground level

cu·vette \kyü-ˈvet\ *noun* [French, diminutive of *cuve* tub, from Latin *cupa* — more at HIVE] (circa 1909)
: a small often transparent laboratory vessel (as a tube)

cwm \ˈküm\ *noun* [Welsh, valley] (1853)
chiefly British **:** CIRQUE 3

-cy *noun suffix* [Middle English *-cie*, from Old French, from Latin *-tia*, partly from *-t-* (final stem consonant) + *-ia* -y, partly from Greek *-tia, -teia*, from *-t-* (final stem consonant) + *-ia, -eia* -y]
1 : action **:** practice ⟨mendican*cy*⟩
2 : rank **:** office ⟨baronet*cy*⟩ ⟨chaplain*cy*⟩
3 : body **:** class ⟨magistra*cy*⟩
4 : state **:** quality ⟨accura*cy*⟩ — often replacing a final *-t* or *-te* of the base word

cy·an \ˈsī-ˌan, -ən\ *noun* [Greek *kyanos*] (circa 1889)
: a greenish blue color — used in photography and color printing of one of the primary colors

cyan- *or* **cyano-** *combining form* [German, from Greek *kyan-, kyano-*, from *kyanos* dark blue enamel]
1 : dark blue **:** blue ⟨*cyano*bacterium⟩
2 : cyanogen ⟨*cyan*ide⟩
3 : cyanide ⟨*cyano*genetic⟩

cy·an·a·mide \sī-ˈa-nə-məd\ *noun* [International Scientific Vocabulary] (1838)
1 : a caustic acidic compound CH_2N_2
2 : CALCIUM CYANAMIDE

cy·a·nate \ˈsī-ə-ˌnāt, -nət\ *noun* [International Scientific Vocabulary] (circa 1846)
: a salt (as ammonium cyanate) or ester of cyanic acid

cy·an·ic acid \sī-ˈa-nik-\ *noun* (1832)
: a strong acid HOCN used to prepare cyanates

cy·a·nide \ˈsī-ə-ˌnīd, -nəd\ *noun* [International Scientific Vocabulary] (1826)
1 : a compound of cyanogen with a more electropositive element or group: as **a :** POTASSIUM CYANIDE **b :** SODIUM CYANIDE
2 : CYANOGEN 1

cy·a·nine \ˈsī-ə-ˌnēn, -nən\ *noun* [International Scientific Vocabulary] (circa 1872)
: any of various dyes that sensitize photographic film to light from the green, yellow, red, and infrared regions of the spectrum

cy·a·no \ˈsī-ə-(ˌ)nō, sī-ˈa-(ˌ)nō\ *adjective* [*cyan-*] (1929)
: relating to or containing the cyanogen group — often used in combination

cy·a·no·ac·ry·late \ˌsī-ə-nō-ˈa-krə-ˌlāt, sī-ˌa-nō-\ *noun* (1963)
: any of several liquid acrylate monomers that readily polymerize as anions and are used as adhesives in industry and in closing wounds in surgery

cy·a·no·bac·te·ri·um \-bak-ˈtir-ē-əm\ *noun* [New Latin] (1974)
: BLUE-GREEN ALGA

cy·a·no·co·bal·a·min \-kō-ˈba-lə-mən\ *also* **cy·a·no·co·bal·a·mine** \-ˌmēn\ *noun* [*cyan-* + *coba*lt + vit*amin*] (1950)
: VITAMIN B_{12} 1

cy·a·no·eth·yl·ate \-ˈe-thə-ˌlāt\ *transitive verb* (1942)
: to introduce a cyano-ethyl group CNC_2H_4 into (a compound) usually by means of acrylonitrile ⟨*cyanoethylate* cotton⟩
— cy·a·no·eth·yl·ation \-ˌe-thə-ˈlā-shən\ *noun*

cy·an·o·gen \sī-ˈa-nə-jən\ *noun* [French *cyanogène*, from *cyan-* + *gène* -gen] (1826)
1 : a univalent group −CN present in cyanides
2 : a colorless flammable poisonous gas $(CN)_2$

cy·a·no·ge·net·ic \ˌsī-ə-nō-jə-ˈne-tik, sī-ˌa-nō-\ *or* **cy·a·no·gen·ic** \-ˈje-nik\ *adjective* (1902)
: capable of producing cyanide (as hydrogen cyanide) ⟨a *cyanogenetic* plant⟩ ⟨a *cyanogenetic* glucoside⟩
— cy·a·no·gen·e·sis \-ˈje-nə-səs\ *noun*

cy·a·no·hy·drin \-ˈhī-drən\ *noun* [International Scientific Vocabulary, from *cyan-* + *hydr-* + [1]*-in*] (1925)
: any of various compounds containing both cyano and hydroxyl groups

cy·a·no·sis \ˌsī-ə-ˈnō-səs\ *noun* [New Latin, from Greek *kyanōsis* dark blue color, from *kyanos*] (1834)
: a bluish or purplish discoloration (as of skin) due to deficient oxygenation of the blood
— cy·a·not·ic \-ˈnä-tik\ *adjective*

cy·an·uric acid \ˌsī-ə-ˈnu̇r-ik-, -ˈnyu̇r-\ *noun* [*cyan-* + *urea*] (1838)
: a crystalline weak acid $C_3N_3(OH)_3$ that yields cyanic acid when heated

Cyb·e·le \ˈsi-bə-(ˌ)lē\ *noun* [Latin, from Greek *Kybelē*]
: a nature goddess of the ancient peoples of Asia Minor

cy·ber·na·tion \ˌsī-bər-ˈnā-shən\ *noun* [*cyber*netics + *-ation*] (1962)
: the automatic control of a process or operation (as in manufacturing) by means of computers
— cy·ber·nat·ed \ˈsī-bər-ˌnā-təd\ *adjective*

cy·ber·ne·ti·cian \ˌsī-(ˌ)bər-nə-ˈti-shən\ *noun* (1951)
: a specialist in cybernetics

cy·ber·net·i·cist \ˌsī-bər-ˈne-tə-sist\ *noun* (1948)
: CYBERNETICIAN

cy·ber·net·ics \ˌsī-bər-ˈne-tiks\ *noun plural but singular in construction* [Greek *kybernētēs* pilot, governor (from *kybernan* to steer, govern) + English *-ics*] (1948)
: the science of communication and control theory that is concerned especially with the comparative study of automatic control systems (as the nervous system and brain and mechanical-electrical communication systems)
— cy·ber·net·ic \-tik\ *also* **cy·ber·net·i·cal** \-ti-kəl\ *adjective*
— cy·ber·net·i·cal·ly \-ti-k(ə-)lē\ *adverb*

cy·ber·punk \ˈsī-bər-ˌpəŋk\ *noun* [*cyber*netic + [1]*punk*] (1984)
1 : science fiction dealing with future urban societies dominated by computer technology
2 : an opportunistic computer hacker

cy·ber·space \-ˌspās\ *noun* [*cyber*netic + *space*] (1986)
: the on-line world of computer networks

cy·borg \ˈsī-ˌbȯrg\ *noun* [*cyb*ernetic + *org*anism] (1960)
: a bionic human

cy·cad \ˈsī-kəd\ *noun* [New Latin *Cycad-, Cycas*, genus name, from Greek *kykas*, manuscript variant of *koïkas*, accusative plural of *koïx*, a kind of palm] (1845)
: any of an order (Cycadales) of dioecious cycadophytes that are represented by a single surviving family (Cycadaceae) of palmlike tropical plants that reproduce by means of spermatozoids

cy·cad·e·oid \sī-ˈka-dē-ˌȯid\ *noun* [New Latin *Cycadeoidales*, group name, ultimately from *Cycad-, Cycas*] (1928)
: any of an extinct order (Cycadeoidales or Bennettitales) of Mesozoic cycadophytes that differ from the cycads chiefly in having the reproductive organs on the trunk embedded in a thick external covering of persistent leaf bases

cy·cad·o·phyte \sī-ˈka-də-ˌfīt\ *noun* [ultimately from New Latin *Cycad-, Cycas* + Greek *phyton* plant — more at -PHYTE] (1911)
: any of a subclass (Cycadophytae) of usually unbranched gymnosperms with pinnate leaves, large pith, little xylem, and a thick cortex that includes the cycads, cycadeoids, and seed ferns

cy·ca·sin \ˈsī-kə-sən\ *noun* [New Latin *Cycas* cycad + International Scientific Vocabulary [1]*-in*] (circa 1965)
: a glucoside $C_8H_{16}N_2O_7$ that occurs in cycads and results in toxic and carcinogenic effects when introduced into mammals

cycl- *or* **cyclo-** *combining form* [New Latin, from Greek *kykl-, kyklo-*, from *kyklos*]
1 : circle ⟨*cyclo*meter⟩
2 : cyclic ⟨*cyclo*hexane⟩

cy·cla·mate \ˈsī-klə-ˌmāt, -mət\ *noun* [*cycl*ohexyl-sulf*amate*] (1954)
: an artificially prepared salt of sodium or calcium used especially formerly as a sweetener

cy·cla·men \ˈsī-klə-mən, ˈsi-\ *noun* [New Latin, genus name, from Greek *kyklaminos*] (circa 1550)
: any of a genus (*Cyclamen*) of plants of the primrose family having showy nodding flowers

cy·clase \ˈsī-ˌklās, -ˌklāz\ *noun* [*cycl-* + *-ase*] (1946)
: an enzyme (as adenylate cyclase) that catalyzes cyclization of a compound

cy·claz·o·cine \sī-ˈkla-zə-ˌsēn, -sən\ *noun* [*cycl-* + *azocine* (C_7H_7N), probably from *az-* + *oc*ta- + [2]*-ine*] (1966)
: an analgesic drug $C_{18}H_{25}NO$ that inhibits the effect of morphine and related addictive drugs and is used in the treatment of drug addiction

[1]**cy·cle** \ˈsī-kəl\ *noun* [Middle English *cicle*, from Late Latin *cyclus*, from Greek *kyklos* circle, wheel, cycle — more at WHEEL] (14th century)
1 : an interval of time during which a sequence of a recurring succession of events or phenomena is completed
2 a : a course or series of events or operations that recur regularly and usually lead back to

the starting point **b :** one complete performance of a vibration, electric oscillation, current alternation, or other periodic process **c :** a permutation of a set of ordered elements in which each element takes the place of the next and the last becomes first
3 : a circular or spiral arrangement: as **a :** an imaginary circle or orbit in the heavens **b :** WHORL **c :** RING 10
4 : a long period of time **:** AGE
5 a : a group of poems, plays, novels, or songs treating the same theme **b :** a series of narratives dealing typically with the exploits of a legendary hero
6 a : BICYCLE **b :** TRICYCLE **c :** MOTORCYCLE
7 : the series of a single, double, triple, and home run hit in any order by one player during one baseball game

²cycle *verb* **cy·cled; cy·cling** \'sī-k(ə-)liŋ\ (1842)
intransitive verb
1 a : to pass through a cycle **b :** to recur in cycles
2 : to ride a cycle; *specifically* **:** BICYCLE
transitive verb
: to cause to go through a cycle
— **cy·cler** \'sī-k(ə-)lər\ *noun*

cy·clic \'sī-klik *also* 'si-\ *or* **cy·cli·cal** \'sī-kli-kəl, 'si-\ *adjective* (1794)
1 a : of, relating to, or being a cycle **b :** moving in cycles ⟨*cyclic* time⟩ **c :** of, relating to, or being a chemical compound containing a ring of atoms
2 *cyclic* **:** being a mathematical group that has an element such that every element of the group can be expressed as one of its powers
— **cy·cli·cal·ly** \-k(ə-)lē\ *also* **cy·clic·ly** \'sī-kli-klē, 'si-\ *adverb*

cyclic AMP *noun* (1966)
: a cyclic mononucleotide of adenosine that is formed from ATP and is responsible for the intracellular mediation of hormonal effects on various cellular processes — called also *adenosine 3',5'-monophosphate*

cyclic GMP \-,jē-(,)em-'pē\ *noun* [*g*uanosine + *m*on- + *p*hosphate] (1969)
: a cyclic mononucleotide of guanosine that acts similarly to cyclic AMP as a secondary messenger in response to hormones

cy·clic·i·ty \sī-'kli-sə-tē, si-\ *noun* (1944)
: the quality or state of being cyclic ⟨estrous *cyclicity*⟩ — called also *cy·cli·cal·i·ty* \,sī-klə-'ka-lə-tē, ,si-\

cy·clist \'sī-k(ə-)ləst\ *noun* (1882)
: one who rides a cycle

cy·cli·tol \'sī-klə-,tȯl, 'si-, -,tōl\ *noun* [*cycl-* + *-itol* (as in *inositol*)] (1943)
: an alicyclic polyhydroxy compound (as inositol)

cy·cli·za·tion \,sī-k(ə-)lə-'zā-shən, ,si-\ *noun* (1909)
: formation of one or more rings in a chemical compound
— **cy·clize** \'sī-kə-,līz, 'si-, -,klīz\ *verb*

cy·clo \'sē-(,)klō, 'si-\ *noun, plural* **cyclos** [French, bicycle, moped, from *cyclo-* (as in *cyclomoteur* moped), from *cycle* two- or three-wheeled vehicle] (1964)
: a 3-wheeled often motor-driven taxi

cy·clo·ad·di·tion \,sī-(,)klō-ə-'di-shən\ *noun* (1963)
: a chemical reaction leading to ring formation in a compound

cy·clo·al·i·phat·ic \,sī-klō-,a-lə-'fa-tik\ *adjective* (1936)
: ALICYCLIC

cy·clo·dex·trin \-'dek-strən\ *noun* (1960)
: any of a class of complex cyclic sugars that are products of the enzymatic decomposition of starch and that can catalyze reactions between simpler molecules which come together within the cylindrical body of the sugar

cy·clo·di·ene \-'dī-,ēn, -dī-'\ *noun* (1942)
: an organic insecticide (as dieldrin or chlordane) with a chlorinated methylene group forming a bridge across a 6-membered carbon ring

cy·clo·gen·e·sis \-'je-nə-səs\ *noun* [*cyclo*ne + *genesis*] (circa 1938)
: the development or intensification of a cyclone

cy·clo·hex·ane \,sī-klō-'hek-,sān\ *noun* [International Scientific Vocabulary] (circa 1909)
: a pungent saturated cyclic hydrocarbon C_6H_{12} found in petroleum or made synthetically and used chiefly as a solvent and in organic synthesis

cy·clo·hex·a·none \-'hek-sə-,nōn\ *noun* (circa 1909)
: a liquid ketone $C_6H_{10}O$ used especially as a solvent and in organic synthesis

cy·clo·hex·i·mide \-'hek-sə-,mīd, -məd\ *noun* [*cyclohex*ane + *imide*] (1950)
: an agricultural fungicide $C_{15}H_{23}NO_4$ that inhibits protein synthesis and is obtained from a soil bacterium (*Streptomyces griseus*)

cy·clo·hex·yl·amine \-hek-'si-lə-,mēn\ *noun* [*cyclohex*ane + *-yl* + *amine*] (1943)
: a colorless liquid amine $C_6H_{11}NH_2$ that is used in organic synthesis and to prevent corrosion in boilers and that is believed to be harmful as a metabolic breakdown product of cyclamate

¹cy·cloid \'sī-,klȯid\ *noun* [French *cycloïde,* from Greek *kykloeidēs* circular, from *kyklos*] (1661)
: a curve that is generated by a point on the circumference of a circle as it rolls along a straight line

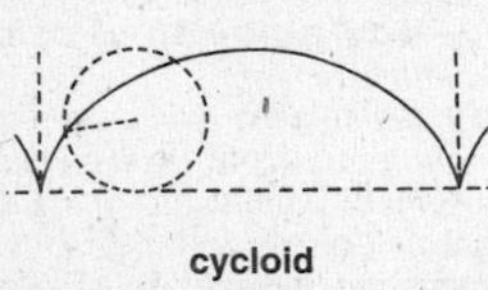
cycloid

— **cy·cloi·dal** \sī-'klȯi-dᵊl\ *adjective*

²cycloid *adjective* (1847)
1 : smooth with concentric lines of growth ⟨*cycloid* scales⟩; *also* **:** having or consisting of cycloid scales
2 : relating to or being a personality characterized by alternating high and low moods — compare CYCLOTHYMIC

cy·clom·e·ter \sī-'klä-mə-tər\ *noun* (1880)
: a device made for recording the revolutions of a wheel and often used for registering distance traversed by a wheeled vehicle

cy·clone \'sī-,klōn\ *noun* [modification of Greek *kyklōma* wheel, coil, from *kykloun* to go around, from *kyklos* circle] (1848)
1 a : a storm or system of winds that rotates about a center of low atmospheric pressure, advances at a speed of 20 to 30 miles (about 30 to 50 kilometers) an hour, and often brings heavy rain **b :** TORNADO **c :** LOW 1b
2 : any of various centrifugal devices for separating materials (as solid particles from gases)
— **cy·clon·ic** \sī-'klä-nik\ *adjective*
— **cy·clon·i·cal·ly** \-ni-k(ə-)lē\ *adverb*

Cy·clone \'sī-,klōn\ *trademark*
— used for a chain-link fence

cyclone cellar *noun* (1887)
: STORM CELLAR

cy·clo·ole·fin \,sī-klō-'ō-lə-fən\ *noun* [International Scientific Vocabulary] (circa 1929)
: a hydrocarbon containing a ring having one or more double bonds
— **cy·clo·ole·fin·ic** \-,ō-lə-'fi-nik\ *adjective*

cy·clo·par·af·fin \-'par-ə-fən\ *noun* (1900)
: a saturated cyclic hydrocarbon of the formula C_nH_{2n}

cy·clo·pe·an \,sī-klə-'pē-ən, sī-'klō-pē-\ *adjective* (1626)
1 *often capitalized* **:** of, relating to, or characteristic of a Cyclops
2 : HUGE, MASSIVE
3 : of or relating to a style of stone construction marked typically by the use of large irregular blocks without mortar

cy·clo·pe·dia *also* **cy·clo·pae·dia** \,sī-klə-'pē-dē-ə\ *noun* (1728)
: ENCYCLOPEDIA
— **cy·clo·pe·dic** \-'pē-dik\ *adjective*

cy·clo·phos·pha·mide \,sī-klō-'fäs-fə-,mīd\ *noun* (1960)
: an immunosuppressive and antineoplastic agent $C_7H_{15}Cl_2N_2O_2P$ used especially in the treatment of lymphomas and some leukemias

cy·clo·pro·pane \,sī-klə-'prō-,pān\ *noun* [International Scientific Vocabulary] (1894)
: a flammable gaseous saturated cyclic hydrocarbon C_3H_6 sometimes used as a general anesthetic

cy·clops \'sī-,kläps\ *noun* [Latin, from Greek *Kyklōps,* from *kykl-* cycl- + *ōps* eye] (1513)
1 *plural* **cy·clo·pes** \sī-'klō-(,)pēz\ *capitalized* **:** any of a race of giants in Greek mythology with a single eye in the middle of the forehead
2 *plural* **cyclops** [New Latin, genus name, from Latin] **:** any of a genus (*Cyclops*) of freshwater predatory copepods having a median eye

cy·clo·ra·ma \,sī-klə-'ra-mə, -'rä-\ *noun* [*cycl-* + *-orama* (as in *panorama*)] (1840)
1 : a large pictorial representation encircling the spectator and often having real objects as a foreground
2 : a curved curtain or wall used as a background of a stage set to suggest unlimited space
— **cy·clo·ram·ic** \-'ra-mik\ *adjective*

cy·clo·ser·ine \,sī-klō-'ser-,ēn\ *noun* (1952)
: an amino antibiotic $C_3H_6N_2O_2$ produced by an actinomycete (*Streptomyces orchidaceus*) and used especially in the treatment of tuberculosis

cy·clo·sis \sī-'klō-səs\ *noun* [New Latin, from Greek *kyklōsis* encirclement, from *kykloun* to go around] (1835)
: the streaming of protoplasm within a cell

cy·clo·spor·ine \,sī-klə-'spȯr-ᵊn, -,ēn\ *noun* [International Scientific Vocabulary *cycl-* + *spor-* + ²*-ine*] (1976)
: an immunosuppressive polypeptide drug $C_{62}H_{111}N_{11}O_{12}$ obtained from various imperfect fungi and used especially to prevent rejection of organ transplants

cy·clo·stome \'sī-klə-,stōm\ *noun* [ultimately from Greek *kykl-* + *stoma* mouth — more at STOMACH] (1835)
: any of a class (Cyclostomata) of jawless fishes having a large sucking mouth and comprising the hagfishes and lampreys

cy·clo·style \-,stīl\ *noun* [from *Cyclostyle,* a trademark] (1883)
: a machine for making multiple copies that utilizes a stencil cut by a graver whose tip is a small rowel
— **cyclostyle** *transitive verb*

cy·clo·thy·mic \,sī-klə-'thī-mik\ *adjective* [New Latin *cyclothymia* (from German *Zyklothymie,* from *zykl-* cycl- + *-thymie* -thymia) + English *-ic*] (1923)
: relating to or being an affective disorder characterized by the alternation of depressed moods with elevated, expansive, or irritable moods without psychotic features — compare CYCLOID 2
— **cy·clo·thy·mia** \-'thī-mē-ə\ *noun*

cy·clo·tom·ic \-'tä-mik\ *adjective* [*cyclotomy* mathematical theory of the division of the circle into equal parts, from *cycl-* + *-tomy*] (1879)
: relating to, being, or containing a polynomial of the form $x^{p-1} + x^{p-2} + \ldots + x + 1$ where p is a prime number

cy·clo·tron \'sī-klə-,trän\ *noun* [*cycl-* + *-tron;* from the circular movement of the particles] (1935)
: an accelerator in which charged particles (as protons, deuterons, or ions) are propelled by an alternating electric field in a constant magnetic field

cy·der *British variant of* CIDER

cyg•net \'sig-nət\ *noun* [Middle English *sygnett,* from Middle French *cygne* swan, from Latin *cycnus, cygnus,* from Greek *kyknos*] (15th century)
: a young swan

Cyg•nus \'sig-nəs\ *noun* [Latin (genitive *Cygni*), literally, swan]
: a northern constellation between Lyra and Pegasus in the Milky Way

cyl•in•der \'si-lən-dər\ *noun* [Middle French or Latin; Middle French *cylindre,* from Latin *cylindrus,* from Greek *kylindros,* from *kylindein* to roll; perhaps akin to Greek *kyklos* wheel — more at WHEEL] (1570)
1 a : the surface traced by a straight line moving parallel to a fixed straight line and intersecting a fixed planar closed curve **b :** the space bounded by a cylinder and two parallel planes cutting all its elements — see VOLUME table
2 : a cylindrical body or space: as **a :** the turning chambered breech of a revolver **b** (1) **:** the piston chamber in an engine (2) **:** a chamber in a pump from which the piston expels the fluid **c :** any of various rotating members in a press (as a printing press); *especially* **:** one that impresses paper on an inked form **d :** a cylindrical clay object inscribed with cuneiform inscriptions
— cyl•in•dered \-dərd\ *adjective*

cylinder head *noun* (1884)
: the closed end of an engine or pump cylinder

cylinder seal *noun* (1887)
: a cylinder (as of stone) engraved in intaglio and used especially in ancient Mesopotamia to roll an impression on wet clay

cy•lin•dri•cal \sə-'lin-dri-kəl\ *also* **cy•lin•dric** \-drik\ *adjective* (1646)
: relating to or having the form or properties of a cylinder
— cy•lin•dri•cal•ly \-dri-k(ə-)lē\ *adverb*

cylindrical coordinate *noun* (circa 1934)
: any of the coordinates in space obtained by constructing in a plane a polar coordinate system and on a line perpendicular to the plane a linear coordinate system

cy•ma \'sī-mə\ *noun* [Greek *kyma,* literally, wave] (1563)
1 : a projecting molding whose profile is an S-shaped curve
2 : an S-shaped curve formed by the union of a concave line and a convex line

cy•ma•tium \sī-'mā-sh(ē-)əm\ *noun, plural* **-tia** \-sh(ē-)ə\ [Latin, from Greek *kymation,* diminutive of *kymat-, kyma*] (1563)
: a crowning molding in classic architecture; *especially* **:** CYMA

cym•bal \'sim-bəl\ *noun* [Middle English, from Old English *cymbal* & Middle French *cymbale,* from Latin *cymbalum,* from Greek *kymbalon,* from *kymbē* bowl, boat] (before 12th century)
: a concave brass plate that produces a brilliant clashing tone and that is struck with a drumstick or is used in pairs struck glancingly together
— cym•bal•ist \-bə-list\ *noun*

cym•bid•i•um \sim-'bi-dē-əm\ *noun* [New Latin, genus name, from Latin *cymba* boat, from Greek *kymbē*] (1815)
: any of a genus (*Cymbidium*) of tropical Old World epiphytic orchids with showy flowers

cyme \'sīm\ *noun* [New Latin *cyma,* from Latin, cabbage sprout, from Greek *kyma* swell, wave, cabbage sprout, from *kyein* to be pregnant; akin to Sanskrit *śvayati* it swells, grows] (1794)
: an inflorescence in which each floral axis terminates in a single flower; *especially* **:** a determinate inflorescence of this type containing several flowers with the first-opening central flower terminating the main axis and subsequent flowers developing from lateral buds — see INFLORESCENCE illustration

cym•ling \'sim-lən, -liŋ\ *noun* [probably alteration of *simnel*] (1779)
: PATTYPAN

cy•mo•phane \'sī-mə-,fān\ *noun* [French, from Greek *kyma* wave + French *-phane* -phane] (circa 1804)
: CHRYSOBERYL; *especially* **:** an opalescent chrysoberyl

cy•mose \'sī-,mōs\ *adjective* (1807)
: of, relating to, being, or bearing a cyme

¹Cym•ric \'kəm-rik, 'kim-\ *adjective* (1839)
: of, relating to, or characteristic of the non-Gaelic Celtic people of Britain or their language; *specifically* **:** WELSH

²Cymric *noun* (circa 1889)
: BRYTHONIC; *specifically* **:** the Welsh language

Cym•ry \-rē\ *noun* [Welsh] (1833)
: WELSH 2

cyn•ic \'si-nik\ *noun* [Middle French or Latin, Middle French *cynique,* from Latin *cynicus,* from Greek *kynikos,* literally, like a dog, from *kyn-, kyōn* dog — more at HOUND] (circa 1564)
1 *capitalized* **:** an adherent of an ancient Greek school of philosophers who held the view that virtue is the only good and that its essence lies in self-control and independence
2 : a faultfinding captious critic; *especially* **:** one who believes that human conduct is motivated wholly by self-interest ◆
— cynic *adjective*

cyn•i•cal \'si-ni-kəl\ *adjective* (1584)
1 : CAPTIOUS, PEEVISH
2 : having or showing the attitude or temper of a cynic; *especially* **:** contemptuously distrustful of human nature and motives ⟨those *cynical* men who say that democracy cannot be honest and efficient —F. D. Roosevelt⟩ ☆
— cyn•i•cal•ly \-k(ə-)lē\ *adverb*

cyn•i•cism \'si-nə-,si-zəm\ *noun* (1672)
1 *capitalized* **:** the doctrine of the Cynics
2 : cynical character, attitude, or quality; *also* **:** an expression of such quality

cy•no•mol•gus monkey \,sī-nə-'mäl-gəs-\ *noun* [New Latin, alteration of *cynamolgus,* from Latin, member of an ancient tribe in Africa, from Greek *Kynamolgoi,* literally, dog milkers] (1936)
: a macaque (*Macaca fascicularis* synonym *M. cynomolgus*) of southeastern Asia, Borneo, and the Philippines that sometimes feeds on marine crustaceans and shellfish and is often used in medical research

cy•no•sure \'sī-nə-,shür, 'si-\ *noun* [Middle French & Latin; Middle French, Ursa Minor, guide, from Latin *cynosura* Ursa Minor, from Greek *kynosoura,* from *kynos oura,* literally, dog's tail]
1 *capitalized* **:** the northern constellation Ursa Minor; *also* **:** NORTH STAR
2 : one that serves to direct or guide
3 : a center of attraction or attention

Cyn•thia \'sin(t)-thē-ə\ *noun* [Latin, from feminine of *Cynthius* of Cynthus, from *Cynthus,* mountain on Delos where she was born, from Greek *Kynthos*]
1 : ARTEMIS
2 : MOON 1a

cy•pher *chiefly British variant of* CIPHER

¹cy pres \,sī-'prā, ,sē-\ *noun* [Anglo-French, so near, as near (as may be)] (1802)
: a rule providing for the interpretation of instruments in equity as nearly as possible in conformity to the intention of the testator when literal construction is illegal, impracticable, or impossible — called also *cy pres doctrine*

²cy pres *adverb* (1885)
: in accordance with the rule of cy pres

¹cy•press \'sī-prəs\ *noun* [Middle English, from Middle French *ciprès,* from Latin *cyparissus,* from Greek *kyparissos*] (14th century)
1 a (1) **:** any of a genus (*Cupressus* of the family Cupressaceae, the cypress family) of evergreen trees and shrubs with small overlapping leaves resembling scales (2) **:** any of several coniferous trees of the cypress family or the bald cypress family; *especially* **:** BALD CYPRESS 1 **b :** the wood of a cypress tree
2 : branches of cypress used as a symbol of mourning

²cypress *noun* [Middle English *ciprus, cipres,* from *Cyprus,* Mediterranean island] (15th century)
: a silk or cotton usually black gauze formerly used for mourning

cypress vine *noun* (1819)
: a tropical American vine (*Ipomoea quamoclit* synonym *Quamoclit pennata*) of the morning-glory family with usually red or white tubular flowers and finely dissected leaves

cyp•ri•an \'si-prē-ən\ *noun, often capitalized* [Latin *cyprius* of Cyprus, from Greek *kyprios,* from *Kypros* Cyprus, birthplace of Aphrodite] (1819)
: PROSTITUTE

cyp•ri•nid \'si-prə-nəd\ *noun* [ultimately from Latin *cyprinus* carp, from Greek *kyprinos*] (circa 1889)
: any of a family (Cyprinidae) of soft-finned freshwater fishes including the carps and minnows
— cyprinid *adjective*

cyp•ri•pe•di•um \,si-prə-'pē-dē-əm\ *noun* [New Latin, genus name, from Late Latin *Cypris,* a name for Venus + Greek *pedilon* sandal] (1813)
1 : any of a genus (*Cypripedium*) of Eurasian and North American terrestrial orchids having large usually showy drooping flowers with the lip inflated or pouched — compare LADY'S SLIPPER
2 : any of a genus (*Paphiopedalum*) of widely cultivated Asian orchids

cy•pro•hep•ta•dine \,sī-prō-'hep-tə-,dēn\ *noun* [cyclic + *pro*pyl + *hepta-* + piper*idine*] (1971)

☆ SYNONYMS
Cynical, misanthropic, pessimistic mean deeply distrustful. CYNICAL implies having a sneering disbelief in sincerity or integrity ⟨*cynical* about politicians' motives⟩. MISANTHROPIC suggests a rooted distrust and dislike of human beings and their society ⟨a solitary and *misanthropic* artist⟩. PESSIMISTIC implies having a gloomy, distrustful view of life ⟨*pessimistic* about the future⟩.

◇ WORD HISTORY
cynic The ancient school of philosophers known collectively as the Cynics was founded by Antisthenes (c. 445–c. 360 B.C.), a contemporary of Plato. Antisthenes is said to have taught at a gymnasium outside Athens called the Kynosarges, from which the Greek name of the school *kynikoi,* literally, "doglike ones," may be derived. On the other hand, Antisthenes' most famous follower, Diogenes of Sinope (c. 400–c. 325 B.C.), is most closely associated with the appellation. Diogenes rejected social conventions as impeding the most convenient satisfaction of one's needs; whatever was natural and easy could not be indecent, and therefore can and should be done in public. This shamelessness earned him the label *ho kyōn* "the dog," *kyōn* being an epithet applied to shameless or audacious people as far back as Homer. What English *cynic* and *cynical* now reflect, however, is not the Cynics' shamelessness, but their biting criticism of conventional mores.

: a drug $C_{21}H_{21}N$ that acts antagonistically to histamine and serotonin and is used especially in the treatment of asthma

cy·prot·er·one \sī-ˈprä-tə-ˌrōn\ *noun* [probably from cycl- + *progesterone*] (1966)
: a synthetic steroid $C_{22}H_{27}ClO_3$ that inhibits androgenic secretions (as testosterone)

Cy·re·na·ic \ˌsir-ə-ˈnā-ik, ˌsī-rə-\ *noun* [Latin *cyrenaicus,* from Greek *kyrēnaikos,* from *Kyrēnē* Cyrene, Africa, home of Aristippus, author of the doctrine] (1586)
: an adherent of the doctrine that pleasure is the chief end of life
— **Cyrenaic** *adjective*
— **Cy·re·na·icism** \-ˈnā-ə-ˌsi-zəm\ *noun*

Cy·ril·lic \sə-ˈri-lik\ *adjective* [Saint *Cyril,* reputed inventor of the Cyrillic alphabet] (1842)
: of, relating to, or constituting an alphabet used for writing Old Church Slavonic and for Russian and a number of other languages of eastern Europe and Asia

cyst \ˈsist\ *noun* [New Latin *cystis,* from Greek *kystis* bladder, pouch; akin to Sanskrit *śvasiti* he blows, snorts — more at WHEEZE] (circa 1720)
1 : a closed sac having a distinct membrane and developing abnormally in a cavity or structure of the body
2 : a body resembling a cyst: as **a :** a resting spore of many algae **b :** a gas-filled vesicle (as of a rockweed or bladderwort) **c :** a capsule formed about a minute organism going into a resting or spore stage; *also* **:** this capsule with its contents **d :** a resistant cover about a parasite produced by the parasite or the host

cyst- *or* **cysti-** *or* **cysto-** *combining form* [French, from Greek *kyst-, kysto-,* from *kystis*]
: bladder ⟨*cyst*itis⟩ **:** sac ⟨*cysto*carp⟩

-cyst *noun combining form* [New Latin *-cystis,* from Greek *kystis*]
: bladder **:** sac ⟨blasto*cyst*⟩

cys·te·amine \sis-ˈtē-ə-mən\ *noun* [*cyste*ine + *amine*] (1943)
: a cysteine derivative C_2H_7NS that has been used experimentally in the prevention of radiation sickness (as of cancer patients)

cys·te·ine \ˈsis-tə-ˌēn\ *noun* [International Scientific Vocabulary, from *cystine* + *-ein*] (1884)
: a crystalline sulfur-containing amino acid $C_3H_7NO_2S$ readily oxidizable to cystine

cys·tic \ˈsis-tik\ *adjective* (1713)
1 : relating to, composed of, or containing cysts
2 : of or relating to the urinary bladder or the gallbladder
3 : enclosed in a cyst

cys·ti·cer·coid \ˌsis-tə-ˈsər-ˌkȯid\ *noun* (1858)
: a tapeworm larva having an invaginated scolex and solid tailpiece

cys·ti·cer·co·sis \ˌsis-tə-(ˌ)sər-ˈkō-səs\ *noun, plural* **-co·ses** \-ˈkō-ˌsēz\ [New Latin] (1905)
: infestation with or disease caused by cysticerci

cys·ti·cer·cus \-ˈsər-kəs\ *noun, plural* **-cer·ci** \-ˈsər-ˌsī, -ˌkī\ [New Latin, from *cyst-* + Greek *kerkos* tail] (circa 1871)
: a tapeworm larva that consists of a fluid-filled sac containing an invaginated scolex and is situated in the tissues of an intermediate host

cystic fibrosis *noun* (1938)
: a common hereditary disease especially among whites that appears usually in early childhood, involves functional disorder of the exocrine glands, and is marked especially by faulty digestion due to a deficiency of pancreatic enzymes, by difficulty in breathing due to mucus accumulation in airways, and by excessive loss of salt in the sweat

cys·tine \ˈsis-ˌtēn\ *noun* [from its discovery in bladder stones] (1843)
: a crystalline amino acid $C_6H_{12}N_2O_4S_2$ that is widespread in proteins (as keratins) and is a major metabolic sulfur source

cys·tin·uria \ˌsis-tə-ˈnu̇r-ē-ə, -ˈnyu̇r-\ *noun* [New Latin] (1853)
: a metabolic defect characterized by excretion of excessive amounts of cystine in the urine and inherited as an autosomal recessive trait

cys·ti·tis \sis-ˈtī-təs\ *noun* [New Latin] (circa 1783)
: inflammation of the urinary bladder

cys·to·carp \ˈsis-tə-ˌkärp\ *noun* [International Scientific Vocabulary] (1875)
: the fruiting structure produced in the red algae after fertilization

cys·to·lith \ˈsis-tə-ˌlith\ *noun* [German *Zystolith,* from *zyst-* cyst- + *-lith*] (1857)
: a calcium carbonate concretion arising from the cellulose wall of cells of higher plants

cys·to·scope \ˈsis-tə-ˌskōp\ *noun* [International Scientific Vocabulary] (1889)
: a medical instrument for the visual examination of the bladder and for the passage of instruments under visual control
— **cys·to·scop·ic** \ˌsis-tə-ˈskä-pik\ *adjective*
— **cys·tos·co·py** \si-ˈstäs-kə-pē\ *noun*

cyt- *or* **cyto-** *combining form* [German *zyt-, zyto-,* from Greek *kytos* hollow vessel — more at HIDE]
1 : cell ⟨*cyto*logy⟩
2 : cytoplasm ⟨*cyto*kinesis⟩

-cyte *noun combining form* [New Latin *-cyta,* from Greek *kytos* hollow vessel]
: cell ⟨leuko*cyte*⟩

Cyth·er·ea \ˌsi-thə-ˈrē-ə\ *noun* [Latin, from Greek *Kythereia,* from *Kythēra* Cythera, island associated with Aphrodite]
: APHRODITE

Cyth·er·e·an \-ˈrē-ən\ *adjective* (1885)
: of or relating to the planet Venus

cy·ti·dine \ˈsi-tə-ˌdēn, ˈsī-\ *noun* [*cyt*osine + *-idine*] (1911)
: a nucleoside containing cytosine

cy·ti·dyl·ic acid \ˌsi-tə-ˈdi-lik-, ˌsī-\ *noun* [*cyti*dine + *-yl* + [1]*-ic*] (1936)
: a nucleotide containing cytosine

cy·to·cha·la·sin \ˌsī-tō-kə-ˈlā-sən\ *noun* [*cyt-* + Greek *chalasis* slackening + English [1]*-in*] (1967)
: any of a group of metabolites isolated from fungi (especially *Helminthosporium dematioideum*) that inhibit various cell processes

cy·to·chem·is·try \-ˈke-mə-strē\ *noun* (circa 1905)
1 : microscopical biochemistry
2 : the chemistry of cells
— **cy·to·chem·i·cal** \-ˈke-mi-kəl\ *adjective*

cy·to·chrome \ˈsī-tə-ˌkrōm\ *noun* (1925)
: any of several intracellular hemoprotein respiratory pigments that are enzymes functioning in electron transport as carriers of electrons

cytochrome c *noun, often 2d c italic* (1940)
: the most abundant and stable of the cytochromes

cytochrome oxidase *noun* (1942)
: an iron-porphyrin enzyme important in cell respiration because of its ability to catalyze the oxidation of reduced cytochrome c in the presence of oxygen

cy·to·dif·fer·en·ti·a·tion \ˈsī-tō-ˌdi-fə-ˌren(t)-shē-ˈā-shən\ *noun* (1959)
: the development of specialized cells (as muscle, blood, or nerve cells) from undifferentiated precursors

cy·to·ge·net·ics \ˌsī-tō-jə-ˈne-tiks\ *noun plural but singular or plural in construction* [International Scientific Vocabulary] (1942)
: a branch of biology that deals with the study of heredity and variation by the methods of both cytology and genetics
— **cy·to·ge·net·ic** \-jə-ˈne-tik\ *or* **cy·to·ge·net·i·cal** \-ti-kəl\ *adjective*
— **cy·to·ge·net·i·cal·ly** \-ti-k(ə-)lē\ *adverb*
— **cy·to·ge·net·i·cist** \-ˈne-tə-sist\ *noun*

cy·to·kine \ˈsī-tə-ˌkīn\ *noun* [*cyt-* + *-kine* (as in *lymphokine*)] (1979)
: any of a class of immunoregulatory substances (as lymphokines) that are secreted by cells of the immune system

cy·to·ki·ne·sis \ˌsī-tō-kə-ˈnē-səs, -kī-\ *noun* [New Latin] (1919)
1 : the cytoplasmic changes accompanying mitosis
2 : cleavage of the cytoplasm into daughter cells following nuclear division
— **cy·to·ki·net·ic** \-ˈne-tik\ *adjective*

cy·to·ki·nin \ˌsī-tə-ˈkī-nən\ *noun* [*cyt-* + *kinin*] (1965)
: any of various plant growth substances (as kinetin) that are usually derivatives of adenine

cy·tol·o·gy \sī-ˈtä-lə-jē\ *noun* [International Scientific Vocabulary] (1889)
1 : a branch of biology dealing with the structure, function, multiplication, pathology, and life history of cells
2 : the cytological aspects of a process or structure
— **cy·to·log·i·cal** \ˌsī-tᵊl-ˈä-ji-kəl\ *or* **cy·to·log·ic** \-ˈä-jik\ *adjective*
— **cy·to·log·i·cal·ly** \-ji-k(ə-)lē\ *adverb*
— **cy·tol·o·gist** \sī-ˈtä-lə-jist\ *noun*

cy·to·ly·sin \ˌsī-tᵊl-ˈī-sᵊn\ *noun* [International Scientific Vocabulary] (circa 1903)
: a substance (as an antibody that lyses bacteria) producing cytolysis

cy·tol·y·sis \sī-ˈtä-lə-səs\ *noun* [New Latin] (1907)
: the usually pathologic dissolution or disintegration of cells
— **cy·to·lyt·ic** \ˌsī-tᵊl-ˈi-tik\ *adjective*

cy·to·me·gal·ic \ˌsī-tō-mi-ˈga-lik\ *adjective* [New Latin *cytomegalia* condition of having enlarged cells, from *cyt-* + *megal-* + *-ia*] (1952)
: characterized by or causing the formation of enlarged cells

cy·to·meg·a·lo·vi·rus \ˌsī-tə-ˌme-gə-lō-ˈvī-rəs\ *noun* [New Latin, from *cytomegalia* + *-o-* + *virus*] (1963)
: any of several viruses that cause cellular enlargement and formation of eosinophilic inclusion bodies especially in the nucleus and include some acting as opportunistic infectious agents in immunosuppressed conditions (as AIDS)

cy·to·mem·brane \ˌsī-tō-ˈmem-ˌbrān\ *noun* (1962)
: one of the cellular membranes including those of the plasmalemma, endoplasmic reticulum, nuclear envelope, and Golgi apparatus; *specifically* **:** UNIT MEMBRANE

cy·to·path·ic \ˌsī-tə-ˈpa-thik\ *adjective* (1952)
: of, relating to, characterized by, or producing pathological changes in cells

cy·to·path·o·gen·ic \-ˌpa-thə-ˈje-nik\ *adjective* (1952)
: pathologic for or destructive to cells
— **cy·to·path·o·ge·nic·i·ty** \-jə-ˈni-sə-tē\ *noun*

cy·to·phil·ic \ˌsī-tə-ˈfi-lik\ *adjective* (circa 1909)
: having an affinity for cells ⟨*cytophilic* antibodies⟩

cy·to·pho·tom·e·try \-fō-ˈtä-mə-trē\ *noun* (1955)
: photometry applied to the study of the cell or its constituents
— **cy·to·pho·to·met·ric** \-ˌfō-tə-ˈme-trik\ *adjective*

cy·to·plasm \ˈsī-tə-ˌpla-zəm\ *noun* [International Scientific Vocabulary] (1874)
: the organized complex of inorganic and organic substances external to the nuclear membrane of a cell and including the cytosol and membrane-bound organelles (as mitochondria or chloroplasts) — see CELL illustration
— **cy·to·plas·mic** \ˌsī-tə-ˈplaz-mik\ *adjective*
— **cy·to·plas·mi·cal·ly** \-mi-k(ə-)lē\ *adverb*

cy·to·sine \'sī-tə-ˌsēn\ *noun* [International Scientific Vocabulary *cyt-* + *-ose* + ²*-ine*] (1894)
: a pyrimidine base $C_4H_5N_3O$ that codes genetic information in the polynucleotide chain of DNA or RNA — compare ADENINE, GUANINE, THYMINE, URACIL

cy·to·skel·e·ton \ˌsī-tə-'ske-lə-tᵊn\ *noun* (1940)
: the network of protein filaments and microtubules in the cytoplasm that controls cell shape, maintains intracellular organization, and is involved in cell movement
— **cy·to·skel·e·tal** \-tᵊl\ *adjective*

cy·to·sol \'sī-tə-ˌsäl, -ˌsȯl\ *noun* (1970)
: the fluid portion of the cytoplasm exclusive of organelles and membranes — called also *hyaloplasm, ground substance*
— **cy·to·sol·ic** \ˌsī-tə-'sä-lik, -'sȯ-\ *adjective*

cy·to·stat·ic \ˌsī-tə-'sta-tik\ *adjective* (1949)
: tending to retard cellular activity and multiplication ⟨*cytostatic* treatment of tumor cells⟩
— **cytostatic** *noun*
— **cy·to·stat·i·cal·ly** \-ti-k(ə-)lē\ *adverb*

cy·to·tax·on·o·my \ˌsī-tō-(ˌ)tak-'sä-nə-mē\ *noun* (1930)
1 : study of the relationships and classification of organisms using both classical systematic techniques and comparative studies of chromosomes
2 : the nuclear cytologic makeup of a kind of organism
— **cy·to·tax·o·nom·ic** \-ˌtak-sə-'nä-mik\ *adjective*
— **cy·to·tax·o·nom·i·cal·ly** \-mi-k(ə-)lē\ *adverb*

cy·to·tech·nol·o·gist \ˌsī-tə-tek-'nä-lə-jist\ *noun* (1961)
: a medical technician trained in the identification of cells and cellular abnormalities (as in cancer)
— **cy·to·tech·nol·o·gy** \-lə-jē\ *noun*

cy·to·tox·ic \ˌsī-tə-'täk-sik\ *adjective* (1904)
1 : of or relating to a cytotoxin
2 : toxic to cells ⟨*cytotoxic* properties of platinum⟩
— **cy·to·tox·ic·i·ty** \-(ˌ)täk-'si-sə-tē\ *noun*

cy·to·tox·in \-'täk-sən\ *noun* (1902)
: a substance (as a toxin or antibody) having a toxic effect on cells

czar \'zär, '(t)sär\ *noun* [New Latin *czar,* from Russian *tsar',* from Old Russian *tsĭsarĭ,* from Gothic *kaisar,* from Greek or Latin; Greek, from Latin *Caesar* — more at CAESAR] (1555)
1 : EMPEROR; *specifically* **:** the ruler of Russia until the 1917 revolution
2 : one having great power or authority ⟨a banking *czar*⟩ ◆
— **czar·dom** \'zär-dəm, '(t)sär-\ *noun*

czar·das \'chär-ˌdash, -ˌdäsh\ *noun, plural* **czardas** [Hungarian *csárdás*] (1860)
: a Hungarian dance to music in duple time in which the dancers start slowly and finish with a rapid whirl

czar·e·vitch \'zär-ə-ˌvich, '(t)sär-\ *noun* [Russian *tsarevich,* from *tsar'* + *-evich,* patronymic suffix] (1710)
: an heir apparent of a Russian czar

cza·ri·na \zä-'rē-nə, (t)sä-\ *noun* [probably modification of German *Zarin,* from *Zar* czar, from Russian *tsar'*] (1717)
: the wife of a czar

czar·ism \'zär-ˌi-zəm, '(t)sär-\ *noun* (1855)
1 : the government of Russia under the czars
2 : autocratic rule
— **czar·ist** \'zär-ist, ˌ'(t)sär-\ *noun or adjective*

Czech \'chek\ *noun* [Czech *Čech*] (1841)
1 : a native or inhabitant of western Czechoslovakia including Bohemia and Moravia
2 : the Slavic language of the Czechs
3 : a native or inhabitant of Czechoslovakia or the Czech Republic
— **Czech** *adjective*
— **Czech·ish** \'che-kish\ *adjective*

◇ WORD HISTORY

czar *Czar* or *tsar* is the English word for a pre-Soviet Russian emperor. *Tsar* is a straightforward transliteration of the Russian word and is now the usual form in Slavic studies; but *czar,* now usually used figuratively (as in "drug czar") has an older and more involved origin. We owe this spelling to the Austrian diplomat Seigmund, Freiherr (Baron) von Herberstein (1486–1566), who was an ambassador to the court at Moscow during the reigns of the Holy Roman Emperors Maximilian I and Charles V. In 1549 he published *Rerum moscovitarum commentarii* ("Commentaries on Muscovite Matters"), one of the first Western books concerning things Russian and long a major source of information about Russia in Western Europe. Herberstein wrote in Latin, but his use of *cz* to represent the sound \ts\ reflected a largely obsolete spelling convention in Hungarian. In English words such as *czardas* and *Czech, cz* represents, in contrast, \ch\, which is a convention continued in modern Polish.

D is the fourth letter of the English alphabet and of most alphabets that are closely related to that of English. It came to Latin through Etruscan from a western Greek alphabet. The Greeks took it from the Phoenician alphabet, naming the letter delta *after Phoenician* dāleth. *In modern English* d *usually represents the voiced alveolar stop, a consonant sound made with the tip of the tongue touching the inner surface of the gums above the upper front teeth (as the* d *sound in English* dog). *Small* d *evolved from a Roman variant of captial* D *in which the top of the curved line swept beyond the straight upright line of the letter (*Ɗ*). The straight line was later rounded on the left to form a lobe, and the original curved line ascended above the lobe from its right side (∂). Eventually, this ascender was straightened to produce the modern form of the letter.*

d \'dē\ *noun, plural* **d's** *or* **ds** \'dēz\ *often capitalized, often attributive* (before 12th century)
1 a : the 4th letter of the English alphabet **b :** a graphic representation of this letter **c :** a speech counterpart of orthographic *d*
2 : 500 — see NUMBER table
3 : the 2d tone of a C-major scale
4 : a graphic device for reproducing the letter *d*
5 : one designated *d* especially as the 4th in order or class
6 a : a grade rating a student's work as poor in quality **b :** one graded or rated with a D
7 : something shaped like the letter D; *specifically* **:** a semicircle on a pool table about 23 inches in diameter for use especially in snooker
8 : DEFENSE 2b, 4b ⟨play tough *D*⟩
d- \ˌdē, 'dē\ *prefix* [International Scientific Vocabulary, from *dextr-*]
1 : dextrorotatory ⟨*d*-tartaric acid⟩
2 : having a similar configuration at a selected carbon atom to the configuration of dextrorotatory glyceraldehyde — usually printed as a small capital ⟨D-fructose⟩
-d *symbol*
— used after the figure 2 or 3 to indicate the ordinal number second or third ⟨2*d*⟩ ⟨53*d*⟩
'd \d, əd, id\ *verb*
1 : HAD
2 : WOULD
3 : DID
DA \ˌdē-'ā\ *noun* [*d*uck's *a*ss; from its resemblance to the tail of a duck] (1951)
: DUCKTAIL
[1]**dab** \'dab\ *noun* [Middle English *dabbe*] (14th century)
1 : a sudden blow or thrust **:** POKE
2 : a small amount
3 : a gentle touch or stroke **:** PAT
4 : DAUB
[2]**dab** *verb* **dabbed; dab·bing** (1562)
transitive verb
1 : to strike or touch lightly **:** PAT
2 : to apply lightly or irregularly **:** DAUB
intransitive verb
: to make a dab
— dab·ber *noun*
[3]**dab** *noun* [Anglo-French *dabbe*] (15th century)
: FLATFISH; *especially* **:** any of several flounders (genus *Limanda*)
[4]**dab** *noun* [origin unknown] (1691)
chiefly British **:** a skillful person
dab·ble \'da-bəl\ *verb* **dab·bled; dab·bling** \-b(ə-)liŋ\ [perhaps frequentative of [2]*dab*] (1557)
transitive verb
: to wet by splashing or by little dips or strokes **:** SPATTER
intransitive verb
1 a : to paddle, splash, or play in or as if in water **b :** to reach with the bill to the bottom of shallow water in order to obtain food
2 : to work or involve oneself superficially or intermittently especially in a secondary activity or interest ⟨*dabbles* in art⟩
dab·bler \-b(ə-)lər\ *noun* (1611)
: one that dabbles: as **a :** one not deeply engaged in or concerned with something **b :** a duck (as a mallard or shoveler) that feeds by dabbling — called also *dabbling duck, puddle duck*
synonym see AMATEUR
dab·bling \-b(ə-)liŋ\ *noun* (circa 1847)
: a superficial or intermittent interest, investigation, or experiment ⟨his *dabblings* in philosophy and art⟩
dab·chick \'dab-ˌchik\ *noun* [probably irregular from obsolete English *dop* (to dive) + English *chick*] (circa 1550)
: any of several small grebes
dab hand *noun* [[4]*dab*] (circa 1828)
chiefly British **:** EXPERT
da ca·po \dä-'kä-(ˌ)pō, də-\ *adverb or adjective* [Italian] (circa 1724)
: from the beginning — used as a direction in music to repeat
dace \'dās\ *noun, plural* **dace** [Middle English *dace, darce,* from Middle French *dars,* from Medieval Latin *darsus*] (15th century)
1 : a small freshwater European cyprinid fish (*Leuciscus leuciscus*)
2 : any of various small North American freshwater cyprinid fishes
da·cha \'dä-chə *also* 'da-\ *noun* [Russian, from Old Russian, land allotted by a prince; akin to Latin *dos* dowry — more at DATE] (1896)
: a Russian country cottage used especially in the summer
dachs·hund \'däks-ˌhu̇nt, -ˌhu̇nd; 'däk-sənt; *especially British* 'dak-sənd\ *noun* [German, from *Dachs* badger + *Hund* dog] (1882)
: any of a breed of long-bodied, short-legged dogs of German origin that occur in short-haired, long-haired, and wirehaired varieties

dachshund

Da·cron \'dā-ˌkrän, 'da-\ *trademark*
— used for a synthetic polyester textile fiber
dac·tyl \'dak-tᵊl, -ˌtil\ *noun* [Middle English *dactile,* from Latin *dactylus,* from Greek *daktylos,* literally, finger; from the fact that the first of three syllables is the longest, like the joints of the finger] (14th century)
: a metrical foot consisting of one long and two short syllables or of one stressed and two unstressed syllables (as in *tenderly*)
— dac·tyl·ic \dak-'ti-lik\ *adjective or noun*
dactyl- *or* **dactylo-** *combining form* [Greek *daktyl-, daktylo-,* from *daktylos*]
: finger **:** toe **:** digit ⟨*dactylo*logy⟩
dac·ty·lol·o·gy \ˌdak-tə-'lä-lə-jē\ *noun* (circa 1656)
: FINGER SPELLING
dad \'dad\ *noun* [probably baby talk] (15th century)
: FATHER 1a
Da·da \'dä-(ˌ)dä\ *noun* [French] (1919)
: a movement in art and literature based on deliberate irrationality and negation of traditional artistic values; *also* **:** the art and literature produced by this movement ◇
da·da·ism \-ˌi-zəm\ *noun, often capitalized* [French *dadaïsme*] (1919)
: DADA
— da·da·ist \-ˌist\ *noun or adjective, often capitalized*
— da·da·is·tic \ˌdä-dä-'is-tik\ *adjective, often capitalized*
dad·dy \'da-dē\ *noun, plural* **daddies** (15th century)
1 : FATHER 1a
2 : GRANDDADDY 2
dad·dy long·legs \ˌda-dē-'lȯŋ-ˌlegz, -ˌlāgz\ *noun, plural* **daddy longlegs** (circa 1814)
1 : CRANE FLY
2 : HARVESTMAN
[1]**da·do** \'dā-(ˌ)dō\ *noun, plural* **da·does** [Italian, die, plinth] (1664)
1 a : the part of a pedestal of a column between the base and the surbase **b :** the lower part of an interior wall when specially decorated or faced; *also* **:** the decoration adorning this part of a wall
2 : a rectangular groove cut to make a joint in woodworking; *specifically* **:** one cut across the grain

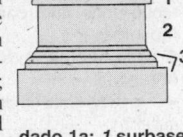

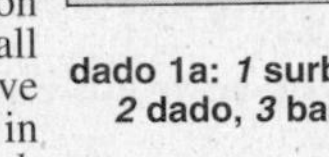

dado 1a: ***1*** **surbase,** ***2*** **dado,** ***3*** **base**

[2]**dado** *transitive verb* **da·doed; da·do·ing** (1881)
1 : to provide with a dado
2 a : to set into a groove **b :** to cut a dado in (as a plank)
dae·dal \'dē-dᵊl\ *adjective* [Latin *daedalus,* from Greek *daidalos*] (1590)
1 a : SKILLFUL, ARTISTIC **b :** INTRICATE ⟨the computer's *daedal* circuitry⟩
2 : adorned with many things ⟨visions of cloud and light and *daedal* earth are the airman's daily scene —Laurence Binyon⟩
Dae·da·lus \'de-dᵊl-əs, 'dē-\ *noun* [Latin, from Greek *Daidalos*]
: the legendary builder of the Cretan labyrinth who makes wings to enable himself and his son Icarus to escape imprisonment

◇ WORD HISTORY

Dada In February 1916, while Europe was engulfed in World War I, a group of aesthetes and war resisters, including the artists Marcel Duchamp and Jean Arp and the poet Tristan Tzara, sought refuge in tranquil, neutral Switzerland. They decided to respond to the madness about them by creating an artistic movement based on irrationality, one that would repudiate traditional values and conventions and would shock bourgeois sensibilities. At a café in Zurich one evening, they hit upon a name for the new movement. Recollections of the movement's founding members differ regarding the arbitrariness of the naming process, but in any case the word chosen was *dada,* in French a child's word for a hobbyhorse; its very childishness was a way the group could mock the world around them. Though the Zurich group soon broke up and Dadaism as a movement lost its steam in the 1920s, it had far-reaching effects on the art of the 20th century, for trends as diverse as surrealism and conceptual art have their roots in Dada.

— **Dae·da·lian** *or* **Dae·da·lean** \di-ˈdāl-yən\ *adjective*

dae·mon *variant of* DEMON

daff \ˈdaf\ *transitive verb* [alteration of *doff*] (1596)
1 *archaic* **:** to thrust aside
2 *obsolete* **:** to put off (as with an excuse)

daf·fo·dil \ˈda-fə-ˌdil\ *noun* [perhaps from Dutch *de affodil* the asphodel, from *de* the (from Middle Dutch) + *affodil* asphodel, from Middle French *afrodille,* from Latin *asphodelus;* akin to Old High German *thaz* the — more at THAT, ASPHODEL] (1548)
: any of various bulbous herbs (genus *Narcissus*); *especially* **:** a plant whose flowers have a large corona elongated into a trumpet — compare JONQUIL

daf·fy \ˈda-fē\ *adjective* **daf·fi·er; -est** [obsolete English *daff,* noun (fool)] (circa 1884)
: CRAZY, FOOLISH
— **daf·fi·ly** \ˈda-fə-lē\ *adverb*

daft \ˈdaft *also* ˈdäft\ *adjective* [Middle English *dafte* gentle, stupid; akin to Old English *gedæfte* mild, gentle, Middle English *defte* deft, Old Church Slavonic po*dobati* to be fitting] (14th century)
1 a : SILLY, FOOLISH **b :** MAD, INSANE
2 *Scottish* **:** frivolously gay
— **daft·ly** *adverb*
— **daft·ness** \ˈdaf(t)-nəs\ *noun*

dag \ˈdag\ *noun* [Middle English *dagge*] (14th century)
1 : a hanging end or shred
2 : matted or manure-coated wool

dag·ger \ˈda-gər\ *noun* [Middle English] (14th century)
1 : a sharp pointed knife for stabbing
2 a : something that resembles a dagger **b :** a character † used as a reference mark or to indicate a death date
— **dag·ger·like** \-ˌlīk\ *adjective*
— **at daggers drawn :** in a state of open hostility or conflict

da·go \ˈdā-(ˌ)gō\ *noun, plural* **dagos** *or* **dagoes** [alteration of earlier *diego,* from *Diego,* a common Spanish given name] (1832)
: a person of Italian or Spanish birth or descent — usually used disparagingly

da·guerre·o·type \də-ˈge-rō-ˌtīp, -rə- *also* də-ˈge-rē-ō-ˌtīp\ *noun* [French *daguerréotype,* from L. J. M. *Daguerre* + French *-o-* + *type*] (1839)
: an early photograph produced on a silver or a silver-covered copper plate; *also* **:** the process of producing such photographs
— **daguerreotype** *transitive verb*
— **da·guerre·o·typ·ist** \-ˌtī-pist\ *noun*
— **da·guerre·o·typy** \-ˌtī-pē\ *noun*

dah \ˈdä\ *noun* [imitative] (1940)
: DASH 7

dahl·ia \ˈdal-yə, ˈdäl-, *U.S. also & British usually* ˈdāl-\ *noun* [New Latin, genus name, from Anders *Dahl* (died 1789) Swedish botanist] (1840)
: any of a genus (*Dahlia*) of American tuberous-rooted composite herbs having opposite pinnate leaves and rayed flower heads and including many that are cultivated as ornamentals

dai·kon \ˈdī-kən\ *noun* [Japanese, from *dai* big + *kon* root] (1876)
: a large long hard white radish used especially in Oriental cuisine; *also* **:** a plant (*Raphanus sativus longipinnatus*) whose root is a daikon

dai·li·ness \ˈdā-lē-nəs, -li-\ *noun* (1596)
: daily or routine quality **:** ORDINARINESS ⟨the *dailiness* of family life⟩

[1]**dai·ly** \ˈdā-lē\ *adjective* (15th century)
1 a : occurring, made, or acted upon every day **b :** issued every day or every weekday **c :** of or providing for every day
2 a : reckoned by the day ⟨average *daily* wage⟩ **b :** covering the period of or based on a day ⟨*daily* statistics⟩

[2]**daily** *adverb* (15th century)
1 : every day
2 : every weekday

[3]**daily** *noun, plural* **dai·lies** (1832)
1 : a newspaper published every weekday
2 *British* **:** a servant who works on a daily basis
3 *plural* **:** RUSH 6

daily double *noun* (1942)
: a system of betting (as on horse races) in which the bettor must pick the winners of two stipulated races in order to win

daily dozen *noun* (1919)
1 : a series of physical exercises to be performed daily **:** WORKOUT
2 : a set of routine duties or tasks

dai·mon \ˈdī-ˌmōn\ *noun, plural* **dai·mo·nes** \ˈdī-mə-ˌnēz\ *or* **daimons** [Greek *daimōn*] (1852)
: DEMON 2, 3
— **dai·mon·ic** \dī-ˈmä-nik\ *adjective*

dai·myo *also* **dai·mio** \ˈdī-mē-ˌō, -(ˌ)myō\ *noun, plural* **-myo** *or* **-myos** *also* **-mio** *or* **-mios** [Japanese *daimyō*] (1727)
: a Japanese feudal baron

[1]**dain·ty** \ˈdān-tē\ *noun, plural* **dain·ties** [Middle English *deinte,* from Old French *deintié,* from Latin *dignitat-, dignitas* dignity, worth] (14th century)
1 a : something delicious to the taste **b :** something choice or pleasing
2 *obsolete* **:** FASTIDIOUSNESS

[2]**dainty** *adjective* **dain·ti·er; -est** (14th century)
1 a : tasting good **:** TASTY **b :** attractively prepared and served
2 : marked by delicate or diminutive beauty, form, or grace
3 *obsolete* **:** CHARY, RELUCTANT
4 a : marked by fastidious discrimination or finical taste **b :** showing avoidance of anything rough
synonym see CHOICE
— **dain·ti·ly** \ˈdān-tə-lē\ *adverb*
— **dain·ti·ness** \ˈdān-tē-nəs\ *noun*

dai·qui·ri \ˈda-kə-rē, ˈdī-\ *noun* [*Daiquirí,* Cuba] (1920)
: a cocktail made of rum, lime juice, and sugar

dairy \ˈder-ē, ˈdar-\ *noun, plural* **dair·ies** *often attributive* [Middle English *deyerie,* from *deye* dairymaid, from Old English *dǣge* kneader of bread; akin to Old English *dāg* dough — more at DOUGH] (14th century)
1 : a room, building, or establishment where milk is kept and butter or cheese is made
2 a : the department of farming or of a farm that is concerned with the production of milk, butter, and cheese **b :** a farm devoted to such production
3 : an establishment for the sale or distribution chiefly of milk and milk products

dairy cattle *noun plural* (1895)
: cattle kept for milk production

dairy·ing \ˈder-ē-iŋ\ *noun* (1649)
: the business of operating a dairy

dairy·maid \-ˌmād\ *noun* (1599)
: a woman employed in a dairy

dairy·man \-mən, -ˌman\ *noun* (circa 1617)
: one who operates a dairy farm or works in a dairy

da·is \ˈdā-əs, ÷ˈdī-\ *noun* [Middle English *deis,* from Old French, from Late Latin *discus* high table, from Latin, dish, quoit — more at DISH] (13th century)
: a raised platform (as in a hall or large room)

dai·shi·ki \dī-ˈshē-kē\ *variant of* DASHIKI

dai·sy \ˈdā-zē\ *noun, plural* **daisies** [Middle English *dayeseye,* from Old English *dægesēage,* from *dæg* day + *ēage* eye] (before 12th century)
1 : a composite plant (as of the genera *Bellis* or *Chrysanthemum*) having a flower head with well-developed ray flowers usually arranged in one or a few whorls: as **a :** a low European herb (*Bellis perennis*) with white or pink ray flowers — called also *English daisy* **b :** a leafy-stemmed perennial herb (*Chrysanthemum leucanthemum*) with long white ray flowers and a yellow disk that was introduced into the U.S. from Europe — called also *ox-eye daisy*
2 : the flower head of a daisy
3 : a first-rate person or thing
4 *capitalized* **:** a member of a program of the Girl Scouts for girls in kindergarten and first grade in school

daisy chain *noun* (1841)
1 : a string of daisies with stems linked to form a chain
2 : an interlinked series

daisy ham *noun* (circa 1938)
: a boned and smoked piece of pork from the shoulder

daisy wheel *noun* [from its resemblance to the flower] (circa 1977)
: a disk with spokes bearing type that serves as the printing element of an electric typewriter or printer; *also* **:** a printer that uses such a disk

Da·ko·ta \də-ˈkō-tə\ *noun, plural* **Dakotas** *also* **Dakota** (1804)
1 : a member of an American Indian people of the northern Mississippi valley
2 : the language of the Dakota people

dal *also* **dahl** \ˈdäl\ *noun* [Hindi *dāl*] (1673)
: a dried legume (as lentils, beans, or peas); *also* **:** an Indian dish made of simmered and usually pureed and spiced legumes

Da·lai La·ma \ˌdä-lī-ˈlä-mə, ˌdä-lā-, ˌda-\ *noun* [Mongolian *dalai* ocean] (1698)
: the spiritual head of Lamaism

da·la·si \dä-ˈlä-sē\ *noun, plural* **dalasi** *or* **dalasis** [Wolof] (1966)
— see MONEY table

dale \ˈdā(ə)l\ *noun* [Middle English, from Old English *dæl;* akin to Old High German *tal* valley, Welsh *dôl*] (before 12th century)
: VALLEY, VALE ⟨went riding over hill and *dale*⟩

dales·man \ˈdā(ə)lz-mən\ *noun* (1769)
British **:** one living or born in a dale

da·leth \ˈdä-ˌleth, -ˌlet *also* -ləd\ *noun* [Hebrew *dāleth,* from *deleth* door] (1823)
: the 4th letter of the Hebrew alphabet — see ALPHABET table

dal·li·ance \ˈda-lē-ən(t)s\ *noun* (14th century)
: an act of dallying: as **a :** PLAY; *especially* **:** amorous play **b :** frivolous action **:** TRIFLING

Dal·lis grass \ˈda-ləs-\ *noun* [A. T. *Dallis,* 19th century American farmer] (1907)
: a tall tufted tropical perennial grass (*Paspalum dilatatum*) introduced as a pasture and forage grass in the southern U.S.

Dall por·poise \ˈdȯl-\ *noun* [William H. *Dall* (died 1927) American naturalist] (1951)
: a black-and-white porpoise (*Phocoenoides dalli*) of temperate and arctic waters of the rim of the North Pacific Ocean — called also *Dall's porpoise* \ˈdȯlz-\

Dall sheep *noun* [W. H. *Dall*] (1887)
: a large white wild sheep (*Ovis dalli*) of Alaska and northern British Columbia — called also *Dall's sheep*

dal·ly \ˈda-lē\ *intransitive verb* **dal·lied; dal·ly·ing** [Middle English *dalyen,* from Anglo-French *dalier*] (15th century)
1 a : to act playfully; *especially* **:** to play amorously **b :** to deal lightly **:** TOY ⟨accused him of *dallying* with a serious problem⟩
2 a : to waste time **b :** LINGER, DAWDLE
synonym see TRIFLE, DELAY
— **dal·li·er** *noun*

dal·ma·tian \dal-ˈmā-shən\ *noun, often capitalized* [from the supposed origin of the breed in Dalmatia] (1824)
: any of a breed of medium-sized dogs having a

white short-haired coat with black or brown spots

dalmatian

dal·mat·ic \dal-ˈma-tik\ *noun* [Middle English *dalmatyk,* from Old English *dalmatice,* from Late Latin *dalmatica,* from Latin, feminine of *dalmaticus* Dalmatian, from *Dalmatia*] (before 12th century)
1 : a wide-sleeved overgarment with slit sides worn by a deacon or prelate
2 : a robe worn by a British sovereign at his or her coronation

dal se·gno \däl-ˈsā-(ˌ)nyō\ *adverb* [Italian, from the sign] (circa 1854)
— used as a direction in music to return to the sign that marks the beginning of a repeat

dal·ton \ˈdȯl-tᵊn\ *noun* [John *Dalton* (died 1844) English chemist] (circa 1928)
: ATOMIC MASS UNIT — used chiefly in biochemistry

¹dam \ˈdam\ *noun* [Middle English *dam, dame* lady, dam — more at DAME] (13th century)
: a female parent — used especially of a domestic animal

²dam *noun* [Middle English, probably from Middle Dutch; akin to Old English for*demman* to stop up] (14th century)
1 : a body of water confined by a barrier
2 a : a barrier preventing the flow of water or of loose solid materials (as soil or snow); *especially* **:** a barrier built across a watercourse for impounding water **b :** a barrier to check the flow of liquid, gas, or air

³dam *transitive verb* **dammed; dam·ming** (15th century)
1 : to provide or restrain with a dam
2 : to stop up **:** BLOCK

¹dam·age \ˈda-mij\ *noun* [Middle English, from Middle French, from *dam* damage, from Latin *damnum*] (14th century)
1 : loss or harm resulting from injury to person, property, or reputation
2 *plural* **:** compensation in money imposed by law for loss or injury
3 : EXPENSE, COST ⟨"What's the *damage*?" he asked the waiter⟩

²damage *transitive verb* **dam·aged; dam·ag·ing** (14th century)
: to cause damage to
synonym see INJURE
— dam·age·abil·i·ty \ˌda-mi-jə-ˈbi-lə-tē\ *noun*
— dam·ag·er *noun*

dam·ag·ing *adjective* (circa 1828)
: causing or able to cause damage **:** INJURIOUS ⟨has a *damaging* effect on wildlife⟩
— dam·ag·ing·ly \ˈda-mi-jiŋ-lē\ *adverb*

dam·ar *or* **dam·mar** \ˈda-mər\ *noun* [Malay] (1698)
: any of various resins used in varnishes and inks and obtained chiefly in Malaya and Indonesia from several timber trees (families Dipterocarpaceae and Burseraceae)

¹dam·a·scene \ˈda-mə-ˌsēn, ˌda-mə-ˈ\ *noun* (14th century)
1 *capitalized* **:** a native or inhabitant of Damascus
2 : the characteristic markings of Damascus steel

²damascene *adjective* (14th century)
1 *capitalized* **:** of, relating to, or characteristic of Damascus or the Damascenes
2 : of or relating to damask or the art of damascening

³damascene *transitive verb* **-scened; -scen·ing** [Middle French *damasquiner,* from *damasquin* of Damascus] (1585)
: to ornament (as iron or steel) with wavy patterns like those of watered silk or with inlaid work of precious metals

Da·mas·cus steel \də-ˈmas-kəs-\ *noun* (circa 1727)
: hard elastic steel ornamented with wavy patterns and used especially for sword blades

¹dam·ask \ˈda-məsk\ *noun* [Middle English *damaske,* from Medieval Latin *damascus,* from *Damascus*] (14th century)
1 : a firm lustrous fabric (as of linen, cotton, silk, or rayon) made with flat patterns in a satin weave on a plain-woven ground on jacquard looms
2 : DAMASCUS STEEL; *also* **:** the characteristic markings of this steel
3 : a grayish red

²damask *adjective* (15th century)
1 : made of or resembling damask
2 : of the color damask

damask rose *noun* [obsolete *Damask* of Damascus, from obsolete *Damask* Damascus] (1540)
: a large hardy fragrant pink rose (*Rosa damascena*) widely introduced from Asia Minor and the major source of attar of roses

dame \ˈdām\ *noun* [Middle English, from Old French, from Latin *domina,* feminine of *dominus* master; akin to Latin *domus* house — more at DOME] (13th century)
1 : a woman of rank, station, or authority: as **a** *archaic* **:** the mistress of a household **b :** the wife or daughter of a lord **c :** a female member of an order of knighthood — used as a title prefixed to the given name
2 a : an elderly woman **b :** WOMAN

dame school *noun* (1817)
: a school in which the rudiments of reading and writing were taught by a woman in her own home

dame's rocket *noun* (circa 1900)
: a Eurasian perennial plant (*Hesperis matronalis*) of the mustard family cultivated for its spikes of showy fragrant white or purple flowers — called also *dame's violet, rocket*

¹damn \ˈdam\ *verb* **damned; damn·ing** \ˈda-miŋ\ [Middle English *dampnen,* from Old French *dampner,* from Latin *damnare,* from *damnum* damage, loss, fine] (13th century)
transitive verb
1 : to condemn to a punishment or fate; *especially* **:** to condemn to hell
2 a : to condemn vigorously and often irascibly for some real or fancied fault or defect ⟨*damned* the storm for their delay⟩ **b :** to condemn as a failure by public criticism
3 : to bring ruin on
4 : to swear at **:** CURSE — often used to express annoyance, disgust, or surprise
intransitive verb
: CURSE, SWEAR
synonym see EXECRATE

²damn *noun* (1619)
1 : the utterance of the word *damn* as a curse
2 : a minimum amount or degree (as of care or consideration) **:** the least bit ⟨don't give a *damn*⟩

³damn *adjective or adverb* (1775)
: DAMNED ⟨a *damn* nuisance⟩ ⟨ran *damn* fast⟩
— damn well : beyond doubt or question **:** CERTAINLY ⟨knew *damn well* what would happen⟩

dam·na·ble \ˈdam-nə-bəl\ *adjective* (14th century)
1 : liable to or deserving condemnation
2 : very bad **:** DETESTABLE ⟨*damnable* weather⟩
— dam·na·ble·ness *noun*
— dam·na·bly \-blē\ *adverb*

dam·na·tion \dam-ˈnā-shən\ *noun* (14th century)
: the act of damning **:** the state of being damned

dam·na·to·ry \ˈdam-nə-ˌtōr-ē, -ˌtȯr-\ *adjective* (1682)
: expressing, imposing, or causing condemnation **:** CONDEMNATORY

¹damned \ˈdam(d)\ *adjective* **damned·er** \ˈdam-dər\; **damned·est** *or* **damnd·est** \-dəst\ (1563)
1 : DAMNABLE ⟨this *damned* smog⟩
2 : COMPLETE, UTTER — often used as an intensive
3 : EXTRAORDINARY — used in the superlative ⟨the *damnedest* contraption you ever saw⟩

²damned \ˈdam(d)\ *adverb* (1757)
: EXTREMELY, VERY ⟨a *damned* good job⟩

damned·est *or* **damnd·est** \ˈdam-dəst\ *noun* (1830)
: UTMOST, BEST — used chiefly in the phrase *do one's damnedest* ⟨doing my *damnedest* to win⟩

dam·ni·fy \ˈdam-nə-ˌfī\ *transitive verb* **-fied; -fy·ing** [Middle French *damnifier,* from Old French, from Late Latin *damnificare,* from Latin *damnificus* injurious, from *damnum* damage] (1512)
: to cause loss or damage to

damn·ing \ˈda-miŋ\ *adjective* (1599)
1 : bringing damnation ⟨a *damning* sin⟩
2 : causing or leading to condemnation or ruin ⟨presented some *damning* testimony⟩
— damn·ing·ly \-miŋ-lē\ *adverb*

Dam·o·cles \ˈda-mə-ˌklēz\ *noun* [Latin, from Greek *Damoklēs*]
: a courtier of ancient Syracuse held to have been seated at a banquet beneath a sword hung by a single hair
— Dam·o·cle·an \ˌda-mə-ˈklē-ən\ *adjective*

Da·mon \ˈdā-mən\ *noun* [Latin, from Greek *Damōn*]
: a legendary Sicilian who pledges his life for his condemned friend Pythias

¹damp \ˈdamp\ *noun* [Middle Dutch or Middle Low German, vapor; akin to Old High German *damph* vapor] (14th century)
1 : a noxious gas — compare BLACK DAMP, FIREDAMP
2 : MOISTURE: **a :** HUMIDITY, DAMPNESS **b** *archaic* **:** FOG, MIST
3 a : DISCOURAGEMENT, CHECK **b** *archaic* **:** DEPRESSION, DEJECTION

²damp (14th century)
transitive verb
1 a : to affect with or as if with a noxious gas **:** CHOKE **b :** to diminish the activity or intensity of ⟨*damping* down the causes of inflation⟩ ⟨liquid *damps* out compass oscillations⟩ **c :** to check the vibration or oscillation of (as a string or voltage)
2 : DAMPEN
intransitive verb
: to diminish progressively in vibration or oscillation

³damp *adjective* (1590)
1 a *archaic* **:** being confused, bewildered, or shocked **:** STUPEFIED **b :** DEPRESSED, DULL
2 : slightly or moderately wet **:** MOIST; *also* **:** HUMID
synonym see WET
— damp·ish \ˈdam-pish\ *adjective*
— damp·ly *adverb*
— damp·ness *noun*

damp·en \ˈdam-pən\ *verb* **damp·ened; damp·en·ing** \ˈdamp-niŋ, ˈdam-pə-\ (1547)
transitive verb
1 : to check or diminish the activity or vigor of **:** DEADEN ⟨the heat *dampened* our spirits⟩
2 : to make damp ⟨the shower barely *dampened* the ground⟩
3 : DAMP 1c
intransitive verb
1 : to become damp
2 : to become deadened or depressed
— damp·en·er \-nər\ *noun*

damp·er \ˈdam-pər\ *noun* (1748)
1 : a dulling or deadening influence ⟨put a *damper* on the celebration⟩
2 : a device that damps: as **a :** a valve or plate (as in the flue of a furnace) for regulating the draft **b :** a small felted block to stop the vibration of a piano string **c** *chiefly British* **:** SHOCK ABSORBER

damp·ing–off \ˌdam-piŋ-ˈȯf\ *noun* (1890)

: a diseased condition of seedlings or cuttings caused by fungi and marked by wilting or rotting

dam·sel \'dam-zəl\ *also* **dam·o·sel** *or* **dam·o·zel** \'da-mə-ˌzel\ *noun* [Middle English *damesel,* from Old French *dameisele,* from (assumed) Vulgar Latin *domnicella* young noblewoman, diminutive of Latin *domina* lady] (13th century)
: a young woman: **a** *archaic* **:** a young unmarried woman of noble birth **b :** GIRL

dam·sel·fish \'dam-zəl-ˌfish\ *noun* (1904)
: any of numerous often brilliantly colored marine fishes (family Pomacentridae) living especially along coral reefs — called also *demoiselle*

dam·sel·fly \-ˌflī\ *noun* (1815)
: any of numerous odonate insects (suborder Zygoptera) distinguished from dragonflies by laterally projecting eyes and petiolate wings folded above the body when at rest

dam·son \'dam-zən\ *noun* [Middle English, from Latin (*prunum*) *damascenum,* literally, plum of Damascus] (14th century)
: the small tart fruit of a widely cultivated Asian plum tree (*Prunus insititia*); *also* **:** this tree

[1]Dan \'dan\ *noun* [Hebrew *Dān*]
: a son of Jacob and the traditional eponymous ancestor of one of the tribes of Israel

[2]Dan \'dan\ *noun* [Middle English, title of members of religious orders, from Middle French, from Latin *dominus* master] (13th century)
archaic **:** MASTER, SIR

Dan·aë \'da-nə-ˌē\ *noun* [Latin, from Greek *Danaē*]
: a princess of Argos visited by Zeus in the form of a shower of gold and by him the mother of Perseus

[1]dance \'dan(t)s, 'dȧn(t)s\ *verb* **danced; danc·ing** [Middle English *dauncen,* from Old French *dancier*] (14th century)
intransitive verb
1 : to engage in or perform a dance
2 : to move or seem to move up and down or about in a quick or lively manner
transitive verb
1 : to perform or take part in as a dancer
2 : to cause to dance
3 : to bring into a specified condition by dancing
— dance·able \'dan(t)-sə-bəl\ *adjective*
— danc·er *noun*

[2]dance *noun, often attributive* (14th century)
1 : an act or instance of dancing
2 : a series of rhythmic and patterned bodily movements usually performed to music
3 : a social gathering for dancing
4 : a piece of music by which dancing may be guided
5 : the art of dancing

dan·de·li·on \'dan-də-ˌlī-ən, -dē-\ *noun* [Middle French *dent de lion,* literally, lion's tooth] (14th century)
: any of a genus (*Taraxacum*) of yellow-flowered composite plants; *especially* **:** an herb (*T. officinale*) sometimes grown as a potherb and nearly cosmopolitan as a weed

dan·der \'dan-dər\ *noun* [alteration of *dandruff*] (1786)
1 : DANDRUFF; *specifically* **:** minute scales from hair, feathers, or skin that may be allergenic
2 : ANGER, TEMPER ⟨don't get your *dander* up⟩

dan·di·a·cal \dan-'dī-ə-kəl\ *adjective* [[1]*dandy* + *-acal* (as in *demoniacal*)] (1831)
: of, relating to, or suggestive of a dandy

Dan·die Din·mont terrier \'dan-dē-'din-ˌmänt-\ *noun* [*Dandie Dinmont,* character owning six such dogs in the novel *Guy Mannering* by Sir Walter Scott] (1875)
: any of a breed of terriers characterized by short legs, a long body, pendulous ears, a rough coat, and a full silky topknot

dan·di·fy \'dan-di-ˌfī\ *transitive verb* **-fied; -fy·ing** (1823)
: to cause to resemble a dandy
— dan·di·fi·ca·tion \ˌdan-di-fə-'kā-shən\ *noun*

Dandie Dinmont terrier

dan·dle \'dan-d^{ə}l\ *transitive verb* **dan·dled; dan·dling** \-(d)liŋ, -d^{ə}l-iŋ\ [origin unknown] (1530)
1 : to move (as a baby) up and down in one's arms or on one's knee in affectionate play
2 : PAMPER, PET

dan·druff \'dan-drəf\ *noun* [probably from *dand-* (origin unknown) + *-ruff,* from Middle English *rove* scabby condition, from Old Norse *hrufa* scab; akin to Old High German *hruf* scurf, Lithuanian *kraupus* rough] (1545)
: a scurf that forms on skin especially of the scalp and comes off in small white or grayish scales
— dan·druffy \-drə-fē\ *adjective*

[1]dan·dy \'dan-dē\ *noun, plural* **dandies** [probably short for *jack-a-dandy,* from [1]*jack* + *a* (of) + *dandy* (origin unknown)] (circa 1780)
1 : a man who gives exaggerated attention to personal appearance
2 : something excellent in its class
— dan·dy·ish \-dē-ish\ *adjective*
— dan·dy·ish·ly *adverb*

[2]dandy *adjective* **dan·di·er; -est** (1792)
1 : of, relating to, or suggestive of a dandy **:** FOPPISH
2 : very good **:** FIRST-RATE ⟨a *dandy* place to stay⟩

dan·dy·ism \'dan-dē-ˌi-zəm\ *noun* (1819)
1 : the style or conduct of a dandy
2 : a literary and artistic style of the latter part of the 19th century marked by artificiality and excessive refinement

Dane \'dān\ *noun* [Middle English *Dan,* from Old Norse *Danr*] (14th century)
1 : a native or inhabitant of Denmark
2 : a person of Danish descent
3 : GREAT DANE

dane·geld \'dān-ˌgeld\ *noun, often capitalized* (before 12th century)
: an annual tax believed to have been imposed originally to buy off Danish invaders in England or to maintain forces to oppose them but continued as a land tax

Dane·law \'dān-ˌlȯ\ *noun* (before 12th century)
1 : the law in force in the part of England held by the Danes before the Norman Conquest
2 : the part of England under the Danelaw

[1]dang \'daŋ\ *transitive verb* [euphemism] (1797)
: DAMN 4

[2]dang *adjective or adverb* (1914)
: DAMNED

[1]dan·ger \'dān-jər\ *noun* [Middle English *daunger,* from Old French *dangier,* alteration of *dongier,* from (assumed) Vulgar Latin *dominiarium,* from Latin *dominium* ownership] (13th century)
1 a *archaic* **:** JURISDICTION **b** *obsolete* **:** REACH, RANGE
2 *obsolete* **:** HARM, DAMAGE
3 : exposure or liability to injury, pain, harm, or loss ⟨could play there without *danger*⟩
4 : a case or cause of danger ⟨the *dangers* of mining⟩

[2]danger *transitive verb* (14th century)
archaic **:** ENDANGER

dan·ger·ous \'dān-jə-rəs; 'dān-jərs, -zhrəs\ *adjective* (15th century)
1 : exposing to or involving danger
2 : able or likely to inflict injury or harm ☆
— dan·ger·ous·ly *adverb*
— dan·ger·ous·ness *noun*

[1]dan·gle \'daŋ-gəl\ *verb* **dan·gled; dan·gling** \-g(ə-)liŋ\ [probably of Scandinavian origin; akin to Danish *dangle* to dangle] (circa 1590)
intransitive verb
1 : to hang loosely and usually so as to be able to swing freely
2 : to be a hanger-on or a dependent
3 : to occur in a sentence without having a normally expected syntactic relation to the rest of the sentence ⟨the word *climbing* in "Climbing the mountain the cabin came into view" is *dangling*⟩
transitive verb
1 : to cause to dangle **:** SWING
2 a : to keep hanging uncertainly **b :** to hold out as an inducement
— dan·gler \-g(ə-)lər\ *noun*

[2]dangle *noun* (1756)
1 : the action of dangling
2 : something that dangles

Dan·iel \'da-nyəl *also* 'da-n^{ə}l\ *noun* [Hebrew *Dānīʼēl*]
1 : the Jewish hero of the Book of Daniel who as an exile in Babylon interprets dreams, gives accounts of apocalyptic visions, and is divinely delivered from a den of lions
2 : a book of narratives, visions, and prophecies in canonical Jewish and Christian Scripture — see BIBLE table

da·nio \'dā-nē-ˌō\ *noun, plural* **da·ni·os** [New Latin, genus name] (circa 1889)
: any of several small brightly colored Asian cyprinid fishes

[1]Dan·ish \'dā-nish\ *adjective* (14th century)
: of, relating to, or characteristic of Denmark, the Danes, or the Danish language

[2]Danish *noun* (15th century)
1 : the Germanic language of the Danes
2 *plural* **Danish :** a piece of Danish pastry

Danish pastry *noun* (circa 1928)
: a pastry made of a rich raised dough

dank \'daŋk\ *adjective* [Middle English *danke*] (1573)
: unpleasantly moist or wet
synonym see WET
— dank·ly *adverb*
— dank·ness *noun*

dan·seur \dä^{n}-'sər, dän-\ *noun* [French, from *danser* to dance] (1828)
: a male ballet dancer

dan·seuse \dä^{n}-'sœ̄z; dä^{n}-'sə(r)z, dän-'süz\ *noun* [French, feminine of *danseur*] (1828)
: a female ballet dancer

Dan·te·an \'dan-tē-ən, 'dän-\ *noun* (circa 1850)
: a student or admirer of Dante

Dao·ism, Dao·ist, Dao·is·tic *variant of* TAOISM, TAOIST, TAOISTIC

daph·ne \'daf-nē\ *noun* [New Latin, genus name, from Latin, laurel, from Greek *daphnē*] (1862)
: any of a genus (*Daphne*) of Eurasian shrubs of the mezereon family with apetalous flowers whose colored calyx resembles a corolla

☆ SYNONYMS
Dangerous, hazardous, precarious, perilous, risky mean bringing or involving the chance of loss or injury. DANGEROUS applies to something that may cause harm or loss unless dealt with carefully ⟨soldiers on a *dangerous* mission⟩. HAZARDOUS implies great and continuous risk of harm or failure ⟨claims that smoking is *hazardous* to your health⟩. PRECARIOUS suggests both insecurity and uncertainty ⟨earned a *precarious* living by gambling⟩. PERILOUS strongly implies the immediacy of danger ⟨*perilous* mountain roads⟩. RISKY often applies to a known and accepted danger ⟨made *risky* decisions⟩.

Daph·ne \'daf-nē\ *noun* [Latin, from Greek *Daphnē*]
: a nymph who is transformed into a laurel tree to escape the pursuing Apollo
daph·nia \'daf-nē-ə\ *noun* [New Latin, genus name] (1847)
: any of a genus (*Daphnia*) of minute freshwater branchiopod crustaceans with biramous antennae used as locomotor organs — compare WATER FLEA
Daph·nis \'daf-nəs\ *noun* [Latin, from Greek]
: a Sicilian shepherd renowned in Greek mythology as the inventor of pastoral poetry
dap·per \'da-pər\ *adjective* [Middle English *dapyr,* from Middle Dutch *dapper* quick, strong; akin to Old High German *tapfar* heavy, Old Church Slavonic *debelŭ* thick] (15th century)
1 a : neat and trim in appearance **b :** excessively spruce and stylish
2 : alert and lively in movement and manners
— dap·per·ly *adverb*
— dap·per·ness *noun*
¹dap·ple \'da-pəl\ *noun* [Middle English *dappel-gray,* adjective, gray marked with spots of another color] (1580)
1 : any of numerous usually cloudy and rounded spots or patches of a color or shade different from their background
2 : the quality or state of being dappled
3 : a dappled animal
²dapple *verb* **dap·pled; dap·pling** \-p(ə-)liŋ\ (1599)
transitive verb
: to mark with dapples
intransitive verb
: to produce a dappled pattern ⟨sun *dappling* through trees⟩
dap·pled *also* **dap·ple** *adjective* (15th century)
: marked with small spots or patches contrasting with the background ⟨a *dappled* fawn⟩
dap·sone \'dap-,sōn\ *noun* [*di*amine + *p*henyl + *sul*f*one*] (1952)
: an antimicrobial agent $C_{12}H_{12}N_2O_2S$ used especially against leprosy
Dar·by and Joan \,där-bē-ən-'jōn\ *noun* [probably from *Darby & Joan,* couple in an 18th century song] (1857)
: a happily married usually elderly couple
Dard \'därd\ *noun* (circa 1885)
: DARDIC
Dar·dan \'där-dᵊn\ *adjective or noun* [Latin *Dardanus,* from Greek *Dardanos*] (1606)
archaic **:** TROJAN
Dar·da·ni·an \där-'dā-nē-ən\ *adjective* (1596)
: TROJAN
Dar·dic \'där-dik\ *noun* (1910)
: a complex of Indic languages spoken in the upper valley of the Indus
¹dare \'dar, 'der\ *verb* **dared; dar·ing; dares** *or (auxiliary)* **dare** [Middle English *dar* (1st & 3d singular present indicative), from Old English *dear;* akin to Old High German *gitar* (1st & 3d singular present indicative) dare, Greek *tharsos* courage] (before 12th century)
verbal auxiliary
: to be sufficiently courageous to ⟨no one *dared* say a word⟩ ⟨she *dare* not let herself love —G. B. Shaw⟩
intransitive verb
: to have sufficient courage ⟨try it if you *dare*⟩
transitive verb
1 a : to challenge to perform an action especially as a proof of courage ⟨*dared* him to jump⟩ **b :** to confront boldly **:** ⟨*dared* the mob⟩
2 : to have the courage to contend against, venture, or try ⟨the actress *dared* a new interpretation of this classic role⟩
— dar·er \'dar-ər, 'der-\ *noun*
²dare *noun* (1594)
1 : an act or instance of daring **:** CHALLENGE ⟨foolishly took a *dare*⟩
2 : imaginative or vivacious boldness **:** DARING
¹dare·dev·il \'dar-,de-vᵊl, 'der-\ *noun* (1794)
: a recklessly bold person
— dare·dev·il·ry \-vᵊl-rē\ *noun*
— dare·dev·il·try \-vᵊl-trē\ *noun*
²daredevil *adjective* (1832)
: recklessly and often ostentatiously daring
synonym see ADVENTUROUS
dareful *adjective* (1605)
obsolete **:** DARING
daren't \'dar-ənt, 'der-\
: dare not **:** dared not
dare·say \,dar-'sā, ,der-\ (13th century)
transitive verb
: venture to say **:** think probable — used in present 1st singular
intransitive verb
: AGREE, SUPPOSE — used in present 1st singular
¹dar·ing *adjective* (1582)
: venturesomely bold in action or thought
synonym see ADVENTUROUS
— dar·ing·ly \-iŋ-lē\ *adverb*
— dar·ing·ness *noun*
²daring *noun* (1651)
: venturesome boldness
Dar·jee·ling \där-'jē-liŋ\ *noun* [*Darjeeling,* India] (1907)
: a tea of high quality grown especially in the mountainous districts of northern India
¹dark \'därk\ *adjective* [Middle English *derk,* from Old English *deorc;* akin to Old High German *tarchannen* to hide] (before 12th century)
1 a : devoid or partially devoid of light **:** not receiving, reflecting, transmitting, or radiating light **b :** transmitting only a portion of light
2 a : wholly or partially black **b** *of a color* **:** of low or very low lightness
3 a : arising from or showing evil traits or desires **:** EVIL ⟨the *dark* powers that lead to war⟩ **b :** DISMAL, GLOOMY ⟨had a *dark* view of the future⟩ **c :** lacking knowledge or culture
4 : not clear to the understanding
5 : not fair in complexion **:** SWARTHY
6 : SECRET ⟨kept his plans *dark*⟩
7 : possessing depth and richness ⟨a *dark* voice⟩
8 : closed to the public ⟨the theater is *dark* in the summer⟩
synonym see OBSCURE
— dark·ish \'där-kish\ *adjective*
— dark·ly *adverb*
— dark·ness *noun*
²dark *noun* (13th century)
1 a : a place or time of little or no light **:** NIGHT, NIGHTFALL **b :** absence of light **:** DARKNESS
2 : a dark or deep color
— in the dark 1 : in secrecy ⟨most of his dealings were done *in the dark*⟩ **2 :** in ignorance ⟨kept the public *in the dark* about the agreement⟩
³dark (14th century)
intransitive verb
obsolete **:** to grow dark
transitive verb
: to make dark
dark adaptation *noun* (1909)
: the phenomena including dilation of the pupil, increase in retinal sensitivity, shift of the region of maximum luminosity toward the blue, and regeneration of visual purple by which the eye adapts to conditions of reduced illumination
— dark–adapt·ed \,därk-ə-'dap-təd\ *adjective*
dark age *noun* (1730)
1 : a time during which a civilization undergoes a decline: as **a** *plural, D&A capitalized* **:** the European historical period from about A.D. 476 to about 1000; *broadly* **:** MIDDLE AGES **b** *often plural, often D&A capitalized* **:** the Greek historical period of three to four centuries from about 1100 B.C.
2 a *often plural, often D&A capitalized* **:** the primitive period in the development of something ⟨in the 1890s, way back in baseball's *Dark Ages* —R. W. Creamer⟩ **b** *often plural, often D&A capitalized* **:** a state of stagnation or decline
dark·en \'där-kən\ *verb* **dark·ened; dark·en·ing** \'där-kə-niŋ\ (14th century)
intransitive verb
1 : to grow dark **:** become obscured
2 : to become gloomy
transitive verb
1 : to make dark
2 : to make less clear **:** OBSCURE ⟨the financial crisis *darkened* the future of the company⟩
3 : TAINT, TARNISH
4 : to cast a gloom over
5 : to make of darker color
— dark·en·er \-kə-nər\ *noun*
dark field *noun* (1865)
: the dark area that serves as the background for objects viewed in an ultramicroscope
dark–field microscope *noun* (1926)
: ULTRAMICROSCOPE
dark horse *noun* (1831)
1 a : a usually little known contender (as a racehorse) that makes an unexpectedly good showing **b :** an entrant in a contest that is judged unlikely to succeed
2 : a political candidate unexpectedly nominated usually as a compromise between factions ◆
dark lantern *noun* (1650)
: a lantern that can be closed to conceal the light
dar·kle \'där-kəl\ *intransitive verb* **dar·kled; dar·kling** \-k(ə-)liŋ\ [back-formation from *darkling*] (1800)
1 a : to become clouded or gloomy **b :** to grow dark
2 : to become concealed in the dark
¹dark·ling \'därk-liŋ\ *adverb* [Middle English *derkelyng,* from *derk* dark + *-lyng* -ling] (15th century)
: in the dark
²dark·ling *adjective* (1739)
1 : DARK
2 : done or taking place in the dark
darkling beetle *noun* (1816)
: TENEBRIONID
dark matter *noun* (1982)
: nonluminous matter not yet directly detected by astronomers that is hypothesized to exist because the visible matter in the universe is insufficient to account for various observed gravitational effects
dark reaction *noun* (1927)
: any of a series of chemical reactions in photosynthesis not requiring the presence of light and involving the reduction of carbon dioxide to form carbohydrate; *also* **:** CALVIN CYCLE
dark·room \'därk-,rüm, -,ru̇m\ *noun* (1841)
: a room with no light or with a safelight for handling and processing light-sensitive photographic materials
dark·some \'därk-səm\ *adjective* (circa 1530)
: gloomily somber **:** DARK
darky *or* **dark·ie** \'där-kē\ *noun, plural* **darkies** (1775)
: a black person — usually taken to be offensive
¹dar·ling \'där-liŋ\ *noun* [Middle English *der-*

◇ WORD HISTORY
dark horse Sometimes in a race a horse whose name and ability are not widely known puts on a surprisingly good show and defeats its more famous rivals. Such a horse is called *dark,* not because of its color (which might be anything), but because of its obscurity. Since the 19th century, the phrase *dark horse* has been extended from racehorses to obscure competitors who do unexpectedly well in contests of other kinds. Now it is perhaps most often applied to candidates for elected office whose chances appear poor.

ling, from Old English *dēorling,* from *dēore* dear] (before 12th century)
1 : a dearly loved person
2 : FAVORITE

²darling *adjective* (1509)
1 : dearly loved **:** FAVORITE
2 : very pleasing **:** CHARMING
— **dar·ling·ly** \-liŋ-lē\ *adverb*
— **dar·ling·ness** *noun*

¹darn \'därn\ *verb* [perhaps from French dialect *darner*] (circa 1600)
transitive verb
1 : to mend with interlacing stitches
2 : to embroider by filling in with long running or interlacing stitches
intransitive verb
: to do darning
— **darn·er** *noun*

²darn *noun* (1720)
: a place that has been darned ⟨a sweater full of *darns*⟩

³darn *adjective or adverb* [euphemism] (1781)
: DAMNED

⁴darn *verb* (1781)
: DAMN
— **darned** \'därn(d)\ *adjective or adverb*

⁵darn *noun* (1840)
: DAMN

dar·nel \'där-nᵊl\ *noun* [Middle English] (14th century)
: any of several usually weedy grasses (genus *Lolium*)

darning needle *noun* (1761)
1 : a long needle with a large eye for use in darning
2 : DRAGONFLY, DAMSELFLY

¹dart \'därt\ *noun* [Middle English, from Middle French, of Germanic origin; akin to Old High German *tart* dart, Old English *daroth*] (14th century)
1 a *archaic* **:** a light spear **b** (1) **:** a small missile usually with a pointed shaft at one end and feathers at the other (2) *plural but singular in construction* **:** a game in which darts are thrown at a target
2 a : something projected with sudden speed; *especially* **:** a sharp glance **b :** something causing sudden pain or distress ⟨*darts* of sarcasm⟩
3 : something with a slender pointed shaft or outline; *specifically* **:** a stitched tapering fold in a garment
4 : a quick movement ⟨made a *dart* for the door⟩

²dart (1580)
transitive verb
1 : to throw with a sudden movement
2 : to thrust or move with sudden speed
3 : to shoot with a dart containing a usually tranquilizing drug
intransitive verb
: to move suddenly or rapidly ⟨*darted* across the street⟩

dart·board \'därt-ˌbōrd, -ˌbȯrd\ *noun* (1901)
: a usually circular board (as of compressed bristles) used as a target in the game of darts

dart·er \'där-tər\ *noun* (1813)
1 : ANHINGA
2 : any of numerous small American freshwater bony fishes (especially genera *Ammocrypta, Etheostoma,* and *Percina* of the family Percidae)

Dar·win·i·an \där-'wi-nē-ən\ *adjective* (1860)
: of or relating to Charles Darwin, his theories especially of evolution, or his followers
— **Darwinian** *noun*

Dar·win·ism \'där-wə-ˌni-zəm\ *noun* (1864)
: a theory of the origin and perpetuation of new species of animals and plants that offspring of a given organism vary, that natural selection favors the survival of some of these variations over others, that new species have arisen and may continue to arise by these processes, and that widely divergent groups of plants and animals have arisen from the same ancestors; *broadly* **:** biological evolutionism ◆
— **Dar·win·ist** \-wə-nist\ *noun or adjective*

Dar·win's finches \ˌdär-wənz-\ *noun plural* [Charles *Darwin*] (1947)
: finches of a subfamily (Geospizinae) having great variation in bill shape and confined mostly to the Galapagos islands

Dar·win tulip \ˌdär-wən-\ *noun* (1889)
: a tall late-flowering tulip with the flowers single and of one color

¹dash \'dash\ *verb* [Middle English *dasshen*] (13th century)
transitive verb
1 : to break by striking or knocking
2 : to knock, hurl, or thrust violently
3 : SPLASH, SPATTER
4 a : RUIN, DESTROY ⟨the news *dashed* his hopes⟩ **b :** DEPRESS, SADDEN **c :** to make ashamed
5 : to affect by mixing in something different ⟨his delight was *dashed* with bitterness over the delay⟩
6 : to complete, execute, or finish off hastily — used with *down* or *off* ⟨*dashed* down a drink⟩ ⟨*dash* off a letter⟩
7 [euphemism] **:** ¹DAMN 4
intransitive verb
1 : to move with sudden speed ⟨*dashed* through the rain⟩
2 : SMASH

²dash *noun* (14th century)
1 a *archaic* **:** BLOW **b** (1) **:** a sudden burst or splash (2) **:** the sound produced by such a burst
2 a : a stroke of a pen **b :** a punctuation mark — that is used especially to indicate a break in the thought or structure of a sentence
3 : a small usually distinctive addition ⟨a *dash* of salt⟩ ⟨a *dash* of humor⟩
4 : flashy display
5 : animation in style and action
6 a : a sudden onset, rush, or attempt **b :** a short fast race
7 : a long click or buzz forming a letter or part of a letter (as in Morse code)
8 : DASHBOARD 2

dash·board \'dash-ˌbōrd, -ˌbȯrd\ *noun* (1846)
1 : a screen on the front of a usually horse-drawn vehicle to intercept water, mud, or snow
2 : a panel extending across the interior of a vehicle (as an automobile) below the windshield and usually containing dials and controls

dashed \'dasht\ *adjective* (circa 1889)
: made up of a series of dashes

da·sheen \da-'shēn, də-\ *noun* [origin unknown] (circa 1899)
: TARO

dash·er \'da-shər\ *noun* (1790)
1 : a dashing person
2 : one that dashes; *specifically* **:** a device having blades for agitating a liquid or semisolid

dashi \'dä-(ˌ)shē\ *noun* [Japanese, broth] (circa 1961)
: a fish broth made from dried bonito

da·shi·ki \də-'shē-kē, dä-, da-\ *noun* [modification of Yoruba *dàǹṣíkí*] (circa 1968)
: a usually brightly colored loose-fitting pullover garment

dash·ing *adjective* (1796)
1 : marked by vigorous action **:** SPIRITED ⟨a *dashing* young horse⟩
2 : marked by smartness especially in dress and manners
— **dash·ing·ly** \-iŋ-lē\ *adverb*

dash·pot \'dash-ˌpät\ *noun* (1861)
: a device for cushioning or damping a movement (as of a mechanical part) to avoid shock

das·sie \'da-sē\ *noun* [Afrikaans] (1786)
: HYRAX

das·tard \'das-tərd\ *noun* [Middle English] (15th century)
1 : COWARD
2 : a person who acts treacherously or underhandedly

das·tard·ly \-lē\ *adjective* (1576)
1 : COWARDLY
2 : characterized by underhandedness or treachery ⟨a *dastardly* attack⟩ ⟨a *dastardly* villain⟩
synonym see COWARDLY
— **das·tard·li·ness** *noun*

da·ta \'dā-tə, 'da- *also* 'dä-\ *noun plural but singular or plural in construction, often attributive* [Latin, plural of *datum*] (1646)
1 : factual information (as measurements or statistics) used as a basis for reasoning, discussion, or calculation ⟨the *data* is plentiful and easily available —H. A. Gleason, Jr.⟩ ⟨comprehensive *data* on economic growth have been published —N. H. Jacoby⟩
2 : information output by a sensing device or organ that includes both useful and irrelevant or redundant information and must be processed to be meaningful
3 : information in numerical form that can be digitally transmitted or processed ■

data bank *noun* (1966)
: DATABASE

da·ta·base \'dā-tə-ˌbās, 'da- *also* 'dä-\ *noun* (circa 1962)
: a usually large collection of data organized especially for rapid search and retrieval (as by a computer)

data processing *noun* (1954)
: the converting of raw data to machine-readable form and its subsequent processing (as storing, updating, combining, rearranging, or printing out) by a computer
— **data processor** *noun*

data structure *noun* (1963)
: any of various methods of organizing data items (as records) in a computer

□ USAGE

data *Data* leads a life of its own quite independent of *datum,* of which it was originally the plural. It occurs in two constructions: as a plural noun (like *earnings*), taking a plural verb and plural modifiers (as *these, many, a few*) but not cardinal numbers, and serving as a referent for plural pronouns (as *they, them*); and as an abstract mass noun (like *information*), taking a singular verb and singular modifiers (as *this, much, little*), and being referred to by a singular pronoun (*it*). Both constructions are standard. The plural construction is more common in print, evidently because the house style of several publishers mandates it.

◇ WORD HISTORY

Darwinism *Darwinism* is the theory initiated by the British naturalist Charles Robert Darwin (1809–1882). Although he was not the first to question the immutability of species in nature or to conceive the notion of evolution, Darwin is credited with providing voluminous evidence for evolution and with developing the principle of natural selection. In 1831 he embarked on a five-year voyage around the world, making stopovers in South America and the Galapagos Islands, where he gained critical insight into the variation between populations of animals. His landmark *On the Origin of Species by Means of Natural Selection* (1859) brought about a revolution in biological thinking and firmly established the study of evolution as part of the science of biology. *The Descent of Man* (1871) was a follow-up work that presented his related theory of sexual selection.

[1]date \'dāt\ *noun* [Middle English, from Old French, ultimately from Latin *dactylus* — more at DACTYL] (14th century)
1 : the oblong edible fruit of a palm (*Phoenix dactylifera*)
2 : the tall palm with pinnate leaves that yields the date ◆

[2]date *noun* [Middle English, from Middle French, from Late Latin *data,* from *data* (as in *data Romae* given at Rome), feminine of Latin *datus,* past participle of *dare* to give; akin to Latin *dos* gift, dowry, Greek *didonai* to give] (14th century)
1 a : the time at which an event occurs ⟨the *date* of his birth⟩ **b :** a statement of the time of execution or making ⟨the *date* on the letter⟩
2 : DURATION
3 : the period of time to which something belongs
4 a : an appointment to meet at a specified time; *especially* **:** a social engagement between two persons that often has a romantic character **b :** a person with whom one has a usually romantic date
5 : an engagement for a professional performance (as of a dance band)
— to date : up to the present moment

[3]date *verb* **dat·ed; dat·ing** (15th century)
transitive verb
1 : to determine the date of ⟨*date* an antique⟩
2 : to record the date of **:** mark with the date
3 a : to mark with characteristics typical of a particular period **b :** to show up plainly the age of
4 : to make or have a date with
intransitive verb
1 : to reckon chronologically
2 : ORIGINATE ⟨a friendship *dating* from college days⟩
3 : to become dated
— dat·able *also* **date·able** \'dā-tə-bəl\ *adjective*
— dat·er \'dā-tər\ *noun*

dat·ed *adjective* (1731)
1 : provided with a date ⟨a *dated* document⟩
2 : OUTMODED, OLD-FASHIONED ⟨*dated* formalities⟩
— dat·ed·ly *adverb*
— dat·ed·ness *noun*

date·less \'dāt-ləs\ *adjective* (1593)
1 : ENDLESS
2 : having no date
3 : too ancient to be dated
4 : TIMELESS ⟨the play's *dateless* theme⟩

date·line \'dāt-,līn\ *noun* (1888)
1 : a line in a written document or a printed publication giving the date and place of composition or issue
2 *usually* **date line :** INTERNATIONAL DATE LINE
— dateline *transitive verb*

date rape *noun* (circa 1983)
: rape committed by someone known to the victim

[1]da·tive \'dā-tiv\ *adjective* [Middle English *datif,* from Latin *dativus,* from *datus*] (15th century)
: of, relating to, or being the grammatical case that marks typically the indirect object of a verb, the object of some prepositions, or a possessor

[2]dative *noun* (15th century)
: a dative case or form

dative bond *noun* [from the donation of electrons by one of the atoms] (circa 1929)
: COORDINATE BOND

da·tum \'dā-təm, 'da-, 'dä-\ *noun* [Latin, from neuter of *datus*] (1646)
1 *plural* **da·ta** \-ə\ **:** something given or admitted especially as a basis for reasoning or inference
2 *plural* **datums :** something used as a basis for calculating or measuring
usage see DATA

da·tu·ra \də-'t(y)u̇r-ə\ *noun* [New Latin, genus name, from Hindi *dhatūrā* jimsonweed] (1598)
: any of a genus (*Datura*) of widely distributed strong-scented herbs, shrubs, or trees of the nightshade family including some used as sources of medicinal alkaloids (as stramonium) or in folk rites or illicitly for their poisonous, narcotic, or hallucinogenic properties — compare JIMSONWEED

[1]daub \'dȯb, 'däb\ *verb* [Middle English, from Middle French *dauber*] (14th century)
transitive verb
1 : to cover or coat with soft adhesive matter **:** PLASTER
2 : to coat with a dirty substance
3 a : to apply coloring material crudely to **b :** to apply (as paint) crudely
intransitive verb
1 *archaic* **:** to put on a false exterior
2 : to apply colors crudely
— daub·er *noun*

[2]daub *noun* (15th century)
1 : material used to daub walls
2 : an act or instance of daubing
3 : something daubed on **:** SMEAR
4 : a crude picture

daube \'dōb\ *noun* [French] (1723)
: a stew of braised meat, vegetables, herbs, and spices

[1]daugh·ter \'dȯ-tər, 'dä-\ *noun* [Middle English, *doughter,* from Old English *dohtor;* akin to Old High German *tohter* daughter, Greek *thygatēr*] (before 12th century)
1 a (1) **:** a human female having the relation of child to parent (2) **:** a female offspring of a lower animal **b :** a human female having a specified ancestor or belonging to a group of common ancestry
2 : something considered as a daughter ⟨the United States is a *daughter* of Great Britain⟩
3 : an atomic species that is the product of the radioactive decay of a given element
— daugh·ter·less \-ləs\ *adjective*

[2]daughter *adjective* (1614)
1 : having the characteristics or relationship of a daughter
2 : belonging to the first generation of offspring, organelles, or molecules produced by reproduction, division, or replication ⟨*daughter* cell⟩ ⟨*daughter* DNA molecules⟩

daugh·ter–in–law \'dȯ-tər-in-,lȯ, 'dȯ-tərn-\ *noun, plural* **daugh·ters–in–law** \-tər-zin-\ (14th century)
: the wife of one's son

dau·no·my·cin \,dȯ-nə-'mī-s^{ə}n, ,dau̇-\ *noun* [International Scientific Vocabulary *dauno-* (from Latin *Daunus* Apulian) + *-mycin*] (1964)
: DAUNORUBICIN

dau·no·ru·bi·cin \-'rü-bə-s^{ə}n\ *noun* [*dauno-* (as in *daunomycin*) + *rubidomycin,* a substance found to be identical with daunomycin (from International Scientific Vocabulary *rubido-* — from Latin *rubidus* red — + *-mycin*)] (circa 1968)
: an antibiotic $C_{27}H_{29}NO_{10}$ that is a nitrogenous glycoside and is used especially in the treatment of leukemia

daunt \'dȯnt, 'dänt\ *transitive verb* [Middle English, from Middle French *danter,* alteration of *donter,* from Latin *domitare* to tame, frequentative of *domare* — more at TAME] (14th century)
: to lessen the courage of **:** COW, SUBDUE
synonym see DISMAY
— daunt·ing·ly \-iŋ-lē\ *adverb*

daunt·less \-ləs\ *adjective* (1593)
: FEARLESS, UNDAUNTED ⟨a *dauntless* hero⟩
— daunt·less·ly *adverb*
— daunt·less·ness *noun*

dau·phin \'dȯ-fən, 'dō-; ,dō-'fa^{n}\ *noun, often capitalized* [Middle English *dolphin,* from Middle French *dalfin,* from Old French, title of lords of the Dauphiné, from *Dalfin,* a surname] (15th century)
: the eldest son of a king of France

dau·phine \dȯ-'fēn, dō-\ *noun, often capitalized* [French] (circa 1864)
: the wife of the dauphin

da·ven \'dä-vən\ *intransitive verb* [Yiddish *davnen*] (circa 1930)
: to recite the prescribed prayers in a Jewish liturgy

dav·en·port \'da-v^{ə}n-,pōrt, 'da-v^{ə}m-, -,pȯrt\ *noun* [probably from the name *Davenport*] (1853)
1 : a small compact writing desk
2 : a large upholstered sofa often convertible into a bed

Da·vid \'dā-vəd\ *noun* [Hebrew *Dāwīdh*]
1 : a Hebrew shepherd who became the second king of Israel in succession to Saul according to biblical accounts
2 : UNDERDOG 1
— Da·vid·ic \də-'vi-dik, dā-\ *adjective*

da·vit \'dā-vət, 'da-\ *noun* [probably from the name *David*] (15th century)
: a crane that projects over the side of a ship or a hatchway and is used especially for boats, anchors, or cargo

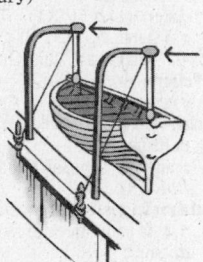

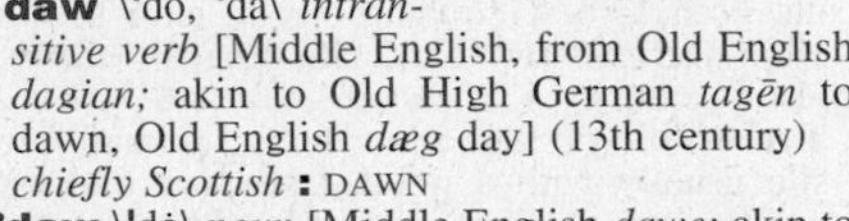
davit

Da·vy Jones \,dā-vē-'jōnz\ *noun* (1751)
: the bottom of the sea personified

Da·vy Jones's lock·er \,dā-vē-,jōnz(-əz)-\ *noun* (circa 1777)
: the bottom of the ocean

[1]daw \'dȯ, 'då\ *intransitive verb* [Middle English, from Old English *dagian;* akin to Old High German *tagēn* to dawn, Old English *dæg* day] (13th century)
chiefly Scottish **:** DAWN

[2]daw \'dȯ\ *noun* [Middle English *dawe;* akin to Old High German *taha* jackdaw] (15th century)
: JACKDAW

daw·dle \'dȯ-d^{ə}l\ *verb* **daw·dled; daw·dling** \'dȯ-dliŋ, -d^{ə}l-iŋ\ [origin unknown] (circa 1656)
intransitive verb
1 : to spend time idly
2 : to move lackadaisically ⟨*dawdled* up the hill⟩
transitive verb
: to spend fruitlessly or lackadaisically ⟨*dawdled* the day away⟩
synonym see DELAY
— daw·dler \'dȯ-dlər, -d^{ə}l-ər\ *noun*

◇ WORD HISTORY

date The English word *date* in its temporal sense has nothing to do etymologically with *day,* but rather is descended from the Latin verb *dare* "to give." In ancient Rome, the date of a letter was written in the following manner: *Dabam Romae Kal. Aprilis* ("I gave [this letter] at Rome on the calends of April [April 1].") A later formula used *data Romae* "given at Rome," instead of *dabam Romae. Data,* past participle of *dare* "to give," had the feminine case ending because it was understood to refer to the unexpressed feminine noun *epistula* "letter." By the 6th century of the modern era, *data* itself had become a noun, used for the formula indicating the date on a letter. Borrowed into French as *date,* it was used not only for the formula on a letter, but also for the time that such a formula indicated, or indeed for any point in time. The English word *date,* borrowed from French in the late Middle Ages, did not come to mean an appointment or engagement until the late 19th century.

¹dawn \'dȯn, 'dän\ *intransitive verb* [Middle English, probably back-formation from *dawning* daybreak, alteration of *dawing,* from Old English *dagung,* from *dagian*] (15th century)
1 : to begin to grow light as the sun rises
2 : to begin to appear or develop
3 : to begin to be perceived or understood ⟨the truth finally *dawned* on us⟩
²dawn *noun* (1599)
1 : the first appearance of light in the morning followed by sunrise
2 : BEGINNING ⟨the *dawn* of the space age⟩
dawn horse *noun* (1930)
: EOHIPPUS
dawn redwood *noun* (1948)
: a metasequoia (*Metasequoia glyptostroboides*) of China resembling the coast redwood but having deciduous foliage
daw•son•ite \'dȯ-sə-ˌnīt\ *noun* [Sir John W. *Dawson*] (1874)
: a mineral consisting of a basic sodium aluminum carbonate
day \'dā\ *noun* [Middle English, from Old English *dæg;* akin to Old High German *tag* day] (before 12th century)
1 a : the time of light between one night and the next **b :** DAYLIGHT 1, 2
2 : the period of rotation of a planet (as earth) or a moon on its axis
3 : the mean solar day of 24 hours beginning at mean midnight
4 : a specified day or date
5 : a specified time or period **:** AGE ⟨in grandfather's *day*⟩
6 : the conflict or contention of the day ⟨played hard and won the *day*⟩
7 : the time established by usage or law for work, school, or business
8 : a period of existence or prominence of a person or thing
— day after day : for an indefinite or seemingly endless number of days
— day in, day out : for an indefinite number of successive days
Day•ak \'dī-ˌak\ *noun* [Malay, literally, upcountry] (1836)
: a member of any of several Indonesian peoples of the interior of Borneo
day•bed \'dā-ˌbed\ *noun* (1818)
1 : a chaise longue of a type made 1680–1780
2 : a couch that can be converted into a bed
day•book \-ˌbu̇k\ *noun* (1580)
: DIARY, JOURNAL
day•break \-ˌbrāk\ *noun* (1530)
: DAWN
day care *noun* (1947)
1 : supervision of and care for children or disabled adults that is provided during the day by a person or organization
2 : a program, facility, or organization offering day care
¹day•dream \'dā-ˌdrēm\ *noun* (1685)
: a pleasant visionary usually wishful creation of the imagination
— day•dream•like \-ˌlīk\ *adjective*
²daydream *intransitive verb* (1820)
: to have a daydream
— day•dream•er *noun*
day•flow•er \'dā-ˌflau̇(-ə)r\ *noun* (circa 1688)
: any of a genus (*Commelina*) of herbs of the spiderwort family having one petal smaller than the other two; *especially* **:** a blue-flowered weed (*C. communis*) with the smaller petal white
Day–Glo \'dā-ˌglō\ *trademark*
— used for fluorescent materials or colors
day•glow \'dā-ˌglō\ *noun* (circa 1960)
: airglow seen during the day
day laborer *noun* (1548)
: one who works for daily wages especially as an unskilled laborer
day letter *noun* (circa 1913)
: a telegram sent during the day that has a lower priority than a regular telegram
day•light \'dā-ˌlīt\ *noun* (13th century)
1 : the light of day
2 : DAYTIME
3 : DAWN
4 a : knowledge or understanding of something that has been obscure ⟨began to see *daylight* on the problem⟩ **b :** the quality or state of being open **:** OPENNESS
5 *plural* **a :** CONSCIOUSNESS **b :** mental soundness or stability **:** WITS ⟨scared the *daylights* out of him⟩
6 : an opening or opportunity especially for action
day•light•ing \'dā-ˌlī-tiŋ\ *noun* (1929)
: illumination of indoor spaces by natural light
daylight saving time *noun* (1919)
: time usually one hour ahead of standard time — called also *daylight saving, daylight time*
day•lily \'dā-ˌli-lē\ *noun* (1597)
: any of various Eurasian plants (genus *Hemerocallis*) of the lily family that have short-lived flowers resembling lilies and are widespread in cultivation and as escapes
day•long \'dā-ˌlȯŋ\ *adjective* (1855)
: lasting all day ⟨a *daylong* tour⟩
day•mare \'dā-ˌmar, -ˌmer\ *noun* [*day* + *-mare* (as in *nightmare*)] (1737)
: a nightmarish fantasy experienced while awake
day–neutral *adjective* (1941)
: developing and maturing regardless of relative length of alternating exposures to light and dark periods — used especially of a plant; compare LONG-DAY, SHORT-DAY
day nursery *noun* (1844)
: a public center for the care and training of young children; *specifically* **:** NURSERY SCHOOL
Day of Atonement (1611)
: YOM KIPPUR
day one *noun, often D&O capitalized* (1971)
: the first day or very beginning of something
day•room \'dā-ˌrüm, -ˌru̇m\ *noun* (1823)
: a room (as in a hospital) equipped for reading, writing, and recreation
days \'dāz\ *adverb* (before 12th century)
: in the daytime repeatedly **:** on any day
day school *noun* (1838)
: an elementary or secondary school held on weekdays; *specifically* **:** a private school without boarding facilities
day•side \'dā-ˌsīd\ *noun* (1963)
: the side of a planet in sunlight
days of grace (1726)
: the days allowed for payment of a note or an insurance premium after it becomes due
day•star \'dā-ˌstär\ *noun* (before 12th century)
1 : MORNING STAR
2 : SUN 1a
day student *noun* (1883)
: a student who attends regular classes at a college or preparatory school but does not live at the institution
day•time \'dā-ˌtīm\ *noun, often attributive* (1535)
: the time during which there is daylight
day•times \'dā-ˌtīmz\ *adverb* (1854)
: during the day or during the workday repeatedly **:** DAYS ⟨has a housekeeper *daytimes*⟩
day–to–day \ˌdā-tə-ˌdā\ *adjective* (1883)
1 : taking place, made, or done in the course of successive days ⟨*day-to-day* problems⟩
2 : providing for a day at a time with little thought for the future ⟨lived an aimless *day-to-day* existence⟩
day–trip•per \'dā-ˌtri-pər\ *noun* (1897)
: one who takes a trip that does not last overnight
daze \'dāz\ *transitive verb* **dazed; daz•ing** [Middle English *dasen,* from (assumed) Old Norse *dasa;* akin to Old Norse *dasask* to become exhausted] (14th century)
1 : to stupefy especially by a blow **:** STUN
2 : to dazzle with light
— daze *noun*
— dazed•ly \'dā-zəd-lē\ *adverb*
— dazed•ness \'dā-zəd-nəs, 'dāzd-\ *noun*
daz•zle \'da-zəl\ *verb* **daz•zled; daz•zling** \-z(ə-)liŋ\ [frequentative of *daze*] (15th century)
intransitive verb
1 : to lose clear vision especially from looking at bright light
2 a : to shine brilliantly **b :** to arouse admiration by an impressive display
transitive verb
1 : to overpower with light
2 : to impress deeply, overpower, or confound with brilliance ⟨*dazzled* us with her wit⟩
— dazzle *noun*
— daz•zler \-z(ə-)lər\ *noun*
— daz•zling•ly \-z(ə-)liŋ-lē\ *adverb*
DBCP \ˌdē-(ˌ)bē-(ˌ)sē-'pē\ *noun* [*d*i- + *b*rom- + *c*hlor- + *p*ropane] (1967)
: a halocarbon compound $C_3H_5Br_2Cl$ used as an agricultural pesticide that is a suspected carcinogen and cause of sterility in human males
D day *noun* [*D,* abbreviation for *day*] (1918)
: a day set for launching an operation; *specifically* **:** June 6, 1944, on which Allied forces began the invasion of France in World War II
DDC \ˌdē-(ˌ)dē-'sē\ *noun* [*d*i- + *d*eoxy + *c*ytidine] (1987)
: an antiviral drug $C_{10}H_{13}N_3O_3$ used to reduce virus growth and reproduction in AIDS and HIV infection
DDD \ˌdē-(ˌ)dē-'dē\ *noun* [*d*ichlor- + *d*iphenyl + *d*ichlor-] (1946)
: an insecticide $(ClC_6H_4)_2CHCHCl_2$ closely related chemically and similar in properties to DDT
DDE \ˌdē-(ˌ)dē-'ē\ *noun* [*d*ichlor- + *d*iphenyl + *e*thylene] (1949)
: a persistent organochlorine $C_{15}H_8Cl_4$ that is produced by the metabolic breakdown of DDT
DDI \ˌdē-(ˌ)dē-'ī\ *noun* [*d*i- + *d*eoxy + *i*nosine, a nucleoside] (1988)
: an antiviral drug $C_{10}H_{12}N_4O_3$ used to reduce virus growth and reproduction in AIDS and HIV infection
DDT \ˌdē-(ˌ)dē-'tē\ *noun* [*d*ichlor- + *d*iphenyl + *t*richlor- (from *tri-* +*chlor-*)] (1943)
: a colorless odorless water-insoluble crystalline insecticide $C_{14}H_9Cl_5$ that tends to accumulate in ecosystems and has toxic effects on many vertebrates
DDVP \ˌdē-(ˌ)dē-ˌvē-'pē\ *noun* [*d*imethyl + *d*ichlor- + *v*inyl + *p*hosphate] (1954)
: DICHLORVOS
de- *prefix* [Middle English, from Old French *de-, des-,* partly from Latin *de-* from, down, away (from *de,* preposition) and partly from Latin *dis-;* Latin *de* akin to Old Irish *di* from, Old English *tō* to — more at TO, DIS-]
1 a : do the opposite of ⟨*de*activate⟩ **b :** reverse of ⟨*de*-emphasis⟩
2 a : remove (a specified thing) from ⟨*de*louse⟩ **b :** remove from (a specified thing) ⟨*de*throne⟩
3 : reduce ⟨*de*value⟩
4 : something derived from (a specified thing) ⟨*de*compound⟩ **:** derived from something (of a specified nature) ⟨*de*nominative⟩
5 : get off of (a specified thing) ⟨*de*train⟩
6 : having a molecule characterized by the removal of one or more atoms (of a specified element) ⟨*de*oxy-⟩
de•acid•i•fy \ˌdē-ə-'si-də-ˌfī\ *transitive verb* (1786)
: to remove acid from **:** reduce the acidity of (as by neutralization)
— de•acid•i•fi•ca•tion \-ˌsi-də-fə-'kā-shən\ *noun*
dea•con \'dē-kən\ *noun* [Middle English *dekene,* from Old English *dēacon,* from Late Latin *diaconus,* from Greek *diakonos,* literally,

servant, from *dia-* + *-konos* (akin to en*konein* to be active); perhaps akin to Latin *conari* to attempt] (before 12th century)
: a subordinate officer in a Christian church: as **a :** a Roman Catholic, Anglican, or Eastern Orthodox cleric ranking next below a priest **b :** one of the laymen elected by a church with congregational polity to serve in worship, in pastoral care, and on administrative committees **c :** a Mormon in the lowest grade of the Aaronic priesthood

dea·con·ess \'dē-kə-nəs\ *noun* (15th century)
: a woman chosen to assist in the church ministry; *specifically* **:** one in a Protestant order

deacon's bench *noun* (1922)
: a bench with usually spindled arms and back

deacon's bench

de·ac·ti·vate \(,)dē-'ak-tə-,vāt\ *transitive verb* (1926)
: to make inactive or ineffective
— de·ac·ti·va·tion \(,)dē-,ak-tə-'vā-shən\ *noun*
— de·ac·ti·va·tor \(,)dē-'ak-tə-,vā-tər\ *noun*

¹dead \'ded\ *adjective* [Middle English *deed,* from Old English *dēad;* akin to Old Norse *dauthr* dead, *deyja* to die, Old High German *tōt* dead — more at DIE] (before 12th century)
1 : deprived of life **:** having died
2 a (1) **:** having the appearance of death **:** DEATHLY ⟨in a *dead* faint⟩ (2) **:** lacking power to move, feel, or respond **:** NUMB **b :** very tired **c** (1) **:** incapable of being stirred emotionally or intellectually **:** UNRESPONSIVE ⟨*dead* to pity⟩
2 : grown cold **:** EXTINGUISHED ⟨*dead* coals⟩
3 a : INANIMATE, INERT ⟨*dead* matter⟩ **b :** BARREN, INFERTILE ⟨*dead* soil⟩ **c :** no longer producing or functioning **:** EXHAUSTED ⟨a *dead* battery⟩
4 a (1) **:** lacking power or effect ⟨a *dead* law⟩ (2) **:** no longer having interest, relevance, or significance ⟨a *dead* issue⟩ **b :** no longer in use **:** OBSOLETE ⟨a *dead* language⟩ **c :** no longer active **:** EXTINCT ⟨a *dead* volcano⟩ **d :** lacking in gaiety or animation ⟨a *dead* party⟩ **e** (1) **:** lacking in commercial activity **:** QUIET (2) **:** commercially idle or unproductive ⟨*dead* capital⟩ **f :** lacking elasticity ⟨a *dead* tennis ball⟩ **g :** being out of action or out of use; *specifically* **:** free from any connection to a source of voltage and free from electric charges **h** (1) **:** being out of play ⟨a *dead* ball⟩ (2) **:** temporarily forbidden to play or to make a certain play in croquet
5 a : not running or circulating **:** STAGNANT ⟨*dead* water⟩ **b :** not turning ⟨a *dead* lathe center⟩ **c :** not imparting motion or power although otherwise functioning ⟨a *dead* rear axle⟩ **d :** lacking warmth, vigor, or taste
6 a : absolutely uniform ⟨a *dead* level⟩ **b** (1) **:** UNERRING (2) **:** EXACT ⟨*dead* center of the target⟩ (3) **:** certain to be doomed ⟨he's *dead* if he's late for curfew⟩ (4) **:** IRREVOCABLE ⟨a *dead* loss⟩ **c :** ABRUPT ⟨brought to a *dead* stop⟩ **d** (1) **:** COMPLETE, ABSOLUTE ⟨a *dead* silence⟩ (2) **:** ALL-OUT ⟨caught it on the *dead* run⟩
7 : devoid of former occupants ⟨*dead* villages⟩ ☆
— dead·ness *noun*
— dead in the water 1 : incapable of being effective **:** STALLED **2 :** ¹DEAD 6b(3)

²dead *noun, plural* **dead** (before 12th century)
1 : one that is dead — usually used collectively
2 : the state of being dead ⟨raised him from the *dead* —Colossians 2:12(Revised Standard Version)⟩
3 : the time of greatest quiet ⟨the *dead* of night⟩

³dead *adverb* (14th century)
1 : ABSOLUTELY, UTTERLY ⟨*dead* certain⟩
2 : suddenly and completely ⟨stopped *dead*⟩
3 : DIRECTLY ⟨*dead* ahead⟩

dead air *noun* (circa 1943)
: a period of silence especially during a broadcast

dead–air space \'ded-'ar-, -'er-\ *noun* (1902)
: an unventilated air space

¹dead·beat \'ded-,bēt\ *noun* (1863)
1 : LOAFER
2 : one who persistently fails to pay personal debts or expenses

²deadbeat *adjective* (circa 1864)
: having a pointer that gives a reading with little or no oscillation

dead bolt *noun* (circa 1902)
: a lock bolt that is moved by turning the knob or key without action of a spring

dead duck *noun* (1943)
: one that is doomed

dead·en \'de-dᵊn\ *verb* **dead·ened; dead·en·ing** \'ded-niŋ, 'de-dᵊn-iŋ\ (1665)
transitive verb
1 : to impair in vigor or sensation **:** BLUNT ⟨*deadened* his enthusiasm⟩
2 a : to deprive of brilliance **b :** to make vapid or spiritless **c :** to make (as a wall) impervious to sound
3 : to deprive of life **:** KILL
intransitive verb
: to become dead **:** lose life or vigor
— dead·en·er \'ded-nər, -dᵊn-ər\ *noun*
— dead·en·ing·ly \-niŋ-lē, -dᵊn-iŋ-\ *adverb*

¹dead–end \'ded-,end\ *adjective* (1919)
1 a : lacking opportunities especially for advancement ⟨a *dead-end* job⟩ **b :** lacking an exit ⟨a *dead-end* street⟩
2 : UNRULY ⟨*dead-end* kids⟩
— dead–end·ed·ness \,ded-'en-dəd-nəs\ *noun*

²dead–end \'ded-'end\ *intransitive verb* (1944)
: to come to a dead end **:** TERMINATE

dead end \'ded-'end\ *noun* (1886)
1 : an end (as of a street) without an exit
2 : a position, situation, or course of action that leads to nothing further

dead·en·ing *noun* (circa 1874)
: material used to soundproof walls or floors

dead·eye \'ded-,ī\ *noun* (1748)
1 : a rounded wood block encircled by a rope or an iron band and having holes to receive the lanyard that is used especially to set up shrouds and stays
2 : an unerring marksman

dead·fall \-,fȯl\ *noun* (1611)
1 : a trap so constructed that a weight (as a heavy log) falls on an animal and kills or disables it
2 : a tangled mass of fallen trees and branches

dead hand *noun* (14th century)
1 : MORTMAIN 1
2 : the oppressive influence of the past

¹dead·head \'ded-,hed\ *noun* (1841)
1 : one who has not paid for a ticket
2 : a dull or stupid person

²deadhead (1911)
intransitive verb
: to make a return trip without a load
transitive verb
: to remove the faded flowers of (a plant) especially to keep a neat appearance and to promote reblooming by preventing seed production

dead heat *noun* (1796)
: a tie with no single winner of a race

dead horse *noun* (1830)
: an exhausted or profitless topic or issue — usually used in the phrases *beat a dead horse* and *flog a dead horse*

dead letter *noun* (1663)
1 : something that has lost its force or authority without being formally abolished
2 : a letter that is undeliverable and unreturnable by the post office

dead lift *noun* (1963)
: a lift in weight lifting in which the weight is lifted from the floor to hip level
— dead·lift \'ded-,lift\ *transitive verb*

dead·light \'ded-,līt\ *noun* (1726)
: a metal cover or shutter fitted to a port to keep out light and water

dead·line \-,līn\ *noun* (1864)
1 : a line drawn within or around a prison that a prisoner passes at the risk of being shot
2 a : a date or time before which something must be done **b :** the time after which copy is not accepted for a particular issue of a publication

dead load *noun* (circa 1888)
: a constant load in a structure (as a bridge, building, or machine) that is due to the weight of the members, the supported structure, and permanent attachments or accessories

dead·lock \'ded-,läk\ *noun* (1779)
1 : a state of inaction or neutralization resulting from the opposition of equally powerful uncompromising persons or factions **:** STANDSTILL
2 : a tie score
— deadlock *verb*

¹dead·ly \'ded-lē\ *adjective* **dead·li·er; -est** (before 12th century)
1 : likely to cause or capable of producing death
2 a : aiming to kill or destroy **:** IMPLACABLE ⟨a *deadly* enemy⟩ **b :** highly effective ⟨a *deadly* exposé⟩ **c :** UNERRING ⟨a *deadly* marksman⟩ **d :** marked by determination or extreme seriousness
3 a : tending to deprive of force or vitality ⟨a *deadly* habit⟩ **b :** suggestive of death especially in dullness or lack of animation ⟨*deadly* bores⟩ ⟨a *deadly* conversation⟩
4 : very great **:** EXTREME ☆
— dead·li·ness *noun*

²deadly *adverb* (before 12th century)
1 *archaic* **:** in a manner to cause death **:** MORTALLY
2 : suggesting death
3 : EXTREMELY ⟨*deadly* serious⟩

deadly nightshade *noun* (1578)
: BELLADONNA 1

deadly sin *noun* (13th century)
: one of seven sins of pride, covetousness, lust, anger, gluttony, envy, and sloth held to be fatal to spiritual progress

dead man's float *noun* (circa 1946)

☆ SYNONYMS

Dead, defunct, deceased, departed, late mean devoid of life. DEAD applies literally to what is deprived of vital force but is used figuratively of anything that has lost any attribute (as energy, activity, radiance) suggesting life ⟨a *dead,* listless performance⟩. DEFUNCT stresses cessation of active existence or operation ⟨a *defunct* television series⟩. DECEASED, DEPARTED, and LATE apply to persons who have died recently. DECEASED is the preferred term in legal use ⟨the estate of the *deceased*⟩. DEPARTED is used usually as a euphemism ⟨our *departed* sister⟩. LATE is used especially with reference to a person in a specific relation or status ⟨the company's *late* president⟩.

Deadly, mortal, fatal, lethal mean causing or capable of causing death. DEADLY applies to an established or very likely cause of death ⟨a *deadly* disease⟩. MORTAL implies that death has occurred or is inevitable ⟨a *mortal* wound⟩. FATAL stresses the inevitability of what has in fact resulted in death or destruction ⟨*fatal* consequences⟩. LETHAL applies to something that is bound to cause death or exists for the destruction of life ⟨*lethal* gas⟩.

: a prone floating position with the arms extended forward

dead march *noun* (1603)
: a solemn march for a funeral

dead metaphor *noun* (1922)
: a word or phrase (as *time is running out*) that has lost its metaphoric force through common usage

dead–on \'ded-'ȯn, -'än\ *adjective* (circa 1889)
: exactly correct or accurate

[1]**dead·pan** \'ded-ˌpan\ *adjective* (circa 1928)
: marked by an impassive matter-of-fact manner, style, or expression ⟨a *deadpan* comedy⟩
— **deadpan** *adverb*

[2]**deadpan** *noun* (circa 1930)
1 : a completely expressionless face
2 : a deadpan manner of behavior or presentation

[3]**deadpan** *transitive verb* (circa 1942)
: to express in a deadpan manner
— **dead·pan·ner** *noun*

dead reckoning *noun* (1613)
1 : the determination without the aid of celestial observations of the position of a ship or aircraft from the record of the courses sailed or flown, the distance made, and the known or estimated drift
2 : GUESSWORK
— **dead reckon** *verb*

dead space *noun* (circa 1923)
: the portion of the respiratory system which is external to the bronchioles and through which air must pass to reach the bronchioles and alveoli

dead–stick landing \'ded-ˌstik-\ *noun* (circa 1917)
: a landing of an airplane or spacecraft made without power

dead·weight \'ded-'wāt\ *noun* (1660)
1 : the unrelieved weight of an inert mass
2 : DEAD LOAD
3 : a ship's load including the total weight of cargo, fuel, stores, crew, and passengers

deadweight ton *noun* (circa 1917)
: a long ton used in indicating a ship's gross capacity — abbreviation *dwt*

dead·wood \-ˌwu̇d\ *noun* (15th century)
1 : wood dead on the tree
2 : useless personnel or material
3 : solid timbers built in at the extreme bow and stern of a ship when too narrow to permit framing
4 : bowling pins that have been knocked down but remain on the alley

de·aer·ate \ˌdē-'ar-ˌāt, -'er-\ *transitive verb* (1791)
: to remove air or gas from
— **de·aer·a·tion** \ˌdē-ˌar-'ā-shən, -ˌer-\ *noun*
— **de·aer·a·tor** \-'ar-ˌā-tər, -'er-\ *noun*

deaf \'def, *dialect* 'dēf\ *adjective* [Middle English *deef,* from Old English *dēaf;* akin to Greek *typhlos* blind, *typhein* to smoke, Latin *fumus* smoke — more at FUME] (before 12th century)
1 : lacking or deficient in the sense of hearing
2 : unwilling to hear or listen **:** not to be persuaded ⟨was overwrought and *deaf* to reason⟩
— **deaf·ish** \'de-fish\ *adjective*
— **deaf·ly** *adverb*
— **deaf·ness** *noun*

deaf·en \'de-fən\ *transitive verb* **deaf·ened; deaf·en·ing** \-fə-niŋ, 'def-niŋ\ (1597)
: to make deaf

deaf·en·ing *adjective* (1597)
1 : that deafens
2 : very loud **:** EARSPLITTING ⟨fell with a *deafening* clap⟩
— **deaf·en·ing·ly** *adverb*

deaf–mute \'def-'myüt\ *noun* (circa 1837)
: a deaf person who cannot speak
— **deaf–mute** *adjective*

[1]**deal** \'dē(ə)l\ *noun* [Middle English *deel,* from Old English *dǣl;* akin to Old English *dāl* division, portion, Old High German *teil* part] (before 12th century)
1 *obsolete* **:** PART, PORTION
2 : a usually large or indefinite quantity or degree ⟨a great *deal* of support⟩ ⟨a good *deal* faster⟩
3 a : the act or right of distributing cards to players in a card game **b :** HAND 9b

[2]**deal** *verb* **dealt** \'delt\; **deal·ing** \'dē-liŋ\ (before 12th century)
transitive verb
1 a : to give as one's portion **:** APPORTION ⟨tried to *deal* justice to all⟩ ⟨*dealt* out three sandwiches apiece⟩ **b :** to distribute (playing cards) to players in a game
2 : ADMINISTER, DELIVER ⟨*dealt* him a blow⟩
3 : SELL ⟨*deals* marijuana⟩
intransitive verb
1 : to distribute the cards in a card game
2 : to concern oneself or itself ⟨the book *deals* with education⟩
3 a : to engage in bargaining **:** TRADE **b :** to sell or distribute something as a business ⟨*deal* in insurance⟩
4 a : to take action with regard to someone or something ⟨*deal* with an offender⟩ **b :** to reach or try to reach a state of acceptance or reconcilement ⟨trying to *deal* with her son's death⟩
synonym see DISTRIBUTE
— **deal·er** \'dē-lər\ *noun*

[3]**deal** *noun* (15th century)
1 a : an act of dealing **:** TRANSACTION **b :** BARGAIN
2 : PACKAGE DEAL
3 : treatment received ⟨a dirty *deal*⟩
4 : an arrangement for mutual advantage
5 : AFFAIR 2 ⟨dinner was an informal *deal*⟩

[4]**deal** *noun* [Middle English *dele,* from Middle Dutch or Middle Low German, plank; akin to Old High German *dili* plank — more at THILL] (14th century)
1 a *British* **:** a board of fir or pine **b :** sawed yellow-pine lumber nine inches (22.5 centimeters) or wider and three, four, or five inches (7.6 to 12.4 centimeters) thick
2 : pine or fir wood
— **deal** *adjective*

de·alat·ed \(ˌ)dē-'ā-ˌlā-təd\ *adjective* (1904)
: divested of the wings — used of postnuptial adults of insects (as ants) that drop their wings after a nuptial flight
— **de·ala·tion** \ˌdē-(ˌ)ā-'lā-shən\ *noun*

deal·er·ship \'dē-lər-ˌship\ *noun* (1916)
: an authorized sales agency ⟨an automobile *dealership*⟩

deal·fish \'dē(ə)l-ˌfish\ *noun* [[4]*deal*] (1845)
: any of several long thin fishes (genus *Trachipterus* especially *T. arcticus* of the family Trachipteridae) inhabiting the deep sea

deal·ing *noun* (15th century)
1 : method of business **:** manner of conduct
2 *plural* **:** friendly or business interactions

dealing box *noun* (1897)
: a case that holds a deck of playing cards so that they may be dealt one by one

de·am·i·nase \(ˌ)dē-'a-mi-ˌnās, -ˌnāz\ *noun* [*de-* + *amino* + *-ase*] (1920)
: an enzyme that hydrolyzes amino compounds (as amino acids) with removal of the amino group

de·am·i·nate \-ˌnāt\ *transitive verb* **-nat·ed; -nat·ing** (1926)
: to remove the amino group from (a compound)
— **de·am·i·na·tion** \(ˌ)dē-ˌa-mi-'nā-shən\ *noun*

dean \'dēn\ *noun* [Middle English *deen,* from Old French *deien,* from Late Latin *decanus* chief of ten, from Greek *dekanos,* from *deka* ten — more at TEN] (13th century)
1 a : the head of the chapter of a collegiate or cathedral church **b :** a Roman Catholic priest who supervises one district of a diocese
2 a : the head of a division, faculty, college, or school of a university **b :** a college or secondary school administrator in charge of counseling and disciplining students
3 : DOYEN 1
— **dean** *intransitive verb*
— **dean·ship** \-ˌship\ *noun*

dean·ery \'dē-n-rē, 'dē-nə-rē\ *noun, plural* **-er·ies** (15th century)
: the office, jurisdiction, or official residence of a clerical dean

dean's list *noun* (circa 1926)
: a list of students receiving special recognition from the dean of a college because of superior scholarship

[1]**dear** \'dir\ *adjective* [Middle English *dere,* from Old English *dēor*] (before 12th century)
: SEVERE, SORE ⟨in our *dear* peril —Shakespeare⟩

[2]**dear** *adjective* [Middle English *dere,* from Old English *dēore;* akin to Old High German *tiuri* distinguished, costly] (before 12th century)
1 *obsolete* **:** NOBLE
2 : highly valued **:** PRECIOUS ⟨a *dear* friend⟩ — often used in a salutation ⟨*dear* Sir⟩
3 : AFFECTIONATE, FOND
4 : high or exorbitant in price **:** EXPENSIVE ⟨eggs are very *dear* just now⟩
5 : HEARTFELT
— **dear** *adverb*
— **dear·ly** *adverb*
— **dear·ness** *noun*

[3]**dear** *noun* (13th century)
1 : a loved one **:** SWEETHEART
2 : a lovable person

[4]**dear** *interjection* (1694)
— used especially to express annoyance or dismay

Dear John \-'jän\ *noun* (1945)
: a letter (as to a soldier) in which a wife asks for a divorce or a girlfriend breaks off an engagement or a friendship

dearth \'dərth\ *noun* [Middle English *derthe,* from (assumed) Old English *dierth,* from *dēore* dear] (13th century)
1 : scarcity that makes dear; *specifically* **:** FAMINE
2 : an inadequate supply **:** LACK

dea·sil \'dē-zəl\ *adverb* [Scottish Gaelic *deiseil;* akin to Latin *dexter* right hand] (1771)
: CLOCKWISE — compare WIDDERSHINS

death \'deth\ *noun* [Middle English *deeth,* from Old English *dēath;* akin to Old Norse *dauthi* death, *deyja* to die — more at DIE] (before 12th century)
1 : a permanent cessation of all vital functions **:** the end of life — compare BRAIN DEATH
2 : the cause or occasion of loss of life ⟨drinking was the *death* of him⟩
3 *capitalized* **:** the destroyer of life represented usually as a skeleton with a scythe
4 : the state of being dead
5 a : the passing or destruction of something inanimate ⟨the *death* of vaudeville⟩ **b :** EXTINCTION
6 : CIVIL DEATH
7 : SLAUGHTER
8 *Christian Science* **:** the lie of life in matter **:** that which is unreal and untrue **:** ILLUSION
— **at death's door :** close to death **:** critically ill
— **to death :** beyond endurance **:** EXCESSIVELY ⟨bored *to death*⟩

death·bed \'deth-ˌbed\ *noun* (before 12th century)
1 : the bed in which a person dies
2 : the last hours of life
— **on one's deathbed :** near the point of death

death benefit *noun* (1921)
: money payable to the beneficiary of a deceased

death·blow \'deth-ˌblō\ *noun* (1795)
: a destructive or killing stroke or event

death camas *noun* (circa 1889)
: any of several plants (genus *Zigadenus*) of the lily family that cause poisoning of livestock in the western U.S.

death camp *noun* (1944)
: a concentration camp in which large numbers of prisoners are systematically killed

death cap *noun* (1925)
: a very poisonous mushroom (*Amanita phalloides*) of deciduous woods of North America and Europe that varies in color from pure white to olive or yellow and has a prominent volva at the base — called also *death cup*

death duty *noun* (1881)
chiefly British **:** DEATH TAX

death instinct *noun* (1922)
: an innate and unconscious tendency toward self-destruction postulated in psychoanalytic theory to explain aggressive and destructive behavior not satisfactorily explained by the pleasure principle — called also *Thanatos*; compare EROS 2

death·less \'deth-ləs\ *adjective* (1589)
: IMMORTAL, IMPERISHABLE ⟨*deathless* fame⟩
— **death·less·ly** *adverb*
— **death·less·ness** *noun*

death·ly \'deth-lē\ *adjective* (before 12th century)
1 : FATAL
2 : of, relating to, or suggestive of death ⟨a *deathly* pallor⟩
— **deathly** *adverb*

death mask *noun* (1877)
: a cast taken from the face of a dead person

death rattle *noun* (1822)
: a rattling or gurgling sound produced by air passing through mucus in the lungs and air passages of a dying person

death ray *noun* (1919)
: a weapon that generates an intense beam of particles or radiation by which it destroys its target

death row *noun* (1950)
: a prison area housing inmates sentenced to death

death's–head \'deths-,hed\ *noun* (1596)
: a human skull or a depiction of a human skull symbolizing death

death's–head hawkmoth *noun* (1879)
: a large dark hawkmoth (*Acherontia atropos*) with markings resembling a human skull on the back of the thorax — called also *death's–head moth*

deaths·man \'deths-mən\ *noun* (1589)
archaic **:** EXECUTIONER

death squad *noun* (1969)
: any of various right-wing vigilante groups in Latin America whose members kill suspected political adversaries and criminals

death tax *noun* (1937)
: a tax arising on the transmission of property after the owner's death; *especially* **:** ESTATE TAX

death trap *noun* (1835)
: a structure or situation that is potentially very dangerous to life

death warrant *noun* (1692)
1 : a warrant for the execution of a death sentence
2 : DEATHBLOW

¹death·watch \'deth-,wäch\ *noun* [*death* + *watch* (timepiece); from the superstition that its ticking presages death] (1646)
: a small insect that makes a ticking sound; *especially* **:** DEATHWATCH BEETLE

²deathwatch *noun* [*death* + *watch* (vigil)] (circa 1890)
1 : a vigil kept over the dead or dying
2 : the guard set over a criminal to be executed

deathwatch beetle *noun* (1877)
: any of various small beetles (family Anobiidae) that are common in old houses where they bore in woodwork and furniture and make a tapping noise as a mating call

death wish *noun* (1913)
: the conscious or unconscious desire for the death of oneself or of another

deb \'deb\ *noun* (1920)
: DEBUTANTE

de·ba·cle \dē-'bä-kəl, -'ba-; ÷'de-bə-kəl\ *also* **dé·bâ·cle** *also* dā-'bäk(lᵊ)\ *noun* [French *débâcle*, from *débâcler* to clear, from Middle French *desbacler*, from *des-* de- + *bacler* to block, perhaps from (assumed) Vulgar Latin *bacculare*, from Latin *baculum* staff] (1802)
1 : a tumultuous breakup of ice in a river
2 : a violent disruption (as of an army) **:** ROUT
3 a : a great disaster **b :** a complete failure **:** FIASCO

de·bar \di-'bär, dē-\ *transitive verb* [Middle English *debarren*, from Middle French *desbarrer* to unbar, from *des-* de- + *barrer* to bar] (15th century)
: to bar from having or doing something **:** PRECLUDE
— **de·bar·ment** \-mənt\ *noun*

¹de·bark \di-'bärk, dē-\ *verb* [French *debarquer*, from *de-* + *barque* bark (ship)] (1654)
: DISEMBARK
— **de·bar·ka·tion** \,dē-,bär-'kā-shən\ *noun*

²de·bark \(,)dē-'bärk\ *transitive verb* (1742)
: to remove bark from

de·base \di-'bās, dē-\ *transitive verb* (1565)
1 : to lower in status, esteem, quality, or character
2 a : to reduce the intrinsic value of (a coin) by increasing the base-metal content **b :** to reduce the exchange value of (a monetary unit) ☆
— **de·base·ment** \-'bās-mənt\ *noun*
— **de·bas·er** \-'bā-sər\ *noun*

de·bat·able \di-'bā-tə-bəl, dē-\ *adjective* (1536)
1 : claimed by more than one country ⟨*debatable* border territory⟩
2 a : open to dispute **:** QUESTIONABLE **b :** open to debate
3 : capable of being debated

¹de·bate \di-'bāt, dē-\ *noun* (13th century)
: a contention by words or arguments: as **a :** the formal discussion of a motion before a deliberative body according to the rules of parliamentary procedure **b :** a regulated discussion of a proposition between two matched sides

²debate *verb* **de·bat·ed; de·bat·ing** [Middle English, from Middle French *debatre*, from Old French, from *de-* + *batre* to beat, from Latin *battuere*] (14th century)
intransitive verb
1 *obsolete* **:** FIGHT, CONTEND
2 a : to contend in words **b :** to discuss a question by considering opposed arguments
3 : to participate in a debate
transitive verb
1 a : to argue about **b :** to engage (an opponent) in debate
2 : to turn over in one's mind
synonym see DISCUSS
— **de·bate·ment** \-'bāt-mənt\ *noun*
— **de·bat·er** *noun*

¹de·bauch \di-'bȯch, -'bäch, dē-\ *transitive verb* [Middle French *debaucher*, from Old French *desbauchier* to scatter, disperse, from *des-* de- + *bauch* beam, of Germanic origin; akin to Old High German *balko* beam — more at BALK] (1595)
1 a *archaic* **:** to make disloyal **b :** to seduce from chastity
2 a : to lead away from virtue or excellence **b :** to corrupt by intemperance or sensuality
synonym see DEBASE
— **de·bauch·er** *noun*

²debauch *noun* (1603)
1 : an act or occasion of debauchery
2 : ORGY

de·bauch·ee \di-,bȯ-'chē, -,bä-; ,de-bə-'shē, -'shā\ *noun* [French *débauché*, from past participle of *débaucher*] (1661)
: one given to debauchery

de·bauch·ery \di-'bȯ-chə-rē, -chrē, -'bä-\ *noun, plural* **-er·ies** (1642)
1 a : extreme indulgence in sensuality **b** *plural* **:** ORGIES
2 *archaic* **:** seduction from virtue or duty

de·beak \(,)dē-'bēk\ *transitive verb* (1937)
: to remove the tip of the upper mandible of (as a chicken) to prevent cannibalism and fighting

de·ben·ture \di-'ben-chər\ *noun* [Middle English *debentur*, from Latin, they are due, 3d plural present passive of *debēre* to owe — more at DEBT] (15th century)
1 *British* **:** a corporate security other than an equity security **:** BOND
2 : a bond backed by the general credit of the issuer rather than a specific lien on particular assets

de·bil·i·tate \di-'bi-lə-,tāt, dē-\ *transitive verb* **-tat·ed; -tat·ing** [Latin *debilitatus*, past participle of *debilitare* to weaken, from *debilis* weak] (1533)
: to impair the strength of **:** ENFEEBLE
synonym see WEAKEN
— **de·bil·i·ta·tion** \-,bi-lə-'tā-shən\ *noun*

de·bil·i·ty \di-'bi-lə-tē, dē-\ *noun, plural* **-ties** [Middle French *debilité*, from Latin *debilitat-, debilitas*, from *debilis*, from *de-* de- + *-bilis*; akin to Sanskrit *bala* strength] (15th century)
: WEAKNESS, INFIRMITY

¹deb·it \'de-bət\ *transitive verb* (1682)
: to enter upon the debit side of an account **:** charge with a debit

²debit *noun* [Latin *debitum* debt] (1776)
1 a : a record of an indebtedness; *specifically* **:** an entry on the left-hand side of an account constituting an addition to an expense or asset account or a deduction from a revenue, net worth, or liability account **b :** the sum of the items entered as debits
2 : a charge against a bank deposit account
3 : DRAWBACK, SHORTCOMING

debit card *noun* (1977)
: a card like a credit card by which money may be withdrawn or the cost of purchases paid directly from the holder's bank account without the payment of interest

deb·o·nair \,de-bə-'nar, -'ner\ *adjective* [Middle English *debonere*, from Old French *debonaire*, from *de bon aire* of good family or nature] (13th century)
1 *archaic* **:** GENTLE, COURTEOUS
2 a : SUAVE, URBANE **b :** LIGHTHEARTED, NONCHALANT
— **deb·o·nair·ly** *adverb*
— **deb·o·nair·ness** *noun*

de·bone \(,)dē-'bōn\ *transitive verb* (1944)
: BONE ⟨*debone* a roast⟩
— **de·bon·er** *noun*

Deb·o·rah \'de-b(ə-)rə\ *noun* [Hebrew *Dĕbhōrāh*]
: a Hebrew prophetess who rallied the Israelites in their struggles against the Canaanites

☆ SYNONYMS
Debase, vitiate, deprave, corrupt, debauch, pervert mean to cause deterioration or lowering in quality or character. DEBASE implies a loss of position, worth, value, or dignity ⟨commercialism has *debased* the holiday⟩. VITIATE implies a destruction of purity, validity, or effectiveness by allowing entrance of a fault or defect ⟨a foreign policy *vitiated* by partisanship⟩. DEPRAVE implies moral deterioration by evil thoughts or influences ⟨the claim that society is *depraved* by pornography⟩. CORRUPT implies loss of soundness, purity, or integrity ⟨the belief that bureaucratese *corrupts* the language⟩. DEBAUCH implies a debasing through sensual indulgence ⟨the long stay on a tropical isle had *debauched* the ship's crew⟩. PERVERT implies a twisting or distorting from what is natural or normal ⟨*perverted* the original goals of the institute⟩.

de·bouch \di-ˈbau̇ch, -ˈbüsh, dē-\ *verb* [French *déboucher,* from *dé-* de- + *bouche* mouth, from Latin *bucca* cheek] (1745)
transitive verb
: to cause to emerge **:** DISCHARGE
intransitive verb
1 : to march out into open ground
2 : EMERGE, ISSUE
— de·bouch·ment \-mənt\ *noun*

de·bride·ment \di-ˈbrēd-mənt, dā-, -ˌmänt, dā-brēd-ˈmä^{n}\ *noun* [French *débridement,* from *débrider* to remove adhesions, literally, to unbridle, from Middle French *desbrider,* from *des-* de- + *bride* bridle, from Middle High German *brīdel* — more at BRIDLE] (circa 1842)
: the surgical removal of lacerated, devitalized, or contaminated tissue
— de·bride \di-ˈbrēd, dā-\ *transitive verb*

de·brief \(ˌ)dē-ˈbrēf\ *transitive verb* (1945)
: to interrogate (as a pilot) usually upon return (as from a mission) in order to obtain useful information

de·bris \də-ˈbrē, dā-ˈ, ˈdā-ˌ, *British usually* ˈde-(ˌ)brē\ *noun, plural* **de·bris** \-ˈbrēz, -ˌbrēz\ [French *débris,* from Middle French, from *debriser* to break to pieces, from Old French *debrisier,* from *de-* + *brisier* to break — more at BRISANCE] (1708)
1 : the remains of something broken down or destroyed
2 : an accumulation of fragments of rock
3 : something discarded **:** RUBBISH

debt \ˈdet\ *noun* [Middle English *dette, debte,* from Old French *dette* something owed, from (assumed) Vulgar Latin *debita,* from Latin, plural of *debitum* debt, from neuter of *debitus,* past participle of *debēre* to owe, from *de-* + *habēre* to have — more at GIVE] (13th century)
1 : SIN, TRESPASS
2 : something owed **:** OBLIGATION
3 : a state of owing
4 : the common-law action for the recovery of money held to be due
— debt·less \-ləs\ *adjective*

debt·or \ˈde tər\ *noun* (13th century)
1 : one guilty of neglect or violation of duty
2 : one who owes a debt

debt service *noun* (1929)
: the amount of interest and sinking fund payments due annually on long-term debt

de·bug \(ˌ)dē-ˈbəg\ *transitive verb* (1944)
1 : to remove insects from
2 : to eliminate errors in or malfunctions of
3 : to remove a concealed microphone or wiretapping device from
— de·bug·ger *noun*

de·bunk \(ˌ)dē-ˈbəŋk\ *transitive verb* (1923)
: to expose the sham or falseness of ⟨*debunk* a legend⟩
— de·bunk·er *noun*

[1]**de·but** *also* **dé·but** \ˈdā-ˌbyü, dā-ˈ\ *noun* [French *début,* from *débuter* to begin, from Middle French *desbuter* to play first, from *des-* de- + *but* starting point, goal — more at BUTT] (1751)
1 : a first appearance ⟨made her singing *debut*⟩
2 : a formal entrance into society

[2]**debut** (1830)
intransitive verb
: to make a debut
transitive verb
: to present to the public for the first time **:** INTRODUCE

deb·u·tant \ˈde-byu̇-ˌtänt\ *noun* [French *débutant,* from present participle of *débuter*] (1821)
: one making a debut

deb·u·tante \ˈde-byu̇-ˌtänt\ *noun* [French *débutante,* feminine of *débutant*] (1801)
: DEBUTANT; *especially* **:** a young woman making her formal entrance into society

deca- *or* **dec-** *or* **deka-** *or* **dek-** *combining form* [Middle English, from Latin, from Greek *deka-, dek-,* from *deka* — more at TEN]
: ten ⟨*deca*syllabic⟩ ⟨*deka*meter⟩

de·cade \ˈde-ˌkād, de-ˈkād; *especially sense 1b* ˈde-kəd\ *noun* [Middle English, from Middle French *décade,* from Late Latin *decad-, decas,* from Greek *dekad-, dekas,* from *deka*] (15th century)
1 : a group or set of 10: as **a :** a period of 10 years **b :** a division of the rosary that consists primarily of 10 Hail Marys
2 : a ratio of 10 to 1 **:** ORDER OF MAGNITUDE
— de·cad·al \ˈde-kə-d^{ə}l\ *adjective*

dec·a·dence \ˈde-kə-dən(t)s *also* di-ˈkā-\ *noun* [Middle French, from Medieval Latin *decadentia,* from Late Latin *decadent-, decadens,* present participle of *decadere* to fall, sink — more at DECAY] (1549)
1 : the process of becoming decadent **:** the quality or state of being decadent
2 : a period of decline
synonym see DETERIORATION

dec·a·den·cy \-d^{ə}n-sē\ *noun* (1632)
: DECADENCE 1

[1]**dec·a·dent** \ˈde-kə-dənt *also* di-ˈkā-\ *adjective* [back-formation from *decadence*] (1837)
1 : marked by decay or decline
2 : of, relating to, or having the characteristics of the decadents
3 : characterized by or appealing to self-indulgence
— dec·a·dent·ly *adverb*

[2]**decadent** *noun* (1886)
1 : one of a group of late 19th century French and English writers tending toward artificial and unconventional subjects and subtilized style
2 : one that is decadent

de·caf \ˈdē-ˌkaf\ *noun* [short for *decaffeinated*] (1984)
: decaffeinated coffee

de·caf·fein·at·ed \(ˌ)dē-ˈka fə-nā-təd, -fē-ə-\ *adjective* (1921)
: having the caffeine removed ⟨*decaffeinated* coffee⟩ ⟨*decaffeinated* tea⟩

deca·gon \ˈde-kə-ˌgän\ *noun* [New Latin *decagonum,* from Greek *dekagōnon,* from *deka-* deca- + *-gōnon* -gon] (1571)
: a plane polygon of 10 angles and 10 sides

decagon

deca·gram \-ˌgram\ *noun* [French *décagramme,* from *déca-* deca- + *gramme* gram] (1810)
: DEKAGRAM

deca·he·dron \ˌde-kə-ˈhē-drən\ *noun* [International Scientific Vocabulary] (circa 1828)
: a polyhedron of 10 faces

de·cal \ˈdē-ˌkal, di-ˈkal; *Canadian usually* ˈde-kəl\ *noun* [short for *decalcomania*] (1937)
: a picture, design, or label made to be transferred (as to glass) from specially prepared paper

de·cal·ci·fi·ca·tion \(ˌ)dē-ˌkal-sə-fə-ˈkā-shən\ *noun* (1859)
: the removal or loss of calcium or calcium compounds (as from bones or soil)
— de·cal·ci·fy \(ˌ)dē-ˈkal-sə-ˌfī\ *transitive verb*

de·cal·co·ma·nia \di-ˌkal-kə-ˈmā-nē-ə\ *noun* [French *décalcomanie,* from *décalquer* to copy by tracing (from *dé-* de- + *calquer* to trace, from Italian *calcare,* literally, to tread, from Latin) + *manie* mania, from Late Latin *mania* — more at CAULK] (1864)
1 : the art or process of transferring pictures and designs from specially prepared paper (as to glass)
2 : DECAL

deca·li·ter \ˈde-kə-ˌlē-tər\ *noun* [French *décalitre,* from *déca-* + *litre* liter] (1810)
: DEKALITER

deca·logue \ˈde-kə-ˌlȯg, -ˌläg\ *noun* [Middle English *decaloge,* from Late Latin *decalogus,* from Greek *dekalogos,* from *deka-* + *logos* word — more at LEGEND] (14th century)
1 *capitalized* **:** TEN COMMANDMENTS
2 : a basic set of rules carrying binding authority

[1]**deca·me·ter** \ˈde-kə-ˌmē-tər\ *noun* [French *décamètre,* from *déca-* + *mètre* meter] (1810)
: DEKAMETER

[2]**de·cam·e·ter** \de-ˈka-mə-tər, də-\ *noun* [Greek *dekametron,* from *deka-* + *metron* measure, meter] (1821)
: a line of verse consisting of 10 metrical feet

deca·me·tho·ni·um \ˌde-kə-mə-ˈthō-nē-əm\ *noun* [*deca-* + *meth*ylene + *-onium*] (circa 1949)
: a synthetic ion used in the form of either its bromide or iodide salts ($C_{16}H_{38}Br_2N_2$ or $C_{16}H_{38}I_2N_2$) as a skeletal muscle relaxant; *also* **:** either of these salts

deca·met·ric \ˌde-kə-ˈme-trik\ *adjective* [*decameter* + *-ic;* from the wavelength range being between 1 and 10 dekameters] (1950)
: of, relating to, or being a radio wave of high frequency

de·camp \di-ˈkamp, dē-\ *intransitive verb* [French *décamper,* from Middle French *descamper,* from *des-* de- + *camper* to camp] (1676)
1 : to break up a camp
2 : to depart suddenly **:** ABSCOND
— de·camp·ment \-mənt\ *noun*

dec·ane \ˈde-ˌkān\ *noun* [International Scientific Vocabulary *deca-*] (circa 1875)
: any of several isomeric liquid alkanes $C_{10}H_{22}$

de·cant \di-ˈkant, dē-\ *transitive verb* [New Latin *decantare,* from Latin *de-* + Medieval Latin *cantus* edge, from Latin, iron ring round a wheel — more at CANT] (1633)
1 : to draw off (a liquid) without disturbing the sediment or the lower liquid layers
2 : to pour from one vessel into another
3 : to pour out, transfer, or unload as if by pouring
— de·can·ta·tion \ˌdē-ˌkan-ˈtā-shən\ *noun*

de·cant·er \di-ˈkan-tər, dē-\ *noun* (1712)
: a vessel used to decant or to receive decanted liquids; *especially* **:** an ornamental glass bottle used for serving wine

decanter

de·cap·i·tate \di-ˈka-pə-ˌtāt, dē-\ *transitive verb* **-tat·ed; -tat·ing** [Late Latin *decapitatus,* past participle of *decapitare,* from Latin *de-* + *capit-, caput* head — more at HEAD] (circa 1611)
1 : to cut off the head of **:** BEHEAD
2 : to make ineffective **:** DESTROY
— de·cap·i·ta·tion \-ˌka-pə-ˈtā-shən\ *noun*
— de·cap·i·ta·tor \-ˈka-pə-ˌtā-tər\ *noun*

deca·pod \ˈde-kə-ˌpäd\ *noun* [New Latin *Decapoda,* order name] (1836)
1 : any of an order (Decapoda) of crustaceans (as shrimps, lobsters, and crabs) with five pairs of thoracic appendages one or more of which are modified into pincers, with stalked eyes, and with the head and thorax fused into a cephalothorax and covered by a carapace
2 : any of the cephalopod mollusks (orders Sepioidea and Teuthoidea) with 10 arms including cuttlefishes, squids, and related forms
— decapod *adjective*
— de·cap·o·dan \di-ˈka-pə-dən\ *adjective or noun*
— de·cap·o·dous \-ə-dəs\ *adjective*

de·car·bon·ate \(ˌ)dē-ˈkär-bə-ˌnāt\ *transitive verb* (1831)
: to remove carbon dioxide or carbonic acid from
— de·car·bon·ation \-ˌkär-bə-ˈnā-shən\ *noun*

\ə\ **abut** \$^{ə}$\ **kitten** \ər\ **further** \a\ **ash** \ā\ **ace** \ä\ **mop, mar** \au̇\ **out** \ch\ **chin** \e\ **bet** \ē\ **easy** \g\ **go** \i\ **hit** \ī\ **ice** \j\ **job** \ŋ\ **sing** \ō\ **go** \ȯ\ **law** \ȯi\ **boy** \th\ **thin** \t͟h\ **the** \ü\ **loot** \u̇\ **foot** \y\ **yet** \zh\ **vision** *see also* Guide to Pronunciation

de·car·bon·ize \(,)dē-ˈkär-bə-ˌnīz\ *transitive verb* [International Scientific Vocabulary] (1825)
: to remove carbon from
— **de·car·bon·iz·er** *noun*

de·car·box·yl·ase \ˌdē-kär-ˈbäk-sə-ˌlās, -ˌlāz\ *noun* (1940)
: any of a group of enzymes that accelerate decarboxylation especially of amino acids

de·car·box·yl·ation \-ˌbäk-sə-ˈlā-shən\ *noun* (1922)
: the removal or elimination of carboxyl from a molecule
— **de·car·box·yl·ate** \-ˈbäk-sə-ˌlāt\ *transitive verb*

de·car·bu·rize \(,)dē-ˈkär-b(y)ə-ˌrīz\ *transitive verb* (1856)
: DECARBONIZE
— **de·car·bu·ri·za·tion** \-ˌkär-b(y)ə-rə-ˈzā-shən\ *noun*

dec·are \ˈde-ˌkar, -ˌker, -ˌkär\ *noun* [French *décare,* from *déca-* deca- + *are*] (1810)
: a metric unit of area equal to 10 ares or 0.2471 acre

de·ca·su·al·iza·tion \(,)dē-ˌka-zh(ə-)wə-lə-ˈzā-shən, -ˌka-zhə-lə-\ *noun* (1892)
: the process of eliminating the employment of casual workers to stabilize the work force

deca·syl·lab·ic \ˌde-kə-sə-ˈla-bik\ *adjective* [probably from French *décasyllabique,* from Greek *dekasyllabos,* from *deka-* deca- + *syllabē* syllable] (1771)
: consisting of 10 syllables or composed of verses of 10 syllables
— **decasyllabic** *noun*
— **deca·syl·la·ble** \ˈde-kə-ˌsi-lə-bəl, ˌde-kə-ˈ\ *noun*

de·cath·lete \di-ˈkath-ˌlēt\ *noun* [blend of *decathlon* and *athlete*] (1968)
: an athlete who competes in the decathlon

de·cath·lon \di-ˈkath-lən, -ˌlän\ *noun* [French *décathlon,* from *déca-* deca- + *-athlon* (as in *pentathlon*)] (1912)
: a 10-event athletic contest; *specifically* **:** a composite contest that consists of the 100-meter, 400-meter, and 1500-meter runs, the 110-meter high hurdles, the javelin and discus throws, shot put, pole vault, high jump, and long jump

[1]de·cay \di-ˈkā\ *verb* [Middle English, from Old North French *decaïr,* from Late Latin *decadere* to fall, sink, from Latin *de-* + *cadere* to fall — more at CHANCE] (15th century)
intransitive verb
1 : to decline from a sound or prosperous condition
2 : to decrease gradually in quantity, activity, or force
3 : to fall into ruin
4 : to decline in health, strength, or vigor
5 : to undergo decomposition
transitive verb
1 *obsolete* **:** to cause to decay **:** IMPAIR ⟨infirmity that *decays* the wise —Shakespeare⟩
2 : to destroy by decomposition ☆
— **de·cay·er** *noun*

[2]decay *noun* (15th century)
1 : gradual decline in strength, soundness, or prosperity or in degree of excellence or perfection
2 : a wasting or wearing away **:** RUIN
3 *obsolete* **:** DESTRUCTION, DEATH
4 a : ROT; *specifically* **:** aerobic decomposition of proteins chiefly by bacteria **b :** the product of decay
5 : a decline in health or vigor
6 : decrease in quantity, activity, or force: as **a :** spontaneous decrease in the number of radioactive atoms in radioactive material **b :** spontaneous disintegration (as of an atom or a particle)

Dec·ca \ˈde-kə\ *noun* [*Decca* Co., British firm which developed it] (1946)
: a system of long-range navigation used chiefly in Europe that utilizes the phase differences of continuous-wave signals from synchronized ground transmitters to establish position

de·cease \di-ˈsēs\ *noun* [Middle English *deces,* from Middle French, from Latin *decessus* departure, death, from *decedere* to depart, die, from *de-* + *cedere* to go] (14th century)
: departure from life **:** DEATH
— **decease** *intransitive verb*

[1]de·ceased \-ˈsēst\ *adjective* (15th century)
: no longer living; *especially* **:** recently dead — used of persons
synonym see DEAD

[2]deceased *noun, plural* **deceased** (1625)
: a dead person ⟨the will of the *deceased*⟩

de·ce·dent \di-ˈsē-d^{ə}nt\ *noun* [Latin *decedent-, decedens,* present participle of *decedere*] (1599)
: a deceased person — used chiefly in law

de·ceit \di-ˈsēt\ *noun*[Middle English *deceite,* from Middle French, from Latin *decepta,* feminine of *deceptus,* past participle of *decipere*] (14th century)
1 : the act or practice of deceiving **:** DECEPTION
2 : an attempt or device to deceive **:** TRICK
3 : the quality of being deceitful **:** DECEITFULNESS

de·ceit·ful \-fəl\ *adjective* (15th century)
: having a tendency or disposition to deceive: **a :** not honest ⟨a *deceitful* child⟩ **b :** DECEPTIVE, MISLEADING
synonym see DISHONEST
— **de·ceit·ful·ly** \-fə-lē\ *adverb*
— **de·ceit·ful·ness** *noun*

de·ceiv·able \di-ˈsē-və-bəl\ *adjective* (14th century)
1 *archaic* **:** DECEITFUL, DECEPTIVE
2 *archaic* **:** capable of being deceived

de·ceive \di-ˈsēv\ *verb* **de·ceived; de·ceiv·ing** [Middle English, from Middle French *deceivre,* from Latin *decipere,* from *de-* + *capere* to take — more at HEAVE] (13th century)
transitive verb
1 *archaic* **:** ENSNARE
2 a *obsolete* **:** to be false to **b** *archaic* **:** to fail to fulfill
3 *obsolete* **:** CHEAT
4 : to cause to accept as true or valid what is false or invalid
5 *archaic* **:** to while away
intransitive verb
: to practice deceit; *also* **:** to give a false impression ⟨appearances can *deceive*⟩ ☆
— **de·ceiv·er** *noun*
— **de·ceiv·ing·ly** \-ˈsē-viŋ-lē\ *adverb*

de·cel·er·ate \(,)dē-ˈse-lə-ˌrāt\ *verb* **-at·ed; -at·ing** [*de-* + *accelerate*] (1899)
transitive verb
1 : to reduce the speed of **:** slow down
2 : to decrease the rate of progress of
intransitive verb
: to move at decreasing speed
— **de·cel·er·a·tion** \(,)dē-ˌse-lə-ˈrā-shən\ *noun*
— **de·cel·er·a·tor** \(,)dē-ˈse-lə-rā-tər\ *noun*

De·cem·ber \di-ˈsem-bər, dē-\ *noun* [Middle English *Decembre,* from Old English or Old French, both from Latin *December* (tenth month), from *decem* ten — more at TEN] (before 12th century)
: the 12th month of the Gregorian calendar
word history see SEPTEMBER

De·cem·brist \-brist\ *noun* (1877)
: one taking part in the unsuccessful uprising against the Russian emperor Nicholas I in December 1825

de·cem·vir \di-ˈsem-vər\ *noun* [Middle English, from Latin, back-formation from *decemviri,* plural, from *decem* + *viri,* plural of *vir* man — more at VIRILE] (1579)
: one of a ruling body of 10; *specifically* **:** one of a body of 10 magistrates in ancient Rome
— **de·cem·vi·ral** \-vər-əl\ *adjective*
— **de·cem·vi·rate** \-vər-ət\ *noun*

de·cen·cy \ˈdē-s^{ə}n-sē\ *noun, plural* **-cies** (1567)
1 *archaic* **a :** FITNESS **b :** ORDERLINESS
2 a : the quality or state of being decent **:** PROPRIETY **b :** conformity to standards of taste, propriety, or quality
3 : standard of propriety — usually used in plural
4 *plural* **:** conditions or services considered essential for a proper standard of living
5 : literary decorum

de·cen·ni·al \di-ˈse-nē-əl\ *adjective* (circa 1656)
1 : consisting of or lasting for 10 years
2 : occurring or being done every 10 years ⟨the *decennial* census⟩
— **decennial** *noun*
— **de·cen·ni·al·ly** \-nē-ə-lē\ *adverb*

de·cen·ni·um \-nē-əm\ *noun, plural* **-ni·ums** *or* **-nia** \-nē-ə\ [Latin, from *decem* + *annus* year — more at ANNUAL] (1685)
: a period of 10 years **:** DECADE

de·cent \ˈdē-s^{ə}nt\ *adjective* [Middle French or Latin; Middle French, from Latin *decent-, decens,* present participle of *decēre* to be fitting; akin to Latin *decus* honor, *dignus* worthy, Greek *dokein* to seem, seem good] (1539)
1 *archaic* **a :** APPROPRIATE **b :** well-formed **:** HANDSOME
2 a : conforming to standards of propriety, good taste, or morality **b :** modestly clothed
3 : free from immodesty or obscenity
4 : fairly good but not excellent **:** ADEQUATE, SATISFACTORY ⟨*decent* wages⟩
5 : marked by moral integrity, kindness, and goodwill ⟨hard-working and *decent* folks⟩
synonym see CHASTE
— **de·cent·ly** *adverb*

de·cen·tral·iza·tion \(,)dē-ˌsen-trə-lə-ˈzā-shən\ *noun* (1846)
1 : the dispersion or distribution of functions and powers; *specifically* **:** the delegation of power from a central authority to regional and local authorities
2 : the redistribution of population and industry from urban centers to outlying areas
— **de·cen·tral·ize** \(,)dē-ˈsen-trə-ˌlīz\ *verb*

de·cep·tion \di-ˈsep-shən\ *noun* [Middle English *decepcioun,* from Middle French *deception,* from Late Latin *deception-, deceptio,* from Latin *decipere* to deceive] (15th century)
1 a : the act of deceiving **b :** the fact or condition of being deceived

☆ SYNONYMS
Decay, decompose, rot, putrefy, spoil mean to undergo destructive dissolution. DECAY implies a slow change from a state of soundness or perfection ⟨a *decaying* mansion⟩. DECOMPOSE stresses a breaking down by chemical change and when applied to organic matter a corruption ⟨the strong odor of *decomposing* vegetation⟩. ROT is a close synonym of DECOMPOSE and often connotes foulness ⟨fruit was left to *rot* in warehouses⟩. PUTREFY implies the rotting of animal matter and offensiveness to sight and smell ⟨corpses *putrefying* on the battlefield⟩. SPOIL applies chiefly to the decomposition of foods ⟨keep the ham from *spoiling*⟩.

Deceive, mislead, delude, beguile mean to lead astray or frustrate usually by underhandedness. DECEIVE implies imposing a false idea or belief that causes ignorance, bewilderment, or helplessness ⟨tried to *deceive* me about the cost⟩. MISLEAD implies a leading astray that may or may not be intentional ⟨I was *misled* by the confusing sign⟩. DELUDE implies deceiving so thoroughly as to obscure the truth ⟨we were *deluded* into thinking we were safe⟩. BEGUILE stresses the use of charm and persuasion in deceiving ⟨was *beguiled* by false promises⟩.

2 : something that deceives **:** TRICK ☆
— **de·cep·tion·al** \-shə-nəl\ *adjective*

de·cep·tive \di-ˈsep-tiv\ *adjective* (circa 1611)
: tending or having power to deceive **:** MISLEADING
— **de·cep·tive·ly** *adverb*
— **de·cep·tive·ness** *noun*

[1]de·cer·e·brate \(ˌ)dē-ˈser-ə-brət, -ˌbrāt\ *adjective* (1897)
1 : characteristic of decerebration ⟨*decerebrate* rigidity⟩
2 : having the cerebrum removed or made inactive

[2]de·cer·e·brate \(ˌ)dē-ˈser-ə-ˌbrāt\ *transitive verb* (circa 1900)
: to remove the cerebrum from; *also* **:** to make incapable of cerebral activity
— **de·cer·e·bra·tion** \(ˌ)dē-ˌser-ə-ˈbrā-shən\ *noun*

de·cer·ti·fy \(ˌ)dē-ˈsər-tə-ˌfī\ *transitive verb* (1918)
: to withdraw or revoke the certification of
— **de·cer·ti·fi·ca·tion** \(ˌ)dē-ˌsər-tə-fə-ˈkā-shən\ *noun*

de·chlo·ri·nate \(ˌ)dē-ˈklōr-ə-ˌnāt, -ˈklȯr-\ *transitive verb* (1941)
: to remove chlorine from ⟨*dechlorinate* water⟩
— **de·chlo·ri·na·tion** \(ˌ)dē-ˌklōr-ə-ˈnā-shən, -ˌklȯr-\ *noun*

deci- *combining form* [French *déci-,* from Latin *decimus* tenth, from *decem* ten — more at TEN]
: tenth part ⟨*decimeter*⟩

deci·bel \ˈde-sə-ˌbel, -bəl\ *noun* [International Scientific Vocabulary *deci-* + *bel*] (1928)
1 a : a unit for expressing the ratio of two amounts of electric or acoustic signal power equal to 10 times the common logarithm of this ratio **b :** a unit for expressing the ratio of the magnitudes of two electric voltages or currents or analogous acoustic quantities equal to 20 times the common logarithm of the voltage or current ratio
2 : a unit for expressing the relative intensity of sounds on a scale from zero for the average least perceptible sound to about 130 for the average pain level
3 : degree of loudness; *also* **:** extremely loud sound — usually used in plural

de·cide \di-ˈsīd, dē-\ *verb* **de·cid·ed; de·cid·ing** [Middle English, from Middle French *decider,* from Latin *decidere,* literally, to cut off, from *de-* + *caedere* to cut] (14th century)
transitive verb
1 a : to arrive at a solution that ends uncertainty or dispute about ⟨*decide* what to do⟩ **b :** to select as a course of action — used with an infinitive ⟨*decided* to go⟩
2 : to bring to a definitive end ⟨one blow *decided* the fight⟩
3 : to induce to come to a choice ⟨her pleas *decided* him to help⟩
intransitive verb
: to make a choice or judgment ☆
— **de·cid·abil·i·ty** \di-ˌsī-də-ˈbi-lə-tē\ *noun*
— **de·cid·able** \di-ˈsī-də-bəl\ *adjective*
— **de·cid·er** *noun*

de·cid·ed *adjective* (1790)
1 : UNQUESTIONABLE ⟨a *decided* advantage⟩
2 : free from doubt or wavering
— **de·cid·ed·ly** *adverb*
— **de·cid·ed·ness** *noun*

de·cid·ing *adjective* (1658)
: that decides **:** DECISIVE ⟨drove in the *deciding* run⟩

de·cid·ua \di-ˈsi-jə-wə\ *noun, plural* **-u·ae** \-jə-ˌwē\ [New Latin, from Latin, feminine of *deciduus*] (1785)
1 : the part of the mucous membrane lining the uterus that in higher placental mammals undergoes special modifications in preparation for and during pregnancy and is cast off at parturition
2 : the part of the mucous membrane of the uterus cast off in the process of menstruation
— **de·cid·u·al** \-wəl\ *adjective*

de·cid·u·ate \-wət\ *adjective* (1868)
: having the fetal and maternal tissues firmly interlocked so that a layer of maternal tissue is torn away at parturition and forms part of the afterbirth

de·cid·u·ous \di-ˈsi-jə-wəs\ *adjective* [Latin *deciduus,* from *decidere* to fall off, from *de-* + *cadere* to fall — more at CHANCE] (1688)
1 : falling off or shed seasonally or at a certain stage of development in the life cycle ⟨*deciduous* leaves⟩ ⟨*deciduous* teeth⟩
2 a : having deciduous parts ⟨*deciduous* trees⟩ **b :** having the dominant plants deciduous ⟨a *deciduous* forest⟩
3 : EPHEMERAL
— **de·cid·u·ous·ness** *noun*

deci·gram \ˈde-sə-ˌgram\ *noun* [French *décigramme,* from *déci-* + *gramme* gram] (1810)
— see METRIC SYSTEM table

dec·ile \ˈde-ˌsīl, -səl\ *noun* [Latin *decem* ten — more at TEN] (1882)
: any one of nine numbers that divide a frequency distribution into 10 classes such that each contains the same number of individuals; *also* **:** any one of these 10 classes
— **decile** *adjective*

deci·li·ter \ˈde-sə-ˌlē-tər\ *noun* [French *décilitre,* from *déci-* + *litre* liter] (1801)
— see METRIC SYSTEM table

de·cil·lion \di-ˈsil-yən\ *noun, often attributive* [Latin *decem* + English *-illion* (as in *million*)] (1847)
— see NUMBER table

[1]dec·i·mal \ˈde-sə-məl, ˈdes-məl\ *adjective* [French *décimal,* from Medieval Latin *decimalis* of a tithe, from Latin *decima* tithe — more at DIME] (1608)
: numbered or proceeding by tens: **a :** based on the number 10; *especially* **:** expressed in or utilizing a decimal system especially with a decimal point **b :** subdivided into 10th or 100th units ⟨*decimal* coinage⟩
— **dec·i·mal·ly** \-mə-lē\ *adverb*

[2]decimal *noun* (1651)
: any real number expressed in base 10; *especially* **:** DECIMAL FRACTION

decimal fraction *noun* (1660)
: a fraction (as .25 = $^{25}/_{100}$ or .025 = $^{25}/_{1000}$) or mixed number (as 3.025 = $3^{25}/_{1000}$) in which the denominator is a power of 10 usually expressed by use of the decimal point

dec·i·mal·i·za·tion \ˌde-sə-mə-lə-ˈzā-shən\ *noun* (1855)
: conversion (as of a currency) to a decimal system
— **dec·i·mal·ize** \ˈde-sə-mə-ˌlīz\ *transitive verb*

decimal point *noun* (circa 1864)
: a period, centered dot, or in some countries a comma at the left of a proper decimal fraction (as .678) or between the parts of a mixed number (as 3.678) expressed by a whole number and a decimal fraction

decimal system *noun* (1864)
1 : a number system that uses a notation in which each number is expressed in base 10 by using one of the first nine integers or 0 in each place and letting each place value be a power of 10
2 : a system of measurement or currency in which the basic units increase by powers of 10

dec·i·mate \ˈde-sə-ˌmāt\ *transitive verb* **-mat·ed; -mat·ing** [Latin *decimatus,* past participle of *decimare,* from *decimus* tenth, from *decem* ten] (1660)
1 : to select by lot and kill every tenth man of
2 : to exact a tax of 10 percent from ⟨poor as a *decimated* Cavalier —John Dryden⟩
3 a : to reduce drastically especially in number ⟨cholera *decimated* the population⟩ **b :** to destroy a large part of ⟨firebombs *decimated* large sections of the city⟩ □

☆ SYNONYMS

Deception, fraud, double-dealing, subterfuge, trickery mean the acts or practices of one who deliberately deceives. DECEPTION may or may not imply blameworthiness, since it may suggest cheating or merely tactical resource ⟨magicians are masters of *deception*⟩. FRAUD always implies guilt and often criminality in act or practice ⟨indicted for *fraud*⟩. DOUBLE-DEALING suggests treachery or at least action contrary to a professed attitude ⟨a go-between suspected of *double-dealing*⟩. SUBTERFUGE suggests the adoption of a stratagem or the telling of a lie in order to escape guilt or to gain an end ⟨obtained the papers by *subterfuge*⟩. TRICKERY implies ingenious acts intended to dupe or cheat ⟨resorted to *trickery* to gain their ends⟩.

Decide, determine, settle, rule, resolve mean to come or cause to come to a conclusion. DECIDE implies previous consideration of a matter causing doubt, wavering, debate, or controversy ⟨she *decided* to sell her house⟩. DETERMINE implies fixing the identity, character, scope, or direction of something ⟨*determined* the cause of the problem⟩. SETTLE implies a decision reached by someone with power to end all dispute or uncertainty ⟨the dean's decision *settled* the campus alcohol policy⟩. RULE implies a determination by judicial or administrative authority ⟨the judge *ruled* that the evidence was inadmissible⟩. RESOLVE often implies a firm decision to do or refrain from doing something ⟨he *resolved* to quit smoking⟩.

□ USAGE

decimate The first sense of *decimate* was born in the Roman practice of disciplining refractory military units by selecting one tenth of the men by lot and executing them. A bit harsh, from a modern standpoint, but presumably the Romans thought it effective. Sense 1 is used only in historical contexts. Sense 2 is also historical; it refers in the main to a tax levied by Oliver Cromwell on the defeated Royalists. Sense 3 is the live sense. It remembers not the arithmetic of the Romans, but the ferocity of their methods, and it has been in continuous use since the second half of the 17th century. Beginning around 1870 and continuing for more than a century, a school of critics—many of them newspaper editors—has emphasized the etymology and disapproved of the use that is regularly met in current writing. All the same, uses like those that follow are standard ⟨though the buffalo herds have been *decimated* . . . this is still the frontier —John Updike⟩ ⟨the curse of scurvy, which *decimated* the crew —Daniel J. Boorstin⟩ ⟨injuries *decimated* the Yankees in the first half of the season —Murray Chass⟩ ⟨tanks and enough weaponry to *decimate* the continent —William Safire⟩ ⟨logrolling that ultimately brought a new tariff compromise bill, the price of which was to *decimate* the distribution act —James MacGregor Burns⟩. This sense is even used occasionally for humorous overstatement ⟨is there a homicidal maniac on the loose, who intends to *decimate* the Detroit priesthood? —Newgate Callendar⟩.

— **dec·i·ma·tion** \ˌde-sə-ˈmā-shən\ *noun*

deci·me·ter \ˈde-sə-ˌmē-tər\ *noun* [French *décimètre,* from *déci-* deci- + *mètre* meter] (1809)
— see METRIC SYSTEM table

de·ci·pher \dē-ˈsī-fər\ *transitive verb* (1545)
1 : DECODE 1a
2 *obsolete* **:** DEPICT
3 a : to make out the meaning of despite indistinctness or obscurity **b :** to interpret the meaning of
— **de·ci·pher·able** \-f(ə-)rə-bəl\ *adjective*
— **de·ci·pher·er** \-fər-ər\ *noun*
— **de·ci·pher·ment** \-fər-mənt\ *noun*

¹de·ci·sion \di-ˈsi-zhən\ *noun* [Middle French, from Latin *decision-, decisio,* from *decidere* to decide] (15th century)
1 a : the act or process of deciding **b :** a determination arrived at after consideration **:** CONCLUSION
2 : a report of a conclusion
3 : promptness and firmness in deciding **:** DETERMINATION
4 a : WIN; *specifically* **:** a victory in boxing decided on points **b :** a win or loss officially credited to a pitcher in baseball
— **de·ci·sion·al** \-ˈsi-zhnəl, -ˈsi-zhə-nᵊl\ *adjective*

²decision *transitive verb* (1943)
: to win a decision over (a boxing opponent)

decision theory *noun* (1961)
: a branch of statistical theory concerned with quantifying the process of making choices between alternatives

decision tree *noun* (1964)
: a tree diagram which is used for making decisions in business or computer programming and in which the branches represent choices with associated risks, costs, results, or probabilities

de·ci·sive \di-ˈsī-siv\ *adjective* (1611)
1 : having the power or quality of deciding
2 : RESOLUTE, DETERMINED
3 : UNMISTAKABLE, UNQUESTIONABLE ⟨a *decisive* superiority⟩
synonym see CONCLUSIVE
— **de·ci·sive·ly** *adverb*
— **de·ci·sive·ness** *noun*

¹deck \ˈdek\ *noun* [Middle English *dekke* covering of a ship, from (assumed) Middle Dutch *dec* covering, probably from Middle Low German *vordeck,* from *vordecken* to cover, from *vor-* for- + *decken* to cover; akin to Old High German *decchen* to cover — more at THATCH] (1509)
1 : a platform in a ship serving usually as a structural element and forming the floor for its compartments
2 : something resembling the deck of a ship: as **a :** a story or tier of a building **b :** the roadway of a bridge **c :** a flat floored roofless area adjoining a house **d :** the lid of the compartment at the rear of the body of an automobile; *also* **:** the compartment **e :** a layer of clouds
3 a : a pack of playing cards **b :** a packet of narcotics
4 : TAPE DECK
— **on deck 1 :** ready for duty **2 :** next in line **:** next in turn

²deck *transitive verb* [Dutch *dekken* to cover; akin to Old High German *decchen*] (1513)
1 *obsolete* **:** COVER
2 a : to clothe elegantly **:** ARRAY ⟨*decked* out in furs⟩ **b :** DECORATE ⟨*deck* the halls with boughs of holly —*English carol*⟩ **c :** to portray or present with embellishments
3 [¹*deck*] **:** to furnish with or as if with a deck
4 [¹*deck*] **:** to knock down forcibly **:** FLOOR ⟨*decked* him with one punch⟩
synonym see ADORN

deck chair *noun* (1884)
: a folding chair often having an adjustable leg rest

deck·er \ˈde-kər\ *noun* (1790)
: something having a specified number of decks, levels, floors, or layers — used in combination ⟨many of the city's buses are double-*deckers*⟩

deck chair

deck·hand \ˈdek-ˌhand\ *noun* (1844)
: a seaman who performs manual duties

deck·house \-ˌhau̇s\ *noun* (1856)
: a superstructure on a ship's upper deck

deck·ing \ˈde-kiŋ\ *noun* (1580)
: DECK; *also* **:** material for a deck

deck·le \ˈde-kəl\ *noun* [German *Deckel,* literally, cover, from *decken* to cover, from Old High German *decchen*] (1816)
: a frame around the edges of a mold used in making paper by hand; *also* **:** either of the bands around the edge of the wire of a papermaking machine that determine the width of the web

deckle edge *noun* (circa 1874)
: the rough untrimmed edge of paper left by a deckle or produced artificially
— **deck·le–edged** \-ˈejd\ *adjective*

deck tennis *noun* [from its being played on the decks of ocean liners] (1927)
: a game in which players toss a ring or quoit back and forth over a net stretched across a small court

de·claim \di-ˈklām, dē-\ *verb* [Middle English *declamen,* from Latin *declamare,* from *de-* + *clamare* to cry out; akin to Latin *calare* to call — more at LOW] (14th century)
intransitive verb
1 : to speak rhetorically; *specifically* **:** to recite something as an exercise in elocution
2 : to speak pompously or bombastically **:** HARANGUE
transitive verb
: to deliver rhetorically; *specifically* **:** to recite in elocution
— **de·claim·er** *noun*
— **dec·la·ma·tion** \ˌde-klə-ˈmā-shən\ *noun*

de·clam·a·to·ry \di-ˈkla-mə-ˌtōr-ē, -ˌtȯr-\ *adjective* (1581)
: of, relating to, or marked by declamation or rhetorical display

de·clar·ant \di-ˈklar-ənt, -ˈkler-\ *noun* (1681)
: a person who makes a statement or declaration especially in connection with a legal proceeding

dec·la·ra·tion \ˌde-klə-ˈrā-shən\ *noun* (15th century)
1 : the act of declaring **:** ANNOUNCEMENT
2 a : the first pleading in a common-law action **b :** a statement made by a party to a legal transaction usually not under oath
3 a : something that is declared **b :** the document containing such a declaration

de·clar·a·tive \di-ˈklar-ə-tiv, -ˈkler-\ *adjective* (1628)
: making a declaration **:** DECLARATORY ⟨a *declarative* sentence⟩
— **de·clar·a·tive·ly** *adverb*

de·clar·a·to·ry \-ə-ˌtōr-ē, -ˌtȯr-\ *adjective* (15th century)
1 : serving to declare, set forth, or explain
2 a : declaring what is the existing law ⟨*declaratory* statute⟩ **b :** declaring a legal right or interpretation ⟨a *declaratory* judgment⟩

de·clare \di-ˈklar, -ˈkler\ *verb* **de·clared; de·clar·ing** [Middle English, from Middle French & Latin; Middle French *declarer,* from Latin *declarare,* from *de-* + *clarare* to make visible, from *clarus* clear — more at CLEAR] (14th century)
transitive verb
1 : to make known formally, officially, or explicitly
2 *obsolete* **:** to make clear
3 : to make evident **:** SHOW
4 : to state emphatically **:** AFFIRM ⟨*declares* his innocence⟩
5 : to make a full statement of (one's taxable or dutiable property)
6 a : to announce (as a trump suit) in a card game **b :** MELD
7 : to make payable ⟨*declare* a dividend⟩
intransitive verb
1 : to make a declaration
2 : to avow one's opinion or support ☆
— **de·clar·able** \-ˈklar-ə-bəl, -ˈkler-\ *adjective*

de·clar·er \di-ˈklar-ər, -ˈkler-\ *noun* (14th century)
: one that declares; *specifically* **:** the bridge player who names the trump and plays both his own hand and that of the dummy

de·class \(ˌ)dē-ˈklas\ *transitive verb* (1888)
: to remove from a class; *especially* **:** to assign to a lower social status

dé·clas·sé \ˌdā-ˌkla-ˈsā, -ˌklä-\ *adjective* [French, from past participle of *déclasser* to declass] (1887)
1 : fallen or lowered in class, rank, or social position
2 : of inferior status

de·clas·si·fy \(ˌ)dē-ˈkla-sə-ˌfī\ *transitive verb* (1945)
: to remove or reduce the security classification of ⟨*declassify* a secret document⟩
— **de·clas·si·fi·ca·tion** \(ˌ)dē-ˌkla-sə-fə-ˈkā-shən\ *noun*

de·claw \-ˈklȯ\ *transitive verb* (1953)
: to remove the claws of (as a cat) surgically

de·clen·sion \di-ˈklen(t)-shən\ *noun* [Middle English *declenson,* modification of Middle French *declinaison,* from Latin *declination-, declinatio* grammatical inflection, turning aside, from *declinare* to inflect, turn aside] (15th century)
1 a : noun, adjective, or pronoun inflection especially in some prescribed order of the forms **b :** a class of nouns or adjectives having the same type of inflectional forms
2 : a falling off or away **:** DETERIORATION
3 : DESCENT, SLOPE
— **de·clen·sion·al** \-ˈklen(t)-shə-nᵊl\ *adjective*

dec·li·na·tion \ˌde-klə-ˈnā-shən\ *noun* [Middle English *declinacioun,* from Middle French *declination,* from Latin *declination-, declinatio* angle of the heavens, turning aside] (14th century)
1 : angular distance north or south from the celestial equator measured along a great circle passing through the celestial poles
2 : a turning aside or swerving
3 : DETERIORATION ⟨moral *declination*⟩
4 : a bending downward **:** INCLINATION
5 : a formal refusal
6 : the angle formed between a magnetic needle and the geographical meridian
— **dec·li·na·tion·al** \-shnəl, -shə-nᵊl\ *adjective*

¹de·cline \di-ˈklīn, dē-\ *verb* **de·clined; de·clin·ing** [Middle English, from Middle French *decliner,* from Latin *declinare* to turn aside, inflect, from *de-* + *clinare* to incline — more at LEAN] (14th century)
intransitive verb

☆ SYNONYMS
Declare, announce, proclaim, promulgate mean to make known publicly. DECLARE implies explicitness and usually formality in making known ⟨the referee *declared* the contest a draw⟩. ANNOUNCE implies the declaration of something for the first time ⟨*announced* their engagement at a party⟩. PROCLAIM implies declaring clearly, forcefully, and authoritatively ⟨the president *proclaimed* a national day of mourning⟩. PROMULGATE implies the proclaiming of a dogma, doctrine, or law ⟨*promulgated* an edict of religious toleration⟩. See in addition ASSERT.

1 *archaic* **:** to turn from a straight course **:** STRAY
2 a : to slope downward **:** DESCEND **b :** to bend down **:** DROOP **c :** to stoop to what is unworthy
3 a *of a celestial body* **:** to sink toward setting **b :** to draw toward a close **:** WANE ⟨the day *declined*⟩
4 : to tend toward an inferior state or weaker condition ⟨his health *declined*⟩
5 : to withhold consent
6 : to become less in amount ⟨prices *declined*⟩
transitive verb
1 : to give in prescribed order the grammatical forms of (a noun, pronoun, or adjective)
2 *obsolete* **a :** AVERT **b :** AVOID
3 : to cause to bend or bow downward
4 a : to refuse to undertake, undergo, engage in, or comply with **b :** to refuse especially courteously ⟨*decline* an invitation⟩ ☆
— de·clin·able \-ˈklī-nə-bəl\ *adjective*
— de·clin·er \-ˈklī-nər\ *noun*

²decline *also* ˈdē-ˌklīn\ *noun* (14th century)
1 : the process of declining: **a :** a gradual physical or mental sinking and wasting away **b :** a change to a lower state or level
2 : the period during which something is deteriorating or approaching its end
3 : a downward slope
4 : a wasting disease; *especially* **:** pulmonary tuberculosis
synonym see DETERIORATION

de·clin·ing \-niŋ\ *adjective* (1593)
: of or relating to the period during which something is deteriorating or nearing its end ⟨her *declining* years⟩

de·cliv·i·tous \di-ˈkliv-ət-əs\ *adjective* (1799)
: moderately steep

de·cliv·i·ty \-ət-ē\ *noun, plural* **-ties** [Latin *declivitat-, declivitas,* from *declivis* sloping down, from *de-* + *clivus* slope, hill; akin to Latin *clinare*] (1612)
1 : downward inclination
2 : a descending slope

de·co \ˈde-(ˌ)kō, dā-ˈkō, ˈdā-ˌ\ *noun, often capitalized* (1969)
: ART DECO

de·coct \di-ˈkäkt\ *transitive verb* [Latin *decoctus,* past participle of *decoquere,* from *de-* + *coquere* to cook — more at COOK] (15th century)
1 : to extract the flavor of by boiling
2 : BOIL DOWN, CONCENTRATE

de·coc·tion \di-ˈkäk-shən\ *noun* (15th century)
1 : an extract obtained by decocting
2 : the act or process of decocting

de·code \(ˌ)dē-ˈkōd\ *transitive verb* (1896)
1 a : to convert (as a coded message) into intelligible form **b :** to recognize and interpret (an electronic signal)
2 a : DECIPHER 3a **b :** to discover the underlying meaning of ⟨*decode* the play's imagery⟩

de·cod·er \-ˈkō-dər\ *noun* (1920)
: one that decodes; *especially* **:** an electronic device for unscrambling a television transmission

de·col·late \dē-ˈkä-ˌlāt\ *transitive verb* **-lat·ed; -lat·ing** [Latin *decollatus,* past participle of *decollare,* from *de-* + *collum* neck — more at COLLAR] (1599)
: BEHEAD
— de·col·la·tion \ˌdē-kä-ˈlā-shən\ *noun*

dé·col·le·tage \(ˌ)dā-ˌkä-lə-ˈtäzh, (ˌ)dā-ˌkȯl-ˈtäzh, ˌde-klə-\ *noun* [French, action of cutting or wearing a low neckline, from *décolleter*] (1894)
1 : the low-cut neckline of a dress
2 : a décolleté dress
3 : ¹BUST 2

¹dé·col·le·té \(ˌ)dā-ˌkäl-ˈtā, -ˌkȯl-, -lə-ˈtā; *also* dā-ˈkȯl-tā\ *adjective* [French, from past participle of *décolleter* to give a low neckline to, from *dé-* de- + *collet* collar, from Old French *colet,* from *col* collar, neck, from Latin *collum* neck] (1831)
1 : wearing a strapless or low-necked dress
2 : having a low-cut neckline

²dé·col·le·té *also* **de·col·le·te** *noun* (1900)
: DÉCOLLETAGE

de·col·o·nize \(ˌ)dē-ˈkä-lə-ˌnīz\ *transitive verb* (1963)
: to free from colonial status
— de·col·o·ni·za·tion \(ˌ)dē-ˌkä-lə-nə-ˈzā-shən\ *noun*

de·col·or·ize \(ˌ)dē-ˈkə-lə-ˌrīz\ *transitive verb* **-ized; -iz·ing** (1839)
: to remove color from ⟨*decolorize* vinegar by adsorption of impurities on activated charcoal⟩
— de·col·or·iza·tion \(ˌ)dē-ˌkə-lə-rə-ˈzā-shən\ *noun*
— de·col·or·iz·er \(ˈ)dē-ˈkə-lə-ˌrī-zər\ *noun*

de·com·mis·sion \ˌdē-kə-ˈmi-shən\ *transitive verb* (1922)
: to remove (as a ship or nuclear power plant) from service

de·com·pen·sa·tion \(ˌ)dē-ˌkäm-pən-ˈsā-shən, -ˌpen-\ *noun* [International Scientific Vocabulary] (circa 1903)
: loss of physiological compensation or psychological balance; *especially* **:** inability of the heart to maintain adequate circulation
— de·com·pen·sate \(ˌ)dē-ˈkäm-pən-ˌsāt, -ˌpen-\ *verb*

de·com·pose \ˌdē-kəm-ˈpōz\ *verb* [French *décomposer,* from *dé-* de + *composer* to compose] (circa 1751)
transitive verb
1 : to separate into constituent parts or elements or into simpler compounds ⟨*decompose* water by electrolysis⟩ ⟨*decompose* a word into its base and affixes⟩
2 : ROT
intransitive verb
: to break up into constituent parts by or as if by a chemical process **:** DECAY, ROT ⟨fruit *decomposes*⟩
synonym see DECAY
— de·com·pos·abil·i·ty \-ˌpō-zə-ˈbi-lə-tē\ *noun*
— de·com·pos·able \-ˈpō-zə-bəl\ *adjective*
— de·com·po·si·tion \(ˌ)dē-ˌkäm-pə-ˈzi-shən\ *noun*

de·com·pos·er \ˌdē-kəm-ˈpō-zər\ *noun* (1959)
: any of various organisms (as many bacteria and fungi) that return constituents of organic substances to ecological cycles by feeding on and breaking down dead protoplasm

de·com·pound \ˌdē-ˈkäm-ˌpau̇nd, ˌdē-kəm-ˈ\ *adjective* (circa 1793)
of a leaf **:** having divisions that are themselves compound

de·com·press \ˌdē-kəm-ˈpres\ (1905)
transitive verb
: to release from pressure or compression
intransitive verb
: to undergo release from pressure; *especially* **:** RELAX ⟨need a week off to *decompress*⟩
— de·com·pres·sion \-ˈpre-shən\ *noun*

decompression sickness *noun* (circa 1941)
: ²BEND 3, AEROEMBOLISM

de·con·cen·trate \(ˌ)dē-ˈkän(t)-sən-ˌtrāt\ *transitive verb* (circa 1889)
: to reduce or abolish the concentration of **:** DECENTRALIZE
— de·con·cen·tra·tion \(ˌ)dē-ˌkän(t)-sən-ˈtrā-shən\ *noun*

de·con·di·tion \ˌdē-kən-ˈdi-shən\ *transitive verb* (1941)
1 : to cause extinction of (a conditioned response)
2 : to cause to lose physical fitness

de·con·ges·tant \ˌdē-kən-ˈjes-tənt\ *noun* (1947)
: an agent that relieves congestion (as of mucous membranes)
— decongestant *adjective*

de·con·ges·tion \ˌdē-kən-ˈjes(h)-chən\ *noun* (1908)
: the process of relieving congestion
— de·con·gest \-ˈjest\ *transitive verb*
— de·con·ges·tive \-ˈjes-tiv\ *adjective*

de·con·se·crate \(ˌ)dē-ˈkän(t)-sə-ˌkrāt\ *transitive verb* (1876)
: to remove the sacred character of ⟨*deconsecrate* a church⟩
— de·con·se·cra·tion \(ˌ)dē-ˌkän(t)-sə-ˈkrā-shən\ *noun*

de·con·struct \ˌdē-kən-ˈstrəkt\ *transitive verb* (1973)
: to discuss (as a work of literature) using the methods of deconstruction
— de·con·struc·tive \-tiv\ *adjective*
— de·con·struc·tor \-tər\ *noun*

de·con·struc·tion \ˌdē-kən-ˈstrək-shən\ *noun* [French *déconstruction,* from *dé-* de- + *construction*] (1973)
: a method of literary criticism that assumes language refers only to itself rather than to an extratextual reality, that asserts multiple conflicting interpretations of a text, and that bases such interpretations on the philosophical, political, or social implications of the use of language in the text rather than on the author's intention
— de·con·struc·tion·ist \-shə-nist\ *noun*

de·con·tam·i·nate \ˌdē-kən-ˈta-mə-ˌnāt\ *transitive verb* (1936)
: to rid of contamination (as radioactive material)
— de·con·tam·i·na·tion \-ˌta-mə-ˈnā-shən\ *noun*
— de·con·tam·i·na·tor \-ˈta-mə-ˌnā-tər\ *noun*

de·con·trol \ˌdē-kən-ˈtrōl\ *transitive verb* (1919)
: to end control of
— decontrol *noun*

de·cor *or* **dé·cor** \dā-ˈkȯr, di-ˈ; ˈde-ˌkȯr, ˈdā-ˌ\ *noun* [French *décor,* from *décorer* to decorate, from Latin *decorare*] (1897)
1 : a stage setting
2 a : DECORATION 2 **b :** the style and layout of interior furnishings

dec·o·rate \ˈde-kə-ˌrāt\ *transitive verb* **-rat·ed; -rat·ing** [Latin *decoratus,* past participle of *decorare,* from *decor-, decus* ornament, honor — more at DECENT] (1530)
1 : to add honor to
2 : to furnish with something ornamental ⟨*decorate* a room⟩
3 : to award a mark of honor to
synonym see ADORN

dec·o·ra·tion \ˌde-kə-ˈrā-shən\ *noun* (1530)
1 : the act or process of decorating
2 : something that adorns, enriches, or beautifies **:** ORNAMENT
3 : a badge of honor (as a U.S. military award)

☆ SYNONYMS
Decline, refuse, reject, repudiate, spurn mean to turn away by not accepting, receiving, or considering. DECLINE often implies courteous refusal especially of offers or invitations ⟨*declined* his party's nomination⟩. REFUSE suggests more positiveness or ungraciousness and often implies the denial of something asked for ⟨*refused* to lend them the money⟩. REJECT implies a peremptory refusal by sending away or discarding ⟨*rejected* the manuscript as unpublishable⟩. REPUDIATE implies a casting off or disowning as untrue, unauthorized, or unworthy of acceptance ⟨teenagers who *repudiate* the values of their parents⟩. SPURN stresses contempt or disdain in rejection or repudiation ⟨*spurned* his overtures of friendship⟩.

Decoration Day *noun* [from the custom of decorating graves on this day] (1871)
: MEMORIAL DAY

dec·o·ra·tive \ˈde-k(ə-)rə-tiv, ˈde-kə-ˌrā-\ *adjective* (1791)
: serving to decorate; *especially* : purely ornamental
— **dec·o·ra·tive·ly** *adverb*
— **dec·o·ra·tive·ness** *noun*

decorative art *noun* (1967)
1 : an art concerned primarily with the creation of useful items (as furniture, ceramics, and textiles) — usually used in plural
2 : objects of decorative art

¹dec·o·ra·tor \ˈde-kə-ˌrā-tər\ *noun* (circa 1755)
: one that decorates; *especially* : one that designs or executes interiors and their furnishings

²decorator *adjective* (1950)
: suitable for interior decoration ⟨*decorator* fabrics⟩

dec·o·rous \ˈde-kər-əs *also* di-ˈkōr-əs *or* -ˈkȯr-\ *adjective* [Latin *decorus,* from *decor* beauty, grace; akin to Latin *decēre* to be fitting — more at DECENT] (1673)
: marked by propriety and good taste : CORRECT ⟨*decorous* conduct⟩
— **dec·o·rous·ly** *adverb*
— **dec·o·rous·ness** *noun*

de·cor·ti·ca·tion \(ˌ)dē-ˌkȯr-tə-ˈkā-shən\ *noun* [Latin *decortication-, decorticatio,* from *decorticare* to remove the bark from, from *de-* + *cortic-, cortex* bark — more at CUIRASS] (circa 1623)
1 : the act or process of removing the outer coverings (as bark or husks) from something (as fiber or seed)
2 : the surgical removal of the cortex of an organ (as the brain), an enveloping membrane, or a constrictive fibrinous covering
— **de·cor·ti·cate** \(ˌ)dē-ˈkȯr-tə-ˌkāt\ *transitive verb*
— **de·cor·ti·cat·or** \-ˌkā-tər\ *noun*

de·co·rum \di-ˈkōr-əm, -ˈkȯr-\ *noun* [Latin, from neuter of *decorus*] (1576)
1 : literary and dramatic propriety : FITNESS
2 : propriety and good taste in conduct or appearance
3 : ORDERLINESS
4 *plural* **:** the conventions of polite behavior

de·cou·page *or* **dé·cou·page** \ˌdā-(ˌ)kü-ˈpäzh\ *noun* [French *découpage,* literally, act of cutting out, from Middle French, from *decouper* to cut out, from *de-* + *couper* to cut — more at COPE] (1946)
1 : the art of decorating surfaces by applying cutouts (as of paper) and then coating with usually several layers of finish (as lacquer or varnish)
2 : work produced by decoupage
— **decoupage** *or* **découpage** *transitive verb*

de·cou·ple \(ˌ)dē-ˈkə-pəl\ *transitive verb* (1953)
: to eliminate the interrelationship of : SEPARATE

¹de·coy \ˈdē-ˌkȯi, di-ˈ\ *noun* [probably from Dutch *de kooi,* literally, the cage, from *de,* masculine definite article (akin to Old English *thæt,* neuter definite article) + *kooi* cage, from Latin *cavea* — more at THAT, CAGE] (1641)
1 : a pond into which wildfowl are lured for capture
2 : someone or something used to lure or lead another into a trap; *especially* : an artificial bird used to attract live birds within shot
3 : someone or something used to draw attention away from another

²de·coy \di-ˈkȯi, ˈdē-ˌ\ *transitive verb* (circa 1674)
: to lure by or as if by a decoy : ENTICE
synonym see LURE

¹de·crease \di-ˈkrēs, ˈdē-ˌ\ *verb* **de·creased; de·creas·ing** [Middle English *decreessen,* from (assumed) Anglo-French *decreistre,* from Latin *decrescere,* from *de-* + *crescere* to grow — more at CRESCENT] (14th century)
intransitive verb
: to grow progressively less (as in size, amount, number, or intensity)
transitive verb
: to cause to decrease ☆
— **de·creas·ing·ly** \di-ˈkrē-siŋ-lē, dē-\ *adverb*

²de·crease \ˈdē-ˌkrēs, di-ˈ\ *noun* (14th century)
1 : the process of decreasing
2 : an amount of diminution : REDUCTION

¹de·cree \di-ˈkrē\ *noun* [Middle English, from Middle French *decré,* from Latin *decretum,* from neuter of *decretus,* past participle of *decernere* to decide, from *de-* + *cernere* to sift, decide — more at CERTAIN] (14th century)
1 : an order usually having the force of law
2 a : a religious ordinance enacted by council or titular head **b :** a foreordaining will
3 a : a judicial decision of the Roman emperor **b :** a judicial decision especially in an equity or probate court

²decree *verb* **de·creed; de·cree·ing** (14th century)
transitive verb
1 : to command or enjoin by or as if by decree ⟨*decree* an amnesty⟩
2 : to determine or order judicially ⟨*decree* a punishment⟩
intransitive verb
: ORDAIN
— **de·cre·er** \-ˈkrē-ər\ *noun*

de·cree–law \di-ˈkrē-ˌlȯ\ *noun* (1926)
: a decree of a ruler or ministry having the force of a law enacted by the legislature

dec·re·ment \ˈde-krə-mənt\ *noun* [Latin *decrementum,* from *decrescere*] (1610)
1 : a gradual decrease in quality or quantity
2 a : the quantity lost by diminution or waste **b :** the amount of decrease (as of a variable)
— **dec·re·men·tal** \ˌde-krə-ˈmen-tᵊl\ *adjective*

de·crep·it \di-ˈkre-pət\ *adjective* [Middle English, from Middle French, from Latin *decrepitus*] (15th century)
1 : wasted and weakened by or as if by the infirmities of old age
2 a : impaired by use or wear : WORN-OUT **b :** fallen into ruin or disrepair
3 : DILAPIDATED, RUN-DOWN
synonym see WEAK
— **de·crep·it·ly** *adverb*

de·crep·i·tate \di-ˈkre-pə-ˌtāt\ *verb* [probably from (assumed) New Latin *decrepitatus,* past participle of *decrepitare,* from Latin *de-* + *crepitare* to crackle — more at CREPITATE] (1646)
transitive verb
: to roast or calcine (as salt) so as to cause crackling or until crackling stops
intransitive verb
: to become decrepitated
— **de·crep·i·ta·tion** \-ˌkre-pə-ˈtā-shən\ *noun*

de·crep·i·tude \di-ˈkre-pə-ˌtüd, -ˌtyüd\ *noun* (1603)
: the quality or state of being decrepit

¹de·cre·scen·do \ˌdā-krə-ˈshen-(ˌ)dō\ *noun, plural* **-dos** [Italian, literally, decreasing, from Latin *decrescendum,* gerund of *decrescere*] (circa 1880)
1 : a gradual decrease in volume of a musical passage
2 : a decrescendo musical passage

mark indicating decrescendo 2

²decrescendo *adverb or adjective* (circa 1890)
: with a decrease in volume — used as a direction in music

de·cres·cent \di-ˈkre-sᵊnt\ *adjective* [alteration of earlier *decressant,* probably from Anglo-French, present participle of (assumed) Anglo-French *decreistre* to decrease] (1610)
: becoming less by gradual diminution : DECREASING, WANING

de·cre·tal \di-ˈkrē-tᵊl, ˈde-kri-tᵊl\ *noun* [Middle English, from Middle French, from Late Latin *decretalis* of a decree, from Latin *decretum* decree] (14th century)
: DECREE; *especially* : a papal letter giving an authoritative decision on a point of canon law

de·cre·tive \-ˈkrē-tiv\ *adjective* (1609)
: having the force of a decree : DECRETORY

de·cre·to·ry \ˈde-krə-ˌtōr-ē, -ˌtȯr-; di-ˈkrē-tər-ē\ *adjective* (circa 1631)
: relating to or fixed by a decree or decision

de·crim·i·nal·ize \(ˌ)dē-ˈkri-mə-nə-ˌlīz, -ˈkrim-nəl-\ *transitive verb* (1969)
: to remove or reduce the criminal classification or status of; *especially* : to repeal a strict ban on while keeping under some form of regulation ⟨*decriminalize* the possession of marijuana⟩
— **de·crim·i·nal·iza·tion** \(ˌ)dē-ˌkri-mə-nə-lə-ˈzā-shən, -ˌkrim-nəl-\ *noun*

de·cry \di-ˈkrī, dē-\ *transitive verb* [French *décrier,* from Old French *descrier,* from *des-* de- + *crier* to cry] (1614)
1 : to depreciate (as a coin) officially or publicly
2 : to express strong disapproval of ⟨*decry* the emphasis on sex⟩ ☆
— **de·cri·er** \-ˈkrī(-ə)r\ *noun*

de·crypt \(ˌ)dē-ˈkript\ *transitive verb* [International Scientific Vocabulary *de-* + *crypto*gram, *crypto*graph] (1935)
: DECODE 1a
— **de·cryp·tion** \-ˈkrip-shən\ *noun*

de·cum·bent \di-ˈkəm-bənt, dē-\ *adjective* [Latin *decumbent-, decumbens,* present participle of *decumbere* to lie down, from *de-* + *-cumbere* to lie down; akin to Latin *cubare* to lie] (1656)
1 : lying down
2 *of a plant* **:** reclining on the ground but with ascending apex or extremity

dec·u·ple \ˈde-kyə-pəl\ *adjective* [French *décuple,* from Middle French, from Late Latin *decuplus,* from Latin *decem* ten + *-uplus* (as in *quadruplus* quadruple)] (1613)
1 : TENFOLD
2 : taken in groups of 10

de·cu·ri·on \di-ˈkyu̇r-ē-ən\ *noun* [Middle English *decurioun,* from Latin *decurion-, decurio,* from *decuria* division of ten, from *decem*] (14th century)
1 : a Roman cavalry officer in command of 10 men
2 : a member of a Roman senate

de·cur·rent \di-ˈkər-ənt, -ˈkə-rənt\ *adjective* [Latin *decurrent-, decurrens,* present participle of *decurrere* to run down, from *de-* + *currere* to run — more at CAR] (circa 1753)

☆ SYNONYMS
Decrease, lessen, diminish, reduce, abate, dwindle mean to grow or make less. DECREASE suggests a progressive decline in size, amount, numbers, or intensity ⟨slowly *decreased* the amount of pressure⟩. LESSEN suggests a decline in amount rather than in number ⟨has been unable to *lessen* her debt⟩. DIMINISH emphasizes a perceptible loss and implies its subtraction from a total ⟨his visual acuity has *diminished*⟩. REDUCE implies a bringing down or lowering ⟨you must *reduce* your caloric intake⟩. ABATE implies a reducing of something excessive or oppressive in force or amount ⟨the storm *abated*⟩. DWINDLE implies progressive lessening and is applied to things growing visibly smaller ⟨their provisions *dwindled* slowly⟩.

: running or extending downward along the stem ⟨*decurrent* leaves⟩

de·curved \(,)dē-ˈkərvd\ *adjective* [part translation of Late Latin *decurvatus,* from Latin *de-* + *curvatus* curved] (1835)
: curved downward : bent down

¹de·cus·sate \ˈde-kə-,sāt, di-ˈkə-,sāt\ *verb* **-sat·ed; -sat·ing** [Latin *decussatus,* past participle of *decussare* to arrange crosswise, from *decussis* the number ten, numeral X, intersection, from *decem* + *ass-, as* unit — more at ACE] (1658)
: INTERSECT, CROSS

²de·cus·sate \ˈde-kə-,sāt, di-ˈkə-sət\ *adjective* (circa 1823)
: arranged in pairs each at right angles to the next pair above or below ⟨*decussate* leaves⟩

de·cus·sa·tion \,de-kə-ˈsā-shən, ,dē-,kə-\ *noun* (circa 1656)
1 : the action of crossing (as of nerve fibers) especially in the form of an X
2 : a crossed tract of nerve fibers passing between centers on opposite sides of the nervous system

¹ded·i·cate \ˈde-di-kət\ *adjective* [Middle English, from Latin *dedicatus,* past participle of *dedicare* to dedicate, from *de-* + *dicare* to proclaim, dedicate — more at DICTION] (14th century)
: DEDICATED

²ded·i·cate \ˈde-di-,kāt *also* ˈde-,dē-\ *transitive verb* **-cat·ed; -cat·ing** (15th century)
1 : to devote to the worship of a divine being; *specifically* **:** to set apart (a church) to sacred uses with solemn rites
2 a : to set apart to a definite use ⟨money *dedicated* to their vacation fund⟩ **b :** to commit to a goal or way of life ⟨ready to *dedicate* his life to public service⟩
3 : to inscribe or address by way of compliment ⟨*dedicate* a book to a friend⟩
4 : to open to public use
synonym see DEVOTE
— ded·i·ca·tor \-,kā-tər\ *noun*

ded·i·cat·ed *adjective* (circa 1600)
1 : devoted to a cause, ideal, or purpose : ZEALOUS ⟨a *dedicated* scholar⟩
2 : given over to a particular purpose ⟨a *dedicated* process control computer⟩
— ded·i·cat·ed·ly *adverb*

ded·i·ca·tee \,de-di-kə-ˈtē\ *noun* (1760)
: one to whom a thing is dedicated

ded·i·ca·tion \,de-di-ˈkā-shən\ *noun* (14th century)
1 : an act or rite of dedicating to a divine being or to a sacred use
2 : a devoting or setting aside for a particular purpose
3 : a name and often a message prefixed to a literary, musical, or artistic production in tribute to a person or cause
4 : self-sacrificing devotion
— ded·i·ca·to·ry \ˈde-di-kə-,tōr-ē, -,tȯr-\ *adjective*

de·dif·fer·en·ti·a·tion \(,)dē-,di-fə-,ren-chē-ˈā-shən\ *noun* (1915)
: reversion of specialized structures (as cells) to a more generalized or primitive condition often as a preliminary to major physiological or structural change
— de·dif·fer·en·ti·ate \-ˈren-chē-,āt\ *intransitive verb*

de·duce \di-ˈdüs, dē-; *chiefly British* -ˈdyüs\ *transitive verb* **de·duced; de·duc·ing** [Middle English, from Latin *deducere,* literally, to lead away, from *de-* + *ducere* to lead — more at TOW] (15th century)
1 : to determine by deduction; *specifically* **:** to infer from a general principle
2 : to trace the course of
synonym see INFER
— de·duc·ible \-ˈd(y)ü-sə-bəl\ *adjective*

de·duct \di-ˈdəkt, dē-\ *transitive verb* [Latin *deductus,* past participle of *deducere*] (15th century)
1 : to take away (an amount) from a total : SUBTRACT
2 : DEDUCE, INFER

¹de·duct·ible \di-ˈdək-tə-bəl, dē-\ *adjective* (1856)
: allowable as a deduction
— de·duct·ibil·i·ty \-,dək-tə-ˈbi-lə-tē\ *noun*

²deductible *noun* (1929)
: a clause in an insurance policy that relieves the insurer of responsibility for an initial specified loss of the kind insured against; *also* **:** the amount of the loss specified in such a clause

de·duc·tion \di-ˈdək-shən, dē-\ *noun* (15th century)
1 a : an act of taking away ⟨*deduction* of legitimate business expenses⟩ **b :** something that is or may be subtracted ⟨*deductions* from his taxable income⟩
2 a : the deriving of a conclusion by reasoning; *specifically* **:** inference in which the conclusion about particulars follows necessarily from general or universal premises — compare INDUCTION **b :** a conclusion reached by logical deduction

de·duc·tive \di-ˈdək-tiv, dē-\ *adjective* (1665)
1 : of, relating to, or provable by deduction
2 : employing deduction in reasoning
— de·duc·tive·ly *adverb*

dee \ˈdē\ *noun* (13th century)
1 : the letter *d*
2 : something shaped like the letter D

¹deed \ˈdēd\ *noun* [Middle English *dede,* from Old English *dǣd;* akin to Old English *dōn* to do] (before 12th century)
1 : something that is done ⟨evil *deeds*⟩
2 : a usually illustrious act or action : FEAT, EXPLOIT
3 : the act of performing : ACTION ⟨righteous in word and in *deed*⟩
4 : a signed and usually sealed instrument containing some legal transfer, bargain, or contract
— deed·less \-ləs\ *adjective*

²deed *transitive verb* (1806)
: to convey or transfer by deed

deed poll \-ˈpōl\ *noun, plural* **deeds poll** [¹*deed* + *poll,* adjective (having the edges cut even rather than indented), from ²*poll*] (1588)
British **:** a deed (as to change one's name) made and executed by only one party

deedy \ˈdē-dē\ *adjective* **deed·i·er; -est** (1615)
dialect chiefly English **:** INDUSTRIOUS

dee·jay \ˈdē-,jā\ *noun* [*d*isc *j*ockey] (circa 1949)
: DISC JOCKEY

deem \ˈdēm\ *verb* [Middle English *demen,* from Old English *dēman;* akin to Old High German *tuomen* to judge, Old English *dōm* doom] (before 12th century)
transitive verb
: to come to think or judge : CONSIDER ⟨*deemed* it wise to go slow⟩
intransitive verb
: to have an opinion : BELIEVE

de–em·pha·size \(,)dē-ˈem(p)-fə-,sīz\ *transitive verb* (1938)
: to reduce in relative importance; *also* **:** PLAY DOWN
— de–em·pha·sis \-fə-səs\ *noun*

de–en·er·gize \,dē-ˈe-nər-,jīz\ *transitive verb* (1925)
: to disconnect from a source of electricity : shut off the power to

¹deep \ˈdēp\ *adjective* [Middle English *dep,* from Old English *dēop;* akin to Old High German *tiof* deep, Old English *dyppan* to dip — more at DIP] (before 12th century)
1 : extending far from some surface or area: as **a :** extending far downward ⟨a *deep* well⟩ **b** (1) **:** extending well inward from an outer surface ⟨a *deep* gash⟩ ⟨a *deep*-chested animal⟩ (2) **:** not located superficially within the body ⟨*deep* pressure receptors in muscles⟩ **c :** extending well back from a surface accepted as front ⟨a *deep* closet⟩ **d :** extending far laterally from the center ⟨*deep* borders of lace⟩ **e :** occurring or located near the outer limits of the playing area ⟨hit to *deep* right field⟩ **f :** thrown deep ⟨a *deep* pass⟩
2 : having a specified extension in an implied direction usually downward or backward ⟨a shelf 20 inches *deep*⟩ ⟨cars parked three-*deep*⟩
3 a : difficult to penetrate or comprehend : RECONDITE ⟨*deep* mathematical problems⟩ **b :** MYSTERIOUS, OBSCURE ⟨a *deep* dark secret⟩ **c :** grave in nature or effect ⟨in *deepest* disgrace⟩ **d :** of penetrating intellect : WISE ⟨a *deep* thinker⟩ **e :** INVOLVED, ENGROSSED ⟨*deep* in debt⟩ **f :** characterized by profundity of feeling or quality ⟨a *deep* sleep⟩; *also* **:** DEEP-SEATED ⟨*deep* religious beliefs⟩
4 a *of color* **:** high in saturation and low in lightness **b :** having a low musical pitch or pitch range ⟨a *deep* voice⟩
5 a : situated well within the boundaries ⟨a house *deep* in the woods⟩ **b :** remote in time or space **c :** being below the level of the conscious ⟨*deep* neuroses⟩ **d :** covered, enclosed, or filled to a specified degree — usually used in combination ⟨ankle-*deep* in mud⟩
6 : LARGE ⟨*deep* discounts⟩
7 : having many good players ⟨a *deep* bull pen⟩
synonym see BROAD
— deep·ly *adverb*
— deep·ness *noun*
— in deep water : in difficulty or distress

²deep *adverb* (before 12th century)
1 : to a great depth : DEEPLY ⟨still waters run *deep*⟩
2 : far on : LATE ⟨danced *deep* into the night⟩
3 a : near the outer limits of the playing area ⟨the shortstop was playing *deep*⟩ **b :** LONG 6

³deep *noun* (before 12th century)
1 a : a vast or immeasurable extent : ABYSS **b** (1) **:** the extent of surrounding space or time (2) **:** OCEAN
2 : any of the deep portions of a body of water; *specifically* **:** a generally long and narrow area in the ocean where the depth exceeds 3000 fathoms (5500 meters)
3 : the middle or most intense part ⟨the *deep* of winter⟩
4 : any of the fathom points on a sounding line other than the marks

deep–dish pie *noun* (1918)
: a pie usually with a fruit filling and no bottom crust that is baked in a deep dish

deep·en \ˈdē-pən\ *verb* **deep·ened; deep·en·ing** \ˈdē-pə-niŋ, ˈdēp-niŋ\ (1598)
transitive verb
: to make deep or deeper
intransitive verb
: to become deeper or more profound

deep fat *noun* (1921)
: hot fat or oil deep enough in a cooking utensil to cover the food to be fried

☆ SYNONYMS
Decry, depreciate, disparage, belittle mean to express a low opinion of. DECRY implies open condemnation with intent to discredit ⟨*decried* their defeatist attitude⟩. DEPRECIATE implies a representing as being of less value than commonly believed ⟨critics *depreciate* his plays for being unabashedly sentimental⟩. DISPARAGE implies depreciation by indirect means such as slighting or invidious comparison ⟨*disparaged* polo as a game for the rich⟩. BELITTLE usually suggests a contemptuous or envious attitude ⟨*belittled* the achievements of others⟩.

Deep·freeze \ˈdēp-ˌfrēz\ *trademark*
— used for a freezer for food storage

deep–freeze \ˈdēp-ˈfrēz\ *transitive verb* **-froze** \-ˈfrōz\; **-fro·zen** \-ˈfrō-zᵊn\ (1943)
1 : QUICK-FREEZE
2 : to store in a frozen state

deep freeze \ˈdēp-ˌfrēz\ *noun* (1948)
1 : COLD STORAGE 2 ⟨a bill presently in *deep freeze* awaiting a new congress —*Newsweek*⟩
2 : intense cold

deep–fry \ˈdēp-ˈfrī\ *transitive verb* (1922)
: to cook in deep fat

deep fryer *noun* (1950)
: a utensil suitable for deep-fat frying

deep pocket *noun* (1976)
1 : a person or an organization having substantial financial resources
2 *plural* **:** substantial financial resources

deep–root·ed \ˈdēp-ˈrü-təd, -ˈru̇-\ *adjective* (15th century)
: deeply implanted or established ⟨a *deep-rooted* loyalty⟩
synonym see INVETERATE

deep–sea \ˈdēp-ˈsē\ *adjective* (1626)
: of, relating to, or occurring in the deeper parts of the sea ⟨*deep-sea* fishing⟩

deep–seat·ed \ˈdēp-ˈsē-təd\ *adjective* (1741)
1 : situated far below the surface ⟨a *deep-seated* inflammation⟩
2 : firmly established ⟨a *deep-seated* tradition⟩
synonym see INVETERATE

deep–six \ˈdēp-ˈsiks\ *transitive verb* (1952)
1 *slang* **:** to throw away **:** DISCARD
2 *slang* **:** to throw overboard

deep six *noun* [from the leadsman's call *by the deep six* for a depth corresponding to the sixth deep on a sounding line] (1944)
slang **:** a place of disposal or abandonment — used especially in the phrase *give it the deep six*

deep–sky \ˈdēp-ˌskī\ *adjective* (1968)
: relating to or existing in space outside the solar system ⟨*deep-sky* objects⟩

deep space *noun* (circa 1952)
: space well outside the earth's atmosphere and especially that part lying beyond the earth-moon system

deep structure *noun* (1964)
: a formal representation of the underlying semantic content of a sentence; *also* **:** the structure which such a representation specifies

deep throat *noun, often D&T capitalized* [from the nickname given to such an informant in the Watergate scandal by Bob Woodward (born 1943) U.S. journalist, from the title of a pornographic film (1972)] (1973)
: an informant who divulges damaging information under cover of anonymity

deep·wa·ter \ˈdēp-ˌwȯ-tər, -ˌwä-\ *adjective* (1795)
: of, relating to, or characterized by water of considerable depth; *especially* **:** able to accommodate oceangoing vessels ⟨*deepwater* ports⟩

deer \ˈdir\ *noun, plural* **deer** *also* **deers** [Middle English, deer, animal, from Old English *dēor* beast; akin to Old High German *tior* wild animal, Lithuanian *dvasia* breath, spirit] (before 12th century)
1 *archaic* **:** ANIMAL; *especially* **:** a small mammal
2 : any of numerous ruminant mammals (family Cervidae, the deer family) having two large and two small hooves on each foot and antlers borne by the males of nearly all and by the females of a few forms ◆
— **deer·like** \-ˌlīk\ *adjective*

deer·ber·ry \-ˌber-ē\ *noun* (1814)
1 : either of two shrubs (*Vaccinium stamineum* or *V. caesium*) of dry woods and scrub of the eastern U.S.
2 : the edible fruit of a deerberry

deer·fly \ˈdir-ˌflī\ *noun* (1853)
: any of numerous small horseflies (as of the genus *Chrysops*) that include important vectors of tularemia

deer·hound \-ˌhau̇nd\ *noun* (1818)
: SCOTTISH DEERHOUND

deer mouse *noun* [from its agility] (1833)
: WHITE-FOOTED MOUSE; *especially* **:** a mouse (*Peromyscus maniculatus*) widely distributed in forests and grasslands of North America and Mexico

deer·skin \ˈdir-ˌskin\ *noun* (14th century)
: leather made from the skin of a deer; *also* **:** a garment of this leather

deer·stalk·er \-ˌstȯ-kər\ *noun* (1870)
: a close-fitting hat with a visor at the front and the back and with earflaps that may be worn up or down — called also *deerstalker cap, deerstalker hat*

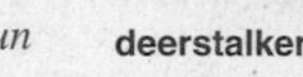
deerstalker

deer tick *noun* (1982)
: an ixodid tick (*Ixodes dammini*) that transmits the bacterium causing Lyme disease

deer·yard \ˈdir-ˌyärd\ *noun* (1849)
: a place where deer herd in winter

de–es·ca·late \(ˌ)dē-ˈes-kə-ˌlāt, ÷-kyə-\ (1964)
transitive verb
: LIMIT 2b
intransitive verb
: to decrease in extent, volume, or scope
— **de–es·ca·la·tion** \(ˌ)dē-ˌes-kə-ˈlā-shən, ÷-kyə-\ *noun*
— **de–es·ca·la·to·ry** \(ˌ)dē-ˈes-kə-lə-ˌtōr-ē, -ˌtȯr-, ÷-kyə-\ *adjective*

deet \ˈdēt\ *noun, often all capitalized* [probably from *d. e. t.*, from *di-* + *ethyl* + *toluamide* (C_8H_9NO)] (1962)
: a colorless oily liquid insect repellent $C_{12}H_{17}NO$

de·face \di-ˈfās, dē-\ *transitive verb* [Middle English, from Middle French *desfacier,* from Old French, from *des-* de- + *face* front, face] (14th century)
1 : to mar the external appearance of **:** injure by effacing significant details ⟨*deface* an inscription⟩
2 : IMPAIR
3 *obsolete* **:** DESTROY
— **de·face·ment** \-ˈfās-mənt\ *noun*
— **de·fac·er** *noun*

¹de fac·to \di-ˈfak-(ˌ)tō, dā-, dē-\ *adverb* [Medieval Latin, literally, from the fact] (1601)
: in reality **:** ACTUALLY

²de facto *adjective* (1696)
1 : ACTUAL; *especially* **:** being such in effect though not formally recognized ⟨a *de facto* state of war⟩
2 : exercising power as if legally constituted ⟨a *de facto* government⟩

de·fal·cate \di-ˈfal-ˌkāt, -ˈfȯl-, dē-; ˈde-fəl-\ *verb* **-cat·ed; -cat·ing** [Medieval Latin *defalcatus,* past participle of *defalcare,* from Latin *de-* + *falc-, falx* sickle] (1540)
transitive verb
archaic **:** DEDUCT, CURTAIL
intransitive verb
: to engage in embezzlement
— **de·fal·ca·tor** \-ˌkā-tər\ *noun*

de·fal·ca·tion \ˌdē-ˌfal-ˈkā-shən, ˌdē-ˌfȯl-, di-; ˌde-fəl-\ *noun* (15th century)
1 *archaic* **:** DEDUCTION
2 : the act or an instance of embezzling
3 : a failure to meet a promise or an expectation

def·a·ma·tion \ˌde-fə-ˈmā-shən\ *noun* (14th century)
: the act of defaming another **:** CALUMNY
— **de·fam·a·to·ry** \di-ˈfa-mə-ˌtōr-ē, dē-, -ˌtȯr-\ *adjective*

de·fame \di-ˈfām, dē-\ *transitive verb* **defamed; de·fam·ing** [Middle English, from Middle French & Medieval Latin; Middle French *defamer,* from Medieval Latin *defamare,* alteration of Latin *diffamare,* from *dis-* + *fama* reputation, fame] (14th century)
1 *archaic* **:** DISGRACE
2 : to harm the reputation of by libel or slander
3 *archaic* **:** ACCUSE
synonym see MALIGN
— **de·fam·er** *noun*

de·fang \(ˌ)dē-ˈfaŋ\ *transitive verb* (1953)
: to make harmless or less powerful

de·fat \(ˌ)dē-ˈfat\ *transitive verb* (1919)
: to remove fat from

¹de·fault \di-ˈfȯlt, dē-; ˈdē-ˌfȯlt\ *noun* [Middle English *defaute, defaulte,* from Old French *defaute,* from *defaillir* to be lacking, fail, from *de-* + *faillir* to fail] (13th century)
1 : failure to do something required by duty or law **:** NEGLECT
2 *archaic* **:** FAULT
3 : a failure to pay financial debts
4 a : failure to appear at the required time in a legal proceeding **b :** failure to compete in or to finish an appointed contest
5 : a selection automatically used by a computer program in the absence of a choice made by the user
— **in default of :** in the absence of

²default (15th century)
intransitive verb
: to fail to fulfill a contract, agreement, or duty: as **a :** to fail to meet a financial obligation **b :** to fail to appear in court **c :** to fail to compete in or to finish an appointed contest; *also* **:** to forfeit a contest by such failure
transitive verb
1 : to fail to perform, pay, or make good
2 a : FORFEIT **b :** to exclude (a player or a team) from a contest by default
— **de·fault·er** *noun*

de·fea·sance \di-ˈfē-zən(t)s\ *noun* [Middle English *defesance,* from Anglo-French, from Old French *deffesant,* present participle of *deffaire*] (15th century)
1 a (1) **:** the termination of a property interest in accordance with stipulated conditions (as in a deed) (2) **:** an instrument stating such conditions of limitation **b :** a rendering null or void
2 : DEFEAT, OVERTHROW

de·fea·si·ble \di-ˈfē-zə-bəl\ *adjective* (15th century)
: capable of being annulled or made void ⟨a *defeasible* claim⟩
— **de·fea·si·bil·i·ty** \-ˌfē-zə-ˈbi-lə-tē\ *noun*

¹de·feat \di-ˈfēt, dē-\ *transitive verb* [Middle English *deffeten,* from Middle French *deffait,* past participle of *deffaire* to destroy, from Old French *desfaire,* from Medieval Latin *disfacere,* from Latin *dis-* + *facere* to do — more at DO] (14th century)
1 *obsolete* **:** DESTROY
2 a : NULLIFY ⟨*defeat* an estate⟩ **b :** FRUSTRATE 2a(1) ⟨*defeat* a hope⟩
3 : to win victory over **:** BEAT ⟨*defeat* the opposing team⟩
synonym see CONQUER

◇ WORD HISTORY

deer In English the development of a word's meaning is sometimes from the general to the specific. For instance, *deer* is used in modern English to refer to several animals, including the white-tailed deer, reindeer, elk, and moose, that belong to the same zoological family. In Old English, however, *dēor* could refer to any animal, tame or wild, or to wild animals collectively. In time, *deer* came to be used only for wild animals that were hunted and then quite specifically for the red deer, which was once the primary object of the hunt in England. From that rather narrow usage, *deer* has expanded to the other members of its family mentioned above, thereby becoming somewhat more general again, though not as general as it once was.

²defeat *noun* (1590)
1 : frustration by nullification or by prevention of success ⟨the bill suffered *defeat* in the Senate⟩
2 *obsolete* **:** DESTRUCTION
3 a : an overthrow especially of an army in battle **b :** the loss of a contest

de·feat·ism \di-ˈfē-ˌti-zəm, dē-\ *noun* (1918)
: acceptance or expectation of or resignation to defeat
— de·feat·ist \-tist\ *noun or adjective*

de·fea·ture \di-ˈfē-chər, dē-\ *noun* [probably from *de-* + *feature*] (1590)
1 *archaic* **:** DISFIGUREMENT
2 *archaic* **:** DEFEAT

def·e·cate \ˈde-fi-ˌkāt\ *verb* **-cat·ed; -cat·ing** [Latin *defaecatus,* past participle of *defaecare,* from *de-* + *faec-, faex* dregs, lees] (1575)
transitive verb
1 : to free from impurity or corruption
2 : to discharge from the anus
intransitive verb
: to discharge feces from the bowels
— def·e·ca·tion \ˌde-fi-ˈkā-shən\ *noun*

¹de·fect \ˈdē-ˌfekt, di-ˈ\ *noun* [Middle English, from Middle French, from Latin *defectus* lack, from *deficere* to desert, fail, from *de-* + *facere* to do — more at DO] (15th century)
1 a : an imperfection that impairs worth or utility **:** SHORTCOMING ⟨the grave *defects* in our foreign policy⟩ **b :** an imperfection (as a vacancy or a foreign atom) in a crystal lattice
2 [Latin *defectus*] **:** a lack of something necessary for completeness, adequacy, or perfection **:** DEFICIENCY ⟨a hearing *defect*⟩

²de·fect \di-ˈfekt\ *intransitive verb* [Latin *defectus,* past participle of *deficere*] (1596)
1 : to forsake one cause, party, or nation for another often because of a change in ideology
2 : to leave one situation (as a job) often to go over to a rival ⟨the reporter *defected* to another network⟩
— de·fec·tor \-ˈfek-tər\ *noun*

de·fec·tion \di-ˈfek-shən\ *noun* (1552)
: conscious abandonment of allegiance or duty (as to a person, cause, or doctrine) **:** DESERTION

¹de·fec·tive \di-ˈfek-tiv\ *adjective* (14th century)
1 a : imperfect in form or function **:** FAULTY ⟨a *defective* pane of glass⟩ **b :** falling below the norm in structure or in mental or physical function ⟨*defective* eyesight⟩
2 : lacking one or more of the usual forms of grammatical inflection ⟨*must* is a *defective* verb⟩
— de·fec·tive·ly *adverb*
— de·fec·tive·ness *noun*

²defective *noun* (1881)
: a person who is subnormal physically or mentally

de·fem·i·nize \(ˌ)dē-ˈfe-mə-ˌnīz\ *transitive verb* (1907)
: to divest of feminine qualities or characteristics **:** MASCULINIZE
— de·fem·i·ni·za·tion \(ˌ)dē-ˌfe-mə-nə-ˈzā-shən\ *noun*

de·fence, de·fence·man *chiefly British variant of* DEFENSE, DEFENSEMAN

de·fend \di-ˈfend\ *verb* [Middle English, from Old French *defendre,* from Latin *defendere,* from *de-* + *-fendere* to strike; akin to Old English *gūth* battle, war, Greek *theinein* to strike] (14th century)
transitive verb
1 a : to drive danger or attack away from **b** (1) **:** to maintain or support in the face of argument or hostile criticism (2) **:** to prove (as a doctoral thesis) valid by answering questions in an oral exam **c :** to attempt to prevent an opponent from scoring at ⟨elects to *defend* the south goal⟩
2 *archaic* **:** PREVENT, FORBID
3 : to act as attorney for
4 : to deny or oppose the right of a plaintiff in regard to (a suit or a wrong charged) **:** CONTEST
5 : to seek to retain (as a title or position) against a challenge in a contest
intransitive verb
1 : to take action against attack or challenge ⟨couldn't fight back, could only *defend*⟩
2 : to play or be on defense ⟨playing deep to *defend* against a pass⟩
3 : to play against the high bidder in a card game ☆
— de·fend·able \ˈfen-də-bəl\ *adjective*

¹de·fen·dant \di-ˈfen-dənt, *in legal circles often* -ˌdant\ *noun* (14th century)
: a person required to make answer in a legal action or suit — compare PLAINTIFF

²defendant *adjective* (15th century)
: being on the defensive **:** DEFENDING

de·fend·er \di-ˈfen-dər\ *noun* (14th century)
1 : one that defends
2 : a player in a sport (as football) assigned to a defensive position

de·fen·es·tra·tion \(ˌ)dē-ˌfe-nə-ˈstrā-shən\ *noun* [*de-* + Latin *fenestra* window] (1620)
: a throwing of a person or thing out of a window
— de·fen·es·trate \(ˌ)dē-ˈfe-nə-ˌstrāt\ *transitive verb*

¹de·fense \di-ˈfen(t)s; *as antonym of "offense," often* ˈdē-ˌ\ *noun* [Middle English, from Middle French, from Late Latin *defensa* vengeance, from Latin, feminine of *defensus,* past participle of *defendere*] (14th century)
1 a : the act or action of defending ⟨the *defense* of our country⟩ ⟨speak out in *defense* of justice⟩ **b :** a defendant's denial, answer, or plea
2 a : capability of resisting attack **b :** defensive play or ability ⟨a player known for good *defense*⟩
3 a : means or method of defending or protecting oneself, one's team, or another; *also* **:** a defensive structure **b :** an argument in support or justification **c :** the collected facts and method adopted by a defendant to protect him self against a plaintiff's action **d :** a sequence of moves available in chess to the second player in the opening
4 a : a defending party or group (as in a court of law) ⟨the *defense* rests⟩ **b :** a defensive team
5 : the military and industrial aggregate that authorizes and supervises arms production ⟨appropriations for *defense*⟩ ⟨*defense* contract⟩
— de·fense·less \-ləs\ *adjective*
— de·fense·less·ly *adverb*
— de·fense·less·ness *noun*

²defense *transitive verb* **de·fensed; de·fens·ing** (1950)
: to take specific defensive action against (an opposing team or player or an offensive play)

de·fense·man \-mən, -ˌman\ *noun* (1895)
: a player in a sport (as hockey) who is assigned to a defensive zone or position

defense mechanism *noun* (1913)
1 : an often unconscious mental process (as repression) that makes possible compromise solutions to personal problems
2 : a defensive reaction by an organism

de·fen·si·ble \di-ˈfen(t) sə bəl\ *adjective* (14th century)
: capable of being defended
— de·fen·si·bil·i·ty \di-ˌfen(t)-sə-ˈbi-lə-tē, ˌdē-\ *noun*
— de·fen·si·bly \-blē\ *adverb*

¹de·fen·sive \di-ˈfen(t)-siv, ˈdē-ˌ\ *adjective* (14th century)
1 : serving to defend or protect
2 a : devoted to resisting or preventing aggression or attack **b :** of or relating to the attempt to keep an opponent from scoring in a game or contest
3 a : valuable in defensive play ⟨a *defensive* card in bridge⟩ **b :** designed to keep an opponent from being the highest bidder ⟨a *defensive* bid⟩
— de·fen·sive·ly *adverb*
— de·fen·sive·ness *noun*

²defensive *noun* (1601)
: a defensive position
— on the defensive : in the state or condition of being prepared for an expected aggression or attack

¹de·fer \di-ˈfər\ *transitive verb* **de·ferred; de·fer·ring** [Middle English *deferren, differren,* from Middle French *differer,* from Latin *differre* to postpone, be different — more at DIFFER] (14th century)
1 : PUT OFF, DELAY
2 : to postpone induction of (a person) into military service ☆
— de·fer·rer *noun*

²defer *verb* **deferred; deferring** [Middle English *deferren, differren,* from Middle French *deferer, defferer,* from Late Latin *deferre,* from Latin, to bring down, bring, from *de-* + *ferre* to carry — more at BEAR] (15th century)
transitive verb
: to delegate to another ⟨he could *defer* his job to no one —J. A. Michener⟩
intransitive verb
: to submit to another's wishes, opinion, or governance usually through deference or respect ⟨*deferred* to her father's wishes⟩
synonym see YIELD

def·er·ence \ˈde-fə-rən(t)s, ˈdef-rən(t)s\ *noun* (1660)
: respect and esteem due a superior or an elder; *also* **:** affected or ingratiating regard for another's wishes
synonym see HONOR
— in deference to : in consideration of

def·er·ent \ˈde-fə-rənt, ˈdef-rənt\ *adjective* [back-formation from *deference*] (1822)
: DEFERENTIAL

def·er·en·tial \ˌde-fə-ˈren-chəl\ *adjective* (1822)
: showing or expressing deference ⟨*deferential* attention⟩

☆ SYNONYMS
Defend, protect, shield, guard, safeguard mean to keep secure from danger or against attack. DEFEND denotes warding off actual or threatened attack ⟨*defend* the country⟩. PROTECT implies the use of something (as a covering) as a bar to the admission or impact of what may attack or injure ⟨a hard hat to *protect* your head⟩. SHIELD suggests protective intervention in imminent danger or actual attack ⟨*shielded* her eyes from the sun with her hand⟩. GUARD implies protecting with vigilance and force against expected danger ⟨White House entrances are well *guarded*⟩. SAFEGUARD implies taking precautionary protective measures against merely possible danger ⟨our civil liberties must be *safeguarded*⟩. See in addition MAINTAIN.

Defer, postpone, suspend, stay mean to delay an action or proceeding. DEFER implies a deliberate putting off to a later time ⟨*deferred* buying a car until spring⟩. POSTPONE implies an intentional deferring usually to a definite time ⟨the game is *postponed* until Saturday⟩. SUSPEND implies temporary stoppage with an added suggestion of waiting until some condition is satisfied ⟨business will be *suspended* while repairs are under way⟩. STAY often suggests the stopping or checking by an intervening agency or authority ⟨the governor *stayed* the execution⟩.

— **def·er·en·tial·ly** \-ˈren-chə-lē\ *adverb*
de·fer·ment \di-ˈfər-mənt\ *noun* (1612)
: the act of delaying or postponing; *specifically* **:** official postponement of military service
de·fer·ra·ble \di-ˈfər-ə-bəl\ *adjective* (1943)
: capable of or suitable or eligible for being deferred
— **deferrable** *noun*
de·fer·ral \di-ˈfər-əl\ *noun* (1895)
: the act of delaying **:** POSTPONEMENT
de·ferred *adjective* (1651)
1 : withheld for or until a stated time ⟨a *deferred* payment⟩
2 : charged in cases of delayed handling ⟨a *deferred* rate⟩
de·fer·ves·cence \ˌdē-(ˌ)fər-ˈve-sᵊn(t)s, ˌde-fər-\ *noun* [German *Deferveszenz,* from Latin *defervescent-, defervescens,* present participle of *defervescere* to stop boiling, from *de-* + *fervescere* to begin to boil — more at EFFERVESCE] (1866)
: the subsidence of a fever
de·fi·ance \di-ˈfī-ən(t)s, dē-\ *noun* (15th century)
1 : the act or an instance of defying **:** CHALLENGE
2 : disposition to resist **:** willingness to contend or fight
— **in defiance of :** contrary to **:** DESPITE
de·fi·ant \-ənt\ *adjective* [French *défiant,* from Old French, present participle of *defier* to defy] (circa 1837)
: full of defiance **:** BOLD
— **de·fi·ant·ly** *adverb*
de·fi·bril·la·tor \(ˌ)dē-ˈfi-brə-ˌlā-tər\ *noun* (1952)
: an electronic device that applies an electric shock to restore the rhythm of a fibrillating heart
— **de·fi·bril·late** \-ˌlāt\ *transitive verb*
— **de·fi·bril·la·tion** \-ˌfi-brə-ˈlā-shən\ *noun*
de·fi·brin·ate \(ˌ)dē-ˈfi-brə-ˌnāt, -ˈfī-\ *transitive verb* **-at·ed; -at·ing** (1845)
: to remove fibrin from (blood)
— **de·fi·brin·ation** \(ˌ)dē-ˌfi-brə-ˈnā-shən, -ˌfī-\ *noun*
de·fi·cien·cy \di-ˈfi-shən-sē\ *noun, plural* **-cies** (1634)
1 : the quality or state of being deficient **:** INADEQUACY
2 : an amount that is lacking or inadequate **:** SHORTAGE: as **a :** a shortage of substances necessary to health **b :** DELETION 2b(1)
deficiency disease *noun* (1912)
: a disease (as scurvy) caused by a lack of essential dietary elements and especially a vitamin or mineral
¹de·fi·cient \di-ˈfi-shənt\ *adjective* [Latin *deficient-, deficiens,* present participle of *deficere* to be wanting — more at DEFECT] (1581)
1 : lacking in some necessary quality or element ⟨*deficient* in judgment⟩
2 : not up to a normal standard or complement **:** DEFECTIVE ⟨*deficient* strength⟩
— **de·fi·cient·ly** *adverb*
²deficient *noun* (1906)
: one that is deficient ⟨a mental *deficient*⟩
def·i·cit \ˈde-fə-sət, *British also* di-ˈfis-ət *or* ˈdē-fə-sət\ *noun* [French *déficit,* from Latin *deficit* it is wanting, 3d singular present indicative of *deficere*] (1782)
1 a (1) **:** deficiency in amount or quality ⟨a *deficit* in rainfall⟩ (2) **:** a lack or impairment in a functional capacity ⟨cognitive *deficits*⟩ ⟨a hearing *deficit*⟩ **b :** DISADVANTAGE ⟨scored two runs to overcome a 2–1 *deficit*⟩
2 a : an excess of expenditure over revenue **b :** a loss in business operations
deficit spending *noun* (1938)
: the spending of public funds raised by borrowing rather than by taxation
de·fi·er \di-ˈfī(-ə)r\ *noun* (1585)
: one that defies
def·i·lade \ˈde-fə-ˌlād, -ˌläd\ *transitive verb* **-lad·ed; -lad·ing** [probably from *de-* + *-filade* (as in *enfilade*)] (1828)
: to arrange (fortifications) so as to protect the lines from frontal or enfilading fire and the interior from fire from above or behind
— **defilade** *noun*
¹de·file \di-ˈfī(ə)l, dē-\ *transitive verb* **de·filed; de·fil·ing** [Middle English, alteration (influenced by Old English *fȳlan* to defile) of *defoulen* to trample, defile, from Old French *defouler* to trample, from *de-* + *fouler* to trample, literally, to full — more at FULL] (14th century)
: to make unclean or impure: as **a :** to corrupt the purity or perfection of **:** DEBASE ⟨the countryside *defiled* by billboards⟩ **b :** to violate the chastity of **:** DEFLOWER **c :** to make physically unclean especially with something unpleasant or contaminating ⟨boots *defiled* with blood⟩ **d :** to violate the sanctity of **:** DESECRATE ⟨*defile* a sanctuary⟩ **e :** SULLY, DISHONOR
synonym see CONTAMINATE
— **de·file·ment** \-ˈfī(ə)l-mənt\ *noun*
— **de·fil·er** \-ˈfī-lər\ *noun*
²de·file \di-ˈfī(ə)l, ˈdē-ˌfīl\ *noun* [French *défilé,* from past participle of *défiler*] (1685)
: a narrow passage or gorge
³de·file \di-ˈfī(ə)l, ˈdē-ˌfīl\ *intransitive verb* **de·filed; de·fil·ing** [French *défiler,* from *dé-* de- + *filer* to move in a column — more at FILE] (1705)
: to march off in a line
de·fin·able \di-ˈfī-nə-bəl\ *adjective* (circa 1660)
1 : able to be defined
2 : able to be specified to have a particular function or operation ⟨*definable* keys⟩
— **de·fin·ably** \-ˈfī-nə-blē\ *adverb*
de·fine \di-ˈfīn\ *verb* **de·fined; de·fin·ing** [Middle English, from Middle French & Latin; Middle French *definer,* from Latin *definire,* from *de-* + *finire* to limit, end, from *finis* boundary, end] (14th century)
transitive verb
1 a : to determine or identify the essential qualities or meaning of ⟨whatever *defines* us as human⟩ **b :** to discover and set forth the meaning of (as a word) **c :** to create on a computer ⟨*define* a window⟩ ⟨*define* a procedure⟩
2 a : to fix or mark the limits of **:** DEMARCATE ⟨rigidly *defined* property lines⟩ **b :** to make distinct, clear, or detailed especially in outline ⟨the issues aren't too well *defined*⟩
3 : CHARACTERIZE, DISTINGUISH ⟨you *define* yourself by the choices you make —*Denison University Bulletin*⟩
intransitive verb
: to make a definition
— **de·fine·ment** \-ˈfīn-mənt\ *noun*
— **de·fin·er** \-ˈfī-nər\ *noun*
de·fin·i·en·dum \di-ˌfi-nē-ˈen-dəm\ *noun, plural* **-da** \-də\ [Latin, something to be defined, neuter of *definiendus,* gerundive of *definire*] (1871)
: an expression that is being defined
de·fin·i·ens \di-ˈfi-nē-ˌenz\ *noun, plural* **de·fin·i·en·tia** \di-ˌfi-nē-ˈen(t)-shē-ə\ [Latin, present participle of *definire*] (1871)
: an expression that defines **:** DEFINITION
def·i·nite \ˈde-fə-nit, ˈdef-nət\ *adjective* [Latin *definitus,* past participle of *definire*] (1553)
1 : having distinct or certain limits ⟨set *definite* standards for pupils to meet⟩
2 a : free of all ambiguity, uncertainty, or obscurity ⟨demanded a *definite* answer⟩ **b :** UNQUESTIONABLE, DECIDED ⟨the quarterback was a *definite* hero today⟩
3 : typically designating an identified or immediately identifiable person or thing ⟨the *definite* article *the*⟩
4 a : being constant in number, usually less than 20, and occurring in multiples of the petal number ⟨stamens *definite*⟩ **b :** CYMOSE ⟨a *definite* inflorescence⟩
synonym see EXPLICIT
— **def·i·nite·ly** *adverb*
— **def·i·nite·ness** *noun*
definite integral *noun* (1860)
: the difference between the values of the integral of a given function $f(x)$ for an upper value b and a lower value a of the independent variable x
def·i·ni·tion \ˌde-fə-ˈni-shən\ *noun* [Middle English *diffinicioun,* from Middle French *definition,* from Latin *definition-, definitio,* from *definire*] (14th century)
1 : an act of determining; *specifically* **:** the formal proclamation of a Roman Catholic dogma
2 a : a statement expressing the essential nature of something **b :** a statement of the meaning of a word or word group or a sign or symbol ⟨dictionary *definitions*⟩ **c :** a product of defining
3 : the action or process of defining
4 a : the action or the power of describing, explaining, or making definite and clear ⟨the *definition* of a telescope⟩ ⟨her comic genius is beyond *definition*⟩ **b** (1) **:** clarity of visual presentation **:** distinctness of outline or detail ⟨high-*definition* television⟩ (2) **:** clarity especially of musical sound in reproduction **c :** sharp demarcation of outlines or limits ⟨a jacket with distinct waist *definition*⟩
— **def·i·ni·tion·al** \-ˈni-shə-nᵊl\ *adjective*
¹de·fin·i·tive \di-ˈfi-nə-tiv\ *adjective* [Middle English *diffinityf,* from Middle French *definitif,* from Latin *definitivus,* from *definitus*] (14th century)
1 : serving to provide a final solution or to end a situation ⟨a *definitive* victory⟩
2 : authoritative and apparently exhaustive ⟨a *definitive* edition⟩
3 a : serving to define or specify precisely ⟨*definitive* laws⟩ **b :** serving as a perfect example **:** QUINTESSENTIAL ⟨a *definitive* bourgeois⟩
4 : fully differentiated or developed ⟨a *definitive* organ⟩
5 *of a postage stamp* **:** issued as a regular stamp for the country or territory in which it is to be used
synonym see CONCLUSIVE
— **de·fin·i·tive·ly** *adverb*
— **de·fin·i·tive·ness** *noun*
²definitive *noun* (1951)
: a definitive postage stamp — compare PROVISIONAL
definitive host *noun* (1901)
: the host in which the sexual reproduction of a parasite takes place — compare INTERMEDIATE HOST 1
de·fi·ni·tize \ˈde-fə-nə-ˌtīz, di-ˈfi-\ *transitive verb* **-tized; -tiz·ing** (1876)
: to make definite
de·fin·i·tude \di-ˈfi-nə-ˌtüd, -ˌtyüd\ *noun* [irregular from *definite*] (1836)
: PRECISION, DEFINITENESS
def·la·grate \ˈdef-lə-ˌgrāt\ *verb* **-grat·ed; -grat·ing** [Latin *deflagratus,* past participle of *deflagrare* to burn down, from *de-* + *flagrare* to burn — more at BLACK] (circa 1727)
transitive verb
: to cause to deflagrate — compare DETONATE 1
intransitive verb
: to burn rapidly with intense heat and sparks being given off
— **def·la·gra·tion** \ˌdef-lə-ˈgrā-shən\ *noun*
de·flate \di-ˈflāt, ˌdē-\ *verb* **de·flat·ed; de·flat·ing** [*de-* + *-flate* (as in *inflate*)] (1891)
transitive verb
1 : to release air or gas from
2 : to reduce in size, importance, or effectiveness ⟨*deflate* his ego with cutting remarks⟩
3 : to reduce (a price level) or cause (a volume of credit) to contract
intransitive verb
: to lose firmness through or as if through the escape of contained gas
synonym see CONTRACT
— **de·fla·tor** *also* **de·fla·ter** \-ˈflā-tər\ *noun*

de·fla·tion \di-ˈflā-shən, ˌdē-\ *noun* (1891)
1 : an act or instance of deflating **:** the state of being deflated
2 : a contraction in the volume of available money or credit that results in a general decline in prices
3 : the erosion of soil by the wind
— de·fla·tion·ary \-shə-ˌner-ē\ *adjective*

de·flect \di-ˈflekt, dē-\ *verb* [Latin *deflectere* to bend down, turn aside, from *de-* + *flectere* to bend] (circa 1555)
transitive verb
: to turn aside especially from a straight course or fixed direction
intransitive verb
: to turn aside **:** DEVIATE
— de·flect·able \-ˈflek-tə-bəl\ *adjective*
— de·flec·tive \-tiv\ *adjective*
— de·flec·tor \-tər\ *noun*

de·flec·tion \di-ˈflek-shən, dē-\ *noun* (1605)
1 : a turning aside or off course **:** DEVIATION
2 : the departure of an indicator or pointer from the zero reading on the scale of an instrument

de·flexed \ˈdē-ˌflekst, di-ˈ\ *adjective* [Latin *deflexus,* past participle of *deflectere*] (1826)
: turned abruptly downward ⟨a *deflexed* leaf⟩

de·flo·ra·tion \ˌdef-lə-ˈrā-shən, ˌdē-flȯ-\ *noun* [Middle English *defloracioun,* from Middle French & Late Latin; Middle French *defloracion,* from Late Latin *defloration-, defloratio,* from *deflorare*] (15th century)
: rupture of the hymen

de·flow·er \(ˌ)dē-ˈflau̇(-ə)r\ *transitive verb* [Middle English *deflouren,* from Middle French or Late Latin; Old French *desflorer,* from Late Latin *deflorare,* from Latin *de-* + *flor-, flos* flower — more at BLOW] (14th century)
1 : to deprive of virginity
2 : to take away the prime beauty of
— de·flow·er·er *noun*

de·fog \(ˌ)dē-ˈfȯg, -ˈfäg\ *transitive verb* (1904)
: to remove fog or condensed moisture from
— de·fog·ger *noun*

de·fo·li·ant \(ˌ)dē-ˈfō-lē-ənt\ *noun* (1943)
: a chemical spray or dust applied to plants in order to cause the leaves to drop off prematurely

de·fo·li·ate \-lē-ˌāt\ *transitive verb* [Late Latin *defoliatus,* past participle of *defoliare,* from Latin *de-* + *folium* leaf — more at BLADE] (1791)
: to deprive of leaves especially prematurely
— de·fo·li·a·tion \(ˌ)dē-ˌfō-lē-ˈā-shən\ *noun*
— de·fo·li·a·tor \(ˌ)dē-ˈfō-lē-ˌā-tər\ *noun*

de·force \(ˌ)dē-ˈfōrs, -ˈfȯrs\ *transitive verb* [Middle English, from Middle French *deforcier,* from *de-* + *forcier* to force] (15th century)
1 : to keep (as lands) by force from the rightful owner
2 : to eject (a person) from possession by force
— de·force·ment \-ˈfōr-smənt, -ˈfȯr-\ *noun*

de·for·es·ta·tion \(ˌ)dē-ˌfȯr-ə-ˈstā-shən, -ˌfär-\ *noun* (1874)
: the action or process of clearing of forests; *also* **:** the state of having been cleared of forests
— de·for·est \(ˈ)dē-ˈfȯr-əst, -ˈfär-\ *transitive verb*

de·form \di-ˈfȯrm, dē-\ *verb* [Middle English, from Old French or Latin; Old French *deformer,* from Latin *deformare,* from *de-* + *formare* to form, from *forma* form] (15th century)
transitive verb
1 : to spoil the form of
2 a : to spoil the looks of **:** DISFIGURE ⟨a face *deformed* by bitterness⟩ **b :** to mar the character of ⟨a marriage *deformed* by jealousy⟩
3 : to alter the shape of by stress
intransitive verb
: to become misshapen or changed in shape ☆
— de·form·able \-mə-bəl\ *adjective*

de·for·mal·ize \(ˌ)dē-ˈfȯr-mə-ˌlīz\ *transitive verb* (1880)
: to make less formal

de·for·ma·tion \ˌdē-ˌfȯr-ˈmā-shən, ˌde-fər-\ *noun* (15th century)
1 : alteration of form or shape; *also* **:** the product of such alteration
2 : the action of deforming **:** the state of being deformed
3 : change for the worse
— de·for·ma·tion·al \-shə-nᵊl\ *adjective*

de·for·ma·tive \di-ˈfȯr-mə-tiv, dē-\ *adjective* (1641)
: tending to deform

de·formed *adjective* (15th century)
: distorted or unshapely in form **:** MISSHAPEN

de·for·mi·ty \di-ˈfȯr-mə-tē, dē-\ *noun, plural* **-ties** [Middle English *deformite,* from Middle French *defformeteit,* from Latin *deformitat-, deformitas,* from *deformis* deformed, from *de-* + *forma*] (15th century)
1 : the state of being deformed
2 : IMPERFECTION, BLEMISH: as **a :** a physical blemish or distortion **:** DISFIGUREMENT **b :** a moral or aesthetic flaw or defect

de·fraud \di-ˈfrȯd, dē-\ *transitive verb* [Middle English, from Middle French *defrauder,* from Latin *defraudare,* from *de-* + *fraudare* to cheat, from *fraud-, fraus* fraud] (14th century)
: to deprive of something by deception or fraud
synonym see CHEAT
— de·fraud·er \di-ˈfrȯ-dər\ *noun*

de·fray \di-ˈfrā, dē-\ *transitive verb* [Middle French *deffrayer,* from *des-* de- + *frayer* to expend, from Old French, from (assumed) Old French *frai* expenditure, literally, damage by breaking, from Latin *fractum,* neuter of *fractus,* past participle of *frangere* to break — more at BREAK] (1536)
1 : to provide for the payment of **:** PAY
2 *archaic* **:** to bear the expenses of
— de·fray·able \-ə-bəl\ *adjective*
— de·fray·al \-ˈfrā(-ə)l\ *noun*

de·frock \(ˌ)dē-ˈfräk\ *transitive verb* (1581)
1 : to deprive (as a priest) of the right to exercise the functions of office
2 : to remove from a position of honor or privilege

de·frost \di-ˈfrȯst, ˈdē-ˌ\ (1895)
transitive verb
1 : to release from a frozen state ⟨*defrost* meat⟩
2 : to free from ice ⟨*defrost* the refrigerator⟩
intransitive verb
: to thaw out especially from a deep-frozen state
— de·frost·er *noun*

deft \ˈdeft\ *adjective* [Middle English *defte* gentle — more at DAFT] (15th century)
: characterized by facility and skill
synonym see DEXTEROUS
— deft·ly *adverb*
— deft·ness \ˈdef(t)-nəs\ *noun*

de·funct \di-ˈfəŋkt, dē-\ *adjective* [Latin *defunctus,* from past participle of *defungi* to finish, die, from *de-* + *fungi* to perform — more at FUNCTION] (1599)
: no longer living, existing, or functioning ⟨the committee is now *defunct*⟩
synonym see DEAD

de·fund \(ˌ)dē-ˈfənd\ *transitive verb* (1948)
: to withdraw funding from

de·fuse \(ˌ)dē-ˈfyüz\ *transitive verb* (1943)
1 : to remove the fuse from (as a mine or bomb)
2 : to make less harmful, potent, or tense ⟨*defuse* the crisis⟩

¹de·fy \di-ˈfī, dē-\ *transitive verb* **de·fied; de·fy·ing** [Middle English, to renounce faith in, challenge, from Middle French *defier,* from *de-* + *fier* to entrust, from (assumed) Vulgar Latin *fidare,* alteration of Latin *fidere* to trust — more at BIDE] (14th century)
1 *archaic* **:** to challenge to combat
2 : to challenge to do something considered impossible **:** DARE
3 : to confront with assured power of resistance **:** DISREGARD ⟨*defy* public opinion⟩
4 : to resist attempts at **:** WITHSTAND ⟨the paintings *defy* classification⟩

²de·fy \di-ˈfī, ˈdē-ˌ\ *noun, plural* **defies** (1580)
: CHALLENGE, DEFIANCE

dé·ga·gé \ˌdā-ˌgä-ˈzhā\ *adjective* [French, from past participle of *dégager* to put at ease, from Old French *desgagier* to redeem a pledge, free, from *des-* de- + *gage* pledge — more at GAGE] (1696)
1 : free of constraint **:** NONCHALANT
2 : being free and easy ⟨clothes with a *dégagé* look⟩
3 : extended with toe pointed in preparation for a ballet step

de·gas \(ˌ)dē-ˈgas\ *transitive verb* (1928)
: to remove gas from ⟨*degas* an electron tube⟩

de Gaull·ism \di-ˈgō-ˌli-zəm, -ˈgȯ-\ *noun* (1943)
: GAULLISM
— de Gaull·ist \-ləst\ *noun*

de·gauss \(ˌ)dē-ˈgau̇s\ *transitive verb* [*de-* + *gauss,* after Karl F. *Gauss*] (circa 1940)
: to remove or neutralize the magnetic field of ⟨*degauss* a ship⟩ ⟨*degauss* a magnetic tape⟩
— de·gauss·er *noun*

de·gen·er·a·cy \di-ˈjen-rə-sē, -ˈje-nə-, dē-\ *noun, plural* **-cies** (1664)
1 : the state of being degenerate
2 : the process of becoming degenerate
3 : sexual perversion
4 : the coding of an amino acid by more than one codon

¹de·gen·er·ate \di-ˈjen-rət, -ˈje-nə-, de-\ *adjective* [Middle English *degenerat,* from Latin *degeneratus,* past participle of *degenerare* to degenerate, from *de-* + *gener-, genus* race, kind — more at KIN] (15th century)
1 a : having declined (as in nature, character, structure, or function) from an ancestral or former state **b :** having sunk to a condition below that which is normal to a type; *especially* **:** having sunk to a lower and usually corrupt and vicious state **c :** DEGRADED 2
2 : being mathematically simpler (as by having a factor or constant equal to zero) than the typical case ⟨a *degenerate* hyperbola⟩
3 : characterized by atoms stripped of their electrons and by very great density ⟨*degenerate* matter⟩; *also* **:** consisting of degenerate matter ⟨a *degenerate* star⟩
4 : having two or more states or subdivisions ⟨*degenerate* energy level⟩
5 : having more than one codon representing an amino acid; *also* **:** being such a codon
synonym see VICIOUS
— de·gen·er·ate·ly *adverb*
— de·gen·er·ate·ness *noun*

☆ SYNONYMS
Deform, distort, contort, warp means to mar or spoil by or as if by twisting. DEFORM may imply a change of shape through stress, injury, or some accident of growth ⟨his face was *deformed* by hatred⟩. DISTORT and CONTORT both imply a wrenching from the natural, normal, or justly proportioned, but CONTORT suggests a more involved twisting and a more grotesque and painful result ⟨the odd camera angle *distorts* the figure in the photograph⟩ ⟨disease had painfully *contorted* her body⟩. WARP indicates physically an uneven shrinking that bends or twists out of a flat plane ⟨*warped* floorboards⟩.

[2]**de·gen·er·ate** \di-ˈje-nə-ˌrāt, dē-\ (1545)
intransitive verb
1 : to pass from a higher to a lower type or condition **:** DETERIORATE
2 : to sink into a low intellectual or moral state
3 : to decline in quality ⟨the poetry gradually *degenerates* into jingles⟩
4 : to decline from a condition or from the standards of a species, race, or breed
5 : to evolve or develop into a less autonomous or less functionally active form ⟨*degenerated* into dependent parasites⟩
transitive verb
: to cause to degenerate

[3]**de·gen·er·ate** \di-ˈjen-rət, -ˈje-nə-, dē-\ *noun* (1555)
: one that is degenerate: as **a :** one degraded from the normal moral standard **b :** a sexual pervert **c :** one showing signs of reversion to an earlier culture stage

de·gen·er·a·tion \di-ˌje-nə-ˈrā-shən, ˌdē-\ *noun* (15th century)
1 : degenerate condition
2 : a lowering of effective power, vitality, or essential quality to an enfeebled and worsened kind or state
3 : intellectual, moral, or artistic decline
4 a : progressive deterioration of physical characters from a level representing the norm of earlier generations or forms **b :** deterioration of a tissue or an organ in which its function is diminished or its structure is impaired
synonym see DETERIORATION

de·gen·er·a·tive \di-ˈje-nə-rə-tiv, -ˈjen-rə-; -ˈje-nə-ˌrā-; dē-\ *adjective* (circa 1846)
: of, relating to, involving, or causing degeneration ⟨a *degenerative* disease⟩

de·gla·ci·a·tion \(ˌ)dē-ˌglā-s(h)ē-ˈā-shən\ *noun* (1895)
: the melting of ice; *specifically* **:** the retreat of a glacier or ice sheet
— de·gla·ci·at·ed \-ˈglā-s(h)ē-ˌā-təd\ *adjective*

de·glam·or·ize \(ˌ)dē-ˈgla-mə-ˌrīz\ *transitive verb* (1938)
: to remove the glamor from
— de·glam·or·iza·tion \(ˌ)dē-ˌgla-mə-rə-ˈzā-shən\ *noun*

[1]**de·glaze** \(ˌ)dē-ˈglāz\ *transitive verb* (circa 1889)
: to remove the glaze from

[2]**deglaze** *transitive verb* [modification of French *déglacer,* literally, to melt the ice from, from *dé-* + *glacer* to freeze — more at GLACÉ] (1968)
: to dissolve the small particles of sautéed meat remaining in (a pan) by adding a liquid and heating

de·glu·ti·tion \ˌdē-glü-ˈti-shən, ˌde-glü-\ *noun* [French *déglutition,* from Latin *deglutire* to swallow down, from *de-* + *glutire, gluttire* to swallow — more at GLUTTON] (1650)
: the act or process of swallowing

de·grad·able \di-ˈgrā-də-bəl, dē-\ *adjective* (circa 1962)
: capable of being chemically degraded ⟨*degradable* detergents⟩ — compare BIODEGRADABLE

deg·ra·da·tion \ˌde-grə-ˈdā-shən\ *noun* (circa 1535)
1 : the act or process of degrading
2 a : decline to a low, destitute, or demoralized state **b :** moral or intellectual decadence **:** DEGENERATION
— deg·ra·da·tive \ˈde-grə-ˌdā-tiv\ *adjective*

de·grade \di-ˈgrād, dē-\ *verb* [Middle English, from Old French *degrader,* from Late Latin *degradare,* from Latin *de-* + *gradus* step, grade — more at GRADE] (14th century)
transitive verb
1 a : to lower in grade, rank, or status **:** DEMOTE **b :** to strip of rank or honors **c :** to lower to an inferior or less effective level **d :** to scale down in desirability or salability
2 a : to bring to low esteem or into disrepute **b :** to drag down in moral or intellectual character **:** CORRUPT
3 : to impair in respect to some physical property
4 : to wear down by erosion
5 : to reduce the complexity of (a chemical compound) **:** DECOMPOSE
intransitive verb
1 : to pass from a higher grade or class to a lower
2 *of a chemical compound* **:** to become reduced in complexity
— de·grad·er *noun*
— de·grad·ing·ly \-ˈgrā-diŋ-lē\ *adverb*

de·grad·ed *adjective* (1643)
1 : reduced far below ordinary standards of civilized life and conduct
2 : characterized by degeneration of structure or function
— de·grad·ed·ly *adverb*

de·gran·u·la·tion \(ˌ)dē-ˌgra-nyə-ˈlā-shən\ *noun* (circa 1941)
: the process of losing granules ⟨*degranulation* of leukocytes⟩

de·grease \(ˌ)dē-ˈgrēs, -ˈgrēz\ *transitive verb* (circa 1889)
: to remove grease from
— de·greas·er \-ˈgrē-sər, -zər\ *noun*

de·gree \di-ˈgrē\ *noun* [Middle English, from Old French *degré,* from (assumed) Vulgar Latin *degradus,* from Latin *de-* + *gradus*] (13th century)
1 : a step or stage in a process, course, or order of classification ⟨advanced by *degrees*⟩
2 a : a rank or grade of official, ecclesiastical, or social position ⟨people of low *degree*⟩ **b** *archaic* **:** a particular standing especially as to dignity or worth **c :** the civil condition or status of a person
3 : a step in a direct line of descent or in the line of ascent to a common ancestor
4 a *obsolete* **:** STEP, STAIR **b** *archaic* **:** a member of a series arranged in steps
5 : a measure of damage to tissue caused by injury or disease — compare FIRST-DEGREE BURN, SECOND-DEGREE BURN, THIRD-DEGREE BURN
6 a : the extent, measure, or scope of an action, condition, or relation ⟨different in *degree* but not in kind⟩ **b :** relative intensity ⟨a high *degree* of stress⟩ **c :** one of the forms or sets of forms used in the comparison of an adjective or adverb **d :** a legal measure of guilt or negligence ⟨found guilty of robbery in the first *degree*⟩
7 a : a title conferred on students by a college, university, or professional school on completion of a program of study **b :** a grade of membership attained in a ritualistic order or society **c :** an academic title conferred to honor distinguished achievement or service **d :** the formal ceremonies observed in the conferral of such a distinction
8 : a unit of measure for angles equal to an angle with its vertex at the center of a circle and its sides cutting off $1/360$ of the circumference; *also* **:** a unit of measure for arcs of a circle equal to the amount of arc that subtends a central angle of one degree
9 *archaic* **:** a position or space on the earth or in the heavens as measured by degrees of latitude
10 a : a step, note, or tone of a musical scale **b :** a line or space of the musical staff
11 : one of the divisions or intervals marked on a scale of a measuring instrument; *specifically* **:** any of various units for measuring temperature
12 a : the sum of the exponents of the variables in the term of highest degree in a polynomial, polynomial function, or polynomial equation **b :** the sum of the exponents of the variable factors of a monomial **c :** the greatest power of the derivative of highest order in a differential equation after the equation has been rationalized and cleared of fractions with respect to the derivative
— de·greed \-ˈgrēd\ *adjective*
— to a degree 1 : to a remarkable extent **2 :** in a small way

de·gree–day \di-ˈgrē-ˌdā\ *noun* (1832)
: a unit that represents one degree of difference from a given point (as 65°) in the mean daily outdoor temperature and that is used especially to measure heat requirements

degree of freedom (1867)
1 : any of a limited number of ways in which a body may move or in which a dynamic system may change
2 : one of the capabilities of a statistic for variation of which there are as many as the number of unrestricted and independent variables determining its value

de·gres·sive \di-ˈgre-siv, ˈdē-\ *adjective* [*de-* + *-gressive* (as in *progressive*)] (1886)
: tending to descend or decrease
— de·gres·sive·ly *adverb*

dé·grin·go·lade \ˌdā-ˌgraⁿ(ŋ)-gə-ˈläd\ *noun* [French, from *dégringoler* to tumble down, from Middle French *desgringueler,* from *des-* de- + *gringueler* to tumble, from Middle Dutch *crinkelen* to make curl, from *crinc, cring* ring, circle] (1883)
: a rapid decline or deterioration (as in strength, position, or condition) **:** DOWNFALL

de·gum \(ˌ)dē-ˈgəm\ *transitive verb* (1887)
: to free from gum, a gummy substance, or sericin

de·gus·ta·tion \ˌdē-gəs-ˈtā-shən, de-\ *noun* [French *dégustation,* from Latin *degustation-, degustatio,* from *degustare* to taste, from *de-* + *gustare* to taste — more at CHOOSE] (circa 1656)
: the action or an instance of tasting especially in a series of small portions
— de·gust \di-ˈgəst, dē-\ *transitive verb*

de haut en bas \də-ō-täⁿ-bä\ *adjective or adverb* [French, literally, from top to bottom] (1696)
: of superiority **:** of or with condescension

de·hisce \di-ˈhis\ *intransitive verb* **de·hisced; de·hisc·ing** [Latin *dehiscere* to split open, from *de-* + *hiscere* to gape; akin to Latin *hiare* to yawn — more at YAWN] (1657)
: to split along a natural line; *also* **:** to discharge contents by so splitting ⟨seedpods *dehiscing* at maturity⟩

de·his·cence \di-ˈhi-sᵊn(t)s\ *noun* [New Latin *dehiscentia,* from Latin *dehiscent-, dehiscens,* present participle of *dehiscere*] (circa 1828)
: an act or instance of dehiscing ⟨pollen freed by *dehiscence* of the anther⟩
— de·his·cent \-sᵊnt\ *adjective*

de·horn \(ˌ)dē-ˈhȯrn\ *transitive verb* (1888)
1 : to deprive of horns
2 : to prevent the growth of the horns of
— de·horn·er *noun*

de·hu·man·ize \(ˌ)dē-ˈhyü-mə-ˌnīz, (ˌ)dē-ˈyü-\ *transitive verb* (1818)
: to deprive of human qualities, personality, or spirit
— de·hu·man·iza·tion \(ˌ)dē-ˌhyü-mə-nə-ˈzā-shən, (ˌ)dē-ˌyü-\ *noun*

de·hu·mid·i·fy \ˌdē-hyü-ˈmi-də-ˌfī, ˌdē-yü-\ *transitive verb* (1927)
: to remove moisture from (as air)
— de·hu·mid·i·fi·ca·tion \-ˌmi-də-fə-ˈkā-shən\ *noun*
— de·hu·mid·i·fi·er \-ˈmi-də-ˌfī(-ə)r\ *noun*

de·hy·drate \(ˌ)dē-ˈhī-ˌdrāt\ (1876)
transitive verb
1 a : to remove bound water or hydrogen and oxygen from (a chemical compound) in the proportion in which they form water **b :** to remove water from (as foods)
2 : to deprive of vitality or savor
intransitive verb
: to lose water or body fluids
— de·hy·dra·tor \-ˌdrā-tər\ *noun*

de·hy·dra·tion \ˌdē-ˌhī-ˈdrā-shən\ *noun* (1854)
: the process of dehydrating; *especially* **:** an abnormal depletion of body fluids

de·hy·dro·chlo·ri·nase \(ˌ)dē-ˌhī-drə-ˈklōr-ə-ˌnās, -ˈklȯr-, -ˌnāz\ *noun* (1956)
: an enzyme that dehydrochlorinates a chlorinated hydrocarbon (as DDT) and is found especially in some DDT resistant insects

de·hy·dro·chlo·ri·na·tion \-ˌklōr-ə-ˈnā-shən, -ˌklȯr-\ *noun* (1936)
: the process of removing hydrogen and chlorine or hydrogen chloride from a compound
— de·hy·dro·chlo·ri·nate \-ˈklōr-ə-ˌnāt, -ˈklȯr-\ *transitive verb*

de·hy·dro·ge·nase \ˌdē-(ˌ)hī-ˈdrä-jə-ˌnās, (ˌ)dē-ˈhī-drə-jə-, -ˌnāz\ *noun* [International Scientific Vocabulary] (1923)
: an enzyme that accelerates the removal of hydrogen from metabolites and its transfer to other substances ⟨succinic *dehydrogenase*⟩

de·hy·dro·ge·na·tion \ˌdē-(ˌ)hī-ˌdrä-jə-ˈnā-shən, (ˌ)dē-ˌhī-drə-jə-\ *noun* (1866)
: the removal of hydrogen from a chemical compound
— de·hy·dro·ge·nate \ˌdē-(ˌ)hī-ˈdrä-jə-ˌnāt, (ˌ)dē-ˈhī-drə-jə-\ *transitive verb*

de·ice \(ˌ)dē-ˈīs\ *transitive verb* (1934)
: to rid or keep free of ice
— de·ic·er *noun*

de·i·cide \ˈdē-ə-ˌsīd, ˈdā-ə-\ *noun* [ultimately from Latin *deus* god + *-cidium, -cida* -cide — more at DEITY] (1611)
1 : the act of killing a divine being or a symbolic substitute of such a being
2 : the killer or destroyer of a god

deic·tic \ˈdīk-tik *also* ˈdāk-\ *adjective* [Greek *deiktikos* able to show, from *deiktos,* verbal of *deiknynai* to show — more at DICTION] (1876)
: showing or pointing out directly ⟨the words *this, that,* and *those* have a *deictic* function⟩

de·i·fi·ca·tion \ˌdē-ə-fə-ˈkā-shən, ˌdā-\ *noun* (14th century)
: the act or an instance of deifying

de·i·fy \ˈdē-ə-ˌfī, ˈdā-\ *transitive verb* **-fied; -fy·ing** [Middle English, from Middle French *deifier,* from Late Latin *deificare,* from Latin *deus* god + *-ficare* -fy] (14th century)
1 a : to make a god of **b :** to take as an object of worship
2 : to glorify as of supreme worth

deign \ˈdān\ *verb* [Middle English, from Old French *deignier,* from Latin *dignare, dignari,* from *dignus* worthy — more at DECENT] (14th century)
intransitive verb
: to condescend reluctantly and with a strong sense of the affront to one's superiority that is involved
transitive verb
: to condescend to give or offer

deil \ˈdē(ə)l\ *noun* [Middle English *devel, del*] (15th century)
Scottish **:** DEVIL

de·in·dus·tri·al·iza·tion \(ˌ)dē-in-ˌdəs-trē-ə-lə-ˈzā-shən\ *noun* (1940)
: the reduction or destruction of a nation's industrial capacity
— de·in·dus·tri·al·ize \(ˌ)dē-in-ˈdəs-trē-ə-ˌlīz\ *verb*

dei·non·y·chus \(ˌ)dī-ˈnä-ni-kəs\ *noun* [New Latin, from Greek *deinos* terrifying + *-onychos* -clawed (from *onych-, onyx* claw, nail) — more at DIRE, NAIL] (1969)
: any of a genus (*Deinonychus*) of small bipedal carnivorous dinosaurs from the Cretaceous

de·in·sti·tu·tion·al·iza·tion \(ˌ)dē-ˌin(t)-stə-ˌtü-shə-nə-lə-ˈzā-shən, -ˌtyü-\ *noun* (1955)
1 : the release of institutionalized individuals (as mental patients) from institutional care to care in the community
2 : the reform or modification of an institution to remove or disguise its institutional character
— de·in·sti·tu·tion·al·ize \-ˈtü-shə-nə-ˌlīz, -ˈtyü-\ *transitive verb*

de·ion·ize \(ˌ)dē-ˈī-ə-ˌnīz\ *transitive verb* (1906)
: to remove ions from ⟨*deionize* water by ion exchange⟩
— de·ion·iza·tion \(ˌ)dē-ˌī-ə-nə-ˈzā-shən\ *noun*
— de·ion·iz·er \-ˌnī-zər\ *noun*

de·ism \ˈdē-ˌi-zəm, ˈdā-\ *noun, often capitalized* (1682)
: a movement or system of thought advocating natural religion, emphasizing morality, and in the 18th century denying the interference of the Creator with the laws of the universe
— de·ist \ˈdē-ist, ˈdā-\ *noun, often capitalized*
— de·is·tic \dē-ˈis-tik, dā-\ *adjective*
— de·is·ti·cal \-ti-kəl\ *adjective*
— de·is·ti·cal·ly \-ti-k(ə-)lē\ *adverb*

de·i·ty \ˈdē-ə-tē, ˈdā-\ *noun, plural* **-ties** [Middle English *deitee,* from Old French *deité,* from Late Latin *deitat-, deitas,* from Latin *deus* god; akin to Old English *Tīw,* god of war, Latin *divus* god, *dies* day, Greek *dios* heavenly, Sanskrit *deva* heavenly, god] (14th century)
1 a : the rank or essential nature of a god **:** DIVINITY **b** *capitalized* **:** GOD 1, SUPREME BEING
2 : a god or goddess ⟨the *deities* of ancient Greece⟩
3 : one exalted or revered as supremely good or powerful

deix·is \ˈdīk-sis *also* ˈdāk-\ *noun* [Greek, literally, display, from *deiknynai* to show — more at DICTION] (1949)
: the pointing or specifying function of some words (as definite articles and demonstrative pronouns) whose denotation changes from one discourse to another

dé·jà vu \ˌdā-ˌzhä-ˈvü, dā-zhà-vᵫ\ *noun* [French, adjective, literally, already seen] (1903)
1 a : the illusion of remembering scenes and events when experienced for the first time **b :** a feeling that one has seen or heard something before
2 : something overly or unpleasantly familiar

¹de·ject \di-ˈjekt, dē-\ *adjective* (15th century)
archaic **:** DEJECTED

²deject *transitive verb* [Middle English, to throw down, from Latin *dejectus,* past participle of *deicere,* from *de-* + *jacere* to throw — more at JET] (1581)
: to make gloomy

de·jec·ta \di-ˈjek-tə, dē-\ *noun plural* [New Latin, from Latin, neuter plural of *dejectus*] (1887)
: FECES, EXCREMENT

de·ject·ed \di-ˈjek-təd, dē-\ *adjective* (1581)
1 : cast down in spirits **:** DEPRESSED
2 a *obsolete, of the eyes* **:** DOWNCAST **b** *archaic* **:** thrown down
3 *obsolete* **:** lowered in rank or condition
— de·ject·ed·ly *adverb*
— de·ject·ed·ness *noun*

de·jec·tion \di-ˈjek-shən, dē-\ *noun* (15th century)
: lowness of spirits

de ju·re \(ˌ)dē-ˈjür-ē, (ˌ)dā-ˈyür-\ *adverb or adjective* [Medieval Latin] (1611)
: by right **:** of right

deka- *or* **dek-** — see DECA-

deka·gram \ˈde-kə-ˌgram\ *noun* (circa 1879)
— see METRIC SYSTEM table

deka·li·ter \-ˌlē-tər\ *noun* (circa 1879)
— see METRIC SYSTEM table

deka·me·ter \-ˌmē-tər\ *noun* (circa 1879)
— see METRIC SYSTEM table

deka·met·ric \ˌde-kə-ˈme-trik\ *adjective* (1968)
: DECAMETRIC

de·lam·i·na·tion \(ˌ)dē-ˌla-mə-ˈnā-shən\ *noun* (1877)
1 : gastrula formation in which the endoderm is split off as a layer from the inner surface of the blastoderm and the archenteron is represented by the space between this endoderm and the yolk mass
2 : separation into constituent layers
— de·lam·i·nate \(ˌ)dē-ˈla-mə-ˌnāt\ *intransitive verb*

de·late \di-ˈlāt, dē-\ *transitive verb* **de·lat·ed; de·lat·ing** [Latin *delatus* (past participle of *deferre* to bring down, report, accuse), from *de-* + *latus,* past participle of *ferre* to bear — more at TOLERATE] (15th century)
1 : ACCUSE, DENOUNCE
2 : REPORT, RELATE
— de·la·tion \-ˈlā-shən\ *noun*
— de·la·tor \-ˈlā-tər\ *noun*

Del·a·ware \ˈde-lə-ˌwar, -ˌwer, -wər\ *noun, plural* **Delaware** *or* **Delawares** [*Delaware* River] (1721)
1 : a member of an American Indian people originally of the Delaware valley
2 : the Algonquian language of the Delaware

¹de·lay \di-ˈlā, dē-\ *noun* (13th century)
1 a : the act of delaying **:** the state of being delayed **b :** an instance of being delayed
2 : the time during which something is delayed

²delay *verb* [Middle English, from Old French *delaier,* from *de-* + *laier* to leave, perhaps alteration of *laissier,* from Latin *laxare* to slacken, from *laxus* loose — more at SLACK] (14th century)
transitive verb
1 : PUT OFF, POSTPONE
2 : to stop, detain, or hinder for a time
intransitive verb
: to move or act slowly; *also* **:** to cause delay
☆ ☆
— de·lay·er *noun*

¹de·le \ˈdē-(ˌ)lē\ *transitive verb* **de·led; de·le·ing** [Latin, imperative singular of *delēre*] (1705)
: to delete especially from typeset matter

☆ SYNONYMS

Delay, retard, slow, slacken, detain mean to cause to be late or behind in movement or progress. DELAY implies a holding back, usually by interference, from completion or arrival ⟨bad weather *delayed* our arrival⟩. RETARD suggests reduction of speed without actual stopping ⟨language barriers *retarded* their progress⟩. SLOW and SLACKEN also imply a reduction of speed, SLOW often suggesting deliberate intention ⟨medication *slowed* the patient's heart rate⟩, SLACKEN an easing up or relaxing of power or effort ⟨on hot days runners *slacken* their pace⟩. DETAIN implies a holding back beyond a reasonable or appointed time ⟨unexpected business had *detained* her⟩.

Delay, procrastinate, lag, loiter, dawdle, dally mean to move or act slowly so as to fall behind. DELAY usually implies a putting off (as a beginning or departure) ⟨we cannot *delay* any longer⟩. PROCRASTINATE implies blameworthy delay especially through laziness or apathy ⟨*procrastinates* about making decisions⟩. LAG implies failure to maintain a speed set by others ⟨*lagging* behind in technology⟩. LOITER and DAWDLE imply delay while in progress, especially in walking, but DAWDLE more clearly suggests an aimless wasting of time ⟨*loitered* at several store windows⟩ ⟨children *dawdling* on their way home from school⟩. DALLY suggests delay through trifling or vacillation when promptness is necessary ⟨stop *dallying* and get to work⟩.

[2]**dele** *noun* (circa 1751)
: a mark indicating that something is to be deled

[1]**de•lec•ta•ble** \di-ˈlek-tə-bəl\ *adjective* [Middle English, from Old French, from Latin *delectabilis,* from *delectare* to delight — more at DELIGHT] (15th century)
1 : highly pleasing **:** DELIGHTFUL
2 : DELICIOUS
— de•lec•ta•bil•i•ty \-ˌlek-tə-ˈbi-lə-tē\ *noun*
— de•lec•ta•bly \-blē\ *adverb*

[2]**delectable** *noun* (1921)
: something that is delectable

de•lec•ta•tion \ˌdē-ˌlek-ˈtā-shən, di-; ˌde-lək-\ *noun* (14th century)
: DELIGHT, ENJOYMENT

del•e•ga•ble \ˈde-li-gə-bəl\ *adjective* (1660)
: capable of being delegated

del•e•ga•cy \-gə-sē\ *noun, plural* **-cies** (15th century)
1 : a body of delegates **:** BOARD
2 a : the act of delegating **b :** appointment as delegate

[1]**del•e•gate** \ˈde-li-gət, -ˌgāt\ *noun* [Middle English *delegat,* from Medieval Latin *delegatus,* from Latin, past participle of *delegare* to delegate, from *de-* + *legare* to send — more at LEGATE] (15th century)
: a person acting for another: as **a :** a representative to a convention or conference **b :** a representative of a U.S. territory in the House of Representatives **c :** a member of the lower house of the legislature of Maryland, Virginia, or West Virginia

[2]**del•e•gate** \-ˌgāt\ *verb* **-gat•ed; -gat•ing** (1530)
transitive verb
1 : to entrust to another ⟨*delegate* authority⟩
2 : to appoint as one's representative
intransitive verb
: to assign responsibility or authority
— del•e•ga•tee \ˌde-li-gə-ˈtē\ *noun*
— del•e•ga•tor \ˈde-li-ˌgā-tər\ *noun*

del•e•ga•tion \ˌde-li-ˈgā-shən\ *noun* (1612)
1 : the act of empowering to act for another
2 : a group of persons chosen to represent others

de•le•git•i•ma•tion \ˌdē-lə-ˌji-tə-ˈmā-shən\ *noun* (1968)
: a decline in or loss of prestige or authority

de•lete \di-ˈlēt, dē-\ *transitive verb* **de•let•ed; de•let•ing** [Latin *deletus,* past participle of *delēre* to wipe out, destroy] (circa 1605)
: to eliminate especially by blotting out, cutting out, or erasing

del•e•te•ri•ous \ˌde-lə-ˈti-rē-əs\ *adjective* [Greek *dēlētērios,* from *dēleisthai* to hurt] (1643)
: harmful often in a subtle or unexpected way ⟨*deleterious* effects⟩ ⟨*deleterious* to health⟩
synonym see PERNICIOUS
— del•e•te•ri•ous•ly *adverb*
— del•e•te•ri•ous•ness *noun*

de•le•tion \di-ˈlē-shən, dē-\ *noun* [Latin *deletion-, deletio* destruction, from *delēre*] (1590)
1 : the act of deleting
2 a : something deleted **b** (1) **:** the absence of a section of genetic material from a chromosome (2) **:** the mutational process that results in a deletion

delft \ˈdelft\ *noun* [*Delft,* Netherlands] (1723)
1 : tin-glazed Dutch earthenware with blue and white or polychrome decoration
2 : a ceramic ware resembling or imitative of Dutch delft

delft•ware \ˈdelft-ˌwar, -ˌwer\ *noun* (1714)
: DELFT

deli \ˈde-lē\ *noun, plural* **del•is** (circa 1954)
: DELICATESSEN

[1]**de•lib•er•ate** \di-ˈli-bə-ˌrāt\ *verb* **-at•ed; -at•ing** (14th century)
intransitive verb
: to think about or discuss issues and decisions carefully
transitive verb
: to think about deliberately and often with formal discussion before reaching a decision
synonym see THINK

[2]**de•lib•er•ate** \di-ˈli-bə-rət, -ˈlib-rət\ *adjective* [Latin *deliberatus,* past participle of *deliberare* to consider carefully, perhaps alteration of (assumed) *delibrare,* from *de-* + *libra* scale, pound] (15th century)
1 : characterized by or resulting from careful and thorough consideration ⟨a *deliberate* decision⟩
2 : characterized by awareness of the consequences ⟨*deliberate* falsehood⟩
3 : slow, unhurried, and steady as though allowing time for decision on each individual action involved ⟨a *deliberate* pace⟩
synonym see VOLUNTARY
— de•lib•er•ate•ly *adverb*
— de•lib•er•ate•ness *noun*

de•lib•er•a•tion \di-ˌli-bə-ˈrā-shən\ *noun* (14th century)
1 a : the act of deliberating **b :** a discussion and consideration by a group of persons of the reasons for and against a measure
2 : the quality or state of being deliberate
— de•lib•er•a•tive \-ˈli-bə-ˌrāt-iv, -ˈli-b(ə-)rət-\ *adjective*
— de•lib•er•a•tive•ly *adverb*
— de•lib•er•a•tive•ness *noun*

del•i•ca•cy \ˈde-li-kə-sē\ *noun, plural* **-cies** (14th century)
1 *obsolete* **a :** the quality or state of being luxurious **b :** INDULGENCE
2 : something pleasing to eat that is considered rare or luxurious ⟨considered caviar a *delicacy*⟩
3 a : the quality or state of being dainty **:** FINENESS ⟨lace of great *delicacy*⟩ **b :** FRAILTY 1
4 : fineness or subtle expressiveness of touch (as in painting or music)
5 a : precise and refined perception and discrimination **b :** extreme sensitivity **:** PRECISION ⟨an electronic instrument of great *delicacy*⟩
6 a : refined sensibility in feeling or conduct **b :** the quality or state of being squeamish
7 : the quality or state of requiring delicate handling

[1]**del•i•cate** \ˈde-li-kət\ *adjective* [Middle English *delicat,* from Latin *delicatus* delicate, addicted to pleasure; akin to Latin *delicere* to allure] (14th century)
1 : pleasing to the senses: **a :** generally pleasant ⟨the climate's *delicate,* the air most sweet —Shakespeare⟩ **b :** pleasing to the sense of taste or smell especially in a mild or subtle way ⟨a *delicate* aroma⟩ ⟨a robust wine will dominate *delicate* dishes⟩ **c :** marked by daintiness or charm of color, lines, or proportions ⟨a *delicate* floral print⟩ ⟨an ample tear trilled down her *delicate* cheek —Shakespeare⟩ **d :** marked by fineness of structure, workmanship, or texture ⟨a *delicate* tracery⟩ ⟨a *delicate* lace⟩
2 a : marked by keen sensitivity or fine discrimination ⟨*delicate* insights⟩ ⟨a more *delicate* syntactic analysis —R. H. Robins⟩ **b :** FASTIDIOUS, SQUEAMISH ⟨a person of *delicate* tastes⟩
3 a : not robust in health or constitution **:** WEAK, SICKLY ⟨had been considered a *delicate* child⟩ **b :** easily torn or damaged **:** FRAGILE ⟨the *delicate* chain of life⟩
4 a : requiring careful handling: (1) **:** easily unsettled or upset ⟨a *delicate* balance⟩ ⟨the *delicate* relationships defined by the Constitution —*New Yorker*⟩ (2) **:** requiring skill or tact ⟨in a *delicate* position⟩ ⟨*delicate* negotiations⟩ ⟨a *delicate* operation⟩ (3) **:** involving matters of a deeply personal nature **:** SENSITIVE ⟨this is a *delicate* matter. Could I possibly speak to you alone —Daphne Du Maurier⟩ **b :** marked by care, skill, or tact ⟨*delicate* handling of a difficult situation⟩
5 : marked by great precision or sensitivity ⟨a *delicate* instrument⟩
synonym see CHOICE
— del•i•cate•ly *adverb*

[2]**delicate** *noun* (15th century)
: something delicate

del•i•ca•tes•sen \ˌde-li-kə-ˈte-sᵊn\ *noun plural* [obsolete German (now *Delikatessen*), plural of *Delicatesse* delicacy, from French *délicatesse,* probably from Old Italian *delicatezza,* from *delicato* delicate, from Latin *delicatus*] (1889)
1 : ready-to-eat food products (as cooked meats and prepared salads)
2 *singular, plural* **delicatessens** [*delicatessen (store)*] **:** a store where delicatessen are sold ◆

de•li•cious \di-ˈli-shəs\ *adjective* [Middle English, from Middle French, from Late Latin *deliciosus,* from Latin *deliciae* delights, from *delicere* to allure] (14th century)
1 : affording great pleasure **:** DELIGHTFUL ⟨*delicious* anecdotes⟩
2 : appealing to one of the bodily senses especially of taste or smell
— de•li•cious•ly *adverb*
— de•li•cious•ness *noun*

Delicious *noun, plural* **De•li•cious•es** *or* **Delicious** (circa 1903)
: an important red or yellow market apple of American origin that has a crown of five rounded prominences on the end opposite the stem

de•lict \di-ˈlikt, dē-\ *noun* [Latin *delictum* fault, from neuter of *delictus,* past participle of *delinquere*] (1523)
: an offense against the law

[1]**de•light** \di-ˈlīt, dē-\ *noun* (13th century)
1 : a high degree of gratification **:** JOY; *also* **:** extreme satisfaction
2 : something that gives great pleasure
3 *archaic* **:** the power of affording pleasure

[2]**delight** *verb* [Middle English *deliten,* from Old French *delitier,* from Latin *delectare,* frequentative of *delicere* to allure, from *de-* + *lacere* to allure] (13th century)
intransitive verb
1 : to take great pleasure ⟨*delighted* in playing the guitar⟩
2 : to give keen enjoyment ⟨a book certain to *delight*⟩
transitive verb
: to give joy or satisfaction to

◇ WORD HISTORY
delicatessen The word *delicatessen,* borrowed from a German plural noun meaning "delicacies," began to appear in U.S. publications toward the end of the 19th century, usually in combination with *store* or *shop.* These collocations are partial translations of German compound words such as *Delikatessenhandlung* or *Delikatessengeschäft.* By the 1920s the noun *store* or *shop* was dropped and the word *delicatessen* was used alone for the store. The delicatessen has become an American institution, especially in large urban centers such as New York City, where the *deli* (a clipping that first appeared in the 1950s) is typically a sparely furnished restaurant that serves traditional American Jewish fare such as pastrami and corned beef sandwiches, knishes, and matzo ball soup. More generally in the U.S., *deli* evokes a store or now more commonly a supermarket department that sells cold cuts and prepared foods. It is worth noting that, contrary to some popular notions, *delicatessen* has no etymological connection with the German verb *essen* "to eat."

— **de·light·er** *noun*

de·light·ed *adjective* (1603)
1 *obsolete* **:** DELIGHTFUL
2 : highly pleased
— **de·light·ed·ly** *adverb*
— **de·light·ed·ness** *noun*

de·light·ful \di-ˈlīt-fəl, dē-\ *adjective* (circa 1530)
: highly pleasing
— **de·light·ful·ly** \-fə-lē\ *adverb*
— **de·light·ful·ness** *noun*

de·light·some \-ˈlīt-səm\ *adjective* (1520)
: very pleasing **:** DELIGHTFUL

De·li·lah \di-ˈlī-lə\ *noun* [Hebrew *Dĕlīlāh*]
: the mistress and betrayer of Samson in the book of Judges

de·lim·it \di-ˈli-mət\ *transitive verb* [French *délimiter,* from Latin *delimitare,* from *de-* + *limitare* to limit, from *limit-, limes* boundary, limit] (1852)
: to fix or define the limits of
— **de·lim·i·ta·tion** \di-ˌli-mə-ˈtā-shən, ˌdē-\ *noun*

de·lim·it·er \di-ˈli-mə-tər\ *noun* (1960)
: a character that marks the beginning or end of a unit of data

de·lin·eate \di-ˈli-nē-ˌāt\ *transitive verb* **-eat·ed; -eat·ing** [Latin *delineatus,* past participle of *delineare,* from *de-* + *linea* line] (1559)
1 a : to indicate or represent by drawn or painted lines **b :** to mark the outline of ⟨lights *delineating* the narrow streets⟩
2 : to describe, portray, or set forth with accuracy or in detail ⟨*delineate* a character in the story⟩ ⟨*delineate* the steps to be taken by the government⟩
— **de·lin·ea·tor** \-ē-ˌā-tər\ *noun*

de·lin·ea·tion \di-ˌli-nē-ˈā-shən\ *noun* (1570)
1 : the act of delineating
2 : something made by delineating
— **de·lin·ea·tive** \-ˈli-nē-ˌā-tiv\ *adjective*

de·lin·quen·cy \di-ˈliŋ-kwən-sē, -ˈlin-\ *noun, plural* **-cies** (1636)
1 a : a delinquent act **b :** conduct that is out of accord with accepted behavior or the law; *especially* **:** JUVENILE DELINQUENCY
2 : a debt on which payment is overdue

¹de·lin·quent \-kwənt\ *noun* (15th century)
: a delinquent person

²delinquent *adjective* [Latin *delinquent-, delinquens,* present participle of *delinquere* to fail, offend, from *de-* + *linquere* to leave — more at LOAN] (1603)
1 : offending by neglect or violation of duty or of law
2 : being overdue in payment ⟨a *delinquent* charge account⟩
3 : of, relating to, or characteristic of delinquents **:** marked by delinquency
— **de·lin·quent·ly** *adverb*

del·i·quesce \ˌde-li-ˈkwes\ *intransitive verb* **-quesced; -quesc·ing** [Latin *deliquescere,* from *de-* + *liquescere,* inchoative of *liquēre* to be fluid — more at LIQUID] (1756)
1 : to dissolve or melt away
2 : to become soft or liquid with age — used of plant structures (as mushrooms)

del·i·ques·cent \-ˈkwe-sᵊnt\ *adjective* [Latin *deliquescent-, deliquescens,* present participle of *deliquescere*] (1791)
1 : tending to melt or dissolve; *especially* **:** tending to undergo gradual dissolution and liquefaction by the attraction and absorption of moisture from the air
2 : having repeated division into branches ⟨elms are *deliquescent* trees⟩ — compare EXCURRENT 2a
— **del·i·ques·cence** \-sᵊn(t)s\ *noun*

de·lir·i·ous \di-ˈlir-ē-əs\ *adjective* (1599)
1 : of, relating to, or characteristic of delirium
2 : affected with or marked by delirium
— **de·lir·i·ous·ly** *adverb*
— **de·lir·i·ous·ness** *noun*

de·lir·i·um \di-ˈlir-ē-əm\ *noun* [Latin, from *delirare* to be crazy, literally, to leave the furrow (in plowing), from *de-* + *lira* furrow — more at LEARN] (circa 1563)
1 : a mental disturbance characterized by confusion, disordered speech, and hallucinations
2 : frenzied excitement ⟨he would stride about his room in a *delirium* of joy —Thomas Wolfe⟩

delirium tre·mens \-ˈtrē-mənz, -ˈtre-\ *noun* [New Latin, literally, trembling delirium] (1865)
: a violent delirium with tremors that is induced by excessive and prolonged use of alcoholic liquors — called also *d.t.'s*

de·list \(ˌ)dē-ˈlist\ *transitive verb* (1933)
: to remove from a list; *especially* **:** to remove (a security) from the list of securities that may be dealt in on a particular exchange

de·liv·er \di-ˈli-vər, dē-\ *verb* **de·liv·ered; de·liv·er·ing** \-v(ə-)riŋ\ [Middle English, from Old French *delivrer,* from Late Latin *deliberare,* from Latin *de-* + *liberare* to liberate] (13th century)
transitive verb
1 : to set free ⟨and lead us not into temptation, but *deliver* us from evil —Matthew 6:13 (Authorized Version)⟩
2 a : to take and hand over to or leave for another **:** CONVEY ⟨*deliver* a package⟩ **b :** HAND OVER, SURRENDER ⟨*delivered* the prisoners to the sheriff⟩ ⟨*delivered* themselves over to God⟩
3 a (1) **:** to assist in giving birth (2) **:** to aid in the birth of **b :** to give birth to **c :** to cause (oneself) to produce as if by giving birth ⟨has *delivered* himself of half an autobiography —H. C. Schonberg⟩
4 : SPEAK, SING, UTTER ⟨*delivered* their lines with style⟩ ⟨*deliver* a song⟩ ⟨*deliver* a speech⟩
5 : to send (something aimed or guided) to an intended target or destination ⟨ability to *deliver* nuclear warheads⟩ ⟨*delivered* a fastball⟩
6 a : to bring (as votes) to the support of a candidate or cause **b :** to come through with **:** PRODUCE ⟨can *deliver* the goods⟩ ⟨new car *delivers* high gas mileage⟩
intransitive verb
: to produce the promised, desired, or expected results **:** COME THROUGH ⟨can't *deliver* on all these promises⟩
synonym see RESCUE
— **de·liv·er·abil·i·ty** \-ˌli-v(ə-)rə-ˈbi-lə-tē\ *noun*
— **de·liv·er·able** \-ˈli-v(ə-)rə-bəl\ *adjective*
— **de·liv·er·er** \-ˈli-vər-ər\ *noun*
— **deliver the goods :** to give results that are promised, expected, or desired

de·liv·er·ance \di-ˈli-v(ə-)rən(t)s, dē-\ *noun* (14th century)
1 : the act of delivering someone or something **:** the state of being delivered; *especially* **:** LIBERATION, RESCUE
2 : something delivered; *especially* **:** an opinion or decision (as the verdict of a jury) expressed publicly

de·liv·ery \di-ˈli-v(ə-)rē, dē-\ *noun, plural* **-er·ies** (15th century)
: the act of delivering something; *also* **:** something delivered

delivery boy *noun* (1920)
: a person employed by a retail store to deliver small orders to customers on call

de·liv·ery·man \-v(ə-)rē-mən, -ˌman\ *noun* (1920)
: a person who delivers wholesale or retail goods to customers usually over a regular local route

dell \ˈdel\ *noun* [Middle English *delle;* akin to Middle High German *telle* ravine, Old English *dæl* valley — more at DALE] (13th century)
: a secluded hollow or small valley usually covered with trees or turf

delly *variant of* DELI

Del·mon·i·co steak \del-ˈmä-ni-(ˌ)kō-\ *noun* [from the *Delmonico* restaurants, New York City, after Lorenzo Delmonico (died 1881) American restaurateur] (1925)
: CLUB STEAK — called also *Delmonico*

de·lo·cal·ize \(ˌ)dē-ˈlō-kə-ˌlīz\ *transitive verb* (1855)
: to free from the limitations of locality; *specifically* **:** to remove (a charge or charge carrier) from a particular position
— **de·lo·cal·iza·tion** \(ˌ)dē-ˌlō-kə-lə-ˈzā-shən\ *noun*

de·louse \(ˌ)dē-ˈlau̇s, -ˈlau̇z\ *transitive verb* (circa 1919)
: to remove lice from

Del·phi·an \ˈdel-fē-ən\ *adjective* (1625)
: DELPHIC

Del·phic \ˈdel-fik\ *adjective* (circa 1599)
1 : of or relating to ancient Delphi or its oracle
2 *often not capitalized* **:** AMBIGUOUS, OBSCURE ⟨*Delphic* utterances⟩ ◆
— **del·phi·cal·ly** \-fi-k(ə-)lē\ *adverb*

delphinium

del·phin·i·um \del-ˈfi-nē-əm\ *noun* [New Latin, genus name, from Greek *delphinion* larkspur, diminutive of *delphin-, delphis* dolphin; probably from the shape of the nectary] (1664)
: any of a large genus (*Delphinium*) of the buttercup family that comprises chiefly perennial erect branching herbs with palmately divided leaves and irregular flowers in showy spikes and includes several that are poisonous — compare LARKSPUR

Del·phi·nus \del-ˈfī-nəs, -ˈfē-\ *noun* [Latin (genitive *Delphini*), literally, dolphin, from Greek *delphin-, delphis*]
: a northern constellation nearly west of Pegasus

¹del·ta \ˈdel-tə\ *noun* [Middle English *deltha,* from Greek *delta,* of Semitic origin; akin to Hebrew *dāleth* daleth] (13th century)
1 : the 4th letter of the Greek alphabet — see ALPHABET table
2 : something shaped like a capital Greek delta; *especially* **:** the alluvial deposit at the mouth of a river
3 : an increment of a variable — symbol Δ
— **del·ta·ic** \del-ˈtā-ik\ *adjective*

◇ WORD HISTORY

Delphic *Delphic* alludes to the ancient Greek town of Delphoi (better known now in its Latin form *Delphi*). Situated on the southern slopes of Mount Parnassus, Delphi was the site of classical Greece's most important oracle. The oracular medium or priestess, known as a Pythia, would chew the leaves of the laurel, the tree sacred to Apollo, as vapors from the earth below engulfed her. While in her divine ecstasy, she would utter incoherent responses to questions posed in advance by those wishing to know what the future might hold or how the gods might be propitiated or evil averted. The Pythia's responses were interpreted by attending priests who recast them into verses that were often highly ambiguous or obscure—so much so that they allowed for just about any interpretation. This ambiguity is the source of a common modern meaning of English *Delphic.*

[2]**delta** *adjective* (circa 1929)
: fourth in position in the structure of an organic molecule from a particular group or atom — symbol δ

Delta (1952)
— a communications code word for the letter *d*

delta ray *noun* (1908)
: an electron ejected by an ionizing particle in its passage through matter

delta wave *noun* (1936)
: a high amplitude electrical rhythm of the brain with a frequency of less than 6 cycles per second that occurs especially in deep sleep, in infancy, and in many diseased conditions of the brain — called also *delta, delta rhythm*

delta wing *noun* [[1]*delta;* from its shape] (1946)
: a triangular swept-back airplane wing with a usually straight trailing edge

[1]**del·toid** \ˈdel-ˌtȯid\ *noun* [New Latin *deltoides,* from Greek *deltoeidēs* shaped like a delta, from *delta*] (circa 1681)
: a large triangular muscle that covers the shoulder joint and serves to raise the arm laterally

[2]**deltoid** *adjective* (circa 1753)
1 : having a triangular shape ⟨a *deltoid* leaf⟩ — see LEAF illustration
2 : relating to, associated with, or supplying the deltoid

del·toi·de·us \del-ˈtȯi-dē-əs\ *noun, plural* **del·toi·dei** \-ē-ˌī\ [New Latin, alteration of *deltoides*] (circa 1860)
: DELTOID

de·lude \di-ˈlüd, dē-\ *transitive verb* **de·lud·ed; de·lud·ing** [Middle English, from Latin *deludere,* from *de-* + *ludere* to play — more at LUDICROUS] (15th century)
1 : to mislead the mind or judgment of **:** DECEIVE, TRICK
2 *obsolete* **a :** FRUSTRATE, DISAPPOINT **b :** EVADE, ELUDE
synonym see DECEIVE
— **de·lud·er** *noun*

[1]**del·uge** \ˈdel-ˌyüj, -ˌyüzh; ÷də-ˈlüj, ˈdā-ˌlüj\ *noun* [Middle English, from Old French, from Latin *diluvium,* from *diluere* to wash away, from *dis-* + *lavere* to wash — more at LYE] (14th century)
1 a : an overflowing of the land by water **b :** a drenching rain
2 : an overwhelming amount or number

[2]**deluge** *transitive verb* **del·uged; del·ug·ing** (1593)
1 : to overflow with water **:** INUNDATE
2 : OVERWHELM, SWAMP

de·lu·sion \di-ˈlü-zhən, dē-\ *noun* [Middle English, from Late Latin *delusion-, delusio,* from *deludere*] (15th century)
1 a : the act of deluding **:** the state of being deluded **b :** an abnormal mental state characterized by the occurrence of psychotic delusions
2 a : something that is falsely or delusively believed or propagated **b :** a persistent false psychotic belief regarding the self or persons or objects outside the self
— **de·lu·sion·al** \-ˈlüzh-nəl, -ˈlü-zhə-nᵊl\ *adjective*
— **de·lu·sion·ary** \-zhə-ˌner-ē\ *adjective*

de·lu·sive \-ˈlü-siv, -ˈlü-ziv\ *adjective* (1605)
1 : likely to delude
2 : constituting a delusion
— **de·lu·sive·ly** *adverb*
— **de·lu·sive·ness** *noun*

de·lu·so·ry \-sə-rē, -zə-\ *adjective* (15th century)
: DECEPTIVE, DELUSIVE

de·lus·ter \(ˌ)dē-ˈləs-tər\ *transitive verb* (1926)
: to reduce the sheen of (as yarn or fabric)

de·luxe \di-ˈləks, dē- *also* -ˈlu̇ks, -ˈlüks\ *adjective* [French *de luxe,* literally, of luxury] (1819)
: notably luxurious, elegant, or expensive ⟨a *deluxe* edition⟩ ⟨*deluxe* hotels⟩

[1]**delve** \ˈdelv\ *verb* **delved; delv·ing** [Middle English, from Old English *delfan;* akin to Old High German *telban* to dig] (before 12th century)
transitive verb
archaic **:** EXCAVATE
intransitive verb
1 : to dig or labor with or as if with a spade
2 : to make a careful or detailed search for information ⟨*delved* into the past⟩
— **delv·er** *noun*

[2]**delve** *noun* (14th century)
archaic **:** CAVE, HOLLOW

de·mag·ne·tize \(ˌ)dē-ˈmag-nə-ˌtīz\ *transitive verb* (1839)
: to deprive of magnetic properties
— **de·mag·ne·ti·za·tion** \(ˌ)dē-ˌmag-nə-tə-ˈzā-shən\ *noun*
— **de·mag·ne·tiz·er** \(ˌ)dē-ˈmag-nə-ˌtī-zər\ *noun*

dem·a·gog·ic \ˌde-mə-ˈgä-gik *also* -ˈgä-jik *or* -ˈgō-jik\ *adjective* (1831)
: of, relating to, or characteristic of a demagogue **:** employing demagoguery
— **dem·a·gog·i·cal·ly** \-gi-k(ə-)lē, -ji-\ *adverb*

[1]**dem·a·gogue** *or* **dem·a·gog** \ˈde-mə-ˌgäg\ *noun* [Greek *dēmagōgos,* from *dēmos* people (perhaps akin to Greek *daiesthai* to divide) + *agōgos* leading, from *agein* to lead — more at TIDE, AGENT] (1648)
1 : a leader who makes use of popular prejudices and false claims and promises in order to gain power
2 : a leader championing the cause of the common people in ancient times
— **dem·a·gogu·ery** \-ˌgä-g(ə-)rē\ *noun*
— **dem·a·gogy** \-ˌgä-gē, -ˌgä-jē, -ˌgō-jē\ *noun*

[2]**demagogue** *or* **demagog** *verb* **-gogued** *or* **-goged; -gogu·ing** *or* **-gog·ing** (1656)
intransitive verb
: to behave like a demagogue
transitive verb
: to treat (as an issue) in a demogogic manner

[1]**de·mand** \di-ˈmand, -ˈmänd, dē-\ *noun* (13th century)
1 a : an act of demanding or asking especially with authority **b :** something claimed as due
2 *archaic* **:** QUESTION
3 a : willingness and ability to purchase a commodity or service **b :** the quantity of a commodity or service wanted at a specified price and time
4 a : a seeking or state of being sought after ⟨in great *demand* as an entertainer⟩ **b :** urgent need
5 : the requirement of work or of the expenditure of a resource ⟨equal to the *demands* of the office⟩ ⟨oxygen *demand* for waste oxidation⟩
— **on demand :** upon presentation and request for payment; *also* **:** when requested or needed

[2]**demand** *verb* [Middle English *demaunden,* from Old French *demander,* from Medieval Latin *demandare,* from Latin, to entrust, charge, from *de-* + *mandare* to enjoin — more at MANDATE] (14th century)
intransitive verb
: to make a demand **:** ASK
transitive verb
1 : to ask or call for with authority **:** claim as due or just ⟨*demanded* to see a lawyer⟩
2 : to call for urgently, peremptorily, or insistently ⟨*demanded* that the rioters disperse⟩
3 a : to ask authoritatively or earnestly to be informed of **b :** to require to come **:** SUMMON
4 : to call for as useful or necessary ☆
— **de·mand·able** \-ˈman-də-bəl\ *adjective*
— **de·mand·er** *noun*

de·man·dant \di-ˈman-dənt\ *noun* (15th century)
1 *archaic* **:** the plaintiff in a real action
2 *archaic* **:** one who makes a demand or claim

demand deposit *noun* (1923)
: a bank deposit that can be withdrawn without advance notice

de·mand·ing *adjective* (1926)
: requiring much time, effort, or attention **:** EXACTING
— **de·mand·ing·ly** \-ˈman-diŋ-lē\ *adverb*
— **de·mand·ing·ness** *noun*

demand loan *noun* (1913)
: CALL LOAN

demand note *noun* (1862)
: a note payable on demand

de·mand–pull \di-ˈman(d)-ˌpu̇l\ *noun* (1952)
: an increase or upward trend in spendable money that tends to result in increased competition for available goods and services and a corresponding increase in consumer prices — compare COST-PUSH
— **demand–pull** *adjective*

de·mand–side \di-ˈmand-ˌsīd\ *adjective* (1980)
: of, relating to, or being an economic theory that advocates use of government spending and growth in the money supply to stimulate the demand for goods and services and therefore expand economic activity — compare SUPPLY-SIDE

dem·an·toid \ˈde-mən-ˌtȯid\ *noun* [German, from obsolete German *Demant* diamond, from Middle High German *diemant,* from Old French *diamant* — more at DIAMOND] (circa 1890)
: a green garnet used as a gem

de·mar·cate \di-ˈmär-ˌkāt, ˈdē-ˌ\ *transitive verb* **-cat·ed; -cat·ing** [back-formation from *demarcation,* from Spanish *demarcación,* from *demarcar* to delimit, from *de-* + *marcar* to mark, probably from Italian *marcare,* of Germanic origin; akin to Old High German *marha* boundary — more at MARK] (1816)
1 : DELIMIT
2 : to set apart **:** SEPARATE
— **de·mar·ca·tion** \ˌdē-ˌmär-ˈkā-shən\ *noun*

dé·marche *or* **de·marche** \dā-ˈmärsh, di-ˈ, ˈdā-ˌ\ *noun* [French *démarche,* literally, gait, from Middle French, from *demarcher* to march, from Old French *demarchier,* from *de-* + *marchier* to march] (1658)
1 a : a course of action **:** MANEUVER **b :** a diplomatic or political initiative or maneuver
2 : a petition or protest presented through diplomatic channels

de·mark \di-ˈmärk\ *transitive verb* (1834)
: DEMARCATE

de·ma·te·ri·al·ize \ˌdē-mə-ˈtir-ē-ə-ˌlīz\ (circa 1864)
transitive verb

☆ SYNONYMS
Demand, claim, require, exact mean to ask or call for something as due or as necessary. DEMAND implies peremptoriness and insistence and often the right to make requests that are to be regarded as commands ⟨*demanded* payment of the debt⟩. CLAIM implies a demand for the delivery or concession of something due as one's own or one's right ⟨*claimed* the right to manage his own affairs⟩. REQUIRE suggests the imperativeness that arises from inner necessity, compulsion of law or regulation, or the exigencies of the situation ⟨the patient *requires* constant attention⟩. EXACT implies not only demanding but getting what one demands ⟨*exacts* absolute loyalty⟩.

: to cause to become or appear immaterial
intransitive verb
: to lose or appear to lose materiality
— **de·ma·te·ri·al·iza·tion** \-ˌtir-ē-ə-lə-ˈzā-shən\ *noun*

deme \ˈdēm\ *noun* [Greek *dēmos* people, deme] (1833)
1 : a unit of local government in ancient Attica
2 : a local population of closely related interbreeding organisms

[1]de·mean \di-ˈmēn\ *transitive verb* **de·meaned; de·mean·ing** [Middle English *demenen,* from Middle French *demener* to conduct, from *de-* + *mener* to lead, from Latin *minare* to drive, from *minari* to threaten — more at MOUNT] (14th century)
: to conduct or behave (oneself) usually in a proper manner

[2]demean *transitive verb* **de·meaned; de·mean·ing** [*de-* + [1]*mean*] (1601)
: to lower in character, status, or reputation

de·mean·or \di-ˈmē-nər\ *noun* [[1]*demean*] (circa 1485)
: behavior toward others **:** outward manner
synonym see BEARING

demeanour *British variant of* DEMEANOR

de·ment·ed \di-ˈmen-təd\ *adjective* (1644)
: MAD, INSANE
— **de·ment·ed·ly** *adverb*
— **de·ment·ed·ness** *noun*

de·men·tia \di-ˈmen(t)-shə, -shē-ə\ *noun* [Latin, from *dement-, demens* mad, from *de-* + *ment-, mens* mind — more at MIND] (1806)
1 : a condition of deteriorated mentality often with emotional apathy
2 : MADNESS, INSANITY ⟨a fanaticism bordering on *dementia*⟩
— **de·men·tial** \-shəl\ *adjective*

dementia prae·cox \-ˈprē-ˌkäks\ *noun* [New Latin, literally, premature dementia] (1899)
: SCHIZOPHRENIA

de·mer·it \di-ˈmer-ət, dē-\ *noun* [Middle English, from Middle French & Medieval Latin; Middle French *demerite,* from Medieval Latin *demeritum,* from neuter of *demeritus,* past participle of *demerēre* to be undeserving of, from Latin, to earn, from *de-* + *merēre* to merit] (15th century)
1 *obsolete* **:** OFFENSE
2 a : a quality that deserves blame or lacks merit **:** FAULT, DEFECT **b :** lack of merit
3 : a mark usually entailing a loss of privilege given to an offender

De·mer·ol \ˈde-mə-ˌrȯl, -ˌrōl\ *trademark*
— used for meperidine

de·mer·sal \di-ˈmər-səl\ *adjective* [Latin *demersus* (past participle of *demergere* to sink, from *de-* + *mergere* to dip, sink) + English [1]*-al* — more at MERGE] (1889)
: living near, deposited on, or sinking to the bottom of the sea ⟨*demersal* fish eggs⟩

de·mesne \di-ˈmān, -ˈmēn\ *noun* [Middle English, from Anglo-French, modification of Old French *demaine* — more at DOMAIN] (14th century)
1 : legal possession of land as one's own
2 : REALM 4b, DOMAIN
3 : manorial land actually possessed by the lord and not held by tenants
4 a : the land attached to a mansion **b :** landed property **:** ESTATE **c :** REGION 2, TERRITORY

De·me·ter \di-ˈmē-tər\ *noun* [Latin, from Greek *Dēmētēr*]
: the Greek goddess of agriculture — compare CERES

demi- *prefix* [Middle English, from *demi,* from Middle French, from (assumed) Vulgar Latin *dimedius,* modification of Latin *dimidius,* from *dis-* + *medius* mid — more at MID]
1 : half ⟨*demi*semiquaver⟩
2 : one that partly belongs to (a specified type or class) ⟨*demi*god⟩

demi·god \ˈde-mē-ˌgäd\ *noun* (1530)
1 : a mythological being with more power than a mortal but less than a god
2 : a person so outstanding as to seem to approach the divine

demi·god·dess \ˈde-mē-ˌgä-dəs\ *noun* (1603)
: a female demigod

demi·john \ˈde-mē-ˌjän\ *noun* [by folk etymology from French *dame-jeanne,* literally, Lady Jane] (1806)
: a large narrow-necked bottle usually enclosed in wickerwork

demijohn

de·mil·i·ta·rize \(ˌ)dē-ˈmi-lə-tə-ˌrīz, di-\ *transitive verb* (1883)
1 a : to do away with the military organization or potential of **b :** to prohibit (as a zone or frontier area) from being used for military purposes
2 : to rid of military characteristics or uses
— **de·mil·i·tar·iza·tion** \(ˌ)dē-ˌmi-lə-t(ə-)rə-ˈzā-shən, di-\ *noun*

demi·mon·daine \ˌde-mi-ˌmän-ˈdān, -ˈmän-ˌ, -mē-\ *noun* [French *demi-mondaine,* from feminine of *demi-mondain,* from *demi-monde*] (1894)
: a woman of the demimonde

demi·monde \ˈde-mi-ˌmänd, -mē-\ *noun* [French *demi-monde,* from *demi-* + *monde* world, from Latin *mundus*] (1855)
1 a : a class of women on the fringes of respectable society supported by wealthy lovers **b :** PROSTITUTES
2 : DEMIMONDAINE
3 : a distinctive class, group, or activity that is often an isolated part of a larger class, group, or activity ⟨the pop music *demimonde*⟩ ⟨the literary *demimonde*⟩; *especially* **:** one having little reputation or prestige

de·min·er·al·iza·tion \(ˌ)dē-ˌmi-nə-rə-lə-ˈzā-shən, di-\ *noun* (1903)
1 : loss of bodily minerals (as calcium salts) especially in disease
2 : the process of removing mineral matter or salts (as from water)
— **de·min·er·al·ize** \(ˌ)dē-ˈmi-nə-rə-ˌlīz\ *transitive verb*
— **de·min·er·al·iz·er** \-ˌlī-zər\ *noun*

demi·rep \ˈde-mi-ˌrep, -mē-\ *noun* [*demi-* + *rep* (reprobate)] (1749)
: DEMIMONDAINE

[1]de·mise \di-ˈmīz\ *verb* **de·mised; de·mis·ing** (15th century)
transitive verb
1 : to convey (as an estate) by will or lease
2 *obsolete* **:** CONVEY, GIVE
3 : to transmit by succession or inheritance
intransitive verb
1 : DIE, DECEASE
2 : to pass by descent or bequest ⟨the property has *demised* to the king⟩

[2]demise *noun* [Middle French, feminine of *demis,* past participle of *demettre* to dismiss, from Latin *demittere* to send down, from *de-* + *mittere* to send] (15th century)
1 : the conveyance of an estate
2 : transfer of the sovereignty to a successor
3 a : DEATH **b :** a cessation of existence or activity **c :** a loss of position or status

demi–sec \ˌde-mi-ˈsek, -mē-\ *adjective* [French] (1926)
of champagne **:** moderately sweet

demi·semi·qua·ver \ˌde-mē-ˈse-mē-ˌkwā-vər\ *noun* (circa 1706)
: THIRTY-SECOND NOTE

de·mis·sion \di-ˈmi-shən\ *noun* [Middle French, from Latin *demission-, demissio* lowering, from *demittere*] (15th century)
: RESIGNATION, ABDICATION

de·mit \di-ˈmit\ *verb* **de·mit·ted; de·mit·ting** [Middle French *demettre*] (15th century)
transitive verb
1 *archaic* **:** DISMISS
2 : RESIGN
intransitive verb
: to withdraw from office or membership

demi·tasse \ˈde-mi-ˌtas, -ˌtäs, -mē-\ *noun* [French *demi-tasse,* from *demi-* + *tasse* cup, from Middle French, from Arabic *ṭass,* from Persian *tast*] (1842)
: a small cup of black coffee; *also* **:** the cup used to serve it

demi·urge \ˈde-mē-ˌərj\ *noun* [Late Latin *demiurgus,* from Greek *dēmiourgos,* literally, artisan, one with special skill, from *dēmios* of the people (from *dēmos* people) + *-ourgos* worker (from *ergon* work) — more at DEMAGOGUE, WORK] (1678)
1 *capitalized* **a :** a Platonic subordinate deity who fashions the sensible world in the light of eternal ideas **b :** a Gnostic subordinate deity who is the creator of the material world
2 : one that is an autonomous creative force or decisive power
— **demi·ur·gic** \-jik\ *also* **demi·ur·gi·cal** \-ji-kəl\ *adjective*

demi·world \ˈde-mē-ˌwərld\ *noun* (1862)
: DEMIMONDE 3

demo \ˈde-(ˌ)mō\ *noun, plural* **dem·os** (1793)
1 *capitalized* **:** DEMOCRAT 2
2 a : DEMONSTRATION 1b **b** *British* **:** DEMONSTRATION 4
3 a : DEMONSTRATOR a **b :** a recording intended to show off a song or performer to a record producer

[1]de·mob \(ˌ)dē-ˈmäb, di-\ *transitive verb* (1919)
chiefly British **:** DEMOBILIZE

[2]demob *noun* (1945)
chiefly British **:** the act or process of demobilizing

de·mo·bi·lize \di-ˈmō-bə-ˌlīz, ˌdē-\ *transitive verb* (1882)
1 : DISBAND
2 : to discharge from military service
— **de·mo·bi·li·za·tion** \di-ˌmō-bə-lə-ˈzā-shən, ˌdē-\ *noun*

de·moc·ra·cy \di-ˈmä-krə-sē\ *noun, plural* **-cies** [Middle French *democratie,* from Late Latin *democratia,* from Greek *dēmokratia,* from *dēmos* + *-kratia* -cracy] (1576)
1 a : government by the people; *especially* **:** rule of the majority **b :** a government in which the supreme power is vested in the people and exercised by them directly or indirectly through a system of representation usually involving periodically held free elections
2 : a political unit that has a democratic government
3 *capitalized* **:** the principles and policies of the Democratic party in the U.S.
4 : the common people especially when constituting the source of political authority
5 : the absence of hereditary or arbitrary class distinctions or privileges

dem·o·crat \ˈde-mə-ˌkrat\ *noun* (1740)
1 a : an adherent of democracy **b :** one who practices social equality
2 *capitalized* **:** a member of the Democratic party of the U.S.

dem·o·crat·ic \ˌde-mə-ˈkra-tik\ *adjective* (1602)
1 : of, relating to, or favoring democracy
2 *often capitalized* **:** of or relating to one of the two major political parties in the U.S. evolving in the early 19th century from the antifederalists and the Democratic-Republican party and associated in modern times with policies of broad social reform and internationalism

3 : relating to, appealing to, or available to the broad masses of the people ⟨*democratic* art⟩
4 : favoring social equality **:** not snobbish
— **dem·o·crat·i·cal·ly** \-ti-k(ə-)lē\ *adverb*

democratic centralism *noun* (1926)
: a principle of Communist party organization by which members take part in policy discussions and elections at all levels but must follow decisions made at higher levels

Democratic–Republican *adjective* (1818)
: of or relating to a major American political party of the early 19th century favoring a strict interpretation of the constitution to restrict the powers of the federal government and emphasizing states' rights

de·moc·ra·tize \di-ˈmä-krə-ˌtīz\ *transitive verb* **-tized; -tiz·ing** (1798)
: to make democratic
— **de·moc·ra·ti·za·tion** \-ˌmä-krə-tə-ˈzā-shən\ *noun*
— **de·moc·ra·tiz·er** \-ˈmä-krə-ˌtī-zər\ *noun*

dé·mo·dé \ˌdā-mō-ˈdā\ *adjective* [French, from *dé-* de- + *mode*] (1873)
: no longer fashionable **:** OUT-OF-DATE

de·mod·ed \(ˌ)dē-ˈmō-dəd\ *adjective* (1887)
: DÉMODÉ

de·mod·u·late \(ˌ)dē-ˈmä-jə-ˌlāt\ *transitive verb* (1927)
: to extract the intelligence from (a modulated signal)
— **de·mod·u·la·tion** \(ˌ)dē-ˌmä-jə-ˈlā-shən\ *noun*
— **de·mod·u·la·tor** \(ˈ)dē-ˈmä-jə-ˌlā-tər\ *noun*

De·mo·gor·gon \ˌdē-mə-ˈgȯr-gən, ˈdē-mə-ˌ\ *noun* [Late Latin]
: a mysterious spirit or deity often explained as a primeval creator god who antedates the gods of Greek mythology

de·mo·graph·ic \ˌde-mə-ˈgra-fik, ˌdē-mə-\ *also* **de·mo·graph·i·cal** \-fi-kəl\ *adjective* (1882)
1 : of or relating to demography or demographics
2 : relating to the dynamic balance of a population especially with regard to density and capacity for expansion or decline
— **de·mo·graph·i·cal·ly** \-fi-k(ə-)lē\ *adverb*

de·mo·graph·ics \-fiks\ *noun plural* (circa 1966)
1 : the statistical characteristics of human populations (as age or income) used especially to identify markets
2 : the demographic profile of a market (as the viewers of a TV show)

de·mog·ra·phy \di-ˈmä-grə-fē\ *noun* [French *démographie,* from Greek *dēmos* people + French *-graphie* -graphy] (1880)
: the statistical study of human populations especially with reference to size and density, distribution, and vital statistics
— **de·mog·ra·pher** \-fər\ *noun*

dem·oi·selle \ˌdem-wə-ˈzel\ *noun* [French, from Old French *dameisele* — more at DAMSEL] (1520)
1 : a young lady
2 : DAMSELFLY
3 : DAMSELFISH

De·Moi·vre's theorem \di-ˈmȯi-vərz-, -ˈmwäv(-rə)z-\ *noun* [Abraham *De Moivre* (died 1754) French mathematician] (circa 1891)
: a theorem of complex numbers: the *n*th power of a complex number has for its absolute value and its argument respectively the *n*th power of the absolute value and *n* times the argument of the complex number

de·mol·ish \di-ˈmä-lish\ *transitive verb* [Middle French *demoliss-,* stem of *demolir,* from Latin *demoliri,* from *de-* + *moliri* to construct, from *moles* mass — more at MOLE] (1570)
1 a : TEAR DOWN, RAZE **b :** to break to pieces **:** SMASH
2 a : to do away with **:** DESTROY **b :** to strip of any pretense of merit or credence
— **de·mol·ish·er** *noun*
— **de·mol·ish·ment** \-lish-mənt\ *noun*

de·mo·li·tion \ˌde-mə-ˈli-shən, ˌdē-mə-\ *noun* (1549)
1 : the act of demolishing; *especially* **:** destruction in war by means of explosives
2 *plural* **:** explosives for destruction in war
— **de·mo·li·tion·ist** \-ˈli-sh(ə-)nəst\ *noun*

demolition derby *noun* (circa 1953)
: a contest in which skilled drivers ram old cars into one another until only one car remains running

de·mon *or* **dae·mon** \ˈdē-mən\ *noun* [Middle English *demon,* from Late Latin & Latin; Late Latin *daemon* evil spirit, from Latin, divinity, spirit, from Greek *daimōn,* probably from *daiesthai* to distribute — more at TIDE] (13th century)
1 a : an evil spirit **b :** a source or agent of evil, harm, distress, or ruin
2 *usually* *daemon* **:** an attendant power or spirit **:** GENIUS
3 *usually* *daemon* **:** a supernatural being of Greek mythology intermediate between gods and men
4 : one that has exceptional enthusiasm, drive, or effectiveness ⟨a *demon* for work⟩
— **de·mo·ni·an** \di-ˈmō-nē-ən\ *adjective*
— **de·mon·iza·tion** \ˌdē-mə-nə-ˈzā-shən\ *noun*
— **de·mon·ize** \ˈdē-mə-ˌnīz\ *transitive verb*

de·mon·e·tize \(ˌ)dē-ˈmä-nə-ˌtīz, -ˈmə-\ *transitive verb* [French *démonétiser,* from *dé-* de- + Latin *moneta* coin — more at MINT] (1852)
1 : to stop using (a metal) as a monetary standard
2 : to deprive of value for official payment
— **de·mon·e·ti·za·tion** \(ˌ)dē-ˌmä-nə-tə-ˈzā-shən, -ˌmə-\ *noun*

¹de·mo·ni·ac \di-ˈmō-nē-ˌak\ *also* **de·mo·ni·a·cal** \ˌdē-mə-ˈnī-ə-kəl\ *adjective* [Middle English *demoniak,* from Late Latin *daemoniacus,* from Greek *daimoniakos,* from *daimon-, daimōn*] (14th century)
1 : possessed or influenced by a demon
2 : of, relating to, or suggestive of a demon **:** FIENDISH ⟨*demoniac* cruelty⟩
— **de·mo·ni·a·cal·ly** \ˌdē-mə-ˈnī-ə-k(ə-)lē\ *adverb*

²demoniac *noun* (14th century)
: one possessed by a demon

de·mon·ic \di-ˈmä-nik, dē-\ *also* **de·mon·i·cal** \-ni-kəl\ *adjective* (1662)
: DEMONIAC 2
— **de·mon·i·cal·ly** \-ni-k(ə-)lē\ *adverb*

de·mon·ol·o·gy \ˌdē-mə-ˈnä-lə-jē\ *noun* (1597)
1 : the study of demons or evil spirits
2 : belief in demons **:** a doctrine of evil spirits
3 : a catalog of enemies ⟨the liberal creed at that time put Big Business in a central place in its *demonology* —Carl Kaysen⟩
— **de·mon·olog·i·cal** \ˌdē-mə-nə-ˈlä-ji-kəl\ *adjective*
— **de·mon·ol·o·gist** \-ˈnä-lə-jist\ *noun*

de·mon·stra·ble \di-ˈmän(t)-strə-bəl\ *adjective* (15th century)
1 : capable of being demonstrated
2 : APPARENT, EVIDENT
— **de·mon·stra·bil·i·ty** \-ˌmän(t)-strə-ˈbi-lə-tē\ *noun*
— **de·mon·stra·bly** \-blē\ *adverb*

dem·on·strate \ˈde-mən-ˌstrāt\ *verb* **-strat·ed; -strat·ing** [Latin *demonstratus,* past participle of *demonstrare,* from *de-* + *monstrare* to show — more at MUSTER] (1552)
transitive verb
1 : to show clearly
2 a : to prove or make clear by reasoning or evidence **b :** to illustrate and explain especially with many examples
3 : to show or prove the value or efficiency of to a prospective buyer
intransitive verb
: to make a demonstration
synonym see SHOW

dem·on·stra·tion \ˌde-mən-ˈstrā-shən\ *noun* (14th century)
1 : an act, process, or means of demonstrating to the intelligence: as **a** (1) **:** conclusive evidence **:** PROOF (2) **:** DERIVATION 5 **b :** a showing of the merits of a product or service to a prospective consumer
2 : an outward expression or display
3 : a show of armed force
4 : a public display of group feelings toward a person or cause
— **dem·on·stra·tion·al** \-shnəl, -shə-nᵊl\ *adjective*

¹de·mon·stra·tive \di-ˈmän(t)-strə-tiv\ *adjective* (14th century)
1 a : demonstrating as real or true **b :** characterized or established by demonstration
2 : pointing out the one referred to and distinguishing it from others of the same class ⟨*demonstrative* pronouns⟩
3 a : marked by display of feeling **b :** inclined to display feelings openly
— **de·mon·stra·tive·ly** *adverb*
— **de·mon·stra·tive·ness** *noun*

²demonstrative *noun* (15th century)
: a demonstrative word or morpheme

dem·on·stra·tor \ˈde-mən-ˌstrā-tər\ *noun* (1611)
: one that demonstrates: **a :** a product (as an automobile) used to demonstrate performance or merits to prospective buyers **b :** someone who engages in a public demonstration

de·mor·al·ize \di-ˈmȯr-ə-ˌlīz, ˌdē-, -ˈmär-\ *transitive verb* (circa 1793)
1 : to corrupt the morals of
2 a : to weaken the morale of **:** DISCOURAGE, DISPIRIT **b :** to upset or destroy the normal functioning of **c :** to throw into disorder
— **de·mor·al·iza·tion** \di-ˌmȯr-ə-lə-ˈzā-shən, ˌdē-, -ˌmär-\ *noun*
— **de·mor·al·iz·er** \di-ˈmȯr-ə-ˌlī-zər, ˌdē-, -ˈmär-\ *noun*
— **de·mor·al·iz·ing·ly** \-ziŋ-lē\ *adverb*

de·mos \ˈdē-ˌmäs\ *noun* [Greek *dēmos* — more at DEMAGOGUE] (1831)
1 : POPULACE
2 : the common people of an ancient Greek state

de·mote \di-ˈmōt, ˌdē-\ *transitive verb* **de·mot·ed; de·mot·ing** [*de-* + *-mote* (as in *promote*)] (circa 1891)
1 : to reduce to a lower grade or rank
2 : to relegate to a less important position
— **de·mo·tion** \-ˈmō-shən\ *noun*

de·mot·ic \di-ˈmä-tik\ *adjective* [Greek *dēmotikos,* from *dēmotēs* commoner, from *dēmos*] (1822)
1 : of, relating to, or written in a simplified form of the ancient Egyptian hieratic writing
2 : POPULAR, COMMON ⟨*demotic* idiom⟩
3 : of or relating to the form of Modern Greek that is based on everyday speech

de·mount \(ˌ)dē-ˈmau̇nt\ *transitive verb* (circa 1930)
1 : to remove from a mounted position
2 : DISASSEMBLE
— **de·mount·able** \-tə-bəl\ *adjective*

¹de·mul·cent \di-ˈməl-sənt\ *adjective* [Latin *demulcent-, demulcens,* present participle of *demulcēre* to soothe, from *de-* + *mulcēre* to soothe] (1732)
: SOOTHING

²demulcent *noun* (1732)
: a usually mucilaginous or oily substance (as tragacanth) that can soothe or protect an abraded mucous membrane

de·mul·ti·plex·er \(ˌ)dē-ˈməl-tə-ˌplek-sər\ *noun* (1963)
: an electronic device that separates a multiplex signal into its component parts

[1]**de·mur** \di-'mər\ *intransitive verb* **de·murred; de·mur·ring** [Middle English *demeoren* to linger, from Old French *demorer,* from Latin *demorari,* from *de-* + *morari* to linger, from *mora* delay — more at MORA] (13th century)
1 *archaic* **:** DELAY, HESITATE
2 : to file a demurrer
3 : to take exception **:** OBJECT — often used with *to* or *at*

[2]**demur** *noun* (13th century)
1 : hesitation (as in doing or accepting) usually based on doubt of the acceptability of something offered or proposed
2 : the act or an instance of objecting **:** PROTEST
synonym see QUALM

de·mure \di-'myu̇r\ *adjective* [Middle English] (14th century)
1 : RESERVED, MODEST
2 : affectedly modest, reserved, or serious **:** COY
— **de·mure·ly** *adverb*
— **de·mure·ness** *noun*

de·mur·rage \di-'mər-ij, -'mə-rij\ *noun* (1641)
1 : the detention of a ship by the freighter beyond the time allowed for loading, unloading, or sailing
2 : a charge for detaining a ship, freight car, or truck

de·mur·ral \di-'mər-əl, -'mə-rəl\ *noun* (1810)
: the act or an instance of demurring

[1]**de·mur·rer** \di-'mər-ər, -'mə-rər\ *noun* [Middle French *demorer,* verb] (circa 1521)
1 : a response in a court proceeding in which the defendant does not dispute the truth of the allegation but claims it is not sufficient grounds to justify legal action
2 : OBJECTION

[2]**de·mur·rer** \-'mər-ər\ *noun* ([1]*demur*) (1711)
: one that demurs

de·my·e·lin·at·ing \(ˌ)dē-'mī-ə-lə-ˌnā-tiŋ\ *adjective* (1939)
: causing or characterized by the loss or destruction of myelin ⟨*demyelinating* diseases⟩ ⟨a *demyelinating* agent⟩
— **de·my·e·lin·ation** \(ˌ)dē-ˌmī-ə-lə-'nā-shən\ *noun*

de·mys·ti·fy \(ˌ)dē-'mis-tə-ˌfī\ *transitive verb* (1963)
: to eliminate the mystifying features of
— **de·mys·ti·fi·ca·tion** \(ˌ)dē-ˌmis-tə-fə-'kā-shən\ *noun*

de·my·thol·o·gize \ˌdē-mi-'thä-lə-ˌjīz\ *transitive verb* (1950)
1 : to divest of mythological forms in order to uncover the meaning underlying them ⟨*demythologize* the Gospels⟩
2 : to divest of mythical elements or associations
— **de·my·thol·o·gi·za·tion** \-ˌthä-lə-jə-'zā-shən\ *noun*
— **de·my·thol·o·giz·er** \-'thä-lə-ˌjī-zər\ *noun*

[1]**den** \'den\ *noun* [Middle English, from Old English *denn;* akin to Old English *denu* valley, Old High German *tenni* threshing floor] (before 12th century)
1 : the lair of a wild usually predatory animal
2 a (1) **:** a hollow or cavern used especially as a hideout (2) **:** a center of secret activity **b :** a small usually squalid dwelling
3 : a comfortable usually secluded room
4 : a subdivision of a Cub Scout pack made up of two or more boys

[2]**den** *verb* **denned; den·ning** (13th century)
intransitive verb
: to live in or retire to a den
transitive verb
: to drive into a den

de·nar·i·us \di-'nar-ē-əs, -'ner-\ *noun, plural* **de·nar·ii** \-ē-ˌī, -ē-ˌē\ [Middle English, from Latin — more at DENIER] (14th century)
1 : a small silver coin of ancient Rome
2 : a gold coin of the Roman Empire equivalent to 25 denarii

de·na·tion·al·ize \(')dē-'nash-nə-ˌlīz, -'na-shə-nᵊl-ˌīz\ *transitive verb* (1807)
1 : to divest of national character or rights
2 : to remove from ownership or control by the national government
— **de·na·tion·al·iza·tion** \(ˌ)dē-ˌnash-nə-lə-'zā-shən, -ˌna-shə-nᵊl-ə-'zā-\ *noun*

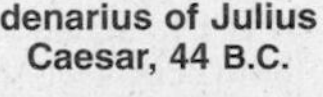
denarius of Julius Caesar, 44 B.C.

de·nat·u·ral·ize \(ˌ)dē-'na-ch(ə-)rə-ˌlīz\ *transitive verb* (1800)
1 : to make unnatural
2 : to deprive of the rights and duties of a citizen
— **de·nat·u·ral·iza·tion** \(ˌ)dē-ˌna-ch(ə-)rə-lə-'zā-shən\ *noun*

de·na·ture \(ˌ)dē-'nā-chər\ *verb* **de·na·tured; de·na·tur·ing** \-'nā-ch(ə-)riŋ\ (1685)
transitive verb
1 : DEHUMANIZE
2 : to deprive of natural qualities **:** change the nature of: as **a :** to make (alcohol) unfit for drinking (as by adding an obnoxious substance) without impairing usefulness for other purposes **b :** to modify the molecular structure of (as a protein or DNA) especially by heat, acid, alkali, or ultraviolet radiation so as to destroy or diminish some of the original properties and especially the specific biological activity
intransitive verb
: to become denatured
— **de·na·tur·ant** \(ˌ)dē-'nā-chər-ənt\ *noun*
— **de·na·tur·ation** \(ˌ)dē-ˌnā-chə-'rā-shən\ *noun*

de·na·zi·fy \(ˌ)dē-'nät-si-ˌfī, -'nat-\ *transitive verb* **-fied; -fy·ing** (1940)
: to rid of Nazism and its influence
— **de·na·zi·fi·ca·tion** \(ˌ)dē-ˌnät-si-fə-'kā-shən, -ˌnat-\ *noun*

dendr- *or* **dendro-** *combining form* [Greek, from *dendron;* akin to Greek *drys* tree — more at TREE]
: tree ⟨*dendrology*⟩ **:** resembling a tree ⟨*dendrite*⟩

den·dri·form \'den-drə-ˌfȯrm\ *adjective* (circa 1847)
: resembling a tree in structure

den·drite \'den-ˌdrīt\ *noun* (1751)
1 : a branching treelike figure produced on or in a mineral by a foreign mineral; *also* **:** the mineral so marked
2 : a crystallized arborescent form
3 : any of the usually branching protoplasmic processes that conduct impulses toward the body of a nerve cell

den·drit·ic \(ˌ)den-'dri-tik\ *adjective* (1816)
: resembling or having dendrites **:** branching like a tree ⟨a *dendritic* drainage system⟩ ⟨*dendritic* cells⟩

den·dro·chro·nol·o·gy \ˌden-(ˌ)drō-krə-'nä-lə-jē\ *noun* (circa 1928)
: the science of dating events and variations in environment in former periods by comparative study of growth rings in trees and aged wood
— **den·dro·chro·no·log·i·cal** \-ˌkrä-nə-'lä-ji-kəl, -ˌkrō-\ *adjective*
— **den·dro·chro·no·log·i·cal·ly** \-ji-k(ə-)lē\ *adverb*
— **den·dro·chro·nol·o·gist** \-krə-'nä-lə-jist\ *noun*

den·dro·gram \'den-drə-ˌgram\ *noun* (circa 1953)
: a branching diagram representing a hierarchy of categories based on degree of similarity or number of shared characteristics especially in biological taxonomy — compare CLADOGRAM

den·droid \'den-ˌdrȯid\ *adjective* [Greek *dendroeidēs,* from *dendron*] (circa 1828)
: resembling a tree in form **:** ARBORESCENT

den·drol·o·gy \den-'drä-lə-jē\ *noun* (circa 1708)
: the study of trees
— **den·dro·log·ic** \ˌden-drə-'lä-jik\ *or* **den·dro·log·i·cal** \-ji-kəl\ *adjective*
— **den·drol·o·gist** \den-'drä-lə-jist\ *noun*

dene \'dēn\ *noun* [Middle English, from Old English *denu*] (before 12th century)
British **:** VALLEY

Dé·né \de-'nā, dā-\ *noun, plural* **Déné** *or* **Dénés** \-'āz\ [Canadian French, from Athabascan; akin to Chipewyan *dene* person] (1891)
: a member of any of the Athabascan-speaking peoples of the interior of Alaska and northwestern Canada; *also* **:** the languages of these peoples

Den·eb \'de-ˌneb, -nəb\ *noun* [Arabic *dhanab al-dajāja,* literally, the tail of the hen]
: a star of the first magnitude in Cygnus

den·e·ga·tion \ˌde-ni-'gā-shən\ *noun* [Middle English *denegacioun,* from Middle French or Latin; Middle French *denegation,* from Latin *denegation-, denegatio,* from *denegare* to deny — more at DENY] (15th century)
: DENIAL

de·ner·vate \'dē-(ˌ)nər-ˌvāt\ *transitive verb* **-vat·ed; -vat·ing** (1905)
: to deprive of a nerve supply (as by cutting a nerve)
— **de·ner·va·tion** \ˌdē-(ˌ)nər-'vā-shən\ *noun*

den·gue \'deŋ-gē, -ˌgā\ *noun* [American Spanish] (1828)
: an acute infectious disease caused by an arbovirus, transmitted by aedes mosquitoes, and characterized by headache, severe joint pain, and a rash — called also *breakbone fever, dengue fever*

de·ni·abil·i·ty \dē-ˌnī-ə-'bi-lə-tē\ *noun* (1973)
: the ability to deny something especially on the basis of being officially uninformed

de·ni·able \di-'nī-ə-bəl, de-\ *adjective* (1548)
: capable of being denied

de·ni·al \di-'nī(-ə)l, dē-\ *noun* (1528)
1 : refusal to satisfy a request or desire
2 a (1) **:** refusal to admit the truth or reality (as of a statement or charge) (2) **:** assertion that an allegation is false **b :** refusal to acknowledge a person or a thing **:** DISAVOWAL
3 : the opposing by the defendant of an allegation of the opposite party in a lawsuit
4 : SELF-DENIAL
5 : negation in logic
6 : a psychological defense mechanism in which confrontation with a personal problem or with reality is avoided by denying the existence of the problem or reality

[1]**de·ni·er** \di-'nī(-ə)r, dē-\ *noun* (15th century)
: one that denies

[2]**de·nier** *noun* [Middle English *denere,* from Middle French *denier,* from Latin *denarius,* coin worth ten asses, from *denarius* containing ten, from *deni* ten each, from *decem* ten — more at TEN] (15th century)
1 \də-'ni(ə)r, də-'nyā\ **:** a small originally silver coin formerly used in western Europe
2 \'de-nyər\ **:** a unit of fineness for yarn equal to the fineness of a yarn weighing one gram for each 9000 meters ⟨100-*denier* yarn is finer than 150 *denier* yarn⟩

den·i·grate \'de-ni-ˌgrāt\ *transitive verb* **-grat·ed; -grat·ing** [Latin *denigratus,* past participle of *denigrare,* from *de-* + *nigrare* to blacken, from *nigr-, niger* black] (1526)
1 : to cast aspersions on **:** DEFAME
2 : to deny the importance or validity of **:** BELITTLE
— **den·i·gra·tion** \ˌde-ni-'grā-shən\ *noun*
— **den·i·gra·tive** \'de-ni-ˌgrā-tiv\ *adjective*
— **den·i·gra·tor** \-ˌgrā-tər\ *noun*

— **de·ni·gra·to·ry** \'de-ni-grə-ˌtōr-ē, -ˌtȯr-\ *adjective*

den·im \'de-nəm\ *noun* [French (*serge*) *de Nîmes* serge of Nîmes, France] (1695)
1 a : a firm durable twilled usually cotton fabric woven with colored warp and white filling threads **b :** a similar fabric woven in colored stripes
2 *plural* **:** overalls or trousers usually of blue denim ◆

de·ni·tri·fi·ca·tion \(ˌ)dē-ˌnī-trə-fə-'kā-shən\ *noun* (1883)
: the loss or removal of nitrogen or nitrogen compounds; *specifically* **:** reduction of nitrates or nitrites commonly by bacteria (as in soil) that usually results in the escape of nitrogen into the air
— **de·ni·tri·fi·er** \-'nī-trə-ˌfī-ər\ *noun*
— **de·ni·tri·fy** \-'nī-trə-ˌfī\ *transitive verb*

den·i·zen \'de-nə-zən\ *noun* [Middle English *denizeine,* from Anglo-French *denzein* inhabitant, inner part, inner, from Old French *denz* within, from Late Latin *deintus,* from Latin *de-* + *intus* within — more at ENT-] (15th century)
1 : INHABITANT
2 : one admitted to residence in a foreign country; *especially* **:** an alien admitted to rights of citizenship
3 : one that frequents a place

den mother *noun* (1946)
: a female adult leader of a Cub Scout den; *also* **:** a person seen in the role of leader or protector of a group

de·nom·i·nal \dē-'nä-mə-nᵊl\ *adjective* (1951)
: derived from a noun

de·nom·i·nate \di-'nä-mə-ˌnāt, dē-\ *transitive verb* [Latin *denominatus,* past participle of *denominare,* from *de-* + *nominare* to name — more at NOMINATE] (circa 1552)
1 : to give a name to **:** DESIGNATE
2 : to express or designate in some denomination ⟨will *denominate* prices in U.S. dollars⟩

de·nom·i·nate number \di-ˌnä-mə-nət-\ *noun* [Latin *denominatus*] (1579)
: a number (as 7 in 7 *feet*) that specifies a quantity in terms of a unit of measurement

de·nom·i·na·tion \di-ˌnä-mə-'nā-shən\ *noun* (15th century)
1 : an act of denominating
2 : a value or size of a series of values or sizes (as of money)
3 : NAME, DESIGNATION; *especially* **:** a general name for a category
4 : a religious organization uniting local congregations in a single legal and administrative body
— **de·nom·i·na·tion·al** \-shnəl, -shə-nᵊl\ *adjective*

de·nom·i·na·tion·al·ism \-shə-nə-ˌli-zəm\ *noun* (circa 1855)
1 : devotion to denominational principles or interests
2 : the emphasizing of denominational differences to the point of being narrowly exclusive **:** SECTARIANISM

de·nom·i·na·tive \di-'nä-mə-nə-tiv\ *adjective* [Latin *de* from + *nomin-, nomen* name] (circa 1783)
: derived from a noun or adjective
— **denominative** *noun*

de·nom·i·na·tor \di-'nä-mə-ˌnā-tər\ *noun* (circa 1542)
1 : the part of a fraction below the line signifying the division that functions as the divisor of the numerator and in fractions with 1 as the numerator indicates into how many parts the unit is divided
2 a : a common trait **b :** the average level (as of taste or opinion) **:** STANDARD

de·no·ta·tion \ˌdē-nō-'tā-shən\ *noun* (circa 1532)
1 : an act or process of denoting
2 : MEANING; *especially* **:** a direct specific meaning as distinct from an implied or associated idea
3 a : a denoting term **:** NAME **b :** SIGN, INDICATION ⟨visible *denotations* of divine wrath⟩
4 : the totality of things to which a term is applicable especially in logic — compare CONNOTATION

de·no·ta·tive \'dē-nō-ˌtā-tiv, di-'nō-tə-tiv\ *adjective* (circa 1611)
1 : denoting or tending to denote
2 : relating to denotation

de·note \di-'nōt, dē-\ *transitive verb* [Middle French *denoter,* from Latin *denotare,* from *de-* + *notare* to note] (1592)
1 : to serve as an indication of **:** BETOKEN ⟨the swollen bellies that *denote* starvation⟩
2 : to serve as an arbitrary mark for ⟨red flares *denoting* danger⟩
3 : to make known **:** ANNOUNCE ⟨his crestfallen look *denoted* his distress⟩
4 a : to serve as a linguistic expression of the notion of **:** MEAN **b :** to stand for **:** DESIGNATE
— **de·note·ment** \-'nōt-mənt\ *noun*

de·noue·ment *also* **dé·noue·ment** \ˌdā-ˌnü-'mäⁿ, dā-'nü-ˌ\ *noun* [French *dénouement,* literally, untying, from Middle French *desnouement,* from *desnouer* to untie, from Old French *desnoer,* from *des-* de- + *noer* to tie, from Latin *nodare,* from *nodus* knot — more at NODE] (1752)
1 : the final outcome of the main dramatic complication in a literary work
2 : the outcome of a complex sequence of events

de·nounce \di-'nau̇n(t)s, dē-\ *transitive verb* **de·nounced; de·nounc·ing** [Middle English, from Old French *denoncier* to proclaim, from Latin *denuntiare,* from *de-* + *nuntiare* to report — more at ANNOUNCE] (13th century)
1 : to pronounce especially publicly to be blameworthy or evil
2 *archaic* **a :** PROCLAIM **b :** to announce threateningly
3 : to inform against **:** ACCUSE
4 *obsolete* **:** PORTEND
5 : to announce formally the termination of (as a treaty)
synonym see CRITICIZE
— **de·nounce·ment** \-'nau̇n(t)-smənt\ *noun*
— **de·nounc·er** *noun*

de no·vo \di-'nō-(ˌ)vō, dā-, dē-\ *adverb or adjective* [Latin] (1536)
: over again **:** ANEW

dense \'den(t)s\ *adjective* **dens·er; dens·est** [Latin *densus;* akin to Greek *dasys* thick with hair or leaves] (15th century)
1 a : marked by compactness or crowding together of parts **b :** having a high mass per unit volume ⟨carbon dioxide is a *dense* gas⟩
2 a : marked by a stupid imperviousness to ideas or impressions **:** THICKHEADED **b :** EXTREME ⟨*dense* ignorance⟩
3 : having between any two elements at least one element ⟨the set of rational numbers is *dense*⟩
4 : demanding concentration to follow or comprehend ⟨*dense* prose⟩
5 : having high or relatively high opacity ⟨a *dense* fog⟩ ⟨a *dense* photographic negative⟩
synonym see STUPID
— **dense·ly** *adverb*
— **dense·ness** \'den(t)-snəs\ *noun*

den·si·fy \'den(t)-sə-ˌfī\ *transitive verb* **-fied; -fy·ing** (1820)
: to make denser **:** COMPRESS
— **den·si·fi·ca·tion** \ˌden(t)-sə-fə-'kā-shən\ *noun*

den·si·tom·e·ter \ˌden(t)-sə-'tä-mə-tər\ *noun* (1901)
: an instrument for determining optical or photographic density
— **den·si·to·met·ric** \ˌden(t)-sə-tə-'me-trik\ *adjective*
— **den·si·tom·e·try** \ˌden(t)-sə-'tä-mə-trē\ *noun*

den·si·ty \'den(t)-sə-tē\ *noun, plural* **-ties** (1603)
1 : the quality or state of being dense
2 : the quantity per unit volume, unit area, or unit length: as **a :** the mass of a substance per unit volume **b :** the distribution of a quantity (as mass, electricity, or energy) per unit usually of space (as length, area, or volume) **c :** the average number of individuals or units per space unit ⟨a population *density* of 500 per square mile⟩ ⟨a housing *density* of 10 houses per acre⟩
3 a : the degree of opacity of a translucent medium **b :** the common logarithm of the opacity

density function *noun* (circa 1962)
: PROBABILITY DENSITY FUNCTION

[1]**dent** \'dent\ *verb* [Middle English, short for *indenten* to make dents in, indent] (14th century)
transitive verb
1 : to make a dent in
2 : to have a weakening effect on
intransitive verb
: to form a dent by sinking inward **:** become dented

[2]**dent** *noun* (1565)
1 : a depression or hollow made by a blow or by pressure
2 a : an impression or effect often made against resistance and usually having a weakening effect **b :** initial progress **:** HEADWAY

[3]**dent** *noun* [French, literally, tooth, from Latin *dent-, dens*] (1703)
: TOOTH 3a

dent- *or* **denti-** *or* **dento-** *combining form* [Middle English *denti-,* from Latin, from *dent-, dens* tooth — more at TOOTH]
: tooth **:** teeth ⟨*denti*form⟩

[1]**den·tal** \'den-tᵊl\ *adjective* [Latin *dentalis,* from *dent-, dens*] (1594)
1 : articulated with the tip or blade of the tongue against or near the upper front teeth
2 : of or relating to the teeth or dentistry
— **den·tal·ly** \-ē\ *adverb*

[2]**dental** *noun* (circa 1727)
: a dental consonant

dental floss *noun* (1910)
: a thread used to clean between the teeth

dental hygienist *noun* (circa 1922)
: a licensed dental professional who cleans and examines teeth

den·ta·li·um \den-'tā-lē-əm\ *noun, plural* **-lia** \-lē-ə\ [New Latin, genus name, from Latin *dentalis*] (1864)
: any of a genus (*Dentalium*) of widely distributed tooth shells; *broadly* **:** TOOTH SHELL

dental technician *noun* (1946)
: one who makes dental appliances

den·tate \'den-ˌtāt\ *adjective* [Latin *dentatus,* from *dent-, dens*] (1810)
: having teeth or pointed conical projections ⟨a *dentate* margin of a leaf⟩

dent corn *noun* (1873)
: an Indian corn having kernels that contain both hard and soft starch and that become indented at maturity

den·ti·cle \'den-ti-kəl\ *noun* [Middle English, from Latin *denticulus,* diminutive of *dent-, dens*] (14th century)
: a conical pointed projection (as a small tooth)

◇ WORD HISTORY

denim It is not unusual for fabrics to be named after the places where they were originally made. Nîmes is a city in France known traditionally for its textiles. A heavy serge produced there became known as *serge de Nîmes,* meaning literally, "serge from Nîmes." The "s" in *Nîmes* is not pronounced in French, so when the name of the fabric came into English in the late 17th century, it was often written *serge de Nim* and later *serge denim.* In time the latter was shortened to *denim.*

den·tic·u·late \den-ˈti-kyə-lət\ *or* **den·tic·u·lat·ed** \-ˌlā-təd\ *adjective* (1661)
1 : finely dentate or serrate ⟨a *denticulate* shell⟩ ⟨a *denticulate* margin of a leaf⟩
2 : cut into dentils
— den·tic·u·la·tion \(ˌ)den-ˌti-kyə-ˈlā-shən\ *noun*

den·ti·form \ˈden-tə-ˌfȯrm\ *adjective* (1708)
: shaped like a tooth

den·ti·frice \ˈden-tə-frəs\ *noun* [Middle French, from Latin *dentifricium,* from *denti-* + *fricare* to rub — more at FRICTION] (15th century)
: a powder, paste, or liquid for cleaning the teeth

den·til \ˈden-tᵊl, -ˌtil\ *noun* [obsolete French *dentille,* from Middle French, diminutive of *dent*] (1663)
: one of a series of small projecting rectangular blocks forming a molding especially under a cornice
— den·tiled \-tᵊld, -ˌtild\ *adjective*

den·tin \ˈden-tᵊn\ *or* **den·tine** \ˈden-ˌtēn, den-ˈ\ *noun* (1845)
: a calcareous material similar to but harder and denser than bone that composes the principal mass of a tooth — see TOOTH illustration
— den·tin·al \den-ˈtē-nᵊl, ˈden-tᵊn-əl\ *adjective*

den·tist \ˈden-təst\ *noun* [French *dentiste,* from *dent*] (1752)
: one who is skilled in and licensed to practice the prevention, diagnosis, and treatment of diseases, injuries, and malformations of the teeth, jaws, and mouth and who makes and inserts false teeth

den·tist·ry \ˈden-tə-strē\ *noun* (1838)
: the art or profession of a dentist

den·ti·tion \den-ˈti-shən\ *noun* [Latin *dentition-, dentitio,* from *dentire* to cut teeth, from *dent-, dens*] (1615)
1 : the development and cutting of teeth
2 : the character of a set of teeth especially with regard to their number, kind, and arrangement — see TOOTH illustration
3 : TEETH

den·tu·lous \ˈden-chə-ləs\ *adjective* [back-formation from *edentulous*] (1926)
: having teeth

den·ture \ˈden-chər\ *noun* [French, from Middle French, from *dent*] (1874)
1 : a set of teeth
2 : an artificial replacement for one or more teeth; *especially* **:** a set of false teeth

den·tur·ist \ˈden-chə-rist\ *noun* (1965)
: a dental technician who makes, fits, and repairs dentures directly for the public

de·nu·cle·ar·ize \(ˈ)dē-ˈn(y)ü-klē-ə-ˌrīz, ÷-kyə-lə-ˌrīz\ *transitive verb* **-ized; -iz·ing** (1958)
: to remove nuclear arms from **:** prohibit the use of nuclear arms in
— de·nu·cle·ar·iza·tion \(ˌ)dē-ˌn(y)ü-klē-ə-rə-ˈzā-shən, ÷-kyə-lə-rə-\ *noun*

de·nude \di-ˈn(y)üd, dē-\ *transitive verb* **de·nud·ed; de·nud·ing** [Latin *denudare,* from *de-* + *nudus* bare — more at NAKED] (15th century)
1 : to deprive of something important
2 a : to strip of all covering or surface layers **b :** to lay bare by erosion **c :** to strip (land) of forests
— de·nu·da·tion \ˌdē-(ˌ)n(y)ü-ˈdā-shən, ˌden-yü-\ *noun*
— de·nude·ment \di-ˈn(y)üd-mənt\ *noun*

de·nu·mer·a·ble \di-ˈn(y)ü-mə-rə-bəl\ *adjective* (1902)
: capable of being put into one-to-one correspondence with the positive integers
— de·nu·mer·a·bil·i·ty \-ˌn(y)ü-mə-rə-ˈbi-lə-tē\ *noun*
— de·nu·mer·a·bly \-ˈn(y)ü-mə-rə-blē\ *adverb*

de·nun·ci·a·tion \di-ˌnən(t)-sē-ˈā-shən\ *noun* (1842)
: an act of denouncing; *especially* **:** a public condemnation
— de·nun·ci·a·tive \-ˈnən(t)-sē-ˌā-tiv\ *adjective*
— de·nun·ci·a·to·ry \-sē-ə-ˌtȯr-ē, -ˌtȯr-\ *adjective*

Den·ver boot \ˈden-vər-\ *noun* [*Denver,* Colo.] (1968)
: a metal clamp that is locked onto one of the wheels of an automobile to immobilize it especially until its owner pays accumulated parking fines

Denver omelet *noun* (1954)
: WESTERN OMELET

Denver sandwich *noun* (1950)
: WESTERN SANDWICH

de·ny \di-ˈnī, dē-\ *transitive verb* **de·nied; de·ny·ing** [Middle English, from Middle French *denier,* from Latin *denegare,* from *de-* + *negare* to deny — more at NEGATE] (14th century)
1 : to declare untrue
2 : to disclaim connection with or responsibility for **:** DISAVOW
3 a : to give a negative answer to **b :** to refuse to grant **c :** to restrain (oneself) from gratification of desires
4 *archaic* **:** DECLINE
5 : to refuse to accept the existence, truth, or validity of ☆
— de·ny·ing·ly \-ˈnī-iŋ-lē\ *adverb*

deoch an dor·is \ˌd(y)ȯk-ᵊn-ˈdōr-əs\ *Scottish & Irish variant of* DOCH-AN-DORRIS

de·o·dar \ˈdē-ə-ˌdär\ *also* **de·o·da·ra** \ˌdē-ə-ˈdär-ə\ *noun* [Hindi *deodār,* from Sanskrit *devadāru,* literally, timber of the gods, from *deva* god + *dāru* wood — more at DEITY, TREE] (1842)
: an East Indian cedar (*Cedrus deodara*)

de·odor·ant \dē-ˈō-də-rənt\ *noun* (1869)
: a preparation that destroys or masks unpleasant odors
— deodorant *adjective*

de·odor·ize \dē-ˈō-də-ˌrīz\ *transitive verb* (1856)
1 : to eliminate or prevent the offensive odor of
2 : to make (something unpleasant or reprehensible) more acceptable ⟨the movie *deodorizes* his scandalous career⟩
— de·odor·iza·tion \-ˌō-də-rə-ˈzā-shən\ *noun*
— de·odor·iz·er *noun*

de·on·tic \(ˌ)dē-ˈän-tik\ *adjective* [Greek *deont-, deon* that which is obligatory, from neuter of present participle of *dein* to lack, be needful — more at DEUTER-] (circa 1866)
: of or relating to moral obligation **:** DEONTOLOGICAL

de·on·tol·o·gy \ˌdē-ˌän-ˈtä-lə-jē\ *noun* (1826)
: the theory or study of moral obligation
— de·on·to·log·i·cal \ˌdē-ˌän-tə-ˈlä-ji-kəl\ *adjective*
— de·on·tol·o·gist \ˌdē-ˌän-ˈtä-lə-jist\ *noun*

Deo vo·len·te \ˌdā-(ˌ)ō-vō-ˈlen-tē, ˌdē-\ [Latin] (1767)
: God being willing

de·ox·i·dize \(ˌ)dē-ˈäk-sə-ˌdīz\ *transitive verb* (1794)
: to remove oxygen from
— de·ox·i·da·tion \(ˌ)dē-ˌäk-sə-ˈdā-shən\ *noun*
— de·ox·i·diz·er \(ˌ)dē-ˈäk-sə-ˌdī-zər\ *noun*

de·oxy \(ˌ)dē-ˈäk-sē\ *also* **des·oxy** \(ˌ)de-ˈzäk-sē, -ˈsäk-\ *adjective* [International Scientific Vocabulary]
: containing less oxygen in the molecule than the compound from which it is derived ⟨*deoxy* sugars⟩ — usually used in combination ⟨*deoxy*ribonucleic acid⟩

de·ox·y·gen·ate \(ˌ)dē-ˈäk-si-jə-ˌnāt, ˌdē-äk-ˈsi-jə-\ *transitive verb* (1799)
: to remove oxygen from
— de·ox·y·gen·ation \(ˌ)dē-ˌäk-si-jə-ˈnā-shən, ˌdē-äk-ˌsi-jə-\ *noun*

de·ox·y·gen·at·ed *adjective* (1799)
: having the hemoglobin in the reduced state

de·oxy·ri·bo·nu·cle·ase \(ˌ)dē-ˈäk-si-ˌrī-bō-ˈn(y)ü-klē-ˌās, -ˌāz\ *noun* (1946)
: an enzyme that hydrolyzes DNA to nucleotides — called also *DNase*

de·oxy·ri·bo·nu·cle·ic acid \(ˌ)dē-ˈäk-si-ˌrī-bō-n(y)ü-ˌklē-ik-, -ˌklā-\ *noun* [*deoxyribose* + *nucleic acid*] (1938)
: DNA

de·oxy·ri·bo·nu·cle·o·tide \-ˈn(y)ü-klē-ə-ˌtīd\ *noun* (1972)
: a nucleotide that contains deoxyribose and is a constituent of DNA

de·oxy·ri·bose \(ˌ)dē-ˌäk-si-ˈrī-ˌbōs, -ˌbōz\ *noun* [International Scientific Vocabulary] (1957)
: a pentose sugar $C_5H_{10}O_4$ that is a structural element of DNA

de·part \di-ˈpärt\ *verb* [Middle English, to divide, part company, from Old French *departir,* from *de-* + *partir* to divide, from Latin *partire,* from *part-, pars* part] (13th century)
intransitive verb
1 a : to go away **:** LEAVE **b :** DIE
2 : to turn aside **:** DEVIATE
transitive verb
: to go away from **:** LEAVE
synonym see SWERVE

de·part·ed *adjective* (14th century)
1 : BYGONE
2 : having died especially recently ⟨mourning our *departed* friend⟩
synonym see DEAD

de·part·ment \di-ˈpärt-mənt\ *noun* [French *département,* from Old French, act of dividing, from *departir*] (1735)
1 a : a distinct sphere **:** PROVINCE ⟨that's not my *department*⟩ **b :** a category consisting especially of a measurable activity or attribute ⟨lacking in the trustworthiness *department* —Garrison Keillor⟩
2 : a functional or territorial division: as **a :** a major administrative division of a government **b :** a major territorial administrative subdivision **c :** a division of a college or school giving instruction in a particular subject **d :** a major division of a business **e :** a section of a department store handling a particular kind of merchandise **f :** a territorial subdivision made for the administration and training of military units
— de·part·men·tal \di-ˌpärt-ˈmen-tᵊl, ˌdē-\ *adjective*
— de·part·men·tal·ly \-tᵊl-ē\ *adverb*

de·part·men·tal·ize \di-ˌpärt-ˈmen-tᵊl-ˌīz, ˌdē-\ *transitive verb* **-ized; -iz·ing** (circa 1895)
: to divide into departments
— de·part·men·tal·iza·tion \-ˌmen-tᵊl-ə-ˈzā-shən\ *noun*

department store *noun* (1887)
: a store selling a wide variety of goods and arranged in several departments

☆ SYNONYMS
Deny, gainsay, contradict, contravene mean to refuse to accept as true or valid. DENY implies a firm refusal to accept as true, to grant or concede, or to acknowledge the existence or claims of ⟨*denied* the charges⟩. GAINSAY implies disputing the truth of what another has said ⟨no one can *gainsay* her claims⟩. CONTRADICT implies an open or flat denial ⟨her account *contradicts* his⟩. CONTRAVENE implies not so much an intentional opposition as some inherent incompatibility ⟨laws that *contravene* tradition⟩.

de·par·ture \di-ˈpär-chər\ *noun* (15th century)
1 a (1) **:** the act or an instance of departing (2) *archaic* **:** DEATH **b :** a setting out (as on a new course)
2 : DIVERGENCE 2

de·pau·per·ate \di-ˈpȯ-pə-rət\ *adjective* [Middle English *depauperat,* from Medieval Latin *depauperatus,* past participle of *depauperare* to impoverish, from Latin *de-* + *pauperare* to impoverish, from *pauper* poor — more at POOR] (1670)
1 : falling short of natural development or size
2 : IMPOVERISHED ⟨a *depauperate* fauna⟩

de·pend \di-ˈpend\ *intransitive verb* [Middle English, from Middle French *dependre,* modification of Latin *dependēre,* from *de-* + *pendēre* to hang — more at PENDANT] (15th century)
1 : to be determined, based, or contingent ⟨life *depends* on food⟩ ⟨the value of *Y depends* on *X*⟩
2 : to be pending or undecided
3 a : to place reliance or trust ⟨you can *depend* on me⟩ **b :** to be dependent especially for financial support
4 : to hang down

de·pend·able \di-ˈpen-də-bəl\ *adjective* (1735)
: capable of being depended on **:** RELIABLE
— **de·pend·abil·i·ty** \-ˌpen-də-ˈbi-lə-tē\ *noun*
— **de·pend·able·ness** *noun*
— **de·pend·ably** \-blē\ *adverb*

de·pen·dence *also* **de·pen·dance** \di-ˈpen-dən(t)s\ *noun* (15th century)
1 : the quality or state of being dependent; *especially* **:** the quality or state of being influenced or determined by or subject to another
2 : RELIANCE, TRUST
3 : one that is relied on
4 a : drug addiction **b :** HABITUATION 2b

de·pen·den·cy \-dən(t)-sē\ *noun, plural* **-cies** (1594)
1 : DEPENDENCE 1
2 : something that is dependent on something else; *especially* **:** a territorial unit under the jurisdiction of a nation but not formally annexed by it
3 : a building (as a stable) that is an adjunct to a main dwelling

[1]**de·pen·dent** \di-ˈpen-dənt\ *adjective* [Middle English *dependant,* from Middle French, present participle of *dependre*] (14th century)
1 : hanging down
2 a : determined or conditioned by another **:** CONTINGENT **b** (1) **:** relying on another for support (2) **:** affected with a drug dependence **c :** subject to another's jurisdiction **d :** SUBORDINATE 3a
3 a : not mathematically or statistically independent ⟨a *dependent* set of vectors⟩ ⟨*dependent* events⟩ **b :** EQUIVALENT 6a ⟨*dependent* equations⟩
— **de·pen·dent·ly** *adverb*

[2]**dependent** *also* **de·pen·dant** \-dənt\ *noun* (1523)
1 *archaic* **:** DEPENDENCY
2 : one that is dependent; *especially* **:** a person who relies on another for support

dependent variable *noun* (circa 1852)
: a mathematical variable whose value is determined by that of one or more other variables in a function

de·per·son·al·iza·tion \(ˌ)dē-ˌpər-snə-lə-ˈzā-shən, -ˌpər-sᵊn-ə-lə-\ *noun* (1906)
1 a : an act or process of depersonalizing **b :** the quality or state of being depersonalized
2 : a psychopathological syndrome characterized by loss of identity and feelings of unreality and strangeness about one's own behavior

de·per·son·al·ize \(ˈ)dē-ˈpər-snə-ˌlīz, -ˈpər-sᵊn-ə-\ *transitive verb* (1866)
1 : to deprive of the sense of personal identity ⟨schools that *depersonalize* students⟩
2 : to make impersonal

de·phos·phor·y·la·tion \(ˌ)dē-ˌfäs-fȯr-ə-ˈlā-shən\ *noun* (1931)
: the process of removing phosphate groups from an organic compound (as ATP) by hydrolysis; *also* **:** the resulting state
— **de·phos·phor·y·late** \(ˌ)dē-fäs-ˈfȯr-ə-ˌlāt\ *transitive verb*

de·pict \di-ˈpikt, dē-\ *transitive verb* [Latin *depictus,* past participle of *depingere,* from *de-* + *pingere* to paint — more at PAINT] (15th century)
1 : to represent by or as if by a picture
2 : DESCRIBE 1
— **de·pic·ter** \-ˈpik-tər\ *noun*
— **de·pic·tion** \-ˈpik-shən\ *noun*

de·pig·men·ta·tion \(ˌ)dē-ˌpig-mən-ˈtā-shən, -ˌmen-\ *noun* (1889)
: loss of normal pigmentation

dep·i·la·tion \ˌde-pə-ˈlā-shən\ *noun* [Middle French or Medieval Latin; Middle French, from Medieval Latin *depilation-, depilatio,* from Latin *depilare,* from *de-* + *pilus* hair — more at PILE] (1547)
: the removal of hair, wool, or bristles by chemical or mechanical methods
— **dep·i·late** \ˈde-pə-ˌlāt\ *transitive verb*

de·pil·a·to·ry \di-ˈpi-lə-ˌtōr-ē, -ˌtȯr-\ *noun, plural* **-ries** (1606)
: an agent for removing hair, wool, or bristles
— **depilatory** *adjective*

de·plane \(ˌ)dē-ˈplān\ *intransitive verb* (1923)
: to disembark from an airplane

de·plete \di-ˈplēt\ *transitive verb* **de·plet·ed; de·plet·ing** [Latin *depletus,* past participle of *deplēre,* from *de-* + *plēre* to fill — more at FULL] (1807)
1 : to empty of a principal substance
2 : to lessen markedly in quantity, content, power, or value ☆
— **de·plet·able** \-ˈplē-tə-bəl\ *adjective*
— **de·ple·tion** \-ˈplē-shən\ *noun*
— **de·ple·tive** \-ˈplē-tiv\ *adjective*

de·plor·able \di-ˈplōr-ə-bəl, -ˈplȯr-\ *adjective* (1612)
1 : LAMENTABLE
2 : deserving censure or contempt **:** WRETCHED
— **de·plor·able·ness** *noun*
— **de·plor·ably** \-blē\ *adverb*

de·plore \di-ˈplōr, -ˈplȯr\ *transitive verb* **de·plored; de·plor·ing** [Middle French or Latin; Middle French *deplorer,* from Latin *deplorare,* from *de-* + *plorare* to wail] (1567)
1 a : to feel or express grief for **b :** to regret strongly
2 : to consider unfortunate or deserving of deprecation ☆
— **de·plor·er** \-ˈplōr-ər\ *noun*
— **de·plor·ing·ly** \-iŋ-lē\ *adverb*

de·ploy \di-ˈplȯi\ *verb* [French *déployer,* literally, to unfold, from Old French *desploier,* from *des-* dis- + *ploier, plier* to fold — more at PLY] (1786)
transitive verb
1 a : to extend (a military unit) especially in width **b :** to place in battle formation or appropriate positions
2 : to spread out, utilize, or arrange especially strategically
intransitive verb
: to move in being deployed
— **de·ploy·able** \-ə-bəl\ *adjective*
— **de·ploy·ment** \-mənt\ *noun*

de·po·lar·ize \(ˌ)dē-ˈpō-lə-ˌrīz\ *transitive verb* (1818)
1 : to cause to become partially or wholly unpolarized
2 : to prevent or remove polarization of (as a dry cell or cell membrane)
— **de·po·lar·iza·tion** \(ˌ)dē-ˌpō-lə-rə-ˈzā-shən\ *noun*
— **de·po·lar·iz·er** \(ˌ)dē-ˈpō-lə-ˌrī-zər\ *noun*

de·po·lit·i·cize \ˌdē-pə-ˈli-tə-ˌsīz\ *transitive verb* (1937)
: to remove the political character of **:** take out of the realm of politics ⟨*depoliticize* foreign aid⟩
— **de·po·lit·i·ci·za·tion** \ˌdē-pə-ˌli-tə-sə-ˈzā-shən\ *noun*

de·po·ly·mer·ize \(ˌ)dē-pə-ˈli-mə-ˌrīz, -ˈpä-lə-mə-\ (circa 1909)
transitive verb
: to decompose (macromolecules) into simpler compounds (as monomers)
intransitive verb
: to undergo decomposition into simpler compounds
— **de·po·ly·mer·iza·tion** \ˌdē-pə-ˌli-mə-rə-ˈzā-shən, (ˌ)dē-ˌpä-lə-mə-rə-\ *noun*

de·pone \di-ˈpōn\ *verb* **de·poned; de·pon·ing** [Medieval Latin *deponere,* from Latin, to put down, from *de-* + *ponere* to put — more at POSITION] (15th century)
: TESTIFY

[1]**de·po·nent** \di-ˈpō-nənt\ *adjective* [Late Latin *deponent-, deponens,* from Latin, present participle of *deponere*] (15th century)
: occurring with passive or middle voice forms but with active voice meaning ⟨the *deponent* verbs in Latin and Greek⟩

[2]**deponent** *noun* (1530)
1 : a deponent verb
2 : one who gives evidence

de·pop·u·late \(ˌ)dē-ˈpä-pyə-ˌlāt\ *transitive verb* [Latin *depopulatus,* past participle of *depopulari,* from *de-* + *populari* to ravage] (1548)
1 *obsolete* **:** RAVAGE
2 : to reduce greatly the population of
— **de·pop·u·la·tion** \(ˌ)dē-ˌpä-pyə-ˈlā-shən\ *noun*

de·port \di-ˈpōrt, -ˈpȯrt, dē-\ *transitive verb* [Middle French *deporter,* from Latin *deportare* to carry away, from *de-* + *portare* to carry — more at FARE] (1598)
1 : to behave or comport (oneself) especially in accord with a code
2 [Latin *deportare*] **a :** to carry away **b :** to send out of the country by legal deportation
synonym see BANISH, BEHAVE

de·port·able \di-ˈpōr-tə-bəl, -ˈpȯr-, dē-\ *adjective* (1891)
1 : punishable by deportation ⟨*deportable* offenses⟩

☆ SYNONYMS
Deplete, drain, exhaust, impoverish, bankrupt mean to deprive of something essential to existence or potency. DEPLETE implies a reduction in number or quantity so as to endanger the ability to function ⟨*depleting* our natural resources⟩. DRAIN implies a gradual withdrawal and ultimate deprivation of what is necessary to an existence ⟨personal tragedy had *drained* him of all spirit⟩. EXHAUST stresses a complete emptying ⟨her lecture *exhausted* the subject⟩. IMPOVERISH suggests a deprivation of something essential to richness or productiveness ⟨*impoverished* soil⟩. BANKRUPT suggests impoverishment to the point of imminent collapse ⟨war had *bankrupted* the nation of resources⟩.

Deplore, lament, bewail, bemoan mean to express grief or sorrow for something. DEPLORE implies regret for the loss or impairment of something of value ⟨*deplores* the breakdown in family values⟩. LAMENT implies a profound or demonstrative expression of sorrow ⟨*lamenting* the loss of their only child⟩. BEWAIL and BEMOAN imply sorrow, disappointment, or protest finding outlet in words or cries, BEWAIL commonly suggesting loudness, and BEMOAN lugubriousness ⟨fans *bewailed* the defeat⟩ ⟨purists *bemoaning* the corruption of the language⟩.

2 : subject to deportation ⟨*deportable* aliens⟩

de·por·ta·tion \\,dē-,pōr-'tā-shən, -,pȯr-, -pər-\\ *noun* (1595)
1 : an act or instance of deporting
2 : the removal from a country of an alien whose presence is unlawful or prejudicial

de·por·tee \\,dē-,pōr-'tē, di-, -,pȯr-\\ *noun* (1865)
: one who has been deported or is under sentence of deportation

de·port·ment \\di-'pōrt-mənt, -'pȯrt-, dē-\\ *noun* (1601)
: the manner in which one conducts oneself **:** BEHAVIOR
synonym see BEARING

de·pos·al \\di-'pō-zəl, dē-\\ *noun* (14th century)
: an act of deposing from office

de·pose \\di-'pōz, dē-\\ *verb* **de·posed; de·pos·ing** [Middle English, from Middle French *deposer,* from Late Latin *deponere* (perfect indicative *deposui*), from Latin, to put down] (14th century)
transitive verb
1 : to remove from a throne or other high position
2 : to put down **:** DEPOSIT
3 a [Middle English, from Medieval Latin *deponere,* from Late Latin] **:** to testify to under oath or by affidavit **b :** AFFIRM, ASSERT
intransitive verb
: to bear witness

[1]de·pos·it \\di-'pä-zət\\ *verb* **de·pos·it·ed** \\-'pä-zə-təd, -'päz-təd\\; **de·pos·it·ing** \\-'pä-zə-tiŋ, -'päz-tiŋ\\ [Latin *depositus,* past participle of *deponere*] (1624)
transitive verb
1 : to place especially for safekeeping or as a pledge; *especially* **:** to put in a bank
2 a : to lay down **:** PLACE **b :** to let fall (as sediment)
intransitive verb
: to become deposited
— **de·pos·i·tor** \\-'pä-zə-tər, -'päz-tər\\ *noun*

[2]deposit *noun* (1624)
1 : the state of being deposited
2 : something placed for safekeeping: as **a :** money deposited in a bank **b :** money given as a pledge or down payment
3 : a place of deposit **:** DEPOSITORY
4 : an act of depositing
5 a : something laid down; *especially* **:** matter deposited by a natural process **b :** a natural accumulation (as of iron ore, coal, or gas)

de·pos·i·tary \\di-'pä-zə-,ter-ē\\ *noun, plural* **-tar·ies** (1605)
1 : a person to whom something is entrusted
2 : DEPOSITORY 2

de·po·si·tion \\,de-pə-'zi-shən, ,dē-pə-\\ *noun* (14th century)
1 : an act of removing from a position of authority
2 a : a testifying especially before a court **b :** DECLARATION; *specifically* **:** testimony taken down in writing under oath
3 : an act or process of depositing
4 : something deposited **:** DEPOSIT
— **de·po·si·tion·al** \\-'zish-nəl, -shə-nᵊl\\ *adjective*

de·pos·i·to·ry \\di-'pä-zə-,tōr-ē, -,tȯr-\\ *noun, plural* **-ries** (1656)
1 : DEPOSITARY 1
2 : a place where something is deposited especially for safekeeping

depository library *noun* (circa 1930)
: a library designated to receive U.S. government publications

de·pot *1 & 2 are* 'de-(,)pō *also* 'dē-, *3 is* 'dē- *sometimes* 'de-\\ *noun* [French *dépôt,* from Middle French *depost,* Medieval Latin *depositum,* from Latin, neuter of *depositus*] (1795)
1 a : a place for storing goods or motor vehicles **b :** STORE, CACHE
2 a : a place for the storage of military supplies **b :** a place for the reception and forwarding of military replacements
3 : a building for railroad or bus passengers or freight

de·prave \\di-'prāv\\ *transitive verb* **de·praved; de·prav·ing** [Middle English, from Middle French *depraver,* from Latin *depravare* to pervert, from *de-* + *pravus* crooked, bad] (14th century)
1 *archaic* **:** to speak ill of **:** MALIGN
2 : to make bad **:** CORRUPT; *especially* **:** to corrupt morally
synonym see DEBASE
— **de·pra·va·tion** \\,de-prə-'vā-shən, ,dē-,prā-\\ *noun*
— **de·prave·ment** \\di-'prāv-mənt\\ *noun*
— **de·prav·er** \\di-'prā-vər\\ *noun*

de·praved \\di-'prāvd\\ *adjective* (14th century)
: marked by corruption or evil; *especially* **:** PERVERTED
— **de·praved·ly** \\-'prā-vəd-lē, -'prāvd-lē\\ *adverb*
— **de·pra·ved·ness** \\-'prā-vəd-nəs, -'prāvd-nəs\\ *noun*

de·prav·i·ty \\di-'pra-və-tē *also* -'prā-\\ *noun, plural* **-ties** (1641)
1 : the quality or state of being depraved
2 : a corrupt act or practice

dep·re·cate \\'de-pri-,kāt\\ *transitive verb* **-cat·ed; -cat·ing** [Latin *deprecatus,* past participle of *deprecari* to avert by prayer, from *de-* + *precari* to pray — more at PRAY] (1628)
1 a *archaic* **:** to pray against (as an evil) **b :** to seek to avert ⟨*deprecate* the wrath . . . of the Roman people —Tobias Smollett⟩
2 : to express disapproval of
3 a : PLAY DOWN **:** make little of ⟨speaks five languages . . . but *deprecates* this facility —*Time*⟩ **b :** BELITTLE, DISPARAGE ⟨the most reluctantly admired and least easily *deprecated* of . . . novelists —*New Yorker*⟩
— **dep·re·cat·ing·ly** \\-,kā-tiŋ-lē\\ *adverb*
— **dep·re·ca·tion** \\,de-pri-'kā-shən\\ *noun*

dep·re·ca·to·ry \\'de-pri-kə-,tōr-ē, -,tȯr-, 'de-prə-,kā-tə-rē\\ *adjective* (1586)
1 : seeking to avert disapproval **:** APOLOGETIC
2 : serving to deprecate **:** DISAPPROVING
— **dep·re·ca·to·ri·ly** \\,de-pri-kə-'tōr-ə-lē, -'tȯr-\\ *adverb*

de·pre·ci·ate \\di-'prē-shē-,āt\\ *verb* **-at·ed; -at·ing** [Late Latin *depretiatus,* past participle of *depretiare,* from Latin *de-* + *pretium* price — more at PRICE] (15th century)
transitive verb
1 : to lower in estimation or esteem
2 : to lower the price or estimated value of
intransitive verb
: to fall in value
synonym see DECRY
— **de·pre·cia·ble** \\-shə-bəl\\ *adjective*
— **de·pre·ci·at·ing·ly** \\-shē-,ā-tiŋ-lē\\ *adverb*
— **de·pre·ci·a·tion** \\-,prē-shē-'ā-shən\\ *noun*
— **de·pre·ci·a·tive** \\-'prē-shə-tiv, -shē-,ā-tiv\\ *adjective*
— **de·pre·ci·a·tor** \\-shē-,ā-tər\\ *noun*
— **de·pre·cia·to·ry** \\-shə-,tōr-ē, -,tȯr-\\ *adjective*

dep·re·date \\'de-prə-,dāt\\ *verb* **-dat·ed; -dat·ing** [Late Latin *depraedatus,* past participle of *depraedari,* from Latin *de-* + *praedari* to plunder — more at PREY] (1626)
transitive verb
: to lay waste **:** PLUNDER, RAVAGE
intransitive verb
: to engage in plunder
— **dep·re·da·tion** \\,de-prə-'dā-shən\\ *noun*
— **dep·re·da·tor** \\'de-prə-,dā-tər, di-'pre-də-\\ *noun*
— **de·pre·da·to·ry** \\di-'pre-də-,tōr-ē, 'de-pri-də-, -,tȯr-\\ *adjective*

de·press \\di-'pres, dē-\\ *transitive verb* [Middle English, from Middle French *depresser,* from Latin *depressus,* past participle of *deprimere* to press down, from *de-* + *premere* to press — more at PRESS] (14th century)
1 *obsolete* **:** REPRESS, SUBJUGATE
2 a : to press down ⟨*depress* a typewriter key⟩ **b :** to cause to sink to a lower position
3 : to lessen the activity or strength of
4 : SADDEN, DISCOURAGE
5 : to decrease the market value or marketability of
— **de·press·ible** \\-'pre-sə-bəl\\ *adjective*

de·pres·sant \\di-'pre-sᵊnt, dē-\\ *noun* (1876)
: one that depresses; *specifically* **:** an agent that reduces a bodily functional activity or an instinctive desire (as appetite)
— **depressant** *adjective*

de·pressed *adjective* (1621)
1 : low in spirits **:** SAD; *especially* **:** affected by psychological depression
2 a : vertically flattened ⟨a *depressed* cactus⟩ **b :** having the central part lower than the margin **c :** lying flat or prostrate **d :** dorsoventrally flattened
3 : suffering from economic depression; *especially* **:** UNDERPRIVILEGED
4 : being below the standard

de·press·ing *adjective* (1789)
: that depresses; *especially* **:** causing emotional depression ⟨a *depressing* story⟩
— **de·press·ing·ly** \\-siŋ-lē\\ *adverb*

de·pres·sion \\di-'pre-shən, dē-\\ *noun* (14th century)
1 a : the angular distance of a celestial object below the horizon **b :** the size of an angle of depression
2 : an act of depressing or a state of being depressed: as **a :** a pressing down **:** LOWERING **b** (1) **:** a state of feeling sad **:** DEJECTION (2) **:** a psychoneurotic or psychotic disorder marked especially by sadness, inactivity, difficulty in thinking and concentration, a significant increase or decrease in appetite and time spent sleeping, feelings of dejection and hopelessness, and sometimes suicidal tendencies **c** (1) **:** a reduction in activity, amount, quality, or force (2) **:** a lowering of vitality or functional activity
3 : a depressed place or part **:** HOLLOW
4 : LOW 1b
5 : a period of low general economic activity marked especially by rising levels of unemployment

Depression glass *noun* [Great *Depression* of 1929 to circa 1939] (1971)
: tinted glassware machine-produced during the 1930s

[1]de·pres·sive \\di-'pre-siv, dē-\\ *adjective* (1620)
1 : tending to depress
2 : of, relating to, marked by, or affected by psychological depression
— **de·pres·sive·ly** *adverb*

[2]depressive *noun* (1937)
: one who is affected with or prone to psychological depression

de·pres·sor \\di-'pre-sər, dē-\\ *noun* [Late Latin, from Latin *deprimere*] (1611)
: one that depresses: as **a :** a muscle that draws down a part — compare LEVATOR **b :** a device for pressing down or aside **c :** a nerve or nerve fiber that decreases the activity or the tone of the organ or part it innervates

de·pres·sur·ize \\(,)dē-'pre-shə-,rīz\\ *transitive verb* (1944)
: to release pressure from
— **de·pres·sur·iza·tion** \\(,)dē-,pre-shə-rə-'zā-shən\\ *noun*

dep·ri·va·tion \\,de-prə-'vā-shən *also* ,dē-,prī-\\ *noun* (15th century)
1 : the state of being deprived **:** PRIVATION; *especially* **:** removal from an office, dignity, or benefice
2 : an act or instance of depriving **:** LOSS

\\ə\\ **abut** \\ᵊ\\ **kitten** \\ər\\ **further** \\a\\ **ash** \\ā\\ **ace** \\ä\\ **mop, mar** \\au̇\\ **out** \\ch\\ **chin** \\e\\ **bet** \\ē\\ **easy** \\g\\ **go** \\i\\ **hit** \\ī\\ **ice** \\j\\ **job** \\ŋ\\ **sing** \\ō\\ **go** \\ȯ\\ **law** \\ȯi\\ **boy** \\th\\ **thin** \\t͟h\\ **the** \\ü\\ **loot** \\u̇\\ **foot** \\y\\ **yet** \\zh\\ **vision** *see also* Guide to Pronunciation

de·prive \di-ˈprīv\ *transitive verb* **de·prived; de·priv·ing** [Middle English *depriven,* from Medieval Latin *deprivare,* from Latin *de-* + *privare* to deprive — more at PRIVATE] (14th century)
1 *obsolete* **:** REMOVE
2 : to take something away from ⟨*deprived* him of his professorship —J. M. Phalen⟩
3 : to remove from office
4 : to withhold something from ⟨*deprived* a citizen of her rights⟩

de·prived *adjective* (circa 1522)
: marked by deprivation especially of the necessities of life or of healthful environmental influences ⟨culturally *deprived* children⟩

de·pro·gram \(ˌ)dē-ˈprō-ˌgram, -grəm\ *transitive verb* (1973)
: to dissuade from convictions usually of a religious nature often by coercive means
— de·pro·gram·mer \-mər\ *noun*

depth \ˈdepth\ *noun, plural* **depths** \ˈdepths, ˈdep(t)s\ [Middle English, probably from *dep* deep] (14th century)
1 a (1) **:** a deep place in a body of water (2) **:** a part that is far from the outside or surface ⟨the *depths* of the woods⟩ (3) **:** ABYSS 2 **b** (1) **:** a profound or intense state (as of thought or feeling) ⟨the *depths* of misery⟩; *also* **:** a reprehensibly low condition ⟨hadn't realized that standards had fallen to such *depths*⟩ (2) **:** the middle of a time (as winter) (3) **:** the worst part
2 a : the perpendicular measurement downward from a surface **b :** the direct linear measurement from front to back
3 : the quality of being deep
4 : the degree of intensity ⟨*depth* of a color⟩; *also* **:** the quality of being profound (as in insight) or full (as of knowledge)
5 : the quality or state of being complete or thorough ⟨a study will be made in *depth*⟩
— depth·less \ˈdepth-ləs\ *adjective*

depth charge *noun* (1917)
: an antisubmarine weapon that consists essentially of a drum filled with explosives which is dropped near a target and descends to a predetermined depth where it explodes — called also *depth bomb*

depth of field (1911)
: the range of distances of the object in front of an image-forming device (as a camera lens) measured along the axis of the device throughout which the image has acceptable sharpness

depth perception *noun* (circa 1911)
: the ability to judge the distance of objects and the spatial relationship of objects at different distances

depth psychology *noun* (1924)
: PSYCHOANALYSIS; *also* **:** psychology concerned especially with the unconscious mind

dep·u·ta·tion \ˌde-pyə-ˈtā-shən\ *noun* (14th century)
1 : the act of appointing a deputy
2 : a group of people appointed to represent others

de·pute \di-ˈpyüt\ *transitive verb* **de·put·ed; de·put·ing** [Middle English, to appoint, from Middle French *deputer,* from Late Latin *deputare* to assign, from Latin, to consider (as), from *de-* + *putare* to consider — more at PAVE] (14th century)
: DELEGATE

dep·u·tize \ˈde-pyə-ˌtīz\ *verb* **-tized; -tiz·ing** (circa 1736)
transitive verb
: to appoint as deputy
intransitive verb
: to act as deputy
— dep·u·ti·za·tion \ˌde-pyə-tə-ˈzā-shən\ *noun*

dep·u·ty \ˈde-pyə-tē\ *noun, plural* **-ties** [Middle English, from Middle French *deputé,* past participle of *deputer*] (15th century)
1 a : a person appointed as a substitute with power to act **b :** a second in command or assistant who usually takes charge when his or her superior is absent
2 : a member of the lower house of some legislative assemblies

de·rac·i·nate \(ˌ)dē-ˈra-sᵊn-ˌāt\ *transitive verb* **-nat·ed; -nat·ing** [Middle French *desraciner,* from *des-* de- + *racine* root, from Late Latin *radicina,* from Latin *radic-, radix* — more at ROOT] (1599)
: UPROOT
— de·rac·i·na·tion \(ˌ)dē-ˌra-sᵊn-ˈā-shən\ *noun*

de·rail \di-ˈrā(ə)l, dē-\ *verb* [French *dérailler* to throw off the track, from *dé-* de- + *rail,* from English] (1850)
transitive verb
1 : to cause to run off the rails
2 : to obstruct the progress of **:** FRUSTRATE
intransitive verb
: to leave the rails
— de·rail·ment \-mənt\ *noun*

de·rail·leur \di-ˈrā-lər\ *noun* [French *dérailleur,* from *dérailler*] (1930)
: a mechanism for shifting gears on a bicycle that operates by moving the chain from one set of exposed gears to another

de·range \di-ˈrānj\ *transitive verb* **deranged; de·rang·ing** [French *déranger,* from Old French *desrengier,* from *des-* de- + *reng* line, row — more at RANK] (1776)
1 : to disturb the operation or functions of
2 : DISARRANGE ⟨hatless, with tie *deranged* —G. W. Stonier⟩
3 : to make insane
— de·range·ment \-mənt\ *noun*

de·rate \(ˌ)dē-ˈrāt\ *transitive verb* (1947)
: to lower the rated capability of (as electrical or mechanical apparatus) because of deterioration or inadequacy

der·by \ˈdər-bē, *especially British* ˈdär-\ *noun, plural* **derbies** [Edward Stanley (died 1834), 12th earl of *Derby*] (1844)
1 : any of several horse races held annually and usually restricted to three-year-olds
2 : a race or contest open to all comers or to a specified category of contestants ⟨bicycle *derby*⟩
3 : a man's stiff felt hat with dome-shaped crown and narrow brim ◆

derby 3

de·re·al·iza·tion \(ˌ)dē-ˌrē-ə-lə-ˈzā-shən, -ˌri-ə-\ *noun* (1942)
: a feeling of altered reality that occurs often in schizophrenia and in some drug reactions

de·reg·u·la·tion \(ˌ)dē-ˌre-gyə-ˈlā-shən\ *noun* (1963)
: the act or process of removing restrictions and regulations
— de·reg·u·late \(ˈ)dē-ˈre-gyə-ˌlāt\ *transitive verb*

[1]**der·e·lict** \ˈder-ə-ˌlikt\ *adjective* [Latin *derelictus,* past participle of *derelinquere* to abandon, from *de-* + *relinquere* to leave — more at RELINQUISH] (1649)
1 : abandoned especially by the owner or occupant **:** RUN-DOWN
2 : lacking a sense of duty **:** NEGLIGENT

[2]**derelict** *noun* (1670)
1 a : something voluntarily abandoned; *specifically* **:** a ship abandoned on the high seas **b :** a tract of land left dry by receding water
2 : a destitute homeless social misfit **:** VAGRANT, BUM

der·e·lic·tion \ˌder-ə-ˈlik-shən\ *noun* (1597)
1 a : an intentional abandonment **b :** the state of being abandoned
2 : a recession of water leaving permanently dry land
3 a : intentional or conscious neglect **:** DELINQUENCY ⟨*dereliction* of duty⟩ **b :** FAULT, SHORTCOMING

de·re·press \ˌdē-ri-ˈpres\ *transitive verb* (1962)
: to activate (a gene or enzyme) by releasing from a blocked state
— de·re·pres·sion \-ˈpre-shən\ *noun*

de·ride \di-ˈrīd, dē-\ *transitive verb* **de·rid·ed; de·rid·ing** [Latin *deridēre,* from *de-* + *ridēre* to laugh] (1530)
1 : to laugh at contemptuously
2 : to subject to usually bitter or contemptuous ridicule
synonym see RIDICULE
— de·rid·er *noun*
— de·rid·ing·ly \-ˈrī-diŋ-lē\ *adverb*

de ri·gueur \də-(ˌ)rē-ˈgər\ *adjective* [French] (1833)
: prescribed or required by fashion, etiquette, or custom **:** PROPER

de·ri·sion \di-ˈri-zhən\ *noun* [Middle English, from Middle French, from Late Latin *derision-, derisio,* from Latin *deridēre*] (14th century)
1 a : the use of ridicule or scorn to show contempt **b :** a state of being derided
2 : an object of ridicule or scorn

de·ri·sive \di-ˈrī-siv, -ziv; -ˈri-ziv, -ˈri-siv\ *adjective* (circa 1662)
: expressing or causing derision
— de·ri·sive·ly *adverb*
— de·ri·sive·ness *noun*

de·ri·so·ry \di-ˈrī-sə-rē, -zə-\ *adjective* (1618)
1 : expressing derision **:** DERISIVE
2 : worthy of derision; *especially* **:** laughably small ⟨a *derisory* sum⟩

de·riv·able \di-ˈrī-və-bəl\ *adjective* (1653)
: capable of being derived

der·i·vate \ˈder-ə-ˌvāt\ *noun* (1660)
: DERIVATIVE

der·i·va·tion \ˌder-ə-ˈvā-shən\ *noun* (15th century)
1 a (1) **:** the formation of a word from another word or base (as by the addition of a usually noninflectional affix) (2) **:** an act of ascertaining or stating the derivation of a word (3) **:** ETYMOLOGY 1 **b :** the relation of a word to its base
2 a : SOURCE, ORIGIN **b :** DESCENT, ORIGINATION
3 : something derived **:** DERIVATIVE
4 : an act or process of deriving
5 : a sequence of statements (as in logic or mathematics) showing that a result is a necessary consequence of previously accepted statements
— der·i·va·tion·al \-shnəl, -shə-nᵊl\ *adjective*

[1]**de·riv·a·tive** \di-ˈri-və-tiv\ *noun* (15th century)
1 : a word formed by derivation
2 : something derived
3 : the limit of the ratio of the change in a function to the corresponding change in its independent variable as the latter change approaches zero

◇ WORD HISTORY
derby The first horse race referred to as a Derby was named after Edward Stanley, the twelfth Earl of Derby (1752–1834). Stanley's inherited title originally derived from the borough and county of Derby in north central England. In 1780 the earl initiated an annual race at Epsom Downs in the southeastern county of Surrey. That race continues to be run up to the present on the first Wednesday of June. The word *derby* has become attached to other horse races usually restricted to three-year-olds, such as the Kentucky Derby, first run in 1875. It is used as well of races that have nothing to do with horses, such as the Soap Box Derby, for small homemade motorless racing cars.

4 a : a chemical substance related structurally to another substance and theoretically derivable from it **b :** a substance that can be made from another substance

[2]derivative *adjective* (circa 1530)
1 : formed by derivation
2 : made up of or marked by derived elements
3 : lacking originality **:** BANAL
— **de·riv·a·tive·ly** *adverb*
— **de·riv·a·tive·ness** *noun*

de·riv·a·ti·za·tion \də-ˌri-və-tə-'zā-shən\ *noun* (1967)
: the conversion of a chemical compound into a derivative (as for identification)
— **de·riv·a·tize** \də-'ri-və-ˌtīz\ *transitive verb*

de·rive \di-'rīv, dē-\ *verb* **de·rived; de·riv·ing** [Middle English, from Middle French *deriver,* from Latin *derivare,* literally, to draw off (water), from *de-* + *rivus* stream — more at RUN] (14th century)
transitive verb
1 a : to take, receive, or obtain especially from a specified source **b :** to obtain (a chemical substance) actually or theoretically from a parent substance
2 : INFER, DEDUCE
3 *archaic* **:** BRING
4 : to trace the derivation of
intransitive verb
: to have or take origin **:** come as a derivative
synonym see SPRING
— **de·riv·er** *noun*

derm- *or* **derma-** *or* **dermo-** *combining form* [New Latin, from Greek *derm-, dermo-,* from *derma,* from *derein* to skin — more at TEAR]
: skin ⟨*derm*al⟩

-derm \ˌdərm\ *noun combining form* [probably from French *-derme,* from Greek *derma*]
: skin **:** covering ⟨ecto*derm*⟩

-der·ma \'dər-mə\ *noun combining form, plural* **-dermas** *or* **-der·ma·ta** \-mə-tə\ [New Latin, from Greek *dermat-, derma* skin]
: skin or skin ailment of a (specified) type ⟨sclero*derma*⟩

derm·abra·sion \ˌdər-mə-'brā-zhən\ *noun* (circa 1954)
: surgical removal of skin blemishes or imperfections (as scars or tattoos) by abrasion (as with sandpaper or wire brushes)

der·mal \'dər-məl\ *adjective* (circa 1803)
1 : of or relating to skin and especially to the dermis **:** CUTANEOUS
2 : EPIDERMAL

dermat- *or* **dermato-** *combining form* [Greek, from *dermat-, derma*]
: skin ⟨*dermat*itis⟩ ⟨*dermato*logy⟩

der·ma·ti·tis \ˌdər-mə-'tī-təs\ *noun* (1876)
: inflammation of the skin

der·mat·o·gen \(ˌ)dər-'ma-tə-jən\ *noun* [International Scientific Vocabulary] (1882)
: PROTODERM

der·ma·to·glyph·ics \ˌdər-mə-tə-'gli-fiks\ *noun plural but singular or plural in construction* [*dermat-* + Greek *glyphein* to carve + English *-ics* — more at CLEAVE] (1926)
1 : skin patterns; *especially* **:** patterns of the specialized skin of the inferior surfaces of the hands and feet
2 : the science of the study of skin patterns
— **der·ma·to·glyph·ic** \-fik\ *adjective*

der·ma·tol·o·gy \ˌdər-mə-'tä-lə-jē\ *noun* (1819)
: a branch of science dealing with the skin, its structure, functions, and diseases
— **der·ma·to·log·ic** \-mə-t^{ə}l-'ä-jik\ *or* **der·ma·to·log·i·cal** \-ji-kəl\ *adjective*
— **der·ma·tol·o·gist** \-mə-'tä-lə-jist\ *noun*

der·ma·tome \'dər-mə-ˌtōm\ *noun* [International Scientific Vocabulary *dermat-* + *-ome*] (1910)
: the lateral wall of a somite from which the dermis is produced
— **der·ma·to·mal** \ˌdər-mə-'tō-məl\ *adjective*

der·ma·to·phyte \(ˌ)dər-'ma-tə-ˌfīt, 'dər-mə-\ *noun* [International Scientific Vocabulary] (1882)
: a fungus parasitic on the skin or skin derivatives (as hair or nails)

der·ma·to·sis \ˌdər-mə-'tō-səs\ *noun, plural* **-to·ses** \-ˌsēz\ (1866)
: a disease of the skin

-der·ma·tous \'dər-mə-təs\ *adjective combining form* [Greek *dermat-, derma* skin]
: having a (specified) type of skin ⟨pachy*dermatous*⟩

der·mes·tid \(ˌ)dər-'mes-təd\ *noun* [ultimately from Greek *dermēstēs,* a leather-eating worm, literally, skin eater, from *derm-* + *edmenai* to eat — more at EAT] (circa 1888)
: any of a family (Dermestidae) of beetles with clubbed antennae that are very destructive to organic material of animal origin (as dried meat, wool, or museum specimens)
— **dermestid** *adjective*

der·mis \'dər-məs\ *noun* [New Latin, from Late Latin *-dermis*] (circa 1830)
: the sensitive vascular inner mesodermic layer of the skin — called also *corium, cutis*

-der·mis \'dər-məs\ *noun combining form* [Late Latin, from Greek, from *derma*]
: layer of skin or tissue ⟨endo*dermis*⟩

der·moid cyst \'dər-ˌmȯid-\ *noun* (1872)
: a cystic tumor often of the ovary that contains skin and skin derivatives (as hair or teeth) — called also *dermoid*

der·nier cri \ˌdern-ˌyā-'krē\ *noun* [French, literally, last cry] (1896)
: the newest fashion

der·o·gate \'der-ə-ˌgāt\ *verb* **-gat·ed; -gat·ing** [Middle English, from Late Latin *derogatus,* past participle of *derogare,* from Latin, to annul (a law), detract, from *de-* + *rogare* to ask, propose (a law) — more at RIGHT] (15th century)
transitive verb
: to cause to seem inferior **:** DISPARAGE
intransitive verb
1 : to take away a part so as to impair **:** DETRACT
2 : to act beneath one's position or character
— **der·o·ga·tion** \ˌder-ə-'gā-shən\ *noun*
— **de·ro·ga·tive** \di-'rä-gə-tiv\ *adjective*

de·rog·a·to·ry \di-'rä-gə-ˌtōr-ē, -ˌtȯr-\ *adjective* (1503)
1 : detracting from the character or standing of something — often used with *to, of,* or *from*
2 : expressive of a low opinion **:** DISPARAGING ⟨*derogatory* remarks⟩
— **de·rog·a·to·ri·ly** \-ˌrä-gə-'tōr-ə-lē, -'tȯr-\ *adverb*

der·rick \'der-ik\ *noun* [obsolete *derrick* hangman, gallows, from *Derick,* name of 17th century English hangman] (circa 1752)
1 : a hoisting apparatus employing a tackle rigged at the end of a beam
2 : a framework or tower over a deep drill hole (as of an oil well) for supporting boring tackle or for hoisting and lowering ◆

derrick 2

der·ri·ere *or* **der·ri·ère** \ˌder-ē-'er\ *noun* [French *derrière,* from Old French *derrier* back part, rear, from *derier,* adverb, behind, from Late Latin *deretro,* from Latin *de* from + *retro* back] (1774)
: BUTTOCKS

der·ring–do \ˌder-iŋ-'dü\ *noun* [Middle English *dorring don* daring to do, from *dorring* (gerund of *dorren* to dare) + *don* to do] (1579)
: daring action **:** DARING ⟨deeds of *derring-do*⟩

der·rin·ger \'der-ən-jər\ *noun* [Henry *Deringer* (died 1869) American inventor] (1853)
: a short-barreled pocket pistol

der·ris \'der-əs\ *noun* [New Latin, genus name, from Greek, skin, from *derein* to skin — more at TEAR] (1919)
1 : a preparation of derris roots and stems used as an insecticide
2 : any of a large genus (*Derris*) of leguminous tropical Old World shrubs and woody vines including sources of poisons and especially commercial sources of rotenone

der·vish \'dər-vish\ *noun* [Turkish *derviş,* literally, beggar, from Persian *darvīsh*] (1585)
1 : a member of a Muslim religious order noted for devotional exercises (as bodily movements leading to a trance)
2 : one that whirls or dances with or as if with the abandonment of a dervish

des- *prefix* [French *dés-,* from Old French *des-* — more at DE-]
: DE- 6 — especially before vowels ⟨*des*oxy⟩

DES \ˌdē-(ˌ)ē-'es\ *noun* (1971)
: DIETHYLSTILBESTROL

de·sa·cral·ize \(ˌ)dē-'sā-krə-ˌlīz, -'sa-\ *transitive verb* **-ized; -iz·ing** (1911)
: to divest of sacred qualities
— **de·sa·cral·iza·tion** \(ˌ)dē-ˌsā-krə-lī-'zā-shən, -ˌsa-\ *noun*

de·sa·li·nate \(ˈ)dē-'sa-lə-ˌnāt *also* -'sā-\ *transitive verb* **-nat·ed; -nat·ing** (1949)
: DESALT
— **de·sa·li·na·tion** \(ˌ)dē-ˌsa-lə-'nā-shən *also* -ˌsā-\ *noun*
— **de·sa·li·na·tor** \(ˌ)dē-'sa-lə-ˌnā-tər *also* -'sā-\ *noun*

de·sa·li·nize \(ˈ)dē-'sa-lə-ˌnīz *also* -'sā-\ *transitive verb* **-nized; -niz·ing** (1963)
: DESALT
— **de·sa·li·ni·za·tion** \(ˌ)dē-ˌsa-lə-nə-'zā-shən *also* -ˌsā-\ *noun*

de·salt \(ˌ)dē-'sȯlt\ *transitive verb* (circa 1904)
: to remove salt from
— **de·salt·er** *noun*

[1]des·cant \'des-ˌkant\ *noun* [Middle English *dyscant,* from Old North French & Medieval Latin; Old North French *descant,* from Medieval Latin *discantus,* from Latin *dis-* + *cantus* song — more at CHANT] (14th century)
1 a : a melody or counterpoint sung above the plainsong of the tenor **b :** the art of composing or improvising contrapuntal part music; *also* **:** the music so composed or improvised **c :** SOPRANO, TREBLE **d :** a superimposed counterpoint to a simple melody sung typically by some or all of the sopranos
2 : discourse or comment on a theme

◇ WORD HISTORY

derrick *Derick* or *Derrick* was apparently the surname of a well-known public executioner active about 1600 at Tyburn, an execution site in what is now central London. Our sole evidence for his existence, however, consists of a few references in plays and popular ballads that circulated in the first decade of the 17th century. According to Thomas Blount's *Glossographia,* an English dictionary published in 1656, *Deric* had become an abusive byword for a hangman. Nothing more is heard of this word for a time, but then *derrick* turns up in the 18th century applied to a kind of block and tackle used aboard ships, and the habitual, if tenuous, assumption is that the famous hangman's name was transferred to a gallows, and the name for a gallows applied to other hoisting apparatuses. In the United States, *derrick* is probably most frequently used for the tower-like structure suspended over an oil well.

²des·cant \'des-ˌkant, des-', dis-'\ *intransitive verb* (15th century)
1 : to sing or play a descant; *broadly* **:** SING
2 : COMMENT, DISCOURSE

de·scend \di-'send, dē-\ *verb* [Middle English, from Old French *descendre,* from Latin *descendere,* from *de-* + *scandere* to climb — more at SCAN] (13th century)
intransitive verb
1 : to pass from a higher place or level to a lower one ⟨*descended* from the platform⟩
2 : to pass in discussion from what is logically prior or more comprehensive
3 a : to come down from a stock or source **:** DERIVE ⟨*descends* from an old merchant family⟩ **b :** to pass by inheritance ⟨a desk that has *descended* in the family⟩ **c :** to pass by transmission ⟨songs *descended* from old ballads⟩
4 : to incline, lead, or extend downward ⟨the road *descends* to the river⟩
5 a : to swoop or pounce down (as in a sudden attack) **b :** to appear suddenly and often disconcertingly as if from above ⟨reporters *descended* on the candidate⟩
6 : to proceed in a sequence or gradation from higher to lower or from more remote to nearer or more recent
7 a : to lower oneself in status or dignity **:** STOOP **b :** to worsen and sink in condition or estimation
transitive verb
1 : to pass, move, or climb down or down along
2 : to extend down along
— de·scend·ible \-'sen-də-bəl\ *adjective*

¹de·scen·dant *or* **de·scen·dent** \di-'sen-dənt\ *adjective* [Middle French & Latin; Middle French *descendant,* from Latin *descendent-, descendens,* present participle of *descendere*] (1572)
1 : moving or directed downward
2 : proceeding from an ancestor or source

²descendant *or* **descendent** *noun* [French & Latin; French *descendant,* from Late Latin *descendent-, descendens,* from Latin] (1600)
1 : one descended from another or from a common stock
2 : one deriving directly from a precursor or prototype

de·scend·er \di-'sen-dər, 'dē-ˌ\ *noun* (1802)
: the part of a lowercase letter (as p) that descends below the main body of the letter; *also* **:** a letter that has such a part

de·scen·sion \di-'sen-chən\ *noun* (15th century)
archaic **:** DESCENT 2

de·scent \di-'sent\ *noun* [Middle English, from Middle French *descente,* from Old French *descendre*] (14th century)
1 a : derivation from an ancestor **:** BIRTH, LINEAGE ⟨of French *descent*⟩ **b :** transmission or devolution of an estate by inheritance usually in the descending line **c :** the fact or process of originating from an ancestral stock **d :** the shaping or development in nature and character by transmission from a source **:** DERIVATION
2 : the act or process of descending
3 : a step downward in a scale of gradation; *specifically* **:** one generation in an ancestral line or genealogical scale
4 a : an inclination downward **:** SLOPE **b :** a descending way (as a downgrade or stairway) **c** *obsolete* **:** the lowest part
5 a : ATTACK, INVASION **b :** a sudden disconcerting appearance (as for a visit)
6 : a downward step (as in station or value) **:** DECLINE ⟨*descent* of the family to actual poverty⟩

de·scribe \di-'skrīb\ *transitive verb* **de·scribed; de·scrib·ing** [Middle English, from Latin *describere,* from *de-* + *scribere* to write — more at SCRIBE] (15th century)
1 : to represent or give an account of in words ⟨*describe* a picture⟩
2 : to represent by a figure, model, or picture **:** DELINEATE
3 *obsolete* **:** DISTRIBUTE
4 : to trace or traverse the outline of ⟨*describe* a circle⟩
5 *archaic* **:** OBSERVE, PERCEIVE
— de·scrib·able \-'skrī-bə-bəl\ *adjective*
— de·scrib·er *noun*

de·scrip·tion \di-'skrip-shən\ *noun* [Middle English *descripcioun,* from Middle French & Latin; Middle French *description,* from Latin *description-, descriptio,* from *describere*] (14th century)
1 a : an act of describing; *specifically* **:** discourse intended to give a mental image of something experienced **b :** a descriptive statement or account
2 : kind or character especially as determined by salient features ⟨opposed to any tax of so radical a *description*⟩
synonym see TYPE

de·scrip·tive \di-'skrip-tiv\ *adjective* (1751)
1 : serving to describe ⟨a *descriptive* account⟩
2 a : referring to, constituting, or grounded in matters of observation or experience ⟨the *descriptive* basis of science⟩ **b :** factually grounded or informative rather than normative, prescriptive, or emotive ⟨*descriptive* cultural studies⟩
3 *of a modifier* **a :** expressing the quality, kind, or condition of what is denoted by the modified term ⟨*hot* in "hot water" is a *descriptive* adjective⟩ **b :** NONRESTRICTIVE
4 : of, relating to, or dealing with the structure of a language at a particular time usually with exclusion of historical and comparative data ⟨*descriptive* linguistics⟩
— de·scrip·tive·ly *adverb*
— de·scrip·tive·ness *noun*

de·scrip·tor \di-'skrip-tər\ *noun* (1933)
: something (as a word or characteristic feature) that serves to describe or identify; *especially* **:** a word or phrase (as an index term) used to identify an item (as a subject or document) in an information retrieval system

¹de·scry \di-'skrī\ *transitive verb* **de·scried; de·scry·ing** [Middle English *descrien,* from Middle French *descrier* to proclaim, decry] (14th century)
1 a : to catch sight of **b :** FIND OUT, DISCOVER
2 *obsolete* **:** to make known **:** REVEAL

²descry *noun* (1605)
obsolete **:** discovery or view from afar

Des·de·mo·na \ˌdez-də-'mō-nə\ *noun*
: the wife of Othello in Shakespeare's *Othello*

des·e·crate \'de-si-ˌkrāt\ *transitive verb* **-crat·ed; -crat·ing** [*de-* + *-secrate* (as in *consecrate*)] (1677)
1 : to violate the sanctity of **:** PROFANE
2 : to treat disrespectfully, irreverently, or outrageously ⟨the kind of shore development . . . that has *desecrated* so many waterfronts —John Fischer⟩
— des·e·crat·er *or* **des·e·cra·tor** \-ˌkrā-tər\ *noun*

des·e·cra·tion \ˌde-si-'krā-shən\ *noun* (circa 1717)
: an act or instance of desecrating **:** the state of being desecrated

de·seg·re·gate \(ˌ)dē-'se-gri-ˌgāt\ (1944)
transitive verb
: to eliminate segregation in; *specifically* **:** to free of any law, provision, or practice requiring isolation of the members of a particular race in separate units
intransitive verb
: to become desegregated

de·seg·re·ga·tion \(ˌ)dē-ˌse-gri-'gā-shən\ *noun* (1951)
1 : the action or an instance of desegregating
2 : the state of being desegregated

de·se·lect \ˌdē-sə-'lekt\ *transitive verb* (1965)
: DISMISS, REJECT

de·sen·si·tize \(ˌ)dē-'sen(t)-sə-ˌtīz\ *transitive verb* (1898)
1 : to make (a sensitized or hypersensitive individual) insensitive or nonreactive to a sensitizing agent
2 : to make emotionally insensitive or callous; *specifically* **:** to extinguish an emotional response (as of fear, anxiety, or guilt) to stimuli that formerly induced it
— de·sen·si·ti·za·tion \(ˌ)dē-ˌsen(t)-sə-tə-'zā-shən\ *noun*
— de·sen·si·tiz·er \(ˌ)dē-'sen-sə-ˌtī-zər\ *noun*

¹des·ert \'de-zərt\ *noun* [Middle English, from Old French, from Late Latin *desertum,* from Latin, neuter of *desertus,* past participle of *deserere* to desert, from *de-* + *serere* to join together — more at SERIES] (13th century)
1 a : arid barren land; *especially* **:** a tract incapable of supporting any considerable population without an artificial water supply **b :** an area of water apparently devoid of life
2 *archaic* **:** a wild uninhabited and uncultivated tract
3 : a desolate or forbidding area ⟨lost in a *desert* of doubt⟩
— de·ser·tic \de-'zər-tik\ *adjective*

²des·ert \'de-zərt\ *adjective* (13th century)
1 : desolate and sparsely occupied or unoccupied ⟨a *desert* island⟩
2 : of or relating to a desert
3 *archaic* **:** FORSAKEN

³de·sert \di-'zərt\ *noun* [Middle English *deserte,* from Old French, from feminine of *desert,* past participle of *deservir* to deserve] (13th century)
1 : the quality or fact of deserving reward or punishment
2 : deserved reward or punishment — usually used in plural ⟨got their just *deserts*⟩
3 : EXCELLENCE, WORTH

⁴de·sert \di-'zərt\ *verb* [French *déserter,* from Late Latin *desertare,* frequentative of Latin *deserere*] (1603)
transitive verb
1 : to withdraw from or leave usually without intent to return
2 a : to leave in the lurch ⟨*desert* a friend in trouble⟩ **b :** to abandon (military service) without leave
intransitive verb
: to quit one's post, allegiance, or service without leave or justification; *especially* **:** to abandon military duty without leave and without intent to return
synonym see ABANDON
— de·sert·er *noun*

de·sert·i·fi·ca·tion \di-ˌzər-tə-fə-'kā-shən\ *noun* (1974)
: the process of becoming desert (as from land mismanagement or climate change)

de·ser·tion \di-'zər-shən\ *noun* (1591)
1 : an act of deserting; *especially* **:** the abandonment without consent or legal justification of a person, post, or relationship and the associated duties and obligations ⟨sued for divorce on grounds of *desertion*⟩
2 : a state of being deserted or forsaken

desert locust *noun* (1944)
: a destructive migratory locust (*Schistocerca gregaria*) of southwestern Asia and parts of northern Africa

desert soil *noun* (circa 1938)
: a soil that develops under sparse shrub vegetation in warm to cool arid climates with a light-colored surface soil usually underlain by calcareous material and a hardpan layer

desert varnish *noun* (circa 1898)
: a dark coating which is found on rocks after long exposure in desert regions and whose color is due to iron and manganese oxides

de·serve \di-'zərv\ *verb* **de·served; de·serv·ing** [Middle English, from Old French *deservir,* from Latin *deservire* to devote oneself to, from *de-* + *servire* to serve] (13th century)
transitive verb
: to be worthy of **:** MERIT ⟨*deserves* another chance⟩
intransitive verb
: to be worthy, fit, or suitable for some reward

or requital ⟨have become recognized as they *deserve* —T. S. Eliot⟩
— **de•serv•er** *noun*

de•served \-ˈzərvd\ *adjective* (circa 1552)
: of, relating to, or being that which one deserves ⟨a *deserved* reputation⟩
— **de•served•ly** \-ˈzər-vəd-lē, -ˈzərvd-lē\ *adverb*
— **de•served•ness** \-ˈzər-vəd-nəs, -ˈzərvd-nəs\ *noun*

¹de•serv•ing \-ˈzər-viŋ\ *noun* (14th century)
: DESERT, MERIT ⟨reward the proud according to their *deservings* —Charles Kingsley⟩

²deserving *adjective* (1576)
: MERITORIOUS, WORTHY; *especially* : meriting financial aid ⟨scholarships for *deserving* students⟩

de•sex \(ˌ)dē-ˈseks\ *transitive verb* (1911)
1 : CASTRATE, SPAY
2 : to eliminate perceived sexism from ⟨*desex* the language of church Bible study programs —R. M. Harley⟩

de•sex•u•al•ize \(ˌ)dē-ˈsek-sh(ə-)wə-ˌlīz, -ˈsek-shə-ˌlīz\ *transitive verb* (1894)
1 : to deprive of sexual characters or power
2 : to divest of sexual quality
— **de•sex•u•al•iza•tion** \(ˌ)dē-ˌsek-sh(ə-)wə-lə-ˈzā-shən, -ˈsek-shə-lə-\ *noun*

des•ha•bille \ˌde-sə-ˈbē(ə)l, -ˈbil, -ˈbē\ *variant of* DISHABILLE

des•ic•cant \ˈde-si-kənt\ *noun* (1676)
: a drying agent (as calcium chloride)

des•ic•cate \ˈde-si-ˌkāt\ *verb* **-cat•ed; -cat•ing** [Latin *desiccatus,* past participle of *desiccare* to dry up, from *de-* + *siccare* to dry, from *siccus* dry — more at SACK] (1575)
transitive verb
1 : to dry up
2 : to preserve (a food) by drying : DEHYDRATE
3 : to drain of emotional or intellectual vitality
intransitive verb
: to become dried up
— **des•ic•ca•tion** \ˌde-si-ˈkā-shən\ *noun*
— **des•sic•ca•tive** \ˈde-si-ˌkā-tiv\ *adjective*
— **des•ic•ca•tor** \ˈde-si-ˌkā-tər\ *noun*

de•sid•er•ate \di-ˈsi-də-ˌrāt, -ˈzi-\ *transitive verb* **-at•ed; -at•ing** [Latin *desideratus,* past participle of *desiderare* to desire] (1645)
: to entertain or express a wish to have or attain
— **de•sid•er•a•tion** \-ˌsi-də-ˈrā-shən, -ˌzi-\ *noun*
— **de•sid•er•a•tive** \-ˈsi-də-ˌrā-tiv, -ˈsi-d(ə-)rət-, -ˈzi-\ *adjective*

de•sid•er•a•tum \di-ˌsi-də-ˈrä-təm, -ˌzi-, -ˈrā-\ *noun, plural* **-ta** \-tə\ [Latin, neuter of *desideratus*] (1652)
: something desired as essential

¹de•sign \di-ˈzīn\ *verb* [Middle English, to outline, indicate, mean, from Middle French & Medieval Latin; Middle French *designer* to designate, from Medieval Latin *designare,* from Latin, to mark out, from *de-* + *signare* to mark — more at SIGN] (14th century)
transitive verb
1 : to create, fashion, execute, or construct according to plan : DEVISE, CONTRIVE
2 a : to conceive and plan out in the mind ⟨he *designed* the perfect crime⟩ **b :** to have as a purpose : INTEND ⟨she *designed* to excel in her studies⟩ **c :** to devise for a specific function or end ⟨a book *designed* primarily as a college textbook⟩
3 *archaic* **:** to indicate with a distinctive mark, sign, or name
4 a : to make a drawing, pattern, or sketch of **b :** to draw the plans for
intransitive verb
1 : to conceive or execute a plan
2 : to draw, lay out, or prepare a design
— **de•sign•ed•ly** \-ˈzī-nəd-lē\ *adverb*

²design *noun* (1588)
1 a : a particular purpose held in view by an individual or group ⟨he has ambitious *designs* for his son⟩ **b :** deliberate purposive planning ⟨more by accident than *design*⟩
2 : a mental project or scheme in which means to an end are laid down
3 a : a deliberate undercover project or scheme : PLOT **b** *plural* **:** aggressive or evil intent — used with *on* or *against* ⟨he has *designs* on the money⟩
4 : a preliminary sketch or outline showing the main features of something to be executed : DELINEATION
5 a : an underlying scheme that governs functioning, developing, or unfolding : PATTERN, MOTIF ⟨the general *design* of the epic⟩ **b :** a plan or protocol for carrying out or accomplishing something (as a scientific experiment); *also* **:** the process of preparing this
6 : the arrangement of elements or details in a product or work of art
7 : a decorative pattern
8 : the creative art of executing aesthetic or functional designs
synonym see INTENTION, PLAN

¹des•ig•nate \ˈde-zig-ˌnāt, -nət\ *adjective* [Latin *designatus,* past participle of *designare*] (1646)
: chosen but not yet installed ⟨ambassador *designate*⟩

²des•ig•nate \-ˌnāt\ *transitive verb* **-nat•ed; -nat•ing** (1791)
1 : to indicate and set apart for a specific purpose, office, or duty ⟨*designate* a group to prepare a plan⟩
2 a : to point out the location of ⟨a marker *designating* the battle⟩ **b :** to distinguish as to class ⟨the area we *designate* as that of spiritual values —J. B. Conant⟩ **c :** SPECIFY, STIPULATE ⟨to be sent by a *designated* shipper⟩
3 : DENOTE ⟨associate names with the people they *designate*⟩
4 : to call by a distinctive title, term, or expression ⟨a particle *designated* the neutron⟩
— **des•ig•na•tive** \-ˌnā-tiv\ *adjective*
— **des•ig•na•tor** \-ˌnā-tər\ *noun*
— **des•ig•na•to•ry** \-nə-ˌtōr-ē, -ˌtȯr-\ *adjective*

designated hitter *noun* (1973)
: a baseball player designated at the start of the game to bat in place of the pitcher without causing the pitcher to be removed from the game

des•ig•na•tion \ˌde-zig-ˈnā-shən\ *noun* (14th century)
1 : the act of indicating or identifying
2 : appointment to or selection for an office, post, or service
3 : a distinguishing name, sign, or title
4 : the relation between a sign and the thing signified

des•ig•nee \ˌde-zig-ˈnē\ *noun* (1925)
: one who is designated

¹de•sign•er \di-ˈzī-nər\ *noun* (1662)
: one that designs: as **a :** one who creates and often executes plans for a project or structure ⟨urban *designers*⟩ ⟨a theater set *designer*⟩ **b :** one that creates and manufactures a new product style or design; *especially* **:** one who designs and manufactures high-fashion clothing ⟨the *designer's* new fall line⟩

²designer *adjective* (1966)
1 : of, relating to, or produced by a designer ⟨*designer* wallpaper⟩ ⟨wearing a *designer* original⟩; *also* **:** displaying the name, signature, or logo of a designer or manufacturer ⟨*designer* jeans⟩
2 : intended to reflect the latest in sophisticated taste or fashion ⟨*designer* ice cream⟩ ⟨a *designer* haircut⟩

designer drug *noun* (1983)
: a synthetic version of a controlled substance (as heroin) that is produced with a slightly altered molecular structure to avoid having it classified as an illicit drug

de•sign•ing \di-ˈzī-niŋ\ *adjective* (1653)
1 : practicing forethought
2 : CRAFTY, SCHEMING

de•sign•ment \di-ˈzīn-mənt\ *noun* (1583)
obsolete **:** PLAN, PURPOSE

de•si•pra•mine \də-ˈzi-prə-ˌmēn\ *noun* [*des*methyl (from *des-* + *methyl*) + *imipramine*] (1965)
: a tricyclic antidepressant $C_{18}H_{22}N_2$

de•sir•abil•i•ty \di-ˌzī-rə-ˈbi-lə-tē\ *noun, plural* **-ties** (1824)
1 *plural* **:** desirable conditions ⟨had understood and studied certain *desirabilities* —D. D. Eisenhower⟩
2 : the quality, fact, or degree of being desirable

¹de•sir•able \di-ˈzī-rə-bəl\ *adjective* (14th century)
1 : having pleasing qualities or properties : ATTRACTIVE ⟨a *desirable* woman⟩
2 : worth seeking or doing as advantageous, beneficial, or wise : ADVISABLE ⟨*desirable* legislation⟩
— **de•sir•able•ness** *noun*
— **de•sir•ably** \-blē\ *adverb*

²desirable *noun* (1645)
: one that is desirable

¹de•sire \di-ˈzīr, dē-\ *verb* **de•sired; de•sir•ing** [Middle English, from Old French *desirer,* from Latin *desiderare,* from *de-* + *sider-, sidus* heavenly body] (13th century)
transitive verb
1 : to long or hope for : exhibit or feel desire for
2 a : to express a wish for : REQUEST **b** *archaic* **:** to express a wish to : ASK
3 *obsolete* **:** INVITE
4 *archaic* **:** to feel the loss of
intransitive verb
: to have or feel desire ☆

²desire *noun* (14th century)
1 : conscious impulse toward something that promises enjoyment or satisfaction in its attainment
2 a : LONGING, CRAVING **b :** sexual urge or appetite
3 : a usually formal request or petition for some action
4 : something desired

de•sir•ous \di-ˈzīr-əs\ *adjective* (14th century)
: impelled or governed by desire ⟨*desirous* of fame⟩
— **de•sir•ous•ly** *adverb*
— **de•sir•ous•ness** *noun*

de•sist \di-ˈzist, -ˈsist, dē-\ *intransitive verb* [Middle English, from Middle French *desister,* from Latin *desistere,* from *de-* + *sistere* to stand, stop; akin to Latin *stare* to stand — more at STAND] (15th century)
: to cease to proceed or act
synonym see STOP
— **de•sis•tance** \-ˈzis-tən(t)s, -ˈsis-\ *noun*

desk \ˈdesk\ *noun* [Middle English *deske,* from Medieval Latin *desca,* modification of Old Italian *desco* table, from Latin *discus* dish, disc — more at DISH] (14th century)
1 a : a table, frame, or case with a sloping or horizontal surface especially for writing and

☆ SYNONYMS
Desire, wish, want, crave, covet mean to have a longing for. DESIRE stresses the strength of feeling and often implies strong intention or aim ⟨*desires* to start a new life⟩. WISH sometimes implies a general or transient longing especially for the unattainable ⟨*wishes* for permanent world peace⟩. WANT specifically suggests a felt need or lack ⟨*wants* to have a family⟩. CRAVE stresses the force of physical appetite or emotional need ⟨*craves* sweets⟩. COVET implies strong envious desire ⟨*covets* his rise to fame⟩.

reading and often with drawers, compartments, and pigeonholes **b :** a reading table or lectern from which a liturgical service is read **c :** a table, counter, stand, or booth at which a person works
2 a : a division of an organization specializing in a particular phase of activity ⟨the Russian *desk* in the Department of State⟩ **b :** a seating position according to rank in an orchestra ⟨a first-*desk* violinist⟩
desk·bound \ˈdesk-ˌbau̇nd\ *adjective* (1944)
: restricted to work at a desk
de–skill \ˌdē-ˈskil\ *transitive verb* (1941)
1 : to reduce the level of skill needed for (a job)
2 : to reduce the level of skill needed for a job by (a worker)
desk·man \ˈdesk-ˌman, -mən\ *noun* (1913)
: one that works at a desk; *specifically* **:** a newspaperman who processes news and prepares copy
1**desk·top** \ˈdesk-ˌtäp\ *noun* (1929)
1 : the top of a desk; *also* **:** an office desktop simulated by a computer program
2 : a desktop computer
2**desktop** *adjective* (1958)
: of a size that can be conveniently used on a desk or table ⟨*desktop* computers⟩ — compare LAPTOP
desktop publishing *noun* (1984)
: the production of printed matter by means of a desktop computer having a layout program that integrates text and graphics
desm- *or* **desmo-** *combining form* [New Latin, from Greek, from *desmos,* from *dein* to bind — more at DIADEM]
: bond **:** ligament ⟨*desmo*some⟩
des·mid \ˈdez-məd\ *noun* [ultimately from Greek *desmos*] (1862)
: any of numerous unicellular or colonial green algae (order Zygnematales)

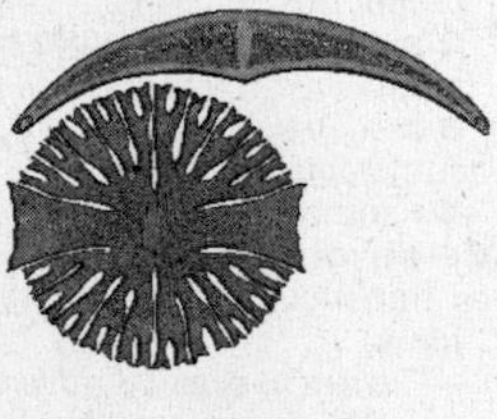

desmid

des·mo·some \ˈdez-mə-ˌsōm\ *noun* (circa 1932)
: a specialized local thickening of the cell membrane of an epithelial cell that serves to anchor contiguous cells together
— des·mo·som·al \-ˌsō-məl\ *adjective*
1**des·o·late** \ˈde-sə-lət, ˈde-zə-\ *adjective* [Middle English *desolat,* from Latin *desolatus,* past participle of *desolare* to abandon, from *de-* + *solus* alone] (14th century)
1 : devoid of inhabitants and visitors **:** DESERTED
2 : joyless, disconsolate, and sorrowful through or as if through separation from a loved one
3 a : showing the effects of abandonment and neglect **:** DILAPIDATED **b :** BARREN, LIFELESS ⟨a *desolate* landscape⟩ **c :** devoid of warmth, comfort, or hope **:** GLOOMY ⟨*desolate* memories⟩
synonym see ALONE
— des·o·late·ly *adverb*
— des·o·late·ness *noun*
2**des·o·late** \-ˌlāt\ *transitive verb* **-lat·ed; -lat·ing** (14th century)
: to make desolate: **a :** to deprive of inhabitants **b :** to lay waste **c :** FORSAKE **d :** to make wretched
— des·o·lat·er *or* **des·o·la·tor** \-ˌlā-tər\ *noun*
— des·o·lat·ing·ly \-ˌlā-tiŋ-lē\ *adverb*
des·o·la·tion \ˌde-sə-ˈlā-shən, ˌde-zə-\ *noun* (14th century)
1 : the action of desolating
2 a : GRIEF, SADNESS **b :** LONELINESS
3 : DEVASTATION, RUIN
4 : barren wasteland

de·sorb \(ˌ)dē-ˈsȯrb, -ˈzȯrb\ *transitive verb* (1924)
: to remove (a sorbed substance) by the reverse of adsorption or absorption
de·sorp·tion \-ˈsȯrp-shən, -ˈzȯrp-\ *noun* (1924)
: the process of desorbing
desoxy- — see DEOXY-
des·oxy·ri·bo·nu·cle·ic acid \de-ˌzäk-sē-ˈrī-bō-n(y)u̇-ˌklē-ik-, -ˌklā-\ *noun* (1931)
: DNA
1**de·spair** \di-ˈspar, -ˈsper\ *verb* [Middle English *despeiren,* from Middle French *desperer,* from Latin *desperare,* from *de-* + *sperare* to hope; akin to Latin *spes* hope — more at SPEED] (14th century)
intransitive verb
: to lose all hope or confidence ⟨*despair* of winning⟩
transitive verb
obsolete **:** to lose hope for
— de·spair·er *noun*
2**despair** *noun* (14th century)
1 : utter loss of hope
2 : a cause of hopelessness ⟨an incorrigible child is the *despair* of his parents⟩
de·spair·ing *adjective* (1589)
: given to, arising from, or marked by despair **:** devoid of hope
synonym see DESPONDENT
— de·spair·ing·ly \-iŋ-lē\ *adverb*
des·patch \dis-ˈpach\ *variant of* DISPATCH
des·per·a·do \ˌdes-pə-ˈrä-(ˌ)dō, -ˈrā-\ *noun, plural* **-does** *or* **-dos** [probably alteration of obsolete *desperate* desperado, from *desperate,* adjective] (1647)
: a bold or violent criminal; *especially* **:** a bandit of the western U.S. in the 19th century
des·per·ate \ˈdes-p(ə-)rət, -pərt\ *adjective* [Latin *desperatus,* past participle of *desperare*] (15th century)
1 a : having lost hope ⟨a *desperate* spirit crying for relief⟩ **b :** giving no ground for hope ⟨the outlook was *desperate*⟩
2 a : moved by despair ⟨victims made *desperate* by abuse⟩ **b :** involving or employing extreme measures in an attempt to escape defeat or frustration ⟨made a *desperate* leap for the rope⟩
3 : suffering extreme need or anxiety ⟨*desperate* for money⟩
4 : involving extreme danger or possible disaster ⟨a *desperate* situation⟩
5 : of extreme intensity
6 : SHOCKING, OUTRAGEOUS
synonym see DESPONDENT
— des·per·ate·ly *adverb*
— des·per·ate·ness *noun*
des·per·a·tion \ˌdes-pə-ˈrā-shən\ *noun* (14th century)
1 : loss of hope and surrender to despair
2 : a state of hopelessness leading to rashness
de·spi·ca·ble \di-ˈspi-kə-bəl, ˈdes-(ˌ)pi-\ *adjective* [Late Latin *despicabilis,* from Latin *despicari* to despise] (1553)
: deserving to be despised **:** so worthless or obnoxious as to rouse moral indignation ⟨*despicable* behavior⟩
synonym see CONTEMPTIBLE
— de·spi·ca·ble·ness *noun*
— de·spi·ca·bly \-blē\ *adverb*
de·spir·i·tu·al·ize \(ˌ)dē-ˈspi-ri-ch(ə-)wə-ˌlīz, -chə-ˌlīz\ *transitive verb* (1868)
: to deprive of spiritual character or influence
de·spise \di-ˈspīz\ *transitive verb* **de·spised; de·spis·ing** [Middle English, from Old French *despis-,* stem of *despire,* from Latin *despicere,* from *de-* + *specere* to look — more at SPY] (14th century)
1 : to look down on with contempt or aversion ⟨*despised* the weak⟩
2 : to regard as negligible, worthless, or distasteful ☆
— de·spise·ment \-ˈspīz-mənt\ *noun*
— de·spis·er \-ˈspī-zər\ *noun*

1**de·spite** \di-ˈspīt\ *noun* [Middle English, from Old French *despit,* from Latin *despectus,* from *despicere*] (13th century)
1 : the feeling or attitude of despising **:** CONTEMPT
2 : MALICE, SPITE
3 a : an act showing contempt or defiance **b :** DETRIMENT, DISADVANTAGE ⟨I know of no government which stands to its obligations, even in its own *despite,* more solidly —Sir Winston Churchill⟩
— in despite of : in spite of
2**despite** *transitive verb* **de·spit·ed; de·spit·ing** (14th century)
1 *archaic* **:** to treat with contempt
2 *obsolete* **:** to provoke to anger **:** VEX
3**despite** *preposition* (15th century)
: in spite of ⟨played *despite* an injury⟩
de·spite·ful \di-ˈspīt-fəl\ *adjective* (15th century)
: expressing malice or hate
— de·spite·ful·ly \-fə-lē\ *adverb*
— de·spite·ful·ness *noun*
de·spit·eous \dis-ˈpi-tē-əs\ *adjective* (14th century)
archaic **:** feeling or showing despite **:** MALICIOUS
— de·spit·eous·ly *adverb, archaic*
de·spoil \di-ˈspȯi(ə)l\ *transitive verb* [Middle English *despoylen,* from Middle French *despoillier,* from Latin *despoliare,* from *de-* + *spoliare* to strip, rob — more at SPOIL] (14th century)
: to strip of belongings, possessions, or value **:** PILLAGE
synonym see RAVAGE
— de·spoil·er *noun*
— de·spoil·ment \-mənt\ *noun*
de·spo·li·a·tion \di-ˌspō-lē-ˈā-shən\ *noun* [Late Latin *despoliation-, despoliatio,* from *despoliare*] (circa 1657)
: the condition of being despoiled **:** SPOLIATION
1**de·spond** \di-ˈspänd\ *intransitive verb* [Latin *despondēre,* from *de-* + *spondēre* to promise solemnly — more at SPOUSE] (1655)
: to become despondent
2**despond** *noun* (1678)
: DESPONDENCY
de·spon·dence \di-ˈspän-dən(t)s\ *noun* (1676)
: DESPONDENCY
de·spon·den·cy \-dən-sē\ *noun* (1653)
: the state of being despondent **:** DEJECTION, HOPELESSNESS
de·spon·dent \-dənt\ *adjective* [Latin *despondent-, despondens,* present participle of *despondēre*] (circa 1699)
: feeling or showing extreme discouragement, dejection, or depression ⟨*despondent* about his health⟩ ☆
— de·spon·dent·ly *adverb*
des·pot \ˈdes-pət, -ˌpät\ *noun* [Middle French *despote,* from Greek *despotēs* master, lord, autocrat, from *des-* (akin to *domos* house) + *-potēs* (akin to *posis* husband); akin to Sanskrit *dampati* lord of the house — more at DOME, POTENT] (1585)

☆ SYNONYMS
Despise, contemn, scorn, disdain mean to regard as unworthy of one's notice or consideration. DESPISE may suggest an emotional response ranging from strong dislike to loathing ⟨*despises* cowards⟩. CONTEMN implies a vehement condemnation of a person or thing as low, vile, feeble, or ignominious ⟨*contemns* the image of women promoted by advertisers⟩. SCORN implies a ready or indignant contempt ⟨*scorns* the very thought of retirement⟩. DISDAIN implies an arrogant or supercilious aversion to what is regarded as unworthy ⟨*disdained* popular music⟩.

1 a : a Byzantine emperor or prince **b :** a bishop or patriarch of the Eastern Orthodox Church **c :** an Italian hereditary prince or military leader during the Renaissance
2 a : a ruler with absolute power and authority **b :** a person exercising power tyrannically

des·pot·ic \des-ˈpä-tik, dis-\ *adjective* (1650)
: of, relating to, or characteristic of a despot
— des·pot·i·cal·ly \-ti-k(ə-)lē\ *adverb*

des·po·tism \ˈdes-pə-ˌti-zəm\ *noun* (circa 1727)
1 a : rule by a despot **b :** despotic exercise of power
2 a : a system of government in which the ruler has unlimited power **:** ABSOLUTISM **b :** a despotic state

des·qua·mate \ˈdes-kwə-ˌmāt\ *intransitive verb* **-mat·ed; -mat·ing** [Latin *desquamatus,* past participle of *desquamare* to scale, from *de-* + *squama* scale] (1828)
: to peel off in scales
— des·qua·ma·tion \ˌdes-kwə-ˈmā-shən\ *noun*

des·sert \di-ˈzərt\ *noun* [Middle French, from *desservir* to clear the table, from *des-* de- + *servir* to serve, from Latin *servire*] (1600)
1 : a usually sweet course or dish (as of pastry or ice cream) served at the end of a meal
2 *British* **:** a fresh fruit served after a sweet course

des·sert·spoon \-ˌspün\ *noun* (1754)
1 : a spoon intermediate in size between a teaspoon and a tablespoon for use in eating dessert
2 : DESSERTSPOONFUL

des·sert·spoon·ful \di-ˈzərt-ˌspün-ˌfu̇l\ *noun* (1875)
1 : as much as a dessertspoon will hold
2 *chiefly British* **:** a unit of measure equal to about 2½ fluidrams

dessert wine *noun* (1773)
: a usually sweet wine typically served with dessert or afterward

de·sta·bi·lize \(ˌ)dē-ˈstā-bə-ˌlīz\ *transitive verb* (1924)
1 : to make unstable
2 : to cause (as a government) to be incapable of functioning or surviving
— de·sta·bi·li·za·tion \(ˌ)dē-ˌstā-bə-lə-ˈzā-shən\ *noun*

de·stain \(ˌ)dē-ˈstān\ *transitive verb* (1927)
: to selectively remove stain from (a specimen for microscopic study)

de–Sta·lin·iza·tion \(ˌ)dē-ˌstä-lə-nə-ˈzā-shən, -ˌsta-\ *noun* (1951)
: the discrediting of Stalin and his policies

de Stijl \də-ˈstī(ə)l, -ˈstā(ə)l\ *noun* [Dutch *De Stijl,* literally, the style, magazine published by members of the school] (1934)
: a school of art founded in Holland in 1917 typically using rectangular forms and the primary colors plus black and white and asymmetric balance

des·ti·na·tion \ˌdes-tə-ˈnā-shən\ *noun* (14th century)
1 : the purpose for which something is destined
2 : an act of appointing, setting aside for a purpose, or predetermining
3 : a place to which one is journeying or to which something is sent ⟨kept their *destination* secret⟩

des·tine \ˈdes-tən\ *transitive verb* **des·tined; des·tin·ing** [Middle English, from Middle French *destiner,* from Latin *destinare,* from *de-* + *-stinare* (akin to Latin *stare* to stand) — more at STAND] (14th century)
1 : to decree beforehand **:** PREDETERMINE
2 a : to designate, assign, or dedicate in advance ⟨the younger son was *destined* for the priesthood⟩ **b :** to direct, devise, or set apart for a specific purpose or place ⟨freight *destined* for European ports⟩

des·ti·ny \ˈdes-tə-nē\ *noun, plural* **-nies** [Middle English *destinee,* from Middle French, from feminine of *destiné,* past participle of *destiner*] (14th century)
1 : something to which a person or thing is destined **:** FORTUNE
2 : a predetermined course of events often held to be an irresistible power or agency
synonym see FATE

des·ti·tute \ˈdes-tə-ˌtüt, -ˌt(y)üt\ *adjective* [Middle English, from Latin *destitutus,* past participle of *destituere* to abandon, deprive, from *de-* + *statuere* to set up — more at STATUTE] (14th century)
1 : lacking something needed or desirable ⟨a lake *destitute* of fish⟩
2 : lacking possessions and resources; *especially* **:** suffering extreme poverty ⟨a *destitute* old man⟩
— des·ti·tute·ness *noun*

des·ti·tu·tion \ˌdes-tə-ˈtü-shən, -ˈtyü-\ *noun* (15th century)
: the state of being destitute; *especially* **:** such extreme want as threatens life unless relieved
synonym see POVERTY

des·trier \ˈdes-trē-ər, də-ˈstri(ə)r\ *noun* [Middle English, from Middle French, from *destre* right hand, from Latin *dextra,* from feminine of *dexter*] (14th century)
archaic **:** WAR-HORSE; *also* **:** a charger used especially in medieval tournaments

de·stroy \di-ˈstrȯi, dē-\ *verb* [Middle English, from Old French *destruire,* from (assumed) Vulgar Latin *destrugere,* alteration of Latin *destruere,* from *de-* + *struere* to build — more at STRUCTURE] (13th century)
transitive verb
1 : to ruin the structure, organic existence, or condition of ⟨*destroyed* the files⟩; *also* **:** to ruin as if by tearing to shreds ⟨their reputation was *destroyed*⟩
2 a : to put out of existence **:** KILL **b :** NEUTRALIZE ⟨the moon *destroys* the light of the stars⟩ **c :** ANNIHILATE, VANQUISH ⟨armies had been crippled but not *destroyed* —W. L. Shirer⟩
intransitive verb
: to cause destruction

de·stroy·er \di-ˈstrȯi(-ə)r, dē-\ *noun* (14th century)
1 : one that destroys
2 : a small fast warship used especially to support larger vessels and usually armed with 5-inch guns, depth charges, torpedoes, and often guided missiles

destroyer escort *noun* (1924)
: a warship similar to but smaller than a destroyer

destroying angel *noun* (circa 1900)
: DEATH CAP; *also* **:** a related poisonous mushroom (*Amanita verna*)

de·struc·ti·ble \di-ˈstrək-tə-bəl\ *adjective* (circa 1755)
: capable of being destroyed
— de·struc·ti·bil·i·ty \di-ˌstrək-tə-ˈbi-lə-tē\ *noun*

de·struc·tion \di-ˈstrək-shən\ *noun* [Middle English *destruccioun,* from Middle French *destruction,* from Latin *destruction-, destructio,* from *destruere*] (14th century)
1 : the state or fact of being destroyed **:** RUIN
2 : the action or process of destroying something
3 : a destroying agency

de·struc·tion·ist \-sh(ə-)nəst\ *noun* (1833)
: one who delights in or advocates destruction

de·struc·tive \di-ˈstrək-tiv\ *adjective* (15th century)
1 : causing destruction **:** RUINOUS ⟨*destructive* storm⟩
2 : designed or tending to destroy ⟨*destructive* criticism⟩
— de·struc·tive·ly *adverb*
— de·struc·tive·ness *noun*

destructive distillation *noun* (circa 1831)
: decomposition of a substance (as wood, coal, or oil) by heat in a closed container and collection of the volatile products produced

de·struc·tiv·i·ty \di-ˌstrək-ˈti-və-tē, ˌdē-\ *noun* (1902)
: capacity for destruction

de·sue·tude \ˈde-swi-ˌtüd, -ˌtyüd, di-ˈsü-ə-, -ˈsyü-\ *noun* [Middle English *dissuetude,* from Latin *desuetudo,* from *desuescere* to become unaccustomed, from *de-* + *suescere* to become accustomed; akin to Latin *sodalis* comrade — more at SIB] (15th century)
: discontinuance from use or exercise **:** DISUSE

de·sul·fur·iza·tion \(ˌ)dē-ˌsəl-fər-ə-ˈzā-shən\ *noun* (1854)
: the removal of sulfur or sulfur compounds (as from coal or flue gas)
— de·sul·fur·ize \(ˌ)dē-ˈsəl-fər-ˌīz\ *transitive verb*

des·ul·to·ry \ˈde-səl-ˌtōr-ē, -ˌtȯr- *also* -zəl-\ *adjective* [Latin *desultorius,* literally, of a circus rider who leaps from horse to horse, from *desilire* to leap down, from *de-* + *salire* to leap — more at SALLY] (1581)
1 : marked by lack of definite plan, regularity, or purpose ⟨a dragged-out ordeal of . . . *desultory* shopping —Herman Wouk⟩
2 : not connected with the main subject
3 : disappointing in progress or performance **:** SLUGGISH
— des·ul·to·ri·ly \ˌde-səl-ˈtōr-ə-lē, -ˈtȯr-\ *adverb*
— des·ul·to·ri·ness \ˈde-səl-ˌtōr-ē-nəs, -ˌtȯr-\ *noun*

de·tach \di-ˈtach, dē-\ *transitive verb* [French *détacher,* from Old French *destachier,* from *des-* de- + *-tachier* (as in *atachier* to attach)] (1686)
1 : to separate especially from a larger mass and usually without violence or damage
2 : DISENGAGE, WITHDRAW
— de·tach·abil·i·ty \-ˌta-chə-ˈbi-lə-tē\ *noun*
— de·tach·able \-ˈta-chə-bəl\ *adjective*
— de·tach·ably \-blē\ *adverb*

de·tached \di-ˈtacht, dē-\ *adjective* (circa 1706)
1 : standing by itself **:** SEPARATE, UNCONNECTED; *especially* **:** not sharing any wall with another building ⟨a *detached* house⟩
2 : exhibiting an aloof objectivity usually free from prejudice or self-interest ⟨a *detached* observer⟩
synonym see INDIFFERENT
— de·tached·ly \-ˈta-chəd-lē, -ˈtacht-lē\ *adverb*
— de·tached·ness \-ˈta-chəd-nəs, -ˈtacht-nəs\ *noun*

detached service *noun* (circa 1889)

☆ SYNONYMS
Despondent, despairing, desperate, hopeless mean having lost all or nearly all hope. DESPONDENT implies a deep dejection arising from a conviction of the uselessness of further effort ⟨*despondent* about yet another rejection⟩. DESPAIRING suggests the slipping away of all hope and often despondency ⟨*despairing* appeals for the return of the kidnapped child⟩. DESPERATE implies despair that prompts reckless action or violence in the face of defeat or frustration ⟨one last *desperate* attempt to turn the tide of battle⟩. HOPELESS suggests despair and the cessation of effort or resistance and often implies acceptance or resignation ⟨the situation of the trapped miners is *hopeless*⟩.

: military service away from one's assigned organization
de·tach·ment \di-ˈtach-mənt, dē-\ *noun* (1669)
1 : the action or process of detaching **:** SEPARATION
2 a : the dispatch of a body of troops or part of a fleet from the main body for a special mission or service **b :** the part so dispatched **c :** a permanently organized separate unit usually smaller than a platoon and of special composition
3 a : indifference to worldly concerns **:** ALOOFNESS **b :** freedom from bias or prejudice
¹de·tail \di-ˈtā(ə)l, ˈdē-ˌtāl\ *noun* [French *détail,* from Old French *detail* slice, piece, from *detaillier* to cut in pieces, from *de-* + *taillier* to cut — more at TAILOR] (1603)
1 : extended treatment of or attention to particular items
2 : a part of a whole: as **a :** a small and subordinate part **:** PARTICULAR; *also* **:** a reproduction of such a part of a work of art **b :** a part considered or requiring to be considered separately from the whole **c :** the small elements that collectively constitute a work of art **d :** the small elements of a photographic image corresponding to those of the subject
3 a : selection of a person or group for a particular task (as in military service) **b** (1) **:** the person or group selected (2) **:** the task to be performed
synonym see ITEM
— in detail : with all the particulars
²detail (1650)
transitive verb
1 : to report minutely and distinctly **:** SPECIFY ⟨*detailed* their grievances⟩
2 : to assign to a particular task
3 : to furnish with the smaller elements of design and finish ⟨trimmings that *detail* slips and petticoats⟩
intransitive verb
: to make detail drawings
— de·tail·er *noun*
de·tailed \di-ˈtā(ə)ld, ˈdē-ˌtāld\ *adjective* (1740)
: marked by abundant detail or by thoroughness in treating small items or parts ⟨the *detailed* study of history⟩
synonym see CIRCUMSTANTIAL
— de·tailed·ly \di-ˈtāld-lē, -ˈtā-ləd-, ˈdē-ˌ\ *adverb*
— de·tailed·ness \di-ˈtā-ləd-nəs, -ˈtāld-, ˈdē-ˌ\ *noun*
detail man *noun* (1928)
: a representative of a drug manufacturer who introduces new drugs especially to physicians and pharmacists
de·tain \di-ˈtān, dē-\ *transitive verb* [Middle English *deteynen,* from Middle French *detenir,* modification of Latin *detinēre,* from *de-* + *tenēre* to hold — more at THIN] (15th century)
1 : to hold or keep in or as if in custody
2 *obsolete* **:** to keep back (as something due) **:** WITHHOLD
3 : to restrain especially from proceeding **:** STOP
synonym see KEEP, DELAY
— de·tain·ment \-mənt\ *noun*
de·tain·ee \di-ˌtā-ˈnē, ˌdē-\ *noun* (circa 1928)
: a person held in custody especially for political reasons
de·tain·er \di-ˈtā-nər\ *noun* [Anglo-French *detener,* from *detener* to detain, from Latin *detinēre*] (1619)
1 : the act of keeping something in one's possession; *specifically* **:** the withholding from the rightful owner of something that has lawfully come into the possession of the holder
2 : detention in custody
3 : a writ authorizing the keeper of a prison to continue to hold a person in custody
de·tect \di-ˈtekt, dē-\ *verb* [Middle English, from Latin *detectus,* past participle of *detegere* to uncover, detect, from *de-* + *tegere* to cover — more at THATCH] (1581)
transitive verb
1 : to discover the true character of
2 : to discover or determine the existence, presence, or fact of ⟨*detect* alcohol in the blood⟩
3 : DEMODULATE
intransitive verb
: to work as a detective
— de·tect·abil·i·ty \-ˌtek-tə-ˈbi-lə-tē\ *noun*
— de·tect·able \-ˈtek-tə-bəl\ *adjective*
de·tec·tion \di-ˈtek-shən\ *noun* (15th century)
1 : the act of detecting **:** the state or fact of being detected
2 : the process of demodulating
¹de·tec·tive \di-ˈtek-tiv\ *adjective* (1843)
1 : fitted for or used in detecting something ⟨had perfected his *detective* sensibilities⟩
2 : of or relating to detectives or their work ⟨a *detective* novel⟩
— de·tec·tive·like \-ˌlīk\ *adjective*
²detective *noun* (1850)
: one employed or engaged in detecting lawbreakers or in getting information that is not readily or publicly accessible
de·tec·tor \di-ˈtek-tər\ *noun* (1541)
: one that detects: as **a :** a device for detecting the presence of electromagnetic waves or of radioactivity **b :** a rectifier of high-frequency current used especially for extracting the intelligence from a radio signal
de·tent \ˈdē-ˌtent, di-ˈ\ *noun* [French *détente,* from Middle French *destente,* from *destendre* to slacken, from Old French, from *des-* de- + *tendre* to stretch, from Latin *tendere* — more at THIN] (1688)
: a device (as a catch, dog, or spring-operated ball) for positioning and holding one mechanical part in relation to another so that the device can be released by force applied to one of the parts
dé·tente *or* **detente** \dā-ˈtänt\ *noun* [French] (1908)
1 : the relaxation of strained relations or tensions (as between nations); *also* **:** a policy promoting this
2 : a period of détente
de·ten·tion \di-ˈten-chən\ *noun* [Middle English *detencion,* from Middle French or Latin; Middle French, from Latin *detention-, detentio,* from *detinēre* to detain] (15th century)
1 : the act or fact of detaining or holding back; *especially* **:** a holding in custody
2 : the state of being detained; *especially* **:** a period of temporary custody prior to disposition by a court
detention home *noun* (circa 1930)
: a house of detention for juvenile delinquents usually under the supervision of a juvenile court
de·ter \di-ˈtər, dē-\ *transitive verb* **de·terred; de·ter·ring** [Latin *deterrēre,* from *de-* + *terrēre* to frighten — more at TERROR] (1579)
1 : to turn aside, discourage, or prevent from acting
2 : INHIBIT
— de·ter·ment \-ˈtər-mənt\ *noun*
— de·ter·ra·bil·i·ty \-ˌtər-ə-ˈbi-lə-tē\ *noun*
— de·ter·ra·ble \-ˈtər-ə-bəl\ *adjective*
de·terge \di-ˈtərj\ *transitive verb* **de·terged; de·terg·ing** [French or Latin; French *déterger,* from Latin *detergēre,* from *de-* + *tergēre* to wipe] (circa 1623)
: to wash off **:** CLEANSE
— de·terg·er *noun*
de·ter·gen·cy \di-ˈtər-jən(t)-sē\ *noun* (1710)
: cleansing quality or power
¹de·ter·gent \-jənt\ *adjective* (1616)
: that cleanses **:** CLEANSING
²detergent *noun* (1676)
: a cleansing agent: as **a :** SOAP **b :** any of numerous synthetic water-soluble or liquid organic preparations that are chemically different from soaps but are able to emulsify oils, hold dirt in suspension, and act as wetting agents **c :** an oil-soluble substance that holds insoluble foreign matter in suspension and is used in lubricating oils and dry-cleaning solvents
de·te·ri·o·rate \di-ˈtir-ē-ə-ˌrāt, dē-\ *verb* **-rat·ed; -rat·ing** [Late Latin *deterioratus,* past participle of *deteriorare,* from Latin *deterior* worse, from *de-* + *-ter* (suffix as in Latin *uter* which of two) + *-ior* (comparative suffix) — more at WHETHER, -ER] (1572)
transitive verb
1 : to make inferior in quality or value **:** IMPAIR
2 : DISINTEGRATE
intransitive verb
: to become impaired in quality, functioning, or condition **:** DEGENERATE ⟨allowed a tradition of academic excellence to *deteriorate*⟩ ⟨his health *deteriorated*⟩
— de·te·ri·o·ra·tive \-ˌrā-tiv\ *adjective*
de·te·ri·o·ra·tion \di-ˌtir-ē-ə-ˈrā-shən, dē-\ *noun* (circa 1658)
: the action or process of deteriorating **:** the state of having deteriorated ☆
de·ter·min·able \-ˈtər-mə-nə-bəl\ *adjective* (15th century)
1 : capable of being determined, definitely ascertained, or decided upon
2 : liable to be terminated **:** TERMINABLE
— de·ter·min·able·ness *noun*
— de·ter·min·ably \-blē\ *adverb*
de·ter·mi·na·cy \di-ˈtər-mə-nə-sē\ *noun, plural* **-cies** (1873)
1 : the quality or state of being determinate
2 a : the state of being definitely and unequivocally characterized **:** EXACTNESS **b :** the state of being determined or necessitated
de·ter·mi·nant \di-ˈtər-mə-nənt\ *noun* (1686)
1 : an element that identifies or determines the nature of something or that fixes or conditions an outcome
2 : a square array of numbers bordered on the left and right by a vertical line and having a value equal to the algebraic sum of all possible products where the number of factors in each product is the same as the number of rows or columns, each factor in a given product is taken from a different row and column, and the sign of a product is positive or negative depending upon whether the number of permutations necessary to place the indices representing each factor's position in its row or column in the order of the natural numbers is odd or even
3 : GENE
4 : EPITOPE
— de·ter·mi·nan·tal \-ˌtər-mə-ˈnan-tᵊl\ *adjective*
de·ter·mi·nate \di-ˈtər-mə-nət\ *adjective* [Middle English, from Latin *determinatus,* past participle of *determinare*] (14th century)
1 : having defined limits
2 : definitely settled

☆ SYNONYMS
Deterioration, degeneration, decadence, decline mean the falling from a higher to a lower level in quality, character, or vitality. DETERIORATION implies generally the impairment of value or usefulness ⟨the *deterioration* of the house through neglect⟩. DEGENERATION stresses physical, intellectual, or especially moral retrogression ⟨the *degeneration* of their youthful idealism into cynicism⟩. DECADENCE presupposes a reaching and passing the peak of development and implies a turn downward with a consequent loss in vitality or energy ⟨cited love of luxury as a sign of cultural *decadence*⟩. DECLINE differs from DECADENCE in suggesting a more markedly downward direction and greater momentum as well as more obvious evidence of deterioration ⟨the meteoric *decline* of his career after the scandal⟩.

3 : conclusively determined **:** DEFINITIVE
4 : characterized by sequential flowering from the central or uppermost bud to the lateral or basal buds; *also* **:** characterized by growth in which the main stem ends in an inflorescence and stops growing with only branches from the main stem having further and similarly restricted growth ⟨*determinate* tomato plants⟩ — compare INDETERMINATE 4
5 : relating to, being, or undergoing egg cleavage in which each division irreversibly separates portions of the zygote with specific potencies for further development
— de·ter·mi·nate·ly *adverb*
— de·ter·mi·nate·ness *noun*

de·ter·mi·na·tion \di-ˌtər-mə-ˈnā-shən\ *noun* (14th century)
1 a : a judicial decision settling and ending a controversy **b :** the resolving of a question by argument or reasoning
2 *archaic* **:** TERMINATION
3 a : the act of deciding definitely and firmly; *also* **:** the result of such an act of decision **b :** the power or habit of deciding definitely and firmly
4 : a fixing or finding of the position, magnitude, value, or character of something: as **a :** the act, process, or result of an accurate measurement **b :** an identification of the taxonomic position of a plant or animal
5 a : the definition of a concept in logic by its essential constituents **b :** the addition of a differentia to a concept to limit its denotation
6 : direction or tendency to a certain end **:** IMPULSION
7 : the fixation of the destiny of undifferentiated embryonic tissue

de·ter·mi·na·tive \-ˈtər-mə-ˌnā-tiv, -ˈtər-mə-nə-\ *adjective* (1655)
: having power or tendency to determine **:** tending to fix, settle, or define something ⟨regard experiments as *determinative* of the principles from which deductions could be made —S. F. Mason⟩
synonym see CONCLUSIVE
— determinative *noun*

de·ter·mi·na·tor \di-ˈtər-mə-ˌnā-tər\ *noun* (1556)
: DETERMINER

de·ter·mine \di-ˈtər-mən, dē-\ *verb* **de·ter·mined; de·ter·min·ing** [Middle English, from Middle French *determiner,* from Latin *determinare,* from *de-* + *terminare* to limit, from *terminus* boundary, limit — more at TERM] (14th century)
transitive verb
1 a : to fix conclusively or authoritatively **b :** to decide by judicial sentence **c :** to settle or decide by choice of alternatives or possibilities **d :** RESOLVE
2 a : to fix the form, position, or character of beforehand **:** ORDAIN ⟨two points *determine* a straight line⟩ **b :** to bring about as a result **:** REGULATE ⟨demand *determines* the price⟩
3 a : to fix the boundaries of **b :** to limit in extent or scope **c :** to put or set an end to **:** TERMINATE ⟨*determine* an estate⟩
4 : to find out or come to a decision about by investigation, reasoning, or calculation ⟨*determine* the answer to the problem⟩ ⟨*determine* a position at sea⟩
5 : to bring about the determination of ⟨*determine* the fate of a cell⟩
intransitive verb
1 : to come to a decision
2 : to come to an end or become void
synonym see DECIDE, DISCOVER

de·ter·mined \-ˈtər-mənd\ *adjective* (1513)
1 : having reached a decision **:** firmly resolved ⟨*determined* to be a pilot⟩
2 a : showing determination ⟨a *determined* effort⟩ **b :** characterized by determination ⟨will deter all but the most *determined* thief —*Security World*⟩
— de·ter·mined·ly \-mənd-lē, -mə-nəd-lē\ *adverb*
— de·ter·mined·ness \-mənd-nəs\ *noun*

de·ter·min·er \-ˈtər-mə-nər\ *noun* (circa 1530)
: one that determines: as **a :** GENE **b :** a word (as an article, possessive, demonstrative, or quantifier) that makes specific the denotation of a noun phrase

de·ter·min·ism \di-ˈtər-mə-ˌni-zəm, dē-\ *noun* (1846)
1 a : a theory or doctrine that acts of the will, occurrences in nature, or social or psychological phenomena are causally determined by preceding events or natural laws **b :** a belief in predestination
2 : the quality or state of being determined
— de·ter·min·ist \-nəst\ *noun or adjective*
— de·ter·min·is·tic \-ˌtər-mə-ˈnis-tik\ *adjective*
— de·ter·min·is·ti·cal·ly \-ti-k(ə-)lē\ *adverb*

de·ter·rence \di-ˈtər-ən(t)s, -ˈter-; -ˈtə-rən(t)s, dē-\ *noun* (1861)
: the inhibition of criminal behavior by fear especially of punishment; *also* **:** the maintenance of military power for the purpose of discouraging attack

de·ter·rent \-ənt, -rənt\ *adjective* [Latin *deterrent-, deterrens,* present participle of *deterrēre* to deter] (1829)
1 : serving to deter
2 : relating to deterrence
— deterrent *noun*
— de·ter·rent·ly *adverb*

de·ter·sive \di-ˈtər-siv, -ziv\ *adjective* [Middle French *detersif,* from Latin *detersus,* past participle of *detergēre* to deterge] (1586)
: DETERGENT
— detersive *noun*

de·test \di-ˈtest, dē-\ *transitive verb* [Middle French *detester* or Latin *detestari;* Middle French *detester,* from Latin *detestari,* literally, to curse while calling a deity to witness, from *de-* + *testari* to call to witness — more at TESTAMENT] (circa 1535)
1 : to feel intense and often violent antipathy toward **:** LOATHE
2 *obsolete* **:** CURSE, DENOUNCE
synonym see HATE
— de·test·er *noun*

de·test·able \di-ˈtes-tə-bəl, dē-\ *adjective* (15th century)
: arousing or meriting intense dislike **:** ABOMINABLE
— de·test·able·ness *noun*
— de·test·ably \-blē\ *adverb*

de·tes·ta·tion \ˌdē-ˌtes-ˈtā-shən, di-\ *noun* (15th century)
1 : extreme hatred or dislike **:** ABHORRENCE, LOATHING
2 : an object of hatred or contempt

de·throne \di-ˈthrōn, dē-\ *transitive verb* (1609)
: to remove from a throne or place of power or prominence **:** DEPOSE
— de·throne·ment \-mənt\ *noun*
— de·thron·er *noun*

de·tick \(ˌ)dē-ˈtik\ *transitive verb* (1925)
: to remove ticks from
— de·tick·er *noun*

det·i·nue \ˈde-tᵊn-ˌ(y)ü\ *noun* [Middle English *detenewe,* from Middle French *detenue* detention, from feminine of *detenu,* past participle of *detenir* to detain] (15th century)
1 : a common-law action for the recovery of a personal chattel wrongfully detained or of its value
2 : detention of something due; *especially* **:** the unlawful detention of a personal chattel from another

det·o·na·ble \ˈde-tᵊn-ə-bəl, -tə-nə-\ *adjective* (1884)
: capable of being detonated
— det·o·na·bil·i·ty \ˌdet-ᵊn-ə-ˈbil-ət-ē, ˌdet-ə-nə-\ *noun*

det·o·nate \ˈde-tᵊn-ˌāt, ˈde-tə-ˌnāt\ *verb* **-nat·ed; -nat·ing** [French *détoner* to explode, from Latin *detonare* to expend thunder, from *de-* + *tonare* to thunder — more at THUNDER] (1729)
intransitive verb
: to explode with sudden violence
transitive verb
1 : to cause to detonate ⟨*detonate* a bomb⟩ — compare DEFLAGRATE
2 : to set off in a burst of activity **:** SPARK ⟨programs that *detonated* controversies⟩
— det·o·nat·able \-ˌā-tə-bəl, -ˌnā-\ *adjective*
— det·o·na·tive \ˈde-tᵊn-ˌā-tiv, ˈde-tə-ˌnā-\ *adjective*

det·o·na·tion \ˌde-tᵊn-ˈā-shən, ˌde-tə-ˈnā-\ *noun* (1686)
1 : the action or process of detonating
2 : rapid combustion in an internal combustion engine that results in knocking

det·o·na·tor \ˈde-tᵊn-ˌā-tər, -tə-ˌnā-\ *noun* (1822)
: a device or small quantity of explosive used for detonating a high explosive

¹de·tour \ˈdē-ˌtu̇r *also* di-ˈtu̇ər\ *noun* [French *détour,* from Old French *destor,* from *destorner* to divert, from *des-* de- + *torner* to turn — more at TURN] (1738)
: a deviation from a direct course or the usual procedure; *especially* **:** a roundabout way temporarily replacing part of a route

²detour (1836)
intransitive verb
: to proceed by a detour ⟨*detour* around road construction⟩
transitive verb
1 : to send by a circuitous route
2 : to avoid by going around **:** BYPASS

de·tox \ˈdē-ˌtäks, di-ˈtäks\ *noun, often attributive* (1973)
: detoxification from an intoxicating or an addictive substance
— detox *verb*

de·tox·i·cate \(ˌ)dē-ˈtäk-sə-ˌkāt\ *transitive verb* **-cat·ed; -cat·ing** [*de-* + ²*in*toxicate] (1867)
: DETOXIFY 1, 2
— de·tox·i·cant \-si-kənt\ *noun*
— de·tox·i·ca·tion \(ˌ)dē-ˌtäk-sə-ˈkā-shən\ *noun*

de·tox·i·fy \(ˌ)dē-ˈtäk-sə-ˌfī\ *transitive verb* **-fied; -fy·ing** (circa 1905)
1 a : to remove a poison or toxin or the effect of such from **b :** to render (a harmful substance) harmless
2 : to free (as a drug user or an alcoholic) from an intoxicating or an addictive substance in the body or from dependence on or addiction to such a substance
3 : NEUTRALIZE 2
— de·tox·i·fi·ca·tion \(ˌ)dē-ˌtäk-sə-fə-ˈkā-shən\ *noun*

de·tract \di-ˈtrakt, dē-\ *verb* [Middle English, from Latin *detractus,* past participle of *detrahere* to pull down, disparage, from *de-* + *trahere* to draw (15th century)
transitive verb
1 *archaic* **:** to speak ill of
2 *archaic* **:** to take away
3 : DIVERT ⟨*detract* attention⟩
intransitive verb
: to diminish the importance, value, or effectiveness of something — often used with *from*
— de·trac·tor \-ˈtrak-tər\ *noun*

de·trac·tion \di-ˈtrak-shən, dē-\ *noun* (14th century)
1 : a lessening of reputation or esteem especially by envious, malicious, or petty criticism **:** BELITTLING, DISPARAGEMENT
2 : a taking away ⟨it is no *detraction* from its dignity or prestige —J. F. Golay⟩

— **de·trac·tive** \-ˈtrak-tiv\ *adjective*
— **de·trac·tive·ly** *adverb*
de·train \(ˌ)dē-ˈtrān\ (1881)
intransitive verb
: to get off a railroad train
transitive verb
: to remove from a railroad train
— **de·train·ment** \-mənt\ *noun*
de·trib·al·ize \(ˌ)dē-ˈtrī-bə-ˌlīz\ *transitive verb* **-ized; -iz·ing** (1920)
: to cause to lose tribal identity **:** ACCULTURATE
— **de·trib·al·iza·tion** \(ˌ)dē-ˌtrī-bə-lə-ˈzā-shən\ *noun*
de·tri·ment \ˈde-trə-mənt\ *noun* [Middle English, from Middle French or Latin; Middle French, from Latin *detrimentum,* from *deterere* to wear away, impair, from *de-* + *terere* to rub — more at THROW] (15th century)
1 : INJURY, DAMAGE ⟨did hard work without *detriment* to his health⟩
2 : a cause of injury or damage ⟨a *detriment* to progress⟩
[1]**det·ri·men·tal** \ˌde-trə-ˈmen-tᵊl\ *adjective* (1590)
: obviously harmful **:** DAMAGING ⟨the *detrimental* effects of pollution⟩
synonym see PERNICIOUS
— **det·ri·men·tal·ly** \-tᵊl-ē\ *adverb*
[2]**detrimental** *noun* (1831)
: an undesirable or harmful person or thing
de·tri·tion \di-ˈtri-shən\ *noun* (1674)
: a wearing off or away
de·tri·tus \di-ˈtrī-təs\ *noun, plural* **de·tri·tus** \-ˈtrī-təs, -ˈtrī-ˌtüs\ [French *détritus,* from Latin *detritus,* past participle of *deterere*] (1802)
1 : loose material (as rock fragments or organic particles) that results directly from disintegration
2 : a product of disintegration, destruction, or wearing away **:** DEBRIS
— **de·tri·tal** \-ˈtrī-tᵊl\ *adjective*
de trop \də-ˈtrō\ *adjective* [French] (1752)
: too much or too many **:** SUPERFLUOUS
de·tu·mes·cence \ˌdē-t(y)ü-ˈme-sᵊn(t)s\ *noun* [Latin *detumescere* to become less swollen, from *de-* + *tumescere* to swell — more at TUMESCENT] (1678)
: subsidence or diminution of swelling or erection
— **de·tu·mes·cent** \-sᵊnt\ *adjective*
Deu·ca·lion \d(y)ü-ˈkāl-yən\ *noun* [Latin, from Greek *Deukaliōn*]
: a survivor with his wife Pyrrha of a great flood by which Zeus destroys the rest of the human race
[1]**deuce** \ˈdüs *also* ˈdyüs\ *noun* [Middle French *deus* two, from Latin *duos,* accusative masculine of *duo* two — more at TWO] (15th century)
1 a (1) **:** the face of a die that bears two spots (2) **:** a playing card bearing an index number two **b :** a throw of the dice yielding two points
2 : a tie in tennis after each side has scored 40 requiring two consecutive points by one side to win
3 [obsolete English *deuce* bad luck] **a :** DEVIL, DICKENS — used chiefly as a mild oath ⟨what the *deuce* is he up to now⟩ **b :** something notable of its kind ⟨a *deuce* of a mess⟩
[2]**deuce** *transitive verb* **deuced; deuc·ing** (1919)
: to bring the score of (a tennis game or set) to deuce
deuc·ed \ˈdü-səd *also* ˈdyü-\ *adjective* (1782)
: DAMNED, CONFOUNDED ⟨in a *deuced* fix⟩
— **deuc·ed** *or* **deuc·ed·ly** *adverb*
deuces wild *noun* (1927)
: a card game (as poker) in which each deuce may represent any card designated by its holder
de·us ex ma·chi·na \ˈdā-əs-ˌeks-ˈmä-ki-nə, -ˈma-, -ˌnä; -mə-ˈshē-nə\ *noun* [New Latin, a god from a machine, translation of Greek *theos ek mēchanēs*] (1697)
1 : a god introduced by means of a crane in ancient Greek and Roman drama to decide the final outcome
2 : a person or thing (as in fiction or drama) that appears or is introduced suddenly and unexpectedly and provides a contrived solution to an apparently insoluble difficulty ◆
deuter- *or* **deutero-** *combining form* [alteration of Middle English *deutro-,* modification of Late Latin *deutero-,* from Greek *deuter-, deutero-,* from *deuteros;* probably akin to Greek *dein* to lack, Sanskrit *doṣa* fault, lack]
: second **:** secondary ⟨*deuter*anopia⟩
deu·ter·ag·o·nist \ˌdü-tə-ˈra-gə-nist *also* ˌdyü-\ *noun* [Greek *deuteragōnistēs,* from *deuter-* + *agōnistēs* combatant, actor — more at PROTAGONIST] (1855)
1 : the actor taking the part of second importance in a classical Greek drama
2 : a person who serves as a foil to another
deu·ter·anom·a·lous \ˌdü-tə-rə-ˈnä-mə-ləs *also* ˌdyü-\ *adjective* [New Latin *deuteranomalia* (from *deuter-* + Latin *anomalia* anomaly) abnormal trichromatism + English *-ous*] (circa 1931)
: exhibiting partial loss of green color vision so that an increased intensity of this color is required in a mixture of red and green to match a given yellow
— **deu·ter·anom·a·ly** \-ˈnä-mə-lē\ *noun*
deu·ter·an·ope \ˈdü-tə-rə-ˌnōp *also* ˈdyü-\ *noun* (1902)
: an individual affected with deuteranopia
deu·ter·an·opia \ˌdü-tə-rə-ˈnō-pē-ə *also* ˌdyü-\ *noun* [New Latin, from *deuter-* + [2]*a-* + *-opia;* from the blindness to green, regarded as the second primary color] (circa 1901)
: color blindness marked by confusion of purplish red and green
— **deu·ter·an·opic** \-ˈnō-pik, -ˈnä-pik\ *adjective*
deu·ter·ate \ˈdü-tə-ˌrāt *also* ˈdyü-\ *transitive verb* **-at·ed; -at·ing** (1947)
: to introduce deuterium into (a compound)
— **deu·ter·a·tion** \ˌdyü-tə-ˈrā-shən\ *noun*
deu·te·ri·um \dü-ˈtir-ē-əm *also* dyü-\ *noun* [New Latin, from Greek *deuteros* second] (1933)
: an isotope of hydrogen that has one proton and one neutron in its nucleus and that has twice the mass of ordinary hydrogen — symbol *D;* called also *heavy hydrogen*
deuterium oxide *noun* (1934)
: HEAVY WATER 1
deu·tero·ca·non·i·cal \ˌdü-tə-rō-kə-ˈnä-ni-kəl *also* ˌdyü-\ *adjective* [New Latin *deuterocanonicus,* from *deuter-* + Late Latin *canonicus* canonical] (1684)
: of, relating to, or constituting the books of Scripture contained in the Septuagint but not in the Hebrew canon
deu·ter·on \ˈdü-tə-ˌrän *also* ˈdyü-\ *noun* (1933)
: a deuterium nucleus
Deu·ter·o·nom·ic \ˌdü-tə-rə-ˈnä-mik *also* ˌdyü-\ *adjective* (1857)
: of or relating to the book of Deuteronomy, its style, or its contents
Deu·ter·on·o·mist \ˌdü-tə-ˈrä-nə-mist *also* ˌdyü-\ *noun* (1862)
: any of the writers or editors of a Deuteronomic body of source material often distinguished in the earlier books of the Old Testament
— **Deu·ter·on·o·mis·tic** \-ˌrä-nə-ˈmis-tik\ *adjective*
Deu·ter·on·o·my \ˌdü-tə-ˈrä-nə-mē *also* ˌdyü-\ *noun* [Middle English *Deutronomie,* from Late Latin *Deuteronomium,* from Greek *Deuteronomion,* from *deuter-* + *nomos* law — more at NIMBLE]
: the fifth book of canonical Jewish and Christian Scripture containing narrative and Mosaic laws — see BIBLE table
deu·tero·stome \ˈdü-tə-rō-ˌstōm *also* ˈdyü-\ *noun* [New Latin *Deuterostomia,* group name, from *deuter-* + Greek *stoma* mouth — more at STOMACH] (1950)
: any of a major division (Deuterostomia) of the animal kingdom that includes the bilaterally symmetrical animals (as the chordates) with indeterminate cleavage and a mouth that does not arise from the blastopore
deu·to·plasm \ˈdü-tō-ˌpla-zəm *also* ˈdyü-\ *noun* [International Scientific Vocabulary *deuter-* + *-plasm*] (1884)
: the nutritive inclusions of protoplasm; *especially* **:** the yolk reserves of an egg
deut·sche mark \ˈdȯi-chə-ˌmärk, ˈdȯi-chə-ˌ\ *noun* [German, German mark] (1948)
— see MONEY table
deut·zia \ˈdüt-sē-ə *also* ˈdyüt-\ *noun* [New Latin, from Jean *Deutz* (died 1784?) Dutch patron of botanical research] (1837)
: any of a genus (*Deutzia*) of the saxifrage family of ornamental shrubs with white or pink flowers
de·val·u·ate \(ˌ)dē-ˈval-yə-ˌwāt\ *verb* (1898)
: DEVALUE
de·val·u·a·tion \(ˌ)dē-ˌval-yə-ˈwā-shən\ *noun* (1914)
1 : an official reduction in the exchange value of a currency by a lowering of its gold equivalency or its value relative to another currency
2 : a lessening especially of status or stature **:** DECLINE
de·val·ue \(ˌ)dē-ˈval-(ˌ)yü, -yə(-w)\ (1918)
transitive verb
1 : to institute the devaluation of (money)
2 : to lessen the value of
intransitive verb
: to institute devaluation
De·va·na·ga·ri \ˌdā-və-ˈnä-gə-rē\ *noun* [Sanskrit *devanāgarī,* from *deva* divine + *nāgarī* script of the city — more at DEITY] (1781)
: an alphabet usually employed for Sanskrit and also used as a literary hand for various modern languages of India — see ALPHABET table
dev·as·tate \ˈde-və-ˌstāt\ *transitive verb* **-tat·ed; -tat·ing** [Latin *devastatus,* past participle of *devastare,* from *de-* + *vastare* to lay waste — more at WASTE] (1638)
1 : to bring to ruin or desolation by violent action
2 : to reduce to chaos, disorder, or helplessness **:** OVERWHELM ⟨*devastated* by grief⟩ ⟨her wisecrack *devastated* the class⟩
synonym see RAVAGE
— **dev·as·tat·ing·ly** \-ˌstā-tiŋ-lē\ *adverb*
— **dev·as·ta·tion** \ˌde-və-ˈstā-shən\ *noun*
— **dev·as·tat·ive** \ˈde-və-ˌstā-tiv\ *adjective*
— **dev·as·ta·tor** \-ˌstā-tər\ *noun*

◇ WORD HISTORY

deus ex machina Since the birth of the drama, playwrights have wrestled with the problem of how to resolve the conflicts that they develop in their dramatic works. In antiquity, even the greatest playwrights often resorted to divine intervention to resolve seemingly irresolvable dramatic situations. In Greek drama of the 4th century B.C., divine intervention took the form of the sudden appearance of a god in the sky. The use of a machine like a crane to achieve this primitive stage effect gave birth to the term *theos ek mēchanēs,* which literally means "a god from a machine." The resolving of a tangled plot by the timely intervention of a god hoisted on a crane became a standard convention in the Greek and Roman theater. The extended application of the term for this stage effect to any unlikely but providential savior, whether person or event, also became established during classical times. It was in this extended sense that *deus ex machina,* the New Latin translation of the Greek phrase, first appeared in English contexts in the 17th century.

de·vein \(ˌ)dē-ˈvān\ *transitive verb* (1953)
: to remove the dark dorsal vein from (shrimp)

de·vel·op \di-ˈvel-əp, dē-\ *verb* [French *développer,* from Old French *desveloper, desvoluper* to unwrap, expose, from *des-* de- + *envoloper* to enclose — more at ENVELOP] (1750)
transitive verb
1 a : to set forth or make clear by degrees or in detail **:** EXPOUND **b :** to make visible or manifest **c :** to treat (as in dyeing) with an agent to cause the appearance of color **d :** to subject (exposed photograph material) especially to chemicals in order to produce a visible image; *also* **:** to make visible by such a method **e :** to elaborate (a musical idea) by the working out of rhythmic and harmonic changes in the theme
2 : to work out the possibilities of
3 a : to make active or promote the growth of ⟨*developed* his muscles⟩ **b** (1) **:** to make available or usable ⟨*develop* natural resources⟩ (2) **:** to make suitable for commercial or residential purposes **c :** to move (as a chess piece) from the original position to one providing more opportunity for effective use
4 a : to cause to unfold gradually ⟨*developed* his argument⟩ **b :** to expand by a process of growth ⟨*developed* a strong organization⟩ **c :** to cause to grow and differentiate along lines natural to its kind ⟨rain and sun *develop* the grain⟩ **d :** to have unfold or differentiate within one ⟨*developed* pneumonia⟩
5 : to acquire gradually ⟨*develop* an appreciation for ballet⟩
intransitive verb
1 a : to go through a process of natural growth, differentiation, or evolution by successive changes ⟨a blossom *develops* from a bud⟩ **b :** to acquire secondary sex characters
2 : to become gradually manifest
3 : to come into being gradually ⟨the situation *developing* in eastern Europe⟩; *also* **:** TURN OUT *intransitive verb* 2a ⟨it *developed* that no one had paid the bill⟩
— de·vel·op·able \-ˈve-lə-pə-bəl\ *adjective*

de·vel·oped \di-ˈve-ləpt\ *adjective* (1945)
: having a relatively high level of industrialization and standard of living ⟨a *developed* country⟩

de·vel·op·er \-lə-pər\ *noun* (1883)
: one that develops: as **a :** a chemical used to develop exposed photographic materials **b :** a person who develops real estate

de·vel·op·ing \-lə-piŋ\ *adjective* (1964)
: UNDERDEVELOPED 2 ⟨*developing* nations⟩

de·vel·op·ment \di-ˈve-ləp-mənt, dē-\ *noun* (1756)
1 : the act, process, or result of developing
2 : the state of being developed
3 : a developed tract of land; *especially* **:** one with houses built on it

de·vel·op·men·tal \dē-ˌve-ləp-ˈmen-t^{ə}l\ *adjective* (1849)
1 a : of, relating to, or being development ⟨*developmental* processes⟩; *broadly* **:** EXPERIMENTAL 2 ⟨*developmental* aircraft⟩ **b :** serving economic development ⟨*developmental* highways⟩
2 : designed to assist growth or bring about improvement (as of a skill) ⟨*developmental* toys⟩
— de·vel·op·men·tal·ly \-t^{ə}l-ē\ *adverb*

developmentally disabled *adjective* (1975)
: having a physical or mental handicap that impedes or prevents normal development

de·verb·al \(ˌ)dē-ˈvər-bəl\ *adjective* (1943)
: DEVERBATIVE

de·verb·a·tive \(ˌ)dē-ˈvər-bə-tiv\ *adjective* (1930)
1 : derived from a verb ⟨the *deverbative* noun *developer* is derived from *develop*⟩
2 : used in derivation from a verb ⟨the *deverbative* suffix *-er* in *developer*⟩
— deverbative *noun*

de·vest \di-ˈvest\ *transitive verb* [Middle French *desvestir,* from Medieval Latin *disvestire,* from Latin *dis-* + *vestire* to clothe — more at VEST] (1563)
: DIVEST

de·vi·ance \ˈdē-vē-ən(t)s\ *noun* (1944)
: deviant quality, state, or behavior

de·vi·an·cy \-ən-sē\ *noun, plural* **-cies** (1947)
: DEVIANCE

de·vi·ant \-ənt\ *adjective* (15th century)
: deviating especially from an accepted norm ⟨*deviant* behavior⟩
— deviant *noun*

[1]**de·vi·ate** \ˈdē-vē-ˌāt\ *verb* **-at·ed; -at·ing** [Late Latin *deviatus,* past participle of *deviare,* from Latin *de-* + *via* way — more at WAY] (circa 1633)
intransitive verb
1 : to stray especially from a standard, principle, or topic
2 : to depart from an established course or norm
transitive verb
: to cause to turn out of a previous course
synonym see SWERVE
— de·vi·a·tor \-ˌā-tər\ *noun*
— de·vi·a·to·ry \-ə-ˌtōr-ē, -ˌtȯr-\ *adjective*

[2]**de·vi·ate** \-vē-ət, -vē-ˌāt\ *noun* (1912)
1 : one that deviates from a norm; *especially* **:** a person who differs markedly from a group norm
2 : a statistical variable that gives the deviation of another variable from a fixed value (as the mean)

[3]**de·vi·ate** \-vē-ət, -vē-ˌāt\ *adjective* (1929)
: departing significantly from the behavioral norms of a particular society

de·vi·a·tion \ˌde-vē-ˈā-shən\ *noun* (15th century)
: an act or instance of deviating: as **a :** deflection of the needle of a compass caused by local magnetic influences (as in a ship) **b :** the difference between a value in a frequency distribution and a fixed number (as the mean) **c :** departure from an established ideology or party line **d :** noticeable or marked departure from accepted norms of behavior
— de·vi·a·tion·ism \-shə-ˌni-zəm\ *noun*
— de·vi·a·tion·ist \-sh(ə-)nist\ *noun or adjective*

de·vice \di-ˈvīs\ *noun* [Middle English *devis, devise,* from Middle French, division, intention, from Old French *deviser* to divide, regulate, tell — more at DEVISE] (14th century)
1 : something devised or contrived: as **a** (1) **:** PLAN, PROCEDURE, TECHNIQUE (2) **:** a scheme to deceive **:** STRATAGEM, TRICK **b :** something fanciful, elaborate, or intricate in design **c :** something (as a figure of speech) in a literary work designed to achieve a particular artistic effect **d** *archaic* **:** MASQUE, SPECTACLE **e :** a conventional stage practice or means (as a stage whisper) used to achieve a particular dramatic effect **f :** a piece of equipment or a mechanism designed to serve a special purpose or perform a special function
2 : DESIRE, INCLINATION ⟨left to my own *devices*⟩
3 : an emblematic design used especially as a heraldic bearing

[1]**dev·il** \ˈde-v^{ə}l, *dialect* ˈdi-\ *noun* [Middle English *devel,* from Old English *dēofol,* from Late Latin *diabolus,* from Greek *diabolos,* literally, slanderer, from *diaballein* to throw across, slander, from *dia-* + *ballein* to throw; probably akin to Sanskrit *gurate* he lifts up] (before 12th century)
1 *often capitalized* **:** the personal supreme spirit of evil often represented in Jewish and Christian belief as the tempter of mankind, the leader of all apostate angels, and the ruler of hell — usually used with *the*; often used as an interjection, an intensive, or a generalized term of abuse ⟨what the *devil* is this?⟩ ⟨the *devil* you say!⟩
2 : an evil spirit **:** DEMON
3 a : an extremely wicked person **:** FIEND **b** *archaic* **:** a great evil
4 : a person of notable energy, recklessness, and dashing spirit; *also* **:** one who is mischievous ⟨those kids are little *devils* today⟩
5 : FELLOW — usually used in the phrases *poor devil, lucky devil*
6 a : something very trying or provoking ⟨having a *devil* of a time with this problem⟩ **b :** severe criticism or rebuke **:** HELL — used with *the* ⟨I'll probably catch the *devil* for this⟩
7 : DUST DEVIL
8 *Christian Science* **:** the opposite of Truth **:** a belief in sin, sickness, and death **:** EVIL, ERROR
— between the devil and the deep blue sea : faced with two equally objectionable alternatives
— devil to pay : severe consequences — used with *the*

[2]**devil** *transitive verb* **-iled** *or* **-illed; -il·ing** *or* **-il·ling** \ˈde-v^{ə}l-iŋ, ˈdev-liŋ\ (1800)
1 : to season highly ⟨*deviled* eggs⟩
2 : TEASE, ANNOY

dev·il·fish \ˈde-v^{ə}l-ˌfish\ *noun* (1709)
1 : any of several extremely large rays (genera *Manta* and *Mobula*) widely distributed in warm seas
2 : OCTOPUS; *broadly* **:** any large cephalopod

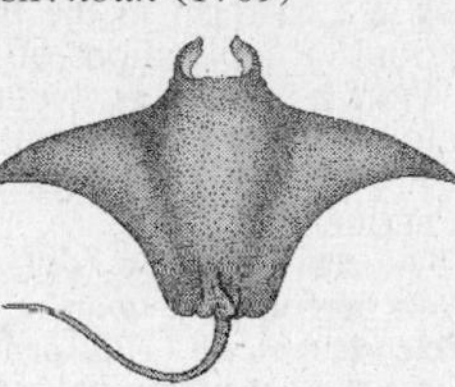
devilfish 1

dev·il·ish \ˈde-v^{ə}l-ish, ˈdev-lish\ *adjective* (15th century)
1 : resembling or befitting a devil: as **a :** EVIL, SINISTER **b :** MISCHIEVOUS, ROGUISH
2 : EXTREME ⟨in a *devilish* hurry⟩
— devilish *adverb*
— dev·il·ish·ly *adverb*
— dev·il·ish·ness *noun*

dev·il–may–care \ˌde-v^{ə}l-(ˌ)mā-ˈker, -ˈkar\ *adjective* (1837)
: EASYGOING, CAREFREE

dev·il·ment \ˈde-v^{ə}l-mənt, -ˌment\ *noun* (1771)
: MISCHIEF

dev·il·ry \ˈde-v^{ə}l-rē\ *or* **dev·il·try** \-v^{ə}l-trē\ *noun, plural* **-ries** *or* **-tries** (14th century)
1 a : action performed with the help of the devil **:** WITCHCRAFT **b :** WICKEDNESS **c :** MISCHIEF
2 : an act of devilry

devil's advocate *noun* [translation of New Latin *advocatus diaboli*] (1760)
1 : a Roman Catholic official whose duty is to examine critically the evidence on which a demand for beatification or canonization rests
2 : a person who champions the less accepted cause for the sake of argument

devil's claw *noun* (circa 1900)
: any of several herbs (genus *Proboscidea* synonym *Martynia* of the family Martyniaceae) of the southwestern U.S. and Mexico that have edible pods yielding a black sewing material used in basket making

devil's darning needle *noun* (1809)
1 : DRAGONFLY
2 : DAMSELFLY

devil's food cake *noun* (1905)
: a rich chocolate cake

devil's paintbrush *noun* (1900)
: ORANGE HAWKWEED; *broadly* **:** any of various hawkweeds that are naturalized weeds in the eastern U.S.

devil theory *noun* (1937)

: a theory of history: political and social crises arise from the deliberate actions of evil or misguided leaders rather than as a natural result of conditions

dev·il·wood \'de-vᵊl-ˌwu̇d\ *noun* (1818)
: a small tree (*Osmanthus americanus*) of the southern U.S. that is related to the olive

de·vi·ous \'dē-vē-əs, -vyəs\ *adjective* [Latin *devius,* from *de* from + *via* way — more at DE-, WAY] (1599)
1 : OUT-OF-THE-WAY, REMOTE
2 a : WANDERING, ROUNDABOUT ⟨a *devious* path⟩ **b :** moving without a fixed course **:** ERRANT ⟨*devious* breezes⟩
3 a : deviating from a right, accepted, or common course **b :** not straightforward **:** CUNNING; *also* **:** DECEPTIVE
— de·vi·ous·ly *adverb*
— de·vi·ous·ness *noun*

¹de·vise \di-'vīz\ *transitive verb* **de·vised; de·vis·ing** [Middle English, from Old French *deviser* to divide, regulate, tell, modification of (assumed) Vulgar Latin *divisare,* frequentative of Latin *dividere* to divide] (13th century)
1 a : to form in the mind by new combinations or applications of ideas or principles **:** INVENT **b** *archaic* **:** CONCEIVE, IMAGINE **c :** to plan to obtain or bring about **:** PLOT
2 : to give (real estate) by will — compare BEQUEATH
— de·vis·able \-'vī-zə-bəl\ *adjective*
— de·vis·er *noun*

²devise *noun* (15th century)
1 : the act of giving or disposing of real property by will
2 : a will or clause of a will disposing of real property
3 : property devised by will

de·vi·see \ˌde-və-'zē, di-ˌvī-'zē\ *noun* (1543)
: one to whom a devise of property is made

de·vi·sor \ˌde-və-'zȯr; di-'vī-zər, -ˌvī-'zȯr\ *noun* (1543)
: one who devises property in a will

de·vi·tal·ize \(ˌ)dē-'vī-tᵊl-ˌīz\ *transitive verb* (1849)
: to deprive of life, vigor, or effectiveness

de·vit·ri·fy \(ˌ)dē-'vi-trə-ˌfī\ *transitive verb* [French *dévitrifier,* from *dé-* de- + *vitrifier* to vitrify] (1832)
: to deprive of glassy luster and transparency; *especially* **:** to change (as a glass) from a vitreous to a crystalline condition
— de·vit·ri·fi·ca·tion \(ˌ)dē-ˌvi-trə-fə-'kā-shən\ *noun*

de·vo·cal·ize \(ˌ)dē-'vō-kə-ˌlīz\ *transitive verb* (1877)
: DEVOICE

de·voice \(ˌ)dē-'vȯis\ *transitive verb* (1932)
: to pronounce (as a sometimes or formerly voiced sound) without vibration of the vocal cords

de·void \di-'vȯid\ *adjective* [Middle English, past participle of *devoiden* to dispel, from Middle French *desvuidier* to empty, from Old French, from *des-* dis- + *vuidier* to empty — more at VOID] (15th century)
: being without a usual, typical, or expected attribute or accompaniment ⟨an argument *devoid* of sense⟩

de·voir \də-'vwär, 'de-ˌ\ *noun* [Middle English, alteration of *dever,* from Anglo-French *deveir,* from Old French *devoir, deveir* to owe, be obliged, from Latin *debēre* — more at DEBT] (14th century)
1 : DUTY, RESPONSIBILITY
2 : a usually formal act of civility or respect

de·vo·lu·tion \ˌde-və-'lü-shən *also* ˌdē-və-\ *noun* [Medieval Latin *devolution-, devolutio,* from Latin *devolvere*] (1545)
1 : transference (as of rights, powers, property, or responsibility) to another; *especially* **:** the surrender of powers to local authorities by a central government
2 : retrograde evolution **:** DEGENERATION
— de·vo·lu·tion·ary \-shə-ˌner-ē\ *adjective*
— de·vo·lu·tion·ist \-sh(ə-)nist\ *noun*

de·volve \di-'välv, -'vȯlv, dē-\ *verb* **de·volved; de·volv·ing** [Middle English, from Latin *devolvere,* from *de-* + *volvere* to roll — more at VOLUBLE] (15th century)
transitive verb
: to pass on (as responsibility, rights, or powers) from one person or entity to another ⟨*devolving* to western Europe full responsibility for its own defense —Christopher Lane⟩
intransitive verb
1 a : to pass by transmission or succession ⟨the estate *devolved* on a distant cousin⟩ **b :** to fall or be passed usually as a responsibility or obligation ⟨the responsibility for breadwinning has *devolved* increasingly upon women —Barbara Ehrenreich⟩
2 : to come by or as if by flowing down ⟨his allegedly subversive campaigns . . . *devolve* from his belief in basic American rights —Frank Deford⟩
3 : to degenerate through a gradual change or evolution ⟨where order *devolves* into chaos —*Johns Hopkins Magazine*⟩

dev·on \'de-vən\ *noun, often capitalized* [*Devon,* England] (1834)
: any of a breed of vigorous red dual-purpose cattle of English origin

De·vo·ni·an \di-'vō-nē-ən\ *adjective* [*Devon,* England] (1612)
1 : of or relating to Devonshire, England
2 : of, relating to, or being the period of the Paleozoic era between the Silurian and the Mississippian or the corresponding system of rocks — see GEOLOGIC TIME table
— Devonian *noun*

Dev·on·shire cream \ˌde-vən-ˌshir-, -shər-\ *noun* (1825)
: CLOTTED CREAM

de·vote \di-'vōt, dē-\ *transitive verb* **de·vot·ed; de·vot·ing** [Latin *devotus,* past participle of *devovēre,* from *de-* + *vovēre* to vow] (1586)
1 : to commit by a solemn act ⟨*devoted* herself to serving God⟩
2 : to give over or direct (as time, money, or effort) to a cause, enterprise, or activity ☆
— de·vote·ment \-'vōt-mənt\ *noun*

de·vot·ed *adjective* (1593)
: characterized by loyalty and devotion
— de·vot·ed·ly *adverb*
— de·vot·ed·ness *noun*

dev·o·tee \ˌde-və-'tē, -'tā; dē-ˌvō-'tē\ *noun* (1645)
: an ardent follower, supporter, or enthusiast (as of a religion, art form, or sport)

de·vo·tion \di-'vō-shən, dē-\ *noun* (13th century)
1 a : religious fervor **:** PIETY **b :** an act of prayer or private worship — usually used in plural **c :** a religious exercise or practice other than the regular corporate worship of a congregation
2 a : the act of devoting **b :** the fact or state of being ardently dedicated and loyal (as to an idea or person)
3 *obsolete* **:** the object of one's devotion
synonym see FIDELITY

¹de·vo·tion·al \-shnəl, -shə-nᵊl\ *adjective* (1648)
: of, relating to, or characterized by devotion
— de·vo·tion·al·ly \-ē\ *adverb*

²devotional *noun* (1659)
: a short worship service

de·vour \di-'vau̇(ə)r, dē-\ *transitive verb* [Middle English, from Middle French *devourer,* from Latin *devorare,* from *de-* + *vorare* to devour — more at VORACIOUS] (14th century)
1 : to eat up greedily or ravenously
2 : to use up or destroy as if by eating ⟨we are *devouring* the world's resources⟩
3 : to prey upon ⟨*devoured* by guilt⟩
4 : to enjoy avidly ⟨*devours* books⟩
— de·vour·er *noun*

de·vout \di-'vau̇t\ *adjective* [Middle English *devot,* from Old French, from Late Latin *devotus,* from Latin, past participle of *devovēre*] (13th century)
1 : devoted to religion or to religious duties or exercises
2 : expressing devotion or piety ⟨a *devout* attitude⟩
3 : devoted to a pursuit, belief, or mode of behavior **:** SERIOUS, EARNEST ⟨a *devout* baseball fan⟩ ⟨born a *devout* coward —G. B. Shaw⟩
— de·vout·ly *adverb*
— de·vout·ness *noun*

dew \'dü *also* 'dyü\ *noun* [Middle English, from Old English *dēaw;* akin to Old High German *tou* dew, Greek *thein* to run] (before 12th century)
1 : moisture condensed upon the surfaces of cool bodies especially at night
2 : something resembling dew in purity, freshness, or power to refresh
3 : moisture especially when appearing in minute droplets: as **a :** TEARS **b :** SWEAT **c :** droplets of water produced by a plant in transpiration
— dew *transitive verb*
— dew·less \-ləs\ *adjective*

dew·ar \'dü-ər *also* 'dyü-\ *noun, often capitalized* [Sir James *Dewar*] (circa 1909)
: a glass or metal container made like a vacuum bottle that is used especially for storing liquefied gases — called also *Dewar flask*

de·wa·ter \(ˌ)dē-'wȯt-ər, -'wät-\ *transitive verb* (circa 1909)
: to remove water from
— de·wa·ter·er *noun*

dew·ber·ry \'dü-ˌber-ē *also* 'dyü-\ *noun* (circa 1578)
1 : any of several sweet edible berries related to and resembling blackberries
2 : a trailing or decumbent bramble (genus *Rubus*) that bears dewberries

dew·claw \'dü-ˌklȯ *also* 'dyü-\ *noun* (1576)
: a vestigial digit not reaching to the ground on the foot of a mammal; *also* **:** a claw or hoof terminating such a digit — see COW illustration

dew·drop \'dü-ˌdräp *also* 'dyü-\ *noun* (13th century)
: a drop of dew

Dew·ey decimal classification \'dü-ē- *also* 'dyü-\ *noun* [Melvil *Dewey*] (1924)
: a system of classifying books and other publications whereby main classes are designated by a three-digit number and subdivisions are shown by numbers after a decimal point — called also *Dewey decimal system*

dew·fall \'dü-ˌfȯl *also* 'dyü-\ *noun* (1622)
: formation of dew; *also* **:** the time when dew begins to deposit

dew·lap \'dü-ˌlap *also* 'dyü-\ *noun* (14th century)
1 : loose skin hanging under the neck of an animal — see COW illustration
2 : loose flesh on the human throat
— dew·lapped \-ˌlapt\ *adjective*

de·worm \(ˌ)dē-'wərm\ *transitive verb* (1926)
: to rid (as a dog) of worms **:** WORM 4
— de·worm·er \(ˌ)dē-'wər-mər\ *noun*

☆ SYNONYMS
Devote, dedicate, consecrate, hallow mean to set apart for a special and often higher end. DEVOTE is likely to imply compelling motives and often attachment to an objective ⟨*devoted* his evenings to study⟩. DEDICATE implies solemn and exclusive devotion to a sacred or serious use or purpose ⟨*dedicated* her life to medical research⟩. CONSECRATE stresses investment with a solemn or sacred quality ⟨*consecrate* a church to the worship of God⟩. HALLOW, often differing little from *dedicate* or *consecrate,* may distinctively imply an attribution of intrinsic sanctity ⟨battlegrounds *hallowed* by the blood of patriots⟩.

dew point *noun* (circa 1833)
: the temperature at which a vapor begins to condense

dew worm *noun* (1599)
: NIGHT CRAWLER

dewy \'dü-ē *also* 'dyü-\ *adjective* **dew·i·er; -est** (before 12th century)
1 : moist with, affected by, or suggestive of dew
2 : INNOCENT, UNSOPHISTICATED ⟨from a *dewy* bride to an ill-mannered, murderous courtesan —Melvin Gussow⟩
— **dew·i·ly** \'dü-ə-lē, 'dyü-\ *adverb*
— **dew·i·ness** \'dü-ē-nəs, 'dyü-\ *noun*

dewy–eyed \'dü-ē-,īd *also* 'dyü-\ *adjective* (1938)
: naively credulous

dex \'deks\ *noun* (circa 1961)
: the sulfate of dextroamphetamine

dexa·meth·a·sone \,dek-sə-'me-thə-,sōn, -,zōn\ *noun* [*dexa-* (blend of *deca-* and *hexa-*) + *methyl* + *-a-* (perhaps from *pregnane,* a parent compound of corticoid hormones) + *-sone* (as in *cortisone*)] (1958)
: a synthetic glucocorticoid $C_{22}H_{29}FO_5$ used especially as an anti-inflammatory agent

Dex·e·drine \'dek-sə-,drēn, -drən\ *trademark*
— used for a preparation of the sulfate of dextroamphetamine

dex·ies \'dek-sēz\ *noun plural* [*Dex*edrine + *-ie* + [1]*-s*] (1956)
: tablets or capsules of the sulfate of dextroamphetamine

dex·ter \'dek-stər\ *adjective* [Latin; akin to Old High German *zeso* situated on the right, Greek *dexios*] (1562)
1 : relating to or situated on the right
2 : being or relating to the side of a heraldic shield at the right of the person bearing it
— **dexter** *adverb*

dex·ter·i·ty \dek-'ster-ə-tē\ *noun, plural* **-ties** [Middle French or Latin; Middle French *dexterité,* from Latin *dexteritat-, dexteritas,* from *dexter*] (1527)
1 : mental skill or quickness **:** ADROITNESS
2 : readiness and grace in physical activity; *especially* **:** skill and ease in using the hands

dex·ter·ous *also* **dex·trous** \'dek-st(ə-)rəs\ *adjective* [Latin *dextr-, dexter* on the right side, skillful] (1622)
1 : mentally adroit and skillful **:** CLEVER
2 : done with dexterity **:** ARTFUL
3 : skillful and competent with the hands ☆
— **dex·ter·ous·ly** *adverb*
— **dex·ter·ous·ness** *noun*

dextr- *or* **dextro-** *combining form* [Latin *dextr-, dexter*]
1 : right **:** on or toward the right ⟨*dextro*rotatory⟩
2 *usually dextro-* **:** dextrorotatory ⟨*dextro*amphetamine⟩

dex·tral \'dek-strəl\ *adjective* (1646)
: of or relating to the right **:** inclined to the right: as **a :** RIGHT-HANDED 3 **b** *of a gastropod shell* **:** having the whorls coiling clockwise down the spire when viewed with the apex toward the observer and having the aperture situated on the right of the axis when held with the spire uppermost and with the aperture opening toward the observer — compare SINISTRAL

dex·tran \'dek-stran, -strən\ *noun* [International Scientific Vocabulary] (1879)
: any of numerous glucose biopolymers of variable molecular weight produced especially by bacteria (as genus *Leuconostoc*), occurring in dental plaque, and used after chemical modification as blood plasma substitutes, as packing materials in chromatography, and as pharmaceutical agents

dex·tran·ase \-strə-,nās, -,nāz\ *noun* (circa 1949)
: a hydrolase that prevents tooth decay by breaking down dextran and eliminating dental plaque

dex·trin \'dek-strən\ *also* **dex·trine** \-,strēn, -strən\ *noun* [French *dextrine,* from *dextr-*] (1838)
: any of various water-soluble gummy polysaccharides $(C_6H_{10}O_5)n$ obtained from starch by the action of heat, acids, or enzymes and used as adhesives, as sizes for paper and textiles, as thickening agents (as in syrups), and in beer

dex·tro \'dek-(,)strō\ *adjective* [*dextr-*] (circa 1929)
: DEXTROROTATORY

dex·tro·am·phet·amine \'dek-(,)strō-am-'fe-tə-,mēn, -mən\ *noun* (1943)
: dextrorotatory amphetamine sulfate

dex·tro·ro·ta·to·ry \-'rō-tə-,tōr-ē, -,tȯr-\ *also* **dex·tro·ro·ta·ry** \-'rō-tə-rē\ *adjective* (1878)
: turning clockwise or toward the right; *especially* **:** rotating the plane of polarization of light toward the right ⟨*dextrorotatory* crystals⟩ — compare LEVOROTATORY

dex·trose \'dek-,strōs, -,strōz\ *noun* (circa 1869)
: dextrorotatory glucose

dey \'dā\ *noun* [French, from Turkish *dayı,* literally, maternal uncle] (1659)
: a ruling official of the Ottoman Empire in northern Africa

[1]DH \,dē-'āch\ *noun* (1973)
: DESIGNATED HITTER

[2]DH \,dē-'āch, 'dē-,āch\ *intransitive verb* **DH–d; DH–ing** (1975)
: to play the position of designated hitter

dhar·ma \'dər-mə, 'där-\ *noun* [Sanskrit; akin to Latin *firmus* firm] (1796)
1 *Hinduism* **:** an individual's duty fulfilled by observance of custom or law
2 *Hinduism & Buddhism* **a :** the basic principles of cosmic or individual existence **:** divine law **b :** conformity to one's duty and nature
— **dhar·mic** \-mik\ *adjective*

dhar·na \'dər-nə, 'där-\ *noun* [Hindi *dharnā,* from Sanskrit *dharaṇa* support, prop; akin to Latin *firmus* firm] (1747)
: a fast held at the door of an offender in India as an appeal for justice

dhole \'dōl\ *noun* [perhaps from Kannada *tōḷa* wolf] (circa 1827)
: a wild dog (*Cuon alpinus*) occurring from India to southern Siberia

dho·ti \'dō-tē\ *noun, plural* **dhotis** [Hindi *dhotī*] (1614)
: a loincloth worn by men in some parts of India

dhow \'dau̇\ *noun* [Arabic *dāwa*] (1785)
: an Arab lateen-rigged boat usually having a long overhang forward, a high poop, and a low waist

dhow

Dhu'l–Hij·ja \,dül-'hi-(,)jä, ,thül-\ *noun* [Arabic *Dhū-l-ḥijjah,* literally, the one of the pilgrimage] (circa 1771)
: the 12th month of the Islamic year — see MONTH table

Dhu'l–Qa'·dah \-'kä-(,)dä\ *noun* [Arabic *Dhū-l-qa'dah,* literally, the one of the sitting] (circa 1771)
: the 11th month of the Islamic year — see MONTH table

dhur·rie \'də-rē, 'dər-ē\ *noun* [Hindi *darī*] (1880)
: a thick flat-woven cotton or wool cloth or rug made in India

di- *combining form* [Middle English, from Middle French, from Latin, from Greek; akin to Old English *twi-*]
1 : twice **:** twofold **:** double ⟨*di*chromatic⟩
2 : containing two atoms, radicals, or groups ⟨*di*oxide⟩

dia- *also* **di-** *prefix* [Middle English, from Old French, from Latin, from Greek, through, apart, from *dia;* akin to Latin *dis-*]
: through ⟨*dia*positive⟩ **:** across ⟨*dia*dromous⟩

di·a·base \'dī-ə-,bās\ *noun* [French, probably from Greek *diabasis* act of crossing over, from *diabainein* to cross over, from *dia-* + *bainein* to go — more at COME] (circa 1816)
1 *archaic* **:** DIORITE
2 *chiefly British* **:** an altered basalt
3 : a fine-grained rock of the composition of gabbro but with an ophitic texture
— **di·a·ba·sic** \,dī-ə-'bā-sik\ *adjective*

di·a·be·tes \,dī-ə-'bē-tēz, -'bē-təs\ *noun* [Latin, from Greek *diabētēs* diabetes insipidus, from *diabainein* to walk with the legs apart, cross over] (15th century)
: any of various abnormal conditions characterized by the secretion and excretion of excessive amounts of urine; *especially* **:** DIABETES MELLITUS ◻

diabetes in·sip·i·dus \-in-'si-pə-dəs\ *noun* [New Latin, literally, bland diabetes] (circa 1860)
: a disorder of the pituitary gland characterized by intense thirst and by the excretion of large amounts of urine

diabetes mel·li·tus \-'me-lə-təs\ *noun* [New Latin, literally, honey-sweet diabetes] (circa 1860)
: a variable disorder of carbohydrate metabolism caused by a combination of hereditary and environmental factors and usually characterized by inadequate secretion or utilization of insulin, by excessive urine production, by excessive amounts of sugar in the blood and urine, and by thirst, hunger, and loss of weight

[1]di·a·bet·ic \,dī-ə-'be-tik\ *adjective* (1799)
1 : of or relating to diabetes or diabetics
2 : affected with diabetes
3 : occurring in or caused by diabetes ⟨*diabetic* coma⟩
4 : suitable for diabetics ⟨*diabetic* food⟩

[2]diabetic *noun* (1840)

☆ SYNONYMS
Dexterous, adroit, deft mean ready and skilled in physical movement. DEXTEROUS implies expertness with consequent facility and quickness in manipulation ⟨unrolled the sleeping bag with a *dexterous* toss⟩. ADROIT implies dexterity but usually also stresses resourcefulness or artfulness or inventiveness ⟨the magician's *adroit* response to the failure of her prop won applause⟩. DEFT emphasizes lightness, neatness, and sureness of touch or handling ⟨a surgeon's *deft* manipulation of the scalpel⟩.

◻ USAGE
diabetes One of the forces in linguistic change is *analogy,* the alteration of one word to make it similar to other words which are perceived to be related. Analogy has probably been the source of the variant pronunciation of *diabetes* in which the last syllable has a reduced form \-təs\, and not the older, fuller \-tēz\. The \-təs\ ending is more like the final syllable of the names of other diseases and conditions: *phlebitis, arthritis, conjunctivitis.* Although *diabetes* shares no word elements with these names, the last syllable has come to be pronounced as if it were parallel in form to these words. The newer pronunciation has been labelled "colloquial" by some, but it is common in all circles today.

: a person affected with diabetes

di·a·be·to·gen·ic \ˌdī-ə-ˌbe-tə-ˈje-nik, -ˌbē-\ *adjective* (circa 1903)
: producing diabetes

di·a·be·tol·og·ist \ˌdī-ə-(ˌ)be-ˈtä-lə-jist, -ˌbē-\ *noun* (1970)
: a specialist in diabetes

di·a·ble·rie \dē-ˈä-blə-(ˌ)rē, -ˈa-blə-\ *noun* [French, from Old French, from *diable* devil, from Late Latin *diabolus* — more at DEVIL] (1751)
1 : black magic **:** SORCERY
2 a : a representation in words or pictures of black magic or of dealings with the devil **b :** demon lore
3 : mischievous conduct or manner

di·a·bol·i·cal \ˌdī-ə-ˈbä-li-kəl\ *or* **di·a·bol·ic** \-ˈbä-lik\ *adjective* [Middle English *deabolik,* from Middle French *diabolique,* from Late Latin *diabolicus,* from *diabolus*] (14th century)
: of, relating to, or characteristic of the devil **:** DEVILISH
— di·a·bol·i·cal·ly \-li-k(ə-)lē\ *adverb*
— di·a·bol·i·cal·ness \-li-kəl-nəs\ *noun*

di·ab·o·lism \dī-ˈa-bə-ˌli-zəm\ *noun* (1614)
1 : dealings with or possession by the devil
2 : belief in or worship of devils
3 : evil character or conduct
— di·ab·o·list \-list\ *noun*

di·ab·o·lize \-ˌlīz\ *transitive verb* **-lized; -liz·ing** (1702)
: to represent as or make diabolic

dia·chron·ic \ˌdī-ə-ˈkrä-nik\ *adjective* (1927)
: of, relating to, or dealing with phenomena (as of language or culture) as they occur or change over a period of time
— dia·chron·i·cal·ly \-ˈkrä-ni-k(ə-)lē\ *adverb*

di·ach·ro·ny \dī-ˈa-krə-nē\ *noun* [International Scientific Vocabulary *dia-* + *-chrony* (as in *synchrony*)] (circa 1939)
1 : diachronic analysis
2 : change extending through time

[1]**di·ac·id** \(ˌ)dī-ˈa-səd\ *or* **di·acid·ic** \ˌdī-ə-ˈsi-dik\ *adjective* (1866)
: able to react with two molecules of a monobasic acid or one of a dibasic acid to form a salt or ester — used especially of bases

[2]**diacid** *noun* [International Scientific Vocabulary] (circa 1929)
: an acid with two acid hydrogen atoms

di·ac·o·nal \dī-ˈa-kə-nᵊl, dē-\ *adjective* [Late Latin *diaconalis,* from *diaconus* deacon — more at DEACON] (circa 1611)
: of or relating to a deacon or deaconess

di·ac·o·nate \-ˈa-kə-nət, -ˌnāt\ *noun* (circa 1751)
1 : the office or period of office of a deacon or deaconess
2 : an official body of deacons

di·a·crit·ic \ˌdī-ə-ˈkri-tik\ *noun* (1866)
: an accent near or through an orthographic or phonetic character or combination of characters indicating a phonetic value different from that given the unmarked or otherwise marked element

di·a·crit·i·cal \ˌdī-ə-ˈkri-ti-kəl\ *also* **di·a·crit·ic** \-ˈkri-tik\ *adjective* [Greek *diakritikos* separative, from *diakrinein* to distinguish, from *dia-* + *krinein* to separate — more at CERTAIN] (1749)
1 : serving as a diacritic
2 a : DISTINCTIVE ⟨the *diacritical* elements in culture —S. F. Nadel⟩ **b :** capable of distinguishing ⟨students of superior *diacritical* powers⟩

di·adel·phous \ˌdī-ə-ˈdel-fəs\ *adjective* [*di-* + *-adelphous*] (1807)
: united by filaments into two fascicles — used of stamens

di·a·dem \ˈdī-ə-ˌdem, -dəm\ *noun* [Middle English *diademe,* from Old French, from Latin *diadema,* from Greek *diadēma,* from *diadein* to bind around, from *dia-* + *dein* to bind; akin to Sanskrit *dāman* rope] (13th century)

DIACRITICS

´	(é)	acute accent
`	(è)	grave accent
^ *or* ˆ *or* ˜	(ô)	circumflex
~	(ñ)	tilde
¯	(ō)	macron
˘	(ŭ)	breve
ˇ	(č)	haček
¨	(oö)	diaeresis
¸	(ç)	cedilla

1 a : CROWN 1; *specifically* **:** a royal headband **b :** CROWN 6a(1)
2 : something that adorns like a crown

di·ad·ro·mous \dī-ˈa-drə-məs\ *adjective* (circa 1949)
of a fish **:** migratory between salt and fresh waters

di·aer·e·sis \dī-ˈer-ə-səs, *British also* -ˈir-\ *noun, plural* **-e·ses** \-ˌsēz\ [Late Latin *diaeresis,* from Greek *diairesis,* literally, division, from *diairein* to divide, from *dia-* + *hairein* to take] (circa 1611)
1 : a mark ¨ placed over a vowel to indicate that the vowel is pronounced in a separate syllable (as in *naïve* or *Brontë*)
2 : the break in a verse caused by the coincidence of the end of a foot with the end of a word
— di·ae·ret·ic \ˌdī-ə-ˈre-tik\ *adjective*

dia·gen·e·sis \ˌdī-ə-ˈje-nə-səs\ *noun* [New Latin] (circa 1886)
1 : recombination or rearrangement of constituents (as of a chemical or mineral) resulting in a new product
2 : the conversion (as by compaction or chemical reaction) of sediment into rock
— dia·ge·net·ic \ˌdī-ə-jə-ˈne-tik\ *adjective*
— dia·ge·net·i·cal·ly \-ˈne-ti-k(ə-)lē\ *adverb*

dia·geo·tro·pic \ˌdī-ə-ˌjē-ə-ˈtrō-pik, -ˈträ-pik\ *adjective* (1880)
: tending to grow at right angles to the line of gravity ⟨*diageotropic* branches and roots⟩

di·ag·nose \ˈdī-ig-ˌnōs, -ˌnōz, ˌdī-ig-ˈ, -əg-\ *verb* **-nosed; -nos·ing** [back-formation from *diagnosis*] (circa 1859)
transitive verb
1 a : to recognize (as a disease) by signs and symptoms **b :** to diagnose a disease or condition in ⟨*diagnosed* the patient⟩
2 : to analyze the cause or nature of ⟨*diagnose* the problem⟩
intransitive verb
: to make a diagnosis
— di·ag·nos·able *or* **di·ag·nose·able** \ˌdī-ig-ˈnō-sə-bəl, -əg-, -zə-\ *adjective*

di·ag·no·sis \ˌdī-ig-ˈnō-səs, -əg-\ *noun, plural* **-no·ses** \-ˌsēz\ [New Latin, from Greek *diagnōsis,* from *diagignōskein* to distinguish, from *dia-* + *gignōskein* to know — more at KNOW] (circa 1681)
1 a : the art or act of identifying a disease from its signs and symptoms **b :** the decision reached by diagnosis
2 : a concise technical description of a taxon
3 a : investigation or analysis of the cause or nature of a condition, situation, or problem ⟨*diagnosis* of engine trouble⟩ **b :** a statement or conclusion from such an analysis

[1]**di·ag·nos·tic** \-ˈnäs-tik\ *also* **di·ag·nos·ti·cal** \-ti-kəl\ *adjective* (1625)
1 a : of, relating to, or used in diagnosis **b :** using the methods of or yielding a diagnosis
2 : serving to distinguish or identify
— di·ag·nos·ti·cal·ly \-ti-k(ə-)lē\ *adverb*

[2]**diagnostic** *noun* (1625)
1 : the art or practice of diagnosis — often used in plural
2 : a distinguishing mark
— di·ag·nos·ti·cian \-(ˌ)näs-ˈti-shən\ *noun*

[1]**di·ag·o·nal** \dī-ˈa-gə-nᵊl, -ˈag-nəl\ *adjective* [Latin *diagonalis,* from Greek *diagōnios* from angle to angle, from *dia-* + *gōnia* angle; akin to Greek *gony* knee — more at KNEE] (1563)
1 a : joining two vertices of a rectilinear figure that are nonadjacent or two vertices of a polyhedral figure that are not in the same face **b :** passing through two nonadjacent edges of a polyhedron ⟨a *diagonal* plane⟩
2 a : inclined obliquely from a reference line (as the vertical) ⟨wood with a *diagonal* grain⟩ **b :** having diagonal markings or parts ⟨a *diagonal* weave⟩

[2]**diagonal** *noun* (1571)
1 : a diagonal straight line or plane
2 a (1) **:** a diagonal direction (2) **:** a diagonal row, arrangement, or pattern **b :** something oriented in diagonal position
3 : a mark / used typically to denote "or" (as in *and/or*), "and or" (as in *straggler/deserter*), or "per" (as in *feet/second*) — called also *solidus, virgule*
— on the diagonal : in an oblique direction **:** DIAGONALLY

di·ag·o·nal·ize \-nə-ˌlīz\ *transitive verb* **-ized; -iz·ing** (1942)
: to put (a matrix) in a form with all the nonzero elements along the diagonal from upper left to lower right
— di·ag·o·nal·iz·able \-ˌlī-zə-bəl\ *adjective*
— di·ag·o·nal·iza·tion \-ˌa-gə-nᵊl-ə-ˈzā-shən, -ˌag-nə-lə-ˈzā-\ *noun*

di·ag·o·nal·ly \dī-ˈa-gə-nᵊl-ē, -ˈag-nə-lē\ *adverb* (1541)
: in a diagonal manner

diagonal matrix *noun* (circa 1928)
: a diagonalized matrix

[1]**di·a·gram** \ˈdī-ə-ˌgram\ *noun* [Greek *diagramma,* from *diagraphein* to mark out by lines, from *dia-* + *graphein* to write — more at CARVE] (1619)
1 : a graphic design that explains rather than represents; *especially* **:** a drawing that shows arrangement and relations (as of parts)
2 : a line drawing made for mathematical or scientific purposes
— di·a·gram·ma·ble \-ˌgra-mə-bəl\ *adjective*
— di·a·gram·mat·ic \ˌdī-ə-grə-ˈma-tik\ *also* **di·a·gram·mat·i·cal** \-ˈma-ti-kəl\ *adjective*
— di·a·gram·mat·i·cal·ly \-ti-k(ə-)lē\ *adverb*

[2]**diagram** *transitive verb* **-grammed** *or* **-gramed** \-ˌgramd\; **-gram·ming** *or* **-gram·ing** \-ˌgra-miŋ\ (1840)
: to represent by or put into the form of a diagram

dia·ki·ne·sis \ˌdī-ə-kə-ˈnē-səs, -(ˌ)kī-\ *noun, plural* **-ne·ses** \-ˌsēz\ [New Latin] (circa 1902)
: the final stage of the meiotic prophase marked by contraction of the bivalents

[1]**di·al** \ˈdī(-ə)l\ *noun* [Middle English *dyal,* from Medieval Latin *dialis* clock wheel revolving daily, from Latin *dies* day — more at DEITY] (15th century)
1 : the face of a sundial
2 *obsolete* **:** TIMEPIECE
3 : the graduated face of a timepiece
4 a : a face upon which some measurement is registered usually by means of graduations and a pointer ⟨the thermometer *dial* reads 70°F⟩ **b :** a device that may be operated to make electrical connections or to regulate the operation of a machine ⟨a radio *dial*⟩ ⟨a telephone *dial*⟩

[2]**dial** *verb* **di·aled** *or* **di·alled; di·al·ing** *or* **di·al·ling** (1821)
transitive verb

1 : to measure with a dial
2 a : to manipulate a device (as a dial) so as to operate, regulate, or select ⟨*dial* your favorite program⟩ ⟨*dialed* the wrong number⟩ **b :** CALL 1m(1) ⟨*dialed* the office⟩
intransitive verb
1 : to manipulate a dial
2 : to make a telephone call or connection
— di·al·er *noun*

di·a·lect \ˈdī-ə-ˌlekt\ *noun, often attributive* [Middle French *dialecte,* from Latin *dialectus,* from Greek *dialektos* conversation, dialect, from *dialegesthai* to converse — more at DIALOGUE] (1577)
1 a : a regional variety of language distinguished by features of vocabulary, grammar, and pronunciation from other regional varieties and constituting together with them a single language ⟨the Doric *dialect* of ancient Greek⟩ **b :** one of two or more cognate languages ⟨French and Italian are Romance *dialects*⟩ **c :** a variety of a language used by the members of a group ⟨such *dialects* as politics and advertising —Philip Howard⟩ **d :** a variety of language whose identity is fixed by a factor other than geography (as social class) ⟨spoke a rough peasant *dialect*⟩ **e :** REGISTER 4c **f :** a version of a computer programming language
2 : manner or means of expressing oneself **:** PHRASEOLOGY
— di·a·lec·tal \ˌdī-ə-ˈlek-tᵊl\ *adjective*
— di·a·lec·tal·ly \-tᵊl-ē\ *adverb*

dialect atlas *noun* (1932)
: LINGUISTIC ATLAS

dialect geography *noun* (1929)
: LINGUISTIC GEOGRAPHY

di·a·lec·tic \ˌdī-ə-ˈlek-tik\ *noun* [Middle English *dialetik,* from Middle French *dialetique,* from Latin *dialectica,* from Greek *dialektikē,* from feminine of *dialektikos* of conversation, from *dialektos*] (14th century)
1 : LOGIC 1a(1)
2 a : discussion and reasoning by dialogue as a method of intellectual investigation; *specifically* **:** the Socratic techniques of exposing false beliefs and eliciting truth **b :** the Platonic investigation of the eternal ideas
3 : the logic of fallacy
4 a : the Hegelian process of change in which a concept or its realization passes over into and is preserved and fulfilled by its opposite; *also* **:** the critical investigation of this process **b** (1) *usually plural but singular or plural in construction* **:** development through the stages of thesis, antithesis, and synthesis in accordance with the laws of dialectical materialism (2) **:** the investigation of this process (3) **:** the theoretical application of this process especially in the social sciences
5 *usually plural but singular or plural in construction* **a :** any systematic reasoning, exposition, or argument that juxtaposes opposed or contradictory ideas and usually seeks to resolve their conflict **b :** an intellectual exchange of ideas
6 : the dialectical tension or opposition between two interacting forces or elements

di·a·lec·ti·cal \ˌdī-ə-ˈlek-ti-kəl\ *also* **di·a·lec·tic** \-tik\ *adjective* (1548)
1 a : of, relating to, or in accordance with dialectic ⟨*dialectical* method⟩ **b :** practicing, devoted to, or employing dialectic ⟨a *dialectical* philosopher⟩
2 : of, relating to, or characteristic of a dialect
— di·a·lec·ti·cal·ly \-ti-k(ə-)lē\ *adverb*

dialectical materialism *noun* (1927)
: the Marxist theory that maintains the material basis of a reality constantly changing in a dialectical process and the priority of matter over mind — compare HISTORICAL MATERIALISM

di·a·lec·ti·cian \ˌdī-ə-ˌlek-ˈti-shən\ *noun* (circa 1693)
1 : one who is skilled in or practices dialectic
2 : a student of dialects

di·a·lec·tol·o·gist \-ˈtä-lə-jist\ *noun* (1883)
: a specialist in dialectology

di·a·lec·tol·o·gy \-jē\ *noun* [International Scientific Vocabulary] (circa 1864)
1 : the systematic study of dialect
2 : the body of data available for study of a dialect
— di·a·lec·to·log·i·cal \-ˌlek-tə-ˈlä-ji-kəl\ *adjective*
— di·a·lec·to·log·i·cal·ly \-k(ə-)lē\ *adverb*

di·al·lel \ˈdī-ə-ˌlel\ *adjective* [Greek *diallēlos* reciprocating, from *dia* through + *allēlōn* one another — more at ALLELO-] (1920)
: relating to or being the crossing of each of several individuals with two or more others in order to determine the relative genetic contribution of each parent to specific characters in the offspring

di·a·log·ic \ˌdī-ə-ˈlä-jik\ *or* **di·a·log·i·cal** \-ji-kəl\ *adjective* (1833)
: of, relating to, or characterized by dialogue ⟨*dialogic* writing⟩
— di·a·log·i·cal·ly \-ji-k(ə-)lē\ *adverb*

di·a·lo·gist \dī-ˈa-lə-jist; ˈdī-ə-ˌlȯ-gist, -ˌlä-\ *noun* (circa 1660)
1 : a writer of dialogues
2 : one who participates in a dialogue
— di·a·lo·gis·tic \(ˌ)dī-ˌa-lə-ˈjis-tik; ˌdī-ə-ˌlȯ-ˈgis-, -ˌlä-ˈgis-\ *adjective*

[1]di·a·logue *also* **di·a·log** \ˈdī-ə-ˌlȯg, -ˌläg\ *noun* [Middle English *dialoge,* from Old French *dialogue,* from Latin *dialogus,* from Greek *dialogos,* from *dialegesthai* to converse, from *dia-* + *legein* to speak — more at LEGEND] (13th century)
1 : a written composition in which two or more characters are represented as conversing
2 a : a conversation between two or more persons; *also* **:** a similar exchange between a person and something else (as a computer) **b :** an exchange of ideas and opinions **c :** a discussion between representatives of parties to a conflict that is aimed at resolution
3 : the conversational element of literary or dramatic composition
4 : a musical composition for two or more parts suggestive of a conversation

[2]dialogue *verb* **-logued; -logu·ing** (1597)
transitive verb
: to express in dialogue
intransitive verb
: to take part in a dialogue

dial tone *noun* (1923)
: a tone emitted by a telephone as a signal that the system is ready for dialing

dial–up \ˈdī(-ə)l-ˌəp\ *adjective* (1972)
: relating to or being a standard telephone line used for computer communications; *also* **:** accessible via a standard telephone line ⟨a *dial-up* information service⟩

di·al·y·sate \dī-ˈa-lə-ˌzāt, -ˌsāt\ *also* **di·al·y·zate** \-ˌzāt\ *noun* (circa 1867)
: the material that passes through the membrane in dialysis; *also* **:** the liquid into which this material passes

di·al·y·sis \dī-ˈa-lə-səs\ *noun, plural* **-y·ses** \-ˌsēz\ [New Latin, from Greek, separation, from *dialyein* to dissolve, from *dia-* + *lyein* to loosen — more at LOSE] (1861)
1 : the separation of substances in solution by means of their unequal diffusion through semipermeable membranes; *especially* **:** such a separation of colloids from soluble substances
2 : HEMODIALYSIS
— di·a·lyt·ic \ˌdī-ə-ˈli-tik\ *adjective*

di·a·lyze \ˈdī-ə-ˌlīz\ *verb* **-lyzed; -lyz·ing** (1861)
transitive verb
: to subject to dialysis
intransitive verb
: to undergo dialysis
— di·a·lyz·able \-ˌlī-zə-bəl\ *adjective*
— di·a·lyz·er \-ˌlī-zər\ *noun*

dia·mag·net·ic \ˌdī-ə-mag-ˈne-tik\ *adjective* (1846)
: having a magnetic permeability less than that of a vacuum **:** slightly repelled by a magnet
— dia·mag·ne·tism \-ˈmag-nə-ˌti-zəm\ *noun*

di·a·man·té \ˌdē-ə-ˌmän-ˈtā\ *noun* [French, adjective, like a diamond, from *diamant* diamond, from Middle French] (1904)
: a sparkling decoration (as of sequins) or material decorated with this ⟨a gown trimmed with *diamanté*⟩

di·am·e·ter \dī-ˈa-mə-tər\ *noun* [Middle English *diametre,* from Middle French, from Latin *diametros,* from Greek, from *dia-* + *metron* measure — more at MEASURE] (14th century)
1 : a chord passing through the center of a figure or body
2 : the length of a straight line through the center of an object
3 : a unit of magnification for an optical instrument equal to the number of times the linear dimensions of an object are apparently increased ⟨a microscope magnifying 60 *diameters*⟩
— di·am·e·tral \-ˈa-mə-trəl\ *adjective*

di·a·met·ric \ˌdī-ə-ˈme-trik\ *or* **di·a·met·ri·cal** \-tri-kəl\ *adjective* (1553)
1 : of, relating to, or constituting a diameter **:** located at the diameter
2 : completely opposed **:** being at opposite extremes ⟨in *diametric* contradiction to his claims⟩
— di·a·met·ri·cal·ly \-tri-k(ə-)lē\ *adverb*

di·amide \ˈdī-ə-ˌmīd, dī-ˈa-məd\ *noun* (1866)
: a compound containing two amido groups

di·amine \ˈdī-ə-ˌmēn, dī-ˈa-mən\ *noun* [International Scientific Vocabulary] (1866)
: a compound containing two amino groups

di·am·mo·ni·um phosphate \ˌdī-ə-ˈmō-nē-əm-\ *noun* (circa 1929)
: a white crystalline compound $(NH_4)_2HPO_4$ used especially as a fertilizer and as a fire retardant

[1]di·a·mond \ˈdī-(ə-)mənd\ *noun, often attributive* [Middle English *diamaunde,* from Middle French *diamant,* from Old French, from Late Latin *diamant-, diamas,* alteration of Latin *adamant-, adamas* hardest metal, diamond, from Greek] (14th century)
1 a : native crystalline carbon that is usually nearly colorless, that when transparent and free from flaws is highly valued as a precious stone, and that is used industrially as an abrasive powder and in rock drills because of its great hardness; *also* **:** a piece of this substance **b :** crystallized carbon produced artificially
2 : something that resembles a diamond (as in brilliance, value, or fine quality)
3 : a square or rhombus-shaped figure usually oriented with the long diagonal vertical
4 a : a playing card marked with a stylized figure of a red diamond **b** *plural but singular or plural in construction* **:** the suit comprising cards marked with diamonds
5 : a baseball infield; *also* **:** the entire playing field

[2]diamond *transitive verb* (1751)
: to adorn with or as if with diamonds

[3]diamond *adjective* (1872)
: of, relating to, or being a 60th or 75th anniversary or its celebration ⟨*diamond* jubilee⟩

di·a·mond·back \ˈdī-(ə-)mən(d)-ˌbak\ *adjective* (1887)
: having marks like diamonds or lozenges on the back

diamondback moth *noun* (1891)
: a nearly cosmopolitan moth (*Plutella xylostella* of the family Plutellidae) whose larva is a pest on cruciferous plants

diamondback rattlesnake *noun* (1894)

: either of two large and deadly rattlesnakes (*Crotalus adamanteus* of the southeastern U.S. and *C. atrox* of the south central and southwestern U.S. and Mexico) — called also *diamondback, diamondback rattler*

diamondback terrapin *noun* (1887)
: any of several terrapins (genus *Malaclemys*) formerly widely distributed in salt marshes along the Atlantic and Gulf coasts but now much restricted

di·a·mond·if·er·ous \ˌdī-(ə-)mən-ˈdi-fə-rəs\ *adjective* (1870)
: containing diamonds ⟨*diamondiferous* earth⟩

diamond in the rough (circa 1947)
: a person of exceptional qualities or potential but lacking refinement or polish

Di·ana \dī-ˈa-nə\ *noun* [Latin]
: an ancient Italian goddess of the forest and of childbirth who was identified with Artemis by the Romans

di·an·thus \dī-ˈan(t)-thəs\ *noun* [New Latin, genus name, from Greek *dios* heavenly + *anthos* flower — more at DEITY, ANTHOLOGY] (1849)
: [2]PINK 1

di·a·pa·son \ˌdī-ə-ˈpā-zᵊn, -sᵊn\ *noun* [Middle English, from Latin, from Greek *(hē) dia pasōn (chordōn symphōnia),* literally, the concord through all the notes, from *dia* through + *pasōn,* genitive feminine plural of *pas* all — more at DIA-, PAN-] (circa 1501)
1 a : a burst of sound ⟨*diapasons* of laughter⟩ **b :** the principal foundation stop in the organ extending through the complete range of the instrument **c** (1) **:** the entire compass of musical tones (2) **:** RANGE, SCOPE
2 a : TUNING FORK **b :** a standard of pitch

dia·pause \ˈdī-ə-ˌpȯz\ *noun* [Greek *diapausis* pause, from *diapauein* to pause, from *dia-* + *pauein* to stop] (1893)
: a period of physiologically enforced dormancy between periods of activity

dia·paus·ing \-ˌpȯ-ziŋ\ *adjective* (1944)
: undergoing diapause

di·a·pe·de·sis \ˌdī-ə-pə-ˈdē-səs\ *noun, plural* **-de·ses** \-ˌsēz\ [New Latin, from Greek *diapēdēsis,* literally, act of leaping through, from *diapēdan* to leap through, from *dia-* + *pēdan* to leap] (1625)
: the passage of blood cells through capillary walls into the tissues

[1]di·a·per \ˈdī-pər *also* ˈdī-ə-\ *noun* [Middle English *diapre,* from Middle French, from Medieval Latin *diasprum*] (14th century)
1 : a fabric with a distinctive pattern: **a :** a rich silk fabric **b :** a soft usually white linen or cotton fabric used for tablecloths or towels
2 : an allover pattern consisting of one or more small repeated units of design (as geometric figures) connecting with one another or growing out of one another with continuously flowing or straight lines
3 : a basic garment for infants consisting of a folded cloth or other absorbent material drawn up between the legs and fastened about the waist

diaper 2

[2]diaper *transitive verb* **di·a·pered; di·a·per·ing** \-p(ə-)riŋ\ (14th century)
1 : to ornament with diaper designs
2 : to put on or change the diaper of (an infant)

diaper rash *noun*
: skin irritation of the diaper-covered area of an infant especially from excessive urinary ammonia

di·a·pha·ne·ity \(ˌ)dī-ˌa-fə-ˈnē-ə-tē, ˌdī-ə-fə-, -ˈnā-\ *noun* (15th century)
: the quality or state of being diaphanous

di·aph·a·nous \dī-ˈa-fə-nəs\ *adjective* [Medieval Latin *diaphanus,* from Greek *diaphanēs,* from *diaphainein* to show through, from *dia-* + *phainein* to show — more at FANCY] (1614)
1 : characterized by such fineness of texture as to permit seeing through
2 : characterized by extreme delicacy of form **:** ETHEREAL ⟨painted *diaphanous* landscapes⟩
3 : INSUBSTANTIAL, VAGUE ⟨had only a *diaphanous* hope of success⟩
— di·aph·a·nous·ly *adverb*
— di·aph·a·nous·ness *noun*

dia·phone \ˈdī-ə-ˌfōn\ *noun* (1906)
: a fog signal similar to a siren but producing a blast of two tones

di·aph·o·rase \dī-ˈa-fə-ˌrās, -ˌrāz\ *noun* [International Scientific Vocabulary *diaphor-* (from Greek *diaphoros* different, from *diapherein* to differ, from *dia-* + *pherein* to carry) + *-ase* — more at BEAR] (1938)
: a flavoprotein enzyme capable of oxidizing the reduced form of NAD

di·a·pho·re·sis \ˌdī-ə-fə-ˈrē-səs, (ˌ)dī-ˌa-fə-\ *noun, plural* **-re·ses** \-ˌsēz\ [Late Latin, from Greek *diaphorēsis,* from *diaphorein* to dissipate by perspiration, from *dia-* + *phorein,* frequentative of *pherein* to carry] (circa 1681)
: PERSPIRATION; *especially* **:** profuse perspiration artificially induced

di·a·pho·ret·ic \-ˈre-tik\ *adjective* (15th century)
: having the power to increase perspiration
— diaphoretic *noun*

di·a·phragm \ˈdī-ə-ˌfram\ *noun* [Middle English *diafragma,* from Late Latin *diaphragma,* from Greek, from *diaphrassein* to barricade, from *dia-* + *phrassein* to enclose] (14th century)
1 : a body partition of muscle and connective tissue; *specifically* **:** the partition separating the chest and abdominal cavities in mammals
2 : a dividing membrane or thin partition especially in a tube
3 a : a more or less rigid partition in the body or shell of an invertebrate **b :** a transverse septum in a plant stem
4 : a device that limits the aperture of a lens or optical system — compare IRIS DIAPHRAGM
5 : a thin flexible disk (as in a microphone or loudspeaker) that vibrates when struck by sound waves or that vibrates to generate sound waves
6 : a molded cap usually of thin rubber fitted over the uterine cervix to act as a mechanical contraceptive barrier
— di·a·phrag·mat·ic \ˌdī-ə-frə(g)-ˈma-tik, -ˌfrag-\ *adjective*
— di·a·phrag·mat·i·cal·ly \-ˈma-ti-k(ə-)lē\ *adverb*

di·aph·y·sis \dī-ˈa-fə-səs\ *noun, plural* **-y·ses** \-ˌsēz\ [New Latin, from Greek, spinous process of the tibia, from *diaphyesthai* to grow between, from *dia-* + *phyein* to bring forth — more at BE] (1831)
: the shaft of a long bone
— di·aph·y·se·al \(ˌ)dī-ˌa-fə-ˈsē-əl\ *or* **di·a·phys·i·al** \ˌdī-ə-ˈfi-zē-əl\ *adjective*

di·a·pir \ˈdī-ə-ˌpir\ *noun* [French, probably from Greek *diapeirein* to drive through, from *dia-* + *peirein* to pierce; akin to Greek *poros* passage — more at FARE] (1918)
: an anticlinal fold in which a mobile core has broken through brittle overlying rocks
— di·a·pir·ic \ˌdī-ə-ˈpir-ik\ *adjective*

dia·pos·i·tive \ˌdī-ə-ˈpä-zə-tiv, -ˈpäz-tiv\ *noun* (1893)
: a positive photographic image on transparent material (as glass or film)

di·ap·sid \dī-ˈap-səd\ *adjective* [ultimately from Greek *di-* + *hapsid-, hapsis* loop, arch — more at APSIS] (circa 1909)
: of, relating to, or including reptiles (as the crocodiles) with two pairs of temporal openings in the skull

di·ar·chy *variant of* DYARCHY

di·a·rist \ˈdī-ə-rəst\ *noun* (circa 1818)
: one who keeps a diary

di·ar·rhea \ˌdī-ə-ˈrē-ə\ *noun* [Middle English *diaria,* from Late Latin *diarrhoea,* from Greek *diarrhoia,* from *diarrhein* to flow through, from *dia-* + *rhein* to flow — more at STREAM] (14th century)
1 : abnormally frequent intestinal evacuations with more or less fluid stools
2 : excessive flow ⟨verbal *diarrhea*⟩
— di·ar·rhe·ic \-ˈrē-ik\ *adjective*
— di·ar·rhe·al \-ˈrē-əl\ *adjective*
— di·ar·rhet·ic \-ˈre-tik\ *adjective*

di·ar·rhoea *chiefly British variant of* DIARRHEA

di·ar·thro·sis \ˌdī-är-ˈthrō-səs\ *noun, plural* **-thro·ses** \-ˌsēz\ [New Latin, from Greek *diarthrōsis,* from *diarthroun* to joint, from *dia-* + *arthroun* to fasten by a joint, from *arthron* joint — more at ARTHR-] (1578)
1 : articulation that permits free movement
2 : a freely movable joint

di·a·ry \ˈdī-(ə-)rē\ *noun, plural* **-ries** [Latin *diarium,* from *dies* day — more at DEITY] (1581)
1 : a record of events, transactions, or observations kept daily or at frequent intervals **:** JOURNAL; *especially* **:** a daily record of personal activities, reflections, or feelings
2 : a book intended or used for a diary

di·as·po·ra \dī-ˈas-p(ə-)rə, dē-\ *noun* [Greek, dispersion, from *diaspeirein* to scatter, from *dia-* + *speirein* to sow] (1881)
1 *capitalized* **a :** the settling of scattered colonies of Jews outside Palestine after the Babylonian exile **b :** the area outside Palestine settled by Jews **c :** the Jews living outside Palestine or modern Israel
2 a : the breaking up and scattering of a people **:** MIGRATION ⟨the black *diaspora* to northern cities⟩ **b :** people settled far from their ancestral homelands ⟨African *diaspora*⟩ **c :** the place where these people live

di·a·spore \ˈdī-ə-ˌspōr, -ˌspȯr\ *noun* [French, from Greek *diaspora*] (1805)
: a mineral consisting of aluminum hydrogen oxide

di·a·stase \ˈdī-ə-ˌstās, -ˌstāz\ *noun* [French, from Greek *diastasis* separation, interval, from *diistanai* to separate, from *dia-* + *histanai* to cause to stand — more at STAND] (1838)
1 : AMYLASE; *especially* **:** a mixture of amylases from malt
2 : ENZYME

di·a·stat·ic \ˌdī-ə-ˈsta-tik\ *adjective* (1881)
: relating to or having the properties of diastase; *especially* **:** converting starch into sugar

di·a·ste·ma \ˌdī-ə-ˈstē-mə\ *noun, plural* **-ma·ta** \-mə-tə\ [New Latin, from Late Latin, interval, from Greek *diastēma,* from *diistanai*] (1854)
: a space between teeth in a jaw

di·a·ste·reo·mer \ˌdī-ə-ˈster-ē-ō-(ˌ)mər, -ˈstir-\ *or* **di·a·ste·reo·iso·mer** \-ˌster-ē-ō-ˈī-sə-mər, -ˌstir-\ *noun* (1936)
: a stereoisomer of a compound having two or more chiral centers that is not a mirror image of another stereoisomer of the same compound — compare ENANTIOMORPH
— di·a·ste·reo·mer·ic \-ˌster-ē-ō-ˈmer-ik, -ˌstir-\ *or* **di·a·ste·reo·iso·mer·ic** \-ˌī-sə-ˈmer-ik\ *adjective*
— di·a·ste·reo·isom·er·ism \-ī-ˈsä-mə-ˌri-zəm\ *noun*

di·as·to·le \dī-ˈas-tə-(ˌ)lē\ *noun* [Greek *diastolē* dilatation, from *diastellein* to expand, from *dia-* + *stellein* to prepare, send] (circa 1578)
: a rhythmically recurrent expansion; *especially* **:** the dilatation of the cavities of the heart during which they fill with blood
— di·a·stol·ic \ˌdī-ə-ˈstä-lik\ *adjective*

di·as·tro·phism \dī-ˈas-trə-ˌfi-zəm\ *noun* [Greek *diastrophē* twisting, from *diastrephein* to distort, from *dia-* + *strephein* to twist] (1890)

: the process of deformation that produces in the earth's crust its continents and ocean basins, plateaus and mountains, folds of strata, and faults
— **di·a·stroph·ic** \ˌdī-ə-ˈsträ-fik\ *adjective*
— **di·a·stroph·i·cal·ly** \-fi-k(ə-)lē\ *adverb*

di·a·tes·sa·ron \ˌdī-ə-ˈte-sə-rən\ *noun* [Greek (*Euangelion*) *dia tessarōn,* literally, Gospel out of four, from *dia* through, out of + *tessarōn,* genitive of *tessares* four — more at DIA-, FOUR] (1803)
: a harmony of the four Gospels edited and arranged into a single connected narrative

dia·ther·ma·nous \ˌdī-ə-ˈthər-mə-nəs\ *adjective* [French *diathermane,* modification of Greek *diathermainein* to heat through, from *dia-* + *thermainein* to heat, from *thermos* hot — more at THERM] (1834)
: DIATHERMIC 1

dia·ther·mic \ˌdī-ə-ˈthər-mik\ *adjective* (1840)
1 : transmitting infrared radiation
2 : of or relating to diathermy ⟨*diathermic* treatment⟩

dia·ther·my \ˈdī-ə-ˌthər-mē\ *noun* [International Scientific Vocabulary] (1909)
: the generation of heat in tissue by electric currents for medical or surgical purposes

di·ath·e·sis \dī-ˈa-thə-səs\ *noun, plural* **-e·ses** \-ˌsēz\ [New Latin, from Greek, literally, arrangement, from *diatithenai* to arrange, from *dia-* + *tithenai* to set — more at DO] (1651)
: a constitutional predisposition toward a particular state or condition and especially one that is abnormal or diseased
— **di·a·thet·ic** \ˌdī-ə-ˈthe-tik\ *adjective*

di·a·tom \ˈdī-ə-ˌtäm\ *noun* [ultimately from Greek *diatomos* cut in half, from *diatemnein* to cut through, from *dia-* + *temnein* to cut — more at TOME] (1845)
: any of a class (Bacillariophyceae) of minute planktonic unicellular or colonial algae with silicified skeletons that form diatomite

di·a·to·ma·ceous \ˌdī-ə-tə-ˈmā-shəs, (ˌ)dī-ˌa-tə-\ *adjective* (1847)
: consisting of or abounding in diatoms or their siliceous remains ⟨*diatomaceous* silica⟩

diatomaceous earth *noun* (1883)
: DIATOMITE

di·atom·ic \ˌdī-ə-ˈtä-mik\ *adjective* [International Scientific Vocabulary] (circa 1859)
: consisting of two atoms **:** having two atoms in the molecule

di·at·o·mite \dī-ˈa-tə-ˌmīt\ *noun* (1887)
: a light friable siliceous material derived chiefly from diatom remains and used especially as a filter

dia·ton·ic \ˌdī-ə-ˈtä-nik\ *adjective* [Late Latin *diatonicus,* from Greek *diatonikos,* from *diatonos* stretching, from *diateinein* to stretch out, from *dia-* + *teinein* to stretch — more at THIN] (1694)
: of or relating to a major or minor musical scale comprising intervals of five whole steps and two half steps
— **dia·ton·i·cal·ly** \-ˈtä-ni-k(ə-)lē\ *adverb*

di·a·tribe \ˈdī-ə-ˌtrīb\ *noun* [Latin *diatriba,* from Greek *diatribē* pastime, discourse, from *diatribein* to spend (time), wear away, from *dia-* + *tribein* to rub — more at THROW] (1581)
1 *archaic* **:** a prolonged discourse
2 : a bitter and abusive speech or writing
3 : ironical or satirical criticism

di·az·e·pam \dī-ˈa-zə-ˌpam\ *noun* [benzo*diaz*epine + *-am* (of unknown origin)] (circa 1961)
: a tranquilizer $C_{16}H_{13}ClN_2O$ used especially to relieve anxiety and tension and as a muscle relaxant — compare VALIUM

Di·az·i·non \dī-ˈa-zi-ˌnän\ *trademark*
— used for a cholinesterase-inhibiting organophosphate insecticide

di·azo \dī-ˈa-(ˌ)zō, -ˈā-\ *adjective* [International Scientific Vocabulary *diaz-, diazo-,* from *di-* + *az-*] (1878)
1 a : relating to or containing the group N_2 composed of two nitrogen atoms united to a single carbon atom of an organic radical — often used in combination **b :** relating to or containing diazonium — often used in combination
2 : of or relating to a photograph or photocopy whose production involves the use of a coating of a diazo compound that is decomposed by exposure to light

di·a·zo·ni·um \ˌdī-ə-ˈzō-nē-əm\ *noun* [International Scientific Vocabulary *di-* + *az-* + *-onium*] (1895)
: the univalent cation N_2^+ that is composed of two nitrogen atoms united to carbon in an organic radical and that usually exists in salts used in the manufacture of azo dyes

di·az·o·tize \dī-ˈa-zə-ˌtīz\ *transitive verb* **-tized; -tiz·ing** [International Scientific Vocabulary *di-* + *azote* nitrogen + *-ize* — more at AZ-] (circa 1889)
: to convert (a compound) into a diazo compound (as a diazonium salt)
— **di·az·o·ti·za·tion** \-ˌa-zə-tə-ˈzā-shən\ *noun*

di·ba·sic \(ˌ)dī-ˈbā-sik\ *adjective* (1868)
: having two replaceable hydrogen atoms — used of acids

dib·ber \ˈdi-bər\ *noun* (circa 1736)
: DIBBLE

[1]**dib·ble** \ˈdi-bəl\ *noun* [Middle English *debyl-le*] (15th century)
: a small hand implement used to make holes in the ground for plants, seeds, or bulbs

[2]**dibble** *transitive verb* **dib·bled; dib·bling** \ˈdi-b(ə-)liŋ\ (1583)
1 : to plant with a dibble
2 : to make holes in (soil) with or as if with a dibble

di·ben·zo·fu·ran \ˌdī-ˌben-zō-ˈfyu̇-ˌran, -fyə-ˈran\ *noun* (1940)
: a highly toxic chemical compound $C_{12}H_8O$ that is used in chemical synthesis and as an insecticide and is a hazardous pollutant in its chlorinated form

dibs \ˈdibz\ *noun plural* [short for *dibstones* (jacks), from obsolete *dib* (to dab)] (1812)
1 *slang* **:** money especially in small amounts
2 : CLAIM, RIGHTS ⟨I have *dibs* on that piece of cake⟩

di·bu·tyl phthal·ate \ˌdī-ˌbyü-tᵊl-ˈtha-ˌlāt\ *noun* [*phthal*ic acid + [1]*-ate*] (1925)
: a colorless oily ester $C_{16}H_{22}O_4$ used chiefly as a solvent, plasticizer, pesticide, and repellent (as for chiggers and mites)

di·cal·ci·um silicate \(ˌ)dī-ˈkal-sē-əm-\ *noun* (1925)
: a calcium silicate $2CaO \cdot SiO_2$ that is an essential ingredient of portland cement

di·car·box·yl·ic \ˌdī-ˌkär-ˌbäk-ˈsi-lik\ *adjective* (circa 1890)
: containing two carboxyl groups in the molecule ⟨*dicarboxylic* acids⟩

di·cast \ˈdī-ˌkast, ˈdi-\ *noun* [Greek *dikastēs,* from *dikazein* to judge, from *dikē* judgment — more at DICTION] (1822)
: an ancient Athenian performing the functions of both judge and juror at a trial

[1]**dice** \ˈdīs\ *noun, plural* **dice** [Middle English *dyce,* from *dees, dyce,* plural of *dee* die — more at DIE] (14th century)
1 a : DIE 1 **b :** a gambling game played with dice
2 *plural also* **dic·es :** a small cubical piece (as of food)
3 : a close contest between two racing-car drivers for position during a race
— **no dice 1 :** of no avail **:** no use **:** FUTILE **2 :** [1]NO 3 ⟨said *no dice* to my request⟩

[2]**dice** *verb* **diced; dic·ing** [Middle English *dycen,* from *dyce*] (14th century)
transitive verb
1 a : to cut into small cubes **b :** to ornament with square markings ⟨*diced* leather⟩
2 a : to bring by playing dice ⟨*dice* himself into debt⟩ **b :** to lose by dicing ⟨*dice* her money away⟩
intransitive verb
1 : to play games with dice ⟨*dice* for drinks in the bar —Malcolm Lowry⟩
2 : to take a chance ⟨the temptation to *dice* with death —*Newsweek*⟩
— **dic·er** *noun*

di·cen·tric \(ˌ)dī-ˈsen-trik\ *adjective* (1937)
: having two centromeres ⟨a *dicentric* chromosome⟩
— **dicentric** *noun*

dic·ey \ˈdī-sē\ *adjective* **dic·i·er; -est** [[1]*dice* + *-y*] (1950)
: RISKY, UNPREDICTABLE ⟨a *dicey* proposition⟩ ⟨*dicey* weather⟩

dich- *or* **dicho-** *combining form* [Late Latin, from Greek, from *dicha;* akin to Greek *di-*]
: in two **:** apart ⟨*dicho*gamous⟩

di·cha·sium \dī-ˈkā-z(h)ē-əm, -zhəm\ *noun, plural* **-sia** \-z(h)ē-ə, -zhə\ [New Latin, from Greek *dichasis* halving, from *dichazein* to halve, from *dicha*] (1875)
: a cymose inflorescence that produces two main axes

dichlor- *or* **dichloro-** *combining form*
: containing two atoms of chlorine ⟨*dichloro*ethane⟩

di·chlo·ro·ben·zene \(ˌ)dī-ˌklōr-ə-ˈben-ˌzēn, -ˌklȯr-, -(ˌ)ben-ˈ\ *noun* (1873)
: any of three isomeric compounds $C_6H_4Cl_2$; *especially* **:** PARADICHLOROBENZENE

di·chlo·ro·di·flu·o·ro·meth·ane \-ˌdī-ˌflu̇r-ə-ˈme-ˌthān\ *noun* (1936)
: a chlorofluoromethane CCl_2F_2

di·chlo·ro·eth·ane \(ˌ)dī-ˌklō-rō-ˈe-ˌthān\ *noun* (1936)
: a colorless toxic liquid compound $C_2H_4Cl_2$ that is used chiefly as a solvent

di·chlor·vos \(ˈ)dī-ˈklōr-ˌväs, -ˈklȯr-, -vəs\ *noun* [*dichlor-* + vinyl + phosphate] (1957)
: an organophosphorus insecticide and anthelmintic $C_4H_7Cl_2O_4P$ used especially in veterinary medicine — called also *DDVP*

di·chog·a·my \dī-ˈkä-gə-mē\ *noun, plural* **-mies** [German *Dichogamie,* from *dich-* + *-gamie* -gamy] (1862)
: the production of male and female reproductive elements at different times by a hermaphroditic organism in order to ensure cross-fertilization
— **di·chog·a·mous** \-gə-məs\ *adjective*

di·chon·dra \dī-ˈkän-drə\ *noun* [New Latin, genus name, from *di-* + Greek *chondros* grain — more at GRIND] (1947)
: any of a genus (*Dichondra*) of chiefly tropical perennial herbs of the morning glory family that includes some (especially *D. repens* or its varieties) used as a ground cover and a substitute for lawn grasses in warmer parts of the U.S.

dich·otic \(ˌ)dī-ˈkō-tik\ *adjective* [*dich-* + [2]*-otic*] (circa 1911)
: affecting or relating to the two ears differently in regard to a conscious aspect (as pitch or loudness) or a physical aspect (as frequency or energy) of sound
— **dich·oti·cal·ly** \-ti-k(ə-)lē\ *adverb*

di·chot·o·mist \dī-ˈkä-tə-məst *also* də-\ *noun* (circa 1592)
: one that dichotomizes

di·chot·o·mize \-ˌmīz\ *verb* **-mized; -miz·ing** [Late Latin *dichotomos*] (1606)
transitive verb
: to divide into two parts, classes, or groups
intransitive verb
: to exhibit dichotomy
— **di·chot·o·mi·za·tion** \-ˌkä-tə-mə-ˈzā-shən\ *noun*

di·chot·o·mous \dī-'kä-tə-məs *also* də-\ *adjective* [Late Latin *dichotomos,* from Greek, from *dich-* + *temnein* to cut — more at TOME] (1752)
1 : dividing into two parts
2 : relating to, involving, or proceeding from dichotomy
— di·chot·o·mous·ly *adverb*
— di·chot·o·mous·ness *noun*

dichotomous key *noun* (circa 1889)
: a key for the identification of organisms based on a series of choices between alternative characters

di·chot·o·my \dī-'kä-tə-mē *also* də-\ *noun, plural* **-mies** [Greek *dichotomia,* from *dichotomos*] (1610)
1 : a division or the process of dividing into two especially mutually exclusive or contradictory groups or entities
2 : the phase of the moon or an inferior planet in which half its disk appears illuminated
3 a : BIFURCATION; *especially* **:** repeated bifurcation (as of a plant's stem) **b :** a system of branching in which the main axis forks repeatedly into two branches **c :** branching of an ancestral line into two equal diverging branches
4 : something with seemingly contradictory qualities

di·chro·ic \dī-'krō-ik\ *adjective* [Greek *dichroos* two-colored, from *di-* + *chrōs* color, literally, skin] (circa 1859)
1 : having the property of dichroism ⟨a *dichroic* crystal⟩ ⟨a *dichroic* mirror⟩
2 : DICHROMATIC

di·chro·ism \'dī-(,)krō-,i-zəm\ *noun* (1819)
: the property of some crystals and solutions of absorbing one of two plane-polarized components of transmitted light more strongly than the other; *also* **:** the property of exhibiting different colors by reflected or transmitted light — compare CIRCULAR DICHROISM

di·chro·mat \'dī-krō-,mat, (,)dī-'\ *noun* [back-formation from *dichromatic*] (circa 1909)
: one affected with dichromatism

di·chro·mate \(,)dī-'krō-,māt, 'dī-krō-\ *noun* [International Scientific Vocabulary] (circa 1864)
: a usually orange to red chromium salt containing the anion $Cr_2O_7^{2-}$ ⟨*dichromate* of potassium⟩ — called also *bichromate*

di·chro·mat·ic \,dī-krō-'ma-tik\ *adjective* (circa 1847)
1 : having or exhibiting two colors
2 : of, relating to, or exhibiting dichromatism

di·chro·ma·tism \dī-'krō-mə-,ti-zəm\ *noun* (circa 1901)
: partial color blindness in which only two colors are perceptible

di·chro·scope \'dī-krə-,skōp\ *noun* (1857)
: an instrument for examining crystals for dichroism

dick \'dik\ *noun* [*Dick,* nickname for *Richard*] (1553)
1 *chiefly British* **:** FELLOW, CHAP
2 : PENIS — usually considered vulgar
3 [by shortening & alteration] **:** DETECTIVE

dick·cis·sel \dik-'si-səl, 'dik-,\ *noun* [imitative] (1886)
: a common migratory black-throated finch (*Spiza americana*) of the central U.S.

dickcissel

dick·ens \'di-kənz\ *noun* [euphemism] (1598)
: DEVIL, DEUCE

[1]dick·er \'di-kər\ *noun* [Middle English *dyker,* from Latin *decuria* quantity of ten, from *decem* ten — more at TEN] (14th century)
: the number or quantity of 10 especially of hides or skins

[2]dicker *intransitive verb* **dick·ered; dick·er·ing** \'di-k(ə-)riŋ\ [origin unknown] (circa 1802)
: BARGAIN ⟨*dickered* over the price⟩

[3]dicker *noun* (1823)
1 : BARTER
2 : an act or session of bargaining

dick·ey *or* **dicky** *also* **dick·ie** \'di-kē\ *noun, plural* **dickeys** *or* **dick·ies** [*Dicky,* nickname for *Richard*] (1753)
1 : any of various articles of clothing: as **a :** a man's separate or detachable shirtfront **b :** a small fabric insert worn to fill in the neckline
2 *chiefly British* **a :** the driver's seat in a carriage **b :** a seat at the back of a carriage or automobile
3 : a small bird

Dick test \'dik-\ *noun* [George French *Dick* and Gladys H. *Dick*] (1925)
: a test to determine susceptibility or immunity to scarlet fever by an injection of scarlet fever toxin

di·cli·nous \(,)dī-'klī-nəs\ *adjective* (1830)
: having the stamens and pistils in separate flowers

di·cot \'dī-,kät\ *noun* (1877)
: DICOTYLEDON

di·cot·y·le·don \,dī-,kä-t^{ə}l-'ē-d^{ə}n\ *noun* [New Latin] (circa 1727)
: any of a class or subclass (Magnoliopsida or Dicotyledoneae) of angiospermous plants that produce an embryo with two cotyledons and usually have floral organs arranged in cycles of four or five and leaves with reticulate venation — compare MONOCOTYLEDON
— di·cot·y·le·don·ous \-d^{ə}n-əs\ *adjective*

di·cou·ma·rin \(,)dī-'kü-mə-rən\ *noun* (1886)
: DICUMAROL

di·crot·ic \dī-'krä-tik\ *adjective* [Greek *dikrotos,* from *di-* + *krotos* rattling noise, beat] (circa 1811)
1 *of the pulse* **:** having a double beat
2 : being or relating to the second part of the arterial pulse occurring during diastole of the heart or of an arterial pressure recording made during the same period
— di·cro·tism \'dī-krə-,ti-zəm\ *noun*

Dic·ta·phone \'dik-tə-,fōn\ *trademark*
— used for a dictating machine

[1]dic·tate \'dik-,tāt, dik-'\ *verb* **dic·tat·ed; dic·tat·ing** [Latin *dictatus,* past participle of *dictare* to assert, dictate, frequentative of *dicere* to say — more at DICTION] (1592)
intransitive verb
1 : to give dictation
2 : to speak or act domineeringly **:** PRESCRIBE
transitive verb
1 : to speak or read for a person to transcribe or for a machine to record
2 a : to issue as an order **b :** to impose, pronounce, or specify authoritatively **c :** to require or determine necessarily ⟨injuries *dictated* the choice of players⟩

[2]dic·tate \'dik-,tāt\ *noun* (1594)
1 a : an authoritative rule, prescription, or injunction **b :** a ruling principle ⟨according to the *dictates* of your conscience⟩
2 : a command by one in authority

dictating machine *noun* (1907)
: a machine used especially for the recording of human speech for transcription

dic·ta·tion \dik-'tā-shən\ *noun* (circa 1656)
1 a : PRESCRIPTION **b :** arbitrary command
2 a (1) **:** the act or manner of uttering words to be transcribed (2) **:** material that is dictated or transcribed **b** (1) **:** the performing of music to be reproduced by a student (2) **:** music so reproduced

dic·ta·tor \'dik-,tā-tər, dik-'\ *noun* [Latin, from *dictare*] (14th century)
1 a : a person granted absolute emergency power; *especially* **:** one appointed by the senate of ancient Rome **b :** one holding complete autocratic control **c :** one ruling absolutely and often oppressively
2 : one that dictates

dic·ta·to·ri·al \,dik-tə-'tōr-ē-əl, -'tȯr-\ *adjective* (1701)
1 a : of, relating to, or befitting a dictator ⟨*dictatorial* power⟩ **b :** ruled by a dictator
2 : oppressive to or arrogantly overbearing toward others ☆
— dic·ta·to·ri·al·ly \-ē-ə-lē\ *adverb*
— dic·ta·to·ri·al·ness *noun*

dic·ta·tor·ship \dik-'tā-tər-,ship, 'dik-,\ *noun* (1542)
1 : the office of dictator
2 : autocratic rule, control, or leadership
3 a : a form of government in which absolute power is concentrated in a dictator or a small clique **b :** a government organization or group in which absolute power is so concentrated **c :** a despotic state

dic·tion \'dik-shən\ *noun* [Latin *diction-, dictio* speaking, style, from *dicere* to say; akin to Old English *tēon* to accuse, Latin *dicare* to proclaim, dedicate, Greek *deiknynai* to show, *dikē* judgment, right] (1581)
1 *obsolete* **:** verbal description
2 : choice of words especially with regard to correctness, clearness, or effectiveness
3 a : vocal expression **:** ENUNCIATION **b :** pronunciation and enunciation of words in singing
— dic·tion·al \-shnəl, -shə-n^{ə}l\ *adjective*
— dic·tion·al·ly \-ē\ *adverb*

dic·tio·nary \'dik-shə-,ner-ē\ *noun, plural* **-nar·ies** [Medieval Latin *dictionarium,* from Late Latin *diction-, dictio* word, from Latin, speaking] (1526)
1 : a reference book containing words usually alphabetically arranged along with information about their forms, pronunciations, functions, etymologies, meanings, and syntactical and idiomatic uses
2 : a reference book listing alphabetically terms or names important to a particular subject or activity along with discussion of their meanings and applications
3 : a reference book giving for words of one language equivalents in another
4 : a list (as of items of data or words) stored in a computer for reference (as for information retrieval or word processing)

dic·tum \'dik-təm\ *noun, plural* **dic·ta** \-tə\ *also* **dictums** [Latin, from neuter of *dictus,* past participle of *dicere*] (1599)
1 : a noteworthy statement: as **a :** a formal pronouncement of a principle, proposition, or opinion **b :** an observation intended or regarded as authoritative
2 : a judicial opinion on a point other than the precise issue involved in determining a case

dicty- *or* **dictyo-** *combining form* [New Latin, from Greek *dikty-, diktyo-,* from *diktyon,* from *dikein* to throw]
: net ⟨*dictyo*stele⟩ ⟨*dictyo*some⟩

☆ SYNONYMS
Dictatorial, magisterial, dogmatic, doctrinaire, oracular mean imposing one's will or opinions on others. DICTATORIAL stresses autocratic, high-handed methods and a domineering manner ⟨exercised *dictatorial* control over the office⟩. MAGISTERIAL stresses assumption or use of prerogatives appropriate to a magistrate or schoolmaster in forcing acceptance of one's opinions ⟨the *magisterial* tone of his pronouncements⟩. DOGMATIC implies being unduly and offensively positive in laying down principles and expressing opinions ⟨*dogmatic* about what is art and what is not⟩. DOCTRINAIRE implies a disposition to follow abstract theories in framing laws or policies affecting people ⟨a *doctrinaire* approach to improving the economy⟩. ORACULAR implies the manner of one who delivers opinions in cryptic phrases or with pompous dogmatism ⟨a designer who is the *oracular* voice of fashion⟩.

dic•tyo•some \'dik-tē-ə-,sōm\ *noun* (circa 1930)
: any of the membranous or vesicular structures making up the Golgi apparatus

dic•tyo•stele \'dik-tē-ə-,stēl, ,dik-tē-ə-'stē-lē\ *noun* (circa 1902)
: a stele in which the vascular cylinder is broken up into a longitudinal series or network of vascular strands around a central pith (as in many ferns)

di•cu•ma•rol *also* **di•cou•ma•rol** \dī-'k(y)ü-mə-,rȯl, -,rōl\ *noun* [*di-* + *coumar*in + [1]*-ol*] (1943)
: a crystalline compound $C_{19}H_{12}O_6$ originally obtained from spoiled sweet clover hay and used to delay clotting of blood especially in preventing and treating thromboembolic disease

di•cyn•o•dont \(,)dī-'si-nə-,dänt, -'sī-\ *noun* [ultimately from Greek *di-* + *kyn-*, *kyōn* dog + *odont-*, *odous* tooth — more at HOUND, TOOTH] (1854)
: any of a suborder (Dicynodontia) of small herbivorous therapsid reptiles with reduced dentition
— **dicynodont** *adjective*

did *past of* DO

di•dact \'dī-,dakt\ *noun* [back-formation from *didactic*] (1954)
: a didactic person

di•dac•tic \dī-'dak-tik, də-\ *adjective* [Greek *didaktikos*, from *didaskein* to teach] (1658)
1 a : designed or intended to teach **b :** intended to convey instruction and information as well as pleasure and entertainment
2 : making moral observations
— **di•dac•ti•cal** \-ti-kəl\ *adjective*
— **di•dac•ti•cal•ly** \-ti-k(ə-)lē\ *adverb*
— **di•dac•ti•cism** \-tə-,si-zəm\ *noun*

di•dac•tics \-tiks\ *noun plural but singular or plural in construction* (1644)
: systematic instruction **:** PEDAGOGY

did•dle \'di-dᵊl\ *verb* **did•dled; did•dling** \'did-liŋ, -dᵊl-iŋ\ [origin unknown] (1786)
transitive verb
1 *chiefly dialect* **:** to move with short rapid motions
2 : to waste (as time) in trifling
3 : HOAX, SWINDLE
4 : to copulate with — often considered vulgar
intransitive verb
1 : DAWDLE, FOOL
2 : FIDDLE, TOY — usually used with *with* ⟨*diddled* with the machine until it broke⟩
— **did•dler** \'did-lər, -dᵊl-ər\ *noun*

did•dly \'di-dᵊl-ē, 'did-lē\ *noun* (1964)
slang **:** DIDDLY-SQUAT

did•dly–squat \-,skwät\ *noun* [probably alteration of *doodly-squat*] (1974)
slang **:** the least amount **:** anything at all ⟨didn't know *diddly-squat* about sports —Sam Toperoff⟩

did•ger•i•doo *also* **did•jer•i•doo** \'di-jə-rē-,dü, ,di-jə-rē-'\ *noun* [Yolngu (Australian aboriginal language group of Arnhem Land) *didjeridu*] (1919)
: a large bamboo or wooden trumpet of the Australian Aborigines

didn't \'di-dᵊnt, -dᵊn, *dialect also* 'dit-ᵊn(t) *or* 'dint\ (1705)
: did not

di•do \'dī-(,)dō\ *noun, plural* **didoes** *or* **didos** [origin unknown] (1807)
1 : a mischievous or capricious act **:** PRANK, ANTIC — often used in the phrase *cut didoes*
2 : something that is frivolous or showy

Di•do \'dī-(,)dō\ *noun* [Latin, from Greek *Didō*]
: a legendary queen of Carthage in Virgil's *Aeneid* who kills herself when Aeneas leaves her

didst \'didst, 'ditst\ *archaic past 2d singular of* DO

di•dym•i•um \dī-'di-mē-əm\ *noun* [New Latin, from Greek *didymos* twin, from *dyo* two — more at TWO] (1842)
: a mixture of rare-earth elements made up chiefly of neodymium and praseodymium and used especially for coloring glass for optical filters

[1]**die** \'dī\ *intransitive verb* **died; dy•ing** \'dī-iŋ\ [Middle English *dien*, from or akin to Old Norse *deyja* to die; akin to Old High German *touwen* to die] (12th century)
1 : to pass from physical life **:** EXPIRE
2 a : to pass out of existence **:** CEASE ⟨their anger *died* at these words⟩ **b :** to disappear or subside gradually — often used with *away, down,* or *out* ⟨the storm *died* down⟩
3 a : SINK, LANGUISH ⟨*dying* from fatigue⟩ **b :** to long keenly or desperately ⟨*dying* to go⟩ **c :** to be overwhelmed by emotion ⟨*die* of embarrassment⟩
4 a : to cease functioning **:** STOP ⟨the motor *died*⟩ **b :** to end in failure ⟨the bill *died* in committee⟩
5 : to become indifferent ⟨*die* to worldly things⟩
— **die hard 1 :** to be long in dying ⟨such rumors *die hard*⟩ **2 :** to continue resistance against hopeless odds ⟨hard-shell conservatism *dies hard*⟩
— **die on the vine :** to fail especially at an early stage through lack of support or enthusiasm ⟨let the proposal *die on the vine*⟩

[2]**die** \'dī\ *noun, plural* **dice** \'dīs\ *or* **dies** \'dīz\ [Middle English *dee*, from Middle French *dé*] (14th century)
1 *plural dice* **:** a small cube marked on each face with from one to six spots and used usually in pairs in various games and in gambling by being shaken and thrown to come to rest at random on a flat surface — often used figuratively in expressions concerning chance or the irrevocability of a course of action ⟨the *die* was cast⟩
2 *plural dies* **:** DADO 1a
3 *plural dies* **:** any of various tools or devices for imparting a desired shape, form, or finish to a material or for impressing an object or material: as **a** (1) **:** the larger of a pair of cutting or shaping tools that when moved toward each other produce a desired form in or impress a desired device on an object by pressure or by a blow (2) **:** a device composed of a pair of such tools **b :** a hollow internally threaded screw-cutting tool used for forming screw threads **c :** a mold into which molten metal or other material is forced **d :** a perforated block through which metal or plastic is drawn or extruded for shaping

die•back \'dī-,bak\ *noun* (circa 1886)
: a condition in woody plants in which peripheral parts are killed especially by parasites

di•ecious *variant of* DIOECIOUS

dief•fen•bach•ia \,dē-fən-'ba-kē-ə, ,di-, -'bä-\ *noun* [New Latin, from Ernst *Dieffenbach* (died 1855) German naturalist] (circa 1900)
: any of a genus (*Dieffenbachia* of the family Araceae) of erect poisonous tropical American plants that have oblong usually variegated leaves and are often used as houseplants

die–hard \'dī-,härd\ *adjective* (1922)
: strongly or fanatically determined or devoted ⟨*die-hard* fans⟩; *especially* **:** strongly resisting change ⟨a *die-hard* conservative⟩
— **die•hard** *noun*
— **die–hard•ism** \-,iz-əm\ *noun*

di•el \'dī-əl, -,el\ *adjective* [irregular from Latin *dies* day + English *-al*] (circa 1935)
: involving a 24-hour period that usually includes a day and the adjoining night ⟨*diel* fluctuations in temperature⟩

diel•drin \'dē(ə)l-drən\ *noun* [*Diels-Alder* reaction (an addition reaction forming a 6-membered ring), after Otto *Diels* & Kurt *Alder*] (circa 1949)
: a white crystalline persistent chlorinated hydrocarbon insecticide $C_{12}H_8Cl_6O$ that accumulates in and becomes toxic to vertebrates

di•elec•tric \,dī-ə-'lek-trik\ *noun* [*dia-* + *electric*] (1837)
: a nonconductor of direct electric current
— **dielectric** *adjective*

dielectric constant *noun* (1875)
: PERMITTIVITY

dielectric heating *noun* (1944)
: the rapid and uniform heating throughout a nonconducting material by means of a high-frequency electromagnetic field

di•en•ceph•a•lon \,dī-ən-'se-fə-,län, ,dī-(,)en-, -lən\ *noun* [New Latin, from *dia-* + *encephalon*] (circa 1883)
: the posterior subdivision of the forebrain — called also *betweenbrain*
— **di•en•ce•phal•ic** \-sə-'fa-lik\ *adjective*

di•ene \'dī-,ēn\ *noun* [*di-* + *-ene*] (1917)
: a compound containing two double bonds between carbon atoms

die–off \'dī-,ȯf\ *noun* (1936)
: a sudden sharp decline of a population (as of rabbits) that is not caused directly by human activity (as hunting)

die off \'dī-'ȯf\ *intransitive verb* (1697)
: to die sequentially either singly or in numbers so that the total number is greatly diminished

die out *intransitive verb* (1853)
: to become extinct

di•er•e•sis *variant of* DIAERESIS

die•sel \'dē-zəl, -səl\ *noun* [Rudolf *Diesel*] (1894)
1 : DIESEL ENGINE
2 : a vehicle driven by a diesel engine

diesel–electric *adjective* (1914)
: of, relating to, or employing the combination of a diesel engine driving an electric generator ⟨a *diesel-electric* locomotive⟩

diesel engine *noun* (1894)
: an internal combustion engine in which air is compressed to a temperature sufficiently high to ignite fuel injected into the cylinder where the combustion and expansion actuate a piston

die•sel•ing \'dē-zə-liŋ, -sə-\ *noun* (circa 1955)
: the continued operation of an internal combustion engine after the ignition is turned off

die•sel•ize \'dē-zə-,līz, -sə-\ *transitive verb* **-ized; -iz•ing** (1925)
: to equip with a diesel engine or with electric locomotives having generators powered by diesel engines
— **die•sel•iza•tion** \,dē-zə-lə-'zā-shən, -sə-\ *noun*

Di•es Irae \,dē-(,)äs-'ē-,rā\ *noun* [Medieval Latin, day of wrath; from the first words of the hymn] (1860)
: a medieval Latin hymn on the Day of Judgment sung in requiem masses

di•esis \'dī-ə-səs\ *noun, plural* **di•eses** \-,sēz\ [probably from Italian, sharp (in music), symbol for a sharp, from Medieval Latin, quarter tone, from Latin, from Greek, from *diiēnai* to send through, from *dia-* + *hienai* to send — more at JET] (circa 1706)
: DOUBLE DAGGER

di•es•ter \'dī-,es-tər\ *noun* (1935)
: a compound containing two ester groups

die•stock \'dī-,stäk\ *noun* (circa 1859)
: a stock to hold dies used for cutting threads

di•es•trus \(,)dī-'es-trəs\ *noun* [New Latin, from *dia-* + *estrus*] (1942)
: a period of sexual quiescence that intervenes between two periods of estrus
— **di•es•trous** \-trəs\ *adjective*

[1]**di•et** \'dī-ət\ *noun* [Middle English *diete*, from Old French, from Latin *diaeta*, from Greek *diaita*, literally, manner of living, from *diaitasthai* to lead one's life] (13th century)
1 a : food and drink regularly provided or consumed **b :** habitual nourishment **c :** the kind and amount of food prescribed for a person or animal for a special reason

2 **:** something provided especially habitually ⟨a *diet* of Broadway shows and nightclubs —Frederick Wyatt⟩

²diet (14th century)
transitive verb
1 : to cause to take food **:** FEED
2 : to cause to eat and drink sparingly or according to prescribed rules
intransitive verb
: to eat sparingly or according to prescribed rules
— di•et•er *noun*

³diet *adjective* (1969)
: reduced in calories ⟨a *diet* soft drink⟩

⁴diet *noun* [Middle English *diete* day's journey, day set for a meeting, from Medieval Latin *dieta,* literally, daily regimen, diet (taken as a derivative of Latin *dies* day), from Latin *diaeta*] (1565)
1 : a formal deliberative assembly of princes or estates
2 : any of various national or provincial legislatures

¹di•e•tary \ˈdī-ə-ˌter-ē\ *noun, plural* **di•e•taries** (1838)
: the kinds and amounts of food available to or eaten by an individual, group, or population

²dietary *adjective* (1614)
: of or relating to a diet or to the rules of a diet
— di•e•tar•i•ly \ˌdī-ə-ˈter-ə-lē\ *adverb*

dietary law *noun* (circa 1930)
: any of the laws observed by Orthodox Jews that permit or prohibit certain foods

di•e•tet•ic \ˌdī-ə-ˈte-tik\ *adjective* (1579)
1 : of or relating to diet
2 : adapted for use in special diets
— di•e•tet•i•cal•ly \-ti-k(ə-)lē\ *adverb*

di•e•tet•ics \-ˈte-tiks\ *noun plural but singular or plural in construction* (1799)
: the science or art of applying the principles of nutrition to the diet

di•ether \(ˌ)dī-ˈē-thər\ *noun* (1950)
: a compound containing two atoms of oxygen with ether linkages

di•eth•yl•car•bam•a•zine \ˌdī-ˌe-thəl-kär-ˈba-mə-ˌzēn, -zən\ *noun* [*di-* + *ethyl* + *carb*oxy- + *am*ide + *azine*] (1948)
: an anthelmintic administered in the form of its crystalline citrate $C_{10}H_{21}N_3O{\cdot}C_6H_8O_7$ especially to control human filariasis and large roundworms in dogs and cats

di•eth•yl ether \(ˌ)dī-ˌe-thəl-\ *noun* (circa 1930)
: ETHER 3a

di•eth•yl•stil•bes•trol \-stil-ˈbes-ˌtrȯl, -ˌtrōl\ *noun* [International Scientific Vocabulary] (1938)
: a colorless crystalline synthetic compound $C_{18}H_{20}O_2$ used as a potent estrogen but contraindicated in pregnancy for its tendency to cause cancer or birth defects in offspring — called also *stilbestrol*

di•eth•yl zinc *noun* (1952)
: a volatile pyrophoric liquid organometallic compound $C_4H_{10}Zn$ used especially to catalyze polymerization reactions and to deacidify paper

di•e•ti•tian *or* **di•e•ti•cian** \ˌdī-ə-ˈti-shən\ *noun* [*dietitian* irregular from ¹*diet* + *-ician*] (circa 1846)
: a specialist in dietetics

dif•fer \ˈdi-fər\ *intransitive verb* **dif•fered; dif•fer•ing** \-f(ə-)riŋ\ [Middle English, from Middle French or Latin; Middle French *differer* to postpone, be different, from Latin *differre,* from *dis-* + *ferre* to carry — more at BEAR] (14th century)
1 a : to be unlike or distinct in nature, form, or characteristics ⟨the law of one state *differs* from that of another⟩ **b :** to change from time to time or from one instance to another **:** VARY ⟨the number of cookies in a box may *differ*⟩
2 : to be of unlike or opposite opinion **:** DISAGREE ⟨they *differ* on religious matters⟩

¹dif•fer•ence \ˈdi-fərn(t)s, ˈdi-f(ə-)rən(t)s\ *noun* (14th century)
1 a : the quality or state of being different **b :** an instance of differing in nature, form, or quality **c** *archaic* **:** a characteristic that distinguishes one from another or from the average **d :** the element or factor that separates or distinguishes contrasting situations
2 : distinction or discrimination in preference
3 a : disagreement in opinion **:** DISSENSION **b :** an instance or cause of disagreement
4 : the degree or amount by which things differ in quantity or measure; *specifically* **:** REMAINDER 2b(1)
5 : a significant change in or effect on a situation

²difference *transitive verb* **-enced; -encing** (1576)
: DIFFERENTIATE, DISTINGUISH

¹dif•fer•ent \ˈdi-fərnt, ˈdi-f(ə-)rənt\ *adjective* [Middle French, from Latin *different-, differens,* present participle of *differre*] (14th century)
1 : partly or totally unlike in nature, form, or quality **:** DISSIMILAR ⟨could hardly be more *different*⟩ — often followed by *from, than,* or chiefly British *to* ⟨small, neat hand, very *different* from the captain's tottery characters —R. L. Stevenson⟩ ⟨vastly *different* in size than it was twenty-five years ago —N. M. Pusey⟩ ⟨a very *different* situation to the . . . one under which we live —Sir Winston Churchill⟩
2 : not the same: as **a :** DISTINCT ⟨*different* age groups⟩ **b :** VARIOUS ⟨*different* members of the class⟩ **c :** ANOTHER ⟨switched to a *different* TV program⟩
3 : UNUSUAL, SPECIAL ⟨she was *different* and superior⟩ ☆ □
— dif•fer•ent•ness *noun*

²different *adverb* (1744)
: DIFFERENTLY

dif•fer•en•tia \ˌdi-fə-ˈren(t)-sh(ē-)ə\ *noun, plural* **-ti•ae** \-shē-ˌē, -shē-ˌī\ [Latin, difference, from *different-, differens*] (1690)
: an element, feature, or factor that distinguishes one entity, state, or class from another; *especially* **:** a characteristic trait distinguishing a species from other species of the same genus

¹dif•fer•en•tial \ˌdi-fə-ˈren(t)-shəl\ *adjective* (1647)
1 a : of, relating to, or constituting a difference **:** DISTINGUISHING **b :** making a distinction between individuals or classes **c :** based on or resulting from a differential **d :** functioning or proceeding differently or at a different rate
2 : being, relating to, or involving a differential or differentiation
3 a : relating to quantitative differences **b :** producing effects by reason of quantitative differences
— dif•fer•en•tial•ly \-ˈren(t)-shə-lē\ *adverb*

²differential *noun* (1704)
1 a : the product of the derivative of a function of one variable by the increment of the independent variable **b :** a sum of products in which each product consists of a partial derivative of a given function of several variables multiplied by the corresponding increment and which contains as many products as there are independent variables in the function
2 : a difference between comparable individuals or classes ⟨a price *differential*⟩; *also* **:** the amount of such a difference
3 a : DIFFERENTIAL GEAR **b :** a case covering a differential gear

differential calculus *noun* (1702)
: a branch of mathematics concerned chiefly with the study of the rate of change of functions with respect to their variables especially through the use of derivatives and differentials

differential equation *noun* (1763)
: an equation containing differentials or derivatives of functions — compare PARTIAL DIFFERENTIAL EQUATION

differential gear *noun* (circa 1859)
: an arrangement of gears forming an epicyclic train for connecting two shafts or axles in the same line, dividing the driving force equally between them, and permitting one shaft to revolve faster than the other — called also *differential gearing*

differential geometry *noun* (circa 1909)
: a branch of mathematics using calculus to study the geometric properties of curves and surfaces

dif•fer•en•ti•ate \ˌdif-ə-ˈren(t)-shē-ˌāt\ *verb* **-at•ed; -at•ing** (1816)
transitive verb
1 : to obtain the mathematical derivative of
2 : to mark or show a difference in **:** constitute a difference that distinguishes
3 : to develop differential characteristics in
4 : to cause differentiation of in the course of development
5 : to express the specific distinguishing quality of **:** DISCRIMINATE
intransitive verb
1 : to recognize or give expression to a difference
2 : to become distinct or different in character
3 : to undergo differentiation
— dif•fer•en•tia•bil•i•ty \-ˌren(t)-sh(ē-)ə-ˈbi-lə-tē\ *noun*
— dif•fer•en•tia•ble \-ˈren(t)-sh(ē-)ə-bəl\ *adjective*

dif•fer•en•ti•a•tion \-ˌren(t)-shē-ˈā-shən\ *noun* (1802)
1 : the act or process of differentiating
2 : development from the one to the many, the simple to the complex, or the homogeneous to the heterogeneous
3 a : modification of body parts for performance of particular functions **b :** the sum of the processes whereby apparently indifferent cells, tissues, and structures attain their adult form and function
4 : the processes by which various rock types are produced from a common magma

dif•fer•ent•ly \ˈdi-fərnt-lē, ˈdi-f(ə-)rənt-\ *adverb* (14th century)
1 : in a different manner
2 : OTHERWISE

dif•fi•cile \ˌdē-fi-ˈsē(ə)l\ *adjective* [French, literally, difficult] (1536)
: STUBBORN, UNREASONABLE

dif•fi•cult \ˈdi-fi-(ˌ)kəlt\ *adjective* [Middle English, back-formation from *difficulty*] (14th century)
1 : hard to do, make, or carry out **:** ARDUOUS ⟨a *difficult* climb⟩

☆ SYNONYMS
Different, diverse, divergent, disparate, various mean unlike in kind or character. DIFFERENT may imply little more than separateness but it may also imply contrast or contrariness ⟨*different* foods⟩. DIVERSE implies both distinctness and marked contrast ⟨such *diverse* interests as dancing and football⟩. DIVERGENT implies movement away from each other and unlikelihood of ultimate meeting or reconciliation ⟨went on to pursue *divergent* careers⟩. DISPARATE emphasizes incongruity or incompatibility ⟨*disparate* notions of freedom⟩. VARIOUS stresses the number of sorts or kinds ⟨tried *various* methods⟩.

□ USAGE
different Numerous commentators have condemned *different than* in spite of its use since the 17th century by many of the best-known names in English literature. It is nevertheless standard and is even recommended in many handbooks when followed by a clause. *Different from,* the generally safe choice, is more common and is even used in constructions where *than* would work more smoothly.

2 a : hard to deal with, manage, or overcome ⟨a *difficult* child⟩ **b :** hard to understand **:** PUZZLING ⟨*difficult* reading⟩
synonym see HARD
— dif·fi·cult·ly *adverb*

dif·fi·cul·ty \-(ˌ)kəl-tē\ *noun, plural* **-ties** [Middle English *difficulte,* from Latin *difficultas,* from *difficilis* not easy, from *dis-* + *facilis* easy — more at FACILE] (14th century)
1 : the quality or state of being difficult
2 : CONTROVERSY, DISAGREEMENT
3 : OBJECTION
4 : something difficult **:** IMPEDIMENT
5 : EMBARRASSMENT, TROUBLE — usually used in plural

dif·fi·dence \ˈdi-fə-dən(t)s, -fə-ˌden(t)s\ *noun* (14th century)
: the quality or state of being diffident

dif·fi·dent \-dənt, -ˌdent\ *adjective* [Middle English, from Latin *diffident-, diffidens,* present participle of *diffidere* to distrust, from *dis-* + *fidere* to trust — more at BIDE] (15th century)
1 : hesitant in acting or speaking through lack of self-confidence
2 *archaic* **:** DISTRUSTFUL
3 : RESERVED, UNASSERTIVE
synonym see SHY
— dif·fi·dent·ly *adverb*

dif·fract \di-ˈfrakt\ *transitive verb* [back-formation from *diffraction*] (1803)
: to cause to undergo diffraction

dif·frac·tion \di-ˈfrak-shən\ *noun* [New Latin *diffraction-, diffractio,* from Latin *diffringere* to break apart, from *dis-* + *frangere* to break — more at BREAK] (1671)
: a modification which light undergoes in passing by the edges of opaque bodies or through narrow slits or in being reflected from ruled surfaces and in which the rays appear to be deflected and to produce fringes of parallel light and dark or colored bands; *also* **:** a similar modification of other waves (as sound waves)

diffraction grating *noun* (1867)
: GRATING 3

dif·frac·tom·e·ter \di-ˌfrak-ˈtä-mə-tər\ *noun* (circa 1909)
: an instrument for analyzing the structure of a usually crystalline substance from the scattering pattern produced when a beam of radiation or particles (as X rays or neutrons) strikes it
— dif·frac·to·met·ric \di-ˌfrak-tə-ˈme-trik\ *adjective*
— dif·frac·tom·e·try \di-ˌfrak-ˈtä-mə-trē\ *noun*

¹dif·fuse \di-ˈfyüs\ *adjective* [Middle English, from Latin *diffusus,* past participle of *diffundere* to spread out, from *dis-* + *fundere* to pour — more at FOUND] (15th century)
1 : being at once verbose and ill-organized
2 : not concentrated or localized ⟨*diffuse* sclerosis⟩
synonym see WORDY
— dif·fuse·ly *adverb*
— dif·fuse·ness *noun*

²dif·fuse \di-ˈfyüz\ *verb* **dif·fused; dif·fus·ing** [Middle English *diffused,* past participle, from Latin *diffusus,* past participle] (14th century)
transitive verb
1 a : to pour out and permit or cause to spread freely **b :** EXTEND, SCATTER **c :** to spread thinly or wastefully
2 : to subject to diffusion; *especially* **:** to break up and distribute (incident light) by reflection
intransitive verb
1 : to spread out or become transmitted especially by contact
2 : to undergo diffusion
— dif·fus·ible \di-ˈfyü-zə-bəl\ *adjective*

dif·fuse–po·rous \di-ˌfyüs-ˈpōr-əs, -ˈpȯr-\ *adjective* [¹*diffuse*] (circa 1902)
: having vessels more or less evenly distributed throughout an annual ring and not varying greatly in size — compare RING-POROUS

dif·fus·er \di-ˈfyü-zər\ *noun* (circa 1679)
1 : one that diffuses: as **a :** a device (as a reflector) for distributing the light of a lamp evenly **b :** a screen (as of cloth or frosted glass) for softening lighting (as in photography) **c :** a device (as slats at different angles) for deflecting air from an outlet in various directions
2 : a device for reducing the velocity and increasing the static pressure of a fluid passing through a system

dif·fu·sion \di-ˈfyü-zhən\ *noun* (14th century)
1 : the action of diffusing **:** the state of being diffused
2 : PROLIXITY, DIFFUSENESS
3 a : the process whereby particles of liquids, gases, or solids intermingle as the result of their spontaneous movement caused by thermal agitation and in dissolved substances move from a region of higher to one of lower concentration **b** (1) **:** reflection of light by a rough reflecting surface (2) **:** transmission of light through a translucent material **:** SCATTERING
4 : the spread of cultural elements from one area or group of people to others by contact
5 : the softening of sharp outlines in a photographic image
— dif·fu·sion·al \-ˈfyü-zhə-nᵊl\ *adjective*

dif·fu·sion·ist \-ˈfyü-zhə-nəst\ *noun* (1938)
: an anthropologist who emphasizes the role of diffusion in the history of culture rather than independent invention or discovery
— dif·fu·sion·ism \-ˈfyü-zhə-ˌni-zəm\ *noun*
— diffusionist *adjective*

dif·fu·sive \di-ˈfyü-siv, -ziv\ *adjective* (1614)
: tending to diffuse **:** characterized by diffusion ⟨*diffusive* motion of atoms⟩
— dif·fu·sive·ly *adverb*
— dif·fu·sive·ness *noun*
— dif·fu·siv·i·ty \di-ˌfyü-ˈsi-və-tē, -ˈzi-\ *noun*

di·func·tion·al \(ˌ)dī-ˈfəŋ(k)-shnəl, -shə-nᵊl\ *adjective* (1943)
: of, relating to, or being a compound with two highly reactive sites in each molecule

¹dig \ˈdig\ *verb* **dug** \ˈdəg\; **dig·ging** [Middle English *diggen*] (13th century)
transitive verb
1 a : to break up, turn, or loosen (earth) with an implement **b :** to prepare the soil of ⟨*dig* a garden⟩
2 a : to bring to the surface by digging **:** UNEARTH **b :** to bring to light or out of hiding ⟨*dig* up facts⟩
3 : to hollow out or form by removing earth **:** EXCAVATE
4 : to drive down so as to penetrate **:** THRUST
5 : POKE, PROD
6 a : to pay attention to **:** NOTICE ⟨*dig* that fancy hat⟩ **b :** UNDERSTAND, APPRECIATE ⟨if you . . . do something subtle . . . only one tenth of the audience will *dig* it —Nat Hentoff⟩ **c :** LIKE, ADMIRE ⟨high school students *dig* short poetry —David Burmester⟩
intransitive verb
1 : to turn up, loosen, or remove earth **:** DELVE
2 : to work hard or laboriously
3 : to advance by or as if by removing or pushing aside material

²dig *noun* (1819)
1 a : THRUST, POKE **b :** a cutting remark
2 *plural* **a :** living accommodations **b** *chiefly British* **:** LODGING, HOTEL
3 : an archaeological excavation site; *also* **:** the excavation itself

dig·a·my \ˈdi-gə-mē\ *noun, plural* **-mies** [Late Latin *digamia,* from Late Greek, from Greek *digamos* married to two people, from *di-* + *-gamos* -gamous] (1635)
: a second marriage after the termination of the first

di·gas·tric \(ˌ)dī-ˈgas-trik\ *adjective* [New Latin *digastricus,* from *di-* + *gastricus* gastric] (circa 1721)
: of, relating to, or being a muscle with two bellies separated by a median tendon

di·ge·net·ic \ˌdī-jə-ˈne-tik\ *adjective* [New Latin *Digenetica,* subclass name (synonym of *Digenea*), from *di-* + *genetica,* neuter plural of *geneticus* genetic] (circa 1883)
: of or relating to a subclass (Digenea) of trematode worms in which sexual reproduction as an internal parasite of a vertebrate alternates with asexual reproduction in a mollusk

¹di·gest \ˈdī-ˌjest\ *noun* [Middle English, systematic arrangement of laws, from Latin *digesta,* from neuter plural of *digestus,* past participle of *digerere* to arrange, distribute, digest, from *dis-* + *gerere* to carry] (14th century)
1 : a summation or condensation of a body of information: as **a :** a systematic compilation of legal rules, statutes, or decisions **b :** a periodical devoted to condensed versions of previously published articles
2 : a product of digestion

²di·gest \dī-ˈjest, də-\ *verb* [Middle English, from Latin *digestus*] (14th century)
transitive verb
1 : to distribute or arrange systematically **:** CLASSIFY
2 : to convert (food) into absorbable form
3 : to take into the mind or memory; *especially* **:** to assimilate mentally
4 a : to soften, decompose, or break down by heat and moisture or chemicals **b :** to extract soluble ingredients from by warming with a liquid
5 : to compress into a short summary
6 : ABSORB 1 ⟨the capacity of the U.S. to *digest* immigrants⟩
intransitive verb
1 : to digest food
2 : to become digested

di·gest·er \-ˈjes-tər\ *noun* (1614)
1 : one that digests or makes a digest
2 : a vessel for digesting especially plant or animal materials

di·gest·ibil·i·ty \-ˌjes-tə-ˈbi-lə-tē\ *noun, plural* **-ties** (1740)
1 : the fitness of something for digestion
2 : the percentage of a foodstuff taken into the digestive tract that is absorbed into the body

di·gest·ible \-ˈjes-tə-bəl\ *adjective* (14th century)
: capable of being digested

di·ges·tion \dī-ˈjes-chən, də-, -ˈjesh-\ *noun* (14th century)
: the action, process, or power of digesting: as **a :** the process of making food absorbable by dissolving it and breaking it down into simpler chemical compounds that occurs in the living body chiefly through the action of enzymes secreted into the alimentary canal **b :** the process in sewage treatment by which organic matter in sludge is decomposed by anaerobic bacteria with the release of a burnable mixture of gases

¹di·ges·tive \-ˈjes-tiv\ *noun* (14th century)
: an aid to digestion especially of food

²digestive *adjective* (15th century)
1 : relating to or functioning in digestion ⟨the *digestive* system⟩
2 : having the power to cause or promote digestion ⟨*digestive* enzymes⟩
— di·ges·tive·ly *adverb*

digestive gland *noun* (1940)
: a gland secreting digestive enzymes

dig·ger \ˈdi-gər\ *noun* (15th century)
1 a : one that digs **b :** a tool or machine for digging

2 *capitalized* **:** a North American Indian (as a Paiute) who dug roots for food — usually used disparagingly especially formerly
3 *often capitalized, chiefly Australian & New Zealand* **:** an Australian or New Zealand soldier

digger wasp *noun* (1880)
: a burrowing wasp; *especially* **:** a usually solitary wasp (superfamily Sphecoidea) that digs nest burrows in the soil and provisions them with insects or spiders paralyzed by stinging

dig·gings *noun plural* (1538)
1 : a place of excavating especially for ore, metals, or precious stones
2 : material dug out
3 a : QUARTERS, PREMISES **b** *chiefly British* **:** lodgings for a student

dight \'dīt\ *transitive verb* **dight·ed** *or* **dight; dight·ing** [Middle English, from Old English *dihtan* to arrange, compose, from Latin *dictare* to dictate, compose] (13th century)
archaic **:** DRESS, ADORN

dig in (1839)
transitive verb
1 : to cover or incorporate by burying ⟨*dig in* compost⟩
2 : to establish in a dug defensive position ⟨the platoon was well *dug in*⟩
intransitive verb
1 : to establish a defensive position especially by digging trenches
2 a : to go resolutely to work **b :** to begin eating
3 : to hold stubbornly to a position
4 : to scuff the ground for better footing while batting (as in baseball)
— dig in one's heels : to take or persist in an uncompromising position or attitude despite opposition

dig·it \'di-jət\ *noun* [Middle English, from Latin *digitus* finger, toe; perhaps akin to Greek *deiknynai* to show — more at DICTION] (14th century)
1 a : any of the Arabic numerals 1 to 9 and usually the symbol 0 **b :** one of the elements that combine to form numbers in a system other than the decimal system
2 : a unit of length based on the breadth of a finger and equal in English measure to ¾ inch
3 : any of the divisions in which the limbs of amphibians and all higher vertebrates terminate, which are typically five in number but may be reduced (as in the horse), and which typically have a series of phalanges bearing a nail, claw, or hoof at the tip — compare FINGER 1, TOE 1a

dig·i·tal \'di-jə-tᵊl\ *adjective* [Latin *digitalis*] (circa 1656)
1 : of or relating to the fingers or toes **:** DIGITATE
2 : done with a finger
3 : of, relating to, or using calculation by numerical methods or by discrete units
4 : of or relating to data in the form of numerical digits
5 : providing a readout in numerical digits ⟨a *digital* voltmeter⟩
6 : relating to an audio recording method in which sound waves are represented digitally (as on magnetic tape) so that in the recording wow and flutter are eliminated and background noise is reduced
— dig·i·tal·ly \-tᵊl-ē\ *adverb*

digital computer *noun* (1947)
: a computer that operates with numbers expressed directly as digits — compare ANALOG COMPUTER, HYBRID COMPUTER

dig·i·tal·in \ˌdi-jə-'ta-lən *also* -'tā-\ *noun* [New Latin *Digitalis*] (1837)
1 : a white crystalline steroid glycoside $C_{36}H_{56}O_{14}$ obtained from seeds especially of the common foxglove
2 : a mixture of the glycosides of digitalis leaves or seeds

dig·i·tal·is \-'ta-ləs *also* -'tā-\ *noun* [New Latin, genus name, from Latin, of a finger, from *digitus;* from its finger-shaped corolla] (1664)
1 : FOXGLOVE
2 : the dried powdered leaf of the common foxglove containing important glycosides and serving as a powerful cardiac stimulant and a diuretic

dig·i·ta·li·za·tion \ˌdi-jə-tᵊl-ə-'zā-shən\ *noun* [*digitalis*] (1882)
: the administration of digitalis until the desired physiological adjustment is attained; *also* **:** the bodily state so produced

¹dig·i·ta·lize \'di-jə-tᵊl-ˌīz\ *transitive verb* **-lized; -liz·ing** [*digitalis*] (1927)
: to subject to digitalization

²dig·i·tal·ize \'di-jə-tᵊl-ˌīz\ *transitive verb* **-ized; -iz·ing** [*digital*] (1962)
: DIGITIZE

dig·i·tate \'di-jə-ˌtāt\ *adjective* (1661)
: having divisions arranged like those of a bird's foot ⟨*digitate* leaves⟩
— dig·i·tate·ly *adverb*

digiti- *combining form* [French, from Latin *digitus*]
: digit **:** finger ⟨*digiti*grade⟩

dig·i·ti·grade \'di-jə-tə-ˌgrād\ *adjective* [French, from *digiti-* + *-grade*] (circa 1833)
: walking on the digits with the posterior of the foot more or less raised

dig·i·tize \'di-jə-ˌtīz\ *transitive verb* **-tized; -tiz·ing** (1953)
: to convert (as data or an image) to digital form
— dig·i·ti·za·tion \ˌdi-jə-tə-'zā-shən\ *noun*
— dig·i·tiz·er \'di-jə-ˌtī-zər\ *noun*

dig·i·to·nin \ˌdi-jə-'tō-nən\ *noun* [International Scientific Vocabulary *digit-* (from New Latin *Digitalis*) + sap*onin*] (1875)
: a steroid saponin $C_{56}H_{92}O_{29}$ occurring in the leaves and seeds of foxglove

digi·toxi·gen·in \ˌdi-jə-ˌtäk-sə-'je-nən\ *noun* [International Scientific Vocabulary, blend of *digitoxin* and *-gen*] (circa 1909)
: a steroid lactone $C_{23}H_{34}O_4$ obtained especially by hydrolysis of digitoxin

digi·tox·in \ˌdi-jə-'täk-sən\ *noun* [International Scientific Vocabulary, blend of New Latin *Digitalis* and International Scientific Vocabulary *toxin*] (circa 1883)
: a poisonous glycoside $C_{41}H_{64}O_{13}$ that is the most active constituent of digitalis; *also* **:** a mixture of digitalis glycosides consisting chiefly of digitoxin

di·glyc·er·ide \ˌdī-'gli-sə-ˌrīd\ *noun* (1918)
: an ester formed from glycerol by reacting two of its alcohol hydroxy groups with fatty acids

dig·ni·fied \'dig-nə-ˌfīd\ *adjective* (1763)
: showing or expressing dignity

dig·ni·fy \'dig-nə-ˌfī\ *transitive verb* **-fied; -fy·ing** [Middle French *dignifier,* from Late Latin *dignificare,* from Latin *dignus* worthy — more at DECENT] (15th century)
1 : to give distinction to **:** ENNOBLE
2 : to confer dignity upon; *also* **:** to give undue attention or status to ⟨won't *dignify* that remark with a reply⟩

dig·ni·tary \'dig-nə-ˌter-ē\ *noun, plural* **-tar·ies** (1673)
: one who possesses exalted rank or holds a position of dignity or honor
— dignitary *adjective*

dig·ni·ty \'dig-nə-tē\ *noun, plural* **-ties** [Middle English *dignete,* from Old French *digneté,* from Latin *dignitat-, dignitas,* from *dignus*] (13th century)
1 : the quality or state of being worthy, honored, or esteemed
2 a : high rank, office, or position **b :** a legal title of nobility or honor
3 *archaic* **:** DIGNITARY
4 : formal reserve or seriousness of manner, appearance, or language

dig out (14th century)
transitive verb
1 : FIND, UNEARTH
2 : to make hollow by digging
intransitive verb
: TAKE OFF 2a

di·gox·in \di-'jäk-sən, -'gäk-\ *noun* [International Scientific Vocabulary *dig-* (from New Latin *Digitalis*) + *toxin*] (circa 1930)
: a poisonous cardiotonic steroid $C_{41}H_{64}O_{14}$ obtained from a foxglove (*Digitalis lanata*) and used similarly to digitalis

di·graph \'dī-ˌgraf\ *noun* (1780)
1 : a group of two successive letters whose phonetic value is a single sound (as *ea* in *bread* or *ng* in *sing*) or whose value is not the sum of a value borne by each in other occurrences (as *ch* in *chin* where the value is \t\ + \sh\)
2 : a group of two successive letters
3 : LIGATURE 4
— di·graph·ic \dī-'gra-fik\ *adjective*
— di·graph·i·cal·ly \-fi-k(ə-)lē\ *adverb*

di·gress \dī-'gres, də-\ *intransitive verb* [Latin *digressus,* past participle of *digredi,* from *dis-* + *gradi* to step — more at GRADE] (1530)
: to turn aside especially from the main subject of attention or course of argument
synonym see SWERVE

di·gres·sion \-'gre-shən\ *noun* (14th century)
1 : the act or an instance of digressing in a discourse or other usually organized literary work
2 *archaic* **:** a going aside
— di·gres·sion·al \-'gresh-nəl, -ən-ᵊl\ *adjective*
— di·gres·sion·ary \-'gre-shə-ˌner-ē\ *adjective*

di·gres·sive \-'gre-siv\ *adjective* (circa 1611)
: characterized by digressions ⟨a *digressive* talk⟩
— di·gres·sive·ly *adverb*
— di·gres·sive·ness *noun*

dig up *transitive verb* (14th century)
: FIND, UNEARTH

di·he·dral \(ˌ)dī-'hē-drəl\ *noun* (circa 1911)
1 : DIHEDRAL ANGLE
2 : the angle between an aircraft supporting surface (as a wing) and a horizontal transverse line

dihedral angle *noun* [*di-* + *-hedral*] (1826)
: a figure formed by two intersecting planes

di·hy·brid \(ˌ)dī-'hī-brəd\ *adjective* [International Scientific Vocabulary] (1907)
: of, relating to, involving, or being an individual or strain that is heterozygous at two genetic loci
— dihybrid *noun*

dihydr- *or* **dihydro-** *combining form*
: combined with two atoms of hydrogen ⟨*dihydro*ergotamine⟩

di·hy·dro·er·got·a·mine \(ˌ)dī-ˌhī-drō-ˌər-'gä-tə-ˌmēn\ *noun* (1945)
: a hydrogenated derivative $C_{33}H_{37}N_5O_5$ of ergotamine that is used in the treatment of migraine

dihydroxy- *combining form*
: containing two hydroxyl groups ⟨*dihydroxy*acetone⟩

di·hy·droxy·ac·e·tone \ˌdī-hī-ˌdräk-sē-'a-sə-ˌtōn\ *noun* (1895)
: a glyceraldehyde isomer $C_3H_6O_3$ used especially to stain the skin to resemble a tan

dik–dik \'dik-ˌdik\ *noun, plural* **dik–diks** *or* **dik–dik** [origin unknown] (1883)
: any of a genus (*Madoqua*) of small eastern African antelopes

dik-dik

¹dike \'dīk\ *noun* [Middle English, probably from Old Norse *dīk* ditch and Middle Low German *dīk* dam; akin to Old

English *dīc* ditch — more at DITCH] (13th century)
1 : an artificial watercourse **:** DITCH
2 a : a bank usually of earth constructed to control or confine water **:** LEVEE **b :** a barrier preventing passage especially of something undesirable
3 a : a raised causeway **b :** a tabular body of igneous rock that has been injected while molten into a fissure
²dike *transitive verb* **diked; dik·ing** (14th century)
1 : to surround or protect with a dike
2 : to drain by a dike
— dik·er *noun*
³dike *variant of* ²DYKE
dik·tat \dik-ˈtät\ *noun* [German, literally, something dictated, from New Latin *dictatum,* from Latin, neuter of *dictatus,* past participle of *dictare* to dictate] (1933)
1 : a harsh settlement unilaterally imposed (as on a defeated nation)
2 : DECREE, ORDER
Di·lan·tin \dī-ˈlan-tᵊn, də-\ *trademark*
— used for phenytoin
di·lap·i·date \də-ˈla-pə-ˌdāt\ *verb* **-dat·ed; -dat·ing** [Latin *dilapidatus,* past participle of *dilapidare* to squander, destroy, from *dis-* + *lapidare* to pelt with stones, from *lapid-, lapis* stone] (circa 1570)
transitive verb
1 : to bring into a condition of decay or partial ruin ⟨furniture is *dilapidated* by use —Janet Flanner⟩
2 *archaic* **:** SQUANDER
intransitive verb
: to become dilapidated
— di·lap·i·da·tion \-ˌla-pə-ˈdā-shən\ *noun*
di·lap·i·dat·ed *adjective* (circa 1806)
: decayed, deteriorated, or fallen into partial ruin especially through neglect or misuse ⟨a *dilapidated* old house⟩
di·lat·an·cy \dī-ˈlā-tᵊn(t)-sē\ *noun* (1885)
: the property of being dilatant
di·lat·ant \-tᵊnt\ *adjective* (1885)
: increasing in viscosity and setting to a solid as a result of deformation by expansion, pressure, or agitation
di·la·ta·tion \ˌdi-lə-ˈtā-shən, ˌdī-\ *noun* (14th century)
1 : amplification in writing or speech
2 a : the condition of being stretched beyond normal dimensions especially as a result of overwork, disease, or abnormal relaxation ⟨*dilatation* of the heart⟩ ⟨*dilatation* of the stomach⟩ **b :** DILATION 2
3 : the action of expanding **:** the state of being expanded
4 : a dilated part or formation
— di·la·ta·tion·al \-shnəl, -shə-nᵊl\ *adjective*
di·late \ˈdī-ˌlāt, dī-ˈ\ *verb* **di·lat·ed; di·lat·ing** [Middle English, from Middle French *dilater,* from Latin *dilatare,* literally, to spread wide, from *dis-* + *latus* wide — more at LATITUDE] (14th century)
transitive verb
1 *archaic* **:** to describe or set forth at length or in detail
2 : to enlarge or expand in bulk or extent **:** DISTEND, WIDEN
intransitive verb
1 : to comment at length **:** DISCOURSE — usually used with *on* or *upon*
2 : to become wide **:** SWELL
synonym see EXPAND
— di·lat·abil·i·ty \(ˌ)dī-ˌlā-tə-ˈbi-lə-tē\ *noun*
— di·lat·able \dī-ˈlā-tə-bəl, ˈdī-ˌ\ *adjective*
— di·la·tor \dī-ˈlā-tər, ˈdī-ˌ\ *noun*
di·lat·ed *adjective* (15th century)
1 : expanded laterally; *especially* **:** being flat and widened ⟨*dilated* leaves⟩
2 : expanded normally or abnormally in all dimensions
di·la·tion \dī-ˈlā-shən\ *noun* (15th century)
1 : the act or action of dilating **:** the state of being dilated **:** EXPANSION, DILATATION
2 : the action of stretching or enlarging an organ or part of the body
di·la·tive \dī-ˈlā-tiv, ˈdī-ˌ\ *adjective* (1634)
: causing dilation **:** tending to dilate
di·la·tom·e·ter \ˌdi-lə-ˈtä-mə-tər, ˌdī-\ *noun* [International Scientific Vocabulary] (circa 1883)
: an instrument for measuring expansion
— di·la·to·met·ric \-tō-ˈme-trik\ *adjective*
— di·la·tom·e·try \-ˈtä-mə-trē\ *noun*
dil·a·to·ry \ˈdi-lə-ˌtōr-ē, -ˌtȯr-\ *adjective* [Middle English, from Late Latin *dilatorius,* from Latin *differre* (past participle *dilatus*) to postpone, differ — more at DIFFER, TOLERATE] (15th century)
1 : tending or intended to cause delay ⟨*dilatory* tactics⟩
2 : characterized by procrastination **:** TARDY ⟨*dilatory* in answering letters⟩
— dil·a·to·ri·ly \ˌdi-lə-ˈtōr-ə-lē, -ˈtȯr-\ *adverb*
— dil·a·to·ri·ness \ˈdi-lə-ˌtōr-ē-nəs, -ˌtȯr-\ *noun*
dil·do \ˈdil-(ˌ)dō\ *noun, plural* **dildos** *also* **dildoes** [origin unknown] (circa 1593)
: an object serving as a penis substitute for vaginal insertion
di·lem·ma \də-ˈle-mə *also* dī-\ *noun* [Late Latin, from Late Greek *dilēmmat-, dilēmma,* probably back-formation from Greek *dilēmmatos* involving two assumptions, from *di-* + *lēmmat-, lēmma* assumption — more at LEMMA] (1523)
1 : an argument presenting two or more equally conclusive alternatives against an opponent
2 a : a usually undesirable or unpleasant choice ⟨faces this *dilemma*: raise interest rates and slow the economy or lower them and risk serious inflation⟩ **b :** a situation involving such a choice ⟨here am I brought to a very pretty *dilemma*; I must commit murder or commit matrimony —George Farquhar⟩; *broadly* **:** PREDICAMENT ⟨lords and bailiffs were in a terrible *dilemma* —G. M. Trevelyan⟩
3 a : a problem involving a difficult choice ⟨the *dilemma* of "liberty versus order" —J. M. Burns⟩ **b :** a difficult or persistent problem ⟨unemployment . . . the great central *dilemma* of our advancing technology —August Heckscher⟩ ■
— dil·em·mat·ic \ˌdi-lə-ˈma-tik *also* ˌdī-\ *adjective*
dil·et·tante \ˈdi-lə-ˌtänt, -ˌtant; ˌdi-lə-ˈ\ *noun, plural* **-tantes** *or* **-tan·ti** \-ˈtän-tē, -ˈtan-tē\ [Italian, from present participle of *dilettare* to delight, from Latin *dilectare* — more at DELIGHT] (1748)
1 : an admirer or lover of the arts
2 : a person having a superficial interest in an art or a branch of knowledge **:** DABBLER
synonym see AMATEUR
— dilettante *adjective*
— dil·et·tant·ish \-ˌtän-tish, -ˌtan-, ˌdi-lə-ˈ\ *adjective*
— dil·et·tan·tism \-ˌtän-ˌti-zəm, -ˌtan-, ˌdi-lə-ˈ\ *noun*
¹dil·i·gence \ˈdi-lə-jən(t)s\ *noun* [Middle English, from Middle French, from Latin *diligentia,* from *diligent-, diligens*] (14th century)
1 a : persevering application **:** ASSIDUITY **b** *obsolete* **:** SPEED, HASTE
2 : the attention and care legally expected or required of a person (as a party to a contract)
²di·li·gence \ˈdi-lə-ˌzhäⁿs, ˈdi-lə-jən(t)s\ *noun* [French, literally, haste, from Middle French, persevering application] (1742)
: STAGECOACH
dil·i·gent \ˈdi-lə-jənt\ *adjective* [Middle English, from Middle French, from Latin *diligent-, diligens,* from present participle of *diligere* to esteem, love, from *di-* (from *dis-* apart) + *legere* to select — more at LEGEND] (14th century)
: characterized by steady, earnest, and energetic effort **:** PAINSTAKING
synonym see BUSY
— dil·i·gent·ly *adverb*
dill \ˈdil\ *noun* [Middle English *dile,* from Old English; akin to Old High German *tilli* dill] (before 12th century)
1 : any of several plants of the carrot family; *especially* **:** a European herb (*Anethum graveolens*) with aromatic foliage and seeds both of which are used in flavoring foods and especially pickles
2 : DILL PICKLE
— dilled *adjective*
dill pickle *noun* (1904)
: a pickle seasoned with fresh dill or dill juice
dil·ly \ˈdi-lē\ *noun, plural* **dillies** [obsolete *dilly,* adjective (delightful), perhaps by shortening & alteration from *delightful*] (1935)
: one that is remarkable or outstanding ⟨had a *dilly* of a storm⟩ ⟨for a practical joke, that was a *dilly*⟩
dil·ly bag \ˈdi-lē-\ *noun* [Jagara (Australian aboriginal language of Queensland) *dili* coarse grass, fiber bag] (1867)
: an Australian mesh bag made of native fibers
dil·ly·dal·ly \ˈdi-lē-ˌda-lē\ *intransitive verb* [reduplication of *dally*] (1741)
: to waste time by loitering or delaying **:** DAWDLE
dil·u·ent \ˈdil-yə-wənt\ *noun* [Latin *diluent-, diluens,* present participle of *diluere*] (circa 1721)
: a diluting agent (as the vehicle in a medicinal preparation)
¹di·lute \dī-ˈlüt, də-\ *transitive verb* **di·lut·ed; di·lut·ing** [Latin *dilutus,* past participle of *diluere* to wash away, dilute, from *di-* + *lavere* to wash — more at LYE] (circa 1555)
1 : ATTENUATE
2 : to make thinner or more liquid by admixture
3 : to diminish the strength, flavor, or brilliance of by admixture
4 : to decrease the per share value of (common stock) by increasing the total number of shares
— di·lut·er *or* **di·lu·tor** \-ˈlü-tər\ *noun*
— di·lu·tive \-ˈlü-tiv\ *adjective*
²dilute *adjective* (1605)
: WEAK, DILUTED
— di·lute·ness *noun*
di·lu·tion \dī-ˈlü-shən, də-\ *noun* (1646)
1 : the action of diluting **:** the state of being diluted
2 : something (as a solution) that is diluted
3 : a lessening of real value (as of equity) by a decrease in relative worth; *specifically* **:** a decrease of per share value of common stock by an increase in the total number of shares

□ USAGE
dilemma Although some commentators insist that *dilemma* be restricted to instances in which the alternatives to be chosen are equally unsatisfactory, their concern is misplaced; the unsatisfactoriness of the options is usually a matter of how the author presents them. What is distressing or painful about a dilemma is having to make a choice one does not want to make. The use of such adjectives as *terrible, painful,* and *irreconcilable* suggests that *dilemma* is losing some of its unpleasant force. There also seems to be a tendency especially in sense 3b toward applying the word to less weighty problems ⟨solved their goaltending *dilemma* —Pat Calabria⟩.

di·lu·vi·al \də-ˈlü-vē-əl, dī-\ *or* **di·lu·vi·an** \-vē-ən\ *adjective* [Late Latin *diluvialis,* from Latin *diluvium* deluge — more at DELUGE] (circa 1656)
: of, relating to, or brought about by a flood

¹dim \ˈdim\ *adjective* **dim·mer; dim·mest** [Middle English, from Old English *dimm;* akin to Old High German *timber* dark] (before 12th century)
1 a : emitting a limited or insufficient amount of light **b :** DULL, LUSTERLESS **c :** lacking pronounced, clear-cut, or vigorous quality or character
2 a : seen indistinctly or without clear outlines or details **b :** perceived by the senses or mind indistinctly or weakly **:** FAINT ⟨had only a *dim* notion of what was going on⟩ **c :** having little prospect of favorable result or outcome ⟨a *dim* future⟩ **d :** characterized by an unfavorable, skeptical, or pessimistic attitude — usually used in the phrase *take a dim view of*
3 : not perceiving clearly and distinctly ⟨*dim* eyes⟩
4 : DIM-WITTED
— dim·ly *adverb*
— dim·ma·ble \ˈdi-mə-bəl\ *adjective*
— dim·ness *noun*

²dim *verb* **dimmed; dim·ming** (before 12th century)
transitive verb
1 : to make dim or lusterless
2 : to reduce the light from
intransitive verb
: to become dim

³dim *noun* (14th century)
1 *archaic* **:** DUSK, DIMNESS
2 : LOW BEAM

dim bulb *noun* (1927)
slang **:** DIMWIT

dime \ˈdīm\ *noun* [Middle English, tenth part, tithe, from Middle French, from Latin *decima,* from feminine of *decimus* tenth, from *decem* ten — more at TEN] (1786)
1 a : a coin of the U.S. worth 1/10 dollar **b :** a petty sum of money
2 : a Canadian 10-cent piece
— a dime a dozen : so plentiful or commonplace as to be of little esteem or slight value
— on a dime 1 : in a very small area ⟨these cars can turn *on a dime*⟩ **2 :** INSTANTLY

di·men·hy·dri·nate \ˌdī-ˌmen-ˈhī-drə-ˌnāt\ *noun* [*dime*thyl + am*ine* + *hydr-* + am*ine* + [1]*-ate*] (circa 1950)
: a crystalline antihistaminic compound $C_{24}H_{28}ClN_5O_3$ used especially to prevent nausea

dime novel *noun* (1864)
: a usually paperback melodramatic novel; *especially* **:** one popular in the U.S. from about the mid-19th century to the early 20th often featuring a Western theme

¹di·men·sion \də-ˈmen(t)-shən *also* dī-\ *noun* [Middle English, from Middle French, from Latin *dimension-, dimensio,* from *dimetiri* to measure out, from *dis-* + *metiri* to measure — more at MEASURE] (14th century)
1 a (1) **:** measure in one direction; *specifically* **:** one of three coordinates determining a position in space or four coordinates determining a position in space and time (2) **:** one of a group of properties whose number is necessary and sufficient to determine uniquely each element of a system of usually mathematical entities (as an aggregate of points in real or abstract space) ⟨the surface of a sphere has two *dimensions*⟩; *also* **:** a parameter or coordinate variable assigned to such a property ⟨the three *dimensions* of momentum⟩ (3) **:** the number of elements in a basis of a vector space **b :** the quality of spatial extension **:** MAGNITUDE, SIZE **c :** a lifelike or realistic quality **d :** the range over which or the degree to which something extends **:** SCOPE — usually used in plural **e :** one of the elements or factors making up a complete personality or entity **:** ASPECT
2 *obsolete* **:** bodily form or proportions
3 : any of the fundamental units (as of mass, length, or time) on which a derived unit is based; *also* **:** the power of such a unit
4 : wood or stone cut to pieces of specified size
5 : a level of existence or consciousness
— di·men·sion·al \-ˈmench-nəl, -ˈmen(t)-shə-nᵊl\ *adjective*
— di·men·sion·al·i·ty \-ˌmen(t)-shə-ˈna-lə-tē\ *noun*
— di·men·sion·al·ly \-ˈmench-nə-lē, -ˈmen(t)-shə-nᵊl-ē\ *adverb*
— di·men·sion·less \-ˈmen(t)-shən-ləs\ *adjective*

²dimension *transitive verb* **di·men·sioned; di·men·sion·ing** \-ˈmen(t)-shə-niŋ\ (1754)
1 : to form to the required dimensions
2 : to indicate the dimensions of (as on a drawing)

di·mer \ˈdī-mər\ *noun* [International Scientific Vocabulary] (circa 1926)
: a compound formed by the union of two radicals or two molecules of a simpler compound; *specifically* **:** a polymer formed from two molecules of a monomer
— di·mer·ic \(ˌ)dī-ˈmer-ik\ *adjective*
— di·mer·iza·tion \ˌdī-mə-rə-ˈzā-shən\ *noun*
— di·mer·ize \ˈdī-mə-ˌrīz\ *transitive verb*

di·mer·cap·rol \ˌdī-mər-ˈka-ˌpról, -ˌprōl\ *noun* [*di-* + *mercap*tan + p*ro*pane + [1]*-ol*] (1947)
: a compound $C_3H_8OS_2$ developed as an antidote against lewisite and used against arsenic, mercury, and gold poisoning — called also *BAL*

dime–store \ˈdīm-ˌstōr, -ˌstór\ *adjective* (1938)
1 : INEXPENSIVE ⟨*dime-store* perfume⟩
2 : TAWDRY, SECOND-RATE ⟨*dime-store* philosophy⟩

dime store *noun* (circa 1928)
: FIVE-AND-TEN

dim·e·ter \ˈdi-mə-tər\ *noun* [Late Latin, from Greek *dimetros,* adjective, being a dimeter, from *di-* + *metron* measure — more at MEASURE] (1589)
: a line of verse consisting of two metrical feet or of two dipodies

di·meth·o·ate \dī-ˈme-thə-ˌwāt\ *noun* [*di*-*me*thyl + *th*io acid + [1]*-ate*] (1960)
: an insecticide $C_5H_{12}NO_3PS_2$ used in agriculture and animal husbandry

di·meth·yl \ˌdī-ˈme-thəl\ *adjective*
: containing two methyl groups in the molecule — often used in combination

di·meth·yl·hy·dra·zine \ˌdī-ˌme-thəl-ˈhī-drə-ˌzēn\ *noun* (1961)
: either of two flammable corrosive isomeric liquids $C_2H_8N_2$ which are methylated derivatives of hydrazine and of which one is used as a rocket fuel

di·meth·yl·ni·tro·sa·mine \-nī-ˈtrō-sə-ˌmēn\ *noun* (1965)
: a carcinogenic nitrosamine $C_2H_6N_2O$ that occurs especially in tobacco smoke

dimethyl sulfoxide *noun* (1964)
: a compound $(CH_3)_2SO$ obtained as a by-product in wood-pulp manufacture and used as a solvent and in veterinary medicine as an anti-inflammatory — called also *DMSO*

di·meth·yl·tryp·ta·mine \-ˈtrip-tə-ˌmēn\ *noun* (1966)
: a naturally occurring or easily synthesized hallucinogenic drug $C_{12}H_{16}N_2$ that is chemically similar to but shorter acting than psilocybin — called also *DMT*

di·min·ish \də-ˈmi-nish\ *verb* [Middle English *deminishen,* alteration of *diminuen,* from Middle French *diminuer,* from Late Latin *diminuere,* alteration of Latin *deminuere,* from *de-* + *minuere* to lessen — more at MINOR] (15th century)
transitive verb
1 : to make less or cause to appear less
2 : to lessen the authority, dignity, or reputation of **:** BELITTLE
3 : to cause to taper
intransitive verb
1 : to become gradually less (as in size or importance) **:** DWINDLE
2 : TAPER
synonym see DECREASE
— di·min·ish·able \-ni-shə-bəl\ *adjective*
— di·min·ish·ment \-mənt\ *noun*

di·min·ished *adjective* (circa 1751)
of a musical interval **:** made one half step less than perfect or minor ⟨a *diminished* fifth⟩

diminishing returns *noun plural* (1815)
: a rate of yield that beyond a certain point fails to increase in proportion to additional investments of labor or capital

di·min·u·en·do \də-ˌmin-yə-ˈwen-(ˌ)dō *also* -ˌmi-nə-\ *adverb or adjective* [Italian, literally, diminishing, from Late Latin *diminuendum,* gerund of *diminuere*] (1775)
: DECRESCENDO
— diminuendo *noun*

dim·i·nu·tion \ˌdi-mə-ˈnü-shən *also* -ˈnyü-\ *noun* [Middle English *diminucioun,* from Middle French *diminution,* from Medieval Latin *diminution-, diminutio,* alteration of Latin *deminution-, deminutio,* from *deminuere*] (14th century)
: the act, process, or an instance of diminishing **:** DECREASE

¹di·min·u·tive \də-ˈmi-nyə-tiv\ *noun* [Middle English *diminutif,* from Medieval Latin *diminutivum,* alteration of Late Latin *deminutivum,* from neuter of *deminutivus,* adjective, from *deminutus,* past participle of *deminuere*] (14th century)
1 : a diminutive word, affix, or name
2 : a diminutive individual

²diminutive *adjective* (14th century)
1 : indicating small size and sometimes the state or quality of being familiarly known, lovable, pitiable, or contemptible — used of affixes (as *-ette, -kin, -ling*) and of words formed with them (as *kitchenette, manikin, duckling*), of clipped forms (as *Jim*), and of altered forms (as *Peggy*); compare AUGMENTATIVE
2 : exceptionally or notably small **:** TINY
synonym see SMALL
— di·min·u·tive·ly *adverb*
— di·min·u·tive·ness *noun*

dim·i·ty \ˈdi-mə-tē\ *noun, plural* **-ties** [alteration of Middle English *demyt,* from Medieval Latin *dimitum,* from Middle Greek *dimitos* of double thread, from Greek *di-* + *mitos* warp thread] (1570)
: a sheer usually corded cotton fabric of plain weave in checks or stripes

dim·mer \ˈdi-mər\ *noun* (circa 1896)
1 : a device for regulating the intensity of an electric lighting unit
2 : LOW BEAM

di·mor·phic \(ˌ)dī-ˈmór-fik\ *adjective* (1859)
1 a : DIMORPHOUS 1 **b :** occurring in two distinct forms ⟨*dimorphic* leaves of emergent plants⟩ ⟨sexually *dimorphic* coloration in birds⟩
2 : combining qualities of two kinds of individuals in one

di·mor·phism \-ˌfi-zəm\ *noun* (1832)
: the condition or property of being dimorphic or dimorphous: as **a :** the existence of two different forms (as of color or size) of a species especially in the same population ⟨sexual *dimorphism*⟩ **b :** the existence of an organ (as the leaves of a plant) in two different forms

di·mor·phous \(ˌ)dī-ˈmór-fəs\ *adjective* [Greek *dimorphos* having two forms, from *di-* + *-morphos* -morphous] (1832)
1 : crystallizing in two different forms
2 : DIMORPHIC 1b

dim·out \ˈdim-ˌau̇t\ *noun* (1942)
: a restriction limiting the use or showing of lights at night especially during the threat of

an air raid; *also* **:** a condition of partial darkness produced by this restriction

[1]dim·ple \ˈdim-pəl\ *noun* [Middle English *dympull;* akin to Old High German *tumphilo* whirlpool, Old English *dyppan* to dip — more at DIP] (15th century)
1 : a slight natural indentation in the surface of some part of the human body
2 : a depression or indentation on a surface (as of a golf ball)
— dim·ply \-p(ə-)lē\ *adjective*

[2]dimple *verb* **dim·pled; dim·pling** \-p(ə-)liŋ\ (1602)
transitive verb
: to mark with dimples
intransitive verb
: to exhibit or form dimples

dim sum \ˈdim-ˈsəm\ *noun* [Chinese (Guangdong) *dímsām,* from *dím* dot, speck + *sām* heart] (1948)
: traditional Chinese food consisting of a variety of items (as steamed or fried dumplings, pieces of cooked chicken, and rice balls) served in small portions

dim·wit \ˈdim-ˌwit\ *noun* (circa 1922)
: a stupid or mentally slow person

dim–wit·ted \-ˈwi-təd\ *adjective* (1934)
: not mentally bright **:** STUPID
— dim–wit·ted·ly *adverb*
— dim–wit·ted·ness *noun*

[1]din \ˈdin\ *noun* [Middle English, from Old English *dyne;* akin to Old Norse *dynr* din, Sanskrit *dhvanati* it roars] (before 12th century)
1 : a loud continued noise; *especially* **:** a welter of discordant sounds
2 : a situation or condition resembling a din

[2]din *verb* **dinned; din·ning** (before 12th century)
intransitive verb
: to make a loud noise
transitive verb
1 : to assail with loud continued noise
2 : to impress by insistent repetition — often used with *into*

di·nar \di-ˈnär, ˈdē-ˌ\ *noun* [Arabic *dīnār,* from Late Greek *dēnarion* denarius, from Latin *denarius* — more at DENIER] (1634)
1 : a gold coin formerly used in countries of southwest Asia and north Africa
2 a — see MONEY table **b** — see *rial* at MONEY table

[1]dine \ˈdīn\ *verb* **dined; din·ing** [Middle English, from Old French *diner,* from (assumed) Vulgar Latin *disjejunare* to break one's fast, from Latin *dis-* + Late Latin *jejunare* to fast, from Latin *jejunus* fasting] (13th century)
intransitive verb
: to take dinner
transitive verb
: to give a dinner to

[2]dine *noun* (15th century)
Scottish **:** DINNER

dine out *intransitive verb* (1816)
: to eat a meal away from home
— dine out on : to use as a subject for dining table conversation

din·er \ˈdī-nər\ *noun* (1815)
1 : one that dines
2 a : DINING CAR **b :** a restaurant usually resembling a dining car in shape

din·er–out \ˌdī-nə-ˈrau̇t\ *noun, plural* **di·ners–out** \-nər-ˈzau̇t\ (1808)
: one who dines out

di·nette \dī-ˈnet\ *noun* (1925)
: a small space usually off a kitchen used for informal dining; *also* **:** furniture for such a space

[1]ding \ˈdiŋ\ *verb* [probably imitative] (1582)
transitive verb
: to dwell on with tiresome repetition ⟨keeps *dinging* it into him that the less he smokes the better —Samuel Butler (died 1902)⟩
intransitive verb
1 : to make a ringing sound **:** CLANG
2 : to speak with tiresome reiteration

[2]ding *noun* [*ding* (to strike), from Middle English *dingen*] (circa 1945)
: an instance of minor surface damage (as a dent)

ding–a–ling \ˈdiŋ-ə-ˌliŋ\ *noun* [reduplication of *[1]ding*] (circa 1935)
: NITWIT, KOOK

ding·bat \ˈdiŋ-ˌbat\ *noun* [origin unknown] (1904)
1 : a typographical symbol or ornament (as *, ¶, or ✠)
2 : NITWIT, KOOK

[1]ding·dong \ˈdiŋ-ˌdȯŋ, -ˌdäŋ\ *noun* [imitative] (1611)
: the ringing sound produced by repeated strokes especially on a bell

[2]dingdong *intransitive verb* (1659)
1 : to make a dingdong sound
2 : to repeat a sound or action tediously or insistently

[3]dingdong *adjective* (1792)
1 : of, relating to, or resembling the ringing sound made by a bell
2 : marked by a rapid exchange or alternation

dinge \ˈdinj\ *noun* [back-formation from *dingy*] (1846)
: the condition of being dingy

ding·er \ˈdiŋ-ər\ *noun* [perhaps from *ding* (to strike) + *[2]-er*] (1974)
: HOME RUN

din·ghy \ˈdiŋ-ē, -gē\ *noun, plural* **dinghies** [Bengali *ḍiṅgi* & Hindi *ḍiṅgī*] (1810)
1 : an East Indian rowboat or sailboat
2 a : a small boat carried on or towed behind a larger boat as a tender or a lifeboat **b :** a small sailboat
3 : a rubber life raft

din·gle \ˈdiŋ-gəl\ *noun* [Middle English, deep hollow] (13th century)
: a small wooded valley **:** DELL

din·gle·ber·ry \ˈdiŋ-gəl-ˌber-ē\ *noun* [origin unknown] (1955)
: a piece of dried fecal matter clinging to the hair around the anus

din·go \ˈdiŋ-(ˌ)gō\ *noun, plural* **dingoes** [Dharuk (Australian aboriginal language of the Port Jackson area) *dingu*] (1789)
: a reddish brown wild dog (*Canis dingo*) of Australia

dingo

din·gus \ˈdiŋ-(g)əs\ *noun* [Dutch or German; Dutch *dinges,* probably from German *Dings,* from genitive of *Ding* thing, from Old High German — more at THING] (1876)
: DOODAD 2

din·gy \ˈdin-jē\ *adjective* **din·gi·er; -est** [origin unknown] (circa 1736)
1 : DIRTY, DISCOLORED
2 : SHABBY, SQUALID
— din·gi·ly \-jə-lē\ *adverb*
— din·gi·ness \-jē-nəs\ *noun*

dining car *noun* (1838)
: a railroad car in which meals are served

dining room *noun* (1601)
: a room used for eating meals

di·ni·tro \(ˌ)dī-ˈnī-trō\ *adjective*
: containing two nitro groups — often used in combination

di·ni·tro·ben·zene \(ˌ)dī-ˌnī-trō-ˈben-ˌzēn, -(ˌ)ben-ˈ\ *noun* [International Scientific Vocabulary] (1873)
: any of three isomeric toxic compounds $C_6H_4(NO_2)_2$

di·ni·tro·phe·nol \-ˈfē-ˌnōl, -fi-ˈ\ *noun* (circa 1897)
: any of six isomeric crystalline compounds $C_6H_4N_2O_5$ some of whose derivatives are pesticides

[1]dink \ˈdiŋk\ *noun* [by shortening & alteration] (1903)
: DINGHY

[2]dink *noun* [*dink* (to hit with a drop shot), probably of imitative origin] (1939)
: DROP SHOT

[3]dink *noun* [perhaps from *dink,* disparaging name for a Vietnamese] (1974)
slang **:** NITWIT, JERK, NERD

Din·ka \ˈdiŋ-kə\ *noun, plural* **Dinkas** *also* **Dinka** (1861)
1 : a member of a pastoral people of the Nile Valley in south central Sudan
2 : the Nilotic language of the Dinkas

din·key *or* **din·ky** \ˈdiŋ-kē\ *noun, plural* **dinkeys** *or* **dinkies** [probably from *dinky*] (1874)
: a small locomotive

[1]din·kum \ˈdiŋ-kəm\ *adjective* [English dialect *dinkum,* noun, work, share of work] (1905)
Australian & New Zealand **:** AUTHENTIC, GENUINE — often used with *fair*

[2]dinkum *adverb* (1915)
Australian & New Zealand **:** TRULY, HONESTLY — often used with *fair;* often used interjectionally

din·ky \ˈdiŋ-kē\ *adjective* **din·ki·er; -est** [Scots *dink* neat] (1880)
: SMALL 1a; *also* **:** INSIGNIFICANT

din·ner \ˈdi-nər\ *noun, often attributive* [Middle English *diner,* from Old French, from *diner* to dine] (13th century)
1 a : the principal meal of the day **b :** a formal feast or banquet
2 : TABLE D'HÔTE 2
3 : the food prepared for a dinner ⟨eat your *dinner*⟩
4 : a packaged meal usually for quick preparation ⟨warmed up a frozen *dinner*⟩
— din·ner·less \-ləs\ *adjective*

dinner jacket *noun* (1891)
: a jacket for formal evening wear

dinner theater *noun* (1960)
: a restaurant in which a play is presented after the meal is over

din·ner·time \ˈdi-nər-ˌtīm\ *noun* (14th century)
: the customary time for dinner

din·ner·ware \-ˌwar, -ˌwer\ *noun* (1895)
: tableware other than flatware

di·no·fla·gel·late \ˌdī-nō-ˈfla-jə-lət, -ˌlāt; -flə-ˈje-lət\ *noun* [ultimately from Greek *dinos* rotation, eddy + New Latin *flagellum*] (1889)
: any of an order (Dinoflagellata) of chiefly marine planktonic usually solitary unicellular phytoflagellates that include luminescent forms, forms important in marine food chains, and forms causing red tide

di·no·saur \ˈdī-nə-ˌsȯr\ *noun* [New Latin *Dinosaurus,* genus name, from Greek *deinos* terrifying + *sauros* lizard — more at DIRE] (1841)
1 : any of a group (Dinosauria) of extinct chiefly terrestrial carnivorous or herbivorous reptiles of the Mesozoic era
2 : any of various large extinct reptiles other than the true dinosaurs
3 : one that is impractically large, out-of-date, or obsolete
— di·no·sau·ri·an \ˌdī-nə-ˈsȯr-ē-ən\ *adjective*

[1]dint \ˈdint\ *noun* [Middle English, from Old English *dynt;* akin to Old Norse *dyntr* noise] (before 12th century)
1 *archaic* **:** BLOW, STROKE
2 : FORCE, POWER
3 : [2]DENT
— by dint of : by force of **:** BECAUSE OF

[2]dint *transitive verb* (1597)
1 : to make a dent in
2 : to impress or drive in with force

di·nu·cle·o·tide \(ˌ)dī-ˈn(y)ü-klē-ə-ˌtīd\ *noun* (circa 1927)
: a nucleotide consisting of two units each composed of a phosphate, a pentose, and a nitrogen base

di·oc·e·san \dī-ˈä-sə-sən *also* ˈdī-ə-ˌsē-sᵊn\ *noun* (15th century)
: a bishop having jurisdiction over a diocese

di·o·cese \ˈdī-ə-səs, -ˌsēs, -ˌsēz\ *noun, plural* **-ces·es** \ˈdī-ə-sə-səz, -ˌsē-zəz, ÷ˈdī-ə-ˌsēz\ [Middle English *diocise,* from Middle French, from Late Latin *diocesis,* alteration of *dioecesis,* from Latin, administrative division, from Greek *dioikēsis* administration, administrative division, from *dioikein* to keep house, govern, from *dia-* + *oikein* to dwell, manage, from *oikos* house — more at VICINITY] (14th century)
: the territorial jurisdiction of a bishop
— **di·oc·e·san** \dī-ˈä-sə-sən *also* ˈdī-ə-ˌsē-sᵊn\ *adjective*

di·ode \ˈdī-ˌōd\ *noun* [International Scientific Vocabulary] (1919)
: an electronic device that has two electrodes or terminals and is used especially as a rectifier

di·oe·cious \(ˌ)dī-ˈē-shəs\ *adjective* [ultimately from Greek *di-* + *oikos*] (1752)
1 : having male reproductive organs in one individual and female in another
2 : having staminate and pistillate flowers borne on different individuals
— **di·oe·cism** \-ˈē-ˌsi-zəm\ *noun*
— **di·oe·cy** \ˈdī-ˌē-sē\ *noun*

di·ol \ˈdī-ˌōl, -ˌȯl\ *noun* [International Scientific Vocabulary] (1923)
: a compound containing two hydroxyl groups

di·ole·fin \dī-ˈō-lə-fən\ *noun* [International Scientific Vocabulary] (circa 1909)
: DIENE

Di·o·me·des \ˌdī-ə-ˈmē-dēz\ *noun* [Latin, from Greek *Diomēdēs*]
: one of the Greek heroes of the Trojan War

Di·o·ny·sia \ˌdī-ə-ˈni-zhē-ə, -ˈnē-, -shē-, -zē-, -sē-\ *noun plural* [Latin, from Greek, from neuter plural of *dionysios* of Dionysus, from *Dionysos*] (circa 1895)
: ancient Greek festival observances held in seasonal cycles in honor of Dionysus; *especially* **:** such observances marked by dramatic performances

Di·o·ny·si·ac \-ˈni-zhē-ˌak, -ˈnē-, -shē-, -zē-, -sē-\ *adjective* [Latin *dionysiacus,* from Greek *dionysiakos,* from *Dionysos*] (1844)
: DIONYSIAN 2
— **Dionysiac** *noun*

Di·o·ny·sian \-ˈni-zhē-ən, -ˈnē-, -shē-, -zē-, -sē-; -zhən, -shən\ *adjective* (1607)
1 a : of or relating to Dionysius **b :** of or related to the theological writings once mistakenly attributed to Dionysius the Areopagite
2 a : devoted to the worship of Dionysus **b :** characteristic of Dionysus or the cult of worship of Dionysus; *especially* **:** being of a frenzied or orgiastic character — compare APOLLONIAN

Di·o·ny·sus \ˌdī-ə-ˈnī-səs, -ˈnē-\ *noun* [Latin, from Greek *Dionysos*]
: BACCHUS

Di·o·phan·tine equation \ˌdī-ə-ˈfan-ˌtīn-, -ˈfan-tᵊn-\ *noun* [*Diophantus,* 3d century A.D. Greek mathematician] (circa 1928)
: an indeterminate polynomial equation with integral coefficients for which it is required to find all integral solutions

di·op·side \dī-ˈäp-ˌsīd\ *noun* [French, from *di-* + Greek *opsis* appearance — more at OPTIC] (circa 1808)
: a green to white mineral that consists of pyroxene containing little or no aluminum
— **di·op·sid·ic** \ˌdī-ˌäp-ˈsi-dik\ *adjective*

di·op·ter \dī-ˈäp-tər, ˈdī-ˌäp-\ *noun* [*diopter* (an optical instrument), from Middle French *dioptre,* from Latin *dioptra,* from Greek, from *dia-* + *opsesthai* to be going to see — more at OPTIC] (circa 1864)
: a unit of measurement of the refractive power of lenses equal to the reciprocal of the focal length in meters

di·op·tric \dī-ˈäp-trik\ *adjective* [Greek *dioptrikos* of a diopter (instrument), from *dioptra*] (1653)
: REFRACTIVE; *specifically* **:** assisting vision by refracting and focusing light

di·ora·ma \ˌdī-ə-ˈra-mə, -ˈrä-\ *noun* [French, from *dia-* + *-orama* (as in *panorama,* from English)] (1823)
1 : a scenic representation in which a partly translucent painting is seen from a distance through an opening
2 a : a scenic representation in which sculptured figures and lifelike details are displayed usually in miniature so as to blend indistinguishably with a realistic painted background **b :** a life-size exhibit (as of a wildlife specimen or scene) with realistic natural surroundings and a painted background
— **di·oram·ic** \-ˈra-mik\ *adjective*

di·o·rite \ˈdī-ə-ˌrīt\ *noun* [French, irregular from Greek *diorizein* to distinguish, from *dia-* + *horizein* to define — more at HORIZON] (1826)
: a granular crystalline igneous rock commonly of acid plagioclase and hornblende, pyroxene, or biotite
— **di·o·rit·ic** \ˌdī-ə-ˈri-tik\ *adjective*

Di·os·cu·ri \ˌdī-əs-ˈkyu̇(ə)r-ˌī, dī-ˈäs-kyə-ˌrī\ *noun plural* [New Latin, from Greek *Dioskouroi,* literally, sons of Zeus, from *Dios* (genitive of *Zeus;* akin to Latin *divus* divine) + *kouroi,* plural of *kouros, koros* boy — more at DEITY, CRESCENT]
: the twins Castor and Pollux reunited as stars in the sky by Zeus after Castor's death and regarded as patrons of athletes and sailors

di·ox·ane \dī-ˈäk-ˌsān\ *also* **di·ox·an** \-ˌsan, -sən\ *noun* [International Scientific Vocabulary] (1912)
: a flammable toxic liquid diether $C_4H_8O_2$ used especially as a solvent

di·ox·ide \(ˌ)dī-ˈäk-ˌsīd\ *noun* [International Scientific Vocabulary] (circa 1847)
: an oxide (as carbon dioxide) containing two atoms of oxygen in the molecule

di·ox·in \(ˌ)dī-ˈäk-sən\ *noun* (circa 1919)
: any of several heterocyclic hydrocarbons that occur especially as persistent toxic impurities in herbicides; *especially* **:** TCDD

¹dip \ˈdip\ *verb* **dipped; dip·ping** [Middle English *dippen,* from Old English *dyppan;* akin to Old High German *tupfen* to wash, Lithuanian *dubus* deep] (before 12th century)
transitive verb
1 a : to plunge or immerse momentarily or partially under the surface (as of a liquid) so as to moisten, cool, or coat ⟨*dip* candles⟩ **b :** to thrust in a way to suggest immersion **c :** to immerse (as a sheep or dog) in an antiseptic or parasiticidal solution
2 a : to lift a portion of by reaching below the surface with something shaped to hold liquid **:** LADLE **b :** to take a portion of (snuff)
3 a *archaic* **:** INVOLVE **b :** MORTGAGE
4 a : to lower and then raise again ⟨*dip* a flag in salute⟩ **b** *chiefly British* **:** DIM 2
intransitive verb
1 a : to plunge into a liquid and quickly emerge **b :** to immerse something into a processing liquid or finishing material
2 a : to suddenly drop down or out of sight **b** *of an airplane* **:** to drop suddenly before climbing **c :** to decline or decrease moderately and usually temporarily ⟨prices *dipped*⟩ **d :** to lower the body momentarily especially as part of an athletic or dancing motion
3 a : to reach down inside or below or as if inside or below a surface especially to withdraw a part of the contents — used with *into* ⟨*dipped* into the family's savings⟩
4 : to examine or read something casually or superficially — used with *into*
5 : to incline downward from the plane of the horizon
— **dip·pa·ble** \ˈdi-pə-bəl\ *adjective*

²dip *noun* (1599)
1 : an act of dipping; *especially* **:** a brief plunge into the water for sport or exercise
2 : inclination downward: **a :** PITCH **b :** a sharp downward course **:** DROP **c :** the angle that a stratum or similar geological feature makes with a horizontal plane
3 : the angle formed with the horizon by a magnetic needle free to rotate in the vertical plane
4 : HOLLOW, DEPRESSION
5 : something obtained by or used in dipping
6 a : a sauce or soft mixture into which food may be dipped **b :** a liquid preparation for the dipping of something; *especially* **:** an insecticide or parasiticide for the dipping of animals — compare SHEEP-DIP
7 *slang* **:** PICKPOCKET
8 : a stupid or unsophisticated person

di·pep·ti·dase \dī-ˈpep-tə-ˌdās, -ˌdāz\ *noun* (1927)
: any of various enzymes that hydrolyze dipeptides but not polypeptides

di·pep·tide \(ˌ)dī-ˈpep-ˌtīd\ *noun* (1903)
: a peptide that yields two molecules of amino acid on hydrolysis

di·pha·sic \(ˌ)dī-ˈfā-zik\ *adjective* (1881)
: having two phases

di·phen·hy·dra·mine \ˌdī-ˌfen-ˈhī-drə-ˌmēn\ *noun* (1948)
: an antihistamine $C_{17}H_{21}NO$ used especially in the form of its hydrochloride to treat allergy symptoms and motion sickness

di·phe·nyl·amine \(ˌ)dī-ˌfe-nᵊl-ə-ˈmēn, -ˌfē-, -nᵊl-ˈa-mən\ *noun* [International Scientific Vocabulary] (1872)
: a crystalline pleasant-smelling compound $(C_6H_5)_2NH$ used chiefly in the manufacture of dyes and as an indicator

di·phe·nyl·hy·dan·to·in \-hī-ˈdan-tə-wən\ *noun* [*di-* + *phenyl* + *hyd*rogen + *allantoin*] (1937)
: PHENYTOIN

di·phos·gene \(ˌ)dī-ˈfäz-ˌjēn\ *noun* [International Scientific Vocabulary] (circa 1922)
: a liquid compound $C_2Cl_4O_2$ used as a poison gas in World War I

di·phos·phate \(ˌ)dī-ˈfäs-ˌfāt\ *noun* (1826)
: a phosphate containing two phosphate groups

di·phos·pho·gly·cer·ic acid \(ˌ)dī-ˈfäs-fō-gli-ˌser-ik-\ *noun* (1959)
: a diphosphate of glyceric acid that is an important intermediate in photosynthesis and in glycolysis and fermentation

di·phos·pho·pyr·i·dine nucleotide \-ˌpir-ə-ˌdēn-\ *noun* (1938)
: NAD

diph·the·ria \dif-ˈthir-ē-ə, ÷dip-\ *noun* [New Latin, from French *diphthérie,* from Greek *diphthera* leather; from the toughness of the false membrane] (circa 1851)
: an acute febrile contagious disease marked by the formation of a false membrane especially in the throat and caused by a bacterium (*Corynebacterium diphtheriae*) that produces a toxin causing inflammation of the heart and nervous system
— **diph·the·ri·al** \-ē-əl\ *adjective*
— **diph·the·rit·ic** \ˌdif-thə-ˈri-tik, ÷ˌdip-\ *adjective*

¹diph·the·roid \ˈdif-thə-ˌrȯid\ *adjective* (1861)
: resembling diphtheria

²diphtheroid *noun* (1908)
: a bacterium (especially genus *Corynebacterium*) that resembles the bacterium of diphtheria but does not produce diphtheria toxin

diph·thong \ˈdif-ˌthȯŋ, ˈdip-\ *noun* [Middle English *diptonge,* from Middle French *diptongue,* from Late Latin *dipthongus,* from Greek *diphthongos,* from *di-* + *phthongos* voice, sound] (15th century)

1 : a gliding monosyllabic speech sound (as the vowel combination at the end of *toy*) that starts at or near the articulatory position for one vowel and moves to or toward the position of another
2 : DIGRAPH
3 : the ligature æ or œ
— diph·thon·gal \dif-ˈthȯŋ-(g)əl, dip-\ *adjective*
diph·thong·i·za·tion \(ˌ)dif-ˌthȯŋ-ə-ˈzā-shən, (ˌ)dip-\ *noun* (1874)
: the act of diphthongizing **:** the state of being diphthongized
diph·thong·ize \ˈdif-ˌthȯŋ-ˌīz, ˈdip-\ *verb* **-ized; -iz·ing** (1867)
intransitive verb
of a simple vowel **:** to change into a diphthong
transitive verb
: to pronounce as a diphthong
diphy- *or* **diphyo-** *combining form* [New Latin, from Greek *diphy-*, from *diphyēs,* from *di-* + *phyein* to bring forth — more at BE]
: double **:** bipartite ⟨*diphy*odont⟩
di·phy·let·ic \ˌdī-fī-ˈle-tik\ *adjective* (1902)
: derived from two lines of evolutionary descent ⟨*diphyletic* dinosaurs⟩
di·phy·odont \(ˌ)dī-ˈfī-ə-ˌdänt\ *adjective* [International Scientific Vocabulary] (1854)
: marked by the successive development of deciduous and permanent sets of teeth
dipl- *or* **diplo-** *combining form* [Greek, from *diploos* — more at DOUBLE]
1 : double **:** twofold ⟨*dipl*opia⟩
2 : diploid ⟨*diplo*phase⟩
di·ple·gia \dī-ˈplē-j(ē-)ə\ *noun* [New Latin] (circa 1881)
: paralysis of corresponding parts on both sides of the body
dip·lo·blas·tic \ˌdip-lō-ˈblas-tik\ *adjective* (circa 1885)
: having two germ layers — used of an embryo or lower invertebrate that lacks a true mesoderm
dip·lo·coc·cus \-ˈkä-kəs\ *noun* [New Latin, genus name] (circa 1881)
: any of various encapsulated bacteria (as *Streptococcus pneumoniae,* a common cause of pneumonia) that usually occur in pairs and that were formerly grouped in a single taxon (genus *Diplococcus*) but are now all assigned to other genera
di·plod·o·cus \də-ˈplä-də-kəs, dī-\ *noun* [New Latin, genus name, from *dipl-* + Greek *dokos* beam, from *dekesthai, dechesthai* to receive; akin to Latin *decēre* to be fitting — more at DECENT] (1928)
: any of a genus (*Diplodocus*) of large herbivorous sauropod dinosaurs of the late Jurassic period in Colorado and Wyoming
dip·loe \ˈdi-plə-ˌwē\ *noun* [New Latin, from Greek *diploē,* from *diploos* double] (circa 1696)
: cancellous bony tissue between the external and internal layers of the skull
— di·plo·ic \də-ˈplō-ik, dī-\ *adjective*
[1]**dip·loid** \ˈdi-ˌplȯid\ *adjective* [International Scientific Vocabulary] (1908)
: having the basic chromosome number doubled
— dip·loi·dy \-ˌplȯi-dē\ *noun*
[2]**diploid** *noun* (1908)
: a single cell, individual, or generation characterized by the diploid chromosome number
di·plo·ma \də-ˈplō-mə\ *noun, plural* **diplomas** [Latin, passport, diploma, from Greek *diplōma* folded paper, passport, from *diploun* to double, from *diploos*] (1622)
1 *plural also* **di·plo·ma·ta** \-mə-tə\ **:** an official or state document **:** CHARTER
2 : a writing usually under seal conferring some honor or privilege
3 : a document bearing record of graduation from or of a degree conferred by an educational institution
di·plo·ma·cy \də-ˈplō-mə-sē\ *noun* (1796)
1 : the art and practice of conducting negotiations between nations
2 : skill in handling affairs without arousing hostility **:** TACT
diploma mill *noun* (1914)
: a usually unregulated institution of higher education granting degrees with few or no academic requirements
dip·lo·mat \ˈdi-plə-ˌmat\ *noun* [French *diplomate,* back-formation from *diplomatique*] (1813)
: one employed or skilled in diplomacy
dip·lo·mate \ˈdi-plə-ˌmāt\ *noun* (1879)
: one who holds a diploma; *especially* **:** a physician qualified to practice in a medical specialty by advanced training and experience in the specialty followed by passing an intensive examination by a national board of senior specialists
dip·lo·mat·ic \ˌdi-plə-ˈma-tik\ *adjective* [in sense 1, from New Latin *diplomaticus,* from Latin *diplomat-, diploma;* in other senses, from French *diplomatique* connected with documents regulating international relations, from New Latin *diplomaticus*] (1711)
1 a : PALEOGRAPHIC **b :** exactly reproducing the original ⟨a *diplomatic* edition⟩
2 : of, relating to, or concerned with diplomacy or diplomats ⟨*diplomatic* relations⟩
3 : employing tact and conciliation especially in situations of stress
synonym see SUAVE
— dip·lo·mat·i·cal·ly \-ti-k(ə-)lē\ *adverb*
di·plo·ma·tist \də ˈplō-mə-tist\ *noun* (1815)
: DIPLOMAT
dip·lont \ˈdi-ˌplänt\ *noun* [International Scientific Vocabulary] (1925)
: an organism with somatic cells having the diploid chromosome number — compare HAPLONT
— dip·lon·tic \di-ˈplän-tik\ *adjective*
dip·lo·phase \ˈdi plə-ˌfāz\ *noun* (circa 1925)
: a diploid phase in a life cycle
dip·lo·pia \di-ˈplō-pē-ə\ *noun* [New Latin] (circa 1811)
: a disorder of vision in which two images of a single object are seen because of unequal action of the eye muscles — called also *double vision*
— dip·lo·pic \-ˈplō-pik, -ˈplä-pik\ *adjective*
dip·lo·pod \ˈdi-plə-ˌpäd\ *noun* [ultimately from Greek *dipl-* + *pod-, pous* foot — more at FOOT] (circa 1864)
: MILLIPEDE
dip·lo·tene \ˈdi-plə-ˌtēn\ *noun* [International Scientific Vocabulary] (1925)
: a stage of meiotic prophase which follows the pachytene and during which the paired homologous chromosomes begin to separate and chiasmata become visible
— diplotene *adjective*
dip net *noun* (1820)
: a bag net with a handle that is used especially to scoop fish from the water
— dip·net \ˈdip-ˌnet\ *transitive verb*
dip·no·an \ˈdip-nə-wən\ *adjective* [ultimately from Greek *dipnoos* having two breathing apertures, from *di-* + *pnoē* breath, from *pnein* to breathe — more at SNEEZE] (1883)
: of or relating to a group (Dipnoi) of fishes with pulmonary circulation, gills, and lungs
— dipnoan *noun*
dip·o·dy \ˈdi-pə-dē\ *noun, plural* **-dies** [Late Latin *dipodia,* from Greek, from *dipod-, dipous* having two feet, from *di-* + *pod-, pous*] (circa 1844)
: a prosodic unit or measure of two feet
— di·pod·ic \dī-ˈpä-dik\ *adjective*
di·pole \ˈdī-ˌpōl\ *noun* [International Scientific Vocabulary] (1912)
1 a : a pair of equal and opposite electric charges or magnetic poles of opposite sign separated especially by a small distance **b :** a body or system (as a molecule) having such charges
2 : a radio antenna consisting of two horizontal rods in line with each other with their ends slightly separated
— di·po·lar \ˈdī-ˌpō-lər, ˌdī-ˈpō-\ *adjective*
dipole moment *noun* (1926)
: the product of the distance between the two poles (as magnetic or electric) of a dipole and the magnitude of either pole
dip·per \ˈdi-pər\ *noun* (1611)
1 : one that dips: as **a :** a worker who dips articles **b :** something (as a long-handled cup) used for dipping **c** *slang* **:** PICKPOCKET
2 : any of a genus (*Cinclus* and especially *C. cinclus* of the Old World and *C. mexicanus* of North America) of birds that comprise an oscine family (Cinclidae) related to the thrushes and including individuals that are not web-footed but dive into swift mountain streams and walk on the bottom in search of food — called also *water ouzel*
3 *capitalized* **:** a group of stars that resembles a dipper: as **a :** BIG DIPPER **b :** LITTLE DIPPER
— dip·per·ful \-ˌfu̇l\ *noun*
dip·py \ˈdi-pē\ *adjective* **dip·pi·er; -est** [origin unknown] (1911)
: FOOLISH
dip·so \ˈdip-(ˌ)sō\ *noun, plural* **dipsos** [by shortening] (1880)
: one affected with dipsomania
dip·so·ma·nia \ˌdip-sə-ˈmā-nē-ə, -nyə\ *noun* [New Latin, from Greek *dipsa* thirst + Late Latin *mania*] (circa 1844)
: an uncontrollable craving for alcoholic liquors
— dip·so·ma·ni·ac \-nē-ˌak\ *noun*
— dip·so·ma·ni·a·cal \ˌdip-sō-mə-ˈnī-ə-kəl\ *adjective*
dip·stick \ˈdip-ˌstik\ *noun* (1927)
: a graduated rod for indicating depth (as of oil in a crankcase)
dip·ter·an \ˈdip-tə-rən\ *adjective* [ultimately from Greek *dipteros* two-winged, from *di-* + *pteron* wing — more at FEATHER] (circa 1842)
: of, relating to, or being a fly (sense 2a)
— dipteran *noun*
— dip·ter·ous \ˈdip-t(ə-)rəs\ *adjective*
dip·tero·carp \ˈdip-tə-rō-ˌkärp\ *noun* [ultimately from Greek *dipteros* + *-karpos* -carpous] (circa 1876)
: any of a family (Dipterocarpaceae) of tall hardwood trees of tropical Asia, Indonesia, and the Philippines that have a 2-winged fruit and are the source of valuable timber, aromatic oils, and resins; *especially* **:** a member of the type genus (*Dipterocarpus*)
dip·ter·on \ˈdip-tə-ˌrän\ *noun, plural* **-tera** \-rə\ [Greek, neuter of *dipteros*] (circa 1891)
: [4]FLY 2a
dip·tych \ˈdip-(ˌ)tik\ *noun* [Late Latin *diptycha,* plural, from Greek, from neuter plural of *diptychos* folded in two, from *di-* + *ptychē* fold] (1622)
1 : a 2-leaved hinged tablet folding together to protect writing on its waxed surfaces
2 : a picture or series of pictures (as an altarpiece) painted or carved on two hinged tablets
3 : a work made up of two matching parts

diptych 2

di·quat \ˈdī-ˌkwät\ *noun* [*di-* + *qua*ternary] (1960)
: a powerful nonpersistent herbicide $C_{12}H_{12}Br_2N_2$ that has been used to control water weeds (as the water hyacinth)

\ə\ **abut** \ᵊ\ **kitten** \ər\ **further** \a\ **ash** \ā\ **ace** \ä\ **mop, mar** \au̇\ **out** \ch\ **chin** \e\ **bet** \ē\ **easy** \g\ **go** \i\ **hit** \ī\ **ice** \j\ **job** \ŋ\ **sing** \ō\ **go** \ȯ\ **law** \ȯi\ **boy** \th\ **thin** \t͟h\ **the** \ü\ **loot** \u̇\ **foot** \y\ **yet** \zh\ **vision** *see also* Guide to Pronunciation

dir·dum \ˈdir-dəm, ˈdər-\ *noun* [Middle English (northern dialect) *durdan, durdum* uproar, from Celtic; akin to Welsh *dwrdd* noise, clamor, Middle Irish *dordán* humming, droning] (circa 1693)
Scottish **:** BLAME

dire \ˈdīr\ *adjective* **dir·er; dir·est** [Latin *dirus;* akin to Greek *deinos* terrifying, Sanskrit *dveṣṭi* he hates] (1567)
1 a : exciting horror ⟨*dire* suffering⟩ **b :** DISMAL, OPPRESSIVE ⟨*dire* days⟩
2 : warning of disaster ⟨a *dire* forecast⟩
3 a : desperately urgent ⟨*dire* need⟩ **b :** EXTREME ⟨*dire* poverty⟩
— dire·ly *adverb*
— dire·ness *noun*

[1]di·rect \də-ˈrekt, dī-\ *verb* [Middle English, from Latin *directus* straight, from past participle of *dirigere* to direct — more at DRESS] (14th century)
transitive verb
1 a *obsolete* **:** to write (a letter) to a person **b :** to mark with the name and address of the intended recipient **c :** to impart orally **d :** to adapt in expression so as to have particular applicability ⟨a lawyer who *directs* his appeals to intelligence⟩
2 a : to regulate the activities or course of **b :** to carry out the organizing, energizing, and supervising of **c :** to dominate and determine the course of **d :** to train and lead performances of
3 : to cause to turn, move, or point undeviatingly or to follow a straight course ⟨X rays are *directed* through the body⟩
4 : to point, extend, or project in a specified line or course
5 : to request or enjoin with authority
6 : to show or point out the way for
intransitive verb
1 : to point out, prescribe, or determine a course or procedure
2 : to act as director
synonym see COMMAND, CONDUCT

[2]direct *adjective* [Middle English, from Latin *directus*] (15th century)
1 *of a celestial body* **:** moving in the general planetary direction from west to east **:** not retrograde
2 a : stemming immediately from a source ⟨*direct* result⟩ **b :** being or passing in a straight line of descent from parent to offspring **:** LINEAL ⟨*direct* ancestor⟩ **c :** having no compromising or impairing element ⟨*direct* insult⟩
3 a : proceeding from one point to another in time or space without deviation or interruption **:** STRAIGHT **b :** proceeding by the shortest way ⟨the *direct* route⟩
4 : NATURAL, STRAIGHTFORWARD ⟨*direct* manner⟩
5 a : marked by absence of an intervening agency, instrumentality, or influence **b :** effected by the action of the people or the electorate and not by representatives ⟨*direct* democracy⟩ **c :** consisting of or reproducing the exact words of a speaker or writer
6 : characterized by close logical, causal, or consequential relationship ⟨*direct* evidence⟩
7 : capable of dyeing without the aid of a mordant

[3]direct *adverb* (14th century)
: in a direct way: as **a :** from point to point without deviation **:** by the shortest way ⟨flew *direct* to Miami⟩ **b :** from the source without interruption or diversion ⟨the writer must take his material *direct* from life —Douglas Stewart⟩ **c :** without an intervening agency or step ⟨buy *direct* from the manufacturer⟩

direct action *noun* (1912)
: action that seeks to achieve an end directly and by the most immediately effective means (as boycott or strike)

direct current *noun* (circa 1889)
: an electric current flowing in one direction only and substantially constant in value — abbreviation *DC*

di·rect·ed *adjective* (1891)
1 : subject to supervision or regulation ⟨a *directed* reading program for students⟩
2 : having a positive or negative sense ⟨*directed* line segment⟩
— di·rect·ed·ness *noun*

direct examination *noun* (circa 1859)
: the first examination of a witness by the party calling the witness — compare CROSS-EXAMINATION

di·rec·tion \də-ˈrek-shən, dī-\ *noun* (15th century)
1 : guidance or supervision of action or conduct **:** MANAGEMENT
2 *archaic* **:** SUPERSCRIPTION
3 a : an explicit instruction **:** ORDER **b :** assistance in pointing out the proper route — usually used in plural ⟨asked for *directions* to the beach⟩
4 : the line or course on which something is moving or is aimed to move or along which something is pointing or facing
5 *archaic* **:** DIRECTORATE 1
6 a : a channel or direct course of thought or action **b :** TENDENCY, TREND **c :** a guiding, governing, or motivating purpose
7 a : the art and technique of directing an orchestra, band, or a show (as for stage or screen) **b :** a word, phrase, or sign indicating the appropriate tempo, mood, or intensity of a passage or movement in music
— di·rec·tion·less \-ləs\ *adjective*
— di·rec·tion·less·ness \-nəs\ *noun*

di·rec·tion·al \-shnəl, -shə-nᵊl\ *adjective* (1881)
1 : of, relating to, or indicating direction in space: **a :** suitable for detecting the direction from which radio signals come or for sending out radio signals in one direction only **b :** operating most effectively in a particular direction
2 : relating to direction or guidance especially of thought or effort
— di·rec·tion·al·i·ty \-ˌrek-shə-ˈna-lə-tē\ *noun*

direction angle *noun* (circa 1909)
: an angle made by a given line with an axis of reference; *specifically* **:** such an angle made by a straight line with the three axes of a rectangular Cartesian coordinate system — usually used in plural

direction cosine *noun* (circa 1889)
: any of the cosines of the three angles between a directed line in space and the positive direction of the axes of a rectangular Cartesian coordinate system — usually used in plural

direction finder *noun* (1913)
: a radio receiving device for determining the direction of incoming radio waves that typically consists of a coil antenna rotating freely on a vertical axis

[1]di·rec·tive \də-ˈrek-tiv, dī-\ *adjective* (15th century)
1 : serving or intended to guide, govern, or influence
2 : serving to point direction; *specifically* **:** DIRECTIONAL 1b
3 : of or relating to psychotherapy or counseling in which the counselor introduces information, content, or attitudes not previously expressed by the client

[2]directive *noun* (1902)
: something that serves to direct, guide, and usually impel toward an action or goal; *especially* **:** an authoritative instrument issued by a high-level body or official

di·rec·tiv·i·ty \də-ˌrek-ˈti-və-tē, (ˌ)dī-\ *noun* (1928)
: the property of being directional

direct lighting *noun* (1928)
: lighting in which the greater part of the light goes directly from the source to the area lit

[1]di·rect·ly \də-ˈrek(t)-lē, dī-, *in sense 2 especially* də-ˈrek-lē *or* ˈdrek-lē\ *adverb* (15th century)
1 a : in a direct manner ⟨*directly* relevant⟩ ⟨the road runs *directly* east and west⟩ **b :** in immediate physical contact **c :** in the manner of direct variation
2 a : without delay **:** IMMEDIATELY **b :** in a little while **:** SHORTLY

[2]di·rect·ly \də-ˈrek(t)-lē, dī-; ˈdrek-lē\ *conjunction* (1795)
chiefly British **:** immediately after **:** AS SOON AS ⟨*directly* I received it I rang up the shipping company —F. W. Crofts⟩

directly proportional *noun* (1796)
: related by direct variation — compare INVERSELY PROPORTIONAL

direct mail *noun* (circa 1923)
: printed matter (as circulars) prepared for soliciting business or contributions and mailed directly to individuals

di·rect·ness \də-ˈrek(t)-nəs, dī-\ *noun* (1598)
1 : the character of being accurate in course or aim
2 : strict pertinence **:** STRAIGHTFORWARDNESS ⟨her *directness* was disarming —Robin Cook⟩

direct object *noun* (1879)
: a word or phrase denoting the goal or the result of the action of a verb

Di·rec·toire \dē-(ˌ)rek-ˈtwär(r), -ˈrek-ˌ\ *adjective* [French, from *Directoire,* the group of five officials who governed France from 1795–99, from *directeur* director] (1878)
: of, relating to, or imitative of the style of clothing, furniture, or decoration prevalent in France during the period of the Directory

di·rec·tor \də-ˈrek-tər, dī-\ *noun* (15th century)
: one that directs: as **a :** the head of an organized group or administrative unit (as a bureau or school) **b :** one of a group of persons entrusted with the overall direction of a corporate enterprise **c :** one that supervises the production of a show (as for stage or screen) usually with responsibility for action, lighting, music, and rehearsals **d :** CONDUCTOR c
— di·rec·tor·ship \-ˌship\ *noun*

di·rec·tor·ate \də-ˈrek-t(ə-)rət, dī-\ *noun* (1837)
1 : the office of director
2 a : a board of directors (as of a corporation) **b :** membership on a board of directors
3 : an executive staff (as of a department)

di·rec·to·ri·al \də-ˌrek-ˈtōr-ē-əl, (ˌ)dī-, -ˈtȯr-\ *adjective* (1770)
1 : serving to direct
2 : of or relating to a director or to theatrical or motion-picture direction
3 : of, relating to, or administered by a directory

director's chair *noun* [from its use by motion-picture directors on the set] (1953)
: a lightweight folding armchair with a back and seat usually of cotton duck

director's chair

[1]di·rec·to·ry \də-ˈrek-t(ə-)rē, dī-\ *adjective* (15th century)
: serving to direct; *specifically* **:** providing advisory but not compulsory guidance

[2]directory *noun, plural* **-ries** [Middle English *directorie* guide, from Medieval Latin *directorium,* from neuter of Late Latin *directorius* directorial, from Latin *dirigere*] (1543)
1 a : a book or collection of directions, rules, or ordinances **b :** an alphabetical or classified list (as of names and addresses)
2 : a body of directors

direct primary *noun* (1900)
: a primary in which nominations of candidates for office are made by direct vote

direct product *noun* (circa 1925)
: CARTESIAN PRODUCT; *especially* **:** a group that is the Cartesian product of two other groups

di·rec·tress \də-ˈrek-trəs, dī-\ *noun* (1580)
: a woman who is a director

di·rec·trice \də-ˌrek-ˈtrēs\ *noun* [French, from Medieval Latin *directric-, directrix*] (1631)
: DIRECTRESS

di·rec·trix \-ˈrek-triks\ *noun, plural* **-trix·es** \-trik-səz\ *also* **-tri·ces** \-trə-ˌsēz\ [Medieval Latin, feminine of Late Latin *director,* from Latin *dirigere*] (1622)
1 *archaic* : DIRECTRESS
2 : a fixed curve with which a generatrix maintains a given relationship in generating a geometric figure; *specifically* : a straight line the distance to which from any point of a conic section is in fixed ratio to the distance from the same point to a focus

direct sum *noun* (circa 1928)
: CARTESIAN PRODUCT — compare DIRECT PRODUCT

direct tax *noun* (1801)
: a tax exacted directly from the taxpayer

direct variation *noun* (1949)
1 : mathematical relationship between two variables that can be expressed by an equation in which one variable is equal to a constant times the other
2 : an equation or function expressing direct variation — compare INVERSE VARIATION

dire·ful \ˈdīr-fəl\ *adjective* (1583)
1 : DREADFUL
2 : OMINOUS
— **dire·ful·ly** \-fə-lē\ *adverb*

dire wolf *noun* (1925)
: a large extinct lupine mammal (*Canis dirus*) known from Pleistocene deposits of North America

dirge \ˈdərj\ *noun* [Middle English *dirige,* the Office of the Dead, from the first word of a Late Latin antiphon, from Latin, imperative of *dirigere* to direct — more at DRESS] (13th century)
1 : a song or hymn of grief or lamentation; *especially* : one intended to accompany funeral or memorial rites
2 : a slow, solemn, and mournful piece of music
3 : something (as a poem) that has the qualities of a dirge ◆
— **dirge·like** \-ˌlīk\ *adjective*

dir·ham \ˈdir-həm\ *noun* [Arabic, from Latin *drachma* drachma] (1788)
1 — see MONEY table
2 — see *dinar, riyal* at MONEY table

¹di·ri·gi·ble \ˈdir-ə-jə-bəl, də-ˈri-jə-\ *adjective* [Latin *dirigere*] (1581)
: capable of being steered

²dirigible *noun* [*dirigible (balloon)*] (1885)
: AIRSHIP

di·ri·gisme \di-ri-ˈzhi-zəm, dē-rē-zhēs-mə\ *noun* [French, from *diriger* to direct (from Latin *dirigere*) + *-isme* -ism] (1947)
: economic planning and control by the state
— **di·ri·giste** \di-ri-ˈzhēst, dē-rē-\ *adjective*

¹dirk \ˈdərk\ *noun* [Scots *durk*] (1557)
: a long straight-bladed dagger

²dirk *transitive verb* (1599)
: to stab with a dirk

dirl \ˈdir(-ə)l, ˈdərl\ *intransitive verb* [perhaps alteration of *thirl*] (1715)
Scottish : TREMBLE, QUIVER

dirndl \ˈdərn-dəl\ *noun* [short for German *Dirndlkleid,* from German dialect *Dirndl* girl + German *Kleid* dress] (1937)
1 : a dress style with tight bodice, short sleeves, low neck, and gathered skirt
2 : a full skirt with a tight waistband

dirt \ˈdərt\ *noun* [Middle English *drit,* from Old Norse; akin to Old English *drītan* to defecate] (13th century)
1 a : EXCREMENT **b** : a filthy or soiling substance (as mud, dust, or grime) **c** *archaic* : something worthless **d** : a contemptible person
2 : loose or packed soil or sand : EARTH
3 a : an abject or filthy state : SQUALOR **b** : CORRUPTION, CHICANERY **c** : licentiousness of language or theme **d** : scandalous or malicious gossip **e** : embarrassing or incriminating information

dirt·bag \ˈdərt-ˌbag\ *noun* (circa 1967)
slang : a dirty, unkempt, or contemptible person

dirt bike *noun* (1970)
: a usually lightweight motorcycle designed for operation on unpaved surfaces

dirt cheap *adjective or adverb* (1821)
: exceedingly cheap

dirt farmer *noun* (1920)
: a farmer who earns his living by farming his own land especially without the help of hired hands or tenants

dirt–poor \ˈdərt-ˈpu̇r\ *adjective* (1937)
: suffering extreme poverty

¹dirty \ˈdər-tē\ *adjective* **dirt·i·er; -est** (14th century)
1 a : not clean or pure ⟨*dirty* clothes⟩ **b** : likely to befoul or defile with dirt ⟨*dirty* jobs⟩ **c** : tedious, disagreeable, and unrecognized or thankless ⟨had to do the *dirty* work⟩ **d** : contaminated with infecting organisms ⟨*dirty* wounds⟩ **e** : containing impurities ⟨*dirty* coal⟩
2 a : morally unclean or corrupt: as (1) : INDECENT, VULGAR ⟨*dirty* jokes⟩ ⟨a *dirty* movie⟩ (2) : DISHONORABLE, BASE ⟨*dirty* tricks⟩ (3) : UNSPORTSMANLIKE ⟨*dirty* players⟩ **b** : acquired by disreputable or illegal means : ILL-GOTTEN ⟨*dirty* money⟩
3 a : ABOMINABLE, HATEFUL ⟨war is a *dirty* business⟩ **b** : highly regrettable ⟨a *dirty* shame⟩
4 : FOGGY, STORMY ⟨*dirty* weather⟩
5 a *of color* : not clear and bright : DULLISH ⟨*dirty* blond⟩ **b** : characterized by a husky, rasping, or raw tonal quality ⟨*dirty* trumpet tones⟩
6 : conveying ill-natured resentment ⟨gave him a *dirty* look⟩
7 : having considerable fallout ⟨*dirty* bombs⟩ ☆
— **dirt·i·ly** \ˈdər-təl-ē\ *adverb*
— **dirt·i·ness** \ˈdər-tē-nəs\ *noun*

²dirty *adverb* **dirt·i·er; -est** (circa 1934)
: in a dirty manner: as **a** : DECEPTIVELY, UNDERHANDEDLY ⟨fight *dirty*⟩ **b** : INDECENTLY ⟨talk *dirty*⟩

³dirty *verb* **dirt·ied; dirty·ing** (1591)
transitive verb
1 : to make dirty
2 a : to stain with dishonor : SULLY **b** : to debase by distorting the real nature of
intransitive verb
: to become soiled

dirty linen *noun* (1946)
: private matters whose public exposure brings distress and embarrassment — called also *dirty laundry*

dirty old man *noun* (1932)
: a lecherous older man

dirty pool *noun* (1940)
: underhanded or unsportsmanlike conduct

dirty rice *noun* (circa 1967)
: a Cajun dish of white rice cooked with chopped or ground giblets

dirty word *noun* (1842)
: a word, expression, or idea that is disagreeable or unpopular in a particular frame of reference

dis \ˈdis\ *transitive verb* **dissed; dis·sing** [perhaps short for *disrespect*] (1986)
1 *slang* : to treat with disrespect or contempt : INSULT
2 *slang* : to find fault with : CRITICIZE

Dis \ˈdis\ *noun* [Latin]
: the Roman god of the underworld — compare PLUTO

dis- *prefix* [Middle English *dis-, des-,* from Old French & Latin; Old French *des-, dis-,* from Latin *dis-,* literally, apart; akin to Old English *te-* apart, Latin *duo* two — more at TWO]
1 a : do the opposite of ⟨*dis*establish⟩ **b** : deprive of (a specified quality, rank, or object) ⟨*dis*franchise⟩ **c** : exclude or expel from ⟨*dis*bar⟩
2 : opposite or absence of ⟨*dis*union⟩ ⟨*dis*affection⟩
3 : not ⟨*dis*agreeable⟩
4 : completely ⟨*dis*annul⟩
5 [by folk etymology] : DYS- ⟨*dis*function⟩

dis·abil·i·ty \ˌdi-sə-ˈbi-lə-tē\ *noun* (1580)
1 a : the condition of being disabled **b** : inability to pursue an occupation because of physical or mental impairment
2 : lack of legal qualification to do something
3 : a disqualification, restriction, or disadvantage

dis·able \di-ˈsā-bəl, di-ˈzā-\ *transitive verb* **dis·abled; dis·abling** \-b(ə-)liŋ\ (15th century)
1 : to deprive of legal right, qualification, or capacity
2 : to make incapable or ineffective; *especially* : to deprive of physical, moral, or intellectual strength
synonym see WEAKEN
— **dis·able·ment** \-bəl-mənt\ *noun*

dis·abled *adjective* (1633)

☆ SYNONYMS

Dirty, filthy, foul, nasty, squalid mean conspicuously unclean or impure. DIRTY emphasizes the presence of dirt more than an emotional reaction to it ⟨a *dirty* littered street⟩. FILTHY carries a strong suggestion of offensiveness and typically of gradually accumulated dirt that begrimes and besmears ⟨a stained greasy floor, utterly *filthy*⟩. FOUL implies extreme offensiveness and an accumulation of what is rotten or stinking ⟨a *foul*-smelling open sewer⟩. NASTY applies to what is actually foul or is repugnant to one expecting freshness, cleanliness, or sweetness ⟨it's a *nasty* job to clean up after a sick cat⟩. In practice, *nasty* is often weakened to the point of being no more than a synonym of *unpleasant* or *disagreeable* ⟨had a *nasty* fall⟩ ⟨his answer gave her a *nasty* shock⟩. SQUALID adds to the idea of dirtiness and filth that of slovenly neglect ⟨*squalid* slums⟩. All these terms are also applicable to moral uncleanness or baseness or obscenity. DIRTY then stresses meanness or despicableness ⟨don't ask me to do your *dirty* work⟩, while FILTHY and FOUL describe disgusting obscenity or loathsome behavior ⟨*filthy* street language⟩ ⟨a *foul* story of lust and greed⟩, and NASTY implies a peculiarly offensive unpleasantness ⟨a stand-up comedian known for *nasty* humor⟩. Distinctively, SQUALID implies sordidness as well as baseness and dirtiness ⟨engaged in a series of *squalid* affairs⟩.

◇ WORD HISTORY

dirge *Dirge* is a borrowed word whose English meaning is derived from its use, as distinct from its meaning, in its original language. *Dirge* and its earlier form *dirige,* meaning "a song or hymn of mourning," come from the first word of a Latin chant used in the church service for the dead: "Dirige, Domine deus meus, in conspectu tuo viam meam." ("Direct, O Lord my God, my way in thy sight.") Because hymns and chants were often referred to by their first words, *dirge* became the common word for this chant. Later it was used for any slow, solemn piece of music.

\ə\ **abut** \ᵊ\ **kitten** \ər\ **further** \a\ **ash** \ā\ **ace** \ä\ **mop, mar** \au̇\ **out** \ch\ **chin** \e\ **bet** \ē\ **easy** \g\ **go** \i\ **hit** \ī\ **ice** \j\ **job** \ŋ\ **sing** \ō\ **go** \ȯ\ **law** \ȯi\ **boy** \th\ **thin** \t͟h\ **the** \ü\ **loot** \u̇\ **foot** \y\ **yet** \zh\ **vision** *see also* Guide to Pronunciation

: incapacitated by illness, injury, or wounds; *broadly* **:** physically or mentally impaired

dis·abuse \ˌdi-sə-ˈbyüz\ *transitive verb* [French *désabuser,* from *dés-* dis- + *abuser* to abuse] (1611)
: to free from error, fallacy, or misconception

di·sac·cha·ri·dase \(ˌ)dī-ˈsa-kə-rə-ˌdās, -ˌdāz\ *noun* (1961)
: an enzyme (as maltase or lactase) that hydrolyzes disaccharides

di·sac·cha·ride \(ˌ)dī-ˈsa-kə-ˌrīd\ *noun* (1892)
: any of a class of sugars (as sucrose) that yields on hydrolysis two monosaccharide molecules

dis·ac·cord \ˌdi-sə-ˈkȯrd\ *intransitive verb* [Middle English *disacorden,* from Middle French *desacorder,* from *desacort* disagreement, from *des-* dis- + *acort* accord] (15th century)
: CLASH, DISAGREE
— disaccord *noun*

dis·ac·cus·tom \ˌdi-sə-ˈkəs-təm\ *transitive verb* [Middle French *desaccoustumer,* from Old French *desacostumer,* from *des-* dis- + *acostumer* to accustom] (1530)
: to free from a habit

[1]dis·ad·van·tage \ˌdi-səd-ˈvan-tij\ *noun* [Middle English *disavauntage,* from Middle French *desavantage,* from Old French, from *des-* dis- + *avantage* advantage] (14th century)
1 : loss or damage especially to reputation, credit, or finances **:** DETRIMENT
2 a : an unfavorable, inferior, or prejudicial condition ⟨we were at a *disadvantage*⟩ **b :** HANDICAP ⟨it put us under a serious *disadvantage*⟩

[2]disadvantage *transitive verb* (circa 1534)
: to place at a disadvantage **:** HARM

dis·ad·van·taged \-tijd\ *adjective* (1648)
: lacking in the basic resources or conditions (as standard housing, medical and educational facilities, and civil rights) believed to be necessary for an equal position in society
— dis·ad·van·taged·ness \-tijd-nəs\ *noun*

dis·ad·van·ta·geous \(ˌ)dis-ˌad-ˌvan-ˈtā-jəs, -vən-\ *adjective* (1603)
1 : constituting a disadvantage
2 : DEROGATORY, DISPARAGING
— dis·ad·van·ta·geous·ly *adverb*
— dis·ad·van·ta·geous·ness *noun*

dis·af·fect \ˌdi-sə-ˈfekt\ *transitive verb* (1641)
: to alienate the affection or loyalty of
synonym see ESTRANGE
— dis·af·fec·tion \-ˈfek-shən\ *noun*

dis·af·fect·ed *adjective* (1632)
: discontented and resentful especially against authority **:** REBELLIOUS

dis·af·fil·i·ate \ˌdi-sə-ˈfi-lē-ˌāt\ (1870)
transitive verb
: DISASSOCIATE
intransitive verb
: to terminate an affiliation
— dis·af·fil·i·a·tion \-ˌfil-ē-ˈā-shən\ *noun*

dis·af·firm \ˌdi-sə-ˈfərm\ *transitive verb* (1531)
1 : to refuse to confirm **:** ANNUL, REPUDIATE
2 : CONTRADICT
— dis·af·fir·mance \-ˈfər-mən(t)s\ *noun*

dis·ag·gre·gate \(ˌ)dis-ˈa-gri-ˌgāt\ (circa 1828)
transitive verb
: to separate into component parts ⟨*disaggregate* sandstone⟩ ⟨*disaggregate* demographic data⟩
intransitive verb
: to break up or apart ⟨the molecules of a gel *disaggregate* to form a sol⟩
— dis·ag·gre·ga·tion \(ˌ)dis-ˌa-gri-ˈgā-shən\ *noun*
— dis·ag·gre·ga·tive \(ˌ)dis-ˈa-gri-ˌgā-tiv\ *adjective*

dis·agree \ˌdi-sə-ˈgrē\ *intransitive verb* [Middle English, from Middle French *desagreer,* from *des-* dis- + *agreer* to agree] (15th century)
1 : to fail to agree ⟨the two accounts *disagree*⟩
2 : to differ in opinion ⟨he *disagreed* with me on every topic⟩
3 : to cause discomfort or distress ⟨fried foods *disagree* with me⟩

dis·agree·able \-ə-bəl\ *adjective* (15th century)
1 : causing discomfort **:** UNPLEASANT, OFFENSIVE
2 : marked by ill temper **:** PEEVISH
— dis·agree·able·ness *noun*
— dis·agree·ably \-blē\ *adverb*

dis·agree·ment \ˌdi-sə-ˈgrē-mənt\ *noun* (15th century)
1 : the act of disagreeing
2 a : the state of being at variance **:** DISPARITY **b :** QUARREL

dis·al·low \ˌdi-sə-ˈlau̇\ *transitive verb* (14th century)
1 : to deny the force, truth, or validity of
2 : to refuse to allow
— dis·al·low·ance \-ən(t)s\ *noun*

dis·am·big·u·ate \ˌdi-sam-ˈbi-gyə-ˌwāt\ *transitive verb* **-at·ed; -at·ing** (1963)
: to establish a single semantic or grammatical interpretation for
— dis·am·big·u·a·tion \-ˌbi-gyə-ˈwā-shən\ *noun*

dis·an·nul \ˌdi-sə-ˈnəl\ *transitive verb* (15th century)
: ANNUL, CANCEL

dis·ap·pear \ˌdi-sə-ˈpir\ (15th century)
intransitive verb
1 : to pass from view
2 : to cease to be **:** pass out of existence or notice
transitive verb
: to cause the disappearance of
— dis·ap·pear·ance \-ˈpir-ən(t)s\ *noun*

dis·ap·point \ˌdi-sə-ˈpȯint\ *verb* [Middle English *disapoynten,* from Middle French *desapointier,* from *des-* dis- + *apointier* to arrange — more at APPOINT] (15th century)
transitive verb
: to fail to meet the expectation or hope of **:** FRUSTRATE
intransitive verb
: to cause disappointment ⟨where the show *disappoints* most is in the work of the younger generation —John Ashbery⟩

dis·ap·point·ed *adjective* (1537)
1 : defeated in expectation or hope
2 *obsolete* **:** not adequately equipped
— dis·ap·point·ed·ly *adverb*

dis·ap·point·ing *adjective* (1530)
: failing to meet expectations
— dis·ap·point·ing·ly \-ˈpȯin-tiŋ-lē\ *adverb*

dis·ap·point·ment \ˌdi-sə-ˈpȯint-mənt\ *noun* (1614)
1 : the act or an instance of disappointing **:** the state or emotion of being disappointed
2 : one that disappoints

dis·ap·pro·ba·tion \(ˌ)di-ˌsa-prə-ˈbā-shən\ *noun* (1647)
: the act or state of disapproving **:** the state of being disapproved **:** CONDEMNATION

dis·ap·prov·al \ˌdi-sə-ˈprü-vəl\ *noun* (1662)
: DISAPPROBATION, CENSURE

dis·ap·prove \-ˈprüv\ (1647)
transitive verb
1 : to pass unfavorable judgment on
2 : to refuse approval to **:** REJECT
intransitive verb
: to feel or express disapproval
— dis·ap·prov·er *noun*
— dis·ap·prov·ing·ly \-ˈprü-viŋ-lē\ *adverb*

dis·arm \di-ˈsärm, -ˈzärm, ˈdi-ˌsärm\ *verb* [Middle English *desarmen,* literally, to divest of arms, from Middle French *desarmer,* from Old French, from *des-* dis- + *armer* to arm] (14th century)
transitive verb
1 a : to deprive of means, reason, or disposition to be hostile **b :** to win over
2 a : to divest of arms **b :** to deprive of a means of attack or defense **c :** to make harmless
intransitive verb
1 : to lay aside arms
2 : to give up or reduce armed forces
— dis·ar·ma·ment \-ˈsär-mə-mənt, -ˈzär-\ *noun*
— dis·arm·er *noun*

dis·arm·ing *adjective* (1839)
: allaying criticism or hostility **:** INGRATIATING
— dis·arm·ing·ly \-ˈsär-miŋ-lē, -ˈzär-\ *adverb*

dis·ar·range \ˌdi-sə-ˈrānj\ *transitive verb* (1744)
: to disturb the arrangement or order of
— dis·ar·range·ment \-mənt\ *noun*

[1]dis·ar·ray \ˌdi-sə-ˈrā\ *noun* (15th century)
1 : a lack of order or sequence **:** CONFUSION, DISORDER
2 : disorderly dress **:** DISHABILLE

[2]disarray *transitive verb* [Middle English *disarayen,* from Middle French *desarroyer,* from Old French *desareer,* from *des-* dis- + *areer* to array] (14th century)
1 : to throw into disorder
2 : UNDRESS

dis·ar·tic·u·late \ˌdi-sär-ˈti-kyə-ˌlāt\ (1830)
intransitive verb
: to become disjointed
transitive verb
: DISJOINT
— dis·ar·tic·u·la·tion \-ˌti-kyə-ˈlā-shən\ *noun*

dis·as·sem·ble \ˌdi-sə-ˈsem-bəl\ (1903)
transitive verb
: to take apart ⟨*disassemble* a watch⟩
intransitive verb
1 : to come apart ⟨the frame *disassembles* into sections⟩
2 : DISPERSE, SCATTER ⟨the crowd began to *disassemble*⟩
— dis·as·sem·bly \-blē\ *noun*

dis·as·so·ci·ate \ˌdi-sə-ˈsō-sē-ˌāt, -shē-\ *transitive verb* (1603)
: to detach from association **:** DISSOCIATE
— dis·as·so·ci·a·tion \-ˌsō-sē-ˈā-shən, -shē-\ *noun*

di·sas·ter \di-ˈzas-tər, -ˈsas-\ *noun* [Middle French & Old Italian; Middle French *desastre,* from Old Italian *disastro,* from *dis-* (from Latin) + *astro* star, from Latin *astrum* — more at ASTRAL] (1591)
1 : a sudden calamitous event bringing great damage, loss, or destruction; *broadly* **:** a sudden or great misfortune or failure
2 *obsolete* **:** an unfavorable aspect of a planet or star ◆

disaster area *noun* (1953)

◇ WORD HISTORY

disaster People who are beset by bad luck are regularly described as "star-crossed." This expression comes from the traditional belief that the position of the stars and planets can have a direct influence on earthly events. The origins of the word *disaster* can likewise be traced to this belief. *Disaster* is ultimately borrowed from *disastro,* an Italian word formed by compounding the negative prefix *dis-* and the noun *astro,* meaning "star," *Disastro* originally referred to an unfortunate event, such as a military defeat, that took place under an unlucky conjunction of heavenly bodies. Eventually the astrological connection became superfluous, and *disastro,* like English *disaster,* was used for any calamity.

: an area officially declared to be the scene of an emergency created by a disaster and therefore qualified to receive certain types of governmental aid (as emergency loans and relief supplies)

di·sas·trous \di-ˈzas-trəs *also* -ˈsas-\ *adjective* (1603)
1 : attended by or causing suffering or disaster **:** CALAMITOUS
2 : TERRIBLE, HORRENDOUS ⟨a *disastrous* score⟩
— di·sas·trous·ly *adverb*

dis·avow \ˌdi-sə-ˈvau̇\ *transitive verb* [Middle English *desavowen,* from Middle French *desavouer,* from Old French, from *des-* dis- + *avouer* to avow] (14th century)
1 : to deny responsibility for **:** REPUDIATE
2 : to refuse to acknowledge **:** DISCLAIM
— dis·avow·able \-ə-bəl\ *adjective*
— dis·avow·al \-ˈvau̇(-ə)l\ *noun*

dis·band \dis-ˈband\ *verb* [Middle French *desbander,* from *des-* dis- + *bande* band] (1591)
transitive verb
: to break up the organization of **:** DISSOLVE
intransitive verb
: to break up as an organization **:** DISPERSE
— dis·band·ment \-ˈban(d)-mənt\ *noun*

dis·bar \dis-ˈbär\ *transitive verb* (1633)
: to expel from the bar or the legal profession **:** deprive (an attorney) of legal status and privileges
— dis·bar·ment \-mənt\ *noun*

dis·be·lief \ˌdis-bə-ˈlēf\ *noun* (1672)
: the act of disbelieving **:** mental rejection of something as untrue

dis·be·lieve \-ˈlēv\ (circa 1644)
transitive verb
: to hold not worthy of belief **:** not believe
intransitive verb
: to withhold or reject belief
— dis·be·liev·er *noun*

dis·ben·e·fit \(ˌ)dis-ˈbe-nə-fit\ *noun* (1968)
: something disadvantageous or objectionable **:** DRAWBACK

dis·bud \(ˌ)dis-ˈbəd\ *transitive verb* (1727)
1 : to thin out flower buds in order to improve the quality of bloom of
2 : to dehorn (cattle) by destroying the undeveloped horn bud

dis·bur·den \(ˌ)dis-ˈbər-dᵊn\ (1532)
transitive verb
1 a : to rid of a burden ⟨*disburden* a pack animal⟩ **b :** UNBURDEN ⟨*disburden* your conscience⟩
2 : UNLOAD ⟨*disburdened* their merchandise in the town square⟩
intransitive verb
: DISCHARGE ⟨the vessels *disburdened* at the dock⟩
— dis·bur·den·ment \-mənt\ *noun*

dis·burse \dis-ˈbərs\ *transitive verb* **dis·bursed; dis·burs·ing** [Middle French *desbourser,* from Old French *desborser,* from *des-* dis- + *borser* to get money, from *borse* burse] (1530)
1 a : to pay out **:** expend especially from a fund **b :** to make a payment in settlement of
2 : DISTRIBUTE
— dis·burs·er *noun*

dis·burse·ment \-ˈbər-smənt\ *noun* (1596)
: the act of disbursing; *also* **:** funds paid out

disc *variant of* DISK

disc- *or* **disci-** *or* **disco-** *combining form* [Latin, from Greek *disk-, disko-,* from *diskos*]
1 : disk ⟨*discoid*⟩
2 : phonograph record ⟨*discophile*⟩

dis·calced \(ˌ)dis-ˈkalst\ *adjective* [part translation of Latin *discalceatus,* from *dis-* + *calceatus,* past participle of *calceare* to put on shoes, from *calceus* shoe, from *calc-, calx* heel] (1631)
: UNSHOD, BAREFOOT ⟨*discalced* friars⟩

dis·cant \ˈdis-ˌkant\ *variant of* DESCANT

¹dis·card \dis-ˈkärd, ˈdis-ˌ\ (circa 1586)
transitive verb
1 : to get rid of especially as useless or unpleasant
2 a : to remove (a playing card) from one's hand **b :** to play (any card except a trump) from a suit different from the one led
intransitive verb
: to discard a playing card ☆
— dis·card·able \-də-bəl\ *adjective*
— dis·card·er *noun*

²dis·card \ˈdis-ˌkärd\ *noun* (1744)
1 a : the act of discarding in a card game **b :** a card discarded
2 : one that is cast off or rejected

dis·car·nate \dis-ˈkär-nət, -ˌnāt\ *adjective* [*dis-* + *-carnate* (as in *incarnate*)] (1895)
: having no physical body **:** INCORPOREAL

disc brake *noun* (1904)
: a brake that operates by the friction of a caliper pressing against the sides of a rotating disc

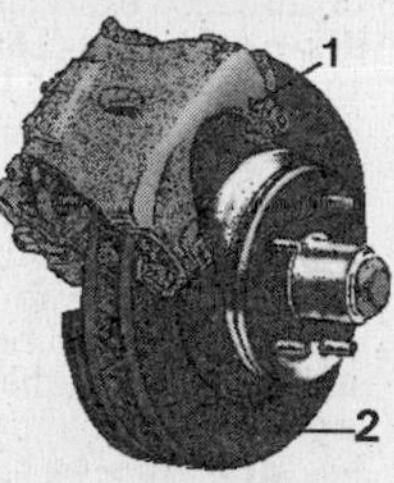

disc brake: *1* caliper, *2* disc

dis·cern \di-ˈsərn, -ˈzərn\ *verb* [Middle English, from Middle French *discerner,* from Latin *discernere* to separate, distinguish between, from *dis-* apart + *cernere* to sift — more at DIS-, CERTAIN] (14th century)
transitive verb
1 a : to detect with the eyes **b :** to detect with senses other than vision
2 : to recognize or identify as separate and distinct **:** DISCRIMINATE
3 : to come to know or recognize mentally
intransitive verb
: to see or understand the difference
— dis·cern·er *noun*
— dis·cern·ible *also* **dis·cern·able** \-ˈsər-nə-bəl, -ˈzər-\ *adjective*
— dis·cern·ibly \-blē\ *adverb*

dis·cern·ing *adjective* (1589)
: showing insight and understanding **:** DISCRIMINATING ⟨a *discerning* critic⟩
— dis·cern·ing·ly \-ˈsər-niŋ-lē, -ˈzər-\ *adverb*

dis·cern·ment \di-ˈsərn-mənt, -ˈzərn-\ *noun* (1586)
1 : the quality of being able to grasp and comprehend what is obscure **:** skill in discerning
2 : an act of discerning ☆

¹dis·charge \dis-ˈchärj, ˈdis-ˌ\ *verb* [Middle English, from Middle French *descharger,* from Late Latin *discarricare,* from Latin *dis-* + Late Latin *carricare* to load — more at CHARGE] (14th century)
transitive verb
1 : to relieve of a charge, load, or burden: **a :** UNLOAD **b :** to release from an obligation **c :** to release electrical energy from (as a battery or capacitor) by a discharge
2 a : to let or put off ⟨*discharge* passengers⟩ ⟨*discharge* cargo⟩ **b :** SHOOT ⟨*discharge* an arrow⟩ **c :** to release from confinement, custody, or care ⟨*discharge* a prisoner⟩ **d :** to give outlet or vent to **:** EMIT
3 a (1) **:** to dismiss from employment (2) **:** to release from service or duty ⟨*discharge* a soldier⟩ **b :** to get rid of (as a debt or obligation) by performing an appropriate action (as payment) **c :** to set aside **:** ANNUL **d :** to order (a legislative committee) to end consideration of a bill in order to bring it before the house for action
4 : to bear and distribute (as the weight of a wall above an opening)
5 : to bleach out or remove (color or dye) in dyeing and printing textiles
6 : to cancel the record of the loan of (a library book) upon return
intransitive verb
1 a : to throw off or deliver a load, charge, or burden **b :** to release electrical energy by a discharge
2 a : GO OFF, FIRE — used of a gun **b :** SPREAD, RUN ⟨some dyes *discharge*⟩ **c :** to pour forth fluid or other contents
synonym see PERFORM
— dis·charge·able \-jə-bəl\ *adjective*
— dis·charg·ee \(ˌ)dis-ˌchär-ˈjē\ *noun*
— dis·charg·er \dis-ˈchär-jər, ˈdis-ˌ\ *noun*

²dis·charge \ˈdis-ˌchärj, dis-ˈ\ *noun* (14th century)
1 a : the act of relieving of something that oppresses **:** RELEASE **b :** something that discharges or releases; *especially* **:** a certification of release or payment
2 : the state of being discharged or relieved
3 : the act of discharging or unloading
4 : legal release from confinement
5 : a firing off
6 a : a flowing or issuing out ⟨a *discharge* of spores⟩; *also* **:** a rate of flow **b :** something that is emitted ⟨a purulent *discharge*⟩
7 : the act of removing an obligation or liability
8 a : release or dismissal especially from an office or employment **b :** complete separation from military service
9 a : the equalization of a difference of electric potential between two points **b :** the conversion of the chemical energy of a battery into electrical energy

discharge lamp *noun* (1936)
: an electric lamp in which an enclosed gas or vapor glows or causes a phosphor coating on the lamp's inner surface to glow

discharge tube *noun* (1898)

☆ SYNONYMS

Discard, cast, shed, slough, scrap, junk mean to get rid of. DISCARD implies the letting go or throwing away of something that has become useless or superfluous though often not intrinsically valueless ⟨*discard* old clothes⟩. CAST, especially when used with *off, away,* or *out,* implies a forceful rejection or repudiation ⟨*cast* off her friends⟩. SHED and SLOUGH imply a throwing off of something both useless and encumbering and often suggest a consequent renewal of vitality or luster ⟨*shed* a bad habit⟩ ⟨finally *sloughed* off the depression⟩. SCRAP and JUNK imply throwing away or breaking up as worthless in existent form ⟨*scrap* all the old ways⟩ ⟨would *junk* our educational system⟩.

Discernment, discrimination, perception, penetration, insight, acumen mean a power to see what is not evident to the average mind. DISCERNMENT stresses accuracy (as in reading character or motives or appreciating art) ⟨the *discernment* to know true friends⟩. DISCRIMINATION stresses the power to distinguish and select what is true or appropriate or excellent ⟨the *discrimination* that develops through listening to a lot of great music⟩. PERCEPTION implies quick and often sympathetic discernment (as of shades of feeling) ⟨a novelist of keen *perception* into human motives⟩. PENETRATION implies a searching mind that goes beyond what is obvious or superficial ⟨lacks the *penetration* to see the scorn beneath their friendly smiles⟩. INSIGHT suggests depth of discernment coupled with understanding sympathy ⟨a documentary providing *insight* into the plight of the homeless⟩. ACUMEN implies characteristic penetration combined with keen practical judgment ⟨a director of reliable box-office *acumen*⟩.

: an electron tube which contains gas or vapor at low pressure and through which conduction takes place when a high voltage is applied

disci- — see DISC-

dis·ci·form \'di-sə-ˌfȯrm\ *adjective* (1830)
: round or oval in shape

dis·ci·ple \di-'sī-pəl\ *noun* [Middle English, from Old English *discipul* & Old French *desciple,* from Late Latin and Latin; Late Latin *discipulus* follower of Jesus Christ in his lifetime, from Latin, pupil] (before 12th century)
1 : one who accepts and assists in spreading the doctrines of another: as **a :** one of the twelve in the inner circle of Christ's followers according to the Gospel accounts **b :** a convinced adherent of a school or individual
2 *capitalized* **:** a member of the Disciples of Christ founded in the U.S. in 1809 that holds the Bible alone to be the rule of faith and practice, usually baptizes by immersion, and has a congregational polity
synonym see FOLLOWER
— dis·ci·ple·ship \-ˌship\ *noun*

dis·ci·plin·able \ˌdi-sə-'pli-nə-bəl; 'di-sə-pli-\ *adjective* (15th century)
1 : DOCILE, TEACHABLE
2 : subject to or deserving discipline ⟨a *disciplinable* offense⟩

dis·ci·pli·nar·i·an \ˌdi-sə-plə-'ner-ē-ən\ *noun* (1639)
: one who disciplines or enforces order
— disciplinarian *adjective*

dis·ci·plin·ary \'di-sə-plə-ˌner-ē, *especially British* ˌdi-sə-'pli-nə-rē\ *adjective* (1598)
1 a : of or relating to discipline **b :** designed to correct or punish breaches of discipline ⟨took *disciplinary* action⟩
2 : of or relating to a particular field of study
— dis·ci·plin·ar·i·ly \ˌdi-sə-plə-'ner-ə-lē\ *adverb*
— dis·ci·plin·ar·i·ty \-'nar-ə-tē\ *noun*

¹dis·ci·pline \'di-sə-plən\ *noun* [Middle English, from Old French & Latin; Old French, from Latin *disciplina* teaching, learning, from *discipulus* pupil] (13th century)
1 : PUNISHMENT
2 *obsolete* **:** INSTRUCTION
3 : a field of study
4 : training that corrects, molds, or perfects the mental faculties or moral character
5 a : control gained by enforcing obedience or order **b :** orderly or prescribed conduct or pattern of behavior **c :** SELF-CONTROL
6 : a rule or system of rules governing conduct or activity
— dis·ci·plin·al \-plə-nᵊl\ *adjective*

²discipline *transitive verb* **-plined; -plin·ing** (14th century)
1 : to punish or penalize for the sake of discipline
2 : to train or develop by instruction and exercise especially in self-control
3 a : to bring (a group) under control ⟨*discipline* troops⟩ **b :** to impose order upon ⟨serious writers *discipline* and refine their writing styles⟩
synonym see PUNISH, TEACH
— dis·ci·plin·er *noun*

dis·ci·plined *adjective* (14th century)
: marked by or possessing discipline ⟨a *disciplined* mind⟩

disc jockey *noun* (1941)
: an announcer of a radio show of popular recorded music; *also* **:** one who plays recorded music for dancing at a nightclub or party

dis·claim \dis-'klām\ *verb* [Middle English, from Anglo-French *disclaimer,* from *dis-* + *claimer* to claim, from Old French *clamer*] (15th century)
intransitive verb
1 : to make a disclaimer
2 a *obsolete* **:** to disavow all part or share **b :** to utter denial
transitive verb
1 : to renounce a legal claim to
2 : DENY, DISAVOW ⟨*disclaimed* any knowledge of the contents of the letter⟩

dis·claim·er \-'klā-mər\ *noun* (15th century)
1 a : a denial or disavowal of legal claim **:** relinquishment of or formal refusal to accept an interest or estate **b :** a writing that embodies a legal disclaimer
2 a : DENIAL, DISAVOWAL **b :** REPUDIATION

dis·cla·ma·tion \ˌdis-klə-'mā-shən\ *noun* (1592)
: RENUNCIATION, DISAVOWAL

dis·cli·max \(ˌ)dis-'klī-ˌmaks\ *noun* (1935)
: a relatively stable ecological community often including kinds of organisms foreign to the region and displacing the climax because of disturbance especially by man

¹dis·close \dis-'klōz\ *transitive verb* [Middle English, from Middle French *desclos-,* stem of *desclore* to disclose, from Medieval Latin *disclaudere* to open, from Latin *dis-* + *claudere* to close — more at CLOSE] (14th century)
1 *obsolete* **:** to open up
2 a : to expose to view **b** *archaic* **:** HATCH **c :** to make known or public ⟨demands that politicians *disclose* the sources of their income⟩
synonym see REVEAL
— dis·clos·er *noun*

²disclose *noun* (1548)
obsolete **:** DISCLOSURE

dis·clos·ing \-'klō-ziŋ\ *adjective* (1965)
: being or using an agent (as a tablet or liquid) that contains a usually red dye that stains dental plaque

dis·clo·sure \dis-'klō-zhər\ *noun* (circa 1598)
1 : the act or an instance of disclosing **:** EXPOSURE
2 : something disclosed **:** REVELATION

¹dis·co \'dis-(ˌ)kō\ *noun, plural* **discos** [short for *discotheque*] (1964)
1 : a nightclub for dancing to live and recorded music
2 : popular dance music characterized by hypnotic rhythm, repetitive lyrics, and electronically produced sounds

²disco *intransitive verb* (1979)
: to dance to disco music

disco- — see DISC-

dis·cog·ra·pher \dis-'kä-grə-fər\ *noun* (1941)
: a person who compiles discographies

dis·cog·ra·phy \-fē\ *noun, plural* **-phies** [French *discographie,* from *disc-* + *-graphie* -graphy] (1933)
1 : a descriptive list of phonograph records by category, composer, performer, or date of release
2 : the history of recorded music
— dis·co·graph·i·cal \ˌdis-kə-'gra-fi-kəl\ *also* **dis·co·graph·ic** \-fik\ *adjective*

dis·coid \'dis-ˌkȯid\ *adjective* [Late Latin *discoides* quoit-shaped, from Greek *diskoeidēs,* from *diskos* disk] (1794)
1 : relating to or having a disk: as **a** *of a composite floret* **:** situated in the floral disk **b** *of a composite flower head* **:** having only tubular florets
2 : flat and circular like a disc

dis·coi·dal \dis-'kȯi-dᵊl\ *adjective* (circa 1706)
: of, resembling, or producing a disk

discoidal cleavage *noun* (circa 1909)
: meroblastic cleavage in which a disk of cells is produced at the animal pole of the zygote (as in bird eggs)

dis·col·or \(ˌ)dis-'kə-lər\ *verb* [Middle English *discolouren,* from Middle French *descolourer,* from Late Latin *discolorari,* from Latin *discolor* of another color, from *dis-* + *color* color] (14th century)
transitive verb
: to alter or change the hue or color of
intransitive verb
: to change color especially for the worse

dis·col·or·ation \(ˌ)dis-ˌkə-lə-'rā-shən\ *noun* (1642)
1 : the act of discoloring **:** the state of being discolored
2 : a discolored spot or formation **:** STAIN

dis·com·bob·u·late \ˌdis-kəm-'bä-b(y)ə-ˌlāt\ *transitive verb* **-lat·ed; -lat·ing** [probably alteration of *discompose*] (circa 1916)
: UPSET, CONFUSE ⟨the offensive had *discombobulated* all the German defensive arrangements —A. J. Liebling⟩
— dis·com·bob·u·la·tion \-ˌbä-b(y)ə-'lā-shən\ *noun*

¹dis·com·fit \dis-'kəm(p)-fət, *especially Southern* ˌdis-kəm-'fit\ *transitive verb* [Middle English, from Old French *desconfit,* past participle of *desconfire,* from *des-* dis- + *confire* to prepare — more at COMFIT] (13th century)
1 a *archaic* **:** to defeat in battle **b :** to frustrate the plans of **:** THWART
2 : to put into a state of perplexity and embarrassment **:** DISCONCERT
synonym see EMBARRASS

²discomfit *noun* (15th century)
: DISCOMFITURE

dis·com·fi·ture \dis-'kəm(p)-fə-ˌchu̇r, -chər, *especially Southern* -ˌt(y)u̇r\ *noun* (14th century)
: the act of discomfiting **:** the state of being discomfited

¹dis·com·fort \dis-'kəm(p)-fərt\ *transitive verb* [Middle English, from Middle French *desconforter,* from Old French, from *des-* dis- + *conforter* to comfort] (14th century)
1 *archaic* **:** DISMAY
2 : to make uncomfortable or uneasy
— dis·com·fort·able \-'kəm(p)-fər-tə-bəl, -'kəm(p)f-tər-bəl\ *adjective*

²discomfort *noun* (14th century)
1 *archaic* **:** DISTRESS, GRIEF
2 : mental or physical uneasiness **:** ANNOYANCE

dis·com·mend \ˌdis-kə-'mend\ *transitive verb* [Middle English *dyscommenden*] (15th century)
1 : DISAPPROVE, DISPARAGE
2 : to cause to be viewed unfavorably

dis·com·mode \ˌdis-kə-'mōd\ *transitive verb* **-mod·ed; -mod·ing** [Middle French *discommoder,* from *dis-* + *commode* convenient — more at COMMODE] (circa 1721)
: to cause inconvenience to **:** TROUBLE

dis·com·pose \ˌdis-kəm-'pōz\ *transitive verb* [Middle English] (15th century)
1 : to destroy the composure of
2 : to disturb the order of ☆
— dis·com·po·sure \-'pō-zhər\ *noun*

dis·con·cert \ˌdis-kən-'sərt\ *transitive verb* [obsolete French *disconcerter,* alteration of Middle French *desconcerter,* from *des-* dis- + *concerter* to concert] (1687)
1 : to throw into confusion

☆ SYNONYMS
Discompose, disquiet, disturb, perturb, agitate, upset, fluster mean to destroy capacity for collected thought or decisive action. DISCOMPOSE implies some degree of loss of self-control or self-confidence especially through emotional stress ⟨*discomposed* by the loss of his beloved wife⟩. DISQUIET suggests loss of sense of security or peace of mind ⟨the *disquieting* news of factories closing⟩. DISTURB implies interference with one's mental processes caused by worry, perplexity, or interruption ⟨the discrepancy in accounts *disturbed* me⟩. PERTURB implies deep disturbance of mind and emotions ⟨*perturbed* by her husband's strange behavior⟩. AGITATE suggests obvious external signs of nervous or emotional excitement ⟨in his *agitated* state we could see he was unable to work⟩. UPSET implies the disturbance of normal or habitual functioning by disappointment, distress, or grief ⟨the family's constant bickering *upsets* the youngest child⟩. FLUSTER suggests bewildered agitation ⟨his declaration of love completely *flustered* her⟩.

2 : to disturb the composure of
synonym see EMBARRASS
— **dis·con·cert·ing** *adjective*
— **dis·con·cert·ing·ly** \-tiŋ-lē\ *adverb*
— **dis·con·cert·ment** \-mənt\ *noun*

dis·con·firm \ˌdis-kən-ˈfərm\ *transitive verb* (1936)
: to deny the validity of

dis·con·for·mi·ty \ˌdis-kən-ˈfȯr-mə-tē\ *noun* (1587)
1 : NONCONFORMITY
2 : a break in a sequence of sedimentary rocks all of which have approximately the same dip

dis·con·nect \ˌdis-kə-ˈnekt\ (1770)
transitive verb
: to sever the connection of or between
intransitive verb
1 : to terminate a connection
2 : to become detached or withdrawn ⟨*disconnects* into dark moods⟩
— **dis·con·nec·tion** \-ˈnek-shən\ *noun*

dis·con·nect·ed *adjective* (1783)
: not connected **:** SEPARATE; *also* **:** INCOHERENT
— **dis·con·nect·ed·ly** *adverb*
— **dis·con·nect·ed·ness** *noun*

dis·con·so·late \dis-ˈkän(t)-sə-lət\ *adjective* [Middle English, from Medieval Latin *disconsolatus,* from Latin *dis-* + *consolatus,* past participle of *consolari* to console] (14th century)
1 : CHEERLESS ⟨a clutch of *disconsolate* houses —D. H. Lawrence⟩
2 : DEJECTED, DOWNCAST ⟨the team returned *disconsolate* from three losses⟩
— **dis·con·so·late·ly** *adverb*
— **dis·con·so·late·ness** *noun*
— **dis·con·so·la·tion** \(ˌ)dis-ˌkän(t)-sə-ˈlā-shən\ *noun*

¹dis·con·tent \ˌdis-kən-ˈtent\ *adjective* (15th century)
: DISCONTENTED

²discontent *transitive verb* (1549)
: to make discontented
— **dis·con·tent·ment** \-mənt\ *noun*

³discontent *noun* (1591)
: lack of contentment: **a :** a sense of grievance **:** DISSATISFACTION ⟨the winter of our *discontent* —Shakespeare⟩ **b :** restless aspiration for improvement

⁴discontent *noun* (1596)
: one who is discontented **:** MALCONTENT

dis·con·tent·ed *adjective* (1525)
: DISSATISFIED, MALCONTENT
— **dis·con·tent·ed·ly** *adverb*
— **dis·con·tent·ed·ness** *noun*

dis·con·tinu·ance \ˌdis-kən-ˈtin-yə-wən(t)s\ *noun* (14th century)
1 : the act or an instance of discontinuing
2 : the interruption or termination of a legal action by the plaintiff's not continuing it

dis·con·tin·ue \ˌdis-kən-ˈtin-(ˌ)yü, -yə(-w)\ *verb* [Middle English, from Middle French *discontinuer,* from Medieval Latin *discontinuare,* from Latin *dis-* + *continuare* to continue] (14th century)
transitive verb
1 : to break the continuity of **:** cease to operate, administer, use, produce, or take
2 : to abandon or terminate by a legal discontinuance
intransitive verb
: to come to an end
synonym see STOP
— **dis·con·tin·u·a·tion** \-kən-ˌtin-yə-ˈwā-shən\ *noun*

dis·con·ti·nu·ity \(ˌ)dis-ˌkän-tə-ˈnü-ə-tē, -ˈnyü-\ *noun* (1570)
1 : lack of continuity or cohesion
2 : GAP 5
3 a : the property of being not mathematically continuous ⟨a point of *discontinuity*⟩ **b :** an instance of being not mathematically continuous; *especially* **:** a value of an independent variable at which a function is not continuous

dis·con·tin·u·ous \ˌdis-kən-ˈtin-yə-wəs\ *adjective* (1718)
1 a (1) **:** not continuous ⟨a *discontinuous* series of events⟩ (2) **:** not continued **:** DISCRETE ⟨*discontinuous* features of terrain⟩ **b :** lacking sequence or coherence
2 : having one or more mathematical discontinuities — used of a variable or a function
— **dis·con·tin·u·ous·ly** *adverb*

dis·co·phile \ˈdis-kə-ˌfīl\ *noun* (1940)
: one who studies and collects phonograph records

¹dis·cord \ˈdis-ˌkȯrd\ *noun* (13th century)
1 a : lack of agreement or harmony (as between persons, things, or ideas) **b :** active quarreling or conflict resulting from discord among persons or factions **:** STRIFE
2 a (1) **:** a combination of musical sounds that strikes the ear harshly (2) **:** DISSONANCE **b :** a harsh or unpleasant sound ☆

²dis·cord \ˈdis-ˌkȯrd, dis-ˈ\ *intransitive verb* [Middle English, from Old French *discorder,* from Latin *discordare,* from *discord-, discors* discordant, from *dis-* + *cord-, cor* heart — more at HEART] (14th century)
: DISAGREE, CLASH

dis·cor·dance \dis-ˈkȯr-dᵊn(t)s\ *noun* (14th century)
1 : the state or an instance of being discordant
2 : DISSONANCE

dis·cor·dan·cy \-dᵊn-sē\ *noun, plural* **-cies** (1607)
: DISCORDANCE

dis·cor·dant \-dᵊnt\ *adjective* (14th century)
1 a : being at variance **:** DISAGREEING **b :** QUARRELSOME
2 : relating to a discord
— **dis·cor·dant·ly** *adverb*

dis·co·theque *or* **discothèque** \ˈdis-kə-ˌtek, ˌdis-kə-ˈ\ *noun* [French *discothèque,* from *disque* disk, record + *-o-* + *-thèque* (as in *bibliothèque* library)] (1954)
: DISCO 1

¹dis·count \ˈdis-ˌkau̇nt\ *noun* (1622)
1 : a reduction made from the gross amount or value of something: as **a** (1) **:** a reduction made from a regular or list price (2) **:** a proportionate deduction from a debt account usually made for cash or prompt payment **b :** a deduction made for interest in advancing money upon or purchasing a bill or note not due
2 : the act or practice of discounting
3 : a deduction taken or allowance made

²dis·count \ˈdis-ˌkau̇nt, dis-ˈ\ *verb* [modification of French *décompter,* from Old French *desconter,* from Medieval Latin *discomputare,* from Latin *dis-* + *computare* to count — more at COUNT] (1629)
transitive verb
1 a : to make a deduction from usually for cash or prompt payment **b :** to sell or offer for sale at a discount
2 : to lend money on after deducting the discount
3 a : to leave out of account **:** DISREGARD **b :** to minimize the importance of **c** (1) **:** to make allowance for bias or exaggeration in (2) **:** to view with doubt **d :** to take into account (as a future event) in present calculations
intransitive verb
: to give or make discounts
— **dis·count·er** \-ˌkau̇n-tər, -ˈkau̇n-\ *noun*

³dis·count \ˈdis-ˌkau̇nt\ *adjective* (1889)
1 a : selling goods or services at a discount ⟨*discount* stores⟩ ⟨a *discount* broker⟩ ⟨*discount* airlines⟩ **b :** offered or sold at a discount
2 : reflecting a discount ⟨*discount* prices⟩

dis·count·able \dis-ˈkau̇n-tə-bəl, ˈdis-ˌ\ *adjective* (1800)
1 : set apart for discounting ⟨within the *discountable* period⟩
2 : subject to being discounted ⟨a *discountable* note⟩

¹dis·coun·te·nance \dis-ˈkau̇n-tᵊn-ən(t)s, -ˈkau̇nt-nən(t)s\ *transitive verb* (1580)
1 : ABASH, DISCONCERT
2 : to look with disfavor on **:** discourage by evidence of disapproval

²discountenance *noun* (1580)
: DISAPPROBATION, DISFAVOR

discount rate *noun* (circa 1927)
1 : the interest on an annual basis deducted in advance on a loan
2 : the charge levied by a central bank for advances and rediscounts

dis·cour·age \dis-ˈkər-ij, -ˈkə-rij\ *transitive verb* **-aged; -ag·ing** [Middle English *discoragen,* from Middle French *descorager,* from Old French *descoragier,* from *des-* dis- + *corage* courage] (15th century)
1 : to deprive of courage or confidence **:** DISHEARTEN
2 a : to hinder by disfavoring **b :** to attempt to dissuade
— **dis·cour·age·able** \-jə-bəl\ *adjective*
— **dis·cour·ag·er** *noun*
— **dis·cour·ag·ing·ly** \-jiŋ-lē\ *adverb*

dis·cour·age·ment \-mənt\ *noun* (1561)
1 : the act of discouraging **:** the state of being discouraged
2 : something that discourages

¹dis·course \ˈdis-ˌkōrs, -ˌkȯrs, dis-ˈ\ *noun* [Middle English *discours,* from Medieval Latin & Late Latin *discursus;* Medieval Latin, argument, from Late Latin, conversation, from Latin, act of running about, from *discurrere* to run about, from *dis-* + *currere* to run — more at CAR] (14th century)
1 *archaic* **:** the capacity of orderly thought or procedure **:** RATIONALITY
2 : verbal interchange of ideas; *especially* **:** CONVERSATION
3 a : formal and orderly and usually extended expression of thought on a subject **b :** connected speech or writing **c :** a linguistic unit (as a conversation or a story) larger than a sentence
4 *obsolete* **:** social familiarity

²dis·course \dis-ˈkōrs, -ˈkȯrs, ˈdis-ˌ\ *verb* **dis·coursed; dis·cours·ing** (1559)
intransitive verb
1 : to express oneself especially in oral discourse

☆ SYNONYMS
Discord, strife, conflict, contention, dissension, variance mean a state or condition marked by a lack of agreement or harmony. DISCORD implies an intrinsic or essential lack of harmony producing quarreling, factiousness, or antagonism ⟨a political party long racked by *discord*⟩. STRIFE emphasizes a struggle for superiority rather than the incongruity or incompatibility of the persons or things involved ⟨during his brief reign the empire was never free of civil *strife*⟩. CONFLICT usually stresses the action of forces in opposition but in static applications implies an irreconcilability as of duties or desires ⟨the *conflict* of freedom and responsibility⟩. CONTENTION applies to strife or competition that shows itself in quarreling, disputing, or controversy ⟨several points of *contention* about the new zoning law⟩. DISSENSION implies strife or discord and stresses a division into factions ⟨religious *dissension* threatened to split the colony⟩. VARIANCE implies a clash between persons or things owing to a difference in nature, opinion, or interest ⟨cultural *variances* that work against a national identity⟩.

2 : TALK, CONVERSE
transitive verb
archaic **:** to give forth **:** UTTER
— dis·cours·er *noun*

discourse analysis *noun* (1952)
: the study of linguistic relations and structures in discourse

dis·cour·te·ous \(ˌ)dis-ˈkər-tē-əs\ *adjective* (1578)
: lacking courtesy **:** RUDE
— dis·cour·te·ous·ly *adverb*
— dis·cour·te·ous·ness *noun*

dis·cour·te·sy \-ˈkər-tə-sē\ *noun* (1555)
1 : RUDENESS
2 : a rude act

dis·cov·er \dis-ˈkə-vər\ *verb* **dis·cov·ered; dis·cov·er·ing** \-ˈkə-v(ə-)riŋ\ [Middle English, from Middle French *descovrir,* from Late Latin *discooperire,* from Latin *dis-* + *cooperire* to cover — more at COVER] (14th century)
transitive verb
1 a : to make known or visible **:** EXPOSE **b** *archaic* **:** DISPLAY
2 a : to obtain sight or knowledge of for the first time **:** FIND ⟨*discover* the solution⟩ **b :** FIND OUT ⟨*discovered* he was out of gas⟩
intransitive verb
: to make a discovery ☆
— dis·cov·er·able \-ˈkə-v(ə-)rə-bəl\ *adjective*
— dis·cov·er·er \-ər-ər\ *noun*

dis·cov·ery \dis-ˈkə-v(ə-)rē\ *noun, plural* **-eries** (1539)
1 a : the act or process of discovering **b** (1) *archaic* **:** DISCLOSURE (2) *obsolete* **:** DISPLAY **c** *obsolete* **:** EXPLORATION
2 : something discovered
3 : the usually pretrial disclosure of pertinent facts or documents by one or both parties to a legal action or proceeding

Discovery Day *noun* (circa 1913)
: COLUMBUS DAY

¹dis·cred·it \(ˌ)dis-ˈkre-dət\ *transitive verb* (1559)
1 : to refuse to accept as true or accurate **:** DISBELIEVE
2 : to cause disbelief in the accuracy or authority of
3 : to deprive of good repute **:** DISGRACE

²discredit *noun* (1565)
1 : loss of credit or reputation ⟨I knew stories to the *discredit* of England —W. B. Yeats⟩
2 : lack or loss of belief or confidence **:** DOUBT ⟨contradictions cast *discredit* on his testimony⟩

dis·cred·it·able \-tə-bəl\ *adjective* (1640)
: injurious to reputation **:** DISGRACEFUL
— dis·cred·it·ably \-blē\ *adverb*

dis·creet \di-ˈskrēt\ *adjective* [Middle English, from Middle French *discret,* from Medieval Latin *discretus,* from Latin, past participle of *discernere* to separate, distinguish between — more at DISCERN] (14th century)
1 : having or showing discernment or good judgment in conduct and especially in speech **:** PRUDENT; *especially* **:** capable of preserving prudent silence
2 : UNPRETENTIOUS, MODEST ⟨the warmth and *discreet* elegance of a civilized home —Joseph Wechsberg⟩
3 : UNOBTRUSIVE, UNNOTICEABLE ⟨followed at a *discreet* distance⟩
— dis·creet·ly *adverb*
— dis·creet·ness *noun*

dis·crep·an·cy \dis-ˈkre-pən-sē\ *noun, plural* **-cies** (circa 1623)
1 : the quality or state of being discrepant
2 : an instance of being discrepant

dis·crep·ant \-pənt\ *adjective* [Middle English *discrepaunt,* from Latin *discrepant-, discrepans,* present participle of *discrepare* to sound discordantly, from *dis-* + *crepare* to rattle, creak — more at RAVEN] (15th century)
: being at variance **:** DISAGREEING ⟨widely *discrepant* conclusions⟩
— dis·crep·ant·ly *adverb*

dis·crete \dis-ˈkrēt, ˈdis-ˌ\ *adjective* [Middle English, from Latin *discretus*] (14th century)
1 : constituting a separate entity **:** individually distinct
2 a : consisting of distinct or unconnected elements **:** NONCONTINUOUS **b :** taking on or having a finite or countably infinite number of values ⟨*discrete* probabilities⟩ ⟨a *discrete* random variable⟩
synonym see DISTINCT
— dis·crete·ly *adverb*
— dis·crete·ness *noun*

dis·cre·tion \dis-ˈkre-shən\ *noun* (14th century)
1 : the quality of being discreet **:** CIRCUMSPECTION; *especially* **:** cautious reserve in speech
2 : ability to make responsible decisions
3 a : individual choice or judgment ⟨left the decision to his *discretion*⟩ **b :** power of free decision or latitude of choice within certain legal bounds ⟨reached the age of *discretion*⟩
4 : the result of separating or distinguishing

dis·cre·tion·ary \-ˈkre-shə-ˌner-ē\ *adjective* (1698)
1 : left to discretion **:** exercised at one's own discretion
2 : available for discretionary use ⟨*discretionary* purchasing power⟩

discretionary account *noun* (circa 1920)
: a security or commodity market account in which an agent (as a broker) is given power of attorney so as to be able to make independent decisions and buy and sell for the principal's account

dis·crim·i·na·bil·i·ty \-ˌkri-mə-nə-ˈbi-lə-tē\ *noun, plural* **-ties** (circa 1901)
1 : the quality of being discriminable
2 : the ability to discriminate

dis·crim·i·na·ble \dis-ˈkri-mə-nə-bəl\ *adjective* (1736)
: capable of being discriminated
— dis·crim·i·na·bly \-blē\ *adverb*

dis·crim·i·nant \-ˈkri-mə-nənt\ *noun* (circa 1948)
: a mathematical expression providing a criterion for the behavior of another more complicated expression, relation, or set of relations

discriminant function *noun* (circa 1936)
: a function of a set of variables (as measurements of taxonomic specimens) that is evaluated for samples of events or objects and used as an aid in discriminating between or classifying them

dis·crim·i·nate \dis-ˈkri-mə-ˌnāt\ *verb* **-nated; -nat·ing** [Latin *discriminatus,* past participle of *discriminare,* from *discrimin-, discrimen* distinction, from *discernere* to distinguish between — more at DISCERN] (1628)
transitive verb
1 a : to mark or perceive the distinguishing or peculiar features of **b :** DISTINGUISH, DIFFERENTIATE ⟨*discriminate* hundreds of colors⟩
2 : to distinguish by discerning or exposing differences; *especially* **:** to distinguish from another like object
intransitive verb
1 a : to make a distinction ⟨*discriminate* among historical sources⟩ **b :** to use good judgment
2 : to make a difference in treatment or favor on a basis other than individual merit ⟨*discriminate* in favor of your friends⟩ ⟨*discriminate* against a certain nationality⟩

dis·crim·i·nat·ing *adjective* (1647)
1 : making a distinction **:** DISTINGUISHING
2 : marked by discrimination: **a :** DISCERNING, JUDICIOUS **b :** DISCRIMINATORY
— dis·crim·i·nat·ing·ly \-ˌnā-tiŋ-lē\ *adverb*

dis·crim·i·na·tion \dis-ˌkri-mə-ˈnā-shən\ *noun* (1648)
1 a : the act of discriminating **b :** the process by which two stimuli differing in some aspect are responded to differently
2 : the quality or power of finely distinguishing
3 a : the act, practice, or an instance of discriminating categorically rather than individually **b :** prejudiced or prejudicial outlook, action, or treatment ⟨racial *discrimination*⟩
synonym see DISCERNMENT
— dis·crim·i·na·tion·al \-shnəl, -shə-nᵊl\ *adjective*

dis·crim·i·na·tive \dis-ˈkri-mə-ˌnā-tiv, -ˈkri-mə-nət-\ *adjective* (1677)
1 : making distinctions
2 : DISCRIMINATORY 2

dis·crim·i·na·tor \dis-ˈkri-mə-ˌnā-tər\ *noun* (1828)
: one that discriminates; *especially* **:** a circuit that can be adjusted to accept or reject signals of different characteristics (as amplitude or frequency)

dis·crim·i·na·to·ry \dis-ˈkri-mə-nə-ˌtōr-ē, -ˌtȯr-, -ˈkrim-nə-\ *adjective* (1828)
1 : DISCRIMINATIVE 1
2 : applying or favoring discrimination in treatment
— dis·crim·i·na·to·ri·ly \-ˌkri-mə-nə-ˈtōr-ə-lē, -ˈtȯr-, -ˌkrim-nə-\ *adverb*

dis·cur·sive \dis-ˈkər-siv\ *adjective* [Medieval Latin *discursivus,* from Latin *discursus,* past participle of *discurrere* to run about — more at DISCOURSE] (1598)
1 a : moving from topic to topic without order **:** RAMBLING **b :** proceeding coherently from topic to topic
2 : marked by analytical reasoning
— dis·cur·sive·ly *adverb*
— dis·cur·sive·ness *noun*

dis·cus \ˈdis-kəs\ *noun, plural* **dis·cus·es** [Latin — more at DISH] (1656)
: a disk (as of wood or plastic) that is thicker in the center than at the perimeter and that is hurled for distance as a track-and-field event; *also* **:** the event

discus

dis·cuss \di-ˈskəs\ *transitive verb* [Middle English, from Latin *discussus,* past participle of *discutere* to disperse, from *dis-* apart + *quatere* to shake — more at DIS-, QUASH] (14th century)
1 *obsolete* **:** DISPEL
2 a : to investigate by reasoning or argument **b :** to present in detail for examination or consideration ⟨*discussed* plans for the party⟩ **c :** to talk about

☆ SYNONYMS
Discover, ascertain, determine, unearth, learn mean to find out what one did not previously know. DISCOVER may apply to something requiring exploration or investigation or to a chance encounter ⟨*discovered* the source of the river⟩. ASCERTAIN implies effort to find the facts or the truth proceeding from awareness of ignorance or uncertainty ⟨attempts to *ascertain* the population of the region⟩. DETERMINE emphasizes the intent to establish the facts definitely or precisely ⟨unable to *determine* the origin of the word⟩. UNEARTH implies bringing to light something forgotten or hidden ⟨*unearth* old records⟩. LEARN may imply acquiring knowledge with little effort or conscious intention (as by simply being told) or it may imply study and practice ⟨I *learned* her name only today⟩ ⟨*learning* Greek⟩.

3 *obsolete* **:** DECLARE ☆
— **dis·cuss·able** *or* **dis·cuss·ible** \-ˈskə-sə-bəl\ *adjective*
— **dis·cuss·er** *noun*

dis·cus·sant \di-ˈskə-sᵊnt\ *noun* (1926)
: one who takes part in a formal discussion or symposium

dis·cus·sion \di-ˈskə-shən\ *noun* (14th century)
1 : consideration of a question in open and usually informal debate
2 : a formal treatment of a topic in speech or writing

¹dis·dain \dis-ˈdān\ *noun* [Middle English *desdeyne,* from Old French *desdeign,* from *desdeignier*] (14th century)
: a feeling of contempt for what is beneath one **:** SCORN

²disdain *transitive verb* [Middle English *desdeynen,* from Middle French *desdeignier,* from Old French, from (assumed) Vulgar Latin *disdignare,* from Latin *dis-* + *dignare* to deign — more at DEIGN] (14th century)
1 : to look on with scorn
2 : to refuse or abstain from because of disdain
3 : to treat as beneath one's notice or dignity
synonym see DESPISE

dis·dain·ful \-fəl\ *adjective* (circa 1542)
: full of or expressing disdain
synonym see PROUD
— **dis·dain·ful·ly** \-fə-lē\ *adverb*
— **dis·dain·ful·ness** *noun*

dis·ease \di-ˈzēz\ *noun* [Middle English *disese,* from Middle French *desaise,* from *des-* dis- + *aise* ease] (14th century)
1 *obsolete* **:** TROUBLE
2 : a condition of the living animal or plant body or of one of its parts that impairs normal functioning **:** SICKNESS, MALADY
3 : a harmful development (as in a social institution)
— **dis·eased** \-ˈzēzd\ *adjective*

dis·econ·o·my \ˌdi-si-ˈkä-nə-mē\ *noun* (1937)
1 : a lack of economy
2 : a factor responsible for an increase in cost

dis·em·bark \ˌdi-səm-ˈbärk\ *verb* [Middle French *desembarquer,* from *des-* dis- + *embarquer* to embark] (1582)
transitive verb
: to remove to shore from a ship
intransitive verb
1 : to go ashore out of a ship
2 : to get out of a vehicle or craft
— **dis·em·bar·ka·tion** \(ˌ)di-ˌsem-ˌbär-ˈkā-shən, -bər-\ *noun*

dis·em·bar·rass \ˌdi-səm-ˈbar-əs\ *transitive verb* (1726)
: to free (as oneself) from something troublesome or superfluous
synonym see EXTRICATE

dis·em·body \ˌdi-səm-ˈbä-dē\ *transitive verb* (1714)
: to divest of a body, of corporeal existence, or of reality

dis·em·bogue \ˌdi-sim-ˈbōg\ *verb* **-bogued; -bogu·ing** [modification of Spanish *desembocar,* from *des-* dis- (from Latin *dis-*) + *embocar* to put into the mouth, from *en* in (from Latin *in*) + *boca* mouth, from Latin *bucca*] (1595)
intransitive verb
: to flow or come forth from or as if from a channel
transitive verb
archaic **:** to pour out from or as if from a container

dis·em·bow·el \ˌdi-səm-ˈbau̇(-ə)l\ *transitive verb* (1618)
1 : to take out the bowels of **:** EVISCERATE
2 : to remove the substance of
— **dis·em·bow·el·ment** \-mənt\ *noun*

dis·en·chant \ˌdi-sᵊn-ˈchant\ *transitive verb* [Middle French *desenchanter,* from *des-* dis- + *enchanter* to enchant] (circa 1586)
: to free from illusion
— **dis·en·chant·er** *noun*
— **dis·en·chant·ing** *adjective*
— **dis·en·chant·ing·ly** \-ˈchan-tiŋ-lē\ *adverb*
— **dis·en·chant·ment** \-mənt\ *noun*

dis·en·chant·ed \-ˈchan-təd\ *adjective* (1838)
: DISAPPOINTED, DISSATISFIED

dis·en·cum·ber \ˌdi-sᵊn-ˈkəm-bər\ *transitive verb* [Middle French *desencombrer,* from *des-* dis- + *encombrer* to encumber] (1598)
: to free from encumbrance **:** DISBURDEN
synonym see EXTRICATE

dis·en·dow \ˌdi-sᵊn-ˈdau̇\ *transitive verb* (1861)
: to strip of endowment
— **dis·en·dow·er** \-ˈdau̇(-ə)r\ *noun*
— **dis·en·dow·ment** \-ˈdau̇-mənt\ *noun*

dis·en·fran·chise \ˌdi-sᵊn-ˈfran-ˌchīz\ *transitive verb* (1664)
: DISFRANCHISE
— **dis·en·fran·chise·ment** \-ˌchīz-mənt, -chəz-\ *noun*

dis·en·gage \ˌdi-sᵊn-ˈgāj\ *verb* [French *désengager,* from Middle French, from *des-* dis- + *engager* to engage] (1611)
transitive verb
: to release from something that engages or involves
intransitive verb
: to release or detach oneself **:** WITHDRAW
— **dis·en·gage·ment** \-mənt\ *noun*

dis·en·tail \ˌdi-sᵊn-ˈtā(ə)l\ *transitive verb* (1641)
: to free from entail

dis·en·tan·gle \ˌdi-sᵊn-ˈtaŋ-gəl\ (1598)
transitive verb
: to free from entanglement **:** UNRAVEL
intransitive verb
: to become disentangled
synonym see EXTRICATE
— **dis·en·tan·gle·ment** \-mənt\ *noun*

dis·en·thrall *also* **dis·en·thral** \ˌdi sᵊn ˈthrȯl\ *transitive verb* (1643)
: to free from bondage **:** LIBERATE

dis·en·ti·tle \ˌdi-sᵊn-ˈtī-tᵊl\ *transitive verb* (1654)
: to deprive of title, claim, or right

dis·equil·i·brate \ˌdi-si-ˈkwi-lə-ˌbrāt\ *transitive verb* (1891)
: to put out of balance
— **dis·equil·i·bra·tion** \-ˌkwi-lə-ˈbrā-shən\ *noun*

dis·equi·lib·ri·um \(ˌ)di-ˌsē-kwə-ˈli-brē-əm, -ˌse-kwə-\ *noun* (1840)
: loss or lack of equilibrium

dis·es·tab·lish \ˌdi-sə-ˈstab-lish\ *transitive verb* (1598)
: to deprive of an established status; *especially* **:** to deprive of the status and privileges of an established church
— **dis·es·tab·lish·ment** \-mənt\ *noun*

dis·es·tab·lish·men·tar·i·an \-ˌstab-lish-ˌmen-ˈter-ē-ən, -mən-\ *noun, often capitalized* [*disestablishment*] (1885)
: one who opposes an established order
— **disestablishmentarian** *adjective, often capitalized*

¹dis·es·teem \ˌdi-sə-ˈstēm\ *transitive verb* (1594)
: to regard with disfavor

²disesteem *noun* (1603)
: DISFAVOR, DISREPUTE

di·seuse \dē-ˈzüz, -ˈzœz\ *noun, plural* **di·seuses** *same*\ [French, feminine of *diseur,* from Old French, from *dire* to say, from Latin *dicere* — more at DICTION] (1896)
: a skilled and usually professional woman reciter

¹dis·fa·vor \(ˌ)dis-ˈfā-vər\ *noun* [probably from Middle French *desfaveur,* from *des-* dis- + *faveur* favor, from Old French *favor*] (circa 1533)
1 : DISAPPROVAL, DISLIKE ⟨practices looked upon with *disfavor*⟩
2 : the state or fact of being no longer favored ⟨fell into *disfavor*⟩
3 : DISADVANTAGE

²disfavor *transitive verb* (1570)
: to withhold or withdraw favor from

dis·fig·ure \dis-ˈfi-gyər, *especially British* -ˈfi-gər\ *transitive verb* [Middle English, from Middle French *desfigurer,* from *des-* dis- + *figure* figure] (14th century)
1 : to impair (as in beauty) by deep and persistent injuries ⟨a face *disfigured* by smallpox⟩
2 *obsolete* **:** DISGUISE
— **dis·fig·ure·ment** \-mənt\ *noun*

dis·fran·chise \(ˌ)dis-ˈfran-ˌchīz\ *transitive verb* (15th century)
: to deprive of a franchise, of a legal right, or of some privilege or immunity; *especially* **:** to deprive of the right to vote
— **dis·fran·chise·ment** \-ˌchīz-mənt, -chəz-\ *noun*

dis·frock \(ˌ)dis-ˈfräk\ *transitive verb* (1837)
: DEFROCK

dis·func·tion *variant of* DYSFUNCTION

dis·fur·nish \(ˌ)dis-ˈfər-nish\ *transitive verb* [Middle French *desfourniss-,* stem of *desfournir,* from *des-* dis- + *fournir* to furnish] (1531)
: to make destitute of possessions **:** DIVEST
— **dis·fur·nish·ment** \-mənt\ *noun*

dis·gorge \(ˌ)dis-ˈgȯrj\ *verb* [Middle English, from Middle French *desgorger,* from *des-* dis- + *gorge* gorge] (15th century)
transitive verb
1 a : to discharge by the throat and mouth **:** VOMIT **b :** to discharge or let go of in a manner suggesting vomiting ⟨the train *disgorged* its passengers⟩ **c :** to give up on request or under pressure ⟨refused to *disgorge* his ill-gotten gains⟩
2 : to discharge the contents of (as the stomach)
intransitive verb
: to discharge contents ⟨where the river *disgorges* into the sea⟩

¹dis·grace \di-ˈskrās, dis-ˈgrās\ *transitive verb* (1580)
1 *archaic* **:** to humiliate by a superior showing
2 : to be a source of shame to ⟨your actions *disgraced* the family⟩
3 : to cause to lose favor or standing ⟨was *disgraced* by the hint of scandal⟩
— **dis·grac·er** *noun*

²disgrace *noun* [Middle French, from Old Italian *disgrazia,* from *dis-* (from Latin) + *grazia* grace, from Latin *gratia* — more at GRACE] (1586)
1 a : the condition of one fallen from grace or honor **b :** loss of grace, favor, or honor

☆ SYNONYMS
Discuss, argue, debate mean to discourse about in order to reach conclusions or to convince. DISCUSS implies a sifting of possibilities especially by presenting considerations pro and con ⟨*discussed* the need for a new highway⟩. ARGUE implies the offering of reasons or evidence in support of convictions already held ⟨*argued* that the project would be too costly⟩. DEBATE suggests formal or public argument between opposing parties ⟨*debated* the merits of the amendment⟩; it may also apply to deliberation with oneself ⟨I'm *debating* whether I should go⟩.

2 : something that disgraces ⟨your manners are a *disgrace*⟩ ☆
dis·grace·ful \-fəl\ *adjective* (1597)
: bringing or involving disgrace
— **dis·grace·ful·ly** \-fə-lē\ *adverb*
— **dis·grace·ful·ness** *noun*
dis·grun·tle \dis-ˈgrən-tᵊl\ *transitive verb* **dis·grun·tled; dis·grun·tling** \-ˈgrənt-liŋ, -ˈgrən-tᵊl-iŋ\ [*dis-* + *gruntle* (to grumble), from Middle English *gruntlen,* frequentative of *grunten* to grunt] (1682)
: to make ill-humored or discontented — usually used as a participial adjective ⟨they were a very *disgruntled* crew —Flannery O'Connor⟩
— **dis·grun·tle·ment** \-tᵊl-mənt\ *noun*
¹dis·guise \də-ˈskīz, dis-ˈgīz *also* diz-\ *transitive verb* **dis·guised; dis·guis·ing** [Middle English *disgisen,* from Middle French *desguiser,* from Old French, from *des-* dis- + *guise* guise] (14th century)
1 a : to change the customary dress or appearance of **b :** to furnish with a false appearance or an assumed identity
2 *obsolete* **:** DISFIGURE
3 : to obscure the existence or true state or character of **:** CONCEAL ☆
— **dis·guised·ly** \-ˈgīz(-ə)d-lē, -ˈkīz(-ə)d-\ *adverb*
— **dis·guise·ment** \-ˈgīz-mənt, -ˈkīz-\ *noun*
— **dis·guis·er** *noun*
²disguise *noun* (14th century)
1 : apparel assumed to conceal one's identity or counterfeit another's
2 : the act of disguising
3 a : form misrepresenting the true nature of something ⟨blessings in *disguise*⟩ **b :** an artificial manner **:** PRETENSE ⟨threw off all *disguise*⟩
¹dis·gust \di-ˈskəst, dis-ˈgəst *also* diz-\ *noun* (1598)
: marked aversion aroused by something highly distasteful **:** REPUGNANCE
²disgust *verb* [Middle French *desgouster,* from *des-* dis- + *goust* taste, from Latin *gustus;* akin to Latin *gustare* to taste — more at CHOOSE] (1616)
transitive verb
1 : to provoke to loathing, repugnance, or aversion **:** be offensive to
2 : to cause (one) to lose an interest or intention
intransitive verb
: to cause disgust
— **dis·gust·ed** *adjective*
— **dis·gust·ed·ly** *adverb*
dis·gust·ful \-fəl\ *adjective* (circa 1616)
1 : provoking disgust
2 : full of or accompanied by disgust
— **dis·gust·ful·ly** \-fə-lē\ *adverb*
dis·gust·ing *adjective* (1754)
: causing disgust
— **dis·gust·ing·ly** \di-ˈskəs-tiŋ-lē, dis-ˈgəs- *also* diz-\ *adverb*
¹dish \ˈdish\ *noun* [Middle English, from Old English *disc* plate, from Latin *discus* quoit, disk, dish, from Greek *diskos,* from *dikein* to throw] (before 12th century)
1 a : a more or less concave vessel from which food is served **b :** the contents of a dish ⟨a *dish* of strawberries⟩
2 a : food prepared in a particular way **b :** something one particularly enjoys **:** CUP OF TEA
3 a (1) **:** any of various shallow concave vessels; *broadly* **:** anything shallowly concave (2) **:** a directional receiver having a concave usually parabolic reflector; *especially* **:** one used as a microwave antenna **b :** the state of being concave or the degree of concavity
4 a : something that is favored ⟨entertainment that is just his *dish*⟩ **b :** an attractive or sexy person
²dish (14th century)
transitive verb
1 : to put (as food for serving) into a dish — often used with *up*
2 : PRESENT — usually used with *up*
3 : to make concave like a dish
intransitive verb
: to pass to a teammate (as in basketball) — usually used with *off*
dis·ha·bille \ˌdis-ə-ˈbē(ə)l, -ˈbil\ *noun* [French *déshabillé,* from past participle of *déshabiller* to undress, from *dés-* dis- + *habiller* to dress — more at HABILIMENT] (1673)
1 a *archaic* **:** NEGLIGEE **b :** the state of being dressed in a casual or careless style
2 : a deliberately careless or casual manner
dis·har·mo·ni·ous \ˌdis-(ˌ)här-ˈmō-nē-əs\ *adjective* (1659)
: lacking in harmony
dis·har·mo·nize \(ˌ)dis-ˈhär-mə-ˌnīz\ *transitive verb* (1801)
: to make disharmonious
dis·har·mo·ny \-nē\ *noun* (circa 1602)
: lack of harmony **:** DISCORD
dish·cloth \ˈdish-ˌklȯth\ *noun* (circa 1828)
: a cloth for washing dishes
dish·clout \ˈdish-ˌklau̇t\ *noun* (circa 1530)
British **:** DISHCLOTH
dis·heart·en \(ˌ)dis-ˈhär-tᵊn\ *transitive verb* (1590)
: to cause to lose spirit or morale
— **dis·heart·en·ing·ly** \-ˈhärt-niŋ-lē, -ˈhär-tᵊn-iŋ-\ *adverb*
— **dis·heart·en·ment** \-ˈhär-tᵊn-mənt\ *noun*
dished \ˈdisht\ *adjective* (1737)
: CONCAVE
di·shev·el \di-ˈshev-əl\ *transitive verb* **di·shev·eled** *or* **di·shev·elled; di·shev·el·ing** *or* **di·shev·el·ling** \-ˈshe-v(ə-)liŋ\ [back-formation from *disheveled*] (1598)
: to throw into disorder or disarray
di·shev·eled *or* **di·shev·elled** *adjective* [Middle English *discheveled* with disordered hair, part translation of Middle French *deschevelé,* from past participle of *descheveler* to disarrange the hair, from *des-* dis- + *chevel* hair, from Latin *capillus*] (1583)
: marked by disorder or disarray
dis·hon·est \(ˌ)di-ˈsä-nəst *also* -ˈzä-\ *adjective* [Middle English, from Middle French *deshoneste,* from *des-* dis- + *honeste* honest] (14th century)
1 *obsolete* **:** SHAMEFUL, UNCHASTE
2 : characterized by lack of truth, honesty, or trustworthiness **:** UNFAIR, DECEPTIVE ☆
— **dis·hon·est·ly** *adverb*
dis·hon·es·ty \-nə-stē\ *noun* (1599)
1 : lack of honesty or integrity **:** disposition to defraud or deceive
2 : a dishonest act **:** FRAUD
¹dis·hon·or \(ˌ)di-ˈsä-nər *also* -ˈzä-\ *noun* [Middle English *dishonour,* from Middle French *deshonor,* from *des-* dis- + *honor* honor] (13th century)
1 : lack or loss of honor or reputation
2 : the state of one who has lost honor or prestige **:** SHAME
3 : a cause of disgrace
4 : the nonpayment or nonacceptance of commercial paper by the party on whom it is drawn
synonym see DISGRACE
— **dis·hon·or·er** \-ˈsän-ər-ər *also* -ˈzä-\ *noun*
²dishonor *transitive verb* (13th century)
1 a : to treat in a degrading manner **b :** to bring shame on
2 : to refuse to accept or pay (as a bill or check)
dis·hon·or·able \(ˌ)di-ˈsä-nə-rə-bəl, -ˈsä-nər-bəl\ *adjective* (1534)
1 : lacking honor **:** SHAMEFUL ⟨*dishonorable* conduct⟩
2 *archaic* **:** not honored
— **dis·hon·or·able·ness** *noun*
— **dis·hon·or·ably** \-blē\ *adverb*
dish out *transitive verb* (1641)
: to give or dispense freely ⟨*dish out* gifts⟩ ⟨*dish out* advice⟩ ⟨*dish out* punishment⟩
dish·pan \ˈdish-ˌpan\ *noun* (1872)
: a large flat-bottomed pan used for washing dishes
dishpan hands *noun plural but singular or plural in construction* (1944)
: a condition of dryness, redness, and scaling of the hands that results typically from repeated exposure to, sensitivity to, or overuse of cleaning materials (as detergents) used in housework
dish·rag \ˈdish-ˌrag\ *noun* (1839)
: DISHCLOTH
dish·ware \ˈdish-ˌwar, -ˌwer\ *noun* (1946)
: tableware (as of china) used in serving food
dish·wash·er \-ˌwȯ-shər, -ˌwä-\ *noun* (15th century)
1 : a worker employed to wash dishes
2 : a machine for washing dishes
dish·wa·ter \-ˌwȯ-tər, -ˌwä-\ *noun* (15th century)
: water in which dishes have been or are to be washed
dishy \ˈdi-shē\ *adjective* **dish·i·er; -est** (1961)
: ATTRACTIVE 2
¹dis·il·lu·sion \ˌdi-sə-ˈlü-zhən\ *noun* (1851)
: the condition of being disenchanted
²disillusion *transitive verb* **dis·il·lu·sioned; dis·il·lu·sion·ing** \-ˈlü-zhə-niŋ\ (1855)
: to leave without illusion or naive faith and trust
— **dis·il·lu·sion·ment** \-ˈlü-zhən-mənt\ *noun*

☆ SYNONYMS

Disgrace, dishonor, disrepute, infamy, ignominy mean the state or condition of suffering loss of esteem and of enduring reproach. DISGRACE often implies humiliation and sometimes ostracism ⟨sent home in *disgrace*⟩. DISHONOR emphasizes the loss of honor that one has enjoyed or the loss of self-esteem ⟨preferred death to life with *dishonor*⟩. DISREPUTE stresses loss of one's good name or the acquiring of a bad reputation ⟨a once proud name fallen into *disrepute*⟩. INFAMY usually implies notoriety as well as exceeding shame ⟨a day that lives in *infamy*⟩. IGNOMINY stresses humiliation ⟨the *ignominy* of being arrested⟩.

Disguise, cloak, mask mean to alter the dress or appearance of so as to conceal the identity or true nature. DISGUISE implies a change in appearance or behavior that misleads by presenting a different apparent identity ⟨*disguised* himself as a peasant⟩. CLOAK suggests a means of hiding a movement or an intention ⟨*cloaked* their maneuvers in secrecy⟩. MASK suggests some often obvious means of hiding or disguising something ⟨smiling to *mask* his discontent⟩.

Dishonest, deceitful, mendacious, untruthful mean unworthy of trust or belief. DISHONEST implies a willful perversion of truth in order to deceive, cheat, or defraud ⟨a swindle usually involves two *dishonest* people⟩. DECEITFUL usually implies an intent to mislead and commonly suggests a false appearance or double-dealing ⟨the secret affairs of a *deceitful* spouse⟩. MENDACIOUS may suggest bland or even harmlessly mischievous deceit and when used of people often suggests a habit of telling untruths ⟨*mendacious* tales of adventure⟩. UNTRUTHFUL stresses a discrepancy between what is said and fact or reality ⟨an *untruthful* account of their actions⟩.

disillusioned *adjective* (1871)
: DISAPPOINTED, DISSATISFIED
dis·in·cen·tive \ˌdi-s^{ə}n-ˈsen-tiv\ *noun* (1946)
: DETERRENT
dis·in·cli·na·tion \(ˌ)di-ˌsin-klə-ˈnā-shən, -ˌsiŋ-\ *noun* (1647)
: a preference for avoiding something **:** slight aversion
dis·in·cline \ˌdi-s^{ə}n-ˈklīn\ *transitive verb* (1647)
: to make unwilling
dis·in·clined *adjective* (1647)
: unwilling because of mild dislike or disapproval ☆
dis·in·fect \ˌdi-s^{ə}n-ˈfekt\ *transitive verb* [Middle French *desinfecter,* from *des-* dis- + *infecter* to infect] (1598)
: to free from infection especially by destroying harmful microorganisms; *broadly* **:** CLEANSE
— dis·in·fec·tion \-ˈfek-shən\ *noun*
dis·in·fec·tant \-ˈfek-tənt\ *noun* (1837)
: an agent that frees from infection; *especially* **:** a chemical that destroys vegetative forms of harmful microorganisms especially on inanimate objects but that may be less effective in destroying bacterial spores
dis·in·fest \ˌdi-s^{ə}n-ˈfest\ *transitive verb* (circa 1920)
: to rid of small animal pests (as insects or rodents)
— dis·in·fes·ta·tion \(ˌ)di-ˌsin-ˌfes-ˈtā-shən\ *noun*
dis·in·fes·tant \ˌdi-s^{ə}n-ˈfes-tənt\ *noun* (1943)
: a disinfesting agent
dis·in·fla·tion \ˌdi-s^{ə}n-ˈflā-shən\ *noun* (1880)
: a reversal of inflationary pressures
— dis·in·fla·tion·ary \-shə-ˌner-ē\ *adjective*
dis·in·for·ma·tion \(ˌ)di-ˌsin-fər-ˈmā-shən\ *noun* (1939)
: false information deliberately and often covertly spread (as by the planting of rumors) in order to influence public opinion or obscure the truth ◆
dis·in·gen·u·ous \ˌdi-s^{ə}n-ˈjen-yə-wəs\ *adjective* (1655)
: lacking in candor; *also* **:** giving a false appearance of simple frankness **:** CALCULATING
— dis·in·gen·u·ous·ly *adverb*
— dis·in·gen·u·ous·ness *noun*
dis·in·her·it \ˌdi-s^{ə}n-ˈher-ət\ *transitive verb* [Middle English] (15th century)
1 : to prevent deliberately from inheriting something (as by making a will)
2 : to deprive of natural or human rights or of previously held special privileges
— dis·in·her·i·tance \-ˈher-ə-tən(t)s\ *noun*
dis·in·hi·bi·tion \(ˌ)di-ˌsin-(h)ə-ˈbi-shən\ *noun* (circa 1927)
: loss or reduction of an inhibition (as by the action of interfering stimuli or events) ⟨*disinhibition* of a reflex⟩ ⟨*disinhibition* of violent tendencies⟩
— dis·in·hib·it \(ˌ)di-s^{ə}n-ˈhi-bət\ *transitive verb*
dis·in·te·grate \(ˌ)di-ˈsin-tə-ˌgrāt\ (1796)
transitive verb
1 : to break or decompose into constituent elements, parts, or small particles
2 : to destroy the unity or integrity of
intransitive verb
1 : to break or separate into constituent elements or parts
2 : to lose unity or integrity by or as if by breaking into parts
3 : to undergo a change in composition ⟨an atomic nucleus that *disintegrates* because of radioactivity⟩
— dis·in·te·gra·tion \(ˌ)di-ˌsin-tə-ˈgrā-shən\ *noun*
— dis·in·te·gra·tive \(ˌ)di-ˈsin-tə-ˌgrā-tiv\ *adjective*
— dis·in·te·gra·tor \-ˌgrā-tər\ *noun*
dis·in·ter \ˌdi-s^{ə}n-ˈtər\ *transitive verb* (1611)
1 : to take out of the grave or tomb
2 : to bring back into awareness or prominence; *also* **:** to bring to light **:** UNEARTH
— dis·in·ter·ment \-mənt\ *noun*
¹dis·in·ter·est \(ˌ)dis-ˈin-trəst; -ˈin-tə-ˌrest, -tə-rəst, -tərst; -ˈin-ˌtrest\ *transitive verb* (1612)
: to cause to regard something with no interest or concern
²disinterest *noun* (1658)
1 : DISINTERESTEDNESS
2 : lack of interest **:** INDIFFERENCE
dis·in·ter·est·ed \-təd\ *adjective* (circa 1612)
1 a : not having the mind or feelings engaged **:** not interested ⟨telling them in a *disinterested* voice —Tom Wicker⟩ ⟨Introverted. Unsocial . . . *Disinterested* in women —J. A. Brussel⟩ **b :** no longer interested ⟨husband and wife become *disinterested* in each other —T. I. Rubin⟩
2 : free from selfish motive or interest **:** UNBIASED ⟨a *disinterested* decision⟩ ⟨*disinterested* intellectual curiosity is the lifeblood of real civilization —G. M. Trevelyan⟩ ▪
synonym see INDIFFERENT
— dis·in·ter·est·ed·ly *adverb*
dis·in·ter·est·ed·ness \-təd-nəs\ *noun* (circa 1682)
: the quality of being objective or impartial
dis·in·ter·me·di·a·tion \ˌdi-ˌsin-tər-ˌmē-dē-ˈā-shən\ *noun* (1967)
: the diversion of savings from accounts with low fixed interest rates to direct investment in high-yielding instruments
dis·in·tox·i·cate \ˌdi-s^{ə}n-ˈtäk-sə-ˌkāt\ *transitive verb* (1685)
: DETOXIFY 2
— dis·in·tox·i·ca·tion \-ˌtäk-sə-ˈkā-shən\ *noun*
dis·in·vest \(ˌ)di-s^{ə}n-ˈvest\ *intransitive verb* (1945)
: to reduce or eliminate capital investment (as in an industry or area)
dis·in·vest·ment \ˌdi-s^{ə}n-ˈves(t)-mənt\ *noun* (1936)
: consumption of capital; *also* **:** the withdrawing of investment
dis·in·vite \(ˌ)di-s^{ə}n-ˈvīt\ *transitive verb* (1580)
: to withdraw an invitation to
dis·join \(ˌ)dis ˈjȯin\ *verb* [Middle English *disjoynen,* from Middle French *desjoindre,* from Latin *disjungere,* from *dis-* + *jungere* to join — more at YOKE] (15th century)
transitive verb
: to end the joining of
intransitive verb
: to become detached
¹dis·joint \-ˈjȯint\ *adjective* [Middle English *disjoynt,* from Middle French *desjoint,* past participle of *desjoindre*] (15th century)
1 *obsolete* **:** DISJOINTED 1a
2 : having no elements in common ⟨*disjoint* mathematical sets⟩
²disjoint (15th century)
transitive verb
1 : to disturb the orderly structure or arrangement of
2 : to take apart at the joints
intransitive verb
: to come apart at the joints
dis·joint·ed *adjective* (circa 1586)
1 a : being thrown out of orderly function ⟨a *disjointed* society⟩ **b :** lacking coherence or orderly sequence ⟨an incomplete and *disjointed* history⟩
2 : separated at or as if at the joint
— dis·joint·ed·ly *adverb*
— dis·joint·ed·ness *noun*
¹dis·junct \dis-ˈjəŋ(k)t\ *adjective* [Middle English, from Latin *disjunctus,* past participle of *disjungere* to disjoin] (15th century)
: marked by separation of or from usually contiguous parts or individuals: as **a :** DISCONTINUOUS **b :** relating to melodic progression by intervals larger than a major second — compare CONJUNCT

☆ SYNONYMS
Disinclined, hesitant, reluctant, loath, averse mean lacking the will or desire to do something indicated. DISINCLINED implies lack of taste for or inclination ⟨*disinclined* to move again⟩ ⟨*disinclined* for reading⟩. HESITANT implies a holding back especially through fear or uncertainty ⟨*hesitant* about asking for a date⟩. RELUCTANT implies a holding back through unwillingness ⟨a *reluctant* witness⟩. LOATH implies hesitancy because of conflict with one's opinions, predilections, or liking ⟨seems *loath* to trust anyone⟩. AVERSE implies a holding back from or avoiding because of distaste or repugnance ⟨*averse* to hard work⟩.

▫ USAGE
disinterested *Disinterested* and *uninterested* have a tangled history. *Uninterested* originally meant impartial, but this sense fell into disuse during the 18th century. About the same time the original sense of *disinterested* also disappeared, with *uninterested* developing a new sense—the present meaning—to take its place. The original sense of *uninterested* is still out of use, but the original sense of *disinterested* revived in the early 20th century. The revival has since been under frequent attack as an illiteracy and a blurring or loss of a useful distinction. Actual usage shows otherwise. Sense 2 of *disinterested* is still its most frequent sense, especially in edited prose; it shows no sign of vanishing. A careful writer may choose sense 1a of *disinterested* in preference to *uninterested* for emphasis ⟨teaching the letters of the alphabet to her wiggling and supremely *disinterested* little daughter —C. L. Sulzberger⟩ Further, *disinterested* has developed a sense (1b), perhaps influenced by sense 1 of the prefix *dis-*, that contrasts with *uninterested* ⟨when I grow tired or *disinterested* in anything, I experience a disgust —Jack London (letter, 1914)⟩ Still, use of senses 1a and 1b will incur the disapproval of some who may not fully appreciate the history of this word or the subtleties of its present use.

◇ WORD HISTORY
disinformation Though its roots are several decades older, *disinformation* became a common word in the 1970s and '80s. Many uses of the word during this period associate *disinformation* with the KGB, the Soviet secret police; and claims have been made that the KGB established a branch in the 1950s called, presumably unofficially, the "Department of Disinformation." It has been said that the word first appeared in English in the *Times* of London in 1955, but actually the U.S. magazine *Time* used it a year earlier to refer to Communist propaganda. The earliest known attestation of *disinformation,* however, is in the 1939 book *In Stalin's Secret Police,* a pioneering exposé of Stalin's purges, written by Walter Krivitsky, a Soviet intelligence officer who defected in the fall of 1937 and—meeting a fate common among Stalin's enemies in those days—was found shot in his hotel room in February 1941. In all likelihood, *disinformation* in Krivitsky's book as well as later uses are translations of Russian *dezinformatsiya,* which began to appear in Russian dictionaries in 1949.

2dis·junct \'dis-ˌjəŋ(k)t, dis-'\ *noun* (1921)
1 : any of the alternatives that make up a logical disjunction
2 : an adverb or adverbial (as *luckily* in "Luckily we had an extra set" or *in short* in "In short, there is nothing we can do") that is loosely connected to a sentence and conveys the speaker's or writer's comment on its content, truth, or manner — compare ADJUNCT 2b

dis·junc·tion \dis-'jəŋ(k)-shən\ *noun* (14th century)
1 : a sharp cleavage **:** DISUNION, SEPARATION ⟨the *disjunction* between theory and practice⟩
2 : a compound sentence in logic formed by joining two simple statements by *or:* **a :** INCLUSIVE DISJUNCTION **b :** EXCLUSIVE DISJUNCTION

1dis·junc·tive \-'jəŋ(k)-tiv\ *adjective* (15th century)
1 a : relating to, being, or forming a logical disjunction **b :** expressing an alternative or opposition between the meanings of the words connected ⟨the *disjunctive* conjunction *or*⟩ **c :** expressed by mutually exclusive alternatives joined by *or* ⟨*disjunctive* pleading⟩
2 : marked by breaks or disunity ⟨a *disjunctive* narrative sequence⟩
3 *of a pronoun form* **:** stressed and not attached to the verb as an enclitic or proclitic
— dis·junc·tive·ly *adverb*

2disjunctive *noun* (1530)
: a disjunctive conjunction

dis·junc·ture \-'jəŋ(k)-chər\ *noun* [Middle English, modification (influenced by Latin *disjunctus*) of Middle French *desjointure,* from *desjoint* disjoint] (14th century)
: DISJUNCTION 1

1disk *or* **disc** \'disk\ *noun, often attributive* [Latin *discus* — more at DISH] (1664)
1 a : the seemingly flat figure of a celestial body ⟨the solar *disk*⟩ **b** *archaic* **:** DISCUS
2 : any of various rounded and flattened animal anatomical structures (as an intervertebral disk) — compare SLIPPED DISK
3 : the central part of the flower head of a typical composite made up of closely packed tubular flowers
4 a : a thin circular object **b** *usually disc* **:** a phonograph record **c :** a round flat plate coated with a magnetic substance on which data for a computer is stored **d** *usually disc* **:** OPTICAL DISK: as (1) **:** VIDEODISC (2) **:** COMPACT DISC
5 *usually disc* **:** one of the concave circular steel tools with sharpened edge making up the working part of a disc harrow or plow; *also* **:** an implement employing such tools
— disk·like \-ˌlīk\ *adjective*

2disk *or* **disc** *transitive verb* (circa 1884)
: to cultivate with an implement (as a harrow or plow) that turns and loosens the soil with a series of disks

dis·kette \ˌdis-'ket\ *noun* (1973)
: FLOPPY DISK

disk flower *noun* (1870)
: one of the tubular flowers in the disk of a composite plant — called also *disk floret*

dis·lik·able *also* **dis·like·able** \(ˌ)dis-'lī-kə-bəl\ *adjective* (1843)
: easy to dislike

1dis·like \(ˌ)dis-'līk, 'dis-ˌ\ *noun* (1577)
1 : a feeling of aversion or disapproval
2 *obsolete* **:** DISCORD

2dislike *transitive verb* (1579)
1 *archaic* **:** DISPLEASE
2 : to regard with dislike **:** DISAPPROVE
3 *obsolete* **:** to show aversion to
— dis·lik·er *noun*

dis·limn \(ˌ)dis-'lim\ *verb* (1606)
: DIM

dis·lo·cate \'dis-lō-ˌkāt, -lə-; (ˌ)dis-'lō-\ *transitive verb* [Medieval Latin *dislocatus,* past participle of *dislocare,* from Latin *dis-* + *locare* to locate] (1605)
1 : to put out of place; *specifically* **:** to displace (a bone) from normal connections with another bone
2 : to force a change in the usual status, relationship, or order of **:** DISRUPT

dis·lo·ca·tion \ˌdis-(ˌ)lō-'kā-shən, -lə-\ *noun* (14th century)
: the act of dislocating **:** the state of being dislocated: as **a :** displacement of one or more bones at a joint **:** LUXATION **b :** a discontinuity in the otherwise normal lattice structure of a crystal **c :** disruption of an established order

dis·lodge \(ˌ)dis-'läj\ *verb* [Middle English *disloggen,* from Middle French *desloger,* from *des-* dis- + *loger* to lodge, from Old French *loge* lodge] (15th century)
transitive verb
1 : to drive from a position of hiding, defense, or advantage
2 : to force out of a secure or settled position ⟨*dislodged* the rock with a shovel⟩
intransitive verb
: to leave a lodging place
— dis·lodg·ment *or* **dis·lodge·ment** *noun*

dis·loy·al \(ˌ)dis-'lȯi(-ə)l\ *adjective* [Middle English, from Middle French *desloial,* from Old French, from *des-* dis- + *loial* loyal] (15th century)
: lacking in loyalty; *also* **:** showing an absence of allegiance, devotion, obligation, faith, or support ⟨his *disloyal* refusal to help his friend⟩
synonym see FAITHLESS
— dis·loy·al·ly \-'lȯi-ə-lē\ *adverb*

dis·loy·al·ty \-'lȯi(-ə)l-tē\ *noun* (15th century)
: lack of loyalty

dis·mal \'diz-məl\ *adjective* [Middle English, from *dismal,* noun, days marked as unlucky in medieval calendars, from Anglo-French, from Medieval Latin *dies mali,* literally, evil days] (15th century)
1 *obsolete* **:** DISASTROUS, DREADFUL
2 : showing or causing gloom or depression
3 : lacking merit **:** particularly bad ◆
— dis·mal·ly \-mə-lē\ *adverb*
— dis·mal·ness *noun*

dis·man·tle \(ˌ)dis-'man-tᵊl\ *transitive verb* **dis·man·tled; dis·man·tling** \-'mant-liŋ, -'man-tᵊl-\ [Middle French *desmanteler,* from *des-* dis- + *mantel* mantle] (1602)
1 : to take to pieces; *also* **:** to destroy the integrity or functioning of
2 : to strip of dress or covering **:** DIVEST
3 : to strip of furniture and equipment
— dis·man·tle·ment \-'man-tᵊl-mənt\ *noun*

dis·mast \(ˌ)dis-'mast\ *transitive verb* (1747)
: to remove or break off the mast of

1dis·may \dis-'mā, diz-\ *transitive verb* **dis·mayed; dis·may·ing** [Middle English, from (assumed) Old French *desmaiier,* from Old French *des-* dis- + *-maiier* (as in *esmaiier* to dismay), from (assumed) Vulgar Latin *-magare,* of Germanic origin; akin to Old High German *magan* to be able — more at MAY] (13th century)
: to deprive of courage, resolution, and initiative through the pressure of sudden fear or anxiety or great perplexity ☆
— dis·may·ing·ly \-iŋ-lē\ *adverb*

2dismay *noun* (14th century)
1 : sudden loss of courage or resolution from alarm or fear
2 a : sudden disappointment **b :** PERTURBATION 1

disme \'dīm\ *noun* [obsolete English, tenth, from obsolete French, from Middle French *disme, dime* — more at DIME] (1792)
: a U.S. 10-cent coin struck in 1792

dis·mem·ber \(ˌ)dis-'mem-bər\ *transitive verb* **-bered; -ber·ing** \-b(ə-)riŋ\ [Middle English *dismembren,* from Old French *desmembrer,* from *des-* dis- + *membre* member] (14th century)
1 : to cut off or disjoin the limbs, members, or parts of
2 : to break up or tear into pieces
— dis·mem·ber·ment \-bər-mənt\ *noun*

dis·miss \dis-'mis\ *transitive verb* [Middle English, modification of Latin *dimissus,* past participle of *dimittere,* from *dis-* apart + *mittere* to send — more at DIS-, SMITE] (15th century)
1 : to permit or cause to leave ⟨*dismissed* the visitors⟩
2 : to remove from position or service **:** DISCHARGE
3 a : to reject serious consideration of ⟨*dismissed* the thought⟩ **b :** to put out of judicial consideration ⟨*dismissed* all charges⟩
— dis·mis·sion \-'mi-shən\ *noun*
— dis·mis·sive \-'mi-siv\ *adjective*
— dis·mis·sive·ly *adverb*

dis·miss·al \-'mi-səl\ *noun* (1816)
: the act of dismissing **:** the fact or state of being dismissed

1dis·mount \(ˌ)dis-'mau̇nt\ *verb* [probably modification of Middle French *desmonter,* from *des-* dis- + *monter* to mount] (1579)
intransitive verb
1 *obsolete* **:** DESCEND
2 : to alight from an elevated position (as on a horse); *also* **:** to get out of an enclosed craft or vehicle
transitive verb
1 : to throw down or remove from a mount or an elevated position; *especially* **:** UNHORSE
2 : DISASSEMBLE

2dismount *noun* (1654)
: the act of dismounting

dis·obe·di·ence \ˌdi-sə-'bē-dē-ən(t)s\ *noun* (15th century)
: refusal or neglect to obey

dis·obe·di·ent \-ənt\ *adjective* [Middle English, from Middle French *desobedient,* from Old French, from *des-* dis- + *obedient* obedient] (15th century)
: refusing or neglecting to obey
— dis·obe·di·ent·ly *adverb*

☆ SYNONYMS
Dismay, appall, horrify, daunt mean to unnerve or deter by arousing fear, apprehension, or aversion. DISMAY implies that one is disconcerted and at a loss as to how to deal with something ⟨*dismayed* at the size of the job⟩. APPALL implies that one is faced with that which perturbs, confounds, or shocks ⟨I am *appalled* by your behavior⟩. HORRIFY stresses a reaction of horror or revulsion ⟨was *horrified* by such wanton cruelty⟩. DAUNT suggests a cowing, disheartening, or frightening in a venture requiring courage ⟨a cliff that would *daunt* the most intrepid climber⟩.

◇ WORD HISTORY
dismal In late antiquity, certain days of each month, called in Latin *dies Aegyptiaci* "Egyptian days," were regarded as unlucky times, when new enterprises of any sort were not to be undertaken. These days of ill omen were probably a relic of ancient Egyptian astrological practices, but their source had been forgotten by the Middle Ages. People then took them to be anniversaries of the plagues visited on Egypt in Moses' time—despite the fact that there were 24 Egyptian days in the year and only ten Biblical plagues. In a 13th century Anglo-French calendar, the Egyptian days were called collectively *dismal* (from Latin *dies mali* "evil days"), and this word was borrowed into Middle English. Any day of the 24 was a *dismal day,* but the original sense "evil days" was forgotten, and *dismal* in the phrase was simply taken as an adjective meaning "disastrous." In Modern English this sense has been weakened to "causing gloom," perhaps by association with *dismay.*

dis·obey \ˌdi-sə-ˈbā\ *verb* [Middle English, from Middle French *desobeir,* from Old French, from *des-* dis- + *obeir* to obey] (14th century)
intransitive verb
: to be disobedient
transitive verb
: to fail to obey
— **dis·obey·er** *noun*

dis·oblige \ˌdi-sə-ˈblīj\ *transitive verb* [French *désobliger,* from Middle French, from *des-* dis- + *obliger* to oblige] (1632)
1 : to go counter to the wishes of
2 : INCONVENIENCE

di·so·di·um phosphate \(ˌ)dī-ˈsō-dē-əm-\ *noun* (circa 1928)
: a sodium phosphate Na_2HPO_4

di·so·mic \(ˌ)dī-ˈsō-mik\ *adjective* [*di-* + *-somic*] (1924)
: having one or more chromosomes present in twice the normal number but not having the entire genome doubled

[1]**dis·or·der** \(ˌ)di-ˈsȯr-dər, -ˈzȯr-\ *transitive verb* (15th century)
1 : to disturb the order of
2 : to disturb the regular or normal functions of

[2]**disorder** *noun* (1530)
1 : lack of order ⟨clothes in *disorder*⟩
2 : breach of the peace or public order ⟨troubled times marked by social *disorders*⟩
3 : an abnormal physical or mental condition **:** AILMENT ⟨a nervous *disorder*⟩

dis·or·dered *adjective* (1548)
1 *obsolete* **a :** morally reprehensible **b :** UNRULY
2 a : marked by disorder **b :** not functioning in a normal orderly healthy way
— **dis·or·dered·ly** *adverb*
— **dis·or·dered·ness** *noun*

[1]**dis·or·der·ly** \-ˈsȯrd-ər-lē\ *adverb* (1564)
archaic **:** in a disorderly manner

[2]**disorderly** *adjective* (1585)
1 : engaged in conduct offensive to public order ⟨charged with being drunk and *disorderly*⟩
2 : characterized by disorder ⟨a *disorderly* pile of clothes⟩
— **dis·or·der·li·ness** *noun*

disorderly conduct *noun* (circa 1845)
: a petty offense chiefly against public order and decency that falls short of an indictable misdemeanor

disorderly house *noun* [euphemism] (circa 1809)
: BORDELLO

dis·or·ga·nize \(ˌ)di-ˈsȯr-gə-ˌnīz\ *transitive verb* [French *désorganiser,* from *dés-* dis- + *organiser* to organize] (1793)
: to destroy or interrupt the orderly structure or function of
— **dis·or·ga·ni·za·tion** \(ˌ)di-ˌsȯr-gə-nə-ˈzā-shən\ *noun*

dis·or·ga·nized *adjective* (1812)
: lacking coherence, system, or central guiding agency ⟨*disorganized* work habits⟩

dis·ori·ent \(ˌ)di-ˈsōr-ē-ˌent, -ˈsȯr-\ *transitive verb* [French *désorienter,* from *dés-* dis- + *orienter* to orient] (1655)
1 a : to cause to lose bearings **:** displace from normal position or relationship **b :** to cause to lose the sense of time, place, or identity
2 : CONFUSE

dis·ori·en·tate \-ē-ən-ˌtāt, -ē-ˌen-\ *transitive verb* (circa 1704)
: DISORIENT
— **dis·ori·en·ta·tion** \(ˌ)di-ˌsōr-ē-ən-ˈtā-shən, -ˌsȯr-, -ē-ˌen-\ *noun*

dis·own \(ˌ)di-ˈsōn\ *transitive verb* (1649)
1 : to refuse to acknowledge as one's own
2 a : to repudiate any connection or identification with **b :** to deny the validity or authority of
— **dis·own·ment** \-mənt\ *noun*

dis·par·age \di-ˈspar-ij\ *transitive verb* **-aged; -ag·ing** [Middle English, to degrade by marriage below one's class, disparage, from Middle French *desparagier* to marry below one's class, from Old French, from *des-* dis- + *parage* extraction, lineage, from *per* peer] (14th century)
1 : to lower in rank or reputation **:** DEGRADE
2 : to depreciate by indirect means (as invidious comparison) **:** speak slightingly about
synonym see DECRY
— **dis·par·age·ment** \-ij-mənt\ *noun*
— **dis·par·ag·er** *noun*
— **dis·par·ag·ing** *adjective*
— **dis·par·ag·ing·ly** \-ij-iŋ-lē\ *adverb*

dis·pa·rate \ˈdis-p(ə-)rət, di-ˈspar-ət\ *adjective* [Latin *disparatus,* past participle of *disparare* to separate, from *dis-* + *parare* to prepare — more at PARE] (15th century)
1 : containing or made up of fundamentally different and often incongruous elements
2 : markedly distinct in quality or character
synonym see DIFFERENT
— **dis·pa·rate·ly** *adverb*
— **dis·pa·rate·ness** *noun*
— **dis·par·i·ty** \di-ˈspar-ə-tē\ *noun*

dis·part \(ˌ)dis-ˈpärt\ *verb* [Italian & Latin; Italian *dispartire,* from Latin, from *dis-* + *partire* to divide — more at PART] (1590)
archaic **:** SEPARATE, DIVIDE

dis·pas·sion \(ˌ)dis-ˈpa-shən\ *noun* (1692)
: absence of passion **:** COOLNESS

dis·pas·sion·ate \-sh(ə-)nət\ *adjective* (1594)
: not influenced by strong feeling; *especially* **:** not affected by personal or emotional involvement ⟨a *dispassionate* critic⟩ ⟨a *dispassionate* approach to an issue⟩
synonym see FAIR
— **dis·pas·sion·ate·ly** *adverb*
— **dis·pas·sion·ate·ness** *noun*

[1]**dis·patch** \di-ˈspach\ *verb* [Spanish *despachar* or Italian *dispacciare,* from Provençal *despachar* to get rid of, from Middle French *despeechier* to set free, from Old French, from *des-* dis- + *-peechier* (as in *empeechier* to hinder) — more at IMPEACH] (1517)
transitive verb
1 : to send off or away with promptness or speed; *especially* **:** to send off on official business
2 a : to kill with quick efficiency **b** *obsolete* **:** DEPRIVE
3 : to dispose of (as a task) rapidly or efficiently
4 : DEFEAT 3
intransitive verb
archaic **:** to make haste **:** HURRY
synonym see KILL
— **dis·patch·er** *noun*

[2]**dis·patch** \di-ˈspach, ˈdis-ˌpach\ *noun* (1537)
1 a : a message sent with speed; *especially* **:** an important official message sent by a diplomatic, military, or naval officer **b :** a news item filed by a correspondent
2 : the act of dispatching: as **a** *obsolete* **:** DISMISSAL **b :** the act of killing **c** (1) **:** prompt settlement (as of an item of business) (2) **:** quick riddance **d :** a sending off **:** SHIPMENT
3 : promptness and efficiency in performance or transmission ⟨done with *dispatch*⟩
synonym see HASTE

dispatch case *noun* (circa 1918)
: a case for carrying papers

dis·pel \di-ˈspel\ *transitive verb* **dis·pelled; dis·pel·ling** [Middle English, from Latin *dispellere,* from *dis-* + *pellere* to drive, beat — more at FELT] (15th century)
: to drive away by or as if by scattering **:** DISSIPATE ⟨*dispel* a rumor⟩
synonym see SCATTER

dis·pens·able \di-ˈspen(t)-sə-bəl\ *adjective* (1649)
: capable of being dispensed with
— **dis·pens·abil·i·ty** \-ˌspen(t)-sə-ˈbi-lə-tē\ *noun*

dis·pen·sa·ry \di-ˈspen(t)s-(ə-)rē\ *noun, plural* **-ries** (1699)
: a place where medicine or medical or dental treatment is dispensed

dis·pen·sa·tion \ˌdis-pən-ˈsā-shən, -ˌpen-\ *noun* (14th century)
1 a : a general state or ordering of things; *specifically* **:** a system of revealed commands and promises regulating human affairs **b :** a particular arrangement or provision especially of providence or nature
2 a : an exemption from a law or from an impediment, vow, or oath **b :** a formal authorization
3 a : the act of dispensing **b :** something dispensed or distributed
— **dis·pen·sa·tion·al** \-shnəl, -shə-nᵊl\ *adjective*

dis·pen·sa·to·ry \di-ˈspen(t)-sə-ˌtōr-ē, -ˌtȯr-\ *noun, plural* **-ries** (1566)
: a medicinal formulary

dis·pense \di-ˈspen(t)s\ *verb* **dis·pensed; dis·pens·ing** [Middle English, from Medieval Latin & Latin; Medieval Latin *dispensare* to exempt, from Latin, to distribute, from *dis-* + *pensare* to weigh, frequentative of *pendere* to weigh, pay out — more at SPIN] (14th century)
transitive verb
1 a : to deal out in portions **b :** ADMINISTER ⟨*dispense* justice⟩
2 : to give dispensation to **:** EXEMPT
3 : to prepare and distribute (medication)
intransitive verb
archaic **:** to grant dispensation
synonym see DISTRIBUTE
— **dispense with 1 :** to set aside **:** DISCARD ⟨*dispensing with* the usual introduction⟩ **2 :** to do without ⟨could *dispense with* such a large staff⟩

dis·pens·er \-ˈspen(t)-sər\ *noun* (14th century)
: one that dispenses: as **a :** a container that extrudes, sprays, or feeds out in convenient units **b :** a usually mechanical device for vending merchandise

dis·peo·ple \(ˌ)dis ˈpē-pəl\ *transitive verb* (15th century)
: DEPOPULATE

dis·pers·al \di-ˈspər-səl\ *noun* (1821)
: the act or result of dispersing; *especially* **:** the process or result of the spreading of organisms from one place to another

dis·per·sant \di-ˈspər-sənt\ *noun* (1941)
: a dispersing agent; *especially* **:** a substance for promoting the formation and stabilization of a dispersion of one substance in another
— **dispersant** *adjective*

dis·perse \di-ˈspərs\ *verb* **dis·persed; dis·pers·ing** [Middle English *dysparsen,* from Middle French *disperser,* from Latin *dispersus,* past participle of *dispergere* to scatter, from *dis-* + *spargere* to scatter — more at SPARK] (14th century)
transitive verb
1 a : to cause to break up ⟨police *dispersed* the crowd⟩ **b :** to cause to become spread widely **c :** to cause to evaporate or vanish ⟨sunlight *dispersing* the mist⟩
2 : to spread or distribute from a fixed or constant source: as **a** *archaic* **:** DISSEMINATE **b :** to subject (as light) to dispersion **c :** to distribute (as fine particles) more or less evenly throughout a medium
intransitive verb
1 : to break up in random fashion ⟨the crowd *dispersed* on request⟩
2 a : to become dispersed **b :** DISSIPATE, VANISH ⟨the fog *dispersed* toward morning⟩
synonym see SCATTER
— **dis·persed·ly** \-ˈspər-səd-lē, -ˈspərst-lē\ *adverb*

\ə\ **abut** \ᵊ\ **kitten** \ər\ **further** \a\ **ash** \ā\ **ace** \ä\ **mop, mar** \au̇\ **out** \ch\ **chin** \e\ **bet** \ē\ **easy** \g\ **go** \i\ **hit** \ī\ **ice** \j\ **job** \ŋ\ **sing** \ō\ **go** \ȯ\ **law** \ȯi\ **boy** \th\ **thin** \t͟h\ **the** \ü\ **loot** \u̇\ **foot** \y\ **yet** \zh\ **vision** *see also* Guide to Pronunciation

— **dis·pers·er** *noun*
— **dis·pers·ible** \-ˈspər-sə-bəl\ *adjective*
dis·per·sion \di-ˈspər-zhən, -shən\ *noun* (14th century)
1 *capitalized* **:** DIASPORA 1a
2 : the act or process of dispersing **:** the state of being dispersed
3 : the scattering of the values of a frequency distribution from an average
4 : the separation of light into colors by refraction or diffraction with formation of a spectrum; *also* **:** the separation of radiation into components in accordance with some varying characteristic (as energy)
5 a : a dispersed substance **b :** a system consisting of a dispersed substance and the medium in which it is dispersed **:** COLLOID 2b
dis·per·sive \-ˈspər-siv, -ziv\ *adjective* (1677)
1 : of or relating to dispersion ⟨a *dispersive* medium⟩ ⟨the *dispersive* power of a lens⟩
2 : tending to disperse
— **dis·per·sive·ly** *adverb*
— **dis·per·sive·ness** *noun*
dis·per·soid \-ˈspər-ˌsȯid\ *noun* (1911)
: finely divided particles of one substance dispersed in another
dispir·it \(ˌ)dis-ˈpir-ət\ *transitive verb* [*dis-* + *spirit*] (1647)
: to deprive of morale or enthusiasm
— **dispir·it·ed** *adjective*
— **dispir·it·ed·ly** *adverb*
— **dispir·it·ed·ness** *noun*
dis·pit·eous \di-ˈspi-tē-əs\ *adjective* [alteration of *despiteous*] (1803)
archaic **:** CRUEL
dis·place \(ˌ)dis-ˈplās\ *transitive verb* [probably from Middle French *desplacer,* from *des-* dis- + *place* place] (1553)
1 a : to remove from the usual or proper place; *specifically* **:** to expel or force to flee from home or homeland **b :** to remove from an office, status, or job **c** *obsolete* **:** to drive out **:** BANISH
2 a : to move physically out of position ⟨a floating object *displaces* water⟩ **b :** to take the place of (as in a chemical reaction) **:** SUPPLANT
synonym see REPLACE
— **dis·place·able** \-ˈplā-sə-bəl\ *adjective*
dis·place·ment \di-ˈsplā-smənt\ *noun* (1611)
1 : the act or process of displacing **:** the state of being displaced
2 a : the volume or weight of a fluid (as water) displaced by a floating body (as a ship) of equal weight **b :** the difference between the initial position of something (as a body or geometric figure) and any later position **c :** the volume displaced by a piston (as in a pump or an engine) in a single stroke; *also* **:** the total volume so displaced by all the pistons in an internal combustion engine (as in an automobile)
3 a : the redirection of an emotion or impulse from its original object (as an idea or person) to another **b :** the substitution of another form of behavior for what is usual or expected especially when the usual response is nonadaptive
dis·plant \di-ˈsplant\ *transitive verb* [Middle French *desplanter,* from *des-* dis- + *planter* to plant, from Late Latin *plantare*] (15th century)
1 : DISPLACE, REMOVE
2 : SUPPLANT
¹dis·play \di-ˈsplā\ *verb* [Middle English, from Anglo-French *despleer, desploier,* literally, to unfold — more at DEPLOY] (14th century)
transitive verb
1 a : to put or spread before the view ⟨*display* the flag⟩ **b :** to make evident ⟨*displayed* great skill⟩ **c :** to exhibit ostentatiously ⟨liked to *display* his erudition⟩
2 *obsolete* **:** DESCRY
intransitive verb
1 *obsolete* **:** SHOW OFF
2 : to make a breeding display ⟨penguins *displayed* and copulated⟩
synonym see SHOW

— **dis·play·able** \-ˈsplā-ə-bəl\ *adjective*
²display *noun, often attributive* (1665)
1 a (1) **:** a setting or presentation of something in open view ⟨a fireworks *display*⟩ (2) **:** a clear sign or evidence **:** EXHIBITION ⟨a *display* of courage⟩ **b :** ostentatious show **c :** type, composition, or printing designed to catch the eye **d :** an eye-catching arrangement by which something is exhibited **e :** an electronic device (as a cathode-ray tube) that temporarily presents information in visual form; *also* **:** the visual information
2 : a pattern of behavior exhibited especially by male birds in the breeding season
dis·please \(ˌ)dis-ˈplēz\ *verb* [Middle English *displesen,* from Middle French *desplaisir,* modification of (assumed) Vulgar Latin *displacēre,* from Latin *dis-* + *placēre* to please — more at PLEASE] (14th century)
transitive verb
1 : to incur the disapproval or dislike of especially by annoying ⟨their gossip *displeases* her⟩
2 : to be offensive to ⟨abstract art *displeases* him⟩
intransitive verb
: to give displeasure ⟨behavior calculated to *displease*⟩
dis·plea·sure \(ˌ)dis-ˈple-zhər, -ˈplā-\ *noun* (15th century)
1 : the feeling of one that is displeased **:** DISFAVOR
2 : DISCOMFORT, UNHAPPINESS
3 *archaic* **:** OFFENSE, INJURY
dis·plode \di-ˈsplōd\ *verb* **dis·plod·ed; dis·plod·ing** [Latin *displodere,* from *dis-* + *plaudere* to clap, applaud] (1667)
archaic **:** EXPLODE
— **dis·plo·sion** \-ˈsplō-zhən\ *noun*
¹dis·port \di-ˈspȯrt, -ˈspȯrt\ *noun* (14th century)
archaic **:** SPORT, PASTIME
²disport *verb* [Middle English, from Middle French *desporter,* from *des-* dis- + *porter* to carry, from Latin *portare* — more at FARE] (14th century)
transitive verb
1 : DIVERT, AMUSE
2 : DISPLAY
intransitive verb
: to amuse oneself in light or lively fashion **:** FROLIC
— **dis·port·ment** \-mənt\ *noun*
¹dis·pos·able \di-ˈspō-zə-bəl\ *adjective* (1643)
1 : subject to or available for disposal; *specifically* **:** remaining to an individual after deduction of taxes ⟨*disposable* income⟩
2 : designed to be used once and then thrown away ⟨*disposable* diapers⟩
— **dis·pos·abil·i·ty** \-ˌspō-zə-ˈbi-lə-tē\ *noun*
²disposable *noun* (1963)
: something that is disposable
dis·pos·al \di-ˈspō-zəl\ *noun* (1630)
1 : the power or authority to dispose or make use of as one chooses ⟨the car was at my *disposal*⟩
2 : the act or process of disposing: as **a :** orderly placement or distribution **b :** REGULATION, ADMINISTRATION **c :** the act or action of presenting or bestowing something ⟨*disposal* of favors⟩ **d :** systematic destruction; *especially* **:** destruction or transformation of garbage
3 [*garbage disposal unit*] **:** a device used to reduce waste matter (as by grinding)
¹dis·pose \di-ˈspōz\ *verb* **dis·posed; dis·pos·ing** [Middle English, from Middle French *disposer,* from Latin *disponere* to arrange (perfect indicative *disposui*), from *dis-* + *ponere* to put — more at POSITION] (14th century)
transitive verb
1 : to give a tendency to **:** INCLINE ⟨faulty diet *disposes* one to sickness⟩

2 a : to put in place **:** set in readiness **:** ARRANGE ⟨*disposing* troops for withdrawal⟩ **b** *obsolete* **:** REGULATE **c :** BESTOW
intransitive verb
1 : to settle a matter finally
2 *obsolete* **:** to come to terms
synonym see INCLINE
— **dis·pos·er** *noun*
— **dispose of 1 :** to place, distribute, or arrange especially in an orderly way **2 a :** to transfer to the control of another ⟨*disposing of* personal property to a total stranger⟩ **b** (1) **:** to get rid of ⟨how to *dispose of* toxic waste⟩ (2) **:** to deal with conclusively ⟨*disposed of* the matter efficiently⟩
²dispose *noun* (1590)
1 *obsolete* **:** DISPOSAL
2 *obsolete* **a :** DISPOSITION **b :** DEMEANOR
dis·po·si·tion \ˌdis-pə-ˈzi-shən\ *noun* [Middle English, from Middle French, from Latin *disposition-, dispositio,* from *disponere*] (14th century)
1 : the act or the power of disposing or the state of being disposed: as **a :** ADMINISTRATION, CONTROL **b :** final arrangement **:** SETTLEMENT ⟨the *disposition* of the case⟩ **c** (1) **:** transfer to the care or possession of another (2) **:** the power of such transferal **d :** orderly arrangement
2 a : prevailing tendency, mood, or inclination **b :** temperamental makeup **c :** the tendency of something to act in a certain manner under given circumstances ☆
— **dis·po·si·tion·al** \-ˈzish-nəl, -ˈzi-shənᵊl\ *adjective*
dis·pos·i·tive \di-ˈspä-zə-tiv\ *adjective* (1613)
: directed toward or effecting disposition (as of a case) ⟨*dispositive* evidence⟩
dis·pos·sess \ˌdis-pə-ˈzes *also* -ˈses\ *transitive verb* [Middle French *despossesser,* from *des-* dis- + *possesser* to possess] (1555)
: to put out of possession or occupancy ⟨*dispossessed* the nobles of their land⟩
— **dis·pos·ses·sion** \-ˈze-shən *also* -ˈse-\ *noun*
— **dis·pos·ses·sor** \-ˈze-sər *also* -ˈse-\ *noun*
dis·pos·sessed *adjective* (15th century)
: deprived of homes, possessions, and security
dis·po·sure \di-ˈspō-zhər\ *noun* (1569)
archaic **:** DISPOSAL, DISPOSITION
dis·praise \(ˌ)dis-ˈprāz\ *transitive verb* [Middle English *dispraisen,* from Middle French *despreisier,* from Old French, from *des-* dis- + *preisier* to praise] (13th century)
: to comment on with disapproval or censure
— **dispraise** *noun*
— **dis·prais·er** *noun*
— **dis·prais·ing·ly** \-ˈprā-ziŋ-lē\ *adverb*
dispread \di-ˈspred\ *transitive verb* (1590)
: to spread abroad or out

☆ SYNONYMS
Disposition, temperament, temper, character, personality mean the dominant quality or qualities distinguishing a person or group. DISPOSITION implies customary moods and attitude toward the life around one ⟨a cheerful *disposition*⟩. TEMPERAMENT implies a pattern of innate characteristics associated with one's specific physical and nervous organization ⟨an artistic *temperament*⟩. TEMPER implies the qualities acquired through experience that determine how a person or group meets difficulties or handles situations ⟨a resilient *temper*⟩. CHARACTER applies to the aggregate of moral qualities by which a person is judged apart from intelligence, competence, or special talents ⟨strength of *character*⟩. PERSONALITY applies to an aggregate of qualities that distinguish one as a person ⟨a somber *personality*⟩.

dis·prize \(ˌ)dis-ˈprīz\ *transitive verb* [Middle French *despriser,* from Old French *despreisier*] (15th century)
archaic **:** UNDERVALUE, SCORN

dis·proof \(ˌ)dis-ˈprüf\ *noun* (15th century)
1 : the action of disproving
2 : evidence that disproves

[1]dis·pro·por·tion \ˌdis-prə-ˈpōr-shən, -ˈpȯr-\ *noun* (1555)
: lack of proportion, symmetry, or proper relation **:** DISPARITY; *also* **:** an instance of such disparity
— dis·pro·por·tion·al \-shnəl, -shə-nᵊl\ *adjective*

[2]disproportion *transitive verb* (1593)
: to make out of proportion **:** MISMATCH

dis·pro·por·tion·ate \-sh(ə-)nət\ *adjective* (1555)
: being out of proportion ⟨a *disproportionate* share⟩
— dis·pro·por·tion·ate·ly *adverb*

dis·pro·por·tion·ation \-ˌpōr-shə-ˈnā-shən, -ˌpȯr-\ *noun* (circa 1929)
: the transformation of a substance into two or more dissimilar substances usually by simultaneous oxidation and reduction
— dis·pro·por·tion·ate \-ˈpōr-shə-ˌnāt, -ˈpȯr-\ *intransitive verb*

dis·prove \(ˌ)dis-ˈprüv\ *transitive verb* [Middle English, from Middle French *desprover,* from *des-* dis- + *prover* to prove] (14th century)
: to prove to be false or wrong **:** REFUTE
— dis·prov·able \-ˈprü-və-bəl\ *adjective*

dis·pu·tant \di-ˈspyü-tᵊnt, ˈdis-pyə-tənt\ *noun* (1593)
: one that is engaged in a dispute

dis·pu·ta·tion \ˌdis-pyə-ˈtā-shən\ *noun* (14th century)
1 : the action of disputing **:** verbal controversy ⟨continuous *disputation* between them⟩ ⟨ideological *disputations*⟩
2 : an academic exercise in oral defense of a thesis by formal logic

dis·pu·ta·tious \-shəs\ *adjective* (1660)
1 a : inclined to dispute **b :** marked by disputation
2 : provoking debate **:** CONTROVERSIAL
— dis·pu·ta·tious·ly *adverb*
— dis·pu·ta·tious·ness *noun*

[1]dis·pute \di-ˈspyüt\ *verb* **dis·put·ed; dis·put·ing** [Middle English, from Old French *desputer,* from Latin *disputare* to discuss, from *dis-* + *putare* to think] (13th century)
intransitive verb
: to engage in argument **:** DEBATE; *especially* **:** to argue irritably or with irritating persistence
transitive verb
1 a : to make the subject of disputation **b :** to call into question ⟨her honesty was never *disputed*⟩
2 a : to struggle against **:** OPPOSE ⟨*disputed* the advance of the invaders⟩ **b :** to contend over ⟨both sides *disputed* the bridgehead⟩
— dis·put·able \di-ˈspyü-tə-bəl, ˈdis-pyə-\ *adjective*
— dis·put·ably \-blē\ *adverb*
— dis·put·er *noun*

[2]dis·pute \di ˈspyüt, ˈdis-ˌpyüt\ *noun* (1608)
1 a : verbal controversy **:** DEBATE **b :** QUARREL
2 *obsolete* **:** physical combat

dis·qual·i·fi·ca·tion \(ˌ)dis-ˌkwä-lə-fə-ˈkā-shən\ *noun* (1714)
1 : something that disqualifies or incapacitates
2 : the act of disqualifying **:** the state of being disqualified ⟨*disqualification* from office⟩

dis·qual·i·fy \(ˌ)dis-ˈkwä-lə-ˌfī\ *transitive verb* (1723)
1 : to deprive of the required qualities, properties, or conditions **:** make unfit
2 : to deprive of a power, right, or privilege
3 : to make ineligible for a prize or for further competition because of violations of the rules

dis·quan·ti·ty \(ˌ)dis-ˈkwän-(t)ə-tē\ *transitive verb* (1605)
obsolete **:** DIMINISH, LESSEN

[1]dis·qui·et \(ˌ)dis-ˈkwī-ət\ *transitive verb* (circa 1530)
: to take away the peace or tranquillity of **:** DISTURB, ALARM
synonym see DISCOMPOSE
— dis·qui·et·ing *adjective*
— dis·qui·et·ing·ly \-ˈkwī-ə-tiŋ-lē\ *adverb*

[2]disquiet *noun* (1581)
: lack of peace or tranquillity **:** ANXIETY

[3]disquiet *adjective* (1587)
archaic **:** UNEASY, DISQUIETED
— dis·qui·et·ly *adverb*

dis·qui·etude \(ˌ)dis-ˈkwī-ə-ˌt(y)üd\ *noun* (1709)
: ANXIETY, AGITATION

dis·qui·si·tion \ˌdis-kwə-ˈzi-shən\ *noun* [Latin *disquisition-, disquisitio,* from *disquirere* to investigate, from *dis-* + *quaerere* to seek] (1647)
: a formal inquiry into or discussion of a subject **:** DISCOURSE

dis·rate \(ˌ)dis-ˈrāt\ *transitive verb* (1811)
: to reduce in rank **:** DEMOTE

[1]dis·re·gard \ˌdis-ri-ˈgärd\ *transitive verb* (1641)
: to pay no attention to **:** treat as unworthy of regard or notice
synonym see NEGLECT

[2]disregard *noun* (1665)
: the act of disregarding **:** the state of being disregarded **:** NEGLECT
— dis·re·gard·ful \-fəl\ *adjective*

dis·re·lat·ed \ˌdis-ri-ˈlā-təd\ *adjective* (1894)
: not related

dis·re·la·tion \-ˈlā-shən\ *noun* (1893)
: lack of a fitting or proportionate connection or relationship

[1]dis·rel·ish \(ˌ)dis-ˈre-lish\ *transitive verb* (1604)
: to find unpalatable or distasteful

[2]disrelish *noun* (circa 1625)
: lack of relish **:** DISTASTE, DISLIKE

dis·re·mem·ber \ˌdis-ri-ˈmem-bər\ *transitive verb* (1815)
: FORGET

dis·re·pair \ˌdis-ri-ˈpar, -ˈper\ *noun* (1798)
: the state of being in need of repair ⟨a building fallen into *disrepair*⟩

dis·rep·u·ta·ble \(ˌ)dis-ˈre-pyə-tə-bəl\ *adjective* (1772)
: not reputable
— dis·rep·u·ta·bil·i·ty \(ˌ)dis-ˌre-pyə-tə-ˈbi-lə-tē\ *noun*
— dis·rep·u·ta·ble·ness \(ˌ)dis-ˈre-pyə-tə-bəl-nəs\ *noun*
— dis·rep·u·ta·bly \-blē\ *adverb*

dis·re·pute \ˌdis-ri-ˈpyüt\ *noun* (1653)
: lack or decline of good reputation **:** a state of being held in low esteem
synonym see DISGRACE

[1]dis·re·spect \ˌdis-ri-ˈspekt\ *transitive verb* (1614)
: to have disrespect for

[2]disrespect *noun* (1631)
: lack of respect
— dis·re·spect·ful \-fəl\ *adjective*
— dis·re·spect·ful·ly \-fə-lē\ *adverb*
— dis·re·spect·ful·ness *noun*

dis·re·spect·able \ˌdis-ri-ˈspek-tə-bəl\ *adjective* (1813)
: not respectable
— dis·re·spect·abil·i·ty \-ˌspek-tə-ˈbi-lə-tē\ *noun*

dis·robe \(ˌ)dis-ˈrōb\ *verb* [Middle French *desrober,* from *des-* dis- + *robe* garment, from Old French] (1581)
intransitive verb
: to take off one's clothing
transitive verb
: to strip of clothing or covering

dis·rupt \dis-ˈrəpt\ *transitive verb* [Latin *disruptus,* past participle of *disrumpere,* from *dis-* + *rumpere* to break — more at REAVE] (1817)
1 a : to break apart **:** RUPTURE **b :** to throw into disorder ⟨agitators trying to *disrupt* the meeting⟩
2 : to interrupt the normal course or unity of
— dis·rupt·er *noun*
— dis·rup·tion \-ˈrəp-shən\ *noun*
— dis·rup·tive \-ˈrəp-tiv\ *adjective*
— dis·rup·tive·ly *adverb*
— dis·rup·tive·ness *noun*

diss *variant of* DIS

dis·sat·is·fac·tion \(ˌ)di(s)-ˌsat-əs-ˈfak-shən\ *noun* (1640)
: the quality or state of being dissatisfied **:** DISCONTENT

dis·sat·is·fac·to·ry \-ˈfak-t(ə-)rē\ *adjective* (circa 1610)
: causing dissatisfaction

dis·sat·is·fied \(ˌ)di(s)-ˈsa-təs-ˌfīd\ *adjective* (1675)
: expressing or showing lack of satisfaction **:** not pleased or satisfied

dis·sat·is·fy \-ˌfī\ *transitive verb* (1666)
: to fail to satisfy **:** DISPLEASE

dis·save \(ˌ)di(s)-ˈsāv\ *intransitive verb* (1936)
: to use savings for current expenses

dis·seat \(ˌ)di(s)-ˈsēt\ *transitive verb* (1612)
archaic **:** UNSEAT

dis·sect \di-ˈsekt; ÷dī-ˈsekt, ÷ˈdī-ˌ\ *verb* [Latin *dissectus,* past participle of *dissecare* to cut apart, from *dis-* + *secare* to cut — more at SAW] (1607)
transitive verb
1 : to separate into pieces **:** expose the several parts of (as an animal) for scientific examination
2 : to analyze and interpret minutely
intransitive verb
: to make a dissection
synonym see ANALYZE
— dis·sec·tor \-ˈsek-tər\ *noun*

dis·sect·ed *adjective* (1652)
1 : cut deeply into fine lobes ⟨a *dissected* leaf⟩
2 : divided into hills and ridges (as by gorges) ⟨a *dissected* plateau⟩

dis·sect·ing microscope *noun* (circa 1897)
: a microscope with low magnification

dis·sec·tion \di-ˈsek-shən; ÷dī-ˈsek-, ÷ˈdī-ˌ\ *noun* (1605)
1 : the act or process of dissecting **:** the state of being dissected
2 : an anatomical specimen prepared by dissecting

dis·seise *or* **dis·seize** \(ˌ)di(s)-ˈsēz\ *transitive verb* **dis·seised** *or* **dis·seized; dis·seis·ing** *or* **dis·seiz·ing** [Middle English *disseisen,* from Medieval Latin *disseisiare* & Anglo-French *disseisir,* from Old French *dessaisir,* from *des-* dis- + *saisir* to put in possession of — more at SEIZE] (14th century)
: to deprive especially wrongfully of seisin **:** DISPOSSESS
— dis·sei·sor \-ˈsē-zər\ *noun*

dis·sei·sin *or* **dis·sei·zin** \-ˈsē-zᵊn\ *noun* [Middle English *dysseysyne,* from Anglo-French *disseisine,* from Old French *dessaisine,* from *des-* dis- + *saisine* seisin] (14th century)
: the act of disseising **:** the state of being disseised

dis·sem·ble \di-ˈsem-bəl\ *verb* **dis·sem·bled; dis·sem·bling** \-b(ə-)liŋ\ [Middle English *dissymblen,* alteration of *dissimulen,* from Middle French *dissimuler,* from Latin *dissimulare* — more at DISSIMULATE] (15th century)
transitive verb
1 : to hide under a false appearance
2 : to put on the appearance of **:** SIMULATE
intransitive verb
: to put on a false appearance **:** conceal facts, intentions, or feelings under some pretense
— dis·sem·bler \-b(ə-)lər\ *noun*

dis·sem·i·nate \di-'se-mə-ˌnāt\ *transitive verb* **-nat·ed; -nat·ing** [Latin *disseminatus,* past participle of *disseminare,* from *dis-* + *seminare* to sow, from *semin-, semen* seed — more at SEMEN] (1603)
1 : to spread abroad as though sowing seed ⟨*disseminate* ideas⟩
2 : to disperse throughout
— dis·sem·i·na·tion \-ˌse-mə-'nā-shən\ *noun*
— dis·sem·i·na·tor \-'se-mə-ˌnā-tər\ *noun*

dis·sem·i·nat·ed *adjective* (1876)
: widely dispersed in a tissue, organ, or the entire body ⟨*disseminated* gonococcal disease⟩

dis·sem·i·nule \di-'sem-ə-ˌnyül\ *noun* (1904)
: a part or organ (as a seed or spore) of a plant that ensures propagation

dis·sen·sion *also* **dis·sen·tion** \di-'sen(t)-shən\ *noun* [Middle English, from Middle French, from Latin *dissension-, dissensio,* from *dissentire*] (14th century)
: DISAGREEMENT; *especially* **:** partisan and contentious quarreling
synonym see DISCORD

dis·sen·sus \(ˌ)di(s)-'sen(t)-səs\ *noun* [*dis-* + *consensus*] (1966)
: difference of opinion

¹dis·sent \di-'sent\ *intransitive verb* [Middle English, from Latin *dissentire,* from *dis-* + *sentire* to feel — more at SENSE] (15th century)
1 : to withhold assent
2 : to differ in opinion

²dissent *noun* (1585)
: difference of opinion: as **a :** religious nonconformity **b :** a justice's nonconcurrence with a decision of the majority — called also *dissenting opinion*

dis·sent·er \di-'sen-tər\ *noun* (1639)
1 : one that dissents
2 *capitalized* **:** an English Nonconformist

dis·sen·tient \di-'sen(t)-sh(ē-)ənt\ *adjective* [Latin *dissentient-, dissentiens,* present participle of *dissentire*] (1651)
: expressing dissent
— dissentient *noun*

dis·sent·ing \di-'sen-tiŋ\ *adjective, often capitalized* (1644)
: of or relating to the English Nonconformists ⟨a *dissenting* church⟩ ⟨*dissenting* merchants⟩

dis·sen·tious \di-'sen(t)-shəs\ *adjective* (1560)
: characterized by dissension or dissent

dis·sep·i·ment \di-'se-pə-mənt\ *noun* [Latin *dissaepimentum* partition, from *dissaepire* to divide, from *dis-* + *saepire* to fence in — more at SEPTUM] (circa 1727)
: a dividing tissue **:** SEPTUM

dis·sert \di-'sərt\ *intransitive verb* [Latin *dissertus,* past participle of *disserere,* from *dis-* + *serere* to join, arrange — more at SERIES] (1657)
: to speak or write at length

dis·ser·tate \'di-sər-ˌtāt\ *intransitive verb* **-tat·ed; -tat·ing** [Latin *dissertatus,* past participle of *dissertare,* from *dissertus*] (1766)
: DISSERT; *also* **:** to write a dissertation
— dis·ser·ta·tor \-ˌtā-tər\ *noun*

dis·ser·ta·tion \ˌdi-sər-'tā-shən\ *noun* (1651)
: an extended usually written treatment of a subject; *specifically* **:** one submitted for a doctorate
— dis·ser·ta·tion·al \-'tāsh-nəl, -'tā-shənᵊl\ *adjective*

dis·serve \(ˌ)di(s)-'sərv\ *transitive verb* (1629)
: to serve badly or falsely **:** HARM

dis·ser·vice \(ˌ)di(s)-'sər-vəs\ *noun* (1599)
: ill service **:** HARM; *also* **:** an unhelpful, unkind, or harmful act

dis·ser·vice·able \di(s)-'sər-və-sə-bəl\ *adjective* (1644)
: COUNTERPRODUCTIVE

dis·sev·er \di-'se-vər\ *verb* [Middle English, from Old French *dessevrer,* from Late Latin *disseparare,* from Latin *dis-* + *separare* to separate] (13th century)
transitive verb
: SEVER, SEPARATE
intransitive verb
: to come apart **:** DISUNITE
— dis·sev·er·ance \-'se-v(ə-)rən(t)s\ *noun*
— dis·sev·er·ment \-'se-vər-mənt\ *noun*

dis·si·dence \'di-sə-dən(t)s\ *noun* (circa 1656)
: DISSENT, DISAGREEMENT

dis·si·dent \-dənt\ *adjective* [Latin *dissident-, dissidens,* present participle of *dissidēre* to sit apart, disagree, from *dis-* + *sedēre* to sit — more at SIT] (1837)
: disagreeing especially with an established religious or political system, organization, or belief
— dissident *noun*

dis·sim·i·lar \(ˌ)di(s)-'si-mə-lər, -'sim-lər\ *adjective* (1599)
: UNLIKE
— dis·sim·i·lar·i·ty \(ˌ)di(s)-ˌsi-mə-'lar-ə-tē\ *noun*
— dis·sim·i·lar·ly \(')di(s)-'si-mə-lər-lē, -'sim-lər-\ *adverb*
— dis·sim·i·lars \-lərz\ *noun plural*

dis·sim·i·late \(ˌ)di-'si-mə-ˌlāt\ *intransitive verb* **-lat·ed; -lat·ing** [*dis-* + *-similate* (as in *assimilate*)] (1841)
: to undergo dissimilation
— dis·sim·i·la·to·ry \-mə-lə-ˌtōr-ē, -ˌtȯr-\ *adjective*

dis·sim·i·la·tion \(ˌ)di-ˌsi-mə-'lā-shən\ *noun* (circa 1874)
: the change or omission of one of two identical or closely related sounds in a word

dis·si·mil·i·tude \ˌdi(s)-sə-'mi-lə-ˌtüd, -ˌtyüd\ *noun* [Latin *dissimilitudo,* from *dissimilis* unlike, from *dis-* + *similis* like — more at SAME] (15th century)
: lack of resemblance

dis·sim·u·late \(ˌ)di-'sim-yə-ˌlāt\ *verb* **-lat·ed; -lat·ing** [Latin *dissimulatus,* past participle of *dissimulare,* from *dis-* + *simulare* to simulate] (15th century)
transitive verb
: to hide under a false appearance ⟨smiled to *dissimulate* her urgency —Alice Glenday⟩
intransitive verb
: DISSEMBLE
— dis·sim·u·la·tion \(ˌ)di-ˌsim-yə-'lā-shən\ *noun*
— dis·sim·u·la·tor \(')di-'sim-yə-ˌlā-tər\ *noun*

dis·si·pate \'di-sə-ˌpāt\ *verb* **-pat·ed; -pat·ing** [Latin *dissipatus,* past participle of *dissipare, dissupare,* from *dis-* + *supare* to throw] (15th century)
transitive verb
1 a : to break up and drive off (as a crowd) **b :** to cause to spread thin or scatter and gradually vanish ⟨one's sympathy is eventually *dissipated* —Andrew Feinberg⟩ **c :** to lose (as heat or electricity) irrecoverably
2 : to spend or use up wastefully or foolishly ⟨lifelong tendency to *dissipate* his gifts in travel and pleasure —Edmund Morris⟩ ⟨his fortune is *dissipated* in imprudent political adventures —John Butt⟩
intransitive verb
1 : to break up and scatter or vanish
2 : to be extravagant or dissolute in the pursuit of pleasure; *especially* **:** to drink to excess
synonym see SCATTER
— dis·si·pat·er *noun*

dis·si·pat·ed *adjective* (1744)
: given to or marked by dissipation **:** DISSOLUTE
— dis·si·pat·ed·ly *adverb*
— dis·si·pat·ed·ness *noun*

dis·si·pa·tion \ˌdi-sə-'pā-shən\ *noun* (15th century)
1 : the action or process of dissipating **:** the state of being dissipated: **a :** DISPERSION, DIFFUSION **b** *archaic* **:** DISSOLUTION, DISINTEGRATION **c :** wasteful expenditure **d :** intemperate living; *especially* **:** excessive drinking
2 : an act of self-indulgence; *especially* **:** one that is not harmful **:** AMUSEMENT

dis·si·pa·tive \'di-sə-ˌpā-tiv\ *adjective* (1684)
: relating to dissipation especially of heat

dis·so·cia·ble \(ˌ)di-'sō-sh(ē-)ə-bəl, -sē-ə-\ *adjective* (1833)
: SEPARABLE
— dis·so·cia·bil·i·ty \(ˌ)di-ˌsō-sh(ē-)ə-'bi-lə-tē, -sē-ə-\ *noun*

dis·so·cial \(ˌ)di(s)-'sō-shəl\ *adjective* (1762)
: UNSOCIAL, SELFISH

dis·so·ci·ate \(ˌ)di-'sō-shē-ˌāt, -sē-\ *verb* **-at·ed; -at·ing** [Latin *dissociatus,* past participle of *dissociare,* from *dis-* + *sociare* to join, from *socius* companion — more at SOCIAL] (1590)
transitive verb
1 : to separate from association or union with another
2 : DISUNITE; *specifically* **:** to subject to chemical dissociation
intransitive verb
1 : to undergo dissociation
2 : to mutate especially reversibly

dis·so·ci·a·tion \(ˌ)di-ˌsō-sē-'ā-shən, -shē-\ *noun* (1611)
1 : the act or process of dissociating **:** the state of being dissociated: as **a :** the process by which a chemical combination breaks up into simpler constituents; *especially* **:** one that results from the action of energy (as heat) on a gas or of a solvent on a dissolved substance **b :** the separation of whole segments of the personality (as in multiple personality) or of discrete mental processes (as in the schizophrenias) from the mainstream of consciousness or of behavior
2 : the property inherent in some biological stocks (as of certain bacteria) of differentiating into two or more distinct and relatively permanent strains; *also* **:** such a strain
— dis·so·cia·tive \(ˌ)di-'sō-shē-ˌā-tiv, -sē-, -shə-tiv\ *adjective*

dis·sol·u·ble \di-'säl-yə-bəl\ *adjective* [Latin *dissolubilis,* from *dissolvere* to dissolve] (1534)
: capable of being dissolved or disintegrated

dis·so·lute \'di-sə-ˌlüt, -lət\ *adjective* [Latin *dissolutus,* from past participle of *dissolvere* to loosen, dissolve] (14th century)
: lacking restraint; *especially* **:** marked by indulgence in things (as drink or promiscuous sex) deemed vices ⟨*dissolute* and degrading aspects of human nature —Wallace Fowlie⟩
— dis·so·lute·ly *adverb*
— dis·so·lute·ness *noun*

dis·so·lu·tion \ˌdi-sə-'lü-shən\ *noun* (14th century)
1 : the act or process of dissolving: as **a :** separation into component parts **b** (1) **:** DECAY, DISINTEGRATION (2) **:** DEATH **c :** termination or destruction by breaking down, disrupting, or dispersing ⟨the *dissolution* of the republic⟩ **d :** the dissolving of an assembly or organization **e :** LIQUEFACTION
2 : a dissolute act or practice

¹dis·solve \di-'zälv, -'zȯlv *also* -'zäv *or* -'zȯv\ *verb* [Middle English, from Latin *dissolvere,* from *dis-* + *solvere* to loosen — more at SOLVE] (14th century)
transitive verb
1 a : to cause to disperse or disappear **:** DESTROY **b :** to separate into component parts **:** DISINTEGRATE **c :** to bring to an end **:** TERMINATE ⟨*dissolve* parliament⟩ **d :** ANNUL ⟨*dissolve* an injunction⟩
2 a : to cause to pass into solution ⟨*dissolve* sugar in water⟩ **b :** MELT, LIQUEFY **c :** to cause to be emotionally moved **d :** to cause to fade out in a dissolve
3 *archaic* **:** DETACH, LOOSEN
4 : to clear up ⟨*dissolve* a problem⟩
intransitive verb
1 a : to become dissipated or decomposed **b :** BREAK UP, DISPERSE **c :** to fade away

2 a : to become fluid **:** MELT **b :** to pass into solution **c :** to be overcome emotionally ⟨*dissolved* into tears⟩ **d :** to resolve itself as if by dissolution ⟨hate *dissolved* into fear⟩ **e :** to change by a dissolve ⟨the scene *dissolves* to a Victorian parlor⟩
— **dis·solv·able** \-ˈzäl-və-bəl, -ˈzȯl-\ *adjective*
— **dis·sol·vent** \-ˈzäl-vənt, -ˈzȯl-\ *noun or adjective*
— **dis·solv·er** *noun*

[2]dissolve *noun* (1916)
: a gradual superimposing of one motion-picture or television shot upon another on a screen

dis·so·nance \ˈdi-sə-nən(t)s\ *noun* (15th century)
1 a : lack of agreement, *especially* **:** inconsistency between the beliefs one holds or between one's actions and one's beliefs — compare COGNITIVE DISSONANCE **b :** an instance of such inconsistency or disagreement
2 : a mingling of discordant sounds; *especially* **:** a clashing or unresolved musical interval or chord

dis·so·nant \-nənt\ *adjective* [Middle French or Latin; Middle French, from Latin *dissonant-, dissonans,* present participle of *dissonare* to be discordant, from *dis-* + *sonare* to sound — more at SOUND] (15th century)
1 : marked by dissonance **:** DISCORDANT
2 : INCONGRUOUS
3 : harmonically unresolved
— **dis·so·nant·ly** *adverb*

dis·suade \di-ˈswād\ *transitive verb* **dis·suad·ed; dis·suad·ing** [Middle French or Latin; Middle French *dissuader,* from Latin *dissuadēre,* from *dis-* + *suadēre* to urge — more at SWEET] (15th century)
1 a : to advise (a person) against something **b** *archaic* **:** to advise against (an action)
2 : to turn from something by persuasion
— **dis·suad·er** *noun*

dis·sua·sion \di-ˈswā-zhən\ *noun* [Middle French or Latin; Middle French, from Latin *dissuasion-, dissuasio,* from *dissuadēre*] (15th century)
: the action of dissuading

dis·sua·sive \di-ˈswā-siv, -ziv\ *adjective* (1609)
: tending to dissuade
— **dis·sua·sive·ly** *adverb*
— **dis·sua·sive·ness** *noun*

dis·syl·la·ble \ˈdi-ˌsi-lə-bəl, (ˌ)di(s)-ˈsi-; ˈdī-ˌsi-, (ˌ)dī-ˈsi-\ *variant of* DISYLLABLE

dis·sym·me·try \(ˌ)di(s)-ˈsi-mə-trē\ *noun* (1845)
: the absence of or the lack of symmetry
— **dis·sym·met·ric** \(ˌ)di(s)-sə-ˈme-trik\ *adjective*

[1]dis·taff \ˈdis-ˌtaf\ *noun, plural* **distaffs** \-ˌtafs, -ˌtavz\ [Middle English *distaf,* from Old English *distæf,* from *dis-* (akin to Middle Low German *dise* bunch of flax) + *stæf* staff] (before 12th century)
1 a : a staff for holding the flax, tow, or wool in spinning **b :** woman's work or domain
2 : the female branch or side of a family ◆

[2]distaff *adjective* (circa 1633)
: MATERNAL ⟨the *distaff* side of the family⟩ — compare SPEAR; *also* **:** FEMALE ⟨*distaff* executives⟩

distaff 1a

dis·tain \dis-ˈtān\ *transitive verb* [Middle English *disteynen,* from Middle French *desteindre* to take away the color of, from Old French, from *des-* dis- + *teindre* to dye, from Latin *tingere* to wet, dye — more at TINGE] (14th century)
1 *archaic* **:** STAIN
2 *archaic* **:** DISHONOR

dis·tal \ˈdis-tᵊl\ *adjective* [*dist*ant + *-al*] (1808)
1 : situated away from the point of attachment or origin or a central point especially of the body — compare PROXIMAL
2 : of, relating to, or being the surface of a tooth that is next to the tooth behind it or that is farthest from the middle of the front of the jaw — compare MESIAL 2
— **dis·tal·ly** \-tᵊl-ē\ *adverb*

distal convoluted tubule *noun* (circa 1901)
: the convoluted portion of the vertebrate nephron that lies between the loop of Henle and the nonsecretory part of the nephron and that is concerned especially with the concentration of urine

[1]dis·tance \ˈdis-tən(t)s\ *noun* (14th century)
1 *obsolete* **:** DISCORD
2 a : separation in time **b :** the degree or amount of separation between two points, lines, surfaces, or objects **c** (1) **:** an extent of area or an advance along a route measured linearly (2) **:** an extent of space measured other than linearly ⟨within walking *distance*⟩ **d :** an extent of advance from a beginning **e :** EXPANSE **f** (1) **:** length of a race or contest ⟨won at all *distances*⟩ (2) **:** the full length (as of a prizefight or ball game) (3) **:** a long race ⟨*distance* training⟩
3 : the quality or state of being distant: as **a :** spatial remoteness **b :** personal and especially emotional separation; *also* **:** RESERVE, COLDNESS **c :** DIFFERENCE, DISPARITY
4 : a distant point or region
5 a : AESTHETIC DISTANCE **b :** capacity to observe dispassionately
— **go the distance** *also* **last the distance :** to complete a course of action

[2]distance *transitive verb* **dis·tanced; dis·tanc·ing** (1578)
1 : to place or keep at a distance
2 : to leave far behind **:** OUTSTRIP

dis·tant \ˈdis-tənt\ *adjective* [Middle English, from Middle French, from Latin *distant-, distans,* present participle of *distare* to stand apart, be distant, from *dis-* + *stare* to stand — more at STAND] (14th century)
1 a : separated in space **:** AWAY **b :** situated at a great distance **:** FAR OFF **c :** separated by a great distance from each other **:** far apart **d :** far behind ⟨finished a *distant* third⟩
2 : separated in a relationship other than spatial ⟨a *distant* cousin⟩ ⟨the *distant* past⟩
3 : different in kind
4 : reserved or aloof in personal relationship **:** COLD ⟨was *distant* and distracted⟩
5 a : going a long distance ⟨*distant* voyages⟩ **b :** concerned with remote things ⟨*distant* thoughts⟩
— **dis·tant·ly** *adverb*
— **dis·tant·ness** *noun*

[1]dis·taste \(ˌ)dis-ˈtāst\ (1592)
transitive verb
1 *archaic* **:** to feel aversion to
2 *archaic* **:** OFFEND, DISPLEASE
intransitive verb
obsolete **:** to have an offensive taste

[2]distaste *noun* (1598)
1 a *archaic* **:** dislike of food or drink **b :** AVERSION, DISINCLINATION ⟨a *distaste* for opera⟩
2 *obsolete* **:** ANNOYANCE, DISCOMFORT

dis·taste·ful \(ˌ)dis-ˈtāst-fəl\ *adjective* (1607)
1 : objectionable because offensive to one's personal taste **:** DISAGREEABLE
2 : unpleasant to the taste **:** LOATHSOME
— **dis·taste·ful·ly** \-fə-lē\ *adverb*
— **dis·taste·ful·ness** *noun*

dis·tel·fink \ˈdish-tᵊl-ˌfiŋk, ˈdis-\ *noun* [Pennsylvania German *dischdelfink,* literally, goldfinch, from *dischdel* thistle + *fink* finch] (1939)
: a traditional Pennsylvania Dutch design motif in the form of a stylized bird

[1]dis·tem·per \dis-ˈtem-pər\ *transitive verb* [Middle English *distempren,* from Late Latin *distemperare* to temper badly, from Latin *dis-* + *temperare* to temper] (14th century)
1 : to throw out of order
2 *archaic* **:** DERANGE, UNSETTLE

[2]distemper *noun* (circa 1555)
1 : bad humor or temper
2 : a disordered or abnormal bodily state especially of quadruped mammals: as **a :** a highly contagious virus disease especially of dogs marked by fever, leukopenia, and respiratory, gastrointestinal, and neurological symptoms **b :** STRANGLES **c :** PANLEUKOPENIA
3 : AILMENT, DISORDER ⟨political *distemper*⟩ ⟨intellectual *distempers*⟩
— **dis·tem·per·ate** \-p(ə-)rət\ *adjective*

[3]distemper *noun* [obsolete *distemper,* verb, to dilute, mix to produce distemper, from Middle English, from Middle French *destemprer,* from Latin *dis-* + *temperare*] (1632)
1 : a process of painting in which the pigments are mixed with an emulsion of egg yolk, with size, or with white of egg as a vehicle and which is used for painting scenery and murals
2 a : the paint or the prepared ground used in the distemper process **b :** a painting done in distemper
3 : any of various water-based paints

[4]distemper *transitive verb* (circa 1873)
: to paint in or with distemper

dis·tem·per·a·ture \di-ˈstem-p(ə-)rə-ˌchu̇r, -pə(r)-, -chər, *chiefly Southern* -ˌt(y)u̇(ə)r\ *noun* (1531)
: a disordered condition

dis·tend \di-ˈstend\ *verb* [Middle English, from Latin *distendere,* from *dis-* + *tendere* to stretch — more at THIN] (15th century)
transitive verb
1 : EXTEND
2 : to enlarge from internal pressure **:** SWELL
intransitive verb
: to become expanded
synonym see EXPAND

dis·ten·si·ble \-ˈsten(t)-sə-bəl\ *adjective* [*distens-* (from Latin *distensus,* past participle of *distendere*) + *-ible*] (circa 1828)
: capable of being distended
— **dis·ten·si·bil·i·ty** \-ˌsten(t)-sə-ˈbi-lə-tē\ *noun*

dis·ten·sion *or* **dis·ten·tion** \di-ˈsten(t)-shən\ *noun* [Latin *distention-, distentio,* from *distendere*] (15th century)
: the act of distending or the state of being distended especially unduly or abnormally

dis·tich \ˈdis-(ˌ)tik\ *noun* [Latin *distichon,* from Greek, from neuter of *distichos* having two rows, from *di-* + *stichos* row, verse; akin to Greek *steichein* to go — more at STAIR] (1553)
: a strophic unit of two lines

dis·ti·chous \ˈdis-ti-kəs\ *adjective* [Late Latin *distichus,* from Greek *distichos*] (circa 1753)

◇ WORD HISTORY

distaff A distaff was originally a short staff that held a bundle of fibers, such as flax or wool, that were drawn and twisted into yarn or thread either by hand or with the aid of a spinning wheel. The job of spinning customarily fell to women, and since it was such a basic daily task, the distaff naturally came to be a symbol for women's work. This symbolic use of the word *distaff* dates back at least to the time of Chaucer. Eventually *distaff* came to be used figuratively for everything relating to the female domain and for women collectively. The women of a family became known as the *distaff* or the *distaff side.*

\ə\ **abut** \ᵊ\ **kitten** \ər\ **further** \a\ **ash** \ā\ **ace** \ä\ **mop, mar** \au̇\ **out** \ch\ **chin** \e\ **bet** \ē\ **easy** \g\ **go** \i\ **hit** \ī\ **ice** \j\ **job** \ŋ\ **sing** \ō\ **go** \ȯ\ **law** \ȯi\ **boy** \th\ **thin** \t͟h\ **the** \ü\ **loot** \u̇\ **foot** \y\ **yet** \zh\ **vision** *see also* Guide to Pronunciation

: disposed in two vertical rows ⟨*distichous* leaves⟩

dis·till *also* **dis·til** \di-ˈstil\ *verb* **dis·tilled; dis·till·ing** [Middle English *distillen,* from Middle French *distiller,* from Late Latin *distillare,* alteration of Latin *destillare,* from *de-* + *stillare* to drip, from *stilla* drop] (14th century)
transitive verb
1 : to let fall, exude, or precipitate in drops or in a wet mist
2 a : to subject to or transform by distillation **b :** to obtain by or as if by distillation **c :** to extract the essence of **:** CONCENTRATE
intransitive verb
1 a : to fall or materialize in drops or in a fine moisture **b :** to appear slowly or in small quantities at a time
2 a : to undergo distillation **b :** to perform distillation

dis·til·late \ˈdis-tə-ˌlāt, -lət; di-ˈsti-lət\ *noun* (circa 1859)
1 : a liquid product condensed from vapor during distillation
2 : something concentrated or extracted as if by distilling

dis·til·la·tion \ˌdis-tə-ˈlā-shən\ *noun* (14th century)
1 a : the process of purifying a liquid by successive evaporation and condensation **b :** a process like distillation; *also* **:** an instance of distilling
2 : something distilled **:** DISTILLATE 2

dis·till·er \di-ˈsti-lər\ *noun* (1577)
: one that distills especially alcoholic liquors

dis·till·ery \di-ˈsti-lə-rē, -ˈstil-rē\ *noun, plural* **-er·ies** (1759)
: the works where distilling (as of alcoholic liquors) is done

dis·tinct \di-ˈstiŋ(k)t\ *adjective* [Middle English, from Middle French, from Latin *distinctus,* from past participle of *distinguere*] (14th century)
1 : distinguishable to the eye or mind as discrete **:** SEPARATE ⟨a *distinct* cultural group⟩ ⟨teaching as *distinct* from research⟩
2 : presenting a clear unmistakable impression ⟨a neat *distinct* handwriting⟩
3 *archaic* **:** notably decorated
4 a : NOTABLE ⟨a *distinct* contribution to scholarship⟩ **b :** readily and unmistakably apprehended ⟨a *distinct* possibility of snow⟩ ⟨a *distinct* British accent⟩ ☆
— dis·tinct·ly \-ˈstiŋ(k)-tlē, -ˈstiŋ-klē\ *adverb*
— dis·tinct·ness \-ˈstiŋ(k)t-nəs, -ˈstiŋk-nəs\ *noun*

dis·tinc·tion \di-ˈstiŋ(k)-shən\ *noun* (13th century)
1 a *archaic* **:** DIVISION **b :** CLASS 4
2 : the distinguishing of a difference ⟨without *distinction* as to race, sex, or religion⟩; *also* **:** the difference distinguished ⟨the *distinction* between *imply* and *infer*⟩
3 : something that distinguishes ⟨regional *distinctions*⟩
4 : the quality or state of being distinguishable
5 a : the quality or state of being distinguished or worthy ⟨a politician of *distinction*⟩ **b :** special honor or recognition ⟨took a law degree with *distinction*⟩ **c :** achievement that sets one apart ⟨the *distinction* of winning the title⟩

dis·tinc·tive \di-ˈstiŋ(k)-tiv\ *adjective* (15th century)
1 a : serving to distinguish **b :** having or giving style or distinction
2 : capable of making a segment of utterance different in meaning as well as in sound from an otherwise identical utterance
synonym see CHARACTERISTIC
— dis·tinc·tive·ly *adverb*
— dis·tinc·tive·ness *noun*

dis·tin·gué \ˌdēs-ˌtaŋ-ˈgā, (ˌ)dis-; di-ˈstaŋ-ˌ\ *adjective* [French, from past participle of *distinguer*] (1813)
: distinguished especially in manner or bearing

dis·tin·guish \di-ˈstiŋ-(g)wish\ *verb* [Middle French *distinguer,* from Latin *distinguere,* literally, to separate by pricking, from *dis-* + *-stinguere* (akin to Latin in*stigare* to urge on) — more at STICK] (1561)
transitive verb
1 : to perceive a difference in **:** mentally separate ⟨so alike they could not be *distinguished*⟩
2 a : to mark as separate or different **b :** to separate into kinds, classes, or categories **c :** to give prominence or distinction to ⟨*distinguished* herself in music⟩ **d :** CHARACTERIZE
3 a : DISCERN ⟨*distinguished* a light in the distance⟩ **b :** to single out **:** take special notice of
intransitive verb
: to perceive a difference
— dis·tin·guish·abil·i·ty \-ˌstiŋ-(g)wi-shə-ˈbi-lə-tē\ *noun*
— dis·tin·guish·able \-ˈstiŋ-(g)wi-shə-bəl\ *adjective*
— dis·tin·guish·ably \-blē\ *adverb*

dis·tin·guished *adjective* (1714)
1 : marked by eminence, distinction, or excellence
2 : befitting an eminent person
synonym see FAMOUS

Distinguished Conduct Medal *noun* (1862)
: a British military decoration awarded for distinguished conduct in the field

Distinguished Flying Cross *noun* (1918)
1 : a British military decoration awarded for acts of gallantry when flying in operations against an enemy
2 : a U.S. military decoration awarded for heroism or extraordinary achievement while participating in an aerial flight

Distinguished Service Cross *noun* (1914)
1 : a British military decoration awarded for distinguished service against the enemy
2 : a U.S. Army decoration awarded for extraordinary heroism during operations against an armed enemy

Distinguished Service Medal *noun* (1914)
1 : a British military decoration awarded for distinguished conduct in war
2 : a U.S. military decoration awarded for exceptionally meritorious service to the government in a wartime duty of great responsibility

Distinguished Service Order *noun* (1886)
: a British military decoration awarded for special services in action

dis·tort \di-ˈstȯrt\ *verb* [Latin *distortus,* past participle of *distorquēre,* from *dis-* + *torquēre* to twist — more at TORTURE] (circa 1586)
transitive verb
1 : to twist out of the true meaning or proportion ⟨*distorted* the facts⟩
2 : to twist out of a natural, normal, or original shape or condition ⟨a face *distorted* by pain⟩; *also* **:** to cause to be perceived unnaturally ⟨the new lights *distorted* colors⟩
3 : PERVERT
intransitive verb
: to become distorted; *also* **:** to cause a twisting from the true, natural, or normal
synonym see DEFORM
— dis·tort·er *noun*

dis·tor·tion \di-ˈstȯr-shən\ *noun* (1581)
1 : the act of distorting
2 : the quality or state of being distorted **:** a product of distorting: as **a :** a lack of proportionality in an image resulting from defects in the optical system **b :** falsified reproduction of an audio or video signal caused by change in the wave form of the original signal
— dis·tor·tion·al \-shnəl, -shə-nᵊl\ *adjective*

1**dis·tract** \di-ˈstrakt, ˈdis-ˌtrakt\ *adjective* (14th century)
archaic **:** INSANE, MAD

2**dis·tract** \di-ˈstrakt\ *transitive verb* [Middle English, from Latin *distractus,* past participle of *distrahere,* literally, to draw apart, from *dis-* + *trahere* to draw] (14th century)
1 a : to turn aside **:** DIVERT **b :** to draw or direct (as one's attention) to a different object or in different directions at the same time
2 : to stir up or confuse with conflicting emotions or motives
synonym see PUZZLE
— dis·tract·i·bil·i·ty \-ˌstrak-tə-ˈbi-lə-tē\ *noun*
— dis·tract·ible *or* **dis·tract·able** \-ˈstrak-tə-bəl\ *adjective*
— dis·tract·ing·ly \-tiŋ-lē\ *adverb*

dis·tract·ed *adjective* (1590)
1 : maddened or deranged especially by grief or anxiety
2 : mentally confused, troubled, or remote
— dis·tract·ed·ly *adverb*

dis·trac·tion \di-ˈstrak-shən\ *noun* (15th century)
1 : the act of distracting or the state of being distracted; *especially* **:** mental confusion
2 : something that distracts; *especially* **:** AMUSEMENT
— dis·trac·tive \-ˈstrak-tiv\ *adjective*

dis·train \di-ˈstrān\ *verb* [Middle English *distreynen,* from Old French *destreindre,* from Medieval Latin *distringere,* from Latin, to draw apart, detain, from *dis-* + *stringere* to bind tight — more at STRAIN] (14th century)
transitive verb
1 : to force or compel to satisfy an obligation by means of a distress
2 : to seize by distress
intransitive verb
: to levy a distress
— dis·train·able \-ˈstrā-nə-bəl\ *adjective*
— dis·train·er \-ˈstrā-nər\ *or* **dis·train·or** \-ˈstrā-nər, -ˌstrā-ˈnȯr\ *noun*

dis·traint \di-ˈstrānt\ *noun* [*distrain* + *-t* (as in *constraint*)] (circa 1736)
: the act or action of distraining

dis·trait \di-ˈstrā\ *adjective* [Middle English, from Old French *destrait,* from Latin *distractus*] (15th century)
: apprehensively divided or withdrawn in attention **:** DISTRACTED 2

dis·traite \di-ˈstrāt\ *adjective* (15th century)
: DISTRAIT — used especially of women

dis·traught \dis-ˈtrȯt\ *adjective* [Middle English, modification of Latin *distractus*] (14th century)
1 : agitated with doubt or mental conflict
2 : INSANE
— dis·traught·ly *adverb*

1**dis·tress** \di-ˈstres\ *noun* [Middle English *destresse,* from Old French, from (assumed) Vulgar Latin *districtia,* from Latin *districtus,* past participle of *distringere*] (13th century)
1 a : seizure and detention of the goods of another as pledge or to obtain satisfaction of a claim by the sale of the goods seized **b :** something that is distrained
2 a : pain or suffering affecting the body, a bodily part, or the mind **:** TROUBLE ⟨gastric *distress*⟩ **b :** a painful situation **:** MISFORTUNE
3 : a state of danger or desperate need ⟨a ship in *distress*⟩ ☆

2**distress** *transitive verb* (14th century)
1 : to subject to great strain or difficulties

> ☆ SYNONYMS
> **Distinct, separate, discrete** mean not being each and every one the same. DISTINCT indicates that something is distinguished by the mind or eye as being apart or different from others ⟨two *distinct* versions⟩. SEPARATE often stresses lack of connection or a difference in identity between two things ⟨*separate* rooms⟩. DISCRETE strongly emphasizes individuality and lack of connection ⟨broke the job down into *discrete* stages⟩. See in addition EVIDENT.

2 *archaic* **:** to force or overcome by inflicting pain
3 : to cause to worry or be troubled **:** UPSET
4 : to mar (as clothing or wood) deliberately to give an effect of age
— dis·tress·ing·ly \-ˈstre-siŋ-lē\ *adverb*

³distress *adjective* (1926)
1 : offered for sale at a loss ⟨*distress* merchandise⟩
2 : involving distress goods ⟨a *distress* sale⟩

dis·tress·ful \di-ˈstres-fəl\ *adjective* (1591)
: causing distress **:** full of distress
— dis·tress·ful·ly \-fə-lē\ *adverb*
— dis·tress·ful·ness *noun*

dis·trib·u·tary \di-ˈstri-byə-ˌter-ē\ *noun, plural* **-tar·ies** (1863)
: a river branch flowing away from the main stream

dis·trib·ute \di-ˈstri-byət, *British also* ˈdis-tri-ˌbyüt\ *verb* **-ut·ed; -ut·ing** [Middle English, from Latin *distributus,* past participle of *distribuere,* from *dis-* + *tribuere* to allot — more at TRIBUTE] (15th century)
transitive verb
1 : to divide among several or many **:** APPORTION ⟨*distribute* expenses⟩
2 a : to spread out so as to cover something **:** SCATTER **b :** to give out or deliver especially to members of a group ⟨*distribute* newspapers⟩ ⟨*distribute* leaflets⟩ **c :** to place or position so as to be properly apportioned over or throughout an area ⟨200 pounds *distributed* on a 6-foot frame⟩ **d :** to use (a term) so as to convey information about every member of the class named ⟨the proposition "all men are mortal" *distributes* "man" but not "mortal"⟩
3 a : to divide or separate especially into kinds **b :** to return the units of (as typeset matter) to storage
4 : to use in or as an operation so as to be mathematically distributive
intransitive verb
: to be mathematically distributive ⟨multiplication *distributes* over addition⟩ ☆
— dis·trib·u·tee \dis-ˌtri-byə-ˈtē\ *noun*

dis·trib·ut·ed *adjective* (1968)
1 : characterized by a statistical distribution of a particular kind ⟨a normally *distributed* random variable⟩
2 : of, relating to, or being a computer network in which at least some of the processing is done by the individual workstations and information is shared by and often stored at the workstations

dis·tri·bu·tion \ˌdis-trə-ˈbyü-shən\ *noun* (14th century)
1 a : the act or process of distributing **b :** the apportionment by a court of the personal property of an intestate
2 a : the position, arrangement, or frequency of occurrence (as of the members of a group) over an area or throughout a space or unit of time **b :** the natural geographic range of an organism
3 a : something distributed **b** (1) **:** FREQUENCY DISTRIBUTION (2) **:** PROBABILITY FUNCTION (3) **:** PROBABILITY DENSITY FUNCTION 2
4 : the pattern of branching and termination of a ramifying structure (as a nerve)
5 : the marketing or merchandising of commodities
— dis·tri·bu·tion·al \-shnəl, -shə-nᵊl\ *adjective*

distribution function *noun* (circa 1909)
: a function that gives the probability that a random variable is less than or equal to the independent variable of the function

dis·trib·u·tive \di-ˈstri-byə-tiv\ *adjective* (15th century)
1 : of or relating to distribution: as **a :** dealing a proper share to each of a group **b :** diffusing more or less evenly
2 *of a word* **:** referring singly and without exception to the members of a group ⟨*each, either,* and *none* are *distributive*⟩
3 a : being an operation (as multiplication in $a(b + c) = ab + ac$) that produces the same result when operating on the whole mathematical expression as when operating on each part and collecting the results **b :** being or relating to a rule or property concerning a distributive operation ⟨the *distributive* property of multiplication with respect to addition⟩
— dis·trib·u·tive·ly *adverb*
— dis·trib·u·tiv·i·ty \-ˌstri-byə-ˈti-və-tē\ *noun*

dis·trib·u·tor \di-ˈstri-byə-tər\ *noun* (1526)
1 : one that distributes
2 : one that markets a commodity; *especially* **:** WHOLESALER
3 : an apparatus for directing the secondary current from the induction coil to the various spark plugs of an engine in their proper firing order

¹dis·trict \ˈdis-(ˌ)trikt\ *noun, often attributive* [French, from Medieval Latin *districtus* jurisdiction, district, from *distringere* to distrain — more at DISTRAIN] (1611)
1 a : a territorial division (as for administrative or electoral purposes) **b :** the basic administrative unit for local government in Northern Ireland
2 : an area, region, or section with a distinguishing character

²district *transitive verb* (1792)
: to divide or organize into districts

district attorney *noun* (1789)
: the prosecuting officer of a judicial district

district court *noun* (1789)
: a trial court that has jurisdiction over certain cases within a specific judicial district

¹dis·trust \(ˌ)dis-ˈtrəst\ *noun* (1513)
: the lack or absence of trust

²distrust *transitive verb* (1548)
: to have no trust or confidence in

dis·trust·ful \-ˈtrəs(t)-fəl\ *adjective* (1589)
: having or showing distrust
— dis·trust·ful·ly \-fə-lē\ *adverb*
— dis·trust·ful·ness *noun*

dis·turb \di-ˈstərb\ *verb* [Middle English *disturben, destourben,* from Old French & Latin; Old French *destourber,* from Latin *disturbare,* from *dis-* + *turbare* to throw into disorder, from *turba* disorder — more at TURBID] (14th century)
transitive verb
1 a : to interfere with **:** INTERRUPT **b :** to alter the position or arrangement of **c :** to upset the natural and especially the ecological balance or relations of ⟨land *disturbed* by dumping⟩
2 a : to destroy the tranquillity or composure of **b :** to throw into disorder **c :** ALARM **d :** to put to inconvenience
intransitive verb
: to cause disturbance
synonym see DISCOMPOSE
— dis·turb·er *noun*
— dis·turb·ing·ly \-ˈstər-biŋ-lē\ *adverb*

dis·tur·bance \di-ˈstər-bən(t)s\ *noun* (13th century)
1 : the act of disturbing **:** the state of being disturbed
2 : a local variation from the average or normal wind conditions

dis·turbed *adjective* (1904)
: showing symptoms of emotional illness ⟨*disturbed* children⟩ ⟨*disturbed* behavior⟩

di·sub·sti·tut·ed \(ˌ)dī-ˈsəb-stə-ˌtü-təd, -ˌtyü-\ *adjective* (circa 1909)
: having two substituent atoms or groups in a molecule

di·sul·fide \(ˌ)dī-ˈsəl-ˌfīd\ *noun* (1872)
1 : a compound containing two atoms of sulfur combined with an element or radical
2 : an organic compound containing the bivalent group SS composed of two sulfur atoms

di·sul·fi·ram \dī-ˈsəl-fə-ˌram\ *noun* [*disulf*ide + *thi*our*e*a + *am*yl] (1952)
: a compound $C_{10}H_{20}N_2S_4$ that causes a severe physiological reaction to alcohol and is used especially in the treatment of alcoholism

di·sul·fo·ton \dī-ˈsəl-fə-ˌtän\ *noun* [*di-* + *sulfo-* + *-ton* (probably from *thion-*)] (1965)
: an organophosphorus systemic insecticide $C_8H_{19}O_2PS_3$

dis·union \(ˌ)dis-ˈyü-nyən, dish-\ *noun* (15th century)
1 : the termination of union **:** SEPARATION
2 : DISUNITY
— dis·union·ist \-nyə-nist\ *noun*

dis·unite \(ˌ)dis-yü-ˈnīt, -yə-, dish-\ *transitive verb* (1598)
: DIVIDE, SEPARATE

dis·uni·ty \(ˌ)dis-ˈyü-nə-tē, dish-\ *noun* (1632)
: lack of unity; *especially* **:** DISSENSION

¹dis·use \(ˌ)dis-ˈyüz, dish-\ *transitive verb* (15th century)
: to discontinue the use or practice of

²dis·use \-ˈyüs\ *noun* (15th century)
: cessation of use or practice

dis·util·i·ty \(ˌ)dis-yü-ˈti-lə tē, -yə-, dish-\ *noun* (1879)
: the state or fact of being counterproductive

¹dis·val·ue \(ˌ)dis-ˈval-(ˌ)yü, -yə(-w)\ *transitive verb* (1603)
1 *archaic* **:** UNDERVALUE, DEPRECIATE
2 : to consider of little value

²disvalue *noun* (1603)
1 *obsolete* **:** DISREGARD, DISESTEEM
2 : a negative value

di·syl·la·ble \ˈdī-ˌsi-lə-bəl, (ˌ)dī-ˈsil-; ˈdi-ˌsil-, (ˌ)di(s)-ˈsil-\ *noun* [part translation of Middle French *dissilabe,* from Latin *disyllabus* having two syllables, from Greek *disyllabos,* from *di-* + *syllabē* syllable] (1589)
: a linguistic form consisting of two syllables
— di·syl·lab·ic \ˌdī-sə-ˈla-bik, ˌdi(s)-sə-\ *adjective*

dit \ˈdit\ *noun* [imitative] (1940)
: a dot in radio or telegraphic code

¹ditch \ˈdich\ *noun* [Middle English *dich,* from Old English *dīc* dike, ditch; akin to Middle High German *tīch* pond, dike] (before 12th century)

☆ SYNONYMS

Distress, suffering, misery, agony mean the state of being in great trouble. DISTRESS implies an external and usually temporary cause of great physical or mental strain and stress ⟨the hurricane put everyone in great *distress*⟩. SUFFERING implies conscious endurance of pain or distress ⟨the *suffering* of famine victims⟩. MISERY stresses the unhappiness attending especially sickness, poverty, or loss ⟨the homeless live with *misery* every day⟩. AGONY suggests pain too intense to be borne ⟨in *agony* over the death of their child⟩.

Distribute, dispense, divide, deal, dole out mean to give out, usually in shares, to each member of a group. DISTRIBUTE implies an apportioning by separation of something into parts, units, or amounts ⟨*distributed* food to the needy⟩. DISPENSE suggests the giving of a carefully weighed or measured portion to each of a group according to due or need ⟨*dispensed* wisdom to the students⟩. DIVIDE stresses the separation of a whole into parts and implies that the parts are equal ⟨three charitable groups *divided* the proceeds⟩. DEAL emphasizes the allotment of something piece by piece ⟨*deal* out equipment and supplies⟩. DOLE OUT implies a carefully measured portion that is often scant or niggardly ⟨*doled out* what little food there was⟩.

: a long narrow excavation dug in the earth (as for drainage)
²ditch (14th century)
transitive verb
1 a : to enclose with a ditch **b :** to dig a ditch in
2 : to make a forced landing of (an airplane) on water
3 : to get rid of **:** DISCARD
intransitive verb
1 : to dig a ditch
2 : to crash-land at sea
ditch·dig·ger \-ˌdi-gər\ *noun* (circa 1897)
1 : one that digs ditches
2 : one employed at menial and usually hard physical labor
dite \ˈdīt\ *noun* [variant of *doit*] (circa 1877)
dialect **:** MITE, BIT
¹dith·er \ˈdi-t͟hər\ *intransitive verb* **dith·ered; dith·er·ing** \-t͟h(ə-)riŋ\ [Middle English *didderen*] (15th century)
1 : SHIVER, TREMBLE
2 : to act nervously or indecisively **:** VACILLATE
— dith·er·er \-t͟hər-ər\ *noun*
²dither *noun* (1819)
: a highly nervous, excited, or agitated state **:** EXCITEMENT, CONFUSION
— dith·ery \ˈdi-t͟hə-rē\ *adjective*
dithi- *or* **dithio-** *combining form* [International Scientific Vocabulary *di-* + *thi-*]
: containing two atoms of sulfur usually in place of two oxygen atoms ⟨*dithio*carbamate⟩
di·thio·car·ba·mate \ˌdī-ˌthī-ō-ˈkär-bə-ˌmāt\ *noun* (1929)
: any of several sulfur analogues of the carbamates including some used as fungicides
dith·y·ramb \ˈdi-thi-ˌram(b)\ *noun, plural* **-rambs** \-ˌramz\ [Greek *dithyrambos*] (1656)
1 : a usually short poem in an inspired wild irregular strain
2 : a statement or writing in an exalted or enthusiastic vein
— dith·y·ram·bic \ˌdi-thi-ˈram-bik\ *adjective*
— dith·y·ram·bi·cal·ly \-bi-k(ə-)lē\ *adverb*
di·tran·si·tive \ˌdī-ˈtran(t)-sə-tiv, -ˈtran-zə-\ *adjective* (1972)
: able to take both a direct and an indirect object ⟨a *ditransitive* verb⟩
— ditransitive *noun*
dit·sy *or* **dit·zy** \ˈdit-sē\ *adjective* **dits·i·er** *or* **ditz·i·er; -est** [origin unknown] (1973)
: eccentrically silly, giddy, or inane **:** DIZZY
dit·ta·ny \ˈdi-tᵊn-ē\ *noun, plural* **-nies** [Middle English *ditoyne,* from Old French *ditayne,* from Latin *dictamnum,* from Greek *diktamnon*] (12th century)
1 : a pink-flowered herb (*Origanum dictamnus*) that is native to Crete
2 : an American herb (*Conila origanoides*) of the mint family that has much-branched stems
¹dit·to \ˈdi-(ˌ)tō\ *noun, plural* **dittos** [Italian *ditto, detto,* past participle of *dire* to say, from Latin *dicere* — more at DICTION] (circa 1678)
1 : a thing mentioned previously or above — used to avoid repeating a word; often symbolized by inverted commas or apostrophes
2 : a ditto mark
²ditto *adjective* (1776)
: having the same characteristics **:** SIMILAR
³ditto *transitive verb* (1837)
1 : to repeat the action or statement of
2 [from *Ditto,* a trademark] **:** to copy (as printed matter) on a duplicator
⁴ditto *adverb* (circa 1864)
: as before or aforesaid **:** in the same manner
dit·ty \ˈdi-tē\ *noun, plural* **ditties** [Middle English *ditee,* from Middle French *ditié* poem, from past participle of *ditier* to compose, from Latin *dictare* to dictate, compose] (14th century)
: an especially simple and unaffected song
dit·ty bag \ˈdi-tē-\ *noun* [origin unknown] (circa 1860)
: a bag used especially by sailors to hold small articles (as needles and thread)
ditty box *noun* (circa 1880)
: a box used for the same purpose as a ditty bag
ditz \ˈdits\ *noun* (1982)
: a ditsy person
di·ure·sis \ˌdī-yə-ˈrē-səs\ *noun, plural* **-ure·ses** \-ˌsēz\ [New Latin] (circa 1681)
: an increased excretion of urine
di·uret·ic \ˌdī-yə-ˈre-tik\ *adjective* [Middle English, from Middle French or Late Latin; Middle French *diuretique,* from Late Latin *diureticus,* from Greek *diourētikos,* from *diourein* to urinate, from *dia-* + *ourein* to urinate — more at URINE] (14th century)
: tending to increase the flow of urine
— diuretic *noun*
— di·uret·i·cal·ly \-ti-k(ə-)lē\ *adverb*
¹di·ur·nal \dī-ˈər-nᵊl\ *adjective* [Middle English, from Latin *diurnalis* — more at JOURNAL] (14th century)
1 a : recurring every day ⟨*diurnal* tasks⟩ **b :** having a daily cycle ⟨*diurnal* tides⟩
2 a : of, relating to, or occurring in the daytime ⟨the city's *diurnal* noises⟩ **b :** active chiefly in the daytime ⟨*diurnal* animals⟩ **c :** opening during the day and closing at night ⟨*diurnal* flowers⟩
— di·ur·nal·ly \-nᵊl-ē\ *adverb*
²diurnal *noun* (1600)
1 *archaic* **:** DIARY, DAYBOOK
2 : JOURNAL 2a
di·u·ron \ˈdī-yə-ˌrän\ *noun* [*di*chlor- + *ur*ea + ¹*-on*] (1957)
: a persistent herbicide $C_9H_{10}Cl_2N_2O$ used especially to control annual weeds
di·va \ˈdē-və\ *noun, plural* **divas** *or* **di·ve** \-(ˌ)vā\ [Italian, literally, goddess, from Latin, feminine of *divus* divine, god — more at DEITY] (1883)
: PRIMA DONNA 1
di·va·gate \ˈdī-və-ˌgāt, ˈdi-\ *intransitive verb* **-gat·ed; -gat·ing** [Late Latin *divagatus,* past participle of *divagari,* from Latin *dis-* + *vagari* to wander — more at VAGARY] (1599)
: to wander or stray from a course or subject **:** DIVERGE, DIGRESS
— di·va·ga·tion \ˌdī-və-ˈgā-shən, ˌdi-\ *noun*
di·va·lent \(ˌ)dī-ˈvā-lənt\ *adjective* (1869)
: having a chemical valence of two; *also* **:** bonded to two other atoms or groups
di·van \di-ˈvan, ˈdī-ˌvan, *especially in senses 1, 2, & 4 also* di-ˈvän, dī-ˈvan\ *noun* [Turkish, from Persian *dīwān* account book] (1586)
1 a : the privy council of the Ottoman Empire **b :** COUNCIL
2 : a council chamber
3 : a large couch usually without back or arms often designed for use as a bed
4 : a collection of poems in Persian or Arabic usually by one author
di·var·i·cate \dī-ˈvar-ə-ˌkāt, də-\ *transitive verb* **-cat·ed; -cat·ing** [Latin *divaricatus,* past participle of *divaricare,* from *dis-* + *varicare* to straddle — more at PREVARICATE] (1673)
: to spread apart **:** branch off **:** DIVERGE
di·var·i·ca·tion \(ˌ)dī-ˌvar-ə-ˈkā-shən, də-\ *noun* (1578)
1 : the action, process, or fact of divaricating
2 : a divergence of opinion
¹dive \ˈdīv\ *verb* **dived** \ˈdīvd\ *or* **dove** \ˈdōv\; **dived** *also* **dove; div·ing** [Middle English *diven, duven,* from Old English *dȳfan* to dip & *dūfan* to dive; akin to Old English *dyppan* to dip — more at DIP] (before 12th century)
intransitive verb
1 a : to plunge into water intentionally and especially headfirst; *also* **:** to execute a dive **b :** SUBMERGE
2 a : to come or drop down precipitously **:** PLUNGE 3 **b :** to plunge one's hand into something **c** *of an airplane* **:** to descend in a dive
3 a : to plunge into some matter or activity **b :** to plunge or dash for some place ⟨*diving* for cover⟩; *also* **:** to lunge especially in order to seize something ⟨*dove* for the ball⟩
transitive verb
1 : to thrust into something
2 : to cause to dive ■
²dive *noun* (1700)
1 : the act or an instance of diving: as **a** (1) **:** a plunge into water executed in a prescribed manner (2) **:** a submerging of a submarine (3) **:** a steep descent of an airplane at greater than the maximum horizontal speed **b :** a sharp decline
2 : a disreputable entertainment establishment
3 : a faked knockout — usually used in the phrase *take a dive*
4 : an offensive play in football in which the ballcarrier plunges into the line for short yardage
dive–bomb \ˈdīv-ˌbäm\ *transitive verb* (1935)
: to bomb from an airplane by making a steep dive toward the target before releasing the bomb
— dive–bomb·er *noun*
div·er \ˈdī-vər\ *noun* (1506)
1 : one that dives
2 a : a person who stays underwater for long periods by having air supplied from the surface or by carrying a supply of compressed air **b :** any of various birds that obtain food by diving in water; *especially* **:** LOON
di·verge \də-ˈvərj, dī-\ *verb* **di·verged; di·verg·ing** [Medieval Latin *divergere,* from Latin *dis-* + *vergere* to incline — more at WRENCH] (1665)
intransitive verb
1 a : to move or extend in different directions from a common point **:** draw apart ⟨*diverging* roads⟩ **b :** to become or be different in character or form **:** differ in opinion
2 : to turn aside from a path or course **:** DEVIATE
3 : to be mathematically divergent
transitive verb
: DEFLECT
synonym see SWERVE
di·ver·gence \-ˈvər-jən(t)s\ *noun* (1656)
1 a : a drawing apart (as of lines extending from a common center) **b :** DIFFERENCE, DISAGREEMENT **c :** the acquisition of dissimilar characters by related organisms in unlike environments
2 : a deviation from a course or standard
3 : the condition of being mathematically divergent
di·ver·gen·cy \-jən(t)-sē\ *noun, plural* **-cies** (1709)
: DIVERGENCE
di·ver·gent \-jənt\ *adjective* [Latin *divergent-, divergens,* present participle of *divergere*] (1696)
1 a : diverging from each other **b :** differing from each other or from a standard **:** DEVIANT ⟨the *divergent* interests of capital and labor⟩

☐ USAGE
dive *Dive,* which was originally a weak verb, developed a past tense *dove,* probably by analogy with verbs like *drive, drove. Dove* exists in some British dialects and has become the standard past tense especially in speech in some parts of Canada. In the U.S. *dived* and *dove* are both widespread in speech as past tense and past participle, with *dove* less common than *dived* in the south Midland area, and *dived* less common than *dove* in the Northern and north Midland areas. In writing, the past tense *dived* is usual in British English and somewhat more common in American English. *Dove* seems relatively rare as a past participle in writing.

2 : relating to or being an infinite sequence that does not have a limit or an infinite series whose partial sums do not have a limit
3 : causing divergence of rays ⟨a *divergent* lens⟩
synonym see DIFFERENT
— di·ver·gent·ly *adverb*

di·vers \'dī-vərz\ *adjective* [Middle English *divers, diverse*] (14th century)
: VARIOUS

di·verse \dī-'vərs, də-', 'dī-,\ *adjective* [Middle English *divers, diverse,* from Old French & Latin; Old French *divers,* from Latin *diversus,* from past participle of *divertere*] (14th century)
1 : differing from one another **:** UNLIKE
2 : composed of distinct or unlike elements or qualities
synonym see DIFFERENT
— di·verse·ly *adverb*
— di·verse·ness *noun*

di·ver·si·fy \də-'vər-sə-,fī, dī-\ *verb* **-fied; -fy·ing** (15th century)
transitive verb
1 : to make diverse **:** give variety to ⟨*diversify* a course of study⟩
2 : to balance (as an investment portfolio) defensively by dividing funds among securities of different industries or of different classes
3 : to increase the variety of the products of
intransitive verb
1 : to produce variety
2 : to engage in varied operations
— di·ver·si·fi·ca·tion \-,vər-sə-fə-'kā-shən\ *noun*
— di·ver·si·fi·er \-'vər-sə-,fī(-ə)r\ *noun*

di·ver·sion \də-'vər-zhən, dī-, -shən\ *noun* (1600)
1 : the act or an instance of diverting from a course, activity, or use **:** DEVIATION
2 : something that diverts or amuses **:** PASTIME
3 : an attack or feint that draws the attention and force of an enemy from the point of the principal operation

di·ver·sion·ary \də-'vər-zhə-,ner-ē, dī-, -shə-\ *adjective* (1846)
: tending to draw attention away from the principal concern **:** being a diversion

di·ver·sion·ist \-zhə-nəst, -shə-\ *noun* (1937)
1 : one engaged in diversionary activities
2 : one characterized by political deviation

di·ver·si·ty \də-'vər-sə-tē, dī-\ *noun, plural* **-ties** (14th century)
1 : the condition of being diverse **:** VARIETY
2 : an instance of being diverse

di·vert \də-'vərt, dī-\ *verb* [Middle English, from Middle French & Latin; Middle French *divertir,* from Latin *divertere* to turn in opposite directions, from *dis-* + *vertere* to turn — more at WORTH] (15th century)
intransitive verb
: to turn aside **:** DEVIATE ⟨studied law but *diverted* to diplomacy⟩
transitive verb
1 a : to turn from one course or use to another **:** DEFLECT **b :** DISTRACT
2 : to give pleasure to especially by distracting the attention from what burdens or distresses
synonym see AMUSE

di·ver·tic·u·li·tis \,dī-vər-,ti-kyə-'lī-təs\ *noun* [New Latin] (circa 1900)
: inflammation of a diverticulum

di·ver·tic·u·lo·sis \,dī-vər-,ti-kyə-'lō-səs\ *noun* [New Latin] (1917)
: an intestinal disorder characterized by the presence of many diverticula

di·ver·tic·u·lum \,dī-vər-'ti-kyə-ləm\ *noun, plural* **-la** \-lə\ [New Latin, from Latin, bypath, probably alteration of *deverticulum,* from *devertere* to turn aside, from *de-* + *vertere*] (circa 1819)
1 : an abnormal pouch or sac opening from a hollow organ (as the intestine or bladder)
2 : a pocket or closed branch opening off a main passage
— di·ver·tic·u·lar \-kyə-lər\ *adjective*

di·ver·ti·men·to \di-,vər-tə-'men-(,)tō, -,ver-\ *noun, plural* **-men·ti** \-'men-(,)tē\ *or* **-mentos** [Italian, literally, diversion, from *divertire* to divert, amuse, from Latin *divertere*] (1823)
1 : an instrumental chamber work in several movements usually light in character
2 : DIVERTISSEMENT 1

di·ver·tisse·ment \di-'vər-təs-mənt, -təz-, *French* dē-ver-tē-smä^{n}\ *noun, plural* **divertissements** \-mən(t)s, -smä^{n}(z)\ [French, literally, diversion, from *divertiss-* (stem of *divertir*)] (1728)
1 : a dance sequence or short ballet usually used as an interlude
2 : DIVERTIMENTO 1
3 : DIVERSION, ENTERTAINMENT

Di·ves \'dī-(,)vēz\ *noun* [Middle English, from Latin, rich, rich man; misunderstood as a proper name in Luke 16:19] (14th century)
: a rich man

di·vest \dī-'vest, də-\ *transitive verb* [alteration of *devest*] (1563)
1 a : to deprive or dispossess especially of property, authority, or title **b :** to undress or strip especially of clothing, ornament, or equipment **c :** RID, FREE
2 : to take away from a person
— di·vest·ment \-'ves(t)-mənt\ *noun*

di·ves·ti·ture \dī-'ves-tə-,chu̇r, -chər, də-, *chiefly Southern* -t(y)u̇(ə)r\ *noun* [*divest* + *-iture* (as in *investiture*)] (1601)
1 : the act of divesting
2 : the compulsory transfer of title or disposal of interests (as stock in a corporation) upon government order

[1]di·vide \də-'vīd\ *verb* **di·vid·ed; di·vid·ing** [Middle English, from Latin *dividere,* from *dis-* + *-videre* to separate — more at WIDOW] (14th century)
transitive verb
1 a : to separate into two or more parts, areas, or groups **b :** to separate into classes, categories, or divisions **c :** CLEAVE, PART
2 a : to separate into portions and give out in shares **:** DISTRIBUTE **b :** to possess, enjoy, or make use of in common **c :** APPORTION
3 a : to cause to be separate, distinct, or apart from one another **b :** to separate into opposing sides or parties **c :** to cause (a parliamentary body) to vote by division
4 a : to subject (a number or quantity) to the operation of finding how many times it contains another number or quantity ⟨*divide* 42 by 14⟩ **b :** to be used as a divisor with respect to (a dividend) ⟨4 *divides* 16 evenly⟩ **c :** to use as a divisor — used with *into* ⟨*divide* 14 into 42⟩
intransitive verb
1 : to perform mathematical division
2 a (1) **:** to undergo replication, multiplication, fission, or separation into parts (2) **:** to branch out **b :** to become separated or disunited especially in opinion or interest
synonym see SEPARATE, DISTRIBUTE
— di·vid·able \-'vī-də-bəl\ *adjective*

[2]divide *noun* (1642)
1 : an act of dividing
2 a : a dividing ridge between drainage areas **:** WATERSHED **b :** a point or line of division

di·vid·ed *adjective* (14th century)
1 a : separated into parts or pieces **b** *of a leaf* **:** cut into distinct parts by incisions extending to the base or to the midrib **c :** having a barrier (as a guardrail) to separate lanes of traffic going in opposite directions ⟨a *divided* highway⟩
2 a : disagreeing with each other **:** DISUNITED **b :** directed or moved toward conflicting interests, states, or objects ⟨*divided* loyalties⟩
3 : separated by distance ⟨familiar objects from which she had never dreamed of being *divided* —James Joyce⟩
— di·vid·ed·ly \də-'vī-dəd-lē\ *adverb*
— di·vid·ed·ness \-nəs\ *noun*

div·i·dend \'di-və-,dend, -dənd\ *noun* [Middle English *divident,* from Latin *dividendus,* gerundive of *dividere*] (15th century)
1 : an individual share of something distributed: as **a :** a share in a pro rata distribution (as of profits) to stockholders **b :** a share of surplus allocated to a policyholder in a participating insurance policy
2 a : a resultant return or reward **b :** BONUS
3 a : a number to be divided **b :** a sum or fund to be divided and distributed
— div·i·dend·less \-ləs\ *adjective*

di·vid·er \də-'vī-dər\ *noun* (1534)
1 : one that divides
2 *plural* **:** an instrument for measuring or marking (as in dividing lines)
3 : something serving as a partition between separate spaces within a larger area

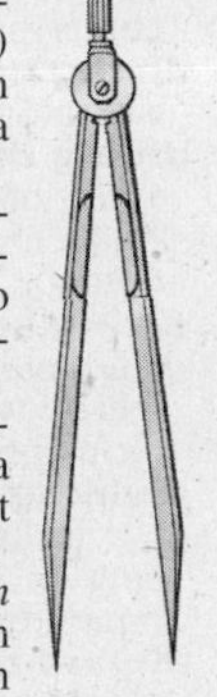
dividers

di·vi–di·vi \,dē vē-'dē-vē, ,di-vē-'di-vē\ *noun* [American Spanish *dividivi,* probably from Cumanagoto (extinct Cariban language of northern Venezuela)] (circa 1837)
: a small leguminous tree (*Caesalpinia coriaria*) of tropical America with twisted astringent pods that contain a large proportion of tannin

div·i·na·tion \,di-və-'nā-shən\ *noun* [Middle English *divinacioun,* from Latin *divination-, divinatio,* from *divinare*] (14th century)
1 : the art or practice that seeks to foresee or foretell future events or discover hidden knowledge usually by the interpretation of omens or by the aid of supernatural powers
2 : unusual insight **:** intuitive perception
— di·vi·na·to·ry \də-'vi-nə-,tōr-ē, də-'vī-nə-, 'di-və-nə-, -,tȯr-\ *adjective*

[1]di·vine \də-'vīn\ *adjective* **di·vin·er; -est** [Middle English *divin,* from Middle French, from Latin *divinus,* from *divus* god — more at DEITY] (14th century)
1 a : of, relating to, or proceeding directly from God or a god ⟨*divine* love⟩ **b :** being a deity ⟨the *divine* Savior⟩ **c :** directed to a deity ⟨*divine* worship⟩
2 a : supremely good **:** SUPERB ⟨the pie was *divine*⟩ **b :** HEAVENLY, GODLIKE
— di·vine·ly *adverb*

[2]divine *noun* [Middle English, from Medieval Latin *divinus,* from Latin, soothsayer, from *divinus,* adjective] (14th century)
1 : CLERGYMAN
2 : THEOLOGIAN

[3]divine *verb* **di·vined; di·vin·ing** [Middle English, from Middle French & Latin; Middle French *diviner,* from Latin *divinare,* from *divinus,* noun] (14th century)
transitive verb
1 : to discover intuitively **:** INFER
2 : to discover or locate (as water or minerals underground) usually by means of a divining rod
intransitive verb
1 : to practice divination **:** PROPHESY
2 : to perceive intuitively
synonym see FORESEE

Divine Liturgy *noun* (1870)
: the eucharistic rite of Eastern churches

Divine Office *noun* (15th century)
: the office for the canonical hours of prayer that priests and religious say daily

di·vin·er \də-'vī-nər\ *noun* (14th century)
1 : one who practices divination **:** SOOTHSAYER
2 : a person who divines the location of water or minerals

divine right *noun* (circa 1600)

: the right of a sovereign to rule as set forth by the theory of government that holds that a monarch receives the right to rule directly from God and not from the people

divine service *noun* (14th century)
: a service of Christian worship; *specifically* : such a service that is not sacramental in character

diving beetle *noun* (circa 1889)
: any of various predatory aquatic beetles (family Dytiscidae) that breathe while submerged using air trapped under their elytra

diving bell *noun* (1661)
: a diving apparatus consisting of a container open only at the bottom and supplied with compressed air by a hose

diving board *noun* (1893)
: SPRINGBOARD 1

diving duck *noun* (1813)
: any of various ducks (as a bufflehead) that frequent deep waters and obtain their food by diving

diving suit *noun* (1908)
: a waterproof suit with a removable helmet that is worn by a diver who is supplied with air pumped through a tube

divining rod *noun* (1751)
: a forked rod believed to indicate the presence of water or minerals especially by dipping downward when held over a vein

di·vin·i·ty \də-ˈvi-nə-tē\ *noun, plural* **-ties** (14th century)
1 : THEOLOGY
2 : the quality or state of being divine
3 *often capitalized* : a divine being: as **a** : GOD 1 **b** (1) : GOD 2 (2) : GODDESS
4 : fudge made of whipped egg whites, sugar, and nuts

divinity school *noun* (circa 1555)
: a professional school having a religious curriculum especially for ministerial candidates

di·vis·i·ble \də-ˈvi-zə-bəl\ *adjective* (15th century)
: capable of being divided
— **di·vis·i·bil·i·ty** \də-ˌvi-zə-ˈbi-lə-tē\ *noun*

di·vi·sion \də-ˈvi-zhən\ *noun* [Middle English, from Middle French, from Latin *division-, divisio,* from *dividere* to divide] (14th century)
1 a : the act or process of dividing : the state of being divided **b** : the act, process, or an instance of distributing among a number : DISTRIBUTION **c** *obsolete* : a method of arranging or disposing (as troops)
2 : one of the parts or groupings into which a whole is divided or is divisible
3 : the condition or an instance of being divided in opinion or interest : DISAGREEMENT, DISUNITY ⟨exploited the *divisions* between the two countries⟩
4 a : something that divides, separates, or marks off **b** : the act, process, or an instance of separating or keeping apart : SEPARATION
5 : the mathematical operation of dividing something
6 a : a self-contained major military unit capable of independent action **b** : a tactical military unit composed of headquarters and usually three to five brigades **c** (1) : the basic naval administrative unit (2) : a tactical subdivision of a squadron of ships **d** : a unit of the U.S. Air Force higher than a wing and lower than an air force
7 a : a portion of a territorial unit marked off for a particular purpose (as administrative or judicial functions) **b** : an administrative or operating unit of a governmental, business, or educational organization
8 : the physical separation into different lobbies of the members of a parliamentary body voting for and against a question
9 : plant propagation by dividing parts and planting segments capable of producing roots and shoots
10 : a group of organisms forming part of a larger group; *specifically* : a primary category of the plant kingdom
11 : a competitive class or category (as in boxing or wrestling)
synonym see PART
— **di·vi·sion·al** \-ˈvizh-nəl, -ˈvi-zhə-nᵊl\ *adjective*

di·vi·sion·ism \-ˈvi-zhə-ˌni-zəm\ *noun, often capitalized* (1901)
: POINTILLISM
— **di·vi·sion·ist** \-ˈvi-zhə-nəst\ *noun or adjective*

division of labor (1776)
: the breakdown of labor into its components and their distribution among different persons, groups, or machines to increase productive efficiency

division sign *noun* (circa 1934)
1 : the symbol ÷ used to indicate division
2 : the diagonal / used to indicate a fraction

di·vi·sive \də-ˈvī-siv *also* -ˈvi- *or* -ziv\ *adjective* (1642)
: creating disunity or dissension
— **di·vi·sive·ly** *adverb*
— **di·vi·sive·ness** *noun*

di·vi·sor \də-ˈvī-zər\ *noun* (15th century)
: the number by which a dividend is divided

¹di·vorce \də-ˈvōrs, -ˈvȯrs *also* dī-\ *noun* [Middle English *divorse,* from Middle French, from Latin *divortium,* from *divertere, divortere* to divert, to leave one's husband] (14th century)
1 : the action or an instance of legally dissolving a marriage
2 : SEPARATION, SEVERANCE

²divorce *verb* **di·vorced; di·vorc·ing** (15th century)
transitive verb
1 a : to end marriage with (one's spouse) by divorce **b** : to dissolve the marriage contract between
2 : to terminate an existing relationship or union : SEPARATE ⟨*divorce* church from state⟩
intransitive verb
: to obtain a divorce
synonym see SEPARATE
— **di·vorce·ment** \-ˈvōr-smənt, -ˈvȯr-\ *noun*

di·vor·cé \də-ˌvōr-ˈsā, -ˌvȯr-, -ˈsē, -ˈvōr-ˌ, -ˈvȯr-ˌ\ *noun* (1877)
: a divorced man

di·vor·cée \də-ˌvōr-ˈsā, -ˌvȯr-, -ˈsē, -ˈvōr-ˌ, -ˈvȯr-ˌ\ *noun* [French, from feminine of *divorcé,* past participle of *divorcer* to divorce, from Middle French *divorse*] (1813)
: a divorced woman

div·ot \ˈdi-vət\ *noun* [alteration of earlier Scots *devat,* from Middle English (Scots dialect) *duvat*] (1586)
1 *Scottish* : a square of turf or sod
2 : a loose piece of turf (as one dug from a golf fairway in making a shot)

di·vulge \də-ˈvəlj, dī-\ *transitive verb* **di·vulged; di·vulg·ing** [Middle English, from Latin *divulgare,* from *dis-* + *vulgare* to make known, from *vulgus* mob] (15th century)
1 *archaic* : to make public : PROCLAIM
2 : to make known (as a confidence or secret)
synonym see REVEAL
— **di·vul·gence** \-ˈvəl-jən(t)s\ *noun*

div·vy \ˈdi-vē\ *transitive verb* **div·vied; div·vy·ing** [by shortening & alteration from *divide*] (1877)
: DIVIDE, SHARE — usually used with *up*

Dix·ie \ˈdik-sē\ *noun* [name for the Southern states in the song *Dixie* (1859) by Daniel D. Emmett] (1859)
: the Southern states of the U.S.

Dix·ie·crat \-ˌkrat\ *noun* (1948)
: a dissident Southern Democrat; *specifically* : a supporter of a 1948 presidential ticket opposing the civil rights stand of the Democrats
— **Dix·ie·crat·ic** \ˌdik-sē-ˈkra-tik\ *adjective*

Dix·ie·land \-ˌland\ *noun* [probably from the *Original Dixieland Jazz Band*] (1927)
: jazz music in duple time usually played by a small band and characterized by ensemble and solo improvisation

DIY \ˌdē-(ˌ)ī-ˈwī\ *noun, often attributive* (1955)
chiefly British : DO-IT-YOURSELF

di·zen \ˈdī-zᵊn, ˈdi-\ *transitive verb* [earlier *disen* to dress a distaff with flax, from Middle Dutch] (1619)
archaic : BEDIZEN

di·zy·got·ic \ˌdī-(ˌ)zī-ˈgä-tik\ *also* **di·zy·gous** \(ˌ)dī-ˈzī-gəs\ *adjective* (1916)
of twins : FRATERNAL 2

¹diz·zy \ˈdi-zē\ *adjective* **diz·zi·er; -est** [Middle English *disy,* from Old English *dysig* stupid; akin to Old High German *tusig* stupid] (before 12th century)
1 : FOOLISH, SILLY
2 a : having a whirling sensation in the head with a tendency to fall **b** : mentally confused
3 a : causing giddiness or mental confusion **b** : caused by or marked by giddiness **c** : extremely rapid
— **diz·zi·ly** \ˈdi-zə-lē\ *adverb*
— **diz·zi·ness** \-zē-nəs\ *noun*

²dizzy *transitive verb* **diz·zied; diz·zy·ing** (1501)
1 : to make dizzy or giddy
2 : BEWILDER ⟨disasters that *dizzy* the mind⟩
— **diz·zy·ing·ly** \-zē-iŋ-lē\ *adverb*

DJ \ˈdē-ˌjā\ *noun, often not capitalized* (1950)
: DISC JOCKEY

djel·la·ba *also* **djel·la·bah** \jə-ˈlä-bə\ *noun* [French *djellaba,* from Arabic *jallābīya*] (1919)
: a long loose garment with full sleeves and a hood

djin·ni \ˈjē-nē\ *or* **djinn** *or* **djin** \ˈjin\ *variant of* JINNI

dl- \(ˌ)dē-ˈel, ˈdē-ˌ\ *prefix*
1 *also* **d,l-** : consisting of equal amounts of the dextro and levo forms of a specified compound ⟨*dl*-tartaric acid⟩
2 : consisting of equal amounts of the D- and L- forms of a specified compound ⟨DL-fructose⟩

D layer *noun* (circa 1934)
: a layer that may exist within the D region of the ionosphere; *also* : D REGION

D–mark \ˈdȯich-ˌmärk, ˈdoi-chə-ˌ\ *noun* (1948)
: DEUTSCHE MARK

DME \ˌdē-(ˌ)em-ˈē\ *noun* [*d*istance *m*easuring *e*quipment] (1947)
: an electronic device that informs the pilot of an airplane of its distance from a particular ground station

DMSO \ˌdē-ˌem-ˌes-ˈō\ *noun* (1964)
: DIMETHYL SULFOXIDE

DMT \ˌdē-(ˌ)em-ˈtē\ *noun* (circa 1966)
: DIMETHYLTRYPTAMINE

DNA \ˌdē-ˌen-ˈā\ *noun* [*d*eoxyribo*n*ucleic *a*cid] (1944)
: any of various nucleic acids that are usually the molecular basis of heredity, are localized especially in cell nuclei, and are constructed of a double helix held together by hydrogen bonds between purine and pyrimidine bases

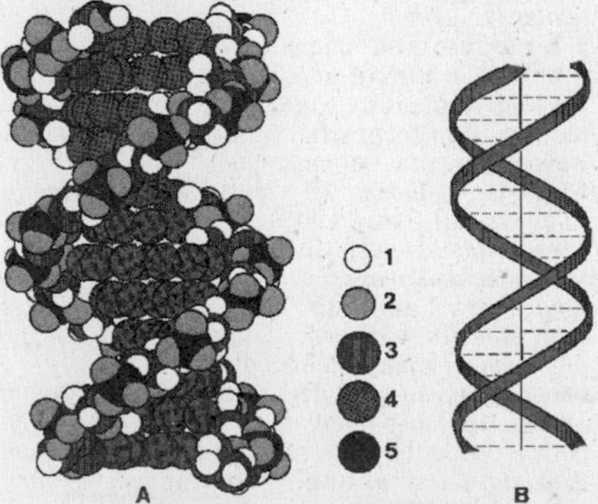

DNA: ***A*** **molecular model:** ***1*** **hydrogen,** ***2*** **oxygen,** ***3*** **carbon in the helical phosphate ester chains,** ***4*** **carbon and nitrogen in the cross-linked purine and pyrimidine bases,** ***5*** **phosphorus;** ***B*** **double helix**

which project inward from two chains containing alternate links of deoxyribose and phosphate — compare RECOMBINANT DNA

DNA fingerprinting *noun* (1984)
: a method of identification (as for forensic purposes) by determining the sequence of base pairs in the DNA especially of a person
— **DNA fingerprint** *noun*

DNA polymerase *noun* (circa 1962)
: any of several polymerases that promote replication or repair of DNA usually using single-stranded DNA as a template

DN•ase \(ˌ)dē-ˈen-ˌās, -ˌāz\ *also* **DNA•ase** \(ˌ)dē-ˌen-ˈā-ˌās, -ˌāz\ *noun* (circa 1956)
: DEOXYRIBONUCLEASE

[1]**do** \ˈdü, də(-w)\ *verb* **did** \ˈdid, dəd\; **done** \ˈdən\; **do•ing** \ˈdü-iŋ\; **does** \ˈdəz\ [Middle English *don,* from Old English *dōn;* akin to Old High German *tuon* to do, Latin *-dere* to put, *facere* to make, do, Greek *tithenai* to place, set] (before 12th century)
transitive verb
1 : to bring to pass **:** CARRY OUT
2 : PUT — used chiefly in *do to death*
3 a : PERFORM, EXECUTE ⟨*do* some work⟩ ⟨*did* his duty⟩ **b :** COMMIT ⟨crimes *done* deliberately⟩
4 a : BRING ABOUT, EFFECT ⟨trying to *do* good⟩ ⟨*do* violence⟩ **b :** to give freely **:** PAY ⟨*do* honor to her memory⟩
5 : to bring to an end **:** FINISH — used in the past participle ⟨the job is finally *done*⟩
6 : to put forth **:** EXERT ⟨*did* her best to win the race⟩
7 a : to wear out especially by physical exertion **:** EXHAUST ⟨at the end of the race they were pretty well *done*⟩ **b** *British* **:** to attack physically **:** BEAT; *also* **:** KILL
8 : to bring into existence **:** PRODUCE ⟨*do* a biography on the general⟩
9 — used as a substitute verb especially to avoid repetition ⟨if you must make such a racket, *do* it somewhere else⟩
10 a : to play the role or character of **b :** MIMIC; *also* **:** to behave like ⟨*do* a Houdini and disappear⟩ **c :** to perform in or serve as producer of ⟨*do* a play⟩
11 : to treat unfairly; *especially* **:** CHEAT ⟨*did* him out of his inheritance⟩
12 : to treat or deal with in any way typically with the sense of preparation or with that of care or attention: **a** (1) **:** to put in order **:** CLEAN ⟨was *doing* the kitchen⟩ (2) **:** WASH ⟨*did* the dishes after supper⟩ **b :** to prepare for use or consumption; *especially* **:** COOK ⟨like my steak *done* rare⟩ **c :** SET, ARRANGE ⟨had her hair *done*⟩ **d :** to apply cosmetics to ⟨wanted to *do* her face before the party⟩ **e :** DECORATE, FURNISH ⟨*did* the living room in Early American⟩
13 : to be engaged in the study or practice of ⟨*do* science⟩; *especially* **:** to work at as a vocation ⟨what to *do* after college⟩
14 a : to pass over (as distance) **:** TRAVERSE **b :** to travel at a speed of ⟨*doing* 55 on the turnpike⟩
15 : TOUR ⟨*doing* 12 countries in 30 days⟩
16 : to spend or serve out (a period of time) ⟨*did* ten years in prison⟩
17 : to serve the needs of **:** SUIT, SUFFICE ⟨worms will *do* us for bait⟩
18 : to approve especially by custom, opinion, or propriety ⟨you oughtn't to say a thing like that . . . it's not *done* —Dorothy Sayers⟩
19 : to treat with respect to physical comforts ⟨*did* themselves well⟩
20 : USE *transitive verb* 3 ⟨doesn't *do* drugs⟩
intransitive verb
1 : ACT, BEHAVE ⟨*do* as I say⟩
2 a : GET ALONG, FARE ⟨*do* well in school⟩ **b :** to carry on business or affairs **:** MANAGE ⟨we can *do* without your help⟩
3 : to take place **:** HAPPEN ⟨what's *doing* across the street⟩
4 : to come to or make an end **:** FINISH — used in the past participle
5 : to be active or busy ⟨let us then be up and *doing* —H. W. Longfellow⟩
6 : to be adequate or sufficient **:** SERVE ⟨half of that will *do*⟩
7 : to be fitting **:** conform to custom or propriety ⟨won't *do* to be late⟩
8 — used as a substitute verb to avoid repetition ⟨wanted to run and play as children *do*⟩; used especially in British English following a modal auxiliary or perfective *have* ⟨a great many people had died, or would *do* —Bruce Chatwin⟩
9 — used in the imperative after an imperative to add emphasis ⟨be quiet *do*⟩
verbal auxiliary
1 a — used with the infinitive without *to* to form present and past tenses in legal and parliamentary language ⟨*do* hereby bequeath⟩ and in poetry ⟨give what she *did* crave —Shakespeare⟩ **b** — used with the infinitive without *to* to form present and past tenses in declarative sentences with inverted word order ⟨fervently *do* we pray —Abraham Lincoln⟩, in interrogative sentences ⟨*did* you hear that⟩, and in negative sentences ⟨we *don't* know⟩ ⟨*don't* go⟩
2 — used with the infinitive without *to* to form present and past tenses expressing emphasis ⟨I *do* say⟩ ⟨*do* be careful⟩
— **do•able** \ˈdü-ə-bəl\ *adjective*
— **do a number on :** to defeat or confound thoroughly especially by indirect or deceptive means
— **do away with 1 :** to put an end to **:** ABOLISH **2 :** to put to death **:** KILL
— **do by :** to deal with **:** TREAT
— **do for** *chiefly British* **1 :** to attend to the wants and needs of **:** take care of **2 :** to bring about the death or ruin of
— **do it :** to have sexual intercourse
— **do justice 1 a :** to act justly **b :** to treat fairly or adequately **c :** to show due appreciation for **2 :** to acquit in a way worthy of one's abilities
— **do proud :** to give cause for pride or gratification
— **do the trick :** to produce a desired result
— **do with :** to make good use of **:** benefit by ⟨could *do with* a cup of coffee⟩
— **to do :** necessary to be done ⟨I've done my best and all's *to do* again —A. E. Housman⟩

[2]**do** \ˈdü\ *noun, plural* **dos** *or* **do's** \ˈdüz\ (1599)
1 *chiefly dialect* **:** FUSS, ADO
2 *archaic* **:** DEED, DUTY
3 a : a festive get-together **:** AFFAIR, PARTY **b** *chiefly British* **:** BATTLE
4 : a command or entreaty to do something ⟨a list of *dos* and don'ts⟩
5 *British* **:** CHEAT, SWINDLE
6 : HAIRDO

[3]**do** \ˈdō\ *noun* [Italian] (circa 1754)
: the 1st tone of the diatonic scale in solmization

dob•bin \ˈdä-bᵊn\ *noun* [*Dobbin,* nickname for *Robert*] (1596)
1 : a farm horse
2 : a quiet plodding horse

dob•by \ˈdä-bē\ *noun, plural* **dobbies** [perhaps from *Dobby,* nickname for *Robert*] (1878)
1 : a loom attachment for weaving small figures
2 : a fabric or figured weave made with a dobby

Dobe \ˈdōb\ *noun* (1946)
: DOBERMAN PINSCHER

Do•ber•man pin•scher \ˌdō-bər-mən-ˈpin-chər\ *noun* [German *Dobermann-pinscher,* from Ludwig *Dobermann,* 19th century German dog breeder + German *Pinscher,* a breed of hunting dog] (1917)
: any of a breed of short-haired medium-sized dogs of German origin — called also *Doberman*

do•bra \ˈdō-brə, ˈdȯ-\ *noun* [Portuguese, from feminine of obsolete *dobro* double, from Latin *duplus* — more at DOUBLE] (1978)
— see MONEY table

Do•bro \ˈdō-(ˌ)brō\ *trademark*
— used for an acoustic guitar having a metal resonator

dob•son•fly \ˈdäb-sən-ˌflī\ *noun* [origin unknown] (circa 1904)
: a winged insect (family Corydalidae) that has very long slender mandibles in the male and a large carnivorous aquatic larva and that is now usually considered a neuropteran — compare HELLGRAMMITE

doc \ˈdäk\ *noun* (circa 1850)
: DOCTOR

do•cent \ˈdō-sᵊnt, dō(t)-ˈsent\ *noun* [obsolete German (now *Dozent*), from Latin *docent-, docens,* present participle of *docēre*] (1880)
1 : a college or university teacher or lecturer
2 : a person who leads guided tours especially through a museum or art gallery

do•ce•tic \dō-ˈsē-tik, -ˈse-\ *adjective, often capitalized* [Late Greek *Dokētai* Docetists, from Greek *dokein* to seem — more at DECENT] (1846)
: of or relating to Docetism or the Docetists

Do•ce•tism \dō-ˈsē-ˌti-zəm, ˈdō-sə-\ *noun* (1846)
: a belief opposed as heresy in early Christianity that Christ only seemed to have a human body and to suffer and die on the cross
— **Do•ce•tist** \-ˈsē-tist, -sə-\ *noun*

doch–an–dor•ris *or* **doch–an–dor•is** \ˌdäk-ən-ˈdȯr-əs\ *noun* [Scottish Gaelic *deoch an doruis* & Irish *deoch an dorais,* literally, drink of the door] (1691)
Scottish & Irish **:** a parting drink **:** STIRRUP CUP

doc•ile \ˈdä-səl *also* -ˌsīl, *especially British* ˈdō-ˌsīl\ *adjective* [Latin *docilis,* from *docēre* to teach; akin to Latin *decēre* to be fitting — more at DECENT] (15th century)
1 : easily taught
2 : easily led or managed **:** TRACTABLE
synonym see OBEDIENT
— **doc•ile•ly** \ˈdä-sə(l)-lē\ *adverb*
— **do•cil•i•ty** \dä-ˈsi-lə-tē, dō-\ *noun*

[1]**dock** \ˈdäk\ *noun* [Middle English, from Old English *docce;* akin to Middle Dutch *docke* dock] (before 12th century)
1 : any of a genus (*Rumex*) of the buckwheat family of coarse weedy plants that have long taproots and are sometimes used as potherbs
2 : any of several usually broad-leaved weedy plants

[2]**dock** *noun* [Middle English *dok,* perhaps from Old English *-docca* (as in *fingirdocca* finger muscle); akin to Old High German *tocka* doll, Old Norse *dokka* bundle] (14th century)
1 : the solid part of an animal's tail as distinguished from the hair
2 : the part of an animal's tail left after it has been shortened

[3]**dock** *transitive verb* (14th century)
1 a : to cut off the end of a body part of; *specifically* **:** to remove part of the tail of **b :** to cut (as ears or a tail) short
2 a : to take away a part of **:** ABRIDGE **b :** to subject (as wages) to a deduction **c :** to penalize by depriving of a benefit ordinarily due; *especially* **:** to fine by a deduction of wages ⟨*docked* him for tardiness⟩

[4]**dock** *noun* [Middle English *dokke,* probably from Middle Dutch *docke*] (15th century)

1 : a usually artificial basin or enclosure for the reception of ships that is equipped with means for controlling the water height
2 : the waterway extending between two piers for the reception of ships
3 : a place (as a wharf or platform) for the loading or unloading of materials ▫

[5]dock (1600)
transitive verb
1 : to haul or guide into a dock
2 : to join (as two spacecraft) mechanically while in space
intransitive verb
1 : to come into dock
2 : to become docked

[6]dock *noun* [Flemish *docke* cage] (1586)
: the place in a criminal court where a prisoner stands or sits during trial
— in the dock : on trial

dock·age \'dä-kij\ *noun* (1648)
1 : a charge for the use of a dock
2 : the docking of ships
3 : docking facilities

[1]dock·er \'dä-kər\ *noun* (1810)
: one that docks the tails of animals

[2]docker *noun* (1887)
: one connected with docks; *especially* **:** LONGSHOREMAN

[1]dock·et \'dä-kət\ *noun* [Middle English *doggette*] (15th century)
1 : a brief written summary of a document **:** ABSTRACT
2 a (1) **:** a formal abridged record of the proceedings in a legal action (2) **:** a register of such records **b** (1) **:** a list of legal causes to be tried (2) **:** a calendar of business matters to be acted on **:** AGENDA
3 : an identifying statement about a document placed on its outer surface or cover

[2]docket *transitive verb* (1615)
1 : to place on the docket for legal action
2 : to make a brief abstract of (as a legal matter) and inscribe it in a list
3 : to inscribe (as a document) with an identifying statement

dock·hand \'däk-ˌhand\ *noun* (1920)
: LONGSHOREMAN

dock·land \-ˌland\ *noun* (1904)
British **:** the part of a port occupied by docks; *also* **:** a residential section adjacent to docks

dock·mas·ter \'däk-ˌmas-tər\ *noun* (1736)
: a person in charge of a dock or marina or of the docking of ships

dock·side \-ˌsīd\ *noun, often attributive* (1887)
: the shore or area adjacent to a dock

dock·work·er \-ˌwər-kər\ *noun* (1921)
: LONGSHOREMAN

dock·yard \-ˌyärd\ *noun* (1704)
1 : SHIPYARD
2 *British* **:** NAVY YARD

[1]doc·tor \'däk-tər\ *noun* [Middle English *doctour* teacher, doctor, from Middle French & Medieval Latin; Middle French, from Medieval Latin *doctor,* from Latin, teacher, from *docēre* to teach — more at DOCILE] (14th century)
1 a : an eminent theologian declared a sound expounder of doctrine by the Roman Catholic Church — called also *doctor of the church* **b :** a learned or authoritative teacher **c :** a person who has earned one of the highest academic degrees (as a PhD) conferred by a university **d :** a person awarded an honorary doctorate (as an LLD or Litt D) by a college or university
2 a : one skilled or specializing in healing arts; *especially* **:** a physician, surgeon, dentist, or veterinarian who is licensed to practice **b :** MEDICINE MAN
3 a : material added (as to food) to produce a desired effect **b :** a blade (as of metal) for spreading a coating or scraping a surface
4 : a person who restores or repairs things
— doc·tor·al \-t(ə-)rəl\ *adjective*
— doc·tor·less \-tər-ləs\ *adjective*
— doc·tor·ship \-ˌship\ *noun*

[2]doctor *verb* **doc·tored; doc·tor·ing** \-t(ə-)riŋ\ (1712)
transitive verb
1 a : to give medical treatment to **b :** to restore to good condition **:** REPAIR ⟨*doctor* an old clock⟩
2 a : to adapt or modify for a desired end by alteration or special treatment ⟨*doctored* the play to suit the audience⟩ ⟨the drink was *doctored*⟩ **b :** to alter deceptively ⟨accused of *doctoring* the election returns⟩
intransitive verb
1 : to practice medicine
2 *dialect* **:** to take medicine

doc·tor·ate \'däk-t(ə-)rət\ *noun* (1676)
: the degree, title, or rank of a doctor

[1]doc·tri·naire \ˌdäk-trə-'nar, -'ner\ *noun* [French, from *doctrine*] (1831)
: one who attempts to put into effect an abstract doctrine or theory with little or no regard for practical difficulties

[2]doctrinaire *adjective* (1834)
: of, relating to, or characteristic of a doctrinaire **:** DOGMATIC
synonym see DICTATORIAL
— doc·tri·nair·ism \-'nar-ˌi-zəm, -'ner-\ *noun*

doc·trin·al \'däk-trə-nᵊl, *especially British* däk-'trī-\ *adjective* (15th century)
: of, relating to, or preoccupied with doctrine
— doc·trin·al·ly \-nᵊl-ē\ *adverb*

doc·trine \'däk-trən\ *noun* [Middle English, from Middle French & Latin; Middle French, from Latin *doctrina,* from *doctor*] (14th century)
1 *archaic* **:** TEACHING, INSTRUCTION
2 a : something that is taught **b :** a principle or position or the body of principles in a branch of knowledge or system of belief **:** DOGMA **c :** a principle of law established through past decisions **d :** a statement of fundamental government policy especially in international relations

docu·dra·ma \'dä-kyə-ˌdrä-mə, -ˌdra-\ *noun* [*documentary* + *drama*] (circa 1961)
: a drama for television, motion pictures, or theater dealing freely with historical events especially of a recent and controversial nature

[1]doc·u·ment \'dä-kyə-mənt\ *noun* [Middle English, from Middle French, from Late Latin & Latin; Late Latin *documentum* official paper, from Latin, lesson, proof, from *docēre* to teach — more at DOCILE] (15th century)
1 a *archaic* **:** PROOF, EVIDENCE **b :** an original or official paper relied on as the basis, proof, or support of something **c :** something (as a photograph or a recording) that serves as evidence or proof
2 a : a writing conveying information **b :** a material substance (as a coin or stone) having on it a representation of thoughts by means of some conventional mark or symbol **c :** DOCUMENTARY
— doc·u·men·tal \ˌdä-kyə-'men-tᵊl\ *adjective*

[2]doc·u·ment \'dä-kyə-ˌment\ *transitive verb* (1711)
1 : to furnish documentary evidence of
2 : to furnish with documents
3 a : to provide with factual or substantial support for statements made or a hypothesis proposed; *especially* **:** to equip with exact references to authoritative supporting information **b** (1) **:** to construct or produce (as a movie or novel) with authentic situations or events (2) **:** to portray realistically
4 : to furnish (a ship) with ship's papers
— doc·u·ment·able \-ˌmen-tə-bəl, ˌdäk-yə-'\ *adjective*
— doc·u·ment·er \'dä-kyə-ˌmen-tər\ *noun*

doc·u·men·tal·ist \ˌdä-kyə-'men-tᵊl-ist\ *noun* (1939)
: a specialist in documentation

doc·u·men·tar·i·an \ˌdä-kyə-mən-'ter-ē-ən, -ˌmen-\ *noun* [[2]*documentary*] (1943)
: one who makes a documentary

doc·u·men·ta·rist \-'men-tə-rist\ *noun* [[2]*documentary*] (1949)
: DOCUMENTARIAN

[1]doc·u·men·ta·ry \ˌdä-kyə-'men-tə-rē, -'men-trē\ *adjective* (1802)
1 : being or consisting of documents **:** contained or certified in writing ⟨*documentary* evidence⟩
2 : of, relating to, or employing documentation in literature or art; *broadly* **:** FACTUAL, OBJECTIVE ⟨a *documentary* film of the war⟩
— doc·u·men·tar·i·ly \-mən-'ter-ə-lē, -ˌmen-\ *adverb*

[2]documentary *noun, plural* **-ries** (1935)
: a documentary presentation (as a film or novel)

doc·u·men·ta·tion \ˌdä-kyə-mən-'tā-shən, -ˌmen-\ *noun* (1884)
1 : the act or an instance of furnishing or authenticating with documents
2 a : the provision of documents in substantiation; *also* **:** documentary evidence **b** (1) **:** the use of historical documents (2) **:** conformity to historical or objective facts (3) **:** the provision of footnotes, appendices, or addenda referring to or containing documentary evidence
3 : INFORMATION SCIENCE
4 : the usually printed instructions, comments, and information for using a particular piece or system of computer software or hardware
— doc·u·men·ta·tion·al \-shnəl, -shə-nᵊl\ *adjective*

[1]dod·der \'dä-dər\ *noun* [Middle English *doder;* akin to Middle High German *toter* dodder, egg yolk] (13th century)
: any of a genus (*Cuscuta*) of parasitic wiry twining vines of the morning-glory family that are deficient in chlorophyll and have tiny scales instead of leaves

[2]dodder *intransitive verb* **dod·dered; dod·der·ing** \'dä-d(ə-)riŋ\ [Middle English *dadiren*] (14th century)
1 : to tremble or shake from weakness or age
2 : to progress feebly and unsteadily ⟨was *doddering* down the walk⟩
— dod·der·er \-dər-ər\ *noun*

dod·dered \'dä-dərd\ *adjective* [probably alteration of *dodded,* from past participle of English dialect *dod* to lop, from Middle English *dodden*] (1697)
1 : deprived of branches through age or decay ⟨a *doddered* oak⟩
2 : INFIRM, ENFEEBLED

dod·der·ing \'dä-d(ə-)riŋ\ *adjective* (1898)
: FEEBLE, SENILE

dod·dery \-d(ə-)rē\ *adjective* (1866)
1 : DODDERED 2
2 : DODDERING

dodeca- *or* **dodec-** *combining form* [Latin, from Greek *dōdeka-, dōdek-,* from *dōdeka, dyōdeka,* from *dyō, dyo* two + *deka* ten]
: twelve ⟨*dodeca*phonic⟩

▫ USAGE
dock It is a nautical-minded newspaperman's tradition to ridicule the use of sense 3 as showing the user's lack of nautical knowledge. But sense 3 is a standard sense, and it may be the only sense of the word used in reference to lakes and inland waterways. It has been used by many an American literary light and even a British visitor or two ⟨Jean was on the *dock* when the ship came in —Mark Twain⟩ ⟨I thought of Gatsby's wonder when he first picked out the green light at the end of Daisy's *dock* —F. Scott Fitzgerald⟩ ⟨she walked across the *dock* and up the steep sandy road —Ernest Hemingway⟩ ⟨staring about me, until we came alongside the *dock* —Charles Dickens (*American Notes,* 1842)⟩.

do·deca·gon \dō-ˈde-kə-ˌgän\ *noun* [Greek *dōdekagōnon,* from *dōdeka-* + *-gōnon* -gon] (circa 1658)
: a polygon of 12 angles and 12 sides

do·deca·he·dron \(ˌ)dō-ˌde-kə-ˈhē-drən\ *noun, plural* **-drons** *or* **-dra** \-drə\ [Greek *dōdekaedron,* from *dōdeka-* + *-edron* -hedron] (circa 1570)
: a solid having 12 plane faces
— do·deca·he·dral \-drəl\ *adjective*

do·deca·phon·ic \(ˌ)dō-ˌde-kə-ˈfä-nik\ *adjective* [*dodeca-* + *phon-* + *-ic*] (1949)
: TWELVE-TONE
— do·deca·phon·i·cal·ly \-ni-k(ə-)lē\ *adverb*
— do·deca·pho·nist \dō-ˈde-kə-fə-nist, -ˌfō-; ˌdō-di-ˈka-fə-nist\ *noun*
— do·deca·pho·ny \-nē\ *noun*

¹dodge \ˈdäj\ *noun* [origin unknown] (1575)
1 a : an artful device to evade, deceive, or trick **b :** EXPEDIENT
2 : an act of evading by sudden bodily movement

²dodge *verb* **dodged; dodg·ing** (1680)
transitive verb
1 : to evade (as a duty) usually indirectly or by trickery ⟨*dodged* the draft by leaving the country⟩ ⟨*dodged* questions⟩
2 a : to evade by a sudden or repeated shift of position **b :** to avoid an encounter with
intransitive verb
1 a : to make a sudden movement in a new direction (as to evade a blow) ⟨*dodged* behind the door⟩ **b :** to move to and fro or from place to place usually in an irregular course ⟨*dodged* through the crowd⟩
2 : to evade a responsibility or duty especially by trickery or deceit

dodge·ball \ˈdäj-ˌbȯl\ *noun* (circa 1922)
: a game in which players stand in a circle and try to hit opponents within the circle with a large inflated ball

dodg·em \ˈdä-jəm\ *noun, often capitalized* (1921)
: an amusement ride featuring bumper cars

dodgem car *noun* (1945)
: BUMPER CAR

dodg·er \ˈdä-jər\ *noun* (1568)
1 : one that dodges; *especially* **:** one who uses tricky devices
2 : a small leaflet **:** CIRCULAR
3 : CORN DODGER
4 : a usually canvas screen on a boat or ship that provides protection from spray

dodg·ery \ˈdä-jə-rē\ *noun, plural* **-er·ies** (1670)
: EVASION, TRICKERY

dodgy \ˈdä-jē\ *adjective* (1861)
1 *chiefly British* **:** EVASIVE, TRICKY
2 *chiefly British* **:** not sound, stable, or reliable **:** SHAKY
3 *British* **:** requiring skill or care in handling; *also* **:** CHANCY, RISKY
— dodg·i·ness \-nəs\ *noun*

do·do \ˈdō-(ˌ)dō\ *noun, plural* **dodoes** *or* **dodos** [Portuguese *doudo,* from *doudo* silly, stupid] (1628)
1 a : an extinct heavy flightless bird (*Raphus cucullatus,* synonym *Didus ineptus*) of the island of Mauritius that is related to the pigeon and larger than a turkey **b :** an extinct bird of the island of Réunion similar to and apparently closely related to the dodo
2 a : one hopelessly behind the times **b :** a stupid person

dodo 1a

do down *transitive verb* (14th century)
British **:** to get the better of (as by trickery)

doe \ˈdō\ *noun, plural* **does** *or* **doe** [Middle English *do,* from Old English *dā;* akin to German dialect *tē* doe] (before 12th century)
: the adult female of various mammals (as a deer, rabbit, or kangaroo) of which the male is called buck

doe–eyed \ˈdō-ˌīd\ *adjective* (1933)
: having large innocent-looking eyes

do·er \ˈdü-ər\ *noun* (14th century)
: one that takes an active part ⟨a thinker or a doer⟩

does *present 3d singular of* DO, *plural of* DOE

doe·skin \ˈdō-ˌskin\ *noun* (15th century)
1 : the skin of does or leather made of it; *also* **:** soft leather from sheep- or lambskins
2 : a compact coating and sportswear fabric napped and felted for a smooth surface

doesn't \ˈdə-zᵊnt\ (1860)
: does not

do·est \ˈdü-əst\ *archaic present 2d singular of* DO

do·eth \ˈdü-əth\ *archaic present 3d singular of* DO

doff \ˈdäf, ˈdȯf\ *transitive verb* [Middle English, from *don* to do + *of* off] (14th century)
1 a : to remove (an article of wear) from the body **b :** to take off (the hat) in greeting or as a sign of respect
2 : to rid oneself of **:** put aside
— doff one's hat to *or* **doff one's cap to :** to show respect to **:** SALUTE

¹dog \ˈdȯg, ˈdäg\ *noun, often attributive* [Middle English, from Old English *docga*] (before 12th century)
1 a : CANID; *especially* **:** a highly variable domestic mammal (*Canis familiaris*) closely related to the common wolf (*Canis lupus*) **b :** a male dog; *also* **:** a male usually carnivorous mammal
2 a : a worthless person **b :** FELLOW, CHAP ⟨a lazy *dog*⟩ ⟨you lucky *dog*⟩
3 a : any of various usually simple mechanical devices for holding, gripping, or fastening that consist of a spike, bar, or hook **b :** ANDIRON
4 : uncharacteristic or affected stylishness or dignity ⟨put on the *dog*⟩
5 *capitalized* **:** either of the constellations Canis Major or Canis Minor
6 *plural* **:** FEET
7 *plural* **:** RUIN ⟨going to the *dogs*⟩
8 : one inferior of its kind: as **a :** an investment not worth its price **b :** an undesirable piece of merchandise
9 : an unattractive person and especially a girl or woman
10 : HOT DOG 1
— dog·like \ˈdȯg-ˌlīk\ *adjective*

dog 1a: *1* **pastern,** *2* **chest,** *3* **flews,** *4* **muzzle,** *5* **stop,** *6* **occiput,** *7* **leather,** *8* **crest,** *9* **withers,** *10* **loin,** *11* **point of rump,** *12* **hock** *or* **tarsus,** *13* **knee** *or* **stifle,** *14* **brisket,** *15* **elbow,** *16* **feathering**

²dog *adjective* (14th century)
1 : CANINE
2 : SPURIOUS; *especially* **:** unlike that used by native speakers or writers ⟨*dog* Latin⟩ ⟨*dog* French⟩

³dog *transitive verb* **dogged** \ˈdȯgd, ˈdägd\; **dog·ging** (1519)
1 a : to hunt or track like a hound **b :** to worry as if by pursuit with dogs **:** PLAGUE
2 : to fasten with a dog
— dog it : to fail to do one's best **:** GOLDBRICK

⁴dog *adverb* (circa 1552)
: EXTREMELY, UTTERLY ⟨*dog*-tired⟩

dog and pony show *noun* (1965)
: an often elaborate public relations or sales presentation

dog·bane \ˈdȯg-ˌbān\ *noun* (1597)
: any of a genus (*Apocynum* of the family Apocynaceae, the dogbane family) comprising chiefly tropical and often poisonous plants with milky juice and usually showy flowers

dog biscuit *noun* (circa 1858)
: a hard dry cracker for dogs

dog·cart \ˈdȯg-ˌkärt\ *noun* (1668)
1 : a cart drawn by a dog
2 : a light two-wheeled carriage with two transverse seats set back to back

dog·catch·er \-ˌka-chər, -ˌke-\ *noun* (1835)
: a community official assigned to catch and dispose of stray dogs

dog collar *noun* (1524)
1 : a collar for a dog
2 *slang* **:** CLERICAL COLLAR
3 : a wide flexible snug-fitting necklace

dog days *noun plural* [from their being reckoned from the heliacal rising of the Dog Star (Sirius)] (1538)
1 : the period between early July and early September when the hot sultry weather of summer usually occurs in the northern hemisphere
2 : a period of stagnation or inactivity ◆

dog·dom \ˈdȯg-dəm\ *noun* (1854)
: the world of dogs or of dog fanciers

doge \ˈdōj\ *noun* [Italian dialect, from Latin *duc-, dux* leader — more at DUKE] (1549)
: the chief magistrate in the republics of Venice and Genoa

dog–ear \ˈdȯ-ˌgir\ *noun* (circa 1725)
: the turned-down corner of a page especially of a book
— dog–ear *transitive verb*

dog–eared \ˈdȯ-ˌgird\ *adjective* (1784)
1 : having dog-ears ⟨a *dog-eared* book⟩
2 : SHABBY, TIMEWORN ⟨a *dog-eared* resort⟩ ⟨*dog-eared* myths⟩

dog–eat–dog \ˌdȯ-ˌgēt-ˈdȯg\ *adjective* (1834)
: marked by ruthless self-interest ⟨*dog-eat-dog* competition⟩

dog·face \ˈdȯg-ˌfās\ *noun* (1941)
: SOLDIER; *especially* **:** INFANTRYMAN

dog fennel *noun* (14th century)
: a strong-scented chamomile (*Anthemis cotula*) of Europe and Asia naturalized along roadsides in the U.S.

dog·fight \ˈdȯg-ˌfīt\ *noun* (1656)

◇ WORD HISTORY

dog days The Greek name for the brightest star in the sky was *Seirios,* a word of uncertain origin; the Latin version of this word is *Sirius,* which gives us our name for the star. Our alternative name, the Dog Star, is ultimately a translation of Greek *Kyōn,* literally, "hound." This epithet was given to Sirius because it was considered the hound of the hunter Orion, whose constellation was nearby. Sirius was regarded by the Greeks as the bringer of scorching heat, because its early-morning rising coincided with the hottest summer days of July and August. The Greek writer Plutarch referred to these days as *hēmerai kynades,* literally, "dog days"—the days of the Dog Star—and a later Latin translation of this expression as *dies caniculares* is the source of the English phrase *dog days.*

1 : a fight between dogs; *broadly* **:** a fiercely disputed contest
2 : a fight between two or more fighter planes usually at close quarters
— **dogfight** *intransitive verb*

dog·fish \-ˌfish\ *noun* (15th century)
: any of various usually small bottom-dwelling sharks (as of the families Squalidae, Carcharhinidae, and Scyliorhinidae) that often appear in schools near shore, prey chiefly on fish and invertebrates, and are a valuable food source

dog·ged \ˈdȯ-gəd\ *adjective* (1653)
: marked by stubborn determination
synonym see OBSTINATE
— **dog·ged·ly** *adverb*
— **dog·ged·ness** *noun*

[1]dog·ger·el \ˈdȯ-g(ə-)rəl, ˈdä-\ *adjective* [Middle English *dogerel,* probably diminutive of *dogge* dog] (14th century)
: loosely styled and irregular in measure especially for burlesque or comic effect; *also* **:** marked by triviality or inferiority

[2]doggerel *noun* (1630)
: doggerel verse

dog·gery \ˈdȯ-gə-rē\ *noun, plural* **-ger·ies** (1830)
: a cheap saloon **:** DIVE

dog·gie bag *or* **doggy bag** \ˈdȯ-gē-\ *noun* [[1]*doggy;* from the presumption that such leftovers are intended for a pet dog] (1963)
: a container for carrying home leftover food and especially meat from a restaurant meal

dog·gish \ˈdȯ-gish\ *adjective* (15th century)
1 : CANINE
2 : stylish in a showy way
— **dog·gish·ly** *adverb*
— **dog·gish·ness** *noun*

dog·go \ˈdȯ-(ˌ)gō\ *adverb* [probably from [1]*dog*] (1893)
: in hiding — used chiefly in the phrase *to lie doggo*

[1]dog·gone \ˈdäg-ˈgän, ˈdȯg-ˈgȯn\ *verb* **dog·goned; dog·gon·ing** [euphemism for *God damn*] (1828)
: DAMN

[2]dog·gone *or* **dog·goned** \ˌdäg-ˌgän(d), ˌdȯg-ˌgȯn(d)\ *adjective or adverb* (1851)
: DAMNED

[3]doggone *noun* (1928)
: DAMN

[1]dog·gy *or* **dog·gie** \ˈdȯ-gē\ *noun, plural* **doggies** (1825)
: a small dog

[2]dog·gy \ˈdȯ-gē\ *adjective* **dog·gi·er; -est** (1859)
1 : concerned with or fond of dogs
2 : resembling or suggestive of a dog ⟨*doggy* odor⟩
3 : STYLISH, SHOWY
4 : not worthy or profitable **:** INFERIOR

dog·house \ˈdȯg-ˌhau̇s\ *noun* (1594)
: a shelter for a dog
— **in the doghouse :** in a state of disfavor

do·gie \ˈdō-gē\ *noun* [origin unknown] (1888)
chiefly West **:** a motherless calf in a range herd

dog in the manger [from the fable of the dog who prevented an ox from eating hay which he did not want himself] (1573)
: a person who selfishly withholds from others something useless to himself

[1]dog·leg \ˈdȯg-ˌleg, -ˌläg\ *adjective* (1889)
: crooked or bent like a dog's hind leg

[2]dogleg *noun* (circa 1909)
1 a : something having an abrupt angle **b :** a sharp bend (as in a road)
2 : a golf hole having an angled fairway

[3]dogleg *intransitive verb* (1947)
: to proceed along a dogleg course ⟨the single narrow street that *doglegs* through town —Russ Leadabrand⟩

dog·ma \ˈdȯg-mə, ˈdäg-\ *noun, plural* **dogmas** *also* **dog·ma·ta** \-mə-tə\ [Latin *dogmat-, dogma,* from Greek, from *dokein* to seem — more at DECENT] (1638)
1 a : something held as an established opinion; *especially* **:** a definite authoritative tenet **b :** a code of such tenets ⟨pedagogical *dogma*⟩ **c :** a point of view or tenet put forth as authoritative without adequate grounds
2 : a doctrine or body of doctrines concerning faith or morals formally stated and authoritatively proclaimed by a church

dog·mat·ic \dȯg-ˈma-tik, däg-\ *also* **dog·mat·i·cal** \-ti-kəl\ *adjective* (circa 1681)
1 : characterized by or given to the use of dogmatism ⟨a *dogmatic* critic⟩
2 : of or relating to dogma
synonym see DICTATORIAL
— **dog·mat·i·cal·ly** \-ti-k(ə-)lē\ *adverb*
— **dog·mat·i·cal·ness** \-ti-kəl-nəs\ *noun*

dog·mat·ics \-tiks\ *noun plural but singular or plural in construction* (1845)
: a branch of theology that seeks to interpret the dogmas of a religious faith

dogmatic theology *noun* (1846)
: DOGMATICS

dog·ma·tism \ˈdȯg-mə-ˌti-zəm, ˈdäg-\ *noun* (1603)
1 : positiveness in assertion of opinion especially when unwarranted or arrogant
2 : a viewpoint or system of ideas based on insufficiently examined premises

dog·ma·tist \-mə-tist\ *noun* (1541)
: one who dogmatizes

dog·ma·tize \ˈdȯg-mə-ˌtīz, ˈdäg-\ *verb* **-tized; -tiz·ing** [French *dogmatiser,* from Late Latin *dogmatizare,* from Greek *dogmatizein,* from *dogmat-, dogma*] (1611)
intransitive verb
: to speak or write dogmatically
transitive verb
: to state as a dogma or in a dogmatic manner
— **dog·ma·ti·za·tion** \ˌdȯg-mə-tə-ˈzā-shən, ˌdäg-\ *noun*
— **dog·ma·tiz·er** *noun*

dog·nap \ˈdȯg-ˌnap\ *transitive verb* **-napped** *or* **-naped** \-ˌnapt\; **-nap·ping** *or* **-nap·ing** \-ˌna-piŋ\ [[1]*dog* + *-nap* (as in *kidnap*)] (1947)
: to steal (a dog) especially to obtain a reward for its return or to sell to a scientific laboratory
— **dog·nap·per** *or* **dog·nap·er** *noun*

Do·gon \ˈdō-ˌgän\ *noun, plural* **Dogon** *or* **Dogons** (circa 1931)
1 : a member of a people of Mali noted for their sculpture
2 : the language of the Dogon

do–good \ˈdü-ˌgu̇d\ *adjective* (1952)
: designed or disposed sometimes impracticably and too zealously toward bettering the conditions under which others live
— **do–good·ism** \-ˌi-zəm\ *noun*

do–good·er \-ˌgu̇-dər\ *noun* (1926)
: an earnest often naive humanitarian or reformer
— **do–good·ing** \-diŋ\ *noun or adjective*

dog paddle *noun* (1904)
: an elementary swimming stroke in which the arms paddle in the water and the legs maintain a kicking motion
— **dog–pad·dle** *intransitive verb*

dog rose *noun* (circa 1597)
: a common European wild rose (*Rosa canina*)

dogs·body \ˈdȯgz-ˌbä-dē\ *noun* [British naval slang *dogsbody* pudding made of peas, junior officer] (1922)
chiefly British **:** DRUDGE

dog's chance *noun* (1902)
: a bare chance in one's favor

dog·sled \ˈdȯg-ˌsled\ *noun* (1810)
: a sled drawn by dogs
— **dogsled** *intransitive verb*
— **dog·sled·der** \-ˌsle-dər\ *noun*

Dog Star *noun*
: SIRIUS

dog tag *noun* (1918)
: an identification tag (as for military personnel or pets)

dog tick *noun* (circa 1552)
: AMERICAN DOG TICK

dog·tooth \ˈdȯg-ˌtüth\ *noun* (1552)
1 : CANINE 1, EYETOOTH
2 : an architectural ornament common in early English Gothic consisting usually of four leaves radiating from a raised point at the center
3 *chiefly British* **:** HOUNDSTOOTH CHECK

dogtooth 2

dogtooth violet *noun* (1629)
: any of a genus (*Erythronium*) of small spring-flowering bulbous herbs of the lily family

[1]dog·trot \ˈdȯg-ˌträt\ *noun* (15th century)
1 : a quick easy gait suggesting that of a dog
2 *chiefly Southern & Midland* **:** a roofed passage similar to a breezeway; *especially* **:** one connecting two parts of a cabin

[2]dogtrot *intransitive verb* (circa 1900)
: to move or progress at a dogtrot

dog·watch \ˈdȯg-ˌwäch\ *noun* (1700)
1 : either of two watches of two hours on shipboard that extend from 4 to 6 and 6 to 8 p.m.
2 : any of various night shifts; *especially* **:** the last shift

dog·wood \ˈdȯg-ˌwu̇d\ *noun* (1617)
: any of various trees and shrubs (genus *Cornus* of the family Cornaceae, the dogwood family) with clusters of small flowers and often large white, pink, or red involucral bracts

doi·ly \ˈdȯi-lē\ *noun, plural* **doilies** [*Doily* or *Doyley* (flourished 1711) London draper] (1711)
1 : a small napkin
2 : a small often decorative mat

do in *transitive verb* (1905)
1 a : to bring about the defeat or destruction of; *also* **:** KILL **b :** EXHAUST, WEAR OUT
2 : CHEAT

do·ing \ˈdü-iŋ\ *noun* (14th century)
1 : the act of performing or executing **:** ACTION ⟨that will take a great deal of *doing*⟩
2 *plural* **a :** things that are done or that occur **:** GOINGS-ON ⟨everyday *doings*⟩ **b :** social activities

doit \ˈdȯit\ *noun* [Dutch *duit;* akin to Old Norse *thveiti* small coin, *thveita* to hew] (1592)
1 : an old Dutch coin equal to about ⅛ stiver
2 : TRIFLE 1

do–it–your·self \ˌdü-ə-chər-ˈself, -tyər-\ *noun, often attributive* (1952)
: the activity of doing or making something (as in woodworking or home repair) without professional training or assistance; *broadly* **:** an activity in which one does something oneself or on one's own initiative
— **do–it–your·self·er** \-ˈsel-fər\ *noun*

do·jo \ˈdō-(ˌ)jō\ *noun, plural* **dojos** [Japanese *dōjō,* from *dō* way, art + *-jō* ground] (1942)
: a school for training in various arts of self-defense (as judo or karate)

Dol·by \ˈdȯl-bē, ˈdōl-\ *trademark*
— used for an electronic device that eliminates noise from recorded sound or sound broadcast on FM radio

dol·ce \ˈdōl-(ˌ)chā\ *adjective or adverb* [Italian, literally, sweet, from Latin *dulcis* — more at DULCET] (circa 1847)
: SOFT, SMOOTH — used as a direction in music

dol·ce far nien·te \ˈdōl-chē-ˌfär-nē-ˈen-tē\ *noun* [Italian, literally, sweet doing nothing] (1814)
: pleasant relaxation in carefree idleness

dol·ce vi·ta \ˌdōl-chā-ˈvē-(ˌ)tä\ *noun* [Italian, literally, sweet life] (1961)
: a life of indolence and self-indulgence

dol·drums \ˈdōl-drəmz, ˈdäl-, ˈdȯl-\ *noun plural* [probably akin to Old English *dol* foolish] (1811)
1 : a spell of listlessness or despondency
2 *often capitalized* **:** a part of the ocean near the equator abounding in calms, squalls, and light shifting winds

3 : a state or period of inactivity, stagnation, or slump

[1]**dole** \'dōl\ *noun* [Middle English, from Old English *dāl* portion — more at DEAL] (before 12th century)
1 *archaic* **:** one's allotted share, portion, or destiny
2 a (1) **:** a giving or distribution of food, money, or clothing to the needy (2) **:** a grant of government funds to the unemployed **b :** something distributed at intervals to the needy; *also* **:** HANDOUT 1 **c :** something portioned out bit by bit

[2]**dole** *transitive verb* **doled; dol•ing** (15th century)
: to give or distribute as a charity — usually used with *out*

[3]**dole** *noun* [Middle English *dol,* from Old French, from Late Latin *dolus,* alteration of Latin *dolor*] (13th century)
archaic **:** GRIEF, SORROW

dole•ful \'dōl-fəl\ *adjective* (13th century)
1 : causing grief or affliction ⟨a *doleful* loss⟩
2 : full of grief **:** CHEERLESS ⟨a *doleful* face⟩
3 : expressing grief **:** SAD ⟨a *doleful* melody⟩
— dole•ful•ly \-fə-lē\ *adverb*
— dole•ful•ness *noun*

dole out *transitive verb* (1749)
1 : to give or deliver in small portions
2 : DISH OUT
synonym see DISTRIBUTE

dol•er•ite \'dä-lə-ˌrīt\ *noun* [French *dolérite,* from Greek *doleros* deceitful, from *dolos* deceit; from its being easily mistaken for diorite] (1838)
1 : any of various coarse basalts
2 *chiefly British* **:** DIABASE 3
— dol•er•it•ic \ˌdä-lə-'ri-tik\ *adjective*

dole•some \'dōl-səm\ *adjective* (1533)
: DOLEFUL

dol•i•cho•ce•phal•ic \ˌdä-li-kō-sə-'fa-lik\ *adjective* [New Latin *dolichocephalus* long-headed, from Greek *dolichos* long + *-kephalos,* from *kephalē* head — more at CEPHALIC] (1852)
: having a relatively long head with cephalic index of less than 75
— dol•i•cho•ceph•a•ly \-'se-fə-lē\ *noun*

doll \'däl, 'dȯl\ *noun* [probably from *Doll,* nickname for *Dorothy*] (circa 1700)
1 : a small-scale figure of a human being used especially as a child's plaything
2 a (1) **:** a pretty but often empty-headed young woman (2) **:** WOMAN **b :** DARLING, SWEETHEART **c :** an attractive person
— doll•ish \'dä-lish, 'dȯ-\ *adjective*
— doll•ish•ly *adverb*
— doll•ish•ness *noun*

dol•lar \'dä-lər\ *noun, often attributive* [Dutch or Low German *daler,* from German *Taler,* short for *Joachimstaler,* from Sankt *Joachimsthal,* Bohemia, where talers were first made] (1553)
1 : TALER
2 : any of numerous coins patterned after the taler (as a Spanish peso)
3 a : any of various basic monetary units (as in the U.S. and Canada) — see MONEY table **b :** a coin, note, or token representing one dollar
4 : RINGGIT ◆

dollar cost averaging *noun* (circa 1957)
: investment in a security at regular intervals of a uniform sum regardless of the price level in order to obtain an overall reduction in cost per unit — called also *dollar averaging*

dollar–a–year *adjective* (1918)
: compensated by a token salary usually for government service ⟨a *dollar-a-year* man⟩

dollar day *noun* (1949)
: a day on which special low prices are offered — often used in plural

dollar diplomacy *noun* (1910)
1 : diplomacy used by a country to promote its financial or commercial interests abroad
2 : diplomacy that seeks to strengthen the power of a country or effect its purposes in foreign relations by the use of its financial resources

dollars–and–cents \'dä-lər-zᵊn-'sen(t)s\ *adjective* (1899)
: dealing with or expressed in terms of money, sales, or profits

dollar sign *noun* (1881)
: a mark $ placed before a number to indicate that it stands for dollars — called also *dollar mark*

doll•house \'däl-ˌhau̇s, 'dȯl-\ *also* **doll's house** *or* **dolls' house** *noun* (1783)
1 : a child's small-scale toy house
2 : a dwelling so small as to suggest a house for dolls

[1]**dol•lop** \'dä-ləp\ *noun* [origin unknown] (circa 1812)
1 *chiefly British* **:** an indefinite often large quantity especially of something liquid ⟨out of heaven, as if a plug had been pulled, fell a jolly *dollop* of rain —E. M. Forster⟩
2 : a lump or glob of something soft or mushy ⟨top it with a *dollop* of jam⟩
3 : an amount given, spooned, or ladled out **:** PORTION ⟨hold out their mess tins for a *dollop* of gruel —Robert Craft⟩
4 : a small lump, portion, or amount ⟨want just a *dollop* of catsup⟩ ⟨a *dollop* of brandy after dinner⟩
5 : something added or served as if in dollops ⟨fantasy with a *dollop* of satire —Lee Rogow⟩ ⟨a delicious *dollop* of gossip —Leon Harris⟩

[2]**dollop** *transitive verb* (circa 1860)
: to serve or dispense in dollops

doll up (1906)
transitive verb
1 : to dress elegantly or extravagantly
2 : to make more attractive (as by addition of decorative details)
intransitive verb
: to get dolled up

[1]**dol•ly** \'dä-lē, 'dȯ-lē\ *noun, plural* **dollies** (1790)
1 : DOLL
2 : a wooden-pronged instrument for beating and stirring clothes in the process of washing them in a tub
3 : a compact narrow-gauge railroad locomotive for moving construction trains and for switching
4 a : a platform on a roller or on wheels or casters for moving heavy objects **b :** a wheeled platform for a television or motion-picture camera

[2]**dolly** *verb* **dol•lied; dol•ly•ing** (1878)
transitive verb
1 : to treat with a dolly
2 : to move or convey on a dolly
intransitive verb
: to move a motion-picture or television camera about on a dolly while shooting a scene; *also of a camera* **:** to be moved on a dolly

dol•ly bird \'dä-lē-ˌbərd, 'dȯ-lē-\ *noun* (1964)
British **:** a pretty young woman

dolly shot *noun* (1933)
: TRACKING SHOT

Dol•ly Var•den trout \ˌdä lē 'vär-dᵊn-\ *noun* [*Dolly Varden,* gaily dressed coquette in *Barnaby Rudge* (1841), novel by Charles Dickens] (circa 1876)
: a large char (*Salvelinus malma*) widespread in streams of western North America and Japan as well as in coastal salt waters — called also *Dolly Varden*

dol•ma \ˌdōl-'mä, 'dōl-mə, 'däl-\ *noun, plural* **dolmas** *or* **dol•ma•des** \dōl-'mä-(ˌ)thēz, dȯl-, däl-\ [Turkish, literally, something stuffed] (circa 1889)
: a stuffed grape leaf or vegetable shell

dolman sleeve *noun* [French *dolman* coat with dolman sleeves, from German *Dolman* or Hungarian *dolmány,* from Turkish *dolama,* a Turkish robe] (1934)
: a sleeve very wide at the armhole and tight at the wrist often cut in one piece with the bodice

dol•men \'dōl-mən, 'dȯl-, 'däl-\ *noun* [French, from Breton *tolmen,* from *tol* table + *men* stone] (1859)
: a prehistoric monument of two or more upright stones supporting a horizontal stone slab found especially in Britain and France and thought to be a tomb

dolmen

do•lo•mite \'dō-lə-ˌmīt, 'dä-\ *noun* [French, from Déodat de *Dolomieu* (died 1801) French geologist] (1794)
1 : a mineral $CaMg(CO_3)_2$ consisting of a calcium magnesium carbonate found in crystals and in extensive beds as a compact limestone
2 : a limestone or marble rich in magnesium carbonate
— do•lo•mit•ic \ˌdō-lə-'mi-tik, ˌdä-\ *adjective*

do•lo•mi•ti•za•tion \ˌdō-lə-mə-tə-'zā-shən, ˌdä-, -ˌmī-\ *noun* (1862)
: the process of converting into dolomite
— do•lo•mi•tize \'dō-lə-mə-ˌtīz, 'dä-\ *transitive verb*

do•lor \'dō-lər *also* 'dä-\ *noun* [Middle English *dolour,* from Middle French, from Latin *dolor* pain, grief, from *dolēre* to feel pain, grieve] (14th century)
: mental suffering or anguish **:** SORROW

do•lor•ous \'dō-lə-rəs *also* 'dä-\ *adjective* (15th century)
: causing, marked by, or expressing misery or grief
— do•lor•ous•ly *adverb*
— do•lor•ous•ness *noun*

do•lour *chiefly British variant of* DOLOR

dol•phin \'däl-fən, 'dȯl-\ *noun* [Middle English, from Middle French *dophin, daufin,* from Old French *dalfin,* from Old Provençal, from Medieval Latin *dalfinus,* alteration of Latin *delphinus,* from Greek *delphin-, delphis;* akin to Greek *delphys* womb, Sanskrit *garbha*] (14th century)

dolphin 1a

◇ WORD HISTORY
dollar The Erzgebirge ("Ore Mountains") on the present-day German-Czech border have long served as a source of metals, as their name might suggest. Beginning about 1515 silver from a mine near the Erzgebirge town of Sankt Joachimstal (now Jáchymov in the Czech Republic) was used to mint a coin for the kings of Bohemia called in German *Joachimstaler,* which was later shortened to *taler.* In the Low German or Dutch form *daler* the word was borrowed into English and applied to similar silver coins, including the Spanish peso or eight-real piece, which circulated widely in Britain's North American colonies. When the newly independent American colonies settled on an official monetary unit in 1785, they chose the name *dollar.* Many nations now call their own official unit the *dollar.*

1 a : any of various small toothed whales (family Delphinidae) with the snout more or less elongated into a beak and the neck vertebrae partially fused **b :** PORPOISE 1
2 : either of two active pelagic bony food fishes (genus *Coryphaena* of the family Coryphaenidae) of tropical and temperate seas that are used for food — called also *dolphinfish*
3 *capitalized* **:** DELPHINUS
4 : a spar or buoy for mooring boats; *also* **:** a cluster of closely driven piles used as a fender for a dock or as a mooring or guide for boats

dolphin striker *noun* (1833)
: a vertical spar under the end of the bowsprit of a sailboat to extend and support the martingale

dolt \'dōlt\ *noun* [probably akin to Old English *dol* foolish] (1553)
: a stupid person
— dolt·ish \'dōl-tish\ *adjective*
— dolt·ish·ly *adverb*
— dolt·ish·ness *noun*

Dom [Latin *dominus* master] (1716)
1 \'däm\ — used as a title for some monks and canons regular
2 \'dōⁿ\ — used as a title prefixed to the Christian name of a Portuguese or Brazilian man of rank

-dom \dəm\ *noun suffix* [Middle English, from Old English *-dōm;* akin to Old High German *-tuom* -dom, Old English *dōm* judgment — more at DOOM]
1 a : dignity **:** office ⟨duke*dom*⟩ **b :** realm **:** jurisdiction ⟨king*dom*⟩
2 : state or fact of being ⟨free*dom*⟩
3 : those having a (specified) office, occupation, interest, or character ⟨official*dom*⟩

do·main \dō-'mān, də-\ *noun* [Middle English *domayne,* from Middle French *domaine, demaine,* from Latin *dominium,* from *dominus*] (15th century)
1 a : complete and absolute ownership of land — compare EMINENT DOMAIN **b :** land so owned
2 : a territory over which dominion is exercised
3 : a region distinctively marked by some physical feature ⟨the *domain* of rushing streams, tall trees, and lakes⟩
4 : a sphere of knowledge, influence, or activity ⟨the *domain* of art⟩
5 : the set of elements to which a mathematical or logical variable is limited; *specifically* **:** the set on which a function is defined
6 : any of the small randomly oriented regions of uniform magnetization in a ferromagnetic substance
7 : INTEGRAL DOMAIN

¹dome \'dōm\ *noun* [French, Italian, & Latin; French *dôme* dome, cathedral, from Italian *duomo* cathedral, from Medieval Latin *domus* church, from Latin, house; akin to Greek *domos* house, Sanskrit *dam*] (1513)
1 *archaic* **:** a stately building **:** MANSION
2 : a large hemispherical roof or ceiling
3 : a natural formation or structure that resembles the dome or cupola of a building
4 : a form of crystal composed of planes parallel to a lateral axis that meet above in a horizontal edge like a roof
5 : an upward fold in rock whose sides dip uniformly in all directions
6 : a roofed sports stadium
— dom·al \'dō-məl\ *adjective*

²dome *verb* **domed; dom·ing** (1876)
transitive verb
1 : to cover with a dome
2 : to form into a dome
intransitive verb
: to swell upward or outward like a dome

Domes·day Book \'dümz-,dā-, 'dōmz-\ *noun* [Middle English, from *domesday* doomsday] (1591)
: a record of a survey of English lands and landholdings made by order of William the Conqueror about 1086

¹do·mes·tic \də-'mes-tik\ *adjective* [Middle English, from Middle French *domestique,* from Latin *domesticus,* from *domus*] (15th century)
1 a : living near or about human habitations **b :** TAME, DOMESTICATED
2 : of, relating to, or originating within a country and especially one's own country ⟨*domestic* politics⟩ ⟨*domestic* wines⟩
3 : of or relating to the household or the family
4 : devoted to home duties and pleasures
5 : INDIGENOUS
— do·mes·ti·cal·ly \-ti-k(ə-)lē\ *adverb*

²domestic *noun* (1613)
1 : a household servant
2 : an article of domestic manufacture — usually used in plural

domestic animal *noun* (circa 1855)
: any of various animals (as the horse or sheep) domesticated so as to live and breed in a tame condition

¹do·mes·ti·cate \də-'mes-ti-,kāt\ *transitive verb* **-cat·ed; -cat·ing** (circa 1639)
1 : to bring into domestic use **:** ADOPT
2 : to adapt (an animal or plant) to life in intimate association with and to the advantage of humans
3 : to make domestic **:** fit for domestic life
4 : to bring to the level of ordinary people **:** FAMILIARIZE
— do·mes·ti·ca·tion \-,mes-ti-'kā-shən\ *noun*

²do·mes·ti·cate \-kət, -,kāt\ *noun* (1951)
: a domesticated animal or plant

do·mes·tic·i·ty \,dō-,mes-'ti-sə-tē, -məs-; ,dä-\ *noun, plural* **-ties** (1721)
1 : the quality or state of being domestic or domesticated
2 : domestic activities or life
3 *plural* **:** domestic affairs

domestic prelate *noun* (1929)
: a priest having permanent honorary membership in the papal household

domestic relations court *noun* (circa 1939)
: COURT OF DOMESTIC RELATIONS

domestic science *noun* (1869)
: HOME ECONOMICS

dom·i·cal \'dō-mi-kəl, 'dä-\ *adjective* (1846)
: relating to, shaped like, or having a dome

¹do·mi·cile \'dä-mə-,sīl, 'dō-; 'dä-mə-sil\ *also* **dom·i·cil** \'däm-ə-səl\ *noun* [Middle English, from Middle French, from Latin *domicilium,* from *domus*] (15th century)
1 : a dwelling place **:** place of residence **:** HOME
2 a : a person's fixed, permanent, and principal home for legal purposes **b :** RESIDENCE 2b

²domicile *transitive verb* **-ciled; -cil·ing** (1809)
: to establish in or provide with a domicile

do·mi·cil·i·ary \,dä-mə-'si-lē-,er-ē, ,dō-\ *adjective* (1790)
: of, relating to, or constituting a domicile: as **a :** provided or taking place in the home ⟨*domiciliary* midwifery⟩ **b :** providing care and living space (as for disabled veterans)

do·mi·cil·i·ate \,dä-mə-'si-lē-,āt, ,dō-\ *verb* **-at·ed; -at·ing** [Latin *domicilium*] (1778)
transitive verb
: DOMICILE
intransitive verb
: RESIDE
— do·mi·cil·i·a·tion \-,si-lē-'ā-shən\ *noun*

dom·i·nance \'dä-mə-nən(t)s, 'däm-nən(t)s\ *noun* (1819)
1 : the fact or state of being dominant: as **a :** dominant position especially in a social hierarchy **b :** the property of one of a pair of alleles or traits that suppresses expression of the other in the heterozygous condition **c :** the influence or control over ecological communities exerted by a dominant
2 : functional asymmetry between a pair of bodily structures (as the right and left hands)

¹dom·i·nant \-nənt\ *adjective* [Middle French or Latin; Middle French, from Latin *dominant-, dominans,* present participle of *dominari*] (circa 1532)
1 : commanding, controlling, or prevailing over all others
2 : overlooking and commanding from a superior position
3 : of, relating to, or exerting ecological or genetic dominance
4 : being the one of a pair of bodily structures that is the more effective or predominant in action ⟨*dominant* eye⟩ ☆
— dom·i·nant·ly *adverb*

²dominant *noun* (1819)
1 : the fifth tone of a diatonic scale
2 a : a dominant genetic character or factor **b :** any of one or more kinds of organism (as a species) in an ecological community that exerts a controlling influence on the environment and thereby largely determines what other kinds of organisms are present **c :** a dominant individual in a social hierarchy

dom·i·nate \'dä-mə-,nāt\ *verb* **-nat·ed; -nat·ing** [Latin *dominatus,* past participle of *dominari,* from *dominus* master; akin to Latin *domus* house — more at DOME] (1611)
transitive verb
1 : RULE, CONTROL
2 : to exert the supreme determining or guiding influence on
3 : to overlook from a superior elevation or command because of superior height or position
4 : to have a commanding or preeminent place or position in ⟨name brands *dominate* the market⟩
intransitive verb
1 : to have or exert mastery, control, or preeminence
2 : to occupy a more elevated or superior position
— dom·i·na·tive \-,nā-tiv\ *adjective*
— dom·i·na·tor \-,nā-tər\ *noun*

dom·i·na·tion \,dä-mə-'nā-shən\ *noun* (14th century)
1 : supremacy or preeminence over another
2 : exercise of mastery or ruling power
3 : exercise of preponderant, governing, or controlling influence
4 *plural* **:** DOMINION 3

do·mi·na·trix \,dä-mi-'nā-triks\ *noun, plural* **-trices** \-'nā-trə-sēz, -nə-'trī-sēz\ [Latin, feminine of *dominator*] (1971)
: a woman who physically and psychologically dominates and abuses her partner in sadomasochistic sex; *broadly* **:** a dominating woman

dom·i·neer \,dä-mə-'nir\ *verb* [Dutch *domineren,* from French *dominer,* from Latin *dominari*] (1591)
intransitive verb
: to exercise arbitrary or overbearing control

☆ SYNONYMS
Dominant, predominant, paramount, preponderant mean superior to all others in influence or importance. DOMINANT applies to something that is uppermost because ruling or controlling ⟨a *dominant* social class⟩. PREDOMINANT applies to something that exerts, often temporarily, the most marked influence ⟨a *predominant* emotion⟩. PARAMOUNT implies supremacy in importance, rank, or jurisdiction ⟨unemployment was the *paramount* issue in the campaign⟩. PREPONDERANT applies to an element or factor that outweighs all others in influence or effect ⟨*preponderant* evidence in her favor⟩.

transitive verb
: to tyrannize over

dom·i·neer·ing *adjective* (1588)
: inclined to domineer
synonym see MASTERFUL
— dom·i·neer·ing·ly \-iŋ-lē\ *adverb*
— dom·i·neer·ing·ness *noun*

do·min·i·cal \də-ˈmi-ni-kəl\ *adjective* [Late Latin *dominicalis,* from *dominicus* (*dies*) the Lord's day, from Latin *dominicus* of a lord, from *dominus* lord, master] (15th century)
1 : of or relating to Jesus Christ as Lord
2 : of or relating to the Lord's day

Do·min·i·can \də-ˈmi-ni-kən\ *noun* [Saint *Dominic*] (circa 1632)
: a member of a mendicant order of friars founded by Saint Dominic in 1215 and dedicated especially to preaching
— Dominican *adjective*

dom·i·nick·er \ˈdä-mə-ˌne-kər, -ˌni-\ *also* **dom·i·nick** \-(ˌ)nik, -nek\ *noun, often capitalized* (1806)
: DOMINIQUE

do·mi·nie *1 usually* ˈdä-mə-nē, *2 usually* ˈdō-\ *noun* [Latin *domine,* vocative of *dominus*] (1612)
1 *chiefly Scottish* **:** SCHOOLMASTER
2 : CLERGYMAN

do·min·ion \də-ˈmi-nyən\ *noun* [Middle English *dominioun,* from Middle French *domin-ion,* modification of Latin *dominium,* from *dominus*] (14th century)
1 : DOMAIN
2 : supreme authority **:** SOVEREIGNTY
3 *plural* **:** an order of angels — see CELESTIAL HIERARCHY
4 *often capitalized* **:** a self-governing nation of the Commonwealth other than the United Kingdom that acknowledges the British monarch as chief of state
5 : absolute ownership
synonym see POWER

Dominion Day *noun* (1867)
: CANADA DAY

dom·i·nique \ˈdä-mə-ˌnēk\ *noun* [*Dominique* (Dominica), one of the Windward islands, West Indies] (1849)
: any of an American breed of domestic fowl with a rose comb, yellow legs, and barred plumage; *broadly* **:** a barred fowl

dom·i·no \ˈdä-mə-ˌnō\ *noun, plural* **-noes** *or* **-nos** [French, probably from Latin (in the ritual formula *benedicamus Domino* let us bless the Lord)] (circa 1694)
1 a (1) **:** a long loose hooded cloak usually worn with a half mask as a masquerade costume (2) **:** a half mask worn over the eyes with a masquerade costume **b :** a person wearing a domino
2 a : a flat rectangular block (as of wood or plastic) whose face is divided into two equal parts that are blank or bear usually from one to six dots arranged as on dice faces **b** *plural but usually singular in construction* **:** any of several games played with a set of usually 28 dominoes
3 : a member of a group (as of nations) expected to behave in accordance with the domino theory ◆

domino effect *noun* (1966)
: a cumulative effect produced when one event initiates a succession of similar events — compare RIPPLE EFFECT

domino theory *noun* [from the fact that if dominoes are stood on end one slightly behind the other, a slight push on the first will topple the others] (1965)
1 : a theory that if one nation becomes Communist-controlled the neighboring nations will also become Communist-controlled
2 : the theory that if one act or event is allowed to take place a series of similar acts or events will follow

¹don \ˈdän\ *transitive verb* **donned; don·ning** [Middle English, contraction of *do on*] (14th century)
1 : to put on (an article of clothing)
2 : to wrap oneself in **:** TAKE ON 3a

²don \ˈdän\ *noun* [Spanish, from Latin *dominus* master — more at DAME] (1523)
1 : a Spanish nobleman or gentleman — used as a title prefixed to the Christian name
2 *archaic* **:** a person of consequence **:** GRANDEE
3 : a head, tutor, or fellow in a college of Oxford or Cambridge University; *broadly* **:** a college or university professor
4 [Italian, title of respect, from *donno,* literally, lord, from Latin *dominus*] **:** a powerful Mafia leader

do·na \ˌdō-nə\ *noun* [Portuguese, from Latin *domina* lady — more at DAME] (circa 1897)
: a Portuguese or Brazilian woman of rank — used as a title prefixed to the Christian name

do·ña \ˈdō-nyə\ *noun* [Spanish, from Latin *domina*] (1622)
: a Spanish woman of rank — used as a title prefixed to the Christian name

do·nate \ˈdō-ˌnāt, dō-ˈ\ *verb* **do·nat·ed; do·nat·ing** [back-formation from *donation*] (1785)
transitive verb
1 : to make a gift of; *especially* **:** to contribute to a public or charitable cause
2 : to transfer (as electrons) to another atom or molecule
intransitive verb
: to make a donation
synonym see GIVE

do·na·tion \dō-ˈnā-shən\ *noun* [Middle English *donatyowne,* from Latin *donation-, donatio,* from *donare* to present, from *donum* gift; akin to Latin *dare* to give — more at DATE] (15th century)
: the act or an instance of donating: as **a :** the making of a gift especially to a charity or public institution **b :** a free contribution **:** GIFT

Do·na·tism \ˈdō-nə-ˌti-zəm, ˈdä-\ *noun* [*Donatus,* 4th century bishop of Carthage] (1588)
: the doctrines of a Christian sect arising in North Africa in 311 and holding that sanctity is essential for the administration of sacraments and church membership
— Do·na·tist \-tist\ *noun*

¹do·na·tive \ˈdō-nə-tiv, ˈdä-\ *noun* (15th century)
: a special gift or donation

²do·na·tive *same or* ˈdō-ˌnā-, dō-ˈ\ *adjective* [Latin *donativus,* from *donatus*] (1559)
: of or relating to donation

do·na·tor \ˈdō-ˌnā-tər, dō-ˈ\ *noun* (15th century)
: DONOR

¹done \ˈdən\ *past participle of* DO

²done *adjective* (14th century)
1 : arrived at or brought to an end
2 : doomed to failure, defeat, or death
3 : gone by **:** OVER
4 : physically exhausted
5 : cooked sufficiently
6 : conformable to social convention

do·nee \ˌdō-ˈnē\ *noun* [*don*or] (1523)
: a recipient of a gift

done for \ˈdən-ˌfȯr\ *adjective* (1803)
1 : sunk in defeat **:** BEATEN
2 : mortally stricken **:** DOOMED

done·ness \ˈdən-nəs\ *noun* (1927)
: the condition of being cooked to the desired degree

¹dong \ˈdȯŋ, ˈdäŋ\ *noun* [origin unknown] (circa 1930)
: PENIS — usually considered vulgar

²dong *noun, plural* **dong** [Vietnamese *đồng*] (1948)
— see MONEY table

don·jon \ˈdän-jən, ˈdən-\ *noun* [Middle English — more at DUNGEON] (14th century)
: a massive inner tower in a medieval castle
word history see DUNGEON

Don Juan \ˈdän-ˈ(h)wän, *chiefly British & in poetry* dän-ˈjü-ən\ *noun* [Spanish]
1 : a legendary Spaniard proverbial for his seduction of women
2 : a captivating man known as a great lover or seducer of women
— Don Juan·ism \-ˈ(h)wän-ˌi-zəm, -ˈjü-ə-ˌni-\ *noun*

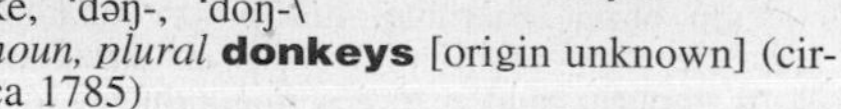

1 donjon

don·key \ˈdäŋ-kē, ˈdəŋ-, ˈdȯŋ-\ *noun, plural* **donkeys** [origin unknown] (circa 1785)
1 : the domestic ass (*Equus asinus*)
2 : a stupid or obstinate person

donkey engine *noun* (1858)
1 : a small usually portable auxiliary engine
2 : a small locomotive used in switching

donkey jacket *noun* (1929)
British **:** a jacket of heavy material worn especially by laborers

donkey's years *noun plural* (1927)
chiefly British **:** a very long time

don·key·work \ˈdäŋ-kē-ˌwərk, ˈdəŋ-, ˈdȯŋ-\ *noun* (1920)
: monotonous and routine work **:** DRUDGERY

don·na \ˌdä-nə, ˌdȯ-\ *noun, plural* **don·ne** \-(ˌ)nā\ [Italian, from Latin *domina*] (1740)
: an Italian woman especially of rank — used as a title prefixed to the Christian name

don·née \dȯ-ˈnā, (ˌ)də-\ *noun, plural* **don·nées** \-ˈnā(z)\ [French, from feminine of *donné,* past participle of *donner* to give, from Latin *donare* to donate — more at DONATION] (1876)
: the set of assumptions on which a work of fiction or drama proceeds

don·nick·er *or* **don·ni·ker** \ˈdä-ni-kər\ *noun* [alteration of English dialect *dunnekin* toilet, cesspool] (circa 1931)
: TOILET 3a

don·nish \ˈdä-nish\ *adjective* (1848)
: of, relating to, or characteristic of a university don
— don·nish·ly *adverb*

◇ WORD HISTORY

domino The origin of dominoes is rather obscure, and that is as true of the word as it is of the game. Both appeared in France shortly after the middle of the 18th century, but details of their history are unrecorded. The word *domino* in both French and English earlier denoted a long hooded cloak worn together with a silk or velvet mask as a costume during carnival season or at a masquerade ball. Dominoes were particularly associated with the carnival in Venice, and masqueraders garbed in domino and mask, together with a three-cornered hat, were often depicted by 18th century Venetian painters. It seems that it is an even earlier sense, however, that links the garment with the playing piece. In French, *domino* originally referred to a cape worn by a priest in winter over his surplice. (Presumably at the source of this sense is some connection with Latin *dominus* "lord," but this is another circumstance that remains conjectural.) Domino pieces were once made with ebony backs and ivory faces, and it has been suggested that the playing piece brought to mind the priest's black domino contrasting with the white surplice.

— **don•nish•ness** *noun*

don•ny•brook \'dä-nē-ˌbrůk\ *noun, often capitalized* [*Donnybrook* Fair, annual Irish event known for its brawls] (1852)
1 : FREE-FOR-ALL, BRAWL
2 : a usually public quarrel or dispute ◆

do•nor \'dō-nər, -ˌnȯr\ *noun* [Middle French *doneur,* from Latin *donator,* from *donare*] (15th century)
1 : one that gives, donates, or presents something
2 : one used as a source of biological material (as blood or an organ)
3 a : a compound capable of giving up a part (as an atom, chemical group, or subatomic particle) for combination with an acceptor **b :** an impurity added to a semiconductor to increase the number of mobile electrons

¹do–noth•ing \'dü-ˌnə-thiŋ\ *noun* (1579)
: a shiftless or lazy person

²do–nothing *adjective* (1832)
: marked by inactivity or failure to make positive progress
— **do–noth•ing•ism** \-ˌi-zəm\ *noun*

Don Qui•xote \ˌdän-kē-'(h)ō-tē, ˌdäŋ-; *chiefly British* dän-'kwik-sət\ *noun* [Spanish, hero of Cervantes' *Don Quixote*]
: an impractical idealist

don•sie *or* **don•sy** \'dän(t)-sē\ *adjective* [Scottish Gaelic *donas* evil, harm + English *-ie*] (1720)
1 *dialect British* **:** UNLUCKY
2 *Scottish* **a :** RESTIVE **b :** SAUCY
3 *chiefly northern Midland* **:** slightly ill

¹don't \'dōnt\ (1639)
1 : do not
2 : does not ▢

²don't \'dōnt\ *noun* (1894)
: a command or entreaty not to do something

do•nut \'dō-(ˌ)nət\ *variant of* DOUGHNUT

doo•dad \'dü-ˌdad\ *noun* [origin unknown] (1888)
1 : an ornamental attachment or decoration
2 : an often small article whose common name is unknown or forgotten **:** GADGET

¹doo•dle \'dü-dᵊl\ *verb* **doo•dled; doo•dling** \'düd-liŋ, 'dü-dᵊl-iŋ\ [perhaps from *doodle* (to ridicule)] (1936)
intransitive verb
1 : to make a doodle
2 : DAWDLE, TRIFLE
transitive verb
: to produce by doodling
— **doo•dler** \'düd-lər, 'dü-dᵊl-ər\ *noun*

²doodle *noun* (1937)
: an aimless or casual scribble, design, or sketch; *also* **:** a minor work

doo•dle•bug \'dü-dᵊl-ˌbəg\ *noun* [probably from *doodle* (fool) + *bug*] (circa 1866)
1 : the larva of an ant lion; *also* **:** any of several other insects
2 : a device (as a divining rod) used in attempting to locate underground gas, water, oil, or ores
3 : BUZZ BOMB

doodley–squat *or* **doodly–squat** \'dü-dᵊl-ē-ˌskwät\ *noun* [perhaps from *doodle* + ⁴-y + *squat,* euphemism for *shit*] (1934)
: DIDDLY-SQUAT

doo–doo \'dü-(ˌ)dü\ *noun* [baby talk] (1948)
: EXCREMENT
— **in deep doo–doo :** in trouble

doo•fus \'dü-fəs, -fis\ *noun, plural* **doo•fus•es** \-fə-siz\ [perhaps alteration of ¹*goof*] (1970)
slang **:** a stupid, incompetent, or foolish person

doo•hick•ey \'dü-ˌhi-kē\ *noun, plural* **-hickeys** *also* **-hickies** [probably from *doo*dad + *hickey*] (1914)
: DOODAD 2

¹doom \'düm\ *noun* [Middle English, from Old English *dōm;* akin to Old High German *tuom* condition, state, Old English *dōn* to do] (before 12th century)
1 : a law or ordinance especially in Anglo-Saxon England
2 a : JUDGMENT, DECISION; *especially* **:** a judicial condemnation or sentence **b** (1) **:** JUDGMENT 3a (2) **:** JUDGMENT DAY 1
3 a : DESTINY; *especially* **:** unhappy destiny **b :** DEATH, RUIN
synonym see FATE

²doom *transitive verb* (15th century)
1 : to give judgment against **:** CONDEMN
2 a : to fix the fate of **:** DESTINE **b :** to make certain the failure or destruction of

doom•ful \'düm-fəl\ *adjective* (1586)
: presaging doom **:** OMINOUS
— **doom•ful•ly** \-fə-lē\ *adverb*

doom•say•er \'düm-ˌsā-ər\ *noun* (1953)
: one given to forebodings and predictions of impending calamity
— **doom•say•ing** \-ˌsā-iŋ\ *noun*

dooms•day \'dümz-ˌdā\ *noun, often attributive* (before 12th century)
: a day of final judgment

dooms•day•er \-ˌdā-ər, -ˌde(ə)r\ *noun* (1972)
: DOOMSAYER

doom•ster \'düm(p)-stər\ *noun* (15th century)
1 : JUDGE
2 : DOOMSAYER

doomy \'dü-mē\ *adjective* (1971)
: suggestive of doom **:** DOOMFUL
— **doom•i•ly** \'dü-mi-lē\ *adverb*

door \'dōr, 'dȯr\ *noun, often attributive* [Middle English *dure, dor,* from Old English *duru* door & *dor* gate; akin to Old High German *turi* door, Latin *fores,* Greek *thyra*] (before 12th century)
1 : a usually swinging or sliding barrier by which an entry is closed and opened; *also* **:** a similar part of a piece of furniture
2 : DOORWAY
3 : a means of access or participation **:** OPPORTUNITY ⟨opens new *doors*⟩ ⟨*door* to success⟩
— **door•less** \-ləs\ *adjective*
— **at one's door :** as a charge against one as being responsible ⟨laid the blame *at our door*⟩
— **on one's doorstep :** close at hand; *especially* **:** too close to be overlooked

door•bell \'dōr-ˌbel, 'dȯr-\ *noun* (circa 1815)
: a bell or set of chimes to be rung usually by a push button at an outer door

do–or–die \'dü-ər-'dī, -ȯr-\ *adjective* (1879)
1 : doggedly determined to reach one's objective **:** INDOMITABLE
2 : presenting as the only alternatives complete success or complete ruin ⟨a *do-or-die* situation⟩

door•jamb \'dōr-ˌjam, 'dȯr-\ *noun* (1837)
: an upright piece forming the side of a door opening

door•keep•er \-ˌkē-pər\ *noun* (1535)
: a person who tends a door

door•knob \-ˌnäb\ *noun* (1846)
: a knob that releases a door latch

door•man \-ˌman, -mən\ *noun* (circa 1897)
: a usually uniformed attendant at the door of a building (as a hotel or apartment house)

door•mat \-ˌmat\ *noun* (1665)
1 : a mat placed before or inside a door for wiping dirt from the shoes
2 : one that submits without protest to abuse or indignities
3 : a team that regularly finishes last

door•nail \-ˌnāl, -'nā(ə)l\ *noun* (14th century)
: a large-headed nail — used chiefly in the phrase *dead as a doornail*

door•plate \-ˌplāt\ *noun* (1823)
: a nameplate on a door

door•post \-ˌpōst\ *noun* (1535)
: DOORJAMB

door prize *noun* (1951)
: a prize awarded to the holder of a winning ticket passed out at the entrance to an entertainment or function

door•sill \'dōr-ˌsil, 'dȯr-\ *noun* (1587)
: SILL 1b

door•step \-ˌstep\ *noun* (1767)
: a step before an outer door

door•stop \-ˌstäp\ *noun* (circa 1895)
1 : a device (as a wedge or weight) for holding a door open
2 : a usually rubber-tipped device attached to a wall or floor to prevent damaging contact between an opened door and the wall

door–to–door \ˌdōr-tə-'dōr, ˌdȯr-tə-'dȯr\ *adjective* (1902)
: going or made by going to each house in a neighborhood ⟨*door-to-door* salesmen⟩ ⟨a *door-to-door* canvass⟩
— **door–to–door** *adverb*

door•way \'dōr-ˌwā, 'dȯr-\ *noun* (1799)
1 : the opening that a door closes; *especially* **:** an entrance into a building or room
2 : DOOR 3

door•yard \-ˌyärd\ *noun* (circa 1764)
: a yard next to the door of a house

doo–wop \'dü-ˌwäp\ *noun* [from nonsense syllables typical of the style] (1969)
: a vocal style of rock and roll characterized by the a cappella singing of nonsense syllables in rhythmical support of the melody

doo•zy *or* **doo•zie** \'dü-zē\ *also* **doo•zer** \-zər\ *noun, plural* **doozies** *or* **doozers** [perhaps alteration of *daisy*] (1916)
: an extraordinary one of its kind

do•pa \'dō-pə, -(ˌ)pä\ *noun* [International Scientific Vocabulary *d*ihydroxy- + *p*henyl*a*lanine] (1917)
: an amino acid $C_9H_{11}NO_4$ that in the levorotatory form is found in the broad bean and is used in the treatment of Parkinson's disease

do•pa•mine \'dō-pə-ˌmēn\ *noun* [*dopa* + *amine*] (1959)

▢ USAGE

don't *Don't* is the earliest attested contraction of *does not* and until about 1900 was the standard spoken form in the U.S. (it survived as spoken standard longer in British English). Dialect surveys find it more common in the speech of the less educated than in that of the educated; in those places (as the Midland and southern Atlantic seaboard regions) where it has lasted in educated speech, it is most common with older informants. Surveys of attitudes toward usage show it more widely disapproved in 1971 than it had been 40 years earlier. Its chief use in edited prose is in fiction for purposes of characterization. It is sometimes used consciously, like *ain't,* to gain an informal effect.

◇ WORD HISTORY

donnybrook Donnybrook, now a suburban district of Dublin, was once a village about a mile and a half from the city's center. From the year 1204 it was the site of a fair held annually in August, where working-class Dubliners would come to find whatever amusement they could afford. Horses were traded, and trinkets and food were sold from crude wattle booths and tents, with entertainment provided mainly by strolling pipers and dancers. By the end of the 18th century, the Donnybrook fair had become notorious for its rowdy brawls, fueled by the consumption of great quantities of liquor. Victorian Dublin became more conservative and temperance-minded, however, and both Catholic and evangelical clergymen campaigned to have the fair abolished. Though the fighting and drunkenness for which it was renowned had greatly subsided, the fair ended in 1855. By that time, nonetheless, *donnybrook* had acquired the generic sense in English that it still maintains.

: a monoamine $C_8H_{11}NO_2$ that is a decarboxylated form of dopa and that occurs especially as a neurotransmitter in the brain

do·pa·mi·ner·gic \ˌdō-pə-ˌmē-ˈnər-jik\ *adjective* (1970)
: relating to, participating in, or activated by dopamine or related substances

dop·ant \ˈdō-pənt\ *noun* [²*dope*] (1962)
: an impurity added usually in minute amounts to a pure substance to alter its properties

¹dope \ˈdōp\ *noun* [Dutch *doop* sauce, from *dopen* to dip; akin to Old English *dyppan* to dip] (1807)
1 a : a thick liquid or pasty preparation **b :** a preparation for giving a desired quality to a substance or surface
2 : absorbent or adsorbent material used in various manufacturing processes (as the making of dynamite)
3 a (1) **:** an illicit, habit-forming, or narcotic drug (2) **:** a preparation given to a racehorse to help or hinder its performance **b** *chiefly Southern* **:** a cola drink **c :** a stupid person
4 : information especially from a reliable source ⟨the inside *dope*⟩

²dope *verb* **doped; dop·ing** (1889)
transitive verb
1 : to treat or affect with dope; *especially* **:** to give a narcotic to
2 : FIGURE OUT — usually used with *out*
3 : to treat with a dopant
intransitive verb
: to take dope
— dop·er *noun*

dope·head \ˈdōp-ˌhed\ *noun* (1903)
: a drug addict

dope·ster \ˈdōp-stər\ *noun* (1907)
: a forecaster of the outcome of future events (as sports contests or elections)

dop·ey *also* **dopy** \ˈdō-pē\ *adjective* **dop·i·er; -est** (1896)
1 a : dulled by alcohol or a narcotic **b :** SLUGGISH, STUPEFIED
2 : STUPID, FATUOUS
— dop·i·ness *noun*

dop·pel·gäng·er *or* **dop·pel·gang·er** \ˈdä-pəl-ˌgaŋ-ər, -ˌgeŋ-, ˌdä-pəl-ˈ\ *n* [German *Doppelgänger,* from *doppel* double + *-gänger* goer] (1851)
1 : a ghostly counterpart of a living person
2 a : DOUBLE 2a **b :** ALTER EGO b **c :** a person who has the same name as another

Dopp·ler \ˈdä-plər\ *adjective* (1905)
: of, relating to, being, or utilizing a shift in frequency in accordance with the Doppler effect; *also* **:** of or relating to Doppler radar

Doppler effect *noun* [Christian J. *Doppler*] (1905)
: a change in the frequency with which waves (as sound or light) from a given source reach an observer when the source and the observer are in motion with respect to each other so that the frequency increases or decreases according to the speed at which the distance is decreasing or increasing

Doppler radar *noun* (1954)
: a radar system that utilizes the Doppler effect for measuring velocity

Dor·cas \ˈdȯr-kəs\ *noun* [Greek *Dorkas*]
: a Christian woman of New Testament times who made clothing for the poor

Do·ri·an \ˈdōr-ē-ən, ˈdȯr-\ *noun* [Latin *Dorius* of Doris, from Greek *dōrios,* from *Dōris,* region of ancient Greece] (1662)
: a member of an ancient Hellenic race that completed the overthrow of Mycenaean civilization and settled especially in the Peloponnisos and Crete
— Dorian *adjective*

¹Dor·ic \ˈdȯr-ik, ˈdär-\ *adjective* (1569)
1 : of, relating to, or characteristic of the Dorians
2 : belonging to the oldest and simplest Greek architectural order — see ORDER illustration
3 : of, relating to, or constituting Doric

²Doric *noun* (1837)
: a dialect of ancient Greek spoken especially in the Peloponnisos, Crete, Sicily, and southern Italy

dork \ˈdȯrk\ *noun* [perhaps alteration of *dick*] (1972)
slang **:** NERD; *also* **:** JERK 4

dorky \ˈdȯr-kē\ *adjective* **dork·i·er; -est** (1983)
slang **:** foolishly stupid **:** INEPT

dorm \ˈdȯrm\ *noun* (1900)
: DORMITORY

dor·man·cy \ˈdȯr-mən(t)-sē\ *noun* (1789)
: the quality or state of being dormant

dor·mant \ˈdȯr-mənt\ *adjective* [Middle English, fixed, stationary, from Middle French, from present participle of *dormir* to sleep, from Latin *dormire;* akin to Sanskrit *drāti* he sleeps] (1500)
1 : represented on a coat of arms in a lying position with the head on the forepaws
2 : marked by a suspension of activity: as **a :** temporarily devoid of external activity ⟨a *dormant* volcano⟩ **b :** temporarily in abeyance yet capable of being activated
3 a : ASLEEP, INACTIVE **b :** having the faculties suspended **:** SLUGGISH **c :** having biological activity suspended: as (1) **:** being in a state of suspended animation (2) **:** not actively growing but protected (as by bud scales) from the environment — used of plant parts
4 : associated with, carried out, or applied during a period of dormancy ⟨*dormant* grafting⟩
synonym see LATENT

dor·mer \ˈdȯr-mər\ *noun* [Middle French *dormeor* dormitory, from Latin *dormitorium*] (1592)
: a window set vertically in a structure projecting through a sloping roof; *also* **:** the roofed structure containing such a window

dor·mie *or* **dor·my** \ˈdȯr-mē\ *adjective* [origin unknown] (1847)
: being ahead by as many holes in golf as remain to be played

dor·mi·to·ry \ˈdȯr-mə-ˌtōr-ē, -ˌtȯr-\ *noun, plural* **-ries** [Middle English, from Latin *dormitorium,* from *dormitus,* past participle of *dormire*] (15th century)
1 : a room for sleeping; *especially* **:** a large room containing numerous beds
2 : a residence hall providing rooms for individuals or for groups usually without private baths
3 *chiefly British* **:** a residential community inhabited chiefly by commuters

dor·mouse \ˈdȯr-ˌmau̇s\ *noun, plural* **dor·mice** \-ˌmīs\ [Middle English *dormowse,* perhaps from Middle French *dormir* + Middle English *mous* mouse] (15th century)
: any of numerous small Old World rodents (families Gliridae and Seleviniidae) that are intermediate in form and behavior between mice and squirrels

dor·nick \ˈdȯr-nik, ˈdä-nik\ *noun* [probably from Irish *dornóg*] (1840)
: a stone small enough to throw; *also* **:** a large piece of rock

do·ron·i·cum \də-ˈrä-ni-kəm\ *noun* [New Latin, genus name, from Arabic *darūnaj,* a plant of this genus] (1892)
: any of a genus (*Doronicum*) of Eurasian perennial composite herbs including several cultivated for their showy yellow flower heads

dorp \ˈdȯrp\ *noun* [Dutch, from Middle Dutch; akin to Old High German *dorf* village — more at THORP] (circa 1576)
: VILLAGE

dors- *or* **dorsi-** *or* **dorso-** *combining form* [Late Latin *dors-,* from Latin *dorsum*]
1 : back ⟨*dors*ad⟩
2 : dorsal and ⟨*dorso*lateral⟩

dor·sad \ˈdȯr-ˌsad\ *adverb* (circa 1803)
: toward the back **:** DORSALLY

¹dorsal *variant of* DOSSAL

²dor·sal \ˈdȯr-səl\ *adjective* [Late Latin *dorsalis,* from Latin *dorsum* back] (1727)
1 : relating to or situated near or on the back especially of an animal or of one of its parts
2 : ABAXIAL
— dor·sal·ly \-sə-lē\ *adverb*

³dorsal *noun* (1834)
: a dorsally located part; *especially* **:** a thoracic vertebra

dorsal lip *noun* (1940)
: the margin of the fold of blastula wall that delineates the dorsal limit of the blastopore, constitutes the primary organizer, and forms the point of origin of chordamesoderm

dorsal root *noun* (circa 1934)
: the one of the two roots of a spinal nerve that passes dorsally to the spinal cord and consists of sensory fibers

Dor·set \ˈdȯr-sət\ *noun* (1891)
: any of a breed of domestic white-faced sheep originally developed in Dorset, England

dor·si·ven·tral \ˌdȯr-si-ˈven-trəl\ *adjective* (circa 1882)
1 : having distinct dorsal and ventral surfaces
2 : DORSOVENTRAL 1
— dor·si·ven·tral·i·ty \-ven-ˈtra-lə-tē\ *noun*
— dor·si·ven·tral·ly \-ˈven-trə-lē\ *adverb*

dor·so·lat·er·al \ˌdȯr-sō-ˈla-tə-rəl, -ˈla-trəl\ *adjective* (1835)
: of, relating to, or involving both the back and the sides

dor·so·ven·tral \-ˈven-trəl\ *adjective* [International Scientific Vocabulary] (1870)
1 : relating to, involving, or extending along the axis joining the dorsal and ventral sides
2 : DORSIVENTRAL 1
— dor·so·ven·tral·i·ty \-ven-ˈtra-lə-tē\ *noun*
— dor·so·ven·tral·ly \-trə-lē\ *adverb*

dor·sum \ˈdȯr-səm\ *noun, plural* **dor·sa** \-sə\ [Latin] (1840)
1 : the upper surface of an appendage or part
2 : BACK; *especially* **:** the entire dorsal surface of an animal

do·ry \ˈdōr-ē, ˈdȯr-\ *noun, plural* **dories** [Miskito *dóri* dugout] (1709)
: a flat-bottomed boat with high flaring sides, sharp bow, and deep V-shaped transom

dos *or* **do's** *plural of* ²DO

dos·age \ˈdō-sij\ *noun* (circa 1867)
1 a : the addition of an ingredient or the application of an agent in a measured dose **b :** the presence and relative representation or strength of a factor or agent
2 a : DOSE 2 **b** (1) **:** the giving of a dose (2) **:** regulation or determination of doses
3 : an exposure to some experience in or as if in measured portions

¹dose \ˈdōs\ *noun* [Middle English, from Middle French, from Late Latin *dosis,* from Greek, literally, act of giving, from *didonai* to give — more at DATE] (15th century)
1 a : the measured quantity of a therapeutic agent to be taken at one time **b :** the quantity of radiation administered or absorbed
2 : a portion of a substance added during a process
3 : something experienced as if in a prescribed or measured amount ⟨a daily *dose* of hard work⟩
4 : a gonorrheal infection

²dose *transitive verb* **dosed; dos·ing** (1654)
1 : to give a dose to; *especially* **:** to give medicine to
2 : to divide (as a medicine) into doses
3 : to treat with an application or agent

do–si–do \ˌdō-(ˌ)sē-ˈdō\ *noun, plural* **do–si–dos** [French *dos-à-dos* back to back] (1926)
: a square-dance figure: **a :** a figure in which the dancers pass each other right shoulder to right shoulder and circle each other back to

back **b :** a figure in which the woman moves in a figure circling first her partner and then the man on her right

do·sim·e·ter \dō-ˈsi-mə-tər\ *noun* [Late Latin *dosis* + International Scientific Vocabulary *-meter*] (1938)
: a device for measuring doses of radiations (as X rays)
— do·si·met·ric \ˌdō-sə-ˈme-trik\ *adjective*
— do·sim·e·try \dō-ˈsi-mə-trē\ *noun*

¹doss \ˈdȯs, ˈdäs\ *intransitive verb* [origin unknown] (circa 1785)
chiefly British **:** to sleep or bed down in a convenient place — usually used with *down*

²doss *noun* (circa 1789)
chiefly British **:** a crude or makeshift bed

dos·sal \ˈdä-səl\ *or* **dor·sal** \ˈdȯr-səl\ *or* **dos·sel** \ˈdä-səl\ *noun* [Medieval Latin *dossale, dorsale,* from neuter of Late Latin *dorsalis* dorsal] (1851)
: an ornamental cloth hung behind and above an altar

doss–house \ˈdȯs-ˌhau̇s, ˈdäs-\ *noun* (1888)
chiefly British **:** a cheap rooming house or hotel

dos·sier \ˈdȯs-ˌyā, ˈdäs-; ˈdȯ-sē-ˌā, ˈdä-\ *noun* [French, bundle of documents labeled on the back, dossier, from *dos* back, from Latin *dorsum*] (1880)
: a file containing detailed records on a particular person or subject

dost \ˈdəst\ *archaic present 2d singular of* DO

¹dot \ˈdät\ *noun* [(assumed) Middle English, from Old English *dott* head of a boil; akin to Old High German *tutta* nipple] (1674)
1 : a small spot **:** SPECK
2 a (1) **:** a small point made with a pointed instrument ⟨a *dot* on the chart marked the ship's position⟩ (2) **:** a small round mark used in orthography or punctuation ⟨put a *dot* over the *i*⟩ **b :** a centered point used as a multiplication sign (as in 6 · 5 = 30) **c** (1) **:** a point after a note or rest in music indicating augmentation of the time value by one half (2) **:** a point over or under a note indicating that it is to be played staccato
3 : a precise point especially in time ⟨arrived at six on the *dot*⟩
4 : a short click or buzz forming a letter or part of a letter (as in the Morse code)

²dot *verb* **dot·ted; dot·ting** (circa 1740)
transitive verb
1 : to mark with a dot
2 : to intersperse with dots or objects scattered at random ⟨boats *dotting* the lake⟩
intransitive verb
: to make a dot
— dot·ter *noun*

³dot \ˈdȯt\ *noun* [French, from Latin *dot-, dos* dowry] (1855)
: DOWRY 2

dot·age \ˈdō-tij\ *noun* [*dote*] (14th century)
: a state or period of senile decay marked by decline of mental poise and alertness

do·tal \ˈdō-tᵊl\ *adjective* [Latin *dotalis,* from *dot-, dos*] (1513)
: of or relating to a woman's marriage dowry

dot·ard \ˈdō-tərd\ *noun* (14th century)
: a person in his or her dotage

dote \ˈdōt\ *intransitive verb* **dot·ed; dot·ing** [Middle English; akin to Middle Low German *dotten* to be foolish] (13th century)
1 : to exhibit mental decline of or like that of old age **:** be in one's dotage
2 : to be lavish or excessive in one's attention, fondness, or affection — used especially with *on* ⟨*doted* on her only grandchild⟩
— dot·er *noun*
— dot·ing·ly \ˈdō-tiŋ-lē\ *adverb*

doth \ˈdəth\ *archaic present 3d singular of* DO

dot matrix *noun* (1963)
: a pattern of dots in a grid from which alphanumeric characters can be formed ⟨a *dot matrix* printer⟩

dot product *noun* [¹*dot;* from its being commonly written *A* · *B*] (1901)
: SCALAR PRODUCT

dotted swiss *noun* (circa 1924)
: a sheer light muslin ornamented with evenly spaced raised dots

dot·ter·el \ˈdä-tə-rəl, ˈdä-trəl\ *noun* [Middle English *dotrelle,* irregular from *doten* to dote] (15th century)
: a Eurasian plover (*Eudromias morinellus*) formerly common in England; *also* **:** any of various related plovers chiefly of eastern Asia, Australia, and South America

dot·tle \ˈdä-tᵊl, ˈdȯ-\ *noun* [Middle English *dottel* plug, from (assumed) Middle English *dot*] (circa 1825)
: unburned and partially burned tobacco in the bowl of a pipe

¹dot·ty \ˈdä-tē\ *adjective* **dot·ti·er; -est** [alteration of Scots *dottle* fool, from Middle English *dotel,* from *doten*] (15th century)
1 a : mentally unbalanced **:** CRAZY **b :** amiably eccentric
2 : being obsessed or infatuated
3 : amusingly absurd **:** RIDICULOUS ⟨some sublimely *dotty* exchanges of letters⟩
— dot·ti·ly \ˈdä-tᵊl-ē\ *adverb*
— dot·ti·ness \ˈdä-tē-nəs\ *noun*

²dotty *adjective* (1812)
: composed of or marked by dots

Dou·ay Version \dü-ˈā-\ *noun* [*Douay,* France] (1837)
: an English translation of the Vulgate used by Roman Catholics

¹dou·ble \ˈdə-bəl\ *adjective* [Middle English, from Old French, from Latin *duplus* (akin to Greek *diploos*), from *duo* two + *-plus* multiplied by; akin to Old English *-feald* -fold — more at TWO, -FOLD] (13th century)
1 : having a twofold relation or character **:** DUAL
2 : consisting of two usually combined members or parts ⟨an egg with a *double* yolk⟩
3 a : being twice as great or as many ⟨*double* the number of expected applicants⟩ **b** *of a coin* **:** worth two of the specified amount ⟨a *double* eagle⟩ ⟨a *double* crown⟩
4 : marked by duplicity **:** DECEITFUL
5 : folded in two
6 : of extra size, strength, or value ⟨a *double* martini⟩
7 : having more than the normal number of floral leaves often at the expense of the sporophylls
8 *of rhyme* **:** involving correspondence of two syllables (as in *exciting* and *inviting*)
9 : designed for the use of two persons ⟨a *double* room⟩ ⟨a *double* bed⟩
— dou·ble·ness *noun*

²double *verb* **dou·bled; dou·bling** \ˈdə-b(ə-)liŋ\ (13th century)
transitive verb
1 : to make twice as great or as many: as **a :** to increase by adding an equal amount **b :** to amount to twice the number of **c :** to make a call in bridge that increases the value of odd tricks or undertricks at (an opponent's bid)
2 a : to make of two thicknesses **:** FOLD **b :** CLENCH ⟨*doubled* his fist⟩ **c :** to cause to stoop
3 : to avoid by doubling **:** ELUDE
4 a : to replace in a dramatic role **b :** to play (dramatic roles) by doubling
5 a (1) **:** to advance or score (a base runner) by a double (2) **:** to bring about the scoring of (a run) by a double **b :** to put out (a base runner) in completing a double play
intransitive verb
1 a : to become twice as much or as many **b :** to double a bid (as in bridge)
2 a : to turn sharply and suddenly; *especially* **:** to turn back on one's course **b :** to follow a circuitous course
3 : to become bent or folded usually in the middle — usually used with *up* ⟨she *doubled* up in pain⟩
4 a : to serve an additional purpose or perform an additional duty **b :** to play a dramatic role as a double
5 : to make a double in baseball
— dou·bler \-b(ə-)lər\ *noun*

³double *adverb* (14th century)
1 : to twice the extent or amount
2 : two together

⁴double *noun* (14th century)
1 : something twice the usual size, strength, speed, quantity, or value: as **a :** a double amount **b :** a base hit that enables the batter to reach second base
2 : one that is the counterpart of another **:** DUPLICATE: as **a :** a living person that closely resembles another living person **b :** WRAITH **c** (1) **:** UNDERSTUDY (2) **:** one who resembles an actor and takes his or her place especially in scenes calling for special skills (3) **:** an actor who plays more than one role in a production
3 a : a sharp turn (as in running) **:** REVERSAL **b :** an evasive shift
4 : something consisting of two paired members: as **a :** FOLD **b :** a combined bet placed on two different contests **c :** two consecutive strikes in bowling
5 *plural* **:** a game between two pairs of players
6 : an act of doubling in a card game
7 : a room (as in a hotel) for two guests — compare SINGLE 4

double agent *noun* (1935)
: a spy pretending to serve one government while actually serving another

double bar *noun* (circa 1674)
: two adjacent vertical lines or a heavy single line separating principal sections of a musical composition

dou·ble–bar·rel \ˌdə-bəl-ˈbar-əl\ *noun* (1811)
: a double-barreled gun

dou·ble–bar·reled \-əld\ *adjective* (1709)
1 *of a firearm* **:** having two barrels mounted side by side or one beneath the other
2 : TWOFOLD; *especially* **:** having a double purpose ⟨asked a *double-barreled* question⟩

double bass *noun* (1752)
: the largest and lowest-pitched of the stringed instruments tuned in fourths
— double bass·ist \-ˈbā-sist\ *noun*

double bass

double bassoon *noun* (circa 1876)
: CONTRABASSOON

double bill *noun* (1917)
: a bill (as at a theater) offering two principal features

double bind *noun* (1956)
: a psychological predicament in which a person receives from a single source conflicting messages that allow no appropriate response to be made; *broadly* **:** DILEMMA 2

dou·ble–blind \ˌdə-bəl-ˈblīnd\ *adjective* (1950)
: of, relating to, or being an experimental procedure in which neither the subjects nor the experimenters know the makeup of the test and control groups during the actual course of the experiments — compare SINGLE-BLIND

double boiler *noun* (1879)
: a cooking utensil consisting of two saucepans fitting together so that the contents of the upper can be cooked or heated by boiling water in the lower

double bond *noun* (1889)
: a chemical bond in which two pairs of electrons are shared by two atoms in a molecule — compare SINGLE BOND, TRIPLE BOND

dou·ble–breast·ed \ˌdə-bəl-ˈbres-təd\ *adjective* (1701)
1 : having one half of the front lapped over the other and usually a double row of buttons

and a single row of buttonholes ⟨a *double-breasted* coat⟩
2 : having a double-breasted coat ⟨a *double-breasted* suit⟩

double–check \ˌdə-bəl-ˈchek, ˈdə-bəl-ˌ\ (1944)
transitive verb
: to subject to a double check ⟨an article *double-checked* for accuracy⟩
intransitive verb
: to make a double check

double check *noun* (1953)
: a careful checking to determine accuracy, condition, or progress especially of something already checked

dou•ble–clutch \ˌdə-bəl-ˈkləch\ *intransitive verb* (1928)
: to shift gears in an automotive vehicle by shifting into neutral and pumping the clutch before shifting to another gear

dou•ble–crop \ˌdə-bəl-ˈkräp\ *intransitive verb* (1918)
: to grow two or more crops on the same land in the same season or at the same time

dou•ble–cross \ˌdə-bəl-ˈkrȯs\ *transitive verb* (1903)
: to deceive by double-dealing **:** BETRAY
— dou•ble–cross•er *noun*

double cross *noun* (1834)
1 a : an act of winning or trying to win a fight or match after agreeing to lose it **b :** an act of betraying or cheating an associate
2 : a cross between first-generation hybrids of four separate inbred lines (as in the production of hybrid seed corn)

double dagger *noun* (1706)
: the character ‡ used as a reference mark — called also *diesis*

double date *noun* (circa 1931)
: a date participated in by two couples
— dou•ble–date *intransitive verb*

¹dou•ble–deal•ing \ˌdə-bəl-ˈdē-liŋ\ *noun* (1529)
: action contradictory to a professed attitude **:** DUPLICITY
synonym see DECEPTION
— dou•ble–deal•er \-ˈdē lər\ *noun*

²double–dealing *adjective* (1587)
: given to or marked by duplicity

dou•ble–deck \ˌdə-bəl-ˌdek\ *or* **dou•ble–decked** \-ˈdekt\ *adjective* (1894)
: having two decks, levels, or layers ⟨a *double-deck* bus⟩ ⟨a *double-deck* sandwich⟩

dou•ble–deck•er \-ˈde-kər\ *noun* (1835)
: something that is double-deck

dou•ble–dig•it \ˌdə-bəl-ˈdi-jət\ *adjective* (1959)
: amounting to 10 percent or more ⟨*double-digit* inflation⟩ ⟨*double-digit* price increases⟩

dou•ble–dip•per \-ˈdi-pər\ *noun* (circa 1974)
: a person who collects both a government pension and a government salary
— dou•ble–dip•ping \-piŋ\ *noun*

dou•ble–dome \ˈdə-bəl-ˌdōm\ *noun* (1938)
: INTELLECTUAL

double door *noun* (1840)
: an opening with two vertical doors that meet in the middle of the opening when closed — compare DUTCH DOOR

double dribble *noun* (circa 1949)
: an illegal action in basketball made when a player dribbles the ball with two hands simultaneously or continues to dribble after allowing the ball to come to rest in one or both hands

double Dutch *noun* (1876)
1 : unintelligible language **:** GIBBERISH
2 : the jumping of two jump ropes rotating in opposite directions simultaneously

dou•ble–edged \ˌdə-bəl-ˈejd\ *adjective* (15th century)
1 : having two cutting edges
2 a : having two components or aspects ⟨a spy with a *double-edged* mission⟩ **b :** capable of being taken in two ways ⟨a *double-edged* remark⟩

double–edged sword *noun* (15th century)
: something that has or can have both favorable and unfavorable consequences ⟨freedom of expression . . . can be a *double-edged sword* —Linda Connors⟩

dou•ble–end•ed \ˌdə-bəl-ˈen-dəd\ *adjective* (circa 1874)
: similar at both ends ⟨a *double-ended* bolt⟩

dou•ble–end•er \-dər\ *noun* (1864)
: a ship or boat with bow and stern of similar shape

dou•ble en•ten•dre \ˈdüb-^{ə}l-ä^{n}-ˈtä^{n}d(-r^{ə}); ˈdə-bəl-än-ˈtänd(-rə)\ *noun, plural* **double en•tendres** *same also* -ˈtä^{n}z; -ˈtän-drəz\ [obsolete French, literally, double meaning] (1673)
1 : ambiguity of meaning arising from language that lends itself to more than one interpretation
2 : a word or expression capable of two interpretations with one usually risqué

double entry *noun* (1741)
: a method of bookkeeping that recognizes both sides of a business transaction by debiting the amount of the transaction to one account and crediting it to another account so the total debits equal the total credits

dou•ble–faced \ˌdə-bəl-ˈfāst\ *adjective* (1577)
1 : HYPOCRITICAL, TWO-FACED
2 a : having two faces or sides designed for use ⟨a *double-faced* bookshelf⟩ **b** *also* **dou•ble–face** \-ˈfās\ **:** finished on both sides **:** REVERSIBLE — used of fabric

double fault *noun* (circa 1909)
: two consecutive serving faults in tennis that result in the loss of a point
— dou•ble–fault \ˌdə-bəl-ˈfȯlt\ *intransitive verb*

double feature *noun* (1928)
: a movie program with two main films

double fertilization *noun* (circa 1909)
: fertilization characteristic of seed plants in which one sperm nucleus fuses with the egg nucleus to form an embryo and another fuses with polar nuclei to form endosperm

double genitive *noun* (1824)
: a syntactic construction in English in which possession is marked both by the preposition *of* and a noun or pronoun in the possessive case (as in "A friend of Bob's is a friend of mine") — called also *double possessive*

double glazing *noun* (1943)
: two layers of glass set in a window to reduce heat flow in either direction

Double Glouces•ter \-ˈgläs-tər, -ˌglȯs-\ *noun* [*Gloucester,* England] (1816)
: a firm mild orange-colored English cheese similar to cheddar

dou•ble•head•er \ˌdə-bəl-ˈhe-dər\ *noun* (1878)
1 : a train pulled by two locomotives
2 : two games, contests, or events held consecutively on the same program

double helix *noun* (1954)
: a helix or spiral consisting of two strands in the surface of a cylinder that coil around its axis; *especially* **:** the structural arrangement of DNA in space that consists of paired polynucleotide strands stabilized by cross-links between purine and pyrimidine bases — compare ALPHA-HELIX, WATSON-CRICK MODEL
— dou•ble–he•li•cal \-ˈhe-li-kəl, -ˈhē-\ *adjective*

double–hung \ˌdə-bəl-ˈhəŋ\ *adjective* (1823)
of a window **:** having an upper and a lower sash that can slide vertically past each other

double hyphen *noun* (1893)
: a punctuation mark ⸗ used in place of a hyphen at the end of a line to indicate that the word so divided is normally hyphenated

double indemnity *noun* (1924)
: a provision in a life-insurance or accident policy whereby the company agrees to pay twice the face of the contract in case of accidental death

double jeopardy *noun* (1910)
: the putting of a person on trial for an offense for which he or she has previously been put on trial under a valid charge **:** two adjudications for one offense

dou•ble–joint•ed \ˌdə-bəl-ˈjȯin-təd\ *adjective* (1831)
: having a joint that permits an exceptional degree of freedom of motion of the parts joined

double knit *noun* (1895)
: a knitted fabric (as wool) made with a double set of needles to produce a double thickness of fabric with each thickness joined by interlocking stitches; *also* **:** an article of clothing made of such fabric

double negative *noun* (1827)
: a now substandard syntactic construction containing two negatives and having a negative meaning ("I didn't hear nothing" is a *double negative*) ■

dou•ble–park \ˌdə-bəl-ˈpärk\ (1936)
transitive verb
: to park (a vehicle) beside a row of vehicles already parked parallel to the curb
intransitive verb
: to double-park a vehicle

double play *noun* (1867)
: a play in baseball by which two players are put out

dou•ble–quick \ˈdə-bəl-ˌkwik\ *noun* (1834)
: DOUBLE TIME 1; *broadly* **:** a rapid pace
— double–quick *adjective or adverb*

double reed *noun* (circa 1876)
: two reeds bound together with a slight separation between them so that air passing through them causes them to beat against one another and that are used as a sound⸗

□ USAGE

double negative The grammarian Otto Jespersen once observed that although negation is very important to the meaning of the sentence, it is often marked in an almost inconspicuous way—perhaps only by an unstressed particle like *-n't.* Hence there has long been a tendency to strengthen the negative idea by adding more negative elements to the sentence. This multiple negation is known popularly as the *double negative.* Multiple negation occurs in Chaucer, and it continued in common use at least through Shakespeare's time ⟨no woman has; nor never none shall mistress be of it —*Twelfth Night*⟩ and the ordinary double negative into the early 18th century ⟨lost no time, nor abated no Diligence —Daniel DeFoe (*Robinson Crusoe,* 1719)⟩. During the 18th century it began to drop out of use in writing but continued in familiar use—conversation and letters. Thus grammarians, seeing it in older literature but not in current literature, decided that it was obsolete. Later grammarians were content to call it wrong. More important than the strictures of the grammarians was the perception that it was associated with the lower classes, both rural and urban, and hence was a construction to be avoided by the socially ambitious. The current status of the double negative remains much the same, confined to the most familiar discourse of those who choose to use it ⟨as you know, Ez hasn't changed none —Archibald MacLeish (letter)⟩ and used as an everyday form only in the speech of less educated people and of some educated people who are comfortable with it. It is not a prestige form, even though it may be used on occasion for effect ⟨the sailplane sure ain't no 747! —Susan Ochshorn (*Saturday Rev.*)⟩.

producing device in certain woodwind instruments (as members of the oboe family)

double refraction *noun* (1831)
: BIREFRINGENCE

dou·ble–ring \ˈdə-bəl-ˌriŋ\ *adjective* (circa 1959)
: of or relating to a wedding ceremony in which each partner ceremonially gives the other a wedding ring while reciting vows

double salt *noun* (circa 1849)
: a salt (as an alum) yielding on hydrolysis two different cations or anions

dou·ble–space \ˌdə-bəl-ˈspās\ (circa 1937)
transitive verb
: to type (text) leaving alternate lines blank
intransitive verb
: to type on every other line

dou·ble·speak \ˈdə-bəl-ˌspēk\ *noun* (1952)
: language used to deceive usually through concealment or misrepresentation of truth; *also* **:** GOBBLEDYGOOK
— dou·ble·speak·er \-ˌspē-kər\ *noun*

double standard *noun* (1894)
1 : BIMETALLISM
2 : a set of principles that applies differently and usually more rigorously to one group of people or circumstances than to another; *especially* **:** a code of morals that applies more severe standards of sexual behavior to women than to men

double star *noun* (1781)
1 : BINARY STAR
2 : two stars in very nearly the same line of sight but actually physically separate

dou·ble–stop \ˌdə-bəl-ˈstäp\ *transitive verb* (circa 1889)
: to produce two or more tones simultaneously on (as a violin)
— double–stop *noun*

double sugar *noun* (1956)
: DISACCHARIDE

dou·blet \ˈdəb-lət\ *noun* [Middle English, from Middle French, from *double*] (14th century)
1 : a man's close-fitting jacket worn in Europe especially during the Renaissance

doublet 1

2 : something consisting of two identical or similar parts: as **a :** a lens consisting of two components; *especially* **:** a handheld magnifier consisting of two lenses in a metal cylinder **b :** a spectrum line having two close components **c :** a domino with the same number of spots on each end
3 : a set of two identical or similar things: as **a :** two thrown dice with the same number of spots on the upper face **b :** one of nine pairs of microtubules found in cilia and flagella
4 : one of a pair; *specifically* **:** one of two or more words (as *guard* and *ward*) in the same language derived by different routes of transmission from the same source

dou·ble take \ˈdə-bəl-ˌtāk\ *noun* (1930)
: a delayed reaction to a surprising or significant situation after an initial failure to notice anything unusual — usually used in the phrase *do a double take*

dou·ble–talk \-ˌtȯk\ *noun* (1936)
1 : language that appears to be earnest and meaningful but in fact is a mixture of sense and nonsense
2 : inflated, involved, and often deliberately ambiguous language
— dou·ble–talk *intransitive verb*
— dou·ble–talk·er *noun*

dou·ble–team \-ˌtēm\ *transitive verb* (1860)
: to block or guard (an opponent) with two players at one time

Double Ten *noun* [translation of Chinese (Beijing) *shuāngshí;* from its being the tenth day of the tenth month] (1940)
: October 10 observed by the Republic of China in commemoration of the revolution of 1911

dou·ble·think \ˈdə-bəl-ˌthiŋk\ *noun* (1949)
: a simultaneous belief in two contradictory ideas

dou·ble–time \ˈdə-bəl-ˌtīm\ *intransitive verb* (1943)
: to move at double time

double time *noun* (1853)
1 : a marching cadence of 180 36-inch steps per minute
2 : payment of a worker at twice the regular wage rate

dou·ble·ton \ˈdə-bəl-tən\ *noun* [*double* + *-ton* (as in *singleton*)] (circa 1894)
: two cards that are the only ones of their suit originally dealt to a player — compare SINGLETON 1, VOID 4

dou·ble–tongue \ˌdə-bəl-ˈtəŋ\ *intransitive verb* (circa 1900)
: to cause the tongue to alternate rapidly between the positions for *t* and *k* so as to produce a fast succession of detached notes on a wind instrument

dou·ble–u *as at* W\ *noun* (1840)
: the letter *w*

double up *intransitive verb* (1789)
: to share accommodations designed for one

double vision *noun* (circa 1860)
: DIPLOPIA

double whammy *noun* (1951)
: a combination of two usually adverse forces, circumstances, or effects

double–wide \ˌdə-bəl-ˈwīd\ *noun* (1970)
: a mobile home consisting of two units which have been fastened together along their length

dou·bloon \ˌdə-ˈblün\ *noun* [Spanish *doblón,* augmentative of *dobla,* an old Spanish coin, from Latin *dupla,* feminine of *duplus* double — more at DOUBLE] (1622)
: an old gold coin of Spain and Spanish America

dou·bly \ˈdə-b(ə-)lē\ *adverb* (15th century)
1 : in a twofold manner
2 : to twice the degree

[1]doubt \ˈdau̇t\ *verb* [alteration of Middle English *douten,* from Old French *douter* to doubt, from Latin *dubitare;* akin to Latin *dubius* dubious] (13th century)
transitive verb
1 *archaic* **a :** FEAR **b :** SUSPECT
2 : to be in doubt about ⟨he *doubts* everyone's word⟩
3 a : to lack confidence in **:** DISTRUST ⟨find myself *doubting* him even when I know that he is honest —H. L. Mencken⟩ **b :** to consider unlikely ⟨I *doubt* if I can go⟩
intransitive verb
: to be uncertain
— doubt·able \ˈdau̇-tə-bəl\ *adjective*
— doubt·er *noun*
— doubt·ing·ly \-tiŋ-lē\ *adverb*

[2]doubt *noun* (13th century)
1 a : uncertainty of belief or opinion that often interferes with decision-making **b :** a deliberate suspension of judgment
2 : a state of affairs giving rise to uncertainty, hesitation, or suspense
3 a : a lack of confidence **:** DISTRUST **b :** an inclination not to believe or accept
synonym see UNCERTAINTY
— no doubt : [1]DOUBTLESS

doubt·ful \ˈdau̇t-fəl\ *adjective* (14th century)
1 : giving rise to doubt **:** open to question ⟨it is *doubtful* that they ever knew what happened⟩
2 a : lacking a definite opinion, conviction, or determination ⟨they were *doubtful* about the advantages of the new system⟩ **b :** uncertain in outcome **:** UNDECIDED ⟨the outcome of the election remains *doubtful*⟩
3 : marked by qualities that raise doubts about worth, honesty, or validity ⟨of *doubtful* repute⟩
☆
— doubt·ful·ly \-fə-lē\ *adverb*
— doubt·ful·ness *noun*

doubting Thom·as \-ˈtä-məs\ *noun* [Saint *Thomas,* apostle who doubted Jesus' resurrection until he had proof of it (John 20:24–29)] (1883)
: a habitually doubtful person

[1]doubt·less \ˈdau̇t-ləs\ *adverb* (14th century)
1 : without doubt
2 : PROBABLY

[2]doubtless *adjective* (14th century)
: free from doubt **:** CERTAIN
— doubt·less·ly *adverb*
— doubt·less·ness *noun*

douce \ˈdüs\ *adjective* [Middle English, sweet, pleasant, from Middle French, from feminine of *douz,* from Latin *dulcis* sweet — more at DULCET] (1721)
chiefly Scottish **:** SOBER, SEDATE ⟨the *douce* faces of the mourners —L. J. A. Bell⟩
— douce·ly *adverb, chiefly Scottish*

dou·ceur \dü-ˈsər\ *noun* [French, pleasantness, from Late Latin *dulcor* sweetness, from Latin *dulcis*] (1763)
: a conciliatory gift

douche \ˈdüsh\ *noun* [French, from Italian *doccia,* from *docciare* to douche, from *doccia* water pipe, probably back-formation from *doccione* conduit, from Latin *duction-, ductio* means of conveying water, from *ducere* to lead — more at TOW] (1766)
1 a : a jet or current especially of water directed against a part or into a cavity of the body **b :** an act of cleansing with a douche
2 : a device for giving douches
3 *British* **:** a rude awakening ⟨the icy *douche* (what he said about my work) —John Fowles⟩
— douche *verb*

douche bag *noun* (circa 1963)
slang **:** an unattractive or offensive person

dough \ˈdō\ *noun* [Middle English *dogh,* from Old English *dāg;* akin to Old High German *teic* dough, Latin *fingere* to shape, Greek *teichos* wall] (before 12th century)
1 : a mixture that consists essentially of flour or meal and a liquid (as milk or water) and is stiff enough to knead or roll
2 : something resembling dough especially in consistency
3 : MONEY
4 : DOUGHBOY
— dough·like \-ˌlīk\ *adjective*

dough box *noun* (circa 1944)
: a rectangular wooden box mounted on legs that is used as a worktable and storage space

dough·boy \-ˌbȯi\ *noun* (1865)
: an American infantryman especially in World War I

dough·face \-ˌfās\ *noun* (1830)
: a Northern congressman not opposed to slavery in the South before or during the Civil War; *also* **:** a Northerner sympathetic to the South during the same period
— dough–faced \-ˈfāst\ *adjective*

dough·nut \-(ˌ)nət\ *noun* (circa 1809)
1 : a small usually ring-shaped cake fried in fat

☆ SYNONYMS
Doubtful, dubious, problematic, questionable mean not affording assurance of the worth, soundness, or certainty of something. DOUBTFUL implies little more than a lack of conviction or certainty ⟨*doubtful* about whether I said the right thing⟩. DUBIOUS stresses suspicion, mistrust, or hesitation ⟨*dubious* about the practicality of the scheme⟩. PROBLEMATIC applies especially to things whose existence, meaning, fulfillment, or realization is highly uncertain ⟨whether the project will ever be finished is *problematic*⟩. QUESTIONABLE may imply no more than the existence of doubt but usually suggests that the suspicions are well-grounded ⟨a man of *questionable* honesty⟩.

2 : something (as a mathematical torus) that resembles a doughnut especially in shape
— **dough·nut·like** \-ˌlīk\ *adjective*

dough·ty \'dau̇-tē\ *adjective* **dough·ti·er; -est** [Middle English, from Old English *dohtig;* akin to Old High German *toug* is useful, Greek *teuchein* to make] (before 12th century)
: marked by fearless resolution **:** VALIANT
— **dough·ti·ly** \'dau̇-tᵊl-ē\ *adverb*
— **dough·ti·ness** \'dau̇-tē-nəs\ *noun*

doughy \'dō-ē\ *adjective* **dough·i·er; -est** (1601)
: resembling dough: as **a :** not thoroughly baked **b :** unhealthily pale **:** PASTY

Doug·las fir \ˌdə-gləs-\ *noun* [David *Douglas* (died 1834) Scottish botanist] (1873)
: any of a genus (*Pseudotsuga*) of tall evergreen timber trees having thick bark, pitchy wood, and pendulous cones; *especially* **:** one (*P. menziesii* synonym *P. taxifolia*) of the western U.S.

Dou·kho·bor \'dü-kə-ˌbȯr\ *noun* [Russian *dukhobor, dukhoborets,* from *dukh* spirit + *borets* wrestler] (1876)
: a member of a Christian sect of 18th century Russian origin emphasizing the duty of obeying the inner light and rejecting church or civil authority

do up *transitive verb* (1666)
1 : to prepare (as by cleaning or repairing) for wear or use ⟨*do up* a shirt⟩
2 a : to wrap up ⟨*do up* a package⟩ **b :** PUT UP, CAN
3 a : to deck out **:** CLOTHE **b :** to furnish with something ornamental **:** DECORATE
4 : EXHAUST, WEAR OUT

dour \'du̇(-ə)r, 'dau̇(-ə)r\ *adjective* [Middle English, from Latin *durus* hard — more at DURING] (14th century)
1 : STERN, HARSH
2 : OBSTINATE, UNYIELDING
3 : GLOOMY, SULLEN
— **dour·ly** *adverb*
— **dour·ness** *noun*

dou·rou·cou·li \ˌdu̇r-ə-'kü-lē\ *noun* [French, from an unidentified American Indian language of Venezuela] (1842)
: any of several small nocturnal monkeys (genus *Aotus*) of tropical America that have round heads, large eyes, and densely furred bodies

1douse \'dau̇s *also* 'dau̇z\ *verb* **doused; dous·ing** [perhaps from obsolete English *douse* (to smite)] (1600)
transitive verb
1 : to plunge into water
2 a : to throw a liquid on **:** DRENCH **b :** SLOSH
3 : EXTINGUISH ⟨*douse* the lights⟩
intransitive verb
: to fall or become plunged into water
— **dous·er** *noun*

2douse \'dau̇s *also* 'dau̇z\ *noun* (1881)
: a heavy drenching

3douse \'düs, 'dau̇s\ *noun* [origin unknown] (circa 1625)
British **:** BLOW, STROKE

4douse \'dau̇s\ *transitive verb* **doused; dous·ing** (1627)
1 a : FURL ⟨*douse* a sail⟩ **b :** SLACKEN ⟨*douse* a rope⟩
2 : TAKE OFF, DOFF

doux \'dü\ *adjective* [French, literally, sweet, from Old French *douz* — more at DOUCE] (circa 1943)
of champagne **:** very sweet

1dove \'dəv\ *noun* [Middle English, from (assumed) Old English *dūfe;* akin to Old High German *tūba* dove] (13th century)
1 : any of numerous pigeons; *especially* **:** a small wild pigeon
2 : a gentle woman or child
3 : one who takes a conciliatory attitude and advocates negotiations and compromise; *especially* **:** an opponent of war — compare HAWK
— **dov·ish** \'də-vish\ *adjective*
— **dov·ish·ness** *noun*

2dove \'dōv\ *past of* DIVE

dove·cote \'dəv-ˌkōt, -ˌkät\ *also* **dove·cot** \-ˌkät\ *noun* (15th century)
1 : a small compartmented raised house or box for domestic pigeons
2 : a settled or harmonious group or organization

dove·kie \'dəv-kē\ *noun* [diminutive of *dove*] (1821)
: a small short-billed auk (*Alle alle*) breeding on arctic coasts and ranging south in winter

doven *variant of* DAVEN

Dover sole *noun* [probably from *Dover,* England] (circa 1911)
1 : a common European sole (*Solea solea*) esteemed as a food fish
2 : a brownish blotched flatfish (*Microstomus pacificus*) of the Pacific coast of North America that is a market fish in California

Do·ver's powder \ˌdō-vərz-\ *noun* [Thomas *Dover* (died 1742) English physician] (1801)
: a powder of ipecac and opium and used as a pain reliever and diaphoretic

1dove·tail \'dəv-ˌtāl\ *noun* (1573)
: something resembling a dove's tail; *especially* **:** a flaring tenon and a mortise into which it fits tightly making an interlocking joint between two pieces (as of wood)

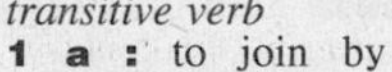

dovetail: *1* mortises, *2* tenons, *3* joint

2dovetail (circa 1656)
transitive verb
1 a : to join by means of dovetails **b :** to cut to a dovetail
2 a : to fit skillfully to form a whole **b :** to fit together with
intransitive verb
: to fit together into a whole

dow \'dau̇\ *intransitive verb* **dought** \'dau̇t\ *or* **dowed** \'dau̇d\; **dow·ing** [Middle English *dow, deih* have worth, am able, from Old English *dēah, dēag,* akin to Old High German *toug* is worthy, is useful — more at DOUGHTY] (before 12th century)
chiefly Scottish **:** to be able or capable

Dow \'dau̇\ *noun* (1949)
: DOW-JONES AVERAGE

dow·a·ger \'dau̇-i-jər\ *noun* [Middle French *douagiere,* from *douage* dower, from *douer* to endow, from Latin *dotare,* from *dot-, dos* gift, dower — more at DATE] (1530)
1 : a widow holding property or a title from her deceased husband
2 : a dignified elderly woman

dowager's hump *noun* (1948)
: an abnormal curvature of the upper back with round shoulders and stooped posture caused especially by bone loss and anterior compression of the vertebrae in osteoporosis

1dowdy \'dau̇-dē\ *noun, plural* **dowd·ies** [diminutive of *dowd* (dowdy), from Middle English *doude*] (1581)
archaic **:** a dowdy woman

2dowdy *adjective* **dowd·i·er; -est** (1676)
1 : not neat or becoming in appearance **:** SHABBY
2 a : lacking smartness or taste **b :** OLD-FASHIONED
— **dowd·i·ly** \'dau̇-dᵊl-ē\ *adverb*
— **dowd·i·ness** \'dau̇-dē-nəs\ *noun*
— **dowdy·ish** \-ish\ *adjective*

3dowdy *noun* [origin unknown] (1936)
: PANDOWDY

1dow·el \'dau̇(-ə)l\ *noun* [Middle English *dowle;* akin to Old High German *tubili* plug, Late Greek *typhos* wedge] (14th century)
1 : a pin fitting into a hole in an abutting piece to prevent motion or slipping; *also* **:** a round rod or stick used especially for cutting up into dowels
2 : a piece of wood driven into a wall so that other pieces can be nailed to it

2dowel *transitive verb* **-elled** *also* **-eled; -elling** *also* **-eling** (1713)
: to fasten by or furnish with dowels

1dow·er \'dau̇(-ə)r\ *noun* [Middle English *dowere,* from Middle French *douaire,* modification of Medieval Latin *dotarium* — more at DOWRY] (14th century)
1 : the part of or interest in the real estate of a deceased husband given by law to his widow during her life — compare CURTESY
2 : DOWRY 2, 3

2dower *transitive verb* (1605)
: to supply with a dower or dowry **:** ENDOW

dow·itch·er \'dau̇-i-chər\ *noun, plural* **dow·itchers** *also* **dowitcher** [probably of Iroquoian origin; akin to Oneida *tawístawis* dowitcher] (1841)
: any of several long-billed wading birds (especially *Limnodromus griseus* and *L. scolopaceus* of the family Scolopacidae) related to the sandpipers

Dow–Jones average \ˌdau̇-'jōnz-\ *noun* [Charles H. *Dow* (died 1902) & Edward D. *Jones* (died 1920) American financial statisticians] (1922)
: an index of the relative price of securities

1down \'dau̇n\ *adverb* [Middle English *doun,* from Old English *dūne,* short for *adūne, of dūne,* from *a-* (from *of*), *of* off, from + *dūne,* dative of *dūn* hill] (before 12th century)
1 a (1) **:** toward or in a lower physical position (2) **:** to a lying or sitting position (3) **:** toward or to the ground, floor, or bottom **b :** as a down payment ⟨paid $10 *down*⟩ **c :** on paper ⟨put *down* what he says⟩
2 : in a direction that is the opposite of up: as **a :** SOUTHWARD **b :** to or toward a point away from the speaker or the speaker's point of reference
3 : to a lesser degree, level, or rate ⟨cool *down* tensions⟩
4 : to or toward a lower position in a series
5 a : to or in a lower or worse condition or status **b** — used to indicate completion ⟨dusted *down* the house⟩
6 : from a past time
7 : to or in a state of less activity or prominence
8 : to a concentrated state ⟨got the report *down* to three pages⟩
9 : into defeat ⟨voted the motion *down*⟩
— **down to the ground :** PERFECTLY, COMPLETELY ⟨that suits me *down to the ground*⟩

2down *preposition* (14th century)
: down along, around, through, toward, in, into, or on ⟨fell *down* the stairs⟩ ⟨*down* the years⟩

3down (1562)
transitive verb
1 : to cause to go or come down
2 : to cause (a football) to be out of play
3 : DEFEAT
intransitive verb
: to go down

4down *adjective* (circa 1565)
1 a (1) **:** occupying a low position; *specifically* **:** lying on the ground ⟨*down* timber⟩ (2) **:** directed or going downward ⟨attendance is *down*⟩ **b :** lower in price **c :** not being in play in football because of wholly stopped progress or because the officials stop the play **d :** defeated or trailing an opponent (as in points scored) ⟨*down* by two runs⟩ **e** *baseball* **:** OUT
2 a : reduced or low in activity or intensity ⟨a *down* economy⟩ **b :** not operating or able to function ⟨the computer is *down*⟩ **c :** DEPRESSED, DEJECTED; *also* **:** DEPRESSING ⟨a *down* movie⟩ **d :** SICK ⟨*down* with flu⟩

3 : DONE, FINISHED ⟨eight *down* and two to go⟩
4 : completely mastered ⟨had her lines *down*⟩ — often used with *pat*
5 *slang* **:** COOL 7 ⟨a *down* dude⟩
6 : being a quark with an electric charge of −⅓, zero charm, and zero strangeness ⟨a *down* quark⟩ — compare ²UP 5
7 : being on record ⟨you're *down* for two tickets⟩
— down on : having a low opinion of or dislike for
⁵down *noun* (1710)
1 : DESCENT, DEPRESSION
2 : an instance of putting down
3 a : a complete play to advance the ball in football **b :** one of a series of four attempts in American football or three attempts in Canadian football to advance the ball 10 yards
4 *chiefly British* **:** DISLIKE, GRUDGE
5 : DOWNER
⁶down *noun* [Middle English *doun* hill, from Old English *dūn*] (14th century)
1 : an undulating usually treeless upland with sparse soil — usually used in plural
2 *often capitalized* **:** a sheep of any breed originating in the downs of southern England
⁷down *noun* [Middle English *doun*, from Old Norse *dūnn*] (14th century)
1 : a covering of soft fluffy feathers; *also* **:** these feathers
2 : something soft and fluffy like down
down and dirty *adjective or adverb* (1967)
1 : UNVARNISHED ⟨the *down and dirty* truth⟩
2 : made or done hastily **:** not revised or polished
3 : marked by fierce competition
4 : BAWDY
5 : SEEDY 2
down–and–out \ˌdau̇-nən-ˈ(d)au̇t\ *adjective* (1901)
1 : DESTITUTE, IMPOVERISHED
2 : physically weakened or incapacitated
— down–and–out *or* **down–and–out•er** \-ˈ(d)au̇t-ər\ *noun*
down–at–the–heels \ˌdau̇-nət-t͟hə-ˈhē(ə)lz, -nat-\ *or* **down–at–heel** \-nət-ˈhē(ə)l\ *also* **down–at–the–heel** \-t͟hə-ˈhē(ə)l\ *or* **down–at–heels** \-nət-ˈhē(ə)lz\ *adjective* (1732)
: SHABBY
¹down•beat \ˈdau̇n-ˌbēt\ *noun* (1876)
1 : the downward stroke of a conductor indicating the principally accented note of a measure of music; *also* **:** the first beat of a measure
2 : a decline in activity or prosperity
²downbeat *adjective* (1950)
: PESSIMISTIC, GLOOMY
down–bow \ˈdau̇n-ˌbō\ *noun* (circa 1889)
: a stroke in playing a bowed instrument (as a violin) in which the bow is drawn across the strings from the frog to the tip
down•burst \-ˌbərst\ *noun* (1978)
: a powerful downdraft usually associated with a thunderstorm that strikes the ground and deflects in all directions and that constitutes a hazard especially for low-flying aircraft; *also* **:** MICROBURST
down•cast \ˈdau̇n-ˌkast\ *adjective* (14th century)
1 : low in spirit **:** DEJECTED
2 : directed downward ⟨with *downcast* eyes⟩
down•court \-ˈkōrt, -ˈkȯrt\ *adverb or adjective* (1952)
: in or into the opposite end of the court (as in basketball)
down•draft \-ˌdraft\ *noun* (1849)
: a downward current of gas (as air during a thunderstorm)
down east *adverb, often D&E capitalized* (1825)
: in or into the northeast coastal section of the U.S. and parts of the Maritime Provinces of Canada; *specifically* **:** in or into coastal Maine
— down east *adjective, often D&E capitalized*
down–east•er \dau̇-ˈnē-stər\ *noun, often D&E capitalized* (1827)
: one born or living down east
down•er \ˈdau̇-nər\ *noun* (1966)
1 : a depressant drug; *especially* **:** BARBITURATE
2 : someone or something depressing
down•fall \ˈdau̇n-ˌfȯl\ *noun* (13th century)
1 a : a sudden fall (as from power) **b :** a fall (as of snow or rain) especially when sudden or heavy
2 : something that causes a downfall (as of a person)
— down•fall•en \-ˌfȯ-lən\ *adjective*
down•field \-ˈfē(ə)ld\ *adverb or adjective* (1944)
: in or into the part of the field toward which the offensive team is headed
¹down•grade \ˈdau̇n-ˌgrād\ *noun* (1858)
1 : a downward grade (as of a road)
2 : a descent toward an inferior state ⟨a career on the *downgrade*⟩
²downgrade *transitive verb* (1930)
1 : MINIMIZE, DEPRECIATE
2 : to lower in quality, value, status, or extent
down•haul \ˈdau̇n-ˌhȯl\ *noun* (1669)
: a rope or line for hauling down or holding down a sail or spar
down•heart•ed \-ˈhär-təd\ *adjective* (circa 1774)
: DOWNCAST, DEJECTED
— down•heart•ed•ly *adverb*
— down•heart•ed•ness *noun*
¹down•hill \ˌdau̇n-ˈhil\ *adverb* (14th century)
1 : toward the bottom of a hill
2 : toward a worsened or inferior state or level — used especially in the phrase *go downhill*
²down•hill \ˈdau̇n-ˌhil\ *noun* (1591)
1 : a descending slope
2 : a skiing race against time down a trail — often used attributively
³down•hill \-ˌhil\ *adjective* (1727)
1 : sloping downhill
2 : closer to the bottom of an incline ⟨your *downhill* ski⟩
3 : not difficult **:** EASY ⟨after that problem the rest was *downhill*⟩
4 : progressively worse
down•hill•er \-hi-lər\ *noun* (1967)
: a downhill skier
down–home \ˈdau̇n-ˌhōm\ *adjective* (1938)
: of, relating to, or having qualities (as informality and simplicity) associated with rural or small-town people especially of the Southern U.S. ⟨*down-home* country cooking⟩; *broadly* **:** SIMPLE, UNPRETENTIOUS
down in the mouth *adjective* (1649)
: DEJECTED 1
down•land \ˈdau̇n-ˌland\ *noun* (before 12th century)
: ⁶DOWN 1
down•link \ˈdau̇n-ˌliŋk\ *noun* (circa 1969)
: a communications channel for receiving transmissions from a spacecraft; *also* **:** such transmissions
down•load \ˈdau̇n-ˌlōd\ *transitive verb* (1980)
: to transfer (data) from a usually large computer to the memory of another device (as a smaller computer)
— down•load•able \-ˌlō-də-bəl\ *adjective*
down–market \ˈdau̇n-ˌmär-kət\ *adjective* (1970)
: relating or appealing to lower-income consumers
down payment *noun* (1926)
: a part of the full price paid at the time of purchase or delivery with the balance to be paid later; *broadly* **:** the first step in a process
down•pipe \ˈdau̇n-ˌpīp\ *noun* (1858)
British **:** DOWNSPOUT
down•play \ˈdau̇n-ˌplā\ *transitive verb* (1954)
: PLAY DOWN, DE-EMPHASIZE
down•pour \-ˌpōr, -ˌpȯr\ *noun* (1811)
: a pouring or streaming downward; *especially* **:** a heavy rain
down•range \-ˈrānj\ *adverb* (1952)
: away from a launching site
¹down•right \-ˌrīt\ *adverb* (13th century)
1 *archaic* **:** straight down
2 : ABSOLUTELY 1 ⟨*downright* handsome⟩ ⟨*downright* mean⟩
3 *obsolete* **:** FORTHRIGHT
²downright *adjective* (1530)
1 *archaic* **:** directed vertically downward
2 : OUTRIGHT, THOROUGH ⟨a *downright* lie⟩
3 : PLAIN, BLUNT ⟨stories he had heard of her *downright* tongue —Angus Wilson⟩
— down•right•ly *adverb*
— down•right•ness *noun*
down•riv•er \ˈdau̇n-ˈri-vər\ *adverb or adjective* (1852)
: toward or at a point nearer the mouth of a river
¹down•scale \ˈdau̇n-ˌskāl\ *transitive verb* **down•scaled; down•scal•ing** (1945)
: to cut back in size or scope ⟨the recession forced us to *downscale* vacation plans⟩
²downscale *adjective* (circa 1966)
: lower in class, income, or quality
down•shift \-ˌshift\ *intransitive verb* (1955)
: to shift an automotive vehicle into a lower gear
— downshift *noun*
down•side \ˈdau̇n-ˌsīd\ *noun* (1946)
1 : a downward trend (as of prices)
2 : a negative aspect ⟨the *downside* of fame⟩
down•size \ˈdau̇n-ˌsīz\ (1975)
transitive verb
: to reduce in size; *especially* **:** to design or produce in smaller size
intransitive verb
: to undergo a reduction in size
down•slide \ˈdau̇n-ˌslīd\ *noun* (1926)
: a downward movement
down•slope \ˈdau̇n-ˌslōp\ *adjective or adverb* (1928)
: toward the bottom of a slope
down•spout \ˈdau̇n-ˌspau̇t\ *noun* (circa 1896)
: a vertical pipe used to drain rainwater from a roof
Down's syndrome \ˈdau̇n(z)-\ *noun* [J. L. H. *Down* (died 1896) English physician] (1961)
: a congenital condition characterized by moderate to severe mental retardation, slanting eyes, a broad short skull, broad hands with short fingers, and trisomy of the human chromosome numbered 21 — called also *Down's, Down syndrome*
¹down•stage \ˈdau̇n-ˈstāj\ *adverb or adjective* (1898)
1 : toward or at the front of a theatrical stage
2 : toward a motion-picture or television camera
²down•stage \-ˌstāj\ *noun* (circa 1931)
: the part of a stage that is nearest the audience or camera
¹down•stairs \ˈdau̇n-ˈstarz, -ˈsterz\ *adverb* (1596)
: down the stairs **:** on or to a lower floor
²down•stairs \ˈdau̇n-ˌstarz, -ˌsterz\ *adjective* (1819)
: situated on the main, lower, or ground floor of a building
³down•stairs \ˈdau̇n-ˈ, ˈdau̇n-ˌ\ *noun plural but singular or plural in construction* (1843)
: the lower floor of a building
down•state \-ˌstāt\ *noun* (1909)
: the chiefly southerly sections of a state; *also* **:** the chiefly rural part of a state when the major metropolitan area is to the north
— down•state \-ˈstāt\ *adjective or adverb*
— down•stat•er \-ˈstā-tər\ *noun*
down•stream \ˈdau̇n-ˈstrēm\ *adverb or adjective* (1706)
1 : in the direction of or nearer to the mouth of a stream

2 : in or toward the latter stages of a usually industrial process or the stages (as marketing) after manufacture
down·stroke \-ˌstrōk\ *noun* (1852)
: a downward stroke
down·swing \-ˌswiŋ\ *noun* (1899)
1 : a downward swing
2 : DOWNTREND
down–the–line *adjective* (1940)
: COMPLETE ⟨a *down-the-line* union supporter⟩
down·time \ˈdau̇n-ˌtīm\ *noun* (1928)
1 : time during which production is stopped especially during setup for an operation or when making repairs
2 : break time between periods of work ⟨napping during our *downtime*⟩
down–to–earth \ˌdau̇n-tə-ˈ(w)ərth\ *adjective* (1932)
1 : PRACTICAL ⟨*down-to-earth* traveling tips⟩
2 : UNPRETENTIOUS ⟨surprised to find the movie star so *down-to-earth*⟩
down–to–the–wire \ˈdau̇n-tə-t͟hə-ˈwīr\ *adjective* (1952)
: full of suspense; *especially* **:** unsettled until the very end
down·town \ˌdau̇n-ˈtau̇n, ˈdau̇n-ˌ\ *noun* (1851)
: the lower part of a city; *especially* **:** the main business district — often used attributively
— downtown \ˌdau̇n-ˈtau̇n\ *adverb*
— down·town·er \-ˈtau̇-nər\ *noun*
down·trend \-ˌtrend\ *noun* (1926)
: a downturn especially in business and economic activity
down·trod·den \ˈdau̇n-ˈträ-d^{ə}n\ *adjective* (1595)
: suffering oppression
down·turn \-ˌtərn\ *noun* (1926)
: a downward turn especially toward a decline in business activity
down under *adverb or adjective, often D&U capitalized* (1886)
: to or in Australia or New Zealand
[1]**down·ward** \ˈdau̇n-wərd\ *or* **down·wards** \-wərdz\ *adverb* (13th century)
1 a : from a higher to a lower place **b :** toward a direction that is the opposite of up
2 : from a higher to a lower condition
3 a : from an earlier time **b :** from an ancestor or predecessor
[2]**downward** *adjective* (circa 1552)
1 : moving or extending downward
2 : descending from a head, origin, or source
— down·ward·ly *adverb*
— down·ward·ness *noun*
down·wash \ˈdau̇n-ˌwȯsh, -ˌwäsh\ *noun* (1915)
: an airstream directed downward (as by an airfoil)
down·wind \ˈdau̇n-ˈwind\ *adverb or adjective* (1855)
: in the direction that the wind is blowing
downy \ˈdau̇-nē\ *adjective* **down·i·er; -est** (1578)
1 : resembling a bird's down
2 : covered with down
3 : made of down
4 : SOFT, SOOTHING
downy mildew *noun* (1886)
1 : any of various parasitic lower fungi (family Peronosporaceae) that produce whitish masses of sporangiophores or conidiophores on the undersurface of the leaves of the host
2 : a plant disease caused by a downy mildew
downy woodpecker *noun* (1808)
: a small black-and-white woodpecker (*Dendrocopos pubescens*) of North America that has a white back and is smaller than the hairy woodpecker
dow·ry \ˈdau̇(-ə)-rē\ *noun, plural* **dowries** [Middle English *dowarie,* from Anglo-French, irregular from Medieval Latin *dotarium,* from Latin *dot-, dos* gift, marriage portion — more at DATE] (14th century)
1 *archaic* **:** DOWER 1
2 : the money, goods, or estate that a woman brings to her husband in marriage
3 : a gift of money or property by a man to or for his bride
4 : a natural talent
dowsabel *noun* [*Dowsabel,* feminine name] (circa 1652)
obsolete **:** SWEETHEART
[1]**dowse** *variant of* DOUSE
[2]**dowse** \ˈdau̇z\ *verb* **dowsed; dows·ing** [origin unknown] (1894)
intransitive verb
: to use a divining rod
transitive verb
: to find (as water) by dowsing
dows·er \ˈdau̇-zər\ *noun* (1838)
: DIVINING ROD; *also* **:** a person who uses it
dowsing rod *noun* (1691)
: DIVINING ROD
dox·ol·o·gy \däk-ˈsä-lə-jē\ *noun, plural* **-gies** [Medieval Latin *doxologia,* from Late Greek, from Greek *doxa* opinion, glory (from *dokein* to seem, seem good) + *-logia* -logy — more at DECENT] (1649)
: a usually liturgical expression of praise to God
doxo·ru·bi·cin \ˈdäk-sə-ˈrü-bə-sən\ *noun* [*deoxy-* + *orubicin* (as in *daunorubicin*)] (circa 1977)
: an antibiotic with broad antitumor activity that is obtained from a bacterium (*Streptomyces peucetius*) and is administered as the hydrochloride $C_{27}H_{29}NO_{11}\cdot HCl$
doxy *also* **dox·ie** \ˈdäk-sē\ *noun, plural* **doxies** [perhaps modification of obsolete Dutch *docke* doll, from Middle Dutch] (circa 1530)
1 : FLOOZY, PROSTITUTE
2 : MISTRESS 4a
doxy·cy·cline \ˌdäk-sə-ˈsī-ˌklēn\ *noun* [*deoxy-* + tetra*cycline*] (1966)
: a broad-spectrum tetracycline antibiotic $C_{22}H_{24}N_2O_8$ with potent antibacterial activity that is often taken orally as a prophylactic against diarrhea by travelers
doy·en \ˈdȯi-ən, -ˌ(y)en; ˈdwä-ˌya^{n}(n)\ *noun* [French, from Late Latin *decanus* dean — more at DEAN] (1670)
1 a : the senior member of a body or group **b :** a person considered to be knowledgeable or uniquely skilled as a result of long experience in some field of endeavor
2 : the oldest example of a category
doy·enne \dȯi-ˈ(y)en, dwä-ˈyen\ *noun* [French, feminine of *doyen*] (circa 1897)
: a woman who is a doyen
doy·ley *chiefly British variant of* DOILY
[1]**doze** \ˈdōz\ *verb* **dozed; doz·ing** [probably of Scandinavian origin; akin to Old Norse *dūsa* to doze] (1693)
transitive verb
: to pass (as time) drowsily
intransitive verb
1 a : to sleep lightly **b :** to fall into a light sleep — usually used with *off*
2 : to be in a dull or stupefied condition
— doze *noun*
— doz·er *noun*
[2]**doze** *transitive verb* **dozed; doz·ing** [probably back formation from *dozer* (bulldozer)] (1945)
: BULLDOZE 2
— doz·er *noun*
doz·en \ˈdə-z^{ə}n\ *noun, plural* **dozens** *or* **dozen** [Middle English *dozeine,* from Middle French *dozaine,* from *doze* twelve, from Latin *duodecim,* from *duo* two + *decem* ten — more at TWO, TEN] (13th century)
1 : a group of 12
2 : an indefinitely large number ⟨*dozens* of times⟩
3 *plural but singular in construction* **:** a ritualized word game that consists of exchanging insults usually about the members of the opponent's family — used with *the*
— dozen *adjective*
— doz·enth \-z^{ə}n(t)th\ *adjective*
dozy \ˈdō-zē\ *adjective* **doz·i·er; -est** (1693)
: DROWSY, SLEEPY
— doz·i·ly \ˈdō-zə-lē\ *adverb*
— doz·i·ness *noun*
DP \ˌdē-ˈpē\ *noun, plural* **DP's** *or* **DPs** (circa 1944)
: a displaced person
DPN \ˌdē-ˌpē-ˈen\ *noun* [*d*iphospho*p*yridine *n*ucleotide] (1938)
: NAD
[1]**drab** \ˈdrab\ *noun* [origin unknown] (circa 1515)
1 : SLATTERN
2 : PROSTITUTE
[2]**drab** *intransitive verb* **drabbed; drab·bing** (1602)
: to associate with prostitutes
[3]**drab** *noun* [Middle French *drap* cloth, from Late Latin *drappus*] (1541)
1 : any of various cloths of a dull brown or gray color
2 a : a light olive brown **b :** a dull, lifeless, or faded appearance or quality
[4]**drab** *adjective* **drab·ber; drab·best** (1686)
1 a : of the dull brown color of drab **b :** of the color drab
2 : characterized by dullness and monotony **:** CHEERLESS
— drab·ly *adverb*
— drab·ness *noun*
[5]**drab** *noun* [probably alteration of *drib*] (1828)
: a small amount — usually used in the phrase *dribs and drabs*
drab·ble \ˈdra-bəl\ *verb* **drab·bled; drab·bling** \-b(ə-)liŋ\ [Middle English *drabelen;* akin to Low German *drabbelen*] (15th century)
transitive verb
: DRAGGLE
intransitive verb
: to become wet and muddy
dra·cae·na \drə-ˈsē-nə\ *noun* [New Latin, from Late Latin, she-serpent, from Greek *drakaina,* feminine of *drakōn* serpent — more at DRAGON] (circa 1823)
: any of two genera (*Dracaena* and *Cordyline*) of Old World tropical shrubs or trees of the agave family that have naked branches ending in tufts of sword-shaped leaves and include some used as houseplants
drachm \ˈdram\ *noun* [alteration of Middle English *dragme* — more at DRAM] (14th century)
1 : DRACHMA
2 *chiefly British* **:** DRAM
drach·ma \ˈdrak-mə\ *noun, plural* **drach·mas** *or* **drach·mai** \-ˌmī\ *or* **drach·mae** \-(ˌ)mē, -ˌmī\ [Latin, from Greek *drachmē* — more at DRAM] (1525)
1 a : any of various ancient Greek units of weight **b :** any of various modern units of weight; *especially* **:** DRAM 1
2 a : an ancient Greek silver coin equivalent to 6 obols **b** — see MONEY table
Dra·co \ˈdrā-(ˌ)kō\ *noun* [Latin (genitive *Draconis*), literally, dragon — more at DRAGON]
: a northern circumpolar constellation within which is the north pole of the ecliptic
dra·co·ni·an \drā-ˈkō-nē-ən, drə-\ *adjective, often capitalized* [Latin *Dracon-, Draco,* from Greek *Drakōn* Draco (Athenian lawgiver)] (1876)
1 : of, relating to, or characteristic of Draco or the severe code of laws held to have been framed by him

2 : CRUEL; *also* **:** SEVERE ⟨*draconian* littering fines⟩ ◇

¹dra·con·ic \drə-ˈkä-nik\ *adjective* [Latin *dracon-, draco*] (1680)
: of or relating to a dragon

²dra·con·ic \drā-ˈkä-nik, drə-\ *adjective* (1708)
: DRACONIAN

¹draft \ˈdraft, ˈdrȧft\ *noun* [Middle English *draght;* akin to Old English *dragan* to draw — more at DRAW] (13th century)
1 : the act of drawing a net; *also* **:** the quantity of fish taken at one drawing
2 a : the act or an instance of drinking or inhaling; *also* **:** the portion drunk or inhaled in one such act **b :** a portion poured out or mixed for drinking **:** DOSE
3 a : the force required to pull an implement **b :** load or load-pulling capacity
4 a : the act of moving loads by drawing or pulling **:** PULL **b :** a team of animals together with what they draw
5 a : DELINEATION, REPRESENTATION **b :** SCHEME, DESIGN **c :** a preliminary sketch, outline, or version ⟨the author's first *draft*⟩ ⟨a *draft* treaty⟩
6 : the act, result, or plan of drawing out or stretching
7 a : the act of drawing (as from a cask) **b :** a portion of liquid so drawn ⟨a *draft* of ale⟩ **c :** draft beer ⟨a glass of *draft*⟩
8 : the depth of water a ship draws especially when loaded
9 a (1) **:** a system for or act of selecting individuals from a group (as for compulsory military service) (2) **:** an act or process of selecting an individual (as for political candidacy) without his expressed consent **b :** a group of individuals selected especially by military draft **c :** a system whereby exclusive rights to select ed new players are apportioned among professional teams
10 a : an order for the payment of money drawn by one person or bank on another **b :** the act or an instance of drawing from or making demands upon something **:** DEMAND
11 a : a current of air in a closed-in space ⟨felt a *draft*⟩ **b :** a device for regulating the flow of air (as in a fireplace)
12 : ANGLE, TAPER; *specifically* **:** the taper given to a pattern or die so that the work can be easily withdrawn
13 : a pocket of reduced air pressure behind a moving object; *also* **:** the use of such a draft to save energy
— on draft : ready to be drawn from a receptacle ⟨beer *on draft*⟩

²draft *adjective* (15th century)
1 : used or adapted for drawing loads ⟨*draft* horses⟩
2 : being or having been on draft ⟨drinking *draft* beer⟩

³draft (1714)
transitive verb
1 : to select for some purpose: as **a :** to conscript for military service **b :** to select (a professional athlete) by draft
2 a : to draw the preliminary sketch, version, or plan of **b :** COMPOSE, PREPARE
3 : to draw off or away ⟨water *drafted* by pumps⟩
4 : to stay close behind (another racer) so as to take advantage of the reduced air pressure created by the leading racer
intransitive verb
1 : to practice draftsmanship
2 : to draft another racer (as in car or bike racing)
— draft·able \ˈdraf-tə-bəl, ˈdrȧf-\ *adjective*
— draft·ee \ˌdraf-ˈtē, ˌdrȧf-\ *noun*
— draft·er \ˈdraf-tər, ˈdrȧf-\ *noun*

draft board *noun* (1953)
: a civilian board that registers, classifies, and selects men for compulsory military service

drafts·man \ˈdraf(t)-smən, ˈdrȧf(t)-\ *noun* (1663)
1 : a person who draws plans and sketches (as of machinery or structures)
2 : a person who draws legal documents or other writings
3 : an artist who excels in drawing
— drafts·man·ship \-ˌship\ *noun*

drafts·person \ˈdraf(t)s-ˌpər-sᵊn\ *noun* (1975)
: DRAFTSMAN 1

drafty \ˈdraf-tē, ˈdrȧf-\ *adjective* **draft·i·er; -est** (1846)
: exposed to or abounding in drafts of air
— draft·i·ly \-tə-lē\ *adverb*
— draft·i·ness \-tē-nəs\ *noun*

¹drag \ˈdrag\ *noun* (14th century)
1 : something used to drag with; *especially* **:** a device for dragging under water to detect or obtain objects
2 : something that is dragged, pulled, or drawn along or over a surface: as **a :** HARROW **b :** a sledge for conveying heavy bodies **c :** CONVEYANCE
3 a : the act or an instance of dragging or drawing: as (1) **:** a drawing along or over a surface with effort or pressure (2) **:** motion effected with slowness or difficulty; *also* **:** the condition of having or seeming to have such motion (3) **:** a draw on a pipe, cigarette, or cigar; *also* **:** a draft of liquid **b :** a movement, inclination, or retardation caused by or as if by dragging **c** *slang* **:** influence securing special favor **:** PULL
4 a : something that retards motion or action **b** (1) **:** the retarding force acting on a body (as an airplane) moving through a fluid (as air) parallel and opposite to the direction of motion (2) **:** friction between engine parts; *also* **:** retardation due to friction **c :** BURDEN, ENCUMBRANCE ⟨the *drag* of population growth on living standards⟩ ⟨maturity is a *drag*⟩ **d :** one that is boring or gets in the way of enjoyment ⟨this sickly kid is going to be a social *drag* —Edmund Morris⟩
5 a : an object drawn over the ground to leave a scented trail **b :** a clog fastened to a trap to prevent the escape of a trapped animal
6 : STREET, ROAD ⟨the main *drag*⟩
7 a : COSTUME, OUTFIT ⟨in Victorian *drag*⟩ **b :** clothing typical of one sex worn by a person of the opposite sex — often used in the phrase *in drag*
8 : DRAG RACE

²drag *verb* **dragged; drag·ging** [Middle English *draggen,* from Old Norse *draga* — more at DRAW] (15th century)
transitive verb
1 a (1) **:** to draw slowly or heavily **:** HAUL (2) **:** to cause to move with painful or undue slowness or difficulty ⟨*dragging* the national anthem⟩ (3) **:** to cause to trail along a surface ⟨wandered off *dragging* the leash⟩ **b** (1) **:** to bring by or as if by force or compulsion ⟨had to *drag* her husband to the opera⟩ (2) **:** to extract by or as if by pulling **c :** PROTRACT ⟨*drag* a story out⟩
2 a : to explore with a drag **b :** to catch with a dragnet
3 : to hit (a drag bunt) while moving toward first base
intransitive verb
1 : to hang or lag behind
2 : to fish or search with a drag
3 : to trail along on the ground
4 : to move or proceed laboriously or tediously ⟨the lawsuit *dragged* on for years⟩ ⟨was *dragging* after the long trip⟩
5 : DRAW 4a ⟨*drag* on a cigarette⟩
6 : to make a plucking or pulling movement
7 : to participate in a drag race
— drag·ging·ly \ˈdra-giŋ-lē\ *adverb*
— drag one's feet *also* **drag one's heels :** to act in a deliberately slow or dilatory manner

³drag *adjective* (1887)
: of, being, involving, or intended for a person in drag ⟨a *drag* ball⟩

drag bunt *noun* (circa 1949)
: a bunt in baseball made by a left-handed batter by trailing the bat while moving toward first base; *broadly* **:** a bunt made with the object of getting on base safely rather than sacrificing

drag coefficient *noun* (circa 1942)
: a factor representing the drag acting on a body (as an automobile or airfoil)

dra·gée \dra-ˈzhā\ *noun* [French, from Middle French *dragie* — more at DREDGE] (1853)
1 : a sugar-coated nut
2 : a small silver-colored ball for decorating cakes

drag·ger \ˈdra-gər\ *noun* (circa 1500)
: one that drags; *specifically* **:** a fishing boat operating a trawl or dragnet

drag·gle \ˈdra-gəl\ *verb* **drag·gled; drag·gling** \-g(ə-)liŋ\ [frequentative of *drag*] (1513)
transitive verb
: to make wet and dirty by dragging
intransitive verb
1 : to trail on the ground
2 : STRAGGLE

drag·gle–tail \ˈdra-gəl-ˌtāl\ *noun* (1596)
: SLATTERN

drag·gy \ˈdra-gē\ *adjective* **drag·gi·er; -est** (15th century)
: SLUGGISH, DULL

drag·line \ˈdrag-ˌlīn\ *noun* (circa 1911)
1 : a line used in or for dragging
2 : an excavating machine in which the bucket is attached by cables and operates by being drawn toward the machine

drag·net \ˈdrag-ˌnet\ *noun* (circa 1541)
1 a : a net drawn along the bottom of a body of water **b :** a net used on the ground (as to capture small game)
2 : a network of measures for apprehension (as of criminals)

drag·o·man \ˈdra-gə-mən\ *noun, plural* **-mans** *or* **-men** \-mən\ [Middle English *drogman,* from Middle French, from Old Italian *dragomanno,* from Middle Greek *dragomanos,* from Arabic *tarjumān,* from Aramaic *tūrgĕmānā*] (14th century)
: an interpreter chiefly of Arabic, Turkish, or Persian employed especially in the Near East

drag·on \ˈdra-gən\ *noun* [Middle English, from Old French, from Latin *dracon-, draco* serpent, dragon, from Greek *drakōn* serpent; akin to Old English *torht* bright, Greek *derkesthai* to see, look at] (13th century)
1 *archaic* **:** a huge serpent
2 : a mythical animal usually represented as a monstrous winged and scaly serpent or saurian with a crested head and enormous claws
3 : a violent, combative, or very strict person
4 *capitalized* **:** DRACO
5 : something or someone formidable or baneful
— drag·on·ish \-gə-nish\ *adjective*

drag·on·et \ˌdra-gə-ˈnet, ˈdra-gə-nət\ *noun* (14th century)
1 : a little dragon

◇ WORD HISTORY
draconian According to ancient Athenian tradition, Draco was a lawgiver of the 7th century B.C., associated with the first commitment of Greek law to writing. Draco's code was famed for its harshness, as most criminal offenses required the death penalty. According to the 4th century orator Demades, the statutes were literally written in blood. Not many decades after Draco supposedly lived, the 6th century statesman Solon abolished all of Draco's punishments except those dealing with homicide. Modern scholars are unsure if Draco or his code ever actually existed, but in any event *draconian* remains in English as a synonym for "severe" in reference to laws or other measures taken by persons in a position of power.

2 : any of various small often brightly colored scaleless marine fishes constituting a family (Callionymidae); *especially* **:** a European fish (*Callionymus lyra*) sometimes used as food

drag·on·fly \ˈdra-gən-ˌflī\ *noun* (1626)
: any of a suborder (Anisoptera) of odonate insects that are larger and stouter than damselflies, hold the wings horizontal in repose, and have rectal gills during the naiad stage; *broadly* **:** ODONATE

dragonfly

drag·on·head \-ˌhed\ *noun* (1784)
: any of several mints (genus *Dracocephalum*) often grown for their showy flower heads; *especially* **:** a North American plant (*D. parviflorum*)

dragon lady *noun* [character in the comic strip "Terry and the Pirates" by Milton Caniff] (1973)
: an overbearing or tyrannical woman; *also* **:** a glamorous often mysterious woman

dragon's blood *noun* (1599)
: any of several resinous mostly dark-red plant products; *specifically* **:** a resin from the fruit of a palm (genus *Daemonorops*) used for coloring varnish and in photoengraving

dragon's teeth *noun plural* [from the dragon's teeth sown by Cadmus which sprang up as armed warriors who killed one another off] (1853)
1 : seeds of strife
2 : wedge-shaped concrete antitank barriers laid in multiple rows

[1]**dra·goon** \drə-ˈgün, dra-\ *noun* [French *dragon* dragon, dragoon, from Middle French] (1622)
: a member of a European military unit formerly composed of heavily armed mounted troops

[2]**dragoon** *transitive verb* (1689)
1 : to subjugate or persecute by harsh use of troops
2 : to force or attempt to force into submission by violent measures **:** COERCE

drag queen *noun* (circa 1941)
: a male homosexual who dresses as a woman

drag race *noun* (1949)
: an acceleration contest between vehicles (as automobiles)
— drag racing *noun*

drag·ster \ˈdrag-stər\ *noun* (circa 1954)
1 : a vehicle built or modified for use in a drag race
2 : one who participates in a drag race

drag strip *noun* (1952)
: the site of a drag race; *specifically* **:** a strip of pavement with a racing area at least ¼ mile long

[1]**drain** \ˈdrān\ *verb* [Middle English *draynen,* from Old English *drēahnian* — more at DRY] (before 12th century)
transitive verb
1 *obsolete* **:** FILTER
2 a : to draw off (liquid) gradually or completely ⟨*drained* all the water out⟩ **b :** to cause the gradual disappearance of **c :** to exhaust physically or emotionally
3 a : to make gradually dry ⟨*drain* a swamp⟩ **b :** to carry away the surface water of ⟨the river that *drains* the valley⟩ **c :** to deplete or empty by or as if by drawing off by degrees or in increments **d :** to empty by drinking the contents of
4 : DROP 7c, SINK ⟨*drained* the putt⟩
intransitive verb
1 a : to flow off gradually **b :** to disappear gradually **:** DWINDLE
2 : to become emptied or freed of liquid by its flowing or dropping
3 : to discharge surface or surplus water
synonym see DEPLETE
— drain·er *noun*

[2]**drain** *noun* (1552)
1 : a means (as a pipe) by which usually liquid matter is drained
2 a : the act of draining **b :** a gradual outflow or withdrawal **:** DEPLETION
3 : something that causes depletion **:** BURDEN
4 : an electrode in a field-effect transistor toward which charge carriers move — compare GATE, SOURCE
— down the drain : to a state of being wasted or irretrievably lost

drain·age \ˈdrā-nij\ *noun* (1652)
1 : the act, process, or mode of draining; *also* **:** something drained off
2 : a device for draining **:** DRAIN; *also* **:** a system of drains
3 : an area or district drained

drain·pipe \ˈdrān-ˌpīp\ *noun* (1857)
: a pipe for drainage

Draize test \ˈdrāz-\ *noun* [John H. *Draize* (born 1900) American pharmacologist] (1980)
: a test for harmfulness of chemicals to the human eye that involves dropping the test substance into one eye of a rabbit without anesthesia using the other eye as a control — called also *Draize eye test*

drake \ˈdrāk\ *noun* [Middle English; akin to Old High German an*trahho* drake] (14th century)
: a male duck

dram \ˈdram\ *noun* [Middle English *dragme,* from Middle French & Late Latin; Middle French, dram, drachma, from Late Latin *dragma,* from Latin *drachma,* from Greek *drachmē,* literally, handful, from *drassesthai* to grasp] (14th century)
1 a — see WEIGHT table **b :** FLUID DRAM
2 a : a small portion of something to drink **b :** a small amount

DRAM \ˈdram *also* ˈdē-ˌram\ *noun* [*d*ynamic + *RAM* (*r*andom-*a*ccess *m*emory)] (1980)
: a computer memory chip that must be continuously supplied with power in order to retain data

dra·ma \ˈdrä-mə, ˈdra-\ *noun* [Late Latin *dramat-, drama,* from Greek, deed, drama, from *dran* to do, act] (1515)
1 : a composition in verse or prose intended to portray life or character or to tell a story usually involving conflicts and emotions through action and dialogue and typically designed for theatrical performance **:** PLAY — compare CLOSET DRAMA
2 : dramatic art, literature, or affairs
3 a : a state, situation, or series of events involving interesting or intense conflict of forces **b :** dramatic state, effect, or quality ⟨the *drama* of the courtroom proceedings⟩

Dram·a·mine \ˈdra-mə-ˌmēn\ *trademark*
— used for dimenhydrinate

dra·mat·ic \drə-ˈma-tik\ *adjective* (1589)
1 : of or relating to the drama
2 a : suitable to or characteristic of the drama **b :** striking in appearance or effect
3 *of an opera singer* **:** having a powerful voice and a declamatory style — compare LYRIC ☆
— dra·mat·i·cal·ly \-ti-k(ə-)lē\ *adverb*

dramatic irony *noun* (circa 1907)
: IRONY 3b

dramatic monologue *noun* (circa 1935)
: a literary work in which a character reveals himself in a monologue usually addressed to a second person

dra·mat·ics \drə-ˈma-tiks\ *noun plural but singular or plural in construction* (1796)
1 : the study or practice of theatrical arts (as acting and stagecraft)
2 : dramatic behavior or expression

dramatic unities *noun plural* (circa 1922)
: the unities of time, place, and action that are observed in classical drama

dramatisation, dramatise *British variant of* DRAMATIZATION, DRAMATIZE

dra·ma·tis per·so·nae \ˌdra-mə-təs-pər-ˈsō-(ˌ)nē, ˌdrä-, -ˌnī\ *noun plural* [New Latin] (1730)
1 : the characters or actors in a drama
2 *singular in construction* **:** a list of the characters or actors in a drama
3 : the participants in an event

dra·ma·tist \ˈdra-mə-tist, ˈdrä-\ *noun* (1678)
: PLAYWRIGHT

dra·ma·ti·za·tion \ˌdra-mə-tə-ˈzā-shən, ˌdrä-\ *noun* (1796)
1 : the action of dramatizing
2 : a dramatized version (as of a novel)

dra·ma·tize \ˈdra-mə-ˌtīz, ˈdrä-\ *verb* **-tized; -tiz·ing** (1783)
transitive verb
1 : to adapt (as a novel) for theatrical presentation
2 : to present or represent in a dramatic manner
intransitive verb
1 : to be suitable for dramatization
2 : to behave dramatically
— dra·ma·tiz·able \-ˌtī-zə-bəl\ *adjective*

dra·ma·turge *or* **dra·ma·turg** \ˈdra-mə-ˌtərj, ˈdrä-\ *noun* (1870)
: a specialist in dramaturgy

dra·ma·tur·gy \ˈdra-mə-ˌtər-jē, ˈdrä-\ *noun* [German *Dramaturgie,* from Greek *dramatourgia* dramatic composition, from *dramat-, drama* + *-ourgia* -urgy] (1801)
: the art or technique of dramatic composition and theatrical representation
— dra·ma·tur·gic \ˌdra-mə-ˈtər-jik, ˌdrä-\ *or* **dra·ma·tur·gi·cal** \-ji-kəl\ *adjective*
— dra·ma·tur·gi·cal·ly \-ji-k(ə-)lē\ *adverb*

dra·me·dy \ˈdrä-mə-dē, ˈdra-\ *noun* [blend of *drama* and *comedy*] (1978)
: a situation comedy having dramatic scenes

dram·mock \ˈdra-mək\ *noun* [Scottish Gaelic *dramag*] (1562)
chiefly Scottish **:** raw oatmeal mixed with cold water

dram·shop \ˈdram-ˌshäp\ *noun* (1725)
: BARROOM

drank *past and past participle of* DRINK

[1]**drape** \ˈdrāp\ *verb* **draped; drap·ing** [probably back-formation from *drapery*] (1847)
transitive verb
1 : to cover or adorn with or as if with folds of cloth
2 : to cause to hang or stretch out loosely or carelessly
3 : to arrange in flowing lines or folds
intransitive verb
: to become arranged in folds ⟨this silk *drapes* beautifully⟩

☆ SYNONYMS
Dramatic, theatrical, histrionic, melodramatic mean having a character or an effect like that of acted plays. DRAMATIC applies to situations in life and literature that stir the imagination and emotions deeply ⟨a *dramatic* meeting of world leaders⟩. THEATRICAL implies a crude appeal through artificiality or exaggeration in gesture or vocal expression ⟨a *theatrical* oration⟩. HISTRIONIC applies to tones, gestures, and motions and suggests a deliberate affectation or staginess ⟨a *histrionic* show of grief⟩. MELODRAMATIC suggests an exaggerated emotionalism or an inappropriate theatricalism ⟨made a *melodramatic* plea⟩.

— **drap·abil·i·ty** *also* **drape·abil·i·ty** \ˌdrā-pə-ˈbi-lə-tē\ *noun*
— **drap·able** *also* **drape·able** \ˈdrā-pə-bəl\ *adjective*

²drape *noun* (1889)
1 : arrangement in or of folds
2 a : a drapery especially for a window **:** CURTAIN **b :** a sterile covering used in an operating room — usually used in plural
3 : the cut or hang of clothing
— **drap·ey** \ˈdrā-pē\ *adjective*

drap·er \ˈdrā-pər\ *noun* [Middle English, weaver, clothier, from Middle French *drapier,* from Old French, from *drap* cloth — more at DRAB] (14th century)
chiefly British **:** a dealer in cloth and sometimes also in clothing and dry goods

drap·ery \ˈdrā-p(ə-)rē\ *noun, plural* **-er·ies** (14th century)
1 *British* **:** DRY GOODS
2 a : a decorative piece of material usually hung in loose folds and arranged in a graceful design **b :** hangings of heavy fabric for use as a curtain
3 : the draping or arranging of materials

dras·tic \ˈdras-tik\ *adjective* [Greek *drastikos,* from *dran* to do] (circa 1691)
1 : acting rapidly or violently ⟨a *drastic* purgative⟩
2 : extreme in effect or action **:** SEVERE ⟨*drastic* measures⟩
— **dras·ti·cal·ly** \-ti-k(ə-)lē\ *adverb*

drat \ˈdrat\ *verb* **drat·ted; drat·ting** [probably euphemistic alteration of *God rot*] (1815)
: DAMN — used as a mild oath

draught \ˈdraft\, **draughty** \ˈdraf-tē\ *chiefly British variant of* DRAFT, DRAFTY

draughts \ˈdraf(t)s\ *noun plural but singular or plural in construction* [Middle English *draghtes,* from plural of *draght* draft, move in chess] (15th century)
British **:** CHECKERS

draughts·man *chiefly British variant of* DRAFTSMAN

Dra·vid·i·an \drə-ˈvi-dē-ən\ *noun* [Sanskrit *Drāviḍa*] (1856)
1 : a member of an ancient dark-skinned people of southern India
2 : DRAVIDIAN LANGUAGES
— **Dravidian** *adjective*

Dravidian languages *noun plural* (1871)
: a language family of India, Sri Lanka, and Pakistan that includes Tamil, Telugu, Kannada, and Malayalam

¹draw \ˈdrȯ\ *verb* **drew** \ˈdrü\; **drawn** \ˈdrȯn, ˈdrän\; **draw·ing** [Middle English *drawen, dragen,* from Old English *dragan;* akin to Old Norse *draga* to draw, drag] (before 12th century)
transitive verb
1 : to cause to move continuously toward or after a force applied in advance **:** PULL ⟨*draw* your chair up by the fire⟩: as **a :** to move (as a covering) over or to one side ⟨*draw* the drapes⟩ **b :** to pull up or out of a receptacle or place where seated or carried ⟨*draw* water from the well⟩ ⟨*drew* a gun⟩; *also* **:** to cause to come out of a container ⟨*draw* water for a bath⟩
2 : to cause to go in a certain direction (as by leading) ⟨*drew* him aside⟩
3 a : to bring by inducement or allure **:** ATTRACT ⟨honey *draws* flies⟩ **b :** to bring in or gather from a specified group or area ⟨a college that *draws* its students from many states⟩ **c :** BRING ON, PROVOKE ⟨*drew* enemy fire⟩ **d :** to bring out by way of response **:** ELICIT ⟨*drew* cheers from the audience⟩ **e :** to receive in the course of play ⟨the batter *drew* a walk⟩ ⟨*draw* a foul⟩
4 : INHALE ⟨*drew* a deep breath⟩
5 a : to extract the essence from ⟨*draw* tea⟩ **b :** EVISCERATE ⟨plucking and *drawing* a goose before cooking⟩ **c :** to derive to one's benefit ⟨*drew* inspiration from the old masters⟩
6 : to require (a specified depth) to float in ⟨a ship that *draws* 12 feet of water⟩
7 a : ACCUMULATE, GAIN ⟨*drawing* interest⟩ **b :** to take (money) from a place of deposit **c :** to use in making a cash demand ⟨*drawing* a check against his account⟩ **d :** to receive regularly or in due course ⟨*draw* a salary⟩
8 a : to take (cards) from a stack or from the dealer **b :** to receive or take at random ⟨*drew* a winning number⟩
9 : to bend (a bow) by pulling back the string
10 : to cause to shrink, contract, or tighten
11 : to strike (a ball) so as to impart a backward spin
12 : to leave (a contest) undecided **:** TIE
13 a (1) **:** to produce a likeness or representation of by making lines on a surface ⟨*draw* a picture⟩ ⟨*draw* a graph with chalk⟩ (2) **:** to give a portrayal of **:** DELINEATE ⟨a writer who *draws* characters well⟩ **b :** to write out in due form ⟨*draw* a will⟩ **c :** to design or describe in detail **:** FORMULATE ⟨*draw* comparisons⟩
14 : to infer from evidence or premises ⟨*draw* a conclusion⟩
15 : to spread or elongate (metal) by hammering or by pulling through dies; *also* **:** to shape (as plastic) by stretching or by pulling through dies
intransitive verb
1 : to come or go steadily or gradually ⟨night *draws* near⟩
2 a : to move something by pulling ⟨*drawing* at the well⟩ **b :** to exert an attractive force ⟨the play is *drawing* well⟩
3 a : to pull back a bowstring **b :** to bring out a weapon ⟨*drew,* aimed, and fired⟩
4 a : to produce a draft ⟨the chimney *draws* well⟩ ⟨*draw* on a cigar⟩ **b :** to swell out in a wind ⟨all sails *drawing*⟩
5 a : to wrinkle or tighten up **:** SHRINK **b :** to change shape by pulling or stretching
6 : to cause blood or pus to localize at one point
7 : to create a likeness or a picture in outlines **:** SKETCH
8 : to come out even in a contest
9 a : to make a written demand for payment of money on deposit **b :** to obtain resources (as of information) ⟨*drawing* from a common fund of knowledge⟩
— **draw·able** \-ə-bəl\ *adjective*
— **draw a bead on :** to take aim at
— **draw a blank :** to fail to gain a desired object (as information sought); *also* **:** to be unable to think of something
— **draw on** *or* **draw upon :** to use as a source of supply ⟨*drawing on* the whole community for support⟩
— **draw straws :** to decide or assign something by lottery in which straws of unequal length are used
— **draw the line** *or* **draw a line 1 :** to fix an arbitrary boundary between things that tend to intermingle **2 :** to fix a boundary excluding what one will not tolerate or engage in

²draw *noun* (1663)
1 : the act or process of drawing: as **a :** a sucking pull on something held with the lips **b :** a removal of a handgun from its holster ⟨quick on the *draw*⟩ **c :** backward spin given to a ball by striking it below center — compare FOLLOW
2 : something that is drawn: as **a :** a card drawn to replace a discard in poker **b :** a lot or chance drawn at random **c :** the movable part of a drawbridge
3 : a contest left undecided or deadlocked **:** TIE
4 : one that draws attention or patronage
5 a : the distance from the string to the back of a drawn bow **b :** the force required to draw a bow fully
6 : a gully shallower than a ravine
7 : the deal in draw poker to improve the players' hands after discarding
8 : a football play that simulates a pass play so a runner can go straight up the middle past the pass rushers

draw away *intransitive verb* (1670)
: to move ahead (as of an opponent in a race)

draw·back \ˈdrȯ-ˌbak\ *noun* (1697)
1 : a refund of duties especially on an imported product subsequently exported or used to produce a product for export
2 : an objectionable feature **:** DISADVANTAGE

draw back \drȯ-ˈbak\ *intransitive verb* (14th century)
: to avoid an issue or commitment

draw·bar \ˈdrȯ-ˌbär\ *noun* (1839)
1 : a railroad coupler
2 : a beam across the rear of a vehicle (as a tractor) to which implements are hitched

draw·bridge \-ˌbrij\ *noun* (14th century)
: a bridge made to be raised up, let down, or drawn aside so as to permit or hinder passage

drawbridge

draw·down \-ˌdau̇n\ *noun* (1918)
1 : a lowering of a water level (as in a reservoir)
2 a : the process of depleting **b :** REDUCTION

draw down \ˌdrȯ-ˈdau̇n\ *transitive verb* (1949)
: to deplete by using or spending

draw·ee \drȯ-ˈē\ *noun* (1766)
: the party on which an order or bill of exchange is drawn

draw·er \ˈdrȯ(-ə)r\ *noun* (14th century)
1 : one that draws: as **a :** a person who draws liquor **b :** DRAFTSMAN **c :** one that draws a bill of exchange or order for payment or makes a promissory note
2 : a sliding box or receptacle opened by pulling out and closed by pushing in
3 *plural* **:** an article of clothing (as underwear) for the lower body
— **drawer·ful** \-ˌfu̇l\ *noun*

draw in (1558)
transitive verb
1 : to cause or entice to enter or participate
2 : to sketch roughly
intransitive verb
1 a : to draw to an end ⟨the day *drew in*⟩ **b :** to shorten seasonally ⟨the evenings are already *drawing in*⟩
2 : to become more cautious or economical

draw·ing \ˈdrȯ(-)iŋ\ *noun* (14th century)
1 : an act or instance of drawing; *especially* **:** the process of deciding something by drawing lots
2 : the art or technique of representing an object or outlining a figure, plan, or sketch by means of lines
3 : something drawn or subject to drawing: as **a :** an amount drawn from a fund **b :** a representation formed by drawing **:** SKETCH

drawing account *noun* (1920)
: an account showing payments made to an employee in advance of actual earnings or for traveling expenses

drawing board *noun* (1725)
1 : a board used as a base for drafting on paper
2 : a planning stage ⟨a project still on the *drawing board*⟩

drawing card *noun* (1887)
: one that attracts attention or patronage

drawing pin *noun* (1859)
British **:** THUMBTACK

drawing room *noun* [short for *withdrawing room*] (1642)
1 a : a formal reception room **b :** a private room on a railroad passenger car with three berths and an enclosed toilet
2 : a formal reception

drawing table *noun* (1706)
: a table with a surface adjustable for elevation and angle of incline

draw•knife \ˈdrȯ-ˌnīf\ *noun* (1703)
: a woodworker's tool consisting of a blade with a handle at each end for use in shaving off surfaces

1**drawl** \ˈdrȯl\ *verb* [probably frequentative of *draw*] (1598)
intransitive verb
: to speak slowly with vowels greatly prolonged
transitive verb
: to utter in a slow lengthened tone
— drawl•er *noun*
— drawl•ing•ly \ˈdrȯ-liŋ-lē\ *adverb*

2**drawl** *noun* (1760)
: a drawling manner of speaking
— drawly \ˈdrȯ-lē\ *adjective*

1**drawn** *past participle of* DRAW

2**drawn** \ˈdrȯn, ˈdrän\ *adjective* (1613)
: showing the effects of tension, pain, or illness **:** HAGGARD

drawn butter *noun* (1826)
: melted clarified butter

drawn•work \ˈdrȯn-ˌwərk\ *noun* (1595)
: decoration on cloth made by drawing out threads according to a pattern

draw off (13th century)
transitive verb
: REMOVE, WITHDRAW
intransitive verb
: to move apart or ahead

draw on (15th century)
intransitive verb
: APPROACH ⟨night *draws on*⟩
transitive verb
: BRING ON, CAUSE

draw out *transitive verb* (14th century)
1 : REMOVE, EXTRACT
2 : to extend beyond a minimum in time **:** PROTRACT 2
3 : to cause to speak freely ⟨a reporter's ability to *draw* a person *out*⟩

draw•plate \ˈdrȯ-ˌplāt\ *noun* (1832)
: a die with holes through which wires are drawn

draw play *noun* (1952)
: DRAW 8

draw poker *noun* (1849)
: poker in which each player is dealt five cards face down and after betting may get replacements for discards

draw•shave \ˈdrȯ-ˌshāv\ *noun* (1828)
: DRAWKNIFE

draw shot *noun* (1897)
: a shot in billiards or pool made by hitting the cue ball with draw so it moves back after striking the object ball

draw•string \ˈdrȯ-ˌstriŋ\ *noun* (1845)
: a string, cord, or tape inserted into hems or casings or laced through eyelets for use in closing a bag or controlling fullness in garments or curtains

draw•tube \-ˌtüb, -ˌtyüb\ *noun* (circa 1891)
: a telescoping tube (as for the eyepiece of a microscope)

draw up (1605)
transitive verb
1 : to bring (as troops) into array
2 : to prepare a draft or version of ⟨*draw up* plans⟩
3 : to bring to a halt
4 : to straighten (oneself) to an erect posture especially as an assertion of dignity or resentment
intransitive verb
: to come to a halt

1**dray** \ˈdrā\ *noun* [Middle English *draye,* a wheelless vehicle; akin to Old English *dræge* dragnet, *dragan* to pull — more at DRAW] (14th century)
: a vehicle used to haul goods; *especially* **:** a strong cart or wagon without sides

2**dray** *transitive verb* (1857)
: to haul on a dray **:** CART

dray•age \ˈdrā-ij\ *noun* (1791)
: the work or cost of hauling by dray

dray horse *noun* (1709)
: a horse adapted for drawing heavy loads

dray•man \ˈdrā-mən\ *noun* (1581)
: one whose work is hauling by dray

1**dread** \ˈdred\ *verb* [Middle English *dreden,* from Old English *drǣdan*] (before 12th century)
transitive verb
1 a : to fear greatly **b** *archaic* **:** to regard with awe
2 : to feel extreme reluctance to meet or face
intransitive verb
: to be apprehensive or fearful

2**dread** *noun* (13th century)
1 a : great fear especially in the face of impending evil **b :** extreme uneasiness in the face of a disagreeable prospect ⟨*dread* of a social blunder⟩ **c** *archaic* **:** AWE
2 : one causing fear or awe
synonym see FEAR

3**dread** *adjective* (15th century)
1 : causing great fear or anxiety
2 : inspiring awe

1**dread•ful** \ˈdred-fəl\ *adjective* (13th century)
1 a : inspiring dread **:** causing great and oppressive fear **b :** inspiring awe or reverence
2 : extremely bad, distasteful, unpleasant, or shocking
3 : EXTREME ⟨*dreadful* disorder⟩
— dread•ful•ly \-f(ə-)lē\ *adverb*
— dread•ful•ness \-fəl-nəs\ *noun*

2**dreadful** *noun* (1873)
: a cheap and sensational story or periodical

dread•lock \ˈdred-ˌläk\ *noun* (1960)
1 : a narrow ropelike strand of hair formed by matting or braiding
2 *plural* **:** a hairstyle consisting of dreadlocks

dread•nought \ˈdred-ˌnȯt, -ˌnät\ *noun* (1806)
1 : a warm garment of thick cloth; *also* **:** the cloth
2 [*Dreadnought,* British battleship] **:** BATTLESHIP
3 : one that is among the largest or most powerful of its kind

1**dream** \ˈdrēm\ *noun, often attributive* [Middle English *dreem,* from Old English *drēam* noise, joy, and Old Norse *draumr* dream; akin to Old High German *troum* dream] (13th century)
1 : a series of thoughts, images, or emotions occurring during sleep — compare REM SLEEP
2 : an experience of waking life having the characteristics of a dream: as **a :** a visionary creation of the imagination **:** DAYDREAM **b :** a state of mind marked by abstraction or release from reality **:** REVERIE **c :** an object seen in a dreamlike state **:** VISION
3 : something notable for its beauty, excellence, or enjoyable quality ⟨the new car is a *dream* to operate⟩
4 a : a strongly desired goal or purpose ⟨a *dream* of becoming president⟩ **b :** something that fully satisfies a wish **:** IDEAL ⟨a meal that was a gourmet's *dream*⟩ ◆
— dream•ful \-fəl\ *adjective*
— dream•ful•ly \-fə-lē\ *adverb*
— dream•ful•ness *noun*
— dream•less *adjective*
— dream•less•ly *adverb*
— dream•less•ness *noun*
— dream•like \ˈdrēm-ˌlīk\ *adjective*

2**dream** *verb* **dreamed** \ˈdrem(p)t, ˈdrēmd\ *or* **dreamt** \ˈdrem(p)t\; **dream•ing** \ˈdrē-miŋ\ (13th century)
intransitive verb
1 : to have a dream
2 : to indulge in daydreams or fantasies ⟨*dreaming* of a better future⟩
3 : to appear tranquil or dreamy ⟨houses *dream* in leafy shadows —Gladys Taber⟩
transitive verb
1 : to have a dream of
2 : to consider as a possibility **:** IMAGINE
3 : to pass (time) in reverie or inaction ⟨*dreaming* the hours away⟩
— dream of : to consider possible or fitting ⟨wouldn't *dream of* disturbing you⟩

dream•er \ˈdrē-mər\ *noun* (14th century)
1 : one that dreams
2 a : one who lives in a world of fancy and imagination **b :** one who has ideas or conceives projects regarded as impractical **:** VISIONARY

dream•land \ˈdrēm-ˌland\ *noun* (circa 1834)
: an unreal delightful country existing only in imagination or in dreams **:** NEVER-NEVER LAND

dream•time \-ˌtīm\ *noun, often capitalized* (1896)
: the time of creation in the mythology of the Australian aborigines

dream up *transitive verb* (1941)
: to form in the mind **:** DEVISE, CONCOCT

dream vision *noun* (1906)
: a usually medieval poem having a framework in which the poet pictures himself as falling asleep and envisioning in his dream a series of allegorical people and events

dream•world \ˈdrēm-ˌwərld\ *noun* (1817)
: a world of illusion or fantasy

dreamy \ˈdrē-mē\ *adjective* **dream•i•er; -est** (1567)
1 a : full of dreams ⟨a *dreamy* night's sleep⟩ **b :** pleasantly abstracted from immediate reality
2 : given to dreaming or fantasy ⟨a *dreamy* child⟩
3 a : suggestive of a dream in vague or visionary quality **b :** quiet and soothing **c :** DELIGHTFUL, IDEAL ⟨he's so handsome . . . real *dreamy* —Greg Foley⟩
— dream•i•ly \-mə-lē\ *adverb*
— dream•i•ness \-mē-nəs\ *noun*

drear \ˈdrir\ *adjective* (1629)
: DREARY
— drear *noun*

drea•ry \ˈdrir-ē\ *adjective* **drea•ri•er; -est** [Middle English *drery,* from Old English *drēorig* sad, bloody, from *drēor* gore; akin to Old High German *trūrēn* to be sad, Gothic *driusan* to fall] (before 12th century)
1 : feeling, displaying, or reflecting listlessness or discouragement
2 : having nothing likely to provide cheer, comfort, or interest **:** GLOOMY, DISMAL
— drea•ri•ly \ˈdrir-ə-lē\ *adverb*
— drea•ri•ness \ˈdrir-ē-nəs\ *noun*

dreck *also* **drek** \ˈdrek\ *noun* [Yiddish *drek* & German *Dreck,* from Middle High German *drec;* akin to Old English *threax* rubbish] (1922)
: TRASH, RUBBISH

◇ WORD HISTORY

dream Not until the 13th century did *dream,* in the Middle English forms *drem* and *dreem,* appear in reference to images seen during sleep; however, the word itself is considerably older. In Old English *drēam* meant "joy, ecstasy" or "melody, music." Yet the shift in sense between Old and Middle English was not simply the result of an adaptation and specialization of the earlier senses. Rather it would appear that after the Scandinavian conquests and settlements in Britain of the 9th and 10th centuries, which had a major impact on the history of English, the Old Norse word *draumr,* referring to a dream during sleep, influenced the meaning of the similar and almost certainly related English word. By the end of the 14th century, the Old English meanings of the word had been entirely replaced.

¹dredge \'drej\ *verb* **dredged; dredg·ing** (1508)
transitive verb
1 a : to dig, gather, or pull out with or as if with a dredge — often used with *up* **b :** to deepen (as a waterway) with a dredging machine
2 : to bring to light by deep searching — often used with *up* ⟨*dredging* up memories⟩
intransitive verb
1 : to use a dredge
2 : to search deeply
— dredg·er *noun*

²dredge *noun* [perhaps from (assumed) Old English *drecge;* akin to Old English *dræge* dragnet, *dragan* to draw] (1602)
1 : an apparatus usually in the form of an oblong iron frame with an attached bag net used especially for gathering fish and shellfish
2 : a machine for removing earth usually by buckets on an endless chain or a suction tube
3 : a barge used in dredging

³dredge *transitive verb* **dredged; dredg·ing** [obsolete *dredge,* noun, sweetmeat, from Middle English *drage, drege,* from Middle French *dragie,* modification of Latin *tragemata* sweetmeats, from Greek *tragēmata,* plural of *tragēma* sweetmeat, from *trōgein* to gnaw] (1596)
: to coat (food) by sprinkling (as with flour)
— dredg·er *noun*

dree \'drē\ *transitive verb* **dreed; dree·ing** [Middle English, from Old English *drēogan;* akin to Gothic *driugan* to perform military service] (before 12th century)
chiefly Scottish **:** ENDURE, SUFFER

dreg \'dreg\ *noun* [Middle English, from Old Norse *dregg;* perhaps akin to Latin *fraces* dregs of oil] (14th century)
1 : sediment contained in a liquid or precipitated from it **:** LEES — usually used in plural
2 : the most undesirable part — usually used in plural
3 : the last remaining part **:** VESTIGE
— dreg·gy \'dre-gē\ *adjective*

D region *noun* (1930)
: the lowest part of the ionosphere occurring approximately between 30 and 55 miles (50 and 90 kilometers) above the surface of the earth

dreich \'drēk\ *adjective* [Middle English, of Scandinavian origin; akin to Old Norse *drjūgr* lasting] (1813)
chiefly Scottish **:** DREARY

drei·del *also* **dreidl** \'drā-dᵊl\ *noun* [Yiddish *dreydl,* from *dreyen* to turn, from Middle High German *drǣjen,* from Old High German *drāen* — more at THROW] (1926)
1 : a 4-sided toy marked with Hebrew letters and spun like a top in a game of chance
2 : a children's game of chance played especially at Hanukkah with a dreidel

dreidel 1

¹drench \'drench\ *noun* (before 12th century)
1 : a poisonous or medicinal drink; *specifically* **:** a large dose of medicine mixed with liquid and put down the throat of an animal
2 a : something that drenches **b :** a quantity sufficient to drench or saturate

²drench *transitive verb* [Middle English, from Old English *drencan;* akin to Old English *drincan* to drink] (before 12th century)
1 a *archaic* **:** to force to drink **b :** to administer a drench to (an animal)
2 : to wet thoroughly (as by soaking or immersing in liquid)
3 : to soak or cover thoroughly with liquid that falls or is precipitated
4 : to fill or cover completely as if by soaking or precipitation ⟨was *drenched* in furs and diamonds —Richard Brautigan⟩
synonym see SOAK
— drench·er *noun*

¹dress \'dres\ *verb* [Middle English, from Middle French *dresser,* from Old French *drecier,* from (assumed) Vulgar Latin *directiare,* from Latin *directus* direct, past participle of *dirigere* to direct, from *dis-* + *regere* to lead straight — more at RIGHT] (14th century)
transitive verb
1 a : to make or set straight **b :** to arrange (as troops) in a straight line and at proper intervals
2 : to prepare for use or service; *specifically* **:** to prepare for cooking or for the table
3 : to add decorative details or accessories to **:** EMBELLISH
4 a : to put clothes on **b :** to provide with clothing
5 *archaic* **:** DRESS DOWN
6 a : to apply dressings or medicaments to **b** (1) **:** to arrange (the hair) by combing, brushing, or curling (2) **:** to groom and curry (an animal) **c :** to kill and prepare for market or for consumption — often used with *out* **d :** CULTIVATE, TEND; *especially* **:** to apply manure or fertilizer to **e :** to put through a finishing process; *especially* **:** to trim and smooth the surface of (as lumber or stone)
intransitive verb
1 a : to put on clothing **b :** to put on or wear formal, elaborate, or fancy clothes ⟨*dress* for dinner⟩
2 *of a food animal* **:** to weigh after being dressed — often used with *out*
3 : to align oneself with the next soldier in a line to make the line straight
— dress ship : to ornament a ship for a celebration by hoisting national ensigns at the mastheads and running a line of signal flags and pennants from bow to stern

²dress *noun* (1606)
1 : APPAREL, CLOTHING
2 : an outer garment (as for a woman or girl) usually consisting of a one-piece bodice and skirt
3 : covering, adornment, or appearance appropriate or peculiar to a particular time
4 : a particular form of presentation **:** GUISE

³dress *adjective* (1767)
1 : suitable for a formal occasion
2 : requiring or permitting formal dress ⟨a *dress* affair⟩
3 : relating to or used for a dress

dres·sage \drə-'säzh, dre-\ *noun, often attributive* [French, from *dresser* to train, drill, from Middle French] (1936)
: the execution by a trained horse of precision movements in response to barely perceptible signals from its rider

dress circle *noun* (1825)
: the first or lowest curved tier of seats above the main floor in a theater or opera house

dress code *noun* (1968)
: formally or socially imposed standards of dress

dress down (circa 1897)
transitive verb
: to reprove severely
intransitive verb
: to dress casually especially for reasons of fashion

¹dress·er \'dre-sər\ *noun* (15th century)
1 *obsolete* **:** a table or sideboard for preparing and serving food
2 : a cupboard to hold dishes and cooking utensils
3 : a chest of drawers or bureau with a mirror

²dresser *noun* (1520)
: one that dresses ⟨a fashionable *dresser*⟩

dresser set *noun* (circa 1934)
: a set of toilet articles including hairbrush, comb, and mirror for use at a dresser or dressing table

dress·ing *noun* (15th century)
1 a : the act or process of one who dresses **b :** an instance of such act or process
2 a : a sauce for adding to a dish (as a salad) **b :** a seasoned mixture usually used as a stuffing (as for poultry)
3 a : material (as ointment or gauze) applied to cover a lesion **b :** fertilizing material (as manure or compost)

dres·sing–down \ˌdre-siŋ-'dau̇n\ *noun* (circa 1890)
: a severe reprimand

dressing glass *noun* (1714)
: a small mirror set to swing in a standing frame and used at a dresser or dressing table

dressing gown *noun* (1777)
: a robe worn especially while dressing or resting

dressing room *noun* (1675)
: a room used chiefly for dressing; *especially* **:** a room in a theater for changing costumes and makeup

dressing table *noun* (1692)
: a table often fitted with drawers and a mirror in front of which one sits while dressing and grooming oneself

¹dress·mak·er \'dres-ˌmā-kər\ *noun* (1803)
: one that makes dresses
— dress·mak·ing \-ˌmā-kiŋ\ *noun*

²dressmaker *adjective* (1904)
of women's clothes **:** having softness, rounded lines, and intricate detailing ⟨a *dressmaker* suit⟩

dress rehearsal *noun* (1828)
1 : a full rehearsal (as of a play) in costume and with stage properties shortly before the first performance
2 : a practice exercise for something to come **:** DRY RUN

dress shield *noun* (1884)
: a pad worn inside a part of the clothing liable to be soiled by perspiration (as at the underarm)

dress shirt *noun* (1892)
: a man's shirt especially for wear with evening dress; *broadly* **:** a shirt suitable for wear with a necktie

dress uniform *noun* (circa 1897)
: a uniform for formal wear

dress up (1674)
transitive verb
1 a : to attire in best or formal clothes **b :** to attire in clothes suited to a particular role
2 a : to present in the most attractive or impressive light ⟨a fiasco *dressed up* as a triumph⟩ **b :** to make more attractive, glamorous, or fancy ⟨*dress up* a plain dessert with a rich chocolate sauce⟩
intransitive verb
: to get dressed up

dressy \'dre-sē\ *adjective* **dress·i·er; -est** (1768)
1 : showy in dress
2 : STYLISH, SMART
— dress·i·ness *noun*

drew *past of* DRAW

Drey·fu·sard \ˌdrī-f(y)ə-'sär(d), ˌdrā-, -'zär(d)\ *noun* [French] (1898)
: a defender or partisan of Alfred Dreyfus

drib \'drib\ *noun* [probably back-formation from *dribble & driblet*] (circa 1730)
: a small amount — usually used in the phrase *dribs and drabs*

¹drib·ble \'dri-bəl\ *verb* **drib·bled; drib·bling** \-(ə-)liŋ\ [frequentative of *drib* (to dribble)] (circa 1589)
transitive verb
1 : to issue sporadically and in small bits
2 : to let or cause to fall in drops little by little
3 a : to propel by successive slight taps or bounces with hand, foot, or stick **b :** to hit (as a baseball) so as to cause a slow bouncing
intransitive verb
1 : to fall or flow in drops or in a thin intermittent stream **:** TRICKLE
2 : to let saliva trickle from the corner of the mouth **:** DROOL
3 : to come or issue in piecemeal or desultory fashion

4 a : to dribble a ball or puck **b :** to proceed by dribbling **c** *of a ball* **:** to move with short bounces
— **drib·bler** \-b(ə-)lər\ *noun*
²dribble *noun* (circa 1680)
1 : a tiny or insignificant bit or quantity
2 : a small trickling stream or flow
3 : an act, instance, or manner of dribbling a ball or puck
— **drib·bly** \'dri-b(ə-)lē\ *adjective*
drib·let \'drib-lət\ *noun* (1678)
1 : a trifling or small sum or part
2 : a drop of liquid
dried–up \'drī-ˌdəp, ˌdrī-'\ *adjective* (1885)
: being wizened and shriveled
¹drier *comparative of* DRY
²dri·er *or* **dry·er** \'drī(-ə)r\ *noun* (1528)
1 : something that extracts or absorbs moisture
2 : a substance that accelerates drying (as of oils, paints, and printing inks)
3 *usually dryer* **:** a device for drying
driest *superlative of* DRY
¹drift \'drift\ *noun* [Middle English; akin to Old English *drīfan* to drive — more at DRIVE] (14th century)
1 a : the act of driving something along **b :** the flow or the velocity of the current of a river or ocean stream
2 : something driven, propelled, or urged along or drawn together in a clump by or as if by a natural agency: as **a :** wind-driven snow, rain, cloud, dust, or smoke usually at or near the ground surface **b** (1) **:** a mass of matter (as sand) deposited together by or as if by wind or water (2) **:** a helter-skelter accumulation **c :** DROVE, FLOCK **d :** something (as driftwood) washed ashore **e :** rock debris deposited by natural agents; *specifically* **:** a deposit of clay, sand, gravel, and boulders transported by a glacier or by running water from a glacier
3 a : a general underlying design or tendency **b :** the underlying meaning, import, or purport of what is spoken or written
4 : something (as a tool) driven down upon or forced into a body
5 : the motion or action of drifting especially spatially and usually under external influence: as **a :** the lateral motion of an aircraft due to air currents **b :** an easy moderate more or less steady flow or sweep along a spatial course **c :** a gradual shift in attitude, opinion, or position **d :** an aimless course; *especially* **:** a foregoing of any attempt at direction or control **e :** a deviation from a true reproduction, representation, or reading
6 a : a nearly horizontal mine passageway driven on or parallel to the course of a vein or rock stratum **b :** a small crosscut in a mine connecting two larger tunnels
7 a : an assumed trend toward a general change in the structure of a language over a period of time **b :** GENETIC DRIFT **c :** a gradual change in the zero reading of an instrument or in any quantitative characteristic that is supposed to remain constant
synonym see TENDENCY
— **drifty** \'drif-tē\ *adjective*
²drift (circa 1600)
intransitive verb
1 a : to become driven or carried along (as by a current of water, wind, or air) **b :** to move or float smoothly and effortlessly
2 a : to move along a line of least resistance **b :** to move in a random or casual way **c :** to become carried along subject to no guidance or control ⟨the talk *drifted* from topic to topic⟩
3 a : to accumulate in a mass or become piled up in heaps by wind or water **b :** to become covered with a drift
4 : to vary or deviate from a set course or adjustment
transitive verb
1 a : to cause to be driven in a current **b** *West* **:** to drive (livestock) slowly especially to allow grazing
2 a : to pile in heaps **b :** to cover with drifts
— **drift·ing·ly** \'drif-tiŋ-lē\ *adverb*
drift·age \'drif-tij\ *noun* (1768)
: drifted material
drift·er \'drif-tər\ *noun* (1897)
: one that drifts; *especially* **:** one that travels or moves about aimlessly
drift fence *noun* (1907)
: a stretch of fence on rangeland especially in the western U.S. for preventing cattle from drifting from their home range
drift net *noun* (1848)
: a fishing net often miles in extent arranged to drift with the tide or current and buoyed up by floats or attached to a boat
drift·wood \'drift-ˌwu̇d\ *noun* (1633)
1 : wood drifted or floated by water
2 : FLOTSAM 2
¹drill \'dril\ *verb* [Dutch *drillen*] (1622)
transitive verb
1 a : to fix something in the mind or habit pattern of by repetitive instruction ⟨*drill* pupils in spelling⟩ **b :** to impart or communicate by repetition ⟨impossible to *drill* the simplest idea into some people⟩ **c :** to train or exercise in military drill
2 a (1) **:** to bore or drive a hole in (2) **:** to make by piercing action **b :** to shoot with or as if with a gun **c** (1) **:** to propel (as a ball) with force or accuracy ⟨*drilled* a single to right field⟩ (2) **:** to hit with force ⟨*drilled* the batter with the first pitch⟩
intransitive verb
1 : to make a hole with a drill
2 : to engage in an exercise
— **drill·abil·i·ty** \ˌdri-lə-'bi-lə-tē\ *noun*
— **drill·able** \-lə-bəl\ *adjective*
— **drill·er** \'dri-lər\ *noun*
²drill *noun* (1611)
1 : an instrument with an edged or pointed end for making holes in hard substances by revolving or by a succession of blows; *also* **:** a machine for operating such an instrument
2 : the act or exercise of training soldiers in marching and in executing prescribed movements with a weapon
3 a : a physical or mental exercise aimed at perfecting facility and skill especially by regular practice **b :** a formal exercise by a team of marchers **c** *chiefly British* **:** the approved or correct procedure for accomplishing something efficiently
4 a : a marine snail (*Urosalpinx cinerea*) destructive to oysters by boring through their shells and feeding on the soft parts **b :** any of several mollusks related to the drill
5 : a drilling sound
³drill *noun* [origin unknown] (1644)
: a western African baboon (*Papio leucophaeus* synonym *Mandrillus leucophaeus*) having a black face and brown coat and closely related to the typical mandrills
⁴drill *noun* [perhaps from *drill* (rill)] (1727)
1 a : a shallow furrow or trench into which seed is sown **b :** a row of seed sown in such a furrow
2 : a planting implement that makes holes or furrows, drops in the seed and sometimes fertilizer, and covers them with earth
⁵drill *transitive verb* (circa 1740)
1 : to sow (seeds) by dropping along a shallow furrow
2 a : to sow with seed or set with seedlings inserted in drills **b :** to distribute seed or fertilizer in by means of a drill
⁶drill *noun* [short for *drilling*] (1743)
: a durable cotton twilled fabric
dril·ling \'dri-liŋ\ *noun* [modification of German *Drillich,* from Middle High German *drilich* fabric woven with a threefold thread, from Old High German *drilīh* made up of three threads, from Latin *trilic-, trilix,* from *tri-* + *licium* thread] (1640)
: ⁶DRILL
drill·mas·ter \'dril-ˌmas-tər\ *noun* (1869)
1 : an instructor in military drill
2 : an instructor or director who maintains severe discipline and often stresses method and detail
drill press *noun* (circa 1864)
: an upright drilling machine in which the drill is pressed to the work by a hand lever or by power
drill team *noun* (1928)
: an exhibition marching team that engages in precision drill
drily *variant of* DRYLY
D ring \'dē-ˌriŋ\ *noun* (circa 1899)
: a usually metal ring having the shape of a capital D
¹drink \'driŋk\ *verb* **drank**\'draŋk\; **drunk** \'drəŋk\ *or* **drank; drink·ing** [Middle English, from Old English *drincan;* akin to Old High German *trinkan* to drink] (before 12th century)
transitive verb
1 a : SWALLOW, IMBIBE **b :** to take in or suck up **:** ABSORB ⟨*drinking* air into his lungs⟩ **c :** to take in or receive avidly — usually used with *in* ⟨*drank* in every word of the lecture⟩
2 : to join in a toast to ⟨I'll *drink* your good health⟩
3 : to bring to a specified state by drinking alcoholic beverages ⟨*drank* himself into oblivion⟩
intransitive verb
1 a : to take liquid into the mouth for swallowing **b :** to receive into one's consciousness
2 : to partake of alcoholic beverages
3 : to make or join in a toast ⟨I'll *drink* to that!⟩
²drink *noun* (before 12th century)
1 a : a liquid suitable for swallowing **b :** alcoholic beverages
2 : a draft or portion of liquid
3 : excessive consumption of alcoholic beverages
4 : a sizable body of water — used with *the*
¹drink·able \'driŋ-kə-bəl\ *adjective* (1611)
: suitable or safe for drinking
— **drink·abil·i·ty** \ˌdriŋ-kə-'bi-lə-tē\ *noun*
²drinkable *noun* (1708)
: a liquid suitable for drinking **:** BEVERAGE
drink·er \'driŋ-kər\ *noun* (before 12th century)
1 a : one that drinks **b :** a person who drinks alcoholic beverages especially to a notable degree ⟨a heavy *drinker*⟩
2 : WATERER b
drinking fountain *noun* (1860)
: a fixture with nozzle that delivers a stream of water for drinking
drinking song *noun* (1597)
: a song on a convivial theme appropriate for a group engaged in social drinking
¹drip \'drip\ *verb* **dripped; drip·ping** [Middle English *drippen,* from Old English *dryppan;* akin to Old English *dropa* drop] (before 12th century)
transitive verb
1 : to let fall in drops
2 : to let out or seem to spill copiously ⟨her voice *dripping* sarcasm⟩
intransitive verb
1 a : to let fall drops of moisture or liquid **b :** to overflow with or as if with moisture ⟨a uniform *dripping* with gold braid⟩
2 : to fall in or as if in drops
3 : to waft or pass gently
— **drip·per** *noun*
²drip *noun* (1664)
1 : a part of a cornice or other member that projects to throw off rainwater; *also* **:** an overlapping metal strip or an underneath groove for the same purpose
2 a : a falling in drops **b :** liquid that falls, overflows, or is extruded in drops

3 : the sound made by or as if by falling drops
4 : a device for the administration of a fluid at a slow rate especially into a vein; *also* **:** a material so administered
5 : a dull or unattractive person

³drip *adjective* (1895)
: of, relating to, or being coffee made by letting boiling water drip slowly through finely ground coffee ⟨*drip* coffee⟩ ⟨a *drip* pot⟩

¹drip–dry \'drip-ˌdrī\ (1953)
intransitive verb
: to dry with few or no wrinkles when hung wet
transitive verb
: to hang (as wet clothing) to drip-dry

²drip–dry *adjective* (1957)
: made of a washable fabric that drip-dries

³drip–dry *noun* (1959)
: a drip-dry garment

drip·less \'dri-pləs\ *adjective* (1887)
: designed not to drip ⟨*dripless* candles⟩

drip·ping \'dri-piŋ\ *noun* (15th century)
: fat and juices drawn from meat during cooking — often used in plural

drip·py \'dri-pē\ *adjective* **drip·pi·er; -est** (circa 1718)
1 : characterized by dripping; *especially* **:** RAINY, DRIZZLY
2 : MAWKISH 2

drip·stone \'drip-ˌstōn\ *noun* (circa 1818)
1 : a stone drip (as over a window)
2 : calcium carbonate in the form of stalactites or stalagmites

¹drive \'drīv\ *verb* **drove**\'drōv\; **driv·en** \'dri-vən\; **driv·ing**\'drī-viŋ\ [Middle English, from Old English *drīfan;* akin to Old High German *trīban* to drive] (before 12th century)
transitive verb
1 a : to frighten or prod (as game or cattle) into moving in a desired direction **b :** to go through (a district) driving game animals
2 : to carry on or through energetically ⟨*drives* a hard bargain⟩
3 a : to impart a forward motion to by physical force ⟨waves *drove* the boat ashore⟩ **b :** to repulse, remove, or cause to go by force, authority, or influence ⟨*drive* the enemy back⟩ **c :** to set or keep in motion or operation ⟨*drive* machinery by electricity⟩
4 a : to direct the motions and course of (a draft animal) **b :** to operate the mechanism and controls and direct the course of (as a vehicle) **c :** to convey in a vehicle **d :** to float (logs) down a stream
5 a : to exert inescapable or coercive pressure on **:** FORCE **b :** to compel to undergo or suffer a change (as in situation or emotional state) ⟨*drove* him crazy⟩ ⟨*drove* her out of business⟩ **c :** to urge relentlessly to continuous exertion ⟨the sergeant *drove* his recruits⟩ **d :** to press or force into an activity, course, or direction ⟨the drug habit *drives* addicts to steal⟩ **e :** to project, inject, or impress incisively ⟨*drove* her point home⟩
6 : to force (a passage) by pressing or digging
7 a : to propel (an object of play) swiftly **b :** to hit (a golf ball) from the tee especially with a driver; *also* **:** to drive a golf ball onto (a green) **c :** to cause (a run or runner) to be scored in baseball — usually used with *in*
8 : to give shape or impulse to ⟨factors that *drive* the business cycle⟩
intransitive verb
1 a : to dash, plunge, or surge ahead rapidly or violently **b :** to progress with strong momentum ⟨the rain was *driving* hard⟩
2 a : to operate a vehicle **b :** to have oneself carried in a vehicle
3 : to drive a golf ball
synonym see MOVE
— driv·abil·i·ty *also* **drive·abil·i·ty** \ˌdrī-və-'bi-lə-tē\ *noun*
— driv·able *also* **drive·able** \'drī-və-bəl\ *adjective*
— drive at : to intend to express, convey, or accomplish ⟨did not understand what she was *driving at* —Eric Goldman⟩

²drive *noun, often attributive* (1785)
1 : an act of driving: **a :** a trip in a carriage or automobile **b :** a collection and driving together of animals; *also* **:** the animals gathered **c :** a driving of cattle or sheep overland **d :** a hunt or shoot in which the game is driven within the hunter's range **e :** the guiding of logs downstream to a mill; *also* **:** the floating logs amassed in a drive **f** (1) **:** the act or an instance of driving an object of play (as a golf ball) (2) **:** the flight of a ball
2 a : a private road **:** DRIVEWAY **b :** a public road for driving (as in a park)
3 : the state of being hurried and under pressure
4 a : a strong systematic group effort ⟨a fund-raising *drive*⟩ **b :** a sustained offensive effort ⟨the *drive* ended in a touchdown⟩
5 a : the means for giving motion to a machine or machine part **b :** the means by which the propulsive power of an automobile is applied to the road ⟨front wheel *drive*⟩ **c :** the means by which the propulsion of an automotive vehicle is controlled and directed ⟨a left-hand *drive*⟩
6 : an offensive, aggressive, or expansionist move; *especially* **:** a strong military attack against enemy-held terrain
7 a : an urgent, basic, or instinctual need **:** a motivating physiological condition of an organism ⟨a sexual *drive*⟩ **b :** an impelling culturally acquired concern, interest, or longing **c :** dynamic quality
8 : a device for reading and writing on magnetic media (as tapes or disks)

drive–by \'drīv-'bī\ *adjective* (1986)
: carried out from a moving vehicle ⟨a *drive-by* shooting⟩

drive–in \'drī-ˌvin\ *noun* (1937)
: an establishment (as a theater or restaurant) so laid out that patrons can be accommodated while remaining in their automobiles
— drive–in *adjective*

¹driv·el \'dri-vəl\ *verb* **-eled** *or* **-elled; -el·ing** *or* **-el·ling**\-v(ə-)liŋ\ [Middle English, from Old English *dreflian;* perhaps akin to Old Norse *draf* malt dregs] (before 12th century)
intransitive verb
1 : to let saliva dribble from the mouth **:** SLAVER
2 : to talk stupidly and carelessly
transitive verb
1 : to utter in an infantile or imbecilic way
2 : to waste or fritter in a childish fashion
— driv·el·er \-v(ə-)lər\ *noun*

²drivel *noun* (14th century)
1 *archaic* **:** DROOL 1
2 : NONSENSE

drive·line \'drīv-ˌlīn\ *noun* (1949)
: DRIVETRAIN

driv·en *adjective* (1925)
: having a compulsive or urgent quality ⟨a *driven* sense of obligation⟩
— driv·en·ness \'dri-vən-nəs\ *noun*

driv·er \'drī-vər\ *noun* (14th century)
: one that drives: as **a :** COACHMAN **b :** the operator of a motor vehicle **c :** an implement (as a hammer) for driving **d :** a mechanical piece for imparting motion to another piece **e :** a golf wood with a nearly straight face used in driving **f :** an electronic circuit that supplies input to another electronic circuit; *also* **:** LOUDSPEAKER **g :** a piece of computer software that controls input and output operations
— driv·er·less \-ləs\ *adjective*

driver ant *noun* (1859)
: ARMY ANT; *specifically* **:** any of various African and Asian ants (*Dorylus* or related genera) that move in vast armies

driver's license *noun* (1926)
: a license issued under governmental authority that permits the holder to operate a motor vehicle

driver's seat *noun* (1923)
: the position of top authority or dominance

drive·shaft \'drīv-ˌshaft\ *noun* (1895)
: a shaft that transmits mechanical power

drive time *noun* (1966)
: a time during rush hour when radio audiences are swelled by commuters listening to car radios

drive·train \'drīv-ˌtrān\ *noun* (1954)
: the parts (as the universal joint and the driveshaft) that connect the transmission with the driving axles of an automobile

drive·way \-ˌwā\ *noun* (1871)
: a private road giving access from a public way to a building on abutting grounds

driv·ing *adjective* (14th century)
1 a : communicating force ⟨a *driving* wheel⟩ **b :** exerting pressure ⟨a *driving* influence⟩
2 a : having great force ⟨a *driving* rain⟩ **b :** acting with vigor **:** ENERGETIC ⟨a hard-*driving* worker⟩

driving range *noun* (circa 1949)
: an area equipped with distance markers, clubs, balls, and tees for practicing golf drives

¹driz·zle \'dri-zəl\ *noun* (1554)
: a fine misty rain
— driz·zly \'driz-(ə-)lē\ *adjective*

²drizzle *verb* **driz·zled; driz·zling** \-z(ə-)liŋ\ [perhaps alteration of Middle English *drysnen* to fall, from Old English *-drysnian* to disappear; akin to Gothic *driusan* to fall] (1584)
transitive verb
1 : to shed or let fall in minute drops or particles
2 : to make wet with minute drops
intransitive verb
: to rain in very small drops or very lightly **:** SPRINKLE
— driz·zling·ly \-z(ə-)liŋ-lē\ *adverb*

drogue \'drōg\ *noun* [probably alteration of ¹*drag*] (1875)
1 : SEA ANCHOR
2 a : a cylindrical or funnel-shaped device towed as a target by an airplane **b :** a small parachute for stabilizing or decelerating something (as an astronaut's capsule) or for pulling a larger parachute out of stowage
3 : a funnel-shaped device which is attached to the end of a long flexible hose suspended from a tanker airplane in flight and into which the probe of another airplane is fitted so as to receive fuel from the tanker

droit \'drȯit, 'drwä\ *noun* [Middle English, from Middle French, from Medieval Latin *directum,* from Late Latin, neuter of *directus* just, from Latin, direct — more at DRESS] (15th century)
: a legal right

droit du sei·gneur \drwä-dœ-se-nʸœr\ *noun* [French, right of the lord] (1825)
: a supposed legal or customary right of a feudal lord to have sexual relations with a vassal's bride on her wedding night

¹droll \'drōl\ *adjective* [French *drôle,* from *drôle* scamp, from Middle French *drolle,* from Middle Dutch, imp] (1623)
: having a humorous, whimsical, or odd quality
— droll·ness *noun*
— drol·ly \'drō(l)-lē\ *adverb*

²droll *noun* (circa 1645)
: an amusing person **:** JESTER, COMEDIAN

³droll *intransitive verb* (1654)
archaic **:** to make fun **:** JEST, SPORT

droll·ery \'drōl-rē, 'drō-lə-\ *noun, plural* **-er·ies** (1597)
1 : something that is droll; *especially* **:** a comic picture or drawing
2 : the act or an instance of jesting or burlesquing
3 : whimsical humor

-drome \ˌdrōm\ *noun combining form* [*hippodrome*]
1 : racecourse ⟨motor*drome*⟩
2 : large specially prepared place ⟨aero*drome*⟩

drom·e·dary \'drä-mə-ˌder-ē *also* 'drə-\ *noun, plural* **-dar·ies** [Middle English *dromedarie,*

from Old French *dromedaire,* from Late Latin *dromedarius,* from Latin *dromad-, dromas,* from Greek, running; akin to Greek *dramein* to run, *dromos* racecourse, Sanskrit *dramati* he runs about] (13th century)
: CAMEL 1a

-d·ro·mous \drə-məs\ *adjective combining form* [New Latin *-dromus,* from Greek *-dromos* (akin to Greek *dramein*)]
: running ⟨cata*dromous*⟩

[1]**drone** \'drōn\ *noun* [Middle English, from Old English *drān;* akin to Old High German *treno* drone, Greek *thrēnos* dirge] (before 12th century)
1 : the male of a bee (as the honeybee) that has no sting and gathers no honey
2 : one that lives on the labors of others **:** PARASITE
3 : an unmanned airplane, helicopter, or ship guided by remote control
4 : DRUDGE

[2]**drone** *verb* **droned; dron·ing** (circa 1520)
intransitive verb
1 a : to make a sustained deep murmuring, humming, or buzzing sound **b :** to talk in a persistently dull or monotonous tone
2 : to pass, proceed, or act in a dull, drowsy, or indifferent manner
transitive verb
1 : to utter or pronounce with a drone
2 : to pass or spend in dull or monotonous activity or in idleness
— dron·er *noun*
— dron·ing·ly \'drō-niŋ-lē\ *adverb*

[3]**drone** *noun* (circa 1520)
1 : a deep sustained or monotonous sound **:** HUM
2 : an instrument or part of an instrument (as one of the fixed-pitch pipes of a bagpipe) that sounds a continuous unvarying tone
3 : PEDAL POINT

[1]**drool** \'drü(-ə)l\ *verb* [perhaps alteration of *drivel*] (1802)
intransitive verb
1 a : to secrete saliva in anticipation of food **b :** DRIVEL 1
2 : to make an effusive show of pleasure or often envious or covetous appreciation
3 : to talk nonsense
transitive verb
: to express sentimentally or effusively

[2]**drool** *noun* (1869)
1 : saliva trickling from the mouth
2 : NONSENSE

[1]**droop** \'drüp\ *verb* [Middle English *drupen,* from Old Norse *drūpa;* akin to Old English *dropa* drop] (13th century)
intransitive verb
1 : to hang or incline downward
2 : to sink gradually
3 : to become depressed or weakened **:** LANGUISH
transitive verb
: to let droop
— droop·ing·ly \'drü-piŋ-lē\ *adverb*

[2]**droop** *noun* (1647)
: the condition or appearance of drooping

droopy \'drü-pē\ *adjective* **droop·i·er; -est** (13th century)
1 : GLOOMY
2 : drooping or tending to droop

[1]**drop** \'dräp\ *noun, often attributive* [Middle English, from Old English *dropa;* akin to Old High German *tropfo* drop] (before 12th century)
1 a (1) **:** the quantity of fluid that falls in one spherical mass (2) *plural* **:** a dose of medicine measured by drops; *especially* **:** a solution for dilating the pupil of the eye **b :** a minute quantity or degree of something nonmaterial or intangible **c :** a small quantity of drink **d :** the smallest practical unit of liquid measure
2 : something that resembles a liquid drop: as **a :** a pendent ornament attached to a piece of jewelry; *also* **:** an earring with such a pendant **b :** a small globular cookie or candy
3 [[2]*drop*] **a :** the act or an instance of dropping **:** FALL **b :** a decline in quantity or quality **c :** a descent by parachute; *also* **:** the people or equipment dropped by parachute **d :** a place or central depository to which something (as mail, money, or stolen property) is brought for distribution or transmission; *also* **:** the act of depositing something at such a place ⟨made the *drop*⟩
4 a : the distance from a higher to a lower level or through which something drops **b :** a fall of electric potential
5 : a slot into which something is to be dropped
6 [[2]*drop*] **:** something that drops, hangs, or falls: as **a :** a movable plate that covers the keyhole of a lock **b :** an unframed piece of cloth stage scenery; *also* **:** DROP CURTAIN **c :** a hinged platform on a gallows **d :** a fallen fruit
7 : the advantage of having an opponent covered with a firearm; *broadly* **:** ADVANTAGE, SUPERIORITY — usually used in the phrase *get the drop on*
— at the drop of a hat : as soon as the slightest provocation is given **:** IMMEDIATELY
— drop in the bucket : a part so small as to be negligible

[2]**drop** *verb* **dropped; drop·ping** (before 12th century)
intransitive verb
1 : to fall in drops
2 a (1) **:** to fall unexpectedly or suddenly (2) **:** to descend from one line or level to another **b :** to fall in a state of collapse or death **c** *of a card* **:** to become played by reason of the obligation to follow suit **d** *of a ball* **:** to fall or roll into a hole or basket
3 : to enter or pass as if without conscious effort of will into some state, condition, or activity ⟨*dropped* into sleep⟩
4 a : to cease to be of concern **:** LAPSE ⟨let the matter *drop*⟩ **b :** to pass from view or notice **:** DISAPPEAR — often used with *out* ⟨*drop* out of sight⟩ **c :** to become less ⟨production *dropped*⟩ — often used with *off*
5 : to move with a favoring wind or current — usually used with *down*
transitive verb
1 : to let fall **:** cause to fall
2 a : GIVE UP 2, ABANDON ⟨*drop* an idea⟩ ⟨*drop* the charges⟩ **b :** DISCONTINUE ⟨*dropped* what she was doing⟩ **c :** to break off an association or connection with **:** DISMISS ⟨*drop* a failing student⟩
3 a : to utter or mention in a casual way ⟨*drop* a suggestion⟩ ⟨*drop* names⟩ **b :** WRITE ⟨*drop* us a line soon⟩
4 a : to lower or cause to descend from one level or position to another **b :** to cause to lessen or decrease **:** REDUCE ⟨*dropped* his speed⟩
5 *of an animal* **:** to give birth to
6 a : LOSE ⟨*dropped* three games⟩ ⟨*dropped* $50 in a poker game⟩ **b :** SPEND ⟨*drop* $20 for lunch⟩
7 a : to bring down with a shot or a blow **b :** to cause (a high card) to fall **c :** to toss or roll into a hole or basket ⟨*drop* a putt⟩
8 a : to deposit or deliver during a usually brief stop — usually used with *off* ⟨*drop* the kids off at school⟩ **b :** AIR DROP
9 : to cause (the voice) to be less loud
10 a : to leave (a letter representing a speech sound) unsounded ⟨*drop* the *g* in *running*⟩ **b :** to leave out in writing **:** OMIT
11 : to draw from an external point ⟨*drop* a perpendicular to the line⟩
12 : to take (a drug) orally **:** SWALLOW ⟨*drop* acid⟩
— drop·pa·ble \'drä-pə-bəl\ *adjective*
— drop behind : to fail to keep up

drop back *intransitive verb* (1927)
1 : RETREAT
2 : to move straight back from the line of scrimmage ⟨the quarterback *drops back* to pass⟩

drop by (circa 1905)
intransitive verb
: to pay a brief casual visit
transitive verb
: to visit casually or unexpectedly ⟨*drop by* a friend's house⟩

drop cloth *noun* (circa 1928)
: a protective sheet (as of cloth or plastic) used especially by painters to cover floors and furniture

drop curtain *noun* (1832)
: a stage curtain that can be lowered and raised

drop–dead *adjective* (1970)
: sensationally striking, attractive, or impressive ⟨a *drop-dead* evening gown⟩
— drop–dead *adverb*

drop–forge \'dräp-ˌfōrj, -ˌfȯrj\ *transitive verb* (1886)
: to forge between dies by means of a drop hammer or punch press
— drop forger *noun*

drop front *noun* (1925)
: a hinged cover on the front of a desk that may be lowered to provide a surface for writing

drop hammer *noun* (circa 1864)
: a power hammer raised and then released to drop (as on metal resting on an anvil or die)

drop·head \'dräp-ˌhed\ *noun, often attributive* (1932)
British **:** a convertible automobile

drop–in \'dräp-ˌin\ *noun* (1819)
1 : a casual visit or brief stop
2 : one who drops in **:** a casual visitor

drop in *intransitive verb* (circa 1600)
: to pay an unexpected or casual visit — often used with *on*

drop·kick \-ˌkik\ *noun* (1857)
: a kick made by dropping a ball to the ground and kicking it at the moment it starts to rebound

drop–kick \-ˌkik\ (circa 1909)
intransitive verb
: to make a dropkick
transitive verb
: to kick by means of a dropkick ⟨*drop-kick* a ball⟩ ⟨*drop-kick* a field goal⟩
— drop·kick·er *noun*

drop leaf *noun* (1882)
: a hinged leaf on the side or end of a table that can be folded down

drop·let \'dräp-lət\ *noun* (1607)
: a tiny drop (as of a liquid)

drop·light \'dräp-ˌlīt\ *noun* (1890)
: an electric light suspended by a cord or on a portable extension

drop–off \'dräp-ˌȯf\ *noun* (1923)
1 : a very steep or perpendicular descent
2 : a marked dwindling or decline ⟨a *drop-off* in attendance⟩

drop off \ˌdräp-'ȯf\ *intransitive verb* (1820)
: to fall asleep

drop·out \'dräp-ˌau̇t\ *noun* (1930)
1 a : one who drops out of school **b :** one who drops out of conventional society **c :** one who abandons an attempt, activity, or chosen path ⟨a corporate *dropout*⟩
2 : a spot on a magnetic tape or disk from which data has disappeared

drop out \ˌdräp-'au̇t\ *intransitive verb* (1883)
: to withdraw from participation or membership **:** QUIT; *especially* **:** to withdraw from conventional society

drop pass *noun* (1949)
: a pass in ice hockey in which the passer skates past the puck leaving it for a teammate following close behind

dropped *adjective* (1953)

: designed to extend or begin lower than normal ⟨a dress with a *dropped* waist⟩ ⟨*dropped* shoulders⟩

dropped egg *noun* (1824)
: a poached egg

drop·per \ˈdrä-pər\ *noun* (circa 1700)
1 : one that drops
2 : a short glass tube fitted with a rubber bulb and used to measure liquids by drops — called also *eyedropper, medicine dropper*
— drop·per·ful \-ˌfu̇l\ *noun*

drop·ping *noun* (14th century)
1 : something dropped
2 *plural* **:** DUNG

drop seat *noun* (1926)
1 : a hinged seat (as in a taxi) that may be dropped down
2 : a seat (as in an undergarment) that can be unbuttoned

drop shot *noun* (1908)
: a delicately hit shot (as in tennis or squash) that drops quickly after crossing the net or dies after hitting a wall

drop·si·cal \ˈdräp-si-kəl\ *adjective* (1678)
1 : relating to or affected with dropsy
2 : TURGID, SWOLLEN

drop·sy \ˈdräp-sē\ *noun* [Middle English *dropesie,* short for *ydropesie,* from Old French, from Latin *hydropisis,* modification of Greek *hydrōps,* from *hydōr* water — more at WATER] (13th century)
: EDEMA

drop tank *noun* (1943)
: an auxiliary fuel tank for airplanes that can be jettisoned when empty

drop volley *noun* (1907)
: a drop shot made on a volley in tennis

drop zone *noun* (circa 1943)
: the area in which troops, supplies, or equipment are to be air-dropped; *also* **:** the target on which a skydiver lands

drosh·ky \ˈdräsh-kē\ *also* **dros·ky** \ˈdräs-kē\ *noun, plural* **droshkies** *also* **droskies** [Russian *drozhki,* from *droga* pole of a wagon] (1808)
: any of various 2- or 4-wheeled carriages used especially in Russia

dro·soph·i·la \drō-ˈsä-fə-lə\ *noun* [New Latin, genus name, from Greek *drosos* dew + New Latin *-phila,* feminine of *-philus* -phil] (1910)
: any of a genus (*Drosophila*) of small two-winged flies used in genetic research

dross \ˈdräs, ˈdrȯs\ *noun* [Middle English *dros,* from Old English *drōs* dregs] (before 12th century)
1 : the scum that forms on the surface of molten metal
2 : waste or foreign matter **:** IMPURITY
3 : something that is base, trivial, or inferior
— drossy \ˈdrä-sē, ˈdrȯ-\ *adjective*

drought \ˈdrau̇t\ *also* **drouth** \ˈdrau̇th\ *noun* [Middle English, from Old English *drūgath,* from *drūgian* to dry up; akin to Old English *drȳge* dry — more at DRY] (before 12th century)
1 : a period of dryness especially when prolonged that causes extensive damage to crops or prevents their successful growth
2 : a prolonged or chronic shortage or lack of something expected or desired
— drought·i·ness *noun*
— droughty \ˈdrau̇-tē\ *adjective*

[1]**drove** \ˈdrōv\ *noun* [Middle English, from Old English *drāf,* from *drīfan* to drive — more at DRIVE] (before 12th century)
1 : a group of animals driven or moving in a body
2 : a large number **:** CROWD — usually used in plural especially with *in* ⟨tourists arriving in *droves*⟩ ⟨stayed away in *droves*⟩

[2]**drove** *past of* DRIVE

drov·er \ˈdrō-vər\ *noun* (15th century)
: one that drives cattle or sheep

drown \ˈdrau̇n\ *verb* **drowned** \ˈdrau̇nd\; **drown·ing** \ˈdrau̇-niŋ\ [Middle English *drounen*] (14th century)
intransitive verb
: to become drowned
transitive verb
1 a : to suffocate by submersion especially in water **b :** to submerge especially by a rise in the water level **c :** to soak, drench, or cover with a liquid
2 : to engage (oneself) deeply and strenuously
3 : to cause (a sound) not to be heard by making a loud noise — usually used with *out*
4 a : to drive out (as a sensation or an idea) ⟨*drowned* his sorrows in liquor⟩ **b :** OVERWHELM

drownd \ˈdrau̇nd\ *nonstandard variant of* DROWN

[1]**drowse** \ˈdrau̇z\ *verb* **drowsed; drows·ing** [probably akin to Gothic *driusan* to fall — more at DREARY] (1573)
intransitive verb
1 : to be inactive
2 : to fall into a light slumber
transitive verb
1 : to make drowsy or inactive
2 : to pass (time) drowsily or in drowsing

[2]**drowse** *noun* (1814)
: the act or an instance of drowsing **:** DOZE

drowsy \ˈdrau̇-zē\ *adjective* **drows·i·er; -est** (1530)
1 a : ready to fall asleep **b :** inducing or tending to induce sleep **c :** INDOLENT, LETHARGIC
2 : giving the appearance of peaceful inactivity
— drows·i·ly \-zə-lē\ *adverb*
— drows·i·ness \-zē-nəs\ *noun*

drub \ˈdrəb\ *verb* **drubbed; drub·bing** [perhaps from Arabic *ḍaraba*] (1634)
transitive verb
1 : to beat severely
2 : to abuse with words **:** BERATE
3 : to defeat decisively
intransitive verb
: DRUM, STAMP
— drub·ber *noun*
— drub·bing *noun*

[1]**drudge** \ˈdrəj\ *verb* **drudged; drudg·ing** [Middle English *druggen*] (14th century)
intransitive verb
: to do hard, menial, or monotonous work
transitive verb
: to force to do hard, menial, or monotonous work
— drudg·er *noun*

[2]**drudge** *noun* (15th century)
1 : one who is obliged to do menial work
2 : one whose work is routine and boring

drudg·ery \ˈdrəj-rē, ˈdrə-jə-rē\ *noun, plural* **-er·ies** (1550)
: dull, irksome, and fatiguing work **:** uninspiring or menial labor
synonym see WORK

drudg·ing \ˈdrə-jiŋ\ *adjective* (1548)
: MONOTONOUS, TIRING
— drudg·ing·ly \-jiŋ-lē\ *adverb*

[1]**drug** \ˈdrəg\ *noun* [Middle English *drogge*] (14th century)
1 a *obsolete* **:** a substance used in dyeing or chemical operations **b :** a substance used as a medication or in the preparation of medication **c** *according to the Food, Drug, and Cosmetic Act* (1) **:** a substance recognized in an official pharmacopoeia or formulary (2) **:** a substance intended for use in the diagnosis, cure, mitigation, treatment, or prevention of disease (3) **:** a substance other than food intended to affect the structure or function of the body (4) **:** a substance intended for use as a component of a medicine but not a device or a component, part, or accessory of a device
2 : a commodity that is not salable or for which there is no demand — used in the phrase *drug on the market*
3 : something and often an illegal substance that causes addiction, habituation, or a marked change in consciousness
— drug·gy *also* **drug·gie** \ˈdrə-gē\ *adjective*

[2]**drug** *verb* **drugged; drug·ging** (1605)
transitive verb
1 : to affect with a drug; *especially* **:** to stupefy by a narcotic drug
2 : to administer a drug to
3 : to lull or stupefy as if with a drug
intransitive verb
: to take drugs for narcotic effect

[3]**drug** *dialect past of* DRAG

drug·get \ˈdrə-gət\ *noun* [Middle French *droguet,* diminutive of *drogue* trash, drug] (1580)
1 : a wool or partly wool fabric formerly used for clothing
2 : a coarse durable cloth used chiefly as a floor covering
3 : a rug having a cotton warp and a wool filling

drug·gie *also* **drug·gy** \ˈdrə-gē\ *noun, plural* **druggies** (1967)
: one who habitually uses drugs

drug·gist \ˈdrə-gist\ *noun* (1611)
: one who sells or dispenses drugs and medicines: as **a :** PHARMACIST **b :** one who owns or manages a drugstore

drug·mak·er \ˈdrəg-ˌmā-kər\ *noun* (1964)
: one that manufactures pharmaceuticals

drug·store \-ˌstōr, -ˌstȯr\ *noun* (1810)
: a retail store where medicines and miscellaneous articles are sold **:** PHARMACY

drugstore cowboy *noun* (1925)
1 : one who wears cowboy clothes but has had no experience as a cowboy
2 : one who loafs on street corners and in drugstores

dru·id \ˈdrü-id\ *noun, often capitalized* [Latin *druides, druidae,* plural, from Gaulish *druides;* akin to Old Irish *druí* druid, and perhaps to Old English *trēow* tree] (1563)
: one of an ancient Celtic priesthood appearing in Irish and Welsh sagas and Christian legends as magicians and wizards
— dru·id·ic \drü-ˈi-dik\ *or* **dru·id·i·cal** \-di-kəl\ *adjective, often capitalized*

dru·id·ism \ˈdrü-ə-ˌdi-zəm\ *noun, often capitalized* (1715)
: the system of religion, philosophy, and instruction of the druids

[1]**drum** \ˈdrəm\ *noun* [probably from Dutch *trom;* akin to Middle High German *trumme* drum] (1539)
1 : a percussion instrument consisting of a hollow shell or cylinder with a drumhead stretched over one or both ends that is beaten with the hands or with some implement (as a stick or wire brush)
2 : TYMPANIC MEMBRANE

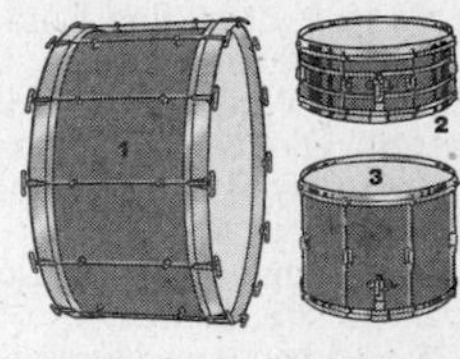

drum 1: *1* **bass,** *2* **snare (orchestra),** *3* **snare (parade)**

3 : the sound of a drum; *also* **:** a sound similar to that of a drum
4 : any of various bony fishes (family Sciaenidae) that make a drumming noise
5 : something resembling a drum in shape: as **a :** a cylindrical machine or mechanical device or part **b :** a cylindrical container; *specifically* **:** a large usually metal container for liquids ⟨a 55-gallon *drum*⟩ **c :** a disk-shaped magazine for an automatic weapon
— drum·like \-ˌlīk\ *adjective*

[2]**drum** *verb* **drummed; drum·ming** (1583)
intransitive verb
1 : to make a succession of strokes or vibrations that produce sounds like drumbeats
2 : to beat a drum
3 : to throb or sound rhythmically
4 : to stir up interest **:** SOLICIT
transitive verb
1 : to summon or enlist by or as if by beating a drum ⟨were *drummed* into service⟩

2 : to dismiss ignominiously **:** EXPEL — usually used with *out*
3 : to drive or force by steady effort or reiteration ⟨*drummed* the speech into her head⟩
4 a : to strike or tap repeatedly **b :** to produce (rhythmic sounds) by such action

[3]drum *noun* [Scottish Gaelic *druim* back, ridge, from Old Irish *druimm*] (1725)
1 *chiefly Scottish* **:** a long narrow hill or ridge
2 : DRUMLIN

drum·beat \'drəm-ˌbēt\ *noun* (1855)
1 : a stroke on a drum or its sound; *also* **:** a series of such strokes
2 : vociferous advocacy of a cause
3 : DRUMFIRE 2
— **drum·beat·er** \-ˌbē-tər\ *noun*
— **drum·beat·ing** \-tiŋ\ *noun*

drum brake *noun* (1950)
: a brake that operates by the friction of usually a pair of shoes pressing against the inner surface of the cylinder of a rotating drum — compare DISC BRAKE

drum·fire \'drəm-ˌfīr\ *noun* (1916)
1 : artillery firing so continuous as to sound like a drumroll
2 : something suggestive of drumfire in intensity **:** BARRAGE ⟨a *drumfire* of publicity⟩

drum·head \-ˌhed\ *noun* (1622)
1 : the material (as skin or plastic) stretched over one or both ends of a drum
2 : the top of a capstan that is pierced with sockets for the levers used in turning it

drumhead court–martial *noun* [from the use of a drumhead as a table] (1835)
: a summary court-martial that tries offenses on the battlefield

drum·lin \'drəm-lən\ *noun* [Irish *druim* back, ridge (from Old Irish *druimm*) + English *-lin* (alteration of *-ling*)] (circa 1833)
: an elongate or oval hill of glacial drift

drum major *noun* (1844)
: the leader of a marching band

drum ma·jor·ette \ˌdrəm-ˌmā-jə-'ret\ *noun* (1938)
1 : a girl or woman who leads a marching band
2 : a baton twirler who accompanies a marching band

drum·mer \'drə-mər\ *noun* (1580)
1 a : one that plays a drum **b** — used figuratively in expressions denoting an unconventional way of behaving or thinking ⟨march to a different *drummer*⟩
2 : TRAVELING SALESMAN

drum·roll \'drəm-ˌrōl\ *noun* (1887)
: a roll on a drum or its sound

drum·stick \-ˌstik\ *noun* (1589)
1 : a stick for beating a drum
2 : the segment of a fowl's leg between the thigh and tarsus

drum up *transitive verb* (1830)
1 : to bring about by persistent effort ⟨*drum up* some business⟩
2 : INVENT, ORIGINATE ⟨*drum up* a new method⟩

[1]drunk *past participle of* DRINK

[2]drunk \'drəŋk\ *adjective* [Middle English *drunke*, alteration of *drunken*] (14th century)
1 a : having the faculties impaired by alcohol **b :** having a level of alcohol in the blood that exceeds a maximum prescribed by law ⟨legally *drunk*⟩
2 : dominated by an intense feeling ⟨*drunk* with rage⟩
3 : of, relating to, or caused by intoxication **:** DRUNKEN

[3]drunk *noun* (1779)
1 : a period of drinking to intoxication or of being intoxicated ⟨a 2-day *drunk*⟩
2 : one who is drunk; *especially* **:** DRUNKARD

drunk·ard \'drəŋ-kərd\ *noun* (15th century)
: one who is habitually drunk

drunk·en \'drəŋ-kən\ *adjective* [Middle English, from Old English *druncen*, from past participle of *drincan* to drink] (before 12th century)
1 : DRUNK 1 ⟨a *drunken* driver⟩
2 *obsolete* **:** saturated with liquid
3 a : given to habitual excessive use of alcohol **b :** of, relating to, or characterized by intoxication ⟨they come from . . . broken homes, *drunken* homes —P. B. Gilliam⟩ **c :** resulting from or as if from intoxication ⟨a *drunken* brawl⟩
4 : unsteady or lurching as if from alcoholic intoxication
— **drunk·en·ly** *adverb*
— **drunk·en·ness** \-kən-nəs\ *noun*

drunk tank *noun* (1947)
: a large detention cell for arrested drunks

dru·pa·ceous \drü-'pā-shəs\ *adjective* (1822)
1 : of or relating to a drupe
2 : bearing drupes

drupe \'drüp\ *noun* [New Latin *drupa*, from Latin, overripe olive, from Greek *dryppa* olive] (circa 1753)
: a one-seeded indehiscent fruit having a hard bony endocarp, a fleshy mesocarp, and a thin exocarp that is flexible (as in the cherry) or dry and almost leathery (as in the almond)

drupe·let \'drü-plət\ *noun* (1880)
: a small drupe; *specifically* **:** one of the individual parts of an aggregate fruit (as the raspberry)

druth·ers \'drə-t͟hərz\ *noun plural* [*druther*, alteration of *would rather*] (1875)
dialect **:** free choice **:** PREFERENCE — used especially in the phrase *if one had one's druthers*

Druze *or* **Druse** \'drüz\ *noun, plural* **Druze** *or* **Druzes** *or* **Druse** *or* **Druses** *often attributive* [Arabic *Durūz*, plural, from Muḥammed ibn-Isma'īl al-*Darazīy* (died 1019) Muslim religious leader] (1786)
: a member of a religious sect originating among Muslims and centered in the mountains of Lebanon and Syria

[1]dry \'drī\ *adjective* **dri·er** *also* **dry·er** \'drī(-ə)r\; **dri·est** *also* **dry·est** \'drī-əst\ [Middle English, from Old English *drȳge*; akin to Old High German *truckan* dry, Old English *drēahnian* to drain] (before 12th century)
1 a : free or relatively free from a liquid and especially water **b :** not being in or under water ⟨*dry* land⟩ **c :** lacking precipitation or humidity ⟨*dry* climate⟩
2 a : characterized by exhaustion of a supply of liquid ⟨a *dry* well⟩ **b :** devoid of running water ⟨a *dry* ravine⟩ **c :** devoid of natural moisture ⟨my throat was *dry*⟩ **d :** no longer sticky or damp ⟨the paint is *dry*⟩ **e :** not giving milk ⟨a *dry* cow⟩ **f :** lacking freshness **:** STALE **g :** ANHYDROUS
3 a : marked by the absence or scantiness of secretions ⟨a *dry* cough⟩ **b :** not shedding or accompanied by tears ⟨a *dry* sob⟩
4 *obsolete* **:** involving no bloodshed or drowning ⟨I would fain die a *dry* death —Shakespeare⟩
5 a : marked by the absence of alcoholic beverages ⟨a *dry* party⟩ **b :** prohibiting the manufacture or distribution of alcoholic beverages
6 : served or eaten without butter or margarine ⟨*dry* toast⟩
7 a : lacking sweetness **:** SEC ⟨*dry* champagne⟩ **b :** having all or most sugar fermented to alcohol ⟨a *dry* wine⟩ ⟨*dry* beer⟩
8 a : solid as opposed to liquid ⟨*dry* groceries⟩ **b :** reduced to powder or flakes **:** DEHYDRATED ⟨*dry* milk⟩
9 : functioning without lubrication ⟨a *dry* clutch⟩
10 *of natural gas* **:** containing no recoverable hydrocarbon (as gasoline)
11 : requiring no liquid in preparation or operation ⟨a *dry* photocopying process⟩
12 a : not showing or communicating warmth, enthusiasm, or tender feeling **:** SEVERE ⟨a *dry* style of painting⟩ **b :** WEARISOME, UNINTERESTING ⟨*dry* passages of description⟩ **c :** lacking embellishment **:** PLAIN ⟨the *dry* facts⟩
13 a : not yielding what is expected or desired **:** UNPRODUCTIVE **b :** having no personal bias or emotional concern ⟨the *dry* light of reason⟩ **c :** RESERVED, ALOOF
14 : marked by matter-of-fact, ironic, or terse manner of expression ⟨*dry* wit⟩
15 : lacking smooth sound qualities ⟨a *dry* rasping voice⟩
16 : being a dry run ⟨a *dry* rehearsal⟩
— **dry·ish** \'drī-ish\ *adjective*
— **dry·ly** *adverb*
— **dry·ness** *noun*

[2]dry *verb* **dried; dry·ing** (before 12th century)
transitive verb
: to make dry
intransitive verb
: to become dry
— **dry·able** \'drī-ə-bəl\ *adjective*

[3]dry *noun, plural* **drys** (13th century)
1 : the condition of being dry **:** DRYNESS
2 : something dry; *especially* **:** a dry place
3 : PROHIBITIONIST

dry·ad \'drī-əd, -ˌad\ *noun* [Latin *dryad-*, *dryas*, from Greek, from *drys* tree — more at TREE] (14th century)
: WOOD NYMPH

dry·as·dust \'drī-əz-ˌdəst\ *adjective* (circa 1872)
: BORING
— **dryasdust** *noun*

dry cell *noun* (1893)
: a voltaic cell whose contents are not spillable — called also *dry battery*

dry–clean \'drī-ˌklēn\ (1817)
transitive verb
: to subject to dry cleaning
intransitive verb
: to undergo dry cleaning
— **dry–clean·able** \-'klē-nə-bəl\ *adjective*

dry cleaner *noun* (1897)
: one whose business is dry cleaning

dry cleaning *noun* (1897)
1 : the cleansing of fabrics with substantially nonaqueous organic solvents
2 : something that is dry-cleaned

dry–dock \'drī-ˌdäk\ *transitive verb* (1884)
: to place in a dry dock

dry dock *noun* (circa 1627)
: a dock that can be kept dry for use during the construction or repairing of ships

dry·er *variant of* DRIER

dry–eyed \'drī-ˌīd\ *adjective* (1667)
1 : not moved to tears or to empathy
2 : marked by the absence of sentimentalism or romanticism

dry farming *noun* (1878)
: farming on nonirrigated land with little rainfall that relies on moisture-conserving tillage and drought-resistant crops
— **dry–farm** *transitive verb*
— **dry farm** *noun*
— **dry farmer** *noun*

dry fly *noun* (1846)
: an artificial angling fly designed to float

Dry·gas \'drī-ˌgas\ *trademark*
— used for fuel-line antifreeze for motor vehicles

dry goods \'drī-ˌgu̇dz\ *noun plural* (1657)
: textiles, ready-to-wear clothing, and notions as distinguished especially from hardware and groceries

dry hole *noun* (1883)
: a well (as for gas or oil) that proves unproductive

dry ice *noun* (1925)
: solidified carbon dioxide

drying oil *noun* (circa 1865)
: an oil (as linseed oil) that changes readily to a hard tough elastic substance when exposed in a thin film to air

dry·land \'drī-ˌland\ *adjective* (1893)
: of, relating to, or being a relatively arid region ⟨a *dryland* wheat state⟩; *also* **:** of, adapted

to, practicing, or being agricultural methods (as dry farming) suited to such a region

dry·lot \'drī-,lät\ *noun* (1924)
: an enclosure of limited size usually bare of vegetation and used for fattening livestock

dry measure *noun* (1688)
: a series of units of capacity for dry commodities — see METRIC SYSTEM table, WEIGHT table

dry mop *noun* (1933)
: a long-handled mop for dusting floors

dry–nurse *transitive verb* (1581)
1 : to act as dry nurse to
2 : to give unnecessary supervision to

dry nurse *noun* (1598)
: a nurse who takes care of but does not breast-feed another woman's baby

dryo·pith·e·cine \,drī-ō-'pi-thə-,sīn\ *noun* [ultimately from Greek *drys* tree + *pithēkos* ape] (1948)
: any of a subfamily (Dryopithecinae) of Miocene and Pliocene Old World anthropoid apes sometimes regarded as ancestors of both man and modern anthropoids
— **dryopithecine** *adjective*

dry out *intransitive verb* (1892)
: to undergo an extended period of withdrawal from alcohol or drug use especially at a special clinic

dry·point \'drī-,pòint\ *noun* (1883)
: an engraving made with a steel or jeweled point directly into the metal plate without the use of acid as in etching; *also* **:** a print made from such an engraving

dry–rot (1870)
transitive verb
: to affect with dry rot
intransitive verb
: to become affected with dry rot

dry rot *noun* (1795)
1 a : a decay of seasoned timber caused by fungi that consume the cellulose of wood leaving a soft skeleton which is readily reduced to powder **b :** a fungal rot of plant tissue in which the affected areas are dry and often firmer than normal or more or less mummified
2 : a fungus causing dry rot
3 : decay from within caused especially by resistance to new forces

dry run *noun* (circa 1941)
1 : a practice exercise **:** REHEARSAL, TRIAL
2 : a practice firing without ammunition

dry·salt·er \'drī-,sòl-tər\ *noun* (1707)
British **:** a dealer in crude dry chemicals and dyes
— **dry·salt·ery** \-tə-rē\ *noun, British*

dry–shod \'drī-,shäd\ *adjective* (15th century)
: having dry shoes or feet

dry sink *noun* (1951)
: a wooden cabinet with a tray top for holding a wash basin

dry·stone \'drī-,stōn\ *adjective* (circa 1702)
chiefly British **:** constructed of stone without the use of mortar as an adhesive ⟨a *drystone* wall⟩

dry suit *noun* (1955)
: a close-fitting air-insulated waterproof suit for divers

dry up (14th century)
transitive verb
: to cut off the supply of
intransitive verb
1 : to disappear as if by evaporation, draining, or cutting off of a source of supply
2 : to wither or die through gradual loss of vitality
3 : to stop talking

dry·wall \'drī-,wòl\ *noun* (1952)
: PLASTERBOARD

dry wash *noun* (1872)
West **:** WASH 1d

dry well *noun* (circa 1942)
: a hole in the ground filled with gravel or rubble to receive drainage water and allow it to percolate away

d.t.'s \,dē-'tēz\ *noun plural, often D&T capitalized* (1858)
: DELIRIUM TREMENS

¹du·al \'dü(-ə)l *also* 'dyü-əl\ *adjective* [Latin *dualis,* from *duo* two — more at TWO] (1607)
1 *of grammatical number* **:** denoting reference to two
2 a : consisting of two parts or elements or having two like parts **:** DOUBLE **b :** having a double character or nature
— **du·al·ly** \-ə(l)-lē\ *adverb*

²dual *noun* (1650)
1 : the dual number of a language
2 : a linguistic form in the dual

dual carriageway *noun* (1933)
chiefly British **:** a divided highway

dual citizenship *noun* (circa 1924)
: the status of an individual who is a citizen of two or more nations

du·al·ism \'dü-ə-,li-zəm *also* 'dyü-\ *noun* (1794)
1 : a theory that considers reality to consist of two irreducible elements or modes
2 : the quality or state of being dual or of having a dual nature
3 a : a doctrine that the universe is under the dominion of two opposing principles one of which is good and the other evil **b :** a view of human beings as constituted of two irreducible elements (as matter and spirit)
— **du·al·ist** \-list\ *noun*
— **du·al·is·tic** \,dü-ə-'lis-tik, ,dyü-\ *adjective*
— **du·al·is·ti·cal·ly** \-ti-k(ə-)lē\ *adverb*

du·al·i·ty \dü-'a-lə-tē *also* dyü-\ *noun, plural* **-ties** (15th century)
: DUALISM 2; *also* **:** DICHOTOMY

du·al·ize \'dü-ə-,līz *also* 'dyü-\ *transitive verb* **-ized; -iz·ing** (1838)
: to make dual

dual–purpose *adjective* (1904)
: having breed characteristics that serve two purposes ⟨*dual-purpose* cattle that supply milk and meat⟩

¹dub \'dəb\ *transitive verb* **dubbed; dub·bing** [Middle English *dubben,* from Old English *dubbian;* akin to Old Norse *dubba* to dub, Old High German *tubili* plug] (before 12th century)
1 a : to confer knighthood on **b :** to call by a distinctive title, epithet, or nickname
2 : to trim or remove the comb and wattles of
3 a : to hit (a golf ball) poorly **b :** to execute poorly
— **dub·ber** *noun*

²dub *noun* (1887)
: one who is inept or clumsy

³dub *noun* [Middle English (Scots dialect) *dubbe*] (15th century)
chiefly Scottish **:** POOL, PUDDLE

⁴dub *transitive verb* **dubbed; dub·bing** [by shortening & alteration from *double*] (1930)
1 : to add (sound effects or new dialogue) to a film or to a radio or television production — usually used with *in*
2 : to provide (a motion-picture film) with a new sound track and especially dialogue in a different language
3 : to make a new recording of (sound or videotape already recorded); *also* **:** to mix (recorded sound or videotape from different sources) into a single recording
— **dubber** *noun*

⁵dub *noun* (1974)
: Jamaican pop music in which audio effects and spoken or chanted words are imposed on an instrumental reggae background

dub·bin \'də-bən\ *also* **dub·bing** \-bən, -biŋ\ *noun* [*dubbing,* gerund of *dub* (to dress leather)] (1781)
: a dressing of oil and tallow for leather

du·bi·ety \dù-'bī-ət-ē *also* dyü-\ *noun, plural* **-eties** [Late Latin *dubietas,* from Latin *dubius*] (1750)
1 : a usually hesitant uncertainty or doubt that tends to cause vacillation
2 : a matter of doubt
synonym see UNCERTAINTY

du·bi·ous \'dü-bē-əs *also* dyü-\ *adjective* [Latin *dubius,* from *dubare* to vacillate; akin to Latin *duo* two — more at TWO] (1548)
1 : giving rise to uncertainty: as **a :** of doubtful promise or outcome ⟨felt that our plan was a little *dubious*⟩ **b :** questionable or suspect as to true nature or quality ⟨the practice is of *dubious* legality⟩ ⟨the *dubious* honor of being the world's biggest polluter⟩
2 : unsettled in opinion **:** DOUBTFUL ⟨I was *dubious* about the plan⟩
synonym see DOUBTFUL
— **du·bi·ous·ly** *adverb*
— **du·bi·ous·ness** *noun*

du·bi·ta·ble \'dü-bə-tə-bəl *also* 'dyü-\ *adjective* [Latin *dubitabilis,* from *dubitare* to doubt — more at DOUBT] (circa 1616)
: open to doubt or question

du·bi·ta·tion \,d(y)ü-bə-'tā-shən\ *noun* (15th century)
archaic **:** DOUBT

du·cal \'dü-kəl *also* 'dyü-\ *adjective* [Middle English, from Middle French, from Late Latin *ducalis* of a leader, from Latin *duc-, dux* leader — more at DUKE] (15th century)
: of or relating to a duke or dukedom
— **du·cal·ly** \-kə-lē\ *adverb*

duc·at \'də-kət\ *noun* [Middle English, from Middle French, from Old Italian *ducato* coin with the doge's portrait on it, from *duca* doge, from Late Greek *douk-, doux* leader, from Latin *duc-, dux*] (14th century)
1 : a usually gold coin formerly used in various European countries
2 : TICKET 2

du·ce \'dü-(,)chā\ *noun* [Italian (*Il*) *Duce,* literally, the leader, title of Benito Mussolini, from Latin *duc-, dux*] (1923)
: LEADER — used especially for the leader of the Italian Fascist party

Du·chenne \dü-'shen, də-\ *also* **Du·chenne's** \-'shenz\ *adjective* [Guillaume Armand *Duchenne* (died 1875) French neurologist] (circa 1882)
: relating to or being a severe form of muscular dystrophy of males that affects the muscles of the pelvic and shoulder girdles and the pectoral muscles first and is inherited as a sex-linked recessive trait

duch·ess \'də-chəs\ *noun* [Middle English *duchesse,* from Middle French, from *duc* duke] (14th century)
1 : the wife or widow of a duke
2 : a woman who holds the rank of duke in her own right
usage see -ESS

duchy \'də-chē\ *noun, plural* **duch·ies** [Middle English *duche,* from Middle French *duché,* from *duc*] (14th century)
1 : the territory of a duke or duchess **:** DUKEDOM
2 : special domain

¹duck \'dək\ *noun, plural* **ducks** *often attributive* [Middle English *duk, doke,* from Old English *dūce*] (before 12th century)
1 *or plural* **duck a :** any of various swimming birds (family Anatidae, the duck family) in which the neck and legs are short, the body more or less depressed, the bill often broad and flat, and the sexes almost always different from each other in plumage **b :** the flesh of any of these birds used as food
2 : a female duck — compare DRAKE
3 *chiefly British* **:** DARLING — often used in plural but sing. in construction
4 : PERSON, CREATURE

²duck *verb* [Middle English *douken;* akin to Old High German *tūhhan* to dive, Old English *dūce* duck] (14th century)
transitive verb
1 : to thrust under water
2 : to lower (as the head) quickly **:** BOW
3 : AVOID, EVADE ⟨*duck* the issue⟩
intransitive verb
1 a : to plunge under the surface of water **b :** to descend suddenly **:** DIP

2 a : to lower the head or body suddenly **:** DODGE **b :** BOW, BOB
3 a : to move quickly **b :** to evade a duty, question, or responsibility
— duck·er *noun*

duck 1a (male): *1* bean, *2* bill, *3* nostril, *4* head, *5* eye, *6* auricular region, *7* neck, *8* cape, *9* shoulder, *10, 11* wing coverts, *12* saddle, *13* secondaries, *14* primaries, *15* rump, *16* drake feathers, *17* tail, *18* tail coverts, *19* down, *20* shank, *21* web, *22* breast, *23* wing front, *24* wing bow

3**duck** *noun* (1554)
: an instance of ducking

4**duck** *noun* [Dutch *doek* cloth; akin to Old High German *tuoh* cloth] (1640)
1 : a durable closely woven usually cotton fabric
2 *plural* **:** light clothes and especially trousers made of duck

duck·bill \ˈdək-ˌbil\ *noun* (1840)
1 : PLATYPUS
2 : HADROSAUR

duck–billed dinosaur *noun* (circa 1928)
: HADROSAUR

duck–billed platypus *also* **duckbill platypus** *noun* (1799)
: PLATYPUS

duck·board \-ˌbōrd, -ˌbȯrd\ *noun* (1917)
: a boardwalk or slatted flooring laid on a wet, muddy, or cold surface — usually used in plural

duck call *noun* (1872)
: a device for imitating the calls of ducks

duck hook *noun* (1973)
: a pronounced and unintended hook in golf

ducking stool *noun* (1597)
: a seat attached to a plank and formerly used to plunge culprits tied to it into water

duck·ling \ˈdək-liŋ, ˈdə-kliŋ\ *noun* (15th century)
: a young duck

duck·pin \-ˌpin\ *noun* (circa 1911)
1 : a small bowling pin shorter than a tenpin but proportionately wider at mid-diameter
2 *plural but singular in construction* **:** a bowling game using duckpins

ducks and drakes *or* **duck and drake** *noun* (1583)
: the pastime of skimming flat stones or shells along the surface of calm water
— play ducks and drakes with *or* **make ducks and drakes of :** to use recklessly **:** SQUANDER ⟨*played ducks and drakes with* his money⟩

duck soup *noun* (1912)
: something easy to do

duck·tail \ˈdək-ˌtāl\ *noun* [from its resemblance to the tail of a duck] (1948)
: a hairstyle in which the hair on each side is slicked back to meet in a ridge at the back of the head

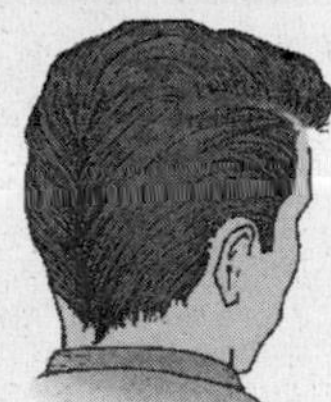
ducktail

duck·walk \ˈdək ˌwȯk\ *intransitive verb* (1950)
: to walk while in a crouch or full squatting position

duck·weed \ˈdək-ˌwēd\ *noun* (15th century)
: a small floating aquatic monocotyledonous plant (family Lemnaceae, the duckweed family)

ducky \ˈdə-kē\ *adjective* **duck·i·er; -est** (1897)
1 : DARLING, CUTE ⟨a *ducky* little tearoom⟩
2 : SATISFACTORY, FINE ⟨everything is just *ducky*⟩

1**duct** \ˈdəkt\ *noun* [New Latin *ductus,* from Medieval Latin, aqueduct, from Latin, act of leading, from *ducere* to lead — more at TOW] (1667)
1 : a bodily tube or vessel especially when carrying the secretion of a gland
2 a : a pipe, tube, or channel that conveys a substance **b :** a pipe or tubular runway for carrying an electric power line, telephone cables, or other conductors
3 a : a continuous tube formed in plant tissue by a row of elongated cells that have lost their intervening end walls **b :** an elongated cavity (as a resin canal of a conifer) formed by disintegration or separation of cells
4 : a layer (as in the atmosphere or the ocean) which occurs under usually abnormal conditions and in which radio or sound waves are confined to a restricted path
— duc·tal \ˈdək-tᵊl\ *adjective*
— duct·less \ˈdək(t)-ləs\ *adjective*

2**duct** *transitive verb* (1936)
1 : to enclose in a duct
2 : to convey (as a gas) through a duct; *also* **:** to propagate (as radio waves) through a duct

duc·tile \ˈdək-tᵊl, -ˌtīl\ *adjective* [Middle French & Latin; Middle French, from Latin *ductilis,* from *ducere*] (14th century)
1 : capable of being drawn out or hammered thin ⟨*ductile* iron⟩
2 : easily led or influenced
3 : capable of being fashioned into a new form
synonym see PLASTIC
— duc·til·i·ty \ˌdək-ˈti-lə-tē\ *noun*

duct·ing \ˈdək-tiŋ\ *noun* (1945)
: a system of ducts; *also* **:** the material composing a duct

ductless gland *noun* (circa 1852)
: ENDOCRINE GLAND

duct tape \ˈdək(t)-\ *noun* (1970)
: a wide silvery cloth adhesive tape designed for sealing joints in heating or air-conditioning ducts

duct·ule \ˈdək-(ˌ)chül\ *noun* (1883)
: a small duct

duc·tus ar·te·ri·o·sus \ˈdək-təs-är-ˌtir-ē-ˈō-səs\ *noun* [New Latin, literally, arterial duct] (1811)
: a short broad vessel in the fetus that connects the pulmonary artery with the aorta and conducts most of the blood directly from the right ventricle to the aorta bypassing the lungs

duct·work \ˈdəkt-ˌwərk\ *noun* (1934)
: DUCTING

1**dud** \ˈdəd\ *noun* [Middle English *dudde*] (1567)
1 *plural* **a :** CLOTHING **b :** personal belongings
2 a : one that is ineffectual; *also* **:** FAILURE ⟨a box-office *dud*⟩ **b :** MISFIT
3 : a bomb or missile that fails to explode

2**dud** *adjective* (1903)
: of little or no worth **:** VALUELESS ⟨*dud* checks⟩

dud·die *or* **dud·dy** \ˈdə-dē\ *adjective* (1718)
Scottish **:** RAGGED, TATTERED

1**dude** \ˈdüd *also* ˈdyüd\ *noun* [origin unknown] (1883)
1 : a man extremely fastidious in dress and manner **:** DANDY
2 : a city dweller unfamiliar with life on the range; *especially* **:** an Easterner in the West
3 : FELLOW, GUY
— dud·ish \ˈd(y)üd-ish\ *adjective*
— dud·ish·ly *adverb*

2**dude** *transitive verb* **dud·ed; dud·ing** (1899)
: DRESS UP — usually used with *up*

du·deen \dü-ˈdēn\ *noun* [Irish *dúidín,* diminutive of *dúd* pipe] (1841)
: a short tobacco pipe made of clay

dude ranch *noun* (1921)
: a vacation resort offering activities (as horseback riding) typical of western ranches

1**dud·geon** \ˈdə-jən\ *noun* [Middle English *dogeon,* from Anglo-French *digeon*] (15th century)
1 *obsolete* **:** a wood used especially for dagger hilts
2 a *archaic* **:** a dagger with a handle of dudgeon **b** *obsolete* **:** a haft made of dudgeon

2**dudgeon** *noun* [origin unknown] (1573)
: a fit or state of indignation — often used in the phrase *in high dudgeon*
synonym see OFFENSE

1**due** \ˈdü, ˈdyü\ *adjective* [Middle English, from Middle French *deu,* past participle of *devoir* to owe, from Latin *debēre* — more at DEBT] (14th century)
1 : owed or owing as a debt
2 a : owed or owing as a natural or moral right ⟨everyone's right to dissent . . . is *due* the full protection of the Constitution —Nat Hentoff⟩ **b :** according to accepted notions or procedures **:** APPROPRIATE
3 a : satisfying or capable of satisfying a need, obligation, or duty **:** ADEQUATE **b :** REGULAR, LAWFUL ⟨*due* proof of loss⟩
4 : capable of being attributed **:** ASCRIBABLE — used with *to* ⟨this advance is partly *due* to a few men of genius —A. N. Whitehead⟩
5 : having reached the date at which payment is required **:** PAYABLE
6 : required or expected in the prescribed, normal, or logical course of events **:** SCHEDULED; *also* **:** expected to give birth
— due·ness *noun*

2**due** *noun* (15th century)
: something due or owed: as **a :** something that rightfully belongs to one **b :** a payment or obligation required by law or custom **:** DEBT **c** *plural* **:** FEES, CHARGES

3**due** *adverb* (1582)
1 : DIRECTLY, EXACTLY ⟨*due* north⟩
2 *obsolete* **:** DULY

due diligence *noun* (1903)
: the care that a reasonable person exercises under the circumstances to avoid harm to other persons or their property

1**du·el** \ˈdü-əl *also* ˈdyü-\ *noun* [Middle English, from Medieval Latin *duellum,* from Old Latin, war] (15th century)
1 : a combat between two persons; *specifically* **:** a formal combat with weapons fought between two persons in the presence of witnesses
2 : a conflict between antagonistic persons, ideas, or forces; *also* **:** a hard-fought contest between two opponents

2**duel** *verb* **du·eled** *or* **du·elled; du·el·ing** *or* **du·el·ling** (circa 1645)
intransitive verb
: to fight a duel
transitive verb
: to encounter (an opponent) in a duel
— du·el·er *or* **du·el·ler** *noun*
— du·el·ist *or* **du·el·list** \ˈdü-ə-list\ *noun*

du·el·lo \d(y)ü-ˈe-(ˌ)lō\ *noun, plural* **-los** [Italian, from Medieval Latin *duellum*] (1588)
1 : the rules or practice of dueling
2 : DUEL

du·en·de \dü-ˈen-(ˌ)dā\ *noun* [Spanish dialect, charm, from Spanish, ghost, goblin, probably from *duen de casa,* from *dueño de casa* owner of a house] (1964)
: the power to attract through personal magnetism and charm

du·en·na \dü-ˈe-nə, ˈdyü-\ *noun* [Spanish *dueña,* from Latin *domina* mistress — more at DAME] (1623)
1 : an elderly woman serving as governess and companion to the younger ladies in a Spanish or a Portuguese family
2 : CHAPERON
— du·en·na·ship \-ˌship\ *noun*

due process *noun* (1791)

\ə\ **abut** \ᵊ\ **kitten** \ər\ **further** \a\ **ash** \ā\ **ace** \ä\ **mop, mar** \au̇\ **out** \ch\ **chin** \e\ **bet** \ē\ **easy** \g\ **go** \i\ **hit** \ī\ **ice** \j\ **job** \ŋ\ **sing** \ō\ **go** \ȯ\ **law** \ȯi\ **boy** \th\ **thin** \t͟h\ **the** \ü\ **loot** \u̇\ **foot** \y\ **yet** \zh\ **vision** *see also* Guide to Pronunciation

1 : a course of formal proceedings (as legal proceedings) carried out regularly and in accordance with established rules and principles — called also *procedural due process*
2 : a judicial requirement that enacted laws may not contain provisions that result in the unfair, arbitrary, or unreasonable treatment of an individual — called also *substantive due process*

¹du·et \dü-'et *also* dyü-\ *noun* [Italian *duetto,* diminutive of *duo*] (circa 1740)
: a composition for two performers

²duet *intransitive verb* **du·et·ted; du·et·ting** (1822)
: to perform a duet

due to *preposition* (1897)
: as a result of **:** BECAUSE OF ⟨*due to* the complaints of uptight parents . . . he lost his job —Herbert Gold⟩ ◻

¹duff \'dəf\ *noun* [English dialect, alteration of *dough*] (1816)
1 : a boiled or steamed pudding often containing dried fruit
2 : the partly decayed organic matter on the forest floor
3 : fine coal **:** SLACK

²duff *noun* [origin unknown] (circa 1888)
: BUTTOCKS ⟨get off your *duff*⟩

³duff *adjective* [*duff,* noun, something worthless, from ¹*duff*] (circa 1889)
British **:** INFERIOR, WORTHLESS

duf·fel *or* **duf·fle** \'də-fəl\ *noun* [Dutch *duffel,* from *Duffel,* Belgium] (1677)
1 : a coarse heavy woolen material with a thick nap
2 : transportable personal belongings, equipment, and supplies
3 : DUFFEL BAG
4 : DUFFLE COAT

duffel bag *noun* (1917)
: a large cylindrical fabric bag for personal belongings

duf·fer \'də-fər\ *noun* [perhaps from *duff,* noun, something worthless] (1756)
1 a : a peddler especially of cheap flashy articles **b :** something counterfeit or worthless
2 : an incompetent, ineffectual, or clumsy person; *especially* **:** a mediocre golfer
3 *Australian* **:** a cattle rustler

duffle coat \'də-fəl-\ *or* **duffel coat** *noun* (1684)
: a heavy usually woolen medium-length coat with toggle fasteners and a hood

¹dug *past and past participle of* DIG

²dug \'dəg\ *noun* [perhaps of Scandinavian origin; akin to Old Swedish *dæggia* to suckle; akin to Old English *delu* nipple — more at FEMININE] (1530)
: UDDER; *also* **:** TEAT — usually used of a suckling animal; usually considered vulgar when used of a woman

du·gong \'dü-ˌgäŋ, -ˌgȯŋ\ *noun* [New Latin, genus name, modification of Malay *duyong* dugong] (1800)
: an aquatic herbivorous mammal of a monotypic genus (*Dugong*) that has a bilobed tail and in the male upper incisors altered into short tusks, is related to the manatee, and inhabits warm coastal regions — called also *sea cow*

dugong

dug·out \'dəg-ˌau̇t\ *noun* (1819)
1 : a boat made by hollowing out a large log
2 a : a shelter dug in a hillside; *also* **:** a shelter dug in the ground and roofed with sod **b :** an area in the side of a trench for quarters, storage, or protection
3 : either of two low shelters on either side of and facing a baseball diamond that contain the players' benches

dui·ker \'dī-kər\ *noun* [Afrikaans, literally, diver, from *duik* to dive, from Middle Dutch *dūken;* akin to Old High German *tūhhan* to dive — more at DUCK] (1777)
: any of several small African antelopes comprising two genera (*Cephalophus* and *Sylvicapra*)

duit *variant of* DOIT

du jour \dü-'zhər, də-, -'zhu̇r, -'zhür\ *adjective* [French, literally, of the day] (1969)
1 : made for a particular day — used of an item not specified on the regular menu ⟨soup *du jour*⟩
2 : popular, fashionable, or prominent at a particular time ⟨the buzzword *du jour*⟩

¹duke \'dük *also* 'dyük\ *noun* [Middle English, from Old French *duc,* from Latin *duc-, dux,* from *ducere* to lead — more at TOW] (12th century)
1 : a sovereign male ruler of a continental European duchy
2 : a nobleman of the highest hereditary rank; *especially* **:** a member of the highest grade of the British peerage
3 [probably from *dukes of York,* rhyming slang for *fork* (hand, fist)] *slang* **:** FIST, HAND — usually used in plural
— duke·dom \-dəm\ *noun*

²duke *intransitive verb* **duked; duk·ing** (circa 1947)
: FIGHT
— duke it out : to engage in a fight and especially a fistfight

Du·kho·bor *variant of* DOUKHOBOR

dul·cet \'dəl-sət\ *adjective* [Middle English *doucet,* from Middle French, from *douz* sweet, from Latin *dulcis;* perhaps akin to Greek *glykys* sweet] (14th century)
1 : sweet to the taste
2 : pleasing to the ear
3 : generally pleasing or agreeable
— dul·cet·ly *adverb*

dul·ci·fy \'dəl-sə-ˌfī\ *transitive verb* **-fied; -fy·ing** [Late Latin *dulcificare,* from Latin *dulcis*] (1599)
1 : to make sweet
2 : to make agreeable **:** MOLLIFY

dul·ci·mer \'dəl-sə-mər\ *noun* [Middle English *dowcemere,* from Middle French *doulcemer,* from Old Italian *dolcimelo,* from *dolce* sweet, from Latin *dulcis*] (15th century)
1 : a stringed instrument of trapezoidal shape played with light hammers held in the hands
2 *or* **dul·ci·more** \-ˌmōr, -ˌmȯr\ **:** an American folk instrument with three or four strings stretched over an elongate fretted sound box that is held on the lap and played by plucking or strumming

dul·ci·nea \ˌdəl-sə-'nē-ə, -'si-nē-ə\ *noun* [Spanish, from *Dulcinea* del Toboso, beloved of Don Quixote] (1748)
: MISTRESS, SWEETHEART

¹dull \'dəl\ *adjective* [Middle English *dul;* akin to Old English *dol* foolish, Old Irish *dall* blind] (13th century)
1 : mentally slow **:** STUPID
2 a : slow in perception or sensibility **:** INSENSIBLE **b :** lacking zest or vivacity **:** LISTLESS
3 : slow in action **:** SLUGGISH
4 a : lacking in force, intensity, or sharpness **b :** not resonant or ringing ⟨a *dull* booming sound⟩
5 : lacking sharpness of edge or point
6 : lacking brilliance or luster
7 *of a color* **:** low in saturation and low in lightness
8 : CLOUDY
9 : TEDIOUS, UNINTERESTING ☆
— dull·ness *or* **dul·ness** \'dəl-nəs\ *noun*
— dul·ly \'də(l)-lē\ *adverb*

²dull (13th century)
transitive verb
: to make dull
intransitive verb
: to become dull

dull·ard \'də-lərd\ *noun* (15th century)
: a stupid or unimaginative person

dull·ish \'də-lish\ *adjective* (14th century)
: somewhat dull
— dull·ish·ly *adverb*

dulls·ville \'dəlz-ˌvil\ *noun* (circa 1960)
: something or some place that is dull or boring; *also* **:** BOREDOM

dulse \'dəls\ *noun* [modification of Scottish Gaelic *duileasg;* akin to Welsh *delysg* dulse] (circa 1698)
: any of several coarse red seaweeds (especially *Palmaria palmata* synonym *Rhodymenia palmata*) found especially in northern latitudes and used as a food condiment

du·ly \'dü-lē *also* 'dyü-\ *adverb* (14th century)
: in a due manner or time **:** PROPERLY

du·ma \'dü-mə, -(ˌ)mä\ *noun* [Russian, from Old Russian, council, thought, probably of Germanic origin; akin to Old English *dōm* judgment — more at DOOM] (circa 1870)
: a representative council in Russia; *especially, often capitalized* **:** the principal legislative assembly in czarist Russia

¹dumb \'dəm\ *adjective* [Middle English, from Old English; akin to Old High German *tumb* mute] (before 12th century)
1 : lacking the power of speech ⟨deaf and *dumb* from birth⟩ ⟨*dumb* animals⟩
2 : temporarily unable to speak (as from shock or astonishment) ⟨struck *dumb* with fear⟩
3 : not expressed in uttered words ⟨*dumb* grief⟩
4 : SILENT; *also* **:** TACITURN
5 : lacking some usual attribute or accompaniment; *especially* **:** having no means of self-propulsion ⟨*dumb* barge⟩
6 a : markedly lacking in intelligence **:** STUPID **b :** showing a lack of intelligence **c :** having little or no meaning — sometimes used in the phrase *dumb luck*
7 : not having the capability to process data ⟨a *dumb* terminal⟩ — compare INTELLIGENT 3a ◻
synonym see STUPID
— dumb·ly \'dəm-lē\ *adverb*
— dumb·ness *noun*

²dumb *transitive verb* (1608)
: to make silent **:** DEADEN ⟨would lie around, *dumbed* by the drugs —Norman Mailer⟩

dumb·bell \'dəm-ˌbel\ *noun* (1785)

☆ SYNONYMS
Dull, blunt, obtuse mean not sharp, keen, or acute. DULL suggests a lack or loss of keenness, zest, or pungency ⟨a *dull* pain⟩ ⟨a *dull* mind⟩. BLUNT suggests an inherent lack of sharpness or quickness of feeling or perception ⟨a person of *blunt* sensibility⟩. OBTUSE implies such bluntness as makes one insensitive in perception or imagination ⟨too *obtuse* to realize the remark was offensive⟩. See in addition STUPID.

◻ USAGE
due to The objection to *due to* as a preposition is only a continuation of disagreements that began in the 18th century over the proper uses of *owing* and *due. Due to* is as grammatically sound as *owing to,* which is frequently recommended in its place. It has been and is used by reputable writers and has been recognized as standard for decades. There is no solid reason to avoid *due to.*

dumb There is evidence that, when applied to persons who cannot speak, *dumb* has come to be considered offensive.

1 : a short bar with weights at each end that is used usually in pairs for exercise
2 : one that is dull and stupid

dumb·cane \'dəm-ˌkān\ *noun* [from the fact that chewing it causes the tongue and throat to swell] (1696)
: DIEFFENBACHIA

dumb down *transitive verb* (1980)
: to lower the level of difficulty and the intellectual content of (as a textbook)

dumb·found *or* **dum·found** \ˌdəm-'faund, 'dəm-ˌ\ *transitive verb* [*dumb* + *-found* (as in *confound*)] (1653)
: to confound briefly and usually with astonishment
synonym see PUZZLE

dumb·foun·der \-'faun-dər\ *transitive verb* (1710)
British **:** DUMBFOUND

dumb·head \'dəm-ˌhed\ *noun* [probably translation of German *dummkopf*] (1887)
slang **:** a stupid person **:** BLOCKHEAD

dumb show *noun* (1561)
1 : a part of a play presented in pantomime
2 : signs and gestures without words **:** PANTOMIME

dumb·struck \'dəm-ˌstrək\ *adjective* (1887)
: made silent by astonishment

dumb·wait·er \'dəm-ˌwā-tər, ˌdəm-'\ *noun* (1749)
1 : a portable serving table or stand
2 : a small elevator used for conveying food and dishes from one story of a building to another

dum·dum \'dəm-ˌdəm\ *noun* [*Dum Dum,* arsenal near Calcutta, India] (circa 1889)
: a bullet (as one with a hollow point) that expands more than usual upon hitting an object

dum–dum *noun* [reduplication of [1]*dumb*] (1928)
: a stupid person **:** DUMMY

dum·ka \'dum-kə\ *noun, plural* **dum·ky** \-kē\ [Czech, elegy, from Ukrainian, diminutive of *duma* narrative folk poem, from Old Russian] (1895)
: a musical composition containing alternating sad and gay passages

dumm·kopf \'dum-ˌkopf\ *noun* [German, from *dumm* stupid + *Kopf* head] (1809)
: BLOCKHEAD

[1]**dum·my** \'də-mē\ *noun, plural* **dummies** [[1]*dumb* + [4]*-y*] (1598)
1 a : one who is incapable of speaking **b :** one who is habitually silent **c :** one who is stupid
2 a : the exposed hand in bridge played by the declarer in addition to his own hand **b :** a bridge player whose hand is a dummy
3 : an imitation, copy, or likeness of something used as a substitute: as **a :** MANNEQUIN **b :** a stuffed figure or cylindrical bag used by football players for tackling and blocking practice **c :** a large puppet usually having movable features (as mouth and arms) manipulated by a ventriloquist **d** *chiefly British* **:** PACIFIER 2
4 : one seeming to act independently but in reality controlled by another
5 a : a mock-up of a proposed publication (as a book or magazine) **b :** a set of pages (as for a newspaper or magazine) with the position of text and artwork indicated for the printer

[2]**dummy** *adjective* (1816)
1 a : having the appearance of being real **:** ARTIFICIAL ⟨*dummy* foods in the display case⟩ **b :** existing in name only **:** FICTITIOUS ⟨*dummy* corporations⟩
2 : apparently acting for oneself while really acting for or at the direction of another ⟨a *dummy* director⟩

[3]**dummy** *transitive verb* **dum·mied; dum·my·ing** (1928)
: to make a dummy of (as a publication) — often used with *up* ⟨*dummied* up the front page⟩

dummy up *intransitive verb* (1926)
: to say nothing **:** CLAM UP

dummy variable *noun* (1957)
: an arbitrary mathematical symbol or variable that can be replaced by another without affecting the value of the expression in which it occurs

du·mor·ti·er·ite \du-'mor-tē-ə-ˌrīt *also* dyu-\ *noun* [French *dumortiérite,* from Eugène *Dumortier* (died 1876) French paleontologist] (1881)
: a bright blue or greenish blue mineral consisting of a silicate of aluminum and used especially for jewelry

[1]**dump** \'dəmp\ *verb* [perhaps from Middle Dutch *dompen* to immerse, topple; akin to Old Norse *dumpa* to thump, fall suddenly] (14th century)
transitive verb
1 a : to let fall in or as if in a heap or mass **b :** to get rid of unceremoniously or irresponsibly **c :** JETTISON ⟨an airplane *dumping* gasoline⟩
2 *slang* **:** to knock down **:** BEAT ⟨the man rushed out and *dumped* him —John Corry⟩
3 : to sell in quantity at a very low price; *specifically* **:** to sell abroad at less than the market price at home
4 : to copy (data in a computer's internal storage) to an external storage or output device
5 : to throw (as a pass) short and softly — often used with *off*
intransitive verb
1 : to fall abruptly **:** PLUNGE
2 : to dump refuse
— dump·er *noun*
— dump on : to treat disrespectfully; *especially* **:** BELITTLE, BAD-MOUTH

[2]**dump** *noun* (1784)
1 a : an accumulation of refuse and discarded materials **b :** a place where such materials are dumped
2 a : a quantity of reserve materials accumulated at one place **b :** a place where such materials are stored ⟨ammunition *dump*⟩
3 : a disorderly, slovenly, or objectionable place
4 : an instance of dumping data stored in a computer
5 : an act of defecation — usually used with *take;* often considered vulgar

dump·ing *noun* (1857)
: the act of one that dumps; *especially* **:** the selling of goods in quantity at below market price

dumping ground *noun* (1857)
: a place to which unwanted people or things are sent

dump·ish \'dəm-pish\ *adjective* [*dumps*] (1519)
: SAD, MELANCHOLY

dump·ling \'dəm-pliŋ\ *noun* [perhaps alteration of *lump*] (circa 1600)
1 a : a small mass of leavened dough cooked by boiling or steaming **b :** a usually baked dessert of fruit wrapped in dough
2 : something soft and rounded like a dumpling; *especially* **:** a short fat person or animal

dumps \'dəm(p)s\ *noun plural* [probably from Dutch *domp* haze, from Middle Dutch *damp* — more at DAMP] (1529)
: a gloomy state of mind **:** DESPONDENCY ⟨in the *dumps*⟩

Dump·ster \'dəm(p)-stər\ *trademark*
— used for a large trash receptacle

dump truck *noun* (1930)
: an automotive truck for the transportation of bulk material that has a body which tilts to dump its contents

dumpy \'dəm-pē\ *adjective* **dump·i·er; -est** [English dialect *dump* (lump)] (1750)
1 : being short and thick in build **:** SQUAT
2 : SHABBY, DINGY
— dump·i·ly \-pə-lē\ *adverb*
— dump·i·ness \-pē-nəs\ *noun*

dumpy level *noun* (1838)
: a surveyor's level with a short telescope rigidly fixed and rotating only in a horizontal plane

[1]**dun** \'dən\ *adjective* [Middle English, from Old English *dunn* — more at DUSK] (before 12th century)
1 a : having the color dun **b** *of a horse* **:** having a grayish yellow coat with black mane and tail
2 : marked by dullness and drabness
— dun·ness \'dən-nəs\ *noun*

[2]**dun** *noun* (14th century)
1 : a dun horse
2 : a variable color averaging a nearly neutral slightly brownish dark gray
3 : a subadult mayfly; *also* **:** an artificial fly tied to imitate such an insect

[3]**dun** *transitive verb* **dunned; dun·ning** [origin unknown] (circa 1626)
1 : to make persistent demands upon for payment
2 : PLAGUE, PESTER

[4]**dun** *noun* (1628)
1 : one who duns
2 : an urgent request; *especially* **:** a demand for payment

Dun·can Phyfe \ˌdəŋ-kən-'fīf\ *adjective* (1926)
: of, relating to, or constituting furniture designed and built by or in the style of Duncan Phyfe

dunce \'dən(t)s\ *noun* [John *Duns* Scotus, whose once accepted writings were ridiculed in the 16th century] (1587)
: one who is slow-witted or stupid ◆

dunce cap *noun* (1840)
: a conical cap formerly used as a punishment for slow learners at school — called also *dunce's cap*

dun·der·head \'dən-dər-ˌhed\ *noun* [perhaps from Dutch *donder* thunder + English *head;* akin to Old High German *thonar* thunder — more at THUNDER] (circa 1625)
: DUNCE, BLOCKHEAD
— dun·der·head·ed \ˌdən-dər-'he-dəd\ *adjective*

dun·drea·ries \ˌdən-'drir-ēz\ *noun plural, often capitalized* [Lord *Dundreary,* character in the play *Our American Cousin* (1858), by Tom Taylor] (circa 1922)
: long flowing sideburns

dune \'dün *also* 'dyün\ *noun* [French, from Old French, from Middle Dutch; akin to Old English *dūn* down — more at DOWN] (1790)
: a hill or ridge of sand piled up by the wind

◇ WORD HISTORY

dunce The irony of *dunce* lies in the fact that this synonym for *dullard* is derived from the name of one of the most brilliant thinkers of the later Middle Ages. *Duns* was the name of a Scottish market town and the birthplace in the mid 13th century of John Duns Scotus, one of the luminaries of Scholasticism. So ingenious were the theological and metaphysical speculations of this thinker that he was given the epithet of *Doctor Subtilis* ("the Subtle Doctor"). His followers, who were known as *Scotists* or *dunsmen,* were a powerful conservative element in English universities through the 16th century. They tended to resist both the new learning of humanism and the complaints of church reformers. As a result, *dunsman* and the shorter form *duns* (later respelled as we have it today) became terms of opprobrium and scorn meaning first "sophist" or "pedant" and gradually taking on the modern sense "slow-witted person."

— **dune·like** \-ˌlīk\ *adjective*

dune buggy *noun* (1956)
: an off-road motor vehicle with oversize tires for use especially on sand

dune·land \ˈdün-ˌland\ *noun* (1922)
: an area having many dunes

¹dung \ˈdəŋ\ *noun* [Middle English, from Old English; akin to Old Norse *dyngja* manure pile] (before 12th century)
1 : the excrement of an animal **:** MANURE
2 : something repulsive
— **dungy** \ˈdəŋ-ē\ *adjective*

²dung (before 12th century)
transitive verb
: to fertilize or dress with manure
intransitive verb
: DEFECATE

dun·ga·ree \ˌdəŋ-gə-ˈrē, ˈdəŋ-gə-ˌ\ *noun* [Hindi *dūgrī*] (1673)
1 : a heavy coarse durable cotton twill woven from colored yarns; *specifically* **:** blue denim
2 *plural* **:** clothes made usually of blue denim

dung beetle *noun* (circa 1634)
: a beetle (as a tumblebug) that rolls balls of dung in which to lay eggs and on which the larvae feed

Dunge·ness crab \ˌdən-jə-ˌnes-\ *noun* [*Dungeness,* village on the Strait of Juan de Fuca, Washington] (1925)
: a large edible crab (*Cancer magister*) of the Pacific coast of North America from Alaska to California

dun·geon \ˈdən-jən\ *noun* [Middle English *donjon,* from Middle French, from (assumed) Vulgar Latin *domnion-, domnio* keep, mastery, from Latin *dominus* lord — more at DOMINATE] (14th century)
1 : DONJON
2 : a dark usually underground prison or vault ◇

dung·hill \ˈdəŋ-ˌhil\ *noun* (14th century)
1 : a heap of dung
2 : something (as a situation or condition) that is repulsive or degraded

du·nite \ˈdü-ˌnīt, ˈdə-\ *noun* [Mount *Dun,* New Zealand] (circa 1868)
: a granitoid igneous rock consisting chiefly of olivine
— **du·nit·ic** \dü-ˈni-tik, ˌdə-\ *adjective*

¹dunk \ˈdəŋk\ *verb* [Pennsylvania German *dunke,* from Middle High German *dunken,* from Old High German *dunkōn* — more at TINGE] (1919)
transitive verb
1 : to dip (as a piece of bread) into a beverage while eating
2 : to dip or submerge temporarily in liquid
3 : to throw (a basketball) into the basket from above the rim
intransitive verb
1 : to submerge oneself in water
2 : to make a dunk shot in basketball

²dunk *noun* (circa 1944)
: the act or action of dunking; *especially* **:** DUNK SHOT

Dun·ker \ˈdəŋ-kər\ *or* **Dun·kard** \-kərd\ *noun* [Pennsylvania German *Dunker,* from *dunke*] (1744)
: a member of the Church of the Brethren or any of several other originally German Baptist denominations practicing trine immersion and love feasts and refusing to take oaths or to perform military service

Dun·kirk \ˈdən-ˌkərk, ˌdən-ˈ\ *noun* [*Dunkirk* or *Dunkerque,* France, scene of the evacuation of Allied forces in 1940] (1941)
1 : a retreat to avoid total defeat
2 : a crisis situation that requires a desperate last effort to forestall certain failure ⟨a *Dunkirk* for U.S. foreign policy —*Time*⟩

dunk shot *noun* (circa 1961)
: a shot in basketball made by jumping high into the air and throwing the ball down through the basket

dun·lin \ˈdən-lən\ *noun, plural* **dunlins** *or* **dunlin** [¹*dun* + *-lin* (alteration of *-ling*)] (circa 1532)
: a small widely distributed sandpiper (*Calidris alpina*) largely cinnamon to rusty brown above and white below

Dun·lop \ˈdən-ˌläp, ˌdən-ˈ\ *noun* [*Dunlop,* Ayrshire, Scotland] (circa 1780)
: a Scottish cheese similar to cheddar

dun·nage \ˈdə-nij\ *noun* [origin unknown] (15th century)
1 : loose materials used to support and protect cargo in a ship's hold; *also* **:** padding in a shipping container
2 : BAGGAGE

duo \ˈdü-(ˌ)ō *also* ˈdyü-\ *noun, plural* **du·os** [Italian, from Latin, two — more at TWO] (1590)
1 : DUET
2 : PAIR 2

duo- *combining form* [Latin *duo*]
: two ⟨*duo*logue⟩

duo·de·cil·lion \ˌdü-ō-di-ˈsil-yən, ˌdyü-\ *noun, often attributive* [Latin *duodecim* twelve + English *-illion* (as in *million*)] (1875)
— see NUMBER table

duo·dec·i·mal \ˌdü-ə-ˈde-sə-məl, ˌdyü-\ *adjective* [Latin *duodecim* — more at DOZEN] (1727)
: of, relating to, or proceeding by twelve or the scale of twelves
— **duodecimal** *noun*

duo·dec·i·mo \-ˌmō\ *noun, plural* **-mos** [Latin, ablative of *duodecimus* twelfth, from *duodecim*] (1658)
: TWELVEMO

du·o·de·num \ˌdü-ə-ˈdē-nəm, du̇-ˈä-dᵊn-əm *also* (ˌ)dyü-\ *noun, plural* **-de·na** \-ˈdē-nə, -dᵊn-ə\ *or* **-denums** [Middle English, from Medieval Latin, from Latin *duodeni* twelve each, from *duodecim* twelve; from its length, about 12 fingers' breadth] (14th century)
: the first part of the small intestine extending from the pylorus to the jejunum
— **du·o·de·nal** \-ˈdē-nᵊl, -dᵊn-əl\ *adjective*

duo·logue \ˈdü-ə-ˌlȯg, -ˌläg *also* ˈdyü-\ *noun* (1864)
: a dialogue between two persons

duo·mo \ˈdwȯ-(ˌ)mō\ *noun, plural* **duomos** [Italian, from Latin *domus* house — more at DOME] (1549)
: CATHEDRAL

du·op·o·ly \du̇-ˈä-pə-lē *also* dyu̇-\ *noun, plural* **-lies** [*duo-* + *-poly* (as in *monopoly*)] (1920)
1 : an oligopoly limited to two sellers
2 : preponderant influence or control by two political powers
— **du·op·o·lis·tic** \-ˌä-pə-ˈlis-tik\ *adjective*

dup \ˈdəp\ *transitive verb* [contraction of *do up*] (1547)
archaic **:** OPEN

¹dupe \ˈdüp *also* ˈdyüp\ *noun* [French, from Middle French *duppe,* probably alteration of *huppe* hoopoe] (1681)
: one that is easily deceived or cheated **:** FOOL

²dupe *transitive verb* **duped; dup·ing** (1704)
: to make a dupe of ☆
— **dup·er** *noun*

³dupe *noun or verb* (circa 1900)
: DUPLICATE

dup·ery \ˈdü-pə-rē *also* ˈdyü-\ *noun, plural* **-er·ies** (1759)
1 : the condition of being duped
2 : the act or practice of duping

du·ple \ˈdü-pəl *also* ˈdyü-\ *adjective* [Latin *duplus* double — more at DOUBLE] (15th century)
1 : having two elements
2 a : marked by two or a multiple of two beats per measure of music ⟨*duple* time⟩ **b** *of rhythm* **:** consisting of a meter based on disyllabic feet

¹du·plex \ˈdü-ˌpleks *also* ˈdyü-\ *adjective* [Latin, from *duo* two + *-plex* -fold — more at TWO, -FOLD] (1567)
1 a : having two principal elements or parts **:** DOUBLE, TWOFOLD **b :** having complementary polynucleotide strands ⟨*duplex* DNA⟩
2 : allowing telecommunication in opposite directions simultaneously

²duplex *transitive verb* (1833)
: to make duplex

³duplex *noun* (1922)
: something duplex: as **a :** a 2-family house **b :** DUPLEX APARTMENT **c :** a duplex molecule of DNA or of RNA and DNA

duplex apartment *noun* (circa 1925)
: an apartment having rooms on two floors

du·plex·er \ˈdü-ˌplek-sər *also* ˈdyü-\ *noun* (circa 1932)
: a switching device that permits alternate transmission and reception with the same radio antenna

¹du·pli·cate \ˈdü-pli-kət *also* ˈdyü-\ *adjective* [Middle English, from Latin *duplicatus,* past participle of *duplicare* to double, from *duplic-, duplex*] (15th century)
1 : consisting of or existing in two corresponding or identical parts or examples ⟨*duplicate* invoices⟩
2 : being the same as another ⟨*duplicate* copies⟩

²du·pli·cate \ˈdü-pli-ˌkāt *also* ˈdyü-\ *verb* **-cat·ed; -cat·ing** (15th century)
transitive verb
1 : to make double or twofold
2 a : to make a copy of ⟨a cell *duplicates* itself when it divides⟩ **b :** to produce something equal to ⟨trying to *duplicate* last year's success⟩ **c :** to do over or again often needlessly ⟨*duplicated* effort⟩
intransitive verb
: to become duplicated; *also* **:** REPEAT
— **du·pli·ca·tive** \-ˌkā-tiv\ *adjective*

³du·pli·cate \-kət\ *noun* (1532)
1 a : either of two things exactly alike and usually produced at the same time or by the

☆ SYNONYMS
Dupe, gull, trick, hoax mean to deceive by underhanded means. DUPE suggests unwariness in the person deluded. GULL stresses credulousness or readiness to be imposed on (as through greed) on the part of the victim. TRICK implies an intent to delude by means of a ruse or fraud but does not always imply a vicious intent. HOAX implies the contriving of an elaborate or adroit imposture in order to deceive.

◇ WORD HISTORY
dungeon The word *dungeon,* in use in English since the 14th century, originally referred to the keep of a castle—the massive inner tower detached from the rest of the structure that was its most securely located and protected part. During its early history this word had about a dozen different spellings, but nowadays, in the sense of a castle's keep, the usual form is *donjon.* The donjon was the stronghold to which the garrison and residents of the castle retreated if the outer walls had been scaled or breached in a siege. The subterranean part of a donjon was called by the same word in the form *dungeon,* the usual spelling for this sense. This dark, damp chamber was used as a cell for the confinement of prisoners. Both *donjon* and *dungeon* are borrowed from Old French *donjon,* most likely the descendant of an unrecorded vernacular Latin form *domnio,* reflected in Medieval Latin renderings such as *dominionus* and *domnionus,* and ultimately a derivative of Latin *dominus* "lord." The underlying sense of *domnio* would have been "dominating tower," reflecting the relation between the keep and the rest of the castle.

same process **b :** an additional copy of something (as a book or stamp) already in a collection
2 : one that resembles or corresponds to another **:** COUNTERPART
3 : two identical copies — used in the phrase *in duplicate*
synonym see REPRODUCTION
duplicate bridge *noun* (1926)
: a tournament form of contract bridge in which identical deals are played in order to compare individual scores
du·pli·ca·tion \ˌdü-pli-ˈkā-shən *also* ˌdyü-\ *noun* (15th century)
1 a : the act or process of duplicating **b :** the quality or state of being duplicated
2 : DUPLICATE, COUNTERPART
3 : a part of a chromosome in which the genetic material is repeated; *also* **:** the process of forming a duplication
du·pli·ca·tor \ˈdü-pli-ˌkā-tər *also* ˈdyü-\ *noun* (1893)
: one that duplicates; *specifically* **:** a machine for making copies of graphic matter
du·plic·i·tous \dü-ˈpli-sə-təs *also* dyü-\ *adjective* (1928)
: marked by duplicity
— **du·plic·i·tous·ly** *adverb*
du·plic·i·ty \dü-ˈpli-sə-tē *also* dyü-\ *noun, plural* **-ties** [Middle English *duplicite,* from Middle French, from Late Latin *duplicitat-, duplicitas,* from Latin *duplex*] (15th century)
1 : contradictory doubleness of thought, speech, or action; *especially* **:** the belying of one's true intentions by deceptive words or action
2 : the quality or state of being double or twofold
3 : the technically incorrect use of two or more distinct items (as claims, charges, or defenses) in a single legal action
du·ra·ble \ˈdür-ə-bəl *also* ˈdyür-\ *adjective* [Middle English, from Middle French, from Latin *durabilis,* from *durare* to last — more at DURING] (14th century)
: able to exist for a long time without significant deterioration; *also* **:** designed to be durable ⟨*durable* goods⟩
synonym see LASTING
— **du·ra·bil·i·ty** \ˌdür-ə-ˈbi-lə-tē, ˌdyür-\ *noun*
— **du·ra·ble·ness** \ˈdür-ə-bəl-nəs, ˈdyür-\ *noun*
— **du·ra·bly** \-blē\ *adverb*
durable press *noun* (1966)
: PERMANENT PRESS
du·ra·bles \ˈdür-ə-bəlz *also* ˈdyür-\ *noun plural* (1941)
: consumer goods (as vehicles and household appliances) that are typically used repeatedly over a period of years — called also *durable goods*
du·ral·u·min \dü-ˈral-yə-mən *also* dyü-\ *noun* [from *Duralumin,* a trademark] (1910)
: a light strong alloy of aluminum, copper, manganese, and magnesium
du·ra ma·ter \ˈdür-ə-ˌmā-tər, ˈdyür-, -ˌmä-\ *noun* [Middle English, from Medieval Latin, literally, hard mother] (14th century)
: the tough fibrous membrane that envelops the brain and spinal cord external to the arachnoid and pia mater
du·rance \ˈdür-ən(t)s *also* ˈdyür-\ *noun* [Middle English, from Middle French, from *durer* to endure, from Latin *durare*] (15th century)
1 *archaic* **:** ENDURANCE
2 : restraint by or as if by physical force — usually used in the phrase *durance vile*
du·ra·tion \dü-ˈrā-shən *also* dyü-\ *noun* (14th century)
1 : continuance in time
2 : the time during which something exists or lasts
du·ra·tive \ˈdür-ə-tiv, ˈdyür-\ *adjective* (1889)
: CONTINUATIVE
— **durative** *noun*
dur·bar \ˈdər-ˌbär, ˌdər-ˈ\ *noun* [Hindi *darbār,* from Persian, from *dar* door + *bār* admission, audience] (1609)
1 : court held by an Indian prince
2 : a formal reception held by an Indian prince or an African ruler
du·ress \dü-ˈres *also* dyü-\ *noun* [Middle English *duresse,* from Middle French *duresce* hardness, severity, from Latin *duritia,* from *durus*] (15th century)
1 : forcible restraint or restriction
2 : compulsion by threat; *specifically* **:** unlawful constraint
Dur·ham \ˈdər-əm, ˈdə-rəm, ˈdür-əm\ *noun* [County *Durham,* England] (1810)
: SHORTHORN
Durham Rule *noun* [Monte *Durham,* 20th century American litigant] (1955)
: a legal hypothesis under which a person is not judged responsible for a criminal act that is attributed to a mental disease or defect
du·ri·an \ˈdür-ē-ən, -ē-ˌän *also* ˈdyür-\ *noun* [Malay] (1588)
1 : a large oval tasty but foul-smelling fruit with a prickly rind
2 : an East Indian tree (*Durio zibethinus*) of the silk-cotton family that bears durians
dur·ing \ˈdür-iŋ *also* ˈdyür-\ *preposition* [Middle English, from present participle of *duren* to last, from Old French *durer,* from Latin *durare* to harden, endure, from *durus* hard; perhaps akin to Sanskrit *dāru* wood — more at TREE] (14th century)
1 : throughout the duration of ⟨swims every day *during* the summer⟩
2 : at a point in the course of ⟨was offered a job *during* a visit to the capital⟩
dur·mast oak \ˈdər-ˌmast-\ *noun* [perhaps alteration of *dun mast,* from [1]*dun* + *mast*] (1791)
: a European oak (*Quercus petraea*) valued especially for its dark heavy tough elastic wood and for its tannin-rich bark
durn \ˈdərn\, **durned** \ˈdərn(d)\ *variant of* DARN, DARNED
du·ro \ˈdür-(ˌ)ō\ *noun, plural* **duros** [Spanish, short for *peso duro* hard peso] (1832)
: a Spanish or Spanish American peso or silver dollar
du·roc \ˈdür-ˌäk *also* ˈdyür-\ *noun, often capitalized* [*Duroc,* 19th century American stallion] (1883)
: any of a breed of large vigorous red American hogs
du·rom·e·ter \dü-ˈrä-mə-tər *also* dyü-\ *noun* [Latin *durus* hard] (circa 1879)
: an instrument for measuring hardness
dur·ra *also* **du·ra** \ˈdür-ə\ *noun* [Arabic *dhurah*] (1798)
: any of several grain sorghums widely grown in warm dry regions
durst \ˈdərst\ *archaic & dialect past of* DARE
du·rum wheat \ˈdür-əm-, ˈdyür-, ˈdər-əm-, ˈdə-rəm-\ *noun* [New Latin *durum,* from Latin, neuter of *durus* hard] (circa 1903)
: a wheat (*Triticum durum*) that yields a glutenous flour used especially in pasta — called also *durum*
[1]**dusk** \ˈdəsk\ *adjective* [Middle English *dosk,* alteration of Old English *dox;* akin to Latin *fuscus* dark brown, Old English *dunn* dun, *dūst* dust] (13th century)
: DUSKY
[2]**dusk** (13th century)
intransitive verb
: to become dark
transitive verb
: to make dark or gloomy
[3]**dusk** *noun* (1622)
1 : the darker part of twilight especially at night
2 : darkness or semidarkness caused by the shutting out of light
dusky \ˈdəs-kē\ *adjective* **dusk·i·er; -est** (1558)
1 : somewhat dark in color; *specifically* **:** having dark skin
2 : marked by slight or deficient light **:** SHADOWY
— **dusk·i·ly** \-kə-lē\ *adverb*
— **dusk·i·ness** \-kē-nəs\ *noun*
[1]**dust** \ˈdəst\ *noun* [Middle English, from Old English *dūst;* akin to Old High German *tunst* storm, and probably to Latin *fumus* smoke — more at FUME] (before 12th century)
1 : fine particles of matter (as of earth)
2 : the particles into which something disintegrates
3 a : something worthless **b :** a state of humiliation
4 a : the earth especially as a place of burial **b :** the surface of the ground
5 a : a cloud of dust **b :** CONFUSION, DISTURBANCE
6 *archaic* **:** a single particle (as of earth)
7 *British* **:** refuse ready for collection
— **dust·less** \-ləs\ *adjective*
— **dust·like** \-ˌlīk\ *adjective*
[2]**dust** (1530)
transitive verb
1 *archaic* **:** to make dusty
2 : to make free of dust
3 a : to sprinkle with fine particles **b :** to sprinkle in the form of dust
intransitive verb
1 *of a bird* **:** to work dust into the feathers
2 : to remove dust
3 : to give off dust
dust·bin \ˈdəs(t)-ˌbin\ *noun* (1848)
1 *British* **:** a can for trash or garbage
2 : DUSTHEAP 2
dust bowl *noun* (1936)
: a region that suffers from prolonged droughts and dust storms
dust·cov·er \-ˌkə-vər\ *noun* (1899)
1 : a cover (as of cloth or plastic) used to protect furniture or equipment from dust
2 : DUST JACKET
dust devil *noun* (1888)
: a small whirlwind containing sand or dust
dust·er \ˈdəs-tər\ *noun* (1576)
1 : one that removes dust
2 a (1) **:** a long lightweight overgarment to protect clothing from dust (2) **:** a long coat cut like a duster — called also *duster coat* **b :** a dress-length housecoat
3 : one that scatters fine particles; *specifically* **:** a device for applying insecticidal or fungicidal dusts to crops
4 : DUST STORM
dust·heap \ˈdəst-ˌ(h)ēp\ *noun* (1599)
1 : a pile of refuse
2 : a category of forgotten items ⟨the *dustheap* of history —*New Republic*⟩
dust jacket *noun* (1926)
: a paper cover for a book
dust·man \ˈdəs(t)-mən\ *noun* (1707)
British **:** a collector of trash or garbage
dust mop *noun* (1953)
: DRY MOP
dust off *transitive verb* (1940)
: to bring out or back to use again
dust·pan \ˈdəs(t)-ˌpan\ *noun* (1783)
: a shovel-shaped pan for sweepings
dust storm *noun* (1879)
1 : a dust-laden whirlwind that moves across an arid region and is usually associated with hot dry air and marked by high electrical tension
2 : strong winds bearing clouds of dust
dust·up \ˈdəs-ˌtəp\ *noun* (1897)
: ROW, FIGHT
dust wrapper *noun* (1932)
: DUST JACKET

dusty \ˈdəs-tē\ *adjective* **dust·i·er; -est** (13th century)
1 : covered or abounding with dust
2 : consisting of dust **:** POWDERY
3 : resembling dust
4 : lacking vitality **:** DRY ⟨*dusty* scholarship⟩
5 *British* **:** UNSATISFACTORY — used especially in the phrases *dusty answer* and *not so dusty*
— dust·i·ly \ˈdəs-tə-lē\ *adverb*
— dust·i·ness \-tē-nəs\ *noun*

dusty miller *noun* (circa 1825)
: any of several plants having ashy-gray or white tomentose leaves; *especially* **:** an herbaceous artemisia (*Artemisia stelleriana*) with greyish foliage found along the eastern coast of the U.S.

dutch \ˈdəch\ *adverb, often capitalized* (1914)
: with each person paying his or her own way

[1]Dutch \ˈdəch\ *adjective* [Middle English *Duch,* from Middle Dutch *duutsch;* akin to Old High German *diutisc* German, Old English *thēod* nation, Gothic *thiudisko* as a gentile, *thiuda* people, Oscan *touto* city] (14th century)
1 a *archaic* **:** of, relating to, or in any of the Germanic languages of Germany, Austria, Switzerland, and the Low Countries **b :** of, relating to, or in the Dutch of the Netherlands
2 a *archaic* **:** of or relating to the Germanic peoples of Germany, Austria, Switzerland, and the Low Countries **b :** of or relating to the Netherlands or its inhabitants **c :** [2]GERMAN
3 : of or relating to the Pennsylvania Dutch or their language
— Dutch·ly *adverb*

[2]Dutch *noun* (14th century)
1 a *archaic* (1) **:** any of the Germanic languages of Germany, Austria, Switzerland, and the Low Countries (2) **:** GERMAN 3 **b :** the Germanic language of the Netherlands
2 Dutch *plural* **a** *archaic* **:** the Germanic peoples of Germany, Austria, Switzerland, and the Low Countries **b :** GERMANS 2a, b **c :** the people of the Netherlands
3 : PENNSYLVANIA DUTCH
4 : DANDER ⟨her *Dutch* is up⟩
5 : DISFAVOR, TROUBLE ⟨in *Dutch* with the boss⟩

Dutch cheese *noun* (1829)
chiefly Northern **:** COTTAGE CHEESE

Dutch clover *noun* (1800)
: WHITE CLOVER

Dutch Colonial *adjective* (1922)
: characterized by a gambrel roof with overhanging eaves

Dutch courage *noun* (1809)
: courage artificially stimulated especially by drink; *also* **:** drink taken for courage

Dutch door *noun* (circa 1890)
: a door divided horizontally so that the lower or upper part can be shut separately

Dutch door

Dutch elm disease *noun* (1927)
: a disease of elms caused by an ascomycetous fungus (*Ceratocystis ulmi*) and characterized by yellowing of the foliage, defoliation, and death

Dutch hoe *noun* (1744)
: SCUFFLE HOE

dutch·man \ˈdəch-mən\ *noun* (14th century)
1 *capitalized* **a** *archaic* **:** a member of any of the Germanic peoples of Germany, Austria, Switzerland, and the Low Countries **b :** a native or inhabitant of the Netherlands **c :** a person of Dutch descent **d :** GERMAN 1a, b
2 : a device for hiding or counteracting structural defects

Dutch·man's–breech·es \ˌdəch-mənz-ˈbri-chəz\ *noun plural but singular or plural in construction* (1837)
: a spring-flowering herb (*Dicentra cucullaria*) of the fumitory family occurring in the eastern U.S. and having finely divided leaves and cream-white double-spurred flowers

Dutchman's-breeches

Dutchman's–pipe \-ˈpīp\ *noun, plural* **Dutchman's–pipes** \-ˈpīps\ (1845)
: a vine (*Aristolochia durior*) with large leaves and early summer flowers having the tube of the calyx curved like the bowl of a pipe

Dutch oven *noun* (1769)
1 : a metal shield for roasting before an open fire
2 : a brick oven in which cooking is done by the preheated walls
3 a : a cast-iron kettle with a tight cover that is used for baking in an open fire **b :** a heavy pot with a tight-fitting domed cover

Dutch roll *noun* (1913)
: a combination of directional and lateral oscillation of an airplane

[1]dutch treat *noun, often D capitalized* (1887)
: a meal or other entertainment for which each person pays his or her own way

[2]dutch treat *adverb, often D capitalized* (1942)
: DUTCH ⟨go *dutch treat*⟩

Dutch uncle *noun* (1837)
: one who admonishes sternly and bluntly

du·te·ous \ˈdü-tē-əs *also* ˈdyü-\ *adjective* [irregular from *duty*] (1593)
: DUTIFUL, OBEDIENT

du·ti·able \ˈdü-tē-ə-bəl *also* ˈdyü-\ *adjective* (1774)
: subject to a duty

du·ti·ful \ˈdü-ti-fəl *also* ˈdyü-\ *adjective* (1552)
1 : filled with or motivated by a sense of duty
2 : proceeding from or expressive of a sense of duty
— du·ti·ful·ly \-f(ə-)lē\ *adverb*
— du·ti·ful·ness \-fəl-nəs\ *noun*

[1]du·ty \ˈdü-tē *also* ˈdyü-\ *noun, plural* **duties** [Middle English *duete,* from Anglo-French *dueté,* from Old French *deu* due] (13th century)
1 : conduct due to parents and superiors **:** RESPECT
2 a : obligatory tasks, conduct, service, or functions that arise from one's position (as in life or in a group) **b** (1) **:** assigned service or business (2) **:** active military service (3) **:** a period of being on duty
3 a : a moral or legal obligation **b :** the force of moral obligation
4 : TAX; *especially* **:** a tax on imports
5 a : WORK 1a **b** (1) **:** the service required (as of an electric machine) under specified conditions (2) **:** functional application **:** USE ⟨got double *duty* out of the trip⟩ (3) **:** use as a substitute ⟨making the word do *duty* for the thing —Edward Sapir⟩
synonym see FUNCTION, TASK
— off duty : free from assignment or responsibility
— on duty : engaged in or responsible for an assigned task or duty

[2]duty *adjective* (1806)
1 : done as a duty
2 : being on duty **:** assigned to specified tasks or functions ⟨the *duty* officer⟩

duty–free \ˌdü-tē-ˈfrē, ˌdyü-, ˈdü-tē-ˌ\ *adjective or adverb* (1689)
1 : without payment of customs duties **:** free from duties ⟨imported *duty-free*⟩ ⟨*duty-free* goods⟩
2 : relating to or selling duty-free goods ⟨a *duty-free* shop⟩

du·um·vir \dü-ˈəm-vər *also* dyü-\ *noun* [Latin, from *duum* (genitive of *duo* two) + *vir* man] (1600)
1 : one of two Roman officers or magistrates constituting a board or court
2 : one of two people jointly holding power
— du·um·vi·rate \-və-rət\ *noun*

du·vet \d(y)ü-ˈvā, ˈd(y)ü-ˌ\ *noun* [French] (1758)
: COMFORTER 2b

duve·tyn \ˈdü-və-ˌtēn, ˈdyü-, ˈdəv-ˌtēn\ *noun* [French *duvetine,* from *duvet* down, from Middle French, alteration of (assumed) Middle French *dumet,* diminutive of Old French *dun, dum* down, from Old Norse *dūnn* — more at DOWN] (1913)
: a smooth lustrous velvety fabric

dux·elles \ˌdük-ˈsel, (ˌ)dü-ˈsel\ *noun* [Marquis Louis Chalon du Blé *d'Uxelles* (died 1658) French nobleman] (1877)
: a garnish or stuffing made especially of finely chopped sautéed mushrooms

[1]dwarf \ˈdwȯrf\ *noun, plural* **dwarfs** \ˈdwȯrfs\ *also* **dwarves** \ˈdwȯrvz\ *often attributive* [Middle English *dwerg, dwerf,* from Old English *dweorg, dweorh;* akin to Old High German *twerg* dwarf] (before 12th century)
1 a : a person of unusually small stature; *especially* **:** one whose bodily proportions are abnormal **b :** an insignificant person
2 : an animal or plant much below normal size
3 : a small legendary manlike being who is usually misshapen and ugly and skilled as an artificer
4 : a star (as the sun) of ordinary or low luminosity and relatively small mass and size
— dwarf·ish \ˈdwȯr-fish\ *adjective*
— dwarf·ish·ly *adverb*
— dwarf·ish·ness *noun*
— dwarf·like \ˈdwȯr-ˌflīk\ *adjective*
— dwarf·ness \ˈdwȯrf-nəs\ *noun*

[2]dwarf (circa 1626)
transitive verb
1 : to restrict the growth of **:** STUNT
2 : to cause to appear smaller or to seem inferior
intransitive verb
: to become smaller

[3]dwarf *adjective* (1664)
of a plant **:** low-growing in habit ⟨*dwarfer* forms of citrus⟩

dwarf·ism \ˈdwȯr-ˌfi-zəm\ *noun* (1865)
: the condition of stunted growth

dweeb \ˈdwēb\ *noun* [origin unknown] (1983)
slang **:** an unattractive, insignificant, or inept person

dwell \ˈdwel\ *intransitive verb* **dwelled** \ˈdweld, ˈdwelt\ *or* **dwelt** \ˈdwelt\; **dwelling** [Middle English, from Old English *dwellan* to go astray, hinder; akin to Old High German *twellen* to tarry] (13th century)
1 : to remain for a time
2 a : to live as a resident **b :** EXIST, LIE
3 a : to keep the attention directed — used with *on* or *upon* ⟨tried not to *dwell* on my fears⟩ **b :** to speak or write insistently — used with *on* or *upon* ⟨leering reviewers *dwelled* on a publicity photograph —James Atlas⟩
— dwell·er *noun*

dwell·ing *noun* (14th century)
: a shelter (as a house) in which people live

dwin·dle \ˈdwin-dᵊl\ *verb* **dwin·dled; dwin·dling** \-(d)liŋ, -dᵊl-iŋ\ [probably frequentative of *dwine* to waste away, from Middle English, from Old English *dwīnan;* akin to Old Norse *dvīna* to pine away, *deyja* to die — more at DIE] (1596)
intransitive verb
: to become steadily less **:** SHRINK
transitive verb
: to make steadily less
synonym see DECREASE

DX \(ˌ)dē-ˈeks\ *noun* (circa 1924)
: DISTANCE — used of long-distance radio transmission

dy- *or* **dyo-** *combining form* [Late Latin, from Greek, from *dyo* — more at TWO]
: two ⟨dyarchy⟩

dy·ad \ˈdī-ˌad, -əd\ *noun* [Late Latin *dyad-, dyas,* from Greek, from *dyo*] (1675)

1 : PAIR; *specifically* **:** two individuals (as husband and wife) maintaining a sociologically significant relationship
2 : a meiotic chromosome after separation of the two homologous members of a tetrad
3 : a mathematical operator indicated by writing the symbols of two vectors without a dot or cross between (as AB)
— **dy·ad·ic** \dī-'a-dik\ *adjective*
— **dy·ad·i·cal·ly** \-di-k(ə-)lē\ *adverb*

dy·ad·ic \dī-'a-dik\ *noun* (1884)
: a mathematical expression formed by addition or subtraction of dyads

Dy·ak *variant of* DAYAK

dy·ar·chy \'dī-,är-kē\ *noun, plural* **-chies** (1640)
: a government in which power is vested in two rulers or authorities

dyb·buk \'di-bək\ *noun, plural* **dyb·bu·kim** \,di-bü-'kēm\ *also* **dybbuks** [Late Hebrew *dibbūq*] (circa 1903)
: a wandering soul believed in Jewish folklore to enter and control a living body until exorcised by a religious rite

[1]**dye** \'dī\ *noun* [Middle English *dehe,* from Old English *dēah, dēag*] (before 12th century)
1 : color from dyeing
2 : a soluble or insoluble coloring matter

[2]**dye** *verb* **dyed; dye·ing** (before 12th century)
transitive verb
1 : to impart a new and often permanent color to especially by impregnating with a dye
2 : to impart (a color) by dyeing ⟨*dyeing* blue on yellow⟩
intransitive verb
: to take up or impart color in dyeing
— **dye·abil·i·ty** \,dī-ə-'bi-lə-tē\ *noun*
— **dye·able** \'dī-ə-bəl\ *adjective*
— **dy·er** \'dī(-ə)r\ *noun*

dyed–in–the–wool \,dīd-ᵊn-th̲ə-'wu̇l\ *adjective* (1580)
: THOROUGHGOING, UNCOMPROMISING ⟨a *dyed-in-the-wool* conservative⟩

dye·stuff \'dī-,stəf\ *noun* (1837)
: DYE 2

dye·wood \-,wu̇d\ *noun* (1699)
: a wood (as logwood or fustic) from which coloring matter is extracted for dyeing

dying *present participle of* DIE

[1]**dyke** *chiefly British variant of* DIKE

[2]**dyke** \'dīk\ *noun* [origin unknown] (circa 1942)
: LESBIAN — often used disparagingly
— **dykey** \'dī-kē\ *adjective*

[1]**dy·nam·ic** \dī-'na-mik\ *adjective* [French *dynamique,* from Greek *dynamikos* powerful, from *dynamis* power, from *dynasthai* to be able] (1827)
1 *also* **dy·nam·i·cal** \-mi-kəl\ **a :** of or relating to physical force or energy **b :** of or relating to dynamics
2 a : marked by usually continuous and productive activity or change ⟨a *dynamic* city⟩ **b :** ENERGETIC, FORCEFUL ⟨a *dynamic* personality⟩
3 *of random-access memory* **:** requiring periodic refreshment of charge in order to retain data
— **dy·nam·i·cal·ly** \-mi-k(ə-)lē\ *adverb*

[2]**dynamic** *noun* (1879)
1 : a dynamic force
2 : DYNAMICS 2; *also* **:** an underlying cause of change or growth

dynamic range *noun* (1949)
: the ratio of the strongest to the weakest sound intensity that can be transmitted or reproduced by an audio or broadcasting system

dy·nam·ics \dī-'na-miks\ *noun plural but singular or plural in construction* (circa 1789)
1 : a branch of mechanics that deals with forces and their relation primarily to the motion but sometimes also to the equilibrium of bodies
2 : a pattern or process of change, growth, or activity ⟨population *dynamics*⟩
3 : variation and contrast in force or intensity (as in music)

dy·na·mism \'dī-nə-,mi-zəm\ *noun* (circa 1857)
1 a : a theory that all phenomena (as matter or motion) can be explained as manifestations of force — compare MECHANISM **b :** DYNAMICS 2
2 : a dynamic or expansionist quality
— **dy·na·mist** \-mist\ *noun*
— **dy·na·mis·tic** \,dī-nə-'mis-tik\ *adjective*

[1]**dy·na·mite** \'dī-nə-,mīt\ *noun* (1867)
1 : an explosive that is made of nitroglycerin absorbed in a porous material and that often contains ammonium nitrate or cellulose nitrate; *also* **:** an explosive (as a mixture of ammonium nitrate and nitrocellulose) that contains no nitroglycerin
2 : one that has a powerful effect; *also* **:** something that has great potential to cause trouble or conflict
— **dy·na·mit·ic** \,dī-nə-'mi-tik\ *adjective*

[2]**dynamite** *transitive verb* **-mit·ed; -mit·ing** (1881)
1 : to blow up with dynamite
2 : to cause the failure or destruction of
— **dy·na·mit·er** *noun*

[3]**dynamite** *adjective* (1940)
: TERRIFIC, WONDERFUL

dy·na·mo \'dī-nə-,mō\ *noun, plural* **-mos** [short for *dynamoelectric machine*] (circa 1882)
1 : GENERATOR 3
2 : a forceful energetic individual

dy·na·mom·e·ter \,dī-nə-'mä-mə-tər\ *noun* [French *dynamomètre,* from Greek *dynamis* power + French *-mètre* -meter] (1810)
1 : an instrument for measuring mechanical force
2 : an apparatus for measuring mechanical power (as of an engine)
— **dy·na·mo·met·ric** \-mō-'me-trik\ *adjective*
— **dy·na·mom·e·try** \-'mä-mə-trē\ *noun*

dy·na·mo·tor \'dī-nə-,mō-tər\ *noun* [*dynamo* + *motor*] (1899)
: a motor generator combining the electric motor and generator

dy·nast \'dī-,nast, -nəst\ *noun* [Latin *dynastes,* from Greek *dynastēs,* from *dynasthai* to be able, have power] (1631)
: RULER 1

dy·nas·ty \'dī-nə-stē *also* -,nas-tē, *especially British* 'di-nə-stē\ *noun, plural* **-ties** (14th century)
1 : a succession of rulers of the same line of descent
2 : a powerful group or family that maintains its position for a considerable time
— **dy·nas·tic** \dī-'nas-tik\ *adjective*
— **dy·nas·ti·cal·ly** \-ti-k(ə-)lē\ *adverb*

dy·na·tron \'dī-nə-,trän\ *noun* [Greek *dynamis* power] (1918)
: a vacuum tube in which the secondary emission of electrons from the plate results in a decrease in the plate current as the plate voltage increases

dyne \'dīn\ *noun* [French, from Greek *dynamis*] (circa 1873)
: the unit of force in the centimeter-gram-second system equal to the force that would give a free mass of one gram an acceleration of one centimeter per second per second

dy·ne·in \'dī-nē-in\ *noun* [*dyne* (force) + *-in*] (1965)
: an ATPase that is associated especially with microtubules involved in the ciliary and flagellar movement of cells

dy·node \'dī-,nōd\ *noun* [Greek *dynamis*] (1939)
: an electrode in an electron tube that functions to produce secondary emission of electrons

dys- *prefix* [Middle English *dis-* bad, difficult, from Middle French & Latin; Middle French *dis-,* from Latin *dys-,* from Greek; akin to Old English *tō-, te-* apart, Sanskrit *dus-* bad, difficult]
1 : abnormal ⟨*dys*plasia⟩
2 : difficult ⟨*dys*phagia⟩ — compare EU-
3 : impaired ⟨*dys*function⟩
4 : bad ⟨*dys*logistic⟩ — compare EU-

dys·ar·thria \di-'sär-thrē-ə\ *noun* [New Latin, from *dys-* + *arthr-* + *-ia*] (1878)
: difficulty in articulating words due to disease of the central nervous system

dys·cra·sia \dis-'krā-zh(ē-)ə\ *noun* [New Latin, from Medieval Latin, bad mixture of humors, from Greek *dyskrasia,* from *dys-* + *krasis* mixture, from *kerannynai* to mix — more at CRATER] (14th century)
: an abnormal condition of the body and especially the blood

dys·en·ter·ic \,di-sᵊn-'ter-ik\ *adjective* (1727)
: of or relating to dysentery

dys·en·tery \'di-sᵊn-,ter-ē\ *noun, plural* **-teries** [Middle English *dissenterie,* from Latin *dysenteria,* from Greek, from *dys-* + *enteron* intestine — more at INTER-] (14th century)
1 : a disease characterized by severe diarrhea with passage of mucus and blood and usually caused by infection
2 : DIARRHEA

dys·func·tion \(,)dis-'fəŋ(k)-shən\ *noun* (circa 1916)
: impaired or abnormal functioning
— **dys·func·tion·al** \-shnəl, -shə-nᵊl\ *adjective*

dys·gen·e·sis \(,)dis-'je-nə-səs\ *noun* [New Latin] (circa 1883)
: defective development especially of the gonads (as in Klinefelter's syndrome)

dys·gen·ic \(,)dis-'je-nik\ *adjective* (1912)
1 : tending to promote survival of or reproduction by less well-adapted individuals (as the weak or diseased) especially at the expense of well-adapted individuals (as the strong or healthy) ⟨the *dysgenic* effect of war⟩
2 : biologically defective or deficient

dys·ki·ne·sia \,dis-kə-'nē-zh(ē-)ə, -kī-\ *noun* [New Latin, from Greek *dyskinēsia* difficulty in moving, from *dys-* + *-kinesia,* from *kinēsis* motion, from *kinein* to move — more at HIGHT] (circa 1706)
: impairment of voluntary movements resulting in fragmented or jerky motions (as in Parkinson's disease) — compare TARDIVE DYSKINESIA
— **dys·ki·net·ic** \-'net-ik\ *adjective*

dys·lex·ia \dis-'lek-sē-ə\ *noun* [New Latin, from *dys-* + Greek *lexis* word, speech, from *legein* to say — more at LEGEND] (circa 1888)
: a disturbance of the ability to read; *broadly* **:** disturbance of the ability to use language
— **dys·lex·ic** \-sik\ *adjective or noun*

dys·lo·gis·tic \,dis-lə-'jis-tik\ *adjective* [*dys-* + *-logistic* (as in *eulogistic*)] (1812)
: UNCOMPLIMENTARY
— **dys·lo·gis·ti·cal·ly** \-ti-k(ə-)lē\ *adverb*

dys·men·or·rhea \(,)dis-,me-nə-'rē-ə\ *noun* [New Latin] (circa 1810)
: painful menstruation
— **dys·men·or·rhe·ic** \-'rē-ik\ *adjective*

dys·pep·sia \dis-'pep-shə, -sē-ə\ *noun* [Latin, from Greek, from *dys-* + *pepsis* digestion, from *peptein, pessein* to cook, digest — more at COOK] (circa 1706)
1 : INDIGESTION
2 : ill humor **:** DISGRUNTLEMENT
— **dys·pep·tic** \-'pep-tik\ *adjective or noun*
— **dys·pep·ti·cal·ly** \-ti-k(ə-)lē\ *adverb*

dys·pha·gia \dis-'fā-j(ē-)ə\ *noun* [New Latin] (1783)

: difficulty in swallowing

dys•pha•sia \dis-ˈfā-zh(ē-)ə\ *noun* [New Latin] (circa 1883)
: loss of or deficiency in the power to use or understand language as a result of injury to or disease of the brain
— **dys•pha•sic** \-ˈfā-zik\ *noun or adjective*

dys•phe•mism \ˈdis-fə-ˌmi-zəm\ *noun* [*dys-* + *-phemism* (as in *euphemism*)] (1884)
: the substitution of a disagreeable, offensive, or disparaging expression for an agreeable or inoffensive one; *also* **:** an expression so substituted
— **dys•phe•mis•tic** \ˌdis-fə-ˈmis-tik\ *adjective*

dys•pho•nia \dis-ˈfō-nē-ə\ *noun* [New Latin] (circa 1706)
: defective use of the voice

dys•pho•ria \dis-ˈfōr-ē-ə, -ˈfȯr-\ *noun* [New Latin, from Greek, from *dysphoros* hard to bear, from *dys-* + *pherein* to bear — more at BEAR] (circa 1842)
: a state of feeling unwell or unhappy
— **dys•phor•ic** \-ˈfȯr-ik, -ˈfär-\ *adjective*

dys•pla•sia \dis-ˈplā-zh(ē-)ə\ *noun* [New Latin] (circa 1923)
: abnormal growth or development (as of organs or cells); *broadly* **:** abnormal anatomic structure due to such growth
— **dys•plas•tic** \-ˈplas-tik\ *adjective*

dys•pnea \ˈdis(p)-nē-ə\ *noun* [Latin *dyspnoea,* from Greek *dyspnoia,* from *dyspnoos* short of breath, from *dys-* + *pnein* to breathe — more at SNEEZE] (circa 1681)
: difficult or labored respiration
— **dys•pne•ic** \-nē-ik\ *adjective*

dyspnoea *chiefly British variant of* DYSPNEA

dys•pro•si•um \dis-ˈprō-zē-əm, -zh(ē-)əm\ *noun* [New Latin, from Greek *dysprositos* hard to get at, from *dys-* + *prositos* approachable, from *prosienai* to approach, from *pros-* + *ienai* to go — more at ISSUE] (1886)
: an element of the rare-earth group that forms highly magnetic compounds — see ELEMENT table

dys•rhyth•mia \dis-ˈrit͟h-mē-ə\ *noun* [New Latin, from *dys-* + Latin *rhythmus* rhythm] (circa 1909)
: an abnormal rhythm; *especially* **:** a disordered rhythm exhibited in a record of electrical activity of the brain or heart
— **dys•rhyth•mic** \-mik\ *adjective*

dys•to•cia \di-ˈstō-sh(ē-)ə\ *noun* [New Latin, from Greek *dystokia,* from *dys-* + *tokos* childbirth; akin to Greek *tiktein* to give birth to — more at THANE] (circa 1706)
: slow or difficult labor or delivery

dys•to•pia \(ˌ)dis-ˈtō-pē-ə\ *noun* [New Latin, from *dys-* + *-topia* (as in *utopia*)] (circa 1950)
1 : an imaginary place where people lead dehumanized and often fearful lives
2 : ANTI-UTOPIA 2
— **dys•to•pi•an** \-pē-ən\ *adjective*

dys•tro•phic \dis-ˈtrō-fik\ *adjective* (1893)
1 a : relating to or caused by faulty nutrition **b :** relating to or affected with a dystrophy ⟨a *dystrophic* patient⟩
2 *of a lake* **:** brownish with much dissolved humic matter, a sparse bottom fauna, and a high oxygen consumption

dys•tro•phy \ˈdis-trə-fē\ *noun, plural* **-phies** [New Latin *dystrophia,* from *dys-* + *-trophia* -trophy] (1901)
1 : a condition produced by faulty nutrition
2 : any myogenic atrophy; *especially* **:** MUSCULAR DYSTROPHY

dys•uria \dis-ˈyu̇r-ē-ə, dish-\ *noun* [New Latin, from Greek *dysouria,* from *dys-* + *-ouria* -uria] (14th century)
: difficult or painful discharge of urine — compare STRANGURY

E *is the fifth letter of the English alphabet and of most alphabets that are closely related to it. It came through Latin from Etruscan and there from Greek, which took it from a Phoenician letter called* he, *representing a sound like English* h. *In Greek the letter, renamed* epsilon *("simple e"), designated a short vowel equivalent to the* e *in English* end. *Although the corresponding long vowel (pronounced like the* a *in English* ale*) was indicated in Greek by the letter* eta *(see* H*), in Latin* E *was used for both the short and the long vowel. English* E *inherited these Latin phonetic values, but in Modern English "long"* e *has come to represent the sound of the letter* e *in* eve. *The small letterform of* e *developed in two main stages: first by a curving of the upright line of capital* E *so as to form a semicircle with the top and bottom bars (ϵ); next by the joining of the top curve and middle bar to form a crescent.*

e \'ē\ *noun, plural* **e's** *or* **es** \'ēz\ *often capitalized, often attributive* (before 12th century)
1 a : the 5th letter of the English alphabet **b :** a graphic representation of this letter **c :** a speech counterpart of orthographic *e*
2 : the 3d tone of a C-major scale
3 : a graphic device for reproducing the letter *e*
4 : one designated *e* especially as the 5th in order or class
5 a : a grade rating a student's work as poor and usually constituting a conditional pass **b :** a grade rating a student's work as failing **c :** one graded or rated with an E
6 : a transcendental number having a value to eight decimal places of 2.71828183 that is the base of natural logarithms
7 : something shaped like the letter E

e- \(')ē, i\ *prefix* [Middle English, from Old French & Latin; Old French, out, forth, away, from Latin, from *ex-*]
1 : missing **:** absent ⟨edentulous⟩
2 : away ⟨eluviation⟩

[1]each \'ēch\ *adjective* [Middle English *ech,* from Old English *ǣlc;* akin to Old High German *iogilīh* each; both from a prehistoric West Germanic compound whose first and second constituents respectively are represented by Old English *ā* always and by Old English *gelīc* alike] (before 12th century)
: being one of two or more distinct individuals having a similar relation and often constituting an aggregate

[2]each *pronoun* (before 12th century)
: each one

[3]each *adverb* (before 12th century)
: to or for each **:** APIECE

each other *pronoun* (before 12th century)
: each of two or more in reciprocal action or relation ⟨looked at *each other* in surprise⟩ ■

ea·ger \'ē-gər\ *adjective* [Middle English *egre,* from Middle French *aigre,* from Latin *acer* — more at EDGE] (14th century)
1 a *archaic* **:** SHARP **b** *obsolete* **:** SOUR
2 : marked by enthusiastic or impatient desire or interest ☆
— ea·ger·ly *adverb*
— ea·ger·ness *noun*

eager beaver *noun* (1943)
: a person who is extremely zealous about performing duties and volunteering for more

ea·gle \'ē-gəl\ *noun* [Middle English *egle,* from Old French *aigle,* from Latin *aquila*] (13th century)
1 : any of various large diurnal birds of prey of the accipiter family noted for their strength, size, keenness of vision, and powers of flight
2 a : the eagle-bearing standard of the ancient Romans **b :** one of a pair of eagle-bearing silver insignia of rank worn by a military colonel or a navy captain
3 : a gold coin of the U.S. bearing an eagle on the reverse and usually having a value of ten dollars
4 : a golf score of two strokes less than par on a hole — compare BIRDIE
5 *capitalized* [Fraternal Order of *Eagles*] **:** a member of a major fraternal order

eagle eye *noun* (1802)
1 : the ability to see or observe with exceptional keenness
2 : one that sees or observes keenly
— eagle–eyed \'ē-gəl-ˌīd\ *adjective*

eagle ray *noun* (circa 1856)
: any of several widely distributed large active stingrays (family Myliobatidae) with broad pectoral fins like wings

Eagle Scout *noun* (1913)
1 : a Boy Scout who has reached the highest level of achievement in scouting
2 : a straight-arrow and self-reliant man

ea·glet \'ē-glət\ *noun* (1572)
: a young eagle

eal·dor·man \'al-dər-mən\ *noun* [Old English — more at ALDERMAN] (before 12th century)
: the chief officer in a district (as a shire) in Anglo-Saxon England

Eames chair \'ēmz-, 'āmz-\ *noun* [Charles *Eames* (died 1978) American designer] (1946)
: any of several chairs designed by Charles Eames to fit the contours of the body and to be made from modern materials

-ean — see -AN

[1]ear \'ir, ēr\ *noun* [Middle English *ere,* from Old English *ēare;* akin to Old High German *ōra* ear, Latin *auris,* Greek *ous*] (before 12th century)
1 a : the characteristic vertebrate organ of hearing and equilibrium consisting in the typical mammal of a sound-collecting outer ear separated by the tympanic membrane from a sound-transmitting middle ear that in turn is separated from a sensory inner ear by membranous fenestrae **b :** any of various organs capable of detecting vibratory motion
2 : the external ear of humans and most mammals
3 a : the sense or act of hearing **b :** acuity of hearing **c :** sensitivity to musical tone and pitch; *also* **:** the ability to retain and reproduce music that has been heard **d :** sensitivity to nuances of language especially as revealed in the command of verbal melody and rhythm or in the ability to render a spoken idiom accurately
4 : something resembling a mammalian ear in shape, position, or function: as **a :** a projecting part (as a lug or handle) **b :** either of a pair of tufts of lengthened feathers on the head of some birds
5 a : sympathetic attention **b :** ATTENTION, AWARENESS
6 : a space in the upper corner of the front page of a periodical (as a newspaper) usually containing advertising for the periodical itself or a weather forecast
— by ear : without reference to or memorization of written music ⟨plays *by ear*⟩
— in one ear and out the other : through one's mind without making an impression ⟨everything you say to him goes *in one ear and out the other*⟩
— on one's ear : in or into a state of irritation, shock, or discord ⟨set the racing world *on its ear* by breaking the record⟩
— up to one's ears : deeply involved **:** heavily implicated ⟨*up to his ears* in shady deals⟩

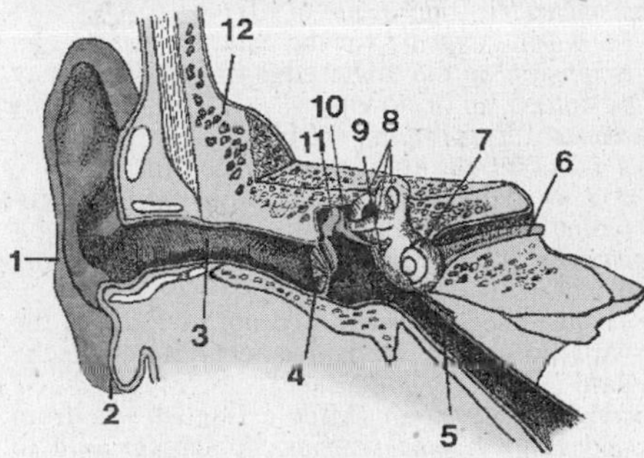

ear 1a: *1* pinna, *2* lobe, *3* auditory meatus, *4* tympanic membrane, *5* eustachian tube, *6* auditory nerve, *7* cochlea, *8* semicircular canals, *9* stapes, *10* incus, *11* malleus, *12* bones of skull

[2]ear *noun* [Middle English *er,* from Old English *ēar;* akin to Old High German *ahir* ear, Old English *ecg* edge — more at EDGE] (before 12th century)
: the fruiting spike of a cereal (as wheat or Indian corn) including both the seeds and protective structures

[3]ear *intransitive verb* (14th century)
: to form ears in growing ⟨the rye should be *earing* up⟩

ear·ache \'ir-ˌāk\ *noun* (1789)
: an ache or pain in the ear

ear candy *noun* (1977)
: music that is pleasing to listen to but lacks depth

☆ SYNONYMS
Eager, avid, keen, anxious, athirst mean moved by a strong and urgent desire or interest. EAGER implies ardor and enthusiasm and sometimes impatience at delay or restraint ⟨*eager* to get started⟩. AVID adds to EAGER the implication of insatiability or greed ⟨*avid* for new thrills⟩. KEEN suggests intensity of interest and quick responsiveness in action ⟨*keen* on the latest fashions⟩. ANXIOUS emphasizes fear of frustration or failure or disappointment ⟨*anxious* not to make a social blunder⟩. ATHIRST stresses yearning but not necessarily readiness for action ⟨*athirst* for adventure⟩.

□ USAGE
each other Some handbooks and textbooks recommend that *each other* be restricted to reference to two and *one another* to reference to three or more. The distinction, while neat, is not observed in actual usage. *Each other* and *one another* are used interchangeably by good writers and have been since at least the sixteenth century.

ear clip *noun* (1945)
: an earring with a clip fastener
ear·drop \-ˌdräp\ *noun* (1720)
: EARRING; *especially* **:** one with a pendant
ear·drum \-ˌdrəm\ *noun* (1645)
: TYMPANIC MEMBRANE
eared \ˈird\ *adjective* (14th century)
: having ears especially of a specified kind or number ⟨a big-*eared* man⟩ ⟨golden-*eared* corn⟩
eared seal *noun* (1883)
: any of a family (Otariidae) of seals including the sea lions and fur seals and having independent mobile hind limbs and small well-developed external ears — compare HAIR SEAL
ear·flap \ˈir-ˌflap\ *noun* (1907)
: a warm covering for the ears; *especially* **:** an extension on the lower edge of a cap that may be folded up or down
ear·ful \ˈir-ˌfu̇l\ *noun* (1911)
1 : an outpouring of news or gossip
2 : an outpouring of anger, abuse, or complaint
ear·ing \ˈir-iŋ\ *noun* [perhaps from [1]*ear*] (1626)
: a line used to fasten a corner of a sail to the yard or gaff or to haul a reef cringle to the yard
earl \ˈər(-ə)l\ *noun* [Middle English *erl,* from Old English *eorl* warrior, nobleman; akin to Old Norse *jarl* warrior, nobleman] (12th century)
: a member of the British peerage ranking below a marquess and above a viscount
— earl·dom \-dəm\ *noun*
earl marshal *noun* (13th century)
: an officer of state in England serving chiefly as a royal attendant on ceremonial occasions, as marshal of state processions, and as head of the College of Arms
ear·lobe \ˈir-ˌlōb\ *noun* (1859)
: the pendent part of the ear of humans or some fowls
ear·lock \-ˌläk\ *noun* (circa 1775)
: a curl of hair hanging in front of the ear
[1]**ear·ly** \ˈər-lē\ *adverb* **ear·li·er; -est** [Middle English *erly,* from Old English *ǣrlīce,* from *ǣr* early, soon — more at ERE] (before 12th century)
1 a : near the beginning of a period of time ⟨awoke *early* in the morning⟩ **b :** near the beginning of a course, process, or series ⟨*early* in his senatorial career⟩
2 a : before the usual or expected time **b** *archaic* **:** SOON **c :** sooner than related forms ⟨these apples bear *early*⟩
[2]**early** *adjective* **ear·li·er; -est** (13th century)
1 a : of, relating to, or occurring near the beginning of a period of time, a development, or a series **b** (1) **:** distant in past time (2) **:** PRIMITIVE
2 a : occurring before the usual or expected time **b :** occurring in the near future ⟨at your *earliest* convenience⟩ **c :** maturing or producing sooner than related forms ⟨an *early* peach⟩
— ear·li·ness *noun*
Early American *noun* (1895)
: a style (as of furniture, architecture, or fabric) originating in or characteristic of colonial America
early bird *noun* [from the proverb, "the early bird catches the worm"] (circa 1922)
1 : an early riser
2 : one that arrives early and especially before possible competitors
early on *adverb* (1928)
: at or during an early point or stage ⟨the reasons were obvious *early on* in the experiment⟩ ◻
ear·ly·wood \ˈər-lē-ˌwu̇d\ *noun* (circa 1914)
: SPRINGWOOD
[1]**ear·mark** \ˈir-ˌmärk\ *noun* (15th century)
1 : a mark of identification on the ear of an animal
2 : a distinguishing mark ⟨all the *earmarks* of poverty⟩
[2]**earmark** *transitive verb* (1591)
1 a : to mark (livestock) with an earmark **b :** to mark in a distinguishing manner
2 : to designate (as funds) for a specific use or owner
ear·muff \ˈir-ˌməf\ *noun* (1889)
: one of a pair of ear coverings connected by a flexible band and worn as protection against cold or noises
[1]**earn** \ˈərn\ *transitive verb* [Middle English *ernen,* from Old English *earnian;* akin to Old High German *arnōn* to reap, Czech *jeseň* autumn] (before 12th century)
1 a : to receive as return for effort and especially for work done or services rendered **b :** to bring in by way of return ⟨bonds *earning* 10% interest⟩
2 a : to come to be duly worthy of or entitled or suited to ⟨she *earned* a promotion⟩ **b :** to make worthy of or obtain for ⟨the suggestion *earned* him a promotion⟩
— earn·er *noun*
[2]**earn** *intransitive verb* [probably alteration of *yearn*] (1599)
obsolete **:** GRIEVE
earned run *noun* (1886)
: a run in baseball that scores without benefit of an error before the fielding team has had a chance to make the third putout of the inning
earned run average *noun* (1947)
: the average number of earned runs per game scored against a pitcher in baseball determined by dividing the total of earned runs scored against him by the total number of innings pitched and multiplying by nine
[1]**ear·nest** \ˈər-nəst\ *noun* [Middle English *ernest,* from Old English *eornost;* akin to Old High German *ernust* earnest] (before 12th century)
: a serious and intent mental state ⟨in *earnest*⟩
[2]**earnest** *adjective* (before 12th century)
1 : characterized by or proceeding from an intense and serious state of mind
2 : GRAVE, IMPORTANT
synonym see SERIOUS
— ear·nest·ly *adverb*
— ear·nest·ness \-nəs(t)-nəs\ *noun*
[3]**earnest** *noun* [Middle English *ernes, ernest,* from Old French *erres,* plural of *erre* earnest, from Latin *arra,* short for *arrabo,* from Greek *arrhabōn,* of Semitic origin; akin to Hebrew *ʿērābhōn* pledge] (13th century)
1 : something of value given by a buyer to a seller to bind a bargain
2 : a token of what is to come **:** PLEDGE ◇
earn·ings \ˈər-niŋz\ *noun plural* (1732)
1 : something (as wages) earned
2 : the balance of revenue after deduction of costs and expenses
ear·phone \ˈir-ˌfōn\ *noun* (1924)
: a device that converts electrical energy into sound waves and is worn over or inserted into the ear
ear pick *noun* (14th century)
: a device often of precious metal for removing wax or foreign bodies from the ear
ear·piece \ˈir-ˌpēs\ *noun* (1853)
1 : a part of an instrument (as a stethoscope or hearing aid) which is applied to the ear; *especially* **:** EARPHONE
2 : one of the two sidepieces that support eyeglasses by passing over or behind the ears
ear·plug \-ˌpləg\ *noun* (1904)
1 : an ornament inserted in the lobe of the ear especially to distend it
2 : a device of pliable material for insertion into the outer opening of the ear (as to keep out water or deaden sound)
ear·ring \ˈir-(ˌ)iŋ, -ˌriŋ\ *noun* (before 12th century)
: an ornament for the earlobe
ear rot *noun* (1926)
: a condition of Indian corn that is characterized by molding and decay of the ears and that is caused by fungi (genera *Diplodia, Fusarium,* or *Gibberella*)
ear shell *noun* (circa 1753)
: ABALONE
ear·shot \ˈir-ˌshät\ *noun* (1607)
: the range within which the unaided voice may be heard
ear·split·ting \-ˌspli-tiŋ\ *adjective* (1884)
: distressingly loud or shrill
synonym see LOUD
[1]**earth** \ˈərth\ *noun* [Middle English *erthe,* from Old English *eorthe;* akin to Old High German *erda* earth, Greek *era*] (before 12th century)
1 : the fragmental material composing part of the surface of the globe; *especially* **:** cultivable soil
2 : the sphere of mortal life as distinguished from spheres of spirit life — compare HEAVEN, HELL
3 a : areas of land as distinguished from sea and air **b :** the solid footing formed of soil **:** GROUND
4 *often capitalized* **:** the planet on which we live that is third in order from the sun — see PLANET table
5 a : the people of the planet Earth **b :** the mortal human body **c :** the pursuits, interests, and pleasures of earthly life as distinguished from spiritual concerns
6 : the lair of a burrowing animal
7 : an excessive amount of money — used with *the* ⟨real suede, which costs the *earth* to clean —Joanne Winship⟩
— earth·like \-ˌlīk\ *adjective*
— on earth — used as an intensive ⟨to find out what *on earth* he was up to —Michael Holroyd⟩
[2]**earth** (1575)

◻ USAGE

early on This adverb is sometimes objected to in American writing as an obtrusive Briticism. It is a relative newcomer to the language, having arisen in British English around 1928. It seems to have filled a need, however. It came into frequent use in American English in the late 1960s and is now well established on both sides of the Atlantic in both speech and writing.

◇ WORD HISTORY

earnest *Earnest money,* which is given to ensure the seriousness of a buyer's intent in a transaction, would seem to be money given "in earnest," but the third homograph *earnest,* the financial word, actually has no historical connection with the first homograph *earnest,* the one meaning "seriousness." As its etymology shows, the *earnest* given in a transaction has reached English through a centuries-long passage from one language to another. English borrowed the word from medieval French, which inherited it from ancient Rome. The Romans acquired the word, perhaps via the Etruscans, from the Greeks, who in turn had borrowed it from Semitic peoples of the eastern Mediterranean with whom they were in contact, such as the Phoenicians. The Semitic source of the Greek word is exactly paralleled by Biblical Hebrew *ʿērābhōn,* a derivative of a verb *ʿārabh* "pledge." The long trail of this word meaning "pledge in a bargain" shows us the remarkable continuity of mercantile customs first devised millennia ago, which have spread from culture to culture as each culture in its turn discovered the usefulness of its more sophisticated neighbor's business practices.

transitive verb
1 : to drive to hiding in the earth
2 : to draw soil about (plants) — often used with *up*
3 *chiefly British* **:** GROUND 3
intransitive verb
of a hunted animal **:** to hide in the ground

earth·born \'ərth-ˌbȯrn\ *adjective* (1667)
1 : born on this earth **:** MORTAL
2 : associated with earthly life ⟨*earthborn* cares⟩

earth·bound \-ˌbau̇nd\ *adjective* (1605)
1 a : fast in or to the soil ⟨*earthbound* roots⟩ **b :** located on or restricted to land or to the surface of the earth
2 a : bound by earthly interests **b :** PEDESTRIAN, UNIMAGINATIVE

earth color *noun* (1931)
: EARTH TONE

earth·en \'ər-thən, -t͟hən\ *adjective* (13th century)
1 : made of earth
2 : EARTHLY

earth·en·ware \-ˌwar, -ˌwer\ *noun* (1648)
: ceramic ware made of slightly porous opaque clay fired at low heat

earth·i·ly \'ər-thə-lē, -t͟hə-\ *adverb* (1953)
: in an earthy manner

earth·light \'ərth-ˌlīt\ *noun* (1833)
: EARTHSHINE

earth·ling \'ərth-liŋ\ *noun* (1593)
1 : an inhabitant of the earth
2 : WORLDLING

earth·ly \'ərth-lē\ *adjective* (before 12th century)
1 a : characteristic of or belonging to this earth **b :** relating to man's actual life on this earth
2 : POSSIBLE ⟨there is no *earthly* reason for such behavior⟩ ☆
— earth·li·ness *noun*

earth mother *noun, often E&M capitalized* (1902)
1 : the earth viewed (as in primitive theology) as the divine source of terrestrial life
2 : the embodiment of the female principle of fertility **:** a nurturing maternal woman

earth·mov·er \'ərth-ˌmü-vər\ *noun* (1941)
: a machine (as a bulldozer) for excavating, pushing, or transporting large quantities of earth (as in road building)
— earth·mov·ing \-ˌmü-viŋ\ *noun*

earth·quake \'ərth-ˌkwāk\ *noun* (14th century)
1 : a shaking or trembling of the earth that is volcanic or tectonic in origin
2 : UPHEAVAL 2

earth·rise \'ərth-ˌrīz\ *noun* (1968)
: the rising of the earth above the horizon of the moon as seen from lunar orbit

earth science *noun* (1939)
: any of the sciences (as geology, meteorology, or oceanography) that deal with the earth or with one or more of its parts — compare GEOSCIENCE
— earth scientist *noun*

earth·shak·er \'ərth-ˌshā-kər\ *noun* (1953)
: one that is earthshaking

earth·shak·ing \-kiŋ\ *adjective* (1948)
: of great importance **:** MOMENTOUS
— earth·shak·ing·ly \-kiŋ-lē\ *adverb*

earth–shat·ter·ing \'ərth-'sha-tə-riŋ\ *adjective* (1970)
: EARTHSHAKING

earth–shel·tered \'ərth-ˌshel-tərd\ *adjective* (1979)
: built partly or mostly underground ⟨an *earth-sheltered* house⟩

earth·shine \'ərth-ˌshīn\ *noun* (1876)
: sunlight reflected by the earth that illuminates the dark part of the moon — called also *earthlight*

earth·star \-ˌstär\ *noun* (1885)
: any of a genus (*Geastrum*) of globose basidiomycetous fungi with an outer peridium that splits into the shape of a star

earth station *noun* (1970)
: DISH 3a(2); *especially* **:** one used primarily for receiving and transmitting television signals

earth tone *noun* (1973)
: any of various rich colors containing some brown

earth·ward \-wərd\ *also* **earth·wards** \-wərdz\ *adverb* (14th century)
: toward the earth

earth·work \'ərth-ˌwərk\ *noun* (1633)
1 : an embankment or other construction made of earth; *especially* **:** one used as a field fortification
2 : the operations connected with excavations and embankments of earth
3 : a work of art consisting of a portion of land modified by an artist

earth·worm \-ˌwərm\ *noun* (14th century)
: a terrestrial annelid worm (class Oligochaeta); *especially* **:** any of a family (Lumbricidae) of numerous widely distributed hermaphroditic worms that move through the soil by means of setae

earthy \'ər-thē, -t͟hē\ *adjective* **earth·i·er; -est** (14th century)
1 a : of, relating to, or consisting of earth ⟨*earthy* creatures like worms⟩ **b :** suggestive of earth (as in texture, odor, or color) ⟨an *earthy* yellow⟩ **c :** rough, coarse, or plain in taste ⟨*earthy* flavors⟩
2 a *archaic* **:** EARTHLY, WORLDLY **b :** characteristic of or associated with mortal life on the earth ⟨prefers *earthy* to ethereal themes⟩
3 : suggestive of plain or poor people or their ways: as **a :** PRACTICAL, DOWN-TO-EARTH ⟨*earthy* problems of daily life⟩ **b :** CRUDE, GROSS ⟨*earthy* humor⟩ **c :** plain and simple in style **:** UNSOPHISTICATED ⟨*earthy* peasant cookery⟩ ⟨*earthy* decor⟩ ⟨*earthy* clothes⟩
— earth·i·ness *noun*

ear·wax \'ir-ˌwaks\ *noun* (14th century)
: CERUMEN

[1]ear·wig \-ˌwig\ *noun* [Middle English *erwigge,* from Old English *ēarwicga,* from *ēare* ear + *wicga* insect] (before 12th century)
: any of numerous insects (order Dermaptera) having slender many-jointed antennae and a pair of cerci resembling forceps at the end of the body

[2]earwig *transitive verb* **ear·wigged; ear·wig·ging** (1837)
: to annoy or attempt to influence by private talk

ear·wit·ness \'ir-'wit-nəs\ *noun* (1594)
: one who overhears something; *especially* **:** one who gives a report on what has been heard

[1]ease \'ēz\ *noun* [Middle English *ese,* from Old French *aise* convenience, comfort, from Latin *adjacent-, adjacens* neighboring — more at ADJACENT] (13th century)
1 : the state of being comfortable: as **a :** freedom from pain or discomfort **b :** freedom from care **c :** freedom from labor or difficulty **d :** freedom from embarrassment or constraint **:** NATURALNESS **e :** an easy fit
2 : relief from discomfort or obligation
3 : FACILITY, EFFORTLESSNESS
4 : an act of easing or a state of being eased
— ease·ful \-fəl\ *adjective*
— ease·ful·ly \-fə-lē\ *adverb*
— at ease 1 : free from pain or discomfort **2 a :** free from restraint or formality **b :** standing silently (as in a military formation) with the feet apart, the right foot in place, and one or both hands behind the body — often used as a command

[2]ease *verb* **eased; eas·ing** (14th century)
transitive verb
1 : to free from something that pains, disquiets, or burdens
2 : to make less painful **:** ALLEVIATE ⟨*ease* his suffering⟩
3 a : to lessen the pressure or tension of especially by slackening, lifting, or shifting **b :** to maneuver gently or carefully **c :** to moderate or reduce especially in amount or intensity
4 : to make less difficult ⟨*ease* credit⟩
5 a : to put the helm of (a ship) alee **b :** to let (a helm or rudder) come back a little after having been put hard over
intransitive verb
1 : to give freedom or relief
2 : to move or pass with freedom or with little resistance
3 : MODERATE, SLACKEN

ea·sel \'ē-zəl\ *noun* [Dutch *ezel,* literally, ass, from Middle Dutch *esel;* akin to Old English *esol* ass; both from a prehistoric East Germanic-West Germanic word borrowed from Latin *asinus* ass] (1596)
: a frame for supporting something (as an artist's canvas) ◆

ease·ment \'ēz-mənt\ *noun* (14th century)
1 : an act or means of easing or relieving (as from discomfort)
2 : an interest in land owned by another that entitles its holder to a specific limited use or enjoyment

eas·i·ly \'ē-zə-lē, 'ēz-lē\ *adverb* (13th century)
1 : in an easy manner **:** without difficulty
2 : without question **:** by far
3 : WELL 10b ⟨it could *easily* have been me⟩

[1]east \'ēst\ *adverb* [Middle English *est,* from Old English *ēast;* akin to Old High German *ōstar* to the east, Latin *aurora* dawn, Greek *ēōs, heōs*] (before 12th century)
: to, toward, or in the east

[2]east *adjective* (before 12th century)
1 : situated toward or at the east ⟨an *east* window⟩
2 : coming from the east ⟨an *east* wind⟩

[3]east *noun* (before 12th century)
1 a : the general direction of sunrise **:** the direction toward the right of one facing north **b :** the compass point directly opposite to west
2 *capitalized* **a :** regions lying to the east of a specified or implied point of orientation **b :** regions having a culture derived from ancient non-European especially Asian areas
3 : the altar end of a church

☆ SYNONYMS
Earthly, worldly, mundane mean belonging to or characteristic of the earth. EARTHLY often implies a contrast with what is heavenly or spiritual ⟨abandoned *earthly* concerns and entered a convent⟩. WORLDLY and MUNDANE both imply a relation to the immediate concerns and activities of human beings, WORLDLY suggesting tangible personal gain or gratification ⟨*worldly* goods⟩ and MUNDANE suggesting reference to the immediate and practical ⟨a *mundane* discussion of finances⟩.

◇ WORD HISTORY
easel The word *easel* was borrowed into English from Middle Dutch *esel* or Dutch *ezel,* alluding to the frame that supports an artist's canvas. This sense was a metaphorical extension of the literal meaning "ass, donkey," probably from the fact that, like a beast of burden, an easel is used to hold things. If this metaphor seems at all odd, we need only recall that since at least the 18th century English has used *horse* in a very similar way in the compound *sawhorse,* a framework used to support wood for cutting.

4 *often capitalized* **a :** the one of four positions at 90-degree intervals that lies to the east or at the right of a diagram **b :** a person (as a bridge player) occupying this position in the course of a specified activity

east·bound \ˈēs(t)-ˌbau̇nd\ *adjective* (1880)
: traveling or heading east

east by north (1594)
: a compass point that is one point north of due east **:** N78°45′E

east by south (14th century)
: a compass point that is one point south of due east **:** S78°45′E

East Caribbean dollar *noun* (circa 1974)
: the basic monetary unit shared by a number of islands of the British West Indies

Eas·ter \ˈē-stər\ *noun* [Middle English *estre,* from Old English *ēastre;* akin to Old High German *ōstarun* (plural) Easter, Old English *ēast* east] (before 12th century)
: a feast that commemorates Christ's resurrection and is observed with variations of date due to different calendars on the first Sunday after the paschal full moon

EASTER DATES

YEAR	ASH WEDNESDAY	EASTER
1993	February 24	April 11
1994	February 16	April 3
1995	March 1	April 16
1996	February 21	April 7
1997	February 12	March 30
1998	February 25	April 12
1999	February 17	April 4
2000	March 8	April 23
2001	February 28	April 15
2002	February 13	March 31
2003	March 5	April 20
2004	February 25	April 11
2005	February 9	March 27
2006	March 1	April 16
2007	February 21	April 8
2008	February 6	March 23
2009	February 25	April 12
2010	February 17	April 4
2011	March 9	April 24
2012	February 22	April 8

Easter egg *noun* (1804)
: an egg that is dyed and sometimes decorated and that is associated with the celebration of Easter

Easter lily *noun* (1877)
: any of several white cultivated lilies (especially *Lilium longiflorum*) that bloom in early spring

[1]**east·er·ly** \ˈē-stər-lē\ *adjective or adverb* [obsolete *easter* (eastern)] (1548)
1 : situated toward or belonging to the east ⟨the *easterly* shore of the lake⟩
2 : coming from the east ⟨an *easterly* storm⟩

[2]**easterly** *noun, plural* **-lies** (1901)
: a wind from the east

Easter Monday *noun* (14th century)
: the Monday after Easter observed as a legal holiday in some nations of the Commonwealth and in North Carolina

east·ern \ˈē-stərn\ *adjective* [Middle English *estern,* from Old English *ēasterne;* akin to Old High German *ōstrōni* eastern, Old English *ēast* east] (before 12th century)
1 *capitalized* **:** of, relating to, or characteristic of a region conventionally designated East
2 *capitalized* **a :** of, relating to, or being the Christian churches originating in the church of the Eastern Roman Empire **b :** EASTERN ORTHODOX
3 a : lying toward the east **b :** coming from the east ⟨an *eastern* wind⟩
— east·ern·most \-ˌmōst\ *adjective*

East·ern·er \ˈē-stə(r)-nər\ *noun* (1840)
: a native or inhabitant of the East; *especially* **:** a native or resident of the eastern part of the U.S.

eastern hemisphere *noun, often E&H capitalized* (1624)
: the half of the earth east of the Atlantic Ocean including Europe, Asia, Australia, and Africa

Eastern Orthodox *adjective* (1909)
: of or consisting of the Eastern churches that form a loose federation according primacy of honor to the patriarch of Constantinople and adhering to the decisions of the first seven ecumenical councils and to the Byzantine rite

eastern time *noun, often E capitalized* (1883)
: the time of the 5th time zone west of Greenwich that includes the eastern U.S. — see TIME ZONE illustration

eastern white pine *noun* (1925)
: WHITE PINE 1a

Eas·ter·tide \ˈē-stər-ˌtīd\ *noun* [Middle English *estertide,* from Old English *ēastortīd,* from *ēastor* + *tīd* time — more at TIDE] (before 12th century)
: the period from Easter to Ascension Day, to Whitsunday, or to Trinity Sunday

East Germanic *noun* (circa 1901)
: a subdivision of the Germanic languages that includes Gothic — see INDO-EUROPEAN LANGUAGES table

east·ing \ˈē-stiŋ\ *noun* (1628)
1 : easterly progress
2 : difference in longitude to the east from the last preceding point of reckoning

east–northeast *noun* (1725)
: a compass point that is two points north of due east **:** N67°30′E

east–southeast *noun* (1555)
: a compass point that is two points south of due east **:** S67°30′E

[1]**east·ward** \ˈēs-twərd\ *adverb or adjective* (before 12th century)
: toward the east
— east·wards \-twərdz\ *adverb*

[2]**eastward** *noun* (1582)
: eastward direction or part ⟨sail to the *eastward*⟩

[1]**easy** \ˈē-zē\ *adjective* **eas·i·er; -est** [Middle English *esy,* from Old French *aaisié,* past participle of *aaisier* to ease, from *a-* ad- (from Latin *ad-*) + *aise* ease] (13th century)
1 a : causing or involving little difficulty or discomfort ⟨within *easy* reach⟩ **b :** requiring or indicating little effort, thought, or reflection ⟨*easy* clichés⟩
2 a : not severe **:** LENIENT **b :** not steep or abrupt ⟨*easy* slopes⟩ **c :** not difficult to endure or undergo ⟨an *easy* penalty⟩ **d :** readily taken advantage of ⟨an *easy* target for takeovers⟩ **e** (1) **:** readily available **:** easily come by ⟨*easy* pickings⟩ (2) **:** plentiful in supply at low or declining interest rates ⟨*easy* money⟩ (3) **:** less in demand and usually lower in price ⟨bonds were *easier*⟩ **f :** PLEASANT ⟨*easy* listening⟩
3 a : marked by peace and comfort ⟨the *easy* life of a courtier⟩ **b :** not hurried or strenuous ⟨an *easy* pace⟩
4 a : free from pain, annoyance, or anxiety ⟨did all she could to make him *easier*⟩ **b :** marked by social ease ⟨an air of *easy* assurance⟩ **c :** EASYGOING ⟨an *easy* disposition⟩
5 a : giving ease, comfort, or relaxation ⟨*easy* chairs⟩ **b :** not burdensome or straitened ⟨bought on *easy* terms⟩ **c :** fitting comfortably **:** allowing freedom of movement ⟨*easy* jackets⟩ **d :** marked by ready facility ⟨an *easy* flowing style⟩ **e :** felt or attained to readily, naturally, and spontaneously ⟨an *easy* smile⟩ ☆
— eas·i·ness *noun*

[2]**easy** *adverb* **eas·i·er; -est** (14th century)
1 : EASILY 1 ⟨promises come *easy*⟩
2 a : without undue speed or excitement ⟨take it *easy*⟩ **b :** in or with moderation ⟨go *easy* on the mustard⟩
3 a : without worry or care ⟨rest *easy*⟩ **b :** without a severe penalty ⟨got off *easy*⟩ **c :** without violent movement ⟨the boat rode *easy*⟩
4 : EASILY 2 ⟨cost $500 *easy*⟩

easy·go·ing \ˌē-zē-ˈgō-iŋ, -ˈgȯ(-)iŋ\ *adjective* (1674)
1 a : relaxed and casual in style or manner ⟨an *easygoing* boss⟩ **b :** morally lax
2 : UNHURRIED, COMFORTABLE ⟨an *easygoing* pace⟩
— easy·go·ing·ness *noun*

easy mark *noun* (circa 1896)
: one easily taken advantage of

easy street *noun* (1900)
: a situation with no worries

easy virtue *noun* (1809)
: sexually promiscuous behavior or habits ⟨ladies of *easy virtue*⟩

[1]**eat** \ˈēt\ *verb* **ate** \ˈāt, *dialect or British* ˈet\; **eat·en** \ˈē-tᵊn\; **eat·ing** [Middle English *eten,* from Old English *etan;* akin to Old High German *ezzan* to eat, Latin *edere,* Greek *edmenai*] (before 12th century)
transitive verb
1 : to take in through the mouth as food **:** ingest, chew, and swallow in turn
2 a : to destroy, consume, or waste by or as if by eating ⟨expenses *ate* up the profits⟩ **b :** to bear the expense of **:** take a loss on
3 a : to consume gradually **:** CORRODE **b :** to consume with vexation **:** BOTHER ⟨what's *eating* you now⟩
intransitive verb
1 : to take food or a meal
2 : to affect something by gradual destruction or consumption — usually used with *into, away,* or *at*
— eat·er *noun*
— eat another's lunch : to deprive of profit, dominance, or success
— eat crow : to accept what one has fought against
— eat humble pie : to apologize or retract under pressure
— eat one's heart out 1 : to grieve bitterly **2 :** to be jealous
— eat one's words : to retract what one has said
— eat out of one's hand : to accept the domination of another

[2]**eat** *noun* [Middle English *et,* from Old English *ǣt;* akin to Old High German *āz* food; derivative from the root of [1]*eat*] (before 12th century)
: something to eat **:** FOOD — usually used in plural

[1]**eat·able** \ˈē-tə-bəl\ *adjective* (14th century)
: fit or able to be eaten

[2]**eatable** *noun* (1672)
1 : something to eat
2 *plural* **:** FOOD

eat·ery \ˈē-tə-rē\ *noun, plural* **-er·ies** (1901)

☆ SYNONYMS
Easy, facile, simple, light, effortless, smooth mean not demanding effort or involving difficulty. EASY is applicable either to persons or things imposing tasks or to activity required by such tasks ⟨an *easy* college course⟩. FACILE often adds to EASY the connotation of undue haste or shallowness ⟨*facile* answers to complex questions⟩. SIMPLE stresses ease in understanding or dealing with because complication is absent ⟨a *simple* problem in arithmetic⟩. LIGHT stresses freedom from what is burdensome ⟨a *light* teaching load⟩. EFFORTLESS stresses the appearance of ease and usually implies the prior attainment of artistry or expertness ⟨moving with *effortless* grace⟩. SMOOTH stresses the absence or removal of all difficulties, hardships, or obstacles ⟨a *smooth* ride⟩. See in addition COMFORTABLE.

: LUNCHEONETTE, RESTAURANT

eath \'ēth\ *adverb or adjective* [Middle English *ethe,* from Old English *ēathe;* akin to Old High German *ōdi* easy] (before 12th century) *Scottish* : EASY

eat·ing \'ē-tiŋ\ *adjective* (15th century)
1 : used for eating ⟨*eating* utensils⟩
2 : suitable to eat ⟨the finest *eating* fish⟩; *also* **:** suitable to eat raw ⟨an *eating* apple⟩

eau de co·logne \,ō-də-kə-'lōn\ *noun, plural* **eaux de cologne** \,ō(z)d-ə-\ [French, literally, Cologne water, from *Cologne,* Germany] (1802)
: COLOGNE

eau–de–vie \,ō-də-'vē\ *noun, plural* **eaux–de–vie** \,ō(z)-də-\ [French, literally, water of life, translation of Medieval Latin *aqua vitae*] (1748)
: a clear brandy distilled from the fermented juice of fruit (as pears or raspberries)

eaves \'ēvz\ *noun plural* [Middle English *eves* (singular), from Old English *efes;* akin to Old High German *obasa* portico, Old English *ūp* up — more at UP] (before 12th century)
1 : the lower border of a roof that overhangs the wall
2 : a projecting edge (as of a hill)

eaves·drop \'ēvz-,dräp\ *intransitive verb* [probably back-formation from *eavesdropper,* literally, one standing under the drip from the eaves] (1606)
: to listen secretly to what is said in private ◇
— eaves·drop·per *noun*

eaves trough *noun* (1878)
: GUTTER 1a

¹ebb \'eb\ *noun* [Middle English *ebbe,* from Old English *ebba;* akin to Middle Dutch *ebbe* ebb, Old English *of* from — more at OF] (before 12th century)
1 : the reflux of the tide toward the sea
2 : a point or condition of decline ⟨our spirits were at a low *ebb*⟩

²ebb *intransitive verb* (before 12th century)
1 : to recede from the flood
2 : to fall from a higher to a lower level or from a better to a worse state
synonym see ABATE

ebb tide *noun* (1823)
1 : the tide while ebbing or at ebb
2 : a period or state of decline

EBCDIC \'ep-sə-,dik, 'eb-\ *noun* [extended *bi*nary *c*oded *d*ecimal *i*nterchange *c*ode] (circa 1966)
: a code for representing alphanumeric information (as on magnetic tape)

Ebo·la virus \ē-'bō-lə, i-, e-\ *noun* [Ebola River, Zaire, site of an outbreak of the virus in 1976] (1977)
: an RNA-containing virus of African origin that causes an often fatal hemorrhagic fever — called also *Ebola*

eb·on \'e-bən\ *adjective* (15th century)
: EBONY

eb·o·nite \'e-bə-,nīt\ *noun* (1861)
: hard rubber especially when black

eb·o·nize \-,nīz\ *transitive verb* **-nized; -niz·ing** (circa 1828)
: to stain black in imitation of ebony

¹eb·o·ny \'e-bə-nē\ *noun, plural* **-nies** [probably from Late Latin *hebeninus* of ebony, from Greek *ebeninos,* from *ebenos* ebony, from Egyptian *hbnj*] (14th century)
1 : a hard heavy wood yielded by various Old World tropical dicotyledonous trees (genus *Diospyros* of the family Ebonaceae, the ebony family)
2 a : a tree yielding ebony **b :** any of several trees yielding wood like ebony

²ebony *adjective* (1598)
1 : made of or resembling ebony
2 : BLACK, DARK

ebul·lience \i-'bu̇l-yən(t)s, -'bəl-\ *noun* (1749)
: the quality of lively or enthusiastic expression of thoughts or feelings : EXUBERANCE

ebul·lien·cy \-yən(t)-sē\ *noun* (1676)
: EBULLIENCE

ebul·lient \-yənt\ *adjective* [Latin *ebullient-, ebulliens,* present participle of *ebullire* to bubble out, from *e-* + *bullire* to bubble, boil — more at BOIL] (1599)
1 : BOILING, AGITATED
2 : characterized by ebullience
— ebul·lient·ly *adverb*

eb·ul·li·tion \,e-bə-'li-shən\ *noun* (1534)
1 : a sudden violent outburst or display
2 : the act, process, or state of boiling or bubbling up

EB virus \,ē-'bē-\ *noun* (1968)
: EPSTEIN-BARR VIRUS

¹ec·cen·tric \ik-'sen-trik, ek-\ *adjective* [Middle English, from Medieval Latin *eccentricus,* from Greek *ekkentros,* from *ex* out of + *kentron* center] (circa 1630)
1 a : deviating from an established or usual pattern or style **b :** deviating from conventional or accepted usage or conduct especially in odd or whimsical ways
2 a : deviating from a circular path; *especially* **:** ELLIPTICAL 1 ⟨an *eccentric* orbit⟩ **b :** located elsewhere than at the geometrical center; *also* **:** having the axis or support so located ⟨an *eccentric* wheel⟩
synonym see STRANGE
— ec·cen·tri·cal·ly \-tri-k(ə-)lē\ *adverb*

²eccentric *noun* (1827)
1 : a mechanical device consisting of a disk through which a shaft is keyed eccentrically and a circular strap which works freely round the rim of the disk for communicating its motion to one end of a rod whose other end is constrained to move in a straight line so as to produce reciprocating motion
2 : an eccentric person

ec·cen·tric·i·ty \,ek-(,)sen-'tri-sə-tē\ *noun, plural* **-ties** (1545)
1 a : the quality or state of being eccentric **b :** deviation from an established pattern or norm; *especially* **:** odd or whimsical behavior
2 a : a mathematical constant that for a given conic section is the ratio of the distances from any point of the conic section to a focus and the corresponding directrix **b :** the eccentricity of an astronomical orbit used as a measure of its deviation from circularity

ec·chy·mo·sis \,e-ki-'mō-səs\ *noun, plural* **-mo·ses** \-,sēz\ [New Latin, from Greek *ekchymōsis,* from *ekchymousthai* to extravasate blood, from *ex-* + *chymos* juice — more at CHYME] (1541)
: the escape of blood into the tissues from ruptured blood vessels
— ec·chy·mot·ic \-'mä-tik\ *adjective*

ecclesi- *or* **ecclesio-** *combining form* [Middle English *ecclesi-,* from Late Latin *ecclesia,* from Greek *ekklēsia* assembly of citizens, church, from *ekkalein* to call forth, summon, from *ex-* + *kalein* to call — more at LOW]
: church ⟨*ecclesio*logy⟩

ec·cle·si·al \i-'klē-zē-əl, e-'klē-\ *adjective* (1641)
: of or relating to a church

Ec·cle·si·as·tes \i-,klē-zē-'as-(,)tēz, e-,klē-\ *noun* [Greek *Ekklēsiastēs,* literally, preacher (translation of Hebrew *qōheleth*), from *ekklēsiastēs* member of an assembly, from *ekklēsia*]
: a book of wisdom literature in canonical Jewish and Christian Scripture — see BIBLE table

¹ec·cle·si·as·tic \-'as-tik\ *adjective* (15th century)
: ECCLESIASTICAL

²ecclesiastic *noun* (1651)
: CLERGYMAN

ec·cle·si·as·ti·cal \-ti-kəl\ *adjective* [Middle English, from Late Latin *ecclesiasticus,* from Late Greek *ekklēsiastikos,* from Greek, of an assembly of citizens, from *ekklēsiastēs*] (15th century)
1 : of or relating to a church especially as an established institution
2 : suitable for use in a church
— ec·cle·si·as·ti·cal·ly \-ti-k(ə-)lē\ *adverb*

ec·cle·si·as·ti·cism \-tə-,si-zəm\ *noun* (circa 1859)
: excessive attachment to ecclesiastical forms and practices

Ec·cle·si·as·ti·cus \-ti-kəs\ *noun* [Late Latin, from *ecclesiasticus*]
: a didactic book included in the Protestant Apocrypha and as Sirach in the Roman Catholic canon of the Old Testament

ec·cle·si·ol·o·gy \i-,klē-zē-'ä-lə-jē, e-,klē-\ *noun, plural* **-gies** (circa 1837)
1 : the study of church architecture and adornment
2 : theological doctrine relating to the church
— ec·cle·si·o·log·i·cal \-zē-ə-'lä-ji-kəl\ *adjective*
— ec·cle·si·ol·o·gist \-zē-'ä-lə-jist\ *noun*

ec·crine gland \'e-krən-, -,krīn-, -,krēn-\ *noun* [Greek *ekkrinein* to secrete, from *ek-, ex-* out + *krinein* to separate — more at CERTAIN] (circa 1927)
: any of the rather small sweat glands in the human skin that produce a fluid secretion without removing cytoplasm from the secretory cells — called also *eccrine sweat gland*

ec·dys·i·ast \ek-'di-zē-,ast, -ē-əst\ *noun* [Greek *ekdysis*] (1940)
: STRIPTEASER

ec·dy·sis \'ek-də-səs\ *noun, plural* **ec·dy·ses** \-də-,sēz\ [New Latin, from Greek *ekdysis* act of getting out, from *ekdyein* to strip, from *ex-* + *dyein* to enter, don] (circa 1854)
: the act of molting or shedding an outer cuticular layer (as in insects and crustaceans)

ec·dy·sone \'ek-də-,sōn\ *also* **ec·dy·son** \-,sän\ *noun* [International Scientific Vocabulary *ecdys*is + horm*one*] (1956)
: any of several arthropod hormones that in insects are produced by the prothoracic gland and that trigger molting and metamorphosis

ece·sis \i-'sē-səs, -'kē-\ *noun* [New Latin, from Greek *oikēsis* inhabitation, from *oikein* to inhabit — more at ECUMENICAL] (circa 1904)
: the establishment of a plant or animal in a new habitat

echelle \ā-'shel\ *noun* [French, literally, ladder, from Old French *eschele*] (1949)

◇ WORD HISTORY

eavesdrop The verb *eavesdrop* first appeared in the 17th century and is probably a back-formation from *eavesdropper,* which is attested in Middle English as *evesdropper* and is apparently a derivative of an earlier noun *evesdrop. Evesdrop,* and a related Old English word *yfesdrype,* occur very rarely and then only in law texts that refer to the ground below the eaves of a house on which rainwater drips. This use comes from an old sanction against building so close to the edge of one's property that water from the eaves falls on a neighbor's land. For a while the word *eavesdropper* too occurred mainly in legal contexts—persons snooping on others' conversations seem to have been considered a sort of public nuisance, like Peeping Toms—but less literal uses of *eavesdropper* and the verb *eavesdrop* appear in the 17th and 18th centuries. Today applications of the words to secret observation by satellite or electronic device are far removed from eaves, rainwater, or stealthy listening outside the walls of someone's house.

: a diffraction grating made by ruling a plane metallic mirror with lines having a relatively wide spacing

[1]**ech·e·lon** \'e-shə-ˌlän\ *noun* [French *échelon,* literally, rung of a ladder, from Old French *eschelon,* from *eschele* ladder, from Late Latin *scala*] (1796)
1 a (1) **:** an arrangement of a body of troops with its units each somewhat to the left or right of the one in the rear like a series of steps (2) **:** a formation of units or individuals resembling such an echelon (3) **:** a flight formation in which each airplane flies at a certain elevation above or below and at a certain distance behind and to the right or left of the airplane ahead **b :** any of several military units in echelon formation
2 a : one of a series of levels or grades (as of leadership or responsibility) in an organization or field of activity **b :** a group of individuals at a particular level or grade in an organization ◆

[2]**echelon** (circa 1860)
transitive verb
: to form or arrange in an echelon
intransitive verb
: to take position in an echelon

ech·e·ve·ria \ˌe-chə-və-'rē-ə\ *noun* [New Latin, genus name, after Atanasio *Echeverría* (flourished 1771) Mexican botanical illustrator] (1883)
: any of a large genus (*Echeveria*) of tropical American succulent plants of the orpine family that have showy rosettes of often plushy basal leaves and axillary clusters of flowers with erect petals and that are often grown as ornamentals

echid·na \i-'kid-nə\ *noun* [New Latin, from Latin, viper, from Greek — more at OPHITIC] (1832)
: an oviparous spiny-coated toothless burrowing nocturnal monotreme mammal (*Tachyglossus aculeatus*) of Australia, Tasmania, and New Guinea that has a long extensile tongue and long heavy claws and that feeds chiefly on ants; *also* **:** a related mammal (*Zaglossus bruijni*) of New Guinea having a longer snout and shorter spines

echidna

echin- *or* **echino-** *combining form* [Latin, from Greek, from *echinos* sea urchin]
1 : prickle ⟨*echino*derm⟩
2 : sea urchin ⟨*echin*oid⟩

echi·no·coc·co·sis \i-ˌkī-nə-kä-'kō-səs\ *noun, plural* **-co·ses** \-ˌsēz\ [New Latin] (1900)
: infestation with or disease caused by an echinococcus (as *Echinococcus granulosus*)

echi·no·coc·cus \i-ˌkī-nə-'kä-kəs\ *noun, plural* **-coc·ci** \-'käk-ˌ(s)ī, -'käk-(ˌ)(s)ē\ [New Latin, genus name] (1839)
: any of a genus (*Echinococcus*) of tapeworms that alternate a minute adult living as a commensal in the intestine of carnivores with a hydatid larva invading tissues especially of the liver of cattle, sheep, swine, and humans and acting as a dangerous pathogen

echi·no·derm \i-'kī-nə-ˌdərm\ *noun* [New Latin *Echinodermata,* phylum name, from *echin-* + *-dermata* (from Greek *derma* skin)] (1835)
: any of a phylum (Echinodermata) of radially symmetrical coelomate marine animals including the starfishes, sea urchins, and related forms
— echi·no·der·ma·tous \-ˌkī-nə-'dər-mə-təs\ *adjective*

echi·noid \i-'kī-ˌnȯid, 'e-kə-ˌnȯid\ *noun* (1864)
: SEA URCHIN

echi·nus \i-'kī-nəs\ *noun, plural* **-ni** \-ˌnī\ [Middle English, from Latin, from Greek *echinos* hedgehog, sea urchin — more at OPHITE] (14th century)
1 : SEA URCHIN
2 a : the rounded molding forming the bell of the capital in the Greek Doric order **b :** a similar member in other orders

echi·uroid \ˌe-ki-'yu̇r-ˌȯid\ *noun* [New Latin *Echiuroidea* or *Echiura,* ultimately from Greek *echis* viper + *oura* tail] (circa 1889)
: any of a taxon (class Echiuroidea or phylum Echiura) of marine worms of uncertain taxonomic affinities that have a sensitive but nonretractile proboscis above the mouth

[1]**echo** \'e-(ˌ)kō\ *noun, plural* **ech·oes** *also* **echos** [Middle English *ecco,* from Middle French & Latin; Middle French *echo,* from Latin, from Greek *ēchō;* akin to Latin *vagire* to wail, Greek *ēchē* sound] (14th century)
1 a : the repetition of a sound caused by reflection of sound waves **b :** the sound due to such reflection
2 a : a repetition or imitation of another **:** REFLECTION **b :** REPERCUSSION, RESULT **c :** TRACE, VESTIGE **d :** RESPONSE
3 : one who closely imitates or repeats another's words, ideas, or acts
4 : a soft repetition of a musical phrase
5 a : the repetition of a received radio signal due especially to reflection of part of the wave from an ionized layer of the atmosphere **b** (1) **:** the reflection of transmitted radar signals by an object (2) **:** the visual indication of this reflection on a radarscope
— echo·ey \'e-ˌkō-ē\ *adjective*

[2]**echo** *verb* **ech·oed; echo·ing** \'e-(ˌ)kō-iŋ, 'e-kə-wiŋ\ (1596)
intransitive verb
1 : to resound with echoes
2 : to produce an echo
transitive verb
1 : REPEAT, IMITATE
2 : to send back (a sound) by the reflection of sound waves

[1]**Echo** *noun* [Greek *Ēchō*]
: a nymph in Greek mythology who pines away for love of Narcissus until nothing is left of her but her voice

[2]**Echo** (circa 1952)
— a communications code word for the letter *e*

echo·car·dio·gram \ˌe-kō-'kär-dē-ə-ˌgram\ *noun* (1967)
: a visual record made by echocardiography

echo·car·di·og·ra·phy \-ˌkär-dē-'ä-grə-fē\ *noun, plural* **-phies** (1965)
: the use of ultrasound to examine the structure and functioning of the heart for abnormalities and disease
— echo·car·di·og·raph·er \-grə-fər\ *noun*
— echo·car·dio·graph·ic \-dē-ə-'gra-fik\ *adjective*

echo chamber *noun* (circa 1937)
: a room with sound-reflecting walls used for producing hollow or echoing sound effects

echo·ic \ə-'kō-ik, e-\ *adjective* (circa 1880)
1 : formed in imitation of some natural sound **:** ONOMATOPOEIC
2 : of or relating to an echo

echo·la·lia \ˌe-kō-'lā-lē-ə\ *noun* [New Latin] (circa 1885)
: the often pathological repetition of what is said by other people as if echoing them
— echo·lal·ic \-'la-lik\ *adjective*

echo·lo·ca·tion \ˌe-kō-lō-'kā-shən\ *noun* (circa 1944)
: a physiological process for locating distant or invisible objects (as prey) by means of sound waves reflected back to the emitter (as a bat) by the objects

echo sounder *noun* (1927)
: an instrument for determining the depth of a body of water or of an object below the surface by sound waves

echo·vi·rus \'e-kō-ˌvī-rəs\ *noun* [*e*nteric *c*ytopathogenic *h*uman *o*rphan + *virus*] (1955)
: any of a group of picornaviruses that are found in the gastrointestinal tract, that cause cytopathic changes in cells in tissue culture, and that are sometimes associated with respiratory ailments and meningitis

éclair \ā-'klar, i-, -'kler, 'ā-ˌ, 'ē-ˌ\ *noun* [French, literally, lightning] (1861)
: a usually chocolate-frosted oblong light pastry with whipped cream or custard filling

éclair·cis·se·ment \ā-kler-sēs-(ə-)mäⁿ\ *noun, plural* **-ments** \-mäⁿ(z)\ [French] (1667)
: a clearing up of something obscure **:** ENLIGHTENMENT

eclamp·sia \i-'klam(p)-sē-ə\ *noun* [New Latin, from Greek *eklampsis* sudden flashing, from *eklampein* to shine forth, from *ex-* out + *lampein* to shine] (circa 1860)
: a convulsive state; *especially* **:** an attack of convulsions during pregnancy or parturition
— eclamp·tic \-'klam(p)-tik\ *adjective*

éclat \ā-'klä, 'ā-ˌ\ *noun* [French, splinter, burst, éclat] (1672)
1 : ostentatious display **:** PUBLICITY
2 : dazzling effect **:** BRILLIANCE
3 a : brilliant or conspicuous success **b :** PRAISE, APPLAUSE

[1]**eclec·tic** \e-'klek-tik, i-\ *adjective* [Greek *eklektikos,* from *eklegein* to select, from *ex-* out + *legein* to gather — more at LEGEND] (1683)
1 : selecting what appears to be best in various doctrines, methods, or styles
2 : composed of elements drawn from various sources; *also* **:** HETEROGENEOUS
— eclec·ti·cal·ly \-ti-k(ə-)lē\ *adverb*

[2]**eclectic** *noun* (1817)
: one who uses an eclectic method or approach

eclec·ti·cism \-'klek-tə-ˌsi-zəm\ *noun* (1798)
1 : the theory or practice of an eclectic method
2 : a mid-19th-century movement in architecture and design that revived the styles of various historical periods

[1]**eclipse** \i-'klips\ *noun* [Middle English, from Old French, from Latin *eclipsis,* from Greek *ekleipsis,* from *ekleipein* to omit, fail, suffer eclipse, from *ex-* + *leipein* to leave — more at LOAN] (13th century)
1 a : the total or partial obscuring of one celestial body by another **b :** the passing into the shadow of a celestial body — compare OCCULTATION, TRANSIT
2 : a falling into obscurity or decline; *also* **:** the state of being eclipsed
3 : the state of being in eclipse plumage

[2]**eclipse** *transitive verb* **eclipsed; eclips·ing** (13th century)
: to cause an eclipse of: as **a :** OBSCURE, DARKEN **b :** to reduce in importance or repute **c :** SURPASS

◇ WORD HISTORY

echelon In Old French *eschele* meant "ladder," and its diminutive form *eschelon* meant "a rung of a ladder." From *eschelon* a number of figurative meanings developed, including the sense referring to a formation of troops in which individual units are arranged like a staggered series of steps. It was this extended sense that English borrowed in the 18th century. After the borrowing, the meaning of *echelon* was expanded to cover similar formations of ships and aircraft. The pyramidal formation and the steplike arrangement of an echelon made it easy to apply the word figuratively to the hierarchical structure of a military organization. Subdivisions of a military organization became known as *echelons.* Then in extended and transferred use *echelon* came to refer to a single level of any hierarchical organization or to the individuals occupying such a level.

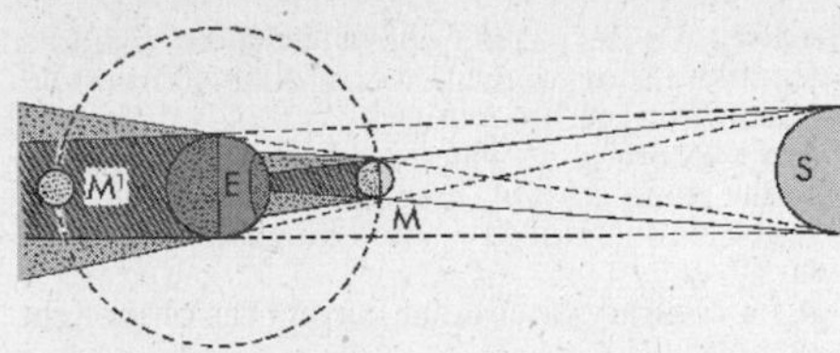

eclipse 1a: *S* sun, *E* earth, *M* moon in solar eclipse, *M*[1] moon in lunar eclipse

eclipse plumage *noun* (1906)
: comparatively dull plumage that is usually of seasonal occurrence in birds exhibiting a distinct breeding plumage

¹eclip·tic \i-'klip-tik\ *adjective* [Middle English *ecliptik,* from Late Latin *ecliptica linea,* literally, line of eclipses] (14th century)
: of or relating to the ecliptic or an eclipse

²ecliptic *noun* (15th century)
: the great circle of the celestial sphere that is the apparent path of the sun among the stars or of the earth as seen from the sun **:** the plane of the earth's orbit extended to meet the celestial sphere

ec·logue \'ek-,lòg, -,läg\ *noun* [Middle English *eclog,* from Latin *Eclogae,* title of Virgil's pastorals, literally, selections, plural of *ecloga,* from Greek *eklogē,* from *eklegein* to select] (15th century)
: a poem in which shepherds converse

eclo·sion \i-'klō-zhən\ *noun* [French *éclosion,* from *éclore* to hatch, from (assumed) Vulgar Latin *exclaudere,* alteration of Latin *excludere* to hatch out, exclude] (circa 1889)
of an insect **:** the act of emerging from the pupal case or hatching from the egg

eco- *combining form* [Late Latin *oeco-* household, from Greek *oik-, oiko-,* from *oikos* house — more at VICINITY]
1 : habitat or environment ⟨*eco*species⟩
2 : ecological or environmental ⟨*eco*catastrophe⟩

eco·ca·tas·tro·phe \,ē-(,)kō-kə-'tas-trə-fē, ,e-(,)kō-\ *noun* (1969)
: a major destructive upset in the balance of nature especially when caused by the intervention of humans

ecol·o·gy \i-'kä-lə-jē, e-\ *noun, plural* **-gies** [German *Ökologie,* from *öko-* eco- + *-logie* -logy] (1873)
1 : a branch of science concerned with the interrelationship of organisms and their environments
2 : the totality or pattern of relations between organisms and their environment
3 : HUMAN ECOLOGY
— eco·log·i·cal \,ē-kə-'lä-ji-kəl, ,e-kə-\ *also* **eco·log·ic** \-jik\ *adjective*
— eco·log·i·cal·ly \-ji-k(ə-)lē\ *adverb*
— ecol·o·gist \i-'kä-lə-jist, e-\ *noun*

econo·box \i-'kä-nō-,bäks, ē-\ *noun* [*economical* + ²*box*] (1980)
: a small economical car

econo·met·rics \i-,kä-nə-'me-triks, ē-,kä-\ *noun plural but singular in construction* [blend of *economics* and *metric*] (1933)
: the application of statistical methods to the study of economic data and problems
— econo·met·ric \-trik\ *adjective*
— econo·met·ri·cal·ly \-tri-k(ə-)lē\ *adverb*
— econo·me·tri·cian \-mə-'tri-shən\ *noun*
— econo·met·rist \-'me-trist\ *noun*

eco·nom·ic \,e-kə-'nä-mik, ,ē-kə-\ *adjective* (1592)
1 *archaic* **:** of or relating to a household or its management
2 : ECONOMICAL 2
3 a : of or relating to economics **b :** of, relating to, or based on the production, distribution, and consumption of goods and services **c :** of or relating to an economy
4 : having practical or industrial significance or uses **:** affecting material resources
5 : PROFITABLE □

eco·nom·i·cal \-'nä-mi-kəl\ *adjective* (15th century)
1 *archaic* **:** ECONOMIC 1
2 : marked by careful, efficient, and prudent use of resources **:** THRIFTY
3 : operating with little waste or at a saving
synonym see SPARING
— eco·nom·i·cal·ly \-mi-k(ə-)lē\ *adverb*

economic rent *noun* (1889)
: the return for the use of a factor in excess of the minimum required to bring forth its service

eco·nom·ics \,e-kə-'nä-miks, ,ē-kə-\ *noun plural but singular or plural in construction* (1792)
1 : a social science concerned chiefly with description and analysis of the production, distribution, and consumption of goods and services
2 : economic aspect or significance

economise *British variant of* ECONOMIZE

econ·o·mist \i-'kä-nə-mist\ *noun* (1586)
1 *archaic* **:** one who practices economy
2 : a specialist in economics

econ·o·mize \-,mīz\ *verb* **-mized; -miz·ing** (1820)
intransitive verb
: to practice economy **:** be frugal
transitive verb
: to use more economically **:** SAVE
— econ·o·miz·er *noun*

¹econ·o·my \i-'kä-nə-mē, ə-, ē-\ *noun, plural* **-mies** [Middle French *yconomie,* from Medieval Latin *oeconomia,* from Greek *oikonomia,* from *oikonomos* household manager, from *oikos* house + *nemein* to manage — more at VICINITY, NIMBLE] (15th century)
1 *archaic* **:** the management of household or private affairs and especially expenses
2 a : thrifty and efficient use of material resources **:** frugality in expenditures; *also* **:** an instance or a means of economizing **:** SAVING **b :** efficient and concise use of nonmaterial resources (as effort, language, or motion)
3 : the arrangement or mode of operation of something **:** ORGANIZATION
4 : the structure of economic life in a country, area, or period; *specifically* **:** an economic system

²economy *adjective* (circa 1906)
: designed to save money ⟨*economy* cars⟩

economy of scale (1957)
: a reduction in unit costs brought about especially by increased size of production facilities — usually used in plural

eco·phys·i·ol·o·gy \,ē-kō-,fi-zē-'ä-lə-jē, ,e-kō-\ *noun* (1962)
: the science of the interrelationships between the physiology of organisms and their environment
— eco·phys·i·o·log·i·cal \-ē-ə-'lä-ji-kəl\ *adjective*

eco·spe·cies \'ē-kō-,spē-(,)shēz, 'e-kō-, -(,)sēz\ *noun, plural* **ecospecies** (1922)
: a subdivision of a cenospecies comparable with a taxonomic species and composed of interbreeding individuals that are less capable of fertile crosses with individuals of other subdivisions

eco·sphere \'ē-kō-,sfir, 'e-kō-\ *noun* (1953)
: the parts of the universe habitable by living organisms; *especially* **:** BIOSPHERE 1

eco·sys·tem \-,sis-təm\ *noun* (1935)
: the complex of a community of organisms and its environment functioning as an ecological unit in nature

eco·tone \'ē-kə-,tōn, 'e-kə-\ *noun* [*ec-* + Greek *tonos* tension — more at TONE] (1904)
: a transition area between two adjacent ecological communities

eco·tour·ism \,ē-kō-'tùr-,i-zəm, ,e-kō-\ *noun* (1989)
: the practice of touring natural habitats in a manner meant to minimize ecological impact
— eco·tour·ist \-'tùr-ist\ *noun*

eco·type \'ē-kə-,tīp, 'e-\ *noun* (1922)
: a subdivision of an ecospecies that survives as a distinct group through environmental selection and isolation and that is comparable with a taxonomic subspecies
— eco·typ·ic \,ē-kə-'ti-pik, ,e-kə-\ *adjective*

ecru \'e-(,)krü, 'ā-(,)krü\ *noun* [French *écru,* literally, unbleached, raw, from Old French *escru,* from *es-* completely (from Latin *ex-*) + *cru* raw, from Latin *crudus* — more at RAW] (1850)
: BEIGE 2
— ecru *adjective*

ec·sta·sy \'ek-stə-sē\ *noun, plural* **-sies** [Middle English *extasie,* from Middle French, from Late Latin *ecstasis,* from Greek *ekstasis,* from *existanai* to derange, from *ex-* out + *histanai* to cause to stand — more at EX-, STAND] (14th century)
1 a : a state of being beyond reason and self-control **b** *archaic* **:** SWOON
2 : a state of overwhelming emotion; *especially* **:** rapturous delight
3 : TRANCE; *especially* **:** a mystic or prophetic trance
4 : a synthetic amphetamine analogue $C_{11}H_{15}NO_2$ used illicitly for its mood-enhancing and hallucinogenic properties ☆

¹ec·stat·ic \ek-'sta-tik, ik-'sta-\ *adjective* [Medieval Latin *ecstaticus,* from Greek *ekstatikos,* from *existanai*] (1590)
: of, relating to, or marked by ecstasy
— ec·stat·i·cal·ly \-'sta-ti-k(ə-)lē\ *adverb*

²ecstatic *noun* (1659)
: one that is subject to ecstasies

ect- *or* **ecto-** *combining form* [New Latin, from Greek *ekto-,* from *ektos,* from *ex* out — more at EX-]
: outside **:** external ⟨*ecto*derm⟩ — compare END-, EXO-

ec·to·derm \'ek-tə-,dərm\ *noun* [International Scientific Vocabulary] (1861)
1 : the outer cellular membrane of a diploblastic animal (as a jellyfish)

☆ SYNONYMS

Ecstasy, rapture, transport mean intense exaltation of mind and feelings. ECSTASY and RAPTURE both suggest a state of trance or near immobility produced by an overpowering emotion. ECSTASY may apply to any strong emotion (as joy, fear, rage, adoration) ⟨religious *ecstasy*⟩. RAPTURE usually implies intense bliss or beatitude ⟨in speechless *rapture*⟩. TRANSPORT applies to any powerful emotion that lifts one out of oneself and usually provokes vehement expression or frenzied action ⟨in a *transport* of rage⟩.

□ USAGE

economic While orthoepists have disagreed for centuries over the first vowel of *economic* and related words, common usage has in that same time established both \ē\ and \e\ variants of *economic* as acceptable. British speakers prefer the initial \ē\ vowel, as they do in *ecumenical.* Americans are split evenly on *economic,* but overwhelmingly prefer initial \e\ in *ecumenical,* as they do also in *estrogen* (spelled *oestrogen* in British English and pronounced most often with initial \ē\). The variable vowel in these words derives ultimately from the Greek diphthong *oi,* which was pronounced \òi\ in ancient times but is \ē\ in modern Greek.

2 a : the outermost of the three primary germ layers of an embryo **b :** a tissue (as neural tissue) derived from this germ layer
— **ec·to·der·mal** \ˌek-tə-ˈdər-məl\ *adjective*
ec·to·morph \ˈek-tə-ˌmȯrf\ *noun* [*ecto*derm + *-morph*] (1940)
: an ectomorphic individual
ec·to·mor·phic \ˌek-tə-ˈmȯr-fik\ *adjective* [*ecto*derm + *-morphic;* from the predominance in such types of structures developed from the ectoderm] (1940)
1 : of or relating to the component in W. H. Sheldon's classification of body types that measures the body's degree of slenderness, angularity, and fragility
2 : characterized by a light body build with slight muscular development
-ec·to·my \ˈek-tə-mē\ *noun combining form* [New Latin *-ectomia,* from Greek *ektemnein* to cut out, from *ec-, ex-* out + *temnein* to cut — more at TOME]
: surgical removal ⟨gastr*ectomy*⟩
ec·to·par·a·site \ˌek-tō-ˈpar-ə-ˌsīt\ *noun* [International Scientific Vocabulary] (1861)
: a parasite that lives on the exterior of its host
— **ec·to·par·a·sit·ic** \-ˌpar-ə-ˈsi-tik\ *adjective*
ec·top·ic \ek-ˈtä-pik\ *adjective* [Greek *ektopos* out of place, from *ex-* out + *topos* place] (1873)
: occurring in an abnormal position or in an unusual manner or form ⟨*ectopic* lesions⟩
— **ec·top·i·cal·ly** \-pi-k(ə-)lē\ *adverb*
ectopic pregnancy *noun* (1929)
: gestation elsewhere than in the uterus (as in a fallopian tube or in the peritoneal cavity)
ec·to·plasm \ˈek-tə-ˌpla-zəm\ *noun* (1883)
1 : the outer relatively rigid granule-free layer of the cytoplasm usually held to be a gel reversibly convertible to a sol
2 : a substance held to produce spirit materialization and telekinesis
— **ec·to·plas·mic** \ˌek-tə-ˈplaz-mik\ *adjective*
ec·to·therm \ˈek-tə-ˌthərm\ *noun* (1945)
: a cold-blooded animal **:** POIKILOTHERM
— **ec·to·ther·mic** \ˌek-tə-ˈthər-mik\ *adjective*
ec·to·tro·phic \ˌek-tə-ˈtrō-fik\ *adjective* (circa 1889)
of a mycorrhiza **:** growing in a close web on the surface of the associated root — compare ENDOTROPHIC
¹ecu \ˈā-ˌkyü, ā-kœ\ *noun, plural* **ecus** \-ˌkyüz, -kœ\ [Middle French, literally, shield, from Old French *escu,* from Latin *scutum;* from the device of a shield on the coin — more at ESQUIRE] (circa 1593)
: any of various old French units of value; *also* **:** a coin representing an ecu
²ecu \ˌē-(ˌ)sē-ˈyü\ *noun, often E&C&U capitalized* [*E*uropean *C*urrency *U*nit (influenced by French *écu* ecu)] (1970)
: a money of account based on the currency units of members of the European Community
ec·u·men·i·cal \ˌe-kyə-ˈme-ni-kəl\ *adjective* [Late Latin *oecumenicus,* from Late Greek *oikoumenikos,* from Greek *oikoumenē* the inhabited world, from feminine of *oikoumenos,* present passive participle of *oikein* to inhabit, from *oikos* house — more at VICINITY] (circa 1587)
1 : worldwide or general in extent, influence, or application
2 a : of, relating to, or representing the whole of a body of churches **b :** promoting or tending toward worldwide Christian unity or cooperation
— **ec·u·men·i·cal·ly** \-k(ə-)lē\ *adverb*
usage see ECONOMIC
ec·u·men·i·cal·ism \-ˈme-ni-kə-ˌli-zəm\ *noun* (1888)
: ECUMENISM
ecumenical patriarch *noun* (1862)
: the patriarch of Constantinople as the dignitary given first honor in the Eastern Orthodox Church
ec·u·men·i·cism \ˌe-kyə-ˈme-nə-ˌsi-zəm\ *noun* (1961)
: ECUMENISM
— **ec·u·men·i·cist** \-sist\ *noun*
ec·u·me·nic·i·ty \ˌe-kyə-mə-ˈni-sə-tē, -me-\ *noun* (1840)
: the quality or state of being drawn close to others especially through Christian ecumenism
ec·u·men·ics \-ˈme-niks\ *noun plural but singular in construction* (circa 1937)
: the study of the nature, mission, problems, and strategy of the Christian church from the perspective of its ecumenical character
ecu·me·nism \e-ˈkyü-mə-ˌni-zəm, i- *also* ˈe-kyə- *or* ˌe-kyə-ˈme-\ *noun* (1948)
: ecumenical principles and practices especially as shown among religious groups (as Christian denominations)
— **ecu·me·nist** \e-ˈkyü-mə-nist, i- *also* ˈe-kyə-mə-nist *or* ˌe-kyə-ˈme-nist\ *noun*
ec·ze·ma \ig-ˈzē-mə, ˈeg-zə-mə, ˈek-sə-\ *noun* [New Latin, from Greek *ekzema,* from *ekzein* to erupt, from *ex-* out + *zein* to boil — more at EX-, YEAST] (circa 1753)
: an inflammatory condition of the skin characterized by redness, itching, and oozing vesicular lesions which become scaly, crusted, or hardened
— **ec·zem·a·tous** \ig-ˈze-mə-təs\ *adjective*
ed \ˈed\ *noun* (1954)
: EDUCATION ⟨driver's *ed*⟩ ⟨adult *ed*⟩
¹-ed \d *after a vowel or* b, g, j, l, m, n, ŋ, r, t͟h, v, z, *or* zh; əd, id *after* d *or* t; t *after other sounds; exceptions are pronounced at their entries*\ *verb suffix or adjective suffix* [Middle English, from Old English *-ed, -od, -ad;* akin to Old High German *-t,* past participle ending, Latin *-tus,* Greek *-tos,* suffix forming verbals]
1 — used to form the past participle of regular weak verbs ⟨end*ed*⟩ ⟨fad*ed*⟩ ⟨tri*ed*⟩ ⟨patt*ed*⟩
2 — used to form adjectives of identical meaning from Latin-derived adjectives ending in *-ate* ⟨crenulat*ed*⟩
3 a : having **:** characterized by ⟨cultur*ed*⟩ ⟨two-legg*ed*⟩ **b :** having the characteristics of ⟨bigot*ed*⟩
²-ed *verb suffix* [Middle English *-ede, -de,* from Old English *-de, -ede, -ode, -ade;* akin to Old High German *-ta,* past ending (1st singular) and probably to Old High German *-t,* past participle ending]
— used to form the past tense of regular weak verbs ⟨judg*ed*⟩ ⟨deni*ed*⟩ ⟨dropp*ed*⟩
eda·cious \i-ˈdā-shəs\ *adjective* [Latin *edac-, edax,* from *edere* to eat — more at EAT] (circa 1798)
1 *archaic* **:** of or relating to eating
2 : VORACIOUS
— **edac·i·ty** \-ˈda-sə-tē\ *noun*
Edam \ˈē-dəm, ˈē-ˌdam\ *noun* [*Edam,* Netherlands] (1836)
: a yellow pressed cheese of Dutch origin usually made in flattened balls and often coated with red wax
edaph·ic \i-ˈda-fik\ *adjective* [Greek *edaphos* bottom, ground] (circa 1900)
1 : of or relating to the soil
2 : resulting from or influenced by the soil rather than the climate — compare CLIMATIC 2
— **edaph·i·cal·ly** \-ˈda-fi-k(ə-)lē\ *adverb*
edaphic climax *noun* (circa 1934)
: an ecological climax resulting from soil factors and commonly persisting through cycles of climatic and physiographic change — compare CLIMATIC CLIMAX
Ed·dic \ˈe-dik\ *adjective* [Old Norse *Edda,* a 13th century collection of mythological, heroic, and aphoristic poetry] (1868)
: of, relating to, or resembling the Old Norse *Edda*
¹ed·dy \ˈe-dē\ *noun, plural* **eddies** [Middle English (Scots dialect) *ydy,* probably from Old Norse *itha*] (15th century)
1 a : a current of water or air running contrary to the main current; *especially* **:** a circular current **:** WHIRLPOOL **b :** something moving similarly
2 : a contrary or circular current (as of thought or policy)
²eddy *verb* **ed·died; ed·dy·ing** (1810)
transitive verb
: to cause to move in an eddy
intransitive verb
: to move in an eddy or in the manner of an eddy
eddy current *noun* (1887)
: an electric current induced by an alternating magnetic field
edel·weiss \ˈā-dᵊl-ˌvīs, -ˌwīs\ *noun* [German, from *edel* noble + *weiss* white] (1862)
: a small alpine perennial composite herb (*Leontopodium alpinum*) of central and southeast Europe that has a dense woolly white pubescence

edelweiss

ede·ma \i-ˈde-mə\ *noun* [New Latin, from Greek *oidēma* swelling, from *oidein* to swell; akin to Old English *ātor* poison] (15th century)
1 : an abnormal infiltration and excess accumulation of serous fluid in connective tissue or in a serous cavity — called also *dropsy*
2 a : watery swelling of plant organs or parts **b :** any of various plant diseases characterized by such swellings
— **edem·a·tous** \-ˈde-mə-təs\ *adjective*
Eden \ˈē-dᵊn\ *noun* [Late Latin, from Hebrew *ʽĒdhen*] (13th century)
1 : PARADISE 2
2 : the garden where according to the account in Genesis Adam and Eve first lived
— **Eden·ic** \i-ˈde-nik, ē-\ *adjective*
¹eden·tate \(ˌ)ē-ˈden-ˌtāt\ *adjective* [Latin *edentatus,* past participle of *edentare* to make toothless, from *e-* + *dent-, dens* tooth — more at TOOTH] (1828)
1 : lacking teeth
2 : being an edentate
²edentate *noun* (1835)
: any of an order (Edentata) of mammals having few or no teeth and including the sloths, armadillos, and New World anteaters and formerly also the pangolins and the aardvark
eden·tu·lous \(ˌ)ē-ˈden-chə-ləs\ *adjective* [Latin *edentulus,* from *e-* + *dent-, dens*] (1782)
: TOOTHLESS
Ed·gar \ˈed-gər\ *noun* [*Edgar* Allan Poe, regarded as father of the detective story] (1947)
: a statuette awarded annually by a professional organization for notable achievement in mystery-novel writing
¹edge \ˈej\ *noun* [Middle English *egge,* from Old English *ecg;* akin to Latin *acer* sharp, Greek *akmē* point] (before 12th century)
1 a : the cutting side of a blade **b :** the sharpness of a blade **c** (1) **:** FORCE, EFFECTIVENESS ⟨blunted the *edge* of the legislation⟩ (2) **:** vigor or energy especially of body ⟨maintains his hard *edge*⟩ **d** (1) **:** incisive or penetrating quality ⟨writing with a satirical *edge*⟩ (2) **:** a noticeably harsh or sharp quality ⟨her voice had an *edge* to it⟩ **e :** keenness of desire or enjoyment ⟨lost my competitive *edge*⟩ ⟨took the *edge* off our appetites⟩
2 a : the line where an object or area begins or ends **:** BORDER ⟨on the *edge* of a plain⟩ **b :** the narrow part adjacent to a border ⟨the *edge* of the deck⟩ **c :** a point near the beginning or the end; *especially* **:** BRINK, VERGE ⟨on the *edge* of disaster⟩ **d :** a favorable margin **:** ADVANTAGE ⟨has an *edge* on the competition⟩

3 : a line or line segment that is the intersection of two plane faces (as of a pyramid) or of two planes
— **edge·less** *adjective*
— **on edge :** ANXIOUS, NERVOUS

[2]**edge** *verb* **edged; edg·ing** (14th century)
transitive verb
1 a : to give an edge to **b :** to be on an edge of ⟨trees *edging* the lake⟩
2 : to move or force gradually ⟨*edged* him off the road⟩
3 : to incline (a ski) sideways so that one edge cuts into the snow
4 : to defeat by a small margin — often used with *out* ⟨*edged* out her opponent⟩
intransitive verb
: to advance by short moves

edged \'ejd, 'e-jid\ *adjective* (before 12th century)
1 : having a specified kind of edge, boundary, or border or a specified number of edges ⟨rough-*edged*⟩ ⟨two-*edged*⟩
2 : SHARP, CUTTING ⟨an *edged* knife⟩ ⟨an *edged* remark⟩

edge effect *noun* (1933)
: the effect of an abrupt transition between two quite different adjoining ecological communities on the numbers and kinds of organisms in the marginal habitat

edge–grain \'ej-,grān\ *or* **edge–grained** \'ej-,grānd\ *adjective* (1906)
: QUARTERSAWN

edge in *transitive verb* (1683)
: to work in **:** INTERPOLATE ⟨*edged in* a few remarks⟩

edg·er \'e-jər\ *noun* (1591)
: one that edges; *especially* **:** a tool used to trim the edge of a lawn along a sidewalk or curb

edge tool *noun* (14th century)
: a tool with a sharp cutting edge

edge·ways \'ej-,wāz\ *adverb* (1566)
chiefly British **:** SIDEWAYS

edge·wise \-,wīz\ *adverb* (1677)
: SIDEWAYS

edg·ing *noun* (1558)
: something that forms an edge or border

edgy \'e-jē\ *adjective* **edg·i·er; -est** (1775)
1 : having an edge **:** SHARP
2 a : being on edge **:** TENSE, IRRITABLE **b :** characterized by tension ⟨*edgy* negotiations⟩
— **edg·i·ly** \'e-jə-lē\ *adverb*
— **edg·i·ness** \'e-jē-nəs\ *noun*

edh \'eth\ *noun* [Icelandic *eth*] (1875)
: the letter ð used in Old English and in Icelandic to represent either of the fricatives \th\ or \th\ and in some phonetic alphabets to represent the fricative \th\

ed·i·ble \'e-də-bəl\ *adjective* [Late Latin *edibilis,* from Latin *edere* to eat — more at EAT] (1594)
: fit to be eaten **:** EATABLE
— **ed·i·bil·i·ty** \,e-də-'bi-lə-tē\ *noun*
— **edible** *noun*
— **ed·i·ble·ness** \'e-də-bəl-nəs\ *noun*

edict \'ē-,dikt\ *noun* [Middle English, from Latin *edictum,* from neuter of *edictus,* past participle of *edicere* to decree, from *e-* + *dicere* to say — more at DICTION] (14th century)
1 : a proclamation having the force of law
2 : ORDER, COMMAND ⟨we held firm to Grandmother's *edict* —M. F. K. Fisher⟩
— **edic·tal** \i-'dik-tᵊl\ *adjective*

ed·i·fi·ca·tion \,e-də-fə-'kā-shən\ *noun* (14th century)
: an act or process of edifying

ed·i·fice \'e-də-fəs\ *noun* [Middle English, from Middle French, from Latin *aedificium,* from *aedificare*] (14th century)
1 : BUILDING; *especially* **:** a large or massive structure
2 : a large abstract structure ⟨holds together the social *edifice* —R. H. Tawney⟩

ed·i·fy \'e-də-,fī\ *transitive verb* **-fied; -fy·ing** [Middle English, from Middle French *edifier,* from Late Latin & Latin; Late Latin *aedificare* to instruct or improve spiritually, from Latin, to erect a house, from *aedes* temple, house; akin to Old English *ād* funeral pyre, Latin *aestas* summer] (14th century)
1 *archaic* **a :** BUILD **b :** ESTABLISH
2 : to instruct and improve especially in moral and religious knowledge **:** UPLIFT; *also* **:** ENLIGHTEN 2, INFORM

[1]**ed·it** \'e-dət\ *transitive verb* [back-formation from *editor*] (1791)
1 a : to prepare (as literary material) for publication or public presentation **b :** to assemble (as a moving picture or tape recording) by cutting and rearranging **c :** to alter, adapt, or refine especially to bring about conformity to a standard or to suit a particular purpose ⟨carefully *edited* the speech⟩
2 : to direct the publication of ⟨*edits* the daily newspaper⟩
3 : DELETE — usually used with *out*
— **ed·it·able** \'e-də-tə-bəl\ *adjective*

[2]**edit** *noun* (1955)
: an instance of editing

edi·tion \i-'di-shən\ *noun* [Middle French, from Latin *edition-, editio* publication, edition, from *edere* to bring forth, publish, from *e-* + *-dere* to put or *-dere* (from *dare* to give) — more at DO, DATE] (1555)
1 a : the form or version in which a text is published ⟨a paperback *edition*⟩ ⟨the German *edition*⟩ **b** (1) **:** the whole number of copies published at one time (2) **:** a usually special issue of a newspaper (as for a particular day or purpose) ⟨Sunday *edition*⟩ ⟨international *edition*⟩ (3) **:** one of the usually several issues of a newspaper in a single day ⟨city *edition*⟩ ⟨late *edition*⟩
2 a : one of the forms in which something is presented ⟨this year's *edition* of the annual charity ball⟩ **b :** the whole number of articles of one style put out at one time ⟨a limited *edition* of collectors' pieces⟩
3 : COPY, VERSION

edi·tio prin·ceps \ā-,di-tē-(,)ō-'priŋ-,keps, i-,di-shē-(,)ō-'prin-,seps\ *noun, plural* **edi·ti·o·nes prin·ci·pes** \ā-,di-tē-'ō-,nās-'priŋ-kə-,pās, i-,di-she-'ō-(,)nēz-'prin(t)-sə-,pēz\ [New Latin, literally, first edition] (1802)
: the first printed edition especially of a work that circulated in manuscript before printing became common

ed·i·tor \'e-də-tər\ *noun* (1649)
1 : someone who edits especially as an occupation
2 : a device used in editing motion-picture film or magnetic tape
3 : a computer program that permits the user to create or modify data (as text or graphics) especially on a display screen
— **ed·i·tor·ship** \-,ship\ *noun*

[1]**ed·i·to·ri·al** \,e-də-'tōr-ē-əl, -'tȯr-\ *adjective* (1744)
1 : of or relating to an editor or editing ⟨an *editorial* office⟩
2 : being or resembling an editorial ⟨an *editorial* statement⟩
— **ed·i·to·ri·al·ly** \-ē-ə-lē\ *adverb*

[2]**editorial** *noun* (1830)
: a newspaper or magazine article that gives the opinions of the editors or publishers; *also* **:** an expression of opinion that resembles such an article ⟨a television *editorial*⟩

ed·i·to·ri·al·ist \-ē-ə-list\ *noun* (1901)
: a writer of editorials

ed·i·to·ri·al·ize \,e-də-'tōr-ē-ə-,līz, -'tȯr-\ *intransitive verb* **-ized; -iz·ing** (1856)
1 : to express an opinion in the form of an editorial
2 : to introduce opinion into the reporting of facts
3 : to express an opinion (as on a controversial issue)
— **ed·i·to·ri·al·iza·tion** \-,tōr-ē-ə-lə-'zā-shən, -,tȯr-\ *noun*
— **ed·i·to·ri·al·iz·er** *noun*

editor in chief *noun* (1873)
: an editor who heads an editorial staff

ed·i·tress \'e-də-trəs\ *noun* (1799)
: a woman who is an editor

Edom·ite \'ē-də-,mīt\ *noun* [*Edom* (Esau), ancestor of the Edomites] (1534)
: a member of a Semitic people living south of the Dead Sea in biblical times

EDTA \,ē-(,)dē-(,)tē-'ā\ *noun* [*e*thylene*d*iamine*t*etra*a*cetic acid] (1954)
: a white crystalline acid $C_{10}H_{16}N_2O_8$ used especially as a chelating agent and in medicine as an anticoagulant and in the treatment of lead poisoning

ed·u·ca·ble \'e-jə-kə-bəl\ *adjective* (1845)
: capable of being educated; *specifically* **:** capable of some degree of learning
— **ed·u·ca·bil·i·ty** \,e-jə-kə-'bi-lə-tē\ *noun*

ed·u·cate \'e-jə-,kāt\ *verb* **-cat·ed; -cat·ing** [Middle English, to rear, from Latin *educatus,* past participle of *educare* to rear, educate, from *educere* to lead forth — more at EDUCE] (15th century)
transitive verb
1 a : to provide schooling for **b :** to train by formal instruction and supervised practice especially in a skill, trade, or profession
2 a : to develop mentally, morally, or aesthetically especially by instruction **b :** to provide with information **:** INFORM
3 : to persuade or condition to feel, believe, or act in a desired way ⟨*educate* the public to support our position⟩
intransitive verb
: to educate a person or thing
synonym see TEACH

ed·u·cat·ed *adjective* (1588)
1 : having an education; *especially* **:** having an education beyond the average
2 a : giving evidence of training or practice **:** SKILLED **b :** befitting one that is educated ⟨*educated* taste⟩ **c :** based on some knowledge of fact ⟨an *educated* guess⟩
— **ed·u·cat·ed·ness** *noun*

ed·u·ca·tion \,e-jə-'kā-shən\ *noun* (1531)
1 a : the action or process of educating or of being educated; *also* **:** a stage of such a process **b :** the knowledge and development resulting from an educational process ⟨a man of little *education*⟩
2 : the field of study that deals mainly with methods of teaching and learning in schools
— **ed·u·ca·tion·al** \-shnəl, -shə-nᵊl\ *adjective*
— **ed·u·ca·tion·al·ly** \-ē\ *adverb*

educational psychology *noun* (1911)
: psychology concerned with human maturation, school learning, teaching methods, guidance, and evaluation of aptitude and progress by standardized tests
— **educational psychologist** *noun*

educational television *noun* (1951)
1 : television that provides instruction especially for students
2 : PUBLIC TELEVISION

ed·u·ca·tion·ese \,e-jə-,kā-shə-'nēz, -'nēs\ *noun* (1954)
: the jargon used especially by educational theorists

ed·u·ca·tion·ist \,e-jə-'kā-sh(ə-)nist\ *also* **ed·u·ca·tion·al·ist** \-shnə-list, -shə-nᵊl-ist\ *noun* (1829)
1 *chiefly British* **:** a professional educator
2 : an educational theorist

ed·u·ca·tive \'e-jə-,kā-tiv\ *adjective* (1856)
1 : tending to educate **:** INSTRUCTIVE
2 : of or relating to education

ed·u·ca·tor \'e-jə-,kā-tər\ *noun* (1673)
1 : one skilled in teaching **:** TEACHER

2 a : a student of the theory and practice of education **:** EDUCATIONIST 2 **b :** an administrator in education

educe \i-ˈdüs *also* -ˈdyüs\ *transitive verb* **educed; educ·ing** [Latin *educere* to draw out, from *e-* + *ducere* to lead — more at TOW] (1603)
1 : to bring out (as something latent)
2 : DEDUCE ☆
— educ·ible \-ˈdü-sə-bəl *also* -ˈdyü-\ *adjective*
— educ·tion \-ˈdək-shən\ *noun*

educ·tor \i-ˈdək-tər\ *noun* [Late Latin, one that leads out, from Latin *educere*] (1796)
: EJECTOR 2

edul·co·rate \i-ˈdəl-kə-ˌrāt\ *transitive verb* **-rat·ed; -rat·ing** [New Latin *edulcoratus,* past participle of *edulcorare,* from Latin *e-* + *dulcor* sweetness, from *dulcis* sweet — more at DULCET] (1641)
: to free from harshness (as of attitude) **:** SOFTEN

ed·u·tain·ment \ˌe-jə-ˈtān-mənt, ˌe-dyü-\ *noun* [*edu*cation + enter*tainment*] (1973)
: a form of entertainment (as by games, films, or shows) that is designed to be educational

Ed·war·di·an \e-ˈdwär-dē-ən, -ˈdwȯr-\ *adjective* (1908)
: of, relating to, or characteristic of Edward VII of England or his age; *especially, of clothing* **:** marked by the hourglass silhouette for women and long narrow fitted suits and high collars for men
— Edwardian *noun*

¹-ee \ˈē, ˌē, ē\ *noun suffix* [Middle English *-e,* from Middle French *-é,* from *-é,* past participle ending, from Latin *-atus*]
1 : recipient or beneficiary of (a specified action) ⟨appoint*ee*⟩ ⟨grant*ee*⟩
2 : person furnished with (a specified thing) ⟨patent*ee*⟩
3 : person that performs (a specified action) ⟨escap*ee*⟩

²-ee *noun suffix* [probably alteration of *-y*]
1 : one associated with ⟨barg*ee*⟩
2 : a particular especially small kind of ⟨boot*ee*⟩
3 : one resembling or suggestive of ⟨goat*ee*⟩

eel \ˈē(ə)l\ *noun* [Middle English *ele,* from Old English *ǣl;* akin to Old High German *āl* eel] (before 12th century)
1 a : any of numerous voracious elongate snakelike bony fishes (order Anguilliformes) that have a smooth slimy skin, lack pelvic fins, and have the median fins confluent around the tail — compare AMERICAN EEL **b :** any of numerous other elongate fishes (as of the order Symbranchii)
2 : any of various nematodes
— eel·like \ˈē(ə)l-ˌlīk\ *adjective*
— eely \ˈē-lē\ *adjective*

eel 1a

eel·grass \ˈē(ə)l-ˌgras\ *noun* (1790)
1 : a submerged long-leaved monocotyledonous marine plant (*Zostera marina*) of the eelgrass family that is abundant along the Atlantic coast and has stems used especially in woven products (as mats and hats)
2 : TAPE GRASS

eel·pout \-ˌpau̇t\ *noun* (before 12th century)
1 : any of various elongate tapered marine fishes (family Zoarcidae) usually living on the bottom of cold seas
2 : BURBOT

eel·worm \-ˌwərm\ *noun* (1888)
: a nematode worm; *especially* **:** any of various small free-living or plant-parasitic roundworms

-een \ˈēn\ *noun suffix* [probably from *ratteen*]
: inferior fabric resembling (a specified fabric) **:** imitation ⟨velvet*een*⟩

e'en \ˈēn\ *adverb* (circa 1553)
: EVEN

-eer \ˈir\ *noun suffix* [Middle French *-ier,* from Latin *-arius* — more at -ARY]
: one that is concerned with professionally, conducts, or produces ⟨auction*eer*⟩ ⟨pamphlet*eer*⟩ — often in words with derogatory meaning ⟨profit*eer*⟩

e'er \ˈer, ˈar\ *adverb* (13th century)
: EVER

ee·rie *also* **ee·ry** \ˈir-ē, ˈēr-\ *adjective* **ee·ri·er; -est** [Middle English (northern dialect) *eri*] (14th century)
1 *chiefly Scottish* **:** affected with fright **:** SCARED
2 : so mysterious, strange, or unexpected as to send a chill up the spine ⟨a coyote's *eerie* howl⟩ ⟨the similarities were *eerie*⟩; *also* **:** seemingly not of earthly origin ⟨the flames cast an *eerie* glow⟩
synonym see WEIRD
— ee·ri·ly \ˈir-ə-lē, ˈēr-\ *adverb*
— ee·ri·ness \ˈir-ē-nəs, ˈēr-\ *noun*

ef \ˈef\ *noun* (before 12th century)
: the letter *f*

ef·face \i-ˈfās, e-\ *transitive verb* **ef·faced; ef·fac·ing** [Middle English, from Middle French *effacer,* from Old French *esfacier,* from *e-* + *face* face] (15th century)
1 : to eliminate or make indistinct by or as if by wearing away a surface ⟨coins with dates *effaced* by wear⟩; *also* **:** to cause to vanish ⟨daylight *effaced* the stars⟩
2 : to make (oneself) modestly or shyly inconspicuous
— ef·face·able \-ˈfā-sə-bəl\ *adjective*
— ef·face·ment \-ˈfā-smənt\ *noun*
— ef·fac·er *noun*

¹ef·fect \i-ˈfekt, e-, ē-\ *noun* [Middle English, from Middle French & Latin; Middle French, from Latin *effectus,* from *efficere* to bring about, from *ex-* + *facere* to make, do — more at DO] (14th century)
1 a : PURPORT, INTENT **b :** basic meaning **:** ESSENCE
2 : something that inevitably follows an antecedent (as a cause or agent)
3 : an outward sign **:** APPEARANCE
4 : ACCOMPLISHMENT, FULFILLMENT
5 : power to bring about a result **:** INFLUENCE ⟨the content itself of television . . . is therefore less important than its *effect* —*Current Biography*⟩
6 *plural* **:** movable property **:** GOODS ⟨personal *effects*⟩
7 a : a distinctive impression ⟨the color gives the *effect* of being warm⟩ **b :** the creation of a desired impression ⟨her tears were purely for *effect*⟩ **c** (1) **:** something designed to produce a distinctive or desired impression — usually used in plural (2) *plural* **:** SPECIAL EFFECTS
8 : the quality or state of being operative **:** OPERATION ⟨the law goes into *effect* next week⟩
— in effect : in substance **:** VIRTUALLY ⟨the . . . committee agreed to what was *in effect* a reduction in the hourly wage —*Current Biography*⟩
— to the effect : with the meaning ⟨issued a statement *to the effect* that he would resign⟩

²effect *transitive verb* (1533)
1 : to cause to come into being
2 a : to bring about often by surmounting obstacles **:** ACCOMPLISH ⟨*effect* a settlement of a dispute⟩ **b :** to put into operation ⟨the duty of the legislature to *effect* the will of the citizens⟩ ▪
synonym see PERFORM

¹ef·fec·tive \i-ˈfek-tiv, e-, ē-\ *adjective* (14th century)
1 a : producing a decided, decisive, or desired effect **b :** IMPRESSIVE, STRIKING ⟨a gold lamé fabric studded with *effective* . . . precious stones —Stanley Marcus⟩
2 : ready for service or action ⟨*effective* manpower⟩
3 : ACTUAL ⟨the need to increase *effective* demand for goods⟩
4 : being in effect **:** OPERATIVE ⟨the tax becomes *effective* next year⟩
5 *of a rate of interest* **:** equal to the rate of simple interest that yields the same amount when the interest is paid once at the end of the interest period as a quoted rate of interest does when calculated at compound interest over the same period — compare NOMINAL 4 ☆
— ef·fec·tive·ness *noun*
— ef·fec·tiv·i·ty \ˌe-ˌfek-ˈti-və-tē, i-, ē-\ *noun*

²effective *noun* (1722)
: one that is effective; *especially* **:** a soldier equipped for duty

ef·fec·tive·ly \-lē\ *adverb* (circa 1536)
1 : in effect **:** VIRTUALLY
2 : in an effective manner

ef·fec·tor \i-ˈfek-tər, -ˌtȯr\ *noun* (1906)
1 : a bodily organ (as a gland or muscle) that becomes active in response to stimulation
2 : a substance (as an inducer or corepressor) that controls protein synthesis by combining allosterically with a genetic repressor

ef·fec·tu·al \i-ˈfek-chə(-wə)l, -ˈfeksh-wəl\ *adjective* (14th century)
: producing or able to produce a desired effect
synonym see EFFECTIVE
— ef·fec·tu·al·i·ty \-ˌfek-chə-ˈwa-lə-tē\ *noun*
— ef·fec·tu·al·ness \-ˈfek-chə(-wə)l-nəs, -ˈfeksh-wəl-\ *noun*

ef·fec·tu·al·ly \i-ˈfek-chə(-wə)-lē, -ˈfek-shwə-\ *adverb* (14th century)

☆ SYNONYMS
Educe, evoke, elicit, extract, extort mean to draw out something hidden, latent, or reserved. EDUCE implies the bringing out of something potential or latent ⟨*educed* order out of chaos⟩. EVOKE implies a strong stimulus that arouses an emotion or an interest or recalls an image or memory ⟨a song that *evokes* warm memories⟩. ELICIT usually implies some effort or skill in drawing forth a response ⟨careful questioning *elicited* the truth⟩. EXTRACT implies the use of force or pressure in obtaining answers or information ⟨*extracted* a confession from him⟩. EXTORT suggests a wringing or wresting from one who resists strongly ⟨*extorted* their cooperation by threatening to inform⟩.

Effective, effectual, efficient, efficacious mean producing or capable of producing a result. EFFECTIVE stresses the actual production of or the power to produce an effect ⟨an *effective* rebuttal⟩. EFFECTUAL suggests the accomplishment of a desired result especially as viewed after the fact ⟨the measures to stop the pilfering proved *effectual*⟩. EFFICIENT suggests an acting or a potential for action or use in such a way as to avoid loss or waste of energy in effecting, producing, or functioning ⟨an *efficient* small car⟩. EFFICACIOUS suggests possession of a special quality or virtue that gives effective power ⟨a detergent that is *efficacious* in removing grease⟩.

□ USAGE
effect The confusion of the verbs *affect* and *effect* is not only quite common but has a long history. *Effect* was used in place of ³*affect* as early as 1494 and in place of ²*affect* as early as 1652. If you think you want to use the verb *effect* but are not certain, check the definitions in this dictionary. The noun *affect* is sometimes mistakenly used for *effect.* Except when your topic is psychology, you will seldom need the noun *affect.*

1 : in an effectual manner
2 : with great effect **:** COMPLETELY

ef·fec·tu·ate \i-ˈfek-chə-ˌwāt\ *transitive verb* **-at·ed; -at·ing** (1580)
: EFFECT 2
— **ef·fec·tu·a·tion** \-ˌfek-chə-ˈwā-shən\ *noun*

ef·fem·i·na·cy \ə-ˈfe-mə-nə-sē\ *noun* (1602)
: the quality of being effeminate

[1]**ef·fem·i·nate** \-nət\ *adjective* [Middle English, from Latin *effeminatus,* from past participle of *effeminare* to make effeminate, from *ex-* + *femina* woman — more at FEMININE] (15th century)
1 : having feminine qualities untypical of a man **:** not manly in appearance or manner
2 : marked by an unbecoming delicacy or overrefinement ⟨*effeminate* art⟩ ⟨an *effeminate* civilization⟩

[2]**effeminate** *noun* (1597)
: an effeminate person

ef·fen·di \e-ˈfen-dē, ə-\ *noun* [Turkish *efendi* master, from New Greek *aphentēs,* alteration of Greek *authentēs* — more at AUTHENTIC] (1614)
: a man of property, authority, or education in an eastern Mediterranean country

ef·fer·ent \ˈe-fər-ənt, -ˌfer-, ˈē-\ *adjective* [French *efférent,* from Latin *efferent-, efferens,* present participle of *efferre* to carry outward, from *ex-* + *ferre* to carry — more at BEAR] (1856)
: conducting outward from a part or organ; *specifically* **:** conveying nervous impulses to an effector — compare AFFERENT
— **efferent** *noun*
— **ef·fer·ent·ly** *adverb*

ef·fer·vesce \ˌe-fər-ˈves\ *intransitive verb* **-vesced; -vesc·ing** [Latin *effervescere,* from *ex-* + *fervescere* to begin to boil, inchoative of *fervēre* to boil — more at BREW] (1784)
1 : to bubble, hiss, and foam as gas escapes
2 : to show liveliness or exhilaration
— **ef·fer·ves·cence** \-ˈve-sᵊn(t)s\ *noun*
— **ef·fer·ves·cent** \-sᵊnt\ *adjective*
— **ef·fer·ves·cent·ly** *adverb*

ef·fete \e-ˈfēt, i-\ *adjective* [Latin *effetus,* from *ex-* + *fetus* fruitful — more at FEMININE] (1660)
1 : no longer fertile
2 a : having lost character, vitality, or strength ⟨the *effete* monarchies . . . of feudal Europe —G. M. Trevelyan⟩ **b :** marked by weakness or decadence ⟨the *effete* East⟩ **c :** soft or delicate from or as if from a pampered existence ⟨peddled . . . trendy tweeds to *effete* Easterners —William Helmer⟩ ⟨*effete* tenderfeet⟩; *also* **:** characteristic of an effete person ⟨a wool scarf . . . a bit *effete* on an outdoorsman —Nelson Bryant⟩
3 : EFFEMINATE 1 ⟨a good-humored, *effete* boy brought up by maiden aunts —Herman Wouk⟩
— **ef·fete·ly** *adverb*
— **ef·fete·ness** *noun*

ef·fi·ca·cious \ˌe-fə-ˈkā-shəs\ *adjective* [Latin *efficac-, efficax,* from *efficere*] (1528)
: having the power to produce a desired effect
synonym see EFFECTIVE
— **ef·fi·ca·cious·ly** *adverb*
— **ef·fi·ca·cious·ness** *noun*

ef·fi·cac·i·ty \ˌe-fə-ˈka-sə-tē\ *noun* (15th century)
: EFFICACY

ef·fi·ca·cy \ˈe-fi-kə-sē\ *noun, plural* **-cies** (13th century)
: the power to produce an effect

ef·fi·cien·cy \i-ˈfi-shən-sē\ *noun, plural* **-cies** (1633)
1 : the quality or degree of being efficient
2 a : efficient operation **b** (1) **:** effective operation as measured by a comparison of production with cost (as in energy, time, and money) (2) **:** the ratio of the useful energy delivered by a dynamic system to the energy supplied to it
3 : EFFICIENCY APARTMENT

efficiency apartment *noun* (1930)
: a small usually furnished apartment with minimal kitchen and bath facilities

efficiency expert *noun* (1913)
: one who analyzes methods, procedures, and jobs in order to secure maximum efficiency — called also *efficiency engineer*

ef·fi·cient \i-ˈfi-shənt\ *adjective* [Middle English, from Middle French or Latin; Middle French, from Latin *efficient-, efficiens,* from present participle of *efficere*] (14th century)
1 : being or involving the immediate agent in producing an effect ⟨the *efficient* action of heat in changing water to steam⟩
2 : productive of desired effects; *especially* **:** productive without waste
synonym see EFFECTIVE
— **ef·fi·cient·ly** *adverb*

ef·fi·gy \ˈe-fə-jē\ *noun, plural* **-gies** [Middle French *effigie,* from Latin *effigies,* from *effingere* to form, from *ex-* + *fingere* to shape — more at DOUGH] (1539)
: an image or representation especially of a person; *especially* **:** a crude figure representing a hated person
— **in effigy :** publicly in the form of an effigy ⟨the football coach was burned *in effigy*⟩

ef·flo·resce \ˌe-flə-ˈres\ *intransitive verb* **-resced; -resc·ing** [Latin *efflorescere,* from *ex-* + *florescere* to begin to blossom — more at FLORESCENCE] (1775)
1 : to burst forth **:** BLOOM
2 a : to change to a powder from loss of water of crystallization **b :** to form or become covered with a powdery crust ⟨bricks may *effloresce* owing to the deposition of soluble salts⟩

ef·flo·res·cence \-ˈre-sᵊn(t)s\ *noun* (1626)
1 a : the action or process of developing and unfolding as if coming into flower **:** BLOSSOMING ⟨periods of . . . intellectual and artistic *efflorescence* —Julian Huxley⟩ **b :** an instance of such development **c :** fullness of manifestation **:** CULMINATION
2 : the period or state of flowering
3 : the process or product of efflorescing chemically
— **ef·flo·res·cent** \-sᵊnt\ *adjective*

ef·flu·ence \ˈe-ˌflü-ən(t)s; e-ˈflü-, ə-ˈ\ *noun* (1603)
1 : something that flows out
2 : an action or process of flowing out

[1]**ef·flu·ent** \-ənt\ *adjective* [Latin *effluent-, effluens,* present participle of *effluere* to flow out, from *ex-* + *fluere* to flow — more at FLUID] (1726)
: flowing out **:** EMANATING, OUTGOING ⟨an *effluent* river⟩

[2]**effluent** *noun* (1859)
: something that flows out: as **a :** an outflowing branch of a main stream or lake **b :** waste material (as smoke, liquid industrial refuse, or sewage) discharged into the environment especially when serving as a pollutant

ef·flu·vi·um \e-ˈflü-vē-əm\ *also* **ef·flu·via** \-vē-ə\ *noun, plural* **-via** *or* **-vi·ums** [Latin *effluvium* act of flowing out, from *effluere*] (1651)
1 : an invisible emanation; *especially* **:** an offensive exhalation or smell
2 : a by-product especially in the form of waste

ef·flux \ˈe-ˌfləks\ *noun* [Medieval Latin *effluxus,* from *effluere*] (1647)
1 : something given off in or as if in a stream
2 a : EFFLUENCE 2 **b :** a passing away **:** EXPIRATION
— **ef·flux·ion** \e-ˈflək-shən\ *noun*

ef·fort \ˈe-fərt, -ˌfȯrt\ *noun* [Middle English, from Middle French, from Old French *esfort,* from *esforcier* to force, from *ex-* + *forcier* to force] (15th century)
1 : conscious exertion of power **:** hard work
2 : a serious attempt **:** TRY
3 : something produced by exertion or trying ⟨the novel was her most ambitious *effort*⟩
4 : effective force as distinguished from the possible resistance called into action by such a force
5 : the total work done to achieve a particular end ⟨the war *effort*⟩

ef·fort·ful \-fərt-fəl\ *adjective* (circa 1895)
: showing or requiring effort
— **ef·fort·ful·ly** \-fə-lē\ *adverb*
— **ef·fort·ful·ness** \-fəl-nəs\ *noun*

ef·fort·less \-fərt-ləs\ *adjective* (1801)
: showing or requiring little or no effort
synonym see EASY
— **ef·fort·less·ly** *adverb*
— **ef·fort·less·ness** *noun*

ef·fron·tery \i-ˈfrən-tə-rē, e-\ *noun, plural* **-ter·ies** [French *effronterie,* ultimately from Medieval Latin *effront-, effrons* shameless, from Latin *ex-* + *front-, frons* forehead] (1697)
: shameless boldness **:** INSOLENCE
synonym see TEMERITY

ef·ful·gence \i-ˈfu̇l-jən(t)s, e-, -ˈfəl-\ *noun* [Late Latin *effulgentia,* from Latin *effulgent-, effulgens,* present participle of *effulgēre* to shine forth, from *ex-* + *fulgēre* to shine — more at FULGENT] (1667)
: radiant splendor **:** BRILLIANCE
— **ef·ful·gent** \-jənt\ *adjective*

[1]**ef·fuse** \i-ˈfyüz, e-\ *verb* **ef·fused; ef·fus·ing** [Latin *effusus,* past participle of *effundere,* from *ex-* + *fundere* to pour — more at FOUND] (1526)
transitive verb
: to pour out (a liquid)
intransitive verb
: to flow out **:** EMANATE

[2]**ef·fuse** \-ˈfyüs\ *adjective* (circa 1530)
: DIFFUSE; *specifically* **:** spread out flat without definite form ⟨*effuse* lichens⟩

ef·fu·sion \i-ˈfyü-zhən, e-\ *noun* (15th century)
1 : an act of effusing
2 : unrestrained expression of words or feelings ⟨greeted her with great *effusion* —Olive H. Prouty⟩
3 a (1) **:** the escape of a fluid from anatomical vessels by rupture or exudation (2) **:** the flow of a gas through an aperture whose diameter is small as compared with the distance between the molecules of the gas **b :** the fluid that escapes

ef·fu·sive \i-ˈfyü-siv, e-, -ziv\ *adjective* (1662)
1 : excessively demonstrative
2 *archaic* **:** pouring freely
3 : characterized or formed by a nonexplosive outpouring of lava ⟨*effusive* rocks⟩
— **ef·fu·sive·ly** *adverb*
— **ef·fu·sive·ness** *noun*

Ef·ik \ˈe-fik\ *noun* (1849)
1 : a member of a people of southeastern Nigeria
2 : the language of the Efik people

eft \ˈeft\ *noun* [Middle English *evete, ewte,* from Old English *efete*] (before 12th century)
: NEWT; *especially* **:** the terrestrial phase of a predominantly aquatic newt

eft·soons \eft-ˈsünz\ *adverb* [Middle English *eftsones,* alteration of Old English *eftsona,* from Old English *eft* after + *sōna* soon; akin to Old English *æfter* after] (before 12th century)
archaic **:** soon after

egad \i-ˈgad\ *or* **egads** \-ˈgadz\ *interjection* [probably euphemism for *oh God*] (1673)
— used as a mild oath

egal \ˈē-gəl\ *adjective* [Middle English, from Middle French, from Latin *aequalis*] (14th century)
obsolete **:** EQUAL

egal·i·tar·i·an \i-ˌga-lə-ˈter-ē-ən\ *adjective* [French *égalitaire,* from *égalité* equality, from Latin *aequalitat-, aequalitas,* from *aequalis* equal] (1885)
: asserting, promoting, or marked by egalitarianism
— **egalitarian** *noun*

egal·i·tar·i·an·ism \-ē-ə-ˌni-zəm\ *noun* (1905)
1 : a belief in human equality especially with respect to social, political, and economic rights and privileges
2 : a social philosophy advocating the removal of inequalities among people

éga·li·té \ā-gà-lē-tā\ *noun* [French] (1794)
: social or political equality

Ege·ria \i-ˈjir-ē-ə\ *noun* [Latin, a nymph who advised the legendary Roman king Numa Pompilius] (1621)
: a woman adviser or companion

eges·ta \i-ˈjes-tə\ *noun plural* [New Latin, from Latin, neuter plural of *egestus*] (1727)
: something egested

eges·tion \i-ˈjes(h)-chən\ *noun* [Middle English *egestioun,* from Middle French or Latin; Middle French *egestion,* from Latin *egestion-, egestio,* from *egerere* to carry outside, discharge, from *e-* + *gerere* to carry] (1547)
: the act or process of discharging undigested or waste material from a cell or organism; *specifically* **:** DEFECATION
— **egest** \i-ˈjest\ *transitive verb*
— **eges·tive** \-ˈjes-tiv\ *adjective*

¹egg \ˈeg, ˈāg\ *transitive verb* [Middle English, from Old Norse *eggja;* akin to Old English *ecg* edge — more at EDGE] (13th century)
: to incite to action — usually used with *on* ⟨*egged* the mob on to riot⟩

²egg *noun, often attributive* [Middle English *egge,* from Old Norse *egg;* akin to Old English *ǣg* egg, Latin *ovum,* Greek *ōion*] (14th century)
1 a : the hard-shelled reproductive body produced by a bird and especially by the common domestic chicken; *also* **:** its contents used as food **b :** an animal reproductive body consisting of an ovum together with its nutritive and protective envelopes and having the capacity to develop into a new individual capable of independent existence **c :** OVUM
2 : something resembling an egg
3 : PERSON, SORT ⟨a good *egg*⟩
— **egg·less** *adjective*
— **eggy** \-gē\ *adjective*
— **egg on one's face :** a state of embarrassment or humiliation

egg 1a: ***1*** **shell,** ***2*** **shell membrane,** ***3*** **egg membrane,** ***4*** **air space,** ***5*** **chalaza,** ***6*** **albumen or white layers,** ***7*** **yolk layers,** ***8*** **blastodisc,** ***9*** **vitelline membrane**

³egg *transitive verb* (1833)
1 : to cover with egg
2 : to pelt with eggs

egg and dart *noun* (circa 1864)
: a carved ornamental design in relief consisting of an egg-shaped figure alternating with a figure somewhat like an elongated javelin or arrowhead

egg·beat·er \ˈeg-ˌbē-tər, ˈāg-\ *noun* (1828)
1 : a hand-operated kitchen utensil used for beating, stirring, or whipping; *especially* **:** a rotary device for these purposes
2 : HELICOPTER

egg case *noun* (1847)
: a protective case enclosing eggs **:** OOTHECA — called also *egg capsule*

egg cell *noun* (1880)
: OVUM

egg cream *noun* (1954)
: a drink consisting of milk, a flavoring syrup, and soda water

egg·cup \ˈeg-ˌkəp, ˈāg-\ *noun* (1773)
: a cup for holding an egg that is to be eaten from the shell

egg·head \-ˌhed\ *noun, often attributive* (1952)
: INTELLECTUAL, HIGHBROW

egg·head·ed \-ˈhe-dəd\ *adjective* (1938)
: having the characteristics of an egghead
— **egg·head·ed·ness** *noun*

egg·nog \-ˌnäg\ *noun* (circa 1775)
: a drink consisting of eggs beaten with sugar, milk or cream, and often alcoholic liquor

egg·plant \-ˌplant\ *noun* (1767)
1 a : a widely cultivated perennial herb (*Solanum melongena*) of the nightshade family yielding edible fruit **b :** the usually smooth ovoid fruit of the eggplant
2 : a dark grayish or blackish purple

egg roll *noun* (1947)
: a thin egg-dough casing filled with minced vegetables and often bits of meat (as shrimp or chicken) and usually deep-fried

eggs Ben·e·dict \-ˈbe-nə-ˌdikt\ *noun plural but singular or plural in construction* [probably from the name *Benedict*] (1898)
: poached eggs and broiled ham placed on toasted halves of English muffin and covered with hollandaise

¹egg·shell \ˈeg-ˌshel, ˈāg-\ *noun* (14th century)
1 : the hard exterior covering of an egg
2 : something resembling an eggshell especially in fragility

²eggshell *adjective* (1835)
1 : thin and fragile
2 : slightly glossy
3 : yellowish white

egg timer *noun* (1884)
: a small sandglass for timing the boiling of eggs

egg tooth *noun* (1893)
: a hard sharp prominence on the beak of an unhatched bird or the nose of an unhatched reptile that is used to break through the eggshell

egis \ˈē-jəs\ *variant of* AEGIS

eg·lan·tine \ˈe-glən-ˌtīn, -ˌtēn\ *noun* [Middle English *eglentyn,* from Middle French *aiglent,* from (assumed) Vulgar Latin *aculentum,* from Latin *acus* needle; akin to Latin *acer* sharp — more at EDGE] (14th century)
: SWEETBRIER

eglo·mi·se *also* **églo·mi·sé** \ˌā-glə-(ˌ)mē-ˈzā, ˌe-; ˈā-glə-(ˌ)mē-ˌ, ˈe-\ *adjective* [French, past participle of *églomiser* to decorate a glass panel by painting on its back, from Jean-Baptiste *Glomy* (died 1786) French decorator] (1877)
: made of glass on the back of which is a painted picture that shows through ⟨a clock with *eglomise* panels⟩

ego \ˈē-(ˌ)gō *also* ˈe-\ *noun, plural* **egos** [New Latin, from Latin, I — more at I] (1789)
1 : the self especially as contrasted with another self or the world
2 a : EGOTISM 2 **b :** SELF-ESTEEM 1
3 : the one of the three divisions of the psyche in psychoanalytic theory that serves as the organized conscious mediator between the person and reality especially by functioning both in the perception of and adaptation to reality — compare ID, SUPEREGO
— **ego·less** *adjective*

ego·cen·tric \ˌē-gō-ˈsen-trik *also* ˌe-\ *adjective* (1894)
1 : concerned with the individual rather than society
2 : taking the ego as the starting point in philosophy
3 a : limited in outlook or concern to one's own activities or needs **b :** SELF-CENTERED, SELFISH
— **egocentric** *noun*
— **ego·cen·tri·cal·ly** \-tri-k(ə-)lē\ *adverb*
— **ego·cen·tric·i·ty** \-ˌsen-ˈtri-sə-tē\ *noun*
— **ego·cen·trism** \-ˈsen-ˌtri-zəm\ *noun*

ego ideal *noun* (1922)
: the standards, ideals, and ambitions that according to psychoanalytic theory are assimilated from the superego

ego·ism \ˈē-gə-ˌwi-zəm *also* ˈe-\ *noun* (1785)
1 a : a doctrine that individual self-interest is the actual motive of all conscious action **b :** a doctrine that individual self-interest is the valid end of all actions
2 : excessive concern for oneself with or without exaggerated feelings of self-importance — compare EGOTISM 2

ego·ist \-wist\ *noun* (1785)
1 : a believer in egoism
2 : an egocentric or egotistic person
— **ego·is·tic** \ˌē-gə-ˈwis-tik *also* ˌe-\ *also* **ego·is·ti·cal** \-ti-kəl\ *adjective*
— **ego·is·ti·cal·ly** \-ti-k(ə-)lē\ *adverb*

egoistic hedonism *noun* (1874)
: the ethical theory that achieving one's own happiness is the proper goal of all conduct

ego·ma·nia \ˌē-gō-ˈmā-nē-ə, -nyə\ *noun* (1825)
: the quality or state of being extremely egocentric
— **ego·ma·ni·ac** \-nē-ˌak\ *noun*
— **ego·ma·ni·a·cal** \-mə-ˈnī-ə-kəl\ *adjective*
— **ego·ma·ni·a·cal·ly** \-k(ə-)lē\ *adverb*

ego·tism \ˈē-gə-ˌti-zəm *also* ˈe-\ *noun* [Latin *ego* + English *-tism* (as in *idiotism*)] (1714)
1 a : excessive use of the first person singular personal pronoun **b :** the practice of talking about oneself too much
2 : an exaggerated sense of self-importance **:** CONCEIT — compare EGOISM 2
— **ego·tist** \-tist\ *noun*
— **ego·tis·tic** \-ˈtis-tik\ *or* **ego·tis·ti·cal** \-ˈtis-ti-kəl\ *adjective*
— **ego·tis·ti·cal·ly** \-k(ə-)lē\ *adverb*

ego trip *noun* (1967)
: an act or course of action that enhances and satisfies one's ego
— **ego–trip** *intransitive verb*
— **ego–trip·per** *noun*

egre·gious \i-ˈgrē-jəs\ *adjective* [Latin *egregius,* from *e-* + *greg-, grex* herd — more at GREGARIOUS] (circa 1534)
1 *archaic* **:** DISTINGUISHED
2 : CONSPICUOUS; *especially* **:** conspicuously bad **:** FLAGRANT ⟨an *egregious* mistake⟩ ◆
— **egre·gious·ly** *adverb*
— **egre·gious·ness** *noun*

¹egress \ˈē-ˌgres\ *noun* [Latin *egressus,* from *egressus,* past participle of *egredi* to go out, from *e-* + *gradi* to go — more at GRADE] (1538)
1 : the action or right of going or coming out
2 : a place or means of going out **:** EXIT

²egress \ē-ˈgres\ *intransitive verb* (1578)
: to go or come out

◇ WORD HISTORY

egregious *Egregious* is one of several English words whose connotations have taken a decided turn for the worse since their appearance in the language. The word is borrowed from *egregius,* a Latin word meaning "outstanding" or "excellent," which was formed from *e* "out of" and *grex* "herd." Thus, a person who was *egregius* stood out from the herd. Originally, English *egregious,* like its Latin source, denoted a positive quality. In the 16th century, however, an opposite sense came into use when *egregious* began to be applied to a person who was conspicuously or outrageously bad. Though it may have been the result of ironic use of the original sense, this pejorative meaning, now applied to things as well as people, is the one that persists in common use today. See in addition GREGARIOUS.

egres·sion \ē-ˈgre-shən\ *noun* (15th century)
: EGRESS, EMERGENCE

egret \ˈē-grət, -ˌgret *also* i-ˈgret, ˈeg-rət\ *noun* [Middle English, from Middle French *aigrette,* from Old Provençal *aigreta,* of Germanic origin; akin to Old High German *heigaro* heron] (14th century)
: any of various herons that bear long plumes during the breeding season

egret

[1]**Egyp·tian** \i-ˈjip-shən\ *adjective* (14th century)
: of, relating to, or characteristic of Egypt or the Egyptians

[2]**Egyptian** *noun* (14th century)
1 : a native or inhabitant of Egypt
2 : the Afro-Asiatic language of the ancient Egyptians from earliest times to about the 3d century A.D.
3 *often not capitalized* **:** a typeface having little contrast between thick and thin strokes and squared serifs

Egyptian alfalfa weevil *noun* (1943)
: an Old World weevil (*Hypera brunneipennis*) established in western North America where it feeds on alfalfa and various clovers

Egyptian clover *noun* (circa 1900)
: BERSEEM

Egyptian cotton *noun* (1877)
: a fine long-staple often somewhat brownish cotton grown chiefly in Egypt

Egyp·tol·o·gy \ˌē-(ˌ)jip-ˈtä-lə-jē\ *noun* (1862)
: the study of Egyptian antiquities
— Egyp·to·log·i·cal \-tə-ˈlä-jə-kəl\ *adjective*
— Egyp·tol·o·gist \-ˈtä-lə-jist\ *noun*

eh \ˈā, ˈe, ˈa(i), *also with* h *preceding and/or with nasalization*\ *interjection* [Middle English *ey*] (13th century)
— used to ask for confirmation or repetition or to express inquiry; used especially in Canadian English in anticipation of the listener's or reader's agreement

ei·cos·a·noid \ī-ˈkō-sə-ˌnȯid\ *noun* [*eicosa-* containing 20 atoms (from Greek *eikosa-* twenty, from *eikosi*) + *-noic,* suffix used in names of fatty acids (from *-ane* + *-oic*) + [1]*-oid* — more at VIGESIMAL] (1980)
: any of a class of compounds (as the prostaglandins) derived from polyunsaturated acids (as arachidonic acid) and involved in cellular activity

ei·der \ˈī-dər\ *noun* [Dutch, German, or Swedish, from Icelandic *æthur,* from Old Norse *æthr*] (1743)
1 : any of several large northern sea ducks (genera *Somateria* and *Polysticta*) having fine soft down that is used by the female for lining the nest — called also *eider duck*
2 : EIDERDOWN 1

ei·der·down \-ˌdau̇n\ *noun* [probably from German *Eiderdaune,* from Icelandic *æthardūnn,* from *æthur* + *dūnn* [7]down] (1774)
1 : the down of the eider
2 : a comforter filled with eiderdown
3 : a soft lightweight clothing fabric knitted or woven and napped on one or both sides

ei·det·ic \ī-ˈde-tik\ *adjective* [Greek *eidētikos* of a form, from *eidos* form — more at WISE] (circa 1924)
: marked by or involving extraordinarily accurate and vivid recall especially of visual images ⟨an *eidetic* memory⟩
— ei·det·i·cal·ly \-ti-k(ə-)lē\ *adverb*

ei·do·lon \ī-ˈdō-lən\ *noun, plural* **-lons** \-lənz\ *or* **-la** \-lə\ [Greek *eidōlon* — more at IDOL] (1828)
1 : an unsubstantial image **:** PHANTOM
2 : IDEAL

ei·gen·mode \ˈī-gən-ˌmōd\ *noun* [*eigen-* (as in *eigenvector*) + [1]*mode*]
: a normal mode of vibration of an oscillating system

ei·gen·val·ue \ˈī-gən-ˌval-(ˌ)yü, -yə(-w)\ *noun* [part translation of German *Eigenwert,* from *eigen* own, peculiar + *Wert* value] (1927)
: a scalar associated with a given linear transformation of a vector space and having the property that there is some nonzero vector which when multiplied by the scalar is equal to the vector obtained by letting the transformation operate on the vector; *especially* **:** a root of the characteristic equation of a matrix

ei·gen·vec·tor \-ˌvek-tər\ *noun* [International Scientific Vocabulary *eigen-* (from German *eigen*) + *vector*] (1941)
: a nonzero vector that is mapped by a given linear transformation of a vector space onto a vector that is the product of a scalar multiplied by the original vector — called also *characteristic vector*

eight \ˈāt\ *noun* [Middle English *eighte,* from *eighte,* adjective, from Old English *eahta;* akin to Old High German *ahto* eight, Latin *octo,* Greek *oktō*] (before 12th century)
1 — see NUMBER table
2 : the eighth in a set or series ⟨the *eight* of spades⟩
3 : something having eight units or members: as **a :** an 8-oared racing boat or its crew **b :** an 8-cylinder engine or automobile
— eight *adjective*
— eight *pronoun, plural in construction*

eight ball *noun* (1932)
1 : a black pool ball numbered 8
2 : MISFIT
— behind the eight ball : in a highly disadvantageous position

eigh·teen \(ˈ)ā(t)-ˈtēn\ *noun* [Middle English *eightetene,* adjective, from Old English *eahtatīene,* from *eahta* + *-tīene;* akin to Old English *tīen* ten] (before 12th century)
— see NUMBER table
— eighteen *adjective*
— eighteen *pronoun, plural in construction*
— eigh·teenth \-ˈtēn(t)th\ *adjective or noun*

18–wheel·er *or* **eigh·teen–wheel·er** \ˌā(t)-(ˌ)tēn-ˈwē-lər\ *noun* (1976)
: a trucking rig consisting of a tractor and a trailer and typically having eighteen wheels

eight·fold \ˈāt-ˌfōld, -ˈfōld\ *adjective* (before 12th century)
1 : having eight units or members
2 : being eight times as great or as many
— eight·fold \-ˈfōld\ *adverb*

eightfold way *noun* (1961)
: a unified theoretical scheme for classifying the relationship among strongly interacting elementary particles on the basis of isospin and hypercharge

eighth \ˈātth, ˈāth\ *noun, plural* **eighths** \ˈāt(th)s, ˈāths\ (1557)
1 — see NUMBER table
2 : OCTAVE
— eighth *adjective or adverb*

eighth note *noun* (circa 1864)
: a musical note with the time value of ⅛ of a whole note — see NOTE illustration

eighth rest *noun* (circa 1890)
: a musical rest corresponding in time value to an eighth note — see REST illustration

800 number \ˌāt-ˈhən-drəd-, -dərd-\ *noun* (1979)
: a toll-free telephone number for long-distance calls (as to a business) that is prefixed by the number 800

eight·pen·ny nail \ˌāt-ˌpe-nē-\ *noun* [from its original price per hundred] (15th century)
: a nail typically 2½ inches (6.35 centimeters) long

eighty \ˈā-tē\ *noun, plural* **eight·ies** [Middle English *eighty,* adjective, from Old English *eahtatig,* short for *hundeahtatig,* noun, group of eighty, from *hund-,* literally, hundred + *eahta* eight + *-tig* group of ten; akin to Old English *tīen* ten] (before 12th century)
1 — see NUMBER table
2 *plural* **:** the numbers 80 to 89; *specifically* **:** the years 80 to 89 in a lifetime or century
— eight·i·eth \ˈā-tē-əth\ *adjective or noun*
— eighty *adjective*
— eighty *pronoun, plural in construction*

eighty–six *or* **86** \ˌā-tē-ˈsiks\ *transitive verb* [probably rhyming slang for [3]*nix*] (1967)
slang **:** to refuse to serve (a customer); *also* **:** to get rid of **:** THROW OUT

ein·korn \ˈīn-ˌkȯrn\ *noun* [German, from Old High German, from *ein* one + *korn* grain — more at ONE, CORN] (circa 1901)
: a one-grained wheat (*Triticum monococcum*) sometimes considered the most primitive wheat and grown especially in poor soils in central Europe — called also *einkorn wheat*

ein·stei·ni·um \īn-ˈstī-nē-əm\ *noun* [New Latin, from Albert *Einstein*] (1955)
: a radioactive element produced artificially — see ELEMENT table

ei·re·nic *chiefly British variant of* IRENIC

eis·ege·sis \ˌī-sə-ˈjē-səs, ˈī-sə-ˌ\ *noun, plural* **-ege·ses** \-ˌsēz\ [Greek *eis* into (akin to Greek *en* in) + English ex*egesis* — more at IN] (1892)
: the interpretation of a text (as of the Bible) by reading into it one's own ideas — compare EXEGESIS

ei·stedd·fod \ī-ˈsteth-ˌvȯd, ā-\ *noun, plural* **-fods** \-ˌvȯdz\ *or* **-fod·au** \-ˌsteth-ˈvȯ-ˌdī\ [Welsh, literally, session, from *eistedd* to sit + *bod* being] (1822)
: a usually Welsh competitive festival of the arts especially in poetry and singing
— ei·stedd·fod·ic \ˌī-ˌsteth-ˈvȯ-dik, ˌā-\ *adjective*

eis·wein \ˈīs-ˌwīn, -ˌvīn\ *noun, often capitalized* [German, from *Eis* ice + *Wein* wine] (1967)
: a sweet German wine made from grapes that have frozen on the vine

[1]**ei·ther** \ˈē-thər *also* ˈī-\ *adjective* [Middle English, from Old English *ǣghwæther* both, each, from *ā* always + *ge-,* collective prefix + *hwæther* which of two, whether — more at AYE, CO-] (before 12th century)
1 : being the one and the other of two **:** EACH ⟨flowers blooming on *either* side of the walk⟩
2 : being the one or the other of two ⟨take *either* road⟩ ■

[2]**either** *pronoun* (before 12th century)
: the one or the other

[3]**either** *conjunction* (before 12th century)
— used as a function word before two or more coordinate words, phrases, or clauses joined usually by *or* to indicate that what immediately follows is the first of two or more alternatives

□ USAGE
either Americans prefer \ē\ as the first vowel of *either* and *neither,* while the British overwhelmingly prefer \ī\. This *ei* has had still other pronunciations through the centuries, including the vowel sound in *eight.* The variant with \ī\ was not at first accepted by British commentators, but now has become standard in British speech. This entails, by the lights of some, that the British \ī\ should be the standard for the American pronunciation of *either* and *neither* as well. However, the weight of long usage has established the older vowel \ē\ for *either* and *neither* in American speech.

[4]**either** *adverb* (15th century)
1 : LIKEWISE, MOREOVER — used for emphasis after a negative ⟨not smart or handsome *either*⟩
2 : for that matter — used for emphasis after an alternative following a question or conditional clause especially where negation is implied ⟨who answers for the Irish parliament? or army *either*? —Robert Browning⟩

[1]**ei·ther–or** \ˌē-t͟hər-ˈȯr *also* ˌī-\ *noun* (1922)
: an unavoidable choice or exclusive division between only two alternatives

[2]**either–or** *adjective* (1926)
: of or marked by either-or **:** BLACK-AND-WHITE

[1]**ejac·u·late** \i-ˈja-kyə-ˌlāt\ *verb* **-lat·ed; -lat·ing** [Latin *ejaculatus,* past participle of *ejaculari* to throw out, from *e-* + *jaculari* to throw, from *jaculum* dart, from *jacere* to throw — more at JET] (1578)
transitive verb
1 : to eject from a living body; *specifically* **:** to eject (semen) in orgasm
2 : to utter suddenly and vehemently
intransitive verb
: to eject a fluid
— ejac·u·la·tor \-ˌlā-tər\ *noun*

[2]**ejac·u·late** \-lət\ *noun* (1927)
: the semen released by one ejaculation

ejac·u·la·tion \i-ˌja-kyə-ˈlā-shən\ *noun* (1603)
1 : an act of ejaculating; *specifically* **:** a sudden discharging of a fluid from a duct
2 : something ejaculated; *especially* **:** a short sudden emotional utterance

ejac·u·la·to·ry \i-ˈja-kyə-lə-ˌtōr-ē, -ˌtȯr-\ *adjective* (1644)
1 : marked by or given to vocal ejaculation
2 : casting or throwing out; *specifically* **:** associated with or concerned in physiological ejaculation ⟨*ejaculatory* vessels⟩

ejaculatory duct *noun* (1751)
: a duct through which semen is ejaculated; *specifically* **:** either of the paired ducts in the human male that are formed by the junction of the duct from the seminal vesicle with the vas deferens and that pass through the prostate to empty into the urethra by means of a small opening

eject \i-ˈjekt\ *transitive verb* [Middle English, from Latin *ejectus,* past participle of *eicere,* from *e-* + *jacere*] (15th century)
1 a : to drive out especially by physical force **b :** to evict from property
2 : to throw out or off from within ⟨*ejects* the empty cartridges⟩ ☆
— eject·able \-ˈjek-tə-bəl\ *adjective*
— ejec·tion \-ˈjek-shən\ *noun*
— ejec·tive \-ˈjek-tiv\ *adjective*

ejec·ta \i-ˈjek-tə\ *noun plural but singular or plural in construction* [New Latin, from Latin, neuter plural of *ejectus*] (1886)
: material thrown out (as from a volcano)

ejection seat *noun* (1945)
: an emergency escape seat for propelling an occupant out and away from an airplane

eject·ment \i-ˈjek(t)-mənt\ *noun* (1523)
1 : the act or an instance of ejecting **:** DISPOSSESSION
2 : an action for the recovery of possession of real property and damages and costs

ejec·tor \i-ˈjek-tər\ *noun* (1640)
1 : one that ejects; *especially* **:** a mechanism of a firearm that ejects an empty cartridge
2 : a jet pump for withdrawing a gas, fluid, or powdery substance from a space

eka- \ˌe-kə, ˌā-kə\ *combining form* [International Scientific Vocabulary, from Sanskrit *eka* one — more at ONE]
: standing or assumed to stand next in order beyond (a specified element) in the same family of the periodic table — in names of chemical elements especially when not yet discovered ⟨*eka*-lead is the hypothetical element 114⟩

[1]**eke** \ˈēk\ *adverb* [Middle English, from Old English *ēac;* akin to Old High German *ouh* also, Latin *aut* or, Greek *au* again] (before 12th century)
archaic **:** ALSO

[2]**eke** *transitive verb* **eked; ek·ing** [Middle English, from Old English *īecan, ēcan;* akin to Old High German *ouhhōn* to add, Latin *augēre* to increase, Greek *auxein*] (before 12th century)
1 *archaic* **:** INCREASE, LENGTHEN
2 : to get with great difficulty — usually used with *out* ⟨*eke* out a living⟩

eke out *transitive verb* (1596)
1 : to make up for the deficiencies of **:** SUPPLEMENT ⟨*eked out* his income by getting a second job⟩
2 : to make (a supply) last by economy

ekis·tics \i-ˈkis-tiks\ *noun plural but singular in construction* [New Greek *oikistikē,* from feminine of *oikistikos* of settlement, from Greek, from *oikizein* to settle, colonize, from *oikos* house — more at VICINITY] (1958)
: a science dealing with human settlements and drawing on the research and experience of professionals in various fields (as architecture, engineering, city planning, and sociology)
— ekis·tic \-tik\ *adjective*

Ek·man dredge \ˌek-mən-\ *noun* [probably from V. W. *Ekman* (died 1954) Swedish oceanographer] (1948)
: a dredge that has opposable jaws operated by a messenger traveling down a cable to release a spring catch and that is used in ecology for sampling the bottom of a body of water

ekue·le \ā-ˈkwā-(ˌ)lā\ *also* **ek·pwe·le** \ek-ˈpwā-\ *noun, plural* **ekuele** *also* **ekpweles** [Fang (Bantu language of western equatorial Africa)] (circa 1973)
: the basic monetary unit of Equatorial Guinea 1975–85

[1]**el** \ˈel\ *noun* (14th century)
: the letter *l*

[2]**el** *noun, often capitalized* (circa 1906)
: an urban railway that operates chiefly on an elevated structure; *also* **:** a train belonging to such a railway

[1]**elab·o·rate** \i-ˈla-b(ə-)rət\ *adjective* [Latin *elaboratus,* from past participle of *elaborare* to work out, acquire by labor, from *e-* + *laborare* to work — more at LABORATORY] (1592)
1 : planned or carried out with great care ⟨took *elaborate* precautions⟩
2 : marked by complexity, fullness of detail, or ornateness ⟨*elaborate* prose⟩
— elab·o·rate·ly *adverb*
— elab·o·rate·ness *noun*

[2]**elab·o·rate** \i-ˈla-bə-ˌrāt\ *verb* **-rat·ed; -rat·ing** (1611)
transitive verb
1 : to produce by labor
2 : to build up (as complex organic compounds) from simple ingredients
3 : to work out in detail **:** DEVELOP
intransitive verb
1 : to become elaborate
2 : to expand something in detail ⟨would you care to *elaborate* on that statement⟩
— elab·o·ra·tion \-ˌla-bə-ˈrā-shən\ *noun*
— elab·o·ra·tive \-ˈla-bə-ˌrā-tiv\ *adjective*

Elaine \i-ˈlān\ *noun*
: any of several women in Arthurian legend; *especially* **:** one who dies for unrequited love of Lancelot

Elam·ite \ˈē-lə-ˌmīt\ *noun* (1874)
: a language of unknown affinities used in Elam approximately from the 25th to the 4th centuries B.C.

élan \ā-ˈläⁿ\ *noun* [French, from Middle French *eslan* rush, from *(s')eslancer* to rush, from *ex-* + *lancer* to hurl — more at LANCE] (1864)
: vigorous spirit or enthusiasm

eland \ˈē-lənd, -ˌland\ *noun, plural* **eland** *also* **elands** [Afrikaans, elk, from Dutch, from obsolete German *Elend,* probably from obsolete Lithuanian *ellenis;* akin to Old High German *elaho* elk — more at ELK] (1600)
: either of two large African antelopes (*Tragelaphus oryx* and *T. derbianus*) bovine in form with short spirally twisted horns in both sexes

eland

élan vi·tal \ā-läⁿ-vē-tàl\ *noun* [French] (1907)
: the vital force or impulse of life; *especially* **:** a creative principle held by Bergson to be immanent in all organisms and responsible for evolution

el·a·pid \ˈe-lə-pəd\ *noun* [New Latin *Elap-, Elaps,* genus of snakes, from Middle Greek, a fish, alteration of Greek *elops*] (1885)
: any of a family (Elapidae) of venomous snakes with grooved fangs

[1]**elapse** \i-ˈlaps\ *intransitive verb* **elapsed; elaps·ing** [Latin *elapsus,* past participle of *elabi,* from *e-* + *labi* to slip — more at SLEEP] (1644)
: to slip or glide away **:** PASS ⟨four years *elapsed* before he returned⟩

[2]**elapse** *noun* (circa 1677)
: PASSAGE ⟨returned after an *elapse* of 15 years⟩

elapsed time *noun* (circa 1909)
: the actual time taken (as by a boat or automobile in traveling over a racecourse)

elas·mo·branch \i-ˈlaz-mə-ˌbraŋk\ *noun, plural* **-branchs** [ultimately from Greek *elasmos* metal plate (from *elaunein*) + *branchia* gills] (1872)
: any of a subclass (Elasmobranchii) of cartilaginous fishes that have five to seven lateral to ventral gill openings on each side and that comprise the sharks, rays, skates, and extinct related fishes
— elasmobranch *adjective*

elas·tase \i-ˈlas-ˌtās, -ˌtāz\ *noun* [*elastin* + *-ase*] (1949)
: an enzyme especially of pancreatic juice that digests elastin

[1]**elas·tic** \i-ˈlas-tik\ *adjective* [New Latin *elasticus,* from Late Greek *elastos* ductile, beaten, from Greek *elaunein* to drive, beat out; probably akin to Greek *ēlythe* he went, Old Irish *luid*] (1674)
1 a *of a solid* **:** capable of recovering size and shape after deformation **b :** being a collision between particles in which the total kinetic energy of the particles remains unchanged
2 : capable of recovering quickly especially from depression or disappointment
3 : capable of being easily stretched or expanded and resuming former shape **:** FLEXIBLE
4 a : capable of ready change or easy expansion or contraction **:** not rigid or constricted **b :** receptive to new ideas **:** ADAPTABLE ☆
— elas·ti·cal·ly \-ti-k(ə-)lē\ *adverb*

☆ SYNONYMS
Eject, expel, oust, evict mean to drive or force out. EJECT carries an especially strong implication of throwing or thrusting out from within as a physical action ⟨*ejected* an obnoxious patron from the bar⟩. EXPEL stresses a thrusting out or driving away especially permanently which need not be physical ⟨a student *expelled* from college⟩. OUST implies removal or dispossession by power of the law or by force or compulsion ⟨got the sheriff to *oust* the squatters⟩. EVICT chiefly applies to turning out of house and home ⟨*evicted* for nonpayment of rent⟩.

[2]elastic *noun* (1847)
1 a : easily stretched rubber usually prepared in cords, strings, or bands **b :** RUBBER BAND
2 a : an elastic fabric usually made of yarns containing rubber **b :** something made from this fabric

elastic fiber *noun* (1849)
: a thick very elastic smooth yellowish anastomosing fiber of connective tissue that contains elastin

elas·tic·i·ty \i-ˌlas-ˈti-sə-tē, ˌē-ˌlas-, -ˈtis-tē\ *noun, plural* **-ties** (1664)
: the quality or state of being elastic: as **a :** the capability of a strained body to recover its size and shape after deformation **:** SPRINGINESS **b :** RESILIENCE 2 **c :** the quality of being adaptable

elas·ti·cized \i-ˈlas-tə-ˌsīzd\ *adjective* (circa 1909)
: made with elastic thread or inserts

elastic limit *noun* (1898)
: the greatest stress that an elastic solid can sustain without undergoing permanent deformation

elastic modulus *noun* (1904)
: the ratio of the stress in a body to the corresponding strain

elastic scattering *noun* (1933)
: a scattering of particles as the result of an elastic collision

elas·tin \i-ˈlas-tən\ *noun* [International Scientific Vocabulary, from New Latin *elasticus*] (1875)
: a protein that is similar to collagen and is the chief constituent of elastic fibers

elas·to·mer \-tə-mər\ *noun* [*elastic* + *-o-* + *-mer*] (circa 1939)
: any of various elastic substances resembling rubber ⟨polyvinyl *elastomers*⟩
— elas·to·mer·ic \i-ˌlas-tə-ˈmer-ik\ *adjective*

[1]elate \i-ˈlāt\ *transitive verb* **elat·ed; elat·ing** [Latin *elatus* (past participle of *efferre* to carry out, elevate), from *e-* + *latus,* past participle of *ferre* to carry — more at TOLERATE, BEAR] (circa 1619)
: to fill with joy or pride

[2]elate *adjective* (1647)
: ELATED

elat·ed *adjective* (circa 1619)
: marked by high spirits **:** EXULTANT
— elat·ed·ly *adverb*
— elat·ed·ness *noun*

el·a·ter \ˈe-lə-tər\ *noun* [New Latin, from Greek *elatēr* driver, from *elaunein* to drive] (1830)
: a plant structure functioning in the distribution of spores: as **a :** one of the elongated filaments among the spores in the capsule of a liverwort **b :** one of the filamentous appendages of the spores in the scouring rushes

elat·er·ite \i-ˈla-tə-ˌrīt\ *noun* [German *Elaterit,* from Greek *elatēr*] (1826)
: a dark brown elastic mineral resin occurring in soft flexible masses

ela·tion \i-ˈlā-shən\ *noun* (14th century)
1 : the quality or state of being elated
2 : pathological euphoria

E layer *noun* (1933)
: a layer of the ionosphere occurring about 65 miles (110 kilometers) above the earth's surface during daylight hours that is capable of reflecting shortwave frequencies

El·ba \ˈel-bə\ *noun* [*Elba* (Mediterranean island), residence of Napoléon Bonaparte after his first abdication May 14, 1814 to Feb. 26, 1815] (1924)
: a place or state of exile

[1]el·bow \ˈel-ˌbō\ *noun* [Middle English *elbowe,* from Old English *elboga,* from *el-* (akin to *eln* ell) + Old English *boga* bow — more at ELL, BOW] (before 12th century)
1 a : the joint of the human arm **b :** a corresponding joint in the anterior limb of a lower vertebrate
2 : something (as macaroni or an angular pipe fitting) resembling an elbow
— at one's elbow : at one's side
— out at elbows *or* **out at the elbows**
1 : shabbily dressed **2 :** short of funds

[2]elbow (1605)
transitive verb
1 a : to push with the elbow **:** JOSTLE **b :** to shove aside by pushing with or as if with the elbow
2 : to force (as one's way) by pushing with or as if with the elbow ⟨*elbowing* our way through the crowd⟩ ⟨*elbows* her way into the best social circles⟩
intransitive verb
1 : to advance by pushing with the elbow
2 : to make an angle **:** TURN

elbow grease *noun* (1672)
: vigorously applied physical labor **:** hard work

el·bow·room \ˈel-ˌbō-ˌrüm, -ˌru̇m\ *noun* (circa 1540)
1 a : room for moving the elbows freely **b :** adequate space for work or operation ⟨the large house gives plenty of *elbowroom*⟩
2 : free scope ⟨*elbowroom* to try new ideas⟩

el cheapo \(ˌ)el-ˈchē-(ˌ)pō, ˈel-\ *adjective* [Spanish *el* the + English *cheap* + Spanish *-o* (masculine noun ending)] (1969)
: CHEAP 3a, b

eld \ˈeld\ *noun* [Middle English, from Old English *ieldo;* akin to Old English *eald* old — more at OLD] (before 12th century)
1 : old age
2 *archaic* **:** old times **:** ANTIQUITY

[1]el·der \ˈel-dər\ *noun* [Middle English *eldre,* from Old English *ellærn;* perhaps akin to Old English *alor* alder — more at ALDER] (before 12th century)
: ELDERBERRY 2

[2]elder *adjective* [Middle English, from Old English *ieldra,* comparative of *eald* old] (before 12th century)
1 : of earlier birth or greater age ⟨his *elder* brother⟩
2 : of or relating to earlier times **:** FORMER
3 *obsolete* **:** of or relating to a more advanced time of life
4 : prior or superior in rank, office, or validity

[3]elder *noun* (before 12th century)
1 : one living in an earlier period
2 a : one who is older **:** SENIOR ⟨a child trying to please her *elders*⟩ **b :** an aged person
3 : one having authority by virtue of age and experience ⟨the village *elders*⟩
4 : any of various officers of religious groups: as **a :** PRESBYTER 1 **b :** a permanent officer elected by a Presbyterian congregation and ordained to serve on the session and assist the pastor at communion **c :** MINISTER 2 **d :** a leader of the Shakers **e :** a Mormon ordained to the Melchizedek priesthood
— el·der·ship \-ˌship\ *noun*

el·der·ber·ry \ˈel-də(r)-ˌber-ē\ *noun* (1589)
1 : the edible black or red berrylike drupe of any of a genus (*Sambucus*) of shrubs or trees of the honeysuckle family bearing flat clusters of small white or pink flowers
2 : a tree or shrub bearing elderberries

[1]el·der·ly \ˈel-dər-lē\ *adjective* (1611)
1 a : rather old; *especially* **:** being past middle age **b :** OLD-FASHIONED
2 : of, relating to, or characteristic of later life or elderly persons
— el·der·li·ness *noun*

[2]elderly *noun, plural* **-ly** *or* **-lies** (1865)
: an elderly person

elder statesman *noun* (1904)
: an eminent senior member of a group or organization; *especially* **:** a retired statesman who unofficially advises current leaders

el·dest \ˈel-dəst\ *adjective* (before 12th century)
: of the greatest age or seniority **:** OLDEST

eldest hand *noun* (1599)
: the card player who first receives cards in the deal

El Do·ra·do \ˌel-də-ˈrä-(ˌ)dō, -ˈrā-\ *noun* [Spanish, literally, the gilded one]
1 : a city or country of fabulous riches held by 16th century explorers to exist in South America
2 : a place of fabulous wealth or opportunity

el·dress \ˈel-drəs\ *noun* (1640)
: a woman elder especially of the Shakers

el·dritch \ˈel-drich\ *adjective* [perhaps from (assumed) Middle English *elfriche* fairyland, from Middle English *elf* + *riche* kingdom, from Old English *rīce* — more at RICH] (1508)
: WEIRD, EERIE

El·e·at·ic \ˌel-ē-ˈa-tik\ *adjective* [Latin *Eleaticus,* from Greek *Eleatikos,* from *Elea* (Velia), ancient town in southern Italy] (1695)
: of or relating to a school of Greek philosophers founded by Parmenides and developed by Zeno and marked by belief in the unity of being and the unreality of motion or change
— Eleatic *noun*
— El·e·at·i·cism \-ˈa-tə-ˌsi-zəm\ *noun*

ele·cam·pane \ˌe-li-ˌkam-ˈpān\ *noun* [Middle English *elena campana,* from Medieval Latin *enula campana,* literally, field elecampane, from *inula, enula* elecampane + *campana* of the field] (14th century)
: a large coarse European composite herb (*Inula helenium*) that has yellow ray flowers and is naturalized in the U.S.

[1]elect \i-ˈlekt\ *adjective* [Middle English, from Latin *electus,* past participle of *eligere* to select, from *e-* + *legere* to choose — more at LEGEND] (15th century)
1 : carefully selected **:** CHOSEN
2 : chosen for salvation through divine mercy
3 a : chosen for office or position but not yet installed ⟨the president-*elect*⟩ **b :** chosen for marriage at some future time ⟨the bride-*elect*⟩

[2]elect *noun, plural* **elect** (15th century)
1 : one chosen or set apart (as by divine favor)
2 *plural* **:** a select or exclusive group of people

[3]elect *verb* [Middle English, from Latin *electus*] (15th century)
transitive verb
1 : to select by vote for an office, position, or membership ⟨*elected* her class president⟩
2 : to make a selection of ⟨will *elect* an academic program⟩
3 : to choose (as a course of action) especially by preference ⟨might *elect* to sell the business⟩
intransitive verb
: to make a selection

elect·able \i-ˈlek-tə-bəl\ *adjective* (1879)
: capable of being elected (as to public office)

☆ SYNONYMS
Elastic, resilient, springy, flexible, supple mean able to endure strain without being permanently injured. ELASTIC implies the property of resisting deformation by stretching ⟨an *elastic* waistband⟩. RESILIENT implies the ability to recover shape quickly when the deforming force or pressure is removed ⟨a *resilient* innersole⟩. SPRINGY stresses both the ease with which something yields to pressure and the quickness of its return to original shape ⟨the cake is done when the top is *springy*⟩. FLEXIBLE applies to something which may or may not be resilient or elastic but which can be bent or folded without breaking ⟨*flexible* plastic tubing⟩. SUPPLE applies to something that can be readily bent, twisted, or folded without any sign of injury ⟨*supple* leather⟩.

— **elect·abil·i·ty** \-ˌlek-tə-ˈbi-lə-tē\ *noun*
elec·tion \i-ˈlek-shən\ *noun* (13th century)
1 a : an act or process of electing **b :** the fact of being elected
2 : predestination to eternal life
3 : the right, power, or privilege of making a choice
Election Day *noun* (15th century)
: a day legally established for the election of public officials; *especially* **:** the first Tuesday after the first Monday in November in an even year designated for national elections in the U.S. and observed as a legal holiday in many states
elec·tion·eer \i-ˌlek-shə-ˈnir\ *intransitive verb* [*election* + *-eer* (as in *privateer,* verb)] (1789)
: to take an active part in an election; *specifically* **:** to work for the election of a candidate or party
— **elec·tion·eer·er** *noun*
[1]**elec·tive** \i-ˈlek-tiv\ *adjective* (1531)
1 a : chosen or filled by popular election ⟨an *elective* official⟩ **b :** of or relating to election **c :** based on the right or principle of election ⟨the presidency is an *elective* office⟩
2 a : permitting a choice **:** OPTIONAL ⟨an *elective* course in school⟩ **b :** beneficial to the patient but not essential for survival ⟨*elective* surgery⟩
3 a : tending to operate on one substance rather than another ⟨*elective* absorption⟩ **b :** favorably inclined to one more than to another **:** SYMPATHETIC ⟨an *elective* affinity⟩
— **elec·tive·ly** *adverb*
— **elec·tive·ness** *noun*
[2]**elective** *noun* (1850)
: an elective course or subject
elec·tor \i-ˈlek-tər, -ˌtȯr\ *noun* (15th century)
1 : one qualified to vote in an election
2 : one entitled to participate in an election: as **a :** any of the German princes entitled to take part in choosing the Holy Roman Emperor **b :** a member of the electoral college in the U.S.
elec·tor·al \i-ˈlek-t(ə-)rəl, ˌē-lek-ˈtȯr-əl\ *adjective* (1675)
1 : of or relating to an elector ⟨the *electoral* vote⟩
2 : of or relating to election ⟨an *electoral* system⟩
— **elec·tor·al·ly** \-t(ə-)rə-lē, -ˈtȯr-ə-lē\ *adverb*
electoral college *noun* (circa 1691)
: a body of electors; *especially* **:** one that elects the president and vice president of the U.S.
elec·tor·ate \i-ˈlek-t(ə-)rət\ *noun* (1675)
1 : the territory, jurisdiction, or dignity of a German elector
2 : a body of people entitled to vote
electr- *or* **electro-** *combining form* [New Latin *electricus*]
1 a : electricity ⟨*electro*meter⟩ **b :** electric ⟨*electro*de⟩ **:** electric and ⟨*electro*chemical⟩ **:** electrically ⟨*electro*positive⟩
2 : electrolytic ⟨*electro*analysis⟩
3 : electron ⟨*electro*philic⟩
Elec·tra \i-ˈlek-trə\ *noun* [Latin, from Greek *Ēlektra*]
: a sister of Orestes who aids him in killing their mother Clytemnestra
Electra complex *noun* (1913)
: the Oedipus complex when it occurs in a female
elec·tress \i-ˈlek-trəs\ *noun* (1618)
: the wife or widow of a German elector
elec·tret \i-ˈlek-trət, -ˌtret\ *noun* [*electr*icity + magn*et*] (1885)
: a dielectric body in which a permanent state of electric polarization has been set up
[1]**elec·tric** \i-ˈlek-trik, ē-\ *adjective* [New Latin *electricus* produced from amber by friction, electric, from Medieval Latin, of amber, from Latin *electrum* amber, electrum, from Greek *ēlektron;* akin to Greek *ēlektōr* beaming sun] (1675)
1 *or* **elec·tri·cal** \-tri-kəl\ **:** of, relating to, or operated by electricity
2 : exciting as if by electric shock ⟨an *electric* performance⟩ ⟨an *electric* personality⟩; *also* **:** charged with strong emotion ⟨the room was *electric* with tension⟩
3 a : ELECTRONIC 3a **b :** amplifying sound by electronic means — used of a musical instrument ⟨an *electric* guitar⟩
4 : very bright ⟨*electric* blue⟩ ⟨*electric* orange⟩
— **elec·tri·cal·ly** \-tri-k(ə-)lē\ *adverb*
[2]**electric** *noun* (1646)
1 *archaic* **:** a nonconductor of electricity used to excite or accumulate electricity
2 : something (as a light, automobile, or train) operated by electricity
electrical storm *noun* (1941)
: THUNDERSTORM — called also *electric storm*
electric chair *noun* (1889)
1 : a chair used in legal electrocution
2 : the penalty of death by electrocution
electric eel *noun* (1794)
: a large eel-shaped fish (*Electrophorus electricus*) of the Orinoco and Amazon basins that is capable of giving a severe shock with its electric organs
electric eye *noun* (1898)
: PHOTOELECTRIC CELL
electric field *noun* (circa 1889)
: a region associated with a distribution of electric charge or a varying magnetic field in which forces due to that charge or field act upon other electric charges
elec·tri·cian \i-ˌlek-ˈtri-shən, ē-ˌ\ *noun* (1869)
: one who installs, maintains, operates, or repairs electrical equipment
elec·tric·i·ty \i-ˌlek-ˈtri-sə-tē, ē-ˌ, -ˈtris-tē\ *noun, plural* **-ties** (1646)
1 a : a fundamental entity of nature consisting of negative and positive kinds, observable in the attractions and repulsions of bodies electrified by friction and in natural phenomena (as lightning or the aurora borealis), and usually utilized in the form of electric currents **b :** electric current or power
2 : a science that deals with the phenomena and laws of electricity
3 : keen contagious excitement
electric organ *noun* (1773)
: a specialized tract of tissue (as in the electric eel) in which electricity is generated
electric ray *noun* (1774)
: any of various round-bodied short-tailed rays (family Torpedinidae) of warm seas with a pair of electric organs
elec·tri·fi·ca·tion \i-ˌlek-trə-fə-ˈkā-shən, ē-ˌ\ *noun* (1748)
1 : an act or process of electrifying
2 : the state of being electrified
elec·tri·fy \i-ˈlek-trə-ˌfī, ē-ˈ\ *transitive verb* **-fied; -fy·ing** (1745)
1 a : to charge with electricity **b** (1) **:** to equip for use of electric power (2) **:** to supply with electric power (3) **:** to amplify (music) electronically
2 : to excite intensely or suddenly as if by electric shock
elec·tro·acous·tics \i-ˌlek-trō-ə-ˈküs-tiks\ *noun plural but singular in construction* (1927)
: a science that deals with the transformation of acoustic energy into electric energy or vice versa
— **elec·tro·acous·tic** \-tik\ *adjective*
elec·tro·anal·y·sis \-ə-ˈna-lə-səs\ *noun* (1903)
: chemical analysis by electrolytic methods
— **elec·tro·an·a·lyt·i·cal** \-ˈlit-i-kəl\ *adjective*
elec·tro·car·dio·gram \-ˈkär-dē-ə-ˌgram\ *noun* (circa 1904)
: the tracing made by an electrocardiograph
elec·tro·car·dio·graph \-ˌgraf\ *noun* (1913)
: an instrument for recording the changes of electrical potential occurring during the heartbeat used especially in diagnosing abnormalities of heart action
— **elec·tro·car·dio·graph·ic** \-ˌkär-dē-ə-ˈgra-fik\ *adjective*
— **elec·tro·car·dio·graph·i·cal·ly** \-fi-k(ə-)lē\ *adverb*
— **elec·tro·car·di·og·ra·phy** \-dē-ˈä-grə-fē\ *noun*
elec·tro·chem·is·try \-ˈke-mə-strē\ *noun* (1814)
: a science that deals with the relation of electricity to chemical changes and with the interconversion of chemical and electrical energy
— **elec·tro·chem·i·cal** \-ˈke-mi-kəl\ *adjective*
— **elec·tro·chem·i·cal·ly** \-k(ə-)lē\ *adverb*
elec·tro·con·vul·sive \i-ˌlek-trō-kən-ˈvəl-siv\ *adjective* (1947)
: of, relating to, or involving convulsive response to electroshock ⟨*electroconvulsive* shocks⟩
electroconvulsive therapy *noun* (1948)
: ELECTROSHOCK THERAPY
elec·tro·cor·ti·co·gram \i-ˌlek-trō-ˈkȯr-ti-kə-ˌgram\ *noun* (1939)
: an electroencephalogram made with the electrodes in direct contact with the brain
elec·tro·cute \i-ˈlek-trə-ˌkyüt\ *transitive verb* **-cut·ed; -cut·ing** [*electr-* + *-cute* (as in *execute*)] (1889)
1 : to execute (a criminal) by electricity
2 : to kill by electric shock
— **elec·tro·cu·tion** \-ˌlek-trə-ˈkyü-shən\ *noun*
elec·trode \i-ˈlek-ˌtrōd\ *noun* (1834)
1 : a conductor used to establish electrical contact with a nonmetallic part of a circuit
2 : an element in a semiconductor device (as a transistor) that emits or collects electrons or holes or controls their movements
[1]**elec·tro·de·pos·it** \i-ˌlek-trō-di-ˈpä-zət\ *noun* (1864)
: a deposit formed in or at an electrode by electrolysis
[2]**electrodeposit** *transitive verb* (1882)
: to deposit (as a metal or rubber) by electrolysis
— **elec·tro·de·po·si·tion** \-ˌde-pə-ˈzi-shən, -ˌdē-pə-\ *noun*
elec·tro·der·mal \i-ˌlek-trō-ˈdər-məl\ *adjective* (1946)
: of or relating to electrical activity in or electrical properties of the skin
elec·tro·des·ic·ca·tion \i-ˌlek-trō-ˌde-si-ˈkā-shən\ *noun* (1919)
: the drying up of tissue by a high-frequency electric current applied with a needle-shaped electrode — called also *fulguration*
elec·tro·di·al·y·sis \i-ˌlek-trō-dī-ˈa-lə-səs\ *noun* (1921)
: dialysis accelerated by an electromotive force applied to electrodes adjacent to the membranes
— **elec·tro·di·a·lyt·ic** \-ˌdī-ə-ˈli-tik\ *adjective*
elec·tro·dy·nam·ics \-dī-ˈna-miks\ *noun plural but singular in construction* (1827)
: a branch of physics that deals with the effects arising from the interactions of electric currents with magnets, with other currents, or with themselves
— **elec·tro·dy·nam·ic** \-mik\ *adjective*
elec·tro·dy·na·mom·e·ter \-ˌdī-nə-ˈmä-mə-tər\ *noun* [International Scientific Vocabulary] (1876)
: an instrument that measures current by indicating the strength of the forces between a current flowing in fixed coils and one flowing in movable coils
elec·tro·en·ceph·a·lo·gram \-in-ˈse-f(ə-)lə-ˌgram\ *noun* [International Scientific Vocabulary] (1934)
: the tracing of brain waves made by an electroencephalograph

elec·tro·en·ceph·a·lo·graph \-ˌgraf\ *noun* [International Scientific Vocabulary] (1936)
: an apparatus for detecting and recording brain waves
— **elec·tro·en·ceph·a·log·ra·pher** \-ˌse-fə-ˈlä-grə-fər\ *noun*
— **elec·tro·en·ceph·a·lo·graph·ic** \-ˌse-f(ə-)lə-ˈgra-fik\ *adjective*
— **elec·tro·en·ceph·a·lo·graph·i·cal·ly** \-fi-k(ə-)lē\ *adverb*
— **elec·tro·en·ceph·a·log·ra·phy** \-ˈlä-grə-fē\ *noun*

elec·tro·fish·ing \i-ˈlek-trō-ˌfi-shiŋ\ *noun* (1950)
: the taking of fish by a system based on their tendency to respond positively to a source of direct electric current

elec·tro·form \i-ˈlek-trə-ˌfȯrm\ *transitive verb* (1931)
: to form (shaped articles) by electrodeposition on a mold
— **electroform** *noun*

elec·tro·gen·ic \i-ˌlek-trə-ˈje-nik\ *adjective* (circa 1891)
: of or relating to the production of electrical activity in living tissue ⟨an *electrogenic* pump⟩
— **elec·tro·gen·e·sis** \-ˈje-nə-sis\ *noun*

elec·tro·gram \i-ˈlek-trə-ˌgram\ *noun* (circa 1935)
: a tracing of the electrical potentials of a tissue (as the brain or heart) made by means of electrodes placed directly in the tissue instead of on the surface of the body

elec·tro·hy·drau·lic \i-ˌlek-trō-hī-ˈdrȯ-lik, -ˈdrä-\ *adjective* (1922)
1 : of or relating to a combination of electric and hydraulic mechanisms
2 : involving or produced by the action of very brief but powerful pulse discharges of electricity under a liquid resulting in the generation of shock waves and highly reactive chemical species ⟨an *electrohydraulic* effect⟩

elec·tro·jet \i-ˈlek-trə-ˌjet\ *noun* (1955)
: an overhead concentration of electric current found in the regions of strong auroral displays and along the magnetic equator

elec·tro·ki·net·ic \i-ˌlek-trō-kə-ˈne-tik, kī-\ *adjective* (1881)
: of or relating to the motion of particles or liquids that results from or produces a difference of electric potential

elec·tro·ki·net·ics \-tiks\ *noun plural but singular in construction* (circa 1925)
: a branch of physics dealing with the motion of electric currents or charged particles

elec·tro·less \i-ˈlek-ˌtrō-ləs, -trə-\ *adjective* (1947)
: being or involving chemical deposition of metal instead of electrodeposition

elec·trol·o·gist \i-ˌlek-ˈträ-lə-jist\ *noun* [blend of *electrolysis* and *-logist* (from *-logy* + *-ist*)] (circa 1902)
: one that removes hair, warts, moles, and birthmarks by means of an electric current applied to the body with a needle-shaped electrode
— **elec·trol·o·gy** \-lə-jē\ *noun*

elec·tro·lu·mi·nes·cent \i-ˌlek-trō-ˌlü-mə-ˈne-sᵊnt\ *adjective* (circa 1909)
: of or relating to luminescence resulting from a high-frequency discharge through a gas or from application of an alternating current to a layer of phosphor
— **elec·tro·lu·mi·nes·cence** \-sᵊn(t)s\ *noun*

elec·trol·y·sis \i-ˌlek-ˈträ-lə-səs\ *noun* (1834)
1 a : the producing of chemical changes by passage of an electric current through an electrolyte **b :** subjection to this action
2 : the destruction of hair roots with an electric current

elec·tro·lyte \i-ˈlek-trə-ˌlīt\ *noun* (1834)
1 : a nonmetallic electric conductor in which current is carried by the movement of ions
2 : a substance that when dissolved in a suitable solvent or when fused becomes an ionic conductor

elec·tro·lyt·ic \i-ˌlek-trə-ˈli-tik\ *adjective* (1842)
: of or relating to electrolysis or an electrolyte; *also* **:** involving, produced by, or used in electrolysis ⟨*electrolytic* cell⟩
— **elec·tro·lyt·i·cal·ly** \-ti-k(ə-)lē\ *adverb*

elec·tro·lyze \i-ˈlek-trə-ˌlīz\ *transitive verb* **-lyzed; -lyz·ing** (1834)
: to subject to electrolysis

elec·tro·mag·net \i-ˌlek-trō-ˈmag-nət\ *noun* (1831)
: a core of magnetic material surrounded by a coil of wire through which an electric current is passed to magnetize the core

elec·tro·mag·net·ic \-mag-ˈne-tik\ *adjective* (1821)
: of, relating to, or produced by electromagnetism
— **elec·tro·mag·net·i·cal·ly** \-ti-k(ə-)lē\ *adverb*

electromagnetic pulse *noun* (1981)
: high-intensity electromagnetic radiation generated by a nuclear blast high above the earth's surface and held to disrupt electronic and electrical systems

electromagnetic radiation *noun* (1939)
: a series of electromagnetic waves

electromagnetic spectrum *noun* (circa 1934)
: the entire range of wavelengths or frequencies of electromagnetic radiation extending from gamma rays to the longest radio waves and including visible light

electromagnetic unit *noun* (circa 1889)
: any of a system of electrical units based primarily on the magnetic properties of electrical currents

electromagnetic wave *noun* (1908)
: one of the waves that are propagated by simultaneous periodic variations of electric and magnetic field intensity and that include radio waves, infrared, visible light, ultraviolet, X rays, and gamma rays

elec·tro·mag·ne·tism \i-ˌlek-tro-ˈmag-nə-ˌti-zəm\ *noun* (1828)
1 : magnetism developed by a current of electricity
2 a : a fundamental physical force that is responsible for interactions between charged particles which occur because of their charge and for the emission and absorption of photons, that is about 100 times weaker than the strong force, and that extends over infinite distances but is dominant over atomic and molecular distances — called also *electromagnetic force*; compare GRAVITY 3a(2), STRONG FORCE, WEAK FORCE **b :** a branch of physical science that deals with the physical relations between electricity and magnetism

elec·tro·me·chan·i·cal \-mə-ˈka-ni-kəl\ *adjective* (1888)
: of, relating to, or being a mechanical process or device actuated or controlled electrically; *especially* **:** being a transducer for converting electrical energy to mechanical energy
— **elec·tro·me·chan·i·cal·ly** \-k(ə-)lē\ *adverb*

elec·tro·met·al·lur·gy \-ˈme-tᵊl-ˌər-jē, *especially British* -mə-ˈta-lər-\ *noun* (1840)
: a branch of metallurgy that deals with the application of electric current either for electrolytic deposition or as a source of heat

elec·trom·e·ter \i-ˌlek-ˈträ-mə-tər\ *noun* (1749)
: any of various instruments for detecting or measuring electric-potential differences or ionizing radiations by means of the forces of attraction or repulsion between charged bodies

elec·tro·mo·tive force \i-ˌlek-trō-ˌmō-tiv-, -trə-\ *noun* (1827)
: something that moves or tends to move electricity **:** the potential difference derived from an electrical source per unit quantity of electricity passing through the source (as a cell or generator)

elec·tro·myo·gram \i-ˌlek-trō-ˈmī-ə-ˌgram\ *noun* (1917)
: a tracing made with an electromyograph

elec·tro·myo·graph \-ˌgraf\ *noun* [*electr-* + *my-* + *-graph*] (1948)
: an instrument that converts the electrical activity associated with functioning skeletal muscle into a visual record or into sound and used to diagnose neuromuscular disorders and in biofeedback training
— **elec·tro·myo·graph·ic** \-ˌmī-ə-ˈgra-fik\ *adjective*
— **elec·tro·myo·graph·i·cal·ly** \-fi-k(ə-)lē\ *adverb*
— **elec·tro·my·og·ra·phy** \-mī-ˈä-grə-fē\ *noun*

elec·tron \i-ˈlek-ˌträn\ *noun* [*electr-* + [2]*-on*] (1891)
: an elementary particle consisting of a charge of negative electricity equal to about 1.602×10^{-19} coulomb and having a mass when at rest of about 9.109534×10^{-28} gram or about $1/1836$ that of a proton

electron cloud *noun* (1926)
: the system of electrons surrounding the nucleus of an atom

elec·tro·neg·a·tive \i-ˌlek-trō-ˈne-gə-tiv\ *adjective* (1834)
: having a tendency to attract electrons
— **elec·tro·neg·a·tiv·i·ty** \-ˌne-gə-ˈti-və-tē\ *noun*

electron gas *noun* (circa 1929)
: a population of free electrons in a vacuum or in a metallic conductor

electron gun *noun* (1924)
: an electron-emitting cathode and its surrounding assembly (as in a cathode-ray tube) for directing, controlling, and focusing a beam of electrons

elec·tron·ic \i-ˌlek-ˈträ-nik\ *adjective* (1902)
1 : of or relating to electrons
2 : of, relating to, or utilizing devices constructed or working by the methods or principles of electronics; *also* **:** implemented on or by means of a computer ⟨*electronic* food stamps⟩ ⟨*electronic* banking⟩
3 a : generating musical tones by electronic means ⟨an *electronic* organ⟩ **b :** of, relating to, or being music that consists of sounds electronically generated or modified
4 : of, relating to, or being a medium (as television) by which information is transmitted electronically ⟨*electronic* journalism⟩
— **elec·tron·i·cal·ly** \-ni-k(ə-)lē\ *adverb*

electronic countermeasure *noun* (1962)
: the disruption of the operation of an enemy's equipment (as by jamming radio or radar signals)

electronic mail *noun* (1977)
: messages sent and received electronically (as between terminals linked by telephone lines or microwave relays)

electronic publishing *noun* (1980)
: publishing in which information is distributed by means of a computer network or is produced in a format for use with a computer

elec·tron·ics \i-ˌlek-ˈträ-niks\ *noun plural* (1910)
1 *singular in construction* **:** a branch of physics that deals with the emission, behavior, and effects of electrons (as in electron tubes and transistors) and with electronic devices
2 : electronic devices or equipment

electron lens *noun* (1931)
: a device for focusing a beam of electrons by means of an electric or a magnetic field

electron micrograph *noun* (1934)

: a micrograph made with an electron microscope
— **electron micrography** \-mī-'krä-grə-fē\ *noun*
electron microscope *noun* (1932)
: an electron-optical instrument in which a beam of electrons focused by means of an electron lens is used to produce an enlarged image of a minute object on a fluorescent screen or photographic plate
— **electron microscopist** *noun*
— **electron microscopy** *noun*
electron multiplier *noun* (1936)
: a device utilizing secondary emission of electrons for amplifying a current of electrons
electron optics *noun plural but singular in construction* (1916)
: a branch of physics in which the principles of optics are applied to beams of electrons
— **elec·tron–op·ti·cal** \i-ˌlek-ˌträn-'äp-ti-kəl\ *adjective*
electron probe *noun* (1962)
: a microprobe that uses an electron beam to induce X-ray emissions in a sample
electron transport *noun* (1951)
: the sequential transfer of electrons especially by cytochromes in cellular respiration from an oxidizable substrate to molecular oxygen by a series of oxidation-reduction reactions
electron tube *noun* (1922)
: an electronic device in which conduction by electrons takes place through a vacuum or a gaseous medium within a sealed glass or metal container and which has various uses based on the controlled flow of electrons
electron volt *noun* (1930)
: a unit of energy equal to the energy gained by an electron in passing from a point of low potential to a point one volt higher in potential : 1.60×10^{-19} joule
elec·tro·oc·u·lo·gram \i-ˌlek-trō-'ä-kyə-lə-ˌgram\ *noun* [*electr-* + Latin *oculus* eye + English *-gram* — more at EYE] (1947)
: a record of the standing voltage between the front and back of the eye that is correlated with eyeball movement (as in REM sleep) and obtained by electrodes placed on the skin near the eye
elec·tro·oc·u·log·ra·phy \-ˌä-kyə-'lä-grə-fē\ *noun, plural* **-phies** (1951)
: the preparation and study of electrooculograms
elec·tro–op·ti·cal \i-ˌlek-trō-'äp-ti-kəl\ *or* **elec·tro–op·tic** \-'äp-tik\ *adjective* (1879)
1 : of or relating to electro-optics
2 a : relating to or being a change in the refractive index of a material due to an electric field ⟨*electro-optical* effect⟩ **b :** using or being a material that exhibits electro-optical properties ⟨an *electro-optical* crystal⟩
3 : relating to or being an electronic device for emitting, modulating, transmitting, or sensing light
— **elec·tro–op·ti·cal·ly** \-k(ə-)lē\ *adverb*
elec·tro–op·tics \-trō-'äp-tiks\ *noun plural* (circa 1889)
1 *singular in construction* **:** a branch of physics that deals with the effects of an electric field on light traversing it
2 : electro-optical devices
elec·tro·os·mo·sis \i-ˌlek-trō-äz-'mō-səs, -äs-\ *noun* (1906)
: the movement of a liquid out of or through a porous material or a biological membrane under the influence of an electric field
— **elec·tro·os·mot·ic** \-'mä-tik\ *adjective*
elec·tro·phe·ro·gram \-trə-'fir-ə-ˌgram, -'fer-\ *noun* [*electr-* + *phero-* (from Greek *pherein* to carry) + *-gram*] (1951)
: ELECTROPHORETOGRAM
elec·tro·phile \i-'lek-trə-ˌfīl\ *noun* (1943)
: an electrophilic substance (as an electron-accepting reagent)
elec·tro·phil·ic \i-ˌlek-trə-'fi-lik\ *adjective* (1936)
1 *of an atom, ion, or molecule* **:** having an affinity for electrons **:** being an electron acceptor
2 : involving an electrophilic species ⟨an *electrophilic* reaction⟩ — compare NUCLEOPHILIC
— **elec·tro·phi·lic·ity** \-trō-fi-'li-sə-tē\ *noun*
elec·tro·pho·re·sis \-trə-fə-'rē-səs\ *noun* [New Latin] (1911)
: the movement of suspended particles through a fluid or gel under the action of an electromotive force applied to electrodes in contact with the suspension
— **elec·tro·pho·rese** \-'rēs, -'rēz\ *transitive verb*
— **elec·tro·pho·ret·ic** \-'re-tik\ *adjective*
— **elec·tro·pho·ret·i·cal·ly** \-ti-k(ə-)lē\ *adverb*
elec·tro·pho·reto·gram \-'re-tə-ˌgram\ *noun* [*electrophoretic* + *-o-* + *-gram*] (1954)
: a record that consists of the separated components of a mixture (as of proteins) produced by electrophoresis in a supporting medium (as filter paper)
elec·troph·o·rus \i-ˌlek-'trä-fə-rəs\ *noun, plural* **-ri** \-ˌrī, -ˌrē\ [New Latin, from *electr-* + *-phorus* -phore] (1778)
: a device for producing electric charges consisting of a disk that is negatively electrified by friction and a metal plate that becomes charged by induction when placed on the disk
elec·tro·pho·tog·ra·phy \i-ˌlek-trō-fə-'tä-grə-fē\ *noun* (1894)
: photography in which images are produced by electrical means (as in xerography)
— **elec·tro·pho·to·graph·ic** \-trə-ˌfō-tə-'gra-fik\ *adjective*
elec·tro·phys·i·ol·o·gy \i-ˌlek-trō-ˌfi-zē-'ä-lə-jē\ *noun* (1838)
1 : physiology that is concerned with the electrical aspects of physiological phenomena
2 : electrical phenomena associated with a physiological process (as the function of a body or bodily part) ⟨*electrophysiology* of the eye⟩
— **elec·tro·phys·i·o·log·i·cal** \-zē-ə-'lä-ji-kəl\ *also* **elec·tro·phys·i·o·log·ic** \-jik\ *adjective*
— **elec·tro·phys·i·o·log·i·cal·ly** \-ji-k(ə-)lē\ *adverb*
— **elec·tro·phys·i·ol·o·gist** \-zē-'äl-ə-jist\ *noun*
elec·tro·plate \i-'lek-trə-ˌplāt\ *transitive verb* (circa 1859)
: to plate with an adherent continuous coating by electrodeposition
elec·tro·pos·i·tive \i-ˌlek-trō-'pä-zə-tiv, -'päz-tiv\ *adjective* (1834)
: having a tendency to release electrons
elec·tro·ret·i·no·gram \-'re-tᵊn-ə-ˌgram\ *noun* (1936)
: a graphic record of electrical activity of the retina used especially in the diagnosis of retinal conditions
elec·tro·ret·i·no·graph \-ˌgraf\ *noun* (1962)
: an instrument for recording electrical activity in the retina
— **elec·tro·ret·i·no·graph·ic** \-ˌre-tᵊn-ə-'gra-fik\ *adjective*
— **elec·tro·ret·i·nog·ra·phy** \-tᵊn-'ä-grə-fē\ *noun*
elec·tro·scope \i-'lek-trə-ˌskōp\ *noun* [probably from French *électroscope*] (1810)
: any of various instruments for detecting the presence of an electric charge on a body, for determining whether the charge is positive or negative, or for indicating and measuring intensity of radiation
elec·tro·shock \-trō-ˌshäk\ *noun* (1941)
1 : ³SHOCK 5
2 : ELECTROSHOCK THERAPY
electroshock therapy *noun* (1942)
: the treatment of mental disorder and especially depression by the induction of unconsciousness and convulsions through the use of an electric current now usually on an anesthetized patient — called also *electroconvulsive therapy*
elec·tro·stat·ic \i-ˌlek-trə-'sta-tik\ *adjective* [International Scientific Vocabulary] (1860)
1 : of or relating to static electricity or electrostatics
2 : of or relating to painting with a spray that utilizes electrically charged particles to ensure complete coating
— **elec·tro·stat·i·cal·ly** \-'sta-ti-k(ə-)lē\ *adverb*
electrostatic generator *noun* (circa 1931)
: VAN DE GRAAFF GENERATOR
electrostatic precipitator *noun* (1949)
: an electrostatic device in chimney flues that removes particles from escaping gases
elec·tro·stat·ics \i-ˌlek-trə-'sta-tiks\ *noun plural but singular in construction* (1827)
: physics that deals with phenomena due to attractions or repulsions of electric charges but not dependent upon their motion
elec·tro·sur·gery \i-ˌlek-trō-'sər-jə-rē\ *noun* (circa 1903)
: surgery by means of diathermy
— **elec·tro·sur·gi·cal** \-'sər-ji-kəl\ *adjective*
elec·tro·ther·a·py \-'ther-ə-pē\ *noun* (1881)
: treatment of disease by means of electricity (as in diathermy)
elec·tro·ther·mal \-'thər-məl\ *adjective* (1884)
: relating to or combining electricity and heat; *specifically* **:** relating to the generation of heat by electricity
— **elec·tro·ther·mal·ly** \-mə-lē\ *adverb*
elec·tro·ton·ic \i-ˌlek-trə-'tä-nik\ *adjective* (1832)
1 : of, induced by, relating to, or constituting electrotonus
2 : of, relating to, or being the spread of electrical activity through living tissue or cells in the absence of repeated action potentials ⟨an *electrotonic* junction between cells⟩
— **elec·tro·ton·i·cal·ly** \-ni-k(ə-)lē\ *adverb*
elec·trot·o·nus \i-ˌlek-'trä-tᵊn-əs\ *noun* [New Latin] (1878)
: the altered sensitivity of a nerve when a constant current of electricity passes through any part of it
elec·tro·type \i-'lek-trə-ˌtīp\ *noun* (1840)
1 : a duplicate printing surface made by an electroplating process
2 : a copy (as of a coin) made by an electroplating process
— **electrotype** *transitive verb*
— **elec·tro·typ·er** \-ˌtī-pər\ *noun*
elec·tro·weak \i-'lek-trō-ˌwēk\ *adjective* (1978)
: of, relating to, or being the unification of electromagnetism and the weak force
elec·tro·win·ning \i-'lek-trō-ˌwi-niŋ\ *noun* (1924)
: the recovery especially of metals from solutions by electrolysis
elec·trum \i-'lek-trəm\ *noun* [Middle English, from Latin — more at ELECTRIC] (14th century)
: a natural pale yellow alloy of gold and silver
elec·tu·ary \i-'lek-chə-ˌwer-ē\ *noun, plural* **-ar·ies** [Middle English *electuarie,* from Late Latin *electuarium,* probably from Greek *ekleikton,* from *ekleichein* to lick up, from *ex-* + *leichein* to lick — more at LICK] (14th century)
: CONFECTION 2b
el·e·doi·sin \ˌe-lə-'dȯi-sᵊn\ *noun* [irregular from New Latin *Eledone,* from Greek *eledōnē,* a kind of octopus] (1963)
: a small protein $C_{54}H_{85}N_{13}O_{15}S$ from the salivary glands of several octopuses (genus *Eledone*) that is a powerful vasodilator and hypotensive agent

el·ee·mo·sy·nary \ˌe-li-ˈmä-sᵊn-ˌer-ē, -ˈmō-; -ˈmä-zᵊn-\ *adjective* [Medieval Latin *eleemosynarius,* from Late Latin *eleemosyna* alms — more at ALMS] (circa 1616)
: of, relating to, or supported by charity

el·e·gance \ˈe-li-gən(t)s\ *noun* (circa 1510)
1 a : refined grace or dignified propriety **:** URBANITY **b :** tasteful richness of design or ornamentation ⟨the sumptuous *elegance* of the furnishings⟩ **c :** dignified gracefulness or restrained beauty of style **:** POLISH ⟨the essay is marked by lucidity, wit, and *elegance*⟩ **d :** scientific precision, neatness, and simplicity ⟨the *elegance* of a mathematical proof⟩
2 : something that is elegant

el·e·gan·cy \-gən(t)-sē\ *noun, plural* **-cies** (15th century)
: ELEGANCE

el·e·gant \ˈe-li-gənt\ *adjective* [Middle French or Latin; Middle French, from Latin *elegant-, elegans;* akin to Latin *eligere* to select — more at ELECT] (15th century)
1 : marked by elegance
2 : of a high grade or quality **:** SPLENDID ⟨*elegant* gems priced at hundreds of thousands of dollars⟩
synonym see CHOICE
— el·e·gant·ly *adverb*

ele·gi·ac \ˌe-lə-ˈjī-ək, -ˌak *also* i-ˈlē jē-ˌak\ *also* **el·e·gi·a·cal** \ˌe-lə-ˈjī-ə-kəl\ *adjective* [Late Latin *elegiacus,* from Greek *elegeiakos,* from *elegeion*] (1542)
1 a : of, relating to, or consisting of two dactylic hexameter lines the second of which lacks the arsis in the third and sixth feet **b** (1) **:** written in or consisting of elegiac couplets (2) **:** noted for having written poetry in such couplets **c :** of or relating to the period in Greece about the seventh century B.C. when poetry written in such couplets flourished
2 : of, relating to, or comprising elegy or an elegy; *especially* **:** expressing sorrow often for something now past ⟨an *elegiac* lament for departed youth⟩
— elegiac *noun*
— el·e·gi·a·cal·ly \ˌe-lə-ˈjī-ə-k(ə-)lē\ *adverb*

ele·git \i-ˈlē-jət\ *noun* [Latin, literally, he has chosen, from *eligere*] (1504)
: a judicial writ of execution by which a defendant's goods and if necessary his or her lands are delivered for debt to the plaintiff until the debt is paid

el·e·gize \ˈe-lə-ˌjīz\ *verb* **-gized; -giz·ing** (1702)
intransitive verb
: to write an elegy
transitive verb
: to write an elegy on

el·e·gy \ˈe-lə-jē\ *noun, plural* **-gies** [Latin *elegia* poem in elegiac couplets, from Greek *elegeia, elegeion,* from *elegos* song of mourning] (1501)
1 : a poem in elegiac couplets
2 a : a song or poem expressing sorrow or lamentation especially for one who is dead **b :** something (as a speech) resembling such a song or poem
3 a : a pensive or reflective poem that is usually nostalgic or melancholy **b :** a short pensive musical composition

el·e·ment \ˈe-lə-mənt\ *noun* [Middle English, from Old French & Latin; Old French, from Latin *elementum*] (13th century)
1 a : any of the four substances air, water, fire, and earth formerly believed to compose the physical universe **b** *plural* **:** weather conditions; *especially* **:** violent or severe weather ⟨battling the *elements*⟩ **c :** the state or sphere natural or suited to a person or thing ⟨at school she was in her *element*⟩
2 : a constituent part: as **a** *plural* **:** the simplest principles of a subject of study **:** RUDIMENTS **b** (1) **:** a part of a geometric magnitude ⟨an infinitesimal *element* of volume⟩ (2) **:** a generator of a geometric figure; *also* **:** a line or line segment contained in the surface of a cone or cylinder (3) **:** a basic member of a mathematical or logical class or set (4) **:** one of the individual entries in a mathematical matrix or determinant **c :** one of a number of distinct groups composing a larger group or community ⟨the criminal *element* in the city⟩ **d** (1) **:** one of the necessary data or values on which calculations or conclusions are based (2) **:** one of the factors determining the outcome of a process **e :** any of more than 100 fundamental substances that consist of atoms of only one kind and that singly or in combination constitute all matter **f :** a distinct part of a composite device **g :** a subdivision of a military unit
3 *plural* **:** the bread and wine used in the Eucharist ☆

CHEMICAL ELEMENTS

ELEMENT	SYMBOL	ATOMIC NUMBER	ATOMIC WEIGHT (C = 12)
actinium	Ac	89	227.0278
aluminum	Al	13	26.98154
americium	Am	95	
antimony	Sb	51	121.75
argon	Ar	18	39.948
arsenic	As	33	74.9216
astatine	At	85	
barium	Ba	56	137.33
berkelium	Bk	97	
beryllium	Be	4	9.01218
bismuth	Bi	83	208.9804
bohrium	Bh	107	
boron	B	5	10.81
bromine	Br	35	79.904
cadmium	Cd	48	112.41
calcium	Ca	20	40.08
californium	Cf	98	
carbon	C	6	12.011
cerium	Ce	58	140.12
cesium	Cs	55	132.9054
chlorine	Cl	17	35.453
chromium	Cr	24	51.996
cobalt	Co	27	58.9332
copper	Cu	29	63.546
curium	Cm	96	
dubnium	Db	105	
dysprosium	Dy	66	162.50
einsteinium	Es	99	
erbium	Er	68	167.26
europium	Eu	63	151.96
fermium	Fm	100	
fluorine	F	9	18.998403
francium	Fr	87	
gadolinium	Gd	64	157.25
gallium	Ga	31	69.72
germanium	Ge	32	72.59
gold	Au	79	196.9665
hafnium	Hf	72	178.49
hassium	Hs	108	
helium	He	2	4.00260
holmium	Ho	67	164.9304
hydrogen	H	1	1.0079
indium	In	49	114.82
iodine	I	53	126.9045
iridium	Ir	77	192.22
iron	Fe	26	55.847
krypton	Kr	36	83.80
lanthanum	La	57	138.9055
lawrencium	Lr	103	
lead	Pb	82	207.2
lithium	Li	3	6.941
lutetium	Lu	71	174.967
magnesium	Mg	12	24.305
manganese	Mn	25	54.9380
meitnerium	Mt	109	
mendelevium	Md	101	
mercury	Hg	80	200.59
molybdenum	Mo	42	95.94
neodymium	Nd	60	144.24
neon	Ne	10	20.179
neptunium	Np	93	237.0482
nickel	Ni	28	58.69
niobium	Nb	41	92.9064
nitrogen	N	7	14.0067
nobelium	No	102	
osmium	Os	76	190.2
oxygen	O	8	15.9994
palladium	Pd	46	106.42
phosphorus	P	15	30.97376
platinum	Pt	78	195.08
plutonium	Pu	94	
polonium	Po	84	
potassium	K	19	39.0983
praseodymium	Pr	59	140.9077
promethium	Pm	61	
protactinium	Pa	91	231.0359
radium	Ra	88	226.0254
radon	Rn	86	
rhenium	Re	75	186.207
rhodium	Rh	45	102.9055
rubidium	Rb	37	85.4678
ruthenium	Ru	44	101.07
rutherfordium	Rf	104	
samarium	Sm	62	150.36
scandium	Sc	21	44.9559
seaborgium	Sg	106	
selenium	Se	34	78.96
silicon	Si	14	28.0855
silver	Ag	47	107.868
sodium	Na	11	22.98977
strontium	Sr	38	87.62
sulfur	S	16	32.06
tantalum	Ta	73	180.9479
technetium	Tc	43	
tellurium	Te	52	127.60
terbium	Tb	65	158.9254
thallium	Tl	81	204.383
thorium	Th	90	232.0381
thulium	Tm	69	168.9342
tin	Sn	50	118.69
titanium	Ti	22	47.88
tungsten	W	74	183.85
uranium	U	92	238.0289
vanadium	V	23	50.9415
xenon	Xe	54	131.29
ytterbium	Yb	70	173.04
yttrium	Y	39	88.9059
zinc	Zn	30	65.38
zirconium	Zr	40	91.22

el·e·men·tal \ˌe-lə-ˈmen-tᵊl\ *adjective* (15th century)
1 a : of, relating to, or being an element; *specifically* **:** existing as an uncombined chemical element **b** (1) **:** of, relating to, or being the basic or essential constituent of something **:** FUNDAMENTAL ⟨*elemental* biological needs⟩ (2) **:** SIMPLE, UNCOMPLICATED ⟨*elemental* food⟩ **c**

☆ SYNONYMS
Element, component, constituent, ingredient mean one of the parts of a compound or complex whole. ELEMENT applies to any such part and often connotes irreducible simplicity ⟨the basic *elements* of geometry⟩. COMPONENT and CONSTITUENT may designate any of the substances (whether elements or compounds) or the qualities that enter into the makeup of a complex product; COMPONENT stresses its separate entity or distinguishable character ⟨the *components* of a stereo system⟩. CONSTITUENT stresses its essential and formative character ⟨the *constituents* of a chemical compound⟩. INGREDIENT applies to any of the substances which when combined form a particular mixture (as a medicine or alloy) ⟨the *ingredients* of a cocktail⟩.

: of, relating to, or dealing with the rudiments of something : ELEMENTARY ⟨taught *elemental* crafts to the children⟩ **d :** forming an integral part : INHERENT ⟨an *elemental* sense of rhythm⟩
2 : of, relating to, or resembling a great force of nature ⟨the rains come with *elemental* violence⟩ ⟨*elemental* passions⟩
— elemental *noun*
— el·e·men·tal·ly \-tᵊl-ē\ *adverb*

el·e·men·ta·ry \ˌe-lə-ˈmen-tə-rē, -ˈmen-trē\ *adjective* (14th century)
1 a : of, relating to, or dealing with the simplest elements or principles of something ⟨avoids the most *elementary* decision-making⟩ **b :** of or relating to an elementary school ⟨an *elementary* curriculum⟩
2 : ELEMENTAL 1a, 1b
3 : ELEMENTAL 2
— el·e·men·ta·ri·ly \-ˌmen-ˈter-ə-lē, -ˈmen-trə-lē\ *adverb*
— el·e·men·ta·ri·ness \-ˈmen-tə-rē-nəs, -ˈmen-trē-\ *noun*

elementary particle *noun* (1934)
: any of the particles of which matter and energy are composed or which mediate the fundamental forces of nature; *especially* : one whose existence has not been attributed to the combination of other more fundamental entities

elementary school *noun* (1841)
: a school including usually the first four to the first eight grades and often a kindergarten

el·e·mi \ˈe-lə-mē\ *noun* [New Latin *elimi*, probably from Arabic *al lāmi* the elemi] (1543)
: any of various fragrant oleoresins obtained from tropical trees (family Burseraceae) and used chiefly in varnishes, lacquers, and printing inks

elen·chus \i-ˈleŋ-kəs\ *noun, plural* **-chi** \-ˌkī, -(ˌ)kē\ [Latin, from Greek *elenchos*] (1663)
: REFUTATION; *especially* : one in syllogistic form

el·e·phant \ˈe-lə-fənt\ *noun, plural* **elephants** *also* **elephant** *often attributive* [Middle English, from Middle French & Latin; Middle French *olifant*, from Latin *elephantus*, from Greek *elephant-, elephas*] (14th century)
1 : any of a family (Elephantidae, the elephant family) of thickset usually extremely large nearly hairless herbivorous mammals that have a snout elongated into a muscular trunk and two incisors in the upper jaw developed especially in the male into large ivory tusks and that include two living forms and various extinct relatives: as **a :** a tall large-eared mammal (*Loxodonta africana*) of tropical Africa — called also *African elephant* **b :** a relatively small-eared mammal (*Elephas maximus*) of forests of southeastern Asia — called also *Asian elephant, Indian elephant*
2 : an animal or fossil related to the elephants
3 : one that is uncommonly large or hard to manage

elephant: ***1*** **African,** ***2*** **Asian**

elephant bird *noun* (circa 1889)
: AEPYORNIS

elephant grass *noun* (1832)
1 : an Old World cattail (*Typha elephantina*) used especially in making baskets
2 : NAPIER GRASS

el·e·phan·ti·a·sis \ˌe-lə-fən-ˈtī-ə-səs, -ˌfan-\ *noun, plural* **-a·ses** \-ˌsēz\ [New Latin, from Latin, a kind of leprosy, from Greek, from *elephant-, elephas*] (1581)
1 : enlargement and thickening of tissues; *specifically* : the enormous enlargement of a limb or the scrotum caused by obstruction of lymphatics by filarial worms (especially *Wuchereria bancrofti*)
2 : an undesirable usually enormous growth, enlargement, or overdevelopment ⟨*elephantiasis* of intellect and atrophy of emotion — Michael Lerner⟩

el·e·phan·tine \ˌe-lə-ˈfan-ˌtēn, -ˌtīn, ˈe-lə-fən-\ *adjective* (1610)
1 a : having enormous size or strength : MASSIVE **b :** CLUMSY, PONDEROUS ⟨*elephantine* verse⟩
2 : of or relating to an elephant

elephant seal *noun* (1841)
: either of two very large seals (genus *Mirounga* of the family Phocidae) characterized by a long inflatable proboscis: **a :** one (*M. angustirostris*) found in Pacific coastal waters from southeastern Alaska to Baja California **b :** one (*M. leonina*) found in coastal waters of subantarctic islands and Patagonia

El·eu·sin·i·an \ˌel-yu̇-ˈsi-nē-ən\ *adjective* (1611)
: of or relating to ancient Eleusis or to the religious mysteries celebrated there in worship of Demeter and Persephone

[1]el·e·vate \ˈe-lə-ˌvāt, -vət\ *adjective* (14th century)
archaic : ELEVATED

[2]el·e·vate \-ˌvāt\ *verb* **-vat·ed; -vat·ing** [Middle English, from Latin *elevatus*, past participle of *elevare*, from *e-* + *levare* to raise — more at LEVER] (15th century)
transitive verb
1 : to lift up : RAISE
2 : to raise in rank or status : EXALT
3 : to improve morally, intellectually, or culturally
4 : to raise the spirits of : ELATE
intransitive verb
: to become elevated : RISE ⟨his voice *elevated* to a shout⟩
synonym see LIFT

[1]el·e·vat·ed \-ˌvā-təd\ *adjective* (1553)
1 a : raised especially above the ground or other surface ⟨an *elevated* highway⟩ **b :** increased especially abnormally (as in degree or amount) ⟨*elevated* blood pressure⟩
2 a : being morally or intellectually on a high plane ⟨*elevated* conversation⟩ **b :** FORMAL, DIGNIFIED ⟨*elevated* diction⟩
3 : exhilarated in mood or feeling

[2]elevated *noun* (1881)
: [2]EL

el·e·va·tion \ˌe-lə-ˈvā-shən\ *noun* (14th century)
1 : the height to which something is elevated: as **a :** the angular distance of something (as a celestial object) above the horizon **b :** the degree to which a gun is aimed above the horizon **c :** the height above the level of the sea : ALTITUDE
2 : a ballet dancer's or a skater's leap and seeming suspension in the air; *also* : the ability to achieve an elevation
3 : an act or instance of elevating
4 : something that is elevated: as **a :** an elevated place **b :** a swelling especially on the skin
5 : the quality or state of being elevated
6 : a geometrical projection (as of a building) on a vertical plane
synonym see HEIGHT

el·e·va·tor \ˈe-lə-ˌvā-tər\ *noun* (1646)
1 : one that raises or lifts something up: as **a :** an endless belt or chain conveyor with cleats, scoops, or buckets for raising material **b :** a cage or platform and its hoisting machinery for conveying people or things to different levels **c :** GRAIN ELEVATOR
2 : a movable auxiliary airfoil usually attached to the tail plane of an airplane for controlling pitch — see AIRPLANE illustration

elevator music *noun* (1979)
: instrumental arrangements of popular songs often piped in (as to an elevator or retail store)

elev·en \i-ˈle-vən\ *noun* [Middle English *enleven*, from *enleven*, adjective, from Old English *endleofan*, from *end-* (alteration of *ān* one) + *-leofan;* akin to Old English *lēon* to lend — more at ONE, LOAN] (before 12th century)
1 — see NUMBER table
2 : the 11th in a set or series
3 : something having 11 units or members; *especially* : a football team
— eleven *adjective*
— eleven *pronoun, plural in construction*
— elev·enth \-vən(t)th\ *adjective or noun*

eleven–plus \i-ˌle-vən-ˈpləs\ *noun* (1955)
British : an examination taken by schoolchildren between the ages of 11 and 12 that determines the type of secondary education to which they are assigned

elev·ens·es \-vən-zəz\ *noun plural but sometimes singular in construction* [double plural of *eleven* (o'clock)] (circa 1819)
British : light refreshment (as a snack) taken in the middle of the morning

eleventh hour *noun* (1826)
: the latest possible time ⟨still making changes at the *eleventh hour*⟩

el·e·von \ˈe-lə-ˌvän\ *noun* [*elev*ator + ailer*on*] (1944)
: an airplane control surface that combines the functions of elevator and aileron

elf \ˈelf\ *noun, plural* **elves** \ˈelvz\ [Middle English, from Old English *ælf;* akin to Old Norse *alfr* elf & perhaps to Latin *albus* white — more at ALB] (before 12th century)
1 : a small often mischievous fairy
2 : a small lively creature; *also* : a usually lively mischievous or malicious person
— elf·ish \ˈel-fish\ *adjective*
— elf·ish·ly *adverb*

elf·in \ˈel-fən\ *adjective* [irregular from *elf*] (1596)
1 a : of, relating to, or produced by an elf **b :** resembling an elf especially in its tiny size ⟨*elfin* portions⟩
2 : having an otherworldly or magical quality or charm

elf·lock \ˈelf-ˌläk\ *noun* (1592)
: hair matted as if by elves — usually used in plural

elf owl *noun* (1887)
: a very small insectivorous owl (*Micrathene whitneyi*) of the Southwestern U.S. and Mexico that often roosts and nests in giant cacti

el-hi \(ˌ)el-ˈhī\ *adjective* [*el*ementary (school) + *hi*gh (school)] (circa 1948)
: of, relating to, or designed for use in grades 1 to 12

Eli \ˈē-ˌlī\ *noun* [Hebrew *ʽĒlī*]
: a judge and priest of Israel who according to the account in I Samuel was entrusted with the care of the boy Samuel

Eli·as \i-ˈlī-əs\ *noun* [Late Latin, from Greek *Ēlias*, from Hebrew *Ēlīyāh*]
: ELIJAH

elic·it \i-ˈli-sət\ *transitive verb* [Latin *elicitus*, past participle of *elicere*, from *e-* + *lacere* to allure] (1605)
1 : to draw forth or bring out (something latent or potential) ⟨hypnotism *elicited* his hidden fears⟩
2 : to call forth or draw out (as information or a response) ⟨her performance *elicited* wild applause⟩
synonym see EDUCE
— elic·i·ta·tion \i-ˌli-sə-ˈtā-shən, ˌē-\ *noun*
— elic·i·tor \i-ˈli-sə-tər\ *noun*

elide \i-ˈlīd\ *transitive verb* **elid·ed; elid·ing** [Latin *elidere* to strike out, from *e-* + *laedere* to injure by striking] (1796)

1 a : to suppress or alter (as a vowel or syllable) by elision **b :** to strike out (as a written word)
2 a : to leave out of consideration **:** OMIT **b :** CURTAIL, ABRIDGE

el·i·gi·ble \ˈe-lə-jə-bəl\ *adjective* [Middle English, from Middle French & Late Latin; Middle French, from Late Latin *eligibilis,* from Latin *eligere* to choose — more at ELECT] (15th century)
1 a : qualified to be chosen **:** ENTITLED ⟨*eligible* to retire⟩ **b :** permitted under football rules to catch a forward pass ⟨an *eligible* receiver⟩
2 : worthy of being chosen **:** DESIRABLE ⟨an *eligible* young bachelor⟩
— el·i·gi·bil·i·ty \ˌe-lə-jə-ˈbi-lə-tē\ *noun*
— eligible *noun*
— el·i·gi·bly \ˈe-lə-jə-blē\ *adverb*

Eli·jah \i-ˈlī-jə\ *noun* [Hebrew *Ēlīyāh*]
: a Hebrew prophet of the 9th century B.C. who according to the account in I Kings championed the worship of Jehovah as against Baal

elim·i·nate \i-ˈli-mə-ˌnāt\ *transitive verb* **-nat·ed; -nat·ing** [Latin *eliminatus,* past participle of *eliminare,* from *e-* + *limin-, limen* threshold] (1568)
1 a : to cast out or get rid of **:** REMOVE, ERADICATE ⟨the need to *eliminate* poverty⟩ **b :** to set aside as unimportant **:** IGNORE
2 : to expel (as waste) from the living body
3 : to cause (as an unknown) to disappear by combining two or more mathematical equations
— elim·i·na·tive \-ˈli-mə-ˌnā-tiv\ *adjective*
— elim·i·na·tor \-ˌnā-tər\ *noun*

elim·i·na·tion \i-ˌli-mə-ˈnā-shən\ *noun, often attributive* (1627)
: the act, process, or an instance of eliminating or discharging: as **a :** the act of discharging or excreting waste products from the body **b :** the removal from a molecule of the constituents of a simpler molecule ⟨ethylene is formed by the *elimination* of water from ethanol⟩ — compare ADDITION 4

ELISA \ē-ˈlī-sə, -zə\ *noun* (1978)
: ENZYME-LINKED IMMUNOSORBENT ASSAY

Eli·sha \i-ˈlī-shə\ *noun* [Hebrew *Ĕlīshāʻ*]
: a Hebrew prophet and disciple and successor of Elijah

eli·sion \i-ˈli-zhən\ *noun* [Late Latin *elision-, elisio,* from Latin *elidere*] (1581)
1 a : the use of a speech form that lacks a final or initial sound which a variant speech form has (as *'s* instead of *is* in *there's*) **b :** the omission of an unstressed vowel or syllable in a verse to achieve a uniform metrical pattern
2 : the act or an instance of omitting something **:** OMISSION

elite \ā-ˈlēt, i-, ē-\ *noun* [French *élite,* from Old French *eslite,* from feminine of *eslit,* past participle of *eslire* to choose, from Latin *eligere*] (1823)
1 a *singular or plural in construction* **:** the choice part **:** CREAM ⟨the *elite* of the entertainment world⟩ **b** *singular or plural in construction* **:** the best of a class ⟨superachievers who dominate the computer *elite* —Marilyn Chase⟩ **c** *singular or plural in construction* **:** the socially superior part of society ⟨how the *elite* live —*A P World*⟩ ⟨how the French-speaking *elite* . . . was changing —*Economist*⟩ **d :** a group of persons who by virtue of position or education exercise much power or influence ⟨members of the ruling *elite*⟩ ⟨the intellectual *elites* of the country⟩ **e :** a member of such an elite — usually used in plural ⟨the *elites* . . ., pursuing their studies in Europe —Robert Wernick⟩
2 : a typewriter type providing 12 characters to the linear inch
— elite *adjective*

élite, élitism *chiefly British variant of* ELITE, ELITISM

elit·ism \ā-ˈlē-ˌti-zəm, i-, ē-\ *noun* (1947)
1 : leadership or rule by an elite
2 : the selectivity of the elite; *especially* **:** SNOBBERY 1 ⟨*elitism* in choosing new members⟩
3 : consciousness of being or belonging to an elite
— elit·ist \-ˈlē-tist\ *noun or adjective*

elix·ir \i-ˈlik-sər\ *noun* [Middle English, from Medieval Latin, from Arabic *al-iksīr* the elixir, from *al* the + *iksīr* elixir, probably from Greek *xērion* desiccative powder, from *xēros* dry] (14th century)
1 a (1) **:** a substance held capable of changing base metals into gold (2) **:** a substance held capable of prolonging life indefinitely **b** (1) **:** CURE-ALL (2) **:** a medicinal concoction
2 : a sweetened liquid usually containing alcohol that is used in medication either for its medicinal ingredients or as a flavoring
3 : the essential principle

Eliz·a·be·than \i-ˌli-zə-ˈbē-thən\ *adjective* (1807)
: of, relating to, or characteristic of Elizabeth I of England or her reign
— Elizabethan *noun*

elk \ˈelk\ *noun, plural* **elks** [Middle English, probably from Old English *eolh;* akin to Old High German *elaho* elk, Greek *elaphos* deer] (before 12th century)
1 *plural usually* **elk a :** MOOSE 1 — used for one of the Old World **b :** a large gregarious deer (*Cervus elaphus*) of North America, Europe, Asia, and northwestern Africa — called also *red deer, wapiti* **c :** any of various large Asian deer
2 : soft tanned rugged leather
3 *capitalized* [Benevolent and Protective Order of *Elks*] **:** a member of a major benevolent and fraternal order

elk·hound \ˈelk-ˌhau̇nd, ˈel-ˌkau̇nd\ *noun* (1889)
: NORWEGIAN ELKHOUND

¹ell \ˈel\ *noun* [Middle English *eln,* from Old English; akin to Old High German *elina* ell, Latin *ulna* forearm, Greek *ōlenē* elbow, Sanskrit *aratni*] (before 12th century)
: a former English unit of length (as for cloth) equal to 45 inches (about 1.14 meters); *also* **:** any of various units of length used similarly

²ell *noun* [alteration of ¹*el*] (1773)
1 : an extension at right angles to the length of a building
2 : an elbow in a pipe or conduit

el·lag·ic acid \ə-ˈla-jik-, e-\ *noun* [French *ellagique,* from *ellag,* anagram of *galle* gall] (1810)
: a crystalline phenolic compound $C_{14}H_6O_8$ with two lactone groupings that is obtained especially from oak galls and some tannins and is used medicinally as a hemostatic

el·lipse \i-ˈlips, e-\ *noun* [Greek *elleipsis*] (circa 1753)
1 a : OVAL **b :** a closed plane curve generated by a point moving in such a way that the sums of its distances from two fixed points is a constant **:** a plane section of a right circular cone that is a closed curve
2 : ELLIPSIS

ellipse 1b: *F, F'* **foci;** *P, P', P''* **any point on the curve;** *FP + PF' = FP'' + P''F' = FP' + P'F'*

el·lip·sis \i-ˈlip-səs, e-\ *noun, plural* **el·lip·ses** \-ˌsēz\ [Latin, from Greek *elleipsis* ellipsis, ellipse, from *elleipein* to leave out, fall short, from *en* in + *leipein* to leave — more at IN, LOAN] (1540)
1 a : the omission of one or more words that are obviously understood but that must be supplied to make a construction grammatically complete **b :** a sudden leap from one topic to another
2 : marks or a mark (as . . . or *** or —) indicating an omission (as of words) or a pause

el·lip·soid \i-ˈlip-ˌsȯid, e-\ *noun* (1721)
: a surface all plane sections of which are ellipses or circles
— el·lip·soi·dal \i-ˌlip-ˈsȯi-dᵊl, (ˌ)e-\ *also* **ellipsoid** *adjective*

el·lip·ti·cal \i-ˈlip-ti-kəl, e-\ *or* **el·lip·tic** \-tik\ *adjective* [Greek *elleiptikos* defective, marked by ellipsis, from *elleipein*] (1656)
1 : of, relating to, or shaped like an ellipse
2 a : of, relating to, or marked by ellipsis or an ellipsis **b** (1) **:** of, relating to, or marked by extreme economy of speech or writing (2) **:** of or relating to deliberate obscurity (as of literary or conversational style)
— el·lip·ti·cal·ly \-ti-k(ə-)lē\ *adverb*

elliptical galaxy *noun* (1948)
: a galaxy that has a generally elliptical shape and that has no apparent internal structure or spiral arms — called also *elliptical;* compare SPIRAL GALAXY

el·lip·tic·i·ty \i-ˌlip-ˈti-sə-tē, (ˌ)e-\ *noun* (1753)
: deviation of an ellipse or a spheroid from the form of a circle or a sphere

elm \ˈelm\ *noun* [Middle English, from Old English; akin to Old High German *elme* elm, Latin *ulmus*] (before 12th century)
1 : any of a genus (*Ulmus* of the family Ulmaceae, the elm family) comprising large trees with alternate stipulate leaves and small apetalous flowers
2 : the wood of an elm

elm bark beetle *noun* (circa 1909)
: either of two beetles (family Scolytidae) that are vectors for the fungus causing Dutch elm disease: **a :** one (*Hylurgopinus rufipes*) native to eastern North America **b :** one (*Scolytus multistriatus*) introduced from Europe into eastern North America

elm leaf beetle *noun* (1881)
: a small orange-yellow black-striped Old World chrysomelid beetle (*Pyrrhalta luteola*) that in the larval and adult stage is a leaf-eating pest of elms in eastern North America

El Ni·ño \el-ˈnē-nyō\ *noun, plural* **El Niños** [Spanish, literally, the child (i.e., the Christ child); from the appearance of the flow at the Christmas season] (1925)
: an irregularly occurring flow of unusually warm surface water along the western coast of South America that is accompanied by abnormally high rainfall in usually arid areas and that prevents upwelling of nutrient-rich cold deep water causing a decline in the regional fish population

el·o·cu·tion \ˌe-lə-ˈkyü-shən\ *noun* [Middle English *elocucioun,* from Latin *elocution-, elocutio,* from *eloqui*] (15th century)
1 : a style of speaking especially in public
2 : the art of effective public speaking
— el·o·cu·tion·ary \-shə-ˌner-ē\ *adjective*
— el·o·cu·tion·ist \-sh(ə-)nist\ *noun*

elo·dea \i-ˈlō-dē-ə\ *noun* [New Latin, genus name, from Greek *helōdēs* marshy, from *helos* marsh; akin to Sanskrit *saras* pond] (circa 1868)
: any of a small American genus (*Elodea*) of submerged aquatic monocotyledonous herbs

eloign \i-ˈlȯin\ *transitive verb* [Middle English *eloynen,* from Middle French *esloigner,* from Old French, from *es-* ex- (from Latin *ex-*) + *loing* (adverb) far, from Latin *longe,* from *longus* long] (15th century)
1 *archaic* **:** to take (oneself) far away
2 *archaic* **:** to remove to a distant or unknown place **:** CONCEAL

¹elon·gate \i-ˈlȯŋ-ˌgāt, (ˌ)ē-, ˈē-ˌ\ *verb* **-gat·ed; -gat·ing** [Late Latin *elongatus,* past participle of *elongare,* to withdraw, from Latin *e-* + *longus*] (1578)
transitive verb

: to extend the length of
intransitive verb
: to grow in length
²elongate *or* **elon·gat·ed** *adjective* (1751)
1 : stretched out
2 : SLENDER
elon·ga·tion \(ˌ)ē-ˌlȯŋ-ˈgā-shən\ *noun* (14th century)
1 : the angular distance of a celestial body from another around which it revolves or from a particular point in the sky
2 a : the state of being elongated or lengthened; *also* **:** the process of growing or increasing in length **b :** something that is elongated
elope \i-ˈlōp\ *intransitive verb* **eloped; elop·ing** [Anglo-French *aloper*] (1628)
1 : to slip away **:** ESCAPE
2 a : to run away from one's husband with a lover **b :** to run away secretly with the intention of getting married usually without parental consent
— elope·ment \-ˈlōp-mənt\ *noun*
— elop·er *noun*
el·o·quence \ˈe-lə-kwən(t)s\ *noun* (14th century)
1 : discourse marked by force and persuasiveness; *also* **:** the art or power of using such discourse
2 : the quality of forceful or persuasive expressiveness
el·o·quent \-kwənt\ *adjective* [Middle English, from Middle French, from Latin *eloquent-, eloquens,* from present participle of *eloqui* to speak out, from *e-* + *loqui* to speak] (14th century)
1 : marked by forceful and fluent expression ⟨an *eloquent* preacher⟩
2 : vividly or movingly expressive or revealing ⟨an *eloquent* monument⟩
— el·o·quent·ly *adverb*
¹else \ˈel(t)s\ *adverb* [Middle English *elles,* from Old English; akin to Latin *alius* other, *alter* other of two, Greek *allos* other] (before 12th century)
1 a : in a different manner or place or at a different time ⟨how *else* could he have acted⟩ ⟨here and nowhere *else*⟩ **b :** in an additional manner or place or at an additional time ⟨where *else* is gold found⟩
2 : if not **:** OTHERWISE ⟨leave or *else* you'll be sorry⟩ — used absolutely to express a threat ⟨do what I tell you or *else*⟩
²else *adjective* (before 12th century)
: OTHER: **a :** being different in identity ⟨it must have been somebody *else*⟩ **b :** being in addition ⟨what *else* did he say⟩
else·where \-ˌ(h)wer, -ˌ(h)war\ *adverb* [Middle English *elleswher,* from Old English *elles hwær*] (before 12th century)
: in or to another place ⟨took my business *elsewhere*⟩
el·u·ant *or* **el·u·ent** \ˈel-yə-wənt\ *noun* [Latin *eluent-, eluens,* present participle of *eluere*] (1941)
: a solvent used in eluting
el·u·ate \ˈel-yə-wət, -ˌwāt\ *noun* [Latin *eluere* + English ¹*-ate*] (1932)
: the washings obtained by eluting
elu·ci·date \i-ˈlü-sə-ˌdāt\ *verb* **-dat·ed; -dat·ing** [Late Latin *elucidatus,* past participle of *elucidare,* from Latin *e-* + *lucidus* lucid] (circa 1568)
transitive verb
: to make lucid especially by explanation or analysis
intransitive verb
: to give a clarifying explanation
synonym see EXPLAIN
— elu·ci·da·tion \-ˌlü-sə-ˈdā-shən\ *noun*
— elu·ci·da·tive \-ˈlü-sə-ˌdā-tiv\ *adjective*
— elu·ci·da·tor \-ˌdā-tər\ *noun*
elu·cu·brate \i-ˈlü-k(y)ə-ˌbrāt\ *transitive verb* **-brat·ed; -brat·ing** [Latin *elucubratus,* past participle of *elucubrare* to work on far into the night, from *e-* + *lucubrare* to work by lamplight — more at LUCUBRATION] (circa 1623)
: to work out or express by studious effort
— elu·cu·bra·tion \-ˌlü-k(y)ə-ˈbrā-shən\ *noun*
elude \ē-ˈlüd\ *transitive verb* **elud·ed; elud·ing** [Latin *eludere,* from *e-* + *ludere* to play — more at LUDICROUS] (1667)
1 : to avoid adroitly **:** EVADE ⟨the mice *eluded* the traps⟩ ⟨managed to *elude* capture⟩
2 : to escape the perception, understanding, or grasp of ⟨subtlety simply *eludes* them⟩ ⟨victory continued to *elude* us⟩
3 : DEFY 4 ⟨it *eludes* explanation⟩
synonym see ESCAPE
Elul \e-ˈlül\ *noun* [Hebrew *Ĕlūl*] (1535)
: the 12th month of the civil year or the 6th month of the ecclesiastical year in the Jewish calendar — see MONTH table
elu·sion \ē-ˈlü-zhən\ *noun* [Medieval Latin *elusion-, elusio,* from Late Latin, deception, from Latin *eludere*] (1617)
: an act of eluding
elu·sive \ē-ˈlü-siv, -ˈlü-ziv\ *adjective* (1719)
: tending to elude: as **a :** tending to evade grasp or pursuit **b :** hard to comprehend or define ⟨an *elusive* concept⟩ **c :** hard to isolate or identify ⟨a haunting *elusive* aroma⟩
— elu·sive·ly *adverb*
— elu·sive·ness *noun*
elute \ē-ˈlüt\ *transitive verb* **elut·ed; elut·ing** [Latin *elutus,* past participle of *eluere* to wash out, from *e-* + *lavere* to wash — more at LYE] (1731)
: EXTRACT; *specifically* **:** to remove (adsorbed material) from an adsorbent by means of a solvent
— elu·tion \-ˈlü-shən\ *noun*
elu·tri·ate \ē-ˈlü-trē-ˌāt\ *transitive verb* **-at·ed; -at·ing** [Latin *elutriatus,* past participle of *elutriare* to put in a vat, perhaps from (assumed) *elutrum* vat, from Greek *elytron* reservoir, literally, covering] (circa 1727)
: to purify, separate, or remove by washing
— elu·tri·a·tion \ē-ˌlü-trē-ˈā-shən\ *noun*
— elu·tri·a·tor \ē-ˈlü-trē-ˌā-tər\ *noun*
elu·vi·a·tion \(ˌ)ē-ˌlü-vē-ˈā-shən\ *noun* [*eluvial* of eluviation (from *e-* + *-luvial*—as in *alluvial*) + *-ation*] (1899)
: the transportation of dissolved or suspended material within the soil by the movement of water when rainfall exceeds evaporation
— elu·vi·al \ē-ˈlü-vē-əl\ *adjective*
— elu·vi·at·ed \-ˈlü-vē-ā-təd\ *adjective*
el·ver \ˈel-vər\ *noun* [alteration of *eelfare* (migration of eels)] (circa 1640)
: a young eel
elves *plural of* ELF
el·vish \ˈel-vish\ *adjective* (13th century)
1 : of or relating to elves
2 : MISCHIEVOUS
ely·sian \i-ˈli-zhən\ *adjective, often capitalized* (1579)
1 : of or relating to Elysium
2 : BLISSFUL, DELIGHTFUL
elysian fields *noun plural, often E capitalized* (1579)
: ELYSIUM
Ely·si·um \i-ˈli-zhē-əm, -zē-\ *noun, plural* **-si·ums** *or* **-sia** \-zhē-ə, -zē-\ [Latin, from Greek *Ēlysion*]
1 : the abode of the blessed after death in classical mythology
2 : PARADISE 2
el·y·tron \ˈe-lə-ˌträn\ *noun, plural* **-tra** \-trə\ [New Latin, from Greek, sheath, wing cover, from *eilyein* to roll, wrap — more at VOLUBLE] (1774)
: one of the anterior wings in beetles and some other insects that serve to protect the posterior pair of functional wings
em \ˈem\ *noun* (13th century)
1 : the letter *m*
2 : the width of a piece of type about as wide as it is tall used as a unit of measure of typeset matter
em- — see EN-
'em \əm; *after* p,b,f, *or* v *often* ᵊm\ *pronoun* [Middle English *hem,* from Old English *heom, him,* dative plural of *hē* he] (before 12th century)
: THEM
ema·ci·ate \i-ˈmā-shē-ˌāt\ *verb* **-at·ed; -at·ing** [Latin *emaciatus,* past participle of *emaciare,* from *e-* + *macies* leanness, from *macer* lean — more at MEAGER] (1646)
intransitive verb
: to waste away physically
transitive verb
1 : to cause to lose flesh so as to become very thin
2 : to make feeble
— ema·ci·a·tion \-ˌmā-s(h)ē-ˈā-shən\ *noun*
E–mail \ˈē-ˌmāl\ *noun* (1982)
: ELECTRONIC MAIL
emalangeni *plural of* LILANGENI
em·a·nate \ˈe-mə-ˌnāt\ *verb* **-nat·ed; -nat·ing** [Latin *emanatus,* past participle of *emanare,* from *e-* + *manare* to flow] (1756)
intransitive verb
: to come out from a source
transitive verb
: EMIT
synonym see SPRING
em·a·na·tion \ˌe-mə-ˈnā-shən\ *noun* (1570)
1 a : the action of emanating **b :** the origination of the world by a series of hierarchically descending radiations from the Godhead through intermediate stages to matter
2 a : something that emanates or is produced by emanation **:** EFFLUENCE **b :** an isotope of radon produced by radioactive disintegration ⟨radium *emanation*⟩
— em·a·na·tive \ˈe-mə-ˌnā-tiv\ *adjective*
eman·ci·pate \i-ˈman(t)-sə-ˌpāt\ *transitive verb* **-pat·ed; -pat·ing** [Latin *emancipatus,* past participle of *emancipare,* from *e-* + *mancipare* to transfer ownership of, from *mancip-, manceps* contractor, from *manus* hand + *capere* to take — more at MANUAL, HEAVE] (1613)
1 : to free from restraint, control, or the power of another; *especially* **:** to free from bondage
2 : to release from paternal care and responsibility and make sui juris
3 : to free from any controlling influence (as traditional mores or beliefs)
synonym see FREE
— eman·ci·pa·tor \-ˌpā-tər\ *noun*
eman·ci·pa·tion \i-ˌman(t)-sə-ˈpā-shən\ *noun* (1631)
: the act or process of emancipating
— eman·ci·pa·tion·ist \-sh(ə-)nist\ *noun*
emar·gin·ate \(ˌ)ē-ˈmär-jə-nət, -ˌnāt\ *adjective* [Latin *emarginatus,* past participle of *emarginare* to deprive of a margin, from *e-* + *margin-, margo* margin] (1794)
: having the margin notched
— emar·gi·na·tion \(ˌ)ē-ˌmär-jə-ˈnā-shən\ *noun*
emas·cu·late \i-ˈmas-kyə-ˌlāt\ *transitive verb* **-lat·ed; -lat·ing** [Latin *emasculatus,* past participle of *emasculare,* from *e-* + *masculus* male — more at MALE] (1607)
1 : to deprive of strength, vigor, or spirit **:** WEAKEN
2 : to deprive of virility or procreative power **:** CASTRATE
3 : to remove the androecium of (a flower) in the process of artificial cross-pollination
synonym see UNNERVE
— emas·cu·late \-lət\ *adjective*
— emas·cu·la·tion \-ˌmas-kyə-ˈlā-shən\ *noun*
— emas·cu·la·tor \-ˈmas-kyə-ˌlā-tər\ *noun*
em·balm \im-ˈbä(l)m, *New England also* -ˈbȧm\ *transitive verb* [Middle English *embaumen,* from Middle French *embaumer,* from Old French *embasmer,* from *en-* + *basme* balm — more at BALM] (14th century)
1 : to treat (a dead body) so as to protect from decay
2 : to fill with sweet odors **:** PERFUME

3 : to protect from decay or oblivion : PRESERVE
4 : to fix in a static condition
— **em·balm·er** *noun*
— **em·balm·ment** \-ˈbä(l)m-mənt, -ˈbȧm-\ *noun*

em·bank \im-ˈbaŋk\ *transitive verb* (1576)
: to enclose or confine by an embankment

em·bank·ment \-mənt\ *noun* (1786)
1 : a raised structure to hold back water or to carry a roadway
2 : the action of embanking

em·bar·ca·de·ro \(ˌ)em-ˌbär-kə-ˈder-(ˌ)ō\ *noun, plural* **-ros** [Spanish, from *embarcado,* past participle of *embarcar* to embark, from *em-* (from Latin *in-*) + *barca* bark, from Late Latin] (1846)
West : a landing place especially on an inland waterway

¹em·bar·go \im-ˈbär-(ˌ)gō\ *noun, plural* **-goes** [Spanish, from *embargar* to bar, from (assumed) Vulgar Latin *imbarricare,* from Latin *in-* + (assumed) Vulgar Latin *barra* bar] (1593)
1 : an order of a government prohibiting the departure of commercial ships from its ports
2 : a legal prohibition on commerce ⟨an *embargo* on arms shipments⟩
3 : STOPPAGE, IMPEDIMENT; *especially* : PROHIBITION ⟨I lay no *embargo* on anybody's words —Jane Austen⟩
4 : an order by a common carrier or public regulatory agency prohibiting or restricting freight transportation

²embargo *transitive verb* **-goed; -go·ing** (1755)
: to place an embargo on

em·bark \im-ˈbärk\ *verb* [Middle French *embarquer,* from Old Provençal *embarcar,* from *em-* (from Latin *in-*) + *barca* bark] (1533)
intransitive verb
1 : to go on board a vehicle for transportation
2 : to make a start ⟨*embarked* on a new career⟩
transitive verb
1 : to cause to go on board (as a boat or airplane)
2 : to engage, enlist, or invest in an enterprise
— **em·bar·ka·tion** \ˌem-ˌbär-ˈkā-shən, -bər-\ *noun*
— **em·bark·ment** \im-ˈbärk-mənt\ *noun*

em·bar·rass \im-ˈbar-əs\ *transitive verb* [French *embarrasser,* from Spanish *embarazar,* from Portuguese *embaraçar,* from *em-* (from Latin *in-*) + *baraça* noose] (1672)
1 a : to place in doubt, perplexity, or difficulties **b** : to involve in financial difficulties **c** : to cause to experience a state of self-conscious distress ⟨bawdy stories *embarrassed* him⟩
2 a : to hamper the movement of **b** : HINDER, IMPEDE
3 : to make intricate : COMPLICATE
4 : to impair the activity of (a bodily function) or the function of (a bodily part) ⟨digestion *embarrassed* by overeating⟩ ☆
— **em·bar·rass·able** \-sə-bəl\ *adjective*

em·bar·rassed·ly \-əst-lē, -ə-səd-lē\ *adverb* (1883)
: with embarrassment

em·bar·rass·ing·ly \-ə-siŋ-lē\ *adverb* (circa 1864)
: to an embarrassing degree : so as to cause embarrassment

em·bar·rass·ment \im-ˈbar-ə-smənt\ *noun* (1729)
1 a : something that embarrasses : IMPEDIMENT **b** : an excessive quantity from which to select — used especially in the phrase *embarrassment of riches*
2 : the state of being embarrassed: as **a** : confusion or disturbance of mind **b** : difficulty arising from the want of money to pay debts **c** : difficulty in functioning as a result of disease ⟨cardiac *embarrassment*⟩

em·bas·sage \ˈem-bə-sij\ *noun* (1526)
1 : the message or commission entrusted to an ambassador
2 *archaic* : EMBASSY

em·bas·sy \ˈem-bə-sē\ *noun, plural* **-sies** [Middle French *ambassee,* ultimately of Germanic origin; akin to Old High German *ambaht* service] (1534)
1 : a body of diplomatic representatives; *specifically* : one headed by an ambassador
2 a : the function or position of an ambassador **b** : a mission abroad undertaken officially especially by an ambassador
3 : EMBASSAGE 1
4 : the official residence and offices of an ambassador

em·bat·tle \im-ˈba-tᵊl\ *transitive verb* **em·bat·tled; em·bat·tling** \-ˈbat-liŋ, -tᵊl-iŋ\ [Middle English *embatailen,* from Middle French *embatailler,* from *en-* + *batailler* to battle] (14th century)
1 : to arrange in order of battle : prepare for battle
2 : FORTIFY

em·bat·tled *adjective* (15th century)
1 a : ready to fight : prepared to give battle ⟨here once the *embattled* farmers stood —R. W. Emerson⟩ **b** : engaged in battle, conflict, or controversy ⟨an *embattled* official accused of extortion⟩
2 a : being a site of battle, conflict, or controversy ⟨the *embattled* capital⟩ **b** : characterized by conflict or controversy ⟨his . . . often *embattled* experience as an educator —Nat Hentoff⟩

em·bat·tle·ment \-ˈba-tᵊl-mənt\ *noun* (15th century)
1 : BATTLEMENT
2 : the state of being embattled

em·bay \im-ˈbā\ *transitive verb* (1600)
: to trap or catch in or as if in a bay ⟨an *embayed* sailing ship⟩

em·bay·ment \-ˈbā-mənt\ *noun* (1815)
1 : formation of a bay
2 : a bay or a conformation resembling a bay

Emb·den \ˈem-dən\ *noun* [*Emden,* Germany] (1903)
: a breed of large white domestic geese with an orange bill and deep orange shanks and toes

em·bed \im-ˈbed\ *verb* **em·bed·ded; em·bed·ding** (circa 1794)
transitive verb
1 a : to enclose closely in or as if in a matrix ⟨fossils *embedded* in stone⟩ **b** : to make something an integral part of ⟨the prejudices *embedded* in our language⟩ **c** : to prepare (a microscopy specimen) for sectioning by infiltrating with and enclosing in a supporting substance
2 : to surround closely ⟨a sweet pulp *embeds* the plum seed⟩
intransitive verb
: to become embedded
— **em·bed·ment** \-ˈbed-mənt\ *noun*

em·bed·ded \im-ˈbe-dəd\ *adjective* (1961)
: occurring as a grammatical constituent (as a verb phrase or clause) within a like constituent
— **em·bed·ding** \-diŋ\ *noun*

em·bel·lish \im-ˈbe-lish\ *transitive verb* [Middle English, from Middle French *embeliss-,* stem of *embelir,* from *en-* + *bel* beautiful — more at BEAUTY] (14th century)
1 : to make beautiful with ornamentation : DECORATE
2 : to heighten the attractiveness of by adding ornamental details : ENHANCE ⟨*embellished* our account of the trip⟩
synonym see ADORN
— **em·bel·lish·er** *noun*

em·bel·lish·ment \-lish-mənt\ *noun* (1591)
1 : the act or process of embellishing
2 : something serving to embellish
3 : ORNAMENT 5

em·ber \ˈem-bər\ *noun* [Middle English *eymere,* from Old Norse *eimyrja;* akin to Old English *ǣmerge* ashes, Latin *urere* to burn] (14th century)
1 : a glowing fragment (as of coal) from a fire; *especially* : one smoldering in ashes
2 *plural* : the smoldering remains of a fire
3 *plural* : slowly dying or fading emotions, memories, ideas, or responses still capable of being revived

ember day \ˈem-bər-\ *noun* [Middle English, from Old English *ymbrendæg,* from *ymbrene* circuit, anniversary + *dæg* day] (before 12th century)
: a Wednesday, Friday, or Saturday following the first Sunday in Lent, Whitsunday, September 14, or December 13 set apart for fasting and prayer in Western churches

em·bez·zle \im-ˈbe-zəl\ *transitive verb* **em·bez·zled; em·bez·zling** \-(ə-)liŋ\ [Middle English *embesilen,* from Anglo-French *embeseiller,* from Middle French *en-* + *besillier* to destroy] (15th century)
: to appropriate (as property entrusted to one's care) fraudulently to one's own use
— **em·bez·zle·ment** \-zəl-mənt\ *noun*
— **em·bez·zler** \-z(ə-)lər\ *noun*

em·bit·ter \im-ˈbi-tər\ *transitive verb* (15th century)
1 : to excite bitter feelings in
2 : to make bitter
— **em·bit·ter·ment** \-mənt\ *noun*

¹em·blaze \im-ˈblāz\ *transitive verb* **em·blazed; em·blaz·ing** (15th century)
1 : to illuminate especially by a blaze
2 : to set ablaze

²emblaze *transitive verb* **em·blazed; em·blaz·ing** [*en-* + *blaze* (to blazon)] (1593)
1 *archaic* : EMBLAZON 1
2 : to adorn sumptuously ⟨with gems and golden luster rich *emblazed* —John Milton⟩

em·bla·zon \im-ˈblā-zᵊn\ *transitive verb* **em·bla·zoned; em·bla·zon·ing** \-ˈblāz-niŋ, -ˈblā-zᵊn-iŋ\ (1589)
1 a : to inscribe or adorn with or as if with heraldic bearings or devices **b** : to inscribe (as heraldic bearings) on a surface
2 : CELEBRATE, EXTOL ⟨have his . . . deeds *emblazoned* by a poet —Thomas Nash⟩
— **em·bla·zon·er** \-ˈblāz-nər, -zᵊn-ər\ *noun*
— **em·bla·zon·ment** \-ˈblā-zᵊn-mənt\ *noun*

em·bla·zon·ry \-zᵊn-rē\ *noun* (1667)
1 : emblazoned figures : brilliant decoration
2 : the act or art of emblazoning

¹em·blem \ˈem-bləm\ *noun* [Middle English, from Latin *emblema* inlaid work, from Greek *emblēmat-, emblēma,* from *emballein* to insert, from *en-* + *ballein* to throw — more at DEVIL] (15th century)

☆ SYNONYMS
Embarrass, discomfit, abash, disconcert, rattle mean to distress by confusing or confounding. EMBARRASS implies some influence that impedes thought, speech, or action ⟨*embarrassed* to admit that she liked the movie⟩. DISCOMFIT implies a hampering or frustrating accompanied by confusion ⟨hecklers *discomfited* the speaker⟩. ABASH presupposes some initial self-confidence that receives a sudden check, producing shyness, shame, or a feeling of inferiority ⟨*abashed* by her swift and cutting retort⟩. DISCONCERT implies an upsetting of equanimity or assurance producing uncertainty or hesitancy ⟨*disconcerted* by finding so many in attendance⟩. RATTLE implies an agitation that impairs thought and judgment ⟨*rattled* by all the television cameras⟩.

1 : a picture with a motto or set of verses intended as a moral lesson
2 : an object or the figure of an object symbolizing and suggesting another object or an idea
3 a : a symbolic object used as a heraldic device **b :** a device, symbol, or figure adopted and used as an identifying mark
²emblem *transitive verb* (1584)
: EMBLEMATIZE
em·blem·at·ic \ˌem-blə-ˈma-tik\ *also* **em·blem·at·i·cal** \-ti-kəl\ *adjective* (1645)
: of, relating to, or constituting an emblem **:** SYMBOLIC, REPRESENTATIVE
— em·blem·at·i·cal·ly \-ti-k(ə-)lē\ *adverb*
em·blem·a·tize \em-ˈble-mə-ˌtīz\ *transitive verb* **-tized; -tiz·ing** (1615)
: to represent by or as if by an emblem **:** SYMBOLIZE
em·ble·ments \ˈem-blə-mən(t)s\ *noun plural* [Middle English *emblayment,* from Middle French *emblaement,* from *emblaer* to sow with grain, from *en-*+ *blee* grain, of Germanic origin; akin to Old English *blǣd* fruit, growth, leaf — more at BLADE] (15th century)
: crops from annual cultivation legally belonging to the tenant
em·bodi·ment \im-ˈbä-di-mənt\ *noun* (1828)
1 : one that embodies something ⟨the *embodiment* of all our hopes⟩
2 : the act of embodying **:** the state of being embodied
em·body \im-ˈbä-dē\ *transitive verb* **em·bod·ied; em·body·ing** (circa 1548)
1 : to give a body to (a spirit) **:** INCARNATE
2 a : to deprive of spirituality **b :** to make concrete and perceptible
3 : to cause to become a body or part of a body **:** INCORPORATE
4 : to represent in human or animal form **:** PERSONIFY ⟨men who greatly *embodied* the idealism of American life —A. M. Schlesinger (born 1917)⟩
— em·bodi·er *noun*
em·bold·en \im-ˈbōl-dən\ *transitive verb* (15th century)
: to instill with boldness or courage
em·bo·lec·to·my \ˌem-bə-ˈlek-tə-mē\ *noun, plural* **-mies** (1923)
: surgical removal of an embolus
em·bol·ic \em-ˈbä-lik, im-\ *adjective* (1866)
: of or relating to an embolus or embolism
em·bo·lism \ˈem-bə-ˌli-zəm\ *noun* [Middle English *embolisme,* from Medieval Latin *embolismus,* from Greek *embol-* (from *emballein* to insert, intercalate) — more at EMBLEM] (14th century)
1 : the insertion of one or more days in a calendar **:** INTERCALATION
2 a : the sudden obstruction of a blood vessel by an embolus **b :** EMBOLUS
— em·bo·lis·mic \ˌem-bə-ˈliz-mik\ *adjective*
em·bo·li·za·tion \ˌem-bə-lə-ˈzā-shən\ *noun* (1942)
: the process or state in which a blood vessel or organ is obstructed by the lodgment of a material mass (as an embolus)
em·bo·lus \ˈem-bə-ləs\ *noun, plural* **-li** \-ˌlī\ [New Latin, from Greek *embolos* wedge-shaped object, stopper, from *emballein*] (1859)
: an abnormal particle (as an air bubble) circulating in the blood — compare THROMBUS
em·bon·point \äⁿ-bōⁿ-pwaⁿ\ *noun* [French, from Middle French, from *en bon point* in good condition] (1670)
: plumpness of person **:** STOUTNESS
em·bos·om \im-ˈbu̇-zəm *also* -ˈbü-\ *transitive verb* (circa 1590)
1 *archaic* **:** to take into or place in the bosom
2 : to shelter closely **:** ENCLOSE ⟨his house *embosomed* in the grove —Alexander Pope⟩
¹em·boss \im-ˈbäs, -ˈbȯs\ *transitive verb* [Middle English *embosen* to become exhausted from being hunted, ultimately from Middle French *bois* woods] (14th century)
archaic **:** to drive (as a hunted animal) to bay or to exhaustion
²emboss *transitive verb* [Middle English *embosen,* from Middle French *embocer,* from *en-* + *boce* boss] (15th century)
1 : to raise the surface of into bosses; *especially* **:** to ornament with raised work
2 : to raise in relief from a surface
3 : ADORN, EMBELLISH
— em·boss·able \-ˈbä-sə-bəl, -ˈbȯ-\ *adjective*
— em·boss·er \-sər\ *noun*
— em·boss·ment \-mənt\ *noun*
em·bou·chure \ˈäm-bu̇-ˌshu̇r, ˌäm-bu̇-ˈ\ *noun* [French, from *(s')emboucher* to flow into, from *en-* + *bouche* mouth — more at DEBOUCH] (1760)
1 : the position and use of the lips, tongue, and teeth in playing a wind instrument
2 : the mouthpiece of a musical instrument
em·bour·geoise·ment \em-ˈbu̇rzh-ˌwäz-mənt, äm-; äⁿ-bu̇rzh-wáz-mäⁿ\ *noun* [French, from *embourgeoiser* to make bourgeois, from *em-* + *bourgeois*] (1937)
: a shift to bourgeois values and practices
em·bowed \im-ˈbōd\ *adjective* (15th century)
: bent like a bow **:** ARCHED
em·bow·el \im-ˈbau̇(-ə)l\ *transitive verb* **-eled** *or* **-elled; -el·ing** *or* **-el·ling** (1521)
1 : DISEMBOWEL
2 *obsolete* **:** ENCLOSE
em·bow·er \im-ˈbau̇(-ə)r\ *transitive verb* (1580)
: to shelter or enclose in or as if in a bower ⟨like a rose *embowered* in its own green leaves —P. B. Shelley⟩
¹em·brace \im-ˈbrās\ *verb* **em·braced; em·brac·ing** [Middle English, from Middle French *embracer,* from Old French *embracier,* from *en-* + *brace* two arms — more at BRACE] (14th century)
transitive verb
1 a : to clasp in the arms **:** HUG **b :** CHERISH, LOVE
2 : ENCIRCLE, ENCLOSE
3 a : to take up especially readily or gladly ⟨*embrace* a cause⟩ **b :** to avail oneself of **:** WELCOME ⟨*embraced* the opportunity to study further⟩
4 a : to take in or include as a part, item, or element of a more inclusive whole ⟨charity *embraces* all acts that contribute to human welfare⟩ **b :** to be equal or equivalent to ⟨his assets *embraced* $10⟩
intransitive verb
: to participate in an embrace
synonym see ADOPT, INCLUDE
— em·brace·able \-ˈbrā-sə-bəl\ *adjective*
— em·brace·ment \-ˈbrā-smənt\ *noun*
— em·brac·er *noun*
— em·brac·ing·ly \-ˈbrā-siŋ-lē\ *adverb*
²embrace *noun* (1592)
1 : a close encircling with the arms and pressure to the bosom especially as a sign of affection **:** HUG
2 : GRIP, ENCIRCLEMENT ⟨in the *embrace* of terror⟩
3 : ACCEPTANCE ⟨her *embrace* of new ideas⟩
em·bra·ceor \im-ˈbrā-sər\ *noun* [Anglo-French, from Middle French *embraseor* instigator, from *embraser* to set on fire, from *en-* + *brase, brese* live coals] (15th century)
: one guilty of embracery
em·brac·ery \im-ˈbrā-sə-rē\ *noun, plural* **-er·ies** [Middle English, from Anglo-French *embraceor*] (15th century)
: an attempt to influence a jury corruptly
em·brac·ive \-ˈbrā-siv\ *adjective* (1855)
1 : disposed to embrace
2 : INCLUSIVE, COMPREHENSIVE
em·bran·gle \im-ˈbraŋ-gəl\ *transitive verb* **-gled; -gling** \-g(ə-)liŋ\ [*en-* + *brangle* (squabble)] (1664)
: EMBROIL
— em·bran·gle·ment \-gəl-mənt\ *noun*
em·bra·sure \im-ˈbrā-zhər\ *noun* [French, from obsolete *embraser* to widen an opening] (1702)
1 : an opening with sides flaring outward in a wall or parapet of a fortification usually for allowing the firing of cannon
2 : a recess of a door or window

E embrasure 2

em·brit·tle \im-ˈbri-tᵊl\ *verb* **-brit·tled; -brit·tling** \-ˈbrit-liŋ, -tᵊl-iŋ\ (1902)
transitive verb
: to make brittle
intransitive verb
: to become brittle
— em·brit·tle·ment \-ˈbri-tᵊl-mənt\ *noun*
em·bro·ca·tion \ˌem-brə-ˈkā-shən\ *noun* [Middle English *embrocacioun,* from Middle French *embrocacion,* from Medieval Latin *embrocation-, embrocatio,* from Late Latin *embrocare* to rub with lotion, from Greek *embroche* lotion, from *en-* + *brechein* to wet] (15th century)
: LINIMENT
em·broi·der \im-ˈbrȯi-dər\ *verb* **em·broi·dered; em·broi·der·ing** \-d(ə-)riŋ\ [Middle English *embroderen,* from Middle French *embroder,* from *en-* + *broder* to embroider, of Germanic origin; akin to Old English *brord* point, *byrst* bristle] (14th century)
transitive verb
1 a : to ornament with needlework **b :** to form with needlework
2 : to elaborate on **:** EMBELLISH
intransitive verb
1 : to make embroidery
2 : to provide embellishments **:** ELABORATE
— em·broi·der·er \-ˈbrȯi-dər-ər\ *noun*
em·broi·dery \im-ˈbrȯi-d(ə-)rē\ *noun, plural* **-der·ies** (14th century)
1 a : the art or process of forming decorative designs with hand or machine needlework **b :** a design or decoration formed by or as if by embroidery **c :** an object decorated with embroidery
2 : elaboration by use of decorative and often fictitious detail
3 : something pleasing or desirable but unimportant ⟨considered the humanities mere educational *embroidery*⟩
em·broil \im-ˈbrȯi(ə)l\ *transitive verb* [French *embrouiller,* from Middle French, from *en-* + *brouiller* to broil] (1603)
1 : to throw into disorder or confusion
2 : to involve in conflict or difficulties
— em·broil·ment \-mənt\ *noun*
em·brown \im-ˈbrau̇n\ *transitive verb* (1667)
1 : DARKEN
2 : to cause to turn brown
embrue *variant of* IMBRUE
embry- *or* **embryo-** *combining form* [Late Latin, from Greek, from *embryon*]
: embryo ⟨*embryo*geny⟩
em·bryo \ˈem-brē-ˌō\ *noun, plural* **em·bry·os** [Medieval Latin *embryon-, embryo,* from Greek *embryon,* from *en-* + *bryein* to swell; akin to Greek *bryon* catkin] (1548)
1 a *archaic* **:** a vertebrate at any stage of development prior to birth or hatching **b :** an animal in the early stages of growth and differentiation that are characterized by cleavage, the laying down of fundamental tissues, and the formation of primitive organs and organ systems; *especially* **:** the developing human individual from the time of implantation to the end of the eighth week after conception
2 : the young sporophyte of a seed plant usually comprising a rudimentary plant with plumule, radicle, and cotyledons
3 a : something as yet undeveloped **b :** a beginning or undeveloped state of something

⟨productions seen in *embryo* during their out-of-town tryout period —Henry Hewes⟩

em·bryo·gen·e·sis \ˌem-brē-ō-ˈje-nə-səs\ *noun* (1830)
: the formation and development of the embryo
— em·bryo·ge·net·ic \-jə-ˈne-tik\ *adjective*

em·bry·og·e·ny \ˌem-brē-ˈä-jə-nē\ *noun, plural* **-nies** (1835)
: EMBRYOGENESIS
— em·bryo·gen·ic \-brē-ō-ˈje-nik\ *adjective*

em·bry·oid \ˈem-brē-ˌȯid\ *noun* (circa 1927)
: a mass of plant or animal tissue that resembles an embryo
— embryoid *adjective*

em·bry·ol·o·gy \ˌem-brē-ˈä-lə-jē\ *noun* [French *embryologie*] (circa 1847)
1 : a branch of biology dealing with embryos and their development
2 : the features and phenomena exhibited in the formation and development of an embryo
— em·bry·o·log·i·cal \-brē-ə-ˈlä-ji-kəl\ *adjective*
— em·bry·o·log·i·cal·ly \-ji-k(ə-)lē\ *adverb*
— em·bry·ol·o·gist \-brē-ˈä-lə-jist\ *noun*

embryon- *or* **embryoni-** *combining form* [Medieval Latin *embryon-, embryo*]
: embryo ⟨*embryon*ic⟩

em·bry·o·nal \em-ˈbrī-ə-nᵊl\ *adjective* (1652)
: EMBRYONIC 1

em·bry·o·nat·ed \ˈem-brē-ə-ˌnā-təd\ *adjective* (1687)
: having an embryo

em·bry·on·ic \ˌem-brē-ˈä-nik\ *adjective* (circa 1841)
1 : of or relating to an embryo
2 : being in an early stage of development **:** INCIPIENT, RUDIMENTARY
— em·bry·on·i·cal·ly \-ni-k(ə-)lē\ *adverb*

embryonic disk *noun* (circa 1938)
1 a : BLASTODISC **b :** BLASTODERM
2 : the part of the inner cell mass of a blastocyst from which the embryo of a placental mammal develops — called also *embryonic shield*

embryonic membrane *noun* (1947)
: a structure (as the amnion) that derives from the fertilized ovum but does not form a part of the embryo

em·bryo·phyte \ˈem-brē-ə-ˌfīt\ *noun* (circa 1909)
: any of a subkingdom (Embryophyta) of plants in which the embryo is retained within maternal tissue and which include the bryophytes and tracheophytes

embryo sac *noun* (1872)
: the female gametophyte of a seed plant consisting of a thin-walled sac within the nucellus that contains the egg nucleus and other nuclei which give rise to endosperm on fertilization

embryo transfer *noun* (1969)
: a procedure used especially in animal breeding in which an embryo from a superovulated female is removed and reimplanted in the uterus of another female — called also *embryo transplant*

¹em·cee \ˌem-ˈsē\ *noun* [*MC*] (circa 1933)
: MASTER OF CEREMONIES

²emcee *verb* **em·ceed; em·cee·ing** (1937)
transitive verb
: to act as master of ceremonies of
intransitive verb
: to act as master of ceremonies

-eme \ˌēm\ *noun suffix* [French *-ème* (from *phonème* speech sound, phoneme)]
: significantly distinctive unit of language structure ⟨tax*eme*⟩

emend \ē-ˈmend\ *transitive verb* [Middle English, from Latin *emendare* — more at AMEND] (15th century)
: to correct usually by textual alterations
synonym see CORRECT
— emend·able \-ˈmen-də-bəl\ *adjective*
— emend·er *noun*

emen·da·tion \ˌē-ˌmen-ˈdā-shən; ˌe-mən-, e-ˌmen-\ *noun* (1536)
1 : the act or practice of emending
2 : an alteration designed to correct or improve

¹em·er·ald \ˈem-rəld, ˈe-mə-\ *noun* [Middle English *emerallde,* from Middle French *esmeralde,* from (assumed) Vulgar Latin *smaralda,* from Latin *smaragdus,* from Greek *smaragdos*] (14th century)
1 : a rich green variety of beryl prized as a gemstone
2 : any of various green gemstones (as synthetic corundum or demantoid)

²emerald *adjective* (1508)
: brightly or richly green

emerald cut *noun* (1926)
: a rectangular cut for a gem having a series of parallel facets on each side and at each corner

emerald green *noun* (1646)
1 : a clear bright green resembling that of the emerald
2 : any of various strong greens

emerge \i-ˈmərj\ *intransitive verb* **emerged; emerg·ing** [Latin *emergere,* from *e-* + *mergere* to plunge — more at MERGE] (1563)
1 : to become manifest
2 : to rise from or as if from an enveloping fluid **:** come out into view
3 : to rise from an obscure or inferior position or condition
4 : to come into being through evolution

emer·gence \i-ˈmər-jən(t)s\ *noun* (1704)
1 : the act or an instance of emerging
2 : any of various superficial outgrowths of plant tissue usually formed from both epidermis and immediately underlying tissues
3 : penetration of the soil surface by a newly germinated plant

emer·gen·cy \i-ˈmər-jənt-sē\ *noun, plural* **-cies** *often attributive* (circa 1631)
1 : an unforeseen combination of circumstances or the resulting state that calls for immediate action
2 : an urgent need for assistance or relief ⟨the governor declared a state of *emergency* after the flood⟩
synonym see JUNCTURE

emergency brake *noun* (1900)
: a brake (as on an automobile) that can be used for stopping in the event of failure of the main brakes and to keep the vehicle from rolling when parked

emergency medical technician *noun* (1980)
: EMT

emergency room *noun* (1964)
: a hospital room or area staffed and equipped for the reception and treatment of persons requiring immediate medical care

¹emer·gent \i-ˈmər-jənt\ *adjective* [Middle English, from Latin *emergent-, emergens,* present participle of *emergere*] (1593)
1 a : arising unexpectedly **b :** calling for prompt action **:** URGENT
2 : rising out of or as if out of a fluid
3 : arising as a natural or logical consequence
4 : newly formed or prominent

²emergent *noun* (1620)
1 : something emergent
2 a : a tree that rises above the surrounding forest **b :** a plant rooted in shallow water and having most of the vegetative growth above water

emergent evolution *noun* (1923)
: evolution that according to some theories involves the appearance of new characters and qualities at complex levels of organization (as the cell or organism) which cannot be predicted solely from the study of less complex levels (as the atom or molecule) — compare CREATIVE EVOLUTION

emerg·ing *adjective* (1646)
: EMERGENT 4 ⟨the *emerging* nations of Africa⟩

emer·i·ta \i-ˈmer-ə-tə\ *adjective* [Latin, feminine of *emeritus*] (1928)
: EMERITUS — used of a woman ⟨Professor *Emerita* Mary Smith⟩

¹emer·i·tus \i-ˈmer-ə-təs\ *noun, plural* **-i·ti** \-ə-ˌtī, -ˌtē\ (1750)
: one retired from professional life but permitted to retain as an honorary title the rank of the last office held

²emeritus *adjective* [Latin, past participle of *emereri* to serve out one's term, from *e-* + *mereri, merēre* to earn, deserve, serve — more at MERIT] (1794)
1 : holding after retirement an honorary title corresponding to that held last during active service
2 : retired from an office or position ⟨professor *emeritus*⟩ — converted to *emeriti* after a plural ⟨professors *emeriti*⟩

emersed \(ˌ)ē-ˈmərst\ *adjective* (1686)
: standing out of or rising above a surface (as of a fluid) ⟨*emersed* aquatic weeds⟩

emer·sion \(ˌ)ē-ˈmər-zhən, -shən\ *noun* [Latin *emersus,* past participle of *emergere*] (1633)
: an act of emerging **:** EMERGENCE

em·ery \ˈem-rē, ˈe-mə-\ *noun, plural* **em·er·ies** *often attributive* [Middle English, from Middle French *emeri,* from Old Italian *smiriglio,* from Medieval Latin *smiriglum,* from Greek *smyrid-, smyris*] (15th century)
: a dark granular mineral that consists essentially of corundum and is used for grinding and polishing; *also* **:** a hard abrasive powder

emery board *noun* (1725)
: a cardboard nail file covered with emery

eme·sis \ˈe-mə-səs, i-ˈmē-\ *noun, plural* **eme·ses** \-ˌsēz\ [New Latin, from Greek, from *emein*] (circa 1847)
: an act or instance of vomiting

emet·ic \i-ˈme-tik\ *noun* [Latin *emetica,* from Greek *emetikē,* from feminine of *emetikos* causing vomiting, from *emein* to vomit — more at VOMIT] (1657)
: an agent that induces vomiting
— emetic *adjective*
— emet·i·cal·ly \-ti-k(ə-)lē\ *adverb*

em·e·tine \ˈe-mə-ˌtēn\ *noun* (1819)
: an amorphous alkaloid $C_{29}H_{40}N_2O_4$ extracted from ipecac root and used as an emetic and expectorant

émeute \ā-mœ̄t\ *noun, plural* **émeutes** *same*\ [French, from Old French *esmeute* act of starting, from feminine of *esmeut,* past participle of *esmovoir* to start — more at EMOTION] (1782)
: UPRISING

emf \ˌē-(ˌ)em-ˈef\ *noun* [*e*lectro*m*otive *f*orce] (1868)
: POTENTIAL DIFFERENCE

-emia \ˈē-mē-ə\ *noun combining form* [New Latin *-emia, -aemia,* from Greek *-aimia,* from *haima* blood]
1 : condition of having (such) blood ⟨leuk*emia*⟩
2 : condition of having (a specified thing) in the blood ⟨ur*emia*⟩

emic \ˈē-mik\ *adjective* [phon*emic*] (1954)
: of, relating to, or involving analysis of linguistic or behavioral phenomena in terms of the internal structural or functional elements of a particular system — compare ETIC

¹em·i·grant \ˈe-mi-grənt\ *noun* (1754)
1 : one who emigrates
2 : a migrant plant or animal

²emigrant *adjective* (1794)
: departing or having departed from a country to settle elsewhere

em·i·grate \ˈe-mə-ˌgrāt\ *intransitive verb* **-grat·ed; -grat·ing** [Latin *emigratus,* past

participle of *emigrare,* from *e-* + *migrare* to migrate] (1778)
: to leave one's place of residence or country to live elsewhere
— **em·i·gra·tion** \ˌe-mə-ˈgrā-shən\ *noun*

émi·gré *also* **emi·gré** \ˈe-mi-ˌgrā, ˌe-mi-ˈ\ *noun, often attributive* [French *émigré,* from past participle of *émigrer* to emigrate, from Latin *emigrare*] (1792)
: EMIGRANT; *especially* **:** a person forced to emigrate for political reasons

em·i·nence \ˈe-mə-nən(t)s\ *noun* (15th century)
1 : a position of prominence or superiority
2 : one that is eminent, prominent, or lofty: as **a :** an anatomical protuberance (as on a bone) **b :** a person of high rank or attainments — often used as a title for a cardinal **c :** a natural elevation

émi·nence grise \ā-mē-nä^{n}s-grēz\ *noun, plural* **éminences grises** *same*\ [French, literally, gray eminence, nickname of Père Joseph (François du Tremblay) (died 1638) French monk and diplomat, confidant of Cardinal Richelieu who was known as *Éminence Rouge* red eminence; from the colors of their respective habits] (1925)
: a confidential agent; *especially* **:** one exercising unsuspected or unofficial power

em·i·nen·cy \ˈe-mə-nən(t)-sē\ *noun, plural* **-cies** (1605)
archaic **:** EMINENCE

em·i·nent \ˈe-mə-nənt\ *adjective* [Middle English, from Middle French or Latin; Middle French, from Latin *eminent-, eminens,* present participle of *eminēre* to stand out, from *e-* + *-minēre;* akin to Latin *mont-, mons* mountain — more at MOUNT] (15th century)
1 : standing out so as to be readily perceived or noted **:** CONSPICUOUS
2 : jutting out **:** PROJECTING
3 : exhibiting eminence especially in standing above others in some quality or position **:** PROMINENT
synonym see FAMOUS

eminent domain *noun* (1783)
: a right of a government to take private property for public use by virtue of the superior dominion of the sovereign power over all lands within its jurisdiction

em·i·nent·ly \-lē\ *adverb* (1641)
: to a high degree **:** VERY ⟨*eminently* worthy⟩ ⟨an *eminently* sensible plan⟩

emir \i-ˈmir, ā-\ *noun* [Arabic *amīr* commander] (1595)
: a ruler, chief, or commander in Islamic countries

emir·ate \ˈe-mə-rət, -ˌrāt\ *noun* (1863)
: the state or jurisdiction of an emir

em·is·sary \ˈe-mə-ˌser-ē\ *noun, plural* **-saries** [Latin *emissarius,* from *emissus,* past participle of *emittere*] (1616)
1 : one designated as the agent of another **:** REPRESENTATIVE
2 : a secret agent

emis·sion \ē-ˈmi-shən\ *noun* (1607)
1 a : an act or instance of emitting **:** EMANATION **b** *archaic* **:** PUBLICATION **c :** a putting into circulation
2 a : something sent forth by emitting: as (1) **:** electromagnetic waves radiated by an antenna or a celestial body (2) **:** substances discharged into the air (as by a smokestack or an automobile gasoline engine) **b :** EFFLUVIUM
— **emis·sive** \-ˈmi-siv\ *adjective*

emis·siv·i·ty \ˌe-mə-ˈsi-və-tē, ˌē-ˌmi-ˈsiv-\ *noun, plural* **-ties** (1880)
: the relative power of a surface to emit heat by radiation **:** the ratio of the radiant energy emitted by a surface to that emitted by a blackbody at the same temperature

emit \ē-ˈmit\ *transitive verb* **emit·ted; emit·ting** [Latin *emittere* to send out, from *e-* + *mittere* to send] (1626)
1 a : to throw or give off or out (as light) **b :** to send out **:** EJECT
2 a : to issue with authority; *especially* **:** to put (as money) into circulation **b** *obsolete* **:** PUBLISH
3 : to give utterance or voice to ⟨*emitted* a groan⟩
— **emit·ter** *noun*

emit·tance \ē-ˈmi-t^{ə}n(t)s\ *noun* (1940)
1 : the energy radiated by the surface of a body per second per unit area
2 : EMISSIVITY

Emmanuel *variant of* IMMANUEL

em·men·a·gogue \ə-ˈme-nə-ˌgäg, e-\ *noun* [Greek *emmēna* menses (from neuter plural of *emmēnos* monthly, from *en-* + *mēn* month) + English *-agogue* — more at MOON] (circa 1732)
: an agent that promotes the menstrual discharge

Em·men·ta·ler *or* **Em·men·tha·ler** \ˈe-mən-ˌtä-lər\ *or* **Em·men·tal** *or* **Em·men·thal** \-ˌtäl\ *noun* [German, from *Emmental,* Switzerland] (1902)
: SWISS CHEESE

em·mer \ˈe-mər\ *noun* [German, from Old High German *amari*] (circa 1900)
: a wheat (*Triticum dicoccum*) having spikelets with two hard red kernels that remain in the glumes after threshing; *broadly* **:** a tetraploid wheat — called also *emmer wheat*

em·met \ˈe-mət\ *noun* [Middle English *emete,* from Old English *æmette* ant — more at ANT] (before 12th century)
chiefly dialect **:** ANT

Em·my \ˈe-mē\ *noun, plural* **Emmys** [from alteration of *Immy,* nickname for *image orthicon* (a camera tube used in television)] (1949)
: a statuette awarded annually by a professional organization for notable achievement in television

em·o·din \ˈe-mə-dən\ *noun* [International Scientific Vocabulary *emodi-* (from New Latin *Rheum emodi,* species of rhubarb) + *-in*] (1858)
: an orange crystalline phenolic compound $C_{15}H_{10}O_5$ that is obtained from plants (as rhubarb and cascara buckthorn) and is used as a laxative

[1]**emol·lient** \i-ˈmäl-yənt\ *adjective* [Latin *emollient-, emolliens,* present participle of *emollire* to soften, from *e-* + *mollis* soft — more at MOLLIFY] (circa 1640)
1 : making soft or supple; *also* **:** soothing especially to the skin or mucous membrane
2 : making less intense or harsh **:** MOLLIFYING ⟨soothe us in our agonies with *emollient* words —H. L. Mencken⟩

[2]**emollient** *noun* (1656)
: something that softens or soothes

emol·u·ment \i-ˈmäl-yə-mənt\ *noun* [Middle English, from Latin *emolumentum* advantage, from *emolere* to produce by grinding, from *e-* + *molere* to grind — more at MEAL] (15th century)
1 : the returns arising from office or employment usually in the form of compensation or perquisites
2 *archaic* **:** ADVANTAGE

emote \i-ˈmōt\ *intransitive verb* **emot·ed; emot·ing** [back-formation from *emotion*] (1917)
: to give expression to emotion especially in or as if in acting

emo·tion \i-ˈmō-shən\ *noun* [Middle French, from *emouvoir* to stir up, from Old French *esmovoir,* from Latin *emovēre* to remove, displace, from *e-* + *movēre* to move] (1579)
1 a *obsolete* **:** DISTURBANCE **b :** EXCITEMENT
2 a : the affective aspect of consciousness **:** FEELING **b :** a state of feeling **c :** a psychic and physical reaction (as anger or fear) subjectively experienced as strong feeling and physiologically involving changes that prepare the body for immediate vigorous action
synonym see FEELING

emo·tion·al \-shnəl, -shə-n^{ə}l\ *adjective* (1834)
1 : of or relating to emotion ⟨an *emotional* disorder⟩
2 : dominated by or prone to emotion ⟨an *emotional* person⟩
3 : appealing to or arousing emotion ⟨an *emotional* sermon⟩
4 : markedly aroused or agitated in feeling or sensibilities ⟨gets *emotional* at weddings⟩
— **emo·tion·al·i·ty** \-ˌmō-shə-ˈna-lə-tē\ *noun*
— **emo·tion·al·ly** \-ˈmō-shnə-lē, -shə-n^{ə}l-ē\

emo·tion·al·ism \i-ˈmō-shnə-ˌli-zəm, -shə-n^{ə}l-ˌiz-\ *noun* (1865)
1 : a tendency to regard things emotionally
2 : undue indulgence in or display of emotion

emo·tion·al·ist \-shnə-list, -shə-n^{ə}l-ist\ *noun* (circa 1866)
1 : one who bases a theory or policy on an emotional conviction
2 : one prone to emotionalism
— **emo·tion·al·is·tic** \-ˌmō-shnə-ˈlis-tik, -shə-n^{ə}l-ˈis-\ *adjective*

emo·tion·al·ize \i-ˈmō-shnə-ˌlīz, -shə-n^{ə}l-ˌīz\ *transitive verb* **-ized; -iz·ing** (1879)
: to give an emotional quality to

emo·tion·less \i-ˈmō-shən-ləs\ *adjective* (1862)
: showing, having, or expressing no emotion
— **emo·tion·less·ness** *noun*
— **emo·tion·less·ly** *adverb*

emo·tive \i-ˈmō-tiv\ *adjective* (1830)
1 : of or relating to the emotions
2 : appealing to or expressing emotion ⟨the *emotive* use of language⟩
— **emo·tive·ly** *adverb*
— **emo·tiv·i·ty** \i-ˌmō-ˈti-və-tē, ˌē-ˌmō-\ *noun*

em·pa·na·da \ˌem-pə-ˈnä-də\ *noun* [American Spanish, from Spanish, feminine of *empanado,* past participle of *empanar* to bread, from *em-* (from Latin *in-*) + *pan* bread, from Latin *panis* — more at FOOD] (circa 1922)
: a turnover with a sweet or savory filling

empanel *variant of* IMPANEL

em·pa·thet·ic \ˌem-pə-ˈthe-tik\ *adjective* (1932)
: EMPATHIC
— **em·pa·thet·i·cal·ly** \-ti-k(ə-)lē\ *adverb*

em·path·ic \em-ˈpa-thik, im-\ *adjective* (1909)
: involving, eliciting, characterized by, or based on empathy
— **em·path·i·cal·ly** \-thi-k(ə-)lē\ *adverb*

em·pa·thise *British variant of* EMPATHIZE

em·pa·thize \ˈem-pə-ˌthīz\ *intransitive verb* **-thized; -thiz·ing** (circa 1921)
: to experience empathy ⟨adults unable to *empathize* with a child's frustrations⟩

em·pa·thy \ˈem-pə-thē\ *noun* [Greek *empatheia,* literally, passion, from *empathēs* emotional, from *em-* + *pathos* feelings, emotion — more at PATHOS] (1904)
1 : the imaginative projection of a subjective state into an object so that the object appears to be infused with it
2 : the action of understanding, being aware of, being sensitive to, and vicariously experiencing the feelings, thoughts, and experience of another of either the past or present without having the feelings, thoughts, and experience fully communicated in an objectively explicit manner; *also* **:** the capacity for this

em·pen·nage \ˌäm-pə-ˈnäzh, ˌem-\ *noun* [French, feathers of an arrow, empennage, from *empenner* to feather an arrow, from *em-* [1]*en-* + *penne* feather, from Middle French — more at PEN] (1909)
: the tail assembly of an airplane

em·per·or \ˈem-pər-ər, -prər\ *noun* [Middle English, from Old French *empereor,* from Latin *imperator,* literally, commander, from *imperare* to command, from *in-* + *parare* to prepare, order — more at PARE] (13th century)
: the sovereign or supreme male monarch of an empire
— **em·per·or·ship** \-ˌship\ *noun*

emperor penguin *noun* (1885)
: a penguin (*Aptenodytes forsteri*) that is the largest known and that is noted for its habit of brooding the egg or young between the feet and a fold of abdominal skin resembling a pouch

em·pery \'em-p(ə-)rē\ *noun, plural* **em·per·ies** [Middle English *emperie,* from Old French, from *emperer* to command, from Latin *imperare*] (13th century)
: wide dominion **:** EMPIRE

em·pha·sise *British variant of* EMPHASIZE

em·pha·sis \'em(p)-fə-səs\ *noun, plural* **-pha·ses** \-ˌsēz\ [Latin, from Greek, exposition, emphasis, from *emphainein* to indicate, from *en-* + *phainein* to show — more at FANCY] (1573)
1 a : force or intensity of expression that gives impressiveness or importance to something **b :** a particular prominence given in reading or speaking to one or more words or syllables
2 : special consideration of or stress or insistence on something

em·pha·size \'em(p)-fə-ˌsīz\ *transitive verb* **-sized; -siz·ing** (circa 1806)
: to place emphasis on **:** STRESS ⟨*emphasized* the need for reform⟩

em·phat·ic \im-'fa-tik, em-\ *adjective* [Greek *emphatikos,* from *emphainein*] (circa 1708)
1 : uttered with or marked by emphasis
2 : tending to express oneself in forceful speech or to take decisive action
3 : attracting special attention
4 : constituting or belonging to a set of tense forms in English consisting of the auxiliary *do* followed by an infinitive without *to* that are used to facilitate rhetorical inversion or to emphasize something
— em·phat·i·cal·ly \-'fa-ti-k(ə-)lē\ *adverb*

em·phy·se·ma \ˌem(p)-fə-'zē-mə, -'sē-\ *noun* [New Latin, from Greek *emphysēma,* from *emphysan* to inflate, from *em-* ²en- + *physan* to blow, from *physa* breath — more at PUSTULE] (1661)
: a condition characterized by air-filled expansions of body tissues; *specifically* **:** a condition of the lung marked by abnormal dilation of its air spaces and distension of its walls and frequently by impairment of heart action
— em·phy·se·ma·tous \-'ze-mə-təs, -'se-, -'zē-, -'sē-\ *adjective*
— em·phy·se·mic \-'zē-mik, -'sē-\ *adjective*

em·pire \'em-ˌpīr\ *noun* [Middle English, from Old French *empire, empirie,* from Latin *imperium* absolute authority, empire, from *imperare*] (14th century)
1 a (1) **:** a major political unit having a territory of great extent or a number of territories or peoples under a single sovereign authority; *especially* **:** one having an emperor as chief of state (2) **:** the territory of such a political unit **b :** something resembling a political empire; *especially* **:** an extensive territory or enterprise under single domination or control
2 : imperial sovereignty, rule, or dominion

Em·pire \'äm-ˌpir, 'em-ˌpīr\ *adjective* [French, from (*le premier*) *Empire* the first Empire of France] (1869)
: of, relating to, or characteristic of a style (as of clothing or furniture) popular in early 19th century France

Em·pire Day \'em-ˌpīr-\ *noun* (1902)
: COMMONWEALTH DAY — used before the official adoption of *Commonwealth Day* in 1958

em·pir·ic \im-'pir-ik, em-\ *noun* [Latin *empiricus,* from Greek *empeirikos* doctor relying on experience alone, from *empeiria* experience, from *em-* ²en- + *peiran* to attempt — more at FEAR] (1562)
1 : CHARLATAN 2
2 : one who relies on practical experience

em·pir·i·cal \-i-kəl\ *also* **em·pir·ic** \-ik\ *adjective* (1569)
1 : originating in or based on observation or experience ⟨*empirical* data⟩
2 : relying on experience or observation alone often without due regard for system and theory
3 : capable of being verified or disproved by observation or experiment ⟨*empirical* laws⟩
4 : of or relating to empiricism
— em·pir·i·cal·ly \-i-k(ə-)lē\ *adverb*

empirical formula *noun* (1885)
: a chemical formula showing the simplest ratio of elements in a compound rather than the total number of atoms in the molecule ⟨CH_2O is the *empirical formula* for glucose⟩

em·pir·i·cism \im-'pir-ə-ˌsi-zəm, em-\ *noun* (1657)
1 a : a former school of medical practice founded on experience without the aid of science or theory **b :** QUACKERY, CHARLATANRY
2 a : the practice of relying on observation and experiment especially in the natural sciences **b :** a tenet arrived at empirically
3 : a theory that all knowledge originates in experience
— em·pir·i·cist \-sist\ *noun*

em·place \im-'plās\ *transitive verb* [back-formation from *emplacement*] (1865)
: to put into position ⟨missiles *emplaced* around the city⟩

em·place·ment \-'plā-smənt\ *noun* [French, from Middle French *emplacer* to emplace, from *en-* + *place*] (1802)
1 : the situation or location of something
2 : a prepared position for weapons or military equipment ⟨radar *emplacements*⟩
3 : a putting into position **:** PLACEMENT

em·plane \im-'plān\ *variant of* ENPLANE

¹em·ploy \im-'plȯi, em-\ *transitive verb* [Middle English *emploien,* from Middle French *emploier,* from Latin *implicare* to enfold, involve, implicate, from *in-* + *plicare* to fold — more at PLY] (15th century)
1 a : to make use of (someone or something inactive) ⟨*employ* a pen for sketching⟩ **b :** to use (as time) advantageously **c** (1) **:** to use or engage the services of (2) **:** to provide with a job that pays wages or a salary
2 : to devote to or direct toward a particular activity or person ⟨*employed* all her energies to help the poor⟩
synonym see USE
— em·ploy·er *noun*

²em·ploy \im-'plȯi, 'im-ˌ, 'em-ˌ\ *noun* (1666)
1 a : USE, PURPOSE **b :** OCCUPATION, JOB
2 : the state of being employed ⟨in the government's *employ*⟩

¹em·ploy·able \im-'plȯi-ə-bəl\ *adjective* (1593)
: capable of being employed
— em·ploy·abil·i·ty \-ˌplȯi-ə-'bi-lə-tē\ *noun*

²employable *noun* (1934)
: one who is employable

em·ploy·ee *or* **em·ploye** \im-ˌplȯ(i)-'ē, (ˌ)em-; im-'plȯ(i)-ˌē, em-\ *noun* (1822)
: one employed by another usually for wages or salary and in a position below the executive level

em·ploy·ment \im-'plȯi-mənt\ *noun* (15th century)
1 : USE, PURPOSE
2 a : activity in which one engages or is employed **b :** an instance of such activity
3 : the act of employing **:** the state of being employed
synonym see WORK

employment agency *noun* (1888)
: an agency whose business is to find jobs for people seeking them or to find people to fill jobs that are open

em·poi·son \im-'pȯi-zᵊn\ *transitive verb* [Middle English *empoysonen,* from Middle French *empoisoner,* from *en-* + *poison* poison, from Old French] (14th century)
1 *archaic* **:** POISON
2 : EMBITTER ⟨a look of *empoisoned* acceptance —Saul Bellow⟩
— em·poi·son·ment \-mənt\ *noun*

em·po·ri·um \im-'pōr-ē-əm, em-, -'pȯr-\ *noun, plural* **-ri·ums** *also* **-ria** \-ē-ə\ [Latin, from Greek *emporion,* from *emporos* traveler, trader, from *em-* ²en- + *poros* passage, journey — more at FARE] (1586)
1 a : a place of trade; *especially* **:** a commercial center **b :** a retail outlet ⟨a hardware *emporium*⟩ ⟨a pizza *emporium*⟩
2 : a store carrying a diversity of merchandise

em·pow·er \im-'pau̇(-ə)r\ *transitive verb* (1648)
1 : to give official authority or legal power to
2 : ENABLE 1a
3 : to promote the self-actualization or influence of ⟨women's movement has been inspiring and *empowering* women —Ron Hansen⟩
— em·pow·er·ment \-mənt\ *noun*

em·press \'em-prəs\ *noun* [Middle English *emperesse,* from Old French, feminine of *empereor* emperor] (12th century)
1 : the wife or widow of an emperor
2 : a woman who is the sovereign or supreme monarch of an empire

em·presse·ment \äⁿ-pres-mäⁿ\ *noun* [French, from (*s'*)*empresser* to hurry, from *en-* + *presser* to press] (1709)
: demonstrative warmth or cordiality

em·prise \em-'prīz\ *noun* [Middle English, from Middle French, from Old French, from *emprendre* to undertake, from (assumed) Vulgar Latin *imprehendere,* from Latin *in-* + *prehendere* to seize — more at GET] (13th century)
: an adventurous, daring, or chivalric enterprise

¹emp·ty \'em(p)-tē\ *adjective* **emp·ti·er; -est** [Middle English, from Old English *ǣmettig* unoccupied, from *ǣmetta* leisure, perhaps from *ǣ-* without + *-metta* (probably akin to *mōtan* to have to) — more at MUST] (before 12th century)
1 a : containing nothing **b :** not occupied or inhabited **c :** UNFREQUENTED **d :** not pregnant ⟨*empty* heifer⟩ **e :** NULL 4a ⟨the *empty* set⟩
2 a : lacking reality, substance, meaning, or value **:** HOLLOW ⟨an *empty* pleasure⟩ **b :** destitute of effect or force **c :** devoid of sense **:** FOOLISH
3 : HUNGRY
4 a : IDLE ⟨*empty* hours⟩ **b :** having no purpose or result **:** USELESS
5 : marked by the absence of human life, activity, or comfort ☆
— emp·ti·ly \-tə-lē\ *adverb*
— emp·ti·ness \-tē-nəs\ *noun*

²empty *verb* **emp·tied; emp·ty·ing** (1555)
transitive verb
1 a : to make empty **:** remove the contents of **b :** DEPRIVE, DIVEST **c :** to discharge (itself) of contents **d :** to fire (a repeating firearm) until empty

☆ SYNONYMS
Empty, vacant, blank, void, vacuous mean lacking contents which could or should be present. EMPTY suggests a complete absence of contents ⟨an *empty* bucket⟩. VACANT suggests an absence of appropriate contents or occupants ⟨a *vacant* apartment⟩. BLANK stresses the absence of any significant, relieving, or intelligible features on a surface ⟨a *blank* wall⟩. VOID suggests absolute emptiness as far as the mind or senses can determine ⟨a statement *void* of meaning⟩. VACUOUS suggests the emptiness of a vacuum and especially the lack of intelligence or significance ⟨a *vacuous* facial expression⟩. See in addition VAIN.

2 : to remove from what holds or encloses ⟨*empty* the grain from sacks⟩
intransitive verb
1 : to become empty
2 : to discharge contents ⟨the river *empties* into the ocean⟩

[3]empty *noun, plural* **emp·ties** (1535)
: something (as a container) that is empty

emp·ty–hand·ed \ˌem(p)-tē-ˈhan-dəd\ *adjective* (1589)
1 : having or bringing nothing
2 : having acquired or gained nothing ⟨came back *empty-handed*⟩

emp·ty–head·ed \-ˈhe-dəd\ *adjective* (1650)
: SCATTERBRAINED

empty nest·er \-ˈnes-tər\ *noun* (1962)
: a parent whose children have grown and moved away from home

emp·ty–nest syndrome \ˌem(p)-tē-ˈnest-\ *noun* (1972)
: the emotional letdown experienced by an empty nester

em·pur·ple \im-ˈpər-pəl\ *verb* **em·pur·pled; em·pur·pling** \-ˈpər-p(ə-)liŋ\ (1590)
transitive verb
: to tinge or color purple
intransitive verb
: to become purple

em·py·ema \ˌem-ˌpī-ˈē-mə\ *noun, plural* **-ema·ta** \-mə-tə\ *or* **-emas** [Late Latin, from Greek *empyēma,* from *empyein* to suppurate, from *em-* [2]en- + *pyon* pus — more at FOUL] (circa 1605)
: the presence of pus in a bodily cavity
— em·py·emic \-mik\ *adjective*

em·py·re·al \ˌem-ˌpī-ˈrē-əl, -pə-; em-ˈpir-ē-əl, -ˈpī-rē-\ *adjective* [Late Latin *empyrius, empyreus,* from Late Greek *empyrios,* from Greek *em-* [2]en- + *pyr* fire] (15th century)
1 : of or relating to the empyrean **:** CELESTIAL
2 : SUBLIME

[1]em·py·re·an \-ən\ *adjective* (15th century)
: EMPYREAL

[2]empyrean *noun* (circa 1610)
1 a : the highest heaven or heavenly sphere in ancient and medieval cosmology usually consisting of fire or light **b :** the true and ultimate heavenly paradise
2 : FIRMAMENT, HEAVENS
3 : an ideal place or state

EMT \ˌē-(ˌ)em-ˈtē\ *noun* [*e*mergency *m*edical *t*echnician] (1972)
: a specially trained medical technician licensed to provide basic emergency services before and during transportation to a hospital — compare PARAMEDIC 2

emu \ˈē-(ˌ)myü\ *noun* [modification of Portuguese *ema* cassowary] (1656)
1 : any of various tall flightless birds (as the rhea)
2 : a swift-running Australian bird (*Dromaius novae-hollandiae*) with undeveloped wings that is related to and smaller than the ostrich

emu 2

[1]em·u·late \ˈem-yə-ˌlāt\ *transitive verb* **-lat·ed; -lat·ing** [Latin *aemulatus,* past participle of *aemulari,* from *aemulus* rivaling] (1582)
1 a : to strive to equal or excel **b :** IMITATE; *especially* **:** to imitate by means of an emulator
2 : to equal or approach equality with

[2]em·u·late \-lət\ *adjective* (1602)
obsolete **:** EMULOUS 1b ⟨pricked on by a most *emulate* pride —Shakespeare⟩

em·u·la·tion \ˌem-yə-ˈlā-shən\ *noun* (1542)
1 *obsolete* **:** ambitious or envious rivalry
2 : ambition or endeavor to equal or excel others (as in achievement)
3 a : IMITATION **b :** the use of or technique of using an emulator
— em·u·la·tive \ˈem-yə-ˌlā-tiv\ *adjective*
— em·u·la·tive·ly *adverb*

em·u·la·tor \ˈem-yə-ˌlā-tər\ *noun* (1589)
1 : one that emulates
2 : hardware or software that permits programs written for one computer to be run on another usually newer computer

em·u·lous \ˈem-yə-ləs\ *adjective* (1535)
1 a : inspired by or deriving from a desire to emulate **b :** ambitious or eager to emulate
2 *obsolete* **:** JEALOUS
— em·u·lous·ly *adverb*
— em·u·lous·ness *noun*

emul·si·fi·er \i-ˈməl-sə-ˌfī(-ə)r\ *noun* (1888)
: one that emulsifies; *especially* **:** a surface-active agent (as a soap) promoting the formation and stabilization of an emulsion

emul·si·fy \-ˌfī\ *transitive verb* **-fied; -fy·ing** (1859)
: to disperse (as an oil) in an emulsion; *also* **:** to convert (two or more immiscible liquids) into an emulsion
— emul·si·fi·able \i-ˈməl-sə-ˌfī-ə-bəl\ *adjective*
— emul·si·fi·ca·tion \i-ˌməl-sə-fə-ˈkā-shən\ *noun*

emul·sion \i-ˈməl-shən\ *noun* [New Latin *emulsion-, emulsio,* from Latin *emulgēre* to milk out, from *e-* + *mulgēre* to milk; akin to Old English *melcan* to milk, Greek *amelgein*] (1612)
1 a : a system (as fat in milk) consisting of a liquid dispersed with or without an emulsifier in an immiscible liquid usually in droplets of larger than colloidal size **b :** the state of such a system
2 : SUSPENSION 2b(3); *especially* **:** a suspension of a sensitive silver salt or a mixture of silver halides in a viscous medium (as a gelatin solution) forming a coating on photographic plates, film, or paper

emul·soid \i-ˈməl-ˌsȯid\ *noun* (circa 1909)
1 : a colloidal system consisting of a liquid dispersed in a liquid
2 : a lyophilic sol (as a gelatin solution)
— emul·soi·dal \-ˌməl-ˈsȯi-dᵊl\ *adjective*

en \ˈen\ *noun* (1792)
1 : the width of a piece of type half the width of an em
2 : the letter *n*

[1]en- *also* **em-** \in *also* en; *sometimes only* in *is shown when* en *is infrequent*\ *prefix* [Middle English, from Old French, from Latin *in-, im-,* from *in*]
1 : put into or onto ⟨*en*throne⟩ **:** cover with ⟨*en*shroud⟩ **:** go into or onto ⟨*en*plane⟩ — in verbs formed from nouns
2 : cause to be ⟨*en*slave⟩ — in verbs formed from adjectives or nouns
3 : provide with ⟨*em*power⟩ — in verbs formed from nouns
4 : so as to cover ⟨*en*wrap⟩ **:** thoroughly ⟨*en*tangle⟩ — in verbs formed from verbs; in all senses usually *em-* before *b, m,* or *p*

[2]en- *also* **em-** *prefix* [Middle English, from Latin, from Greek, from *en* in — more at IN]
: in **:** within ⟨*en*zootic⟩ — usually *em-* before *b, m,* or *p* ⟨*em*pathy⟩

[3]en- *combining form* [International Scientific Vocabulary, from *-ene*]
: chemically unsaturated; *especially* **:** having one double bond ⟨*en*amine⟩

[1]-en \ən, ᵊn\ *also* **-n** \n\ *adjective suffix* [Middle English, from Old English; akin to Old High German *-īn* made of, Latin *-īnus* of or belonging to, Greek *-inos* made of, of or belonging to]
: made of **:** consisting of ⟨earth*en*⟩ ⟨leather*n*⟩

[2]-en *verb suffix* [Middle English *-nen,* from Old English *-nian;* akin to Old High German *-inōn* -en]
1 a : cause to be ⟨sharp*en*⟩ **b :** cause to have ⟨length*en*⟩
2 a : come to be ⟨steep*en*⟩ **b :** come to have ⟨length*en*⟩

en·able \i-ˈnā-bəl\ *transitive verb* **en·abled; en·abling** \-b(ə-)liŋ\ (15th century)
1 a : to provide with the means or opportunity ⟨training that *enables* people to earn a living⟩ **b :** to make possible, practical, or easy **c :** to cause to operate ⟨software that *enables* the keyboard⟩
2 : to give legal power, capacity, or sanction to ⟨a law *enabling* admission of a state⟩

en·act \i-ˈnakt\ *transitive verb* (15th century)
1 : to establish by legal and authoritative act; *specifically* **:** to make (as a bill) into law
2 : ACT OUT ⟨*enact* a role⟩
— en·ac·tor \-ˈnak-tər\ *noun*

en·act·ment \-ˈnak(t)-mənt\ *noun* (1817)
1 : the act of enacting **:** the state of being enacted
2 : something (as a law) that has been enacted

[1]enam·el \i-ˈna-məl\ *transitive verb* **-eled** *or* **-elled; -el·ing** *or* **-el·ling** \-ˈnam-liŋ, -ˈna-mə-\ [Middle English, from Middle French *enamailler,* from *en-* + *esmail* enamel, of Germanic origin; akin to Old High German *smelzan* to melt — more at SMELT] (14th century)
1 : to cover, inlay, or decorate with enamel
2 : to beautify with a colorful surface
3 : to form a glossy surface on (as paper, leather, or cloth)
— enam·el·er *noun*
— enam·el·ist \-mə-list\ *noun*

[2]enamel *noun* (15th century)
1 : a usually opaque vitreous composition applied by fusion to the surface of metal, glass, or pottery
2 : a surface or outer covering that resembles enamel
3 a : something that is enameled **b :** ENAMELWARE
4 : a cosmetic intended to give a smooth or glossy appearance
5 : a hard calcareous substance that forms a thin layer capping the teeth — see TOOTH illustration
6 : a paint that flows out to a smooth coat when applied and that dries with a glossy appearance

enam·el·ware \i-ˈna-məl-ˌwar, -ˌwer\ *noun* (1903)
: metalware (as kitchen utensils) coated with enamel

en·amine \ˈe-nə-ˌmēn, ˈē-\ *noun* (1942)
: an amine containing the double bond linkage C=C−N

en·am·or \i-ˈna-mər\ *transitive verb* **-ored; -or·ing** \-mə-riŋ, -ˈnam-riŋ\ [Middle English *enamouren,* from Middle French *enamourer,* from *en-* + *amour* love — more at AMOUR] (14th century)
1 : to inflame with love — usually used in the passive with *of*
2 : FASCINATE 2b — usually used in the passive with *of* or *with*

en·am·our *chiefly British variant of* ENAMOR

en·an·tio·mer \i-ˈnan-tē-ə-mər\ *noun* [Greek *enantios* + English *-mer*] (circa 1929)
: either of a pair of chemical compounds whose molecular structures have a mirror-image relationship to each other
— en·an·tio·mer·ic \-ˌnan-tē-ə-ˈmer-ik\ *adjective*

en·an·tio·morph \i-ˈnan-tē-ə-ˌmȯrf\ *noun* [Greek *enantios* opposite (from *enanti* facing, from *en* in + *anti* against) + International Scientific Vocabulary *-morph* -morph] (1885)
: ENANTIOMER; *also* **:** either of a pair of crystals (as of quartz) that are structural mirror images
— en·an·tio·mor·phic \-ˌnan-tē-ə-ˈmȯr-fik\ *adjective*
— en·an·tio·mor·phism \-ˈmȯr-ˌfi-zəm\ *noun*
— en·an·tio·mor·phous \-ˈmȯr-fəs\ *adjective*

ena·tion \i-ˈnā-shən\ *noun* [Latin *enatus,* past participle of *enasci* to rise out of, from *e-* + *nasci* to be born — more at NATION] (circa 1842)

: an outgrowth from the surface of an organ ⟨a plant virus forming *enations* on leaves⟩

en banc \ä^n-'bä^n\ *adverb or adjective* [French, on the bench] (1863)
: in full court **:** with full judiciary authority

en bloc \ä^n-'bläk\ *adverb or adjective* [French] (1861)
: as a whole **:** in a mass

en bro·chette \ˌä^n-brō-'shet\ *adjective* [French] (circa 1909)
of food **:** cooked or served on a skewer ⟨shrimp *en brochette*⟩

en·cae·nia \en-'sē-nyə\ *noun plural but singular or plural in construction, often capitalized* [New Latin, from Latin, dedication festival, from Greek *enkainia,* from *en* + *kainos* new — more at RECENT] (1691)
: an annual university ceremony (as at Oxford) of commemoration with recital of poems and essays and conferring of degrees

en·cage \in-'kāj, en-\ *transitive verb* (1593)
: CAGE 1

en·camp \in-'kamp, en-\ (1568)
transitive verb
: to place or establish in a camp
intransitive verb
: to set up or occupy a camp

en·camp·ment \-mənt\ *noun* (1598)
1 a : the place where a group (as a body of troops) is encamped **b :** the individuals that make up an encampment
2 : the act of encamping **:** the state of being encamped

en·cap·su·late \in-'kap-sə-ˌlāt, en-\ *verb* **-lat·ed; -lat·ing** (1876)
transitive verb
1 : to enclose in or as if in a capsule ⟨a pilot *encapsulated* in the cockpit⟩
2 : EPITOMIZE, SUMMARIZE ⟨*encapsulate* an era in an aphorism⟩
intransitive verb
: to become encapsulated
— en·cap·su·la·tion \-ˌkap-sə-'lā-shən\ *noun*

en·cap·su·lat·ed *adjective* (1894)
1 : surrounded by a gelatinous or membranous envelope ⟨*encapsulated* water bacteria⟩
2 : CONDENSED

en·cap·sule \in-'kap-səl, -(ˌ)sül, en-\ *transitive verb* **-suled; -sul·ing** (1877)
: ENCAPSULATE

en·case \in-'kās, en-\ *transitive verb* (1633)
: to enclose in or as if in a case

en·case·ment \in-'kā-smənt, en-\ *noun* (1741)
: the act or process of encasing **:** the state of being encased; *also* **:** CASE, COVERING

en·cash \in-'kash, en-\ *transitive verb* (1861)
British **:** CASH
— en·cash·able \-'ka-shə-bəl\ *adjective, chiefly British*
— en·cash·ment \-mənt\ *noun, chiefly British*

en·caus·tic \in-'kȯ-stik\ *noun* [*encaustic,* adjective, from Latin *encausticus,* from Greek *enkaustikos,* from *enkaiein* to burn in, from *en-* + *kaiein* to burn] (1601)
1 : a paint made from pigment mixed with melted beeswax and resin and after application fixed by heat
2 : the method involving the use of encaustic; *also* **:** a work produced by this method
— encaustic *adjective*

-ence \ən(t)s, ᵊn(t)s\ *noun suffix* [Middle English, from Old French, from Latin *-entia,* from *-ent-, -ens,* present participle ending + *-ia* ²-y]
1 : action or process ⟨emerg*ence*⟩ **:** instance of an action or process ⟨refer*ence*⟩
2 : quality or state ⟨despond*ence*⟩

¹en·ceinte \ä^n(n)-'sant\ *adjective* [Middle French, perhaps from (assumed) Vulgar Latin *incenta,* alteration of Latin *incient-, inciens* being with young, modification of Greek *enkyos* pregnant, from *en-* + *kyein* to be pregnant — more at CYME] (1599)
: PREGNANT 3

²enceinte *noun* [French, from Old French, enclosing wall, from *enceindre* to surround, from Latin *incingere,* from *in-* + *cingere* to gird — more at CINCTURE] (circa 1708)
: a line of fortification enclosing a castle or town; *also* **:** the area so enclosed

encephal- *or* **encephalo-** *combining form* [French *encéphal-,* from Greek *enkephal-,* from *enkephalos,* from *en-* + *kephalē* head — more at CEPHALIC]
: brain ⟨*encephal*itis⟩ ⟨*encephalo*myocarditis⟩

en·ceph·a·li·tis \in-ˌse-fə-'lī-təs\ *noun, plural* **-lit·i·des** \-'li-tə-ˌdēz\ (1843)
: inflammation of the brain
— en·ceph·a·lit·ic \-'li-tik\ *adjective*

en·ceph·a·li·to·gen·ic \-ˌlī-tə-'je-nik\ *adjective* (1923)
: tending to cause encephalitis ⟨an *encephalitogenic* virus⟩
— en·ceph·a·li·to·gen \-'lī-tə-jən, -ˌjen\ *noun*

en·ceph·a·lo·gram \in-'se-fə-lə-ˌgram\ *noun* (1928)
: an X-ray picture of the brain made by encephalography

en·ceph·a·lo·graph \-ˌgraf\ *noun* (1928)
1 : ENCEPHALOGRAM
2 : ELECTROENCEPHALOGRAPH

en·ceph·a·log·ra·phy \in-ˌse-fə-'lä-grə-fē\ *noun* (1922)
: roentgenography of the brain after the cerebrospinal fluid has been replaced by a gas (as air)

en·ceph·a·lo·my·eli·tis \in-ˌse-fə-lō-ˌmī-ə-'lī-təs\ *noun, plural* **-elit·i·des** \-ə-'li-tə-ˌdēz\ [New Latin] (1908)
: concurrent inflammation of the brain and spinal cord; *specifically* **:** any of several virus diseases of horses

en·ceph·a·lo·myo·car·di·tis \-ˌmī-ə-kär-'dī-təs\ *noun* [New Latin] (1947)
: an acute febrile virus disease characterized by degeneration and inflammation of skeletal and cardiac muscle and lesions of the central nervous system

en·ceph·a·lon \in-'se-fə-ˌlän, -lən\ *noun, plural* **-la** \-lə\ [New Latin, from Greek *enkephalos*] (1741)
: the vertebrate brain

en·ceph·a·lop·a·thy \in-ˌse-fə-'lä-pə-thē\ *noun* (1866)
: a disease of the brain; *especially* **:** one involving alterations of brain structure
— en·ceph·a·lo·path·ic \-lə-'pa-thik\ *adjective*

en·chain \in-'chān\ *transitive verb* [Middle English *encheynen,* from Middle French *enchainer,* from Old French, from *en-* + *chaeine* chain] (14th century)
: to bind or hold with or as if with chains
— en·chain·ment \-mənt\ *noun*

en·chant \in-'chant, en-\ *transitive verb* [Middle English, from Middle French *enchanter,* from Latin *incantare,* from *in-* + *cantare* to sing — more at CHANT] (14th century)
1 : to influence by or as if by charms and incantation **:** BEWITCH
2 : to attract and move deeply **:** rouse to ecstatic admiration ⟨the scene *enchanted* her to the point of tears —Elinor Wylie⟩
synonym see ATTRACT

en·chant·er *noun* (13th century)
: one that enchants; *especially* **:** SORCERER

en·chant·ing *adjective* (1606)
: CHARMING
— en·chant·ing·ly \-'chan-tiŋ-lē\ *adverb*

en·chant·ment \in-'chant-mənt, en-\ *noun* (13th century)
1 a : the act or art of enchanting **b :** the quality or state of being enchanted
2 : something that enchants

en·chant·ress \in-'chan-trəs, en-\ *noun* (14th century)
1 : a woman who practices magic **:** SORCERESS
2 : a fascinating woman

en·chase \in-'chās\ *transitive verb* [Middle English, to emboss, from Middle French *enchasser* to enshrine, set, from *en-* + *chasse* reliquary, from Latin *capsa* case — more at CASE] (15th century)
1 : ORNAMENT: as **a :** to cut or carve in relief **b :** INLAY
2 : SET ⟨*enchase* a gem⟩

en·chi·la·da \ˌen-chə-'lä-də\ *noun* [American Spanish, from feminine of *enchilado,* past participle of *enchilar* to season with chili, from Spanish *en-* ¹en- + *chile* chili] (1887)
1 : a rolled filled tortilla covered with chili sauce and usually baked
2 : SCHMEAR, BALL OF WAX ⟨the whole *enchilada*⟩

en·chi·rid·i·on \ˌen-ˌkī-'ri-dē-ən, -ˌki-'\ *noun, plural* **-rid·ia** \-dē-ə\ [Late Latin, from Greek *encheiridion,* from *en* in + *cheir* hand — more at IN, CHIR-] (15th century)
: HANDBOOK, MANUAL

-en·chy·ma \'eŋ-kə-mə\ *noun combining form, plural* **-en·chy·ma·ta** \en-'ki-mə-tə, -'kī-mə-\ *or* **-enchymas** [New Latin, from *parenchyma*]
: cellular tissue ⟨coll*enchyma*⟩

en·ci·pher \in-'sī-fər, en-\ *transitive verb* (1577)
: to convert (a message) into cipher
— en·ci·pher·er \-fər-ər\ *noun*
— en·ci·pher·ment \-fər-mənt\ *noun*

en·cir·cle \in-'sər-kəl, en-\ *transitive verb* [Middle English *enserclen*] (15th century)
1 : to form a circle around **:** SURROUND
2 : to pass completely around
— en·cir·cle·ment \-mənt\ *noun*

en clair \ä^n-'kler\ *adverb or adjective* [French, in clear] (circa 1897)
: in plain language ⟨a message sent *en clair*⟩

en·clasp \in-'klasp, en-\ *transitive verb* (1596)
: to seize and hold **:** EMBRACE

en·clave \'en-ˌklāv; ÷'än-ˌklāv, ÷'äŋ-\ *noun* [French, from Middle French, from *enclaver* to enclose, from (assumed) Vulgar Latin *inclavare* to lock up, from Latin *in-* + *clavis* key — more at CLAVICLE] (1868)
: a distinct territorial, cultural, or social unit enclosed within or as if within foreign territory ⟨ethnic *enclaves*⟩

en·clit·ic \en-'kli-tik\ *noun* [Late Latin *encliticus,* from Greek *enklitikos,* from *enklinesthai* to lean on, from *en-* + *klinein* to lean — more at LEAN] (circa 1663)
: a clitic that is associated with a preceding word
— enclitic *adjective*

en·close \in-'klōz, en-\ *transitive verb* [Middle English, probably from *enclos* enclosed, from Middle French, past participle of *enclore* to enclose, from (assumed) Vulgar Latin *inclaudere,* alteration of Latin *includere* — more at INCLUDE] (14th century)
1 a (1) **:** to close in **:** SURROUND ⟨*enclose* a porch with glass⟩ (2) **:** to fence off (common land) for individual use **b :** to hold in **:** CONFINE
2 : to include along with something else in a parcel or envelope ⟨a check is *enclosed* herewith⟩

en·clo·sure \in-'klō-zhər, en-\ *noun* (15th century)
1 : the act or action of enclosing **:** the quality or state of being enclosed
2 : something that encloses
3 : something enclosed ⟨a letter with two *enclosures*⟩

en·code \in-'kōd, en-\ *transitive verb* (circa 1919)

1 : to convert (as a body of information) from one system of communication into another; *especially* **:** to convert (a message) into code
2 : to specify the genetic code for
— en·cod·er *noun*

en·co·mi·ast \en-ˈkō-mē-ˌast, -mē-əst\ *noun* [Greek *enkōmiastēs,* from *enkōmiazein* to praise, from *enkōmion*] (1610)
: one that praises **:** EULOGIST
— en·co·mi·as·tic \-ˌkō-mē-ˈas-tik\ *adjective*

en·co·mi·um \en-ˈkō-mē-əm\ *noun, plural* **-mi·ums** *or* **-mia** \-mē-ə\ [Latin, from Greek *enkōmion,* from *en* in + *kōmos* revel, celebration] (1589)
: glowing and warmly enthusiastic praise; *also* **:** an expression of this ☆

en·com·pass \in-ˈkəm-pəs, en- *also* -ˈkäm-\ *transitive verb* [Middle English] (14th century)
1 a : to form a circle about **:** ENCLOSE **b** *obsolete* **:** to go completely around
2 a : ENVELOP **b :** INCLUDE ⟨a plan that *encompasses* a number of aims⟩
3 : BRING ABOUT, ACCOMPLISH ⟨*encompass* a task⟩
— en·com·pass·ment \-pə-smənt\ *noun*

¹en·core \ˈän-ˌkōr, -ˌkȯr\ *noun* [French, still, again] (1712)
: a demand for repetition or reappearance made by an audience; *also* **:** a reappearance or additional performance in response to such a demand

²encore *transitive verb* **en·cored; en·cor·ing** (1748)
: to request an encore of or by

¹en·coun·ter \in-ˈkau̇n-tər, en-\ *verb* **en·coun·tered; en·coun·ter·ing** \-ˈkau̇n-t(ə-)riŋ\ [Middle English *encountren,* from Middle French *encontrer,* from Medieval Latin *incontrare,* from Late Latin *incontra* toward, from Latin *in-* + *contra* against — more at COUNTER] (14th century)
transitive verb
1 a : to meet as an adversary or enemy **b :** to engage in conflict with
2 : to come upon face-to-face
3 : to come upon especially unexpectedly
intransitive verb
: to meet especially by chance

²encounter *noun* (14th century)
1 a : a meeting especially between hostile factions or persons **b :** a sudden often violent clash **:** COMBAT
2 a : a chance meeting **b :** a direct often momentary meeting
3 : a coming into the vicinity of a celestial body ⟨the Martian *encounter* of a spacecraft⟩

encounter group *noun* (1967)
: a usually unstructured group that seeks to develop the capacity of the individual to express feelings and to form emotional ties by unrestrained confrontation of individuals

en·cour·age \in-ˈkər-ij, -ˈkə-rij, en-\ *transitive verb* **-aged; -ag·ing** [Middle English *encoragen,* from Middle French *encoragier,* from Old French, from *en-* + *corage* courage] (15th century)
1 : to inspire with courage, spirit, or hope **:** HEARTEN
2 : to spur on **:** STIMULATE
3 : to give help or patronage to **:** FOSTER ☆
synonym see HEARTEN
— en·cour·ag·er *noun*

en·cour·age·ment \-ij-mənt, -rij-\ *noun* (1568)
1 : the act of encouraging **:** the state of being encouraged
2 : something that encourages

en·cour·ag·ing \-i-jiŋ, -ri-jiŋ\ *adjective* (1593)
: giving hope or promise **:** INSPIRITING
— en·cour·ag·ing·ly \-jiŋ-lē\ *adverb*

en·crim·son \in-ˈkrim-zən\ *transitive verb* (1597)
: to make or dye crimson

en·croach \in-ˈkrōch, en-\ *intransitive verb* [Middle English *encrochen* to get, seize, from Middle French *encrochier,* from Old French, from *en-* + *croc, croche* hook — more at CROCHET] (circa 1534)
1 : to enter by gradual steps or by stealth into the possessions or rights of another
2 : to advance beyond the usual or proper limits ⟨the gradually *encroaching* sea⟩
synonym see TRESPASS
— en·croach·er *noun*
— en·croach·ment \-ˈkrōch-mənt\ *noun*

en·crust \in-ˈkrəst, iŋ-\ *verb* [probably from Latin *incrustare,* from *in-* + *crusta* crust] (1641)
transitive verb
: to cover, line, or overlay with or as if with a crust
intransitive verb
: to form a crust

en·crus·ta·tion \(ˌ)in-ˌkrəs-ˈtā-shən, ˌen-\ *variant of* INCRUSTATION

en·crypt \in-ˈkript, en-\ *transitive verb* [*en-* + *crypt-* (as in *cryptogram*)] (1944)
1 : ENCIPHER
2 : ENCODE 1
— en·cryp·tion \-ˈkrip-shən\ *noun*

en·cum·ber \in-ˈkəm-bər\ *transitive verb* **-cumbered; -cum·ber·ing** \-b(ə-)riŋ\ [Middle English *encombren,* from Middle French *encombrer,* from Old French, from *en-* + (assumed) Old French *combre* dam, weir] (14th century)
1 : WEIGH DOWN, BURDEN
2 : to impede or hamper the function or activity of **:** HINDER
3 : to burden with a legal claim (as a mortgage) ⟨*encumber* an estate⟩

en·cum·brance \in-ˈkəm-brən(t)s\ *noun* (1535)
1 : something that encumbers **:** IMPEDIMENT
2 : a claim (as a mortgage) against property

en·cum·branc·er \-brən(t)-sər\ *noun* (1858)
: one that holds an encumbrance

-en·cy \ən(t)-sē, ᵊn(t)-\ *noun suffix* [Middle English *-encie,* from Latin *-entia* — more at -ENCE]
: quality or state ⟨despond*ency*⟩

¹en·cyc·li·cal \in-ˈsi-kli-kəl, en-\ *adjective* [Late Latin *encyclicus,* from Greek *enkyklios* circular, general, from *en* in + *kyklos* circle — more at IN, WHEEL] (1647)
: addressed to all the individuals of a group **:** GENERAL

²encyclical *noun* (1837)
: an encyclical letter; *specifically* **:** a papal letter to the bishops of the church as a whole or to those in one country

en·cy·clo·pe·dia *also* **en·cy·clo·pae·dia** \in-ˌsī-klə-ˈpē-dē-ə\ *noun* [Medieval Latin *encyclopaedia* course of general education, from Greek *enkyklios* + *paideia* education, child rearing, from *paid-, pais* child — more at FEW] (1644)
: a work that contains information on all branches of knowledge or treats comprehensively a particular branch of knowledge usually in articles arranged alphabetically often by subject

en·cy·clo·pe·dic *also* **en·cy·clo·pae·dic** \-ˈpē-dik\ *adjective* (1824)
: of, relating to, or suggestive of an encyclopedia or its methods of treating or covering a subject **:** COMPREHENSIVE ⟨an *encyclopedic* mind⟩ ⟨an *encyclopedic* collection of armor⟩
— en·cy·clo·pe·di·cal·ly \-di-k(ə-)lē\ *adverb*

en·cy·clo·pe·dism \-ˈpē-ˌdi-zəm\ *noun* (1833)
: the quality or state of being encyclopedic

en·cy·clo·pe·dist \-ˈpē-dist\ *noun* (1651)
1 : one who compiles or writes for an encyclopedia
2 *often capitalized* **:** one of the writers of a French encyclopedia (1751–80) who were identified with the Enlightenment and advocated deism and scientific rationalism

en·cyst \in-ˈsist, en-\ (1845)
transitive verb
: to enclose in a cyst
intransitive verb
: to form or become enclosed in a cyst
— en·cyst·ment \-ˈsis(t)-mənt\ *noun*

¹end \ˈend\ *noun* [Middle English *ende,* from Old English; akin to Old High German *enti* end, Latin *ante* before, Greek *anti* against] (before 12th century)
1 a : the part of an area that lies at the boundary **b** (1) **:** a point that marks the extent of something (2) **:** the point where something ceases to exist ⟨world without *end*⟩ **c :** the extreme or last part lengthwise **:** TIP **d :** the terminal unit of something spatial that is marked off by units **e :** a player stationed at the extremity of a line (as in football)
2 a : cessation of a course of action, pursuit, or activity **b :** DEATH, DESTRUCTION **c** (1) **:** the ultimate state (2) **:** RESULT, ISSUE
3 : something incomplete, fragmentary, or undersized **:** REMNANT
4 a : an outcome worked toward **:** PURPOSE ⟨the *end* of poetry is to be poetry —R. P. Warren⟩ **b :** the object by virtue of or for the sake of which an event takes place
5 a : a share in an undertaking ⟨kept your *end* up⟩ **b :** a particular operation or aspect of an undertaking or organization ⟨the sales *end* of the business⟩
6 : something that is extreme **:** ULTIMATE — used with *the*
7 : a period of action or turn in any of various sports events (as archery or lawn bowling)
synonym see INTENTION
— end·ed \ˈen-dəd\ *adjective*
— in the end : AFTER ALL, ULTIMATELY
— no end : EXCEEDINGLY
— on end : without a stop or letup ⟨it rained for days *on end*⟩

²end (before 12th century)
transitive verb
1 a : to bring to an end **b :** DESTROY
2 : to make up the end of
intransitive verb

☆ SYNONYMS
Encomium, eulogy, panegyric, tribute, citation mean a formal expression of praise. ENCOMIUM implies enthusiasm and warmth in praising a person or a thing. EULOGY applies to a prepared speech or writing extolling the virtues and services of a person. PANEGYRIC suggests an elaborate often poetic compliment. TRIBUTE implies deeply felt praise conveyed either through words or through a significant act. CITATION applies to the formal praise accompanying the mention of a person in a military dispatch or in awarding an honorary degree.

Encourage, inspirit, hearten, embolden mean to fill with courage or strength of purpose. ENCOURAGE suggests the raising of one's confidence especially by an external agency ⟨the teacher's praise *encouraged* the students to greater efforts⟩. INSPIRIT, somewhat literary, implies instilling life, energy, courage, or vigor into something ⟨patriots *inspirited* the people to resist⟩. HEARTEN implies the lifting of dispiritedness or despondency by an infusion of fresh courage or zeal ⟨a hospital patient *heartened* by good news⟩. EMBOLDEN implies the giving of courage sufficient to overcome timidity or reluctance ⟨*emboldened* by her first success, she tried an even more difficult climb⟩.

1 a : to come to an end **b :** to reach a specified ultimate rank or situation — usually used with *up* ⟨*ended* up as a colonel⟩
2 : DIE
synonym see CLOSE

[3]end *adjective* (13th century)
: FINAL, ULTIMATE ⟨*end* results⟩ ⟨*end* markets⟩ ⟨*end* product⟩

[4]end *transitive verb* [probably alteration of English dialect *in* (to harvest)] (1607)
dialect English **:** to put (grain or hay) into a barn or stack

end- *or* **endo-** *combining form* [French, from Greek, from *endon* within; akin to Greek *en* in, Old Latin *indu,* Hittite *andan* within — more at IN]
1 : within **:** inside ⟨*endo*skeleton⟩ — compare ECT-, EXO-
2 : taking in ⟨*endo*thermic⟩

en·dam·age \in-'da-mij\ *transitive verb* [Middle English] (14th century)
: to cause loss or damage to

end·amoe·ba \ˌen-də-'mē-bə\ *noun* [New Latin, genus name] (circa 1879)
: any of a genus (*Endamoeba*) comprising amoebas parasitic in the intestines of insects — compare ENTAMOEBA

en·dan·ger \in-'dān-jər\ *verb* **-dan·gered; -dan·ger·ing** \-'dānj-riŋ, -'dān-jə-\ (1509)
transitive verb
: to bring into danger or peril
intransitive verb
: to create a dangerous situation ⟨driving to *endanger*⟩
— en·dan·ger·ment \-'dān-jər-mənt\ *noun*

en·dan·gered *adjective* (1964)
: being or relating to an endangered species ⟨an *endangered* bird⟩ ⟨put on the *endangered* list⟩

endangered species *noun* (1964)
: a species threatened with extinction; *broadly* **:** anyone or anything whose continued existence is threatened

en·darch \'en-ˌdärk\ *adjective* (circa 1900)
: formed or taking place from inner cells outward ⟨*endarch* xylem⟩

end around *noun* (1926)
: a football play in which an offensive end comes behind the line of scrimmage to take a handoff and attempts to carry the ball around the opposite flank

end·ar·ter·ec·to·my \ˌen-ˌdär-tə-'rek-tə-mē\ *noun, plural* **-mies** [New Latin *endarter*ium intima of an artery (from *end-* + *arteria* artery) + English *-ectomy*] (1950)
: surgical removal of the inner layer of an artery when thickened and atheromatous or occluded (as by intimal plaques)

end·brain \'en(d)-ˌbrān\ *noun* (1927)
: TELENCEPHALON

end brush *noun* (circa 1891)
: END PLATE

en·dear \in-'dir\ *transitive verb* (1580)
1 *obsolete* **:** to make higher in cost, value, or estimation
2 : to cause to become beloved or admired
— en·dear·ing·ly \-iŋ-lē\ *adverb*

en·dear·ment \in-'dir-mənt\ *noun* (1610)
1 : a word or an act (as a caress) expressing affection
2 : the act or process of endearing

[1]en·deav·or \in-'de-vər\ *verb* **en·deav·ored; en·deav·or·ing** \-v(ə-)riŋ\ [Middle English *endeveren* to exert oneself, from *en-* + *dever* duty — more at DEVOIR] (15th century)
transitive verb
1 *archaic* **:** to strive to achieve or reach
2 : to attempt (as the fulfillment of an obligation) by exertion of effort ⟨*endeavors* to finish the race⟩
intransitive verb
: to work with set purpose
synonym see ATTEMPT

[2]endeavor *noun* (15th century)
1 : serious determined effort
2 : activity directed toward a goal **:** ENTERPRISE ⟨fields of *endeavor*⟩

en·deav·our *chiefly British variant of* ENDEAVOR

[1]en·dem·ic \en-'de-mik, in-\ *adjective* [French *endémique,* from *endémie* endemic disease, from Greek *endēmia* action of dwelling, from *endēmos* endemic, from *en* in + *dēmos* people, populace — more at DEMAGOGUE] (1759)
1 a : belonging or native to a particular people or country **b :** characteristic of or prevalent in a particular field, area, or environment ⟨problems *endemic* to translation⟩ ⟨the self-indulgence *endemic* in the film industry⟩
2 : restricted or peculiar to a locality or region ⟨*endemic* diseases⟩ ⟨an *endemic* species⟩
synonym see NATIVE
— en·dem·i·cal·ly \-'de-mi-k(ə-)lē\ *adverb*
— en·de·mic·i·ty \ˌen-ˌde-'mi-sə-tē, -də-'mi-\ *noun*
— en·de·mism \'en-də-ˌmi-zəm\ *noun*

[2]endemic *noun* (1926)
: an endemic organism

end·er·gon·ic \ˌen-ˌdər-'gä-nik\ *adjective* [*end-* + Greek *ergon* work — more at WORK] (1940)
: ENDOTHERMIC 1 ⟨an *endergonic* biochemical reaction⟩

end·ex·ine \(ˌ)en-'dek-ˌsēn, -ˌsīn\ *noun* (1947)
: an inner membranous layer of the exine

end·game \'en(d)-ˌgām\ *noun* (1884)
: the stage of a chess game after major reduction of forces; *also* **:** the final stages of some action

end·ing \'en-diŋ\ *noun* (before 12th century)
: something that constitutes an end: as **a :** CONCLUSION **b :** one or more letters or syllables added to a word base especially in inflection

endite *variant of* INDITE

en·dive \'en-ˌdīv, ˌän-'dēv\ *noun* [Middle English, from Middle French, from Late Latin *endivia,* from Late Greek *entybion,* from Latin *intubus*] (14th century)
1 : an annual or biennial composite herb (*Cichorium endivia*) widely cultivated as a salad plant — called also *escarole*
2 : the developing crown of chicory when blanched for use as a vegetable or in salads by growing in darkness or semidarkness

end·leaf \'end-ˌlēf\ *noun* (1888)
: ENDPAPER

end·less \'en(d)-ləs\ *adjective* (before 12th century)
1 : being or seeming to be without end ⟨an *endless* speech⟩
2 : extremely numerous ⟨all the multiplied, *endless,* nameless iniquities —Edmund Burke⟩
3 : joined at the ends ⟨an *endless* chain⟩
— end·less·ly *adverb*
— end·less·ness *noun*

end line *noun* (1893)
: a line marking an end or boundary especially of a playing area: as **a :** a line at either end of a football field 10 yards beyond and parallel to the goal line **b :** a line at either end of a court (as in basketball or tennis) perpendicular to the sidelines

end·long \'end-ˌlȯŋ\ *adverb* [Middle English *endelong,* alteration of *andlong,* from Old English *andlang* along, from *andlang,* preposition — more at ALONG] (13th century)
archaic **:** LENGTHWISE

end man *noun* (1865)
: a man at each end of the line of performers in a minstrel show who engages in comic repartee with the interlocutor

end·most \'en(d)-ˌmōst\ *adjective* (before 12th century)
: situated at the very end

end·note \'en(d)-ˌnōt\ *noun* (1926)
: a note placed at the end of the text

en·do·bi·ot·ic \ˌen-dō-ˌbī-'ä-tik, -bē-\ *adjective* [International Scientific Vocabulary] (circa 1900)
: dwelling within the cells or tissues of a host ⟨*endobiotic* fungi⟩

en·do·car·di·al \ˌen-dō-'kär-dē-əl\ *adjective* (circa 1849)
1 : situated within the heart
2 : of or relating to the endocardium

en·do·car·di·tis \-ˌkär-'dī-təs\ *noun* [New Latin] (circa 1839)
: inflammation of the lining of the heart and its valves

en·do·car·di·um \-'kär-dē-əm\ *noun, plural* **-dia** \-dē-ə\ [New Latin, from *end-* + Greek *kardia* heart] (circa 1864)
: a thin serous membrane lining the cavities of the heart

en·do·carp \'en-də-ˌkärp\ *noun* [French *endocarpe*] (1830)
: the inner layer of the pericarp of a fruit (as an apple or orange) when it consists of two or more layers of different texture or consistency

endocarp (cross section of a cherry): *1* exocarp, *2* mesocarp, *3* endocarp, *4* seed; *1, 2,* and *3* together form the pericarp

en·do·cast \'en-dō-ˌkast\ *noun* (1949)
: ENDOCRANIAL CAST

en·do·chon·dral \ˌen-də-'kän-drəl\ *adjective* (1882)
: relating to, formed by, or being ossification that takes place from centers arising in cartilage and involves deposition of lime salts in the cartilage matrix followed by secondary absorption and replacement by true bony tissue

en·do·cra·ni·al cast \ˌen-də-ˌkrā-nē-əl-\ *noun* (1923)
: a cast of the cranial cavity showing the approximate shape of the brain

[1]en·do·crine \'en-də-krən, -ˌkrīn, -ˌkrēn\ *adjective* [International Scientific Vocabulary *end-* + Greek *krinein* to separate — more at CERTAIN] (circa 1911)
1 : secreting internally; *specifically* **:** producing secretions that are distributed in the body by way of the bloodstream ⟨an *endocrine* system⟩
2 : of, relating to, affecting, or resembling an endocrine gland or secretion ⟨*endocrine* tumors⟩

[2]endocrine *noun* (1922)
1 : HORMONE
2 : ENDOCRINE GLAND

endocrine gland *noun* (1914)
: a gland (as the thyroid or the pituitary) that produces an endocrine secretion — called also *ductless gland*

en·do·cri·no·log·ic \ˌen-də-ˌkri-nᵊl-'ä-jik, -ˌkrī-, -ˌkrē-\ *or* **en·do·cri·no·log·i·cal** \-ji-kəl\ *adjective* (circa 1934)
: involving or relating to the endocrine glands or secretions or to endocrinology

en·do·cri·nol·o·gy \ˌen-də-kri-'nä-lə-jē, -ˌkrī-\ *noun* [International Scientific Vocabulary] (circa 1913)
: a science dealing with the endocrine glands
— en·do·cri·nol·o·gist \-jist\ *noun*

en·do·cy·to·sis \-sī-'tō-səs\ *noun* [New Latin, from *end-* + *-cytosis* (as in *phagocytosis*)] (1963)
: incorporation of substances into a cell by phagocytosis or pinocytosis
— en·do·cy·tot·ic \-'tä-tik\ *adjective*

en·do·derm \'en-də-ˌdərm\ *noun* [French *endoderme,* from *end-* + Greek *derma* skin — more at DERM-] (1861)
: the innermost of the germ layers of an embryo that is the source of the epithelium of the

digestive tract and its derivatives; *also* **:** a tissue that is derived from this germ layer — compare HYPOBLAST
— **en·do·der·mal** \ˌen-də-ˈdər-məl\ *adjective*

en·do·der·mis \ˌen-də-ˈdər-məs\ *noun* [New Latin] (1884)
: the innermost tissue of the cortex in many roots and stems

end·odon·tics \-ˈdän-tiks\ *noun plural but singular in construction* [*end-* + *odont-* + *-ics*] (1946)
: a branch of dentistry concerned with diseases of the pulp
— **end·odon·tic** \-ˈdän-tik\ *adjective*
— **end·odon·ti·cal·ly** \-ˈdän-ti-k(ə-)lē\ *adverb*
— **end·odon·tist** \-ˈdän-tist\ *noun*

en·do·en·zyme \ˌen-dō-ˈen-ˌzīm\ *noun* [International Scientific Vocabulary] (circa 1909)
: an enzyme that functions inside the cell

en·do·er·gic \ˌen-dō-ˈər-jik\ *adjective* (1940)
: absorbing energy **:** ENDOTHERMIC ⟨*endoergic* nuclear reactions⟩

en·dog·a·my \en-ˈdä-gə-mē\ *noun* (1865)
: marriage within a specific group as required by custom or law
— **en·dog·a·mous** \-məs\ *adjective*

en·do·gen·ic \ˌen-də-ˈje-nik\ *adjective* (circa 1904)
1 : of or relating to metamorphism taking place within a planet or moon
2 : ENDOGENOUS

en·dog·e·nous \en-ˈdä-jə-nəs\ *adjective* (1830)
1 : growing or produced by growth from deep tissue ⟨*endogenous* plant roots⟩
2 a : caused by factors inside the organism or system ⟨an *endogenous* psychic depression⟩ ⟨*endogenous* business cycles⟩ **b :** produced or synthesized within the organism or system ⟨an *endogenous* hormone⟩
— **en·dog·e·nous·ly** *adverb*

en·do·lith·ic \ˌen-də-ˈli-thik\ *adjective* (1886)
: living within or penetrating deeply into stony substances (as rocks, coral, or mollusk shells) ⟨*endolithic* lichens⟩

en·do·lymph \ˈen-də-ˌlim(p)f\ *noun* [International Scientific Vocabulary] (circa 1839)
: the watery fluid in the membranous labyrinth of the ear
— **en·do·lym·phat·ic** \ˌen-də-lim-ˈfa-tik\ *adjective*

en·do·me·tri·osis \ˌen-dō-ˌmē-trē-ˈō-səs\ *noun* [New Latin] (1925)
: the presence and growth of functioning endometrial tissue in places other than the uterus that often results in severe pain and infertility

en·do·me·tri·tis \ˌen-dō-mə-ˈtrī-təs\ *noun* [New Latin] (1872)
: inflammation of the endometrium

en·do·me·tri·um \-ˈmē-trē-əm\ *noun, plural* **-tria** \-trē-ə\ [New Latin, from *end-* + Greek *mētra* uterus, from *mētr-, mētēr* mother — more at MOTHER] (circa 1882)
: the mucous membrane lining the uterus
— **en·do·me·tri·al** \-trē-əl\ *adjective*

en·do·mi·to·sis \-mī-ˈtō-səs\ *noun* [New Latin] (1942)
: division of chromosomes not followed by nuclear division that results in an increased number of chromosomes in the cell
— **en·do·mi·tot·ic** \-mī-ˈtä-tik\ *adjective*

en·do·mix·is \-ˈmik-səs\ *noun* [New Latin, from *end-* + Greek *mixis* act of mixing, from *mignynai* to mix — more at MIX] (1914)
: a periodic nuclear reorganization in ciliated protozoans

en·do·morph \ˈen-də-ˌmȯrf\ *noun* [*endo*derm + *-morph*] (1940)
: an endomorphic individual

en·do·mor·phic \ˌen-də-ˈmȯr-fik\ *adjective* [*endo*derm + *-morphic;* from the predominance in such types of structures developed from the endoderm] (1940)
1 : of or relating to the component in W. H. Sheldon's classification of body types that measures the massiveness of the digestive viscera and the body's degree of roundedness and softness
2 : having a heavy rounded body build often with a marked tendency to become fat
— **en·do·mor·phy** \ˈen-də-ˌmȯr-fē\ *noun*

en·do·mor·phism \ˌen-də-ˈmȯr-ˌfi-zəm\ *noun* (1909)
: a homomorphism that maps a mathematical set into itself — compare ISOMORPHISM

en·do·nu·cle·ase \ˌen-dō-ˈnü-klē-ˌās, -ˌāz, -ˈnyü-\ *noun* (1962)
: an enzyme that breaks down a nucleotide chain into two or more shorter chains by cleaving it at points not adjacent to the end — compare EXONUCLEASE

en·do·nu·cleo·lyt·ic \-ˌnü-klē-ō-ˈli-tik, -ˌnyü-\ *adjective* [*end-* + *nucleo-* + *-lytic*] (1967)
: cleaving a nucleotide chain at an internal point ⟨*endonucleolytic* nicks⟩

en·do·par·a·site \-ˈpar-ə-ˌsīt\ *noun* [International Scientific Vocabulary] (circa 1882)
: a parasite that lives in the internal organs or tissues of its host
— **en·do·par·a·sit·ic** \-ˌpar-ə-ˈsi-tik\ *adjective*
— **en·do·par·a·sit·ism** \-ˈpar-ə-ˌsī-ˌti-zəm, -sə-ˌti-\ *noun*

en·do·pep·ti·dase \-ˈpep-tə-ˌdās, -ˌdāz\ *noun* (1936)
: any of a group of enzymes that hydrolyze peptide bonds within the long chains of protein molecules **:** PROTEASE — compare EXOPEPTIDASE

en·do·per·ox·ide \-pə-ˈräk-ˌsīd\ *noun* (1962)
: any of various biosynthetic intermediates in the formation of prostaglandins

en·do·phyte \ˈen-də-ˌfīt\ *noun* [International Scientific Vocabulary] (1854)
: a plant living within another plant
— **en·do·phyt·ic** \ˌen-də-ˈfi-tik\ *adjective*

en·do·plasm \ˈen-də-ˌpla-zəm\ *noun* [International Scientific Vocabulary] (1882)
: the inner relatively fluid part of the cytoplasm
— **en·do·plas·mic** \ˌen-də-ˈplaz-mik\ *adjective*

endoplasmic reticulum *noun* (1947)
: a system of interconnected vesicular and lamellar cytoplasmic membranes that functions especially in the transport of materials within the cell and that is studded with ribosomes in some places — see CELL illustration

en·dop·o·dite \en-ˈdä-pə-ˌdīt\ *noun* [International Scientific Vocabulary] (1870)
: the mesial or internal branch of a typical limb of a crustacean

en·do·poly·ploidy \ˌen-dō-ˈpä-li-ˌplȯi-dē\ *noun* (1945)
: a polyploid state in which the chromosomes have divided repeatedly without subsequent division of the nucleus or cell
— **en·do·poly·ploid** \-ˌplȯid\ *adjective*

end organ *noun* (1878)
: a structure forming the end of a neural path and consisting of an effector or a receptor with its associated nerve terminations

en·dor·phin \en-ˈdȯr-fən\ *noun* [International Scientific Vocabulary *endo*genous + *morphine*] (1976)
: any of a group of proteins with potent analgesic properties that occur naturally in the brain — compare ENKEPHALIN

en·dorse \in-ˈdȯrs, en-\ *transitive verb* **en·dorsed; en·dors·ing** [alteration of obsolete *endoss,* from Middle English *endosen,* from Middle French *endosser,* from Old French, to put on the back, from *en-* + *dos* back, from Latin *dorsum*] (1581)
1 a : to write on the back of; *especially* **:** to sign one's name as payee on the back of (a check) in order to obtain the cash or credit represented on the face **b :** to inscribe (one's signature) on a check, bill, or note **c :** to inscribe (as an official document) with a title or memorandum **d :** to make over to another (the value represented in a check, bill, or note) by inscribing one's name on the document **e :** to acknowledge receipt of (a sum specified) by one's signature on a document
2 : to approve openly ⟨*endorse* an idea⟩; *especially* **:** to express support or approval of publicly and definitely ⟨*endorse* a mayoral candidate⟩
synonym see APPROVE
— **en·dors·able** \-ˈdȯr-sə-bəl\ *adjective*
— **en·dors·ee** \in-ˌdȯr-ˈsē, ˌen-\ *noun*
— **en·dors·er** \in-ˈdȯr-sər\ *noun*

en·dorse·ment \in-ˈdȯr-smənt, en-\ *noun* (1547)
1 : the act or process of endorsing
2 a : something that is written in the process of endorsing **b :** a provision added to an insurance contract altering its scope or application
3 : SANCTION, APPROVAL

en·do·scope \ˈen-də-ˌskōp\ *noun* [International Scientific Vocabulary] (1861)
: an instrument for visualizing the interior of a hollow organ (as the rectum or urethra)
— **en·dos·co·py** \en-ˈdäs-kə-pē\ *noun*

en·do·scop·ic \ˌen-də-ˈskä-pik\ *adjective* (1861)
: of, relating to, or performed by means of an endoscope or endoscopy
— **en·do·scop·i·cal·ly** \-pi-k(ə-)lē\ *adverb*

en·do·skel·e·ton \ˌen-dō-ˈske-lə-tᵊn\ *noun* (circa 1847)
: an internal skeleton or supporting framework in an animal
— **en·do·skel·e·tal** \-lə-tᵊl\ *adjective*

en·do·sperm \ˈen-dō-ˌspərm\ *noun* [French *endosperme,* from *end-* + Greek *sperma* seed — more at SPERM] (circa 1850)
: a nutritive tissue in seed plants formed within the embryo sac

endosperm nucleus *noun* (circa 1902)
: the triploid nucleus formed in the embryo sac of a seed plant by fusion of a sperm nucleus with two polar nuclei or with a nucleus formed by the prior fusion of the polar nuclei

en·do·spore \ˈen-dō-ˌspōr, -ˌspȯr\ *noun* [International Scientific Vocabulary] (1875)
: an asexual spore developed within the cell especially in bacteria

end·os·te·al \en-ˈdäs-tē-əl\ *adjective* (circa 1868)
1 : of or relating to the endosteum
2 : located within bone or cartilage
— **end·os·te·al·ly** *adverb*

end·os·te·um \en-ˈdäs-tē-əm\ *noun, plural* **-tea** \-tē-ə\ [New Latin, from *end-* + Greek *osteon* bone — more at OSSEOUS] (circa 1881)
: the layer of vascular connective tissue lining the medullary cavities of bone

en·do·style \ˈen-dō-ˌstīl\ *noun* [International Scientific Vocabulary *end-* + Greek *stylos* pillar — more at STEER] (1854)
: a pair of parallel longitudinal folds projecting into the pharyngeal cavity and bounding a furrow lined with glandular ciliated cells in lower chordates (as the tunicates)

en·do·sul·fan \ˌen-də-ˈsəl-fən, -ˌfan\ *noun* [*endo-* + *sulf-* + ³*-an*] (1962)
: a brownish crystalline insecticide $C_9H_6Cl_6O_3S$ that is used in the control of numerous crop insects and some mites

en·do·sym·bi·o·sis \ˌen-dō-ˌsim-bī-ˈō-səs, -bē-\ *noun* (circa 1940)
: symbiosis in which a symbiont dwells within the body of its symbiotic partner
— **en·do·sym·bi·ont** \-ˈsim-bī-ˌänt, -bē-\ *noun*
— **en·do·sym·bi·ot·ic** \-ˌsim-bī-ˈä-tik, -bē-\ *adjective*

en·do·the·ci·um \ˌen-dō-ˈthē-s(h)ē-əm\ *noun, plural* **-cia** \-s(h)ē-ə\ [New Latin] (1832)

: the inner lining of a mature anther

en·do·the·li·o·ma \-ˌthē-lē-ˈō-mə\ *noun, plural* **-o·mas** *or* **-o·ma·ta** \-mə-tə\ [New Latin] (1880)
: a tumor developing from endothelial tissue

en·do·the·li·um \ˌen-də-ˈthē-lē-əm\ *noun, plural* **-lia** \-lē-ə\ [New Latin, from *end-* + *-thelium* (as in *epithelium*)] (1872)
1 : an epithelium of mesodermal origin composed of a single layer of thin flattened cells that lines internal body cavities
2 : the inner layer of the seed coat of some plants
— en·do·the·li·al \-lē-əl\ *adjective*

en·do·therm \ˈen-dō-ˌthərm\ *noun* (1946)
: a warm-blooded animal

en·do·ther·mic \ˌen-də-ˈthər-mik\ *adjective* [International Scientific Vocabulary] (1884)
1 : characterized by or formed with absorption of heat
2 : WARM-BLOODED

en·do·ther·my \ˈen-də-ˌthər-mē\ *noun* (1922)
: physiological regulation of body temperature by metabolic means; *especially* **:** the property or state of being warm-blooded

en·do·tox·in \ˌen-dō-ˈtäk-sən\ *noun* [International Scientific Vocabulary] (1904)
: a toxin of internal origin; *specifically* **:** a poisonous substance present in bacteria (as the causative agent of typhoid fever) but separable from the cell body only on its disintegration
— en·do·tox·ic \-sik\ *adjective*

en·do·tra·che·al \-ˈtrā-kē-əl\ *adjective* (1910)
1 : placed within the trachea ⟨an *endotracheal* tube⟩
2 : applied or effected through the trachea

en·do·tro·phic \ˌen-də-ˈtrō-fik\ *adjective* (1899)
of a mycorrhiza **:** penetrating into the associated root and ramifying between the cells — compare ECTOTROPHIC

en·dow \in-ˈdau̇, en-\ *transitive verb* [Middle English, from Anglo-French *endouer,* from Middle French *en-* + *douer* to endow, from Latin *dotare,* from *dot-, dos* gift, dowry — more at DATE] (15th century)
1 : to furnish with an income; *especially* **:** to make a grant of money providing for the continuing support or maintenance of ⟨*endow* a hospital⟩
2 : to furnish with a dower
3 : to provide with something freely or naturally ⟨*endowed* with a good sense of humor⟩

en·dow·ment \-mənt\ *noun* (15th century)
1 : the act or process of endowing
2 : something that is endowed; *specifically* **:** the part of an institution's income derived from donations
3 : natural capacity, power, or ability

end·pa·per \ˈen(d)-ˌpā-pər\ *noun* (1818)
: a once-folded sheet of paper having one leaf pasted flat against the inside of the front or back cover of a book and the other pasted at the base to the first or last page

end plate *noun* (1878)
: a complex terminal treelike branching of a motor nerve cell

end point *noun* (1899)
1 : a point marking the completion of a process or stage of a process; *especially* **:** a point in a titration at which a definite effect (as a color change) is observed
2 *usually* **end·point :** either of two points or values that mark the ends of a line segment or interval; *also* **:** a point that marks the end of a ray

en·drin \ˈen-drən\ *noun* [*end-* + dield*rin*] (1952)
: a chlorinated hydrocarbon insecticide $C_{12}H_8Cl_6O$ that is a stereoisomer of dieldrin and resembles dieldrin in toxicity

end run *noun* (1902)
1 : a football play in which the ballcarrier attempts to run wide around the end of the line; *specifically* **:** SWEEP 3e
2 : an evasive trick or maneuver

end–stopped \ˈen(d)-ˌstäpt\ *adjective* (1877)
: marked by a logical or rhetorical pause at the end ⟨an *end-stopped* line of verse⟩ — compare RUN-ON

end table *noun* (1851)
: a small table usually about the height of the arm of a chair that is used beside a larger piece of furniture (as a sofa)

en·due \in-ˈdü, -ˈdyü, en-\ *transitive verb* **en·dued; en·du·ing** [Middle English, from Middle French *enduire* to bring in, introduce, from Latin *inducere* — more at INDUCE] (15th century)
1 : PROVIDE, ENDOW
2 : IMBUE, TRANSFUSE
3 [Middle English *induen;* influenced by Latin *induere* to put on] **:** PUT ON, DON

en·dur·able \in-ˈdu̇r-ə-bəl, -ˈdyu̇r-, en-\ *adjective* (1800)
: capable of being endured **:** BEARABLE
— en·dur·ably \-blē\ *adverb*

en·dur·ance \in-ˈdu̇r-ən(t)s, -ˈdyu̇r-, en-\ *noun* (15th century)
1 : PERMANENCE, DURATION
2 : the ability to withstand hardship or adversity; *especially* **:** the ability to sustain a prolonged stressful effort or activity ⟨a marathon runner's *endurance*⟩
3 : the act or an instance of enduring or suffering ⟨*endurance* of many hardships⟩

en·dure \in-ˈdu̇r, -ˈdyu̇r, en-\ *verb* **en·dured; en·dur·ing** [Middle English, from Middle French *endurer,* from (assumed) Vulgar Latin *indurare,* from Latin, to harden, from *in-* + *durare* to harden, endure — more at DURING] (14th century)
transitive verb
1 : to undergo (as a hardship) especially without giving in **:** SUFFER
2 : to regard with acceptance or tolerance ⟨could not *endure* noisy children⟩
intransitive verb
1 : to continue in the same state **:** LAST
2 : to remain firm under suffering or misfortune without yielding
synonym see BEAR, CONTINUE

en·dur·ing *adjective* (15th century)
: LASTING, DURABLE
— en·dur·ing·ly \-ˈd(y)u̇r-iŋ-lē\ *adverb*
— en·dur·ing·ness *noun*

en·duro \in-ˈdu̇r-(ˌ)ō, -ˈdyu̇r-\ *noun, plural* **en·dur·os** [*endurance* + *-o* (Italian or Spanish masculine noun ending)] (1935)
: a long race (as for automobiles or motorcycles) stressing endurance rather than speed

end user *noun* (circa 1945)
: the ultimate consumer of a finished product

end·ways \ˈen-ˌdwāz\ *adverb or adjective* (circa 1608)
1 : in or toward the direction of the ends **:** LENGTHWISE ⟨*endways* pressure⟩
2 : with the end forward (as toward the observer)
3 : on end **:** UPRIGHT ⟨boxes set *endways*⟩

end·wise \ˈen-ˌdwīz\ *adverb or adjective* (1657)
: ENDWAYS

En·dym·i·on \en-ˈdi-mē-ən\ *noun* [Latin, from Greek *Endymiōn*]
: a beautiful youth loved by Selene in Greek mythology

end zone *noun* (circa 1916)
: the area at either end of a football field between the goal line and the end line

-ene \ˌēn\ *noun suffix* [International Scientific Vocabulary, from Greek *-ēnē,* feminine of *-ēnos,* adjective suffix]
: unsaturated carbon compound ⟨benz*ene*⟩; *especially* **:** carbon compound with one double bond ⟨ethyl*ene*⟩

en·e·ma \ˈe-nə-mə\ *noun, plural* **enemas** *also* **ene·ma·ta** \ˌe-nə-ˈmä-tə, ˈe-nə-mə-tə\ [Late Latin, from Greek, from *enienai* to inject, from *en-* + *hienai* to send — more at JET] (15th century)
1 : the injection of liquid into the intestine by way of the anus
2 : material for injection as an enema

en·e·my \ˈe-nə-mē\ *noun, plural* **-mies** [Middle English *enemi,* from Old French, from Latin *inimicus,* from *in-* [1]in- + *amicus* friend — more at AMIABLE] (13th century)
1 : one that is antagonistic to another; *especially* **:** one seeking to injure, overthrow, or confound an opponent
2 : something harmful or deadly
3 a : a military adversary **b :** a hostile unit or force

en·er·get·ic \ˌe-nər-ˈje-tik\ *adjective* [Greek *energētikos,* from *energein* to be active, from *energos*] (1651)
1 : operating with or marked by vigor or effect
2 : marked by energy **:** STRENUOUS
3 : of or relating to energy ⟨*energetic* equation⟩
synonym see VIGOROUS
— en·er·get·i·cal·ly \-ti-k(ə-)lē\ *adverb*

en·er·get·ics \-tiks\ *noun plural but singular in construction* (1855)
1 : a branch of mechanics that deals primarily with energy and its transformations
2 : the total energy relations and transformations of a physical, chemical, or biological system ⟨the *energetics* of an ecological community⟩

en·er·gise *British variant of* ENERGIZE

en·er·gize \ˈe-nər-ˌjīz\ *verb* **-gized; -giz·ing** (1752)
intransitive verb
: to put forth energy **:** ACT
transitive verb
1 : to make energetic, vigorous, or active
2 : to impart energy to
3 : to apply voltage to
— en·er·gi·za·tion \ˌe-nər-ˌjī-ˈzā shən\ *noun*
— en·er·giz·er *noun*

en·er·gy \ˈe-nər-jē\ *noun, plural* **-gies** [Late Latin *energia,* from Greek *energeia* activity, from *energos* active, from *en* in + *ergon* work — more at WORK] (1599)
1 a : dynamic quality ⟨narrative *energy*⟩ **b :** the capacity of acting or being active ⟨intellectual *energy*⟩
2 : vigorous exertion of power **:** EFFORT ⟨investing time and *energy*⟩
3 : the capacity for doing work
4 : usable power (as heat or electricity); *also* **:** the resources for producing such power
synonym see POWER

energy level *noun* (1910)
: one of the stable states of constant energy that may be assumed by a physical system — used especially of the quantum states of electrons in atoms and of nuclei; called also *energy state*

[1]**ener·vate** \i-ˈnər-vət\ *adjective* (1603)
: lacking physical, mental, or moral vigor **:** ENERVATED

[2]**en·er·vate** \ˈe-nər-ˌvāt\ *transitive verb* **-vat·ed; -vat·ing** [Latin *enervatus,* past participle of *enervare,* from *e-* + *nervus* sinew — more at NERVE] (1614)
1 : to reduce the mental or moral vigor of
2 : to lessen the vitality or strength of
synonym see UNNERVE
— en·er·va·tion \ˌe-nər-ˈvā-shən\ *noun*

en·fant ter·ri·ble \ä^{n}-fä^{n}-te-rēbl^{ə}\ *noun, plural* **enfants terribles** *same*\ [French, literally, terrifying child] (1851)
1 a : a child whose inopportune remarks cause embarrassment **b :** a person known for shocking remarks or outrageous behavior

2 : a usually young and successful person who is strikingly unorthodox, innovative, or avant-garde

en·fee·ble \in-ˈfē-bəl, en-\ *transitive verb* **en·fee·bled; en·fee·bling** \-b(ə-)liŋ\ [Middle English *enfeblen,* from Middle French *enfeblir,* from Old French, from *en-* + *feble* feeble] (14th century)
: to make feeble **:** deprive of strength
synonym see WEAKEN
— en·fee·ble·ment \-bəl-mənt\ *noun*

en·feoff \in-ˈfef, -ˈfēf, en-\ *transitive verb* [Middle English *enfeoffen,* from Anglo-French *enfeoffer,* from Old French *en-* + *fief* fief] (15th century)
: to invest with a fief or fee
— en·feoff·ment \-mənt\ *noun*

en·fet·ter \in-ˈfe-tər, en-\ *transitive verb* (1599)
: to bind in fetters **:** ENCHAIN

en·fe·ver \in-ˈfē-vər, en-\ *transitive verb* (1647)
: FEVER

En·field rifle \ˈen-ˌfēld-\ *noun* [*Enfield,* England] (1854)
: a .30 caliber bolt-action repeating rifle used by U.S. and British troops in World War I

[1]**en·fi·lade** \ˈen-fə-ˌlād, -ˌläd\ *noun* [French, from *enfiler* to thread, enfilade, from Old French, to thread, from *en-* + *fil* thread — more at FILE] (circa 1730)
1 : an interconnected group of rooms arranged usually in a row with each room opening into the next
2 : gunfire directed from a flanking position along the length of an enemy battle line

[2]**enfilade** *transitive verb* **-lad·ed; -lad·ing** (1706)
: to rake or be in a position to rake with gunfire in a lengthwise direction

enflame *variant of* INFLAME

en·fleu·rage \ˌä^{n}-ˌflər-ˈäzh\ *noun* [French, from *enfleurer* to saturate with the perfume of flowers, from *en-* [1]en- + *fleur* flower, from Old French *flor* — more at FLOWER] (1855)
: a process of extracting perfumes by exposing absorbents to the exhalations of flowers

en·fold \in-ˈfōld, en-\ *transitive verb* (circa 1674)
1 a : to cover with or as if with folds **:** ENVELOP **b :** to surround with a covering **:** CONTAIN
2 : to clasp within the arms **:** EMBRACE

en·force \in-ˈfōrs, -ˈfȯrs, en-\ *transitive verb* [Middle English, from Middle French *enforcier,* from Old French, from *en-* + *force* force] (14th century)
1 : to give force to **:** STRENGTHEN
2 : to urge with energy
3 : CONSTRAIN, COMPEL
4 *obsolete* **:** to effect or gain by force
5 : to carry out effectively ⟨*enforce* laws⟩
— en·force·abil·i·ty \-ˌfōr-sə-ˈbi-lə-tē, -ˌfȯr-\ *noun*
— en·force·able \-ˈfōr-sə-bəl, -ˈfȯr-\ *adjective*
— en·force·ment \-ˈfōr-smənt, -ˈfȯr-\ *noun*

en·forc·er \in-ˈfōr-sər, -ˈfȯr-\ *noun* (1580)
1 : one that enforces
2 a : a violent criminal employed by a crime syndicate; *especially* **:** HIT MAN 1 **b :** an aggressive player (as in ice hockey) known for rough play and fighting

en·frame \in-ˈfrām\ *transitive verb* (1848)
: FRAME 6
— en·frame·ment \-ˈfrām-mənt\ *noun*

en·fran·chise \in-ˈfran-ˌchīz, en-\ *transitive verb* **-chised; -chis·ing** [Middle English, from Middle French *enfranchiss-,* stem of *enfranchir,* from Old French, from *en-* + *franc* free — more at FRANK] (15th century)
1 : to set free (as from slavery)
2 : to endow with a franchise: as **a :** to admit to the privileges of a citizen and especially to the right of suffrage **b :** to admit (a municipality) to political privileges or rights
— en·fran·chise·ment \-ˌchīz-mənt, -chəz-\ *noun*

en·gage \in-ˈgāj, en-\ *verb* **en·gaged; en·gag·ing** [Middle English, from Middle French *engagier,* from Old French, from *en-* + *gage* token, gage] (15th century)
transitive verb
1 : to offer (as one's word) as security for a debt or cause
2 a *obsolete* **:** to entangle or entrap in or as if in a snare or bog **b :** to attract and hold by influence or power **c :** to interlock with **:** MESH; *also* **:** to cause (mechanical parts) to mesh
3 : to bind (as oneself) to do something; *especially* **:** to bind by a pledge to marry
4 a : to provide occupation for **:** INVOLVE ⟨*engage* him in a new project⟩ **b :** to arrange to obtain the use or services of **:** HIRE
5 a : to hold the attention of **:** ENGROSS ⟨her work *engages* her completely⟩ **b :** to induce to participate ⟨*engaged* the shy boy in conversation⟩
6 a : to enter into contest with **b :** to bring together or interlock (weapons)
7 : to deal with especially at length
intransitive verb
1 a : to pledge oneself **:** PROMISE **b :** GUARANTEE ⟨he *engages* for the honesty of his brother⟩
2 a : to begin and carry on an enterprise or activity ⟨*engaged* in trade for a number of years⟩ **b :** to take part **:** PARTICIPATE ⟨at college she *engaged* in gymnastics⟩
3 : to enter into conflict
4 : to come together and interlock (as of machinery parts) **:** be or become in gear

en·ga·gé \ˌä^{n}-ˌgä-ˈzhā\ *adjective* [French, past participle of *engager* to engage, from Middle French *engagier*] (1946)
: committed to or supportive of a cause

en·gaged \in-ˈgājd, en-\ *adjective* (1665)
1 : involved in activity **:** OCCUPIED
2 : pledged to be married **:** BETROTHED
3 : greatly interested **:** COMMITTED
4 : involved especially in a hostile encounter
5 : partly embedded in a wall ⟨an *engaged* column⟩
6 : being in gear **:** MESHED

en·gage·ment \in-ˈgāj-mənt, en-\ *noun* (1601)
1 a : an arrangement to meet or be present at a specified time and place ⟨a dinner *engagement*⟩ **b :** a job or period of employment especially as a performer
2 : something that engages **:** PLEDGE
3 a : the act of engaging **:** the state of being engaged **b :** emotional involvement or commitment ⟨seesaws between obsessive *engagement* and ambiguous detachment —Gary Taylor⟩ **c :** BETROTHAL
4 : the state of being in gear
5 : a hostile encounter between military forces

en·gag·ing *adjective* (1673)
: tending to draw favorable attention **:** ATTRACTIVE
— en·gag·ing·ly \-ˈgā-jiŋ-lē\ *adverb*

en·gar·land \in-ˈgär-lənd, en-\ *transitive verb* (1581)
: to adorn with or as if with a garland

En·gel·mann spruce \ˌeŋ-gəl-mən-\ *noun* [George *Engelmann* (died 1884) American botanist] (1908)
: a large spruce (*Picea engelmannii*) of the Rocky mountain region and British Columbia that yields a light-colored wood

en·gen·der \in-ˈjen-dər, en-\ *verb* **en·gen·dered; en·gen·der·ing** \-d(ə-)riŋ\ [Middle English *engendren,* from Middle French *engendrer,* from Latin *ingenerare,* from *in-* + *generare* to generate] (14th century)
transitive verb
1 : BEGET, PROCREATE
2 : to cause to exist or to develop **:** PRODUCE ⟨angry words *engender* strife⟩
intransitive verb
: to assume form **:** ORIGINATE

en·gild \in-ˈgild, en-\ *transitive verb* (15th century)
: to make bright with or as if with light

[1]**en·gine** \ˈen-jən\ *noun* [Middle English *engin,* from Middle French, from Latin *ingenium* natural disposition, talent, from *in-* + *gignere* to beget — more at KIN] (13th century)
1 *obsolete* **a :** INGENUITY **b :** evil contrivance **:** WILE
2 : something used to effect a purpose **:** AGENT, INSTRUMENT ⟨mournful and terrible *engine* of horror and of crime —E. A. Poe⟩
3 a : a mechanical tool: as (1) **:** an instrument or machine of war (2) *obsolete* **:** a torture implement **b :** MACHINERY **c :** any of various mechanical appliances — often used in combination ⟨fire *engine*⟩
4 : a machine for converting any of various forms of energy into mechanical force and motion; *also* **:** a mechanism or object that serves as an energy source ⟨black holes may be the *engines* for quasars⟩
5 : a railroad locomotive

[2]**engine** *transitive verb* **en·gined; en·gin·ing** (1868)
: to equip with engines

engine driver *noun* (1828)
British **:** ENGINEER 4

[1]**en·gi·neer** \ˌen-jə-ˈnir\ *noun* [alteration of earlier *enginer,* from Middle English, alteration of *enginour,* from Middle French *engigneur,* from Old French *engignier* to contrive, from *engin*] (14th century)
1 : a member of a military group devoted to engineering work
2 *obsolete* **:** a crafty schemer **:** PLOTTER
3 a : a designer or builder of engines **b :** a person who is trained in or follows as a profession a branch of engineering **c :** a person who carries through an enterprise by skillful or artful contrivance
4 : a person who runs or supervises an engine or an apparatus

[2]**engineer** *transitive verb* (1843)
1 : to lay out, construct, or manage as an engineer
2 a : to contrive or plan out usually with more or less subtle skill and craft **b :** to guide the course of
3 : to modify or produce by genetic engineering ⟨grain crops *engineered* to require fewer nutrients and produce higher yields⟩
synonym see GUIDE

en·gi·neer·ing *noun* (1720)
1 : the activities or function of an engineer
2 a : the application of science and mathematics by which the properties of matter and the sources of energy in nature are made useful to people **b :** the design and manufacture of complex products ⟨software *engineering*⟩
3 : calculated manipulation or direction (as of behavior) ⟨social *engineering*⟩ — compare GENETIC ENGINEERING

en·gine·ry \ˈen-jən-rē\ *noun* (1641)
: instruments of war

en·gird \in-ˈgərd, en-\ *transitive verb* (1566)
archaic **:** GIRD, ENCOMPASS

en·gir·dle \in-ˈgər-d^{ə}l, en-\ *transitive verb* (1602)
: to encircle with or as if with a girdle

[1]**En·glish** \ˈiŋ-glish, ˈiŋ-lish\ *adjective* [Middle English, from Old English *englisc,* from *Engle* (plural) Angles] (before 12th century)
: of, relating to, or characteristic of England, the English people, or the English language
— En·glish·ness *noun*

[2]**English** *noun* (before 12th century)
1 a : the language of the people of England and the U.S. and many areas now or formerly under British control **b :** a particular variety of English distinguished by peculiarities (as of pronunciation) **c :** English language, literature, or composition when a subject of study
2 *plural in construction* **:** the people of England

3 a : an English translation **b :** idiomatic or intelligible English
4 : spin around the vertical axis deliberately imparted to a ball that is driven or rolled — compare DRAW, FOLLOW; BODY ENGLISH

³English *transitive verb* (14th century)
1 : to translate into English
2 : to adopt into English **:** ANGLICIZE

English breakfast *noun* (1807)
1 : a substantial breakfast (as of eggs, ham or bacon, toast, and cereal)
2 : CONGOU; *broadly* **:** any similar black tea

English cocker spaniel *noun* (1948)
: any of a breed of spaniels that have square muzzles, wide well-developed noses, and distinctive heads which are ideally half muzzle and half skull with the forehead and skull arched and slightly flattened

English cocker spaniel

English daisy *noun* (circa 1890)
: DAISY 1a

English foxhound *noun* (1929)
: any of a breed of medium-sized foxhounds developed in England and characterized by a muscular body, bi- or tri-colored short coat, and lightly fringed tail

English horn *noun* [translation of Italian *corno inglese*] (1838)
: a double-reed woodwind instrument resembling the oboe in design but having a longer tube and a range a fifth lower than that of the oboe

English ivy *noun* (1624)
: IVY 1

En·glish·man \ˈiŋ-glish-mən, ˈiŋ-lish-\ *noun* (before 12th century)
: a native or inhabitant of England

English muffin *noun* (1902)
: bread dough rolled and cut into rounds, baked on a griddle, and split and toasted just before eating

English pea *noun* (1634)
Southern **:** PEA 1a, b

En·glish·ry \ˈiŋ-glish-rē, ˈiŋ-lish-\ *noun* (1620)
: the state, fact, or quality of being English **:** ENGLISHNESS

English saddle *noun* (1817)
: a saddle with long side bars, steel cantle and pommel, no horn, and a leather seat supported by webbing stretched between the saddlebow and cantle — see SADDLE illustration

English setter *noun* (1859)
: any of a breed of dogs often trained as bird dogs and characterized by a moderately long flat silky coat of white or white with color and by feathering on the tail and legs

English shepherd *noun* (1950)
: any of a breed of vigorous medium-sized working dogs with a long and glossy black coat usually with tan to brown markings that was developed in England for herding sheep and cattle

English sonnet *noun* (circa 1903)
: a sonnet consisting of three quatrains and a couplet with a rhyme scheme of *abab cdcd efef gg* — called also *Shakespearean sonnet*

English sparrow *noun* (1876)
: HOUSE SPARROW

English springer spaniel *noun* (1929)
: any of a breed of springer spaniels having a muscular build and a moderately long silky coat usually of black and white or liver and white hair — called also *English springer*

English system *noun* (1927)
: the foot-pound-second system of units

English toy spaniel *noun* (circa 1934)
: any of a breed of small blocky spaniels with well-rounded upper skull projecting forward toward the short turned-up nose

English walnut *noun* (1772)
: a Eurasian walnut (*Juglans regia*) valued for its large edible nut and its hard richly figured wood; *also* **:** its nut

English toy spaniel

En·glish·wom·an \ˈiŋ-glish-ˌwu̇-mən *also* ˈiŋ-lish-\ *noun* (15th century)
: a woman of English birth, nationality, or origin

English yew *noun* (circa 1930)
: YEW 1a(1)

en·gorge \in-ˈgȯrj, en-\ *verb* [Middle French *engorgier,* from Old French, to devour, from *en-* + *gorge* throat — more at GORGE] (1515)
transitive verb
: GORGE, GLUT; *especially* **:** to fill with blood to the point of congestion
intransitive verb
: to suck blood to the limit of body capacity
— en·gorge·ment \-mənt\ *noun*

en·graft \in-ˈgraft, en-\ *transitive verb* (1585)
1 : to join or fasten as if by grafting
2 : GRAFT 1, 3 ⟨*engrafted* embryonic gill tissue into the back⟩
— en·graft·ment \-ˈgraf(t)-mənt\ *noun*

en·grailed \in-ˈgrāld, en-\ *adjective* [Middle English *engreled,* from Middle French *engreslé,* from *en-* + *gresle* slender, from Latin *gracilis*] (15th century)
1 : indented with small concave curves ⟨an *engrailed* heraldic bordure⟩
2 : made of or bordered by a circle of raised dots ⟨an *engrailed* coin⟩

en·grain \in-ˈgrān\ *transitive verb* (circa 1641)
: INGRAIN

en·gram *also* **en·gramme** \ˈen-ˌgram\ *noun* [International Scientific Vocabulary] (1908)
: a hypothetical change in neural tissue postulated in order to account for persistence of memory

en·grave \in-ˈgrāv, en-\ *transitive verb* **en·graved; en·grav·ing** [Middle French *engraver,* from *en-* + *graver* to grave, of Germanic origin; akin to Old English *grafan* to grave] (1509)
1 a : to form by incision (as on wood or metal) **b :** to impress deeply as if with a graver ⟨the incident was *engraved* in his memory⟩
2 a : to cut figures, letters, or designs on for printing; *also* **:** to print from an engraved plate **b :** PHOTOENGRAVE
— en·grav·er *noun*

en·grav·ing *noun* (1601)
1 : the act or process of one that engraves
2 : something that is engraved: as **a :** an engraved printing surface **b :** engraved work
3 : an impression from an engraved printing surface

en·gross \in-ˈgrōs, en-\ *transitive verb* [Middle English, from Anglo-French *engrosser,* probably from Medieval Latin *ingrossare,* from Latin *in* + Medieval Latin *grossa* large handwriting, from Latin, feminine of *grossus* thick] (15th century)
1 a : to copy or write in a large hand **b :** to prepare the usually final handwritten or printed text of (an official document)
2 [Middle English, from Middle French *en gros* in large quantities] **a :** to purchase large quantities of (as for speculation) **b** *archaic* **:** AMASS, COLLECT **c :** to take or engage the whole attention of **:** occupy completely ⟨ideas that have *engrossed* the minds of scholars for generations⟩
— en·gross·er *noun*

en·gross·ing \-ˈgrō-siŋ\ *adjective* (1820)
: taking up the attention completely **:** ABSORBING
— en·gross·ing·ly \-siŋ-lē\ *adverb*

en·gross·ment \in-ˈgrō-smənt, en-\ *noun* (1526)
1 : the act of engrossing
2 : the state of being absorbed or occupied **:** PREOCCUPATION

en·gulf \in-ˈgəlf, en-\ *transitive verb* (1555)
1 : to flow over and enclose **:** OVERWHELM ⟨the mounting seas threatened to *engulf* the island⟩
2 : to take in (food) by or as if by flowing over and enclosing
— en·gulf·ment \-mənt\ *noun*

en·ha·lo \in-ˈhā-(ˌ)lō, en-\ *transitive verb* (1842)
: to surround with or as if with a halo

en·hance \in-ˈhan(t)s, en-\ *transitive verb* **en·hanced; en·hanc·ing** [Middle English *enhauncen,* from Anglo-French *enhauncer,* alteration of Old French *enhaucier,* from (assumed) Vulgar Latin *inaltiare,* from Latin *in* + *altus* high — more at OLD] (13th century)
1 *obsolete* **:** RAISE
2 : HEIGHTEN, INCREASE; *especially* **:** to increase or improve in value, quality, desirability, or attractiveness
— en·hance·ment \-ˈhan(t)-smənt\ *noun*

enhanced recovery *noun* (1970)
: the extraction of oil from a nearly exhausted well by methods more costly and complex than waterflooding alone

en·hanc·er \in-ˈhan(t)-sər, en-\ *noun* (14th century)
1 : one that enhances
2 : a nucleotide sequence that increases the rate of genetic transcription by preferentially increasing the activity of the nearest promoter on the same DNA molecule

en·har·mon·ic \ˌen-(ˌ)här-ˈmä-nik\ *adjective* [French *enharmonique,* from Middle French, of a scale employing quarter tones, from Greek *enarmonios,* from *en* in + *harmonia* harmony, scale] (1794)
: of, relating to, or being notes that are written differently (as A flat and G sharp) but sound the same in the tempered scale
— en·har·mon·i·cal·ly \-ni-k(ə-)lē\ *adverb*

enig·ma \i-ˈnig-mə, e-\ *noun* [Latin *aenigma,* from Greek *ainigmat-, ainigma,* from *ainissesthai* to speak in riddles, from *ainos* fable] (15th century)
1 : an obscure speech or writing
2 : something hard to understand or explain
3 : an inscrutable or mysterious person
synonym see MYSTERY

enig·mat·ic \ˌe-(ˌ)nig-ˈma-tik *also* ˌē-(ˌ)nig-\ *also* **enig·mat·i·cal** \-ti-kəl\ *adjective* (1677)
: of, relating to, or resembling an enigma **:** MYSTERIOUS
synonym see OBSCURE
— enig·mat·i·cal·ly \-ti-k(ə-)lē\ *adverb*

en·isle \in-ˈī(ə)l, en-\ *transitive verb* (1612)
1 : to place apart **:** ISOLATE
2 : to make an island of

en·jamb·ment \in-ˈjam-mənt\ *or* **en·jambe·ment** *same or* äⁿ-zhäⁿb-mäⁿ\ *noun* [French *enjambement,* from Middle French, encroachment, from *enjamber* to straddle, encroach on, from *en-* + *jambe* leg — more at JAMB] (1839)
: the running over of a sentence from one verse or couplet into another so that closely related words fall in different lines — compare RUN-ON

en·join \in-ˈjȯin, en-\ *transitive verb* [Middle English, from Old French *enjoindre,* from Latin *injungere,* from *in-* + *jungere* to join — more at YOKE] (13th century)
1 : to direct or impose by authoritative order or with urgent admonition ⟨*enjoined* us to be careful⟩

2 a : FORBID, PROHIBIT ⟨was *enjoined* by conscience from telling a lie⟩ **b :** to prohibit by a judicial order **:** put an injunction on ⟨a book had been *enjoined* prior to publication —David Margolick⟩
synonym see COMMAND

en·joy \in-ˈjȯi, en-\ *verb* [Middle English *enjoie,* from Middle French *enjoir,* from Old French, from *en-* + *joir* to enjoy, from Latin *gaudēre* to rejoice — more at JOY] (15th century)
intransitive verb
: to have a good time
transitive verb
1 : to have for one's use, benefit, or lot **:** EXPERIENCE ⟨*enjoyed* great success⟩
2 : to take pleasure or satisfaction in
— en·joy·able \-ə-bəl\ *adjective*
— en·joy·able·ness *noun*
— en·joy·ably \-blē\ *adverb*
— en·joy·er *noun*
— enjoy oneself : to have a good time

en·joy·ment \in-ˈjȯi-mənt\ *noun* (1553)
1 a : the action or state of enjoying **b :** possession and use ⟨the *enjoyment* of civic rights⟩
2 : something that gives keen satisfaction

en·keph·a·lin \in-ˈke-fə-lən, -(ˌ)lin, en-\ *noun* [*enkephal-* (alteration of *encephal-*) + *-in*] (1975)
: either of two pentapeptides with opiate and analgesic activity that occur naturally in the brain and have a marked affinity for opiate receptors — compare ENDORPHIN

en·kin·dle \in-ˈkin-dᵊl, en-\ (1542)
transitive verb
1 : to set (as fuel) on fire
2 : to make bright and glowing
intransitive verb
: to take fire **:** FLAME

en·lace \in-ˈlās, en-\ *transitive verb* [Middle English, from Middle French *enlacier,* from Old French, from *en-* + *lacier* to lace] (14th century)
1 : ENCIRCLE, ENFOLD
2 : ENTWINE, INTERLACE

en·lace·ment \in-ˈlā-smənt, en-\ *noun* (1830)
1 : the process or result of interlacing
2 : a pattern of interlacing elements

en·large \in-ˈlärj, en-\ *verb* **en·larged; en·larg·ing** [Middle English, from Middle French *enlargier,* from Old French, from *en-* + *large* large, abundant] (14th century)
transitive verb
1 : to make larger **:** EXTEND
2 : to give greater scope to **:** EXPAND
3 : to set free (as a captive)
intransitive verb
1 : to grow larger
2 : to speak or write at length **:** ELABORATE ⟨let me *enlarge* upon that point⟩
synonym see INCREASE
— en·large·able \-ˈlär-jə-bəl\ *adjective*
— en·larg·er *noun*

en·large·ment \in-ˈlärj-mənt, en-\ *noun* (1540)
1 : an act or instance of enlarging **:** the state of being enlarged
2 : a photographic print larger than the negative that is made by projecting the negative image through a lens onto a photographic printing surface

en·light·en \in-ˈlī-tᵊn, en-\ *transitive verb* **en·light·ened; en·light·en·ing** \-ˈlīt-niŋ, -tᵊn-iŋ\ (1587)
1 *archaic* **:** ILLUMINATE
2 a : to furnish knowledge to **:** INSTRUCT **b :** to give spiritual insight to

en·light·ened *adjective* (1652)
1 : freed from ignorance and misinformation ⟨an *enlightened* people⟩
2 : based on full comprehension of the problems involved ⟨issued an *enlightened* ruling⟩

en·light·en·ment \in-ˈlī-tᵊn-mənt, en-\ *noun* (1669)
1 : the act or means of enlightening **:** the state of being enlightened
2 *capitalized* **:** a philosophic movement of the 18th century marked by a rejection of traditional social, religious, and political ideas and an emphasis on rationalism — used with *the*
3 *Buddhism* **:** a final blessed state marked by the absence of desire or suffering

en·list \in-ˈlist, en-\ (1599)
transitive verb
1 a : to secure the support and aid of **:** employ in advancing an interest ⟨*enlist* all the available resources⟩ ⟨*enlist* the community in an experiment⟩ **b :** to win over **:** ATTRACT ⟨trying to *enlist* my sympathies⟩
2 : to engage (a person) for duty in the armed forces
intransitive verb
1 : to enroll oneself in the armed forces
2 : to participate heartily (as in a cause, drive, or crusade)
— en·list·ee \-ˌlis-ˈtē, -ˈlis-tē\ *noun*
— en·list·ment \-ˈlis(t)-mənt\ *noun*

en·list·ed \-ˈlis-təd\ *adjective* (1724)
: of, relating to, or constituting the part of a military or naval force below commissioned or warrant officers

enlisted man *noun* (1724)
: a man or woman in the armed forces ranking below a commissioned or warrant officer; *specifically* **:** one ranking below a noncommissioned officer or petty officer

en·liv·en \in-ˈlī-vən, en-\ *transitive verb* (1604)
: to give life, action, or spirit to **:** ANIMATE
synonym see QUICKEN

en masse \äⁿ(n)-ˈmas, -ˈmäs, en-\ *adverb* [French] (1795)
: in a body **:** as a whole

en·mesh \in-ˈmesh, en-\ *transitive verb* (1604)
: to catch or entangle in or as if in meshes
— en·mesh·ment \-mənt\ *noun*

en·mi·ty \ˈen-mə-tē\ *noun, plural* **-ties** [Middle English *enmite,* from Middle French *enemité,* from Old French *enemisté,* from *enemi* enemy] (13th century)
: positive, active, and typically mutual hatred or ill will ☆

en·ne·ad \ˈe-nē-ˌad\ *noun* [Greek *ennead-, enneas,* from *ennea* nine — more at NINE] (1550)
: a group of nine

en·no·ble \i-ˈnō-bəl, e-ˈnō-\ *transitive verb* **en·no·bled; en·no·bling** \-b(ə-)liŋ\ [Middle English *ennobelen,* from Middle French *ennoblir,* from Old French, from *en-* + *noble* noble] (15th century)
1 : to make noble **:** ELEVATE ⟨seemed *ennobled* by suffering⟩
2 : to raise to the rank of nobility
— en·no·ble·ment \-bəl-mənt\ *noun*

en·nui \ˌän-ˈwē\ *noun* [French, from Old French *enui* annoyance, from *enuier* to annoy — more at ANNOY] (1732)
: a feeling of weariness and dissatisfaction **:** BOREDOM

Enoch \ˈē-nək, -nik\ *noun* [Greek *Enōch,* from Hebrew *Ḥănōkh*]
: an Old Testament patriarch and father of Methuselah

eno·ki·da·ke \e-ˌnō-kē-ˈdä-kē\ *noun* [Japanese *enokitake,* from *enoki* Chinese hackberry (*Celtis sinensis*) + *take* mushroom] (1983)
: ENOKI MUSHROOM

enoki mushroom \e-ˈnō-kē-\ *noun* [Japanese *enoki*] (circa 1977)
: a small white edible mushroom (*Flammulina velutipes* synonym *Collybia velutipes* of the family Agaricaceae) — called also *enoki*

enol \ˈē-ˌnȯl, -ˌnōl\ *noun* [International Scientific Vocabulary *ene-* (from *-ene*) + *-ol*] (1904)
: an organic compound that contains a hydroxyl group bonded to a carbon atom having a double bond and that is usually characterized by the grouping C=C(OH)
— eno·lic \ē-ˈnō-lik, -ˈnä-lik\ *adjective*

eno·lase \ˈē-nə-ˌlās, -ˌlāz\ *noun* [International Scientific Vocabulary *enol* + *-ase*] (1942)
: a crystalline enzyme that is found especially in muscle and yeast and is important in the metabolism of carbohydrates

enol·o·gy \ē-ˈnä-lə-jē\ *noun* [Greek *oinos* wine + English *-logy* — more at WINE] (1814)
: a science that deals with wine and wine making
— eno·log·i·cal \ˌē-nə-ˈlä-ji-kəl\ *adjective*
— enol·o·gist \ē-ˈnä-lə-jist\ *noun*

enor·mi·ty \i-ˈnȯr-mə-tē\ *noun, plural* **-ties** (15th century)
1 : an outrageous, improper, vicious, or immoral act ⟨the *enormities* of state power —Susan Sontag⟩ ⟨other *enormities* too juvenile to mention —Richard Freedman⟩
2 : the quality or state of being immoderate, monstrous, or outrageous; *especially* **:** great wickedness ⟨the *enormity* of the crimes committed during the Third Reich —G. A. Craig⟩
3 : the quality or state of being huge **:** IMMENSITY
4 : a quality of momentous importance ■

enor·mous \i-ˈnȯr-məs, ē-\ *adjective* [Latin *enormis,* from *e, ex* out of + *norma* rule] (1531)
1 a *archaic* **:** ABNORMAL, INORDINATE **b :** exceedingly wicked **:** SHOCKING ⟨an *enormous* sin⟩

☆ SYNONYMS
Enmity, hostility, antipathy, antagonism, animosity, rancor, animus mean deep-seated dislike or ill will. ENMITY suggests positive hatred which may be open or concealed ⟨an unspoken *enmity*⟩. HOSTILITY suggests an enmity showing itself in attacks or aggression ⟨*hostility* between the two nations⟩. ANTIPATHY and ANTAGONISM imply a natural or logical basis for one's hatred or dislike, ANTIPATHY suggesting repugnance, a desire to avoid or reject, and ANTAGONISM suggesting a clash of temperaments leading readily to hostility ⟨a natural *antipathy* for self-seekers⟩ ⟨*antagonism* between the brothers⟩. ANIMOSITY suggests intense ill will and vindictiveness that threaten to kindle hostility ⟨*animosity* that led to revenge⟩. RANCOR is especially applied to bitter brooding over a wrong ⟨*rancor* filled every line of his letters⟩. ANIMUS adds to animosity the implication of strong prejudice ⟨objections devoid of personal *animus*⟩.

□ USAGE
enormity *Enormity,* some people insist, is improperly used to denote large size. They insist on *enormousness* for this meaning, and would limit *enormity* to the meaning "great wickedness." Those who urge such a limitation may not recognize the subtlety with which *enormity* is actually used. It regularly denotes a considerable departure from the expected or normal ⟨they awakened; they sat up; and then the *enormity* of their situation burst upon them. "How did the fire start?" —John Steinbeck⟩. When used to denote large size, either literal or figurative, it usually suggests something so large as to seem overwhelming ⟨no intermediate zone of study. Either the *enormity* of the desert or the sight of a tiny flower —Paul Theroux⟩ ⟨the *enormity* of the task of teachers in slum schools —J. B. Conant⟩ and may even be used to suggest both great size and deviation from morality ⟨the *enormity* of existing stockpiles of atomic weapons —*New Republic*⟩. It can also emphasize the momentousness of what has happened ⟨the sombre *enormity* of the Russian Revolution —George Steiner⟩ or of its consequences ⟨perceived as no one in the family could the *enormity* of the misfortune —E. L. Doctorow⟩.

2 : marked by extraordinarily great size, number, or degree; *especially* **:** exceeding usual bounds or accepted notions ☆
— enor·mous·ly *adverb*
— enor·mous·ness *noun*

eno·sis \i-ˈnō-səs\ *noun* [New Greek *henōsis,* from Greek, union, from *henoun* to unite, from *hen-, heis* one — more at SAME] (1928)
: a movement to secure the political union of Greece and Cyprus

[1]**enough** \i-ˈnəf; *after* t, d, s, z *often* ᵊn-ˈəf\ *adjective* [Middle English *ynough,* from Old English *genōg* (akin to Old High German *ginuog* enough), from *ge-* (perfective prefix) + *-nōg;* akin to Latin *nancisci* to get, Greek *enenkein* to carry — more at CO-] (before 12th century)
: occurring in such quantity, quality, or scope as to fully meet demands, needs, or expectations
synonym see SUFFICIENT

[2]**enough** *adverb* (before 12th century)
1 : in or to a degree or quantity that satisfies or that is sufficient or necessary for satisfaction **:** SUFFICIENTLY
2 : FULLY, QUITE
3 : in a tolerable degree

[3]**enough** *pronoun* (before 12th century)
: a sufficient number, quantity, or amount ⟨*enough* were present to constitute a quorum⟩ ⟨had *enough* of their foolishness⟩

enounce \ē-ˈnau̇n(t)s\ *transitive verb* **enounced; enounc·ing** [French *énoncer,* from Latin *enuntiare* to report — more at ENUNCIATE] (1805)
1 : to set forth or state (as a proposition)
2 : to pronounce distinctly **:** ARTICULATE

enow \i-ˈnau̇, i-ˈnō\ *adverb or adjective* [Middle English *inow,* from Old English *genōg*] (before 12th century)
archaic **:** ENOUGH

en pas·sant \ˌäⁿ-ˌpä-ˈsäⁿ, -pə-\ *adverb* [French] (1665)
1 : in passing
2 — used in chess of the capture of a pawn as it makes a first move of two squares by an enemy pawn that threatens the first of these squares

en·plane \in-ˈplān, en-\ *intransitive verb* (1941)
: to board an airplane

en prise \äⁿ-prēz\ *adjective* [French, literally, engaged, within grasp] (1899)
of a chess piece **:** exposed to capture

en·quire \in-ˈkwīr\, **en·qui·ry** \ˈin-ˌkwīr-ē, in-ˈ; ˈin-kwə-rē, ˈiŋ-\ *variant of* INQUIRE, INQUIRY

en·rage \in-ˈrāj, en-\ *transitive verb* [Middle French *enrager* to become mad, from Old French *enragier,* from *en-* + *rage* rage] (1589)
: to fill with rage **:** ANGER

en·rapt \in-ˈrapt, en-\ *adjective* (1606)
: wholly absorbed with rapture

en·rap·ture \in-ˈrap-chər, en-\ *transitive verb* **en·rap·tured; en·rap·tur·ing** \-ˈrap-chə-riŋ, -ˈrap-shriŋ\ (1740)
: to fill with delight

en·reg·is·ter \in-ˈre-jə-stər, en-\ *transitive verb* [Middle French *enregistrer,* from Old French, from *en-* + *registre* register] (1523)
: to put on record **:** REGISTER

en·rich \in-ˈrich, en-\ *transitive verb* [Middle English, from Middle French *enrichir,* from Old French, from *en-* + *riche* rich] (14th century)
: to make rich or richer especially by the addition or increase of some desirable quality, attribute, or ingredient ⟨the experience will *enrich* your life⟩: as **a :** to add beauty to **:** ADORN **b :** to enhance the taste of ⟨butter will *enrich* the sauce⟩ **c :** to make (a soil) more fertile **d :** to improve the nutritive value of (a food) by adding nutrients (as vitamins or amino acids) and especially by restoring part of the nutrients lost in processing ⟨*enriched* flour⟩ **e :** to process so as to add or increase the proportion of a desirable ingredient ⟨*enriched* uranium⟩ ⟨*enriched* natural gas⟩
— en·rich·er *noun*
— en·rich·ment \-ˈrich-mənt\ *noun*

en·robe \in-ˈrōb, en-\ *transitive verb* (1593)
1 : to cover with or as if with a robe
2 : COAT 2

en·roll *or* **en·rol** \in-ˈrōl, en-\ *verb* **en·rolled; en·roll·ing** [Middle English, from Middle French *enroller,* from *en-* + *rolle* roll, register] (14th century)
transitive verb
1 : to insert, register, or enter in a list, catalog, or roll ⟨the school *enrolls* about 800 pupils⟩
2 : to prepare a final perfect copy of (a bill passed by a legislature) in written or printed form
3 : to roll or wrap up
intransitive verb
: to enroll oneself or cause oneself to be enrolled ⟨we *enrolled* in the history course⟩
— en·roll·ee \-rō-ˈlē\ *noun*
— en·roll·ment \-ˈrōl-mənt\ *noun*

en·root \in-ˈrüt, -ˈru̇t\ *transitive verb* [Middle English] (15th century)
: ESTABLISH, IMPLANT

en route \äⁿ(n)-ˈrüt, en-, in-, -ˈrau̇t\ *adverb or adjective* [French] (1779)
: on or along the way ⟨he reads *en route*⟩ ⟨arrived early despite *en route* delays⟩

en·sam·ple \in-ˈsam-pəl\ *noun* [Middle English, from Middle French *ensample, example*] (13th century)
archaic **:** EXAMPLE, INSTANCE

en·san·guine \in-ˈsaŋ-gwən\ *transitive verb* **-guined; -guin·ing** (1667)
1 : to make bloody
2 : CRIMSON

en·sconce \in-ˈskän(t)s\ *transitive verb* **en·sconced; en·sconc·ing** [*en-* + [2]*sconce*] (1598)
1 : SHELTER, CONCEAL
2 : ESTABLISH, SETTLE ⟨*ensconced* in a new job⟩

[1]**en·sem·ble** \än-ˈsäm-bəl, äⁿ-\ *noun* [French, from *ensemble* together, from Latin *insimul* at the same time, from *in-* + *simul* at the same time — more at SAME] (1750)
: a group producing a single effect: as **a :** concerted music of two or more parts **b :** a complete costume of harmonizing or complementary clothing and accessories **c** (1) **:** the musicians engaged in the performance of a musical ensemble (2) **:** a group of supporting players, singers, or dancers; *especially* **:** CORPS DE BALLET

[2]**ensemble** *adjective* (circa 1911)
: emphasizing the roles of all performers as a whole rather than a star performance ⟨*ensemble* acting⟩

en·serf \in-ˈsərf, en-\ *transitive verb* (1882)
: to make a serf of **:** deprive of liberty and personal rights
— en·serf·ment \-mənt\ *noun*

en·sheathe \in-ˈshēth, en-\ *transitive verb* (1593)
: to cover with or as if with a sheath

en·shrine \in-ˈshrīn, en-, *especially Southern* -ˈsrīn\ *transitive verb* [Middle English] (14th century)
1 : to enclose in or as if in a shrine
2 : to preserve or cherish as sacred
— en·shrine·ment \-mənt\ *noun*

en·shri·nee \in-ˌshrī-nē, -ˌshrī-ˈnē, en-\ *noun* (1968)
: a person inducted into a Hall of Fame

en·shroud \in-ˈshrau̇d, en-, *especially Southern* -ˈsrau̇d\ *transitive verb* (1583)
: to cover or enclose with or as if with a shroud

en·si·form \ˈen(t)-sə-ˌfȯrm\ *adjective* [French *ensiforme,* from Latin *ensis* sword + French *-forme* -form; akin to Sanskrit *asi* sword] (1541)
: having sharp edges and tapering to a slender point ⟨*ensiform* leaves of the gladiolus⟩ — see LEAF illustration

en·sign \ˈen(t)-sən, *also* ˈen-ˌsīn *for 1, 2, & 3a*\ *noun* [Middle English *ensigne,* from Middle French *enseigne,* from Latin *insignia* insignia, flags] (15th century)
1 : a flag that is flown (as by a ship) as the symbol of nationality and that may also be flown with a distinctive badge added to its design
2 a : a badge of office, rank, or power **b :** EMBLEM, SIGN
3 a : an infantry officer of what was formerly the lowest commissioned rank **b :** a commissioned officer in the navy or coast guard ranking above a chief warrant officer and below a lieutenant junior grade

en·si·lage \ˈen(t)-s(ə-)lij, *for 1 also* in-ˈsī-lij\ *noun* [French, from *ensiler* to ensile, from *en-* + *silo* silo, from Spanish] (1876)
1 : the process of preserving fodder by ensiling
2 : SILAGE

en·sile \en-ˈsī(ə)l, in-\ *transitive verb* **en·siled; en·sil·ing** (1883)
: to prepare and store (fodder) for silage

en·sky \in-ˈskī, en-\ *transitive verb* (1603)
: EXALT

en·slave \in-ˈslāv, en-\ *transitive verb* (1630)
: to reduce to or as if to slavery **:** SUBJUGATE
— en·slave·ment \-mənt\ *noun*
— en·slav·er *noun*

en·snare \in-ˈsnar, -ˈsner, en-\ *transitive verb* (1576)
: to take in or as if in a snare
synonym see CATCH

en·snarl \in-ˈsnär(-ə)l, en-\ *transitive verb* (1593)
: to involve in a snarl

en·sor·cell *or* **en·sor·cel** \in-ˈsȯr-səl\ *transitive verb* **-celled** *or* **-celed; -cell·ing** *or* **-cel·ing** [Middle French *ensorceler,* alteration of Old French *ensorcerer,* from *en-* + *-sorcerer,* from *sorcier* sorcerer — more at SORCERY] (circa 1541)
: BEWITCH, ENCHANT
— en·sor·cell·ment \-mənt\ *noun*

en·soul \in-ˈsōl, en-\ *transitive verb* (1605)
: to endow or imbue with a soul

en·sphere \in-ˈsfir, en-\ *transitive verb* (1612)
: to enclose in or as if in a sphere

en·sue \in-ˈsü, en-\ *verb* **en·sued; en·su·ing** [Middle English, from Middle French *ensuivre,* from Old French, from *en-* + *suivre* to follow — more at SUE] (14th century)
transitive verb
: to strive to attain **:** PURSUE ⟨I wander, seeking peace, and *ensuing* it —Rupert Brooke⟩
intransitive verb
: to take place afterward or as a result
synonym see FOLLOW

☆ SYNONYMS
Enormous, immense, huge, vast, gigantic, colossal, mammoth mean exceedingly large. ENORMOUS and IMMENSE both suggest an exceeding of all ordinary bounds in size or amount or degree, but ENORMOUS often adds an implication of abnormality or monstrousness ⟨an *enormous* expense⟩ ⟨an *immense* shopping mall⟩. HUGE commonly suggests an immensity of bulk or amount ⟨incurred a *huge* debt⟩. VAST usually suggests immensity of extent ⟨the *vast* Russian steppes⟩. GIGANTIC stresses the contrast with the size of others of the same kind ⟨a *gigantic* sports stadium⟩. COLOSSAL applies especially to a human creation of stupendous or incredible dimensions ⟨a *colossal* statue of Lincoln⟩. MAMMOTH suggests both hugeness and ponderousness of bulk ⟨a *mammoth* boulder⟩.

en suite \ä^{n}-'swēt\ *adverb or adjective* [French] (1818)
: so as to form a suite **:** CONNECTED ⟨bathroom *en suite*⟩; *also* **:** so as to make a matching set

en·sure \in-'shůr\ *transitive verb* **en·sured; en·sur·ing** [Middle English, from Anglo-French *enseurer,* probably alteration of Old French *aseürer* — more at ASSURE] (circa 1704)
: to make sure, certain, or safe **:** GUARANTEE ☆

en·swathe \in-'swäth, -'swȯth, -'swāth, en-\ *transitive verb* (1597)
: to enfold or enclose with or as if with a covering **:** SWATHE

ent- *or* **ento-** *combining form* [New Latin, from Greek *entos* within; akin to Latin *intus* within, Greek *en* in — more at IN]
: inner **:** within ⟨*ento*derm⟩

en·tab·la·ture \in-'ta-blə-,chůr, -chər, -,t(y)ůr\ *noun* [obsolete French, modification of Italian *intavolatura,* from *intavolare* to put on a board or table, from *in-* (from Latin) + *tavola* board, table, from Latin *tabula*] (1611)
: a horizontal part in classical architecture that rests on the columns and consists of architrave, frieze, and cornice

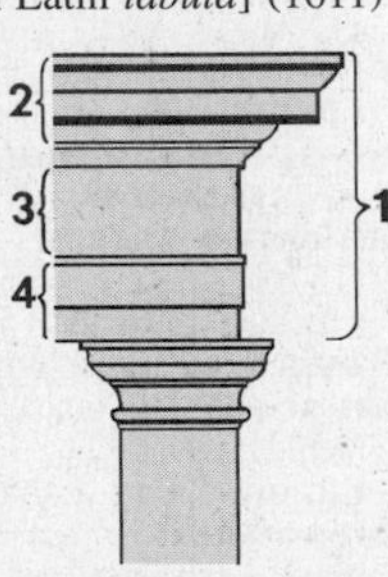

1 **entablature,** *2* **cornice,** *3* **frieze,** *4* **architrave**

[1]en·tail \in-'tā(ə)l, en-\ *transitive verb* [Middle English *entailen, entaillen,* from [1]*en-* + *taile, taille* limitation — more at TAIL] (14th century)
1 : to restrict (property) by limiting the inheritance to the owner's lineal descendants or to a particular class thereof
2 a : to confer, assign, or transmit as if by entail **:** FASTEN ⟨*entailed* on them indelible disgrace —Robert Browning⟩ **b :** to fix (a person) permanently in some condition or status ⟨*entail* him and his heirs unto the crown —Shakespeare⟩
3 : to impose, involve, or imply as a necessary accompaniment or result ⟨the project will *entail* considerable expense⟩
— en·tail·er \-'tā-lər\ *noun*
— en·tail·ment \-'tāl-mənt\ *noun*

[2]en·tail \'en-,tāl, in-'tā(ə)l\ *noun* (14th century)
1 a : an entailing especially of lands **b :** an entailed estate
2 : something transmitted as if by entail

ent·amoe·ba \,en-tə-'mē-bə\ *noun* [New Latin] (1914)
: any of a genus (*Entamoeba*) comprising various amoebas parasitic in vertebrates and including one (*E. histolytica*) that causes amebic dysentery in humans — compare ENDAMOEBA

en·tan·gle \in-'taŋ-gəl, en-\ *transitive verb* (15th century)
1 a : to wrap or twist together **:** INTERWEAVE **b :** ENSNARE
2 a : to involve in a perplexing or troublesome situation ⟨became *entangled* in a lawsuit⟩ **b :** to make complicated ⟨the story is *entangled* with legends⟩
— en·tan·gler \-g(ə-)lər\ *noun*

en·tan·gle·ment \in-'taŋ-gəl-mənt, en-\ *noun* (1535)
1 a : the action of entangling **:** the state of being entangled **b :** something that entangles, confuses, or ensnares
2 : the condition of being deeply involved

en·ta·sis \'en-tə-sis\ *noun, plural* **-ta·ses** \-tə-,sēz\ [Greek, literally, distension, stretching, from *enteinen* to stretch tight (from *en-* [2]en- + *teinein* to stretch) — more at THIN] (1664)
: a slight convexity especially in the shaft of a column

en·tel·e·chy \en-'te-lə-kē, in-\ *noun, plural* **-chies** [Late Latin *entelechia,* from Greek *entelecheia,* from *entelēs* complete (from *en-* [2]en- + *telos* end) + *echein* to have — more at WHEEL, SCHEME] (1603)
1 : the actualization of form-giving cause as contrasted with potential existence
2 : a hypothetical agency not demonstrable by scientific methods that in some vitalist doctrines is considered an inherent regulating and directing force in the development and functioning of an organism

en·tente \än-'tänt\ *noun* [French, from Old French, intent, understanding — more at INTENT] (1854)
1 : an international understanding providing for a common course of action
2 [French *entente cordiale*] **:** a coalition of parties to an entente

entente cor·diale \(')än-'tä^{n}t-,kȯr-'dyäl\ *noun* [French, literally, cordial understanding] (1844)
1 : ENTENTE 1
2 : a friendly agreement or working relationship

en·ter \'en-tər\ *verb* **en·tered; en·ter·ing** \'en-t(ə-)riŋ\ [Middle English *entren,* from Old French *entrer,* from Latin *intrare,* from *intra* within; akin to Latin *inter* between — more at INTER-] (13th century)
intransitive verb
1 : to go or come in
2 : to come or gain admission into a group **:** JOIN — often used with *into*
3 a : to make a beginning ⟨*entering* upon a career⟩ **b :** to begin to consider a subject — usually used with *into* or *upon*
4 : to go upon land for the purpose of taking possession
5 : to play a part **:** be a factor ⟨other considerations *enter* when money is involved⟩
transitive verb
1 : to come or go into ⟨*enter* a room⟩
2 : INSCRIBE, REGISTER ⟨*enter* the names of qualified voters⟩
3 : to cause to be received or admitted ⟨*enter* a child at a school⟩
4 : to put in **:** INSERT ⟨*enter* the new data into the computer⟩
5 a : to make a beginning in ⟨*enter* politics⟩ **b :** to go into (a particular period of time) ⟨*enter* middle age⟩
6 : to become a member of or an active participant in ⟨*enter* the university⟩ ⟨*enter* a race⟩
7 : to make report of (a ship or its cargo) to customs authorities
8 : to place in proper form before a court of law or upon record ⟨*enter* a writ⟩
9 : to go into or upon and take actual possession of (as land)
10 : to put formally on record ⟨*entering* a complaint⟩ ☆
— en·ter·able \'en-t(ə-)rə-bəl\ *adjective*
— enter into 1 : to make oneself a party to or in ⟨*enter into* an agreement⟩ **2 :** to form or be part of ⟨your prejudices shouldn't *enter into* it⟩ **3 :** to participate or share in ⟨*enter into* the spirit of the occasion⟩
— enter the lists : to engage in a fight or struggle

enter- *or* **entero-** *combining form* [Greek, from *enteron* — more at INTER-]
: intestine ⟨*enter*itis⟩

en·ter·al \'en-tə-rəl\ *adjective* (1903)
: ENTERIC
— en·ter·al·ly \-rə-lē\ *adverb*

en·ter·ic \en-'ter-ik, in-\ *adjective* (circa 1859)
1 : of or relating to the intestines; *broadly* **:** ALIMENTARY
2 : of, relating to, or being a medicinal preparation treated to pass through the stomach unaltered and disintegrate in the intestines

en·ter·i·tis \,en-tə-'rī-təs\ *noun* (1808)
1 : inflammation of the intestines and especially of the human ileum
2 : a disease of domestic animals (as panleucopenia of cats) marked by enteritis and diarrhea

en·tero·bac·te·ri·um \,en-tə-rō-bak-'tir-ē-əm\ *noun* [New Latin] (circa 1951)
: any of a family (Enterobacteriaceae) of gram-negative straight rod bacteria (as a salmonella or a colon bacillus) that ferment glucose and include saprophytes as well as some serious plant and animal pathogens
— en·tero·bac·te·ri·al \-ē-əl\ *adjective*

en·tero·bi·a·sis \-'bī-ə-səs\ *noun, plural* **-a·ses** \-,sēz\ [New Latin, from *Enterobius,* genus name (from Greek *enter-* + *bios* mode of life) + *-iasis*] (circa 1927)
: infestation with or disease caused by pinworms (genus *Enterobius*) that occurs especially in children

en·tero·chro·maf·fin \-'krō-mə-fən\ *adjective* (circa 1941)
: of or relating to epithelial cells of the intestinal mucosa that stain especially with chromium salts and usually contain serotonin

en·tero·coc·cus \-'kä-kəs\ *noun, plural* **-coc·ci** \-'käk-,(s)ī, -'käk-(,)(s)ē\ [New Latin, genus name] (1908)
: STREPTOCOCCUS; *especially* **:** a streptococcus (as *Streptococcus faecalis*) normally present in the intestine
— en·tero·coc·cal \-'kä-kəl\ *adjective*

en·tero·coele *or* **en·tero·coel** \'en-tə-rō-,sēl\ *noun* (1877)
: a coelom originating by outgrowth from the archenteron
— en·tero·coe·lous \,en-tə-rō-'sē-ləs\ *adjective*
— en·tero·coe·lic \-lik\ *adjective*

en·tero·co·li·tis \,en-tə-rō-kə-'lī-təs\ *noun* [New Latin] (circa 1857)
: enteritis affecting both the large and small intestine

en·tero·gas·trone \-'gas-,trōn\ *noun* [*enter-* + *gastr-* + *-one* (as in *hormone*)] (circa 1930)
: a hormone that is produced by the duodenal mucosa and has an inhibitory action on gastric motility and secretion

en·tero·ki·nase \,en-tə-rō-'kī-,nās, -,nāz\ *noun* [International Scientific Vocabulary] (circa 1902)
: an enzyme especially of the upper intestinal mucosa that activates trypsinogen by converting it to trypsin

en·ter·on \'en-tə-,rän, -rən\ *noun* [New Latin, from Greek, intestine — more at INTER-] (circa 1842)

☆ SYNONYMS

Ensure, insure, assure, secure mean to make a thing or person sure. ENSURE, INSURE, and ASSURE are interchangeable in many contexts where they indicate the making certain or inevitable of an outcome, but INSURE sometimes stresses the taking of necessary measures beforehand, and ASSURE distinctively implies the removal of doubt and suspense from a person's mind. SECURE implies action taken to guard against attack or loss.

Enter, penetrate, pierce, probe mean to make way into something. ENTER is the most general of these and may imply either going in or forcing a way in ⟨*entered* the city in triumph⟩. PENETRATE carries a strong implication of an impelling force or compelling power that achieves entrance ⟨the enemy *penetrated* the fortress⟩. PIERCE means an entering or cutting through with a sharp pointed instrument ⟨*pierced* the boil with a lancet⟩. PROBE implies penetration to investigate or explore something hidden from sight or knowledge ⟨*probed* the depths of the sea⟩.

: the alimentary canal or system — used especially of the embryo

en·tero·patho·gen·ic \ˌen-tə-rō-ˌpa-thə-ˈje-nik\ *adjective* (1961)
: tending to produce disease in the intestinal tract ⟨*enteropathogenic* bacteria⟩

en·ter·op·a·thy \ˌen-tə-ˈrä-pə-thē\ *noun* (circa 1889)
: a disease of the intestinal tract

en·ter·os·to·my \ˌen-tə-ˈräs-tə-mē\ *noun, plural* **-mies** [International Scientific Vocabulary] (1878)
: a surgical formation of an opening into the intestine through the abdominal wall
— **en·ter·os·to·mal** \-tə-məl\ *adjective*

en·tero·tox·in \ˌen-tə-rō-ˈtäk-sən\ *noun* (circa 1928)
: a toxin that is produced by microorganisms (as some staphylococci) and causes gastrointestinal symptoms (as in some forms of food poisoning or cholera)

en·tero·vi·rus \-ˈvī-rəs\ *noun* [New Latin] (1957)
: any of a group of picornaviruses (as the poliomyelitis virus) that typically occur in the gastrointestinal tract but may be involved in respiratory ailments, meningitis, and neurological disorders
— **en·tero·vi·ral** \-rəl\ *adjective*

en·ter·prise \ˈen-tə(r)-ˌprīz\ *noun* [Middle English, from Middle French, from Old French *entreprendre* to undertake, from *entre-* inter- + *prendre* to take — more at PRIZE] (15th century)
1 : a project or undertaking that is especially difficult, complicated, or risky
2 : readiness to engage in daring action **:** INITIATIVE
3 a : a unit of economic organization or activity; *especially* **:** a business organization **b :** a systematic purposeful activity ⟨agriculture is the main economic *enterprise* among these people⟩

en·ter·pris·er \-ˌprī-zər\ *noun* (1523)
: ENTREPRENEUR

en·ter·pris·ing \-ˌprī ziŋ\ *adjective* (1611)
: marked by an independent energetic spirit and by readiness to undertake or experiment

en·ter·tain \ˌen-tər-ˈtān\ *verb* [Middle English *entertinen,* from Middle French *entretenir,* from *entre-* inter- + *tenir* to hold — more at TENABLE] (15th century)
transitive verb
1 a *archaic* **:** MAINTAIN **b** *obsolete* **:** RECEIVE
2 : to show hospitality to
3 a : to keep, hold, or maintain in the mind ⟨I *entertain* grave doubts about her sincerity⟩ **b :** to receive and take into consideration ⟨refused to *entertain* our plea⟩
4 : to provide entertainment for
5 : to play against (an opposing team) on one's home field or court
intransitive verb
: to provide entertainment especially for guests
synonym see AMUSE
— **en·ter·tain·er** *noun*

en·ter·tain·ing *adjective* (1568)
: providing entertainment **:** DIVERTING
— **en·ter·tain·ing·ly** \-ˈtā-niŋ-lē\ *adverb*

en·ter·tain·ment \ˌen tər ˈtān mənt\ *noun* (15th century)
1 : the act of entertaining
2 a *archaic* **:** MAINTENANCE, PROVISION **b** *obsolete* **:** EMPLOYMENT
3 : something diverting or engaging: as **a :** a public performance **b :** a usually light comic or adventure novel

en·thal·py \ˈen-ˌthal-pē, en-ˈ\ *noun* [*en-* + Greek *thalpein* to heat] (circa 1924)
: the sum of the internal energy of a body and the product of its volume multiplied by the pressure

en·thrall *or* **en·thral** \in-ˈthrȯl, en-\ *transitive verb* **en·thralled; en·thrall·ing** [Middle English] (15th century)
1 : to hold in or reduce to slavery
2 : to hold spellbound **:** CHARM
— **en·thrall·ment** \-ˈthrȯl-mənt\ *noun*

en·throne \in-ˈthrōn, en-\ *transitive verb* (1606)
1 a : to seat ceremonially on a throne **b :** to seat in a place associated with a position of authority or influence
2 : to assign supreme virtue or value to **:** EXALT
— **en·throne·ment** \-mənt\ *noun*

en·thuse \in-ˈthüz, en-, *also* -ˈthyüz\ *verb* **en·thused; en·thus·ing** [back-formation from *enthusiasm*] (1827)
transitive verb
1 : to make enthusiastic
2 : to express with enthusiasm
intransitive verb
: to show enthusiasm ⟨a splendid performance, and I was *enthusing* over it —Julian Huxley⟩
□

en·thu·si·asm \in-ˈthü-zē-ˌa-zəm, en-, *also* -ˈthyü-\ *noun* [Greek *enthousiasmos,* from *enthousiazein* to be inspired, irregular from *entheos* inspired, from *en-* + *theos* god] (1603)
1 a : belief in special revelations of the Holy Spirit **b :** religious fanaticism
2 a : strong excitement of feeling **:** ARDOR **b :** something inspiring zeal or fervor ◇
synonym see PASSION

en·thu·si·ast \-ˌast, -əst\ *noun* (1570)
: a person filled with enthusiasm: as **a :** one who is ardently attached to a cause, object, or pursuit ⟨a sports car *enthusiast*⟩ **b :** one who tends to become ardently absorbed in an interest

en·thu·si·as·tic \in-ˌthü-zē-ˈas-tik, en-, *also* -ˌthyü-\ *adjective* (1603)
: filled with or marked by enthusiasm
— **en·thu·si·as·ti·cal·ly** \-ti-k(ə-)lē\ *adverb*

en·thy·meme \ˈen(t)-thi-ˌmēm\ *noun* [Latin *enthymema,* from Greek *enthymēma,* from *enthymeisthai* to keep in mind, from *en-* + *thymos* mind, soul] (1552)
: a syllogism in which one of the premises is implicit

en·tice \in-ˈtīs, en-\ *transitive verb* **en·ticed; en·tic·ing** [Middle English, from Middle French *enticier,* from (assumed) Vulgar Latin *intitiare,* from Latin *in-* + *titio* firebrand] (14th century)
: to attract artfully or adroitly or by arousing hope or desire **:** TEMPT
synonym see LURE
— **en·tice·ment** \-ˈtī-smənt\ *noun*
— **en·tic·ing·ly** \-ˈtī-siŋ-lē\ *adverb*

[1]en·tire \in-ˈtīr, ˈen-ˌ\ *adjective* [Middle English, from Middle French *entir,* from Latin *integer,* literally, untouched, from *in-* + *tangere* to touch — more at TANGENT] (14th century)
1 : having no element or part left out **:** WHOLE ⟨was alone the *entire* day⟩
2 : complete in degree **:** TOTAL ⟨their *entire* devotion to their family⟩
3 a : consisting of one piece **b :** HOMOGENEOUS, UNMIXED **c :** INTACT ⟨strove to keep the collection *entire*⟩
4 : not castrated
5 : having the margin continuous or free from indentations ⟨an *entire* leaf⟩
synonym see WHOLE, PERFECT
— **entire** *adverb*
— **en·tire·ness** *noun*

[2]entire *noun* (1597)
1 *archaic* **:** the whole **:** ENTIRETY
2 : STALLION

en·tire·ly *adverb* (14th century)
1 : to the full or entire extent **:** COMPLETELY ⟨I agree *entirely*⟩ ⟨you are *entirely* welcome⟩
2 : to the exclusion of others **:** SOLELY ⟨*entirely* by my own efforts⟩

en·tire·ty \in-ˈtī-rə-tē, -ˈtī(-ə)r-tē\ *noun, plural* **-ties** (1548)
1 : the state of being entire or complete
2 : SUM TOTAL, WHOLE

en·ti·tle \in-ˈtī-tᵊl, en-\ *transitive verb* **en·ti·tled; en·ti·tling** \-ˈtīt-liŋ, -ᵊl-iŋ\ [Middle English, from Middle French *entituler,* from Late Latin *intitulare,* from Latin *in-* + *titulus* title] (14th century)
1 : to give a title to **:** DESIGNATE
2 : to furnish with proper grounds for seeking or claiming something ⟨this ticket *entitles* the bearer to free admission⟩

en·ti·tle·ment \-ˈtī-tᵊl-mənt\ *noun* (1944)
1 a : the state or condition of being entitled **:** RIGHT **b :** a right to benefits specified especially by law or contract
2 : a government program providing benefits to members of a specified group; *also* **:** funds supporting or distributed by such a program

en·ti·ty \ˈen-tə-tē, ˈe-nə-\ *noun, plural* **-ties** [Medieval Latin *entitas,* from Latin *ent-, ens* existing thing, from coined present participle of *esse* to be — more at IS] (1596)
1 a : BEING, EXISTENCE; *especially* **:** independent, separate, or self-contained existence **b :** the existence of a thing as contrasted with its attributes
2 : something that has separate and distinct existence and objective or conceptual reality

ento- — see ENT-

en·to·derm \ˈen-tə-ˌdərm\ *noun* (1879)
: ENDODERM
— **en·to·der·mal** \ˌen-tə-ˈdər-məl\ *adjective*
— **en·to·der·mic** \-mik\ *adjective*

en·toil \in-ˈtȯi(ə)l\ *transitive verb* (1581)
: ENTRAP, ENMESH

entom- *or* **entomo-** *combining form* [French, from Greek *entomon*]
: insect ⟨*entomo*phagous⟩

en·tomb \in-ˈtüm, en-\ *transitive verb* [Middle English *entoumben,* from Middle French *entomber,* from *en-* + *tombe* tomb] (1576)

□ USAGE
enthuse *Enthuse* is apparently American in origin, although the earliest known example of its use occurs in a letter written in 1827 by a young Scotsman who spent about two years in the Pacific Northwest. It has been disapproved since about 1870. Current evidence shows it to be flourishing nonetheless on both sides of the Atlantic especially in journalistic prose.

◇ WORD HISTORY
enthusiasm In Greek *theos* means "god" and *entheos* or *enthous* means "inspired or possessed by a god." Hence, the noun *enthousiasmos,* the source of our *enthusiasm,* literally means "inspiration or possession by a god." It was in this sense that *enthusiasm* was first used when it made its appearance in English in the 17th century. Shortly thereafter, the word also took on a secular dimension, coming to be used of any very strong emotion, such as the fervor that the greatest poetry can inspire. For a time *enthusiasm* took on negative connotations, being applied specifically to the display of religious emotion associated with dissenters from the Church of England such as the Methodists, a display considered unseemly by members of the established church. By the 18th century *enthusiasm* was increasingly being used in nonreligious contexts. In the process it lost its suggestion of excess and acquired the neutral or positive sense of "ardent zeal for a cause or subject."

1 : to deposit in a tomb **:** BURY
2 : to serve as a tomb for
— **en·tomb·ment** \-'tüm-mənt\ *noun*

en·to·mo·fau·na \ˌen-tə-mō-'fö-nə, -'fä-\ *noun* [New Latin] (1951)
: a fauna of insects **:** the insects of an environment or region

en·to·mol·o·gy \ˌen-tə-'mä-lə-jē\ *noun* [French *entomologie,* from Greek *entomon* insect (from neuter of *entomos* cut up, from *en-* + *temnein* to cut) + French *-logie* -logy — more at TOME] (1766)
: a branch of zoology that deals with insects
— **en·to·mo·log·i·cal** \-mə-'lä-ji-kəl\ *adjective*
— **en·to·mo·log·i·cal·ly** \-k(ə-)lē\ *adverb*
— **en·to·mol·o·gist** \ˌen-tə-'mä-lə-jist\ *noun*

en·to·moph·a·gous \ˌen-tə-'mä-fə-gəs\ *adjective* (1847)
: feeding on insects

en·to·moph·i·lous \ˌen-tə-'mä-fə-ləs\ *adjective* (1880)
: normally pollinated by insects — compare ZOOPHILIC
— **en·to·moph·i·ly** \-lē\ *noun*

en·to·proct \'en-tə-ˌpräkt\ *noun* [ultimately from *ent-* + Greek *prōktos* anus] (1940)
: any of a phylum (Entoprocta) of animals that are very similar to bryozoans but lack a true coelom and have the anus located near the mouth inside a crown of tentacles

en·tou·rage \ˌän-tü-'räzh\ *noun* [French, from Middle French, from *entourer* to surround, from *entour* around, from *en* in (from Latin *in*) + *tour* circuit — more at TURN] (1834)
1 : one's attendants or associates
2 : SURROUNDINGS

en·tr'acte \'ä^{n}(n)-ˌtrakt, -ˌträkt, ä^{n}(n)-'\ *noun* [French, from *entre-* inter- + *acte* act] (circa 1842)
1 : a dance, piece of music, or interlude performed between two acts of a play
2 : the interval between two acts of a play

en·trails \'en-ˌtrālz, -trəlz\ *noun plural* [Middle English *entrailles,* from Middle French, from Medieval Latin *intralia,* alteration of Latin *interanea,* plural of *interaneum* intestine, from neuter of *interaneus* interior] (14th century)
1 : BOWELS, VISCERA; *broadly* **:** internal parts
2 : the inner workings of something ⟨the *entrails* of the movie industry⟩

^{1}**en·train** \in-'trān\ *transitive verb* [Middle French *entrainer,* from *en-* + *trainer* to draw, drag — more at TRAIN] (1568)
1 : to draw along with or after oneself
2 : to draw in and transport (as solid particles or gas) by the flow of a fluid
3 : to incorporate (air bubbles) into concrete
4 : to determine or modify the phase or period of ⟨circadian rhythms *entrained* by a light cycle⟩
— **en·train·er** *noun*
— **en·train·ment** \-'trān-mənt\ *noun*

^{2}**entrain** (1881)
transitive verb
: to put aboard a train
intransitive verb
: to go aboard a train

^{1}**en·trance** \'en-trən(t)s\ *noun* (15th century)
1 : power or permission to enter **:** ADMISSION
2 : the act of entering
3 : the means or place of entry
4 : the point at which a voice or instrument part begins in ensemble music
5 : the first appearance of an actor in a scene

^{2}**en·trance** \in-'tran(t)s, en-\ *transitive verb* **en·tranced; en·tranc·ing** (1593)
1 : to carry away with delight, wonder, or rapture
2 : to put into a trance
— **en·trance·ment** \-'tran(t)-smənt\ *noun*

en·trance·way \'en-trən(t)s-ˌwā\ *noun* (1865)
: ENTRYWAY

en·trant \'en-trənt\ *noun* (1635)
: one that enters; *especially* **:** one that enters a contest

en·trap \in-'trap, en-\ *transitive verb* [Middle French *entraper,* from *en-* + *trape* trap] (1534)
1 : to catch in or as if in a trap
2 : to lure into a compromising statement or act
synonym see CATCH

en·trap·ment \-mənt\ *noun* (1597)
1 a : the action or process of entrapping **b :** the condition of being entrapped
2 : the action of luring an individual into committing a crime in order to prosecute the person for it

en·treat \in-'trēt, en-\ *verb* [Middle English *entreten,* from Middle French *entraitier,* from *en-* + *traitier* to treat] (14th century)
intransitive verb
1 *obsolete* **a :** NEGOTIATE **b :** INTERCEDE
2 : to make an earnest request **:** PLEAD
transitive verb
1 : to plead with especially in order to persuade **:** ask urgently ⟨*entreated* his boss for another chance⟩
2 *archaic* **:** to deal with **:** TREAT
synonym see BEG
— **en·treat·ing·ly** \-trē-tiŋ-lē\ *adverb*
— **en·treat·ment** \-mənt\ *noun*

en·treaty \-'trē-tē\ *noun, plural* **-treat·ies** (15th century)
: an act of entreating **:** PLEA

en·tre·chat \'ä^{n}(n)-trə-ˌshä\ *noun* [French, modification of Italian (*capriola*) *intrecciata,* literally, intertwined caper] (1775)
: a leap in which a ballet dancer repeatedly crosses the legs and sometimes beats them together

en·tre·côte *also* **en·tre·cote** \'ä^{n}(n)-trə-ˌkōt\ *noun* [French *entrecôte,* from *entre-* inter- + *côte* rib, from Latin *costa* — more at INTER-, COAST] (1841)
: a steak cut from between the ribs

en·trée *or* **en·tree** \'än-ˌtrā *also* än-'\ *noun* [French *entrée,* from Old French — more at ENTRY] (1761)
1 a : the act or manner of entering **:** ENTRANCE **b :** freedom of entry or access
2 : the main course of a meal in the U.S. ◆

en·tre·mets *as singular* ˌä^{n}(n)-trə-'mā, *as plural* -'mā(z)\ *noun plural but singular or plural in construction* [Middle English, from Middle French, from Old French *entremes,* from Latin *intermissus,* past participle of *intermittere* to intermit] (15th century)
: dishes served in addition to the main course of a meal; *especially* **:** DESSERT

en·trench \in-'trench, en-\ (1555)
transitive verb
1 a : to place within or surround with a trench especially for defense **b :** to place (oneself) in a strong defensive position **c :** to establish solidly ⟨*entrenched* themselves in the business⟩
2 : to cut into **:** FURROW; *specifically* **:** to erode downward so as to form a trench
intransitive verb
1 : to dig or occupy a trench for defensive purposes
2 : to enter upon or take over something unfairly, improperly, or unlawfully **:** ENCROACH — used with *on* or *upon*
— **en·trench·ment** \-mənt\ *noun*

en·tre·pôt \'ä^{n}(n)-trə-ˌpō\ *noun* [French, from Middle French *entrepost,* from *entreposer* to put between, from *entre-* inter- + *poser* to pose, put] (1758)
: an intermediary center of trade and transshipment

en·tre·pre·neur \ˌä^{n}n-trə-p(r)ə-'nər, -'n(y)u̇r\ *noun* [French, from Old French, from *entreprendre* to undertake — more at ENTERPRISE] (1852)
: one who organizes, manages, and assumes the risks of a business or enterprise
— **en·tre·pre·neur·ial** \-'n(y)u̇r-ē-əl, -'nər-\ *adjective*
— **en·tre·pre·neur·ial·ism** \-ē-ə-li-zəm\ *noun*
— **en·tre·pre·neur·ial·ly** \-ē-ə-lē\ *adverb*
— **en·tre·pre·neur·ship** \-'nər-ˌship, -'n(y)u̇r-\ *noun*

en·tre·sol \'ä^{n}(n)-trə-ˌsäl, -ˌsȯl\ *noun* [French, from Spanish *entresuelo,* from *entre* between + *suelo* level, from (assumed) Vulgar Latin *sola,* from Latin *solea* sandal, sole, sill — more at SOLE] (1711)
: MEZZANINE

en·tro·pi·on \en-'trō-pē-ˌän, -pē-ən\ *noun* [New Latin, from *en-* ^{2}en- + *ectropion* turning out of the eyelid, from Greek *ektropion,* from *ektrepein* to turn out, from *ex-* out + *trepein* to turn] (circa 1860)
: the inversion or turning inward of the border of the eyelid against the eyeball

en·tro·py \'en-trə-pē\ *noun, plural* **-pies** [International Scientific Vocabulary ^{2}*en-* + Greek *tropē* change, literally, turn, from *trepein* to turn] (1875)
1 : a measure of the unavailable energy in a closed thermodynamic system that is also usually considered to be a measure of the system's disorder and that is a property of the system's state and is related to it in such a manner that a reversible change in heat in the system produces a change in the measure which varies directly with the heat change and inversely with the absolute temperature at which the change takes place; *broadly* **:** the degree of disorder or uncertainty in a system
2 a : the degradation of the matter and energy in the universe to an ultimate state of inert uniformity **b :** a process of degradation or running down or a trend to disorder
3 : CHAOS, DISORGANIZATION, RANDOMNESS
— **en·tro·pic** \en-'trō-pik, -'trä-pik\ *adjective*
— **en·tro·pi·cal·ly** \-pi-k(ə-)lē\ *adverb*

en·trust \in-'trəst, en-\ *transitive verb* (1602)
1 : to confer a trust on; *especially* **:** to deliver something in trust to
2 : to commit to another with confidence
synonym see COMMIT
— **en·trust·ment** \-'trəs(t)-mənt\ *noun*

en·try \'en-trē\ *noun, plural* **entries** [Middle English *entre,* from Old French *entree,* from feminine of *entré,* past participle of *entrer* to enter] (13th century)
1 : the right or privilege of entering **:** ENTRÉE
2 : the act of entering **:** ENTRANCE
3 : a place of entrance: as **a :** VESTIBULE, PASSAGE **b :** DOOR, GATE

◇ WORD HISTORY

entrée *Entrée,* which signifies the main course of a meal in the U.S., is in fact the French word for "entry." The culinary sense of *entrée* can be traced back to 18th century Britain. In those days a formal dinner could be a gastronomic marathon. In addition to the principal courses of soup, fish, meat, and dessert, there would be an impressive array of side dishes, not to mention the salad and cheese courses. Between the fish and meat courses would come a small dish concocted of several ingredients and often garnished and sauced. Because this secondary dish immediately preceded the centerpiece of the whole meal—typically a roast—it was called the *entrée,* being, in effect, the "entrance" to the really important part of the meal. As Anglo-American dining habits changed, meals gradually diminished in their elaborateness; fewer and simpler courses were served. In the U.S. the course following the appetizer course continued to be known as the *entrée,* even if it did turn out to be a roast. Perhaps the continued preference for *entrée* in menu terminology lay in the fact that it was obviously French, and anything French was considered to have prestige.

4 a : the act of making or entering a record **b :** something entered: as (1) **:** a record or notation of an occurrence, transaction, or proceeding (2) **:** a descriptive record (as in a card catalog or an index) (3) **:** HEADWORD (4) **:** a headword with its definition or identification (5) **:** VOCABULARY ENTRY
5 : a person, thing, or group entered in a contest

en·try–lev·el \'en-trē-ˌle-vəl\ *adjective* (1975)
: of or being at the lowest level of a hierarchy ⟨*entry-level* jobs⟩

en·try·way \-trē-ˌwā\ *noun* (1746)
: a passage for entrance

entry word *noun* (circa 1908)
: HEADWORD

en·twine \in-'twīn, en-\ (1590)
transitive verb
: to twine together or around
intransitive verb
: to become twisted or twined

en·twist \in-'twist, en-\ *transitive verb* (1590)
: ENTWINE

enu·cle·ate \(ˌ)ē-'nü-klē-ˌāt, -'nyü-\ *transitive verb* **-at·ed; -at·ing** [Latin *enucleatus,* past participle of *enucleare,* literally, to remove the kernel from, from *e-* + *nucleus* kernel — more at NUCLEUS] (1548)
1 *archaic* **:** EXPLAIN
2 : to deprive of a nucleus
3 : to remove without cutting into ⟨*enucleate* a tumor⟩ ⟨*enucleate* the eyeball⟩
— enu·cle·ation \(ˌ)ē-ˌn(y)ü-klē-'ā-shən\ *noun*

enu·mer·a·ble \i-'n(y)üm-rə-bəl, -'n(y)ü-mə-\ *adjective* (circa 1889)
: DENUMERABLE
— enu·mer·a·bil·i·ty \-ˌn(y)üm-rə-'bi-lə-tē, -'n(y)ü-mə-\ *noun*

enu·mer·ate \i-'n(y)ü-mə-ˌrāt\ *transitive verb* **-at·ed; -at·ing** [Latin *enumeratus,* past participle of *enumerare,* from *e-* + *numerare* to count, from *numerus* number] (1616)
1 : to ascertain the number of **:** COUNT
2 : to specify one after another **:** LIST
— enu·mer·a·tion \-ˌn(y)ü-mə-'rā-shən\ *noun*
— enu·mer·a·tive \-'n(y)ü-mə-ˌrā-tiv, -'n(y)üm-rə , -'n(y)ü-mə-rə-\ *adjective*

enu·mer·a·tor \-'n(y)ü-mə-ˌrā-tər\ *noun* (1856)
: one that enumerates; *especially* **:** a census taker

enun·ci·ate \ē-'nən(t)-sē-ˌāt\ *verb* **-at·ed; -at·ing** [Latin *enuntiatus,* past participle of *enuntiare* to report, declare, from *e-* + *nuntiare* to report — more at ANNOUNCE] (1623)
transitive verb
1 a : to make a definite or systematic statement of **b :** ANNOUNCE, PROCLAIM ⟨*enunciated* the new policy⟩
2 : ARTICULATE, PRONOUNCE ⟨*enunciate* all the syllables⟩
intransitive verb
: to utter articulate sounds
— enun·cia·ble \-'nən(t)-sē-ə-bəl, -'nən-ch(ē-)ə-\ *adjective*
— enun·ci·a·tion \-ˌnən(t)-sē-'ā-shən\ *noun*
— enun·ci·a·tor \-'nən(t)-sē-ˌā-tər\ *noun*

enure *variant of* INURE

en·ure·sis \ˌen-yu̇-'rē-səs\ *noun* [New Latin, from Greek *enourein* to urinate in, wet the bed, from *en-* + *ourein* to urinate — more at URINE] (circa 1800)
: the involuntary discharge of urine **:** incontinence of urine
— en·uret·ic \-'re-tik\ *adjective or noun*

en·vel·op \in-'ve-ləp, en-\ *transitive verb* [Middle English *envolupen,* from Middle French *envoluper, enveloper,* from Old French *envoloper,* from *en-* + *voloper* to wrap] (14th century)
1 : to enclose or enfold completely with or as if with a covering
2 : to mount an attack on (an enemy's flank)
— en·vel·op·ment \-mənt\ *noun*

en·ve·lope \'en-və-ˌlōp, ÷'än-\ *noun* (circa 1714)
1 : a flat usually paper container (as for a letter)
2 : something that envelops **:** WRAPPER ⟨the *envelope* of air around the earth⟩
3 a : the outer covering of an aerostat **b :** the bag containing the gas in a balloon or airship
4 : a natural enclosing covering (as a membrane, shell, or integument)
5 a : a curve tangent to each of a family of curves **b :** a surface tangent to each of a family of surfaces
6 : a set of performance limits (as of an aircraft) that may not be safely exceeded; *also* **:** the set of operating parameters that exists within these limits ▪

en·ven·om \in-'ve-nəm, en-\ *transitive verb* [Middle English *envenimen,* from Old French *envenimer,* from *en-* + *venim* venom] (13th century)
1 : to make poisonous
2 : EMBITTER

en·ven·om·iza·tion \in-ˌve-nə-mə-'zā-shən, en-\ *noun* (1960)
: a poisoning caused by a bite or sting

en·vi·able \'en-vē-ə-bəl\ *adjective* (1602)
: highly desirable
— en·vi·able·ness *noun*
— en·vi·ably \-blē\ *adverb*

en·vi·er \'en-vē-ər\ *noun* (15th century)
: one that envies

en·vi·ous \'en-vē-əs\ *adjective* (13th century)
1 : feeling or showing envy ⟨*envious* of their neighbor's new car⟩ ⟨*envious* looks⟩
2 *archaic* **a :** EMULOUS **b :** ENVIABLE
— en·vi·ous·ly *adverb*
— en·vi·ous·ness *noun*

en·vi·ron \in-'vī-rən, -'vī(-ə)rn\ *transitive verb* [Middle English *envirounen,* from Middle French *environner,* from *environ* around, from *en* in (from Latin *in*) + *viron* circle, from *virer* to turn — more at VEER] (14th century)
: ENCIRCLE, SURROUND

en·vi·ron·ment \in-'vī-rə(n)-mənt, -'vī(-ə)r(n)-\ *noun* (1827)
1 : the circumstances, objects, or conditions by which one is surrounded
2 a : the complex of physical, chemical, and biotic factors (as climate, soil, and living things) that act upon an organism or an ecological community and ultimately determine its form and survival **b :** the aggregate of social and cultural conditions that influence the life of an individual or community
3 : the position or characteristic position of a linguistic element in a sequence
synonym see BACKGROUND
— en·vi·ron·men·tal \-ˌvī-rə(n)-'men-tᵊl, -ˌvī(-ə)r(n)-\ *adjective*
— en·vi·ron·men·tal·ly \-tᵊl-ē\ *adverb*

en·vi·ron·men·tal·ism \-ˌvī-rə(n)-'men-tᵊl-ˌi-zəm, -ˌvī(-ə)r(n)-\ *noun* (circa 1922)
1 : a theory that views environment rather than heredity as the important factor in the development and especially the cultural and intellectual development of an individual or group
2 : advocacy of the preservation or improvement of the natural environment; *especially* **:** the movement to control pollution

en·vi·ron·men·tal·ist \-tᵊl-əst\ *noun* (1916)
1 : an advocate of environmentalism
2 : one concerned about environmental quality especially of the human environment with respect to the control of pollution

en·vi·rons \in-'vī-rənz, -'vī(-ə)rnz\ *noun plural* (1665)
1 : the districts around a city
2 a : environing things **:** SURROUNDINGS **b :** an adjoining region or space **:** VICINITY

en·vis·age \in-'vi-zij, en-\ *transitive verb* **-aged; -ag·ing** [French *envisager,* from *en-* + *visage* face] (1837)
1 : to view or regard in a certain way ⟨*envisages* the slum as a hotbed of crime⟩
2 : to have a mental picture of especially in advance of realization ⟨*envisages* an entirely new system of education⟩
synonym see THINK

en·vi·sion \in-'vi-zhən, en-\ *transitive verb* (1919)
: to picture to oneself ⟨*envisions* a career dedicated to promoting peace⟩
synonym see THINK

en·voi *or* **en·voy** \'en-ˌvoi, 'än-\ *noun* [Middle English *envoye,* from Middle French *envoi,* literally, message, from Old French *envei,* from *envoier* to send on one's way, from (assumed) Vulgar Latin *inviare,* from Latin *in-* + *via* way — more at WAY] (14th century)
: the usually explanatory or commendatory concluding remarks to a poem, essay, or book; *especially* **:** a short final stanza of a ballade serving as a summary or dedication

en·voy \'en-ˌvȯi, ÷'än-\ *noun* [French *envoyé,* from past participle of *envoyer* to send, from Old French *envoier*] (1635)
1 a : a minister plenipotentiary accredited to a foreign government who ranks between an ambassador and a minister resident — called also *envoy extraordinary* **b :** a person delegated to represent one government in its dealings with another
2 : MESSENGER, REPRESENTATIVE

¹en·vy \'en-vē\ *noun, plural* **envies** [Middle English *envie,* from Old French, from Latin *invidia,* from *invidus* envious, from *invidēre* to look askance at, envy, from *in-* + *vidēre* to see — more at WIT] (13th century)
1 : painful or resentful awareness of an advantage enjoyed by another joined with a desire to possess the same advantage
2 *obsolete* **:** MALICE
3 : an object of envious notice or feeling ⟨his new car made him the *envy* of his friends⟩

²envy *verb* **en·vied; en·vy·ing** (14th century)
transitive verb
1 : to feel envy toward or on account of
2 *obsolete* **:** BEGRUDGE
intransitive verb
obsolete **:** to feel or show envy
— en·vy·ing·ly \-vē-iŋ-lē\ *adverb*

en·wheel \in-'hwē(ə)l, -'wē(ə)l, en-\ *transitive verb* (1604)
obsolete **:** ENCIRCLE

en·wind \in-'wīnd, en-\ *transitive verb* **en·wound** \-'wau̇nd\; **en·wind·ing** (1850)
: to wind in or about **:** ENFOLD

en·womb \in-'wüm, en-\ *transitive verb* (circa 1591)
: to shut up as if in a womb

en·wrap \in-'rap, en-\ *transitive verb* (14th century)
1 : to wrap in a covering **:** ENFOLD
2 a : ENVELOP **b :** to preoccupy or absorb mentally **:** ENGROSS

en·wreathe \in-'rēth, en-\ *transitive verb* (15th century)

□ USAGE
envelope The \'en-\ and \'än-\ pronunciations are used with about equal frequency, and both are fully acceptable, though the \'än-\ version is sometimes decried as "pseudo-French." Actually \'än-\ is exactly what one would expect to hear when a French word like *entrepreneur* is becoming anglicized. *Envelope,* however, has been in English for nearly 300 years, plenty of time for it to become completely anglicized and for both of its pronunciations to win respectability.

: to encircle with or as if with a wreath : ENVELOP

en·zo·ot·ic \ˌen-zə-ˈwä-tik\ *adjective* [*en-* + epi*zootic*] (1882)
of animal diseases : peculiar to or constantly present in a locality
— **enzootic** *noun*

en·zy·mat·ic \ˌen-zə-ˈma-tik\ *also* **en·zy·mic** \en-ˈzī-mik\ *adjective* (1900)
: of, relating to, or produced by an enzyme
— **en·zy·mat·i·cal·ly** \ˌen-zə-ˈma-ti-k(ə-)lē\ *also* **en·zy·mi·cal·ly** \en-ˈzī-mi-k(ə-)lē\ *adverb*

en·zyme \ˈen-ˌzīm\ *noun* [German *Enzym,* from Middle Greek *enzymos* leavened, from Greek *en-* + *zymē* leaven — more at JUICE] (1881)
: any of numerous complex proteins that are produced by living cells and catalyze specific biochemical reactions at body temperatures

enzyme–linked immunosorbent assay *noun* (1977)
: a quantitative in vitro test for an antibody or antigen in which the test material is adsorbed on a surface and exposed to a complex of an enzyme linked to an antibody specific for the substance being tested for with a positive result indicated by a treatment yielding a color in proportion to the amount of antigen or antibody in the test material — called also *ELISA*

en·zy·mol·o·gy \ˌen-ˌzī-ˈmä-lə-jē, -zə-\ *noun* [International Scientific Vocabulary] (circa 1900)
: a branch of biochemistry that deals with enzymes, their nature, activity, and significance
— **en·zy·mol·o·gist** \-jist\ *noun*

eo- *combining form* [Greek *ēō-* dawn, from *ēōs* — more at EAST]
: earliest : oldest ⟨*eo*lithic⟩

Eo·cene \ˈē-ə-ˌsēn\ *adjective* (1831)
: of, relating to, or being an epoch of the Tertiary between the Paleocene and the Oligocene or the corresponding system of rocks — see GEOLOGIC TIME table
— **Eocene** *noun*

eo·hip·pus \ˌē-ō-ˈhi-pəs\ *noun* [New Latin, from *eo-* + Greek *hippos* horse — more at EQUINE] (circa 1879)
: any of a genus (*Hyracotherium* synonym *Eohippus*) of very small primitive horses from the Lower Eocene having four-toed forefeet and three-toed hind feet — called also *dawn horse*

eo·lian \ē-ˈō-lē-ən, -ˈōl-yən\ *adjective* [Latin *Aeolus,* Aeolus] (1853)
: borne, deposited, produced, or eroded by the wind

eo·lith \ˈē-ə-ˌlith\ *noun* (1895)
: a very crudely chipped flint

Eo·lith·ic \ˌē-ə-ˈli-thik\ *adjective* (1890)
: of or relating to the early period of the Stone Age marked by the use of eoliths

eon \ˈē-ən, ˈē-ˌän\ *variant of* AEON

eo no·mi·ne \ˌē-ō-ˈnä-mə-nē\ [Latin] (1627)
: by or under that name

Eos \ˈē-ˌäs\ *noun* [Greek *Ēōs*]
: the Greek goddess of dawn — compare AURORA

eo·sin \ˈē-ə-sən\ *also* **eo·sine** \-sən, -ˌsēn\ *noun* [International Scientific Vocabulary, from Greek *ēōs* dawn] (1866)
1 : a red fluorescent dye $C_{20}H_8Br_4O_5$ obtained by the action of bromine on fluorescein and used especially in cosmetics and as a toner; *also* : its red to brown sodium or potassium salt used especially as a biological stain for cytoplasmic structures
2 : any of several dyes related to eosin

[1]**eo·sin·o·phil** \ˌē-ə-ˈsi-nə-ˌfil\ *adjective* (circa 1882)
: EOSINOPHILIC 1

[2]**eosinophil** \-ˈfil\ *noun* (circa 1900)
: a white blood cell or other granulocyte with cytoplasmic inclusions readily stained by eosin

eo·sin·o·phil·ia \-ˌsi-nə-ˈfi-lē-ə\ *noun* [New Latin] (circa 1903)
: abnormal increase in the number of eosinophils in the blood that is characteristic of allergic states and various parasitic infections

eo·sin·o·phil·ic \-ˌsi-nə-ˈfi-lik\ *adjective* (circa 1900)
1 : staining readily with eosin
2 : of, relating to, or characterized by eosinophilia

epact \ˈē-ˌpakt, ˈe-ˌpakt\ *noun* [Middle French *epacte,* from Late Latin *epacta,* from Greek *epaktē,* from *epagein* to bring in, intercalate, from *epi-* + *agein* to drive — more at AGENT] (1588)
: a period added to harmonize the lunar with the solar calendar

ep·ar·chy \ˈe-ˌpär-kē\ *noun, plural* **-chies** [Greek *eparchia* province, from *eparchos* prefect, from *epi-* + *archos* ruler — more at ARCH-] (1796)
: a diocese of an Eastern church

ep·au·let *also* **ep·au·lette** \ˌe-pə-ˈlet; ˈe-pə-ˌlet, -lət\ *noun* [French *épaulette,* diminutive of *épaule* shoulder, from Late Latin *spatula* shoulder blade, spoon, diminutive of Latin *spatha* spoon, sword — more at SPADE] (1783)
: something that ornaments or protects the shoulder: as **a** : an ornamental fringed shoulder pad formerly worn as part of a military uniform **b** : an ornamental strip or loop sewn across the shoulder of a dress or coat
— **ep·au·let·ted** \ˌe-pə-ˈle-təd, ˈe-pə-ˌ\ *adjective*

E epaulet

ep·a·zote \ˈe-pə-ˌzōt\ *noun* [Mexican Spanish, from Nahuatl *epazōtl*] (1946)
: WORMSEED b

épée \ˈe-ˌpā, ā-ˈpā\ *noun* [French, from Latin *spatha*] (1889)
1 : a fencing or dueling sword having a bowl-shaped guard and a rigid blade of triangular section with no cutting edge that tapers to a sharp point blunted for fencing — compare [4]FOIL 1, SABER
2 : the art or sport of fencing with the épée

épée·ist \-ist\ *noun* (1910)
: one who fences with an épée

ep·ei·rog·e·ny \ˌe-ˌpī-ˈrä-jə-nē\ *noun, plural* **-nies** [Greek *ēpeiros* mainland, continent + English *-geny*] (1890)
: the deformation of the earth's crust by which the broader features of relief are produced
— **epei·ro·gen·ic** \i-ˌpī-rə-ˈje-nik\ *adjective*
— **epei·ro·gen·i·cal·ly** \-ni-k(ə-)lē\ *adverb*

epen·the·sis \i-ˈpen(t)-thə-səs, e-\ *noun, plural* **-the·ses** \-ˌsēz\ [Late Latin, from Greek, from *epentithenai* to insert a letter, from *epi-* + *entithenai* to put in, from *en-* + *tithenai* to put — more at DO] (1543)
: the insertion or development of a sound or letter in the body of a word (as \ə\ in \ˈa-thə-ˌlēt\ *athlete*)
— **ep·en·thet·ic** \ˌe-pən-ˈthe-tik\ *adjective*

epergne \i-ˈpərn, ā-\ *noun* [probably from French *épargne* saving] (1761)
: an often ornate tiered centerpiece consisting typically of a frame of wrought metal (as silver or gold) bearing dishes, vases, or candle holders or a combination of these

ep·ex·e·ge·sis \ˌe-ˌpek-sə-ˈjē-səs\ *noun, plural* **-ge·ses** \-ˌsēz\ [Greek *epexēgēsis,* from *epi-* + *exēgēsis* exegesis] (circa 1577)
: additional explanation or explanatory matter
— **ep·ex·e·get·i·cal** \-ˈje-ti-kəl\ *or* **ep·ex·e·get·ic** \-ˈje-tik\ *adjective*
— **ep·ex·e·get·i·cal·ly** \-ˈje-ti-k(ə-)lē\ *adverb*

ephah \ˈē-fə, ˈe-fə\ *noun* [Middle English *ephi,* from Late Latin, from Hebrew *ēphāh,* from Egyptian *'pt*] (1611)
: an ancient Hebrew unit of dry measure equal to 1/10 homer or a little over a bushel

ephebe \ˈe-ˌfēb, i-ˈfēb\ *noun* [Latin *ephebus*] (1880)
: EPHEBUS; *also* : a young man : YOUTH

ephe·bic \i-ˈfē-bik\ *adjective* (1865)
: of, relating to, or characteristic of an ephebe or ephebus

ephe·bus \i-ˈfē-bəs, e-\ *noun, plural* **-bi** \-ˌbī\ [Latin, from Greek *ephēbos,* from *epi-* + *hēbē* youth, puberty] (1697)
: a youth of ancient Greece; *especially* : an Athenian 18 or 19 years old in training for full citizenship

ephe·dra \i-ˈfe-drə, ˈe-fə-drə\ *noun* [New Latin, genus name, from Latin, equisetum, from Greek, from *ephedros* sitting upon, from *epi-* + *hedra* seat — more at SIT] (circa 1889)
: any of a large genus (*Ephedra* of the family Gnetaceae) of jointed nearly leafless desert shrubs with the leaves reduced to scales at the nodes

ephed·rine \i-ˈfe-drən, *British also* ˈe-fə-drən\ *noun* [New Latin *Ephedra*] (1889)
: a crystalline alkaloid $C_{10}H_{15}NO$ extracted from Chinese ephedras or synthesized and used in the form of a salt for relief of hay fever, asthma, and nasal congestion

ephem·era \i-ˈfe-mər-ə, -ˈfem-rə\ *noun, plural* **ephemera** *also* **ephem·er·ae** \-mər-ē, -rē\ *or* **ephemeras** [New Latin, from Greek *ephēmera,* neuter plural of *ephēmeros*] (1751)
1 : something of no lasting significance — usually used in plural
2 *ephemera plural* : collectibles (as posters, broadsides, and tickets) not intended to have lasting value

[1]**ephem·er·al** \i-ˈfem-rəl, -ˈfēm-; -ˈfe-mə-, -ˈfē-\ *adjective* [Greek *ephēmeros* lasting a day, daily, from *epi-* + *hēmera* day] (1576)
1 : lasting one day only ⟨an *ephemeral* fever⟩
2 : lasting a very short time ⟨*ephemeral* pleasures⟩
synonym see TRANSIENT
— **ephem·er·al·ly** \-rə-lē\ *adverb*

[2]**ephemeral** *noun* (1817)
: something ephemeral; *specifically* : a plant that grows, flowers, and dies in a few days

ephem·er·al·i·ty \i-ˌfe-mə-ˈra-lə-tē, -ˌfē-\ *noun, plural* **-ties** (1822)
1 : the quality or state of being ephemeral
2 *plural* : ephemeral things

ephem·er·id \i-ˈfe-mə-rəd\ *noun* [ultimately from Greek *ephēmeron*] (1872)
: MAYFLY

ephem·er·is \-mə-rəs\ *noun, plural* **eph·e·mer·i·des** \ˌe-fə-ˈmer-ə-ˌdēz\ [Latin, diary, ephemeris, from Greek *ephēmeris,* from *ephēmeros*] (1508)
: a tabular statement of the assigned places of a celestial body for regular intervals

ephemeris time *noun* (1950)
: a uniform measure of time defined by the orbital motions of the planets

Ephe·sians \i-ˈfē-zhənz\ *noun plural but singular in construction* [short for *Epistle to the Ephesians*]
: a letter addressed to early Christians and included as a book in the New Testament — see BIBLE table

eph·od \ˈe-ˌfäd, ˈē-ˌfäd\ *noun* [Middle English, from Late Latin, from Hebrew *ēphōdh*] (14th century)
1 : a linen apron worn in ancient Hebrew rites; *especially* : a vestment for the high priest
2 : an ancient Hebrew instrument of priestly divination

eph·or \ˈe-fər, -ˌfȯr\ *noun* [Latin *ephorus,* from Greek *ephoros,* from *ephoran* to oversee, from *epi-* + *horan* to see — more at WARY] (1579)
1 : one of five ancient Spartan magistrates having power over the king

2 : a government official in modern Greece; *especially* **:** one who oversees public works
— **eph·or·ate** \'e-fə-ˌrāt\ *noun*

Ephra·im \'ē-frē-əm\ *noun* [Hebrew *Ephrayim*]
: a son of Joseph and the traditional eponymous ancestor of one of the tribes of Israel

Ephra·im·ite \-frē-ə-ˌmīt\ *noun* (1611)
1 : a member of the Hebrew tribe of Ephraim
2 : a native or inhabitant of the biblical northern kingdom of Israel

epi- *or* **ep-** *prefix* [Middle English, from Middle French & Latin; Middle French, from Latin, from Greek, from *epi* on, at, besides, after; akin to Old English *eofot* crime]
1 : upon ⟨*epi*phyte⟩ **:** besides ⟨*epi*phenomenon⟩ **:** attached to ⟨*epi*didymis⟩ **:** over ⟨*epi*center⟩ **:** outer ⟨*epi*blast⟩ **:** after ⟨*epi*genesis⟩
2 a : chemical entity related to (such) another ⟨*epi*mer⟩ **b :** chemical entity distinguished from (such) another by having a bridge connection ⟨*epi*chlorohydrin⟩

epi·blast \'e-pə-ˌblast\ *noun* (1875)
: the outer layer of the blastoderm **:** ECTODERM
— **epi·blas·tic** \ˌe-pə-'blas-tik\ *adjective*

epib·o·ly \i-'pi-bə-lē\ *noun, plural* **-lies** [Greek *epibolē* addition, from *epiballein* to throw on, from *epi-* + *ballein* to throw — more at DEVIL] (1875)
: the growing of one part about another; *especially* **:** such growth of the dorsal lip area during gastrulation
— **ep·i·bol·ic** \ˌe-pə-'bä-lik\ *adjective*

[1]ep·ic \'e-pik\ *adjective* [Latin *epicus,* from Greek *epikos,* from *epos* word, speech, poem — more at VOICE] (1589)
1 : of, relating to, or having the characteristics of an epic
2 a : extending beyond the usual or ordinary especially in size or scope ⟨his genius was *epic —Times Literary Supplement*⟩ **b :** HEROIC
— **ep·i·cal** \-pi-kəl\ *adjective*
— **ep·i·cal·ly** \-pi-k(ə-)lē\ *adverb*

[2]epic *noun* (1706)
1 : a long narrative poem in elevated style recounting the deeds of a legendary or historical hero ⟨the *Iliad* and the *Odyssey* are *epics*⟩
2 : a work of art (as a novel or drama) that resembles or suggests an epic
3 : a series of events or body of legend or tradition thought to form the proper subject of an epic ⟨the winning of the West was a great American *epic*⟩

epi·ca·lyx \ˌe-pi-'kā-liks *also* -'ka-liks\ *noun* (1870)
: an involucre resembling the calyx but consisting of a whorl of bracts that is exterior to the calyx or results from the union of the sepal appendages

epi·can·thic fold \ˌe-pə-'kan(t)-thik-\ *noun* [New Latin *epicanthus* epicanthic fold, from *epi-* + *canthus* canthus] (1913)
: a prolongation of a fold of the skin of the upper eyelid over the inner angle or both angles of the eye

epi·car·di·um \ˌe-pə-'kar-dē-əm\ *noun, plural* **-dia** \-dē-ə\ [New Latin] (circa 1865)
: the visceral part of the pericardium that closely envelops the heart
— **epi·car·di·al** \-dē-əl\ *adjective*

epi·carp \'e-pi-ˌkärp\ *noun* [French *épicarpe,* from *épi-* epi- + *-carpe* -carp] (1835)
: EXOCARP

ep·i·cene \'e-pə-ˌsēn\ *adjective* [Middle English, from Latin *epicoenus,* from Greek *epikoinos,* from *epi-* + *koinos* common — more at CO-] (15th century)
1 *of a noun* **:** having but one form to indicate either sex
2 a : having characteristics typical of the other sex **:** INTERSEXUAL **b :** EFFEMINATE
3 : lacking characteristics of either sex
— **epicene** *noun*
— **ep·i·cen·ism** \-ˌsē-ˌni-zəm, ˌe-pə-'\ *noun*

epi·cen·ter \'e-pi-ˌsen-tər\ *noun* [New Latin *epicentrum,* from *epi-* + Latin *centrum* center] (1887)
1 : the part of the earth's surface directly above the focus of an earthquake — compare HYPOCENTER 1
2 : CENTER 2a, b, c ⟨the *epicenter* of world finance⟩
— **epi·cen·tral** \ˌe-pi-'sen-trəl\ *adjective*

epi·chlo·ro·hy·drin \ˌe-pi-ˌklōr-ə-'hī-drən, -ˌklȯr-\ *noun* (circa 1891)
: a volatile liquid toxic epoxide C_3H_5ClO having a chloroform odor and used especially in making epoxy resins and rubbers

epi·con·ti·nen·tal \ˌe-pi-ˌkän-tᵊn-'en-tᵊl\ *adjective* (1900)
: lying upon a continent or a continental shelf ⟨*epicontinental* seas⟩

epi·cot·yl \'e-pi-ˌkä-tᵊl\ *noun* [*epi-* + *cotyledon*] (1880)
: the portion of the axis of a plant embryo or seedling above the cotyledonary node

ep·i·crit·ic \ˌe-pə-'kri-tik\ *adjective* [Greek *epikritikos* determinative, from *epikrinein* to decide, from *epi-* + *krinein* to judge — more at CERTAIN] (circa 1905)
: of, relating to, being, or mediating cutaneous sensory reception marked by accurate discrimination between small degrees of sensation

epic simile *noun* (1931)
: an extended simile that is used typically in epic poetry to intensify the heroic stature of the subject

ep·i·cure \'e-pi-ˌkyu̇r\ *noun* [*Epicurus*] (1565)
1 *archaic* **:** one devoted to sensual pleasure **:** SYBARITE
2 : one with sensitive and discriminating tastes especially in food or wine ☆ ◆

ep·i·cu·re·an \ˌe-pi-kyu̇-'rē-ən, -'kyu̇r-ē-\ *adjective* (1586)
1 *capitalized* **:** of or relating to Epicurus or Epicureanism
2 : of, relating to, or suited to an epicure

Epicurean *noun* (14th century)
1 : a follower of Epicurus
2 *often not capitalized* **:** EPICURE 2

ep·i·cu·re·an·ism \-ə-ˌni-zəm\ *noun* (circa 1751)
1 *capitalized* **a :** the philosophy of Epicurus who subscribed to a hedonistic ethics that considered an imperturbable emotional calm the highest good and whose followers held intellectual pleasures superior to transient sensualism **b :** a mode of life in consonance with Epicureanism
2 : EPICURISM

ep·i·cur·ism \'e-pi-ˌkyu̇r-ˌi-zəm, ˌe-pi-'\ *noun* (1586)
: the practices or tastes of an epicure or an epicurean

epi·cu·ti·cle \ˌe-pi-'kyü-ti-kəl\ *noun* (1929)
: an outermost waxy layer of the insect exoskeleton
— **epi·cu·tic·u·lar** \-kyu̇-'ti-kyə-lər\ *adjective*

epi·cy·cle \'e-pə-ˌsī-kəl\ *noun* [Middle English *epicicle,* from Late Latin *epicyclus,* from Greek *epikyklos,* from *epi-* + *kyklos* circle — more at WHEEL] (14th century)
1 *in Ptolemaic astronomy* **:** a circle in which a planet moves and which has a center that is itself carried around at the same time on the circumference of a larger circle
2 : a process going on within a larger one
— **epi·cy·clic** \ˌe-pə-'sī-klik, -'si-klik\ *adjective*

epicyclic train *noun* (circa 1890)
: a train (as of gear wheels) designed to have one or more parts travel around the circumference of another fixed or revolving part

epi·cy·cloid \ˌe-pə-'sī-ˌklȯid\ *noun* (circa 1755)
: a curve traced by a point on a circle that rolls on the outside of a fixed circle
— **epi·cy·cloi·dal** \-sī-'klȯi-dᵊl\ *adjective*

[1]ep·i·dem·ic \ˌe-pə-'de-mik\ *adjective* [French *épidémique,* from Middle French, from *epidemie,* noun, epidemic, from Late Latin *epidemia,* from Greek *epidēmia* visit, epidemic, from *epidēmos* visiting, epidemic, from *epi-* + *dēmos* people — more at DEMAGOGUE] (1603)

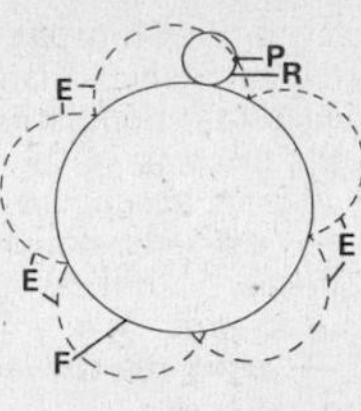

epicycloid *E*, traced by point *P*, on circle *R*, rolling on fixed circle *F*

1 : affecting or tending to affect a disproportionately large number of individuals within a population, community, or region at the same time ⟨typhoid was *epidemic*⟩
2 a : excessively prevalent **b :** CONTAGIOUS 4 ⟨*epidemic* laughter⟩
3 : of, relating to, or constituting an epidemic ⟨the practice had reached *epidemic* proportions⟩
— **ep·i·dem·i·cal** \-'de-mi-kəl\ *adjective*
— **ep·i·dem·i·cal·ly** \-'de-mi-k(ə-)lē\ *adverb*
— **ep·i·de·mic·i·ty** \-də-'mi-sə-tē\ *noun*

[2]epidemic *noun* (1799)
1 : an outbreak of epidemic disease
2 : an outbreak or product of sudden rapid spread, growth, or development; *specifically* **:** a natural population suddenly and greatly enlarged

ep·i·de·mi·ol·o·gy \ˌe-pə-ˌdē-mē-'ä-lə-jē, -ˌde-mē-\ *noun* [Late Latin *epidemia* + International Scientific Vocabulary *-logy*] (circa 1860)

☆ SYNONYMS
Epicure, gourmet, gourmand, gastronome mean one who takes pleasure in eating and drinking. EPICURE implies fastidiousness and voluptuousness of taste. GOURMET implies being a connoisseur in food and drink and the discriminating enjoyment of them. GOURMAND implies a hearty appetite for good food and drink, not without discernment, but with less than a gourmet's. GASTRONOME implies that one has studied extensively the history and rituals of haute cuisine.

◇ WORD HISTORY
epicure Although *epicure* currently suggests hedonism, nothing could be more remote from the lifestyle and teachings of the man to whom we owe the word. The Greek philosopher Epicurus (341– 270 B.C.) articulated a philosophy of simple pleasure, friendship, and a secluded, private life. At the heart of his teachings was the pursuit of pleasure, but pleasure equated with tranquility of mind and freedom from pain—not the indulgence of the senses. Indeed, at his school of philosophy, the usual drink was water and the staple food, barley bread. Detractors of Epicurus in his own time and later, however, reduced his lofty notion of pleasure and happiness to material and sensual gratification and corrupted his reputation in the popular mind. When *epicure* entered the English language in the 16th century, his philosophy had long been trivialized, and so the word became synonymous with "hedonist" and "bon vivant."

1 : a branch of medical science that deals with the incidence, distribution, and control of disease in a population
2 : the sum of the factors controlling the presence or absence of a disease or pathogen
— ep·i·de·mi·o·log·i·cal \-ˌdē-mē-ə-ˈlä-ji-kəl, -ˌde-mē-\ *also* **ep·i·de·mi·o·log·ic** \-jik\ *adjective*
— ep·i·de·mi·o·log·i·cal·ly \-ji-k(ə-)lē\ *adverb*
— ep·i·de·mi·ol·o·gist \-ˌdē-mē-ˈä-lə-jist, -ˌde-mē-\ *noun*

ep·i·den·drum \ˌe-pə-ˈden-drəm\ *noun* [New Latin, from Greek *epi-* + *dendron* tree — more at DENDR-] (1791)
: any of a large genus (*Epidendrum*) of chiefly epiphytic orchids found especially in tropical America

epi·der·mal \ˌe-pə-ˈdər-məl\ *also* **epi·der·mic** \-mik\ *adjective* (1816)
: of, relating to, or arising from the epidermis

epidermal growth factor *noun* (1966)
: a polypeptide hormone that stimulates cell proliferation

epi·der·mis \ˌe-pə-ˈdər-məs\ *noun* [Late Latin, from Greek, from *epi-* + *derma* skin — more at DERM-] (1626)
1 a : the outer epithelial layer of the external integument of the animal body that is derived from the embryonic epiblast; *specifically* **:** the outer nonsensitive and nonvascular layer of the skin of a vertebrate that overlies the dermis **b :** any of various animal integuments
2 : a thin surface layer of tissue in higher plants formed by growth of a primary meristem

epi·der·moid \-ˌmȯid\ *adjective* (1836)
: resembling epidermis or epidermal cells **:** made up of elements like those of epidermis ⟨*epidermoid* cancer of the lung⟩

epi·dia·scope \ˌe-pə-ˈdī-ə-ˌskōp\ *noun* [International Scientific Vocabulary] (1903)
1 : a projector for images of both opaque objects and transparencies
2 : EPISCOPE

ep·i·did·y·mis \ˌe-pə-ˈdi-də-məs\ *noun, plural* **-mi·des** \-mə-ˌdēz\ [New Latin, from Greek, from *epi-* + *didymos* testicle, twin, from *dyo* two — more at TWO] (1610)
: a system of ductules emerging posteriorly from the testis that holds sperm during maturation and that forms a tangled mass before uniting into a single coiled duct which is continuous with the vas deferens
— ep·i·did·y·mal \-məl\ *adjective*

epi·did·y·mi·tis \ˌe-pə-ˌdi-də-ˈmī-təs\ *noun* [New Latin] (1852)
: inflammation of the epididymis

ep·i·dote \ˈe-pə-ˌdōt\ *noun* [French *épidote,* from Greek *epididonai* to give in addition, from *epi-* + *didonai* to give — more at DATE] (1808)
: a yellowish green mineral $Ca_2(Al,Fe)_3Si_3O_{12}OH$ usually occurring in grains or columnar masses and sometimes used as a gemstone

epi·du·ral \ˌe-pi-ˈd(y)u̇r-əl\ *adjective* (1882)
: situated upon or administered outside the dura mater ⟨*epidural* anesthesia⟩ ⟨*epidural* structures⟩

epi·fau·na \-ˈfȯ-nə, -ˈfä-\ *noun* [New Latin] (circa 1914)
: benthic fauna living on the substrate (as a hard sea floor) or on other organisms — compare INFAUNA
— epi·fau·nal \-ˈfȯ-nᵊl, -ˈfä-\ *adjective*

epi·gas·tric \ˌe-pi-ˈgas-trik\ *adjective* (circa 1678)
1 : lying upon or over the stomach
2 a : of, relating to, supplying, or draining the anterior walls of the abdomen **b :** of or relating to the abdominal region

epi·ge·al \ˌe-pi-ˈjē-əl\ *or* **epi·ge·an** \-ˈjē-ən\ *also* **epi·ge·ous** \-ˈjē-əs\ *or* **epi·ge·ic** \-ˈjē-ik\ *adjective* [Greek *epigaios* upon the earth, from *epi-* + *gaia* earth] (1861)
1 *of a cotyledon* **:** forced above ground by elongation of the hypocotyl
2 : marked by the production of epigeal cotyledons ⟨*epigeal* germination⟩
3 : living on or near the surface of the ground; *also* **:** relating to or being the environment near the surface of the ground

epi·gen·e·sis \ˌe-pə-ˈje-nə-səs\ *noun* [New Latin] (1798)
1 : development of a plant or animal from an egg or spore through a series of processes in which unorganized cell masses differentiate into organs and organ systems; *also* **:** the theory that plant and animal development proceeds in this way — compare PREFORMATION 2
2 : change in the mineral character of a rock owing to outside influences

epi·ge·net·ic \-jə-ˈne-tik\ *adjective* (1883)
1 : of, relating to, or produced by the chain of developmental processes in epigenesis that lead from genotype to phenotype after the initial action of the genes
2 *of a deposit or structure* **:** formed after the laying down of the enclosing rock
— epi·ge·net·i·cal·ly \-ˈne-ti-k(-ə)lē\ *adverb*

epi·glot·tal \ˌe-pə-ˈglä-tᵊl\ *also* **epi·glot·tic** \-ˈglä-tik\ *adjective* (1926)
: of, relating to, or produced with the aid of the epiglottis

epi·glot·tis \-ˈglä-təs\ *noun* [New Latin, from Greek *epiglōttis,* from *epi-* + *glōttis* glottis] (1615)
: a thin plate of flexible cartilage in front of the glottis that folds back over and protects the glottis during swallowing

ep·i·gone \ˈe-pə-ˌgōn\ *noun* [German, from Latin *epigonus* successor, from Greek *epigonos,* from *epigignesthai* to be born after, from *epi-* + *gignesthai* to be born — more at KIN] (1865)
: FOLLOWER, DISCIPLE; *also* **:** an inferior imitator
— ep·i·gon·ic \ˌe-pə-ˈgä-nik\ *or* **epig·o·nous** \i-ˈpi-gə-nəs, e-\ *adjective*
— epig·o·nism \-ˈpi-gə-ˌni-zəm\ *noun*

epig·o·nus \i-ˈpi-gə-nəs, e-\ *noun, plural* **-ni** \-ˌnī, -ˌnē\ [Latin] (1922)
: EPIGONE — usually used in plural

ep·i·gram \ˈe-pə-ˌgram\ *noun* [Middle English *epigrame,* from Latin *epigrammat-, epigramma,* from Greek, from *epigraphein* to write on, inscribe, from *epi-* + *graphein* to write — more at CARVE] (15th century)
1 : a concise poem dealing pointedly and often satirically with a single thought or event and often ending with an ingenious turn of thought
2 : a terse, sage, or witty and often paradoxical saying
3 : epigrammatic expression
— ep·i·gram·ma·tism \ˌe-pə-ˈgra-mə-ˌti-zəm\ *noun*
— ep·i·gram·ma·tist \-ˈgra-mə-tist\ *noun*

ep·i·gram·mat·ic \ˌe-pə-grə-ˈma-tik\ *adjective* (circa 1704)
1 : of, relating to, or resembling an epigram
2 : marked by or given to the use of epigrams
— ep·i·gram·mat·i·cal·ly \-ˈma-ti-k(ə-)lē\ *adverb*

ep·i·gram·ma·tize \-ˈgra-mə-ˌtīz\ *verb* **-tized; -tiz·ing** (1691)
transitive verb
1 : to express in the form of an epigram
2 : to make an epigram about
intransitive verb
: to make an epigram
— ep·i·gram·ma·tiz·er *noun*

ep·i·graph \ˈe-pə-ˌgraf\ *noun* [Greek *epigraphē,* from *epigraphein*] (1624)
1 : an engraved inscription
2 : a quotation set at the beginning of a literary work or one of its divisions to suggest its theme

epig·ra·pher \i-ˈpi-grə-fər, e-\ *noun* (1887)
: EPIGRAPHIST

ep·i·graph·ic \ˌe-pə-ˈgra-fik\ *also* **ep·i·graph·i·cal** \-fi-kəl\ *adjective* (1858)
: of or relating to epigraphs or epigraphy
— ep·i·graph·i·cal·ly \-fi-k(ə-)lē\ *adverb*

epig·ra·phist \i-ˈpi-grə-fist, e-\ *noun* (circa 1864)
: a specialist in epigraphy

epig·ra·phy \-fē\ *noun* (1851)
1 : EPIGRAPHS, INSCRIPTIONS
2 : the study of inscriptions; *especially* **:** the deciphering of ancient inscriptions

epig·y·nous \i-ˈpi-jə-nəs, e-\ *adjective* (1830)
1 *of a floral organ* **:** adnate to the surface of the ovary and appearing to grow from the top of it
2 : having epigynous floral organs
— epig·y·ny \-nē\ *noun*

ep·i·la·tion \ˌe-pə-ˈlā-shən\ *noun* [French *épilation,* from *épiler* to remove hair, from *é-* e- + Latin *pilus* hair — more at PILE] (1878)
: the loss or removal of hair

ep·i·lep·sy \ˈe-pə-ˌlep-sē\ *noun, plural* **-sies** [Middle French *epilepsie,* from Late Latin *epilepsia,* from Greek *epilēpsia,* from *epilambanein* to seize, from *epi-* + *lambanein* to take, seize — more at LATCH] (1543)
: any of various disorders marked by disturbed electrical rhythms of the central nervous system and typically manifested by convulsive attacks usually with clouding of consciousness

epilept- *or* **epilepti-** *or* **epilepto-** *combining form* [Greek *epilēpt-,* from *epilēptos* seized by epilepsy, from *epilambanein*]
: epilepsy ⟨*epilept*oid⟩

ep·i·lep·tic \ˌe-pə-ˈlep-tik\ *adjective* (1605)
: relating to, affected with, or having the characteristics of epilepsy
— epileptic *noun*
— ep·i·lep·ti·cal·ly \-ti-k(ə-)lē\ *adverb*

ep·i·lep·ti·form \-ˈlep-tə-ˌfȯrm\ *adjective* (circa 1859)
: resembling that of epilepsy ⟨an *epileptiform* convulsion⟩

ep·i·lep·to·gen·ic \-ˌlep-tə-ˈje-nik\ *adjective* (circa 1882)
: inducing or tending to induce epilepsy ⟨an *epileptogenic* drug⟩

ep·i·lep·toid \-ˈlep-ˌtȯid\ *adjective* (circa 1860)
1 : EPILEPTIFORM
2 : exhibiting symptoms resembling those of epilepsy ⟨the *epileptoid* person⟩

epi·lim·ni·on \ˌe-pə-ˈlim-nē-ˌän, -nē-ən\ *noun* [New Latin, from *epi-* + Greek *limnion,* diminutive of *limnē* marshy lake — more at LIMNETIC] (circa 1910)
: the water layer overlying the thermocline of a lake

ep·i·logue *also* **ep·i·log** \ˈe-pə-ˌlȯg, -ˌläg\ *noun* [Middle English *epiloge,* from Middle French *epilogue,* from Latin *epilogus,* from Greek *epilogos,* from *epilegein* to say in addition, from *epi-* + *legein* to say — more at LEGEND] (15th century)
1 : a concluding section that rounds out the design of a literary work
2 a : a speech often in verse addressed to the audience by an actor at the end of a play; *also* **:** the actor speaking such an epilogue **b :** the final scene of a play that comments on or summarizes the main action
3 : the concluding section of a musical composition **:** CODA

epi·mer \ˈe-pi-mər\ *noun* [*epi-* + isomer] (circa 1911)
: either of the stereoisomers of a sugar or sugar derivative that differ in the arrangement of the hydrogen atom and the hydroxyl group on the first asymmetric carbon atom of a chain
— epi·mer·ic \ˌe-pi-ˈmer-ik\ *adjective*

epim·er·ase \i-ˈpi-mə-ˌrās, e-, -ˌrāz\ *noun* (1960)
: any of various isomerases that catalyze the inversion of asymmetric groups in a substrate with several centers of asymmetry

epi·my·si·um \ˌe-pə-ˈmi-zhē-əm, -zē-\ *noun, plural* **-sia** \-zhē-ə, -zē-ə\ [New Latin, irregular from *epi-* + Greek *mys* mouse, muscle — more at MOUSE] (1900)
: the external connective-tissue sheath of a muscle

epi·nas·ty \ˈe-pə-ˌnas-tē\ *noun* [International Scientific Vocabulary *epi-* + Greek *nastos* close-pressed (from *nassein* to press) + International Scientific Vocabulary ²-y] (1880)
: a nastic movement in which a plant part (as a flower petal) is bent outward and often downward

epi·neph·rine *also* **epi·neph·rin** \ˌe-pə-ˈne-frən\ *noun* [International Scientific Vocabulary *epi-* + Greek *nephros* kidney — more at NEPHRITIS] (1899)
: a colorless crystalline feebly basic sympathomimetic hormone $C_9H_{13}NO_3$ that is the principal blood-pressure raising hormone secreted by the adrenal medulla and is used medicinally especially as a heart stimulant, a vasoconstrictor in controlling hemorrhages of the skin, and a muscle relaxant in bronchial asthma — called also *adrenaline*

epi·neu·ri·um \ˌe-pə-ˈn(y)ůr-ē-əm\ *noun* [New Latin] (circa 1882)
: the external connective-tissue sheath of a nerve trunk

epi·pe·lag·ic \ˌe-pi-pə-ˈla-jik\ *adjective* (1940)
: of, relating to, or constituting the part of the oceanic zone into which enough light penetrates for photosynthesis

ep·i·phan·ic \ˌe-pə-ˈfa-nik\ *adjective* (1951)
: of or having the character of an epiphany

epiph·a·nous \i-ˈpi-fə-nəs\ *adjective* (1823)
: EPIPHANIC

epiph·a·ny \i-ˈpi-fə-nē\ *noun, plural* **-nies** [Middle English *epiphanie,* from Middle French, from Late Latin *epiphania,* from Late Greek, plural, probably alteration of Greek *epiphaneia* appearance, manifestation, from *epiphainein* to manifest, from *epi-* + *phainein* to show — more at FANCY] (14th century)
1 *capitalized* **:** January 6 observed as a church festival in commemoration of the coming of the Magi as the first manifestation of Christ to the Gentiles or in the Eastern Church in commemoration of the baptism of Christ
2 : an appearance or manifestation especially of a divine being
3 a (1) **:** a usually sudden manifestation or perception of the essential nature or meaning of something (2) **:** an intuitive grasp of reality through something (as an event) usually simple and striking (3) **:** an illuminating discovery **b :** a revealing scene or moment

epi·phe·nom·e·nal \ˌe-pi-fi-ˈnä-mə-nᵊl\ *adjective* (1899)
: of or relating to an epiphenomenon **:** DERIVATIVE
— epi·phe·nom·e·nal·ly \-nᵊl-ē\ *adverb*

epi·phe·nom·e·nal·ism \-nᵊl-ˌi-zəm\ *noun* (1899)
: a doctrine that mental processes are epiphenomena of brain processes

epi·phe·nom·e·non \-ˈnä-mə-ˌnän, -nən\ *noun, plural* **-na** \-nə, -ˌnä\ (circa 1706)
: a secondary phenomenon accompanying another and caused by it

ep·i·phragm \ˈe-pə-ˌfram\ *noun* [Greek *epiphragma* covering] (circa 1854)
: a closing membrane or septum (as of a snail shell or a moss capsule)

epiph·y·se·al \i-ˌpi-fə-ˈsē-əl\ *also* **ep·i·phys·i·al** \ˌe-pə-ˈfi-zē-əl\ *adjective* (1842)
: of or relating to an epiphysis

epiph·y·sis \i-ˈpi-fə-səs\ *noun, plural* **-y·ses** \-ˌsēz\ [New Latin, from Greek, growth, from *epiphyesthai* to grow on, from *epi-* + *phyesthai* to grow, middle voice of *phyein* to bring forth — more at BE] (1634)
1 : a part or process of a bone that ossifies separately and later becomes ankylosed to the main part of the bone; *especially* **:** an end of a long bone
2 : PINEAL GLAND

epi·phyte \ˈe-pə-ˌfīt\ *noun* (circa 1847)
: a plant that derives its moisture and nutrients from the air and rain and grows usually on another plant

epi·phyt·ic \ˌe-pə-ˈfi-tik\ *adjective* (1830)
1 : of, relating to, or being an epiphyte
2 : living on the surface of plants
— epi·phyt·i·cal·ly \-ˈfi-ti-k(ə-)lē\ *adverb*
— epi·phyt·ism \ˈe-pə-ˌfī-ˌti-zəm\ *noun*

ep·i·phy·tol·o·gy \ˌe-pə-ˌfī-ˈtä-lə-jē\ *noun* [*epiphyto*tic + *-logy*] (1940)
1 : a science that deals with character, ecology, and causes of outbreak of plant diseases
2 : the sum of the factors controlling the occurrence of a disease or pathogen of plants

ep·i·phy·tot·ic \-ˈtä-tik\ *adjective* [*epi-* + *-phyte* + *-otic* (as in *epizootic*)] (circa 1899)
: of, relating to, or being a plant disease that tends to recur sporadically and to affect large numbers of susceptible plants
— epiphytotic *noun*

epi·scia \i-ˈpi-sh(ē-)ə\ *noun* [New Latin, from Greek *episkios* shaded, from *epi-* + *skia* shadow — more at SHINE] (circa 1868)
: any of a genus (*Episcia*) of tropical American herbs that have hairy foliage and are related to the African violet

epis·co·pa·cy \i-ˈpis-kə-pə-sē\ *noun, plural* **-cies** (1647)
1 : government of the church by bishops or by a hierarchy
2 : EPISCOPATE

epis·co·pal \i-ˈpis-kə-pəl, -bəl\ *adjective* [Middle English, from Late Latin *episcopalis,* from *episcopus* bishop — more at BISHOP] (15th century)
1 : of or relating to a bishop
2 : of, having, or constituting government by bishops
3 *capitalized* **:** of or relating to the Protestant Episcopal Church representing the Anglican communion in the U.S.
— epis·co·pal·ly \-p(ə-)lē\ *adverb*

Episcopal *noun* (1752)
: EPISCOPALIAN

Epis·co·pa·lian \i-ˌpis-kə-ˈpāl-yən\ *noun* (1690)
1 : an adherent of the episcopal form of church government
2 : a member of an episcopal church (as the Protestant Episcopal Church)
— Episcopalian *adjective*
— Epis·co·pa·lian·ism \-yə-ˌni-zəm\ *noun*

epis·co·pate \i-ˈpis-kə-pət, -ˌpāt\ *noun* (1641)
1 : the rank, office, or term of bishop
2 : DIOCESE
3 : the body of bishops (as in a country)

epi·scope \ˈe-pə-ˌskōp\ *noun* [International Scientific Vocabulary] (circa 1909)
: a projector for images of opaque objects (as photographs)

epi·si·ot·o·my \i-ˌpi-zē-ˈä-tə-mē, -ˌpē-\ *noun* [International Scientific Vocabulary *episio-* vulva (from Greek *epision* pubic region) + *-tomy*] (1878)
: surgical enlargement of the vulval orifice for obstetrical purposes during parturition

ep·i·sode \ˈe-pə-ˌsōd *also* -ˌzōd\ *noun* [Greek *epeisodion,* from neuter of *epeisodios* coming in besides, from *epi-* + *eisodios* coming in, from *eis* into (akin to Greek *en* in) + *hodos* road, journey — more at IN] (1678)
1 : a usually brief unit of action in a dramatic or literary work: as **a :** the part of an ancient Greek tragedy between two choric songs **b :** a developed situation that is integral to but separable from a continuous narrative **:** INCIDENT **c :** one of a series of loosely connected stories or scenes **d :** the part of a serial presented at one performance
2 : an event that is distinctive and separate although part of a larger series
3 : a digressive subdivision in a musical composition
synonym see OCCURRENCE

ep·i·sod·ic \ˌe-pə-ˈsä-dik *also* -ˈzä-\ *also* **ep·i·sod·i·cal** \-di-kəl\ *adjective* (1711)
1 : made up of separate especially loosely connected episodes
2 : having the form of an episode
3 : of or limited in duration or significance to a particular episode **:** TEMPORARY ⟨may be able to establish whether the sea-floor spreading is continuous or *episodic* —A. I. Hammond⟩
4 : occurring, appearing, or changing at usually irregular intervals **:** OCCASIONAL ⟨an *episodic* illness⟩
— ep·i·sod·i·cal·ly \-di-k(ə-)lē\ *adverb*

epi·some \ˈe-pə-ˌsōm, -ˌzōm\ *noun* (circa 1931)
: a genetic determinant (as the DNA of some bacteriophages) that can replicate autonomously in bacterial cytoplasm or as an integral part of the chromosomes
— epi·som·al \ˌe-pə-ˈsō-məl, -ˈzō-\ *adjective*
— epi·som·al·ly \-mə-lē\ *adverb*

epis·ta·sis \i-ˈpis-tə-səs\ *noun, plural* **-ta·ses** \-ˌsēz\ [New Latin, from Greek, act of stopping, from *ephistanai* to stop, from *epi-* + *histanai* to cause to stand — more at STAND] (circa 1917)
: suppression of the effect of a gene by a nonallelic gene
— ep·i·stat·ic \ˌe-pə-ˈsta-tik\ *adjective*

ep·i·stax·is \ˌe-pə-ˈstak-səs\ *noun, plural* **-stax·es** \-ˌsēz\ [New Latin, from Greek, from *epistazein* to drip on, to bleed at the nose again, from *epi-* + *stazein* to drip] (1793)
: NOSEBLEED

ep·i·ste·mic \ˌe-pə-ˈstē-mik, -ˈste-mik\ *adjective* (1922)
: of or relating to knowledge or knowing **:** COGNITIVE
— ep·i·ste·mi·cal·ly \-mi-k(ə-)lē\ *adverb*

epis·te·mol·o·gy \i-ˌpis-tə-ˈmä-lə-jē\ *noun* [Greek *epistēmē* knowledge, from *epistanai* to understand, know, from *epi-* + *histanai* to cause to stand — more at STAND] (circa 1856)
: the study or a theory of the nature and grounds of knowledge especially with reference to its limits and validity
— epis·te·mo·log·i·cal \-mə-ˈlä-ji-kəl\ *adjective*
— epis·te·mo·log·i·cal·ly \-k(ə-)lē\ *adverb*
— epis·te·mol·o·gist \-ˈmä-lə-jist\ *noun*

epis·tle \i-ˈpi-səl\ *noun* [Middle English, letter, Epistle, from Old French, from Latin *epistula, epistola* letter, from Greek *epistolē* message, letter, from *epistellein* to send to, from *epi-* + *stellein* to send] (13th century)
1 *capitalized* **a :** one of the letters adopted as books of the New Testament **b :** a liturgical lection usually from one of the New Testament Epistles
2 a : LETTER; *especially* **:** a formal or elegant letter **b :** a composition in the form of a letter
— epis·tler \-ˈpi-sə-lər\ *noun*

¹epis·to·lary \i-ˈpis-tə-ˌler-ē, ˌe-pi-ˈstȯ-lə-rē\ *adjective* (circa 1656)
1 : of, relating to, or suitable to a letter
2 : contained in or carried on by letters ⟨an endless sequence of . . . *epistolary* love affairs —*Times Literary Supplement*⟩
3 : written in the form of a series of letters ⟨*epistolary* novel⟩

[2]**epistolary** *noun, plural* **-lar·ies** (circa 1900)
: a lectionary containing a body of liturgical epistles

epis·to·ler \i-'pis-tə-lər\ *noun* (1530)
: the reader of the liturgical Epistle especially in Anglican churches

ep·i·stome \'e-pə-ˌstōm\ *noun* [New Latin *epistoma*] (1852)
: any of several structures or regions situated above or covering the mouth of various invertebrates

epis·tro·phe \i-'pis-trə-(ˌ)fē\ *noun* [Greek *epistrophē*, literally, turning about, from *epi-* + *strophē* turning — more at STROPHE] (circa 1584)
: repetition of a word or expression at the end of successive phrases, clauses, sentences, or verses especially for rhetorical or poetic effect (as Lincoln's "of the people, by the people, for the people") — compare ANAPHORA

ep·i·taph \'e-pə-ˌtaf\ *noun* [Middle English *epitaphe*, from Middle French & Medieval Latin; Middle French, from Medieval Latin *epi taphium*, from Latin, funeral oration, from Greek *epitaphion*, from *epi-* + *taphos* tomb, funeral] (14th century)
1 : an inscription on or at a tomb or a grave in memory of the one buried there
2 : a brief statement commemorating or epitomizing a deceased person or something past
— **ep·i·taph·ial** \ˌe-pə-'ta-fē-əl\ *adjective*
— **ep·i·taph·ic** \-'ta-fik\ *adjective*

epit·a·sis \i-'pi-tə-səs\ *noun, plural* **-a·ses** \-ˌsēz\ [Greek, increased intensity, from *epiteinein* to stretch tighter, from *epi-* + *teinein* to stretch — more at THIN] (1589)
: the part of a play developing the main action and leading to the catastrophe

ep·i·taxy \'e-pə-ˌtak-sē\ *noun* [International Scientific Vocabulary] (circa 1931)
: the growth on a crystalline substrate of a crystalline substance that mimics the orientation of the substrate
— **ep·i·tax·i·al** \ˌe-pə-'tak-sē-əl\ *adjective*
— **ep·i·tax·i·al·ly** \-sē-ə-lē\ *adverb*

ep·i·tha·la·mi·um \ˌe-pə-thə-'lā-mē-əm\ *or* **ep·i·tha·la·mi·on** \-mē-ən\ *noun, plural* **-mi·ums** *or* **-mia** \-mē-ə\ [Latin & Greek; Latin *epithalamium*, from Greek *epithalamion*, from *epi-* + *thalamos* room, bridal chamber; perhaps akin to Greek *tholos* rotunda] (circa 1589)
: a song or poem in honor of a bride and bridegroom
— **ep·i·tha·lam·ic** \-'la-mik\ *adjective*

ep·i·the·li·al \ˌe-pə-'thē-lē-əl\ *adjective* (1845)
: of or relating to epithelium

ep·i·the·li·oid \-lē-ˌȯid\ *adjective* (1878)
: resembling epithelium ⟨*epithelioid* cells⟩

ep·i·the·li·o·ma \-ˌthē-lē-'ō-mə\ *noun* (1872)
: a tumor derived from epithelial tissue
— **ep·i·the·li·o·ma·tous** \-mə-təs\ *adjective*

ep·i·the·li·um \ˌe-pə-'thē-lē-əm\ *noun, plural* **-lia** \-lē-ə\ [New Latin, from *epi-* + Greek *thēlē* nipple — more at FEMININE] (1748)
1 : a membranous cellular tissue that covers a free surface or lines a tube or cavity of an animal body and serves especially to enclose and protect the other parts of the body, to produce secretions and excretions, and to function in assimilation
2 : a usually thin layer of parenchyma that lines a cavity or tube of a plant

ep·i·the·li·za·tion \ˌe-pə-ˌthē-lə-'zā-shən\ *or* **ep·i·the·li·al·iza·tion** \-ˌthē-lē-ə-\ *noun* (circa 1934)
: the process of becoming covered with or converted to epithelium
— **ep·i·the·lize** \ˌe-pə-'thē-ˌlīz\ *or* **ep·i·the·li·al·ize** \-'thē-lē-ə-ˌlīz\ *transitive verb*

ep·i·thet \'e-pə-ˌthet *also* -thət\ *noun* [Latin *epitheton*, from Greek, from neuter of *epithetos* added, from *epitithenai* to put on, add, from *epi-* + *tithenai* to put — more at DO] (1579)
1 a : a characterizing word or phrase accompanying or occurring in place of the name of a person or thing **b :** a disparaging or abusive word or phrase **c :** the part of a taxonomic name identifying a subordinate unit within a genus
2 *obsolete* **:** EXPRESSION
— **ep·i·thet·ic** \ˌe-pə-'the-tik\ *or* **ep·i·thet·i·cal** \-ti-kəl\ *adjective*

epit·o·me \i-'pi-tə-mē\ *noun* [Latin, from Greek *epitomē*, from *epitemnein* to cut short, from *epi-* + *temnein* to cut — more at TOME] (1520)
1 a : a summary of a written work **b :** a brief presentation or statement of something
2 : a typical or ideal example **:** EMBODIMENT ⟨the British monarchy itself is the *epitome* of tradition —Richard Joseph⟩
3 : brief or miniature form — usually used with *in*
— **ep·i·tom·ic** \ˌe-pə-'tä-mik\ *or* **ep·i·tom·i·cal** \-mi-kəl\ *adjective*

epit·o·mise *British variant of* EPITOMIZE

epit·o·mize \-ˌmīz\ *transitive verb* **-mized; -miz·ing** (1594)
1 : to make or give an epitome of
2 : to serve as the typical or ideal example of

epi·tope \'e-pə-ˌtōp\ *noun* [International Scientific Vocabulary, from *epi-* + Greek *topos* place] (1960)
: a molecular region on the surface of an antigen capable of eliciting an immune response and of combining with the specific antibody produced by such a response — called also *determinant, antigenic determinant*

epi·zo·ic \ˌe-pə-'zō-ik\ *adjective* (circa 1857)
: living upon the body of an animal ⟨an *epizoic* plant⟩
— **epi·zo·ite** \-ˌīt\ *noun*

epi·zo·ot·ic \ˌe-pə-zə-'wä-tik\ *noun* [French *épizootique*, from *épizootie* such an outbreak, from *épi-* (as in *épidemie* epidemic) + Greek *zōiotēs* animal nature, from *zōē* life — more at QUICK] (1748)
: an outbreak of disease affecting many animals of one kind at the same time; *also* **:** the disease itself
— **epizootic** *adjective*

epi·zo·ot·i·ol·o·gy \ˌe-pə-zə-ˌwä-tē-'ä-lə-jē\ *noun* (1910)
1 : the sum of the factors controlling the occurrence of a disease or pathogen of animals
2 : a science that deals with the character, ecology, and causes of outbreaks of animal diseases
— **epi·zo·oti·o·log·i·cal** \-zə-ˌwä-tē-ə-'lä-ji-kəl\ *also* **epi·zo·oti·o·log·ic** \-jik\ *adjective*

ep·och \'e-pək, 'e-ˌpäk, *US also & British usually* 'ē-ˌpäk\ *noun* [Medieval Latin *epocha*, from Greek *epochē* cessation, fixed point, from *epechein* to pause, hold back, from *epi-* + *echein* to hold — more at SCHEME] (1614)
1 a : an event or a time marked by an event that begins a new period or development **b :** a memorable event or date
2 a : an extended period of time usually characterized by a distinctive development or by a memorable series of events **b :** a division of geologic time less than a period and greater than an age
3 : an instant of time or a date selected as a point of reference (as in astronomy)
synonym see PERIOD

ep·och·al \'e-pə-kəl, 'e-ˌpä-kəl\ *adjective* (1685)
1 : of or relating to an epoch
2 : uniquely or highly significant **:** MOMENTOUS ⟨during his three *epochal* years in the assembly —C. G. Bowers⟩; *also* **:** UNPARALLELED ⟨*epochal* stupidity⟩
— **ep·och·al·ly** *adverb*

ep·ode \'e-ˌpōd\ *noun* [Latin *epodos*, from Greek *epōidos*, from *epōidos* sung or said after, from *epi-* + *aidein* to sing — more at ODE] (1598)
1 : a lyric poem in which a long verse is followed by a shorter one
2 : the third part of a triadic Greek ode following the strophe and the antistrophe

ep·onym \'e-pə-ˌnim\ *noun* [Greek *epōnymos*, from *epōnymos* eponymous, from *epi-* + *onyma* name — more at NAME] (1846)
1 : one for whom or which something is or is believed to be named
2 : a name (as of a drug or a disease) based on or derived from an eponym
— **ep·onym·ic** \ˌe-pə-'ni-mik\ *adjective*

epon·y·mous \i-'pä-nə-məs, e-\ *adjective* (1846)
: of, relating to, or being an eponym

epon·y·my \-mē\ *noun, plural* **-mies** (1865)
: the explanation of a proper name (as of a town or tribe) by supposing a fictitious eponym

ep·o·pee \'e-pə-ˌpē\ *noun* [French *épopée*, from Greek *epopoiia*, from *epos* + *poiein* to make — more at POET] (1697)
: EPIC; *especially* **:** an epic poem

ep·os \'e-ˌpäs\ *noun* [Greek, word, epic poem — more at VOICE] (circa 1828)
1 : EPIC
2 : a number of poems that treat an epic theme but are not formally united

ep·ox·i·da·tion \(ˌ)e-ˌpäk-sə-'dā-shən\ *noun* (1944)
: a conversion of a usually unsaturated compound into an epoxide

ep·ox·ide \(ˌ)e-'päk-ˌsīd\ *noun* (1930)
: an epoxy compound

ep·ox·i·dize \(ˌ)e-'päk-sə-ˌdīz\ *transitive verb* **-dized; -diz·ing** (1945)
: to convert into an epoxide ⟨*epoxidized* esters⟩

[1]**ep·oxy** \i-'päk-sē\ *adjective* [*epi-* + *oxy*] (1916)
1 : containing oxygen attached to two different atoms already united in some other way; *specifically* **:** containing a 3-membered ring consisting of one oxygen and two carbon atoms
2 : of or relating to an epoxide

[2]**epoxy** *transitive verb* **ep·ox·ied** *or* **ep·oxyed; ep·oxy·ing** (1966)
: to glue with epoxy resin

epoxy resin *noun* (1950)
: a flexible usually thermosetting resin made by copolymerization of an epoxide with another compound having two hydroxyl groups and used chiefly in coatings and adhesives — called also *epoxy*

EPROM \'ē-ˌpräm\ *noun* [erasable *p*rogrammable *r*ead-*o*nly *m*emory] (1977)
: a programmable read-only memory that can be erased usually by exposure to ultraviolet radiation

ep·si·lon \'ep-sə-ˌlän, -lən\ *noun* [Greek *e psilon*, literally, simple e] (15th century)
1 : the 5th letter of the Greek alphabet — see ALPHABET table
2 : an arbitrarily small positive quantity in mathematical analysis
— **ep·si·lon·ic** \ˌep-sə-'lä-nik\ *adjective*

Ep·som salt \'ep-səm-\ *noun* (1770)
: EPSOM SALTS

Epsom salts *noun plural but singular in construction* [*Epsom*, England] (1876)
: a bitter colorless or white crystalline salt $MgSO_4 \cdot 7H_2O$ that is a hydrated magnesium sulfate with cathartic properties

Ep·stein–Barr virus \ˌep-ˌstīn-'bär-\ *noun* [Michael Anthony *Epstein* (born 1921) and Y. M. *Barr* (born 1932) English pathologists] (1968)
: a herpesvirus that causes infectious mononucleosis and is associated with Burkitt's lymphoma and nasopharyngeal carcinoma

equa·ble \ˈe-kwə-bəl, ˈē-\ *adjective* [Latin *aequabilis,* from *aequare* to make level or equal, from *aequus*] (1677)
1 : marked by lack of variation or change **:** UNIFORM
2 : marked by lack of noticeable, unpleasant, or extreme variation or inequality
synonym see STEADY
— equa·bil·i·ty \ˌe-kwə-ˈbi-lə-tē, ˌē-\ *noun*
— equa·ble·ness \ˈe-kwə-bəl-nəs, ˈē-\ *noun*
— equa·bly \-blē\ *adverb*

[1]equal \ˈē-kwəl\ *adjective* [Middle English, from Latin *aequalis,* from *aequus* level, equal] (14th century)
1 a (1) **:** of the same measure, quantity, amount, or number as another (2) **:** identical in mathematical value or logical denotation **:** EQUIVALENT **b :** like in quality, nature, or status **c :** like for each member of a group, class, or society ⟨provide *equal* employment opportunities⟩
2 : regarding or affecting all objects in the same way **:** IMPARTIAL
3 : free from extremes: as **a :** tranquil in mind or mood **b :** not showing variation in appearance, structure, or proportion
4 a : capable of meeting the requirements of a situation or a task **b :** SUITABLE ⟨bored with work not *equal* to his abilities⟩
synonym see SAME

[2]equal *transitive verb* **equaled** *or* **equalled; equal·ing** *or* **equal·ling** (1590)
1 : to be equal to; *especially* **:** to be identical in value to
2 *archaic* **:** EQUALIZE
3 : to make or produce something equal to

[3]equal *noun* (1753)
1 : one that is equal ⟨insists that women can be absolute *equals* with men —Anne Bernays⟩
2 : an equal quantity

equal–area *adjective* (circa 1929)
of a map projection **:** maintaining constant ratio of size between quadrilaterals formed by the meridians and parallels and the quadrilaterals of the globe thereby preserving true areal extent of forms represented

equal·ise, equal·is·er *British variant of* EQUALIZE, EQUALIZER

equal·i·tar·i·an \i-ˌkwä-lə-ˈter-ē-ən\ *adjective or noun* (1799)
: EGALITARIAN
— equal·i·tar·i·an·ism \-ē-ə-ˌni-zəm\ *noun*

equal·i·ty \i-ˈkwä-lə-tē\ *noun, plural* **-ties** (15th century)
1 : the quality or state of being equal
2 : EQUATION 2a

equal·ize \ˈē-kwə-ˌlīz\ *verb* **-ized; -iz·ing** (1622)
transitive verb
1 : to make equal
2 a : to compensate for **b :** to make uniform; *especially* **:** to distribute evenly or uniformly ⟨*equalize* the tax burden⟩ **c :** to adjust or correct the frequency characteristics of (an electronic signal) by restoring to their original level high frequencies that have been attenuated
intransitive verb
chiefly British **:** to tie the score
— equal·iza·tion \ˌē-kwə-lə-ˈzā-shən\ *noun*

equal·iz·er \-ˌlī-zər\ *noun* (1792)
: one that equalizes: as **a :** a score that ties a game **b :** an electronic device (as in a sound-reproducing system) used to adjust response to different audio frequencies

equal·ly \ˈē-kwə-lē\ *adverb* (14th century)
1 : in an equal or uniform manner **:** EVENLY
2 : to an equal degree ⟨respected *equally* by young and old⟩

equal opportunity employer *noun* (1963)
: an employer who agrees not to discriminate against any employee or job applicant because of race, color, religion, national origin, sex, physical or mental handicap, or age

equal sign *noun* (circa 1909)
: a sign = indicating mathematical or logical equivalence — called also *equality sign, equals sign*

equa·nim·i·ty \ˌē-kwə-ˈni-mə-tē, ˌe-kwə-\ *noun, plural* **-ties** [Latin *aequanimitas,* from *aequo animo* with even mind] (circa 1616)
1 : evenness of mind especially under stress
2 : right disposition **:** BALANCE ☆

equate \i-ˈkwāt, ˈē-ˌ\ *verb* **equat·ed; equat·ing** [Middle English, from Latin *aequatus,* past participle of *aequare*] (15th century)
transitive verb
1 a : to make equal **:** EQUALIZE **b :** to make such an allowance or correction in as will reduce to a common standard or obtain a correct result
2 : to treat, represent, or regard as equal, equivalent, or comparable ⟨*equates* disagreement with disloyalty⟩
intransitive verb
: to correspond as equal

equa·tion \i-ˈkwā-zhən *also* -shən\ *noun* (14th century)
1 a : the act or process of equating **b** (1) **:** an element affecting a process **:** FACTOR (2) **:** a complex of variable factors **c :** a state of being equated; *specifically* **:** a state of close association or identification ⟨bring governmental enterprises and payment for them into immediate *equation* —R. G. Tugwell⟩
2 a : a usually formal statement of the equality or equivalence of mathematical or logical expressions **b :** an expression representing a chemical reaction quantitatively by means of chemical symbols

equa·tion·al \i-ˈkwāzh-nəl, -ˈkwā-zhə-nᵊl *also* -ˈkwāsh-\ *adjective* (1864)
1 : of, using, or involving equation or equations
2 : dividing into two equal parts — used especially of the mitotic cell division usually following reduction in meiosis
— equa·tion·al·ly *adverb*

equation of time (1726)
: the difference between apparent time and mean time usually expressed as a correction which is to be added to apparent time to give local mean time

equa·tor \i-ˈkwā-tər, ˈē-ˌ\ *noun* [Middle English, from Medieval Latin *aequator,* literally, equalizer, from Latin *aequare*] (14th century)
1 : the great circle of the celestial sphere whose plane is perpendicular to the axis of the earth
2 : a great circle of the earth or a celestial body that is everywhere equally distant from the two poles and divides the surface into the northern and southern hemispheres
3 a : a circle or circular band dividing the surface of a body into two usually equal and symmetrical parts **b :** EQUATORIAL PLANE ⟨the *equator* of a dividing cell⟩
4 : GREAT CIRCLE

equa·to·ri·al \ˌē-kwə-ˈtōr-ē-əl, ˌe-kwə-, -ˈtȯr-\ *adjective* (1664)
1 a : of, relating to, or located at the equator or an equator; *also* **:** being in the plane of the equator ⟨a satellite in *equatorial* orbit⟩ **b :** of, originating in, or suggesting the region around the geographic equator
2 a : being or having a support that includes two axles at right angles to each other with one parallel to the earth's axis of rotation ⟨an *equatorial* telescope⟩ **b :** extending in a direction essentially in the plane of a cyclic structure (as of cyclohexane) ⟨*equatorial* hydrogens⟩ — compare AXIAL

equatorial plane *noun* (circa 1892)
: the plane perpendicular to the spindle of a dividing cell and midway between the poles

equatorial plate *noun* (1887)
1 : METAPHASE PLATE
2 : EQUATORIAL PLANE

equa·tor·ward \i-ˈkwā-tər-wərd\ *adverb or adjective* (1875)
: toward or near the equator ⟨currents flowing *equatorward*⟩ ⟨*equatorward* winds⟩

equer·ry \ˈe-kwə-rē, i-ˈkwer-ē\ *noun, plural* **-ries** [obsolete *escuirie, equerry* stable, from Middle French *escuirie* office of a squire, stable, from *escuier* squire — more at ESQUIRE] (1591)
1 : an officer of a prince or noble charged with the care of horses
2 : an officer of the British royal household in personal attendance on the sovereign or a member of the royal family

[1]eques·tri·an \i-ˈkwes-trē-ən\ *adjective* [Latin *equestr-, equester* of a horseman, from *eques* horseman, from *equus* horse — more at EQUINE] (circa 1681)
1 a : of, relating to, or featuring horseback riding **b** *archaic* **:** riding on horseback **:** MOUNTED **c :** representing a person on horseback ⟨an *equestrian* statue⟩
2 : of, relating to, or composed of knights

[2]equestrian *noun* (1791)
: one who rides on horseback

eques·tri·enne \i-ˌkwes-trē-ˈen\ *noun* [[2]*equestrian* + *-enne* (as in *tragedienne*)] (circa 1864)
: a female rider on horseback

equi- *combining form* [Middle English, from Middle French, from Latin *aequi-,* from *aequus* equal]
: equal ⟨*equi*poise⟩ **:** equally ⟨*equi*probable⟩

equi·an·gu·lar \ˌē-kwi-ˈaŋ-gyə-lər, ˌe-kwi-\ *adjective* (1660)
: having all or corresponding angles equal ⟨mutually *equiangular* parallelograms⟩

equi·ca·lor·ic \ˌē-kwə-kə-ˈlȯr-ik, ˌe-kwə-, -ˈlär-\ *adjective* (1940)
: capable of yielding equal amounts of energy in the body ⟨*equicaloric* diets⟩

equid \ˈe-ˌkwid, ˈē-\ *noun* [New Latin *Equidae,* family name, from *Equus,* genus name, from Latin, horse] (circa 1889)
: any of a family (Equidae) of perissodactyl mammals consisting of the horses, asses, zebras, and extinct related animals

equi·dis·tant \ˌē-kwə-ˈdis-tənt, ˌe-kwə-\ *adjective* [Middle French or Late Latin; Middle French, from Late Latin *aequidistant-, aequidistans,* from Latin *aequi-* + *distant-, distans,* present participle of *distare* to stand apart — more at DISTANT] (1593)
1 : equally distant
2 : representing map distances true to scale in all directions
— equi·dis·tant·ly *adverb*

equi·lat·er·al \ˌē-kwə-ˈla-tə-rəl, ˌe-kwə-, -ˈla-trəl\ *adjective* [Late Latin *aequilateralis,* from Latin *aequi-* + *later-, latus* side — more at LATERAL] (1570)
1 : having all sides equal ⟨an *equilateral* triangle⟩ ⟨an *equilateral* polygon⟩ — see TRIANGLE illustration

☆ SYNONYMS
Equanimity, composure, sangfroid mean evenness of mind under stress. EQUANIMITY suggests a habit of mind that is only rarely disturbed under great strain ⟨accepted her troubles with *equanimity*⟩. COMPOSURE implies the controlling of emotional or mental agitation by an effort of will or as a matter of habit ⟨maintaining his *composure* even under hostile questioning⟩. SANGFROID implies great coolness and steadiness under strain ⟨handled the situation with professional *sangfroid*⟩.

2 : having all the faces equal ⟨an *equilateral* polyhedron⟩

equilateral hyperbola *noun* (1880)
: a hyperbola with its asymptotes at right angles

equil·i·brant \i-ˈkwi-lə-brənt, ē-, *also* ˌē-kwə-ˈli-brənt\ *noun* (1883)
: a force that will balance one or more unbalanced forces

equil·i·brate \i-ˈkwi-lə-ˌbrāt\ *verb* **-brat·ed; -brat·ing** (1635)
transitive verb
: to bring into or keep in equilibrium **:** BALANCE
intransitive verb
: to bring about, come to, or be in equilibrium
— equil·i·bra·tion \-ˌkwi-lə-ˈbrā-shən\ *noun*
— equil·i·bra·tor \-ˈkwi-lə-ˌbrā-tər\ *noun*
— equil·i·bra·to·ry \-brə-ˌtōr-ē, -ˌtȯr-\ *adjective*

equi·li·brist \ˌē-kwə-ˈli-brist, ˌe-kwə-; i-ˈkwi-lə-brist\ *noun* (1760)
: one (as a rope dancer) who performs difficult feats of balancing
— equil·i·bris·tic \i-ˌkwi-lə-ˈbris-tik\ *adjective*

equi·lib·ri·um \ˌē-kwə-ˈli-brē-əm, ˌe-kwə-\ *noun, plural* **-ri·ums** *or* **-ria** \-brē-ə\ [Latin *aequilibrium,* from *aequilibris* being in equilibrium, from *aequi-* + *libra* weight, balance] (1608)
1 a : a state of intellectual or emotional balance **:** POISE **b :** a state of adjustment between opposing or divergent influences or elements
2 : a state of balance between opposing forces or actions that is either static (as in a body acted on by forces whose resultant is zero) or dynamic (as in a reversible chemical reaction when the velocities in both directions are equal)
3 : BALANCE 6a

equilibrium constant *noun* (1929)
: a number that expresses the relationship between the amounts of products and reactants present at equilibrium in a reversible chemical reaction at a given temperature

equi·mo·lar \-ˈmō-lər\ *adjective* (circa 1909)
1 : of or relating to an equal number of moles ⟨an *equimolar* mixture⟩
2 : having equal molar concentration

equine \ˈē-ˌkwīn, ˈe-ˌkwīn\ *adjective* [Latin *equinus,* from *equus* horse; akin to Old English *eoh* horse, Greek *hippos,* Sanskrit *aśva*] (1778)
: of, relating to, or resembling a horse or the horse family
— equine *noun*
— equine·ly *adverb*

[1]equi·noc·tial \ˌē-kwə-ˈnäk-shəl, ˌe-kwə-\ *adjective* (1545)
1 : relating to an equinox or to a state or the time of equal day and night
2 : relating to the regions or climate on or near the equator
3 : relating to the time when the sun passes an equinoctial point

[2]equinoctial *noun* (1527)
1 : EQUATOR 1
2 : an equinoctial storm

equi·nox \ˈē-kwə-ˌnäks, ˈe-kwə-\ *noun* [Middle English, from Middle French or Medieval Latin; Middle French *equinoxe,* from Medieval Latin *equinoxium,* alteration of Latin *aequinoctium,* from *aequi-* equi- + *noct-, nox* night — more at NIGHT] (14th century)
1 : either of the two points on the celestial sphere where the celestial equator intersects the ecliptic
2 : either of the two times each year (as about March 21 and September 23) when the sun crosses the equator and day and night are everywhere of equal length

equip \i-ˈkwip\ *transitive verb* **equipped; equip·ping** [Middle French *equiper,* from Old French *esciper,* of Germanic origin; akin to Old English *scip* ship] (1523)
1 : to furnish for service or action by appropriate provisioning
2 : DRESS, ARRAY
3 : to make ready **:** PREPARE
synonym see FURNISH

eq·ui·page \ˈe-kwə-pij\ *noun* (1579)
1 a : material or articles used in equipment **:** OUTFIT **b** *archaic* (1) **:** a set of small articles (as for table service) (2) **:** ETUI **c :** TRAPPINGS
2 *archaic* **:** RETINUE
3 : a horse-drawn carriage with its servants; *also* **:** such a carriage alone

equip·ment \i-ˈkwip-mənt\ *noun* (1717)
1 a : the set of articles or physical resources serving to equip a person or thing: as (1) **:** the implements used in an operation or activity **:** APPARATUS (2) **:** all the fixed assets other than land and buildings of a business enterprise (3) **:** the rolling stock of a railway **b :** a piece of such equipment
2 a : the equipping of a person or thing **b :** the state of being equipped
3 : mental or emotional traits or resources **:** ENDOWMENT

[1]equi·poise \ˈe-kwə-ˌpȯiz, ˈē-\ *noun* (1658)
1 : a state of equilibrium
2 : COUNTERBALANCE

[2]equipoise *transitive verb* (1664)
1 : to serve as an equipoise to
2 : to put or hold in equipoise

equi·pol·lence \ˌē-kwə-ˈpä-lən(t)s, ˌe-kwə-\ *noun* (15th century)
: the quality of being equipollent

equi·pol·lent \-lənt\ *adjective* [Middle English, from Middle French, from Latin *aequipollent-, aequipollens,* from *aequi-* equi- + *pollent-, pollens,* present participle of *pollēre* to be able] (15th century)
1 : equal in force, power, or validity
2 : the same in effect or signification
— equipollent *noun*
— equi·pol·lent·ly *adverb*

equi·pon·der·ant \-ˈpän-d(ə-)rənt\ *adjective* [Medieval Latin *aequiponderant-, aequiponderans,* present participle of *aequiponderare,* from Latin *aequi-* + *ponderare* to weigh — more at PONDER] (1630)
: evenly balanced

equi·po·ten·tial \ˌē-kwə-pə-ˈten(t)-shəl, ˌe-kwə-\ *adjective* (circa 1865)
: having the same potential **:** of uniform potential throughout ⟨*equipotential* points⟩

equi·prob·a·ble \-ˈprä-bə-bəl\ *adjective* (1921)
: having the same degree of logical or mathematical probability ⟨*equiprobable* alternatives⟩

eq·ui·se·tum \ˌe-kwə-ˈsē-təm\ *noun, plural* **-se·tums** *or* **-se·ta** \-ˈsē-tə\ [New Latin, from Latin *equisaetum* horsetail plant, from *equus* horse + *saeta* bristle] (1830)
: any of a genus (*Equisetum*) of lower tracheophytes comprising perennial plants that spread by creeping rhizomes and have leaves reduced to nodal sheaths on the hollow jointed ribbed shoots — called also *horsetail, scouring rush*

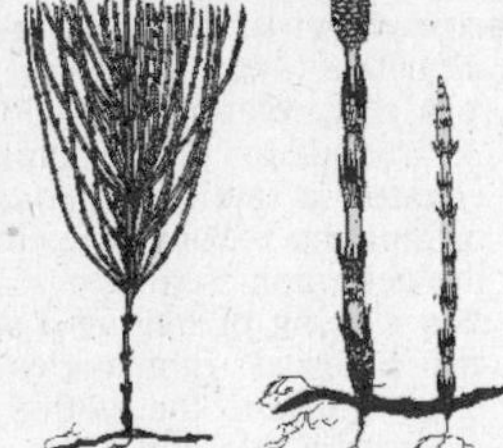

equisetum: ***1*** **vegetative plant,** ***2*** **fertile plant**

eq·ui·ta·ble \ˈe-kwə-tə-bəl\ *adjective* (1646)
1 : having or exhibiting equity **:** dealing fairly and equally with all concerned
2 : existing or valid in equity as distinguished from law
synonym see FAIR
— eq·ui·ta·bil·i·ty \ˌe-kwə-tə-ˈbi-lə-tē\ *noun*
— eq·ui·ta·ble·ness \ˈe-kwə-tə-bəl-nəs\ *noun*
— eq·ui·ta·bly \-blē\ *adverb*

eq·ui·ta·tion \ˌe-kwə-ˈtā-shən\ *noun* [Middle French, from Latin *equitation-, equitatio,* from *equitare* to ride on horseback, from *equit-, eques* horseman, from *equus* horse] (1562)
: the act or art of riding on horseback

eq·ui·ty \ˈe-kwə-tē\ *noun, plural* **-ties** [Middle English *equite,* from Middle French *equité,* from Latin *aequitat-, aequitas,* from *aequus* equal, fair] (14th century)
1 a : justice according to natural law or right; *specifically* **:** freedom from bias or favoritism **b :** something that is equitable
2 a : a system of law originating in the English chancery and comprising a settled and formal body of legal and procedural rules and doctrines that supplement, aid, or override common and statute law and are designed to protect rights and enforce duties fixed by substantive law **b :** trial or remedial justice under or by the rules and doctrines of equity **c :** a body of legal doctrines and rules developed to enlarge, supplement, or override a narrow rigid system of law
3 a : a right, claim, or interest existing or valid in equity **b :** the money value of a property or of an interest in a property in excess of claims or liens against it **c :** a risk interest or ownership right in property **d :** the common stock of a corporation

equity capital *noun* (1942)
: VENTURE CAPITAL

equiv·a·lence \i-ˈkwiv-lən(t)s, -ˈkwi-və-\ *noun* (circa 1541)
1 a : the state or property of being equivalent **b :** the relation holding between two statements if they are either both true or both false so that to affirm one and to deny the other would result in a contradiction
2 : a presentation of terms as equivalent
3 : equality in metrical value of a regular foot and one in which there are substitutions

equivalence class *noun* (1952)
: a set for which an equivalence relation holds between every pair of elements

equivalence relation *noun* (circa 1949)
: a relation (as equality) between elements of a set (as the real numbers) that is symmetric, reflexive, and transitive and for any two elements either holds or does not hold

equiv·a·len·cy \i-ˈkwiv-lən(t)-sē, -ˈkwi-və-\ *noun, plural* **-cies** (1535)
1 : EQUIVALENCE
2 : a level of achievement equivalent to completion of an educational or training program ⟨a high school *equivalency* certificate⟩

equiv·a·lent \-lənt\ *adjective* [Middle English, from Middle French or Late Latin; Middle French, from Late Latin *aequivalent-, aequivalens,* present participle of *aequivalēre* to have equal power, from Latin *aequi-* + *valēre* to be strong — more at WIELD] (15th century)
1 : equal in force, amount, or value; *also* **:** equal in area or volume but not admitting of superposition ⟨a square *equivalent* to a triangle⟩
2 a : like in signification or import **b :** having logical equivalence ⟨*equivalent* statements⟩
3 : corresponding or virtually identical especially in effect or function
4 *obsolete* **:** equal in might or authority
5 : having the same chemical combining capacity ⟨*equivalent* quantities of two elements⟩
6 a : having the same solution set ⟨*equivalent* equations⟩ **b :** capable of being placed in one-to-one correspondence ⟨*equivalent* sets⟩ **c :** related by an equivalence relation
synonym see SAME
— equivalent *noun*
— equiv·a·lent·ly *adverb*

equivalent weight *noun* (1904)
: the mass of a substance especially in grams that combines with or is chemically equivalent to eight grams of oxygen or one gram of hydrogen **:** the atomic or molecular weight divided by the valence

equiv•o•cal \i-ˈkwi-və-kəl\ *adjective* [Late Latin *aequivocus,* from *aequi-* equi- + *voc-, vox* voice — more at VOICE] (1599)
1 a : subject to two or more interpretations and usually used to mislead or confuse **b :** uncertain as an indication or sign
2 a : of uncertain nature or classification **b :** of uncertain disposition toward a person or thing **:** UNDECIDED **c :** of doubtful advantage, genuineness, or moral rectitude ⟨*equivocal* behavior⟩
synonym see OBSCURE
— equiv•o•cal•i•ty \-ˌkwi-və-ˈka-lə-tē\ *noun*
— equiv•o•cal•ly \-ˈkwi-və-k(ə-)lē\ *adverb*
— equiv•o•cal•ness \-kəl-nəs\ *noun*

equiv•o•cate \i-ˈkwi-və-ˌkāt\ *intransitive verb* **-cat•ed; -cat•ing** (1590)
1 : to use equivocal language especially with intent to deceive
2 : to avoid committing oneself in what one says
synonym see LIE
— equiv•o•ca•tion \-ˌkwi-və-ˈkā-shən\ *noun*
— equiv•o•ca•tor \-ˈkwi-və-ˌkā-tər\ *noun*

equi•voque *also* **equi•voke** \ˈe-kwə-ˌvōk, ˈē-\ *noun* [French *équivoque,* from *équivoque* equivocal, from Late Latin *aequivocus*] (1599)
1 : an equivocal word or phrase; *specifically* **:** PUN
2 a : double meaning **b :** WORDPLAY

[1]-er \ər; *after some vowels, often* r; *after* ŋ, *usually* gər\ *adjective suffix or adverb suffix* [Middle English *-er, -ere, -re,* from Old English *-ra* (in adjectives), *-or* (in adverbs); akin to Old High German *-iro,* adjective comparative suffix, Latin *-ior,* Greek *-iōn*]
— used to form the comparative degree of adjectives and adverbs of one syllable ⟨hott*er*⟩ ⟨dri*er*⟩ and of some adjectives and adverbs of two syllables ⟨complet*er*⟩ and sometimes of longer ones ⟨beautiful*er*⟩

[2]-er \ər; *after some vowels, often* r\ *also* **-ier** \ē-ər, yər\ *or* **-yer** \yər\ *noun suffix* [Middle English *-er, -ere, -ier, -iere;* partly from Old English *-ere* (from Latin *-arius*); partly from Old French *-ier, -iere,* from Latin *-arius, -aria, -arium* -ary; partly from Middle French *-ere,* from Latin *-ator* -or — more at -ARY, -OR]
1 a : person occupationally connected with ⟨hatt*er*⟩ ⟨furri*er*⟩ ⟨lawy*er*⟩ **b :** person or thing belonging to or associated with ⟨head*er*⟩ ⟨old-tim*er*⟩ **c :** native of **:** resident of ⟨cottag*er*⟩ ⟨New York*er*⟩ **d :** one that has ⟨three-deck*er*⟩ **e :** one that produces or yields ⟨pork*er*⟩
2 a : one that does or performs (a specified action) ⟨report*er*⟩ — sometimes added to both elements of a compound ⟨build*er*-upp*er*⟩ **b :** one that is a suitable object of (a specified action) ⟨broil*er*⟩
3 : one that is ⟨foreign*er*⟩ — in all senses *-yer* in a few words after *w, ier* in a few other words, otherwise *-er*

era \ˈir-ə, ˈer-ə, ˈē-rə\ *noun* [Late Latin *aera,* from Latin, counters, plural of *aer-, aes* copper, money — more at ORE] (1615)
1 a : a fixed point in time from which a series of years is reckoned **b :** a memorable or important date or event; *especially* **:** one that begins a new period in the history of a person or thing
2 : a system of chronological notation computed from a given date as basis
3 a : a period identified by some prominent figure or characteristic feature **b :** a stage in development; *especially* **:** one of the four major divisions of geologic time ⟨Paleozoic *era*⟩
synonym see PERIOD

erad•i•cate \i-ˈra-də-ˌkāt\ *transitive verb* **-cat•ed; -cat•ing** [Latin *eradicatus,* past participle of *eradicare,* from *e-* + *radic-, radix* root — more at ROOT] (1578)
1 : to pull up by the roots
2 : to do away with as completely as if by pulling up by the roots ⟨programs to *eradicate* illiteracy⟩
synonym see EXTERMINATE
— erad•i•ca•ble \-ˈra-di-kə-bəl\ *adjective*
— erad•i•ca•tion \-ˌra-də-ˈkā-shən\ *noun*
— erad•i•ca•tor \-ˌkā-tər\ *noun*

erase \i-ˈrās, *British* -ˈrāz\ *verb* **erased; eras•ing** [Latin *erasus,* past participle of *eradere,* from *e-* + *radere* to scratch, scrape — more at RODENT] (1605)
transitive verb
1 a : to rub or scrape out (as written, painted, or engraved letters) **b :** to remove (recorded matter) from a magnetic medium; *also* **:** to remove recorded matter from ⟨*erase* a videotape⟩ **c :** to delete from a computer storage device
2 a : to remove from existence or memory as if by erasing **b :** to nullify the effect or force of
intransitive verb
: to yield to being erased
— eras•abil•i•ty \-ˌrā-sə-ˈbi-lə-tē\ *noun*
— eras•able \-ˈrā-sə-bəl\ *adjective*

eras•er \i-ˈrā-sər\ *noun* (1790)
: one that erases; *especially* **:** a device (as a piece of rubber, or a felt pad) used to erase marks (as of ink or chalk)

Eras•tian \i-ˈras-tē-ən, -ˈras-chən\ *adjective* [Thomas *Erastus* (died 1583) Swiss physician and Zwinglian theologian] (1839)
: of, characterized by, or advocating the doctrine of state supremacy in ecclesiastical affairs
— Erastian *noun*
— Eras•tian•ism \-ˌni-zəm\ *noun*

era•sure \i-ˈrā-shər *also* -zhər\ *noun* (1734)
: an act or instance of erasing

Er•a•to \ˈer-ə-ˌtō\ *noun* [Greek *Eratō*]
: the Greek Muse of lyric and love poetry

er•bi•um \ˈər-bē-əm\ *noun* [New Latin, from *Ytterby,* Sweden] (1843)
: a metallic element of the rare-earth group that occurs with yttrium — see ELEMENT table

[1]ere \ˈer, ˈar\ *preposition* [Middle English *er,* from Old English *ǣr,* from *ǣr,* adverb, early, soon; akin to Old High German *ēr* earlier, Greek *ēri* early] (before 12th century)
: [2]BEFORE 2 ⟨contrived *ere* the beginning of the world —Norman Douglas⟩

[2]ere *conjunction* (before 12th century)
: [3]BEFORE

Er•e•bus \ˈer-ə-bəs\ *noun* [Latin, from Greek *Erebos*]
1 : a personification of darkness in Greek mythology
2 : a place of darkness in the underworld on the way to Hades

[1]erect \i-ˈrekt\ *adjective* [Middle English, from Latin *erectus,* past participle of *erigere* to erect, from *e-* + *regere* to lead straight, guide — more at RIGHT] (14th century)
1 a : vertical in position; *also* **:** not spread out or decumbent ⟨an *erect* plant stem⟩ **b :** standing up or out from the body ⟨*erect* hairs⟩ **c :** characterized by firm or rigid straightness in bodily posture ⟨an *erect* bearing⟩
2 *archaic* **:** directed upward
3 *obsolete* **:** ALERT, WATCHFUL
4 : being in a state of physiological erection
— erect•ly \-ˈrek-(t)lē\ *adverb*
— erect•ness \-ˈrek(t)-nəs\ *noun*

[2]erect *transitive verb* (15th century)
1 a (1) **:** to put up by the fitting together of materials or parts **:** BUILD (2) **:** to fix in an upright position (3) **:** to cause to stand up or stand out **b** *archaic* **:** to direct upward **c :** to change (an image) from an inverted to a normal position
2 : to elevate in status
3 : SET UP, ESTABLISH
4 *obsolete* **:** ENCOURAGE, EMBOLDEN
5 : to draw or construct (as a perpendicular or figure) upon a given base
— erect•able \-ˈrek-tə-bəl\ *adjective*

erec•tile \i-ˈrek-tᵊl, -ˌtīl\ *adjective* (1830)
: capable of becoming erect ⟨*erectile* tissue⟩ — compare CAVERNOUS 1b
— erec•til•i•ty \-ˌrek-ˈti-lə-tē\ *noun*

erec•tion \i-ˈrek-shən\ *noun* (15th century)
1 a : the state marked by firm turgid form and erect position of a previously flaccid bodily part containing cavernous tissue when that tissue becomes dilated with blood **b :** an occurrence of such a state in the penis or clitoris
2 : the act or process of erecting something **:** CONSTRUCTION
3 : something erected

erec•tor \i-ˈrek-tər\ *noun* (1538)
: one that erects; *especially* **:** a muscle that raises or keeps a part erect

Erector *trademark*
— used for a metal toy construction set

E region *noun* (1930)
: the part of the ionosphere occurring between 55 and 80 miles (90 and 130 kilometers) above the surface of the earth and containing the daytime E layer and the sporadic E layer

ere•long \er-ˈlȯŋ, ar-\ *adverb* (1577)
archaic **:** before long **:** SOON

er•e•mite \ˈer-ə-ˌmīt\ *noun* [Middle English — more at HERMIT] (13th century)
: HERMIT; *especially* **:** a religious recluse
— er•e•mit•ic \ˌer-ə-ˈmi-tik\ *or* **er•e•mit•i•cal** \-ti-kəl\ *adjective*
— er•e•mit•ism \ˈer-ə-ˌmī-ˌti-zəm\ *noun*

er•em•urus \ˌer-ə-ˈmyu̇r-əs\ *noun, plural* **-uri** \-ˈmyu̇r-ˌī\ [New Latin, from Greek *erēmos* solitary + *oura* tail — more at ASS] (1829)
: any of a genus (*Eremurus*) of perennial herbs of the lily family that produce tall racemes of showy blooms — called also *foxtail lily*

ere•now \er-ˈnau̇, ar-\ *adverb* (14th century)
: before now **:** HERETOFORE

erep•sin \i-ˈrep-sən\ *noun* [International Scientific Vocabulary *er-* (probably from Latin *eripere* to snatch away, from *e-* + *rapere* to seize) + *pepsin* — more at RAPID] (1902)
: a proteolytic fraction obtained especially from the intestinal juice and known to be a mixture of exopeptidases

er•e•thism \ˈer-ə-ˌthi-zəm\ *noun* [French *éréthisme,* from Greek *erethismos* irritation, from *erethizein* to irritate; akin to Greek *ornynai* to rouse — more at ORIENT] (1800)
: abnormal irritability or responsiveness to stimulation

ere•while \er-ˈ(h)wī(ə)l, ar-\ *also* **ere•whiles** \-ˈ(h)wī(ə)lz\ *adverb* (13th century)
archaic **:** HERETOFORE

erg \ˈərg\ *noun* [Greek *ergon* work — more at WORK] (circa 1873)
: a centimeter-gram-second unit of work equal to the work done by a force of one dyne acting through a distance of one centimeter and equivalent to 10^{-7} joule

erg- *or* **ergo-** *combining form* [Greek, from *ergon*]
: work ⟨*ergo*motor⟩

er•gas•tic \(ˌ)ər-ˈgas-tik\ *adjective* [Greek *ergastikos* able to work, from *ergazesthai* to work, from *ergon* work] (circa 1896)
: constituting the nonliving by-products of protoplasmic activity ⟨*ergastic* substances⟩

er•gas•to•plasm \-tə-ˌpla-zəm\ *noun* [International Scientific Vocabulary] (1902)
: ribosome-studded endoplasmic reticulum
— er•gas•to•plas•mic \-ˌgas-tə-ˈplaz-mik\ *adjective*

\ə\ **abut** \ᵊ\ kitt**e**n \ər\ f**ur**th**er** \a\ **a**sh \ā\ **a**ce
\ä\ m**o**p, m**a**r \au̇\ **ou**t \ch\ **ch**in \e\ b**e**t \ē\ **ea**sy
\g\ **g**o \i\ h**i**t \ī\ **i**ce \j\ **j**ob \ŋ\ si**ng** \ō\ g**o**
\ȯ\ l**aw** \ȯi\ b**oy** \th\ **th**in \t͟h\ **th**e \ü\ l**oo**t \u̇\ f**oo**t
\y\ **y**et \zh\ vi**si**on *see also* Guide to Pronunciation

er·ga·tive \ˈər-gə-tiv\ *adjective* [Greek *ergatēs* worker, from *ergon* work] (1939)
: of, relating to, or being a language (as Inuit or Georgian) in which the objects of transitive verbs and subjects of intransitive verbs are typically marked by the same linguistic forms; *also* **:** being an inflectional morpheme that typically marks the subject of a transitive verb in an ergative language

-er·gic \(ˌ)ər-jik\ *adjective combining form* [*-ergy* work (from Late Latin *-ergia,* from Greek, from *ergon* work) + *-ic* — more at WORK]
: exhibiting or stimulating activity of ⟨dopamin*ergic*⟩

er·go \ˈer-(ˌ)gō, ˈər-\ *adverb* [Middle English, from Latin, from Old Latin, because of, from (assumed) Old Latin *e rogo* from the direction (of)] (14th century)
: THEREFORE, HENCE

ergo- *combining form* [French, from *ergot*]
: ergot ⟨*ergo*sterol⟩

er·go·dic \(ˌ)ər-ˈgä-dik, -ˈgō-\ *adjective* [International Scientific Vocabulary *erg-* + *-ode*] (1926)
1 : of or relating to a process in which every sequence or sizable sample is equally representative of the whole (as in regard to a statistical parameter)
2 : involving or relating to the probability that any state will recur; *especially* **:** having zero probability that any state will never recur
— er·go·dic·i·ty \ˌər-gə-ˈdi-sə-tē\ *noun*

er·go·graph \ˈər-gə-ˌgraf\ *noun* [International Scientific Vocabulary] (1892)
: an apparatus for measuring the work capacity of a muscle

er·gom·e·ter \(ˌ)ər-ˈgä-mə-tər\ *noun* (circa 1879)
: an apparatus for measuring the work performed (as by a person exercising); *also* **:** an exercise machine equipped with an ergometer
— er·go·met·ric \ˌər-gə-ˈme-trik\ *adjective*

er·go·nom·ics \ˌər-gə-ˈnä-miks\ *noun plural but singular or plural in construction* [*erg-* + *-nomics* (as in *economics*)] (1949)
: an applied science concerned with designing and arranging things people use so that the people and things interact most efficiently and safely — called also *human engineering*
— er·go·nom·ic \-mik\ *adjective*
— er·go·nom·i·cal·ly \-mi-k(ə-)lē\ *adverb*
— er·gon·o·mist \(ˌ)ər-ˈgä-nə-mist\ *noun*

er·go·no·vine \ˌər-gə-ˈnō-ˌvēn\ *noun* [*ergo-* + Latin *nov*us new — more at NEW] (circa 1936)
: an alkaloid $C_{19}H_{23}N_3O_2$ from ergot with similar pharmacological action but reduced toxicity

er·gos·ter·ol \(ˌ)ər-ˈgäs-tə-ˌrȯl, -ˌrōl\ *noun* [International Scientific Vocabulary] (1906)
: a crystalline steroid alcohol $C_{28}H_{44}O$ that occurs especially in yeast, molds, and ergot and is converted by ultraviolet irradiation ultimately into vitamin D_2

er·got \ˈər-gət, -ˌgät\ *noun* [French, literally, cock's spur] (1683)
1 : the black or dark purple sclerotium of fungi (genus *Claviceps*) that occurs as a club-shaped body replacing the seed of a grass (as rye); *also* **:** a fungus bearing ergots
2 : a disease of rye and other cereals caused by an ergot fungus
3 a : the dried sclerotia of an ergot fungus grown on rye and containing several alkaloids (as ergonovine and ergotamine) **b :** any of such alkaloids used medicinally for their contractile effect on smooth muscle (as of peripheral arterioles)
— er·got·ic \(ˌ)ər-ˈgä-tik\ *adjective*

er·got·a·mine \(ˌ)ər-ˈgä-tə-ˌmēn\ *noun* [International Scientific Vocabulary] (1921)
: an alkaloid $C_{33}H_{35}N_5O_5$ from ergot that is used chiefly in the form of its tartrate especially in treating migraine

er·got·ism \ˈər-gə-ˌti-zəm\ *noun* (circa 1841)
: a toxic condition produced by eating grain, grain products (as rye bread), or grasses infected with ergot fungus or by chronic excessive use of an ergot drug

er·got·ized \-ˌtīzd\ *adjective* (1860)
: infected with ergot ⟨*ergotized* grain⟩; *also* **:** poisoned by ergot ⟨*ergotized* cattle⟩

er·i·ca \ˈer-i-kə\ *noun* [New Latin, from Latin *erice* heather, from Greek *ereikē* — more at BRIER] (1826)
: any of a large genus (*Erica*) of the heath family of low much-branched evergreen shrubs

er·i·ca·ceous \ˌer-ə-ˈkā-shəs\ *adjective* (circa 1859)
: of, relating to, or being a heath or the heath family

er·i·coid \ˈer-ə-ˌkȯid\ *adjective* (circa 1900)
: resembling heath ⟨*ericoid* foliage⟩

Erie \ˈir-ē, ˈēr-ē\ *noun* (circa 1909)
1 : a member of an American Indian people living south of Lake Erie in the 17th century
2 : the extinct and probably Iroquoian language of the Erie people

erig·er·on \i-ˈri-jə-ˌrän\ *noun* [New Latin, from Latin, groundsel, from Greek *ērigerōn,* from *ēri* early + *gerōn* old man; from the hoary down of some species — more at ERE, GERONT-] (1601)
: any of a widely distributed genus (*Erigeron*) of composite herbs with flower heads that resemble asters but have fewer and narrower involucral bracts

Eri·nys \i-ˈri-nəs, -ˈrī-\ *noun, plural* **Eriny·es** \-ˈri-nē-ˌēz\ [Greek] (1590)
: FURY 2a

er·i·o·phy·id \ˌer-ē-ˈä-fē-əd, -ē-ə-ˈfī-əd\ *noun* [ultimately from Greek *erion* wool + *phyē* growth; akin to Greek *physis* growth — more at PHYSICS] (1942)
: any of a large family (Eriophyidae) of minute plant-feeding mites that have two pairs of legs placed far anterior and lack a respiratory system
— eriophyid *adjective*

[1]eris·tic \i-ˈris-tik, e-\ *also* **eris·ti·cal** \-ti-kəl\ *adjective* [Greek *eristikos* fond of wrangling, from *erizein* to wrangle, from *eris* strife] (1637)
: characterized by disputatious and often subtle and specious reasoning
— eris·ti·cal·ly \-ti-k(ə-)lē\ *adverb*

[2]eristic *noun* (1659)
1 : a person devoted to logical disputation
2 : the art or practice of disputation and polemics

Er·len·mey·er flask \ˌər-lən-ˌmī(-ə)r-, ˌer-lən-\ *noun* [Emil *Erlenmeyer*] (circa 1890)
: a flat-bottomed conical laboratory flask

er·mine \ˈər-mən\ *noun, plural* **ermines** [Middle English, from Old French, of Germanic origin; akin to Old High German *harmo* weasel] (12th century)
1 *or plural* **ermine a :** any of several weasels whose coats become white in winter usually with black on the tip of the tail; *especially* **:** a short-tailed weasel (*Mustela erminea*) of the forests and tundra of Eurasia and North America **b :** the white fur of the ermine
2 : a rank or office whose ceremonial or official robe is ornamented with ermine

ermine 1a: winter coat

er·mined \-mənd\ *adjective* (15th century)
: clothed or adorned with ermine

erne \ˈərn, ˈern\ *noun* [Middle English, from Old English *earn;* akin to Old High German *arn* eagle, Greek *ornis* bird] (before 12th century)
: EAGLE; *especially* **:** a long-winged sea eagle (*Haliäetus albicilla*) with a short white wedge-shaped tail

erode \i-ˈrōd\ *verb* **erod·ed; erod·ing** [Latin *erodere* to eat away, from *e-* + *rodere* to gnaw — more at RODENT] (1612)
transitive verb
1 : to diminish or destroy by degrees: **a :** to eat into or away by slow destruction of substance (as by acid, infection, or cancer) **b :** to wear away by the action of water, wind, or glacial ice **c :** to cause to deteriorate or disappear as if by eating or wearing away ⟨inflation *eroding* buying power⟩
2 : to produce or form by eroding ⟨glaciers *erode* U-shaped valleys⟩
intransitive verb
: to undergo erosion
— erod·ibil·i·ty \-ˌrō-də-ˈbi-lə-tē\ *noun*
— erod·ible \-ˈrō-də-bəl\ *adjective*

erog·e·nous \i-ˈrä-jə-nəs\ *adjective* [Greek *erōs* + English *-genous, -genic*] (circa 1889)
1 : producing sexual excitement or libidinal gratification when stimulated **:** sexually sensitive
2 : of, relating to, or arousing sexual feelings

Eros \ˈer-ˌäs, ˈir-\ *noun* [Greek *Erōs,* from *erōs* sexual love; akin to Greek *erasthai* to love, desire]
1 : the Greek god of erotic love — compare CUPID
2 : the sum of life-preserving instincts that are manifested as impulses to gratify basic needs, as sublimated impulses, and as impulses to protect and preserve the body and mind — compare DEATH INSTINCT
3 a : love conceived in the philosophy of Plato as a fundamental creative impulse having a sensual element **b** *often not capitalized* **:** erotic love or desire

erose \i-ˈrōs\ *adjective* [Latin *erosus,* past participle of *erodere*] (1793)
: IRREGULAR, UNEVEN; *specifically* **:** having the margin irregularly notched as if gnawed ⟨an *erose* leaf⟩

ero·sion \i-ˈrō-zhən\ *noun* (1541)
1 a : the action or process of eroding **b :** the state of being eroded
2 : an instance or product of erosive action
— ero·sion·al \-ˈrōzh-nəl, -ˈrō-zhə-nᵊl\ *adjective*
— ero·sion·al·ly \-ē\ *adverb*

ero·sive \i-ˈrō-siv, -ziv\ *adjective* (1830)
: tending to erode or to induce or permit erosion
— ero·sive·ness *noun*
— ero·siv·i·ty \i-ˌrō-ˈsi-və-tē\ *noun*

erot·ic \i-ˈrä-tik\ *also* **erot·i·cal** \-ti-kəl\ *adjective* [Greek *erōtikos,* from *erōt-, erōs*] (1651)
1 : of, devoted to, or tending to arouse sexual love or desire ⟨*erotic* art⟩
2 : strongly marked or affected by sexual desire ◆

◇ WORD HISTORY

erotic Eros was the ancient Greek god of sexual love. In early classical art and literature, he was depicted as a handsome young man of athletic build. The love that he symbolized was frankly sexual, hence the meaning of our adjective *erotic.* Later, in Hellenistic times, the concept of love became romanticized, and the image of Eros became accordingly sentimentalized. Eros gradually evolved into a figure of a playful infant or child sporting wings and equipped with a bow and a quiver full of arrows. The Romans called him Amor ("Love") or Cupido ("Desire"). In Roman art and literature this embodiment of the god became the dominant image, giving rise to the chubby cherub of a *cupid* that traditionally adorns St. Valentine's Day cards.

— **erotic** *noun*
— **erot·i·cal·ly** \i-tik(ə-)lē\ *adverb*
erot·i·ca \i-'rä-ti-kə\ *noun plural but singular or plural in construction* [New Latin, from Greek *erōtika,* neuter plural of *erōtikos*] (1854)
1 : literary or artistic works having an erotic theme or quality
2 : depictions of things erotic
erot·i·cism \i-'rä-tə-,si-zəm\ *noun* (1881)
1 : an erotic theme or quality
2 : a state of sexual arousal
3 : insistent sexual impulse or desire
— **erot·i·cist** \-sist\ *noun*
erot·i·cize \-,sīz\ *transitive verb* **-cized; -ciz·ing** (circa 1914)
: to make erotic
— **erot·i·ci·za·tion** \i-,rä-tə-sə-'zā-shən\ *noun*
er·o·tism \'er-ə-,ti-zəm\ *noun* (1849)
: EROTICISM
er·o·tize \'er-ə-,tīz\ *transitive verb* **-tized; -tiz·ing** (1936)
: to invest with erotic significance or sexual feeling
— **er·o·ti·za·tion** \,er-ə-tə-'zā-shən\ *noun*
eroto- *combining form* [New Latin, from Greek *erōto-,* from *erōt-, erōs*]
: sexual desire ⟨*eroto*genic⟩
ero·to·gen·ic \i-,rō-tə-'je-nik, -,rä-\ *adjective* (circa 1909)
: EROGENOUS
err \'er, 'ər\ *intransitive verb* [Middle English, from Middle French *errer,* from Latin *errare* to wander, err; akin to Old English *ierre* wandering, angry, Old Norse *rās* race — more at RACE] (14th century)
1 *archaic* **:** STRAY
2 a : to make a mistake **b :** to violate an accepted standard of conduct ■
er·ran·cy \'er-ən(t)-sē\ *noun, plural* **-cies** (1621)
: the state or an instance of erring
er·rand \'er-ənd\ *noun* [Middle English *erend* message, business, from Old English *ǣrend;* akin to Old High German *ārunti* message] (before 12th century)
1 *archaic* **a :** an oral message entrusted to a person **b :** EMBASSY, MISSION
2 a : a short trip taken to attend to some business often for another ⟨was on an *errand* for his mother⟩ **b :** the object or purpose of such a trip
er·rant \'er-ənt\ *adjective* [Middle English *erraunt,* from Middle French *errant,* present participle of *errer* to err & *errer* to travel, from Late Latin *iterare,* from Latin *iter* road, journey — more at ITINERANT] (14th century)
1 : traveling or given to traveling ⟨an *errant* knight⟩
2 a : straying outside the proper path or bounds ⟨an *errant* calf⟩ **b :** moving about aimlessly or irregularly ⟨an *errant* breeze⟩ **c :** deviating from a standard (as of truth or propriety) ⟨an *errant* child⟩ **d :** FALLIBLE
— **errant** *noun*
— **er·rant·ly** *adverb*
er·rant·ry \'er-ən-trē\ *noun, plural* **-ries** (1654)
: the quality, condition, or fact of wandering; *especially* **:** a roving in search of chivalrous adventure
er·ra·ta \e-'rä-tə, -'rā-, -'ra-\ *noun* [Latin, plural of *erratum*] (1589)
: a list of corrigenda; *also* **:** a page bearing such a list
[1]**er·rat·ic** \ir-'a-tik\ *adjective* [Middle English, from Middle French or Latin; Middle French *erratique,* from Latin *erraticus,* from *erratus,* past participle of *errare*] (14th century)
1 a : having no fixed course **:** WANDERING ⟨an *erratic* comet⟩ **b** *archaic* **:** NOMADIC
2 : transported from an original resting place especially by a glacier ⟨an *erratic* boulder⟩
3 a : characterized by lack of consistency, regularity, or uniformity **b :** deviating from what is ordinary or standard **:** ECCENTRIC ⟨an *erratic* genius⟩
synonym see STRANGE
— **er·rat·i·cal** \-ti-kəl\ *adjective*
— **er·rat·i·cal·ly** \-ti-k(ə-)lē\ *adverb*
— **er·rat·i·cism** \-'a-tə-,si-zəm\ *noun*
[2]**erratic** *noun* (circa 1623)
: one that is erratic; *especially* **:** an erratic boulder or block of rock
er·ra·tum \e-'rä-təm, -'rā-, -'ra-\ *noun, plural* **-ta** \-tə\ [Latin, from neuter of *erratus*] (1589)
: ERROR; *especially* **:** CORRIGENDUM
er·ro·ne·ous \i-'rō-nē-əs, e-'rō-\ *adjective* [Middle English, from Latin *erroneus,* from *erron-, erro* wanderer, from *errare*] (15th century)
1 : containing or characterized by error **:** MISTAKEN ⟨*erroneous* assumptions⟩
2 *archaic* **:** WANDERING
— **er·ro·ne·ous·ly** *adverb*
— **er·ro·ne·ous·ness** *noun*
er·ror \'er-ər\ *noun* [Middle English *errour,* from Middle French, from Latin *error,* from *errare*] (13th century)
1 a : an act or condition of ignorant or imprudent deviation from a code of behavior **b :** an act involving an unintentional deviation from truth or accuracy **c :** an act that through ignorance, deficiency, or accident departs from or fails to achieve what should be done: as (1) **:** a defensive misplay other than a wild pitch or passed ball made by a baseball player when normal play would have resulted in an out or prevented an advance by a base runner (2) **:** the failure of a player (as in tennis) to make a successful return of a ball during play **d :** a mistake in the proceedings of a court of record in matters of law or of fact
2 a : the quality or state of erring **b** *Christian Science* **:** illusion about the nature of reality that is the cause of human suffering **:** the contradiction of truth **c :** an instance of false belief
3 : something produced by mistake; *especially* **:** a postage stamp exhibiting a consistent flaw (as a wrong color) in its manufacture
4 a : the difference between an observed or calculated value and a true value; *specifically* **:** variation in measurements, calculations, or observations of a quantity due to mistakes or to uncontrollable factors **b :** the amount of deviation from a standard or specification
5 : a deficiency or imperfection in structure or function ⟨an *error* of metabolism⟩ ☆
— **er·ror·less** \'er-ər-ləs\ *adjective*
er·satz \'er-,säts, -,zäts, er-'; 'ər-,sats\ *adjective* [German *ersatz-,* from *Ersatz,* noun, substitute] (1875)
: being a usually artificial and inferior substitute or imitation ⟨*ersatz* turf⟩ ⟨*ersatz* intellectuals⟩
— **ersatz** *noun*
Erse \'ərs\ *noun* [Middle English (Scots) *Erisch,* adjective, Irish, alteration of *Irish*] (15th century)
1 : SCOTTISH GAELIC
2 : IRISH GAELIC
— **Erse** *adjective*
erst \'ərst\ *adverb* [Middle English *erest* earliest, formerly, from Old English *ǣrest,* superlative of *ǣr* early — more at ERE] (12th century)
archaic **:** ERSTWHILE
[1]**erst·while** \'ərst-,(h)wīl\ *adverb* (1569)
: in the past **:** FORMERLY ⟨cultures, *erstwhile* unknown to each other —Robert Plank⟩
[2]**erstwhile** *adjective* (1903)
: FORMER, PREVIOUS ⟨her *erstwhile* students⟩
eru·cic acid \i-'rü-sik-\ *noun* [New Latin *Eruca,* genus of herbs, from Latin, colewort] (1869)
: a crystalline fatty acid $C_{22}H_{42}O_2$ found in the form of glycerides especially in rapeseed oil
eruct \i-'rəkt\ *verb* [Latin *eructare,* frequentative of *erugere* to belch, disgorge; akin to Old English *rocettan* to belch, Greek *ereugesthai*] (1666)
: BELCH
eruc·ta·tion \i-,rək-'tā-shən, ,ē-\ *noun* (15th century)
: an act or instance of belching
er·u·dite \'er-ə-,dīt, 'er-yə-\ *adjective* [Middle English *erudit,* from Latin *eruditus,* from past participle of *erudire* to instruct, from *e-* + *rudis* rude, ignorant] (15th century)
: possessing or displaying erudition **:** LEARNED ⟨an *erudite* scholar⟩
— **er·u·dite·ly** *adverb*
er·u·di·tion \,er-ə-'di-shən, ,er-yə-\ *noun* (15th century)
: extensive knowledge acquired chiefly from books **:** profound, recondite, or bookish learning
synonym see KNOWLEDGE
erum·pent \i-'rəm-pənt\ *adjective* [Latin *erumpent-, erumpens,* present participle of *erumpere*] (1650)
: bursting forth ⟨*erumpent* fungi⟩
erupt \i-'rəpt\ *verb* [Latin *eruptus,* past participle of *erumpere* to burst forth, from *e-* + *rumpere* to break — more at REAVE] (1657)
intransitive verb
1 a (1) **:** to burst from limits or restraint (2) *of a tooth* **:** to emerge through the gum **b :** to force out or release suddenly and often violently something (as lava or steam) that is pent up **c :** to become active or violent especially suddenly
2 : to break out with or as if with a skin eruption

☆ SYNONYMS
Error, mistake, blunder, slip, lapse mean a departure from what is true, right, or proper. ERROR suggests the existence of a standard or guide and a straying from the right course through failure to make effective use of this ⟨procedural *errors*⟩. MISTAKE implies misconception or inadvertence and usually expresses less criticism than *error* ⟨dialed the wrong number by *mistake*⟩. BLUNDER regularly imputes stupidity or ignorance as a cause and connotes some degree of blame ⟨diplomatic *blunders*⟩. SLIP stresses inadvertence or accident and applies especially to trivial but embarrassing mistakes ⟨a *slip* of the tongue⟩. LAPSE stresses forgetfulness, weakness, or inattention as a cause ⟨a *lapse* in judgment⟩.

□ USAGE
err The sound of the letter *r* often colors a preceding vowel in English, so that the originally distinct vowels of *curt, word, bird,* and *were* are now pronounced the same. Originally *err* and *error* had the same first vowel, but over time *err* developed the pronunciation \'ər\ as well. Commentators have expressed a visceral dislike for the original pronunciation \'er\; perhaps they believe that once usage has established a new pronunciation for a word that there can be no going back. By this reasoning, though, we should embrace the once established innovative pronunciations of *gold* \'güld\ and *Rome* \'rüm\ (as seen in Shakespeare's pun on *Rome* and *room* in *Julius Caesar* I.ii.156). For these two words the English language has returned to the older forms, and no sound reason prevents us from accepting again the \'er\ pronunciation of *err,* which is today also the more common variant in American speech.

transitive verb
: to force out or release usually suddenly and violently
— **erupt·ible** \-ˈrəp-tə-bəl\ *adjective*
— **erup·tive** \-tiv\ *adjective*
— **erup·tive·ly** *adverb*

erup·tion \i-ˈrəp-shən\ *noun* (1555)
1 a : an act, process, or instance of erupting **b :** the breaking out of a rash on the skin or mucous membrane
2 : a product of erupting (as a skin rash)

-ery \(ə-)rē\ *noun suffix, plural* **-er·ies** [Middle English *-erie,* from Old French, from *-ier* -er + *-ie* -y]
1 : qualities collectively **:** character **:** -NESS ⟨snobb*ery*⟩
2 : art **:** practice ⟨quack*ery*⟩
3 : place of doing, keeping, producing, or selling (the thing specified) ⟨fish*ery*⟩ ⟨bak*ery*⟩
4 : collection **:** aggregate ⟨fin*ery*⟩
5 : state or condition ⟨slav*ery*⟩

eryn·go \i-ˈriŋ-(ˌ)gō\ *noun, plural* **-goes** *or* **-gos** [modification of Latin *eryngion* sea holly, from Greek *ēryngion*] (1543)
1 : any of various plants (genus *Eryngium*) of the carrot family that have elongate spinulose-margined leaves and flowers in dense bracted heads
2 *obsolete* **:** candied root of the sea holly made to be used as an aphrodisiac

ery·sip·e·las \ˌer-ə-ˈsi-p(ə-)ləs, ˌir-\ *noun* [Middle English *erisipila,* from Latin *erysipelas,* from Greek, from *erysi-* (probably akin to Greek *erythros* red) + *-pelas* (probably akin to Latin *pellis* skin) — more at RED, FELL] (14th century)
: an acute febrile disease associated with intense edematous local inflammation of the skin and subcutaneous tissues caused by a hemolytic streptococcus

er·y·the·ma \ˌer-ə-ˈthē-mə\ *noun* [New Latin, from Greek *erythēma,* from *erythainein* to redden, from *erythros*] (circa 1783)
: abnormal redness of the skin due to capillary congestion
— **er·y·them·a·tous** \-ˈthe-mə-təs\ *adjective*

er·y·thor·bate \ˌer-ə-ˈthȯr-ˌbāt\ *noun* (1963)
: a salt of erythorbic acid that is used in foods as an antioxidant

er·y·thor·bic acid \-ˌthȯr-bik-\ *noun* [*erythr-* + *ascorbic acid*] (1963)
: a diastereoisomer of ascorbic acid with optical activity

erythr- *or* **erythro-** *combining form* [Greek, from *erythros* — more at RED]
1 : red ⟨*erythro*cyte⟩
2 : erythrocyte ⟨*erythr*oid⟩

er·y·thre·mia \ˌer-ə-ˈthrē-mē-ə\ *noun* [New Latin] (1908)
: POLYCYTHEMIA VERA

er·y·thrism \ˈer-ə-ˌthri-zəm\ *noun* (1864)
: a condition marked by exceptional prevalence of red pigmentation (as in skin or hair)
— **er·y·thris·tic** \ˌer-ə-ˈthris-tik\ *also* **er·y·thris·mal** \-ˈthriz-məl\ *adjective*

er·y·thrite \ˈer-ə-ˌthrīt\ *noun* (1844)
: a usually rose-colored mineral consisting of a hydrous cobalt arsenate occurring especially in monoclinic crystals

eryth·ro·blast \i-ˈrith-rə-ˌblast\ *noun* [International Scientific Vocabulary] (circa 1890)
: a polychromatic nucleated cell of red marrow that synthesizes hemoglobin and that is an intermediate in the initial stage of red blood cell formation; *broadly* **:** a cell ancestral to red blood cells
— **eryth·ro·blas·tic** \-ˌrith-rə-ˈblas-tik\ *adjective*

eryth·ro·blas·to·sis \i-ˌrith-rə-ˌblas-ˈtō-səs\ *noun, plural* **-to·ses** \-ˌsēz\ [New Latin] (circa 1923)
: abnormal presence of erythroblasts in the circulating blood; *especially* **:** ERYTHROBLASTOSIS FETALIS

erythroblastosis fe·ta·lis \-fi-ˈta-ləs\ *noun* [New Latin, fetal erythroblastosis] (circa 1934)
: a hemolytic disease of the fetus and newborn that occurs when the system of an Rh-negative mother produces antibodies to an antigen in the blood of an Rh-positive fetus which cross the placenta and destroy fetal erythrocytes and that is characterized by an increase in circulating erythroblasts and by jaundice

eryth·ro·cyte \i-ˈrith-rə-ˌsīt\ *noun* [International Scientific Vocabulary] (circa 1894)
: RED BLOOD CELL
— **eryth·ro·cyt·ic** \-ˌrith-rə-ˈsi-tik\ *adjective*

ery·throid \i-ˈrith-ˌrȯid, ˈer-ə-ˌthrȯid\ *adjective* (1927)
: relating to erythrocytes or their precursors

eryth·ro·my·cin \i-ˌrith-rə-ˈmī-sᵊn\ *noun* (1952)
: a broad-spectrum antibiotic $C_{37}H_{67}NO_{13}$ produced by an actinomycete (*Streptomyces erythreus*)

eryth·ro·poi·e·sis \i-ˌrith-rō-pȯi-ˈē-səs\ *noun* [New Latin, from *erythr-* + Greek *poiēsis* creation — more at POESY] (1918)
: the production of red blood cells (as from the bone marrow)
— **eryth·ro·poi·et·ic** \-ˈe-tik\ *adjective*

eryth·ro·poi·e·tin \-ˈpȯi-ə-tᵊn\ *noun* (1948)
: a hormonal substance that is formed especially in the kidney and stimulates red blood cell formation

eryth·ro·sin \i-ˈrith-rə-sən\ *also* **eryth·ro·sine** \-sən, -ˌsēn\ *noun* [International Scientific Vocabulary *erythr-* + *eosin*] (circa 1882)
: any of several dyes made from fluorescein that yield reddish shades

¹-es \əz, iz *after* s, z, sh, ch; z *after* v *or a vowel*\ *noun plural suffix* [Middle English *-es, -s* — more at ¹-s]
— used to form the plural of most nouns that end in *s* ⟨glass*es*⟩, *z* ⟨fuzz*es*⟩, *sh* ⟨bush*es*⟩, *ch* ⟨peach*es*⟩, or a final *y* that changes to *i* ⟨lad*ies*⟩ and of some nouns ending in *f* that changes to *v* ⟨loav*es*⟩; compare ¹-s

²-es *verb suffix* [Middle English — more at ³-s]
— used to form the third person singular present of most verbs that end in *s* ⟨bless*es*⟩, *z* ⟨fizz*es*⟩, *sh* ⟨hush*es*⟩, *ch* ⟨catch*es*⟩, or a final *y* that changes to *i* ⟨def*ies*⟩; compare ³-s

Esau \ˈē-(ˌ)sȯ\ *noun* [Latin, from Greek *Ēsau,* from Hebrew *ʽĒsāw*]
: the elder son of Isaac and Rebekah who sold his birthright to his twin brother Jacob

es·ca·drille \ˈes-kə-ˌdril, -ˌdrē\ *noun* [French, flotilla, escadrille, from Spanish *escuadrilla,* diminutive of *escuadra* squadron, squad — more at SQUAD] (1912)
: a unit of a European air command containing usually six airplanes

es·ca·lade \ˈes-kə-ˌlād, -ˌläd\ *noun* [French, from Italian *scalata,* from *scalare* to scale, from *scala* ladder, from Late Latin — more at SCALE] (1598)
: an act of scaling especially the walls of a fortification
— **escalade** *transitive verb*
— **es·ca·lad·er** *noun*

es·ca·late \ˈes-kə-ˌlāt, ÷-kyə-\ *verb* **-lat·ed; -lat·ing** [back-formation from *escalator*] (1944)
intransitive verb
: to increase in extent, volume, number, amount, intensity, or scope ⟨a little war threatens to *escalate* into a huge ugly one —Arnold Abrams⟩
transitive verb
: EXPAND 2
— **es·ca·la·tion** \ˌes-kə-ˈlā-shən, ÷-kyə-\ *noun*
— **es·ca·la·to·ry** \ˈes-kə-lə-ˌtōr-ē, -ˌtȯr-, ÷-kyə-\ *adjective*

¹es·ca·la·tor \ˈes-kə-ˌlā-tər, ÷-kyə-\ *noun* [from *Escalator,* a trademark] (1900)
1 a : a power-driven set of stairs arranged like an endless belt that ascend or descend continuously **b :** an upward course suggestive of an escalator ⟨a never-stopping *escalator* of economic progress —D. W. Brogan⟩
2 : an escalator clause or provision

²escalator *adjective* (1930)
: providing for a periodic proportional upward or downward adjustment (as of prices or wages) ⟨an *escalator* arrangement tying the base pay . . . to living costs —*N. Y. Times*⟩

es·cal·lop \is-ˈkä-ləp, -ˈka-\ *variant of* SCALLOP

es·ca·pade \ˈes-kə-ˌpād\ *noun* [French, action of escaping, from Spanish *escapada,* from *escapar* to escape, from (assumed) Vulgar Latin *excappare*] (1672)
: a usually adventurous action that runs counter to approved or conventional conduct

¹es·cape \is-ˈkāp, es-, *dialect* iks-ˈkāp\ *verb* **es·caped; es·cap·ing** [Middle English, from Old North French *escaper,* from (assumed) Vulgar Latin *excappare,* from Latin *ex-* + Late Latin *cappa* head covering, cloak] (13th century)
intransitive verb
1 a : to get away (as by flight) ⟨*escaped* from prison⟩ **b :** to issue from confinement ⟨gas is *escaping*⟩ **c** *of a plant* **:** to run wild from cultivation
2 : to avoid a threatening evil
transitive verb
1 : to get free of **:** break away from ⟨*escape* the jungle⟩ ⟨*escape* the solar system⟩
2 : to get or stay out of the way of **:** AVOID
3 : to fail to be noticed or recallable by ⟨his name *escapes* me⟩
4 a : to issue from **b :** to be uttered involuntarily by ☆
— **es·cap·er** *noun*

²escape *noun* (14th century)
1 : an act or instance of escaping: as **a :** flight from confinement **b :** evasion of something undesirable **c :** leakage or outflow especially of a fluid **d :** distraction or relief from routine or reality
2 : a means of escape
3 : a cultivated plant run wild

³escape *adjective* (1817)
1 : providing a means of escape ⟨*escape* literature⟩
2 : providing a means of evading a regulation, claim, or commitment ⟨an *escape* clause in a contract⟩

escape artist *noun* (1943)
: one (as a performer or criminal) unusually adept at escaping from confinement

es·cap·ee \is-ˌkā-ˈpē, ˌes-(ˌ)kā-, ˌes-kə-\ *noun* (circa 1866)
: one that has escaped; *especially* **:** an escaped prisoner

☆ SYNONYMS
Escape, avoid, evade, elude, shun, eschew mean to get away or keep away from something. ESCAPE stresses the fact of getting away or being passed by not necessarily through effort or by conscious intent ⟨nothing *escapes* her sharp eyes⟩. AVOID stresses forethought and caution in keeping clear of danger or difficulty ⟨try to *avoid* past errors⟩. EVADE implies adroitness, ingenuity, or lack of scruple in escaping or avoiding ⟨*evaded* the question by changing the subject⟩. ELUDE implies a slippery or baffling quality in the person or thing that escapes ⟨what she sees in him *eludes* me⟩. SHUN often implies an avoiding as a matter of habitual practice or policy and may imply repugnance or abhorrence ⟨you have *shunned* your responsibilities⟩. ESCHEW implies an avoiding or abstaining from as unwise or distasteful ⟨a playwright who *eschews* melodrama⟩.

escape hatch *noun* (1925)
1 : a hatch providing an emergency exit from an enclosed space
2 : a means of evading a difficulty, dilemma, or responsibility

escape mechanism *noun* (1927)
: a mode of behavior or thinking adopted to evade unpleasant facts or responsibilities

es·cape·ment \is-ˈkāp-mənt\ *noun* (1779)
1 a : a device in a timepiece which controls the motion of the train of wheelwork and through which the energy of the power source is delivered to the pendulum or balance by means of impulses that permit a tooth to escape from a pallet at regular intervals **b :** a ratchet device (as the spacing mechanism of a typewriter) that permits motion in one direction only in equal steps
2 a : the act of escaping **b :** a way of escape **:** VENT

escapement 1a

escape velocity *noun* (1934)
: the minimum velocity that a moving body (as a rocket) must have to escape from the gravitational field of a celestial body (as the earth) and move outward into space

es·cap·ism \is-ˈkā-ˌpi-zəm\ *noun* (1933)
: habitual diversion of the mind to purely imaginative activity or entertainment as an escape from reality or routine
— es·cap·ist \-pist\ *adjective or noun*

es·cap·ol·o·gy \is-ˌkā-ˈpä-lə-jē, ˌes-(ˌ)\ *noun* (1939)
: the art or practice of escaping
— es·cap·ol·o·gist \-jist\ *noun*

es·car·got \ˌes-ˌkär-ˈgō\ *noun, plural* **-gots** \-ˈgō(z)\ [French, snail, from Middle French, from Old Provençal *escaragol*] (circa 1892)
: a snail prepared for use as food

es·ca·role \ˈes-kə-ˌrōl\ *noun* [French, from Late Latin *escariola,* from Latin *escarius* of food, from *esca* food, from *edere* to eat — more at EAT] (1897)
: ENDIVE 1

es·carp·ment \i-ˈskärp-mənt\ *noun* [French *escarpement,* from *escarper* to scarp, from Middle French, from *escarpe* scarp, from Old Italian *scarpa* — more at SCARP] (circa 1802)
1 : a steep slope in front of a fortification
2 : a long cliff or steep slope separating two comparatively level or more gently sloping surfaces and resulting from erosion or faulting

-es·cence \ˈe-s^{ə}n(t)s\ *noun suffix* [Middle French, from Latin *-escentia,* from *-escent-, -escens* + *-ia* -y]
: state or process of becoming ⟨obsol*escence*⟩

-es·cent \ˈe-s^{ə}nt\ *adjective suffix* [Middle French, from Latin *-escent-, -escens,* present participle suffix of inchoative verbs in *-escere*]
1 : beginning **:** beginning to be **:** slightly ⟨frut*escent*⟩
2 : reflecting or emitting light (in a specified way) ⟨opal*escent*⟩

es·char \ˈes-ˌkär\ *noun* [Middle English *escare* — more at SCAR] (1543)
: a scab formed especially after a burn

es·cha·rot·ic \ˌes-kə-ˈrä-tik\ *adjective* [French or Late Latin; French *escharotique,* from Late Latin *escharoticus,* from Greek *escharōtikos,* from *escharoun* to form an eschar, from *eschara* eschar] (1612)
: producing an eschar
— escharotic *noun*

es·cha·to·log·i·cal \(ˌ)es-ˌka-t^{ə}l-ˈä-ji-kəl, ˌes-kə-\ *adjective* (1854)
1 : of or relating to eschatology or an eschatology
2 : of or relating to the end of the world or the events associated with it in eschatology
— es·cha·to·log·i·cal·ly \-ji-k(ə-)lē\ *adverb*

es·cha·tol·o·gy \ˌes-kə-ˈtä-lə-jē\ *noun, plural* **-gies** [Greek *eschatos* last, farthest] (1844)
1 : a branch of theology concerned with the final events in the history of the world or of mankind
2 : a belief concerning death, the end of the world, or the ultimate destiny of mankind; *specifically* **:** any of various Christian doctrines concerning the Second Coming, the resurrection of the dead, or the Last Judgment

[1]es·cheat \is-ˈchēt, ish-ˈchēt\ *noun* [Middle English *eschete,* from Old French, reversion of property, from *escheoir* to fall, devolve, from (assumed) Vulgar Latin *excadēre,* from Latin *ex-* + (assumed) Vulgar Latin *cadēre* to fall, from Latin *cadere* — more at CHANCE] (14th century)
1 : escheated property
2 a : the reversion of lands in English feudal law to the lord of the fee when there are no heirs capable of inheriting under the original grant **b :** the reversion of property to the crown in England or to the state in the U.S. when there are no legal heirs

[2]escheat (14th century)
transitive verb
: to cause to revert by escheat
intransitive verb
: to revert by escheat
— es·cheat·able \-ˈchē-tə-bəl\ *adjective*

es·chew \e-ˈshü, i-; es-ˈchü, is-; *also* e-ˈskyü\ *transitive verb* [Middle English, from Middle French *eschiuver,* of Germanic origin; akin to Old High German *sciuhen* to frighten off — more at SHY] (14th century)
: to avoid habitually especially on moral or practical grounds **:** SHUN
synonym see ESCAPE
— es·chew·al \-əl\ *noun*

es·co·lar \ˌes-kə-ˈlär\ *noun, plural* **escolar** *or* **escolars** [Spanish, literally, scholar, from Medieval Latin *scholaris* — more at SCHOLAR] (circa 1890)
: a large widely distributed rough-scaled fish (*Lepidocybium flavobrunneum*) that resembles a mackerel

[1]es·cort \ˈes-ˌkȯrt\ *noun* [French *escorte,* from Italian *scorta,* from *scorgere* to guide, from (assumed) Vulgar Latin *excorrigere,* from Latin *ex-* + *corrigere* to make straight, correct — more at CORRECT] (1745)
1 a (1) **:** a person or group of persons accompanying another to give protection or as a courtesy (2) **:** the man who goes on a date with a woman **b :** a protective screen of warships or fighter planes or a single ship or plane used to fend off enemy attack from one or more vulnerable craft
2 : accompaniment by a person or an armed protector (as a ship)

[2]es·cort \is-ˈkȯrt, es-, ˈes-ˌ\ *transitive verb* (1708)
: to accompany as an escort

es·cot \is-ˈkät\ *transitive verb* [Middle French *escoter,* from *escot* contribution, of Germanic origin; akin to Old Norse *skot* contribution, shot — more at SHOT] (1602)
obsolete **:** SUPPORT, MAINTAIN

es·cri·toire \ˈes-krə-ˌtwär\ *noun* [obsolete French, writing desk, scriptorium, from Medieval Latin *scriptorium*] (1694)
: a writing table or desk; *specifically* **:** SECRETARY 4b

[1]es·crow \ˈes-ˌkrō, es-ˈ\ *noun* [Middle French *escroue* scroll — more at SCROLL] (1594)
1 : a deed, a bond, money, or a piece of property held in trust by a third party to be turned over to the grantee only upon fulfillment of a condition
2 : a fund or deposit designed to serve as an escrow
— in escrow : in trust as an escrow ⟨had $1000 *in escrow* to pay taxes⟩

[2]es·crow \es-ˈkrō, ˈes-ˌ\ *transitive verb* (1949)
: to place in escrow

es·cu·do \is-ˈkü-(ˌ)dō\ *noun, plural* **-dos** [Spanish & Portuguese, literally, shield, from Latin *scutum*] (circa 1821)
1 : any of various former gold or silver coins of Hispanic countries
2 — see MONEY table
3 : the basic monetary unit of Chile between 1960 and 1975
4 : the peso of Guinea-Bissau

es·cu·lent \ˈes-kyə-lənt\ *adjective* [Latin *esculentus,* from *esca* food, from *edere* to eat — more at EAT] (1626)
: EDIBLE
— esculent *noun*

es·cutch·eon \is-ˈkə-chən\ *noun* [Middle English *escochon,* from Middle French *escuchon,* from (assumed) Vulgar Latin *scution-, scutio,* from Latin *scutum* shield — more at ESQUIRE] (15th century)
1 : a defined area on which armorial bearings are displayed and which usually consists of a shield
2 : a protective or ornamental plate or flange (as around a keyhole)
3 : the part of a ship's stern on which the name is displayed

Es·dras \ˈez-drəs\ *noun* [Late Latin, from Greek, from Hebrew *ʿEzrā*]
1 : either of two books of the Roman Catholic canon of the Old Testament: **a :** EZRA 2 **b :** NEHEMIAH 2
2 : either of two uncanonical books of Scripture included in the Protestant Apocrypha — see BIBLE table

[1]-ese \ˈēz, ˈēs\ *adjective suffix* [Portuguese *-ês* & Italian *-ese,* from Latin *-ensis*]
: of, relating to, or originating in (a certain place or country) ⟨Japan*ese*⟩

[2]-ese *noun suffix, plural* **-ese**
1 : native or resident of (a specified place or country) ⟨Chin*ese*⟩
2 a : language of (a particular place, country, or nationality) ⟨Siam*ese*⟩ **b :** speech, literary style, or diction peculiar to (a specified place, person, or group) — usually in words applied in depreciation ⟨journal*ese*⟩

es·em·plas·tic \ˌe-ˌsem-ˈplas-tik, -səm-\ *adjective* [Greek *es hen* into one + English *plastic*] (1817)
: shaping or having the power to shape disparate things into a unified whole ⟨the *esemplastic* power of the poetic imagination —W. H. Gardner⟩

es·er·ine \ˈe-sə-ˌrēn\ *noun* [French *ésérine*] (1879)
: PHYSOSTIGMINE

es·ker \ˈes-kər\ *noun* [Irish *eiscir* ridge] (1848)
: a long narrow ridge or mound of sand, gravel, and boulders deposited by a stream flowing on, within, or beneath a stagnant glacier

Es·ki·mo \ˈes-kə-ˌmō\ *noun* [obsolete *Esquimawe,* probably from Spanish *esquimao,* from Montagnais (Algonquian language of eastern Canada) *aiachkime8* Micmac, Eskimo; probably akin to modern Montagnais *assimeˑw* she laces a snowshoe, Ojibwa *aškime·*] (1584)
1 *plural* **Eskimo** *or* **Eskimos :** a member of a group of peoples of northern Canada, Greenland, Alaska, and eastern Siberia
2 : any of the languages of the Eskimo peoples
— Es·ki·mo·an \ˌes-kə-ˈmō-ən\ *adjective*

Eskimo curlew *noun* (1813)
: an extremely rare New World curlew (*Numenius borealis*) that breeds in northern North America and winters in South America

Eskimo curlew

\ə\ **abut** \$^{ə}$\ **kitten** \ər\ **further** \a\ **ash** \ā\ **ace** \ä\ **mop, mar** \au̇\ **out** \ch\ **chin** \e\ **bet** \ē\ **easy** \g\ **go** \i\ **hit** \ī\ **ice** \j\ **job** \ŋ\ **sing** \ō\ **go** \ȯ\ **law** \ȯi\ **boy** \th\ **thin** \t͟h\ **the** \ü\ **loot** \u̇\ **foot** \y\ **yet** \zh\ **vision** *see also* Guide to Pronunciation

Eskimo dog *noun* (1774)
: a sled dog of American origin
ESOP \ˌē-ˌes-(ˌ)ō-ˈpē, ˈē-ˌsäp\ *noun* [employee stock ownership plan] (1975)
: a program by which a corporation's employees acquire its capital stock
esoph·a·gus \i-ˈsä-fə-gəs\ *noun, plural* **-gi** \-ˌgī, -ˌjī\ [Middle English *ysophagus,* from Greek *oisophagos,* from *oisein* to be going to carry + *phagein* to eat — more at BAKSHEESH] (14th century)
: a muscular tube that in humans is about nine inches long and passes from the pharynx down the neck between the trachea and the spinal column and behind the left bronchus where it pierces the diaphragm slightly to the left of the middle line and joins the cardiac end of the stomach
— esoph·a·ge·al \i-ˌsä-fə-ˈjē-əl\ *adjective*
es·o·ter·ic \ˌe-sə-ˈter-ik\ *adjective* [Late Latin *esotericus,* from Greek *esōterikos,* from *esōterō,* comparative of *eisō, esō* within, from *eis* into; akin to Greek *en* in — more at IN] (circa 1660)
1 a : designed for or understood by the specially initiated alone ⟨a body of *esoteric* legal doctrine —B. N. Cardozo⟩ **b :** of or relating to knowledge that is restricted to a small group
2 a : limited to a small circle ⟨*esoteric* pursuits⟩ **b :** PRIVATE, CONFIDENTIAL ⟨an *esoteric* purpose⟩
— es·o·ter·i·cal·ly \-i-k(ə-)lē\ *adverb*
es·o·ter·i·ca \-i-kə\ *noun plural* [New Latin, from Greek *esōterika,* neuter plural of *esōterikos*] (circa 1929)
: esoteric items
es·o·ter·i·cism \-ˈter-ə-ˌsi-zəm\ *noun* (1846)
1 : esoteric doctrines or practices
2 : the quality or state of being esoteric
ESP \ˌē-ˌes-ˈpē\ *noun* [extrasensory perception] (1934)
: EXTRASENSORY PERCEPTION
es·pa·drille \ˈes-pə-ˌdril\ *noun* [French, alteration of *espardille,* ultimately from Latin *spartum*] (1892)
: a sandal usually having a fabric upper and a flexible sole
[1]**es·pal·ier** \is-ˈpal-yər, -ˌyā\ *noun* [French, ultimately from Italian *spalla* shoulder, from Late Latin *spatula* shoulder blade — more at EPAULET] (1662)
1 : a plant (as a fruit tree) trained to grow flat against a support (as a wall)
2 : a railing or trellis on which fruit trees or shrubs are trained to grow flat
[2]**espalier** *transitive verb* (1810)
1 : to train as an espalier
2 : to furnish with an espalier
es·par·to \is-ˈpär-(ˌ)tō\ *noun, plural* **-tos** [Spanish, from Latin *spartum,* from Greek *sparton* — more at SPIRE] (1845)
1 : either of two Spanish and Algerian grasses (*Stipa tenacissima* and *Lygeum spartum*) used especially to make cordage, shoes, and paper — called also *esparto grass*
2 : the fiber of esparto
es·pe·cial \is-ˈpe-shəl\ *adjective* [Middle English, from Middle French — more at SPECIAL] (14th century)
: being distinctive: as **a :** directed toward a particular individual, group, or end ⟨sent *especial* greetings to his son⟩ ⟨took *especial* care to speak clearly⟩ **b :** of special note or importance **:** unusually great or significant ⟨a decision of *especial* relevance⟩ **c :** highly distinctive or personal **:** PECULIAR ⟨had an *especial* dislike for music⟩ **d :** CLOSE, INTIMATE ⟨his *especial* crony⟩ **e :** SPECIFIC, PARTICULAR ⟨had no *especial* destination in mind⟩
synonym see SPECIAL
— in especial : in particular
es·pe·cial·ly \is-ˈpesh-lē, -ˈpe-shə-\ *adverb* (15th century)
1 : SPECIALLY 1
2 a : in particular **:** PARTICULARLY ⟨food seems cheaper, *especially* meats⟩ **b :** for a particular purpose ⟨built *especially* for research⟩
3 — used as an intensive ⟨an *especially* good essay⟩ ⟨nothing *especially* radical in the remarks⟩
es·per·ance \ˈes-p(ə-)rən(t)s\ *noun* [Middle English *esperaunce,* from Middle French *esperance*] (15th century)
obsolete **:** HOPE, EXPECTATION
Es·pe·ran·to \ˌes-pə-ˈrän-(ˌ)tō, -ˈran-(ˌ)tō\ *noun* [Dr. *Esperanto,* pseudonym of L. L. Zamenhof (died 1917) Polish oculist, its inventor] (1892)
: an artificial international language based as far as possible on words common to the chief European languages
— Es·pe·ran·tist \-ˈrän-tist, -ˈran-\ *noun or adjective*
es·pi·al \is-ˈpī(-ə)l\ *noun* (14th century)
1 : OBSERVATION
2 : an act of noticing **:** DISCOVERY
es·piè·gle \es-pyeglᵊ\ *adjective* [French, after *Ulespiegle* (Till Eulenspiegel), peasant prankster] (1816)
: FROLICSOME, ROGUISH
es·piè·gle·rie \es-pyeg-lə-rē\ *noun* [French, from *espiègle*] (1816)
: the quality or state of being roguish or frolicsome
es·pi·o·nage \ˈes-pē-ə-ˌnäzh, -ˌnäj, -nij, *Canadian also* -ˌnazh; ˌes-pē-ə-ˈnäzh; is-ˈpē-ə-nij\ *noun* [French *espionnage,* from Middle French, from *espionner* to spy, from *espion* spy, from Old Italian *spione,* from *spia,* of Germanic origin; akin to Old High German *spehōn* to spy — more at SPY] (1793)
: the practice of spying or using spies to obtain information about the plans and activities especially of a foreign government or a competing company ⟨industrial *espionage*⟩
es·pla·nade \ˈes-plə-ˌnäd, ˌes-plə-ˈ *also* -ˈnād *or* -ˌnād\ *noun* [Middle French, from Italian *spianata,* from *spianare* to level, from Latin *explanare* — more at EXPLAIN] (1591)
: a level open stretch of paved or grassy ground; *especially* **:** one designed for walking or driving along a shore
es·pous·al \is-ˈpau̇-zəl *also* -səl\ *noun* (14th century)
1 a : BETROTHAL **b :** WEDDING **c :** MARRIAGE
2 : a taking up or adopting of a cause or belief
es·pouse \is-ˈpau̇z *also* -ˈpau̇s\ *transitive verb* **es·poused; es·pous·ing** [Middle English, from Middle French *espouser,* from Late Latin *sponsare* to betroth, from Latin *sponsus* betrothed — more at SPOUSE] (15th century)
1 : MARRY
2 : to take up and support as a cause **:** become attached to
synonym see ADOPT
— es·pous·er *noun*
espres·so \e-ˈspre-(ˌ)sō\ *noun, plural* **-sos** [Italian (*caffè*) *espresso,* literally, pressed out coffee] (1945)
1 : coffee brewed by forcing steam through finely ground darkly roasted coffee beans
2 : a cup of espresso
es·prit \is-ˈprē\ *noun* [French, from Latin *spiritus* spirit] (1591)
1 : vivacious cleverness or wit
2 : ESPRIT DE CORPS
es·prit de corps \is-ˌprē-də-ˈkōr, -ˈkȯr\ *noun* [French] (1780)
: the common spirit existing in the members of a group and inspiring enthusiasm, devotion, and strong regard for the honor of the group
es·py \is-ˈpī\ *transitive verb* **es·pied; es·py·ing** [Middle English *espien,* from Old French *espier* — more at SPY] (14th century)
: to catch sight of ⟨among the several horses . . . she *espied* the white mustang —Zane Grey⟩
-esque \ˈesk\ *adjective suffix* [French, from Italian *-esco,* of Germanic origin; akin to Old High German *-isc* -ish — more at -ISH]
: in the manner or style of **:** like ⟨statu*esque*⟩
Es·qui·mau \ˈes-kə-ˌmō\ *noun, plural* **Esquimau** *or* **Es·qui·maux** \-ˌmō(z)\ [French, from Montagnais (Algonquian language)] (1744)
: ESKIMO
es·quire \ˈes-ˌkwīr, is-ˈ\ *noun* [Middle English, from Middle French *escuier* squire, from Late Latin *scutarius,* from Latin *scutum* shield; akin to Old Irish *sciath* shield] (15th century)
1 : a member of the English gentry ranking below a knight
2 : a candidate for knighthood serving as shield bearer and attendant to a knight
3 — used as a title of courtesy usually placed in its abbreviated form after the surname ⟨John R. Smith, *Esq.*⟩
4 *archaic* **:** a landed proprietor ◆
ess \ˈes\ *noun* (1540)
1 : the letter *s*
2 : something resembling the letter *S* in shape; *especially* **:** an S-shaped curve in a road
-ess \əs, is *also* ˌes\ *noun suffix* [Middle English *-esse,* from Old French, from Late Latin *-issa,* from Greek]
: female ⟨giant*ess*⟩ ■
[1]**es·say** \e-ˈsā, ə-ˈsā, ˈe-ˌsā\ *transitive verb* (14th century)
1 : to put to a test
2 : to make an often tentative or experimental effort to perform **:** TRY
synonym see ATTEMPT
— es·say·er *noun*
[2]**es·say** \ˈe-ˌsā; *senses 1, 2 & 4 also* e-ˈsā\ *noun* [Middle English, from Middle French *essai,* ultimately from Late Latin *exagium* act of weighing, from *ex-* + *agere* to drive — more at AGENT] (14th century)
1 : TRIAL, TEST
2 a : EFFORT, ATTEMPT; *especially* **:** an initial tentative effort **b :** the result or product of an attempt
3 a : an analytic or interpretative literary composition usually dealing with its subject from a limited or personal point of view **b :** something resembling such a composition ⟨a photographic *essay*⟩
4 : a proof of an unaccepted design for a stamp or piece of paper money
es·say·ist \ˈe-ˌsā-ist\ *noun* (1601)
: a writer of essays
es·say·is·tic \ˌe-(ˌ)sā-ˈis-tik\ *adjective* (1862)
1 : of or relating to an essay or an essayist
2 : resembling an essay in quality or character
essay question *noun* (1947)

◇ WORD HISTORY
esquire The word *esquire* in its oldest senses, "young candidate for knighthood" and "member of the English gentry ranking below a knight in social precedence," has been less common than the shortened form *squire* since Middle English. It is only as an abbreviated title placed after a person's name that *esquire* maintains a curious sort of afterlife in English, even in the United States, where there have never been knights or gentry. This peculiar usage came about because the technical stipulations for members of the gentry who could be styled *esquire,* as established by 17th century lawyers and antiquaries, were so complex and so loosely observed that the title became practically equivalent to "gentleman." (The title *knight* never became similarly nebulous because since the 16th century knighthood has been an honor specifically bestowed on individuals by the British sovereign, and not hereditary.) In America, *esquire* is less common than in Britain, persisting mainly among attorneys, who use it, chiefly in the form of its abbreviation *Esq.,* when referring to or addressing each other in writing.

: an examination question that requires an answer in a sentence, paragraph, or short composition

es·sence \'e-s^{ə}n(t)s\ *noun* [Middle English, from Middle French & Latin; Middle French, from Latin *essentia,* from *esse* to be — more at IS] (14th century)
1 a : the permanent as contrasted with the accidental element of being **b :** the individual, real, or ultimate nature of a thing especially as opposed to its existence **c :** the properties or attributes by means of which something can be placed in its proper class or identified as being what it is
2 : something that exists **:** ENTITY
3 a (1) **:** a volatile substance or constituent (as of perfume) (2) **:** a constituent or derivative possessing the special qualities (as of a plant or drug) in concentrated form; *also* **:** a preparation of such an essence or a synthetic substitute **b :** ODOR, PERFUME
4 : one that possesses or exhibits a quality in abundance as if in concentrated form ⟨she was the *essence* of punctuality⟩
— in essence : in or by its very nature **:** ESSENTIALLY, BASICALLY ⟨was *in essence* an honest person⟩
— of the essence : of the utmost importance ⟨time is *of the essence*⟩

Es·sene \i-'sēn, 'e-ˌsēn\ *noun* [Greek *Essēnos*] (1553)
: a member of a monastic brotherhood of Jews in Palestine from the 2d century B.C. to the 2d century A.D.
— Es·se·ni·an \i-'sē-nē-ən, e-'se-\ *or* **Es·se·nic** \-'se-nik, -'sē-nik\ *adjective*
— Es·se·nism \-'sē-ˌni-zəm\ *noun*

[1]es·sen·tial \i-'sen(t)-shəl\ *adjective* (14th century)
1 : of, relating to, or constituting essence **:** INHERENT
2 : of the utmost importance **:** BASIC, INDISPENSABLE, NECESSARY ⟨*essential* foods⟩ ⟨an *essential* requirement for admission to college⟩
3 : IDIOPATHIC ⟨*essential* disease⟩ ☆
— es·sen·tial·ly \-'sench-lē, -'sen chə-\ *adverb*
— es·sen·tial·ness \-'sen-chəl-nəs\ *noun*

[2]essential *noun* (15th century)
1 : something basic ⟨the *essentials* of astronomy⟩
2 : something necessary, indispensable, or unavoidable

essential amino acid *noun* (1935)
: an amino acid (as lysine) required for normal health and growth, manufactured in the body in insufficient quantities or not at all, and usually supplied by dietary protein

es·sen·tial·ism \-ˌli-zəm\ *noun* (1927)
1 : an educational theory that ideas and skills basic to a culture should be taught to all alike by time-tested methods — compare PROGRESSIVISM
2 : a philosophical theory ascribing ultimate reality to essence embodied in a thing perceptible to the senses — compare NOMINALISM
— es·sen·tial·ist \-list\ *adjective or noun*

es·sen·ti·al·i·ty \i-ˌsen(t)-shē-'a-lə-tē\ *noun, plural* **-ties** (1616)
1 a : essential nature **:** ESSENCE **b :** an essential quality, property, or aspect
2 : the quality or state of being essential ⟨the *essentiality* of freedom and justice —P. G. Hoffman⟩

es·sen·tial·ize \i-'sen(t)-shə-ˌlīz\ *transitive verb* **-ized; -iz·ing** (1913)
: to express or formulate in essential form **:** reduce to essentials

essential oil *noun* (1674)
: any of a class of volatile oils that give plants their characteristic odors and are used especially in perfumes and flavorings — compare FIXED OIL

es·soin \i-'sȯin\ *noun* [Middle English *essoine,* from Middle French, from Old French, from *essoinier* to offer an essoin, from *es-* ex- + *soine* legal excuse, of Germanic origin; akin to Old Saxon *sunnea* denial, Old English *sōth* truth — more at SOOTH] (14th century)
1 : an excuse for not appearing in an English law court at the appointed time
2 *obsolete* **:** EXCUSE, DELAY

es·so·nite \'e-s^{ə}n-ˌīt\ *noun* [French, from Greek *hēsson* inferior; from its being less hard than true hyacinth] (1820)
: a yellow to brown garnet

[1]-est \əst, ist\ *adjective suffix or adverb suffix* [Middle English, from Old English *-st, -est, -ost;* akin to Old High German *-isto* (adjective superlative suffix), Greek *-istos*]
— used to form the superlative degree of adjectives and adverbs of one syllable ⟨fatt*est*⟩ ⟨lat*est*⟩, of some adjectives and adverbs of two syllables ⟨lucki*est*⟩ ⟨often*est*⟩, and less often of longer ones ⟨beggarli*est*⟩

[2]-est \əst, ist\ *or* **-st** \st\ *verb suffix* [Middle English, from Old English *-est, -ast, -st;* akin to Old High German *-ist, -ōst, -ēst,* 2d singular ending]
— used to form the archaic 2d person singular of English verbs (with *thou*) ⟨did*st*⟩

es·tab·lish \is-'ta-blish\ *transitive verb* [Middle English *establissen,* from Middle French *establiss-,* stem of *establir,* from Latin *stabilire,* from *stabilis* stable] (14th century)
1 : to institute (as a law) permanently by enactment or agreement
2 *obsolete* **:** SETTLE 7
3 a : to make firm or stable **b :** to introduce and cause to grow and multiply ⟨*establish* grass on pasturelands⟩
4 a : to bring into existence **:** FOUND ⟨*established* a republic⟩ **b :** BRING ABOUT, EFFECT ⟨*established* friendly relations⟩
5 a : to put on a firm basis **:** SET UP ⟨*establish* his son in business⟩ **b :** to put into a favorable position **c :** to gain full recognition or acceptance of ⟨the role *established* her as a star⟩
6 : to make (a church) a national or state institution
7 : to put beyond doubt **:** PROVE ⟨*established* my innocence⟩
— es·tab·lish·able \-shə-bəl\ *adjective*
— es·tab·lish·er \-shər\ *noun*

established church *noun* (1731)
: a church recognized by law as the official church of a nation or state and supported by civil authority

es·tab·lish·ment \is-'ta-blish-mənt\ *noun* (15th century)
1 : something established: as **a :** a settled arrangement; *especially* **:** a code of laws **b :** ESTABLISHED CHURCH **c :** a permanent civil or military organization **d :** a place of business or residence with its furnishings and staff **e :** a public or private institution
2 : an established order of society: as **a** *often capitalized* **:** a group of social, economic, and political leaders who form a ruling class (as of a nation) **b** *often capitalized* **:** a controlling group ⟨the literary *establishment*⟩
3 a : the act of establishing **b :** the state of being established

es·tab·lish·men·tar·i·an \is-ˌta-blish-mən-'ter-ē-ən, -ˌmen-\ *adjective* (1847)
: of, relating to, or favoring the social or political establishment
— establishmentarian *noun*
— es·tab·lish·men·tar·i·an·ism \-ē-ə-ˌni-zəm\ *noun*

es·ta·mi·net \e-stȧ-mē-nā\ *noun, plural* **-nets** \-nā(z)\ [French] (1814)
: a small café

es·tate \is-'tāt\ *noun* [Middle English *estat,* from Old French — more at STATE] (13th century)
1 : STATE, CONDITION
2 : social standing or rank especially of a high order
3 : a social or political class; *specifically* **:** one of the great classes (as the nobility, the clergy, and the commons) formerly vested with distinct political powers
4 a : the degree, quality, nature, and extent of one's interest in land or other property **b** (1) **:** POSSESSIONS, PROPERTY; *especially* **:** a person's property in land and tenements ⟨a man of small *estate*⟩ (2) **:** the assets and liabilities left by a person at death **c :** a landed property usually with a large house on it **d** *British* **:** PROJECT 4
5 *British* **:** STATION WAGON

estate agent *noun* (1880)
British **:** a real estate broker or manager

estate car *noun* (1950)
British **:** STATION WAGON

estate tax *noun* (1928)
: an excise in the form of a percentage of the net estate that is levied on the privilege of an

☆ SYNONYMS

Essential, fundamental, vital, cardinal mean so important as to be indispensable. ESSENTIAL implies belonging to the very nature of a thing and therefore being incapable of removal without destroying the thing itself or its character ⟨conflict is *essential* in drama⟩. FUNDAMENTAL applies to something that is a foundation without which an entire system or complex whole would collapse ⟨*fundamental* principles of algebra⟩. VITAL suggests something that is necessary to a thing's continued existence or operation ⟨cut off from *vital* supplies⟩. CARDINAL suggests something on which an outcome turns or depends ⟨a *cardinal* rule in buying a home⟩.

☐ USAGE

-ess In 1865 Sara Josepha Hale, self-styled editress of *Godey's Lady's Book,* proposed for use a list of more than 50 terms ending in *-ess* that she thought would lend dignity to women and their work. We are not sure if her plea for the use of these words had any effect on late 19th century usage, but the male commentators of that time mostly disapproved the forms. In the 20th century, attitudes have become no more nearly unanimous. A few male commentators have looked on the forms with approval, if not with the evangelistic fervor of Mrs. Hale, but more, both men and women, find them objectionable in some way. Of course some words with this ending, like *lioness* or *duchess* or *goddess,* have never been in dispute. But actual users always have the final say. If *stewardess* is falling into disuse, it is probably due as much to the increased number of male flight attendants as to language attitudes. *Poetess,* which has been dismissed as archaic or disused since the 18th century, is still used, but in historical reference; it does not appear to be applied to contemporary writers. *Mayoress* and *manageress,* which look a bit strange to American eyes, are quite common in British English. *Actress* and *waitress* are still in common use. *Priestess* in its first sense generally refers to ancient or pagan religions; its second sense, "a woman regarded as a leader," is still in common use, often in such phrases as *high priestess of fashion.* These are a few of the *-ess* words that have been commented on. You can assume that any *-ess* word entered in this dictionary is in current acceptable use unless it bears a warning label or usage note.

owner of property of transmitting the property to others after his or her death — compare INHERITANCE TAX 1

[1]**es·teem** \is-'tēm\ *noun* (14th century)
1 *archaic* **:** WORTH, VALUE
2 *archaic* **:** OPINION, JUDGMENT
3 : the regard in which one is held; *especially* **:** high regard ⟨the *esteem* we all feel for her⟩

[2]**esteem** *transitive verb* [Middle English *estemen* to estimate, from Middle French *estimer,* from Latin *aestimare*] (15th century)
1 *archaic* **:** APPRAISE
2 a : to view as **:** CONSIDER ⟨*esteem* it a privilege⟩ **b :** THINK, BELIEVE
3 : to set a high value on **:** regard highly and prize accordingly
synonym see REGARD

es·ter \'es-tər\ *noun* [German, from *Essigäther* ethyl acetate, from *Essig* vinegar + *Äther* ether] (circa 1852)
: any of a class of often fragrant compounds that can be represented by the formula RCOOR' and that are usually formed by the reaction between an acid and an alcohol with elimination of water

es·ter·ase \'es-tə-ˌrās, -ˌrāz\ *noun* (1910)
: an enzyme that accelerates the hydrolysis or synthesis of esters

es·ter·i·fy \e-'ster-ə-ˌfī\ *transitive verb* **-fied; -fy·ing** (circa 1905)
: to convert into an ester
— es·ter·i·fi·ca·tion \-ˌster-ə-fə-'kā-shən\ *noun*

Es·ther \'es-tər\ *noun* [Latin, from Hebrew *Estēr*]
1 : the Jewish heroine of the Old Testament book of Esther
2 : a narrative book of canonical Jewish and Christian Scripture — see BIBLE table

es·thete, es·thet·ic, es·the·ti·cian, es·the·ti·cism *variant of* AESTHETE, AESTHETIC, AESTHETICIAN, AESTHETICISM

es·ti·ma·ble \'es-tə-mə-bəl\ *adjective* (15th century)
1 : capable of being estimated
2 *archaic* **:** VALUABLE
3 : worthy of esteem
— es·ti·ma·ble·ness *noun*
— es·ti·ma·bly \-blē\ *adverb*

[1]**es·ti·mate** \'es-tə-ˌmāt\ *transitive verb* **-mat·ed; -mat·ing** [Latin *aestimatus,* past participle of *aestimare* to value, estimate] (circa 1532)
1 *archaic* **a :** ESTEEM **b :** APPRAISE
2 a : to judge tentatively or approximately the value, worth, or significance of **b :** to determine roughly the size, extent, or nature of **c :** to produce a statement of the approximate cost of
3 : JUDGE, CONCLUDE ☆
— es·ti·ma·tive \-ˌmā-tiv\ *adjective*

[2]**es·ti·mate** \'es-tə-mət\ *noun* (1563)
1 : the act of appraising or valuing **:** CALCULATION
2 : an opinion or judgment of the nature, character, or quality of a person or thing ⟨had a high *estimate* of his abilities⟩
3 a : a rough or approximate calculation **b :** a numerical value obtained from a statistical sample and assigned to a population parameter
4 : a statement of the cost of work to be done

es·ti·ma·tion \ˌes-tə-'mā-shən\ *noun* (14th century)
1 : JUDGMENT, OPINION
2 a : the act of estimating something **b :** the value, amount, or size arrived at in an estimate
3 : ESTEEM, HONOR

es·ti·ma·tor \'es-tə-ˌmā-tər\ *noun* (1611)
1 : one that estimates
2 : ESTIMATE 3b; *also* **:** a statistical function whose value for a sample furnishes an estimate of a population parameter

es·ti·val \'es-tə-vəl\ *adjective* [Middle English, from Middle French or Latin; Middle French, from Latin *aestivalis,* from *aestivus* of summer, from *aestas* summer — more at EDIFY] (14th century)
: of or relating to the summer

es·ti·vate \-ˌvāt\ *intransitive verb* **-vat·ed; -vat·ing** (1626)
1 : to spend the summer usually at one place
2 : to pass the summer in a state of torpor — compare HIBERNATE

es·ti·va·tion \ˌes-tə-'vā-shən\ *noun* (1625)
: the state of one that estivates

Es·to·nian \e-'stō-nē-ən, -nyən\ *noun* (1795)
1 : a native or inhabitant of Estonia
2 : the Finno-Ugric language of the Estonian people
— Estonian *adjective*

es·top \e-'stäp\ *transitive verb* **es·topped; es·top·ping** [Middle English *estoppen,* from Middle French *estouper,* from (assumed) Vulgar Latin *stuppare* to stop with a tow — more at STOP] (15th century)
1 *archaic* **:** to stop up
2 : BAR; *specifically* **:** to impede by estoppel

es·top·pel \e-'stä-pəl\ *noun* [probably from Middle French *estoupail* bung, from *estouper*] (1531)
: a legal bar to alleging or denying a fact because of one's own previous actions or words to the contrary

estr- *or* **estro-** *combining form*
: estrus ⟨*estro*gen⟩

es·tra·di·ol \ˌes-trə-'dī-ˌȯl, -ˌōl\ *noun* [International Scientific Vocabulary *estra-* (from *estrane* parent compound of estradiol, from New Latin *estrus* + English *-ane*) + *di-* + [1]*-ol*] (1934)
: an estrogenic hormone that is a phenolic steroid alcohol $C_{18}H_{24}O_2$ usually made synthetically and that is often used combined as an ester especially in treating menopausal symptoms

es·tral cycle \'es-trəl-\ *noun* (1941)
: ESTROUS CYCLE

es·trange \is-'trānj\ *transitive verb* **estranged; es·trang·ing** [Middle English, from Middle French *estranger,* from Medieval Latin *extraneare,* from Latin *extraneus* strange — more at STRANGE] (15th century)
1 : to remove from customary environment or associations
2 : to arouse especially mutual enmity or indifference in where there had formerly been love, affection, or friendliness **:** ALIENATE ☆
— es·trange·ment \-'trānj-mənt\ *noun*
— es·trang·er *noun*

[1]**es·tray** \is-'trā\ *noun* (circa 1523)
: STRAY 1

[2]**estray** *intransitive verb* [Middle French *estraier*] (1572)
archaic **:** STRAY

es·tri·ol \'es-ˌtrī-ˌȯl, e-'strī-, -ˌōl\ *noun* [*estrane* + *tri-* + [1]*-ol*] (1933)
: a crystalline estrogenic hormone that is a glycol $C_{18}H_{24}O_3$ usually obtained from the urine of pregnant women

es·tro·gen \'es-trə-jən\ *noun* [New Latin *estrus* + International Scientific Vocabulary *-o-* + *-gen*] (1927)
: a substance (as a sex hormone) tending to promote estrus and stimulate the development of female secondary sex characteristics •
usage see ECONOMIC

es·tro·gen·ic \ˌes-trə-'je-nik\ *adjective* (1930)
1 : promoting estrus
2 : of, relating to, caused by, or being an estrogen
— es·tro·gen·i·cal·ly \-ni-k(ə-)lē\ *adverb*

es·trone \'es-ˌtrōn\ *noun* [*estrane*] (1933)
: an estrogenic hormone that is a ketone $C_{18}H_{22}O_2$ usually obtained from the urine of pregnant females and used similarly to estradiol

es·trous \'es-trəs\ *adjective* (1900)
1 : of, relating to, or characteristic of estrus
2 : being in heat

estrous cycle *noun* (1900)
: the correlated phenomena of the endocrine and generative systems of a female mammal from the beginning of one period of estrus to the beginning of the next — called also *estral cycle, estrus cycle*

es·tru·al \'es-trə-wəl\ *adjective* (circa 1857)
: ESTROUS

es·trus \'es-trəs\ *noun* [New Latin, from Latin *oestrus* gadfly, frenzy, from Greek *oistros* — more at IRE] (circa 1890)
: a regularly recurrent state of sexual excitability during which the female of most mammals will accept the male and is capable of conceiving **:** HEAT; *also* **:** a single occurrence of this state

es·tu·ar·i·al \ˌes-chə-'wer-ē-əl, ˌesh-\ *adjective* (1883)
: ESTUARINE

es·tu·a·rine \'es-chə-(wə-)ˌrīn, -ˌrēn, -ˌrin, 'esh-\ *adjective* (1846)
: of, relating to, or formed in an estuary ⟨*estuarine* currents⟩ ⟨*estuarine* animals⟩

es·tu·ary \'es-chə-ˌwer-ē, 'esh-\ *noun, plural* **-ar·ies** [Latin *aestuarium,* from *aestus* boiling, tide; akin to Latin *aestas* summer — more at EDIFY] (1538)
: a water passage where the tide meets a river current; *especially* **:** an arm of the sea at the lower end of a river

esu·ri·ence \i-'su̇r-ē-ən(t)s, -'zu̇r-\ *noun* (1825)
: the quality or state of being esurient

esu·ri·ent \-ənt\ *adjective* [Latin *esurient-, esuriens,* present participle of *esurire* to be hungry; akin to Latin *edere* to eat — more at EAT] (circa 1672)
: HUNGRY, GREEDY
— esu·ri·ent·ly *adverb*

et \'et\ *dialect past and past participle of* EAT

[1]**-et** \'et, ˌet, ət, it\ *noun suffix* [Middle English, from Old French *-et,* masculine, & *-ete,* feminine, from Late Latin *-itus & -ita*]
: small one ⟨baron*et*⟩ ⟨cellar*et*⟩

☆ SYNONYMS

Estimate, appraise, evaluate, value, rate, assess mean to judge something with respect to its worth or significance. ESTIMATE implies a judgment, considered or casual, that precedes or takes the place of actual measuring or counting or testing out ⟨*estimated* the crowd at two hundred⟩. APPRAISE commonly implies the fixing by an expert of the monetary worth of a thing, but it may be used of any critical judgment ⟨having their house *appraised*⟩. EVALUATE suggests an attempt to determine relative or intrinsic worth in terms other than monetary ⟨*evaluate* a student's work⟩. VALUE equals APPRAISE but without implying expertness of judgment ⟨a watercolor *valued* by the donor at $500⟩. RATE adds to ESTIMATE the notion of placing a thing according to a scale of values ⟨a highly *rated* restaurant⟩. ASSESS implies a critical appraisal for the purpose of understanding or interpreting, or as a guide in taking action ⟨officials are trying to *assess* the damage⟩.

Estrange, alienate, disaffect mean to cause one to break a bond of affection or loyalty. ESTRANGE implies the development of indifference or hostility with consequent separation or divorcement ⟨his *estranged* wife⟩. ALIENATE may or may not suggest separation but always implies loss of affection or interest ⟨managed to *alienate* all his coworkers⟩. DISAFFECT refers especially to those from whom loyalty is expected and stresses the effects (as rebellion or discontent) of alienation without actual separation ⟨conservatives were *disaffected* by the new tax⟩.

[2]-et *noun suffix* [duet]
: group ⟨oct*et*⟩

eta \ˈā-tə, *chiefly British* ˈē-tə\ *noun* [Middle English, from Late Latin, from Greek *ēta,* of Semitic origin; akin to Hebrew *ḥēth* heth] (15th century)
: the 7th letter of the Greek alphabet — see ALPHABET table

éta·gère *or* **eta·gere** \ˌā-ˌtä-ˈzher, -tə-\ *noun* [French, from Middle French *estagiere,* from *estage* floor of a building, station, from Old French — more at STAGE] (1851)
: a piece of furniture consisting of a set of open shelves for displaying small objects and sometimes having an enclosed cabinet as a base

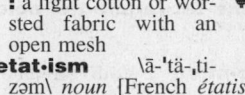

étagère

et alia \(ˌ)et-ˈä-l(ē-)yə, -ˈā-, -ˈa-\ [Latin] (1953)
: and others

eta·mine \ˈā-tə-ˌmēn\ *noun* [French *étamine*] (1714)
: a light cotton or worsted fabric with an open mesh

etat·ism \ā-ˈtä-ˌti-zəm\ *noun* [French *étatisme,* from *état* state, from Old French *estat* — more at STATE] (1923)
: STATE SOCIALISM
— etat·ist \-ˈtä-tist\ *adjective*

et·cet·era \et-ˈse-tə-rə, -ˈse-trə *also* it-, ÷ek-, ÷ik-\ *noun* (1656)
1 : a number of unspecified additional persons or things
2 *plural* **:** unspecified additional items **:** ODDS AND ENDS

et cet·era *same*\ [Latin] (12th century)
: and others especially of the same kind **:** and so forth

[1]etch \ˈech\ *verb* [Dutch *etsen,* from German *ätzen* to etch, corrode, from Old High German *azzen* to feed; akin to Old High German *ezzan* to eat — more at EAT] (1634)
transitive verb
1 a : to produce (as a pattern or design) on a hard material by eating into the material's surface (as by acid or laser beam) **b :** to subject to such etching
2 : to delineate or impress clearly ⟨scenes *etched* in our minds⟩ ⟨pain was *etched* on his features⟩
intransitive verb
: to practice etching
— etch·er *noun*

[2]etch *noun* (1896)
1 : the action or effect of etching a surface
2 : a chemical agent used in etching

etch·ant \ˈe-chənt\ *noun* (1904)
: ETCH 2

etch·ing *noun* (1634)
1 a : the action or process of etching **b :** the art of producing pictures or designs by printing from an etched metal plate
2 a : an etched design **b :** an impression from an etched plate

[1]eter·nal \i-ˈtər-nᵊl\ *adjective* [Middle English, from Middle French, from Late Latin *aeternalis,* from Latin *aeternus* eternal, from *aevum* age, eternity — more at AYE] (14th century)
1 a : having infinite duration **:** EVERLASTING **b :** of or relating to eternity **c :** characterized by abiding fellowship with God ⟨good teacher, what must I do to inherit *eternal* life? —Mark 10:17 (Revised Standard Version)⟩
2 a : continued without intermission **:** PERPETUAL **b :** seemingly endless
3 *archaic* **:** INFERNAL ⟨some *eternal* villain . . . devised this slander —Shakespeare⟩
4 : valid or existing at all times **:** TIMELESS ⟨*eternal* verities⟩
— eter·nal·ize \-nᵊl-ˌīz\ *transitive verb*
— eter·nal·ly \-nᵊl-ē\ *adverb*
— eter·nal·ness *noun*

[2]eternal *noun* (1582)
1 *capitalized* **:** GOD 1 — used with *the*
2 : something eternal

eterne \i-ˈtərn\ *adjective* [Middle English, from Middle French, from Latin *aeternus*] (14th century)
archaic **:** ETERNAL

eter·ni·ty \i-ˈtər-nə-tē\ *noun, plural* **-ties** [Middle English *eternite,* from Middle French *eternité,* from Latin *aeternitat-, aeternitas,* from *aeternus*] (14th century)
1 : the quality or state of being eternal
2 : infinite time
3 *plural* **:** AGE 3b
4 : the state after death **:** IMMORTALITY
5 : a seemingly endless or immeasurable time

eter·nize \i-ˈtər-ˌnīz\ *transitive verb* **-nized; -niz·ing** (1580)
1 a : to make eternal **b :** to prolong indefinitely
2 : IMMORTALIZE
— eter·ni·za·tion \-ˌtər-nə-ˈzā-shən\ *noun*

Ete·sian \i-ˈtē-zhən\ *adjective* [Latin *etesius,* from Greek *etēsios,* from *etos* year — more at WETHER] (1601)
: recurring annually — used of summer winds that blow over the Mediterranean
— Etesian *noun*

eth \ˈeth\ *variant of* EDH

eth- *combining form* [International Scientific Vocabulary]
: ethyl ⟨*eth*ene⟩

[1]-eth \əth, ith\ *or* **-th** \th\ *verb suffix* [Middle English, from Old English *-eth, -ath, -th;* akin to Old High German *-it, -ōt, -ēt,* 3d singular ending, Latin *-t, -it*]
— used to form the archaic third person singular present of verbs ⟨do*th*⟩

[2]-eth — see -TH

eth·a·cryn·ic acid \ˌe-thə-ˌkri-nik-\ *noun* [perhaps from *eth-* + *ac*etic + butyryl + phenol] (1964)
: a potent synthetic diuretic $C_{13}H_{12}Cl_2O_4$ used especially in the treatment of edema

eth·am·bu·tol \e-ˈtham-byu̇-ˌtȯl, -ˌtōl\ *noun* [*eth*ylene + *am*ine + *but*anol] (1965)
: a compound $C_{10}H_{24}N_2O_2$ used especially in the treatment of tuberculosis

eth·ane \ˈe-ˌthān, *British usually* ˈē-\ *noun* [International Scientific Vocabulary, from *ethyl*] (1873)
: a colorless odorless gaseous alkane C_2H_6 found in natural gas and used as a fuel

eth·a·nol \ˈe-thə-ˌnȯl, -ˌnōl, *British also* ˈē-\ *noun* (1900)
: a colorless volatile flammable liquid C_2H_5OH that is the intoxicating agent in liquors and is also used as a solvent — called also *ethyl alcohol, grain alcohol*

eth·a·nol·amine \ˌe-thə-ˈnä-lə-ˌmēn, -ˈnō-, *British also* ˌē-\ *noun* (1897)
: a colorless liquid amino alcohol C_2H_7NO used especially as a solvent in the synthesis of detergents and in gas purification

eth·ene \ˈe-ˌthēn\ *noun* (1873)
: ETHYLENE

eth·e·phon \ˈe-thə-ˌfän\ *noun* [*ethyl* + *phos-phonic* acid (a dibasic organic acid)] (1971)
: a synthetic plant growth regulator $C_2H_6ClO_3P$ that induces flowering and abscission by promoting the release of ethylene and has been used to cause early ripening (as of apples on the tree)

ether \ˈē-thər\ *noun* [Middle English, from Latin *aether,* from Greek *aithēr,* from *aithein* to ignite, blaze; akin to Old English *ād* pyre — more at EDIFY] (14th century)
1 a : the rarefied element formerly believed to fill the upper regions of space **b :** the upper regions of space **:** HEAVENS
2 a : a medium that in the wave theory of light permeates all space and transmits transverse waves **b :** AIRWAVES
3 a : a light volatile flammable liquid $C_4H_{10}O$ used chiefly as a solvent and anesthetic **b :** any of various organic compounds characterized by an oxygen atom attached to two carbon atoms
— ethe·ric \i-ˈther-ik, -ˈthir-\ *adjective*

ethe·re·al \i-ˈthir-ē-əl\ *adjective* (1513)
1 a : of or relating to the regions beyond the earth **b :** CELESTIAL, HEAVENLY **c :** UNWORLDLY, SPIRITUAL
2 a : lacking material substance **:** IMMATERIAL, INTANGIBLE **b :** marked by unusual delicacy or refinement ⟨this smallest, most *ethereal,* and daintiest of birds —William Beebe⟩
3 : relating to, containing, or resembling a chemical ether
— ethe·re·al·i·ty \-ˌthir-ē-ˈa-lə-tē\ *noun*
— ethe·re·al·iza·tion \-ē-ə-lə-ˈzā-shən\ *noun*
— ethe·re·al·ize \-ˈthir-ē-ə-ˌlīz\ *transitive verb*
— ethe·re·al·ly \-ē-ə-lē\ *adverb*
— ethe·re·al·ness *noun*

ether extract *noun* (circa 1900)
: the part of a complex organic material that is soluble in ether and consists chiefly of fats and fatty acids

ether·ize \ˈē-thə-ˌrīz\ *transitive verb* **-ized; -iz·ing** (1853)
1 : to treat or anesthetize with ether
2 : to make numb as if by anesthetizing
— ether·iza·tion \ˌē-thə-rə-ˈzā-shən\ *noun*
— ether·iz·er *noun*

eth·ic \ˈe-thik\ *noun* [Middle English *ethik,* from Middle French *ethique,* from Latin *ethice,* from Greek *ēthikē,* from *ēthikos*] (14th century)
1 *plural but singular or plural in construction* **:** the discipline dealing with what is good and bad and with moral duty and obligation
2 a : a set of moral principles or values **b :** a theory or system of moral values ⟨the present-day materialistic *ethic*⟩ **c** *plural but singular or plural in construction* **:** the principles of conduct governing an individual or a group ⟨professional *ethics*⟩ **d :** a guiding philosophy

eth·i·cal \ˈe-thi-kəl\ *also* **eth·ic** \-thik\ *adjective* [Middle English *etik,* from Latin *ethicus,* from Greek *ēthikos,* from *ethos* character — more at SIB] (1607)
1 : of or relating to ethics
2 : involving or expressing moral approval or disapproval
3 : conforming to accepted professional standards of conduct
4 *of a drug* **:** restricted to sale only on a doctor's prescription
synonym see MORAL
— eth·i·cal·i·ty \ˌe-thə-ˈka-lə-tē\ *noun*
— eth·i·cal·ly \ˈe-thi-k(ə-)lē\ *adverb*
— eth·i·cal·ness \-kəl-nəs\ *noun*

ethi·cian \e-ˈthi-shən\ *noun* (1629)
: ETHICIST

eth·i·cist \ˈe-thə-sist\ *noun* (circa 1890)
: a specialist in ethics

ethid·ium bromide \e-ˈthi-dē-əm-\ *noun* [*eth*yl + *-id* + *-ium*] (1963)
: a biological dye used especially to block nucleic acid synthesis (as in mitochondria) and to destroy trypanosomes

ethinyl *variant of* ETHYNYL
— used especially in pharmacology

ethinyl estradiol *noun* (1939)
: a very potent synthetic estrogen $C_{20}H_{24}O_2$ used orally

eth·i·on \ˈe-thē-ˌän\ *noun* [blend of *eth-* and *thion-*] (circa 1960)
: an organophosphate $C_9H_{22}O_4P_2S_4$ used as a pesticide

eth·i·on·amide \\ˌe-thē-ˈä-nə-ˌmīd\\ *noun* (1960)
: a compound $C_8H_{10}N_2S$ used against mycobacteria (as in tuberculosis and leprosy)
ethi·o·nine \\e-ˈthī-ə-ˌnēn\\ *noun* (1938)
: an amino acid $C_6H_{13}NO_2S$ that is the ethyl homologue of methionine and is biologically antagonistic to methionine
Ethi·op \\ˈē-thē-ˌäp\\ *or* **Ethi·ope** \\-ˌōp\\ *noun* [Middle English *Ethiope,* from Latin *Aethiops,* from Greek *Aithiops*] (13th century)
archaic **:** ETHIOPIAN
¹Ethi·o·pi·an \\ˌē-thē-ˈō-pē-ən\\ *noun* (13th century)
1 : a member of any of the mythical or actual peoples usually described by the ancient Greeks as dark-skinned and living far to the south
2 *archaic* **:** a black person
3 : a native or inhabitant of Ethiopia
²Ethiopian *adjective* (1578)
1 : of, relating to, or characteristic of the inhabitants or the country of Ethiopia
2 : of, relating to, or being the biogeographic region that includes Africa south of the Sahara, southern Arabia, and sometimes Madagascar and the adjacent islands
¹Ethi·o·pic \\-ˈä-pik, -ˈō-pik\\ *adjective* (1654)
1 : ETHIOPIAN
2 a : of, relating to, or constituting Ethiopic **b :** of, relating to, or constituting a group of related Semitic languages spoken in Ethiopia
²Ethiopic *noun* (1677)
1 : a Semitic language formerly spoken in Ethiopia and still used as the liturgical language of the Christian church in Ethiopia
2 : the Ethiopic group of Semitic languages
eth·moid \\ˈeth-ˌmȯid\\ *noun* [French *ethmoïde,* from Greek *ēthmoeidēs,* literally, like a strainer, from *ēthmos* strainer] (1842)
: a light spongy cubical bone forming much of the walls of the nasal cavity and part of those of the orbits
— ethmoid *same*\\ *or* **eth·moi·dal** \\(ˌ)eth-ˈmȯi-dᵊl\\ *adjective*
¹eth·nic \\ˈeth-nik\\ *adjective* [Middle English, from Late Latin *ethnicus,* from Greek *ethnikos* national, gentile, from *ethnos* nation, people; akin to Greek *ēthos* custom — more at SIB] (15th century)
1 : HEATHEN
2 a : of or relating to large groups of people classed according to common racial, national, tribal, religious, linguistic, or cultural origin or background ⟨*ethnic* minorities⟩ ⟨*ethnic* enclaves⟩ **b :** being a member of an ethnic group **c :** of, relating to, or characteristic of ethnics ⟨*ethnic* neighborhoods⟩ ⟨*ethnic* foods⟩
²ethnic *noun* (1941)
: a member of an ethnic group; *especially* **:** a member of a minority group who retains the customs, language, or social views of the group
eth·ni·cal \\ˈeth-ni-kəl\\ *adjective* (15th century)
1 : ETHNIC
2 : of or relating to ethnology **:** ETHNOLOGIC
— eth·ni·cal·ly \\-k(ə-)lē\\ *adverb*
ethnic cleansing *noun* (1992)
: the expulsion, imprisonment, or killing of ethnic minorities by a dominant majority group
eth·nic·i·ty \\eth-ˈni-sə-tē\\ *noun* (1950)
: ethnic quality or affiliation
ethno- *combining form* [French, from Greek *ethno-, ethn-,* from *ethnos*]
: race **:** people **:** cultural group ⟨*ethno*centric⟩
eth·no·bot·a·ny \\ˌeth-nō-ˈbä-tᵊn-ē, -ˈbät-nē\\ *noun* (1890)
: the plant lore of a race or people; *also* **:** the systematic study of such lore
— eth·no·bo·tan·i·cal \\-bə-ˈta-ni-kəl\\ *adjective*
— eth·no·bot·a·nist \\-ˈbä-tᵊn-ist, -ˈbät-nist\\ *noun*
eth·no·cen·tric \\ˌeth-nō-ˈsen-trik\\ *adjective* (1900)
: characterized by or based on the attitude that one's own group is superior
— eth·no·cen·tric·i·ty \\-sen-ˈtri-sə-tē\\ *noun*
— eth·no·cen·trism \\-ˈsen-ˌtri-zəm\\ *noun*
eth·nog·ra·phy \\eth-ˈnä-grə-fē\\ *noun* [French *ethnographie,* from *ethno-* + *-graphie* -graphy] (1834)
: the study and systematic recording of human cultures; *also* **:** a descriptive work produced from such research
— eth·nog·ra·pher \\-fər\\ *noun*
— eth·no·graph·ic \\ˌeth-nə-ˈgra-fik\\ *or* **eth·no·graph·i·cal** \\-fi-kəl\\ *adjective*
— eth·no·graph·i·cal·ly \\-fi-k(ə-)lē\\ *adverb*
eth·no·his·to·ry \\ˌeth-nō-ˈhis-t(ə-)rē\\ *noun* (1943)
: a study of the development of cultures
— eth·no·his·to·ri·an \\-(h)is-ˈtōr-ē-ən, -ˈtȯr-, -ˈtär-\\ *noun*
— eth·no·his·to·ric \\-(h)is-ˈtȯr-ik, -ˈtär-\\ *or* **eth·no·his·to·ri·cal** \\-i-kəl\\ *adjective*
eth·nol·o·gy \\eth-ˈnä-lə-jē\\ *noun* (circa 1828)
1 : a science that deals with the division of human beings into races and their origin, distribution, relations, and characteristics
2 : anthropology dealing chiefly with the comparative and analytical study of cultures **:** CULTURAL ANTHROPOLOGY
— eth·no·log·i·cal \\ˌeth-nə-ˈlä-ji-kəl\\ *also* **eth·no·log·ic** \\-jik\\ *adjective*
— eth·nol·o·gist \\eth-ˈnä-lə-jist\\ *noun*
eth·no·meth·od·ol·o·gy \\ˌeth-nō-ˌme-thə-ˈdä-lə-jē\\ *noun* (1967)
: a branch of sociology dealing with nonspecialists' commonsense understanding of the structure and organization of society
— eth·no·meth·od·ol·o·gist \\-ˈdä-lə-jist\\ *noun*
eth·no·mu·si·col·o·gy \\ˌeth-nō-ˌmyü-zi-ˈkä-lə-jē\\ *noun* (1950)
1 : the study of music that is outside the European art tradition
2 : the study of music in a sociocultural context
— eth·no·mu·si·co·log·i·cal \\-kə-ˈlä-ji-kəl\\ *adjective*
— eth·no·mu·si·col·o·gist \\-ˈkä-lə-jist\\ *noun*
eth·no·sci·ence \\ˈeth-nō-ˌsī-ən(t)s\\ *noun* (1961)
: the nature lore (as folk taxonomy of plants and animals) of primitive peoples
ethol·o·gy \\ē-ˈthä-lə-jē\\ *noun* (circa 1843)
1 : a branch of knowledge dealing with human ethos and with its formation and evolution
2 : the scientific and objective study of animal behavior especially under natural conditions
— etho·log·i·cal \\ˌē-thə-ˈlä-ji-kəl, ˌe-thə-\\ *adjective*
— ethol·o·gist \\ē-ˈthä-lə-jist\\ *noun*
ethos \\ˈē-ˌthäs\\ *noun* [New Latin, from Greek *ēthos* custom, character — more at SIB] (1851)
: the distinguishing character, sentiment, moral nature, or guiding beliefs of a person, group, or institution
eth·oxy \\e-ˈthäk-sē\\ *adjective* (circa 1909)
: relating to or containing the univalent group CH_3CH_2O- composed of ethyl united with oxygen
eth·yl \\ˈe-thəl\\ *noun* [German *Ethyl* (now *Äthyl*), from *Äther* ether + *-yl*] (1838)
: a univalent hydrocarbon group CH_3CH_2-
— eth·yl·ic \\e-ˈthi-lik\\ *adjective*
ethyl acetate *noun* (1874)
: a colorless fragrant volatile flammable liquid ester $C_4H_8O_2$ used especially as a solvent
ethyl alcohol *noun* (1869)
: ETHANOL
eth·yl·ben·zene \\ˌe-thil-ˈben-ˌzēn\\ *noun* [International Scientific Vocabulary] (1873)
: a liquid hydrocarbon C_8H_{10} that is made usually from benzene and ethylene and is used chiefly in the manufacture of styrene
ethyl cellulose *noun* (1936)
: any of various thermoplastic substances used especially in plastics and lacquers
ethyl chloride *noun* (circa 1891)
: a colorless pungent flammable gaseous or volatile liquid C_2H_5Cl used especially as a local surface anesthetic
eth·yl·ene \\ˈe-thə-ˌlēn\\ *noun* (circa 1852)
1 : a colorless flammable gaseous unsaturated hydrocarbon C_2H_4 that is found in coal gas, can be produced by pyrolysis of petroleum hydrocarbons, and occurs in plants functioning especially as a natural growth regulator that promotes the ripening of fruit
2 : a divalent hydrocarbon group C_2H_4 derived from ethane
— eth·yl·enic \\ˌe-thə-ˈlē-nik, -ˈle-nik\\ *adjective*
eth·yl·ene·di·amine·tet·ra·ac·e·tate \\ˌe-thə-ˌlēn-ˌdī-ə-ˌmēn-ˌte-trə-ˈa-sə-ˌtāt, -dī-ˌa-mən-\\ *noun* (1954)
: a salt of EDTA
eth·yl·ene·di·amine·tet·ra·ace·tic acid \\-(ə-)ˈsē-tik-\\ *noun* (1942)
: EDTA
ethylene di·bro·mide \\-(ˌ)dī-ˈbrō-ˌmīd\\ *noun* (circa 1929)
: a colorless toxic liquid compound $C_2H_4Br_2$ that is used chiefly as a fuel additive in leaded gasolines, that has been found to be strongly carcinogenic in laboratory animals, and that was used formerly in the U.S. as an agricultural pesticide — abbreviation *EDB*
ethylene glycol *noun* (1901)
: a thick liquid diol $C_2H_6O_2$ used especially as an antifreeze and in making polyester fibers
ethylene oxide *noun* (1898)
: a colorless flammable toxic compound C_2H_4O used especially in synthesis (as of ethylene glycol) and in sterilization and fumigation
ethyl ether *noun* (1878)
: ETHER 3a
eth·yne \\ˈe-ˌthīn, e-ˈ\\ *noun* [*ethyl* + ²*-ine*] (circa 1929)
: ACETYLENE
ethy·nyl \\e-ˈthī-nᵊl, ˈe-thə-ˌnil\\ *noun* (1929)
: a univalent unsaturated group $HC{\equiv}C-$ derived from acetylene by removal of one hydrogen atom
et·ic \\ˈe-tik\\ *adjective* [phon*etic*] (1954)
: of, relating to, or having linguistic or behavioral characteristics considered without regard to their structural significance — compare EMIC
-et·ic \\ˈe-tik\\ *adjective suffix* [Latin & Greek; Latin *-eticus,* from Greek *-etikos, -ētikos,* from *-etos, -ētos,* ending of certain verbals]
: -IC ⟨limn*etic*⟩ — often in adjectives corresponding to nouns ending in *-esis* ⟨*genetic*⟩
eti·o·late \\ˈē-tē-ə-ˌlāt\\ *transitive verb* **-lat·ed; -lat·ing** [French *étioler*] (1791)
1 : to bleach and alter the natural development of (a green plant) by excluding sunlight
2 a : to make pale **b :** to deprive of natural vigor **:** make feeble
— eti·o·la·tion \\ˌē-tē-ə-ˈlā-shən\\ *noun*
eti·o·log·ic \\ˌē-tē-ə-ˈlä-jik\\ *or* **eti·o·log·i·cal** \\-ji-kəl\\ *adjective* (circa 1753)
1 : assigning or seeking to assign a cause
2 : of or relating to etiology
— eti·o·log·i·cal·ly \\-ji-k(ə-)lē\\ *adverb*
eti·ol·o·gy \\ˌē-tē-ˈä-lə-jē\\ *noun, plural* **-gies** [Medieval Latin *aetiologia* statement of causes, from Greek *aitiologia,* from *aitia* cause] (circa 1555)
1 : CAUSE, ORIGIN; *specifically* **:** all of the causes of a disease or abnormal condition
2 : a branch of knowledge concerned with causes; *specifically* **:** a branch of medical science concerned with the causes and origins of diseases

et·i·quette \'e-ti-kət, -ˌket\ *noun* [French *étiquette,* literally, ticket — more at TICKET] (1750)
: the conduct or procedure required by good breeding or prescribed by authority to be observed in social or official life ◇

Eton collar \ˌē-tᵊn-\ *noun* [*Eton* College, English public school] (1887)
: a large stiff turnover collar

Eto·ni·an \ē-'tō-nē-ən\ *noun* (circa 1770)
: a student or former student of Eton College
— Etonian *adjective*

Eton jacket *noun* (1881)
: a short black jacket with long sleeves, wide lapels, and an open front

étouf·fée *also* **etouf·fee** \ˌā-tü-'fā\ *noun* [Louisiana French, from French *à l'étouffée* braised, from feminine of *étouffer,* past participle of *étouffer* to smother, from Old French *estofer,* alteration of *estoper* to plug up — more at ESTOP] (circa 1933)
: a Cajun stew of shellfish or chicken served over rice

Etru·ri·an \i-'tru̇r-ē-ən\ *noun* (1623)
: ETRUSCAN
— Etrurian *adjective*

[1]**Etrus·can** \i-'trəs-kən\ *adjective* [Latin *etruscus;* akin to Latin *Etruria,* ancient country] (1706)
: of, relating to, or characteristic of Etruria, the Etruscans, or the Etruscan language

[2]**Etruscan** *noun* (1773)
1 : the language of the Etruscans which is of unknown affiliation
2 : a native or inhabitant of ancient Etruria

-ette \'et, ˌet, ət, it\ *noun suffix* [Middle English, from Middle French, feminine diminutive suffix, from Old French *-ete* — more at -ET]
1 : little one ⟨kitchen*ette*⟩
2 : female ⟨farmer*ette*⟩

étude \'ā-ˌtüd, -ˌtyüd\ *noun* [French, literally, study, from Middle French *estude, estudie,* from Old French — more at STUDY] (circa 1837)
1 : a piece of music for the practice of a point of technique
2 : a composition built on a technical motive but played for its artistic value

etui \ā-'twē, 'ā-ˌ\ *noun, plural* **etuis** [French *étui*] (1611)
: a small ornamental case

et·y·mol·o·gise *British variant of* ETYMOLOGIZE

et·y·mol·o·gist \ˌe-tə-'mä-lə-jist\ *noun* (1635)
: a specialist in etymology

et·y·mol·o·gize \-ˌjīz\ *verb* **-gized; -giz·ing** (1530)
transitive verb
: to discover, formulate, or state an etymology for
intransitive verb
: to study or formulate etymologies

et·y·mol·o·gy \-jē\ *noun, plural* **-gies** [Middle English *ethimologie,* from Latin *etymologia,* from Greek, from *etymon* + *-logia* -logy] (14th century)
1 : the history of a linguistic form (as a word) shown by tracing its development since its earliest recorded occurrence in the language where it is found, by tracing its transmission from one language to another, by analyzing it into its component parts, by identifying its cognates in other languages, or by tracing it and its cognates to a common ancestral form in an ancestral language
2 : a branch of linguistics concerned with etymologies
— et·y·mo·log·i·cal \-mə-'lä-ji-kəl\ *adjective*
— et·y·mo·log·i·cal·ly \-k(ə-)lē\ *adverb*

et·y·mon \'e-tə-ˌmän\ *noun, plural* **-ma** \-mə\ *also* **-mons** [Latin, from Greek, literal meaning of a word according to its origin, from *etymos* true; akin to Greek *eteos* true — more at SOOTH] (circa 1576)
1 a : an earlier form of a word in the same language or an ancestral language **b :** a word in a foreign language that is the source of a particular loanword
2 : a word or morpheme from which words are formed by composition or derivation

eu- *combining form* [Middle English, from Latin, from Greek, from *eu* well, from neuter of *eys* good; perhaps akin to Latin *esse* to be]
1 a : well **:** easily ⟨*eu*clase⟩ — compare DYS- **b :** good ⟨*eu*daemonism⟩ — compare DYS-
2 : true ⟨*eu*chromatin⟩

eu·ca·lypt \'yü-kə-ˌlipt\ *noun* (1885)
: EUCALYPTUS

eu·ca·lyp·tol *also* **eu·ca·lyp·tole** \ˌyü-kə-'lip-ˌtȯl, -ˌtōl\ *noun* (1879)
: CINEOLE

eu·ca·lyp·tus \ˌyü-kə-'lip-təs\ *noun, plural* **-ti** \-ˌtī, -ˌtē\ *or* **-tus·es** [New Latin, genus name, from *eu-* + Greek *kalyptos* covered, from *kalyptein* to conceal; from the conical covering of the buds — more at HELL] (1801)
: any of a genus (*Eucalyptus*) of mostly Australian evergreen trees or rarely shrubs of the myrtle family that have rigid entire leaves and umbellate flowers and are widely cultivated for their gums, resins, oils, and woods

eu·cary·ote *variant of* EUKARYOTE

Eu·cha·rist \'yü-k(ə-)rəst\ *noun* [Middle English *eukarist,* from Middle French *eucharistie,* from Late Latin *eucharistia,* from Greek, Eucharist, gratitude, from *eucharistos* grateful, from *eu-* + *charizesthai* to show favor, from *charis* favor, grace, gratitude; akin to Greek *chairein* to rejoice — more at YEARN] (14th century)
1 : COMMUNION 2a
2 *Christian Science* **:** spiritual communion with God
— eu·cha·ris·tic \ˌyü-kə-'ris-tik\ *adjective, often capitalized*

[1]**eu·chre** \'yü-kər\ *noun* [origin unknown] (1841)
: a card game in which each player is dealt five cards and the player making trump must take three tricks to win a hand

[2]**euchre** *transitive verb* **eu·chred; eu·chring** \-k(ə-)riŋ\ (1847)
1 : to prevent from winning three tricks in euchre
2 : CHEAT, TRICK ⟨*euchred* out of their life savings —Pete Martin⟩

eu·chro·ma·tin \(ˌ)yü-'krō-mə-tən\ *noun* [International Scientific Vocabulary] (1932)
: the genetically active portion of chromatin that is largely composed of genes
— eu·chro·mat·ic \ˌyü-krō-'ma-tik\ *adjective*

eu·clase \'yü-ˌklās, -ˌklāz\ *noun* [French, from *eu-* (from Latin) + Greek *klasis* breaking, from *klan* to break] (1804)
: a mineral that consists of a brittle silicate of beryllium and aluminum in pale-yellow, green, or blue prismatic crystals and is used especially as a gemstone

eu·clid·e·an *also* **eu·clid·i·an** \yu̇-'kli-dē-ən\ *adjective, often capitalized* (1660)
: of, relating to, or based on the geometry of Euclid or a geometry with similar axioms

Euclidean algorithm *noun* (circa 1955)
: a method of finding the greatest common divisor of two numbers by dividing the larger by the smaller, the smaller by the remainder, the first remainder by the second remainder, and so on until exact division is obtained whence the greatest common divisor is the exact divisor — called also *Euclid's algorithm*

euclidean geometry *noun, often E capitalized* (circa 1865)
1 : geometry based on Euclid's axioms
2 : the geometry of a euclidean space

euclidean space *noun, often E capitalized* (1883)
: a space in which Euclid's axioms and definitions (as of straight and parallel lines and angles of plane triangles) apply

eu·crite \'yü-ˌkrīt\ *noun* [German *Eukrit,* from Greek *eukritos* easily discerned, from *eu-* + *kritos* separated, from *krinein* to separate — more at CERTAIN] (circa 1899)
1 : a stony meteorite composed essentially of plagioclase and pigeonite
2 : a rock consisting of a very basic gabbro
— eu·crit·ic \yü-'kri-tik\ *adjective*

eu·dae·mo·nism \yu̇-'dē-mə-ˌni-zəm\ *or* **eu·dai·mo·nism** \-'dī-\ *noun* [Greek *eudaimonia* happiness, from *eudaimōn* having a good attendant spirit, happy, from *eu-* + *daimōn* spirit — more at DEMON] (1827)
: a theory that the highest ethical goal is happiness and personal well-being
— eu·dae·mo·nist \-nist\ *noun*
— eu·dae·mo·nis·tic \-ˌdē-mə-'nis-tik\ *adjective*

eu·di·om·e·ter \ˌyü-dē-'ä-mə-tər\ *noun* [modification of Italian *eudiometro,* from Greek *eudia* fair weather (from *eu-* + *-dia* weather— akin to Latin *dies* day) + Italian *-metro* -meter, from Greek *metron* measure] (1777)
: an instrument for the volumetric measurement and analysis of gases
— eu·dio·met·ric \ˌyü-dē-ə-'me-trik\ *adjective*
— eu·dio·met·ri·cal·ly \-tri-k(ə-)lē\ *adverb*

eu·gen·ic \yu̇-'je-nik\ *adjective* [Greek *eugenēs* wellborn, from *eu-* + *-genēs* born — more at -GEN] (1883)
1 : relating to or fitted for the production of good offspring
2 : of or relating to eugenics
— eu·gen·i·cal·ly \-ni-k(ə-)lē\ *adverb*

eu·gen·i·cist \-'je-nə-sist\ *noun* (circa 1909)
: a student or advocate of eugenics

eu·gen·ics \yu̇-'je-niks\ *noun plural but singular or plural in construction* (1883)
: a science that deals with the improvement (as by control of human mating) of hereditary qualities of a race or breed

eu·gen·ist \yü-'je-nist\ *noun* (1908)
: EUGENICIST

◇ WORD HISTORY
etiquette The Middle French word *estiquette* or *etiquette* denoted a label attached to something, such as a bundle of documents, for description or identification; it was a derivative of a medieval French verb *estichier* or *estiquier* "to stick or attach" (itself ultimately borrowed from a verb akin to Modern English *stick* and *stitch* in the old Germanic language called Frankish). Middle French *etiquette* is the source of the English word *ticket,* and Modern French *étiquette* still retains the primary meaning "label." In the 16th century, the court of the Spanish king and Holy Roman Emperor Charles V borrowed Middle French *etiquette* as Spanish *etiqueta* to denote the written protocols describing order of precedence and other aspects of correct courtly behavior. The *etiqueta* eventually became the court ceremonies themselves as well as the documents that described them. Under the influence of Spanish, French *étiquette* also acquired the meaning "proper court behavior" and in the 18th century was borrowed into English. It then expanded in sense to cover all sorts of proper behavior required by society.

eu·ge·nol \ˈyü-jə-ˌnȯl, -ˌnōl\ *noun* [International Scientific Vocabulary *eugen-*, from New Latin *Eugenia,* genus of tropical trees] (1886) **:** a colorless aromatic liquid phenol $C_{10}H_{12}O_2$ found especially in clove oil and used commercially in flavors and perfumes and in dentistry as an analgesic

eu·geo·syn·cline \(ˌ)yü-ˌjē-ō-ˈsin-ˌklīn\ *noun* (1944) **:** a narrow rapidly subsiding geosyncline usually with volcanic materials mingled with clastic sediments
— **eu·geo·syn·cli·nal** \-(ˌ)sin-ˈklī-nᵊl\ *adjective*

eu·gle·na \yu̇-ˈglē-nə\ *noun* [New Latin, genus name, from *eu-* + Greek *glēnē* eyeball, socket of a joint] (circa 1889) **:** any of a genus (*Euglena*) of green freshwater flagellates often classed as algae

eu·gle·noid \-ˌnȯid\ *noun* (1885) **:** any of a taxon (Euglenoidina or Euglenophyta) of varied flagellates (as a euglena) that are typically green or colorless stigma-bearing solitary organisms with one or two flagella emerging from a well-defined gullet
— **euglenoid** *adjective*

euglenoid movement *noun* (1940) **:** writhing usually nonprogressive protoplasmic movement of plastic-bodied euglenoid flagellates

eu·glob·u·lin \yü-ˈglä-byə-lən\ *noun* [International Scientific Vocabulary] (1904) **:** a simple protein that does not dissolve in pure water

eu·he·mer·ism \yu̇-ˈhē-mə-ˌri-zəm, -ˈhe-mə-\ *noun* [*Euhemerus,* 4th century B.C. Greek mythographer] (1846) **:** interpretation of myths as traditional accounts of historical persons and events
— **eu·he·mer·ist** \-rist\ *noun*
— **eu·he·mer·is·tic** \-ˌhē-mə-ˈris-tik, -ˌhe-mə-\ *adjective*

eu·kary·ote \(ˌ)yü-ˈkar-ē-ˌōt, -ē-ət\ *noun* [New Latin *Eukaryotes,* proposed subdivision of protists, from *eu-* + *kary-* + *-otes,* plural noun suffix, from Greek *-ōtos* — more at -OTIC] (1943) **:** an organism composed of one or more cells containing visibly evident nuclei and organelles — compare PROKARYOTE
— **eu·kary·ot·ic** \-ˌkar-ē-ˈä-tik\ *adjective*

eu·la·chon \ˈyü-lə-ˌkän, -li-kən\ *noun, plural* **eulachon** *or* **eulachons** [Chinook Jargon *ulâkân*] (1807) **:** an anadromous marine food fish (*Thaleichthys pacificus*) of the north Pacific coast related to the smelt — called also *candlefish*

eulogise *British variant of* EULOGIZE

eu·lo·gist \ˈyü-lə-jist\ *noun* (1808) **:** one who eulogizes

eu·lo·gi·um \yu̇-ˈlō-jē-əm\ *noun, plural* **-gia** \-jē-ə\ *or* **-gi·ums** [Medieval Latin] (1621) **:** EULOGY

eu·lo·gize \ˈyü-lə-ˌjīz\ *transitive verb* **-gized; -giz·ing** (1810) **:** to speak or write in high praise of **:** EXTOL
— **eu·lo·giz·er** *noun*

eu·lo·gy \ˈyü-lə-jē\ *noun, plural* **-gies** [Middle English *euloge,* from Medieval Latin *eulogium,* from Greek *eulogia* praise, from *eu-* + *-logia* -logy] (15th century)
1 : a commendatory formal statement or set oration
2 : high praise
synonym see ENCOMIUM
— **eu·lo·gis·tic** \ˌyü-lə-ˈjis-tik\ *adjective*
— **eu·lo·gis·ti·cal·ly** \-ti-k(ə-)lē\ *adverb*

Eu·men·i·des \yu̇-ˈme-nə-ˌdēz\ *noun plural* [Latin, from Greek, literally, the gracious ones] **:** the Furies in Greek mythology

eu·nuch \ˈyü-nək, -nik\ *noun* [Middle English *eunuk,* from Latin *eunuchus,* from Greek *eunouchos,* from *eunē* bed + *echein* to have, have charge of — more at SCHEME] (15th century)
1 : a castrated man placed in charge of a harem or employed as a chamberlain in a palace
2 : a man or boy deprived of the testes or external genitals
3 : one that lacks virility or power ⟨political *eunuchs*⟩
— **eu·nuch·ism** \-nə-ˌki-zəm, -ni-\ *noun*

eu·nuch·oid \-nə-ˌkȯid, -ni-\ *noun* (1906) **:** a sexually deficient individual; *especially* **:** one lacking in sexual differentiation and tending toward the intersex state
— **eunuchoid** *adjective*

eu·on·y·mus \yu̇-ˈä-nə-məs\ *noun* [New Latin, genus name, from Latin *euonymos* spindle tree, from Greek *euōnymos,* from *euōnymos* having an auspicious name, from *eu-* + *onyma* name — more at NAME] (1767) **:** SPINDLE TREE

eu·pa·trid \yu̇-ˈpa-trəd, ˈyü-pə-\ *noun, plural* **eu·pat·ri·dae** \yu̇-ˈpa-trə-ˌdē\ *often capitalized* [Greek *eupatridēs,* from *eu-* + *patr-, patēr* father — more at FATHER] (1836) **:** one of the hereditary aristocrats of ancient Athens

eu·pep·tic \yu̇-ˈpep-tik\ *adjective* (1831)
1 : of, relating to, or having good digestion
2 : CHEERFUL, OPTIMISTIC

eu·phau·si·id \yu̇-ˈfȯ-zē-əd\ *noun* [New Latin *Euphausia,* genus of crustaceans] (circa 1893) **:** any of an order (Euphausiacea) of small usually luminescent malacostracan crustaceans that resemble shrimps and in some areas form an important element in marine plankton
— **euphausiid** *adjective*

euphemise *British variant of* EUPHEMIZE

eu·phe·mism \ˈyü-fə-ˌmi-zəm\ *noun* [Greek *euphēmismos,* from *euphēmos* auspicious, sounding good, from *eu-* + *phēmē* speech, from *phanai* to speak — more at BAN] (circa 1681) **:** the substitution of an agreeable or inoffensive expression for one that may offend or suggest something unpleasant; *also* **:** the expression so substituted
— **eu·phe·mist** \-mist\ *noun*
— **eu·phe·mis·tic** \ˌyü-fə-ˈmis-tik\ *adjective*
— **eu·phe·mis·ti·cal·ly** \-ti-k(ə-)lē\ *adverb*

eu·phe·mize \ˈyü-fə-ˌmīz\ *transitive verb* **-mized; -miz·ing** (1857) **:** to express or describe euphemistically ⟨*euphemize* death⟩
— **eu·phe·miz·er** \-ˌmī-zər\ *noun*

eu·phen·ics \yu̇-ˈfe-niks\ *noun plural but singular in construction* [*eu-* + *phen-* (from *phenotype*) + *-ics;* after English *genotype : eugenics*] (1963) **:** a science that deals with the biological improvement of human beings after birth
— **eu·phen·ic** \-nik\ *adjective*

eu·pho·ni·ous \yu̇-ˈfō-nē-əs\ *adjective* (1774) **:** pleasing to the ear
— **eu·pho·ni·ous·ly** *adverb*
— **eu·pho·ni·ous·ness** *noun*

eu·pho·ni·um \yu̇-ˈfō-nē-əm\ *noun* [Greek *euphōnos* + English *-ium* (as in *harmonium*)] (1865) **:** a brass instrument smaller than but resembling a tuba and having a range from B flat below the bass staff upward for three octaves

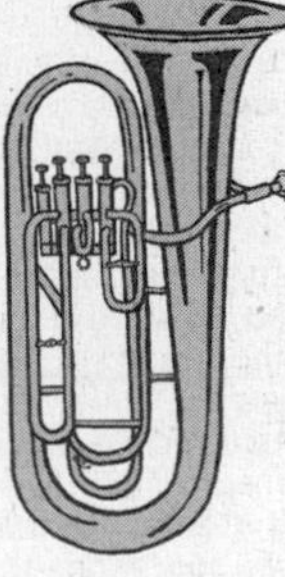
euphonium

eu·pho·ny \ˈyü-fə-nē\ *noun, plural* **-nies** [French *euphonie,* from Late Latin *euphonia,* from Greek *euphōnia,* from *euphōnos* sweet-voiced, musical, from *eu-* + *phōnē* voice — more at BAN] (circa 1616)
1 : pleasing or sweet sound; *especially* **:** the acoustic effect produced by words so formed or combined as to please the ear
2 : a harmonious succession of words having a pleasing sound
— **eu·phon·ic** \yu̇-ˈfä-nik\ *adjective*
— **eu·phon·i·cal·ly** \-ni-k(ə-)lē\ *adverb*

eu·phor·bia \yu̇-ˈfȯr-bē-ə\ *noun* [New Latin, alteration of Latin *euphorbea,* from *Euphorbus,* 1st century A.D. Greek physician] (14th century) **:** any of a large genus (*Euphorbia* of the family Euphorbiaceae) of plants that have a milky juice and flowers lacking a calyx and included in an involucre which surrounds a group of several staminate flowers and a central pistillate flower with 3-lobed pistils; *broadly* **:** SPURGE

eu·pho·ria \yu̇-ˈfȯr-ē-ə, -ˈfȯr-\ *noun* [New Latin, from Greek, from *euphoros* healthy, from *eu-* + *pherein* to bear — more at BEAR] (circa 1751) **:** a feeling of well-being or elation
— **eu·phor·ic** \-ˈfȯr-ik, -ˈfär-\ *adjective*
— **eu·phor·i·cal·ly** \-i-k(ə-)lē\ *adverb*

eu·pho·ri·ant \yu̇-ˈfȯr-ē-ənt, -ˈfȯr-\ *noun* (1947) **:** a drug that tends to induce euphoria
— **euphoriant** *adjective*

eu·pho·tic \yü-ˈfō-tik\ *adjective* [International Scientific Vocabulary] (1909) **:** of, relating to, or constituting the upper layers of a body of water into which sufficient light penetrates to permit growth of green plants

Eu·phros·y·ne \yu̇-ˈfrä-sᵊn-(ˌ)ē, -zᵊn-\ *noun* [Latin, from Greek *Euphrosynē*] **:** one of the three Graces

eu·phu·ism \ˈyü-fyə-ˌwi-zəm\ *noun* [*Euphues,* character in prose romances by John Lyly] (1592)
1 : an elegant Elizabethan literary style marked by excessive use of balance, antithesis, and alliteration and by frequent use of similes drawn from mythology and nature
2 : artificial elegance of language
— **eu·phu·ist** \-wist\ *noun*
— **eu·phu·is·tic** \ˌyü-fyə-ˈwis-tik\ *adjective*
— **eu·phu·is·ti·cal·ly** \-ti-k(ə-)lē\ *adverb*

eu·ploid \ˈyü-ˌplȯid\ *adjective* [International Scientific Vocabulary] (1926) **:** having a chromosome number that is an exact multiple of the monoploid number — compare ANEUPLOID
— **euploid** *noun*
— **eu·ploi·dy** \-ˌplȯi-dē\ *noun*

eup·nea *also* **eup·noea** \yüp-ˈnē-ə\ *noun* [New Latin, from Greek *eupnoia,* from *eupnous* breathing freely, from *eu-* + *pnein* to breathe — more at SNEEZE] (circa 1706) **:** normal respiration
— **eup·ne·ic** \-ˈnē-ik\ *adjective*

Eur- *or* **Euro-** *combining form* [*Europe*] **:** European and ⟨*Eur*american⟩ **:** European ⟨*Euro*centric⟩ **:** western European ⟨*Euro*communism⟩ **:** of the European Communities ⟨*Eu*rocrat⟩

Eur·amer·i·can \ˌyu̇r-ə-ˈmer-ə-kən\ *or* **Eu·ro–Amer·i·can** \ˌyu̇r-ō-ə-ˈmer-\ *adjective* (1941) **:** common to Europe and America

Eur·asian \yu̇-ˈrā-zhən, -shən\ *adjective* (1844)
1 : of a mixed European and Asian origin
2 : of or relating to Europe and Asia
— **Eurasian** *noun*

eu·re·ka \yu̇-ˈrē-kə\ *interjection* [Greek *heurēka* I have found, from *heuriskein* to find; from the exclamation attributed to Archimedes on discovering a method for determining the purity of gold — more at HEURISTIC] (1603)

— used to express triumph on a discovery ◆

eu·ro \'yůr-(,)ō\ *noun, plural* **euros** [Adnyamadhanha (Australian aboriginal language of South Australia) *yuru*] (1855)
: WALLAROO

Eu·ro·bond \'yůr-ō-,bänd\ *noun* (1966)
: a bond of a U.S. corporation that is sold outside the U.S. and that is denominated and paid for in dollars and yields interest in dollars

Eu·ro·cen·tric \,yůr-ə-'sen-trik\ *adjective* (1963)
: centered on Europe or the Europeans; *especially* **:** reflecting a tendency to interpret the world in terms of western and especially European values and experiences
— **Eu·ro·cen·trism** \-,tri-zəm\ *noun*

Eu·ro·com·mu·nism \,yůr-ō-'käm-yə-,ni-zəm\ *noun* (1976)
: the communism especially of western European Communist parties that is marked by a willingness to reach power through coalitions and by independence from Soviet leadership
— **Eu·ro·com·mu·nist** *noun or adjective*

Eu·ro·crat \'yůr-ə-,krat\ *noun* (1961)
: a staff member of the administrative commission of the European Communities

Eu·ro·cur·ren·cy \,yůr-ō-'kər-ən-sē, -'kə-rən-\ *noun* (1963)
: moneys (as of the U.S. and Japan) held outside their countries of origin and used in the money markets of Europe

Eu·ro·dol·lar \'yůr-ō-,dä-lər\ *noun* (1960)
: a U.S. dollar held as Eurocurrency

Eu·ro·pa \yů-'rō-pə\ *noun* [Latin, from Greek *Eurōpē*]
: a Phoenician princess carried off by Zeus in the form of a white bull and by him mother of Minos, Rhadamanthus, and Sarpedon

[1]**Eu·ro·pe·an** \,yůr-ə-'pē-ən, -'pēn\ *adjective* (1603)
: of, relating to, or characteristic of the continent of Europe or its people
— **Eu·ro·pe·an·iza·tion** \-,pē-ə-nə-'zā-shən\ *noun*
— **Eu·ro·pe·an·ize** \-'pē-ə-,nīz\ *transitive verb*

[2]**European** *noun* (1632)
1 : a native or inhabitant of Europe
2 : a person of European descent

European bison *noun* (1860)
: WISENT

European chafer *noun* (1947)
: an Old World beetle (*Rhizotragus majalis* synonym *Amphimallon majalis*) established in parts of eastern North America where its larva is a destructive pest on the roots of turf grasses

European corn borer *noun* (1920)
: an Old World moth (*Ostrinia nubilalis*) that is widespread in eastern North America where its larva is a major pest especially in the stems and crowns of Indian corn, dahlias, and potatoes

European plan *noun* (1834)
: a hotel plan whereby the daily rates cover only the cost of the room — compare AMERICAN PLAN

European red mite *noun* (1940)
: a small bright or brownish red oval mite (*Panonychus ulmi*) that is a destructive orchard pest

eu·ro·pi·um \yů-'rō-pē-əm\ *noun* [New Latin, from *Europa* Europe] (1901)
: a bivalent and trivalent metallic element of the rare-earth group found in monazite sand — see ELEMENT table

Eu·ro·po·cen·tric \yů-,rō-pə-'sen-trik\ *adjective* (1926)
: EUROCENTRIC
— **Eu·ro·po·cen·trism** \-,tri-zəm\ *noun*

eury- *combining form* [New Latin, from Greek, from *eurys;* akin to Sanskrit *uru* broad, wide]
: broad **:** wide ⟨*eury*haline⟩

eu·ry·bath·ic \,yůr-i-'ba-thik\ *adjective* [*eury-* + Greek *bathos* depth] (1902)
: capable of living on the bottom in both deep and shallow water

Eu·ryd·i·ce \yů-'ri-də-(,)sē\ *noun* [Latin, from Greek *Eurydikē*]
: the wife of Orpheus whom he attempts to bring back from Hades

eu·ry·ha·line \,yůr-i-'hā-,līn, -'ha-\ *adjective* [International Scientific Vocabulary *eury-* + Greek *halinos* of salt, from *hals* salt — more at SALT] (1888)
: able to live in waters of a wide range of salinity ⟨*euryhaline* crabs⟩

eu·ryp·ter·id \yů-'rip-tə-rəd\ *noun* [ultimately from Greek *eury-* + *pteron* wing — more at FEATHER] (1871)
: any of an order (Eurypterida) of usually large aquatic Paleozoic arthropods related to the horseshoe crabs
— **eurypterid** *adjective*

eu·ry·ther·mal \,yůr-i-'thər-məl\ *adjective* [International Scientific Vocabulary] (1881)
: tolerating a wide range of temperature ⟨*eurythermal* animals⟩

eu·ry·ther·mic \-mik\ *adjective* [International Scientific Vocabulary] (1903)
: EURYTHERMAL

eu·ry·ther·mous \-məs\ *adjective* [International Scientific Vocabulary] (1940)
: EURYTHERMAL

eu·ryth·mic *or* **eu·rhyth·mic** \yů-'rith-mik\ *adjective* (1921)
1 : HARMONIOUS
2 : of or relating to eurythmy or eurythmics

eu·ryth·mics *or* **eu·rhyth·mics** \-miks\ *noun plural but singular or plural in construction* (1912)
: the art of harmonious bodily movement especially through expressive timed movements in response to improvised music

eu·ryth·my *or* **eu·rhyth·my** \-mē\ *noun* [German *Eurhythmie,* from Latin *eurythmia* rhythmical movement, from Greek, from *eurythmos* rhythmical, from *eu-* + *rhythmos* rhythm] (1949)
: a system of harmonious body movement to the rhythm of spoken words

eu·ry·top·ic \,yůr-i-'tä-pik\ *adjective* [International Scientific Vocabulary *eury-* + Greek *topos* place] (circa 1945)
: tolerant of wide variation in one or more physical factors of the environment

eu·sta·chian tube \yů-'stā-sh(ē-)ən- *also* -'stā-kē-ən-\ *noun, often E capitalized* [Bartolommeo *Eustachio*] (1741)
: a bony and cartilaginous tube connecting the middle ear with the nasopharynx and equalizing air pressure on both sides of the tympanic membrane — called also *auditory tube;* see EAR illustration

eu·stat·ic \yů-'sta-tik\ *adjective* [International Scientific Vocabulary] (1906)
: relating to or characterized by worldwide change of sea level

eu·stele \'yü-,stēl, yů-'stē-lē\ *noun* (circa 1920)
: a stele typical of dicotyledonous plants that consists of vascular bundles of xylem and phloem strands with parenchymal cells between the bundles

eu·tec·tic \yů-'tek-tik\ *adjective* [Greek *eutēktos* easily melted, from *eu-* + *tēktos* melted, from *tēkein* to melt — more at THAW] (1884)
1 *of an alloy or solution* **:** having the lowest melting point possible
2 : of or relating to a eutectic alloy or solution or its melting or freezing point
— **eutectic** *noun*
— **eu·tec·toid** \-,tȯid\ *adjective or noun*

Eu·ter·pe \yů-'tər-pē\ *noun* [Latin, from Greek *Euterpē*]
: the Greek Muse of music

eu·tha·na·sia \,yü-thə-'nā-zh(ē-)ə\ *noun* [Greek, easy death, from *euthanatos,* from *eu-* + *thanatos* death — more at THANATOS] (1869)
: the act or practice of killing or permitting the death of hopelessly sick or injured individuals (as persons or domestic animals) in a relatively painless way for reasons of mercy
— **eu·tha·na·sic** \-zik, -sik\ *adjective*

eu·than·a·tize \yů-'tha-nə-,tīz\ *also* **eu·tha·nize** \'yü-thə-,nīz\ *transitive verb* **-tized** *also* **-nized; -tiz·ing** *also* **-niz·ing** [Greek *euthanatos*] (1873)
: to subject to euthanasia

eu·then·ics \yů-'the-niks\ *noun plural but singular or plural in construction* [Greek *euthenein* to thrive, from *eu-* + *-thenein*] (1905)
: a science that deals with development of human well-being by improvement of living conditions
— **eu·the·nist** \yů-'the-nist, 'yü-thə-\ *noun*

eu·the·ri·an \yů-'thir-ē-ən\ *adjective* [ultimately from New Latin *eu-* + Greek *thērion* beast — more at TREACLE] (1880)
: of or relating to a major division (Eutheria) of mammals comprising the placental mammals
— **eutherian** *noun*

eu·thy·roid \(,)yü-'thī-,rȯid\ *adjective* (1924)
: characterized by normal thyroid function

eu·tro·phic \yů-'trō-fik\ *adjective* [probably from German *Eutroph* eutrophic, from Greek *eutrophos* well-nourished, nourishing, from *eu-* + *trephein* to nourish] (1928)
of a body of water **:** characterized by the state resulting from eutrophication — compare MESOTROPHIC, OLIGOTROPHIC
— **eu·tro·phy** \'yü-trə-fē\ *noun*

eu·tro·phi·ca·tion \yů-,trō-fə-'kā-shən, ,yů-trə-fə-\ *noun* (1946)
: the process by which a body of water becomes enriched in dissolved nutrients (as phosphates) that stimulate the growth of aquatic plant life usually resulting in the depletion of dissolved oxygen

evac·u·ate \i-'va-kyə-,wāt\ *verb* **-at·ed; -at·ing** [Middle English, to draw off morbid humors, from Latin *evacuatus,* past participle of *evacuare* to empty, from *e-* + *vacuus* empty] (15th century)
transitive verb
1 : to remove the contents of **:** EMPTY
2 : to discharge from the body as waste **:** VOID
3 : to remove something (as gas or water) from especially by pumping

◇ WORD HISTORY

eureka When people make an unexpected discovery or have a sudden flash of inspiration, a traditional response is an exclamation of "Eureka!" Whether they know it or not, the elated discoverers are reenacting a legendary event in the life of the Greek mathematician and mechanical inventor Archimedes (ca. 287–212 B.C.). While wrestling with the problem of how to determine the purity of gold, he made the sudden realization that the buoyancy of an object placed in water is equal in magnitude to the weight of the water displaced by the object. According to the most popular and elaborate version of the legend, he made this discovery while at a public bathhouse, whereupon he leapt out of his bath, exclaimed "*Heurēka! Heurēka!*" ("I have found (it)") and ran naked through the streets of Syracuse all the way home. The absence of a contemporary source for the anecdote has done nothing to diminish its popularity over the centuries.

4 a : to remove especially from a military zone or dangerous area **b :** to withdraw from military occupation of **c :** VACATE ⟨were ordered to *evacuate* the building⟩
intransitive verb
1 : to withdraw from a place in an organized way especially for protection
2 : to pass urine or feces from the body
— evac·u·a·tive \-ˌwā-tiv\ *adjective*
evac·u·a·tion \i-ˌva-kyə-ˈwā-shən\ *noun* (14th century)
1 : the act or process of evacuating
2 : something evacuated or discharged
evac·u·ee \i-ˌva-kyə-ˈwē\ *noun* (1918)
: an evacuated person
evade \i-ˈvād\ *verb* **evad·ed; evad·ing** [Middle French & Latin; Middle French *evader,* from Latin *evadere,* from *e-* + *vadere* to go, walk — more at WADE] (1513)
intransitive verb
1 : to slip away
2 : to take refuge in evasion
transitive verb
1 : to elude by dexterity or stratagem
2 a : to avoid facing up to ⟨*evaded* the real issues⟩ **b :** to avoid the performance of **:** DODGE, CIRCUMVENT; *especially* **:** to fail to pay (taxes) **c :** to avoid answering directly **:** turn aside
3 : to be elusive to **:** BAFFLE ⟨the simple, personal meaning *evaded* them —C. D. Lewis⟩
synonym see ESCAPE
— evad·able \-ˈvā-də-bəl\ *adjective*
— evad·er *noun*
evag·i·na·tion \i-ˌva-jə-ˈnā-shən\ *noun* [Late Latin *evagination-, evaginatio,* act of unsheathing, from Latin *evaginatus,* past participle of *evaginare* to unsheathe, from *e-* + *vagina* sheath] (circa 1676)
1 : an act or instance of everting
2 : a product of eversion **:** OUTGROWTH
eval·u·ate \i-ˈval-yə-ˌwāt\ *transitive verb* **-at·ed; -at·ing** [back-formation from *evaluation*] (1842)
1 : to determine or fix the value of
2 : to determine the significance, worth, or condition of usually by careful appraisal and study
synonym see ESTIMATE
— eval·u·a·tion \-ˌval-yə-ˈwā-shən\ *noun*
— eval·u·a·tive \-ˈval-yə-ˌwā-tiv\ *adjective*
— eval·u·a·tor \-ˌwā-tər\ *noun*
ev·a·nesce \ˌe-və-ˈnes\ *intransitive verb* **-nesced; -nesc·ing** [Latin *evanescere* — more at VANISH] (1822)
: to dissipate like vapor
ev·a·nes·cence \ˌe-və-ˈne-sᵊn(t)s\ *noun* (1751)
1 : the process or fact of evanescing
2 : evanescent quality
ev·a·nes·cent \-sᵊnt\ *adjective* [Latin *evanescent-, evanescens,* present participle of *evanescere*] (1717)
: tending to vanish like vapor
synonym see TRANSIENT
¹evan·gel \i-ˈvan-jəl\ *noun* [Middle English *evangile,* from Middle French, from Late Latin *evangelium,* from Greek *euangelion* good news, gospel, from *euangelos* bringing good news, from *eu-* + *angelos* messenger] (14th century)
: GOSPEL
²evangel *noun* (1614)
: EVANGELIST
evan·gel·i·cal \ˌē-ˌvan-ˈje-li-kəl, ˌe-vən-\ *also* **evan·gel·ic** \-ik\ *adjective* (1531)
1 : of, relating to, or being in agreement with the Christian gospel especially as it is presented in the four Gospels
2 : PROTESTANT
3 : emphasizing salvation by faith in the atoning death of Jesus Christ through personal conversion, the authority of Scripture, and the importance of preaching as contrasted with ritual
4 a *capitalized* **:** of or relating to the Evangelical Church in Germany **b** *often capitalized* **:** of, adhering to, or marked by fundamentalism **:** FUNDAMENTALIST **c** *often capitalized* **:** LOW CHURCH
5 : marked by militant or crusading zeal **:** EVANGELISTIC ⟨the *evangelical* ardor of the movement's leaders —Amos Vogel⟩
— Evan·gel·i·cal·ism \-li-kə-ˌli-zəm\ *noun*
— evan·gel·i·cal·ly \-li-k(ə-)lē\ *adverb*
Evangelical *noun* (1532)
: one holding evangelical principles or belonging to an evangelical party or church
evan·ge·lism \i-ˈvan-jə-ˌli-zəm\ *noun* (circa 1626)
1 : the winning or revival of personal commitments to Christ
2 : militant or crusading zeal
— evan·ge·lis·tic \-ˌvan-jə-ˈlis-tik\ *adjective*
— evan·ge·lis·ti·cal·ly \-ti-k(ə-)lē\ *adverb*
evan·ge·list \i-ˈvan-jə-list\ *noun* (13th century)
1 *often capitalized* **:** a writer of any of the four Gospels
2 : a person who evangelizes; *specifically* **:** a Protestant minister or layman who preaches at special services
evan·ge·lize \i-ˈvan-jə-ˌlīz\ *verb* **-lized; -liz·ing** (14th century)
transitive verb
1 : to preach the gospel to
2 : to convert to Christianity
intransitive verb
: to preach the gospel
— evan·ge·li·za·tion \-ˌvan-jə-lə-ˈzā-shən\ *noun*
evap·o·rate \i-ˈva-p(ə-)ˌrāt\ *verb* **-rat·ed; -rat·ing** [Middle English, from Latin *evaporatus,* past participle of *evaporare,* from *e-* + *vapor* steam, vapor] (15th century)
transitive verb
1 a : to convert into vapor; *also* **:** to dissipate or draw off in vapor or fumes **b :** to deposit (as a metal) in the form of a film by sublimation
2 a : to expel moisture from **b :** EXPEL ⟨*evaporate* electrons from a hot wire⟩
intransitive verb
1 a : to pass off in vapor or in invisible minute particles **b** (1) **:** to pass off or away **:** DISAPPEAR ⟨my despair *evaporated* —J. F. Wharton⟩ (2) **:** to diminish quickly
2 : to give forth vapor
— evap·o·ra·tion \-ˌva-pə-ˈrā-shən\ *noun*
— evap·o·ra·tive \-ˈva-pə-ˌrā-tiv\ *adjective*
— evap·o·ra·tor \-ˈva-pə-ˌrā-tər\ *noun*
evaporated milk *noun* (1870)
: unsweetened milk concentrated by partial evaporation
evap·o·rite \i-ˈva-pə-ˌrīt\ *noun* [*evapor*ation + ¹*-ite*] (1924)
: a sedimentary rock (as gypsum) that originates by evaporation of seawater in an enclosed basin
— evap·o·rit·ic \-ˌva-pə-ˈri-tik\ *adjective*
evapo·trans·pi·ra·tion \i-ˈva-pō-ˌtran(t)-spə-ˈrā-shən\ *noun* [*evapo*ration + *transpiration*] (1938)
: loss of water from the soil both by evaporation and by transpiration from the plants growing thereon
eva·sion \i-ˈvā-zhən\ *noun* [Middle English, from Middle French or Late Latin; Middle French, from Late Latin *evasion-, evasio,* from Latin *evadere* to evade] (15th century)
1 : a means of evading **:** DODGE
2 : the act or an instance of evading **:** ESCAPE ⟨suspected of tax *evasion*⟩
eva·sive \i-ˈvā-siv, -ziv\ *adjective* (1725)
: tending or intended to evade **:** EQUIVOCAL ⟨*evasive* answers⟩
— eva·sive·ly *adverb*
— eva·sive·ness *noun*
eve \ˈēv\ *noun* [Middle English *eve, even*] (13th century)
1 : EVENING
2 : the evening or the day before a special day
3 : the period immediately preceding
Eve \ˈēv\ *noun* [Old English *Ēfe,* from Late Latin *Eva,* from Hebrew *Ḥawwāh*]
: the first woman and wife of Adam
¹even \ˈē-vən\ *noun* [Middle English *even, eve,* from Old English *ǣfen*] (before 12th century)
archaic **:** EVENING
²even \ˈē-vən, -bᵊm\ *adjective* [Middle English, from Old English *efen;* akin to Old High German *eban* even] (before 12th century)
1 a : having a horizontal surface **:** FLAT ⟨*even* ground⟩ **b :** being without break, indentation, or irregularity **:** SMOOTH **c :** being in the same plane or line
2 a : free from variation **:** UNIFORM ⟨his disposition was *even*⟩ **b :** LEVEL 4
3 a : EQUAL, FAIR ⟨an *even* exchange⟩ **b** (1) **:** leaving nothing due on either side **:** SQUARE ⟨we will not be *even* until you repay my visit⟩ (2) **:** fully revenged **c :** being in equilibrium **:** BALANCED; *specifically* **:** showing neither profit nor loss **d** *obsolete* **:** CANDID
4 a : being any of the integers (as −2, 0, and +2) that are exactly divisible by two **b :** marked by an even number **c :** being a mathematical function such that $f(x) = f(-x)$ where the value remains unchanged if the sign of the independent variable is reversed
5 : EXACT, PRECISE ⟨an *even* dollar⟩
6 : as likely as not **:** FIFTY-FIFTY ⟨an *even* chance of winning⟩
synonym see LEVEL, STEADY
— even·ly *adverb*
— even·ness \-vən-nəs\ *noun*
— on an even keel *also* **on even keel :** in a sound or stable condition
³even *adverb* [Middle English, from Old English *efne,* from *efen,* adjective] (before 12th century)
1 a : EXACTLY, PRECISELY **b :** to a degree that extends **:** FULLY, QUITE ⟨faithful *even* unto death⟩ **c :** at the very time
2 a — used as an intensive to emphasize the identity or character of something ⟨he looked content, *even* happy⟩ **b** — used as an intensive to stress an extreme or highly unlikely condition or instance ⟨so simple *even* a child can do it⟩ **c** — used as an intensive to stress the comparative degree ⟨he did *even* better⟩ **d** — used as an intensive to indicate a small or minimum amount ⟨didn't *even* try⟩
⁴even *verb* **evened; even·ing** \ˈēv-niŋ, ˈē-və-\ (13th century)
transitive verb
: to make even
intransitive verb
: to become even
— even·er \-nər\ *noun*
even·fall \ˈē-vən-ˌfȯl\ *noun* (1814)
: the beginning of evening **:** DUSK
even·hand·ed \ˌē-vən-ˈhan-dəd\ *adjective* (1605)
: FAIR, IMPARTIAL
— even·hand·ed·ly *adverb*
— even·hand·ed·ness *noun*
¹eve·ning \ˈēv-niŋ\ *noun, often attributive* [Middle English, from Old English *ǣfnung,* from *ǣfnian* to grow toward evening, from *ǣfen* evening; akin to Old High German *āband* evening and perhaps to Greek *epi* on] (before 12th century)
1 a : the latter part and close of the day and early part of the night **b** *chiefly Southern & Midland* **:** AFTERNOON **c :** the period from sunset or the evening meal to bedtime
2 : the latter portion
3 : the period of an evening's entertainment
²evening *adjective* (1797)
: suitable for formal or semiformal evening social occasions ⟨*evening* dress⟩ ⟨*evening* clothes⟩

evening prayer *noun, often E&P capitalized* (1598)
: the daily evening office of the Anglican liturgy

evening primrose *noun* (1806)
: any of several dicotyledonous plants of a family (Onagraceae, the evening-primrose family) and especially of the type genus (*Oenothera*); *especially* **:** a coarse biennial herb (*O. biennis*) with yellow flowers that open in the evening

eve·nings \'ēv-niŋz\ *adverb* (1652)
: in the evening repeatedly **:** on any evening ⟨goes bowling *evenings*⟩

evening star *noun* (1535)
1 : a bright planet (as Venus) seen especially in the western sky at or after sunset
2 : a planet that rises before midnight

even money *noun* (1891)
: a situation in wagering in which the odds are even

even permutation *noun* (circa 1932)
: a permutation that is produced by the successive application of an even number of interchanges of pairs of elements

even so *adverb* (1948)
: in spite of that **:** NEVERTHELESS

even·song \'ē-vən-,sȯŋ\ *noun, often capitalized* [Middle English, from Old English *ǣfensang,* from *ǣfen* evening + *sang* song] (before 12th century)
1 : VESPERS 1
2 : EVENING PRAYER

event \i-'vent\ *noun* [Middle French or Latin; Middle French, from Latin *eventus,* from *evenire* to happen, from *e-* + *venire* to come — more at COME] (1573)
1 a *archaic* **:** OUTCOME **b :** the final outcome or determination of a legal action **c :** a postulated outcome, condition, or eventuality ⟨in the *event* that I am not there, call the house⟩
2 a : something that happens **:** OCCURRENCE **b :** a noteworthy happening **c :** a social occasion or activity
3 : any of the contests in a program of sports
4 : the fundamental entity of observed physical reality represented by a point designated by three coordinates of place and one of time in the space-time continuum postulated by the theory of relativity
5 : a subset of the possible outcomes of an experiment
synonym see OCCURRENCE
— event·less \-ləs\ *adjective*
— at all events : in any case
— in any event : in any case
— in the event *British* **:** as it turns out

event·ful \i-'vent-fəl\ *adjective* (1600)
1 : full of or rich in events
2 : MOMENTOUS
— event·ful·ly \-fə-lē\ *adverb*
— event·ful·ness *noun*

event horizon *noun* (1969)
: the surface of a black hole **:** the boundary of a black hole at which the escape velocity equals the speed of light and beyond which nothing can escape from within it

even·tide \'ē-vən-,tīd\ *noun* (before 12th century)
: the time of evening **:** EVENING

even·tu·al \i-'ven(t)-sh(ə-)wəl, -'ven-chəl\ *adjective* (1683)
1 *archaic* **:** CONTINGENT, CONDITIONAL
2 : taking place at an unspecified later time **:** ultimately resulting ⟨they counted on our *eventual* success⟩
synonym see LAST

even·tu·al·i·ty \i-,ven-chə-'wa-lə-tē\ *noun, plural* **-ties** (1759)
: a possible event or outcome **:** POSSIBILITY

even·tu·al·ly \-ē\ *adverb* (1680)
: at an unspecified later time **:** in the end

even·tu·ate \i-'ven-chə-,wāt\ *intransitive verb* **-at·ed; -at·ing** (1789)
: to come out finally **:** RESULT, COME ABOUT

ev·er \'e-vər\ *adverb* [Middle English, from Old English *ǣfre*] (before 12th century)
1 : ALWAYS ⟨*ever* striving to improve⟩ ⟨the *ever*-increasing population⟩
2 a : at any time ⟨more than *ever* before⟩ **b :** in any way ⟨how can I *ever* thank you⟩
3 — used as an intensive ⟨looks *ever* so angry⟩ ⟨am I *ever* happy to see you⟩

ev·er·bloom·ing \,e-vər-'blü-miŋ\ *adjective* (circa 1891)
: blooming more or less continuously throughout the growing season

ev·er·dur·ing \,e-vər-'du̇r-iŋ, -'dyu̇r-\ *adjective* (15th century)
archaic **:** EVERLASTING

ev·er·glade \'e-vər-,glād\ *noun* [the *Everglades,* Fla.] (1823)
: a swampy grassland especially in southern Florida usually containing saw grass and at least seasonally covered by slowly moving water — usually used in plural

[1]ev·er·green \'e-vər-,grēn\ *noun* (1644)
1 : an evergreen plant; *also* **:** CONIFER
2 *plural* **:** twigs and branches of evergreen plants used for decoration
3 : something that retains its freshness, interest, or popularity

[2]evergreen *adjective* (1671)
1 : having foliage that remains green and functional through more than one growing season — compare DECIDUOUS 1
2 : retaining freshness or interest **:** PERENNIAL

evergreen oak *noun* (circa 1682)
: any of various oaks (as a live oak, a holm oak, or a tan oak) with foliage that persists for two years so that the plant is more or less continuously green

[1]ev·er·last·ing \,e-vər-'las-tiŋ\ *adjective* (13th century)
1 : lasting or enduring through all time **:** ETERNAL
2 a (1) **:** continuing long or indefinitely (2) **:** having or being flowers or foliage that retain form or color for a long time when dried **b :** tediously persistent ⟨the *everlasting* sympathy-seeker who demands attention —H. A. Overstreet⟩
3 : wearing indefinitely
— ev·er·last·ing·ly \-tiŋ-lē\ *adverb*
— ev·er·last·ing·ness *noun*

[2]everlasting *noun* (14th century)
1 : ETERNITY ⟨from *everlasting*⟩
2 *capitalized* **:** GOD 1 — used with *the*
3 a : any of several chiefly composite plants (as cudweed) with flowers that can be dried without loss of form or color — compare PEARLY EVERLASTING **b :** the flower of an everlasting

ev·er·more \,e-vər-'mōr, -'mȯr\ *adverb* (13th century)
1 : FOREVER, ALWAYS
2 : in the future

ever·sion \i-'vər-zhən, -shən\ *noun* (1751)
1 : the act of turning inside out **:** the state of being turned inside out ⟨*eversion* of the bladder⟩
2 : the condition (as of the foot) of being turned or rotated outward
— ever·si·ble \-'vər-sə-bəl\ *adjective*

evert \i-'vərt\ *transitive verb* [Latin *evertere,* from *e-* + *vertere* to turn — more at WORTH] (1538)
1 : OVERTHROW, UPSET
2 : to subject to eversion

ev·ery \'ev-rē\ *adjective* [Middle English *everich, every,* from Old English *ǣfre ǣlc,* from *ǣfre* ever + *ǣlc* each] (before 12th century)
1 a : being each individual or part of a group without exception **b :** being each in a series or succession ⟨*every* few days⟩
2 *obsolete* **:** being all taken severally
3 : being each within a range of possibilities ⟨was given *every* chance⟩
4 : COMPLETE, ENTIRE ⟨we have *every* confidence in her⟩
— every now and then *or* **every now and again** *or* **every so often :** at intervals **:** OCCASIONALLY

ev·ery·body \'ev-ri-,bə-dē, -,bä-\ *pronoun* (1530)
: every person **:** EVERYONE ■

ev·ery·day \,ev-rē-,dā\ *adjective* (circa 1623)
: encountered or used routinely or typically **:** ORDINARY ⟨*everyday* clothes⟩
— ev·ery·day·ness \-'dā-nəs\ *noun*

ev·ery·man \'ev-rē-,man\ *noun, often capitalized* [*Everyman,* allegorical character in *The Summoning of Everyman,* 15th century English morality play] (1906)
: the typical or ordinary person

ev·ery·one \-(,)wən\ *pronoun* (13th century)
: EVERYBODY
usage see EVERYBODY

ev·ery·place \-,plās\ *adverb* (circa 1917)
: EVERYWHERE

ev·ery·thing \'ev-rē-,thiŋ\ *pronoun* (14th century)
1 a : all that exists **b :** all that relates to the subject
2 : all that is important ⟨you mean *everything* to me⟩
3 : all sorts of other things — used to indicate related but unspecified events, facts, or conditions ⟨all the pains and colds and *everything* —E. B. White⟩

ev·ery·where \'ev-rē-,(h)wer, -,(h)war\ *adverb* (13th century)
: in every place or part

every which way *adverb* [probably by folk etymology from Middle English *everich way* every way] (1824)
1 : in every direction
2 : in a disorderly manner **:** IRREGULARLY ⟨toys scattered about *every which way*⟩

ev·ery·wom·an \'ev-rē-,wu̇-mən\ *noun, often capitalized* [after *everyman*] (1945)
: the typical or ordinary woman

evict \i-'vikt\ *transitive verb* [Middle English, from Late Latin *evictus,* past participle of *evincere,* from Latin, to vanquish, win a point — more at EVINCE] (15th century)

□ USAGE
everybody The indefinite pronouns *everybody* and *everyone* share with other indefinite pronouns the characteristic of taking a singular verb and, more often than not, a plural pronoun in reference. The use of the plural pronoun—*they, their, them*—has traditionally been disapproved by those who insist that agreement in number is more important than avoidance of gender-specific third-person singular pronouns. But the use of the plural pronoun has long been established ⟨*everybody* should marry as soon as they can do it to advantage —Jane Austen (1814)⟩ ⟨*everybody* endeavouring to excuse themselves —Samuel Pepys (1668)⟩. The context of the Jane Austen quotation shows why there has been less resistance to the plural pronoun with *everybody* than with the other indefinite pronouns—the meaning is so clearly plural that a singular pronoun can seem silly. For instance, an unthinking copy editor changed Margaret Mitchell's *they* in this sentence from *Gone With the Wind:* "Everyone was very polite and kind to her because he felt sorry for her . . ." *He* simply sounds out of place in this context. Various kinds of mixed usage still occur ⟨*everyone* allegedly being entitled to his ignorance —John Simon⟩ ⟨*everybody* seems to be enjoying themselves —Truman Capote⟩.

1 a : to recover (property) from a person by legal process **b :** to put (a tenant) out by legal process
2 : to force out **:** EXPEL
synonym see EJECT
— **evic·tion** \-ˈvik-shən\ *noun*
— **evic·tor** \-ˈvik-tər\ *noun*

evict·ee \i-ˌvik-ˈtē\ *noun* (1879)
: an evicted person

[1]ev·i·dence \ˈe-və-dən(t)s, -və-ˌden(t)s\ *noun* (14th century)
1 a : an outward sign **:** INDICATION **b :** something that furnishes proof **:** TESTIMONY; *specifically* **:** something legally submitted to a tribunal to ascertain the truth of a matter
2 : one who bears witness; *especially* **:** one who voluntarily confesses a crime and testifies for the prosecution against his accomplices
— **in evidence 1 :** to be seen **:** CONSPICUOUS ⟨trim lawns . . . are everywhere *in evidence* —*American Guide Series: North Carolina*⟩ **2 :** as evidence

[2]evidence *transitive verb* **-denced; -denc·ing** (circa 1610)
: to offer evidence of **:** PROVE, EVINCE
synonym see SHOW

ev·i·dent \ˈe-və-dənt, -və-ˌdent\ *adjective* [Middle English, from Middle French, from Latin *evident-, evidens,* from *e-* + *vident-, videns,* present participle of *vidēre* to see — more at WIT] (14th century)
: clear to the vision or understanding ☆

ev·i·den·tial \ˌe-və-ˈden(t)-shəl\ *adjective* (1641)
: EVIDENTIARY 1
— **ev·i·den·tial·ly** *adverb*

ev·i·den·tia·ry \ˌe-və-ˈden-chə-rē, -chē-ˌer-ē\ *adjective* (1810)
1 : being, relating to, or affording evidence ⟨photographs of *evidentiary* value⟩
2 : conducted so that evidence may be presented ⟨an *evidentiary* hearing⟩

ev·i·dent·ly \ˈe-və-dənt-lē, -ə-ˌdent-, *especially for 2 often* ˌev-ə-ˈdent-\ *adverb* (1690)
1 : in an evident manner **:** CLEARLY, OBVIOUSLY ⟨any style . . . so *evidently* bad or second-rate —T. S. Eliot⟩
2 : on the basis of available evidence ⟨he was born . . . *evidently* in Texas —Robert Coughlan⟩

[1]evil \ˈē-vəl, *British often & US also* ˈē-(ˌ)vil\ *adjective* **evil·er** *or* **evil·ler; evil·est** *or* **evil·lest** [Middle English, from Old English *yfel;* akin to Old High German *ubil* evil] (before 12th century)
1 a : morally reprehensible **:** SINFUL, WICKED ⟨an *evil* impulse⟩ **b :** arising from actual or imputed bad character or conduct ⟨a man of *evil* reputation⟩
2 a *archaic* **:** INFERIOR **b :** causing discomfort or repulsion **:** OFFENSIVE ⟨an *evil* odor⟩ **c :** DISAGREEABLE ⟨woke late and in an *evil* temper⟩
3 a : causing harm **:** PERNICIOUS ⟨the *evil* institution of slavery⟩ **b :** marked by misfortune **:** UNLUCKY
— **evil** *adverb, archaic*
— **evil·ly** \-(l)ē\ *adverb*
— **evil·ness** \-nəs\ *noun*

[2]evil *noun* (before 12th century)
1 a : the fact of suffering, misfortune, and wrongdoing **b :** a cosmic evil force
2 : something that brings sorrow, distress, or calamity

evil·do·er \ˌē-vəl-ˈdü-ər\ *noun* (14th century)
: one who does evil

evil·do·ing \-ˈdü-iŋ\ *noun* (14th century)
: the act or action of doing evil

evil eye *noun* (before 12th century)
: an eye or glance held capable of inflicting harm; *also* **:** a person believed to have such an eye or glance

evil–mind·ed \ˌē-vəl-ˈmīn-dəd, -vil-\ *adjective* (1531)
: having an evil disposition or evil thoughts
— **evil–mind·ed·ly** *adverb*
— **evil–mind·ed·ness** *noun*

evince \i-ˈvin(t)s\ *transitive verb* **evinced; evinc·ing** [Latin *evincere* to vanquish, win a point, from *e-* + *vincere* to conquer — more at VICTOR] (1621)
1 : to constitute outward evidence of
2 : to display clearly **:** REVEAL
synonym see SHOW
— **evinc·ible** \-ˈvin(t)-sə-bəl\ *adjective*

evis·cer·ate \i-ˈvi-sə-ˌrāt\ *verb* **-at·ed; -at·ing** [Latin *evisceratus,* past participle of *eviscerare,* from *e-* + *viscera* viscera] (1621)
transitive verb
1 a : to take out the entrails of **:** DISEMBOWEL **b :** to deprive of vital content or force
2 : to remove an organ from (a patient) or the contents of (an organ)
intransitive verb
: to protrude through a surgical incision or suffer protrusion of a part through an incision
— **evis·cer·a·tion** \-ˌvi-sə-ˈrā-shən\ *noun*

ev·i·ta·ble \ˈe-və-tə-bəl\ *adjective* [Latin *evitabilis,* from *evitare* to avoid, from *e-* + *vitare* to shun] (1502)
: capable of being avoided

evo·ca·ble \ˈe-və-kə-bəl, i-ˈvō-kə-\ *adjective* (1886)
: capable of being evoked

evo·ca·tion \ˌē-vō-ˈkā-shən, ˌe-və-\ *noun* [Latin *evocation-, evocatio,* from *evocare*] (1633)
1 : the act or fact of evoking **:** SUMMONING: as **a :** the summoning of a spirit **b :** imaginative recreation ⟨an *evocation* of the past⟩
2 : INDUCTION 4e
— **evo·ca·tor** \ˈē-vō-ˌkā-tər, ˈe-və-\ *noun*

evoc·a·tive \i-ˈvä-kə-tiv\ *adjective* (1657)
: evoking or tending to evoke an especially emotional response ⟨settings . . . so *evocative* that they bring tears to the eyes —Eric Malpass⟩
— **evoc·a·tive·ly** *adverb*
— **evoc·a·tive·ness** *noun*

evoke \i-ˈvōk\ *transitive verb* **evoked; evok·ing** [French *évoquer,* from Latin *evocare,* from *e-* + *vocare* to call — more at VOCATION] (circa 1626)
1 : to call forth or up: as **a :** CONJURE 2a ⟨*evoke* evil spirits⟩ **b :** to cite especially with approval or for support **:** INVOKE **c :** to bring to mind or recollection ⟨this place *evokes* memories⟩
2 : to recreate imaginatively
synonym see EDUCE

evo·lute \ˈe-və-ˌlüt *also* ˈē-və-\ *noun* (circa 1736)
: the locus of the center of curvature or the envelope of the normals of a curve

evo·lu·tion \ˌe-və-ˈlü-shən, ˌē-və-\ *noun* [Latin *evolution-, evolutio* unrolling, from *evolvere*] (1622)
1 : one of a set of prescribed movements
2 a : a process of change in a certain direction **:** UNFOLDING **b :** the action or an instance of forming and giving something off **:** EMISSION **c** (1) **:** a process of continuous change from a lower, simpler, or worse to a higher, more complex, or better state **:** GROWTH (2) **:** a process of gradual and relatively peaceful social, political, and economic advance **d :** something evolved
3 : the process of working out or developing
4 a : the historical development of a biological group (as a race or species) **:** PHYLOGENY **b :** a theory that the various types of animals and plants have their origin in other preexisting types and that the distinguishable differences are due to modifications in successive generations
5 : the extraction of a mathematical root
6 : a process in which the whole universe is a progression of interrelated phenomena
— **evo·lu·tion·ari·ly** \-shə-ˌner-ə-lē\ *adverb*
— **evo·lu·tion·ary** \-shə-ˌner-ē\ *adjective*
— **evo·lu·tion·ism** \-shə-ˌni-zəm\ *noun*
— **evo·lu·tion·ist** \-sh(ə-)nist\ *noun or adjective*

evolve \i-ˈvälv, -ˈvȯlv, ē- *also* -ˈväv *or* -ˈvȯv\ *verb* **evolved; evolv·ing** [Latin *evolvere* to unroll, from *e-* + *volvere* to roll — more at VOLUBLE] (1641)
transitive verb
1 : EMIT
2 a : DERIVE, EDUCE **b :** to produce by natural evolutionary processes **c :** DEVELOP, WORK OUT ⟨*evolve* social, political, and literary philosophies —L. W. Doob⟩
intransitive verb
: to undergo evolutionary change
— **evolv·able** \-ˈväl-və-bəl, -ˈvȯl- *also* -ˈvä-və- *or* -ˈvȯ-və-\ *adj*
— **evolve·ment** \-ˈvälv-mənt, -ˈvȯlv- *also* -ˈväv- *or* -ˈvȯv-\ *noun*

evul·sion \i-ˈvəl-shən\ *noun* [Latin *evulsion-, evulsio,* from *evellere* to pluck out, from *e-* + *vellere* to pluck — more at VULNERABLE] (circa 1611)
: EXTRACTION

ev·zone \ˈev-ˌzōn\ *noun* [New Greek *euzōnos,* from Greek, active, literally, well girt, from *eu-* + *zōnē* girdle — more at ZONE] (1897)
: a member of a select Greek infantry unit often serving as a palace guard

ewe \ˈyü, *in rural dials also* ˈyō\ *noun* [Middle English, from Old English *ēowu;* akin to Old High German *ouwi* ewe, Latin *ovis* sheep, Greek *ois*] (before 12th century)
: the female of the sheep especially when mature; *also* **:** the female of various related animals

Ewe \ˈā-ˌwā, ˈā-ˌvā\ *noun, plural* **Ewe** *also* **Ewes** (1861)
: a people of Ghana and Togo speaking a Kwa language; *also* **:** the language itself

ewe–neck \-ˈnek\ *noun* (1820)
: a thin neck with a concave arch occurring as a defect in dogs and horses
— **ewe–necked** \-ˈnekt\ *adjective*

ew·er \ˈyü-ər, ˈyu̇(-ə)r\ *noun* [Middle English, from Anglo-French, from Old French *evier,* from Latin *aquarium* water source, neuter of *aquarius* of water, from *aqua* water — more at ISLAND] (14th century)
: a vase-shaped pitcher or jug

ewer

Ew·ing's sarcoma \ˈyü-iŋz-\ *noun* [James Ewing (died 1943) American pathologist] (1951)

☆ SYNONYMS
Evident, manifest, patent, distinct, obvious, apparent, plain, clear mean readily perceived or apprehended. EVIDENT implies presence of visible signs that lead one to a definite conclusion ⟨an *evident* fondness for sweets⟩. MANIFEST implies an external display so evident that little or no inference is required ⟨*manifest* hostility⟩. PATENT applies to a cause, effect, or significant feature that is clear and unmistakable once attention has been directed to it ⟨*patent* defects⟩. DISTINCT implies such sharpness of outline or definition that no unusual effort to see or hear or comprehend is required ⟨a *distinct* refusal⟩. OBVIOUS implies such ease in discovering that it often suggests conspicuousness or little need for perspicacity in the observer ⟨the *obvious* solution⟩. APPARENT is very close to EVIDENT except that it may imply more conscious exercise of inference ⟨for no *apparent* reason⟩. PLAIN suggests lack of intricacy, complexity, or elaboration ⟨her feelings about him are *plain*⟩. CLEAR implies an absence of anything that confuses the mind or obscures the pattern ⟨a *clear* explanation⟩.

: a tumor that invades the shaft of a long bone and that tends to recur but rarely metastasizes

¹ex \(ˈ)eks\ *preposition* [Latin] (circa 1755)
1 : out of **:** FROM: as **a :** from a specified place or source **b :** from a specified dam ⟨a promising calf by Eric XVI *ex* Heatherbell⟩
2 : free from **:** WITHOUT: as **a :** without an indicated value or right — used especially of securities **b :** free of charges precedent to removal from the specified place with purchaser to provide means of subsequent transportation ⟨*ex* dock⟩ ◆

²ex *noun* [¹*ex*-] (1827)
: one that formerly held a specified position or place; *especially* **:** a former spouse

³ex *noun* (circa 1889)
: the letter *x*

¹ex- \e *also occurs in this prefix where only* i *is shown below (as in "express") and* ks *sometimes occurs where only* gz *is shown (as in "exact")*\ *prefix* [Middle English, from Old French & Latin; Old French, from Latin (also, prefix with perfective and causative value), from *ex* out of, from; akin to Greek *ex, ex-* out of, from Old Church Slavonic *iz*]
1 : out of **:** outside ⟨*ex*clave⟩
2 : not ⟨*ex*stipulate⟩
3 \(ˌ)eks, ˈeks\ [Middle English, from Late Latin, from Latin] **:** former ⟨*ex*-president⟩

²ex- — see EXO-

exa- *combining form* [International Scientific Vocabulary, modification of Greek *hexa-* hexa-]
: quintillion ⟨*exa*joules⟩

ex·ac·er·bate \ig-ˈza-sər-ˌbāt\ *transitive verb* **-bat·ed; -bat·ing** [Latin *exacerbatus,* past participle of *exacerbare,* from *ex-* + *acerbus* harsh, bitter, from *acer* sharp — more at EDGE] (1660)
: to make more violent, bitter, or severe ⟨the proposed shutdown . . . would *exacerbate* unemployment problems —*Science*⟩
— ex·ac·er·ba·tion \-ˌza-sər-ˈbā-shən\ *noun*

¹ex·act \ig-ˈzakt\ *transitive verb* [Middle English, to require as payment, from Latin *exactus,* past participle of *exigere* to drive out, demand, measure, from *ex-* + *agere* to drive — more at AGENT] (1564)
1 : to call for forcibly or urgently and obtain ⟨from them has been *exacted* the ultimate sacrifice —D. D. Eisenhower⟩
2 : to call for as necessary or desirable
synonym see DEMAND
— ex·act·able \-ˈzak-tə-bəl\ *adjective*
— ex·ac·tor *also* **ex·act·er** \-ˈzak-tər\ *noun*

²exact *adjective* [Latin *exactus*] (1533)
1 : exhibiting or marked by strict, particular, and complete accordance with fact or a standard
2 : marked by thorough consideration or minute measurement of small factual details
synonym see CORRECT
— ex·act·ness \-ˈzak(t)-nəs\ *noun*

ex·ac·ta \ig-ˈzak-tə\ *noun* [American Spanish (*quiniela*) *exacta* exact quiniela] (1964)
: PERFECTA

exact differential *noun* (1825)
: a differential expression of the form $X_1dx_1 + \ldots + X_ndx_n$ where the *X*'s are the partial derivatives of a function $f(x_1, \ldots, x_n)$ with respect to $x_1, \ldots, x_n$ respectively

ex·act·ing \ig-ˈzak-tiŋ\ *adjective* (1634)
1 : tryingly or unremittingly severe in making demands
2 : requiring careful attention and precision
synonym see ONEROUS
— ex·act·ing·ly \-tiŋ-lē\ *adverb*
— ex·act·ing·ness *noun*

ex·ac·tion \ig-ˈzak-shən\ *noun* (15th century)
1 a : the act or process of exacting **b :** EXTORTION
2 : something exacted; *especially* **:** a fee, reward, or contribution demanded or levied with severity or injustice

ex·ac·ti·tude \ig-ˈzak-tə-ˌtüd, -ˌtyüd\ *noun* (1734)
: the quality or an instance of being exact **:** EXACTNESS

ex·act·ly \ig-ˈzak-(t)lē\ *adverb* (1612)
1 a : in a manner or measure or to a degree or number that strictly conforms to a fact or condition ⟨it's *exactly* 3 o'clock⟩ ⟨these two pieces are *exactly* the same size⟩ **b :** in every respect **:** ALTOGETHER, ENTIRELY ⟨that was *exactly* the wrong thing to do⟩ ⟨not *exactly* what I had in mind⟩
2 : quite so — used to express agreement

exact science *noun* (1843)
: a science (as physics, chemistry, or astronomy) whose laws are capable of accurate quantitative expression

ex·ag·ger·ate \ig-ˈza-jə-ˌrāt\ *verb* **-at·ed; -at·ing** [Latin *exaggeratus,* past participle of *exaggerare,* literally, to heap up, from *ex-* + *agger* heap, from *aggerere* to carry toward, from *ad-* + *gerere* to carry] (circa 1587)
transitive verb
1 : to enlarge beyond bounds or the truth **:** OVERSTATE ⟨a friend *exaggerates* a man's virtues —Joseph Addison⟩
2 : to enlarge or increase especially beyond the normal **:** OVEREMPHASIZE
intransitive verb
: to make an overstatement
— ex·ag·ger·at·ed·ly *adverb*
— ex·ag·ger·at·ed·ness *noun*
— ex·ag·ger·a·tion \-ˌza-jə-ˈrā-shən\ *noun*
— ex·ag·ger·a·tive \-ˈza-jə-ˌrā-tiv, -ˈzaj-rə-, -ˈza-jə-\ *adjective*
— ex·ag·ger·a·tor \-ˈza-jə-ˌrā-tər\ *noun*
— ex·ag·ger·a·to·ry \-ˈzaj-rə-ˌtōr-ē, -ˌtȯr-, -ˈza-jə-\ *adjective*

ex·alt \ig-ˈzȯlt\ *verb* [Middle English, from Middle French & Latin; Middle French *exalter,* from Latin *exaltare,* from *ex-* + *altus* high — more at OLD] (15th century)
transitive verb
1 : to raise in rank, power, or character
2 : to elevate by praise or in estimation **:** GLORIFY
3 *obsolete* **:** ELATE
4 : to raise high **:** ELEVATE
5 : to enhance the activity of **:** INTENSIFY ⟨rousing and *exalting* the imagination —George Eliot⟩
intransitive verb
: to induce exaltation
— ex·alt·ed·ly *adverb*
— ex·alt·er *noun*

ex·al·ta·tion \ˌeg-ˌzȯl-ˈtā-shən, ˌek-ˌsȯl-\ *noun* (14th century)
1 : an act of exalting **:** the state of being exalted
2 : an excessively intensified sense of well-being, power, or importance
3 : an increase in degree or intensity ⟨*exaltation* of virulence of a virus⟩

ex·am \ig-ˈzam\ *noun* (1877)
: EXAMINATION

ex·a·men \ig-ˈzā-mən\ *noun* [Latin, tongue of a balance, examination, from *exigere* — more at EXACT] (1606)
1 : EXAMINATION
2 : a critical study

ex·am·i·nant \-ˈza-mə-nənt\ *noun* (1588)
1 : EXAMINEE
2 : one who examines **:** EXAMINER

ex·am·i·na·tion \ig-ˌza-mə-ˈnā-shən\ *noun* (14th century)
1 : the act or process of examining **:** the state of being examined
2 : an exercise designed to examine progress or test qualification or knowledge
3 : a formal interrogation
— ex·am·i·na·tion·al \-shnəl, -shə-nᵊl\ *adjective*

ex·am·ine \ig-ˈza-mən\ *verb* **ex·am·ined; ex·am·in·ing** \-ˈzam-niŋ, -ˈza-mə-\ [Middle English, from Middle French *examiner,* from Latin *examinare,* from *examen*] (14th century)
transitive verb
1 a : to inspect closely **b :** to test the condition of **c :** to inquire into carefully **:** INVESTIGATE
2 a : to interrogate closely ⟨*examine* a prisoner⟩ **b :** to test by questioning in order to determine progress, fitness, or knowledge
intransitive verb
: to make or give an examination
synonym see SCRUTINIZE
— ex·am·in·able \-ˈza-mə-nə-bəl\ *adjective*
— ex·am·in·er \-ˈzam-nər, -ˈza-mə-\ *noun*

ex·am·in·ee \ig-ˌza-mə-ˈnē\ *noun* (1788)
: a person who is examined

¹ex·am·ple \ig-ˈzam-pəl\ *noun* [Middle English, from Middle French, from Latin *exemplum,* from *eximere* to take out, from *ex-* + *emere* to take — more at REDEEM] (14th century)
1 : one that serves as a pattern to be imitated or not to be imitated ⟨a good *example*⟩ ⟨a bad *example*⟩
2 : a punishment inflicted on someone as a warning to others; *also* **:** an individual so punished
3 : one (as an item or incident) that is representative of all of a group or type
4 : a parallel or closely similar case especially when serving as a precedent or model
5 : an instance (as a problem to be solved) serving to illustrate a rule or precept or to act as an exercise in the application of a rule
synonym see INSTANCE, MODEL
— for example \fər-ig-ˈzam-pəl, frig-\ **:** as an example ⟨there are many sources of air pollution; exhaust fumes, *for example*⟩

²example *transitive verb* **ex·am·pled; ex·am·pling** \-p(ə-)liŋ\ (15th century)
1 : to serve as an example of
2 *archaic* **:** to be or set an example to

ex·an·i·mate \eg-ˈza-nə-mət\ *adjective* [Latin *exanimatus,* past participle of *exanimare* to deprive of life or spirit, from *ex-* + *anima* breath, soul — more at ANIMATE] (circa 1534)
1 : lacking animation **:** SPIRITLESS
2 : being or appearing lifeless

ex·an·them \eg-ˈzan(t)-thəm, ˈek-ˌsan-ˌthem\ *also* **ex·an·the·ma** \ˌeg-ˌzan-ˈthē-mə\ *noun, plural* **-thems** *also* **-them·a·ta** \ˌeg-ˌzan-ˈthe-mə-tə\ *or* **-themas** [Late Latin *exanthema,* from Greek *exanthēma,* from *exanthein* to bloom, break out, from *ex-* + *anthos* flower — more at ANTHOLOGY] (1656)

◇ WORD HISTORY

ex The word *ex* meaning "former spouse" is clearly a shortening of *ex-wife* or *ex-husband,* but the origin of the prefix *ex-* that forms these words is less straightforward. In Latin, both the prefix *ex-* and the preposition *ex* meant "out of, from." The preposition took on the meaning "from being, having formerly been" in phrases describing a man's office and the office he had previously held. For example, *ex assessore praefectus praetorii* meant "prefect of the Praetorian Guard, having formerly held the office of assessor" (literally, "out of assessor, prefect of the Praetorian Guard"). Because of phrases of this type, *ex* in the sense "having formerly been" came to be used in Late Latin simply as a prefix, in words like *exconsul* "person who was formerly consul." Although *ex-* in this meaning has been used sporadically in English since the late Middle Ages, coinages such as *ex-president* and *ex-wife* did not become common until the 19th century.

: an eruptive disease (as measles) or its symptomatic eruption
— **ex•an•them•a•tous** \ˌeg-ˌzan-ˈthe-mə-təs\ *or* **ex•an•the•mat•ic** \-ˌzan-thə-ˈma-tik\ *adjective*

[1]**ex•arch** \ˈek-ˌsärk\ *noun* [Late Latin *exarchus,* from Late Greek *exarchos,* from Greek, leader, from *exarchein* to begin, take the lead, from *ex-* + *archein* to rule, begin — more at ARCH-] (1588)
1 : a Byzantine viceroy
2 : an Eastern bishop ranking below a patriarch and above a metropolitan; *specifically* **:** the head of an independent church
— **ex•ar•chal** \ek-ˈsär-kəl\ *adjective*
— **ex•arch•ate** \ˈek-ˌsär-kət\ *noun*
— **ex•ar•chy** \ˈek-ˌsär-kē\ *noun*

[2]**exarch** *adjective* [*exo-* + *-arch*] (1891)
: formed or taking place from the periphery toward the center ⟨*exarch* xylem⟩

[1]**ex•as•per•ate** \ig-ˈzas-pə-ˌrāt\ *transitive verb* **-at•ed; -at•ing** [Latin *exasperatus,* past participle of *exasperare,* from *ex-* + *asper* rough — more at ASPERITY] (1534)
1 a : to excite the anger of **:** ENRAGE **b :** to cause irritation or annoyance to
2 *obsolete* **:** to make more grievous **:** AGGRAVATE
synonym see IRRITATE
— **ex•as•per•at•ed•ly** *adverb*
— **ex•as•per•at•ing•ly** \-ˌrā-tiŋ-lē\ *adverb*

[2]**ex•as•per•ate** \-p(ə-)rət\ *adjective* (1541)
1 : irritated or annoyed especially to the point of injudicious action **:** EXASPERATED
2 : roughened with irregular prickles or elevations ⟨*exasperate* seed coats⟩

ex•as•per•a•tion \ig-ˌzas-pə-ˈrā-shən\ *noun* (1547)
1 : the state of being exasperated
2 : the act or an instance of exasperating

Ex•cal•i•bur \ek-ˈska-lə-bər\ *noun* [Middle English *Excalaber,* from Old French *Escalibor,* from Medieval Latin *Caliburnus*]
: the sword of King Arthur

ex ca•the•dra \ˌeks-kə-ˈthē-drə\ *adverb or adjective* [New Latin, literally, from the chair] (1818)
: by virtue of or in the exercise of one's office or position ⟨*ex cathedra* pronouncements⟩

ex•ca•vate \ˈek-skə-ˌvāt\ *verb* **-vat•ed; -vat•ing** [Latin *excavatus,* past participle of *excavare,* from *ex-* + *cavare* to make hollow — more at CAVATINA] (1599)
transitive verb
1 : to form a cavity or hole in
2 : to form by hollowing out
3 : to dig out and remove
4 : to expose to view by or as if by digging away a covering ⟨*excavate* the remains of a temple⟩ ⟨another writer whose work I *excavated* —William Zinsser⟩
intransitive verb
: to make excavations

ex•ca•va•tion \ˌek-skə-ˈvā-shən\ *noun* (circa 1611)
1 : the action or process of excavating
2 : a cavity formed by cutting, digging, or scooping
— **ex•ca•va•tion•al** \-shnəl, -shə-nᵊl\ *adjective*

ex•ca•va•tor \ˈek-skə-ˌvā-tər\ *noun* (circa 1815)
: one that excavates; *especially* **:** a power-operated shovel

ex•ceed \ik-ˈsēd\ *verb* [Middle English *exceden,* from Middle French *exceder,* from Latin *excedere,* from *ex-* + *cedere* to go] (14th century)
transitive verb
1 : to extend outside of ⟨the river will *exceed* its banks⟩
2 : to be greater than or superior to
3 : to go beyond a limit set by ⟨*exceeded* his authority⟩
intransitive verb
1 *obsolete* **:** OVERDO
2 : PREDOMINATE ☆

ex•ceed•ing *adjective* (15th century)
: exceptional in amount, quality, or degree

ex•ceed•ing•ly \-ˈsē-diŋ-lē\ *also* **ex•ceed•ing** *adverb* (1535)
: to an extreme degree **:** EXTREMELY

ex•cel \ik-ˈsel\ *verb* **ex•celled; ex•cel•ling** [Middle English *excellen,* from Latin *excellere,* from *ex-* + *-cellere* to rise, project; akin to Latin *collis* hill — more at HILL] (15th century)
transitive verb
: to be superior to **:** surpass in accomplishment or achievement
intransitive verb
: to be distinguishable by superiority **:** surpass others ⟨*excel* in sports⟩ ⟨*excelled* at lipreading⟩
synonym see EXCEED

ex•cel•lence \ˈek-s(ə-)lən(t)s\ *noun* (14th century)
1 : the quality of being excellent
2 : an excellent or valuable quality **:** VIRTUE
3 : EXCELLENCY 2

ex•cel•len•cy \-s(ə-)lən(t)-sē\ *noun, plural* **-cies** (15th century)
1 : EXCELLENCE; *especially* **:** outstanding or valuable quality — usually used in plural ⟨so crammed, as he thinks, with *excellencies* —Shakespeare⟩
2 — used as a title for high dignitaries of state (as a governor or an ambassador) or church (as a Roman Catholic archbishop or bishop)

ex•cel•lent \ˈek-s(ə-)lənt\ *adjective* [Middle English, from Middle French, from Latin *excellent-, excellens,* from present participle of *excellere*] (14th century)
1 *archaic* **:** SUPERIOR
2 : very good of its kind **:** eminently good **:** FIRST-CLASS
— **ex•cel•lent•ly** *adverb*

ex•cel•si•or \ik-ˈsel-sē-ər, -ˌȯr\ *noun* [trade name, from Latin, higher, comparative of *excelsus* high, from past participle of *excellere*] (1868)
: fine curled wood shavings used especially for packing fragile items

[1]**ex•cept** \ik-ˈsept\ *also* **ex•cept•ing** \-ˈsep-tiŋ\ *preposition* (14th century)
: with the exclusion or exception of ⟨daily *except* Sundays⟩

[2]**except** *verb* [Middle English, from Middle French *excepter,* from Latin *exceptare,* from *exceptus,* past participle of *excipere* to take out, except, from *ex-* + *capere* to take — more at HEAVE] (14th century)
transitive verb
: to take or leave out from a number or a whole **:** EXCLUDE
intransitive verb
: to take exception **:** OBJECT
— **ex•cep•tive** \-ˈsep-təv\ *adjective*

[3]**except** *also* **excepting** *conjunction* (15th century)
1 : on any other condition than that **:** UNLESS ⟨*except* you repent⟩
2 : with this exception, namely ⟨was inaccessible *except* by boat⟩
3 : ONLY — often followed by *that* ⟨I would go *except* that it's too far⟩

except for *preposition* (1842)
1 : with the exception of ⟨everyone was gone *except for* me⟩
2 : were it not for ⟨*except for* you I would be dead⟩

ex•cep•tion \ik-ˈsep-shən\ *noun* (14th century)
1 : the act of excepting **:** EXCLUSION
2 : one that is excepted; *especially* **:** a case to which a rule does not apply
3 : QUESTION, OBJECTION ⟨witnesses whose authority is beyond *exception* —T. B. Macaulay⟩
4 : an oral or written legal objection

ex•cep•tion•able \ik-ˈsep-sh(ə-)nə-bəl\ *adjective* (1691)
: being likely to cause objection **:** OBJECTIONABLE ⟨visitors even drink the *exceptionable* beer —W. D. Howells⟩
— **ex•cep•tion•abil•i•ty** \-ˌsep-sh(ə-)nə-ˈbi-lə-tē\ *noun*
— **ex•cep•tion•ably** \-ˈsep-sh(ə-)nə-blē\ *adverb*

ex•cep•tion•al \ik-ˈsep-shnəl, -shə-nᵊl\ *adjective* (circa 1846)
1 : forming an exception **:** RARE ⟨an *exceptional* number of rainy days⟩
2 : better than average **:** SUPERIOR ⟨*exceptional* skill⟩
3 : deviating from the norm: as **a :** having above or below average intelligence **b :** physically handicapped
— **ex•cep•tion•al•i•ty** \-ˌsep-shə-ˈna-lə-tē\ *noun*
— **ex•cep•tion•al•ness** *noun*

ex•cep•tion•al•ism \ik-ˈsep-shnə-ˌli-zəm, -shə-nᵊl-ˌi-\ *noun* (1937)
: the condition of being different from the norm; *also* **:** a theory expounding the exceptionalism especially of a nation or region

ex•cep•tion•al•ly \ik-ˈsep-shnə-lē, -shə-nᵊl-ē\ *adverb* (1848)
: in an exceptional manner **:** to an exceptional degree; *especially* **:** more than average or usual ⟨an *exceptionally* difficult task⟩

[1]**ex•cerpt** \ek-ˈsərpt, eg-ˈzərpt, ˈek-ˌ, ˈeg-ˌ\ *transitive verb* [Latin *excerptus,* past participle of *excerpere,* from *ex-* + *carpere* to gather, pluck — more at HARVEST] (15th century)
1 : to select (a passage) for quoting **:** EXTRACT
2 : to take or publish extracts from (as a book)
— **ex•cerp•tor** *or* **ex•cerpt•er** *noun*
— **ex•cerp•tion** \ek-ˈsərp-shən, eg-ˈzərp-\ *noun*

[2]**ex•cerpt** \ˈek-ˌsərpt, ˈeg-ˌzərpt\ *noun* (1627)
: a passage (as from a book or musical composition) selected, performed, or copied **:** EXTRACT

[1]**ex•cess** \ik-ˈses, ˈek-ˌ\ *noun* [Middle English, from Middle French or Late Latin; Middle French *exces,* from Late Latin *excessus,* from Latin, departure, projection, from *excedere* to exceed] (14th century)
1 a : the state or an instance of surpassing usual, proper, or specified limits **:** SUPERFLUITY **b :** the amount or degree by which one thing or quantity exceeds another ⟨an *excess* of 10 bushels⟩
2 : undue or immoderate indulgence **:** INTEMPERANCE; *also* **:** an act or instance of intemperance ⟨prevent *excesses* and abuses by newly created local powers —Albert Shanker⟩
— **in excess of :** to an amount or degree beyond **:** OVER

[2]**excess** *adjective* (15th century)
: more than the usual, proper, or specified amount

[3]**excess** *transitive verb* (1974)
: to eliminate the position of ⟨*excessed* several teachers because of budget cutbacks⟩

☆ SYNONYMS
Exceed, surpass, transcend, excel, outdo, outstrip mean to go or be beyond a stated or implied limit, measure, or degree. EXCEED implies going beyond a limit set by authority or established by custom or by prior achievement ⟨*exceed* the speed limit⟩. SURPASS suggests superiority in quality, merit, or skill ⟨the book *surpassed* our expectations⟩. TRANSCEND implies a rising or extending notably above or beyond ordinary limits ⟨*transcended* the values of their culture⟩. EXCEL implies preeminence in achievement or quality and may suggest superiority to all others ⟨*excels* in mathematics⟩. OUTDO applies to a bettering or exceeding what has been done before ⟨*outdid* herself this time⟩. OUTSTRIP suggests surpassing in a race or competition ⟨*outstripped* other firms in sales⟩.

ex·ces·sive \ik-'se-siv\ *adjective* (14th century)
: exceeding what is usual, proper, necessary, or normal ☆
— **ex·ces·sive·ly** *adverb*
— **ex·ces·sive·ness** *noun*

[1]**ex·change** \iks-'chānj, 'eks-,\ *noun, often attributive* [Middle English *exchaunge,* from Middle French *eschange,* from *eschangier* to exchange, from (assumed) Vulgar Latin *excambiare,* from Latin *ex-* + *cambiare* to exchange — more at CHANGE] (14th century)
1 : the act of giving or taking one thing in return for another **:** TRADE ⟨an *exchange* of prisoners⟩
2 a : the act or process of substituting one thing for another **b :** reciprocal giving and receiving
3 : something offered, given, or received in an exchange
4 a : funds payable currently at a distant point either in a foreign currency or in domestic currency **b** (1) **:** interchange or conversion of the money of two countries or of current and uncurrent money with allowance for difference in value (2) **:** EXCHANGE RATE (3) **:** the amount of the difference in value between two currencies or between values of a particular currency at two places **c :** instruments (as checks or bills of exchange) presented in a clearinghouse for settlement
5 : a place where things or services are exchanged: as **a :** an organized market or center for trading in securities or commodities **b :** a store or shop specializing in merchandise usually of a particular type **c :** a cooperative store or society **d :** a central office in which telephone lines are connected to permit communication
— **in exchange :** as a substitute

[2]**exchange** *verb* **ex·changed; ex·chang·ing** (15th century)
transitive verb
1 a : to part with, give, or transfer in consideration of something received as an equivalent **b :** to have replaced by other merchandise ⟨*exchanged* the shirt for one in a larger size⟩
2 : to part with for a substitute ⟨*exchanging* future security for immediate pleasure⟩
3 : to give and receive reciprocally ⟨*exchange* gifts⟩
intransitive verb
1 : to pass or become received in exchange
2 : to engage in an exchange
— **ex·change·abil·i·ty** \iks-,chān-jə-'bi-lə-tē\ *noun*
— **ex·change·able** \iks-'chān-jə-bəl\ *adjective*
— **ex·chang·er** \iks-'chān-jər, eks-,\ *noun*

exchange rate *noun* (1896)
: the ratio at which the principal unit of two currencies may be traded

exchange student *noun* (circa 1930)
: a student from one country received into an institution in another country in exchange for one sent to an institution in the home country of the first

Ex·chang·ite \iks-'chān-,jīt\ *noun* [*(National) Exchange (club)*] (circa 1934)
: a member of a major national service club

ex·che·quer \'eks-,che-kər, iks-'\ *noun* [Middle English *escheker,* from Anglo-French, from Old French *eschequier* chessboard, counting table — more at CHECKER] (14th century)
1 *capitalized* **:** a department or office of state in medieval England charged with the collection and management of the royal revenue and judicial determination of all revenue causes
2 *capitalized* **:** a former superior court having jurisdiction in England and Wales primarily over revenue matters and now merged with King's Bench
3 *often capitalized* **a :** the department or office of state in Great Britain and Northern Ireland charged with the receipt and care of the national revenue **b :** the national banking account of this realm
4 : TREASURY; *especially* **:** a national or royal treasury
5 : pecuniary resources **:** FUNDS

ex·ci·mer laser \'ek-si-,mər-\ *noun* [*exci*ted + di*mer*] (1973)
: a laser that uses a noble-gas halide to generate radiation usually in the ultraviolet region of the spectrum

ex·cip·i·ent \ik-'si-pē-ənt\ *noun* [Latin *excipient-, excipiens,* present participle of *excipere* to take out, take up — more at EXCEPT] (1753)
: a usually inert substance (as gum arabic or starch) that forms a vehicle (as for a drug)

ex·ci·ple \'ek-sə-pəl\ *noun* [New Latin *excipulum,* from Latin, receptacle, from *excipere*] (circa 1866)
: a saucer-shaped rim around the hymenium of various lichens

ex·cis·able \'ek-,sī-zə-bəl, -,sī-sə-, ek-'\ *adjective* (1689)
: subject to excise

[1]**ex·cise** \'ek-,sīz, -,sīs\ *noun* [obsolete Dutch *excijs* (now *accijus*), from Middle Dutch, probably modification of Old French *assise* session, assessment — more at ASSIZE] (15th century)
1 : an internal tax levied on the manufacture, sale, or consumption of a commodity
2 : any of various taxes on privileges often assessed in the form of a license or fee

[2]**ex·cise** \'ek-,sīz, -,sīs, ik-'sīz\ *transitive verb* **ex·cised; ex·cis·ing** (1652)
: to impose an excise on

[3]**ex·cise** \ik-'sīz\ *transitive verb* **ex·cised; ex·cis·ing** [Latin *excisus,* past participle of *excidere,* from *ex-* + *caedere* to cut] (1634)
: to remove by or as if by excision

ex·cise·man \'ek-,sīz-mən, -,sīs-, -,man, ek-'\ *noun* (1647)
: an officer who inspects and rates articles liable to excise under British law

ex·ci·sion \ik-'si-zhən\ *noun* [Middle English *excysion,* from Middle French *excision,* from Latin *excision-, excisio,* from *excidere*] (1541)
: the act or procedure of removing by or as if by cutting out; *especially* **:** surgical removal or resection
— **ex·ci·sion·al** \-'sizh-nəl, -zh°n-əl\ *adjective*

ex·cit·able \ik-'sī-tə-bəl\ *adjective* (1609)
1 : capable of being readily roused into action or a state of excitement or irritability
2 : capable of being activated by and reacting to stimuli
— **ex·cit·abil·i·ty** \-,sī-tə-'bi-lə-tē\ *noun*
— **ex·cit·able·ness** \-'sī-tə-bəl-nəs\ *noun*

ex·ci·tant \ik-'sī-t°nt, 'ek-sə-tənt\ *adjective* (1607)
: tending to excite or augment ⟨*excitant* drugs⟩
— **excitant** *noun*

ex·ci·ta·tion \,ek-,sī-'tā-shən, ,ek-sə-\ *noun* (14th century)
: EXCITEMENT; *especially* **:** the disturbed or altered condition resulting from stimulation of an individual organ, tissue, or cell

ex·cit·ative \ik-'sī-tə-tiv\ *adjective* (15th century)
: tending to induce excitation (as of a neuron) ⟨*excitative* substances⟩

ex·cit·ato·ry \ik-'sī-tə-,tōr-ē, -,tòr-\ *adjective* (1803)
: exhibiting, resulting from, related to, or producing excitement or excitation ⟨*excitatory* nerve fibers⟩

ex·cite \ik-'sīt, ek-\ *transitive verb* **ex·cit·ed; ex·cit·ing** [Middle English, from Middle French *exciter,* from Latin *excitare,* from *ex-* + *citare* to rouse — more at CITE] (14th century)
1 a : to call to activity **b :** to rouse to an emotional response ⟨scenes to *excite* the hardest man to pity⟩ **c :** to arouse (as a strong emotional response) by appropriate stimuli ⟨*excite* enthusiasm for the new regime —Arthur Knight⟩
2 a : ENERGIZE ⟨*excite* an electromagnet⟩ **b :** to produce a magnetic field in ⟨*excite* a dynamo⟩
3 : to increase the activity of (as a living organism) **:** STIMULATE
4 : to raise (as an atomic nucleus, an atom, or a molecule) to a higher energy level
synonym see PROVOKE
— **ex·cit·ed·ly** \-'sī-təd-lē\ *adverb*

excited state *noun* (1927)
: a state of a physical system (as an atomic nucleus, an atom, or a molecule) that is higher in energy than the ground state

ex·cite·ment \ik-'sīt-mənt\ *noun* (1604)
1 : something that excites or rouses
2 : the action of exciting **:** the state of being excited

ex·cit·er \ik-'sī-tər\ *noun* (14th century)
1 : one that excites
2 a : a generator or battery that supplies the electric current used to produce the magnetic field in another generator or motor **b :** an electrical oscillator that generates the carrier frequency (as for a radio transmitter)

ex·cit·ing \ik-'sī-tiŋ\ *adjective* (circa 1811)
: producing excitement
— **ex·cit·ing·ly** \-iŋ-lē\ *adverb*

ex·ci·ton \'ek-sə-,tän, -,sī-\ *noun* [International Scientific Vocabulary *excit*ation + [2]*-on*] (1936)
: a mobile combination of an electron and a hole in an excited crystal (as of a semiconductor)
— **ex·ci·ton·ic** \,ek-sə-'tä-nik, -,sī-\ *adjective*

ex·ci·tor \ik-'sī-tər\ *noun* (1816)
: an afferent nerve arousing increased action of the part that it supplies

ex·claim \iks-'klām\ *verb* [Middle French *exclamer,* from Latin *exclamare,* from *ex-* + *clamare* to cry out — more at CLAIM] (circa 1570)
intransitive verb
1 : to cry out or speak in strong or sudden emotion ⟨*exclaimed* in delight⟩
2 : to speak loudly or vehemently ⟨*exclaimed* against immorality⟩
transitive verb
: to utter sharply, passionately, or vehemently **:** PROCLAIM
— **ex·claim·er** *noun*

ex·cla·ma·tion \,eks-klə-'mā-shən\ *noun* (14th century)
1 : a sharp or sudden utterance

☆ SYNONYMS
Excessive, immoderate, inordinate, extravagant, exorbitant, extreme mean going beyond a normal limit. EXCESSIVE implies an amount or degree too great to be reasonable or acceptable ⟨*excessive* punishment⟩. IMMODERATE implies lack of desirable or necessary restraint ⟨*immoderate* spending⟩. INORDINATE implies an exceeding of the limits dictated by reason or good judgment ⟨*inordinate* pride⟩. EXTRAVAGANT implies an indifference to restraints imposed by truth, prudence, or good taste ⟨*extravagant* claims for the product⟩. EXORBITANT implies a departure from accepted standards regarding amount or degree ⟨*exorbitant* prices⟩. EXTREME may imply an approach to the farthest limit possible or conceivable but commonly means only to a notably high degree ⟨*extreme* shyness⟩.

2 : vehement expression of protest or complaint

exclamation point *noun* (1824)
: a mark ! used especially after an interjection or exclamation to indicate forceful utterance or strong feeling — called also *exclamation mark*

ex·clam·a·to·ry \iks-ˈkla-mə-ˌtōr-ē, -ˌtȯr-\ *adjective* (1593)
: containing, expressing, using, or relating to exclamation ⟨an *exclamatory* phrase⟩

ex·clave \ˈeks-ˌklāv, -ˌkläv\ *noun* [*ex-* + *-clave* (as in *enclave*)] (1888)
: a portion of a country separated from the main part and constituting an enclave in respect to the surrounding territory

ex·clud·able *or* **ex·clud·ible** \iks-ˈklü-də-bəl\ *adjective* (1916)
: subject to exclusion ⟨*excludable* income⟩
— ex·clud·abil·i·ty \-ˌklü-də-ˈbi-lə-tē\ *noun*

ex·clude \iks-ˈklüd\ *transitive verb* **ex·clud·ed; ex·clud·ing** [Middle English, from Latin *excludere,* from *ex-* + *claudere* to close — more at CLOSE] (14th century)
1 a : to prevent or restrict the entrance of **b :** to bar from participation, consideration, or inclusion
2 : to expel or bar especially from a place or position previously occupied
— ex·clud·er *noun*

ex·clu·sion \iks-ˈklü-zhən\ *noun* [Latin *exclusion-, exclusio,* from *excludere*] (15th century)
1 : the act or an instance of excluding
2 : the state of being excluded
— ex·clu·sion·ary \-zhə-ˌner-ē\ *adjective*

exclusionary rule *noun* (1964)
: a legal rule that bars unlawfully obtained evidence from being used in court proceedings

ex·clu·sion·ist \iks-ˈklüzh-nist, -ˈklü-zhə-\ *noun* (1822)
: one who would exclude another from some right or privilege
— exclusionist *adjective*

exclusion principle *noun* (1926)
: a principle in physics: no two particles (as electrons) in an atom or molecule can have the same set of quantum numbers

[1]**ex·clu·sive** \iks-ˈklü-siv, -ziv\ *adjective* (1515)
1 a : excluding or having power to exclude **b :** limiting or limited to possession, control, or use by a single individual or group
2 a : excluding others from participation **b :** snobbishly aloof
3 a : accepting or soliciting only a socially restricted patronage (as of the upper class) **b :** STYLISH, FASHIONABLE **c :** restricted in distribution, use, or appeal because of expense
4 a : SINGLE, SOLE ⟨*exclusive* jurisdiction⟩ **b :** WHOLE, UNDIVIDED ⟨his *exclusive* attention⟩
— ex·clu·sive·ly *adverb*
— ex·clu·sive·ness *noun*

[2]**exclusive** *noun* (15th century)
: something exclusive: as **a :** a newspaper story at first released to or printed by only one newspaper **b :** an exclusive right (as to sell a particular product in a certain area)

exclusive disjunction *noun* (1942)
: a compound proposition in logic that is true when one and only one of its constituent statements is true — see TRUTH TABLE table

exclusive of *preposition* (1762)
: not taking into account ⟨there were four of us *exclusive of* the guide⟩

ex·clu·siv·ism \iks-ˈklü-sə-ˌvi-zəm, -ˈklü-zə-\ *noun* (circa 1834)
: the practice of excluding or of being exclusive
— ex·clu·siv·ist \-vist\ *noun or adjective*

ex·clu·siv·i·ty \ˌeks-ˌklü-ˈsi-və-tē, iks-, -ˈzi-\ *noun, plural* **-ties** (1926)
1 : the quality or state of being exclusive
2 : exclusive rights or services

ex·cog·i·tate \ek-ˈskä-jə-ˌtāt\ *transitive verb* [Latin *excogitatus,* past participle of *excogitare,* from *ex-* + *cogitare* to cogitate] (circa 1530)
: to think out **:** DEVISE
— ex·cog·i·ta·tion \(ˌ)ek-ˌskä-jə-ˈtā-shən\ *noun*
— ex·cog·i·ta·tive \ek-ˈskä-jə-ˌtā-tiv\ *adjective*

[1]**ex·com·mu·ni·cate** \ˌek-skə-ˈmyü-nə-ˌkāt\ *transitive verb* [Middle English, from Late Latin *excommunicatus,* past participle of *excommunicare,* from Latin *ex-* + Late Latin *communicare* to communicate] (15th century)
: to subject to excommunication
— ex·com·mu·ni·ca·tor \-ˌkā-tər\ *noun*

[2]**ex·com·mu·ni·cate** \-ni-kət\ *adjective* (1551)
: excluded from the rites of the church **:** EXCOMMUNICATED
— excommunicate *noun*

ex·com·mu·ni·ca·tion \-ˌmyü-nə-ˈkā-shən\ *noun* (15th century)
1 : an ecclesiastical censure depriving a person of the rights of church membership
2 : exclusion from fellowship in a group or community
— ex·com·mu·ni·ca·tive \-ˈmyü-nə-ˌkā-tiv, -ni-kət-\ *adjective*

ex·co·ri·ate \ek-ˈskōr-ē-ˌāt, -ˈskȯr-\ *transitive verb* **-at·ed; -at·ing** [Middle English, from Late Latin *excoriatus,* past participle of *excoriare,* from Latin *ex-* + *corium* skin, hide — more at CUIRASS] (15th century)
1 : to wear off the skin of **:** ABRADE
2 : to censure scathingly
— ex·co·ri·a·tion \(ˌ)ek-ˌskōr-ē-ˈā-shən, -ˌskȯr-\ *noun*

ex·cre·ment \ˈek-skrə-mənt\ *noun* [Latin *excrementum,* from *excernere*] (1533)
: waste matter discharged from the body; *especially* **:** waste (as feces) discharged from the alimentary canal
— ex·cre·men·tal \ˌek-skrə-ˈmen-tᵊl\ *adjective*
— ex·cre·men·ti·tious \-ˌmen-ˈti-shəs, -mən-\ *adjective*

ex·cres·cence \ik-ˈskre-sᵊn(t)s, ek-\ *noun* (15th century)
1 : a projection or outgrowth especially when abnormal ⟨warty *excrescences* in the colon⟩
2 : a disfiguring, extraneous, or unwanted mark or part **:** BLOT

ex·cres·cen·cy \-sᵊn-sē\ *noun, plural* **-cies** (1545)
: EXCRESCENCE

ex·cres·cent \-sᵊnt\ *adjective* [Latin *excrescent-, excrescens,* present participle of *excrescere* to grow out, from *ex-* + *crescere* to grow — more at CRESCENT] (1633)
1 : forming an abnormal, excessive, or useless outgrowth
2 : of, relating to, or constituting epenthesis
— ex·cres·cent·ly *adverb*

ex·cre·ta \ik-ˈskrē-tə\ *noun plural* [New Latin, from Latin, neuter plural of *excretus*] (1857)
: waste matter eliminated or separated from an organism; *especially* **:** EXCRETIONS
— ex·cre·tal \-ˈskrē-tᵊl\ *adjective*

ex·crete \ik-ˈskrēt\ *transitive verb* **ex·cret·ed; ex·cret·ing** [Latin *excretus,* past participle of *excernere* to sift out, discharge, from *ex-* + *cernere* to sift — more at CERTAIN] (1620)
: to separate and eliminate or discharge (waste) from the blood or tissues or from the active protoplasm
— ex·cret·er *noun*

ex·cre·tion \ik-ˈskrē-shən\ *noun* (1603)
1 : the act or process of excreting
2 : something excreted; *especially* **:** metabolic waste products (as urea and carbon dioxide) that are eliminated from the body and differ from a secretion in not being produced to perform a useful function

ex·cre·to·ry \ˈek-skrə-ˌtōr-ē, -ˌtȯr-\ *adjective* (circa 1681)
: of, relating to, or functioning in excretion ⟨*excretory* ducts⟩

ex·cru·ci·ate \ik-ˈskrü-shē-ˌāt\ *transitive verb* **-at·ed; -at·ing** [Latin *excruciatus,* past participle of *excruciare,* from *ex-* + *cruciare* to crucify, from *cruc-, crux* cross] (circa 1570)
1 : to inflict intense pain on **:** TORTURE
2 : to subject to intense mental distress
— ex·cru·ci·a·tion \-ˌskrü-shē-ˈā-shən, -sē-\ *noun*

ex·cru·ci·at·ing *adjective* (1599)
1 : causing great pain or anguish **:** AGONIZING ⟨the nation's most *excruciating* dilemma —W. H. Ferry⟩
2 : very intense **:** EXTREME ⟨*excruciating* pain⟩
— ex·cru·ci·at·ing·ly \-ˌā-tiŋ-lē\ *adverb*

ex·cul·pate \ˈek-(ˌ)skəl-ˌpāt, (ˌ)ek-ˈ\ *transitive verb* **-pat·ed; -pat·ing** [Medieval Latin *exculpatus,* past participle of *exculpare,* from Latin *ex-* + *culpa* blame] (circa 1681)
: to clear from alleged fault or guilt ☆
— ex·cul·pa·tion \ˌek-(ˌ)skəl-ˈpā-shən\ *noun*

ex·cul·pa·to·ry \ek-ˈskəl-pə-ˌtōr-ē, -ˌtȯr-\ *adjective* (1781)
: tending or serving to exculpate

ex·cur·rent \(ˌ)ek-ˈskər-ənt, -ˈskə-rənt\ *adjective* [Latin *excurrent-, excurrens,* present participle of *excurrere* to run out, extend, from *ex-* + *currere* to run — more at CAR] (1826)
1 : characterized by a current that flows outward ⟨*excurrent* canals of a sponge⟩
2 a : having the axis prolonged to form an undivided main stem or trunk (as in conifers) — compare DELIQUESCENT 2 **b :** projecting beyond the apex — used especially of the midrib of a mucronate leaf

ex·cur·sion \ik-ˈskər-zhən\ *noun* [Latin *excursion-, excursio,* from *excurrere*] (1587)
1 a : a going out or forth **:** EXPEDITION **b** (1) **:** a usually brief pleasure trip (2) **:** a trip at special reduced rates
2 : deviation from a direct, definite, or proper course; *especially* **:** DIGRESSION ⟨needless *excursions* into abstruse theory⟩
3 a : a movement outward and back or from a mean position or axis; *also* **:** the distance traversed **:** AMPLITUDE ⟨the *excursion* of a piston⟩ **b :** one complete movement of expansion and contraction of the lungs and their membranes (as in breathing)

ex·cur·sion·ist \-ˈskərzh-nist, -ˈskər-zhə-\ *noun* (1830)
: a person who goes on an excursion

ex·cur·sive \-ˈskər-siv\ *adjective* (1749)
: constituting a digression **:** characterized by digression
— ex·cur·sive·ly *adverb*

☆ SYNONYMS
Exculpate, absolve, exonerate, acquit, vindicate mean to free from a charge. EXCULPATE implies a clearing from blame or fault often in a matter of small importance ⟨*exculpating* himself from the charge of overenthusiasm⟩. ABSOLVE implies a release either from an obligation that binds the conscience or from the consequences of disobeying the law or committing a sin ⟨cannot be *absolved* of blame⟩. EXONERATE implies a complete clearance from an accusation or charge and from any attendant suspicion of blame or guilt ⟨*exonerated* by the investigation⟩. ACQUIT implies a formal decision in one's favor with respect to a definite charge ⟨voted to *acquit* the defendant⟩. VINDICATE may refer to things as well as persons that have been subjected to critical attack or imputation of guilt, weakness, or folly, and implies a clearing effected by proving the unfairness of such criticism or blame ⟨her judgment was *vindicated*⟩.

— **ex·cur·sive·ness** *noun*
ex·cur·sus \ik-ˈskər-səs\ *noun, plural* **ex·cur·sus·es** *also* **ex·cur·sus** \-səs, -ˌsüs\ [Latin, digression, from *excurrere*] (1803)
: an appendix or digression that contains further exposition of some point or topic
ex·cu·sa·to·ry \ik-ˈskyü-zə-ˌtōr-ē, -ˌtȯr-\ *adjective* (1535)
: making or containing excuse
¹ex·cuse \ik-ˈskyüz, *imperatively often* ˈskyüz\ *transitive verb* **ex·cused; ex·cus·ing** [Middle English, from Old French *excuser,* from Latin *excusare,* from *ex-* + *causa* cause, explanation] (13th century)
1 a : to make apology for **b :** to try to remove blame from
2 : to forgive entirely or disregard as of trivial import **:** regard as excusable ⟨graciously *excused* his tardiness⟩
3 a : to grant exemption or release to ⟨was *excused* from jury duty⟩ **b :** to allow to leave ⟨*excused* the class⟩
4 : to serve as excuse for **:** JUSTIFY ⟨nothing can *excuse* such neglect⟩ ☆
— **ex·cus·able** \ik-ˈskyü-zə-bəl\ *adjective*
— **ex·cus·able·ness** *noun*
— **ex·cus·ably** \-blē\ *adverb*
— **ex·cus·er** *noun*
²ex·cuse \ik-ˈskyüs\ *noun* (14th century)
1 : the act of excusing
2 a : something offered as justification or as grounds for being excused **b** *plural* **:** an expression of regret for failure to do something **c :** a note of explanation of an absence
3 : JUSTIFICATION, REASON
synonym see APOLOGY
ex–di·rec·to·ry \ˌeks-də-ˈrek-t(ə-)rē, -dī-\ *adjective* [Latin *ex* out of — more at EX-] (1936)
British **:** not listed in a telephone book **:** UNLISTED
ex·ec \ig-ˈzek\ *noun* (1896)
1 : EXECUTIVE OFFICER
2 : EXECUTIVE
ex·e·cra·ble \ˈek-si-krə-bəl\ *adjective* (14th century)
1 : deserving to be execrated **:** DETESTABLE ⟨*execrable* crimes⟩
2 : very bad **:** WRETCHED ⟨*execrable* hotel food⟩
— **ex·e·cra·ble·ness** *noun*
— **ex·e·cra·bly** \-blē\ *adverb*
ex·e·crate \ˈek-sə-ˌkrāt\ *transitive verb* **-crat·ed; -crat·ing** [Latin *exsecratus,* past participle of *exsecrari* to put under a curse, from *ex* + *sacr-, sacer* sacred] (1561)
1 : to declare to be evil or detestable **:** DENOUNCE
2 : to detest utterly
— **ex·e·cra·tive** \-ˌkrā-tiv\ *adjective*
— **ex·e·cra·tor** \-ˌkrā-tər\ *noun*
ex·e·cra·tion \ˌek-sə-ˈkrā-shən\ *noun* (14th century)
1 : the act of cursing or denouncing; *also* **:** the curse so uttered
2 : an object of curses **:** something detested
ex·ec·u·tant \ig-ˈze-k(y)ə-tənt\ *noun* (1858)
: one who executes or performs; *especially* **:** one skilled in the technique of an art **:** PERFORMER
ex·e·cute \ˈek-si-ˌkyüt\ *verb* **-cut·ed; -cut·ing** [Middle English, from Middle French *executer,* back-formation from *execution*] (14th century)
transitive verb
1 : to carry out fully **:** put completely into effect ⟨*execute* a command⟩
2 : to do what is provided or required by ⟨*execute* a decree⟩
3 : to put to death especially in compliance with a legal sentence
4 : to make or produce (as a work of art) especially by carrying out a design
5 : to perform what is required to give validity to ⟨*execute* a deed⟩
6 : PLAY ⟨*execute* a piece of music⟩
intransitive verb
1 : to perform properly or skillfully the fundamentals of a sport or of a particular play ⟨never had a team *execute* better —Bobby Knight⟩
2 : to perform indicated tasks according to encoded instructions — used of a computer program or routine
synonym see KILL, PERFORM
— **ex·e·cut·able** \-ˌkyü-tə-bəl\ *adjective*
ex·e·cu·tion \ˌek-si-ˈkyü-shən\ *noun* [Middle English, from Middle French, from Latin *execution-, exsecutio,* from *exsequi* to execute, from *ex-* + *sequi* to follow — more at SUE] (14th century)
1 : the act or process of executing **:** PERFORMANCE
2 : a putting to death especially as a legal penalty
3 : the process of enforcing a legal judgment (as against a debtor); *also* **:** a judicial writ directing such enforcement
4 : the act or mode or result of performance
5 *archaic* **:** effective or destructive action ⟨his brandished steel, which smoked with bloody *execution* —Shakespeare⟩ — usually used with *do* ⟨as soon as day came, we went out to see what *execution* we had done —Daniel Defoe⟩
ex·e·cu·tion·er \-sh(ə-)nər\ *noun* (1536)
: one who executes; *especially* **:** one who puts to death
¹ex·ec·u·tive \ig-ˈze-k(y)ə-tiv\ *adjective* (1649)
1 a : of or relating to the execution of the laws and the conduct of public and national affairs **b :** belonging to the branch of government that is charged with such powers as diplomatic representation, superintendence of the execution of the laws, and appointment of officials and that usually has some power over legislation (as through veto) — compare JUDICIAL, LEGISLATIVE
2 a : designed for or relating to execution or carrying into effect ⟨*executive* board⟩ **b :** having administrative or managerial responsibility ⟨*executive* director⟩
3 : of or relating to an executive ⟨the *executive* offices⟩
²executive *noun* (1774)
1 : the executive branch of a government; *also* **:** the person or persons who constitute the executive magistracy of a state
2 : a directing or controlling office of an organization
3 : one that exercises administrative or managerial control
executive agreement *noun* (1942)
: an agreement between the U.S. and a foreign government made by the executive branch of the government either alone or with Congressional approval and dealing usually with routine matters
executive council *noun* (1778)
1 : a council constituted to advise or share in the functions of a political executive
2 : a council that exercises supreme executive power
executive officer *noun* (1881)
: the officer second in command of a military or naval organization
executive order *noun* (1883)
: REGULATION 2b
executive privilege *noun* (1940)
: exemption from legally enforced disclosure of communications within the executive branch of government when such disclosure would adversely affect the functions and decision-making processes of the executive branch
executive secretary *noun* (1950)
: a secretary having administrative duties; *especially* **:** an official responsible for administering the activities and business affairs of an organization
executive session *noun* (1840)
: a usually closed session (as of a legislative body) that functions as an executive council (as of the U.S. Senate when considering appointments or the ratification of treaties)
ex·ec·u·tor \ig-ˈze-k(y)ə-tər *or in sense 1* ˈek-sə-ˌkyü-\ *noun* [Middle English, from Old French, from Latin *exsecutor,* from *exsequi*] (13th century)
1 a : one who executes something **b** *obsolete* **:** EXECUTIONER
2 a : the person appointed by a testator to execute a will **b :** LITERARY EXECUTOR
— **ex·ec·u·to·ri·al** \ig-ˌze-k(y)ə-ˈtōr-ē-əl, -ˈtȯr-\ *adjective*
ex·ec·u·to·ry \ig-ˈze-k(y)ə-ˌtōr-ē, -ˌtȯr-\ *adjective* (1592)
1 : designed or of such a nature as to be executed in time to come or to take effect on a future contingency
2 : relating to administration
ex·ec·u·trix \ig-ˈze-k(y)ə-(ˌ)triks\ *noun, plural* **ex·ec·u·tri·ces** \-ˌze-k(y)ə-ˈtrī-(ˌ)sēz\ *or* **ex·ec·u·trix·es** \-ˈze-k(y)ə-ˌtrik-səz\ (15th century)
: a woman who is an executor
ex·e·dra \ˈek-sə-drə\ *noun, plural* **-drae** \-ˌdrē, -ˌdrī\ [Latin, from Greek, from *ex-* + *hedra* seat — more at SIT] (1706)
1 : a room (as in a temple or house) in ancient Greece and Rome used for conversation and formed by an open or columned recess often semicircular in shape and furnished with seats
2 : a large outdoor nearly semicircular seat with a solid back
ex·e·ge·sis \ˌek-sə-ˈjē-səs, ˈek-sə-ˌ\ *noun, plural* **-ge·ses** \-ˈjē-(ˌ)sēz\ [New Latin, from Greek *exēgēsis,* from *exēgeisthai* to explain, interpret, from *ex-* + *hēgeisthai* to lead — more at SEEK] (1619)
: EXPOSITION, EXPLANATION; *especially* **:** an explanation or critical interpretation of a text
ex·e·gete \ˈek-sə-ˌjēt\ *noun* [Greek *exēgētēs,* from *exēgeisthai*] (circa 1736)
: one who practices exegesis
ex·e·get·i·cal \ˌek-sə-ˈje-ti-kəl\ *also* **ex·e·get·ic** \-tik\ *adjective* [Greek *exēgētikos,* from *exēgeisthai*] (1623)
: of or relating to exegesis **:** EXPLANATORY
ex·e·get·ist \-ˈjē-tist, -ˈje-\ *noun* (1848)
: EXEGETE
ex·em·plar \ig-ˈzem-ˌplär, -plər, eg-\ *noun* [Middle English, from Latin, from *exemplum* example] (15th century)

☆ SYNONYMS
Excuse, condone, pardon, forgive mean to exact neither punishment nor redress. EXCUSE may refer to specific acts especially in social or conventional situations or the person responsible for these ⟨*excuse* an interruption⟩ ⟨*excused* them for interrupting⟩. Often the term implies extenuating circumstances ⟨injustice *excuses* strong responses⟩. CONDONE implies that one overlooks without censure behavior (as dishonesty or violence) that involves a serious breach of a moral, ethical, or legal code, and the term may refer to the behavior or to the agent responsible for it ⟨a society that *condones* alcohol but not narcotics⟩. PARDON implies that one remits a penalty due for an admitted or established offense ⟨*pardon* a criminal⟩. FORGIVE implies that one gives up all claim to requital and to resentment or vengeful feelings ⟨could not *forgive* their rudeness⟩.

: one that serves as a model or example: as **a** : an ideal model **b** : a typical or standard specimen **c** : a copy of a book or writing **d** : IDEA 1a

synonym see MODEL

ex•em•pla•ry \ig-ˈzem-plə-rē\ *adjective* (1589)
1 a : serving as a pattern **b** : deserving imitation : COMMENDABLE ⟨his courage was *exemplary*⟩; *also* : deserving imitation because of excellence ⟨they serve *exemplary* pastries —G. V. Higgins⟩
2 : serving as a warning : MONITORY
3 : serving as an example, instance, or illustration
— **ex•em•plar•i•ly** \ˌeg-ˌzem-ˈpler-ə-lē\ *adverb*
— **ex•em•pla•ri•ness** \ig-ˈzem-plə-rē-nəs\ *noun*
— **ex•em•plar•i•ty** \ˌeg-ˌzem-ˈplar-ə-tē\ *noun*

ex•em•pli•fi•ca•tion \ig-ˌzem-plə-fə-ˈkā-shən\ *noun* (1510)
1 a : the act or process of exemplifying **b** : EXAMPLE, CASE IN POINT
2 : an exemplified copy of a document

ex•em•pli•fy \ig-ˈzem-plə-ˌfī\ *transitive verb* **-fied; -fy•ing** [Middle English *exemplifien*, from Middle French *exemplifier*, from Medieval Latin *exemplificare*, from Latin *exemplum*] (15th century)
1 : to show or illustrate by example
2 : to make an attested copy or transcript of (a document) under seal
3 a : to be an instance of or serve as an example : EMBODY **b** : to be typical of

ex•em•pli gra•tia \ig-ˌzem-(ˌ)plē-ˈgrä-tē-ˌä, -ˈgrā-sh(ē-)ə\ *adverb* [Latin] (1602)
: for example

ex•em•plum \ig-ˈzem-pləm, eg-\ *noun, plural* **-pla** \-plə\ [Latin] (1890)
1 : EXAMPLE, MODEL
2 : an anecdote or short narrative used to point a moral or sustain an argument

¹ex•empt \ig-ˈzem(p)t\ *adjective* [Middle English, from Latin *exemptus*, past participle of *eximere* to take out — more at EXAMPLE] (14th century)
1 *obsolete* : set apart
2 : free or released from some liability or requirement to which others are subject ⟨was *exempt* from jury duty⟩ ⟨the estate was *exempt* from taxes⟩

²exempt *transitive verb* (15th century)
1 : to release or deliver from some liability or requirement to which others are subject : EXCUSE ⟨a man *exempted* from military service⟩
2 *obsolete* : to set apart

³exempt *noun* (1670)
: one exempted or freed from duty

ex•emp•tion \ig-ˈzem(p)-shən\ *noun* (14th century)
1 : the act of exempting or state of being exempt : IMMUNITY
2 : one that exempts or is exempted; *especially* : a source or amount of income exempted from taxation

ex•en•ter•ate \ig-ˈzen-tə-ˌrāt\ *transitive verb* **-at•ed; -at•ing** [Latin *exenteratus*, past participle of *exenterare* to disembowel, modification of Greek *exenterizein*, from *ex-* + *enteron* intestine — more at INTER-] (1607)
: to remove the contents of (as the orbit or pelvis)
— **ex•en•ter•a•tion** \-ˌzen-tə-ˈrā-shən\ *noun*

¹ex•er•cise \ˈek-sər-ˌsīz\ *noun* [Middle English, from Middle French *exercice*, from Latin *exercitium*, from *exercitare* to train, exercise, frequentative of *exercēre* to train, occupy, from *ex-* + *arcēre* to enclose, hold off — more at ARK] (14th century)
1 a : the act of bringing into play or realizing in action : USE **b** : the discharge of an official function or professional occupation **c** : the act or an instance of carrying out the terms of an agreement (as an option)
2 a : regular or repeated use of a faculty or bodily organ **b** : bodily exertion for the sake of developing and maintaining physical fitness
3 : something performed or practiced in order to develop, improve, or display a specific power or skill ⟨arithmetic *exercises*⟩
4 : a performance having a strongly marked secondary or ulterior aspect ⟨party politics has always been an *exercise* in compromise —H. S. Ashmore⟩
5 a : a maneuver, operation, or drill carried out for training and discipline **b** *plural* : a program including speeches, announcements of awards and honors, and various traditional practices of secular or religious character ⟨commencement *exercises*⟩

²exercise *verb* **-cised; -cis•ing** (14th century)
transitive verb
1 a : to make effective in action : USE ⟨didn't *exercise* good judgment⟩ **b** : to bring to bear : EXERT ⟨*exercise* influence⟩ **c** : to implement the terms of (as an option)
2 a : to use repeatedly in order to strengthen or develop **b** : to train (as troops) by drills and maneuvers **c** : to put through exercises ⟨*exercise* the horses⟩
3 a : to engage the attention and effort of **b** : to cause anxiety, alarm, or indignation in ⟨the issues *exercising* voters this year⟩
intransitive verb
: to take exercise
— **ex•er•cis•able** \-ˌsī-zə-bəl\ *adjective*

ex•er•cis•er \ˈek-sər-ˌsī-zər\ *noun* (1552)
1 : one that exercises
2 : an apparatus for use in physical exercise

ex•er•ci•ta•tion \ig-ˌzər-sə-ˈtā-shən\ *noun* [Middle English *exercitacioun*, from Latin *exercitation-*, *exercitatio*, from *exercitare*] (14th century)
: EXERCISE

Ex•er•cy•cle \ˈek-sər-ˌsī-kəl\ *trademark*
— used for a stationary bicycle

ex•er•gon•ic \ˌek-(ˌ)sər-ˈgä-nik\ *adjective* [*exo-* + Greek *ergon* work — more at WORK] (1940)
: EXOTHERMIC ⟨an *exergonic* biochemical reaction⟩

ex•ergue \ˈek-ˌsərg, ˈeg-ˌzərg\ *noun* [French, from New Latin *exergum*, from Greek *ex* out of + *ergon* work] (1697)
: a space on a coin, token, or medal usually on the reverse below the central part of the design

ex•ert \ig-ˈzərt\ *transitive verb* [Latin *exsertus*, past participle of *exserere* to thrust out, from *ex-* + *serere* to join — more at SERIES] (1681)
1 a : to put forth (as strength) **b** : to put (oneself) into action or to tiring effort
2 : to bring to bear especially with sustained effort or lasting effect
3 : EMPLOY, WIELD ⟨*exerted* her leadership abilities intelligently⟩

ex•er•tion \ig-ˈzər-shən\ *noun* (1677)
: the act or an instance of exerting; *especially* : a laborious or perceptible effort

ex•e•unt \ˈek-sē-(ˌ)ənt, -ˌu̇nt\ [Latin, they go out, from *exire* to go out — more at EXIT] (15th century)
— used as a stage direction to specify that all or certain named characters leave the stage

ex•fo•li•ate \(ˌ)eks-ˈfō-lē-ˌāt\ *verb* **-at•ed; -at•ing** [Late Latin *exfoliatus*, past participle of *exfoliare* to strip of leaves, from Latin *ex-* + *folium* leaf — more at BLADE] (1612)
transitive verb
1 : to cast off in scales, laminae, or splinters
2 : to remove the surface of in scales or laminae
3 : to spread or extend by or as if by opening out leaves
intransitive verb
1 : to split into or give off scales, laminae, or body cells
2 : to come off in thin layers or scales
3 : to grow by or as if by producing or unfolding leaves
— **ex•fo•li•a•tion** \(ˌ)eks-ˌfō-lē-ˈā-shən\ *noun*
— **ex•fo•li•a•tive** \eks-ˈfō-le-ˌā-tiv\ *adjective*

ex gra•tia \(ˌ)eks-ˈgrā-sh(ē-)ə\ *adjective or adverb* [New Latin] (1769)
: as a favor : not compelled by legal right ⟨*ex gratia* pension payments⟩

ex•hal•ant *or* **ex•hal•ent** \eks-ˈhā-lənt, ek-ˈsā-\ *adjective* (1771)
: bearing out or outward : EMISSIVE ⟨an *exhalant* siphon of a clam⟩

ex•ha•la•tion \ˌeks-hə-ˈlā-shən, ˌek-sə-\ *noun* (14th century)
1 : something exhaled or given off : EMANATION
2 : an act of exhaling

ex•hale \eks-ˈhā(ə)l, ek-ˈsā(ə)l\ *verb* **exhaled; ex•hal•ing** [Middle English *exalen*, from Latin *exhalare*, from *ex-* + *halare* to breathe] (14th century)
intransitive verb
1 : to rise or be given off as vapor
2 : to emit breath or vapor
transitive verb
1 a : to breathe out **b** : to give forth (gaseous matter) : EMIT
2 *archaic* : to cause to be emitted in vapor

¹ex•haust \ig-ˈzȯst\ *verb* [Latin *exhaustus*, past participle of *exhaurire*, from *ex-* + *haurire* to draw; akin to Middle High German *œsen* to empty, Greek *auein* to take] (1533)
transitive verb
1 a : to consume entirely : USE UP ⟨*exhausted* our funds in a week⟩ **b** : to tire extremely or completely ⟨*exhausted* by overwork⟩ **c** : to deprive of a valuable quality or constituent ⟨*exhaust* a photographic developer⟩ ⟨*exhaust* a soil of fertility⟩
2 a : to draw off or let out completely **b** : to empty by drawing off the contents; *specifically* : to create a vacuum in
3 a : to develop (a subject) completely **b** : to try out the whole number of ⟨*exhausted* all the possibilities⟩
intransitive verb
: DISCHARGE, EMPTY ⟨the engine *exhausts* through the muffler⟩
synonym see DEPLETE, TIRE
— **ex•haust•er** *noun*
— **ex•haust•ibil•i•ty** \-ˌzȯ-stə-ˈbi-lə-tē\ *noun*
— **ex•haust•ible** \-ˈzȯ-stə-bəl\ *adjective*

²exhaust *noun* (1848)
1 a : the escape of used gas or vapor from an engine **b** : the gas or vapor thus escaping
2 a : the conduit through which used gases escape **b** : an arrangement for removing fumes, dusts, or odors from an enclosure
3 : EXHAUSTION

ex•haus•tion \ig-ˈzȯs-chən\ *noun* (1646)
: the act or process of exhausting : the state of being exhausted

ex•haus•tive \ig-ˈzȯ-stiv\ *adjective* (1786)
: testing all possibilities or considering all elements : THOROUGH ⟨conducted an *exhaustive* investigation⟩
— **ex•haus•tive•ly** *adverb*
— **ex•haus•tive•ness** *noun*
— **ex•haus•tiv•i•ty** \-ˌzȯ-ˈsti-və-tē\ *noun*

ex•haust•less \ig-ˈzȯst-ləs\ *adjective* (1712)
: not to be exhausted : INEXHAUSTIBLE
— **ex•haust•less•ly** *adverb*
— **ex•haust•less•ness** *noun*

¹ex•hib•it \ig-ˈzi-bət\ *verb* [Middle English, from Latin *exhibitus*, past participle of *exhibēre*, from *ex-* + *habēre* to have, hold — more at GIVE] (15th century)
transitive verb

1 : to submit (as a document) to a court or officer in course of proceedings; *also* **:** to present or offer officially or in legal form
2 : to present to view: as **a :** to show or display outwardly especially by visible signs or actions ⟨*exhibited* no fear⟩ **b :** to have as a readily discernible quality or feature ⟨in all cultures we know, men *exhibit* an aesthetic sense —H. J. Muller⟩ **c :** to show publicly especially for purposes of competition or demonstration
intransitive verb
: to display something for public inspection
synonym see SHOW
— ex·hib·i·tive \-bə-tiv\ *adjective*
— ex·hib·i·tor \-bə-tər\ *noun*
— ex·hib·i·to·ry \-bə-ˌtōr-ē, -ˌtȯr-\ *adjective*

²exhibit *noun* (1626)
1 : a document or material object produced and identified in court or before an examiner for use as evidence
2 : something exhibited
3 : an act or instance of exhibiting

ex·hi·bi·tion \ˌek-sə-ˈbi-shən\ *noun* (14th century)
1 : an act or instance of exhibiting
2 *British* **:** a grant drawn from the funds of a school or university to help maintain a student
3 : a public showing (as of works of art, objects of manufacture, or athletic skill) ⟨a one-man *exhibition*⟩ ⟨an *exhibition* game⟩

ex·hi·bi·tion·er \-ˈbish-nər, -ˈbi-shə-\ *noun* (1679)
British **:** one who holds a grant from a school or university

ex·hi·bi·tion·ism \-ˈbi-shə-ˌni-zəm\ *noun* (1893)
1 a : a perversion marked by a tendency to indecent exposure **b :** an act of such exposure
2 : the act or practice of behaving so as to attract attention to oneself
— ex·hi·bi·tion·ist \-ˈbish-nist, -ˈbi-shə-\ *noun or adjective*
— ex·hi·bi·tion·is·tic \ˌbi-shə-ˈnis-tik\ *adjective*
— ex·hi·bi·tion·is·ti·cal·ly \-ti-k(ə-)lē\ *adverb*

ex·hil·a·rate \ig-ˈzi-lə-ˌrāt\ *transitive verb* **-rat·ed; -rat·ing** [Latin *exhilaratus,* past participle of *exhilarare,* from *ex-* + *hilarare* to gladden, from *hilarus* cheerful — more at HILARIOUS] (1540)
1 a : to make cheerful **b :** ENLIVEN, EXCITE
2 : REFRESH, STIMULATE
— ex·hil·a·rat·ing·ly \ig-ˈzi-lə-ˌrā-tiŋ-lē\ *adverb*
— ex·hil·a·ra·tive \-ˌrā-tiv\ *adjective*

ex·hil·a·ra·tion \ig-ˌzi-lə-ˈrā-shən\ *noun* (circa 1626)
1 : the action of exhilarating
2 : the feeling or the state of being exhilarated

ex·hort \ig-ˈzȯrt\ *verb* [Middle English, from Middle French *exhorter,* from Latin *exhortari,* from *ex-* + *hortari* to incite — more at YEARN] (15th century)
transitive verb
: to incite by argument or advice **:** urge strongly
intransitive verb
: to give warnings or advice **:** make urgent appeals
— ex·hort·er *noun*

ex·hor·ta·tion \ˌek-ˌsȯr-ˈtā-shən, -sər-; ˌeg-ˌzȯr-, -zər-\ *noun* (14th century)
1 : an act or instance of exhorting
2 : language intended to incite and encourage

ex·hor·ta·tive \ig-ˈzȯr-tə-tiv\ *adjective* (15th century)
: serving to exhort

ex·hor·ta·to·ry \-tə-ˌtōr-ē, -ˌtȯr-\ *adjective* (15th century)
: using exhortation **:** EXHORTATIVE ⟨an *exhortatory* appeal⟩

ex·hume \ig-ˈzüm, igz-ˈyüm, iks-ˈ(h)yüm\ *transitive verb* **ex·humed; ex·hum·ing** [French or Medieval Latin; French *exhumer,* from Medieval Latin *exhumare,* from Latin *ex* out of + *humus* earth — more at EX-, HUMBLE] (15th century)
1 : DISINTER
2 : to bring back from neglect or obscurity
— ex·hu·ma·tion \ˌeks-(h)yü-ˈmā-shən, ˌeg-zü-, ˌegz-yü-\ *noun*
— ex·hum·er \ig-ˈzü-mər, igz-ˈyü-, iks-ˈ(h)yü-\ *noun*

ex hy·po·the·si \ˌeks-hī-ˈpä-thə-ˌsī\ *adverb* [New Latin, from a hypothesis] (1603)
: according to assumptions made **:** by hypothesis ⟨regard . . . all elites as *ex hypothesi* incompatible with democracy —P. G. J. Pulzer⟩

ex·i·gence \ˈek-sə-jən(t)s\ *noun* (15th century)
: EXIGENCY

ex·i·gen·cy \ˈek-sə-jən(t)-sē, ig-ˈzi-jən(t)-\ *noun, plural* **-cies** (1581)
1 : that which is required in a particular situation — usually used in plural
2 a : the quality or state of being exigent **b :** a state of affairs that makes urgent demands ⟨a leader must act in any sudden *exigency*⟩
synonym see JUNCTURE

ex·i·gent \ˈek-sə-jənt\ *adjective* [Latin *exigent-, exigens,* present participle of *exigere* to demand — more at EXACT] (1629)
1 : requiring immediate aid or action
2 : requiring or calling for much **:** DEMANDING
— ex·i·gent·ly *adverb*

ex·i·gu·ity \ˌeg-zi-ˈgyü-ə-tē\ *noun, plural* **-ities** (circa 1626)
: the quality or state of being exiguous **:** SCANTINESS

ex·ig·u·ous \ig-ˈzi-gyə-wəs\ *adjective* [Latin *exiguus,* from *exigere*] (1651)
: excessively scanty **:** INADEQUATE ⟨wrest an *exiguous* existence from the land⟩ ⟨*exiguous* evidence⟩
— ex·ig·u·ous·ly *adverb*
— ex·ig·u·ous·ness *noun*

¹ex·ile \ˈeg-ˌzīl, ˈek-ˌsīl\ *noun* [Middle English *exil,* from Middle French, from Latin *exilium* from *exul, exsul* an exile] (14th century)
1 a : the state or a period of forced absence from one's country or home **b :** the state or a period of voluntary absence from one's country or home
2 : a person who is in exile
— ex·il·ic \eg-ˈzi-lik\ *adjective*

²exile *transitive verb* **ex·iled; ex·il·ing** (14th century)
: to banish or expel from one's own country or home
synonym see BANISH

ex·im·i·ous \eg-ˈzi-mē-əs\ *adjective* [Latin *eximius,* from *eximere* to take out — more at EXAMPLE] (1547)
archaic **:** CHOICE, EXCELLENT

ex·ine \ˈek-ˌsēn, -ˌsīn\ *noun* [International Scientific Vocabulary *ex-* + *in-* fibrous tissue, from Greek *in-, is* tendon] (circa 1884)
: the outer of the two major layers forming the walls of some spores and especially pollen grains

ex·ist \ig-ˈzist\ *intransitive verb* [Latin *exsistere* to come into being, exist, from *ex-* + *sistere* to stand, stop; akin to Latin *stare* to stand — more at STAND] (1602)
1 a : to have real being whether material or spiritual ⟨did unicorns *exist*⟩ ⟨the largest galaxy known to *exist*⟩ **b :** to have being in a specified place or with respect to understood limitations or conditions ⟨strange ideas *existed* in his mind⟩
2 : to continue to be ⟨racism still *exists* in society⟩
3 a : to have life or the functions of vitality ⟨we cannot *exist* without oxygen⟩ **b :** to live at an inferior level or under adverse circumstances ⟨the hungry *existing* from day to day⟩

ex·is·tence \ig-ˈzis-tən(t)s\ *noun* (14th century)
1 a *obsolete* **:** reality as opposed to appearance **b :** reality as presented in experience **c** (1) **:** the totality of existent things (2) **:** a particular being ⟨all the fair *existences* of heaven —John Keats⟩ **d :** sentient or living being **:** LIFE
2 a : the state or fact of having being especially independently of human consciousness and as contrasted with nonexistence **b :** the manner of being that is common to every mode of being **c :** being with respect to a limiting condition or under a particular aspect
3 : continued or repeated manifestation

ex·is·tent \-tənt\ *adjective* [Latin *exsistent-, exsistens,* present participle of *exsistere*] (1561)
1 : having being **:** EXISTING
2 : existing now **:** PRESENT
— existent *noun*

ex·is·ten·tial \ˌeg-(ˌ)zis-ˈten(t)-shəl, ˌek-(ˌ)sis-\ *adjective* (1693)
1 : of, relating to, or affirming existence ⟨*existential* propositions⟩
2 a : grounded in existence or the experience of existence **:** EMPIRICAL **b :** having being in time and space
3 [translation of Danish *eksistentiel* & German *existential*] **:** EXISTENTIALIST
— ex·is·ten·tial·ly *adverb*

ex·is·ten·tial·ism \-ˈten(t)-shə-ˌli-zəm\ *noun* (1941)
: a chiefly 20th century philosophical movement embracing diverse doctrines but centering on analysis of individual existence in an unfathomable universe and the plight of the individual who must assume ultimate responsibility for his acts of free will without any certain knowledge of what is right or wrong or good or bad

¹ex·is·ten·tial·ist \-list\ *noun* (1942)
: an adherent of existentialism

²existentialist *adjective* (1946)
: of or relating to existentialism or existentialists
— ex·is·ten·tial·is·tic \-ˌten(t)-shə-ˈlis-tik\ *adjective*
— ex·is·ten·tial·is·ti·cal·ly \-ti-k(ə-)lē\ *adverb*

existential quantifier *noun* (1940)
: a quantifier (as *for some* in "for some x, $2x + 5 = 8$") that asserts that there exists at least one value of a variable — called also *existential operator*

¹ex·it \ˈeg-zət, ˈek-sət\ [Latin, he goes out, from *exire* to go out, from *ex-* + *ire* to go — more at ISSUE] (1538)
— used as a stage direction to specify who goes off stage

²exit *noun* [Latin *exitus,* from *exire*] (1588)
1 [¹*exit*] **:** a departure from a stage
2 a : the act of going out or going away **b :** DEATH
3 : a way out of an enclosed place or space
4 : one of the designated points of departure from an expressway
— ex·it·less *adjective*

³exit (1607)
intransitive verb
1 : to go out or away **:** DEPART
2 : DIE
transitive verb
: LEAVE 3a

exit poll *noun* (1980)
: a poll taken (as by news media) of voters leaving the voting place that is usually used for predicting the winners
— exit polling *noun*

ex li·bris \eks-ˈlē-brəs, -ˌbrēs\ *noun, plural* **ex libris** [New Latin, from the books; used before the owner's name on bookplates] (1880)
: BOOKPLATE

Ex·moor \'ek-ˌsmu̇r, -ˌsmōr, -ˌsmȯr\ *noun* [*Exmoor,* England] (1808)
1 : any of a breed of horned sheep of Devonshire in England valued especially for mutton
2 : any of a breed of hardy ponies native to the Exmoor district and having a mealy-colored muzzle

Exmoor 1

ex ni·hi·lo \(ˌ)eks-'nē-(h)ə-ˌlō, -'ni-, -'nī-\ *adverb or adjective* [Latin] (1580)
: from or out of nothing ⟨creation *ex nihilo*⟩
exo- *or* **ex-** *combining form* [Greek *exō* out, outside, from *ex* out of — more at EX-]
1 : outside ⟨*exo*gamy⟩ **:** outer ⟨*exo*skeleton⟩ — compare ECT-, END-
2 : turning out ⟨*exo*ergic⟩
exo·bi·ol·o·gy \ˌek-sō-bī-'ä-lə-jē\ *noun* (1960)
: a branch of biology concerned with the search for life outside the earth and with the effects of extraterrestrial environments on living organisms
— **exo·bi·o·log·i·cal** \-ˌbī-ə-'lä-ji-kəl\ *adjective*
— **exo·bi·ol·o·gist** \-bī-'ä-lə-jist\ *noun*
exo·carp \'ek-sō-ˌkärp\ *noun* [International Scientific Vocabulary] (circa 1845)
: the outermost layer of the pericarp of a fruit — see ENDOCARP illustration
exo·crine \'ek-sə-krən, -ˌkrīn, -ˌkrēn\ *adjective* [International Scientific Vocabulary *exo-* + Greek *krinein* to separate — more at CERTAIN] (circa 1911)
: producing, being, or relating to a secretion that is released outside its source ⟨*exocrine* pancreatic cells⟩
exocrine gland *noun* (circa 1927)
: a gland (as a salivary gland or part of the pancreas) that releases a secretion external to or at the surface of an organ by means of a canal or duct
exo·cy·clic \ˌek-sō-'sī-klik, -'si-\ *adjective* (1888)
: situated outside of a ring in a chemical structure
exo·cy·to·sis \ˌek-sō-sī-'tō-səs\ *noun, plural* **-to·ses** \-ˌsēz\ [New Latin, from *exo-* + *cyt-* + *-osis*] (1963)
: the release of cellular substances (as secretory products) contained in cell vesicles by fusion of the vesicular membrane with the plasma membrane and subsequent release of the contents to the exterior of the cell
— **exo·cy·tot·ic** \-'tä-tik\ *adjective*
exo·der·mis \ˌek-sō-'dər-məs\ *noun* [New Latin] (circa 1900)
: a layer of the outer living cortical cells of plants that takes over the functions of the epidermis in roots lacking secondary thickening
ex·odon·tia \ˌek-sə-'dän(t)-sh(ē-)ə\ *noun* [New Latin, from *ex-* + *-odontia*] (1913)
: a branch of dentistry that deals with the extraction of teeth
— **ex·odon·tist** \-'dän-tist\ *noun*
ex·o·dus \'ek-sə-dəs, 'eg-zə-\ *noun* [Latin, from Greek *Exodos,* literally, road out, from *ex-* + *hodos* road]
1 *capitalized* **:** the mainly narrative second book of canonical Jewish and Christian Scripture — see BIBLE table
2 : a mass departure **:** EMIGRATION
exo·en·zyme \ˌek-sō-'en-ˌzīm\ *noun* [International Scientific Vocabulary] (circa 1923)
: an extracellular enzyme
exo·er·gic \ˌek-sō-'ər-jik\ *adjective* (1942)
: releasing energy **:** EXOTHERMIC
exo·eryth·ro·cyt·ic \ˌek-sō-i-ˌri-thrə-'si-tik\ *adjective* (1942)
: occurring outside the red blood cells — used especially of stages of malaria parasites
ex of·fi·cio \ˌek-sə-'fi-shē-ˌō, -sē-\ *adverb or adjective* [Late Latin] (1533)
: by virtue or because of an office ⟨the Vice President serves *ex officio* as president of the Senate⟩
ex·og·a·my \ek-'sä-gə-mē\ *noun, plural* **-mies** (1865)
: marriage outside of a specific group especially as required by custom or law
— **ex·og·a·mous** \ek-'sä-gə-məs\ *or* **exo·gam·ic** \ˌek-sə-'ga-mik\ *adjective*
ex·og·e·nous \ek-'sä-jə-nəs\ *adjective* [French *exogène* exogenous, from *exo-* + *-gène* (from Greek *-genēs* born) — more at -GEN] (1830)
1 : produced by growth from superficial tissue ⟨*exogenous* roots produced by leaves⟩
2 a : caused by factors (as food or a traumatic factor) or an agent (as a disease-producing organism) from outside the organism or system ⟨*exogenous* obesity⟩ ⟨*exogenous* psychic depression⟩ **b :** introduced from or produced outside the organism or system; *specifically* **:** not synthesized within the organism or system
— **ex·og·e·nous·ly** *adverb*
ex·on \'ek-ˌsän\ *noun* [*ex*pressed sequence + [2]*-on*] (circa 1978)
: a polynucleotide sequence in a nucleic acid that codes information for protein synthesis and that is copied and spliced together with other such sequences to form messenger RNA — compare INTRON
— **ex·on·ic** \ek-'sä-nik\ *adjective*
ex·on·er·ate \ig-'zä-nə-ˌrāt, eg-\ *transitive verb* **-at·ed; -at·ing** [Middle English, from Latin *exoneratus,* past participle of *exonerare* to unburden, from *ex-* + *oner-, onus* load] (1524)
1 : to relieve of a responsibility, obligation, or hardship
2 : to clear from accusation or blame
synonym see EXCULPATE
— **ex·on·er·a·tion** \-ˌzä-nə-'rā-shən\ *noun*
— **ex·on·er·a·tive** \-'zä-nə-ˌrā-tiv\ *adjective*
exo·nu·cle·ase \ˌek-sō-'nü-klē-ˌās, -'nyü-, -ˌāz\ *noun* (1963)
: an enzyme that breaks down a nucleic acid by removing nucleotides one by one from the end of a chain — compare ENDONUCLEASE
exo·nu·mia \ˌek-sə-'nü-mē-ə, -'nyü-\ *noun plural* [New Latin, from *exo-* + English *numismatic* + New Latin *-ia*] (1966)
: numismatic items (as tokens, medals, or scrip) other than coins and paper money
exo·pep·ti·dase \-'pep-tə-ˌdās, -ˌdāz\ *noun* (1936)
: any of a group of enzymes that hydrolyze peptide bonds formed by the terminal amino acids of peptide chains **:** PEPTIDASE — compare ENDOPEPTIDASE
ex·oph·thal·mos *also* **ex·oph·thal·mus** \ˌek-säf-'thal-məs, -səf-, -säp-\ *noun* [New Latin, from Greek *exophthalmos* having prominent eyes, from *ex* out + *ophthalmos* eye; akin to Greek *ōps* eye — more at EYE] (1872)
: abnormal protrusion of the eyeball
— **ex·oph·thal·mic** \-mik\ *adjective*
ex·or·bi·tance \ig-'zȯr-bə-tən(t)s\ *noun* (1611)
1 : an exorbitant action or procedure; *especially* **:** excessive or gross deviation from rule, right, or propriety
2 : the tendency or disposition to be exorbitant
ex·or·bi·tant \-tənt\ *adjective* [Middle English, from Middle French, from Late Latin *exorbitant-, exorbitans,* present participle of *exorbitare* to deviate, from Latin *ex-* + *orbita* track of a wheel, rut, from *orbis* disk, hoop] (15th century)
1 : not coming within the scope of the law
2 : exceeding in intensity, quality, amount, or size the customary or appropriate limits
synonym see EXCESSIVE
— **ex·or·bi·tant·ly** *adverb*
ex·or·cise *also* **ex·or·cize** \'ek-ˌsȯr-ˌsīz, -sər-\ *transitive verb* **-cised** *also* **-cized; -cis·ing** *also* **-ciz·ing** [Middle English, from Middle French *exorciser,* from Late Latin *exorcizare,* from Greek *exorkizein,* from *ex-* + *horkizein* to bind by oath, adjure, from *horkos* oath] (1539)
1 a : to expel (an evil spirit) by adjuration **b :** to get rid of (something troublesome, menacing, or oppressive)
2 : to free of an evil spirit
— **ex·or·cis·er** *noun*
ex·or·cism \-ˌsi-zəm\ *noun* (14th century)
1 : the act or practice of exorcising
2 : a spell or formula used in exorcising
— **ex·or·cist** \-ˌsist, -səst\ *noun*
— **ex·or·cis·tic** \ˌek-ˌsȯr-'sis-tik, -sər-\ *or* **ex·or·cis·ti·cal** \-ti-kəl\ *adjective*
ex·or·di·um \eg-'zȯr-dē-əm\ *noun, plural* **-diums** *or* **-dia** \-dē-ə\ [Latin, from *exordiri* to begin, from *ex-* + *ordiri* to begin — more at ORDER] (1577)
: a beginning or introduction especially to a discourse or composition
— **ex·or·di·al** \-dē-əl\ *adjective*
exo·skel·e·ton \ˌek-sō-'ske-lə-tᵊn\ *noun* (1847)
1 : an external supportive covering of an animal (as an arthropod)
2 : bony or horny parts of a vertebrate produced from epidermal tissues
— **exo·skel·e·tal** \-lə-tᵊl\ *adjective*
exo·sphere \'ek-sō-ˌsfir\ *noun* [International Scientific Vocabulary] (1949)
: the outer fringe region of the atmosphere of the earth or a planet
— **exo·spher·ic** \ˌek-sō-'sfir-ik, -'sfer-\ *adjective*
exo·spore \'ek-sə-ˌspōr, -ˌspȯr\ *noun* [International Scientific Vocabulary] (1859)
: an asexual spore cut off from a parent sporophore by the formation of septa
ex·os·to·sis \ˌek-(ˌ)säs-'tō-səs\ *noun, plural* **-to·ses** \-ˌsēz\ [New Latin, from Greek *exostōsis,* from *ex* out of + *osteon* bone — more at EX-, OSSEOUS] (1736)
: a spur or bony outgrowth from a bone or the root of a tooth
ex·o·ter·ic \ˌek-sə-'ter-ik\ *adjective* [Latin & Greek; Latin *exotericus,* from Greek *exōterikos,* literally, external, from *exōterō* more outside, comparative of *exō* outside — more at EXO-] (1660)
1 a : suitable to be imparted to the public ⟨the *exoteric* doctrine⟩ — compare ESOTERIC **b :** belonging to the outer or less initiate circle
2 : relating to the outside **:** EXTERNAL
— **ex·o·ter·i·cal·ly** \-i-k(ə-)lē\ *adverb*
exo·ther·mal \ˌek-sō-'thər-məl\ (1906)
: EXOTHERMIC
— **exo·ther·mal·ly** \-mə-lē\ *adverb*
exo·ther·mic \-mik\ *adjective* [International Scientific Vocabulary] (1884)
: characterized by or formed with evolution of heat
— **exo·ther·mi·cal·ly** \-mi-k(ə-)lē\ *adverb*
— **exo·ther·mi·ci·ty** \-ˌthər-'mis-ə-tē\ *noun*
[1]**ex·ot·ic** \ig-'zä-tik\ *adjective* [Latin *exoticus,* from Greek *exōtikos,* from *exō*] (1599)
1 : introduced from another country **:** not native to the place where found
2 *archaic* **:** FOREIGN, ALIEN
3 : strikingly, excitingly, or mysteriously different or unusual
4 : of or relating to striptease ⟨*exotic* dancing⟩
— **ex·ot·i·cal·ly** \-ti-k(ə-)lē\ *adverb*
— **ex·ot·ic·ness** \-tik-nəs\ *noun*
[2]**exotic** *noun* (1645)
1 : one (as a plant or animal) that is exotic
2 : STRIPTEASER
ex·ot·i·ca \ig-'zä-ti-kə\ *noun plural* [New Latin, from Latin, neuter plural of *exoticus*] (1876)

: things excitingly different or unusual; *especially* : literary or artistic items having an exotic theme or nature
ex·ot·i·cism \ig-'zä-tə-ˌsi-zəm\ *also* **ex·o·tism** \'eg-zə-ˌti-zəm, 'ek-sə-\ *noun* (1827)
: the quality or state of being exotic
exo·tox·in \ˌek-sō-'täk-sən\ *noun* [International Scientific Vocabulary] (1920)
: a soluble poisonous substance produced during growth of a microorganism and released into the surrounding medium
ex·pand \ik-'spand\ *verb* [Middle English *expaunden,* from Latin *expandere,* from *ex-* + *pandere* to spread — more at FATHOM] (15th century)
transitive verb
1 : to open up **:** UNFOLD
2 : to increase the extent, number, volume, or scope of **:** ENLARGE
3 a : to express at length or in greater detail **b :** to write out in full ⟨*expand* all abbreviations⟩ **c :** to subject to mathematical expansion ⟨*expand* a function in a power series⟩
intransitive verb
1 : to open out **:** SPREAD
2 : to increase in extent, number, volume, or scope
3 : to speak or write fully or in detail ⟨*expanded* on the theme⟩
4 : to feel generous or optimistic ☆
— ex·pand·abil·i·ty \-ˌspan-də-'bi-lə-tē\ *noun*
— ex·pand·able \-'span-də-bəl\ *adjective*
ex·pand·ed *adjective* (1875)
of a typeface **:** EXTENDED
expanded metal *noun* (1890)
: sheet metal cut and expanded into a lattice and used especially as lath
expanded plastic *noun* (1945)
: lightweight cellular plastic used especially as insulation and protective packing material — called also *foamed plastic, plastic foam*
ex·pand·er \ik-'span-dər\ *noun* (1862)
: one that expands; *specifically* **:** any of several colloidal substances (as dextran) of high molecular weight used as a blood or plasma substitute for increasing the blood volume
ex·panse \ik-'span(t)s\ *noun* [New Latin *expansum,* from Latin, neuter of *expansus,* past participle of *expandere*] (1667)
1 : FIRMAMENT
2 : great extent of something spread out ⟨an *expanse* of calm ocean⟩
ex·pan·si·ble \ik-'span(t)-sə-bəl\ *adjective* (circa 1691)
: capable of being expanded
— ex·pan·si·bil·i·ty \-ˌspan(t)-sə-'bi-lə-tē\ *noun*
ex·pan·sion \ik-'span(t)-shən\ *noun* (1611)
1 : EXPANSE
2 : the act or process of expanding ⟨territorial *expansion*⟩
3 : the quality or state of being expanded
4 : the increase in volume of working fluid (as steam) in an engine cylinder after cutoff or in an internal combustion engine after explosion
5 a : an expanded part **b :** something that results from an act of expanding ⟨the book is an *expansion* of a lecture series⟩
6 : the result of carrying out an indicated mathematical operation **:** the expression of a function in the form of a series
— ex·pan·sion·al \-'panch-nəl, -chə-nᵊl\ *adjective*
ex·pan·sion·ary \ik-'span(t)-shə-ˌner-ē\ *adjective* (1936)
: tending toward expansion ⟨an *expansionary* economy⟩
ex·pan·sion·ism \ik-'span-(t)-shə-ˌni-zəm\ *noun* (1899)
: a policy or practice of expansion and especially of territorial expansion by a nation
— ex·pan·sion·ist \-'span(t)-sh(ə-)nist\ *noun*
— expansionist *also* **ex·pan·sion·is·tic** \-ˌspan(t)-shə-'nis-tik\ *adjective*
ex·pan·sive \ik-'span(t)-siv\ *adjective* (1651)
1 : having a capacity or a tendency to expand
2 : causing or tending to cause expansion
3 a : characterized by high spirits, generosity, or readiness to talk **:** OPEN ⟨grew *expansive* after dinner⟩ **b :** marked by or indicative of exaggerated euphoria and delusions of self-importance ⟨an *expansive* patient⟩
4 : marked by expansion; *especially* **:** having a great expanse or extent **:** SIZABLE, EXTENSIVE
5 : characterized by richness, abundance, or magnificence ⟨*expansive* living⟩
— ex·pan·sive·ly *adverb*
— ex·pan·sive·ness *noun*
ex·pan·siv·i·ty \ˌek-ˌspan-'si-və-tē, ik-\ *noun* (1837)
: the quality or state of being expansive; *especially* **:** the capacity to expand
ex par·te \(ˌ)eks-'pär-tē\ *adverb or adjective* [Medieval Latin] (1672)
1 : on or from one side or party only — used of legal proceedings
2 : from a one-sided or partisan point of view
ex·pat \'eks-ˌpat\ *noun* (1962)
chiefly British **:** an expatriate person **:** EXPATRIATE
ex·pa·ti·ate \ek-'spā-shē-ˌāt\ *intransitive verb* **-at·ed; -at·ing** [Latin *exspatiatus,* past participle of *exspatiari* to wander, digress, from *ex-* + *spatium* space, course] (1538)
1 : to move about freely or at will **:** WANDER
2 : to speak or write at length or in detail ⟨*expatiating* upon the value of the fabric —Thomas Hardy⟩
— ex·pa·ti·a·tion \(ˌ)ek-ˌspā-shē-'ā-shən\ *noun*
¹ex·pa·tri·ate \ek-'spā-trē-ˌāt\ *verb* **-at·ed; -at·ing** [Medieval Latin *expatriatus,* past participle of *expatriare* to leave one's own country, from Latin *ex-* + *patria* native country, from feminine of *patrius* of a father, from *patr-, pater* father — more at FATHER] (1784)
transitive verb
1 : to withdraw (oneself) from residence in or allegiance to one's native country
2 : BANISH, EXILE
intransitive verb
: to leave one's native country to live elsewhere; *also* **:** to renounce allegiance to one's native country
— ex·pa·tri·ate \-ˌāt, -ət\ *noun*
— ex·pa·tri·a·tion \(ˌ)ek-ˌspā-trē-'ā-shən\ *noun*
²ex·pa·tri·ate \ek-'spā-trē-ət, -trē-ˌāt\ *adjective* (1812)
: living in a foreign land
ex·pa·tri·a·tism \ek-'spā-trē-ə-ˌti-zəm\ *noun* (1937)
: the fact or state of being an expatriate
ex·pect \ik-'spekt\ *verb* [Latin *exspectare* to look forward to, from *ex-* + *spectare* to look at, from *spectus,* past participle of *specere* to look — more at SPY] (1560)
intransitive verb
1 *archaic* **:** WAIT, STAY
2 : to look forward
3 : to be pregnant **:** await the birth of one's child — used in progressive tenses ⟨she's *expecting* next month⟩
transitive verb
1 *archaic* **:** AWAIT
2 : SUPPOSE, THINK
3 : to anticipate or look forward to the coming or occurrence of ⟨we *expect* them any minute now⟩ ⟨*expected* a telephone call⟩
4 a : to consider probable or certain ⟨*expect* to be forgiven⟩ ⟨*expect* that things will improve⟩ **b :** to consider reasonable, due, or necessary ⟨*expected* hard work from the students⟩ **c :** to consider bound in duty or obligated ⟨they *expect* you to pay your bills⟩ ☆
— ex·pect·able \-'spek-tə-bəl\ *adjective*
— ex·pect·ably \-blē\ *adverb*
— ex·pect·ed·ly *adverb*
— ex·pect·ed·ness *noun*
ex·pec·tance \ik-'spek-tən(t)s\ *noun* (1603)
: EXPECTANCY
ex·pec·tan·cy \-tən(t)-sē\ *noun, plural* **-cies** (1600)
1 a : the act, action, or state of expecting **b :** the state of being expected
2 a : something expected **b :** the expected amount (as of the number of years of life) based on statistical probability ⟨life *expectancy*⟩
¹ex·pec·tant \-tənt\ *adjective* (14th century)
1 : characterized by expectation
2 : expecting the birth of a child ⟨*expectant* mothers⟩
— ex·pec·tant·ly *adverb*
²expectant *noun* (1628)
: one who is looking forward to something
ex·pec·ta·tion \ˌek-ˌspek-'tā-shən, ik-\ *noun* (1540)
1 : the act or state of expecting **:** ANTICIPATION
2 a : something expected ⟨not up to *expectations*⟩ **b :** basis for expecting **:** ASSURANCE ⟨they have every *expectation* of success⟩ **c :** prospects of inheritance — usually used in plural
3 : the state of being expected
4 a : EXPECTANCY 2b **b :** EXPECTED VALUE
— ex·pec·ta·tion·al \-'tā-shə-nᵊl, -shnəl\ *adjective*
ex·pec·ta·tive \ik-'spek-tə-tiv\ *adjective* (15th century)
: of, relating to, or constituting an object of expectation
expected value *noun* (1947)
1 : the sum of the values of a random variable with each value multiplied by its probability of occurrence
2 : the integral of the product of a probability density function of a continuous random variable and the random variable itself when taken over all possible values of the variable
ex·pec·to·rant \ik-'spek-t(ə-)rənt\ *noun* (1782)

☆ SYNONYMS
Expand, amplify, swell, distend, inflate, dilate mean to increase in size or volume. EXPAND may apply whether the increase comes from within or without and regardless of manner (as growth, unfolding, addition of parts). AMPLIFY implies the extension or enlargement of something inadequate. SWELL implies gradual expansion beyond a thing's original or normal limits. DISTEND implies outward extension caused by pressure from within. INFLATE implies expanding by introduction of air or something insubstantial and suggests a resulting vulnerability and liability to sudden collapse. DILATE applies especially to expansion of circumference.

Expect, hope, look mean to await some occurrence or outcome. EXPECT implies a high degree of certainty and usually involves the idea of preparing or envisioning ⟨*expects* to be finished by Tuesday⟩. HOPE implies little certainty but suggests confidence or assurance in the possibility that what one desires or longs for will happen ⟨*hopes* to find a job soon⟩. LOOK, with *to,* implies assurance that expectations will be fulfilled ⟨*looks* to a tidy profit from the sale⟩; with *for* it implies less assurance and suggests an attitude of expectancy and watchfulness ⟨*look* for rain when the wind shifts to the northeast⟩.

: an agent that promotes the discharge or expulsion of mucus from the respiratory tract; *broadly* **:** an antitussive agent
— **expectorant** *adjective*

ex·pec·to·rate \-tə-ˌrāt\ *verb* **-rat·ed; -rat·ing** [Latin *expectoratus,* past participle of *expectorare* to banish from the mind (taken to mean literally "to expel from the chest"), from *ex-* + *pector-, pectus* breast, soul — more at PECTORAL] (1601)
transitive verb
1 : to eject from the throat or lungs by coughing or hawking and spitting
2 : SPIT
intransitive verb
1 : to discharge matter from the throat or lungs by coughing or hawking and spitting
2 : SPIT
— **ex·pec·to·ra·tion** \-ˌspek-tə-ˈrā-shən\ *noun*

ex·pe·di·ence \ik-ˈspē-dē-ən(t)s\ *noun* (1593)
: EXPEDIENCY

ex·pe·di·en·cy \-dən-sē\ *noun, plural* **-cies** (1612)
1 : the quality or state of being suited to the end in view **:** SUITABILITY, FITNESS
2 *obsolete* **a :** HASTE, DISPATCH **b :** an enterprise requiring haste or caution
3 : adherence to expedient means and methods ⟨put more emphasis on *expediency* than on principle —W. H. Jones⟩
4 : a means of achieving a particular end **:** EXPEDIENT
— **ex·pe·di·en·tial** \-ˌspē-dē-ˈen-chəl\ *adjective*

¹ex·pe·di·ent \ik-ˈspē-dē-ənt\ *adjective* [Middle English, from Middle French or Latin; Middle French, from Latin *expedient-, expediens,* present participle of *expedire* to extricate, prepare, be useful, from *ex-* + *ped-, pes* foot — more at FOOT] (14th century)
1 : suitable for achieving a particular end in a given circumstance
2 : characterized by concern with what is opportune; *especially* **:** governed by self-interest ☆
— **ex·pe·di·ent·ly** *adverb*

²expedient *noun* (1653)
: something expedient **:** a temporary means to an end
synonym see RESOURCE

ex·pe·dite \ˈek-spə-ˌdīt\ *transitive verb* **-dit·ed; -dit·ing** [Latin *expeditus,* past participle of *expedire*] (15th century)
1 : to execute promptly
2 : to accelerate the process or progress of **:** speed up
3 : ISSUE, DISPATCH

ex·pe·dit·er *also* **ex·pe·di·tor** \-ˌdī-tər\ *noun* (1891)
: one that expedites; *specifically* **:** one employed to ensure efficient movement of goods or supplies in a business

ex·pe·di·tion \ˌek-spə-ˈdi-shən\ *noun* (15th century)
1 a : a journey or excursion undertaken for a specific purpose **b :** the group of persons making such a journey
2 : efficient promptness **:** SPEED
3 : a sending or setting forth
synonym see HASTE

ex·pe·di·tion·ary \-ˈdi-shə-ˌner-ē\ *adjective* (1817)
: of, relating to, or being an expedition; *also* **:** sent on military service abroad ⟨an *expeditionary* force⟩

ex·pe·di·tious \ˌek-spə-ˈdi-shəs\ *adjective* (1559)
: characterized by or acting promptly and efficiently
synonym see FAST
— **ex·pe·di·tious·ly** *adverb*
— **ex·pe·di·tious·ness** *noun*

ex·pel \ik-ˈspel\ *transitive verb* **ex·pelled; ex·pel·ling** [Middle English *expellen,* from Latin *expellere,* from *ex-* + *pellere* to drive — more at FELT] (14th century)
1 : to force out **:** EJECT ⟨*expelled* the smoke from her lungs⟩
2 : to force to leave (as a place or organization) by official action **:** take away rights or privileges of membership ⟨*expelled* from college⟩
synonym see EJECT
— **ex·pel·la·ble** \-ˈspe-lə-bəl\ *adjective*

ex·pel·lee \ˌek-ˌspe-ˈlē, ik-\ *noun* (1888)
: a person who is expelled especially from a native or adopted country

ex·pend \ik-ˈspend\ *transitive verb* [Middle English, from Latin *expendere* to weigh out, expend, from *ex-* + *pendere* to weigh — more at SPIN] (15th century)
1 : to pay out **:** SPEND ⟨the social services upon which public revenue is *expended* —J. A. Hobson⟩
2 : to make use of for a specific purpose **:** UTILIZE ⟨projects on which they *expended* great energy⟩; *also* **:** USE UP
— **ex·pend·er** *noun*

¹ex·pend·able \ik-ˈspen-də-bəl\ *adjective* (1805)
: that may be expended: as **a :** normally used up or consumed in service ⟨*expendable* supplies like pencils and paper⟩ **b :** more easily or economically replaced than rescued, salvaged, or protected
— **ex·pend·abil·i·ty** \-ˌspen-də-ˈbi-lə-tē\ *noun*

²expendable *noun* (1942)
: one that is expendable — usually used in plural

ex·pen·di·ture \ik-ˈspen-di-chər, -də-ˌchu̇r, -də-ˌt(y)u̇r\ *noun* [irregular from *expend*] (1769)
1 : the act or process of expending ⟨an *expenditure* of energy⟩
2 : something expended **:** DISBURSEMENT, EXPENSE ⟨income should exceed *expenditures*⟩

¹ex·pense \ik-ˈspen(t)s\ *noun* [Middle English, from Anglo-French or Late Latin; Anglo-French, from Late Latin *expensa,* from Latin, feminine of *expensus,* past participle of *expendere*] (14th century)
1 *archaic* **:** the act or an instance of expending **:** EXPENDITURE
2 a : something expended to secure a benefit or bring about a result **b :** financial burden or outlay **:** COST ⟨built the monument at their own *expense*⟩ **c :** an item of business outlay chargeable against revenue for a specific period
3 : a cause or occasion of expenditure ⟨an estate is a great *expense*⟩
4 : a loss, detriment, or embarrassment that results from some action or gain **:** SACRIFICE ⟨everyone had a good laugh at my *expense*⟩ — usually used in the phrase *at the expense of* ⟨develop a boy's physique at the *expense* of his intelligence —Bertrand Russell⟩

²expense *transitive verb* **ex·pensed; ex·pens·ing** (circa 1909)
1 : to charge with expenses
2 a : to charge to an expense account **b :** to write off as an expense

expense account *noun* (1872)
: an account of expenses reimbursable to an employee; *also* **:** the right of charging expenses to such an account

ex·pen·sive \ik-ˈspen(t)-siv\ *adjective* (1634)
1 : involving high cost or sacrifice ⟨an *expensive* hobby⟩
2 a : commanding a high price and especially one that is not based on intrinsic worth or is beyond a prospective buyer's means **b :** characterized by high prices ⟨*expensive* shops⟩
— **ex·pen·sive·ly** *adverb*
— **ex·pen·sive·ness** *noun*

¹ex·pe·ri·ence \ik-ˈspir-ē-ən(t)s\ *noun* [Middle English, from Middle French, from Latin *experientia* act of trying, from *experient-, experiens,* present participle of *experiri* to try, from *ex-* + *-periri* (akin to *periculum* attempt) — more at FEAR] (14th century)
1 a : direct observation of or participation in events as a basis of knowledge **b :** the fact or state of having been affected by or gained knowledge through direct observation or participation
2 a : practical knowledge, skill, or practice derived from direct observation of or participation in events or in a particular activity **b :** the length of such participation ⟨has 10 years *experience* in the job⟩
3 a : the conscious events that make up an individual life **b :** the events that make up the conscious past of a community or nation or mankind generally
4 : something personally encountered, undergone, or lived through
5 : the act or process of directly perceiving events or reality

²experience *transitive verb* **-enced; -enc·ing** (1580)
1 : to learn by experience ⟨I have *experienced* that a landscape and the sky unfold the deepest beauty —Nathaniel Hawthorne⟩
2 : to have experience of **:** UNDERGO ⟨*experienced* severe hardships as a child⟩

ex·pe·ri·enced \-ən(t)st\ *adjective* (1576)
: made skillful or wise through experience **:** PRACTICED ⟨an *experienced* driver⟩

ex·pe·ri·en·tial \ik-ˌspir-ē-ˈen(t)-shəl\ *adjective* (1816)
: relating to, derived from, or providing experience **:** EMPIRICAL ⟨*experiential* knowledge⟩ ⟨*experiential* lessons⟩
— **ex·pe·ri·en·tial·ly** \-ˈen(t)-sh(ə-)lē\ *adverb*

¹ex·per·i·ment \ik-ˈsper-ə-mənt *also* -ˈspir-\ *noun* [Middle English, from Middle French, from Latin *experimentum,* from *experiri*] (14th century)
1 a : TEST, TRIAL ⟨make another *experiment* of his suspicion —Shakespeare⟩ **b :** a tentative procedure or policy **c :** an operation carried out under controlled conditions in order to discover an unknown effect or law, to test or establish a hypothesis, or to illustrate a known law
2 *obsolete* **:** EXPERIENCE
3 : the process of testing **:** EXPERIMENTATION

²ex·per·i·ment \-ˌment\ *intransitive verb* (1787)
: to carry out experiments **:** try out a new procedure, idea, or activity
— **ex·per·i·men·ta·tion** \ik-ˌsper-ə-mən-ˈtā-shən, -ˌmen- *also* -ˌspir-\ *noun*
— **ex·per·i·ment·er** \-ˈsper-ə-ˌmen-tər *also* -ˈspir-\ *noun*

ex·per·i·men·tal \ik-ˌsper-ə-ˈmen-tᵊl *also* -ˌspir-\ *adjective* (15th century)
1 : of, relating to, or based on experience or experiment
2 a : serving the ends of or used as a means of experimentation ⟨an *experimental* school⟩ **b :** relating to or having the characteristics of experiment **:** TENTATIVE ⟨still in the *experimental* stage⟩
— **ex·per·i·men·tal·ly** \-tᵊl-ē\ *adverb*

ex·per·i·men·tal·ism \-tᵊl-ˌi-zəm\ *noun* (circa 1834)

☆ SYNONYMS
Expedient, politic, advisable mean dictated by practical or prudent motives. EXPEDIENT usually implies what is immediately advantageous without regard for ethics or consistent principles ⟨a politically *expedient* decision⟩. POLITIC stresses judiciousness and tactical value but usually implies some lack of candor or sincerity ⟨a *politic* show of interest⟩. ADVISABLE applies to what is practical, prudent, or advantageous but lacks the derogatory implication of EXPEDIENT and POLITIC ⟨sometimes it's *advisable* to say nothing⟩.

: reliance on or advocacy of experimental or empirical principles and procedures; *specifically* **:** INSTRUMENTALISM

ex·per·i·men·tal·ist \-tᵊl-əst\ *noun* (1762)
: one who experiments; *specifically* **:** a person conducting scientific experiments

experiment station *noun* (1874)
: an establishment for scientific research (as in agriculture) where experiments are carried out, studies of practical application are made, and information is disseminated

¹ex·pert \'ek-ˌspərt, ik-'\ *adjective* [Middle English, from Middle French & Latin; Middle French, from Latin *expertus,* from past participle of *experiri*] (14th century)
1 *obsolete* **:** EXPERIENCED
2 : having, involving, or displaying special skill or knowledge derived from training or experience
synonym see PROFICIENT
— ex·pert·ly *adverb*
— ex·pert·ness *noun*

²ex·pert \'ek-ˌspərt\ *noun* [French, from *expert,* adjective] (1535)
: one with the special skill or knowledge representing mastery of a particular subject

³ex·pert \'ek-ˌspərt\ *intransitive verb* (circa 1889)
: to serve as an expert

ex·per·tise \ˌek-(ˌ)spər-'tēz, -'tēs\ *noun* [French, from Middle French, expertness, from *expert*] (1868)
1 : expert opinion or commentary
2 : the skill of an expert

ex·pert·ism \'ek-ˌspər-ˌti-zəm\ *noun* (1886)
: EXPERTISE 2

ex·pert·ize \'ek-spər-ˌtīz\ *verb* **-ized; -iz·ing** (1889)
intransitive verb
: to give a professional opinion usually after careful study
transitive verb
: to examine and give expert judgment on

expert system *noun* (1977)
: computer software that attempts to mimic the reasoning of a human specialist

ex·pi·ate \'ek-spē-ˌāt\ *verb* **-at·ed; -at·ing** [Latin *expiatus,* past participle of *expiare* to atone for, from *ex-* + *piare* to atone for, appease, from *pius* faithful, pious] (1594)
transitive verb
1 *obsolete* **:** to put an end to
2 a : to extinguish the guilt incurred by **b :** to make amends for ⟨permission to *expiate* their offences by their assiduous labours —Francis Bacon⟩
intransitive verb
: to make expiation
— ex·pi·a·ble \'ek-spē-ə-bəl\ *adjective*
— ex·pi·a·tor \-spē-ˌā-tər\ *noun*

ex·pi·a·tion \ˌek-spē-'ā-shən\ *noun* (15th century)
1 : the act of making atonement
2 : the means by which atonement is made

ex·pi·a·to·ry \'ek-spē-ə-ˌtōr-ē, -ˌtȯr-\ *adjective* (15th century)
: serving to expiate

ex·pi·ra·tion \ˌek-spə-'rā-shən\ *noun* (1526)
1 a : the last emission of breath **:** DEATH **b** (1) **:** the act or process of releasing air from the lungs through the nose or mouth (2) **:** the escape of carbon dioxide from the body protoplasm (as through the blood and lungs or by diffusion)
2 : the fact of coming to an end or the point at which something ends **:** TERMINATION

ex·pi·ra·to·ry \ik-'spī-rə-ˌtōr-ē, ek-', -ˌtȯr-; 'ek-sp(ə-)rə-\ *adjective* (circa 1847)
: of, relating to, or employed in the expiration of air from the lungs

ex·pire \ik-'spīr, *oftenest for intransitive verb 3 and transitive verb 2* ek-\ *verb* **ex·pired; ex·pir·ing** [Middle English, from Middle French or Latin; Middle French *expirer,* from Latin *exspirare,* from *ex-* + *spirare* to breathe] (14th century)
intransitive verb
1 : to breathe one's last breath **:** DIE
2 : to come to an end
3 : to emit the breath
transitive verb
1 *obsolete* **:** CONCLUDE
2 : to breathe out from or as if from the lungs
3 *archaic* **:** EMIT

ex·pi·ry \ik-'spīr-ē, 'ek-spə-rē\ *noun, plural* **-ries** (1752)
: EXPIRATION: as **a :** exhalation of breath **b :** DEATH **c :** TERMINATION; *especially* **:** the termination of a time or period fixed by law, contract, or agreement

ex·plain \ik-'splān\ *verb* [Middle English *explanen,* from Latin *explanare,* literally, to make level, from *ex-* + *planus* level, flat — more at FLOOR] (15th century)
transitive verb
1 a : to make known **b :** to make plain or understandable ⟨footnotes that *explain* the terms⟩
2 : to give the reason for or cause of
3 : to show the logical development or relationships of
intransitive verb
: to make something plain or understandable ☆
— ex·plain·able \-'splā-nə-bəl\ *adjective*
— ex·plain·er *noun*
— explain oneself : to clarify one's statements or the reasons for one's conduct

explain away *transitive verb* (1704)
1 : to get rid of by or as if by explanation
2 : to minimize the significance of by or as if by explanation

ex·pla·na·tion \ˌek-splə-'nā-shən\ *noun* (14th century)
1 : the act or process of explaining
2 : something that explains ⟨gave no *explanation*⟩

ex·plan·a·tive \ik-'spla-nə-tiv\ *adjective* (circa 1616)
: EXPLANATORY
— ex·plan·a·tive·ly *adverb*

ex·plan·a·to·ry \ik-'spla-nə-ˌtōr-ē, -ˌtȯr-\ *adjective* (1618)
: serving to explain ⟨*explanatory* notes⟩
— ex·plan·a·to·ri·ly \-ˌspla-nə-'tōr-ə-lē, -'tȯr-\ *adverb*

¹ex·plant \(ˌ)ek-'splant\ *transitive verb* [*ex-* + *-plant* (as in *implant*)] (1915)
: to remove (living tissue) especially to a medium for tissue culture
— ex·plan·ta·tion \ˌek-ˌsplan-'tā-shən\ *noun*

²ex·plant \'ek-ˌsplant\ *noun* (1917)
: living tissue removed from an organism and placed in a medium for tissue culture

¹ex·ple·tive \'ek-splə-tiv\ *noun* (1612)
1 a : a syllable, word, or phrase inserted to fill a vacancy (as in a sentence or a metrical line) without adding to the sense; *especially* **:** a word (as *it* in "make it clear which you prefer") that occupies the position of the subject or object of a verb in normal English word order and anticipates a subsequent word or phrase that supplies the needed meaningful content **b :** an exclamatory word or phrase; *especially* **:** one that is obscene or profane
2 : one that serves to fill out or as a filling

²expletive *adjective* [Late Latin *expletivus,* from Latin *expletus,* past participle of *explēre* to fill out, from *ex-* + *plēre* to fill — more at FULL] (1666)
1 : serving to fill up ⟨*expletive* phrases⟩
2 : marked by the use of expletives

ex·ple·to·ry \'ek-splə-ˌtōr-ē, -ˌtȯr-\ *adjective* (1672)
: EXPLETIVE

ex·pli·ca·ble \ek-'spli-kə-bəl, 'ek-(ˌ)splik-\ *adjective* (1556)
: capable of being explained
— ex·pli·ca·bly \-blē\ *adverb*

ex·pli·cate \'ek-splə-ˌkāt\ *transitive verb* **-cat·ed; -cat·ing** [Latin *explicatus,* past participle of *explicare,* literally, to unfold, from *ex-* + *plicare* to fold — more at PLY] (1531)
1 : to give a detailed explanation of
2 : to develop the implications of **:** analyze logically
synonym see EXPLAIN
— ex·pli·ca·tion \ˌek-splə-'kā-shən\ *noun*
— ex·pli·ca·tor \'ek-splə-ˌkā-tər\ *noun*

ex·pli·ca·tion de texte \ek-splē-kä-syōⁿ-də-tekst\ *noun, plural* **explications de texte** *same*\ [French, literally, explanation of text] (1935)
: a method of literary criticism involving a detailed analysis of a work

ex·pli·ca·tive \ek-'spli-kə-tiv, 'ek-splə-ˌkāt-\ *adjective* (1649)
: serving to explicate; *specifically* **:** serving to explain logically what is contained in the subject ⟨an *explicative* proposition⟩
— ex·pli·ca·tive·ly *adverb*

ex·pli·ca·to·ry \ek-'spli-kə-ˌtōr-ē, 'ek-(ˌ)spli-, -ˌtȯr-\ *adjective* (1625)
: EXPLICATIVE

ex·plic·it \ik-'spli-sət\ *adjective* [French or Medieval Latin; French *explicite,* from Medieval Latin *explicitus,* from Latin, past participle of *explicare*] (circa 1609)
1 a : fully revealed or expressed without vagueness, implication, or ambiguity **:** leaving no question as to meaning or intent ⟨*explicit* instructions⟩ **b :** open in the depiction of nudity or sexuality ⟨*explicit* books and films⟩
2 : fully developed or formulated ⟨an *explicit* plan⟩ ⟨an *explicit* notion of our objective⟩
3 : unambiguous in expression ⟨was very *explicit* on how we are to behave⟩
4 *of a mathematical function* **:** defined by an expression containing only independent variables — compare IMPLICIT 1c ☆
— ex·plic·it·ly *adverb*
— ex·plic·it·ness *noun*

ex·plode \ik-'splōd\ *verb* **ex·plod·ed; ex·plod·ing** [Latin *explodere* to drive off the

☆ SYNONYMS
Explain, expound, explicate, elucidate, interpret mean to make something clear or understandable. EXPLAIN implies a making plain or intelligible what is not immediately obvious or entirely known ⟨*explain* the rules⟩. EXPOUND implies a careful often elaborate explanation ⟨*expounding* a scientific theory⟩. EXPLICATE adds the idea of a developed or detailed analysis ⟨*explicate* a poem⟩. ELUCIDATE stresses the throwing of light upon as by offering details or motives previously unclear or only implicit ⟨*elucidate* an obscure passage⟩. INTERPRET adds to EXPLAIN the need for imagination or sympathy or special knowledge in dealing with something ⟨*interpreting* a work of art⟩.

Explicit, definite, express, specific mean perfectly clear in meaning. EXPLICIT implies such verbal plainness and distinctness that there is no need for inference and no room for difficulty in understanding ⟨*explicit* instructions⟩. DEFINITE stresses precise, clear statement or arrangement that leaves no doubt or indecision ⟨the law is *definite* in such cases⟩. EXPRESS implies both explicitness and direct and positive utterance ⟨her *express* wishes⟩. SPECIFIC applies to what is precisely and fully treated in detail or particular ⟨two *specific* criticisms⟩.

stage by clapping, from *ex-* + *plaudere* to clap] (circa 1611)
transitive verb
1 *archaic* **:** to drive from the stage by noisy disapproval
2 : to bring into disrepute or discredit ⟨*explode* a theory⟩
3 : to cause to explode or burst noisily ⟨*explode* a bomb⟩
intransitive verb
1 : to burst forth with sudden violence or noise from internal energy: as **a :** to undergo a rapid chemical or nuclear reaction with the production of noise, heat, and violent expansion of gases ⟨dynamite *explodes*⟩ **b :** to burst violently as a result of pressure from within
2 a : to give forth a sudden strong and noisy outburst of emotion ⟨*exploded* in anger⟩ **b :** to move with sudden speed and force ⟨*exploded* from the starting gate⟩
3 : to increase rapidly ⟨the population of the city *exploded*⟩ ◆
— ex·plod·er *noun*

ex·plod·ed *adjective* (1944)
: showing the parts separated but in correct relationship to each other ⟨an *exploded* view of a carburetor⟩

¹ex·ploit \ˈek-ˌsplȯit, ik-ˈ\ *noun* [Middle English, outcome, success, from Middle French, from Latin *explicitum,* neuter of *explicitus,* past participle] (circa 1538)
: DEED, ACT; *especially* **:** a notable or heroic act
synonym see FEAT

²ex·ploit \ik-ˈsplȯit, ˈek-ˌ\ *transitive verb* (1838)
1 : to make productive use of **:** UTILIZE ⟨*exploiting* your talents⟩ ⟨*exploit* your opponent's weakness⟩
2 : to make use of meanly or unjustly for one's own advantage ⟨*exploiting* migrant farm workers⟩
— ex·ploit·able \-ˈsplȯi-tə-bəl\ *adjective*
— ex·ploit·er *noun*

ex·ploi·ta·tion \ˌek-ˌsplȯi-ˈtā-shən\ *noun* (1803)
: an act or instance of exploiting ⟨*exploitation* of natural resources⟩ ⟨*exploitation* of immigrant laborers⟩ ⟨clever *exploitation* of the system⟩

ex·ploit·ative \ik-ˈsplȯi-tə-tiv, ˈek-ˌsplȯi-\ *adjective* (1885)
: exploiting or tending to exploit; *especially* **:** unfairly or cynically using another person or group for profit or advantage ⟨*exploitative* terms of employment⟩ ⟨an *exploitative* film⟩
— ex·ploit·ative·ly *adverb*

ex·ploit·ive \ik-ˈsplȯi-tiv\ *adjective* (1921)
: EXPLOITATIVE

ex·plo·ra·tion \ˌek-splə-ˈrā-shən, -ˌsplȯ-\ *noun* (1537)
: the act or an instance of exploring
— ex·plo·ra·tion·al \-shnəl, -shə-nᵊl\ *adjective*

ex·plor·ative \ik-ˈsplōr-ə-tiv, -ˈsplȯr-\ *adjective* (1738)
: EXPLORATORY
— ex·plor·ative·ly *adverb*

ex·plor·ato·ry \-ə-ˌtōr-ē, -ˌtȯr-\ *adjective* (1620)
: of, relating to, or being exploration ⟨*exploratory* surgery⟩ ⟨*exploratory* drilling for oil⟩

ex·plore \ik-ˈsplōr, -ˈsplȯr\ *verb* **ex·plored; ex·plor·ing** [Latin *explorare,* from *ex-* + *plorare* to cry out] (1585)
transitive verb
1 a : to investigate, study, or analyze **:** look into ⟨*explore* the relationship between social class and learning ability⟩ — sometimes used with indirect questions ⟨to *explore* where ethical issues arise —R. T. Blackburn⟩ **b :** to become familiar with by testing or experimenting ⟨*explore* new cuisines⟩
2 : to travel over (new territory) for adventure or discovery
3 : to examine minutely especially for diagnostic purposes ⟨*explore* the wound⟩
intransitive verb
: to make or conduct a systematic search ⟨*explore* for oil⟩

ex·plor·er \ik-ˈsplōr-ər, -ˈsplȯr-\ *noun* (1685)
1 : one that explores; *especially* **:** a person who travels in search of geographical or scientific information
2 *capitalized* **:** a member of a coed scouting program of the Boy Scouts of America for young people ages 14 to 20 focusing on career awareness

ex·plo·sion \ik-ˈsplō-zhən\ *noun* [Latin *explosion-, explosio* act of driving off by clapping, from *explodere*] (circa 1616)
1 : the act or an instance of exploding ⟨injured in a laboratory *explosion*⟩
2 : a large-scale, rapid, or spectacular expansion or bursting out or forth ⟨the *explosion* of suburbia⟩ ⟨an *explosion* of red hair⟩
3 : the release of occluded breath that occurs in one kind of articulation of stop consonants

¹ex·plo·sive \ik-ˈsplō-siv, -ziv\ *adjective* (1667)
1 a : relating to, characterized by, or operated by explosion ⟨an *explosive* hatch⟩ **b :** resulting from or as if from an explosion ⟨*explosive* population growth⟩
2 a : tending to explode ⟨an *explosive* person⟩ **b :** likely to erupt in or produce hostile reaction or violence ⟨an *explosive* ghetto situation⟩
— ex·plo·sive·ly *adverb*
— ex·plo·sive·ness *noun*

²explosive *noun* (1874)
1 : an explosive substance
2 : a consonant characterized by explosion in its articulation when it occurs in certain environments **:** STOP

ex·po \ˈek-(ˌ)spō\ *noun, plural* **expos** (1913)
: EXPOSITION 3

ex·po·nent \ik-ˈspō-nənt, ˈek-ˌ\ *noun* [Latin *exponent-, exponens,* present participle of *exponere*] (1706)
1 : a symbol written above and to the right of a mathematical expression to indicate the operation of raising to a power
2 a : one that expounds or interprets **b :** one that champions, practices, or exemplifies

ex·po·nen·tial \ˌek-spə-ˈnen-chəl\ *adjective* (1704)
1 : of or relating to an exponent
2 : involving a variable in an exponent ⟨10^x is an *exponential* expression⟩
3 : expressible or approximately expressible by an exponential function; *especially* **:** characterized by or being an extremely rapid increase (as in size or extent) ⟨an *exponential* growth rate⟩
— ex·po·nen·tial·ly \-ˈnench-(ə-)lē\ *adverb*

exponential function *noun* (circa 1890)
: a mathematical function in which an independent variable appears in one of the exponents — called also *exponential*

ex·po·nen·ti·a·tion \ˌek-spə-ˌnen(t)-shē-ˈā-shən\ *noun* (1903)
: the act or process of raising a quantity to a power — called also *involution*

¹ex·port \ek-ˈspōrt, -ˈspȯrt, ˈek-ˌ\ *verb* [Middle English, from Latin *exportare,* from *ex-* + *portare* to carry — more at FARE] (15th century)
transitive verb
1 : to carry away **:** REMOVE
2 : to carry or send (as a commodity) to some other place (as another country)
intransitive verb
: to export something abroad
— ex·port·abil·i·ty \(ˌ)ek-ˌspōr-tə-ˈbi-lə-tē, -ˌspȯr-\ *noun*
— ex·port·able \ek-ˈspōr-tə-bəl, -ˈspȯr-, ˈek-ˌ\ *adjective*

²ex·port \ˈek-ˌspōrt, -ˌspȯrt\ *noun* (1690)
1 : something exported; *specifically* **:** a commodity conveyed from one country or region to another for purposes of trade
2 : the act of exporting **:** EXPORTATION ⟨the *export* of wheat⟩

³ex·port \ˈek-ˌ\ *adjective* (1795)
: of or relating to exportation or exports ⟨*export* duties⟩

ex·por·ta·tion \ˌek-ˌspōr-ˈtā-shən, -ˌspȯr-, -spər-\ *noun* (1641)
: the act of exporting; *also* **:** a commodity exported

ex·port·er \ek-ˈspōr-tər, -ˈspȯr-, ˈek-ˌ\ *noun* (1691)
: one that exports; *specifically* **:** a wholesaler who sells to merchants or industrial consumers in foreign countries

ex·pose \ik-ˈspōz\ *transitive verb* **ex·posed; ex·pos·ing** [Middle English, from Middle French *exposer,* from Latin *exponere* to set forth, explain (perfect indicative *exposui*), from *ex-* + *ponere* to put, place — more at POSITION] (15th century)
1 a : to deprive of shelter, protection, or care **:** subject to risk from a harmful action or condition ⟨*expose* troops needlessly⟩ ⟨has not yet been *exposed* to measles⟩ **b :** to submit or make accessible to a particular action or influence ⟨*expose* children to good books⟩; *especially* **:** to subject (a sensitive photographic film, plate, or paper) to radiant energy **c :** to abandon (an infant) especially by leaving in the open
2 a : to make known **:** bring to light (as something shameful) **b :** to disclose the faults or crimes of ⟨*expose* a murderer⟩
3 : to cause to be visible or open to view **:** DISPLAY: as **a :** to offer publicly for sale **b :** to exhibit for public veneration **c :** to reveal the face of (a playing card) or the cards of (a player's hand) **d :** to engage in indecent exposure of (oneself)
synonym see SHOW
— ex·pos·er *noun*

ex·po·sé *also* **ex·po·se** \ˌek-spō-ˈzā, -spə-\ *noun* [French *exposé,* from past participle of *exposer*] (1803)
1 : a formal statement of facts
2 : an exposure of something discreditable ⟨a newspaper *exposé* of government corruption⟩

ex·posed \ik-ˈspōzd\ *adjective* (circa 1623)
1 : open to view
2 : not shielded or protected; *also* **:** not insulated ⟨an *exposed* electric wire⟩
synonym see LIABLE

ex·pos·it \ik-ˈspä-zət\ *transitive verb* [Latin *expositus,* past participle of *exponere*] (1882)
: EXPOUND

ex·po·si·tion \ˌek-spə-ˈzi-shən\ *noun* (14th century)
1 : a setting forth of the meaning or purpose (as of a writing)
2 a : discourse or an example of it designed to convey information or explain what is difficult to understand **b** (1) **:** the first part of a musical composition in sonata form in which the thematic material of the movement is presented (2) **:** the opening section of a fugue
3 : a public exhibition or show

◇ WORD HISTORY
explode A modern audience often expresses its disapproval of a performer or performance by hissing. But in ancient Rome a customary expression of audience disapproval was loud clapping, sometimes expressly intended to drive the performer from the stage. The Latin verb meaning "to drive off by clapping" was *explodere,* a compound of *ex-,* meaning "out of, from" and *plaudere,* meaning "to clap." *Explodere* was, of course, the source of English *explode,* which initially also meant "to drive from the stage by a noisy expression of dislike." This sense is no longer current, but the senses that have replaced it have kept the notion either of dismissal and rejection or of a loud, violent burst.

— **ex·po·si·tion·al** \-ˈzish-nəl, -ˈzi-shə-nᵊl\ *adjective*

ex·pos·i·tive \ik-ˈspä-zə-tiv\ *adjective* (15th century)
: DESCRIPTIVE, EXPOSITORY

ex·pos·i·tor \-zə-tər\ *noun* [Middle English *expositour,* from Middle French *expositeur,* from Late Latin *expositor,* from Latin *exponere*] (14th century)
: a person who explains : COMMENTATOR

ex·pos·i·to·ry \-zə-ˌtōr-ē, -ˌtȯr-\ *adjective* (1628)
: of, relating to, or containing exposition ⟨*expository* writing⟩

[1]**ex post fac·to** \ˌeks-ˌpōs(t)-ˈfak-(ˌ)tō\ *adverb* [Late Latin, literally, from a thing done afterward] (1621)
: after the fact : RETROACTIVELY

[2]**ex post facto** *adjective* (1787)
: done, made, or formulated after the fact : RETROACTIVE ⟨*ex post facto* approval⟩ ⟨*ex post facto* laws⟩

ex·pos·tu·late \ik-ˈspäs-chə-ˌlāt\ *verb* [Latin *expostulatus,* past participle of *expostulare* to demand, dispute, from *ex-* + *postulare* to ask for — more at POSTULATE] (1573)
transitive verb
obsolete : DISCUSS, EXAMINE
intransitive verb
: to reason earnestly with a person for purposes of dissuasion or remonstrance

ex·pos·tu·la·tion \-ˌspäs-chə-ˈlā-shən\ *noun* (1540)
: an act or an instance of expostulating
— **ex·pos·tu·la·to·ry** \-ˈspäs-chə-lə-ˌtōr-ē, -ˌtȯr-\ *adjective*

ex·po·sure \ik-ˈspō-zhər\ *noun* (1606)
1 : the fact or condition of being exposed: as **a :** the condition of being presented to view or made known ⟨a politician seeks a lot of *exposure*⟩ **b :** the condition of being unprotected especially from severe weather ⟨died of *exposure*⟩ **c :** the condition of being subject to some effect or influence ⟨risk *exposure* to the flu⟩ **d :** the condition of being at risk of financial loss; *also* **:** an amount at risk
2 : the act or an instance of exposing: as **a :** disclosure of something secret ⟨tried to prevent *exposure* of their past⟩ **b :** the treating of sensitized material (as film) to controlled amounts of radiant energy; *also* **:** the amount of such energy or length of such treatment ⟨a 3-second *exposure*⟩
3 a : the manner of being exposed **b :** the position (as of a house) with respect to weather influences or compass points ⟨a room with a southern *exposure*⟩
4 : a piece or section of sensitized material (as film) on which an exposure is or can be made ⟨36 *exposures* per roll⟩

exposure meter *noun* (1891)
: a device for indicating correct photographic exposure under varying conditions of illumination

ex·pound \ik-ˈspau̇nd\ *verb* [Middle English, from Middle French *expondre,* from Latin *exponere* to explain — more at EXPOSE] (14th century)
transitive verb
1 a : to set forth : STATE **b :** to defend with argument
2 : to explain by setting forth in careful and often elaborate detail ⟨*expound* a law⟩
intransitive verb
: to make a statement : COMMENT
synonym see EXPLAIN
— **ex·pound·er** *noun*

[1]**ex·press** \ik-ˈspres\ *adjective* [Middle English, from Middle French *expres,* from Latin *expressus,* past participle of *exprimere* to press out, express, from *ex-* + *premere* to press — more at PRESS] (14th century)
1 a : directly, firmly, and explicitly stated ⟨my *express* orders⟩ **b :** EXACT, PRECISE
2 a : designed for or adapted to its purpose **b :** of a particular sort : SPECIFIC ⟨for that *express* purpose⟩
3 a : traveling at high speed; *specifically* **:** traveling with few or no stops along the way ⟨*express* train⟩ **b :** adapted or suitable for travel at high speed ⟨an *express* highway⟩ **c** *British* **:** designated to be delivered without delay by special messenger
synonym see EXPLICIT

[2]**express** *adverb* (14th century)
1 *obsolete* : EXPRESSLY
2 : by express ⟨delivered *express*⟩

[3]**express** *noun* (1619)
1 a *British* **:** a messenger sent on a special errand **b** *British* **:** a dispatch conveyed by a special messenger **c** (1) **:** a system for the prompt and safe transportation of parcels, money, or goods at rates higher than standard freight charges (2) **:** a company operating such a merchandise freight service **d** *British* : SPECIAL DELIVERY
2 : an express vehicle

[4]**express** *transitive verb* [Middle English, from Middle French & Latin; Middle French *expresser,* from Old French, from *expres,* adjective, from Latin *expressus,* past participle] (14th century)
1 a : DELINEATE, DEPICT **b :** to represent in words : STATE **c :** to give or convey a true impression of : SHOW, REFLECT **d :** to make known the opinions or feelings of (oneself) **e :** to give expression to the artistic or creative impulses or abilities of (oneself) **f :** to represent by a sign or symbol : SYMBOLIZE
2 a : to force out (as the juice of a fruit) by pressure **b :** to subject to pressure so as to extract something
3 : to send by express
4 : to cause (a gene) to manifest its effects in the phenotype ☆
— **ex·press·er** *noun*
— **ex·press·ible** \-ə-bəl\ *adjective*

ex·press·age \ik-ˈspre-sij\ *noun* (1857)
: a carrying of parcels by express; *also* **:** a charge for such carrying

ex·pres·sion \ik-ˈspre-shən\ *noun* (15th century)
1 a : an act, process, or instance of representing in a medium (as words) : UTTERANCE ⟨freedom of *expression*⟩ **b** (1) **:** something that manifests, embodies, or symbolizes something else ⟨this gift is an *expression* of my admiration for you⟩ (2) **:** a significant word or phrase (3) **:** a mathematical or logical symbol or a meaningful combination of symbols (4) **:** the detectable effect of a gene; *also* : EXPRESSIVITY 1
2 a : a mode, means, or use of significant representation or symbolism; *especially* **:** felicitous or vivid indication or depiction of mood or sentiment ⟨read the poem with *expression*⟩ **b** (1) **:** the quality or fact of being expressive (2) **:** facial aspect or vocal intonation as indicative of feeling
3 : an act or product of pressing out
— **ex·pres·sion·al** \-ˈspresh-nəl, -ˈspre-shə-nᵊl\ *adjective*

ex·pres·sion·ism \ik-ˈspre-shə-ˌni-zəm\ *noun, often capitalized* (circa 1901)
: a theory or practice in art of seeking to depict the subjective emotions and responses that objects and events arouse in the artist
— **ex·pres·sion·ist** \-ˈspre-shə-nist\ *noun or adjective, often capitalized*
— **ex·pres·sion·is·tic** \-ˌspre-shə-ˈnis-tik\ *adjective*
— **ex·pres·sion·is·ti·cal·ly** \-ti-k(ə-)lē\ *adverb*

ex·pres·sion·less \ik-ˈspre-shən-ləs\ *adjective* (1831)
: lacking expression ⟨an *expressionless* face⟩
— **ex·pres·sion·less·ly** *adverb*
— **ex·pres·sion·less·ness** *noun*

ex·pres·sive \ik-ˈspre-siv\ *adjective* (15th century)
1 : of or relating to expression ⟨the *expressive* function of language⟩
2 : serving to express, utter, or represent ⟨foul and novel terms *expressive* of rage —H. G. Wells⟩
3 : effectively conveying meaning or feeling ⟨an *expressive* silence⟩ ⟨*expressive* line drawings⟩
— **ex·pres·sive·ly** *adverb*
— **ex·pres·sive·ness** *noun*

ex·pres·siv·i·ty \ˌek-ˌspre-ˈsi-və-tē\ *noun, plural* **-ties** (1934)
1 : the relative capacity of a gene to affect the phenotype of the organism of which it is a part
2 : the quality of being expressive

ex·press·ly \ik-ˈspres-lē\ *adverb* (14th century)
1 : in an express manner : EXPLICITLY ⟨*expressly* rejected the proposal⟩
2 : for the express purpose : PARTICULARLY, SPECIFICALLY ⟨made *expressly* for me⟩

Express Mail *service mark*
— used for overnight delivery of mail

ex·press·man \ik-ˈspres-ˌman, -mən\ *noun* (1839)
: a person employed in the express business

ex·pres·so \ik-ˈspre-(ˌ)sō\ *variant of* ESPRESSO

ex·press·way \ik-ˈspres-ˌwā\ *noun* (1944)
: a high-speed divided highway for through traffic with access partially or fully controlled

ex·pro·pri·ate \ek-ˈsprō-prē-ˌāt\ *transitive verb* **-at·ed; -at·ing** [Medieval Latin *expropriatus,* past participle of *expropriare,* from Latin *ex-* + *proprius* own] (1611)
1 : to deprive of possession or proprietary rights
2 : to transfer (the property of another) to one's own possession
— **ex·pro·pri·a·tor** \-ˌā-tər\ *noun*

ex·pro·pri·a·tion \(ˌ)ek-ˌsprō-prē-ˈā-shən\ *noun* (15th century)
: the act of expropriating or the state of being expropriated; *specifically* **:** the action of the state in taking or modifying the property rights of an individual in the exercise of its sovereignty

ex·pulse \ik-ˈspəls\ *transitive verb* **expulsed; ex·puls·ing** (15th century)
: EXPEL

ex·pul·sion \ik-ˈspəl-shən\ *noun* [Middle English, from Latin *expulsion-, expulsio,* from *expellere* to expel] (15th century)
: the act of expelling : the state of being expelled
— **ex·pul·sive** \-ˈspəl-siv\ *adjective*

☆ SYNONYMS
Express, vent, utter, voice, broach, air mean to make known what one thinks or feels. EXPRESS suggests an impulse to reveal in words, gestures, actions, or what one creates or produces ⟨*expressed* her feelings in music⟩. VENT stresses a strong inner compulsion to express especially in words ⟨a tirade *venting* his frustration⟩. UTTER implies the use of the voice not necessarily in articulate speech ⟨*utter* a groan⟩. VOICE does not necessarily imply vocal utterance but does imply expression or formulation in words ⟨an editorial *voicing* their concerns⟩. BROACH adds the implication of disclosing for the first time something long thought over or reserved for a suitable occasion ⟨*broached* the subject of a divorce⟩. AIR implies an exposing or parading of one's views often in order to gain relief or sympathy or attention ⟨publicly *airing* their differences⟩.

ex·punc·tion \ik-ˈspəŋ(k)-shən\ *noun* [Latin *expungere*] (1606)
: the act of expunging : the state of being expunged : ERASURE

ex·punge \ik-ˈspənj\ *transitive verb* **ex·punged; ex·pung·ing** [Latin *expungere* to mark for deletion by dots, from *ex-* + *pungere* to prick — more at PUNGENT] (1602)
1 : to strike out, obliterate, or mark for deletion
2 : to efface completely **:** DESTROY
— **ex·pung·er** *noun*

ex·pur·gate \ˈek-spər-ˌgāt\ *transitive verb* **-gat·ed; -gat·ing** [Latin *expurgatus,* past participle of *expurgare,* from *ex-* + *purgare* to purge] (1678)
: to cleanse of something morally harmful, offensive, or erroneous; *especially* **:** to expunge objectionable parts from before publication or presentation ⟨an *expurgated* edition of the letters⟩
— **ex·pur·ga·tion** \ˌek-spər-ˈgā-shən\ *noun*
— **ex·pur·ga·tor** \ˈek-spər-ˌgā-tər\ *noun*

ex·pur·ga·to·ri·al \(ˌ)ek-ˌspər-gə-ˈtōr-ē-əl, -ˈtȯr-\ *adjective* (1807)
: relating to expurgation or an expurgator **:** EXPURGATORY

ex·pur·ga·to·ry \ek-ˈspər-gə-ˌtōr-ē, -ˌtȯr-\ *adjective* (1625)
: serving to purify from something morally harmful, offensive, or erroneous

[1]**ex·qui·site** \ek-ˈskwi-zət, ˈek-(ˌ)\ *adjective* [Middle English *exquisit,* from Latin *exquisitus,* past participle of *exquirere* to search out, from *ex-* + *quaerere* to seek] (15th century)
1 : carefully selected **:** CHOICE
2 *archaic* **:** ACCURATE
3 a : marked by flawless craftsmanship or by beautiful, ingenious, delicate, or elaborate execution **b :** marked by nice discrimination, deep sensitivity, or subtle understanding ⟨*exquisite* taste⟩ **c :** ACCOMPLISHED, PERFECTED ⟨an *exquisite* gentleman⟩
4 a : pleasing through beauty, fitness, or perfection ⟨an *exquisite* white blossom⟩ **b :** ACUTE, INTENSE ⟨*exquisite* pain⟩ **c :** having uncommon or esoteric appeal ◆
synonym see CHOICE
— **ex·qui·site·ly** *adverb*
— **ex·qui·site·ness** *noun*

[2]**exquisite** *noun* (1819)
: one who is overly fastidious in dress or ornament

ex·san·gui·na·tion \(ˌ)ek(s)-ˌsaŋ-gwə-ˈnā-shən\ *noun* [Latin *exsanguinatus* drained of blood, from *ex-* + *sanguin-, sanguis* blood] (circa 1909)
: the action or process of draining or losing blood
— **ex·san·gui·nate** \ek(s)-ˈsaŋ-gwə-ˌnāt\ *transitive verb*

ex·scind \ek-ˈsind\ *transitive verb* [Latin *exscindere,* from *ex-* + *scindere* to cut, tear — more at SHED] (1662)
: to cut off or out **:** EXCISE

ex·sert \ek-ˈsərt\ *transitive verb* [Latin *exsertus,* past participle of *exserere* — more at EXERT] (1836)
: to thrust out
— **ex·ser·tile** \-ˈsər-tᵊl, -ˈsər-ˌtīl\ *adjective*
— **ex·ser·tion** \-ˈsər-shən\ *noun*

ex·sert·ed *adjective* (1816)
: projecting beyond an enclosing organ or part

ex·sic·cate \ˈek-si-ˌkāt\ *transitive verb* **-cat·ed; -cat·ing** [Middle English, from Latin *exsiccatus,* past participle of *exsiccare,* from *ex-* + *siccare* to dry, from *siccus* dry — more at SACK] (15th century)
: to remove moisture from **:** DRY
— **ex·sic·ca·tion** \ˌek-si-ˈkā-shən\ *noun*

ex·so·lu·tion \ˌek-sə-ˈlü-shən\ *noun* (1929)
: the process of separating or precipitating from a solid crystalline phase

ex·tant \ˈek-stənt; ek-ˈstant, ˈek-ˌ\ *adjective* [Latin *exstant-, exstans,* present participle of *exstare* to stand out, be in existence, from *ex-* + *stare* to stand — more at STAND] (1545)
1 *archaic* **:** standing out or above
2 a : currently or actually existing ⟨the most charming writer *extant* —G. W. Johnson⟩ **b :** not destroyed or lost ⟨*extant* manuscripts⟩

ex·tem·po·ral \ek-ˈstem-p(ə-)rəl\ *adjective* [Latin *extemporalis,* from *ex tempore*] (1570) *archaic* **:** EXTEMPORANEOUS
— **ex·tem·po·ral·ly** \-ē\ *adverb*

ex·tem·po·ra·ne·ity \(ˌ)ek-ˌstem-pə-rə-ˈnē-ə-tē, -ˈnā-\ *noun* (1937)
: the quality or state of being extemporaneous

ex·tem·po·ra·ne·ous \(ˌ)ek-ˌstem-pə-ˈrā-nē-əs\ *adjective* [Late Latin *extemporaneus,* from Latin *ex tempore*] (1673)
1 a (1) **:** composed, performed, or uttered on the spur of the moment **:** IMPROMPTU (2) **:** carefully prepared but delivered without notes or text **b :** skilled at or given to extemporaneous utterance **c :** happening suddenly and often unexpectedly and usually without clearly known causes or relationships ⟨a great deal of criminal and delinquent behavior is . . . *extemporaneous* —W. C. Reckless⟩
2 : provided, made, or put to use as an expedient **:** MAKESHIFT
— **ex·tem·po·ra·ne·ous·ly** *adverb*
— **ex·tem·po·ra·ne·ous·ness** *noun*

ex·tem·po·rary \ik-ˈstem-pə-ˌrer-ē\ *adjective* (1596)
: EXTEMPORANEOUS
— **ex·tem·po·rar·i·ly** \-ˌstem-pə-ˈrer-ə-lē\ *adverb*

ex·tem·po·re \ik-ˈstem-pə-(ˌ)rē\ *adverb or adjective* [Latin *ex tempore,* from *ex* + *tempore,* ablative of *tempus* time] (1553)
: in an extemporaneous manner ⟨speaking *extempore*⟩

ex·tem·por·i·sa·tion, ex·tem·po·rise *British variant of* EXTEMPORIZATION, EXTEMPORIZE

ex·tem·po·ri·za·tion \ik-ˌstem-pə-rə-ˈzā-shən\ *noun* (circa 1860)
1 : the act of extemporizing **:** IMPROVISATION
2 : something extemporized

ex·tem·po·rize \ik-ˈstem-pə-ˌrīz\ *verb* **-rized; -riz·ing** (1644)
intransitive verb
1 : to do something extemporaneously **:** IMPROVISE; *especially* **:** to speak extemporaneously
2 : to get along in a makeshift manner
transitive verb
: to compose, perform, or utter extemporaneously **:** IMPROVISE
— **ex·tem·po·riz·er** *noun*

ex·tend \ik-ˈstend\ *verb* [Middle English, from Middle French or Latin; Middle French *estendre,* from Latin *extendere,* from *ex-* + *tendere* to stretch — more at THIN] (14th century)
transitive verb
1 : to spread or stretch forth **:** UNBEND ⟨*extended* both her arms⟩
2 a : to stretch out to fullest length **b :** to cause (as a horse) to move at full stride **c :** to exert (oneself) to full capacity ⟨could work long and hard without seeming to *extend* himself⟩ **d** (1) **:** to increase the bulk of (as by adding a cheaper substance or a modifier) (2) **:** ADULTERATE
3 [Middle English, from Medieval Latin *extendere* (from Latin) or Anglo-French *estendre,* from Old French] **a** *British* **:** to take possession of (as lands) by a writ of extent **b** *obsolete* **:** to take by force
4 a : to make the offer of **:** PROFFER ⟨*extending* aid to the needy⟩ **b :** to make available ⟨*extending* credit to customers⟩
5 a : to cause to reach (as in distance or scope) ⟨national authority was *extended* over new territories⟩ **b :** to cause to be longer **:** PROLONG ⟨*extend* the side of a triangle⟩ ⟨*extended* their visit another day⟩; *also* **:** to prolong the time of payment of **c :** ADVANCE, FURTHER ⟨*extending* her potential through job training⟩
6 a : to cause to be of greater area or volume **:** ENLARGE **b :** to increase the scope, meaning, or application of **:** BROADEN ⟨beauty, I suppose, opens the heart, *extends* the consciousness —Algernon Blackwood⟩ **c** *archaic* **:** EXAGGERATE
intransitive verb
1 : to stretch out in distance, space, or time **:** REACH ⟨their jurisdiction *extended* over the whole area⟩
2 : to reach in scope or application ⟨his concern *extends* beyond mere business to real service to his customers⟩ ☆
— **ex·tend·abil·i·ty** \-ˌsten-də-ˈbi-lə-tē\ *noun*
— **ex·tend·able** *also* **ex·tend·ible** \-ˈsten-də-bəl\ *adjective*

ex·tend·ed *adjective* (15th century)
1 : drawn out in length especially of time ⟨an *extended* visit⟩
2 a : fully stretched out ⟨an *extended* battle line⟩ **b** *of a horse's gait* **:** performed with a greatly lengthened stride but without a break — compare COLLECTED **c :** INTENSIVE ⟨*extended* efforts⟩
3 : having spatial magnitude **:** being larger than a point ⟨an *extended* source of light⟩
4 : EXTENSIVE 1 ⟨made available *extended* information —Ruth G. Strickland⟩
5 : DERIVATIVE 1, SECONDARY 2a ⟨an *extended* sense of a word⟩
6 *of a typeface* **:** having a wider face than that of a standard typeface
— **ex·tend·ed·ly** *adverb*
— **ex·tend·ed·ness** *noun*

extended family *noun* (circa 1935)
: a family that includes in one household near relatives in addition to a nuclear family

ex·tend·er \ik-ˈsten-dər\ *noun* (1611)
: one that extends: as **a :** a substance added to a product especially in the capacity of a diluent, adulterant, or modifier **b :** an added ingredient used to increase the bulk of a food (as soup or meat)

ex·ten·si·ble \ik-ˈsten(t)-sə-bəl\ *adjective* (1611)
: capable of being extended
— **ex·ten·si·bil·i·ty** \-ˌsten(t)-sə-ˈbi-lə-tē\ *noun*

ex·ten·sile \ik-ˈsten(t)-səl, -ˈsten-ˌsīl\ *adjective* (1744)
: EXTENSIBLE

ex·ten·sion \ik-ˈsten(t)-shən\ *noun* [Middle English, from Middle French or Late Latin; Middle French, from Late Latin *extension-, extensio,* from Latin *extendere*] (15th century)

☆ SYNONYMS
Extend, lengthen, prolong, protract mean to draw out or add to so as to increase in length. EXTEND and LENGTHEN imply a drawing out in space or time but EXTEND may also imply increase in width, scope, area, or range ⟨*extend* a vacation⟩ ⟨*extend* welfare services⟩ ⟨*lengthen* a skirt⟩ ⟨*lengthen* the workweek⟩. PROLONG suggests chiefly increase in duration especially beyond usual limits ⟨*prolonged* illness⟩. PROTRACT adds to PROLONG implications of needlessness, vexation, or indefiniteness ⟨*protracted* litigation⟩.

◇ WORD HISTORY
exquisite The source of English *exquisite* is the Latin verb *exquirere,* which meant "to ask about" or "to seek." The past participle of this verb, *exquisitus,* came to be used as an adjective, eventually acquiring not just the meaning of "sought out" but also "sought out with care"—in other words, "choice." This was the sense that was adopted into Middle English as *exquisit* and from which the modern senses of *exquisite* have developed.

1 a : the action of extending **:** state of being extended **b :** an enlargement in scope or operation ⟨tools are *extensions* of human hands⟩
2 a : the total range over which something extends **:** COMPASS **b :** DENOTATION 4
3 a : the stretching of a fractured or dislocated limb so as to restore it to its natural position **b :** an unbending movement around a joint in a limb (as the knee or elbow) that increases the angle between the bones of the limb at the joint — compare FLEXION 4a
4 : a property whereby something occupies space
5 : an increase in length of time; *specifically* **:** an increase in time allowed under agreement or concession ⟨was granted an *extension*⟩
6 : a program that geographically extends the educational resources of an institution by special arrangements (as correspondence courses) to persons otherwise unable to take advantage of such resources
7 a : a part constituting an addition **b :** a section or line segment forming an additional length **c :** an extra telephone connected to the principal line
8 : a mathematical set (as a field or group) that includes a given and similar set as a subset

extension agent *noun* (1949)
: COUNTY AGENT

ex·ten·sion·al \ik-ˈstench-nəl, -ˈsten(t)-shə-nᵊl\ *adjective* (1647)
1 : of, relating to, or marked by extension; *specifically* **:** DENOTATIVE
2 : concerned with objective reality
— ex·ten·sion·al·i·ty \-ˌsten(t)-shə-ˈna-lə-tē\ *noun*
— ex·ten·sion·al·ly \-ˈstench-nə-lē, -ˈsten(t)-shə-nᵊl-ē\ *adverb*

extension cord *noun* (1946)
: an electric cord fitted with a plug at one end and a receptacle at the other

ex·ten·si·ty \ik-ˈsten(t)-sə-tē\ *noun, plural* **-ties** (circa 1834)
1 a : the quality of having extension **b :** degree of extension **:** RANGE
2 : an attribute of sensation whereby space or size is perceived

ex·ten·sive \ik-ˈsten(t)-siv\ *adjective* (1605)
1 : having wide or considerable extent ⟨*extensive* reading⟩
2 : EXTENSIONAL
3 : of, relating to, or constituting farming in which large areas of land are utilized with minimum outlay and labor
— ex·ten·sive·ly *adverb*
— ex·ten·sive·ness *noun*

ex·ten·som·e·ter \ˌek-ˌsten-ˈsä-mə-tər\ *noun* [*extension* + *-o-* + *-meter*] (1887)
: an instrument for measuring minute deformations of test specimens caused by tension, compression, bending, or twisting

ex·ten·sor \ik-ˈsten(t)-sər\ *noun* (1713)
: a muscle serving to extend a bodily part (as a limb)

ex·tent \ik-ˈstent\ *noun* [Middle English, from Anglo-French & Middle French; Anglo-French *extente* land valuation, from Middle French, area, surveying of land, from *extendre, estendre* to extend] (14th century)
1 *archaic* **:** valuation (as of land) in Great Britain especially for taxation
2 a : seizure (as of land) in execution of a writ of extent in Great Britain **:** the condition of being so seized **b :** a writ giving to a creditor temporary possession of his debtor's property
3 a : the range over which something extends **:** SCOPE ⟨the *extent* of her jurisdiction⟩ **b :** the point, degree, or limit to which something extends ⟨using talents to the greatest *extent*⟩ **c :** the amount of space or surface that something occupies or the distance over which it extends **:** MAGNITUDE ⟨the *extent* of the forest⟩

ex·ten·u·ate \ik-ˈsten-yə-ˌwāt\ *transitive verb* **-at·ed; -at·ing** [Latin *extenuatus,* past participle of *extenuare,* from *ex-* + *tenuis* thin — more at THIN] (1529)
1 a *archaic* **:** to make light of **b :** to lessen or to try to lessen the seriousness or extent of by making partial excuses **:** MITIGATE **c** *obsolete* **:** DISPARAGE
2 a *archaic* **:** to make thin or emaciated **b :** to lessen the strength or effect of
— ex·ten·u·a·tor \-ˌwā-tər\ *noun*
— ex·ten·u·a·to·ry \-wə-ˌtōr-ē, -ˌtȯr-\ *adjective*

ex·ten·u·a·tion \ik-ˌsten-yə-ˈwā-shən\ *noun* (circa 1543)
1 : the act of extenuating or state of being extenuated; *especially* **:** partial justification
2 : something extenuating; *especially* **:** a partial excuse

[1]ex·te·ri·or \ek-ˈstir-ē-ər\ *adjective* [Latin, comparative of *exter, exterus* being on the outside, foreign, from *ex*] (1528)
1 : being on an outside surface **:** situated on the outside
2 : observable by outward signs ⟨his *exterior* quietness is belied by an occasional nervous twitch —*Current Biography*⟩
3 : suitable for use on outside surfaces
— ex·te·ri·or·ly *adverb*

[2]exterior *noun* (1591)
1 a : an exterior part or surface **:** OUTSIDE **b :** outward manner or appearance
2 : a representation (as on stage or film) of an outdoor scene; *also* **:** a scene filmed outdoors

exterior angle *noun* (circa 1864)
1 : the angle between a side of a polygon and an extended adjacent side
2 : an angle formed by a transversal as it cuts one of two lines and situated on the outside of the line

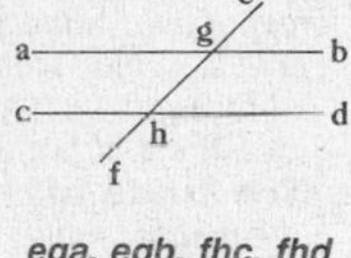

ega, egb, fhc, fhd
exterior angle 2

ex·te·ri·or·ise *British variant of* EXTERIORIZE

ex·te·ri·or·i·ty \(ˌ)ek-ˌstir-ē-ˈȯr-ə-tē, -ˈär-\ *noun* (1611)
: the quality or state of being exterior or exteriorized **:** EXTERNALITY

ex·te·ri·or·ize \ek-ˈstir-ē-ə-ˌrīz\ *transitive verb* **-ized; -iz·ing** (1879)
1 : EXTERNALIZE
2 : to bring out of the body (as for surgery)
— ex·te·ri·or·iza·tion \-ˌstir-ē-ə-rə-ˈzā-shən\ *noun*

ex·ter·mi·nate \ik-ˈstər-mə-ˌnāt\ *transitive verb* **-nat·ed; -nat·ing** [Latin *exterminatus,* past participle of *exterminare,* from *ex-* + *terminus* boundary — more at TERM] (1591)
: to get rid of completely usually by killing off ⟨*exterminate* crabgrass from a lawn⟩ ☆
— ex·ter·mi·na·tion \-ˌstər-mə-ˈnā-shən\ *noun*
— ex·ter·mi·na·tor \-ˈstər-mə-ˌnā-tər\ *noun*

ex·ter·mi·na·to·ry \ik-ˈstər-mə-nə-ˌtōr-ē, -ˌtȯr-\ *adjective* (1790)
: of, relating to, or marked by extermination

ex·ter·mine \ik-ˈstər-mən\ *transitive verb* **-mined; -min·ing** (1539)
obsolete **:** EXTERMINATE

[1]ex·tern \ek-ˈstərn, ˈek-ˌ\ *adjective* [Middle French or Latin; Middle French *externe,* from Latin *externus*] (1530)
archaic **:** EXTERNAL

[2]ex·tern *also* **ex·terne** \ˈek-ˌstərn\ *noun* (circa 1610)
: a person connected with an institution but not living or boarding in it; *specifically* **:** a nonresident doctor or medical student at a hospital

[1]ex·ter·nal \ek-ˈstər-nᵊl\ *adjective* [Latin *externus* external, from *exter*] (1556)
1 a : capable of being perceived outwardly ⟨*external* signs of a disease⟩ ⟨*external* reality⟩ **b** (1) **:** having merely the outward appearance of something **:** SUPERFICIAL (2) **:** not intrinsic or essential ⟨*external* circumstances⟩
2 a : of, relating to, or connected with the outside or an outer part **b :** applied or applicable to the outside
3 a (1) **:** situated outside, apart, or beyond; *specifically* **:** situated near or toward the surface of the body (2) **:** arising or acting from outside ⟨*external* force⟩ **b :** of or relating to dealings or relationships with foreign countries **c :** having existence independent of the mind ⟨*external* reality⟩
— ex·ter·nal·ly \-nᵊl-ē\ *adverb*

[2]external *noun* (circa 1635)
: something that is external: as **a :** an outer part **b :** an external feature or aspect — usually used in plural

external combustion engine *noun* (1915)
: a heat engine (as a steam engine) that derives its heat from fuel consumed outside the engine cylinder

external degree *noun* (1928)
: a degree conferred on a student who has not attended the university but has passed the qualifying examination

ex·ter·nal·isa·tion, ex·ter·nal·ise *British variant of* EXTERNALIZATION, EXTERNALIZE

ex·ter·nal·ism \ek-ˈstər-nᵊl-ˌi-zəm\ *noun* (1856)
1 : attention to externals; *especially* **:** excessive preoccupation with externals
2 : EXTERNALITY 1

ex·ter·nal·i·ty \ˌek-ˌstər-ˈna-lə-tē\ *noun, plural* **-ties** (1673)
1 : the quality or state of being external or externalized
2 : something that is external
3 : a secondary or unintended consequence ⟨pollution and other *externalities* of manufacturing⟩

ex·ter·nal·iza·tion \ek-ˌstər-nᵊl-ə-ˈzā-shən\ *noun* (1803)
1 a : the action or process of externalizing **b :** the quality or state of being externalized
2 : something externalized **:** EMBODIMENT

ex·ter·nal·ize \ek-ˈstər-nᵊl-ˌīz\ *transitive verb* **-ized; -iz·ing** (1852)
1 : to make external or externally manifest
2 : to attribute to causes outside the self **:** RATIONALIZE

external respiration *noun* (1940)
: exchange of gases between the external environment and a distributing system of the animal body (as the lungs of higher vertebrates or the tracheal tubes of insects) or between the alveoli of the lungs and the blood — compare INTERNAL RESPIRATION

ex·tern·ship \ˈek-ˌstərn-ˌship\ *noun* (1945)
: a training program that is part of a course of study of an educational institution and is taken in private business

☆ SYNONYMS
Exterminate, extirpate, eradicate, uproot mean to effect the destruction or abolition of something. EXTERMINATE implies complete and immediate extinction by killing off all individuals ⟨*exterminate* cockroaches⟩. EXTIRPATE implies extinction of a race, family, species, or sometimes an idea or doctrine by destruction or removal of its means of propagation ⟨many species have been *extirpated* from the area⟩. ERADICATE implies the driving out or elimination of something that has established itself ⟨a campaign to *eradicate* illiteracy⟩. UPROOT implies a forcible or violent removal and stresses displacement or dislodgment rather than immediate destruction ⟨the war *uprooted* thousands⟩.

ex·tero·cep·tive \ˌek-stə-rō-ˈsep-tiv\ *adjective* [*exterior* + *-o-* + *-ceptive* (as in *receptive*)] (circa 1921)
: activated by, relating to, or being stimuli received by an organism from outside

ex·tero·cep·tor \-tər\ *noun* (1906)
: a sense organ excited by exteroceptive stimuli

ex·ter·ri·to·ri·al \ˌek-ˌster-ə-ˈtōr-ē-əl, -ˈtȯr-\ *adjective* (circa 1880)
: EXTRATERRITORIAL
— **ex·ter·ri·to·ri·al·i·ty** \-ˌtōr-ē-ˈa-lə-tē, -ˌtȯr-\ *noun*

[1]ex·tinct \ik-ˈstiŋ(k)t, ˈek-ˌ\ *adjective* [Middle English, from Latin *exstinctus*, past participle of *exstinguere*] (15th century)
1 a : no longer burning **b :** no longer active ⟨an *extinct* volcano⟩
2 : no longer existing ⟨an *extinct* animal⟩
3 a : gone out of use **:** SUPERSEDED **b :** having no qualified claimant ⟨an *extinct* title⟩

[2]extinct *transitive verb* (15th century)
archaic **:** EXTINGUISH

ex·tinc·tion \ik-ˈstiŋ(k)-shən\ *noun* (15th century)
1 : the act of making extinct or causing to be extinguished
2 : the condition or fact of being extinct or extinguished; *also* **:** the process of becoming extinct ⟨*extinction* of a species⟩
3 : the process of eliminating or reducing a conditioned response by not reinforcing it

extinction coefficient *noun* (1902)
: a measure of the rate of diminution of transmitted light via scattering and absorption for a medium

ex·tinc·tive \ik-ˈstiŋ(k)-tiv\ *adjective* (1600)
: tending or serving to extinguish or make extinct

ex·tin·guish \ik-ˈstiŋ-(g)wish\ *transitive verb* [Latin *exstinguere* (from *ex-* + *stinguere* to extinguish) + English *-ish* (as in *abolish*); akin to Latin in*stigare* to incite — more at STICK] (1551)
1 a : to cause to cease burning **:** QUENCH **b** (1) **:** to bring to an end **:** make an end of ⟨hope for their safety was slowly *extinguished*⟩ (2) **:** to reduce to silence or ineffectiveness **c :** to cause extinction of (a conditioned response) **d :** to dim the brightness of **:** ECLIPSE
2 a : to cause to be void **:** NULLIFY ⟨*extinguish* a claim⟩ **b :** to get rid of usually by payment ⟨*extinguish* a debt⟩
— **ex·tin·guish·able** \-(g)wi-shə-bəl\ *adjective*
— **ex·tin·guish·er** \-shər\ *noun*
— **ex·tin·guish·ment** \-mənt\ *noun*

ex·tir·pate \ˈek-stər-ˌpāt\ *transitive verb* **-pat·ed; -pat·ing** [Latin *exstirpatus*, past participle of *exstirpare*, from *ex-* + *stirp-*, *stirps* trunk, root — more at TORPID] (1535)
1 a : to destroy completely **:** WIPE OUT **b :** to pull up by the root
2 : to cut out by surgery
synonym see EXTERMINATE
— **ex·tir·pa·tion** \ˌek-stər-ˈpā-shən\ *noun*
— **ex·tir·pa·tor** \ˈek-stər-ˌpā-tər\ *noun*

ex·tol *also* **ex·toll** \ik-ˈstōl\ *transitive verb* **ex·tolled; ex·tol·ling** [Middle English, from Latin *extollere*, from *ex-* + *tollere* to lift up — more at TOLERATE] (15th century)
: to praise highly **:** GLORIFY
— **ex·tol·ler** *noun*
— **ex·tol·ment** \-ˈstōl-mənt\ *noun*

ex·tort \ik-ˈstȯrt\ *transitive verb* [Latin *extortus*, past participle of *extorquēre* to wrench out, extort, from *ex-* + *torquēre* to twist — more at TORTURE] (1529)
: to obtain from a person by force, intimidation, or undue or illegal power **:** WRING; *also* **:** to gain especially by ingenuity or compelling argument
synonym see EDUCE
— **ex·tort·er** *noun*
— **ex·tor·tive** \-ˈstȯr-tiv\ *adjective*

ex·tor·tion \ik-ˈstȯr-shən\ *noun* (14th century)
1 : the act or practice of extorting especially money or other property; *especially* **:** the offense committed by an official engaging in such practice
2 : something extorted; *especially* **:** a gross overcharge
— **ex·tor·tion·er** \-sh(ə-)nər\ *noun*
— **ex·tor·tion·ist** \-sh(ə-)nist\ *noun*

ex·tor·tion·ary \-shə-ˌner-ē\ *adjective* (1805)
archaic **:** EXTORTIONATE 1

ex·tor·tion·ate \ik-ˈstȯr-sh(ə-)nət\ *adjective* (1789)
1 : characterized by extortion
2 : EXCESSIVE, EXORBITANT ⟨*extortionate* prices⟩
— **ex·tor·tion·ate·ly** *adverb*

[1]ex·tra \ˈek-strə\ *adjective* [probably short for *extraordinary*] (1776)
1 a : more than is due, usual, or necessary **:** ADDITIONAL ⟨*extra* work⟩ **b :** subject to an additional charge ⟨room service is *extra*⟩
2 : SUPERIOR ⟨*extra* quality⟩

[2]extra *noun* (1793)
1 : one that is extra or additional: as **a :** a special edition of a newspaper **b :** an added charge **c :** an additional worker; *specifically* **:** one hired to act in a group scene in a motion picture or stage production
2 : something of superior quality or grade

[3]extra *adverb* (1823)
: beyond the usual size, extent, or degree ⟨*extra* large⟩

extra- *prefix* [Middle English, from Latin, from *extra*, adverb & preposition, outside, except, beyond, from *exter* being on the outside — more at EXTERIOR]
: outside **:** beyond ⟨*extra*judicial⟩

extra–base hit *noun* (circa 1949)
: a hit in baseball that lets the batter take more than one base

ex·tra·cel·lu·lar \ˌek-strə-ˈsel-yə-lər\ *adjective* (1867)
: situated or occurring outside a cell or the cells of the body ⟨*extracellular* digestion⟩ ⟨*extracellular* enzymes⟩
— **ex·tra·cel·lu·lar·ly** *adverb*

ex·tra·chro·mo·som·al \-ˌkrō-mə-ˈsō-məl, -ˈzō-\ *adjective* (1940)
: situated or controlled by factors outside the chromosome ⟨*extrachromosomal* inheritance⟩

ex·tra·cor·po·re·al \-kȯr-ˈpōr-ē-əl, -ˈpȯr-\ *adjective* (1865)
: occurring or based outside the living body ⟨heart surgery employing *extracorporeal* circulation⟩
— **ex·tra·cor·po·re·al·ly** \-ē-ə-lē\ *adverb*

ex·tra·cra·ni·al \-ˈkrā-nē-əl\ *adjective* (circa 1884)
: situated or occurring outside the cranium

[1]ex·tract \ik-ˈstrakt, *oftenest in sense* 5 ˈek-ˌ\ *transitive verb* [Middle English, from Latin *extractus*, past participle of *extrahere*, from *ex-* + *trahere* to draw] (15th century)
1 a : to draw forth (as by research) ⟨*extract* data⟩ **b :** to pull or take out forcibly ⟨*extracted* a wisdom tooth⟩ **c :** to obtain by much effort from someone unwilling ⟨*extracted* a confession⟩
2 : to withdraw (as a juice or fraction) by physical or chemical process; *also* **:** to treat with a solvent so as to remove a soluble substance
3 : to separate (a metal) from an ore
4 : to determine (a mathematical root) by calculation
5 : to select (excerpts) and copy out or cite
synonym see EDUCE
— **ex·tract·abil·i·ty** \ik-ˌstrak-tə-ˈbi-lə-tē, (ˌ)ek-\ *noun*
— **ex·tract·able** \ik-ˈstrak-tə-bəl, ˈek-ˌ\ *adjective*

[2]ex·tract \ˈek-ˌstrakt\ *noun* (15th century)
1 : a selection from a writing or discourse **:** EXCERPT
2 : a product (as an essence or concentrate) prepared by extracting; *especially* **:** a solution (as in alcohol) of essential constituents of a complex material (as meat or an aromatic plant)

ex·trac·tion \ik-ˈstrak-shən\ *noun* (15th century)
1 : the act or process of extracting something
2 : ANCESTRY, ORIGIN
3 : something extracted

[1]ex·trac·tive \ik-ˈstrak-tiv, ˈek-ˌ\ *adjective* (1599)
1 a : of, relating to, or involving extraction **b :** tending toward or resulting in withdrawal of natural resources by extraction with no provision for replenishment ⟨*extractive* agriculture⟩
2 : capable of being extracted
— **ex·trac·tive·ly** *adverb*

[2]extractive *noun* (1847)
: something extracted or extractable **:** EXTRACT

ex·trac·tor \ik-ˈstrak-tər\ *noun* (1611)
: one that extracts; *especially* **:** the mechanism in a firearm that dislodges a spent cartridge from the chamber

ex·tra·cur·ric·u·lar \ˌek-strə-kə-ˈri-kyə-lər\ *adjective* (1925)
1 : not falling within the scope of a regular curriculum; *specifically* **:** of or relating to officially or semiofficially approved and usually organized student activities (as athletics) connected with school and usually carrying no academic credit
2 a : lying outside one's regular duties or routine **b :** EXTRAMARITAL
— **extracurricular** *noun*

ex·tra·dit·able \ˈek-strə-ˌdī-tə-bəl\ *adjective* (1881)
1 : subject or liable to extradition
2 : making one liable to extradition ⟨an *extraditable* offense⟩

ex·tra·dite \ˈek-strə-ˌdīt\ *transitive verb* **-dit·ed; -dit·ing** [back-formation from *extradition*] (1864)
1 : to deliver up to extradition
2 : to obtain the extradition of

ex·tra·di·tion \ˌek-strə-ˈdi-shən\ *noun* [French, from *ex-* + Latin *tradition-*, *traditio* act of handing over — more at TREASON] (1839)
: the surrender of an alleged criminal usually under the provisions of a treaty or statute by one authority (as a state) to another having jurisdiction to try the charge

ex·tra·dos \ˈek-strə-ˌdäs, -ˌdō; ek-ˈstrā-ˌdäs\ *noun, plural* **-dos·es** \-ˌdä-səz\ *or* **-dos** \-ˌdōz, -ˌdäs\ [French, from Latin *extra* + French *dos* back — more at DOSSIER] (1772)
: the exterior curve of an arch — see ARCH illustration

ex·tra·em·bry·on·ic \ˌek-strə-ˌem-brē-ˈä-nik\ *adjective* (circa 1903)
: situated outside the embryo; *especially* **:** developed from the zygote but not part of the embryo ⟨*extraembryonic* membranes⟩

ex·tra·ga·lac·tic \ˌek-strə-gə-ˈlak-tik\ *adjective* [International Scientific Vocabulary] (1851)
: lying or coming from outside the Milky Way; *also* **:** of or relating to extragalactic space ⟨*extragalactic* astronomy⟩

ex·tra·he·pat·ic \-hi-ˈpa-tik\ *adjective* (circa 1923)
: situated or originating outside the liver

ex·tra·ju·di·cial \-jü-ˈdi-shəl\ *adjective* (1630)
1 a : not forming a valid part of regular legal proceedings ⟨an *extrajudicial* investigation⟩ **b :** delivered without legal authority **:** PRIVATE 2a(2) ⟨the judge's *extrajudicial* statements⟩
2 : done in contravention of due process of law ⟨an *extrajudicial* execution⟩
— **ex·tra·ju·di·cial·ly** \-ˈdish-lē, -ˈdi-shə-\ *adverb*

ex·tra·le·gal \ˌek-strə-ˈlē-gəl\ *adjective* (1644)
: not regulated or sanctioned by law

— **ex·tra·le·gal·ly** \-gə-lē\ *adverb*

ex·tra·lim·it·al \-'li-mə-tᵊl\ *adjective* (1874)
: not present in a given area — used of kinds of organisms (as species)

ex·tra·lin·guis·tic \-liŋ-'gwis-tik\ *adjective* (1927)
: lying outside the province of linguistics
— **ex·tra·lin·guis·ti·cal·ly** \-ti-kə-lē\ *adverb*

ex·tra·lit·er·ary \-'li-tə-ˌrer-ē\ *adjective* (1945)
: lying outside the field of literature

ex·tral·i·ty \ek-'stra-lə-tē\ *noun* [by contraction] (1925)
: EXTRATERRITORIALITY

ex·tra·log·i·cal \ˌek-strə-'lä-ji-kəl\ *adjective* (1833)
: not guided or determined by considerations of logic

ex·tra·mar·i·tal \ˌek-strə-'mar-ə-tᵊl\ *adjective* (1925)
: of, relating to, or being sexual intercourse between a married person and someone other than his or her spouse : ADULTEROUS

ex·tra·mun·dane \ˌek-strə-ˌmən-'dān, -'mən-ˌdān\ *adjective* [Late Latin *extramundanus*, from Latin *extra* + *mundus* the world] (1665)
: situated in or relating to a region beyond the material world

ex·tra·mu·ral \-'myu̇r-əl\ *adjective* (1854)
1 : existing or functioning outside or beyond the walls, boundaries, or precincts of an organized unit (as a school or hospital)
2 *chiefly British* **:** of, relating to, or taking part in extension courses or facilities
— **ex·tra·mu·ral·ly** \-ə-lē\ *adverb*

ex·tra·mu·si·cal \-'myü-zi-kəl\ *adjective* (1923)
: lying outside the province of music

ex·tra·ne·ous \ek-'strā-nē-əs\ *adjective* [Latin *extraneus* — more at STRANGE] (1638)
1 : existing on or coming from the outside
2 a : not forming an essential or vital part ⟨*extraneous* ornamentation⟩ **b :** having no relevance ⟨an *extraneous* digression⟩
3 : being a number obtained in solving an equation that is not a solution of the equation ⟨*extraneous* roots⟩
synonym see EXTRINSIC
— **ex·tra·ne·ous·ly** *adverb*
— **ex·tra·ne·ous·ness** *noun*

ex·tra·nu·cle·ar \ˌek-strə-'nü-klē-ər, -'nyü-, ÷-kyə-lər\ *adjective* (1887)
1 : situated in or affecting the parts of a cell external to the nucleus : CYTOPLASMIC
2 : situated outside the nucleus of an atom

ex·tra·oc·u·lar muscle \ˌek-strə-'ä-kyə-lər-\ *noun* (1939)
: any of six small voluntary muscles that pass between the eyeball and the orbit and control the movement of the eyeball in relation to the orbit

ex·traor·di·naire \ik-ˌstrȯ(r)-dᵊn-'er, ek-\ *adjective* [French] (1940)
: EXTRAORDINARY — used postpositively ⟨a chef *extraordinaire*⟩

ex·traor·di·nary \ik-'strȯr-dᵊn-ˌer-ē, ˌek-strə-'ȯr-\ *adjective* [Middle English *extraordinarie*, from Latin *extraordinarius*, from *extra ordinem* out of course, from *extra* + *ordinem*, accusative of *ordin-*, *ordo* order] (15th century)
1 a : going beyond what is usual, regular, or customary ⟨*extraordinary* powers⟩ **b :** exceptional to a very marked extent ⟨*extraordinary* beauty⟩ **c** *of a financial transaction* **:** NONRECURRING
2 : employed for or sent on a special function or service ⟨an ambassador *extraordinary*⟩
— **ex·traor·di·nari·ly** \ik-ˌstrȯr-dᵊn-'er-ə-lē, ˌek-strə-ˌȯr-\ *adverb*
— **ex·traor·di·nari·ness** \ik-'strȯr-dᵊn-ˌer-ē-nəs, ˌek-strə-'ȯr-\ *noun*

extra point *noun* (circa 1949)
: a point gained on a conversion in football

ex·trap·o·late \ik-'stra-pə-ˌlāt\ *verb* **-lat·ed; -lat·ing** [Latin *extra* outside + English *-polate* (as in *interpolate*) — more at EXTRA-] (1874)
transitive verb
1 : to infer (values of a variable in an unobserved interval) from values within an already observed interval
2 a : to project, extend, or expand (known data or experience) into an area not known or experienced so as to arrive at a usually conjectural knowledge of the unknown area ⟨*extrapolates* present trends to construct an image of the future⟩ **b :** to predict by projecting past experience or known data ⟨*extrapolate* public sentiment on one issue from known public reaction on others⟩
intransitive verb
: to perform the act or process of extrapolating
— **ex·trap·o·la·tion** \-ˌstra-pə-'lā-shən\ *noun*
— **ex·trap·o·la·tive** \-'stra-pə-ˌlā-tiv\ *adjective*
— **ex·trap·o·la·tor** \-ˌlā-tər\ *noun*

ex·tra·py·ra·mi·dal \ˌek-strə-pə-'ra-mə-dᵊl, -ˌpir-ə-'mi-dᵊl\ *adjective* (circa 1902)
: situated outside of and especially involving descending nerve tracts other than the pyramidal tracts

ex·tra·sen·so·ry \ˌek-strə-'sen(t)s-rē, -'sen(t)-sə-\ *adjective* (1934)
: residing beyond or outside the ordinary senses

extrasensory perception *noun* (1934)
: perception (as in telepathy, clairvoyance, and precognition) that involves awareness of information about events external to the self not gained through the senses and not deducible from previous experience — called also *ESP*

ex·tra·sys·to·le \-'sis-tə-(ˌ)lē\ *noun* [New Latin] (circa 1900)
: a premature beat of one of the chambers of the heart that leads to momentary arrhythmia

¹ex·tra·ter·res·tri·al \-tə-'res-trē-əl, -'res(h)-chəl\ *adjective* (1868)
: originating, existing, or occurring outside the earth or its atmosphere ⟨*extraterrestrial* life⟩

²extraterrestrial *noun* (1960)
: an extraterrestrial being

ex·tra·ter·ri·to·ri·al \-ˌter-ə-'tōr-ē-əl, -'tȯr-\ *adjective* (1869)
: existing or taking place outside the territorial limits of a jurisdiction

ex·tra·ter·ri·to·ri·al·i·ty \-ˌtōr-ē-'a-lə-tē, -ˌtȯr-\ *noun* (1836)
: exemption from the application or jurisdiction of local law or tribunals

ex·tra·tex·tu·al \ˌek-strə-'teks-chə-wəl, -chəl\ *adjective* (1960)
: of, relating to, or being something outside a literary text

ex·tra·trop·i·cal cyclone \ˌek-strə-ˌträ-pi-kəl-\ *noun* (1923)
: a cyclone in the middle latitudes often being 1500 miles (2400 kilometers) in diameter and usually containing a cold front that extends toward the equator for hundreds of miles

ex·tra·uter·ine \ˌek-strə-'yü-tə-rən, -ˌrīn\ *adjective* (1709)
: situated or occurring outside the uterus ⟨*extrauterine* pregnancy⟩

ex·trav·a·gance \ik-'stra-vi-gən(t)s\ *noun* (1650)
1 a : an instance of excess or prodigality; *specifically* **:** an excessive outlay of money **b :** something extravagant
2 : the quality or fact of being extravagant

ex·trav·a·gan·cy \-gən(t)-sē\ *noun, plural* **-cies** (1625)
: EXTRAVAGANCE

ex·trav·a·gant \ik-'stra-vi-gənt\ *adjective* [Middle English, from Middle French, from Medieval Latin *extravagant-*, *extravagans*, present participle of *extravagari* to wander outside, from Latin *extra-* + *vagant-*, *vagans*, present participle of *vagari* to wander about, from *vagus* wandering] (15th century)
1 a *obsolete* **:** STRANGE, CURIOUS **b** *archaic* **:** WANDERING
2 a : exceeding the limits of reason or necessity ⟨*extravagant* claims⟩ **b :** lacking in moderation, balance, and restraint ⟨*extravagant* praise⟩ **c :** extremely or excessively elaborate
3 a : spending much more than necessary **b :** PROFUSE, LAVISH
4 : unreasonably high in price ◆
synonym see EXCESSIVE
— **ex·trav·a·gant·ly** *adverb*

ex·trav·a·gan·za \ik-ˌstra-və-'gan-zə\ *noun* [Italian *estravaganza*, literally, extravagance, from *estravagante* extravagant, from Medieval Latin *extravagant-*, *extravagans*] (1754)
1 : a literary or musical work marked by extreme freedom of style and structure and usually by elements of burlesque or parody
2 : a lavish or spectacular show or event
3 : something extravagant

ex·trav·a·gate \ik-'stra-və-ˌgāt\ *intransitive verb* **-gat·ed; -gat·ing** (circa 1755)
archaic **:** to go beyond proper limits

ex·trav·a·sate \ik-'stra-və-ˌsāt, -ˌzāt\ *verb* **-sat·ed; -sat·ing** [Latin *extra* + *vas* vessel] (1668)
transitive verb
: to force out or cause to escape from a proper vessel or channel
intransitive verb
: to pass by infiltration or effusion from a proper vessel or channel (as a blood vessel) into surrounding tissue
— **ex·trav·a·sa·tion** \-ˌstra-və-'sā-shən, -'zā-\ *noun*

ex·tra·vas·cu·lar \ˌek-strə-'vas-kyə-lər\ *adjective* (1804)
: not occurring or contained in body vessels ⟨*extravascular* tissue fluids⟩

ex·tra·ve·hic·u·lar \-vē-'hi-kyə-lər\ *adjective* (1965)
: taking place outside a vehicle (as a spacecraft) ⟨*extravehicular* activity⟩

¹ex·treme \ik-'strēm\ *adjective* [Middle English, from Middle French, from Latin *extremus*, superlative of *exter*, *exterus* being on the outside — more at EXTERIOR] (15th century)
1 a : existing in a very high degree ⟨*extreme* poverty⟩ **b :** going to great or exaggerated lengths **:** RADICAL ⟨went on an *extreme* diet⟩ **c :** exceeding the ordinary, usual, or expected ⟨*extreme* weather conditions⟩
2 *archaic* **:** LAST
3 : situated at the farthest possible point from a center ⟨the country's *extreme* north⟩
4 a : most advanced or thoroughgoing ⟨the *extreme* political left⟩ **b :** MAXIMUM
synonym see EXCESSIVE

◇ WORD HISTORY

extravagant A person who buys an extravagant gift has in all likelihood strayed quite far from the amount originally budgeted for the purchase. The notion of wandering far afield actually is inherent in the word *extravagant*, which comes from a word in Medieval Latin. *Extravagans* is formed from the Latin prefix *extra-* "outside" or "beyond" and the verb *vagari* "to wander about." Originally *extravagant* was used literally of things that wandered beyond their usual confines. In Shakespeare's *Hamlet* the restless ghost of Hamlet's father is spoken of as an "extravagant and erring spirit." From this literal use of *extravagant*, the extended meanings "exceeding the limits of reason or necessity" and "lacking in moderation, balance, and restraint" developed.

— **ex·treme·ness** *noun*
[2]**extreme** *noun* (1555)
1 a : something situated at or marking one end or the other of a range ⟨*extremes* of heat and cold⟩ **b :** the first term or the last term of a mathematical proportion **c :** the major term or minor term of a syllogism
2 a : a very pronounced or excessive degree **b :** highest degree **:** MAXIMUM
3 : an extreme measure or expedient ⟨going to *extremes*⟩
— **in the extreme :** to the greatest possible extent
ex·treme·ly *adverb* (1531)
1 : in an extreme manner
2 : to an extreme extent
extremely high frequency *noun* (1952)
: a radio frequency in the highest range of the radio spectrum — see RADIO FREQUENCY table
extremely low frequency *noun* (1966)
: a radio frequency in the lowest range of the radio spectrum — see RADIO FREQUENCY table
extreme unction \ik-ˌstrēm-ˈəŋ(k)-shən, ˌek-(ˌ)strēm-\ *noun* (15th century)
: a sacrament in which a priest anoints and prays for the recovery and salvation of a critically ill or injured person
ex·trem·ism \ik-ˈstrē-ˌmi-zəm\ *noun* (1865)
1 : the quality or state of being extreme
2 : advocacy of extreme political measures **:** RADICALISM
— **ex·trem·ist** \-mist\ *noun or adjective*
ex·trem·i·ty \ik-ˈstre-mə-tē\ *noun, plural* **-ties** (14th century)
1 a : the farthest or most remote part, section, or point **b :** a limb of the body; *especially* **:** a human hand or foot
2 a : extreme danger or critical need **b :** a moment marked by imminent destruction or death
3 a : an intense degree ⟨the *extremity* of his participation —*Saturday Review*⟩ **b :** the utmost degree (as of emotion or pain)
4 : a drastic or desperate act or measure ⟨driven to *extremities*⟩
ex·tre·mum \ik-ˈstrē-məm\ *noun, plural* **-ma** \-mə\ [New Latin, from Latin, neuter of *extremus*] (1904)
: a maximum or a minimum of a mathematical function — called also *extreme value*
ex·tri·cate \ˈek-strə-ˌkāt\ *transitive verb* **-cat·ed; -cat·ing** [Latin *extricatus,* past participle of *extricare,* from *ex-* + *tricae* trifles, perplexities] (1614)
1 a *archaic* **:** UNRAVEL **b :** to distinguish from a related thing
2 : to free or remove from an entanglement or difficulty ☆
— **ex·tri·ca·ble** \ik-ˈstri-kə-bəl, ek-ˈ, ˈek-(ˌ)\ *adjective*
— **ex·tri·ca·tion** \ˌek-strə-ˈkā-shən\ *noun*
ex·trin·sic \ek-ˈstrin-zik, -ˈstrin(t)-sik\ *adjective* [French & Late Latin; French *extrinsèque,* from Late Latin *extrinsecus,* from Latin, adverb, from without; akin to Latin *exter* outward and to Latin *sequi* to follow — more at EXTERIOR, SUE] (1613)
1 a : not forming part of or belonging to a thing **:** EXTRANEOUS **b :** originating from or on the outside; *especially* **:** originating outside a part and acting upon the part as a whole ⟨*extrinsic* muscles of the tongue⟩
2 : EXTERNAL ☆
— **ex·trin·si·cal·ly** \-zi-k(ə-)lē, -si-\ *adverb*
extrinsic factor *noun* (1938)
: VITAMIN B_{12}
extro- *prefix* [alteration of Latin *extra-*]
: outward ⟨*extro*vert⟩ — compare INTRO-
ex·trorse \ˈek-ˌstrórs\ *adjective* [probably from (assumed) New Latin *extrorsus,* from Late Latin, adverb, outward, from Latin *extra-* + *-orsus* (as in *introrsus*) — more at INTRORSE] (1858)
: facing outward ⟨an *extrorse* anther⟩
ex·tro·ver·sion *or* **ex·tra·ver·sion** \ˌek-strə-ˈvər-zhən, -shən\ *noun* [German *Extraversion,* from Latin *extra-* + *versus,* past participle of *vertere* to turn — more at WORTH] (1915)
: the act, state, or habit of being predominantly concerned with and obtaining gratification from what is outside the self
ex·tro·vert *also* **ex·tra·vert** \ˈek-strə-ˌvərt\ *noun* [modification of German *extravertiert,* from Latin *extra-* + *vertere*] (1918)
: one whose personality is characterized by extroversion; *broadly* **:** a gregarious and unreserved person
— **extrovert** *also* **extravert** *adjective*
— **ex·tro·vert·ed** *also* **ex·tra·vert·ed** \-ˌvər-təd, ˌek-strə-ˈvər-\ *adjective*
ex·trude \ik-ˈstrüd\ *verb* **ex·trud·ed; ex·trud·ing** [Latin *extrudere,* from *ex-* + *trudere* to thrust — more at THREAT] (1566)
transitive verb
1 : to force, press, or push out
2 : to shape (as metal or plastic) by forcing through a die
intransitive verb
: to become extruded
— **ex·trud·abil·i·ty** \-ˌstrü-də-ˈbi-lə-tē\ *noun*
— **ex·trud·able** \-ˈstrü-də-bəl\ *adjective*
— **ex·trud·er** \-ˈstrü-dər\ *noun*
ex·tru·sion \ik-ˈstrü-zhən\ *noun* [Medieval Latin *extrusion-, extrusio,* from Latin *extrudere*] (1540)
: the act or process of extruding; *also* **:** a form or product produced by this process
ex·tru·sive \ik-ˈstrü-siv, -ziv\ *adjective* (1816)
: relating to or formed by geological extrusion from the earth in a molten state or as volcanic ash
ex·u·ber·ance \ig-ˈzü-b(ə-)rən(t)s\ *noun* (1631)
1 : the quality or state of being exuberant
2 : an exuberant act or expression
ex·u·ber·ant \-b(ə-)rənt\ *adjective* [Middle English, from Middle French, from Latin *exuberant-, exuberans,* present participle of *exuberare* to be abundant, from *ex-* + *uber* fruitful, from *uber* udder — more at UDDER] (15th century)
1 : extreme or excessive in degree, size, or extent
2 a : joyously unrestrained and enthusiastic **b :** unrestrained or elaborate especially in style **:** FLAMBOYANT ⟨*exuberant* architecture⟩
3 : produced in extreme abundance **:** PLENTIFUL
synonym see PROFUSE
— **ex·u·ber·ant·ly** *adverb*
ex·u·ber·ate \-bə-ˌrāt\ *intransitive verb* **-at·ed; -at·ing** (15th century)
1 *archaic* **:** to have something in abundance **:** OVERFLOW
2 : to become exuberant **:** show exuberance ⟨*exuberated* over his victory⟩
ex·u·date \ˈek-s(y)ü-ˌdāt, -shü-\ *noun* (1876)
: exuded matter
ex·u·da·tion \ˌek-s(y)ü-ˈdā-shən, -shü-\ *noun* (1612)
1 : the process of exuding
2 : EXUDATE
— **ex·u·da·tive** \ig-ˈzü-də-tiv; ˈek-s(y)ü-ˌdā-tiv, -shü-\ *adjective*
ex·ude \ig-ˈzüd\ *verb* **ex·ud·ed; ex·ud·ing** [Latin *exsudare,* from *ex-* + *sudare* to sweat — more at SWEAT] (1574)
intransitive verb
1 : to ooze out
2 : to undergo diffusion
transitive verb
1 : to cause to ooze or spread out in all directions
2 : to display conspicuously or abundantly ⟨*exudes* charm⟩
ex·ult \ig-ˈzəlt\ *intransitive verb* [Middle French *exulter,* from Latin *exsultare,* literally, to leap up, from *ex-* + *saltare* to leap — more at SALTATION] (1570)
1 *obsolete* **:** to leap for joy
2 : to be extremely joyful **:** REJOICE
— **ex·ult·ing·ly** \-ˈzəl-tiŋ-lē\ *adverb*
ex·ul·tance \ig-ˈzəl-tᵊn(t)s\ *noun* (1650)
: EXULTATION
ex·ul·tan·cy \-ˈzəl-tᵊn(t)-sē\ *noun* (1621)
: EXULTATION
ex·ul·tant \ig-ˈzəl-tᵊnt\ *adjective* (1653)
: filled with or expressing great joy or triumph **:** JUBILANT
— **ex·ul·tant·ly** *adverb*
ex·ul·ta·tion \ˌek-(ˌ)səl-ˈtā-shən, ˌeg-(ˌ)zəl-\ *noun* (15th century)
: the act of exulting **:** the state of being exultant
ex·urb \ˈek-ˌsərb, ˈeg-ˌzərb\ *noun* [*ex-* + *suburb*] (1955)
: a region or settlement that lies outside a city and usually beyond its suburbs and that often is inhabited chiefly by well-to-do families
— **ex·ur·ban** \ek-ˈsər-bən; eg-ˈzər-, ig-\ *adjective*
ex·ur·ban·ite \ek-ˈsər-bə-ˌnīt; eg-ˈzər-, ig-\ *noun* (1955)
: one who lives in an exurb
ex·ur·bia \-bē-ə\ *noun* (1955)
: the generalized region of exurbs
ex·u·vi·ae \ig-ˈzü-vē-ˌē, -vē-ˌī\ *noun plural* [Latin, from *exuere* to take off, from *ex-* + *-uere* to put on; akin to Old Church Slavonic *obuti* to put on (footwear)] (1653)
: sloughed off natural animal coverings (as the skins of snakes)
— **ex·u·vi·al** \-vē-əl\ *adjective*
ex·u·vi·a·tion \-ˌzü-vē-ˈā-shən\ *noun* (1839)
: the process of molting
[1]**ex-vo·to** \(ˌ)eks-ˈvō-(ˌ)tō\ *noun, plural* **ex-votos** [Latin *ex voto* according to a vow] (1787)
: a votive offering
[2]**ex–voto** *adjective* (1823)
: VOTIVE
-ey — see -Y
ey·as \ˈī-əs\ *noun* [Middle English, alteration (by incorrect division of *a neias*) of *neias,* from Middle French *niais* fresh from the nest,

☆ SYNONYMS
Extricate, disentangle, untangle, disencumber, disembarrass mean to free from what binds or holds back. EXTRICATE implies the use of care or ingenuity in freeing from a difficult position or situation ⟨*extricated* himself from financial difficulties⟩. DISENTANGLE and UNTANGLE suggest painstaking separation of a thing from other things ⟨*disentangling* fact from fiction⟩ ⟨*untangle* a web of deceit⟩. DISENCUMBER implies a release from something that clogs or weighs down ⟨an article *disencumbered* of jargon⟩. DISEMBARRASS suggests a release from something that impedes or hinders ⟨*disembarrassed* herself of her advisers⟩.

Extrinsic, extraneous, foreign, alien mean external to a thing, its essential nature, or its original character. EXTRINSIC applies to what is distinctly outside the thing in question or is not contained in or derived from its essential nature ⟨sentimental value that is *extrinsic* to the house's market value⟩. EXTRANEOUS applies to what is on or comes from the outside and may or may not be capable of becoming an essential part ⟨arguments *extraneous* to the issue⟩. FOREIGN applies to what is so different as to be rejected or repelled or to be incapable of becoming assimilated ⟨techniques *foreign* to French cuisine⟩. ALIEN is stronger than FOREIGN in suggesting opposition, repugnance, or irreconcilability ⟨a practice totally *alien* to her nature⟩.

from (assumed) Vulgar Latin *nidax* nestling, from Latin *nidus* nest — more at NEST] (15th century)
: an unfledged bird; *specifically* **:** a nestling hawk

[1]**eye** \ˈī\ *noun* [Middle English, from Old English *ēage;* akin to Old High German *ouga* eye, Latin *oculus,* Greek *ōps* eye, face, Sanskrit *akṣi* eye] (before 12th century)
1 a : an organ of sight; *especially* **:** a nearly spherical hollow organ that is lined with a sensitive retina, is lodged in a bony orbit in the skull, is the vertebrate organ of sight, and is normally paired **b :** all the visible structures within and surrounding the orbit and including eyelids, eyelashes, and eyebrows **c** (1) **:** the faculty of seeing with eyes (2) **:** the faculty of intellectual or aesthetic perception or appreciation ⟨an *eye* for beauty⟩ **d :** LOOK, GLANCE ⟨cast an eager *eye*⟩ **e** (1) **:** an attentive look ⟨kept an *eye* on his valuables⟩ (2) **:** ATTENTION, NOTICE ⟨caught his *eye*⟩ (3) **:** close observation **:** SCRUTINY ⟨works under the *eye* of her boss⟩ ⟨in the public *eye*⟩ **f :** POINT OF VIEW, JUDGMENT ⟨beauty is in the *eye* of the beholder⟩ — often used in plural ⟨an offender in the *eyes* of the law⟩ **g :** VIEW 5 ⟨with an *eye* to the future⟩

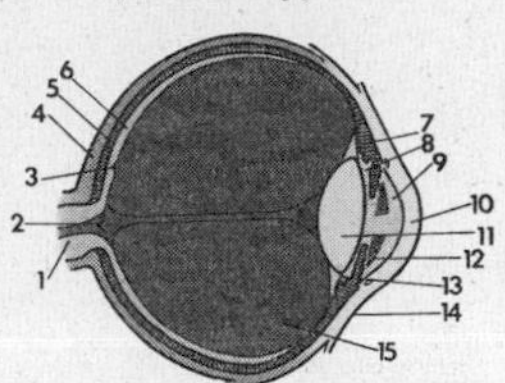

eye 1a: *1* **optic nerve,** *2* **blind spot,** *3* **fovea,** *4* **sclera,** *5* **choroid,** *6* **retina,** *7* **ciliary body,** *8* **posterior chamber,** *9* **anterior chamber,** *10* **cornea,** *11* **lens,** *12* **iris,** *13* **suspensory ligament,** *14* **conjunctiva,** *15* **vitreous humor**

2 : something having an appearance suggestive of an eye: as **a :** the hole through the head of a needle **b :** a usually circular marking (as on a peacock's tail) **c :** LOOP; *especially* **:** a loop or catch to receive a hook **d :** an undeveloped bud (as on a potato) **e :** an area like a hole in the center of a tropical cyclone marked by only light winds or complete calm with no precipitation **f :** the center of a flower especially when differently colored or marked; *specifically* **:** the disk of a composite **g** (1) **:** a triangular piece of beef cut from between the top and bottom of a round (2) **:** the chief muscle of a chop (3) **:** a compact mass of muscular tissue usually embedded in fat in a rib or loin cut of meat **h :** a device (as a photoelectric cell) that functions in a manner analogous to human vision
3 : something central **:** CENTER ⟨the *eye* of the problem —Norman Mailer⟩
4 : the direction from which the wind is blowing
— eye·less \ˈī-ləs\ *adjective*
— eye·like \-ˌlīk\ *adjective*
— an eye for an eye : retribution in kind
— my eye — used to express mild disagreement or sometimes surprise ⟨a diamond, *my eye!* That's glass⟩

[2]**eye** *verb* **eyed; eye·ing** *or* **ey·ing** (15th century)
transitive verb
1 a : to fix the eyes on **:** look at **b :** to watch closely
2 : to furnish with an eye
intransitive verb
obsolete **:** SEEM, LOOK
— ey·er \ˈī(-ə)r\ *noun*

[1]**eye·ball** \ˈī-ˌbȯl\ *noun* (1590)
: the more or less globular capsule of the vertebrate eye formed by the sclera and cornea together with their contained structures

[2]**eyeball** *transitive verb* (1901)
: to look at intently

[3]**eyeball** *adjective* (1971)
: based on observation ⟨*eyeball* judgment⟩

eyeball–to–eyeball *adverb or adjective* (1962)
: FACE-TO-FACE

eye bank *noun* (1944)
: a storage place for human corneas from the newly dead for transplanting to the eyes of those blind through corneal defects

eye·bar \ˈī-ˌbär\ *noun* (circa 1889)
: a metal bar having a closed loop at one or both ends

eye·bolt \ˈī-ˈbōlt\ *noun* (1769)
: a bolt with a looped head

eye·bright \ˈī-ˌbrīt\ *noun* (1533)
: any of a genus (*Euphrasia*) of herbs of the snapdragon family with opposite toothed or cut leaves

eye·brow \ˈī-ˌbrau̇\ *noun* (15th century)
: the ridge over the eye or the hair growing on it

eyebrow pencil *noun* (1881)
: a cosmetic pencil for the eyebrows

eye–catch·er \ˈī-ˌka-chər, -ˌke-\ *noun* (1923)
: something that arrests the eye
— eye–catch·ing \-chiŋ\ *adjective*
— eye–catch·ing·ly \-chiŋ-lē\ *adverb*

eye chart *noun* (1943)
: a chart read at a fixed distance for purposes of testing sight; *especially* **:** one with rows of letters or objects of decreasing size

eye contact *noun* (1955)
: visual contact with another person's eyes

eye·cup \ˈī-ˌkəp\ *noun* (circa 1874)
1 : a small oval cup with a rim curved to fit the orbit of the eye used for applying liquid remedies to the eyes
2 : OPTIC CUP
3 : a usually rubber cup at the eyepiece of an optical instrument (as binoculars) for keeping out extraneous light

eyed \ˈīd\ *adjective* (14th century)
: having an eye or eyes especially of a specified kind or number — often used in combination ⟨an almond-*eyed* girl⟩

eyed·ness \ˈīd-nəs\ *noun* [*-eyed* (as in *right-eyed, left-eyed*)] (1924)
: preference for the use of one eye instead of the other (as in using a monocular microscope)

eye·drop·per \ˈī-ˌdrä-pər\ *noun* (1937)
: DROPPER 2

eye·drops \-ˌdräps\ *noun plural* (1926)
: a medicated solution for the eyes applied in drops

eye·ful \ˈī-ˌfu̇l\ *noun* (circa 1864)
1 : a full or completely satisfying view
2 : one that is visually attractive; *especially* **:** a strikingly beautiful woman

eye·glass \ˈī-ˌglas\ *noun* (1664)
1 a : EYEPIECE **b :** a lens worn to aid vision; *specifically* **:** MONOCLE **c** *plural* **:** GLASSES, SPECTACLES
2 : EYECUP 1

eye·hole \ˈī-ˌhōl\ *noun* (1637)
1 : [1]ORBIT
2 : PEEPHOLE

eye·lash \ˈī-ˌlash\ *noun* (1752)
1 : the fringe of hair edging the eyelid — usually used in plural
2 : a single hair of the eyelashes

eye lens *noun* (1871)
: the lens nearest the eye in an eyepiece

eye·let \ˈī-lət\ *noun* [alteration of Middle English *oilet,* from Middle French *oillet,* diminutive of *oil* eye, from Latin *oculus*] (14th century)
1 a : a small hole designed to receive a cord or used for decoration (as in embroidery) **b :** a small typically metal ring to reinforce an eyelet **:** GROMMET
2 : PEEPHOLE, LOOPHOLE

eye·lid \ˈī-ˌlid\ *noun* (13th century)
: either of the movable folds of skin and muscle that can be closed over the eyeball

eye·lin·er \ˈī-ˌlī-nər\ *noun* (1947)
: makeup used to emphasize the contour of the eyes

ey·en \ˈī(-ə)n\ *archaic plural of* EYE

eye–open·er \ˈī-ˌōp-nər, -ˌō-pə-\ *noun* (1818)
1 : a drink intended to wake one up
2 : something startling, surprising, or enlightening
— eye–open·ing \-niŋ\ *adjective*

eye·piece \ˈī-ˌpēs\ *noun* (1790)
: the lens or combination of lenses at the eye end of an optical instrument

eye·pop·per \ˈī-ˌpä-pər\ *noun* (1941)
: something that excites, astonishes, or attracts the eye
— eye–pop·ping \-ˌpä-piŋ\ *adjective*

eye rhyme *noun* (1871)
: an imperfect rhyme that appears to have identical vowel sounds from similarity of spelling (as *move* and *love*)

eye·shade \ˈī-ˌshād\ *noun* (1845)
: a visor that shields the eyes from strong light and is fastened on with a headband

eye shadow *noun* (1930)
: a cosmetic cream or powder in one of various colors that is applied to the eyelids to accent the eyes

eye·shot \ˈī-ˌshät\ *noun* (1599)
: the range of the eye **:** VIEW

eye·sight \ˈī-ˌsīt\ *noun* (13th century)
1 : SIGHT 4a
2 *archaic* **:** OBSERVATION 1

eye socket *noun* (circa 1844)
: [1]ORBIT

eyes only *adjective* (1972)
: to be read by only the person addressed

eye·sore \ˈī-ˌsōr, -ˌsȯr\ *noun* (1530)
: something offensive to view

eye·spot \ˈī-ˌspät\ *noun* (1877)
1 a : a simple visual organ of pigment or pigmented cells covering a sensory termination **:** OCELLUS **b :** a small pigmented body of various unicellular algae
2 : a spot of color
3 : any of several fungal diseases of plants characterized by yellowish oval lesions on the leaves and stem; *especially* **:** a disease of various grasses (as sugarcane) caused by a fungus (*Helminthosporium sacchari*)

eye·stalk \ˈī-ˌstȯk\ *noun* (1854)
: one of the movable peduncles bearing an eye at the tip in a decapod crustacean

eye·strain \ˈī-ˌstrān\ *noun* (1874)
: weariness or a strained state of the eye

eye·strings \ˈī-ˌstriŋz\ *noun plural* (1590)
obsolete **:** organic eye attachments formerly believed to break at death or blindness

eye·tooth \ˈī-ˈtüth\ *noun* (circa 1545)
: a canine tooth of the upper jaw

eye view *noun* (circa 1771)
: POINT OF VIEW ⟨an alien *eye view*⟩

eye·wash \ˈī-ˌwȯsh, -ˌwäsh\ *noun* (circa 1859)
1 : an eye lotion
2 : misleading or deceptive statements, actions, or procedures

eye·wear \ˈī-ˌwar, -ˌwer\ *noun* (1926)
: corrective or protective devices (as glasses or contact lenses) for the eyes

eye·wink \ˈī-ˌwiŋk\ *noun* (1598)
: LOOK, GLANCE

eye·wit·ness \ˈī-ˈwit-nəs\ *noun* (1539)
: one who sees an occurrence or an object; *especially* **:** one who gives a report on what he or she has seen

eyre \ˈar, ˈer\ *noun* [Middle English *eire,* from Anglo-French, from Old French *erre* trip, from *errer* to travel — more at ERRANT] (14th century)
: a circuit traveled by an itinerant justice in medieval England or the court he presided over

ey·rie \ˈīr-ē, *or same as* AERIE\ *variant of* AERIE

ey•rir \'ā-ˌrir\ *noun, plural* **au•rar** \'au̇-ˌrär, 'œi-\ [Icelandic, from Old Norse, money (in plural), probably from Latin *aureus* a gold coin] (circa 1927)
— see *krona* at MONEY table

Eze•chiel \i-'zē-kyəl, -kē-əl\ *noun* [Late Latin] **:** EZEKIEL

Eze•kiel \i-'zē-kyəl, -kē-əl\ *noun* [Late Latin *Ezechiel,* from Hebrew *Yĕḥezqēl*]
1 : a Hebrew priest and prophet of the 6th century B.C.
2 : a prophetic book of canonical Jewish and Christian Scripture written by Ezekiel — see BIBLE table

Ez•ra \'ez-rə\ *noun* [Late Latin, from Hebrew *'Ezrā*]
1 : a Hebrew priest, scribe, and reformer of Judaism of the 5th century B.C. in Babylon and Jerusalem
2 : a narrative book of canonical Jewish and Christian Scripture — see BIBLE table

F is the sixth letter of the English, Latin, and early Greek alphabets. Corresponding to the Phoenician wāw, *which represented a bilabial fricative, a consonant with a sound somewhat like that of a breathy English* w, *it came into Greek as* digamma *(written "ϝ"). This letter dropped out of Classical Greek but was preserved in western Greek alphabets, from which it passed into Etruscan, taking there the value of a voiceless labiodental fricative with the sound of* f *in English* flag. *Latin adopted both the Etruscan letterform and pronunciation of* f, *which has the same phonetic value in English, except in the word* of *and its compounds, where it is voiced (pronounced with vibrating vocal cords). Small* f *evolved from the capital form through lengthening of the upright stem, shortening of the top and middle horizontal bars, and rounding of the top bar.*

f \ˈef\ *noun, plural* **f's** *or* **fs** \ˈefs\ *often capitalized, often attributive* (before 12th century)
1 a : the 6th letter of the English alphabet **b :** a graphic representation of this letter **c :** a speech counterpart of orthographic *f*
2 : the 4th tone of a C-major scale
3 : a graphic device for reproducing the letter *f*
4 : one designated *f* especially as the 6th in order or class
5 a : a grade rating a student's work as failing **b :** one graded or rated with an F
6 : something shaped like the letter F

fa \ˈfä\ *noun* [Middle English, from Medieval Latin, from the syllable sung to this note in a medieval hymn to Saint John the Baptist] (13th century)
: the 4th tone of the diatonic scale in solmization

Fa·bi·an \ˈfā-bē-ən\ *adjective* (1777)
1 a : of, relating to, or in the manner of the Roman general Quintus Fabius Maximus known for his defeat of Hannibal in the Second Punic War by the avoidance of decisive contests **b :** CAUTIOUS, DILATORY
2 [the *Fabian* Society; from the members' belief in slow rather than revolutionary change in government] **:** of, relating to, or being a society of socialists organized in England in 1884 to spread socialist principles gradually
— Fabian *noun*
— Fa·bi·an·ism \-ə-ˌni-zəm\ *noun*

¹fa·ble \ˈfā-bəl\ *noun* [Middle English, from Middle French, from Latin *fabula* conversation, story, play, from *fari* to speak — more at BAN] (14th century)
: a fictitious narrative or statement: as **a :** a legendary story of supernatural happenings **b :** a narration intended to enforce a useful truth; *especially* **:** one in which animals speak and act like human beings **c :** FALSEHOOD, LIE

²fable *verb* **fa·bled; fa·bling** \-b(ə-)liŋ\ (14th century)
intransitive verb
archaic **:** to tell fables
transitive verb
: to talk or write about as if true
— fa·bler \-b(ə-)lər\ *noun*

fa·bled \ˈfā-bəld\ *adjective* (1606)
1 : FICTITIOUS
2 : told or celebrated in fables
3 : RENOWNED, FAMOUS

fab·li·au \ˈfa-blē-ˌō\ *noun, plural* **-aux** \-ˌō(z)\ [French, from Old French, diminutive of *fable*] (1804)
: a short, usually comic, frankly coarse, and often cynical tale in verse popular especially in the 12th and 13th centuries

fab·ric \ˈfa-brik\ *noun* [Middle French *fabrique,* from Latin *fabrica* workshop, structure] (15th century)
1 a : STRUCTURE, BUILDING **b :** underlying structure **:** FRAMEWORK ⟨the *fabric* of society⟩
2 : an act of constructing **:** ERECTION; *specifically* **:** the construction and maintenance of a church building
3 a : structural plan or style of construction **b :** TEXTURE, QUALITY — used chiefly of textiles **c :** the arrangement of physical components (as of soil) in relation to each other
4 a : CLOTH 1a **b :** a material that resembles cloth
5 : the appearance or pattern produced by the shapes and arrangement of the crystal grains in a rock

fab·ri·cant \ˈfa-bri-kənt\ *noun* (1757)
: MANUFACTURER

fab·ri·cate \ˈfa-bri-ˌkāt\ *transitive verb* **-cat·ed; -cat·ing** [Middle English, from Latin *fabricatus,* past participle of *fabricari,* from *fabrica*] (15th century)
1 a : INVENT, CREATE **b :** to make up for the purpose of deception
2 : CONSTRUCT, MANUFACTURE; *specifically* **:** to construct from diverse and usually standardized parts
— fab·ri·ca·tor \ˈfa-bri-ˌkā-tər\ *noun*

fab·ri·ca·tion \ˌfa-bri-ˈkā-shən\ *noun* (15th century)
1 : the act or process of fabricating
2 : a product of fabrication; *especially* **:** LIE, FALSEHOOD

fabric softener *noun* (1965)
: a product used to make laundered fabrics softer and fluffier

fab·u·lar \ˈfa-byə-lər\ *adjective* (1684)
: of, relating to, or having the form of a fable

fab·u·list \ˈfa-byə-list\ *noun* (1593)
1 : a creator or writer of fables
2 : LIAR
— fabulist *or* **fab·u·lis·tic** \ˌfa-byə-ˈlis-tik\ *adjective*

fab·u·lous \ˈfa-byə-ləs\ *adjective* [Middle English, from Latin *fabulosus,* from *fabula*] (15th century)
1 a : resembling or suggesting a fable **:** of an incredible, astonishing, or exaggerated nature ⟨*fabulous* wealth⟩ **b :** WONDERFUL, MARVELOUS ⟨had a *fabulous* time⟩
2 : told in or based on fable
synonym see FICTITIOUS
— fab·u·lous·ly *adverb*
— fab·u·lous·ness *noun*

fa·cade *also* **fa·çade** \fə-ˈsäd\ *noun* [French *façade,* from Italian *facciata,* from *faccia* face, from (assumed) Vulgar Latin *facia*] (circa 1681)
1 : the front of a building; *also* **:** any face of a building given special architectural treatment ⟨a museum's east *facade*⟩
2 : a false, superficial, or artificial appearance or effect

facade 1

¹face \ˈfās\ *noun, often attributive* [Middle English, from Old French, from (assumed) Vulgar Latin *facia,* from Latin *facies* make, form, face, from *facere* to make, do — more at DO] (13th century)
1 a : the front part of the human head including the chin, mouth, nose, cheeks, eyes, and usually the forehead **b :** the face as a means of identification **:** COUNTENANCE ⟨would know that *face* anywhere⟩
2 *archaic* **:** PRESENCE, SIGHT
3 a : facial expression **b :** GRIMACE **c :** MAKEUP 3a(1)
4 a : outward appearance ⟨suspicious on the *face* of it⟩ **b :** DISGUISE, PRETENSE **c** (1) **:** ASSURANCE, CONFIDENCE ⟨maintaining a firm *face* in spite of adversity⟩ (2) **:** EFFRONTERY ⟨how anyone could have the *face* to ask that question⟩ **d :** DIGNITY, PRESTIGE ⟨afraid to lose *face*⟩
5 : SURFACE: **a** (1) **:** a front, upper, or outer surface (2) **:** the front of something having two or four sides (3) **:** FACADE (4) **:** an exposed surface of rock (5) **:** any of the plane surfaces that bound a geometric solid **b :** a surface specially prepared: as (1) **:** the principal dressed surface (as of a disk) (2) **:** the right side (as of cloth or leather) (3) **:** an inscribed, printed, or marked side **c** (1) **:** the surface (as of type) that receives the ink and transfers it to the paper (2) **:** a style of type
6 : the end or wall of a mine tunnel, drift, or excavation at which work is progressing
7 : FACE VALUE
8 : PERSON ⟨lots of new *faces* around here⟩
— in the face of *also* **in face of :** face-to-face with **:** DESPITE ⟨succeed *in the face of* great difficulties⟩
— to one's face : in one's presence or so that one is fully aware of what is going on **:** FRANKLY

²face *verb* **faced; fac·ing** (15th century)
transitive verb
1 : to confront impudently
2 a : to line near the edge especially with a different material **b :** to cover the front or surface of ⟨*faced* the building with marble⟩
3 : to meet face-to-face or in competition
4 a : to stand or sit with the face toward **b :** to have the front oriented toward ⟨a house *facing* the park⟩
5 a : to recognize and deal with straightforwardly ⟨*face* the facts⟩ **b :** to master by confronting with determination — used with *down* ⟨*faced* down his critics⟩
6 a : to have as a prospect **:** be confronted by ⟨*face* a grim future⟩ **b :** to be a prospect or a source of concern for ⟨the problems that *face* us⟩ **c :** to bring face-to-face ⟨he was *faced* with ruin⟩
7 : to make the surface of (as a stone) flat or smooth
8 : to cause (troops) to face in a particular direction on command
intransitive verb
1 : to have the face or front turned in a specified direction
2 : to turn the face in a specified direction
— face the music : to meet an unpleasant situation, a danger, or the consequences of one's actions

face angle *noun* (1913)
: an angle formed by two edges of a polyhedral angle

face card *noun* (1826)
: a king, queen, or jack in a deck of cards

face–cen·tered \'fās-ˌsen-tərd\ *adjective* (1913)
: relating to or being a crystal space lattice in which each cubic unit cell has an atom at the center and at the corners of each face — compare BODY-CENTERED

face·cloth \'fās-ˌklȯth\ *noun* (1602)
: WASHCLOTH

face cord *noun* (circa 1926)
: a unit of wood cut for fuel equal to a stack 4 × 8 feet with lengths of pieces from about 12 to 16 inches

-faced \ˌfāst\ *adjective combining form*
: having (such) a face or (so many) faces ⟨rosy-*faced*⟩ ⟨two-*faced*⟩

face·down \ˌfās-'dau̇n\ *adverb* (1949)
: with the face down ⟨sliding *facedown*⟩

face fly *noun* (1961)
: a European fly (*Musca autumnalis*) that is similar to the housefly, is widely established in North America, and causes distress to livestock by clustering about the face

face·less \'fās-ləs\ *adjective* (1596)
1 : lacking a face
2 a : lacking character or individuality **:** NONDESCRIPT ⟨the *faceless* masses⟩ **b :** not identified **:** ANONYMOUS ⟨a *faceless* accuser⟩
— face·less·ness *noun*

face–lift \'fās-ˌlift\ *noun* (1934)
1 : a cosmetic operation for removal of facial defects (as wrinkles) typical of aging
2 : an alteration, restoration, or restyling (as of a building) intended especially to modernize
— face–lift *transitive verb*

face mask *noun* (1906)
: a mask covering the face (as in football or scuba diving) **:** MASK 3

face–off \'fās-ˌȯf\ *noun* (1896)
1 : a method of beginning play (as in hockey or lacrosse) in which two opponents face each other and attempt to gain control of a puck or ball dropped between them
2 : CONFRONTATION

face·plate \'fās-ˌplāt\ *noun* (1841)
1 : a disk fixed with its face at right angles to the live spindle of a lathe for the attachment of the work
2 a : a protective plate for a machine or device **b :** a protective cover for the human face (as of a diver)
3 : the glass front of a cathode-ray tube on which the image is seen

fac·er \'fā-sər\ *noun* (15th century)
1 : one that faces
2 *British* **:** a sudden often stunning check or obstacle

face–sav·er \'fā(s)-ˌsā-vər\ *noun* (1923)
: something (as a compromise) that saves face
— face–sav·ing \-ˌsā-viŋ\ *adjective or noun*

fac·et \'fa-sət\ *noun* [French *facette,* diminutive of *face*] (1625)
1 : a small plane surface (as on a cut gem)
2 : any of the definable aspects that make up a subject (as of contemplation) or an object (as of consideration)
3 : the external corneal surface of an ommatidium
4 : a smooth flat circumscribed anatomical surface (as of a bone)
— fac·et·ed *or* **fac·et·ted** \'fa-sə-təd\ *adjective*

fa·cete \fə-'sēt\ *adjective* [Latin *facetus*] (1603)
archaic **:** FACETIOUS, WITTY

fa·ce·ti·ae \fə-'sē-shē-ˌē, -ˌī\ *noun plural* [Latin, from plural of *facetia* jest, from *facetus* elegant, witty] (1529)
: witty or humorous writings or sayings

fa·ce·tious \fə-'sē-shəs\ *adjective* [Middle French *facetieux,* from *facetie* jest, from Latin *facetia*] (1599)
1 : joking or jesting often inappropriately **:** WAGGISH ⟨just being *facetious*⟩
2 : meant to be humorous or funny **:** not serious ⟨a *facetious* remark⟩
synonym see WITTY
— fa·ce·tious·ly *adverb*
— fa·ce·tious·ness *noun*

face–to–face *adverb or adjective* (14th century)
1 : within each other's sight or presence ⟨met and talked *face-to-face*⟩ ⟨a *face-to-face* consultation⟩
2 : in or into direct contact or confrontation ⟨came *face-to-face* with the problem⟩

face·up \'fā-'səp\ *adverb* (1897)
: with the face up

face up *intransitive verb* (1920)
: to confront or deal directly with someone or something previously avoided — usually used with *to* ⟨*faced up* to the situation⟩

face value *noun* (1876)
1 : the value indicated on the face (as of a postage stamp or a stock certificate)
2 : the apparent value or significance ⟨if their remarks may be taken at *face value*⟩

facia *variant of* FASCIA

[1]fa·cial \'fā-shəl\ *adjective* (circa 1818)
1 : of or relating to the face
2 : concerned with or used in improving the appearance of the face
— fa·cial·ly \-shə-lē\ *adverb*

[2]facial *noun* (1914)
: a facial treatment

facial index *noun* (circa 1889)
: the ratio of the breadth of the face to its length multiplied by 100

facial nerve *noun* (circa 1818)
: either of the 7th pair of cranial nerves that supply motor fibers especially to the muscles of the face and jaw and sensory and parasympathetic fibers to the tongue, palate, and fauces

-fa·cient \'fā-shənt\ *adjective combining form* [Latin *-facient-, -faciens* (as in *calefacient-, calefaciens* making warm, present participle of *calefacere* to warm)]
: making **:** causing ⟨somni*facient*⟩

fa·cies \'fā-sh(ē-ˌ)ēz\ *noun, plural* **facies** [New Latin, from Latin, face] (circa 1736)
1 : an appearance and expression of the face characteristic of a particular condition especially when abnormal ⟨adenoid *facies*⟩
2 : general appearance ⟨a plant species with a particularly distinct *facies*⟩
3 : a part of a rock or group of rocks that differs from the whole formation (as in composition, age, or fossil content)

fac·ile \'fa-səl\ *adjective* [Middle French, from Latin *facilis,* from *facere* to do — more at DO] (15th century)
1 a (1) **:** easily accomplished or attained ⟨a *facile* victory⟩ (2) **:** SPECIOUS, SUPERFICIAL ⟨I am not concerned . . . with offering any *facile* solution for so complex a problem —T. S. Eliot⟩ **b :** used or comprehended with ease **c :** readily manifested and often lacking sincerity or depth ⟨*facile* tears⟩
2 *archaic* **:** mild or pleasing in manner or disposition
3 a : READY, FLUENT ⟨*facile* prose⟩ **b :** POISED, ASSURED
synonym see EASY
— fac·ile·ly \-sə(l)-lē\ *adverb*
— fac·ile·ness \-səl-nəs\ *noun*

fa·cil·i·tate \fə-'si-lə-ˌtāt\ *transitive verb* **-tat·ed; -tat·ing** (1611)
: to make easier **:** help bring about ⟨*facilitate* economic recovery⟩
— fa·cil·i·ta·tive \-ˌtā-tiv\ *adjective*
— fa·cil·i·ta·tor \-ˌtā-tər\ *noun*

fa·cil·i·ta·tion \fə-ˌsi-lə-'tā-shən\ *noun* (1619)
1 : the act of facilitating **:** the state of being facilitated
2 a : the lowering of the threshold for reflex conduction along a particular neural pathway especially from repeated use of that pathway **b :** the increasing of the ease or intensity of a response by repeated stimulation

fa·cil·i·ta·to·ry \fə-'si-lə-tə-ˌtōr-ē, -ˌtȯr-\ *adjective* (1944)
: inducing or involved in facilitation especially of a reflex action

fa·cil·i·ty \fə-'si-lə-tē\ *noun, plural* **-ties** (1531)
1 : the quality of being easily performed
2 : ease in performance **:** APTITUDE
3 : readiness of compliance
4 a (1) **:** something that makes an action, operation, or course of conduct easier — usually used in plural ⟨*facilities* for study⟩ (2) **:** LAVATORY 2 — often used in plural **b :** something (as a hospital) that is built, installed, or established to serve a particular purpose

fac·ing \'fā-siŋ\ *noun* (1566)
1 a : a lining at the edge especially of a garment **b** *plural* **:** the collar, cuffs, and trimmings of a uniform coat
2 : an ornamental or protective layer
3 : material for facing

fac·sim·i·le \fak-'si-mə-lē\ *noun* [Latin *fac simile* make similar] (1691)
1 : an exact copy
2 : a system of transmitting and reproducing graphic matter (as printing or still pictures) by means of signals sent over telephone lines
synonym see REPRODUCTION

fact \'fakt\ *noun* [Latin *factum,* from neuter of *factus,* past participle of *facere*] (15th century)
1 : a thing done: as **a** *obsolete* **:** FEAT **b :** CRIME ⟨accessory after the *fact*⟩ **c** *archaic* **:** ACTION
2 *archaic* **:** PERFORMANCE, DOING
3 : the quality of being actual **:** ACTUALITY ⟨a question of *fact* hinges on evidence⟩
4 a : something that has actual existence ⟨space exploration is now a *fact*⟩ **b :** an actual occurrence ⟨prove the *fact* of damage⟩
5 : a piece of information presented as having objective reality
— in fact : in truth

fact finder *noun* (1926)
: one that tries to determine the realities of a case, situation, or relationship; *especially* **:** an impartial examiner designated by a government agency to appraise the facts underlying a particular matter (as a labor dispute)
— fact–find·ing *noun or adjective*

fac·tic·i·ty \fak-'ti-sə-tē\ *noun* [French or German; French *facticité,* from German *Faktizität,* from *Factum* fact, from Latin *factum*] (1945)
: the quality or state of being a fact

fac·tion \'fak-shən\ *noun* [Middle French & Latin; Middle French, from Latin *faction-, factio* act of making, faction — more at FASHION] (1509)
1 : a party or group (as within a government) that is often contentious or self-seeking **:** CLIQUE
2 : party spirit especially when marked by dissension
— fac·tion·al \-shnəl, -shə-nᵊl\ *adjective*
— fac·tion·al·ism \-shnə-ˌli-zəm, -shə-nᵊl-ˌiz-\ *noun*
— fac·tion·al·ly \-ē\ *adverb*

-fac·tion \'fak-shən\ *noun combining form* [Middle English *-faccioun,* from Middle French & Latin; Middle French *-faction,* from Latin *-faction-, -factio* (as in *satisfaction-, satisfactio* satisfaction)]
: making **:** -FICATION ⟨petri*faction*⟩

fac·tious \'fak-shəs\ *adjective* [Middle French or Latin; Middle French *factieux,* from Latin *factiosus,* from *factio*] (1532)
: of or relating to faction: as **a :** caused by faction ⟨*factious* disputes⟩ **b :** inclined to faction or the formation of factions **c :** SEDITIOUS
— fac·tious·ly *adverb*
— fac·tious·ness *noun*

fac·ti·tious \fak-'ti-shəs\ *adjective* [Latin *facticius,* from *factus,* past participle of *facere* to make, do — more at DO] (1646)
1 : produced by humans rather than by natural forces

2 a : formed by or adapted to an artificial or conventional standard **b :** produced by special effort **:** SHAM ⟨created a *factitious* demand by spreading rumors of shortage⟩
— **fac·ti·tious·ly** *adverb*
— **fac·ti·tious·ness** *noun*

fac·ti·tive \ˈfak-tə-tiv\ *adjective* [probably from Latin *factitare* to do habitually, frequentative of *facere*] (1846)
: of, relating to, or being a transitive verb that in some constructions requires an objective complement as well as an object
— **fac·ti·tive·ly** *adverb*

-fac·tive \ˈfak-tiv\ *adjective combining form* [Middle French *-factif,* from *-faction*]
: making **:** causing ⟨putre*factive*⟩

fact of life (1854)
1 : something that exists and must be taken into consideration
2 *plural* **:** the fundamental physiological processes and behavior involved in sex and reproduction

fac·toid \ˈfak-ˌtȯid\ *noun* (1973)
1 : an invented fact believed to be true because of its appearance in print
2 : a brief and usually trivial news item

¹fac·tor \ˈfak-tər\ *noun* [Middle English, from Middle French *facteur,* from Latin *factor* doer, from *facere*] (15th century)
1 : one who acts or transacts business for another: as **a :** BROKER 1b **b :** one that lends money to producers and dealers (as on the security of accounts receivable)
2 a (1) **:** one that actively contributes to the production of a result **:** INGREDIENT ⟨price wasn't a *factor* in the decision⟩ (2) **:** a substance that functions in or promotes the function of a particular physiological process or bodily system **b :** a good or service used in the process of production
3 : GENE
4 a : any of the numbers or symbols in mathematics that when multiplied together form a product; *also* **:** a number or symbol that divides another number or symbol **b :** a quantity by which a given quantity is multiplied or divided in order to indicate a difference in measurement ⟨costs increased by a *factor* of 10⟩
— **fac·tor·ship** \-ˌship\ *noun*

²factor *verb* **fac·tored; fac·tor·ing** \-t(ə-)riŋ\ (1621)
intransitive verb
: to work as a factor
transitive verb
1 : to resolve into factors
2 a : to include or admit as a factor — used with *in* or *into* ⟨*factor* inflation into our calculations⟩ **b :** to exclude as a factor — used with *out*
— **fac·tor·able** \-t(ə-)rə-bəl\ *adjective*

fac·tor·age \-t(ə-)rij\ *noun* (1613)
1 : the charges made by a factor for services
2 : the business of a factor

factor analysis *noun* (1931)
: the analytical process of transforming statistical data (as measurements) into linear combinations of usually independent variables
— **factor analytic** *adjective*

factor VIII \-ˈāt\ *noun* (1954)
: a glycoprotein of blood plasma that is essential for blood clotting and is absent or inactive in hemophilia — called also *antihemophilic factor*

factor group *noun* (1897)
: QUOTIENT GROUP

¹fac·to·ri·al \fak-ˈtōr-ē-əl, -ˈtȯr-\ *adjective* (1837)
: of, relating to, or being a factor or a factorial

²factorial *noun* (1869)
1 : the product of all the positive integers from 1 to *n* — symbol *n!*
2 : the quantity *0!* arbitrarily defined as equal to 1

fac·tor·iza·tion \ˌfak-tə-rə-ˈzā-shən\ *noun* (1886)
: the operation of resolving a quantity into factors; *also* **:** a product obtained by factorization
— **fac·tor·ize** \ˈfak-tə-ˌrīz\ *transitive verb*

fac·to·ry \ˈfak-t(ə-)rē\ *noun, plural* **-ries** (1582)
1 : a station where resident factors trade
2 a : a building or set of buildings with facilities for manufacturing **b :** the seat of some kind of production ⟨the vice *factories* of the slums⟩
— **fac·to·ry·like** \-ˌlīk\ *adjective*

factory ship *noun* (1927)
: a ship equipped to process a whale or fish catch at sea

fac·to·tum \fak-ˈtō-təm\ *noun* [New Latin, literally, do everything, from Latin *fac* (imperative of *facere* do) + *totum* everything] (1566)
1 : a person having many diverse activities or responsibilities
2 : a general servant

fac·tu·al \ˈfak-chə-wəl, -chəl, ˈfaksh-wəl\ *adjective* [*fact* + *-ual* (as in *actual*)] (circa 1834)
1 : of or relating to facts
2 : restricted to or based on fact
— **fac·tu·al·i·ty** \ˌfak-chə-ˈwa-lə-tē\ *noun*
— **fac·tu·al·ly** \ˈfak-chə-wə-lē, -chə-lē, ˈfaksh-wə-\ *adverb*
— **fac·tu·al·ness** *noun*

fac·tu·al·ism \ˈfak-chə-wə-ˌli-zəm, -chə-ˌli-, ˈfaksh-wə-\ *noun* (1936)
: adherence or dedication to facts
— **fac·tu·al·ist** \-list\ *noun*

fac·ture \ˈfak-chər\ *noun* [Middle English, from Middle French, from Latin *factura* action of making, from *factus*] (15th century)
: the manner in which something (as an artistic work) is made **:** EXECUTION

fac·u·la \ˈfa-kyə-lə\ *noun, plural* **-lae** \-ˌlē, -ˌlī\ [New Latin, from Latin, diminutive of *fac-, fax* torch] (1706)
: any of the bright regions of the sun's photosphere seen most easily near the sun's edge

fac·ul·ta·tive \ˈfa-kəl-ˌtā-tiv, *British* -tə-tiv\ *adjective* (1820)
1 a : of or relating to the grant of permission, authority, or privilege ⟨*facultative* legislation⟩ **b :** OPTIONAL
2 : of or relating to a mental faculty
3 a : taking place under some conditions but not under others ⟨*facultative* diapause⟩ **b :** exhibiting an indicated lifestyle under some environmental conditions but not under others ⟨*facultative* anaerobes⟩
— **fac·ul·ta·tive·ly** *adverb*

fac·ul·ty \ˈfa-kəl-tē\ *noun, plural* **-ties** [Middle English *faculte,* from Middle French *faculté,* from Medieval Latin & Latin; Medieval Latin *facultat-, facultas* branch of learning or teaching, from Latin, ability, abundance, from *facilis* facile] (14th century)
1 : ABILITY, POWER: as **a :** innate or acquired ability to act or do **b :** an inherent capability, power, or function ⟨the *faculty* of hearing⟩ **c :** any of the powers of the mind formerly held by psychologists to form a basis for the explanation of all mental phenomena **d :** natural aptitude ⟨has a *faculty* for saying the right things⟩
2 a : a branch of teaching or learning in an educational institution **b** *archaic* **:** something in which one is trained or qualified
3 a : the members of a profession **b :** the teaching and administrative staff and those members of the administration having academic rank in an educational institution **c faculty** *plural* **:** faculty members ⟨many *faculty* were present⟩
4 : power, authority, or prerogative given or conferred
synonym see GIFT

fad \ˈfad\ *noun* [origin unknown] (1867)
: a practice or interest followed for a time with exaggerated zeal **:** CRAZE
synonym see FASHION
— **fad·dish** \ˈfa-dish\ *adjective*
— **fad·dish·ness** *noun*
— **fad·dism** \ˈfa-ˌdi-zəm\ *noun*
— **fad·dist** \ˈfa-dist\ *noun*
— **fad·dy** \-dē\ *adjective*

FAD \ˌef-ˌā-ˈdē\ *noun* (1944)
: FLAVIN ADENINE DINUCLEOTIDE

¹fade \ˈfād\ *verb* **fad·ed; fad·ing** [Middle English, from Middle French *fader,* from *fade* feeble, insipid, from (assumed) Vulgar Latin *fatidus,* alteration of Latin *fatuus* fatuous, insipid] (14th century)
intransitive verb
1 : to lose freshness, strength, or vitality **:** WITHER
2 : to lose freshness or brilliance of color
3 : to sink away **:** VANISH
4 : to change gradually in loudness, strength, or visibility — used of a motion-picture image or of an electronics signal and usually with *in* or *out*
5 *of an automobile brake* **:** to lose braking power gradually
6 : to move back from the line of scrimmage — used of a quarterback
transitive verb
: to cause to fade
— **fad·er** *noun*

²fade *noun* (1918)
1 a : FADE-OUT **b :** a gradual changing of one picture to another in a motion-picture or television sequence
2 : a fading of an automobile brake
3 : a slight to moderate and usually intentional slice in golf

³fade \ˈfäd\ *adjective* [Middle English, from Middle French] (15th century)
: INSIPID, COMMONPLACE

fade·away \ˈfā-də-ˌwā\ *noun* (1909)
1 a : SCREWBALL 1 **b :** a slide in which a base runner throws his body sideways to avoid the tag
2 : an act or instance of fading away

fade–in \ˈfā-ˌdin\ *noun* (1917)
: a gradual increase in a motion-picture or television image's visibility at the beginning of a sequence

fade·less \ˈfād-ləs\ *adjective* (1652)
: not susceptible to fading

fade–out \ˈfā-ˌdau̇t\ *noun* (1917)
: an act or instance of fading out; *especially* **:** a gradual decrease in a motion-picture or television image's visibility at the end of a sequence

fa·do \ˈfä-(ˌ)t͟hü, ˈfa-\ *noun, plural* **fados** [Portuguese, literally, fate, from Latin *fatum*] (1902)
: a plaintive Portuguese folk song

fae·cal, fae·ces *variant of* FECAL, FECES

fa·e·na \fä-ˈā-(ˌ)nä\ *noun* [Spanish, literally, task, from obsolete Catalan, from Latin *facienda* things to be done, from *facere* to do — more at DO] (1927)
: a series of final passes leading to the kill made by the matador in a bullfight

fa·er·ie *also* **fa·ery** \ˈfā-(ə-)rē, ˈfar-ē, ˈfer-ē\ *noun, plural* **fa·er·ies** [Middle French *faerie* — more at FAIRY] (1590)
1 : FAIRYLAND
2 : FAIRY
— **faery** *adjective*

Faero·ese \ˌfar-ə-ˈwēz, ˌfer-, -ˈwēs\ *noun, plural* **Faeroese** (1855)
1 : a member of the people inhabiting the Faeroes
2 : the North Germanic language of the Faeroese people
— **Faeroese** *adjective*

Faf·nir \ˈfäv-nər, ˈfäf-, -ˌnir\ *noun* [Old Norse *Fāfnir*]
: a dragon in Norse mythology that guards the Nibelungs' gold hoard until slain by Sigurd

\ə\ **a**but \ᵊ\ kitt**e**n \ər\ f**ur**th**er** \a\ **a**sh \ā\ **a**ce \ä\ m**o**p, m**a**r \au̇\ **ou**t \ch\ **ch**in \e\ b**e**t \ē\ **ea**sy \g\ **g**o \i\ h**i**t \ī\ **i**ce \j\ **j**ob \ŋ\ si**ng** \ō\ g**o** \ȯ\ l**aw** \ȯi\ b**oy** \th\ **th**in \t͟h\ **th**e \ü\ l**oo**t \u̇\ f**oo**t \y\ **y**et \zh\ vi**si**on *see also* Guide to Pronunciation

[1]fag \ˈfag\ *verb* **fagged; fag·ging** [obsolete *fag* to droop, perhaps from *fag* (fag end)] (1772)
intransitive verb
1 : to work hard **:** TOIL
2 : to act as a fag especially in an English public school ⟨*fagging* for older boys during his first year⟩
transitive verb
: to tire by strenuous activity **:** EXHAUST
synonym see TIRE

[2]fag *noun* (1780)
1 *chiefly British* **:** TOIL, DRUDGERY
2 a : an English public-school boy who acts as servant to an older schoolmate **b :** DRUDGE

[3]fag *noun* [*fag end*] (circa 1888)
: CIGARETTE

[4]fag *noun* [probably by shortening] (circa 1931)
: FAGGOT — usually used disparagingly
— fag·gy \ˈfa-gē\ *adjective*

fag end *noun* [earlier *fag,* from Middle English *fagge* flap] (1613)
1 a : a poor or worn-out end **:** REMNANT **b :** the extreme end
2 a : the last part or coarser end of a web of cloth **b :** the untwisted end of a rope

fag·got \ˈfa-gət\ *noun* [origin unknown] (1914)
: a male homosexual — usually used disparagingly
— fag·got·ry \-gə-trē\ *noun*
— fag·goty \-gə-tē\ *adjective*

fa·gin \ˈfā-gən\ *noun, often capitalized* [*Fagin,* character in Charles Dickens' *Oliver Twist* (1839)] (1847)
: an adult who instructs others (as children) in crime

[1]fag·ot *or* **fag·got** \ˈfa-gət\ *noun* [Middle English *fagot,* from Middle French] (14th century)
: BUNDLE: as **a :** a bundle of sticks **b :** a bundle of pieces of wrought iron to be shaped by rolling or hammering at high temperature

[2]fagot *or* **faggot** *transitive verb* (circa 1598)
: to make a fagot of **:** bind together into a bundle ⟨*fagoted* sticks⟩

fag·ot·ing *or* **fag·got·ing** *noun* (1885)
1 : an embroidery produced by pulling out horizontal threads from a fabric and tying the remaining cross threads into groups of an hourglass shape

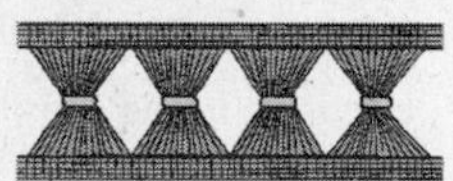
fagoting 1

2 : an openwork stitch joining hemmed edges

Fahr·en·heit \ˈfar-ən-ˌhīt, ˈfer-\ *adjective* [Gabriel D. *Fahrenheit*] (1753)
: relating or conforming to a thermometric scale on which under standard atmospheric pressure the boiling point of water is at 212 degrees above the zero of the scale, the freezing point is at 32 degrees above zero, and the zero point approximates the temperature produced by mixing equal quantities by weight of snow and common salt — abbreviation *F* ◆

fa·ience *or* **fa·ïence** \fā-ˈän(t)s, fī-, -ˈä^{n}s\ *noun* [French, from *Faenza,* Italy] (1714)
: earthenware decorated with opaque colored glazes

[1]fail \ˈfā(ə)l\ *verb* [Middle English *failen,* from Old French *faillir,* from (assumed) Vulgar Latin *fallire,* alteration of Latin *fallere* to deceive, disappoint] (13th century)
intransitive verb
1 a : to lose strength **:** WEAKEN ⟨her health was *failing*⟩ **b :** to fade or die away ⟨until our family line *fails*⟩ **c :** to stop functioning ⟨the patient's heart *failed*⟩
2 a : to fall short ⟨*failed* in his duty⟩ **b :** to be or become absent or inadequate ⟨the water supply *failed*⟩ **c :** to be unsuccessful (as in passing an examination) **d :** to become bankrupt or insolvent
transitive verb
1 a : to disappoint the expectations or trust of ⟨her friends *failed* her⟩ **b :** to miss performing an expected service or function for ⟨his wit *failed* him⟩
2 : to be deficient in **:** LACK ⟨never *failed* an invincible courage —Douglas MacArthur⟩
3 : to leave undone **:** NEGLECT ⟨*fail* to lock the door⟩
4 a : to be unsuccessful in passing (as a test) **b :** to grade (as a student) as not passing
— fail·ing·ly \ˈfā-liŋ-lē\ *adverb*

[2]fail *noun* (13th century)
1 : FAILURE — usually used in the phrase *without fail*
2 : a failure (as by a security dealer) to deliver or receive securities within a prescribed period after purchase or sale

[1]fail·ing \ˈfā-liŋ\ *noun* (1590)
: a usually slight or insignificant defect in character, conduct, or ability
synonym see FAULT

[2]failing *preposition* (1810)
: in absence or default of ⟨*failing* specific instructions, use your own judgment⟩

faille \ˈfī(ə)l\ *noun* [French, from Old French] (1869)
: a somewhat shiny closely woven silk, rayon, or cotton fabric characterized by slight ribs in the weft

fail–safe \ˈfā(ə)l-ˌsāf\ *adjective* (1946)
1 : incorporating some feature for automatically counteracting the effect of an anticipated possible source of failure
2 : being or relating to a safeguard that prevents continuing on a bombing mission according to a preconceived plan
3 : having no chance of failure **:** infallibly problem-free

fail·ure \ˈfā(ə)l-yər\ *noun* [alteration of earlier *failer,* from Anglo-French, from Old French *faillir* to fail] (1643)
1 a : omission of occurrence or performance; *specifically* **:** a failing to perform a duty or expected action **b :** a state of inability to perform a normal function ⟨kidney *failure*⟩ — compare HEART FAILURE **c :** a fracturing or giving way under stress ⟨structural *failure*⟩
2 a : lack of success **b :** a failing in business **:** BANKRUPTCY
3 a : a falling short **:** DEFICIENCY ⟨a crop *failure*⟩ **b :** DETERIORATION, DECAY
4 : one that has failed

[1]fain \ˈfān\ *adjective* [Middle English *fagen, fayn,* from Old English *fægen;* akin to Old Norse *fegiun* happy, Old English *fæger* fair] (before 12th century)
1 *archaic* **:** HAPPY, PLEASED
2 *archaic* **:** INCLINED, DESIROUS
3 a : WILLING ⟨he was very *fain,* for the young widow was "altogether fair and lovely . . . " —Amy Kelly⟩ **b :** being obliged or constrained **:** COMPELLED ⟨Great Britain was *fain* to devote its whole energy . . . to the business of slaying and being slain —G. M. Trevelyan⟩

[2]fain *adverb* (12th century)
1 : with pleasure **:** GLADLY ⟨a speech of fire that *fain* would blaze —Michael Billington⟩
2 a : by preference ⟨knew it, too, though he would *fain* not admit it publicly —John Lukacs⟩ **b :** by desire ⟨I would *fain* consult you —W. S. Gilbert⟩

[1]fai·né·ant \fā-nā-ˈä^{n}\ *noun, plural* **fainéants** \-ˈä^{n}(z)\ [French, from Middle French *fait-nient,* literally, does nothing, by folk etymology from *faignant,* from present participle of *faindre, feindre* to feign] (1619)
: an irresponsible idler

[2]fai·né·ant \fā-nā-ˈä^{n}\ *or* **fai·ne·ant** \ˈfā-nē-ənt\ *adjective* (1855)
: idle and ineffectual **:** INDOLENT

[1]faint \ˈfānt\ *adjective* [Middle English *faint, feint,* from Middle French, from past participle of *faindre, feindre* to feign, shirk — more at FEIGN] (14th century)
1 : lacking courage and spirit **:** COWARDLY
2 : weak, dizzy, and likely to faint
3 : lacking strength or vigor **:** performed, offered, or accomplished weakly or languidly
4 : producing a sensation of faintness **:** OPPRESSIVE ⟨the *faint* atmosphere of a tropical port⟩
5 : lacking distinctness **:** DIM
— faint·ish \ˈfān-tish\ *adjective*
— faint·ish·ness *noun*
— faint·ly *adverb*
— faint·ness *noun*

[2]faint *intransitive verb* (14th century)
1 *archaic* **:** to lose courage or spirit
2 *archaic* **:** to become weak
3 : to lose consciousness because of a temporary decrease in the blood supply to the brain

[3]faint *noun* (1808)
: the physiological action of fainting; *also* **:** the resulting condition **:** SYNCOPE 1

faint·heart·ed \ˈfānt-ˈhär-təd\ *adjective* (15th century)
: lacking courage or resolution **:** TIMID
— faint·heart·ed·ly *adverb*
— faint·heart·ed·ness *noun*

[1]fair \ˈfar, ˈfer\ *adjective* [Middle English *fager, fair,* from Old English *fæger;* akin to Old High German *fagar* beautiful] (before 12th century)
1 : pleasing to the eye or mind especially because of fresh, charming, or flawless quality
2 : superficially pleasing **:** SPECIOUS ⟨she trusted his *fair* promises⟩
3 a : CLEAN, PURE ⟨*fair* sparkling water⟩ **b :** CLEAR, LEGIBLE
4 : not stormy or foul **:** FINE ⟨*fair* weather⟩
5 : AMPLE ⟨a *fair* estate⟩
6 a : marked by impartiality and honesty **:** free from self-interest, prejudice, or favoritism ⟨a very *fair* person to do business with⟩ **b** (1) **:** conforming with the established rules **:** ALLOWED (2) **:** consonant with merit or importance **:** DUE ⟨a *fair* share⟩ **c :** open to legitimate pursuit, attack, or ridicule ⟨*fair* game⟩
7 a : PROMISING, LIKELY ⟨in a *fair* way to win⟩ **b :** favorable to a ship's course ⟨a *fair* wind⟩
8 *archaic* **:** free of obstacles
9 : not dark **:** BLOND
10 : sufficient but not ample **:** ADEQUATE ⟨a *fair* understanding of the work⟩
11 : being such to the utmost **:** UTTER ⟨a *fair* treat to watch him —*New Republic*⟩ ☆
— fair·ness *noun*

[2]fair *noun* (before 12th century)
1 *obsolete* **:** BEAUTY, FAIRNESS
2 : something that is fair or fortunate
3 *archaic* **:** WOMAN; *especially* **:** SWEETHEART
— for fair : to the greatest extent or degree **:** FULLY ⟨the rush is on *for fair*⟩
— no fair : something that is not according to the rules ⟨that's *no fair*⟩

[3]fair *adverb* (before 12th century)
chiefly British **:** FAIRLY 5

[4]fair (1836)
intransitive verb
of the weather **:** CLEAR

◇ WORD HISTORY

Fahrenheit The Fahrenheit scale memorializes the German physicist Daniel Gabriel Fahrenheit (1686–1736). Fahrenheit devoted his career to the making of precision scientific instruments. In 1709 he developed an alcohol thermometer, superseding it with a mercury thermometer in 1714. His most important achievement was his development, from the work of Olaus Roemer, of the thermometric scale that bears his name.

transitive verb
: to join so that the external surfaces blend smoothly

⁵fair *noun* [Middle English *feire,* from Old French, from Medieval Latin *feria* weekday, fair, from Late Latin, festal day, from Latin *feriae* (plural) holidays — more at FEAST] (13th century)
1 : a gathering of buyers and sellers at a particular place and time for trade
2 a : a competitive exhibition (as of farm products) usually with accompanying entertainment and amusements **b :** an exhibition designed to acquaint prospective buyers or the general public with a product
3 : a sale of assorted articles usually for a charitable purpose

fair ball *noun* (1856)
: a batted baseball that lands within the foul lines or that is within the foul lines when bounding to the outfield past first or third base or when going beyond the outfield for a home run

fair catch *noun* (circa 1876)
: a catch of a kicked football by a player who gives a prescribed signal, may not advance the ball, and may not be tackled

fair copy *noun* (1709)
: a neat and exact copy especially of a corrected draft

fair·ground \'far-ˌgrau̇nd, 'fer-\ *noun* (1741)
: an area where outdoor fairs, circuses, or exhibitions are held — often used in plural with singular construction ⟨what a spot for a *fairgrounds* —W. L. Gresham⟩

fair–haired \'far-'hard, 'fer-'herd\ *adjective* (1909)
: specially favored : WHITE-HEADED — used especially in the phrase *fair-haired boy*

¹fair·ing \'far-iŋ, 'fer-\ *noun* (1574)
1 *British* **a :** a present bought or given at a fair **b :** GIFT
2 *British* **:** ³DESERT 2

²fairing *noun* (1914)
: a member or structure whose primary function is to produce a smooth outline and to reduce drag (as on an airplane)

fair·ish \'far-ish, 'fer-\ *adjective* (1611)
: fairly good ⟨a *fairish* wage for those days⟩
— fair·ish·ly *adverb*

Fair Isle *noun* (1851)
: a style of knitting originating in the Shetland Islands that is characterized by bands of multicolored geometric patterns; *also* **:** an article of clothing knitted in this style

fair·lead \'far-ˌlēd, 'fer-\ *noun* (circa 1841)
1 *also* **fair·lead·er** \-ˌlē-dər\ **:** a block, ring, or strip of plank with holes that serves as a guide for the running rigging or any ship's rope and keeps it from chafing
2 : a course of running ship's rope that avoids all chafing

fair·ly \'far-lē, 'fer-\ *adverb* (12th century)
1 : in a handsome manner ⟨a table *fairly* set⟩
2 *obsolete* **a :** in a gentle manner **:** QUIETLY **b :** in a courteous manner
3 : in a manner of speaking ⟨*fairly* bursting with pride⟩
4 a : in a proper or legal manner ⟨*fairly* priced stocks⟩ **b :** without bias or distortion **:** IMPARTIALLY ⟨a story told *fairly* and objectively⟩
5 : to a full degree or extent **:** PLAINLY, DISTINCTLY ⟨had *fairly* caught sight of him⟩
6 : RATHER 5, MODERATELY ⟨a ***fairly*** easy job⟩

fair market value *noun* (1901)
: a price at which both buyers and sellers are willing to do business

fair–mind·ed \'far-ˌmīn-dəd, 'fer-\ *adjective* (1874)
: marked by impartiality and honesty **:** JUST, UNPREJUDICED
— fair–mind·ed·ness *noun*

fairness doctrine *noun* (1967)
: a tenet of licensed broadcasting that ensures a reasonable opportunity for the airing of conflicting viewpoints on controversial issues

fair play *noun* (1595)
: equitable or impartial treatment **:** JUSTICE

fair shake *noun* (1830)
: a fair chance or fair treatment ⟨give the negative side a *fair shake* —S. L. Payne⟩

fair–spo·ken \'far-ˌspō-kən, 'fer-\ *adjective* (15th century)
: pleasant and courteous in speech ⟨a *fair-spoken* youth⟩

fair–trade \'far-'trād, 'fer-\ *transitive verb* (1947)
: to market (a commodity) in compliance with the provisions of a fair-trade agreement
— fair trade *noun*
— fair trader *noun*

fair–trade agreement *noun* (1937)
: an agreement between a producer and a seller that commodities bearing a trademark, label, or brand name belonging to the producer be sold at or above a specified price

fair·way \'far-ˌwā, 'fer-\ *noun* (1584)
1 a : a navigable part of a river, bay, or harbor **b :** an open path or space
2 : the closely mowed part of a golf course between a tee and a green

fair–weather *adjective* (1736)
1 : loyal only during a time of success ⟨a *fair-weather* friend⟩
2 : suitable for or done during fair weather ⟨a *fair-weather* sail⟩

fairy \'far-ē, 'fer-\ *noun, plural* **fair·ies** [Middle English *fairie* fairyland, fairy people, from Old French *faerie,* from *feie, fee* fairy, from Latin *Fata,* goddess of fate, from *fatum* fate] (14th century)
1 : a mythical being of folklore and romance usually having diminutive human form and magic powers
2 : a male homosexual — usually used disparagingly
— fairy *adjective*
— fairy·like \-ˌlīk\ *adjective*

fairy godmother *noun* (1851)
: a generous friend or benefactor

fairy·ism \-ˌi-zəm\ *noun* (1715)
archaic **:** the power to enchant

fairy·land \-ˌland\ *noun* (1590)
1 : the land of fairies
2 : a place of delicate beauty or magical charm

fairy ring *noun* [from the folk belief that such rings were dancing places of the fairies] (1599)
1 : a ring of basidiomycetous mushrooms produced at the periphery of a body of mycelium which has grown outward from an initial growth point; *also* **:** a ring of luxuriant vegetation especially when associated with these mushrooms
2 : a mushroom (especially *Marasmius oreades*) that commonly grows in fairy rings

fairy shrimp *noun* (1857)
: any of several delicate transparent freshwater branchiopod crustaceans (order Anostraca)

fairy–tale *adjective* (1924)
: characteristic of or suitable to a fairy tale; *especially* **:** marked by seemingly unreal beauty, perfection, luck, or happiness

fairy tale *noun* (1749)
1 : a story (as for children) involving fantastic forces and beings (as fairies, wizards, and goblins) — called also *fairy story*
2 : a made-up story usually designed to mislead

fait ac·com·pli \'fā-tə-ˌkäm-'plē, 'fe-, 'fe-ˌta-, -ˌkōⁿ(m)-, *British usually* -'käm-(ˌ)plē\ *noun, plural* **faits accomplis** *same, or* -'plēz\ [French, accomplished fact] (1845)
: a thing accomplished and presumably irreversible

¹faith \'fāth\ *noun, plural* **faiths** \'fāths, *sometimes* 'fāt͟hz\ [Middle English *feith,* from Old French *feid, foi,* from Latin *fides;* akin to Latin *fidere* to trust — more at BIDE] (13th century)
1 a : allegiance to duty or a person **:** LOYALTY **b** (1) **:** fidelity to one's promises (2) **:** sincerity of intentions
2 a (1) **:** belief and trust in and loyalty to God (2) **:** belief in the traditional doctrines of a religion **b** (1) **:** firm belief in something for which there is no proof (2) **:** complete trust
3 : something that is believed especially with strong conviction; *especially* **:** a system of religious beliefs ◆
synonym see BELIEF
— in faith : without doubt or question **:** VERILY

²faith *transitive verb* (15th century)
archaic **:** BELIEVE, TRUST

¹faith·ful \'fāth-fəl\ *adjective* (14th century)
1 *obsolete* **:** full of faith
2 : steadfast in affection or allegiance **:** LOYAL
3 : firm in adherence to promises or in observance of duty **:** CONSCIENTIOUS
4 : given with strong assurance **:** BINDING ⟨*faithful* promise⟩

☆ SYNONYMS
Fair, just, equitable, impartial, unbiased, dispassionate, objective mean free from favor toward either or any side. FAIR implies an elimination of one's own feelings, prejudices, and desires so as to achieve a proper balance of conflicting interests ⟨a *fair* decision⟩. JUST implies an exact following of a standard of what is right and proper ⟨a *just* settlement of territorial claims⟩. EQUITABLE implies a less rigorous standard than JUST and usually suggests equal treatment of all concerned ⟨the *equitable* distribution of the property⟩. IMPARTIAL stresses an absence of favor or prejudice ⟨an *impartial* third party⟩. UNBIASED implies even more strongly an absence of all prejudice ⟨your *unbiased* opinion⟩. DISPASSIONATE suggests freedom from the influence of strong feeling and often implies cool or even cold judgment ⟨a *dispassionate* summation of the facts⟩. OBJECTIVE stresses a tendency to view events or persons as apart from oneself and one's own interest or feelings ⟨I can't be *objective* about my own child⟩. See in addition BEAUTIFUL.

◇ WORD HISTORY
faith *Faith* exemplifies how a language can retain in a word borrowed from another language a sound which that source language has lost or modified drastically. The final \th\ sound of *faith* existed in the northern variety of French from which English borrowed the word, probably not long after the Norman Conquest of 1066. Subsequently, all dialects of French lost this sound entirely, and the usual form of the Old French word was *fei* (later *foi,* leading to the identically spelled Modern French word, pronounced \fwä\). English, on the other hand, from its remote Germanic past to the present day, has retained the sound \th\, and so has kept it in *faith* (perhaps with encouragement from the *-th* suffix of abstract nouns such as *truth*). Like archaeological relics protected by the earth until they are uncovered, words such as *faith* are valuable instruments for studying the history of language change.

5 : true to the facts, to a standard, or to an original ⟨a *faithful* copy⟩ ☆
— **faith·ful·ly** \-fə-lē\ *adverb*
— **faith·ful·ness** *noun*

²faithful *noun* (1558)
1 *plural in construction* **a :** church members in full communion and good standing — used with *the* **b :** the body of believers in Islam — used with *the*
2 *plural* **faithful** *or* **faithfuls :** one who is faithful; *especially* **:** a loyal follower, member, or fan ⟨party *faithfuls*⟩

faith healing *noun* (1885)
: a method of treating diseases by prayer and exercise of faith in God
— **faith healer** *noun*

faith·less \ˈfāth-ləs\ *adjective* (14th century)
1 : not true to allegiance or duty **:** TREACHEROUS, DISLOYAL ⟨a *faithless* servant⟩
2 : not to be relied on **:** UNTRUSTWORTHY ⟨a *faithless* tool⟩ ☆
— **faith·less·ly** *adverb*
— **faith·less·ness** *noun*

fai·tour \ˈfā-tər\ *noun* [Middle English, from Anglo-French, from Old French *faitor* perpetrator, from Latin *factor* doer — more at FACTOR] (14th century)
archaic **:** CHEAT, IMPOSTER

fa·ji·ta \fə-ˈhē-tə, fä-\ *noun* [American Spanish, diminutive of Spanish *faja* sash, belt, probably from Catalan *faixa*, from Latin *fascia* band — more at FASCIA] (1984)
: a marinated strip usually of beef or chicken grilled or broiled and served usually with a flour tortilla and various savory fillings — usually used in plural

¹fake \ˈfāk\ *transitive verb* **faked; fak·ing** [Middle English] (15th century)
: to coil in fakes

²fake *noun* (1627)
: one loop of a coil (as of ship's rope or a fire hose) coiled free for running

³fake *adjective* [origin unknown] (1775)
: COUNTERFEIT, SHAM

⁴fake *noun* (1827)
: one that is not what it purports to be: as **a :** a worthless imitation passed off as genuine **b :** IMPOSTOR, CHARLATAN **c :** a simulated movement in a sports contest (as a pretended kick, pass, or jump or a quick movement in one direction before going in another) designed to deceive an opponent **d :** a device or apparatus used by a magician to achieve the illusion of magic in a trick
synonym see IMPOSTURE

⁵fake *verb* **faked; fak·ing** (1851)
transitive verb
1 : to alter, manipulate, or treat so as to give a spuriously genuine appearance to **:** DOCTOR ⟨*faked* the lab results⟩
2 : COUNTERFEIT, SIMULATE, CONCOCT
3 : to deceive (an opponent) in a sports contest by means of a fake
4 : IMPROVISE, AD-LIB ⟨whistle a few bars . . . and I'll *fake* the rest —Robert Sylvester⟩
intransitive verb
1 : to engage in faking something **:** PRETEND
2 : to give a fake to an opponent
— **fak·er** *noun*
— **fak·ery** \ˈfā-k(ə-)rē\ *noun*

fake out *transitive verb* (1968)
: to deliberately mislead **:** FOOL, TRICK

fa·kir *noun* [Arabic *faqīr*, literally, poor man] (1609)
1 \fə-ˈkir, fä-, fa-; ˈfā-kər\ **a :** a Muslim mendicant **:** DERVISH **b :** an itinerant Hindu ascetic or wonder-worker
2 \ˈfā-kər\ **:** IMPOSTOR; *especially* **:** SWINDLER

fa la \fä-ˈlä\ *noun* [*fa-la*, meaningless syllables often occurring in its refrain] (1597)
: a 16th and 17th century part-song

fa·la·fel \fə-ˈlä-fəl\ *noun, plural* **falafel** [Arabic *falāfil*] (1950)
: a spicy mixture of ground vegetables (as chick-peas or fava beans) formed into balls or patties and then fried

Fa·lan·gist \fə-ˈlan-jist, ˈfā-ˌ\ *noun* [Spanish *falangista*, from *Falange española* Spanish Phalanx, a fascist organization] (1936)
: a member of the fascist political party governing Spain after the civil war of 1936–39

Fa·la·sha \fə-ˈlä-shə\ *noun, plural* **-sha** *or* **-shas** [Amharic *fälaša*] (1710)
: a member of a people of highland Ethiopia who practice a variety of Judaism

fal·cate \ˈfal-ˌkāt, ˈfȯl-\ *adjective* [Latin *falcatus*, from *falc-, falx* sickle, scythe] (1826)
: hooked or curved like a sickle

fal·chion \ˈfȯl-chən\ *noun* [Middle English *fauchoun*, from Middle French *fauchon*, from *fauchier* to mow, from (assumed) Vulgar Latin *falcare*, from Latin *falc-, falx*] (14th century)
1 : a broad-bladed slightly curved sword of medieval times
2 *archaic* **:** SWORD

fal·ci·form \ˈfal-sə-ˌfȯrm, ˈfȯl-\ *adjective* [Latin *falc-, falx* + English *-iform*] (1766)
: having the shape of a scythe or sickle

fal·con \ˈfal-kən, ˈfȯl- *also* ˈfȯ-kən\ *noun* [Middle English, from Old French, from Late Latin *falcon-, falco*, probably from Latin *falc-, falx*] (13th century)
1 : any of various hawks trained for use in falconry; *especially* **:** PEREGRINE — used technically only of a female; compare TIERCEL
2 : any of various hawks (family Falconidae) that have long pointed wings, a long tail, and a notched beak and that usually inhabit open areas

falcon 1: *1* hood, *2* jess, *3* gauntlet

fal·con·er \-kə-nər\ *noun* (14th century)
: a person who breeds, trains, or hunts with hawks

fal·con·et \ˌfal-kə-ˈnet, ˌfȯl- *also* ˌfȯ-\ *noun* (1559)
1 : a very small cannon used in the 16th and 17th centuries
2 : any of several very small falcons (genera *Microhierax, Prolihierax,* and *Spiziapteryx*)

fal·con–gen·tle \-kən-ˈjen-tᵊl\ *noun* [Middle English *faucon gentil* peregrine falcon, from Middle French, literally, noble falcon] (15th century)
: the female peregrine falcon

fal·co·nine \ˈfal-kə-ˌnīn, ˈfȯl- *also* ˈfȯ-\ *adjective* (circa 1889)
: of or resembling a falcon ⟨a *falconine* face⟩

fal·con·ry \ˈfal-kən-rē, ˈfȯl- *also* ˈfȯ-kən-\ *noun* (1575)
1 : the art of training hawks to hunt in cooperation with a person
2 : the sport of hunting with hawks

fal·de·ral \ˈfäl-də-ˌräl\ *variant of* FOLDEROL

fald·stool \ˈfȯl(d)-ˌstül\ *noun* [Medieval Latin *faldistolium*, of Germanic origin; akin to Old High German *faltistuol* folding chair, from *falt* (akin to Old High German *faldan* to fold) + *stuol* chair — more at FOLD, STOOL] (1603)
1: a foldingstool or chair; *specifically* **:** one used by a bishop
2 : a folding stool or small desk at which one kneels during devotions; *especially* **:** one used by the sovereigns of England at their coronation
3 : the desk from which the litany is read in Anglican churches

faldstool 1

¹fall \ˈfȯl\ *verb* **fell** \ˈfel\; **fall·en** \ˈfȯ-lən\; **fall·ing** [Middle English, from Old English *feallan;* akin to Old High German *fallan* to fall and perhaps to Lithuanian *pulti*] (before 12th century)
intransitive verb
1 a : to descend freely by the force of gravity **b :** to hang freely ⟨her hair *falls* over her shoulders⟩ **c :** to drop oneself to a lower position ⟨*fell* to his knees⟩ **d :** to come or go as if by falling ⟨darkness *falls* early in the winter⟩
2 : to become born — usually used of lambs
3 a : to become lower in degree or level ⟨the temperature *fell* 10°⟩ **b :** to drop in pitch or volume ⟨their voices *fell* to a whisper⟩ **c :** ISSUE 1a,b ⟨wisdom that *fell* from his lips⟩ **d :** to become lowered ⟨her eyes *fell*⟩
4 a : to leave an erect position suddenly and involuntarily ⟨slipped and *fell* on the ice⟩ **b :** to enter as if unawares **:** STUMBLE, STRAY ⟨*fell* into error⟩ **c :** to drop down wounded or dead; *especially* **:** to die in battle **d :** to suffer military capture ⟨after a long siege the city *fell*⟩ **e :** to lose office ⟨the party *fell* from power⟩ **f :** to suffer ruin, defeat, or failure ⟨the deal *fell* through⟩
5 : to commit an immoral act; *especially* **:** to lose one's chastity
6 a : to move or extend in a downward direction ⟨the land *falls* away to the east⟩ **b :** SUBSIDE, ABATE ⟨the wind is *falling*⟩ **c :** to decline in quality, activity, or quantity ⟨production *fell* off⟩ **d :** to lose weight — used with *off* or *away* **e :** to assume a look of shame, disappointment, or dejection ⟨his face *fell*⟩ **f :** to decline in financial value or price ⟨stocks *fell* sharply⟩
7 a : to occur at a certain time **b :** to come by chance ⟨*fell* in with a fast crowd⟩ **c :** to come or pass by lot, assignment, or inheritance **:** DEVOLVE ⟨it *fell* to him to break the news⟩ **d :** to have a certain or proper position, place, or station ⟨the accent *falls* on the second syllable⟩

☆ SYNONYMS

Faithful, loyal, constant, staunch, steadfast, resolute mean firm in adherence to whatever one owes allegiance. FAITHFUL implies unswerving adherence to a person or thing or to the oath or promise by which a tie was contracted ⟨*faithful* to her promise⟩. LOYAL implies a firm resistance to any temptation to desert or betray ⟨remained *loyal* to the czar⟩. CONSTANT stresses continuing firmness of emotional attachment without necessarily implying strict obedience to promises or vows ⟨*constant* friends⟩. STAUNCH suggests fortitude and resolution in adherence and imperviousness to influences that would weaken it ⟨a *staunch* defender of free speech⟩. STEADFAST implies a steady and unwavering course in love, allegiance, or conviction ⟨*steadfast* in their support⟩. RESOLUTE implies firm determination to adhere to a cause or purpose ⟨a *resolute* ally⟩.

Faithless, false, disloyal, traitorous, treacherous, perfidious mean untrue to what should command one's fidelity or allegiance. FAITHLESS applies to any failure to keep a promise or pledge or any breach of allegiance or loyalty ⟨*faithless* allies⟩. FALSE stresses the fact of failing to be true in any manner ranging from fickleness to cold treachery ⟨betrayed by *false* friends⟩. DISLOYAL implies a lack of complete faithfulness to a friend, cause, leader, or country ⟨*disloyal* to their country⟩. TRAITOROUS implies either actual treason or a serious betrayal of trust ⟨*traitorous* acts punishable by death⟩. TREACHEROUS implies readiness to betray trust or confidence ⟨a *treacherous* adviser⟩. PERFIDIOUS adds to FAITHLESS the implication of an incapacity for fidelity or reliability ⟨a *perfidious* double-crosser⟩.

8 : to come within the limits, scope, or jurisdiction of something ⟨this word *falls* into the class of verbs⟩
9 : to pass suddenly and passively into a state of body or mind or a new state or condition ⟨*fall* asleep⟩ ⟨*fall* in love⟩
10 : to set about heartily or actively ⟨*fell* to work⟩
11 : STRIKE, IMPINGE ⟨music *falling* on the ear⟩
transitive verb
: FELL 1
— fall apart 1 : DISINTEGRATE **2 :** to succumb to mental or emotional stress **:** BREAK DOWN
— fall behind 1 : to lag behind **2 :** to be in arrears
— fall between two stools : to fail because of inability to choose between or reconcile two alternative or conflicting courses of action
— fall flat : to produce no response or result ⟨the joke *fell flat*⟩
— fall for 1 : to fall in love with **2 :** to become a victim of ⟨*fell for* the trick⟩
— fall foul : to have a quarrel **:** CLASH — often used with *of*
— fall from grace : BACKSLIDE 1
— fall home : to curve inward; used of the timbers or upper parts of a ship's side
— fall into line : to comply with a certain course of action
— fall on *or* **fall upon :** to meet with ⟨*fell on* hard times⟩
— fall over oneself *or* **fall over backward :** to display excessive eagerness
— fall short 1 : to be deficient **2 :** to fail to attain something (as a goal or target)

[2]fall *noun* (13th century)
1 : the act of falling by the force of gravity
2 a : a falling out, off, or away **:** DROPPING ⟨the *fall* of leaves⟩ ⟨a *fall* of snow⟩ **b :** the season when leaves fall from trees **:** AUTUMN **c :** a thing or quantity that falls or has fallen ⟨a *fall* of rock at the base of the cliff⟩; *especially* **:** one or more meteorites or their fragments that have fallen together **d** (1) **:** BIRTH (2) **:** the quantity born — usually used of lambs
3 a : a costume decoration of lace or thin fabric arranged to hang loosely and gracefully **b :** a very wide turned-down collar worn in the 17th century **c :** the part of a turnover collar from the crease to the outer edge **d :** a wide front flap on trousers (as those worn by sailors) **e :** the freely hanging lower edge of the skirt of a coat **f :** one of the three outer and often drooping segments of the flower of an iris **g :** long hair overhanging the face of dogs of some breeds **h :** a usually long straight portion of hair that is attached to a person's own hair
4 : a hoisting-tackle rope or chain; *especially* **:** the part of it to which the power is applied
5 a : loss of greatness **:** COLLAPSE ⟨the *fall* of the Roman Empire⟩ **b :** the surrender or capture of a besieged place ⟨the *fall* of Troy⟩ **c :** lapse or departure from innocence or goodness **d :** loss of a woman's chastity
6 a : the downward slope (as of a hill) **:** DECLIVITY **b :** a precipitous descent of water **:** WATERFALL — usually used in plural but sing. or plural in construction **c :** a musical cadence **d :** a falling-pitch intonation in speech
7 : a decrease in size, quantity, degree, or value
8 a : the distance which something falls **b :** INCLINATION, PITCH
9 a : the act of felling something **b :** the quantity of trees cut down **c** (1) **:** an act of forcing a wrestler's shoulders to the mat for a specified time (as one second) (2) **:** a bout of wrestling
10 *Scottish* **:** DESTINY, LOT

[3]fall *adjective* (1677)
: of, relating to, or suitable for autumn ⟨a new *fall* coat⟩

fal·la·cious \fə-ˈlā-shəs\ *adjective* (1509)
1 : embodying a fallacy
2 : tending to deceive or mislead **:** DELUSIVE
— fal·la·cious·ly *adverb*
— fal·la·cious·ness *noun*

fal·la·cy \ˈfa-lə-sē\ *noun, plural* **-cies** [Latin *fallacia,* from *fallac-, fallax* deceitful, from *fallere* to deceive] (14th century)
1 a *obsolete* **:** GUILE, TRICKERY **b :** deceptive appearance **:** DECEPTION
2 a : a false or mistaken idea ⟨popular *fallacies*⟩ **b :** erroneous character **:** ERRONEOUSNESS
3 : an often plausible argument using false or invalid inference

fal–lal \fa-ˈlal, ˈfa(l)-ˌlal\ *noun* [perhaps alteration of *falbala* furbelow, from French, from French dialect *ferbelà, farbélla*] (circa 1706)
: a fancy ornament especially in dress
— fal·lal·ery \fa-ˈla-lə-rē\ *noun*

fall armyworm *noun* (1881)
: a migratory American moth (*Spodoptera frugiperda*) that is especially destructive to small grains and grasses as a larva

fall·away \ˈfȯ-lə-ˌwā\ *adjective* (1966)
: made while moving away from the basket in basketball ⟨a *fallaway* jumper⟩
— fallaway *noun*

fall away *intransitive verb* (1535)
1 a : to withdraw friendship or support **b :** to renounce one's faith
2 a : to diminish gradually in size **b :** to drift off a course

fall·back \ˈfȯl-ˌbak\ *noun* (1851)
1 : something on which one can fall back **:** RESERVE
2 : a falling back **:** RETREAT
3 : something that falls back ⟨the *fallback* from an explosion⟩

fall back *intransitive verb* (1607)
: RETREAT, RECEDE
— fall back on *or* **fall back upon :** to have recourse to ⟨had to *fall back on* their reserves⟩

fall down *intransitive verb* (1873)
: to fail to meet expectations or requirements ⟨*fell down* on the job⟩

fall·er \ˈfȯ-lər\ *noun* (1677)
1 : a machine part that acts by falling
2 : a logger who fells trees

fall·fish \ˈfȯl-ˌfish\ *noun* (circa 1811)
: a common cyprinid fish (*Semotilus corporalis*) of the streams of northeastern North America

fall guy *noun* (1906)
1 : one that is easily duped
2 : SCAPEGOAT

fal·li·bil·i·ty \ˌfa-lə-ˈbi-lə-tē\ *noun* (1634)
: liability to err

fal·li·ble \ˈfa-lə-bəl\ *adjective* [Middle English, from Medieval Latin *fallibilis,* from Latin *fallere*] (15th century)
1 : liable to be erroneous ⟨a *fallible* generalization⟩
2 : capable of making a mistake ⟨all men are *fallible*⟩
— fal·li·bly \-blē\ *adverb*

fall in *intransitive verb* (1719)
1 : to sink inward ⟨the roof *fell in*⟩
2 : to take one's proper place in a military formation
— fall in with 1 : to concur with ⟨had to *fall in with* her wishes⟩ **2 :** to harmonize with ⟨it *falls in* exactly *with* my views⟩

falling diphthong *noun* (1888)
: a diphthong (as \ȯi\ in \ˈnȯiz\ *noise*) composed of a vowel followed by a less sonorous glide

fall·ing–out \ˌfȯ-liŋ-ˈau̇t\ *noun, plural* **fallings–out** *or* **falling–outs** (1568)
: an instance of falling out **:** QUARREL

falling rhythm *noun* (1918)
: rhythm with stress occurring regularly on the first syllable of each foot — compare RISING RHYTHM

falling star *noun* (1563)
: METEOR 2a

fall line *noun* (1882)
1 : a line joining the waterfalls on numerous rivers that marks the point where each river descends from the upland to the lowland and the limit of the navigability of each river
2 : the natural downhill course (as for skiing) between two points on a slope

fall·off \ˈfȯ-ˌlȯf\ *noun* (1880)
: a decline especially in quantity or quality ⟨a *falloff* in exports⟩ ⟨a *falloff* of light intensity⟩

fall off *intransitive verb* (1613)
1 : TREND 1b
2 *of a ship* **:** to deviate to leeward of the point to which the bow was directed

fal·lo·pi·an tube \fə-ˈlō-pē-ən-\ *noun, often F capitalized* [Gabriel *Fallopius* (died 1562) Italian anatomist] (circa 1706)
: either of the pair of tubes conducting the egg from the ovary to the uterus

fall·out \ˈfȯ-ˌlau̇t\ *noun* (1949)
1 a : the often radioactive particles stirred up by or resulting from a nuclear explosion and descending through the atmosphere; *also* **:** other polluting particles (as volcanic ash) descending likewise **b :** descent (as of fallout) through the atmosphere
2 : a secondary and often lingering effect, result, or set of consequences ⟨have to take a position and accept the political *fallout* —Andy Logan⟩

fall out *intransitive verb* (15th century)
1 : QUARREL; *also* **:** to cut off relations over a quarrel ⟨former friends who have *fallen out*⟩
2 : TURN OUT, HAPPEN ⟨expected to be in the States . . . , but things *fell out* otherwise —Mark Twain⟩
3 a : to leave one's place in the ranks **b :** to leave a building in order to take one's place in a military formation

[1]fal·low \ˈfa-(ˌ)lō, -lə(-w)\ *adjective* [Middle English *falow,* from Old English *fealu;* akin to Old High German *falo* pale, fallow, Latin *pallēre* to be pale, Greek *polios* gray] (before 12th century)
: of a light yellowish brown color

[2]fallow *noun* [Middle English *falwe, falow,* from Old English *fealg* — more at FELLY] (before 12th century)
1 : usually cultivated land that is allowed to lie idle during the growing season
2 *obsolete* **:** plowed land
3 : the state or period of being fallow
4 : the tilling of land without sowing it for a season

[3]fallow *transitive verb* (15th century)
: to plow, harrow, and break up (land) without seeding to destroy weeds and conserve soil moisture

[4]fallow *adjective* (15th century)
1 : left untilled or unsown after plowing
2 : DORMANT, INACTIVE — used especially in the phrase *to lie fallow* ⟨at this very moment there are probably important inventions lying *fallow* —*Harper's*⟩
— fal·low·ness *noun*

fallow deer *noun* [[1]*fallow*] (15th century)
: a deer (*Cervus dama* synonym *Dama dama*) of variable color with broad antlers and typically a yellow-brown coat spotted with white in the summer that was originally found in Europe and Asia Minor but has been introduced elsewhere

fallow deer

fall to *intransitive verb* (1593)

\ə\ **abut** \ᵊ\ **kitten** \ər\ **further** \a\ **ash** \ā\ **ace** \ä\ **mop, mar** \au̇\ **out** \ch\ **chin** \e\ **bet** \ē\ **easy** \g\ **go** \i\ **hit** \ī\ **ice** \j\ **job** \ŋ\ **sing** \ō\ **go** \ȯ\ **law** \ȯi\ **boy** \th\ **thin** \t͟h\ **the** \ü\ **loot** \u̇\ **foot** \y\ **yet** \zh\ **vision** *see also* Guide to Pronunciation

: to begin doing something (as working or eating) especially vigorously — often used in invitation or command

[1]**false** \'fȯls\ *adjective* **fals·er; fals·est** [Middle English *fals,* from Old French & Latin; Old French, from Latin *falsus,* from past participle of *fallere* to deceive] (12th century)
1 : not genuine ⟨*false* documents⟩ ⟨*false* teeth⟩
2 a : intentionally untrue ⟨*false* testimony⟩ **b :** adjusted or made so as to deceive ⟨*false* scales⟩ ⟨a trunk with a *false* bottom⟩ **c :** intended or tending to mislead ⟨a *false* promise⟩
3 : not true ⟨*false* concepts⟩
4 a : not faithful or loyal **:** TREACHEROUS ⟨a *false* friend⟩ **b :** lacking naturalness or sincerity ⟨*false* sympathy⟩
5 a : not essential or permanent — used of parts of a structure that are temporary or supplemental **b :** fitting over a main part to strengthen it, to protect it, or to disguise its appearance ⟨a *false* ceiling⟩
6 : inaccurate in pitch ⟨a *false* note⟩
7 a : based on mistaken ideas ⟨*false* pride⟩ **b :** inconsistent with the facts ⟨a *false* position⟩ ⟨a *false* sense of security⟩
8 : threateningly sudden or deceptive ⟨don't make a *false* move⟩
synonym see FAITHLESS
— false·ly *adverb*
— false·ness *noun*

[2]**false** *adverb* (13th century)
: in a false or faithless manner **:** TREACHEROUSLY ⟨his friends played him *false*⟩

false alarm *noun* (1579)
1 : one causing alarm or excitement that proves to be unfounded
2 : an alarm (as a fire or burglar alarm) that is set off needlessly

false arrest *noun* (1926)
: an arrest not justifiable under law

false color *noun* (1968)
: color in an image (as a photograph) of an object that does not actually appear in the object but is used to enhance, contrast, or distinguish details which are evident solely or chiefly from differences in the absorption and reflection of electromagnetic radiation at wavelengths outside the visual spectrum

false·hood \'fȯls-,hu̇d\ *noun* (13th century)
1 : an untrue statement **:** LIE
2 : absence of truth or accuracy
3 : the practice of lying **:** MENDACITY

false imprisonment *noun* (14th century)
: imprisonment of a person contrary to law

false mi·ter·wort \-'mī-tər-,wərt, -,wȯrt\ *noun* [*miterwort* from the resemblance of the plant's capsule to a bishop's miter] (1868)
: FOAMFLOWER

false morel *noun* (1942)
: any of a genus (*Gyromitra*) of fungi that are often poisonous and have a cap with convolutions resembling a brain

false pregnancy *noun* (circa 1860)
: PSEUDOCYESIS, PSEUDOPREGNANCY

false rib *noun* (15th century)
: a rib whose cartilages unite indirectly or not at all with the sternum — compare FLOATING RIB

false Solomon's seal *noun* (circa 1856)
: any of a genus (*Smilacina*) of herbs of the lily family that differ from Solomon's seal in having flowers in a terminal raceme or panicle — called also *false Solomonseal*

false start *noun* (1815)
1 : a premature start (as of a race)
2 : an unsuccessful attempt to begin something (as a career)

[1]**fal·set·to** \fȯl-'se-(,)tō\ *noun, plural* **-tos** [Italian, from diminutive of *falso* false, from Latin *falsus*] (1774)
1 : an artificially high voice; *especially* **:** an artificially produced singing voice that overlaps and extends above the range of the full voice especially of a tenor
2 : a singer who uses falsetto

[2]**falsetto** *adverb* (1940)
: in falsetto

false·work \'fȯls-,wərk\ *noun* (circa 1874)
: temporary construction work on which a main work is wholly or partly built and supported until the main work is strong enough to support itself

fals·ie \'fȯl-sē\ *noun* (circa 1943)
: an artificial addition to a bodily part worn to enhance appearance; *specifically* **:** a breast-shaped usually fabric or rubber cup used to pad a brassiere — usually used in plural

fal·si·fy \'fȯl-sə-,fī\ *verb* **-fied; -fy·ing** [Middle English *falsifien,* from Middle French *falsifier,* from Medieval Latin *falsificare,* from Latin *falsus*] (15th century)
transitive verb
1 : to prove or declare false
2 : to make false: as **a :** to make false by mutilation or addition ⟨the accounts were *falsified* to conceal a theft⟩ **b :** to represent falsely **:** MISREPRESENT
3 : to prove unsound by experience
intransitive verb
: to tell lies **:** LIE
— fal·si·fi·abil·i·ty \,fȯl-sə-,fī-ə-'bi-lə-tē\ *noun*
— fal·si·fi·able \-'fī-ə-bəl\ *adjective*
— fal·si·fi·ca·tion \,fȯl-sə-fə-'kā-shən\ *noun*
— fal·si·fi·er \'fȯl-sə-,fī(-ə)r\ *noun*

fal·si·ty \'fȯl-sə-tē\ *noun, plural* **-ties** (13th century)
1 : something false **:** LIE
2 : the quality or state of being false

Fal·staff \'fȯl-,staf\ *noun*
: a fat, convivial, roguish character in Shakespeare's *Merry Wives of Windsor* and *Henry IV*
— Fal·staff·ian \fȯl-'sta-fē-ən\ *adjective*

[1]**fal·ter** \'fȯl-tər\ *verb* **fal·tered; fal·ter·ing** \-t(ə-)riŋ\ [Middle English] (14th century)
intransitive verb
1 a : to walk unsteadily **:** STUMBLE **b :** to give way **:** TOTTER ⟨could feel my legs *faltering*⟩ **c :** to move waveringly or hesitatingly
2 : to speak brokenly or weakly **:** STAMMER
3 a : to hesitate in purpose or action **:** WAVER **b :** to lose drive or effectiveness ⟨the business was *faltering*⟩
transitive verb
: to utter hesitatingly or brokenly
synonym see HESITATE
— fal·ter·er \-tər-ər\ *noun*
— fal·ter·ing·ly \-t(ə-)riŋ-lē\ *adverb*

[2]**falter** *noun* (1834)
: an act or instance of faltering

[1]**fame** \'fām\ *noun* [Middle English, from Old French, from Latin *fama* report, fame; akin to Latin *fari* to speak — more at BAN] (13th century)
1 a : public estimation **:** REPUTATION **b :** popular acclaim **:** RENOWN
2 *archaic* **:** RUMOR

[2]**fame** *transitive verb* **famed; fam·ing** (14th century)
1 *archaic* **:** REPORT, REPUTE
2 : to make famous

famed \'fāmd\ *adjective* (circa 1533)
: known widely and well **:** FAMOUS

fa·mil·ial \fə-'mil-yəl, -'mi-lē-əl\ *adjective* [French, from Latin *familia*] (circa 1900)
1 : tending to occur in more members of a family than expected by chance alone ⟨a *familial* disorder⟩
2 : of, relating to, or suggestive of a family

[1]**fa·mil·iar** \fə-'mil-yər\ *noun* (13th century)
1 : a member of the household of a high official
2 : one that is familiar; *especially* **:** an intimate associate **:** COMPANION
3 : a spirit often embodied in an animal and held to attend and serve or guard a person
4 a : one who is well acquainted with something **b :** one who frequents a place

[2]**familiar** *adjective* [Middle English *familier,* from Middle French, from Latin *familiaris,* from *familia*] (14th century)
1 : closely acquainted **:** INTIMATE ⟨a *familiar* family friend⟩
2 *obsolete* **:** AFFABLE, SOCIABLE
3 a : of or relating to a family ⟨remembering past *familiar* celebrations⟩ **b :** frequented by families ⟨a *familiar* resort⟩
4 a : being free and easy ⟨the *familiar* association of old friends⟩ **b :** marked by informality ⟨a *familiar* essay⟩ **c :** overly free and unrestrained **:** PRESUMPTUOUS ⟨grossly *familiar* behavior⟩ **d :** moderately tame ⟨*familiar* animals⟩
5 a : frequently seen or experienced **:** easily recognized ⟨a *familiar* theme⟩ **b :** of everyday occurrence **c :** possibly known but imperfectly remembered ⟨her face looked *familiar*⟩
6 : having personal or intimate knowledge — used with *with* ⟨*familiar* with the facts of the case⟩
synonym see COMMON
— fa·mil·iar·ly *adverb*
— fa·mil·iar·ness *noun*

fa·mil·iar·ise *British variant of* FAMILIARIZE

fa·mil·iar·i·ty \fə-,mi-lē-'(y)ar-ə-tē, -,mil-'yar-\ *noun, plural* **-ties** (13th century)
1 a : the quality or state of being familiar **b :** a state of close relationship **:** INTIMACY
2 a : absence of ceremony **:** INFORMALITY **b :** an unduly informal act or expression **:** IMPROPRIETY **c :** a sexual liberty
3 : close acquaintance with something ⟨my *familiarity* with American history⟩

fa·mil·iar·ize \fə-'mil-yə-,rīz\ *transitive verb* **-ized; -iz·ing** (1608)
1 : to make known or familiar ⟨Shakespeare . . . *familiarizes* the wonderful —Samuel Johnson⟩
2 : to make well acquainted ⟨*familiarize* students with good literature⟩
— fa·mil·iar·iza·tion \-,mil-yə-rə-'zā-shən\ *noun*

familiar spirit *noun* (1565)
1 : a spirit or demon that serves or prompts an individual
2 : the spirit of a dead person invoked by a medium to advise or prophesy

fam·i·lism \'fa-mə-,li-zəm\ *noun* (1925)
: a social pattern in which the family assumes a position of ascendance over individual interests
— fam·i·lis·tic \,fa-mə-'lis-tik\ *adjective*

fa·mille rose \fə-'mē-\ *noun* [French, literally, rose family] (circa 1898)
: Chinese porcelain in the decoration of which a rose color predominates

fa·mille verte \fə-'mē-'vert\ *noun* [French, literally, green family] (1872)
: Chinese porcelain in the decoration of which green predominates

[1]**fam·i·ly** \'fam-lē, 'fa-mə-\ *noun, plural* **-lies** [Middle English *familie,* from Latin *familia* household (including servants as well as kin of the householder), from *famulus* servant] (15th century)
1 : a group of individuals living under one roof and usually under one head **:** HOUSEHOLD
2 a : a group of persons of common ancestry **:** CLAN **b :** a people or group of peoples regarded as deriving from a common stock **:** RACE
3 a : a group of people united by certain convictions or a common affiliation **:** FELLOWSHIP **b :** the staff of a high official (as the President)
4 : a group of things related by common characteristics: as **a :** a closely related series of elements or chemical compounds **b :** a group of soils that have similar profiles and include one or more series **c :** a group of related languages descended from a single ancestral language
5 a : the basic unit in society traditionally consisting of two parents rearing their own or adopted children; *also* **:** any of various social units differing from but regarded as equivalent to the traditional family ⟨a single-parent *family*⟩ **b :** spouse and children ⟨want to spend more time with my *family*⟩

6 a : a group of related plants or animals forming a category ranking above a genus and below an order and usually comprising several to many genera **b** *in livestock breeding* (1) **:** the descendants or line of a particular individual especially of some outstanding female (2) **:** an identifiable strain within a breed
7 : a set of curves or surfaces whose equations differ only in parameters
8 : a unit of a crime syndicate (as the Mafia) operating within a geographical area

[2]family *adjective* (1602)
1 : of or relating to a family
2 : designed or suitable for both children and adults ⟨*family* restaurants⟩ ⟨*family* movies⟩

family Bible *noun* (1740)
: a large Bible usually having special pages for recording births, marriages, and deaths

family court *noun* (circa 1931)
: COURT OF DOMESTIC RELATIONS

family doctor *noun* (1846)
1 : a doctor regularly consulted by a family
2 : a doctor specializing in family practice

family jewels *noun plural* (circa 1946)
slang **:** a man's testicles

family man *noun* (1856)
1 : a man with a wife and children dependent on him
2 : a responsible man of domestic habits

family name *noun* (1699)
: SURNAME 2

family physician *noun* (1807)
: FAMILY DOCTOR

family planning *noun* (1939)
: planning intended to determine the number and spacing of one's children through birth control

family practice *noun* (1969)
: a medical practice or specialty which provides continuing general medical care for the individual and family — called also *family medicine*

family practitioner *noun* (1846)
: FAMILY DOCTOR

family room *noun* (1853)
: a large room designed as a recreation center for members of a family

family style *adverb or adjective* (1932)
: with the food placed on the table in serving dishes from which those eating may help themselves

family tree *noun* (1807)
1 : a genealogical diagram
2 : GENEALOGY

family way *noun* (1796)
: condition of being pregnant — used with *in* and *the* or *a* ⟨she is in a *family way*⟩

fam·ine \ˈfa-mən\ *noun* [Middle English, from Middle French, from *faim* hunger, from Latin *fames*] (14th century)
1 : an extreme scarcity of food
2 *archaic* **:** STARVATION
3 *archaic* **:** a ravenous appetite
4 : a great shortage

fam·ish \ˈfa-mish\ *verb* [Middle English, probably alteration of *famen,* from Middle French *afamer,* from (assumed) Vulgar Latin *affamare,* from Latin *ad-* + *fames*] (15th century)
transitive verb
1 : to cause to suffer severely from hunger
2 *archaic* **:** to cause to starve to death
intransitive verb
1 *archaic* **:** STARVE
2 : to suffer for lack of something necessary ⟨a moment when French poetry in particular was *famishing* for such invention —T. S. Eliot⟩
— fam·ish·ment \-mənt\ *noun*

fam·ished \ˈfa-misht\ *adjective* (15th century)
: intensely hungry; *also* **:** NEEDY

fa·mous \ˈfā-məs\ *adjective* [Middle English, from Middle French *fameux,* from Latin *famosus,* from *fama* fame] (14th century)
1 a : widely known **b :** honored for achievement
2 : EXCELLENT, FIRST-RATE ⟨*famous* weather for a walk⟩ ☆
— fa·mous·ness *noun*

fa·mous·ly *adverb* (1579)
1 : in a celebrated manner
2 : in a superlative fashion
3 : to an unusual degree **:** VERY

fam·u·lus \ˈfam-yə-ləs\ *noun, plural* **-li** \-ˌlī, -ˌlē\ [German, assistant to a professor, from Latin, servant] (1837)
: a private secretary or attendant

[1]fan \ˈfan\ *noun* [Middle English, from Old English *fann,* from Latin *vannus* — more at WINNOW] (before 12th century)
1 : any of various devices for winnowing grain
2 : an instrument for producing a current of air: as **a :** a device for cooling the person that is usually shaped like a segment of a circle and is composed of material (as feathers or paper) mounted on thin rods or slats moving about a pivot so that the device may be closed compactly when not in use **b :** a device that consists of a series of vanes radiating from a hub rotated on its axle by a motor **c** *slang* **:** an airplane propeller
3 a : something resembling an open fan **b :** a gently sloping fan-shaped body of detritus; *especially* **:** ALLUVIAL FAN
— fan·like \-ˌlīk\ *adjective*

[2]fan *verb* **fanned; fan·ning** (before 12th century)
transitive verb
1 a : to drive away the chaff of (grain) by means of a current of air **b :** to eliminate (as chaff) by winnowing
2 : to move or impel (air) with a fan
3 : to blow or breathe upon ⟨the breeze *fanning* her hair⟩
4 a : to direct a current of air upon with a fan **b :** to stir up to activity as if by fanning **:** STIMULATE ⟨*fanning* the fires of nationalism⟩
5 *archaic* **:** WAVE
6 *slang* **:** SPANK
7 : to spread like a fan ⟨the peacock *fanned* his tail⟩
8 : to strike (a batter) out in baseball
9 : to fire a series of shots from (a single-action revolver) by holding the trigger back and successively striking the hammer to the rear with the free hand
intransitive verb
1 : to move like a fan **:** FLUTTER
2 : to spread like a fan — often used with *out* ⟨deputies *fanning* out on the hunt⟩
3 : STRIKE OUT 3
— fan·ner \ˈfa-nər\ *noun*

[3]fan *noun* [probably short for *fanatic*] (1682)
1 : an enthusiastic devotee (as of a sport or a performing art) usually as a spectator
2 : an ardent admirer or enthusiast (as of a celebrity or a pursuit) ⟨science-fiction *fans*⟩
word history see FANATIC

fa·nat·ic \fə-ˈna-tik\ *or* **fa·nat·i·cal** \-ti-kəl\ *adjective* [Latin *fanaticus* inspired by a deity, frenzied, from *fanum* temple — more at FEAST] (1550)
: marked by excessive enthusiasm and often intense uncritical devotion ⟨they're *fanatic* about politics⟩ ◆
— fanatic *noun*
— fa·nat·i·cal·ly \fə-ˈna-ti-k(ə-)lē\ *adverb*
— fa·nat·i·cal·ness \-kəl-nəs\ *noun*

fa·nat·i·cism \fə-ˈna-tə-ˌsi-zəm\ *noun* (1652)
: fanatic outlook or behavior

fa·nat·i·cize \-ˌsīz\ *transitive verb* **-cized; -ciz·ing** (1812)
: to cause to become fanatic

fan·ci·er \ˈfan(t)-sē-ər\ *noun* (1765)
1 : one that has a special liking or interest
2 : a person who breeds or grows a particular animal or plant for points of excellence ⟨a pigeon *fancier*⟩

fan·ci·ful \ˈfan(t)-si-fəl\ *adjective* (circa 1627)
1 : marked by fancy or unrestrained imagination rather than by reason and experience ⟨a *fanciful* person⟩
2 : existing in fancy only ⟨a *fanciful* notion⟩
3 : marked by or as if by fancy or whim ⟨gave their children *fanciful* names⟩
synonym see IMAGINARY
— fan·ci·ful·ly \-f(ə-)lē\ *adverb*
— fan·ci·ful·ness \-fəl-nəs\ *noun*

fan·ci·fy \ˈfan(t)-sə-ˌfī\ *transitive verb* **-fied; -fy·ing** (1890)
: to make ornate, elaborate, or fancy ⟨a *fancified* hamburger⟩

[1]fan·cy \ˈfan(t)-sē\ *transitive verb* **fan·cied; fan·cy·ing** (14th century)
1 : to have a fancy for **:** LIKE
2 : to form a conception of **:** IMAGINE ⟨*fancy* our embarrassment⟩
3 a : to believe mistakenly or without evidence **b :** to believe without being certain ⟨she *fancied* she had met him before⟩
4 : to visualize or interpret as ⟨*fancied* myself a child again⟩
synonym see THINK

[2]fancy *noun, plural* **fancies** [Middle English *fantasie, fantsy* fantasy, fancy, from Middle French *fantasie,* from Latin *phantasia,* from

☆ SYNONYMS

Famous, renowned, celebrated, noted, notorious, distinguished, eminent, illustrious mean known far and wide. FAMOUS implies little more than the fact of being, sometimes briefly, widely and popularly known ⟨a *famous* actress⟩. RENOWNED implies more glory and acclamation ⟨one of the most *renowned* figures in sports history⟩. CELEBRATED implies notice and attention especially in print ⟨the most *celebrated* beauty of her day⟩. NOTED suggests well-deserved public attention ⟨the *noted* mystery writer⟩. NOTORIOUS frequently adds to FAMOUS an implication of questionableness or evil ⟨a *notorious* gangster⟩. DISTINGUISHED implies acknowledged excellence or superiority ⟨a *distinguished* scientist who won the Nobel Prize⟩. EMINENT implies even greater conspicuousness for outstanding quality or character ⟨the country's most *eminent* writers⟩. ILLUSTRIOUS stresses enduring honor and glory attached to a deed or person ⟨our *illustrious* national heroes⟩.

◇ WORD HISTORY

fanatic In Latin the adjective *fanaticus,* a derivative of *fanum* "consecrated space, temple," meant literally "pertaining to a temple," though more common was the sense "inspired by a deity, frenzied." With the latter meaning the word was borrowed into English as *fanatic* in the 16th century. In the following century the word as both adjective and noun was applied to adherents of more radical Protestant sects who espoused their beliefs—in the view of the conservative mainstream who belonged to the Church of England—with excessive zeal, acting as if they were divinely inspired. Eventually *fanatic* became a more general negative label for extreme partisans of dubious causes, and in the 19th and 20th centuries has been associated as much with politics as religion. The word *fan* is probably a shortening of *fanatic;* it occurs in a single text in the 17th century (in the curious spellings *phann* and *fann*), but was in all likelihood formed anew in 19th century American English, where it first appears in reference to enthusiastic devotees of local sports teams—still a frequent usage of the word.

Greek, appearance, imagination, from *phantazein* to present to the mind (middle voice, to imagine), from *phainein* to show; akin to Old English *gebōned* polished, Greek *phōs* light] (15th century)
1 a : a liking formed by caprice rather than reason **:** INCLINATION ⟨took a *fancy* to the strange little animal⟩ **b :** amorous fondness **:** LOVE
2 a : NOTION, WHIM **b :** an image or representation of something formed in the mind
3 *archaic* **:** fantastic quality or state
4 a : imagination especially of a capricious or delusive sort **b :** the power of conception and representation used in artistic expression (as by a poet)
5 : TASTE, JUDGMENT
6 a : devotees of some particular art, practice, or amusement **b :** the object of interest of such a fancy; *especially* **:** [2]BOXING

[3]fancy *adjective* **fan·ci·er; -est** (1646)
1 : dependent or based on fancy **:** WHIMSICAL
2 a (1) **:** not plain **:** ORNAMENTAL ⟨a *fancy* hairdo⟩ (2) **:** SWANKY 2, POSH ⟨a *fancy* restaurant⟩ **b** (1) **:** of particular excellence or highest grade ⟨*fancy* tuna⟩ (2) **:** IMPRESSIVE ⟨posted some *fancy* numbers⟩ **c** *of an animal or plant* **:** bred especially for bizarre or ornamental qualities that lack practical utility
3 : based on conceptions of the fancy ⟨*fancy* sketches⟩
4 a : dealing in fancy goods **b :** EXTRAVAGANT ⟨paying *fancy* prices⟩
5 : executed with technical skill and style ⟨*fancy* footwork⟩ ⟨*fancy* diving⟩
6 : PARTI-COLOR ⟨*fancy* carnations⟩
— fan·ci·ly \ˈfan(t)-sə-lē\ *adverb*
— fan·ci·ness \-sē-nəs\ *noun*

fancy–dan \ˈfan(t)-sē-ˈdan, -ˌdan\ *adjective, often D capitalized* (1938)
: SHOWY 2, FANCY

fancy Dan *noun, often F capitalized* (circa 1943)
: one given to flamboyant display especially of technique or dress

fancy dress *noun* (1770)
: a costume (as for a masquerade) chosen to suit the wearer's fancy

fan·cy–free \ˌfan(t)-sē-ˈfrē\ *adjective* (1590)
1 : free from amorous attachment or engagement
2 : free to imagine or fancy

fancy man *noun* (circa 1811)
: a woman's paramour; *also* **:** PIMP

fancy–pants *adjective* (1945)
: overly elegant or refined **:** LA-DI-DA

fancy up *transitive verb* (1934)
: to add superficial adornment to

fancy woman *noun* (1812)
: a woman of questionable morals; *specifically* **:** PROSTITUTE

fan·cy·work \ˈfan(t)-sē-ˌwərk\ *noun* (1810)
: decorative needlework

fan·dan·go \fan-ˈdaŋ-(ˌ)gō\ *noun, plural* **-gos** [Spanish] (1774)
1 : a lively Spanish or Spanish-American dance in triple time that is usually performed by a man and a woman to the accompaniment of guitar and castanets; *also* **:** music for this dance
2 : TOMFOOLERY

fan·dom \ˈfan-dəm\ *noun* (1903)
: all the fans (as of a sport)

fane \ˈfān\ *noun* [Middle English, from Latin *fanum* — more at FEAST] (15th century)
1 : TEMPLE
2 : CHURCH

fan·fare \ˈfan-ˌfar, -ˌfer\ *noun* [French] (1676)
1 : a showy outward display
2 : a short and lively sounding of trumpets

fan·far·o·nade \ˌfan-ˌfar-ə-ˈnād, -ˈnäd\ *noun* [French *fanfaronnade,* from Spanish *fanfarronada,* from *fanfarrón* braggart] (1652)
: empty boasting **:** BLUSTER

fan·fold \ˈfan-ˌfōld\ *noun* (1925)
: paper (as business forms or tape) made from a web and folded like a fan lengthwise and sometimes crosswise
— fanfold *transitive verb*

fang \ˈfaŋ\ *noun* [Middle English, that which is taken, from Old English; akin to Old High German *fang* seizure, Old English *fōn* to seize — more at PACT] (1555)
1 a : a long sharp tooth: as (1) **:** one by which an animal's prey is seized and held or torn (2) **:** one of the long hollow or grooved and often erectile teeth of a venomous snake **b :** one of the chelicerae of a spider at the tip of which a poison gland opens
2 : the root of a tooth or one of the processes or prongs into which a root divides
3 : a projecting tooth or prong
— fanged \ˈfaŋd\ *adjective*

Fang \ˈfaŋ, ˈfäŋ\ *also* **Fan** \ˈfan, ˈfän\ *noun, plural* **Fang** *or* **Fangs** *also* **Fan** *or* **Fans** (1861)
1 : a member of a Bantu-speaking people of northern Gabon, mainland Equatorial Guinea, and southern Cameroon
2 : the language of the Fang people

fan–jet \ˈfan-ˌjet\ *noun* (1962)
: a jet engine having a fan that operates in a duct and draws in extra air whose compression and expulsion provide extra thrust; *also* **:** an airplane powered by a fan-jet engine

fan letter *noun* (1932)
: a letter sent to a public figure by an admirer

fan·light \ˈfan-ˌlīt\ *noun* (1819)
: a semicircular window with radiating bars like the ribs of a fan that is placed over a door or window

fan mail *noun* (1924)
: FAN LETTERS

fan·ny \ˈfa-nē\ *noun, plural* **fannies** [perhaps from *Fanny,* nickname of *Frances*] (1928)
1 : BUTTOCKS
2 *slang British* **:** VULVA

fanny pack *noun* (1967)
: a pack for carrying personal articles that straps to the waist

fan·tab·u·lous \fan-ˈta-byə-ləs\ *adjective* [blend of *fantastic* and *fabulous*] (1959)
slang **:** marvelously good

fan·tail \ˈfan-ˌtāl\ *noun* (1728)
1 : a fan-shaped tail or end
2 : a domestic pigeon having a broad rounded tail often with 30 or 40 feathers
3 : an architectural part resembling a fan
4 : a counter or after overhang of a ship shaped like a duck's bill

fan–tan \ˈfan-ˌtan\ *noun* [Chinese (Guangdong) *fāantāan*] (1878)
1 : a Chinese gambling game in which the banker divides a pile of objects (as beans) into fours and players bet on what number will be left at the end of the count
2 : a card game in which players must build in sequence upon sevens and attempt to be the first one out of cards

fan·ta·sia \fan-ˈtā-zhə, -z(h)ē-ə; ˌfan-tə-ˈzē-ə\ *noun* [Italian, literally, fancy, from Latin *phantasia* — more at FANCY] (1724)
1 : a free usually instrumental composition not in strict form
2 a : a work (as a poem or play) in which the author's fancy roves unrestricted **b :** something possessing grotesque, bizarre, or unreal qualities

fan·ta·sie \ˌfan-tə-ˈzē, ˌfän-\ *noun* [German *Phantasie,* from Latin *phantasia*] (circa 1859)
: FANTASIA

fan·ta·sied \ˈfan-tə-sēd, -zēd\ *adjective* (1561)
1 : existing only in the imagination **:** FANCIED
2 *obsolete* **:** full of fancies or strange whims

fan·ta·sise *British variant of* FANTASIZE

fan·ta·sist \-sist, -zist\ *noun* (1896)
: one who creates fantasias or fantasies

fan·ta·size \-ˌsīz\ *verb* **-sized; -siz·ing** (1926)
intransitive verb
: to indulge in reverie **:** create or develop imaginative and often fantastic views or ideas ⟨doing things I'd *fantasized* about in my sheltered childhood —Diane Arbus⟩
transitive verb
: to portray in the mind **:** FANCY ⟨likes to *fantasize* herself as very wealthy⟩
— fan·ta·siz·er \-ˌsī-zər\ *noun*

fan·tasm *variant of* PHANTASM

fan·tast \ˈfan-ˌtast\ *noun* [German, from Medieval Latin *fantasta,* probably back-formation from Late Latin *phantasticus*] (1588)
1 : VISIONARY
2 : a fantastic or eccentric person
3 : FANTASIST

[1]fan·tas·tic \fan-ˈtas-tik, fən-\ *also* **fan·tas·ti·cal** \-ti-kəl\ *adjective* [Middle English *fantastic, fantastical,* from Middle French & Late Latin; Middle French *fantastique,* from Late Latin *phantasticus,* from Greek *phantastikos* producing mental images, from *phantazein* to present to the mind] (14th century)
1 a : based on fantasy **:** not real **b :** conceived or seemingly conceived by unrestrained fancy **c :** so extreme as to challenge belief **:** UNBELIEVABLE; *broadly* **:** exceedingly large or great
2 : marked by extravagant fantasy or extreme individuality **:** ECCENTRIC
3 *fantastic* **:** EXCELLENT, SUPERLATIVE ⟨a *fantastic* dinner⟩ ☆
— fan·tas·ti·cal·i·ty \(ˌ)fan-ˌtas-tə-ˈka-lə-tē, fən-\ *noun*
— fan·tas·ti·cal·ness \-ˈtas-tə-kəl-nəs\ *noun*

[2]fantastic *noun* (1598)
: ECCENTRIC 2

fan·tas·ti·cal·ly \fan-ˈtas-ti-k(ə-)lē, fən-\ *adverb* (1543)
1 : in a fantastic manner
2 : to a fantastic degree **:** EXTREMELY ⟨*fantastically* expensive clothes⟩

fan·tas·ti·cate \fan-ˈtas-tə-ˌkāt, fən-\ *transitive verb* **-cat·ed; -cat·ing** (1600)
: to make fantastic
— fan·tas·ti·ca·tion \(ˌ)fan-ˌtas-tə-ˈkā-shən, fən-\ *noun*

fan·tas·ti·co \fan-ˈtas-ti-ˌkō, fən-\ *noun, plural* **-coes** [Italian, fantastic (adjective), from Late Latin *phantasticus*] (1596)
: a ridiculously fantastic individual

[1]fan·ta·sy \ˈfan-tə-sē, -zē\ *noun, plural* **-sies** [Middle English *fantasie* — more at FANCY] (14th century)
1 *obsolete* **:** HALLUCINATION
2 : FANCY; *especially* **:** the free play of creative imagination
3 : a creation of the imaginative faculty whether expressed or merely conceived: as **a :** a fanciful design or invention **b :** a chimerical or fantastic notion **c :** FANTASIA 1 **d :** imaginative fiction featuring especially strange settings and grotesque characters — called also *fantasy fiction*
4 : CAPRICE

☆ SYNONYMS
Fantastic, bizarre, grotesque mean conceived, made, or carried out without adherence to truth or reality. FANTASTIC may connote unrestrained extravagance in conception or merely ingenuity of decorative invention ⟨dreamed up *fantastic* rumors to spread⟩. BIZARRE applies to the sensationally queer or strange and implies violence of contrast or incongruity of combination ⟨a *bizarre* medieval castle built in the heart of a modern city⟩. GROTESQUE may apply to what is conventionally ugly but artistically effective or it may connote ludicrous awkwardness or incongruity often with sinister or tragic overtones ⟨*grotesque* statues adorn the cathedral⟩ ⟨though grief-stricken, she made a *grotesque* attempt at a smile⟩. See in addition IMAGINARY.

5 : the power or process of creating especially unrealistic or improbable mental images in response to psychological need ⟨an object of *fantasy*⟩; *also* **:** a mental image or a series of mental images (as a daydream) so created ⟨sexual *fantasies* of adolescence⟩
6 : a coin usually not intended for circulation as currency and often issued by a dubious authority (as a government-in-exile)

[2]fantasy *verb* **-sied; -sy·ing** (15th century)
: FANTASIZE

fan·ta·sy·land \-ˌland\ *noun* (1967)
: an imaginary or ideal place or situation

fan·toc·ci·ni \ˌfän-tə-ˈchē-nē, ˌfan-\ *noun plural* [Italian, plural of *fantoccino,* diminutive of *fantoccio* doll, augmentative of *fante* child, from Latin *infant-, infans* infant] (1771)
: a puppet show using puppets operated by strings or mechanical devices; *also* **:** such puppets

fan·tod \ˈfan-ˌtäd\ *noun* [perhaps alteration of English dialect *fantique, fanteeg*] (1839)
1 *plural* **a :** a state of irritability and tension **b :** FIDGETS
2 : an emotional outburst **:** FIT

fan·tom *variant of* PHANTOM

fan vault *noun* (circa 1901)
: a Gothic vault in which the ribs from each springer spread out like the vanes of a fan
— fan vaulting *noun*

fan·wise \ˈfan-ˌwīz\ *adverb or adjective* (1882)
: in the manner or position of the slats of an open fan ⟨boats anchored *fanwise* at the pier⟩

fan·zine \ˈfan-ˌzēn\ *noun* [[3]*fan* + maga*zine*] (1949)
: a magazine written by and for fans especially of science fiction or fantasy writing

[1]far \ˈfär\ *adverb* **far·ther** \-t͟hər\ *or* **fur·ther** \ˈfər-\; **far·thest** *or* **fur·thest** \-t͟həst\ [Middle English *fer,* from Old English *feorr;* akin to Old High German *ferro* far, Old English *faran* to go — more at FARE] (before 12th century)
1 : at or to a considerable distance in space ⟨wandered *far* from home⟩
2 a : to a great extent **:** MUCH ⟨*far* better methods⟩ **b :** by a broad interval **:** WIDELY ⟨the *far* distant future⟩
3 : to or at a definite distance, point, or degree ⟨as *far* as I know⟩
4 : to an advanced point or extent ⟨a bright student will go *far*⟩ ⟨worked *far* into the night⟩
5 : at a considerable distance in time ⟨not *far* from the year 1870⟩
— by far : far and away ⟨is *by far* the best runner⟩
— far be it from : it would be inappropriate or impossible for ⟨*far be it from* God, that he should do wickedness —Job 34:10 (Authorized Version)⟩
— far from : of a distinctly different and especially opposite quality than ⟨the trip was *far from* a failure⟩
— how far : to what extent, degree, or distance ⟨didn't know *how far* to trust them⟩
— so far 1 : to a certain extent, degree, or distance ⟨when the water rose *so far,* the villagers sought higher ground⟩ **2 :** up to the present ⟨has written just one novel *so far*⟩
— thus far : so far ⟨*thus far* our findings have been negative⟩

[2]far *adjective* **farther** *or* **further; farthest** *or* **furthest** (before 12th century)
1 a : remote in space **b :** distinctly different in quality or relationship **c :** remote in time
2 a : LONG ⟨a *far* journey⟩ **b :** of notable extent **:** COMPREHENSIVE ⟨a man of *far* vision⟩
3 : the more distant of two
4 : EXTREME ⟨the *far* left⟩ ⟨a *far* right political organization⟩

far·ad \ˈfar-ˌad, ˈfar-əd\ *noun* [Michael *Faraday*] (1873)
: the unit of capacitance equal to the capacitance of a capacitor between whose plates there appears a potential of one volt when it is charged by one coulomb of electricity

far·a·day \ˈfar-ə-ˌdā, -ə-dē\ *noun* [Michael *Faraday*] (1904)
: the quantity of electricity transferred in electrolysis per equivalent weight of an element or ion equal to about 96,500 coulombs

fa·rad·ic \fə-ˈra-dik, fa-ˈra-\ *also* **far·a·da·ic** \ˌfar-ə-ˈdā-ik\ *adjective* (1875)
: of or relating to an asymmetric alternating current of electricity produced by an induction coil ⟨*faradic* stimulation of the muscles⟩

far·a·dism \ˈfar-ə-ˌdi-zəm\ *noun* (1876)
: the application of a faradic current of electricity (as for therapeutic purposes)

far and away *adverb* (1852)
: by a considerable margin ⟨was *far and away* the better team⟩

far·an·dole \ˈfar-ən-ˌdōl\ *noun* [French *farandole,* from Provençal *farandoulo*] (1863)
1 : a lively Provençal dance in which men and women hold hands, form a chain, and follow a leader through a serpentine course
2 : music in sextuple time for a farandole

far and wide *adverb* (before 12th century)
: in every direction **:** EVERYWHERE ⟨searched *far and wide*⟩

far·away \ˈfär-ə-ˌwā\ *adjective* (1735)
1 : lying at a great distance **:** REMOTE
2 : DREAMY, ABSTRACTED ⟨a *faraway* look in her eyes⟩

[1]farce \ˈfärs\ *transitive verb* **farced; farc·ing** [Middle English *farsen,* from Middle French *farcir,* from Latin *farcire*] (14th century)
1 : STUFF
2 : to improve as if by stuffing ◆

[2]farce *noun* [Middle English *farse,* from Middle French *farce,* from (assumed) Vulgar Latin *farsa,* from Latin, feminine of *farsus,* past participle of *farcire*] (14th century)
1 : a savory stuffing **:** FORCEMEAT
2 : a light dramatic composition marked by broadly satirical comedy and improbable plot
3 : the broad humor characteristic of farce or pretense
4 a : ridiculous or empty show **b :** MOCKERY ⟨the enforcement of this law became a *farce*⟩

far·ceur \fär-ˈsər\ *noun* [French, from Middle French, from *farcer* to joke, from Old French, from *farce*] (1781)
1 : JOKER, WAG
2 : a writer or actor of farce

far·ci *or* **far·cie** \fär-ˈsē\ *adjective* [French, from past participle of *farcir*] (1903)
: stuffed especially with forcemeat ⟨oysters *farci*⟩

far·ci·cal \ˈfär-si-kəl\ *adjective* (1716)
1 : of, relating to, or resembling farce **:** LUDICROUS
2 : laughably inept **:** ABSURD
— far·ci·cal·i·ty \ˌfär-sə-ˈka-lə-tē\ *noun*
— far·ci·cal·ly \ˈfär-si-k(ə-)lē\ *adverb*

far·cy \ˈfär-sē\ *noun* [Middle English *farsin, farsi,* from Middle French *farcin,* from Late Latin *farcimen,* from Latin, sausage, from *farcire*] (15th century)
: GLANDERS; *especially* **:** cutaneous glanders

fard \ˈfärd\ *transitive verb* [Middle English, from Middle French *farder,* of Germanic origin; akin to Old High German *faro* colored — more at PERCH] (15th century)
1 : to paint (the face) with cosmetics
2 *archaic* **:** to gloss over
— fard *noun, archaic*

far·del \ˈfär-d^{ə}l\ *noun* [Middle English, from Middle French, probably from Arabic *fardah*] (14th century)
1 : BUNDLE
2 : BURDEN

[1]fare \ˈfar, ˈfer\ *intransitive verb* **fared; far·ing** [Middle English *faren,* from Old English *faran;* akin to Old High German *faran* to go, Latin *portare* to carry, Greek *peran* to pass through, *poros* passage, journey] (before 12th century)
1 : GO, TRAVEL
2 : GET ALONG, SUCCEED ⟨how did you *fare* on your exam?⟩
3 : EAT, DINE

[2]fare *noun* [Middle English, journey, passage, supply of food, from Old English *faru, fær;* akin to Old English *faran* to go] (15th century)
1 a : range of food **:** DIET **b :** material provided for use, consumption, or enjoyment
2 a : the price charged to transport a person **b :** a paying passenger on a public conveyance

fare–thee–well \ˈfar-(ˌ)t͟hē-ˌwel, ˈfer-\ *also* **fare–you–well** \-yə-, -yü-, -yē-\ *noun* (1884)
1 : the utmost degree ⟨researched the story to a *fare-thee-well*⟩
2 : a state of perfection ⟨imitated the speaker's pompous manner to a *fare-thee-well*⟩

[1]fare·well \far-ˈwel, fer-\ *verb imperative* (14th century)
: get along well — used interjectionally to or by one departing

[2]farewell *noun* (14th century)
1 : a wish of well-being at parting **:** GOOD-BYE
2 a : an act of departure **:** LEAVE-TAKING **b :** a formal occasion honoring a person about to leave or retire

[3]fare·well *transitive verb* (1580)
chiefly Australian & New Zealand **:** to bid farewell to

[4]fare·well \ˈfar-ˌwel, ˈfer-\ *adjective* (1669)
: of or relating to leave-taking **:** FINAL ⟨a *farewell* appearance⟩

far·fel *or* **far·fal** \ˈfär-fəl\ *noun* [Yiddish *farfl* (plural), from Middle High German *varveln* noodles, noodle soup] (1892)
: noodles in the form of small pellets or granules

far–fetched \ˈfär-ˈfecht\ *adjective* (1583)
1 : brought from a remote time or place
2 : not easily or naturally deduced or introduced **:** IMPROBABLE ⟨a *far-fetched* story⟩
— far–fetched·ness \-ˈfech(t)-nəs, -ˈfe-chəd-nəs\ *noun*

far–flung \-ˈfləŋ\ *adjective* (1895)
1 : widely spread or distributed ⟨a *far-flung* empire⟩
2 : REMOTE ⟨a *far-flung* correspondent⟩

fa·ri·na \fə-ˈrē-nə\ *noun* [Latin, meal, flour, from *far* spelt — more at BARLEY] (14th century)

◇ WORD HISTORY

farce When *farce* first appeared in English, its connection was with cooking, not comedy. In the 14th century English adopted *farce* (spelling it *farse* until the 18th century) from French, retaining its original meaning of "stuffing or forcemeat." The comedic sense of *farce* in English dates back to the 16th century, when England imported a kind of knockabout comedy that was already well-established in France and Italy. This dramatic genre had its origins in the 13th century practice of interlarding or "stuffing" Latin liturgical texts with explanatory passages in the vernacular. By the 15th century the practice arose of inserting unscripted buffoonery into performances of religious plays. Such farces, which included clowning, acrobatics, reversal of social roles, and indecency, soon developed into a distinct dramatic genre, a kind of broad comedy with a slapstick element. Since its introduction into England, farce in one guise or another has continued its popularity into the age of television.

1 **:** a fine meal of vegetable matter (as cereal grains) used chiefly for puddings or as a breakfast cereal
2 **:** any of various powdery or mealy substances

far·i·na·ceous \ˌfar-ə-ˈnā-shəs\ *adjective* (1646)
1 **:** having a mealy texture or surface
2 **:** containing or rich in starch

far–in·fra·red \ˈfär-ˌin-frə-ˈred\ *adjective* (1923)
: of or relating to the longer wavelengths of radiation in the infrared spectrum and especially to those between 10 and 1000 micrometers

fa·ri·nha \fə-ˈrēn-yə\ *noun* [Portuguese, flour, cassava meal, from Latin *farina*] (1726)
: cassava meal

far·kle·ber·ry \ˈfär-kəl-ˌber-ē\ *noun* [origin unknown] (1765)
: a shrub or small tree (*Vaccinium arboreum*) of the heath family of the southeastern U.S. having a black berry with stony seeds

farl \ˈfär(-ə)l\ *noun* [contraction of Scots *fardel,* literally, fourth part, from Middle English (Scots), from *ferde del;* from *ferde* fourth + *del* part] (1686)
Scottish **:** a small thin triangular cake or biscuit made especially with oatmeal or wheat flour

¹farm \ˈfärm\ *noun, often attributive* [Middle English *ferme* rent, lease, from Old French, lease, from *fermer* to fix, make a contract, from Latin *firmare* to make firm, from *firmus* firm] (14th century)
1 *obsolete* **:** a sum or due fixed in amount and payable at fixed intervals
2 **:** a letting out of revenues or taxes for a fixed sum to one authorized to collect and retain them
3 **:** a district or division of a country leased out for the collection of government revenues
4 **:** a tract of land devoted to agricultural purposes
5 **a :** a plot of land devoted to the raising of animals and especially domestic livestock **b :** a tract of water reserved for the artificial cultivation of some aquatic life form
6 **:** a minor-league team (as in baseball) associated with a major-league team as a subsidiary
7 **:** an area containing a number of similar structures (as radio antennas or storage tanks)

²farm (15th century)
transitive verb
1 *obsolete* **:** RENT
2 **:** to collect and take the fees or profits of (an occupation or business) on payment of a fixed sum
3 **:** to give up (as an estate or a business) to another on condition of receiving in return a fixed sum
4 **a :** to devote to agriculture **b :** to manage and cultivate as a farm **c :** to grow or cultivate in quantity ⟨*farm* trees for fuel⟩ ⟨*farm* salmon⟩
intransitive verb
: to engage in raising crops, animals, or fish

farm·er \ˈfär-mər\ *noun* (14th century)
1 **:** a person who pays a fixed sum for some privilege or source of income
2 **:** a person who cultivates land or crops or raises animals or fish
3 **:** YOKEL, BUMPKIN

farmer cheese *noun* (1949)
: a pressed unripened cheese similar to but drier and firmer than cottage cheese

farm·er·ette \ˌfär-mə-ˈret\ *noun* (1902)
: a woman who is a farmer or farmhand

farmer's lung *noun* (1945)
: an acute pulmonary disorder characterized by sudden onset, fever, cough, expectoration, and breathlessness that results from the inhalation of dust from moldy hay or straw

farm·hand \ˈfärm-ˌhand\ *noun* (1843)
1 **:** a farm laborer; *especially* **:** a hired laborer on a farm
2 **:** a player on a farm team

farm·house \-ˌhau̇s\ *noun* (1598)
: a dwelling on a farm

farm·ing *noun* (1733)
: the practice of agriculture or aquaculture

farm·land \ˈfärm-ˌland\ *noun* (1638)
: land used or suitable for farming

farm out *transitive verb* (1607)
1 **:** to turn over (as a job) for performance by another usually under contract
2 **a :** to put (as children) into the hands of another for care **b :** to send (as a baseball player) to a farm team
3 **:** to exhaust (land) by farming especially by continuously raising one crop

farm·stead \ˈfärm-ˌsted\ *noun* (1807)
: the buildings and adjacent service areas of a farm; *broadly* **:** a farm with its buildings

farm·wife \ˈfärm-ˌwīf\ *noun* (1880)
: a farmer's wife

farm·work·er \ˈfärm-ˌwər-kər\ *noun* (1946)
: FARMHAND 1
— farm·work \-ˌwərk\ *noun*

farm·yard \-ˌyärd\ *noun* (1748)
: land around or enclosed by farm buildings; *especially* **:** BARNYARD

faro \ˈfar-(ˌ)ō, ˈfer-\ *noun, plural* **far·os** [probably alteration of earlier *pharaoh,* translation of French *pharaon*] (circa 1735)
: a gambling game in which players bet on cards drawn from a dealing box

Faro·ese *variant of* FAEROESE

far–off \ˈfär-ˌȯf\ *adjective* (15th century)
: remote in time or space

fa·rouche \fə-ˈrüsh\ *adjective* [French, wild, shy, from Late Latin *forasticus* living outside, from Latin *foras* outdoors; akin to Latin *fores* door — more at DOOR] (1765)
1 **:** WILD
2 **:** marked by shyness and lack of social graces

far–out \ˈfär-ˌau̇t\ *adjective* (1954)
: marked by a considerable departure from the conventional or traditional ⟨*far-out* clothes⟩
— far–out·ness *noun*

far·rag·i·nous \fə-ˈra-jə-nəs\ *adjective* (1615)
: consisting of a farrago

far·ra·go \fə-ˈrä-(ˌ)gō, -ˈrā-\ *noun, plural* **-goes** [Latin *farragin-, farrago* mixed fodder, mixture, from *far* spelt — more at BARLEY] (1632)
: a confused mixture **:** HODGEPODGE

far–reach·ing \ˈfär-ˌrē-chiŋ\ *adjective* (1824)
: having a wide range or effect

far–red \-ˈred\ *adjective* (1951)
: NEAR-INFRARED

far·ri·er \ˈfar-ē-ər\ *noun* [alteration of Middle English *ferrour,* from Middle French *ferrour* blacksmith, from Old French *ferreor,* from *ferrer* to fit with iron, from (assumed) Vulgar Latin *ferrare,* from Latin *ferrum* iron] (15th century)
: a person who shoes horses

¹far·row \ˈfar-(ˌ)ō, -ə(-w)\ *verb* [Middle English *farwen,* from (assumed) Old English *feargian,* from Old English *fearh* young pig; akin to Old High German *farah* young pig, Latin *porcus* pig] (13th century)
transitive verb
: to give birth to (a farrow)
intransitive verb
of swine **:** to bring forth young — often used with *down*

²farrow *noun* (1577)
1 **:** a litter of pigs
2 **:** an act of farrowing

³farrow *adjective* [Middle English (Scots) *ferow*] (15th century)
of a cow **:** not pregnant

far·see·ing \ˈfär-ˌsē-iŋ\ *adjective* (1837)
: FARSIGHTED 1

Far·si \ˈfär-sē\ *noun* [Persian *fārsī,* from *Fārs* Persia] (1878)
: PERSIAN 2b

far·side *noun* (15th century)
: the farther side; *especially* **:** the side of the moon away from the earth
— on the far side of : BEYOND ⟨just *on the far side of* 40⟩

far·sight·ed \ˈfär-ˌsī-təd\ *adjective* (1609)
1 **a :** seeing or able to see to a great distance **b :** having or showing foresight or good judgment **:** SAGACIOUS
2 **:** affected with hyperopia
— far·sight·ed·ly *adverb*

far·sight·ed·ness *noun* (circa 1829)
1 **:** the quality or state of being farsighted
2 **:** HYPEROPIA

¹fart \ˈfärt\ *intransitive verb* [Middle English *ferten, farten;* akin to Old High German *ferzan* to break wind, Old Norse *freta,* Greek *perdesthai,* Sanskrit *pardate* he breaks wind] (13th century)
: to expel intestinal gas from the anus — often considered vulgar

²fart *noun* (14th century)
1 **:** an expulsion of intestinal gas — often considered vulgar
2 **:** a foolish or contemptible person ⟨couldn't stand the old *fart*⟩ — seldom in polite use

¹far·ther \ˈfär-t͟hər\ *adverb* [Middle English *ferther,* alteration of *further*] (14th century)
1 **:** at or to a greater distance or more advanced point ⟨got no *farther* than the first page⟩ ⟨nothing could be *farther* from the truth⟩
2 **:** to a greater degree or extent ⟨see to it that I do not have to act any *farther* in the matter —Bernard DeVoto⟩ ■

²farther *adjective* (14th century)
1 **:** more distant **:** REMOTER ⟨the *farther* side of town⟩
2 **:** FURTHER 2 ⟨clearing his throat preparatory to *farther* revelations —Edith Wharton⟩

far·ther·most \-ˌmōst\ *adjective* (15th century)
: most distant **:** FARTHEST

¹far·thest \ˈfär-t͟həst\ *adjective* (14th century)
: most distant especially in space or time

²farthest *adverb* (15th century)
1 **:** to or at the greatest distance in space or time ⟨who can jump the *farthest*⟩
2 **:** to the most advanced point ⟨goes *farthest* toward answering the question⟩
3 **:** by the greatest degree or extent **:** MOST ⟨the painting *farthest* removed from reality⟩

far·thing \ˈfär-t͟hiŋ\ *noun* [Middle English *ferthing,* from Old English *fēorthung* (akin to Middle High German *vierdunc* fourth part), from Old English *fēortha* fourth] (before 12th century)
1 **a :** a former British monetary unit equal to ¼ of a penny **b :** a coin representing this unit
2 **:** something of small value **:** MITE

far·thin·gale \ˈfär-t͟hən-ˌgāl, -t͟hiŋ-\ *noun* [modification of Middle French *verdugale,* from Old Spanish *verdugado,* from *verdugo* young shoot of a tree, from *verde* green, from Latin *viridis* — more at VERDANT] (1552)

□ USAGE
farther *Farther* and *further* have been used more or less interchangeably throughout most of their history, but currently they are showing signs of diverging. As adverbs they continue to be used interchangeably whenever spatial, temporal, or metaphorical distance is involved. But where there is no notion of distance, *further* is used ⟨our techniques can be *further* refined⟩. *Further* is also used as a sentence modifier ⟨*further,* the workshop participants were scarcely optimistic —L. B. Mayhew⟩, but *farther* is not. A polarizing process appears to be taking place in their adjective use. *Farther* is taking over the meaning of distance ⟨the *farther* shore⟩ and *further* the meaning of addition ⟨needed no *further* invitation⟩.

: a support (as of hoops) worn especially in the 16th century beneath a skirt to expand it at the hipline
far ultraviolet *adjective* (1947)
: of, relating to, or being the shortest wavelengths of radiation in the ultraviolet spectrum and especially those between 100 and 300 nanometers
fas·ces \'fas-ˌēz\ *noun plural but singular or plural in construction* [Latin, from plural of *fascis* bundle; akin to Latin *fascia*] (1598)
: a bundle of rods and among them an ax with projecting blade borne before ancient Roman magistrates as a badge of authority

fasces

fas·cia *1 & 3 are usually* 'fā-sh(ē-)ə, *2 is usually* 'fa-\ *noun, plural* **-ci·ae** \-shē-ˌē\ *or* **-cias** [Italian, from Latin, band, bandage; akin to Middle Irish *basc* necklace] (1563)
1 : a flat usually horizontal member of a building having the form of a flat band or broad fillet: as **a :** a flat piece used as a molding **b :** a horizontal piece (as a board) covering the joint between the top of a wall and the projecting eaves — called also *fascia board* **c :** a nameplate over the front of a shop
2 : a sheet of connective tissue covering or binding together body structures; *also* **:** tissue of this character
3 *or* **fa·cia** \'fā-sh(ē-)ə\ *British* **:** the dashboard of an automobile
— **fas·cial** \'fa-sh(ē-)əl\ *adjective*
fas·ci·at·ed \'fa-shē-ˌā-təd\ *adjective* (circa 1835)
1 : exhibiting fasciation
2 : arranged in fascicles
fas·ci·a·tion \ˌfa-s(h)ē-'ā-shən\ *noun* (1677)
: a malformation of plant stems commonly manifested as enlargement and flattening as if several stems were fused
fas·ci·cle \'fa-si-kəl\ *noun* [Latin *fasciculus,* diminutive of *fascis*] (15th century)
1 : a small or slender bundle (as of pine needles or nerve fibers)
2 : one of the divisions of a book published in parts
— **fas·ci·cled** \ kəld\ *adjective*
fas·cic·u·lar \fə-'si-kyə-lər, fa-\ *adjective* (1816)
: of, relating to, or consisting of fascicles or fasciculi
— **fas·cic·u·lar·ly** *adverb*
fas·cic·u·late \-lət\ *also* **fas·cic·u·lat·ed** \-ˌlā-təd\ *adjective* (1794)
: FASCICULAR
fas·cic·u·la·tion \fə-ˌsi-kyə-'lā-shən, fa-\ *noun* (1938)
: muscular twitching involving the simultaneous contraction of contiguous groups of muscle fibers
fas·ci·cule \'fa-si-ˌkyü(ə)l\ *noun* [French, from Latin *fasciculus*] (1880)
: FASCICLE 2
fas·cic·u·lus \fə-'si-kyə-ləs, fa-\ *noun, plural* **-li** \-ˌlī\ [New Latin, from Latin] (1713)
1 : a slender bundle of anatomical fibers
2 : FASCICLE 2
fas·ci·nate \'fa-sᵊn-ˌāt\ *verb* **fas·ci·nat·ed; fas·ci·nat·ing** \'fas-ˌnā-tiŋ, 'fa-sᵊn-ˌā-\ [Latin *fascinatus,* past participle of *fascinare,* from *fascinum* evil spell] (1598)
transitive verb
1 *obsolete* **:** BEWITCH
2 a : to transfix and hold spellbound by an irresistible power ⟨believed that the serpent could *fascinate* its prey⟩ **b :** to command the interest of **:** ALLURE ⟨was *fascinated* by carnivals⟩
intransitive verb
: to be irresistibly attractive ⟨the novel's flamboyant cover *fascinates*⟩ ◆
synonym see ATTRACT
fas·ci·nat·ing *adjective* (1648)
: extremely interesting or charming **:** CAPTIVATING
— **fas·ci·nat·ing·ly** \-ˌā-tiŋ-lē\ *adverb*
fas·ci·na·tion \ˌfa-sᵊn-'ā-shən\ *noun* (1605)
1 a : the quality or power of fascinating **b :** something fascinating
2 : the state of being fascinated
fas·ci·na·tor \'fa-sᵊn-ˌā-tər\ *noun* (1750)
1 : one that fascinates
2 : a woman's lightweight head scarf usually of crochet or lace
fas·cine \fa-'sēn, fə-\ *noun* [French, from Latin *fascina,* from *fascis*] (circa 1688)
: a long bundle of sticks of wood bound together and used for such purposes as filling ditches and making revetments for riverbanks
fa·sci·o·li·a·sis \fə-ˌsē-ə-'lī-ə-səs, -ˌsī-\ *noun, plural* **-a·ses** \-ˌsēz\ [New Latin, from *Fasciola,* genus of flukes + *-iasis*] (1890)
: infestation with or disease caused by liver flukes (genus *Fasciola*)
fas·cism \'fa-ˌshi-zəm *also* 'fa-ˌsi-\ *noun* [Italian *fascismo,* from *fascio* bundle, fasces, group, from Latin *fascis* bundle & *fasces* fasces] (1921)
1 *often capitalized* **:** a political philosophy, movement, or regime (as that of the Fascisti) that exalts nation and often race above the individual and that stands for a centralized autocratic government headed by a dictatorial leader, severe economic and social regimentation, and forcible suppression of opposition
2 : a tendency toward or actual exercise of strong autocratic or dictatorial control ⟨early instances of army *fascism* and brutality —J. W. Aldridge⟩ ◆
— **fas·cist** \-shist *also* -sist\ *noun or adjective, often capitalized*
— **fas·cis·tic** \fa-'shis-tik *also* -'sis-\ *adjective, often capitalized*
— **fas·cis·ti·cal·ly** \-ti-k(ə)lē\ *adverb, often capitalized*
Fa·sci·sta \fä-'shē-(ˌ)stä\ *noun, plural* **-sti** \-(ˌ)stē\ [Italian, from *fascio*] (1921)
: a member of an Italian political organization under Mussolini governing Italy 1922–1943 according to the principles of fascism
fash \'fash\ *verb* [Middle French *fascher,* from (assumed) Vulgar Latin *fastidiare* to disgust, from Latin *fastidium* disgust — more at FASTIDIOUS] (1533)
chiefly Scottish **:** VEX
— **fash** *noun, chiefly Scottish*
¹fash·ion \'fa-shᵊn\ *noun* [Middle English *facioun, fasoun* shape, manner, from Middle French *façon,* from Latin *faction-, factio* act of making, faction, from *facere* to make — more at DO] (14th century)
1 a : the make or form of something **b** *archaic* **:** KIND, SORT
2 a : a distinctive or peculiar and often habitual manner or way ⟨he will, after his sour *fashion,* tell you —Shakespeare⟩ **b :** mode of action or operation ⟨assembled in an orderly *fashion*⟩
3 a : a prevailing custom, usage, or style **b** (1) **:** the prevailing style (as in dress) during a particular time (2) **:** a garment in such a style ⟨always wears the latest *fashions*⟩ **c :** social standing or prominence especially as signalized by dress or conduct ⟨men and women of *fashion*⟩ ☆
— **after a fashion :** in an approximate or rough way ⟨became an artist *after a fashion*⟩
²fashion *transitive verb* **fash·ioned; fash·ion·ing** \'fash-niŋ, 'fa-shᵊn-iŋ\ (15th century)
1 a : to give shape or form to **:** MOLD **b :** ALTER, TRANSFORM **c :** to mold into a particular character by influencing or training **d :** to make or construct usually with the use of imagination and ingenuity ⟨*fashion* a lamp from an old churn⟩
2 : FIT, ADAPT
3 *obsolete* **:** CONTRIVE
— **fash·ion·er** \'fash-nər, 'fa-shᵊn-ər\ *noun*

☆ SYNONYMS
Fashion, style, mode, vogue, fad, rage, craze mean the usage accepted by those who want to be up-to-date. FASHION is the most general term and applies to any way of dressing, behaving, writing, or performing that is favored at any one time or place ⟨the current *fashion*⟩. STYLE often implies a distinctive fashion adopted by people of taste ⟨a media baron used to traveling in *style*⟩. MODE suggests the fashion of the moment among those anxious to appear elegant and sophisticated ⟨slim bodies are the *mode* at this resort⟩. VOGUE stresses the wide acceptance of a fashion ⟨short skirts are back in *vogue*⟩. FAD suggests caprice in taking up or in dropping a fashion ⟨last year's *fad* is over⟩. RAGE and CRAZE stress intense enthusiasm in adopting a fad ⟨Cajun food was the *rage* nearly everywhere for a time⟩ ⟨crossword puzzles once seemed just a passing *craze* but have lasted⟩. See in addition METHOD.

◇ WORD HISTORY
fascinate The word *fascinate* has its roots in a world where witchcraft and enchantment were taken much more seriously than they are in our time. Synonyms such as *bewitch* and *charm* have a magical side that is still apparent enough from the literal meanings of the words themselves, but this aspect of *fascinate* is clear only if we look at its etymology. In Latin the verb *fascinare* meant "to cast a spell on, bewitch" and was a derivative of a noun *fascinum.* The original meaning of *fascinum* seems to have been "bewitchment," though this usage is very sparsely attested. More frequently *fascinum* referred to a sort of counter charm, an amulet worn around the neck to avert the evil eye. *Fascinum* is similar in sound to Greek *baskanos* "malicious" (with derivatives such as *baskanein* "to avert the evil eye" and *baskanion* "charm, amulet"), but the ultimate source of both words is uncertain.

fascism The words *fascism* and *fascist* have long been associated with the Fascisti of Benito Mussolini and the bundle of rods called *fasces,* which the Fascisti adopted as a symbol of authoritarianism. However, Mussolini did not introduce the word *fascista* when his black-shirted organization, the *Fasci di combattimento,* was organized in 1919, nor did the fasces have any direct connection with the origin of *fascista.* In Italian, the word *fascio* (plural *fasci*) means literally "bundle," and figuratively "group." From at least 1872 *fascio* was used in the names of labor and agrarian unions, and in October 1914 a political coalition was formed called the *Fascio rivoluzionario d'azione internazionalista* ("revolutionary group for international action"), which advocated Italian participation in World War I on the side of the Allies. Members of this group were first called *fascisti* in January 1915. Although Mussolini was closely associated with this interventionist movement, it had no direct link with the postwar *Fasci di combattimento,* and in 1919 the word *fascista* was already in political circulation. Of course, it is to the Fascisti in their 1919 incarnation—who seized power in Italy three years later—that we owe the customary meanings of our *fascism* and *fascist.*

[1]fash•ion•able \'fash-nə-bəl, 'fa-shᵊn-ə-\ *adjective* (1606)
1 : conforming to the custom, fashion, or established mode
2 : of or relating to the world of fashion
— fash•ion•abil•i•ty \ˌfash-nə-'bi-lə-tē, ˌfa-shᵊn-ə-\ *noun*
— fash•ion•able•ness \'fash-nə-bəl-nəs, 'fa-shə-\ *noun*
— fash•ion•ably \-blē\ *adverb*

[2]fashionable *noun* (circa 1800)
: a fashionable person

fash•ion•mon•ger \'fa-shᵊn-ˌmäŋ-gər, -ˌməŋ-\ *noun* (1599)
: one that studies, imitates, or sets the fashion

fashion plate *noun* (1851)
1 : an illustration of a clothing style
2 : a person who dresses in the latest fashions

[1]fast \'fast\ *adjective* [Middle English, from Old English *fæst;* akin to Old High German *festi* firm, Old Norse *fastr,* Armenian *hast*] (before 12th century)
1 a : firmly fixed ⟨roots *fast* in the ground⟩ **b :** tightly shut ⟨the drawers were *fast*⟩ **c :** adhering firmly **d :** not easily freed **:** STUCK ⟨a ball *fast* in the mouth of the cannon⟩ **e :** STABLE ⟨movable items were made *fast* to the deck⟩
2 : firmly loyal ⟨became *fast* friends⟩
3 a : characterized by quick motion, operation, or effect: (1) **:** moving or able to move rapidly **:** SWIFT (2) **:** taking a comparatively short time (3) **:** imparting quickness of motion ⟨a *fast* bowler⟩ (4) **:** accomplished quickly (5) **:** agile of mind; *especially* **:** quick to learn ⟨a class for *fast* students⟩ **b :** conducive to rapidity of play or action **c** (1) *of a timepiece or weighing device* **:** indicating in advance of what is correct (2) **:** according to or being daylight saving time **d :** contributing to a shortening of exposure time ⟨*fast* film⟩ **e :** acquired with unusually little effort and often by shady or dishonest methods ⟨had a keen eye for a *fast* buck —R. A. Keith⟩
4 a : securely attached ⟨a rope *fast* to the wharf⟩ **b :** TENACIOUS ⟨a *fast* hold on her purse⟩
5 a *archaic* **:** sound asleep **b** *of sleep* **:** not easily disturbed
6 : not fading or changing color readily
7 a : WILD ⟨a pretty *fast* crowd⟩ **b :** sexually promiscuous
8 : resistant to change (as from destructive action or fading) ⟨*fast* dyes⟩ — often used in combination ⟨sun*fast*⟩ ⟨acid-*fast* bacteria⟩ ☆

[2]fast *adverb* (before 12th century)
1 : in a firm or fixed manner ⟨stuck *fast*⟩
2 : in a sound manner **:** DEEPLY ⟨*fast* asleep⟩
3 a : in a rapid manner **:** QUICKLY **b :** in quick succession ⟨kaleidoscopic impressions that come so thick and *fast* —M. B. Tucker⟩
4 : in a reckless or dissipated manner
5 : ahead of a correct time or schedule
6 *archaic* **:** CLOSE, NEAR

[3]fast *intransitive verb* [Middle English, from Old English *fæstan*] (before 12th century)
1 : to abstain from food
2 : to eat sparingly or abstain from some foods

[4]fast *noun* (before 12th century)
1 : the practice of fasting
2 : a time of fasting

[5]fast *noun* [alteration of Middle English *fest,* from Old Norse *festr* rope, mooring cable, from *fastr* firm] (15th century)
: something that fastens (as a mooring line) or holds a fastening

fast and loose *adverb* (1557)
1 : in a reckless or irresponsible manner ⟨played *fast and loose* with the public purse strings —Paul Stuewe⟩
2 : in a craftily deceitful way ⟨manipulated evidence . . . and played *fast and loose* with the truth —C. V. Woodward⟩

fast•back \'fas(t)-ˌbak\ *noun* (1954)
: an automobile with a roof having a long curving downward slope to the rear; *also* **:** the back of such an automobile

fast•ball \'fas(t)-ˌbȯl\ *noun* (1912)
: a baseball pitch thrown at full speed and often rising slightly as it nears the plate
— fast•ball•er \-ˌbȯ-lər\ *noun*

fast break *noun* (circa 1949)
: a quick offensive drive toward a goal (as in basketball) in an attempt to score before the opponent's defense is set up
— fast–break *intransitive verb*

fas•ten \'fa-sᵊn\ *verb* **fas•tened; fas•ten•ing** \'fas-niŋ, 'fa-sᵊn-iŋ\ [Middle English *fastnen,* from Old English *fæstnian* to make fast; akin to Old High German *festinōn* to make fast, Old English *fæst* fast] (before 12th century)
transitive verb
1 a : to attach especially by pinning, tying, or nailing **b :** to make fast and secure **c :** to fix firmly or securely **d :** to secure against opening
2 : to fix or set steadily ⟨*fastened* her attention on the main problem⟩
3 : to take a firm grip with ⟨the dog *fastened* its teeth in the shoe⟩
4 a : to attach (oneself) persistently and usually objectionably **b :** IMPOSE ⟨*fastened* the blame on the wrong person⟩
intransitive verb
1 : to become fast or fixed
2 a : to take a firm grip or hold **b :** to focus attention ☆
— fas•ten•er \'fas-nər, 'fa-sᵊn-ər\ *noun*

fas•ten•ing *noun* (12th century)
: something that fastens **:** FASTENER

fast–food \'fas(t)-ˌfüd\ *adjective* (1951)
1 : of, relating to, or specializing in food that can be prepared and served quickly ⟨a *fast-food* restaurant⟩
2 : designed for ready availability, use, or consumption and with little consideration given to quality or significance ⟨*fast-food* TV programming⟩
— fast food *noun*

[1]fast–for•ward \ˌfas(t)-'fȯr-wərd\ *noun, often attributive* (1948)
1 : a function of a tape player by which the tape is advanced at a higher speed than when it is playing normally
2 : a state or an instance of rapid advancement ⟨put her career in *fast-forward*⟩

[2]fast–forward (1974)
transitive verb
1 : to advance (a magnetic tape) using the fast-forward of a tape player
2 : to bypass (as a commercial) by fast-forwarding
intransitive verb
1 : to advance a magnetic tape using the fast-forward
2 : to proceed rapidly forward especially in time ⟨*fast-forward* to the future⟩

fas•tid•i•ous \fa-'sti-dē-əs, fə-\ *adjective* [Middle English, from Latin *fastidiosus,* from *fastidium* disgust, probably from *fastus* arrogance (probably akin to Latin *fastigium* top) + *taedium* irksomeness — more at TEDIUM] (15th century)
1 *archaic* **:** SCORNFUL
2 a : having high and often capricious standards **:** difficult to please ⟨critics . . . so *fastidious* that they can talk only to a small circle of initiates —Granville Hicks⟩ **b :** showing or demanding excessive delicacy or care **c :** reflecting a meticulous, sensitive, or demanding attitude ⟨*fastidious* workmanship⟩
3 : having complex nutritional requirements ⟨*fastidious* microorganisms⟩
— fas•tid•i•ous•ly *adverb*
— fas•tid•i•ous•ness *noun*

fas•ti•gi•ate \fa-'sti-jē-ət\ *adjective* [probably from (assumed) New Latin *fastigiatus,* from Latin *fastigium* — more at BRISTLE] (1662)
: narrowing toward the top; *especially* **:** having upright usually clustered branches ⟨*fastigiate* trees⟩

fast lane *noun* (1966)
1 : a traffic lane intended for vehicles traveling at higher speeds
2 : a way of life marked by a fast pace and usually the pursuit of immediate gratification
3 : FAST TRACK
— fast–lane *adjective*

fast•ness \'fas(t)-nəs\ *noun* (before 12th century)
1 : the quality or state of being fast: as **a :** the quality or state of being fixed **b :** the quality or state of being swift **c :** colorfast quality **d :** resistance (as of an organism) to the action of a usually toxic substance
2 a : a fortified or secure place **b :** a remote and secluded place ⟨vacationed in their mountain *fastness*⟩

Fast of Esther (1887)
: a Jewish fast day observed the day before Purim in commemoration of a fast proclaimed by Queen Esther

fast–talk \'fas(t)-ˌtȯk\ *transitive verb* (1946)
: to influence or persuade by fluent, facile, and usually deceptive or tricky talk ⟨*fast-talked* him into buying a lemon⟩

[1]fast–track \'fas(t)-ˌtrak\ *adjective* (1967)
1 : of, relating to, or moving along a fast track ⟨*fast-track* executives⟩
2 : of, relating to, or being a construction procedure in which work on a building begins before designs are completed

[2]fast–track *transitive verb* (1971)
: to speed up the processing, production, or construction of in order to meet a goal
— fast–track•er *noun*

fast track *noun* (1980)
: a course leading to rapid advancement or success

fast–twitch \'fas(t)-ˌtwich\ *adjective* (1970)
: of, relating to, or being muscle fiber that contracts quickly especially during brief high-intensity physical activity requiring strength — compare SLOW-TWITCH

fas•tu•ous \'fas-chə-wəs\ *adjective* [Latin *fastuosus,* from *fastus* arrogance] (1638)
1 : HAUGHTY, ARROGANT ⟨a *fastuous* air of finality —Carl Van Vechten⟩

☆ SYNONYMS

Fast, rapid, swift, fleet, quick, speedy, hasty, expeditious mean moving, proceeding, or acting with celerity. FAST and RAPID are very close in meaning, but FAST applies particularly to the thing that moves ⟨*fast* horses⟩ and RAPID to the movement itself ⟨*rapid* current⟩. SWIFT suggests great rapidity coupled with ease of movement ⟨returned the ball with one *swift* stroke⟩. FLEET adds the implication of lightness and nimbleness ⟨*fleet* runners⟩. QUICK suggests promptness and the taking of little time ⟨a *quick* wit⟩. SPEEDY implies quickness of successful accomplishment ⟨*speedy* delivery of mail⟩ and may also suggest unusual velocity. HASTY suggests hurry and precipitousness and often connotes carelessness ⟨a *hasty* inspection⟩. EXPEDITIOUS suggests efficiency together with rapidity of accomplishment ⟨the *expeditious* handling of an order⟩.

Fasten, fix, attach, affix mean to make something stay firmly in place. FASTEN implies an action such as tying, buttoning, nailing, locking, or otherwise securing ⟨*fasten* the reins to a post⟩. FIX usually implies a driving in, implanting, or embedding ⟨*fixed* the stake in the ground⟩. ATTACH suggests a connecting or uniting by a bond, link, or tie in order to keep things together ⟨*attach* the W-2 form here⟩. AFFIX implies an imposing of one thing on another by gluing, impressing, or nailing ⟨*affix* your address label here⟩.

2 : OSTENTATIOUS, SHOWY ⟨disdained *fastuous* ceremonies⟩

¹fat \'fat\ *adjective* **fat·ter; fat·test** [Middle English, from Old English *fǣtt,* past participle of *fǣtan* to cram; akin to Old High German *feizit* fat] (before 12th century)
1 : notable for having an unusual amount of fat: **a :** PLUMP **b :** OBESE **c** *of a meat animal* **:** fattened for market **d** *of food* **:** OILY, GREASY
2 a : well filled out **:** THICK, BIG ⟨a *fat* book⟩ **b :** full in tone and quality **:** RICH ⟨a gorgeous *fat* bass voice —*Irish Digest*⟩ **c :** well stocked ⟨a *fat* larder⟩ **d :** PROSPEROUS, WEALTHY ⟨grew *fat* on the war —*Time*⟩ **e :** being substantial and impressive ⟨a *fat* bank account⟩
3 a : richly rewarding or profitable ⟨a *fat* part in a movie⟩ ⟨a *fat* contract⟩ **b :** practically nonexistent ⟨a *fat* chance⟩
4 : PRODUCTIVE, FERTILE ⟨a *fat* year for crops⟩
5 : STUPID, FOOLISH
6 : being swollen ⟨got a *fat* lip from the fight⟩
7 *of a baseball pitch* **:** easy to hit
— fat·ness *noun*

²fat *transitive verb* **fat·ted; fat·ting** (before 12th century)
: to make fat **:** FATTEN

³fat *noun* (14th century)
1 : animal tissue consisting chiefly of cells distended with greasy or oily matter
2 a : oily or greasy matter making up the bulk of adipose tissue and often abundant in seeds **b :** any of numerous compounds of carbon, hydrogen, and oxygen that are glycerides of fatty acids, are the chief constituents of plant and animal fat, are a major class of energy-rich food, and are soluble in organic solvents but not in water **c :** a solid or semisolid fat as distinguished from an oil
3 : the best or richest part
4 : OBESITY
5 : something in excess **:** SUPERFLUITY ⟨trim the *fat* from the news operation —Ray Olson⟩
— fat·less \-ləs\ *adjective*

fa·tal \'fā-tᵊl\ *adjective* [Middle English, from Middle French & Latin; Middle French, from Latin *fatalis,* from *fatum*] (14th century)
1 *obsolete* **:** FATED
2 : FATEFUL ⟨a *fatal* hour⟩
3 a : of or relating to fate **b :** resembling fate in proceeding according to a fixed sequence **c :** determining one's fate
4 a : causing death **b :** bringing ruin ▢
synonym see DEADLY

fa·tal·ism \-ˌi-zəm\ *noun* (1678)
: a doctrine that events are fixed in advance so that human beings are powerless to change them; *also* **:** a belief in or attitude determined by this doctrine
— fa·tal·ist \-ist\ *noun*
— fa·tal·is·tic \ˌfā-tᵊl-'is-tik\ *adjective*
— fa·tal·is·ti·cal·ly \-ti-k(ə-)lē\ *adverb*

fa·tal·i·ty \fā-'ta-lə-tē, fə-\ *noun, plural* **-ties** [Middle English, from Middle French *fatalité,* from Late Latin *fatalitat-, fatalitas,* from Latin *fatalis*] (15th century)
1 a : the quality or state of causing death or destruction **b :** the quality or condition of being destined for disaster
2 : something established by fate
3 a : FATE 1 **b :** FATALISM
4 : the agent or agency of fate
5 a : death resulting from a disaster **b :** one that experiences a fatal outcome

fa·tal·ly \'fā-tᵊl-ē\ *adverb* (15th century)
1 : in a way determined by fate
2 : in a manner suggesting fate or an act of fate: as **a :** in a manner resulting in death **:** MORTALLY ⟨*fatally* wounded⟩ **b :** beyond repair **:** IRREVOCABLY **c :** in a manner resulting in ruin or evil ⟨it is *fatally* easy to pass off our prejudices as our opinions —W. F. Hambly⟩ **d :** in a manner that cannot be easily resisted ⟨thinks she is *fatally* attractive —J. W. Krutch⟩

fa·ta mor·ga·na \ˌfä-tə-mȯr-'gä-nə, -'ga-\ *noun* [Italian, literally, Morgan le Fay (sorceress of Arthurian legend)] (1818)
: MIRAGE

fat·back \'fat-ˌbak\ *noun* (1903)
: the strip of fat from the back of a hog carcass usually cured by drying and salting

fat body *noun* (1869)
: a fatty tissue especially of nearly mature insect larvae that serves as a food reserve

fat cat *noun* (1928)
1 a : a wealthy contributor to a political campaign fund **b :** a wealthy and privileged person **c :** BIG SHOT
2 : a lethargic complacent person
— fat–cat \'fat-ˌkat\ *adjective*

fat cell *noun* (1845)
: one of the fat-laden cells making up adipose tissue

fat depot *noun* (1946)
: ADIPOSE TISSUE

¹fate \'fāt\ *noun* [Middle English, from Middle French or Latin; Middle French, from Latin *fatum,* literally, what has been spoken, from neuter of *fatus,* past participle of *fari* to speak — more at BAN] (14th century)
1 : the principle or determining cause or will by which things in general are believed to come to be as they are or events to happen as they do **:** DESTINY
2 a : an inevitable and often adverse outcome, condition, or end **b :** DISASTER; *especially* **:** DEATH
3 a : final outcome **b :** the expected result of normal development ⟨prospective *fate* of embryonic cells⟩
4 *plural, capitalized* **:** the three goddesses who determine the course of human life in classical mythology ☆

²fate *transitive verb* **fat·ed; fat·ing** (1601)
: DESTINE; *also* **:** DOOM

fat·ed *adjective* (circa 1606)
: decreed, controlled, or marked by fate

fate·ful \'fāt-fəl\ *adjective* (circa 1720)
1 : having a quality of ominous prophecy ⟨a *fateful* remark⟩
2 a : involving momentous consequences **:** DECISIVE ⟨made his *fateful* decision to declare war —W. L. Shirer⟩ **b :** DEADLY, CATASTROPHIC
3 : controlled by fate **:** FOREORDAINED
synonym see OMINOUS
usage see FATAL
— fate·ful·ly \-fə-lē\ *adverb*
— fate·ful·ness *noun*

fat farm *noun* (1969)
: a health spa that specializes in weight reduction

fat·head \'fat-ˌhed\ *noun* (1842)
: a stupid person
— fat·head·ed \-'he-dəd\ *adjective*
— fat·head·ed·ly *adverb*
— fat·head·ed·ness *noun*

¹fa·ther \'fä-t͟hər, 'fȧ-\ *noun* [Middle English *fader,* from Old English *fæder;* akin to Old High German *fater* father, Latin *pater,* Greek *patēr*] (before 12th century)
1 a : a man who has begotten a child; *also* **:** SIRE 3 **b** *capitalized* (1) **:** GOD 1 (2) **:** the first person of the Trinity
2 : FOREFATHER
3 a : one related to another in a way suggesting that of father to child **b :** an old man — used as a respectful form of address
4 *often capitalized* **:** a pre-Scholastic Christian writer accepted by the church as an authoritative witness to its teaching and practice — called also *church father*
5 a : one that originates or institutes ⟨the *father* of modern science⟩ **b :** SOURCE ⟨the sun, the *father* of warmth and light —Lena M. Whitney⟩ **c :** PROTOTYPE
6 : a priest of the regular clergy; *broadly* **:** PRIEST — used especially as a title
7 : one of the leading men (as of a city) — usually used in plural
— fa·ther·hood \-ˌhu̇d\ *noun*
— fa·ther·less \-ləs\ *adjective*
— fa·ther·like \-ˌlīk\ *adjective or adverb*

²father *verb* **fa·thered; fa·ther·ing** \'fät͟h-riŋ, 'fȧt͟h-; 'fä-t͟hə-, 'fȧ-\ (15th century)
transitive verb
1 a : BEGET **b :** to make oneself the founder, producer, or author of ⟨*fathered* the improvement plan⟩ **c :** to accept responsibility for
2 a : to fix the paternity or origin of **b :** to place responsibility for the origin or cause of ⟨collected gossip and *fathered* it on responsible men —J. A. Williamson⟩
3 : FOIST, IMPOSE
intransitive verb
: to care for or look after someone as a father might

Father Christmas *noun*
British **:** SANTA CLAUS

father figure *noun* (1934)
: one often of particular power or influence who serves as an emotional substitute for a father

father image *noun* (1937)
: an idealization of one's father often projected onto someone to whom one looks for guidance and protection

father–in–law \'fä-t͟hə-rən-ˌlȯ, -t͟hərn-ˌlȯ, 'fȧ-\ *noun, plural* **fa·thers–in–law** \-t͟hər-zən-\ (14th century)

☆ SYNONYMS
Fate, destiny, lot, portion, doom mean a predetermined state or end. FATE implies an inevitable and usually an adverse outcome ⟨the *fate* of the submarine is unknown⟩. DESTINY implies something foreordained and often suggests a great or noble course or end ⟨the country's *destiny* to be a model of liberty to the world⟩. LOT and PORTION imply a distribution by fate or destiny, LOT suggesting blind chance ⟨it was her *lot* to die childless⟩, PORTION implying the apportioning of good and evil ⟨remorse was his daily *portion*⟩. DOOM distinctly implies a grim or calamitous fate ⟨if the rebellion fails, his *doom* is certain⟩.

▢ USAGE
fatal *Fatal* and *fateful* have long shared a meaning ⟨a *fatal* or *fateful* day⟩ and in the second half of the 20th century a curious opinion has grown up that *fatal* ought to be restricted to its fourth sense and its second sense abandoned to *fateful.* The purpose of this newly imposed distinction is presumably to tidy up the language and to reduce confusion. But these uses seem plain enough ⟨at Pearl Harbor on the *fatal* day —David McCullough⟩ ⟨weeping when she thinks of the *fatal* moment when they first met —Samuel Eliot Morison⟩. Here, as in most similar uses of *fatal,* the direction of the portent is toward evil or the undesirable. *Fateful* may be used in a less ominous way ⟨the day when the *fateful* letter from the college admission office is due —James B. Conant⟩. Sometimes *fatal* is similarly used ⟨when the *fatal* call comes over the public address system—"Jones, city desk" Jones can feel his colleagues thinking as he walks past them, "I hope he gets a lousy assignment . . . " —Robert Darnton⟩. *Fatal* doesn't share its sense "causing death or destruction" ⟨a *fatal* shooting⟩. This sense is also used frequently with weakened force ⟨tediousness is the most *fatal* of all faults —Samuel Johnson⟩ ⟨to assume that the listeners know all about it already . . . is a *fatal* mistake —George Bernard Shaw⟩.

1 : the father of one's spouse
2 *archaic* **:** STEPFATHER

fa·ther·land \'fä-t͟hər-ˌland, 'fȧ-\ *noun* (12th century)
1 : the native land or country of one's father or ancestors
2 : one's native land or country

fa·ther·ly \'fä-t͟hər-lē, 'fȧ-\ *adjective* (15th century)
1 : of, relating to, or befitting a father ⟨*fatherly* responsibilities⟩
2 : resembling a father (as in affection or care) ⟨a *fatherly* old man⟩
— fa·ther·li·ness \-lē-nəs\ *noun*
— fatherly *adverb*

Father's Day *noun* (1927)
: the third Sunday in June appointed for the honoring of fathers

[1]fath·om \'fa-t͟həm\ *noun* [Middle English *fadme,* from Old English *fæthm* outstretched arms, length of the outstretched arms; akin to Old Norse *fathmr* fathom, Latin *patēre* to be open, *pandere* to spread out, Greek *petannynai*] (before 12th century)
1 : a unit of length equal to six feet (1.83 meters) used especially for measuring the depth of water
2 : COMPREHENSION

[2]fathom (1607)
intransitive verb
1 : PROBE
2 : to take soundings
transitive verb
1 : to measure by a sounding line
2 : to penetrate and come to understand ⟨couldn't *fathom* the problem⟩
— fath·om·able \'fa-t͟hə-mə-bəl\ *adjective*

Fa·thom·e·ter \fa-'t͟hä-mə-tər, 'fa-t͟hə-ˌmē-\ *trademark*
— used for a sonic depth finder

fath·om·less \'fa-t͟həm-ləs\ *adjective* (1638)
: incapable of being fathomed
— fath·om·less·ly *adverb*
— fath·om·less·ness *noun*

fa·tid·ic \fā-'ti-dik, fə-\ *or* **fa·tid·i·cal** \-di-kəl\ *adjective* [Latin *fatidicus,* from *fatum* fate + *dicere* to say — more at DICTION] (1607)
: of or relating to prophecy

fa·ti·ga·ble \fə-'tē-gə-bəl, 'fa-ti-gə-\ *adjective* (1608)
: susceptible to fatigue
— fa·ti·ga·bil·i·ty \fə-ˌtē-gə-'bi-lə-tē, ˌfa-ti-gə-\ *noun*

[1]fa·tigue \fə-'tēg\ *noun* [French, from Middle French, from *fatiguer* to fatigue, from Latin *fatigare;* akin to Latin *affatim* sufficiently] (1669)
1 a : LABOR **b :** manual or menial work performed by military personnel **c** *plural* **:** the uniform or work clothing worn on fatigue and in the field
2 a : weariness or exhaustion from labor, exertion, or stress **b :** the temporary loss of power to respond induced in a sensory receptor or motor end organ by continued stimulation
3 : the tendency of a material to break under repeated stress

[2]fatigue *verb* **fa·tigued; fa·tigu·ing** (1693)
transitive verb
1 : to weary with labor or exertion
2 : to induce a condition of fatigue in
intransitive verb
: to suffer fatigue
synonym see TIRE
— fa·tigu·ing·ly \-'tē-giŋ-lē\ *adverb*

[3]fatigue *adjective* (1774)
1 : consisting of, done, or used in fatigue ⟨*fatigue* detail⟩
2 : belonging to fatigues ⟨a *fatigue* cap⟩

fat·ling \'fat-liŋ\ *noun* (circa 1534)
: a young animal fattened for slaughter

fat·ly *adverb* (15th century)
1 : RICHLY
2 : in the manner of one that is fat ⟨waddled *fatly*⟩
3 : in a smug manner **:** COMPLACENTLY ⟨snickered *fatly*⟩

fats·hed·era \fats-'he-d(ə-)rə, 'fat-'se-\ *noun* [New Latin *Fatsia,* genus of shrubs + *Hedera,* genus of vines (from Latin, ivy)] (1948)
: a vigorous upright hybrid ornamental foliage plant (*Hedera helix* × *Aralia elata*) with glossy deeply lobed palmate leaves

fat·so \'fat-(ˌ)sō\ *noun, plural* **fat·soes** [probably from *Fats,* nickname for a fat person + *-o*] (1944)
: a fat person — often used disparagingly

fat·stock \-ˌstäk\ *noun* (1880)
chiefly British **:** livestock that is fat and ready for market

fat–tailed sheep \'fat-ˌtāld-\ *noun* (1842)
: a coarse-wooled mutton sheep that has great quantities of fat on each side of the tail bones

fat·ten \'fa-tᵊn\ *verb* **fat·tened; fat·ten·ing** \'fat-niŋ, 'fa-tᵊn-iŋ\ (1552)
transitive verb
1 a : to make fat, fleshy, or plump; *especially* **:** to feed (as a stock animal) for slaughter **b :** to make more substantial
2 : to make fertile
intransitive verb
: to become fat
— fat·ten·er \'fat-nər, 'fa-tᵊn-ər\ *noun*

fat·tish \'fa-tish\ *adjective* (14th century)
: somewhat fat

[1]fat·ty \'fa-tē\ *adjective* **fat·ti·er; -est** (14th century)
1 : containing fat especially in unusual amounts; *also* **:** unduly stout **:** CORPULENT
2 : GREASY
3 : derived from or chemically related to fat
— fat·ti·ness *noun*

[2]fatty *noun, plural* **fat·ties** (1797)
: one that is fat; *especially* **:** an overweight person

fatty acid *noun* (circa 1872)
1 : any of numerous saturated aliphatic monocarboxylic acids $C_nH_{2n+1}COOH$ (as acetic acid) including many that occur naturally usually in the form of esters in fats, waxes, and essential oils
2 : any of the saturated or unsaturated monocarboxylic acids (as palmitic acid) usually with an even number of carbon atoms that occur naturally in the form of glycerides in fats and fatty oils

fa·tu·ity \fə-'tü-ət-ē, fa-, -'chü-, -'tyü-\ *noun, plural* **-ities** [Middle French *fatuité* foolishness, from Latin *fatuitat-, fatuitas,* from *fatuus*] (1538)
1 a : something foolish or stupid **b :** STUPIDITY, FOOLISHNESS
2 *archaic* **:** IMBECILITY, DEMENTIA

fat·u·ous \'fa-chü-əs, -tyü-\ *adjective* [Latin *fatuus* foolish] (1633)
: complacently or inanely foolish **:** SILLY
synonym see SIMPLE
— fat·u·ous·ly *adverb*
— fat·u·ous·ness *noun*

fat·wa \'fət-wə\ *noun* [Arabic *fatwā*] (1947)
: a legal opinion or decree handed down by an Islamic religious leader

fat–wit·ted \'fat-ˌwi-təd\ *adjective* (1596)
: STUPID, IDIOTIC

fat·wood \'fat-ˌwu̇d\ *noun* (1904)
chiefly Southern **:** LIGHTWOOD

fau·bourg \fō-'bu̇r\ *noun* [Middle English *fabour,* from Middle French *fauxbourg,* alteration of *forsbourg,* from Old French *forsborc,* from *fors* outside + *borc* town — more at BOURG] (15th century)
1 : SUBURB; *especially* **:** a suburb of a French city
2 : a city quarter

fau·ces \'fȯ-ˌsēz\ *noun plural but singular or plural in construction* [Latin, plural, throat, fauces] (1541)
: the narrow passage from the mouth to the pharynx situated between the soft palate and the base of the tongue
— fau·cial \'fȯ-shəl\ *adjective*

fau·cet \'fȯ-sət, 'fäs-\ *noun* [Middle English, bung, faucet, from Middle French *fausset* bung, perhaps from *fausser* to damage, from Late Latin *falsare* to falsify, from Latin *falsus* false] (15th century)
: a fixture for drawing or regulating the flow of liquid especially from a pipe

faugh *a strong* p-*sound or lip trill; often read as* 'fȯ(k̲)\ *interjection* (1542)
— used to express contempt, disgust, or abhorrence

[1]fault \'fȯlt, *in poetry also* 'fȯt\ *noun* [Middle English *faute,* from Middle French, from (assumed) Vulgar Latin *fallita,* from feminine of *fallitus,* past participle of Latin *fallere* to deceive, disappoint] (14th century)
1 *obsolete* **:** LACK
2 a : WEAKNESS, FAILING; *especially* **:** a moral weakness less serious than a vice **b :** a physical or intellectual imperfection or impairment **:** DEFECT **c :** an error especially in service in a net or racket game

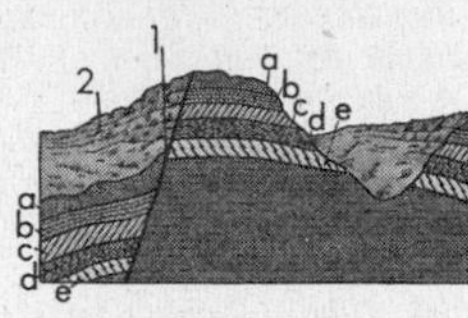

fault 5: *1* **fault with displaced strata** ***a, b, c, d, e; 2*** **scarp**

3 a : MISDEMEANOR **b :** MISTAKE
4 : responsibility for wrongdoing or failure ⟨the accident was the driver's *fault*⟩
5 : a fracture in the crust of a planet (as the earth) or moon accompanied by a displacement of one side of the fracture with respect to the other usually in a direction parallel to the fracture ☆
— at fault 1 : unable to find the scent and continue chase **2 :** open to blame **:** RESPONSIBLE ⟨couldn't determine who was really *at fault*⟩
— to a fault : to an excessive degree ⟨precise *to a fault*⟩

[2]fault (15th century)
intransitive verb
1 : to commit a fault **:** ERR
2 : to fracture so as to produce a geologic fault
transitive verb
1 : to find a fault in ⟨easy to praise this book and to *fault* it —H. G. Roepke⟩
2 : to produce a geologic fault in
3 : BLAME, CENSURE ⟨can't *fault* them for not coming⟩

fault·find·er \'fȯlt-ˌfīn-dər\ *noun* (1561)
: one given to faultfinding

[1]fault·find·ing \-diŋ\ *adjective* (1622)
: disposed to find fault **:** captiously critical
synonym see CRITICAL

[2]faultfinding *noun* (1626)
: petty, nagging, or unreasonable criticism

fault·less \'fȯlt-ləs\ *adjective* (14th century)
: having no fault **:** IRREPROACHABLE ⟨*faultless* workmanship⟩
— fault·less·ly *adverb*
— fault·less·ness *noun*

☆ SYNONYMS
Fault, failing, frailty, foible, vice mean an imperfection or weakness of character. FAULT implies a failure, not necessarily culpable, to reach some standard of perfection in disposition, action, or habit ⟨a writer of many virtues and few *faults*⟩. FAILING suggests a minor shortcoming in character ⟨being late is a *failing* of mine⟩. FRAILTY implies a general or chronic proneness to yield to temptation ⟨human *frailties*⟩. FOIBLE applies to a harmless or endearing weakness or idiosyncrasy ⟨an eccentric's charming *foibles*⟩. VICE can be a general term for any imperfection or weakness, but it often suggests violation of a moral code or the giving of offense to the moral sensibilities of others ⟨compulsive gambling was his *vice*⟩.

fault line *noun* (1869)
: something resembling a fault : SPLIT, RIFT ⟨a major conceptual *fault line* in foreign policy —Morton Kondracke⟩

faulty \'fȯl-tē\ *adjective* **fault·i·er; -est** (14th century)
: marked by fault or defect : IMPERFECT
— **fault·i·ly** \-tə-lē\ *adverb*
— **fault·i·ness** \-tē-nəs\ *noun*

faun \'fȯn, 'fän\ *noun* [Middle English, from Latin *faunus,* from *Faunus*] (14th century)
: a figure in Roman mythology similar to but gentler than the satyr

fau·na \'fȯ-nə, 'fä-\ *noun, plural* **faunas** *also* **fau·nae** \-ˌnē, -ˌnī\ [New Latin, from Latin *Fauna,* sister of Faunus] (1771)
: animal life; *especially* : the animals characteristic of a region, period, or special environment — compare FLORA
— **fau·nal** \-nᵊl\ *adjective*
— **fau·nal·ly** \-nᵊl-ē\ *adverb*

fau·nis·tic \fȯ-'nis-tik, fä-\ *adjective* (1881)
: of or relating to zoogeography : FAUNAL
— **fau·nis·ti·cal·ly** \-ti-k(ə-)lē\ *adverb*

Fau·nus \'fȯ-nəs, 'fä-\ *noun* [Latin]
: the Roman god of animals

Faust \'fau̇st\ *or* **Fau·stus** \'fau̇-stəs, 'fȯ-\ *noun* [German]
: a magician of German legend who enters into a compact with the devil

Faust·ian \'fau̇-stē-ən, 'fȯ-\ *adjective* (1876)
: of, relating to, resembling, or suggesting Faust; *especially* : made or done for present gain without regard for future cost or consequences ⟨a *Faustian* bargain⟩

faute de mieux \ˌfōt-də-'myə(r), -'myœ̄\ *adverb* [French] (1766)
: for lack of something better or more desirable ⟨sherry made him dopey but he drank it *faute de mieux* —F. T. Marsh⟩

fau·teuil \fō-'tœy; 'fō-ˌtil\ *noun, plural* **fauteuils** \fō-'tœy; -ˌtilz\ [French, from Old French *faudestuel,* of Germanic origin; akin to Old High German *faltistuol* folding chair — more at FALDSTOOL] (1744)
: ARMCHAIR; *especially* : an upholstered chair with open arms

¹fauve \'fōv\ *noun, often capitalized* [French, literally, wild animal, from *fauve* tawny, wild, from Old French *falve* tawny, of Germanic origin; akin to Old High German *falo* fallow — more at FALLOW] (1931)
: a painter practicing fauvism : FAUVIST

²fauve *adjective* (1945)
1 *often capitalized* : of or relating to the fauves
2 : vivid in color

fau·vism \'fō-ˌvi-zəm\ *noun, often capitalized* (1922)
: a movement in painting typified by the work of Matisse and characterized by vivid colors, free treatment of form, and a resulting vibrant and decorative effect
— **fau·vist** \-vist\ *noun, often capitalized*

faux \'fō\ *adjective* [French, false] (1975)
: IMITATION, ERSATZ ⟨*faux* marble⟩

faux pas \'fō-ˌpä, fō-'\ *noun, plural* **faux pas** \-ˌpä(z), -'pä(z)\ [French, literally, false step] (1676)
: BLUNDER; *especially* : a social blunder

fa·va \'fä-və\ *noun* [Italian, from Latin *faba* bean] (1928)
: BROAD BEAN — called also *fava bean*

fave \'fāv\ *noun* (1938)
: FAVORITE 1
— **fave** *adjective*

fa·ve·la *also* **fa·vel·la** \fə-'ve-lə\ *noun* [Brazilian Portuguese *favela,* perhaps from *Favela,* hill outside Rio de Janeiro] (1946)
: a settlement of jerry-built shacks lying on the outskirts of a Brazilian city

fa·vism \'fä-ˌvi-zəm\ *noun* (circa 1927)
: a hereditary condition especially of males of Mediterranean descent that involves a severe allergic reaction to the broad bean or its pollen which is characterized by hemolytic anemia, fever, and jaundice

fa·vo·ni·an \fə-'vō-nē-ən\ *adjective* [Latin *favonianus,* from *Favonius,* the west wind] (circa 1681)
: of or relating to the west wind : MILD

¹fa·vor \'fā-vər\ *noun* [Middle English, friendly regard, attractiveness, from Old French *favor* friendly regard, from Latin, from *favēre* to be favorable; perhaps akin to Old High German *gouma* attention, Old Church Slavonic *gověti* to revere] (14th century)
1 a (1) : friendly regard shown toward another especially by a superior (2) : approving consideration or attention : APPROBATION **b** : PARTIALITY **c** *archaic* : LENIENCY **d** *archaic* : PERMISSION **e** : POPULARITY
2 *archaic* **a** : APPEARANCE **b** (1) : FACE (2) : a facial feature
3 a : gracious kindness; *also* : an act of such kindness **b** *archaic* : AID, ASSISTANCE **c** *plural* : effort in one's behalf or interest : ATTENTION
4 a : a token of love (as a ribbon) usually worn conspicuously **b** : a small gift or decorative item given out at a party **c** : BADGE
5 a : a special privilege or right granted or conceded **b** : sexual privileges — usually used in plural
6 *archaic* : LETTER
7 : BEHALF, INTEREST
— **in favor of 1 a** : in accord or sympathy with **b** : for the acquittal of ⟨a verdict *in favor of* the accused⟩ **c** : in support of **2** : to the order of **3** : in order to choose : out of preference for ⟨turned down the scholarship *in favor of* a pro career⟩
— **in one's favor 1** : in one's good graces **2** : to one's advantage ⟨the odds were *in my favor*⟩
— **out of favor** : UNPOPULAR, DISLIKED ⟨was *out of favor* with his neighbors⟩

²favor *transitive verb* **fa·vored; fa·vor·ing** \'fā-v(ə-)riŋ\ (14th century)
1 a : to regard or treat with favor **b** (1) : to do a kindness for : OBLIGE (2) : ENDOW **c** : to treat gently or carefully ⟨*favored* his injured leg⟩
2 : to show partiality toward : PREFER
3 a : to give support or confirmation to : SUSTAIN **b** : to afford advantages for success to : FACILITATE ⟨good weather *favored* the outing⟩
4 : to bear a resemblance to ⟨he *favors* his father⟩
— **fa·vor·er** \'fā-vər-ər\ *noun*

fa·vor·able \'fā-v(ə-)rə-bəl, 'fā-vər-bəl\ *adjective* (14th century)
1 a : disposed to favor : PARTIAL **b** : expressing approval : COMMENDATORY **c** : giving a result that is in one's favor ⟨a *favorable* comparison⟩ **d** : AFFIRMATIVE
2 : winning approval : PLEASING
3 a : tending to promote or facilitate : ADVANTAGEOUS ⟨*favorable* wind⟩ **b** : marked by success ☆
— **fa·vor·able·ness** *noun*
— **fa·vor·ably** \-blē\ *adverb*

fa·vored \'fā-vərd\ *adjective* (15th century)
1 : having an appearance or features of a particular kind ⟨hard-*favored*⟩
2 : endowed with special advantages or gifts
3 : providing preferential treatment

¹fa·vor·ite \'fā-v(ə-)rət, 'fā-vərt, *chiefly dialect* 'fā-və-ˌrīt\ *noun* [Italian *favorito,* past participle of *favorire* to favor, from *favore* favor, from Latin *favor*] (1583)
1 : one that is treated or regarded with special favor or liking; *especially* : one unusually loved, trusted, or provided with favors by a person of high rank or authority
2 : a competitor judged most likely to win

²favorite *adjective* (1711)
: constituting a favorite; *especially* : markedly popular

favorite son *noun* (1788)
1 : one favored by the delegates of his state as presidential candidate at a national political convention
2 : a famous person who is popular with hometown people

fa·vor·it·ism \'fā-v(ə-)rə-ˌti-zəm, 'fā-vər-\ *noun* (1763)
1 : the showing of special favor : PARTIALITY
2 : the state or fact of being a favorite

fa·vour *chiefly British variant of* FAVOR

fa·vus \'fā-vəs\ *noun* [New Latin, from Latin, honeycomb] (circa 1543)
: a contagious skin disease caused by a fungus (as *Trichophyton schoenleinii*) and occurring in humans and many domestic animals and fowls

¹fawn \'fȯn, 'fän\ *intransitive verb* [Middle English *faunen,* from Old English *fagnian* to rejoice, from *fægen, fagan* glad — more at FAIN] (13th century)
1 : to show affection — used especially of a dog
2 : to court favor by a cringing or flattering manner ☆
— **fawn·er** *noun*
— **fawn·ing·ly** \'fȯ-niŋ-lē, 'fä-\ *adverb*

²fawn *noun* [Middle English *foun,* from Middle French *feon, faon* young of an animal, from (assumed) Vulgar Latin *feton-, feto,* from Latin *fetus* offspring — more at FETUS] (14th century)
1 : a young deer; *especially* : one still unweaned or retaining a distinctive baby coat
2 : KID 1
3 : a light grayish brown

fawn lily *noun* (circa 1894)
: DOGTOOTH VIOLET

fawny \'fȯ-nē, 'fä-\ *adjective* (1849)
: of a color approximating fawn

fax \'faks\ *noun* [by shortening & alteration] (1948)
1 : FACSIMILE 2
2 : a machine used to send or receive facsimile communications
3 : a facsimile communication
— **fax** *transitive verb*

¹fay \'fā\ *verb* [Middle English *feien,* from Old English *fēgan;* akin to Old High German *fuogen* to fit, Latin *pangere* to fasten — more at PACT] (before 12th century)

☆ SYNONYMS

Favorable, auspicious, propitious mean pointing toward a happy outcome. FAVORABLE implies that the persons involved are approving or helpful or that the circumstances are advantageous ⟨*favorable* weather conditions⟩. AUSPICIOUS applies to something taken as a sign or omen promising success before or at the start of an event ⟨an *auspicious* beginning⟩. PROPITIOUS may also apply to beginnings but often implies a continuing favorable condition ⟨a *propitious* time for starting a business⟩.

Fawn, toady, truckle, cringe, cower mean to behave abjectly before a superior. FAWN implies seeking favor by servile flattery or exaggerated attention ⟨waiters *fawning* over a celebrity⟩. TOADY suggests the attempt to ingratiate oneself by an abjectly menial or subservient attitude ⟨*toadying* to his boss⟩. TRUCKLE implies the subordination of oneself and one's desires or judgment to those of a superior ⟨*truckling* to a powerful lobbyist⟩. CRINGE suggests a bowing or shrinking in fear or servility ⟨a *cringing* sycophant⟩. COWER suggests a display of abject fear in the company of threatening or domineering people ⟨*cowering* before a bully⟩.

: to fit or join closely or tightly

²fay *noun* [Middle English *fai, fei,* from Old French *feid, fei* — more at FAITH] (13th century)
obsolete **:** FAITH

³fay *noun* [Middle English *faie,* from Middle French *feie, fee* — more at FAIRY] (14th century)
: FAIRY, ELF

⁴fay *adjective* (14th century)
: resembling an elf

⁵fay *noun* (1927)
: OFAY

faze \'fāz\ *transitive verb* **fazed; faz·ing** [alteration of *feeze* (to drive away, frighten), from Middle English *fesen,* from Old English *fēsian* to drive away] (1830)
: to disturb the composure of **:** DISCONCERT, DAUNT ⟨nothing *fazed* her⟩

F clef *noun* (1596)
: BASS CLEF

F distribution *noun* [Sir Ronald *F*isher (died 1962) English geneticist and statistician] (1947)
: a probability density function that is used especially in analysis of variance and is a function of the ratio of two independent random variables each of which has a chi-square distribution and is divided by its number of degrees of freedom

fe·al·ty \'fē(-ə)l-tē\ *noun, plural* **-ties** [alteration of Middle English *feute,* from Old French *feelté, fealté,* from Latin *fidelitat-, fidelitas* — more at FIDELITY] (14th century)
1 a : the fidelity of a vassal or feudal tenant to his lord **b :** the obligation of such fidelity
2 : intense fidelity
synonym see FIDELITY

¹fear \'fir\ (before 12th century)
transitive verb
1 *archaic* **:** FRIGHTEN
2 *archaic* **:** to feel fear in (oneself)
3 : to have a reverential awe of ⟨*fear* God⟩
4 : to be afraid of **:** expect with alarm
intransitive verb
: to be afraid or apprehensive
— fear·er *noun*

²fear *noun* [Middle English *fer,* from Old English *fǣr* sudden danger; akin to Latin *periculum* attempt, peril, Greek *peiran* to attempt] (12th century)
1 a : an unpleasant often strong emotion caused by anticipation or awareness of danger **b** (1) **:** an instance of this emotion (2) **:** a state marked by this emotion
2 : anxious concern **:** SOLICITUDE
3 : profound reverence and awe especially toward God
4 : reason for alarm **:** DANGER ☆

fear·ful \'fir-fəl\ *adjective* (14th century)
1 : causing or likely to cause fear, fright, or alarm especially because of dangerous quality ⟨a *fearful* storm⟩
2 a : full of fear **b :** indicating or arising from fear ⟨a *fearful* glance⟩ **c :** inclined to fear **:** TIMOROUS
3 : very great or bad — used as an intensive ⟨a *fearful* waste⟩ ⟨*fearful* slum conditions⟩ ☆
— fear·ful·ly \-f(ə-)lē\ *adverb*
— fear·ful·ness \-fəl-nəs\ *noun*

fear·less \'fir-ləs\ *adjective* (1591)
: free from fear **:** BRAVE
— fear·less·ly *adverb*
— fear·less·ness *noun*

fear·some \'fir-səm\ *adjective* (1768)
1 : causing fear
2 : TIMID, TIMOROUS
— fear·some·ly *adverb*
— fear·some·ness *noun*

fea·si·ble \'fē-zə-bəl\ *adjective* [Middle English *faisible,* from Middle French, from *fais-,* stem of *faire* to make, do, from Latin *facere* — more at DO] (15th century)
1 : capable of being done or carried out ⟨a *feasible* plan⟩
2 : capable of being used or dealt with successfully **:** SUITABLE
3 : REASONABLE, LIKELY
synonym see POSSIBLE
— fea·si·bil·i·ty \ˌfē-zə-'bi-lə-tē\ *noun*
— fea·si·bly \'fē-zə-blē\ *adverb*

¹feast \'fēst\ *noun* [Middle English *feste* festival, feast, from Middle French, festival, from Latin *festa,* plural of *festum* festival, from neuter of *festus* solemn, festal; akin to Latin *feriae* holidays, *fanum* temple] (13th century)
1 a : an elaborate meal often accompanied by a ceremony or entertainment **:** BANQUET **b :** something that gives unusual or abundant pleasure
2 : a periodic religious observance commemorating an event or honoring a deity, person, or thing

²feast (14th century)
intransitive verb
1 : to take part in a feast
2 : to enjoy some unusual pleasure or delight
transitive verb
1 : to give a feast for
2 : DELIGHT, GRATIFY
— feast·er *noun*

Feast of Tabernacles (14th century)
: SUKKOTH

¹feat \'fēt\ *noun* [Middle English *fait,* from Middle French, from Latin *factum,* from neuter of *factus,* past participle of *facere* to make, do — more at DO] (14th century)
1 : ACT, DEED
2 a : a deed notable especially for courage **b :** an act or product of skill, endurance, or ingenuity ☆

²feat *adjective* [Middle English *fete, fayt,* from Middle French *fait,* past participle of *faire*] (15th century)
1 *archaic* **:** BECOMING, NEAT
2 *archaic* **:** SMART, DEXTEROUS

¹feath·er \'fe-t͟hər\ *noun* [Middle English *fether,* from Old English; akin to Old High German *federa* wing, Latin *petere* to go to, seek, Greek *petesthai* to fly, *piptein* to fall, *pteron* wing] (before 12th century)
1 a : any of the light horny epidermal outgrowths that form the external covering of the body of birds and that consist of a shaft bearing on each side a series of barbs which bear barbules which in turn bear barbicels commonly ending in hooked hamuli and interlocking with the barbules of an adjacent barb to link the barbs into a continuous vane **b :** PLUME **c :** the vane of an arrow

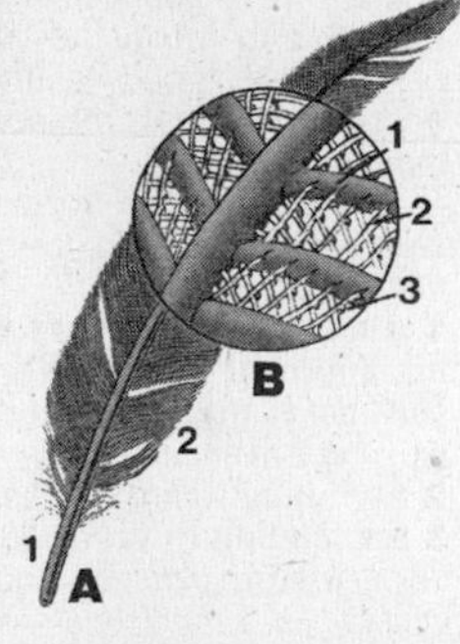

feather 1a: *A: 1* **quill,** *2* **vane;** *B: 1* **barb,** *2* **barbule,** *3* **barbicel with hamulus**

2 a : PLUMAGE **b :** KIND, NATURE ⟨birds of a *feather* flock together⟩ **c :** ATTIRE, DRESS **d :** CONDITION, MOOD
3 : FEATHERING 2
4 : a projecting strip, rib, fin, or flange
5 : a feathery flaw in the eye or in a precious stone
6 : the act of feathering an oar
— feath·ered \-t͟hərd\ *adjective*
— feath·er·less *adjective*
— a feather in one's cap : a mark of distinction **:** HONOR

²feather *verb* **feath·ered; feath·er·ing** \'fet͟h-riŋ, 'fe-t͟hə-\ (13th century)
transitive verb
1 a : to furnish (as an arrow) with a feather **b :** to cover, clothe, or adorn with or as if with feathers
2 a : to turn (an oar blade) almost horizontal when lifting from the water at the end of a stroke to reduce air resistance **b** (1) **:** to change the angle of (airplane propeller blades) so that the chords become approximately parallel to the line of flight; *also* **:** to change the angle of airplane propeller blades of (an engine) in such a manner (2) **:** to change the angle of (a rotor blade of a rotorcraft) periodically in forward flight
3 : to reduce the edge of to a featheredge
4 : to cut (as air) with or as if with a wing
5 : to join by a tongue and groove
intransitive verb
1 : to grow or form feathers
2 : to have or take on the appearance of a feather or something feathered
3 : to soak in and spread **:** BLUR — used of ink or a printed impression
4 : to feather an oar or an airplane propeller blade
— feather one's nest : to provide for oneself especially while in a position of trust

¹feath·er·bed \'fe-t͟hər-ˌbed\ *adjective* (1938)
: calling for, sanctioning, or resulting from featherbedding

²featherbed (1949)
intransitive verb
1 a : to require that more workers be hired than are needed **b :** to limit production under a featherbed rule
2 : to do featherbed work or put in time under a featherbed rule
transitive verb
1 : to bring under a featherbed rule
2 : to assist (as an industry) by government aid

feather bed *noun* (before 12th century)
1 : a feather mattress
2 : a bed having a feather mattress

feath·er·bed·ding *noun* (1921)

☆ SYNONYMS

Fear, dread, fright, alarm, panic, terror, trepidation mean painful agitation in the presence or anticipation of danger. FEAR is the most general term and implies anxiety and usually loss of courage ⟨*fear* of the unknown⟩. DREAD usually adds the idea of intense reluctance to face or meet a person or situation and suggests aversion as well as anxiety ⟨faced the meeting with *dread*⟩. FRIGHT implies the shock of sudden, startling fear ⟨*fright* at being awakened suddenly⟩. ALARM suggests a sudden and intense awareness of immediate danger ⟨view the situation with *alarm*⟩. PANIC implies unreasoning and overmastering fear causing hysterical activity ⟨the news caused widespread *panic*⟩. TERROR implies the most extreme degree of fear ⟨immobilized with *terror*⟩. TREPIDATION adds to DREAD the implications of timidity, trembling, and hesitation ⟨raised the subject with *trepidation*⟩.

Fearful, apprehensive, afraid mean disturbed by fear. FEARFUL implies often a timorous or worrying temperament ⟨the child is *fearful* of loud noises⟩. APPREHENSIVE suggests a state of mind and implies a premonition of evil or danger ⟨*apprehensive* of being found out⟩. AFRAID often suggests weakness or cowardice and regularly implies inhibition of action or utterance ⟨*afraid* to speak the truth⟩.

Feat, exploit, achievement mean a remarkable deed. FEAT implies strength or dexterity or daring ⟨an acrobatic *feat*⟩. EXPLOIT suggests an adventurous or heroic act ⟨his *exploits* as a spy⟩. ACHIEVEMENT implies hard-won success in the face of difficulty or opposition ⟨her *achievements* as a chemist⟩.

: the requiring of an employer usually under a union rule or safety statute to hire more employees than are needed or to limit production

feath·er·brain \-ˌbrān\ *noun* (1839)
: a foolish scatterbrained person
— feath·er·brained \-ˌbrānd\ *adjective*

feath·er·edge \ˈfe-t͟hə-ˌrej, ˌfe-t͟hə-ˈ\ *noun* (1616)
: a very thin sharp edge; *especially* **:** one that is easily broken or bent over
— featheredge *transitive verb*

feath·er·head \ˈfe-t͟hər-ˌhed\ *noun* (1831)
: FEATHERBRAIN
— feath·er·head·ed \ˌfe-t͟hər-ˈhe-dəd\ *adjective*

feath·er·ing \ˈfet͟h-riŋ, ˈfe-t͟hə-\ *noun* (1721)
1 : a covering of feathers **:** PLUMAGE
2 : a fringe of hair (as on the legs of a dog) — see DOG illustration

feath·er·light \ˈfe-t͟hər-ˈlīt\ *adjective* (circa 1837)
: extremely light

feather star *noun* (1862)
: any of an order (Comatulida) of free-swimming stalkless crinoids

feath·er·stitch \ˈfe-t͟hər-ˌstich\ *noun* (1835)
: an embroidery stitch consisting of a line of diagonal blanket stitches worked alternately to the left and right
— featherstitch *verb*

featherstitch

feath·er·weight \-ˌwāt\ *noun, often attributive* (1812)
1 : one that is very light in weight; *especially* **:** a boxer in a weight division having a maximum limit of 126 pounds for professionals and 125 pounds for amateurs — compare BANTAMWEIGHT, LIGHTWEIGHT
2 : LIGHTWEIGHT 2

feath·ery \ˈfet͟h-rē, ˈfe-t͟hə-\ *adjective* (1580)
: resembling, suggesting, or covered with feathers; *especially* **:** extremely light

¹feat·ly \ˈfēt-lē\ *adverb* [Middle English *fetly,* from *fete* feat (adjective)] (14th century)
1 : in a graceful manner **:** NIMBLY
2 : in a suitable manner **:** PROPERLY
3 : with skill and ingenuity

²featly *adjective* (1801)
: GRACEFUL, NEAT

¹fea·ture \ˈfē-chər\ *noun* [Middle English *feture,* from Middle French, from Latin *factura* act of making, from *factus,* past participle of *facere* to make — more at DO] (14th century)
1 a : the structure, form, or appearance especially of a person **b** *obsolete* **:** physical beauty
2 a : the makeup or appearance of the face or its parts **b :** a part of the face **:** LINEAMENT
3 a : a prominent part or characteristic **b :** any of the properties (as voice or gender) that are characteristic of a grammatical element (as a phoneme or morpheme); *especially* **:** one that is distinctive
4 : a special attraction: as **a :** the principal motion picture shown on a program with other pictures **b :** a featured article, story, or department in a newspaper or magazine **c :** something offered to the public or advertised as particularly attractive
— fea·ture·less \-ləs\ *adjective*

²feature *verb* **fea·tured; fea·tur·ing** \ˈfēch-riŋ, ˈfē-chə-\ (circa 1755)
transitive verb
1 *chiefly dialect* **:** to resemble in features
2 : to picture or portray in the mind **:** IMAGINE
3 a : to give special prominence to **b :** to have as a characteristic or feature
intransitive verb
: to play an important part

fea·tured \ˈfē-chərd\ *adjective* (15th century)
1 : having facial features of a particular kind — used in combination ⟨a heavy-*featured* lout⟩
2 : displayed, advertised, or presented as a special attraction

fea·tur·ette \ˈfē-chə-ˌret, ˌfē-chə-ˈ\ *noun* (1940)
: a short film

feaze \ˈfēz, ˈfāz\ *variant of* FAZE

febri- *combining form* [Late Latin, from Latin *febris*]
: fever ⟨*febri*fic⟩

fe·brif·ic \fi-ˈbrif-ik\ *adjective* (1710)
archaic **:** FEVERISH

feb·ri·fuge \ˈfe-brə-ˌfyüj\ *noun* [French *fébrifuge,* probably from (assumed) New Latin *febrifuga,* from Late Latin *febrifuga, febrifugia* centaury, from *febri-* + Late Latin *-fuga* -fuge] (1686)
: ANTIPYRETIC
— febrifuge *adjective*

fe·brile \ˈfe-ˌbrīl *also* ˈfē-\ *adjective* [Medieval Latin *febrilis,* from Latin *febris* fever] (1651)
: FEVERISH

Feb·ru·ary \÷ˈfe-b(y)ə-ˌwer-ē, ˈfe-brə-\ *noun, plural* **-ar·ies** *or* **-ar·ys** [Middle English *Februarie,* from Old English *Februarius,* from Latin, from *Februa,* plural, feast of purification] (before 12th century)
: the 2d month of the Gregorian calendar ▢

fe·cal \ˈfē-kəl\ *adjective* (1541)
: of, relating to, or constituting feces

fe·ces \ˈfē-(ˌ)sēz\ *noun plural* [Middle English, from Latin *faec-, faex* (singular) dregs] (14th century)
: bodily waste discharged through the anus **:** EXCREMENT

feck·less \ˈfek-ləs\ *adjective* [Scots, from *feck* effect, majority, from Middle English (Scots) *fek,* alteration of Middle English *effect*] (circa 1585)
1 : WEAK, INEFFECTIVE
2 : WORTHLESS, IRRESPONSIBLE
— feck·less·ly *adverb*
— feck·less·ness *noun*

feck·ly \ˈfek-lē\ *adverb* [*feck* + *-ly*] (1768)
chiefly Scottish **:** ALMOST, NEARLY

fec·u·lent \ˈfe-kyə-lənt\ *adjective* [Middle English, from Latin *faeculentus,* from *faec-, faex*] (15th century)
: foul with impurities **:** FECAL
— fec·u·lence \-lən(t)s\ *noun*

fe·cund \ˈfe-kənd, ˈfē-\ *adjective* [Middle English, from Middle French *fecond,* from Latin *fecundus* — more at FEMININE] (15th century)
1 : fruitful in offspring or vegetation **:** PROLIFIC
2 : intellectually productive or inventive to a marked degree
synonym see FERTILE
— fe·cun·di·ty \fi-ˈkən-də-tē, fe-\ *noun*

fe·cun·date \ˈfe-kən-ˌdāt, ˈfē-\ *transitive verb* **-dat·ed; -dat·ing** [Latin *fecundatus,* past participle of *fecundare,* from *fecundus*] (circa 1631)
1 : to make fecund
2 : IMPREGNATE
— fe·cun·da·tion \ˌfe-kən-ˈdā-shən, ˌfē-\ *noun*

fed \ˈfed\ *noun, often capitalized* (1916)
: a federal agent, officer, or official — usually used in plural

fe·da·yee \fi-ˌda-ˈ(y)ē, -ˌdä-\ *noun, plural* **fe·da·yeen** \-ˈ(y)ēn\ [Arabic *fidā'ī,* literally, one who sacrifices himself] (1955)
: a member of an Arab commando group operating especially against Israel — usually used in plural

fed·er·al \ˈfe-d(ə-)rəl\ *adjective* [Latin *foeder-, foedus* compact, league; akin to Latin *fidere* to trust — more at BIDE] (1660)
1 *archaic* **:** of or relating to a compact or treaty
2 a : formed by a compact between political units that surrender their individual sovereignty to a central authority but retain limited residuary powers of government **b :** of or constituting a form of government in which power is distributed between a central authority and a number of constituent territorial units **c :** of or relating to the central government of a federation as distinguished from the governments of the constituent units
3 *capitalized* **:** advocating or friendly to the principle of a federal government with strong centralized powers; *especially* **:** of or relating to the American Federalists
4 *often capitalized* **:** of, relating to, or loyal to the federal government or the Union armies of the U.S. in the American Civil War
5 *capitalized* **:** being or belonging to a style of architecture and decoration current in the U.S. following the Revolution
— fed·er·al·ly \-d(ə-)rə-lē\ *adverb*

Federal *noun* (1861)
1 : a supporter of the U.S. government in the Civil War; *especially* **:** a soldier in the federal armies
2 : FED — usually used in plural

federal case *noun* (1955)
: BIG DEAL ⟨don't make a *federal case* out of it⟩

federal court *noun* (1789)
: a court established by a federal government; *especially* **:** one established under the constitution and laws of the U.S.

federal district *noun* (circa 1934)
: a district set apart as the seat of the central government of a federation

federal district court *noun* (1948)
: a district trial court of law and equity that hears cases under federal jurisdiction

fed·er·al·ese \ˌfe-d(ə-)rə-ˈlēz, -ˈlēs; ˈfe-d(ə-)rə-ˌ\ *noun* (1944)
: BUREAUCRATESE

federal funds *noun plural* (1950)
: reserve funds lent overnight by one Federal Reserve bank to another

fed·er·al·ism \ˈfe-d(ə-)rə-ˌli-zəm\ *noun* (1789)
1 a *often capitalized* **:** the distribution of power in an organization (as a government) between a central authority and the constituent units — compare CENTRALISM **b :** support or advocacy of this principle
2 *capitalized* **:** Federalist principles

fed·er·al·ist \-list\ *noun* (1787)
1 : an advocate of federalism: as **a** *often capitalized* **:** an advocate of a federal union between the American colonies after the Revolution and of the adoption of the U.S. Constitution **b** *often capitalized* **:** WORLD FEDERALIST
2 *capitalized* **:** a member of a major political party in the early years of the U.S. favoring a strong centralized national government
— federalist *adjective, often capitalized*

fed·er·al·iza·tion \ˌfe-d(ə-)rə-lə-ˈzā-shən\ *noun* (circa 1860)
1 : the act of federalizing
2 : the state of being federalized

fed·er·al·ize \ˈfe-d(ə-)rə-ˌlīz\ *transitive verb* **-ized; -iz·ing** (1801)
1 : to unite in or under a federal system
2 : to bring under the jurisdiction of a federal government

▢ USAGE
February Dissimilation may occur when a word contains two identical or closely related sounds, resulting in the change or loss of one of them. This happens regularly in *February,* which is more often pronounced \ˈfe-b(y)ə-ˌwer-ē\ than \ˈfe-brə-ˌwer-ē\, though all of these variants are in frequent use and widely accepted. The \y\ heard from many speakers is not an intrusion but rather a common pronunciation of the vowel *u* after a consonant, as in *January* and *annual.*

Federal Reserve bank *noun* (1914)
: one of 12 reserve banks set up under the Federal Reserve Act to hold reserves and discount commercial paper for affiliated banks in their respective districts

¹fed·er·ate \'fe-d(ə-)rət\ *adjective* [Latin *foederatus,* from *foeder-, foedus*] (1710)
: united in an alliance or federation : FEDERATED

²fed·er·ate \'fe-də-ˌrāt\ *transitive verb* **-at·ed; -at·ing** (1837)
: to join in a federation

federated church *noun* (circa 1926)
: a local church uniting two or more congregations that maintain different denominational ties — compare UNION CHURCH

fed·er·a·tion \ˌfe-də-'rā-shən\ *noun* (1791)
1 : something formed by federation: as **a :** a federal government **b :** a union of organizations
2 : the act of federating; *especially* **:** the forming of a federal union

fed·er·a·tive \'fe-də-ˌrā-tiv, 'fe-d(ə-)rə-\ *adjective* (1690)
: FEDERAL
— fed·er·a·tive·ly *adverb*

fe·do·ra \fi-'dōr-ə, -'dȯr-\ *noun* [*Fédora* (1882), drama by V. Sardou] (1895)
: a low soft felt hat with the crown creased lengthwise ◆

fed up *adjective* (1900)
: tired, sated, or disgusted beyond endurance

¹fee \'fē\ *noun* [Middle English, from Middle French *fé, fief,* from Old French, of Germanic origin; akin to Old English *feoh* cattle, property, Old High German *fihu* cattle; akin to Latin *pecus* cattle, *pecunia* money] (14th century)
1 a (1) **:** an estate in land held in feudal law from a lord on condition of homage and service (2) **:** a piece of land so held **b :** an inherited or heritable estate in land
2 a : a fixed charge **b :** a sum paid or charged for a service
— in fee : in absolute and legal possession

²fee *transitive verb* **feed; fee·ing** (15th century)
1 *chiefly Scottish* **:** HIRE
2 : ⁹TIP 2

fee·ble \'fē-bəl\ *adjective* **fee·bler** \-b(ə-)lər\; **fee·blest** \-b(ə-)ləst\ [Middle English *feble,* from Old French, from Latin *flebilis* lamentable, wretched, from *flēre* to weep — more at BLEAT] (12th century)
1 a : markedly lacking in strength **b :** indicating weakness
2 a : deficient in qualities or resources that indicate vigor, authority, force, or efficiency **b :** INADEQUATE, INFERIOR
synonym see WEAK
— fee·ble·ness \-bəl-nəs\ *noun*
— fee·bly \-blē\ *adverb*

fee·ble·mind·ed \ˌfē-bəl-'mīn-dəd\ *adjective* (1534)
1 *obsolete* **:** IRRESOLUTE, VACILLATING
2 : mentally deficient
3 : FOOLISH, STUPID
— fee·ble·mind·ed·ly *adverb*
— fee·ble·mind·ed·ness *noun*

fee·blish \'fē-b(ə-)lish\ *adjective* (1674)
: somewhat feeble

¹feed \'fēd\ *verb* **fed** \'fed\; **feed·ing** [Middle English *feden,* from Old English *fēdan;* akin to Old English *fōda* food — more at FOOD] (before 12th century)
transitive verb
1 a : to give food to **b :** to give as food
2 a : to furnish something essential to the growth, sustenance, maintenance, or operation of **b :** to supply (material to be operated on) to a machine
3 : to produce or provide food for
4 a : SATISFY, GRATIFY **b :** SUPPORT, ENCOURAGE
5 a (1) **:** to supply for use or consumption (2) **:** CHANNEL, ROUTE **b** (1) **:** to supply (a signal) to an electronic circuit (2) **:** to send (as by wire or satellite) to a transmitting station for broadcast
6 : to supply (a fellow actor) with cues and situations that make a role more effective
7 : to pass a ball or puck to (a teammate) especially for a shot at the goal
intransitive verb
1 a : to consume food **:** EAT **b :** PREY — used with *on, upon,* or *off*
2 : to become nourished or satisfied or sustained as if by food
3 a : to become channeled or directed **b :** to move into a machine or opening in order to be used or processed

²feed *noun* (1576)
1 a : an act of eating **b :** MEAL; *especially* **:** a large meal
2 a : food for livestock; *specifically* **:** a mixture or preparation for feeding livestock **b :** the amount given at each feeding
3 a : material supplied (as to a furnace or machine) **b :** a mechanism by which the action of feeding is effected **c :** the motion or process of carrying forward the material to be operated upon (as in a machine) **d :** the process of feeding a television program (as to a local station)
4 : the action of passing a ball or puck to a team member who is in position to score

feed·back \'fēd-ˌbak\ *noun* (1920)
1 : the return to the input of a part of the output of a machine, system, or process (as for producing changes in an electronic circuit that improve performance or in an automatic control device that provide self-corrective action)
2 a : the partial reversion of the effects of a process to its source or to a preceding stage **b :** the transmission of evaluative or corrective information to the original or controlling source about an action, event, or process; *also* **:** the information so transmitted

feedback inhibition *noun* (1960)
: inhibition of an enzyme controlling an early stage of a series of biochemical reactions by the end product when it reaches a critical concentration

feed dog *noun* (circa 1939)
: a notched piece of metal on a sewing machine that feeds material into position under the needle

feed·er \'fē-dər\ *noun* (14th century)
1 : one that feeds: as **a :** one that fattens livestock for slaughter **b :** a device or apparatus for supplying food
2 : one that eats or takes nourishment; *especially* **:** an animal being fattened or one suitable for fattening
3 a : one that supplies, replenishes, or connects **b :** TRIBUTARY **c :** a heavy wire conductor supplying electricity at some point of an electric distribution system (as from a substation) **d :** BRANCH; *especially* **:** a branch transportation line **e :** FEEDER ROAD

feeder road *noun* (1938)
: a road that provides access to a major artery

feeding frenzy *noun* (1973)
: a frenzy of eating; *also* **:** the excited pursuit of something by a group

feed·lot \'fēd-ˌlät\ *noun* (1889)
: a plot of land on which livestock are fattened for market

feed·stock \-ˌstäk\ *noun* (1932)
: raw material supplied to a machine or processing plant

feed·stuff \-ˌstəf\ *noun* (1856)
: FEED 2a; *also* **:** any of the constituent nutrients of an animal ration

¹feel \'fē(ə)l\ *verb* **felt** \'felt\; **feel·ing** [Middle English *felen,* from Old English *fēlan;* akin to Old High German *fuolen* to feel, Latin *palpare* to caress] (before 12th century)
transitive verb
1 a : to handle or touch in order to examine, test, or explore some quality **b :** to perceive by a physical sensation coming from discrete end organs (as of the skin or muscles)
2 a : to undergo passive experience of **b :** to have one's sensibilities markedly affected by
3 : to ascertain by cautious trial — usually used with *out*
4 a : to be aware of by instinct or inference **b :** BELIEVE, THINK
intransitive verb
1 a : to receive or be able to receive a tactile sensation **b :** to search for something by using the sense of touch
2 a : to be conscious of an inward impression, state of mind, or physical condition **b :** to have a marked sentiment or opinion ⟨*feels* strongly about it⟩
3 : SEEM ⟨it *feels* like spring today⟩
4 : to have sympathy or pity ⟨I *feel* for you⟩
— feel like : to have an inclination for ⟨*feel like* a walk?⟩

²feel *noun* (13th century)
1 : SENSATION, FEELING
2 : the sense of touch
3 a : the quality of a thing as imparted through or as if through touch **b :** typical or peculiar quality or atmosphere; *also* **:** an awareness of such a quality or atmosphere
4 : intuitive knowledge or ability

feel·er \'fē-lər\ *noun* (1526)
: one that feels: as **a :** a tactile process (as a tentacle) of an animal **b :** something (as a proposal) ventured to ascertain the views of others

feeler gauge *noun* (1925)
: a thin metal strip or wire of known thickness used as a gauge

feel–good \'fē(ə)l-ˌgu̇d\ *adjective* (1978)
: relating to or promoting an often specious sense of satisfaction or well-being

¹feel·ing \'fē-liŋ\ *noun* (12th century)
1 a (1) **:** the one of the basic physical senses of which the skin contains the chief end organs and of which the sensations of touch and temperature are characteristic **:** TOUCH (2) **:** a sensation experienced through this sense **b :** generalized bodily consciousness or sensation **c :** appreciative or responsive awareness or recognition
2 a : an emotional state or reaction ⟨had a kindly *feeling* toward the child⟩ **b** *plural* **:** susceptibility to impression **:** SENSITIVITY ⟨the remark hurt her *feelings*⟩
3 a : the undifferentiated background of one's awareness considered apart from any identifiable sensation, perception, or thought **b :** the overall quality of one's awareness **c :** conscious recognition **:** SENSE
4 a : often unreasoned opinion or belief **:** SENTIMENT **b :** PRESENTIMENT

◇ WORD HISTORY

fedora The hat known as a *fedora* is generally thought to have some connection with *Fédora,* a melodrama by the French playwright Victorien Sardou (1831–1908), though the exact nature of the connection has yet to be established. Sardou, a popular dramatist in the 19th century, though now nearly forgotten except by literary and theater historians, premiered *Fédora* in 1882, with the celebrated actress Sarah Bernhardt in the title role of Fédora Romanoff, a Russian princess. There is some evidence that the long plume worn by Bernhardt in the play made an impression on Parisian fashion, but the word *fédora,* if it was used for the plume in French, did not survive. Sardou's play ran successfully in New York in 1883 with the American actress Fanny Davenport in the title role. Sometime afterward, *fedora* found its way into American English as a name for a man's soft felt hat, with no obvious link to either Sarah Bernhardt's or Fanny Davenport's headgear. By 1895 the word appeared in a Montgomery Ward catalog—clearly a sign that the hat style had entered mainstream American fashion.

5 : capacity to respond emotionally especially with the higher emotions
6 : the character ascribed to something **:** ATMOSPHERE
7 a : the quality of a work of art that conveys the emotion of the artist **b :** sympathetic aesthetic response
8 : FEEL 4 ☆

²feeling *adjective* (14th century)
1 a : SENTIENT, SENSITIVE **b :** easily moved emotionally
2 *obsolete* **:** deeply felt
3 : expressing emotion or sensitivity
— **feel·ing·ly** \-liŋ-lē\ *adverb*
— **feel·ing·ness** *noun*

feel up *transitive verb* (1930)
: to touch or fondle (someone) for sexual pleasure

fee simple *noun, plural* **fees simple** (14th century)
: a fee without limitation to any class of heirs or restrictions on transfer of ownership

fee splitting *noun* (1943)
: payment by a specialist (as a doctor or a lawyer) of a part of his fee to the person who made the referral

feet *plural of* FOOT

fee tail *noun, plural* **fees tail** (15th century)
: a fee limited to a particular class of heirs

feet·first \ˌfēt-ˈfərst\ *adverb* (circa 1901)
: with the feet foremost

feet of clay [from the feet of the idol in Daniel 2:33] (1814)
: a flaw of character that is usually not readily apparent

feeze \ˈfēz, ˈfāz\ *noun* [Middle English *veze,* from *fesen, vesen* to drive away — more at FAZE] (14th century)
1 *chiefly dialect* **:** RUSH
2 *dialect* **:** a state of alarm or excitement

Feh·ling's solution \ˈfā-liŋ(z)-\ *noun* [Hermann *Fehling* (died 1885) German chemist] (1873)
: a blue solution of Rochelle salt and copper sulfate used as an oxidizing agent in a test for sugars and aldehydes

feign \ˈfān\ *verb* [Middle English, from Old French *feign-,* stem of *feindre,* from Latin *fingere* to shape, feign — more at DOUGH] (13th century)
intransitive verb
: PRETEND, DISSEMBLE
transitive verb
1 a : to give a false appearance of **:** induce as a false impression ⟨*feign* death⟩ **b :** to assert as if true **:** PRETEND
2 *archaic* **a :** INVENT, IMAGINE **b :** to give fictional representation to
3 *obsolete* **:** DISGUISE, CONCEAL
synonym see ASSUME
— **feign·er** *noun*

feigned *adjective* (14th century)
1 : FICTITIOUS
2 : not genuine or real

fei·joa \fā-ˈyō-ə, -ˈhō-ə\ *noun* [New Latin, genus name, from João da Silva *Feijó* (died 1824) Brazilian naturalist] (1898)
: the green edible fruit of a shrub or small tree (*Feijoa sellowiana*) of the myrtle family that is native to South America and is grown commercially especially in New Zealand; *also* **:** the tree or shrub

¹feint \ˈfānt\ *noun* [French *feinte,* from Old French, from *feint,* past participle of *feindre*] (1679)
: something feigned; *specifically* **:** a mock blow or attack on or toward one part in order to distract attention from the point one really intends to attack
synonym see TRICK

²feint (1810)
intransitive verb
: to make a feint
transitive verb
1 : to lure or deceive with a feint
2 : to make a pretense of

fei·rie \ˈfē-rē\ *adjective* [Middle English (Scots) *fery,* from Middle English *fere* strong, from Old English *fēre* able to go; akin to Old English *faran* to travel, fare] (before 12th century)
Scottish **:** NIMBLE, STRONG

feist \ˈfīst\ *noun* [obsolete *fisting hound,* from obsolete *fist* (to break wind)] (1770)
chiefly dialect **:** a small dog

feisty \ˈfī-stē\ *adjective* **feist·i·er; -est** (1896)
1 *chiefly Southern & Midland* **a :** full of nervous energy **:** FIDGETY **b :** being touchy and quarrelsome **c :** being frisky and exuberant
2 : having or showing a lively aggressiveness **:** SPUNKY ◆
— **feist·i·ness** *noun*

fe·la·fel \fə-ˈlä-fəl\ *variant of* FALAFEL

feld·sher \ˈfel(d)-shər\ *noun* [Russian *fel'dsher,* from German *Feldscher, Feldscherer* field surgeon, from *Feld* field + *Scherer* barber, surgeon] (1877)
: a medical or surgical practitioner without full professional qualifications or status in some east European countries and especially Russia

feld·spar \ˈfel(d)-ˌspär\ *noun* [modification of obsolete German *Feldspath* (now *Feldspat*), from German *Feld* field + obsolete German *Spath* (now *Spat*) spar] (1772)
: any of a group of crystalline minerals that consist of aluminum silicates with either potassium, sodium, calcium, or barium and that are an essential constituent of nearly all crystalline rocks

feld·spath·ic \fel(d)-ˈspa-thik\ *adjective* [*feldspath* (variant of *feldspar*), from obsolete German] (circa 1828)
: relating to or containing feldspar — used especially of a porcelain glaze

fe·li·cif·ic \ˌfē-lə-ˈsi-fik\ *adjective* [Latin *felic , felix*] (1865)
: causing or intended to cause happiness

felicific calculus *noun* (1945)
: a method of determining the rightness of an action by balancing the probable pleasures and pains that it would produce

¹fe·lic·i·tate \fi-ˈli-sə-ˌtāt\ *adjective* [Late Latin *felicitatus,* past participle of *felicitare* to make happy, from Latin *felicitas*] (1605)
obsolete **:** made happy

²felicitate *transitive verb* **-tat·ed; -tat·ing** (1628)
1 *archaic* **:** to make happy
2 a : to consider happy or fortunate **b :** to offer congratulations to
— **fe·lic·i·ta·tion** \-ˌli-sə-ˈtā-shən\ *noun*
— **fe·lic·i·ta·tor** \-ˈli-sə-ˌtā-tər\ *noun*

fe·lic·i·tous \fi-ˈli-sə-təs\ *adjective* (1789)
1 : very well suited or expressed **:** APT ⟨a *felicitous* remark⟩
2 : PLEASANT, DELIGHTFUL
synonym see FIT
— **fe·lic·i·tous·ly** *adverb*
— **fe·lic·i·tous·ness** *noun*

fe·lic·i·ty \fi-ˈli-sə-tē\ *noun, plural* **-ties** [Middle English *felicite,* from Middle French *felicité,* from Latin *felicitat-, felicitas,* from *felic-, felix* fruitful, happy — more at FEMININE] (14th century)
1 a : the quality or state of being happy; *especially* **:** great happiness **b :** an instance of happiness
2 : something that causes happiness
3 : a pleasing manner or quality especially in art or language
4 : an apt expression

fe·lid \ˈfē-ləd\ *noun* [New Latin *Felidae,* family name, from *Felis,* genus of cats, from Latin, cat] (circa 1889)
: CAT 1b
— **felid** *adjective*

fe·line \ˈfē-ˌlīn\ *adjective* [Latin *felinus,* from *felis*] (1681)
1 : of, relating to, or affecting cats or the cat family
2 : resembling a cat: as **a :** sleekly graceful **b :** SLY, TREACHEROUS **c :** STEALTHY
— **feline** *noun*
— **fe·line·ly** *adverb*
— **fe·lin·i·ty** \fē-ˈli-nə-tē\ *noun*

feline distemper *noun* (1942)
: PANLEUKOPENIA

feline panleukopenia *noun* (circa 1943)
: PANLEUKOPENIA

¹fell \ˈfel\ *noun* [Middle English, from Old English; akin to Old High German *fel* skin, Latin *pellis*] (before 12th century)
1 : SKIN, HIDE, PELT
2 : a thin tough membrane covering a carcass directly under the hide

²fell *transitive verb* [Middle English, from Old English *fellan;* akin to Old English *feallan* to fall — more at FALL] (before 12th century)
1 a : to cut, knock, or bring down **b :** KILL
2 : to sew (a seam) by folding one raw edge under the other and sewing flat on the wrong side
— **fell·able** \ˈfe-lə-bəl\ *adjective*
— **fell·er** *noun*

³fell *past of* FALL

⁴fell *adjective* [Middle English *fel,* from Middle French, from Old French — more at FELON] (14th century)

☆ SYNONYMS
Feeling, emotion, affection, sentiment, passion mean a subjective response to a person, thing, or situation. FEELING denotes any partly mental, partly physical response marked by pleasure, pain, attraction, or repulsion; it may suggest the mere existence of a response but imply nothing about the nature or intensity of it ⟨the *feelings* that once moved me are gone⟩. EMOTION carries a strong implication of excitement or agitation but, like FEELING, encompasses both positive and negative responses ⟨the drama portrays the *emotions* of adolescence⟩. AFFECTION applies to feelings that are also inclinations or likings ⟨a memoir of childhood filled with *affection* for her family⟩. SENTIMENT often implies an emotion inspired by an idea ⟨her feminist *sentiments* are well known⟩. PASSION suggests a very powerful or controlling emotion ⟨revenge became his ruling *passion*⟩.

◇ WORD HISTORY
feisty As *feisty* is a word now generally used with approval, its malodorous origins may come as a surprise to both user and subject. *Feisty* is ultimately derivative of *fist* (not the same word as *fist* denoting a clenched hand), which is attested in late Middle English as both a noun meaning "the act of breaking wind" and a verb meaning "to break wind." By the 16th century the participle *fisting* had become common in contemptuous expressions for a small dog, such as "fisting cur" and "fisting hound." Although *fist* and *fisting* became obsolete in Britain, the word resurfaced in the 19th century U.S., now spelled *feist* and meaning "a small dog of mixed breed," a usage that still survives, mostly in South Midland and Southern states. In the same area an adjective derivative *feisty* was formed, meaning variously "fidgety" and "irritable," supposed qualities of such dogs. Growing out of regional use, *feisty* has entered mainstream American English in the sense "spunky."

1 a : FIERCE, CRUEL, TERRIBLE **b :** SINISTER, MALEVOLENT ⟨a *fell* purpose⟩ **c :** very destructive **:** DEADLY ⟨a *fell* disease⟩
2 *Scottish* **:** SHARP, PUNGENT
— fell·ness *noun*
— fel·ly \'fel-lē\ *adverb*
— at one fell swoop *or* **in one fell swoop :** all at once; *also* **:** with a single concentrated effort

⁵fell *noun* [Middle English, from Old Norse *fell, fjall* mountain; akin to Old High German *felis* rock] (14th century)
dialect British **:** a high barren field or moor

fel·lah \'fe-lə, fə-'lä\ *noun, plural* **fel·la·hin** *or* **fel·la·heen** \,fel-ə-'hēn, fə-,lä-'hēn\ [Arabic *fallāḥ*] (1743)
: a peasant or agricultural laborer in an Arab country (as Egypt)

fel·late \'fe-,lāt, fə-'lāt\ *verb* **fel·lat·ed; fel·lat·ing** [back-formation from *fellatio*] (1948)
transitive verb
: to perform fellatio on
intransitive verb
: to fellate someone
— fel·la·tor \-,lā-tər, -'lā-tər\ *noun*

fel·la·tio \fə-'lā-shē-,ō, fe-\ *also* **fel·la·tion** \-'lā-shən\ *noun* [New Latin *fellation-, fellatio,* from Latin *felare, fellare,* literally, to suck — more at FEMININE] (circa 1893)
: oral stimulation of the penis

fell·mon·ger \'fel-,məŋ-gər, -,mäŋ-\ *noun* [¹*fell*] (1530)
British **:** one who removes hair or wool from hides in preparation for leather making
— fell·mon·gered \-gərd\ *adjective, British*
— fell·mon·ger·ing \-g(ə-)riŋ\ *noun, British*
— fell·mon·gery \-g(ə-)rē\ *noun, British*

fel·low \'fe-(,)lō, -lə(-w)\ *noun, often attributive* [Middle English *felawe,* from Old English *fēolaga,* from Old Norse *fēlagi,* from *fēlag* partnership, from *fē* cattle, money + *lag* act of laying] (before 12th century)
1 : COMRADE, ASSOCIATE
2 a : an equal in rank, power, or character **:** PEER **b :** one of a pair **:** MATE
3 : a member of a group having common characteristics; *specifically* **:** a member of an incorporated literary or scientific society
4 a *obsolete* **:** a person of one of the lower social classes **b :** a worthless man or boy **c :** MAN, BOY **d :** BOYFRIEND, BEAU
5 : an incorporated member of a college or collegiate foundation especially in a British university
6 : a person appointed to a position granting a stipend and allowing for advanced study or research ◇

fellow feeling *noun* (1712)
: a feeling of community of interest or of mutual understanding

fel·low·ly \-lō-lē, -lə-lē\ *adjective* (13th century)
: SOCIABLE
— fellowly *adverb*

fel·low·man \,fe-lō-'man, -lə-\ *noun* (1667)
: a kindred human being

fellow servant *noun* (1900)
: an employee working with another employee under such circumstances that each one if negligent may expose the other to harm which the employer cannot reasonably be expected to guard against or be held legally liable for

¹fel·low·ship \'fe-lə-,ship, -lō-\ *noun* (before 12th century)
1 : COMPANIONSHIP, COMPANY
2 a : community of interest, activity, feeling, or experience **b :** the state of being a fellow or associate
3 : a company of equals or friends **:** ASSOCIATION
4 : the quality or state of being comradely
5 *obsolete* **:** MEMBERSHIP, PARTNERSHIP
6 a : the position of a fellow (as of a university) **b :** the stipend of a fellow **c :** a foundation for the providing of such a stipend

²fellowship *verb* **-shipped** *also* **-shiped** \-,shipt\; **-ship·ping** *also* **-ship·ing** \-,shi-piŋ\ (14th century)
intransitive verb
: to join in fellowship especially with a church member
transitive verb
: to admit to fellowship (as in a church)

fellow traveler *noun* [translation of Russian *poputchik*] (1925)
: one that sympathizes with and often furthers the ideals and program of an organized group (as the Communist party) without membership in the group or regular participation in its activities
— fel·low–trav·el·ing *adjective*

fel·ly \'fe-lē\ *or* **fel·loe** \-(,)lō\ *noun, plural* **fellies** *or* **felloes** [Middle English *fely, felive,* from Old English *felg;* akin to Old High German *felga* felly, Old English *fealg* piece of plowed land] (before 12th century)
: the exterior rim or a segment of the rim of a wheel supported by the spokes

felo–de–se \,fe-lō-də-'sā, -'sē\ *noun, plural* **fe·lo·nes–de–se** \fə-,lō-(,)nēz-də-\ *or* **felos–de–se** \,fe-lōz-də-\ [Medieval Latin *felo de se, fello de se,* literally, evildoer in respect to oneself] (1607)
1 : one who deliberately kills himself or who dies from the effects of his commission of an unlawful malicious act
2 : an act of deliberate self-destruction **:** SUICIDE

¹fel·on \'fe-lən\ *noun* [Middle English, from Middle French *felon, fel* evildoer, probably of Germanic origin; akin to Old High German *fillen* to beat, whip, *fel* skin — more at FELL] (13th century)
1 : one who has committed a felony
2 *archaic* **:** VILLAIN
3 : WHITLOW

²felon *adjective* (13th century)
1 *archaic* **a :** CRUEL **b :** EVIL
2 *archaic* **:** WILD

fe·lo·ni·ous \fə-'lō-nē-əs\ *adjective* (1575)
1 *archaic* **:** very evil **:** VILLAINOUS
2 : of, relating to, or having the nature of a felony
— fe·lo·ni·ous·ly *adverb*
— fe·lo·ni·ous·ness *noun*

fel·on·ry \'fe-lən-rē\ *noun* (1837)
: FELONS; *especially* **:** the convict population of a penal colony

fel·o·ny \'fe-lə-nē\ *noun, plural* **-nies** (14th century)
1 : an act on the part of a feudal vassal involving the forfeiture of his fee
2 a : a grave crime formerly differing from a misdemeanor under English common law by involving forfeiture in addition to any other punishment **b :** a grave crime declared to be a felony by the common law or by statute regardless of the punishment actually imposed **c :** a crime declared a felony by statute because of the punishment imposed **d :** a crime for which the punishment in federal law may be death or imprisonment for more than one year

fel·site \'fel-,sīt\ *noun* [*felspar* + *-ite*] (1794)
: a dense igneous rock consisting almost entirely of feldspar and quartz
— fel·sit·ic \fel-'si-tik\ *adjective*

fel·spar *chiefly British variant of* FELDSPAR

¹felt \'felt\ *noun* [Middle English, from Old English; akin to Old High German *filz* felt, Latin *pellere* to drive, beat] (before 12th century)
1 a : a cloth made of wool and fur often mixed with natural or synthetic fibers through the action of heat, moisture, chemicals, and pressure **b :** a firm woven cloth of wool or cotton heavily napped and shrunk
2 : an article made of felt
3 : a material resembling felt: as **a :** a heavy paper of organic or asbestos fibers impregnated with asphalt and used in building construction **b :** semirigid pressed fiber insulation used in building
— felt·like *adjective*

²felt *transitive verb* (14th century)
1 : to make out of or cover with felt
2 : to cause to adhere and mat together
3 : to make into felt or a similar substance

³felt *past and past participle of* FEEL

felt·ing \'fel-tiŋ\ *noun* (1686)
1 : the process of making felt
2 : FELT

fe·luc·ca \fə-'lü-kə, -'lə-kə\ *noun* [Italian *feluca*] (1615)
: a narrow fast lateen-rigged sailing vessel chiefly of the Mediterranean area

felucca

¹fe·male \'fē-,māl\ *adjective* (14th century)
1 a : of, relating to, or being the sex that bears young or produces eggs **b :** PISTILLATE
2 : having some quality (as gentleness) associated with the female sex
3 : designed with a hollow or groove into which a corresponding male part fits ⟨*female* coupling of a hose⟩ ◇
— fe·male·ness *noun*

²female *noun* [Middle English, alteration of *femel, femelle,* from Middle French & Medieval Latin; Middle French *femelle,* from Medieval Latin *femella,* from Latin, girl, diminutive of *femina*] (14th century)
1 a : a female person **:** WOMAN, GIRL **b :** an individual that bears young or produces large usually immobile gametes (as eggs) that are fertilized by small usually motile gametes of a male
2 : a pistillate plant

◇ WORD HISTORY

fellow *Fellow* is one of a significant number of words—from *law* and *outlaw* in the social sphere to *egg* and *sky* in the everyday world—that English owes to the Scandinavian conquests and settlement of Britain in the 9th and 10th centuries. *Fēolaga,* the late Old English ancestor of *fellow,* was borrowed from Old Norse *fēlagi,* literally, a "fee-layer," that is, someone who put down property, typically in the form of livestock or money in a joint venture with another person, for trade or some other common goal. This sense, which we might render more loosely as "partner" or "business associate," has been lost in Modern English, but *fellow* over the course of centuries has acquired a wide variety of other meanings, one of the most common now being simply "male person."

female In the 14th century *female* first appears in Middle English as one spelling alongside others such as *femel* and *femelle.* All are borrowed, through medieval French, from Latin *femella* "young woman, girl," a diminutive of *femina* "woman." French *femelle* had no etymological connection to English *male,* but the phonetic similarity of the words and the fact that "male" and "female" are naturally paired concepts led English speakers to reshape the French word to make it look more similar to *male.* This process reflects the simple operation of folk etymology, that is, the alteration of less familiar words in the direction of more familiar ones.

¹fem·i·nine \'fe-mə-nən\ *adjective* [Middle English, from Middle French *feminin,* from Latin *femininus,* from *femina* woman; akin to Old English *delu* nipple, Latin *filius* son, *felix, fetus,* & *fecundus* fruitful, *felare* to suck, Greek *thēlē* nipple] (14th century)
1 : FEMALE 1a
2 : characteristic of or appropriate or unique to women
3 : of, relating to, or constituting the gender that ordinarily includes most words or grammatical forms referring to females ⟨a *feminine* noun⟩
4 a : being an unstressed and usually hypermetric final syllable ⟨*feminine* ending⟩ **b** *of rhyme* **:** having an unstressed final syllable **c :** having the final chord occurring on a weak beat ⟨music in *feminine* cadences⟩
— **fem·i·nine·ly** *adverb*
— **fem·i·nine·ness** \-nə(n)-nəs\ *noun*

²feminine *noun* (15th century)
1 a : a noun, pronoun, adjective, or inflectional form or class of the feminine gender **b :** the feminine gender
2 : the female principle ⟨eternal *feminine*⟩

fem·i·nin·i·ty \ˌfe-mə-'ni-nə-tē\ *noun* (14th century)
1 : the quality or nature of the female sex
2 : EFFEMINACY
3 : WOMEN, WOMANKIND

fem·i·nise *British variant of* FEMINIZE

fem·i·nism \'fe-mə-ˌni-zəm\ *noun* (1895)
1 : the theory of the political, economic, and social equality of the sexes
2 : organized activity on behalf of women's rights and interests
— **fem·i·nist** \-nist\ *noun or adjective*
— **fem·i·nis·tic** \ˌfe-mə-'nis-tik\ *adjective*

fe·min·i·ty \fe-'mi-nə-tē, fə-\ *noun* (14th century)
: FEMININITY

fem·i·nize \'fe-mə-ˌnīz\ *transitive verb* **-nized; -niz·ing** (1652)
1 : to give a feminine quality to
2 : to cause (a male or castrate) to take on feminine characters (as by implantation of ovaries or administration of estrogenic substances)
— **fem·i·ni·za·tion** \ˌfe-mə-nə-'zā-shən\ *noun*

femme *also* **fem** \'fem\ *noun* [probably from French *femme* woman, from Latin *femina*] (1958)
: a lesbian who plays the female role in a homosexual relationship

femme fa·tale \ˌfem-fə-'tal, ˌfam-, -'täl\ *noun, plural* **femmes fa·tales** \-'tal(z), -'täl(z)\ [French, literally, disastrous woman] (1912)
1 : a seductive woman who lures men into dangerous or compromising situations
2 : a woman who attracts men by an aura of charm and mystery

fem·o·ral \'fe-mə-rəl, 'fem-rəl\ *adjective* (circa 1771)
: of or relating to the femur or thigh

femoral artery *noun* (circa 1771)
: the chief artery of the thigh lying in its anterior inner part

fem·to- \ˌfem(p)-tō\ *combining form* [International Scientific Vocabulary, from Danish or Norwegian *femten* fifteen, from Old Norse *fimmtān;* akin to Old English *fīftēne* fifteen]
: one quadrillionth (10^{-15}) part of ⟨*femto*second⟩

fem·to·sec·ond \'fem(p)-tə-ˌse-kənd, -kənt, -tō-\ *noun* (1976)
: one quadrillionth of a second

fe·mur \'fē-mər\ *noun, plural* **fe·murs** *or* **fem·o·ra** \'fe-mə-rə, 'fem-rə\ [New Latin *femor-, femur,* from Latin, thigh] (circa 1771)
1 : the proximal bone of the hind or lower limb — called also *thighbone*
2 : the segment of an insect's leg that is third from the body

¹fen \'fen\ *noun* [Middle English, from Old English *fenn;* akin to Old High German *fenna* fen, Sanskrit *paṅka* mud] (before 12th century)
: low land covered wholly or partly with water unless artificially drained

²fen \'fən\ *noun, plural* **fen** [Chinese (Beijing) *fēn*] (1916)
— see *yuan* at MONEY table

¹fence \'fen(t)s\ *noun, often attributive* [Middle English *fens,* short for *defens* defense] (14th century)
1 *archaic* **:** a means of protection **:** DEFENSE
2 : a barrier intended to prevent escape or intrusion or to mark a boundary; *especially* **:** such a barrier made of posts and wire or boards
3 : FENCING 1
4 a : a receiver of stolen goods **b :** a place where stolen goods are bought
— **fence·less** \-ləs\ *adjective*
— **fence·less·ness** *noun*
— **on the fence :** in a position of neutrality or indecision

²fence *verb* **fenced; fenc·ing** (15th century)
transitive verb
1 a : to enclose with a fence **b** (1) **:** to keep in or out with a fence (2) **:** to ward off
2 : to provide a defense for
3 : to sell (stolen property) to a fence
intransitive verb
1 a : to practice fencing **b** (1) **:** to use tactics of attack and defense resembling those of fencing (2) **:** to parry arguments by shifting ground
2 *archaic* **:** to provide protection
— **fenc·er** *noun*

fence–mend·ing \'fen(t)s-ˌmen-diŋ\ *noun* (1947)
: the rehabilitating of a deteriorated political relationship

fence·row \'fen(t)s-ˌrō\ *noun* (1842)
: the land occupied by a fence including the uncultivated area on each side

fence–sit·ting \'fen(t)s-ˌsi-tiŋ\ *noun* (1904)
: a state of indecision or neutrality with respect to conflicting positions
— **fence–sit·ter** *noun*

fenc·ing *noun* (1581)
1 : the art or practice of attack and defense with the foil, épée, or saber
2 a (1) **:** FENCE 2 (2) **:** the fences of a property or region **b :** material used for building fences

¹fend \'fend\ *verb* [Middle English *fenden,* short for *defenden*] (14th century)
transitive verb
1 : DEFEND
2 : to keep or ward off **:** REPEL — often used with *off*
3 *dialect British* **:** to provide for **:** SUPPORT
intransitive verb
1 *dialect British* **:** to make an effort **:** STRUGGLE
2 a : to try to get along without help **:** SHIFT ⟨had to *fend* for themselves⟩ **b :** to provide a livelihood

²fend *noun* (1721)
chiefly Scottish **:** an effort or attempt especially for oneself

fend·er \'fen-dər\ *noun* (15th century)
: a device that protects: as **a** (1) **:** a cushion (as foam rubber or a wood float) between a boat and a dock or between two boats that lessens shock and prevents chafing (2) **:** a pile or a row or cluster of piles placed to protect a dock or bridge pier from damage by ships or floating objects **b :** RAILING **c :** a device in front of locomotives and streetcars to lessen injury to animals or pedestrians in case of collision **d :** a guard over the wheel of a motor vehicle **e :** a low metal frame or a screen before an open fireplace **f :** an oblong or triangular shield of leather attached to the stirrup leather of a saddle to protect a rider's legs
— **fend·ered** \-dərd\ *adjective*
— **fend·er·less** *adjective*

fender bender *noun* (circa 1962)
: a minor automobile accident

fe·nes·tra \fə-'nes-trə\ *noun, plural* **-trae** \-ˌtrē, -ˌtrī\ [New Latin, from Latin, window] (circa 1737)
1 : a small anatomical opening (as in a bone): as **a :** an oval opening between the middle ear and the vestibule having the base of the stapes or columella attached to its membrane — called also *fenestra ova·lis* \-ō-'vā-ləs\, *fenestra ves·tib·u·li* \-ves-'ti-byə-(ˌ)lē\, *oval window* **b :** a round opening between the middle ear and the cochlea — called also *fenestra cochleae, fenestra ro·tun·da* \-rō-'tən-də\, *round window*
2 : an opening cut in bone
— **fe·nes·tral** \-trəl\ *adjective*

fe·nes·trate \fə-'nes-ˌtrāt, 'fe-nə-ˌstrāt\ *adjective* [Latin *fenestratus,* from *fenestra*] (1835)
: FENESTRATED

fen·es·trat·ed \'fe-nə-ˌstrā-təd\ *adjective* (circa 1852)
: having one or more openings or pores ⟨*fenestrated* blood capillaries⟩

fen·es·tra·tion \ˌfe-nə-'strā-shən\ *noun* (1846)
1 : the arrangement, proportioning, and design of windows and doors in a building
2 : an opening in a surface (as a wall or membrane)
3 : the operation of cutting an opening in the bony labyrinth between the inner ear and tympanum to replace natural fenestrae that are not functional

Fe·ni·an \'fē-nē-ən\ *noun* [Irish *féinne,* plural of *fiann,* legendary band of Irish warriors] (1816)
1 : a member of a legendary band of warriors defending Ireland in the 2d and 3d centuries A.D.
2 : a member of a secret 19th century Irish and Irish-American organization dedicated to the overthrow of British rule in Ireland
— **Fenian** *adjective*
— **Fe·ni·an·ism** \-ə-ˌni-zəm\ *noun*

fen·land \'fen-ˌland\ *noun* (before 12th century)
: an area of low often marshy ground

fen·nec \'fe-nik\ *noun* [Arabic *fanak*] (1790)
: a small pale-fawn fox (*Vulpes zerda* synonym *Fennecus zerda*) with large ears that inhabits the deserts of northern Africa and Arabia

fen·nel \'fe-nᵊl\ *noun* [Middle English *fenel,* from Old English *finugl,* from (assumed) Vulgar Latin *fenuculum,* from Latin *feniculum* fennel, irregular diminutive of *fenum* hay] (before 12th century)
: a perennial European herb (*Foeniculum vulgare*) of the carrot family introduced into North America and cultivated for its foliage and aromatic seeds

fen·ny \'fe-nē\ *adjective* (before 12th century)
1 : having the characteristics of a fen **:** BOGGY
2 *archaic* **:** peculiar to or found in a fen

fenu·greek \'fen-yə-ˌgrēk\ *noun* [Middle English *fenugrek,* from Middle French *fenugrec,* from Latin *fenum Graecum,* literally, Greek hay] (14th century)
: a leguminous annual Asian herb (*Trigonella foenumgraecum*) with aromatic seeds

feoff·ee \fe-'fē, fē-'fē\ *noun* (15th century)
: one to whom a feoffment is made

feoff·ment \'fef-mənt, 'fēf-\ *noun* [Middle English *feoffement,* from Anglo-French, from *feoffer* to invest with a fee, from Old French *fief* fee] (14th century)
: the granting of a fee

\ə\ **abut** \ᵊ\ kitten \ər\ **further** \a\ **ash** \ā\ **ace** \ä\ **mop, mar** \au̇\ **out** \ch\ **chin** \e\ **bet** \ē\ **easy** \g\ **go** \i\ **hit** \ī\ **ice** \j\ **job** \ŋ\ **sing** \ō\ **go** \ȯ\ **law** \ȯi\ **boy** \th\ **thin** \t͟h\ **the** \ü\ **loot** \u̇\ **foot** \y\ **yet** \zh\ **vision** *see also* Guide to Pronunciation

feof·for \'fe-fər, 'fē-; fe-'fȯr, fē-\ *or* **feoff·er** \'fef-ər, 'fēf-\ *noun* (15th century)
: one who makes a feoffment

-fer \fər\ *noun combining form* [French & Latin; French *-fère,* from Latin *-fer* bearing, one that bears, from *ferre* to carry — more at BEAR]
: one that bears ⟨aqui*fer*⟩

fe·rae na·tu·rae \'fer-,ī-nə-'tu̇r-,ī\ *adjective* [Latin, of a wild nature] (circa 1661)
: wild by nature and not usually tamed

fe·ral \'fir-əl, 'fer-\ *adjective* [Medieval Latin *feralis,* from Latin *fera* wild animal, from feminine of *ferus* wild — more at FIERCE] (1604)
1 : of, relating to, or suggestive of a wild beast **:** SAVAGE
2 a : not domesticated or cultivated **:** WILD **b :** having escaped from domestication and become wild
synonym see BRUTAL

fer–de–lance \'fer-dᵊl-'an(t)s, -'än(t)s\ *noun, plural* **fer–de–lance** [French, literally, lance iron, spearhead] (1880)
: a large extremely venomous pit viper (*Bothrops atrox*) of Central and South America

fere \'fir\ *noun* [Middle English, from Old English *gefēra;* akin to Old English *faran* to go, travel — more at FARE] (before 12th century)
1 *archaic* **:** COMPANION, COMRADE
2 *archaic* **:** SPOUSE

¹fe·ria \'fir-ē-ə, 'fer-\ *noun* [Medieval Latin — more at FAIR] (15th century)
: a weekday of a church calendar on which no feast falls
— fe·ri·al \-ē-əl\ *adjective*

²fe·ria \'fer-ē-ə, -ē-,ä\ *noun* [Spanish, fair, market, from Medieval Latin — more at FAIR] (1844)
: an Hispanic market festival often in observance of a religious holiday

fe·rine \'fir-,īn\ *adjective* [Latin *ferinus,* from *fera*] (1640)
: FERAL

fer·i·ty \'fer-ət-ē\ *noun, plural* **-ties** [Latin *feritas,* from *ferus*] (1534)
archaic **:** the quality or state of being feral; *also* **:** BARBARITY

fer·lie *also* **fer·ly** \'fer-lē\ *noun, plural* **ferlies** [Middle English, from *ferly* strange, from Old English *fǣrlīc* unexpected, from *fǣr* sudden danger — more at FEAR] (13th century)
Scottish **:** WONDER

fer·ma·ta \fer-'mä-tə\ *noun* [Italian, literally, stop, from *fermare* to stop, from Latin *firmare* to make firm] (circa 1859)
: a prolongation at the discretion of the performer of a musical note, chord, or rest beyond its given time value; *also* **:** the sign 𝄐 denoting such a prolongation — called also *hold*

¹fer·ment \(,)fər-'ment\ (14th century)
intransitive verb
1 : to undergo fermentation
2 : to be in a state of agitation or intense activity
transitive verb
1 : to cause to undergo fermentation
2 : to work up (as into a state of agitation) **:** FOMENT ⟨quick-spreading rumors *fermented* the city and violence soon broke out⟩
— fer·ment·able \-'men-tə-bəl\ *adjective*

²fer·ment \'fər-,ment *also* (,)fər-'\ *noun* [Middle English, from Latin *fermentum* yeast — more at BARM] (15th century)
1 a : a living organism (as a yeast) that causes fermentation by virtue of its enzymes **b :** ENZYME
2 a : a state of unrest **:** AGITATION **b :** a process of active often disorderly development ⟨the great period of creative *ferment* in literature —William Barrett⟩

fer·men·ta·tion \,fər-mən-'tā-shən, -,men-\ *noun* (1601)
1 a : a chemical change with effervescence **b :** an enzymatically controlled anaerobic breakdown of an energy-rich compound (as a carbohydrate to carbon dioxide and alcohol or to an organic acid); *broadly* **:** an enzymatically controlled transformation of an organic compound
2 : FERMENT 2

fer·men·ta·tive \(,)fər-'men-tə-tiv\ *adjective* (1661)
1 : causing or producing a substance that causes fermentation ⟨*fermentative* organisms⟩
2 : of, relating to, or produced by fermentation

fer·men·ter \(,)fər-'men-tər\ *noun* (1918)
1 : an organism that causes fermentation
2 *or* **fer·men·tor :** an apparatus for carrying out fermentation

fer·mi \'fer-(,)mē, 'fər-\ *noun* [Enrico *Fermi*] (1955)
: a unit of length equal to 10^{-13} centimeter

fer·mi·on \'fer-mē-,än, 'fər-\ *noun* [Enrico *Fermi* + English ²*-on*] (1947)
: a particle (as an electron, proton, or neutron) whose spin quantum number is an odd multiple of ½ — compare BOSON

fer·mi·um \'fer-mē-əm, 'fər-\ *noun* [Enrico *Fermi*] (1955)
: a radioactive metallic element artificially produced (as by bombardment of plutonium with neutrons) — see ELEMENT table

fern \'fərn\ *noun* [Middle English, from Old English *fearn;* akin to Old High German *farn* fern, Sanskrit *parṇa* wing, leaf] (before 12th century)
: any of a large class (Filicopsida) of flowerless spore-producing vascular plants; *especially* **:** any of an order (Filicales) of homosporous plants possessing roots, stems, and leaflike fronds
— fern·like \-,līk\ *adjective*
— ferny \'fər-nē\ *adjective*

fern bar *noun* (1976)
: a bar or restaurant fashionably decorated with green plants and especially ferns

fern·ery \'fər-nə-rē\ *noun, plural* **-er·ies** (1840)
1 : a place or stand where ferns grow
2 : a collection of growing ferns

fern seed *noun* (1596)
: the dustlike asexual spores of ferns formerly thought to be seeds and believed to make the possessor invisible

fe·ro·cious \fə-'rō-shəs\ *adjective* [Latin *feroc-, ferox,* literally, fierce looking, from *ferus* + *-oc-, -ox* (akin to Greek *ōps* eye) — more at EYE] (1646)
1 : exhibiting or given to extreme fierceness and unrestrained violence and brutality
2 : extremely intense ⟨*ferocious* heat⟩
synonym see FIERCE
— fe·ro·cious·ly *adverb*
— fe·ro·cious·ness *noun*

fe·roc·i·ty \fə-'rä-sə-tē\ *noun* (1606)
: the quality or state of being ferocious

-ferous *adjective combining form* [Middle English, from Middle French & Latin; Middle French *-fere* -fer, from Latin *-fer*]
: bearing **:** producing ⟨carboni*ferous*⟩

fer·re·dox·in \,fer-ə-'däk-sən\ *noun* [Latin *ferrum* + English *redox* + *-in*] (1962)
: any of a group of iron-containing plant proteins that function as electron carriers in photosynthetic organisms and in some anaerobic bacteria

¹fer·ret \'fer-ət\ *noun* [Middle English *furet, ferret,* from Middle French *furet,* from (assumed) Vulgar Latin *furittus,* literally, little thief, diminutive of Latin *fur* thief — more at FURTIVE] (14th century)
1 : a partially domesticated usually albino European polecat that is sometimes classed as a separate species (*Mustela furo*)
2 : an active and persistent searcher ◆
— fer·rety \-ə-tē\ *adjective*

²ferret (15th century)
intransitive verb
1 : to hunt with ferrets
2 : to search about
transitive verb
1 a (1) **:** to hunt (as rabbits) with ferrets (2) **:** to force out of hiding **:** FLUSH **b :** to find and bring to light by searching — usually used with *out* ⟨*ferret* out the answers⟩
2 : HARRY, WORRY
— fer·ret·er *noun*

³ferret *noun* [probably modification of Italian *fioretti* floss silk, from plural of *fioretto,* diminutive of *fiore* flower, from Latin *flor-, flos* — more at BLOW] (1649)
: a narrow cotton, silk, or wool tape — called also *ferreting*

ferri- *combining form* [Latin, from *ferrum*]
1 : iron ⟨*ferri*ferous⟩
2 : ferric iron ⟨*ferri*cyanide⟩

fer·ri·age \'fer-ē-ij\ *noun* [Middle English] (14th century)
1 : the fare paid for a ferry passage
2 : the act or business of transporting by ferry

fer·ric \'fer-ik\ *adjective* (1799)
1 : of, relating to, or containing iron
2 : being or containing iron usually with a valence of three

ferric ammonium citrate *noun* (circa 1924)
: a complex salt containing varying amounts of iron and used especially for making blueprints

ferric chloride *noun* (1885)
: a deliquescent dark salt $FeCl_3$ that readily hydrates to the yellow-orange form and that is used in sewage treatment and as an astringent

ferric hydroxide *noun* (1885)
: a hydrate $Fe_2O_3{\cdot}nH_2O$ of ferric oxide that is capable of acting both as a base and as a weak acid

ferric oxide *noun* (1882)
: the red or black oxide of iron Fe_2O_3 found in nature as hematite and as rust and also obtained synthetically and used as a pigment and for polishing

fer·ri·cy·a·nide \,fer-,ī-'sī-ə-,nīd, ,fer-i-\ *noun* [International Scientific Vocabulary] (1845)
1 : the trivalent anion $Fe(CN)_6^{3-}$
2 : a compound containing the ferricyanide anion; *especially* **:** the red salt $K_3Fe(CN)_6$ used in making blue pigments

fer·rif·er·ous \fə-'ri-f(ə-)rəs, fe-\ *adjective* (1811)
: containing or yielding iron

fer·ri·mag·net·ic \'fer-,ī-mag-'ne-tik, ,fer-i-\ *adjective* (1951)
: of or relating to a substance (as ferrite) characterized by magnetization in which one group of magnetic ions is polarized in a direction opposite to the other
— fer·ri·mag·net \'fer-,ī-,mag-nət, 'fer-i-\ *noun*
— fer·ri·mag·net·i·cal·ly \'fer-,ī-mag-'ne-ti-k(ə-)lē, ,fer-i-\ *adverb*
— fer·ri·mag·ne·tism \-'mag-nə-,ti-zəm\ *noun*

Fer·ris wheel \'fer-əs-\ *noun* [G. W. G. *Ferris* (died 1896) American engineer] (1893)

◇ WORD HISTORY
ferret The *ferret* is a partially domesticated breed of the European polecat. For centuries the ferret has been used in Europe in hunting rats and sometimes rabbits. But both the wild polecat and his domestic relative have rather bad reputations, especially as poultry thieves. The polecat's name (Middle English *polcat*) is of uncertain origin, but one suggestion is that it may be a compound of Middle French *pol* (or *poul*) "cock," and Middle English *cat.* And the ferret is a "little thief": Middle English *ferret* (or *furet*) comes from Middle French *furet,* which ultimately goes back to Latin *fur* "thief." The sport of ferreting out rodents and rabbits gives us the common modern sense of the verb *ferret,* "to find and bring to light by searching."

: an amusement device consisting of a large upright power-driven wheel carrying seats that remain horizontal around its rim

fer·rite \'fer-,īt\ *noun* (1851)
1 : any of several magnetic substances that consist essentially of ferric oxide combined with the oxides of one or more metals (as manganese, nickel, or zinc), have high magnetic permeability and high electrical resistivity, and are used especially in electronic devices
2 : a solid solution in which alpha iron is the solvent
— fer·rit·ic \fə-'ri-tik, fe-\ *adjective*

fer·ri·tin \'fer-ə-tᵊn\ *noun* [International Scientific Vocabulary, alteration of *ferratin,* iron-containing protein, from Latin *ferrat*us bound with iron (from *ferrum*) + International Scientific Vocabulary *-in*] (1937)
: a crystalline iron-containing protein that functions in the storage of iron and is found especially in the liver and spleen

ferro- *combining form* [Medieval Latin, from Latin *ferrum*]
1 : iron ⟨*ferro*concrete⟩
2 : ferrous iron ⟨*ferro*cyanide⟩

fer·ro–ce·ment \,fer-ō-si-'ment\ *noun* (1956)
: a building material made of thin cement slabs reinforced with steel mesh

fer·ro·cene \'fer-ō-,sēn\ *noun* [*ferro-* + cycl- + *-ene*] (1952)
: a crystalline stable organometallic coordination compound $(C_5H_5)_2Fe$; *also* **:** an analogous compound with a heavy metal (as chromium)

fer·ro·con·crete \,fer-ō-'kän-,krēt, -kän-'\ *noun* (1900)
: REINFORCED CONCRETE

fer·ro·cy·a·nide \-'sī-ə-,nīd\ *noun* (circa 1826)
1 : the tetravalent anion $Fe(CN)_6^{4-}$
2 : a compound containing the ferrocyanide anion; *especially* **:** the salt $K_4Fe(CN)_6$ used in making blue pigments (as Prussian blue)

fer·ro·elec·tric \,fer-ō-i-'lek-trik\ *adjective* (1935)
: of or relating to crystalline substances having spontaneous electric polarization reversible by an electric field
— ferroelectric *noun*
— fer·ro·elec·tric·i·ty \-,lek-'tri-sə-tē, 'tris-tē\ *noun*

fer·ro·mag·ne·sian \-mag-'nē-zhən, -shən\ *adjective* (1899)
: containing iron and magnesium ⟨*ferromagnesian* minerals⟩

fer·ro·mag·net·ic \-'ne-tik\ *adjective* (1896)
: of or relating to substances with an abnormally high magnetic permeability, a definite saturation point, and appreciable residual magnetism and hysteresis
— fer·ro·mag·net \'fer-ə-,mag-nət\ *noun*
— fer·ro·mag·ne·tism \,fer-ō-'mag-nə-,ti-zəm\ *noun*

fer·ro·man·ga·nese \-'maŋ-gə-,nēz, -,nēs\ *noun* (1864)
: an alloy of iron and manganese containing usually about 80 percent manganese and used in the manufacture of steel

fer·ro·sil·i·con \-'si-li-kən, -lə-,kän\ *noun* (1882)
: an alloy of iron and silicon containing 15 to 95 percent silicon and used for deoxidizing molten steel and making silicon steel and high-silicon cast iron

fer·ro·type \'fer-ə-,tīp\ *noun* (1844)
1 : a positive photograph made by a collodion process on a thin iron plate having a darkened surface
2 : the process by which a ferrotype is made

fer·rous \'fer-əs\ *adjective* [New Latin *ferrosus,* from Latin *ferrum*] (1865)
1 : of, relating to, or containing iron
2 : being or containing divalent iron

ferrous oxide *noun* (1873)
: a black easily oxidizable powder FeO that is the monoxide of iron

ferrous sulfate *noun* (1865)
: a salt $FeSO_4$; *especially* **:** COPPERAS

fer·ru·gi·nous \fə-'rü-jə-nəs, fe-\ *adjective* [Latin *ferrugineus, ferruginus,* from *ferrugin-, ferrugo* iron rust, from *ferrum*] (circa 1661)
1 : of, relating to, or containing iron ⟨a *ferruginous* soil⟩
2 : resembling iron rust in color

fer·rule \'fer-əl\ *noun* [alteration of Middle English *virole,* from Middle French, from Latin *viriola,* diminutive of *viria* bracelet, of Celtic origin; akin to Old Irish *fiar* oblique] (1611)
1 : a ring or cap usually of metal put around a slender shaft (as a cane or a tool handle) to strengthen it or prevent splitting
2 : a usually metal sleeve used especially for joining or binding one part to another (as pipe sections or the bristles and handle of a brush)
— fer·ruled *adjective*

[1]fer·ry \'fer-ē\ *verb* **fer·ried; fer·ry·ing** [Middle English *ferien,* from Old English *ferian* to carry, convey; akin to Old English *faran* to go — more at FARE] (before 12th century)
transitive verb
1 a : to carry by boat over a body of water **b :** to cross by a ferry
2 a : to convey (as by aircraft or motor vehicle) from one place to another **:** TRANSPORT **b :** to fly (an airplane) from the factory or other shipping point to a designated delivery point or from one base to another
intransitive verb
: to cross water in a boat

[2]ferry *noun, plural* **fer·ries** (15th century)
1 : a place where persons or things are carried across a body of water (as a river) in a boat
2 : FERRYBOAT
3 : a franchise or right to operate a ferry service across a body of water
4 : an organized service and route for flying airplanes especially across a sea or continent for delivery to the user

fer·ry·boat \'fer-ē-,bōt\ *noun* (15th century)
: a boat used to ferry passengers, vehicles, or goods

fer·ry·man \-mən\ *noun* (15th century)
: a person who operates a ferry

fer·tile \'fər-tᵊl\ *adjective* [Middle English, from Middle French & Latin; Middle French, from Latin *fertilis,* from *ferre* to carry, bear — more at BEAR] (15th century)
1 a : producing or bearing fruit in great quantities **:** PRODUCTIVE **b :** characterized by great resourcefulness of thought or imagination **:** INVENTIVE ⟨a *fertile* mind⟩ **c** *obsolete* **:** PLENTIFUL
2 a (1) **:** capable of sustaining abundant plant growth ⟨*fertile* soil⟩ (2) **:** affording abundant possibilities for development ⟨a *fertile* area for research⟩ **b :** capable of growing or developing ⟨*fertile* egg⟩ **c** (1) **:** capable of producing fruit (2) *of an anther* **:** containing pollen (3) **:** developing spores or spore-bearing organs **d :** capable of breeding or reproducing
3 : capable of being converted into fissionable material ⟨*fertile* uranium 238⟩ ☆
— fer·tile·ly \-tᵊl-(l)ē\ *adverb*
— fer·tile·ness \-tᵊl-nəs\ *noun*

fer·til·i·ty \(,)fər-'ti-lə-tē\ *noun* (15th century)
1 : the quality or state of being fertile
2 : the birthrate of a population

fer·til·iza·tion \,fər-tᵊl-ə-'zā-shən\ *noun* (circa 1787)
: an act or process of making fertile: as **a :** the application of fertilizer **b** (1) **:** an act or process of fecundation, insemination, or pollination — not used technically (2) **:** the process of union of two gametes whereby the somatic chromosome number is restored and the development of a new individual is initiated

fertilization membrane *noun* (1931)
: a resistant membranous layer in eggs of many animals that forms following fertilization by the thickening and separation of the vitelline membrane from the cell surface and that prevents multiple fertilization

fer·til·ize \'fər-tᵊl-,īz\ *transitive verb* **-ized; -iz·ing** (1648)
: to make fertile: as **a :** to apply a fertilizer to ⟨*fertilize* land⟩ **b :** to cause the fertilization of
— fer·til·iz·able \-,ī-zə-bəl\ *adjective*

fer·til·iz·er \-,ī-zər\ *noun* (circa 1661)
: one that fertilizes; *specifically* **:** a substance (as manure or a chemical mixture) used to make soil more fertile

fer·ule \'fer-əl\ *also* **fer·u·la** \'fer-(y)ə-lə\ *noun* [Latin *ferula* giant fennel, ferule] (1580)
1 : an instrument (as a flat piece of wood like a ruler) used to punish children
2 : school discipline

fe·ru·lic acid \fə-'rü-lik-\ *noun* [*ferula*] (1876)
: a white crystalline acid that is structurally related to vanillin and is obtained especially from plant sources (as aspen bark)

fer·ven·cy \'fər-vən(t)-sē\ *noun, plural* **-cies** (15th century)
: FERVOR

fer·vent \'fər-vənt\ *adjective* [Middle English, from Middle French & Latin; Middle French, from Latin *fervent-, fervens,* present participle of *fervēre* to boil, froth — more at BARM] (14th century)
1 : very hot **:** GLOWING
2 : exhibiting or marked by great intensity of feeling **:** ZEALOUS ⟨*fervent* prayers⟩
synonym see IMPASSIONED
— fer·vent·ly *adverb*

fer·vid \'fər-vəd\ *adjective* [Latin *fervidus,* from *fervēre*] (1599)
1 : very hot **:** BURNING
2 : marked by often extreme fervor ⟨a *fervid* crusader⟩
synonym see IMPASSIONED
— fer·vid·ly *adverb*
— fer·vid·ness *noun*

fer·vor \'fər-vər\ *noun* [Middle English *fervour,* from Middle French & Latin; Middle French *ferveur,* from Latin *fervor,* from *fervēre*] (14th century)
1 : intensity of feeling or expression ⟨booing and cheering with almost equal *fervor* —Alan Rich⟩
2 : intense heat
synonym see PASSION

fer·vour *chiefly British variant of* FERVOR

fes·cen·nine \'fe-sᵊn-,īn, -,ēn\ *adjective* [Latin *fescennini (versus),* ribald songs sung at rustic weddings, probably from *Fescinninus* of Fescennium, from *Fescennium,* town in Etruria] (1601)
: SCURRILOUS, OBSCENE

fes·cue \'fes-(,)kyü\ *noun* [Middle English *festu* stalk, straw, from Middle French, from Late Latin *festucum,* from Latin *festuca*] (1513)
1 : a small pointer (as a stick) used to point out letters to children learning to read

☆ SYNONYMS
Fertile, fecund, fruitful, prolific mean producing or capable of producing offspring or fruit. FERTILE implies the power to reproduce in kind or to assist in reproduction and growth ⟨*fertile* soil⟩; applied figuratively, it suggests readiness of invention and development ⟨a *fertile* imagination⟩. FECUND emphasizes abundance or rapidity in bearing fruit or offspring ⟨a *fecund* herd⟩. FRUITFUL adds to FERTILE and FECUND the implication of desirable or useful results ⟨*fruitful* research⟩. PROLIFIC stresses rapidity of spreading or multiplying by or as if by natural reproduction ⟨a *prolific* writer⟩.

2 : any of a genus (*Festuca*) of tufted perennial grasses with panicled spikelets

fescue foot *noun* (1949)
: a disease of the feet of cattle resembling ergotism that is associated with feeding on fescue grass and especially tall fescue

¹fess *also* **fesse** \'fes\ *noun* [Middle English *fesse,* from Middle French *faisse,* from Latin *fascia* band — more at FASCIA] (15th century)
1 : a broad horizontal bar across the middle of a heraldic field
2 : the center point of an armorial escutcheon

²fess \'fes\ *intransitive verb* [short for *confess*] (1840)
: to own up **:** CONFESS — usually used with *up*

-fest \,fest\ *noun combining form* [German, from *Fest* celebration, from Latin *festum* — more at FEAST]
: meeting or occasion marked by (such) activity ⟨song*fest*⟩

fes·tal \'fes-tᵊl\ *adjective* [Latin *festum*] (15th century)
: of or relating to a feast or festival **:** FESTIVE
— fes·tal·ly \-tᵊl-ē\ *adverb*

¹fes·ter \'fes-tər\ *noun* [Middle English, from Middle French *festre,* from Latin *fistula* pipe, fistulous ulcer] (14th century)
: a suppurating sore **:** PUSTULE

²fester *verb* **fes·tered; fes·ter·ing** \-t(ə-)riŋ\ (14th century)
intransitive verb
1 : to generate pus
2 : PUTREFY, ROT
3 a : to cause increasing poisoning, irritation, or bitterness **:** RANKLE ⟨dissent *festered* unchecked⟩ **b :** to undergo or exist in a state of progressive deterioration ⟨allowed slums to *fester*⟩
transitive verb
: to make inflamed or corrupt

¹fes·ti·nate \'fes-tə-,nāt\ *verb* **-nat·ed; -nat·ing** (1596)
: HASTEN

²fes·ti·nate \-nət, -,nāt\ *adjective* [Latin *festinatus,* past participle of *festinare* to hasten; perhaps akin to Middle Irish *bras* forceful, Welsh *brys* haste] (1605)
: HASTY
— fes·ti·nate·ly *adverb*

¹fes·ti·val \'fes-tə-vəl\ *adjective* [Middle English, from Middle French, from Latin *festivus* festive] (14th century)
: of, relating to, appropriate to, or set apart as a festival

²festival *noun* (1589)
1 a : a time of celebration marked by special observances **b :** FEAST 2
2 : a periodic season or program of cultural events or entertainment
3 : GAIETY, CONVIVIALITY

fes·ti·val·go·er \-,gō(-ə)r\ *noun* (1959)
: one who attends a festival

fes·tive \'fes-tiv\ *adjective* [Latin *festivus,* from *festum*] (1651)
1 : of, relating to, or suitable for a feast or festival
2 : JOYFUL, GAY
— fes·tive·ly *adverb*
— fes·tive·ness *noun*

fes·tiv·i·ty \fes-'ti-və-tē, fəs-\ *noun, plural* **-ties** (14th century)
1 : FESTIVAL 1
2 : the quality or state of being festive **:** GAIETY
3 : festive activity

¹fes·toon \fes-'tün\ *noun* [French *feston,* from Italian *festone,* from *festa* festival, from Latin — more at FEAST] (1630)
1 : a decorative chain or strip hanging between two points
2 : a carved, molded, or painted ornament representing a decorative chain

²festoon *transitive verb* (1800)
1 : to hang or form festoons on
2 : to shape into festoons
3 : DECORATE, ADORN

fes·toon·ery \fes-'tü-nə-rē\ *noun* (1836)
: an arrangement of festoons

Fest·schrift \'fes(t)-,shrift\ *noun, plural* **Fest·schrif·ten** \-,shrif-tən\ *or* **Fest·schrifts** [German, from *Fest* celebration + *Schrift* writing] (1898)
: a volume of writings by different authors presented as a tribute or memorial especially to a scholar

fe·ta \'fe-tə, 'fe-,tä\ *noun, often capitalized* [New Greek (*tyri*) *pheta,* from *tyri* cheese + *pheta* slice, from Italian *fetta* — more at FETTUCCINE] (1940)
: a white moderately hard and crumbly Greek cheese made from sheep's or goat's milk and cured in brine

fe·tal \'fē-tᵊl\ *adjective* (1811)
: of, relating to, or being a fetus

fetal alcohol syndrome *noun* (1974)
: a highly variable group of birth defects including mental retardation, deficient growth, and defects of the skull, face, and brain that tend to occur in the infants of women who consume large amounts of alcohol during pregnancy

fetal hemoglobin *noun* (1950)
: a hemoglobin variant that predominates in the blood of a newborn and persists in increased proportions in some forms of anemia (as thalassemia)

fetal position *noun* (1963)
: a position (as of a sleeping person) in which the body lies on one side, curled up with the arms and legs drawn up toward the chest and the head bowed forward

¹fetch \'fech\ *verb* [Middle English *fecchen,* from Old English *fetian, feccan;* perhaps akin to Old English *fōt* foot — more at FOOT] (before 12th century)
transitive verb
1 a : to go or come after and bring or take back **b :** DERIVE, DEDUCE
2 a : to cause to come **b :** to bring in (as a price) **:** REALIZE **c :** INTEREST, ATTRACT
3 a : to give (a blow) by striking **:** DEAL **b** *chiefly dialect* **:** BRING ABOUT, ACCOMPLISH **c** (1) **:** to take in (as a breath) **:** DRAW (2) **:** to bring forth (as a sound) **:** HEAVE ⟨*fetch* a sigh⟩
4 a : to reach by sailing especially against the wind or tide **b :** to arrive at **:** REACH
intransitive verb
1 : to get and bring something; *specifically* **:** to retrieve killed game
2 : to take a roundabout way **:** CIRCLE
3 a : to hold a course on a body of water **b :** VEER
— fetch·er *noun*

²fetch *noun* (circa 1530)
1 : TRICK, STRATAGEM
2 : an act or instance of fetching
3 a : the distance along open water or land over which the wind blows **b :** the distance traversed by waves without obstruction

³fetch *noun* [origin unknown] (circa 1787)
: DOPPELGÄNGER 1

fetch·ing \'fe-chiŋ\ *adjective* (1880)
: ATTRACTIVE, PLEASING
— fetch·ing·ly \-chiŋ-lē\ *adverb*

fetch up (1599)
transitive verb
1 : to bring up or out **:** PRODUCE
2 : to make up (as lost time)
3 : to bring to a stop
intransitive verb
: to reach a standstill, stopping place, or goal **:** end up ⟨may have *fetched up* running a village store —Geoffrey Household⟩

¹fete *or* **fête** \'fāt, 'fet\ *noun* [Middle English *fete,* from Middle French, from Old French *feste* — more at FEAST] (15th century)
1 : FESTIVAL
2 a : a lavish often outdoor entertainment **b :** a large elaborate party

²fete *or* **fête** *transitive verb* **fet·ed** *or* **fêt·ed; fet·ing** *or* **fêt·ing** (1819)
1 : to honor or commemorate with a fete
2 : to pay high honor to

fête cham·pê·tre \,fāt-,shäⁿ(m)-'petrᵊ, ,fet-\ *noun, plural* **fêtes champêtres** *same*\ [French, literally, rural festival] (1774)
: an outdoor entertainment

fe·ti·cide \'fē-tə-,sīd\ *noun* (circa 1844)
: the act of causing the death of a fetus

fet·id \'fe-təd, *especially British* 'fē-tid\ *adjective* [Middle English, from Latin *foetidus,* from *foetēre* to stink] (15th century)
: having a heavy offensive smell
synonym see MALODOROUS
— fet·id·ly *adverb*
— fet·id·ness *noun*

fe·tish *also* **fe·tich** \'fe-tish *also* 'fē-\ *noun* [French & Portuguese; French *fétiche,* from Portuguese *feitiço,* from *feitiço* artificial, false, from Latin *facticius* factitious] (1613)
1 a : an object (as a small stone carving of an animal) believed to have magical power to protect or aid its owner; *broadly* **:** a material object regarded with superstitious or extravagant trust or reverence **b :** an object of irrational reverence or obsessive devotion **:** PREPOSSESSION **c :** an object or bodily part whose real or fantasied presence is psychologically necessary for sexual gratification and that is an object of fixation to the extent that it may interfere with complete sexual expression
2 : a rite or cult of fetish worshipers
3 : FIXATION ◆

fe·tish·ism *also* **fe·tich·ism** \-tish-,i-zəm\ *noun* (1801)
1 : belief in magical fetishes
2 : extravagant irrational devotion
3 : the pathological displacement of erotic interest and satisfaction to a fetish
— fe·tish·ist \-tish-ist\ *noun*
— fe·tish·is·tic \,fe-ti-'shis-tik *also* ,fē-\ *adjective*
— fe·tish·is·ti·cal·ly \-ti-k(ə-)lē\ *adverb*

fet·lock \'fet-,läk\ *noun* [Middle English *fitlok, fetlak;* akin to Old English *fōt* foot] (14th century)
1 a : a projection bearing a tuft of hair on the back of the leg above the hoof of a horse or similar animal — see HORSE illustration **b :** the tuft of hair itself
2 : the joint of the limb at the fetlock

feto- *or* **feti-** *combining form* [New Latin *fetus*]
: fetus ⟨*feti*cide⟩

fe·tol·o·gy \fē-'tä-lə-jē\ *noun* (1965)

◇ WORD HISTORY

fetish The Portuguese word *feitiço* "charm, amulet" is descended from Latin *facticius* "made, artificial," a derivative of *facere* "to make, do." These Latin words are also the source of English *factitious.* When in the 15th century the Portuguese began to encounter and soon enslave peoples of West Africa, they applied *feitiço* to African religious objects, whether natural or man-made. English borrowed this word in the 17th century, at first directly from Portuguese (in spellings such as *fetisso*). Later it was borrowed again through French *fétiche,* itself ultimately from Portuguese, which gave rise to the current English form *fetish.* Both in English and other languages that borrowed it, the meaning of the word was extended to any object thought to be irrationally or obsessively worshiped. *Fetish* was given new life when it was used at the end of the 19th century by Havelock Ellis, a pioneering student of human sexuality, to refer to objects of displaced sexual obsession, and the term was adopted by psychoanalytic theorists such as Sigmund Freud, who first used the German equivalent *Fetisch* in 1905.

: a branch of medical science concerned with the study and treatment of the fetus in the uterus
— **fe·tol·o·gist** \-jist\ *noun*

fe·to·pro·tein \ˌfē-tō-ˈprō-ˌtēn, -ˈprō-tē-ən\ *noun* (1965)
: any of several fetal antigens present in the adult in some abnormal conditions

fe·tor \ˈfē-tər, ˈfē-ˌtȯr\ *noun* [Middle English *fetoure,* from Latin *foetor,* from *foetēre*] (15th century)
: a strong offensive smell : STENCH

fe·tos·co·py \fē-ˈtäs-kə-pē\ *noun, plural* **-pies** (1971)
: examination of the pregnant uterus by means of a fiber-optic tube
— **fe·to·scope** \ˈfē-tə-ˌskōp\ *noun*

[1]**fet·ter** \ˈfe-tər\ *noun* [Middle English *feter,* from Old English; akin to Old English *fōt* foot] (before 12th century)
1 : a chain or shackle for the feet
2 : something that confines : RESTRAINT

[2]**fetter** *transitive verb* (before 12th century)
1 : to put fetters on : SHACKLE
2 : to restrain from motion, action, or progress
synonym see HAMPER

[1]**fet·tle** \ˈfe-tᵊl\ *noun* (1740)
: state or condition of health, fitness, wholeness, spirit, or form — often used in the phrase *in fine fettle*

[2]**fettle** *transitive verb* **fet·tled; fet·tling** \ˈfet-liŋ, ˈfe-tᵊl-iŋ\ [Middle English *fetlen* to shape, prepare; perhaps akin to Old English *fæt* vessel — more at VAT] (1881)
: to cover or line the hearth of (as a reverberatory furnace) with loose material (as sand or gravel)

fet·tuc·ci·ne *or* **fet·tuc·ci·ni** *or* **fet·tu·ci·ne** *or* **fet·tu·ci·ni** \ˌfe-tə-ˈchē-nē\ *noun plural but singular or plural in construction* [Italian, plural of *fettuccina,* diminutive of *fettuccia* small slice, ribbon, diminutive of *fetta* slice, probably alteration of (assumed) *offetta,* from *offa* flour cake, from Latin] (1912)
: pasta in the form of narrow ribbons; *also* : a dish of which fettuccine forms the base

fettuccine Al·fre·do \-(ˌ)al-ˈfrā-(ˌ)dō, -(ˌ)äl-\ *noun* [from *Alfredo all'Augusteo,* restaurant in Rome where it originated] (1961)
: a dish consisting of fettuccine with butter, Parmesan cheese, cream, and seasonings — called also *fettuccine all'Alfredo* \-ˌal-(ˌ)al-, -ˌäl-(ˌ)äl-\

fe·tus \ˈfē-təs\ *noun* [Middle English, from Latin, act of bearing young, offspring; akin to Latin *fetus* newly delivered, fruitful — more at FEMININE] (14th century)
: an unborn or unhatched vertebrate especially after attaining the basic structural plan of its kind; *specifically* : a developing human from usually three months after conception to birth

[1]**feud** \ˈfyüd\ *noun* [alteration of Middle English *feide,* from Middle French, of Germanic origin; akin to Old High German *fēhida* hostility, feud, Old English *fāh* hostile — more at FOE] (15th century)
: a mutual enmity or quarrel that is often prolonged or inveterate; *especially* : a lasting state of hostilities between families or clans marked by violent attacks for revenge
— **feud** *intransitive verb*

[2]**feud** *noun* [Medieval Latin *feodum, feudum,* of Germanic origin; akin to Old English *feoh* cattle, property — more at FEE] (1614)
: FEE 1a

feu·dal \ˈfyü-dᵊl\ *adjective* (1612)
1 : of, relating to, or having the characteristics of a medieval fee
2 : of, relating to, or suggestive of feudalism ⟨*feudal* law⟩
— **feu·dal·ly** \-dᵊl-ē\ *adverb*

feu·dal·ism \ˈfyü-dᵊl-ˌi-zəm\ *noun* (circa 1818)
1 : the system of political organization prevailing in Europe from the 9th to about the 15th centuries having as its basis the relation of lord to vassal with all land held in fee and as chief characteristics homage, the service of tenants under arms and in court, wardship, and forfeiture
2 : any of various political or social systems similar to medieval feudalism
— **feu·dal·ist** \-dᵊl-ist\ *noun*
— **feu·dal·is·tic** \ˌfyü-dᵊl-ˈis-tik\ *adjective*

feu·dal·i·ty \fyü-ˈda-lə-tē\ *noun, plural* **-ties** (1790)
1 : the quality or state of being feudal
2 : a feudal holding, domain, or concentration of power

feu·dal·ize \ˈfyü-dᵊl-ˌīz\ *transitive verb* **-ized; -iz·ing** (1828)
: to make feudal
— **feu·dal·iza·tion** \ˌfyü-dᵊl-ə-ˈzā-shən\ *noun*

[1]**feu·da·to·ry** \ˈfyü-də-ˌtōr-ē, -ˌtȯr-\ *adjective* [Medieval Latin *feudatorius,* from *feudatus,* past participle of *feudare* to enfeoff, from *feudum*] (1592)
1 : owing feudal allegiance
2 : being under the overlordship of a foreign state

[2]**feudatory** *noun, plural* **-ries** (1644)
1 : a dependent lordship : FEE
2 : one holding lands by feudal tenure

[1]**feud·ist** \ˈfyü-dist\ *noun* (1607)
: a specialist in feudal law

[2]**feudist** *noun* (1901)
: one who feuds

feuil·le·ton \ˌfə-yə-ˈtōⁿ, ˌfər-, ˌfœ-\ *noun* [French, from *feuillet* sheet of paper, from Old French *foillet,* diminutive of *foille* leaf — more at FOIL] (1845)
1 : a part of a European newspaper or magazine devoted to material designed to entertain the general reader
2 : something (as an installment of a novel) printed in a feuilleton
3 a : a novel printed in installments **b :** a work of fiction catering to popular taste
4 : a short literary composition often having a familiar tone and reminiscent content
— **feuil·le·ton·ism** \-ˈtō(ⁿ)-ˌni-zəm\ *noun*
— **feuil·le·ton·ist** \-nist\ *noun*

Feul·gen \ˈfȯil-gən\ *adjective* (1928)
: of, relating to, utilizing, or staining by the Feulgen reaction ⟨positive *Feulgen* mitochondria⟩

Feulgen reaction *noun* [Robert *Feulgen* (died 1955) German physiologist] (1928)
: the development of a brilliant purple color by DNA in a microscopic preparation stained with a modified Schiff's reagent

[1]**fe·ver** \ˈfē-vər\ *noun* [Middle English, from Old English *fēfer,* from Latin *febris*] (before 12th century)
1 a : a rise of body temperature above the normal **b :** any of various diseases of which fever is a prominent symptom
2 a : a state of heightened or intense emotion or activity **b :** a contagious usually transient enthusiasm : CRAZE

[2]**fever** *verb* **fe·vered; fe·ver·ing** \ˈfē-v(ə-)riŋ\ (1606)
transitive verb
: to throw into a fever : AGITATE
intransitive verb
: to contract or be in a fever : become feverish

fever blister *noun* (circa 1860)
: COLD SORE

fe·ver·few \ˈfē-vər-ˌfyü\ *noun* [Middle English, from Old English *feferfuge,* from Late Latin *febrifugia* centaury — more at FEBRIFUGE] (15th century)
: a perennial European composite herb (*Chrysanthemum parthenium*) widely cultivated as an ornamental

fe·ver·ish \ˈfē-v(ə-)rish\ *adjective* (14th century)
1 a : tending to cause fever **b :** having the symptoms of a fever **c :** indicating or relating to fever
2 : marked by intense emotion, activity, or instability
— **fe·ver·ish·ly** *adverb*
— **fe·ver·ish·ness** *noun*

fe·ver·ous \ˈfē-v(ə-)rəs\ *adjective* (14th century)
: FEVERISH

fever pitch *noun* (1915)
: a state of intense excitement and agitation

fever tree *noun* (1868)
: any of several shrubs or trees that are thought to indicate regions free from fever or that yield remedies for fever; *especially* : an African acacia (*Acacia xanthlophloea*)

fe·ver·wort \ˈfē-vər-ˌwərt, -ˌwȯrt\ *noun* (circa 1814)
: a coarse American herb (*Triosteum perfoliatum*) of the honeysuckle family — called also *horse gentian*

[1]**few** \ˈfyü\ *pronoun, plural in construction* [Middle English *fewe,* pronoun & adjective, from Old English *fēawa;* akin to Old High German *fō* little, Latin *paucus* little, *pauper* poor, Greek *paid-, pais* child, Sanskrit *putra* son] (before 12th century)
: not many persons or things ⟨*few* were present⟩ ⟨*few* of his stories are true⟩

[2]**few** *adjective* (before 12th century)
1 : consisting of or amounting to only a small number ⟨one of our *few* pleasures⟩
2 : at least some but indeterminately small in number — used with *a* ⟨caught a *few* fish⟩
— **few·ness** *noun*
— **few and far between :** few in number and infrequently met : RARE

[3]**few** *noun, plural in construction* (before 12th century)
1 : a small number of units or individuals ⟨a *few* of them⟩
2 : a special limited number ⟨the discriminating *few*⟩

[1]**few·er** \ˈfyü-ər, ˈfyu̇r\ *pronoun, plural in construction* (before 12th century)
: a smaller number of persons or things

[2]**fewer** *adjective, comparative of* FEW
usage see LESS

few·trils \ˈfyü-trəlz\ *noun plural* [origin unknown] (circa 1750)
dialect English : things of little value : TRIFLES

fey \ˈfā\ *adjective* [Middle English *feye,* from Old English *fǣge;* akin to Old High German *feigi* fey and perhaps to Old English *fāh* hostile, outlawed — more at FOE] (before 12th century)
1 a *chiefly Scottish* **:** fated to die : DOOMED **b :** marked by a foreboding of death or calamity
2 a : able to see into the future : VISIONARY **b :** marked by an otherworldly air or attitude **c :** CRAZY, TOUCHED
3 a : PRECIOUS 3 **b :** UNCONVENTIONAL, CAMPY
— **fey·ly** *adverb*
— **fey·ness** *noun*

fez \ˈfez\ *noun, plural* **fez·zes** *also* **fez·es** [French, from *Fez,* Morocco] (circa 1803)
: a brimless cone-shaped flat-crowned hat that usually has a tassel, is usually made of red felt, and is worn especially by men in eastern Mediterranean countries
— **fezzed** \ˈfezd\ *adjective*

fez

fi·acre \fē-ˈäkrᵊ\ *noun, plural* **fi·acres** *same, or* -ˈä-krəz\ [French, from the Hotel Saint *Fiacre,* Paris] (1699)
: a small hackney coach

fi·an·cé \ˌfē-ˌän-ˈsā, fē-ˈän-ˌ\ *noun* [French, from Middle French, from past participle of *fiancer* to promise, betroth, from

Old French *fiancier,* from *fiance* promise, trust, from *fier* to trust, from (assumed) Vulgar Latin *fidare,* alteration of Latin *fidere* — more at BIDE] (1864)
: a man engaged to be married

fi·an·cée \ˌfē-ˌän-ˈsā, fē-ˈän-ˌ\ *noun* [French, feminine of *fiancé*] (1853)
: a woman engaged to be married

fi·an·chet·to \ˌfē-ən-ˈke-(ˌ)tō, -ˈche-\ *verb* [*fianchetto* (an opening in chess), from Italian, diminutive of *fianco* side, flank, from Old French *flanc* — more at FLANK] (1927)
transitive verb
: to develop (a bishop) in a chess game to the second square on the adjacent knight's file
intransitive verb
: to fianchetto a bishop in a chess game

¹fi·as·co \fē-ˈas-(ˌ)kō *also* -ˈäs-\ *noun, plural* **-coes** [French, from Italian, from *fare fiasco,* literally, to make a bottle] (circa 1854)
: a complete failure ◆

²fi·as·co \fē-ˈäs-(ˌ)kō, -ˈas-\ *noun, plural* **-coes** *also* **fi·as·chi** \-(ˌ)kē\ [Italian, from Late Latin *flasco* bottle — more at FLASK] (1887)
: BOTTLE, FLASK; *especially* : a bulbous long-necked straw-covered bottle for wine

fi·at \ˈfē-ət, -ˌat, -ˌät; ˈfī-ət, -ˌat\ *noun* [Latin, let it be done, 3d singular present subjunctive of *fieri* to become, be done — more at BE] (circa 1631)
1 : a command or act of will that creates something without or as if without further effort
2 : an authoritative determination : DICTATE ⟨a *fiat* of conscience⟩
3 : an authoritative or arbitrary order : DECREE ⟨government by *fiat*⟩

fiat money *noun* (1876)
: money (as paper currency) not convertible into coin or specie of equivalent value

¹fib \ˈfib\ *noun* [perhaps by shortening & alteration from *fable*] (1611)
: a trivial or childish lie

²fib *intransitive verb* **fibbed; fib·bing** (1690)
: to tell a fib
synonym see LIE
— fib·ber *noun*

³fib *verb* **fibbed; fib·bing** [origin unknown] (circa 1665)
British : BEAT, PUMMEL

fi·ber *or* **fi·bre** \ˈfī-bər\ *noun* [French *fibre,* from Latin *fibra*] (1540)
1 : a thread or a structure or object resembling a thread: as **a** (1) **:** a slender root (as of a grass) (2) **:** an elongated tapering thick-walled plant cell void at maturity that imparts elasticity, flexibility, and tensile strength **b** (1) **:** a strand of nerve tissue : AXON, DENDRITE (2) **:** one of the filaments composing most of the intercellular matrix of connective tissue (3) **:** one of the elongated contractile cells of muscle tissue **c :** a slender and greatly elongated natural or synthetic filament (as of wool, cotton, asbestos, gold, glass, or rayon) typically capable of being spun into yarn **d :** indigestible material in human food that stimulates the intestine to peristalsis — called also *bulk, roughage*
2 : material made of fibers; *especially* : VULCANIZED FIBER
3 a : an element that gives texture or substance **b :** basic toughness : STRENGTH, FORTITUDE **c :** essential structure or character ⟨the very *fiber* of a person's being⟩
— fi·bered \-bərd\ *adjective*

fi·ber·board *also* **fi·bre·board** \-ˌbōrd, -ˌbȯrd\ *noun* (1897)
: a material made by compressing fibers (as of wood) into stiff sheets; *also* : PAPERBOARD

fi·ber·fill *also* **fi·bre·fill** \-ˌfil\ *noun* (1962)
: synthetic fibers used as a filling material (as for cushions)

¹fi·ber·glass *also* **fi·bre·glass** \-ˌglas\ *noun* (1937)
1 : glass in fibrous form used in making various products (as glass wool for insulation)
2 : a composite structural material of plastic and fiberglass

²fiberglass *transitive verb* (1967)
: to protect or repair by the application of fiberglass

fi·ber·ize \ˈfī-bə-ˌrīz\ *transitive verb* **-ized; -iz·ing** (1925)
: to make or break down into fibers
— fi·ber·iza·tion \ˌfī-b(ə-)rə-ˈzā-shən\ *noun*

fi·ber–op·tic \ˈfī-bə-ˌräp-tik\ *adjective* (1961)
: of, relating to, or using fiber optics

fiber optics *noun plural* (1956)
1 : thin transparent fibers of glass or plastic that are enclosed by material of a lower index of refraction and that transmit light throughout their length by internal reflections; *also* : a bundle of such fibers used in an instrument (as for viewing body cavities)
2 *singular in construction* **:** the technique of the use of fiber optics

fi·ber·scope \ˈfī-bər-ˌskōp\ *noun* (1954)
: a flexible instrument utilizing fiber optics and used for examination of inaccessible areas

Fi·bo·nac·ci number \ˌfē-bə-ˌnä-chē-, ˌfi-bə-\ *noun* [Leonardo *Fibonacci* (died *about* 1250) Italian mathematician] (1914)
: an integer in the infinite sequence 1, 1, 2, 3, 5, 8, 13, . . . of which the first two terms are 1 and 1 and each succeeding term is the sum of the two immediately preceding

fibr- *or* **fibro-** *combining form* [Latin *fibra*]
: fiber : fibrous tissue ⟨*fibr*oid⟩ : fibrous and ⟨*fibro*vascular⟩

fi·branne \ˈfī-ˌbran, fī-ˈ\ *noun* [French, viscose rayon, from *fibre*] (1941)
: a fabric made of spun-rayon yarn

fi·bril \ˈfī-brəl, ˈfi-\ *noun* [New Latin *fibrilla,* diminutive of Latin *fibra*] (1664)
: a small filament or fiber: as **a :** ROOT HAIR **b** (1) **:** one of the fine threads into which a striated muscle fiber can be longitudinally split (2) : NEUROFIBRIL
— fi·bril·lar \ˈfi-brə-lər, ˈfī-\ *adjective*

fi·bril·late \ˈfi-brə-ˌlāt, ˈfī-\ *verb* **-lat·ed; -lat·ing** (circa 1847)
intransitive verb
: to undergo or exhibit fibrillation
transitive verb
: to cause to undergo fibrillation

fi·bril·la·tion \ˌfi-brə-ˈlā-shən, ˌfī-\ *noun* (1845)
1 : an act or process of forming fibers or fibrils
2 a : a muscular twitching involving individual muscle fibers acting without coordination **b :** very rapid irregular contractions of the muscle fibers of the heart resulting in a lack of synchronism between heartbeat and pulse

fi·brin \ˈfī-brən\ *noun* (1800)
: a white insoluble fibrous protein formed from fibrinogen by the action of thrombin especially in the clotting of blood

fi·brin·o·gen \fī-ˈbri-nə-jən\ *noun* [International Scientific Vocabulary] (1872)
: a plasma protein that is produced in the liver and is converted into fibrin during blood clot formation

fi·bri·noid \ˈfi-brə-ˌnȯid, ˈfī-\ *noun, often attributive* (1910)
: a homogeneous acidophilic refractile material that somewhat resembles fibrin and is formed in the walls of blood vessels and in connective tissue in some pathological conditions and normally in the placenta

fi·bri·no·ly·sin \ˌfī-brə-nᵊl-ˈī-sᵊn\ *noun* [International Scientific Vocabulary] (1915)
: any of several proteolytic enzymes that promote the dissolution of blood clots; *especially* : PLASMIN

fi·bri·no·ly·sis \-ˈī-səs, -brə-ˈnä-lə-səs\ *noun* [New Latin] (1907)
: the usually enzymatic breakdown of fibrin
— fi·bri·no·lyt·ic \-brə-nᵊl-ˈi-tik\ *adjective*

fi·bri·no·pep·tide \ˌfī-brə-nō-ˈpep-ˌtīd\ *noun* (1960)
: any of the vertebrate polypeptides that are cleaved from fibrinogen by thrombin during blood clot formation

fi·bro·blast \ˈfī-brə-ˌblast, ˈfi-\ *noun* [International Scientific Vocabulary] (1876)
: a connective-tissue cell of mesenchymal origin that secretes proteins and especially molecular collagen from which the extracellular fibrillar matrix of connective tissue forms
— fi·bro·blas·tic \ˌfī-brə-ˈblas-tik, ˌfi-\ *adjective*

fi·bro·cys·tic \ˌfī-brə-ˈsis-tik, ˌfi-\ *adjective* (1854)
: characterized by the presence or development of fibrous tissue and cysts

¹fi·broid \ˈfī-brȯid, ˈfi-\ *adjective* (1852)
: resembling, forming, or consisting of fibrous tissue

²fibroid *noun* (circa 1860)
: a benign tumor that consists of fibrous and muscular tissue and occurs especially in the uterine wall

fi·bro·in \ˈfī-brə-wən, ˈfi-\ *noun* [French *fibroïne,* from *fibr-* + *-ine* -in] (1878)
: an insoluble protein comprising the filaments of the raw silk fiber

fi·bro·ma \fī-ˈbrō-mə\ *noun, plural* **-mas** *also* **-ma·ta** \-mə-tə\ (circa 1849)
: a benign tumor consisting mainly of fibrous tissue
— fi·bro·ma·tous \-mə-təs\ *adjective*

fi·bro·nec·tin \ˌfī-brə-ˈnek-tən\ *noun* [*fibr-* + Latin *nectere* to tie, bind + English *-in*] (1976)
: any of a group of glycoproteins of cell surfaces, blood plasma, and connective tissue that promote cellular adhesion and migration

fi·bro·sar·co·ma \ˌfī-brə-sär-ˈkō-mə, ˌfi-\ *noun* (1878)
: a sarcoma of relatively low malignancy consisting chiefly of spindle-shaped cells that tend to form collagenous fibrils

fi·bro·sis \fī-ˈbrō-səs\ *noun* [New Latin] (1873)
: a condition marked by increase of interstitial fibrous tissue
— fi·brot·ic \-ˈbrä-tik\ *adjective*

fi·bro·si·tis \ˌfī-brə-ˈsī-təs, ˌfi-\ *noun* [New Latin, from *fibrosus* fibrous, from International Scientific Vocabulary *fibrous*] (1904)

◇ WORD HISTORY

fiasco The Italian word *fiasco* refers literally to a kind of bulbous, straw-covered bottle traditionally used for inexpensive wines in Tuscany. The source of English *fiasco* in the sense "failure" is the Italian phrase *fare fiasco,* literally, "to make a bottle," which as an idiom means "to fail." The source of this peculiar sense, first attested in Italian in 1808, remains uncertain. Possibly there is some connection with the much earlier idioms *appiccare il fiasco a qualcuno* "to trick someone" (literally, "to hang the bottle on someone") or *vestir fiaschi* "to accomplish something insignificant" (literally, "to dress bottles," that is, to cover them with straw), but these seem to show only that *fiasco* had many idiomatic possibilities. Attempts have been made to relate *fare fiasco* to some incident in Italian theatrical history, in particular to a comic monologue about a bottle, improvised by a famous 17th century comic actor in Bologna. The monologue fell so flat with the audience that it became proverbial, or so the story goes, but supporting documentation is lacking. With *fare fiasco,* as with other idioms, the loss of original context has made something once self-evident completely opaque.

: a rheumatic disorder of fibrous tissue

fi·brous \ˈfī-brəs\ *adjective* [modification of French *fibreux,* from *fibre* fiber] (1626)
1 a : containing, consisting of, or resembling fibers **b :** characterized by fibrosis **c :** capable of being separated into fibers ⟨a *fibrous* mineral⟩
2 : TOUGH, SINEWY ⟨*fibrous* texture⟩

fibrous root *noun* (1626)
: a root (as in most grasses) that has no prominent central axis and that branches in all directions

fi·bro·vas·cu·lar \ˈfī-brō-ˈvas-kyə-lər, ˌfi-\ *adjective* (1845)
: having or consisting of fibers and conducting cells

fibrovascular bundle *noun* (circa 1889)
: VASCULAR BUNDLE

fib·u·la \ˈfi-byə-lə\ *noun, plural* **-lae** \-lē, -lī\ *or* **-las** [Latin, pin, clasp; akin to Latin *figere* to fasten] (1615)
1 : the outer and usually smaller of the two bones between the knee and ankle in the hind or lower limbs of vertebrates
2 : a clasp resembling a safety pin used especially by the ancient Greeks and Romans
— fib·u·lar \-lər\ *adjective*

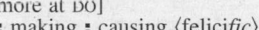

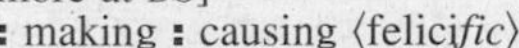
fibula 2

-fic \fik\ *adjective suffix* [Middle French & Latin; Middle French *-fique,* from Latin *-ficus,* from *facere* to make — more at DO]
: making : causing ⟨felici*fic*⟩

-fi·ca·tion \fə-ˈkā-shən\ *noun combining form* [Middle English *-ficacioun,* from Middle French & Latin; Middle French *-fication,* from Latin *-fication-, -ficatio,* from *-ficare* to make, from *-ficus*]
: making : production ⟨rei*fication*⟩

fice \ˈfīs\ *variant of* FEIST

fiche \ˈfēsh *also* ˈfish\ *noun, plural* **fiche** *also* **fiches** (1951)
: MICROFICHE

fi·chu \ˈfi-(ˌ)shü, ˈfē-\ *noun* [French, from past participle of *ficher* to stick in, throw on, from (assumed) Vulgar Latin *figicare,* from Latin *figere* to fasten, pierce] (1803)
: a woman's light triangular scarf that is draped over the shoulders and fastened in front or worn to fill in a low neckline

fi·cin \ˈfī-sᵊn\ *noun* [Latin *ficus* fig] (1930)
: a protease obtained from the latex of fig trees and used as an anthelmintic and protein digestive

fick·le \ˈfi-kəl\ *adjective* [Middle English *fikel* deceitful, inconstant, from Old English *ficol* deceitful; akin to Old English be*fician* to deceive, and probably to Old English *fāh* hostile — more at FOE] (13th century)
: marked by lack of steadfastness, constancy, or stability : given to erratic changeableness
synonym see INCONSTANT
— fick·le·ness *noun*
— fick·ly \ˈfi-k(ə-)lē\ *adverb*

fi·co \ˈfē-(ˌ)kō\ *noun, plural* **ficoes** [obsolete *fico,* obscene gesture of contempt, modification of Italian *fica* fig, vulva, gesture of contempt, from (assumed) Vulgar Latin *fica* fig — more at FIG] (1598)
archaic : FIG 2

fic·tile \ˈfik-tᵊl, -ˌtīl\ *adjective* [Latin *fictilis* molded of clay, from *fingere*] (1626)
1 *archaic* **:** PLASTIC 2a
2 : of or relating to pottery
3 : MALLEABLE 2a

fic·tion \ˈfik-shən\ *noun* [Middle English *ficcioun,* from Middle French *fiction,* from Latin *fiction-, fictio* act of fashioning, fiction, from *fingere* to shape, fashion, feign — more at DOUGH] (14th century)
1 a : something invented by the imagination or feigned; *specifically* **:** an invented story **b :** fictitious literature (as novels or short stories) **c :** a work of fiction; *especially* **:** NOVEL
2 a : an assumption of a possibility as a fact irrespective of the question of its truth ⟨a legal *fiction*⟩ **b :** a useful illusion or pretense
3 : the action of feigning or of creating with the imagination
— fic·tion·al \-shnəl, -shə-nᵊl\ *adjective*
— fic·tion·al·i·ty \ˌfik-shə-ˈna-lə-tē\ *noun*
— fic·tion·al·ly \ˈfik-shnə-lē, -shə-nᵊl-ē\ *adverb*

fictionalise *British variant of* FICTIONALIZE

fic·tion·al·ize \ˈfik-shnə-ˌlīz, -shə-nᵊl-ˌīz\ *transitive verb* **-ized; -iz·ing** (1918)
: to make into or treat in the manner of fiction ⟨*fictionalize* a biography⟩
— fic·tion·al·iza·tion \ˌfik-shnə-lə-ˈzā-shən, -shə-nᵊl-ə-\ *noun*

fic·tion·eer \ˌfik-shə-ˈnir\ *noun* (1923)
: one who writes fiction especially in quantity and without high standards
— fic·tion·eer·ing *noun*

fic·tion·ist \ˈfik-sh(ə-)nist\ *noun* (1829)
: a writer of fiction; *especially* **:** NOVELIST

fic·tion·ize \ˈfik-shə-ˌnīz\ *transitive verb* **-ized; -iz·ing** (1831)
: FICTIONALIZE
— fic·tion·iza·tion \ˌfik-shə-nə-ˈzā-shən\ *noun*

fic·ti·tious \fik-ˈti-shəs\ *adjective* [Latin *ficticius* artificial, feigned, from *fictus*] (1634)
1 : of, relating to, or characteristic of fiction : IMAGINARY
2 a : conventionally or hypothetically assumed or accepted ⟨a *fictitious* concept⟩ **b** *of a name* **:** FALSE, ASSUMED
3 : not genuinely felt ☆
— fic·ti·tious·ly *adverb*
— fic·ti·tious·ness *noun*

fic·tive \ˈfik tiv\ *adjective* (1612)
1 : not genuine **:** FEIGNED
2 : of, relating to, or capable of imaginative creation
3 : of, relating to, or having the characteristics of fiction **:** FICTIONAL
— fic·tive·ly *adverb*
— fic·tive·ness *noun*

fi·cus \ˈfī-kəs\ *noun, plural* **ficus** *or* **fi·cus·es** [New Latin, from Latin, fig] (1864)
: FIG 1b

fid \ˈfid\ *noun* [origin unknown] (1615)
: a tapered usually wooden pin used in opening the strands of a rope

-fid \fəd, ˌfid\ *adjective combining form* [Latin *-fidus,* from *findere* to split — more at BITE]
: divided into (so many) parts or (such) parts ⟨pinnati*fid*⟩

[1]fid·dle \ˈfi-dᵊl\ *noun* [Middle English *fidel,* from (assumed) Old English *fithele,* probably from Medieval Latin *vitula*] (13th century)
1 : VIOLIN
2 : a device (as a slat, rack, or light railing) to keep dishes from sliding off a table aboard ship
3 : FIDDLESTICKS — used as an interjection
4 [[2]*fiddle*] *chiefly British* **:** SWINDLE

[2]fiddle *verb* **fid·dled; fid·dling** \ˈfid-liŋ, ˈfi-dᵊl-iŋ\ (14th century)
intransitive verb
1 : to play on a fiddle
2 a : to move the hands or fingers restlessly **b :** to spend time in aimless or fruitless activity **:** PUTTER, TINKER ⟨*fiddled* around with the engine for hours⟩ **c :** MEDDLE, TAMPER
transitive verb
1 : to play (as a tune) on a fiddle
2 : CHEAT, SWINDLE
3 : to alter or manipulate deceptively for fraudulent gain ⟨accountants *fiddling* the books —Stanley Cohen⟩
— fid·dler \ˈfid-lər, ˈfi-dᵊl-ər\ *noun*

fiddle away *transitive verb* (1667)
: to fritter away ⟨*fiddling away* the time⟩

fid·dle·back \ˈfi-dᵊl-ˌbak\ *noun* (1890)
: something resembling a fiddle

fid·dle–fad·dle \ˈfi-dᵊl-ˌfa-dᵊl\ *noun* [reduplication of *fiddle* (fiddlesticks)] (1577)
: NONSENSE — often used as an interjection

fid·dle–foot·ed \ˌfi-dᵊl-ˈfu̇-təd\ *adjective* (1941)
1 : SKITTISH, JUMPY ⟨a *fiddle-footed* horse⟩
2 : prone to wander ⟨the nameless *fiddle-footed* drifters, the shifty riders who traveled the back trails —Luke Short⟩

fid·dle·head \ˈfi-dᵊl-ˌhed\ *noun* (1882)
: one of the young unfurling fronds of some ferns that are often eaten as greens

fiddler crab *noun* (1843)
: any of a genus (*Uca*) of burrowing crabs in which the male has one claw that is greatly enlarged

fiddler crab

fid·dle·stick \ˈfi-dᵊl-ˌstik\ *noun* (15th century)
1 : a violin bow
2 a : something of little value **:** TRIFLE ⟨didn't care a *fiddlestick* for that⟩ **b** *plural* **:** NONSENSE — used as an interjection

fid·dling \ˈfid-liŋ, -lən\ *adjective* (1652)
: TRIFLING, PETTY ⟨a *fiddling* excuse⟩

fid·dly \ˈfi-dᵊl-ē\ *adjective* (1926)
chiefly British **:** requiring close attention to detail **:** FUSSY

fi·de·ism \ˈfē-(ˌ)dā-ˌi-zəm\ *noun* [probably from French *fidéisme,* from Latin *fides*] (1885)
: reliance on faith rather than reason in pursuit of religious truth
— fi·de·ist \-ˌdā-ist\ *noun*
— fi·de·is·tic \ˌfē-(ˌ)dā-ˈis-tik\ *adjective*

Fi·del·is·ta \ˌfē-dᵊl-ˈē-stə\ *noun* [American Spanish, from *Fidel* Castro + *-ista* -ist] (1960)
: an adherent of Castroism

fi·del·i·ty \fə-ˈde-lə-tē, fī-\ *noun, plural* **-ties** [Middle English *fidelite,* from Middle French *fidelité,* from Latin *fidelitat-, fidelitas,* from *fidelis* faithful, from *fides* faith, from *fidere* to trust — more at BIDE] (15th century)
1 a : the quality or state of being faithful **b :** accuracy in details **:** EXACTNESS
2 : the degree to which an electronic device (as a record player, radio, or television) accu-

> ☆ SYNONYMS
> **Fictitious, fabulous, legendary, mythical, apocryphal** mean having the nature of something imagined or invented. FICTITIOUS implies fabrication and suggests artificiality or contrivance more than deliberate falsification or deception ⟨*fictitious* characters⟩. FABULOUS stresses the marvelous or incredible character of something without necessarily implying impossibility or actual nonexistence ⟨a land of *fabulous* riches⟩. LEGENDARY suggests the elaboration of invented details and distortion of historical facts produced by popular tradition ⟨the *legendary* exploits of Davy Crockett⟩. MYTHICAL implies a purely fanciful explanation of facts or the creation of beings and events out of the imagination ⟨*mythical* creatures⟩. APOCRYPHAL implies an unknown or dubious source or origin or may imply that the thing itself is dubious or inaccurate ⟨a book that repeats many *apocryphal* stories⟩.

rately reproduces its effect (as sound or picture) ☆

fidge \'fij\ *intransitive verb* **fidged; fidg·ing** [probably alteration of English dialect *fitch,* from Middle English *fichen*] (1575)
chiefly Scottish **:** FIDGET

[1]fidg·et \'fi-jət\ *noun* [irregular from *fidge*] (1674)
1 : uneasiness or restlessness as shown by nervous movements — usually used in plural
2 [[2]*fidget*] **:** one that fidgets

[2]fidget (1754)
intransitive verb
: to move or act restlessly or nervously
transitive verb
: to cause to move or act nervously

fidg·ety \'fi-jə-tē\ *adjective* (circa 1736)
1 : inclined to fidget
2 : making unnecessary fuss **:** FUSSY
— fidg·et·i·ness *noun*

fi·do \'fī-(,)dō\ *noun, plural* **fidos** [*f*reaks + *i*rregulars + *d*efects + *o*ddities] (1967)
: a coin having a minting error

fi·du·cial \fə-'dü-shəl, -'dyü-, fī-\ *adjective* (1571)
1 : taken as standard of reference ⟨a *fiducial* mark⟩
2 : founded on faith or trust
3 : having the nature of a trust **:** FIDUCIARY
— fi·du·cial·ly \-shə-lē\ *adverb*

[1]fi·du·cia·ry \-'dü-shē-,er-ē, -shə-rē, -'dyü-\ *noun, plural* **-ries** (1631)
: one that holds a fiduciary relation or acts in a fiduciary capacity

[2]fiduciary *adjective* [Latin *fiduciarius,* from *fiducia* confidence, trust, from *fidere*] (circa 1641)
: of, relating to, or involving a confidence or trust: as **a :** held or founded in trust or confidence **b :** holding in trust **c :** depending on public confidence for value or currency ⟨*fiduciary* fiat money⟩

fie \'fī\ *interjection* [Middle English *fi,* from Old French] (14th century)
— used to express disgust or disapproval

fief \'fēf\ *noun* [French, from Old French — more at FEE] (circa 1611)
1 : a feudal estate **:** FEE
2 : something over which one has rights or exercises control ⟨a politician's *fief*⟩
— fief·dom \-dəm\ *noun*

[1]field \'fē(ə)ld\ *noun* [Middle English, from Old English *feld;* akin to Old High German *feld* field, Old English *flōr* floor — more at FLOOR] (before 12th century)
1 a (1) **:** an open land area free of woods and buildings (2) **:** an area of land marked by the presence of particular objects or features ⟨dune *field*⟩ **b** (1) **:** an area of cleared enclosed land used for cultivation or pasture ⟨a *field* of wheat⟩ (2) **:** land containing a natural resource (3) **:** AIRFIELD **c :** the place where a battle is fought; *also* **:** BATTLE **d :** a large unbroken expanse (as of ice)
2 a : an area or division of an activity **b :** the sphere of practical operation outside a base (as a laboratory, office, or factory) ⟨geologists working in the *field*⟩ **c :** an area for military exercises or maneuvers **d** (1) **:** an area constructed, equipped, or marked for sports (2) **:** the portion of an indoor or outdoor sports area within the running track on which field events are held (3) **:** any of the three sections of a baseball outfield ⟨hits to all *fields*⟩
3 : a space on which something is drawn or projected: as **a :** the space on the surface of a coin, medal, or seal that does not contain the design **b :** the ground of each division in a flag **c :** the whole surface of an escutcheon
4 : the individuals that make up all or part of the participants in a sports activity; *especially* **:** all participants with the exception of the favorite or the winner in a contest where more than two are entered
5 : the area visible through the lens of an optical instrument
6 a : a region or space in which a given effect (as magnetism) exists **b :** a region of embryonic tissue capable of a particular type of differentiation ⟨a morphogenetic *field*⟩
7 : a set of mathematical elements that is subject to two binary operations the second of which is distributive relative to the first and that constitutes a commutative group under the first operation and also under the second if the zero or unit element under the first is omitted
8 : a complex of forces that serve as causative agents in human behavior
9 : a series of drain tiles and an absorption area for septic-tank overflow
10 : a particular area (as of a record in a database) in which the same type of information is regularly recorded

[2]field *adjective* (12th century)
: of or relating to a field: as **a :** growing in or inhabiting the fields or open country **b :** made, conducted, or used in the field **c :** operating or active in the field

[3]field (1823)
transitive verb
1 a : to catch or pick up (a batted ball) and usually throw to a teammate **b :** to take care of or respond to (as a telephone call or a request) **c :** to give an impromptu answer or solution to ⟨the senator *fielded* the reporters' questions⟩
2 : to put into the field ⟨*field* an army⟩ ⟨*field* a team⟩; *also* **:** to enter in competition
intransitive verb
: to play as a fielder

field artillery *noun* (1644)
: artillery other than antiaircraft artillery used with armies in the field

field bed *noun* (1926)
: a four-poster with a canopy arched at the center

field corn *noun* (1856)
: an Indian corn (as dent corn or flint corn) with starchy kernels grown for feeding stock or for market grain

field crop *noun* (1860)
: an agricultural crop (as hay, grain, or cotton) grown on large areas

field day *noun* (1747)
1 a : a day for military exercises or maneuvers **b :** an outdoor meeting or social gathering **c :** a day of sports and athletic competition
2 : a time of extraordinary pleasure or opportunity ⟨TV had a *field day* with the scandal⟩

field–ef·fect transistor \'fēl-də-,fekt-\ *noun* (1959)
: a transistor in which the output current is controlled by a variable electric field

field·er \'fēl-dər\ *noun* (1832)
: one that fields; *especially* **:** a defensive player stationed in the field (as in baseball)

fielder's choice *noun* (1902)
: a situation in baseball in which a batter reaches base safely because the fielder attempts to put out another base runner on the play

field event *noun* (1899)
: an event (as weight-throwing or jumping) in a track-and-field meet other than a race

field·fare \'fē(ə)l(d)-,far, -,fer\ *noun* [Middle English *feldefare,* from Old English, from *feld* + *-fare;* probably akin to Old English *fara* companion; akin to Old English *faran* to go — more at FARE] (before 12th century)
: a medium-sized Eurasian thrush (*Turdus pilaris*) with an ash-colored head and chestnut wings

field glass *noun* (1836)
: a binocular without prisms especially for use outdoors — usually used in plural

field goal *noun* (1902)
1 : a score of three points in football made by drop-kicking or place-kicking the ball over the crossbar from ordinary play
2 : a goal in basketball made while the ball is in play

field grade *noun* (1944)
: the rank of a field officer

field guide *noun* (1934)
: a manual for identifying natural objects, flora, or fauna in the field

field hand *noun* (1826)
: an outdoor farm laborer

field hockey *noun* (1903)
: a game played on a turfed field between two teams of 11 players each whose object is to direct a ball into the opponent's goal with a hockey stick

field house *noun* (1895)
1 : a building at an athletic field for housing equipment or providing dressing facilities
2 : a building enclosing a large area suitable for various forms of athletics and usually providing seats for spectators

fielding average *noun* (1947)
: the average (as of a baseball fielder) determined by dividing the number of putouts and assists by the number of chances — compare BATTING AVERAGE

field judge *noun* (circa 1929)
: a football official whose duties include covering action on kicks and forward passes and timing intermission periods and time-outs

field lens *noun* (1837)
: the lens in a compound eyepiece that is nearer the objective

field magnet *noun* (1883)
: a magnet for producing and maintaining a magnetic field especially in a generator or electric motor

field marshal *noun* (1614)
: the highest ranking military officer (as in the British army)

field mouse *noun* (15th century)
: any of various mice and voles that inhabit fields

field officer *noun* (1656)
: a commissioned officer in the army, air force, or marine corps of the rank of colonel, lieutenant colonel, or major — compare COMPANY OFFICER, GENERAL OFFICER

field of force (1850)
: FIELD 6a

field of honor (1824)
1 : BATTLEFIELD
2 : a place where a duel is fought

field of view (circa 1816)
: FIELD 5

field of vision (1862)
: VISUAL FIELD

field pea *noun* (1709)
: a small-seeded pea (*Pisum sativum* variety *arvense*) widely grown for forage and food

field·piece \'fē(ə)l(d)-,pēs\ *noun* (1590)
: a gun or howitzer for use in the field

field spaniel *noun* (1867)
: any of a breed of medium-sized hunting and retrieving spaniels that have a flat dense usually black, liver, red, or roan coat

field·stone \'fē(ə)l(d)-,stōn\ *noun* (1799)
: stone (as in building) in usually unaltered form as taken from the field

☆ SYNONYMS
Fidelity, allegiance, fealty, loyalty, devotion, piety mean faithfulness to something to which one is bound by pledge or duty. FIDELITY implies strict and continuing faithfulness to an obligation, trust, or duty ⟨marital *fidelity*⟩. ALLEGIANCE suggests an adherence like that of citizens to their country ⟨pledging *allegiance*⟩. FEALTY implies a fidelity acknowledged by the individual and as compelling as a sworn vow ⟨*fealty* to the truth⟩. LOYALTY implies a faithfulness that is steadfast in the face of any temptation to renounce, desert, or betray ⟨valued the *loyalty* of his friends⟩. DEVOTION stresses zeal and service amounting to self-dedication ⟨a painter's *devotion* to her art⟩. PIETY stresses fidelity to obligations regarded as natural and fundamental ⟨filial *piety*⟩.

field·strip \-ˌstrip\ *transitive verb* (1947)
: to take apart (a weapon) to the extent authorized for routine cleaning, lubrication, and minor repairs

field–test \-ˌtest\ *transitive verb* (1948)
: to test (as a procedure or product) in a natural environment
— **field test** *noun*

field theory *noun* (1901)
: a detailed mathematical description of the assumed physical properties of a region under some influence (as gravitation)

field trial *noun* (1895)
1 : a trial of sporting dogs in actual performance
2 : a trial of a new product in actual situations for which it is intended

field trip *noun* (1926)
: a visit (as to a factory, farm, or museum) made (as by students and a teacher) for purposes of firsthand observation

field winding *noun* (1893)
: the winding of a field magnet

field·work \ˈfē(ə)l-ˌdwərk\ *noun* (1819)
1 : a temporary fortification thrown up by an army in the field
2 : work done in the field (as by students) to gain practical experience and knowledge through firsthand observation
3 : the gathering of anthropological or sociological data through the interviewing and observation of subjects in the field
— **field–work·er** *noun*

fiend \ˈfēnd\ *noun* [Middle English, from Old English *fīend;* akin to Old High German *fīant* enemy, Sanskrit *pīyati* he scorns] (before 12th century)
1 a : DEVIL 1 **b :** DEMON **c :** a person of great wickedness or maliciousness
2 : a person extremely devoted to a pursuit or study **:** FANATIC ⟨a golf *fiend*⟩
3 : ADDICT 1 ⟨a dope *fiend*⟩
4 : WIZARD 3 ⟨a *fiend* at mathematics⟩

fiend·ish \ˈfēn-dish\ *adjective* (1529)
1 : perversely diabolical ⟨took a *fiendish* pleasure in hurting people⟩
2 : extremely cruel or wicked
3 : excessively bad, unpleasant, or difficult ⟨*fiendish* weather⟩
— **fiend·ish·ly** *adverb*
— **fiend·ish·ness** *noun*

fierce \ˈfirs\ *adjective* **fierc·er; fierc·est** [Middle English *fiers,* from Middle French, from Latin *ferus* wild, savage; akin to Greek *thēr* wild animal] (14th century)
1 a : violently hostile or aggressive in temperament **b :** given to fighting or killing **:** PUGNACIOUS
2 a : marked by unrestrained zeal or vehemence ⟨a *fierce* argument⟩ **b :** extremely vexatious, disappointing, or intense ⟨*fierce* pain⟩
3 : furiously active or determined ⟨make a *fierce* effort⟩
4 : wild or menacing in appearance ☆
— **fierce·ly** *adverb*
— **fierce·ness** *noun*

fi·eri fa·cias \ˌfī-(ə-)rē-ˈfā-sh(ē-)əs\ *noun* [Latin, cause (it) to be done] (15th century)
: a writ authorizing the sheriff to obtain satisfaction of a judgment in debt or damages from the goods and chattels of the defendant

fi·ery \ˈfī(-ə)-rē\ *adjective* **fi·er·i·er; -est** [Middle English, from *fire, fier* fire] (13th century)
1 a : consisting of fire **b :** BURNING, BLAZING **c :** using or carried out with fire **d :** liable to catch fire or explode **:** FLAMMABLE ⟨a *fiery* vapor⟩
2 a : hot like a fire **b** (1) **:** being in an inflamed state or condition ⟨a *fiery* boil⟩ (2) **:** feverish and flushed ⟨a *fiery* forehead⟩
3 : of the color of fire **:** RED ⟨a *fiery* sunset⟩
4 a : full of or exuding emotion or spirit ⟨a *fiery* sermon⟩ **b :** easily provoked **:** IRRITABLE
— **fi·eri·ly** \ˈfī-rə-lē\ *adverb*
— **fi·eri·ness** \ˈfī-rē-nəs\ *noun*
— **fiery** *adverb*

fi·es·ta \fē-ˈes-tə\ *noun* [Spanish, from Latin *festa* — more at FEAST] (1844)
: FESTIVAL; *specifically* **:** a saint's day celebrated in Spain and Latin America with processions and dances

fife \ˈfīf\ *noun* [German *Pfeife* pipe, fife, from Old High German *pfīfa,* from (assumed) Vulgar Latin *pipa* pipe — more at PIPE] (1539)
: a small transverse flute with six to eight finger holes and usually no keys

fife rail *noun* (circa 1800)
: a rail about the mast near the deck to which rigging is belayed

fif·teen \ˌfif-ˈtēn, ˈfif-ˌ\ *noun* [Middle English *fiftene,* adjective, from Old English *fīftēne,* from *fīf* five + *-tīene* (akin to Old English *tīen* ten) — more at FIVE, TEN] (before 12th century)
1 — see NUMBER table
2 : the first point scored by a side in a game of tennis — called also *five*
— **fifteen** *adjective*
— **fifteen** *pronoun, plural in construction*
— **fif·teenth** \-ˈtēn(t)th, -ˌtēn(t)th\ *adjective or noun*

fifth \ˈfith, ˈfif(t)th, ˈfift\ *noun, plural* **fifths** \ˈfiths, ˈfif(t)ths, ˈfif(t)s\ (before 12th century)
1 — see NUMBER table
2 a : the musical interval embracing five diatonic degrees **b :** a tone at this interval; *specifically* **:** DOMINANT 1 **c :** the harmonic combination of two tones at this interval
3 : a unit of measure for liquor equal to one fifth of a U.S. gallon (0.757 liter)
4 *capitalized* **:** the Fifth Amendment of the U.S. Constitution
— **fifth** *adjective or adverb*
— **fifth·ly** *adverb*

fifth column *noun* [name applied to rebel sympathizers in Madrid in 1936 when four rebel columns advanced on the city] (1936)
: a group of secret sympathizers or supporters of an enemy that engage in espionage or sabotage within defense lines or national borders ◆
— **fifth col·um·nism** \-ˈkä-ləm-ˌ(n)i zəm\ *noun*
— **fifth col·um·nist** \-(n)ist\ *noun*

fifth wheel *noun* (circa 1874)
1 a : a horizontal wheel or segment of a wheel that consists of two parts rotating on each other above the fore axle of a carriage and that forms support to prevent tipping **b :** a similar coupling between tractor and trailer of a semitrailer
2 : one that is superfluous, unnecessary, or burdensome

fif·ty \ˈfif-tē\ *noun, plural* **fifties** [Middle English, from *fifty,* adjective, from Old English *fīftig,* from *fīftig,* noun, group of 50, from *fīf* five + *-tig* group of ten; akin to *tīen* ten] (before 12th century)
1 — see NUMBER table
2 *plural* **:** the numbers 50 to 59; *specifically* **:** the years 50 to 59 in a lifetime or century
3 : a 50-dollar bill
— **fif·ti·eth** \-tē-əth\ *adjective or noun*
— **fifty** *adjective*
— **fifty** *pronoun, plural in construction*
— **fif·ty·ish** \-tē-ish\ *adjective*

fif·ty–fif·ty \ˌfif-tē-ˈfif-tē\ *adjective* (1913)
1 : shared, assumed, or borne equally ⟨a *fifty-fifty* proposition⟩
2 : half favorable and half unfavorable ⟨a *fifty-fifty* chance⟩
— **fifty–fifty** *adverb*

¹fig \ˈfig\ *noun* [Middle English *fige,* from Old French, from Old Provençal *figa,* from (assumed) Vulgar Latin *fica,* from Latin *ficus* fig tree, fig] (13th century)
1 a : an oblong or pear-shaped fruit that is a syconium **b :** any of a genus (*Ficus*) of trees of the mulberry family bearing fruits that are syconia; *especially* **:** a widely cultivated tree (*F. carica*) that produces edible figs
2 : a worthless trifle **:** the least bit ⟨doesn't care a *fig*⟩

fig 1: leaves and fruit

²fig *noun* [*fig,* verb (to adorn)] (1835)
: DRESS, ARRAY ⟨a young woman in dazzling royal full *fig* —Mollie Panter-Downes⟩

¹fight \ˈfīt\ *verb* **fought** \ˈfȯt\; **fight·ing** [Middle English, from Old English *feohtan;* akin to Old High German *fehtan* to fight and perhaps to Latin *pectere* to comb — more at PECTINATE] (before 12th century)
intransitive verb
1 a : to contend in battle or physical combat; *especially* **:** to strive to overcome a person by blows or weapons **b :** to engage in boxing
2 : to put forth a determined effort
transitive verb
1 a (1) **:** to contend against in or as if in battle or physical combat (2) **:** to box against in the

☆ SYNONYMS
Fierce, ferocious, barbarous, savage, cruel mean showing fury or malignity in looks or actions. FIERCE applies to humans and animals that inspire terror because of their wild and menacing aspect or fury in attack ⟨*fierce* warriors⟩. FEROCIOUS implies extreme fierceness and unrestrained violence and brutality ⟨a *ferocious* dog⟩. BARBAROUS implies a ferocity or mercilessness regarded as unworthy of civilized people ⟨*barbarous* treatment of prisoners⟩. SAVAGE implies the absence of inhibitions restraining civilized people filled with rage, lust, or other violent passion ⟨a *savage* criminal⟩. CRUEL implies indifference to suffering and even positive pleasure in inflicting it ⟨the *cruel* jokes of schoolboys⟩.

◇ WORD HISTORY
fifth column *Fifth column* apparently first appeared in English in *New York Times* dispatches by William Carney, a correspondent reporting on the Spanish Civil War. On October 16, 1936, Carney mentioned raids in Republican-held Madrid that "apparently were instigated by a recent broadcast over the Rebel radio station by General Emilio Mola. He stated he was counting on four columns of troops outside Madrid and another column of persons hiding within the city who would join the invaders as soon as they entered the capital." Carney was accurately reflecting the fact that Spanish *quinta columna* "fifth column" was an epithet for subversive Nationalist ("Rebel") elements in the city. But subsequent literature on the war makes no mention of a radio broadcast by Mola, which Carney himself seems not to have heard, and Nationalist biographies of Mola do not attribute the phrase to him. In fact, the first documented use of *quinta columna* was on October 2, in a newspaper article by the Communist leader "La Pasionaria" (Dolores Ibárruri). The phrase may actually have been a propaganda invention of the Republicans attributed to Mola in order to justify reprisals against supposed Nationalist supporters in Madrid. In any event, *fifth column* soon became an international catchphrase for a conspiracy of traitors within one's own camp.

ring **b** (1) **:** to attempt to prevent the success or effectiveness of ⟨the company *fought* the takeover attempt⟩ (2) **:** to oppose the passage or development of ⟨*fight* a bill in Congress⟩
2 a : WAGE, CARRY ON ⟨*fight* a battle⟩ **b :** to take part in (as a boxing match)
3 : to struggle to endure or surmount ⟨*fight* a cold⟩
4 a : to gain by struggle ⟨*fights* his way through⟩ **b :** to resolve by struggle ⟨*fought* out their differences in court⟩
5 a : to manage (a ship) in a battle or storm **b :** to cause to struggle or contend **c :** to manage in an unnecessarily rough or awkward manner
— fight shy of : to avoid facing or meeting

²fight *noun* (before 12th century)
1 a : a hostile encounter **:** BATTLE, COMBAT **b :** a boxing match **c :** a verbal disagreement **:** ARGUMENT
2 : a struggle for a goal or an objective ⟨a *fight* for justice⟩
3 : strength or disposition for fighting **:** PUGNACITY ⟨still full of *fight*⟩

fight·er \'fī-tər\ *noun* (13th century)
: one that fights: as **a** (1) **:** WARRIOR, SOLDIER (2) **:** a pugnacious or game individual (3) **:** ¹BOXER 1 **b :** an airplane of high speed and maneuverability with armament designed to destroy enemy aircraft

fight·er–bomb·er \-'bä-mər\ *noun* (1936)
: a fighter aircraft fitted to carry bombs and rockets in addition to its normal armament

fighting chair *noun* (1950)
: a chair from which a salt-water angler plays a hooked fish

fighting chance *noun* (1889)
: a chance that may be realized by a struggle ⟨the patient had a *fighting chance* to live⟩

fighting word *noun* (1917)
: a word likely to provoke a fight

fight song *noun* (1954)
: a song used to inspire enthusiasm usually during an athletic competition

fig leaf *noun* (14th century)
1 : the leaf of a fig tree
2 [from the use by Adam and Eve of fig leaves to cover their nakedness after eating the forbidden fruit (Genesis 3:7)] **:** something that conceals or camouflages usually inadequately or dishonestly

fig marigold *noun* (1731)
: any of several herbs (genus *Mesembryanthemum*) of the carpetweed family with showy white or pink flowers

fig·ment \'fig-mənt\ *noun* [Middle English, from Latin *figmentum,* from *fingere* to shape — more at DOUGH] (15th century)
: something made up or contrived

fig·ur·al \'fi-g(y)ə-rəl\ *adjective* (1813)
: of, relating to, or consisting of human or animal figures ⟨a *figural* composition⟩

fig·u·ra·tion \,fi-g(y)ə-'rā-shən\ *noun* (14th century)
1 : FORM, OUTLINE
2 : the act or process of creating or providing a figure
3 : an act or instance of representation in figures and shapes ⟨cubism was explained as a synthesis of colored *figurations* of objects —Janet Flanner⟩
4 : ornamentation of a musical passage by using decorative and usually repetitive figures

fig·u·ra·tive \'fi-g(y)ə-rə-tiv\ *adjective* (14th century)
1 a : representing by a figure or resemblance **:** EMBLEMATIC **b :** of or relating to representation of form or figure in art ⟨*figurative* sculpture⟩
2 a : expressing one thing in terms normally denoting another with which it may be regarded as analogous **:** METAPHORICAL ⟨*figurative* language⟩ **b :** characterized by figures of speech ⟨a *figurative* description⟩
— fig·u·ra·tive·ly *adverb*
— fig·u·ra·tive·ness *noun*

¹fig·ure \'fi-gyər, *British & often US* 'fi-gər\ *noun* [Middle English, from Old French, from Latin *figura,* from *fingere*] (13th century)
1 a : a number symbol **:** NUMERAL, DIGIT ⟨a salary running into six *figures*⟩ **b** *plural* **:** arithmetical calculations ⟨good at *figures*⟩ **c :** a written or printed character **d :** value especially as expressed in numbers **:** SUM, PRICE ⟨sold at a low *figure*⟩
2 a : a geometric form (as a line, triangle, or sphere) especially when considered as a set of geometric elements (as points) in space of a given number of dimensions ⟨a square is a plane *figure*⟩ **b :** bodily shape or form especially of a person ⟨a slender *figure*⟩ **c :** an object noticeable only as a shape or form ⟨*figures* moving in the dusk⟩
3 a : the graphic representation of a form especially of a person or geometric entity **b :** a diagram or pictorial illustration of textual matter
4 : a person, thing, or action representative of another
5 a : FIGURE OF SPEECH **b :** an intentional deviation from the ordinary form or syntactical relation of words
6 : the form of a syllogism with respect to the relative position of the middle term
7 : an often repetitive pattern or design in a manufactured article (as cloth) or natural product (as wood) ⟨a polka-dot *figure*⟩
8 : appearance made **:** impression produced ⟨the couple cut quite a *figure*⟩
9 a : a series of movements in a dance **b :** an outline representation of a form traced by a series of evolutions (as with skates on an ice surface or by an airplane in the air)
10 : a prominent personality **:** PERSONAGE ⟨great *figures* of history⟩
11 : a short coherent group of notes or chords that may constitute part of a phrase, theme, or composition

²figure *verb* **fig·ured; fig·ur·ing** \'fi-gyə-riŋ, 'fi-g(ə-)riŋ\ (14th century)
transitive verb
1 : to represent by or as if by a figure or outline
2 : to decorate with a pattern; *also* **:** to write figures over or under (the bass) in order to indicate the accompanying chords
3 : to indicate or represent by numerals
4 a : CALCULATE **b :** CONCLUDE, DECIDE ⟨*figured* there was no use in further effort⟩ **c :** REGARD, CONSIDER **d :** to appear likely ⟨*figures* to win⟩
intransitive verb
1 a : to be or appear important or conspicuous **b :** to be involved or implicated ⟨*figured* in a robbery⟩
2 : to perform a figure in dancing
3 : COMPUTE, CALCULATE
4 : to seem rational, normal, or expected ⟨that *figures*⟩
— fig·ur·er \-g(y)ər-ər\ *noun*
— figure on 1 : to take into consideration ⟨*figuring on* the extra income⟩ **2 :** to rely on **3 :** PLAN ⟨I *figure on* going into town⟩

fig·ured \-g(y)ərd\ *adjective* (15th century)
1 : adorned with, formed into, or marked with a figure ⟨*figured* muslin⟩ ⟨*figured* wood⟩
2 : being represented **:** PORTRAYED
3 : indicated by figures

figured bass *noun* (1801)
: CONTINUO

figure eight *noun* (1887)
: something resembling the Arabic numeral eight in form or shape: as **a :** a small knot — see KNOT illustration **b :** an embroidery stitch **c :** a dance pattern **d :** a skater's figure — called also *figure-of-eight*

fig·ure·head \'fi-g(y)ər-,hed\ *noun* (1765)
1 : the figure on a ship's bow
2 : a head or chief in name only

figure in *transitive verb* (circa 1934)
: to include especially in a reckoning

figure of merit (circa 1865)
: a numerical quantity based on one or more characteristics of a system or device that represents a measure of efficiency or effectiveness

figure of speech (1824)
: a form of expression (as a simile or metaphor) used to convey meaning or heighten effect often by comparing or identifying one thing with another that has a meaning or connotation familiar to the reader or listener

figure out *transitive verb* (1833)
1 : DISCOVER, DETERMINE ⟨try to *figure out* a way to do it⟩
2 : SOLVE, FATHOM ⟨*figure out* a problem⟩

figure skating *noun* (1852)
: skating characterized by the performance of various jumps, spins, and dance movements and formerly by the tracing of prescribed figures
— figure skater *noun*

fig·u·rine \,fi-g(y)ə-'rēn\ *noun* [French, from Italian *figurina,* diminutive of *figura* figure, from Latin — more at FIGURE] (1854)
: a small carved or molded figure **:** STATUETTE

fig wasp *noun* (1883)
: a minute wasp (*Blastophaga psenes* of the family Agaontidae) that breeds in the caprifig and is the agent of caprification; *broadly* **:** any wasp of the same family

fig·wort \'fig-,wərt, -,wȯrt\ *noun* (1548)
: any of a genus (*Scrophularia*) of chiefly herbaceous plants of the snapdragon family with leaves having no stipules, an irregular bilabiate corolla, and a 2-celled ovary

Fi·ji·an \'fē-(,)jē-ən, fi-'\ *noun* (1809)
1 : a member of a Melanesian people of Fiji
2 : the Austronesian language of the Fijians
— Fijian *adjective*

fil·a·ment \'fi-lə-mənt\ *noun* [Middle French, from Medieval Latin *filamentum,* from Late Latin *filare* to spin — more at FILE] (1594)
: a single thread or a thin flexible threadlike object, process, or appendage: as **a :** a tenuous conductor (as of carbon or metal) made incandescent by the passage of an electric current; *specifically* **:** a cathode in the form of a metal wire in an electron tube **b** (1) **:** a thin and fine elongated constituent part of a gill (2) **:** an elongated thin series of cells attached one to another or a very long thin cylindrical single cell (as of some algae, fungi, or bacteria) **c :** the anther-bearing stalk of a stamen — see FLOWER illustration
— fil·a·men·ta·ry \,fil-ə-'men-t(ə-)rē\ *adjective*
— fil·a·men·tous \-'men-təs\ *adjective*

fi·lar \'fī-lər\ *adjective* [Latin *filum* thread] (circa 1859)
: of or relating to a thread or line; *especially* **:** having threads across the field of view ⟨a *filar* eyepiece⟩

fi·lar·ia \fə-'lar-ē-ə, -'ler-\ *noun, plural* **-i·ae** \-ē-,ē, -ē-,ī\ [New Latin, genus name, from Latin *filum*] (1834)
: any of numerous slender filamentous nematodes (*Filaria* and related genera) that as adults are parasites in the blood or tissues of mammals and as larvae usually develop in biting insects
— fi·lar·i·al \-ē-əl\ *adjective*
— fi·lar·i·id \-ē-əd\ *adjective or noun*

fil·a·ri·a·sis \,fi-lə-'rī-ə-səs\ *noun, plural* **-a·ses** \-,sēz\ [New Latin] (1879)
: infestation with or disease caused by filariae

fil·a·ture \'fil-ə-,chu̇(ə)r, -chər, -,t(y)u̇(ə)r\ *noun* [French, from Late Latin *filatus,* past participle of *filare*] (1759)
: a factory where silk is reeled

fil·bert \'fil-bərt\ *noun* [Middle English, from Anglo-French *philber,* from Saint *Philibert* (died 684) Frankish abbot whose feast day falls in the nutting season] (14th century)
1 : either of two European hazels (*Corylus avellana* and *C. maxima*); *also* **:** the sweet thick-shelled nut of the filbert

2 : HAZELNUT ◆

filch \'filch\ *transitive verb* [Middle English] (1561)
: to appropriate furtively or casually ⟨*filch* a cookie⟩
synonym see STEAL

[1]**file** \'fī(ə)l\ *noun* [Middle English, from Old English *fēol;* akin to Old High German *fīla* file] (before 12th century)
1 : a tool usually of hardened steel with cutting ridges for forming or smoothing surfaces especially of metal
2 : a shrewd or crafty person

[2]**file** *transitive verb* **filed; fil·ing** (13th century)
: to rub, smooth, or cut away with or as if with a file

[3]**file** *transitive verb* **filed; fil·ing** [Middle English, from Old English *fȳlan,* from *fūl* foul] (before 12th century)
chiefly dialect **:** DEFILE, CORRUPT

[4]**file** *verb* **filed; fil·ing** [Middle English, from Middle French *filer* to string documents on a string or wire, from *fil* thread, from Old French, from Latin *filum;* akin to Armenian *ǰil* sinew] (15th century)
transitive verb
1 : to arrange in order for preservation and reference ⟨*file* letters⟩
2 a : to place among official records as prescribed by law ⟨*file* a mortgage⟩ **b :** to send (copy) to a newspaper ⟨*filed* a story⟩ **c :** to return to the office of the clerk of a court without action on the merits
3 : to initiate (as a legal action) through proper formal procedure ⟨threatened to *file* charges⟩
intransitive verb
1 : to register as a candidate especially in a primary election
2 : to place items in a file
— fil·er \'fī-lər\ *noun*

[5]**file** *noun* (1525)
1 : a device (as a folder, case, or cabinet) by means of which papers are kept in order
2 a *archaic* **:** ROLL, LIST **b :** a collection of papers or publications usually arranged or classified **c** (1) **:** a collection of related data records (as for a computer) (2) **:** a complete collection of data (as text or a program) treated by a computer as a unit especially for purposes of input and output
— on file : in or as if in a file for ready reference

[6]**file** *noun* [Middle French, from *filer* to spin, from Late Latin *filare,* from Latin *filum*] (1598)
1 : a row of persons, animals, or things arranged one behind the other
2 : any of the rows of squares that extend across a chessboard from one player's side to the other player's side

[7]**file** *intransitive verb* **filed; fil·ing** (1616)
: to march or proceed in file

filé *also* **file** \fə-'lā, (ˌ)fē-'lā, 'fē-(ˌ)lā\ *noun* [Louisiana French, from French, past participle of *filer* to twist, spin] (1806)
: powdered young leaves of sassafras used to thicken soups or stews

file clerk *noun* (1919)
: a clerk who works on files

file·fish \'fī(ə)l-ˌfish\ *noun* (1814)
: any of various bony fishes (order Tetraodontiformes and especially genera *Aluterus, Cantherhines,* and *Monacanthus* of the family Balistidae) with rough granular leathery skins

fi·let \fi-'lā\ *noun* [French, literally, net] (1881)
: a lace with a square mesh and geometric designs

fi·let mi·gnon \ˌfi-(ˌ)lā-mēn-'yō^{n}, fi-ˌlā-\ *noun, plural* **filets mignons** \ˌfi-(ˌ)lā-mēn-'yō^{n}z, fi-ˌlā-\ [French, literally, dainty fillet] (1906)
: a thick slice of beef cut from the narrow end of a beef tenderloin

fili- *or* **filo-** *combining form* [Latin *filum*]
: thread ⟨*fili*form⟩

fil·ial \'fi-lē-əl, 'fil-yəl\ *adjective* [Middle English, from Late Latin *filialis,* from Latin *filius* son — more at FEMININE] (14th century)
1 : of, relating to, or befitting a son or daughter ⟨*filial* obedience⟩
2 : having or assuming the relation of a child or offspring
— fil·ial·ly \-lē-ə-lē, -yə-lē\ *adverb*

filial generation *noun* (1909)
: a generation in a breeding experiment that is successive to a mating between parents of two distinctively different but usually relatively pure genotypes

fil·i·a·tion \ˌfi-lē-'ā-shən\ *noun* (15th century)
1 a : filial relationship especially of a son to his father **b :** the adjudication of paternity
2 a : descent or derivation especially from a culture or language **b :** the act or process of determining such relationship

[1]**fil·i·bus·ter** \'fi-lə-ˌbəs-tər\ *noun* [Spanish *filibustero,* literally, freebooter] (1851)
1 : an irregular military adventurer; *specifically* **:** an American engaged in fomenting insurrections in Latin America in the mid-19th century
2 [[2]*filibuster*] **a :** the use of extreme dilatory tactics in an attempt to delay or prevent action especially in a legislative assembly **b :** an instance of this practice ◆

[2]**filibuster** *verb* **-tered; -ter·ing** \-t(ə-)riŋ\ (1853)
intransitive verb
1 : to carry out insurrectionist activities in a foreign country
2 : to engage in a filibuster
transitive verb
: to subject to a filibuster
— fil·i·bus·ter·er \-tər-ər\ *noun*

fi·li·form \'fi-lə-ˌfȯrm, 'fī-\ *adjective* (1757)
: shaped like a filament

[1]**fil·i·gree** \'fi-lə-ˌgrē\ *noun* [modification of French *filigrane,* from Italian *filigrana,* from Latin *filum* + *granum* grain — more at CORN] (1693)
1 : ornamental work especially of fine wire of gold, silver, or copper applied chiefly to gold and silver surfaces

filigree 2a

2 a : ornamental openwork of delicate or intricate design **b :** a pattern or design resembling such openwork ⟨a *filigree* of frost⟩ **c :** ORNAMENTATION, EMBELLISHMENT ⟨writings . . . heavy with late Victorian *filigree* —Jack Beatty⟩

[2]**filigree** *transitive verb* **fil·i·greed; fil·i·gree·ing** (1831)
: to adorn with or as if with filigree

fil·ing \'fī-liŋ\ *noun* (14th century)
1 : an act or instance of using a file
2 : a fragment rubbed off in filing ⟨iron *filings*⟩

fil·io·pi·etis·tic \ˌfi-lē-ō-ˌpī-ə-'tis-tik\ *adjective* [*fil*ial + *-o-* + *pietistic*] (1893)
: of or relating to an often excessive veneration of ancestors or tradition

Fil·i·pi·na \ˌfi-lə-'pē-nə\ *noun* [Spanish] (1899)
: a Filipino girl or woman

Fil·i·pi·no \-'pē-(ˌ)nō\ *noun, plural* **Filipinos** [Spanish] (circa 1889)
1 : a native of the Philippine islands
2 : a citizen of the Republic of the Philippines
— Filipino *adjective*

[1]**fill** \'fil\ *verb* [Middle English, from Old English *fyllan;* akin to Old English *full* full] (before 12th century)
transitive verb
1 a : to put into as much as can be held or conveniently contained ⟨*fill* a cup with water⟩ **b :** to supply with a full complement ⟨the class is already *filled*⟩ **c** (1) **:** to cause to swell or billow ⟨wind *filled* the sails⟩ (2) **:** to trim (a sail) to catch the wind **d :** to raise the level of with fill ⟨*filled* land⟩ **e :** to repair the cavities of (teeth) **f :** to stop up **:** OBSTRUCT, PLUG ⟨wreckage *filled* the channel⟩ **g :** to stop up the interstices, crevices, or pores of (as cloth, wood, or leather) with a foreign substance
2 a : FEED, SATIATE **b :** SATISFY, FULFILL ⟨*fills* all requirements⟩ **c :** MAKE OUT, COMPLETE — used with *out* or *in* ⟨*fill* out a form⟩ ⟨*fill* in the blanks⟩ **d :** to draw the playing cards necessary to complete (as a straight or flush in poker)
3 a : to occupy the whole of ⟨smoke *filled* the room⟩ **b :** to spread through **c :** to make full ⟨a mind *filled* with fantasies⟩
4 a : to possess and perform the duties of **:** HOLD ⟨*fill* an office⟩ **b :** to place a person in ⟨*fill* a vacancy⟩
5 : to supply as directed ⟨*fill* a prescription⟩
6 : to cover the surface of with a layer of precious metal
intransitive verb
: to become full
— fill one's shoes : to take over one's job, position, or responsibilities

[2]**fill** *noun* (before 12th century)
1 : a full supply; *especially* **:** a quantity that satisfies or satiates ⟨eat your *fill*⟩
2 : something that fills: as **a :** material used to fill a receptacle, cavity, passage, or low place

◇ WORD HISTORY

filbert In all likelihood people who enjoy filberts are unaware that the nut is named after a 7th century Frankish abbot. Saint Philibert was neither a known grower nor a connoisseur of hazelnuts. His connection with the nut is based on the fact that his feast day falls on August 20th, which happens to be the peak of the nutting season in France and England. The Anglo-French word *philber* was borrowed into Middle English as *filbert.*

filibuster The Dutch word *vrijbuiter,* meaning "pirate, brigand" has left its mark on the English vocabulary not once but twice. It first entered English as *freebooter* in the 16th century. From English the word passed, probably by way of French, into Spanish as *filibustero.* In the 1850s, after the Mexican War had resulted in U.S. seizure of vast Mexican territories, soldiers of fortune from the southern states were inspired to launch freelance incursions into a few Latin-American countries. Ostensibly they meant to support one or another side in civil conflicts, but their real long-term goal was to establish new Anglo-American fiefdoms where slavery would be tolerated. Americans adopted the Spanish word *filibustero,* usually Anglicized to *filibuster,* to refer to these men. Before a new decade had arrived, the word *filibuster* was being applied to members of Congress who engaged in delaying tactics to prevent vote on a bill that the legislative majority favored. The analogy apparently was that these politicians, like the filibusters in Latin America, were attempting to usurp legitimate rule. Soon after its adoption in this sense, *filibuster* was used as a verb, and then turned back into a noun to refer to the act of legislative filibustering.

b : a bit of instrumental music that fills the pauses between phrases (as of a vocalist or soloist)

fill away *intransitive verb* (1840)
1 : to trim a sail to catch the wind
2 : to proceed on the course especially after being brought up in the wind

¹fill·er \ˈfi-lər\ *noun* (15th century)
: one that fills: as **a :** a substance added to a product (as to increase bulk, weight, viscosity, opacity, or strength) **b :** a composition used to fill the pores and grain especially of a wood surface before painting or varnishing **c :** a piece used to cover or fill in a space between two parts of a structure **d :** tobacco used to form the core of a cigar **e :** material used to fill extra space in a column or page of a newspaper or magazine **f :** a pack of paper for a loose-leaf notebook **g :** a sound, word, or phrase (as "you know?") used to fill pauses in speaking

²fil·ler \ˈfi-ˌler\ *noun, plural* **fillers** *or* **filler** [Hungarian *fillér*] (1904)
— see *forint* at MONEY table

¹fil·let \ˈfi-lət, *in sense 2b also* fi-ˈlā, ˈfi-(ˌ)lā\ *also* **fi·let** \fi-ˈlā, ˈfi-(ˌ)lā\ *noun* [Middle English *filet,* from Middle French, diminutive of *fil* thread — more at FILE] (14th century)
1 : a ribbon or narrow strip of material used especially as a headband
2 a : a thin narrow strip of material **b :** a piece or slice of boneless meat or fish; *especially* **:** the tenderloin of beef
3 a : a concave junction formed where two surfaces meet (as at an angle) **b :** a strip that gives a rounded appearance to such a junction; *also* **:** a strip to reinforce the corner where two surfaces meet
4 : a narrow flat architectural member: **a :** a flat molding separating others — see BASE illustration **b :** the space between two flutings in a shaft

²fil·let \ˈfi-lət, *in sense 2 also* fi-ˈlā, ˈfi-(ˌ)lā\ *transitive verb* (1604)
1 : to bind, furnish, or adorn with or as if with a fillet
2 : to cut into fillets

fillet weld *noun* (1926)
: a weld of approximately triangular cross section used to join two pieces especially perpendicularly

fill–in \ˈfil-ˌin\ *noun* (1917)
: someone or something that fills in

fill in (1840)
transitive verb
1 : to give necessary or recently acquired information to ⟨I'll *fill* you *in*⟩
2 : to enrich (as a design) with detail
intransitive verb
: to fill a vacancy usually temporarily

fill·ing \ˈfi-liŋ\ *noun* (14th century)
1 : an act or instance of filling
2 : something used to fill a cavity, container, or depression
3 : something that completes: as **a :** the yarn interlacing the warp in a fabric; *also* **:** yarn for the shuttle **b :** a food mixture used to fill pastry or sandwiches

filling station *noun* (1921)
: SERVICE STATION 1

¹fil·lip \ˈfi-ləp\ *transitive verb* [probably of imitative origin] (15th century)
1 a : to make a filliping motion with **b :** to strike or tap with a fillip ⟨*filliped* him on the nose⟩
2 : to project quickly by or as if by a fillip ⟨*fillip* crumbs off the table⟩
3 : STIMULATE ⟨with this to *fillip* his spirits —Robert Westerby⟩

²fillip *noun* (1519)
1 a : a blow or gesture made by the sudden forcible straightening of a finger curled up against the thumb **b :** a short sharp blow **:** BUFFET
2 : something tending to arouse or excite: as **a :** STIMULUS ⟨just the *fillip* my confidence needed⟩ ⟨lent a *fillip* of danger to the sport⟩ **b :** a trivial addition **:** EMBELLISHMENT ⟨showy *fillips* of language⟩ **c :** a significant and often unexpected development **:** WRINKLE ⟨plot twists and *fillips*⟩

fill out *intransitive verb* (1888)
: to put on flesh

fill–up \ˈfil-ˌəp\ *noun* (1853)
: an action or instance of filling up something (as a gas tank)

fil·ly \ˈfi-lē\ *noun, plural* **fillies** [Middle English *fyly,* from Old Norse *fylja;* akin to Old English *fola* foal] (15th century)
1 : a young female horse usually of less than four years
2 : a young woman **:** GIRL

¹film \ˈfilm, *Southern also* ˈfi(ə)m\ *noun, often attributive* [Middle English *filme,* from Old English *filmen;* akin to Greek *pelma* sole of the foot, Old English *fell* skin — more at FELL] (before 12th century)
1 a : a thin skin or membranous covering **:** PELLICLE **b :** an abnormal growth on or in the eye
2 : a thin covering or coating ⟨a *film* of ice⟩
3 a : an exceedingly thin layer **:** LAMINA **b** (1) **:** a thin flexible transparent sheet (as of plastic) used especially as a wrapping (2) **:** such a sheet of cellulose acetate or nitrocellulose coated with a radiation-sensitive emulsion for taking photographs
4 : MOTION PICTURE

²film (1602)
transitive verb
1 : to cover with or as if with a film
2 : to make a motion picture of or from ⟨*film* a scene⟩
intransitive verb
1 : to become covered or obscured with or as if with a film
2 : to make a motion picture
— film·able \ˈfil-mə-bəl\ *adjective*

film badge *noun* (1945)
: a small pack of sensitive photographic film worn as a badge for indicating exposure to radiation

film·dom \ˈfilm-dəm\ *noun* (1914)
: the motion-picture industry

film·go·er \-ˌgō(-ə)r\ *noun* (1919)
: one who frequently attends films

film·ic \ˈfil-mik\ *adjective* (circa 1930)
: of, relating to, or resembling motion pictures
— film·i·cal·ly \-mi-k(ə-)lē\ *adverb*

film·land \ˈfilm-ˌland\ *noun* (1913)
: FILMDOM

film·mak·er \ˈfilm-ˌmā-kər\ *noun* (1908)
: one who makes motion pictures

film·mak·ing \-ˌmā-kiŋ\ *noun* (1913)
: the making of motion pictures

film noir \-ˈnwär\ *noun* [French, literally, black film] (1958)
: a type of crime film featuring cynical malevolent characters in a sleazy setting and an ominous atmosphere that is conveyed by shadowy photography and foreboding background music; *also* **:** a film of this type

film·og·ra·phy \fil-ˈmä-grə-fē\ *noun, plural* **-phies** [*film* + *-ography* (as in *bibliography*)] (1962)
: a list of motion pictures featuring the work of a prominent film figure or relating to a particular topic

film·set·ting \ˈfilm-ˌse-tiŋ\ *noun* (1954)
: PHOTOCOMPOSITION
— film·set *adjective*
— filmset *transitive verb*
— film·set·ter *noun*

film·strip \ˈfilm-ˌstrip\ *noun* (1930)
: a strip of film bearing a sequence of images for projection as still pictures

filmy \ˈfil-mē\ *adjective* **film·i·er; -est** (1604)
1 : of, resembling, or composed of film **:** GAUZY ⟨*filmy* draperies⟩
2 : covered with a haze or film
— film·i·ly \-mə-lē\ *adverb*
— film·i·ness \-mē-nəs\ *noun*

filo *also* **fil·lo** *variant of* PHYLLO

filo- — see FILI-

fils \ˈfils\ *noun, plural* **fils** [Arabic] (1931)
— see *dinar, dirham, rial* at MONEY table

¹fil·ter \ˈfil-tər\ *noun* [Middle English *filtre,* from Medieval Latin *filtrum* piece of felt used as a filter, of Germanic origin; akin to Old High German *filz* felt — more at FELT] (1563)
1 a : a porous article or mass (as of paper or sand) through which a gas or liquid is passed to separate out matter in suspension **b :** something that has the effect of a filter
2 : an apparatus containing a filter medium
3 a : a device or material for suppressing or minimizing waves or oscillations of certain frequencies (as of electricity, light, or sound) **b :** a transparent material (as colored glass) that absorbs light of certain wavelengths or colors selectively and is used for modifying light that reaches a sensitized photographic material — called also *color filter*

²filter *verb* **fil·tered; fil·ter·ing** \-t(ə-)riŋ\ (1576)
transitive verb
1 : to subject to the action of a filter
2 : to remove by means of a filter
intransitive verb
1 : to pass or move through or as if through a filter
2 : to come or go in small units over a period of time ⟨people began *filtering* in⟩

fil·ter·able *also* **fil·tra·ble** \ˈfil-t(ə-)rə-bəl\ *adjective* (1908)
: capable of being filtered or of passing through a filter
— fil·ter·abil·i·ty \ˌfil-t(ə-)rə-ˈbi-lə-tē\ *noun*

filterable virus *noun* (1911)
: any of the infectious agents that pass through a filter of diatomite or unglazed porcelain with the filtrate and remain virulent and that include the viruses as presently understood and various other groups (as the mycoplasmas and rickettsias) which were originally considered viruses before their cellular nature was established

filter bed *noun* (circa 1874)
: a sand or gravel bed for filtering water or sewage

filter feeder *noun* (1928)
: an animal that obtains its food by filtering organic matter or minute organisms from a current of water that passes through some part of its system

filter paper *noun* (circa 1846)
: porous unsized paper used especially for filtering

filter tip *noun* (1932)
: a cigar or cigarette tip designed to filter the smoke before it enters the smoker's mouth; *also* **:** a cigar or cigarette provided with such a tip
— fil·ter–tipped \ˌfil-tər-ˈtipt\ *adjective*

filth \ˈfilth\ *noun* [Middle English, from Old English *fȳlth,* from *fūl* foul] (before 12th century)
1 : foul or putrid matter; *especially* **:** loathsome dirt or refuse
2 a : moral corruption or defilement **b :** something that tends to corrupt or defile

filthy \ˈfil-thē\ *adjective* **filth·i·er; -est** (14th century)
1 : covered with or containing filth **:** offensively dirty
2 a : UNDERHAND, VILE **b :** OBSCENE
synonym see DIRTY
— filth·i·ly \-thə-lē\ *adverb*
— filth·i·ness \-thē-nəs\ *noun*

fil·trate \ˈfil-ˌtrāt\ *noun* (circa 1846)
: material that has passed through a filter

fil·tra·tion \fil-ˈtrā-shən\ *noun* (1605)
1 : the process of filtering
2 : the process of passing through or as if through a filter; *also* **:** DIFFUSION

fim·bria \'fim-brē-ə\ *noun, plural* **-bri·ae** \-brē-,ē, -,ī\ [New Latin, from Latin, fringe] (1752)
: a bordering fringe especially at the entrance of the fallopian tubes
— **fim·bri·al** \-brē-əl\ *adjective*

fim·bri·at·ed \'fim-brē-,ā-təd\ *also* **fim·bri·ate** \-,āt\ *adjective* (15th century)
: having the edge or extremity bordered by slender processes : FRINGED
— **fim·bri·a·tion** \,fim-brē-'ā-shən\ *noun*

[1]fin \'fin\ *noun* [Middle English *finn,* from Old English] (before 12th century)
1 : an external membranous process of an aquatic animal (as a fish) used in propelling or guiding the body — see FISH illustration
2 : something resembling a fin: as **a :** HAND, ARM **b** (1) **:** an appendage of a boat (as a submarine) (2) **:** an airfoil attached to an airplane for directional stability **c :** FLIPPER 1b **d :** any of the projecting ribs on a radiator or an engine cylinder
— **fin·like** \-,līk\ *adjective*
— **finned** \'find\ *adjective*

[2]fin *verb* **finned; fin·ning** (1933)
transitive verb
: to equip with fins
intransitive verb
1 : to show the fins above the water
2 : to move through the water propelled by fins

[3]fin *noun* [Yiddish *finf* five, from Middle High German, from Old High German — more at FIVE] (1925)
slang **:** a 5-dollar bill

fi·na·gle \fə-'nā-gəl\ *verb* **fi·na·gled; fi·na·gling** \-g(ə-)liŋ\ [perhaps alteration of *fainaigue* (to renege)] (circa 1924)
transitive verb
1 : to obtain by indirect or involved means
2 : to obtain by trickery
intransitive verb
: to use devious or dishonest methods to achieve one's ends
— **fi·na·gler** \-g(ə-)lər\ *noun*

[1]fi·nal \'fī-n^{ə}l\ *adjective* [Middle English, from Middle French, from Latin *finalis,* from *finis* boundary, end] (14th century)
1 a : not to be altered or undone ⟨all sales are *final*⟩ **b :** of or relating to a concluding court action or proceeding ⟨*final* decree⟩
2 : coming at the end **:** being the last in a series, process, or progress ⟨the *final* chapter⟩
3 : of or relating to the ultimate purpose or result of a process ⟨our *final* goal⟩
synonym see LAST
— **fi·nal·ly** \'fī-n^{ə}l-ē, 'fīn-lē\ *adverb*

[2]final *noun* (1609)
: something that is final: as **a :** a deciding match, game, heat, or trial — usually used in plural **b :** the last examination in a course — often used in plural

fi·na·le \fə-'na-lē, fi-'nä-\ *noun* [Italian, from *finale,* adjective, final, from Latin *finalis*] (1783)
: the close or termination of something: as **a :** the last section of an instrumental musical composition **b :** the closing part, scene, or number in a public performance **c :** the last and often climactic event or item in a sequence

fi·nal·ise *British variant of* FINALIZE

fi·nal·ist \'fī-n^{ə}l-ist\ *noun* (1898)
: a contestant in the finals of a competition

fi·nal·i·ty \fī-'na-lə-tē, fə-\ *noun, plural* **-ties** (1833)
1 a : the character or condition of being final, settled, irrevocable, or complete **b :** the condition of being at an ultimate point especially of development or authority
2 : something final; *especially* **:** a fundamental fact, action, or belief

fi·nal·ize \'fī-n^{ə}l-,īz\ *transitive verb* **-ized; -iz·ing** (1901)
1 : to put in final or finished form ⟨soon my conclusion will be *finalized* —D. D. Eisenhower⟩
2 : to give final approval to ⟨*finalizing* the papers prepared . . . by his staff —*Newsweek*⟩ ■
— **fi·nal·iza·tion** \,fī-n^{ə}l-ə-'zā-shən\ *noun*

final solution *noun, often F&S capitalized* (1947)
: the Nazi program for extermination of all Jews in Europe

[1]fi·nance \fə-'nan(t)s, 'fī-,, fī-'\ *noun* [Middle English, payment, ransom, from Middle French, from *finer* to end, pay, from *fin* end — more at FINE] (1739)
1 *plural* **:** money or other liquid resources of a government, business, group, or individual
2 : the system that includes the circulation of money, the granting of credit, the making of investments, and the provision of banking facilities
3 : the science or study of the management of funds
4 : the obtaining of funds or capital **:** FINANCING ■

[2]finance *transitive verb* **fi·nanced; fi·nanc·ing** (1866)
1 a : to raise or provide funds or capital for ⟨*finance* a new house⟩ **b :** to furnish with necessary funds ⟨*finance* a son through college⟩
2 : to sell something to on credit

finance company *noun* (circa 1924)
: a company that makes usually small short-term loans usually to individuals

fi·nan·cial \fə-'nan(t)-shəl, fī-\ *adjective* (1769)
: relating to finance or financiers
— **fi·nan·cial·ly** \-'nan(t)-sh(ə-)lē\ *adverb*

fi·nan·cier \,fi-nən-'sir; fə-,nan-, ,fī-\ *noun* (1618)
1 : one who specializes in raising and expending public moneys
2 : one who deals with finance and investment on a large scale

fi·nanc·ing *noun* (1827)
: the act or process or an instance of raising or providing funds; *also* **:** the funds thus raised or provided

fin·back \'fin-,bak\ *noun* (1725)
: FIN WHALE

finch \'finch\ *noun* [Middle English, from Old English *finc;* akin to Old High German *fincho* finch and perhaps to Greek *spiza* chaffinch] (before 12th century)
: any of numerous songbirds (especially families Fringillidae, Estrildidae, and Emberizidae) having a short stout usually conical bill adapted for crushing seeds — compare BUNTING, CROSSBILL, GOLDFINCH, GROSBEAK, LINNET, SPARROW

[1]find \'fīnd\ *verb* **found** \'faund\; **find·ing** [Middle English, from Old English *findan;* akin to Old High German *findan* to find, Latin *pont-, pons* bridge, Greek *pontos* sea, Sanskrit *patha* way, course] (before 12th century)
transitive verb
1 a : to come upon often accidentally **:** ENCOUNTER **b :** to meet with (a particular reception) ⟨hoped to *find* favor⟩
2 a : to come upon by searching or effort ⟨must *find* a suitable person for the job⟩ **b :** to discover by study or experiment ⟨*find* an answer⟩ **c :** to obtain by effort or management ⟨*find* the time to study⟩ **d :** ATTAIN, REACH ⟨the bullet *found* its mark⟩
3 a : to discover by the intellect or the feelings **:** EXPERIENCE ⟨*find* much pleasure in your company⟩ **b :** to perceive (oneself) to be in a certain place or condition **c :** to gain or regain the use or power of ⟨trying to *find* his tongue⟩ **d :** to bring (oneself) to a realization of one's powers or of one's proper sphere of activity ⟨must help the student to *find* himself as an individual —N. M. Pusey⟩
4 a : PROVIDE, SUPPLY **b :** to furnish (room and board) especially as a condition of employment
5 : to determine and make a statement about ⟨*find* a verdict⟩ ⟨*found* her guilty⟩
intransitive verb
: to determine a case judicially by a verdict ⟨*find* for the defendant⟩
— **find·able** \'fīn-də-bəl\ *adjective*
— **find fault :** to criticize unfavorably

[2]find *noun* (1825)
1 : an act or instance of finding
2 : something found: as **a :** a valuable discovery ⟨an archaeological *find*⟩ **b :** a person whose ability proves to be unexpectedly good

find·er \'fīn-dər\ *noun* (14th century)
1 : one that finds
2 : a small astronomical telescope of low power and wide field attached to a larger telescope for finding an object
3 : VIEWFINDER

fin de siè·cle \,fan-də-sē-'e-k^{ə}l; fa^{n}-də-syekl^{ə}\ *adjective* [French, end of the century] (1890)
: of, relating to, or characteristic of the close of the 19th century and especially its literary and artistic climate of sophistication, world-weariness, and fashionable despair

find·ing \'fīn-diŋ\ *noun* (14th century)
1 a : the act of one that finds **b :** FIND 2
2 *plural* **:** small tools and supplies used by an artisan (as a dressmaker, jeweler, or shoemaker)
3 a : the result of a judicial examination or inquiry **b :** the results of an investigation — usually used in plural

find out (13th century)
transitive verb
1 : to learn by study, observation, or search **:** DISCOVER
2 a : to catch in an offense (as a crime) ⟨the culprits were soon *found out*⟩ **b :** to ascertain the true character or identity of ⟨the informer was *found out*⟩
intransitive verb
: to discover, learn, or verify something ⟨I don't know, but I'll *find out* for you⟩

[1]fine \'fīn\ *noun* [Middle English, from Old French *fin,* from Latin *finis* boundary, end] (13th century)
1 *obsolete* **:** END, CONCLUSION

□ USAGE

finalize *Finalize* has been frequently castigated as an unnecessary neologism or as U.S. government gobbledygook. It appears to have first gained currency in Australia (where it has been acceptable all along) in the early 1920s. The U.S. Navy picked it up in the late 20s, and from there it came into widespread use. It is a standard formation (see -IZE). Our current evidence indicates it is most frequently used in government, business dealings, and child adoption; it usually is not found in belles-lettres.

finance Some critics have decreed that the American variant of *finance* with stress on the first syllable is to be avoided, perhaps because it is not the original stress pattern. By this reasoning, however, the original but now obsolete pronunciation of *balcony* and *abdomen,* with stress on the second syllable, should be revived and prescribed, while the British pronunciation of *garage* with stress on the first syllable should be condemned as an innovation. Long usage has proved both stressings of *finance* to be acceptable.

2 : a compromise of a fictitious suit used as a form of conveyance of lands
3 a : a sum imposed as punishment for an offense **b :** a forfeiture or penalty paid to an injured party in a civil action
— in fine : in short
²fine *transitive verb* **fined; fin·ing** (1559)
: to impose a fine on **:** punish by a fine
³fine *adjective* **fin·er; fin·est** [Middle English *fin,* from Old French, from Latin *finis,* noun, end, limit] (13th century)
1 a : free from impurity **b** *of a metal* **:** having a stated proportion of pure metal in the composition expressed in parts per thousand ⟨a gold coin .9166 *fine*⟩
2 a (1) **:** very thin in gauge or texture ⟨*fine* thread⟩ (2) **:** not coarse ⟨*fine* sand⟩ (3) **:** very small ⟨*fine* print⟩ (4) **:** KEEN ⟨a knife with a *fine* edge⟩ (5) **:** very precise or accurate ⟨a *fine* adjustment⟩ ⟨trying to be too *fine* with his pitches⟩ **b :** physically trained or hardened close to the limit of efficiency — used of an athlete or animal
3 : delicate, subtle, or sensitive in quality, perception, or discrimination ⟨a *fine* distinction⟩
4 : superior in kind, quality, or appearance **:** EXCELLENT ⟨a *fine* job⟩ ⟨a *fine* day⟩ ⟨*fine* wines⟩
5 a : ORNATE 1 ⟨*fine* writing⟩ **b :** marked by or affecting elegance or refinement ⟨*fine* manners⟩
6 a : very well ⟨feel *fine*⟩ **b :** ALL RIGHT ⟨that's *fine* with me⟩
7 — used as an intensive ⟨the leader, in a *fine* frenzy, beheaded one of his wives —Brian Crozier⟩
— fine·ness \'fīn-nəs\ *noun*
⁴fine *adverb* (14th century)
1 : FINELY: as **a :** very well **b :** ALL RIGHT
2 : with a very narrow margin of time or space ⟨she had not intended to cut her escape so *fine* —Melinda Beck *et al.*⟩
⁵fine *verb* **fined; fin·ing** (14th century)
transitive verb
1 : PURIFY, CLARIFY ⟨*fine* and filter wine⟩
2 : to make finer in quality or size
intransitive verb
1 : to become pure or clear ⟨the ale will *fine*⟩
2 : to become smaller in lines or proportions
⁶fi·ne \'fē-(,)nā\ *noun* [Italian, from Latin *finis* end] (circa 1798)
: END — used as a direction in music to mark the closing point after a repeat
fine art *noun* (1767)
1 a : art (as painting, sculpture, or music) concerned primarily with the creation of beautiful objects — usually used in plural **b :** objects of fine art
2 : an activity requiring a fine skill
fine·ly \'fīn-lē\ *adverb* (14th century)
: in a fine manner: as **a :** extremely well **:** EXCELLENTLY ⟨plays the hero very *finely* —*New Yorker*⟩ **b :** with close discrimination **:** PRECISELY **c :** with delicacy or subtlety **:** SENSITIVELY ⟨a leader *finely* attuned to the needs of the people⟩ **d :** MINUTELY ⟨*finely* ground meal⟩
fine print *noun* (1951)
: something thoroughly and often deliberately obscure; *especially* **:** a part of an agreement or document spelling out restrictions and limitations often in small type or obscure language
fin·ery \'fīn-rē, 'fī-nə-\ *noun, plural* **-er·ies** (1680)
: ORNAMENT, DECORATION; *especially* **:** dressy or showy clothing and jewels
fines \'fīnz\ *noun plural* [³*fine*] (circa 1903)
: finely crushed or powdered material (as ore); *also* **:** very small particles in a mixture of various sizes
fines herbes \fēn-'zerb, fē-'nerb\ *noun plural* [French, literally, fine herbs] (1846)
: a mixture of herbs used as a seasoning or garnish
fine·spun \'fīn-'spən\ *adjective* (1647)
: developed with extreme care or delicacy; *also* **:** developed in excessively fine or subtle detail
¹fi·nesse \fə-'nes\ *noun* [Middle English, from Middle French, from *fin*] (1528)
1 : refinement or delicacy of workmanship, structure, or texture
2 : skillful handling of a situation **:** adroit maneuvering
3 : the withholding of one's highest card or trump in the hope that a lower card will take the trick because the only opposing higher card is in the hand of an opponent who has already played
²finesse *verb* **fi·nessed; fi·ness·ing** (1746)
intransitive verb
: to make a finesse in playing cards
transitive verb
1 : to play (a card) in a finesse
2 a : to bring about or manage by adroit maneuvering ⟨*finesse* his way through tight places —Marquis James⟩ **b :** EVADE, TRICK
fin·est \'fī-nəst\ *noun, plural in construction* [superlative of ³*fine*] (1951)
: POLICE OFFICERS — usually used with the possessive form of a city or area ⟨the city's *finest*⟩
fine structure *noun* (1935)
: microscopic structure of a biological entity or one of its parts especially as studied in preparations for the electron microscope
— fine structural *adjective*
fine–tooth comb \'fīn-,tüth-\ *noun* (1839)
1 : a comb with close-set teeth used especially for clearing parasites or foreign matter from the hair
2 : an attitude or system of thorough searching or scrutinizing ⟨went over the report with a *fine-tooth comb*⟩
fine–tune \'fīn-'tün\ *transitive verb* (1967)
1 a : to adjust precisely so as to bring to the highest level of performance or effectiveness ⟨*fine-tune* a TV set⟩ ⟨*fine-tune* the format⟩ **b :** to improve through minor alteration or revision ⟨*fine-tune* the temperature of the room⟩
2 : to stabilize (an economy) by small-scale fiscal and monetary manipulations
fin·fish \'fin-,fish\ *noun* (circa 1890)
: FISH 1b — compare SHELLFISH
¹fin·ger \'fiŋ-gər\ *noun* [Middle English, from Old English; akin to Old High German *fingar* finger] (before 12th century)
1 : any of the five terminating members of the hand **:** a digit of the forelimb; *especially* **:** one other than the thumb
2 a : something that resembles a finger ⟨a narrow *finger* of land⟩ **b :** a part of a glove into which a finger is inserted **c :** a projecting piece (as a pawl for a ratchet) brought into contact with an object to affect its motion
3 : the breadth of a finger
4 : INTEREST, SHARE — often used in the phrase *have a finger in the pie*
5 : BIRD 10 — usually used with *the*
— fin·ger·like \-,līk\ *adjective*
²finger *verb* **fin·gered; fin·ger·ing** \-g(ə-)riŋ\ (15th century)
transitive verb
1 : to touch or feel with the fingers
2 a : to play (a musical instrument) with the fingers **b :** to play (as notes or chords) with a specific fingering **c :** to mark the notes of (a music score) as a guide in playing
3 : to point out **:** IDENTIFY
4 : to extend into or penetrate in the shape of a finger
intransitive verb
1 : to touch or handle something
2 a : to use the fingers in playing a musical instrument **b :** to have a certain fingering — used of a musical instrument
3 : to extend in the shape or manner of a finger
fin·ger·board \'fiŋ-gər-,bōrd, -,bȯrd\ *noun* (circa 1672)
: the part of a stringed instrument against which the fingers press the strings to vary the pitch — see VIOLIN illustration
finger bowl *noun* (circa 1860)
: a small water bowl for rinsing the fingers at the table
fin·gered \'fiŋ-gərd\ *adjective* (circa 1529)
1 : having fingers especially of a specified kind or number — used in combination ⟨stubby-*fingered*⟩ ⟨five-*fingered*⟩
2 : having projections or processes like fingers ⟨the *fingered* roots of giant trees⟩
finger food *noun* (1928)
: a food that is to be held with the fingers for eating
fin·ger·hold \'fiŋ-gər-,hōld\ *noun* (1909)
1 : a hold or place of support for the fingers
2 : a tenuous hold or support
finger hole *noun* (1854)
1 : any of several holes in the side of a wind instrument (as a recorder) which may be covered or left open by the fingers to change the pitch of the tone
2 : a hole (as in a telephone dial or a bowling ball) into which the finger is placed to provide a grip
fin·ger·ing \'fiŋ-g(ə-)riŋ\ *noun* (14th century)
1 a : the act or method of using the fingers in playing an instrument **b :** the marking (as by figures on a musical score) of the method of fingering
2 : the act or process of handling or touching with the fingers
fin·ger·ling \'fiŋ-gər-liŋ\ *noun* (1836)
: a small fish especially up to one year of age
fin·ger·nail \'fiŋ-gər-,nāl, ,fiŋ-gər-'nā(ə)l\ *noun* (13th century)
: the nail of a finger
finger painting *noun* (1937)
: a technique of spreading pigment on wet paper chiefly with the fingers; *also* **:** a picture so produced
fin·ger·pick·ing \'fiŋ-gər-,pi-kiŋ\ *noun* (1969)
: a method of playing a stringed instrument (as a guitar) with the thumb and tips of the fingers rather than with a pick
— fin·ger·pick *verb*
fin·ger–point·ing \-,pȯin-tiŋ\ *noun* (1949)
: the act of making explicit and often unfair accusations of blame
fin·ger·post \'fiŋ-gər-,pōst\ *noun* (1785)
1 : a post bearing one or more signs often terminating in a pointing finger
2 : something serving as a guide to understanding or knowledge
fin·ger·print \-,print\ *noun* (1859)
1 : the impression of a fingertip on any surface; *also* **:** an ink impression of the lines upon the fingertip taken for the purpose of identification
2 : something that identifies: as **a :** a trait, trace, or characteristic revealing origin or responsibility **b :** analytical evidence (as a spectrogram) that characterizes an object or substance; *especially* **:** the chromatogram or electropherogram obtained by cleaving a protein by enzymatic action and subjecting the resulting collection of peptides to two-dimensional chromatography or electrophoresis
— fingerprint *transitive verb*
— fin·ger·print·ing *noun*

fingerprint 1: *1* **arch,** *2* **loop,** *3* **whorl,** *4* **composite**

finger spelling *noun* (1918)
: communication by signs made with the fingers — called also *dactylology*

1fin·ger·tip \-ˌtip\ *noun* (1842)
1 : the tip of a finger
2 : a protective covering for the end of a finger
— at one's fingertips : instantly or readily available

2fingertip *adjective* (1926)
1 : readily accessible **:** being within easy reach ⟨*fingertip* information⟩ ⟨*fingertip* controls⟩
2 : extending from head or shoulders to mid-thigh — used of clothing

finger wave *noun* (circa 1934)
: a method of setting hair by dampening with water or wave solution and forming waves or curls with the fingers and a comb

fin·i·al \ˈfi-nē-əl\ *noun* [Middle English, from *final, finial* final] (15th century)
1 : a usually foliated ornament forming an upper extremity especially in Gothic architecture
2 : a crowning ornament or detail (as a decorative knob)

fin·i·cal \ˈfi-ni-kəl\ *adjective* [probably from 3*fine*] (1592)
: FINICKY
— fin·i·cal·ly \-k(ə-)lē\ *adverb*
— fin·i·cal·ness \-kəl-nəs\ *noun*

fin·ick·ing \-kiŋ, -kən\ *adjective* [alteration of *finical*] (1661)
: FINICKY

fin·icky \ˈfi-ni-kē\ *adjective* [alteration of *finicking*] (circa 1825)
1 : extremely or excessively nice, exacting, or meticulous in taste or standards
2 : requiring much care, precision, or attentive effort ⟨a *finicky* recipe⟩
— fin·ick·i·ness *noun*

fi·nis \ˈfi-nəs; ˈfī-nəs; fə-ˈnē\ *noun* [Middle English, from Latin] (15th century)
: END, CONCLUSION

1fin·ish \ˈfi-nish\ *verb* [Middle English *finisshen,* from Middle French *finiss-,* stem of *finir,* from Latin *finire,* from *finis*] (14th century)
intransitive verb
1 a : to come to an end **:** TERMINATE **b :** END 1b
2 a : to come to the end of a course, task, or undertaking **b :** to end relations — used with *with* ⟨decided to *finish* with him for good⟩
3 : to end a competition in a specified manner or position ⟨*finished* third in the race⟩
transitive verb
1 a : to bring to an end **:** TERMINATE ⟨*finished* the speech and sat down⟩ **b :** to use or dispose of entirely ⟨her sandwich *finished* the loaf⟩
2 a : to bring to completion or issue ⟨hope to *finish* their new home before winter⟩ **b :** to provide with a finish; *especially* **:** to put a final coat or surface on ⟨*finish* a table with varnish⟩
3 a : to defeat or ruin utterly and finally ⟨the scandal *finished* his career⟩ **b :** to bring about the death of
synonym see CLOSE
— fin·ish·er *noun*

2finish *noun* (1779)
1 : something that completes or perfects: as **a :** the fine or decorative work required for a building or one of its parts **b :** a finishing material used in painting **c :** the final treatment or coating of a surface **d :** the taste in the mouth after swallowing a beverage (as wine)
2 a : final stage **:** END **b :** the cause of one's ruin
3 : the result or product of a finishing process
4 : the quality or state of being perfected

fin·ished *adjective* (1709)
: marked by the highest quality **:** CONSUMMATE

finishing school *noun* (circa 1837)
: a private school for girls that emphasizes cultural studies and prepares students especially for social activities

finish line *noun* (1899)
: a line marking the end of a racecourse

fi·nite \ˈfī-ˌnīt\ *adjective* [Middle English *finit,* from Latin *finitus,* past participle of *finire*] (15th century)
1 a : having definite or definable limits ⟨*finite* number of possibilities⟩ **b :** having a limited nature or existence ⟨*finite* beings⟩
2 : completely determinable in theory or in fact by counting, measurement, or thought ⟨the *finite* velocity of light⟩
3 a : less than an arbitrary positive integer and greater than the negative of that integer **b :** having a finite number of elements ⟨a *finite* set⟩
4 : of, relating to, or being a verb or verb form that can function as a predicate or as the initial element of one and that is limited (as in tense, person, and number)
— finite *noun*
— fi·nite·ly *adverb*
— fi·nite·ness *noun*

finite difference *noun* (1807)
: any of a sequence of differences obtained by incrementing successively the dependent variable of a function by a fixed amount; *especially* **:** any of such differences obtained from a polynomial function using successive integral values of its dependent variable

fi·ni·tude \ˈfī-nə-ˌtüd, -ˌtyüd, ˈfin-ə-\ *noun* (1644)
: finite quality or state

1fink \ˈfiŋk\ *noun* [origin unknown] (1903)
1 : one who is disapproved of or is held in contempt
2 : STRIKEBREAKER
3 : INFORMER 2

2fink *intransitive verb* (circa 1925)
: to turn informer **:** SQUEAL

fink out *intransitive verb* (1956)
: BACK OUT, COP OUT

Fin·land·isa·tion *British variant of* FINLANDIZATION

Fin·land·iza·tion \ˌfin-lən-də-ˈzā-shən, (ˌ)fin-ˌlan-\ *noun* [*Finland*] (1969)
: a foreign policy of neutrality under the influence of the Soviet Union; *also* **:** the conversion to such a policy
— Fin·land·ize \ˈfin-lən-ˌdīz, (ˌ)fin-ˈlan-\ *transitive verb*

Finn \ˈfin\ *noun* [Swedish *Finne*] (before 12th century)
1 : a member of a people speaking Finnish or a Finnic language
2 a : a native or inhabitant of Finland **b :** one who is of Finnish descent

fin·nan had·die \ˌfi-nən-ˈha-dē\ *noun* [alteration of *findon haddock,* from *Findon,* village in Scotland] (1811)
: smoked haddock — called also *finnan haddock*

Finn·ic \ˈfi-nik\ *adjective* (1668)
1 : of or relating to the Finns
2 : of, relating to, or constituting the branch of the Finno-Ugric subfamily that includes Finnish, Estonian, and Lapp

1Finn·ish \ˈfi-nish\ *adjective* (1699)
: of, relating to, or characteristic of Finland, the Finns, or Finnish

2Finnish *noun* (circa 1845)
: a Finno-Ugric language spoken in Finland, Karelia, and small areas of Sweden and Norway

Fin·no–Ugri·an \ˌfi-(ˌ)nō-ˈ(y)ü-grē-ən\ *adjective* (1880)
: FINNO-UGRIC
— Finno–Ugrian *noun*

Fin·no–Ugric \ˌfi-nō-ˈ(y)ü-grik\ *adjective* (1879)
1 : of, relating to, or constituting a subfamily of the Uralic family of languages comprising various languages spoken in Hungary, Lapland, Finland, Estonia, and parts of western Russia
2 : of or relating to any of the peoples speaking Finno-Ugric languages
— Finno–Ugric *noun*

fin·ny \ˈfi-nē\ *adjective* (1590)
1 : provided with or characterized by fins
2 : relating to or being fish

fino \ˈfē-(ˌ)nō\ *noun, plural* **finos** [Spanish, from *fino* fine, from *fin* end, from Latin *finis*] (1846)
: a very dry Spanish sherry

fin whale *noun* (1885)
: a baleen whale (*Balaenoptera physalus*) that attains a length of over 60 feet (18.3 meters) and is found in arctic to tropical waters worldwide — called also *finback*

fin whale

fiord *variant of* FJORD

fio·ri·tu·ra \fē-ˌȯr-ə-ˈtu̇r-ə\ *noun, plural* **-tu·re** \-ˈtu̇r-ē\ [Italian, literally, flowering, from *fiorito,* past participle of *fiorire* to flower, from (assumed) Vulgar Latin *florire* — more at FLOURISH] (1841)
: ORNAMENT 5

fip·ple flute \ˈfi-pəl-\ *noun* [origin unknown] (1911)
: any of a group of wind instruments (as a flageolet or recorder) having a straight tubular shape, a whistle mouthpiece, and finger holes

fir \ˈfər\ *noun* [Middle English, from Old English *fyrh;* akin to Old High German *forha* fir, Latin *quercus* oak] (before 12th century)
1 : any of a genus (*Abies*) of north temperate evergreen trees of the pine family that have flattish leaves, circular leaf scars, and erect female cones and are valued for their wood; *also* **:** any of various conifers (as the Douglas fir) of other genera
2 : the wood of a fir

1fire \ˈfīr\ *noun, often attributive* [Middle English, from Old English *fȳr;* akin to Old High German *fiur* fire, Greek *pyr*] (before 12th century)
1 a (1) **:** the phenomenon of combustion manifested in light, flame, and heat (2) **:** one of the four elements of the alchemists **b** (1) **:** burning passion **:** ARDOR (2) **:** liveliness of imagination **:** INSPIRATION
2 a : fuel in a state of combustion (as on a hearth) **b** *British* **:** a small gas or electric space heater
3 a : a destructive burning (as of a building) **b** (1) **:** death or torture by fire (2) **:** severe trial or ordeal
4 : BRILLIANCY, LUMINOSITY ⟨the *fire* of a gem⟩
5 a : the firing of weapons (as firearms, artillery, or missiles) **b :** intense verbal attack or criticism **c :** a rapidly delivered series (as of remarks)
— fire·less \-ləs\ *adjective*
— on fire 1 : being consumed by fire **:** AFLAME **2 :** EAGER, BURNING
— under fire 1 : exposed to fire from an enemy's weapons **2 :** under attack

2fire *verb* **fired; fir·ing** (13th century)
transitive verb
1 a : to set on fire **:** KINDLE; *also* **:** IGNITE ⟨*fire* a rocket engine⟩ **b** (1) **:** to give life or spirit to **:** INSPIRE (2) **:** to fill with passion or enthusiasm — often used with *up* **c :** to light up as if by fire **d :** to cause to start operating — usually used with *up*
2 a : to drive out or away by or as if by fire **b :** to dismiss from a position

3 a (1) **:** to cause to explode **:** DETONATE (2) **:** to propel from or as if from a gun **:** DISCHARGE, LAUNCH ⟨*fire* a rocket⟩ (3) **:** SHOOT 1b ⟨*fire* a gun⟩ (4) **:** to score (a number) in a game or contest **b :** to throw with speed or force ⟨*fired* the ball to first base⟩ ⟨*fire* a left jab⟩ **c :** to utter with force and rapidity
4 : to apply fire or fuel to: as **a :** to process by applying heat **b :** to feed or serve the fire of
intransitive verb
1 a : to take fire **:** KINDLE, IGNITE **b :** to begin operation **:** START ⟨the engine *fired*⟩ **c :** to operate especially as the result of the application of an electrical impulse ⟨the spark plug *fires*⟩
2 a : to become irritated or angry — often used with *up* **b :** to become filled with excitement or enthusiasm
3 a : to discharge a firearm **b :** to emit or let fly an object
4 : to tend a fire
5 : to transmit a nerve impulse
— **fire·able** \ˈfīr-ə-bəl, ˈfī-rə-\ *adjective*
— **fir·er** *noun*

fire and brimstone *noun* (13th century)
: the torments suffered by sinners in hell
— **fire–and–brimstone** *adjective*

fire ant *noun* (1796)
: any of a genus (*Solenopsis*) of fiercely stinging omnivorous ants; *especially* **:** IMPORTED FIRE ANT

fire·arm \ˈfīr-ˌärm\ *noun* (1646)
: a weapon from which a shot is discharged by gunpowder — usually used of small arms

fire away *intransitive verb* (1840)
: to speak without hesitation — usually used as an imperative

fire·back \-ˌbak\ *noun* (1847)
: an often decorated cast-iron plate lining the back wall of a fireplace

fire·ball \ˈfīr-ˌbȯl\ *noun* (1555)
1 : a ball of fire; *also* **:** something resembling such a ball ⟨the primordial *fireball* associated with the beginning of the universe —*Scientific American*⟩
2 : a brilliant meteor that may trail bright sparks
3 : the highly luminous cloud of vapor and dust created by a nuclear explosion
4 : a highly energetic person

fire·ball·er \-ˌbȯ-lər\ *noun* (1946)
: a baseball pitcher known for throwing fastballs
— **fire·ball·ing** \-liŋ\ *adjective*

fire·base \-ˌbās\ *noun* (1968)
: a secured site from which field artillery can lay down interdicting fire

fire blight *noun* (1817)
: a destructive infectious disease especially of apples, pears, and related fruits caused by a bacterium (*Erwinia amylovora*)

fire·boat \ˈfīr-ˌbōt\ *noun* (1849)
: a ship equipped with fire-fighting apparatus

fire·bomb \-ˌbäm\ *noun* (1895)
: an incendiary bomb
— **firebomb** *transitive verb*

fire·box \-ˌbäks\ *noun* (1791)
1 : a chamber (as of a furnace or steam boiler) that contains a fire
2 : a box containing an apparatus for transmitting an alarm to a fire station

fire·brand \-ˌbrand\ *noun* (13th century)
1 : a piece of burning wood
2 : one that creates unrest or strife (as in aggressively promoting a cause) **:** AGITATOR

fire·brat \-ˌbrat\ *noun* (1895)
: a wingless insect (*Thermobia domestica*) related to the silverfish and found in warm moist places

fire·break \-ˌbrāk\ *noun* (1841)
: a barrier of cleared or plowed land intended to check a forest or grass fire

fire–breath·ing \-ˌbrē-t͟hiŋ\ *adjective* (1933)
: intimidatingly or violently aggressive in speech and manner ⟨a *fire-breathing* orator⟩

fire·brick \-ˌbrik\ *noun* (1793)
: a refractory brick capable of sustaining high temperatures that is used especially for lining furnaces or fireplaces

fire brigade *noun* (1838)
: a body of firefighters: as **a :** a usually private or temporary fire-fighting organization **b** *British* **:** FIRE DEPARTMENT

fire·bug \ˈfīr-ˌbəg\ *noun* (1872)
: INCENDIARY, PYROMANIAC

fire chief *noun* (1889)
: the head of a fire department

fire·clay \ˈfīr-ˌklā\ *noun* (1819)
: clay capable of withstanding high temperatures that is used especially for firebrick and crucibles

fire control *noun* (1864)
: the planning, preparation, and delivery of fire on targets

fire·crack·er \ˈfīr-ˌkra-kər\ *noun* (1829)
: a usually paper cylinder containing an explosive and a fuse and set off to make a noise

fired \ˈfīrd\ *adjective* (1889)
: using a specified fuel — usually used in combination ⟨oil-*fired* power plant⟩

fire·damp \ˈfīr-ˌdamp\ *noun* (1677)
: a combustible mine gas that consists chiefly of methane; *also* **:** the explosive mixture of this gas with air

fire department *noun* (1825)
1 : an organization for preventing or extinguishing fires; *especially* **:** a government division (as in a municipality) having these duties
2 : the members of a fire department

fire·dog \-ˌdȯg\ *noun* (1792)
chiefly Southern & Midland **:** ANDIRON

fire·drake \ˈfīr-ˌdrāk\ *noun* [Middle English *firdrake,* from Old English *fȳrdraca,* from *fȳr* + *draca* dragon, from Latin *draco* — more at DRAGON] (before 12th century)
: a fire-breathing dragon especially in Germanic mythology

fire drill *noun* (circa 1890)
: a practice drill in extinguishing fires or in the conduct and manner of exit in case of fire

fire–eat·er \ˈfīr-ˌē-tər\ *noun* (1672)
1 : a performer who pretends to eat fire
2 a : a violent or pugnacious person **b :** a person who displays very militant or aggressive partisanship

fire–eat·ing \-ˌē-tiŋ\ *adjective* (1819)
: violent or highly militant in disposition, bearing, or policy ⟨a *fire-eating* radical⟩

fire engine *noun* (circa 1680)
: a usually mobile apparatus for directing an extinguishing agent upon fires; *especially* **:** FIRE TRUCK

fire–en·gine red \ˈfīr-ˌen-jən-\ *noun* (1954)
: a bright red

fire escape *noun* (1788)
: a device for escape from a burning building; *especially* **:** a metal stairway attached to the outside of a building

fire extinguisher *noun* (1849)
: a portable or wheeled apparatus for putting out small fires by ejecting extinguishing chemicals

fire·fight \ˈfīr-ˌfīt\ *noun* (1899)
1 a : a usually brief intense exchange of fire between opposing military units **b :** a hostile confrontation that involves gunfire
2 : SKIRMISH 2b

fire·fight·er \-ˌfī-tər\ *noun* (1903)
: a person who fights fires **:** FIREMAN 2
— **fire fighting** *noun*

fire·fly \-ˌflī\ *noun* (1658)
: any of various winged nocturnal beetles (especially family Lampyridae) that produce a bright soft intermittent light by oxidation of luciferin especially for courtship purposes

fire·guard \-ˌgärd\ *noun* (1833)
1 : a person who watches for the outbreak of fire; *also* **:** a person whose duty is to extinguish fires
2 : FIRE SCREEN
3 : FIREBREAK

fire hall *noun* (1881)
chiefly Canadian **:** FIRE STATION

fire·house \ˈfīr-ˌhau̇s\ *noun* (1901)
: FIRE STATION

fire irons *noun plural* (1648)
: utensils (as tongs) for tending a fire especially in a fireplace

fire·light \ˈfīr-ˌlīt\ *noun* (before 12th century)
: the light of a fire (as in a fireplace)
— **fire·lit** \-ˌlit\ *adjective*

fire·lock \-ˌläk\ *noun* (1547)
1 : a gun's lock employing a slow match to ignite the powder charge; *also* **:** a gun having such a lock
2 a : FLINTLOCK **b :** WHEEL LOCK

fire·man \-mən\ *noun* (14th century)
1 : a person who tends or feeds fires **:** STOKER
2 : a member of a fire department **:** FIREFIGHTER
3 : an enlisted man in the navy who works with engineering machinery
4 : a relief pitcher in baseball

fire·man·ic \-ˌma-nik\ *adjective* [*fireman* + [1]*-ic*] (1902)
: of or relating to firefighters or fire fighting

fire off *transitive verb* (1888)
: to write and send usually in haste or anger ⟨*fired off* a memo⟩

fire opal *noun* (1816)
: GIRASOLE 2

fire·place \ˈfīr-ˌplās\ *noun* (1698)
1 : a framed opening made in a chimney to hold an open fire **:** HEARTH; *also* **:** a metal container with a smoke pipe used for the same purpose
2 : an outdoor structure of brick, stone, or metal for an open fire
— **fire·placed** \-ˌplāst\ *adjective*

fire·plug \-ˌpləg\ *noun* (1713)
: HYDRANT

fire·pot \-ˌpät\ *noun* (1627)
1 : a clay pot filled with combustibles formerly used as a missile in war
2 : a vessel used in Oriental cuisine for cooking foods in broth at the table; *also* **:** the food cooked in it

fire·pow·er \-ˌpau̇(-ə)r\ *noun* (1913)
1 a : the capacity (as of a military unit) to deliver effective fire on a target **b :** effective fire
2 a : effective power or force ⟨intellectual *firepower*⟩ **b :** the scoring action or potential of a team or player

[1]**fire·proof** \-ˌprüf\ *adjective* (circa 1638)
: proof against or resistant to fire

[2]**fireproof** *transitive verb* (1867)
: to make fireproof

fire sale *noun* (1891)
: a sale of merchandise damaged in a fire; *also* **:** a sale at very low prices

fire screen *noun* (15th century)
: a protective screen before a fireplace

fire ship *noun* (1588)
: a ship carrying combustibles or explosives sent among the enemy's ships or works to set them on fire

[1]**fire·side** \ˈfīr-ˌsīd\ *noun* (1563)
1 : a place near the fire or hearth
2 : HOME

[2]**fireside** *adjective* (1740)
: having an informal or intimate quality ⟨a *fireside* chat⟩

fire station *noun* (1877)
: a building housing fire apparatus and usually firefighters

fire·stone \ˈfīr-ˌstōn\ *noun* (before 12th century)
1 : pyrite formerly used for striking fire; *also* **:** FLINT
2 : a stone that will endure high heat

fire–stop \-ˌstäp\ *noun* (1897)
: material used to close open parts especially of a building for preventing the spread of fire
— **fire–stop** *transitive verb*

fire·storm \-ˌstȯrm\ *noun* (1945)
1 : a large usually stationary fire characterized by very high temperatures in which the central

column of rising heated air induces strong inward winds which supply oxygen to the fire
2 : a sudden or violent outburst ⟨a *firestorm* of public protest⟩
fire·thorn \-,thȯrn\ *noun* (circa 1900)
: any of a genus (*Pyracantha*) of usually thorny ornamental shrubs of the rose family; *especially* **:** a European semievergreen shrub (*P. coccinea*) with orange-red berries
fire tower *noun* (1827)
: a tower (as in a forest) from which a watch for fires is maintained
fire·trap \'fīr-,trap\ *noun* (1881)
: a place (as a building) apt to catch on fire or difficult to escape from in case of fire
fire truck *noun* (1935)
: an automotive vehicle equipped with firefighting apparatus
fire wall *noun* (1759)
: a wall constructed to prevent the spread of fire
fire·wa·ter \'fīr-,wȯ-tər, -,wä-\ *noun* (1817)
: strong alcoholic liquor
fire·weed \-,wēd\ *noun* (1784)
: any of several plants that grow especially in clearings or burned districts: as **a :** a weedy composite (*Erechtites hieracifolia*) of the eastern U.S. and Canada that has clusters of brush-shaped flower heads with no ray flowers **b :** a tall perennial (*Epilobium angustifolium*) of the evening-primrose family that has long spikes of pinkish purple flowers and is an important honey plant in some areas — called also *willow herb*
fire·wood \-,wu̇d\ *noun* (14th century)
: wood used for fuel
fire·work \-,wərk\ *noun* (1575)
1 : a device for producing a striking display by the combustion of explosive or flammable compositions
2 *plural* **:** a display of fireworks
3 *plural* **a :** a display of temper or intense conflict **b :** a spectacular display ⟨the *fireworks* of autumn leaves⟩
fir·ing \'fīr-iŋ\ *noun* (14th century)
1 : the act or process of one that fires
2 : the process of maturing ceramic products by the application of heat
firing line *noun* (1881)
1 : a line from which fire is delivered against a target
2 : the forefront of an activity — used especially in the phrase *on the firing line*
firing pin *noun* (1874)
: the pin that strikes the cartridge primer in the breech mechanism of a firearm
firing squad *noun* (1904)
1 : a detachment detailed to fire volleys over the grave of one buried with military honors
2 : a detachment detailed to carry out a sentence of death by shooting
fir·kin \'fər-kən\ *noun* [Middle English, ultimately from Middle Dutch *veerdel* fourth, from *veer* four; akin to Old English *fēower* — more at FOUR] (14th century)
1 : a small wooden vessel or cask
2 : any of various British units of capacity usually equal to ¼ barrel
[1]**firm** \'fərm\ *adjective* [Middle English *ferm* from Middle French, from Latin *firmus;* akin to Greek *thronos* chair, throne] (14th century)
1 a : securely or solidly fixed in place **b :** not weak or uncertain **:** VIGOROUS **c :** having a solid or compact structure that resists stress or pressure
2 a (1) **:** not subject to change or revision (2) **:** not subject to price weakness **:** STEADY **b :** not easily moved or disturbed **:** STEADFAST **c :** WELL-FOUNDED
3 : indicating firmness or resolution ⟨a *firm* mouth⟩
— firm·ly *adverb*
— firm·ness *noun*
[2]**firm** *adverb* (14th century)
: in a firm manner **:** STEADFASTLY, FIXEDLY
[3]**firm** (14th century)
transitive verb
1 a : to make secure or fast **:** TIGHTEN ⟨*firming* her grip on the racquet⟩ — often used with *up* **b :** to make solid or compact ⟨*firm* the soil⟩
2 : to put into final form **:** SETTLE ⟨*firm* a contract⟩ ⟨*firm* up plans⟩
3 : to give additional support to **:** STRENGTHEN — usually used with *up*
intransitive verb
1 : to become firm **:** HARDEN — often used with *up*
2 : to recover from a decline **:** IMPROVE ⟨the market is *firming*⟩
[4]**firm** *noun* [German *Firma,* from Italian, signature, ultimately from Latin *firmare* to make firm, confirm, from *firmus*] (1744)
1 : the name or title under which a company transacts business
2 : a partnership of two or more persons that is not recognized as a legal person distinct from the members composing it
3 : a business unit or enterprise
fir·ma·ment \'fər-mə-mənt\ *noun* [Middle English, from Late Latin & Latin; Late Latin *firmamentum,* from Latin, support, from *firmare*] (13th century)
1 : the vault or arch of the sky **:** HEAVENS
2 *obsolete* **:** BASIS
3 : the field or sphere of an interest or activity ⟨the international fashion *firmament*⟩
— fir·ma·men·tal \,fər-mə-'men-tᵊl\ *adjective*
fir·mer chisel \'fər-mər-\ *noun* [French *fermoir* chisel, alteration of Middle French *formoir,* from *former* to form, from Old French *forme* form] (1823)
: a woodworking chisel with a thin flat blade
firm·ware \'fərm-,war, -,wer\ *noun* (1967)
: computer programs contained permanently in a hardware device (as a read-only memory)
firn \'firn\ *noun* [German, from Old High German *firni* old; akin to Old English *fyrn* ... *faran*] (1853)
: NÉVÉ
[1]**first** \'fərst\ *adjective* [Middle English, from Old English *fyrst;* akin to Old High German *furist* first, Old English *faran* to go — more at FARE] (before 12th century)
: preceding all others in time, order, or importance: as **a :** EARLIEST **b :** being the lowest forward gear or speed of a motor vehicle **c :** having the highest or most prominent part among a group of similar voices or instruments ⟨*first* tenor⟩ ⟨*first* violins⟩
[2]**first** *adverb* (before 12th century)
1 a : before another in time, space, or importance **b :** in the first place — often used with *of all* **c :** for the first time
2 : in preference to something else **:** SOONER
[3]**first** *noun* (13th century)
1 — see NUMBER table
2 : something that is first: as **a :** the first occurrence or item of a kind **b :** the first forward gear or speed of a motor vehicle **c :** the highest or chief voice or instrument of a group **d :** an article of commerce of the finest grade **e :** the winning or highest place in a competition, examination, or contest
3 : FIRST BASE
— at first : at the beginning **:** INITIALLY
first aid *noun* (1882)
: emergency care or treatment given to an ill or injured person before regular medical aid can be obtained
first base *noun* (1845)
1 : the base that must be touched first by a base runner in baseball
2 : the player position for defending the area around first base
3 : the first step or stage in a course of action ⟨plans never got to *first base*⟩
— first base·man \-'bā-smən\ *noun*
first·born \'fərs(t)-'bȯrn\ *adjective* (14th century)
: first brought forth **:** ELDEST
— firstborn *noun*
first cause *noun* (14th century)
: the self-created ultimate source of all being
first-class \'fərs(t)-'klas\ *adjective* (circa 1838)
1 : of or relating to first class
2 : of the highest quality ⟨a *first-class* meal⟩
— first-class *adverb*
first class *noun* (1750)
: the first or highest group in a classification: as **a :** the highest of usually three classes of travel accommodations **b :** a class of mail that comprises letters, postcards, or matter sealed against inspection
first cousin *noun* (1661)
: COUSIN 1a
first day cover *noun* (1932)
: a philatelic cover franked with a newly issued postage stamp and postmarked on the first day of issue at a city officially chosen for first day sale
first-degree burn *noun* (circa 1929)
: a mild burn characterized by heat, pain, and reddening of the burned surface but not exhibiting blistering or charring of tissues
first down *noun* (1897)
1 : the first of a series of usually four downs in which a football team must net a 10-yard gain to retain possession of the ball
2 : a gain of a total of 10 or more yards within usually four downs giving the team the right to start a new series of downs
first edition *noun* (circa 1828)
: the copies of a literary work first printed from the same type and issued at the same time; *also* **:** a single copy from a first edition
first estate *noun, often F&E capitalized* (1935)
: the first of the traditional political estates; *specifically* **:** CLERGY
first floor *noun* (15th century)
1 : GROUND FLOOR 1
2 *chiefly British* **:** the floor next above the ground floor
first·fruits \'fərs(t)-'früts\ *noun plural* (14th century)
1 : the earliest gathered fruits offered to the Deity in acknowledgment of the gift of fruitfulness
2 : the earliest products or results of an endeavor
first·hand \'fərst-'hand\ *adjective* (1748)
: obtained by, coming from, or being direct personal observation or experience ⟨a *firsthand* account of the war⟩
— firsthand *adverb*
first lady *noun, often F&L capitalized* (1834)
1 : the wife or hostess of the chief executive of a country or jurisdiction
2 : the leading woman of an art or profession
first lieutenant *noun* (1782)
1 : a commissioned officer in the army, air force, or marine corps ranking above a second lieutenant and below a captain
2 : a naval officer responsible for a ship's upkeep
first·ling \'fərst-liŋ\ *noun* (1535)
1 : the first of a class or kind
2 : the first produce or result of something
first·ly \-lē\ *adverb* (circa 1532)
: in the first place **:** FIRST
first mortgage *noun* (1855)
: a mortgage that has priority as a lien over all mortgages and liens except those imposed by law
first name *noun* (13th century)
: the name that stands first in one's full name
first night *noun* (1711)

\ə\ **abut** \ᵊ\ **kitten** \ər\ **further** \a\ **ash** \ā\ **ace** \ä\ **mop, mar** \au̇\ **out** \ch\ **chin** \e\ **bet** \ē\ **easy** \g\ **go** \i\ **hit** \ī\ **ice** \j\ **job** \ŋ\ **sing** \ō\ **go** \ȯ\ **law** \ȯi\ **boy** \th\ **thin** \th̲\ **the** \ü\ **loot** \u̇\ **foot** \y\ **yet** \zh\ **vision** *see also* Guide to Pronunciation

: the night on which a theatrical production is first performed at a given place; *also* **:** the performance itself
first-night·er \'fərs(t)-'nī-tər\ *noun* (1882)
: a spectator at a first-night performance
first off *adverb* (1880)
: in the first place **:** before anything else
first offender *noun* (1849)
: one convicted of an offense for the first time
first papers *noun plural* (1912)
: papers declaring intention filed by an applicant for citizenship as the first step in the naturalization process
first person *noun* (1520)
1 a : a set of linguistic forms (as verb forms, pronouns, and inflectional affixes) referring to the speaker or writer of the utterance in which they occur **b :** a linguistic form belonging to such a set **c :** reference of a linguistic form to the speaker or writer of the utterance in which it occurs
2 : a style of discourse marked by general use of verbs and pronouns of the first person
¹first-rate \'fərs(t)-'rāt\ *adjective* (1671)
: of the first order of size, importance, or quality
— first-rate·ness *noun*
— first-rat·er \-'rā-tər\ *noun*
²first-rate *adverb* (1844)
: very well ⟨is getting along *first-rate,* now —Mark Twain⟩
First Reader *noun* (1895)
: a Christian Scientist chosen to conduct meetings for a specified time and specifically to read aloud from the writings of Mary Baker Eddy
first reading *noun* (circa 1703)
: the first submitting of a bill before a quorum of a legislative assembly usually by title or number only
first-run \'fərs(t)-'rən\ *adjective* (1912)
: available for public viewing for the first time ⟨*first-run* movies⟩; *also* **:** exhibiting first-run movies ⟨*first-run* theaters⟩
first sergeant *noun* (circa 1860)
1 : a noncommissioned officer serving as the chief assistant to the commander of a military unit (as a company or squadron)
2 : the rank of a first sergeant; *specifically* **:** a rank in the army above a platoon sergeant and below a command sergeant major and in the marine corps above a gunnery sergeant and below a sergeant major
first strike *noun* (1960)
: a preemptive nuclear attack
— first-strike *adjective*
first-string \'fərs(t)-'striŋ\ *adjective* (1917)
1 : being a regular as distinguished from a substitute (as on a team)
2 : FIRST-RATE
first water *noun* (1753)
1 : the purest luster — used of gems
2 : the highest grade, degree, or quality
first world *noun, often F&W capitalized* [after *third world*] (1967)
: the highly developed industrialized nations often considered the westernized countries of the world
firth \'fərth\ *noun* [Middle English, from Old Norse *fjǫrthr* — more at FORD] (14th century)
: ESTUARY
fisc \'fisk\ *noun* [Latin *fiscus*] (1598)
: a state or royal treasury
¹fis·cal \'fis-kəl\ *adjective* [Latin *fiscalis,* from *fiscus* basket, treasury] (1563)
1 : of or relating to taxation, public revenues, or public debt ⟨*fiscal* policy⟩
2 : of or relating to financial matters
— fis·cal·ly \-kə-lē\ *adverb*
²fiscal *noun* (1929)
1 : REVENUE STAMP
2 : FISCAL YEAR
fiscal year *noun* (1843)
: an accounting period of 12 months

¹fish \'fish\ *noun, plural* **fish** *or* **fish·es** *often attributive* [Middle English, from Old English *fisc;* akin to Old High German *fisc* fish, Latin *piscis*] (before 12th century)
1 a : an aquatic animal — usually used in combination ⟨star*fish*⟩ ⟨cuttle*fish*⟩ **b :** any of numerous cold-blooded strictly aquatic craniate vertebrates that include the bony fishes and usually the cartilaginous and jawless fishes and that have typically an elongated somewhat spindle-shaped body terminating in a broad caudal fin, limbs in the form of fins when present at all, and a 2-chambered heart by which blood is sent through thoracic gills to be oxygenated
2 : the flesh of fish used as food
3 a : a person who is caught or is wanted (as in a criminal investigation) **b :** FELLOW, PERSON ⟨an odd *fish*⟩ **c :** SUCKER 5a
4 : something that resembles a fish: as **a** *plural, capitalized* **:** PISCES 1, 2a **b :** TORPEDO 2b
— fish·less \'fish-ləs\ *adjective*
— fish·like \-,līk\ *adjective*
— fish out of water : a person who is in an unnatural or uncomfortable sphere or situation
— fish to fry : concerns or interests to pursue — usually used with *other*
— neither fish nor fowl : one that does not belong to a particular class or category

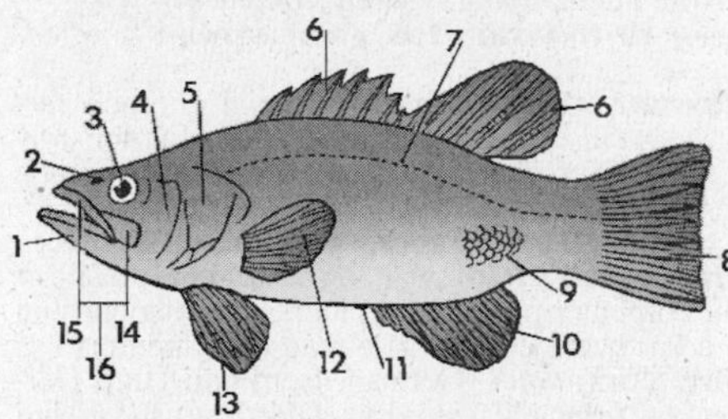

fish 1b: *1* **mandible,** *2* **nasal opening,** *3* **eye,** *4* **cheek,** *5* **operculum,** *6* **dorsal fins,** *7* **lateral line,** *8* **caudal fin,** *9* **scales,** *10* **anal fin,** *11* **anus,** *12* **pectoral fin,** *13* **pelvic fin,** *14* **maxilla,** *15* **premaxilla,** *16* **upper jaw**

²fish (before 12th century)
intransitive verb
1 : to attempt to catch fish
2 : to seek something by roundabout means ⟨*fishing* for a compliment⟩
3 a : to search for something underwater ⟨*fish* for pearls⟩ **b :** to engage in a search by groping or feeling ⟨*fishing* around in her purse for her keys⟩
transitive verb
1 a : to try to catch fish in **b :** to fish with **:** use (as a boat, net, or bait) in fishing
2 a : to go fishing for ⟨*fish* salmon⟩ **b :** to pull or draw as if fishing ⟨*fished* the ball from under the car⟩ ⟨*fish* wires through a conduit⟩
— fish·abil·i·ty \,fi-shə-'bi-lə-tē\ *noun*
— fish·able \'fi-shə-bəl\ *adjective*
— fish or cut bait : to make a choice between alternatives
fish-and-chips \,fi-shᵊn-'chips\ *noun plural* (1876)
: fried fish and french fried potatoes
fish·bowl \'fish-,bōl\ *noun* (1906)
1 : a bowl for the keeping of live fish
2 : a place or condition that affords no privacy
fish cake *noun* (1854)
: a round fried cake made of shredded fish and mashed potato
fish duck *noun* (1858)
: MERGANSER
fish·er \'fi-shər\ *noun* (before 12th century)
1 : one that fishes
2 a : a large dark brown North American carnivorous mammal (*Martes pennanti*) related to the weasels **b :** the fur or pelt of this animal
fish·er·folk \'fi-shər-,fōk\ *noun plural* (1854)
: people who fish especially for a living

fish·er·man \-mən\ *noun* (15th century)
1 : one who engages in fishing as an occupation or for pleasure
2 : a ship used in commercial fishing
fisherman's bend *noun* (1823)
: a knot made by passing the end twice round a spar or through a ring and then back under both turns — see KNOT illustration
fish·er·wo·man \'fi-shər-,wu̇-mən\ *noun* (1816)
: a woman who fishes as an occupation or for pleasure
fish·ery \'fi-shə-rē\ *noun, plural* **-er·ies** (1528)
1 : the occupation, industry, or season of taking fish or other sea animals (as sponges, shrimp, or seals) **:** FISHING
2 : a place for catching fish or taking other sea animals
3 : a fishing establishment; *also* **:** its fishermen
4 : the legal right to take fish at a particular place or in particular waters
5 : the technology of fishery — usually used in plural
fish-eye \'fish-,ī\ *adjective* (1943)
: being, having, or produced by a wide-angle photographic lens that has a highly curved protruding front, that covers an angle of about 180 degrees, and that gives a circular image
fish farm *noun* (1865)
: a commercial facility for raising aquatic animals for human food
— fish-farm \'fish-,färm\ *transitive verb*
fish finger *noun* (1962)
British **:** FISH STICK
fish fry *noun* (1824)
1 : a picnic or supper featuring fried fish
2 : fried fish
fish hawk *noun* (1709)
: OSPREY 1
fish·hook \'fish-,hu̇k\ *noun* (14th century)
: a usually barbed hook for catching fish
fish·ing *noun* (13th century)
1 : the sport or business of catching fish
2 : a place for catching fish
fishing expedition *noun* (circa 1925)
1 : a legal interrogation or examination to discover information for a later proceeding
2 : an investigation that does not stick to a stated objective but hopes to uncover incriminating or newsworthy evidence
fish ladder *noun* (1865)
: a series of pools arranged like steps by which fish can pass over a dam in going upstream
fish meal *noun* (1854)
: ground dried fish and fish waste used as fertilizer and animal food
fish·mong·er \'fish-,mäŋ-gər, -,məŋ-\ *noun* (15th century)
chiefly British **:** a fish dealer
fish·net \-,net\ *noun* (before 12th century)
1 : netting fitted with floats and weights or a supporting frame for catching fish
2 : a coarse open-mesh fabric
fish out *transitive verb* (1892)
: to exhaust the supply of fish in by fishing
fish·plate \-,plāt\ *noun* (1855)
: a steel plate used to lap a butt joint
fish·pond \-,pänd\ *noun* (14th century)
: a pond stocked with fish
fish protein concentrate *noun* (1961)
: a protein-rich food additive made from ground whole fish
fish stick *noun* (1953)
: a small elongated breaded fillet of fish
fish story *noun* [from the traditional exaggeration by fishermen of the size of fish almost caught] (1819)
: an extravagant or incredible story
fish·tail \'fish-,tāl\ *intransitive verb* (1927)
1 : to swing the tail of an airplane from side to side to reduce speed especially when landing
2 : to have the rear end slide from side to side out of control while moving forward ⟨the car *fishtailed* on the icy curve⟩

fish·way \-ˌwā\ *noun* (1845)
: a contrivance for enabling fish to pass around a fall or dam in a stream; *specifically* **:** FISH LADDER

fish·wife \-ˌwīf\ *noun* [Middle English] (15th century)
1 : a woman who sells fish
2 : a vulgar abusive woman

fishy \ˈfi-shē\ *adjective* **fish·i·er; -est** (15th century)
1 : of or resembling fish especially in taste or odor
2 : creating doubt or suspicion **:** QUESTIONABLE

fis·sile \ˈfi-səl, ˈfi-ˌsīl\ *adjective* [Latin *fissilis,* from *findere*] (1661)
1 : capable of being split or divided in the direction of the grain or along natural planes of cleavage ⟨*fissile* wood⟩ ⟨*fissile* crystals⟩
2 : capable of undergoing fission
— fis·sil·i·ty \fi-ˈsi-lə-tē\ *noun*

[1]fis·sion \ˈfi-shᵊn, -zhᵊn\ *noun* [Latin *fission-, fissio,* from *findere* to split — more at BITE] (circa 1617)
1 : a splitting or breaking up into parts
2 : reproduction by spontaneous division of the body into two or more parts each of which grows into a complete organism
3 : the splitting of an atomic nucleus resulting in the release of large amounts of energy
— fis·sion·al \-ᵊl\ *adjective*

[2]fission (1929)
intransitive verb
: to undergo fission
transitive verb
: to cause to undergo fission

fis·sion·able \ˈfi-shᵊn-ə-bəl, -zhᵊn-; ˈfish-nə-, ˈfizh-\ *adjective* (1945)
: FISSILE 2
— fis·sion·abil·i·ty \ˌfi-shᵊn-ə-ˈbi-lə-tē, -zhᵊn-; ˌfish-nə-, ˌfizh-\ *noun*
— fissionable *noun*

fis·sip·a·rous \fi-ˈsi p(ə-)rəs\ *adjective* [Latin *fissus,* past participle of *findere* + English *-parous*] (1874)
: tending to break up into parts **:** DIVISIVE
— fis·sip·a·rous·ness *noun*

[1]fis·sure \ˈfi-shər\ *noun* [Middle English, from Middle French, from Latin *fissura,* from *fissus*] (14th century)
1 : a narrow opening or crack of considerable length and depth usually occurring from some breaking or parting
2 a : a natural cleft between body parts or in the substance of an organ **b :** a break or slit in tissue usually at the junction of skin and mucous membrane
3 : a separation or disagreement in thought or viewpoint **:** SCHISM ⟨*fissures* in a political party⟩

[2]fissure *verb* **fis·sured; fis·sur·ing** (1656)
transitive verb
: to break into fissures **:** CLEAVE
intransitive verb
: CRACK, DIVIDE

[1]fist \ˈfist\ *noun* [Middle English, from Old English *fȳst;* akin to Old High German *fūst* fist, Polish *pięść,* and probably to Old English *fīf* five] (before 12th century)
1 : the hand clenched with the fingers doubled into the palm and the thumb doubled inward across the fingers
2 : the hand closed as in grasping **:** CLUTCH
3 : INDEX 5

[2]fist *transitive verb* (1607)
1 : to grip with the fist **:** HANDLE
2 : to clench into a fist

-fist·ed \ˈfis-təd\ *combining form*
: having (such or so many) fists ⟨two-*fisted*⟩ ⟨tight*fisted*⟩

fist·fight \ˈfist-ˌfīt\ *noun* (1603)
: a usually spontaneous fight with bare fists

fist·ful \-ˌfu̇l\ *noun* (1611)
1 : HANDFUL ⟨a *fistful* of coins⟩
2 : a considerable number or amount ⟨a whole *fistful* of musicians —Thomas Lask⟩

fist·ic \ˈfis-tik\ *adjective* (1806)
: of or relating to boxing or to fighting with the fists

fist·i·cuffs \ˈfis-ti-ˌkəfs\ *noun plural* [alteration of *fisty cuff,* from *fisty* (fistic) + *cuff*] (1605)
: a fight with the fists

fist·note \ˈfis(t)-ˌnōt\ *noun* (circa 1934)
: matter in a text to which attention is directed by means of an index mark

fis·tu·la \ˈfis(h)-chə-lə\ *noun, plural* **-las** *or* **-lae** \-ˌlē, -lī\ [Middle English, from Latin, pipe, fistula] (14th century)
: an abnormal passage leading from an abscess or hollow organ to the body surface or from one hollow organ to another and permitting passage of fluids or secretions

fis·tu·lous \-ləs\ *adjective* (15th century)
1 : of, relating to, or having the form or nature of a fistula
2 : hollow like a pipe or reed

fistulous withers *noun plural but singular or plural in construction* (circa 1900)
: a deep-seated chronic inflammation of the withers of the horse in which bloody fluid is discharged

[1]fit \ˈfit\ *noun* [Middle English, from Old English *fitt;* akin to Old Saxon *fittea* division of a poem, Old High German *fizza* skein] (before 12th century)
archaic **:** a division of a poem or song

[2]fit *adjective* **fit·ter; fit·test** [Middle English; akin to Middle English *fitten*] (14th century)
1 a (1) **:** adapted to an end or design **:** suitable by nature or by art (2) **:** adapted to the environment so as to be capable of surviving **b :** acceptable from a particular viewpoint (as of competence or morality) **:** PROPER ⟨a movie *fit* for the whole family⟩
2 a : put into a suitable state **:** made ready ⟨get the house *fit* for company⟩ **b :** being in such a state as to be or seem ready to do or suffer something ⟨fair *fit* to cry I was —Bryan MacMahon⟩ ⟨laughing *fit* to burst⟩
3 : sound physically and mentally **:** HEALTHY ☆
— fit·ly *adverb*
— fit to be tied : extremely angry or irritated
— fit to kill : in a striking manner ⟨dressed *fit to kill*⟩

[3]fit *noun* [Middle English, from Old English *fitt* strife] (circa 1541)
1 a : a sudden violent attack of a disease (as epilepsy) especially when marked by convulsions or unconsciousness **:** PAROXYSM **b :** a sudden but transient attack of a physical disturbance
2 : a sudden burst or flurry (as of activity) ⟨cleaned the whole house in a *fit* of efficiency⟩
3 : an emotional reaction (as in anger or frustration) ⟨has a *fit* when I show up late⟩
— by fits *or* **by fits and starts** *or* **in fits and starts :** in an impulsive and irregular manner

[4]fit *verb* **fit·ted** *also* **fit; fit·ting** [Middle English *fitten* to marshal troops, from or akin to Middle Dutch *vitten* to be suitable; akin to Old High German *fizza* skein] (15th century)
transitive verb
1 a : to be suitable for or to **:** harmonize with **b** *archaic* **:** to be seemly or proper for ⟨it *fits* us then to be as provident as fear may teach us —Shakespeare⟩
2 a : to conform correctly to the shape or size of ⟨it doesn't *fit* me anymore⟩ **b** (1) **:** to insert or adjust until correctly in place ⟨*fit* the mechanism into the box⟩ (2) **:** to make or adjust to the right shape and size ⟨*fitting* the jacket to the customer⟩ (3) **:** to measure for determining the specifications of something to be worn by ⟨*fitted* him for a new suit⟩ **c :** to make a place or room for **:** ACCOMMODATE
3 : to be in agreement or accord with ⟨the theory *fits* all the facts⟩
4 a : to put into a condition of readiness **b :** to cause to conform to or suit something
5 : SUPPLY, EQUIP ⟨*fitted* the ship with new engines⟩ — often used with *out*
6 : to adjust (a smooth curve of a specified type) to a given set of points
intransitive verb
1 *archaic* **:** to be seemly, proper, or suitable
2 : to conform to a particular shape or size; *also* **:** to be accommodated ⟨will we all *fit* into the car?⟩
3 : to be in harmony or accord **:** BELONG — often used with *in*
— fit·ter *noun*

[5]fit *noun* (1823)
: the fact, condition, or manner of fitting or being fitted: as **a :** the way clothing fits the wearer **b :** the degree of closeness between surfaces in an assembly of parts **c :** GOODNESS OF FIT

[6]fit *dialect past and past participle of* FIGHT

fitch \ˈfich\ *or* **fitch·ew** \ˈfich-(ˌ)ü\ *noun* [Middle English *fiche, ficheux,* from Middle French or Middle Dutch; Middle French *fichau,* from Middle Dutch *vitsau*] (15th century)
1 : POLECAT 1
2 : the fur or pelt of the polecat

fitch·et \ˈfi-chət\ *noun* (1535)
: POLECAT 1

fit·ful \ˈfit-fəl\ *adjective* (1592)
1 *obsolete* **:** characterized by fits or paroxysms
2 : having an erratic or intermittent character **:** IRREGULAR ☆
— fit·ful·ly \-fə-lē\ *adverb*
— fit·ful·ness *noun*

fit·ment \ˈfit-mənt\ *noun* [[4]*fit*] (1851)
chiefly British **:** FURNISHING 2, FIXTURE, CABINETRY — usually used in plural

fit·ness \ˈfit-nəs\ *noun* (1580)
1 : the quality or state of being fit
2 : the capacity of an organism to survive and transmit its genotype to reproductive offspring as compared to competing organisms; *also* **:** the contribution of an allele or genotype to

☆ SYNONYMS

Fit, suitable, meet, proper, appropriate, fitting, apt, happy, felicitous mean right with respect to some end, need, use, or circumstance. FIT stresses adaptability and sometimes special readiness for use or action ⟨*fit* for battle⟩. SUITABLE implies an answering to requirements or demands ⟨clothes *suitable* for camping⟩. MEET suggests a just proportioning ⟨*meet* payment⟩. PROPER suggests a suitability through essential nature or accordance with custom ⟨*proper* acknowledgement⟩. APPROPRIATE implies eminent or distinctive fitness ⟨an *appropriate* gift⟩. FITTING implies harmony of mood or tone ⟨a *fitting* end⟩. APT connotes a fitness marked by nicety and discrimination ⟨*apt* quotations⟩. HAPPY suggests what is effectively or successfully appropriate ⟨a *happy* choice of words⟩. FELICITOUS suggests an aptness that is opportune, telling, or graceful ⟨a *felicitous* phrase⟩.

Fitful, spasmodic, convulsive mean lacking steadiness or regularity in movement. FITFUL implies intermittence, a succession of starts and stops or risings and fallings ⟨*fitful* sleep⟩. SPASMODIC adds to FITFUL the implication of rapid or violent activity alternating with inactivity ⟨*spasmodic* growth⟩. CONVULSIVE suggests the breaking of regularity or quiet by uncontrolled movement ⟨*convulsive* shocks⟩.

the gene pool of subsequent generations as compared to that of other alleles or genotypes

fit·ted \'fi-təd\ *adjective* (circa 1666)
1 : FIT, SUITABLE
2 : shaped for a precise fit ⟨a *fitted* sheet⟩; *especially* **:** shaped to conform to the lines of the body ⟨a *fitted* shirt⟩

¹fit·ting \'fi-tiŋ\ *adjective* (15th century)
: of a kind appropriate to the situation **:** SUITABLE
synonym see FIT
— fit·ting·ly \-tiŋ-lē\ *adverb*
— fit·ting·ness *noun*

²fitting *noun* (1607)
1 : an action or act of one that fits; *specifically* **:** a trying on of clothes which are in the process of being made or altered
2 : a small often standardized part ⟨an electrical *fitting*⟩

five \'fīv\ *noun* [Middle English, from *five,* adjective, from Old English *fīf;* akin to Old High German *finf* five, Latin *quinque,* Greek *pente*] (before 12th century)
1 — see NUMBER table
2 *plural* **:** a British handball game
3 : the fifth in a set or series ⟨the *five* of clubs⟩
4 : something having five units or members; *especially* **:** a basketball team
5 : a 5-dollar bill
6 : FIFTEEN 2
— five *adjective*
— five *pronoun, plural in construction*

five–and–ten \ˌfī-vən-'ten\ *noun* [from the fact that all articles in such stores were formerly priced at either 5 or 10 cents] (1880)
: a retail store that carries chiefly inexpensive merchandise (as notions and household goods) — called also *five-and-dime*

five–fin·ger \'fīv-ˌfiŋ-gər\ *noun* (before 12th century)
: CINQUEFOIL 1

five·fold \'fīv-ˌfōld, -'fōld\ *adjective* (before 12th century)
1 : having five units or members
2 : being five times as great or as many
— five·fold \-'fōld\ *adverb*

five of a kind (1897)
: four cards of the same rank plus a wild card in one hand — see POKER illustration

fiv·er \'fī-vər\ *noun* (1843)
1 *slang* **:** a 5-dollar bill
2 *British* **:** a 5-pound note

five–spice powder \'fīv-'spīs-\ *noun* (1970)
: a blend of spices typically including anise, pepper, fennel, cloves, and cinnamon that is used in Chinese cooking

five–star \'fīv-'stär\ *adjective* (1913)
: of first class or quality

¹fix \'fiks\ *verb* [Middle English, from Latin *fixus,* past participle of *figere* to fasten; akin to Lithuanian *dygti* to sprout, break through] (14th century)
transitive verb
1 a : to make firm, stable, or stationary **b :** to give a permanent or final form to: as (1) **:** to change into a stable compound or available form ⟨bacteria that *fix* nitrogen⟩ (2) **:** to kill, harden, and preserve for microscopic study (3) **:** to make the image of (a photographic film) permanent by removing unused salts **c :** AFFIX, ATTACH
2 a : to hold or direct steadily ⟨*fixes* his eyes on the horizon⟩ **b :** to capture the attention of ⟨*fixed* her with a stare⟩
3 a : to set or place definitely **:** ESTABLISH **b :** to make an accurate determination of **:** DISCOVER ⟨*fixing* our location on the chart⟩ **c :** ASSIGN ⟨*fix* the blame⟩
4 : to set in order **:** ADJUST
5 : to get ready **:** PREPARE ⟨*fix* lunch⟩
6 a : REPAIR, MEND ⟨*fix* the clock⟩ **b :** RESTORE, CURE ⟨the doctor *fixed* him up⟩ **c :** SPAY, CASTRATE
7 a : to get even with **b :** to influence the actions, outcome, or effect of by improper or illegal methods ⟨the race had been *fixed*⟩
intransitive verb
1 : to become firm, stable, or fixed
2 : to get set **:** be on the verge ⟨we're *fixing* to leave soon⟩
3 : to direct one's attention or efforts **:** FOCUS; *also* **:** DECIDE, SETTLE — usually used with *on*
synonym see FASTEN
— fix·able \'fik-sə-bəl\ *adjective*

²fix *noun* (1809)
1 : a position of difficulty or embarrassment **:** PREDICAMENT
2 a : the position (as of a ship) determined by bearings, observations, or radio; *also* **:** a determination of one's position **b :** an accurate determination or understanding especially by observation or analysis
3 : an act or instance of improper or illegal fixing ⟨the *fix* was in⟩
4 : a supply or dose of something strongly desired or craved; *especially* **:** a shot of a narcotic
5 : FIXATION
6 : something that fixes or restores ⟨a quick *fix*⟩

fix·ate \'fik-ˌsāt\ *verb* **fix·at·ed; fix·at·ing** (1885)
transitive verb
1 : to make fixed, stationary, or unchanging
2 : to focus one's gaze on
3 : to direct (the libido) toward an infantile form of gratification
intransitive verb
1 : to focus or concentrate one's gaze or attention intently or obsessively
2 : to undergo arrestment at a stage of development

fix·at·ed *adjective* (1926)
: arrested in development or adjustment; *especially* **:** arrested at a pregenital level of psychosexual development

fix·a·tion \fik-'sā-shən\ *noun* (14th century)
: the act, process, or result of fixing, fixating, or becoming fixated: as **a :** a persistent concentration of libidinal energies upon objects characteristic of psychosexual stages of development preceding the genital stage **b :** stereotyped behavior (as in response to frustration) **c :** an obsessive or unhealthy preoccupation or attachment

fix·a·tive \'fik-sə-tiv\ *noun* (circa 1859)
: something that fixes or sets: as **a :** a substance added to a perfume especially to prevent too rapid evaporation **b :** a substance used to fix living tissue **c :** a varnish used especially for the protection of drawings (as in pastel or charcoal)
— fixative *adjective*

fixed \'fikst\ *adjective* (14th century)
1 a : securely placed or fastened **:** STATIONARY **b** (1) **:** NONVOLATILE (2) **:** formed into a chemical compound **c** (1) **:** not subject to change or fluctuation ⟨a *fixed* income⟩ (2) **:** firmly set in the mind ⟨a *fixed* idea⟩ (3) **:** having a final or crystallized form or character (4) **:** recurring on the same date from year to year ⟨*fixed* holidays⟩ **d :** IMMOBILE, CONCENTRATED ⟨a *fixed* stare⟩
2 : supplied with something (as money) needed ⟨comfortably *fixed*⟩
— fixed·ly \'fik-səd-lē, 'fiks-tlē\ *adverb*
— fixed·ness \'fik-səd-nəs, 'fiks(t)-nəs\ *noun*

fixed charge *noun* (circa 1901)
: a regularly recurring expense (as rent, taxes, or interest) that must be met when due

fixed oil *noun* (circa 1800)
: a nonvolatile oil; *especially* **:** a fatty oil — compare ESSENTIAL OIL

fixed–point *adjective* (1960)
: involving or being a mathematical notation (as in a decimal system) in which the point separating whole numbers and fractions is fixed — compare FLOATING-POINT

fixed star *noun* (1561)
: a star so distant that its motion can be measured only by very precise observations over long periods

fix·er \'fik-sər\ *noun* (1849)
: one that fixes: as **a :** one who intervenes to enable a person to circumvent the law or obtain a political favor **b :** one who adjusts matters or disputes by negotiation **c :** ²HYPO

fix·ing \-siŋ\ *noun* (1605)
1 : the act or process of one that fixes
2 *plural* *often* -sənz\ **:** customary accompaniments **:** TRIMMINGS ⟨a turkey dinner with all the *fixings*⟩

fix·i·ty \'fik-sə-tē\ *noun, plural* **-ties** (1666)
1 : the quality or state of being fixed or stable
2 : something that is fixed

fix·ture \'fiks-chər\ *noun* [modification of Late Latin *fixura,* from Latin *fixus*] (1598)
1 : the act or process of fixing **:** the state of being fixed
2 a : something that is fixed or attached (as to a building) as a permanent appendage or as a structural part ⟨a plumbing *fixture*⟩ **b :** a device for supporting work during machining **c :** an item of movable property so incorporated into real property that it may be regarded as legally a part of it
3 : a familiar or invariably present element or feature in some particular setting; *especially* **:** a person long associated with a place or activity
4 : a settled date or time especially for a sporting or festive event; *also* **:** such an event especially as a regularly scheduled affair

fix up *transitive verb* (1831)
1 : REFURBISH ⟨*fix up* the attic⟩
2 : to set right **:** SETTLE ⟨*fixed up* their dispute⟩
3 : to provide with something needed or wanted; *especially* **:** to arrange a date for

¹fizz \'fiz\ *intransitive verb* [probably of imitative origin] (1685)
1 : to make a hissing or sputtering sound **:** EFFERVESCE
2 : to show excitement or exhilaration

²fizz *noun* (1842)
1 a : a hissing sound **b :** SPIRIT, LIVELINESS
2 : an effervescent beverage
— fizzy \'fi-zē\ *adjective*

¹fiz·zle \'fi-zəl\ *intransitive verb* **fiz·zled; fiz·zling** \-z(ə-)liŋ\ [probably alteration of *fist* (to break wind)] (circa 1841)
1 : FIZZ
2 : to fail or end feebly especially after a promising start — often used with *out*

²fizzle *noun* (1842)
: an abortive effort **:** FAILURE

fjord \fē-'ȯrd, 'fē-ˌ; 'fyȯrd\ *noun* [Norwegian *fjord,* from Old Norse *fjǫrthr* — more at FORD] (1674)
: a narrow inlet of the sea between cliffs or steep slopes

fjord

flab \'flab\ *noun* [back-formation from *flabby*] (1951)
: soft flabby body tissue

flab·ber·gast \'fla-bər-ˌgast\ *transitive verb* [origin unknown] (1772)
: to overwhelm with shock, surprise, or wonder **:** DUMBFOUND
synonym see SURPRISE
— flab·ber·gast·ing·ly \-ˌgas-tiŋ-lē\ *adverb*

flab·by \'fla-bē\ *adjective* **flab·bi·er; -est** [alteration of *flappy*] (1697)
1 : lacking resilience or firmness **:** FLACCID
2 : weak and ineffective **:** FEEBLE
— flab·bi·ly \'fla-bə-lē\ *adverb*
— flab·bi·ness \'fla-bē-nəs\ *noun*

fla·bel·late \flə-ˈbe-lət, ˈfla-bə-ˌlāt\ *adjective* [Latin *flabellum* fan] (1819)
: shaped like a fan

fla·bel·li·form \flə-ˈbe-lə-ˌfȯrm\ *adjective* (1777)
: FLABELLATE

flac·cid \÷ˈfla-səd *also* ˈflak-səd\ *adjective* [Latin *flaccidus,* from *flaccus* flabby] (1620)
1 a : not firm or stiff; *also* **:** lacking normal or youthful firmness ⟨*flaccid* muscles⟩ **b** *of a plant part* **:** deficient in turgor
2 : lacking vigor or force ⟨*flaccid* leadership⟩
— **flac·cid·i·ty** \fla(k)-ˈsi-də-tē\ *noun*
— **flac·cid·ly** \ˈfla(k)-səd-lē\ *adverb*

[1]**flack** \ˈflak\ *noun* [origin unknown] (1939)
: one who provides publicity; *especially* **:** PRESS AGENT
— **flack·ery** \ˈfla-k(ə-)rē\ *noun*

[2]**flack** *variant of* FLAK

[3]**flack** *intransitive verb* (1965)
: to provide publicity **:** engage in press-agentry

fla·con \ˈfla-kən, -ˌkän; fla-ˈkōⁿ\ *noun* [French, from Middle French, bottle — more at FLAGON] (1824)
: a small usually ornamental bottle with a tight cap

[1]**flag** \ˈflag *also* ˈflāg\ *noun* [Middle English *flagge* reed, rush] (14th century)
: any of various monocotyledonous plants with long ensiform leaves: as **a :** IRIS; *especially* **:** a wild iris **b :** SWEET FLAG

[2]**flag** *noun, often attributive* [perhaps from [1]*flag*] (1530)
1 : a usually rectangular piece of fabric of distinctive design that is used as a symbol (as of a nation), as a signaling device, or as a decoration
2 a : the tail of some dogs (as a setter or hound); *also* **:** the long hair fringing a dog's tail **b :** the tail of a deer
3 a : something used like a flag to signal or attract attention **b :** one of the cross strokes of a musical note less than a quarter note in value
4 : something represented by a flag: as **a :** FLAGSHIP **b :** an admiral functioning in his office of command **c :** NATIONALITY; *especially* **:** the nationality of registration of a ship or aircraft

[3]**flag** *transitive verb* **flagged; flag·ging** (1856)
1 : to signal with or as if with a flag; *especially* **:** to signal to stop ⟨*flagged* the train⟩ — often used with *down*
2 : to put a flag on (as for identification)
3 : to call a penalty on **:** PENALIZE ⟨a lineman *flagged* for being offside⟩

[4]**flag** *intransitive verb* **flagged; flag·ging** [probably from [2]*flag*] (1545)
1 : to hang loose without stiffness
2 a : to become unsteady, feeble, or spiritless **b :** to decline in interest or attraction

[5]**flag** *noun* [Middle English *flagge* turf, from Old Norse *flaga* slab; akin to Old English *flōh* chip] (1604)
: a hard evenly stratified stone that splits into flat pieces suitable for paving; *also* **:** a piece of such stone

[6]**flag** *transitive verb* **flagged; flag·ging** (1615)
: to lay (as a pavement) with flags

flag day *noun* (1894)
1 *F&D capitalized* **:** June 14 observed in various states in commemoration of the adoption in 1777 of the official U.S. flag
2 *British* **:** a day on which charitable contributions are solicited in exchange for small flags

fla·gel·lant \ˈfla-jə-lənt, flə-ˈje-lənt\ *noun* (circa 1587)
1 : a person who scourges himself as a public penance
2 : a person who responds sexually to being beaten by or to beating another person
— **flagellant** *adjective*
— **fla·gel·lant·ism** \-lən-ˌti-zəm\ *noun*

fla·gel·lar \flə-ˈje-lər, ˈfla-jə-lər\ *adjective* (circa 1889)
: of or relating to a flagellum

[1]**flag·el·late** \ˈfla-jə-ˌlāt\ *transitive verb* **-lat·ed; -lat·ing** [Latin *flagellatus,* past participle of *flagellare,* from *flagellum,* diminutive of *flagrum* whip; perhaps akin to Old Norse *blaka* to wave] (circa 1623)
1 : WHIP, SCOURGE
2 : to drive or punish as if by whipping

[2]**fla·gel·late** \ˈfla-jə-lət, -ˌlāt; flə-ˈje-lət\ *adjective* [New Latin *flagellatus,* from *flagellum*] (circa 1859)
1 a *or* **flag·el·lat·ed** \ˈfla-jə-ˌlā-təd\ **:** having flagella **b :** shaped like a flagellum
2 [[3]*flagellate*] **:** of, relating to, or caused by flagellates ⟨*flagellate* diarrhea⟩

[3]**flagellate** *same as* [2]\ *noun* [New Latin *Flagellata,* class of unicellular organisms, from neuter plural of *flagellatus*] (1879)
: a flagellate protozoan or alga

flag·el·la·tion \ˌfla-jə-ˈlā-shən\ *noun* (15th century)
: the act or practice of flagellating; *especially* **:** the practice of a flagellant

fla·gel·lin \flə-ˈje-lən\ *noun* [*flagell*um + [1]*-in*] (1955)
: a polymeric protein that is the chief constituent of bacterial flagella and is responsible for the specificity of their flagellar antigens

fla·gel·lum \flə-ˈje-ləm\ *noun, plural* **-la** \-lə\ *also* **-lums** [New Latin, from Latin, whip, shoot of a plant] (1852)
: any of various elongated filiform appendages of plants or animals: as **a :** the slender distal part of an antenna **b :** a long tapering process that projects singly or in groups from a cell and is the primary organ of motion of many microorganisms

[1]**fla·geo·let** \ˌfla-jə-ˈlet, -ˈlā\ *noun* [French, from Old French *flajolet,* from *flajol* flute, from (assumed) Vulgar Latin *flabeolum,* from Latin *flare* to blow — more at BLOW] (1659)
: a small fipple flute resembling the treble recorder

[2]**flageolet** *noun* [French, diminutive of *flageolle* kidney bean, modification of Italian *fagiolo,* from Latin *phaseolus* — more at FRIJOLE] (1877)
: a green kidney bean used in French cuisine

flag football *noun* (1954)
: a variation of football in which a player must remove a flag attached to the ballcarrier's clothing to stop the play

flag·ging \ˈfla-giŋ, *also* ˈflā-\ *adjective* (1545)
1 : LANGUID, WEAK
2 : becoming progressively less **:** DWINDLING
— **flag·ging·ly** \-giŋ-lē\ *adverb*

fla·gi·tious \flə-ˈji-shəs\ *adjective* [Middle English *flagicious,* from Latin *flagitiosus,* from *flagitium* shameful thing] (14th century)
: marked by scandalous crime or vice **:** VILLAINOUS
— **fla·gi·tious·ly** *adverb*
— **fla·gi·tious·ness** *noun*

flag·man \ˈflag-mən\ *noun* (1832)
: a person who signals with a flag

flag of convenience (1956)
: registry of a merchant ship under a foreign flag in order to profit from less restrictive regulations

flag officer *noun* [from his being entitled to display a flag with one or more stars indicating his rank] (1665)
: any of the officers in the navy or coast guard above captain — compare GENERAL OFFICER

flag of truce (1627)
: a white flag carried or displayed to an enemy as an invitation to conference or parley

flag·on \ˈfla-gən\ *noun* [Middle English, from Middle French *flascon, flacon* bottle, from Late Latin *flascon-, flasco* — more at FLASK] (15th century)
1 a : a large usually metal or pottery vessel (as for wine) with handle and spout and often a lid **b :** a large bulging short-necked bottle
2 : the contents of a flagon

flag·pole \ˈflag-ˌpōl\ *noun* (1884)
: a pole on which to raise a flag

flagon 1a

fla·grance \ˈflā-grən(t)s *also* ˈfla-\ *noun* (circa 1615)
: FLAGRANCY

fla·gran·cy \ˈflā-grən(t)-sē *also* ˈfla-\ *noun* (1599)
: the quality or state of being flagrant

flag rank *noun* (1894)
: the rank of a flag officer

fla·grant \ˈflā-grənt *also* ˈfla-\ *adjective* [Latin *flagrant-, flagrans,* present participle of *flagrare* to burn — more at BLACK] (1513)
1 *archaic* **:** fiery hot **:** BURNING
2 : conspicuously offensive ⟨*flagrant* errors⟩; *especially* **:** so obviously inconsistent with what is right or proper as to appear to be a flouting of law or morality ⟨*flagrant* violations of human rights⟩ ☆
— **fla·grant·ly** *adverb*

fla·gran·te de·lic·to \flə-ˌgrän-tē-di-ˈlik-(ˌ)tō, -ˌgran-\ *adverb or adjective* [Medieval Latin, literally, while the crime is blazing] (1826)
: in the very act of committing a misdeed **:** RED-HANDED; *also* **:** in the midst of sexual activity

flag·ship \ˈflag-ˌship\ *noun* (1672)
1 : the ship that carries the commander of a fleet or subdivision of a fleet and flies his flag
2 : the finest, largest, or most important one of a series, network, or chain

flag·staff \-ˌstaf\ *noun* (circa 1613)
: a staff on which a flag is hoisted

flag·stick \-ˌstik\ *noun* (1926)
: a staff for a flag marking the location of the cup on a golf putting green

flag·stone \-ˌstōn\ *noun* (1730)
: [5]FLAG

flag stop *noun* (1941)
: a point at which a vehicle in public transportation stops only on prearrangement or signal

flag–wav·er \ˈflag-ˌwā-vər\ *noun* (1894)
1 : one who is intensely and conspicuously patriotic
2 : one who waves a flag in signaling
3 : a song intended to rouse patriotic sentiment

flag–wav·ing \-viŋ\ *noun* (1892)
: passionate appeal to patriotic or partisan sentiment **:** CHAUVINISM

[1]**flail** \ˈflā(ə)l\ *noun* [Middle English *fleil, flail,* partly from (assumed) Old English *flegel* (whence Old English *fligel*), from Late Latin *flagellum* flail, from Latin, whip & partly from Middle French *flaiel,* from Late Latin *flagellum* — more at FLAGELLATE] (before 12th century)

☆ SYNONYMS
Flagrant, glaring, gross, rank mean conspicuously bad or objectionable. FLAGRANT applies usually to offenses or errors so bad that they can neither escape notice nor be condoned ⟨*flagrant* abuse of the office of president⟩. GLARING implies painful or damaging obtrusiveness of something that is conspicuously wrong, faulty, or improper ⟨*glaring* errors⟩. GROSS implies the exceeding of reasonable or excusable limits ⟨*gross* carelessness⟩. RANK applies to what is openly and extremely objectionable and utterly condemned ⟨*rank* heresy⟩.

: a hand threshing implement consisting of a wooden handle at the end of which a stouter and shorter stick is so hung as to swing freely

[2]flail (15th century)
transitive verb
1 a : to strike with or as if with a flail ⟨arms *flailing* the water⟩ **b :** to move, swing, or beat as if wielding a flail ⟨*flailing* a club to drive away the insects⟩
2 : to thresh (grain) with a flail
intransitive verb
: to move, swing, or beat like a flail

flair \ˈflar, ˈfler\ *noun* [French, literally, sense of smell, from Old French, odor, from *flairier* to give off an odor, from Late Latin *flagrare,* alteration of Latin *fragrare*] (1881)
1 : a skill or instinctive ability to appreciate or make good use of something **:** TALENT ⟨a *flair* for color⟩; *also* **:** INCLINATION, TENDENCY ⟨a *flair* for the dramatic⟩
2 : a uniquely attractive quality **:** STYLE ⟨fashionable dresses with a *flair* all their own⟩

flak \ˈflak\ *noun, plural* **flak** [German, from *F*l*iegera*bwehr*k*anonen, from *Flieger* flyer + *Abwehr* defense + *Kanonen* cannons] (1938)
1 : antiaircraft guns
2 : the bursting shells fired from flak
3 : CRITICISM, OPPOSITION

[1]flake \ˈflāk\ *noun* [Middle English, of Scandinavian origin; akin to Norwegian *flak* disk] (14th century)
1 : a small loose mass or bit ⟨snow *flakes*⟩
2 : a thin flattened piece or layer **:** CHIP
3 *slang* **:** COCAINE

[2]flake *verb* **flaked; flak·ing** (1627)
intransitive verb
: to separate into flakes; *also* **:** to peel in flakes
transitive verb
1 : to form or break into flakes **:** CHIP
2 : to cover with or as if with flakes
— flak·er *noun*

[3]flake *noun* [Middle English, hurdle, from Old Norse *flaki*] (1623)
: a stage, platform, or tray for drying fish or produce

[4]flake *noun* [perhaps from *flake out*] (1964)
: a person who is flaky **:** ODDBALL

flake out *intransitive verb* [probably from dialect *flake* to lie, bask] (1939)
1 *slang* **:** to fall asleep
2 *slang* **:** to be overcome especially by exhaustion

flake tool *noun* (circa 1947)
: a Stone-Age tool that is a flake of stone struck off from a larger piece

flak jacket *noun* (1950)
: a jacket containing metal plates for protection against flak; *broadly* **:** a bulletproof vest — called also *flak vest*

[1]flaky *also* **flak·ey** \ˈflā-kē\ *adjective* **flak·i·er; -est** (1580)
1 : consisting of flakes ⟨*flaky* snow⟩
2 : tending to flake ⟨a *flaky* crust⟩
— flak·i·ness *noun*

[2]flaky *adjective* **flak·i·er; -est** [[4]*flake*] (circa 1963)
: markedly odd or unconventional **:** OFFBEAT, WACKY
— flak·i·ness *noun*

flam \ˈflam\ *noun* [probably imitative] (1819)
: a drumbeat of two strokes of which the first is a very quick grace note

[1]flam·bé \fläm-ˈbā, flä^{n}-\ *adjective* [French *flambé,* from past participle of *flamber* to flame, singe, from Old French, from *flambe* flame] (1914)
: dressed or served covered with flaming liquor — usually used postpositively ⟨crepes suzette *flambé*⟩

[2]flambé *transitive verb* **flam·béed; flam·bé·ing** (circa 1946)
: to douse with a liquor (as brandy, rum, or cognac) and ignite

flam·beau \ˈflam-ˌbō\ *noun, plural* **flam·beaux** \-ˌbōz\ *or* **flambeaus** [French, from Middle French, from *flambe* flame] (1632)
: a flaming torch; *broadly* **:** TORCH

flam·boy·ance \flam-ˈbȯi-ən(t)s\ *noun* (1891)
: the quality or state of being flamboyant

flam·boy·an·cy \-ən(t)-sē\ *noun* (circa 1889)
: FLAMBOYANCE

[1]flam·boy·ant \-ənt\ *adjective* [French, from present participle of *flamboyer* to flame, from Old French, from *flambe*] (1832)
1 *often capitalized* **:** characterized by waving curves suggesting flames ⟨*flamboyant* tracery⟩ ⟨*flamboyant* architecture⟩
2 : marked by or given to strikingly elaborate or colorful display or behavior
— flam·boy·ant·ly *adverb*

[2]flamboyant *noun* (1879)
: ROYAL POINCIANA

[1]flame \ˈflām\ *noun* [Middle English *flaume, flaumbe,* from Middle French *flamme* (from Latin *flamma*) & *flambe,* from Old French, from *flamble,* from Latin *flammula,* diminutive of *flamma* flame; akin to Latin *flagrare* to burn — more at BLACK] (14th century)
1 : the glowing gaseous part of a fire
2 a : a state of blazing combustion ⟨the car burst into *flame*⟩ **b :** a condition or appearance suggesting a flame or burning: as (1) **:** burning zeal or passion (2) **:** a strong reddish orange color
3 : BRILLIANCE, BRIGHTNESS
4 : SWEETHEART
5 : an angry, hostile, or abusive electronic message

[2]flame *verb* **flamed; flam·ing** (14th century)
intransitive verb
1 : to burn with a flame **:** BLAZE
2 a : to burst or break out violently or passionately ⟨*flaming* with indignation⟩ **b :** to send an angry, hostile, or abusive electronic message
3 : to shine brightly **:** GLOW ⟨color *flaming* up in her cheeks⟩
transitive verb
1 : to send or convey by means of flame ⟨*flame* a message by signal fires⟩
2 : to treat or affect with flame: as **a :** to sear, sterilize, or destroy by fire **b :** FLAMBÉ
3 : to send an angry, hostile, or abusive electronic message to or about
— flam·er *noun*

flame cell *noun* (1888)
: a hollow cell that has a tuft of vibratile cilia and is part of some lower invertebrate excretory systems (as of a flatworm)

fla·men \ˈflā-mən\ *noun, plural* **flamens** *or* **flam·i·nes** \ˈfla-mə-ˌnēz\ [Middle English *flamin,* from Latin *flamin-, flamen*] (14th century)
: a priest especially in ancient Rome

fla·men·co \flə-ˈmeŋ-(ˌ)kō\ *noun, plural* **-cos** [Spanish, from *flamenco* of the Gypsies, literally, Flemish, from Middle Dutch *Vlaminc* Fleming] (1896)
1 : a vigorous rhythmic dance style of the Andalusian Gypsies; *also* **:** a dance in flamenco style
2 : music or song suitable to accompany a flamenco dance ◆

flame·out \ˈflā-ˌmau̇t\ *noun* (1950)
1 : the unintentional cessation of operation of a jet airplane engine
2 : a sudden or spectacular failure
3 : a person whose successful career ends abruptly

flame photometer *noun* (1945)
: a spectrophotometer in which a spray of metallic salts in solution is vaporized in a very hot flame and subjected to quantitative analysis by measuring the intensities of the spectral lines of the metals present
— flame photometric *adjective*
— flame photometry *noun*

flame·proof \ˈflām-ˌprüf\ *adjective* (1886)
: resistant to damage or burning on contact with flame
— flameproof *transitive verb*
— flame·proof·er *noun*

flame-retardant *adjective* (1947)
: made or treated so as to resist burning

flame stitch *noun* (1936)
: a needlepoint stitch that produces a pattern resembling flames

flame·throw·er \-ˌthrō(-ə)r\ *noun* (1917)
1 : a device that expels from a nozzle a burning stream of liquid or semiliquid fuel under pressure
2 : a pitcher who throws hard **:** a fastball pitcher

flame tree *noun* (1860)
: any of several trees or shrubs with showy scarlet or yellow flowers: as **a :** a tree (*Brachychiton acerifolium* of the family Sterculiaceae) of southern Australia with panicles of brilliant scarlet flowers **b :** ROYAL POINCIANA

flam·ing \ˈflā-miŋ\ *adjective* (14th century)
1 : resembling or suggesting a flame in color, brilliance, or wavy outline ⟨the *flaming* sunset sky⟩ ⟨*flaming* red hair⟩
2 : being on fire **:** BLAZING
3 : INTENSE, PASSIONATE ⟨*flaming* youth⟩
4 — used as an intensive ⟨you *flaming* idiot⟩
— flam·ing·ly \-miŋ-lē\ *adverb*

fla·min·go \flə-ˈmiŋ-(ˌ)gō\ *noun, plural* **-gos** *also* **-goes** [obsolete Spanish *flamengo* (now *flamenco*), literally, Fleming, German (conventionally thought of as ruddy-complexioned)] (1565)
: any of several large aquatic birds (family Phoenicopteridae) with long legs and neck, webbed feet, a broad lamellate bill resembling that of a duck but abruptly bent downward, and usually rosy-white plumage with scarlet wing coverts and black wing quills

flamingo

flam·ma·bil·i·ty \ˌfla-mə-ˈbi-lə-tē\ *noun* (1646)
: ability to support combustion; *especially* **:** a high capacity for combustion

flam·ma·ble \ˈfla-mə-bəl\ *adjective* [Latin *flammare* to flame, set on fire, from *flamma*] (1813)

◇ WORD HISTORY

flamenco The Spanish word *flamenco* means literally "Flemish," and its usage in the sense "Gypsy-like," especially in reference to a distinctive song, dance, and guitar-music style of Andalusia, has inspired a number of hypotheses. It has been speculated that Gypsies were called *flamenco* because they reached Spain in the 16th century via Germany or Flanders, or that *flamenco* became a derogatory epithet for any foreigner because certain Flemish courtiers of the Spanish king and Holy Roman Emperor Charles V were much disliked in Spain. These notions of such an early origin for the "Gypsy-like" sense of *flamenco* all look implausible in light of the fact that this meaning is unattested in Spanish before 1870. In the later 19th century *flamenco* also meant "jaunty, cocky" and, in reference to women, "provocatively attractive," and the suggestion has been made that "Gypsy-like" is a secondary development from these senses. In any event, a direct connection between the senses "Flemish" and "Gypsy-like" has not been demonstrated. The ordinary Spanish word for "Gypsy" is *gitano,* which, like English *Gypsy,* is altered from a word meaning "Egyptian."

: capable of being easily ignited and of burning quickly
— **flammable** *noun*

flan \'flan, 'flän(n)\ *noun* [French, from Old French *flaon,* from Late Latin *fladon-, flado* flat cake, of Germanic origin; akin to Old High German *flado* flat cake] (1846)
1 a : an open pie containing any of various sweet or savory fillings **b :** custard baked with a caramel glaze
2 : the metal disk of a coin, token, or medal as distinguished from the design and lettering stamped on it

fla·neur \flä-'nər\ *noun* [French *flâneur*] (1854)
: an idle man-about-town

¹flange \'flanj\ *noun* [perhaps alteration of *flanch* (a curving charge on a heraldic shield)] (circa 1735)
1 : a rib or rim for strength, for guiding, or for attachment to another object ⟨a *flange* on a pipe⟩ ⟨a *flange* on a wheel⟩
2 : a projecting edge of cloth used for decoration on clothing ⟨a jacket with *flange* shoulders⟩

²flange *transitive verb* **flanged; flang·ing** (circa 1859)
: to furnish with a flange

¹flank \'flaŋk\ *noun* [Middle English, from Old French *flanc,* of Germanic origin; akin to Old High German *hlanca* loin, flank — more at LANK] (before 12th century)
1 a : the fleshy part of the side between the ribs and the hip; *broadly* **:** the side of a quadruped **b :** a cut of meat from this part of an animal — see BEEF illustration
2 a : SIDE **b :** the right or left of a formation
3 : the area along either side of a heraldic shield

²flank *transitive verb* (1596)
1 : to protect a flank of
2 : to attack or threaten the flank of (as a body of troops)
3 a : to be situated at the side of; *especially* **:** to be situated on both sides of ⟨a road *flanked* with linden trees⟩ **b :** to place something on each side of

flan·ken \'fläŋ-kən\ *noun* [Yiddish, plural of *flank,* literally, flank, ultimately from Old French *flanc*] (1950)
: beef flank cooked especially by boiling

flank·er \'flaŋ-kər\ *noun* (1940)
: a football player stationed wide of the formation slightly behind the line of scrimmage as a pass receiver — called also *flanker back*

flank steak *noun* (1902)
: a pear-shaped muscle of the beef flank; *also* **:** a steak cut from this muscle — see BEEF illustration

flan·nel \'fla-nᵊl\ *noun* [Middle English *flaunneol* woolen cloth or garment] (1503)
1 a : a soft twilled wool or worsted fabric with a loose texture and a slightly napped surface **b :** a napped cotton fabric of soft yarns simulating the texture of wool flannel **c :** a stout cotton fabric usually napped on one side
2 *plural* **a :** flannel underwear **b :** outer garments of flannel; *especially* **:** men's trousers
3 *British* **:** WASHCLOTH
4 *British* **:** flattering or evasive talk; *also* **:** NONSENSE, RUBBISH
— **flannel** *adjective*
— **flan·nel·ly** \-nᵊl-ē\ *adjective*

flan·nel·ette \,fla-nᵊl-'et\ *noun* (circa 1882)
: a lightweight cotton flannel

flan·nel·mouthed \'fla-nᵊl-,mau̇tht, -,mau̇thd\ *adjective* (circa 1893)
1 : speaking indistinctly
2 : speaking in a tricky or ingratiating way

¹flap \'flap\ *noun* [Middle English *flappe*] (14th century)
1 : a stroke with something broad **:** SLAP
2 *obsolete* **:** something broad and flat used for striking
3 : something that is broad, limber, or flat and usually thin and that hangs loose or projects freely: as **a :** a piece on a garment that hangs free **b :** a part of a book jacket that folds under the book's cover **c :** a piece of tissue partly severed from its place of origin for use in surgical grafting **d :** an extended part forming the closure (as of an envelope or carton)
4 : the motion of something broad and limber (as a sail or wing)
5 : a movable auxiliary airfoil usually attached to an airplane wing's trailing edge to increase lift or drag — see AIRPLANE illustration
6 a : a state of excitement or agitation **:** TIZZY, UPROAR **b :** something (as an incident or remark) that generates an uproar
7 : a consonant (as the sound \d\ in *ladder* and \t\ in *latter*) characterized by a single rapid contact of the tongue or lower lip against another point in the mouth — called also *tap*

²flap *verb* **flapped; flap·ping** (14th century)
transitive verb
1 : to beat with or as if with a flap
2 : to toss sharply **:** FLING
3 : to move or cause to move in flaps
intransitive verb
1 : to sway loosely usually with a noise of striking and especially when moved by wind
2 a : to beat or pulsate wings or something suggesting wings **b :** to progress by flapping **c :** to flutter ineffectively
3 : to talk foolishly and persistently

flap·doo·dle \'flap-,dü-dᵊl\ *noun* [origin unknown] (1878)
: NONSENSE

flap·jack \-,jak\ *noun* (circa 1600)
: PANCAKE

flap·pa·ble \'fla-pə-bəl\ *adjective* (1968)
: easily upset

flap·per \'fla-pər\ *noun* (circa 1570)
1 a : one that flaps **b :** something used in flapping or striking **c :** FLIPPER 1
2 : a young woman; *specifically* **:** a young woman of the period of World War I and the following decade who showed freedom from conventions (as in conduct)

flap·py \'fla pē\ *adjective* (1905)
: flapping or tending to flap

¹flare \'flar, 'fler\ *verb* **flared; flar·ing** [origin unknown] (circa 1700)
intransitive verb
1 a : to burn with an unsteady flame **b :** to stream in the wind
2 a : to shine with a sudden light ⟨a match *flares* in the darkness⟩ **b** (1) **:** to become suddenly excited or angry — usually used with *up* (2) **:** to break out or intensify usually suddenly or violently — often used with *up* ⟨fighting *flared* up after a two-week lull⟩ **c :** to express strong emotion (as anger) ⟨*flaring* out at such abuses⟩
3 : to open or spread outward ⟨the pants *flare* at the bottom⟩
transitive verb
1 : to display conspicuously ⟨*flaring* her scarf to attract attention⟩
2 : to cause to flare ⟨the breeze *flares* the candle⟩
3 : to signal with a flare or by flaring
4 : to burn (a jet of waste gas) in the open air

²flare *noun* (1814)
1 : an unsteady glaring light
2 a : a fire or blaze of light used especially to signal, illuminate, or attract attention; *also* **:** a device or composition used to produce such a flare **b :** SOLAR FLARE; *also* **:** a sudden increase and decrease in the brightness of a star often amounting to a difference of several magnitudes
3 : a sudden outburst (as of excitement or anger)
4 a : a spreading outward; *also* **:** a place or part that spreads **b :** an area of skin flush
5 : light resulting from reflection (as between lens surfaces) or an effect of this light (as a fogged or dense area in a photographic negative)

flare–up \-,əp\ *noun* (1839)
1 : a sudden outburst or intensification
2 : a sudden bursting (as of a smoldering fire) into flame or light

flar·ing \'flar-iŋ, 'fler-\ *adjective* (1593)
1 : opening or spreading outward ⟨*flaring* nostrils⟩
2 a : flaming or as if flaming brightly or unsteadily **b :** GAUDY ⟨a *flaring* resort hotel⟩
— **flar·ing·ly** \-iŋ-lē\ *adverb*

¹flash \'flash\ *verb* [Middle English *flaschen,* of imitative origin] (13th century)
intransitive verb
1 : RUSH, DASH — used of flowing water
2 : to break forth in or like a sudden flame or flare
3 a : to appear suddenly ⟨an idea *flashes* into her mind⟩ **b :** to move with great speed ⟨the days *flash* by⟩
4 a : to break forth or out so as to make a sudden display ⟨the sun *flashed* from behind a cloud⟩ **b :** to act or speak vehemently and suddenly especially in anger
5 a : to give off light suddenly or in transient bursts **b :** to glow or gleam especially with animation or passion ⟨her eyes *flashed* with anger⟩
6 : to change suddenly or violently into vapor ⟨hot water *flashing* to steam under reduced pressure⟩
7 : to expose one's genitals usually suddenly and briefly in public
8 : to have sudden insight — often used with *on*
transitive verb
1 a *archaic* **:** SPLASH **b :** to fill by a sudden inflow of water
2 a : to cause the sudden appearance of (light) **b :** to cause to burst violently into flame **c** (1) **:** to cause (light) to reflect (2) **:** to cause (as a mirror) to reflect light (3) **:** to cause (a lamp) to flash **d :** to convey by means of flashes of light
3 a : to make known or cause to appear with great speed ⟨*flash* a message on the screen⟩ **b :** to display obtrusively and ostentatiously ⟨always *flashing* a roll of bills⟩ **c :** to expose to view suddenly and briefly ⟨*flashed* a badge⟩
4 : to cover with or form into a thin layer: as **a :** to protect against rain by covering with sheet metal or a substitute **b :** to coat (as glass) with a thin layer (as of metal or a differently colored glass)
5 : to subject (an exposed photographic negative or positive) to a supplementary uniform exposure to light before development in order to modify detail or tone ☆

²flash *noun* (1566)

☆ SYNONYMS
Flash, gleam, glint, sparkle, glitter, glisten, glimmer, shimmer mean to send forth light. FLASH implies a sudden and transient outburst of bright light ⟨lightning *flashed*⟩. GLEAM suggests a steady light seen through an obscuring medium or against a dark background ⟨lights *gleamed* in the valley⟩. GLINT implies a cold glancing light ⟨*glinting* steel⟩. SPARKLE suggests innumerable moving points of bright light ⟨the *sparkling* waters of the gulf⟩. GLITTER connotes a brilliant sparkling or gleaming ⟨*glittering* diamonds⟩. GLISTEN applies to the soft sparkle from a wet or oily surface ⟨*glistening* rain-drenched sidewalks⟩. GLIMMER suggests a faint or wavering gleam ⟨a distant *glimmering* light⟩. SHIMMER implies a soft tremulous gleaming or a blurred reflection ⟨a *shimmering* satin dress⟩.

1 a : a sudden burst of light **b :** a movement of a flag in signaling
2 : a sudden and often brilliant burst ⟨a *flash* of wit⟩
3 : a brief time ⟨I'll be back in a *flash*⟩
4 a : SHOW, DISPLAY; *especially* **:** a vulgar ostentatious display **b** *archaic* **:** a showy ostentatious person **c :** one that attracts notice; *especially* **:** an outstanding athlete **d :** PIZZAZZ
5 *obsolete* **:** thieves' slang
6 : something flashed: as **a :** GLIMPSE, LOOK **b :** SMILE **c :** a first brief news report **d :** FLASHLIGHT 1,2 **e :** a quick-spreading flame or momentary intense outburst of radiant heat **f :** a device for producing a brief and very bright flash of light for taking photographs
7 : RUSH 7a
8 : the rapid conversion of a liquid into vapor

[3]flash *adjective* (circa 1700)
1 a : FLASHY, SHOWY **b :** of, relating to, or characteristic of flashy people or things ⟨*flash* behavior⟩ **c :** of, relating to, or characteristic of persons considered social outcasts ⟨*flash* language⟩
2 a : of sudden origin and short duration ⟨a *flash* fire⟩ **b :** involving very brief exposure to an intense altering agent (as heat or cold) ⟨*flash* drying of milk⟩ ⟨*flash* freezing of food⟩

flash•back \'flash-ˌbak\ *noun* (1903)
1 : a recession of flame to an unwanted position (as into a blowpipe)
2 a : interruption of chronological sequence (as in a film or literary work) by interjection of events of earlier occurrence; *also* **:** an instance of flashback **b :** a past incident recurring vividly in the mind

flash back *intransitive verb* (1944)
1 : to focus one's mind on or vividly remember a past time or incident — usually used with *to*
2 : to employ a flashback (as in a film) — usually used with *to*

flash•board \-ˌbōrd, -ˌbȯrd\ *noun* (circa 1774)
: one or more boards projecting above the top of a dam to increase the depth of the water

flash•bulb \-ˌbəlb\ *noun* (1935)
: an electric bulb that can be used only once to produce a brief and very bright flash of light for taking photographs

flash card *noun* (1923)
: a card bearing words, numbers, or pictures that is briefly displayed (as by a teacher to a class) usually as a learning aid

flash•cube \'flash-ˌkyüb\ *noun* (1965)
: a cubical device incorporating four flashbulbs

flash•er \'fla-shər\ *noun* (1686)
: one that flashes: as **a :** a light (as a traffic signal or automobile light) that catches the attention by flashing **b :** a device for automatically flashing a light **c :** an exhibitionist who flashes

flash flood *noun* (1940)
: a local flood of great volume and short duration generally resulting from heavy rainfall in the immediate vicinity
— flash flood *verb*

flash–for•ward \'flash-'fȯr-wərd\ *noun* (1949)
: interruption of chronological sequence (as in a film or literary work) by interjection of events of future occurrence; *also* **:** an instance of flash-forward

flash•gun \-ˌgən\ *noun* (1925)
: a device for producing a bright flash of light for photography

flash•ing \'fla-shiŋ\ *noun* (1742)
: sheet metal used in waterproofing (as at roof valleys or hips or the angle between a chimney and a roof)

flash in the pan [from the firing of the priming in the pan of a flintlock musket without discharging the piece] (1901)
1 : a sudden spasmodic effort that accomplishes nothing
2 : one that appears promising but turns out to be disappointing or worthless

flash•lamp \'flash-ˌlamp\ *noun* (1890)
: a lamp for producing a brief but intense flash of light (as for taking photographs)

flash•light \'flash-ˌlīt\ *noun* (1886)
1 a : a sudden bright artificial light used in taking photographic pictures **b :** a photograph taken by such a light
2 : a small battery-operated portable electric light

flash•over \-ˌō-vər\ *noun* (1892)
1 : an abnormal electrical discharge (as through the air to the ground from a high potential source or between two conducting portions of a structure)
2 : the sudden spread of flame over an area when it becomes heated to the flash point

flash point *noun* (1878)
1 : the lowest temperature at which vapors above a volatile combustible substance ignite in air when exposed to flame
2 : a point at which someone or something bursts suddenly into action or being
3 : TINDERBOX 2

flash•tube \'flash-ˌt(y)üb\ *noun* (1945)
: a gas discharge tube that produces very brief intense flashes of light and is used especially in photography

flashy \'fla-shē\ *adjective* **flash•i•er; -est** (circa 1598)
1 *chiefly dialect* **:** lacking in substance or flavor **:** INSIPID
2 : momentarily dazzling
3 a : superficially attractive or impressive **b :** ostentatious or showy often beyond the bounds of good taste; *especially* **:** marked by gaudy brightness
synonym see GAUDY
— flash•i•ly \'fla-shə-lē\ *adverb*
— flash•i•ness \'fla-shē-nəs\ *noun*

flask \'flask, flȧsk\ *noun* [Middle French *flasque* powder flask, ultimately from Late Latin *flascon-, flasco* bottle, probably of Germanic origin; akin to Old High German *flaska* bottle] (1549)
: a container often somewhat narrowed toward the outlet and often fitted with a closure: as **a :** a broad flattened necked vessel used especially to carry alcoholic beverages on the person **b** *British* **:** VACUUM BOTTLE

[1]flat \'flat\ *adjective* **flat•ter; flat•test** [Middle English, from Old Norse *flatr;* akin to Old High German *flaz* flat, and probably Greek *platys* broad — more at PLACE] (14th century)
1 a : lying at full length or spread out upon the ground **:** PROSTRATE **b :** utterly ruined or destroyed **c :** resting with a surface against something
2 a : having a continuous horizontal surface **b :** being or characterized by a horizontal line or tracing without peaks or depressions ⟨a *flat* EEG⟩
3 : having a relatively smooth or even surface
4 : arranged or laid out so as to be level or even
5 a : having the major surfaces essentially parallel and distinctly greater than the minor surfaces ⟨a *flat* piece of wood⟩ **b** *of a heel* **:** very low and broad
6 a : clearly unmistakable **:** DOWNRIGHT ⟨a *flat* denial⟩ **b** (1) **:** not varying **:** FIXED ⟨a *flat* rate⟩ (2) **:** having no fraction either lacking or in excess **:** EXACT ⟨in a *flat* 10 seconds⟩ (3) *of a frequency response* **:** not varying significantly throughout its range
7 a : lacking in animation, zest, or vigor **:** DULL ⟨life seemed *flat* without her⟩ **b :** lacking flavor **:** TASTELESS **c :** lacking effervescence or sparkle ⟨*flat* ginger ale⟩ **d :** commercially inactive; *also* **:** characterized by no significant rise or decline from one period to another ⟨sales were *flat*⟩ **e** *of a tire* **:** lacking air **:** DEFLATED **f** *chiefly British, of a battery* **:** DEAD 3c, DISCHARGED
8 a (1) *of a tone* **:** lowered a half step in pitch (2) **:** lower than the proper pitch **b** *of the vowel a* **:** pronounced as in *bad* or *bat*
9 a : having a low trajectory **b** *of a tennis stroke* **:** made so as to give little or no spin to the ball
10 *of a sail* **:** TAUT
11 a : uniform in hue or shade **b :** having little or no illusion of depth **c** *of a photograph or negative* **:** lacking contrast **d** *of lighting conditions* **:** lacking shadows or contours **e :** free from gloss ⟨a *flat* paint⟩ **f :** TWO-DIMENSIONAL 2
12 : of, relating to, or used in competition on the flat ⟨a *flat* horse⟩
synonym see LEVEL, INSIPID
— flat•ly *adverb*
— flat•ness *noun*
— flat•tish \'fla-tish\ *adjective*

[2]flat *noun* (14th century)
1 a : a level surface of land — usually used in plural ⟨sagebrush *flats*⟩ ⟨tidal *flats*⟩ **b :** a stretch of land without obstacles; *especially* **:** a track or course for a flat race — usually used with *the* ⟨has won twice on the *flat*⟩
2 : a flat part or surface ⟨the *flat* of one's hand⟩
3 a : a musical note or tone one half step lower than a specified note or tone **b :** a character ♭ on a line or space of the musical staff indicating a half step drop in pitch
4 : something flat: as **a :** a shallow container for shipping produce **b :** a shallow box in which seedlings are started **c :** a flat piece of theatrical scenery **d :** a shoe or slipper having a flat heel or no heel
5 *chiefly British* **:** an apartment on one floor
6 : a deflated tire
7 : the area to either side of an offensive football formation

[3]flat *adverb* (1531)
1 : in a flat manner **:** DIRECTLY, POSITIVELY
2 : in a complete manner **:** ABSOLUTELY ⟨*flat* broke⟩
3 : below the proper musical pitch
4 : without interest charge; *especially* **:** without allowance or charge for accrued interest ⟨bonds sold *flat*⟩

[4]flat *verb* **flat•ted; flat•ting** (circa 1604)
transitive verb
1 : FLATTEN
2 : to lower in pitch especially by a half step
intransitive verb
: to sing or play below the true pitch

[1]flat•bed \'flat-ˌbed\ *adjective* (1892)
: having a horizontal bed on which the work rests ⟨a *flatbed* printing press⟩ ⟨a *flatbed* plotter⟩

[2]flatbed *noun* (1944)
: a motortruck or trailer with a body in the form of a platform or shallow box

flat•boat \-ˌbōt\ *noun* (1660)
: a boat with a flat bottom and square ends used for transportation of bulky freight especially in shallow waters

flat•car \-ˌkär\ *noun* (1881)
: a railroad freight car without permanent raised sides, ends, or covering

flat–coated retriever \'flat-ˌkō-təd-\ *noun* (1929)
: any of an English breed of medium-sized sporting dogs that have a dense smooth black or liver-colored coat

flat–earth•er \ˌflat-'ər-thər\ *also* **flat–earth•ist** \-thist\ *noun* (1926)
: a person who believes that the Earth is flat

flat•fish \'flat-ˌfish\ *noun* (1710)
: any of an order (Heterosomata) of marine typically bottom-dwelling bony fishes (as the halibuts, flounders, turbots, and soles) that as adults swim on one side of the laterally compressed body and have both eyes on the upper side

flat•foot \-ˌfu̇t (*always so in sense 3*), ˌflat-'\ *noun, plural* **flat•feet** \-ˌfēt, -'fēt\ (1860)

1 : a condition in which the arch of the instep is flattened so that the entire sole rests upon the ground
2 : a foot affected with flatfoot
3 a *or plural* **flatfoots** *slang* **:** POLICE OFFICER; *especially* **:** a patrolman walking a regular beat **b** *slang* **:** SAILOR

[1]**flat–foot•ed** \ˈflat-ˌfu̇-təd, ˌflat-ˈ\ *adjective* (1601)
1 : affected with flatfoot; *broadly* **:** walking with a dragging or shambling gait
2 a : firm and well balanced on the feet **b :** free from reservation **:** FORTHRIGHT ⟨had an honest *flat-footed* way of saying a thing⟩
3 : not ready **:** UNPREPARED — used chiefly in the phrase *catch one flat-footed*
4 : proceeding in a plodding or unimaginative way **:** PEDESTRIAN ⟨*flat-footed* prose⟩
— **flat–foot•ed•ly** *adverb*
— **flat–foot•ed•ness** *noun*

[2]**flat–footed** *adverb* (1828)
1 : in an open and determined manner **:** FLATLY
2 : with the feet flat on a surface (as the ground)

flat–hat \ˈflat-ˌhat\ *intransitive verb* [from an alleged incident in which a pedestrian's hat was crushed by a low-flying plane] (1940)
: to fly low in an airplane in a reckless manner **:** HEDGEHOP
— **flat–hat•ter** *noun*

Flat•head \-ˌhed\ *noun, plural* **Flatheads** *or* **Flathead** (1709)
1 : a member of any of several North American Indian peoples that practiced head-flattening
2 : an American Indian people of Montana
3 *not capitalized* **:** any of a family (Platycephalidae) of mostly Australian and East Indian marine food fishes that resemble sculpins

flat•head catfish \ˈflat-ˌhed-\ *noun* (1945)
: a large yellowish brown-mottled catfish (*Pylodictis olivaris*) of the central and Gulf states of the U.S.

flat•iron \ˈflat-ˌī(-ə)rn\ *noun* (1744)
: IRON 2c

flat•land \ˈflat-ˌland\ *noun* (1735)
1 : a region in which the land is predominantly flat — usually used in plural
2 : land that lacks significant variation in elevation
— **flat•land•er** \-ˌlan-dər\ *noun*

flat•let \ˈflat-lət\ *noun* (1925)
British **:** EFFICIENCY APARTMENT

flat•ling \ˈflat-liŋ\ *or* **flat•lings** \-liŋz\ *adverb* (15th century)
dialect British **:** with a flat side or edge

flat•mate \ˈflat-ˌmāt\ *noun* (1955)
chiefly British **:** one of two or more persons sharing the same flat

flat–out \ˈflat-ˌau̇t\ *adjective* (1906)
1 : being or going at maximum effort or speed
2 : OUT-AND-OUT, DOWNRIGHT ⟨it was a *flat-out* lie⟩

flat out *adverb* (1932)
1 : in a blunt and direct manner **:** OPENLY ⟨called *flat out* for revolution —*National Review*⟩
2 : at top speed or peak performance ⟨the car does 180 m.p.h. *flat out*⟩
3 *usually* **flat–out :** to the greatest degree **:** COMPLETELY — usually used as an intensive ⟨is just *flat-out* confusing⟩

flat–picking \ˈflat-ˌpi-kiŋ\ *noun* (1970)
: a method of playing a stringed instrument (as a guitar) with a plectrum held between the thumb and forefinger

flat race *noun* (1848)
: a race (as for horses) on a level course without obstacles (as hurdles) — compare STEEPLECHASE
— **flat racing** *noun*

flat•ten \ˈfla-tᵊn\ *verb* **flat•tened; flat•ten•ing** \ˈflat-niŋ, ˈfla-tə-\ (1630)
transitive verb
: to make flat: as **a :** to make level or smooth **b :** to knock down; *also* **:** to defeat decisively **c :** to make dull or uninspired — often used with *out* **d :** to make (as paint) lusterless **e :** to stabilize especially at a lower level
intransitive verb
: to become flat or flatter: as **a :** to become dull or spiritless **b :** to extend in or into a flat position or form **c :** to become uniform or stabilized often at a new lower level — usually used with *out*
— **flat•ten•er** \-nər\ *noun*

[1]**flat•ter** \ˈfla-tər\ *verb* [Middle English *flateren,* from Old French *flater* to lick, flatter, of Germanic origin; akin to Old High German *flaz* flat] (13th century)
transitive verb
1 : to praise excessively especially from motives of self-interest
2 a *archaic* **:** BEGUILE 4 **b :** to encourage or gratify especially with the assurance that something is right ⟨I *flatter* myself that my interpretation is correct⟩
3 a : to portray too favorably ⟨the portrait *flatters* him⟩ **b :** to display to advantage ⟨candlelight often *flatters* the face⟩
intransitive verb
: to use flattery
— **flat•ter•er** \-tər-ər\ *noun*
— **flat•ter•ing•ly** \-tə-riŋ-lē\ *adverb*

[2]**flatter** *noun* (1714)
: one that flattens; *especially* **:** a flat-faced swage used in smithing

flat•tery \ˈfla-tə-rē\ *noun, plural* **-ter•ies** (14th century)
1 a : the act or practice of flattering **b** (1) **:** something that flatters (2) **:** insincere or excessive praise
2 *obsolete* **:** a pleasing self-deception

flat•top \ˈflat-ˌtäp\ *noun* (1940)
: something with a flat or flattened upper surface: as **a :** AIRCRAFT CARRIER **b :** a modified crew cut

flat•u•lence \ˈfla-chə-lən(t)s\ *noun* (1711)
: the quality or state of being flatulent

flat•u•len•cy \-lən(t)-sē\ *noun* (1660)
: FLATULENCE

flat•u•lent \-lənt\ *adjective* [Middle French, from Latin *flatus* act of blowing, wind, from *flare* to blow — more at BLOW] (1599)
1 a : marked by or affected with gas generated in the intestine or stomach **b :** likely to cause gas
2 : pompously or portentously overblown **:** INFLATED
— **flat•u•lent•ly** *adverb*

fla•tus \ˈflā-təs\ *noun* [Latin, act of blowing, act of breaking wind] (1651)
: gas generated in the stomach or bowels

flat•ware \ˈflat-ˌwar, -ˌwer\ *noun* (1851)
: tableware more or less flat and usually formed or cast in a single piece; *especially* **:** eating and serving utensils (as knives, forks, and spoons) — compare HOLLOWARE

flat•ways \-ˌwāz\ *adverb* (1692)
: FLATWISE

flat•wise \-ˌwīz\ *adverb* (1601)
: with the flat surface presented in some expressed or implied position

flat•work \-ˌwərk\ *noun* (1925)
: laundry that can be finished mechanically and does not require hand ironing

flat•worm \-ˌwərm\ *noun* (circa 1889)
: PLATYHELMINTH; *especially* **:** TURBELLARIAN

flaunt \ˈflȯnt, ˈflänt\ *verb* [perhaps of Scandinavian origin; akin to Old Norse *flana* to rush around — more at PLANET] (1566)
intransitive verb
1 : to display or obtrude oneself to public notice
2 : to wave or flutter showily ⟨the flag *flaunts* in the breeze⟩
transitive verb
1 : to display ostentatiously or impudently **:** PARADE ⟨*flaunting* his superiority⟩
2 : to treat contemptuously ⟨*flaunted* the rules —Louis Untermeyer⟩ ■
synonym see SHOW
— **flaunt** *noun*
— **flaunt•ing•ly** \ˈflȯn-tiŋ-lē, ˈflän-\ *adverb*
— **flaunty** \-tē\ *adjective*

flau•tist \ˈflȯ-tist, ˈflau̇-\ *noun* [Italian *flautista,* from *flauto* flute, from Old Provençal *flaut*] (1860)
: FLUTIST

fla•va•none \ˈflā-və-ˌnōn\ *noun* [Latin *flavus* + International Scientific Vocabulary *-ane* + *-one*] (1949)
: a colorless crystalline ketone $C_{15}H_{12}O_2$; *also* **:** any of the derivatives of this ketone many of which occur in plants often in the form of glycosides

fla•vin \ˈflā-vən\ *noun* [International Scientific Vocabulary, from Latin *flavus* yellow — more at BLUE] (1933)
: any of a class of yellow water-soluble nitrogenous pigments derived from isoalloxazine and occurring in the form of nucleotides as coenzymes of flavoproteins; *especially* **:** RIBOFLAVIN

flavin adenine dinucleotide *noun* (1960)
: a coenzyme $C_{27}H_{33}N_9O_{15}P_2$ of some flavoproteins

fla•vine \ˈflā-ˌvēn\ *noun* [International Scientific Vocabulary, from Latin *flavus*] (circa 1853)
: any of a series of yellow acridine dyes (as acriflavine) often used medicinally for their antiseptic properties

flavin mononucleotide *noun* (circa 1953)
: FMN

fla•vone \ˈflā-ˌvōn\ *noun* [International Scientific Vocabulary, from Latin *flavus*] (1897)
: a colorless crystalline ketone $C_{15}H_{10}O_2$ found in the leaves, stems, and seed capsules of many primroses; *also* **:** any of the derivatives of this ketone many of which occur as yellow plant pigments in the form of glycosides and are used as dyestuffs

fla•vo•noid \ˈflā-və-ˌnȯid\ *noun* [*flavone* + *-oid*] (1947)
: any of a group of aromatic compounds that includes many common pigments (as the anthocyanins and flavones)

fla•vo•nol \ˈflā-və-ˌnȯl, -ˌnōl\ *noun* (1898)
: any of various hydroxy derivatives of flavone

fla•vo•pro•tein \ˌflā-vō-ˈprō-ˌtēn, -ˈprō-tē-ən\ *noun* [International Scientific Vocabulary *flavin* + *-o-* + *protein*] (1934)

□ USAGE
flaunt Although transitive sense 2 of *flaunt* undoubtedly arose from confusion with *flout,* the contexts in which it appears cannot be called substandard ⟨meting out punishment to the occasional mavericks who operate rigged games, tolerate rowdyism, or otherwise *flaunt* the law —Oscar Lewis⟩ ⟨observed with horror the *flaunting* of their authority in the suburbs, where men . . . put up buildings that had no place at all in a Christian commonwealth —Marchette Chute⟩ ⟨in our profession . . . very rarely do we publicly chastise a colleague who has *flaunted* our most basic principles —R. T. Blackburn, *AAUP Bulletin*⟩. If you use it, however, you should be aware that many people will consider it a mistake. Use of *flout* in the sense of *flaunt* 1 is found occasionally ⟨"The proper pronunciation," the blonde said, *flouting* her refined upbringing, "is pree feeks" —Mike Royko⟩.

: a dehydrogenase that contains a flavin and often a metal and plays a major role in biological oxidations

¹fla·vor \ˈflā-vər\ *noun* [Middle English, from Middle French *flaor, flavor,* from (assumed) Vulgar Latin *flator,* alteration of Latin *flatus* breath, act of blowing — more at FLATULENT] (14th century)
1 a *archaic* **:** ODOR, FRAGRANCE **b :** the quality of something that affects the sense of taste **c :** the blend of taste and smell sensations evoked by a substance in the mouth ⟨the *flavor* of apples⟩
2 : a substance that flavors ⟨artificial *flavors*⟩
3 : characteristic or predominant quality ⟨the ethnic *flavor* of a neighborhood⟩
4 a : VARIETY 3a **b :** a property that distinguishes different types of elementary particles (as quarks and neutrinos); *also* **:** any of the different types of particles that are distinguished by flavor
— fla·vored \-vərd\ *adjective*
— fla·vor·ful \-vər-fəl\ *adjective*
— fla·vor·ful·ly \-fə-lē\ *adverb*
— fla·vor·less \-vər-ləs\ *adjective*
— fla·vor·some \-səm\ *adjective*

²flavor *transitive verb* **fla·vored; fla·vor·ing** \ˈflā-v(ə-)riŋ\ (1542)
: to give or add flavor to

fla·vor·ing *noun* (1845)
: FLAVOR 2

fla·vor·ist \ˈflā-vər-ist\ *noun* (1964)
: a specialist in the creation of artificial flavors

fla·vour *chiefly British variant of* FLAVOR

¹flaw \ˈflȯ\ *noun* [of Scandinavian origin; akin to Norwegian *flaga* gust] (1513)
1 : a sudden brief burst of wind; *also* **:** a spell of stormy weather
2 *obsolete* **:** an outburst especially of passion

²flaw *noun* [Middle English, flake, probably of Scandinavian origin; akin to Swedish *flaga* flake, flaw; akin to Old English *flōh* flat stone] (1586)
1 a : a defect in physical structure or form **b :** an imperfection or weakness and especially one that detracts from the whole or hinders effectiveness ⟨vanity was the *flaw* in his character⟩ ⟨a *flaw* in the book's plot⟩
2 *obsolete* **:** FRAGMENT
— flawed \ˈflȯd\ *adjective*
— flaw·less \-ləs\ *adjective*
— flaw·less·ly *adverb*
— flaw·less·ness *noun*

³flaw (1610)
transitive verb
: to make flaws in **:** MAR
intransitive verb
: to become defective

flax \ˈflaks\ *noun, often attributive* [Middle English, from Old English *fleax;* akin to Old High German *flahs* flax, Latin *plectere* to braid — more at PLY] (before 12th century)
1 : any of a genus (*Linum* of the family Linaceae, the flax family) of herbs; *especially* **:** a slender erect annual (*Latin usitatissimum*) with blue flowers commonly cultivated for its bast fiber and seed
2 : the fiber of the flax plant especially when prepared for spinning
3 : any of several plants resembling flax

flax·en \ˈflak-sən\ *adjective* (15th century)
1 : made of flax
2 : resembling flax especially in pale soft strawy color ⟨*flaxen* hair⟩

flax·seed \ˈflak(s)-ˌsēd\ *noun* (1562)
: the seed of flax used as a source of oil and medicinally as a demulcent and emollient

flaxy \ˈflak-sē\ *adjective* **flax·i·er; -est** (1634)
: resembling flax especially in texture

flay \ˈflā\ *transitive verb* [Middle English *flen,* from Old English *flēan;* akin to Old Norse *flā* to flay, Lithuanian *plėšti* to tear] (before 12th century)
1 : to strip off the skin or surface of **:** SKIN
2 : to criticize harshly **:** EXCORIATE
3 : LASH 1b ⟨the wind whipped up to gale fury, *flaying* his face —Richard Kent⟩

F layer *noun* (1928)
: the highest and most densely ionized regular layer of the ionosphere occurring at night within the F region

flea \ˈflē\ *noun* [Middle English *fle,* from Old English *flēa;* akin to Old High German *flōh* flea] (before 12th century)
: any of an order (Siphonaptera) of small wingless bloodsucking insects that have a hard laterally compressed body and legs adapted to leaping and that feed on warm-blooded animals
— flea in one's ear : REBUKE ⟨sent him away with a *flea in his ear*⟩

flea·bag \ˈflē-ˌbag\ *noun* (1839)
: an inferior hotel or rooming house

flea·bane \-ˌbān\ *noun* (1548)
: any of various composite plants (as of the genus *Erigeron*) that were once believed to drive away fleas

flea beetle *noun* (1842)
: any of a subfamily (Alticinae, especially genera *Alticia* and *Epitrix*) of small chrysomelid beetles with legs adapted for leaping that feed on foliage and sometimes serve as vectors of virus diseases of plants

flea·bite \ˈflē-ˌbīt\ *noun* (circa 1570)
1 : the bite of a flea; *also* **:** the red spot caused by such a bite
2 : a trifling pain or annoyance

flea–bit·ten \-ˌbi-tᵊn\ *adjective* (1570)
1 *of a horse* **:** having a white or gray coat flecked with a darker color
2 : bitten by or infested with fleas

flea collar *noun* (1953)
: a collar for animals (as dogs and cats) that contains insecticide for killing fleas

flea–flick·er \-ˌfli-kər\ *noun* (1927)
: any of various deceptive football plays in which the ball is quickly transferred between players (as by a lateral) before or after a forward pass

flea·hop·per \-ˌhä-pər\ *noun* (1902)
: any of various small jumping bugs that feed on cultivated plants

flea market *noun* [translation of French *Marché aux Puces,* a market in Paris] (1922)
: a usually open-air market for secondhand articles and antiques

flea·pit \-ˌpit\ *noun* (1937)
British **:** a dilapidated building usually housing a movie theater

flea·wort \ˈflē-ˌwərt, -ˌwȯrt\ *noun* (before 12th century)
: any of three Old World plantains (especially *Plantago psyllium*) whose seeds are sometimes used as a mild laxative — compare PSYLLIUM SEED

flèche \ˈflāsh, ˈflesh\ *noun* [French, literally, arrow, from Old French *fleche,* of Germanic origin; akin to Middle Dutch *vlieke* arrow, Old English *flēogan* to fly] (1848)
: SPIRE; *especially* **:** a slender spire above the intersection of the nave and transepts of a church

flé·chette \flā-ˈshet, fle-\ *noun* [French, from diminutive of *flèche* arrow] (1915)
: a small dart-shaped projectile that is clustered in an explosive warhead, dropped as a missile from an airplane, or fired from a handheld gun

¹fleck \ˈflek\ *transitive verb* [back-formation from *flecked* spotted, from Middle English, probably from Old Norse *flekkōttr,* from *flekkr* spot] (14th century)
1 : STREAK, SPOT ⟨whitecaps *flecked* the blue sea⟩
2 : to color as if by sprinkling with flecks ⟨his wit is *flecked* with sarcasm —James Atlas⟩

²fleck *noun* (1598)
1 : SPOT, MARK ⟨a brown tweed with *flecks* of yellow⟩
2 : FLAKE, PARTICLE ⟨*flecks* of snow drifted down⟩

fledge \ˈflej\ *verb* **fledged; fledg·ing** [*fledge* capable of flying, from Middle English *flegge,* from Old English *-flycge;* akin to Old High German *flucki* capable of flying, Old English *flēogan* to fly — more at FLY] (1566)
intransitive verb
of a bird **:** to acquire the feathers necessary for flight or independent activity
transitive verb
1 : to rear until ready for flight or independent activity
2 : to cover with or as if with feathers or down
3 : to furnish (as an arrow) with feathers

fledg·ling \ˈflej-liŋ\ *noun* (1830)
1 : a young bird just fledged
2 : an immature or inexperienced person
3 : one that is new ⟨a *fledgling* company⟩

flee \ˈflē\ *verb* **fled** \ˈfled\; **flee·ing** [Middle English *flen,* from Old English *flēon;* akin to Old High German *fliohan* to flee] (before 12th century)
intransitive verb
1 a : to run away often from danger or evil **:** FLY **b :** to hurry toward a place of security
2 : to pass away swiftly **:** VANISH
transitive verb
: to run away from **:** SHUN

¹fleece \ˈflēs\ *noun* [Middle English *flees,* from Old English *flēos;* akin to Middle High German *vlius* fleece and perhaps to Latin *pluma* feather, down] (before 12th century)
1 a : the coat of wool covering a wool-bearing animal (as a sheep) **b :** the wool obtained from a sheep at one shearing
2 a : any of various soft or woolly coverings **b :** a soft bulky deep-piled knitted or woven fabric used chiefly for clothing

²fleece *transitive verb* **fleeced; fleec·ing** (1537)
1 a : to strip of money or property by fraud or extortion **b :** to charge excessively for goods or services
2 : to remove the fleece from **:** SHEAR
3 : to dot or cover with fleecy masses

fleeced \ˈflēst\ *adjective* (1580)
1 : covered with or as if with a fleece
2 *of a textile* **:** having a soft nap

fleech \ˈflēch\ *verb* [Middle English (Scots) *flechen*] (14th century)
dialect **:** COAX, WHEEDLE

fleecy \ˈflē-sē\ *adjective* **fleec·i·er; -est** (1590)
: covered with, made of, or resembling fleece ⟨a *fleecy* winter coat⟩

¹fleer \ˈflir\ *intransitive verb* [Middle English *fleryen,* of Scandinavian origin; akin to Norwegian *flire* to giggle] (15th century)
: to laugh or grimace in a coarse derisive manner **:** SNEER
synonym see SCOFF
— fleer·ing·ly \-iŋ-lē\ *adverb*

²fleer *noun* (1604)
: a word or look of derision or mockery

¹fleet \ˈflēt\ *verb* [Middle English *fleten,* from Old English *flēotan;* akin to Old High German *fliozzan* to float, Old English *flōwan* to flow] (before 12th century)
intransitive verb
1 *obsolete* **:** DRIFT
2 a *archaic* **:** FLOW **b :** to fade away **:** VANISH
3 [³*fleet*] **:** to fly swiftly
transitive verb
: to cause (time) to pass usually quickly or imperceptibly

²fleet *noun* [Middle English *flete,* from Old English *flēot* ship, from *flēotan*] (13th century)
1 : a number of warships under a single command; *specifically* **:** an organization of ships and aircraft under the command of a flag officer
2 : GROUP 2a, b; *especially* **:** a group (as of ships, planes, or trucks) operated under unified control

³fleet *adjective* [probably from ¹*fleet*] (circa 1529)
1 : swift in motion **:** NIMBLE

2 : EVANESCENT, FLEETING
synonym see FAST
— fleet·ly *adverb*
— fleet·ness *noun*

fleet admiral *noun* (1946)
: an admiral of the highest rank in the navy whose insignia is five stars

fleet–foot·ed \-ˌfu̇-təd\ *adjective* (circa 1743)
: able to run fast

fleet·ing *adjective* (1563)
: passing swiftly **:** TRANSITORY
synonym see TRANSIENT
— fleet·ing·ly \ˈflē-tiŋ-lē\ *adverb*
— fleet·ing·ness *noun*

Fleet Street \ˈflēt-\ *noun* [*Fleet Street,* London, England, center of the London newspaper district] (1882)
: the London press

flei·shig \ˈflā-shik\ *adjective* [Yiddish *fleyshik,* from Middle High German *vleischic* meaty, from *vleisch* flesh, meat, from Old High German *fleisk*] (1943)
: made of, prepared with, or used for meat or meat products — compare MILCHIG, PAREVE

Flem·ing \ˈfle-miŋ\ *noun* [Middle English, from Middle Dutch *Vlaminc* (akin to Middle Dutch *Vlander* Flanders)] (12th century)
: a member of the Germanic people inhabiting northern Belgium and a small section of northern France

[1]**Flem·ish** \ˈfle-mish\ *adjective* (14th century)
: of, relating to, or characteristic of Flanders or the Flemings or their language

[2]**Flemish** *noun* (circa 1741)
1 : the Germanic language of the Flemings that is made up of dialects of Dutch
2 *plural in construction* **:** FLEMINGS

Flemish giant *noun* (1898)
: any of a breed of very large solid-colored rabbits probably of Belgian origin

flense \ˈflen(t)s\ *transitive verb* **flensed; flens·ing** [Dutch *flensen* or Danish & Norwegian *flense*] (1820)
: to strip (as a whale) of blubber or skin

[1]**flesh** \ˈflesh\ *noun* [Middle English, from Old English *flǣsc;* akin to Old High German *fleisk* flesh and perhaps to Old Norse *flā* to flay — more at FLAY] (before 12th century)
1 a : the soft parts of the body of an animal and especially of a vertebrate; *especially* **:** the parts composed chiefly of skeletal muscle as distinguished from visceral structures, bone, and integuments **b :** sleek well-fatted condition of body **c :** SKIN
2 a : edible parts of an animal **b :** flesh of a mammal or fowl eaten as food
3 a : the physical nature of human beings ⟨the spirit indeed is willing, but the *flesh* is weak —Matthew 26:41 (Authorized Version)⟩ **b :** HUMAN NATURE
4 a : human beings **:** MANKIND **b :** living beings **c :** STOCK, KINDRED
5 : a fleshy plant part used as food; *also* **:** the fleshy part of a fruit
6 *Christian Science* **:** an illusion that matter has sensation
7 : SUBSTANCE ⟨insights buried in the *flesh* of the narrative —Jan Carew⟩
— in the flesh : in person and alive

[2]**flesh** (1530)
transitive verb
1 : to initiate or habituate especially by giving a foretaste
2 *archaic* **:** GRATIFY
3 : to clothe or cover with or as if with flesh; *broadly* **:** to give substance to — usually used with *out*
4 : to free from flesh
intransitive verb
: to become fleshy — often used with *up* or *out*

flesh and blood *noun* (before 12th century)
1 : corporeal nature as composed of flesh and of blood
2 : near kindred — used chiefly in the phrase *one's own flesh and blood*
3 : SUBSTANCE, REALITY

fleshed \ˈflesht\ *adjective* (15th century)
: having flesh especially of a specified kind — often used in combination ⟨pink-*fleshed*⟩ ⟨thick-*fleshed*⟩

flesh fly *noun* (14th century)
: a dipteran fly whose maggots feed on flesh; *especially* **:** any of a family (Sarcophagidae) of flies some of which cause myiasis

flesh·ings \ˈfle-shiŋz\ *noun plural* (1860)
: material removed in fleshing a hide

flesh·ly \ˈflesh-lē\ *adjective* (before 12th century)
1 a : CORPOREAL, BODILY **b :** of, relating to, or characterized by indulgence of bodily appetites; *especially* **:** LASCIVIOUS ⟨*fleshly* desires⟩ **c :** not spiritual **:** WORLDLY
2 : FLESHY, PLUMP
3 : having a sensuous quality ⟨*fleshly* art⟩
synonym see CARNAL

flesh·ment \ˈflesh-mənt\ *noun* [[2]*flesh*] (1605)
obsolete **:** excitement associated with a successful beginning

flesh out *transitive verb* (1886)
: to make fuller or more nearly complete ⟨museums *fleshing out* their collections with borrowed works⟩
— fleshed–out *adjective*

flesh·pot \ˈflesh-ˌpät\ *noun* (1592)
1 *plural* **:** bodily comfort **:** LUXURY
2 : a place of lascivious entertainment — usually used in plural

flesh wound *noun* (circa 1674)
: an injury involving penetration of the body musculature without damage to bones or internal organs

fleshy \ˈfle-shē\ *adjective* **flesh·i·er; -est** (14th century)
1 a : marked by, consisting of, or resembling flesh **b :** marked by abundant flesh; *especially* **:** CORPULENT
2 a : SUCCULENT, PULPY ⟨the *fleshy* texture of a melon⟩ **b :** not thin, dry, or membranous ⟨*fleshy* fungi⟩
— flesh·i·ness *noun*

fleshy fruit *noun* (1929)
: a fruit (as a berry, drupe, or pome) consisting largely of soft succulent tissue

fletch \ˈflech\ *transitive verb* [back-formation from *fletcher*] (circa 1656)
: FEATHER ⟨*fletch* an arrow⟩

fletch·er \ˈfle-chər\ *noun* [Middle English *fleccher,* from Old French *flechier,* from *fleche* arrow — more at FLÈCHE] (14th century)
: a maker of arrows

fletch·ing \-chiŋ\ *noun* (circa 1930)
: the feathers on an arrow; *also* **:** the arrangement of such feathers

fleur de coin \ˌflər-də-ˈkwä^{n}\ *adjective* [French *à fleur de coin,* literally, with the bloom of the die] (circa 1889)
: being in the preserved mint condition

fleur–de–lis *or* **fleur–de–lys** \ˌflər-d^{ə}l-ˈē, ˌflu̇r-\ *noun, plural* **fleurs–de–lis** *or* **fleur–de–lis** *or* **fleurs–de–lys** *or* **fleur–de–lys** \ˌflər-d^{ə}l-ˈē(z), ˌflu̇r-\ [Middle English *flourdelis,* from Middle French *flor de lis,* literally, lily flower] (14th century)
1 : IRIS 3
2 : a conventionalized iris in artistic design and heraldry

fleur-de-lis 2

fleu·ry \ˈflu̇r-ē\ *adjective* [alteration of Middle English *flory,* from Old French *floré,* from *flor* flower — more at FLOWER] (15th century)
of a heraldic cross **:** having the ends of the arms broadening out into the heads of fleurs-de-lis — see CROSS illustration

flew *past of* FLY

flews \ˈflüz\ *noun plural* [origin unknown] (1575)
: the pendulous lateral parts of a dog's upper lip — see DOG illustration

[1]**flex** \ˈfleks\ *verb* [Latin *flexus,* past participle of *flectere* to bend] (circa 1521)
transitive verb
1 : to bend especially repeatedly
2 a : to move muscles so as to cause flexion of (a joint) **b :** to move or tense (a muscle) by contraction
intransitive verb
: BEND
— flex one's muscles : to demonstrate one's strength ⟨an exaggerated need to *flex his* political *muscles* —J. P. Lash⟩

[2]**flex** *noun* [short for *flexible cord*] (1905)
chiefly British **:** an electric cord

[3]**flex** *noun* (circa 1934)
: an act or instance of flexing

flex·i·ble \ˈflek-sə-bəl\ *adjective* (15th century)
1 : capable of being flexed **:** PLIANT
2 : yielding to influence **:** TRACTABLE
3 : characterized by a ready capability to adapt to new, different, or changing requirements ⟨a *flexible* foreign policy⟩ ⟨a *flexible* schedule⟩
synonym see ELASTIC
— flex·i·bil·i·ty \ˌflek-sə-ˈbi-lə-tē\ *noun*
— flex·i·bly \ˈflek-sə-blē\ *adverb*

flex·ile \ˈflek-səl, -ˌsīl\ *adjective* (1633)
: FLEXIBLE

flex·ion \ˈflek-shən\ *noun* [Latin *flexion-, flexio,* from *flectere*] (1615)
1 : the act of flexing or bending
2 : a part bent **:** BEND
3 : INFLECTION 3
4 a : a bending movement around a joint in a limb (as the knee or elbow) that decreases the angle between the bones of the limb at the joint — compare EXTENSION 3b **b :** a forward raising of the arm or leg by a movement at the shoulder or hip joint

flex·og·ra·phy \flek-ˈsä-grə-fē\ *noun* [*flex*ible + *-o-* + *-graphy*] (1954)
: a process of rotary letterpress printing using flexible plates and fast-drying inks
— flexo·graph·ic \ˌflek-sə-ˈgra-fik\ *adjective*
— flexo·graph·i·cal·ly \-fi-k(ə-)lē\ *adverb*

flex·or \ˈflek-sər, -ˌsȯr\ *noun* (1615)
: a muscle serving to bend a body part (as a limb)

flex·time \ˈflek-ˌstīm\ *or* **flexi·time** \ˈflek-si-ˌtīm\ *noun* (1972)
: a system that allows employees to choose their own times for starting and finishing work within a broad range of available hours

flex·u·ous \ˈflek-sh(ə-)wəs\ *adjective* [Latin *flexuosus,* from *flexus* bend, from *flectere*] (1605)
1 : having curves, turns, or windings
2 : lithe or fluid in action or movement

flex·ur·al \ˈflek-sh(ə-)rəl\ *adjective* (1879)
1 : of, relating to, or resulting from flexure
2 : characterized by flexure

flex·ure \ˈflek-shər\ *noun* (1592)
1 : the quality or state of being flexed **:** FLEXION
2 : TURN, BEND, FOLD

fley \ˈflā\ *transitive verb* [Middle English *flayen,* from Old English *āflēgan,* from *ā-,* perfective prefix + *-flēgan* to put to flight] (13th century)
Scottish **:** FRIGHTEN

flib·ber·ti·gib·bet \ˌfli-bər-tē-ˈji-bət\ *noun* [Middle English *flepergebet*] (15th century)
: a silly flighty person
— flib·ber·ti·gib·bety \-bə-tē\ *adjective*

flic \ˈflēk\ *noun* [French] (1899)
: a French police officer

[1]flick \'flik\ *noun* [imitative] (15th century)
1 : a light sharp jerky stroke or movement
2 : a sound produced by a flick
3 : FLICKER 1

[2]flick (1816)
transitive verb
1 a : to move or propel with or as if with a flick ⟨*flicked* her hair back over her shoulder⟩ ⟨*flick* a switch⟩ **b :** to activate, deactivate, or change by or as if by flicking a switch ⟨*flick* on a cigarette lighter⟩ ⟨*flick* off the radio⟩
2 a : to strike lightly with a quick sharp motion ⟨*flicked* the horse with a whip⟩ **b :** to remove with light blows ⟨*flicked* an ash off her sleeve⟩
intransitive verb
1 : to go or pass quickly or abruptly ⟨a bird *flicked* by⟩ ⟨*flicking* through some papers⟩
2 : to direct flicks at something

[3]flick *noun* [short for [2]*flicker*] (1926)
: MOVIE

[1]flick·er \'fli-kər\ *verb* **flick·ered; flick·er·ing** \-k(ə-)riŋ\ [Middle English *flikeren,* from Old English *flicorian*] (before 12th century)
intransitive verb
1 : to move irregularly or unsteadily **:** FLUTTER
2 : to burn or shine fitfully or with a fluctuating light
3 : to appear briefly
transitive verb
1 : to cause to flicker
2 : to produce by flickering
— flick·er·ing·ly \-k(ə-)riŋ-lē\ *adverb*

[2]flicker *noun* (1849)
1 a : an act of flickering **b :** a sudden brief movement **c :** a momentary quickening ⟨a *flicker* of anger⟩ **d :** a slight indication **:** HINT ⟨a *flicker* of recognition⟩
2 : a wavering light
3 : MOVIE — often used in plural
— flick·ery \'fli-k(ə-)rē\ *adjective*

[3]flicker *noun* [probably imitative of its call] (1809)
: a large barred and spotted North American woodpecker (*Colaptes auratus*) with a brown back that commonly forages on the ground for ants — compare RED-SHAFTED FLICKER, YELLOW-SHAFTED FLICKER

flick–knife \'flik-ˌnīf\ *noun* (1957)
British **:** SWITCHBLADE

flied *past and past participle of* [3]FLY

fli·er \'flī(-ə)r\ *noun* (15th century)
1 : one that flies; *specifically* **:** AIRMAN
2 : a reckless or speculative venture — usually used in the phrase *take a flier*
3 *usually* **fly·er :** an advertising circular

[1]flight \'flīt\ *noun, often attributive* [Middle English, from Old English *flyht;* akin to Middle Dutch *vlucht* flight, Old English *flēogan* to fly] (before 12th century)
1 a : an act or instance of passing through the air by the use of wings ⟨the *flight* of a bee⟩ **b :** the ability to fly ⟨*flight* is natural to birds⟩
2 a : a passing through the air or through space outside the earth's atmosphere ⟨*flight* of an arrow⟩ ⟨*flight* of a rocket to the moon⟩ **b :** the distance covered in such a flight **c :** swift movement
3 a : a trip made by or in an airplane or spacecraft **b :** a scheduled airplane trip
4 : a group of similar beings or objects flying through the air together
5 : a brilliant, imaginative, or unrestrained exercise or display ⟨a *flight* of fancy⟩
6 a : a continuous series of stairs from one landing or floor to another **b :** a series (as of terraces or conveyors) resembling a flight of stairs
7 : a unit of the U.S. Air Force below a squadron
— flight·less \-ləs\ *adjective*

[2]flight (1873)
intransitive verb
: to rise, settle, or fly in a flock ⟨geese *flighting* on the marsh⟩
transitive verb
: [1]FLUSH

[3]flight *noun* [Middle English *fluht, fliht;* akin to Old High German *fluht* flight, Old English *flēon* to flee] (13th century)
: an act or instance of running away

flight attendant *noun* (1947)
: a person who attends passengers on an airplane

flight bag *noun* [[1]*flight*] (1943)
1 : a lightweight traveling bag with zippered outside pockets
2 : a small canvas satchel

flight deck *noun* (1924)
1 : the uppermost complete deck of an aircraft carrier
2 : the forward compartment in some airplanes

flight engineer *noun* (1938)
: a flight crewman responsible for mechanical operation

flight feather *noun* (1735)
: one of the quills of a bird's wing or tail that support it in flight — compare CONTOUR FEATHER

flight lieutenant *noun* (1914)
: a commissioned officer in the British air force who ranks with a captain in the army

flight line *noun* (1943)
: a parking and servicing area for airplanes

flight path *noun* (1911)
: the path in the air or space made or followed by something (as a particle, an airplane, or a spacecraft) in flight

flight pay *noun* (1928)
: an additional allowance paid to military personnel who take part in regular authorized aircraft flights

flight plan *noun* (circa 1936)
: a usually written statement (as by a pilot) of the details of an intended flight (as of an airplane or spacecraft) usually filed with an authority

flight recorder *noun* (1939)
: a crashworthy instrument for recording flight data (as airspeed and altitude)

flight suit *noun* (1944)
: a usually one-piece garment especially of fire-resistant fabric worn especially by military aircrews

flight surgeon *noun* (1925)
: a military medical officer trained in aerospace medicine

flight–test \'flīt-ˌtest\ *transitive verb* (1930)
: to test (as an airplane or spacecraft) in flight

flighty \'flī-tē\ *adjective* **flight·i·er; -est** (1552)
1 : SWIFT
2 : lacking stability or steadiness: **a :** easily upset **:** VOLATILE ⟨a *flighty* temper⟩ **b :** easily excited **:** SKITTISH ⟨a *flighty* horse⟩ **c :** CAPRICIOUS, SILLY
— flight·i·ly \'flī-tᵊl-ē\ *adverb*
— flight·i·ness \'flī-tē-nəs\ *noun*

[1]flim–flam \'flim-ˌflam\ *noun* [probably of Scandinavian origin; akin to Old Norse *flim* mockery] (circa 1538)
1 : DECEPTION, FRAUD
2 : deceptive nonsense

[2]flimflam *transitive verb* **flim·flammed; flim·flam·ming** (1660)
: to subject to a flimflam
— flim·flam·mer *noun*
— flim·flam·mery \-ˌfla-mə-rē\ *noun*

[1]flim·sy \'flim-zē\ *adjective* **flim·si·er; -est** [perhaps alteration of [1]*film* + *-sy* (as in *tricksy*)] (1702)
1 a : lacking in physical strength or substance ⟨*flimsy* silks⟩ **b :** of inferior materials and workmanship
2 : having little worth or plausibility
— flim·si·ly \-zə-lē\ *adverb*
— flim·si·ness \-zē-nəs\ *noun*

[2]flimsy *noun, plural* **flim·sies** (1857)
chiefly British **:** a lightweight paper used especially for multiple copies; *also* **:** a document printed on flimsy

flinch \'flinch\ *intransitive verb* [Middle French *flenchir* to bend, of Germanic origin; akin to Middle High German *lenken* to bend, Old High German *hlanca* flank — more at LANK] (1579)
: to withdraw or shrink from or as if from pain **:** WINCE; *also* **:** to tense the muscles involuntarily in anticipation of discomfort
synonym see RECOIL
— flinch *noun*
— flinch·er *noun*

flin·ders \'flin-dərz\ *noun plural* [Middle English *flendris*] (15th century)
: SPLINTERS, FRAGMENTS

[1]fling \'fliŋ\ *verb* **flung** \'fləŋ\; **fling·ing** \'fliŋ-iŋ\ [Middle English, perhaps of Scandinavian origin; akin to Old Norse *flengja* to whip] (14th century)
intransitive verb
1 : to move in a brusque or headlong manner ⟨*flung* out of the room in a rage⟩
2 *of an animal* **:** to kick or plunge vigorously
3 *Scottish* **:** CAPER
transitive verb
1 a : to throw forcefully, impetuously, or casually ⟨*flung* herself down on the sofa⟩ ⟨clothes were *flung* on the floor⟩ **b :** to cast as if by throwing ⟨*flung* off all restraint⟩
2 : to place or send suddenly and unceremoniously ⟨was arrested and *flung* into prison⟩
3 : to give unrestrainedly ⟨*flung* himself into music⟩
synonym see THROW
— fling·er \'fliŋ-ər\ *noun*

[2]fling *noun* (1589)
1 : an act or instance of flinging
2 a : a casual try or involvement **b :** a casual or brief love affair
3 : a period devoted to self-indulgence

flint \'flint\ *noun* [Middle English, from Old English; akin to Old High German *flins* pebble, hard stone] (before 12th century)
1 : a massive hard quartz that produces a spark when struck by steel
2 : an implement of flint used in prehistoric cultures
3 a : a piece of flint **b :** a material used for producing a spark; *especially* **:** an alloy (as of iron and cerium) used in lighters
4 : something resembling flint in hardness
— flint·like \-ˌlīk\ *adjective*

flint corn *noun* (1705)
: an Indian corn (*Zea mays indurata*) having hard horny usually rounded kernels with the soft endosperm enclosed by a hard outer layer

flint glass *noun* (1683)
: heavy brilliant glass that contains lead oxide, has a relatively high index of refraction, and is used in lenses and prisms

flint·lock \'flint-ˌläk\ *noun* (1683)
1 : a lock for a gun or pistol having a flint in the hammer for striking a spark to ignite the charge
2 : a firearm fitted with a flintlock

flinty \'flin-tē\ *adjective* **flint·i·er; -est** (1536)
1 : resembling flint; *especially* **:** STERN, UNYIELDING
2 : composed of or covered with flint
— flint·i·ly \'flin-tᵊl-ē\ *adverb*
— flint·i·ness \'flin-tē-nəs\ *noun*

[1]flip \'flip\ *verb* **flipped; flip·ping** [probably imitative] (1616)
transitive verb
1 : to toss so as to cause to turn over in the air ⟨*flip* a coin⟩; *also* **:** TOSS ⟨*flip* me the ball⟩ ⟨*flip* one end of the scarf over your shoulder⟩
2 a : to cause to turn and especially to turn over ⟨*flipped* the car⟩ ⟨*flipping* the pages of⟩ **b :** to move with a small quick motion ⟨*flip* a switch⟩
intransitive verb

1 : to make a twitching or flicking movement ⟨the fish *flipped* and flopped on the deck⟩; *also* **:** to change from one position to another and especially turn over ⟨the car *flipped*⟩
2 : LEAF 2 ⟨*flipped* through the pages⟩
3 *slang* **a :** to lose one's mind or composure — often used with *out* **b :** to become very enthusiastic
²flip *noun* (1695)
1 : a mixed drink usually consisting of a sweetened spiced liquor with beaten eggs
2 : an act or instance of flipping
3 : the motion used in flipping
4 : a somersault especially in the air
³flip *adjective* (circa 1847)
: FLIPPANT, IMPERTINENT
flip–flop \ˈflip-ˌfläp\ *noun* (1600)
1 : the sound or motion of something flapping loosely
2 a : a backward handspring **b :** a sudden reversal (as of direction or point of view)
3 : a usually electronic device or a circuit (as in a computer) capable of assuming either of two stable states
4 : a rubber sandal loosely fastened to the foot by a thong
— flip–flop *intransitive verb*
flip·pan·cy \ˈfli-pən(t)-sē\ *noun, plural* **-cies** (1746)
: unbecoming levity or pertness especially in respect to grave or sacred matters
flip·pant \ˈfli-pənt\ *adjective* [probably from ¹*flip*] (1605)
1 *archaic* **:** GLIB, TALKATIVE
2 : lacking proper respect or seriousness
— flip·pant·ly *adverb*
flip·per \ˈfli-pər\ *noun* (1822)
1 a : a broad flat limb (as of a seal or cetacean) adapted for swimming **b :** a flat rubber shoe with the front expanded into a paddle used in skin diving
2 : one that flips
flip·py \ˈfli-pē\ *adjective* (1967)
: loose and flaring at the bottom ⟨a *flippy* skirt⟩
flip side *noun* (1949)
1 : the reverse and usually less popular side of a phonograph record
2 : a reverse or opposite side, aspect, or result ⟨the *flip side* of deficient saving . . . is overconsumption —R. S. Gay⟩
¹flirt \ˈflərt\ *verb* [origin unknown] (1583)
transitive verb
1 : FLICK
2 : to move in a jerky manner
intransitive verb
1 : to move erratically **:** FLIT
2 a : to behave amorously without serious intent **b :** to show superficial or casual interest or liking ⟨*flirted* with the idea⟩; *also* **:** EXPERIMENT ⟨a novelist *flirting* with poetry⟩
3 : to come close to — used with *with* ⟨the temperature *flirted* with 100°⟩
synonym see TRIFLE
— flir·ta·tion \ˌflər-ˈtā-shən\ *noun*
— flirt·er *noun*
— flirty \ˈflər-tē\ *adjective*
²flirt *noun* (circa 1590)
1 : an act or instance of flirting
2 : a person who flirts
flir·ta·tious \ˌflər-ˈtā-shəs\ *adjective* (1834)
: inclined to flirt **:** COQUETTISH
— flir·ta·tious·ly *adverb*
— flir·ta·tious·ness *noun*
flit \ˈflit\ *intransitive verb* **flit·ted; flit·ting** [Middle English *flitten,* of Scandinavian origin; akin to Old Norse *flytjask* to move, Old English *flēotan* to float] (13th century)
1 : to pass quickly or abruptly from one place or condition to another
2 *archaic* **:** ALTER, SHIFT
3 : to move in an erratic fluttering manner
— flit *noun*
flitch \ˈflich\ *noun* [Middle English *flicche,* from Old English *flicce;* akin to Old High German *fleisk* flesh — more at FLESH] (before 12th century)
1 : a side of cured meat; *especially* **:** a side of bacon
2 a : a longitudinal section of a log **b :** a bundle of sheets of veneer laid together in sequence
¹flit·ter \ˈfli-tər\ *intransitive verb* [frequentative of *flit*] (1534)
: FLUTTER, FLICKER
²flitter *noun* (1554)
: one that flits
fliv·ver \ˈfli-vər\ *noun* [origin unknown] (1910)
: a small cheap usually old automobile
¹float \ˈflōt\ *noun* [Middle English *flote* boat, float, from Old English *flota* ship; akin to Old High German *flōz* raft, stream, Old English *flēotan* to float — more at FLEET] (before 12th century)
1 : an act or instance of floating
2 : something that floats in or on the surface of a fluid: as **a :** a device (as a cork) buoying up the baited end of a fishing line **b :** a floating platform anchored near a shoreline for use by swimmers or boats **c :** a hollow ball that floats at the end of a lever in a cistern, tank, or boiler and regulates the liquid level **d :** a sac containing air or gas and buoying up the body of a plant or animal — compare PNEUMATOPHORE 1 **e :** a watertight structure giving an airplane buoyancy on water
3 : a tool or apparatus for smoothing a surface (as of wet concrete)
4 : a government grant of a fixed amount of land not yet located by survey out of a larger specific tract
5 a : a vehicle with a platform used to carry an exhibit in a parade **b :** the vehicle and exhibit together
6 a : an amount of money represented by checks outstanding and in process of collection **b :** the time between a transaction (as the writing of a check or a purchase on credit) and the actual withdrawal of funds to cover it **c :** the volume of a company's shares available for active trading in the auction market
7 : a soft drink with ice cream floating in it
— floaty \ˈflō-tē\ *adjective*
²float (before 12th century)
intransitive verb
1 : to rest on the surface of or be suspended in a fluid
2 a : to drift on or through or as if on or through a fluid ⟨yellow leaves *floated* down⟩ **b :** WANDER
3 *of a currency* **:** to find a level in the international exchange market in response to the law of supply and demand and without any restrictive effect of artificial support or control
transitive verb
1 a : to cause to float in or on the surface of a fluid **b :** to cause to float as if in a fluid
2 : FLOOD ⟨*float* a cranberry bog⟩
3 : to smooth (as plaster or cement) with a float
4 a : to put forth (as a proposal) for acceptance **b :** to place (an issue of securities) on the market **c :** to obtain money for the establishment or development of (an enterprise) by issuing and selling securities **d :** NEGOTIATE ⟨*float* a loan⟩
floa·ta·tion *variant of* FLOTATION
float·er \ˈflō-tər\ *noun* (1717)
1 a : one that floats **b :** a person who floats something
2 : a person who votes illegally in various polling places
3 a : a person without a permanent residence or regular employment **b :** a worker who moves from job to job; *especially* **:** one without fixed duties
4 : a policy insuring specific items of personal property (as jewelry or art)
float glass *noun* (1959)
: flat glass produced by solidifying molten glass on the surface of a bath of molten tin
float·ing *adjective* (1600)
1 : buoyed on or in a fluid
2 : located out of the normal position ⟨a *floating* kidney⟩
3 a : continually drifting or changing position ⟨the *floating* population⟩ **b :** not presently committed or invested ⟨*floating* capital⟩ **c :** short-term and usually not funded ⟨*floating* debt⟩ **d :** having no fixed value or rate ⟨*floating* currencies⟩ ⟨*floating* interest rates⟩
4 : connected or constructed so as to operate and adjust smoothly ⟨a *floating* axle⟩
floating dock *noun* (1866)
: a dock that floats on the water and can be partly submerged to permit entry of a ship and raised to keep the ship high and dry — called also *floating drydock*
floating island *noun* (1771)
: a dessert consisting of custard with floating masses of beaten egg whites
floating–point *adjective* (1948)
: expressed in, using, or being a mathematical notation in which a number is represented (as in a computer display) by an integer or a decimal fraction multiplied by a power of the number base indicated by an exponent (as in 4.52E2 for 452) — compare FIXED-POINT
floating rib *noun* (1831)
: a rib (as one of either of the last two pairs in humans) that has no attachment to the sternum — compare FALSE RIB
float·plane \ˈflōt-ˌplān\ *noun* (1922)
: a seaplane supported on the water by one or more floats
floc \ˈfläk\ *noun* [short for *floccule*] (1921)
: a flocculent mass formed by the aggregation of a number of fine suspended particles
floc·cu·late \ˈflä-kyə-ˌlāt\ *verb* **-lat·ed; -lat·ing** (1877)
transitive verb
: to cause to aggregate into a flocculent mass ⟨*flocculate* clay⟩
intransitive verb
: to become flocculent
— floc·cu·lant \-lənt\ *noun*
— floc·cu·la·tion \ˌflä-kyə-ˈlā-shən\ *noun*
— floc·cu·la·tor \ˈflä-kyə-ˌlā-tər\ *noun*
floc·cule \ˈflä-(ˌ)kyü(ə)l\ *noun* [Late Latin *flocculus*] (circa 1846)
: FLOC
floc·cu·lent \-lənt\ *adjective* [Latin *floccus* + English *-ulent*] (1800)
1 : resembling wool especially in loose fluffy organization
2 : made up of flocs or floccules ⟨a *flocculent* precipitate⟩
floc·cu·lus \-ləs\ *noun, plural* **-li** \-ˌlī, -ˌlē\ [Late Latin, diminutive of Latin *floccus* tuft of wool] (1799)
1 : a small loosely aggregated mass
2 : a bright or dark patch on the sun
¹flock \ˈfläk\ *noun* [Middle English, from Old English *flocc* crowd, band; akin to Old Norse *flokkr* crowd, band] (13th century)
1 : a group of birds or mammals assembled or herded together
2 : a group under the guidance of a leader; *especially* **:** a church congregation
3 : a large number ⟨a *flock* of tourists⟩
²flock *intransitive verb* (14th century)
: to gather or move in a flock ⟨they *flocked* to the beach⟩
³flock *noun* [Middle English *flok,* from Old French *floc,* from Latin *floccus*] (13th century)
1 : a tuft of wool or cotton fiber
2 : woolen or cotton refuse used for stuffing furniture and mattresses
3 : very short or pulverized fiber used especially to form a velvety pattern on cloth or paper or a protective covering on metal
4 : FLOC

[4]**flock** *transitive verb* (1530)
1 : to fill with flock
2 : to decorate with flock

flock·ing \ˈflä-kiŋ\ *noun* (circa 1874)
: a design in flock

floe \ˈflō\ *noun* [probably from Norwegian *flo* flat layer] (1817)
1 : floating ice formed in a large sheet on the surface of a body of water
2 : ICE FLOE

flog \ˈfläg\ *verb* **flogged; flog·ging** [perhaps modification of Latin *flagellare* to whip — more at FLAGELLATE] (circa 1676)
transitive verb
1 a : to beat with or as if with a rod or whip **b :** to criticize harshly
2 : to force or urge into action **:** DRIVE
3 a *chiefly British* **:** to sell (as stolen goods) illegally ⟨*flogged* their employers' petrol to ordinary motorists —*Economist*⟩ **b :** SELL 7 ⟨traveled by horse, *flogging* encyclopedias —Robert Darnton⟩ **c :** to promote aggressively **:** PLUG ⟨flying around the world *flogging* your movies —Peter Bogdanovich⟩
4 *British* **:** STEAL 1
intransitive verb
1 : FLAP, FLUTTER ⟨sails *flogging*⟩
2 *British* **:** to move along with difficulty **:** SLOG
— flog·ger *noun*

flo·ka·ti \flō-ˈkä-tē\ *noun* [New Greek *phlokatē*] (1967)
: a hand-woven Greek woolen rug with a thick shaggy pile

[1]**flood** \ˈfləd\ *noun* [Middle English, from Old English *flōd;* akin to Old High German *fluot* flood, Old English *flōwan* to flow] (before 12th century)
1 a : a rising and overflowing of a body of water especially onto normally dry land; *also* **:** a condition of overflowing ⟨rivers in *flood*⟩ **b** *capitalized* **:** a flood described in the Bible as covering the earth in the time of Noah
2 : the flowing in of the tide
3 : an overwhelming quantity or volume; *also* **:** a state of abundant flow or volume ⟨a debate in full *flood*⟩
4 : FLOODLIGHT

[2]**flood** (1663)
transitive verb
1 : to cover with a flood **:** INUNDATE
2 a : to fill abundantly or excessively ⟨*flood* the market⟩ **b :** to supply to (the carburetor of an internal combustion engine) an excess of fuel so that engine operation is hampered
intransitive verb
1 : to pour forth, go, or come in a flood
2 : to become filled with a flood
— flood·er *noun*

flood·gate \ˈfləd-ˌgāt\ *noun* [Middle English *flodgate*] (13th century)
1 : a gate for shutting out, admitting, or releasing a body of water **:** SLUICE
2 : something serving to restrain an outburst ⟨opened the *floodgates*⟩

[1]**flood·light** \-ˌlīt\ *transitive verb* **-lit** \-ˌlit\ *also* **-light·ed; -light·ing** (1923)
: to illuminate by means of one or more floodlights

[2]**floodlight** *noun* (1924)
1 a : artificial illumination in a broad beam **b :** a source of such illumination
2 : a lighting unit for projecting a beam of light

flood·plain \ˈfləd-ˌplān\ *noun* (1873)
1 : level land that may be submerged by floodwaters
2 : a plain built up by stream deposition

flood tide *noun* (1719)
1 : a rising tide
2 a : an overwhelming quantity **b :** a high point **:** PEAK

flood·wa·ter \-ˌwȯ-tər, -ˌwä-\ *noun* (1791)
: the water of a flood

flood·way \-ˌwā\ *noun* (1928)
: a channel for diverting floodwaters

floo·ey \ˈflü-ē\ *adjective* [origin unknown] (1920)
: AWRY, ASKEW ⟨go *flooey*⟩

[1]**floor** \ˈflōr, ˈflȯr\ *noun, often attributive* [Middle English *flor,* from Old English *flōr;* akin to Old High German *fluor* meadow, Latin *planus* level, and perhaps to Greek *planasthai* to wander] (before 12th century)
1 : the level base of a room
2 a : the lower inside surface of a hollow structure (as a cave or bodily part) **b :** a ground surface ⟨the ocean *floor*⟩
3 a : a structure dividing a building into stories; *also* **:** STORY **b :** the occupants of such a floor
4 : the surface of a structure on which one travels ⟨the *floor* of a bridge⟩
5 a : a main level space (as in a stock exchange or legislative chamber) distinguished from a platform or gallery **b :** the members of an assembly ⟨took questions from the *floor*⟩ **c :** the right to address an assembly ⟨the senator from Utah has the *floor*⟩
6 : a lower limit **:** BASE
— floored *adjective*

[2]**floor** *transitive verb* (15th century)
1 : to cover with a floor or flooring
2 a : to knock or bring down **b :** FLABBERGAST, DUMBFOUND
3 : to press (the accelerator of a vehicle) to the floorboard; *also* **:** to accelerate rapidly ⟨*floored* the van⟩
— floor·er *noun*

floor·board \ˈflōr-ˌbōrd, ˈflȯr-ˌbȯrd\ *noun* (1881)
1 : a board in a floor
2 : the floor of an automobile

floor·cloth \-ˌklȯth\ *noun, plural* **-cloths** \-ˌklȯt͟hz, -ˌklȯths\ (1746)
: a usually decorated heavy cloth (as of canvas) used for a floor covering

floor exercise *noun* (1961)
: an event in gymnastics competition consisting of various ballet and tumbling movements (as jumps, somersaults, and handstands) performed without apparatus

floor·ing \ˈflōr-iŋ, ˈflȯr-\ *noun* (1624)
1 : FLOOR, BASE
2 : material for floors

floor lamp *noun* (1892)
: a tall lamp that stands on the floor

floor leader *noun* (1899)
: a member of a legislative body chosen by a party to have charge of its organization and strategy on the floor

floor–length *adjective* (1939)
: reaching to the floor ⟨a *floor-length* gown⟩

floor manager *noun* (1887)
: a person who directs something from the floor (as of a nominating convention)

floor show *noun* (1927)
: a series of acts presented in a nightclub

floor–through \ˈflōr-ˌthrü, ˈflȯr-\ *noun* (1967)
: an apartment that occupies an entire floor of a building

floor·walk·er \ˈflōr-ˌwȯ-kər, ˈflȯr-\ *noun* (1876)
: a person employed in a retail store to oversee the salespeople and aid customers

floo·zy *or* **floo·zie** \ˈflü-zē\ *noun, plural* **floozies** [origin unknown] (1911)
: a usually young woman of loose morals

[1]**flop** \ˈfläp\ *verb* **flopped; flop·ping** [alteration of [2]*flap*] (1602)
intransitive verb
1 : to swing or move loosely **:** FLAP
2 : to throw or move oneself in a heavy, clumsy, or relaxed manner ⟨*flopped* into the chair⟩
3 : to change or turn suddenly
4 : to go to bed ⟨a place to *flop* at night⟩
5 : to fail completely ⟨the play *flopped*⟩
transitive verb
: to move or drop heavily or noisily **:** cause to flop ⟨*flopped* the bundles down⟩ ⟨*flop* the rag over⟩
— flop·per *noun*

[2]**flop** *adverb* (1728)
: RIGHT, SQUARELY ⟨fell *flop* on my face⟩

[3]**flop** *noun* (1823)
1 : an act or sound of flopping
2 : a complete failure
3 *slang* **:** a place to sleep; *especially* **:** FLOPHOUSE
4 : DUNG ⟨cow *flop*⟩; *also* **:** a piece of dung

flop·house \ˈfläp-ˌhau̇s\ *noun* (1916)
: a cheap rooming house or hotel

[1]**flop·py** \ˈflä-pē\ *adjective* **flop·pi·er; -est** (1858)
: tending to flop; *especially* **:** being both soft and flexible
— flop·pi·ly \ˈflä-pə-lē\ *adverb*
— flop·pi·ness \ˈflä-pē-nəs\ *noun*

[2]**floppy** *noun, plural* **-pies** (1974)
: FLOPPY DISK

floppy disk *noun* (1972)
: a small flexible plastic disk coated with magnetic material on which data for a computer can be stored

flo·ra \ˈflōr-ə, ˈflȯr-\ *noun, plural* **floras** *also* **flo·rae** \ˈflōr-ˌē, ˈflȯr-, -ˌī\ [New Latin, from Latin *Flora,* Roman goddess of flowers, from Latin *flor-, flos*] (1777)
1 : a treatise on or list of the plants of an area or period
2 : plant or bacterial life; *especially* **:** such life characteristic of a region, period, or special environment ⟨fossil *flora*⟩ ⟨intestinal *flora*⟩ — compare FAUNA

[1]**flo·ral** \ˈflōr-əl, ˈflȯr-\ *adjective* [Latin *flor-, flos* flower — more at BLOW] (1753)
: of or relating to flowers or a flora

[2]**floral** *noun* (1897)
: a design, pattern, or picture in which flowers predominate

floral envelope *noun* (circa 1829)
: PERIANTH

Flor·ence flask \ˈflȯr-ən(t)s-, ˈflär-\ *noun* [*Florence,* Italy; from the use of flasks of this shape for certain Italian wines] (1744)
: a round usually flat-bottomed laboratory vessel with a long neck

Flor·en·tine \ˈflȯr-ən-ˌtēn, ˈflär-, -ˌtīn\ *adjective* [Medieval Latin *Florentinus*] (1597)
1 a : of or relating to Florence, Italy **b :** MACHIAVELLIAN ⟨*Florentine* politics⟩
2 : served or dressed with spinach ⟨poached eggs *Florentine*⟩
3 : having a matte brushed finish ⟨*Florentine* gold⟩

flo·res·cence \flȯ-ˈre-sᵊn(t)s, flə-\ *noun* [New Latin *florescentia,* from Latin *florescent-, florescens,* present participle of *florescere,* inchoative of *florēre* to blossom, flourish — more at FLOURISH] (1793)
: a state or period of flourishing
— flo·res·cent \-sᵊnt\ *adjective*

flo·ret \ˈflōr-ət, ˈflȯr-\ *noun* [Middle English *flourette,* from Middle French *flouret,* diminutive of *flour* flower] (1671)
1 : a small flower; *especially* **:** one of the small flowers forming the head of a composite plant
2 : a cluster of flower buds separated from a head especially when used as food ⟨cauliflower *florets*⟩

flori- *combining form* [Latin, from *flor-, flos*]
: flower or flowers ⟨*floriculture*⟩

flo·ri·at·ed \ˈflōr-ē-ˌā-təd, ˈflȯr-\ *adjective* (1845)
: having floral ornaments or a floral form
— flo·ri·a·tion \ˌflōr-ē-ˈā-shən, ˌflȯr-\ *noun*

flo·ri·bun·da \ˌflōr-ə-ˈbən-də, ˌflȯr-\ *noun* [New Latin, feminine of *floribundus* flowering freely] (1898)
: any of various bush roses with large flowers in open clusters that derive from crosses of polyantha and tea roses

flo·ri·cul·ture \ˈflōr-ə-ˌkəl-chər, ˈflȯr-\ *noun* (1822)
: the cultivation and management of ornamental and especially flowering plants

— **flo·ri·cul·tur·al** \ˌflōr-ə-ˈkəlch-rəl, -ˈkəl-chə-, ˌflȯr-\ *adjective*
— **flo·ri·cul·tur·ist** \-rist\ *noun*
flor·id \ˈflȯr-əd, ˈflär-\ *adjective* [Latin *floridus* blooming, flowery, from *florēre*] (circa 1656)
1 a *obsolete* **:** covered with flowers **b :** very flowery in style **:** ORNATE ⟨*florid* prose⟩ ⟨*florid* declamations⟩; *also* **:** having a florid style ⟨a *florid* writer⟩ **c :** elaborately decorated ⟨a *florid* interior⟩
2 a : tinged with red **:** RUDDY ⟨a *florid* complexion⟩ **b :** marked by emotional or sexual fervor ⟨a *florid* secret life⟩ ⟨a *florid* sensibility⟩
3 *archaic* **:** HEALTHY
4 : fully developed **:** manifesting a complete and typical clinical syndrome ⟨the *florid* stage of a disease⟩
— **flo·rid·i·ty** \flə-ˈri-də-tē, flȯ-\ *noun*
— **flor·id·ly** \ˈflȯr-əd-lē, ˈflär-\ *adverb*
— **flor·id·ness** \-nəs\ *noun*
flo·rif·er·ous \flȯ-ˈri-f(ə-)rəs\ *adjective* [Latin *florifer,* from *flori-* + *-fer* fer] (1678)
: bearing flowers; *especially* **:** blooming freely
— **flo·rif·er·ous·ness** *noun*
flo·ri·gen \ˈflōr-ə-jən, ˈflȯr-\ *noun* [International Scientific Vocabulary] (1936)
: a hormone or hormonal agent that promotes flowering
— **flo·ri·gen·ic** \ˌflōr-ə-ˈje-nik, ˌflȯr-\ *adjective*
flo·ri·le·gium \ˌflōr-ə-ˈlē-j(ē-)əm, ˌflȯr-\ *noun, plural* **-gia** \-j(ē-)ə\ [New Latin, from Latin *florilegus* culling flowers, from *flori-* + *legere* to gather — more at LEGEND] (1647)
: a volume of writings **:** ANTHOLOGY
flo·rin \ˈflȯr-ən, ˈflär-, ˈflōr-\ *noun* [Middle English, from Middle French, from Old Italian *fiorino,* from *fiore* flower, from Latin *flor-, flos;* from the lily on the coins] (14th century)
1 a : an old gold coin first struck at Florence in 1252 **b :** any of various European gold coins patterned after the Florentine florin
2 a : a British silver coin worth two shillings **b :** any of several similar coins issued in Commonwealth countries
3 : GULDEN
4 : FORINT
flo·rist \ˈflōr-ist, ˈflȯr-, ˈflär-\ *noun* (1623)
: one who sells or grows for sale flowers and ornamental plants
— **flo·rist·ry** \-ə-strē\ *noun*
flo·ris·tic \flō-ˈris-tik\ *adjective* (1898)
: of or relating to flowers, a flora, or the phytogeographical study of plants and plant groups
— **flo·ris·ti·cal·ly** \-ti-k(ə-)lē\ *adverb*
flo·ru·it \ˈflȯr-(y)ə-wət, ˈflär-\ *noun* [Latin, he flourished, from *florēre* to flourish] (1843)
: a period of flourishing (as of a person or movement)
¹floss \ˈfläs, ˈflȯs\ *noun* [probably modification of French *floche* soft, weak (of silk fiber), from Gascon, from Latin *fluxus,* literally, loose, flowing, past participle of *fluere* to flow — more at FLUID] (1759)
1 a : soft thread of silk or mercerized cotton for embroidery **b :** DENTAL FLOSS
2 : fluffy fibrous material
²floss (1974)
transitive verb
: to use dental floss on
intransitive verb
: to use dental floss
flossy \ˈflä-sē, ˈflȯ-\ *adjective* **floss·i·er; -est** (1839)
1 : of, relating to, or having the characteristics of floss
2 : stylish or glamorous especially at first impression ⟨*flossy* new hotels⟩
— **floss·i·ly** \ˈflä-sə-lē\ *adverb*
flo·ta \ˈflō-tə\ *noun* [Spanish] (1527)
: a fleet of Spanish ships
flo·ta·tion \flō-ˈtā-shən\ *noun* [²*float*] (1806)
1 : the act, process, or state of floating
2 : an act or instance of financing (as an issue of stock)
3 : the separation of the particles of a mass of pulverized ore according to their relative capacity for floating on a given liquid; *also* **:** any of various similar processes involving the relative capacity of materials for floating
4 : the ability (as of a tire or snowshoes) to stay on the surface of soft ground or snow
flo·til·la \flō-ˈti-lə\ *noun* [Spanish, diminutive of *flota* fleet, from Old French *flote,* from Old Norse *floti;* akin to Old English *flota* ship, fleet — more at FLOAT] (1711)
1 : a fleet of ships or boats; *especially* **:** a navy organizational unit consisting of two or more squadrons of small warships
2 : an indefinite large number ⟨a *flotilla* of changes⟩
flot·sam \ˈflät-səm\ *noun* [Anglo-French *floteson,* from Old French *floter* to float, of Germanic origin; akin to Old English *flotian* to float, *flota* ship] (circa 1607)
1 : floating wreckage of a ship or its cargo; *broadly* **:** floating debris
2 a : a floating population (as of emigrants or castaways) **b :** an accumulation of miscellaneous or unimportant stuff
¹flounce \ˈflaůn(t)s\ *intransitive verb* **flounced; flounc·ing** [perhaps of Scandinavian origin; akin to Norwegian *flunsa* to hurry] (1542)
1 a : to move with exaggerated jerky or bouncy motions ⟨*flounced* about the room, jerking her shoulders, gesticulating —Agatha Christie⟩; *also* **:** to move so as to draw attention to oneself ⟨*flounced* into the lobby⟩ **b :** to go with sudden determination ⟨*flounced* out of the room in a huff⟩
2 : FLOUNDER, STRUGGLE
²flounce *noun* (1583)
: an act or instance of flouncing
— **flouncy** \ˈflaůn(t)-sē\ *adjective*
³flounce *transitive verb* **flounced; flounc·ing** [alteration of earlier *frounce,* from Middle English *frouncen* to curl] (1711)
: to trim with flounces
⁴flounce *noun* (1713)
: a strip of fabric attached by one edge; *also* **:** a wide ruffle
— **flouncy** \ˈflaůn(t)-sē\ *adjective*
flounc·ing \ˈflaůn(t)-siŋ\ *noun* (1865)
: material used for flounces
¹floun·der \ˈflaůn-dər\ *noun, plural* **flounder** *or* **flounders** [Middle English, from Anglo-French *floundre,* of Scandinavian origin; akin to Old Norse *flythra* flounder] (15th century)
: FLATFISH; *especially* **:** a fish of either of two families (Pleuronectidae and Bothidae) that include important marine food fishes
²flounder *intransitive verb* **floun·dered; floun·der·ing** \-d(ə-)riŋ\ [probably alteration of *founder*] (1592)
1 : to struggle to move or obtain footing **:** thrash about wildly
2 : to proceed or act clumsily or ineffectually
¹flour \ˈflaůr\ *noun* [Middle English — more at FLOWER] (13th century)
1 : finely ground meal of wheat usually largely freed from bran; *also* **:** a similar meal of another material (as a cereal grain, an edible seed, or dried processed fish)
2 : a fine soft powder
— **flour·less** *adjective*
— **floury** \-ē\ *adjective*
²flour (circa 1657)
transitive verb
: to coat with or as if with flour
intransitive verb
: to break up into particles
¹flour·ish \ˈflər-ish, ˈflə-rish\ *verb* [Middle English *florisshen,* from Middle French *floriss-,* stem of *florir,* from (assumed) Vulgar Latin *florire,* alteration of Latin *florēre,* from *flor-, flos* flower] (14th century)
intransitive verb
1 : to grow luxuriantly **:** THRIVE
2 a : to achieve success **:** PROSPER **b :** to be in a state of activity or production ⟨*flourished* around 1850⟩ **c :** to reach a height of development or influence
3 : to make bold and sweeping gestures
transitive verb
: to wield with dramatic gestures **:** BRANDISH
synonym see SWING
— **flour·ish·er** *noun*
— **flour·ish·ing·ly** \-i-shiŋ-lē\ *adverb*
²flourish *noun* (1597)
1 a : a period of thriving **b :** a luxuriant growth or profusion ⟨a *flourish* of white hair⟩ ⟨a springtime *flourish* of color⟩
2 a : a florid bit of speech or writing ⟨rhetorical *flourishes*⟩ **b :** an ornamental stroke in writing or printing **c :** a decorative or finishing detail ⟨a house with clever little *flourishes*⟩
3 : FANFARE
4 : an act or instance of brandishing or waving
5 : showiness in the doing of something ⟨opened the door with a *flourish*⟩
6 : a sudden burst (as of activity) ⟨the week ends with a *flourish* of tests⟩
¹flout \ˈflaůt\ *verb* [probably from Middle English *flouten* to play the flute, from *floute* flute] (1551)
transitive verb
: to treat with contemptuous disregard **:** SCORN ⟨*flouting* the rules⟩
intransitive verb
: to indulge in scornful behavior
synonym see SCOFF
usage see FLAUNT
— **flout·er** *noun*
²flout *noun* (circa 1570)
: JEER
¹flow \ˈflō\ *verb* [Middle English, from Old English *flōwan;* akin to Old High German *flouwen* to rinse, wash, Latin *pluere* to rain, Greek *plein* to sail, float] (before 12th century)
intransitive verb
1 a (1) **:** to issue or move in a stream (2) **:** CIRCULATE **b :** to move with a continual change of place among the constituent particles ⟨molasses *flows* slowly⟩
2 : RISE ⟨the tide ebbs and *flows*⟩
3 : ABOUND
4 a : to proceed smoothly and readily ⟨conversation *flowed* easily⟩ **b :** to have a smooth continuity
5 : to hang loose and billowing
6 : to derive from a source **:** COME ⟨the wealth that *flows* from trade⟩
7 : to deform under stress without cracking or rupturing — used especially of minerals and rocks
8 : MENSTRUATE
transitive verb
1 : to cause to flow
2 : to discharge in a flow
synonym see SPRING
— **flow·ing·ly** \-iŋ-lē\ *adverb*
²flow *noun* (15th century)
1 : an act of flowing
2 : FLOOD 1a, 2
3 a : a smooth uninterrupted movement or progress ⟨a *flow* of information⟩ **b :** STREAM; *also* **:** a mass of material which has flowed when molten ⟨an old lava *flow*⟩ **c :** the direction of movement or development ⟨go with the *flow*⟩
4 : the quantity that flows in a certain time
5 : MENSTRUATION
6 a : the motion characteristic of fluids **b :** a continuous transfer of energy
flow·age \ˈflō-ij\ *noun* (1830)
1 a : an overflowing onto adjacent land **b :** a body of water formed by overflowing or damming **c :** floodwater especially of a stream

2 : gradual deformation of a body of plastic solid (as rock) by intermolecular shear
flow•chart \-ˌchärt\ *noun* (1920)
: a diagram that shows step-by-step progression through a procedure or system especially using connecting lines and a set of conventional symbols
— flow•chart•ing \-ˌchär-tiŋ\ *noun*
flow cy•tom•e•try \-sī-ˈtä-mə-trē\ *noun* (1978)
: a technique for identifying and sorting cells and their components (as DNA) by staining with a fluorescent dye and detecting the fluorescence usually by laser beam illumination
flow diagram *noun* (1943)
: FLOWCHART
¹flow•er \ˈflau̇(-ə)r\ *noun* [Middle English *flour* flower, best of anything, flour, from Old French *flor, flour,* from Latin *flor-, flos* — more at BLOW] (13th century)
1 a : BLOSSOM, INFLORESCENCE **b :** a shoot of the sporophyte of a higher plant that is modified for reproduction and consists of a shortened axis bearing modified leaves; *especially* **:** one of a seed plant differentiated into a calyx, corolla, stamens, and carpels **c :** a plant cultivated for its blossoms

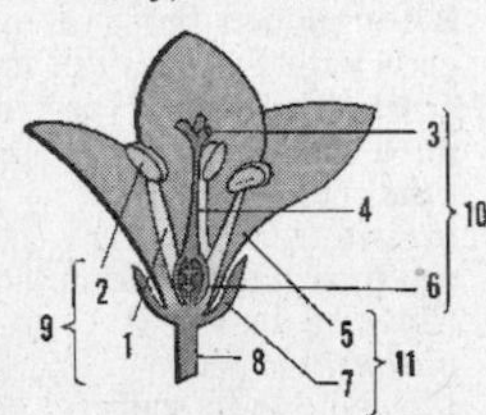

cross section of flower 1b: *1* **filament,** *2* **anther,** *3* **stigma,** *4* **style,** *5* **petal,** *6* **ovary,** *7* **sepal,** *8* **pedicel,** *9* **stamen,** *10* **pistil,** *11* **perianth**

2 a : the best part or example ⟨the *flower* of our youth⟩ **b :** the finest most vigorous period **c :** a state of blooming or flourishing ⟨in full *flower*⟩
3 *plural* **:** a finely divided powder produced especially by condensation or sublimation ⟨*flowers* of sulfur⟩
— flow•ered \ˈflau̇(-ə)rd\ *adjective*
— flow•er•ful \ˈflau̇(-ə)r-fəl\ *adjective*
— flow•er•less \-ləs\ *adjective*
— flow•er•like \-ˌlīk\ *adjective*
²flower (13th century)
intransitive verb
1 a : DEVELOP ⟨*flowered* into young womanhood⟩ **b :** FLOURISH
2 : to produce flowers **:** BLOSSOM
transitive verb
1 : to cause to bear flowers
2 : to decorate with flowers or floral designs
— flow•er•er \ˈflau̇(-ə)r-ər\ *noun*
flow•er•age \ˈflau̇(-ə)r-ij\ *noun* (1840)
: a flowering process, state, or condition
flower bud *noun* (1828)
: a plant bud that produces only a flower
flower bug *noun* (circa 1889)
: any of various small mostly black-and-white predaceous bugs (family Anthocoridae) that frequent flowers and feed on pest insects (as aphids and thrips)
flower child *noun* (1967)
: a hippie who advocates love, beauty, and peace
flow•er•et *also* **flow•er•ette** \ˈflau̇(-ə)r-ət\ *noun* (15th century)
: FLORET 1
flower girl *noun* (1925)
: a little girl who carries flowers at a wedding
flower head *noun* (1845)
: a capitulum (as of a composite) having sessile flowers so arranged that the whole inflorescence looks like a single flower
flowering dogwood *noun* (1843)
: a common spring-flowering white-bracted dogwood (*Cornus florida*)
flowering plant *noun* (1745)
: ANGIOSPERM
flower people *noun plural* (1967)
: FLOWER CHILDREN
flow•er•pot \ˈflau̇(-ə)r-ˌpät\ *noun* (1598)
: a pot in which to grow plants
flow•ery \ˈflau̇(-ə)r-ē\ *adjective* (14th century)
1 : of, relating to, or resembling flowers
2 : marked by or given to rhetorical elegance
— flow•er•i•ly \ˈflau̇(-ə)r-ə-lē\ *adverb*
— flow•er•i•ness *noun*
flow•me•ter \ˈflō-ˌmē-tər\ *noun* (1915)
: an instrument for measuring one or more properties (as velocity or pressure) of a flow (as of a liquid in a pipe)
¹flown \ˈflōn\ *past participle of* FLY
²flown *adjective* [archaic past participle of ¹*flow*] (1626)
: filled to excess
flow sheet *noun* (1912)
: FLOWCHART
flow•stone \ˈflō-ˌstōn\ *noun* (1925)
: calcite deposited by a thin sheet of flowing water usually along the walls or floor of a cave
flu \ˈflü\ *noun* [by shortening] (1839)
1 : INFLUENZA
2 : any of several virus diseases marked especially by respiratory symptoms
¹flub \ˈfləb\ *verb* **flubbed; flub•bing** [origin unknown] (1904)
transitive verb
: to make a mess of **:** BOTCH ⟨*flubbed* my lines⟩
intransitive verb
: BLUNDER
²flub *noun* (1948)
: an act or instance of flubbing
flub•dub \ˈfləb-ˌdəb\ *noun* [origin unknown] (1888)
: BUNKUM, BALDERDASH
fluc•tu•ant \ˈflək-chə-wənt\ *adjective* (1560)
1 : moving in waves
2 : VARIABLE, UNSTABLE
3 : being movable and compressible ⟨a *fluctuant* abscess⟩
fluc•tu•ate \ˈflək-chə-ˌwāt\ *verb* **-at•ed; -at•ing** [Latin *fluctuatus,* past participle of *fluctuare,* from *fluctus* flow, wave, from *fluere* — more at FLUID] (1634)
intransitive verb
1 : to shift back and forth uncertainly
2 : to ebb and flow in waves
transitive verb
: to cause to fluctuate
synonym see SWING
— fluc•tu•a•tion \ˌflək-chə-ˈwā-shən\ *noun*
— fluc•tu•a•tion•al \-ˈwā-shnəl, -shə-nᵊl\ *adjective*
flue \ˈflü\ *noun* [origin unknown] (1582)
: an enclosed passageway for directing a current: as **a :** a channel in a chimney for conveying flame and smoke to the outer air **b :** a pipe for conveying flame and hot gases around or through water in a steam boiler **c :** an air channel leading to the lip of a wind instrument **d :** FLUE PIPE
flue–cured \-ˌkyu̇rd\ *adjective* (1905)
: cured with heat transmitted through a flue without exposure to smoke or fumes ⟨*flue-cured* tobacco⟩
flu•en•cy \ˈflü-ən(t)-sē\ *noun* (1636)
: the quality or state of being fluent
flu•ent \ˈflü-ənt\ *adjective* [Latin *fluent-, fluens,* present participle of *fluere*] (1599)
1 a : capable of flowing **:** FLUID **b :** capable of moving with ease and grace ⟨the *fluent* body of a dancer⟩
2 a : ready or facile in speech ⟨*fluent* in Spanish⟩ **b :** effortlessly smooth and rapid **:** POLISHED ⟨a *fluent* performance⟩
— flu•ent•ly *adverb*
flue pipe *noun* (1852)
: an organ pipe whose tone is produced by an air current striking the lip and causing the air within to vibrate — compare REED PIPE
flue stop *noun* (1855)
: an organ stop made up of flue pipes
¹fluff \ˈfləf\ *noun* [perhaps blend of *flue* (fluff) and *puff*] (1790)
1 : NAP, DOWN
2 : something fluffy
3 : something inconsequential
4 : BLUNDER; *especially* **:** an actor's lapse of memory
²fluff (1875)
intransitive verb
1 : to become fluffy
2 : to make a mistake; *especially* **:** to forget or bungle one's lines in a play
transitive verb
1 : to make fluffy
2 a : to spoil by a mistake **:** BOTCH **b :** to deliver badly or forget (one's lines) in a play
fluffy \ˈflə-fē\ *adjective* **fluff•i•er; -est** (circa 1825)
1 a : covered with or resembling fluff **b :** being light and soft or airy ⟨a *fluffy* omelet⟩
2 : lacking in meaning or substance **:** SUPERFICIAL 2c
— fluff•i•ly \ˈflə-fə-lē\ *adverb*
— fluff•i•ness \ˈflə-fē-nəs\ *noun*
flü•gel•horn *or* **flue•gel•horn** \ˈflü-gəl-ˌhȯrn, ˈflᵫ-\ *noun* [German, from *Flügel* wing, flank + *Horn* horn; from its use to signal the flanking drivers in a battue] (1854)
: a valved brass instrument resembling a cornet but having a larger bore
— flü•gel•horn•ist \-ˌhȯr-nist\ *noun*
¹flu•id \ˈflü-əd\ *adjective* [French or Latin; French *fluide,* from Latin *fluidus,* from *fluere* to flow; akin to Greek *phlyzein* to boil over] (1603)
1 a : having particles that easily move and change their relative position without a separation of the mass and that easily yield to pressure **:** capable of flowing **b :** subject to change or movement ⟨boundaries became *fluid*⟩
2 : characterized by or employing a smooth easy style ⟨the ballerina's *fluid* movements⟩
3 a : available for a different use **b :** LIQUID 4 ⟨*fluid* assets⟩
— flu•id•ly *adverb*
— flu•id•ness *noun*
²fluid *noun* (1661)
: a substance (as a liquid or gas) tending to flow or conform to the outline of its container
— flu•id•al \ˈflü-ə-dᵊl\ *adjective*
— flu•id•al•ly \-dᵊl-ē\ *adverb*
fluid dram *or* **flu•i•dram** \ˌflü-ə(d)-ˈdram\ *noun* (circa 1860)
: a unit of liquid capacity equal to ⅛ fluid ounce — see WEIGHT table
flu•id•ex•tract \ˌflü-ə-ˈdek-ˌstrakt\ *noun* (1851)
: an alcohol preparation of a vegetable drug containing the active constituents of one gram of the dry drug in each milliliter
flu•id•ic \flü-ˈi-dik\ *adjective* (1960)
: of, relating to, or being a device (as an amplifier or control) that depends for operation on the pressures and flows of a fluid in precisely shaped channels
— fluidic *noun*
— flu•id•ics \-diks\ *noun plural but singular in construction*
flu•id•i•ty \flü-ˈi-də-tē\ *noun* (1603)
1 : the quality or state of being fluid
2 : the physical property of a substance that enables it to flow
flu•id•ize \ˈflü-ə-ˌdīz\ *transitive verb* **-ized; -iz•ing** (circa 1855)
1 : to cause to flow like a fluid
2 : to suspend (as solid particles) in a rapidly moving stream of gas or vapor to induce flowing motion of the whole
— flu•id•iza•tion \ˌflü-ə-də-ˈzā-shən\ *noun*
— flu•id•iz•er \ˈflü-ə-ˌdī-zər\ *noun*
fluidized bed *noun* (1949)
: a bed of small solid particles (as in a coal burning furnace) suspended and kept in motion by an upward flow of a fluid (as a gas) — called also *fluid bed*
fluid mechanics *noun plural but singular or plural in construction* (1937)
: a branch of mechanics dealing with the properties of liquids and gases

fluid ounce *noun* (circa 1860)
1 : a U.S. unit of liquid capacity equal to 1/16 pint — see WEIGHT table
2 : a British unit of liquid capacity equal to 1/20 pint — see WEIGHT table

¹fluke \'flük\ *noun* [Middle English, from Old English *flōc;* akin to Old English *flōh* chip, Old High German *flah* smooth, Greek *plax* flat surface, and probably to Old English *flōr* floor — more at FLOOR] (before 12th century)
1 : FLATFISH
2 : a flattened digenetic trematode worm; *broadly* **:** TREMATODE — compare LIVER FLUKE

²fluke *noun* [perhaps from ¹*fluke*] (1561)
1 : the part of an anchor that fastens in the ground — see ANCHOR illustration
2 : one of the lobes of a whale's tail

³fluke *noun* [origin unknown] (1857)
1 : an accidentally successful stroke at billiards or pool
2 : a stroke of luck ⟨the discovery was a *fluke*⟩

fluky *also* **fluk•ey** \'flü-kē\ *adjective* **fluk•i•er; -est** (1867)
1 : happening by or depending on chance
2 : being unsteady or uncertain — used especially of wind

flume \'flüm\ *noun* [probably from Middle English *flum* river, from Old French, from Latin *flumen,* from *fluere* — more at FLUID] (1748)
1 : an inclined channel for conveying water (as for power)
2 : a ravine or gorge with a stream running through it

flum•mery \'flə-mə-rē, 'flə-mə-\ *noun, plural* **-mer•ies** [Welsh *llymru*] (1623)
1 a : a soft jelly or porridge made with flour or meal **b :** any of several sweet desserts
2 : MUMMERY, MUMBO JUMBO

flum•mox \'flə-məks, -miks\ *transitive verb* [origin unknown] (1837)
: CONFUSE

¹flump \'fləmp\ [imitative] (1816)
intransitive verb
: to move or fall suddenly and heavily ⟨*flumped* down into the chair⟩
transitive verb
: to place or drop with a flump

²flump *noun* (1832)
: a dull heavy sound (as of a fall)

flung *past and past participle of* FLING

¹flunk \'fləŋk\ *verb* [perhaps blend of *flinch* and *funk*] (1823)
intransitive verb
: to fail especially in an examination or course
transitive verb
1 : to give a failing grade to
2 : to get a failing grade in
— flunk•er *noun*

²flunk *noun* (1846)
: an act or instance of flunking

flunk out (1920)
intransitive verb
: to be dismissed from a school or college for failure
transitive verb
: to dismiss from a school or college for failure

flun•ky *or* **flun•key** \'fləŋ-kē\ *noun, plural* **flunkies** *or* **flunkeys** [Scots, of unknown origin] (circa 1782)
1 a : a liveried servant **b :** one performing menial or miscellaneous duties
2 : YES-MAN

flu•o•cin•o•lone ac•e•to•nide \ˌflü-ə-'si-nᵊl-ˌōn-ˌa-sə-'tō-ˌnīd\ *noun* [*fluor-* + *-cinolone,* probably alteration of *-nisolone* (as in *prednisolone*)] (1963)
: a glucocorticoid steroid $C_{24}H_{30}F_2O_6$ used especially as an anti-inflammatory agent in the treatment of skin diseases

flu•or \'flü-ˌȯr, 'flü-ər\ *noun* [New Latin, mineral belonging to a group used as fluxes and including fluorite, from Latin, flow, from *fluere* — more at FLUID] (1661)
: FLUORITE

fluor- *or* **fluoro-** *combining form* [French]
1 : fluorine ⟨*fluor*ide⟩
2 *also* **fluori- :** fluorescence ⟨*fluoro*scope⟩ ⟨*fluori*meter⟩

fluo•resce \flu̇-'res, flō-, flȯ-\ *intransitive verb* **-resced; -resc•ing** [back-formation from *fluorescence*] (1874)
: to produce, undergo, or exhibit fluorescence
— fluo•resc•er *noun*

fluo•res•ce•in \-'re-sē-ən\ *noun* (1876)
: a yellow or red crystalline dye $C_{20}H_{12}O_5$ with a bright yellow-green fluorescence in alkaline solution

fluo•res•cence \-'re-sᵊn(t)s\ *noun* [*fluor*spar + opal*escence*] (1852)
: luminescence that is caused by the absorption of radiation at one wavelength followed by nearly immediate reradiation usually at a different wavelength and that ceases almost immediately when the incident radiation stops; *also* **:** the radiation emitted — compare PHOSPHORESCENCE

fluo•res•cent \-sᵊnt\ *adjective* (1853)
1 : having or relating to fluorescence
2 : bright and glowing as a result of fluorescence ⟨*fluorescent* inks⟩; *broadly* **:** very bright in color
— fluorescent *noun*

fluorescent lamp *noun* (1896)
: a tubular electric lamp having a coating of fluorescent material on its inner surface and containing mercury vapor whose bombardment by electrons from the cathode provides ultraviolet light which causes the material to emit visible light

fluo•ri•date \'flu̇r-ə-ˌdāt, 'flōr-, 'flȯr-\ *transitive verb* **-dat•ed; -dat•ing** (1949)
: to add a fluoride to (as drinking water) to reduce tooth decay
— fluo•ri•da•tion \ˌflu̇r-ə-'dā-shən, ˌflōr-, ˌflȯr-\ *noun*

fluo•ride \'flȯr-ˌīd, 'flu̇r-\ *noun, often attributive* (1826)
1 : a compound of fluorine
2 : the monovalent anion of flourine

fluo•ri•nate \'flȯr-ə-ˌnāt, 'flōr-, 'flu̇r-\ *transitive verb* **-nat•ed; -nat•ing** (circa 1929)
: to treat or cause to combine with fluorine or a compound of fluorine
— fluo•ri•na•tion \ˌflȯr-ə-'nā-shən, ˌflōr-, ˌflu̇r-\ *noun*

fluo•rine \'flu̇r-ˌēn, 'flȯr-, 'flōr-\ *noun* [French, from New Latin *fluor*] (1813)
: a nonmetallic halogen element that is isolated as a pale yellowish flammable irritating toxic diatomic gas — see ELEMENT table

fluo•rite \'flu̇r-ˌīt, 'flȯr-, 'flōr-\ *noun* [Italian, from New Latin *fluor*] (1868)
: a transparent or translucent mineral of different colors that consists of the fluoride of calcium and is used as a flux and in the making of opalescent and opaque glasses

fluo•ro•car•bon \ˌflu̇r-ō-'kär-bən, ˌflȯr-, ˌflōr-\ *noun* (1937)
: any of various chemically inert compounds containing carbon and fluorine used chiefly as lubricants, refrigerants, nonstick coatings, and formerly aerosol propellants and in making resins and plastics; *also* **:** CHLOROFLUOROCARBON

fluo•ro•chrome \'flu̇r-ə-ˌkrōm, ˌflȯr-, ˌflōr-\ *noun* (1943)
: any of various fluorescent substances used in biological staining to produce fluorescence in a specimen

fluo•rog•ra•phy \flu̇-'rä-grə-fē, flȯ-, flō-\ *noun* (1941)
: PHOTOFLUOROGRAPHY
— flu•o•ro•graph•ic \ˌflu̇r-ə-'gra-fik, ˌflȯr-, flōr-\ *adjective*

fluo•rom•e•ter \flu̇-'rä-mə-tər, flȯ-, flō-\ *or* **fluo•rim•e•ter** \-'ri-\ *noun* (1897)
: an instrument for measuring fluorescence and related phenomena (as intensity of radiation)
— fluo•ro•met•ric *or* **fluo•ri•met•ric** \ˌflu̇r-ə-'me-trik, ˌflȯr-, ˌflōr-\ *adjective*
— fluo•rom•e•try \flu̇-'rä-mə-trē, flȯ-, flō-\ *or* **fluo•rim•e•try** \-'ri-mə-trē\ *noun*

¹fluo•ro•scope \'flu̇r-ə-ˌskōp, 'flȯr-, 'flōr-\ *noun* [International Scientific Vocabulary] (1896)
: an instrument used for observing the internal structure of an opaque object (as the living body) by means of X rays
— fluo•ro•scop•ic \ˌflu̇r-ə-'skä-pik, ˌflȯr-, ˌflōr-\ *adjective*
— fluo•ro•scop•i•cal•ly \-pi-k(ə-)lē\ *adverb*
— fluo•ros•co•pist \flu̇-'räs-kə-pist, flȯ-, flō-\ *noun*
— fluo•ros•co•py \-pē\ *noun*

²fluoroscope *transitive verb* **-scoped; -scop•ing** (1898)
: to examine by fluoroscopy

fluo•ro•sis \flu̇-'rō-səs, flȯ-\ *noun* [New Latin] (1927)
: an abnormal condition (as mottling of the teeth) caused by fluorine or its compounds
— fluo•rot•ic \-'rä-tik\ *adjective*

fluo•ro•ura•cil \ˌflu̇r-ō-'yu̇r-ə-ˌsil, -ˌsəl, ˌflȯr-, ˌflōr-\ *noun* [*fluor-* + *uracil*] (circa 1958)
: a fluorine-containing pyrimidine base $C_4H_3FN_2O_2$ used to treat some kinds of cancer

fluor•spar \'flu̇r-ˌspär, 'flȯr-, 'flōr-\ *noun* (1794)
: FLUORITE

flu•phen•azine \flü-'fe-nə-ˌzēn\ *noun* [*fluor-* + *phenazine*] (circa 1960)
: a tranquilizing compound $C_{22}H_{26}F_3N_3OS$ used especially combined as a salt

¹flur•ry \'flər-ē, 'flə-rē\ *noun, plural* **flurries** [probably from *flurr* (to throw scatteringly)] (1686)
1 a : a gust of wind **b :** a brief light snowfall
2 a : a brief period of commotion or excitement **b :** a sudden occurrence of many things at once **:** BARRAGE 2 ⟨a *flurry* of insults⟩
3 : a brief advance or decline in prices **:** a short-lived outburst of trading activity

²flurry *verb* **flur•ried; flur•ry•ing** (1757)
transitive verb
: to cause to become agitated and confused
intransitive verb
: to move in an agitated or confused manner

¹flush \'fləsh\ *verb* [Middle English *flusshen*] (13th century)
intransitive verb
: to take wing suddenly
transitive verb
1 : to cause (a bird) to flush
2 : to expose or chase from a place of concealment ⟨*flushed* the boys from their hiding place⟩

²flush *noun* [Middle French *flus, fluz,* from Latin *fluxus* flow, flux] (circa 1529)
1 : a hand of playing cards all of the same suit; *specifically* **:** a poker hand containing five cards of the same suit but not in sequence — see POKER illustration
2 : a series of three or more slalom gates set vertically on a slope

³flush *noun* [perhaps modification of Latin *fluxus*] (1529)
1 : a sudden flow (as of water); *also* **:** a rinsing or cleansing with or as if with a flush of water
2 a : a sudden increase or expansion; *especially* **:** sudden and usually abundant new plant growth **b :** a surge of emotion ⟨felt a *flush* of anger at the insult⟩
3 a : a tinge of red **:** BLUSH **b :** a fresh and vigorous state ⟨in the first *flush* of womanhood⟩
4 : a transitory sensation of extreme heat — compare HOT FLASH

⁴flush (1548)
intransitive verb
1 : to flow and spread suddenly and freely
2 a : to glow brightly **b :** BLUSH
3 : to produce new growth ⟨the plants *flush* twice during the year⟩
transitive verb
1 a : to cause to flow **b :** to pour liquid over or through; *especially* **:** to cleanse or wash out with or as if with a rush of liquid ⟨*flush* the toilet⟩ ⟨*flush* the lungs with air⟩
2 : INFLAME, EXCITE — usually used passively ⟨*flushed* with pride⟩
3 : to cause to blush
⁵flush *adjective* (1594)
1 a : of a ruddy healthy color **b :** full of life and vigor **:** LUSTY
2 a : filled to overflowing **b :** AFFLUENT
3 : readily available **:** ABUNDANT
4 a : having or forming a continuous plane or unbroken surface ⟨*flush* paneling⟩ **b :** directly abutting or immediately adjacent: as (1) **:** set even with an edge of a type page or column **:** having no indention (2) **:** arranged edge to edge so as to fit snugly
— **flush·ness** *noun*
⁶flush *adverb* (1700)
1 : in a flush manner
2 : SQUARELY ⟨hit him *flush* on the chin⟩
⁷flush *transitive verb* (circa 1842)
: to make flush ⟨*flush* the headings on a page⟩
flush·able \ˈflə-shə-bəl\ *adjective* (1973)
: suitable for disposal by flushing down a toilet
¹flus·ter \ˈfləs-tər\ *verb* **flus·tered; flus·ter·ing** \-t(ə-)riŋ\ [probably of Scandinavian origin; akin to Icelandic *flaustur* hurry] (1604)
transitive verb
1 : to make tipsy
2 : to put into a state of agitated confusion **:** UPSET
intransitive verb
: to move or behave in an agitated or confused manner
synonym see DISCOMPOSE
— **flus·tered·ly** *adverb*
²fluster *noun* (1728)
: a state of agitated confusion
¹flute \ˈflüt\ *noun* [Middle English *floute,* from Middle French *fleute,* from Old French *flaüte,* probably from Old Provençal *flaut*] (14th century)
1 a : RECORDER 3 **b :** a keyed woodwind instrument consisting of a cylindrical tube which is stopped at one end and which has a side hole over which air is blown to produce the tone and having a range from middle C upward for three octaves

flute 1b

2 : something long and slender: as **a :** a tall slender wineglass **b :** a grooved pleat (as on a hat brim)
3 : a rounded groove; *specifically* **:** one of the vertical parallel grooves on a classical architectural column
— **flute·like** \-ˌlīk\ *adjective*
— **fluty** *or* **flut·ey** \ˈflü-tē\ *adjective*
²flute *verb* **flut·ed; flut·ing** (14th century)
intransitive verb
1 : to play a flute
2 : to produce a flutelike sound
transitive verb
1 : to utter with a flutelike sound
2 : to form flutes in
— **flut·er** *noun*
flut·ed *adjective* (1611)
: having or marked by grooves
flut·ing \ˈflü-tiŋ\ *noun* (1611)
1 : a series of flutes ⟨the *fluting* of a column⟩
2 : fluted material
flut·ist \ˈflü-tist\ *noun* (1603)
: one who plays a flute
¹flut·ter \ˈflə-tər\ *verb* [Middle English *floteren* to float, flutter, from Old English *floterian,* frequentative of *flotian* to float; akin to Old English *flēotan* to float — more at FLEET] (before 12th century)
intransitive verb
1 : to flap the wings rapidly
2 a : to move with quick wavering or flapping motions **b :** to vibrate in irregular spasms
3 : to move about or behave in an agitated aimless manner
transitive verb
: to cause to flutter
— **flut·ter·er** \-tər-ər\ *noun*
— **flut·tery** \-tə-rē\ *adjective*
²flutter *noun* (1641)
1 : an act of fluttering
2 a : a state of nervous confusion or excitement **b :** FLURRY, COMMOTION **c :** abnormal spasmodic fluttering of a body part ⟨treatment of atrial *flutter*⟩
3 a : a distortion in reproduced sound similar to but of a higher pitch than wow **b :** fluctuation in the brightness of a television image
4 : an unwanted oscillation (as of an aileron or a bridge) set up by natural forces
5 *chiefly British* **:** a small speculative venture or gamble
flut·ter·board \ˈflə-tər-ˌbōrd, -ˌbȯrd\ *noun* (1950)
: a rectangular board used by swimmers in practicing leg strokes
flutter kick *noun* (circa 1934)
: an alternating whipping motion of the legs used in various swimming styles (as the crawl)
flutter sleeve *noun* (1973)
: a loose-fitting tapered sleeve falling in folds over the upper arm
flu·vi·al \ˈflü-vē-əl\ *adjective* [Latin *fluvialis,* from *fluvius* river, from *fluere*] (14th century)
1 : of, relating to, or living in a stream or river
2 : produced by the action of a stream
flu·vi·a·tile \ˈflü-vē-ə-ˌtīl\ *adjective* [Middle French, from Latin *fluviatilis,* from *fluvius*] (1599)
: FLUVIAL
¹flux \ˈfləks\ *noun* [Middle English, from Middle French & Medieval Latin; Middle French, from Medieval Latin *fluxus,* from Latin, flow, from *fluere* to flow — more at FLUID] (14th century)
1 : a flowing of fluid from the body; *especially* **:** an excessive abnormal discharge from the bowels
2 : a continuous moving on or passing by (as of a stream)
3 : a continued flow **:** FLOOD
4 a : INFLUX **b :** CHANGE, FLUCTUATION ⟨in a state of *flux*⟩
5 : a substance used to promote fusion (as of metals or minerals); *especially* **:** one (as rosin) applied to surfaces to be joined by soldering, brazing, or welding to clean and free them from oxide and promote their union
6 : the rate of transfer of fluid, particles, or energy across a given surface
²flux (15th century)
transitive verb
1 : to cause to become fluid
2 : to treat with a flux
intransitive verb
: to become fluid **:** FUSE
flux·gate \ˈfləks-ˌgāt\ *noun* (1944)
: a device used to indicate the direction and intensity of the magnetic field (as on a planet)
flux·ion \ˈflək-shən\ *noun* (1599)
1 : the action of flowing or changing; *also* **:** something subjected to such action
2 : DERIVATIVE 3 — compare METHOD OF FLUXIONS
— **flux·ion·al** \-shnəl, -shə-nᵊl\ *adjective*
¹fly \ˈflī\ *verb* **flew** \ˈflü\; **flown** \ˈflōn\; **fly·ing** [Middle English *flien,* from Old English *flēogan;* akin to Old High German *fliogan* to fly and probably Old English *flōwan* to flow] (before 12th century)
intransitive verb
1 a : to move in or pass through the air with wings **b :** to move through the air or before the wind or through outer space **c :** to float, wave, or soar in the air ⟨flags *flying* at half-mast⟩
2 a : to take flight **:** FLEE **b :** to fade and disappear **:** VANISH
3 a : to move, pass, or spread quickly ⟨rumors were *flying*⟩ **b :** to be moved with sudden extreme emotion ⟨*flew* into a rage⟩ **c :** to seem to pass quickly ⟨the time simply *flew*⟩
4 : to become expended or dissipated rapidly
5 : to operate or travel in an airplane or spacecraft
6 : to work successfully **:** win popular acceptance ⟨knew . . . a pure human-rights approach would not *fly* —Charles Brydon⟩
transitive verb
1 a : to cause to fly, float, or hang in the air ⟨*flying* a kite⟩ **b :** to operate (as a balloon, aircraft, rocket, or spacecraft) in flight **c :** to journey over or through by flying
2 a : to flee or escape from **b :** AVOID, SHUN
3 : to transport by aircraft or spacecraft
— **fly at :** to assail suddenly and violently
— **fly blind :** to fly an airplane solely by instruments
— **fly high :** to be elated
— **fly in the face of** *or* **fly in the teeth of :** to stand or act forthrightly or brazenly in defiance or contradiction of
²fly *noun, plural* **flies** (before 12th century)
1 : the action or process of flying **:** FLIGHT
2 a : a device consisting of two or more radial vanes capable of rotating on a spindle to act as a fan or to govern the speed of clockwork or very light machinery **b :** FLYWHEEL
3 *plural* **:** the space over a theater stage where scenery and equipment can be hung
4 : something attached by one edge: as **a :** a garment closing concealed by a fold of cloth extending over the fastener **b** (1) **:** the length of an extended flag from its staff or support (2) **:** the outer or loose end of a flag
5 : a baseball hit high into the air
6 : FLYLEAF
7 : a sheet of material (as canvas) that is attachable to a tent for use as a double top or as a rooflike extension
8 : a football pass pattern in which the receiver runs straight downfield
— **on the fly 1 :** in motion **:** BUSY **2 :** while still in the air **:** without the ball bouncing ⟨the home run carried 450 feet *on the fly*⟩
³fly *intransitive verb* **flied; fly·ing** (1893)
: to hit a fly in baseball
⁴fly *noun, plural* **flies** [Middle English *flie,* from Old English *flēoge;* akin to Old High German *flioga* fly, Old English *flēogan* to fly] (before 12th century)
1 : a winged insect — used chiefly in combination ⟨may*flies*⟩ ⟨butter*fly*⟩
2 a : any of a large order (Diptera) of winged or rarely wingless insects (as the housefly, mosquito, or gnat) that have segmented often headless, eyeless, and legless larvae, the anterior wings functional, and the posterior wings reduced to halteres **b :** a large stout-bodied fly
3 : a fishhook dressed (as with feathers or tinsel) to suggest an insect
— **fly in the ointment :** a detracting factor or element
⁵fly *adjective* [probably from ¹*fly*] (1811)
chiefly British **:** KEEN, ARTFUL
fly·able \ˈflī-ə-bəl\ *adjective* (circa 1909)
: suitable for flying or for being flown
fly agaric *noun* (1788)
: a poisonous mushroom (*Amanita muscaria*) with a usually bright red cap
fly ash *noun* (1931)
: fine solid particles of ashes, dust, and soot carried out from burning fuel (as coal or oil) by the draft

fly·away \ˈflī-ə-ˌwā\ *adjective* (1844)
1 : loose and flowing especially because of unconfined fullness at the back ⟨a *flyaway* jacket⟩
2 : of, relating to, or being an aircraft that is ready to fly ⟨a *flyaway* price⟩
fly ball *noun* (1865)
: [2]FLY 5
[1]**fly·blow** \-ˌblō\ *transitive verb* **-blew; -blown** [[4]*fly* + [1]*blow*] (1603)
1 : TAINT, CONTAMINATE
2 : to deposit eggs or young larvae of a flesh fly or blowfly in
[2]**flyblow** *noun* (1825)
: FLY-STRIKE
fly·blown \ˈflī-ˌblōn\ *adjective* (circa 1529)
1 a : not pure **:** TAINTED ⟨a world *flyblown* with the vices of irresponsible power —V. L. Parrington⟩ **b :** not bright and new **:** SEEDY, MOTH-EATEN **c :** TRITE, HACKNEYED ⟨a long list of *flyblown* metaphors —*Horizon*⟩
2 a : infested with eggs or young larvae of a flesh fly or blowfly **b :** covered with flyspecks
fly·boat \-ˌbōt\ *noun* [modification of Dutch *vlieboot,* from *Vlie,* channel between North Sea & Wadden Zee + *boot* boat] (1577)
: any of various fast boats
fly·boy \ˈflī-ˌbȯi\ *noun* (1946)
: a member of the air force; *broadly* **:** an aircraft pilot
fly·bridge \ˈflī-ˌbrij\ *noun* (1965)
: an open deck on a cabin cruiser located above the bridge on the cabin roof and usually having a duplicate set of navigating equipment
fly·by \ˈflī-ˌbī\ *noun, plural* **flybys** (1953)
1 : a prearranged usually low-altitude flight by one or more airplanes over a public gathering (as an air show)
2 a : a flight of a spacecraft past a celestial body (as Mars) close enough to obtain scientific data **b :** a spacecraft that makes a flyby
[1]**fly–by–night** \ˈflī-bī-ˌnīt\ *noun* (1822)
1 : one that seeks to evade responsibilities and especially creditors by flight
2 : one without established reputation or standing; *especially* **:** a shaky business enterprise
[2]**fly–by–night** *adjective* (1914)
1 : given to making a quick profit usually by shady or irresponsible acts
2 : TRANSITORY, PASSING ⟨*fly-by-night* fashions⟩
fly–by–night·er \ˌflī-bī-ˈnī-tər\ *noun* (1946)
: FLY-BY-NIGHT
fly–by–wire \ˈflī-bī-ˌwīr\ *adjective* (1968)
: of, relating to, being, or utilizing a flight-control system in which controls are operated electrically rather than mechanically
fly casting *noun* (circa 1889)
: the casting of artificial flies in fly-fishing or as a competitive sport
fly·catch·er \ˈflī-ˌka-chər, -ˌke-\ *noun* (1678)
: any of various passerine birds (families Muscicapidae and Tyrannidae) that feed on insects taken on the wing
fly dope *noun* (1897)
: an insect repellent
fly·er *variant of* FLIER
fly fisherman *noun* (1886)
: an angler who uses the technique of fly-fishing
fly fish·ing \ˈflī-ˌfi-shiŋ\ *noun* (1653)
: a method of fishing in which an artificial fly is cast by use of a fly rod, a reel, and a relatively heavy oiled or treated line
fly front *noun* (1893)
: a concealed closing on the front of a coat, skirt, shirt, dress, or pants
— fly–front \ˈflī-ˌfrənt\ *adjective*
fly gallery *noun* (1888)
: a narrow raised platform at the side of a theater stage from which flying scenery lines are operated
[1]**fly·ing** \ˈflī-iŋ\ *adjective* (before 12th century)
1 a : moving or capable of moving in the air **b :** moving or made by moving rapidly ⟨*flying* feet⟩ ⟨a *flying* leap⟩ **c :** very brief
2 : intended for ready movement or action ⟨a *flying* squad car⟩
3 : having stylized wings — used especially of livestock brand marks
4 : of or relating to the operation of aircraft ⟨a *flying* club⟩
5 : traversed or to be traversed (as in speed-record trials) after a running start ⟨a *flying* kilometer⟩
— with flying colors : with complete or eminent success
[2]**flying** *noun* (1548)
1 : travel by air
2 : the operation of an aircraft or spacecraft
flying boat *noun* (1913)
: a seaplane with a hull designed for floating
flying bomb *noun* (1944)
chiefly British **:** BUZZ BOMB
flying bridge *noun* (circa 1909)
1 : the highest navigational bridge on a ship
2 : FLYBRIDGE
flying buttress *noun* (1669)
: a masonry structure that typically consists of a straight inclined bar carried on an arch and a solid pier or buttress against which it abuts and that receives the thrust of a roof or vault

1 **flying buttress**

Flying Dutchman *noun*
1 : a legendary Dutch mariner condemned to sail the seas until Judgment Day
2 : a spectral ship that according to legend haunts the seas near the Cape of Good Hope
flying fish *noun* (circa 1511)
: any of numerous fishes (family Exocoetidae) chiefly of tropical and warm seas that are capable of long gliding flights out of water by spreading their large pectoral fins like wings
flying fox *noun* (1759)
: FRUIT BAT
flying gurnard *noun* (1884)
: any of several marine fishes (family Dactylopteridae) that resemble gurnards and have large pectoral fins allowing them to glide above the water for short distances
flying jib *noun* (1711)
: a sail outside the jib on an extension of the jibboom — see SAIL illustration
flying lemur *noun* (1883)
: either of two East Indian or Philippine arboreal nocturnal mammals (*Cynocephalus volans* and *C. variegatus*) that are about the size of a cat, that make long gliding leaps using a broad fold of skin on each side attached to and extending between the limbs, and that are placed in a separate order (Dermoptera)
flying machine *noun* (1736)
: an apparatus for navigating the air
flying mare *noun* (1754)
: a wrestling maneuver in which the aggressor seizes his opponent's wrist, turns about, and jerks him over his back
flying officer *noun* (1913)
: a commissioned officer in the British air force who ranks with a first lieutenant in the army
flying saucer *noun* (1947)
: any of various unidentified flying objects usually described as being saucer-shaped or disk-shaped
flying spot *noun* (1933)
: a spot of light moved over a surface (as one bearing printing or an image) so that light reflected from or transmitted by different parts of the surface is translated into electrical signals for transmission (as in television or computers)
flying squad *noun* (1927)
: a usually small standby group of people ready to move or act swiftly; *especially* **:** a police unit formed to respond quickly in an emergency
flying squirrel *noun* (1624)
: either of two small nocturnal North American squirrels (*Glaucomys volans* and *G. sabrinus*) with folds of skin connecting the forelegs and hind legs that enable it to make long gliding leaps; *also* **:** any of various squirrels that possess a patagium
flying start *noun* (1851)
1 : a start in racing in which the participants are already moving when they cross the starting line or receive the starting signal
2 : a favorable start of something
flying wedge *noun* (1909)
: a moving formation (as of guards or police) resembling a wedge
fly·leaf \ˈflī-ˌlēf\ *noun, plural* **fly·leaves** \-ˌlēvz\ (1832)
: one of the free endpapers of a book
fly·man \-mən, -ˌman\ *noun* (1883)
: a worker in the flies of a theater who manipulates curtains and scenery
fly net *noun* (before 12th century)
: a net to exclude or keep off insects (as from a harness horse)
fly·over \ˈflī-ˌō-vər\ *noun* (1931)
1 : FLYBY 1
2 *British* **:** OVERPASS
fly·pa·per \-ˌpā-pər\ *noun* (1847)
: paper coated with a sticky often poisonous substance for killing flies
fly·past \-ˌpast\ *noun* (1914)
chiefly British **:** FLYBY 1
fly rod *noun* (1684)
: a light springy fishing rod used in fly casting
flysch \ˈflish\ *noun* [German dialect] (1853)
: a thick and extensive deposit largely of sandstone that is formed in a geosyncline adjacent to a rising mountain belt and is especially common in the Alpine region of Europe
fly sheet *noun* (1833)
1 : a small loose advertising sheet **:** HANDBILL
2 : a sheet of a folder, booklet, or catalog giving directions for the use of or information about the material that follows
fly·speck \ˈflī-ˌspek\ *noun* (circa 1847)
1 : a speck made by fly excrement
2 : something small and insignificant
— flyspeck *transitive verb*
fly–strike \-ˌstrīk\ *noun* (1940)
: infestation with fly maggots
— fly–struck \-ˌstrək\ *adjective*
fly·swat·ter \-ˌswä-tər\ *noun* (1917)
: a device for killing insects that consists of a flat piece of perforated rubber or plastic or fine-mesh wire netting attached to a handle
fly·ti·er \ˈflī-ˌtī(-ə)r\ *noun* [*fly* + *tier* (one that ties)] (1881)
: a person who makes flies for fishing
flyt·ing \ˈflī-tiŋ\ *noun* [Scots, literally, contention, gerund of *flyte* to contend, argue, from Middle English *fliten,* from Old English *flītan;* akin to Old High German *flīzan* to argue] (1508)
: a dispute or exchange of personal abuse in verse form
fly·way \ˈflī-ˌwā\ *noun* (1891)
: an established air route of migratory birds
fly·weight \-ˌwāt\ *noun* (1911)
: a boxer in a weight division having a maximum limit of 112 pounds — compare BANTAMWEIGHT
fly·wheel \-ˌhwēl, -ˌwēl\ *noun* (1784)
: a heavy wheel for opposing and moderating by its inertia any fluctuation of speed in the machinery with which it revolves; *also* **:** a

similar wheel used for storing kinetic energy (as for motive power)

fly whisk *noun* (1841)
: a whisk for brushing away flies

FM \'ef-,em\ *noun, often attributive* [*f*requency *m*odulation] (1940)
: a broadcasting system using frequency modulation; *also* **:** a radio receiver of such a system

FMN \,ef-,em-'en\ *noun* [*f*lavin *m*ono*n*ucleotide] (circa 1953)
: a yellow crystalline phosphoric ester $C_{17}H_{21}N_4O_9P$ of riboflavin that is a coenzyme of several flavoprotein enzymes

f–num·ber \'ef-,nəm-bər\ *noun* [*f*ocal length] (circa 1903)
1 : the ratio of the focal length to the aperture in an optical system
2 : a number following the symbol f/ that expresses the effectiveness of the aperture of a camera lens in relation to brightness of image so that the smaller the number the brighter the image and therefore the shorter the exposure required

¹foal \'fōl\ *noun* [Middle English *fole,* from Old English *fola;* akin to Latin *pullus* young of an animal, Greek *pais* child — more at FEW] (before 12th century)
: a young animal of the horse family; *especially* **:** one under one year
— in foal : PREGNANT 3

²foal *intransitive verb* (14th century)
: to give birth to a foal

¹foam \'fōm\ *noun* [Middle English *fome,* from Old English *fām;* akin to Old High German *feim* foam, Latin *spuma* foam, *pumex* pumice] (before 12th century)
1 : a light frothy mass of fine bubbles formed in or on the surface of a liquid: as **a :** a frothy mass formed in salivating or sweating **b :** a stabilized froth produced chemically or mechanically and used especially in fighting oil fires **c :** a material in a lightweight cellular form resulting from introduction of gas bubbles during manufacture
2 : SEA
3 : something resembling foam
— foam·less \-ləs\ *adjective*

²foam (before 12th century)
intransitive verb
1 a : to produce or form foam **b :** to froth at the mouth especially in anger; *broadly* **:** to be angry
2 : to gush out in foam
3 : to become covered with or as if with foam ⟨streets . . . *foaming* with life —Thomas Wolfe⟩
transitive verb
1 : to cause to foam; *specifically* **:** to cause air bubbles to form in
2 : to convert (as a plastic) into a foam
— foam·able \'fō-mə-bəl\ *adjective*
— foam·er \fō-mər\ *noun*

foamed plastic *noun* (1945)
: EXPANDED PLASTIC

foam·flow·er \'fōm-,flau̇(-ə)r\ *noun* (1895)
: an American woodland spring-flowering herb (*Tiarella cordifolia*) that has white flowers with very long stamens and no stem leaves — called also *false miterwort*

foam rubber *noun* (circa 1939)
: spongy rubber of fine texture made from latex by foaming (as by whipping) before vulcanization

foamy \'fō-mē\ *adjective* **foam·i·er; -est** (before 12th century)
1 : covered with foam **:** FROTHY
2 : full of, consisting of, or resembling foam
— foam·i·ly \-mə-lē\ *adverb*
— foam·i·ness \-mē-nəs\ *noun*

¹fob \'fäb\ *transitive verb* **fobbed; fob·bing** [Middle English *fobben*] (14th century)
archaic **:** DECEIVE, CHEAT

²fob *noun* [perhaps akin to German dialect *Fuppe* pocket] (1653)
1 : WATCH POCKET
2 : a short strap, ribbon, or chain attached especially to a pocket watch
3 : an ornament attached to a fob chain

fob off *transitive verb* (1597)
1 : to put off with a trick, excuse, or inferior substitute
2 : to pass or offer (something spurious) as genuine
3 : to put aside ⟨now *fob off* what once they would have welcomed eagerly —Walter Lippmann⟩

fo·cac·cia \fō-'kä-ch(ē-)ə\ *noun* [Italian, from Late Latin *focacia* (neuter plural), from Latin *focus* hearth] (1969)
: a flat Italian bread typically seasoned with herbs and olive oil

fo·cal \'fō-kəl\ *adjective* (1693)
: of, relating to, being, or having a focus
— fo·cal·ly \-kə-lē\ *adverb*

focal infection *noun* (circa 1923)
: a persistent bacterial infection of some organ or region; *especially* **:** one causing symptoms elsewhere in the body

fo·cal·ize \'fō-kə-,līz\ *verb* **-ized; -iz·ing** (1845)
transitive verb
1 : to bring to a focus
2 : LOCALIZE
intransitive verb
1 : to come to a focus **:** CONCENTRATE
2 : LOCALIZE
— fo·cal·iza·tion \,fō-kə-lə-'zā-shən\ *noun*

focal length *noun* (1753)
: the distance of a focus from the surface of a lens or curved mirror

focal plane *noun* (1889)
: a plane that is perpendicular to the axis of a lens or mirror and passes through the focus

focal point *noun* (1713)
: FOCUS 1a, 5a

focal ratio *noun* (1926)
: F-NUMBER

fo'·c'sle *variant of* FORECASTLE

¹fo·cus \'fō-kəs\ *noun, plural* **fo·ci** \'fō-,sī *also* -,kī\ *also* **fo·cus·es** [New Latin, from Latin, hearth] (1644)
1 a : a point at which rays (as of light, heat, or sound) converge or from which they diverge or appear to diverge; *specifically* **:** the point where the geometrical lines or their prolongations conforming to the rays diverging from or converging toward another point intersect and give rise to an image after reflection by a mirror or refraction by a lens or optical system **b :** a point of convergence of a beam of particles (as electrons)
2 a : FOCAL LENGTH **b :** adjustment for distinct vision; *also* **:** the area that may be seen distinctly or resolved into a clear image **c :** a state or condition permitting clear perception or understanding ⟨tried to bring the issues into *focus*⟩ **d :** DIRECTION 6c
3 : one of the fixed points that with the corresponding directrix defines a conic section
4 : a localized area of disease or the chief site of a generalized disease or infection
5 a : a center of activity, attraction, or attention ⟨the *focus* of the meeting was drug abuse⟩ **b :** a point of concentration
6 : the place of origin of an earthquake or moonquake
7 : directed attention **:** EMPHASIS ◆
— fo·cus·less \-ləs\ *adjective*
— in focus : having or giving the proper sharpness of outline due to good focusing
— out of focus : not in focus

²focus *verb* **fo·cused** *also* **fo·cussed; fo·cus·ing** *also* **fo·cus·sing** (1775)
transitive verb
1 a : to bring into focus **b :** to adjust the focus of (as the eye or a lens)
2 : to cause to be concentrated ⟨*focused* their attention on the most urgent problems⟩
3 : to bring (as light rays) to a focus **:** CONCENTRATE
intransitive verb
1 : to come to a focus **:** CONVERGE
2 : to adjust one's eye or a camera to a particular range
3 : to concentrate attention or effort
— fo·cus·able \-kə-sə-bəl\ *adjective*
— fo·cus·er *noun*

fod·der \'fä-dər\ *noun* [Middle English, from Old English *fōdor;* akin to Old High German *fuotar* food — more at FOOD] (before 12th century)
1 : something fed to domestic animals; *especially* **:** coarse food for cattle, horses, or sheep
2 : inferior or readily available material used to supply a heavy demand ⟨routine entertainment *fodder*⟩ ⟨*fodder* for gossip columnists⟩
— fodder *transitive verb*

fod·gel \'fä-jəl\ *adjective* [origin unknown] (1724)
Scottish **:** BUXOM

foe \'fō\ *noun* [Middle English *fo,* from Old English *fāh,* from *fāh,* adjective, hostile; akin to Old High German *gifēh* hostile] (before 12th century)
1 : one who has personal enmity for another
2 a : an enemy in war **b :** ADVERSARY, OPPONENT
3 : one who opposes on principle ⟨a *foe* of needless expenditures⟩
4 : something prejudicial or injurious

foehn *or* **föhn** \'fə(r)n, 'fœn, 'fān\ *noun* [German *Föhn*] (1861)
: a warm dry wind blowing down the side of a mountain

foe·man \'fō-mən\ *noun* (before 12th century)
: FOE 2

foe·tal, foe·tus *chiefly British variant of* FETAL, FETUS

foe·tid *variant of* FETID

foeto- *or* **foeti-** *chiefly British variant of* FETO-

¹fog \'fȯg, fäg\ *noun* [probably of Scandinavian origin; akin to Danish *fog* spray, shower] (1544)
1 a : vapor condensed to fine particles of water suspended in the lower atmosphere that differs from cloud only in being near the ground **b :** a fine spray or a foam for fire fighting
2 : a murky condition of the atmosphere or a substance causing it
3 a : a state of confusion or bewilderment **b :** something that confuses or obscures ⟨hid behind a *fog* of rhetoric⟩
4 : cloudiness or partial opacity in a developed photographic image caused by chemical action or stray radiation
— fog·less \-ləs\ *adjective*

²fog *verb* **fogged; fog·ging** (1599)
transitive verb

◇ WORD HISTORY

focus The Latin word *focus* meant "hearth, fireplace." In the spoken Latin of the Roman Empire, the word must have shifted to the more general meaning "fire" and then ousted the literary Latin word *ignis.* It is the sense "fire" that we see reflected in the Romance languages, where the offspring of *focus* include Italian *fuoco,* Spanish *fuego,* French *feu,* and Romanian *foc.* In the scientific Latin of treatises written in the 17th century, the classical Latin word *focus* reappears, used to refer to the point at which rays of light refracted by a lens converge. Because rays of sunlight when directed by a magnifying glass can produce enough heat to burn paper, a word meaning "fireplace" is quite appropriate metaphorically to describe their convergence point. From this optical sense of *focus* have arisen more extended senses in English such as "point of concentration" or "emphasis."

1 : to cover, envelop, or suffuse with or as if with fog ⟨*fog* the barns with pesticide⟩
2 : to make obscure or confusing ⟨accusations which *fogged* the real issues⟩
3 : to make confused
4 : to produce fog on (as a photographic film) during development
intransitive verb
1 : to become covered or thick with fog
2 a : to become blurred by a covering of fog or mist **b :** to become indistinct through exposure to light or radiation

fog·bound \'fȯg-ˌbau̇nd, 'fäg-\ *adjective* (circa 1855)
1 : unable to move because of fog ⟨*fogbound* ship⟩
2 : covered with or surrounded by fog ⟨*fogbound* coast⟩

fog·bow \-ˌbō\ *noun* (1831)
: a nebulous arc or circle of white or yellowish light sometimes seen in fog

fog·gage \'fó-gij, 'fä-\ *noun* [Scots, from Middle English (Scots) *fogage,* from Anglo Latin *fogagium,* from Middle English *fogge* second growth of grass + Medieval Latin *-agium* -age] (1786)
Scottish **:** a second growth of grass

fog·gy \'fó-gē, 'fä-\ *adjective* **fog·gi·er; -est** (15th century)
1 a : filled or abounding with fog **b :** covered or made opaque by moisture or grime
2 : blurred or obscured as if by fog ⟨hadn't the *foggiest* notion⟩
— fog·gi·ly \'fó-gə-lē, 'fä-\ *adverb*
— fog·gi·ness \'fó-gē-nəs, 'fä-\ *noun*

Foggy Bottom *noun* [*Foggy Bottom,* district in Washington, D.C.] (1951)
: the U.S. Department of State

fog·horn \'fȯg-ˌhȯrn, 'fäg-\ *noun* (1858)
1 : a horn (as on a ship) sounded in a fog to give warning
2 : a loud hoarse voice

fo·gy *also* **fo·gey** \'fō-gē\ *noun, plural* **fogies** *also* **fogeys** [origin unknown] (1780)
: a person with old-fashioned ideas — usually used with *old*
— fo·gy·ish \-gē-ish\ *adjective*
— fo·gy·ism \-gē-ˌi-zəm\ *noun*

foi·ble \'fȯi-bəl\ *noun* [obsolete French (now *faible*), from obsolete *foible* weak, from Old French *feble* feeble] (circa 1648)
1 : the part of a sword or foil blade between the middle and point
2 : a minor flaw or shortcoming in character or behavior **:** WEAKNESS
synonym see FAULT

foie gras \'fwä-'grä\ *noun* [French, literally, fat liver] (1818)
: the fatted liver of an animal and especially of a goose usually served as a pâté

¹foil \'fȯi(ə)l\ *transitive verb* [Middle English, to trample, full cloth, from Middle French *fouler* — more at FULL] (14th century)
1 *obsolete* **:** TRAMPLE
2 a : to prevent from attaining an end **:** DEFEAT **b :** to bring to naught **:** THWART
synonym see FRUSTRATE

²foil *noun* [Middle English, leaf, from Middle French *foille* (from Latin *folia,* plural of *folium*) & *foil,* from Latin *folium* — more at BLADE] (14th century)
1 : very thin sheet metal ⟨aluminum *foil*⟩
2 : a thin piece of material (as metal) put under an inferior or paste stone to add color or brilliance
3 : someone or something that serves as a contrast to another ⟨acted as a *foil* for a comedian⟩
4 a : an indentation between cusps in Gothic tracery **b :** one of several arcs that enclose a complex figure
5 : HYDROFOIL 1

³foil *noun* (15th century)
1 *archaic* **:** DEFEAT
2 *archaic* **:** the track or trail of an animal

⁴foil *noun* [origin unknown] (1594)
1 : a light fencing sword having a usually circular guard and a flexible blade of rectangular section tapering to a blunted point — compare ÉPÉE, SABER
2 : the art or sport of fencing with the foil — often used in plural

⁵foil *transitive verb* (1611)
1 : to back or cover with foil
2 : to enhance by contrast

foiled \'fȯi(ə)ld\ *adjective* (1835)
: ornamented with foils ⟨a *foiled* arch⟩

foils·man \'fȯi(ə)lz-mən\ *noun* (1927)
: one who fences with a foil

¹foin \'fȯin\ *intransitive verb* [Middle English, from *foin* fork for spearing fish, from Middle French *foisne*] (14th century)
archaic **:** to thrust with a pointed weapon **:** LUNGE

²foin *noun* (15th century)
archaic **:** a pass in fencing **:** LUNGE

foi·son \'fȯi-zᵊn\ *noun* [Middle English *foisoun,* from Middle French *foison,* from Latin *fusion-, fusio* pouring, effusion — more at FUSION] (14th century)
1 *archaic* **:** rich harvest
2 *chiefly Scottish* **:** physical energy or strength
3 *plural, obsolete* **:** RESOURCES

foist \'fȯist\ *transitive verb* [probably from obsolete Dutch *vuisten* to take into one's hand, from Middle Dutch *vuysten,* from *vuyst* fist; akin to Old English *fȳst* fist] (circa 1587)
1 a : to introduce or insert surreptitiously or without warrant **b :** to force another to accept especially by stealth or deceit
2 : to pass off as genuine or worthy ⟨*foist* costly and valueless products on the public —Jonathan Spivak⟩

fo·la·cin \'fō-lə-sən\ *noun* [*fol*ic *ac*id + *-in*] (1949)
: FOLIC ACID

fo·late \'fō-ˌlāt\ *noun* (1951)
: FOLIC ACID

¹fold \'fōld\ *noun* [Middle English, from Old English *falod;* akin to Old Saxon *faled* enclosure] (before 12th century)
1 : an enclosure for sheep
2 a : a flock of sheep **b :** a group of people or institutions that share a common faith, belief, activity, or enthusiasm

²fold *transitive verb* (before 12th century)
: to pen up or confine (as sheep) in a fold

³fold *verb* [Middle English, from Old English *fealdan;* akin to Old High German *faldan* to fold, Greek di*plasios* twofold] (before 12th century)
transitive verb
1 : to lay one part over another part of ⟨*fold* a letter⟩
2 : to reduce the length or bulk of by doubling over ⟨*fold* a tent⟩
3 : to clasp together **:** ENTWINE ⟨*fold* the hands⟩
4 : to clasp or enwrap closely **:** EMBRACE
5 : to bend (as a layer of rock) into folds
6 a : to incorporate (a food ingredient) into a mixture by repeated gentle overturnings without stirring or beating **b :** to incorporate closely
7 a : to concede defeat by withdrawing (one's cards) from play (as in poker) **b :** to bring to an end
intransitive verb
1 : to become doubled or pleated
2 : to fail completely **:** COLLAPSE; *especially* **:** to go out of business
3 : to fold one's cards (as in poker)
— fold·able \'fōl-də-bəl\ *adjective*

⁴fold *noun* (13th century)
1 : a part doubled or laid over another part **:** PLEAT
2 : something that is folded together or that enfolds
3 a : a bend or flexure produced in rock by forces operative after the depositing or consolidation of the rock **b** *chiefly British* **:** an undulation in the landscape
4 : a margin apparently formed by the doubling upon itself of a flat anatomical structure (as a membrane)
5 : a crease made by folding something (as a newspaper)

-fold \ˌfōld, 'fōld\ *suffix* [Middle English, from Old English *-feald;* akin to Old High German *-falt* -fold, Latin *-plex, -plus,* Old English *fealdan*]
1 : multiplied by (a specified number) **:** times — in adjectives ⟨a six*fold* increase⟩ and adverbs ⟨repay you ten*fold*⟩
2 : having (so many) parts ⟨three*fold* aspect of the problem⟩

fold·away \'fōl-də-ˌwā\ *adjective* (1948)
: designed to be folded for storage or portability ⟨*foldaway* doors⟩ ⟨*foldaway* bed⟩ ⟨a *foldaway* table⟩

fold·boat \'fōl(d)-ˌbōt\ *noun* [translation of German *Faltboot*] (1938)
: a small collapsible canoe made of rubberized sailcloth stretched over a framework

fold·er \'fōl-dər\ *noun* (1552)
1 : one that folds
2 : a folded printed circular
3 : a folded cover or large envelope for holding or filing loose papers

fol·de·rol \'fäl-də-ˌräl\ *noun* [*fol-de-rol,* a nonsense refrain in songs] (1820)
1 : a useless ornament or accessory **:** TRIFLE
2 : NONSENSE

fold·ing \'fōl-diŋ\ *adjective* (15th century)
: capable of being folded into a more compact shape ⟨*folding* chairs⟩ ⟨a *folding* door⟩

folding money *noun* (circa 1930)
: PAPER MONEY

fold·out \'fōl-ˌdau̇t\ *noun, often attributive* (1950)
: a folded leaf in a publication (as a book) that is larger in some dimension than the page ⟨a *foldout* map⟩

fo·li·a·ceous \ˌfō-lē-'ā-shəs\ *adjective* [Latin *foliaceus,* from *foli*um leaf + *-aceus* -aceous] (1658)
: of, relating to, or resembling an ordinary green leaf as distinguished from a modified leaf (as a petal, bract, or scale)

fo·liage \'fō-lē-ij *also* -lyij; ÷'fō-lij, ÷'fȯi-lij\ *noun* [Middle French *fuellage,* from *foille* leaf — more at FOIL] (1598)
1 : a representation of leaves, flowers, and branches for architectural ornamentation
2 : the aggregate of leaves of one or more plants
3 : a cluster of leaves, flowers, and branches ▫
— fo·liaged \-lē-ijd *also* -lyijd; ÷'fō-lijd, ÷'fȯi-\ *adjective*

foliage plant *noun* (1862)
: a plant grown primarily for its decorative foliage

fo·li·ar \'fō-lē-ər\ *adjective* [French *foliaire,* from Latin *folium* leaf + French *-aire* -ar] (circa 1859)

▫ USAGE
foliage The disyllabic pronunciation \'fō-lij\ is very common. Some commentators insist that *foliage* requires a trisyllabic pronunciation because of its spelling, but words of a similar pattern such as *carriage* and *marriage* do not fall under their prescription. The pronunciation \'fȯi-lij\ is disapproved because it suggests the transposition of the *l* and *i* in the spelling. It is not as common as \'fō-lij\ and may be associated with the nonstandard spelling *foilage.*

: of, relating to, or applied to leaves ⟨*foliar* sprays⟩

fo·li·ate \'fō-lē-ət, -ˌāt\ *adjective* [Latin *foliatus* leafy, from *folium* leaf — more at BLADE] (circa 1658)
1 : shaped like a leaf ⟨a *foliate* sponge⟩
2 : FOLIATED

fo·li·at·ed \-ˌā-təd\ *adjective* (1650)
1 : composed of or separable into layers ⟨a *foliated* rock⟩
2 : ornamented with foils or a leaf design

fo·li·a·tion \ˌfō-lē-'ā-shən\ *noun* (circa 1623)
1 a : the process of forming into a leaf **b :** the state of being in leaf **c :** VERNATION
2 : the numbering of the leaves of a manuscript or early printed book
3 a : ornamentation with foliage **b :** a decoration resembling a leaf
4 : the enrichment of an opening by foils
5 : foliated texture

fo·lic acid \ˌfō-lik-\ *noun* [Latin *folium*] (1941)
: a crystalline vitamin $C_{19}H_{19}N_7O_6$ of the B complex that is used especially in the treatment of nutritional anemias — called also *pteroylglutamic acid*

fo·lie à deux \fō-lē-ä-dœ, ˌfä-lē-ˌä-'də(r)\ *noun* [French, literally, double madness] (circa 1892)
: the presence of the same or similar delusional ideas in two persons closely associated with one another

¹fo·lio \'fō-lē-ˌō\ *noun, plural* **fo·li·os** [Middle English, from Latin, ablative of *folium*] (1447)
1 a : a leaf especially of a manuscript or book **b :** a leaf number **c :** a page number **d :** an identifying reference in accounting used in posting
2 a : a sheet of paper folded once **b :** a case or folder for loose papers
3 a : the size of a piece of paper cut two from a sheet; *also* **:** paper or a page of this size **b :** a book printed on folio pages **c :** a book of the largest size
4 : a certain number of words taken as a unit or division in a document for purposes of measurement or reference

²folio *transitive verb* (1858)
: to put a serial number on each leaf or page of

fo·li·ose \'fō-lē-ˌōs\ *adjective* [Latin *foliosus* leafy] (1758)
: having a flat, thin, and usually lobed thallus attached to the substratum ⟨*foliose* lichens⟩ — compare CRUSTOSE, FRUTICOSE

¹folk \'fōk\ *noun, plural* **folk** *or* **folks** [Middle English, from Old English *folc;* akin to Old High German *folc* people] (before 12th century)
1 *archaic* **:** a group of kindred tribes forming a nation **:** PEOPLE
2 : the great proportion of the members of a people that determines the group character and that tends to preserve its characteristic form of civilization and its customs, arts and crafts, legends, traditions, and superstitions from generation to generation
3 *plural* **:** a certain kind, class, or group of people ⟨old *folks*⟩ ⟨just plain *folk*⟩ ⟨country *folk*⟩ ⟨media *folk*⟩
4 *folks plural* **:** people generally
5 *folks plural* **:** the persons of one's own family; *especially* **:** PARENTS

²folk *adjective* (before 12th century)
1 : originating or traditional with the common people of a country or region and typically reflecting their lifestyle ⟨*folk* hero⟩ ⟨*folk* music⟩
2 : of or relating to the common people or to the study of the common people ⟨*folk* sociology⟩

folk etymology *noun* (1882)
: the transformation of words so as to give them an apparent relationship to other better-known or better-understood words (as in the change of Spanish *cucaracha* to English *cockroach*)

¹folk·ie *also* **folky** \'fō-kē\ *noun, plural* **folkies** (1965)
: a folk singer or instrumentalist

²folkie *or* **folky** *adjective* (1965)
: of or relating to folk music

folk·ish \'fō-kish\ *adjective* (1938)
: FOLKLIKE
— folk·ish·ness *noun*

folk·life \'fōk-ˌlīf\ *noun* (1864)
: the traditions, activities, skills, and products (as handicrafts) of a particular people or group

folk·like \'fō-ˌklīk\ *adjective* (1939)
: having a folk character

folk·lore \'fō-ˌklōr, -ˌklȯr\ *noun* (1846)
1 : traditional customs, tales, sayings, dances, or art forms preserved among a people
2 : a branch of knowledge that deals with folklore
3 : an often unsupported notion, story, or saying that is widely circulated
— folk·lor·ic \-ˌklōr-ik, -ˌklȯr-\ *adjective*
— folk·lor·ish \-ish\ *adjective*
— folk·lor·ist \-ist\ *noun*
— folk·lor·is·tic \ˌfō-ˌklōr-'is-tik, -ˌklȯr-\ *adjective*

folk mass *noun* (1966)
: a mass in which traditional liturgical music is replaced by folk music

folk medicine *noun* (1878)
: traditional medicine as practiced nonprofessionally especially by people isolated from modern medical services and usually involving the use of plant-derived remedies on an empirical basis

folk·moot \'fōk-ˌmüt\ *or* **folk·mote** \-ˌmōt\ *noun* [alteration of Old English *folcmōt, folcgemōt,* from *folc* people + *mōt, gemōt* meeting — more at MOOT] (before 12th century)
: a general assembly of the people (as of a shire) in early England

folk·sing·er \-ˌsiŋ-ər\ *noun* (1884)
: one who sings folk songs or sings in a style associated with folk songs
— folk·sing·ing \-ˌsiŋ-iŋ\ *noun*

folk song *noun* (1847)
: a traditional or composed song typically characterized by stanzaic form, refrain, and simplicity of melody

folksy \'fōk-sē\ *adjective* **folks·i·er; -est** (1852)
1 : SOCIABLE, FRIENDLY
2 : informal, casual, or familiar in manner or style ⟨*folksy* humor⟩
— folks·i·ly \-sə-lē\ *adverb*
— folks·i·ness \-sē-nəs\ *noun*

folk·tale \'fōk-ˌtāl\ *noun* (1852)
: a characteristically anonymous, timeless, and placeless tale circulated orally among a people

folk·way \'fō-ˌkwā\ *noun* (circa 1906)
: a mode of thinking, feeling, or acting common to a given group of people; *especially* **:** a traditional social custom

fol·li·cle \'fä-li-kəl\ *noun* [New Latin *folliculus,* from Latin, diminutive of *follis* bag — more at FOOL] (1646)
1 a : a small anatomical cavity or deep narrow-mouthed depression **b :** a small lymph node **c :** a vesicle in the mammalian ovary that contains a developing egg surrounded by a covering of cells; *especially* **:** GRAAFIAN FOLLICLE
2 : a dry dehiscent one-celled many-seeded fruit (as of the milkweed) that has a single carpel and opens along one suture
— fol·lic·u·lar \fə-'li-kyə-lər, fä-\ *adjective*

follicle mite *noun* (1925)
: any of several minute mites (genus *Demodex*) parasitic in hair follicles

follicle–stimulating hormone *noun* (circa 1943)
: a hormone produced by the anterior lobe of the pituitary gland that stimulates the growth of the ovum-containing follicles in the ovary and activates sperm-forming cells

fol·lic·u·li·tis \fə-ˌli-kyə-'lī-təs\ *noun* [New Latin, from *folliculus* + *-itis*] (circa 1860)
: inflammation of one or more follicles especially of the hair

¹fol·low \'fä-(ˌ)lō, -lə(-w)\ *verb* [Middle English *folwen,* from Old English *folgian;* akin to Old High German *folgēnoun* to follow] (before 12th century)
transitive verb
1 : to go, proceed, or come after ⟨*followed* the guide⟩
2 a : to engage in as a calling or way of life **:** PURSUE ⟨wheat-growing is generally *followed* here⟩ **b :** to walk or proceed along ⟨*follow* a path⟩
3 a : to be or act in accordance with ⟨*follow* directions⟩ **b :** to accept as authority **:** OBEY ⟨*followed* his conscience⟩
4 a : to pursue in an effort to overtake **b :** to seek to attain ⟨*follow* knowledge⟩
5 : to come into existence or take place as a result or consequence of ⟨disaster *followed* the blunder⟩
6 a : to come or take place after in time, sequence, or order **b :** to cause to be followed ⟨*followed* dinner with a liqueur⟩
7 : to copy after **:** IMITATE
8 a : to watch steadily ⟨*followed* the flight of the ball⟩ **b :** to keep the mind on ⟨*follow* a speech⟩ **c :** to attend closely to **:** keep abreast of ⟨*followed* his career with interest⟩ **d :** to understand the sense or logic of (as a line of thought)
intransitive verb
1 : to go or come after a person or thing in place, time, or sequence
2 : to result or occur as a consequence, effect, or inference ☆
— as follows : as comes next — used impersonally
— follow one's nose 1 : to go in a straight or obvious course **2 :** to proceed without plan or reflection **:** obey one's instincts
— follow suit 1 : to play a card of the same suit as the card led **2 :** to follow an example set

²follow *noun* (1870)
1 : the act or process of following
2 : forward spin given to a ball by striking it above center — compare DRAW, ENGLISH

fol·low·er \'fä-lə-wər\ *noun* (before 12th century)
1 a : one in the service of another **:** RETAINER **b :** one that follows the opinions or teachings of another **c :** one that imitates another
2 *archaic* **:** one that chases
3 : a sheet added to the first sheet of an indenture or other deed
4 : a machine part that receives motion from another part
5 : a spring-loaded plate at the bottom of a firearm's magazine that angles cartridges for proper insertion into the chamber

☆ SYNONYMS
Follow, succeed, ensue, supervene mean to come after something or someone. FOLLOW may apply to a coming after in time, position, or logical sequence ⟨speeches *followed* the dinner⟩. SUCCEED implies a coming after immediately in a sequence determined by natural order, inheritance, election, or laws of rank ⟨she *succeeded* her father as head of the business⟩. ENSUE commonly suggests a logical consequence or naturally expected development ⟨after the talk a general discussion *ensued*⟩. SUPERVENE suggests the following or beginning of something unforeseen or unpredictable ⟨*supervening* events brought unhappiness to him⟩. See in addition CHASE.

6 : FAN, DEVOTEE ☆
fol·low·er·ship \-ˌship\ *noun* (circa 1928)
1 : FOLLOWING
2 : the capacity or willingness to follow a leader
¹fol·low·ing \ˈfä-lə-wiŋ\ *adjective* (15th century)
1 : being next in order or time ⟨the *following* day⟩
2 : listed or shown next ⟨trains will leave at the *following* times⟩
²following *noun* (15th century)
: a group of followers, adherents, or partisans
³following *preposition* (circa 1926)
: SUBSEQUENT TO ⟨*following* the lecture tea was served⟩
fol·low–on \ˈfä-lə-ˌwȯn, -ˌwän\ *adjective* (1960)
: being or relating to something that follows as a natural or logical consequence, development, or progression
— follow–on *noun*
follow out *transitive verb* (1762)
1 : to follow to the end or to a conclusion
2 : CARRY OUT, EXECUTE ⟨*followed out* their orders⟩
follow shot *noun* (circa 1909)
1 : a shot in billiards or pool made by striking the cue ball above its center to cause it to continue forward after striking the object ball
2 : a camera shot in which the camera follows the movement of the subject
fol·low–through \ˈfä-lō-ˌthrü, ˌfä-lō-ˈ, -lə-\ *noun* (circa 1899)
1 : the act or an instance of following through
2 : the part of the stroke following the striking of a ball
follow through *intransitive verb* (1895)
1 : to continue a stroke or motion to the end of its arc
2 : to press on in an activity or process especially to a conclusion
¹fol·low–up \ˈfä-lə-ˌwəp\ *noun* (1916)
1 a : the act or an instance of following up **b :** something that follows up
2 : maintenance of contact with or reexamination of a person (as a patient) especially following treatment
3 : a news story presenting new information on a story published earlier
²follow–up *adjective* (1912)
1 : of, relating to, or being something that follows up ⟨*follow-up* action by the police —Frank Faulkner⟩
2 : done, conducted, or administered in the course of following up persons ⟨*follow-up* care for discharged hospital patients⟩
follow up (1792)
transitive verb
1 : to follow with something similar, related, or supplementary ⟨*following up* his convictions with action —G. P. Merrill⟩
2 : to maintain contact with (a person) so as to monitor the effects of earlier activities or treatments
3 : to pursue in an effort to take further action ⟨the police *follow up* leads⟩
intransitive verb
: to take appropriate action ⟨*follow up* on complaints⟩
fol·ly \ˈfä-lē\ *noun, plural* **follies** [Middle English *folie,* from Old French, from *fol* fool] (13th century)
1 : lack of good sense or normal prudence and foresight
2 a : criminally or tragically foolish actions or conduct **b** *obsolete* **:** EVIL, WICKEDNESS; *especially* **:** lewd behavior
3 : a foolish act or idea
4 : an excessively costly or unprofitable undertaking
5 : an often extravagant picturesque building erected to suit a fanciful taste
Fol·som \ˈfōl-səm\ *adjective* [*Folsom,* town in New Mexico] (1928)
: of, relating to, or characteristic of a prehistoric culture of North America on the east side of the Rocky Mountains that is characterized by flint projectile points having a concave base with side projections and a longitudinal groove on each face
fo·ment \ˈfō-ˌment, fō-ˈ\ *transitive verb* [Middle English *fomenten* to apply a warm substance to, from Late Latin *fomentare,* from Latin *fomentum* compress, from *fovēre* to heat, soothe; akin to Lithuanian *degti* to burn, Sanskrit *dahati* it burns] (circa 1613)
: to promote the growth or development of **:** ROUSE, INCITE ⟨*foment* a rebellion⟩
synonym see INCITE
— fo·ment·er *noun*
fo·men·ta·tion \ˌfō-mən-ˈtā-shən, -ˌmen-\ *noun* (14th century)
1 a : the application of hot moist substances to the body to ease pain **b :** the material so applied
2 : the act of fomenting **:** INSTIGATION
¹fond \ˈfänd\ *adjective* [Middle English, from *fonne* fool] (14th century)
1 : FOOLISH, SILLY ⟨*fond* pride⟩
2 a : prizing highly **:** DESIROUS ⟨*fond* of praise⟩ **b :** having an affection or liking — used with *of* ⟨*fond* of music⟩
3 a : foolishly tender **:** INDULGENT ⟨a *fond* mother⟩ **b :** AFFECTIONATE, LOVING ⟨absence makes the heart grow *fonder*⟩
4 : cherished with great affection **:** doted on ⟨our *fondest* hopes⟩
²fond *intransitive verb* (1530)
obsolete **:** to lavish affection **:** DOTE
³fond \ˈfōⁿ\ *noun, plural* **fonds** \ˈfōⁿ(z)\ [French, from Latin *fundus* bottom, piece of property — more at BOTTOM] (1664)
1 : BACKGROUND, BASIS
2 *obsolete* **:** FUND
fon·dant \ˈfän-dənt\ *noun* [French, from present participle of *fondre* to melt — more at FOUND] (1877)
1 : a soft creamy preparation of sugar, water, and flavorings that is used as a basis for candies or icings
2 : a candy consisting chiefly of fondant
fon·dle \ˈfän-dᵊl\ *verb* **fon·dled; fon·dling** \-(d)liŋ, -dᵊl-iŋ\ [frequentative of obsolete *fond* to fondle] (1694)
transitive verb
1 *obsolete* **:** PAMPER
2 : to handle tenderly, lovingly, or lingeringly **:** CARESS
intransitive verb
: to show affection or desire by caressing
— fon·dler \-(d)lər, -dᵊl-ər\ *noun*
fond·ly \ˈfän-(d)lē\ *adverb* (14th century)
1 *archaic* **:** in a foolish manner **:** FOOLISHLY
2 : in a fond manner **:** AFFECTIONATELY
3 : in a willingly credulous manner ⟨it would stun, I *fondly* hoped, the reader —Annie Dillard⟩
fond·ness \ˈfän(d)-nəs\ *noun* (14th century)
1 *obsolete* **:** FOOLISHNESS, FOLLY
2 : tender affection
3 : APPETITE, RELISH ⟨had a *fondness* for argument⟩
fon·due *also* **fon·du** \fän-ˈdü, -ˈdyü, ˈfän-ˌ\ *noun* [French *fondue,* from feminine of *fondu,* past participle of *fondre* to melt] (1878)
1 a (1) **:** a preparation of melted cheese (as Swiss cheese and Gruyère) usually flavored with white wine and kirsch (2) **:** a dish that consists of small pieces of food (as meat or fruit) cooked in or dipped into a hot liquid ⟨beef *fondue*⟩ ⟨chocolate *fondue*⟩ **b :** a chafing dish in which fondue is made
2 : a dish similar to a soufflé usually made with cheese and bread crumbs
F_1 layer \ˈef-ˈwən-\ *noun* (1933)
: the lower of the two layers into which the F region of the ionosphere splits in the daytime that occurs at varying heights from about 80 to 120 miles (130 to 200 kilometers) above the earth's surface
¹font \ˈfänt\ *noun* [Middle English, from Old English, from Late Latin *font-, fons,* from Latin, fountain] (before 12th century)
1 a : a receptacle for baptismal water **b :** a receptacle for holy water **c :** a receptacle for various liquids
2 : SOURCE, FOUNTAIN ⟨a *font* of information⟩
— font·al \ˈfän-tᵊl\ *adjective*
²font *noun* [Middle French *fonte* act of founding, from (assumed) Vulgar Latin *fundita,* feminine of *funditus,* past participle of Latin *fundere* to found, pour — more at FOUND] (circa 1688)
: an assortment or set of type all of one size and style
fon·ta·nel *or* **fon·ta·nelle** \ˌfän-tᵊn-ˈel, ˈfän-tᵊn-ˌ\ *noun* [Middle English *fontinelle* a bodily hollow or pit, from Middle French *fontenele,* diminutive of *fontaine* fountain] (1741)
: a membrane-covered opening in bone or between bones; *specifically* **:** any of the spaces closed by membranous structures between the uncompleted angles of the parietal bones and the neighboring bones of a fetal or young skull
fon·ti·na \fän-ˈtē-nə\ *noun, often capitalized* [Italian] (1938)
: a cheese that is semisoft to hard in texture and mild to medium sharp in flavor
food \ˈfüd\ *noun, often attributive* [Middle English *fode,* from Old English *fōda;* akin to Old High German *fuotar* food, fodder, Latin *panis* bread, *pascere* to feed] (before 12th century)
1 a : material consisting essentially of protein, carbohydrate, and fat used in the body of an organism to sustain growth, repair, and vital processes and to furnish energy; *also* **:** such food together with supplementary substances (as minerals, vitamins, and condiments) **b :** inorganic substances absorbed by plants in gaseous form or in water solution
2 : nutriment in solid form
3 : something that nourishes, sustains, or supplies ⟨*food* for thought⟩
— food·less \-ləs\ *adjective*
— food·less·ness *noun*
food chain *noun* (1926)
: an arrangement of the organisms of an ecological community according to the order of predation in which each uses the next usually lower member as a food source
food·ie \ˈfü-dē\ *noun* (1982)
: a person having an avid interest in the latest food fads
foo dog \ˈfü-\ *noun, often F capitalized* [Chinese (Beijing) *fó* Buddha; from the use of such figures in ceramic or stone as guardians of Buddhist temples] (1953)
: a mythical lion-dog used as a decorative motif in Far Eastern art
food poisoning *noun* (1887)
: an acute gastrointestinal disorder caused by bacteria or their toxic products or by chemical residues in food
food processor *noun* (1974)

☆ SYNONYMS
Follower, adherent, disciple, partisan mean one who gives full loyalty and support to another. FOLLOWER may apply to people who attach themselves either to the person or beliefs of another ⟨an evangelist and his *followers*⟩. ADHERENT suggests a close and persistent attachment ⟨*adherents* to Marxism⟩. DISCIPLE implies a devoted allegiance to the teachings of one chosen as a master ⟨*disciples* of Gandhi⟩. PARTISAN suggests a zealous often prejudiced attachment ⟨*partisans* of the President⟩.

: an electric kitchen appliance with a set of interchangeable blades revolving inside a container

food pyramid *noun* (1949)
: an ecological hierarchy of food relationships in which a chief predator is at the top, each level preys on the next lower level, and usually green plants are at the bottom

food stamp *noun* (1940)
: a government-issued coupon that is sold or given to low-income persons and is redeemable for food

food·stuff \'füd-ˌstəf\ *noun* (1872)
: a substance with food value; *specifically* : the raw material of food before or after processing

food vacuole *noun* (circa 1889)
: a membrane-bound vacuole (as in an amoeba) in which ingested food is digested — see AMOEBA illustration

food·ways \'füd-ˌwāz\ *noun plural* (1946)
: the eating habits and culinary practices of a people, region, or historical period

food web *noun* (1949)
: the totality of interacting food chains in an ecological community

foo·fa·raw \'fü-fə-ˌrȯ\ *noun* [origin unknown] (1934)
1 : frills and flashy finery
2 : a disturbance or to-do over a trifle : FUSS

¹fool \'fül\ *noun* [Middle English, from Old French *fol,* from Late Latin *follis,* from Latin, bellows, bag; akin to Old High German *bolla* blister, *balg* bag — more at BELLY] (13th century)
1 : a person lacking in judgment or prudence
2 a : a retainer formerly kept in great households to provide casual entertainment and commonly dressed in motley with cap, bells, and bauble **b** : one who is victimized or made to appear foolish : DUPE
3 a : a harmlessly deranged person or one lacking in common powers of understanding **b** : one with a marked propensity or fondness for something ⟨a dancing *fool*⟩ ⟨a *fool* for candy⟩
4 : a cold dessert of pureed fruit mixed with whipped cream or custard

²fool *adjective* (13th century)
: FOOLISH, SILLY ⟨barking its *fool* head off⟩

³fool (1593)
intransitive verb
1 a : to behave foolishly ⟨why old men *fool* and children calculate —Shakespeare⟩ — often used with *around* **b** : to meddle, tamper, or experiment especially thoughtlessly or ignorantly ⟨don't *fool* with that gun⟩ — often used with *around*
2 a : to play or improvise a comic role **b** : to speak in jest : JOKE ⟨I was only *fooling*⟩
3 : to contend or fight without serious intent or with less than full strength : TOY ⟨a dangerous man to *fool* with⟩
transitive verb
1 : to make a fool of : DECEIVE
2 *obsolete* : INFATUATE
3 : to spend on trifles or without advantage : FRITTER — used with *away*

fool around *intransitive verb* (1837)
1 : to spend time idly, aimlessly, or frivolously
2 : to engage in casual sexual activity

fool·ery \'fül-rē, 'fü-lə-\ *noun, plural* **-er·ies** (1552)
1 : a foolish act, utterance, or belief
2 : foolish behavior

fool·har·dy \'fül-ˌhär-dē\ *adjective* (13th century)
: foolishly adventurous and bold : RASH
synonym see ADVENTUROUS
— **fool·har·di·ly** \-ˌhär-dᵊl-ē\ *adverb*
— **fool·har·di·ness** \-ˌhär-dē-nəs\ *noun*

fool·ish \'fü-lish\ *adjective* (13th century)
1 : lacking in sense, judgment, or discretion
2 a : ABSURD, RIDICULOUS **b** : marked by a loss of composure : NONPLUSSED, ABASHED
3 : INSIGNIFICANT, TRIFLING, HUMBLE
synonym see SIMPLE
— **fool·ish·ly** *adverb*

fool·ish·ness *noun* (15th century)
1 : foolish behavior : FOLLY
2 : a foolish act or idea

fool·proof \'fül-ˌprüf\ *adjective* (1902)
: so simple, plain, or reliable as to leave no opportunity for error, misuse, or failure ⟨a *foolproof* plan⟩

fools·cap *also* **fool's cap** \'fül-ˌskap\ *noun* (1632)
1 : a cap or hood usually with bells worn by jesters
2 : a conical cap for slow or lazy students
3 *usually foolscap* [from the watermark of a foolscap formerly applied to such paper] : a size of paper formerly standard in Great Britain; *broadly* : a piece of writing paper

foolscap 1

fool's gold *noun* (1872)
: PYRITE; *broadly* : any of various pyritic minerals resembling gold

fool's paradise *noun* (15th century)
: a state of delusory happiness

fool's parsley *noun* (1755)
: a poisonous European weed (*Aethusa cynapium*) of the carrot family that resembles parsley and is naturalized in the northern U.S. and southern Canada

¹foot \'fu̇t\ *noun, plural* **feet** \'fēt\ *also* **foot** [Middle English *fot,* from Old English *fōt;* akin to Old High German *fuot* foot, Latin *ped-, pes,* Greek *pod-, pous*] (before 12th century)
1 : the terminal part of the vertebrate leg upon which an individual stands
2 : an invertebrate organ of locomotion or attachment; *especially* : a ventral muscular surface or process of a mollusk — see CLAM illustration
3 : any of various units of length based on the length of the human foot; *especially* : a unit equal to ⅓ yard and comprising 12 inches — plural *foot* used between a number and a noun ⟨a 10-*foot* pole⟩; plural *feet* or *foot* used between a number and an adjective ⟨6 *feet* tall⟩; see WEIGHT table
4 : the basic unit of verse meter consisting of any of various fixed combinations or groups of stressed and unstressed or long and short syllables
5 a : motion or power of walking or running : STEP ⟨fleet of *foot*⟩ **b** : SPEED, SWIFTNESS ⟨showed early *foot*⟩
6 : something resembling a foot in position or use: as **a** : the lower end of the leg of a chair or table **b** (1) : the basal portion of the sporogonium in mosses (2) : a specialized outgrowth by which the embryonic sporophyte of many ferns and related plants and some seed plants absorbs nourishment from the gametophyte **c** : a piece on a sewing machine that presses the cloth against the feed
7 *foot plural, chiefly British* : INFANTRY
8 : the lower edge (as of a sail)
9 : the lowest part : BOTTOM ⟨the *foot* of the hill⟩
10 a : the end that is lower or opposite the head ⟨the *foot* of the bed⟩ **b** : the part (as of a stocking) that covers the foot
11 foots *plural but singular or plural in construction* : material deposited especially in aging or refining : DREGS
12 foots *plural* : FOOTLIGHTS
— **at one's feet** : under one's spell or influence
— **foot in the door** : the initial step toward a goal
— **off one's feet** : in a sitting or lying position
— **on foot** : by walking or running ⟨tour the campus *on foot*⟩
— **on one's feet 1** : in a standing position **2** : in an established position or state **3** : in a recovered condition (as from illness) ⟨back *on my feet*⟩ **4** : in an extemporaneous manner : while in action ⟨good debaters can think *on their feet*⟩
— **to one's feet** : to a standing position ⟨brought the crowd *to its feet*⟩

²foot (15th century)
intransitive verb
1 : DANCE
2 : to go on foot
3 *of a sailboat* : to make speed : MOVE
transitive verb
1 a : to perform the movements of (a dance) **b** : to walk, run, or dance on, over, or through
2 *archaic* **a** : KICK **b** : REJECT
3 *archaic* : ESTABLISH
4 a : to add up **b** : to pay or stand credit for ⟨*foot* the bill⟩
5 : to make or renew the foot of (as a stocking)

foot·age \'fu̇-tij\ *noun* (1892)
: length or quantity expressed in feet: as **a** : BOARD FEET **b** : the total number of running feet of motion-picture film used (as for a scene or subject); *also* : the material contained on such footage

foot–and–mouth disease *noun* (1862)
: an acute contagious febrile virus disease especially of cloven-footed animals marked by ulcerating vesicles in the mouth, about the hooves, and on the udder and teats — called also *foot-and-mouth, hoof-and-mouth disease*

foot·ball \'fu̇t-ˌbȯl\ *noun* (15th century)
1 : any of several games played between two teams on a rectangular field having two goalposts at each end and whose object is to get the ball over a goal line or between goalposts by running, passing, or kicking: as **a** *British* : SOCCER **b** *British* : RUGBY **c** : an American game played between two teams of 11 players each in which the ball is in possession of one side at a time and is advanced by running or passing **d** *Australian* : AUSTRALIAN RULES FOOTBALL **e** *Canadian* : CANADIAN FOOTBALL
2 a : an inflated oval ball used in the game of football **b** *British* : a soccer ball
3 : something treated roughly especially as the subject of a prolonged dispute ⟨the issue became a political *football* in Congress⟩
— **foot·ball·er** \-ˌbȯ-lər\ *noun*

foot·bath \'fu̇t-ˌbath, -ˌbȧth\ *noun* (1599)
: a bath (as at the entrance to an indoor swimming pool) for cleansing, warming, or disinfecting the feet

foot·board \'fu̇t-ˌbōrd, -ˌbȯrd\ *noun* (1766)
1 : a narrow platform on which to stand or brace the feet
2 : a board forming the foot of a bed

foot·boy \-ˌbȯi\ *noun* (1590)
: a serving boy : PAGE, ATTENDANT

foot·bridge \'fu̇t-ˌbrij\ *noun* (14th century)
: a bridge for pedestrians

foot–can·dle \-'kan-dᵊl\ *noun* (1906)
: a unit of illuminance on a surface that is everywhere one foot from a uniform point source of light of one candle and equal to one lumen per square foot

foot·cloth \ˌklȯth\ *noun* (14th century)
1 *archaic* : an ornamental cloth draped over the back of a horse to reach the ground on each side
2 : CARPET

foot·drag·ger \-ˌdra-gər\ *noun* (1957)
: one who engages in foot-dragging

foot–drag·ging \-ˌdra-giŋ\ *noun* (1952)
: failure to act with the necessary promptness or vigor

foot·ed \'fu̇-təd\ *adjective* (14th century)
: having a foot or feet especially of a specified kind or number — often used in combination ⟨a four-*footed* animal⟩

foot·er \'fu̇-tər\ *noun* (1608)
archaic : PEDESTRIAN

-foot·er \'fu̇-tər\ *noun combining form*
: one that is a specified number of feet in height, length, or breadth ⟨a six-*footer*⟩

foot·fall \ˈfu̇t-ˌfȯl\ *noun* (1610)
: the sound of a footstep

foot fault *noun* (1886)
: an infraction of the service rules (as in tennis, racquetball, or volleyball) that results from illegal placement of the server's feet
— **foot-fault** \ˈfu̇t-ˌfȯlt\ *intransitive verb*

foot·gear \ˈfu̇t-ˌgir\ *noun* (1837)
: FOOTWEAR

foot·hill \-ˌhil\ *noun* (1850)
1 : a hill at the foot of higher hills
2 *plural* **:** a hilly region at the base of a mountain range

foot·hold \-ˌhōld\ *noun* (1625)
1 : a hold for the feet **:** FOOTING
2 : a position usable as a base for further advance

foot·ing \ˈfu̇-tiŋ\ *noun* (14th century)
1 : a stable position or placing of the feet
2 : a surface or its condition with respect to one walking or running on it; *especially* **:** the condition of a racetrack
3 : the act of moving on foot **:** STEP, TREAD
4 a : a place or space for standing **:** FOOTHOLD **b :** established position **:** STATUS; *especially* **:** position or rank in relation to others ⟨they all started off on an equal *footing*⟩
5 : BASIS
6 : terms of social intercourse
7 : an enlargement at the lower end of a foundation wall, pier, or column to distribute the load
8 : the sum of a column of figures

foot·lam·bert \ˈfu̇t-ˌlam-bərt\ *noun* (1925)
: a unit of luminance equal to the luminance of a perfectly diffusing surface that emits or reflects one lumen per square foot

foo·tle \ˈfü-tᵊl\ *intransitive verb* **foo·tled; foo·tling** \-tᵊl-iŋ, ˈfüt-liŋ\ [alteration of *footer* (to footle)] (1892)
1 : to talk or act foolishly
2 : to waste time **:** TRIFLE, FOOL
— **footle** *noun*
— **foo·tler** \-tᵊl-ər, ˈfüt-lər\ *noun*

foot·less \ˈfu̇t-ləs\ *adjective* (14th century)
1 a : having no feet **b :** lacking foundation **:** UNSUBSTANTIAL
2 : STUPID, INEPT
— **foot·less·ly** *adverb*
— **foot·less·ness** *noun*

foot·lights \-ˌlīts\ *noun plural* (circa 1839)
1 : a row of lights set across the front of a stage floor
2 : the stage as a profession

foo·tling \ˈfü-tᵊl-iŋ, ˈfüt-liŋ\ *adjective* [*footle*] (circa 1897)
1 : lacking judgment or ability **:** INEPT ⟨*footling* amateurs who understand nothing —E. R. Bentley⟩
2 : lacking use or value **:** TRIVIAL

foot·lock·er \ˈfu̇t-ˌlä-kər\ *noun* (circa 1942)
: a small trunk designed to be placed at the foot of a bed (as in a barracks)

foot·loose \-ˌlüs\ *adjective* (1873)
: having no ties **:** free to move about

foot·man \-mən\ *noun* (14th century)
1 a *archaic* **:** a traveler on foot **:** PEDESTRIAN **b :** INFANTRYMAN
2 a : a servant in livery formerly attending a rider or required to run in front of his master's carriage **b :** a servant who serves at table, tends the door, and runs errands

foot·mark \-ˌmärk\ *noun* (1826)
: FOOTPRINT

[1]**foot·note** \-ˌnōt\ *noun* (1822)
1 : a note of reference, explanation, or comment usually placed below the text on a printed page
2 : something that is subordinately related to a larger event or work **:** COMMENTARY ⟨that biography is an illuminating *footnote* to our times⟩

[2]**footnote** *transitive verb* (1864)
: to furnish with a footnote **:** ANNOTATE

foot·pace \ˈfu̇t-ˌpās\ *noun* (1538)
1 : a walking pace
2 : PLATFORM, DAIS

[1]**foot·pad** \-ˌpad\ *noun* [*foot* + *pad* (highwayman)] (1683)
: one who robs a pedestrian

[2]**footpad** *noun* [*foot* + [1]*pad*] (1966)
: a flattish foot on the leg of a spacecraft for distributing weight to minimize sinking into a surface

foot·path \ˈfu̇t-ˌpath, -ˌpȧth\ *noun* (1526)
: a narrow path for pedestrians

foot–pound \-ˈpau̇nd\ *noun, plural* **foot–pounds** (1850)
: a unit of work equal to the work done by a force of one pound acting through a distance of one foot in the direction of the force

foot–pound–second *adjective* (1892)
: being or relating to a system of units based upon the foot as the unit of length, the pound as the unit of weight and the second as the unit of time — abbreviation *fps*

foot·print \ˈfu̇t-ˌprint\ *noun* (1552)
1 : an impression of the foot on a surface
2 : the area on a surface covered by something ⟨a tire with a wide *footprint*⟩ ⟨the *footprint* of a laser beam on the surface of a planet⟩

foot·race \-ˌrās\ *noun* (1663)
: a race run by humans on foot

foot·rest \-ˌrest\ *noun* (1861)
: a support for the feet

foot·rope \-ˌrōp\ *noun* (1772)
1 : the part of a boltrope sewed to the lower edge of a sail
2 : a rope rigged below a yard for men to stand on

foot rot *noun* (1807)
1 : a progressive inflammation of the feet of sheep or cattle that is associated with bacterial infection
2 : a plant disease marked by rot of the stem near the ground

foot·sie *or* **foot·sy** \ˈfu̇t-sē\ *noun* [diminutive of [1]*foot*] (1944)
1 : a furtive flirtatious caressing with the feet (as under a table)
2 : a usually surreptitious cooperation or negotiation with someone supposed hostile to one's own interests — usually used with *play*

foot·slog \ˈfu̇t-ˌsläg\ *intransitive verb* (1899)
: to march or tramp through mud
— **foot·slog·ger** *noun*

foot soldier *noun* (1622)
: INFANTRYMAN

foot·sore \ˈfu̇t-ˌsōr, -ˌsȯr\ *adjective* (1719)
: having sore or tender feet (as from much walking)
— **foot·sore·ness** *noun*

foot·step \-ˌstep\ *noun* (13th century)
1 : the mark of the foot **:** TRACK
2 a : TREAD **b :** distance covered by a step **:** PACE
3 : a step on which to ascend or descend
4 : a way of life, conduct, or action ⟨followed in his father's *footsteps*⟩

foot·stone \-ˌstōn\ *noun* (1724)
: a stone placed at the foot of a grave

foot·stool \-ˌstül\ *noun* (1530)
: a low stool used to support the feet

foot·wall \-ˌwȯl\ *noun* (1860)
1 : the lower underlying wall of a vein, ore deposit, or coal seam in a mine
2 : the lower wall of an inclined fault

foot·way \-ˌwā\ *noun* (15th century)
: a narrow way or path for pedestrians

foot·wear \-ˌwar, -ˌwer\ *noun* (1881)
: wearing apparel (as shoes or boots) for the feet

foot·work \-ˌwərk\ *noun* (1895)
1 : the management of the feet (as in boxing); *also* **:** the work done with them
2 : the activity of moving from place to place ⟨the investigation entailed a lot of *footwork*⟩
3 : active and adroit maneuvering to achieve an end ⟨fancy political *footwork*⟩

[1]**foo·zle** \ˈfü-zəl\ *noun* (1890)
: an act of foozling; *especially* **:** a bungling golf stroke

[2]**foozle** *transitive verb* **foo·zled; foo·zling** \ˈfüz-liŋ, ˈfü-zə-\ [perhaps from German dialect *fuseln* to work carelessly] (1892)
: to manage or play awkwardly **:** BUNGLE

[1]**fop** \ˈfäp\ *noun* [Middle English; akin to Middle English *fobben* to deceive, Middle High German *voppen*] (15th century)
1 *obsolete* **:** a foolish or silly person
2 : a man who is devoted to or vain about his appearance or dress **:** COXCOMB, DANDY

[2]**fop** *transitive verb* **fopped; fop·ping** (1602)
obsolete **:** FOOL, DUPE

fop·pery \ˈfä-p(ə-)rē\ *noun, plural* **-per·ies** (1546)
1 : foolish character or action **:** FOLLY
2 : the behavior or dress of a fop

fop·pish \ˈfä-pish\ *adjective* (1599)
1 *obsolete* **:** FOOLISH, SILLY
2 a : characteristic of a fop ⟨a *foppish* dressing gown⟩ **b :** behaving or dressing in the manner of a fop
— **fop·pish·ly** *adverb*
— **fop·pish·ness** *noun*

[1]**for** \fər, (ˈ)fȯr, *Southern also* (ˈ)fär\ *preposition* [Middle English, from Old English; akin to Latin *per* through, *prae* before, *pro* before, for, ahead, Greek *pro*, Old English *faran* to go — more at FARE] (before 12th century)
1 a — used as a function word to indicate purpose ⟨a grant *for* studying medicine⟩ **b** — used as a function word to indicate an intended goal ⟨left *for* home⟩ ⟨acted *for* the best⟩ **c** — used as a function word to indicate the object or recipient of a perception, desire, or activity ⟨now *for* a good rest⟩ ⟨run *for* your life⟩ ⟨an eye *for* a bargain⟩
2 a : as being or constituting ⟨taken *for* a fool⟩ ⟨eggs *for* breakfast⟩ **b** — used as a function word to indicate an actual or implied enumeration or selection ⟨*for* one thing, the price is too high⟩
3 : because of ⟨can't sleep *for* the heat⟩
4 — used as a function word to indicate suitability or fitness ⟨it is not *for* you to choose⟩ ⟨ready *for* action⟩
5 a : in place of ⟨go to the store *for* me⟩ **b** (1) **:** on behalf of **:** REPRESENTING ⟨speaks *for* the court⟩ (2) **:** in favor of ⟨all *for* the plan⟩
6 : in spite of — usually used with *all* ⟨*for* all his large size, he moves gracefully⟩
7 : with respect to **:** CONCERNING ⟨a stickler *for* detail⟩ ⟨heavy *for* its size⟩
8 a — used as a function word to indicate equivalence in exchange ⟨$10 *for* a hat⟩, equality in number or quantity ⟨point *for* point⟩, or correspondence or correlation ⟨*for* every one that works, you'll find five that don't⟩ **b** — used as a function word to indicate number of attempts ⟨0 *for* 4⟩
9 — used as a function word to indicate duration of time or extent of space ⟨gone *for* two days⟩
10 : in honor of **:** AFTER ⟨named *for* her grandmother⟩

[2]**for** *conjunction* (12th century)
: for the reason that **:** on this ground **:** BECAUSE

for- *prefix* [Middle English, from Old English; akin to Old High German *far-* for-, Old English *for*]
1 : so as to involve prohibition, exclusion, omission, failure, neglect, or refusal ⟨*for*bid⟩
2 : destructively or detrimentally ⟨*for*do⟩
3 : completely **:** excessively **:** to exhaustion **:** to pieces ⟨*for*spent⟩

fora *plural of* FORUM

[1]**for·age** \ˈfȯr-ij, ˈfär-\ *noun* [Middle English, from Old French, from *forre* fodder, of Germanic origin; akin to Old High German *fuotar* food, fodder — more at FOOD] (14th century)

1 : food for animals especially when taken by browsing or grazing
2 [[2]*forage*] **:** the act of foraging **:** search for provisions

[2]**forage** *verb* **for·aged; for·ag·ing** (15th century)
transitive verb
1 : to strip of provisions **:** collect forage from
2 : to secure by foraging ⟨*foraged* a chicken for the feast⟩
intransitive verb
1 : to wander in search of forage or food
2 : to secure forage (as for horses) by stripping the country
3 : RAVAGE, RAID
4 : to make a search **:** RUMMAGE
— for·ag·er *noun*

fo·ram \'fōr-əm, 'fȯr-\ *noun* (1927)
: FORAMINIFER

fo·ra·men \fə-'rā-mən\ *noun, plural* **fo·ram·i·na** \-'ra-mə-nə\ *or* **fo·ra·mens** \-'rā-mənz\ [Latin *foramin-, foramen,* from *forare* to bore — more at BORE] (1671)
: a small opening, perforation, or orifice **:** FENESTRA
— fo·ram·i·nal \fə-'ra-mə-nᵊl\ *or* **fo·ram·i·nous** \-mə-nəs\ *adjective*

fo·ra·men mag·num \fə-ˌrā-mən-'mag-nəm\ *noun* [New Latin, literally, large opening] (1860)
: the opening in the skull through which the spinal cord passes to become the medulla oblongata

foramen ova·le \-ō-'va-lē, -'vä-, -'vā-\ *noun* [New Latin, literally, oval opening] (circa 1860)
: an opening in the septum between the two atria of the heart that is normally present only in the fetus

for·a·min·i·fer \ˌfȯr-ə-'mi-nə-fər, ˌfär-\ *noun* (circa 1842)
: any of an order (Foraminifera) of large chiefly marine rhizopod protozoans usually having calcareous shells that often are perforated with minute holes for protrusion of slender pseudopodia and form the bulk of chalk and nummulitic limestone
— fo·ra·mi·nif·er·al \fə-ˌra-mə-'ni-f(ə-)rəl; ˌfȯr-ə-mə-'ni-, ˌfär-\ *adjective*

fo·ra·mi·nif·era \fə-ˌra-mə-'ni-f(ə-)rə; ˌfȯr-ə-mə-'ni-, ˌfär-\ *noun plural* [New Latin, from Latin *foramin-, foramen* + *-fera,* neuter plural of *-fer* -fer] (circa 1836)
: organisms that are foraminifers

fo·ra·mi·nif·er·an \-f(ə-)rən\ *noun* (1920)
: FORAMINIFER

for and *conjunction* (circa 1529)
obsolete **:** and also

for·as·much as \'fȯr-əz-ˌmə-chəz\ *conjunction* (13th century)
: in view of the fact that

[1]**for·ay** \'fȯr-ˌā, 'fōr-, 'fär- *also* fō-'rā *or* fə-'\ *verb* [Middle English *forrayen,* from Middle French *forrer,* from *forre* fodder — more at FORAGE] (14th century)
transitive verb
archaic **:** to ravage in search of spoils **:** PILLAGE
intransitive verb
: to make a raid or brief invasion ⟨*forayed* into enemy territory⟩
— for·ay·er *noun*

[2]**foray** *noun* (14th century)
1 : a sudden or irregular invasion or attack for war or spoils **:** RAID
2 : a brief excursion or attempt especially outside one's accustomed sphere ⟨the novelist's *foray* into nonfiction⟩

forb \'fȯrb\ *noun* [Greek *phorbē* fodder, food, from *pherbein* to graze] (1924)
: an herb other than grass

[1]**for·bear** \fȯr-'bar, fər-, -'ber\ *verb* **-bore** \-'bōr, -'bȯr\; **-borne** \-'bōrn, -'bȯrn\; **-bear·ing** [Middle English *forberen,* from Old English *forberan* to endure, do without, from *for-* + *beran* to bear] (before 12th century)
transitive verb
1 *obsolete* **:** to do without
2 : to hold oneself back from especially with an effort
3 *obsolete* **:** to leave alone **:** SHUN ⟨*forbear* his presence —Shakespeare⟩
intransitive verb
1 : HOLD BACK, ABSTAIN ⟨can write with ease what I *forbear* to read —Flannery O'Connor⟩
2 : to control oneself when provoked **:** be patient
— for·bear·er *noun*

[2]**forbear** *variant of* FOREBEAR

for·bear·ance \fȯr-'bar-ən(t)s, fər-, -'ber-\ *noun* (1576)
1 : a refraining from the enforcement of something (as a debt, right, or obligation) that is due
2 : the act of forbearing **:** PATIENCE
3 : the quality of being forbearing **:** LENIENCY

[1]**for·bid** \fər-'bid, fȯr-\ *transitive verb* **-bade** \-'bad, -'bād\ *also* **-bad** \-'bad\; **-bid·den** \-'bi-dᵊn\; **-bid·ding** [Middle English *forbidden,* from Old English *forbēodan,* from *for-* + *bēodan* to bid — more at BID] (before 12th century)
1 : to proscribe from or as if from the position of one in authority **:** command against ⟨the law *forbids* stores to sell liquor to minors⟩ ⟨her mother *forbids* her to go⟩
2 : to hinder or prevent as if by an effectual command ⟨space *forbids* further treatment here⟩ ☆
— for·bid·der *noun*

[2]**forbid** *adjective* (1606)
archaic **:** ACCURSED ⟨he shall live a man *forbid* —Shakespeare⟩

for·bid·dance \fər-'bi-dᵊn(t)s, fȯr-\ *noun* (circa 1611)
: the act of forbidding

for·bid·den \-'bi-dᵊn\ *adjective* (13th century)
1 : not permitted or allowed
2 : not conforming to the usual selection principles — used of quantum phenomena ⟨*forbidden* transition⟩ ⟨*forbidden* radiation⟩ ⟨*forbidden* spectral line⟩

forbidden fruit *noun* [from the forbidden fruit of the Garden of Eden in Genesis 3:2–19] (1662)
: an immoral or illegal pleasure

for·bid·ding *adjective* (1712)
1 : such as to make approach or passage difficult or impossible ⟨*forbidding* walls⟩
2 : DISAGREEABLE, REPELLENT ⟨a *forbidding* task⟩
3 : GRIM, MENACING ⟨a dark *forbidding* sky⟩
— for·bid·ding·ly \-'bi-diŋ-lē\ *adverb*

forbode *variant of* FOREBODE

[1]**for·by** *or* **for·bye** \fȯr-'bī\ *preposition* [Middle English *forby,* preposition & adverb, from *fore-* + *by*] (14th century)
1 *archaic* **a :** PAST **b :** NEAR
2 *chiefly Scottish* **:** BESIDES

[2]**forby** *or* **forbye** *adverb* (1590)
chiefly Scottish **:** BESIDES **:** in addition

[1]**force** \'fōrs, 'fȯrs\ *noun* [Middle English, from Middle French, from (assumed) Vulgar Latin *fortia,* from Latin *fortis* strong] (14th century)
1 a (1) **:** strength or energy exerted or brought to bear **:** cause of motion or change **:** active power ⟨the *forces* of nature⟩ ⟨the motivating *force* in her life⟩ (2) *usually capitalized* — used with a number to indicate the strength of the wind according to the Beaufort scale ⟨a *Force* 10 hurricane⟩ **b :** moral or mental strength **c :** capacity to persuade or convince ⟨the *force* of the argument⟩
2 a : military strength **b** (1) **:** a body (as of troops or ships) assigned to a military purpose (2) *plural* **:** the whole military strength (as of a nation) **c :** a body of persons or things available for a particular end ⟨a labor *force*⟩ ⟨the missile *force*⟩ **d :** an individual or group having the power of effective action ⟨join *forces* to prevent violence⟩ ⟨a *force* in politics⟩ **e** *often capitalized* **:** POLICE FORCE — usually used with *the*
3 : violence, compulsion, or constraint exerted upon or against a person or thing
4 a : an agency or influence that if applied to a free body results chiefly in an acceleration of the body and sometimes in elastic deformation and other effects **b :** any of the natural influences (as electromagnetism, gravity, the strong force, and the weak force) that exist especially between particles and determine the structure of the universe
5 : the quality of conveying impressions intensely in writing or speech
synonym see POWER
— force·less \-ləs\ *adjective*
— in force 1 : in great numbers ⟨picnickers were out *in force*⟩ **2 :** VALID, OPERATIVE ⟨the ban remains *in force*⟩

[2]**force** *transitive verb* **forced; forc·ing** (14th century)
1 : to do violence to; *especially* **:** RAPE
2 : to compel by physical, moral, or intellectual means
3 : to make or cause especially through natural or logical necessity ⟨*forced* to admit my error⟩
4 a : to press, drive, attain to, or effect against resistance or inertia ⟨*force* your way through⟩ **b :** to impose or thrust urgently, importunately, or inexorably ⟨*force* unwanted attentions on a woman⟩
5 : to achieve or win by strength in struggle or violence: as **a :** to win one's way into ⟨*force* a castle⟩ ⟨*forced* the mountain passes⟩ **b :** to break open or through ⟨*force* a lock⟩
6 a : to raise or accelerate to the utmost ⟨*forcing* the pace⟩ **b :** to produce only with unnatural or unwilling effort ⟨*forced* a smile⟩ **c :** to wrench, strain, or use (language) with marked unnaturalness and lack of ease
7 a : to hasten the rate of progress or growth of **b :** to bring (as plants) to maturity out of the normal season ⟨*forcing* lilies for Easter⟩
8 : to induce (as a particular bid or play by another player) in a card game by some conventional act, play, bid, or response
9 a : to cause (a runner in baseball) to be put out on a force-out **b :** to cause (a run) to be scored in baseball by giving a base on balls when the bases are full ☆
— forc·er *noun*
— force one's hand : to cause one to act precipitously **:** force one to reveal one's purpose or intention

forced \'fōrst, 'fȯrst\ *adjective* (circa 1537)
1 : compelled by force or necessity **:** INVOLUNTARY ⟨a *forced* landing⟩
2 : done or produced with effort, exertion, or pressure ⟨a *forced* laugh⟩
— forced·ly \'fōr-səd-lē, 'fȯr-\ *adverb*

force–feed *transitive verb* **-fed; -feed·ing** (1901)
1 : to feed (as an animal) by forcible administration of food

☆ SYNONYMS
Forbid, prohibit, interdict, inhibit mean to debar one from doing something or to order that something not be done. FORBID implies that the order is from one in authority and that obedience is expected ⟨smoking is *forbidden* in the building⟩. PROHIBIT suggests the issuing of laws, statutes, or regulations ⟨*prohibited* the sale of liquor⟩. INTERDICT implies prohibition by civil or ecclesiastical authority usually for a given time or a declared purpose ⟨practices *interdicted* by the church⟩. INHIBIT implies the imposition of restraints or restrictions that amount to prohibitions, not only by authority but also by the exigencies of the time or situation ⟨conditions *inhibiting* the growth of free trade⟩.

2 : to force to take in ⟨*force-feed* students the classics⟩ — also used with a single object ⟨*force-feed* the classics to students⟩ ⟨*force-feed* students with the classics⟩

force·ful \'fōrs-fəl, 'fȯrs-\ *adjective* (1571)
: possessing or filled with force **:** EFFECTIVE
— force·ful·ly \-fə-lē\ *adverb*
— force·ful·ness *noun*

force ma·jeure \ˌfōr-smä-'zhər, ˌfȯr-, -smə-\ *noun* [French, superior force] (1883)
1 : superior or irresistible force
2 : an event or effect that cannot be reasonably anticipated or controlled — compare ACT OF GOD

force·meat \'fōrs-ˌmēt, 'fȯrs-\ *noun* [*force* (alteration of ²*farce*) + *meat*] (circa 1688)
: finely chopped and highly seasoned meat or fish that is either served alone or used as a stuffing — called also *farce*

force of habit (circa 1925)
: behavior made involuntary or automatic by repeated practice

force of nature (1981)
: FORCE 4b

force–out \'fōrs-ˌau̇t, 'fȯrs-\ *noun* (1896)
: a play in baseball in which a runner is put out by being forced to advance to the next base but failing to do so safely

force play *noun* (1912)
: FORCE-OUT

for·ceps \'fȯr-səps, -ˌseps\ *noun, plural* **for·ceps** [Latin, tongs, perhaps from *formus* warm + *capere* to take — more at THERM, HEAVE] (1634)
: an instrument for grasping, holding firmly, or exerting traction upon objects especially for delicate operations (as by jewelers or surgeons)
— for·ceps·like \-ˌlīk\ *adjective*

force pump *noun* (1659)
: a pump with a solid piston for drawing and forcing through valves a liquid (as water) to a considerable height above the pump or under a considerable pressure

forc·ible \'fōr-sə-bəl, 'fȯr-\ *adjective* (15th century)
1 : effected by force used against opposition or resistance
2 : characterized by force, efficiency, or energy **:** POWERFUL
— forc·ible·ness *noun*
— forc·ibly \-blē\ *adverb*

¹ford \'fōrd, 'fȯrd\ *noun* [Middle English, from Old English; akin to Old Norse *fjǫrthr* fjord, Latin *portus* port, Old English *faran* to go — more at FARE] (before 12th century)
: a shallow part of a body of water that may be crossed by wading

²ford *transitive verb* (1614)
: to cross (a body of water) by wading
— ford·able \'fōr-də-bəl, 'fȯr-\ *adjective*

for·do \fȯr-'dü, fōr-\ *transitive verb* **-did** \-'did\; **-done** \-'dən\; **-do·ing** \-'dü-iŋ\ [Middle English *fordon,* from Old English *fordōn,* from *for-* + *dōn* to do] (before 12th century)
1 *archaic* **:** to do away with **:** DESTROY
2 : to overcome with fatigue — used only as past participle ⟨quite *fordone* with the heat⟩

¹fore \'fōr, 'fȯr\ *adverb* [Middle English, from Old English; akin to Old English *for*] (before 12th century)
1 *obsolete* **:** at an earlier time or period
2 : in, toward, or adjacent to the front **:** FORWARD

²fore *also* **'fore** *preposition* (before 12th century)
1 *chiefly dialect* **:** BEFORE
2 : in the presence of

³fore *adjective* [*fore-*] (15th century)
1 : situated in front of something else **:** FORWARD
2 : prior in order of occurrence **:** FORMER

⁴fore *noun* (1842)
: something that occupies a front position
— to the fore : in or into a position of prominence **:** FORWARD

⁵fore *interjection* [probably short for *before*] (circa 1878)
— used by a golfer to warn anyone within range of the probable line of flight of the ball

fore- *combining form* [Middle English *for-, fore-,* from Old English *fore-,* from *fore,* adverb]
1 a : earlier **:** beforehand ⟨*fore*see⟩ **b :** occurring earlier **:** occurring beforehand ⟨*fore*shock⟩
2 a : situated at the front **:** in front ⟨*fore*leg⟩ **b :** front part of (something specified) ⟨*fore*arm⟩ **c :** foremast ⟨*fore*top⟩

fore–and–aft \ˌfōr-ə-'naft, ˌfȯr-\ *adjective* (1820)
1 : lying, running, or acting in the general line of the length of a construction (as a ship or a house) **:** LONGITUDINAL
2 : having no square sails

fore and aft *adverb* (circa 1618)
1 : lengthwise of a ship **:** from stem to stern
2 : in, at, or toward both the bow and stern
3 : in or at the front and back or the beginning and end

fore–and–aft·er \-'naf-tər\ *noun* (1823)
: a ship with a fore-and-aft rig; *especially* **:** SCHOONER

fore–and–aft rig *noun* (1879)
: a sailing-ship rig in which most or all of the sails are not attached to yards but are bent to gaffs or set on the masts or on stays in a fore-and-aft line

¹fore·arm \(ˌ)fōr-'ärm, (ˌ)fȯr-\ *transitive verb* (1592)
: to arm in advance **:** PREPARE

²fore·arm \'fōr-ˌärm, 'fȯr-\ *noun* (1741)
: the part of the arm between the elbow and the wrist; *also* **:** the corresponding part in other vertebrates

fore·bay \'fōr-ˌbā, 'fȯr-\ *noun* (1770)
: a reservoir or canal from which water is taken to run equipment (as a waterwheel or turbine)

fore·bear \-ˌbar, -ˌber\ *noun* [Middle English (Scots), from *fore-* + *-bear* (from *been* to be)] (15th century)
: ANCESTOR, FOREFATHER; *also* **:** PRECURSOR — usually used in plural

fore·bode \(ˌ)fōr-'bōd, (ˌ)fȯr-\ (1603)
transitive verb
1 : to have an inward conviction of (as coming ill or misfortune)
2 : FORETELL, PORTEND
intransitive verb
: AUGUR, PREDICT
— fore·bod·er *noun*

¹fore·bod·ing \-'bō-diŋ\ *noun* (14th century)
: the act of one who forebodes; *also* **:** an omen, prediction, or presentiment especially of coming evil **:** PORTENT

²foreboding *adjective* (1679)
: indicative of or marked by foreboding
— fore·bod·ing·ly \-diŋ-lē\ *adverb*
— fore·bod·ing·ness *noun*

fore·brain \'fōr-ˌbrān, 'fȯr-\ *noun* (1879)
: the anterior of the three primary divisions of the developing vertebrate brain or the corresponding part of the adult brain that includes especially the cerebral hemispheres, the thalamus, and the hypothalamus and that especially in higher vertebrates is the main control center for sensory and associative information processing, visceral functions, and voluntary motor functions — called also *prosencephalon;* compare DIENCEPHALON, TELENCEPHALON

fore·cad·die \-ˌka-dē\ *noun* (1792)
: a golf caddie who is stationed in the fairway and who indicates the position of balls on the course

¹fore·cast \-ˌkast; fōr-'kast, fȯr-'\ *verb* **fore·cast** *also* **fore·cast·ed; fore·cast·ing** (15th century)
transitive verb
1 a : to calculate or predict (some future event or condition) usually as a result of study and analysis of available pertinent data; *especially* **:** to predict (weather conditions) on the basis of correlated meteorological observations **b :** to indicate as likely to occur
2 : to serve as a forecast of **:** PRESAGE ⟨such events may *forecast* peace⟩
intransitive verb
: to calculate the future
synonym see FORETELL
— fore·cast·able \-ˌkas-tə-bəl\ *adjective*
— fore·cast·er *noun*

²fore·cast \'fōr-ˌkast, 'fȯr-\ *noun* (circa 1541)
1 *archaic* **:** foresight of consequences and provision against them **:** FORETHOUGHT
2 : a prophecy, estimate, or prediction of a future happening or condition

fore·cas·tle \'fōk-səl; 'fōr-ˌka-səl, 'fȯr-\ *noun* (15th century)
1 : the forward part of the upper deck of a ship
2 : the crew's quarters usually in a ship's bow

fore·check \'fōr-ˌchek, 'fȯr-\ *intransitive verb* (1951)
: to check an opponent in ice hockey in his own defensive zone
— fore·check·er \-ˌche-kər\ *noun*

fore·close \(ˌ)fōr-'klōz, (ˌ)fȯr-\ *verb* [Middle English, from Middle French *forclos,* past participle of *forclore,* from *fors* outside (from Latin *foris*) + *clore* to close — more at FORUM] (15th century)
transitive verb
1 : to shut out **:** PRECLUDE
2 : to hold exclusively
3 : to deal with or close in advance
4 : to subject to foreclosure proceedings
intransitive verb
: to foreclose a mortgage

fore·clo·sure \-'klō-zhər\ *noun* (1728)
: an act or instance of foreclosing; *specifically* **:** a legal proceeding that bars or extinguishes a mortgagor's right of redeeming a mortgaged estate

fore·court \'fōr-ˌkōrt, 'fȯr-, -ˌkȯrt\ *noun* (1535)
1 : an open court in front of a building
2 : the area near the net in a court game

fore·deck \'fōr-ˌdek, 'fȯr-\ *noun* (1565)
: the forepart of a ship's main deck

fore·do *variant of* FORDO

fore·doom \(ˌ)fōr-'düm, (ˌ)fȯr-\ *transitive verb* (1603)
: DOOM 2

fore·face \'fōr-ˌfās, 'fȯr-\ *noun* (1545)
: the part of the head of a quadruped that is in front of the eyes

fore·fa·ther \-ˌfä-t͟hər, -ˌfȧ-\ *noun* (14th century)
1 : ANCESTOR 1a
2 : a person of an earlier period and common heritage

☆ SYNONYMS
Force, compel, coerce, constrain, oblige mean to make someone or something yield. FORCE is the general term and implies the overcoming of resistance by the exertion of strength, power, or duress ⟨*forced* to flee for their lives⟩. COMPEL typically suggests overcoming of resistance or unwillingness by an irresistible force ⟨*compelled* to admit my mistake⟩. COERCE suggests overcoming resistance or unwillingness by actual or threatened violence or pressure ⟨*coerced* into signing over the rights⟩. CONSTRAIN suggests the effect of a force or circumstance that limits freedom of action or choice ⟨*constrained* by conscience⟩. OBLIGE implies the constraint of necessity, law, or duty ⟨felt *obliged* to go⟩.

fore·feel \(ˌ)fōr-ˈfēl, (ˌ)fȯr-\ *transitive verb* **-felt** \-ˈfelt\; **-feel·ing** (1580)
: to have a presentiment of

fore·fend *variant of* FORFEND

fore·fin·ger \ˈfōr-ˌfiŋ-gər, ˈfȯr-\ *noun* (15th century)
: the finger next to the thumb — called also *index finger*

fore·foot \-ˌfu̇t\ *noun* (14th century)
1 a : one of the anterior feet especially of a quadruped **b :** the front part of the human foot
2 : the forward part of a ship where the stem and keel meet

fore·front \-ˌfrənt\ *noun* (15th century)
: the foremost part or place

fore·gath·er *variant of* FORGATHER

¹fore·go \fōr-ˈgō, fȯr-\ *transitive verb* **-went** \-ˈwent\; **-gone** \-ˈgȯn *also* -ˈgän\; **-go·ing** \-ˈgō-iŋ, -ˈgȯ(-)iŋ\ (before 12th century)
: to go before : PRECEDE
— **fore·go·er** \-ˈgō(-ə)r\ *noun*

²forego *variant of* FORGO

fore·go·ing \-ˈgō-iŋ, -ˈgȯ(-)iŋ\ *adjective* (15th century)
: listed, mentioned, or occurring before ⟨the *foregoing* statement can be proven⟩
synonym see PRECEDING

fore·gone \ˈfōr-ˌgȯn, ˈfȯr- *also* -ˌgän\ *adjective* (circa 1600)
: PREVIOUS, PAST

foregone conclusion *noun* (1604)
1 : a conclusion that has preceded argument or examination
2 : an inevitable result : CERTAINTY ⟨the victory was a *foregone conclusion*⟩

¹fore·ground \ˈfōr-ˌgrau̇nd, ˈfȯr-\ *noun* (1695)
1 : the part of a scene or representation that is nearest to and in front of the spectator
2 : a position of prominence : FOREFRONT

²foreground *transitive verb* (1892)
: to bring to the foreground; *especially* : to give prominence or emphasis to

fore·gut \ˈfōr-ˌgət, ˈfȯr-\ *noun* (circa 1889)
: the anterior part of the alimentary canal of a vertebrate embryo that develops into the pharynx, esophagus, stomach, and extreme anterior part of the intestine

¹fore·hand \-ˌhand\ *noun* (1557)
1 *archaic* **:** superior position : ADVANTAGE
2 : the part of a horse that is before the rider
3 : a forehand stroke (as in tennis or racquets); *also* : the side on which such strokes are made

forehand 3

²forehand *adjective* (1599)
1 *obsolete* **:** done or given in advance : PRIOR
2 : made with the palm of the hand turned in the direction in which the hand is moving ⟨a *forehand* tennis stroke⟩

³forehand *adverb* (1925)
: with a forehand stroke

fore·hand·ed \(ˌ)fōr-ˈhan-dəd, (ˌ)fȯr-\ *adjective* (1650)
1 a : mindful of the future : PRUDENT **b :** WELL-TO-DO
2 : FOREHAND 2
— **fore·hand·ed·ly** *adverb*
— **fore·hand·ed·ness** *noun*

fore·head \ˈfär-əd, ˈfȯr-; ˈfōr-ˌhed, ˈfȯr- *also* -ˌed\ *noun* (before 12th century)
1 : the part of the face above the eyes
2 : the front or forepart of something ⟨flames in the *forehead* of the morning sky —John Milton⟩

fore·hoof \ˈfōr-ˌhu̇f, ˈfȯr-, -ˌhüf\ *noun* (1770)
: the hoof of a forefoot

for·eign \ˈfȯr-ən, ˈfär-\ *adjective* [Middle English *forein,* from Old French, from Late Latin *foranus* on the outside, from Latin *foris* outside — more at FORUM] (13th century)
1 : situated outside a place or country; *especially* : situated outside one's own country
2 : born in, belonging to, or characteristic of some place or country other than the one under consideration
3 : of, relating to, or proceeding from some other person or material thing than the one under consideration
4 : alien in character : not connected or pertinent
5 : related to or dealing with other nations
6 : occurring in an abnormal situation in the living body and often introduced from outside
7 : not being within the jurisdiction of a political unit (as a state)
synonym see EXTRINSIC
— **for·eign·ness** \-ən-nəs\ *noun*

foreign affairs *noun plural* (1611)
: matters having to do with international relations and with the interests of the home country in foreign countries

foreign aid *noun* (1949)
: assistance (as economic aid) provided by one nation to another

foreign bill *noun* (1682)
: a bill of exchange that is not both drawn and payable within a particular jurisdiction

for·eign–born \ˌfȯr-ən-ˈbȯrn, ˌfär-\ *adjective* (1856)
: foreign by birth

foreign correspondent *noun* (1948)
: a correspondent employed to send news or comment from a foreign country

for·eign·er \ˈfȯr-ə-nər, ˈfär-\ *noun* (15th century)
1 : a person belonging to or owing allegiance to a foreign country
2 *chiefly dialect* **:** one not native to a place or community : STRANGER 1c

foreign exchange *noun* (1691)
1 : a process of settling accounts or debts between persons residing in different countries
2 : foreign currency or current short-term credit instruments payable in such currency

for·eign·ism \ˈfȯr-ə-ˌni-zəm, ˈfär-\ *noun* (1855)
: something peculiar to a foreign language or people; *specifically* : a foreign idiom or custom

foreign minister *noun* (1709)
: a governmental minister for foreign affairs

foreign office *noun* (1859)
: a government office (as a ministry) that deals with foreign affairs

foreign policy *noun* (1859)
: the policy of a sovereign state in its interaction with other sovereign states

foreign service *noun* (1927)
: the field force of a foreign office comprising diplomatic and consular personnel

¹fore·judge \fər-ˈjəj, fōr-, fȯr-\ *transitive verb* [Middle English *forjuggen,* from Middle French *forjugier,* from *fors* outside (from Latin *foris*) + *jugier* to judge] (15th century)
: to expel, oust, or put out by judgment of a court

²fore·judge \(ˌ)fōr-ˈjəj, (ˌ)fȯr-\ *transitive verb* (1561)
: PREJUDGE

fore·know \(ˌ)fōr-ˈnō, (ˌ)fȯr-\ *transitive verb* **-knew** \-ˈnü, -ˈnyü\; **-known** \-ˈnōn\; **-know·ing** (14th century)
: to have previous knowledge of : known beforehand especially by paranormal means or by revelation
synonym see FORESEE
— **fore·knowl·edge** \-ˈnä-lij\ *noun*

fore·la·dy \ˈfōr-ˌlā-dē, ˈfȯr-\ *noun* (circa 1889)
: FOREWOMAN

fore·land \ˈfōr-lənd, ˈfȯr-\ *noun* (14th century)
: PROMONTORY, HEADLAND

fore·leg \ˈfōr-ˌleg, ˈfȯr-, -ˌlāg\ *noun* (15th century)
: a front leg

fore·limb \-ˌlim\ *noun* (circa 1796)
: a limb (as an arm, wing, fin, or leg) that is situated anteriorly ⟨the *forelimb* of a bat⟩

fore·lock \-ˌläk\ *noun* (1589)
: a lock of hair growing from the front of the head

fore·man \ˈfōr-mən, ˈfȯr-\ *noun* (15th century)
: a first or chief person: as **a :** a member of a jury who acts as chairman and spokesman **b** (1) **:** a chief and often specially trained worker who works with and commonly leads a gang or crew (2) **:** a person in charge of a group of workers, a particular operation, or a section of a plant
— **fore·man·ship** \-ˌship\ *noun*

fore·mast \ˈfōr-ˌmast, ˈfȯr-, -məst\ *noun* (1582)
: the mast nearest the bow of a ship

¹fore·most \-ˌmōst\ *adjective* [Middle English *formest,* from Old English, superlative of *forma* first; akin to Old High German *fruma* advantage, Old English *fore* fore] (before 12th century)
1 : first in a series or progression
2 : of first rank or position : PREEMINENT

²foremost *adverb* (before 12th century)
1 : in the first place
2 : most importantly ⟨first and *foremost*⟩

fore·moth·er \ˈfōr-ˌmə-t͟hər, ˈfȯr-\ *noun* (1582)
: a female ancestor

fore·name \-ˌnām\ *noun* (1533)
: a name that precedes one's surname

fore·named \-ˌnāmd\ *adjective* (13th century)
: named previously : AFORESAID

fore·noon \ˈfōr-ˌnün, ˈfȯr-ˌ, fōr-ˈ, fȯr-ˈ\ *noun* (15th century)
: the early part of the day ending with noon : MORNING

¹fo·ren·sic \fə-ˈren(t)-sik, -ˈren-zik\ *adjective* [Latin *forensis* public, forensic, from *forum* forum] (1659)
1 : belonging to, used in, or suitable to courts of judicature or to public discussion and debate
2 : ARGUMENTATIVE, RHETORICAL
3 : relating to or dealing with the application of scientific knowledge to legal problems ⟨*forensic* medicine⟩ ⟨*forensic* science⟩ ⟨*forensic* pathologist⟩ ⟨*forensic* experts⟩
— **fo·ren·si·cal·ly** \-si-k(ə-)lē, -zi-\ *adverb*

²forensic *noun* (1814)
1 : an argumentative exercise
2 *plural but singular or plural in construction* **:** the art or study of argumentative discourse

fore·or·dain \ˌfōr-ȯr-ˈdān, ˌfȯr-\ *transitive verb* (15th century)
: to dispose or appoint in advance : PREDESTINE
— **fore·or·di·na·tion** \-ˌȯr-dᵊn-ˈā-shən\ *noun*

fore·part \ˈfōr-ˌpärt, ˈfȯr-\ *noun* (14th century)
1 : the anterior part of something
2 : the earlier part of a period of time

fore·passed *or* **fore·past** \-ˌpast\ *adjective* (1557)
: BYGONE

fore·paw \-ˌpȯ\ *noun* (1825)
: the paw of a foreleg

fore·peak \-ˌpēk\ *noun* (1693)
: the extreme forward lower compartment or tank usually used for trimming or storage in a ship

fore·play \-ˌplā\ *noun* (1929)
1 : erotic stimulation preceding sexual intercourse
2 : action or behavior that precedes an event

fore·quar·ter \-ˌkwȯ(r)-tər, -ˌkȯr-\ *noun* (15th century)
: the front half of a lateral half of the body or carcass of a quadruped ⟨a *forequarter* of beef⟩

fore·reach \fōr-ˈrēch, fȯr-\ (1644)
intransitive verb
of a ship **:** to gain ground in tacking
transitive verb
: to gain on or go ahead of (a ship) when close-hauled

fore·run \-ˈrən\ *transitive verb* **-ran** \-ˈran\; **-run; -run·ning** (before 12th century)
1 : to run before
2 : to come before as a token of something to follow
3 : FORESTALL, ANTICIPATE

fore·run·ner \ˈfōr-ˌrə-nər, ˈfȯr-\ *noun* (13th century)
1 : one that precedes and indicates the approach of another: as **a :** a premonitory sign or symptom **b :** a skier who runs the course before the start of a race
2 : PREDECESSOR, ANCESTOR ☆

fore·said \-ˌsed\ *adjective* (before 12th century)
archaic **:** AFORESAID

fore·sail \ˈfōr-ˌsāl, ˈfȯr-, -səl\ *noun* (15th century)
1 : the lowest sail set on the foremast of a square-rigged ship or schooner — see SAIL illustration
2 : the sole or principal headsail (as of a sloop, cutter, or schooner)

fore·see \fōr-ˈsē, fȯr-\ *transitive verb* **-saw** \-ˈsȯ\; **-seen** \-ˈsēn\; **-see·ing** (before 12th century)
: to see (as a development) beforehand ☆
— fore·seer \fōr-ˈsē-ər, fȯr-, -ˈsi(-ə)r\ *noun*

fore·see·able \-ˈsē-ə-bəl\ *adjective* (1804)
1 : being such as may be reasonably anticipated ⟨*foreseeable* problems⟩
2 : lying within the range for which forecasts are possible ⟨in the *foreseeable* future⟩
— fore·see·abil·i·ty \-ˌsē-ə-ˈbi-lə-tē\ *noun*

fore·shad·ow \-ˈsha-(ˌ)dō, -də(-w)\ *transitive verb* (1577)
: to represent, indicate, or typify beforehand **:** PREFIGURE
— fore·shad·ow·er \-də-wər\ *noun*

fore·shank \ˈfōr-ˌshaŋk, ˈfȯr-\ *noun* (1924)
: the upper part of the foreleg of cattle; *also* **:** meat cut from this part

fore·sheet \-ˌshēt\ *noun* (1667)
1 : one of the sheets of a foresail
2 *plural* **:** the forward part of an open boat

fore·shock \-ˌshäk\ *noun* (1902)
: any of the usually minor tremors commonly preceding the principal shock of an earthquake

fore·shore \-ˌshōr, -ˌshȯr\ *noun* (1764)
1 : a strip of land margining a body of water
2 : the part of a seashore between high-water and low-water marks

fore·short·en \fōr-ˈshȯr-tᵊn, fȯr-\ *transitive verb* (1606)
1 : to shorten by proportionately contracting in the direction of depth so that an illusion of projection or extension in space is obtained
2 : to make more compact **:** ABRIDGE, SHORTEN

fore·side \ˈfōr-ˌsīd, ˈfȯr-\ *noun* (14th century)
: the front side or part **:** FRONT

fore·sight \ˈfōr-ˌsīt, ˈfȯr-\ *noun* (14th century)
1 : an act or the power of foreseeing **:** PRESCIENCE
2 : provident care **:** PRUDENCE ⟨had the *foresight* to invest his money wisely⟩
3 : an act of looking forward; *also* **:** a view forward
— fore·sight·ed \-ˌsī-təd\ *adjective*
— fore·sight·ed·ly *adverb*
— fore·sight·ed·ness *noun*
— fore·sight·ful \-ˌsīt-fəl\ *adjective*

fore·skin \-ˌskin\ *noun* (1535)
: a fold of skin that covers the glans of the penis — called also *prepuce*

fore·speak \fōr-ˈspēk, fȯr-\ *transitive verb* **-spoke** \-ˈspōk\; **-spo·ken** \-ˈspō-kən\; **-speak·ing** (14th century)
1 : FORETELL, PREDICT
2 : to arrange for in advance

[1]for·est \ˈfȯr-əst, ˈfär-\ *noun, often attributive* [Middle English, from Old French, from Late Latin *forestis* (*silva*) unenclosed (woodland), from Latin *foris* outside — more at FORUM] (13th century)
1 : a dense growth of trees and underbrush covering a large tract
2 : a tract of wooded land in England formerly owned by the sovereign and used for game
3 : something resembling a forest especially in profusion ⟨a *forest* of microphones⟩
— for·est·al \-əs-tᵊl\ *or* **fo·res·tial** \fə-ˈres-tē-əl, fō-, -ˈres(h)-chəl\ *adjective*
— for·est·ed \ˈfȯr-ə-stəd, ˈfär-\ *adjective*

[2]forest *transitive verb* (circa 1828)
: to cover with trees or forest
— for·es·ta·tion \ˌfȯr-ə-ˈstā-shən, ˌfär-\ *noun*

fore·stage \ˈfōr-ˌstāj, ˈfȯr-\ *noun* (1923)
: APRON 2e

fore·stall \fōr-ˈstȯl, fȯr-\ *transitive verb* [Middle English, from *forstall* act of waylaying, from Old English *foresteall,* from *fore-* + *steall* position, stall] (before 12th century)
1 : to prevent the normal trading in by buying or diverting goods or by persuading persons to raise prices
2 *archaic* **:** INTERCEPT
3 *obsolete* **:** OBSTRUCT, BESET
4 : to exclude, hinder, or prevent by prior occupation or measures
5 : to get ahead of **:** ANTICIPATE
synonym see PREVENT
— fore·stall·er *noun*
— fore·stall·ment \-ˈstȯl-mənt\ *noun*

fore·stay \ˈfōr-ˌstā, ˈfȯr-\ *noun* (13th century)
: a stay from the foremast to the foredeck or bow of a ship

fore·stay·sail \-ˌsāl, -səl\ *noun* (1742)
: the triangular aftermost headsail of a schooner, ketch, or yawl set on the forestay — see SAIL illustration

for·est·er \ˈfȯr-ə-stər, ˈfär-\ *noun* [Middle English *forster, forester,* from Old French *forestier,* from *forest*] (14th century)
1 : a person trained in forestry
2 : an inhabitant of a forest
3 : any of various woodland moths (family Agaristidae)
4 *capitalized* **:** a member of a major benevolent and fraternal order

forest floor *noun* (1849)
: the richly organic layer of soil and debris characteristic of forested land

forest green *noun* (1810)
: a dark yellowish or moderate olive green

for·est·land \ˈfȯr-əst-ˌland, ˈfär-\ *noun* (1649)
: land covered with forest or reserved for the growth of forests

forest ranger *noun* (1830)
: an officer charged with the patrolling and guarding of a forest; *especially* **:** one in charge of the management and protection of a portion of a public forest

for·est·ry \ˈfȯr-ə-strē, ˈfär-\ *noun* (1823)
1 : FORESTLAND
2 a : the science of developing, caring for, or cultivating forests **b :** the management of growing timber

forest tent caterpillar *noun* (1854)
: a moth (*Malacosoma disstria* of the family Lasiocampidae) whose orange-marked larva is a tent caterpillar and a serious defoliator of deciduous trees

foreswear, foresworn *variant of* FORSWEAR, FORSWORN

[1]fore·taste \ˈfōr-ˌtāst, ˈfȯr-\ *noun* (15th century)
1 : a small anticipatory sample
2 : an advance indication or warning
synonym see PROSPECT

[2]fore·taste \fōr-ˈtāst, fȯr-ˈ, ˈfōr-ˌ, ˈfȯr-ˌ\ *transitive verb* (15th century)
: to taste beforehand **:** ANTICIPATE

fore·tell \fōr-ˈtel, fȯr-\ *transitive verb* **-told** \-ˈtōld\; **-tell·ing** (14th century)
: to tell beforehand **:** PREDICT ☆
— fore·tell·er *noun*

[1]fore·thought \ˈfōr-ˌthȯt, ˈfȯr-\ *noun* (14th century)
1 : a thinking or planning out in advance **:** PREMEDITATION
2 : consideration for the future

[2]forethought *adjective* (15th century)
archaic **:** AFORETHOUGHT

fore·thought·ful \-fəl\ *adjective* (1809)
: full of or having forethought
— fore·thought·ful·ly \-fə-lē\ *adverb*
— fore·thought·ful·ness *noun*

fore·time \ˈfōr-ˌtīm, ˈfȯr-\ *noun* (circa 1540)
: former or past time **:** the time before the present

[1]fore·to·ken \ˈfōr-ˌtō-kən, ˈfȯr-\ *noun* (before 12th century)
: a premonitory sign

[2]fore·to·ken \fōr-ˈtō-kən, fȯr-\ *transitive verb* **fore·to·kened; fore·to·ken·ing** \-ˈtōk-niŋ, -ˈtō-kə-\ (15th century)
: to indicate or warn of in advance

fore·top \ˈfōr-ˌtäp, ˈfȯr-; -təp\ *noun* (1509)
: the platform at the head of a ship's foremast

fore·top·man \ˈfōr-ˌtäp-mən, ˈfȯr-; -təp-\ *noun* (1816)
: a sailor on duty on the foremast and above

fore–top·mast \ˈfōr-ˌtäp-məst, ˈfȯr-; -təp-ˌmast\ *noun* (1626)

☆ SYNONYMS

Forerunner, precursor, harbinger, herald mean one that goes before or announces the coming of another. FORERUNNER is applicable to anything that serves as a sign or presage ⟨the blockade was the *forerunner* of war⟩. PRECURSOR applies to a person or thing paving the way for the success or accomplishment of another ⟨18th-century poets like Burns were *precursors* of the Romantics⟩. HARBINGER and HERALD both apply, chiefly figuratively, to one that proclaims or announces the coming or arrival of a notable event ⟨their early victory was the *harbinger* of a winning season⟩ ⟨the *herald* of a new age in medicine⟩.

Foresee, foreknow, divine, anticipate mean to know beforehand. FORESEE implies nothing about how the knowledge is derived and may apply to ordinary reasoning and experience ⟨economists should have *foreseen* the recession⟩. FOREKNOW usually implies supernatural assistance, as through revelation ⟨if only we could *foreknow* our own destinies⟩. DIVINE adds to FORESEE the suggestion of exceptional wisdom or discernment ⟨was able to *divine* Europe's rapid recovery from the war⟩. ANTICIPATE implies taking action about or responding emotionally to something before it happens ⟨the waiter *anticipated* our every need⟩.

Foretell, predict, forecast, prophesy, prognosticate mean to tell beforehand. FORETELL applies to the telling of the coming of a future event by any procedure or any source of information ⟨seers *foretold* the calamity⟩. PREDICT commonly implies inference from facts or accepted laws of nature ⟨astronomers *predicted* an eclipse⟩. FORECAST adds the implication of anticipating eventualities and differs from PREDICT in being usually concerned with probabilities rather than certainties ⟨*forecast* snow⟩. PROPHESY connotes inspired or mystic knowledge of the future especially as the fulfilling of divine threats or promises ⟨*prophesying* a new messiah⟩. PROGNOSTICATE is used less often than the other words; it may suggest learned or skilled interpretation, but more often it is simply a colorful substitute for PREDICT or PROPHESY ⟨*prognosticating* the future⟩.

: a mast next above the foremast

[1]for·ev·er \fə-ˈrev-ər, fȯ-; *Southern often* fə-ˈe-və\ *adverb* (1629)
1 : for a limitless time ⟨wants to live *forever*⟩
2 : at all times **:** CONTINUALLY ⟨is *forever* making bad puns⟩

[2]forever *noun* (1858)
: a seemingly interminable time **:** excessively long ⟨it took her *forever* to find the answer⟩

for·ev·er·more \-ˌre-və(r)-ˈmōr, -ˈmȯr\ *adverb* (1837)
: FOREVER 1

for·ev·er·ness \-ˈre-vər-nəs\ *noun* (1945)
: ETERNITY

fore·warn \fōr-ˈwȯrn, fȯr-\ *transitive verb* (14th century)
: to warn in advance

fore·wing \ˈfōr-ˌwiŋ, ˈfȯr-\ *noun* (circa 1889)
: either of the anterior wings of a 4-winged insect

fore·wom·an \ˈfōr-ˌwu̇-mən, ˈfȯr-\ *noun* (1709)
: a woman who is a foreman

fore·word \ˈfōr-(ˌ)wərd, ˈfȯr-\ *noun* (1842)
: prefatory comments (as for a book) especially when written by someone other than the author

fore·worn *archaic variant of* FORWORN

[1]for·feit \ˈfȯr-fət\ *noun* [Middle English *forfait,* from Middle French, from past participle of *forfaire* to commit a crime, forfeit, from *fors* outside (from Latin *foris*) + *faire* to do, from Latin *facere* — more at FORUM, DO] (14th century)
1 : something forfeited or subject to being forfeited (as for a crime, offense, or neglect of duty) **:** PENALTY
2 : forfeiture especially of civil rights
3 a : something deposited (as for making a mistake in a game) and then redeemed on payment of a fine **b** *plural* **:** a game in which forfeits are exacted

[2]forfeit *transitive verb* (14th century)
1 : to lose or lose the right to by some error, offense, or crime
2 : to subject to confiscation as a forfeit
— for·feit·able \-fə-tə-bəl\ *adjective*
— for·feit·er *noun*

[3]forfeit *adjective* (14th century)
: forfeited or subject to forfeiture

for·fei·ture \ˈfȯr-fə-ˌchu̇r, -chər, -ˌt(y)u̇r\ *noun* (14th century)
1 : the act of forfeiting **:** the loss of property or money because of a breach of a legal obligation
2 : something (as money or property) that is forfeited **:** PENALTY

for·fend \fȯr-ˈfend, fōr-\ *transitive verb* (14th century)
1 a *archaic* **:** FORBID **b :** to ward off **:** PREVENT
2 : PROTECT, PRESERVE

for·gath·er \fȯr-ˈga-t͟hər, fōr-\ *intransitive verb* (1513)
1 : to come together **:** ASSEMBLE
2 : to meet someone usually by chance

[1]forge \ˈfōrj, ˈfȯrj\ *noun* [Middle English, from Old French, from Latin *fabrica,* from *fabr-, faber* smith] (13th century)
1 : a furnace or a shop with its furnace where metal is heated and wrought **:** SMITHY
2 : a workshop where wrought iron is produced or where iron is made malleable

[2]forge *verb* **forged; forg·ing** (14th century)
transitive verb
1 a : to form (as metal) by heating and hammering **b :** to form (metal) by a mechanical or hydraulic press with or without heat
2 : to make or imitate falsely especially with intent to defraud **:** COUNTERFEIT
3 : to form or bring into being especially by an expenditure of effort ⟨working to *forge* party unity⟩
intransitive verb
1 : to work at a forge
2 : to commit forgery
— forge·abil·i·ty \ˌfōr-jə-ˈbi-lə-tē, ˌfȯr-\ *noun*
— forge·able \ˈfōr-jə-bəl, ˈfȯr-\ *adjective*

[3]forge *intransitive verb* **forged; forg·ing** [origin unknown] (1611)
1 : to move forward slowly and steadily ⟨the ship *forged* ahead through heavy seas⟩
2 : to move with a sudden increase of speed and power ⟨*forged* into the lead⟩ ⟨*forged* ahead in marketing the product⟩

forg·er \ˈfōr-jər, ˈfȯr-\ *noun* [[2]*forge*] (14th century)
1 : one that forges metals
2 a : one that falsifies; *especially* **:** a creator of false tales **b :** a person guilty of forgery

forg·ery \ˈfōrj-rē, ˈfȯrj-; ˈfōr-jə-, ˈfȯr-\ *noun, plural* **-er·ies** (1583)
1 *archaic* **:** INVENTION
2 : something forged
3 : an act of forging; *especially* **:** the crime of falsely and fraudulently making or altering a document (as a check)

for·get \fər-ˈget, fȯr-\ *verb* **-got** \-ˈgät\; **-got·ten** \-ˈgä-tᵊn\ *or* **-got; -get·ting** [Middle English, from Old English *forgietan,* from *for-* + *-gietan* (akin to Old Norse *geta* to get)] (before 12th century)
transitive verb
1 a : to lose the remembrance of **:** be unable to think of or recall ⟨I *forget* his name⟩ **b** *obsolete* **:** to cease from doing
2 : to treat with inattention or disregard ⟨*forgot* their old friends⟩
3 : to disregard intentionally **:** OVERLOOK — usually used in the imperative ⟨*forget* it⟩
intransitive verb
1 : to cease remembering or noticing ⟨forgive and *forget*⟩
2 : to fail to become mindful at the proper time ⟨*forget* about paying the bill⟩
synonym see NEGLECT
— for·get·ter *noun*
— forget oneself : to lose one's dignity, temper, or self-control

for·get·ful \-ˈget-fəl\ *adjective* (14th century)
1 : likely to forget
2 : characterized by negligent failure to remember **:** NEGLECTFUL
3 : inducing oblivion ⟨*forgetful* sleep⟩
— for·get·ful·ly \-fə-lē\ *adverb*
— for·get·ful·ness *noun*

for·ge·tive \ˈfōr-jə-tiv, ˈfȯr-\ *adjective* [probably from [2]*forge* + *-tive* (as in *inventive*)] (1597)
archaic **:** INVENTIVE, IMAGINATIVE

for·get–me–not \fər-ˈget-mē-ˌnät, fȯr-\ *noun* (1532)
: any of a genus (*Myosotis*) of small herbs of the borage family having bright-blue or white flowers usually arranged in a curving spike

for·get·ta·ble \fər-ˈge-tə-bəl, fȯr-\ *adjective* (1845)
: fit or likely to be forgotten ⟨a *forgettable* movie⟩

forg·ing \ˈfōr-jiŋ, ˈfȯr-\ *noun* (14th century)
1 : the art or process of forging
2 : a piece of forged work
3 : FORGERY 3

for·give \fər-ˈgiv, fȯr-\ *verb* **-gave** \-ˈgāv\; **-giv·en** \-ˈgi-vən\; **-giv·ing** [Middle English, from Old English *forgifan,* from *for-* + *gifan* to give] (before 12th century)
transitive verb
1 a : to give up resentment of or claim to requital for ⟨*forgive* an insult⟩ **b :** to grant relief from payment of ⟨*forgive* a debt⟩
2 : to cease to feel resentment against (an offender) **:** PARDON ⟨*forgive* one's enemies⟩
intransitive verb
: to grant forgiveness
synonym see EXCUSE
— for·giv·able \-ˈgi-və-bəl\ *adjective*
— for·giv·ably \-blē\ *adverb*
— for·giv·er *noun*

for·give·ness \-ˈgiv-nəs\ *noun* (before 12th century)
: the act of forgiving

for·giv·ing *adjective* (1690)
1 : willing or able to forgive
2 : allowing room for error or weakness ⟨designed to be a *forgiving* tennis racquet⟩
— for·giv·ing·ly \-ˈgi-viŋ-lē\ *adverb*
— for·giv·ing·ness *noun*

for·go \fȯr-ˈgō, fōr-\ *transitive verb* **-went** \-ˈwent\; **-gone** \-ˈgȯn *also* -ˈgän\; **-go·ing** \-ˈgō-iŋ, -ˈgȯ(-)iŋ\ [Middle English, from Old English *forgān* to pass by, forgo, from *for-* + *gān* to go] (before 12th century)
1 : to give up the enjoyment or advantage of **:** do without
2 *archaic* **:** FORSAKE
— for·go·er \-ˈgō(-ə)r\ *noun*

forgotten man *noun* (1925)
: a person or category of persons that receives less attention than is merited

for instance \fə-ˈrin(t)-stənts, ˈfrin(t)-\ *noun* (1959)
: EXAMPLE ⟨I'll give you a *for instance*⟩

fo·rint \ˈfȯr-ˌint\ *noun, plural* **forints** *also* **forint** [Hungarian] (circa 1916)
— see MONEY table

for·judge *variant of* FOREJUDGE

[1]fork \ˈfȯrk\ *noun* [Middle English *forke,* from Old English & Old North French; Old English *forca* & Old North French *forque,* from Latin *furca*] (before 12th century)
1 : an implement with two or more prongs used especially for taking up (as in eating), pitching, or digging
2 : a forked part, tool, or piece of equipment
3 a : a division into branches or the place where something divides into branches **b :** CONFLUENCE
4 : one of the branches into which something forks
5 : an attack by one chess piece (as a knight) on two pieces simultaneously
— fork·ful \-ˌfu̇l\ *noun*

[2]fork (15th century)
intransitive verb
1 : to divide into two or more branches ⟨where the road *forks*⟩
2 a : to use or work with a fork **b :** to turn into a fork
transitive verb
1 : to give the form of a fork to ⟨*forking* her fingers⟩
2 : to attack (two chessmen) simultaneously
3 : to raise, pitch, dig, or work with a fork ⟨*fork* hay⟩
4 : PAY, CONTRIBUTE — used with *over, out,* or *up* ⟨had to *fork* over $5000⟩
— fork·er *noun*

fork·ball \ˈfȯrk-ˌbȯl\ *noun* (1936)
: a baseball pitch in which the ball is gripped between the forked index and middle fingers

forked \ˈfȯrkt, ˈfȯr-kəd\ *adjective* (13th century)
1 : resembling a fork especially in having one end divided into two or more branches or points ⟨*forked* lightning⟩
2 : shaped like a fork or having a forked part ⟨a *forked* road⟩

forked tongue *noun* (1836)
: intent to mislead or deceive — usually used in the phrase *to speak with forked tongue*

fork·lift \ˈfȯr-ˌklift\ *noun* (1944)
: a self-propelled machine for hoisting and transporting heavy objects by means of steel fingers inserted under the load

fork–ten·der \ˈfȯrk-ˈten-dər\ *adjective* (1973)
: tender enough to be easily pierced or cut with a fork ⟨*fork-tender* filet mignon⟩

forky \ˈfȯr-kē\ *adjective* **fork·i·er; -est** (1697)
: FORKED ⟨a *forky* beard⟩

for·lorn \fər-ˈlȯrn, fȯr-\ *adjective* [Middle English *forloren,* from Old English, past participle of *forlēosan* to lose, from *for-* + *lēosan* to lose — more at LOSE] (before 12th century)

1 a : BEREFT, FORSAKEN ⟨left quite *forlorn* of hope⟩ **b :** sad and lonely because of isolation or desertion **:** DESOLATE
2 : being in poor condition **:** MISERABLE, WRETCHED ⟨*forlorn* tumbledown buildings⟩
3 : nearly hopeless ⟨a *forlorn* attempt⟩
synonym see ALONE
— for·lorn·ly *adverb*
— for·lorn·ness \-ˈlȯrn-nəs\ *noun*

forlorn hope *noun* [by folk etymology from Dutch *verloren hoop,* literally, lost band] (1579)
1 : a body of men selected to perform a perilous service
2 : a desperate or extremely difficult enterprise

[1]form \ˈfȯrm\ *noun* [Middle English *forme,* from Old French, from Latin *forma* form, beauty] (13th century)
1 a : the shape and structure of something as distinguished from its material **b :** a body (as of a person) especially in its external appearance or as distinguished from the face **:** FIGURE **c** *archaic* **:** BEAUTY
2 : the essential nature of a thing as distinguished from its matter: as **a :** IDEA 1a **b :** the component of a thing that determines its kind
3 a : established method of expression or proceeding **:** procedure according to rule or rote **b :** a prescribed and set order of words **:** FORMULA ⟨the *form* of the marriage service⟩
4 : a printed or typed document with blank spaces for insertion of required or requested information ⟨tax *forms*⟩
5 a (1) **:** conduct regulated by extraneous controls (as of custom or etiquette) **:** CEREMONY (2) **:** show without substance **b :** manner or conduct as tested by a prescribed or accepted standard ⟨rudeness is simply bad *form*⟩ **c :** manner or style of performing or accomplishing according to recognized standards of technique ⟨a strong swimmer but weak on *form*⟩
6 a : the resting place or nest of a hare **b :** a long seat **:** BENCH
7 a : a supporting frame model of the human figure or part (as the torso) of the human figure usually used for displaying apparel **b :** a proportioned and often adjustable model for fitting clothes **c :** a mold in which concrete is placed to set
8 : the printing type or other matter arranged and secured in a chase ready for printing
9 a : one of the different modes of existence, action, or manifestation of a particular thing or substance **:** KIND ⟨one *form* of respiratory disorder⟩ ⟨a *form* of art⟩ **b :** a distinguishable group of organisms **c :** LINGUISTIC FORM **d :** one of the different aspects a word may take as a result of inflection or change of spelling or pronunciation ⟨verbal *forms*⟩ **e :** a mathematical expression of a particular type ⟨a bilinear *form*⟩ ⟨a polynomial *form*⟩
10 a (1) **:** orderly method of arrangement (as in the presentation of ideas) **:** manner of coordinating elements (as of an artistic production or course of reasoning) (2) **:** a particular kind or instance of such arrangement ⟨the sonnet is a poetical *form*⟩ **b :** PATTERN, SCHEMA ⟨arguments of the same logical *form*⟩ **c :** the structural element, plan, or design of a work of art — compare CONTENT 2c **d :** a visible and measurable unit defined by a contour **:** a bounded surface or volume
11 : a grade in a British school or in some American private schools
12 a (1) **:** the past performance of a race horse (2) **:** RACING FORM **b :** known ability to perform ⟨a singer at the top of her *form*⟩ **c :** condition suitable for performing (as in athletic competition) ⟨back on *form*⟩

[2]form (13th century)
transitive verb
1 a : to give a particular shape to **:** shape or mold into a certain state or after a particular model ⟨*form* the dough into a ball⟩ ⟨a state *formed* along republican lines⟩ **b :** to arrange themselves in ⟨the dancers *formed* a line⟩ **c :** to model by instruction and discipline ⟨a mind *formed* by classical education⟩
2 : to give form or shape to **:** FASHION, CONSTRUCT
3 : to serve to make up or constitute **:** be an essential or basic element of
4 : DEVELOP, ACQUIRE ⟨*form* a habit⟩
5 : to arrange in order **:** DRAW UP
6 a : to assume an inflection so as to produce (as a tense) ⟨*forms* the past in *-ed*⟩ **b :** to combine to make (a compound word)
intransitive verb
1 : to become formed or shaped
2 : to take form **:** come into existence **:** ARISE
3 : to take on a definite form, shape, or arrangement
— form·abil·i·ty \ˌfȯr-mə-ˈbi-lə-tē\ *noun*
— form·able \ˈfȯr-mə-bəl\ *adjective*
— form on : to take up a formation next to

form- *or* **formo-** *combining form* [*formic*]
: formic acid ⟨*form*ate⟩

-form \ˌfȯrm\ *adjective combining form* [Middle French & Latin; Middle French *-forme,* from Latin *-formis,* from *forma*]
: in the form or shape of **:** resembling ⟨fili*form*⟩

[1]for·mal \ˈfȯr-məl\ *adjective* [Middle English, from Middle French or Latin; Middle French, from Latin *formalis,* from *forma*] (14th century)
1 a : belonging to or constituting the form or essence of a thing ⟨*formal* cause⟩ **b :** relating to or involving the outward form, structure, relationships, or arrangement of elements rather than content ⟨*formal* logic⟩ ⟨*formal* style of painting⟩ ⟨*formal* approach to comparative linguistics⟩
2 a : following or according with established form, custom, or rule ⟨lacked *formal* schooling⟩ ⟨a *formal* dinner party⟩ ⟨*formal* attire⟩ **b :** done in due or lawful form ⟨a *formal* contract⟩ ⟨received *formal* recognition⟩
3 a : characterized by punctilious respect for form **:** METHODICAL ⟨very *formal* in all his dealings⟩ **b :** rigidly ceremonious **:** PRIM
4 : having the appearance without the substance ⟨*formal* Christians who go to church only at Easter⟩
synonym see CEREMONIAL
— for·mal·ly \-mə-lē\ *adverb*
— for·mal·ness *noun*

[2]formal *noun* (1605)
: something (as a dance or a dress) formal in character

[3]formal *adjective* [*form*ula + [1]*-al*] (circa 1934)
: [3]MOLAR

form·al·de·hyde \fȯr-ˈmal-də-ˌhīd, fər-\ *noun* [International Scientific Vocabulary *form-* + *aldehyde*] (1872)
: a colorless pungent irritating gas CH_2O used chiefly as a disinfectant and preservative and in chemical synthesis

for·ma·lin \ˈfȯr-mə-lən, -ˌlēn\ *noun* [*Formalin,* a trademark] (1893)
: a clear aqueous solution of formaldehyde containing a small amount of methanol

for·mal·ise *British variant of* FORMALIZE

for·mal·ism \ˈfȯr-mə-ˌli-zəm\ *noun* (circa 1840)
1 : the practice or the doctrine of strict adherence to prescribed or external forms (as in religion or art); *also* **:** an instance of this
2 : marked attention to arrangement, style, or artistic means (as in art or literature) usually with corresponding de-emphasis of content
— for·mal·ist \-list\ *noun or adjective*
— for·mal·is·tic \ˌfȯr-mə-ˈlis-tik\ *adjective*

for·mal·i·ty \fȯr-ˈma-lə-tē\ *noun, plural* **-ties** (1597)
1 : compliance with formal or conventional rules **:** CEREMONY
2 : the quality or state of being formal
3 : an established form or procedure that is required or conventional

for·mal·ize \ˈfȯr-mə-ˌlīz\ *transitive verb* **-ized; -iz·ing** (1646)
1 : to give a certain or definite form to **:** SHAPE
2 a : to make formal **b :** to give formal status or approval to
— for·mal·iz·able \-ˌlī-zə-bəl\ *adjective*
— for·mal·iza·tion \ˌfȯr-mə-lə-ˈzā-shən\ *noun*
— for·mal·iz·er \ˈfȯr-mə-ˌlī-zər\ *noun*

form·am·ide \fȯr-ˈma-ˌmīd; ˈfȯr-mə-ˌmīd, -məd\ *noun* (1852)
: a colorless hygroscopic liquid $CHONH_2$ used chiefly as a solvent

for·mant \ˈfȯr-mənt, -ˌmant\ *noun* (1901)
: a characteristic component of the quality of a speech sound; *specifically* **:** any of several resonance bands held to determine the phonetic quality of a vowel

[1]for·mat \ˈfȯr-ˌmat\ *noun* [French or German; French, from German, from Latin *formatus,* past participle of *formare* to form, from *forma*] (1840)
1 : the shape, size, and general makeup (as of something printed)
2 : general plan of organization, arrangement, or choice of material (as for a television show)

[2]format *transitive verb* **for·mat·ted; for·mat·ting** (1964)
: to arrange (as material to be printed or stored data) in a particular format
— for·mat·ter *noun*

for·mate \ˈfȯr-ˌmāt\ *noun* (1807)
: a salt or ester of formic acid

for·ma·tion \fȯr-ˈmā-shən\ *noun* (15th century)
1 : an act of giving form or shape to something or of taking form **:** DEVELOPMENT
2 : something that is formed ⟨new word *formations*⟩
3 : the manner in which a thing is formed **:** STRUCTURE ⟨the peculiar *formation* of the heart⟩
4 : the largest unit in an ecological community comprising two or more associations and their precursors ⟨grassland *formation*⟩
5 a : any igneous, sedimentary, or metamorphic rock represented as a unit **b :** any sedimentary bed or consecutive series of beds sufficiently homogeneous or distinctive to be a unit
6 : an arrangement of a body or group of persons or things in some prescribed manner or for a particular purpose

[1]for·ma·tive \ˈfȯr-mə-tiv\ *adjective* (15th century)
1 a : giving or capable of giving form **:** CONSTRUCTIVE ⟨a *formative* influence⟩ **b :** used in word formation or inflection
2 : capable of alteration by growth and development; *also* **:** producing new cells and tissues
3 : of, relating to, or characterized by formative effects or formation ⟨*formative* years⟩
— for·ma·tive·ly *adverb*

[2]formative *noun* (1816)
: the element (as a suffix) in a word that serves to give the word appropriate form and is not part of the base

form class *noun* (1921)
: a class of linguistic forms that can be used in the same position in a construction and that have one or more morphological or syntactical features in common

form–critical *adjective* (1933)
: based on or applying form criticism

form criticism *noun* (1928)
: a method of criticism for determining the sources and historicity of biblical writings through analysis of the writings in terms of ancient literary forms and oral traditions (as love poems, parables, and proverbs)

\ə\ abut \ᵊ\ kitten \ər\ further \a\ ash \ā\ ace \ä\ mop, mar \au̇\ out \ch\ chin \e\ bet \ē\ easy \g\ go \i\ hit \ī\ ice \j\ job \ŋ\ sing \ō\ go \ȯ\ law \ȯi\ boy \th\ thin \t͟h\ the \ü\ loot \u̇\ foot \y\ yet \zh\ vision *see also* Guide to Pronunciation

— **form critic** *noun*
forme \'fȯm, 'fȯrm\ *noun* (15th century)
British **:** FORM 8
formed \'fȯrmd\ *adjective* (1605)
: organized in a way characteristic of living matter ⟨mitochondria are *formed* bodies of the cell⟩
for·mée \'fȯr-ˌmā, fȯr-'\ *adjective* [French, feminine past participle of *former* to form, from Latin *formare*] (15th century)
of a heraldic cross **:** having the arms narrow at the center and expanding toward the ends — see CROSS illustration
¹for·mer \'fȯr-mər\ *adjective* [Middle English, from *forme* first, from Old English *forma* — more at FOREMOST] (12th century)
1 a : coming before in time **b :** of, relating to, or occurring in the past ⟨*former* correspondence⟩
2 : preceding in place or arrangement **:** FOREGOING ⟨*former* part of the chapter⟩
3 : first in order of two or more things cited or understood ⟨of the two given, the *former* spelling is more common⟩ ⟨of the two spellings, the *former* is more common⟩
4 : having been previously **:** ONETIME ⟨a *former* athlete⟩
synonym see PRECEDING
²form·er \'fȯr-mər\ *noun* (14th century)
1 : one that forms
2 *chiefly British* **:** a member of a school form — usually used in combination ⟨sixth *former*⟩
for·mer·ly \'fȯr-mə(r)-lē\ *adverb* (1534)
1 : at an earlier time **:** PREVIOUSLY
2 *obsolete* **:** just before
form·fit·ting \'fȯrm-ˌfi-tiŋ\ *adjective* (1897)
: conforming to the outline of the body **:** fitting snugly ⟨a *formfitting* sweater⟩
form·ful \'fȯrm-fəl\ *adjective* (1832)
: exhibiting or notable for form
form genus *noun* (1873)
: an artificial taxonomic category established for organisms (as imperfect fungi) of obscure true relationships
For·mi·ca \fȯr-'mī-kə, fər-\ *trademark*
— used for any of various laminated plastic products used especially for surface finish
for·mic acid \'fȯr-mik-\ *noun* [irregular from Latin *formica* ant — more at PISMIRE] (1791)
: a colorless pungent fuming vesicant liquid acid CH_2O_2 found especially in ants and in many plants and used chiefly in dyeing and finishing textiles
for·mi·cary \'fȯr-mə-ˌker-ē\ *noun, plural* **-car·ies** [Medieval Latin *formicarium,* from Latin *formica*] (1816)
: an ant nest
for·mi·da·ble \'fȯr-mə-də-bəl; fȯr-'mi-, fər-'mi-\ *adjective* [Middle English, from Latin *formidabilis,* from *formidare* to fear, from *formido* terror, bogey; akin to Greek *mormō* bogey] (15th century)
1 : causing fear, dread, or apprehension ⟨a *formidable* prospect⟩
2 : having qualities that discourage approach or attack
3 : tending to inspire awe or wonder
— **for·mi·da·bil·i·ty** \ˌfȯr-mə-də-'bi-lə-tē; fȯr-ˌmi-, fər-ˌ\ *noun*
— **for·mi·da·ble·ness** \'fȯr-mə-də-bəl-nəs; fȯr-'mi-, fər-'\ *noun*
— **for·mi·da·bly** \-blē\ *adverb*
form·less \'fȯrm-ləs\ *adjective* (1591)
1 : having no regular form or shape
2 : lacking order or arrangement
3 : having no physical existence
— **form·less·ly** *adverb*
— **form·less·ness** *noun*
form letter *noun* (1909)
1 : a letter on a subject of frequent recurrence that can be sent to different people without essential change except in the address
2 : a letter for mass circulation that is printed in many copies and has a very general salutation (as *Dear Friend*)
formo- — see FORM-
¹for·mu·la \'fȯr-myə-lə\ *noun, plural* **-las** *or* **-lae** \-ˌlē, -ˌlī\ [Latin, diminutive of *forma* form] (1618)
1 a : a set form of words for use in a ceremony or ritual **b :** a conventionalized statement intended to express some fundamental truth or principle especially as a basis for negotiation or action
2 a (1) **:** RECIPE (2) **:** PRESCRIPTION **b :** a milk mixture or substitute for feeding an infant
3 a : a general fact, rule, or principle expressed in usually mathematical symbols **b :** a symbolic expression of the chemical composition or constitution of a substance **c :** a group of symbols (as letters and numbers) associated to express concisely facts or data (as the number and kinds of teeth in the jaw) **d :** a combination of signs in a logical calculus
4 : a customary or set form or method allowing little room for originality
— **for·mu·la·ic** \ˌfȯr-myə-'lā-ik\ *adjective*
— **for·mu·la·ical·ly** \-'lā-ə-k(ə-)lē\ *adverb*
²formula *adjective* (1927)
: of, relating to, or being an open-wheel open-cockpit rear-engine racing car conforming to prescribed specifications as to size, weight, and engine displacement
for·mu·la·rize \'fȯr-myə-lə-ˌrīz\ *transitive verb* **-rized; -riz·ing** (1852)
: to state in or reduce to a formula **:** FORMULATE
— **for·mu·la·ri·za·tion** \ˌfȯr-myə-lə-rə-'zā-shən\ *noun*
— **for·mu·la·riz·er** \'fȯr-myə-lə-ˌrī-zər\ *noun*
for·mu·lary \'fȯr-myə-ˌler-ē\ *noun, plural* **-lar·ies** (1541)
1 : a collection of prescribed forms (as oaths or prayers)
2 : FORMULA 1
3 : a book listing medicinal substances and formulas
— **formulary** *adjective*
for·mu·late \'fȯr-myə-ˌlāt\ *transitive verb* **-lat·ed; -lat·ing** (1860)
1 a : to reduce to or express in a formula **b :** to put into a systematized statement or expression **c :** DEVISE ⟨*formulate* a policy⟩
2 a : to develop a formula for the preparation of (as a soap or plastic) **b :** to prepare according to a formula
— **for·mu·la·tor** \-ˌlā-tər\ *noun*
for·mu·la·tion \ˌfȯr-myə-'lā-shən\ *noun* (1876)
: an act or the product of formulating
formula weight *noun* (circa 1920)
: MOLECULAR WEIGHT — used especially of ionic compounds
for·mu·lize \'fȯr-myə-ˌlīz\ *transitive verb* **-lized; -liz·ing** (1842)
: FORMULATE 1
form word *noun* (1875)
: FUNCTION WORD
form·work \'fȯrm-ˌwərk\ *noun* (1918)
: a set of forms in place to hold wet concrete until it sets
for·myl \'fȯr-ˌmil\ *noun* [International Scientific Vocabulary] (circa 1859)
: the radical HCO of formic acid that is also characteristic of aldehydes
for·ni·cate \'fȯr-nə-ˌkāt\ *verb* **-cat·ed; -cat·ing** [Late Latin *fornicatus,* past participle of *fornicare* to have intercourse with prostitutes, from Latin *fornic-, fornix* arch, vault, brothel] (1552)
intransitive verb
: to commit fornication
transitive verb
: to commit fornication with
— **for·ni·ca·tor** \-ˌkā-tər\ *noun*
for·ni·ca·tion \ˌfȯr-nə-'kā-shən\ *noun* (14th century)
: consensual sexual intercourse between two persons not married to each other — compare ADULTERY
for·nix \'fȯr-niks\ *noun, plural* **for·ni·ces** \-nə-ˌsēz\ [New Latin, from Latin] (1681)
: an anatomical arch or fold
for·rad·er *also* **for·rard·er** \'fär-ə-dər\ *adverb* [English dialect, comparative of English *forward*] (1888)
chiefly British **:** further ahead
for·sake \fər-'sāk, fȯr-\ *transitive verb* **for·sook** \-'su̇k\; **for·sak·en** \-'sā-kən\; **for·sak·ing** [Middle English, from Old English *forsacan,* from *for-* + *sacan* to dispute; akin to Old English *sacu* action at law — more at SAKE] (before 12th century)
: to renounce or turn away from entirely ⟨friends have *forsaken* her⟩ ⟨*forsook* the theater for politics⟩
synonym see ABANDON
for·sooth \fər-'süth\ *adverb* [Middle English *for soth,* from Old English *forsōth,* from *for* + *sōth* sooth] (before 12th century)
: in truth **:** INDEED — often used to imply contempt or doubt
for·spent \fər-'spent, fȯr-\ *adjective* (1563)
archaic **:** WORN-OUT, EXHAUSTED
for·swear \fȯr-'swar, fōr-, -'swer\ *verb* **-swore** \-'swōr, -'swȯr\; **-sworn** \-'swōrn, -'swȯrn\; **-swear·ing** [Middle English *forsweren,* from Old English *forswerian,* from *for-* + *swerian* to swear] (before 12th century)
transitive verb
1 : to make a liar of (oneself) under or as if under oath
2 a : to reject or renounce under oath **b :** to renounce earnestly
3 : to deny under oath
intransitive verb
: to swear falsely
synonym see ABJURE
for·sworn \-'swōrn, -'swȯrn\ *adjective* (before 12th century)
1 : guilty of perjury
2 : marked by perjury
for·syth·ia \fər-'si-thē-ə, *chiefly British* -'sī-\ *noun* [New Latin, genus name, from William *Forsyth* (died 1804) British botanist] (circa 1814)
: any of a genus (*Forsythia*) of ornamental shrubs of the olive family with opposite leaves and yellow bell-shaped flowers appearing before the leaves in early spring

forsythia

fort \'fōrt, 'fȯrt\ *noun* [Middle English *forte,* from Middle French *fort,* from *fort,* adjective, strong, from Latin *fortis*] (15th century)
1 : a strong or fortified place; *especially* **:** a fortified place occupied only by troops and surrounded with such works as a ditch, rampart, and parapet **:** FORTIFICATION
2 : a permanent army post — often used in place names
for·ta·lice \'fȯr-tᵊl-əs\ *noun* [Middle English, from Medieval Latin *fortalitia* — more at FORTRESS] (15th century)
1 *archaic* **:** FORTRESS
2 *archaic* **:** a small fort
¹forte \'fōrt, 'fȯrt; *2 is often* 'fȯr-ˌtā *or* fȯr-'tā *or* 'fȯr-tē\ *noun* [French *fort,* from *fort,* adjective, strong] (circa 1648)
1 : the part of a sword or foil blade that is between the middle and the hilt and that is the strongest part of the blade
2 : one's strong point ■
²for·te \'fȯr-ˌtā, 'fȯr-tē\ *adverb or adjective* [Italian, from *forte* strong, from Latin *fortis*] (circa 1724)
: LOUD — used as a direction in music
³for·te \'fȯr-ˌtā, 'fȯr-tē\ *noun* (1759)
: a tone or passage played forte
for·te·pi·a·no \ˌfȯr-ˌtā-pē-'a-(ˌ)nō *also* -'ä-(ˌ)nō\ *noun* [French or Italian; French, from Italian, from *forte* loud + *piano* soft] (1771)

: an early form of the piano originating in the 18th and early 19th centuries and having a smaller range and softer timbre than a modern piano
word history see PIANO

for·te–pi·a·no \ˌfȯr-ˌtā-pē-ˈä-(ˌ)nō, ˌfȯr-tē-\ *adverb or adjective* (circa 1889)
: loud then immediately soft — used as a direction in music

¹forth \ˈfōrth, ˈfȯrth\ *adverb* [Middle English, from Old English; akin to Old English *for*] (before 12th century)
1 : onward in time, place, or order **:** FORWARD ⟨from that day *forth*⟩
2 : out into notice or view ⟨put *forth* leaves⟩
3 *obsolete* **:** AWAY, ABROAD

²forth *preposition* (circa 1577)
archaic **:** forth from **:** OUT OF

forth·com·ing \fōrth-ˈkə-miŋ, fȯrth-; ˈfōrth-ˌ, ˈfȯrth-\ *adjective* [obsolete *forthcome* (to come forth)] (circa 1532)
1 : being about to appear or to be produced or made available ⟨the *forthcoming* holidays⟩ ⟨your *forthcoming* novel⟩ ⟨funds are *forthcoming*⟩
2 a : RESPONSIVE, OUTGOING ⟨a *forthcoming* and courteous man⟩ **b :** characterized by openness, candidness, and forthrightness ⟨not *forthcoming* about his memories of medical school —Mark Kramer⟩

forth of *preposition* (13th century)
archaic **:** out from **:** OUT OF

¹forth·right \ˈfōr-ˌthrīt, ˈfȯr-\ *adverb* [Middle English, from Old English *forthriht,* from *forth* + *riht* right] (before 12th century)
1 *archaic* **a :** directly forward **b :** without hesitation **:** FRANKLY
2 *archaic* **:** at once

²forthright *adjective* (before 12th century)
1 *archaic* **:** proceeding straight on
2 : free from ambiguity or evasiveness **:** going straight to the point ⟨a *forthright* critic⟩ ⟨was *forthright* in appraising the problem⟩
3 : notably simple in style or quality ⟨*forthright* furniture⟩
— forth·right·ly *adverb*
— forth·right·ness *noun*

³forthright *noun* (1606)
archaic **:** a straight path

forth·with \(ˌ)fōrth-ˈwith, (ˌ)fȯrth- *also* -ˈwith\ *adverb* (14th century)
: IMMEDIATELY

for·ti·fi·ca·tion \ˌfȯr-tə-fə-ˈkā-shən\ *noun* (15th century)
1 : an act or process of fortifying
2 : something that fortifies, defends, or strengthens; *especially* **:** works erected to defend a place or position

fortified wine *noun* (1906)
: a wine (as sherry) to which alcohol usually in the form of grape brandy has been added during or after fermentation

for·ti·fi·er \ˈfȯr-tə-ˌfī(-ə)r\ *noun* (circa 1552)
: one that fortifies

for·ti·fy \-ˌfī\ *verb* **-fied; -fy·ing** [Middle English *fortifien,* from Middle French *fortifier,* from Late Latin *fortificare,* from Latin *fortis* strong] (15th century)
transitive verb
: to make strong: as **a :** to strengthen and secure (as a town) by forts or batteries **b :** to give physical strength, courage, or endurance to **c :** to add mental or moral strength to **:** ENCOURAGE ⟨*fortified* by prayer⟩ **d :** to add material to for strengthening or enriching
intransitive verb
: to erect fortifications

for·tis \ˈfȯr-təs\ *adjective* [New Latin, from Latin, strong] (1897)
: produced with relatively great articulatory tenseness and strong expiration (\t\ in *toe* is *fortis,* \d\ in *doe* is lenis)

¹for·tis·si·mo \fȯr-ˈti-sə-ˌmō\ *adverb or adjective* [Italian, superlative of *forte*] (1724)
: very loud — used especially as a direction in music

²fortissimo *noun, plural* **-mos** *or* **-mi** \-ˌmē\ (1856)
: a very loud passage, sound, or tone

for·ti·tude \ˈfȯr-tə-ˌtüd, -ˌtyüd\ *noun* [Middle English, from Latin *fortitudin-, fortitudo,* from *fortis*] (12th century)
1 : strength of mind that enables a person to encounter danger or bear pain or adversity with courage
2 *obsolete* **:** STRENGTH

fort·night \ˈfōrt-ˌnīt, ˈfȯrt-\ *noun* [Middle English *fourtenight,* alteration of *fourtene night,* from Old English *fēowertȳne niht* fourteen nights] (before 12th century)
: a period of 14 days **:** two weeks

¹fort·night·ly \-lē\ *adjective* (1800)
: occurring or appearing once in a fortnight

²fortnightly *adverb* (1820)
: once in a fortnight **:** every fortnight

³fortnightly *noun, plural* **-lies** (1940)
: a publication issued fortnightly

FOR·TRAN *or* **For·tran** \ˈfȯr-ˌtran\ *noun* [*for*mula *tran*slation] (1956)
: a computer programming language that resembles algebra in its notation and is widely used for scientific applications

for·tress \ˈfȯr-trəs\ *noun* [Middle English *forteresse,* from Middle French *forteresce,* from Medieval Latin *fortalitia,* from Latin *fortis* strong] (14th century)
: a fortified place **:** STRONGHOLD; *especially* **:** a large and permanent fortification sometimes including a town
— for·tress·like \-ˌlīk\ *adjective*

for·tu·itous \fȯr-ˈtü-ə-təs, -ˈtyü-, fər-\ *adjective* [Latin *fortuitus;* akin to Latin *fort-, fors*] (1653)
1 : occurring by chance
2 a : FORTUNATE, LUCKY ⟨from a cost standpoint, the company's timing is *fortuitous* —*Business Week*⟩ **b :** coming or happening by a lucky chance ⟨belted down the stairs, and there was a *fortuitous* train —Doris Lessing⟩ ■
synonym see ACCIDENTAL
— for·tu·itous·ly *adverb*
— for·tu·itous·ness *noun*

for·tu·ity \-ə-tē\ *noun, plural* **-ities** (circa 1747)
1 : the quality or state of being fortuitous
2 : a chance event or occurrence

for·tu·nate \ˈfȯrch-nət, ˈfȯr-chə-\ *adjective* (14th century)
1 : bringing some good thing not forseen as certain **:** AUSPICIOUS
2 : receiving some unexpected good
synonym see LUCKY
— for·tu·nate·ness *noun*

for·tu·nate·ly \-lē\ *adverb* (1548)
1 : in a fortunate manner
2 : it is fortunate that

¹for·tune \ˈfȯr-chən\ *noun* [Middle English, from Middle French, from Latin *fortuna;* akin to Latin *fort-, fors* chance, luck, and perhaps to *ferre* to carry — more at BEAR] (14th century)
1 *often capitalized* **:** a hypothetical force or personified power that unpredictably determines events and issues favorably or unfavorably
2 *obsolete* **:** ACCIDENT, INCIDENT
3 a : prosperity attained partly through luck **:** SUCCESS **b :** LUCK 1 **c** *plural* **:** the turns and courses of luck accompanying one's progress (as through life) ⟨her *fortunes* varied but she never gave up⟩
4 : DESTINY, FATE ⟨can tell your *fortune*⟩; *also* **:** a prediction of fortune
5 a : RICHES, WEALTH ⟨a man of *fortune*⟩ **b :** a store of material possessions ⟨the family *fortune*⟩ **c :** a very large sum of money ⟨spent a *fortune* redecorating⟩

²fortune *verb* **for·tuned; for·tun·ing** (14th century)
transitive verb
1 *obsolete* **:** to give good or bad fortune to
2 *archaic* **:** to endow with a fortune
intransitive verb
archaic **:** HAPPEN, CHANCE

fortune cookie *noun* (1962)
: a thin cookie folded to contain a slip of paper on which is printed a fortune, proverb, or humorous statement

fortune hunter *noun* (1689)
: a person who seeks wealth especially by marriage

for·tune–tell·er \-ˌte-lər\ *noun* (1590)
: one that professes to foretell future events
— for·tune–tell·ing \-liŋ\ *noun or adjective*

for·ty \ˈfȯr-tē\ *noun, plural* **forties** [Middle English *fourty,* adjective, from Old English *fēowertig,* from *fēowertig* group of 40, from *fēower* four + *-tig* group of 10; akin to Old English *tīen* ten] (before 12th century)
1 — see NUMBER table
2 *plural* **:** the numbers 40 to 49; *specifically* **:** the years 40 to 49 in a lifetime or century
3 : the third point scored by a side in a game of tennis
— for·ti·eth \ˈfȯr-tē-əth\ *adjective or noun*
— forty *adjective*
— forty *pronoun, plural in construction*
— for·ty·ish \-tē-ish\ *adjective*

for·ty–five \ˌfȯr-tē-ˈfīv\ *noun* (1904)
1 — see NUMBER table
2 : a .45 caliber handgun — usually written .45
3 : a phonograph record designed to be played at 45 revolutions per minute — usually written 45
— forty–five *adjective*
— forty–five *pronoun, plural in construction*

Forty Hours *noun plural but singular or plural in construction* (1759)
: a Roman Catholic devotion in which the churches of a diocese in two-day turns have the Blessed Sacrament exposed on the altar for continuous daytime veneration

for·ty–nin·er \ˌfȯr-tē-ˈnī-nər\ *noun* (1853)
: one taking part in the rush to California for gold in 1849

forty winks *noun plural but singular or plural in construction* (1872)
: a short sleep **:** NAP

□ USAGE

forte In *forte* we have a word derived from French that in its "strong point" sense has no entirely satisfactory pronunciation. Usage writers have denigrated \ˈfȯr-ˌtā\ and \ˈfȯr-tē\ because they reflect the influence of the Italian-derived ²*forte.* Their recommended pronunciation \ˈfȯrt\, however, does not exactly reflect French either: the French would write the word *le fort* and would rhyme it with English *for.* So you can take your choice, knowing that someone somewhere will dislike whichever variant you choose. All are standard, however. In British English \ˈfȯ-ˌtā\ and \ˈfȯt\ predominate; \ˈfȯr-ˌtā\ and \fȯr-ˈtā\ are probably the most frequent pronunciations in American English.

fortuitous Sense 2a has been influenced in meaning by *fortunate.* It has been in standard if not elevated use for some 70 years, but is still disdained by some critics. Sense 2b, a blend of 1 and 2a, is virtually unnoticed by the critics. Sense 1 is the only sense commonly used in negative constructions.

fo·rum \ˈfōr-əm, ˈfȯr-\ *noun, plural* **forums** *also* **fo·ra** \-ə\ [Latin; akin to Latin *foris* outside, *fores* door — more at DOOR] (15th century)
1 a : the marketplace or public place of an ancient Roman city forming the center of judicial and public business **b :** a public meeting place for open discussion **c :** a medium (as a newspaper) of open discussion or expression of ideas
2 : a judicial body or assembly **:** COURT
3 a : a public meeting or lecture involving audience discussion **b :** a program (as on radio or television) involving discussion of a problem usually by several authorities

¹for·ward \ˈfȯr-wərd, *also* ˈfō- *or* ˈfȯ-, *Southern also* ˈfär-\ *adjective* [Middle English, from Old English *foreweard,* from *fore-* + *-weard* -ward] (before 12th century)
1 a : near, being at, or belonging to the forepart **b :** situated in advance
2 a : strongly inclined **:** READY **b :** lacking modesty or reserve **:** BRASH
3 : notably advanced or developed **:** PRECOCIOUS
4 : moving, tending, or leading toward a position in front; *also* **:** moving toward an opponent's goal
5 a : advocating an advanced policy in the direction of what is considered progress **b :** EXTREME, RADICAL
6 : of, relating to, or getting ready for the future ⟨*forward* buying of produce⟩
— for·ward·ly *adverb*
— for·ward·ness *noun*

²forward *adverb* (before 12th century)
: to or toward what is ahead or in front ⟨from that time *forward*⟩ ⟨moved slowly *forward*⟩

³forward *transitive verb* (1596)
1 : to help onward **:** PROMOTE ⟨*forwarded* his friend's career⟩
2 a : to send forward **:** TRANSMIT ⟨will *forward* the goods on receipt of your check⟩ **b :** to send or ship onward from an intermediate post or station in transit ⟨*forward* mail⟩
synonym see ADVANCE

⁴forward *noun* (1879)
: a player who plays at the front of his team's formation near the goal at which his team is attempting to score

for·ward·er \-wər-dər\ *noun* (1549)
: one that forwards; *especially* **:** an agent who performs services (as receiving, transshipping, or delivering) designed to move goods to their destination

for·ward–look·ing \ˈfȯr-wərd-ˌlu̇-kiŋ\ *adjective* (1800)
: concerned with or planning for the future

forward pass *noun* (1903)
: a pass (as in football) made in the direction of the opponents' goal

for·wards \ˈfȯr-wərdz\ *adverb* (15th century)
: FORWARD

for·worn \fər-ˈwōrn, -ˈwȯrn\ *adjective* (1528)
archaic **:** WORN-OUT

for·zan·do \fȯrt-ˈsän-(ˌ)dō\ *adjective or adverb* [Italian] (circa 1828)
: SFORZANDO

¹fos·sa \ˈfä-sə\ *noun, plural* **fos·sae** \-ˌsē, -ˌsī\ [New Latin, from Latin, ditch] (1771)
: an anatomical pit, groove, or depression

²fossa *noun* [Malagasy] (1838)
: a slender lithe carnivorous mammal (*Cryptoprocta ferox*) of Madagascar that resembles a cat especially in having retractile claws but is usually considered a viverrid

fosse *or* **foss** \ˈfäs\ *noun* [Middle English *fosse,* from Old French, from Latin *fossa,* from feminine of *fossus*] (15th century)
: DITCH, MOAT

fos·sick \ˈfä-sik\ *verb* [English dialect *fossick* to ferret out] (1852)
intransitive verb
1 *Australian & New Zealand* **:** to search for gold or gemstones typically by picking over abandoned workings
2 *chiefly Australian & New Zealand* **:** to search about **:** RUMMAGE
transitive verb
chiefly Australian & New Zealand **:** to search for by or as if by rummaging **:** ferret out
— fos·sick·er *noun, chiefly Australian & New Zealand*

¹fos·sil \ˈfä-səl\ *adjective* [Latin *fossilis* obtained by digging, from *fodere* to dig — more at BED] (1665)
1 : preserved from a past geologic age ⟨*fossil* plants⟩ ⟨*fossil* water in an underground reservoir⟩
2 : being or resembling a fossil
3 : of or relating to fossil fuel

²fossil *noun* (1736)
1 : a remnant, impression, or trace of an organism of past geologic ages that has been preserved in the earth's crust — compare LIVING FOSSIL
2 a : one whose views are outmoded **:** FOGY **b :** something (as a theory) that has become rigidly fixed
3 : an old word or word element preserved only by idiom (as *fro* in *to and fro*)

fossil fuel *noun* (1835)
: a fuel (as coal, oil, or natural gas) that is formed in the earth from plant or animal remains
— fos·sil–fueled \-ˌfyü(-ə)ld\ *adjective*

fos·sil·if·er·ous \ˌfä-sə-ˈli-f(ə-)rəs\ *adjective* (circa 1846)
: containing fossils

fos·sil·ise *chiefly British variant of* FOSSILIZE

fos·sil·ize \ˈfä-sə-ˌlīz\ *verb* **-ized; -iz·ing** (1794)
transitive verb
1 : to convert into a fossil
2 : to make outmoded, rigid, or fixed
intransitive verb
: to become changed into a fossil
— fos·sil·iza·tion \ˌfä-sə-lə-ˈzā-shən\ *noun*

fos·so·ri·al \fä-ˈsōr-ē-əl, -ˈsȯr-\ *adjective* [Medieval Latin *fossorius* used for digging, from Latin *fossor* digger, from *fodere*] (1837)
: adapted to digging ⟨a *fossorial* foot⟩

¹fos·ter \ˈfȯs-tər, ˈfäs-\ *adjective* [Middle English, from Old English *fōstor-,* from *fōstor* food, feeding; akin to Old English *fōda* food] (before 12th century)
: affording, receiving, or sharing nurture or parental care though not related by blood or legal ties

²foster *transitive verb* **fos·tered; fos·ter·ing** \-t(ə-)riŋ\ (12th century)
1 : to give parental care to **:** NURTURE
2 : to promote the growth or development of **:** ENCOURAGE
— fos·ter·er \-tər-ər\ *noun*

fos·ter·age \-tə-rij\ *noun* (1614)
1 : the act of fostering
2 : a custom once prevalent in Ireland, Wales, and Scotland of entrusting one's child to foster parents to be brought up

foster home *noun* (1886)
: a household in which an orphaned, neglected, or delinquent child is placed for care

fos·ter·ling \-tər-liŋ\ *noun* (before 12th century)
: a foster child

fou \ˈfü\ *adjective* [Middle English (Scots) *fow* full, from Middle English *full*] (1535)
Scottish **:** DRUNK 1a

Fou·cault pendulum \ˌfü-ˈkō-\ *noun* [J.B.L. *Foucault*] (1931)
: a freely swinging pendulum that consists of a heavy weight hung by a long wire and that swings in a constant direction which appears to change showing that the earth rotates

fouet·té \(ˌ)fwe-ˈtā\ *noun* [French, from past participle of *fouetter* to whip, from Middle French, from *fouet* whip, from Old French, from *fou* beech, from Latin *fagus* — more at BEECH] (1830)
: a quick whipping movement of the raised leg in ballet usually accompanying a pirouette

fought *past and past participle of* FIGHT

¹foul \ˈfau̇(ə)l\ *adjective* [Middle English, from Old English *fūl;* akin to Old High German *fūl* rotten, Latin *pus* pus, *putēre* to stink, Greek *pyon* pus] (before 12th century)
1 a : offensive to the senses **:** LOATHSOME **b :** filled or covered with offensive matter
2 : full of dirt or mud
3 a : morally or spiritually odious **:** DETESTABLE ⟨a *foul* crime⟩ **b :** notably unpleasant or distressing **:** WRETCHED, HORRID ⟨in a *foul* mood⟩
4 : OBSCENE, ABUSIVE ⟨*foul* language⟩
5 a : being wet and stormy **b :** obstructive to navigation ⟨a *foul* tide⟩
6 *dialect British* **:** HOMELY, UGLY
7 a : TREACHEROUS, DISHONORABLE ⟨fair means or *foul*⟩ **b :** constituting an infringement of rules in a game or sport ⟨a *foul* blow in boxing⟩
8 : containing marked-up corrections ⟨*foul* manuscript⟩ ⟨*foul* proofs⟩
9 : encrusted, clogged, or choked with a foreign substance ⟨the chimney was *foul* and smoked badly⟩
10 : being odorous and impure **:** POLLUTED ⟨*foul* air⟩
11 : placed in a situation that impedes physical movement **:** ENTANGLED
12 : being outside the foul lines in baseball
synonym see DIRTY
— foul·ly \ˈfau̇(l)-lē\ *adverb*
— foul·ness *noun*

²foul *noun* (before 12th century)
1 *archaic* **:** something foul
2 : an entanglement or collision especially in angling or sailing
3 a : an infringement of the rules in a game or sport **b :** FREE THROW
4 : FOUL BALL

³foul (before 12th century)
intransitive verb
1 : to become or be foul: as **a :** DECOMPOSE, ROT **b :** to become encrusted, clogged, or choked with a foreign substance **c :** to become entangled or come into collision
2 : to commit a violation of the rules in a sport or game
3 : to hit a foul ball
transitive verb
1 : to make foul: as **a :** to make dirty **:** POLLUTE **b :** to tangle or come into collision with **c :** to encrust with a foreign substance ⟨a ship's bottom *fouled* with barnacles⟩ **d :** OBSTRUCT, BLOCK
2 : DISHONOR, DISCREDIT
3 : to commit a foul against
4 : to hit (a baseball) foul

⁴foul *adverb* (13th century)
: in a foul manner **:** so as to be foul

fou·lard \fü-ˈlärd\ *noun* [French] (1830)
1 a : a lightweight plain-woven or twilled silk usually decorated with a printed pattern **b :** an imitation of this fabric
2 : an article of clothing made of foulard

foul ball *noun* (1860)
: a baseball batted into foul territory

foul·brood \ˈfau̇l-ˌbrüd\ *noun* (1863)
: a destructive disease of honeybee larvae caused by bacteria (as *Bacillus larvae*)

foul·ing *noun* (14th century)
: DEPOSIT, INCRUSTATION ⟨*fouling* on a ship's bottom⟩

foul line *noun* (1878)
1 : either of two straight lines extending from the rear corner of home plate through the outer corners of first and third base respectively and prolonged to the boundary of a baseball field
2 : a line across a bowling alley that a player must not step over when delivering the ball
3 : either of two lines on a basketball court parallel to and 15 feet from the backboards behind which a player must stand while shooting a free throw

foul·mouthed \ˈfau̇l-ˌmau̇t͟hd, -ˌmau̇tht\ *adjective* (1596)

: given to the use of obscene, profane, or abusive language
foul out *intransitive verb* (1948)
: to be put out of a basketball game for exceeding the number of fouls permitted
foul play *noun* (15th century)
: VIOLENCE; *especially* : MURDER
foul shot *noun* (circa 1949)
: FREE THROW
foul tip *noun* [[1]*foul* + *tip* (tap)] (1870)
: a pitched ball in baseball that is slightly deflected by the bat; *specifically* : a tipped pitch legally caught by the catcher and counting as a full strike with the ball remaining in play
foul–up \ˈfau̇l-ˌəp\ *noun* (1950)
1 : a state of confusion or an error caused by ineptitude, carelessness, or mismanagement ⟨*foul-ups* in transportation⟩
2 : a mechanical difficulty
foul up (1947)
transitive verb
1 : to make dirty : CONTAMINATE
2 : to spoil by making mistakes or using poor judgment : CONFUSE
3 : ENTANGLE, BLOCK ⟨*fouled up* communications⟩
intransitive verb
: to cause a foul-up : BUNGLE ⟨it was his fault. He had *fouled up* —Pat Frank⟩
[1]**found** \ˈfau̇nd\ *past and past participle of* FIND
[2]**found** *adjective* (1793)
1 : having all usual, standard, or reasonably expected equipment ⟨the boat comes fully *found*, ready to go —*Holiday*⟩
2 : presented as or incorporated into an artistic work essentially as found ⟨sculpture of fabric, wood, and other *found* materials —Hilton Kramer⟩
[3]**found** *noun* (1830)
: free food and lodging in addition to wages ⟨they're paid $175 a month and *found* —*New Yorker*⟩
[4]**found** *transitive verb* [Middle English, from Old French *fonder,* from Latin *fundare,* from *fundus* bottom — more at BOTTOM] (13th century)
1 : to take the first steps in building
2 : to set or ground on something solid : BASE
3 : to establish (as an institution) often with provision for future maintenance
[5]**found** *transitive verb* [Middle French *fondre* to pour, melt, from Latin *fundere;* akin to Old English *gēotan* to pour, Greek *chein*] (1562)
: to melt (as metal) and pour into a mold
foun·da·tion \fau̇n-ˈdā-shən\ *noun* (14th century)
1 : the act of founding
2 : a basis (as a tenet, principle, or axiom) upon which something stands or is supported ⟨the *foundations* of geometry⟩ ⟨the rumor is without *foundation* in fact⟩
3 a : funds given for the permanent support of an institution : ENDOWMENT **b :** an organization or institution established by endowment with provision for future maintenance
4 : an underlying base or support; *especially* : the whole masonry substructure of a building
5 a : a body or ground upon which something is built up or overlaid **b :** a woman's supporting undergarment : CORSET **c :** a cosmetic usually used as a base for makeup
— **foun·da·tion·al** \-shnəl, -shə-nᵊl\ *adjective*
— **foun·da·tion·al·ly** *adverb*
— **foun·da·tion·less** \-shən-ləs\ *adjective*
foundation stone *noun* (1651)
1 : BASIS, GROUNDWORK
2 : a stone in the foundation of a building; *especially* : such a stone laid with public ceremony — compare CORNERSTONE
[1]**found·er** \ˈfau̇n-dər\ *noun* [[4]*found*] (14th century)
: one that founds or establishes
[2]**foun·der** \ˈfau̇n-dər\ *verb* **foun·dered; foun·der·ing** \-d(ə-)riŋ\ [Middle English *foundren* to send to the bottom, collapse, from Middle French *fondrer,* ultimately from Latin *fundus*] (14th century)
intransitive verb
1 : to become disabled; *especially* : to go lame
2 : to give way : COLLAPSE
3 : to sink below the surface of the water
4 : to come to grief : FAIL
transitive verb
: to disable (an animal) especially by excessive feeding
[3]**foun·der** *noun* (circa 1547)
: LAMINITIS
[4]**found·er** *noun* [[5]*found*] (15th century)
: one that founds metal; *especially* : TYPEFOUNDER
founding father *noun* (1914)
1 : an originator of an institution or movement : FOUNDER
2 *often both Fs capitalized* **:** a leading figure in the founding of the U.S.; *specifically* : a member of the American Constitutional Convention of 1787
found·ling \ˈfau̇n(d)-liŋ\ *noun* (14th century)
: an infant found after its unknown parents have abandoned it
found object *noun* (1950)
: OBJET TROUVÉ
found poem *noun* (1966)
: a poem consisting of words found in a nonpoetic context (as a product label) and usually broken into lines that convey a verse rhythm
found·ry \ˈfau̇n-drē\ *noun, plural* **foundries** (1536)
1 : an establishment where founding is carried on
2 : the act, process, or art of casting metals
[1]**fount** \ˈfau̇nt\ *noun* [Middle English, from Middle French *font,* from Latin *font-, fons*] (15th century)
: FOUNTAIN, SOURCE
[2]**fount** \ˈfänt, ˈfau̇nt\ *noun* [French *fonte,* from Middle French — more at FONT] (circa 1683)
British : a type font
[1]**foun·tain** \ˈfau̇n-tᵊn\ *noun* [Middle English, from Middle French *fontaine,* from Late Latin *fontana,* from Latin, feminine of *fontanus* of a spring, from *font-, fons*] (14th century)
1 : the source from which something proceeds or is supplied
2 : a spring of water issuing from the earth
3 : an artificially produced jet of water; *also* : the structure from which it rises
4 : a reservoir containing a liquid that can be drawn off as needed
5 : SODA FOUNTAIN 2
[2]**fountain** (1903)
intransitive verb
: to flow or spout like a fountain
transitive verb
: to cause to flow like a fountain
foun·tain·head \-ˌhed\ *noun* (1585)
1 : a spring that is the source of a stream
2 : principal source : ORIGIN
fountain pen *noun* (1710)
: a pen containing a reservoir that automatically feeds the writing point with ink
four \ˈfōr, ˈfȯr\ *noun* [Middle English, from *four* adjective, from Old English *fēower;* akin to Old High German *fior* four, Latin *quattuor,* Greek *tessares, tettares*] (before 12th century)
1 — see NUMBER table
2 : the fourth in a set or series ⟨the *four* of hearts⟩
3 : something having four units or members: as **a :** a 4-oared racing shell or its crew **b :** a 4-cylinder engine or automobile
— **four** *adjective*
— **four** *pronoun, plural in construction*
four–bag·ger \-ˈba-gər\ *noun* (1926)
: HOME RUN
four–ball \-ˌbȯl\ *adjective* (1904)
: relating to or being a golf match in which the best individual score of one partnership is matched against the best individual score of another partnership for each hole
four·chée \fu̇r-ˈshā\ *adjective* [French (feminine), literally, forked] (1706)
of a heraldic cross : having the end of each arm forked — see CROSS illustration
four–dimensional *adjective* (1880)
: relating to or having four dimensions ⟨*four-dimensional* space-time continuum⟩; *especially* : consisting of or relating to elements requiring four coordinates to determine them
four·dri·nier \ˌfōr-drə-ˈnir, ˌfȯr-; fu̇r-ˈdri-nē-ər, fōr-, fȯr-\ *noun, often capitalized* [Henry & Sealy *Fourdrinier*] (1839)
: a machine for making paper in an endless web
four–eyed \ˈfōr-ˌīd, ˈfȯr-\ *adjective* (1926)
: wearing glasses
4–F \ˈfōr-ˈef, ˈfȯr-\ *noun* (1944)
: classification as unfit for military service; *also* : a person having this classification
four–flush *intransitive verb* (1896)
: to bluff in poker holding a four flush; *broadly* : to make a false claim : BLUFF
— **four–flush·er** \-ˈflə-shər\ *noun*
four flush *noun* (1887)
: four cards of the same suit in a 5-card poker hand
four·fold \ˈfōr-ˌfōld, ˈfȯr-, -ˈfōld\ *adjective* [Middle English, from Old English *fēowerfeald,* from *fēower* + *-feald* -fold] (before 12th century)
1 : being four times as great or as many
2 : having four units or members
— **four·fold** \-ˈfōld\ *adverb*
four–foot·ed \-ˈfu̇-təd\ *adjective* (14th century)
: having four feet : QUADRUPED
four·gon \fu̇r-ˈgōⁿ\ *noun, plural* **fourgons** \-ˈgōⁿ(z)\ [French] (1848)
: a wagon for carrying baggage
4–H \ˈfōr-ˈāch, ˈfȯr-\ *adjective* [from the fourfold aim of improving the head, heart, hands, and health] (1926)
: of or relating to a program set up by the U.S. Department of Agriculture originally in rural areas to help young people become productive citizens by instructing them in useful skills (as in agriculture, animal husbandry, and carpentry), community service, and personal development
— **4–H'·er** *also* **4–H·er** \-ˈā-chər\ *noun*
four–hand \ˈfōr-ˌhand, ˈfȯr-\ *adjective* (circa 1909)
: FOUR-HANDED
four–hand·ed \-ˈhan-dəd\ *adjective* (1824)
1 : engaged in by four persons ⟨a *four-handed* card game⟩
2 : designed for four hands ⟨a *four-handed* musical composition⟩
Four Horsemen *noun plural* [from the apocalyptic vision in Revelation 6:2–8] (1918)
: war, famine, pestilence, and death personified as the four major plagues of mankind
Four Hundred *or* **400** *noun* (1888)
: the exclusive social set of a community — used with *the*
Fou·ri·er analysis \ˈfu̇r-ē-ˌā-\ *noun* [Baron J.B.J. *Fourier* (died 1830) French geometrician & physicist] (circa 1928)
: the process of using the terms of a Fourier series to find a function that approximates periodic data
Fou·ri·er·ism \ˈfu̇r-ē-ə-ˌri-zəm, -ē-ˌā-ˌi-\ *noun* [French *fouriérisme,* from F.M.C. *Fourier*] (1843)
: a system for reorganizing society into cooperative communities of small self-sustaining groups
— **Fou·ri·er·ist** \-ē-ə-rist, -ē-ˌā-ist\ *noun*
Fou·ri·er series \ˈfu̇r-ē-ˌā-\ *noun* [Baron J.B.J. *Fourier*] (1877)

: an infinite series in which the terms are constants multiplied by sine or cosine functions of integer multiples of the variable and which is used in the analysis of periodic functions

Fourier's theorem *noun* (1834)
: a theorem in mathematics: under suitable conditions any periodic function can be represented by a Fourier series

Fourier transform *noun* (1923)
: any of various functions (as $F(u)$) that under suitable conditions can be obtained from given functions (as $f(x)$) by multiplying by e^{iux} and integrating over all values of x and that in scientific instrumentation describe the dependence of the average of a series of measurements (as of a spectrum) on a quantity of interest (as brightness) especially of a very small magnitude — called also *Fourier transformation*

four–in–hand \ˈfōr-ən-ˌhand, ˈfȯr-\ *noun* (1793)
1 a : a team of four horses driven by one person **b :** a vehicle drawn by such a team
2 : a necktie tied in a slipknot with long ends overlapping vertically in front

four–letter *adjective* (1897)
: of, relating to, or being four-letter words

four–letter word *noun* (1897)
: any of a group of vulgar or obscene words typically made up of four letters

four–line octave *noun* [from the four accent marks appended to the letters representing its notes] (1931)
: the musical octave that begins on the third C above middle C — see PITCH illustration

four–o'clock \ˈfōr-ə-ˌkläk, ˈfȯr-\ *noun* (1756)
: any of a genus (*Mirabilis*) of chiefly American annual or perennial herbs (family Nyctaginaceae, the four-o'clock family) having apetalous flowers with a showy involucre simulating a calyx; *especially* **:** a garden plant (*M. jalapa*) with fragrant yellow, red, or white flowers opening late in the afternoon

four of a kind (circa 1934)
: four cards of the same rank in one hand — see POKER illustration

four·plex \ˈfōr-ˌpleks, ˈfȯr-\ *noun* (1952)
: a building that contains four separate apartments

four–post·er \ˌfōr-ˈpō-stər, ˌfȯr-\ *noun* (1836)
: a bed with tall often carved corner posts originally designed to support curtains or a canopy

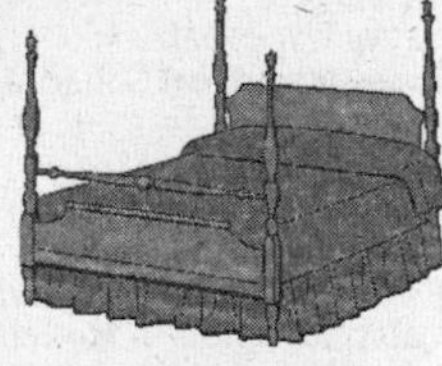
four-poster

four·ra·gère \ˌfu̇r-ə-ˈzher\ *noun* [French, from feminine of *fourrager* of forage, from *fourrage* forage] (1919)
: a braided cord worn usually around the left shoulder; *especially* **:** such a cord awarded as a decoration to a military unit

four·score \ˈfōr-ˈskōr, ˈfȯr-ˈskȯr\ *adjective* (13th century)
: being four times twenty **:** EIGHTY

four·some \ˈfōr-səm, ˈfȯr-\ *noun* (14th century)
1 a : a group of four persons or things **:** QUARTET **b :** two couples
2 : a golf match in which two players compete against two others with players on each side taking turns playing one ball; *broadly* **:** any golf match involving four players

four·square \-ˈskwar, -ˈskwer\ *adjective* (14th century)
1 : SQUARE
2 : marked by boldness and conviction **:** FORTHRIGHT
— foursquare *adverb*

four–star \-ˈstär\ *adjective* [from the number of asterisks used to denote relative excellence in guidebooks] (1921)
: of a superior degree of excellence ⟨a *four-star* French restaurant⟩

four·teen \fōr-ˈtēn, fȯr-; ˈfōr(t)-ˌtēn, ˈfȯr(t)-\ *noun* [Middle English *fourtene,* from Old English *fēowertīene,* from *fēowertīene,* adjective, from *fēower* + *-tīene;* akin to Old English *tīen* ten] (before 12th century)
— see NUMBER table
— fourteen *adjective*
— fourteen *pronoun, plural in construction*
— four·teenth \-ˈtēn(t)th; -ˌtēn(t)th\ *adjective or noun*

four·teen·er \-ˈtē-nər\ *noun* (1884)
: a verse consisting of 14 syllables or especially of 7 iambic feet

fourth \ˈfōrth, ˈfȯrth\ *noun, plural* **fourths** \ˈfōr(th)s, ˈfȯr(th)s\ (before 12th century)
1 — see NUMBER table
2 a : a musical interval embracing four tones of the diatonic scale **b :** a tone at this interval; *specifically* **:** SUBDOMINANT 1 **c :** the harmonic combination of two tones a fourth apart
3 : the 4th forward gear or speed of a motor vehicle
4 *capitalized* **:** INDEPENDENCE DAY — used with *the*
— fourth *adjective or adverb*
— fourth·ly *adverb*

fourth class *noun* (1862)
1 : a class or group ranking fourth in a series
2 : a class of mail in the U.S. that comprises merchandise and non-second-class printed matter and is not sealed against inspection

fourth dimension *noun* (1875)
1 : a dimension in addition to length, breadth, and depth; *specifically* **:** a coordinate in addition to three rectangular coordinates especially when interpreted as the time coordinate in a space-time continuum
2 : something outside the range of ordinary experience
— fourth–dimensional *adjective*

fourth estate *noun, often F&E capitalized* (1752)
: the public press ◆

Fourth of July (1779)
: INDEPENDENCE DAY

fourth world *noun, often F&W capitalized* (1974)
: a group of nations especially in Africa and Asia characterized by extremely low per capita income and an absence of valuable natural resources

four–way \ˈfōr-ˈwā, ˈfȯr-\ *adjective* (1824)
1 : allowing or affecting passage in any of four directions
2 : including four participants

four–wheel \ˈfōr-ˌhwēl, ˈfȯr-, -ˌwēl\ *or* **four–wheeled** \-ˌhwē(ə)ld, -ˌwē(ə)ld\ *adjective* (1740)
1 : having four wheels
2 : acting on or by means of four wheels of an automotive vehicle ⟨*four-wheel* drive⟩

four–wheel·er \-ˌhwē-lər, -ˌwē-\ *noun* (1846)
: a vehicle with four wheels

fo·vea \ˈfō-vē-ə\ *noun, plural* **fo·ve·ae** \-vē-ˌē, -vē-ˌī\ [New Latin, from Latin, pit] (1849)
1 : a small fossa
2 : a small rodless area of the retina that affords acute vision — see EYE illustration
— fo·ve·al \-vē-əl\ *adjective*
— fo·ve·ate \-vē-ˌāt, -ət\ *adjective*

fovea cen·tra·lis \-sen-ˈtra-ləs, -ˈträ-, -ˈtrā-\ *noun* [New Latin, central fovea] (1858)
: FOVEA 2

¹fowl \ˈfau̇(ə)l\ *noun, plural* **fowl** *or* **fowls** [Middle English *foul,* from Old English *fugel;* akin to Old High German *fogal* bird, and probably to Old English *flēogan* to fly — more at FLY] (before 12th century)
1 : a bird of any kind — compare WATERFOWL, WILDFOWL
2 a : a cock or hen of the domestic chicken (*Gallus gallus*); *especially* **:** an adult hen **b :** any of several domesticated or wild gallinaceous birds — compare GUINEA FOWL, JUNGLE FOWL
3 : the meat of fowls used as food

²fowl *intransitive verb* (before 12th century)
: to seek, catch, or kill wildfowl
— fowl·er *noun*

fowling piece *noun* (1596)
: a shotgun for shooting birds or small quadrupeds

¹fox \ˈfäks\ *noun, plural* **fox·es** *also* **fox** *often attributive* [Middle English, from Old English; akin to Old High German *fuhs* fox and perhaps to Sanskrit *puccha* tail] (before 12th century)
1 a : any of various carnivorous mammals (especially genus *Vulpes*) of the dog family related to but smaller than wolves with shorter legs, more pointed muzzle, large erect ears, and long bushy tail **b :** the fur of a fox
2 : a clever crafty person
3 *archaic* **:** SWORD
4 *capitalized* **:** a member of an American Indian people formerly living in what is now Wisconsin
5 : a good-looking young woman or man
word history see VIXEN

²fox *transitive verb* (1611)
1 *obsolete* **:** INTOXICATE
2 a : to trick by ingenuity or cunning **:** OUTWIT **b :** BAFFLE

foxed \ˈfäkst\ *adjective* (1847)
: discolored with yellowish brown stains

fox fire *noun* (15th century)
: an eerie phosphorescent light (as of decaying wood); *also* **:** a luminous fungus (as *Armillaria mellea*) that causes decaying wood to glow

fox·glove \ˈfäks-ˌgləv\ *noun* (before 12th century)
: any of a genus (*Digitalis*) of erect herbs of the snapdragon family; *especially* **:** a common European biennial or perennial (*D. purpurea*) cultivated for its showy racemes of dotted white or purple tubular flowers and as a source of digitalis

fox grape *noun* (1657)
: any of several native grapes (especially *Vitis labrusca*) of eastern North America with sour or musky fruit

fox·hole \ˈfäks-ˌhōl\ *noun* (1919)
: a pit dug usually hastily for individual cover from enemy fire

fox·hound \-ˌhau̇nd\ *noun* (circa 1763)
: any of various large swift powerful hounds of great endurance used in hunting foxes and developed to form several breeds and many distinctive strains — compare AMERICAN FOXHOUND, ENGLISH FOXHOUND

fox·hunt·er \-ˌhən-tər\ *noun* (1692)
1 : one who engages in foxhunting
2 : HUNTER 1c

fox·hunt·ing \-ˌhən-tiŋ\ *noun* (1674)
: a pastime in which participants on horseback ride over the countryside following a pack of hounds on the trail of a fox
— fox·hunt \-ˌhənt\ *intransitive verb*

fox·tail \ˈfäks-ˌtāl\ *noun* (14th century)

◇ WORD HISTORY

fourth estate In Europe, in former times, the people who participated in the political life of a country were generally divided into three classes or estates. In England the three traditional estates were the nobility, the clergy, and the common people. Some other group, like the mob or the public press, that had an unofficial but often great influence on public affairs was called the fourth estate. In the 19th century, *fourth estate* came to be reserved exclusively for the press—or the news media, to use the more contemporary term.

1 a : the tail of a fox **b :** something resembling the tail of a fox
2 : any of several grasses (especially genera *Alopecurus, Hordeum,* and *Setaria*) with spikes resembling brushes — called also *foxtail grass*

foxtail lily *noun* (1946)
: EREMURUS

foxtail millet *noun* (circa 1899)
: a coarse drought-resistant but frost-sensitive annual grass (*Setaria italica*) grown for grain, hay, and forage

fox terrier *noun* (1823)
: any of a smooth-haired or a wirehaired breed of small lively terriers formerly used to dig out foxes

fox terrier

Fox·trot \'fäks-ˌträt\ (1952)
— a communications code word for the letter *f*

1fox–trot \'fäks-ˌträt\ *noun* (1872)
1 : a short broken slow trotting gait in which the hind foot of the horse hits the ground a trifle before the diagonally opposite forefoot
2 : a ballroom dance in duple time that includes slow walking steps, quick running steps, and the step of the two-step

2fox–trot *intransitive verb* (1916)
: to dance the fox-trot

foxy \'fäk-sē\ *adjective* **fox·i·er; -est** (1528)
1 : resembling or suggestive of a fox ⟨a narrow *foxy* face⟩: as **a :** cunningly shrewd **b :** of a warm reddish brown color ⟨*foxy* eyebrows⟩
2 : having a sharp brisk flavor ⟨*foxy* grapes⟩
3 : physically attractive ⟨a *foxy* lady⟩
synonym see SLY
— fox·i·ly \'fäk-sə-lē\ *adverb*
— fox·i·ness \-sē-nəs\ *noun*

foy \'fȯi\ *noun* [Dutch dialect *fooi* feast at end of the harvest] (circa 1645)
chiefly Scottish **:** a farewell feast or gift

foy·er \'fȯi(-ə)r, 'fȯi-ˌ(y)ā *also* 'fwä-ˌyā\ *noun* [French, literally, fireplace, from (assumed) Vulgar Latin *focarium,* from Latin *focus* hearth] (1859)
: an anteroom or lobby especially of a theater; *also* **:** an entrance hallway **:** VESTIBULE

Fra \'frä\ *noun* [Italian, short for *frate,* from Latin *frater* — more at BROTHER] (circa 1890)
— used as a title equivalent to *brother* preceding the name of an Italian monk or friar

fra·cas \'frā-kəs, 'fra-, *British* 'fra-ˌkä\ *noun, plural* **fra·cas·es** \-kə-səz\ *or British* **fracas** \-ˌkäz\ [French, din, row, from Italian *fracasso,* from *fracassare* to shatter] (1727)
: a noisy quarrel **:** BRAWL

frac·tal \'frak-tᵊl\ *noun* [French *fractale,* from Latin *fractus* broken, uneven (past participle of *frangere* to break) + French *-ale* -al (noun suffix)] (1975)
: any of various extremely irregular curves or shapes for which any suitably chosen part is similar in shape to a given larger or smaller part when magnified or reduced to the same size
— fractal *adjective*

fract·ed \'frak-təd\ *adjective* [Latin *fractus*] (1547)
obsolete **:** BROKEN

frac·tion \'frak-shən\ *noun* [Middle English *fraccioun,* from Late Latin *fraction-, fractio* act of breaking, from Latin *frangere* to break — more at BREAK] (14th century)
1 a : a numerical representation (as ¾, ⅝, 3.234) indicating the quotient of two numbers **b** (1) **:** a piece broken off **:** FRAGMENT (2) **:** a discrete unit **:** PORTION
2 : one of several portions (as of a distillate) separable by fractionation
3 : BIT, LITTLE ⟨a *fraction* closer⟩

frac·tion·al \-shnəl, -shə-nᵊl\ *adjective* (1675)
1 : of, relating to, or being a fraction
2 : of, relating to, or being fractional currency
3 : relatively small **:** INCONSIDERABLE
4 : of, relating to, or involving a process for separating components of a mixture through differences in physical or chemical properties ⟨*fractional* distillation⟩
— frac·tion·al·ly *adverb*

fractional currency *noun* (1862)
1 : paper money in denominations of less than one dollar issued by the U.S. 1863–76
2 : currency in denominations less than the basic monetary unit

frac·tion·al·ize \'frak-shnə-ˌlīz, -shə-nᵊl-ˌīz\ *transitive verb* **-ized; -iz·ing** (1924)
: to break up into parts or sections
— frac·tion·al·iza·tion \ˌfrak-shnə-lə-'zā-shən, -shə-nᵊl-ə-'zā-\ *noun*

frac·tion·ate \'frak-shə-ˌnāt\ *transitive verb* **-at·ed; -at·ing** (1867)
1 : to separate (as a mixture) into different portions
2 : to divide or break up
— frac·tion·ation \ˌfrak-shə-'nā-shən\ *noun*
— frac·tion·ator \'frak-shə-ˌnā-tər\ *noun*

frac·tious \'frak-shəs\ *adjective* [*fraction* (discord) + *-ous*] (1725)
1 : tending to be troublesome **:** UNRULY
2 : QUARRELSOME, IRRITABLE
— frac·tious·ly *adverb*
— frac·tious·ness *noun*

1frac·ture \'frak-chər, -shər\ *noun* [Middle English, from Latin *fractura,* from *fractus*] (15th century)
1 : the result of fracturing **:** BREAK
2 a : the act or process of breaking or the state of being broken; *especially* **:** the breaking of hard tissue (as bone) **b :** the rupture (as by tearing) of soft tissue ⟨kidney *fracture*⟩
3 : the general appearance of a freshly broken surface of a mineral

2fracture *verb* **frac·tured; frac·tur·ing** \-chə-riŋ, -shriŋ\ (1612)
transitive verb
1 a : to cause a fracture in **:** BREAK ⟨*fracture* a rib⟩ **b :** RUPTURE, TEAR
2 a : to damage or destroy as if by rupturing **b :** to cause great disorder in **c :** to break up **:** FRACTIONATE **d :** to go beyond the limits of (as rules) **:** VIOLATE ⟨*fractured* the English language with malaprops —Goodman Ace⟩
intransitive verb
: to undergo fracture

fracture zone *noun* (1946)
: an area of suboceanic crust characterized by fractures

frae \'frā\ *preposition* [Middle English (northern) *fra, frae,* from Old Norse *frā;* akin to Old English *fram* from] (1700)
Scottish **:** FROM

frag·ile \'fra-jəl, -ˌjīl\ *adjective* [Middle French, from Latin *fragilis* — more at FRAIL] (1607)
1 a : easily broken or destroyed **b :** constitutionally delicate **:** lacking in physical vigor
2 : TENUOUS, SLIGHT ☆
— fra·gil·i·ty \frə-'ji-lə-tē\ *noun*

1frag·ment \'frag-mənt\ *noun* [Middle English, from Latin *fragmentum,* from *frangere* to break — more at BREAK] (15th century)
: a part broken off, detached, or incomplete
synonym see PART

2frag·ment \-ˌment\ (1818)
intransitive verb
: to fall to pieces
transitive verb
: to break up or apart into fragments

frag·men·tal \frag-'men-tᵊl\ *adjective* (1798)
: FRAGMENTARY
— frag·men·tal·ly \-tᵊl-ē\ *adverb*

frag·men·tary \'frag-mən-ˌter-ē\ *adjective* (1611)
: consisting of fragments **:** INCOMPLETE
— frag·men·tari·ly \ˌfrag-mən-'ter-ə-lē\ *adverb*
— frag·men·tari·ness \-ˌter-ē-nəs\ *noun*

frag·men·ta·tion \ˌfrag-mən-'tā-shən, -ˌmen-\ *noun* (1881)
1 : the act or process of fragmenting or making fragmentary
2 : the state of being fragmented or fragmentary
— frag·men·tate \'frag-mən-ˌtāt\ *verb*

fragmentation bomb *noun* (1918)
: a bomb or shell whose relatively thick casing is splintered upon explosion and thrown in fragments in all directions

frag·men·tize \'frag-mən-ˌtīz\ *transitive verb* **-tized; -tiz·ing** (1815)
: FRAGMENT

fra·grance \'frā-grən(t)s\ *noun* (1667)
1 a : a sweet or delicate odor (as of fresh flowers, pine trees, or perfume) **b :** something (as a perfume) compounded to give off a sweet or pleasant odor
2 : the quality or state of having a sweet odor ☆

fra·gran·cy \-grən(t)-sē\ *noun* (1578)
: FRAGRANCE

fra·grant \'frā-grənt\ *adjective* [Middle English, from Latin *fragrant-, fragrans,* from present participle of *fragrare* to be fragrant] (15th century)
: marked by fragrance
synonym see ODOROUS
— fra·grant·ly *adverb*

frail \'frā(ə)l\ *adjective* [Middle English, from Middle French *fraile,* from Latin *fragilis* fragile, from *frangere*] (14th century)
1 : easily led into evil ⟨*frail* humanity⟩
2 : easily broken or destroyed **:** FRAGILE
3 a : physically weak **b :** SLIGHT, UNSUBSTANTIAL
synonym see WEAK

☆ SYNONYMS
Fragile, frangible, brittle, crisp, friable mean breaking easily. FRAGILE implies extreme delicacy of material or construction and need for careful handling ⟨a *fragile* antique chair⟩. FRANGIBLE implies susceptibility to being broken without implying weakness or delicacy ⟨*frangible* stone used for paving⟩. BRITTLE implies hardness together with lack of elasticity or flexibility or toughness ⟨patients with *brittle* bones⟩. CRISP implies a firmness and brittleness desirable especially in some foods ⟨*crisp* lettuce⟩. FRIABLE applies to substances that are easily crumbled or pulverized ⟨*friable* soil⟩. See in addition WEAK.

Fragrance, perfume, scent, redolence mean a sweet or pleasant odor. FRAGRANCE suggests the odors of flowers or other growing things ⟨the *fragrance* of pine⟩. PERFUME may suggest a stronger or heavier odor ⟨the *perfume* of lilacs⟩. SCENT is very close to PERFUME but of wider application because more neutral in connotation ⟨*scent*-free soaps⟩. REDOLENCE implies a mixture of fragrant or pungent odors ⟨the *redolence* of a forest after a rain⟩.

— **frail·ly** \'frā(ə)(l)-lē\ *adverb*
— **frail·ness** *noun*

frail·ty \'frā(-ə)l-tē\ *noun, plural* **frailties** (14th century)
1 : the quality or state of being frail
2 : a fault due to weakness especially of moral character
synonym see FAULT

fraise \'frāz\ *noun* [French] (1775)
: an obstacle of pointed stakes driven into the ramparts of a fortification in a horizontal or inclined position

Frak·tur *also* **Frac·tur** \fräk-'tu̇r\ *noun* [German, from Latin *fractura* fracture] (1904)
1 : a German style of black letter
2 *often not capitalized* **:** a Pennsylvania German document (as a birth or wedding certificate) that is written in calligraphy and illuminated with decorative motifs (as tulips, birds, and scrolls)

fram·be·sia \fram-'bē-zh(ē-)ə\ *noun* [New Latin, from French *framboise* raspberry; from the appearance of the lesions] (1803)
: YAWS

fram·boise \fräⁿ-'bwäz\ *noun* [French, literally, raspberry, from Old French, of Germanic origin; akin to Dutch *braambes* blackberry, literally, bramble berry, Old High German *brāmberi* — more at BROOM, BERRY] (circa 1945)
: a brandy or liqueur made from raspberries

¹frame \'frām\ *verb* **framed; fram·ing** [Middle English, to benefit, construct, from Old English *framian* to benefit, make progress; akin to Old Norse *fram* forward, Old English *fram* from] (14th century)
transitive verb
1 : to construct by fitting and uniting the parts of the skeleton of (a structure)
2 a : PLAN, CONTRIVE ⟨*framed* a new method of achieving their purpose⟩ **b :** SHAPE, CONSTRUCT **c :** to give expression to **:** FORMULATE **d :** to draw up (as a document)
3 a : to devise falsely (as a criminal charge) **b :** to contrive the evidence against (an innocent person) so that a verdict of guilty is assured **c :** FIX 7b
4 : to fit or adjust especially to something or for an end **:** ARRANGE
5 *obsolete* **:** PRODUCE
6 : to enclose in a frame; *also* **:** to enclose as if in a frame
intransitive verb
1 *archaic* **:** PROCEED, GO
2 *obsolete* **:** MANAGE
— **fram·able** *or* **frame·able** \'frā-mə-bəl\ *adjective*
— **fram·er** *noun*

²frame *noun* (14th century)
1 a : something composed of parts fitted together and united **b :** the physical makeup of an animal and especially a human body **:** PHYSIQUE, FIGURE
2 a : the constructional system that gives shape or strength (as to a building); *also* **:** a frame dwelling **b :** such a skeleton not filled in or covered
3 *obsolete* **:** the act or manner of framing
4 a : a machine built upon or within a framework ⟨a spinning *frame*⟩ **b :** an open case or structure made for admitting, enclosing, or supporting something ⟨a window *frame*⟩ **c** (1) **:** a part of a pair of glasses that holds one of the lenses (2) *plural* **:** that part of a pair of glasses other than the lenses **d :** a structural unit in an automobile chassis supported on the axles and supporting the rest of the chassis and the body
5 a : an enclosing border **b :** the matter or area enclosed in such a border: as (1) **:** one of the squares in which scores for each round are recorded (as in bowling); *also* **:** a round in bowling (2) **:** an individual drawing in a comic strip usually enclosed by a bordering line (3) **:** one picture of the series on a length of film (4) **:** a complete image for display (as on a television set) **c :** an inning in baseball **d** (1) **:** FRAMEWORK 1a (2) **:** CONTEXT, FRAME OF REFERENCE **e :** an event that forms the background for the action of a novel or play
6 : FRAME-UP

³frame *adjective* (1775)
: having a wood frame ⟨*frame* houses⟩

frame of mind (1665)
: mental attitude or outlook **:** MOOD

frame of reference (1897)
1 : an arbitrary set of axes with reference to which the position or motion of something is described or physical laws are formulated
2 : a set of ideas, conditions, or assumptions that determine how something will be approached, perceived, or understood ⟨a Marxian *frame of reference*⟩

frame·shift \'frām-ˌshift\ *adjective* (1967)
: relating to, being, or causing a mutation in which a number of nucleotides not divisible by three is inserted or deleted so as to change the reading frame of some triplet codons during genetic translation
— **frameshift** *noun*

frame–up \'frā-ˌməp\ *noun* (1889)
1 : an act or series of actions in which someone is framed
2 : an action that is framed

frame·work \'frām-ˌwərk\ *noun* (1644)
1 a : a basic conceptional structure (as of ideas) ⟨the *framework* of the constitution⟩ **b :** a skeletal, openwork, or structural frame
2 : FRAME OF REFERENCE
3 : the larger branches of a tree that determine its shape

fram·ing \'frā-miŋ\ *noun* (1703)
: FRAME, FRAMEWORK

franc \'fraŋk\ *noun* [French] (14th century)
— see MONEY table

¹fran·chise \'fran-ˌchīz\ *noun* [Middle English, from Middle French, from *franchir* to free, from Old French *franc* free — more at FRANK] (14th century)
1 : freedom or immunity from some burden or restriction vested in a person or group
2 a : a special privilege granted to an individual or group; *especially* **:** the right to be and exercise the powers of a corporation **b :** a constitutional or statutory right or privilege; *especially* **:** the right to vote **c** (1) **:** the right or license granted to an individual or group to market a company's goods or services in a particular territory; *also* **:** a business granted such a right or license (2) **:** the territory involved in such a right
3 a : the right of membership in a professional sports league **b :** a team and its operating organization having such membership

²franchise *transitive verb* **fran·chised; fran·chis·ing** (14th century)
1 *archaic* **:** FREE
2 : to grant a franchise to

fran·chi·see \ˌfran-ˌchī-'zē, -chə-\ *noun* (1954)
: one granted a franchise

fran·chis·er \'fran-ˌchī-zər\ *noun* [in sense 1, from ¹*franchise;* in sense 2, from ²*franchise*] (1843)
1 : FRANCHISEE
2 : FRANCHISOR

fran·chi·sor \ˌfran-ˌchī-'zȯr, -chə-\ *noun* [²*franchise* + ¹*-or*] (1967)
: one that grants a franchise

Fran·cis·can \fran-'sis-kən\ *noun* [Medieval Latin *Franciscus* Francis] (1599)
: a member of the Order of Friars Minor founded by Saint Francis of Assisi in 1209 and dedicated especially to preaching, missions, and charities
— **Franciscan** *adjective*

fran·ci·um \'fran(t)-sē-əm\ *noun* [New Latin, from *France*] (1946)
: a radioactive element of the alkali-metal group discovered as a disintegration product of actinium and obtained artificially by the bombardment of thorium with protons — see ELEMENT table

Franco- *combining form* [Medieval Latin, from *Francus* Frenchman, from Late Latin, Frank]
1 : French and ⟨*Franco*-American⟩
2 : French ⟨*Franco*phile⟩

Fran·co–Amer·i·can \ˌfraŋ-kō-ə-'mer-ə-kən\ *noun* (1859)
: an American of French or especially French-Canadian descent
— **Franco–American** *adjective*

fran·co·lin \'fraŋ-k(ə-)lən\ *noun* [French, from Italian *francolino*] (1653)
: any of a genus (*Francolinus*) of partridges of southern Asia and Africa

Fran·co·phile \'fraŋ-kə-ˌfīl, -kō-\ *or* **Fran·co·phil** \-ˌfil\ *adjective* (1889)
: markedly friendly to France or French culture
— **Francophile** *noun*
— **Fran·co·phil·ia** \ˌfraŋ-kə-'fi-lē-ə, -lyə, -kō-\ *noun*

Fran·co·phobe \-ˌfōb\ *adjective* (1891)
: marked by a fear or strong dislike of France or French culture or customs
— **Francophobe** *noun*
— **Fran·co·pho·bia** \ˌfraŋ-kə-'fō-bē-ə, -kō-\ *noun*

fran·co·phone \-ˌfōn\ *adjective, often capitalized* (1962)
: of, having, or belonging to a population using French as its first or sometimes second language
— **Francophone** *noun*

franc–ti·reur \ˌfräⁿ-(ˌ)tē-'rər\ *noun* [French, from *franc* free + *tireur* shooter] (1808)
: a civilian fighter or sniper

fran·gi·ble \'fran-jə-bəl\ *adjective* [Middle English, from Middle French & Medieval Latin; Middle French, from Medieval Latin *frangibilis,* from Latin *frangere* to break — more at BREAK] (15th century)
: readily or easily broken
synonym see FRAGILE
— **fran·gi·bil·i·ty** \ˌfran-jə-'bi-lə-tē\ *noun*

fran·gi·pane \'fran-jə-ˌpān, fräⁿ-zhē-pȧn\ *noun* [French, frangipani (perfume), fragipane, from Italian] (1858)
: a custard usually flavored with almonds

fran·gi·pa·ni *also* **fran·gi·pan·ni** \ˌfran-jə-'pa-nē, -'pä-\ *noun, plural* **-pani** *also* **-panni** [modification of Italian *frangipane,* from Muzio *Frangipane,* 16th century Italian nobleman] (1676)
1 : a perfume derived from or imitating the odor of the flower of a frangipani (*Plumeria rubra*)
2 : any of a genus (*Plumeria*) of shrubs or small trees of the dogbane family that are native to the American tropics and widely cultivated as ornamentals

fran·glais \fräⁿ-'glā\ *noun, often capitalized* [French, blend of *français* French and *anglais* English] (1964)
: French marked by a considerable number of borrowings from English

¹frank \'fraŋk\ *adjective* [Middle English, free, from Old French *franc,* from Medieval Latin *francus,* from Late Latin *Francus* Frank] (1548)
1 : marked by free, forthright, and sincere expression ⟨a *frank* reply⟩
2 a : unmistakably evident ⟨*frank* materialism⟩ **b :** clinically evident and unmistakable ⟨*frank* pus⟩ ☆ ◆
— **frank·ness** *noun*

²frank *transitive verb* (1708)
1 a : to mark (a piece of mail) with an official signature or sign indicating the right of the sender to free mailing **b :** to mail free **c :** to affix to (mail) a stamp or a marking indicating the payment of postage
2 : to enable to pass or go freely or easily
— **frank·able** \'fraŋ-kə-bəl\ *adjective*
— **frank·er** *noun*

[3]**frank** *noun* (1713)
1 a : the signature of the sender on a piece of franked mail serving in place of a postage stamp **b :** a mark or stamp on a piece of mail indicating postage paid **c :** a franked envelope
2 : the privilege of sending mail free of charge

[4]**frank** *noun* (1904)
: FRANKFURTER

Frank \ˈfraŋk\ *noun* [Middle English, partly from Old English *Franca;* partly from Old French *Franc,* from Late Latin *Francus,* of Germanic origin; akin to Old High German *Franko* Frank, Old English *Franca*] (before 12th century)
: a member of a West Germanic people that entered the Roman provinces in A.D. 253, occupied the Netherlands and most of Gaul, and established themselves along the Rhine

Fran•ken•stein \ˈfraŋ-kən-ˌstīn *also* -ˌstēn\ *noun*
1 a : the title character in Mary W. Shelley's novel *Frankenstein* who creates a monster by which he is eventually killed **b :** a monster in the shape of a man especially in popularized versions of the Frankenstein story
2 : a monstrous creation; *especially* **:** a work or agency that ruins its originator
— Fran•ken•stein•ian \ˌfraŋ-kən-ˈstī-nē-ən, -ˈstē-\ *adjective*

frank•furt•er \ˈfraŋk-fə(r)t-ər, -ˌfərt-\ *or* **frank•furt** \-fərt\ *noun* [German *Frankfurter* of Frankfurt, from *Frankfurt am Main,* Germany] (1894)
: a cured cooked sausage (as of beef or beef and pork) that may be skinless or stuffed in a casing

frank•in•cense \ˈfraŋ-kən-ˌsen(t)s\ *noun* [Middle English *fraunk encense,* from Old French *franc encens,* from *franc* (perhaps in sense "of high quality") + *encens* incense] (14th century)
: a fragrant gum resin from trees of a genus (*Boswellia* of the family Burseraceae) of Somalia and southern coastal Arabia that is an important incense resin and was used in ancient times in religious rites and in embalming

[1]**Frank•ish** \ˈfraŋ-kish\ *adjective* (14th century)
: of or relating to the Franks

[2]**Frankish** *noun* (1863)
: the Germanic language of the Franks

frank•lin \ˈfraŋ-klən\ *noun* [Middle English *frankeleyn,* from Anglo-French *fraunclein,* from Old French *franc*] (14th century)
: a medieval English landowner of free but not noble birth

frank•lin•ite \-klə-ˌnīt\ *noun* [*Franklin,* N.J.] (1820)
: a black slightly magnetic mineral consisting of an oxide of iron and zinc

Frank•lin stove \ˈfraŋ-klən-\ *noun* [Benjamin *Franklin,* its inventor] (1787)
: a metal heating stove resembling an open fireplace but designed to be set out in a room

Franklin stove

frank•ly \ˈfraŋ-klē\ *adverb* (1537)
1 : in a frank manner
2 : in truth **:** INDEED
usage see HOPEFULLY

frank•pledge \ˈfraŋk-ˌplej\ *noun* [Middle English *frankeplegge,* from Anglo-French *fraunc plege* (probably translation of Middle English *friborg* peace pledge), from *fraunc* free (from Old French *franc*) + *plege* pledge] (15th century)
: an Anglo-Saxon system under which each adult male member of a tithing was responsible for the good conduct of the others; *also* **:** the member himself or the tithing

fran•tic \ˈfran-tik\ *adjective* [Middle English *frenetik, frantik* — more at FRENETIC] (14th century)
1 a *archaic* **:** mentally deranged **b :** emotionally out of control ⟨*frantic* with anger and frustration⟩
2 : marked by fast and nervous, disordered, or anxiety-driven activity ⟨made a *frantic* search for the lost child⟩
— fran•ti•cal•ly \-ti-k(ə-)lē\ *adverb*
— fran•tic•ness \-tik-nəs\ *noun*

frap \ˈfrap\ *transitive verb* **frapped; frap•ping** [Middle English, to strike, beat, from Middle French *fraper*] (1548)
: to draw tight (as with ropes or cables) ⟨*frap* a sail⟩

[1]**frap•pé** \fra-ˈpā\ *adjective* [French, from past participle of *frapper* to strike, chill, from Middle French *fraper* to strike] (1848)
: chilled or partly frozen

[2]**frap•pé** \fra-ˈpā\ *or* **frappe** \ˈfrap, fra-ˈpā\ *noun* (1903)
1 a : a partly frozen drink (as of fruit juice) **b :** a liqueur served over shaved ice
2 : a thick milk shake

Fra•ser fir \ˈfrā-zər-\ *noun* [John *Fraser* (died 1811) British botanist] (1897)
: a southern Appalachian fir (*Abies fraseri*) that resembles the balsam fir

frass \ˈfras\ *noun* [German, insect damage, literally, eating away, from Old High German *vrāz* food, from *frezzan* to devour — more at FRET] (1854)
: debris or excrement produced by insects

frat \ˈfrat\ *noun* (circa 1895)
: FRATERNITY 1c

fra•ter•nal \frə-ˈtər-nᵊl\ *adjective* [Middle English, from Medieval Latin *fraternalis,* from Latin *fraternus,* from *frater* brother — more at BROTHER] (15th century)
1 a : of, relating to, or involving brothers **b :** of, relating to, or being a fraternity or society ⟨a *fraternal* order⟩
2 : derived from two ova **:** DIZYGOTIC ⟨*fraternal* twins⟩
3 : FRIENDLY, BROTHERLY
— fra•ter•nal•ism \-nᵊl-ˌiz-əm\ *noun*
— fra•ter•nal•ly \-nᵊl-ē\ *adverb*

fra•ter•ni•ty \frə-ˈtər-nə-tē\ *noun, plural* **-ties** (14th century)
1 : a group of people associated or formally organized for a common purpose, interest, or pleasure: as **a :** a fraternal order **b :** GUILD **c :** a men's student organization formed chiefly for social purposes having secret rites and a name consisting of Greek letters **d :** a student organization for scholastic, professional, or extracurricular activities ⟨a debating *fraternity*⟩
2 : the quality or state of being brothers **:** BROTHERLINESS
3 : persons of the same class, profession, character, or tastes ⟨the racetrack *fraternity*⟩

frat•er•nize \ˈfra-tər-ˌnīz\ *intransitive verb* **-nized; -niz•ing** (1611)
1 : to associate or mingle as brothers or on fraternal terms
2 a : to associate on close terms with members of a hostile group especially when contrary to military orders **b :** to be friendly or amiable
— frat•er•ni•za•tion \ˌfra-tər-nə-ˈzā-shən\ *noun*
— frat•er•niz•er \ˈfra-tər-ˌnī-zər\ *noun*

frat•ri•cide \ˈfra-trə-ˌsīd\ *noun* [in sense 1, from Middle English, from Middle French or Latin; Middle French, from Latin *fratricida,* from *fratr-, frater* brother + *-cida* -cide; in sense 2, from Middle French or Latin; Middle French, from Latin *fratricidium,* from *fratr-, frater* + *-cidium* -cide] (15th century)
1 : one that murders or kills his or her own brother or sister or an individual (as a countryman) having a relationship like that of a brother or sister
2 : the act of a fratricide
— frat•ri•cid•al \ˌfra-trə-ˈsī-dᵊl\ *adjective*

Frau \ˈfrau̇\ *noun, plural* **Frau•en** \ˈfrau̇(-ə)n\ [German, woman, wife, from Old High German *frouwa* mistress, lady; akin to Old English *frēa* lord, Old High German *fruma* advantage — more at FOREMOST] (circa 1813)
: a German married woman **:** WIFE — used as a title equivalent to *Mrs.*

fraud \ˈfrȯd\ *noun* [Middle English *fraude,* from Middle French, from Latin *fraud-, fraus*] (14th century)
1 a : DECEIT, TRICKERY; *specifically* **:** intentional perversion of truth in order to induce another to part with something of value or to surrender a legal right **b :** an act of deceiving or misrepresenting **:** TRICK
2 a : a person who is not what he or she pretends to be **:** IMPOSTOR; *also* **:** one who defrauds **:** CHEAT **b :** one that is not what it seems or is represented to be
synonym see DECEPTION, IMPOSTURE

fraud•u•lence \ˈfrȯ-jə-lən(t)s\ *noun* (1601)
: the quality or state of being fraudulent

fraud•u•lent \-lənt\ *adjective* (15th century)
: characterized by, based on, or done by fraud **:** DECEITFUL
— fraud•u•lent•ly *adverb*
— fraud•u•lent•ness *noun*

[1]**fraught** \ˈfrȯkt\ *noun* [Middle English, freight, load, from Middle Dutch or Middle Low German *vracht, vrecht*] (14th century)
chiefly Scottish **:** LOAD, CARGO

☆ SYNONYMS
Frank, candid, open, plain mean showing willingness to tell what one feels or thinks. FRANK stresses lack of shyness or secretiveness or of evasiveness from considerations of tact or expedience ⟨*frank* discussions⟩. CANDID suggests expression marked by sincerity and honesty especially in offering unwelcome criticism or opinion ⟨a *candid* appraisal⟩. OPEN implies frankness but suggests more indiscretion than FRANK and less earnestness than CANDID ⟨*open* in saying what they think⟩. PLAIN suggests outspokenness and freedom from affectation or subtlety in expression ⟨*plain* talk⟩.

◇ WORD HISTORY
frank In the 5th century A.D. members of a group of Germanic tribes known collectively as the Franks migrated from the lower Rhine into the northern part of Roman Gaul. The region they settled, which was expanded and unified under their king Clovis (died around 511), was by the 7th century called *Francia* and would eventually become modern France. The Franks tended to merge with existing Gallo-Roman culture, and by the 8th century the Medieval Latin word *Francus* "Frank" had largely lost its ethnic sense and meant any inhabitant of *Francia,* whether of Frankish or Gallo-Roman ancestry, who was free, that is, not a slave or bondsman. From the noun *francus* "freeman" a corresponding adjective *francus* was formed with the meaning "free." Assimilated into the vernacular of Frankish Gaul, the adjective became *franc* in Old French and subsequently developed a wide range of extended senses, including "free of further charges or duties" (hence, ultimately, the English verb *frank*) and "expressing oneself freely, sincere" (hence the English adjective *frank*).

²fraught *transitive verb* **fraught·ed** *or* **fraught; fraught·ing** [Middle English *fraughten,* from ¹*fraught*] (14th century)
chiefly Scottish **:** LOAD, FREIGHT
³fraught \'frȯt\ *adjective* [Middle English, from past participle of *fraughten*] (14th century)
1 *archaic* **a :** LADEN **b :** well supplied or provided
2 : full of or accompanied by something specified — used with *with* ⟨a situation *fraught* with danger⟩
3 *chiefly British* **:** causing or characterized by emotional distress or tension **:** UNEASY
fräu·lein \'frȯi-,līn\ *noun* [German, diminutive of *Frau*] (circa 1689)
1 *capitalized* **:** an unmarried German woman — used as a title equivalent to *Miss*
2 : a German governess
frax·i·nel·la \,frak-sə-'ne-lə\ *noun* [New Latin, diminutive of Latin *fraxinus* ash tree — more at BIRCH] (1664)
: a Eurasian perennial herb (*Dictamnus albus*) of the rue family with flowers that exhale a flammable vapor in hot weather — called also *gas plant*
¹fray \'frā\ *transitive verb* [Middle English *fraien,* short for *affraien* to affray] (14th century)
archaic **:** SCARE; *also* **:** to frighten away
²fray *noun* (14th century)
: a usually disorderly or protracted fight, struggle, or dispute
³fray *verb* [Middle English *fraien,* from Middle French *froyer, frayer* to rub, from Latin *fricare* — more at FRICTION] (15th century)
transitive verb
1 a : to wear (as an edge of cloth) by or as if by rubbing **:** FRET **b :** to separate the threads at the edge of
2 : STRAIN, IRRITATE ⟨tempers became a bit *frayed*⟩
intransitive verb
1 : to wear out or into shreds
2 : to show signs of strain ⟨*fraying* nerves⟩
⁴fray *noun* (1630)
: a raveled place or worn spot (as on fabric)
fray·ing *noun* (1637)
: something rubbed or worn off by fraying
¹fraz·zle \'fra-zəl\ *verb* **fraz·zled; fraz·zling** \'fraz-liŋ, 'fra-zə-\ [alteration of English dialect *fazle* (to tangle, fray)] (circa 1825)
transitive verb
1 : ³FRAY
2 a : to put in a state of extreme physical or nervous fatigue **b :** UPSET
intransitive verb
: to become frazzled
²frazzle *noun* (1865)
1 : the state of being frazzled
2 : a condition of fatigue or nervous exhaustion ⟨worn to a *frazzle*⟩
¹freak \'frēk\ *noun* [origin unknown] (1563)
1 a : a sudden and odd or seemingly pointless idea or turn of the mind **b :** a seemingly capricious action or event
2 *archaic* **:** a whimsical quality or disposition
3 : one that is markedly unusual or abnormal: as **a :** a person or animal with a physical oddity who appears in a circus sideshow **b** *slang* (1) **:** a sexual deviate (2) **:** a person who uses an illicit drug **c :** HIPPIE **d :** an atypical postage stamp usually caused by a unique defect in paper (as a crease) or a unique event in the manufacturing process (as a speck of dirt on the plate) that does not produce a constant or systematic effect
4 : an ardent enthusiast ⟨film *freaks*⟩
²freak *adjective* (circa 1887)
: having the character of a freak ⟨a *freak* accident⟩
³freak (1965)
intransitive verb
1 : to withdraw from reality especially by taking drugs — often used with *out*
2 : to experience nightmarish hallucinations as a result of taking drugs — often used with *out*
3 a : to behave irrationally or unconventionally under the influence of drugs — often used with *out* **b :** to react with extreme or irrational distress or discomposure — often used with *out*
transitive verb
1 : to put under the influence of a psychedelic drug — often used with *out*
2 : to make greatly distressed, astonished, or discomposed — often used with *out* ⟨the news *freaked* them out⟩
— freaked *adjective*
— freaked–out *adjective*
⁴freak *transitive verb* [perhaps from or akin to ¹*freckle*] (1637)
: to streak especially with color ⟨silver and mother-of-pearl *freaking* the intense azure —Robert Bridges (died 1930)⟩
freak·ing \'frē-kᵊn, -kiŋ\ *adjective or adverb* [euphemism for *frigging* or *fucking*] (1963)
: DAMNED — used as an intensive
freak·ish \'frē-kish\ *adjective* (1653)
1 : WHIMSICAL, CAPRICIOUS
2 : markedly strange or abnormal
— freak·ish·ly *adverb*
— freak·ish·ness *noun*
freak of nature (1883)
: FREAK 3a
freak–out \'frē-,kau̇t\ *noun* (1966)
1 : an act or instance of freaking out
2 : a gathering of hippies
freak show *noun* (1887)
: an exhibition (as a sideshow) featuring freaks of nature
freaky \'frē-kē\ *adjective* **freak·i·er; -est** (1824)
: FREAKISH
— freak·i·ness *noun*
¹freck·le \'fre-kəl\ *noun* [Middle English *freken, frekel,* of Scandinavian origin; akin to Old Norse *freknōttr* freckled] (14th century)
: any of the small brownish spots in the skin usually due to precipitation of pigment that increase in number and intensity on exposure to sunlight
— freck·ly \'fre-k(ə-)lē\ *adjective*
²freckle *verb* **freck·led; freck·ling** \'fre-k(ə-)liŋ\ (1613)
transitive verb
: to sprinkle or mark with freckles or small spots
intransitive verb
: to become marked with freckles
¹free \'frē\ *adjective* **fre·er; fre·est** [Middle English, from Old English *frēo;* akin to Old High German *frī* free, Welsh *rhydd,* Sanskrit *priya* own, dear] (before 12th century)
1 a : having the legal and political rights of a citizen **b :** enjoying civil and political liberty ⟨*free* citizens⟩ **c :** enjoying political independence or freedom from outside domination **d :** enjoying personal freedom **:** not subject to the control or domination of another
2 a : not determined by anything beyond its own nature or being **:** choosing or capable of choosing for itself **b :** determined by the choice of the actor or performer ⟨*free* actions⟩ **c :** made, done, or given voluntarily or spontaneously
3 a : relieved from or lacking something unpleasant or burdensome ⟨*free* from pain⟩ ⟨a speech *free* of political rhetoric⟩ **b :** not bound, confined, or detained by force
4 a : having no trade restrictions **b :** not subject to government regulation **c** *of foreign exchange* **:** not subject to restriction or official control
5 a : having no obligations (as to work) or commitments ⟨I'll be *free* this evening⟩ **b :** not taken up with commitments or obligations ⟨a *free* evening⟩
6 : having a scope not restricted by qualification ⟨a *free* variable⟩
7 a (1) **:** not obstructed or impeded **:** CLEAR (2) **:** not being used or occupied ⟨waved with his *free* hand⟩ **b :** not hampered or restricted in its normal operation
8 a : not fastened ⟨the *free* end of the rope⟩ **b :** not confined to a particular position or place; *also* **:** not having a specific opponent to cover in football ⟨a *free* safety⟩ **c :** capable of moving or turning in any direction ⟨a *free* particle⟩ **d :** performed without apparatus ⟨*free* tumbling⟩ **e :** done with artificial aids (as pitons) used only for protection against falling and not for support ⟨a *free* climb⟩
9 a : not parsimonious ⟨*free* spending⟩ **b :** OUTSPOKEN **c :** availing oneself of something without stint **d :** FRANK, OPEN **e :** overly familiar or forward in action or attitude **f :** LICENTIOUS
10 : not costing or charging anything
11 a (1) **:** not united with, attached to, combined with, or mixed with something else **:** SEPARATE ⟨*free* ores⟩ ⟨a *free* surface of a bodily part⟩ (2) **:** FREESTANDING ⟨a *free* column⟩ **b :** chemically uncombined ⟨*free* oxygen⟩ ⟨*free* acids⟩ **c :** not permanently attached but able to move about ⟨a *free* electron in a metal⟩ **d :** capable of being used alone as a meaningful linguistic form ⟨the word *hats* is a *free* form⟩ — compare ⁵BOUND 7
12 a : not literal or exact ⟨*free* translation⟩ **b :** not restricted by or conforming to conventional forms ⟨*free* skating⟩
13 : FAVORABLE — used of a wind blowing from a direction more than six points from dead ahead
14 : not allowing slavery
15 : open to all comers ☆ ■
— free·ness \-nəs\ *noun*
— for free : without charge

☆ SYNONYMS
Free, independent, sovereign, autonomous mean not subject to the rule or control of another. FREE stresses the complete absence of external rule and the full right to make all of one's own decisions ⟨you're *free* to do as you like⟩. INDEPENDENT implies a standing alone; applied to a state it implies lack of connection with any other having power to interfere with its citizens, laws, or policies ⟨the colony's struggle to become *independent*⟩. SOVEREIGN stresses the absence of a superior power and implies supremacy within a thing's own domain or sphere ⟨separate and *sovereign* armed services⟩. AUTONOMOUS stresses independence in matters pertaining to self-government ⟨in this denomination each congregation is regarded as *autonomous*⟩.

□ USAGE
free The idiomatic phrase *for free* apparently came into common use only in the early 1940s—our earliest printed evidence is from 1942—and it was no sooner popular than it was censured. The earliest complaint appeared barely six months after the earliest printed citation. Ever since, the critics have been denigrating the phrase, and others have kept on using it. One reason for the phrase's usefulness is that the preposition provides a little space between *free* and the preceding text ⟨it seemed best to find a way to live *for free* then —Jane Harriman⟩. In this example the *for* eliminates the ambiguous *live free.* But in many instances no potential ambiguity is involved; people seem to use the phrase simply because it sounds right to them. This American idiom is well established in speech and in general prose, but it is not used in writing of high solemnity.

[2]**free** *transitive verb* **freed; free•ing** (before 12th century)
1 a : to cause to be free **b :** to relieve or rid of what restrains, confines, restricts, or embarrasses ⟨*free* a person from debt⟩ **c :** DISENTANGLE, CLEAR
2 *obsolete* **:** BANISH ☆
— fre•er *noun*

[3]**free** *adverb* (1559)
1 : in a free manner
2 : without charge
3 : with the wind more than six points from dead ahead ⟨sailing *free*⟩

free agent *noun* (1955)
: a professional athlete (as a baseball player) who is free to negotiate a contract with any team
— free agency *noun*

free alongside ship *adverb or adjective* (circa 1903)
: with delivery at the side of the ship free of charges and the buyer's liability then beginning

free and easy *adjective* (1699)
1 : marked by informality and lack of constraint ⟨the *free and easy,* open-air life of the plains —Allan Murray⟩
2 : not observant of strict demands ⟨too *free and easy* in accepting political contributions⟩
— free–and–eas•i•ness \ˌfrē-ən-ˈ(d)ē-zē-nəs\ *noun*
— free and easy *adverb*

free association *noun* (1899)
1 a : the expression (as by speaking or writing) of the content of consciousness without censorship as an aid in gaining access to unconscious processes especially in psychoanalysis **b :** the reporting of the first thought that comes to mind in response to a given stimulus (as a word)
2 : an idea or image elicited by free association
3 : a method using free association
— free–as•so•ci•ate \ˌfrē-ə-ˈsō-s(h)ē-ˌāt\ *intransitive verb*
— free–as•so•ci•at•ive \-s(h)ē-ˌā-tiv, -shə-tiv\ *adjective*

[1]**free•base** \ˈfrē-ˌbās\ (1980)
intransitive verb
: to prepare or use freebase cocaine
transitive verb
: to prepare or use (cocaine) as freebase
— free•bas•er *noun*

[2]**freebase** *noun* (1980)
: cocaine freed from impurities by treatment (as with ether) and heated to produce vapors for inhalation or smoked as crack

free beach *noun* (1975)
: a beach at which nudity is permitted

free•bie *or* **free•bee** \ˈfrē-bē\ *noun* [by alteration from obsolete *freeby* gratis, irregular from *free*] (1942)
: something (as a theater ticket) given without charge

free•board \ˈfrē-ˌbōrd, -ˌbȯrd\ *noun* (1726)
1 : the distance between the waterline and the main deck or weather deck of a ship or between the level of the water and the upper edge of the side of a small boat
2 : the height above the recorded high-water mark of a structure (as a dam) associated with the water

free•boo•ter \ˈfrē-ˌbü-tər\ *noun* [by folk etymology from Dutch *vrijbuiter,* from *vrijbuit* plunder, from *vrij* free + *buit* booty] (1570)
: PIRATE, PLUNDERER
— free•boot \-ˌbüt\ *intransitive verb*

free•born \ˈfrē-ˈbȯrn\ *adjective* (13th century)
1 : not born in vassalage or slavery
2 : of, relating to, or befitting one that is freeborn

free diver *noun* (1953)
: one who engages in skin diving
— free diving *noun*

freed•man \ˈfrēd-mən, -ˌman\ *noun* (1601)
: a man freed from slavery

free•dom \ˈfrē-dəm\ *noun* (before 12th century)
1 : the quality or state of being free: as **a :** the absence of necessity, coercion, or constraint in choice or action **b :** liberation from slavery or restraint or from the power of another **:** INDEPENDENCE **c :** the quality or state of being exempt or released usually from something onerous ⟨*freedom* from care⟩ **d :** EASE, FACILITY ⟨spoke the language with *freedom*⟩ **e :** the quality of being frank, open, or outspoken ⟨answered with *freedom*⟩ **f :** improper familiarity **g :** boldness of conception or execution **h :** unrestricted use ⟨gave him the *freedom* of their home⟩
2 a : a political right **b :** FRANCHISE, PRIVILEGE ☆

freedom of the seas (1917)
: the right of a merchant ship to travel any waters except territorial waters either in peace or war

freedom ride *noun, often F&R capitalized* (1961)
: a ride made by civil rights workers through states of the southern U.S. to ascertain whether public facilities (as bus terminals) are desegregated
— freedom rider *noun*

freed•wom•an \ˈfrēd-ˌwu̇-mən\ *noun* (1866)
: a woman freed from slavery

free–elec•tron laser \ˈfrē-i-ˈlek-ˌträn-\ *noun* (1979)
: a laser that can be tuned over a wide range of frequencies and that produces electromagnetic radiation by the motion of electrons moving at relativistic velocities in a magnetic field

free enterprise *noun* (1890)
: freedom of private business to organize and operate for profit in a competitive system without interference by government beyond regulation necessary to protect public interest and keep the national economy in balance

free enterpriser *noun* (1943)
: a supporter or advocate of free enterprise

free fall *noun* (1919)
1 : the condition of unrestrained motion in a gravitational field; *also* **:** such motion
2 a : the part of a parachute jump before the parachute opens **b :** a rapid and continuing drop or decline ⟨a *free fall* in stock prices⟩
— free–fall *intransitive verb*

free–fire zone \ˈfrē-ˌfīr-\ *noun* (1967)
: a combat area in which any moving thing is a legitimate target

free–float•ing \-ˈflō-tiŋ\ *adjective* (1921)
1 a : floating freely ⟨*free-floating* vegetation⟩ **b :** lacking specific attachment, direction, or purpose ⟨*free-floating* ideas⟩
2 : felt as an emotion without apparent cause ⟨*free-floating* anxiety⟩

free–flow•ing \-ˌflō-iŋ\ *adjective* (1920)
: characterized by easy freedom in movement, progression, or style ⟨a *free-flowing* essay⟩

free–for–all \ˈfrē-fə-ˌrȯl\ *noun* (1881)
: a competition, dispute, or fight open to all comers and usually with no rules **:** BRAWL
— free–for–all *adjective*

free–form \ˈfrē-ˈfȯrm\ *adjective* (1950)
1 : having or being an irregular or asymmetrical shape or design ⟨*free-form* furniture⟩
2 : FREE 12b ⟨*free-form* dancing⟩

free•hand \ˈfrē-ˌhand\ *adjective* (circa 1862)
: done without mechanical aids or devices ⟨*freehand* drawing⟩
— freehand *adverb*

free hand \-ˈhand\ *noun* (1929)
: freedom of action or decision

free•hand•ed \ˈfrē-ˈhan-dəd\ *adjective* (circa 1656)
: GENEROUS, OPENHANDED
— free•hand•ed•ly *adverb*
— free•hand•ed•ness *noun*

free•heart•ed \-ˈhär-təd\ *adjective* (14th century)
1 : FRANK, UNRESERVED
2 : GENEROUS
— free•heart•ed•ly *adverb*

free•hold \ˈfrē-ˌhōld\ *noun* (15th century)
1 : a tenure of real property by which an estate of inheritance in fee simple or fee tail or an estate for life is held; *also* **:** an estate held by such tenure — compare FEE 1
2 *British* **:** an estate held in fee simple
— freehold *adjective or adverb*
— free•hold•er \-ˌhōl-dər\ *noun*

free kick *noun* (1882)
: a kick (as in football, soccer, or rugby) with which an opponent may not interfere; *especially* **:** such a kick in any direction awarded because of an infraction of the rules by an opponent

[1]**free•lance** \ˈfrē-ˌlan(t)s\ *noun* (1820)
1 a *usually* **free lance :** a mercenary soldier especially of the Middle Ages **:** CONDOTTIERE **b :** a person who acts independently without being affiliated with or authorized by an organization
2 : a person who pursues a profession without a long-term commitment to any one employer

[2]**freelance** *adjective* (1901)
1 a : of, relating to, or being a freelance **:** INDEPENDENT ⟨a *freelance* photographer⟩ ⟨*freelance* fees⟩ **b :** done by a freelance ⟨*freelance* reviewing⟩ ⟨*freelance* jobs⟩
2 : not sponsored by an organization ⟨*freelance* terrorists⟩ ⟨a *freelance* demonstration⟩
— freelance *adverb*

[3]**freelance** (1902)
intransitive verb
: to act or work as a freelance
transitive verb
: to produce as a freelance ⟨*freelancing* magazine articles⟩

free•lanc•er \ˈfrē-ˌlan(t)-sər\ *noun* (1937)
: FREELANCE 1b, 2

free–liv•ing \ˈfrē-ˈli-viŋ\ *adjective* (1818)
1 : marked by more than usual freedom in the gratification of appetites

☆ SYNONYMS

Free, release, liberate, emancipate, manumit mean to set loose from restraint or constraint. FREE implies a usually permanent removal from whatever binds, confines, entangles, or oppresses ⟨*freed* the animals from their cages⟩. RELEASE suggests a setting loose from confinement, restraint, or a state of pressure or tension, often without implication of permanent liberation ⟨*released* his anger on a punching bag⟩. LIBERATE stresses particularly the resulting state of liberty ⟨*liberated* their country from the tyrant⟩. EMANCIPATE implies the liberation of a person from subjection or domination ⟨labor-saving devices *emancipated* us from household drudgery⟩. MANUMIT implies emancipation from slavery ⟨the document *manumitted* the slaves⟩.

Freedom, liberty, license mean the power or condition of acting without compulsion. FREEDOM has a broad range of application from total absence of restraint to merely a sense of not being unduly hampered or frustrated ⟨*freedom* of the press⟩. LIBERTY suggests release from former restraint or compulsion ⟨the released prisoner had difficulty adjusting to his new *liberty*⟩. LICENSE implies freedom specially granted or conceded and may connote an abuse of freedom ⟨freedom without responsibility may degenerate into *license*⟩.

2 a : not fixed to the substrate but capable of motility ⟨a *free-living* protozoan⟩ **b :** being metabolically independent **:** neither parasitic nor symbiotic

free·load \-ˌlōd\ *intransitive verb* (circa 1934) **:** to impose upon another's generosity or hospitality without sharing in the cost or responsibility involved **:** SPONGE
— **free·load·er** *noun*

free love *noun* (1822)
1 : the practice of living openly with one of the opposite sex without marriage
2 : sexual relations without any commitments by either partner

free lunch *noun* (1975)
: something one does not have to pay for; *also* **:** FREE RIDE

free·ly \ˈfrē-lē\ *adverb* (before 12th century)
: in a free manner: as **a :** of one's own accord ⟨left home *freely*⟩ **b :** with freedom from external control ⟨a *freely* elected government⟩ **c :** without restraint or reservation ⟨spent *freely* on clothes⟩ **d :** without hindrance ⟨a gate swinging *freely*⟩ ⟨currencies are *freely* convertible⟩ **e :** not strictly following a model, convention, or rule ⟨*freely* translated⟩

free·man \ˈfrē-mən, -ˌman\ *noun* (before 12th century)
1 : one enjoying civil or political liberty
2 : one having the full rights of a citizen

free market *noun* (1907)
: an economic market operating by free competition

free–mar·ke·teer \ˈfrē-ˌmär-kə-ˈtir\ *noun* (1954)
: a proponent of a free-market economy

free·mar·tin \ˈfrē-ˌmär-tᵊn\ *noun* [origin unknown] (1681)
: a sexually imperfect usually sterile female calf twinborn with a male

Free·ma·son \-ˈmā-sᵊn\ *noun* (1646)
: a member of a major fraternal organization called Free and Accepted Masons or Ancient Free and Accepted Masons that has certain secret rituals

free·ma·son·ry \-rē\ *noun* (1741)
1 *capitalized* **:** the principles, institutions, or practices of Freemasons — called also *Masonry*
2 : natural fellowship based on some common experience

free on board *adverb or adjective* (1924)
: without charge for delivery to and placing on board a carrier at a specified point

free port *noun* (1711)
: an enclosed port or section of a port where goods are received and shipped free of customs duty

freer *comparative of* FREE

free radical *noun* (1900)
: an especially reactive atom or group of atoms that has one or more unpaired electrons

free–range \ˈfrē-ˌrānj\ *adjective* (1912)
: allowed to range and forage with relative freedom ⟨*free-range* chickens⟩; *also* **:** of, relating to, or produced by free-range poultry ⟨*free-range* eggs⟩

free reed *noun* (1855)
: a reed in a musical instrument (as a harmonium) that vibrates in an air opening just large enough to allow the reed to move freely — compare BEATING REED

free rein *noun* (1952)
: unrestricted liberty of action or decision

free ride *noun* (1899)
: a benefit obtained at another's expense or without the usual cost or effort
— **free ride** *intransitive verb*
— **free rider** *noun*

free safety *noun* (1973)
: a safety in football who has no particular receiver to cover in a man-to-man defense

free·sia \ˈfrē-zh(ē-)ə, -zē-ə\ *noun* [New Latin, from F. H. T. *Freese* (died 1876) German physician] (circa 1882)
: any of a genus (*Freesia*) of the iris family of sweet-scented African herbs with red, pink, white, or yellow flowers

free–soil *adjective* (1846)
1 : characterized by free soil ⟨*free-soil* states⟩
2 *F&S capitalized* **:** opposing the extension of slavery into U.S. territories and the admission of slave states into the Union prior to the Civil War; *specifically* **:** of, relating to, or constituting a minor U.S. political party having these aims
— **Free–Soil·er** \-ˈsȯi-lər\ *noun*
— **Free–Soil·ism** \-ˈsȯi(ə)-ˌli-zəm\ *noun*

free soil *noun* (1848)
: U.S. territory where prior to the Civil War slavery was prohibited

free–spo·ken \ˈfrē-ˈspō-kən\ *adjective* (1625)
: speaking freely **:** OUTSPOKEN

freest *superlative of* FREE

free·stand·ing \ˈfrē-ˈstan-diŋ\ *adjective* (1876)
1 : standing alone or on its own foundation free of support or attachment ⟨a *freestanding* wall⟩
2 : being independent; *especially* **:** not part of or affiliated with another organization ⟨a *freestanding* clinic⟩ ⟨a *freestanding* city⟩ ⟨a *freestanding* computer store⟩

Free State *noun* (1819)
: a state of the U.S. in which slavery was prohibited before the Civil War

free·stone \ˈfrē-ˌstōn\ *noun* (13th century)
1 : a stone that may be cut freely without splitting
2 a : a fruit stone to which the flesh does not cling **b :** a fruit having such a stone

free·style \ˈfrē-ˌstīl\ *noun, often attributive* (circa 1934)
1 : a competition in which the contestant is given more latitude than in related events; *especially* **:** swimming competition in which the swimmer may use any stroke
2 : CRAWL 2
— **free·styl·er** *noun*

free–swim·ming \-ˈswi-miŋ\ *adjective* (circa 1890)
: able to swim about **:** not attached ⟨the *free-swimming* larva of the barnacle⟩

free–swing·ing \-ˈswiŋ-iŋ\ *adjective* (1949)
: bold, forthright, and heedless of personal consequences ⟨a *free-swinging* soldier of fortune —Will Herberg⟩

free–tailed bat \ˈfrē-ˌtāld-\ *noun* (1895)
: any of a family (Molossidae) of bats characterized by a tail that projects beyond the posterior part of the flight membrane and found in warm regions of the world

free·think·er \-ˈthiŋ-kər\ *noun* (1692)
: one that forms opinions on the basis of reason independently of authority; *especially* **:** one who doubts or denies religious dogma
— **free·think·ing** \-kiŋ\ *noun or adjective*

free thought *noun* (1711)
: unorthodox attitudes or beliefs; *specifically* **:** 18th century deism

free throw *noun* (circa 1929)
: an unhindered shot in basketball made from behind a set line and awarded because of a foul by an opponent

free throw lane *noun* (circa 1929)
: a 12 or 16 foot wide lane on a basketball court that extends from underneath the goal to a line 15 feet in front of the backboard and that players may not enter during a free throw

free trade *noun* (1823)
: trade based on the unrestricted international exchange of goods with tariffs used only as a source of revenue

free trader *noun* (1832)
: one that practices or advocates free trade

free verse *noun* (1908)
: verse whose meter is irregular in some respect or whose rhythm is not metrical

free·way \ˈfrē-ˌwā\ *noun* (1930)
1 : an expressway with fully controlled access
2 : a toll-free highway

¹free·wheel \-ˈ(h)wē(ə)l\ *noun* (1930)
1 : a clutch fitted in the rear hub of a bicycle that permits the rear wheel to run on free from the rear sprocket when the pedals are stopped
2 : a power-transmission system in a motor vehicle with a device that permits the propeller shaft to run freely when its speed is greater than that of the engine shaft

²freewheel *intransitive verb* (1903)
1 : to roll along freely independent of a gear
2 : to move, live, or play freely or irresponsibly
— **free·wheel·er** *noun*

free·wheel·ing \ˌfrē-ˈhwē-liŋ, -ˈwē-\ *adjective* (1931)
: free and loose in form or manner: as **a :** heedless of social norms or niceties ⟨the raider style of his *freewheeling* father —Garry Wills⟩ **b :** not repressed or restrained ⟨*freewheeling* promiscuity⟩ ⟨a *freewheeling* competitive spirit⟩ ⟨a *freewheeling* vocabulary⟩ **c :** not bound by formal rules, procedures, or guidelines ⟨a *freewheeling* investigation⟩ ⟨*freewheeling* improvisation⟩ **d :** loose and undisciplined **:** not defensive ⟨a *freewheeling* style of hockey⟩
— **free·wheel·ing·ly** *adverb*

free·will \ˈfrē-ˌwil\ *adjective* (1535)
: VOLUNTARY, SPONTANEOUS

free will *noun* (13th century)
1 : voluntary choice or decision ⟨I do this of my own *free will*⟩
2 : freedom of humans to make choices that are not determined by prior causes or by divine intervention

Freewill Baptist *noun* (1732)
: a member of a Baptist group holding to Arminian doctrine and practicing open communion

free world *noun, often F&W capitalized* (1949)
: the part of the world where democracy and capitalism or moderate socialism rather than totalitarian or Communist political and economic systems prevail

free·writ·ing \ˈfrē-ˈrī-tiŋ\ *noun* (1980)
: automatic writing done especially as a classroom exercise

¹freeze \ˈfrēz\ *verb* **froze** \ˈfrōz\; **fro·zen** \ˈfrō-zᵊn\; **freez·ing** [Middle English *fresen*, from Old English *frēosan;* akin to Old High German *friosan* to freeze, Latin *pruina* hoarfrost, Old English *frost* frost] (before 12th century)
intransitive verb
1 a : to become congealed into ice by cold **b :** to solidify as a result of abstraction of heat **c :** to withstand freezing ⟨the bread *freezes* well⟩
2 : to become chilled with cold ⟨almost *froze* to death⟩
3 : to adhere solidly by or as if by freezing ⟨pressure caused the metals to *freeze*⟩
4 : to become fixed or motionless; *especially* **:** to become incapable of acting or speaking
5 : to become clogged with ice ⟨the water pipes *froze*⟩
transitive verb
1 a : to harden into ice **b :** to convert from a liquid to a solid by cold
2 : to make extremely cold **:** CHILL
3 a : to act on usually destructively by frost **b :** to anesthetize by cold
4 : to cause to grip tightly or remain in immovable contact
5 a : to cause to become fixed, immovable, unavailable, or unalterable ⟨*freeze* interest rates⟩ **b :** to immobilize by governmental regulation the expenditure, withdrawal, or exchange of ⟨*freeze* foreign assets⟩ **c :** to render motionless ⟨a fake *froze* the defender⟩
6 : to attempt to retain continuous possession of (a ball or puck) without an attempt to score usually in order to protect a small lead
— **freez·ing·ly** *adverb*

²freeze *noun* (15th century)
1 a : an act or instance of freezing **b :** the state of being frozen

2 : a state of weather marked by low temperature especially when below the freezing point
3 : a halt in the production, testing, and deployment of military weapons ⟨a nuclear *freeze*⟩

freeze–dried \-ˈdrīd\ *adjective* (1946)
: being in a state produced by or as if by freeze-drying

freeze–dry \-ˈdrī\ *transitive verb* (1949)
: to dry (as food) in a frozen state under high vacuum especially for preservation

freeze–etch·ing \ˈfrē-ˌze-chiŋ\ *noun* (1968)
: preparation of a specimen (as of tissue) for electron microscopic examination by freezing, fracturing along natural structural lines, and preparing a replica (as by simultaneous vapor deposition of carbon and platinum)
— **freeze–etch** \-ˈzech\ *adjective*
— **freeze–etched** \-ˈzecht\ *adjective*

freeze fracture *noun* (1973)
: FREEZE-ETCHING

freeze–frame \ˈfrēz-ˈfrām\ *noun* (1948)
1 a : a frame of a motion-picture film that is repeated so as to give the illusion of a static picture **b :** a static picture produced from a videodisc or videotape recording
2 : something resembling a freeze-frame especially in unchanging quality

freeze out *transitive verb* (1861)
: EXCLUDE
— **freeze–out** \ˈfrē-ˌzau̇t\ *noun*

freez·er \ˈfrē-zər\ *noun* (1847)
: one that freezes or keeps cool; *especially* **:** a compartment, room, or device for freezing food or keeping it frozen

freezer burn *noun* (1926)
: light-colored spots developed in frozen foods as a result of surface evaporation and drying when inadequately wrapped or packaged

freezing point *noun* (1747)
: the temperature at which a liquid solidifies

free zone *noun* (1900)
: an area within which goods may be received and stored without payment of duty

F region *noun* (1923)
: the highest region of the ionosphere occurring from 80 miles (130 kilometers) to more than 300 miles (500 kilometers)

[1]freight \ˈfrāt\ *noun, often attributive* [Middle English, from Middle Dutch or Middle Low German *vracht, vrecht*] (15th century)
1 a : the compensation paid for the transportation of goods **b :** COST ⟨help pay the *freight*⟩
2 a : goods to be shipped **:** CARGO **b :** LOAD, BURDEN **c :** MEANING 3, SIGNIFICANCE
3 a : the ordinary transportation of goods by a common carrier and distinguished from express **b :** a train designed or used for such transportation

[2]freight *transitive verb* (15th century)
1 a : to load with goods for transportation **b :** BURDEN, CHARGE ⟨*freighted* with memories⟩
2 : to transport or ship by freight

freight·age \ˈfrā-tij\ *noun* (1694)
: FREIGHT

freight·er \-tər\ *noun* (1622)
1 : one that loads or charters and loads a ship
2 : SHIPPER
3 : a ship or airplane used chiefly to carry freight

frem·i·tus \ˈfre-mə-təs\ *noun* [New Latin, from Latin, murmur, from *fremere* to murmur, akin to Old English *bremman* to roar] (1879)
: a sensation felt by a hand placed on a part of the body (as the chest) that vibrates during speech

french \ˈfrench\ *transitive verb, often capitalized* (1941)
1 : to trim the meat from the end of the bone of (as a chop)
2 : to cut (green beans) in thin lengthwise strips before cooking

[1]French \ˈfrench\ *adjective* [Middle English, from Old English *frencisc,* from *Franca* Frank] (before 12th century)
1 : of, relating to, or characteristic of France, its people, or their language
2 : of or relating to the overseas descendents of the French people
— **French·ness** *noun*

[2]French *noun* (12th century)
1 : a Romance language that developed out of the Vulgar Latin spoken in northern and central Transalpine Gaul and that became the literary and official language of France
2 *plural in construction* **:** the French people
3 : strong language ⟨pardon my *French*⟩

French bean *noun* (1632)
1 *chiefly British* **:** a bean (as a green bean) of which the whole young pod is eaten
2 *chiefly British* **:** KIDNEY BEAN 2

French bread *noun* (15th century)
: a crusty white bread baked usually in long thin loaves

French bulldog *noun* (1875)
: any of a breed of small compact heavy-boned dogs developed in France and having erect ears

French Canadian *noun* (1758)
: one of the descendants of French settlers in Lower Canada
— **French–Canadian** *adjective*

French chalk *noun* (circa 1728)
: a soft white granular variety of steatite used especially for drawing lines on cloth and for removing grease in dry cleaning

French cuff *noun* (1916)
: a soft double cuff that is made by turning back half of a wide cuff band and fastening with cuff links

french curve *noun, often F capitalized* (1885)
: a curved piece of flat often plastic material used as a guide in drawing curves

French door *noun* (1923)
: a door with rectangular glass panes extending the full length; *also* **:** FRENCH WINDOW

French dressing *noun* (1876)
1 : a salad dressing made with oil and vinegar or lemon juice, and spices
2 : a commercial salad dressing that is tomato-flavored and of creamy consistency

[1]french fry *noun, often 1st F capitalized* (1918)
: a strip of potato fried in deep fat — usually used in plural

[2]french fry *transitive verb, often 1st F capitalized* (circa 1930)
: to fry (as strips of potato) in deep fat until brown

French horn *noun* (1682)
: a circular valved brass instrument having a conical bore, a funnel-shaped mouthpiece, and a usual range from B below the bass staff upward for more than three octaves

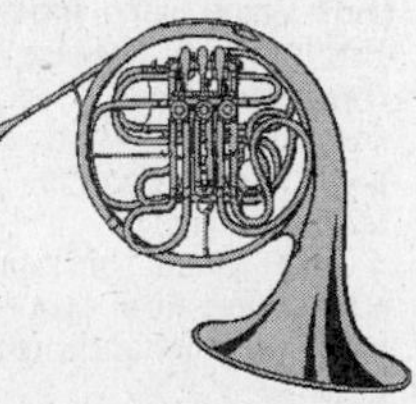
French horn

french·ify \ˈfren-chə-ˌfī\ *transitive verb* **-ified; -ify·ing** *often capitalized* (1592)
: to make French in qualities, traits, or typical practices
— **french·i·fi·ca·tion** \ˌfren-chə-fə-ˈkā-shən\ *noun, often capitalized*

French kiss *noun* (circa 1923)
: an open-mouth kiss usually involving tongue-to-tongue contact
— **French–kiss** *verb*

French leave *noun* [from an 18th century French custom of leaving a reception without taking leave of the host or hostess] (1771)
: an informal, hasty, or secret departure

French letter *noun* (circa 1856)
chiefly British **:** CONDOM

French·man \ˈfrench-mən\ *noun* (before 12th century)
1 : a native or inhabitant of France
2 : one who is of French descent

French pastry *noun* (1922)
: a rich pastry filled especially with custard or fruit

French provincial *noun, often P capitalized* (1945)
: a style of furniture, architecture, or fabric originating in or characteristic of the 17th and 18th century French provinces

French seam *noun* (circa 1890)
: a strong seam stitched on both sides of the fabric to enclose all raw edges

French telephone *noun* (1932)
: HANDSET

French toast *noun* (1871)
: bread dipped in a mixture of egg and milk and sautéed

French window *noun* (1801)
: a pair of casement windows that reaches to the floor, opens in the middle, and is placed in an exterior wall

French·wom·an \ˈfrench-ˌwu̇-mən\ *noun* (1593)
1 : a woman who is a native or inhabitant of France
2 : a woman of French descent

fre·net·ic \fri-ˈne-tik\ *adjective* [Middle English *frenetik* insane, from Middle French *frenetique,* from Latin *phreneticus,* modification of Greek *phrenitikos,* from *phrenitis* inflammation of the brain, from *phren-, phrēn* diaphragm, mind] (14th century)
: FRENZIED, FRANTIC
— **fre·net·i·cal·ly** \-ti-k(ə-)lē\ *adverb*
— **fre·net·i·cism** \-ˈne-tə-ˌsi-zəm\ *noun*

fren·u·lum \ˈfren-yə-ləm\ *noun, plural* **-la** \-lə\ [New Latin, diminutive of Latin *frenum*] (circa 1706)
1 : a connecting fold of membrane serving to support or restrain a part (as the tongue)
2 : a bristle or group of bristles on the front edge of the posterior wings of some lepidoptera that unites the wings by interlocking with the retinaculum of the forewings

fre·num \ˈfrē-nəm\ *noun, plural* **frenums** *or* **fre·na** \-nə\ [New Latin, from Latin, bridle, reins, and bit; probably akin to Latin *frendere* to grind — more at GRIND] (1741)
: FRENULUM 1

fren·zied \ˈfren-zēd\ *adjective* (1796)
: marked by frenzy
— **fren·zied·ly** *adverb*

[1]fren·zy \ˈfren-zē\ *noun, plural* **frenzies** [Middle English *frenesie,* from Middle French, from Medieval Latin *phrenesia,* alteration of Latin *phrenesis,* from *phreneticus*] (14th century)
1 a : a temporary madness **b :** a violent mental or emotional agitation
2 : intense usually wild and often disorderly compulsive or agitated activity

[2]frenzy *transitive verb* **fren·zied; fren·zy·ing** (1791)
: to affect with frenzy

Fre·on \ˈfrē-ˌän\ *trademark*
— used for any of various nonflammable fluorocarbons used as refrigerants and as propellants for aerosols

fre·quence \ˈfrē-kwən(t)s\ *noun* (1603)
: FREQUENCY

fre·quen·cy \ˈfrē-kwən(t)-sē\ *noun, plural* **-cies** (1600)
1 : the fact or condition of occurring frequently
2 a : the number of times that a periodic function repeats the same sequence of values during a unit variation of the independent variable **b :** the number of individuals in a single class when objects are classified according to variations in a set of one or more specified attributes

\ə\ **abut** \ᵊ\ **kitten** \ər\ **further** \a\ **ash** \ā\ **ace** \ä\ **mop, mar** \au̇\ **out** \ch\ **chin** \e\ **bet** \ē\ **easy** \g\ **go** \i\ **hit** \ī\ **ice** \j\ **job** \ŋ\ **sing** \ō\ **go** \ȯ\ **law** \ȯi\ **boy** \th\ **thin** \t͟h\ **the** \ü\ **loot** \u̇\ **foot** \y\ **yet** \zh\ **vision** *see also* Guide to Pronunciation

3 : the number of repetitions of a periodic process in a unit of time: as **a :** the number of complete alternations per second of an alternating current **b :** the number of complete oscillations per second of energy (as sound or electromagnetic radiation) in the form of waves

frequency distribution *noun* (1895)
: an arrangement of statistical data that exhibits the frequency of the occurrence of the values of a variable

frequency modulation *noun* (1922)
: modulation of the frequency of the carrier wave in accordance with speech or a signal; *also* **:** FM

frequency response *noun* (1926)
: the ability of a device (as an audio amplifier) to handle the frequencies applied to it; *also* **:** a graph representing this ability

[1]**fre·quent** \frē-ˈkwent, ˈfrē-kwənt\ *transitive verb* (15th century)
1 : to associate with, be in, or resort to often or habitually
2 *archaic* **:** to read systematically or habitually
— fre·quen·ta·tion \ˌfrē-ˌkwen-ˈtā-shən, -kwən-\ *noun*
— fre·quent·er *noun*

[2]**fre·quent** \ˈfrē-kwənt\ *adjective* [Middle English, ample, from Middle French or Latin; Middle French, crowded, from Latin *frequent-, frequens*] (1531)
1 a : COMMON, USUAL **b :** happening at short intervals **:** often repeated or occurring
2 *obsolete* **:** FULL, THRONGED
3 : HABITUAL, PERSISTENT
4 *archaic* **:** INTIMATE, FAMILIAR
— fre·quent·ness *noun*

fre·quen·ta·tive \frē-ˈkwen-tə-tiv\ *adjective* (1533)
: denoting repeated or recurrent action or state — used of a verb aspect, verb form, or meaning
— frequentative *noun*

fre·quent·ly \ˈfrē-kwənt-lē\ *adverb* (1531)
: at frequent or short intervals

fres·co \ˈfres-(ˌ)kō\ *noun, plural* **frescoes** [Italian, from *fresco* fresh, of Germanic origin; akin to Old High German *frisc* fresh] (1598)
1 : the art of painting on freshly spread moist lime plaster with water-based pigments
2 : a painting executed in fresco ◆
— fresco *transitive verb*

[1]**fresh** \ˈfresh\ *adjective* [Middle English, from Old French *freis,* of Germanic origin; akin to Old High German *frisc* fresh, Old English *fersc* fresh] (13th century)
1 a : having its original qualities unimpaired: as (1) **:** full of or renewed in vigor **:** REFRESHED ⟨rose *fresh* from a good night's sleep⟩ (2) **:** not stale, sour, or decayed ⟨*fresh* bread⟩ (3) **:** not faded ⟨the lessons remain *fresh* in her memory⟩ (4) **:** not worn or rumpled ⟨a *fresh* white shirt⟩ **b :** not altered by processing ⟨*fresh* vegetables⟩
2 a : not salt **b** (1) **:** free from taint **:** PURE ⟨*fresh* air⟩ (2) *of wind* **:** moderately strong
3 a (1) **:** experienced, made, or received newly or anew ⟨form *fresh* friendships⟩ (2) **:** ADDITIONAL, ANOTHER ⟨a *fresh* start⟩ **b :** ORIGINAL, VIVID ⟨a *fresh* portrayal⟩ **c :** lacking experience **:** RAW **d :** newly or just come or arrived ⟨*fresh* from school⟩ **e :** having the milk flow recently established ⟨a *fresh* cow⟩
4 [probably by folk etymology from German *frech*] **:** disposed to take liberties **:** IMPUDENT
synonym see NEW
— fresh·ly *adverb*
— fresh·ness *noun*

[2]**fresh** *adverb* (14th century)
: just recently **:** NEWLY ⟨we're *fresh* out of eggs⟩ ⟨*fresh* caught fish⟩

[3]**fresh** *noun* (1538)
1 : an increased flow or rush (as of water) **:** FRESHET
2 *archaic* **:** a stream, spring, or pool of freshwater

fresh breeze *noun* (circa 1881)
: wind having a speed of 19 to 24 miles per hour (30 to 38 kilometers per hour) — see BEAUFORT SCALE table

fresh·en \ˈfre-shən\ *verb* **fresh·ened; fresh·en·ing** \-sh(ə-)niŋ\ (1697)
intransitive verb
1 : to grow or become fresh: as **a** *of wind* **:** to increase in strength **b :** to become fresh in appearance or vitality — usually used with *up* ⟨*freshen* up with a shower⟩
2 *of a milk animal* **:** to begin lactating
transitive verb
: to make fresh; *also* **:** REFRESH, REVIVE
— fresh·en·er \-sh(ə-)nər\ *noun*

fresh·et \ˈfre-shət\ *noun* (1596)
1 *archaic* **:** STREAM 1
2 a : a great rise or overflowing of a stream caused by heavy rains or melted snow **b :** something resembling or suggesting a freshet ⟨a *freshet* of visitors⟩

fresh gale *noun* (1582)
: wind having a speed of 39 to 46 miles per hour (62 to 74 kilometers per hour) — see BEAUFORT SCALE table

fresh·man \ˈfresh-mən\ *noun, often attributive* (1550)
1 : a first-year student
2 : BEGINNER, NEWCOMER

[1]**fresh·wa·ter** \ˈfresh-ˈwȯ-tər, -ˈwä-\ *noun* (14th century)
: water that is not salty especially when considered as a natural resource

[2]**freshwater** *adjective* (1528)
1 : of, relating to, being, or living in freshwater
2 : accustomed to navigating only in inland waters ⟨a *freshwater* sailor⟩; *also* **:** UNSKILLED
3 : inland and usually provincial ⟨a *freshwater* college⟩

freshwater drum *noun* (1879)
: a croaker (*Aplodinotus grunniens*) of the Great Lakes and Mississippi valley that may attain a weight of 50 pounds (23 kilograms) or more — called also *sheepshead, white perch*

freshwater pearl *noun* (1918)
: a usually very small pearl produced by a freshwater mollusk

Fres·nel lens \ˈfrez-nəl-, frā-ˈnel-\ *noun* [Augustin J. *Fresnel*] (circa 1884)
: a lens that has a surface consisting of a concentric series of simple lens sections so that a thin lens with a short focal length and large diameter is possible and that is used especially for spotlights

[1]**fret** \ˈfret\ *verb* **fret·ted; fret·ting** [Middle English, to devour, fret, from Old English *fretan* to devour; akin to Old High German *frezzan* to devour, *ezzan* to eat — more at EAT] (12th century)
transitive verb
1 a : to eat or gnaw into **:** CORRODE; *also* **:** FRAY **b :** RUB, CHAFE **c :** to make by wearing away a substance ⟨the stream *fretted* a channel⟩
2 : to cause to suffer emotional strain **:** VEX
3 : to pass (as time) in fretting
4 : AGITATE, RIPPLE
intransitive verb
1 a : to eat into something **b :** to affect something as if by gnawing or biting **:** GRATE
2 a : WEAR, CORRODE **b :** CHAFE **c :** FRAY 1
3 a : to become vexed or worried **b** *of running water* **:** to become agitated

[2]**fret** *noun* (15th century)
1 a : the action of wearing away **:** EROSION **b :** a worn or eroded spot
2 : an agitation of mind **:** IRRITATION

[3]**fret** *transitive verb* **fret·ted; fret·ting** [Middle English, back-formation from *fret, fretted* adorned, interwoven, from Middle French *freté,* from Old French, past participle of *freter, ferter* to tie, bind, probably from (assumed) Vulgar Latin *firmitare,* from Latin *firmus* firm] (14th century)
1 a : to decorate with interlaced designs **b :** to form a pattern upon
2 : to enrich with embossed or pierced carved patterns

[4]**fret** *noun* (14th century)
1 : an ornamental network; *especially* **:** a medieval metallic or jeweled net for a woman's headdress
2 : an ornament or ornamental work often in relief consisting of small straight bars intersecting one another in right or oblique angles

fret 2

[5]**fret** *noun* [perhaps from Middle French *frete* ferrule, from *freter*] (circa 1500)
: one of a series of ridges fixed across the fingerboard of a stringed musical instrument (as a guitar)
— fret·less *adjective*
— fret·ted *adjective*

[6]**fret** *transitive verb* **fret·ted; fret·ting** (1602)
: to press (the strings of a stringed instrument) against the frets

fret·ful \ˈfret-fəl\ *adjective* (1602)
: disposed to fret **:** IRRITABLE
— fret·ful·ly \-fə-lē\ *adverb*
— fret·ful·ness *noun*

fret·saw \ˈfret-ˌsȯ\ *noun* (1865)
: a narrow-bladed fine-toothed saw held under tension in a frame and used for cutting curved outlines

fret·work \-ˌwərk\ *noun* (1601)
1 : decoration consisting of work adorned with frets
2 : ornamental openwork or work in relief

Freud·ian \ˈfrȯi-dē-ən\ *adjective* (1910)
: of, relating to, or according with the psychoanalytic theories or practices of Freud ◆
— Freudian *noun*
— Freud·ian·ism \-ə-ˌni-zəm\ *noun*

Freudian slip *noun* (1953)

◇ WORD HISTORY

fresco The Italian word *fresco* literally means "fresh" and is ultimately descended from a Germanic word akin to the source of English *fresh* and German *frisch.* In the Renaissance, *pittura a fresco,* which can be translated as "painting fresh(ly)," meant painting applied while the plaster on a wall was still wet, and was opposed to *pittura a secco* "painting dry(ly)," referring to painting applied when the plaster was dry. The advantage of the "fresh" technique was somewhat greater durability, though it required that the painter work quickly and did not allow for easy alteration. In English, *fresco* appears earliest in the phrase *in fresco,* rendering Italian *a fresco,* and is not attested as a noun referring to a painting in fresco until 1670. A different sense of Italian *fresco* as a noun, "fresh air," appears in the phrase *al fresco* "outdoors," borrowed into English as *alfresco,* and used particularly in the context of dining outdoors.

Freudian The Austrian neurologist and psychiatrist Sigmund Freud (1856–1939) is renowned as the founder of psychoanalysis. He began his study of hysteria under Josef Breuer, whose established method of treatment was hypnosis. From 1892 to 1895 Freud developed his own technique of having the patient freely associate ideas. He came to believe that a complex of repressed and forgotten impressions underlies all abnormal mental states such as hysteria, and that a cure could be effected by a revelation of these impressions. He also developed a theory that dreams are an unconscious representation of repressed desires, especially sexual desires.

: a slip of the tongue that is motivated by and reveals some unconscious aspect of the mind
Frey \\'frā\\ *noun* [Old Norse *Freyr*]
: the Norse god of fertility, crops, peace, and prosperity
Freya \\'frā-ə\\ *noun* [Old Norse *Freyja*]
: the Norse goddess of love and beauty
fri·a·ble \\'frī-ə-bəl\\ *adjective* [Middle French or Latin; Middle French, from Latin *friabilis,* from *friare* to crumble — more at FRICTION] (1563)
: easily crumbled or pulverized ⟨*friable* soil⟩
synonym see FRAGILE
— fri·a·bil·i·ty \\ˌfrī-ə-'bi-lə-tē\\ *noun*
fri·ar \\'frī(-ə)r\\ *noun* [Middle English *frere, fryer,* from Old French *frere,* literally, brother, from Latin *fratr-, frater* — more at BROTHER] (13th century)
: a member of a mendicant order
fri·ar·ly \\-lē\\ *adjective* (1549)
: resembling a friar **:** relating to friars
friar's lantern *noun* (1632)
obsolete **:** IGNIS FATUUS
fri·ary \\'frī(-ə)r-ē\\ *noun, plural* **-ar·ies** (1538)
: a monastery of friars
[1]frib·ble \\'fri-bəl\\ *verb* **frib·bled; frib·bling** \\-b(ə-)liŋ\\ [origin unknown] (1633)
intransitive verb
1 : TRIFLE
2 *obsolete* **:** DODDER
transitive verb
: to trifle or fool away
[2]fribble *noun* (1664)
: a frivolous person, thing, or idea **:** TRIFLER
— frib·ble *adjective*
fric·an·deau \\'fri-kən-ˌdō\\ *noun* [French, from Middle French, probably from *fric*asser + *-ande* (as in *viande* meat) + *-eau,* noun suffix] (1706)
: larded veal roasted and glazed in its own juices
[1]fric·as·see *also* **fric·as·sée** \\'fri-kə-ˌsē, ˌfri-kə-'\\ *noun* [Middle French, from feminine of *fricassé,* past participle of *fricasser* to fricassee] (1568)
: a dish of cut-up pieces of meat (as chicken) stewed in stock and served in a white sauce
[2]fricassee *transitive verb* **-seed; -see·ing** (1657)
: to cook as a fricassee
fric·a·tive \\'fri-kə-tiv\\ *noun* [Latin *fricatus,* past participle of *fricare*] (1863)
: a consonant characterized by frictional passage of the expired breath through a narrowing at some point in the vocal tract
— fricative *adjective*
fric·tion \\'frik-shən\\ *noun* [Middle English, from Middle French or Latin; Middle French, from Latin *friction-, frictio,* from *fricare* to rub; akin to Latin *friare* to crumble, and perhaps to Sanskrit *bhrīṇanti* they injure] (1704)
1 a : the rubbing of one body against another **b :** the force that resists relative motion between two bodies in contact
2 : the clashing between two persons or parties of opposed views **:** DISAGREEMENT
— fric·tion·less \\-ləs\\ *adjective*
— fric·tion·less·ly *adverb*
fric·tion·al \\'frik-shnəl, -shə-nᵊl\\ *adjective* (1850)
1 : of or relating to friction
2 : moved or produced by friction
— fric·tion·al·ly *adverb*
friction clutch *noun* (circa 1842)
: a clutch in which connection is made through sliding friction
friction drive *noun* (1907)
: a power-transmission system that transmits motion by surface friction instead of teeth
friction tape *noun* (1920)
: a usually cloth tape impregnated with water-resistant insulating material and an adhesive and used especially to protect, insulate, and support electrical conductors
Fri·day \\'frī-dē, -(ˌ)dā\\ *noun* [Middle English, from Old English *frīgedæg* (akin to Old High German *frīatag* Friday), from (assumed) *Frīg* Frigga + *dæg* day, prehistoric translation of Latin *dies Veneris* Venus' day] (before 12th century)
: the sixth day of the week
— Fri·days \\-dēz, -(ˌ)dāz\\ *adverb*
fridge *also* **frig** \\'frij\\ *noun* [by shortening & alteration] (1926)
: REFRIGERATOR
fried \\'frīd\\ *adjective* (1926)
: INTOXICATED
fried·cake \\'frīd-ˌkāk\\ *noun* (1839)
: DOUGHNUT, CRULLER
[1]friend \\'frend\\ *noun* [Middle English *frend,* from Old English *frēond;* akin to Old High German *friunt* friend, Old English *frēon* to love, *frēo* free] (before 12th century)
1 a : one attached to another by affection or esteem **b :** ACQUAINTANCE
2 a : one that is not hostile **b :** one that is of the same nation, party, or group
3 : one that favors or promotes something (as a charity)
4 : a favored companion
5 *capitalized* **:** a member of a Christian sect that stresses Inner Light, rejects sacraments and an ordained ministry, and opposes war — called also *Quaker*
— friend·less \\'fren(d)-ləs\\ *adjective*
— friend·less·ness *noun*
— be friends with : to have a friendship or friendly relationship with
— make friends with : to establish a friendship or friendly relations with
[2]friend *transitive verb* (13th century)
: to act as the friend of **:** BEFRIEND
[1]friend·ly \\'fren(d)-lē\\ *adjective* **friend·li·er; -est** (before 12th century)
1 : of, relating to, or befitting a friend: as **a :** showing kindly interest and goodwill **b :** not hostile ⟨a *friendly* merger offer⟩; *also* **:** involving or coming from actions of one's own forces ⟨*friendly* fire⟩ **c :** CHEERFUL, COMFORTING ⟨the *friendly* glow of the fire⟩
2 : serving a beneficial or helpful purpose
3 : easy to use or understand ⟨*friendly* computer software⟩
synonym see AMICABLE
— friend·li·ly \\'fren(d)-lə-lē\\ *adverb*
— friend·li·ness *noun*
[2]friendly *adverb* (before 12th century)
: in a friendly manner **:** AMICABLY
[3]friendly *noun, plural* **friendlies** (1861)
1 : one that is friendly; *especially* **:** a native who is friendly to settlers or invaders
2 *British* **:** a match between sports teams and especially international teams that has no connection with league or championship play
friendly society *noun* (1703)
British **:** a mutual association for providing life and health insurance and old-age pension benefits to members
friend of the court (1944)
: AMICUS CURIAE
friend·ship \\'fren(d)-ˌship\\ *noun* (before 12th century)
1 : the state of being friends
2 : the quality or state of being friendly **:** FRIENDLINESS
3 *obsolete* **:** AID
fri·er *variant of* FRYER
Frie·sian \\'frē-zhən\\ *noun* [variant of *Frisian*] (1923)
chiefly British **:** HOLSTEIN
[1]frieze \\'frēz *or* frē-'zā\\ *noun* [Middle English *frise,* from Middle French, from Middle Dutch *vriese*] (15th century)
1 : a heavy durable coarse wool and shoddy fabric with a rough surface
2 : a pile surface of uncut loops or of patterned cut and uncut loops
[2]frieze \\'frēz\\ *noun* [Middle French *frise,* perhaps from Medieval Latin *phrygium, frisium* embroidered cloth, from Latin *phrygium,* from neuter of *Phrygius* Phrygian, from *Phrygia*] (1563)
1 : the part of an entablature between the architrave and the cornice — see ENTABLATURE illustration
2 : a sculptured or richly ornamented band (as on a building or piece of furniture)
3 : a band, line, or series suggesting a frieze ⟨a constant *frieze* of visitors wound its way around the . . . ruins —Mollie Panter-Downes⟩
— frieze·like *adjective*
frig \\'frig\\ *intransitive verb* **frigged; frig·ging** [Middle English *fryggen* to wriggle] (1598)
: COPULATE — usually considered vulgar; sometimes used in present participle as a meaningless intensive
frig·ate \\'fri-gət\\ *noun* [Middle French, from Old Italian *fregata*] (1585)
1 : a light boat propelled originally by oars but later by sails
2 : a square-rigged war vessel intermediate between a corvette and a ship of the line
3 : a warship that is smaller than a destroyer
frigate bird *noun* (1738)
: any of a family (Fregatidae) of tropical seabirds having a forked tail and large wingspans that are noted for aggressively taking food from other birds
Frig·ga \\'fri-gə\\ *noun* [Old Norse *Frigg*]
: the wife of Odin and Norse goddess of married love and of the hearth
[1]fright \\'frīt\\ *noun* [Middle English, from Old English *fyrhto, fryhto;* akin to Old High German *forhta* fear] (before 12th century)
1 : fear excited by sudden danger **:** ALARM
2 : something strange, ugly, or shocking
synonym see FEAR
[2]fright *transitive verb* (before 12th century)
: to alarm suddenly **:** FRIGHTEN
fright·en \\'frī-tᵊn\\ *verb* **fright·ened; fright·en·ing** \\'frī-tᵊn-iŋ, 'frīt-niŋ\\ (1666)
transitive verb
1 : to make afraid **:** TERRIFY
2 : to drive or force by frightening ⟨*frightened* the boy into confessing⟩
intransitive verb
: to become frightened
— fright·en·ing·ly \\-tᵊn-iŋ-lē, -niŋ-lē\\ *adverb*
fright·ful \\'frīt-fəl\\ *adjective* (1607)
1 : causing intense fear or alarm **:** TERRIFYING
2 : startling especially in being bad or objectionable ⟨a *frightful* novel⟩
3 : EXTREME ⟨*frightful* thirst⟩
— fright·ful·ly \\-fə-lē\\ *adverb*
— fright·ful·ness *noun*
fright wig *noun* (1886)
: a wig with hair that stands out from the head
frig·id \\'fri-jəd\\ *adjective* [Latin *frigidus,* from *frigēre* to be cold; akin to Latin *frigus* frost, cold, Greek *rhigos*] (1622)
1 a : intensely cold **b :** lacking warmth or ardor **:** INDIFFERENT
2 : lacking imaginative qualities **:** INSIPID
3 a : abnormally averse to sexual intercourse — used especially of women **b** *of a female* **:** unable to achieve orgasm during sexual intercourse
— frig·id·ly *adverb*
— frig·id·ness *noun*
Frig·i·daire \\ˌfri-jə-'dar, -'der\\ *trademark*
— used for a mechanical refrigerator
fri·gid·i·ty \\fri-'ji-də-tē\\ *noun* (15th century)
: the quality or state of being frigid; *specifically* **:** marked or abnormal sexual indifference especially in a woman
frigid zone *noun* (1622)
: the area or region between the arctic circle and the north pole or between the antarctic circle and the south pole

\\ə\\ **abut** \\ᵊ\\ **kitten** \\ər\\ **further** \\a\\ **ash** \\ā\\ **ace** \\ä\\ **mop, mar** \\au̇\\ **out** \\ch\\ **chin** \\e\\ **bet** \\ē\\ **easy** \\g\\ **go** \\i\\ **hit** \\ī\\ **ice** \\j\\ **job** \\ŋ\\ **sing** \\ō\\ **go** \\ȯ\\ **law** \\ȯi\\ **boy** \\th\\ **thin** \\t͟h\\ **the** \\ü\\ **loot** \\u̇\\ **foot** \\y\\ **yet** \\zh\\ **vision** *see also* Guide to Pronunciation

frig·o·rif·ic \ˌfri-gə-ˈri-fik\ *adjective* [Latin *frigorificus,* from *frigor-, frigus* frost] (1667)
: causing cold **:** CHILLING

fri·jo·le \frē-ˈhō-lē\ *also* **fri·jol** \frē-ˈhōl, ˈfrē-ˌ\ *noun, plural* **fri·jo·les** \frē-ˈhō-lēz, ˈfrē-ˌ\ [American Spanish *frijol,* from Spanish, kidney bean, from earlier *fesol, fresol,* probably modification of Galician *feijoo,* from Latin *phaseolus,* diminutive of *phaselus* cowpea, from Greek *phasēlos*] (1577)
: any of various beans used in Mexican style cooking — usually used in plural

[1]frill \ˈfril\ *transitive verb* (1574)
: to provide or decorate with a frill

[2]frill *noun* [perhaps from Flemish *frul*] (1591)
1 a : a gathered, pleated, or bias-cut fabric edging used on clothing **b :** a strip of paper curled at one end and rolled to be slipped over the bone end (as of a chop) in serving
2 : a ruff of hair or feathers or a bony projection about the neck of an animal
3 a : AFFECTATION, AIR — usually used in plural ⟨an honest . . . man who had no *frills,* . . . no nonsense about him —W. A. White⟩ **b :** something decorative or useful and desirable but not essential **:** LUXURY
— **frilly** \ˈfri-lē\ *adjective*

[1]fringe \ˈfrinj\ *noun, often attributive* [Middle English *frenge,* from Middle French, from (assumed) Vulgar Latin *frimbia,* from Latin *fimbriae* (plural)] (14th century)
1 : an ornamental border consisting of short straight or twisted threads or strips hanging from cut or raveled edges or from a separate band
2 a : something resembling a fringe **:** EDGE, PERIPHERY — often used in plural ⟨operated on the *fringes* of the law⟩ **b :** one of various light or dark bands produced by the interference or diffraction of light **c :** an area bordering a putting green on a golf course with grass trimmed longer than on the green itself
3 a : something that is marginal, additional, or secondary to some activity, process, or subject **b :** a group with marginal or extremist views **c :** FRINGE BENEFIT
— **fringy** \ˈfrin-jē\ *adjective*

[2]fringe *transitive verb* **fringed; fring·ing** \ˈfrin-jiŋ\ (15th century)
1 : to furnish or adorn with a fringe
2 : to serve as a fringe for **:** BORDER

fringe area *noun* (1950)
: a region in which reception from a given broadcasting station is weak or subject to serious distortion

fringe benefit *noun* (1948)
1 : an employment benefit (as a pension, a paid holiday, or health insurance) granted by an employer that has a monetary value but does not affect basic wage rates
2 : any additional benefit

fringe tree *noun* (circa 1730)
: a small tree (*Chionanthus virginicus*) of the olive family that has clusters of white flowers and is native to the southern U.S. but is widely cultivated as an ornamental

frip·pery \ˈfri-p(ə-)rē\ *noun, plural* **-per·ies** [Middle French *friperie,* alteration of Old French *freperie,* from *frepe* old garment] (1568)
1 *obsolete* **a :** cast-off clothes **b** *archaic* **:** a place where old clothes are sold
2 a : FINERY; *also* **:** an elegant or showy garment **b :** something showy, frivolous, or nonessential **:** LUXURY, TRIFLE **c :** OSTENTATION; *especially* **:** something foolish or affectedly elegant

Fris·bee \ˈfriz-bē\ *trademark*
— used for a plastic disk several inches in diameter sailed between players by a flip of the wrist

fri·sé \frē-ˈzā\ *noun* [French, from past participle of *friser* to curl] (1884)
: [1]FRIEZE

Frise aileron \ˈfrēz-\ *noun* [Leslie George *Frise* (born 1897) English engineer] (circa 1934)
: an aileron having a nose portion projecting ahead of the hinge axis and a lower surface in line with the lower surface of the wing

fri·seur \frē-ˈzər\ *noun* [French, from *friser*] (1750)
: HAIRDRESSER

[1]Fri·sian \ˈfri-zhən, ˈfrē-\ *adjective* [Latin *Frisius* Frisian; akin to Old English *Frīsa, Frēsa* a Frisian] (1598)
: of, relating to, or characteristic of Friesland, the Frisians, or Frisian

[2]Frisian *noun* (1601)
1 : a member of a people that inhabit principally the Netherlands province of Friesland and the Frisian islands in the North Sea
2 : the Germanic language of the Frisian people

[1]frisk \ˈfrisk\ *verb* [obsolete *frisk* (lively)] (1519)
intransitive verb
: to leap, skip, or dance in a lively or playful way **:** GAMBOL
transitive verb
: to search (a person) for something (as a concealed weapon) by running the hand rapidly over the clothing and through the pockets
— **frisk·er** *noun*

[2]frisk *noun* (1525)
1 a *archaic* **:** CAPER **b :** GAMBOL, ROMP **c :** DIVERSION
2 : an act of frisking

fris·ket \ˈfris-kət\ *noun* [French *frisquette,* from Middle French] (circa 1898)
: a masking device or material used especially in printing or graphic arts

frisky \ˈfris-kē\ *adjective* **frisk·i·er; -est** (circa 1500)
: inclined to frisk **:** PLAYFUL; *also* **:** LIVELY
— **frisk·i·ly** \ˈfris-kə-lē\ *adverb*
— **frisk·i·ness** \-kē-nəs\ *noun*

fris·son \frē-ˈsō^{n}\ *noun, plural* **frissons** \-ˈsō^{n}(z)\ [French, shiver, from Old French *friçon,* from Late Latin *friction-, frictio,* from Latin, literally, friction (taken in Late Latin as derivative of *frigēre* to be cold)] (1777)
: a brief moment of emotional excitement **:** SHUDDER, THRILL

[1]frit \ˈfrit\ *noun* [Italian *fritta,* from feminine of *fritto,* past participle of *friggere* to fry, from Latin *frigere* to roast — more at FRY] (1662)
1 : the calcined or partly fused materials of which glass is made
2 : any of various chemically complex glasses used ground especially to introduce soluble or unstable ingredients into glazes or enamels

[2]frit *transitive verb* **frit·ted; frit·ting** (1832)
1 : to prepare (materials for glass) by heat **:** FUSE
2 : to convert into a frit

frith \ˈfrith\ *noun* (14th century)
archaic **:** ESTUARY

frit·il·lar·ia \ˌfri-tᵊl-ˈer-ē-ə, -ˈar-\ *noun* [New Latin, from Latin *fritillus* dice cup; from the markings of the petals] (1664)
: any of a widespread genus (*Fritillaria*) of bulbous herbs of the lily family with variably colored and often mottled or checkered flowers

fritillaria

frit·il·lary \ˈfri-tᵊl-ˌer-ē\ *noun, plural* **-lar·ies** [New Latin *fritillaria*] (1633)
1 : FRITILLARIA
2 : any of numerous nymphalid butterflies (*Argynnis, Speyeria,* and related genera) that usually are orange with black spots on the upper side of both wings and silver spotted on the underside of the hind wing

frit·ta·ta \frē-ˈtä-tə\ *noun* [Italian, from *fritto*] (1931)
: an unfolded omelet often containing chopped vegetables or meats

frit·ted *adjective* [[2]*frit*] (1879)
: being porous glass made of sintered powdered glass or fiberglass

[1]frit·ter \ˈfri-tər\ *noun* [Middle English *fritour,* from Middle French *friture,* from (assumed) Vulgar Latin *frictura,* from Latin *frictus,* past participle of *frigere* to roast] (14th century)
: a small mass of fried or sautéed batter often containing fruit or meat

[2]fritter *verb* [*fritter,* noun (fragment, shred)] (1728)
transitive verb
1 : to spend or waste bit by bit, on trifles, or without commensurate return — usually used with *away*
2 : to break into small fragments
intransitive verb
: DISSIPATE, DWINDLE
— **frit·ter·er** \-tər-ər\ *noun*

frit·to mi·sto \ˈfrē-(ˌ)tō-ˈmē-(ˌ)stō\ *noun* [Italian, literally, mixed fried (food)] (1903)
: small morsels of meat, seafood, or vegetables coated with batter and deep fried

fritz \ˈfritz\ *noun* [origin unknown] (1902)
: a state of disorder or disrepair — used in the phrase *on the fritz*

friv·ol \ˈfri-vᵊl\ *intransitive verb* **-oled** *or* **-olled; -ol·ing** *or* **-ol·ling** \-vᵊl-iŋ, -və-liŋ\ [back-formation from *frivolous*] (1866)
: to act frivolously **:** TRIFLE
— **friv·ol·er** *or* **friv·ol·ler** \-vᵊl-ər, -və-lər\ *noun*

fri·vol·i·ty \fri-ˈvä-lə-tē\ *noun, plural* **-ties** (1796)
1 : the quality or state of being frivolous
2 : a frivolous act or thing

friv·o·lous \ˈfri-vᵊl-əs, -və-ləs\ *adjective* [Middle English, from Latin *frivolus*] (15th century)
1 a : of little weight or importance **b :** having no sound basis (as in fact or law) ⟨a *frivolous* lawsuit⟩
2 a : lacking in seriousness **b :** marked by unbecoming levity
— **friv·o·lous·ly** *adverb*
— **friv·o·lous·ness** *noun*

[1]frizz \ˈfriz\ *verb* [French *friser*] (1660)
transitive verb
: to form into small tight curls
intransitive verb
of hair **:** to form a mass of tight curls

[2]frizz *noun* (1668)
1 : a tight curl
2 : hair that is tightly curled

[3]frizz *verb* [alteration of [1]*fry*] (1835)
transitive verb
: to fry or sear with a sizzling noise
intransitive verb
: SIZZLE

[1]friz·zle \ˈfri-zᵊl\ *verb* **friz·zled; friz·zling** \-zᵊl-iŋ, -zə-liŋ\ [probably akin to Old Frisian *frīsle* curl] (1573)
: FRIZZ, CURL

[2]frizzle *noun* (1613)
: a crisp curl

[3]frizzle *verb* **friz·zled; friz·zling** [[1]*fry* + *sizzle*] (1839)
transitive verb
1 : to fry until crisp and curled
2 : BURN, SCORCH
intransitive verb
: to cook with a sizzling noise

frizzy \ˈfri-zē\ *adjective* **frizz·i·er; -est** (circa 1864)
: tightly curled
— **frizz·i·ness** *noun*

[1]fro \frə, ˈfrō\ *preposition* [Middle English, from Old Norse *frā;* akin to Old English *fram* from] (13th century)
dialect British **:** FROM

[2]fro \ˈfrō\ *adverb* (14th century)
: BACK, AWAY — used in the phrase *to and fro*

[1]frock \ˈfräk\ *noun* [Middle English *frok,* from Middle French *froc,* of Germanic origin; akin to Old High German *hroch* mantle, coat] (14th century)
1 : an outer garment worn by monks and friars **:** HABIT
2 : an outer garment worn chiefly by men: **a :** a long loose mantle **b :** a workman's outer shirt; *especially* **:** SMOCK FROCK **c :** a woolen jersey worn especially by sailors
3 : a woman's dress

[2]frock *transitive verb* (1828)
1 : to clothe in a frock
2 : to make a cleric of

frock coat *noun* (1823)
: a man's knee-length usually double-breasted coat

froe \ˈfrō\ *noun* [perhaps alteration of obsolete *froward* turned away, from Middle English; from the position of the handle] (1574)
: a cleaving tool for splitting cask staves and shingles from the block

frog \ˈfrȯg, ˈfräg\ *noun* [Middle English *frogge,* from Old English *frogga;* akin to Old High German *frosk* frog; senses 2, 3, 5, 7, 8 unclearly derived & perhaps of distinct origin] (before 12th century)
1 : any of various smooth-skinned web-footed largely aquatic tailless agile leaping amphibians (as a ranid) — compare TOAD
2 : the triangular elastic horny pad in the middle of the sole of the foot of a horse — see HOOF illustration
3 a : a loop attached to a belt to hold a weapon or tool **b :** an ornamental braiding for fastening the front of a garment that consists of a button and a loop through which it passes
4 *often capitalized* **:** FRENCHMAN — usually taken to be offensive
5 : a device permitting the wheels on one rail of a track to cross an intersecting rail
6 : a condition in the throat that produces hoarseness ⟨had a *frog* in his throat⟩
7 : the nut of a violin bow
8 : a small holder (as of metal, glass, or plastic) with perforations or spikes for holding flowers in place in a bowl or vase

frog·eye \-ˌī\ *noun* (circa 1909)
: any of numerous leaf diseases characterized by concentric rings about the diseased spots

frog·hop·per \-ˌhä-pər\ *noun* (1711)
: SPITTLEBUG

frog kick *noun* (1940)
: a breaststroke kick executed with the knees turned outward and the legs separated and then swung together

frog·man \ˈfrȯg-ˌman, ˈfräg-, -mən\ *noun* (1945)
: a person equipped (as with face mask, flippers, and air supply) for extended periods of underwater swimming; *especially* **:** a person so equipped for military reconnaissance and demolition

frog–march \-ˌmärch\ *transitive verb* (1923)
: to seize from behind roughly and forcefully propel forward ⟨*frog-marched* him out the door⟩

frog spit *noun* (circa 1825)
: CUCKOO SPIT 1

[1]frol·ic \ˈfrä-lik\ *adjective* [Dutch *vroolijk,* from Middle Dutch *vrolijc,* from *vro* happy; akin to Old High German *frō* happy] (1538)
: full of fun **:** MERRY

[2]frolic *intransitive verb* **frol·icked; frol·ick·ing** (1593)
1 : to amuse oneself **:** make merry
2 : to play and run about happily **:** ROMP

[3]frolic *noun* (1616)
1 : a playful or mischievous action
2 a : an occasion or scene of fun **:** PARTY **b :** FUN, MERRIMENT

frol·ic·some \ˈfrä-lik-səm\ *adjective* (1699)
: full of gaiety **:** SPORTIVE, PLAYFUL

from \ˈfrəm, ˈfräm *also* fəm\ *preposition* [Middle English, from Old English *from, fram;* akin to Old High German *fram,* adverb, forth, away, Old English *faran* to go — more at FARE] (before 12th century)
1 — used as a function word to indicate a starting point of a physical movement or a starting point in measuring or reckoning or in a statement of limits ⟨came here *from* the city⟩ ⟨a week *from* today⟩ ⟨cost *from* $5 to $10⟩
2 — used as a function word to indicate physical separation or an act or condition of removal, abstention, exclusion, release, subtraction, or differentiation ⟨protection *from* the sun⟩ ⟨relief *from* anxiety⟩
3 — used as a function word to indicate the source, cause, agent, or basis ⟨we conclude *from* this⟩ ⟨a call *from* my lawyer⟩ ⟨inherited a love of music *from* his father⟩ ⟨worked hard *from* necessity⟩

frond \ˈfränd\ *noun* [Latin *frond-, frons* foliage] (1785)
1 : a large leaf (especially of a palm or fern) usually with many divisions
2 : a thallus or thalloid shoot (as of a lichen or seaweed) resembling a leaf
— frond·ed \ˈfrän-dəd\ *adjective*

fron·deur \frō^{n}-ˈdər\ *noun* [French, literally, slinger, participant in a 17th century revolt in which the rebels were compared to schoolboys using slings only when the teacher was not looking] (1798)
: REBEL, MALCONTENT

[1]front \ˈfrənt\ *noun* [Middle English, from Old French, from Latin *front-, frons*] (13th century)
1 a : FOREHEAD; *also* **:** the whole face **b :** external and often feigned appearance especially in the face of danger or adversity
2 a (1) **:** VANGUARD (2) **:** a line of battle (3) **:** a zone of conflict between armies **b** (1) **:** a stand on an issue **:** POLICY (2) **:** an area of activity (as study or debate) ⟨progress on the educational *front*⟩ (3) **:** a movement linking divergent elements to achieve common objectives; *especially* **:** a political coalition
3 : a side of a building; *especially* **:** the side that contains the principal entrance
4 a : the forward part or surface **b** (1) **:** FRONTAGE (2) **:** a beach promenade at a seaside resort **c :** DICKEY 1a **d :** the boundary between two dissimilar air masses
5 *archaic* **:** BEGINNING
6 a (1) **:** a position ahead of a person or of the foremost part of a thing (2) — used as a call by a hotel desk clerk in summoning a bellhop **b :** a position of leadership or superiority
7 a : a person, group, or thing used to mask the identity or true character or activity of the actual controlling agent **b :** a person who serves as the nominal head or spokesman of an enterprise or group to lend it prestige
— in front of : directly before or ahead of
— out front : in the audience

[2]front (1523)
intransitive verb
1 : to have the front or principal side adjacent to something; *also* **:** to have frontage on something ⟨a ten-acre plot *fronting* on a lake —*Current Biography*⟩
2 : to serve as a front ⟨*fronting* for special interests⟩
transitive verb
1 a : CONFRONT ⟨went to the woods because I wished . . . to *front* only the essential facts of life —H. D. Thoreau⟩ **b :** to appear before ⟨daily *fronted* him in some fresh splendor —Alfred Tennyson⟩
2 a : to be in front of ⟨lawn *fronting* the house⟩ **b :** to be the leader of (a musical group) ⟨appeared as a soloist and *fronted* bands⟩
3 : to face toward or have frontage on ⟨the house *fronts* the street⟩
4 : to supply a front to ⟨*fronted* the building with bricks⟩
5 a : to articulate (a sound) with the tongue farther forward **b :** to move (a word or phrase) to the beginning of a sentence
6 *basketball* **:** to play in front of (an opposing player) rather than between the player and the basket

[3]front *adjective* (1600)
1 a : of, relating to, or situated at the front **b :** acting as a front ⟨*front* company⟩
2 : articulated at or toward the front of the oral passage ⟨*front* vowels⟩
3 : constituting the first nine holes of an 18-hole golf course
— front *adverb*

front·age \ˈfrən-tij\ *noun* (1622)
1 a : a piece of land that lies adjacent (as to a street or the ocean) **b :** the land between the front of a building and the street **c :** the length of a frontage
2 : the act or fact of facing a given way
3 : the front side of a building

frontage road *noun* (1949)
: a local street that parallels an expressway or through street and that provides access to property near the expressway — called also *service road*

[1]fron·tal \ˈfrən-t^{ə}l\ *noun* (14th century)
1 [Middle English *frontel,* from Medieval Latin *frontellum,* diminutive of Latin *front-, frons*] **:** a cloth hanging over the front of an altar
2 : FACADE 1

[2]frontal *adjective* [New Latin *frontalis,* from Latin *front-, frons*] (1656)
1 : of, relating to, or adjacent to the forehead or the frontal bone
2 a : of, relating to, or situated at the front **b :** directed against the front or at the main point or issue **:** DIRECT ⟨*frontal* assault⟩
3 : parallel to the main axis of the body and at right angles to the sagittal plane
4 : of or relating to a meteorological front
— fron·tal·ly \-t^{ə}l-ē\ *adverb*

frontal bone *noun* (1741)
: a bone that forms the forehead and roofs over most of the orbits and nasal cavity and that at birth consists of two halves separated by a suture

fron·tal·i·ty \ˌfrən-ˈta-lə-tē\ *noun* (1905)
1 *sculpture* **:** a schematic composition of the front view that is complete without lateral movement
2 *painting* **:** the depiction of an object, figure, or scene in a plane parallel to the plane of the picture surface

frontal lobe *noun* (1879)
: the anterior division of each cerebral hemisphere

front and center *adverb* (1951)
: in or to the forefront of activity or consideration

front bench *noun* (circa 1889)
: either of the two benches nearest the chair in a British legislature (as the House of Commons) occupied by government and opposition leaders; *also* **:** the leaders themselves — compare BACKBENCH
— front–bench·er \-ˈben-chər\ *noun*

front burner *noun* (1973)
: the condition of being in active consideration or development — compare BACK BURNER

front·court \ˈfrənt-ˈkōrt, -ˈkȯrt\ *noun* (circa 1949)
: a basketball team's offensive half of the court; *also* **:** the positions of forward and center on a basketball team

front dive *noun* (circa 1934)
: a dive from a position facing the water

front–end *adjective* (1962)
: relating to or required at the beginning of an undertaking ⟨no *front-end* charge at the time of investment⟩

front end *noun* (1973)

: a unit in a computer system devoted to controlling the data communications link between terminals and the main computer and often to the preliminary processing of data

front–end load *noun* (1962)
: the part of the total commission and expenses taken out of early payments under a contract plan for the periodic purchase of investment-company shares

front–end loader *noun* (1954)
: a usually wheeled vehicle with a hydraulically operated scoop in front for excavating and loading loose material — called also *front loader*

fron·tier \ˌfrən-ˈtir, ˈfrən-ˌ, frän-ˈ, ˈfrän-ˌ\ *noun* [Middle English *fronter,* from Middle French *frontiere,* from *front*] (15th century)
1 a : a border between two countries **b** *obsolete* **:** a stronghold on a frontier
2 a : a region that forms the margin of settled or developed territory **b :** the farthermost limits of knowledge or achievement in a particular subject **c :** a line of division between different or opposed things ⟨the *frontiers* separating science and the humanities —R. W. Clark⟩ **d :** a new field for exploitative or developmental activity
— frontier *adjective*

fron·tiers·man \ˌfrən-ˈtirz-mən, frän-\ *noun* (1782)
: one who lives or works on a frontier

fron·tis·piece \ˈfrən-tə-ˌspēs\ *noun* [Middle French *frontispice,* from Late Latin *frontispicium* facade, from Latin *front-, frons* + *-i-* + *specere* to look at — more at SPY] (circa 1598)
1 a : the principal front of a building **b :** a decorated pediment over a portico or window
2 : an illustration preceding and usually facing the title page of a book or magazine ◆

front·less \ˈfrənt-ləs\ *adjective* (1605)
archaic **:** SHAMELESS

front·let \-lət\ *noun* [Middle English *frontlette,* from Middle French *frontelet,* diminutive of *frontel,* from Latin *frontale,* from *front-, frons*] (15th century)
1 : a band or phylactery worn on the forehead
2 : the forehead especially of an animal

front·line \ˈfrənt-ˌlīn\ *adjective* (1915)
1 : relating to, being, or involved in a front line ⟨*frontline* ambulances⟩
2 : FIRST-RATE ⟨*frontline* teachers⟩; *also* **:** FIRST-STRING ⟨*frontline* goalie⟩

front line *noun* (1917)
1 a : a military line formed by the most advanced tactical combat units; *also* **:** FRONT 2a(2) **b :** an area of potential or actual conflict or struggle
2 : the most advanced, responsible, or visible position in a field or activity

front–load *transitive verb* (1976)
: to assign costs or benefits to the early stages of (as a contract, project, or time period)

front man *noun* (1927)
1 : a person serving as a front or figurehead
2 : the lead performer in a musical group

front matter *noun* (circa 1909)
: matter preceding the main text of a book

front money *noun* (circa 1928)
: money that is paid in advance for a promised service or product

fronto- *combining form* [[1]front]
: boundary of an air mass ⟨*fronto*genesis⟩

front office *noun, often attributive* (1900)
: the policy-making officials of an organization

front·o·gen·e·sis \ˌfrən-tō-ˈje-nə-səs\ *noun* [New Latin] (1931)
: the coming together into a distinct front of two dissimilar air masses that commonly react upon each other to induce cloud and precipitation

front·ol·y·sis \ˌfrən-ˈtä-lə-səs\ *noun* [New Latin] (circa 1938)
: a process tending to destroy a meteorological front

fron·ton \ˈfrän-ˌtän\ *noun* [Spanish *frontón* gable, wall of a pelota court, fronton, from augmentative of *frente* forehead, from Latin *front-, frons*] (1896)
: a jai alai arena

[1]front–page \ˈfrənt-ˈpāj\ *adjective* (1917)
: printed on the front page of a newspaper; *also* **:** very newsworthy

[2]front–page *transitive verb* (1929)
: to print or report on the front page

front room *noun* (1853)
: LIVING ROOM, PARLOR

front–run·ner \ˈfrənt-ˌrə-nər\ *noun* (1914)
1 : a contestant who runs best when in the lead
2 : a leading contestant in or as if in a rivalry or competition ⟨a political *front-runner*⟩ ⟨dubbed . . . *front-runner* for the title of "worst Senator" —Edward Roeder⟩

front·ward \ˈfrənt-wərd\ *or* **front·wards** \-wərdz\ *adverb or adjective* (1865)
: toward the front

frore \ˈfrōr, ˈfrȯr\ *adjective* [Middle English *froren,* from Old English, past participle of *frēosan* to freeze] (13th century)
: FROSTY, FROZEN

frosh \ˈfräsh\ *noun, plural* **frosh** [by shortening & alteration] (1915)
: FRESHMAN

[1]frost \ˈfrȯst\ *noun* [Middle English, from Old English; akin to Old High German *frost* — more at FREEZE] (before 12th century)
1 a : the process of freezing **b :** a covering of minute ice crystals on a cold surface; *also* **:** ice particles formed from a gas **c :** the temperature that causes freezing
2 a : coldness of deportment or temperament **:** an indifferent, reserved, or unfriendly manner **b :** FAILURE ⟨the play was . . . a most dreadful *frost* —Arnold Bennett⟩

[2]frost (1635)
transitive verb
1 a : to cover with or as if with frost; *especially* **:** to put icing on (cake) **b :** to produce a fine-grained slightly roughened surface on (as metal or glass)
2 : to injure or kill (as plants) by frost
3 : to make angry or irritated ⟨your attitude really *frosts* me⟩
intransitive verb
: to become frosted

[1]frost·bite \ˈfrȯs(t)-ˌbīt\ *transitive verb* **-bit** \-ˌbit\; **-bit·ten** \-ˌbi-tᵊn\; **-bit·ing** \-ˌbī-tiŋ\ (1601)
: to affect or injure by frost or frostbite

[2]frostbite *noun* (1813)
: the freezing or the local effect of a partial freezing of some part of the body

[3]frostbite *adjective* (1953)
: done in cold weather ⟨*frostbite* sailing⟩; *also* **:** of or relating to cold-weather sailing ⟨*frostbite* sailors⟩

frost·bit·ing \-ˌbī-tiŋ\ *noun* (1965)
: the sport of sailing in cold weather

frost·ed \ˈfrȯ-stəd\ *adjective* (1947)
: having undergone frosting ⟨*frosted* hair⟩

frost heave *noun* (1941)
: an upthrust of ground or pavement caused by freezing of moist soil — called also *frost heaving*

frost·ing \ˈfrȯ-stiŋ\ *noun* (1858)
1 a : ICING **b :** TRIMMING, ORNAMENTATION
2 : lusterless finish of metal or glass **:** MAT; *also* **:** a white finish produced on glass (as by etching)
3 : the lightening (as by chemicals) of small strands of hair throughout the entire head to produce a two-tone effect — compare STREAKING

frost·work \ˈfrȯs(t)-ˌwərk\ *noun* (1729)
: the figures that moisture sometimes forms in freezing (as on a windowpane)

frosty \ˈfrȯ-stē\ *adjective* **frost·i·er; -est** (before 12th century)
1 a : attended with or producing frost **:** FREEZING **b :** briskly cold **:** CHILLY
2 : covered or appearing as if covered with frost **:** HOARY ⟨a man of 65, with *frosty* eyebrows and hair —Nan Robertson⟩
3 : marked by coolness or extreme reserve in manner ⟨his smile was distinctly *frosty* —Erle Stanley Gardner⟩
— frost·i·ly \-stə-lē\ *adverb*
— frost·i·ness \-stē-nəs\ *noun*

[1]froth \ˈfrȯth\ *noun, plural* **froths** \ˈfrȯths, ˈfrȯt͟hz\ [Middle English, from Old Norse *frotha;* akin to Old English *āfrēothan* to froth] (14th century)
1 a : bubbles formed in or on a liquid **:** FOAM **b :** a foamy slaver sometimes accompanying disease or exhaustion
2 : something resembling froth (as in being unsubstantial, worthless, or light and airy)

[2]froth \ˈfrȯth, ˈfrȯt͟h\ (14th century)
intransitive verb
1 : to foam at the mouth
2 : to throw froth out or up
3 : to become covered with or as if with froth ⟨whole groves *froth* with nodding blossoms —Amy Lovejoy⟩
transitive verb
1 : to cause to foam
2 : to cover with froth
3 : VENT, VOICE

frothy \ˈfrȯ-thē, -t͟hē\ *adjective* **froth·i·er; -est** (15th century)
1 : full of or consisting of froth
2 a : gaily frivolous or light in content or treatment **:** INSUBSTANTIAL ⟨a *frothy* comedy⟩ **b :** made of light thin material
— froth·i·ly \-thə-lē, -t͟hə-\ *adverb*
— froth·i·ness \-thē-nəs, -t͟hē-\ *noun*

frot·tage \frȯ-ˈtäzh\ *noun* [French, from *frotter* to rub] (1935)
: the technique of creating a design by rubbing (as with a pencil) over an object placed underneath the paper; *also* **:** a composition so made

frou·frou \ˈfrü-(ˌ)frü\ *noun* [French, of imitative origin] (1870)
1 : a rustling especially of a woman's skirts
2 : showy or frilly ornamentation

frow \ˈfrō\ *variant of* FROE

fro·ward \ˈfrō-(w)ərd\ *adjective* [Middle English, turned away, froward, from *fro* from + *-ward* -ward] (13th century)
1 : habitually disposed to disobedience and opposition
2 *archaic* **:** ADVERSE
— fro·ward·ly *adverb*
— fro·ward·ness *noun*

[1]frown \ˈfrau̇n\ *verb* [Middle English *frounen,* from Middle French *froigner* to snort, frown, of Celtic origin; akin to Welsh *ffroen* nostril, Old Irish *srón* nose] (14th century)
intransitive verb

◇ WORD HISTORY
frontispiece Folk etymology is a process whereby unfamiliar words are altered to give them an apparent relationship to more familiar ones. One word that reflects the workings of folk etymology is *frontispiece,* which has nothing at all to do with the word *piece.* The earliest form of the word in English, *frontispice,* is a loan, via French, from Late Latin *frontispicium,* meaning "façade of a building." Virtually as soon as the word was introduced, English writers began to spell the final syllable as *piece. Frontispiece* was applied early on not only to building fronts and decorated pediments, but also to the title page of a book, which in the 17th century often had columns, pediments, and other architectural detail framing the title itself. As the original architectural sense of the word became secondary in users' minds, associations evoked by *front* ("the foremost part") and *piece* ("an artistic composition") made it natural to apply the word to an illustration preceding the title page.

1 : to contract the brow in displeasure or concentration
2 : to give evidence of displeasure or disapproval by or as if by facial expression ⟨critics *frown* on the idea⟩
transitive verb
: to show displeasure with or disapproval of especially by facial expression
— frown·er *noun*
— frown·ing·ly \'fraù-niŋ-lē\ *adverb*

[2]**frown** *noun* (1581)
1 : an expression of displeasure
2 : a wrinkling of the brow in displeasure or concentration

frows·ty \'fraù-stē\ *adjective* **frowst·i·er; -est** [alteration of *frowsy*] (1865)
1 *chiefly British* **:** MUSTY
2 *chiefly British* **:** FROWSY 2

frow·sy *or* **frow·zy** \'fraù-zē\ *adjective* **frow·si·er** *or* **frow·zi·er; -est** [origin unknown] (1681)
1 : MUSTY, STALE ⟨a *frowsy* smell of stale beer and stale smoke —W. S. Maugham⟩
2 : having a slovenly or uncared-for appearance ⟨a couple of *frowsy* stuffed chairs —R. M. Williams⟩

froze *past of* FREEZE

fro·zen \'frō-zᵊn\ *adjective* (14th century)
1 a : treated, affected, or crusted over by freezing **b :** subject to long and severe cold ⟨*frozen* north⟩
2 a : incapable of being changed, moved, or undone **:** FIXED; *specifically* **:** debarred by official action from movement or from change in status ⟨*frozen* wages⟩ **b :** not available for present use ⟨*frozen* capital⟩ **c** (1) **:** drained or incapable of emotion (2) **:** expressing or characterized by cold unfriendliness
— fro·zen·ly *adverb*
— fro·zen·ness \-zᵊn-(n)əs\ *noun*

fruc·ti·fi·ca·tion \ˌfrək-tə-fə-'kā-shən, ˌfrùk-\ *noun* (1764)
: the reproductive organs or fruit of a plant; *especially* **:** SPOROPHORE

fruc·ti·fy \'frək-tə-ˌfī, 'frùk-\ *verb* **-fied; -fy·ing** [Middle English *fructifien,* from Middle French *fructifier,* from Latin *fructificare,* from *fructus* fruit] (14th century)
intransitive verb
: to bear fruit ⟨its seeds shall *fructify* —Amy Lowell⟩ ⟨no partnership can *fructify* without candor on both sides —D. M. Ogilvy⟩
transitive verb
: to make fruitful or productive

fruc·tose \'frək-ˌtōs, 'frük-, 'frùk-, -ˌtōz\ *noun* (circa 1864)
1 : an optically active sugar $C_6H_{12}O_6$ that differs from glucose in having a ketonic rather than aldehydic carbonyl group
2 : the very sweet soluble levorotatory D-form of fructose that occurs especially in fruit juices and honey — called also *levulose*

fruc·tu·ous \'frək-chə-wəs, 'frùk-\ *adjective* (14th century)
: FRUITFUL ⟨a *fructuous* land⟩

fru·gal \'frü-gəl\ *adjective* [Middle French or Latin; Middle French, from Latin *frugalis* virtuous, frugal, from *frug-, frux* fruit, value; akin to Latin *frui* to enjoy] (1598)
: characterized by or reflecting economy in the use of resources
synonym see SPARING
— fru·gal·i·ty \frü-'ga-lə-tē\ *noun*
— fru·gal·ly \'frü-gə-lē\ *adverb*

fru·giv·o·rous \frü-'ji-və-rəs\ *adjective* [Latin *frug-, frux* + English *-vorous*] (1713)
: feeding on fruit
— fru·gi·vore \'frü-ji-ˌvōr, -ˌvòr\ *noun*

[1]**fruit** \'früt\ *noun, often attributive* [Middle English, from Old French, from Latin *fructus* fruit, use, from *frui* to enjoy, have the use of — more at BROOK] (12th century)
1 a : a product of plant growth (as grain, vegetables, or cotton) ⟨the *fruits* of the field⟩ **b** (1) **:** the usually edible reproductive body of a seed plant; *especially* **:** one having a sweet pulp associated with the seed ⟨the *fruit* of the tree⟩ (2) **:** a succulent plant part (as the petioles of a rhubarb plant) used chiefly in a dessert or sweet course **c :** a dish, quantity, or diet of fruits ⟨live on *fruit*⟩ **d :** a product of fertilization in a plant with its modified envelopes or appendages; *specifically* **:** the ripened ovary of a seed plant and its contents **e :** the flavor or aroma of fresh fruit in mature wine
2 : OFFSPRING, PROGENY
3 a : the state of bearing fruit ⟨a tree in *fruit*⟩ **b :** the effect or consequence of an action or operation **:** PRODUCT, RESULT ⟨the *fruits* of our labor⟩
4 : a male homosexual — often used disparagingly
— fruit·ed \'frü-təd\ *adjective*

[2]**fruit** (14th century)
intransitive verb
: to bear fruit
transitive verb
: to cause to bear fruit

fruit·age \'frü-tij\ *noun* (15th century)
1 a : FRUIT **b :** the condition or process of bearing fruit
2 : the product or result of an action

frui·tar·i·an \(ˌ)frü(t)-'ter-ē-ən\ *noun* (1893)
: a person who lives on fruit

fruit bat *noun* (1877)
: any of a family (Pteropodidae of the suborder Megachiroptera) of often large tropical and subtropical Old World bats that feed on ripe fruit, pollen, and nectar and that usually use visual navigation rather than echolocation — called also *flying fox*

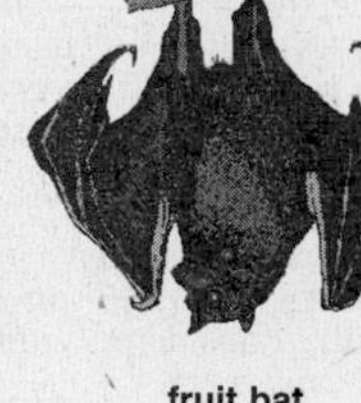
fruit bat

fruit·cake \'früt-ˌkāk\ *noun* (1848)
1 : a rich cake containing nuts, dried or candied fruits, and spices
2 : NUT 6a

fruit·er·er \'frü-tər-ər\ *noun* [Middle English, modification of Middle French *fruitier,* from *fruit*] (15th century)
chiefly British **:** a person who deals in fruit

fruit fly *noun* (circa 1753)
: any of various small dipteran flies (as a drosophila) whose larvae feed on fruit or decaying vegetable matter

fruit·ful \'früt-fəl\ *adjective* (14th century)
1 a : yielding or producing fruit **b :** conducive to an abundant yield
2 : abundantly productive
synonym see FERTILE
— fruit·ful·ly \-fə-lē\ *adverb*
— fruit·ful·ness *noun*

fruiting body *noun* (1918)
: a plant organ specialized for producing spores; *especially* **:** SPOROPHORE

fru·ition \frü-'i-shən\ *noun* [Middle English *fruicioun,* from Middle French or Late Latin; Middle French *fruition,* from Late Latin *fruition-, fruitio,* from Latin *frui*] (15th century)
1 : pleasurable use or possession **:** ENJOYMENT
2 a : the state of bearing fruit **b :** REALIZATION

fruit·less \'früt-ləs\ *adjective* (14th century)
1 : UNSUCCESSFUL
2 : lacking or not bearing fruit
synonym see FUTILE
— fruit·less·ly *adverb*
— fruit·less·ness *noun*

fruit·let \-lət\ *noun* (1882)
1 : a small fruit
2 : a unit of a collective fruit

fruit machine *noun* (1933)
British **:** SLOT MACHINE 2

fruit sugar *noun* (circa 1889)
: FRUCTOSE 2

fruit·wood \'früt-ˌwùd\ *noun, often attributive* (1927)
: the wood of a fruit tree (as the apple, cherry, or pear) ⟨*fruitwood* furniture⟩

fruity \'frü-tē\ *adjective* **fruit·i·er; -est** (1657)
1 a : relating to, made with, or resembling fruit **b :** having the flavor or aroma of ripe fruit
2 a : extremely effective, interesting, or enjoyable **b :** sweet or sentimental especially to excess **c** *of a voice* **:** rich and deep
3 a *slang* **:** CRAZY, SILLY **b :** HOMOSEXUAL — often used disparagingly
— fruit·i·ness *noun*

fru·men·ty \'frü-mən-tē\ *noun, plural* **-ties** [Middle English, from Middle French *frumentee,* from *frument* grain, from Latin *frumentum,* from *frui* to enjoy — more at BROOK] (14th century)
: a dish of wheat boiled in milk and usually sweetened and spiced

frump \'frəmp\ *noun* [probably from *frumple* (to wrinkle)] (1817)
1 : a dowdy unattractive girl or woman
2 : a staid, drab, old-fashioned person

frump·ish \'frəm-pish\ *adjective* (circa 1847)
: DOWDY, DRAB

frumpy \'frəm-pē\ *adjective* **frump·i·er; -est** (circa 1840)
: DOWDY, DRAB

[1]**frus·trate** \'frəs-ˌtrāt\ *transitive verb* **frus·trat·ed; frus·trat·ing** [Middle English, from Latin *frustratus,* past participle of *frustrare* to deceive, frustrate, from *frustra* in error, in vain] (15th century)
1 a : to balk or defeat in an endeavor **b :** to induce feelings of discouragement in
2 a (1) **:** to make ineffectual **:** bring to nothing (2) **:** IMPEDE, OBSTRUCT **b :** to make invalid or of no effect ☆

[2]**frustrate** *adjective* (15th century)
: characterized by frustration

frus·trat·ing \-ˌtrā-tiŋ\ *adjective* (1871)
: tending to produce or characterized by frustration
— frus·trat·ing·ly \-tiŋ-lē\ *adverb*

frus·tra·tion \(ˌ)frəs-'trā-shən\ *noun* (1555)
1 : the act of frustrating
2 a : the state or an instance of being frustrated **b :** a deep chronic sense or state of insecurity and dissatisfaction arising from unresolved problems or unfulfilled needs
3 : something that frustrates

frus·tule \'frəs-(ˌ)chü(ə)l, -(ˌ)t(y)ü(ə)l\ *noun* [French, from Latin *frustulum,* diminutive of *frustum*] (1857)
: the 2-valved siliceous shell of a diatom

frus·tum \'frəs-təm\ *noun, plural* **frustums** *or* **frus·ta** \-tə\ [New Latin, from Latin, piece, bit — more at BRUISE] (1658)

☆ SYNONYMS
Frustrate, thwart, foil, baffle, balk mean to check or defeat another's plan or block achievement of a goal. FRUSTRATE implies making vain or ineffectual all efforts however vigorous or persistent ⟨*frustrated* attempts at government reform⟩. THWART suggests frustration or checking by crossing or opposing ⟨the army *thwarted* his attempt at a coup⟩. FOIL implies checking or defeating so as to discourage further effort ⟨*foiled* by her parents, he stopped trying to see her⟩. BAFFLE implies frustration by confusing or puzzling ⟨*baffled* by the maze of rules and regulations⟩. BALK suggests the interposing of obstacles or hindrances ⟨officials felt that legal restrictions had *balked* their efforts to control crime⟩.

: the basal part of a solid cone or pyramid formed by cutting off the top by a plane parallel to the base; *also* **:** the part of a solid intersected between two usually parallel planes

fru·tes·cent \frü-ˈte-sᵊnt\ *adjective* [Latin *frutex* + English *-escent*] (1709)
: having or approaching the habit or appearance of a shrub **:** SHRUBBY

fru·ti·cose \ˈfrü-ti-ˌkōs\ *adjective* [Latin *fruticosus,* from *frutic-, frutex* shrub] (1882)
: having a shrubby often branched thallus that grows perpendicular to the substrate ⟨*fruticose* lichens⟩ — compare CRUSTOSE, FOLIOSE

¹fry \ˈfrī\ *verb* **fried; fry·ing** [Middle English *frien,* from Old French *frire,* from Latin *frigere* to roast; akin to Greek *phrygein* to roast, fry, Sanskrit *bhṛjjati* he roasts] (13th century)
transitive verb
: to cook in a pan or on a griddle over heat especially with the use of fat
intransitive verb
1 : to undergo frying
2 : to get very hot or burn as if being fried ⟨bodies *frying* on the beach⟩

²fry *noun, plural* **fries** (1833)
1 : a social gathering or picnic where food is fried and eaten ⟨a fish *fry*⟩
2 a : a dish of something fried **b :** FRENCH FRY — usually used in plural

³fry *noun, plural* **fry** [Middle English, probably from Old North French *fri,* from Old French *frier, froyer* to rub, spawn — more at FRAY] (14th century)
1 a : recently hatched or juvenile fishes **b :** the young of other animals
2 : very small adult fishes
3 : members of a group or class **:** INDIVIDUALS ⟨small *fry*⟩ ⟨a great part of the earth is peopled with these *fry* —Katherine Mansfield⟩

fry bread *noun* (1950)
: quick bread cooked (as by American Indians) by deep-frying

fry·er \ˈfrī(-ə)r\ *noun* (1851)
: something intended for or used in frying: as **a :** a young chicken; *especially* **:** one weighing 2½ to 4 pounds (1.1 to 1.8 kilograms) when dressed **b :** a deep utensil for frying foods

frying pan *noun* (14th century)
: a metal pan with a handle that is used for frying foods — called also *fry pan*
— out of the frying pan into the fire
: clear of one difficulty only to fall into a greater one

fry–up \ˈfrī-ˌəp\ *noun* (1967)
British **:** a dish or meal of fried food

f–stop \ˈef-ˌstäp\ *noun* (1946)
: a camera lens aperture setting indicated by an f-number

F_2 layer \ˈef-ˈtü-\ *noun* (1933)
: the upper of the two layers into which the F region of the ionosphere splits in the daytime at varying heights from about 120 miles (200 kilometers) to more than 300 miles (500 kilometers) above the earth

fubsy \ˈfəb-zē\ *adjective* [obsolete English *fubs* (chubby person)] (1780)
: chubby and somewhat squat

fuch·sia \ˈfyü-shə\ *noun* [New Latin, from Leonhard *Fuchs* (died 1566) German botanist] (1846)
1 : any of a genus (*Fuchsia*) of decorative shrubs of the evening-primrose family having showy nodding flowers usually in deep pinks, reds, and purples
2 : a vivid reddish purple

fuchsia 1

fuch·sin *or* **fuch·sine** \ˈfyük-sən, -ˌsēn\ *noun* [French *fuchsine,* probably from New Latin *Fuchsia;* from its color] (1865)
: a dye that is produced by oxidation of a mixture of aniline and toluidines and yields a brilliant bluish red

¹fuck \ˈfək\ *verb* [akin to Dutch *fokken* to breed (cattle), Swedish dialect *fokka* to copulate] (15th century)
intransitive verb
1 : COPULATE — usually considered obscene; sometimes used in the present participle as a meaningless intensive
2 : MESS 3 — used with *with;* usually considered vulgar
transitive verb
1 : to engage in coitus with — usually considered obscene; sometimes used interjectionally with an object (as a personal or reflexive pronoun) to express anger, contempt, or disgust
2 : to deal with unfairly or harshly **:** CHEAT, SCREW — usually considered vulgar

²fuck *noun* (1680)
1 : an act of copulation — usually considered obscene
2 : a sexual partner — usually considered obscene
3 a : DAMN 2 — usually considered vulgar **b** — used especially with *the* as a meaningless intensive; usually considered vulgar ⟨what the *fuck* do they want from me⟩

fucked–up \ˈfək-ˈtəp\ *adjective* (1966)
: thoroughly confused or disordered — usually considered vulgar

fuck·er \ˈfə-kər\ *noun* (1598)
: one that fucks — often used as a generalized term of abuse; usually considered vulgar

fuck off *intransitive verb* (1929)
: SCRAM — usually used as a command; usually considered vulgar

fuck up (1951)
intransitive verb
: to act foolishly or stupidly **:** BLUNDER — usually considered vulgar
transitive verb
: to ruin or spoil especially through stupidity or carelessness **:** BUNGLE — usually considered vulgar
— fuck·up \ˈfək-ˌəp\ *noun*

¹fu·coid \ˈfyü-ˌkȯid\ *adjective* [New Latin *Fucus,* from Latin] (1839)
: relating to or resembling the rockweeds

²fucoid *noun* (circa 1841)
: a fucoid seaweed or fossil

fu·cose \ˈfyü-ˌkōs, -ˌkōz\ *noun* [International Scientific Vocabulary *fuc-* (from Latin *fucus*) + *-ose*] (circa 1909)
: an aldose sugar that occurs in bound form in the dextrorotatory form in various glycosides and in the levorotatory form in some brown algae and in mammalian polysaccharides typical of some blood groups

fu·co·xan·thin \ˌfyü-kō-ˈzan-thən\ *noun* (1873)
: a brown carotenoid pigment $C_{40}H_{60}O_6$ occurring especially in the chloroplasts of brown algae

fu·cus \ˈfyü-kəs\ *noun* [Latin, seaweed, archil, dye obtained from archil, from Greek *phykos*] (1599)
1 *obsolete* **:** a face paint
2 [New Latin, genus name, from Latin] **:** any of a genus (*Fucus*) of cartilaginous brown algae used in the kelp industry; *broadly* **:** any of various brown algae

fud \ˈfəd\ *noun* (1913)
: FUDDY-DUDDY

fud·dle \ˈfə-dᵊl\ *verb* **fud·dled; fud·dling** \ˈfəd-liŋ, ˈfə-dᵊl-iŋ\ [origin unknown] (1588)
intransitive verb
: BOOZE, TIPPLE
transitive verb
1 : to make drunk **:** INTOXICATE
2 : to make confused **:** MUDDLE

fud·dy–dud·dy \ˈfə-dē-ˌdə-dē\ *noun, plural* **-dies** [origin unknown] (circa 1904)
: one that is old-fashioned, unimaginative, or conservative
— fuddy–duddy *adjective*

¹fudge \ˈfəj\ *verb* **fudged; fudg·ing** [origin unknown] (1674)
transitive verb
1 a : to devise as a substitute **:** FAKE **b :** FALSIFY ⟨*fudged* the figures⟩
2 : to fail to come to grips with **:** DODGE ⟨*fudged* the issue⟩
intransitive verb
1 : to exceed the proper bounds or limits of something ⟨feel that the author has *fudged* a little on the . . . rules for crime fiction —*Newsweek*⟩; *also* **:** CHEAT ⟨*fudging* on an exam⟩
2 : to fail to perform as expected
3 : to avoid commitment **:** HEDGE ⟨the government's tendency to *fudge* on delicate matters of policy —Claire Sterling⟩

²fudge *noun* (1766)
1 : foolish nonsense — often used interjectionally to express annoyance, disappointment, or disbelief
2 : a soft creamy candy made typically of sugar, milk, butter, and flavoring
3 : something that is fudged; *especially* **:** a bending of rules or a compromise

fudge factor *noun* (1962)
: an arbitrary mathematical term inserted into a calculation in order to arrive at an expected solution or to allow for errors especially of underestimation; *broadly* **:** any arbitrary unspecified factor

fu dog *often F capitalized, variant of* FOO DOG

¹fu·el \ˈfyü(-ə)l\ *noun, often attributive* [Middle English *fewel,* from Old French *fouaille,* from (assumed) Vulgar Latin *focalia,* from Latin *focus* hearth] (13th century)
1 a : a material used to produce heat or power by burning **b :** nutritive material **c :** a material from which atomic energy can be liberated especially in a reactor
2 : a source of sustenance or incentive **:** REINFORCEMENT

²fuel *verb* **-eled** *or* **-elled; -el·ing** *or* **-el·ling** (1592)
transitive verb
1 : to provide with fuel
2 : SUPPORT, STIMULATE ⟨movement is *fueled* by massive grants-in-aid —Allen Schick⟩
intransitive verb
: to take in fuel — often used with *up*

fuel cell *noun* (1922)
: a device that continuously changes the chemical energy of a fuel (as hydrogen) and an oxidant directly into electrical energy

fuel oil *noun* (1893)
: an oil that is used for fuel and that usually has a higher flash point than kerosene

fuel·wood \ˈfyü(-ə)l-ˌwu̇d\ *noun* (14th century)
: wood grown or used for fuel

¹fug \ˈfəg\ *noun* [perhaps alteration of ¹*fog*] (1888)
: the stuffy atmosphere of a poorly ventilated space; *also* **:** a stuffy or malodorous emanation
— fug·gy \ˈfə-gē\ *adjective*

²fug *verb* **fugged; fug·ging** (circa 1889)
intransitive verb
: to loll indoors in a stuffy atmosphere
transitive verb
: to make stuffy and odorous

fu·ga·cious \fyü-ˈgā-shəs\ *adjective* [Latin *fugac-, fugax,* from *fugere*] (1634)
1 : lasting a short time **:** EVANESCENT
2 : disappearing before the usual time — used chiefly of plant parts (as stipules) other than floral organs

fu·gac·i·ty \fyü-ˈga-sə-tē\ *noun* [*fugacious*] (circa 1929)
: the vapor pressure of a vapor assumed to be an ideal gas obtained by correcting the determined vapor pressure and useful as a measure of the escaping tendency of a substance from a heterogeneous system

fu·gal \ˈfyü-gəl\ *adjective* (1854)
: of, relating to, or being in the style of a musical fugue
— fu·gal·ly \-gə-lē\ *adverb*

-fuge *noun combining form* [French, from Late Latin *-fuga,* from Latin *fugare* to put to flight, from *fuga*]
: one that drives away ⟨febri*fuge*⟩

¹fu·gi·tive \ˈfyü-jə-tiv\ *adjective* [Middle English, from Middle French & Latin; Middle French *fugitif,* from Latin *fugitivus,* from *fugitus,* past participle of *fugere* to flee; akin to Greek *pheugein* to flee] (14th century)
1 : running away or intending flight ⟨*fugitive* slave⟩ ⟨*fugitive* debtor⟩
2 : moving from place to place **:** WANDERING
3 a : being of short duration **b :** difficult to grasp or retain **:** ELUSIVE **c :** likely to evaporate, deteriorate, change, fade, or disappear ⟨dyed with *fugitive* colors⟩
4 : being of transient interest ⟨*fugitive* essays⟩
synonym see TRANSIENT
— fu·gi·tive·ly *adverb*
— fu·gi·tive·ness *noun*

²fugitive *noun* (14th century)
1 : a person who flees or tries to escape; *especially* **:** REFUGEE
2 : something elusive or hard to find

fu·gle·man \ˈfyü-gəl-mən\ *noun* [modification of German *Flügelmann,* from *Flügel* wing + *Mann* man] (1804)
1 : a trained soldier formerly posted in front of a line of soldiers at drill to serve as a model in their exercises
2 : one at the head or forefront of a group or movement

fu·gu \ˈf(y)ü-(ˌ)gü\ *noun* [Japanese] (1909)
: any of various very poisonous puffers (sense 2a) that contain tetrodotoxin and that are used as food in Japan after the toxin-containing organs are removed

fugue \ˈfyüg\ *noun* [probably from Italian *fuga* flight, fugue, from Latin, flight, from *fugere*] (1597)
1 a : a musical composition in which one or two themes are repeated or imitated by successively entering voices and contrapuntally developed in a continuous interweaving of the voice parts **b :** something that resembles a fugue especially in interweaving repetitive elements
2 : a disturbed state of consciousness in which the one affected seems to perform acts in full awareness but upon recovery cannot recollect the deeds
— fugue *verb*
— fugu·ist \ˈfyü-gist\ *noun*

füh·rer *or* **fueh·rer** \ˈfyu̇r-ər, ˈfir-\ *noun* [German (*der*) *Führer,* literally, the leader (title assumed by Adolf Hitler), from Middle High German *vüerer,* from *vüeren* to lead, bear, from Old High German *fuoren* to lead; akin to Old English *faran* to go — more at FARE] (1934)
: LEADER 2; *especially* **:** TYRANT

fu·ji \ˈfü-(ˌ)jē\ *noun* [*Fuji,* mountain in Japan] (1925)
: a spun silk clothing fabric in plain weave originally made in Japan

¹-ful *adjective suffix, sometimes* **-ful·ler** *sometimes* **-ful·lest** [Middle English, from Old English, from *full,* adjective]
1 : full of ⟨pride*ful*⟩
2 : characterized by ⟨peace*ful*⟩
3 : having the qualities of ⟨master*ful*⟩
4 : tending, given, or liable to ⟨help*ful*⟩

²-ful *noun suffix*
: number or quantity that fills or would fill ⟨room*ful*⟩ ■

Fu·la *or* **Fu·lah** \ˈfü-lə\ *noun, plural* **Fula** *or* **Fulas** *or* **Fulah** *or* **Fulahs** (1832)
1 : a member of a mainly pastoral African people dispersed over savanna and desert from Senegal to eastern Sudan
2 : the language of the Fula people

Fu·la·ni \ˈfü-ˌlä-nē, fü-ˈ\ *noun, plural* **-ni** *or* **-nis** (1860)
1 : FULA 1; *especially* **:** the Fula of northern Nigeria and adjacent areas
2 : FULA 2

ful·crum \ˈfu̇l-krəm, ˈfəl-\ *noun, plural* **ful·crums** *or* **ful·cra** \-krə\ [Late Latin, from Latin, bedpost, from *fulcire* to prop — more at BALK] (1674)
1 a : PROP; *specifically* **:** the support about which a lever turns **b :** one that supplies capability for action
2 : a part of an animal that serves as a hinge or support

ful·fill *or* **ful·fil** \fu̇(l)-ˈfil *also* fə(l)-\ *transitive verb* **ful·filled; ful·fill·ing** [Middle English *fulfillen,* from Old English *fullfyllan,* from *full* + *fyllan* to fill] (before 12th century)
1 *archaic* **:** to make full **:** FILL ⟨her subtle, warm, and golden breath . . . *fulfills* him with beatitude —Alfred Tennyson⟩
2 a : to put into effect **:** EXECUTE **b :** to bring to an end **c :** to measure up to **:** SATISFY
3 a : to convert into reality **b :** to develop the full potentialities of
synonym see PERFORM
— ful·fill·er *noun*
— ful·fill·ment \-mənt\ *noun*

ful·gent \ˈfu̇l-jənt, ˈfəl-\ *adjective* [Middle English, from Latin *fulgent-, fulgens,* present participle of *fulgēre* to shine; akin to Latin *flagrare* to burn — more at BLACK] (15th century)
: dazzlingly bright **:** RADIANT
— ful·gent·ly *adverb*

ful·gu·rant \ˈfu̇l-g(y)ə-rənt, ˈfu̇l-jə-, ˈfəl-\ *adjective* (1647)
: flashing like lightning; *also* **:** BRILLIANT

ful·gu·ra·tion \ˌfu̇l-g(y)ə-ˈrā-shən, ˌfu̇l-jə-, fəl-\ *noun* [Latin *fulguration-, fulguratio* sheet lightning, from *fulgurare* to flash with lightning, from *fulgur* lightning, from *fulgēre*] (1633)
1 : the act or process of flashing like lightning
2 : ELECTRODESICCATION
— ful·gu·rate \ˈfu̇l-g(y)ə-ˌrāt, ˈfu̇l-jə-, ˈfəl-\ *transitive verb*

ful·gu·rite \ˈfu̇l-g(y)ə-ˌrīt, ˈfu̇l-jə-, ˈfəl-\ *noun* [International Scientific Vocabulary, from Latin *fulgur*] (1834)
: an often tubular vitrified crust produced by the fusion of sand or rock by lightning

ful·gu·rous \-rəs\ *adjective* [Latin *fulgur*] (1865)
: flashing with lightning

ful·ham \ˈfu̇-ləm\ *noun* [alteration of earlier *fullan,* perhaps from ¹*full* + ³*one*] (circa 1592)
archaic **:** a loaded die

fu·lig·i·nous \fyu̇-ˈli-jə-nəs\ *adjective* [Late Latin *fuliginosus,* from Latin *fuligin-, fuligo* soot; akin to Lithuanian *dūlis* cloud, vapor, and probably to Latin *fumus* smoke — more at FUME] (1621)
1 a : SOOTY **b :** OBSCURE, MURKY
2 : having a dark or dusky color
— fu·lig·i·nous·ly *adverb*

¹full \ˈfu̇l *also* ˈfəl\ *adjective* [Middle English, from Old English; akin to Old High German *fol* full, Latin *plenus* full, *plēre* to fill, Greek *plērēs* full, *plēthein* to be full] (before 12th century)
1 : containing as much or as many as is possible or normal ⟨a bin *full* of corn⟩
2 a : complete especially in detail, number, or duration ⟨a *full* report⟩ ⟨my *full* share⟩ ⟨gone a *full* hour⟩ **b :** lacking restraint, check, or qualification ⟨*full* retreat⟩ ⟨*full* support⟩ **c :** having all distinguishing characteristics **:** enjoying all authorized rights and privileges ⟨*full* member⟩ ⟨*full* professor⟩ **d :** not lacking in any essential **:** PERFECT ⟨in *full* control of your senses⟩
3 a : being at the highest or greatest degree **:** MAXIMUM ⟨*full* speed⟩ ⟨*full* strength⟩ **b :** being at the height of development ⟨*full* bloom⟩
4 : rounded in outline ⟨a *full* figure⟩
5 a : possessing or containing a great number or amount — used with *of* ⟨a room *full* of pictures⟩ ⟨*full* of hope⟩ **b :** having an abundance of material especially in the form of gathered, pleated, or flared parts ⟨a *full* skirt⟩ **c :** rich in experience ⟨a *full* life⟩
6 a : satisfied especially with food or drink **b :** large enough to satisfy ⟨a *full* meal⟩
7 *archaic* **:** completely weary
8 : having both parents in common ⟨*full* sisters⟩
9 : having volume or depth of sound ⟨*full* tones⟩
10 : completely occupied especially with a thought or plan ⟨*full* of their own concerns⟩
11 : possessing a rich or pronounced quality ⟨a food of *full* flavor⟩ ☆
— full·ness *also* **ful·ness** \ˈfu̇l-nəs\ *noun*
— full of it : not to be believed

²full *adverb* (before 12th century)
1 a : VERY, EXTREMELY ⟨knew *full* well they had lied to me⟩ **b :** ENTIRELY ⟨swung *full* around —Morley Callaghan⟩
2 : STRAIGHT, SQUARELY ⟨got hit *full* in the face⟩
3 — used as an intensive ⟨wound up winning by a *full* four strokes —William Johnson⟩

³full *noun* (14th century)
1 a : the highest or fullest state or degree ⟨the *full* of the moon⟩ **b :** the utmost extent ⟨enjoy to the *full*⟩
2 : the requisite or complete amount ⟨paid in *full*⟩

⁴full (1794)
intransitive verb
of the moon **:** to become full
transitive verb
: to make full in sewing

⁵full *transitive verb* [Middle English, from Middle French *fouler* to trample under foot, from Medieval Latin *fullare* to walk, trample, full, from Latin *fullo* fuller] (14th century)
: to shrink and thicken (woolen cloth) by moistening, heating, and pressing

full·back \ˈfu̇l-ˌbak\ *noun* (1887)
1 : an offensive football back used primarily for line plunges and blocking

☆ SYNONYMS
Full, complete, plenary, replete mean containing all that is wanted or needed or possible. FULL implies the presence or inclusion of everything that is wanted or required by something or that can be held, contained, or attained by it ⟨a *full* schedule⟩. COMPLETE applies when all that is needed is present ⟨a *complete* picture of the situation⟩. PLENARY adds to COMPLETE the implication of fullness without qualification ⟨given *plenary* power⟩. REPLETE implies being filled to the brim or to satiety ⟨*replete* with delightful details⟩.

□ USAGE
-ful Nouns ending in *-ful,* like *cupful* and *spoonful,* regularly form the plural by adding *-s* at the end: *cupfuls, spoonfuls.* But these words have long been a puzzle to the public. Part of the problem stems from the fact that most were originally two words with the noun pluralized—many people will recall the nursery rhyme and its "yes sir, yes sir, three bags full." Another factor is that many people remember having been taught that internal pluralization was correct, although we have not yet found a schoolbook giving such a prescription. The result of the continuing uncertainty is the existence of less frequent variants such as *cupsful* and *teaspoonsful.* These variants are not wrong, but most people use *cupfuls* and *teaspoonfuls.*

2 : a primarily defensive player usually stationed nearest the defended goal (as in soccer or field hockey)

full blast *adverb* (1909)
: at full capacity **:** with great intensity

¹full–blood \'fu̇l-ˌbləd\ *adjective* (1812)
: FULL-BLOODED 1

²full–blood *noun* (1846)
: a full-blooded individual

full–blood•ed \'fu̇l-ˌblə-dəd\ *adjective* (1774)
1 : of unmixed ancestry **:** PUREBRED
2 : FLORID, RUDDY ⟨of *full-blooded* face⟩
3 : FORCEFUL ⟨*full-blooded* prose style⟩
4 a : lacking no particulars **:** GENUINE **b :** containing fullness of substance **:** RICH
— full–blood•ed•ness *noun*

full–blown \-'blōn\ *adjective* (1601)
1 a : fully mature **b :** being at the height of bloom **c :** FULL-FLEDGED
2 : possessing all the usual or necessary features ⟨a general philosophy, if not a *full-blown* ideology, is emerging —W. H. Jones⟩

full–bod•ied \-ˌbä-dēd\ *adjective* (1686)
1 : having a large body
2 *of a beverage* **:** imparting to the palate the general impression of substantial weight and rich texture
3 : having importance, significance, or meaningfulness ⟨*full-bodied* study of literature⟩

full circle *adverb* (1879)
: through a series of developments that lead back to the original source, position, or situation or to a complete reversal of the original position — usually used in the phrase *come full circle*

full–dress *adjective* (1761)
: involving attention to every detail in preparation or execution ⟨a *full-dress* rehearsal⟩ ⟨a *full-dress* investigation⟩

full dress *noun* (1790)
: the style of dress prescribed for ceremonial or formal social occasions

¹full•er \'fu̇-lər\ *noun* [Middle English, from Old English *fullere,* from Latin *fullo*] (before 12th century)
: one that fulls cloth

²ful•ler \'fu̇-lər\ *noun* [*fuller* (to form a groove in)] (circa 1864)
: a blacksmithing hammer for grooving and spreading iron

ful•ler•ene \ˌfu̇-lə-'rēn\ *noun* [R. Buckminster *Fuller;* from the resemblance of the molecules to the geodesic domes designed by Fuller] (1988)
: any of a class of closed hollow aromatic carbon compounds that are made up of twelve pentagonal and differing numbers of hexagonal faces

fuller's earth *noun* [¹*fuller;* from its earlier use as fulling agent] (15th century)
: an earthy substance that consists chiefly of clay mineral but lacks plasticity and that is used as an adsorbent, a filter medium, and a carrier for catalysts

ful•ler's teasel *noun* (15th century)
: TEASEL 1a

full–fash•ioned \'fu̇l-'fa-shənd\ *adjective* (1883)
: employing or produced by a knitting process for shaping to conform to body lines ⟨*full-fashioned* hosiery⟩

full–fledged \-'flejd\ *adjective* (1883)
1 : fully developed **:** TOTAL, COMPLETE ⟨a *full-fledged* war⟩
2 : having attained complete status ⟨*full-fledged* lawyer⟩

full house *noun* (1887)
: a poker hand containing three of a kind and a pair — see POKER illustration

full–length \'fu̇l-ˌleŋ(k)th\ *adjective* (1760)
1 : showing or adapted to the entire length especially of the human figure ⟨a *full-length* mirror⟩ ⟨a *full-length* dress⟩
2 : having a length as great as that which is normal or standard for an object of its kind ⟨a *full-length* play⟩

full marks *noun plural* (1916)
chiefly British **:** due credit or commendation

full moon *noun* (before 12th century)
: the moon with its whole apparent disk illuminated

full•mouthed \'fu̇l-'mau̇t͟hd, -'mau̇tht\ *adjective* (1577)
1 : having a full complement of teeth ⟨*full-mouthed* ewes⟩
2 : uttered loudly

full nelson *noun* (circa 1922)
: a wrestling hold gained from behind an opponent by thrusting the arms under the opponent's arms and clasping the hands behind the opponent's head — compare HALF NELSON

full–out \-ˌau̇t\ *adjective* (14th century)
: COMPLETE, TOTAL

full–scale \-ˌskā(ə)l\ *adjective* (1933)
1 : identical to an original in proportion and size ⟨*full-scale* drawing⟩
2 a : involving full use of available resources ⟨a *full-scale* biography⟩ ⟨*full-scale* war⟩ **b :** TOTAL, COMPLETE ⟨a *full-scale* musical renaissance —*Current Biography*⟩

full–ser•vice \-'sər-vəs\ *adjective* (1957)
: providing comprehensive service of a particular kind ⟨a *full-service* bank⟩

full–size \-ˌsīz\ *adjective* (1832)
1 : having the usual or normal size of its kind
2 : having the dimensions 54 inches by 75 inches (about 1.4 by 1.9 meters) — used of a bed; compare KING-SIZE, QUEEN-SIZE, TWIN-SIZE

full stop *noun* (1596)
: PERIOD 5a

full tilt *adverb* [³*tilt*] (1600)
: at high speed

full–time *adjective* (1898)
1 : employed for or involving full time ⟨*full-time* employees⟩
2 : devoting one's full attention and energies to something ⟨a *full-time* gambler⟩
— full–time *adverb*

full time *noun* (1898)
: the amount of time considered the normal or standard amount for working during a given period

full–tim•er \'fu̇l-'tī-mər\ *noun* (1864)
: a person who works full-time

ful•ly \'fu̇(l)-lē\ *adverb* (before 12th century)
1 : in a full manner or degree **:** COMPLETELY
2 : at least ⟨*fully* nine tenths of us⟩
usage see PLENTY

ful•mar \'fu̇l-mər, -ˌmär\ *noun* [of Scandinavian origin; akin to Old Norse *fūlmār* fulmar, from *fūll* foul + *mār* gull — more at MEW] (1698)
: a seabird (*Fulmarus glacialis*) of colder northern seas closely related to the petrels; *also* **:** a related bird (*F. glacialoides*) of circumpolar distribution in colder southern seas

ful•mi•nant \'fu̇l-mə-nənt, 'fəl-\ *adjective* (1602)
: FULMINATING 3

¹ful•mi•nate \-ˌnāt\ *verb* **-nat•ed; -nat•ing** [Middle English, from Medieval Latin *fulminatus,* past participle of *fulminare,* from Latin, to strike (of lightning), from *fulmin-, fulmen* lightning; akin to Latin *flagrare* to burn — more at BLACK] (15th century)
transitive verb
: to utter or send out with denunciation
intransitive verb
: to send forth censures or invectives
— ful•mi•na•tion \ˌfu̇l-mə-'nā-shən, ˌfəl-\ *noun*

²fulminate *noun* [*fulminic acid,* from Latin *fulmin-, fulmen*] (1826)
: an often explosive salt (as mercury fulminate) containing the group −CNO

ful•mi•nat•ing *adjective* (1626)
1 : hurling denunciations or menaces
2 : EXPLOSIVE
3 : coming on suddenly with great severity

ful•mine \'fu̇l-mən, 'fəl-\ *verb* (1590)
archaic **:** FULMINATE

ful•some \'fu̇l-səm\ *adjective* [Middle English *fulsom* copious, cloying, from *full* + *-som* -some] (13th century)
1 a : characterized by abundance **:** COPIOUS ⟨describes in *fulsome* detail —G. N. Shuster⟩ ⟨*fulsome* bird life. The feeder overcrowded —Maxine Kumin⟩ **b :** generous in amount, extent, or spirit ⟨the passengers were *fulsome* in praise of the plane's crew —Don Oliver⟩ ⟨a *fulsome* victory for the far left —Bruce Rothwell⟩ ⟨the greetings have been *fulsome*, the farewells tender —Simon Gray⟩ **c :** being full and well developed ⟨she was in generally *fulsome*, limpid voice —Thor Eckert, Jr.⟩
2 : aesthetically, morally, or generally offensive ⟨*fulsome* lies and nauseous flattery —William Congreve⟩ ⟨the devil take thee for a . . . *fulsome* rogue —George Villiers⟩
3 : exceeding the bounds of good taste **:** OVERDONE ⟨the *fulsome* chromium glitter of the escalators dominating the central hall —Lewis Mumford⟩
4 : excessively complimentary or flattering **:** EFFUSIVE ⟨an admiration whose extent I did not express, lest I be thought *fulsome* —A. J. Liebling⟩ ■
— ful•some•ly *adverb*
— ful•some•ness *noun*

ful•vous \'fu̇l-vəs, 'fəl-\ *adjective* [Latin *fulvus;* perhaps akin to Latin *flavus* yellow — more at BLUE] (1664)
: of a dull brownish yellow **:** TAWNY

Fu Man•chu mustache \ˌfü-(ˌ)man-'chü-\ *noun* [*Fu Manchu,* Chinese villain in stories by "Sax Rohmer" (A. S. Ward) (died 1955)] (1968)
: a long mustache with ends that turn down to the chin

fu•ma•rase \'fyü-mə-ˌrās, -ˌrāz\ *noun* (1936)
: an enzyme that catalyzes the interconversion (as in the Krebs cycle) of fumaric acid and malic acid or their salts

fu•ma•rate \-ˌrāt\ *noun* (1864)
: a salt or ester of fumaric acid

fu•mar•ic acid \fyü-'mar-ik-\ *noun* [International Scientific Vocabulary, from New Latin *Fumaria,* genus of herbs, from Late Latin, fumitory, from Latin *fumus*] (circa 1864)
: a crystalline acid $C_4H_4O_4$ found in various plants or made synthetically and used especially in making resins

fu•ma•role \'fyü-mə-ˌrōl\ *noun* [Italian *fumarola,* from Italian dialect (Neapolitan), from Late Latin *fumariolum* vent, from Latin *fumarium* smoke chamber for aging wine, from *fumus*] (1811)
: a hole in a volcanic region from which hot gases and vapors issue
— fu•ma•rol•ic \ˌfyü-mə-'rō-lik\ *adjective*

¹fum•ble \'fəm-bəl\ *verb* **fum•bled; fum•bling** \-b(ə-)liŋ\ [probably of Scandinavian origin; akin to Swedish *fumla* to fumble] (1534)
intransitive verb

□ USAGE

fulsome The senses shown above are the chief living senses of *fulsome.* Sense 2, which was a generalized term of disparagement in the late 17th century, is the least common of these. *Fulsome* became a point of dispute when sense 1, thought to be obsolete in the 19th century, began to be revived in the 20th. The dispute was exacerbated by the fact that the large dictionaries of the first half of the century missed the beginnings of the revival. Sense 1 has not only been revived but has spread in its application and continues to do so. The chief danger for the user of *fulsome* is ambiguity. Unless the context is made very clear, the reader or hearer can't be sure whether such an expression as "fulsome praise" or "a fulsome tribute" is meant in sense 1b or in sense 4.

1 a : to grope for or handle something clumsily or aimlessly **b :** to make awkward attempts to do or find something ⟨*fumbled* in his pocket for a coin⟩ **c :** to search by trial and error **d :** BLUNDER
2 : to feel one's way or move awkwardly
3 a : to drop or juggle or fail to play cleanly a grounder **b :** to lose hold of a football while handling or running with it
transitive verb
1 : to bring about by clumsy manipulation
2 a : to feel or handle clumsily **b :** to deal with in a blundering way **:** BUNGLE
3 : to make (one's way) in a clumsy manner
4 a : MISPLAY ⟨*fumble* a grounder⟩ **b :** to lose hold of (a football) while handling or running
— **fum·bler** \-b(ə-)lər\ *noun*
— **fum·bling·ly** \-b(ə-)liŋ-lē\ *adverb*

[2]**fumble** *noun* (1634)
1 : an act or instance of fumbling
2 : a fumbled ball

[1]**fume** \'fyüm\ *noun* [Middle English, from Middle French *fum,* from Latin *fumus;* akin to Old High German *toumen* to be fragrant, Sanskrit *dhūma* smoke, Old Church Slavonic *dymŭ*] (14th century)
1 a : a smoke, vapor, or gas especially when irritating or offensive ⟨engine exhaust *fumes*⟩ **b :** an often noxious suspension of particles in a gas (as air)
2 : something (as an emotion) that impairs one's reasoning ⟨sometimes his head gets a little hot with the *fumes* of patriotism —Matthew Arnold⟩
3 : a state of excited irritation or anger — usually used in the phrase *in a fume*
— **fumy** \'fyü-mē\ *adjective*

[2]**fume** *verb* **fumed; fum·ing** (14th century)
transitive verb
1 : to expose to or treat with fumes
2 : to give off in fumes ⟨*fuming* thick black smoke⟩
3 : to utter while in a state of excited irritation or anger
intransitive verb
1 a : to emit fumes **b :** to be in a state of excited irritation or anger ⟨fretted and *fumed* over the delay⟩
2 : to rise in or as if in fumes

fu·met \fyü-'mā, 'fyü-mət\ *noun* [French, literally, pleasant aroma (of meat cooking), from Middle French, from *fumer* to give off smoke or steam, from Latin *fumare,* from *fumus*] (1906)
: a reduced and seasoned fish, meat, or vegetable stock

fu·mi·gant \'fyü-mi-gənt\ *noun* (1890)
: a substance used in fumigating

fu·mi·gate \'fyü-mə-ˌgāt\ *transitive verb* **-gat·ed; -gat·ing** [Latin *fumigatus,* past participle of *fumigare,* from *fumus* + *-igare* (akin to Latin *agere* to drive) — more at AGENT] (1781)
: to apply smoke, vapor, or gas to especially for the purpose of disinfecting or of destroying pests
— **fu·mi·ga·tion** \ˌfyü-mə-'gā-shən\ *noun*
— **fu·mi·ga·tor** \'fyü-mə-ˌgā-tər\ *noun*

fu·mi·to·ry \'fyü-mə-ˌtōr-ē, -ˌtȯr-\ *noun* [Middle English *fumeterre,* from Middle French, from Medieval Latin *fumus terrae,* literally, smoke of the earth, from Latin *fumus* + *terrae,* genitive of *terra* earth — more at TERRACE] (14th century)
: any of a genus (*Fumaria* of the family Fumariaceae, the fumitory family) of erect or climbing herbs; *especially* **:** a common European herb (*F. officinalis*)

[1]**fun** \'fən\ *noun* [English dialect *fun* to hoax, perhaps alteration of Middle English *fonnen,* from *fonne* dupe] (1727)
1 : what provides amusement or enjoyment; *specifically* **:** playful often boisterous action or speech ⟨full of *fun*⟩
2 : a mood for finding or making amusement ⟨all in *fun*⟩
3 a : AMUSEMENT, ENJOYMENT ⟨sickness takes all the *fun* out of life⟩ **b :** derisive jest **:** SPORT, RIDICULE ⟨a figure of *fun*⟩
4 : violent or excited activity or argument ⟨let a snake loose in the classroom; then the *fun* began⟩ ☆

[2]**fun** *intransitive verb* **funned; fun·ning** (1833)
: to indulge in banter or play **:** JOKE

[3]**fun** *adjective, sometimes* **fun·ner;** *sometimes* **fun·nest** (circa 1846)
1 : providing entertainment, amusement, or enjoyment ⟨a *fun* party⟩ ⟨a *fun* person to be with⟩
2 : full of fun **:** PLEASANT ⟨a *fun* night⟩ ⟨have a *fun* time⟩

fu·nam·bu·lism \fyu̇-'nam-byə-ˌli-zəm\ *noun* [Latin *funambulus* ropewalker, from *funis* rope + *ambulare* to walk] (1824)
1 : tightrope walking
2 : a show especially of mental agility
— **fu·nam·bu·list** \-list\ *noun*

fun and games *noun plural but singular or plural in construction* (1920)
: light amusement

[1]**func·tion** \'fəŋ(k)-shən\ *noun* [Latin *function-, functio* performance, from *fungi* to perform; probably akin to Sanskrit *bhuṅkte* he enjoys] (1533)
1 : professional or official position **:** OCCUPATION
2 : the action for which a person or thing is specially fitted or used or for which a thing exists **:** PURPOSE
3 : any of a group of related actions contributing to a larger action; *especially* **:** the normal and specific contribution of a bodily part to the economy of a living organism
4 : an official or formal ceremony or social gathering
5 a : a mathematical correspondence that assigns exactly one element of one set to each element of the same or another set **b :** a variable (as a quality, trait, or measurement) that depends on and varies with another ⟨height is a *function* of age⟩; *also* **:** RESULT ⟨illnesses that are a *function* of stress⟩
6 : characteristic behavior of a chemical compound due to a particular reactive unit; *also* **:** FUNCTIONAL GROUP
7 : a computer subroutine; *specifically* **:** one that performs a calculation with variables provided by a program and supplies the program with a single result ☆
— **func·tion·less** \-ləs\ *adjective*

[2]**function** *intransitive verb* **func·tioned; func·tion·ing** \-sh(ə-)niŋ\ (1856)
1 : to have a function **:** SERVE ⟨an attributive noun *functions* as an adjective⟩
2 : to carry on a function or be in action **:** OPERATE ⟨a government *functions* through numerous divisions⟩

func·tion·al \'fəŋ(k)-shnəl, -shə-nᵊl\ *adjective* (1631)
1 a : of, connected with, or being a function **b :** affecting physiological or psychological functions but not organic structure ⟨*functional* heart disease⟩
2 : used to contribute to the development or maintenance of a larger whole ⟨*functional* and practical school courses⟩; *also* **:** designed or developed chiefly from the point of view of use ⟨*functional* clothing⟩
3 : performing or able to perform a regular function
— **func·tion·al·ity** \ˌfəŋ(k)-shə-'na-lə-tē\ *noun*
— **func·tion·al·ly** \'fəŋ(k)-shnə-lē, -shə-nᵊl-ē\ *adverb*

functional calculus *noun* (1933)
: PREDICATE CALCULUS

functional group *noun* (1943)
: a characteristic reactive unit of a chemical compound especially in organic chemistry

functional illiterate *noun* (1946)
: a person having had some schooling but not meeting a minimum standard of literacy
— **functionally illiterate** *adjective*

func·tion·al·ism \'fəŋ(k)-shnə-ˌli-zəm, -shə-nᵊl-ˌi-\ *noun* (1914)
1 : a philosophy of design (as in architecture) holding that form should be adapted to use, material, and structure
2 : a theory that stresses the interdependence of the patterns and institutions of a society and their interaction in maintaining cultural and social unity
3 : a doctrine or practice that emphasizes practical utility or functional relations
— **func·tion·al·ist** \-shnə-list, -shə-nᵊl-ist\ *noun*
— **functionalist** *or* **func·tion·al·is·tic** \ˌfəŋ(k)-shnə-'lis-tik, -shə-nᵊl-'is-\ *adjective*

functional shift *noun* (1942)
: the process by which a word or form comes to be used in another grammatical function

func·tion·ary \'fəŋ(k)-shə-ˌner-ē\ *noun, plural* **-ar·ies** (1791)
1 : one who serves in a certain function
2 : one holding office in a government or political party

function word *noun* (1940)
: a word (as a preposition, auxiliary verb, or conjunction) expressing primarily grammatical relationship

func·tor \'fəŋ(k)-tər\ *noun* (1935)
: something that performs a function or an operation

[1]**fund** \'fənd\ *noun* [Latin *fundus* bottom, country estate — more at BOTTOM] (1694)

☆ SYNONYMS
Fun, jest, sport, game, play mean action or speech that provides amusement or arouses laughter. FUN usually implies laughter or gaiety but may imply merely a lack of serious or ulterior purpose ⟨played cards just for *fun*⟩. JEST implies lack of earnestness in what is said or done and may suggest a hoaxing or teasing ⟨hurt by remarks said only in *jest*⟩. SPORT applies especially to the arousing of laughter against someone ⟨teasing begun in *sport* led to anger⟩. GAME is close to SPORT, and often stresses mischievous or malicious fun ⟨made *game* of their poor relations⟩. PLAY stresses the opposition to *earnest* without implying any element of malice or mischief ⟨pretended to strangle his brother in *play*⟩.

Function, office, duty, province mean the acts or operations expected of a person or thing. FUNCTION implies a definite end or purpose that the one in question serves or a particular kind of work it is intended to perform ⟨the *function* of language is two-fold: to communicate emotion and to give information —Aldous Huxley⟩. OFFICE is typically applied to the function or service expected of a person by reason of his trade or profession or his special relationship to others ⟨they exercise the *offices* of the judge, the priest, the counsellor —W. E. Gladstone⟩. DUTY applies to a task or responsibility imposed by one's occupation, rank, status, or calling ⟨it is the judicial *duty* of the court, to examine the whole case —R. B. Taney⟩. PROVINCE applies to a function, office, or duty that naturally or logically falls to one ⟨nursing does not belong to a man; it is not his *province* —Jane Austen⟩.

1 a : a sum of money or other resources whose principal or interest is set apart for a specific objective **b :** money on deposit on which checks or drafts can be drawn — usually used in plural **c :** CAPITAL **d** *plural* **:** the stock of the British national debt — usually used with *the*
2 : an available quantity of material or intangible resources **:** SUPPLY
3 *plural* **:** available pecuniary resources
4 : an organization administering a special fund

²fund *transitive verb* (1789)
1 a : to make provision of resources for discharging the interest or principal of **b :** to provide funds for ⟨a science program federally *funded*⟩
2 : to place in a fund **:** ACCUMULATE
3 : to convert into a debt that is payable either at a distant date or at no definite date and that bears a fixed interest ⟨*fund* a floating debt⟩

fun·da·ment \'fən-də-mənt\ *noun* [Middle English, from Old French *fondement,* from Latin *fundamentum,* from *fundare* to found, from *fundus*] (13th century)
1 : an underlying ground, theory, or principle
2 a : BUTTOCKS **b :** ANUS
3 : the part of a land surface that has not been altered by human activities

¹fun·da·men·tal \ˌfən-də-'men-tᵊl\ *adjective* (15th century)
1 a : serving as an original or generating source **:** PRIMARY ⟨a discovery *fundamental* to modern computers⟩ **b :** serving as a basis supporting existence or determining essential structure or function **:** BASIC
2 a : of or relating to essential structure, function, or facts **:** RADICAL ⟨*fundamental* change⟩; *also* **:** of or dealing with general principles rather than practical application ⟨*fundamental* science⟩ **b :** adhering to fundamentalism
3 : of, relating to, or produced by the lowest component of a complex vibration
4 : of central importance **:** PRINCIPAL ⟨*fundamental* purpose⟩
5 : belonging to one's innate or ingrained characteristics **:** DEEP-ROOTED ⟨her *fundamental* good humor⟩
synonym see ESSENTIAL
— fun·da·men·tal·ly \-tᵊl-ē\ *adverb*

²fundamental *noun* (1637)
1 : something fundamental; *especially* **:** one of the minimum constituents without which a thing or a system would not be what it is
2 a : the principal musical tone produced by vibration (as of a string or column of air) on which a series of higher harmonics is based **b :** the root of a chord
3 : the harmonic component of a complex wave that has the lowest frequency and commonly the greatest amplitude

fundamental group *noun* (1957)
: a set that is a subset of all paths defined on a set of points each pair of which is joined by a path and that is the quotient group of the group of all paths beginning and ending with a given point

fun·da·men·tal·ism \-tᵊl-ˌi-zəm\ *noun* (1922)
1 a *often capitalized* **:** a movement in 20th century Protestantism emphasizing the literally interpreted Bible as fundamental to Christian life and teaching **b :** the beliefs of this movement **c :** adherence to such beliefs
2 : a movement or attitude stressing strict and literal adherence to a set of basic principles
— fun·da·men·tal·ist \-tᵊl-ist\ *noun*
— fundamentalist *or* **fun·da·men·tal·is·tic** \-ˌmen-tᵊl-'is-tik\ *adjective*

fundamental law *noun* (circa 1914)
: the organic or basic law of a political unit as distinguished from legislative acts; *specifically* **:** CONSTITUTION

fundamental particle *noun* (1947)
: ELEMENTARY PARTICLE

fun·dic \'fən-dik\ *adjective* (circa 1927)
: of or relating to a fundus

fund–rais·er \'fən-ˌdrā-zər\ *noun* (1957)
1 : a person employed to raise funds
2 : a social event (as a cocktail party) held for the purpose of raising funds

fund–rais·ing \-ziŋ\ *noun, often attributive* (1940)
: the organized activity of raising funds (as for an institution or political cause)

fun·dus \'fən-dəs\ *noun, plural* **fun·di** \-ˌdī, -ˌdē\ [New Latin, from Latin, bottom] (1764)
: the bottom of or part opposite the aperture of the internal surface of a hollow organ: as **a :** the greater curvature of the stomach **b :** the lower back part of the bladder **c :** the large upper end of the uterus **d :** the part of the eye opposite the pupil

¹fu·ner·al \'fyün-rəl, 'fyü-nə-\ *adjective* [Middle English, from Late Latin *funeralis,* from Latin *funer-, funus* funeral (noun)] (14th century)
1 : of, relating to, or constituting a funeral
2 : FUNEREAL 2

²funeral *noun* [Middle English *funerelles* (plural), from Middle French *funerailles* (plural), from Medieval Latin *funeralia* (plural), from Late Latin, neuter plural of *funeralis,* adjective] (circa 1512)
1 : the observances held for a dead person usually before burial or cremation
2 *chiefly dialect* **:** a funeral sermon
3 : a funeral procession
4 : an end of something's existence
5 : a matter of concern to one **:** WORRY ⟨if you flunk, that's your *funeral*⟩

funeral director *noun* (1886)
: one whose profession is the management of funerals and who is usually an embalmer

funeral home *noun* (1926)
: an establishment with facilities for the preparation of the dead for burial or cremation, for the viewing of the body, and for funerals — called also *funeral parlor*

fu·ner·ary \'fyü-nə-ˌrer-ē\ *adjective* (circa 1693)
: of, used for, or associated with burial ⟨a pharaoh's *funerary* chamber⟩

fu·ne·re·al \fyu̇-'nir-ē-əl\ *adjective* [Latin *funereus,* from *funer-, funus*] (1725)
1 : of or relating to a funeral
2 : befitting or suggesting a funeral (as in solemnity)
— fu·ne·re·al·ly \-ə-lē\ *adverb*

fun·fair \'fən-ˌfar, -ˌfer\ *noun* (1925)
chiefly British **:** AMUSEMENT PARK

fun·gal \'fəŋ-gəl\ *adjective* (1835)
1 : of, relating to, or having the characteristics of fungi
2 : caused by a fungus

fungi- *combining form* [Latin *fungus*]
: fungus ⟨*fungi*form⟩

¹fun·gi·ble \'fən-jə-bəl\ *noun* (circa 1765)
: something that is fungible — usually used in plural

²fungible *adjective* [New Latin *fungibilis,* from Latin *fungi* to perform — more at FUNCTION] (1818)
1 : being of such a nature that one part or quantity may be replaced by another equal part or quantity in the satisfaction of an obligation ⟨oil, wheat, and lumber are *fungible* commodities⟩
2 : INTERCHANGEABLE
— fun·gi·bil·i·ty \ˌfən-jə-'bi-lə-tē\ *noun*

fun·gi·cid·al \ˌfən-jə-'sī-dᵊl, ˌfəŋ-gə-\ *adjective* (1905)
: ANTIFUNGAL
— fun·gi·cid·al·ly \-dᵊl-ē\ *adverb*

fun·gi·cide \'fən-jə-ˌsīd, 'fəŋ-gə-\ *noun* [International Scientific Vocabulary] (1889)
: an agent that destroys fungi or inhibits their growth

fun·gi·form \'fən-jə-ˌfȯrm, 'fəŋ-gə-\ *adjective* (1823)
: shaped like a mushroom

fun·gi·stat·ic \ˌfən-jə-'sta-tik\ *adjective* (1922)
: inhibiting the growth of fungi without destroying them

fun·go \'fəŋ-(ˌ)gō\ *noun, plural* **fungoes** [origin unknown] (circa 1867)
1 : a fly ball hit especially for practice fielding by a player who tosses a ball in the air and hits it as it comes down
2 : FUNGO BAT

fungo bat *noun* (1926)
: a long thin bat used for hitting fungoes

fun·goid \'fəŋ-ˌgȯid\ *adjective* (circa 1836)
: resembling, characteristic of, caused by, or being a fungus ⟨a *fungoid* growth⟩
— fungoid *noun*

fun·gous \'fəŋ-gəs\ *adjective* (15th century)
: FUNGAL

fun·gus \'fəŋ-gəs\ *noun, plural* **fun·gi** \'fən-ˌjī, 'fəŋ-ˌgī\ *also* **fun·gus·es** \'fəŋ-gə-səz\ *often attributive* [Latin] (1527)
: any of a major group (Fungi) of saprophytic and parasitic spore-producing organisms usually classified as plants that lack chlorophyll and include molds, rusts, mildews, smuts, mushrooms, and yeasts

fun house *noun* (1948)
: a building in an amusement park that contains various devices designed to startle or amuse

¹fu·nic·u·lar \fyu̇-'ni-kyə-lər, fə-\ *adjective* [Latin *funiculus*] (1664)
1 : having the form of or associated with a cord usually under tension
2 [New Latin *funiculus*] **:** of, relating to, or being a funiculus
3 : of, relating to, or being a funicular ⟨a *funicular* system⟩

²funicular *noun* (1911)
: a cable railway ascending a mountain; *especially* **:** one in which an ascending car counterbalances a descending car

fu·nic·u·lus \-ləs\ *noun, plural* **-li** \-ˌlī, -ˌlē\ [New Latin, from Latin, diminutive of *funis* rope] (1826)
1 : a bodily structure suggesting a cord; *especially* **:** a bundle of nerve fibers
2 : the stalk of a plant ovule

¹funk \'fəŋk\ (circa 1739)
intransitive verb
: to become frightened and shrink back
transitive verb
1 : to be afraid of **:** DREAD
2 : to shrink from undertaking or facing

²funk *noun* [probably from obsolete Flemish *fonck*] (1743)
1 a : a state of paralyzing fear **b :** a depressed state of mind
2 : one that funks **:** COWARD

³funk *noun* [back-formation from ²*funky*] (1959)
1 : music that combines traditional forms of black music (as blues, gospel, or soul) and is characterized by a strong backbeat
2 : the quality or state of being funky ⟨jeans . . . have lost much of their *funk* —Tom Wolfe⟩

funk hole *noun* (1900)
1 : DUGOUT 2
2 : a place of safe retreat

fun·kia \'fəŋ-kē-ə, 'fu̇ŋ-\ *noun* [New Latin, genus name, from C. H. Funck (died 1839) German botanist] (1839)
: PLANTAIN LILY

¹funky \'fəŋ-kē\ *adjective* (1845)
: being in a funk **:** PANICKY

²funky *adjective* **funk·i·er; -est** [*funk* (offensive odor)] (circa 1899)
1 : having an offensive odor **:** FOUL
2 : having an earthy unsophisticated style and feeling; *especially* **:** having the style and feeling of older black American music (as blues or gospel) or of funk ⟨a slick, heavy beat that is unmistakably contemporary and irresistibly *funky* —Jay Cocks⟩

3 a : odd or quaint in appearance or feeling **b :** lacking style or taste **c :** unconventionally stylish **:** HIP ◆

— **funk·i·ness** *noun*

¹fun·nel \ˈfə-nᵊl\ *noun* [Middle English *fonel,* from Old Provençal *fonilh,* from Medieval Latin *fundibulum,* short for Latin *infundibulum,* from *infundere* to pour in, from *in-* + *fundere* to pour — more at FOUND] (15th century)

1 a : a utensil that is usually a hollow cone with a tube extending from the smaller end and that is designed to catch and direct a downward flow **b :** something shaped like a funnel

2 : a stack or flue for the escape of smoke or for ventilation (as on a ship)

²funnel *verb* **-neled** *also* **-nelled; -nel·ing** *also* **-nel·ling** (1594)

intransitive verb

1 : to have or take the shape of a funnel

2 : to pass through or as if through a funnel or conduit

transitive verb

1 : to form in the shape of a funnel ⟨*funneled* his hands and shouted through them⟩

2 : to move to a focal point or into a conduit or central channel ⟨contributions were *funneled* into one account⟩

funnel cloud *noun* (circa 1909)

: a funnel-shaped cloud that projects from the base of a thundercloud and that often betokens the formation of a tornado; *also* **:** TORNADO 2b

fun·nel·form \ˈfə-nᵊl-ˌfȯrm\ *adjective* (circa 1828)

: INFUNDIBULIFORM

¹fun·ny \ˈfə-nē\ *adjective* **fun·ni·er; -est** (1756)

1 a : affording light mirth and laughter **:** AMUSING **b :** seeking or intended to amuse **:** FACETIOUS

2 : differing from the ordinary in a suspicious, perplexing, quaint, or eccentric way **:** PECULIAR — often used as a sentence modifier ⟨*funny,* things didn't turn out the way we planned⟩

3 : involving trickery or deception ⟨told his prisoner not to try anything *funny*⟩

— **fun·ni·ly** \ˈfə-nᵊl-ē\ *adverb*

— **fun·ni·ness** \ˈfə-nē-nəs\ *noun*

— **funny** *adverb*

²funny *noun, plural* **funnies** (1852)

1 : one that is funny; *especially* **:** JOKE

2 *plural* **:** comic strips or the comic section of a periodical — usually used with *the*

funny bone *noun* [from the tingling felt when it is struck] (1840)

1 : the place at the back of the elbow where the ulnar nerve rests against a prominence of the humerus

2 : a sense of humor

funny book *noun* (1947)

: COMIC BOOK

funny car *noun* (1969)

: a specialized dragster that has a one-piece molded body resembling the body of a mass-produced car

funny farm *noun* (1963)

slang **:** a psychiatric hospital

fun·ny·man \ˈfə-nē-ˌman\ *noun* (1852)

: COMEDIAN 2, HUMORIST

funny money *noun* (1943)

1 : artificially inflated currency

2 : counterfeit money

funny paper *noun* (1924)

: a comic section of a newspaper

¹fur \ˈfər\ *verb* **furred; fur·ring** [Middle English *furren,* from Middle French *fourrer,* from Old French *forrer,* from *fuerre* sheath, of Germanic origin; akin to Old High German *fuotar* sheath; akin to Greek *pōma* lid, cover, Sanskrit *pāti* he protects] (14th century)

transitive verb

1 : to cover, line, trim, or clothe with fur

2 : to coat or clog as if with fur

3 : to apply furring to

intransitive verb

: to become coated or clogged as if with fur ◆

²fur *noun, often attributive* (14th century)

1 : a piece of the dressed pelt of an animal used to make, trim, or line wearing apparel

2 : an article of clothing made of or with fur

3 : the hairy coat of a mammal especially when fine, soft, and thick; *also* **:** such a coat with the skin

4 : a coating resembling fur: as **a :** a coat of epithelial debris on the tongue **b :** the thick pile of a fabric (as chenille)

— **fur·less** \ˈfər-ləs\ *adjective*

fu·ran \ˈfyu̇r-ˌan, fyu̇-ˈran\ *also* **fu·rane** \ˈfyu̇r-ˌān, fyu̇-ˈrān\ *noun* [International Scientific Vocabulary, from *furfural*] (1894)

: a flammable liquid C_4H_4O that is obtained from wood oils of pines or made synthetically and is used especially in organic synthesis; *also* **:** a derivative of furan

fu·ra·nose \ˈfyu̇r-ə-ˌnōs, -ˌnōz\ *noun* (1927)

: a sugar having an oxygen-containing ring of five atoms

fu·ran·o·side \fyu̇-ˈra-nə-ˌsīd\ *noun* (1932)

: a glycoside containing the ring characteristic of furanose

fu·ra·zol·i·done \ˌfyu̇r-ə-ˈzä-lə-ˌdōn\ *noun* [*fur*fural + *azole* + *-ide* + *-one*] (1955)

: an antimicrobial drug $C_8H_7N_3O_5$ used against bacteria and some protozoa especially in infections of the gastrointestinal tract

fur·bear·er \ˈfər-ˌbar-ər, -ˌber-\ *noun* (1875)

: an animal that bears fur especially of a commercially desired quality

fur·be·low \ˈfər-bə-ˌlō\ *noun* [by folk etymology from French dialect *farbella*] (1706)

1 : a pleated or gathered piece of material; *especially* **:** a flounce on women's clothing

2 : something that suggests a furbelow especially in being showy or superfluous

— **furbelow** *transitive verb*

fur·bish \ˈfər-bish\ *transitive verb* [Middle English *furbisshen,* from Middle French *fourbiss-,* stem of *fourbir,* of Germanic origin; akin to Old High German *furben* to polish] (14th century)

1 : to make lustrous **:** POLISH

2 : to give a new look to **:** RENOVATE — often used with *up*

— **fur·bish·er** *noun*

fur·ca·tion \ˌfər-ˈkā-shən\ *noun* [Medieval Latin *furcation-, furcatio,* from *furcare* to branch, from Latin *furca* fork] (1646)

1 : something that is branched **:** FORK

2 : the act or process of branching

fur·cu·la \ˈfər-kyə-lə\ *noun, plural* **-lae** \-ˌlē, -ˌlī\ [New Latin, from Latin, forked prop, diminutive of *furca*] (1859)

: a forked process or part: as **a :** WISHBONE **b :** the forked leaping appendage arising from the fourth abdominal segment of a springtail

fur·fu·ral \ˈfər-f(y)ə-ˌral\ *noun* [Latin *furfur* bran + International Scientific Vocabulary ³*-al*] (1879)

: a liquid aldehyde $C_5H_4O_2$ of penetrating odor that is usually made from plant materials and used especially in making furan or phenolic resins and as a solvent

fu·ri·o·so \ˌfyu̇r-ē-ˈō-(ˌ)sō, (ˌ)zō\ *adverb or adjective* [Italian, literally, furious] (circa 1823)

: with great force or vigor — used as a direction in music

fu·ri·ous \ˈfyu̇r-ē-əs\ *adjective* [Middle English, from Middle French *furieus,* from Latin *furiosus,* from *furia* madness, fury] (14th century)

1 a (1) **:** exhibiting or goaded by anger (2) **:** indicative of or proceeding from anger **b :** giving a stormy or turbulent appearance ⟨*furious* bursts of flame⟩ **c :** marked by noise, excitement, activity, or rapidity

2 : INTENSE 1a ⟨the *furious* growth of tropical vegetation⟩

— **fu·ri·ous·ly** *adverb*

¹furl \ˈfər(-ə)l\ *verb* [Middle French *ferler,* from Old North French *ferlier* to tie tightly, from Old French *fer, ferm* tight (from Latin *firmus* firm) + *lier* to tie, from Latin *ligare* — more at LIGATURE] (1556)

transitive verb

: to wrap or roll (as a sail or a flag) close to or around something

intransitive verb

: to curl or fold as in being furled

²furl *noun* (1643)

1 : a furled coil

2 : the act of furling

fur·long \ˈfər-ˌlȯŋ\ *noun* [Middle English, from Old English *furlang,* from *furh* furrow + *lang* long] (14th century)

: a unit of distance equal to 220 yards (about 201 meters)

¹fur·lough \ˈfər-(ˌ)lō\ *noun* [Dutch *verlof,* literally, permission, from Middle Dutch, from *ver-* for- + *lof* permission; akin to Middle High German *loube* permission — more at FOR-, LEAVE] (1625)

: a leave of absence from duty granted especially to a soldier; *also* **:** a document authorizing such a leave of absence

²furlough *transitive verb* (1781)

1 : to grant a furlough to

◇ WORD HISTORY

funky The word *funk,* meaning "foul odor," first appears in the language of 17th century seamen—as near as we can judge from the small amount of evidence available—and in the 18th century it is found in dictionaries of London street talk. Henceforth it vanishes on the far side of the Atlantic but begins to surface in records of regional speech in the U.S. during the 19th century. *Funk* has been well-documented in African-American vernacular, and it may be part of the common lexical heritage of African varieties of English in the New World, to judge by the parallel record of *funk,* meaning "inferior (and thus, foul-smelling) tobacco" in Jamaican English. The derivative *funky* "foul-smelling" took on new life when its use was extended by black jazzmen to an earthy, blues-based style that developed in the 1950s, and *funky* in African-American vernacular began to be used as a term of approbation. In the 1960s *funky* entered mainstream American English, usually with the meaning "odd, eccentric," in either a positive or negative way, though its application has fluctuated in a way that makes the word difficult to define and its future difficult to predict.

fur Though we think of fur as physically part of the animal that wears it—and as something the animal has its entire life with any luck—the history of the word *fur* begins with clothing rather than animals. Middle English *furre* "fur trim or lining for a garment" was shortened from the synonymous word *furrour* or derived from a verb *furren* "to trim or line (a garment) with animal skin." This verb was borrowed from medieval French *fourrer,* with the identical meaning, itself a derivative of *fuerre* "sheath, wrapper," which referred to the lining of a garment providing a sort of warm wrapper for its wearer. The odyssey from "wrapper" to "line with something warm" to "lining material" to "pelt cut from a dead animal" to "coat of a live animal" is a long one, though each step along the way is easily grasped.

2 : to lay off from work

fur·mi·ty \'fər-mə-tē\ *variant of* FRUMENTY

fur·nace \'fər-nəs\ *noun* [Middle English *furnas,* from Old French *fornaise,* from Latin *fornac-, fornax;* akin to Latin *formus* warm — more at THERM] (13th century)
: an enclosed structure in which heat is produced (as for heating a house or for reducing ore)

fur·nish \'fər-nish\ *transitive verb* [Middle English *furnisshen,* from Middle French *fourniss-,* stem of *fournir* to complete, equip, of Germanic origin; akin to Old High German *frummen* to further, *fruma* advantage — more at FOREMOST] (15th century)
1 : to provide with what is needed; *especially* **:** to equip with furniture
2 : SUPPLY, GIVE ⟨*furnished* food and shelter for the refugees⟩ ☆
— fur·nish·er *noun*

fur·nish·ing *noun* (1594)
1 : an article or accessory of dress — usually used in plural
2 : an object that tends to increase comfort or utility; *especially* **:** an article of furniture for the interior of a building — usually used in plural

fur·ni·ture \'fər-ni-chər\ *noun* [Middle French *fourniture,* from *fournir*] (1542)
: equipment that is necessary, useful, or desirable: as **a** *archaic* **:** the trappings of a horse **b :** movable articles used in readying an area (as a room or patio) for occupancy or use

furniture beetle *noun* (1925)
: a widespread deathwatch beetle (*Anobium punctatum*) noted for boring in and damaging furniture and seasoned wood

fu·ror \'fyu̇r-ˌȯr, -ˌōr, -ər\ *noun* [Middle French & Latin; Middle French, from Latin, from *furere* to rage] (15th century)
1 : an angry or maniacal fit **:** RAGE
2 : FURY 4
3 : a fashionable craze **:** VOGUE
4 a : furious or hectic activity **b :** an outburst of public excitement or indignation **:** UPROAR

fu·rore \'fyu̇r-ˌōr, -ˌȯr, -ər, *especially British* fyu̇-'rō-ri\ *noun* [Italian, from Latin *furor*] (1790)
1 : FUROR 4b
2 : FUROR 3

fu·ro·se·mide \fyu̇-'rō-sə-ˌmīd\ *noun* [*fur*fural + *-o-* + sulf- + *-emide,* probably alteration of *amide*] (1965)
: a powerful diuretic $C_{12}H_{11}ClN_2O_5S$ used especially to treat edema

furred \'fərd\ *adjective* [Middle English] (14th century)
1 : lined, trimmed, or faced with fur
2 : coated as if with fur; *specifically* **:** having a coating consisting chiefly of mucus and dead epithelial cells ⟨a *furred* tongue⟩
3 : bearing or wearing fur
4 : provided with furring ⟨*furred* wall⟩

fur·ri·er \'fər-ē-ər, 'fə-rē-\ *noun* [alteration of Middle English *furrer,* from Anglo-French *furrere,* from Old French *forrer* to fur — more at FUR] (14th century)
1 : a fur dealer
2 a : one that dresses furs **b :** one that makes, repairs, alters, or cleans fur garments

fur·ri·ery \-ə-rē\ *noun* (circa 1864)
1 : the fur business
2 : fur craftsmanship

fur·rin·er \'fər-ə-nər\ *noun* [alteration of *foreigner*] (1849)
: FOREIGNER 2 — used to represent a dialectal pronunciation

fur·ring \'fər-iŋ\ *noun* (14th century)
1 : a fur trimming or lining
2 a : the application of thin wood, brick, or metal to joists, studs, or walls to form a level surface (as for attaching wallboard) or an air space **b :** the material used in this process

¹fur·row \'fər-(ˌ)ō, 'fə-(ˌ)rō\ *noun* [Middle English *furgh, forow,* from Old English *furh;* akin to Old High German *furuh* furrow, Latin *porca*] (before 12th century)
1 a : a trench in the earth made by a plow **b :** plowed land **:** FIELD
2 : something that resembles the track of a plow: as **a :** a marked narrow depression **:** GROOVE **b :** a deep wrinkle ⟨*furrows* in his brow⟩

²furrow (15th century)
transitive verb
: to make furrows, grooves, wrinkles, or lines in
intransitive verb
: to make or form furrows, grooves, wrinkles, or lines

fur·ry \'fər-ē\ *adjective* **fur·ri·er; -est** (circa 1674)
1 : consisting of or resembling fur ⟨animals with *furry* coats⟩
2 : covered with fur
3 : thick in quality ⟨spoke with a *furry* voice⟩

fur seal *noun* (1775)
: any of two genera (*Callorhinus* and *Arctocephalus*) of eared seals that have a double coat with a dense soft underfur

¹fur·ther \'fər-t͟hər\ *adverb* [Middle English, from Old English *furthor* (akin to Old High German *furthar* further), comparative, from the base of Old English *forth* forth] (before 12th century)
1 : FARTHER 1 ⟨my ponies are tired, and I have *further* to go —Thomas Hardy⟩
2 : in addition **:** MOREOVER
3 : to a greater degree or extent ⟨*further* annoyed by a second intrusion⟩
usage see FARTHER

²further *transitive verb* **fur·thered; fur·ther·ing** \'fərt͟h-riŋ, 'fər-t͟hə-\ (before 12th century)
: to help forward **:** PROMOTE
synonym see ADVANCE
— fur·ther·er \'fər-t͟hər-ər\ *noun*

³further *adjective* (13th century)
1 : FARTHER 1 ⟨rode . . . across the valley and up the *further* slopes —T. E. Lawrence⟩
2 : going or extending beyond **:** ADDITIONAL ⟨*further* volumes⟩ ⟨*further* education⟩
usage see FARTHER

fur·ther·ance \'fərt͟h-rən(t)s, 'fər-t͟hə-\ *noun* (15th century)
: the act of furthering **:** ADVANCEMENT

further education *noun* (1937)
British **:** ADULT EDUCATION

fur·ther·more \'fər-t͟hə(r)-ˌmōr, -ˌmȯr\ *adverb* (13th century)
: in addition to what precedes **:** BESIDES

fur·ther·most \-t͟hər-ˌmōst\ *adjective* (15th century)
: most distant **:** FARTHEST

fur·thest \'fər-t͟həst\ *adverb or adjective* (14th century)
: FARTHEST

fur·tive \'fər-tiv\ *adjective* [French or Latin; French *furtif,* from Latin *furtivus,* from *furtum* theft, from *fur* thief, from or akin to Greek *phōr* thief, *pherein* to carry — more at BEAR] (1612)
1 a : done by stealth **:** SURREPTITIOUS **b :** expressive of stealth **:** SLY ⟨had a *furtive* look about him⟩
2 : obtained underhandedly **:** STOLEN
synonym see SECRET
— fur·tive·ly *adverb*
— fur·tive·ness *noun*

fu·run·cle \'fyu̇r-ˌəŋ-kəl\ *noun* [Latin *furunculus* petty thief, boil, diminutive of *furon-, furo* ferret, thief, from *fur*] (1676)
: ²BOIL

fu·run·cu·lo·sis \fyu̇-ˌrəŋ-kyə-'lō-səs\ *noun, plural* **-lo·ses** \-ˌsēz\ [New Latin] (1886)
1 : the condition of having or tending to develop multiple furuncles
2 : a highly infectious disease of various salmonoid fishes (as trout) that is caused by a bacterium (*Bacterium salmonicida*) and is especially virulent in dense fish populations (as in hatcheries)

fu·ry \'fyu̇r-ē\ *noun, plural* **furies** [Middle English *furie,* from Middle French & Latin; Middle French, from Latin *furia,* from *furere* to rage] (14th century)
1 : intense, disordered, and often destructive rage
2 a *capitalized* **:** any of the avenging deities in Greek mythology who torment criminals and inflict plagues **b :** an avenging spirit **c :** one who resembles an avenging spirit; *especially* **:** a spiteful woman
3 : extreme fierceness or violence
4 : a state of inspired exaltation **:** FRENZY ◆
synonym see ANGER

furze \'fərz\ *noun* [Middle English *firse,* from Old English *fyrs;* akin to Russian *pyreĭ* quack grass, Greek *pyros* wheat] (before 12th century)
: GORSE
— furzy \'fər-zē\ *adjective*

fus·cous \'fəs-kəs\ *adjective* [Latin *fuscus* — more at DUSK] (1662)
: of any of several colors averaging a brownish gray

¹fuse \'fyüz\ *verb* **fused; fus·ing** [Latin *fusus,* past participle of *fundere* to pour, melt — more at FOUND] (1592)
transitive verb
1 : to reduce to a liquid or plastic state by heat
2 : to blend thoroughly by or as if by melting together **:** COMBINE ⟨in her richest work she *fuses* comedy and tragedy —T. A. Gullason⟩
3 : to stitch by applying heat and pressure with or without the use of an adhesive
intransitive verb
1 a : to become fluid with heat **b** *British* **:** to fail because of the blowing of a fuse
2 : to become blended or joined by or as if by melting together
synonym see MIX

²fuse *noun* (1884)
: an electrical safety device consisting of or including a wire or strip of fusible metal that

☆ SYNONYMS
Furnish, equip, outfit, appoint, accoutre mean to supply one with what is needed. FURNISH implies the provision of any or all essentials for performing a function ⟨a sparsely *furnished* apartment⟩. EQUIP suggests the provision of something making for efficiency in action or use ⟨a fully *equipped* kitchen⟩. OUTFIT implies provision of a complete list or set of articles as for a journey, an expedition, or a special occupation ⟨*outfitted* the family for a ski trip⟩. APPOINT implies provision of complete and usually elegant or elaborate equipment or furnishings ⟨a lavishly *appointed* apartment⟩. ACCOUTRE suggests the supplying of personal dress or equipment for a special activity ⟨fully *accoutred* members of a polar expedition⟩.

◇ WORD HISTORY
fury No more fearsome figures darkened Greek mythology than the Erinyes. Born of the blood drops from the emasculation of Uranus, with snakes coiled in their hair, they roamed the land avenging perjury and murder and carrying out the curses of parent against son. Neither prayer nor tears could sway them, nor sacrifice stave off their wrath. To the Romans they were known as the *Dirae* or the *Furiae.* The latter name is a personified form of the plural noun *furiae* "frenzy," a derivative of *furere* "to rage." English *fury* is borrowed via French from the singular form *furia.*

melts and interrupts the circuit when the current exceeds a particular amperage

[3]fuse *noun* [Italian *fuso* spindle, from Latin *fusus,* of unknown origin] (1644)
1 : a continuous train of a combustible substance enclosed in a cord or cable for setting off an explosive charge by transmitting fire to it
2 *usually* **fuze :** a mechanical or electrical detonating device for setting off the bursting charge of a projectile, bomb, or torpedo

[4]fuse *or* **fuze** \'fyüz\ *transitive verb* **fused** *or* **fuzed; fus·ing** *or* **fuz·ing** (1802)
: to equip with a fuse

fused quartz *noun* (1925)
: QUARTZ GLASS — called also *fused silica*

fu·see \fyü-'zē\ *noun* [French *fusée,* literally, spindleful of yarn, from Old French, from *fus* spindle, from Latin *fusus*] (1622)
1 : a conical spirally grooved pulley in a timepiece from which a cord or chain unwinds onto a barrel containing the spring and which by its increasing diameter compensates for the lessening power of the spring
2 : a red signal flare used especially for protecting stalled trains and trucks

fu·se·lage \'fyü-sə-,läzh, -zə-\ *noun* [French, from *fuselé* spindle-shaped, from Middle French, from *fusel,* diminutive of *fus*] (1909)
: the central body portion of an aircraft designed to accommodate the crew and the passengers or cargo — see AIRPLANE illustration

fu·sel oil \'fyü-zəl-\ *noun* [German *Fusel* bad liquor] (1850)
: an acrid oily liquid occurring in insufficiently distilled alcoholic liquors, consisting chiefly of amyl alcohol, and used especially as a source of alcohols and as a solvent

fus·ible \'fyü-zə-bəl\ *adjective* (14th century)
: capable of being fused and especially liquefied by heat ⟨*fusible* alloy⟩
— fus·ibil·i·ty \,fyü-zə-'bi-lə-tē\ *noun*

fu·si·form \'fyü-zə-,fȯrm\ *adjective* [Latin *fusus* spindle] (1746)
: tapering toward each end ⟨*fusiform* bacteria⟩

[1]fu·sil \'fyü-zəl\ *or* **fu·sile** \'fyü-zəl, -,zīl\ *adjective* [Middle English, from Latin *fusilis,* from *fundere*] (14th century)
1 *archaic* **a :** made by melting and pouring into forms **:** CAST **b :** liquefied by heat
2 *archaic* **:** FUSIBLE

[2]fusil *noun* [French, literally, steel for striking fire, from Old French *foisil,* from (assumed) Vulgar Latin *focilis,* from Late Latin *focus* fire — more at FUEL] (1680)
: a light flintlock musket

fu·sil·ier *or* **fu·sil·eer** \,fyü-zə-'lir\ *noun* [French *fusilier,* from *fusil*] (1680)
1 : a soldier armed with a fusil
2 : a member of a British regiment formerly armed with fusils

fu·sil·lade \'fyü-sə-,läd, -,lād, ,fyü-sə-', -zə-\ *noun* [French, from *fusiller* to shoot, from *fusil*] (1801)
1 a : a number of shots fired simultaneously or in rapid succession **b :** something that gives the effect of a fusillade ⟨a *fusillade* of rocks and bottles⟩
2 : a spirited outburst especially of criticism

fu·sil·li \fyü-'si-lē, -'sē-\ *noun* [Italian, plural of *fusillo,* from Italian dialect (southern Italy), diminutive of *fuso* spindle, from Latin *fusus*] (1948)
: spiral-shaped pasta

fu·sion \'fyü-zhᵊn\ *noun, often attributive* [Latin *fusion-, fusio,* from *fundere*] (1555)
1 : the act or process of liquefying or rendering plastic by heat
2 : a union by or as if by melting: as **a :** a merging of diverse, distinct, or separate elements into a unified whole **b :** a political partnership **:** COALITION **c :** popular music combining different styles (as jazz and rock)
3 : the union of atomic nuclei to form heavier nuclei resulting in the release of enormous quantities of energy when certain light elements unite

fu·sion·ist \'fyü-zhᵊn-ist\ *noun* (1851)
: a person involved in a political fusion or in nuclear or musical fusion

[1]fuss \'fəs\ *noun* [origin unknown] (1701)
1 a : needless bustle or excitement **:** COMMOTION **b :** a show of flattering attention ⟨made a big *fuss* over his favorite niece⟩
2 a : a state of agitation especially over a trivial matter **b :** OBJECTION, PROTEST **c :** an often petty controversy or quarrel ⟨ended up having a pretty good *fuss* with my wife —Mac Hyman⟩

[2]fuss (1792)
intransitive verb
1 a : to create or be in a state of restless activity; *especially* **:** to shower flattering attentions ⟨*fussing* over the grandchildren⟩ **b :** to pay close or undue attention to small details ⟨*fussed* with her hair⟩
2 a : to become upset **:** WORRY **b :** to express annoyance or pique **:** COMPLAIN
transitive verb
: AGITATE, UPSET
— fuss·er *noun*

fuss·bud·get \'fəs-,bə-jət\ *noun* (circa 1904)
: one who fusses or is fussy especially about trifles
— fuss·bud·gety \-jə-tē\ *adjective*

fuss·pot \'fəs-,pät\ *noun* (1921)
: FUSSBUDGET

fussy \'fə-sē\ *adjective* **fuss·i·er; -est** (1831)
1 : easily upset **:** IRRITABLE
2 : overly decorative ⟨a *fussy* wallpaper pattern⟩
3 a : requiring or giving close attention to details ⟨*fussy* bookkeeping procedures⟩ **b :** revealing a sometimes extreme concern for niceties **:** FASTIDIOUS, PICKY
— fuss·i·ly \'fə-sə-lē\ *adverb*
— fuss·i·ness \'fə-sē-nəs\ *noun*

fus·tian \'fəs-chən\ *noun* [Middle English, from Old French *fustaine,* from Medieval Latin *fustaneum,* perhaps from *fustis* tree trunk, from Latin, club] (13th century)
1 a : a strong cotton and linen fabric **b :** a class of cotton fabrics usually having a pile face and twill weave
2 : highflown or affected writing or speech; *broadly* **:** anything highflown or affected in style
— fus·tian *adjective*

fus·tic \'fəs-tik\ *noun* [Middle English *fustyk* smoke tree, from Middle French *fustoc,* from Arabic *fustuq,* from Greek *pistakē* pistachio tree — more at PISTACHIO] (15th century)
: the wood of a tropical American tree (*Chlorophora tinctoria*) of the mulberry family that yields a yellow dye; *also* **:** any of several similar dyewoods

fus·ti·gate \'fəs-tə-,gāt\ *transitive verb* **-gat·ed; -gat·ing** [Late Latin *fustigatus,* past participle of *fustigare,* from Latin *fustis* + *-igare* (as in *fumigare* to fumigate)] (circa 1661)
1 : CUDGEL
2 : to criticize severely
— fus·ti·ga·tion \,fəs-tə-'gā-shən\ *noun*

fus·ty \'fəs-tē\ *adjective* **fus·ti·er; -est** [Middle English, from *fust* wine cask, from Middle French, club, cask, from Latin *fustis*] (14th century)
1 *British* **:** impaired by age or dampness **:** MOLDY
2 : saturated with dust and stale odors **:** MUSTY
3 : rigidly old-fashioned or reactionary
synonym see MALODOROUS
— fus·ti·ly \-tə-lē\ *adverb*
— fus·ti·ness \-tē-nəs\ *noun*

fu·su·li·nid \,fyü-zə-'lī-nid, -'lē-, -'li-\ *noun* [New Latin *Fusulinidae,* from *Fusulina,* a genus, from Latin *fusus* spindle + *-ulus* -ule + New Latin *-ina,* diminutive suffix] (1941)
: any of a family (Fusulinidae) of extinct marine foraminifers

fu·thark \'fü-,thärk\ *also* **fu·thorc** *or* **fu·thork** \-,thȯrk\ *noun* [from the first six letters, *f, u, þ* (*th*), *o* (or *a*), *r, c* (=k)] (1851)
: the runic alphabet — see RUNE illustration

fu·tile \'fyü-tᵊl, 'fyü-,tīl\ *adjective* [Middle French or Latin; Middle French, from Latin *futilis* brittle, pointless, probably from *fu-* (akin to *fundere* to pour) — more at FOUND] (circa 1555)
1 : serving no useful purpose **:** completely ineffective ⟨efforts to convince him were *futile*⟩
2 : occupied with trifles **:** FRIVOLOUS ☆
— fu·tile·ly \-tᵊl-(l)ē, -,tīl-lē\ *adverb*
— fu·tile·ness \-tᵊl-nəs, -,tīl-nəs\ *noun*

fu·til·i·tar·i·an \fyü-,ti-lə-'ter-ē-ən, ,fyü-\ *noun* [blend of *futile* and *utilitarian*] (1827)
: one who believes that human striving is futile
— futilitarian *adjective*
— fu·til·i·tar·i·an·ism \-ē-ə-,ni-zəm\ *noun*

fu·til·i·ty \fyü-'ti-lə-tē\ *noun, plural* **-ties** (1623)
1 : the quality or state of being futile **:** USELESSNESS
2 : a useless act or gesture ⟨the *futilities* of debate for its own sake —W. A. White⟩

fu·ton \'fü-,tän\ *noun, plural* **futons** *also* **futon** [Japanese] (1876)
: a usually cotton-filled mattress used on the floor or in a frame as a bed

fut·tock \'fə-tək\ *noun* [Middle English *votek, futtek*] (13th century)
: one of the curved timbers scarfed together to form the lower part of the compound rib of a ship

futtock shroud *noun* (1840)
: a short iron rod connecting the topmast rigging with the lower mast

[1]fu·ture \'fyü-chər\ *adjective* [Middle English, from Middle French & Latin; Middle French *futur,* from Latin *futurus* about to be — more at BE] (14th century)
1 : that is to be; *specifically* **:** existing after death
2 : of, relating to, or constituting a verb tense expressive of time yet to come
3 : existing or occurring at a later time ⟨met his *future* wife⟩

[2]future *noun* (15th century)
1 a : time that is to come **b :** what is going to happen
2 : an expectation of advancement or progressive development
3 : something (as a bulk commodity) bought for future acceptance or sold for future delivery — usually used in plural ⟨grain *futures*⟩
4 a : the future tense of a language **b :** a verb form in the future tense

fu·ture·less \'fyü-chər-ləs\ *adjective* (1863)
: having no future
— fu·ture·less·ness *noun*

future perfect *adjective* (circa 1898)

☆ SYNONYMS
Futile, vain, fruitless mean producing no result. FUTILE may connote completeness of failure or unwisdom of undertaking ⟨resistance had proved so *futile* that surrender was the only choice left⟩. VAIN usually implies simple failure to achieve a desired result ⟨a *vain* attempt to get the car started⟩. FRUITLESS comes close to VAIN but often suggests long and arduous effort or severe disappointment ⟨*fruitless* efforts to obtain a lasting peace⟩.

: of, relating to, or constituting a verb tense that is traditionally formed in English with *will have* and *shall have* and that expresses completion of an action by a specified time that is yet to come
— **future perfect** *noun*

future shock *noun* (1965)
: the physical and psychological distress suffered by one who is unable to cope with the rapidity of social and technological changes

fu·tur·ism \'fyü-chə-ˌri-zəm\ *noun* (1909)
1 : a movement in art, music, and literature begun in Italy about 1909 and marked especially by an effort to give formal expression to the dynamic energy and movement of mechanical processes
2 : a point of view that finds meaning or fulfillment in the future rather than in the past or present

fu·tur·ist \'fyü-chə-rist\ *noun* (1911)
1 : one who advocates or practices futurism
2 : one who studies and predicts the future especially on the basis of current trends
— **futurist** *adjective*

fu·tur·is·tic \ˌfyü-chə-'ris-tik\ *adjective* (1915)
: of, relating to, or characteristic of the future, futurism, or futurology; *also* **:** very modern
— **fu·tur·is·ti·cal·ly** \-ti-k(ə-)lē\ *adverb*

fu·tur·is·tics \-tiks\ *noun plural but singular in construction* (1969)
: FUTUROLOGY

fu·tu·ri·ty \fyü-'tu̇r-ə-tē, -'tyu̇r-, -'chu̇r-\ *noun, plural* **-ties** (1604)
1 : time to come **:** FUTURE
2 : the quality or state of being future
3 *plural* **:** future events or prospects
4 a : a horse race usually for two-year-olds in which the competitors are nominated at birth or before **b :** a race or competition for which entries are made well in advance of the event

fu·tur·olo·gy \ˌfyü-chə-'rä-lə-jē\ *noun* (1946)
: a study that deals with future possibilities based on current trends
— **fu·tur·olog·i·cal** \-rə-'lä-ji-kəl\ *adjective*
— **fu·tur·olo·gist** \-'rä-lə-jist\ *noun*

futz \'fəts\ *intransitive verb* [perhaps part modification, part translation of Yiddish *arumfartsn zikh,* literally, to fart around] (1932)
slang **:** FOOL AROUND 1 — often used with *around* ⟨*futz* around without producing any worthwhile music —John Koegel⟩

fuze, fu·zee *variant of* FUSE, FUSEE

¹fuzz \'fəz\ *noun* [probably back-formation from *fuzzy*] (1674)
1 : fine light particles or fibers (as of down or fluff)
2 : a blurred effect

²fuzz (circa 1702)
intransitive verb
1 : to fly off in or become covered with fluffy particles
2 : to become blurred ⟨her frame of reference *fuzzing* at the edges —Jane O'Reilly⟩
transitive verb
1 : to make fuzzy
2 : to envelop in a haze **:** BLUR

³fuzz *noun* [origin unknown] (1927)
: POLICE; *also* **:** a police officer

fuzzy \'fə-zē\ *adjective* **fuzz·i·er; -est** [perhaps from Low German *fussig* loose, spongy] (1713)
1 : marked by or giving a suggestion of fuzz ⟨a *fuzzy* covering of felt⟩
2 : lacking in clarity or definition ⟨moving the camera causes *fuzzy* photos⟩ ⟨*fuzzy* thinking⟩
— **fuzz·i·ly** \'fə-zə-lē\ *adverb*
— **fuzz·i·ness** \'fə-zē-nəs\ *noun*

-fy \ˌfī\ *verb suffix* [Middle English *-fien,* from Old French *-fier,* from Latin *-ficare,* from *-ficus* -fic]
1 : make **:** form into ⟨dandi*fy*⟩
2 : invest with the attributes of **:** make similar to ⟨citi*fy*⟩

fyce \'fīs\ *variant of* FEIST

fyke \'fīk\ *noun* [Dutch *fuik*] (1832)
: a long bag net kept open by hoops

fyl·fot \'fil-ˌfät\ *noun* [Middle English, device used to fill the lower part of a painted glass window (from a conjectural manuscript reading)] (1842)
: SWASTIKA

G is the seventh letter of the English alphabet and of most modern alphabets closely related to that of English. It came from Latin, where it was introduced to distinguish the voiced velar stop (the sound of g *in* go*) from the corresponding voiceless stop (the sound of* k *in* king*), both of which had until then been represented by C. In Modern English,* g *represents two chief sounds, commonly called "hard* g*" and "soft* g*." The former, as in* go*, was the earlier sound, the latter, equivalent to* j*, as in* gem*, appears chiefly in words from Latin and Old French. The small* g *developed from a late Roman form of the capital letter that had a short curved flourish extending below the line (ɢ). The upper half of the letter closed to form a full oval, while the tail below the line also closed to form a loop (g or g).*

---

g \'jē\ *noun, plural* **g's** *or* **gs** \'jēz\ *often capitalized, often attributive* (before 12th century)
1 a : the 7th letter of the English alphabet **b :** a graphic representation of this letter **c :** a speech counterpart of orthographic *g*
2 : the 5th tone of a C-major scale
3 : a graphic device for reproducing the letter *g*
4 : one designated *g* especially as the 7th in order or class
5 [gravity] **:** ACCELERATION OF GRAVITY; *also* **:** a unit of force equal to the force exerted by gravity on a body at rest and used to indicate the force to which a body is subjected when accelerated
6 [grand] *slang* **:** a sum of $1000
7 : something shaped like the letter G

G *certification mark*
— used to certify that a motion picture is of such a nature that persons of all ages may be allowed admission; compare NC-17, PG, PG-13, R

[1]gab \'gab\ *intransitive verb* **gabbed; gab·bing** [probably short for *gabble*] (1786)
: to talk in a rapid or thoughtless manner **:** CHATTER
— **gab·ber** *noun*

[2]gab *noun* (1790)
: TALK; *especially* **:** idle talk

[3]gab *noun* (1939)
: GABARDINE 2

gab·ar·dine \'ga-bər-ˌdēn\ *noun* [Middle French *gaverdine*] (1520)
1 : GABERDINE 1
2 a : a firm hard-finish durable fabric (as of wool or rayon) twilled with diagonal ribs on the right side **b :** a garment of gabardine

gab·ble \'ga-bəl\ *verb* **gab·bled; gab·bling** \-b(ə-)liŋ\ [probably of imitative origin] (1577)
intransitive verb
1 : to talk fast or foolishly **:** JABBER
2 : to utter inarticulate or animal sounds
transitive verb
: to say with incoherent rapidity **:** BABBLE
— **gabble** *noun*
— **gab·bler** \-b(ə-)lər\ *noun*

gab·bro \'ga-(ˌ)brō\ *noun, plural* **gabbros** [Italian, probably modification of Latin *glaber* smooth — more at GLAD] (circa 1828)
: a granular igneous rock composed essentially of calcic plagioclase, a ferromagnesian mineral, and accessory minerals
— **gab·bro·ic** \ga-'brō-ik\ *adjective*

gab·by \'ga-bē\ *adjective* **gab·bi·er; -est** (1719)
: TALKATIVE, GARRULOUS

ga·belle \gə-'bel\ *noun* [Middle English, from Middle French, from Old Italian *gabella* tax, from Arabic *qabālah*] (15th century)
: a tax on salt levied in France prior to 1790

gab·er·dine \'ga-bər-ˌdēn\ *noun* [Middle French *gaverdine*] (1520)
1 : a long loose outer garment worn in medieval times and associated especially with Jews since the 16th century
2 : GABARDINE 2

gab·fest \'gab-ˌfest\ *noun* (1897)
1 : an informal gathering for general talk ⟨political *gabfests*⟩
2 : an extended conversation

ga·bi·on \'gā-bē-ən, 'ga-\ *noun* [Middle French, from Old Italian *gabbione,* literally, large cage, augmentative of *gabbia* cage, from Latin *cavea* — more at CAGE] (1579)
: a basket or cage filled with earth or rocks and used especially in building a support or abutment

ga·ble \'gā-bəl\ *noun* [Middle English, from Middle French, of Germanic origin; akin to Old Norse *gafl* gable — more at CEPHALIC] (14th century)
1 a : the vertical triangular end of a building from cornice or eaves to ridge **b :** the similar end of a gambrel roof **c :** the end wall of a building
2 : a triangular part or structure
— **ga·bled** \-bəld\ *adjective*

1 gable 1a

gable roof *noun* (1850)
: a double-sloping roof that forms a gable at each end

gab·oon \gä-'bün, gə-\ *noun* [alteration of [1]*gob* + *-oon* (as in *spittoon*)] (1929)
dialect **:** SPITTOON

Ga·bri·el \'gā-brē-əl\ *noun* [Hebrew *Gabhrī'ēl*]
: one of the four archangels named in Hebrew tradition

ga·by \'gā-bē\ *noun, plural* **gabies** [perhaps of Scandinavian origin; akin to Old Norse *gapa* to gape — more at GAPE] (circa 1796)
dialect chiefly English **:** SIMPLETON

[1]gad \'gad\ *noun* [Middle English, spike, from Old Norse *gaddr;* akin to Old English *geard* rod — more at YARD] (1671)
1 : a chisel or pointed iron or steel bar for loosening ore or rock
2 *chiefly dialect* **:** a long stick

[2]gad *intransitive verb* **gad·ded; gad·ding** [Middle English *gadden*] (15th century)
: to be on the go without a specific aim or purpose — usually used with *about*

[3]gad *interjection* [euphemism for *God*] (1608)
— used as a mild oath

Gad \'gad\ *noun* [Hebrew *Gādh*]
: a son of Jacob and the traditional eponymous ancestor of one of the tribes of Israel
— **Gad·ite** \'ga-ˌdīt\ *noun*

gad·about \'ga-də-ˌbau̇t\ *noun* (1837)
: a person who flits about in social activity
— **gadabout** *adjective*

gad·a·rene \'ga-də-ˌrēn\ *adjective, often capitalized* [from the demon-possessed *Gadarene* swine in Matthew 8:28 that rushed into the sea] (1922)
: HEADLONG, PRECIPITATE ⟨a *gadarene* rush to the cities⟩

gad·fly \'gad-ˌflī\ *noun* [[1]*gad*] (1626)
1 : any of various flies (as a horsefly, botfly, or warble fly) that bite or annoy livestock
2 : a person who stimulates or annoys especially by persistent criticism

gad·get \'ga-jət\ *noun* [origin unknown] (1886)
: an often small mechanical or electronic device with a practical use but often thought of as a novelty
— **gad·ge·teer** \ˌga-jə-'tir\ *noun*
— **gad·get·ry** \'ga-jə-trē\ *noun*
— **gad·gety** \-jə-tē\ *adjective*

ga·doid \'gā-ˌdȯid, 'ga-\ *adjective* [New Latin *Gadus,* genus of fishes, from Greek *gados,* a fish] (circa 1842)
: resembling or related to the cods
— **gadoid** *noun*

gad·o·lin·ite \'ga-dᵊl-ə-ˌnīt\ *noun* [German *Gadolinit,* from Johann *Gadolin* (died 1852) Finnish chemist] (1802)
: a black or brown mineral that is a source of rare earths and consists of a silicate of iron, beryllium, yttrium, cerium, and erbium

gad·o·lin·i·um \ˌga-də-'li-nē-əm\ *noun* [New Latin, from J. *Gadolin*] (1886)
: a magnetic metallic element of the rare-earth group occurring in combination in gadolinite and several other minerals — see ELEMENT table

ga·droon \gə-'drün\ *noun* [French *godron* round plait, gadroon] (circa 1724)
1 : the ornamental notching or carving of a rounded molding
2 : a short often oval fluting or reeding used in decoration
— **gadroon** *transitive verb*
— **ga·droon·ing** *noun*

gad·wall \'gad-ˌwȯl\ *noun, plural* **gadwalls** *or* **gadwall** [origin unknown] (1666)
: a grayish brown medium-sized dabbling duck (*Anas strepera*)

gad·zook·ery \gad-'zü-kə-rē, -'zu̇-\ *noun* (1955)
British **:** the use of archaisms (as in a historical novel)

gad·zooks \gad-'züks, -'zu̇ks\ *interjection, often capitalized* [perhaps from *God's hooks,* the nails of the Crucifixion] (1694)
archaic — used as a mild oath

Gaea \'jē-ə\ *noun* [Greek *Gaia*]
: the Greek earth goddess and mother of the Titans

Gael \'gā(ə)l\ *noun* [Scottish Gaelic *Gàidheal* & Irish *Gaedheal*] (1753)
1 : a Scottish Highlander
2 : a Celtic especially Gaelic-speaking inhabitant of Ireland, Scotland, or the Isle of Man
— **Gael·dom** \-dəm\ *noun*

Gael·ic \'gā-lik, 'ga-, 'gä-\ *adjective* [Scottish Gaelic *Gàidhlig* the Scottish Gaelic language, from *Gàidheal* Gaelic] (1741)
1 : of or relating to the Gaels and especially the Celtic Highlanders of Scotland
2 : of, relating to, or constituting the Goidelic speech of the Celts in Ireland, the Isle of Man, and the Scottish Highlands
— **Gaelic** *noun*

Gael·tacht \'gā(ə)l-təkt\ *noun* [Irish, from *Gaelic,* spelling variant of *Gaedheal* Irishman, Gael] (1929)
: any of the Irish-speaking regions remaining in Ireland

\ə\ **abut** \ᵊ\ **kitten** \ər\ **further** \a\ **ash** \ā\ **ace** \ä\ **mop, mar** \au̇\ **out** \ch\ **chin** \e\ **bet** \ē\ **easy** \g\ **go** \i\ **hit** \ī\ **ice** \j\ **job** \ŋ\ **sing** \ō\ **go** \ȯ\ **law** \ȯi\ **boy** \th\ **thin** \t͟h\ **the** \ü\ **loot** \u̇\ **foot** \y\ **yet** \zh\ **vision** *see also* Guide to Pronunciation

[1]**gaff** \ˈgaf\ *noun* [French *gaffe,* from Provençal *gaf*] (circa 1656)
1 a : a spear or spearhead for taking fish or turtles **b :** a handled hook for holding or lifting heavy fish **c :** a metal spur for a gamecock **d :** a butcher's hook **e :** a climbing iron or its steel point used by a telephone lineman
2 : the spar on which the head of a fore-and-aft sail is extended
3 : GAFFE

[2]**gaff** *transitive verb* (1844)
1 : to strike or secure with a gaff
2 : to fit (a gamecock) with a gaff

[3]**gaff** *noun* [origin unknown] (1812)
British **:** a cheap theater or music hall

[4]**gaff** *noun* [origin unknown] (1896)
1 a : something painful or difficult to bear **:** ORDEAL — usually used in the phrase *stand the gaff; especially* **:** persistent raillery or criticism **b :** rough treatment **:** ABUSE
2 a : HOAX, FRAUD **b :** GIMMICK, TRICK

[5]**gaff** *transitive verb* (1933)
1 : DECEIVE, TRICK; *also* **:** FLEECE
2 : to fix for the purpose of cheating ⟨*gaff* the dice⟩

gaffe \ˈgaf\ *noun* [French, gaff, gaffe] (1909)
: a social or diplomatic blunder

gaf·fer \ˈga-fər\ *noun* [alteration of *godfather*] (1575)
1 : an old man — compare GAMMER
2 *British* **a :** FOREMAN, OVERSEER **b :** EMPLOYER
3 : a head glassblower
4 : a lighting electrician on a motion-picture or television set

gaff–top·sail \ˈgaf-ˈtäp-ˌsāl, -səl\ *noun* (1794)
: a usually triangular topsail with its foot extended upon the gaff — see SAIL illustration

[1]**gag** \ˈgag\ *verb* **gagged; gag·ging** [Middle English *gaggen* to strangle, of imitative origin] (1509)
transitive verb
1 a : to restrict use of the mouth of by inserting a gag **b :** to prevent from exercising freedom of speech or expression **c :** to pry or hold open with a gag
2 : to provide or write quips or pranks for ⟨*gag* a show⟩
3 : to choke or cause to retch
intransitive verb
1 a : CHOKE; *also* **:** to suffer a throat spasm that makes swallowing or breathing difficult **b :** RETCH
2 : to be unable to endure something **:** BALK
3 : to make quips
— gag·ger *noun*

[2]**gag** *noun* (1553)
1 : something thrust into the mouth to keep it open or to prevent speech or outcry
2 : an official check or restraint on debate or free speech ⟨a *gag* order⟩ ⟨a *gag* rule⟩
3 : a laugh-provoking remark or act
4 : PRANK, TRICK

ga·ga \ˈgä-(ˌ)gä\ *adjective* [French, from *gaga* fool, of imitative origin] (1920)
1 : CRAZY, FOOLISH
2 : marked by wild enthusiasm **:** INFATUATED, DOTING

ga·ga·ku \gä-ˈgä-(ˌ)kü\ *noun* [Japanese, from *ga* elegance + *gaku* music] (1929)
: the ancient court music of Japan

[1]**gage** \ˈgāj\ *noun* [Middle English, from Middle French, of Germanic origin; akin to Old High German *wetti* pledge — more at WED] (14th century)
1 : a token of defiance; *specifically* **:** a glove or cap cast on the ground to be taken up by an opponent as a pledge of combat
2 : something deposited as a pledge of performance

[2]**gage** *transitive verb* (15th century)
1 *archaic* **:** PLEDGE
2 *archaic* **:** STAKE, RISK

[3]**gage** *variant of* GAUGE

[4]**gage** *noun* (1847)
: GREENGAGE

gag·gle \ˈga-gəl\ *noun* [Middle English *gagyll,* from *gagelen* to cackle] (15th century)
1 : FLOCK; *especially* **:** a flock of geese when not in flight — compare SKEIN
2 : AGGREGATION, CLUSTER ⟨a *gaggle* of reporters and photographers⟩

gag·man \ˈgag-ˌman\ *noun* (1928)
1 : a gag writer
2 : COMEDIAN 2

gag·ster \ˈgag-stər\ *noun* (1935)
: GAGMAN; *also* **:** one who plays practical jokes

gahn·ite \ˈgä-ˌnīt\ *noun* [German *Gahnit,* from J. G. *Gahn* (died 1818) Swedish chemist] (circa 1808)
: a usually dark green mineral consisting of an oxide of zinc and aluminum

gai·ety \ˈgā-ə-tē\ *noun, plural* **-eties** [French *gaieté*] (1634)
1 : MERRYMAKING; *also* **:** festive activity — often used in plural
2 : high spirits **:** MERRIMENT
3 : ELEGANCE, FINERY

gai·jin \ˈgī-(ˌ)jēn, -(ˌ)jin\ *noun, plural* **gaijin** [Japanese, from *gai-* outer, foreign + *-jin* person] (1964)
: a foreigner in Japan

gail·lar·dia \gə-ˈlär-d(ē-)ə\ *noun* [New Latin, from *Gaillard* de Marentonneau, 18th century French botanist] (1888)
: any of a genus (*Gaillardia*) of chiefly western American composite herbs with showy flower heads

gai·ly \ˈgā-lē\ *adverb* (14th century)
: in a gay manner **:** marked by gaiety

[1]**gain** \ˈgān\ *noun* [Middle English *gayne,* from Middle French *gaigne, gain,* from Old French *gaaigne, gaaing,* from *gaaignier* to till, earn, gain, of Germanic origin; akin to Old High German *weidanōn* to hunt for food, Old English *wāth* pursuit, hunt] (14th century)
1 : resources or advantage acquired or increased **:** PROFIT ⟨made substantial *gains* last year⟩
2 : the act or process of gaining
3 a : an increase in amount, magnitude, or degree ⟨a *gain* in efficiency⟩ **b :** the increase (as of voltage or signal intensity) caused by an amplifier; *especially* **:** the ratio of output over input **c :** the signal-gathering ability of an antenna

[2]**gain** (14th century)
transitive verb
1 a : to acquire or get possession of usually by industry, merit, or craft ⟨*gain* an advantage⟩ ⟨he stood to *gain* a fortune⟩ **b :** to win in competition or conflict ⟨the troops *gained* enemy territory⟩ **c** (1) **:** to arrive at **:** REACH, ATTAIN ⟨*gained* the river that night⟩ (2) **:** TRAVERSE, COVER ⟨*gained* 10 yards on the play⟩ **d :** to get by a natural development or process ⟨*gain* strength⟩ **e :** to establish a specific relationship with ⟨*gain* a friend⟩
2 a : to make an increase of (a specified amount) ⟨*gained* 3% in the past month⟩ **b :** to increase in (a particular quality) ⟨*gain* momentum⟩
3 : to win to one's side **:** PERSUADE ⟨*gain* adherents to a cause⟩
4 : to cause to be obtained or given **:** ATTRACT ⟨*gain* attention⟩
5 *of a timepiece* **:** to run fast by the amount of ⟨the clock *gains* a minute a day⟩
intransitive verb
1 : to get advantage **:** PROFIT ⟨hoped to *gain* by the deal⟩
2 a : INCREASE ⟨the day was *gaining* in warmth⟩ **b :** to increase in weight **c :** to improve in health or ability
3 *of a timepiece* **:** to run fast
4 : to get closer to something pursued — usually used with *on* or *upon*
— gain·er *noun*
— gain ground : to make progress

gain·ful \ˈgān-fəl\ *adjective* (1555)
: productive of gain **:** PROFITABLE ⟨*gainful* employment⟩
— gain·ful·ly \-fə-lē\ *adverb*
— gain·ful·ness *noun*

gain·giv·ing \ˈgān-ˌgi-viŋ, ˌgān-ˈ\ *noun* [*gain-* (against) + *giving*] (1602)
archaic **:** MISGIVING

gain·say \ˌgān-ˈsā\ *transitive verb* **-said** \-ˈsād, -ˈsed\; **-say·ing** \-ˈsā-iŋ\; **-says** \-ˈsāz, -ˈsez\ [Middle English *gainsayen,* from *gain-* against (from Old English *gēan-*) + *sayen* to say — more at AGAIN] (14th century)
1 : to declare to be untrue or invalid
2 : CONTRADICT, OPPOSE
synonym see DENY
— gain·say·er *noun*

[1]**gait** \ˈgāt\ *noun* [Middle English *gait, gate* gate, way] (1509)
1 : a manner of walking or moving on foot
2 : a sequence of foot movements (as a walk, trot, pace, or canter) by which a horse or a dog moves forward
3 : a manner or rate of movement or progress ⟨the leisurely *gait* of summer⟩

[2]**gait** (circa 1900)
transitive verb
1 : to train (a horse or a dog) to use a particular gait or set of gaits
2 : to lead (a show dog) before a judge to display carriage and movement
intransitive verb
: to walk with a particular gait

gait·ed \ˈgā-təd\ *adjective* (1588)
: having a particular gait or so many gaits ⟨slow-*gaited*⟩ ⟨a *gaited* horse⟩

gai·ter \ˈgā-tər\ *noun* [French *guêtre*] (1775)
1 : a cloth or leather leg covering reaching from the instep to above the ankle or to midcalf or knee
2 a : an overshoe with fabric upper **b :** an ankle-high shoe with elastic gores in the sides

[1]**gal** \ˈgal\ *noun* [by alteration] (1795)
: GIRL, WOMAN

[2]**gal** *noun* [*Galileo* Galilei] (1914)
: a unit of acceleration equivalent to one centimeter per second per second — used especially for values of gravity

ga·la \ˈgā-lə, ˈga-, ˈgä-\ *noun* [Italian, from Middle French *gale* festivity, pleasure — more at GALLANT] (1777)
: a festive celebration; *especially* **:** a public entertainment marking a special occasion
— gala *adjective*

ga·la·bia *or* **ga·la·bi·eh** *or* **ga·la·bi·ya** \gə-ˈlä-b(ē-)ə\ *noun* [Arabic dialect (Egyptian) *gallābīyah*] (1725)
: DJELLABA

galact- *or* **galacto-** *combining form* [Latin *galact-,* from Greek *galakt-, galakto-,* from *galakt-, gala*]
1 : milk ⟨*galactor*rhea⟩
2 : related to galactose ⟨*galacto*semia⟩

ga·lac·tic \gə-ˈlak-tik\ *adjective* (1839)
1 : of or relating to a galaxy and especially the Milky Way galaxy
2 : HUGE ⟨a *galactic* sum of money⟩

ga·lac·tor·rhea \gə-ˌlak-tə-ˈrē-ə\ *noun* (circa 1860)
: a spontaneous flow of milk from the nipple

ga·lac·tos·amine \gə-ˌlak-ˈtō-sə-ˌmēn, -zə-\ *noun* (1900)
: an amino derivative $C_6H_{13}O_5N$ of galactose that occurs in cartilage

ga·lac·tose \gə-ˈlak-ˌtōs, -ˌtōz\ *noun* [French, from *galact-*] (1869)
: a sugar $C_6H_{12}O_6$ less soluble and less sweet than glucose

ga·lac·to·semia \gə-ˌlak-tə-ˈsē-mē-ə\ *noun* (1934)
: an inherited metabolic disorder in which galactose accumulates in the blood due to deficiency of an enzyme catalyzing its conversion to glucose
— ga·lac·to·semic \-mik\ *adjective*

ga·lac·to·si·dase \gə-ˌlak-ˈtō-sə-ˌdās, -zə-ˌdāz\ *noun* (1917)
: an enzyme (as lactase) that hydrolyzes a galactoside

ga·lac·to·side \gə-ˈlak-tə-ˌsīd\ *noun* (1862)
: a glycoside that yields galactose on hydrolysis

ga·lac·to·syl \gə-ˈlak-tə-ˌsil\ *noun* (1950)
: a glycosyl radical $C_6H_{11}O_5$ that is derived from galactose

ga·lact·uron·ic acid \gə-ˌlak-tü-ˈrä-nik-, -tyü-\ *noun* [International Scientific Vocabulary] (1917)
: a crystalline aldehyde-acid $C_6H_{10}O_7$ that occurs especially in polymerized form in pectin

ga·la·go \gə-ˈlā-(ˌ)gō, -ˈlä-\ *noun, plural* **-gos** [New Latin, perhaps from Wolof *golokh* monkey] (circa 1848)
: any of several small active nocturnal arboreal African primates (*Galago* and related genera) with large eyes, long ears, a long tail, and elongated hind limbs that enable them to leap with great agility — called also *bush baby*

galago

ga·lah \gə-ˈlä\ *noun* [Yuwaalaraay (Australian aboriginal language of northern New South Wales) *gilaa*] (1862)
: a showy Australian cockatoo (*Eolophus roseicapillus*) that is a pest in wheat-growing areas and is often kept as a cage bird

Gal·a·had \ˈga-lə-ˌhad\ *noun*
1 : the knight of the Round Table who successfully seeks the Holy Grail
2 : one who is pure, noble, and unselfish

gal·an·tine \ˈga-lən-ˌtēn\ *noun* [French, from Old French *galentine, galatine* fish sauce, from Medieval Latin *galatina,* probably from Latin *gelatus,* past participle of *gelare* to congeal, freeze — more at COLD] (1725)
: a cold dish consisting of boned meat or fish that has been stuffed, poached, and covered with aspic

Gal·a·tea \ˌga-lə-ˈtē-ə\ *noun* [Latin, from Greek *Galateia*]
: a female figure sculpted by Pygmalion and given life by Aphrodite in fulfillment of his prayer

Ga·la·tians \gə-ˈlā-shənz\ *noun plural but singular in construction*
: an argumentative letter of Saint Paul written to the Christians of Galatia and included as a book in the New Testament — see BIBLE table

gal·a·vant *variant of* GALLIVANT

ga·lax \ˈgā-ˌlaks\ *noun* [New Latin] (circa 1753)
: an evergreen herb (*Galax urceolata* synonym *G. aphylla* of the order Diapensiales) of the southeastern U.S. that has glossy leaves and is related to the heaths (order Ericales)

gal·axy \ˈga-lək-sē\ *noun, plural* **-ax·ies** [Middle English *galaxie, galaxias,* from Late Latin *galaxias,* from Greek, from *galakt-, gala* milk; akin to Latin *lac* milk] (14th century)
1 a *often capitalized* **:** MILKY WAY GALAXY — used with *the* **b :** any of the very large groups of stars and associated matter that are found throughout the universe
2 : an assemblage of brilliant or notable persons or things ⟨a *galaxy* of artists⟩ ◆

gal·ba·num \ˈgal-bə-nəm, ˈgȯl-\ *noun* [Middle English, from Latin, from Greek *chalbanē,* of Semitic origin; akin to Hebrew *ḥelbĕnāh* galbanum] (14th century)
: a yellowish to green or brown aromatic bitter gum resin derived from several Asian plants (as *Ferula galbaniflua*) and used in incense

gale \ˈgā(ə)l\ *noun* [origin unknown] (circa 1547)
1 a : a strong current of air: (1) **:** a wind from 32 to 63 miles per hour (about 51 to 101 kilometers per hour) (2) **:** FRESH GALE — see BEAUFORT SCALE table **b** *archaic* **:** BREEZE
2 : an emotional outburst ⟨*gales* of laughter⟩

ga·lea \ˈgā-lē-ə\ *noun* [New Latin, from Latin, helmet] (1834)
: an anatomical part suggesting a helmet: as **a :** the upper lip of the corolla of a mint **b :** the outer or lateral lobe of the maxilla in mandibulate insects
— ga·le·ate \-lē-ˌāt\ *adjective*

ga·le·na \gə-ˈlē-nə\ *noun* [Latin, lead ore] (1671)
: a bluish gray mineral with metallic luster consisting of lead sulfide, showing highly perfect cubic cleavage, and constituting the principal ore of lead

ga·len·i·cal \gə-ˈle-ni-kəl, gā-\ *noun* [*Galen* + [1]*-ic* + [1]*-al*] (1768)
: a medicine prepared by extracting one or more active constituents of a plant

ga·lère \ga-ˈler\ *noun* [French, galley, from Middle French, from Catalan *galera,* from Middle Greek *galea*] (1756)
: a group of people having an attribute in common

gal Friday *noun* (1958)
: GIRL FRIDAY

Gal·i·le·an \ˌga-lə-ˈlē-ən, -ˈlā-\ *adjective* (circa 1751)
: of, relating to, or discovered by Galileo Galilei

gal·i·lee \ˈga-lə-ˌlē\ *noun* [Anglo-French, from Medieval Latin *galilaea,* probably from *Galilaea* Galilee, from Latin] (15th century)
: a chapel or porch at the entrance of an English church

gal·in·gale \ˈga-lən-ˌgāl, -liŋ-\ *noun* [Middle English, a kind of ginger, from Middle French *galingal,* from Arabic *khalanjān*] (1578)
: an Old World sedge (*Cyperus longus*) that is used for papermaking and basket-weaving and has an aromatic root; *broadly* **:** any of various other sedges of the same genus

gal·i·ot *variant of* GALLIOT

[1]**gall** \ˈgȯl\ *noun* [Middle English, from Old English *gealla;* akin to Greek *cholē, cholos* gall, wrath, Old English *geolu* yellow — more at YELLOW] (before 12th century)
1 a : BILE; *especially* **:** bile obtained from an animal and used in the arts or medicine **b :** something bitter to endure **c :** bitterness of spirit **:** RANCOR
2 : brazen boldness coupled with impudent assurance and insolence
synonym see TEMERITY

[2]**gall** *noun* [Middle English *galle,* from Old English *gealla,* from Latin *galla* gallnut] (before 12th century)
1 a : a skin sore caused by chronic irritation **b :** a cause or state of exasperation
2 *archaic* **:** FLAW

[3]**gall** (14th century)
transitive verb
1 : to fret and wear away by friction **:** CHAFE ⟨the loose saddle *galled* the horse's back⟩ ⟨the *galling* of a metal bearing⟩
2 : IRRITATE, VEX ⟨sarcasm *galls* her⟩
intransitive verb
1 : to become sore or worn by rubbing
2 : SEIZE 2

[4]**gall** *noun* [Middle English *galle,* from Middle French, from Latin *galla*] (14th century)
: a swelling of plant tissue usually due to fungi or insect parasites and sometimes forming an important source of tannin

Gal·la \ˈga-lə\ *noun, plural* **Galla** *or* **Gallas** (1875)
: OROMO

gal·la·mine tri·eth·io·dide \ˈga-lə-ˌmēn-ˌtrī-ˌe-ˈthī-ə-ˌdīd\ *noun* [pyrogall*ol* + *amine* + *triethyl* + *iodide*] (1951)
: a substituted ammonium salt $C_{30}H_{60}I_3N_3O_3$ that is used to produce muscle relaxation especially during anesthesia — called also *gallamine*

[1]**gal·lant** \gə-ˈlant, gə-ˈlänt, ˈga-lənt\ *noun* (14th century)
1 : a young man of fashion
2 a : LADIES' MAN **b :** SUITOR **c :** PARAMOUR

[2]**gal·lant** \ˈga-lənt *(usually in sense 2);* gə-ˈlant, gə-ˈlänt *(usually in sense 3)*\ *adjective* [Middle English *galaunt,* from Middle French *galant,* from present participle of *galer* to have a good time, from *gale* pleasure, of Germanic origin; akin to Old English *wela* weal — more at WEAL] (15th century)
1 : showy in dress or bearing **:** SMART
2 a : SPLENDID, STATELY ⟨a *gallant* ship⟩ **b :** SPIRITED, BRAVE ⟨*gallant* efforts against the enemy⟩ **c :** nobly chivalrous and often self-sacrificing
3 : courteously and elaborately attentive especially to ladies
synonym see CIVIL
— gal·lant·ly *adverb*

[3]**gal·lant** \gə-ˈlant, -ˈlänt\ (1672)
transitive verb
1 : to pay court to (a lady) **:** ATTEND ⟨used to *gallant* her in his youth —Washington Irving⟩
2 *obsolete* **:** to manipulate (a fan) in a modish manner
intransitive verb
: to pay court to ladies

gal·lant·ry \ˈga-lən-trē\ *noun, plural* **-ries** (1613)
1 *archaic* **:** gallant appearance
2 a : an act of marked courtesy **b :** courteous attention to a lady **c :** amorous attention or pursuit
3 : spirited and conspicuous bravery

gal·late \ˈga-ˌlāt, ˈgȯ-\ *noun* (1794)
: a salt or ester of gallic acid

gall·blad·der \ˈgȯl-ˌbla-dər\ *noun* (1676)
: a membranous muscular sac in which bile from the liver is stored

gal·le·ass \ˈga-lē-əs\ *noun* [Middle French *galeasse,* from Old French *galie* galley] (1544)
: a large fast galley used especially as a warship by Mediterranean countries in the 16th and 17th centuries and having both sails and oars but usually propelled chiefly by rowing

gal·le·on \ˈga-lē-ən\ *noun* [Old Spanish *galeón,* from Middle French *galion,* from Old French *galie*] (1529)
: a heavy square-rigged sailing ship of the 15th to early 18th centuries used for war or commerce especially by the Spanish

galleon

◇ WORD HISTORY

galaxy The earliest known use in English of the words *galaxy* and *Milky Way* is in the following lines from a poem by Geoffrey Chaucer called *The House of Fame,* written sometime before 1385: "Se yonder, loo, the Galaxie, /Which men clepeth the Milky Wey,/ For hit is whit." (See yonder, lo, the galaxy, which they call the Milky Way, because it is white.) These lines make it clear that *Milky Way,* at least, was a term popular in England before that time, though how early it was in use is impossible to say. The idea of the whiteness of the Milky Way being similar to that of milk is considerably older than the English language, however, for *galaxy* is borrowed through Latin from Greek *galaxias,* a derivative of *gala* "milk." Not until the 19th century was *galaxy* used as a generic term for other star systems as well as the one in which we live, and to distinguish our galaxy from others the name *Milky Way* has proved useful in English.

gal·le·ria \ˌga-lə-ˈrē-ə\ *noun* [Italian, gallery, from Medieval Latin *galeria*] (circa 1901)
: a roofed and usually glass-enclosed promenade or court (as at a mall)

gal·lery \ˈga-lə-rē, ˈgal-rē\ *noun, plural* **-ler·ies** [Middle English *galerie,* from Medieval Latin *galeria,* probably alteration of *galilaea* galilee] (15th century)
1 a : a roofed promenade **:** COLONNADE **b :** CORRIDOR
2 a : an outdoor balcony **b** *Southern & Midland* **:** PORCH, VERANDA **c :** a platform at the quarters or stern of a ship
3 a : a long and narrow passage, apartment, or corridor **b :** a subterranean passageway in a cave or military mining system; *also* **:** a working drift or level in mining **c :** an underground passage made by a mole or ant or a passage made in wood by an insect
4 a : a room or building devoted to the exhibition of works of art **b :** an institution or business exhibiting or dealing in works of art **c :** COLLECTION, AGGREGATION ⟨the rich *gallery* of characters in this novel —H. S. Canby⟩
5 a : a structure projecting from one or more interior walls of an auditorium to accommodate additional people; *especially* **:** the highest balcony in a theater commonly having the cheapest seats **b :** the part of a theater audience seated in the top gallery **c :** the undiscriminating general public **d :** the spectators at a tennis or golf match
6 : a small ornamental barrier or railing (as along the edge of a table or shelf)
7 : a photographer's studio
— gal·ler·ied \-rēd\ *adjective*
— gal·lery·ite \-rē-ˌīt\ *noun*

gallery forest *noun* (1920)
: a forest growing along a watercourse in a region otherwise devoid of trees

gal·lery·go·er \ˈga-lə-rē-ˌgō(-ə)r, ˈgal-rē-\ *noun* (1888)
: one who frequently goes to art galleries

ga·lle·ta \gə-ˈye-tə, gī-ˈe-tə\ *noun* [Spanish, hardtack] (1872)
: either of two perennial forage grasses (*Hilaria rigida* and *H. jamesii*) used for hay in the southwestern U.S. and in Mexico

gal·ley \ˈga-lē\ *noun, plural* **galleys** [Middle English *galeie,* from Old French *galie, galee,* ultimately from Middle Greek *galea*] (13th century)
1 : a ship or boat propelled solely or chiefly by oars: as **a :** a long low ship used for war and trading especially in the Mediterranean Sea from the Middle Ages to the 19th century; *also* **:** GALLEASS **b :** a warship of classical antiquity — compare BIREME, TRIREME **c :** a large open boat (as a gig) formerly used in England
2 : the kitchen and cooking apparatus especially of a ship or airplane
3 a : an oblong tray to hold especially a single column of set type **b :** a proof of typeset matter especially in a single column before being made into pages

gal·ley–west \ˌga-lē-ˈwest\ *adverb* [probably alteration of English dialect *collywest* (badly askew)] (1875)
: into destruction or confusion ⟨was knocked *galley-west*⟩

gall·fly \ˈgȯl-ˌflī\ *noun* (circa 1834)
: an insect (as a gall wasp) that deposits its eggs in plants causing the formation of galls in which the larvae feed

[1]**gal·liard** \ˈgal-yərd\ *adjective* [Middle English *gaillard,* from Middle French] (14th century)
archaic **:** GAY, LIVELY

[2]**galliard** *noun* (1533)
: a sprightly dance with five steps to a phrase popular in the 16th and 17th centuries

Gal·lic \ˈga-lik\ *adjective* [Latin *Gallicus,* from *Gallia* Gaul] (1672)
: of or relating to Gaul or France

gal·lic acid \ˈga-lik-, ˈgȯ-lik-\ *noun* [French *gallique,* from *galle* gall] (1791)
: a white crystalline acid $C_7H_6O_5$ found widely in plants or combined in tannins and used especially in dyes and as a photographic developer

Gal·li·can \ˈga-li-kən\ *adjective* (14th century)
1 : GALLIC
2 *often not capitalized* **:** of or relating to Gallicanism
— Gallican *noun*

Gal·li·can·ism \-kə-ˌni-zəm\ *noun* (1858)
: a movement originating in France and advocating administrative independence from papal control for the Roman Catholic Church in each nation

gal·li·cism \ˈga-lə-ˌsi-zəm\ *noun, often capitalized* (circa 1656)
1 : a characteristic French idiom or expression appearing in another language
2 : a French trait

gal·li·cize \-ˌsīz\ *transitive verb* **-cized; -ciz·ing** *often capitalized* (1773)
: to cause to conform to a French mode or idiom
— gal·li·ci·za·tion \ˌga-lə-sə-ˈzā-shən\ *noun, often capitalized*

gal·li·gas·kins \ˌga-li-ˈgas-kənz\ *noun plural* [probably modification of Middle French *garguesques,* from Old Spanish *gregüescos,* from *griego* Greek, from Latin *Graecus*] (1577)
1 a : loose wide hose or breeches worn in the 16th and 17th centuries **b :** very loose trousers
2 *dialect chiefly British* **:** LEGGINGS

gal·li·mau·fry \ˌga-lə-ˈmȯ-frē\ *noun, plural* **-fries** [Middle French *galimafree* stew] (circa 1556)
: HODGEPODGE ⟨a *gallimaufry* of opinions⟩

gal·li·na·ceous \ˌga-lə-ˈnā-shəs\ *adjective* [Latin *gallinaceus* of domestic fowl, from *gallina* hen, from *gallus* cock] (1783)
: of or relating to an order (Galliformes) of heavy-bodied largely terrestrial birds including the pheasants, turkeys, grouse, and the common domestic fowl

gall·ing \ˈgȯ-liŋ\ *adjective* [[3]*gall*] (1583)
: markedly irritating **:** VEXING ⟨a most *galling* defeat⟩
— gall·ing·ly \-liŋ-lē\ *adverb*

gal·li·nip·per \ˈga-lə-ˌni-pər\ *noun* [origin unknown] (1709)
chiefly Southern & Midland **:** any of various insects (as a large mosquito or crane fly)

gal·li·nule \ˈga-lə-ˌnü(ə)l, -ˌnyü(ə)l\ *noun* [New Latin *Gallinula,* genus of birds, from Latin, pullet, diminutive of *gallina*] (1776)
: any of several aquatic birds of the rail family with long thin feet and a platelike frontal area on the head; *especially* **:** one (*Gallinula chloropus*) widespread in the New World, Eurasia, and Africa that has a largely red bill, red frontal area on the head, and a white band on the flanks

gal·li·ot \ˈga-lē-ət\ *noun* [Middle English *galiote,* from Middle French, from Medieval Latin *galeota,* diminutive of *galea* galley, from Middle Greek] (14th century)
1 : a small swift galley formerly used in the Mediterranean
2 [Dutch *galjoot,* from Middle French *galiote*] **:** a long narrow shallow-draft Dutch merchant sailing ship

gal·li·pot \ˈga-li-ˌpät\ *noun* [Middle English *galy pott*] (15th century)
1 : a small usually ceramic vessel
2 *archaic* **:** DRUGGIST

gal·li·um \ˈga-lē-əm\ *noun* [New Latin, from Latin *gallus* cock (intended as translation of surname of Paul *Lecoq* de Boisbaudran (died 1912) French chemist)] (1875)
: a rare bluish white metallic element that is hard and brittle at low temperatures but melts just above room temperature and expands on freezing — see ELEMENT table

gallium arsenide *noun* (circa 1961)
: a synthetic compound GaAs used especially as a semiconducting material

gal·li·vant \ˈga-lə-ˌvant\ *intransitive verb* [perhaps alteration of [3]*gallant*] (1823)
1 : to go about usually ostentatiously or indiscreetly with members of the opposite sex
2 : to travel, roam, or move about for pleasure

gall midge *noun* (circa 1889)
: any of numerous minute dipteran flies (family Cecidomyiidae) most of which cause gall formation in plants

gall mite *noun* (1881)
: any of various minute 4-legged mites (family Eriophyidae) that form galls on plants

gall·nut \ˈgȯl-ˌnət\ *noun* [[4]*gall*] (1572)
: a gall resembling a nut

gal·lon \ˈga-lən\ *noun* [Middle English *galon,* a liquid measure, from Old North French, from Medieval Latin *galeta* pail, a liquid measure] (13th century)
: a unit of liquid capacity equal to 231 cubic inches or four quarts — see WEIGHT table

gal·lon·age \ˈga-lə-nij\ *noun* (circa 1909)
: amount in gallons

gal·loon \gə-ˈlün\ *noun* [French *galon*] (1604)
: a narrow trimming (as of lace, embroidery, or braid with metallic threads) having both edges scalloped

[1]**gal·lop** \ˈga-ləp\ (15th century)
intransitive verb
1 : to progress or ride at a gallop
2 : to run fast
transitive verb
1 : to cause to gallop
2 : to transport at a gallop
— gal·lop·er *noun*

[2]**gallop** *noun* [Middle French *galop*] (1523)
1 : a bounding gait of a quadruped; *specifically* **:** a fast natural 3-beat gait of the horse — compare [3]CANTER, RUN
2 : a ride or run at a gallop
3 : a stretch of land suitable for galloping horses
4 : a rapid or hasty progression or pace

gal·lo·pade \ˌga-lə-ˈpād, -ˈpäd\ *noun* (1831)
: GALOP

Gal·lo·phile \ˈga-lə-ˌfīl\ *adjective* [Latin *Gallus* Gaul + English *-phile*] (1923)
: FRANCOPHILE
— Gallophile *noun*

gal·lop·ing *adjective* (1642)
: progressing, developing, or increasing rapidly ⟨*galloping* inflation⟩ ⟨a *galloping* farce⟩ ⟨*galloping* alcoholism⟩

Gal·lo·way \ˈga-lə-ˌwā\ *noun* [*Galloway,* Scotland] (1805)
: any of a breed of hardy medium-sized hornless chiefly black beef cattle native to southwestern Scotland

gal·low·glass \ˈga-lō-ˌglas\ *noun* [alteration of Irish *gallóglach,* from *gall* foreigner + *óglach* young man, warrior] (circa 1515)
1 : a mercenary or retainer of an Irish chief
2 : an armed Irish foot soldier

[1]**gal·lows** \ˈga-(ˌ)lōz, -ləz, *in sense 3 also* -ləs\ *noun, plural* **gallows** *or* **gal·lows·es** [Middle English *galwes,* plural of *galwe,* from Old English *gealga;* akin to Old Norse *gelgja* pole, stake, Armenian *jałk* twig] (before 12th century)
1 a : a frame usually of two upright posts and a transverse beam from which criminals are hanged — called also *gallows tree* **b :** the punishment of hanging
2 : a structure consisting of an upright frame with a crosspiece
3 : SUSPENDER 2a

[2]**gallows** *adjective* (15th century)
: deserving the gallows

gallows bird *noun* (circa 1785)
: a person who deserves hanging

gallows humor *noun* (1901)
: humor that makes fun of a life-threatening, disastrous, or terrifying situation

gall·stone \ˈgȯl-ˌstōn\ *noun* (1758)

: a calculus formed in the gallbladder or biliary passages

gal·lus \'ga-ləs\ *noun* [alteration of [1]*gallows*] (1836)
: SUSPENDER 2a — usually used in plural

gal·lused \'ga-ləst\ *adjective* (1927)
: wearing galluses

gall wasp *noun* (1879)
: any of a family (Cynipidae) of hymenopterous gallflies

gal·ly \'ga-lē\ *transitive verb* **gal·lied; gal·ly·ing** [origin unknown] (1605)
chiefly dialect **:** FRIGHTEN, TERRIFY

Ga·lois theory \(ˌ)gal-'wä-, 'gal-ˌwä-\ *noun* [Évariste *Galois*] (1893)
: a part of the theory of mathematical groups concerned especially with the conditions under which a solution to a polynomial equation with coefficients in a given mathematical field can be obtained in the field by the repetition of operations and the extraction of nth roots

ga·loot \gə-'lüt\ *noun* [origin unknown] (1866)
slang **:** FELLOW; *especially* **:** one who is strange or foolish

ga·lop \'ga-ləp, ga-'lō\ *noun* [French] (1831)
: a lively dance in duple measure; *also* **:** the music of a galop

ga·lore \gə-'lōr, -'lȯr\ *adjective* [Irish *go leor* enough] (1628)
: ABUNDANT, PLENTIFUL — used postpositively ⟨bargains *galore*⟩

ga·losh \gə-'läsh\ *noun* [Middle English *galoche,* from Middle French] (14th century)
1 *obsolete* **:** a shoe with a heavy sole
2 : a high overshoe worn especially in snow and slush
— ga·loshed \-'läsht\ *adjective*

ga·lumph \gə-'ləm(p)f\ *intransitive verb* [probably alteration of [1]*gallop*] (1872)
: to move with a clumsy heavy tread

gal·van·ic \gal-'va-nik\ *adjective* (1797)
1 : of, relating to, or producing a direct current of electricity ⟨a *galvanic* cell⟩
2 a : having an electric effect **:** intensely exciting ⟨a *galvanic* performance⟩ **b :** produced as if by an electric shock ⟨had a *galvanic* effect on the audience⟩
— gal·van·i·cal·ly \-ni-k(ə-)lē\ *adverb*

galvanic skin response *noun* (1942)
: a change in the electrical resistance of the skin that is a physiochemical response to an emotional change

gal·va·nise *British variant of* GALVANIZE

gal·va·nism \'gal-və-ˌni-zəm\ *noun* [French or Italian; French *galvanisme,* from Italian *galvanismo,* from Luigi *Galvani*] (1797)
1 : a direct current of electricity especially when produced by chemical action
2 : the therapeutic use of direct electric current
3 : vital or forceful activity

gal·va·nize \'gal-və-ˌnīz\ *verb* **-nized; -niz·ing** (1802)
transitive verb
1 a : to subject to the action of an electric current especially for the purpose of stimulating physiologically ⟨*galvanize* a muscle⟩ **b :** to stimulate or excite as if by an electric shock ⟨an issue that would *galvanize* public opinion⟩
2 : to coat (iron or steel) with zinc; *especially* **:** to immerse in molten zinc to produce a coating of zinc-iron alloy
intransitive verb
: to react as if stimulated by an electric shock ⟨they *galvanized* into action⟩
— gal·va·ni·za·tion \ˌgal-və-nə-'zā-shən\ *noun*
— gal·va·niz·er \'gal-və-ˌnī-zər\ *noun*

galvano- *combining form* [*galvanic*]
: galvanic current ⟨*galvano*meter⟩

gal·va·nom·e·ter \ˌgal-və-'nä-mə-tər\ *noun* (1802)
: an instrument for detecting or measuring a small electric current by movements of a magnetic needle or of a coil in a magnetic field
— gal·va·no·met·ric \-nō-'me-trik\ *adjective*

gal·va·no·scope \gal-'va-nə-ˌskōp, 'gal-və-nə-\ *noun* (1832)
: an instrument for detecting the presence and direction of an electric current by the deflection of a magnetic needle

[1]gam \'gam\ *noun* [probably ultimately from Lingua Franca *gamba* leg, from Italian, from Late Latin] (1781)
slang **:** LEG

[2]gam *noun* [perhaps short for obsolete *gammon* (talk)] (1846)
1 : a visit or friendly conversation at sea or ashore especially between whalers
2 : a school of whales

[3]gam *verb* **gammed; gam·ming** (1849)
intransitive verb
: to engage in a gam
transitive verb
1 : to have a gam with
2 : to spend or pass (as time) talking

gam- *or* **gamo-** *combining form* [New Latin, from Greek, marriage, from *gamos,* from *gamein* to marry]
: united **:** joined ⟨*gamo*petalous⟩

gama grass \'ga-mə-\ *noun* [probably alteration of *grama*] (1833)
: a tall coarse American grass (*Tripsacum dactyloides*) valuable for forage

ga·may \ga-'mā, 'ga-ˌmā\ *noun, often capitalized* [French, from *Gamay,* village in Burgundy] (circa 1941)
: a light dry red table wine made from the same grape used for French Beaujolais ⟨*gamay* rosé⟩

gam·ba \'gäm-bə, 'gam-\ *noun* (1598)
: VIOLA DA GAMBA

[1]gam·ba·do \gam-'bā-(ˌ)dō\ *noun, plural* **-does** *also* **-dos** [perhaps modification of Italian *gambale,* from *gamba* leg] (circa 1656)
: a horseman's legging

[2]gambado *noun, plural* **-does** *also* **-dos** [modification of French *gambade* — more at GAMBOL] (1820)
1 : a spring of a horse
2 : CAPER, GAMBOL

gam·bier *also* **gam·bir** \'gam-ˌbir\ *noun* [Malay *gambir*] (1830)
: a yellowish catechu that is obtained from a Malayan woody vine (*Uncaria gambir*) of the madder family and is used for chewing with the betel nut and for tanning and dyeing

gam·bit \'gam-bət\ *noun* [Italian *gambetto,* literally, act of tripping someone, from *gamba* leg, from Late Latin *gamba, camba,* from Greek *kampē* bend; probably akin to Gothic *hamfs* maimed, Lithuanian *kampas* corner] (1656)
1 : a chess opening in which a player risks one or more minor pieces to gain an advantage in position
2 a (1) **:** a remark intended to start a conversation or make a telling point (2) **:** TOPIC **b :** a calculated move **:** STRATAGEM

[1]gam·ble \'gam-bəl\ *verb* **gam·bled; gam·bling** \-b(ə-)liŋ\ [probably back-formation from *gambler,* probably alteration of obsolete *gamner,* from obsolete *gamen* (to play)] (circa 1775)
intransitive verb
1 a : to play a game for money or property **b :** to bet on an uncertain outcome
2 : to stake something on a contingency **:** take a chance
transitive verb
1 : to risk by gambling **:** WAGER
2 : VENTURE, HAZARD
— gam·bler \-blər\ *noun*

[2]gamble *noun* (1823)
1 a : an act having an element of risk **b :** something chancy
2 : the playing of a game of chance for stakes

gam·boge \gam-'bōj, -'büzh\ *noun* [New Latin *gambogium,* alteration of *cambugium,* from or akin to Portuguese *Camboja* Cambodia] (1712)
1 : an orange to brown gum resin from southeast Asian trees (genus *Garcinia*) of the Saint-John's-wort family that is used as a yellow pigment and cathartic
2 : a strong yellow

[1]gam·bol \'gam-bəl\ *intransitive verb* **-boled** *or* **-bolled; -bol·ing** *or* **-bol·ling** \-bə-liŋ *also* -bliŋ\ (1508)
: to skip about in play **:** FRISK, FROLIC

[2]gambol *noun* [modification of Middle French *gambade* spring of a horse, gambol, probably from Old Provençal *camba* leg, from Late Latin] (circa 1510)
: a skipping or leaping about in play

gam·brel \'gam-brəl\ *noun* [Old North French *gamberel,* from *gambe* leg, from Late Latin *gamba*] (1547)
: a stick or iron for suspending slaughtered animals

gambrel roof *noun* (1765)
: a roof with a lower steeper slope and an upper less steep one on each of its two sides — see ROOF illustration

gam·bu·sia \gam-'bü-zh(ē-)ə, -'byü-\ *noun* [New Latin, modification of American Spanish *gambusino* gambusia] (circa 1889)
: any of a genus (*Gambusia*) of live-bearers (family Poeciliidae) introduced as exterminators of mosquito larvae in warm fresh waters

[1]game \'gām\ *noun* [Middle English, from Old English *gamen;* akin to Old High German *gaman* amusement] (before 12th century)
1 a (1) **:** activity engaged in for diversion or amusement **:** PLAY (2) **:** the equipment for a game **b :** often derisive or mocking jesting **:** FUN, SPORT ⟨make *game* of a nervous player⟩
2 a : a procedure or strategy for gaining an end **:** TACTIC **b :** an illegal or shady scheme or maneuver **:** RACKET
3 a (1) **:** a physical or mental competition conducted according to rules with the participants in direct opposition to each other (2) **:** a division of a larger contest (3) **:** the number of points necessary to win (4) **:** points scored in certain card games (as in all fours) by a player whose cards count up the highest (5) **:** the manner of playing in a contest (6) **:** the set of rules governing a game (7) **:** a particular aspect or phase of play in a game or sport ⟨a football team's kicking *game*⟩ **b** *plural* **:** organized athletics **c** (1) **:** a field of gainful activity **:** LINE ⟨the newspaper *game*⟩ (2) **:** any activity undertaken or regarded as a contest involving rivalry, strategy, or struggle ⟨the dating *game*⟩ ⟨the *game* of politics⟩; *also* **:** the course or period of such an activity ⟨got into aviation early in the *game*⟩ (3) **:** area of expertise **:** SPECIALTY 3 ⟨comedy is not my *game*⟩
4 a (1) **:** animals under pursuit or taken in hunting; *especially* **:** wild animals hunted for sport or food (2) **:** the flesh of game animals **b** *archaic* **:** PLUCK **c :** a target or object especially of ridicule or attack — often used in the phrase *fair game*
synonym see FUN
— game·like \-ˌlīk\ *adjective*

[2]game *verb* **gamed; gam·ing** (1529)
intransitive verb
: to play for a stake
transitive verb
archaic **:** to lose or squander by gambling

[3]game *adjective* (1610)
1 a : having or showing a resolute unyielding spirit ⟨*game* to the end⟩ **b :** willing or ready to proceed ⟨were *game* for anything⟩
2 : of or relating to game ⟨*game* laws⟩
— game·ly *adverb*
— game·ness *noun*

⁴game *adjective* [origin unknown] (circa 1787)
: LAME ⟨a *game* leg⟩

game ball *noun* (1966)
: a ball (as a football) presented to a player or coach in recognition of an outstanding contribution to a team victory

game bird *noun* (1866)
: a bird that may be legally hunted according to the laws especially of a state of the U.S.

game•cock \'gām-,käk\ *noun* (1677)
: a rooster of the domestic chicken trained for fighting

game fish *noun* (1862)
1 : a fish of a family (Salmonidae) including salmons, trouts, chars, and whitefishes
2 : SPORT FISH; *especially* **:** a fish made a legal catch by law

game•keep•er \'gām-,kē-pər\ *noun* (circa 1671)
: a person in charge of the breeding and protection of game animals or birds on a private preserve

gam•elan \'ga-mə-,lan\ *noun* [Javanese] (1817)
: an Indonesian orchestra made up especially of percussion instruments (as gongs, xylophones, and drums)

game of chance (circa 1925)
: a game (as a dice game) in which chance rather than skill determines the outcome

game plan *noun* (1941)
: a strategy for achieving an objective

game point *noun* (circa 1949)
: a situation (as in tennis) in which one player will win the game by winning the next point; *also* **:** the point itself

gam•er \'gā-mər\ *noun* (circa 1630)
: a player who is game; *especially* **:** an athlete who relishes competition

game show *noun* (1958)
: a television program on which contestants compete for prizes in a game (as a quiz)

games•man \'gāmz-mən, -,man\ *noun* (1947)
: one who practices gamesmanship; *also* **:** one who plays games

games•man•ship \'gāmz-mən-,ship\ *noun* (1947)
1 : the art or practice of winning games by questionable expedients without actually violating the rules
2 : the use of ethically dubious methods to gain an objective ◆

game•some \'gām-səm\ *adjective* [Middle English] (14th century)
: MERRY, FROLICSOME
— game•some•ly *adverb*
— game•some•ness *noun*

game•ster \'gām-stər\ *noun* (1553)
: one who plays games; *especially* **:** GAMBLER

gamet- *or* **gameto-** *combining form* [New Latin, from *gameta*]
: gamete ⟨*gameto*phore⟩

gam•etan•gi•um \,ga-mə-'tan-jē-əm\ *noun, plural* **-gia** \-jē-ə\ [New Latin, from *gamet-* + Greek *angeion* vessel — more at ANGI-] (1886)
: a cell or organ in which gametes are developed

gam•ete \'ga-,mēt *also* gə-'mēt\ *noun* [New Latin *gameta,* from Greek *gametēs* husband, from *gamein* to marry] (1886)
: a mature male or female germ cell usually possessing a haploid chromosome set and capable of initiating formation of a new diploid individual by fusion with a gamete of the opposite sex
— ga•met•ic \gə-'me-tik, -'mē-\ *adjective*
— ga•met•i•cal•ly \-ti-k(ə-)lē\ *adverb*

game theory *noun* (1949)
: the analysis of a situation involving conflicting interests (as in business or military strategy) in terms of gains and losses among opposing players
— game theorist *noun*

ga•me•to•cyte \gə-'mē-tə-,sīt\ *noun* [International Scientific Vocabulary] (1899)
: a cell (as of a protozoan causing malaria) that divides to produce gametes

ga•me•to•gen•e•sis \gə-,mē-tə-'je-nə-səs, ,ga-mə-tə-\ *noun* [New Latin] (circa 1900)
: the production of gametes
— ga•me•to•gen•ic \-'je-nik\ *or* **gam•etog•e•nous** \,ga-mə-'tä-jə-nəs\ *adjective*

ga•me•to•phore \gə-'mē-tə-,fōr, -,fȯr\ *noun* (1895)
: a modified branch (as of a moss) bearing gametangia

ga•me•to•phyte \gə-'mē-tə-,fīt\ *noun* [International Scientific Vocabulary] (circa 1889)
: the individual or generation of a plant exhibiting alternation of generations that bears sex organs — compare SPOROPHYTE
— ga•me•to•phyt•ic \-,mē-tə-'fi-tik\ *adjective*

-gamic *adjective combining form* [International Scientific Vocabulary, from Greek *-gamos* -gamous]
: having (such) reproductive organs ⟨cleisto*gamic*⟩

gam•in \'ga-mən\ *noun* [French] (1840)
1 : a boy who hangs around on the streets **:** URCHIN
2 : GAMINE 2

¹ga•mine \ga-'mēn, 'ga-,mēn\ *noun* [French, feminine of *gamin*] (1889)
1 : a girl who hangs around on the streets
2 : a small playfully mischievous girl

²gamine *adjective* (1925)
: of, relating to, or suggesting a gamine

gam•ing \'gā-miŋ\ *noun* (1501)
1 : the practice of gambling
2 : the playing of games that simulate actual conditions (as of business or war) especially for training or testing purposes

¹gam•ma \'ga-mə\ *noun* [Middle English, from Late Latin, from Greek, of Semitic origin; akin to Hebrew *gīmel* gimel] (15th century)
1 : the 3d letter of the Greek alphabet — see ALPHABET table
2 : the degree of contrast of a developed photographic image or of a television image
3 : a unit of magnetic flux density equal to one nanotesla
4 : GAMMA RAY — usually used as the attributive form of *gamma ray* ⟨*gamma* counter⟩
5 : MICROGRAM

²gamma *adjective* (1896)
1 : of, relating to, or being one of three or more closely related chemical substances
2 : third in position in the structure of an organic molecule from a particular group or atom — symbol γ

gam•ma–ami•no•bu•tyr•ic acid \,ga-mə-ə-,mē-(,)nō-byü-'tir-ik-, ,ga-mə-,a-mə-(,)nō-\ *noun* (1957)
: an amino acid $C_4H_9NO_2$ that is a neurotransmitter in the central nervous system

gamma camera *noun* (1964)
: a camera that detects the radiation from a radioactive tracer injected into the body and is used especially in medical diagnostic scanning

gamma globulin *noun* (1937)
1 a : a protein fraction of blood rich in antibodies **b :** a sterile solution of gamma globulin from pooled human blood administered especially for passive immunity against measles, German measles, infectious hepatitis, or poliomyelitis
2 : any of numerous globulins of blood plasma or serum that have less electrophoretic mobility at alkaline pH than serum albumins, alpha globulins, or beta globulins and that include most antibodies — compare ANTIBODY

gamma radiation *noun* (1904)
: radiation composed of gamma rays

gamma ray *noun* (1903)
: a photon emitted spontaneously by a radioactive substance; *also* **:** a high-energy photon — usually used in plural

gam•mer \'ga-mər\ *noun* [alteration of *godmother*] (1575)
archaic **:** an old woman — compare GAFFER

¹gam•mon \'ga-mən\ *noun* [Middle English, from Old North French *gambon* ham, from *gambe* leg — more at GAM] (15th century)
1 *chiefly British* **:** HAM 2
2 *chiefly British* **a :** a side of bacon **b :** the lower end of a side of bacon

²gammon *noun* [perhaps alteration of Middle English *gamen* game] (circa 1734)
1 *archaic* **:** BACKGAMMON
2 : the winning of a backgammon game before the loser removes any men from the board

³gammon *transitive verb* (1735)
: to beat by scoring a gammon

⁴gammon (1789)
intransitive verb
1 : to talk gammon
2 : PRETEND, FEIGN
transitive verb
: DECEIVE, FOOL

⁵gammon *noun* [obsolete *gammon* (talk)] (1805)
: talk intended to deceive **:** HUMBUG

gam•my \'ga-mē\ *adjective* [perhaps from ⁴*game*] (1870)
British **:** LAME, GAME

gamo- — see GAM-

gamo•pet•al•ous \,ga-mə-'pe-tᵊl-əs\ *adjective* (1830)
: having the corolla composed of united petals ⟨the morning glory is *gamopetalous*⟩

-gamous *adjective combining form* [Greek *-gamos,* from *gamos* marriage, from *gamein* to marry]
1 : characterized by having or practicing (such) a marriage or (such or so many) marriages ⟨exo*gamous*⟩
2 : -GAMIC ⟨hetero*gamous*⟩

gamp \'gamp\ *noun* [Sarah *Gamp,* nurse with a large umbrella in *Martin Chuzzlewit* by Charles Dickens] (1864)
British **:** a large umbrella

gam•ut \'ga-mət\ *noun* [Medieval Latin *gamma,* lowest note of a medieval scale (from Late Latin, 3d letter of the Greek alphabet) + *ut* ut] (15th century)
1 : the whole series of recognized musical notes

◇ WORD HISTORY

gamesmanship In 1947 the British writer and critic Stephen Potter (1900–69) published a short, drolly witty book called *Gamesmanship*—a word that Potter defined as "the art of winning games without actually cheating." In the book, Potter introduced more *-manship* coinages ("winmanship," "losemanship"), and *Gamesmanship* was followed by further titles: in 1950 *Lifemanship* ("how to make the other man feel that something has gone wrong, however slightly"), *One-Upmanship* in 1952, and *Supermanship* in 1958. The coinages *gamesmanship* and *one-upmanship* have outlived the success of Potter's books—as well as Potter himself—and popularized the use of *-manship* as a suffix on its own, with the approximate meaning "skill in a competitive activity," whether or not there is a corresponding noun ending in *-man.* In 1956 appeared *brinkmanship,* originally coined to describe the foreign policy of John Foster Dulles, Dwight Eisenhower's secretary of state, but now a standard term for a kind of international negotiating tactic. A more recent coinage is *grantsmanship,* first recorded in 1961.

2 : an entire range or series ⟨ran the *gamut* from praise to contempt⟩ ◇
synonym see RANGE

gamy *or* **gam·ey** \'gā-mē\ *adjective* **gam·i·er; -est** (1844)
1 : BRAVE, PLUCKY — used especially of animals
2 a : having the flavor of game; *especially* **:** having the flavor of game near tainting **b :** SMELLY
3 a : SORDID, SCANDALOUS ⟨gave us all the *gamy* details⟩ **b :** CORRUPT, DISREPUTABLE ⟨a *gamy* character⟩ **c :** sexually suggestive **:** RACY ⟨*gamy* witticisms⟩
— **gam·i·ly** \-mə-lē\ *adverb*
— **gam·i·ness** \-mē-nəs\ *noun*

-gamy *noun combining form* [Middle English *-gamie,* from Late Latin *-gamia,* from Greek, from *gamein* to marry]
1 : marriage ⟨exo*gamy*⟩
2 : union for propagation or reproduction ⟨allo*gamy*⟩
3 : possession of (such) reproductive organs or (such) a mode of fertilization ⟨cleisto*gamy*⟩

gan *past of* GIN

ga·nache \(,)gä-'näsh, gə-\ *noun* [French, literally, jowl, from Italian *ganascia,* modification of Greek *gnathos* jaw — more at -GNATHOUS] (1977)
: a sweet creamy chocolate mixture used especially as a filling or frosting

Gan·da \'gan-də, 'gän-\ *noun, plural* **Ganda** *or* **Gandas** (1934)
1 : a member of a Bantu-speaking people of Uganda
2 : LUGANDA

¹gan·der \'gan-dər\ *noun* [Middle English, from Old English *gandra;* akin to Old English *gōs* goose] (before 12th century)
1 : the adult male goose
2 : SIMPLETON

²gander *noun* [probably from ¹*gander;* from the outstretched neck of a person craning to look at something] (circa 1914)
: LOOK, GLANCE ⟨take a *gander*⟩

gan·dy dancer \'gan-dē-\ *noun* [origin unknown] (1923)
1 : a laborer in a railroad section gang
2 : an itinerant or seasonal laborer ◇

ga·nef \'gä-nəf\ *noun* [Yiddish, from Hebrew *gannābh* thief] (1923)
slang **:** THIEF, RASCAL

Ga·ne·lon \,ga-nᵊl-'ōⁿ\ *noun* [French]
: the traitor in the Charlemagne romances who is responsible for the death of Roland

¹gang \'gaŋ\ *intransitive verb* [Middle English, from Old English *gangan;* akin to Lithuanian *žengti* to stride] (before 12th century)
Scottish **:** GO

²gang *noun* [Middle English, walking, journey, from Old English; akin to Old English *gangan*] (15th century)
1 a (1) **:** a set of articles **:** OUTFIT ⟨a *gang* of oars⟩ (2) **:** a combination of similar implements or devices arranged for convenience to act together ⟨a *gang* of saws⟩ **b :** GROUP: as (1) **:** a group of persons working together (2) **:** a group of persons working to unlawful or antisocial ends; *especially* **:** a band of antisocial adolescents
2 : a group of persons having informal and usually close social relations

³gang (1856)
transitive verb
1 a : to assemble or operate simultaneously as a group **b :** to arrange in or produce as a gang
2 : to attack in a gang
intransitive verb
: to move or act as a gang

gang bang *noun* (1950)
1 : copulation by several persons in succession with the same passive partner — often considered vulgar
2 : GANG RAPE — often considered vulgar
— **gang–bang** *verb*

gang·bang·er \'gaŋ-,baŋ-ər\ *noun* (circa 1972)
: a member of a street gang

gang·bust·er \'gaŋ-,bəs-tər\ *noun* (1940)
: one engaged in the aggressive breakup of organized criminal gangs
— **like gangbusters :** with great or excessive force or aggressiveness ⟨came on *like gangbusters*⟩; *also* **:** with great speed or success ⟨making money *like gangbusters* —Martha Hume⟩

gangbusters *also* **gangbuster** *adjective* (1971)
: outstandingly excellent or successful ⟨doesn't make her a *gangbusters* . . . player —Tim Allis⟩
— **gangbusters** *adverb*

gang·er \'gaŋ-ər\ *noun* (1849)
British **:** the foreman of a gang of workers

gang hook *noun* (circa 1934)
: two or three fishhooks with their shanks joined together

gang·land \'gaŋ-,land, -lənd\ *noun, often attributive* (1912)
: the world of organized crime

gan·gling \'gaŋ-gliŋ, -glən\ *adjective* [perhaps alteration of Scots *gangrel* vagrant, lanky person] (circa 1825)
: loosely and awkwardly built **:** LANKY

gan·gli·on \'gaŋ-glē-ən\ *noun, plural* **-glia** \-glē-ə\ *also* **-gli·ons** [Late Latin, from Greek] (circa 1681)
1 : a small cystic tumor connected either with a joint membrane or tendon sheath
2 a : a mass of nerve tissue containing nerve cells external to the brain or spinal cord; *also* **:** NUCLEUS 2b **b :** something likened to a nerve ganglion
— **gan·gli·on·at·ed** \'gaŋ-glē-ə-,nā-təd\ *adjective*
— **gan·gli·on·ic** \,gaŋ-glē-'ä-nik\ *adjective*

gan·gli·o·side \'gaŋ-gle-ə-,sīd\ *noun* [International Scientific Vocabulary *ganglion* + ²*-ose* + *-ide*] (1943)
: any of a group of glycolipids that yield a hexose sugar on hydrolysis and are found especially in the plasma membrane of cells of the gray matter

gan·gly \'gaŋ-glē\ *adjective* **gan·gli·er; -est** (1872)
: GANGLING

gang·plank \'gaŋ-,plaŋk\ *noun* (1846)
: a movable bridge used in boarding or leaving a ship at a pier

gang·plow \-,plau̇\ *noun* (1850)
: a plow designed to turn two or more furrows at one time

gang rape *noun* (1968)
: rape of one person by several attackers in succession
— **gang–rape** *transitive verb*

gang·rel \'gaŋ-(ə-)rəl\ *noun* [Middle English, from *gangen* to go, from Old English *gangan*] (14th century)
Scottish **:** VAGRANT

¹gan·grene \'gaŋ-,grēn, gaŋ-', 'gan-,, gan-'\ *noun* [Latin *gangraena,* from Greek *gangraina;* akin to Greek *gran* to gnaw] (1543)
1 : local death of soft tissues due to loss of blood supply
2 : pervasive decay or corruption **:** ROT ⟨moral *gangrene*⟩
— **gan·gre·nous** \'gaŋ-grə-nəs\ *adjective*

²gangrene *verb* **gan·grened; gan·gren·ing** (1607)
transitive verb
: to make gangrenous
intransitive verb
: to become gangrenous

gang·ster \'gaŋ-stər\ *noun* (1896)
: a member of a gang of criminals **:** RACKETEER
— **gang·ster·dom** \-dəm\ *noun*
— **gang·ster·ish** \-stə-rish\ *adjective*
— **gang·ster·ism** \-stə-,ri-zəm\ *noun*

gang–tackle \'gaŋ-,ta-kəl\ *transitive verb* (1951)
: to bring down (a ballcarrier in football) with several tacklers

gangue \'gaŋ\ *noun* [French, from German *Gang* vein of metal, from Old High German, act of going] (1809)
: the worthless rock or vein matter in which valuable metals or minerals occur

gang up *intransitive verb* (1925)
1 : to make a joint assault ⟨*ganged up* on him and beat him up⟩
2 : to combine for a specific purpose ⟨*ganged up* to raise prices⟩
3 : to exert group pressure ⟨the class *ganged up* against the teacher⟩

gang·way \'gaŋ-,wā\ *noun* (before 12th century)
1 : PASSAGEWAY; *especially* **:** a temporary way of planks
2 a : either of the sides of the upper deck of a ship **b :** the opening by which a ship is boarded **c :** GANGPLANK
3 *British* **:** AISLE
4 a : a cross aisle dividing the front benches from the backbenches in the British House of Commons **b :** an aisle in the British House of Commons that separates government and opposition benches
5 : a clear passage through a crowd — often used as an interjection

◇ WORD HISTORY

gamut In the 11th century, Guido d'Arezzo, a musician and former Benedictine monk, devised a system of musical notation that was later adopted throughout Europe. Guido's system consisted of groups of six notes, which he named *ut, re, mi, fa, sol,* and *la,* from the initial syllables of six lines of a Latin hymn to St. John the Baptist. Guido called the first line of the bass staff *gamma,* and we can assume that *gamma ut* was the term his followers used for the note falling on this line, that is, the first note of the lowest group of six. This was later contracted to *gamut* and used for the whole scale as well as the lowest note. The term was further generalized to mean the whole range of a voice or instrument. Eventually *gamut* came to be used for an entire range of any sort.

gandy dancer *Gandy dancer* is a colorful and now perhaps vanishing expression for a laborer who helped lay and maintain railroad tracks. Such work was hard, relatively unskilled, low-paying, and often taken up temporarily by homeless drifters, so that *gandy dancer* was also applied on occasion to itinerant laborers in general, as well as to hoboes. Though *dancer* plausibly alludes to the repetitive motion involved in tamping ballast around ties, the origin of *gandy* remains mysterious. It has been claimed that the word derives from the Gandy Manufacturing Company of Chicago, which supposedly made tools for section gangs, though no such business appears in Chicago city directories of the early part of the century, nor in railroad trade publications of the period. The suggestion that *gandy* is altered from *gander,* alluding to the rocking motion of a worker as he moved along the track, is phonetically implausible. *Gandy dancer,* like other originally slang and argot expressions, arose in a social milieu that left behind few written records to illuminate the context in which the phrase first was used.

gan·is·ter *also* **gan·nis·ter** \'ga-nə-stər\ *noun* [origin unknown] (1811)
: a fine-grained quartzite used in the manufacture of refractory brick

gan·ja \'gän-jə, 'gan-\ *noun* [Hindi *gā̃jā,* from Sanskrit *gañjā* hemp] (1689)
: a potent and selected preparation of marijuana used especially for smoking; *broadly* **:** MARIJUANA

gan·net \'ga-nət\ *noun, plural* **gannets** *also* **gannet** [Middle English *ganet,* from Old English *ganot;* akin to Old English *gōs* goose] (before 12th century)
: any of a genus (*Morus* of the family Sulidae) of large fish-eating seabirds that breed in colonies chiefly on offshore islands

¹gan·oid \'ga-,nȯid\ *adjective* [ultimately from Greek *ganos* brightness; akin to Greek *gēthein* to rejoice — more at JOY] (circa 1847)
: of, having, or being fish scales consisting of bone and an outer shiny layer resembling enamel; *also* **:** relating to or being fish with ganoid scales

²ganoid *noun* (circa 1839)
: a fish (as a sturgeon or gar) with ganoid scales

gante·lope *or* **gant·lope** \'gant-,lōp\ *noun* [modification of Swedish *gatlopp,* from Old Swedish *gatulop,* from *gata* road + *lop* course] (1646)
archaic **:** ²GAUNTLET

gant·let \'gȯnt-lət, 'gänt-\ *variant of* GAUNTLET

gan·try \'gan-trē\ *noun, plural* **gantries** [perhaps modification of Old North French *gantier,* from Latin *cantherius* trellis] (circa 1574)
1 : a frame for supporting barrels
2 : a frame structure raised on side supports so as to span over or around something: as **a :** a platform made to carry a traveling crane and supported by towers or side frames running on parallel tracks; *also* **:** a movable structure with platforms at different levels used for erecting and servicing rockets before launching **b :** a structure spanning several railroad tracks and displaying signals for each

Gan·y·mede \'ga-ni-,mēd\ *noun* [Latin *Ganymedes,* from Greek *Ganymēdēs*]
: a beautiful youth in classical mythology carried off to Olympus to be the cupbearer of the gods

gaol \'jā(ə)l\, **gaol·er** \'jā-lər\ *chiefly British variant of* JAIL, JAILER

¹gap \'gap\ *noun* [Middle English, from Old Norse, chasm, hole; akin to Old Norse *gapa* to gape] (14th century)
1 a : a break in a barrier (as a wall, hedge, or line of military defense) **b :** an assailable position
2 a : a mountain pass **b :** RAVINE
3 : SPARK GAP
4 a : a separation in space **b :** an incomplete or deficient area ⟨a *gap* in her knowledge⟩
5 : a break in continuity **:** HIATUS
6 : a break in the vascular cylinder of a plant where a vascular trace departs from the central cylinder
7 : lack of balance **:** DISPARITY ⟨the *gap* between imports and exports⟩
8 : a wide difference in character or attitude ⟨the generation *gap*⟩
9 : a problem caused by some disparity ⟨a communication *gap*⟩ ⟨credibility *gap*⟩
— gap·py \'ga-pē\ *adjective*

²gap *verb* **gapped; gap·ping** (1879)
transitive verb
1 : to make an opening in
2 : to adjust the space between the electrodes of (a spark plug)
intransitive verb
: to fall or stand open

¹gape \'gāp *sometimes* 'gap\ *intransitive verb* **gaped; gap·ing** [Middle English, from Old Norse *gapa;* perhaps akin to Latin *hiare* to gape, yawn — more at YAWN] (13th century)
1 a : to open the mouth wide **b :** to open or part widely ⟨holes *gaped* in the pavement⟩
2 : to gaze stupidly or in openmouthed surprise or wonder
3 : YAWN
— gap·ing·ly \'gā-piŋ-lē, 'ga-piŋ-\ *adverb*

²gape *noun* (1535)
1 : an act of gaping: **a :** YAWN **b :** an openmouthed stare
2 : an unfilled space or extent
3 a : the median margin-to-margin length of the open mouth **b :** the line along which the mandibles of a bird close **c :** the width of an opening
4 *plural but singular in construction* **a :** a disease of birds and especially young birds in which gapeworms invade and irritate the trachea **b :** a fit of yawning

gap·er \'gā-pər *sometimes* 'ga-pər\ *noun* (circa 1637)
1 : one that gapes
2 : any of several large sluggish burrowing clams (families Myacidae and Mactridae) including several used for food

gape·worm \'gāp-,wərm *sometimes* 'gap-\ *noun* (1873)
: a nematode worm (*Syngamus trachea*) that causes gapes in birds

gap·ing \'gā-piŋ\ *adjective* (1588)
: wide open ⟨a *gaping* hole⟩

gap junction *noun* (1967)
: an area of contact between adjacent cells characterized by modification of the cell membranes for intercellular communication or transfer of low molecular-weight substances

gapped scale *noun* (1910)
: a musical scale derived from a larger system of tones by omitting certain tones

gap–toothed \'gap-,tütht\ *adjective* (1567)
: having gaps between the teeth

¹gar \'gär\ *interjection* [euphemism for *God*] (1598)
— used as a mild oath in the phrase *by gar*

²gar *noun* [short for *garfish*] (1765)
: any of various fishes that have an elongate body resembling that of a pike and long and narrow jaws: as **a :** NEEDLEFISH 1 **b :** any of several predaceous North American freshwater bony fishes (family Lepisosteidae) with heavy ganoid scales

¹ga·rage \gə-'räzh, -'räj; *chiefly New England* -'rȧzh, -'rȧj; *Canadian also* -'razh, -'raj; *British usually* 'gar-(,)äzh, -(,)äj, -ij\ *noun* [French, act of docking, garage, from *garer* to dock, of Germanic origin; akin to Old High German *biwarōn* to protect — more at WARE] (1902)
: a shelter or repair shop for automotive vehicles

²garage *transitive verb* **ga·raged; ga·rag·ing** (1905)
: to keep or put in a garage

ga·rage·man \-,man\ *noun* (1919)
: one who works in a garage

garage sale *noun* (1964)
: a sale of used household or personal articles (as furniture, tools, or clothing) held on the seller's own premises

ga·ram ma·sa·la \gä-'räm-mə-'sä-lə\ *noun* [Hindi *garam masālā,* literally, hot spices] (1970)
: a pungent and aromatic mixture of ground spices used in Indian cooking

Ga·rand rifle \gə-'rand-, 'gar-ənd-\ *noun* [John C. *Garand*] (1931)
: M1 RIFLE

¹garb \'gärb\ *noun* [Middle French or Old Italian; Middle French *garbe* graceful contour, grace, from Old Italian *garbo* grace] (1599)
1 *obsolete* **:** FASHION, MANNER
2 a : a style of apparel **b :** outward form **:** APPEARANCE

²garb *transitive verb* (1846)
: to cover with or as if with clothing ⟨*garbed* in T-shirt and blue jeans⟩

gar·bage \'gär-bij\ *noun* [Middle English, offal] (15th century)
1 a : food waste **b :** discarded or useless material
2 a : TRASH 1b **b :** inaccurate or useless data

gar·bage·man \-,man\ *noun* (1888)
: one who collects and hauls away garbage

gar·ban·zo \gär-'bän-(,)zō, *also* -'ban-\ *noun, plural* **-zos** [Spanish] (1759)
: CHICKPEA

garbanzo bean *noun* (1944)
: CHICKPEA

¹gar·ble \'gär-bəl\ *transitive verb* **gar·bled; gar·bling** \-b(ə-)liŋ\ [Middle English *garbelen,* from Old Italian *garbellare* to sift, from Arabic *gharbala,* from Late Latin *cribellare,* from *cribellum* sieve; akin to Latin *cernere* to sift — more at CERTAIN] (15th century)
1 *archaic* **:** CULL
2 : to sift impurities from
3 a : to so alter or distort as to create a wrong impression or change the meaning ⟨*garble* a story⟩ **b :** to introduce textual error into (a message) by inaccurate encipherment, transmission, or decipherment
— gar·bler \-b(ə-)lər\ *noun*

²garble *noun* (1502)
1 : the impurities removed from spices in sifting
2 : an act or an instance of garbling

gar·board \'gär-,bōrd, -,bȯrd\ *noun* [obsolete Dutch *gaarboord*] (1627)
: the strake next to a ship's keel

gar·boil \-,bȯil\ *noun* [Middle French *garbouil,* from Old Italian *garbuglio*] (1548)
archaic **:** a confused disordered state **:** TURMOIL

gar·çon \gär-'sōⁿ\ *noun, plural* **garçons** \-'sōⁿ(z)\ [French, boy, servant, from Old French, of Germanic origin; akin to Old High German *hrechjo* fugitive — more at WRETCH] (1788)
: WAITER

garde–man·ger \,gärd-,mäⁿ-'zhā\ *noun, plural* **garde–mangers** \-'zhā(z)\ [French, literally, one who keeps food] (1928)
: a cook who specializes in the preparation of cold foods (as meats, fish, and salads)

¹gar·den \'gär-dᵊn\ *noun* [Middle English *gardin,* from Old North French, of Germanic origin; akin to Old High German *gart* enclosure — more at YARD] (13th century)
1 a : a plot of ground where herbs, fruits, flowers, or vegetables are cultivated **b :** a rich well-cultivated region **c :** a container (as a window box) planted with usually a variety of small plants
2 a : a public recreation area or park usually ornamented with plants and trees ⟨a botanical *garden*⟩ **b :** an open-air eating or drinking place **c :** a large hall for public entertainment
— gar·den·ful \-,fül\ *noun*

²garden *verb* **gar·dened; gar·den·ing** \'gär-'dᵊn-iŋ, 'gärd-niŋ\ (1577)
intransitive verb
: to lay out or work in a garden
transitive verb
1 : to make into a garden
2 : to ornament with gardens
— gar·den·er \'gär-dᵊn-ər, 'gärd-nər\ *noun*

³garden *adjective* (1622)
1 : of, relating to, used in, or frequenting a garden
2 a : of a kind grown in the open as distinguished from one more delicate ⟨*garden* plant⟩ **b :** commonly found **:** GARDEN-VARIETY

garden apartment *noun* (1946)
: a multiple-unit low-rise dwelling having considerable lawn or garden space

garden city *noun* (1898)
: a planned residential community with park and planted areas

garden cress *noun* (1577)
: an annual herb (*Lepidium sativum*) of the mustard family sometimes cultivated for its pungent basal leaves

garden heliotrope *noun* (circa 1902)

: a tall rhizomatous Old World valerian (*Valeriana officinalis*) widely cultivated for its fragrant tiny flowers and for its roots which yield the drug valerian

gar·de·nia \gär-ˈdē-nyə\ *noun* [New Latin, from Alexander *Garden* (died 1791) Scottish naturalist] (1760)
: any of a large genus (*Gardenia*) of Old World tropical trees and shrubs of the madder family with showy fragrant white or yellow flowers

Garden of Eden (1535)
: EDEN

garden rocket *noun* (1832)
: ARUGULA

garden–variety *adjective* (1928)
: ORDINARY, COMMONPLACE

garde·robe \ˈgär-ˌdrōb\ *noun* [Middle English, from Middle French; akin to Old North French *warderobe* wardrobe] (15th century)
1 : a wardrobe or its contents
2 : a private room **:** BEDROOM
3 : PRIVY 1

gar·dy·loo \ˌgär-dē-ˈlü\ *interjection* [perhaps from French *garde à l'eau!* look out for the water!] (1622)
— used in Edinburgh as a warning cry when it was customary to throw slops from the windows into the streets

Gar·eth \ˈgar-əth\ *noun*
: a knight of the Round Table and nephew of King Arthur

gar·fish \ˈgär-ˌfish\ *noun* [Middle English *garfysshe*] (15th century)
: GAR

Gar·gan·tua \gär-ˈgan(t)-sh(ə-)wə\ *noun* [French]
: a gigantic king in Rabelais' *Gargantua* having a great capacity for food and drink

gar·gan·tuan \-wən\ *adjective, often capitalized* [*Gargantua*] (1596)
: of tremendous size or volume **:** GIGANTIC, COLOSSAL ⟨*gargantuan* waterfalls⟩

¹gar·gle \ˈgär-gəl\ *verb* **gar·gled; gar·gling** \-g(ə-)liŋ\ [Middle French *gargouiller,* of imitative origin] (1527)
transitive verb
1 a : to hold (a liquid) in the mouth or throat and agitate with air from the lungs **b :** to cleanse or disinfect (the oral cavity) in this manner
2 : to utter with a gargling sound
intransitive verb
1 : to use a gargle
2 : to speak or sing as if gargling

²gargle *noun* (1657)
1 : a liquid used in gargling
2 : a sound of or like that of gargling

gar·goyle \ˈgär-ˌgȯil\ *noun* [Middle English *gargoyl,* from Middle French *gargouille;* akin to Middle French *gargouiller*] (13th century)
1 a : a spout in the form of a grotesque human or animal figure projecting from a roof gutter to throw rainwater clear of a building **b :** a grotesquely carved figure
2 : a person with an ugly face
— **gar·goyled** \-ˌgȯild\ *adjective*

gargoyle 1a

gar·i·bal·di \ˌgar-ə-ˈbȯl-dē\ *noun* (1862)
: a woman's blouse copied from the red shirt worn by the Italian patriot Garibaldi

ga·rigue \gə-ˈrēg\ *noun* [French] (1896)
: a low open scrubland with many evergreen shrubs, low trees, aromatic herbs, and bunchgrasses found in poor or dry soil in the Mediterranean region

gar·ish \ˈgar-ish, ˈger-\ *adjective* [origin unknown] (1545)
1 : clothed in vivid colors
2 a : excessively vivid **:** FLASHY **b :** offensively or distressingly bright **:** GLARING
3 : tastelessly showy
synonym see GAUDY
— **gar·ish·ly** *adverb*
— **gar·ish·ness** *noun*

¹gar·land \ˈgär-lənd\ *noun* [Middle English, from Middle French *garlande*] (14th century)
1 : WREATH, CHAPLET
2 : ANTHOLOGY, COLLECTION

²garland *transitive verb* (15th century)
1 : to form into a garland
2 : to adorn with or as if with a garland

gar·lic \ˈgär-lik\ *noun* [Middle English *garlek,* from Old English *gārlēac,* from *gār* spear + *lēac* leek — more at GORE] (before 12th century)
1 : a European bulbous herb (*Allium sativum*) of the lily family widely cultivated for its pungent compound bulbs much used in cookery; *broadly* **:** any plant of the same genus
2 : a bulb of garlic
— **gar·licky** \-li-kē\ *adjective*

gar·licked \ˈgär-likt\ *adjective* (1950)
: containing or prepared with garlic

garlic salt *noun* (1927)
: a seasoning of ground dried garlic and salt

¹gar·ment \ˈgär-mənt\ *noun* [Middle English, from Middle French *garnement,* from Old French, from *garnir* to equip — more at GARNISH] (14th century)
: an article of clothing

²garment *transitive verb* (1547)
: to clothe with or as if with a garment

garment bag *noun* (1927)
: a bag used by travelers that folds in half and has a center handle for easy carrying

¹gar·ner \ˈgär-nər\ *noun* [Middle English, from Old French *gernier, grenier,* from Latin *granarium,* from *granum* grain — more at CORN] (12th century)
1 a : GRANARY **b :** a grain bin
2 : something that is collected **:** ACCUMULATION

²garner *transitive verb* **gar·nered; gar·ner·ing** \ˈgärn-riŋ, ˈgär-nə-\ (14th century)
1 a : to gather into storage **b :** to deposit as if in a granary ⟨volumes in which he has *garnered* the fruits of his lifetime labors —Reinhold Niebuhr⟩
2 a : to acquire by effort **:** EARN **b :** ACCUMULATE, COLLECT

gar·net \ˈgär-nət\ *noun* [Middle English *grenat,* from Middle French, from *grenat,* adjective, red like a pomegranate, from (*pomme*) *grenate* pomegranate] (14th century)
1 : a brittle and more or less transparent usually red silicate mineral that has a vitreous luster, occurs mainly in crystals but also in massive form and in grains, is found commonly in gneiss and mica schist, and is used as a semiprecious stone and as an abrasive
2 : a variable color averaging a dark red
— **gar·net·if·er·ous** \ˌgär-nə-ˈti-f(ə-)rəs\ *adjective*

garnet paper *noun* (circa 1902)
: an abrasive paper with crushed garnet as the abrasive

gar·ni·er·ite \ˈgär-nē-ə-ˌrīt\ *noun* [Jules *Garnier* (died 1904) French geologist] (1875)
: a soft mineral consisting of hydrous nickel magnesium silicate and constituting an important ore of nickel

¹gar·nish \ˈgär-nish\ *transitive verb* [Middle English, from Middle French *garniss-,* stem of *garnir* to warn, equip, garnish, of Germanic origin; akin to Old High German *warnōn* to take heed — more at WARN] (14th century)
1 a : DECORATE, EMBELLISH **b :** to add decorative or savory touches to (food or drink)
2 : to equip with accessories **:** FURNISH
3 : GARNISHEE
synonym see ADORN

²garnish *noun* (1596)
1 : EMBELLISHMENT, ORNAMENT
2 : something (as lemon wedges or parsley) used to garnish food or drink
3 a : an unauthorized fee formerly extorted from a new inmate of an English jail **b :** a similar payment required of a new worker

¹gar·nish·ee \ˌgär-nə-ˈshē\ *noun* (1627)
: a person who is served with a legal process of garnishment

²garnishee *transitive verb* **-eed; -ee·ing** (circa 1876)
1 : to serve with a garnishment
2 : to take (as a debtor's wages) by legal authority

gar·nish·ment \ˈgär-nish-mənt\ *noun* (1550)
1 : GARNISH
2 : a legal summons or warning concerning the attachment of property to satisfy a debt
3 : a stoppage of a specified sum from wages to satisfy a creditor

gar·ni·ture \ˈgär-ni-chər, -nə-ˌchu̇r\ *noun* [Middle French, equipment, alteration of Old French *garnesture,* from *garnir*] (1667)
1 : EMBELLISHMENT, TRIMMING
2 : a set of decorative objects (as vases, urns, or clocks)

gar·pike \ˈgär-ˌpīk\ *noun* (1776)
: GAR b

gar·ret \ˈgar-ət\ *noun* [Middle English *garette* watchtower, from Middle French *garite* watchtower, refuge, from Old French *garir*] (14th century)
: a room or unfinished part of a house just under the roof

¹gar·ri·son \ˈgar-ə-sən\ *noun* [Middle English *garisoun* protection, from Old French *garison,* from *garir* to protect, of Germanic origin; akin to Old High German *werien* to defend — more at WEIR] (15th century)
1 : a military post; *especially* **:** a permanent military installation
2 : the troops stationed at a garrison

²garrison *transitive verb* **gar·ri·soned; gar·ri·son·ing** \ˈgar-ə-s(ə-)niŋ\ (1569)
1 : to station troops in
2 a : to assign as a garrison **b :** to occupy with troops

garrison cap *noun* (1944)
: a visorless folding cap worn as part of a military uniform — compare SERVICE CAP

Gar·ri·son finish \ˈgar-ə-sən-\ *noun* [probably from Edward "Snapper" *Garrison,* 19th century American jockey] (1935)
: a finish in which the winner comes from behind at the end

garrison house *noun* (1676)
1 : a house fortified against Indian attack
2 : BLOCKHOUSE
3 : a house having the second story overhanging the first in the front

garrison state *noun* (1937)
: a state organized to serve primarily its own need for military security; *also* **:** a state maintained by military power

gar·ron \ˈgar-ən, gə-ˈrōn\ *noun* [Irish *gearrán* & Scottish Gaelic *gearran,* gelding] (1540)
Scottish & Irish **:** a small sturdy workhorse

¹gar·rote *or* **ga·rotte** \gə-ˈrät, -ˈrōt; ˈgar-ət\ *noun* [Spanish *garrote*] (1622)
1 a : a method of execution by strangulation **b :** the apparatus used
2 : an implement (as a wire with a handle at each end) for strangulation

²garrote *or* **garotte** *transitive verb* **gar·rot·ed** *or* **ga·rott·ed; gar·rot·ing** *or* **ga·rott·ing** (1851)
: to strangle with or as if with a garrote
— **gar·rot·er** *noun*

gar·ru·li·ty \gə-ˈrü-lə-tē, ga-\ *noun* (1581)
: the quality or state of being garrulous

gar·ru·lous \ˈgar-ə-ləs *also* ˈgar-yə-\ *adjective* [Latin *garrulus,* from *garrire* to chatter — more at CARE] (circa 1611)

1 : given to prosy, rambling, or tedious loquacity **:** pointlessly or annoyingly talkative
2 : WORDY 1 ⟨*garrulous* speeches⟩
synonym see TALKATIVE
— **gar·ru·lous·ly** *adverb*
— **gar·ru·lous·ness** *noun*

[1]gar·ter \ˈgär-tər\ *noun* [Middle English, from Old North French *gartier,* from *garet* bend of the knee, of Celtic origin; akin to Welsh *gar* shank] (14th century)
1 a : a band worn to hold up a stocking or sock **b :** a band worn to hold up a shirt sleeve **c :** a strap hanging from a girdle or corset to support a stocking
2 *capitalized* **a :** the British Order of the Garter; *also* **:** the blue velvet garter that is its badge **b :** membership in the order

[2]garter *transitive verb* (15th century)
: to support with or as if with a garter

garter snake *noun* (1769)
: any of a genus (*Thamnophis*) of harmless viviparous American snakes with longitudinal stripes on the back

garth \ˈgärth\ *noun* [Middle English, from Old Norse *garthr* yard; akin to Old High German *gart* enclosure — more at YARD] (14th century)
: a small yard or enclosure **:** CLOSE

gar·vey \ˈgär-vē\ *noun, plural* **garveys** [probably from the name *Garvey*] (circa 1896)
: a small scow especially of the New Jersey coast

[1]gas \ˈgas\ *noun, plural* **gas·es** *also* **gas·ses** [New Latin, alteration of Latin *chaos* space, chaos] (1779)
1 : a fluid (as air) that has neither independent shape nor volume but tends to expand indefinitely
2 a : a combustible gas or gaseous mixture for fuel or lighting; *especially* **:** NATURAL GAS **b :** a gaseous product of digestion; *also* **:** discomfort from this **c :** a gas or gaseous mixture used to produce anesthesia **d :** a substance that can be used to produce a poisonous, asphyxiating, or irritant atmosphere
3 : empty talk **:** BOMBAST
4 : GASOLINE; *also* **:** the accelerator pedal of an automotive vehicle
5 : driving force **:** ENERGY ⟨I was young, and full of *gas* —H. L. Mencken⟩ ⟨ran out of *gas* in the seventh inning⟩
6 *slang* **:** something that gives pleasure **:** DELIGHT ⟨the party was a *gas*⟩

[2]gas *verb* **gassed; gas·sing** (1852)
intransitive verb
1 : to talk idly or garrulously
2 : to give off gas
3 : to fill the tank (as of an automobile) with gasoline — usually used with *up*
transitive verb
1 : to supply with gas or especially gasoline ⟨*gas* up the car⟩
2 a : to treat chemically with gas **b :** to poison or otherwise affect adversely with gas
3 *slang* **:** to please greatly

gas·bag \ˈgas-ˌbag\ *noun* (1827)
1 : a bag for holding gas
2 : an idle or garrulous talker

gas chamber *noun* (1945)
: a chamber in which prisoners are executed by poison gas

gas chromatograph *noun* (1958)
: an instrument used to separate a sample into components in gas chromatography

gas chromatography *noun* (1952)
: chromatography in which the sample mixture is vaporized and injected into a stream of carrier gas (as nitrogen or helium) moving through a column containing a stationary phase composed of a liquid or particulate solid and is separated into its component compounds according to their affinity for the stationary phase
— **gas chromatographic** *adjective*

gas·con \ˈgas-kən\ *noun* (14th century)
1 *capitalized* **a :** a native of Gascony **b :** the Romance speech of Gascony
2 : a boastful swaggering person
— **Gascon** *adjective*

gas·co·nade \ˌgas-kə-ˈnād\ *noun* [French *gasconnade,* from *gasconner* to boast, from *gascon* gascon, boaster] (1709)
: BRAVADO, BOASTING
— **gasconade** *intransitive verb*
— **gas·con·ad·er** *noun*

gas·eous \ˈga-sē-əs, ˈga-shəs\ *adjective* (1799)
1 : having the form of or being gas; *also* **:** of or relating to gases
2 a : lacking substance or solidity **b :** GASSY 3 ⟨trick phrases and *gaseous* circumlocutions —Edwin Newman⟩
— **gas·eous·ness** *noun*

gas fitter *noun* (circa 1858)
: a worker who installs or repairs gas pipes and appliances

gas gangrene *noun* (1914)
: progressive gangrene marked by impregnation of the dead and dying tissue with gas and caused by one or more toxin-producing clostridia

gas–guz·zler \ˈgas-ˈgəz-lər, -ˈgə-zə-\ *noun* (1973)
: a usually large automobile that gets relatively poor mileage
— **gas–guz·zling** \-liŋ\ *adjective*

[1]gash \ˈgash\ *noun* (1548)
1 : a deep long cut in flesh
2 : a deep narrow depression or cut ⟨cut a *gash* through the forest⟩ ⟨a *gash* in the hull⟩

[2]gash *verb* [alteration of Middle English *garsen,* from Old North French *garser,* from (assumed) Vulgar Latin *charissare,* from Greek *charassein* to scratch, engrave] (1570)
transitive verb
: to make a gash in
intransitive verb
: to make a gash **:** CUT

[3]gash *adjective* [origin unknown] (1706)
1 *chiefly Scottish* **:** KNOWING, WITTY
2 *chiefly Scottish* **:** well-dressed **:** TRIM

gas·hold·er \ˈgas-ˌhōl-dər\ *noun* (1802)
: a container for gas; *especially* **:** a huge cylindrical tank for storing fuel gas under pressure

gas·house \-ˌhau̇s\ *noun* (1880)
: GASWORKS

gas·i·fi·ca·tion \ˌga-sə-fə-ˈkā-shən\ *noun* (1812)
: conversion into gas; *especially* **:** conversion of coal into natural gas

gas·i·fy \ˈga-sə-ˌfī\ *verb* **-i·fied; -i·fy·ing** (circa 1828)
transitive verb
: to convert into gas ⟨*gasify* coal⟩
intransitive verb
: to become gaseous
— **gas·ifi·er** \ˈga-sə-ˌfī(-ə)r\ *noun*

gas·ket \ˈgas-kət\ *noun* [perhaps modification of French *garcette*] (circa 1889)
: a material (as rubber) or a member (as an O-ring) used to make a joint fluid-tight

gas·kin \ˈgas-kən\ *noun* [probably short for *galligaskins*] (1573)
1 *plural, obsolete* **:** HOSE, BREECHES
2 : a part of the hind leg of a quadruped between the stifle and the hock — see HORSE illustration

gas·light \ˈgas-ˌlīt, -ˈlīt\ *noun* (1808)
1 : light made by burning illuminating gas
2 : a gas flame or gas lighting fixture

gas–liquid chromatography *noun* (1952)
: gas chromatography in which the stationary phase is a liquid
— **gas–liquid chromatographic** *adjective*

gas·lit \-ˌlit, -ˈlit\ *adjective* (1837)
: illuminated by gaslight

gas log *noun* (1885)
: a hollow perforated imitation log used as a gas burner in a fireplace

gas mask *noun* (1915)
: a mask connected to a chemical air filter and used to protect the face and lungs from toxic gases; *broadly* **:** RESPIRATOR 1

gas·o·gene \ˈga-sə-ˌjēn\ *noun* [French *gazogène,* from *gaz* gas (from New Latin *gas*) + *-o-* + *-gène* -gen] (circa 1853)
1 : a portable apparatus for carbonating liquids
2 : an apparatus carried by a vehicle to produce gas for fuel by partial burning of charcoal or wood

gas·o·hol \ˈga-sə-ˌhȯl\ *noun* [blend of *gasoline* and *alcohol*] (1977)
: a fuel consisting of a blend usually of 10 percent ethyl alcohol and 90 percent gasoline

gas oil *noun* (1901)
: a hydrocarbon oil used as a fuel oil; *especially* **:** a petroleum distillate intermediate in boiling range and viscosity between kerosene and lubricating oil

gas·olier \ˌga-sə-ˈlir\ *noun* [alteration of *gaselier,* from *gas* + *-elier* (as in *chandelier*)] (1905)
: a gaslight chandelier

gas·o·line *also* **gas·o·lene** \ˈga-sə-ˌlēn, ˌga-sə-ˈ *also* -zə-\ *noun* [[1]*gas* + [2]*-ol* + [2]*-ine* or *-ene*] (1865)
: a volatile flammable liquid hydrocarbon mixture used as a fuel especially for internal combustion engines and usually blended from several products of natural gas and petroleum
— **gas·o·lin·ic** \ˌga-sə-ˈlē-nik, -ˈli-\ *adjective*

gas·om·e·ter \ga-ˈsä-mə-tər\ *noun* [French *gazomètre,* from *gaz* + *-o-* + *-mètre* -meter] (1808)
1 : GASHOLDER
2 : a laboratory apparatus for holding and measuring gases

gas–operated *adjective* (1944)
of a firearm **:** using part of the force of expanding propellant gases to operate the action

gasp \ˈgasp\ *verb* [Middle English; akin to Old Norse *geispa* to yawn] (14th century)
intransitive verb
1 : to catch the breath convulsively and audibly (as with shock)
2 : to breathe laboriously
transitive verb
: to utter in a gasping manner
— **gasp** *noun*

gasp·er \ˈgäs-pə(r)\ *noun* (1914)
slang British **:** CIGARETTE

gas plant *noun* (circa 1909)
: FRAXINELLA

gas ring *noun* (1901)
: a ring-shaped portable gas burner for cooking

gassed \ˈgast\ *adjective* (circa 1925)
: DRUNK 1

gas·ser \ˈga-sər\ *noun* (circa 1944)
slang **:** something outstanding

gas station *noun* (1925)
: SERVICE STATION 1

gas·sy \ˈga-sē\ *adjective* **gas·si·er; -est** (1757)
1 : full of or containing gas
2 : having the characteristics of gas
3 : characterized by many words but little content **:** emptily verbose
— **gas·si·ly** \ˈga-sə-lē\ *adverb*
— **gas·si·ness** *noun*

gast \ˈgast\ *transitive verb* [Middle English, from *gast, gost* ghost — more at GHOST] (14th century)
obsolete **:** SCARE ⟨*gasted* by the noise I made, full suddenly he fled —Shakespeare⟩

Gast·ar·beit·er \ˈgast-(ˌ)är-ˌbī-tər\ *noun, plural* **Gastarbeiter** *also* **Gastarbeiters** [German, from *Gast* guest + *Arbeiter* worker] (1970)
: a foreign worker especially in Germany

gas·ter \ˈgas-tər\ *noun* [Greek *gastēr*] (circa 1909)
: the enlarged part of the abdomen behind the pedicel in hymenopterous insects (as ants)

gas·tight \ˈgas-ˈtīt\ *adjective* (1831)
: impervious to gas
— **gas·tight·ness** *noun*

gast·ness \ˈgas(t)-nəs\ *noun* (14th century)
obsolete **:** FRIGHT, TERROR

gastr- *or* **gastro-** *also* **gastri-** *combining form* [Greek, from *gastr-, gastēr*]
1 : stomach ⟨*gastr*ectomy⟩
2 : gastric and ⟨*gastro*intestinal⟩

gas·trea *also* **gas·traea** \ga-ˈstrē-ə\ *noun* [New Latin, from Greek *gastr-, gastēr*] (1879)
: a hypothetical metazoan ancestral form corresponding in organization to a simple invaginated gastrula

gas·trec·to·my \ga-ˈstrek-tə-mē\ *noun, plural* **-mies** [International Scientific Vocabulary] (1886)
: surgical removal of all or part of the stomach

gas·tric \ˈgas-trik\ *adjective* [Greek *gastr-, gastēr,* alteration of (assumed) Greek *grastēr,* from Greek *gran* to gnaw, eat] (1656)
: of or relating to the stomach

gastric gland *noun* (1900)
: any of various glands in the walls of the stomach that secrete gastric juice

gastric juice *noun* (circa 1736)
: a thin watery acid digestive fluid secreted by glands in the mucous membrane of the stomach

gastric ulcer *noun* (circa 1910)
: a peptic ulcer situated in the stomach

gas·trin \ˈgas-trən\ *noun* (1905)
: any of various polypeptide hormones that are secreted by the gastric mucosa and induce secretion of gastric juice

gas·tri·tis \ga-ˈstrī-təs\ *noun* (1806)
: inflammation especially of the mucous membrane of the stomach

gas·troc·ne·mi·us \ˌgas-(ˌ)träk-ˈnē-mē-əs, -trək-\ *noun, plural* **-mii** \-mē-ˌī\ [New Latin, from Greek *gastroknēmē* calf of the leg, from *gastr-* + *knēmē* shank — more at HAM] (1676)
: the largest and most superficial muscle of the calf of the leg arising by two heads from the condyles of the femur and attaching to a tendon that becomes part of the Achilles tendon

gas·tro·du·o·de·nal \ˌgas-trō-ˌd(y)ü-ə-ˈdē-nᵊl, -d(y)u̇-ˈä-dᵊn-əl\ *adjective* (1854)
: of, relating to, or involving the stomach and the duodenum

gas·tro·en·ter·i·tis \ˌgas-trō-ˌen-tə-ˈrī-təs\ *noun* (circa 1829)
: inflammation of the lining membrane of the stomach and the intestines

gas·tro·en·ter·ol·o·gy \-ˌen-tə-ˈrä-lə-jē\ *noun* [International Scientific Vocabulary] (circa 1904)
: a branch of medicine concerned with the structure, functions, diseases, and pathology of the stomach and intestines
— **gas·tro·en·ter·o·log·i·cal** \-rə-ˈlä-ji-kəl\ *adjective*
— **gas·tro·en·ter·ol·o·gist** \-ˌen-tə-ˈrä-lə-jist\ *noun*

gas·tro·esoph·a·ge·al \ˈgas-trō-i-ˌsä-fə-ˈjē-əl\ *adjective* (circa 1909)
: of, relating to, or involving the stomach and esophagus

gas·tro·in·tes·ti·nal \ˌgas-trō-in-ˈtes-tə-nᵊl, -ˈtes(t)-nəl\ *adjective* (1831)
: of, relating to, affecting, or including both stomach and intestine ⟨*gastrointestinal* tract⟩

gas·tro·lith \ˈgas-trə-ˌlith\ *noun* (1854)
: a stone or pebble ingested by an animal and functioning to grind food in gastric digestion

gas·tro·nome \ˈgas-trə-ˌnōm\ *noun* [French, back-formation from *gastronomie*] (1823)
: a lover of good food; *especially* **:** one with a serious interest in gastronomy
synonym see EPICURE

gas·tro·nom·i·cal·ly \ˌgas-trə-ˈnä-mi-k(ə-)lē\ *adverb* (1875)
1 : from the standpoint of gastronomy ⟨*gastronomically* fashionable⟩
2 : toward gastronomy ⟨*gastronomically* inclined⟩

gas·tron·o·mist \ga-ˈsträ-nə-mist\ *noun* (1825)
: GASTRONOME

gas·tron·o·my \-mē\ *noun* [French *gastronomie,* from Greek *Gastronomia,* title of a 4th century B.C. poem, from *gastro-* gastr- + *-nomia* -nomy] (1814)
1 : the art or science of good eating
2 : culinary customs or style
— **gas·tro·nom·ic** \ˌgas-trə-ˈnä-mik\ *also* **gas·tro·nom·i·cal** \-mi-kəl\ *adjective*

gas·tro·pod \ˈgas-trə-ˌpäd\ *noun* [New Latin *Gastropoda,* class name] (1826)
: any of a large class (Gastropoda) of mollusks (as snails and slugs) usually with a univalve shell or none and a distinct head bearing sensory organs
— **gastropod** *adjective*

gas·tro·scope \ˈgas-trə-ˌskōp\ *noun* [International Scientific Vocabulary] (1888)
: an instrument for viewing the interior of the stomach
— **gas·tro·scop·ic** \ˌgas-trə-ˈskä-pik\ *adjective*
— **gas·tros·co·pist** \ga-ˈsträs-kə-pist\ *noun*
— **gas·tros·co·py** \-pē\ *noun*

gas·tro·trich \ˈgas-trə-ˌtrik\ *noun* [ultimately from Greek *gastr-* + *trich-, thrix* hair — more at TRICH-] (1940)
: any of a phylum (Gastrotricha) of minute aquatic pseudocoelomate animals that usually have a spiny or scaly cuticle and cilia on the ventral surface

gas·tro·vas·cu·lar \ˌgas-trō-ˈvas-kyə-lər\ *adjective* [International Scientific Vocabulary] (1876)
: functioning in both digestion and circulation ⟨the *gastrovascular* cavity of a coelenterate⟩

gas·tru·la \ˈgas-trə-lə\ *noun, plural* **-las** *or* **-lae** \-ˌlē, -ˌlī\ [New Latin, from *gastr-*] (1877)
: an early metazoan embryo in which the ectoderm, mesoderm, and endoderm are established either by invagination of the blastula (as in fish and amphibians) to form a multilayered cellular cup with a blastopore opening into the archenteron or by differentiation of the blastodisc (as in reptiles, birds, and mammals) and inward cellular migration — compare BLASTULA, MORULA
— **gas·tru·lar** \-lər\ *adjective*

gas·tru·la·tion \ˌgas-trə-ˈlā-shən\ *noun* (1879)
: the process of becoming or of forming a gastrula
— **gas·tru·late** \ˈgas-trə-ˌlāt\ *intransitive verb*

gas turbine *noun* (1904)
: an internal combustion engine in which expanding gases from the combustion chamber drive the blades of a turbine

gas·works \ˈgas-ˌwərks\ *noun plural but singular or plural in construction* (1819)
: a plant for manufacturing gas and especially illuminating gas

[1]gat \ˈgat\ *archaic past of* GET

[2]gat *noun* [probably from Dutch, literally, hole; akin to Old English *geat* gate] (1723)
: a natural or artificial channel or passage

[3]gat *noun* [short for *Gatling gun*] (1904)
slang **:** HANDGUN

[1]gate \ˈgāt\ *noun* [Middle English, from Old English *geat;* akin to Old Norse *gat* opening] (before 12th century)
1 : an opening in a wall or fence
2 : a city or castle entrance often with defensive structures (as towers)
3 a : the frame or door that closes a gate **b :** a movable barrier (as at a grade crossing)
4 a : a means of entrance or exit **b :** STARTING GATE **c :** an area (as at a railroad station or an airport) for departure or arrival **d :** a space between two markers through which a competitor must pass in the course of a slalom race
5 a : a door, valve, or other device for controlling the passage especially of a fluid **b** (1) **:** an electronic switch that allows or prevents the flow of current in a circuit (2) **:** an electrode in a field-effect transistor that modulates the current flowing through the transistor according to the voltage applied to the electrode — compare DRAIN, SOURCE **c :** a device (as in a computer) that outputs a signal when specified input conditions are met ⟨logic *gate*⟩ **d :** a molecular mechanism controlling the flow of a fluid, ion, or molecule through a channel especially in a cell membrane
6 *slang* **:** DISMISSAL ⟨gave him the *gate*⟩
7 : the total admission receipts or the number of spectators (as at a sports event)

[2]gate *transitive verb* **gat·ed; gat·ing** (1835)
1 *British* **:** to punish by confinement to a campus or dormitory
2 : to supply with a gate
3 : to control by means of a gate

[3]gate *noun* [Middle English, from Old Norse *gata* road; akin to Old High German *gazza* road] (13th century)
1 *archaic* **:** WAY, PATH
2 *dialect* **:** METHOD, STYLE

-gate \ˌgāt\ *noun combining form* [Water*gate*]
: usually political scandal often involving the concealment of wrongdoing ⟨Iran*gate*⟩

gâ·teau *or* **ga·teau** \(ˌ)gä-ˈtō\ *noun, plural* **gâ·teaux** *or* **ga·teaux** \-ˈtō(z)\ *also* **ga·teaus** [French, from Old French *gastel,* probably of Germanic origin; akin to Old English *wist* sustenance, food] (1845)
1 : food baked or served in the form of a cake ⟨eggplant *gâteau*⟩
2 : a rich or fancy cake

gate–crash·er \ˈgāt-ˌkra-shər\ *noun* (1921)
: a person who enters, attends, or participates without ticket or invitation
— **gate–crash** *verb*

gate·fold \-ˌfōld\ *noun* (1946)
: FOLDOUT; *especially* **:** one with a single fold that opens out like a gate

gate·house \-ˌhau̇s\ *noun* (14th century)
: a building or house connected or associated with a gate

gate·keep·er \-ˌkē-pər\ *noun* (1572)
1 : one that tends or guards a gate
2 : a person who controls access
— **gate·keep·ing** \-piŋ\ *adjective*

gate·leg table \ˈgāt-ˌleg-, -ˌlāg-\ *noun* (1926)
: a table with drop leaves supported by movable paired legs

gateleg table

gate·post \ˈgāt-ˌpōst\ *noun* (1522)
: the post to which a gate is hung or the one against which it closes

gate·way \-ˌwā\ *noun* (1707)
1 : an opening for a gate
2 : GATE 4a

[1]gath·er \ˈga-t͟hər *also* ˈge-\ *verb* **gath·ered; gath·er·ing** \ˈgat͟h-riŋ, ˈga-t͟hə-\ [Middle English *gaderen,* from Old English *gaderian;* akin to Sanskrit *gadhya* what one clings to — more at GOOD] (before 12th century)
transitive verb
1 : to bring together **:** COLLECT ⟨tried to *gather* a crowd⟩ ⟨*gathered* firewood⟩

\ə\ **abut** \ᵊ\ **kitten** \ər\ **further** \a\ **ash** \ā\ **ace** \ä\ **mop, mar** \au̇\ **out** \ch\ **chin** \e\ **bet** \ē\ **easy** \g\ **go** \i\ **hit** \ī\ **ice** \j\ **job** \ŋ\ **sing** \ō\ **go** \ȯ\ **law** \ȯi\ **boy** \th\ **thin** \t͟h\ **the** \ü\ **loot** \u̇\ **foot** \y\ **yet** \zh\ **vision** *see also* Guide to Pronunciation

2 a : PICK, HARVEST ⟨*gather* flowers⟩ **b :** to pick up or accumulate as if by harvesting ⟨*gathering* ideas for the project⟩ **c :** to assemble (the signatures of a book) in sequence for binding
3 : to serve as a center of attraction for
4 : to effect the collection of (as tax)
5 a : to summon up ⟨*gathered* his courage⟩ **b :** to gain by gradual increase **:** ACCUMULATE ⟨*gather* speed⟩ **c :** to prepare (as oneself) by mustering strength
6 : to reach a conclusion often intuitively from hints or through inferences ⟨I *gather* that you want to leave⟩
7 a : to pull (fabric) along a line of stitching so as to draw into puckers **b :** to draw about or close to something ⟨*gathering* her cloak about her⟩ **c :** to bring together the parts of **d :** to haul in
intransitive verb
1 a : to come together in a body **b :** to cluster around a focus of attraction
2 a : to swell and fill with pus **b :** GROW, INCREASE ☆
— gath·er·er \-t͟hər-ər\ *noun*

²gather *noun* (1555)
1 : something gathered: as **a :** a puckering in cloth made by gathering **b :** a mass of molten glass collected for use in glassblowing
2 : an act or instance of gathering

gath·er·ing *noun* (before 12th century)
1 : ASSEMBLY, MEETING
2 : a suppurating swelling **:** ABCESS
3 : the collecting of food or raw materials from the wild
4 : COLLECTION, COMPILATION
5 : a gather in cloth

Gat·ling gun \ˈgat-liŋ-\ *noun* [Richard J. *Gatling* (died 1903) American inventor] (1867)
: a machine gun with a revolving cluster of barrels fired once each per revolution

ga·tor \ˈgā-tər\ *noun* (1844)
: ALLIGATOR

gauche \ˈgōsh\ *adjective, sometimes* **gauch·er** *sometimes* **gauch·est** [French, literally, left] (1751)
1 a : lacking social experience or grace; *also* **:** not tactful **:** CRUDE ⟨it would be *gauche* to mention the subject⟩ **b :** crudely made or done ⟨a *gauche* turn of phrase⟩
2 : not planar ⟨*gauche* conformation of molecules⟩
synonym see AWKWARD
— gauche·ly *adverb*
— gauche·ness *noun*

gau·che·rie \ˌgō-sh(ə-)ˈrē\ *noun* [French] (1826)
: a tactless or awkward act

Gau·cher's disease \(ˌ)gō-ˈshāz-\ *noun* [Philippe C. E. *Gaucher* (died 1918) French physician] (1902)
: a rare hereditary disorder of lipid metabolism caused by an enzyme deficiency and characterized by enlargement of the spleen and neurological impairment

gau·cho \ˈgau̇-(ˌ)chō\ *noun, plural* **gauchos** [American Spanish] (1824)
: a cowboy of the South American pampas

gaud \ˈgȯd, ˈgäd\ *noun* [Middle English *gaude*] (15th century)
: ORNAMENT, TRINKET

gaud·ery \ˈgȯ-də-rē, ˈgä-\ *noun* (circa 1598)
: showy ornamentation; *especially* **:** personal finery

¹gau·dy \ˈgȯ-dē, ˈgä-\ *adjective* **gaud·i·er; -est** (1582)
1 : ostentatiously or tastelessly ornamented
2 : marked by dazzling brilliance, showiness, or extravagance ⟨*gaudy* praise⟩ ☆
— gaud·i·ly \ˈgȯ-dᵊl-ē, ˈgä-\ *adverb*
— gaud·i·ness \ˈgȯ-dē-nəs, ˈgä-\ *noun*

²gaudy *noun, plural* **gaudies** [probably from Latin *gaudium* joy — more at JOY] (1651)
: a feast or entertainment especially in the form of an annual college dinner at a British university

gauf·fer \ˈgä-fər, ˈgȯ-, ˈgō-\ *variant of* GOFFER

¹gauge \ˈgāj\ *noun* [Middle English *gauge,* from Old North French] (15th century)
1 a : a measurement (as of linear dimension) according to some standard or system: as (1) **:** the distance between the rails of a railroad (2) **:** the size of a shotgun barrel's inner diameter nominally expressed as the number of lead balls each just fitting that diameter required to make a pound ⟨a 12-*gauge* shotgun⟩ (3) **:** the thickness of a thin material (as sheet metal or plastic film) (4) **:** the diameter of a slender object (as wire or a hypodermic needle) (5) **:** the fineness of a knitted fabric expressed by the number of loops per unit width **b :** DIMENSIONS, SIZE **c :** MEASURE 1 ⟨surveys are a *gauge* of public sentiment⟩
2 : an instrument for or a means of measuring or testing: as **a :** an instrument for measuring a dimension or for testing mechanical accuracy **b :** an instrument with a graduated scale or dial for measuring or indicating quantity
3 : relative position of a ship with reference to another ship and the wind
4 : a function introduced into a field equation to produce a convenient form of the equation but having no observable physical consequences
synonym see STANDARD

²gauge *transitive verb* **gauged; gaug·ing** (15th century)
1 a : to measure precisely the size, dimensions, or other measurable quantity of **b :** to determine the capacity or contents of **c :** ESTIMATE, JUDGE ⟨hard to *gauge* his moods⟩
2 a : to check for conformity to specifications or limits **b :** to measure off or set out

gauged \ˈgājd\ *adjective* (1823)
of masonry **:** dressed to size

gaug·er \ˈgā-jər\ *noun* (15th century)
1 : one that gauges
2 *chiefly British* **:** an exciseman who inspects dutiable bulk goods

gauge theory *noun* (1925)
: any of several theories in physics that explain the transmission of a fundamental force between two interacting particles by the exchange of an elementary particle

Gaul \ˈgȯl\ *noun* (1625)
1 : a Celt of ancient Gaul
2 : FRENCHMAN

¹Gaul·ish \ˈgȯ-lish\ *adjective* (1659)
: of or relating to the Gauls or their language or land

²Gaulish *noun* (1668)
: the Celtic language of the ancient Gauls — see INDO-EUROPEAN LANGUAGES table

Gaull·ism \ˈgō-ˌli-zəm, ˈgȯ-\ *noun* [Charles de *Gaulle*] (1943)
1 : a French political movement during World War II led by Charles de Gaulle in opposition to the Vichy regime
2 : a postwar French political movement led by Charles de Gaulle
— Gaull·ist \-list\ *adjective or noun*

gault \ˈgȯlt\ *noun* [probably of Scandinavian origin; akin to Old Norse *gald* hard-packed snow] (1575)
chiefly British **:** a heavy thick clay soil

gaum \ˈgȯm, ˈgäm\ *transitive verb* [origin unknown] (1796)
dialect **:** SMUDGE, SMEAR

gaunt \ˈgȯnt, ˈgänt\ *adjective* [Middle English] (15th century)
1 : excessively thin and angular ⟨a long *gaunt* face⟩
2 : BARREN, DESOLATE
synonym see LEAN
— gaunt·ly *adverb*
— gaunt·ness *noun*

¹gaunt·let \ˈgȯnt-lət, ˈgänt-\ *noun* [Middle English, from Middle French *gantelet,* diminutive of *gant* glove, of Germanic origin; akin to Middle Dutch *want* glove, Old Norse *vǫttr*] (15th century)
1 : a glove worn with medieval armor to protect the hand
2 : any of various protective gloves used especially in industry
3 : an open challenge (as to combat) — used in phrases like *throw down the gauntlet*
4 : a dress glove extending above the wrist
— gaunt·let·ed \-lə-təd\ *adjective*

²gauntlet *noun* [by folk etymology from *gantelope*] (1661)
1 : a severe trial **:** ORDEAL ⟨ran the *gauntlet* of criticism and censure⟩
2 a : a double file of men facing each other and armed with clubs or other weapons with which to strike at an individual who is made to run between them — used with *run* **b :** a long line (as of guards or well-wishers)

gaur \ˈgau̇r\ *noun* [Hindi, from Sanskrit *gaura;* akin to Sanskrit *go* bull, cow — more at COW] (1806)
: a large wild ox (*Bos gaurus*) of India and southeast Asia with a broad forehead and short thick curved horns

gaur

gauss \ˈgau̇s\ *noun, plural* **gauss** *also* **gauss·es** [Karl F. *Gauss*] (1882)
: the centimeter-gram-second unit of magnetic flux density that is equal to 1×10^{-4} tesla

Gauss·ian \ˈgau̇-sē-ən\ *adjective* [Karl F. *Gauss*] (1905)
: being or having the shape of a Gaussian curve or a Gaussian distribution

Gaussian curve *noun* (1905)
: NORMAL CURVE

Gaussian distribution *noun* (1905)
: NORMAL DISTRIBUTION

gauze \ˈgȯz\ *noun* [Middle French *gaze*] (1561)
1 a : a thin often transparent fabric used chiefly for clothing or draperies **b :** a loosely woven cotton surgical dressing **c :** a firm woven fabric of metal or plastic filaments
2 : HAZE

☆ SYNONYMS
Gather, collect, assemble, congregate mean to come or bring together into a group, mass, or unit. GATHER is the most general term for bringing or coming together from a spread-out or scattered state ⟨a crowd quickly *gathered*⟩. COLLECT often implies careful selection or orderly arrangement ⟨*collected* books on gardening⟩. ASSEMBLE implies an ordered union or organization of persons or things often for a definite purpose ⟨experts *assembled* for a conference⟩. CONGREGATE implies a spontaneous flocking together into a crowd or huddle ⟨*congregating* under shelter in a storm⟩. See in addition INFER.

Gaudy, tawdry, garish, flashy, meretricious mean vulgarly or cheaply showy. GAUDY implies a tasteless use of overly bright, often clashing colors or excessive ornamentation ⟨circus performers in *gaudy* costumes⟩. TAWDRY applies to what is at once gaudy and cheap and sleazy ⟨*tawdry* saloons⟩. GARISH describes what is distressingly or offensively bright ⟨*garish* neon signs⟩. FLASHY implies an effect of brilliance quickly and easily seen to be shallow or vulgar ⟨a *flashy* nightclub act⟩. MERETRICIOUS stresses falsity and may describe a tawdry show that beckons with a false allure or promise ⟨a *meretricious* wasteland of casinos and bars⟩.

— **gauze·like** \-ˌlīk\ *adjective*
— **gauz·i·ly** \ˈgȯ-zə-lē\ *adverb*
— **gauzy** \-zē\ *adjective*

ga·vage \gə-ˈväzh, gä-\ *noun* [French, from *gaver* to stuff, force-feed, from Middle French dialect (Picardy) *gave* gullet, crop] (1889)
: introduction of material into the stomach by a tube

gave *past of* GIVE

[1]**gav·el** \ˈga-vəl\ *noun* [Middle English, from Old English *gafol;* akin to Old English *giefan* to give] (before 12th century)
: rent or tribute in medieval England

[2]**gavel** *noun* [origin unknown] (circa 1859)
: a mallet used (as by a presiding officer or auctioneer) for commanding attention or confirming an action (as a vote or sale)

[3]**gavel** *transitive verb* **-eled** *or* **-elled; -el·ing** *or* **-el·ling** \ˈgav-liŋ, ˈga-və-\ (1925)
: to bring or force by use of a gavel ⟨*gaveled* the audience to silence⟩

gav·el·kind \ˈga-vəl-ˌkīnd\ *noun* [Middle English *gavelkynde,* from [1]*gavel* + *kinde* kind] (14th century)
: a tenure of land existing chiefly in Kent from Anglo-Saxon times until 1925 and providing for division of an intestate's estate equally among the sons or other heirs

gavel–to–gavel *adjective* (1968)
: extending from the beginning to the end of a meeting or session ⟨*gavel-to-gavel* television coverage⟩

ga·vi·al \ˈgā-vē-əl\ *noun* [French, modification of Bengali *ghāṛiyal* or Hindi *ghaṛyāl,* from Sanskrit *ghaṇṭika* crocodilian] (circa 1825)
: a large long-snouted crocodilian (*Gavialis gangeticus*) of India

ga·votte \gə-ˈvät\ *noun* [French, from Middle French, from Old Provençal *gavato,* from *gavot* Alpine dweller] (1696)
1 : a dance of French peasant origin marked by the raising rather than sliding of the feet
2 : a tune for the gavotte in moderately quick 4/4 time
— **gavotte** *intransitive verb*

Ga·wain \gə-ˈwän, ˈgä-ˌwän, ˈgau̇-ən\ *noun*
: a knight of the Round Table and nephew of King Arthur

[1]**gawk** \ˈgȯk\ *noun* [probably from English dialect *gawk* (left-handed)] (1757)
: a clumsy stupid person **:** LOUT

[2]**gawk** *intransitive verb* [perhaps alteration of obsolete *gaw* (to stare)] (1785)
: to gape or stare stupidly
— **gawk·er** *noun*

gawk·ish \ˈgȯ-kish\ *adjective* (1876)
: GAWKY
— **gawk·ish·ly** *adverb*
— **gawk·ish·ness** *noun*

gawky \ˈgȯ-kē\ *adjective* **gawk·i·er; -est** (1759)
: AWKWARD, CLUMSY ⟨a *gawky* adolescent⟩
— **gawk·i·ly** \-kə-lē\ *adverb*
— **gawky** *noun*

gawp \ˈgȯp, ˈgäp\ *intransitive verb* [English dialect *gawp* to yawn, gape, from obsolete *galp,* from Middle English] (1855)
chiefly British **:** GAWK
— **gawp·er** \ˈgȯ-pər, ˈgä-\ *noun*

[1]**gay** \ˈgā\ *adjective* [Middle English, from Middle French *gai*] (14th century)
1 a : happily excited **:** MERRY **b :** keenly alive and exuberant **:** having or inducing high spirits ⟨abandoned a sober traditional style for one more timely and *gay*⟩
2 a : BRIGHT, LIVELY ⟨*gay* sunny meadows⟩ **b :** brilliant in color
3 : given to social pleasures; *also* **:** LICENTIOUS
4 a : HOMOSEXUAL **b :** of, relating to, or used by homosexuals ⟨the *gay* rights movement⟩ ⟨a *gay* bar⟩
synonym see LIVELY
— **gay** *adverb*
— **gay·ness** *noun*

[2]**gay** *noun* (1953)
: HOMOSEXUAL

gay·ety, gay·ly *variant of* GAIETY, GAILY

ga·za·nia \gə-ˈzā-nē-ə, -nyə\ *noun* [New Latin, from Theodorus *Gaza* (died 1478) Greek scholar] (1813)
: any of a genus (*Gazania*) of South African composite herbs often cultivated for their brightly colored flowers

ga·zar \gə-ˈzär\ *noun* [origin unknown] (1967)
: a silk organza

[1]**gaze** \ˈgāz\ *intransitive verb* **gazed; gaz·ing** [Middle English] (14th century)
: to fix the eyes in a steady and intent look and often with eagerness or studious attention
— **gaz·er** *noun*

[2]**gaze** *noun* (1566)
: a fixed intent look

ga·ze·bo \gə-ˈzē-(ˌ)bō *also* -ˈzā-\ *noun, plural* **-bos** [perhaps from [1]*gaze* + Latin *-ebo* (as in *videbo* I shall see)] (1752)
1 : BELVEDERE
2 : a freestanding roofed structure usually open on the sides

gaze·hound \ˈgāz-ˌhau̇nd\ *noun* (1610)
: a dog that hunts by sight rather than by scent; *especially* **:** GREYHOUND

ga·zelle \gə-ˈzel\ *noun, plural* **gazelles** *also* **gazelle** [French, from Middle French, from Arabic *ghazāl*] (1600)
: any of numerous small to medium graceful and swift African and Asian antelopes (*Gazella* and related genera)

gazelle

[1]**ga·zette** \gə-ˈzet\ *noun* [French, from Italian *gazzetta*] (circa 1598)
1 : NEWSPAPER
2 : an official journal
3 *British* **:** an announcement in an official gazette

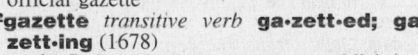

[2]**gazette** *transitive verb* **ga·zett·ed; ga·zett·ing** (1678)
1 *chiefly British* **:** to announce or publish in a gazette
2 *British* **:** to announce the appointment or status of in an official gazette

gaz·et·teer \ˌga-zə-ˈtir\ *noun* (1611)
1 *archaic* **:** JOURNALIST, PUBLICIST
2 [*The Gazetteer's: or, Newsman's Interpreter,* a geographical index edited by Laurence Echard] **:** a geographical dictionary; *also* **:** a book in which a subject is treated especially in regard to geographic distribution and regional specialization

gaz·o·gene \ˈga-zə-ˌjēn\ *variant of* GASOGENE

gaz·pa·cho \gəz-ˈpä-(ˌ)chō, gə-ˈspä-\ *noun, plural* **-chos** [Spanish] (1845)
: a spicy soup that is usually made from chopped raw vegetables (as tomato, onion, pepper, and cucumber) and that is served cold

GB \ˌjē-ˈbē\ *noun* [code name] (1961)
: SARIN

G clef *noun* (1596)
: TREBLE CLEF

ge- *or* **geo-** *combining form* [Middle English *geo-,* from Middle French & Latin; Middle French, from Latin, from Greek *gē-, geō-,* from *gē*]
1 : earth **:** ground **:** soil ⟨*ge*anticline⟩ ⟨*geo*phyte⟩
2 : geographic **:** geography and ⟨*geo*politics⟩

ge·an·ti·cline \jē-ˈan-ti-ˌklīn\ *noun* (1889)
: a great upward flexure of the earth's crust — compare GEOSYNCLINE

[1]**gear** \ˈgir\ *noun* [Middle English *gere,* from Old Norse *gervi, gǫrvi;* akin to Old English *gearwe* equipment, clothing, *gearu* ready — more at YARE] (14th century)
1 a : CLOTHING, GARMENTS **b :** movable property **:** GOODS
2 : EQUIPMENT, PARAPHERNALIA ⟨fishing *gear*⟩
3 a : the rigging of a ship or boat **b :** the harness especially of horses
4 *dialect chiefly British* **:** absurd talk **:** NONSENSE
5 *dialect chiefly British* **:** DOINGS
6 a (1) **:** a mechanism that performs a specific function in a complete machine ⟨steering *gear*⟩ (2) **:** a toothed wheel (3) **:** working relation, position, order, or adjustment ⟨got her career in *gear*⟩ **b :** one of two or more adjustments of a transmission (as of a bicycle or motor vehicle) that determine mechanical advantage, relative speed, and direction of travel
— **gear·less** \-ləs\ *adjective*

[2]**gear** (1851)
transitive verb
1 a : to provide (as machinery) with gearing **b :** to connect by gearing
2 a : to make ready for effective operation **b :** to adjust so as to match, blend with, or satisfy something ⟨*gearing* wages to productivity⟩
intransitive verb
1 a *British, of machinery* **:** to be in gear **:** MESH **b :** SHIFT 1c ⟨*gear* down⟩
2 : to become adjusted so as to match, blend, or harmonize

gear·box \ˈgir-ˌbäks\ *noun* (1887)
1 : GEARING 2
2 : TRANSMISSION 3

gear·change \-ˌchānj\ *noun* (1927)
British **:** GEARSHIFT

gear·ing \ˈgir-iŋ\ *noun* (1833)
1 : the act or process of providing or fitting with gears
2 : the parts by which motion is transmitted from one portion of machinery to another; *especially* **:** a train of gears

gear·shift \ˈgir-ˌshift\ *noun* (1926)
: a mechanism by which the transmission gears in a power-transmission system are engaged and disengaged; *also* **:** a lever for controlling such a mechanism

gear up *intransitive verb* (1951)
: to get ready ⟨the team is *gearing up* for the big game⟩

gear·wheel \ˈgir-ˌhwēl, -ˌwēl\ *noun* (circa 1874)
: GEAR 6a(2)

Geat \ˈgēt, ˈgā-ət, ˈyaət\ *noun* [Old English *Gēat*] (before 12th century)
: a member of a Scandinavian people of southern Sweden to which the legendary hero Beowulf belonged
— **Geat·ish** \ˈgē-tish, ˈgā-, ˈyaə-\ *adjective*

gecko \ˈge-(ˌ)kō\ *noun, plural* **geck·os** *or* **geck·oes** [perhaps from Malay dialect *geʔkok*] (1774)
: any of numerous small harmless chiefly tropical and nocturnal insectivorous lizards (family Gekkonidae)

gecko

ge·dank·en·ex·per·i·ment \gə-ˈdäŋ-kən-ik-ˌsper-ə-mənt *also* -ˌspir-\ *noun* [German, from *Gedanke* thought + *Experiment* experiment] (1941)
: an experiment carried out in thought only

[1]**gee** \ˈjē\ *verb* [origin unknown] (1628)
verb imperative
— used as a direction to turn to the right or move ahead; compare [5]HAW
intransitive verb
geed; gee·ing : to turn to the right side

[2]**gee** *interjection* [euphemism for *Jesus*] (1895)
— used as an introductory expletive or to express surprise or enthusiasm

\ə\ **abut** \ᵊ\ **kitten** \ər\ **further** \a\ **ash** \ā\ **ace** \ä\ **mop, mar** \au̇\ **out** \ch\ **chin** \e\ **bet** \ē\ **easy** \g\ **go** \i\ **hit** \ī\ **ice** \j\ **job** \ŋ\ **sing** \ō\ **go** \ȯ\ **law** \ȯi\ **boy** \th\ **thin** \t͟h\ **the** \ü\ **loot** \u̇\ **foot** \y\ **yet** \zh\ **vision** *see also* Guide to Pronunciation

³gee *noun* (1926)
1 : the letter *g*
2 [grand] *slang* **:** a thousand dollars
gee·gaw \'jē-(ˌ)gȯ, 'gē-\ *variant of* GEWGAW
geek \'gēk\ *noun* [probably from English dialect *geek, geck* fool, from Low German *geck,* from Middle Low German] (1914)
1 : a person often of an intellectual bent who is disapproved of
2 : a carnival performer often billed as a wild man whose act usually includes biting the head off a live chicken or snake
— geeky \'gē-kē\ *adjective*
geese *plural of* GOOSE
gee–whiz \'jē-ˌhwiz, -ˌwiz\ *adjective* (1934)
1 : designed to arouse wonder or excitement or to amplify the merits or significance of something especially by the use of clever or sensational language ⟨play-by-play specialists who wallow in *gee-whiz* banality —Jack Gould⟩
2 : marked by spectacular or astonishing qualities or achievement ⟨*gee-whiz* technology⟩
3 : characterized by wide-eyed enthusiasm, excitement, and wonder
gee whiz \(ˌ)jē-'\ *interjection* (1885)
: ²GEE
geez *variant of* JEEZ
Ge·ez \gē-'ez; 'gē-(ˌ)ez, 'gā-\ *noun* (1790)
: a Semitic language formerly spoken in Ethiopia and still used as the liturgical language of the Christian church in Ethiopia
gee·zer \'gē-zər\ *noun* [probably alteration of Scots *guiser* (one in disguise)] (1885)
: a queer, odd, or eccentric person — used especially of elderly men
ge·fil·te fish \gə-'fil-tə-\ *noun* [Yiddish, literally, stuffed fish] (1892)
: balls or cakes of seasoned minced fish usually simmered in a fish stock or baked in a tomato sauce
ge·gen·schein \'gā-gən-ˌshīn\ *noun, often capitalized* [German, from *gegen* against, counter- + *Schein* shine] (1880)
: a faint light about 20° across on the celestial sphere opposite the sun probably caused by backscatter of sunlight by solar-system dust
Ge·hen·na \gi-'he-nə\ *noun* [Late Latin, from Greek *Geenna,* from Hebrew *Gē' Hinnōm,* literally, valley of Hinnom] (1594)
1 : a place or state of misery
2 : HELL 1a(2)
Gei·ger counter \'gī-gər-\ *noun* [Hans *Geiger* (died 1945) German physicist] (1924)
: an instrument for detecting the presence and intensity of radiations (as cosmic rays or particles from a radioactive substance) by means of the ionizing effect on an enclosed gas which results in a pulse that is amplified and fed to a device giving a visible or audible indication ◆
Gei·ger–Mül·ler counter \-'myü-lər, -'mi-, -'mə-\ *noun* [W. *Müller,* 20th century German physicist] (1932)
: GEIGER COUNTER
gei·sha \'gā-shə, *also* 'gē-\ *noun, plural* **geisha** *or* **geishas** [Japanese, from *gei* art + *-sha* person] (1887)
: a Japanese girl or woman who is trained to provide entertaining and lighthearted company especially for a man or a group of men
¹gel \'jel\ *noun* [*gelatin*] (1899)
1 : a colloid in a more solid form than a sol; *broadly* **:** JELLY 2
2 : a thin colored transparent sheet used over a stage light to color it
3 : a gelatinous preparation used in styling hair
²gel *intransitive verb* **gelled; gel·ling** (1917)
: to change into or take on the form of a gel
: SET
— gel·able \'je-lə-bəl\ *adjective*
gel·a·da baboon \'je-lə-də-, 'ge-; jə-'lä-də-\ *noun* [Amharic *č̣ällada*] (1878)
: a large long-haired primate (*Theropithecus gelada*) of Ethiopia related to the baboon — called also *gelada*
ge·län·de·sprung \gə-'len-də-ˌshprůŋ, -ˌsprůŋ\ *noun* [German, from *Gelände* open fields + *Sprung* jump] (1931)
: a jump usually over an obstacle in skiing that is made from a low crouch with the aid of both ski poles
gel·ate \'je-ˌlāt\ *intransitive verb* **gel·at·ed; gel·at·ing** (1915)
: GEL
gel·a·tin *also* **gel·a·tine** \'je-lə-tᵊn\ *noun* [French *gélatine* edible jelly, gelatin, from Italian *gelatina,* from *gelato,* past participle of *gelare* to freeze, from Latin — more at COLD] (1800)
1 : glutinous material obtained from animal tissues by boiling; *especially* **:** a colloidal protein used as a food, in photography, and in medicine
2 a : any of various substances (as agar) resembling gelatin **b :** an edible jelly made with gelatin
3 : GEL 2
ge·la·ti·ni·za·tion \jə-ˌla-tᵊn-ə-'zā-shən, ˌje-lə-tᵊn-\ *noun* (1843)
: the process of converting into a gelatinous form or into a jelly
— ge·la·ti·nize \jə-'la-tᵊn-ˌīz, 'je-lə-tᵊn-\ *verb*
ge·lat·i·nous \jə-'lat-nəs, -'la-tᵊn-əs\ *adjective* (1766)
1 : resembling gelatin or jelly **:** VISCOUS ⟨a *gelatinous* precipitate⟩
2 : of, relating to, or containing gelatin
— ge·lat·i·nous·ly *adverb*
— ge·lat·i·nous·ness *noun*
¹ge·la·tion \ji-'lā-shən\ *noun* [Latin *gelation-, gelatio,* from *gelare*] (1854)
: the action or process of freezing
²gel·ation \je-'lā-shən\ *noun* [¹*gel* + *-ation*] (1915)
: the formation of a gel from a sol
ge·la·to \je-'lä-(ˌ)tō\ *noun, plural* **-ti** \-tē\ *also* **-tos** [Italian, literally, frozen] (1929)
: a soft rich ice cream containing little or no air
¹geld \'geld\ *transitive verb* [Middle English, from Old Norse *gelda;* akin to Old English *gelde* sterile] (14th century)
1 : CASTRATE
2 : to deprive of a natural or essential part ⟨sick of workingmen being *gelded* of their natural expression —*Atlantic*⟩
²geld *noun* [Old English *gield, geld* service, tribute; akin to Old English *gieldan* to pay, yield — more at YIELD] (1610)
: the crown tax paid under Anglo-Saxon and Norman kings
geld·ing \'gel-diŋ\ *noun* [Middle English, from Old Norse *geldingr,* from *gelda*] (14th century)
1 : a castrated animal; *specifically* **:** a castrated male horse
2 *archaic* **:** EUNUCH
ge·lée \zhə-'lā\ *noun* [French, jelly, from Middle French — more at JELLY] (1966)
: a cosmetic gel
gel electrophoresis *noun* (1960)
: electrophoresis in which molecules (as proteins and nucleic acids) migrate through a gel and especially a polyacrylamide gel and separate into bands according to size
gel·id \'je-ləd\ *adjective* [Latin *gelidus,* from *gelu* frost, cold — more at COLD] (1599)
: extremely cold **:** ICY ⟨*gelid* water⟩ ⟨a man of *gelid* reserve —*New Yorker*⟩
— ge·lid·i·ty \jə-'li-də-tē, je-\ *noun*
— gel·id·ly \'je-ləd-lē\ *adverb*
gel·ig·nite \'je-lig-ˌnīt\ *noun* [*gel*atin + Latin *ignis* fire + English *-ite* — more at IGNEOUS] (1889)
: a dynamite in which the adsorbent base is largely potassium nitrate or a similar nitrate usually with some wood pulp
gel·lant \'je-lənt\ *noun* (1956)
: a substance used to produce gelling
gel permeation chromatography *noun* (1966)
: chromatography in which macromolecules (as polymers) in a solution are separated by size on a column packed with a gel (as of polystyrene)
gelt \'gelt\ *noun* [Dutch & German *geld* & Yiddish *gelt;* all akin to Old English *geld* ²geld] (circa 1529)
: MONEY
¹gem \'jem\ *noun* [Middle English *gemme,* from Middle French, from Latin *gemma* bud, gem] (14th century)
1 a : JEWEL **b :** a precious or sometimes semiprecious stone cut and polished for ornament
2 a : something prized especially for great beauty or perfection **b :** a highly prized or well-beloved person
3 : MUFFIN
²gem *transitive verb* **gemmed; gem·ming** (1610)
: to adorn with or as if with gems
Ge·ma·ra \gə-'mär-ə, -'mȯr-\ *noun* [Aramaic *gĕmārā* completion] (1613)
: a commentary on the Mishnah forming the second part of the Talmud
— Ge·ma·ric \-ik\ *adjective*
— Ge·ma·rist \-ist\ *noun*
ge·mein·schaft \gə-'mīn-ˌshäft\ *noun* [German, community, from *gemein* common, general (from Old High German *gimeini*) + *-schaft* -ship — more at MEAN] (1937)
: a spontaneously arising organic social relationship characterized by strong reciprocal bonds of sentiment and kinship within a common tradition; *also* **:** a community or society characterized by this relationship — compare GESELLSCHAFT
gem·i·nal \'je-mə-nᵊl\ *adjective* [Latin *geminus* twin] (1967)
: relating to or characterized by two usually similar substituents on the same atom
— gem·i·nal·ly \-nᵊl-ē\ *adverb*
¹gem·i·nate \'je-mə-nət, -ˌnāt\ *adjective* [Latin *geminatus,* past participle of *geminare* to double, from *geminus* twin] (15th century)
1 : arranged in pairs **:** DUPLICATE
2 : being a sequence of identical speech sounds (as in *meanness* or Italian *notte* \'nȯt-te\ "night")
²gem·i·nate \-ˌnāt\ *verb* **-nat·ed; -nat·ing** (1637)
transitive verb
: to make geminate
intransitive verb
: to be or become geminate
— gem·i·na·tion \ˌje-mə-'nā-shən\ *noun*
Gem·i·ni \'je-mə-(ˌ)nē, -ˌnī; 'ge-mə-ˌnē\ *noun plural but singular in construction* [Latin (genitive *Geminorum*), literally, the twins (Castor and Pollux)]
1 : the 3d zodiacal constellation pictorially represented as the twins Castor and Pollux sitting together and located on the opposite side of the Milky Way from Taurus and Orion

◇ WORD HISTORY

Geiger counter The Geiger counter is named after the German physicist Hans Wilhelm Geiger (1882–1945). Geiger is credited with significant research on atomic theory, radioactivity, and cosmic rays. He developed a number of techniques and instruments for the detection of individual charged particles. With the British physicist Ernest Rutherford he developed in 1908 the first radiation counter, an instrument that counted alpha particles. Aided by fellow German physicist Walther Müller, he introduced an improved version of the Geiger counter in 1928. The Geiger-Müller counter marked the introduction of modern electrical devices into radiation research.

2 a : the 3d sign of the zodiac in astrology — see ZODIAC table **b :** one born under the sign of Gemini

gem·ma \ˈje-mə\ *noun, plural* **gem·mae** \-ˌmē\ [Latin] (1830)
: BUD; *broadly* **:** an asexual reproductive body that becomes detached from a parent plant

gem·ma·tion \je-ˈmā-shən\ *noun* (circa 1839)
: reproduction by gemmae

gem·mule \ˈjem-(ˌ)yü(ə)l\ *noun* [French, from Latin *gemmula,* diminutive of *gemma*] (circa 1841)
: a small bud: **a :** a theoretical particle proposed in the theory of pangenesis that is shed by a somatic cell and contains all the information necessary to reproduce that cell type (as in an offspring) **b :** a reproductive bud produced by freshwater and some marine sponges that consists of a usually hardened aggregate of cells

gem·my \ˈje-mē\ *adjective* (15th century)
1 : having the characteristics desired in a gemstone
2 : BRIGHT, GLITTERING

gem·ol·o·gist *also* **gem·mol·o·gist** \je-ˈmä-lə-jist, jə-\ *noun* (1931)
: a specialist in gems; *specifically* **:** one who appraises gems

gem·ol·o·gy *or* **gem·mol·o·gy** \-jē\ *noun* [Latin *gemma* gem] (1811)
: the science of gems
— gem·olog·i·cal \ˌje-mə-ˈlä-ji-kəl\ *adjective*

ge·mot *or* **ge·mote** \gə-ˈmōt, yə-\ *noun* [Old English *gemōt,* from *ge-* (perfective prefix) + *mōt* assembly — more at CO-, MOOT] (before 12th century)
: a judicial or legislative assembly in Anglo-Saxon England

gems·bok \ˈgemz-ˌbäk\ *noun* [Afrikaans, from German *Gemsbock* male chamois, from *Gems* chamois + *Bock* male goat] (1777)
: a large and strikingly marked oryx (*Oryx gazella*) formerly abundant in southern Africa

gemsbok

gem·stone \ˈjem-ˌstōn\ *noun* (before 12th century)
: a mineral or petrified material that when cut and polished can be used in jewelry

ge·müt·lich \gə-ˈmu̇et-lik, -ˈmüt-lik\ *adjective* [German, from Middle High German *gemüetlich* pleasant, from *gemüete* mentality, mind] (1852)
: agreeably pleasant **:** COMFORTABLE

ge·müt·lich·keit \gə-ˈmu̇et-lik-ˌkīt, -ˈmüt-lik-\ *noun* [German, from *gemütlich* + *-keit,* alteration of *-heit* -hood] (1892)
: CORDIALITY, FRIENDLINESS

gen \ˈjen\ *noun* [perhaps from *gen*eral information] (1940)
chiefly British **:** INFORMATION 2a

¹gen- *or* **geno-** *combining form* [Greek *genos* birth, race, kind — more at KIN]
1 : race (*geno*cide)
2 : genus **:** kind (*geno*type)

²gen- *or* **geno-** *combining form*
: gene (*geno*me)

-gen *also* **-gene** *noun combining form* [French *-gène,* from Greek *-genēs* born; akin to Greek *genos* birth]
1 : producer (andro*gen*)
2 : one that is (so) produced (culti*gen*)

gen·darme \ˈzhän-ˌdärm *also* ˈjän-\ *noun* [French, from Middle French, back-formation from *gensdarmes,* plural of *gent d'armes,* literally, armed people] (1796)
1 : a member of a body of soldiers especially in France serving as an armed police force for the maintenance of public order
2 : POLICE OFFICER

gen·dar·mer·ie *or* **gen·dar·mery** \zhän-ˈdär-mə-rē, jän-\ *noun, plural* **-mer·ies** [French *gendarmerie,* from *gendarme*] (circa 1796)
: a body of gendarmes

¹gen·der \ˈjen-dər\ *noun* [Middle English *gendre,* from Middle French *genre, gendre,* from Latin *gener-, genus* birth, race, kind, gender — more at KIN] (14th century)
1 a : a subclass within a grammatical class (as noun, pronoun, adjective, or verb) of a language that is partly arbitrary but also partly based on distinguishable characteristics (as shape, social rank, manner of existence, or sex) and that determines agreement with and selection of other words or grammatical forms **b :** membership of a word or a grammatical form in such a subclass **c :** an inflectional form showing membership in such a subclass
2 a : SEX (the feminine *gender*) **b :** the behavioral, cultural, or psychological traits typically associated with one sex ◼

²gender *verb* **gen·dered; gen·der·ing** \-d(ə-)riŋ\ [Middle English *gendren,* from Middle French *gendrer,* from Latin *generare* — more at GENERATE] (14th century)
: ENGENDER

gender bender *noun* (1980)
: a person who dresses and behaves like a member of the opposite sex
— gender–bending *adjective or noun*

gen·dered \ˈjen-dərd\ *adjective* (1972)
: reflecting the experience, prejudices, or orientations of one sex more than the other (*gendered* language)

gene \ˈjēn\ *noun* [German *Gen,* short for *Pangen,* from *pan-* + *-gen*] (1911)
: a specific sequence of nucleotides in DNA or RNA that is located in the germ plasm usually on a chromosome and that is the functional unit of inheritance controlling the transmission and expression of one or more traits by specifying the structure of a particular polypeptide and especially a protein or controlling the function of other genetic material

ge·ne·al·o·gist \ˌjē-nē-ˈä-lə-jist *also* -ˈa-lə- *also* ˌje-nē-\ *noun* (1605)
: a person who traces or studies the descent of persons or families

ge·ne·al·o·gy \-jē\ *noun, plural* **-gies** [Middle English *genealogie,* from Middle French, from Late Latin *genealogia,* from Greek, from *genea* race, family + *-logia* -logy; akin to Greek *genos* race] (14th century)
1 : an account of the descent of a person, family, or group from an ancestor or from older forms
2 : regular descent of a person, family, or group of organisms from a progenitor or older form **:** PEDIGREE
3 : the study of family pedigrees
— ge·ne·a·log·i·cal \ˌjē-nē-ə-ˈlä-ji-kəl, ˌje-nē-\ *adjective*
— ge·ne·a·log·i·cal·ly \-k(ə-)lē\ *adverb*

gene amplification *noun* (1968)
: replication and especially massive replication (as in the polymerase chain reaction) of the genetic material in part of a genome

gene conversion *noun* (1955)
: a genetic process in which a heterozygote with one damaged strand of DNA produces gametes in an aberrant allelic ratio (as 3:1) instead of the normal 1:1 Mendelian ratio due to repair of the damaged strand with genetic material complementary to the other strand

gene flow *noun* (1947)
: the passage and establishment of genes typical of one breeding population into the gene pool of another by hybridization and backcrossing

gene frequency *noun* (1930)
: the ratio of the number of a specified allele in a population to the total of all alleles at its genetic locus

gene mutation *noun* (1927)
: POINT MUTATION

gene pool *noun* (1946)
: the collection of genes of all the individuals in an interbreeding population

genera *plural of* GENUS

gen·er·a·ble \ˈjen-rə-bəl, ˈje-nə-\ *adjective* (15th century)
: capable of being generated

¹gen·er·al \ˈjen-rəl, ˈje-nə-\ *adjective* [Middle English, from Middle French, from Latin *generalis,* from *gener-, genus* kind, class — more at KIN] (14th century)
1 : involving, applicable to, or affecting the whole
2 : involving, relating to, or applicable to every member of a class, kind, or group (the *general* equation of a straight line)
3 : not confined by specialization or careful limitation
4 : belonging to the common nature of a group of like individuals **:** GENERIC
5 a : applicable to or characteristic of the majority of individuals involved **:** PREVALENT **b :** concerned or dealing with universal rather than particular aspects
6 : relating to, determined by, or concerned with main elements rather than limited details (bearing a *general* resemblance to the original)
7 : holding superior rank or taking precedence over others similarly titled (the *general* manager)

²general *noun* (14th century)
1 : something (as a concept, principle, or statement) that involves or is applicable to the whole
2 : SUPERIOR GENERAL
3 *archaic* **:** the general public **:** PEOPLE
4 a : GENERAL OFFICER **b :** a commissioned officer in the army, air force, or marine corps who ranks above a lieutenant general and whose insignia is four stars — compare ADMIRAL
— in general : for the most part **:** GENERALLY

☐ USAGE
gender "Until very recently, this was a purely grammatical term" writes a 1986 commentator. In fact, sense 2 is of just about the same age as sense 1—they both go back to the 1380s. But sense 2 has not had a great deal of use until recently. The 19th century examples that we have do not show the sense used in contexts of high seriousness (what a pity nature had not been of the masculine instead of the feminine *gender*—Washington Irving (*Salmagundi*)) (black divinities of the feminine *gender*—Charles Dickens (*A Tale of Two Cities*)) (a timid tender feminine *gender*—W. S. Gilbert (*The Grand Duke*)). Turn-of-the-century dictionaries labeled such use *jocular* or *humorous.* But by the middle of the 20th century, usage had changed and the old labels were no longer appropriate. Since the 1970s there has been a great increase in use, particularly in political and social reference, as opposed to biological reference (escapes the restrictions of *gender* by achieving power as a newspaper publisher —Justin Kaplan) (*gender* knows no double standard in Chinese sport —Frank Deford). It can be used for purposes of contrast (writers of either *gender* and any sexual preference —George Garrett). Sense 2 is often used attributively (*gender* identity) (*gender* differences) (the *gender* gap—the pattern of women voting differently from men —Nancy J. Walker) (the view that *gender* domination is as inevitable as tyranny —Virginia Held).

general admission *noun* (circa 1949)
: a fee paid for admission to a usually unreserved seating area (as in an auditorium or stadium)

general agent *noun* (1835)
1 : one employed to transact generally all legal business entrusted to him by his principal
2 : an insurance company agent working within a specified area

general assembly *noun* (1619)
1 : the highest governing body in a religious denomination (as the United Presbyterian Church)
2 : a legislative assembly; *especially* **:** a U.S. state legislature
3 *G&A capitalized* **:** the supreme deliberative body of the United Nations

general aviation *noun* (1966)
: the operation of civilian aircraft not under the control of a common carrier; *also* **:** such aircraft collectively

General Court *noun* (1629)
: a legislative assembly; *specifically* **:** the state legislature in Massachusetts and New Hampshire

general delivery *noun* (1846)
: a department of a post office that handles the delivery of mail at a post office window to persons who call for it

general election *noun* (1716)
: an election usually held at regular intervals in which candidates are elected in all or most constituencies of a nation or state

gen·er·al·isa·tion, gen·er·al·ise, gen·er·al·ised *British variant of* GENERALIZATION, GENERALIZE, GENERALIZED

gen·er·a·lis·si·mo \ˌjen-rə-ˈli-sə-ˌmō, ˌje-nə-\ *noun, plural* **-mos** [Italian, from *generale* general] (1621)
: the chief commander of an army

gen·er·al·ist \ˈjen-rə-list, ˈje-nə-\ *noun* (1611)
: one whose skills, interests, or habits are varied or unspecialized

gen·er·al·i·ty \ˌje-nə-ˈra-lə-tē\ *noun, plural* **-ties** (15th century)
1 : the quality or state of being general
2 a : GENERALIZATION 2 **b :** a vague or inadequate statement
3 : the greatest part **:** BULK ⟨the *generality* of the population⟩

gen·er·al·iza·tion \ˌjen-rə-lə-ˈzā-shən, ˌje-nə-\ *noun* (1761)
1 : the act or process of generalizing
2 : a general statement, law, principle, or proposition
3 : the act or process whereby a response is made to a stimulus similar to but not identical with a reference stimulus

gen·er·al·ize \ˈjen-rə-ˌlīz, ˈje-nə-\ *verb* **-ized; -iz·ing** (circa 1751)
transitive verb
1 : to give a general form to
2 a : to derive or induce (a general conception or principle) from particulars **b :** to draw a general conclusion from
3 : to give general applicability to ⟨*generalize* a law⟩; *also* **:** to make indefinite
intransitive verb
1 : to form generalizations; *also* **:** to make vague or indefinite statements
2 : to spread or extend throughout the body
— gen·er·al·iz·abil·i·ty \ˌjen-rə-ˌlī-zə-ˈbi-lə-tē, ˌje-nə-\ *noun*
— gen·er·al·iz·able \-ˌlī-zə-bəl\ *adjective*
— gen·er·al·iz·er *noun*

gen·er·al·ized *adjective* (circa 1843)
: made general; *especially* **:** not highly differentiated biologically nor strictly adapted to a particular environment

gen·er·al·ly \ˈjen-rə-lē, ˈje-nə-, ˈje-nər-lē\ *adverb* (14th century)
: in a general manner: as **a :** in disregard of specific instances and with regard to an overall picture ⟨*generally* speaking⟩ **b :** as a rule **:** USUALLY

general officer *noun* (1681)
: any of the officers in the army, air force, or marine corps above colonel — compare COMPANY OFFICER, FIELD OFFICER, FLAG OFFICER

general of the air force (1949)
: a general of the highest rank in the air force whose insignia is five stars

general of the army (1945)
: a general of the highest rank in the army whose insignia is five stars

general paresis *noun* (1874)
: insanity caused by syphilitic alteration of the brain that leads to dementia and paralysis — called also *general paralysis of the insane*

general partner *noun* (1887)
: a partner whose liability for partnership debts and obligations is unlimited

general practitioner *noun* (circa 1885)
: a physician or veterinarian whose practice is not limited to a specialty; *broadly* **:** GENERALIST

general–purpose *adjective* (1894)
: suitable to be used for two or more basic purposes

general quarters *noun plural* (1902)
: a condition of maximum readiness of a warship for action

general relativity *noun* (1916)
: RELATIVITY 3b

general semantics *noun plural but singular or plural in construction* (1933)
: a doctrine and educational discipline intended to improve habits of response of human beings to their environment and one another especially by training in the more critical use of words and other symbols

gen·er·al·ship \ˈjen-rəl-ˌship, ˈje-nə-\ *noun* (1610)
1 : office or tenure of office of a general
2 : LEADERSHIP
3 : military skill in a high commander

general store *noun* (1835)
: a retail store located usually in a small or rural community that carries a wide variety of goods including groceries but is not divided into departments

general theory of relativity (1921)
: RELATIVITY 3b

general will *noun* (circa 1902)
: the collective will of a community that is the embodiment or expression of its common interest

gen·er·ate \ˈje-nə-ˌrāt\ *transitive verb* **-at·ed; -at·ing** [Latin *generatus,* past participle of *generare,* from *gener-, genus* descent, birth — more at KIN] (1509)
1 : to bring into existence: as **a :** PROCREATE, BEGET **b :** to originate by a vital, chemical, or physical process **:** PRODUCE ⟨*generate* electricity⟩
2 : to be the cause of (a situation, action, or state of mind) ⟨these stories . . . *generate* a good deal of psychological suspense —*Atlantic*⟩
3 : to define or originate (as a mathematical or linguistic set or structure) by the application of one or more rules or operations; *especially* **:** to trace out (as a curve) by a moving point or trace out (as a surface) by a moving curve

gen·er·a·tion \ˌje-nə-ˈrā-shən\ *noun* (14th century)
1 a : a body of living beings constituting a single step in the line of descent from an ancestor **b :** a group of individuals born and living contemporaneously **c :** a group of individuals having contemporaneously a status (as that of students in a school) which each one holds only for a limited period **d :** a type or class of objects usually developed from an earlier type ⟨first of the . . . new *generation* of powerful supersonic fighters —Kenneth Koyen⟩
2 a : the action or process of producing offspring **:** PROCREATION **b :** the process of coming or bringing into being ⟨*generation* of income⟩ **c :** origination by a generating process **:** PRODUCTION; *especially* **:** formation of a geometric figure by motion of another
3 : the average span of time between the birth of parents and that of their offspring
— gen·er·a·tion·al \-shnəl, -shə-nᵊl\ *adjective*
— gen·er·a·tion·al·ly \-shnə-lē, shə-nᵊl-ē\ *adverb*

gen·er·a·tive \ˈjen-rə-tiv, ˈje-nə-, -ˌrā-tiv\ *adjective* (14th century)
: having the power or function of generating, originating, producing, or reproducing

generative cell *noun* (circa 1892)
: a sexual reproductive cell **:** GAMETE

generative grammar *usually* ˈje-nə-rə-tiv-\ *noun* (1959)
1 : a description in the form of an ordered set of rules for producing the grammatical sentences of a language
2 : TRANSFORMATIONAL GRAMMAR

generative nucleus *noun* (circa 1892)
: the one of the two nuclei resulting from the first division in the pollen grain of a seed plant that gives rise to sperm nuclei — compare TUBE NUCLEUS

generative semantics *usually* ˈje-nə-rə-tiv-\ *noun plural but usually singular in construction* (1970)
: a description of a language emphasizing a semantic deep structure that is logical in form, that provides syntactic structure, and that is related to surface structure by transformations

gen·er·a·tor \ˈje-nə-ˌrā-tər\ *noun* (1646)
1 : one that generates
2 : an apparatus in which vapor or gas is formed
3 : a machine by which mechanical energy is changed into electrical energy
4 : a mathematical entity that when subjected to one or more operations yields another mathematical entity or its elements; *specifically* **:** GENERATRIX

gen·er·a·trix \ˌje-nə-ˈrā-triks\ *noun, plural* **-er·a·tri·ces** \-trə-ˌsēz, -ə-rə-ˈtrī-(ˌ)sēz\ (1840)
: a point, line, or surface whose motion generates a line, surface, or solid

¹ge·ner·ic \jə-ˈner-ik\ *adjective* [French *générique,* from Latin *gener-, genus* birth, kind, class] (1676)
1 a : relating to or characteristic of a whole group or class **:** GENERAL **b :** being or having a nonproprietary name **c :** having no particularly distinctive quality or application
2 : relating to or having the rank of a biological genus
— ge·ner·i·cal·ly \-i-k(ə-)lē\ *adverb*
— ge·ner·ic·ness *noun*

²generic *noun* (1967)
: a generic product (as a drug)

gen·er·os·i·ty \ˌje-nə-ˈrä-sə-tē, -ˈräs-tē\ *noun, plural* **-ties** (circa 1616)
1 a : the quality or fact of being generous **b :** a generous act
2 : ABUNDANCE

gen·er·ous \ˈjen-rəs, ˈje-nə-\ *adjective* [Middle French or Latin; Middle French *genereus,* from Latin *generosus,* from *gener-, genus*] (1588)
1 *archaic* **:** HIGHBORN
2 a : characterized by a noble or forbearing spirit **:** MAGNANIMOUS, KINDLY **b :** liberal in giving **:** OPENHANDED **c :** marked by abundance or ample proportions **:** COPIOUS
synonym see LIBERAL
— gen·er·ous·ly *adverb*
— gen·er·ous·ness *noun*

gen·e·sis \ˈje-nə-səs\ *noun, plural* **-e·ses** \-ˌsēz\ [Latin, from Greek, from *gignesthai* to be born — more at KIN] (circa 1604)
: the origin or coming into being of something

Genesis *noun* [Greek]
: the mainly narrative first book of canonical Jewish and Christian Scriptures — see BIBLE table

gene–splic·ing \'jēn-ˌsplī-siŋ\ *noun* (circa 1978)
: any of various techniques by which recombinant DNA is produced and made to function in an organism

gen·et \'je-nət\ *noun* [Middle English *genete,* from Middle French, from Arabic *jarnayt*] (15th century)
: any of a genus (*Genetta*) of small Old World usually carnivorous mammals related to the civets and having retractile claws, spotted or striped fur, and a ringed tail

gene therapy *noun* (1974)
: the insertion of normal or genetically altered genes into cells usually to replace defective genes especially in the treatment of genetic disorders

ge·net·ic \jə-'ne-tik\ *also* **ge·net·i·cal** \-ti-kəl\ *adjective* [*genesis*] (1831)
1 : relating to or determined by the origin, development, or causal antecedents of something
2 a : of, relating to, or involving genetics **b :** GENIC
— ge·net·i·cal·ly \-ti-k(ə-)lē\ *adverb*

-ge·net·ic *adjective combining form*
: -GENIC 1, 2 ⟨psycho*genetic*⟩

genetic code *noun* (1961)
: the biochemical basis of heredity consisting of codons in DNA and RNA that determine the specific amino acid sequence in proteins and appear to be uniform for all known forms of life

genetic counseling *noun* (1968)
: medical education of affected individuals and the general public concerning inherited disorders

genetic drift *noun* (1945)
: random changes in gene frequency especially in small populations when leading to preservation or extinction of particular genes

genetic engineering *noun* (1966)
: the directed alteration of genetic material by intervention in genetic processes; *especially* **:** GENE-SPLICING
— genetically engineered *adjective*
— genetic engineer *noun*

genetic map *noun* (circa 1960)
: MAP 3

genetic marker *noun* (1950)
: a usually dominant gene or trait that serves especially to identify genes or traits linked with it

ge·net·ics \jə-'ne-tiks\ *noun plural but singular in construction* (1905)
1 : a branch of biology that deals with the heredity and variation of organisms
2 : the genetic makeup and phenomena of an organism, type, group, or condition
— ge·net·i·cist \-'ne-tə-sist\ *noun*

ge·ne·va \jə-'nē-və\ *noun* [modification of obsolete Dutch *genever* (now *jenever*), literally, juniper, from Middle Dutch, from Old French *geneivre,* ultimately from Latin *juniperus*] (1706)
: a highly aromatic bitter gin originally made in the Netherlands

Ge·ne·va bands \jə-'nē-və-\ *noun plural* [*Geneva,* Switzerland; from their use by the Calvinist clergy of Geneva] (1882)
: two strips of white cloth suspended from the front of a clerical collar and sometimes used by Protestant clergymen — called also *Geneva tabs*

Geneva bands

Geneva convention *noun* (1880)
: one of a series of agreements concerning the treatment of prisoners of war and of the sick, wounded, and dead in battle first made at Geneva, Switzerland in 1864 and subsequently accepted in later revisions by most nations

Geneva cross *noun* [from its adoption by the Geneva convention] (circa 1889)
: RED CROSS

Geneva gown *noun* [from its use by the Calvinist clergy of Geneva] (1820)
: a loose large-sleeved black academic gown widely used as a vestment by members of the Protestant clergy

Ge·ne·van \jə-'nē-vən\ *adjective* (1573)
1 : of or relating to Geneva, Switzerland
2 : of or relating to Calvinism
— Genevan *noun*

¹ge·nial \'jēn-yəl, 'jē-nē-əl\ *adjective* [Latin *genialis,* from *genius*] (1566)
1 *obsolete* **:** of or relating to marriage or generation ⟨the *genial* bed —John Milton⟩
2 *obsolete* **:** INBORN, NATIVE
3 a : favorable to growth or comfort **:** MILD ⟨*genial* sunshine⟩ **b :** marked by or diffusing sympathy or friendliness ⟨your *genial* host⟩
4 : displaying or marked by genius
synonym see GRACIOUS
— ge·nial·i·ty \ˌjē-nē-'a-lə-tē, jēn-'yal-\ *noun*
— ge·nial·ly \'jēn-yə-lē\ *adverb*

²ge·ni·al \ji-'nī(-ə)l\ *adjective* [Greek *geneion* chin, from *genys* jaw — more at CHIN] (1831)
: of or relating to the chin

gen·ic \'jē-nik, 'je-\ *adjective* (1918)
: of, relating to, or being a gene
— gen·i·cal·ly \-ni-k(ə-)lē\ *adverb*

-gen·ic *adjective combining form* [International Scientific Vocabulary *-gen* & *-geny* + *-ic*]
1 : producing **:** forming ⟨carcino*genic*⟩
2 : produced by **:** formed from ⟨bio*genic*⟩
3 [*photogenic*] **:** suitable for production or reproduction by (such) a medium ⟨tele*genic*⟩

ge·nic·u·late \jə-'ni-kyə-lət\ *or* **ge·nic·u·lat·ed** \-ˌlā-təd\ *adjective* [Latin *geniculatus,* from *geniculum,* diminutive of *genu* knee — more at KNEE] (1657)
: bent abruptly at an angle like a bent knee

ge·nie \'jē-nē\ *noun, plural* **ge·nies** *also* **ge·nii** \'jē-nē(-ˌī)\ [French *génie,* from Arabic *jinnīy*] (1748)
1 : JINNI 1
2 : a magic spirit believed to take human form and serve the person who calls it

gen·i·tal \'je-nə-tᵊl\ *adjective* [Middle English, from Latin *genitalis,* from *genitus,* past participle of *gignere* to beget — more at KIN] (14th century)
1 : GENERATIVE
2 : of, relating to, or being a sexual organ
3 : of, relating to, or characterized by the stage of psychosexual development in psychoanalytic theory during which oral and anal impulses are subordinated to adaptive interpersonal mechanisms
— gen·i·tal·ly \-tə-lē\ *adverb*

genital herpes *noun* (1968)
: herpes simplex of the type typically affecting the genitalia

gen·i·ta·lia \ˌje-nə-'tāl-yə\ *noun plural* [Latin, from neuter plural of *genitalis*] (1876)
: the organs of the reproductive system; *especially* **:** the external genital organs
— gen·i·ta·lic \-'ta-lik, -'tā-\ *adjective*

gen·i·tals \'je-nə-tᵊlz\ *noun plural* (14th century)
: GENITALIA

gen·i·ti·val \ˌje-nə-'tī-vəl\ *adjective* (1818)
: of, relating to, or formed with or from the genitive case
— gen·i·ti·val·ly \-və-lē\ *adverb*

gen·i·tive \'je-nə-tiv\ *adjective* [Middle English, from Latin *genetivus, genitivus,* literally, of generation (erroneous translation of Greek *genikos* genitive), from *genitus*] (14th century)
1 : of, relating to, or constituting a grammatical case marking typically a relationship of possessor or source — compare POSSESSIVE
2 : expressing a relationship that in some inflected languages is often marked by a genitive case — used especially of English prepositional phrases introduced by *of*
— genitive *noun*

genito- *combining form* [*genital*]
: genital and ⟨*genito*urinary⟩

gen·i·to·uri·nary \ˌje-nə-tō-'yůr-ə-ˌner-ē\ *adjective* (circa 1836)
: of or relating to the genital and urinary organs or functions

gen·i·ture \'je-nə-ˌchůr, -chər, -ˌt(y)ůr\ *noun* (15th century)
: NATIVITY, BIRTH

ge·nius \'jēn-yəs, 'jē-nē-əs\ *noun, plural* **ge·nius·es** *or* **ge·nii** \-nē-ˌī\ [Latin, tutelary spirit, natural inclinations, from *gignere* to beget] (1513)
1 a *plural genii* **:** an attendant spirit of a person or place **b** *plural usually genii* **:** a person who influences another for good or bad
2 : a strong leaning or inclination **:** PENCHANT
3 a : a peculiar, distinctive, or identifying character or spirit **b :** the associations and traditions of a place **c :** a personification or embodiment especially of a quality or condition
4 *plural usually genii* **:** SPIRIT, JINNI
5 *plural usually geniuses* **a :** a single strongly marked capacity or aptitude ⟨had a *genius* for getting along with boys —Mary Ross⟩ **b :** extraordinary intellectual power especially as manifested in creative activity **c :** a person endowed with transcendent mental superiority; *especially* **:** a person with a very high intelligence quotient ◇
synonym see GIFT

genius lo·ci \-'lō-ˌsī, -ˌkē\ *noun* [Latin] (1605)
1 : the pervading spirit of a place
2 : a tutelary deity of a place

geno- — see GEN-

gen·oa \'je-nə-wə\ *noun* [*Genoa,* Italy] (1932)
: a large jib that overlaps the mainsail and is used especially in racing

geno·cide \'je-nə-ˌsīd\ *noun* (1944)
: the deliberate and systematic destruction of a racial, political, or cultural group
— geno·cid·al \ˌje-nə-'sī-dᵊl\ *adjective*

ge·noise *or* **gé·noise** \zhā-'nwäz\ *noun* [French *génoise,* from feminine of *génois* of Genoa, Italy] (1931)

◇ WORD HISTORY

genius For the Romans, *genius* was something between "soul" and "life," a spirit resident in every male human during his lifetime, with a counterpart in women called the *juno*. Originally, the *genius* seems to have resided only in the *paterfamilias,* the head of an extended family, for whom it was the expression of his procreative potency, in line with the etymological link between *genius* and the verbal base *gen-* (as in *gignere* "to beget" and *genitor* "parent"). The *genius* was thought of as the personification of a person's natural desires and so was used in phrases such as *indulgere genio* "to indulge oneself" (literally, "to indulge one's genius"). Corporate entities, such as military units, as well as places, were thought of as having their own *genius*. In late Roman religious belief, *genius* was detached from the individual and regarded as a tutelary spirit who presided over a person's well-being. The English word *genius,* borrowed from Latin in the 16th century, has drawn to some degree on several nuances of the Latin word and added complexities of its own. The sense "extraordinary intellectual power" arose in the 18th century, influenced in part by the meanings of French *génie,* itself descended from Latin *genius*.

: a sponge cake containing butter and leavened by stiffly beaten eggs

ge•nome \'jē-ˌnōm\ *noun* [German *Genom,* from *Gen* gene + *-om* (as in *Chromosom* chromosome)] (1930)
: one haploid set of chromosomes with the genes they contain; *broadly* **:** the genetic material of an organism
— **ge•no•mic** \ji-'nō-mik, -'nä-\ *adjective*

ge•no•type \'jē-nə-ˌtīp, 'je-\ *noun* (1897)
1 [International Scientific Vocabulary [1]*gen-*] **:** TYPE SPECIES
2 [International Scientific Vocabulary [2]*gen-*] **:** all or part of the genetic constitution of an individual or group — compare PHENOTYPE
— **ge•no•typ•ic** \ˌjē-nə-'ti-pik, ˌje-\ *also* **ge•no•typ•i•cal** \-pi-kəl\ *adjective*
— **ge•no•typ•i•cal•ly** \-pi-k(ə-)lē\ *adverb*

-genous *adjective combining form* [*-gen* + *-ous*]
1 : producing **:** yielding ⟨ero*genous*⟩
2 : having (such) an origin ⟨terri*genous*⟩

genre \'zhän-rə, 'zhäⁿ-; 'zhäⁿr; 'jän-rə\ *noun* [French, from Middle French, kind, gender — more at GENDER] (1770)
1 : a category of artistic, musical, or literary composition characterized by a particular style, form, or content
2 : KIND, SORT
3 : painting that depicts scenes or events from everyday life usually realistically

gen•ro \'gen-'rō\ *noun plural, often capitalized* [Japanese *genrō*] (1876)
: the elder statesmen of Japan who formerly advised the emperor

gens \'jenz, 'gen(t)s\ *noun, plural* **gen•tes** \'jen-ˌtēz, 'gen-ˌtās\ [Latin *gent-, gens* — more at GENTLE] (1847)
1 : a Roman clan embracing the families of the same stock in the male line with the members having a common name and worshiping a common ancestor
2 : CLAN; *especially* **:** a patrilineal clan
3 : a distinguishable group of related organisms

[1]**gent** \'jent\ *adjective* [Middle English, noble, graceful, from Old French, from Latin *genitus,* past participle of *gignere* to beget — more at KIN] (13th century)
archaic **:** PRETTY, GRACEFUL

[2]**gent** *noun* (1564)
: GENTLEMAN

gen•ta•mi•cin \ˌjen-tə-'mī-sᵊn\ *noun* [alteration of earlier *gentamycin,* from *gen*tian violet + kan*amycin;* from the color of the actinomycete] (1963)
: a broad-spectrum antibiotic that is derived from an actinomycete (*Micromonospora purpurea* or *M. echinospora*) and is extensively used as the sulfate in treating infections especially of the urinary tract

gen•teel \jen-'tē(ə)l\ *adjective* [Middle French *gentil* gentle] (1599)
1 a : having an aristocratic quality or flavor **:** STYLISH **b :** of or relating to the gentry or upper class **c :** elegant or graceful in manner, appearance, or shape **d :** free from vulgarity or rudeness **:** POLITE
2 a : maintaining or striving to maintain the appearance of superior or middle-class social status or respectability **b** (1) **:** marked by false delicacy, prudery, or affectation (2) **:** conventionally or insipidly pretty ⟨timid and *genteel* artistic style⟩
— **gen•teel•ly** \-'tē(ə)(l)-lē\ *adverb*
— **gen•teel•ness** *noun*

gen•teel•ism \-'tē(ə)(l)-ˌli-zəm\ *noun* (1926)
: a word believed by its user to be more polite or less vulgar than a common synonym; *also* **:** the use of genteelisms

gen•tian \'jen(t)-shən\ *noun* [Middle English *gencian,* from Middle French *gentiane,* from Latin *gentiana*] (14th century)
1 : any of numerous herbs (family Gentianaceae, the gentian family, and especially genus *Gentiana*) with opposite smooth leaves and showy usually blue flowers
2 : the rhizome and roots of a yellow-flowered gentian (*Gentiana lutea*) of southern Europe that is used as a tonic, stomachic, and flavoring in vermouth

gentian violet *noun, often G&V capitalized* (1897)
: any of several dyes or dye mixtures consisting of one or more methyl derivatives of pararosaniline; *especially* **:** a dark green or greenish mixture used especially as a bactericide, fungicide, and anthelmintic

[1]**gen•tile** \'jen-ˌtīl\ *noun* [Middle English, from Late Latin *gentilis,* from Latin *gent-, gens* nation] (14th century)
1 *often capitalized* **:** a person of a non-Jewish nation or of non-Jewish faith; *especially* **:** a Christian as distinguished from a Jew
2 : HEATHEN, PAGAN
3 *often capitalized* **:** a non-Mormon ◆

[2]**gentile** *adjective* (15th century)
1 *often capitalized* **a :** of or relating to the nations at large as distinguished from the Jews; *also* **:** of or relating to Christians as distinguished from the Jews **b :** of or relating to non-Mormons
2 : HEATHEN, PAGAN
3 [Latin *gentilis*] **:** relating to a tribe or clan

gen•ti•lesse \ˌjen-tᵊl-'es\ *noun* [Middle English, from Middle French, from *gentil*] (14th century)
: decorum of conduct befitting a member of the gentry

gen•til•i•ty \jen-'ti-lə-tē\ *noun, plural* **-ties** (14th century)
1 a : the condition of belonging to the gentry **b :** GENTLEFOLK, GENTRY
2 a (1) **:** decorum of conduct **:** COURTESY (2) **:** attitudes or activity marked by false delicacy, prudery, or affectation **b :** superior social status or prestige evidenced by manners, possessions, or mode of life

[1]**gen•tle** \'jen-tᵊl\ *adjective* **gen•tler** \'jent-lər, -tᵊl-ər\; **gen•tlest** \'jent-ləst, -tᵊl-əst\ [Middle English *gentil,* from Old French, from Latin *gentilis* of a gens, of one's family, from *gent-, gens* gens, nation; akin to Latin *gignere* to beget — more at KIN] (13th century)
1 a : belonging to a family of high social station **b** *archaic* **:** CHIVALROUS **c :** HONORABLE, DISTINGUISHED; *specifically* **:** of or relating to a gentleman **d :** KIND, AMIABLE — used especially in address as a complimentary epithet ⟨*gentle* reader⟩ **e :** suited to a person of high social station
2 a : TRACTABLE, DOCILE **b :** free from harshness, sternness, or violence
3 : SOFT, DELICATE
4 : MODERATE
— **gent•ly** \'jent-lē\ *adverb*

[2]**gentle** *noun* (14th century)
: a person of gentle birth or status

[3]**gentle** *verb* **gen•tled; gen•tling** \'jent-liŋ, 'jen-tᵊl-iŋ\ (14th century)
transitive verb
1 : to raise from the commonalty **:** ENNOBLE
2 a : to make gentler **b :** to make (an animal) tame and docile **c :** MOLLIFY, PLACATE **d :** to stroke soothingly **:** PET
intransitive verb
: to become gentle ⟨the wind *gentled*⟩

gentle breeze *noun* (circa 1881)
: wind having a speed of 8 to 12 miles per hour (about 13 to 19 kilometers per hour) — see BEAUFORT SCALE table

gen•tle•folk \'jen-tᵊl-ˌfōk\ *also* **gen•tle•folks** \-ˌfōks\ *noun plural* (1594)
: persons of gentle or good family and breeding

gen•tle•man \'jen-tᵊl-mən, 'je-nᵊl-, *in rapid speech also* 'jen-tə-mən, 'je-nə-\ *noun, often attributive* [Middle English *gentilman*] (12th century)
1 a : a man of noble or gentle birth **b :** a man belonging to the landed gentry **c** (1) **:** a man who combines gentle birth or rank with chivalrous qualities (2) **:** a man whose conduct conforms to a high standard of propriety or correct behavior **d** (1) **:** a man of independent means who does not engage in any occupation or profession for gain (2) **:** a man who does not engage in a menial occupation or in manual labor for gain
2 : VALET — often used in the phrase *gentleman's gentleman*
3 : a man of any social class or condition — often used in a courteous reference ⟨show this *gentleman* to a seat⟩ or usually in the plural in address ⟨ladies and *gentlemen*⟩
— **gen•tle•man•like** \-mən-ˌlīk\ *adjective*
— **gen•tle•man•like•ness** *noun*

gentleman–at–arms *noun, plural* **gentlemen–at–arms** (1859)
: one of a military corps of 40 gentlemen who attend the British sovereign on state occasions

gentleman–commoner *noun, plural* **gentlemen–commoners** (1687)
: any of a privileged class of commoners formerly required to pay higher fees than ordinary commoners at the universities of Oxford and Cambridge

gentleman farmer *noun, plural* **gentlemen farmers** (1749)
: a man who farms mainly for pleasure rather than for profit

gen•tle•man•ly \-lē\ *adjective* (15th century)
: characteristic of or having the character of a gentleman
— **gen•tle•man•li•ness** *noun*

gentleman of fortune (1883)
: ADVENTURER

gentleman's agreement *or* **gentlemen's agreement** *noun* (1886)
: an agreement secured only by the honor of the participants

gen•tle•ness \'jen-tᵊl-nəs\ *noun* (14th century)
: the quality or state of being gentle; *especially* **:** mildness of manners or disposition

gen•tle•per•son \'jen-tᵊl-ˌpər-sᵊn\ *noun* (1943)
: a gentleman or lady

gentle sex *noun* (1583)
: the female sex **:** women in general

gen•tle•wom•an \'jen-tᵊl-ˌwu̇-mən\ *noun* (13th century)
1 a : a woman of noble or gentle birth **b :** a woman attendant upon a lady of rank
2 : a woman of refined manners or good breeding **:** LADY

◇ WORD HISTORY

gentile The Latin word *gens,* from the same Indo-European root as English *kin,* originally denoted a Roman clan, that is, a group of people sharing—either as a matter of legend or in reality—a common male ancestor. Early on, however, *gens,* especially in the plural *gentes,* was extended to any non-Roman people who, like the members of the clan, were ultimately all related to each other. Thus, it also meant simply "people, nation." The derived adjective *gentilis* meant accordingly both "of a clan" and "of a people, native." In the later Latin of the Christian church, *gentes* was used to translate Greek *ta ethnē* "the pagan peoples, those who were not the people of God." *Ta ethnē* in turn was a translation of Hebrew *goyim* in the Septuagint, the Greek translation of Jewish scripture. Correspondingly, the adjective *gentilis* was used in church Latin to mean "pagan" or "non-Jewish," the latter being the usual sense of English *gentile,* borrowed from Latin in the 14th century.

Gen·too \ˈjen-(ˌ)tü\ *noun, plural* **Gentoos** [Portuguese *gentio,* literally, gentile, from Late Latin *gentilis*] (1638)
archaic **:** HINDU

gen·too penguin \ˈjen-(ˌ)tü-\ *noun* [perhaps from *Gentoo*] (1860)
: a penguin (*Pygoscelis papua*) of Antarctica and nearby islands with a gray back and throat, white underparts, and white spots above the eyes — called also *gentoo*

gen·trice \ˈjen-trəs\ *noun* [Middle English *gentrise,* from Old French *genterise,* alteration of *gentelise,* from *gentil* gentle] (14th century)
archaic **:** gentility of birth **:** RANK

gen·tri·fi·ca·tion \ˌjen-trə-fə-ˈkā-shən\ *noun* (1964)
: the process of renewal and rebuilding accompanying the influx of middle-class or affluent people into deteriorating areas that often displaces earlier usually poorer residents

gen·tri·fy \ˈjen-trə-ˌfī\ *verb* **-fied; -fying** (1972)
transitive verb
: to attempt or accomplish the gentrification of
intransitive verb
: to become gentrified
— gen·tri·fi·er \-ˌfī(-ə)r\ *noun*

gen·try \ˈjen-trē\ *noun, plural* **gentries** [Middle English *gentrie,* alteration of *gentrise*] (14th century)
1 a *obsolete* **:** the qualities appropriate to a person of gentle birth; *especially* **:** COURTESY **b :** the condition or rank of a gentleman
2 a : upper or ruling class **:** ARISTOCRACY **b :** a class whose members are entitled to bear a coat of arms though not of noble rank; *especially* **:** the landed proprietors having such status
3 : people of a specified class or kind **:** FOLKS ⟨no real heroes or heroines among the academic *gentry* — R. G. Hanvey⟩

gents \ˈjen(t)s\ *noun, often capitalized* (1938)
chiefly British **:** MEN'S ROOM

gen·u·flect \ˈjen-yə-ˌflekt\ *intransitive verb* [Late Latin *genuflectere,* from Latin *genu* knee + *flectere* to bend — more at KNEE] (1630)
1 a : to bend the knee **b :** to touch the knee to the floor or ground especially in worship
2 : to be servilely obedient or respectful
— gen·u·flec·tion \ˌjen-yə-ˈflek-shən\ *noun*

gen·u·ine \ˈjen-yə-wən, -(ˌ)win, ÷-ˌwīn\ *adjective* [Latin *genuinus* innate, genuine; akin to Latin *gignere* to beget — more at KIN] (circa 1639)
1 a : actually having the reputed or apparent qualities or character ⟨*genuine* vintage wines⟩ **b :** actually produced by or proceeding from the alleged source or author ⟨the signature is *genuine*⟩ **c :** sincerely and honestly felt or experienced ⟨a deep and *genuine* love⟩ **d :** ACTUAL, TRUE ⟨a *genuine* improvement⟩
2 : free from hypocrisy or pretense **:** SINCERE ■
synonym see AUTHENTIC
— gen·u·ine·ly *adverb*
— gen·u·ine·ness \-wə(n)-nəs\ *noun*

ge·nus \ˈjē-nəs, ˈje-\ *noun, plural* **gen·era** \ˈje-nə-rə\ [Latin *gener-, genus* birth, race, kind — more at KIN] (1551)
1 : a class, kind, or group marked by common characteristics or by one common characteristic; *specifically* **:** a category of biological classification ranking between the family and the species, comprising structurally or phylogenetically related species or an isolated species exhibiting unusual differentiation, and being designated by a Latin or latinized capitalized singular noun
2 : a class of objects divided into several subordinate species
word history see SPICE

-geny *noun combining form* [Greek *-geneia* act of being born, from *-genēs* born — more at -GEN]
: generation **:** production ⟨phylo*geny*⟩

geo- — see GE-

geo·bot·a·ny \ˌjē-ō-ˈbä-tᵊn-ē, -ˈbät-nē\ *noun* (1904)
: PHYTOGEOGRAPHY
— geo·bo·tan·i·cal \-bə-ˈta-ni-kəl\ *also* **geo·bo·tan·ic** \-nik\ *adjective*
— geo·bot·a·nist \-ˈbä-tᵊn-ist, -ˈbät-nist\ *noun*

geo·cen·tric \ˌjē-ō-ˈsen-trik\ *adjective* (1686)
1 a : relating to, measured from, or as if observed from the earth's center — compare TOPOCENTRIC **b :** having or relating to the earth as center — compare HELIOCENTRIC
2 : taking or based on the earth as the center of perspective and valuation
— geo·cen·tri·cal·ly \-tri-k(ə-)lē\ *adverb*

geo·chem·is·try \ˌjē-ō-ˈke-mə-strē\ *noun* (1902)
1 : a science that deals with the chemical composition of and chemical changes in the solid matter of the earth or a celestial body (as the moon)
2 : the related chemical and geological properties of a substance
— geo·chem·i·cal \-ˈke-mi-kəl\ *adjective*
— geo·chem·i·cal·ly \-k(ə-)lē\ *adverb*
— geo·chem·ist \-ˈke-mist\ *noun*

geo·chro·nol·o·gy \-krə-ˈnä-lə-jē\ *noun* (1893)
1 : the chronology of the past as indicated by geologic data
2 : the study of geochronology
— geo·chro·no·log·i·cal \-ˌkrä-nᵊl-ˈä-ji-kəl, -ˌkrō-\ *also* **geo·chro·no·log·ic** \-ˈä-jik\ *adjective*
— geo·chro·no·log·i·cal·ly \-ji-k(ə-)lē\ *adverb*
— geo·chro·nol·o·gist \-krə-ˈnä-lə-jist\ *noun*

ge·ode \ˈjē-ˌōd\ *noun* [Latin *geodes,* a gem, from Greek *geōdēs* earthlike, from *gē* earth] (circa 1732)
1 : a nodule of stone having a cavity lined with crystals or mineral matter
2 : the cavity in a geode

¹geo·de·sic \ˌjē-ə-ˈde-sik, -ˈdē-, -zik\ *adjective* (1821)
1 : GEODETIC
2 : made of light straight structural elements mostly in tension ⟨a *geodesic* dome⟩

²geodesic *noun* (1883)
: the shortest line between two points that lies in a given surface

ge·od·e·sy \jē-ˈä-də-sē\ *noun* [Greek *geōdaisia,* from *geō-* ge- + *daiesthai* to divide — more at TIDE] (1570)
: a branch of applied mathematics concerned with the determination of the size and shape of the earth and the exact positions of points on its surface and with the description of variations of its gravity field
— ge·od·e·sist \-də-sist\ *noun*

geo·det·ic \ˌjē-ə-ˈde-tik\ *also* **geo·det·i·cal** \-ti-kəl\ *adjective* [*geodesy;* after such pairs as *heresy : heretic*] (circa 1828)
: of, relating to, or determined by geodesy

geodetic survey *noun* (1880)
: a survey of a large land area in which corrections are made for the curvature of the earth's surface

Geo·dim·e·ter \ˌjē-ə-ˈdi-mə-tər\ *trademark*
— used for an electronic-optical device that measures distance using the velocity of light

geo·duck \ˈgü-ē-ˌdək\ *noun* [Lushootseed (Salishan language of the Puget Sound region) *gʷídəq*] (1883)
: a large edible clam (*Panopea abrupta* synonym *P. generosa*) of the Pacific coast that sometimes weighs over five pounds

ge·og·ra·pher \jē-ˈä-grə-fər\ *noun* (1542)
: a specialist in geography

geo·graph·ic \ˌjē-ə-ˈgra-fik\ *or* **geo·graph·i·cal** \-fi-kəl\ *adjective* (1559)
1 : of or relating to geography
2 : belonging to or characteristic of a particular region
— geo·graph·i·cal·ly \-fi-k(ə-)lē\ *adverb*

geographical mile *noun* (1823)
: NAUTICAL MILE a

ge·og·ra·phy \jē-ˈä-grə-fē\ *noun, plural* **-phies** [Latin *geographia,* from Greek *geōgraphia,* from *geōgraphein* to describe the earth's surface, from *geō-* + *graphein* to write — more at CARVE] (15th century)
1 : a science that deals with the description, distribution, and interaction of the diverse physical, biological, and cultural features of the earth's surface
2 : the geographic features of an area
3 : a treatise on geography
4 : a delineation or systematic arrangement of constituent elements **:** CONFIGURATION ⟨the philosophers . . . have tried to construct *geographies* of human reason —*Times Literary Supplement*⟩

geo·hy·drol·o·gy \ˌjē-ō-hī-ˈdrä-lə-jē\ *noun* (circa 1909)
: a science that deals with the character, source, and mode of occurrence of underground water
— geo·hy·dro·log·ic \-ˌhī-drə-ˈlä-jik\ *adjective*
— geo·hy·drol·o·gist \-hī-ˈdrä-lə-jist\ *noun*

ge·oid \ˈjē-ˌȯid\ *noun* [German, from Greek *geoeidēs* earthlike, from *gē*] (1881)
: the surface within or around the earth that is everywhere normal to the direction of gravity and coincides with mean sea level in the oceans
— ge·oi·dal \jē-ˈȯi-dᵊl\ *adjective*

geo·log·ic \ˌjē-ə-ˈlä-jik\ *or* **geo·log·i·cal** \-ji-kəl\ *adjective* (1791)
: of, relating to, or based on geology
— geo·log·i·cal·ly \-ji-k(ə-)lē\ *adverb*

geologic time *noun* (1861)
: the long period of time occupied by the earth's geologic history
▶The Geologic Time table is on page 768.

ge·ol·o·gize \jē-ˈä-lə-ˌjīz\ *intransitive verb* **-gized; -giz·ing** (1831)
: to study geology or make geologic investigations

ge·ol·o·gy \jē-ˈä-lə-jē\ *noun, plural* **-gies** [New Latin *geologia,* from *ge-* + *-logia* -logy] (1735)
1 a : a science that deals with the history of the earth and its life especially as recorded in rocks **b :** a study of the solid matter of a celestial body (as the moon)
2 : geologic features
3 : a treatise on geology
— ge·ol·o·gist \-jist\ *noun*

geo·mag·net·ic \ˌjē-ō-mag-ˈne-tik\ *adjective* (1904)
: of or relating to terrestrial magnetism
— geo·mag·net·i·cal·ly \-ti-k(ə-)lē\ *adverb*
— geo·mag·ne·tism \-ˈmag-nə-ˌti-zəm\ *noun*

geomagnetic storm *noun* (1941)
: MAGNETIC STORM

□ USAGE
genuine The objection which some commentators make to the pronunciation \ˈjen-yə-ˌwīn\ is perhaps occasioned by the fact that it is more frequent among those with less schooling. However, this variant is heard in the speech of cultured or highly educated speakers sufficiently frequently for it to be recognized as a widespread pronunciation at all social levels. This variant was recorded as early as 1890 and appears to be simply a long-standing spelling pronunciation.

\ə\ **abut** \ᵊ\ **kitten** \ər\ **further** \a\ **ash** \ā\ **ace** \ä\ **mop, mar** \au̇\ **out** \ch\ **chin** \e\ **bet** \ē\ **easy** \g\ **go** \i\ **hit** \ī\ **ice** \j\ **job** \ŋ\ **sing** \ō\ **go** \ȯ\ **law** \ȯi\ **boy** \th\ **thin** \t͟h\ **the** \ü\ **loot** \u̇\ **foot** \y\ **yet** \zh\ **vision** *see also* Guide to Pronunciation

GEOLOGIC TIME

EON	ERA	PERIODS AND SYSTEMS	EPOCHS AND SERIES	BEGINNING OF INTERVAL*	BIOLOGICAL FORMS
Phanerozoic	Cenozoic	Quaternary	Holocene	0.01	
			Pleistocene	1.6	Earliest humans
		Tertiary	Pliocene	5	
			Miocene	24	Earliest hominids
			Oligocene	37	
			Eocene	58	Earliest grasses
			Paleocene	65	Earliest large mammals
	Cretaceous-Tertiary boundary (65 million years ago): extinction of dinosaurs				
	Mesozoic	Cretaceous	Upper	98	
			Lower	144	Earliest flowering plants; dinosaurs in ascendance
		Jurassic		208	Earliest birds & mammals
		Triassic		245	Age of Dinosaurs begins
	Paleozoic	Permian		286	
		Carboniferous			
		Pennsylvanian		320	Earliest reptiles
		Mississippian		360	Earliest winged insects
		Devonian		408	Earliest vascular plants (as ferns & mosses) & amphibians
		Silurian		438	Earliest land plants & insects
		Ordovician		505	Earliest corals
		Cambrian		570	Earliest fish
Proterozoic	Precambrian			2500	Earliest colonial algae & soft-bodied invertebrates
Archean				4000	Life appears: earliest algae & primitive bacteria

*In millions of years before the present

geo·man·cy \ˈjē-ə-ˌman(t)-sē\ *noun* [Middle English *geomancie,* from Middle French, from Medieval Latin *geomantia,* from Late Greek *geōmanteia,* from Greek *geō-* + *-manteia* -mancy] (14th century)
: divination by means of figures or lines or geographic features
— **geo·man·cer** \-sər\ *noun*
— **geo·man·tic** \ˌjē-ə-ˈman-tik\ *adjective*

ge·om·e·ter \jē-ˈä-mə-tər\ *noun* (15th century)
1 : a specialist in geometry
2 : GEOMETRID

geo·met·ric \ˌjē-ə-ˈme-trik\ *or* **geo·met·ri·cal** \-ˈme-tri-kəl\ *adjective* (14th century)
1 a : of, relating to, or according to the methods or principles of geometry **b :** increasing in a geometric progression ⟨*geometric* population growth⟩
2 *capitalized* **:** of or relating to a style of ancient Greek pottery characterized by geometric decorative motifs
3 a : utilizing rectilinear or simple curvilinear motifs or outlines in design **b :** of or relating to art based on simple geometric shapes (as straight lines, circles, or squares) ⟨*geometric* abstractions⟩
— **geo·met·ri·cal·ly** \-tri-k(ə-)lē\ *adverb*

geo·me·tri·cian \(ˌ)jē-ˌä-mə-ˈtri-shən, ˌjē-ə-mə-\ *noun* (15th century)
: GEOMETER 1

geometric mean *noun* (circa 1901)
: the nth root of the product of *n* numbers; *specifically* **:** a number that is the second term of three consecutive terms of a geometric progression ⟨the *geometric mean* of 9 and 4 is 6⟩

geometric progression *noun* (circa 1856)
: a sequence (as 1, ½ , ¼) in which the ratio of a term to its predecessor is always the same — called also *geometrical progression, geometric sequence*

geo·met·rics \ˌjē-ə-ˈme-triks\ *noun plural* (1977)
: decorative patterns or designs based on geometric shapes

geometric series *noun* (circa 1909)
: a series (as $1 + x + x^2 + x^3 + \ldots$) whose terms form a geometric progression

geo·me·trid \jē-ˈä-mə-trəd, ˌjē-ə-ˈme-trəd\ *noun* [ultimately from Greek *geōmetrēs* geometer, from *geōmetrein*] (1876)
: any of a family (Geometridae) of usually medium-sized moths with large wings and larvae that are loopers
— **geometrid** *adjective*

ge·om·e·trise *British variant of* GEOMETRIZE

ge·om·e·trize \jē-ˈä-mə-ˌtrīz\ *verb* **-trized; -triz·ing** (1603)
intransitive verb
: to work by or as if by geometric methods or laws
transitive verb
1 : to represent geometrically
2 : to make conform to geometric principles and laws
— **ge·om·e·tri·za·tion** \-ˌä-mə-trə-ˈzā-shən\ *noun*

ge·om·e·try \jē-ˈä-mə-trē\ *noun, plural* **-tries** [Middle English *geometrie,* from Middle French, from Latin *geometria,* from Greek *geōmetria,* from *geōmetrein* to measure the earth, from *geō-* ge- + *metron* measure — more at MEASURE] (14th century)
1 a : a branch of mathematics that deals with the measurement, properties, and relationships of points, lines, angles, surfaces, and solids; *broadly* **:** the study of properties of given elements that remain invariant under specified transformations **b :** a particular type or system of geometry
2 a : CONFIGURATION **b :** surface shape
3 : an arrangement of objects or parts that suggests geometric figures

geo·mor·phic \ˌjē-ə-ˈmȯr-fik\ *adjective* (1893)
: GEOMORPHOLOGICAL

geo·mor·pho·log·i·cal \-mȯr-fə-ˈlä-ji-kəl\ *adjective* (1896)
: of or relating to the form or surface features of the earth or other celestial body (as the moon)

geo·mor·phol·o·gy \-mȯr-ˈfä-lə-jē\ *noun, plural* **-gies** [International Scientific Vocabulary] (1893)
1 : a science that deals with the relief features of the earth or of another celestial body (as the moon) and seeks a genetic interpretation of them
2 : the features dealt with in geomorphology
— **geo·mor·phol·o·gist** \-mȯr-ˈfä-lə-jist\ *noun*

ge·oph·a·gy \jē-ˈä-fə-jē\ *noun* [International Scientific Vocabulary] (1850)
: a practice in rural or preindustrial societies of eating earthy substances (as clay) to augment a scanty or mineral-deficient diet — compare [1]PICA

geo·phone \ˈjē-ə-ˌfōn\ *noun* (1919)
: an instrument for detecting vibrations passing through rocks, soil, or ice

geo·phys·ics \ˌjē-ə-ˈfi-ziks\ *noun plural but singular or plural in construction* [International Scientific Vocabulary] (circa 1889)
: a branch of earth science dealing with the physical processes and phenomena occurring especially in the earth and in its vicinity
— **geo·phys·i·cal** \-zi-kəl\ *adjective*
— **geo·phys·i·cal·ly** \-zi-k(ə-)lē\ *adverb*
— **geo·phys·i·cist** \-ˈfi-zə-sist\ *noun*

geo·phyte \ˈjē-ə-ˌfīt\ *noun* (circa 1900)
: a perennial plant that bears its perennating buds below the surface of the soil

geo·pol·i·ti·cian \ˌjē-ō-ˌpä-lə-ˈti-shən\ *noun* (1941)
: a specialist in geopolitics

geo·pol·i·tics \-ˈpä-lə-ˌtiks\ *noun plural but singular in construction* (1904)
1 : a study of the influence of such factors as geography, economics, and demography on the politics and especially the foreign policy of a state
2 : a governmental policy guided by geopolitics
3 : a combination of political and geographic factors relating to something (as a state or particular resources)
— geo·po·lit·i·cal \-pə-ˈli-ti-kəl\ *adjective*
— geo·po·lit·i·cal·ly \-ti-k(ə-)lē\ *adverb*

geo·pres·sured \ˌjē-ō-ˈpre-shərd\ *adjective* (1968)
: subjected to great pressure from geologic forces ⟨*geopressured* methane⟩

Geor·die \ˈjȯr-dē\ *noun* [from *Geordie,* diminutive of the name *George*] (1866)
chiefly British **:** an inhabitant of Newcastle upon Tyne or its environs; *also* **:** the dialect of English spoken by Geordies

George \ˈjȯrj\ *noun* [Saint *George*] (1506)
1 : either of two of the insignia of the British Order of the Garter
2 : a British coin bearing the image of Saint George

geor·gette \jȯr-ˈjet\ *noun* [from *Georgette,* a trademark] (1915)
: a sheer crepe woven from hard-twisted yarns to produce a dull pebbly surface

¹Geor·gian \ˈjȯr-jən\ *noun* (15th century)
1 : a native or inhabitant of Georgia in the Caucasus
2 : the language of the Georgian people

²Georgian *adjective* (1607)
: of, relating to, or constituting Georgia in the Caucasus, the Georgians, or Georgian

³Georgian *noun* (1741)
: a native or resident of the state of Georgia

⁴Georgian *adjective* (1762)
: of, relating to, or characteristic of the state of Georgia or its people

⁵Georgian *adjective* (circa 1855)
1 : of, relating to, or characteristic of the reigns of the first four Georges of Great Britain
2 : of, relating to, or characteristic of the reign of George V of Great Britain

⁶Georgian *noun* (1901)
1 : one belonging to either of the Georgian periods; *especially* **:** a poet of the second decade of the 20th century
2 : Georgian taste or style especially in architecture

¹geor·gic \ˈjȯr-jik\ *noun* [the *Georgics,* poem by Virgil, from Latin *georgicus*] (1513)
: a poem dealing with agriculture

²georgic *adjective* [Latin *georgicus,* from Greek *geōrgikos,* from *geōrgos* farmer, from *geō-* ge- + *ergon* work — more at WORK] (circa 1720)
: of or relating to agriculture

geo·sci·ence \ˌjē-ō-ˈsī-ən(t)s\ *noun* (1942)
1 : the sciences (as geology, geophysics, and geochemistry) dealing with the earth
2 : any of the geosciences — compare EARTH SCIENCE
— geo·sci·en·tist \-ən-tist\ *noun*

geo·sta·tion·ary \-ˈstā-shə-ˌner-ē\ *adjective* (1961)
: being or having an equatorial orbit requiring an angular velocity the same as that of the earth so that the position of a satellite in such an orbit is fixed with respect to the earth

geo·strat·e·gy \-ˈstra-tə-jē\ *noun* (1942)
1 : a branch of geopolitics that deals with strategy
2 : the combination of geopolitical and strategic factors characterizing a particular geographic region
3 : the use by a government of strategy based on geopolitics
— geo·stra·te·gic \-strə-ˈtē-jik\ *adjective*
— geo·strat·e·gist \-ˈstra-tə-jist\ *noun*

geo·stroph·ic \ˌjē-ə-ˈsträ-fik\ *adjective* [*ge-* + Greek *strophikos* turned, from *strophē* turning — more at STROPHE] (1916)
: of, relating to, or arising from the Coriolis force
— geo·stroph·i·cal·ly \-fi-k(ə-)lē\ *adverb*

geo·syn·chro·nous \ˌjē-ō-ˈsiŋ-krə-nəs, -ˈsin-\ *adjective* (1968)
: GEOSTATIONARY

geo·syn·cline \-ˈsin-ˌklīn\ *noun* (1895)
: a great downward flexure of the earth's crust — compare GEANTICLINE
— geo·syn·cli·nal \-ˌsin-ˈklī-nəl\ *adjective*

geo·tac·tic \ˌjē-ō-ˈtak-tik\ *adjective* (1899)
: of or relating to geotaxis

geo·tax·is \-ˈtak-səs\ *noun* [New Latin] (1899)
: a taxis in which the force of gravity is the directive factor

geo·tech·ni·cal \-ˈtek-ni-kəl\ *adjective* (1967)
: of or relating to geotechnical engineering

geotechnical engineering *noun* (1974)
: a science that deals with the application of geology to engineering

geo·tec·ton·ic \-tek-ˈtä-nik\ *adjective* (1882)
: of or relating to the form, arrangement, and structure of rock masses of the earth's crust resulting from folding or faulting
— geo·tec·ton·i·cal·ly \-ni-k(ə-)lē\ *adverb*

geo·ther·mal \-ˈthər-məl\ *adjective* [International Scientific Vocabulary] (1875)
: of, relating to, or utilizing the heat of the earth's interior; *also* **:** produced or permeated by such heat ⟨*geothermal* steam⟩ ⟨*geothermal* region⟩
— geo·ther·mal·ly \-mə-lē\ *adverb*

geo·tro·pic \ˌjē-ə-ˈtrō-pik, -ˈträ-\ *adjective* (1875)
: of or relating to geotropism
— geo·tro·pi·cal·ly \-ˈtrō-pi-k(ə-)lē, -ˈträ-\ *adverb*

ge·ot·ro·pism \jē-ˈä-trə-ˌpi-zəm\ *noun* [International Scientific Vocabulary] (1875)
: a tropism (as of plant roots) in which gravity is the orienting factor

ge·rah \ˈgir-ə\ *noun* [Hebrew *gērāh*] (1530)
: an ancient Hebrew unit of weight equal to 1⁄20 shekel

ge·ra·ni·ol \jə-ˈrā-nē-ˌȯl, -ˌōl\ *noun* [International Scientific Vocabulary, from New Latin *Geranium*] (1871)
: a fragrant liquid unsaturated alcohol $C_{10}H_{18}O$ used chiefly in perfumes and soap

ge·ra·ni·um \jə-ˈrā-nē-əm, -nyəm\ *noun* [New Latin, from Latin, geranium, from Greek *geranion,* from diminutive of *geranos* crane — more at CRANE] (1548)
1 : any of a widely distributed genus (*Geranium* of the family Geraniaceae, the geranium family) of plants having regular usually white, pink, or purple flowers with elongated styles and glands that alternate with the petals
2 : PELARGONIUM
3 : a vivid or strong red

ge·rar·dia \jə-ˈrär-dē-ə\ *noun* [New Latin, from John *Gerard* (died 1612) English botanist] (1851)
: any of a genus (*Agalinis* synonym *Gerardia*) of often root-parasitic herbs of the snapdragon family having pink, purple, or yellow flowers

ger·bera \ˈgər-bə-rə, ˈjər-\ *noun* [New Latin, from Traugott *Gerber* (died 1743) German naturalist] (1889)
: any of a genus (*Gerbera*) of Old World composite herbs having basal tufted leaves and showy heads of yellow, pink, or orange flowers with prominent rays

ger·bil *also* **ger·bille** \ˈjər-bəl\ *noun* [French *gerbille,* from New Latin *Gerbillus,* diminutive of *gerboa, jerboa* jerboa] (1849)
: any of numerous Old World burrowing desert rodents (*Gerbillus* and related genera) with long hind legs adapted for leaping

ge·rent \ˈjir-ənt\ *noun* [Latin *gerent-, gerens,* present participle of *gerere* to bear, carry on] (1576)
: one that rules or manages

ge·re·nuk \ˈger-ə-ˌnu̇k, gə-ˈre-nək\ *noun, plural* **gerenuk** *or* **gerenuks** [Somali *gáránúug*] (1895)
: a large-eyed antelope (*Litocranius walleri*) of eastern Africa with a long neck and limbs

gerenuk

ger·fal·con *variant of* GYRFALCON

¹ge·ri·at·ric \ˌjer-ē-ˈa-trik, ˌjir-\ *noun* (1909)
1 *plural but singular in construction* **:** a branch of medicine that deals with the problems and diseases of old age and aging people — compare GERONTOLOGY
2 : an aged person

²geriatric *adjective* [Greek *gēras* old age + English *-iatric*] (1926)
1 a : of or relating to geriatrics or the process of aging **b :** of, relating to, or appropriate for elderly people ⟨the *geriatric* set⟩
2 a : OLD, ELDERLY ⟨a *geriatric* dachshund⟩ **b :** being old and outmoded ⟨*geriatric* airplanes⟩

ger·i·a·tri·cian \ˌjer-ē-ə-ˈtri-shən, ˌjir-\ *noun* (1926)
: a specialist in geriatrics

germ \ˈjərm\ *noun* [French *germe,* from Latin *germin-, germen,* from *gignere* to beget — more at KIN] (1644)
1 a : a small mass of living substance capable of developing into an organism or one of its parts **b :** the embryo with the scutellum of a cereal grain that is usually separated from the starchy endosperm during milling
2 : something that initiates development or serves as an origin **:** RUDIMENTS, BEGINNING
3 : MICROORGANISM; *especially* **:** a microorganism causing disease

¹ger·man \ˈjər-mən\ *adjective* [Middle English *germain,* from Middle French, from Latin *germanus* having the same parents, from *germen*] (14th century)
: having the same parents or the same grandparents on either the maternal or paternal side — usually used after the noun which it modifies and joined to it by a hyphen ⟨brother-german⟩ ⟨cousin-german⟩

²german *noun* (15th century)
obsolete **:** a near relative

¹Ger·man \ˈjər-mən\ *noun* [Middle English, from Medieval Latin *Germanus,* from Latin] (14th century)
1 : a member of any of the Germanic peoples inhabiting western Europe in Roman times
2 a : a native or inhabitant of Germany **b :** a person of German descent **c :** one whose native language is German and who is a native of a country other than Germany
3 a : the Germanic language spoken mainly in Germany, Austria, and parts of Switzerland **b :** the literary and official language of Germany
4 *often not capitalized* **a :** a dance consisting of intricate figures that are improvised and intermingled with waltzes **b** *chiefly Midland* **:** a dancing party; *specifically* **:** one at which the german is danced

²German *adjective* (1552)
: of, relating to, or characteristic of Germany, the Germans, or German

German cockroach *noun* (1896)
: a small active winged cockroach (*Blattella germanica*) probably of African origin that is a

common household pest in the U.S. — called also *Croton bug*

ger·man·der \(ˌ)jər-ˈman-dər\ *noun* [ultimately from Greek *chamaidrys,* from *chamai* on the ground + *drys* tree — more at HUMBLE, TREE] (15th century)
: any of a genus (*Teucrium*) of plants of the mint family with flowers having four exserted stamens, a short corolla tube, and a prominent lower lip

ger·mane \(ˌ)jər-ˈmān\ *adjective* [Middle English *germain,* literally, having the same parents, from Middle French] (14th century)
1 *obsolete* **:** closely akin
2 : being at once relevant and appropriate **:** FITTING ⟨omit details that are not *germane* to the discussion⟩
synonym see RELEVANT
— ger·mane·ly *adverb*

¹Ger·man·ic \(ˌ)jər-ˈma-nik\ *adjective* (1633)
1 : GERMAN
2 : of, relating to, or characteristic of the Germanic-speaking peoples
3 : of, relating to, or constituting Germanic

²Germanic *noun* (1892)
: a branch of the Indo-European language family containing English, German, Dutch, Afrikaans, Flemish, Frisian, the Scandinavian languages, and Gothic — see INDO-EUROPEAN LANGUAGES table

Ger·man·ism \ˈjər-mə-ˌni-zəm\ *noun* (1611)
1 : a characteristic feature of German occurring in another language
2 : partiality for Germany or German customs
3 : the practices or objectives characteristic of the Germans

Ger·man·ist \-nist\ *noun* (1831)
: a specialist in German or Germanic language, literature, or culture

ger·ma·ni·um \(ˌ)jər-ˈmā-nē-əm\ *noun* [New Latin, from Medieval Latin *Germania* Germany] (1886)
: a grayish white hard brittle metalloid element that resembles silicon and is used as a semiconductor — see ELEMENT table

ger·man·ize \ˈjər-mə-ˌnīz\ *verb* **-ized; -iz·ing** *often capitalized* (1598)
transitive verb
1 *archaic* **:** to translate into German
2 : to cause to acquire German characteristics
intransitive verb
: to have or acquire German customs or leanings
— ger·man·iza·tion \ˌjər-mə-nə-ˈzā-shən\ *noun, often capitalized*

German measles *noun plural but singular or plural in construction* (circa 1875)
: an acute contagious virus disease that is milder than typical measles but is damaging to the fetus when occurring early in pregnancy

Ger·mano- *combining form*
: German ⟨*Germano*phile⟩

¹Ger·mano·phile \(ˌ)jər-ˈma-nə-ˌfīl\ *adjective* (1898)
: approving or favoring the German people and their institutions and customs

²Germanophile *noun* (1911)
: one that is Germanophile

German shepherd *noun* (1926)
: any of a breed of working dogs of German origin that are intelligent and responsive and are often used in police work and as guide dogs for the blind — called also *Alsatian*

German shorthaired pointer *noun* (1931)
: any of a breed of gundogs of German origin that have a liver or liver and white coat

German silver *noun* (1830)
: a silver-white alloy of copper, zinc, and nickel

German wirehaired pointer *noun* (circa 1964)
: any of a breed of gundogs of German origin that have a liver or liver and white flat-lying wiry coat

germ cell *noun* (circa 1855)
: a gamete (as an egg or sperm cell) or one of its antecedent cells

ger·men \ˈjər-mən\ *noun* [Latin] (1605)
archaic **:** GERM 1a, 2

germ·free \ˈjərm-ˌfrē\ *adjective* (1926)
: free of microorganisms **:** AXENIC

ger·mi·cid·al \ˌjər-mə-ˈsī-dᵊl\ *adjective* (circa 1888)
: of or relating to a germicide; *also* **:** destroying germs

ger·mi·cide \ˈjər-mə-ˌsīd\ *noun* (1881)
: an agent that destroys germs

ger·mi·na·bil·i·ty \ˌjər-mə-nə-ˈbi-lə-tē\ *noun* (1896)
: the capacity to germinate

ger·mi·nal \ˈjərm-nəl, ˈjer-mə-nᵊl\ *adjective* [French, from Latin *germin-, germen* — more at GERM] (1808)
1 a : being in the earliest stage of development **b :** CREATIVE, PRODUCTIVE
2 : of, relating to, or having the characteristics of a germ cell or early embryo
— ger·mi·nal·ly *adverb*

germinal vesicle *noun* (circa 1856)
: the enlarged nucleus of the egg before completion of meiosis

ger·mi·nate \ˈjər-mə-ˌnāt\ *verb* **-nat·ed; -nat·ing** [Latin *germinatus,* past participle of *germinare* to sprout, from *germin-, germen* bud, germ] (1610)
transitive verb
: to cause to sprout or develop
intransitive verb
1 : to come into being **:** EVOLVE ⟨before Western civilization began to *germinate* —A. L. Kroeber⟩
2 : to begin to grow **:** SPROUT
— ger·mi·na·tion \ˌjər-mə-ˈnā-shən\ *noun*
— ger·mi·na·tive \ˈjər-mə-ˌnā-tiv, -mə-nə-\ *adjective*

germ layer *noun* (1879)
: any of the three primary layers of cells differentiated in most embryos during and immediately following gastrulation

germ plasm *noun* (1889)
1 : germ cells and their precursors serving as the bearers of heredity and being fundamentally independent of other cells
2 : the hereditary material of the germ cells **:** GENES

germ·proof \ˈjərm-ˌprüf\ *adjective* (1902)
: impervious to the penetration or action of germs

germ theory *noun* (1871)
: a theory in medicine: infections, contagious diseases, and various other conditions result from the action of microorganisms

germ warfare *noun* (1938)
: the use of harmful microorganisms (as bacteria) as weapons in war

germy \ˈjər-mē\ *adjective* **germ·i·er; -est** (1912)
: full of germs

geront- *or* **geronto-** *combining form* [French *géront-, géronto-,* from Greek *geront-, geronto-,* from *geront-, gerōn* old man; akin to Greek *gēras* old age, Sanskrit *jarati* he grows old]
: aged one **:** old age ⟨*geronto*logy⟩

ge·ron·tic \jə-ˈrän-tik\ *adjective* (1885)
: of or relating to decadence or old age

ger·on·toc·ra·cy \ˌjer-ən-ˈtä-krə-sē\ *noun, plural* **-cies** [French *gérontocratie,* from *géronto-* geront- + *-cratie* -cracy] (1830)
: rule by elders; *specifically* **:** a form of social organization in which a group of old men or a council of elders dominates or exercises control
— ge·ron·to·crat \jə-ˈrän-tə-ˌkrat\ *noun*
— ge·ron·to·crat·ic \-ˌrän-tə-ˈkra-tik\ *adjective*

ger·on·tol·o·gy \ˌjer-ən-ˈtä-lə-jē\ *noun* [International Scientific Vocabulary] (1903)
: the comprehensive study of aging and the problems of the aged — compare GERIATRIC 1
— ge·ron·to·log·i·cal \jə-ˌrän-tᵊl-ˈä-ji-kəl\ *also* **ge·ron·to·log·ic** \-jik\ *adjective*
— ger·on·tol·o·gist \ˌjer-ən-ˈtä-lə-jist\ *noun*

ge·ron·to·mor·phic \jə-ˌrän-tə-ˈmȯr-fik\ *adjective* (1939)
: characterized by physical specialization most fully developed in the aged male of a species ⟨*gerontomorphic* traits⟩

¹ger·ry·man·der \ˈjer-ē-ˌman-dər, *also* ˈger-; *originally* ˈger-\ *noun* [Elbridge *Gerry* + salamander; from the shape of an election district formed during Gerry's governorship of Mass.] (1812)
1 : the act or method of gerrymandering
2 : a district or pattern of districts varying greatly in size or population as a result of gerrymandering ◆

²gerrymander *transitive verb* **-dered; -der·ing** \-d(ə-)riŋ\ (1812)
1 : to divide (a territorial unit) into election districts to give one political party an electoral majority in a large number of districts while concentrating the voting strength of the opposition in as few districts as possible
2 : to divide (an area) into political units to give special advantages to one group ⟨*gerrymander* a school district⟩

ger·und \ˈjer-ənd\ *noun* [Late Latin *gerundium,* from Latin *gerundus,* gerundive of *gerere* to bear, carry on] (1513)
1 : a verbal noun in Latin that expresses generalized or uncompleted action
2 : any of several linguistic forms analogous to the Latin gerund in languages other than Latin; *especially* **:** the English verbal noun in *-ing* that has the function of a substantive and at the same time shows the verbal features of tense, voice, and capacity to take adverbial qualifiers and to govern objects ■

ge·run·dive \jə-ˈrən-div\ *noun* (15th century)
1 : the Latin future passive participle that functions as the verbal adjective, that expresses the fitness or necessity of the action to be performed, and that has the same suffix as the gerund
2 : a verbal adjective in a language other than Latin analogous to the gerundive

ge·sell·schaft \gə-ˈzel-ˌshäft\ *noun* [German, companionship, society] (1887)
: a rationally developed mechanistic type of social relationship characterized by impersonally contracted associations between persons;

◇ WORD HISTORY

gerrymander Elbridge Gerry (1744–1814), a Massachusetts politician who signed the Declaration of Independence and was a U.S. vice president, is now chiefly remembered for a bit of political flimflam. As a Democratic-Republican, Gerry was twice elected to the governorship of Massachusetts. His second administration achieved notoriety because of a measure of February 11, 1812, that has since become known as the "Gerrymander Bill." Its purpose was to redistrict the state in such a way as to give the Democratic-Republicans a majority in the state senate. During the ensuing election campaign a member of the opposition Federalists sketched an animal's head, wings, and claws onto an outline of one irregularly shaped senatorial district in Essex County. The fancied resemblance of the resulting image to the mythical salamander inspired the coinage *gerrymander.* This caricature of the district as a winged monster—with Gerry's profile superimposed upon its back—was widely reproduced, and *gerrymander* immediately caught on, at first applied to the caricature itself and then to the actual act of redistricting in such a way as to favor the party in power. Before long it was being used as a verb, too.

also **:** a community or society characterized by this relationship — compare GEMEINSCHAFT

ges·ne·ri·ad \ges-ˈnir-ē-ˌad\ *noun* [New Latin *Gesneria,* genus name, from Konrad *Gesner* (died 1565) Swiss naturalist] (circa 1925)
: any of a family (Gesneriaceae) of tropical herbs (as an African violet or gloxinia) with chiefly opposite leaves and highly zygomorphic flowers

ges·so \ˈje-(ˌ)sō\ *noun, plural* **gessoes** [Italian, literally, gypsum, from Latin *gypsum*] (1596)
1 : plaster of paris or gypsum prepared with glue for use in painting or making bas-reliefs
2 : a paste prepared by mixing whiting with size or glue and spread upon a surface to fit it for painting or gilding
— ges·soed \-(ˌ)sōd\ *adjective*

gest *or* **geste** \ˈjest\ *noun* [Middle English *geste* — more at JEST] (13th century)
1 : a tale of adventures; *especially* **:** a romance in verse
2 : ADVENTURE, EXPLOIT

ge·stalt \gə-ˈshtält, -ˈstält, -ˈstȯlt, -ˈshtȯlt\ *noun, plural* **ge·stalt·en** \-ˈstäl-tᵊn, -ˈshtäl-, -ˈstȯl-, -ˈshtȯl-\ *or* **gestalts** [German, literally, shape, form] (1922)
: a structure, configuration, or pattern of physical, biological, or psychological phenomena so integrated as to constitute a functional unit with properties not derivable by summation of its parts

ge·stalt·ist \-ˈstäl-tist, -ˈshtäl-, -ˈstȯl-, -ˈshtȯl-\ *noun, often capitalized* (1931)
: a specialist in Gestalt psychology

Gestalt psychology *noun* (1927)
: the study of perception and behavior from the standpoint of an individual's response to configurational wholes with stress on the uniformity of psychological and physiological events and rejection of analysis into discrete events of stimulus, percept, and response

ge·sta·po \gə-ˈstä-(ˌ)pō\ *noun, plural* **-pos** [German, from *Ge*heime *Sta*ats*po*lizei, literally, secret state police] (1934)
: a secret-police organization employing underhanded and terrorist methods against persons suspected of disloyalty

ges·tate \ˈjes-ˌtāt\ *verb* **ges·tat·ed; ges·tat·ing** [back-formation from *gestation*] (1866)
transitive verb
1 : to carry in the uterus during pregnancy
2 : to conceive and gradually develop in the mind
intransitive verb
: to be in the process of gestation

ges·ta·tion \je-ˈstā-shən\ *noun* [Latin *gestation-, gestatio,* from *gestare* to bear, frequentative of *gerere* to bear] (1615)
1 : the carrying of young in the uterus **:** PREGNANCY
2 : conception and development especially in the mind
— ges·ta·tion·al \-shnəl, -shə-nᵊl\ *adjective*

geste *also* **gest** \ˈjest\ *noun* [Middle French *geste,* from Latin *gestus,* from *gerere*] (14th century)
1 *archaic* **:** DEPORTMENT
2 *archaic* **:** GESTURE

ges·tic \ˈjes-tik\ *adjective* (1764)
: relating to or consisting of bodily movements or gestures

ges·tic·u·lant \je-ˈsti-kyə-lənt\ *adjective* (1877)
: making gesticulations ⟨the little wiry man *gesticulant* and wild —William Faulkner⟩

ges·tic·u·late \je-ˈsti-kyə-ˌlāt\ *intransitive verb* **-lat·ed; -lat·ing** [Latin *gesticulatus,* past participle of *gesticulari,* from (assumed) Latin *gesticulus,* diminutive of Latin *gestus*] (circa 1609)
: to make gestures especially when speaking
— ges·ti·cu·la·tive \je-ˈsti-kyə-ˌlā-tiv\ *adjective*
— ges·tic·u·la·tor \-ˌlā-tər\ *noun*
— ges·tic·u·la·to·ry \-lə-ˌtōr-ē, -ˌtȯr-\ *adjective*

ges·tic·u·la·tion \je-ˌsti-kyə-ˈlā-shən\ *noun* (15th century)
1 : the act of making gestures
2 : GESTURE; *especially* **:** an expressive gesture made in showing strong feeling or in enforcing an argument

[1]ges·ture \ˈjes-chər, ˈjesh-\ *noun* [Middle English, from Medieval Latin *gestura* mode of action, from Latin *gestus,* past participle of *gerere*] (15th century)
1 *archaic* **:** CARRIAGE, BEARING
2 : a movement usually of the body or limbs that expresses or emphasizes an idea, sentiment, or attitude
3 : the use of motions of the limbs or body as a means of expression
4 : something said or done by way of formality or courtesy, as a symbol or token, or for its effect on the attitudes of others ⟨a political *gesture* to draw popular support —V. L. Parrington⟩
— ges·tur·al \-chə-rəl\ *adjective*
— ges·tur·al·ly \-chə-rə-lē\ *adverb*

[2]gesture *verb* **ges·tured; ges·tur·ing** (1542)
intransitive verb
: to make a gesture
transitive verb
: to express or direct by a gesture

gesture language *noun* (1865)
: communication by gestures; *especially* **:** SIGN LANGUAGE — called also *gestural language*

ge·sund·heit \gə-ˈzu̇nt-ˌhīt\ *interjection* [German, literally, health, from *gesund* healthy (from Old High German *gisunt*) + *-heit* -hood — more at SOUND] (1914)
— used to wish good health especially to one who has just sneezed

[1]get \ˈget, ÷ˈgit\ *verb* **got** \ˈgät\; **got** *or* **got·ten** \ˈgä-tᵊn\; **get·ting** [Middle English, from Old Norse *geta* to get, beget; akin to Old English *bigietan* to beget, Latin *prehendere* to seize, grasp, Greek *chandanein* to hold, contain] (13th century)
transitive verb
1 a : to gain possession of **b :** to receive as a return **:** EARN ⟨he *got* a bad reputation for carelessness⟩
2 a : to obtain by concession or entreaty ⟨*get* your mother's permission to go⟩ **b :** to become affected by (a disease or bodily condition) **:** CATCH ⟨*got* measles from his sister⟩
3 a : to seek out and obtain ⟨hoped to *get* dinner at the inn⟩ **b :** to obtain and bring where wanted or needed ⟨*get* a pencil from the desk⟩
4 : BEGET
5 a : to cause to come or go ⟨quickly *got* his luggage through customs⟩ **b :** to cause to move ⟨*get* it out of the house⟩ **c :** to cause to be in a certain position or condition ⟨*got* his feet wet⟩ **d :** to make ready **:** PREPARE ⟨*get* breakfast⟩
6 a : to be subjected to ⟨*got* a bad fall⟩ **b :** to receive by way of punishment **c :** to suffer a specified injury to ⟨*got* my nose broken⟩
7 a : to achieve as a result of military activity **b :** to obtain or receive by way of benefit or advantage ⟨he *got* little for his trouble⟩ ⟨*get* the better of an enemy⟩
8 a : SEIZE **b :** OVERCOME **c :** to have an emotional effect on ⟨the final scene always *gets* me⟩ **d :** IRRITATE ⟨the delays were starting to *get* her⟩ **e :** PUZZLE **f :** to take vengeance on; *specifically* **:** KILL **g :** HIT
9 : to prevail on **:** CAUSE ⟨finally *got* them to tidy up their room⟩
10 a : HAVE — used in the present perfect tense form with present meaning ⟨I've *got* no money⟩ **b :** to have as an obligation or necessity — used in the present perfect tense form with present meaning ⟨you have *got* to come⟩
11 a : to find out by calculation ⟨*get* the answer to a problem⟩ **b :** MEMORIZE ⟨*got* the verse by heart⟩ **c :** HEAR **d :** UNDERSTAND
12 : to establish communication with
13 : to put out in baseball
14 : DELIVER 6b ⟨the car *gets* 20 miles to the gallon⟩
intransitive verb

☐ USAGE
gerund A question discussed in many grammars and handbooks is whether the possessive case of a noun or pronoun must be used before a gerund: must it be "I approve of Pat's going" only or is "I approve of Pat going" also acceptable? Actually both forms have been used side by side for many centuries. When grammarians first discovered this fact in the 18th century, they were perplexed; it never occurred to them that there could be two right ways of doing anything—one of them must be wrong. The 18th century grammarians disagreed as to which was right, but 20th century handbooks opt for the possessive and disapprove the objective pronoun and the common-case noun. Historical grammarians have in the meantime been studying the evidence, and they have discovered that certain circumstances influence the choice of the common (or objective) case or the possessive case. If the gerund is certainly a noun, as in Samuel Pepys' "and thence to Jacob Hall's dancing on the ropes" —where *dancing* could be replaced by the noun *performance*—the possessive is always used. Unfortunately, clear-cut noun gerunds are relatively uncommon; most gerunds seem to be more verblike. Verblike gerunds can take either the possessive case or the common (or objective) case, but again subtle influences are at work. Pronouns, for instance, are more often possessive ⟨if you don't mind *my saying* so⟩ ⟨no chance of *its being* ignored⟩, but if the pronoun is linked to a noun in the common case, the objective is usual ⟨they joked about my brother and *me liking* rice pudding⟩. Words intervening between the pronoun and the gerund can also trigger the objective case ⟨I don't understand *you,* of all people, *complaining*⟩. The objective case may also be felt to be more emphatic ⟨I can't imagine *me letting* that happen⟩. These same considerations can bring on the common case of nouns ⟨imagine *John,* of all people, *being* frightened⟩. Nouns appear to be somewhat more evenly divided between possessive and common case ⟨afraid of the *money's being* wasted⟩ ⟨seemed pleased at the *book being* out of print⟩. But the possessive is usually not used after nouns ending in an \s\ sound ⟨afraid of the *house being* robbed⟩. Some plurals without final *-s* are used in the common case because the possessive simply sounds wrong ⟨whatever is responsible for my *feet swelling*⟩. Using the common case permits the addition of a parallel construction with the gerund omitted ⟨was sorry to see *Tom getting* all the credit and Paul none⟩. Some words lack a possessive form ⟨never dreamed of *there being* a different way to do it⟩. And the more informal, the more conversational the context, the less likely is the possessive. All of these constructions are standard English, and when there is no reason prompting one form or the other, the choice is open.

1 a : to succeed in coming or going **:** to bring or move oneself ⟨*get* away to the country⟩ ⟨*got* into the car⟩ **b :** to reach or enter into a certain condition ⟨*got* to sleep after midnight⟩ **c :** to make progress ⟨hasn't *gotten* far with the essay⟩
2 : to acquire wealth
3 a : to be able ⟨never *got* to go to college⟩ **b :** to come to be — often used with following present participle ⟨*got* talking about old times⟩
4 a : to succeed in becoming **:** BECOME ⟨how to *get* clear of all the debts I owe —Shakespeare⟩ **b :** to become involved ⟨people who *get* into trouble with the law⟩
5 : to leave immediately ⟨told them to *get*⟩
verbal auxiliary
— used with the past participle of transitive verbs as a passive voice auxiliary ⟨they *got* caught in the act⟩ ▫
— get a bead on : to gain a precise and telling knowledge or understanding of ⟨*got a bead on* his motives⟩
— get after : to pursue with exhortation, reprimand, or attack
— get ahead : to achieve success ⟨determined to *get ahead* in life⟩
— get a move on : HURRY
— get at 1 : to reach effectively **2 :** to influence corruptly **:** BRIBE **3 :** to turn one's attention to **4 :** to try to prove or make clear ⟨what is he *getting at*⟩
— get away with : to avoid criticism or punishment for or the consequences of (as a reprehensible act)
— get cracking : to make a start **:** get going ⟨ought to *get cracking* on that assignment⟩
— get even : to get revenge
— get even with : to repay in kind
— get going : to make a start
— get into : to become strongly involved with or deeply interested in
— get it : to receive a scolding or punishment
— get it on 1 : to become enthusiastic, energetic, or excited **2 :** to engage in sexual intercourse
— get on : to produce an unfortunate effect on **:** UPSET ⟨the noise *got on* my nerves⟩
— get one's act together 1 : to put one's life, thoughts, or emotions in order **:** cease to be confused or misdirected **2 :** to begin to function in a skillful or efficient manner ⟨the company finally *got its act together*⟩
— get one's goat : to make one angry or annoyed
— get over 1 a : OVERCOME, SURMOUNT **b :** to recover from **c :** to reconcile oneself to **:** become accustomed to **2 :** to move or travel across
— get somewhere : to be successful
— get there : to be successful
— get through : to reach the end of **:** COMPLETE
— get to 1 a : BEGIN ⟨*gets to* worrying over nothing at all⟩ **b :** to be ready to begin or deal with ⟨I'll *get to* the accounts as soon as I can⟩ **2 :** to have an effect on: as **a :** INFLUENCE **b :** BOTHER
— get together 1 : to bring together **:** ACCUMULATE **2 :** to come together **:** ASSEMBLE, MEET **3 :** to reach agreement
— get wind of : to become aware of
— get with it : to become alert or aware **:** show sophisticated consciousness

²get \ˈget\ *noun* (14th century)
1 a : something begotten: (1) **:** OFFSPRING (2) **:** the entire progeny of a male animal **b :** LINEAGE
2 : a return of a difficult shot in a game (as tennis)

³get \ˈget\ *noun, plural* **git·tin** \gē-ˈtēn, ˈgi-tin\ [Late Hebrew *gēṭ*] (1892)
1 : a document of release from obligation in Jewish law; *specifically* **:** a bill of divorce
2 : a religious divorce by Jewish law

ge·ta \ˈge-(ˌ)tä, -tə\ *noun, plural* **geta** *or* **ge·tas** [Japanese] (1884)
: a Japanese wooden clog for outdoor wear

geta

get about *intransitive verb* (1816)
1 : to become current **:** CIRCULATE
2 : to be up and about **:** begin to walk ⟨able to *get about* again⟩

get across (1913)
intransitive verb
: to become clear or convincing
transitive verb
: to make clear or convincing ⟨we couldn't *get* our point *across*⟩

get along *intransitive verb* (1768)
1 a : to proceed toward a destination **:** PROGRESS **b :** to approach an advanced stage; *especially* **:** to approach old age
2 : to meet one's needs **:** MANAGE ⟨we *got along* on a minimum of clothing⟩
3 : to be or remain on congenial terms

get around (1875)
transitive verb
1 : CIRCUMVENT, EVADE
2 : to get the better of
intransitive verb
1 a : to find or take the necessary time or effort — used with *to* **b :** to give attention or consideration — used with *to*
2 : to go from place to place
3 : to become known or current ⟨word *got around* that he was resigning⟩

get·at·able \get-ˈa-tə-bəl\ *adjective* (1799)
: ACCESSIBLE, APPROACHABLE

get·away \ˈge-tə-ˌwā\ *noun* (1890)
1 : an act or instance of getting away: as **a :** ESCAPE **b :** START
2 : a place suitable for a vacation
3 : a vacation especially of brief duration

get back (1605)
intransitive verb
1 : to come or go again to a person, place, or condition **:** RETURN, REVERT ⟨*getting back* to the main topic of the lecture⟩
2 : to gain revenge **:** RETALIATE — used with *at*
transitive verb
: to regain possession of **:** RECOVER

get by *intransitive verb* (1908)
1 : to succeed with the least possible effort or accomplishment
2 : to make ends meet **:** SURVIVE
3 : to proceed without being discovered, criticized, or punished

get down (1757)
intransitive verb
1 : to alight especially from a vehicle **:** DESCEND
2 : to give one's attention or consideration — used with *to* ⟨*get down* to business⟩
transitive verb
1 : to cause to be physically, mentally, or emotionally exhausted **:** DEPRESS ⟨the weather was *getting* her *down*⟩
2 : to manage to swallow
3 : to commit to writing **:** DESCRIBE

get–go \ˈgit-ˌgō, ˈget-\ *noun* (1980)
: the very beginning — used in the phrase *from the get-go* ⟨didn't like me from the *get-go*⟩

Geth·sem·a·ne \geth-ˈse-mə-nē\ *noun* [Greek *Gethsēmanē*]
1 : the garden outside Jerusalem mentioned in Mark 14 as the scene of the agony and arrest of Jesus
2 : a place or occasion of great mental or spiritual suffering

get in (circa 1533)
intransitive verb
1 a : ENTER **b :** ARRIVE
2 a : to become friendly **b :** to become involved
3 : to become accepted for membership or chosen for office
transitive verb
1 : to succeed in doing, making, or delivering
2 : to include in one's schedule
3 : INVOLVE 2a, b

get off (1640)
intransitive verb
1 : to avoid the most serious consequences of a dangerous situation or punishment ⟨*got off* with a light sentence⟩
2 : START, LEAVE ⟨*got off* on the trip early⟩
3 : to leave work with permission or as scheduled
4 : to get high on a drug
5 : to experience orgasm
6 : to experience great pleasure — often used with *on*
transitive verb
1 : to secure the release of or procure a modified penalty for ⟨his lawyers *got* him *off*⟩
2 a : UTTER ⟨*get off* a joke⟩ **b :** to write and send
3 : to succeed in doing, making, or delivering
4 : to cause to get off

get on *intransitive verb* (1816)
1 : GET ALONG ⟨was *getting on* in years⟩ ⟨*got on* well with the boss⟩ ⟨*get on* with the game⟩
2 : to gain knowledge or understanding ⟨*got on* to the racket⟩

get out (14th century)
intransitive verb
1 : LEAVE, ESCAPE ⟨doubted that he would *get out* alive⟩
2 : to become known **:** leak out ⟨their secret *got out*⟩
transitive verb
1 : to cause to leave or escape
2 : to bring before the public; *especially* **:** PUBLISH

get round *intransitive verb* (1748)
: GET AROUND

get·ter \ˈge-tər\ *noun* (15th century)
1 : one that gets
2 : a substance introduced into a vacuum tube or electric lamp to remove traces of gas

get through *intransitive verb* (1694)
1 : to reach a destination
2 : to gain approval or a desired outcome
3 a : to make something clear **b :** to complete a communications connection

get–to·geth·er \ˈget-tə-ˌge-t͟hər\ *noun* (1911)
: MEETING; *especially* **:** an informal social gathering

get·up \ˈget-ˌəp\ *noun* (1847)
1 : OUTFIT, COSTUME
2 : general composition or structure

get up (14th century)
intransitive verb
1 a : to arise from bed **b :** to rise to one's feet **c :** CLIMB, ASCEND
2 : to go ahead or faster — used in the imperative as a command especially to driven animals
transitive verb
1 : to make preparations for **:** ORGANIZE ⟨*got up* a party for the newcomers⟩
2 : to arrange as to external appearance **:** DRESS
3 : to acquire a knowledge of
4 : to create in oneself ⟨cannot *get up* the courage to tell them⟩

▫ USAGE
get The pronunciation \ˈgit\ has been noted as a feature of some British and American dialects since the 16th century. In the phonetic spelling of his own speech Benjamin Franklin records *git.* However, since at least 1687 some grammarians and teachers have disapproved this pronunciation. It nonetheless remains in widespread and unpredictable use in many dialects, often, but not exclusively, when *get* is a passive auxiliary (as in *get married*) or an imperative (as in *get up!*).

get–up–and–go \ˌget-ˌəp-ᵊn-ˈgō, ˌgit-, -ᵊm-, -ᵊŋ-\ *noun* (1906)
: ENERGY, DRIVE

ge·um \ˈjē-əm\ *noun* [Latin] (circa 1548)
: AVENS

gew·gaw \ˈg(y)ü-(ˌ)gȯ\ *noun* [origin unknown] (circa 1529)
: a showy trifle **:** BAUBLE, TRINKET

ge·würz·tra·mi·ner \gə-ˈvu̇rt-ˌstra-mə-nər, -ˈvərt-, -ˌsträ; -strə-ˈmē-nər\ *noun, often capitalized* [German, variety of grape, from *Gewürz* spice + *Traminer,* variety of grape, from *Tramin* (Termeno, Italy)] (circa 1950)
: a light dry Alsatian white wine with a spicy bouquet; *also* **:** a similar wine made elsewhere

gey \ˈgā\ *adverb* [alteration of *gay,* adverb] (1796)
chiefly Scottish **:** VERY, QUITE

gey·ser \ˈgī-zər, *British sometimes* ˈgā- *or* ˈgē- *for 1 & usually* ˈgē- *for 2*\ *noun* [Icelandic *Geysir,* hot spring in Iceland, from *geysa* to rush forth, from Old Norse; akin to Old English *gēotan* to pour — more at FOUND] (1780)
1 : a spring that throws forth intermittent jets of heated water and steam
2 *British* **:** an apparatus for heating water rapidly with a gas flame (as for a bath)

gey·ser·ite \-zə-ˌrīt\ *noun* [French *geysérite,* from *geyser,* from Icelandic *Geysir*] (circa 1814)
: a hydrous silica that constitutes one variety of opal and is deposited around some hot springs and geysers in white or grayish concretions

g–fac·tor \ˈjē-ˌfak-tər\ *noun* (1942)
: GYROMAGNETIC RATIO

gha·ri·al \ˈger-ē-əl\ *noun* [Hindi *gharyāl*] (circa 1809)
: GAVIAL

ghar·ry \ˈgar-ē, ˈgär \ *noun, plural* **gharries** [Hindi *gāṛī*] (1810)
: a horse-drawn cab used especially in India and Egypt

ghast \ˈgast\ *adjective* (1622)
archaic **:** GHASTLY

ghast·ful \-fəl\ *adjective* (14th century)
archaic **:** FRIGHTFUL
— ghast·ful·ly *adverb, archaic*

ghast·ly \ˈgast-lē\ *adjective* **ghast·li·er; -est** [Middle English *gastly,* from *gasten* to terrify — more at GAST] (14th century)
1 a : terrifyingly horrible to the senses **:** FRIGHTENING ⟨a *ghastly* crime⟩ **b :** intensely unpleasant, disagreeable, or objectionable ⟨such a life seems *ghastly* in its emptiness and sterility —Aldous Huxley⟩
2 : resembling a ghost
3 *obsolete* **:** filled with fear
4 : very great ⟨*ghastly* mistake⟩ ☆
— ghast·li·ness *noun*
— ghastly *adverb*

ghat \ˈgȯt, ˈgät\ *noun* [Hindi *ghāṭ,* from Sanskrit *ghaṭṭa*] (1783)
: a broad flight of steps that is situated on an Indian riverbank and that provides access to the water especially for bathing

ghee *or* **ghi** \ˈgē\ *noun* [Hindi *ghī,* from Sanskrit *ghṛta*] (1665)
: a semifluid clarified butter made especially in India

gher·kin \ˈgər-kən\ *noun* [Dutch *gurken,* plural of *gurk* cucumber, ultimately from Middle Greek *agouros*] (1661)
1 : a small prickly fruit used for pickling; *also* **:** the slender annual vine (*Cucumis anguria*) of the gourd family that bears it
2 : the immature fruit of the cucumber

¹ghet·to \ˈge-(ˌ)tō\ *noun, plural* **ghettos** *also* **ghettoes** [Italian, from Venetian dialect *ghèto* island where Jews were forced to live, literally, foundry (located on the island), from *ghetàr* to cast, from Latin *jactare* to throw — more at JET] (1611)
1 : a quarter of a city in which Jews were formerly required to live
2 : a quarter of a city in which members of a minority group live especially because of social, legal, or economic pressure
3 a : an isolated group ⟨a geriatric *ghetto*⟩ **b :** a situation that resembles a ghetto especially in conferring inferior status or limiting opportunity ⟨stuck in daytime TV's *ghetto*⟩ ◇

²ghetto *transitive verb* (1936)
: GHETTOIZE

ghetto blaster *noun* (1983)
: BOOM BOX

ghet·to·ize \ˈge-tō-ˌīz\ *transitive verb* **-ized; -iz·ing** (1939)
: to isolate in or as if in a ghetto
— ghet·to·iza·tion \ˌge-tō-ə-ˈzā-shən\ *noun*

Ghib·el·line \ˈgi-bə-ˌlēn, -ˌlīn, -lən\ *noun* [Italian *Ghibellino*] (1573)
: a member of an aristocratic political party in medieval Italy supporting the authority of the German emperors — compare GUELF

ghib·li \ˈgib-lē\ *noun* [Arabic dialect *giblī* south wind, from Arabic *qiblī* southern] (1942)
: a hot desert wind of northern Africa

ghil·lie *variant of* GILLIE

¹ghost \ˈgōst\ *noun* [Middle English *gost, gast,* from Old English *gāst;* akin to Old High German *geist* spirit, Sanskrit *heḍa* anger] (before 12th century)
1 : the seat of life or intelligence **:** SOUL ⟨give up the *ghost*⟩
2 : a disembodied soul; *especially* **:** the soul of a dead person believed to be an inhabitant of the unseen world or to appear to the living in bodily likeness
3 : SPIRIT, DEMON
4 a : a faint shadowy trace ⟨a *ghost* of a smile⟩ **b :** the least bit ⟨not a *ghost* of a chance⟩
5 : a false image in a photographic negative or on a television screen caused especially by reflection
6 : one who ghostwrites
7 : a red blood cell that has lost its hemoglobin ◇
— ghost·like \-ˌlīk\ *adjective*
— ghosty \ˈgō-stē\ *adjective*

²ghost (1606)
transitive verb
1 : to haunt like a ghost
2 : GHOSTWRITE
intransitive verb
1 a : to move silently like a ghost **b :** to sail quietly in light winds
2 : GHOSTWRITE

Ghost Dance *noun* (1890)
: a group dance of a late 19th century American Indian messianic cult believed to promote the return of the dead and the restoration of traditional ways of life

ghost·ing *noun* (circa 1957)
: a false image on a television screen; *also* **:** the formation of such images

ghost·ly \ˈgōst-lē\ *adjective* **ghost·li·er; -est** (before 12th century)
1 : of or relating to the soul **:** SPIRITUAL
2 : of, relating to, or having the characteristics of a ghost **:** SPECTRAL
— ghost·li·ness *noun*
— ghostly *adverb*

ghost story *noun* (1819)
1 : a story about ghosts
2 : a tale based on imagination rather than fact

ghost town *noun* (1931)
: a once-flourishing town wholly or nearly deserted usually as a result of the exhaustion of some natural resource

ghost word *noun* (1886)
: a word form never in established usage

ghost·write \ˈgōs(t)-ˌrīt\ *verb* **-wrote** \-ˌrōt\; **-writ·ten** \-ˌri-tᵊn\ [back-formation from *ghostwriter*] (1927)
intransitive verb
: to write for and in the name of another

☆ SYNONYMS

Ghastly, grisly, gruesome, macabre, lurid mean horrifying and repellent in appearance or aspect. GHASTLY suggests the terrifying aspects of corpses and ghosts ⟨a *ghastly* accident⟩. GRISLY and GRUESOME suggest additionally the results of extreme violence or cruelty ⟨an unusually *grisly* murder⟩ ⟨the *gruesome* history of the Nazi death camps⟩. MACABRE implies a morbid preoccupation with the physical aspects of death ⟨a *macabre* tale of premature burial⟩. LURID adds to GRUESOME the suggestion of shuddering fascination with violent death and especially with murder ⟨the *lurid* details of a crime⟩.

◇ WORD HISTORY

ghetto The etymology of Italian *ghetto* was formerly the subject of much speculation; among many hypotheses were suggestions that it had been shortened from Italian *borghetto,* a diminutive of *borgo* "small town, suburb," or that it had developed from Talmudic Hebrew *gēṭ* "document of release." Today there is little doubt on historical and linguistic grounds that the word comes from an Italian dialect form *ghèto* "foundry." A foundry for cannons was once located on an island forming part of Venice, where in 1516 the Venetians restricted Jewish residence. The word *ghèto* became the name for the quarter and was borrowed into standard Tuscan-based Italian as *ghetto,* with the generic meaning "quarter of a city where Jews are forced to live." From there it passed into most of the languages of Europe. The word took on its darkest connotations during World War II, when Nazi imprisonment of Jews in sectors of Warsaw, Łódź, and other Polish cities led to thousands of deaths through malnutrition and disease. Since the late 19th century, the reference of *ghetto* has been extended to crowded urban districts where other ethnic or racial groups have been confined by poverty and prejudice, and this is probably its most frequent current use.

ghost English *ghost* is descended directly from Old English *gāst,* which meant "soul" or "spirit." These senses now survive mainly in the idiom *to give up the ghost,* and in the Christian epithet *Holy Ghost* for the third person of the Trinity. Not untypically for English, a loanword, *spirit* (from Latin *spiritus*), fills in for the primary sense of an old native word that later developed a very restricted sense. *Ghost* now usually alludes to the spirit of a dead person that supposedly appears to the living in its bodily likeness—a meaning not attested in English before the 14th century. This situation contrasts with the history of German *Geist,* an exact cognate of *ghost,* which from Old High German to the present has retained its primary sense, "spirit." The peculiar spelling of English *ghost* with an initial *gh,* which did not become general until the end of the 16th century, was apparently introduced by the 15th century English printer and translator William Caxton. Having lived in the Low Countries for many years, Caxton most likely adopted the *gh* from Middle Dutch *gheest,* another Germanic cognate of *ghost.*

transitive verb
: to write (as a speech) for another who is the presumed author
— **ghost·writ·er** *noun*

ghoul \'gül\ *noun* [Arabic *ghūl*] (1786)
1 : a legendary evil being that robs graves and feeds on corpses
2 : one suggestive of a ghoul
— **ghoul·ish** \'gü-lish\ *adjective*
— **ghoul·ish·ly** *adverb*
— **ghoul·ish·ness** *noun*

ghou·lie \'gü-lē\ *noun* (1928)
: GHOUL 1

1GI \,jē-'ī\ *adjective* [galvanized *i*ron; from abbreviation used in listing such articles as garbage cans, but taken as abbreviation for *government issue*] (circa 1935)
1 : provided by an official U.S. military supply department ⟨*GI* shoes⟩
2 : of, relating to, or characteristic of U.S. military personnel
3 : conforming to military regulations or customs ⟨a *GI* haircut⟩

2GI *noun, plural* **GIs** *or* **GI's** \-'īz\ (1943)
: a member or former member of the U.S. armed forces; *especially* **:** a man enlisted in the army

3GI *adverb* (1949)
: in a strictly regulation manner

4GI *transitive verb* **GI'd** \-'īd\; **GI'·ing** \-'ī-iŋ\ (1951)
: to clean thoroughly (as floors) in preparation for or as if for a military inspection

1gi·ant \'jī-ənt\ *noun* [Middle English *giaunt,* from Old French *geant,* from Latin *gigant-, gigas,* from Greek] (14th century)
1 : a legendary humanlike being of great stature and strength
2 a : a living being of great size **b :** a person of extraordinary powers
3 : something unusually large or powerful
— **gi·ant·like** \-,līk\ *adjective*

2giant *adjective* (15th century)
: having extremely large size, proportion, or power

giant anteater *noun* (1940)
: a large bushy-tailed anteater (*Myrmecophaga tridactyla*) of Central and South America

giant cactus *noun* (circa 1884)
: SAGUARO

giant clam *noun* (circa 1889)
: a very large clam (*Tridacna gigas*) found on the coral reefs of the Indian and Pacific oceans that sometimes weighs more than 500 pounds (227 kilograms)

gi·ant·ess \'jī-ən-təs\ *noun* (14th century)
: a female giant

gi·ant·ism \'jī-ən-,ti-zəm\ *noun* (1639)
1 : the quality or state of being a giant ⟨*giantism* in industry⟩
2 : GIGANTISM 2

giant panda *noun* (1920)
: a large black-and-white mammal (*Ailuropoda melanoleuca*) of chiefly central China now usually classified with the bears (family Ursidae) — see PANDA illustration

giant reed *noun* (1851)
: a tall European grass (*Arundo donax*) with woody stems used especially in making organ reeds, baskets, and shelters

giant schnauzer *noun* (circa 1934)
: any of a breed of powerful heavyset schnauzers that attain a height of 23½ to 27½ inches (60 to 70 centimeters)

giant sequoia *noun* (circa 1931)
: BIG TREE

giant slalom *noun* (1952)
: a slalom race for skiers on a longer and steeper course than that used for the regular slalom

giant squid *noun* (circa 1890)
: any of a genus (*Architeuthis*) of extremely large squids that include the largest mollusks known

giant star *noun* (1912)
: a star of great luminosity and of large mass

giant tortoise *noun* (circa 1909)
: any of various large long-lived herbivorous land tortoises (genus *Geochelone*) formerly common on the islands of the western Indian Ocean and on the Galapagos Islands

giant water bug *noun* (1901)
: any of a family (Belostomatidae and especially genus *Lethocerus*) of very large predatory bugs capable of inflicting a painful bite

giaour \'jau̇(-ə)r\ *noun* [French, from Italian dialect (Venetian) *giaur,* from Turkish *gâvur,* from Persian *gawr*] (1564)
: one outside the Islamic faith **:** INFIDEL 2a

giar·di·a·sis \(,)jē-,är-'dī-ə-səs, jē-ər-, jär-\ *noun, plural* **-a·ses** \-,sēz\ [New Latin, from *Giardia,* from Alfred M. *Giard* (died 1908) French biologist] (1919)
: infestation with or disease caused by a flagellate protozoan (genus *Giardia* and especially *G. lamblia* in humans) that is often characterized by diarrhea

1gib \'gib\ *noun* [Middle English, from *Gib,* nickname for *Gilbert*] (1561)
: a male cat; *specifically* **:** a castrated male cat

2gib *noun* [origin unknown] (1795)
: a plate of metal or other material machined to hold other parts in place, to afford a bearing surface, or to provide means for overcoming looseness

gib·ber \'ji-bər\ *intransitive verb* **gib·bered; gib·ber·ing** \-b(ə-)riŋ\ [imitative] (1604)
: to speak rapidly, inarticulately, and often foolishly
— **gibber** *noun*

gib·ber·el·lic acid \,ji-bə-'re-lik-\ *noun* (1954)
: a crystalline acid $C_{19}H_{22}O_6$ that is a gibberellin used especially in the malting of barley

gib·ber·el·lin \-'re-lən\ *noun* [New Latin *Gibberella fujikoroi,* fungus from which it was first isolated] (1939)
: any of several growth-regulating plant hormones that promote cell elongation and activity of the cambium, induce parthenocarpy, and stimulate synthesis of nucleic acids and proteins

gib·ber·ish \'ji-b(ə-)rish, 'gi-\ *noun* [probably from *gibber*] (circa 1554)
: unintelligible or meaningless language: **a :** a technical or esoteric language **b :** pretentious or needlessly obscure language

1gib·bet \'ji-bət\ *noun* [Middle English *gibet,* from Old French] (13th century)
1 : GALLOWS
2 : an upright post with a projecting arm for hanging the bodies of executed criminals as a warning

2gibbet *transitive verb* (1646)
1 a : to expose to infamy or public scorn **b :** to hang on a gibbet
2 : to execute by hanging on a gibbet

gib·bon \'gi-bən\ *noun* [French] (1774)
: any of a genus (*Hylobates*) of brachiating tailless apes of southeastern Asia that are the smallest and most arboreal anthropoid apes

gibbon

gib·bos·i·ty \ji-'bä-sə-tē, gi-\ *noun, plural* **-ties** (14th century)
: PROTUBERANCE, SWELLING

gib·bous \'ji-bəs, 'gi-\ *adjective* [Middle English, from Middle French *gibbeux,* from Late Latin *gibbosus* humpbacked, from Latin *gibbus* hump] (14th century)
1 a : marked by convexity or swelling **b** *of the moon or a planet* **:** seen with more than half but not all of the apparent disk illuminated
2 : having a hump **:** HUMPBACKED

gibe \'jīb\ *verb* **gibed; gib·ing** [perhaps from Middle French *giber* to shake, handle roughly] (1567)
intransitive verb
: to utter taunting words
transitive verb
: to deride or tease with taunting words
synonym see SCOFF
— **gibe** *noun*
— **gib·er** *noun*

gib·lets \'jib-ləts *also* 'gib-\ *noun plural* [Middle English *gibelet* entrails, garbage, from Middle French, stew of wildfowl] (15th century)
: the edible viscera of a fowl

Gi·bral·tar \jə-'bröl-tər\ *noun* [*Gibraltar,* fortress in the British colony of Gibraltar] (1856)
: an impregnable stronghold

Gib·son \'gib-sən\ *noun* [from the name *Gibson*] (1948)
: a martini garnished with a cocktail onion

Gibson girl *adjective* [Charles D. *Gibson*] (1936)
: of or relating to a style especially in women's clothing characterized by high necks, full sleeves, and wasp waists

gid \'gid\ *noun* [back-formation from *giddy*] (1601)
: a disease especially of sheep caused by the larva of a tapeworm (*Multiceps multiceps*) in the brain

gid·dap \gi-'dap, -'dəp\ *also* **gid·dy·ap** \,gi-dē-'ap, -'əp\ *verb imperative* [alteration of *get up*] (circa 1897)
— a command to a horse to go ahead or go faster

1gid·dy \'gi-dē\ *adjective* **gid·di·er; -est** [Middle English *gidy* mad, foolish, from Old English *gydig* possessed, mad; akin to Old English *god* god] (14th century)
1 a : DIZZY ⟨*giddy* from the unaccustomed exercise⟩ **b :** causing dizziness ⟨a *giddy* height⟩ **c :** whirling rapidly
2 a : lightheartedly silly **:** FRIVOLOUS **b :** joyfully elated **:** EUPHORIC ◆
— **gid·di·ly** \'gi-dᵊl-ē\ *adverb*
— **gid·di·ness** \'gi-dē-nəs\ *noun*

2giddy *verb* **gid·died; gid·dy·ing** (1602)
transitive verb
: to make giddy
intransitive verb
: to become giddy

Gid·e·on \'gi-dē-ən\ *noun* [Hebrew *Gidh'ōn*]
1 : an early Hebrew hero noted for his defeat of the Midianites
2 : a member of an interdenominational organization whose activities include the placing of Bibles in hotel rooms

gie \'gē\ *chiefly Scottish variant of* GIVE

◇ WORD HISTORY

giddy Before the nature of mental illness began to be understood, the cause of insanity was long thought to be possession by spirits or demons. Old English *gidig* or *gydig* "insane" meant literally "possessed by a spirit" and was an adjectival derivative in prehistoric Old English of the noun that would become Old English and Modern English *god. Gidy,* the Middle English outcome of Old English *gydig,* retained overtones of madness, but its meaning tended more toward "foolish, stupid." Only a small semantic sidestep was then required to produce the sense "dizzy," which was first recorded at the end of the 14th century. By the 17th century, the new meaning of *giddy,* as it had come to be spelled, had produced several extended senses, including "causing dizziness" and "rapidly whirling." These meanings are still in use, as is the sense "lightheartedly or exuberantly silly," which first occurred in the middle of the 16th century and which combines elements of both the "foolish" and "dizzy" senses.

Gi·em·sa stain \gē-ˈem-zə-\ *noun* [Gustav *Giemsa* (died 1948) German chemotherapist] (circa 1909)
: a stain consisting of eosin and a blue dye and used chiefly in the differential staining of blood films — called also *Giemsa, Giemsa's stain*

[1]**gift** \ˈgift\ *noun* [Middle English, from Old Norse, something given, talent; akin to Old English *giefan* to give] (12th century)
1 : a notable capacity, talent, or endowment
2 : something voluntarily transferred by one person to another without compensation
3 : the act, right, or power of giving ☆

[2]**gift** *transitive verb* (circa 1550)
1 : to endow with some power, quality, or attribute
2 : PRESENT ⟨generously *gifted* us with a copy —*Saturday Review*⟩

gift certificate *noun* (1942)
: a certificate entitling the recipient to purchase goods or services in the establishment of the issuer to the amount specified

gift·ed \ˈgif-təd\ *adjective* (1644)
1 : having great natural ability **:** TALENTED ⟨*gifted* children⟩
2 : revealing a special gift ⟨*gifted* voices⟩
— gift·ed·ly *adverb*
— gift·ed·ness *noun*

gift of gab (circa 1890)
: the ability to talk glibly and persuasively

gift of tongues [from the gifts of the Spirit in 1 Corinthians 12:1–13] (1560)
: the charismatic gift of ecstatic speech

gift·ware \ˈgift-ˌwar, -ˌwer\ *noun* (1904)
: wares or goods suitable for gifts

gift wrap *transitive verb* (1936)
: to wrap (merchandise intended as a gift) decoratively

[1]**gig** \ˈgig\ *noun* [Middle English *-gyge* (in *whyrlegyge* whirligig), of unknown origin] (1570)
1 : something that whirls or is whirled: as **a** *obsolete* **:** TOP, WHIRLIGIG **b :** a 3-digit selection in a numbers game
2 : a person of odd or grotesque appearance
3 a : a long light ship's boat **b :** a rowboat designed for speed rather than for work
4 : a light two-wheeled one-horse carriage

[2]**gig** *intransitive verb* **gigged; gig·ging** (1807)
: to travel in a gig

[3]**gig** *noun* [short for earlier *fizgig, fishgig,* of unknown origin] (1722)
1 : a pronged spear for catching fish
2 : an arrangement of hooks to be drawn through a school of fish in order to hook their bodies

[4]**gig** *verb* **gigged; gig·ging** (1803)
transitive verb
1 : to spear with a gig
2 a *chiefly West* **:** SPUR, JAB **b :** GOAD, PROVOKE
intransitive verb
: to fish with a gig

[5]**gig** *noun* [origin unknown] (1926)
: a job usually for a specified time; *especially* **:** an entertainer's engagement

[6]**gig** *intransitive verb* **gigged; gig·ging** (1939)
: to work as a musician ⟨*gigged* with various bands —*Downbeat*⟩

[7]**gig** *noun* [origin unknown] (circa 1941)
: a military demerit

[8]**gig** *transitive verb* **gigged; gig·ging** (circa 1941)
: to give a military gig to

giga- \ˈji-gə, ˈgi-\ *combining form* [International Scientific Vocabulary, from Greek *gigas* giant]
: billion ⟨*giga*hertz⟩ ⟨*giga*watt⟩

giga·bit \-ˌbit\ *noun* (1970)
: a unit of information equal to one billion bits

giga·byte \-ˌbīt\ *noun* (1975)
: 1,073,741,824 bytes

giga·hertz \-ˌhərts, -ˌherts\ *noun* (1964)
: a unit of frequency equal to one billion hertz

gi·gan·tesque \ˌjī-ˌgan-ˈtesk, -gən-\ *adjective* (1821)
: of enormous or grotesquely large proportions

gi·gan·tic \jī-ˈgan-tik, jə-\ *adjective* [Greek *gigantikos,* from *gigant-, gigas* giant] (1651)
: exceeding the usual or expected (as in size, force, or prominence) ⟨a man of *gigantic* stature⟩ ⟨made a last *gigantic* effort⟩ ⟨*gigantic* industrial combines⟩
synonym see ENORMOUS
— gi·gan·ti·cal·ly \-ti-k(ə-)lē\ *adverb*

gi·gan·tism \jī-ˈgan-ˌti-zəm, jə-; ˈjī-gən-\ *noun* (circa 1885)
1 : GIANTISM 1
2 : development to abnormally large size
3 : excessive vegetative growth often accompanied by the inhibiting of reproduction

gi·gas \ˈjī-gəs\ *adjective* [New Latin, from Latin, giant, from Greek] (1915)
of a polyploid plant **:** having thicker stem, taller growth, darker thicker leaves, and larger flowers and seeds than a corresponding diploid

giga·watt \ˈji-gə-ˌwät, ˈgi-\ *noun* (circa 1962)
: a unit of power equal to one billion watts

[1]**gig·gle** \ˈgi-gəl\ *verb* **gig·gled; gig·gling** \-g(ə-)liŋ\ [imitative] (1509)
intransitive verb
: to laugh with repeated short catches of the breath
transitive verb
: to utter with a giggle
— gig·gler \-g(ə-)lər\ *noun*
— gig·gling·ly \-g(ə-)liŋ-lē\ *adverb*
— gig·gly \-g(ə-)lē\ *adjective*

[2]**giggle** *noun* (circa 1677)
1 : the act of giggling
2 *chiefly British* **:** a source of amusement

gig·o·lo \ˈji-gə-ˌlō, ˈzhi-\ *noun, plural* **-los** [French] (1922)
1 : a man supported by a woman usually in return for his attentions
2 : a professional dancing partner or male escort

gi·got \ˈji-gət, zhē-ˈgō\ *noun, plural* **gi·gots** \-gəts, -ˈgō(z)\ [Middle French, diminutive of *gigue* fiddle; from its shape — more at JIG] (1526)
1 : a leg of meat (as lamb) especially when cooked
2 : a leg-of-mutton sleeve

gigue \ˈzhēg\ *noun* [French, from English *jig*] (1685)
: a lively dance movement (as of a suite) having compound triple rhythm and composed in fugal style

Gi·la monster \ˈhē-lə-\ *noun* [*Gila* River, Ariz.] (1877)
: a large orange and black venomous lizard (*Heloderma suspectum*) of the southwestern U.S.; *also* **:** a related lizard (*H. horridum*) of Mexico

Gila monster

gil·bert \ˈgil-bərt\ *noun* [William *Gilbert*] (1893)
: the centimeter-gram-second unit of magnetomotive force equivalent to $10\div4\pi$ ampere-turn

[1]**gild** \ˈgild\ *transitive verb* **gild·ed** \ˈgil-dəd\ *or* **gilt** \ˈgilt\; **gild·ing** [Middle English, from Old English *gyldan;* akin to Old English *gold* gold] (14th century)
1 : to overlay with or as if with a thin covering of gold
2 a : to give money to **b :** to give an attractive but often deceptive appearance to **c** *archaic* **:** to make bloody
— gild·er *noun*
— gild the lily : to add unnecessary ornamentation to something beautiful in its own right

[2]**gild** *variant of* GUILD

Gil·ga·mesh \ˈgil-gə-ˌmesh\
: a legendary Sumerian king and hero of the *Gilgamesh Epic*

[1]**gill** \ˈjil\ *noun* [Middle English *gille,* from Medieval Latin *gillus,* from Late Latin *gillo, gello* water pot] (14th century)
— see WEIGHT table

[2]**gill** \ˈgil\ *noun* [Middle English *gile,* of Scandinavian origin; akin to Swedish *gäl* gill, Old Norse *gjǫlnar* lips; akin to Greek *chelynē* lip, jawbone] (14th century)
1 : an organ (as of a fish) for obtaining oxygen from water
2 a : WATTLE 2 **b :** the flesh under or about the chin or jaws — usually used in plural **c :** one of the radiating plates forming the undersurface of the cap of a mushroom fungus
— gilled \ˈgild\ *adjective*
— to the gills : as full as possible

[3]**gill** \ˈgil\ (1884)
transitive verb
: GILLNET
intransitive verb
of fish **:** to become entangled in a gill net
— gill·er *noun*

[4]**gill** \ˈgil\ *noun* [Middle English *gille,* from Old Norse *gil;* akin to Old High German *gil* hernia] (14th century)
1 *British* **:** RAVINE
2 *British* **:** a narrow stream or rivulet

[5]**gill** \ˈjil\ *noun, often capitalized* [Middle English, from *Gill,* nickname for *Gillian*] (15th century)
: GIRL, SWEETHEART

gill arch *noun* (1879)
1 : any of the bony or cartilaginous arches or curved bars extending dorsoventrally and placed one behind the other on each side of the pharynx and supporting the gills of fishes and amphibians
2 : any of the rudimentary ridges in the embryos of all higher vertebrates that correspond to the gill arches

gill cleft *noun* (circa 1889)
: GILL SLIT 1, 2

gill cover *noun* (1776)
: OPERCULUM 1b

[1]**gil·lie** \ˈgi-lē\ *noun* [Scottish Gaelic *gille* & Irish *giolla* youth, gillie] (1705)
1 : a male attendant on a Scottish Highland chief; *broadly* **:** ATTENDANT
2 *chiefly Scottish & Irish* **:** a fishing and hunting guide
3 : a shoe with a low top and decorative lacing

[2]**gillie** *intransitive verb* **gil·lied; gil·ly·ing** (1900)

☆ SYNONYMS
Gift, faculty, aptitude, bent, talent, genius, knack mean a special ability for doing something. GIFT often implies special favor by God or nature ⟨the *gift* of singing beautifully⟩. FACULTY applies to an innate or less often acquired ability for a particular accomplishment or function ⟨a *faculty* for remembering names⟩. APTITUDE implies a natural liking for some activity and the likelihood of success in it ⟨a mechanical *aptitude*⟩. BENT is nearly equal to APTITUDE but it stresses inclination perhaps more than specific ability ⟨a family with an artistic *bent*⟩. TALENT suggests a marked natural ability that needs to be developed ⟨has enough *talent* to succeed⟩. GENIUS suggests impressive inborn creative ability ⟨has no great *genius* for poetry⟩. KNACK implies a comparatively minor but special ability making for ease and dexterity in performance ⟨the *knack* of getting along⟩.

: to serve as a gillie
gill·net \'gil-ˌnet\ *transitive verb* (1949)
: to catch (fish) with a gill net
gill net *noun* (1796)
: a flat net suspended vertically in the water with meshes that allow the head of a fish to pass but entangle it as it seeks to withdraw
gill·net·ter \-ˌne-tər\ *noun* (circa 1889)
: a boat equipped for or engaged in fishing with gill nets; *also* **:** a person who fishes with a gill net
gill raker *noun* (1880)
: any of the bony processes on a gill arch that divert solid substances away from the gills
gill slit *noun* (1854)
1 : any of the openings or clefts between the gill arches in vertebrates that breathe by gills through which water taken in at the mouth passes to the exterior and so bathes the gills
2 : any of the rudimentary grooves in the neck region of the embryos of air-breathing vertebrates that correspond to the gill slits
3 : the external opening to the cavity containing the gills when a protective covering of the gills is present
gil·ly·flow·er \'ji-lē-ˌflau̇(-ə)r\ *noun* [by folk etymology from Middle English *gilofre* clove, from Middle French *girofle, gilofre,* from Latin *caryophyllum,* from Greek *karyophyllon,* from *karyon* nut + *phyllon* leaf — more at CAREEN, BLADE] (1551)
: CARNATION 2
Gil·son·ite \'gil-sə-ˌnīt\ *trademark*
— used for uintaite
¹gilt \'gilt\ *adjective* [Middle English, from past participle of *gilden* to gild] (14th century)
: covered with gold or gilt **:** of the color of gold
²gilt *noun* (15th century)
1 : gold or something that resembles gold laid on a surface
2 *slang* **:** MONEY
3 : superficial brilliance
4 : a bond issued by the government of the United Kingdom
³gilt *noun* [Middle English *gylte,* from Old Norse *gyltr;* akin to Old English *gelde* sterile — more at GELD] (14th century)
: a young female swine
gilt–edged \'gilt-ˌejd\ *or* **gilt–edge** \-'ej\ *adjective* (1818)
1 : having a gilt edge
2 : of the best quality or rating ⟨*gilt-edged* securities⟩
¹gim·bal \'gim-bəl, 'jim-\ *noun* [alteration of obsolete *gemel* (double ring)] (1780)
: a device that permits a body to incline freely in any direction or suspends it so that it will remain level when its support is tipped — usually used in plural; called also *gimbal ring*
²gimbal *transitive verb* **-balled** *or* **-baled; -bal·ling** *or* **-bal·ing** (1875)
: to provide with or support on gimbals
gim·crack \'jim-ˌkrak\ *noun* [origin unknown] (1676)
: a showy object of little use or value **:** GEWGAW
— **gimcrack** *adjective*
— **gim·crack·ery** \-ˌkra-k(ə-)rē\ *noun*
gim·el \'gi-məl\ *noun* [Hebrew *gīmel*] (1828)
: the 3d letter of the Hebrew alphabet — see ALPHABET table
¹gim·let \'gim-lət\ *noun* [Middle English, from Middle French *guimbelet,* probably modification of Middle Dutch *wimmelkijn,* from *wimmel* wimble] (14th century)
: a small tool with a screw point, grooved shank, and cross handle for boring holes

gimlet

²gimlet *adjective* (1752)
: having a piercing or penetrating quality
³gimlet *transitive verb* (1840)
: to pierce as if with a gimlet
⁴gimlet *noun* [probably from ¹*gimlet*] (1928)
: a drink consisting of sweetened lime juice and gin or vodka and sometimes carbonated or plain water
gim·let–eyed \-'īd\ *adjective* (1752)
: SHARP-SIGHTED
gim·mal \'gi-məl, 'ji-\ *noun* [alteration of obsolete *gemel* (double ring)] (1598)
1 *plural* **:** joined work (as in a clock) whose parts move within each other
2 : a pair or series of interlocked rings
gim·me \'gi-mē\ *noun, plural* **gimmes** [from *gimme,* contraction of *give me*] (1982)
: something easily achieved or won especially in a contest
¹gim·mick \'gi-mik\ *noun* [origin unknown] (circa 1926)
1 a : a mechanical device for secretly and dishonestly controlling gambling apparatus **b :** an ingenious or novel mechanical device **:** GADGET
2 a : an important feature that is not immediately apparent **:** CATCH ⟨what's the *gimmick* . . . what's in it for you —Maxwell Griffith⟩ **b :** an ingenious and usually new scheme or angle
— **gim·micky** \-mi-kē\ *adjective*
²gimmick *transitive verb* (1928)
1 : to alter or influence by means of a gimmick
2 : to provide with a gimmick — often used with *up*
gim·mick·ry \'gi-mi-krē\ *noun, plural* **-ries** (1948)
: an array or profusion of gimmicks; *also* **:** use of gimmicks
¹gimp \'gimp\ *noun* [perhaps from Dutch] (1664)
: an ornamental flat braid or round cord used as a trimming
²gimp *noun* [perhaps from *gimp* (fish line strengthened with wire)] (1901)
: SPIRIT, VIM
³gimp *noun* [origin unknown] (1929)
1 : CRIPPLE
2 : LIMP ⟨walks with a *gimp* —Damon Runyon⟩
— **gimpy** \'gim-pē\ *adjective*
⁴gimp *intransitive verb* (1948)
: LIMP, HOBBLE ⟨*gimping* up the stairs⟩
¹gin \'gin\ *verb* **gan** \'gan\; **gin·ning** [Middle English *ginnen,* short for *beginnen*] (13th century)
archaic **:** BEGIN
²gin \'jin\ *noun* [Middle English *gin,* modification of Old French *engin* — more at ENGINE] (13th century)
: any of various tools or mechanical devices: as **a :** a snare or trap for game **b :** COTTON GIN
³gin \'jin\ *transitive verb* **ginned; gin·ning** (circa 1625)
1 : SNARE
2 : to separate (cotton fiber) from seeds and waste material
— **gin·ner** *noun*
⁴gin \'gin\ *conjunction* [perhaps by contraction from dialect *gif* if + *an* if] (1580)
dialect **:** IF
⁵gin \'jin\ *noun* [by shortening & alteration from *geneva*] (1714)
1 : a colorless alcoholic beverage made from distilled or redistilled neutral grain spirits flavored with juniper berries and aromatics (as anise and caraway seeds)
2 a : GIN RUMMY **b :** the act of laying down a full hand of matched cards in gin rummy ◆
— **gin·ny** \'jin-ē\ *adjective*
¹gin·ger \'jin-jər\ *noun* [Middle English, from Old English *gingifer,* from Medieval Latin *gingiber,* alteration of Latin *zingiber,* from Greek *zingiberi,* ultimately from Pali *siṅgivera*] (before 12th century)
1 a (1) **:** a thickened pungent aromatic rhizome that is used as a spice and sometimes medicinally (2) **:** the spice usually prepared by drying and grinding ginger **b :** any of a genus (*Zingiber* of the family Zingiberaceae, the ginger family) of herbs with pungent aromatic rhizomes; *especially* **:** a widely cultivated tropical herb (*Z. officinale*) that supplies most of the ginger of commerce
2 : high spirit **:** PEP ⟨the *ginger* to care hard and work hard —Willa Cather⟩
3 : a strong brown
— **gin·gery** \'jinj-rē, 'jin-jə-\ *adjective*
²ginger *transitive verb* **gin·gered; gin·ger·ing** \'jinj-riŋ, 'jin-jə-\ (1849)
: to make lively **:** pep up ⟨*ginger* up the tourist trade —*N.Y. Times*⟩
ginger ale *noun* (1886)
: a sweetened carbonated nonalcoholic beverage flavored mainly with ginger extract
ginger beer *noun* (1809)
: a sweetened carbonated nonalcoholic beverage heavily flavored with ginger or capsicum or both
gin·ger·bread \'jin-jər-ˌbred\ *noun* (15th century)
1 : a cake whose ingredients include molasses and ginger
2 [from the fancy shapes and gilding formerly often applied to gingerbread] **:** lavish or superfluous ornament especially in architecture
— **gingerbread** *adjective*
— **gin·ger·bread·ed** \-ˌbre-dəd\ *adjective*
— **gin·ger·bready** \-dē\ *adjective*
ginger group *noun* (1925)
chiefly British **:** a group that serves as an energizing force within a larger body (as a political party)
gin·ger·ly \'jin-jər-lē\ *adjective* [perhaps from ¹*ginger*] (1594)
: very cautious or careful
— **gin·ger·li·ness** *noun*
— **gingerly** *adverb*
ginger nut *noun* (1842)
British **:** GINGERSNAP
gin·ger·root \'jin-jər-ˌ(r)üt, -ˌ(r)u̇t\ *noun* (1831)
: GINGER 1a(1)
gin·ger·snap \-ˌsnap\ *noun* (1805)
: a thin brittle cookie sweetened with molasses and flavored with ginger
ging·ham \'giŋ-əm\ *noun* [modification of Malay *genggang* striped cloth] (1615)
: a clothing fabric usually of yarn-dyed cotton in plain weave
gin·gi·va \'jin-jə-və, jin-'jī-\ *noun, plural* **-vae** \-ˌvē\ [Latin] (circa 1889)
: ¹GUM
— **gin·gi·val** \'jin-jə-vəl\ *adjective*

◇ WORD HISTORY

gin *Genever* (now spelled with an initial *j*) was the Dutch word for a drink made of distilled spirits and flavored with juniper berries. Ultimately it came from Latin *juniperus* by way of Middle Dutch and Old French. Both the word and the beverage were introduced into Britain around the end of the 17th century, probably by soldiers returning from wars in the Low Countries. In its new land, however, the word was influenced by the similar-sounding name of a city in Switzerland, and thus the earliest name for the drink is *geneva,* which appeared in print in 1706. The creation of new words by clipping longer words to a monosyllable was as popular in 18th century English as it is today, and so a shortened form *gin* arose, to designate not only the Dutch drink but also a similar liquor made in Britain, not necessarily flavored with juniper but remarkably popular. So popular did it become, in fact, that gin drinking among the poorer classes became a major social problem. The epidemic proportions of this menace are long past, but the liquor and its clipped name are still with us.

gin·gi·vec·to·my \ˌjin-jə-ˈvek-tə-mē\ *noun, plural* **-mies** (circa 1927)
: the excision of a portion of the gingiva

gin·gi·vi·tis \ˌjin-jə-ˈvī-təs\ *noun* (1874)
: inflammation of the gums

gink \ˈgiŋk\ *noun* [origin unknown] (1908)
slang : PERSON, GUY

gink·go *also* **ging·ko** \ˈgiŋ-(ˌ)kō *also* ˈgiŋk-(ˌ)gō\ *noun, plural* **ginkgoes** *or* **ginkgos** [New Latin *Ginkgo,* from Japanese *ginkyō*] (1773)
: a gymnospermous tree (*Ginkgo biloba*) of eastern China that is widely grown as an ornamental or shade tree and has fan-shaped leaves and yellow fruit

gin mill *noun* (1865)
: BAR, SALOON

gin rummy *noun* [[5]*gin*] (1941)
: a rummy game for two players in which each player is dealt 10 cards and in which a player may win a hand by matching all the cards in it or may end play when the unmatched cards count up to 10 points or less

gin·seng \ˈjin-ˌseŋ, -ˌsiŋ\ *noun* [Chinese (Beijing) *rénshēn*] (1654)
1 a : a Chinese perennial herb (*Panax schinseng* of the family Araliaceae, the ginseng family) having 5-foliolate leaves, scarlet berries, and an aromatic root valued especially locally as a medicine **b :** any of several plants related to ginseng; *especially* **:** a North American herb (*P. quinquefolius*)
2 : the root of a ginseng

Gip·sy *chiefly British variant of* GYPSY

gi·raffe \jə-ˈraf\ *noun, plural* **giraffes** [Italian *giraffa,* from Arabic *zirāfah*] (circa 1600)
1 *or plural* **giraffe :** a large fleet African ruminant mammal (*Giraffa camelopardalis*) that is the tallest of living quadrupeds and has a very long neck and a short coat with dark blotches separated by pale lines
2 *capitalized* **:** CAMELOPARDALIS
— gi·raff·ish \-ˈra-fish\ *adjective*

giraffe 1

gir·an·dole \ˈjir-ən-ˌdōl\ *noun* [French & Italian; French, from Italian *girandola,* from *girare* to turn, from Late Latin *gyrare,* from Latin *gyrus* gyre] (1749)
1 : a radiating and showy composition (as a cluster of skyrockets fired together)
2 : an ornamental branched candlestick
3 : a pendant earring usually with three ornaments hanging from a central piece

girandole 2

gir·a·sole *also* **gir·a·sol** \ˈjir-ə-ˌsōl, -ˌsōl, -ˌsäl\ *noun* [Italian *girasole* sun flower, from *girare* + *sole* sun, from Latin *sol* — more at SOLAR] (circa 1586)
1 : JERUSALEM ARTICHOKE
2 *usually* **girasol :** an opal of varying color that gives out fiery reflections in bright light

[1]gird \ˈgərd\ *verb* **gird·ed** \ˈgər-dəd\ *or* **girt** \ˈgərt\; **gird·ing** [Middle English, from Old English *gyrdan;* akin to Old English *geard* yard — more at YARD] (before 12th century)
transitive verb
1 a : to encircle or bind with a flexible band (as a belt) **b :** to make fast (as a sword by a belt or clothing with a cord) **c :** SURROUND
2 : PROVIDE, EQUIP; *especially* **:** to invest with the sword of knighthood
3 : to prepare (oneself) for action
intransitive verb
: to prepare for action
— gird one's loins : to prepare for action **:** muster up one's resources

[2]gird *verb* [Middle English, to strike, thrust] (1546)
intransitive verb
: GIBE, RAIL
transitive verb
: to sneer at **:** MOCK

[3]gird *noun* (1566)
: a sarcastic remark

gird·er \ˈgər-dər\ *noun* [[1]*gird*] (1611)
: a horizontal main structural member (as in a building or bridge) that supports vertical loads and that consists of a single piece or of more than one piece bound together

[1]gir·dle \ˈgər-dᵊl\ *noun* [Middle English *girdel,* from Old English *gyrdel;* akin to Old High German *gurtil* girdle, Old English *gyrdan* to gird] (before 12th century)
1 : something that encircles or confines: as **a :** an article of dress encircling the body usually at the waist **b :** a woman's close-fitting undergarment often boned and usually elasticized that extends from the waist to below the hips **c** (1) **:** PECTORAL GIRDLE (2) **:** PELVIC GIRDLE
2 : the edge of a brilliant that is grasped by the setting

[2]girdle *transitive verb* **gir·dled; gir·dling** \ˈgər-dᵊl-iŋ\ (1582)
1 : to encircle with or as if with a girdle
2 : to cut away the bark and cambium in a ring around (a plant) usually to kill by interrupting the circulation of water and nutrients
3 : to move around **:** CIRCLE ⟨*girdled* the world⟩

girl \ˈgər(-ə)l\ *noun* [Middle English *gurle, girle* young person of either sex] (14th century)
1 a : a female child **b :** a young unmarried woman **c :** a single or married woman of any age — sometimes taken to be offensive
2 a : SWEETHEART **b :** a female servant or employee — sometimes taken to be offensive **c :** DAUGHTER ◇
— girl·hood \-ˌhu̇d\ *noun*

girl Friday *noun* [*girl* + *Friday* (as in *man Friday*)] (1940)
: a female assistant (as in an office) entrusted with a wide variety of tasks

girl·friend \ˈgər(-ə)l-ˌfrend\ *noun* (1859)
1 : a female friend
2 : a frequent or regular female companion of a boy or man
3 : MISTRESS 4a

Girl Guide *noun* (1909)
: a member of a worldwide scouting movement for girls 7 to 18 years of age that is equivalent to the Girl Scouts in the U.S.

girl·ie *also* **girly** \ˈgər-lē\ *adjective* (1942)
: featuring scantily clothed women ⟨*girlie* magazines⟩ ⟨*girlie* show⟩

girl·ish \ˈgər-lish\ *adjective* (1565)
: of, relating to, or having the characteristics of a girl or girlhood
— girl·ish·ly *adverb*
— girl·ish·ness *noun*

Girl Scout *noun* (1909)
: a member of any of the scouting programs of the Girl Scouts of the United States of America for girls ages 5 through 17

girn \ˈgirn\ *intransitive verb* [Middle English, alteration of *grinnen* to grin, snarl] (12th century)
chiefly Scottish **:** SNARL
— girn *noun, chiefly Scottish*

gi·ro \ˈjir-(ˌ)ō, ˈzhir-; ˈjē-(ˌ)rō, ˈzhē-, *especially British* ˈjī-\ *noun* [German, circulation (of currency), from Italian, from Latin *gyrus* gyre] (1890)
: a service of many European banks that permits authorized direct transfer of funds among account holders as well as conventional transfers by check

gi·ron *variant of* GYRON

Gi·ron·din \jə-ˈrän-din, zhə-\ *noun* [French, from *girondin* of Gironde] (1837)
: GIRONDIST

Gi·rond·ist \-dist\ *noun* [French *girondiste,* from *Gironde,* a political party, from *Gironde,* department of France represented by its leaders] (1795)
: a member of the moderate republican party in the French legislative assembly in 1791

girt \ˈgərt\ *verb* [Middle English *girten,* alteration of *girden*] (15th century)
transitive verb
1 : GIRD
2 : to fasten by means of a girth
intransitive verb
: to measure in girth

[1]girth \ˈgərth\ *noun* [Middle English, from Old Norse *gjǫrth;* akin to Old English *gyrdan* to gird] (13th century)
1 : a band or strap that encircles the body of an animal to fasten something (as a saddle) on its back
2 a : a measure around a body ⟨a man of more than average *girth*⟩ **b :** SIZE, DIMENSIONS

[2]girth *transitive verb* (15th century)
1 : ENCIRCLE
2 : to bind or fasten with a girth
3 : to measure the girth of

gi·sarme \gi-ˈzärm\ *noun* [Middle English, from Old French] (13th century)
: a medieval weapon consisting of a blade mounted on a long staff and carried by foot soldiers

gist \ˈjist\ *noun* [Anglo-French, it lies, from Middle French, from *gesir* to lie, ultimately from Latin *jacēre* — more at ADJACENT] (circa 1711)
1 : the ground of a legal action
2 : the main point or part **:** ESSENCE ⟨the *gist* of an argument⟩

git \ˈgit\ *noun* [variant of *get,* term of abuse, from [2]*get*] (1943)
British **:** a foolish or worthless person

◇ WORD HISTORY

girl Words in European languages that correspond to "boy" or "girl" are often of relatively recent origin and of obscure etymology, having floated upwards into standard use from the most colloquial and innovative varieties of speech. The English word *girl* is no exception. It first appeared in Middle English, about the end of the 13th century, in the forms *gurle, gerle,* and *girle,* denoting a child of either sex. Toward the end of the 14th century it was applied specifically to a female, a sense that became dominant by Shakespeare's time. Attempts have been made to relate the Middle English word to Old English *gierela, gerela* "apparel, banner." The connection may be possible on semantic grounds—witness the now fading English use of *flapper* to mean "young woman"—but it offers phonetic difficulties, as the expected Old English ancestor of *girl* would more plausibly be an unattested *gyrela.* The word *girl* shows a variety of dialect pronunciations in Modern English, often with a dropped \r\ even in accents that do not normally lose \r\ before a consonant. One such form, *gal,* is well-established in American English, though now somewhat old-fashioned and restricted in context.

git–go \ˈgit-ˌgō\ *variant of* GET-GO

git·tern \ˈgi-tərn\ *noun* [Middle English *giterne,* from Middle French *guiterne,* modification of Old Spanish *guitarra* guitar] (14th century)
: a medieval guitar

¹give \ˈgiv\ *verb* **gave** \ˈgāv\; **giv·en** \ˈgi-vən\; **giv·ing** [Middle English, of Scandinavian origin; akin to Old Swedish *giva* to give; akin to Old English *giefan, gifan* to give, and perhaps to Latin *habēre* to have, hold] (13th century)
transitive verb
1 : to make a present of ⟨*give* a doll to a child⟩
2 a : to grant or bestow by formal action ⟨the law *gives* citizens the right to vote⟩ **b :** to accord or yield to another ⟨*gave* him her confidence⟩
3 a : to put into the possession of another for his use **b** (1) **:** to administer as a sacrament (2) **:** to administer as a medicine **c :** to commit to another as a trust or responsibility and usually for an expressed reason **d :** to transfer from one's authority or custody ⟨the sheriff *gave* the prisoner to the warden⟩ **e :** to execute and deliver ⟨all employees must *give* bond⟩ **f :** to convey to another ⟨*give* them my regards⟩
4 a : to offer to the action of another **:** PROFFER ⟨*gave* her his hand⟩ **b :** to yield (oneself) to a man in sexual intercourse
5 a : to present in public performance ⟨*give* a concert⟩ **b :** to present to view or observation ⟨*gave* the signal to start⟩
6 : to provide by way of entertainment ⟨*give* a party⟩
7 : to propose as a toast
8 a : to designate as a share or portion **:** ALLOT ⟨all the earth to thee and to thy race I *give* —John Milton⟩ **b :** to make assignment of (a name) **c :** to set forth as an actual or hypothetical datum ⟨*give* the dimensions of the room⟩ **d :** to attribute in thought or utterance **:** ASCRIBE ⟨*gave* the credit to you⟩
9 a : to yield as a product, consequence, or effect **:** PRODUCE ⟨cows *give* milk⟩ ⟨84 divided by 12 *gives* 7⟩ **b :** to bring forth **:** BEAR
10 a : to yield possession of by way of exchange **:** PAY **b :** to dispose of for a price **:** SELL
11 a : to deliver by some bodily action ⟨*gave* him a push⟩ **b :** to carry out (as a bodily movement) ⟨*gave* a cynical smile⟩ **c :** to inflict as punishment **d :** to award by formal verdict ⟨judgment was *given* against the plaintiff⟩
12 : to offer for consideration, acceptance, or use ⟨*gives* no reason for his absence⟩
13 a : to suffer the loss of **:** SACRIFICE **b :** to offer as appropriate or due especially to something higher or more worthy ⟨*gave* his spirit to God⟩ **c :** to apply freely or fully **:** DEVOTE ⟨*gave* themselves to their work⟩ **d :** to offer as a pledge ⟨I *give* you my word⟩
14 a : to cause one to have or receive ⟨mountains always *gave* him pleasure⟩ **b :** to cause a person to catch by contagion, infection, or exposure
15 a : to allow one to have or take ⟨*give* me time⟩ **b :** to lead or attempt to lead — used with an infinitive ⟨you *gave* me to understand you'd be late⟩
16 : to care to the extent of ⟨didn't *give* a hang⟩
intransitive verb
1 : to make gifts or presents
2 a : to yield to physical force or strain **b :** to collapse from the application of force or pressure **c :** to undergo or submit to change ⟨for the strike to be settled, something has to *give*⟩
3 : to afford a view or passage **:** OPEN
4 : to enter wholeheartedly into an activity
5 *slang* **:** GO ON, HAPPEN ⟨wants to know what *gives*⟩ ☆
— give a good account of : to acquit (oneself) well
— give birth : to have a baby ⟨*gave birth* last Thursday⟩
— give birth to 1 : to produce as offspring ⟨*gave birth to* a son⟩ **2 :** to be the source of
— give chase : to set off in pursuit
— give ground : to withdraw before superior force **:** RETREAT
— give of : to make available **:** provide generously ⟨freely *gave of* their time⟩
— give or take : as an estimate accurate within (an amount to be added or subtracted)
— give place to : to be replaced or succeeded by ⟨optimism *gave place to* worry⟩
— give rise to : to be the cause or source of **:** PRODUCE
— give the gun : to open the throttle of **:** speed up
— give the lie to 1 : to accuse of falsehood **2 :** to show to be false, inaccurate, or invalid
— give tongue *of hounds* **:** to begin barking on the scent
— give way 1 a : RETREAT **b :** to yield the right of way **2 :** to yield oneself without restraint or control **3 a :** to yield to or as if to physical stress **b :** to yield to entreaty or insistence **4 :** to yield place **5 :** to begin to row

²give *noun* (1868)
1 : capacity or tendency to yield to force or strain **:** FLEXIBILITY
2 : the quality or state of being springy

give–and–go \ˌgiv-ən-ˈgō\ *noun* (1965)
: a play (as in basketball or hockey) in which a player passes to a teammate and immediately cuts toward the net or goal to receive a return pass

give–and–take \ˌgiv-ən-ˈtāk\ *noun* (1778)
1 : the practice of making mutual concessions **:** COMPROMISE
2 : a usually good-natured exchange (as of ideas or comments)

give·away \ˈgi-və-ˌwā\ *noun* (1882)
1 : an unintentional revelation or betrayal
2 a : something given away free; *specifically* **:** PREMIUM 1d **b :** the act of giving something away free ⟨staging a promotional *giveaway*⟩
3 : a radio or television program on which prizes are given away

give away *transitive verb* (14th century)
1 : to make a present of
2 : to deliver (a bride) ceremonially to the bridegroom at a wedding
3 a : BETRAY **b :** DISCLOSE, REVEAL
4 : to give (as weight) by way of a handicap

give·back \ˈgiv-ˌbak\ *noun* (1978)
: a previous gain (as an increase in wages or benefits) given back to management by workers (as in a labor contract)

give back (1548)
intransitive verb
: RETIRE, RETREAT
transitive verb
: to send in return or reply **:** RESTORE, RETURN

give in (1602)
transitive verb
: DELIVER, SUBMIT ⟨*gave in* his resignation⟩
intransitive verb
: to yield under insistence or entreaty **:** SURRENDER

¹giv·en \ˈgi-vən\ *adjective* (14th century)
1 : PRONE, DISPOSED ⟨*given* to swearing⟩
2 : presented as a gift **:** bestowed without compensation
3 : PARTICULAR, SPECIFIED ⟨at a *given* time⟩
4 *of an official document* **:** having been executed **:** DATED
5 : immediately present in experience

²given *noun* (1879)
: something given; *especially* **:** something taken for granted **:** a basic condition or assumption

³given *preposition* (1904)
: in view of **:** CONSIDERING ⟨*given* what she knew about others' lives, how could she complain about her own? —Marilyn French⟩

given name *noun* (1827)
: a name that precedes one's surname; *especially* **:** FIRST NAME

give off (1831)
transitive verb
1 : to send out as a branch
2 : EMIT ⟨*gave off* an unpleasant smell⟩
intransitive verb
: to branch off

give out (14th century)
transitive verb
1 a : DECLARE, PUBLISH ⟨*giving out* that the doctor . . . required a few days of complete rest —Charles Dickens⟩ **b :** to read aloud the words of (a hymn or psalm) for congregational singing
2 : EMIT ⟨*gave out* a constant hum⟩
3 : ISSUE ⟨*gave out* new uniforms⟩
intransitive verb
1 : BREAK DOWN, FAIL
2 : to become exhausted **:** COLLAPSE

give over (14th century)
transitive verb
1 : CEASE
2 : ENTRUST
3 a : to yield without restraint or control **:** ABANDON ⟨*gave* themselves *over* to laughter⟩ **b :** to set apart for a particular purpose or use
4 *archaic* **:** to pronounce incurable
intransitive verb
British **:** to cease an activity **:** STOP ⟨told him to *give over* and let me alone —Brendan Behan⟩

giv·er \ˈgi-vər\ *noun* (14th century)
: one that gives **:** DONOR

give up (13th century)
transitive verb
1 : to yield control or possession of **:** SURRENDER ⟨forced to *give up* his job⟩
2 : to desist from **:** ABANDON ⟨refused to *give up* her efforts⟩
3 : to declare incurable or insoluble
4 a : to abandon (oneself) to a particular feeling, influence, or activity ⟨*gave* himself *up* to despair⟩ **b :** to devote to a particular purpose or use
5 : to despair of seeing ⟨we'd *given* you *up*⟩
6 : to allow (a hit or run in baseball) while pitching
intransitive verb
: to cease doing or attempting something especially as an admission of defeat **:** QUIT — often used with *on* ⟨don't *give up* on the project⟩
— give up the ghost : to cease to live or function **:** DIE

giz·mo *also* **gis·mo** \ˈgiz-(ˌ)mō\ *noun, plural* **gizmos** *also* **gismos** [origin unknown] (1943)
: GADGET

giz·zard \ˈgi-zərd\ *noun* [alteration of Middle English *giser,* from Old French *guisier, giser,* from Latin *gigeria* (plural) giblets] (1565)
1 a : the muscular enlargement of the alimentary canal of birds that has usually thick muscular walls and a tough horny lining for grinding the food and when the crop is present

☆ SYNONYMS
Give, present, donate, bestow, confer, afford mean to convey to another as a possession. GIVE, the general term, is applicable to any passing over of anything by any means ⟨*give* alms⟩ ⟨*gave* her a ride on a pony⟩ ⟨*give* my love to your mother⟩. PRESENT carries a note of formality and ceremony ⟨*present* an award⟩. DONATE is likely to imply a publicized giving (as to charity) ⟨*donate* a piano to the orphanage⟩. BESTOW implies the conveying of something as a gift and may suggest condescension on the part of the giver ⟨*bestow* unwanted advice⟩. CONFER implies a gracious giving (as of a favor or honor) ⟨*confer* an honorary degree⟩. AFFORD implies a giving or bestowing usually as a natural or legitimate consequence of the character of the giver ⟨the trees *afford* shade⟩ ⟨a development that *affords* us some hope⟩.

follows it and the proventriculus **b :** a thickened part of the alimentary canal in some animals (as an insect or an earthworm) that is similar in function to the crop of a bird
2 : INNARDS

gla·bel·la \glə-ˈbe-lə\ *noun, plural* **-bel·lae** \-ˈbe-(ˌ)lē, -ˌlī\ [New Latin, from Latin, feminine of *glabellus* hairless, diminutive of *glaber*] (circa 1823)
: the smooth prominence between the eyebrows
— gla·bel·lar \-ˈbe-lər\ *adjective*

gla·bres·cent \glā-ˈbre-sᵊnt\ *adjective* (1857)
1 : somewhat glabrous
2 : tending to become glabrous

gla·brous \ˈglā-brəs\ *adjective* [Latin *glabr-, glaber* smooth, bald — more at GLAD] (1640)
: SMOOTH; *especially* **:** having a surface without hairs or projections ⟨*glabrous* skin⟩ ⟨*glabrous* leaves⟩

gla·cé \gla-ˈsā\ *adjective* [French, from past participle of *glacer* to freeze, ice, glaze, from Latin *glaciare,* from *glacies*] (1847)
1 : made or finished so as to have a smooth glossy surface ⟨*glacé* silk⟩
2 *also* **gla·céed** \-ˈsād\ **:** coated with a glaze **:** CANDIED ⟨*glacé* cherries⟩

gla·cial \ˈglā-shəl\ *adjective* [Latin *glacialis,* from *glacies*] (1656)
1 a : extremely cold **:** FRIGID ⟨a *glacial* wind⟩ **b :** devoid of warmth and cordiality ⟨a *glacial* handshake⟩ **c :** coldly imperturbable ⟨maintained a *glacial* calm⟩
2 : resembling ice in appearance ⟨*glacial* acetic acid⟩
3 a (1) **:** of, relating to, or being any of those parts of geologic time from Precambrian onward when a much larger portion of the earth was covered by glaciers than at present (2) *capitalized* **:** PLEISTOCENE **b :** of, relating to, or produced by glaciers **c :** suggestive of the very slow movement of glaciers ⟨progress on the bill has been *glacial*⟩
— gla·cial·ly \-shə-lē\ *adverb*

gla·ci·ate \ˈglā-shē-ˌāt, -sē-\ *transitive verb* **-at·ed; -at·ing** (circa 1623)
1 : FREEZE
2 a : to subject to glacial action; *also* **:** to produce glacial effects in or on **b :** to cover with a glacier
— gla·ci·a·tion \ˌglā-shē-ˈā-shən, -sē-\ *noun*

gla·cier \ˈglā-shər *also* -zhər, *especially British* ˈgla-sē-ər *or* ˈglā-sē-\ *noun* [French dialect, from Middle French dialect, from Middle French *glace* ice, from Latin *glacies;* akin to Latin *gelu* frost — more at COLD] (1744)
: a large body of ice moving slowly down a slope or valley or spreading outward on a land surface

gla·ci·ol·o·gy \ˌglā-shē-ˈä-lə-jē, -sē-\ *noun* [Latin *glacies* + International Scientific Vocabulary *-logy*] (1889)
: any of the branches of science dealing with snow or ice accumulation, glaciation, or glacial epochs
— gla·ci·o·log·i·cal \-ə-ˈlä-ji-kəl\ *adjective*
— gla·ci·ol·o·gist \-ˈä-lə-jist\ *noun*

gla·cis \gla-ˈsē, ˈgla-sē\ *noun, plural* **glacis** \-ˈsēz, -sēz\ [French, from *glacer* to freeze, slide] (1672)
1 a : a gentle slope **:** INCLINE **b :** a slope that runs downward from a fortification
2 : BUFFER STATE; *also* **:** BUFFER ZONE

[1]glad \ˈglad\ *adjective* **glad·der; glad·dest** [Middle English, shining, glad, from Old English *glæd;* akin to Old High German *glat* shining, smooth, Latin *glaber* smooth, bald] (before 12th century)
1 *archaic* **:** having a cheerful or happy disposition by nature
2 a : experiencing pleasure, joy, or delight **:** made happy **b :** made pleased, satisfied, or grateful — often used with *of* ⟨was *glad* of their help⟩ **c :** very willing ⟨*glad* to do it⟩
3 a : marked by, expressive of, or caused by happiness and joy ⟨a *glad* shout⟩ **b :** causing happiness and joy **:** PLEASANT ⟨*glad* tidings⟩
4 : full of brightness and cheerfulness ⟨a *glad* spring morning⟩
— glad·ly *adverb*
— glad·ness *noun*

[2]glad *verb* **glad·ded; glad·ding** (before 12th century)
archaic **:** GLADDEN

[3]glad *noun* (1923)
: GLADIOLUS 1

glad·den \ˈgla-dᵊn\ *verb* **glad·dened; glad·den·ing** \ˈgla-dᵊn-iŋ\ (13th century)
intransitive verb
archaic **:** to be glad
transitive verb
: to make glad

glade \ˈglād\ *noun* [perhaps from [1]*glad*] (1529)
: an open space surrounded by woods
— glady \ˈglā-dē\ *adjective*

glad–hand \ˈglad-ˌhand\ (1903)
transitive verb
: to extend a glad hand to ⟨candidates *glad-handing* everyone they meet⟩
intransitive verb
: to extend a glad hand ⟨*glad-handing* as if he were running for mayor⟩
— glad–hand·er \ˈglad-ˌhan-dər\ *noun*

glad hand *noun* (circa 1895)
: a warm welcome or greeting often prompted by ulterior reasons

glad·i·a·tor \ˈgla-dē-ˌā-tər\ *noun* [Latin, from *gladius* sword, of Celtic origin; akin to Welsh *cleddyf* sword] (15th century)
1 : a person engaged in a fight to the death as public entertainment for ancient Romans
2 : a person engaging in a public fight or controversy
3 : a trained fighter; *especially* **:** a professional boxer
— glad·i·a·to·ri·al \ˌgla-dē-ə-ˈtōr-ē-əl, -ˈtȯr-\ *adjective*

glad·i·o·la \ˌgla-dē-ˈō-lə\ *noun* [back-formation from *gladiolus,* taken as a plural] (1926)
: GLADIOLUS 1

glad·i·o·lus \ˌgla-dē ˈō-ləs\ *noun, plural* **-li** \-(ˌ)lē, -ˌlī\ [New Latin, from Latin, gladiolus, from diminutive of *gladius*] (15th century)
1 *or plural* **gladiolus** *also* **glad·i·o·lus·es :** any of a genus (*Gladiolus*) of chiefly African plants of the iris family with erect sword-shaped leaves and spikes of brilliantly colored irregular flowers arising from corms
2 : the large middle portion of the sternum

gladiolus 1

glad rags *noun plural* (1896)
: dressy clothes

glad·some \ˈglad-səm\ *adjective* (14th century)
: giving or showing joy **:** CHEERFUL
— glad·some·ly *adverb*
— glad·some·ness *noun*

glad·stone \ˈglad-ˌstōn, *chiefly British* -stən\ *noun often capitalized* [W. E. *Gladstone*] (1889)
: a suitcase with flexible sides on a rigid frame that opens flat into two equal compartments — called also *gladstone bag*

glai·kit *or* **glai·ket** \ˈglā-kət\ *adjective* [Middle English (Scots) *glaikit*] (15th century)
chiefly Scottish **:** FOOLISH, GIDDY

glair *or* **glaire** \ˈglar, ˈgler\ *noun* [Middle English *gleyre* egg white, from Middle French *glaire,* modification of (assumed) Vulgar Latin *claria,* from Latin *clarus* clear — more at CLEAR] (13th century)
1 : a sizing liquid made from egg white
2 : a viscid substance suggestive of an egg white

glairy \-ē\ *adjective* **glair·i·er; -est** (1662)
: having the characteristics of or overlaid with glair

glaive \ˈglāv\ *noun* [Middle English, from Middle French, javelin, sword, modification of Latin *gladius* sword] (15th century)
archaic **:** SWORD; *especially* **:** BROADSWORD

glam·or·ise *British variant of* GLAMORIZE

glam·or·ize *also* **glam·our·ize** \ˈgla-mə-ˌrīz\ *transitive verb* **-ized; -iz·ing** (1936)
1 : to look upon or depict as glamorous **:** ROMANTICIZE ⟨the novel *glamorizes* war⟩
2 : to make glamorous ⟨*glamorize* the living room⟩
— glam·or·iza·tion \ˌgla-mə-rə-ˈzā-shən\ *noun*
— glam·or·iz·er \ˈgla-mə-ˌrī-zər\ *noun*

glam·or·ous *also* **glam·our·ous** \ˈglam-rəs, ˈgla-mə-\ *adjective* (1882)
: full of glamour
— glam·or·ous·ly *adverb*
— glam·or·ous·ness *noun*

glam·our *or* **glam·or** \ˈgla-mər\ *noun* [Scots *glamour,* alteration of English *grammar;* from the popular association of erudition with occult practices] (1715)
1 : a magic spell ⟨the girls appeared to be under a *glamour* —Llewelyn Powys⟩
2 : an exciting and often illusory and romantic attractiveness; *especially* **:** alluring or fascinating personal attraction — often used attributively ⟨*glamour* stock⟩ ⟨*glamour* boy⟩ ⟨*glamour* girl⟩ ◆
— glamour *transitive verb*
— glam·our·less \-ləs\ *adjective*

glam·our–puss \-ˌpu̇s\ *noun* (1941)
: a glamorously attractive person

[1]glance \ˈglan(t)s\ *verb* **glanced; glanc·ing** [Middle English *glencen, glenchen*] (15th century)
intransitive verb
1 : to strike a surface obliquely so as to go off at an angle ⟨the bullet *glanced* off the wall⟩
2 a : to make sudden quick movements ⟨dragonflies *glancing* over the pond⟩ **b :** to flash or gleam with quick intermittent rays of light ⟨brooks *glancing* in the sun⟩

◇ WORD HISTORY

glamour In classical antiquity the Greek and Latin ancestors of the English word *grammar* were used in reference not only to the study of language but also to the study of literature. In the medieval period, Latin *grammatica* and its outcomes in other languages were extended to include learning in general. Since almost all learning was couched in a language not spoken or understood by the unschooled populace, it was commonly believed that subjects such as magic and astrology were included in *grammatica.* Scholars were sometimes regarded with awe and more than a little suspicion, and the Faust legend illustrates the popular intertwining of learning and alchemical studies with necromancy and the devil. This connection between grammar and the black arts was evident in a number of languages, and in Scotland by the 18th century a form of *grammar,* altered by dissimilation to *glamer* or *glamour,* meant "a magic spell." As *glamour* passed into more general English usage, it came to have its current meanings "an exciting and often deceptive attractiveness" and particularly "alluring personal attractiveness."

3 : to touch on a subject or refer to it briefly or indirectly ⟨the work *glances* at the customs of ancient cultures⟩
4 a *of the eyes* **:** to move swiftly from one thing to another **b :** to take a quick look at something ⟨*glanced* at his watch⟩
transitive verb
1 *archaic* **a :** to take a quick look at **b :** to catch a glimpse of
2 : to give an oblique path of direction to: **a :** to throw or shoot so that the object glances from a surface **b** *archaic* **:** to aim (as an innuendo) indirectly **:** INSINUATE
— glanc·er *noun*

[2]glance *noun* (1503)
1 a : a quick intermittent flash or gleam **b** *archaic* **:** a sudden quick movement
2 a *archaic* **:** a rapid oblique movement **b :** a deflected impact or blow
3 a : a swift movement of the eyes **b :** a quick or cursory look
4 *archaic* **a :** a brief satirical reference to something **:** GIBE **b :** ALLUSION
— at first glance : on first consideration ⟨*at first glance* the subject seems harmless enough⟩

glanc·ing \'glan(t)-siŋ\ *adjective* (circa 1541)
1 : hitting so as to glance off ⟨a *glancing* blow⟩
2 : INCIDENTAL, INDIRECT ⟨made *glancing* allusions to her past⟩
— glanc·ing·ly \-siŋ-lē\ *adverb*

[1]gland \'gland\ *noun* [French *glande,* from Old French, glandular swelling on the neck, gland, ultimately from Latin *gland-, glans* acorn; akin to Greek *balanos* acorn] (1692)
1 a : a cell, group of cells, or organ of endothelial origin that selectively removes materials from the blood, concentrates or alters them, and secretes them for further use in the body or for elimination from the body **b :** any of various animal structures suggestive of glands though not secretory in function
2 : any of various secreting organs (as a nectary) of plants
— gland·less \'gland-ləs\ *adjective*

[2]gland *noun* [origin unknown] (1839)
1 : a device for preventing leakage of fluid past a joint in machinery
2 : the movable part of a stuffing box by which the packing is compressed

glan·dered \'glan-dərd\ *adjective* (1667)
: affected with glanders

glan·ders \-dərz\ *noun plural but singular or plural in construction* [Middle French *glandre* glandular swelling on the neck, from Latin *glandulae,* from plural of *glandula,* diminutive of *gland-, glans*] (1523)
: a contagious and destructive disease especially of horses caused by a bacterium (*Pseudomonas mallei* synonym *Actinobacillus mallei*) and characterized by caseating nodular lesions especially on the respiratory mucosae and lungs that tend to break down and form ulcers

glan·du·lar \'glan-jə-lər\ *adjective* (circa 1740)
1 a : of, relating to, or involving glands, gland cells, or their products **b :** having the characteristics or function of a gland
2 a : INNATE, INHERENT ⟨the almost *glandular* . . . instinct for adventure and romance —*Newsweek*⟩ **b :** PHYSICAL, SEXUAL
— glan·du·lar·ly *adverb*

glandular fever *noun* (1902)
: INFECTIOUS MONONUCLEOSIS

glans \'glanz\ *noun, plural* **glan·des** \'glan-ˌdēz\ [Latin *gland-, glans,* literally, acorn] (1650)
1 : a conical vascular body forming the extremity of the penis — called also *glans penis*
2 : a structure of the clitoris similar to the glans penis — called also *glans cli·to·ri·dis* \-klə-'tȯr-ə-dəs\

[1]glare \'glar, 'gler\ *verb* **glared; glar·ing** [Middle English *glaren;* akin to Old English *glæs* glass] (13th century)
intransitive verb
1 a : to shine with a harsh uncomfortably brilliant light **b :** STAND OUT, OBTRUDE
2 : to stare angrily or fiercely
transitive verb
1 : to express (as hostility) by staring angrily
2 *archaic* **:** to cause to be sharply reflected

[2]glare *noun* (15th century)
1 a : a harsh uncomfortably bright light ⟨the *glare* of a neon sign⟩ ⟨the *glare* of publicity⟩; *especially* **:** painfully bright sunlight **b :** cheap showy brilliance **:** GARISHNESS
2 : an angry or fierce stare
3 : a surface or sheet of smooth and slippery ice

glar·ing \'glar-iŋ, 'gler-\ *adjective* (14th century)
1 : having a fixed look of hostility, fierceness, or anger
2 a : shining with or reflecting an uncomfortably bright light **b** (1) **:** GARISH (2) **:** vulgarly ostentatious
3 : obtrusively and often painfully obvious ⟨a *glaring* error⟩
synonym see FLAGRANT
— glar·ing·ly \-iŋ-lē\ *adverb*
— glar·ing·ness *noun*

glary \'glar-ē, 'gler-\ *adjective* **glar·i·er; -est** (1632)
: having a dazzling brightness **:** GLARING

glas·nost \'glaz-(ˌ)nōst, 'glas-, 'gläz-, 'gläs-\ *noun* [Russian *glasnost',* literally, publicity, from *glasnyĭ* public, from *glas* voice, from Old Church Slavonic *glasŭ* — more at CALL] (1986)
: a Soviet policy permitting open discussion of political and social issues and freer dissemination of news and information

[1]glass \'glas, 'glȧs\ *noun, often attributive* [Middle English *glas,* from Old English *glæs;* akin to Old English *geolu* yellow — more at YELLOW] (before 12th century)
1 : any of various amorphous materials formed from a melt by cooling to rigidity without crystallization: as **a :** a usually transparent or translucent material consisting especially of a mixture of silicates **b :** a material (as obsidian) produced by fast cooling of magma
2 a : something made of glass: as (1) **:** TUMBLER; *also* **:** GLASSWARE (2) **:** MIRROR (3) **:** BAROMETER (4) **:** HOURGLASS (5) **:** BACKBOARD **b** (1) **:** an optical instrument or device that has one or more lenses and is designed to aid in the viewing of objects not readily seen (2) **:** FIELD GLASSES, BINOCULARS — usually used in plural **c** *plural* **:** a device used to correct defects of vision or to protect the eyes that consists typically of a pair of glass or plastic lenses and the frame by which they are held in place — called also *eyeglasses, spectacles*
3 : the quantity held by a glass container
4 : FIBERGLASS
— glass·ful \'glas-ˌfu̇l\ *noun*
— glass·less \-ləs\ *adjective*

[2]glass (14th century)
transitive verb
1 a : to provide with glass **:** GLAZE 1 **b :** to enclose, case, or wall with glass ⟨the sunroom was *glassed* in⟩
2 : to make glassy
3 a : REFLECT **b :** to see mirrored
4 : to look at through an optical instrument (as binoculars)
intransitive verb
: [1]GLAZE 1

glass·blow·ing \-ˌblō-iŋ\ *noun* (circa 1829)
: the art of shaping a mass of glass that has been softened by heat by blowing air into it through a tube
— glass·blow·er \-ˌblō(-ə)r\ *noun*

glass ceiling *noun* (1986)
: an intangible barrier within the hierarchy of a company that prevents women or minorities from obtaining upper-level positions

glass eye *noun* (1687)
1 : an artificial eye made of glass
2 : an eye having a pale, whitish, or colorless iris
— glass–eyed \-ˌīd\ *adjective*

glass fiber *noun* (1882)
: FIBERGLASS

glass harmonica *noun* (circa 1909)
: a musical instrument consisting of a series of rotating glass bowls of differing sizes played by touching the dampened edges with a finger

glass·house \'glas-ˌhau̇s\ *noun* (14th century)
1 : a place where glass is made
2 *chiefly British* **:** GREENHOUSE
3 *British* **:** a military prison

glass·ie \'gla-sē\ *or* **glassy** *noun, plural* **glass·ies** (1887)
: a playing marble made of glass

glass·ine \gla-'sēn\ *noun* (1916)
: a thin dense transparent or semitransparent paper highly resistant to the passage of air and grease

glass jaw *noun* (1940)
: vulnerability (as of a boxer) to knockout punches

glass·mak·er \'glas-ˌmā-kər\ *noun* (1576)
: one that makes glass
— glass·mak·ing \-kiŋ\ *noun*

glass·pa·per \'gläs-ˌpā-pər\ *noun* (1815)
British **:** abrasive paper coated with pulverized glass and used like sandpaper
— glasspaper *verb*

glass snake *noun* (1709)
: any of a genus (*Ophisaurus*) of limbless snakelike lizards of the southern U.S., Eurasia, and Africa with a fragile tail that readily breaks into pieces

glass sponge *noun* (1875)
: any of a class (Hexactinellida synonym Hyalospongiae) of chiefly deep-water siliceous sponges with 6-rayed spicules and a skeleton often resembling glass when dried

glass·ware \'glas-ˌwar, -ˌwer\ *noun* (1745)
: articles made of glass

glass wool *noun* (1879)
: glass fibers in a mass resembling wool and being used especially for thermal insulation and air filters

glass·work \'glas-ˌwərk\ *noun* (1611)
1 a : the manufacture of glass or glassware; *also* **:** glaziers' work **b** *plural* **:** GLASSHOUSE 1
2 : GLASSWARE
— glass·work·er \-ˌwər-kər\ *noun*

glass·wort \-ˌwərt, -ˌwȯrt\ *noun* [from its former use in the manufacture of glass] (1597)
: any of a genus (*Salicornia*) of woody jointed succulent herbs of the goosefoot family with leaves reduced to fleshy sheaths

glassy \'gla-sē\ *adjective* **glass·i·er; -est** (14th century)
1 : resembling or made of glass
2 : having little animation **:** DULL, LIFELESS ⟨*glassy* eyes⟩
— glass·i·ly \'gla-sə-lē\ *adverb*
— glass·i·ness \'gla-sē-nəs\ *noun*

glassy–eyed \-ˌīd\ *adjective* (1895)
: marked by or having glassy eyes

Glau·ber's salt \'glau̇-bər(z)-\ *noun* [Johann R. *Glauber* (died 1668) German chemist] (1736)
: a colorless crystalline sulfate of sodium $Na_2SO_4 \cdot 10H_2O$ used especially in dyeing, as a cathartic, and in solar energy systems — called also *Glauber salt*; sometimes used in plural

glau·co·ma \glau̇-'kō-mə, glȯ-\ *noun* [Latin, cataract, from Greek *glaukōma,* from *glaukoun* to have a cataract, from *glaukos*] (1885)
: a disease of the eye marked by increased pressure within the eyeball that can result in damage to the optic disk and gradual loss of vision

glau·co·nite \'glȯ-kə-ˌnīt\ *noun* [German *Glaukonit,* irregular from Greek *glaukos*] (1836)
: a mineral consisting of a dull green earthy iron potassium silicate occurring in greensand
— **glau·co·nit·ic** \ˌglȯ-kə-'ni-tik\ *adjective*

glau·cous \'glȯ-kəs\ *adjective* [Latin *glaucus,* from Greek *glaukos* gleaming, gray] (1671)
1 a : of a pale yellow-green color **b :** of a light bluish gray or bluish white color
2 : having a powdery or waxy coating that gives a frosted appearance and tends to rub off
— **glau·cous·ness** *noun*

[1]glaze \'glāz\ *verb* **glazed; glaz·ing** [Middle English *glasen,* from *glas* glass] (14th century)
transitive verb
1 : to furnish or fit with glass
2 a : to coat with or as if with a glaze ⟨the storm *glazed* trees with ice⟩ **b :** to apply a glaze to ⟨*glaze* doughnuts⟩
3 : to give a smooth glossy surface to
intransitive verb
1 : to become glazed or glassy ⟨my eyes *glazed* over⟩
2 : to form a glaze
— **glaz·er** *noun*

[2]glaze *noun* (1752)
1 : a smooth slippery coating of thin ice
2 a (1) **:** a liquid preparation applied to food on which it forms a firm glossy coating (2) **:** a mixture mostly of oxides (as silica and alumina) applied to the surface of ceramic wares to form a moisture-impervious and often lustrous or ornamental coating **b :** a transparent or translucent color applied to modify the effect of a painted surface **c :** a smooth glossy or lustrous surface or finish
3 : a glassy film

[3]glaze *intransitive verb* **glazed; glaz·ing** [probably blend of *glare* and *gaze*] (1601)
archaic **:** STARE

glazed \'glāzd\ *adjective* (15th century)
1 : covered with or as if with a glassy film ⟨*glazed* eyes⟩
2 : marked by lack of expression

gla·zier \'glā-zhər, -zē-ər\ *noun* (14th century)
: one who sets glass
— **gla·ziery** \'glā-zh(ə-)rē, 'glā-zē ə-rē\ *noun*

glaz·ing \'glā-ziŋ\ *noun* (1677)
1 : the action, process, or trade of fitting windows with glass
2 a : GLASSWORK **b :** GLAZE
3 : transparent material (as glass) used for windows

[1]gleam \'glēm\ *noun* [Middle English *gleem,* from Old English *glǣm;* akin to Old English *geolu* yellow — more at YELLOW] (15th century)
1 a : a transient appearance of subdued or partly obscured light ⟨the *gleam* of dawn in the east⟩ **b** (1) **:** a small bright light ⟨the *gleam* of a match⟩ (2) **:** GLINT ⟨a *gleam* in his eyes⟩
2 : a brief or faint appearance ⟨a *gleam* of hope⟩
— **gleamy** \'glē-mē\ *adjective*

[2]gleam (1508)
intransitive verb
1 : to shine with or as if with subdued steady light or moderate brightness
2 : to appear briefly or faintly ⟨a light *gleamed* in the distance⟩
transitive verb
: to cause to gleam
synonym see FLASH

glean \'glēn\ *verb* [Middle English *glenen,* from Middle French *glener,* from Late Latin *glennare,* of Celtic origin; akin to Old Irish do*glenn* he selects] (14th century)
intransitive verb
1 : to gather grain or other produce left by reapers
2 : to gather information or material bit by bit
transitive verb
1 a : to pick up after a reaper **b :** to strip (as a field) of the leavings of reapers
2 a : to gather (as information) bit by bit **b :** to pick over in search of relevant material ⟨*gleaning* old files for information⟩
3 : FIND OUT
— **glean·able** \'glē-nə-bəl\ *adjective*
— **glean·er** *noun*

glean·ings \'glē-niŋz\ *noun plural* (15th century)
: things acquired by gleaning

glebe \'glēb\ *noun* [Latin *gleba* clod, land] (14th century)
1 *archaic* **:** LAND; *specifically* **:** a plot of cultivated land
2 : land belonging or yielding revenue to a parish church or ecclesiastical benefice

glede \'glēd\ *noun* [Middle English, from Old English *glida;* akin to Old English *glīdan* to glide] (before 12th century)
: any of several birds of prey (as a kite of Europe)

glee \'glē\ *noun* [Middle English, from Old English *glēo* entertainment, music; akin to Old Norse *glȳ* joy, and perhaps to Greek *chleuē* joke] (before 12th century)
1 : exultant high-spirited joy **:** MERRIMENT
2 : a part-song for usually male voices

glee club *noun* (1844)
: a chorus organized for singing usually short pieces

gleed \'glēd\ *noun* [Middle English, from Old English *glēd;* akin to Old English *glōwan* to glow] (before 12th century)
archaic **:** a glowing coal

glee·ful \'glē-fəl\ *adjective* (1586)
: full of glee **:** MERRY
— **glee·ful·ly** \-fə-lē\ *adverb*
— **glee·ful·ness** *noun*

gleek \'glēk\ *intransitive verb* [origin unknown] (1590)
archaic **:** GIBE, JOKE

glee·man \'glē-mən\ *noun* [Middle English *gleman,* from Old English *glēoman,* from *glēo* + *man* man] (before 12th century)
: JONGLEUR

glee·some \'glē-səm\ *adjective* (1603)
archaic **:** GLEEFUL

gleet \'glēt\ *noun* [Middle English *glet* slimy or mucous matter, from Middle French *glete,* from Latin *glittus* viscous; akin to Latin *gluten* glue — more at CLAY] (14th century)
: a chronic inflammation (as gonorrhea) of a bodily orifice usually accompanied by an abnormal discharge; *also* **:** the discharge itself

gleg \'gleg\ *adjective* [Middle English, from Old Norse *glǫggr* clear-sighted] (14th century)
Scottish **:** marked by quickness of perception or movement

glei·za·tion \glā-'zā-shən\ *noun* (1938)
: development of or conversion into gley

glen \'glen\ *noun* [Middle English (Scots), valley, from (assumed) Scottish Gaelic *glenn;* akin to Middle Irish *glend* valley] (15th century)
: a secluded narrow valley

glen·gar·ry \glen-'gar-ē\ *noun, plural* **-ries** *often capitalized* [*Glengarry,* valley in Scotland] (1845)
: a woolen cap of Scottish origin — called also *glengarry bonnet*

glen plaid \'glen-\ *noun* [short for *glenurquhart plaid,* from *Glen Urquhart,* valley in Inverness-shire, Scotland] (1926)
: a twill pattern of broken checks; *also* **:** a fabric woven in this pattern — called also *glen check*

gley \'glā\ *noun, often attributive* [Ukrainian *gleĭ* clayey earth; akin to Old English *clǣg* clay — more at CLAY] (1927)
: a sticky clay soil or soil layer formed under the surface of some waterlogged soils
— **gleyed** *adjective*

gley·ing \'glā-iŋ\ *noun* (1949)
: GLEIZATION

glia \'glē-ə, 'glī-ə\ *noun* [New Latin, from Middle Greek, glue — more at CLAY] (1891)
: NEUROGLIA
— **gli·al** \-əl\ *adjective*

gli·a·din \'glī-ə-dən\ *noun* [Italian *gliadina,* from Middle Greek *glia*] (circa 1828)
: PROLAMIN; *especially* **:** one obtained by alcoholic extraction of gluten from wheat and rye

glib \'glib\ *adjective* **glib·ber; glib·best** [probably modification of Low German *glibberig* slippery] (1593)
1 a : marked by ease and informality **:** NONCHALANT **b :** showing little forethought or preparation **:** OFFHAND ⟨*glib* answers⟩ **c :** lacking depth and substance **:** SUPERFICIAL ⟨*glib* solutions to knotty problems⟩
2 *archaic* **:** SMOOTH, SLIPPERY
3 : marked by ease and fluency in speaking or writing often to the point of being insincere or deceitful ⟨a *glib* politician⟩
— **glib·ly** *adverb*
— **glib·ness** *noun*

[1]glide \'glīd\ *verb* **glid·ed; glid·ing** [Middle English, from Old English *glīdan;* akin to Old High German *glītan* to glide] (before 12th century)
intransitive verb
1 : to move smoothly, continuously, and effortlessly ⟨swans *gliding* over the lake⟩
2 : to go or pass imperceptibly ⟨hours *glided* by⟩
3 a *of an airplane* **:** to descend gradually in controlled flight **b :** to fly in a glider
4 : to produce a glide (as in music or speech)
transitive verb
: to cause to glide

[2]glide *noun* (1596)
1 : the act or action of gliding
2 : a calm stretch of shallow water flowing smoothly
3 : PORTAMENTO
4 a : a less prominent vowel sound produced by the passing of the vocal organs to or from the articulatory position of a speech sound — compare DIPHTHONG **b :** SEMIVOWEL
5 : a device for facilitating movement of something; *especially* **:** a circular usually metal button attached to the bottom of furniture legs to provide a smooth surface

glide path *noun* (1936)
: GLIDE SLOPE

glid·er \'glī-dər\ *noun* (15th century)
1 : one that glides: as **a :** an aircraft similar to an airplane but without an engine **b :** a porch seat suspended from an upright framework
2 : something that aids gliding

glide slope *noun* (circa 1949)
1 : the proper path of descent for an aircraft preparing to land; *especially* **:** such a path indicated by a radio beam
2 : the radio beam that marks a glide slope

glim \'glim\ *noun* [perhaps short for [2]*glimmer*] (circa 1700)
: something that furnishes light (as a lantern or candle); *also* **:** illumination given off by such a source

[1]glim·mer \'gli-mər\ *intransitive verb* **glim·mered; glim·mer·ing** \'glim-riŋ, 'gli-mə-\ [Middle English *glimeren;* akin to Old English *glǣm* gleam] (15th century)
1 a : to shine faintly or unsteadily **b :** to give off a subdued unsteady reflection
2 : to appear indistinctly with a faintly luminous quality
synonym see FLASH

[2]glimmer *noun* (1590)
1 a : a feeble or intermittent light **b :** a subdued unsteady shining or sparkle

\ə\ **abut** \ᵊ\ **kitten** \ər\ **further** \a\ **ash** \ā\ **ace** \ä\ **mop, mar** \au̇\ **out** \ch\ **chin** \e\ **bet** \ē\ **easy** \g\ **go** \i\ **hit** \ī\ **ice** \j\ **job** \ŋ\ **sing** \ō\ **go** \ȯ\ **law** \ȯi\ **boy** \th\ **thin** \t͟h\ **the** \ü\ **loot** \u̇\ **foot** \y\ **yet** \zh\ **vision** *see also* Guide to Pronunciation

2 a : a dim perception or faint idea **:** INKLING **b :** HINT, SPARK ⟨a *glimmer* of intelligence⟩

glim·mer·ing *noun* (15th century)
: GLIMMER

¹glimpse \'glim(p)s\ *verb* **glimpsed; glimps·ing** [Middle English *glimsen;* akin to Middle High German *glimsen* to glimmer, Old English *glǣm* gleam] (14th century)
intransitive verb
1 *archaic* **:** GLIMMER
2 : to look briefly
transitive verb
: to get a brief look at
— glimps·er *noun*

²glimpse *noun* (circa 1540)
1 *archaic* **:** GLIMMER
2 : a fleeting view or look

¹glint \'glint\ *verb* [Middle English, to dart obliquely, glint, alteration of *glenten,* of Scandinavian origin; akin to Swedish dialect *glänta* to clear up; akin to Old High German *glanz* bright, Old English *geolu* yellow — more at YELLOW] (14th century)
intransitive verb
1 a *archaic* **:** to glance off an object **b** *of rays of light* **:** to be reflected at an angle from a surface
2 : to give off reflection in brilliant flashes; *also* **:** GLEAM
3 : to look quickly or briefly **:** GLANCE
4 : to appear briefly or faintly
transitive verb
: to cause to glint
synonym see FLASH

²glint *noun* (14th century)
1 : a tiny bright flash of light
2 : a brief or faint manifestation **:** GLIMMER ⟨a *glint* of recognition⟩; *also* **:** a trace of emotion expressed through the eyes ⟨a steely *glint* in his eye⟩

glio·blas·to·ma \,glī-ō-,bla-'stō-mə\ *noun, plural* **-mas** *or* **-ma·ta** \-mə-tə\ [New Latin, from *glia* neuroglia + *blast-* + *-oma*] (circa 1923)
: a malignant tumor of the central nervous system and usually of a cerebral hemisphere

gli·o·ma \glē-'ō-mə, glī-\ *noun, plural* **-mas** *or* **-ma·ta** \-mə-tə\ [New Latin, from *glia*] (1870)
: a tumor arising from neuroglia

¹glis·sade \gli-'säd, -'sād\ *intransitive verb* **glis·sad·ed; glis·sad·ing** [French, noun, slide, glissade, from *glisser* to slide, from Old French *glicier,* alteration of *glier,* of Germanic origin; akin to Old High German *glītan* to glide] (1837)
1 : to perform a ballet glissade
2 : to slide in a standing or squatting position down a snow-covered slope without the aid of skis
— glis·sad·er *noun*

²glissade *noun* (1843)
1 : a gliding step in ballet
2 : the action of glissading

glis·san·do \gli-'sän-(,)dō\ *noun, plural* **-di** \-(,)dē\ *or* **-dos** [probably modification of French *glissade*] (circa 1854)
: a rapid sliding up or down the musical scale

¹glis·ten \'gli-sᵊn\ *intransitive verb* **glis·tened; glis·ten·ing** \'gli-sᵊn-iŋ\ [Middle English *glistnen,* from Old English *glisnian;* akin to Old English *glisian* to glitter, *geolu* yellow — more at YELLOW] (before 12th century)
: to give off a sparkling or lustrous reflection of or as if of a moist or polished surface
synonym see FLASH

²glisten *noun* (1840)
: GLITTER, SPARKLE

glis·ter \'glis-tər\ *intransitive verb* **glis·tered; glis·ter·ing** \-t(ə-)riŋ\ [Middle English *glistren;* akin to Old English *glisian*] (14th century)
: GLITTER
— glister *noun*

glitch \'glich\ *noun* [perhaps from Yiddish *glitsh* slippery place, from *glitshn* (*zikh*) to slide, glide; akin to Old High German *glītan* to glide — more at GLIDE] (1962)
1 a : a usually minor malfunction ⟨a *glitch* in a spacecraft's fuel cell⟩; *also* **:** ²BUG 2 **b :** a minor problem that causes a temporary setback **:** SNAG
2 : a false or spurious electronic signal

¹glit·ter \'gli-tər\ *intransitive verb* [Middle English *gliteren,* from Old Norse *glitra;* akin to Old English *geolu* yellow] (14th century)
1 a : to shine by reflection with many small flashes of brilliant light **:** SPARKLE ⟨sequins *glittered* in the spotlight⟩ **b :** to shine with strong emotion **:** FLASH ⟨eyes *glittering* in anger⟩
2 : to be brilliantly attractive, lavish, or spectacular; *also* **:** to be superficially attractive or exciting
synonym see FLASH
— glit·ter·ing·ly \-tə-riŋ-lē\ *adverb*

²glitter *noun* (1602)
1 a : sparkling brilliance of something that glitters **b :** a bright usually superficial attractiveness **c :** the quality of being spectacular
2 : small glittering objects used for ornamentation
— glit·tery \'gli-tə-rē\ *adjective*

glit·te·ra·ti \,gli-tə-'rä-tē\ *noun plural* [blend of ²*glitter* and *literati*] (1940)
: CELEBRITIES, BEAUTIFUL PEOPLE

glitter rock *noun* (1973)
: rock music characterized by performers wearing glittering costumes and bizarre often grotesque makeup

glitz \'glits\ *noun* [perhaps modification of German *glitzern* to glitter, from Middle High German *glitzen;* akin to Old Norse *glitra* to glitter] (1971)
: extravagant showiness **:** GLITTER, OSTENTATION
— glitzy \'glit-sē\ *adjective*

gloam \'glōm\ *noun* [Scots *gloam* to become twilight, back-formation from *gloaming*] (circa 1821)
archaic **:** TWILIGHT

gloam·ing \'glō-miŋ\ *noun* [Middle English (Scots) *gloming,* from Old English *glōming,* from *glōm* twilight; akin to Old English *glōwan* to glow] (before 12th century)
: TWILIGHT, DUSK

¹gloat \'glōt\ *intransitive verb* [probably of Scandinavian origin; akin to Old Norse *glotta* to grin scornfully] (1676)
1 *obsolete* **:** to look or glance admiringly or amorously
2 : to observe or think about something with triumphant and often malicious satisfaction, gratification, or delight ⟨*gloat* over an enemy's misfortune⟩
— gloat·er *noun*
— gloat·ing·ly *adverb*

²gloat *noun* (1899)
: the act or feeling of one who gloats

glob \'gläb\ *noun* [perhaps blend of *globe* and *blob*] (1900)
1 : a small drop **:** BLOB
2 : a usually large and rounded mass
— glob·by \'glä-bē\ *adjective*

glob·al \'glō-bəl\ *adjective* (1676)
1 : SPHERICAL
2 : of, relating to, or involving the entire world **:** WORLDWIDE ⟨*global* warfare⟩ ⟨a *global* system of communication⟩; *also* **:** of or relating to a celestial body (as the moon)
3 : of, relating to, or applying to a whole (as a mathematical function or a computer program) ⟨a *global* search of a file⟩
— glob·al·ly \'glō-bə-lē\ *adverb*

glob·al·ise *British variant of* GLOBALIZE

glob·al·ism \'glō-bə-,li-zəm\ *noun* (1943)
: a national policy of treating the whole world as a proper sphere for political influence — compare IMPERIALISM, INTERNATIONALISM
— glob·al·ist \-list\ *noun*

glob·al·ize \'glō-bə-,līz\ *transitive verb* **-ized; -iz·ing** (1944)
: to make global; *especially* **:** to make worldwide in scope or application
— glob·al·iza·tion \,glō-bə-lə-'zā-shən\ *noun*

global village *noun* (1960)
: the world viewed as a community in which distance and isolation have been dramatically reduced by electronic media (as television)

¹globe \'glōb\ *noun* [Middle French, from Latin *globus*] (15th century)
: something spherical or rounded: as **a :** a spherical representation of the earth, a celestial body, or the heavens **b :** EARTH 4 **c :** ORB 5

²globe *transitive verb* **globed; glob·ing** (1641)
archaic **:** to form into a globe

globe artichoke *noun* (circa 1858)
: ARTICHOKE 1

globe·fish \'glōb-,fish\ *noun* (1668)
: PUFFER 2a

globe·flow·er \-,flau̇(-ə)r\ *noun* (1597)
: any of a genus (*Trollius*) of plants of the buttercup family with globose yellow flowers

globe thistle *noun* (1597)
: any of a genus (*Echinops*) of widely cultivated Asian and Mediterranean composite herbs with spiky globose blue or white flowers

globe–trot·ter \-,trä-tər\ *noun* (1875)
: a person who travels widely
— globe–trot·ting \-,trä-tiŋ\ *noun or adjective*

glo·bin \'glō-bən\ *noun* [International Scientific Vocabulary, from *hemoglobin*] (1877)
: a colorless protein obtained by removal of heme from a conjugated protein and especially hemoglobin

glo·boid \'glō-,bȯid\ *adjective* (1887)
: shaped like a sphere

glo·bose \'glō-,bōs\ *adjective* (15th century)
: GLOBULAR 1a(1) ⟨*globose* pollen⟩

glob·u·lar \'glä-byə-lər, *1b is also* 'glō-\ *adjective* [partly from Latin *globus* + English *-ular;* partly from Latin *globulus* + English *-ar*] (1654)
1 a (1) **:** having the shape of a globe or globule (2) **:** having a compact folded molecular structure ⟨*globular* proteins⟩ **b :** GLOBAL
2 : having or consisting of globules

globular cluster *noun* (1859)
: any of various approximately spherical clusters of gravitationally associated stars that typically populate galactic halos

glob·ule \'glä-(,)byü(ə)l\ *noun* [French, from Latin *globulus,* diminutive of *globus*] (1661)
: a tiny globe or ball especially of a liquid ⟨*globules* of mercury⟩

glob·u·lin \'glä-byə-lən\ *noun* (1845)
: any of a class of simple proteins (as myosin) that are insoluble in pure water but are soluble in dilute salt solutions and that occur widely in plant and animal tissues — compare ALPHA GLOBULIN, BETA GLOBULIN, GAMMA GLOBULIN

glo·chid·i·um \glō-'ki-dē-əm\ *noun, plural* **-ia** \-dē-ə\ [New Latin, from Greek *glōchis* projecting point + New Latin *-idium*] (1882)
: the larva of a freshwater mussel (family Unionidae) that develops as an external parasite on fish

glock·en·spiel \'glä-kən-,spēl, -,shpēl\ *noun* [German, from *Glocke* bell + *Spiel* play] (circa 1834)
: a percussion instrument consisting of a series of graduated metal bars tuned to the chromatic scale and played with two hammers

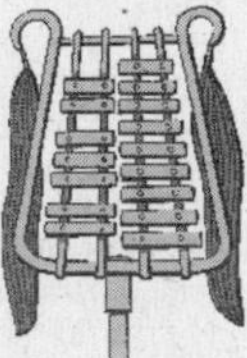
glockenspiel

glogg *or* **glögg** \'glu̇g, 'glüg\ *noun* [Swedish *glögg,* from *glödga* to burn, mull, from *glöd* glowing coal, from Old Norse *glōth;* akin to Old English *glēd* glowing coal — more at GLEED] (1927)

: a hot spiced wine and liquor punch served in Scandinavian countries as a Christmas drink

glom \ˈgläm\ *transitive verb* **glommed; glom·ming** [alteration of English dialect *glaum* to grab] (1907)
1 : TAKE, STEAL
2 : SEIZE, CATCH
— glom on to : to grab hold of **:** appropriate to oneself

glo·mer·u·lar \glə-ˈmer-(y)ə-lər, glō-\ *adjective* (1885)
: of, relating to, or produced by a glomerulus ⟨*glomerular* nephritis⟩ ⟨*glomerular* capillaries⟩

glom·er·ule \ˈglä-mə-ˌrül, -mər-ˌyü(ə)l\ *noun* [New Latin *glomerulus*] (1793)
: a compacted cyme of almost sessile and usually small flowers

glo·mer·u·lo·ne·phri·tis \glə-ˌmer-(y)ə-lō-ni-ˈfrī-təs\ *noun, plural* **-phrit·i·des** \-ˈfri-tə-ˌdēz\ (circa 1886)
: nephritis marked by inflammation of the capillaries of the renal glomeruli

glo·mer·u·lus \glə-ˈmer-(y)ə-ləs, glō-\ *noun, plural* **-li** \-ˌlī, -ˌlē\ [New Latin, glomerulus, glomerule, diminutive of Latin *glomer-, glomus* ball; akin to Latin *globus* globe] (1856)
: a small convoluted or intertwined mass; *especially* **:** a tuft of capillaries at the point of origin of each vertebrate nephron that passes a protein-free filtrate to the surrounding Bowman's capsule

glo·mus \ˈglō-məs\ *noun, plural* **glo·mera** \ˈglä-mə-rə, ˈglō-\ [New Latin, from Latin *glomer-, glomus*] (1903)
: a small arteriovenous anastomosis together with its supporting structures

[1]gloom \ˈglüm\ *verb* [Middle English *gloumen*] (14th century)
intransitive verb
1 : to look, feel, or act sullen or despondent
2 : to be or become overcast
3 : to loom up dimly
transitive verb
: to make dark, murky, or somber **:** make gloomy

[2]gloom *noun* (1629)
1 a : partial or total darkness **b :** a dark or shadowy place
2 a : lowness of spirits **:** DEJECTION **b :** an atmosphere of despondency ⟨a *gloom* fell over the household⟩

gloomy \ˈglü-mē\ *adjective* **gloom·i·er; -est** (1588)
1 a : partially or totally dark; *especially* **:** dismally and depressingly dark ⟨*gloomy* weather⟩ **b :** having a frowning or scowling appearance **:** FORBIDDING ⟨a *gloomy* countenance⟩ **c :** low in spirits **:** MELANCHOLY
2 a : causing gloom **:** DEPRESSING ⟨a *gloomy* story⟩ ⟨a *gloomy* landscape⟩ **b :** lacking in promise or hopefulness **:** PESSIMISTIC ⟨*gloomy* prophecies⟩ ⟨a *gloomy* future⟩
synonym see SULLEN
— gloom·i·ly \-mə-lē\ *adverb*
— gloom·i·ness \-mē-nəs\ *noun*

glop \ˈgläp\ *noun* [origin unknown] (circa 1944)
1 : a thick semiliquid substance (as food) that is usually unattractive in appearance
2 : tasteless or worthless material
— glop·py \ˈglä-pē\ *adjective*

Glo·ria \ˈglōr-ē-ə, ˈglȯr-\ *noun* [Latin, glory] (13th century)
1 : GLORIA IN EXCELSIS
2 : GLORIA PATRI

Gloria in Ex·cel·sis \-ˌin-eks-ˈchel-səs, -ek-ˈshel-\ [Late Latin, glory (be to God) on high] (14th century)
: a Christian liturgical hymn having the verse form of the Psalms

Gloria Pa·tri \-ˈpä-(ˌ)trē\ *noun* [Late Latin, glory (be) to the Father] (13th century)
: a 2-verse doxology to the Trinity

glo·ri·fy \ˈglōr-ə-ˌfī, ˈglȯr-\ *transitive verb* **-fied; -fy·ing** [Middle English *glorifien,* from Middle French *glorifier,* from Late Latin *glorificare,* from *gloria*] (14th century)
1 a : to make glorious by bestowing honor, praise, or admiration **b :** to elevate to celestial glory
2 : to light up brilliantly ⟨a large chandelier *glorifies* the whole room⟩
3 a : to represent as glorious **:** EXTOL ⟨a song *glorifying* romantic love⟩ **b :** to cause to be or seem to be better than the actual condition ⟨the new position is just a *glorified* version of the old stockroom job⟩
4 : to give glory to (as in worship)
— glo·ri·fi·ca·tion \ˌglōr-ə-fə-ˈkā-shən, ˌglȯr-\ *noun*
— glo·ri·fi·er \ˈglōr-ə-ˌfī(-ə)r, ˈglȯr-\ *noun*

glo·ri·ous \ˈglōr-ē-əs, ˈglȯr-\ *adjective* [Middle English, from Old French & Latin; Old French *glorieus, glorios,* from Latin *gloriosus* glorious, vainglorious, from *gloria*] (13th century)
1 a : possessing or deserving glory **:** ILLUSTRIOUS **b :** entitling one to glory ⟨a *glorious* victory⟩
2 : marked by great beauty or splendor **:** MAGNIFICENT
3 : DELIGHTFUL, WONDERFUL ⟨had a *glorious* weekend⟩
synonym see SPLENDID
— glo·ri·ous·ly *adverb*
— glo·ri·ous·ness *noun*

[1]glo·ry \ˈglōr-ē, ˈglȯr-\ *noun, plural* **glories** [Middle English *glorie,* from Middle French & Latin; Middle French, from Latin *gloria*] (14th century)
1 a : praise, honor, or distinction extended by common consent **:** RENOWN **b :** worshipful praise, honor, and thanksgiving ⟨giving *glory* to God⟩
2 a : something that secures praise or renown ⟨the *glory* of a brilliant career⟩ **b :** a distinguished quality or asset
3 a (1) **:** great beauty and splendor **:** MAGNIFICENCE ⟨the *glory* that was Greece and the grandeur that was Rome —E. A. Poe⟩ (2) **:** something marked by beauty or resplendence ⟨a perfect *glory* of a day⟩ **b :** the splendor and beatific happiness of heaven; *broadly* **:** ETERNITY
4 a : a state of great gratification or exaltation ⟨when she's acting she's in her *glory*⟩ **b :** a height of prosperity or achievement
5 : a ring or spot of light: as **a :** AUREOLE **b :** CORONA 2a, b

[2]glory *intransitive verb* **glo·ried; glo·ry·ing** (14th century)
: to rejoice proudly — used with *in*

[3]glory *or* **glory be** *interjection* (1816)
— used to express surprise or delight

glory–of–the–snow *noun* (circa 1890)
: any of a genus (*Chionodoxa*) of hardy spring-flowering bulbous herbs of the lily family with narrow basal leaves and terminal racemes of blue, white, or pink flowers

[1]gloss \ˈgläs, ˈglȯs\ *noun* [probably of Scandinavian origin; akin to Icelandic *glossa* to glow; akin to Old English *geolu* yellow] (1538)
1 : a surface luster or brightness **:** SHINE
2 a : a deceptively attractive appearance ⟨selfishness that had a *gloss* of humanitarianism about it⟩ **b :** bright often superficial attractiveness ⟨show-biz *gloss*⟩
3 : a transparent cosmetic preparation for adding shine and usually color to the lips

[2]gloss *transitive verb* (1656)
1 a : to mask the true nature of **:** give a deceptively attractive appearance to — used with *over* ⟨the misery was general, where not *glossed* over by liberal application of alcohol —Marston Bates⟩ **b :** to deal with (a subject or problem) too lightly or not at all — used with *over* ⟨*glosses* over scholarly controversies rather than confronting them head-on —John Israel⟩
2 : to give a gloss to

[3]gloss *noun* [alteration of *gloze,* from Middle English *glose,* from Middle French, from Medieval Latin *glosa, glossa,* from Greek *glōssa, glōtta* tongue, language, obscure word; akin to Greek *glōchis* projecting point] (1548)
1 a : a brief explanation (as in the margin or between the lines of a text) of a difficult or obscure word or expression **b :** a false and often willfully misleading interpretation (as of a text)
2 a : GLOSSARY **b :** an interlinear translation **c :** a continuous commentary accompanying a text
3 : COMMENTARY, INTERPRETATION

[4]gloss *transitive verb* (1603)
1 a : to provide a gloss for **:** EXPLAIN, DEFINE **b :** INTERPRET
2 : to dispose of by false or perverse interpretation ⟨trying to *gloss* away the irrationalities of the universe —Irwin Edman⟩

gloss- *or* **glosso-** *combining form* [Latin, from Greek *glōss-, glōsso-,* from *glōssa*]
1 : tongue ⟨*gloss*itis⟩
2 : language ⟨*glosso*lalia⟩

glos·sa \ˈglä-sə, ˈglȯ-\ *noun, plural* **glos·sae** \-ˌsē, -ˌsī\ *also* **glossas** [New Latin, from Greek *glōssa*] (circa 1852)
: a tongue or lingual structure especially in an insect; *especially* **:** the median distal lobe of the labium of an insect

glos·sa·rist \ˈglä-sə-rist, ˈglȯ-\ *noun* (1774)
: GLOSSATOR

glos·sa·ry \-sə-rē\ *noun, plural* **-ries** (14th century)
: a collection of textual glosses or of specialized terms with their meanings
— glos·sar·i·al \glä-ˈsar-ē-əl, glȯ-, -ˈser-\ *adjective*

glos·sa·tor \ˈglä-ˌsā-tər, ˈglȯ-\ *noun* (14th century)
1 : one that makes textual glosses
2 : a compiler of a glossary

glos·si·tis \glä-ˈsī-təs, ˈglȯ-\ *noun* (circa 1834)
: inflammation of the tongue

glos·sog·ra·pher \glä-ˈsä-grə-fər, glȯ-\ *noun* [Greek *glōssographos,* from *glōssa* + *graphein* to write — more at CARVE] (1607)
: GLOSSATOR

glos·so·la·lia \ˌglä-sə-ˈlā-lē-ə, ˌglȯ-\ *noun* [New Latin] (1879)
: TONGUE 4c
— glos·so·la·list \-ˈlā-list\ *noun*

glos·so·pha·ryn·geal nerve \ˌglä-sō-ˌfar-ən-ˈjē-əl-, ˌglȯ-, -fə-ˈrin-j(ē-)əl-\ *noun* (circa 1823)
: either of the 9th pair of cranial nerves that are mixed nerves and supply chiefly the pharynx, posterior tongue, and parotid gland — called also *glossopharyngeal*

[1]glossy \ˈglä-sē, ˈglȯ-\ *adjective* **gloss·i·er; -est** (1556)
1 : having a surface luster or brightness ⟨rich *glossy* leather⟩ ⟨*glossy* paper⟩
2 : attractive in an artificially opulent, sophisticated, or smoothly captivating manner **:** SLICK ⟨lots of *glossy* and phony chatter⟩
synonym see SLEEK
— gloss·i·ly \-sə-lē\ *adverb*
— gloss·i·ness \-sē-nəs\ *noun*

[2]glossy *noun, plural* **gloss·ies** (1928)
1 : a photograph printed on smooth shiny paper
2 *chiefly British* **:** SLICK 3

glott- *or* **glotto-** *combining form* [Greek *glōtt-, glōtto-* tongue, from *glōssa, glōtta*]
: language ⟨*glotto*chronology⟩

glot·tal \ˈglä-t^{ə}l\ *adjective* (circa 1846)
: of, relating to, or produced in or by the glottis ⟨*glottal* constriction⟩

glottal stop *noun* (1888)
: the interruption of the breath stream during speech by closure of the glottis

glot·tis \'glä-təs\ *noun, plural* **glot·tis·es** *or* **glot·ti·des** \-tə-ˌdēz\ [Greek *glōttid-, glōttis,* from *glōtta* tongue — more at GLOSS] (1578)
: the elongated space between the vocal cords; *also* : the structures that surround this space — compare EPIGLOTTIS

glot·to·chro·nol·o·gy \ˌglä-tō-krə-'nä-lə-jē\ *noun* (1953)
: a linguistic method that uses the rate of vocabulary replacement to estimate the date of divergence for distinct but genetically related languages
— **glot·to·chro·no·log·i·cal** \-ˌkrä-nᵊl-'ä-ji-kəl, -ˌkrō-\ *adjective*

glout \'glüt, 'glau̇t\ *intransitive verb* [Middle English, probably of Scandinavian origin; akin to Old Norse *glotta* to grin scornfully] (14th century)
archaic : FROWN, SCOWL

[1]**glove** \'gləv\ *noun* [Middle English, from Old English *glōf;* akin to Old Norse *glōfi* glove] (before 12th century)
1 a : a covering for the hand having separate sections for each of the fingers and the thumb and often extending part way up the arm **b :** [1]GAUNTLET 1, 3
2 a (1) **:** a padded leather covering for the hand used in baseball to catch a thrown or batted ball; *specifically* **:** one having individual thumb and finger sections usually connected with a lacing or webbing — compare MITT (2) **:** fielding ability ⟨he's got a good *glove* at three positions and can pinch-hit —Casey Stengel⟩ **b :** BOXING GLOVE

[2]**glove** *transitive verb* **gloved; glov·ing** (15th century)
1 a : to cover with or as if with a glove **b :** to furnish with gloves
2 : to catch (a baseball) in one's gloved hand

glove box *noun* (1946)
1 : GLOVE COMPARTMENT
2 : a sealed protectively lined compartment having holes to which are attached gloves for use in handling dangerous materials inside the compartment

glove compartment *noun* (1939)
: a small storage cabinet in the dashboard of an automobile

glove leather *noun* (1721)
: a soft lightweight leather

glov·er \'glə-vər\ *noun* (14th century)
: one that makes or sells gloves

[1]**glow** \'glō\ *intransitive verb* [Middle English, from Old English *glōwan;* akin to Old English *geolu* yellow — more at YELLOW] (before 12th century)
1 a : to shine with or as if with an intense heat ⟨embers *glowing* in the darkness⟩ **b** (1) **:** to have a rich warm typically ruddy color ⟨cheeks *glowing* with health⟩ (2) **:** FLUSH, BLUSH ⟨the children *glowed* with excitement⟩
2 a : to experience a sensation of or as if of heat ⟨*glowing* with rage⟩ **b :** to show exuberance or elation ⟨*glow* with pride⟩
— **glow·ing·ly** \-iŋ-lē\ *adverb*

[2]**glow** *noun* (1600)
1 : brightness or warmth of color; *especially* **:** REDNESS
2 a : warmth of feeling or emotion **b :** a sensation of warmth ⟨the drug produces a sustained *glow*⟩
3 a : the state of glowing with heat and light **b :** light such as is emitted by a solid body heated to luminosity **:** INCANDESCENCE

glow discharge *noun* (1844)
: a luminous electrical discharge without sparks through a gas

[1]**glow·er** \'glau̇(-ə)r, ÷'glō(-ə)r\ *intransitive verb* [Middle English (Scots) *glowren;* perhaps of Scandinavian origin; akin to Norwegian dialect *glyra* to look askance, Icelandic *glossa* to glow — more at GLOSS] (15th century)
: to look or stare with sullen annoyance or anger

[2]**glower** *noun* (1715)
: a sullen brooding look of annoyance or anger

glow lamp *noun* (1884)
: a gas-discharge electric lamp in which most of the light proceeds from the glow of the gas near the cathode

glow plug *noun* (circa 1941)
: a heating element in a diesel-engine cylinder to preheat the air and facilitate starting; *also* : a similar element for ignition in other internal-combustion engines

glow·worm \'glō-ˌwərm\ *noun* [Middle English] (14th century)
: any of various luminous insect larvae or adults with wings rudimentary or lacking; *especially* **:** a larva or wingless female of a firefly (family Lampyridae) that emits light from the abdomen

glox·in·ia \gläk-'si-nē-ə\ *noun* [New Latin, from B. P. *Gloxin* 18th century German botanist] (circa 1820)
: any of a genus (*Sinningia*) of Brazilian tuberous herbaceous gesneriads; *especially* **:** a plant (*S. speciosa*) widely cultivated for its showy bell-shaped or slipper-shaped flowers

gloxinia

[1]**gloze** \'glōz\ *transitive verb* **glozed; gloz·ing** [Middle English *glosen* to gloss, flatter, from *glose* gloss] (14th century)
archaic **:** [4]GLOSS 1

[2]**gloze** *transitive verb* **glozed; gloz·ing** (14th century)
: [2]GLOSS 1 — often used with *over*

gluc- *or* **gluco-** *combining form* [International Scientific Vocabulary]
1 : glucose ⟨*gluco*kinase⟩
2 : related to or containing glucose ⟨*glucos*amine⟩

glu·ca·gon \'glü-kə-ˌgän\ *noun* [*gluc-* + *-agon* (perhaps from Greek *agōn,* present participle of *agein* to lead, drive) — more at AGENT] (1923)
: a protein hormone that is produced especially by the islets of Langerhans and that promotes an increase in the sugar content of the blood by increasing the rate of glycogen breakdown in the liver

glu·can \'glü-ˌkan, -kən\ *noun* (1941)
: a polysaccharide (as glycogen or cellulose) that is a polymer of glucose

glu·co·cor·ti·coid \ˌglü-kō-'kȯr-ti-ˌkȯid\ *noun* (1950)
: any of a group of corticoids (as hydrocortisone) that are involved especially in carbohydrate, protein, and fat metabolism, that are anti-inflammatory and immunosuppressive, and that are used widely in medicine (as in the alleviation of the symptoms of rheumatoid arthritis) — compare MINERALOCORTICOID

glu·co·ki·nase \-'kī-ˌnās, -ˌnāz\ *noun* (1950)
: a hexokinase found especially in the liver that catalyzes the phosphorylation of glucose

glu·co·nate \'glü-kə-ˌnāt\ *noun* (1884)
: a salt or ester of gluconic acid

glu·co·neo·gen·e·sis \ˌglü-kə-ˌnē-ə-'je-nə-səs\ *noun* [New Latin] (1912)
: formation of glucose within the animal body especially by the liver from substances (as fats and proteins) other than carbohydrates

glu·con·ic acid \(ˌ)glü-'kä-nik-\ *noun* [International Scientific Vocabulary, irregular from *glucose* + *-ic*] (1871)
: a crystalline acid $C_6H_{12}O_7$ obtained by oxidation of glucose and used chiefly in cleaning metals

glu·cos·amine \glü-'kō-sə-ˌmēn, -zə-\ *noun* (1884)
: an amino derivative $C_6H_{13}NO_5$ of glucose that occurs especially as a constituent of polysaccharides in animal supporting structures and some plant cell walls

glu·cose \'glü-ˌkōs, -ˌkōz\ *noun* [French, modification of Greek *gleukos* must, sweet wine; akin to Greek *glykys* sweet — more at DULCET] (1840)
1 : an optically active sugar $C_6H_{12}O_6$ that has an aldehydic carbonyl group; *especially* **:** the sweet colorless soluble dextrorotatory form that occurs widely in nature and is the usual form in which carbohydrate is assimilated by animals
2 : a light-colored syrup made from cornstarch

glucose–1–phosphate *noun* [from the position at which the phosphate group is attached] (1938)
: an ester $C_6H_{13}O_9P$ that reacts in the presence of a phosphorylase with aldoses and ketoses to yield disaccharides or with itself in liver and muscle to yield glycogen and phosphoric acid

glucose phosphate *noun* (1912)
: a phosphate ester of glucose: as **a :** GLUCOSE-1-PHOSPHATE **b :** GLUCOSE-6-PHOSPHATE

glucose–6–phosphate *noun* [from the position at which the phosphate group is attached] (1954)
: an ester $C_6H_{13}O_9P$ that is formed from glucose and ATP in the presence of a glucokinase and that is an essential early stage in glucose metabolism

glucose–6–phosphate dehydrogenase *noun* (1954)
: an enzyme found especially in red blood cells that dehydrogenates glucose-6-phosphate in a glucose degradation pathway alternative to the Krebs cycle

glu·co·si·dase \glü-'kō-sə-ˌdās, -zə-ˌdāz\ *noun* (circa 1926)
: an enzyme (as maltase) that hydrolyzes a glucoside

glu·co·side \'glü-kə-ˌsīd\ *noun* (1855)
: GLYCOSIDE; *especially* **:** a glycoside that yields glucose on hydrolysis
— **glu·co·sid·ic** \ˌglü-kə-'si-dik\ *adjective*

glu·cu·ron·ic acid \ˌglü-kyə-'rä-nik-\ *noun* (1911)
: a compound $C_6H_{10}O_7$ that occurs especially as a constituent of mucopolysaccharides (as hyaluronic acid) and combined as a glucuronide

glu·cu·ron·i·dase \-'rä-nə-ˌdās, -ˌdāz\ *noun* (1945)
: an enzyme that hydrolyzes a glucuronide; *especially* **:** one that occurs widely (as in liver and spleen) and hydrolyzes the beta form of a glucuronide

glu·cu·ro·nide \glü-'kyu̇r-ə-ˌnīd\ *noun* (1934)
: any of various derivatives of glucuronic acid that are formed especially as combinations with often toxic aromatic hydroxyl compounds and are excreted in the urine

[1]**glue** \'glü\ *noun* [Middle English *glu,* from Middle French, from Late Latin *glut-, glus;* akin to Latin *gluten* glue — more at CLAY] (14th century)
1 a : any of various strong adhesive substances; *especially* **:** a hard protein chiefly gelatinous substance that absorbs water to form a viscous solution with strong adhesive properties and that is obtained by cooking down collagenous materials (as hides or bones) **b :** a solution of glue used for sticking things together
2 : something that binds together ⟨enough social *glue* . . . to satisfy the human desire for community —E. D. Hirsch, Jr.⟩
— **glu·ey** \'glü-ē\ *adjective*
— **glu·i·ly** \'glü-ə-lē\ *adverb*

[2]**glue** *transitive verb* **glued; glu·ing** *also* **glue·ing** (14th century)
1 : to cause to stick tightly with or as if with glue ⟨*gluing* the wings onto the model airplane⟩ ⟨used that war to *glue* together a frail story —Gloria Emerson⟩
2 : to fix (as the eyes) on an object steadily or with deep concentration ⟨kept his eyes *glued* to the TV screen⟩

glum \ˈgləm\ *adjective* **glum·mer; glum·mest** [akin to Middle English *gloumen* to gloom] (1547)
1 : broodingly morose ⟨became *glum* when they heard the news⟩
2 : DREARY, GLOOMY ⟨a *glum* countenance⟩
synonym see SULLEN
— glum·ly *adverb*
— glum·ness *noun*

glume \ˈglüm\ *noun* [New Latin *gluma,* from Latin, hull, husk; akin to Latin *glubere* to peel — more at CLEAVE] (1789)
: a chaffy bract; *specifically* **:** either of two empty bracts at the base of the spikelet in grasses

glu·on \ˈglü-ˌän\ *noun* [[1]*glue* + [2]*-on*] (1971)
: a hypothetical neutral massless particle held to bind together quarks to form hadrons

[1]**glut** \ˈglət\ *verb* **glut·ted; glut·ting** [Middle English *glouten,* probably from Middle French *gloutir* to swallow, from Latin *gluttire* — more at GLUTTON] (14th century)
transitive verb
1 : to fill especially with food to satiety
2 : to flood (the market) with goods so that supply exceeds demand
intransitive verb
: to eat gluttonously
synonym see SATIATE

[2]**glut** *noun* (circa 1546)
1 : an excessive quantity **:** OVERSUPPLY
2 *archaic* **:** the act or process of glutting

[3]**glut** *transitive verb* **glut·ted; glut·ting** [probably from obsolete *glut,* noun, swallow] (1600)
archaic **:** to swallow greedily

glu·ta·mate \ˈglü-tə-ˌmāt\ *noun* (1876)
: a salt or ester of glutamic acid — compare MONOSODIUM GLUTAMATE

glu·tam·ic acid \(ˌ)glü-ˈta-mik-\ *noun* [International Scientific Vocabulary *glu*ten + *am*ino + *-ic*] (1871)
: a crystalline amino acid $C_5H_9NO_4$ widely distributed in plant and animal proteins

glu·ta·min·ase \ˈglü-tə-mə-ˌnās, glü-ˈta-mə-, -ˌnāz\ *noun* (1938)
: an enzyme that hydrolyzes glutamine to glutamic acid and ammonia

glu·ta·mine \ˈglü-tə-ˌmēn\ *noun* [International Scientific Vocabulary *glut*en + *amine*] (circa 1885)
: a crystalline amino acid $C_5H_{10}N_2O_3$ that is found both free and in proteins in plants and animals and that yields glutamic acid and ammonia on hydrolysis

glu·tar·al·de·hyde \ˌglü-tə-ˈral-də-ˌhīd\ *noun* [*glutar*ic acid + *aldehyde*] (1951)
: a compound $C_5H_8O_2$ that contains two aldehyde groups and is used especially in tanning leather and in the fixation of biological tissues

glu·tar·ic acid \glü-ˈtar-ik-\ *noun* [probably from *glut*en + *-aric* (as in *tartaric acid*)] (circa 1885)
: a crystalline acid $C_5H_8O_4$ used especially in organic synthesis

glu·ta·thi·one \ˌglü-tə-ˈthī-ˌōn\ *noun* [International Scientific Vocabulary *gluta-* (from *glutamic acid*) + *thi-* + *-one*] (1921)
: a peptide $C_{10}H_{17}N_3O_6S$ that contains one amino-acid residue each of glutamic acid, cysteine, and glycine, that occurs widely in plant and animal tissues, and that plays an important role in biological oxidation-reduction processes and as a coenzyme

glu·te·al \ˈglü-tē-əl, glü-ˈtē-\ *adjective* (1804)
: of or relating to the gluteus muscles

glu·ten \ˈglü-tᵊn\ *noun* [Latin *glutin-, gluten* glue — more at CLAY] (1803)
: a tenacious elastic protein substance especially of wheat flour that gives cohesiveness to dough
— glu·ten·ous \ˈglüt-nəs, ˈglü-tᵊn-əs\ *adjective*

glu·teth·i·mide \glü-ˈte-thə-ˌmīd, -məd\ *noun* [*glu*ten + *eth-* + *imide*] (1955)
: a sedative-hypnotic drug $C_{13}H_{15}NO_2$ that induces sleep with less depression of respiration than occurs with comparable doses of barbiturates

glu·te·us \ˈglü-tē-əs, glü-ˈtē-\ *noun, plural* **glu·tei** \ˈglü-tē-ˌī, -tē-ˌē; glü-ˈtē-ˌī\ [New Latin *glutaeus, gluteus,* from Greek *gloutos* buttock — more at CLOUD] (circa 1681)
: any of the large muscles of the buttocks; *especially* **:** GLUTEUS MAXIMUS

gluteus max·i·mus \-ˈmak-sə-məs\ *noun, plural* **glutei max·i·mi** \-ˈmak-sə-ˌmī\ [New Latin, literally, largest gluteus] (1886)
: the outermost muscle of the three glutei found in each of the human buttocks

glu·ti·nous \ˈglüt-nəs, ˈglü-tᵊn-əs\ *adjective* [Middle French or Latin; Middle French *glutineux,* from Latin *glutinosus,* from *glutin-, gluten*] (15th century)
: having the quality of glue **:** GUMMY
— glu·ti·nous·ly *adverb*

glut·ton \ˈglə-tᵊn\ *noun* [Middle English *glotoun,* from Old French *gloton,* from Latin *glutton-, glutto;* akin to Latin *gluttire* to swallow, *gula* throat, Old English *ceole*] (13th century)
1 a : one given habitually to greedy and voracious eating and drinking **b :** one that has a great capacity for accepting or enduring something ⟨a *glutton* for punishment⟩
2 : WOLVERINE 1a

glut·ton·ous \ˈglət-nəs, ˈglə-tᵊn-əs\ *adjective* (14th century)
: marked by or given to gluttony
synonym see VORACIOUS
— glut·ton·ous·ly *adverb*
— glut·ton·ous·ness *noun*

glut·tony \ˈglət-nē, ˈglə-tᵊn-ē\ *noun, plural* **-ton·ies** (13th century)
1 : excess in eating or drinking
2 : greedy or excessive indulgence

glyc- *or* **glyco-** *combining form* [International Scientific Vocabulary, from Greek *glyk-* sweet, from *glykys*]
1 : carbohydrate and especially sugar ⟨*glyco*protein⟩
2 : glycine ⟨*glyc*yl⟩

gly·can \ˈglī-ˌkan\ *noun* (1953)
: POLYSACCHARIDE

glyc·er·al·de·hyde \ˌgli-sə-ˈral-də-ˌhīd\ *noun* [*glycer*ic acid + *aldehyde*] (1882)
: a sweet crystalline compound $C_3H_6O_3$ that is formed as an intermediate in carbohydrate metabolism by the breakdown of sugars and that yields glycerol on reduction

gly·cer·ic acid \gli-ˈser-ik-\ *noun* [International Scientific Vocabulary, from *glycerin*] (circa 1864)
: a syrupy acid $C_3H_6O_4$ obtainable by oxidation of glycerol or glyceraldehyde

glyc·er·ide \ˈgli-sə-ˌrīd\ *noun* (circa 1864)
: an ester of glycerol especially with fatty acids
— glyc·er·id·ic \ˌgli-sə-ˈri-dik\ *adjective*

glyc·er·in *or* **glyc·er·ine** \ˈglis-rən, ˈgli-sə-\ *noun* [French *glycérine,* from Greek *glykeros* sweet; akin to Greek *glykys*] (1838)
: GLYCEROL

glyc·er·in·ate \ˈglis-rə-ˌnāt, ˈgli-sə-\ *transitive verb* **-at·ed; -at·ing** (1897)
: to treat with or preserve in glycerin

glyc·er·ol \ˈgli-sə-ˌrȯl, -ˌrōl\ *noun* [*glycer*in + *-ol*] (1884)
: a sweet syrupy hygroscopic trihydroxy alcohol $C_3H_8O_3$ usually obtained by the saponification of fats and used especially as a solvent and plasticizer

glyc·er·yl \ˈglis-rəl, ˈgli-sə-\ *noun* (1845)
: a radical derived from glycerol by removal of hydroxide; *especially* **:** a trivalent radical CH_2CHCH_2

gly·cine \ˈglī-ˌsēn, ˈglī-sᵊn\ *noun* (1851)
: a sweet crystalline amino acid $C_2H_5NO_2$ obtained especially by hydrolysis of proteins

gly·co·gen \ˈglī-kə-jən\ *noun* (circa 1864)
: a white amorphous tasteless polysaccharide $(C_6H_{10}O_5)_x$ that is the principal form in which carbohydrate is stored in animal tissues and especially muscle and liver tissue

gly·co·gen·e·sis \ˌglī-kə-ˈje-nə-səs\ *noun* [New Latin] (circa 1886)
: the formation and storage of glycogen

gly·co·gen·ol·y·sis \ˌglī-kə-jə-ˈnä-lə-səs\ *noun, plural* **-y·ses** \-ˌsēz\ [New Latin] (1909)
: the breakdown of glycogen especially to glucose in the animal body
— gly·co·gen·o·lyt·ic \-jə-nᵊl-ˈi-tik, -ˌje-\ *adjective*

gly·col \ˈglī-ˌkȯl, -ˌkōl\ *noun* [International Scientific Vocabulary *glyc-* + *-ol*] (1858)
: DIOL; *especially* **:** ETHYLENE GLYCOL

gly·col·ic acid *or* **gly·col·lic acid** \(ˌ)glī-ˈkä-lik-\ *noun* [International Scientific Vocabulary *glycol* + [1]*-ic*] (1852)
: a translucent crystalline compound $C_2H_4O_3$ found especially in unripe grapes and sugar beets and used especially in textile and leather processing

gly·co·lip·id \ˌglī-kō-ˈli-pəd\ *noun* (1936)
: a lipid (as a ganglioside or a cerebroside) that contains a carbohydrate radical

gly·col·y·sis \glī-ˈkä-lə-səs\ *noun* [New Latin] (1892)
: the enzymatic breakdown of a carbohydrate (as glucose) by way of phosphate derivatives with the production of pyruvic or lactic acid and energy stored in high-energy phosphate bonds of ATP
— gly·co·lyt·ic \ˌglī-kə-ˈli-tik\ *adjective*

gly·co·pep·tide \ˌglī-kō-ˈpep-ˌtīd\ *noun* (1959)
: GLYCOPROTEIN

gly·co·pro·tein \-ˈprō-ˌtēn, -ˈprō-tē-ən\ *noun* (circa 1908)
: a conjugated protein in which the nonprotein group is a carbohydrate

gly·cos·ami·no·gly·can \ˌglī-kō-sə-ˌmē-nō-ˈglī-ˌkan, -kō-ˌsa-mə-nō-\ *noun* [*glyc* + *hexosamine* (amine derived from a hexose) + *-o-* + *glycan*] (1962)
: any of various polysaccharides derived from an amino hexose that are constituents of mucoproteins, glycoproteins, and blood-group substances — called also *mucopolysaccharide*

gly·co·si·dase \glī-ˈkō-sə-ˌdās, -zə-ˌdāz\ *noun* (1944)
: an enzyme that catalyzes the hydrolysis of a bond joining a sugar of a glycoside to an alcohol or another sugar unit

gly·co·side \ˈglī-kə-ˌsīd\ *noun* [alteration of *glucoside*] (circa 1889)
: any of numerous sugar derivatives that contain a nonsugar group bonded to an oxygen or nitrogen atom and that on hydrolysis yield a sugar (as glucose)
— gly·co·sid·ic \ˌglī-kə-ˈsi-dik\ *adjective*
— gly·co·sid·i·cal·ly \-di-k(ə-)lē\ *adverb*

gly·cos·uria \ˌglī-kō-ˈshu̇r-ē-ə, ˌglī-kəs-ˈyu̇r-\ *noun* [New Latin, from International Scientific Vocabulary *glycose* glucose + New Latin *-uria*] (1860)
: the presence in the urine of abnormal amounts of sugar

gly·co·syl \ˈglī-kə-ˌsil\ *noun* [International Scientific Vocabulary *glycose* glucose] (1945)
: a univalent radical derived from a cyclic form of glucose by removal of the hemiacetal hydroxyl group

gly·co·syl·a·tion \ˌglī-kō-sə-ˈlā-shən\ *noun* (1946)
: the process of adding glycosyl groups to a protein to form a glycoprotein
— gly·co·sy·late \ˌglī-kō-ˈsī-ˌlāt\ *transitive verb*

\ə\ abut \ᵊ\ kitten \ər\ further \a\ ash \ā\ ace \ä\ mop, mar \au̇\ out \ch\ chin \e\ bet \ē\ easy \g\ go \i\ hit \ī\ ice \j\ job \ŋ\ sing \ō\ go \ȯ\ law \ȯi\ boy \th\ thin \th\ the \ü\ loot \u̇\ foot \y\ yet \zh\ vision *see also* Guide to Pronunciation

gly·cyl \'glī-səl\ *noun* (1901)
: the univalent acyl radical NH_2CH_2CO– of glycine

glyph \'glif\ *noun* [Greek *glyphē* carved work, from *glyphein* to carve — more at CLEAVE] (1775)
1 : an ornamental vertical groove especially in a Doric frieze
2 : a symbolic figure or a character (as in the Mayan system of writing) usually incised or carved in relief
3 : a symbol (as a curved arrow on a road sign) that conveys information nonverbally
— glyph·ic \'gli-fik\ *adjective*

Glyp·tal \'glip-tᵊl\ *trademark*
— used for an alkyd

glyp·tic \'glip-tik\ *noun* [probably from French *glyptique,* from Greek *glyptikē,* from *glyphein*] (circa 1818)
: the art or process of carving or engraving especially on gems

G–man \'jē-ˌman\ *noun* [probably from government *man*] (1928)
: a special agent of the Federal Bureau of Investigation

gnar *or* **gnarr** \'när\ *intransitive verb* **gnarred; gnar·ring** [imitative] (15th century)
: SNARL, GROWL

¹gnarl \'när(-ə)l\ *intransitive verb* [probably frequentative of *gnar*] (1593)
: SNARL, GROWL

²gnarl *transitive verb* [back-formation from *gnarled*] (1814)
: to twist into a state of deformity

³gnarl *noun* (1824)
: a hard protuberance with twisted grain on a tree

gnarled \'när(-ə)ld\ *adjective* [probably alteration of *knurled*] (1603)
1 : full of knots or gnarls **:** KNOTTY ⟨*gnarled* hands⟩
2 : crabbed in disposition, aspect, or character

gnarly \'när-lē\ *adjective* (1829)
1 : GNARLED
2 *slang* — used as a generalized term of approval or disparagement

gnash \'nash\ *transitive verb* [alteration of Middle English *gnasten*] (15th century)
: to strike or grind (as the teeth) together
— gnash *noun*

gnat \'nat\ *noun* [Middle English, from Old English *gnætt;* akin to Old English *gnagan* to gnaw] (before 12th century)
: any of various small usually biting dipteran flies
— gnat·ty \-ē\ *adjective*

gnat·catch·er \'nat-ˌka-chər, -ˌke-\ *noun* (1839)
: any of a genus (*Polioptila*) of several small North and South American insectivorous oscine birds

gnath·ic \'na-thik\ *or* **gna·thal** \'nā-thəl, 'na-\ *adjective* [Greek *gnathos* jaw] (1882)
: of or relating to the jaw

-gnathous *adjective combining form* [New Latin *-gnathus,* from Greek *gnathos;* akin to Greek *genys* jaw — more at CHIN]
: having (such) a jaw ⟨pro*gnathous*⟩

gnaw \'nȯ\ *verb* [Middle English *gnawen,* from Old English *gnagan;* akin to Old High German *gnagan* to gnaw] (before 12th century)
transitive verb
1 a : to bite or chew on with the teeth; *especially* **:** to wear away by persistent biting or nibbling ⟨a dog *gnawing* a bone⟩ **b :** to make by gnawing ⟨rats *gnawed* a hole⟩
2 a : to be a source of vexation to **:** PLAGUE ⟨anxiety always *gnawing* him⟩ **b :** to affect like gnawing ⟨hunger *gnawing* her vitals⟩
3 : ERODE, CORRODE
intransitive verb
1 : to bite or nibble persistently ⟨*gnawing* at his underlip⟩
2 : to produce an effect of or as if of gnawing ⟨waves *gnawing* away at the cliffs⟩
— gnaw·er \'nȯ(-ə)r\ *noun*

gneiss \'nīs\ *noun* [German *Gneis,* alteration of Middle High German *gneiste* spark, from Old High German *gneisto;* akin to Old English *fȳrgnāst* spark] (1757)
: a foliated metamorphic rock corresponding in composition to a feldspathic plutonic rock (as granite)
— gneiss·ic \'nī-sik\ *adjective*
— gneiss·oid \-ˌsȯid\ *adjective*
— gneiss·ose \-ˌsōs\ *adjective*

gnoc·chi \'nȯ-kē, 'nyȯ-, 'nä-\ *noun plural* [Italian, plural of *gnocco,* from Italian dialect (Veneto), probably of Germanic origin; akin to Middle High German *knöchel* knuckle, *knoche* bone — more at KNUCKLE] (1891)
: dumplings usually made with potato or semolina and served with sauce

¹gnome \'nōm\ *noun* [Greek *gnōmē,* from *gignōskein* to know — more at KNOW] (1577)
: MAXIM, APHORISM

²gnome *noun* [French, from New Latin *gnomus*] (1712)
1 : an ageless and often deformed dwarf of folklore who lives in the earth and usually guards treasure
2 : an elemental being in the theory of Paracelsus that inhabits earth
— gnome·like \-ˌlīk\ *adjective*
— gnom·ish \'nō-mish\ *adjective*

gno·mic \'nō-mik\ *adjective* (1815)
1 : characterized by aphorism ⟨*gnomic* utterances⟩
2 : given to the composition of gnomic writing ⟨a *gnomic* poet⟩

gno·mon \'nō-mən, -ˌmän\ *noun* [Latin, from Greek *gnōmōn* interpreter, pointer on a sundial, from *gignōskein*] (1546)
1 : an object that by the position or length of its shadow serves as an indicator especially of the hour of the day: as **a :** the pin of a sundial **b :** a column or shaft erected perpendicular to the horizon

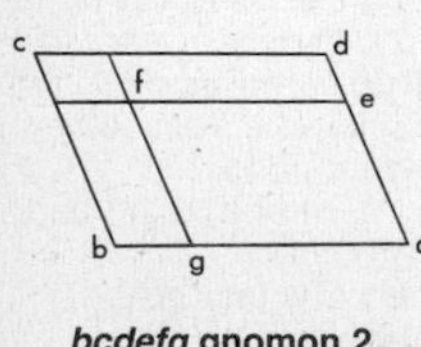

bcdefg **gnomon 2**

2 : the remainder of a parallelogram after the removal of a similar parallelogram containing one of its corners

gno·sis \'nō-səs\ *noun* [Greek *gnōsis,* literally, knowledge, from *gignōskein*] (1703)
: esoteric knowledge of spiritual truth held by the ancient Gnostics to be essential to salvation

gnos·tic \'näs-tik\ *noun, often capitalized* [Late Latin *gnosticus,* from Greek *gnōstikos* of knowledge, from *gignōskein*] (circa 1587)
: an adherent of gnosticism
— gnostic *adjective, often capitalized*

gnos·ti·cism \'näs-tə-ˌsi-zəm\ *noun, often capitalized* (1664)
: the thought and practice especially of various cults of late pre-Christian and early Christian centuries distinguished by the conviction that matter is evil and that emancipation comes through gnosis

gno·to·bi·ot·ic \ˌnō-tō-bī-'ä-tik, -bē-\ *adjective* [Greek *gnōtos* known (from *gignōskein* to know) + *biotē* life, way of life — more at KNOW, BIOTA] (1949)
: of, relating to, living in, or being a controlled environment containing one or a few kinds of organisms; *also* **:** AXENIC
— gno·to·bi·ot·i·cal·ly \-ti-k(ə-)lē\ *adverb*

gnu \'nü *also* 'nyü\ *noun, plural* **gnu** *or* **gnus** [Khoikhoi *t'gnu*] (1777)
: either of two large African antelopes (*Connochaetes gnou* and *C. taurinus*) with a head like that of an ox, short mane, long tail, and horns in both sexes that curve downward and outward

gnu

¹go \'gō\ *verb* **went** \'went\; **gone** \'gȯn *also* 'gän\; **go·ing** \'gō-iŋ, 'gȯ(-)iŋ; *"going to" in sense 13 is often* 'gō-nə *or* 'gȯ-nə *or* 'gə-nə\; **goes** \'gōz\ [Middle English *gon,* from Old English *gān;* akin to Old High German *gān* to go, Greek *kichanein* to reach, attain] (before 12th century)
intransitive verb
1 : to move on a course **:** PROCEED ⟨*go* slow⟩ ⟨*went* by train⟩ — compare STOP
2 : to move out of or away from a place expressed or implied **:** LEAVE, DEPART ⟨*went* from school to the party⟩ ⟨*going* away for vacation⟩
3 a : to take a certain course or follow a certain procedure ⟨reports *go* through channels to the president⟩ **b :** to pass by means of a process like journeying ⟨the message *went* by wire⟩ **c :** to proceed without delay and often in a thoughtless or reckless manner — used especially to intensify a complementary verb ⟨why did you *go* and spoil it⟩ ⟨*go* jump in a lake⟩ **d** (1) **:** to extend from point to point or in a certain direction ⟨the road *goes* to the lake⟩ (2) **:** to give access **:** LEAD ⟨that door *goes* to the cellar⟩
4 *obsolete* **:** WALK
5 : to be habitually in a certain state or condition ⟨*go* bareheaded⟩
6 a : to become lost, consumed, or spent ⟨our time has *gone*⟩ **b :** DIE **c :** to slip away **:** ELAPSE ⟨the evening *went* quickly⟩ **d :** to come to be given up or discarded ⟨these slums have to *go*⟩ **e :** to pass by sale ⟨*went* for a good price⟩ **f :** to become impaired or weakened ⟨his hearing started to *go*⟩ **g :** to give way especially under great force or pressure **:** BREAK ⟨the roof *went*⟩
7 a : to move along in a specified manner **:** FARE ⟨everything was *going* well⟩ **b :** to be in general or on an average ⟨cheap, as yachts *go*⟩ **c :** to be or become especially as the result of a contest ⟨the election *went* in her favor⟩ **d :** to turn out well **:** SUCCEED ⟨worked hard to make the party *go*⟩
8 a : to apply oneself ⟨*went* to work⟩ **b :** to put or subject oneself ⟨*went* to unnecessary expense⟩ **c** *chiefly Southern & Midland* **:** INTEND ⟨I didn't *go* to do it⟩
9 : to have recourse to another for corroboration, vindication, or decision **:** RESORT ⟨*go* to court to recover damages⟩
10 a : to begin an action or motion ⟨here *goes*⟩ **b :** to maintain or perform a certain action or motion ⟨still *going* strong⟩ **c :** to function in the proper or expected manner **:** RUN ⟨the motor won't *go*⟩
11 : to be known ⟨*goes* by an alias⟩
12 a : to act in accordance or harmony ⟨a good rule to *go* by⟩ **b :** to come to be determined ⟨dreams *go* by contraries⟩ **c :** to come to be applied or appropriated ⟨all proceeds *go* to charity⟩ **d :** to pass by award, assignment, or lot ⟨the prize *went* to a sophomore⟩ **e** (1) **:** to contribute to an end or result ⟨qualities that *go* to make a hero⟩ (2) **:** to be of advantage ⟨has a lot *going* for her⟩
13 : to be about, intending, or expecting something — used in a progressive tense before an infinitive ⟨is *going* to leave town⟩
14 a : EXTEND ⟨his knowledge fails to *go* very deep⟩ **b :** to come or arrive at a certain state or condition ⟨*go* to sleep⟩ **c :** to come to be **:** BECOME ⟨the tire *went* flat⟩ — often used to express conversion to specified values or a specified state ⟨*gone* Hollywood⟩ ⟨*go* condo⟩ **d :** to undergo a change ⟨leaves *go* from green to red⟩
15 a : to be in phrasing or expression **:** READ ⟨as the story *goes*⟩ **b :** to be capable of being sung or played ⟨the tune *goes* like this⟩

16 : to be compatible, suitable, or becoming **:** HARMONIZE ⟨the tie *goes* with his suit⟩

17 a : to be capable of passing, extending, or being contained or inserted ⟨will these clothes *go* in your suitcase⟩ **b :** to have a usual or proper place or position **:** BELONG ⟨these books *go* on the top shelf⟩

18 : to have a tendency **:** CONDUCE ⟨it *goes* to show⟩

19 a (1) **:** to carry authority ⟨what she said *went*⟩ (2) **:** to be acceptable, satisfactory, or adequate ⟨anything *goes* here⟩ **b :** to hold true **:** be valid ⟨the rule *goes* for you, too⟩

20 : to empty the bladder or bowels

transitive verb

1 : to proceed along or according to **:** FOLLOW ⟨if I were *going* his way⟩ ⟨*went* the conventional route⟩

2 : to travel through or along **:** TRAVERSE ⟨*went* the length of the street⟩

3 a : to make a wager of **:** BET ⟨*go* a dollar on the outcome⟩ **b :** to make an offer of **:** BID ⟨willing to *go* $50 for the clock⟩

4 a : to assume the function or obligation of ⟨promised to *go* bail for his friend⟩ **b :** to participate to the extent of ⟨decided to *go* halves on the winnings⟩

5 : YIELD, WEIGH ⟨this fish *goes* ten pounds⟩

6 a : to put up with **:** TOLERATE ⟨couldn't *go* the noise⟩ **b :** AFFORD ⟨can't *go* the price⟩ **c :** ENJOY ⟨I could *go* a soda⟩

7 a : to cause (a characteristic sound) to occur ⟨the gun *went* bang⟩ **b :** SAY — used chiefly in oral narration of speech

8 : to engage in ⟨don't *go* telling everyone⟩

9 *of a sports team or player* **:** to have a record of ⟨*went* 11-0 last season⟩

— **go·er** \'gō(-ə)r\ *noun*

— **go about :** to set about

— **go after :** to try to get **:** SEEK

— **go all the way 1 :** to enter into complete agreement **2 :** to engage in sexual intercourse

— **go at 1 a :** to make an attack on **b :** to make an approach to **2 :** UNDERTAKE

— **go back on 1 :** ABANDON **2 :** BETRAY **3 :** FAIL

— **go begging :** to be in little demand

— **go by the board 1 :** to be carried over a ship's side **2 :** to be discarded

— **go down the line :** to give wholehearted support

— **go easy :** to be sparing ⟨*go easy* with the sugar⟩ ⟨*go easy* on the kid⟩

— **go fly a kite :** to stop being an annoyance or disturbance ⟨told him to *go fly a kite*⟩

— **go for 1 :** to pass for or serve as **2 :** to try to secure or attain (as a goal) ⟨*go for* the prize⟩ **3 a :** FAVOR, ACCEPT ⟨cannot *go for* your idea⟩ **b :** to have an interest in or liking for ⟨she *went for* him in a big way —Chandler Brossard⟩ **4 :** ATTACK, ASSAIL ⟨my dog *went for* the intruder⟩

— **go for broke :** to put forth all one's strength or resources

— **go great guns :** to achieve great success

— **go hang :** to cease to be of interest or concern

— **go into :** to be contained in ⟨5 *goes into* 60 12 times⟩

— **go it 1 :** to behave in a reckless, excited, or impromptu manner **2 :** to proceed in a rapid or furious manner **3 :** to conduct one's affairs **:** ACT ⟨insists on *going it* alone⟩

— **go missing** *chiefly British* **:** to become lost **:** DISAPPEAR

— **go one better :** OUTDO, SURPASS

— **go over 1 :** EXAMINE **2 a :** REPEAT **b :** STUDY, REVIEW

— **go places :** to be on the way to success

— **go public 1** *of a close corporation* **:** to offer shares for sale to the general public **2 :** to make a public disclosure

— **go steady :** to date one person exclusively and frequently

— **go the vole :** to risk all for great gains

— **go through 1 :** to subject to thorough examination, consideration, or study **2 :** EXPERIENCE, UNDERGO **3 :** CARRY OUT, PERFORM ⟨*went through* his work in a daze⟩

— **go to bat for :** to give active support or assistance to **:** DEFEND, CHAMPION

— **go to bed with :** to have sexual intercourse with

— **go to one's head 1 :** to cause one to become confused, excited, or dizzy **2 :** to cause one to become conceited or overconfident

— **go to pieces :** to become shattered (as in nerves or health)

— **go to the mat :** to make an all-out combative effort (as in support of a position)

— **go to town 1 :** to work or act rapidly or efficiently **2 :** to be markedly successful **3 :** to indulge oneself excessively

— **go with 1 :** DATE **2 :** CHOOSE 2 ⟨*went with* an iron off the tee⟩

— **go without saying :** to be self-evident

— **go with the flow :** CONFORM 2b

— **to go 1 :** still remaining ⟨ten minutes *to go*⟩ **2** *of prepared food* **:** sold for consumption off the premises

[2]go \'gō\ *noun, plural* **goes** (1727)

1 : the act or manner of going

2 : the height of fashion **:** RAGE ⟨elegant shawls labeled . . . "quite the *go*" —R. S. Surtees⟩

3 : an often unexpected turn of affairs **:** OCCURRENCE

4 : the quantity used or furnished at one time ⟨you can obtain a *go* of brandy for sixpence —C. B. Fairbanks⟩

5 : ENERGY, VIGOR

6 a : a turn in an activity (as a game) ⟨it's your *go*⟩ **b :** ATTEMPT, TRY ⟨have a *go* at painting⟩

7 : a spell of activity ⟨finished the job at one *go*⟩

8 : SUCCESS ⟨made a *go* of the business⟩

9 : permission to proceed **:** GO-AHEAD ⟨gave the astronauts a *go* for another orbit⟩

— **no go :** to no avail **:** USELESS

— **on the go :** constantly or restlessly active

[3]go *adjective* (1961)

: functioning properly **:** being in good and ready condition ⟨declared all systems *go*⟩

[4]go *noun, often capitalized* [Japanese] (1890)

: a Japanese game played between two players who alternately place black and white stones on a board checkered by 19 vertical lines and 19 horizontal lines in an attempt to enclose the larger area on the board

[1]goad \'gōd\ *noun* [Middle English *gode,* from Old English *gād* spear, goad; akin to Langobardic *gaida* spear, and perhaps to Sanskrit *hinoti* he urges on] (before 12th century)

1 a : something that pricks like a goad **:** THORN **b :** something that urges or stimulates into action **:** SPUR

2 : a pointed rod used to urge on an animal

synonym see MOTIVE

[2]goad *transitive verb* (1579)

1 : to incite or rouse as if with a goad

2 : to drive (as cattle) with a goad

[1]go–ahead \'gō-ə-ˌhed\ *adjective* (1834)

1 : marked by energy and enterprise **:** PROGRESSIVE ⟨a vigorous *go-ahead* company⟩

2 : indicating that one may proceed ⟨*go-ahead* signal⟩

3 : being a score that gives a team the lead in a game ⟨drove in the *go-ahead* run⟩

[2]go–ahead *noun* (1840)

1 a : ENERGY, SPIRIT ⟨had a great deal of courage and *go-ahead*⟩ **b :** one possessing go-ahead

2 : a sign, signal, or authority to proceed **:** GREEN LIGHT

goal \'gōl, *chiefly Northern especially in 1b & 3a also* 'gül\ *noun* [Middle English *gol* boundary, limit] (1531)

1 a : the terminal point of a race **b :** an area to be reached safely in children's games

2 : the end toward which effort is directed **:** AIM

3 a : an area or object toward which players in various games attempt to advance a ball or puck and usually through or into which it must go to score points **b :** the act or action of causing a ball or puck to go through or into such a goal **c :** the score resulting from such an act

synonym see INTENTION

— **goal** *intransitive verb*

— **goal·less** \'gō(l)-ləs\ *adjective*

goal·ie \'gō-lē\ *noun* (1921)

: GOALKEEPER

goal·keep·er \'gōl-ˌkē-pər\ *noun* (1658)

: a player who defends the goal in any of various games (as hockey, lacrosse, or soccer)

goal kick *noun* (1891)

: a free kick in soccer awarded to a defensive player when the ball is driven out of bounds over the end line by an opposing player

goal line *noun* (1867)

: a line at either end and usually running the width of a playing area on which a goal or goalpost is situated

goal·mouth \'gōl-ˌmau̇th\ *noun* (1882)

: the area directly in front of the goal (as in soccer or hockey)

go along *intransitive verb* (1602)

1 : to move along **:** PROCEED

2 : to go or travel as a companion

3 : to act in cooperation or express agreement ⟨*go along* with the crowd⟩

goal·post \'gōl-ˌpōst\ *noun* (1857)

: one of usually two vertical posts that with or without a crossbar constitute the goal in various games

goal·tend·er \'gōl-ˌten-dər\ *noun* (circa 1909)

: GOALKEEPER

goal·tend·ing \-diŋ\ *noun* (1968)

1 : the act of guarding a goal (as in hockey)

2 : a violation in basketball that involves touching or deflecting a ball that is on its downward path toward the basket or on or within the rim of the basket

goal·ward \'gōl-wərd\ *adverb or adjective* (1949)

: toward a goal

go·an·na \gō-'a-nə\ *noun* [alteration of *iguana*] (1831)

: any of several large Australian monitor lizards (genus *Varanus* of the family Varanidae)

go–around \'gō-ə-ˌrau̇nd\ *noun* (circa 1929)

1 : RUNAROUND ⟨gave me the *go-around*⟩

2 a : ROUND ⟨reached an agreement during the first *go-around*⟩ **b :** a heated argument or struggle ⟨had a real *go-around* with her about it⟩

3 : an act or instance of going around (as in an air traffic pattern)

go around *intransitive verb* (circa 1520)

1 a : to pass from place to place **:** go here and there **b :** to have currency **:** CIRCULATE ⟨an amusing story is *going around*⟩

2 : to satisfy demand **:** fill the need ⟨not enough jobs to *go around*⟩

goat \'gōt\ *noun, plural* **goats** [Middle English *gote,* from Old English *gāt;* akin to Old High German *geiz* goat, Old Norse *geit,* Latin *haedus* kid] (before 12th century)

1 a *or plural* **goat :** any of various hollow-horned ruminant mammals (especially of the genus *Capra*) related to the sheep but of lighter build and with backwardly arching horns, a short tail, and usually straight hair **b** *capitalized* **:** CAPRICORN

2 : a licentious man **:** LECHER

3 : SCAPEGOAT 2

— **goat·ish** \'gō-tish\ *adjective*

— **goat·like** \-ˌlīk\ *adjective*
goat cheese *noun* (1893)
: any of various cheeses made from goat's milk
goa·tee \gō-ˈtē\ *noun* [from its resemblance to the beard of a he-goat] (1844)
: a small pointed or tufted beard on a man's chin
— **goa·teed** \-ˈtēd\ *adjective*
goat·fish \ˈgōt-ˌfish\ *noun* (circa 1639)
: any of a family (Mullidae) of moderate-sized usually red or golden fishes with two barbels on the chin — called also *red mullet*
goat·herd \-ˌhərd\ *noun* [Middle English *goteherd,* from Old English *gāthyrd*] (before 12th century)
: a person who tends goats
goat·skin \-ˌskin\ *noun* (14th century)
1 : the skin of a goat
2 : leather made from goatskin
goat·suck·er \-ˌsə-kər\ *noun* (1611)
: NIGHTJAR
[1]**gob** \ˈgäb\ *noun* [Middle English *gobbe,* from Middle French *gobe* large piece of food, back-formation from *gobet*] (14th century)
1 : LUMP
2 : a large amount — usually used in plural ⟨*gobs* of money⟩
[2]**gob** \ˈgȯb, ˈgäb\ *noun* [probably from Scottish Gaelic, mouth & Irish, beak, pursed mouth] (circa 1550)
chiefly British **:** MOUTH
[3]**gob** \ˈgäb, ˈgȯb\ *noun* [origin unknown] (1915)
: SAILOR
gob·bet \ˈgä-bət\ *noun* [Middle English *gobet,* from Middle French, mouthful, piece] (14th century)
1 : a piece or portion (as of meat)
2 : LUMP, MASS
3 : a small fragment or extract ⟨a *gobbet* of information⟩
4 : a small quantity of liquid **:** DROP
[1]**gob·ble** \ˈgä-bəl\ *transitive verb* **gob·bled; gob·bling** \-b(ə-)liŋ\ [probably irregular from [1]*gob*] (1601)
1 : to swallow or eat greedily
2 : to take eagerly — usually used with *up*
3 : to read rapidly or greedily — usually used with *up*
[2]**gobble** *intransitive verb* **gob·bled; gob·bling** \-b(ə-)liŋ\ [imitative] (1680)
1 : to make the natural guttural noise of a male turkey
2 : to make a sound resembling a gobble
— **gobble** *noun*
gob·ble·dy·gook *also* **gob·ble·de·gook** \ˈgä-bəl-dē-ˌgu̇k, -ˌgük\ *noun* [irregular from *gobble,* noun] (1944)
: wordy and generally unintelligible jargon
[1]**gob·bler** \ˈgä-blər\ *noun* (circa 1737)
: a male turkey
[2]**gobbler** *noun* (circa 1755)
: one that gobbles
Go·be·lin \ˈgō-bə-lən, ˌgō-bə-ˈla^{n}\ *adjective* [*Gobelin* dye and tapestry works, Paris, France] (1788)
: of, relating to, or characteristic of tapestry produced at the Gobelin works in Paris
— **Gobelin** *noun*
go–be·tween \ˈgō-bə-ˌtwēn\ *noun* (1598)
: an intermediate agent **:** BROKER
gob·let \ˈgäb-lət\ *noun* [Middle English *gobelet,* from Middle French] (14th century)
1 *archaic* **:** a bowl-shaped drinking vessel without handles
2 : a drinking vessel (as of glass) with a foot and stem — compare TUMBLER
goblet cell *noun* [from its shape] (1878)
: a mucus-secreting epithelial cell (as of intestinal columnar epithelium) that is distended at the free end
gob·lin \ˈgäb-lən\ *noun* [Middle English *gobelin,* from Middle French, from Medieval Latin *gobelinus,* ultimately from Greek *kobalos* rogue] (14th century)
: an ugly or grotesque sprite that is mischievous and sometimes evil and malicious
go·bo \ˈgō-(ˌ)bō\ *noun, plural* **gobos** *also* **goboes** [origin unknown] (circa 1930)
1 : a dark strip (as of wallboard) to shield a motion-picture or television camera from light
2 : a device to shield a microphone from sound
go·by \ˈgō-bē\ *noun, plural* **gobies** *also* **goby** [Latin *gobius* gudgeon, from Greek *kōbios*] (1769)
: any of numerous spiny-finned fishes (family Gobiidae) that usually have the pelvic fins united to form a ventral sucking disk

goby

go by *intransitive verb* (1508)
: PASS 3b ⟨as time *goes by*⟩
go–cart \ˈgō-ˌkärt\ *noun* (1689)
1 a : WALKER **b :** STROLLER
2 : HANDCART
3 : a light open carriage
[1]**god** \ˈgäd *also* ˈgȯd\ *noun* [Middle English, from Old English; akin to Old High German *got* god] (before 12th century)
1 *capitalized* **:** the supreme or ultimate reality: as **a :** the Being perfect in power, wisdom, and goodness who is worshipped as creator and ruler of the universe **b** *Christian Science* **:** the incorporeal divine Principle ruling over all as eternal Spirit **:** infinite Mind
2 : a being or object believed to have more than natural attributes and powers and to require human worship; *specifically* **:** one controlling a particular aspect or part of reality
3 : a person or thing of supreme value
4 : a powerful ruler
[2]**god** *transitive verb* **god·ded; god·ding** (1595)
: to treat as a god **:** IDOLIZE, DEIFY
god–aw·ful \gä-ˈdȯ-fəl\ *adjective* [*god*damned + *awful*] (1878)
: extremely unpleasant or disagreeable **:** ABOMINABLE ⟨*god-awful* weather⟩
god·child \ˈgäd-ˌchīld *also* ˈgȯd-\ *noun* (13th century)
: a person for whom another person becomes sponsor at baptism
[1]**god·damn** *or* **god·dam** \ˈgä(d)-ˈdam\ *noun, often capitalized* (1640)
: DAMN ⟨they were in no mood to give a good *goddamn* about anything —Robert Lowry⟩
[2]**goddamn** *or* **goddam** *verb, often capitalized* (1928)
: DAMN
god·damned \ˌgä(d)-ˌdamd\ *or* **god·damn** *or* **god·dam** \-ˌdam\ *adjective or adverb* (1918)
: DAMNED
god·daugh·ter \ˈgäd-ˌdȯ-tər *also* ˈgȯd-\ *noun* (before 12th century)
: a female godchild
god·dess \ˈgä-dəs *also* ˈgȯ-\ *noun* (14th century)
1 : a female god
2 : a woman whose great charm or beauty arouses adoration
usage see -ESS
Go·del's theorem \ˈgō-dəlz-, ˈgər-, ˈgœ-\ *noun* [Kurt *Gödel* (died 1978) American mathematician] (1933)
: a theorem in advanced logic: in any logical system as complex as or more complex than the arithmetic of the integers there can always be found either a statement which can be shown to be both true and false or a statement whose truth or falsity cannot be deduced from other statements in the system — called also *Godel's incompleteness theorem*
go·det \gō-ˈdet, ˈgō-ˌdā\ *noun, often attributive* [French, from *goder* to form creases, probably from *goderon* rounded pleat on a ruffle, gadroon] (1872)
: an inset of cloth placed in a seam to give fullness (as at the bottom of a skirt)
go–dev·il \ˈgō-ˌde-vəl\ *noun* (1852)
: any of various devices: as **a :** a cultivator with wooden runners **b :** a weight formerly dropped in a bored hole (as of an oil well) to set off an explosive **c :** a cleaning scraper propelled through a pipeline **d :** a handcar or small gasoline car used on a railroad **e :** a child's sled
[1]**god·fa·ther** \ˈgäd-ˌfä-t͟hər, -ˌfà-, *also* ˈgȯd-\ *noun* (before 12th century)
1 : a man who sponsors a person at baptism
2 : one having a relation to someone or something analogous to that of a male sponsor to his godchild: as **a :** one that founds, supports, or inspires ⟨made him the *godfather* of a whole generation of rebels —*Times Literary Supplement*⟩ **b :** the leader of an organized crime syndicate
[2]**godfather** *transitive verb* (1780)
: to act as godfather to
God–fear·ing \-ˌfir-iŋ\ *adjective* (1835)
: having a reverent feeling toward God **:** DEVOUT
god·for·sak·en \-fər-ˌsā-kən\ *adjective* (1860)
1 : REMOTE, DESOLATE
2 : neglected and miserable in appearance or circumstances
god·head \-ˌhed\ *noun* [Middle English *godhed,* from *god* + *-hed* -hood; akin to Middle English *-hod* -hood] (13th century)
1 : divine nature or essence **:** DIVINITY
2 *capitalized* **a :** GOD 1 **b :** the nature of God especially as existing in three persons — used with *the*
god·hood \-ˌhu̇d\ *noun* [Middle English *godhod,* from Old English *godhād,* from *god* + *-hād* -hood] (1563)
: DIVINITY
Go·di·va \gə-ˈdī-və\ *noun*
: an English earl's wife who in legend rode naked through Coventry to save its citizens from a tax
god·less \ˈgäd-ləs *also* ˈgȯd-\ *adjective* (1528)
: not acknowledging a deity or divine law
— **god·less·ness** *noun*
god·like \-ˌlīk\ *adjective* (before 12th century)
: resembling or having the qualities of God or a god **:** DIVINE
— **god·like·ness** *noun*
god·ling \-liŋ\ *noun* (circa 1500)
: an inferior or local god
god·ly \-lē\ *adjective* **god·li·er; -est** (14th century)
1 : DIVINE
2 : PIOUS, DEVOUT
— **god·li·ness** *noun*
— **godly** *adverb, archaic*
god·moth·er \-ˌmə-t͟hər\ *noun* (before 12th century)
: a woman who sponsors a person at baptism
go·down \ˈgō-ˌdau̇n\ *noun* [by folk etymology from Malay *gudang*] (1552)
: a warehouse in an oriental country
go down *intransitive verb* (14th century)
1 a : to go below the horizon **:** SET ⟨the sun *went down*⟩ **b :** to fall to or as if to the ground ⟨the plane *went down* in flames⟩ **c :** to become submerged ⟨the ship *went down* with all hands⟩
2 : to admit of being swallowed ⟨the medicine *went down* easily⟩
3 a : to find acceptance ⟨will the plan *go down* with the farmers⟩ **b :** to come to be remembered especially in posterity ⟨will he *go down* in history as a great president⟩
4 *British* **:** to leave a university
5 a : to undergo defeat or failure **b** *chiefly British* **:** to become incapacitated ⟨*went down* with . . . acute tonsillitis —Helen Cathcart⟩
6 *slang* **:** to take place **:** HAPPEN
— **go down on :** to perform fellatio or cunnilingus on
god·par·ent \ˈgäd-ˌpar-ənt, -ˌper- *also* ˈgȯd-\ *noun* (1865)

: a sponsor at baptism

God's acre *noun* (1617)
: CHURCHYARD

god·send \'gäd-ˌsend *also* 'gȯd-\ *noun* [back-formation from *god-sent*] (1820)
: a desirable or needed thing or event that comes unexpectedly

god·son \-ˌsən\ *noun* (before 12th century)
: a male godchild

God·speed \-'spēd\ *noun* [Middle English *god speid,* from the phrase *God spede you* God prosper you] (15th century)
: a prosperous journey **:** SUCCESS ⟨bade him *Godspeed*⟩

god·wit \'gäd-ˌwit\ *noun* [origin unknown] (1552)
: any of a genus (*Limosa*) of wading birds that are related to the curlews and sandpipers and have a long slender slightly upturned or straight bill

goes *present 3d singular of* GO, *plural of* GO

goe·thite \'gə(r)-ˌtīt\ *noun* [German *Göthit,* from J. W. von *Goethe*] (circa 1823)
: a mineral that consists of an iron hydrogen oxide and is the commonest constituent of many forms of natural rust

go·fer \'gō-fər\ *noun* [alteration of *go for*] (1967)
: an employee whose duties include running errands

gof·fer \'gä-fər, 'gȯ-\ *transitive verb* [French *gaufrer,* from *gaufre* honeycomb, waffle, from Old French, of Germanic origin; akin to Middle Dutch *wafel* waffle] (1706)
: to crimp, plait, or flute (as linen or lace) especially with a heated iron
— goffer *noun*

go–get·ter \'gō-'ge-tər, -ˌge-\ *noun* (1921)
: an aggressively enterprising person
— go–get·ting \-ˌge-tiŋ\ *adjective or noun*

¹gog·gle \'gä-gəl\ *intransitive verb* **gog·gled; gog·gling** [Middle English *gogelen* to squint] (1616)
: to stare with wide or protuberant eyes
— gog·gler \'gä-gə-lər\ *noun*

²goggle *adjective* (1540)
: PROTUBERANT, STARING ⟨*goggle* eyes⟩
— gog·gly \'gä-gə-lē\ *adjective*

gog·gle–eye \'gä-gəl-ˌī\ *noun* (1840)
1 : ROCK BASS
2 : WARMOUTH

gog·gle–eyed \-ˌīd\ *adjective* (1711)
: having bulging or rolling eyes

gog·gles \'gä-gəlz\ *noun plural* (1715)
: protective glasses set in a flexible frame (as of rubber or plastic) that fits snugly against the face
— gog·gled \-gəld\ *adjective*

go–go \'gō-(ˌ)gō\ *adjective* [partly from *a-go-go,* partly from reduplication of ¹*go*] (1965)
1 a : employed to entertain in a disco ⟨*go-go* dancers⟩ **b :** of, relating to, or being a disco or the music or dances performed there
2 : marked by spirited or aggressive action ⟨*go-go* baseball⟩
3 a : relating to or dealing in popular often speculative investment expected to yield high returns ⟨*go-go* mutual funds⟩ **b :** relating to, involved in, or marked by business growth and prosperity and aggressive efforts to turn a quick profit ⟨*go-go* bankers and entrepreneurs . . . who put together the megabuck deals —Ken Auletta⟩

¹Goi·del·ic \gȯi-'de-lik\ *adjective* [Middle Irish *Goídel* Gael, Irishman] (1896)
1 : of, relating to, or characteristic of the Gaels
2 : of, relating to, or constituting Goidelic

²Goidelic *noun* (1882)
: the branch of the Celtic languages that includes Irish, Scottish Gaelic, and Manx — see INDO-EUROPEAN LANGUAGES table

go in *intransitive verb* (1812)
1 : to make an approach (as in attacking)
2 a : to take part in a game or contest **b :** to call the opening bet in poker **:** STAY
3 *of a celestial body* **:** to become obscured by a cloud
4 : to form a union or alliance **:** JOIN — often used with *with* ⟨asked the rest of us to *go in* with them⟩
— go in for 1 : to give support to **:** ADVOCATE **2 :** to have or show an interest in or a liking for **3 :** to engage in **:** take part in

¹go·ing \'gō-iŋ, 'gȯ(-)iŋ\ *noun* (14th century)
1 : an act or instance of going
2 *plural* **:** BEHAVIOR, ACTIONS ⟨for his eyes are upon the ways of man, and he seeth all his *goings* —Job 34:21 (Authorized Version)⟩
3 : the condition of the ground (as for walking)
4 : advance toward an objective ⟨when the *going* gets tough⟩

²going *adjective* (14th century)
1 a : that goes — often used in combination ⟨easy*going*⟩ ⟨out*going*⟩ **b :** WORKING, MOVING ⟨everything was in *going* order⟩
2 : LIVING, EXISTING ⟨the best novelist *going*⟩
3 : CURRENT, PREVAILING ⟨*going* price⟩
4 : conducting business with the expectation of indefinite continuance ⟨*going* concern⟩
— going on : drawing near to **:** APPROACHING ⟨is six years old *going on* seven⟩

go·ing–over \ˌgō-iŋ-'ō-vər, ˌgȯ(-)iŋ-\ *noun, plural* **go·ings–over** (1872)
1 a : a severe scolding **b :** BEATING
2 : a thorough examination or investigation

go·ings–on \ˌgō-iŋ-'zȯn, ˌgȯ(-)iŋ-, -'zän\ *noun plural* (1775)
1 : ACTIONS, EVENTS
2 : irregular or reprehensible happenings or conduct ⟨titillating stories about the *goings-on* of the carefree millionaires —Eleanor Early⟩

goi·ter \'gȯi-tər\ *noun* [French *goitre,* from Middle French, back formation from *goitron* throat, from (assumed) Vulgar Latin *guttrion-, guttrio,* from Latin *guttur*] (1625)
: an enlargement of the thyroid gland visible as a swelling of the front of the neck — compare HYPERTHYROIDISM, HYPOTHYROIDISM
— goi·trous \'gȯi-trəs, 'gȯi-tə rəs\ *adjective*

goi·tre *chiefly British variant of* GOITER

goi·tro·gen \'gȯi-trə-jən\ *noun* [*goiter* + *-o-* + *-gen*] (1946)
: a substance (as thiourea or thiouracil) that induces goiter formation

goi·tro·gen·ic \ˌgȯi-trə-'je-nik\ *adjective* (1929)
: producing or tending to produce goiter
— goi·tro·ge·nic·i·ty \ˌgȯi-trə-jə-'ni-sə-tē\ *noun*

Gol·con·da \gäl-'kän-də\ *noun* [*Golconda,* India, famous for its diamonds] (1884)
: a rich mine; *broadly* **:** a source of great wealth

gold \'gōld\ *noun, often attributive* [Middle English, from Old English; akin to Old High German *gold* gold, Old English *geolu* yellow — more at YELLOW] (before 12th century)
1 : a malleable ductile yellow metallic element that occurs chiefly free or in a few minerals and is used especially in coins, jewelry, and dentures — see ELEMENT table
2 a (1) **:** gold coins (2) **:** a gold piece **b :** MONEY **c :** GOLD STANDARD
3 : a variable color averaging deep yellow
4 : something resembling gold; *especially* **:** something valued as the finest of its kind ⟨a heart of *gold*⟩
5 : a medal awarded as the top prize in a competition **:** a gold medal

¹gold·brick \'gōl(d)-ˌbrik\ *noun* (1881)
1 a : a worthless brick that appears to be of gold **b :** something that appears to be valuable but is actually worthless
2 : a person who shirks assigned work

²goldbrick (1902)
transitive verb
: SWINDLE
intransitive verb
: to shirk duty or responsibility

gold·bug \-ˌbəg\ *noun* (1878)
1 : a supporter of the gold standard
2 : a person who invests in or hoards gold

gold coast *noun, often G&C capitalized* (1877)
: an exclusive residential district

gold digger *noun* (1830)
1 : one who digs for gold
2 : a person who uses charm to extract money or gifts from others

gold·en \'gōl-dən\ *adjective* [Middle English] (13th century)
1 : consisting of, relating to, or containing gold
2 a : being or having the color gold or the color of gold **b :** BLOND 1a
3 : LUSTROUS, SHINING
4 : of a high degree of excellence **:** SUPERB
5 : PROSPEROUS, FLOURISHING ⟨*golden* days⟩
6 a : radiantly youthful and vigorous **b :** possessing talents that promise worldly success — often used with *boy* **c :** highly favored **:** POPULAR
7 : FAVORABLE, ADVANTAGEOUS ⟨a *golden* opportunity⟩
8 : of, relating to, or being a 50th anniversary or its celebration
9 : MELLOW, RESONANT ⟨a smooth *golden* tenor⟩
— gold·en·ly *adverb*
— gold·en·ness \-də(n)-nəs\ *noun*

golden age *noun* (1555)
: a period of great happiness, prosperity, and achievement

gold·en–ag·er \'gōl-dən-'ā-jər\ *noun* (1961)
: an elderly and often retired person usually engaging in club activities

golden al·ex·an·ders \-ˌa-lig-'zan-dərz, -ˌe-lig-\ *noun plural but singular or plural in construction, often A capitalized* [Medieval Latin *alexandrum*] (circa 1923)
: a North American yellow-flowered perennial herb (*Zizia aurea*) of the carrot family occurring in moist woods and meadows; *also* **:** any of several related herbs

golden–brown alga *noun* (circa 1957)
: any of a division (Chrysophyta) of marine and freshwater algae (as diatoms) containing yellowish green to golden brown pigments that obscure the chlorophyll — called also *chrysophyte, golden alga*

golden club *noun* (1837)
: an American aquatic plant (*Orontium aquaticum*) of the arum family with a spadix of tiny yellow flowers

golden eagle *noun* (1839)
: a brown eagle (*Aquila chrysaetos*) of the northern hemisphere with gold-colored feathers on the back of the head and neck

golden eagle

gold·en·eye \'gōl-dən-ˌī\ *noun* (circa 1678)
1 : either of two diving ducks (genus *Bucephala*); *especially* **:** a large-headed swift-flying Holarctic diving duck (*B. clangula*) with the male having a green head and striking black and white markings
2 : a lacewing (family Chrysopidae) with yellow eyes

Golden Fleece *noun* (14th century)

: a fleece of gold placed by the king of Colchis in a dragon-guarded grove and recovered by the Argonauts

golden glow *noun* (1902)
: a tall branching composite herb (*Rudbeckia laciniata hortensia*) with showy yellow flower heads

golden hamster *noun* (1939)
: a small tawny hamster (*Mesocricetus auratus*) native to Asia Minor — called also *Syrian hamster*

golden hamster

golden handshake *noun* (1960)
: a generous severance agreement given especially as an inducement to early retirement

Golden Horde *noun* (1863)
: a body of Mongols that overran eastern Europe in the 13th century and dominated Russia until 1486

golden lion tamarin *noun* (1975)
: a marmoset (*Leontopithecus rosalia rosalia*) with a reddish gold coat and long mane that occurs in remnants of tropical forest in southeastern Brazil

golden mean *noun* (1587)
: the medium between extremes **:** MODERATION

golden nematode *noun* (1946)
: a small yellowish Old World nematode worm (*Globodera rostochiensis* synonym *Heterodera rostochiensis*) established locally as a pest of potatoes in eastern North America

golden oldie *noun* (1970)
: one that was a hit or favorite in the past

golden parachute *noun* (1981)
: a generous severance agreement for an executive in the event of a sudden dismissal (as because of a merger)

golden plover *noun* (1785)
: either of two gregarious plovers (genus *Pluvialis*); *especially* **:** one (*P. dominica*) that breeds in arctic America and Siberia and winters in Hawaii and the southern hemisphere

gold·en·rain tree \ˈgōl-dən-ˌrān-\ *noun* (1923)
: an Asian tree (*Koelreuteria paniculata* of the family Sapindaceae) that has a rounded crown with very long showy clusters of yellow flowers

golden retriever *noun* (1919)
: any of a breed of medium-sized retrievers having a flat moderately long golden coat

gold·en·rod \ˈgōl-dən-ˌräd\ *noun* (1568)
: any of numerous chiefly North American composite biennial or perennial plants (especially of the genus *Solidago*) with alternate leaves and heads of small yellow or sometimes white flowers often clustered in panicles

golden rule *noun* (1861)
1 : a guiding principle
2 *G&R capitalized* **:** a rule of ethical conduct referring to Matthew 7:12 and Luke 6:31: do to others as you would have them do to you

gold·en·seal \ˈgōl-dən-ˌsēl\ *noun* (1839)
: a perennial American herb (*Hydrastis canadensis*) of the buttercup family with large rounded leaves and a thick knotted yellow rootstock sometimes used in pharmacy

golden section *noun* (1875)
: a proportion (as one involving a line divided into two segments or the length and width of a rectangle and their sum) in which the ratio of the whole to the larger part is the same as the ratio of the larger part to the smaller

golden shiner *noun* (circa 1889)
: a common cyprinid fish (*Notemigonus crysoleucas*) of eastern North America having silvery sides with bright golden reflections

golden syrup *noun* (1860)
chiefly British **:** TREACLE 2b

gold·field \ˈgōl(d)-ˌfēld\ *noun* (1851)
: a gold-mining district

gold–filled \-ˈfild\ *adjective* (circa 1903)
: consisting of a base metal covered with a layer of gold ⟨a *gold-filled* bracelet⟩

gold·finch \-ˌfinch\ *noun* (before 12th century)
1 : a small Palearctic finch (*Carduelis carduelis*) with a red, white, and black head and yellow and black wings
2 : any of three small American finches (genus *Carduelis* and especially *C. tristis*) having the breeding plumage of the male variably yellow with black markings on the wings, tail, and crown

gold·fish \-ˌfish\ *noun* (1791)
: a small usually golden yellow or orange cyprinid fish (*Carassius auratus*) often kept as an aquarium and pond fish

goldfish bowl *noun* (1935)
: a place or situation offering no privacy

gold leaf *noun* (circa 1741)
: an extremely thin sheet of gold that is used especially for gilding

gold mine *noun* (1833)
: a rich source of something desired

gold of pleasure (1597)
: a European herb (*Camelina sativa*) of the mustard family that is cultivated for its oil-rich seeds

gold rush *noun* (1876)
1 : a rush to newly discovered goldfields in pursuit of riches
2 : the headlong pursuit of sudden wealth in a new or lucrative field
— gold rush·er \-ˈrə-shər\ *noun*

gold·smith \ˈgōl(d)-ˌsmith\ *noun* (before 12th century)
: one who makes or deals in articles of gold

gold standard *noun* (1831)
: a monetary standard under which the basic unit of currency is defined by a stated quantity of gold and which is usually characterized by the coinage and circulation of gold, unrestricted convertibility of other money into gold, and the free export and import of gold for the settlement of international obligations

gold·stone \ˈgōl(d)-ˌstōn\ *noun* (circa 1889)
: aventurine glass spangled close and fine with particles of gold-colored material

go·lem \ˈgō-ləm, ˈgȯi-, ˈgā-\ *noun* [Yiddish *goylem,* from Hebrew *gōlem* shapeless mass] (1897)
1 : an artificial human being in Hebrew folklore endowed with life
2 : something or someone resembling a golem: as **a :** AUTOMATON **b :** BLOCKHEAD

golf \ˈgälf, ˈgȯlf, ˈgäf, ˈgȯf *sometimes* ˈgəlf\ *noun, often attributive* [Middle English (Scots)] (15th century)
: a game in which a player using special clubs attempts to sink a ball with as few strokes as possible into each of the 9 or 18 successive holes on a course ■
— golf *intransitive verb*
— golf·er *noun*

Golf (1952)
— a communications code word for the letter *g*

golf ball *noun* (1545)
1 : a small hard dimpled ball used in golf
2 : the spherical printing element of an electric typewriter or printer

golf cart *noun* (1899)
1 : a small cart for wheeling a golf bag around a golf course
2 : a motorized cart for carrying golfers and their equipment over a golf course — called also *golf car*

golf course *noun* (1890)
: an area of land laid out for golf with a series of 9 or 18 holes each including tee, fairway, and putting green and often one or more natural or artificial hazards — called also *golf links*

Gol·gi \ˈgȯl-(ˌ)jē\ *adjective* (1891)
: of or relating to the Golgi apparatus, Golgi bodies, or a method of staining nerve tissue ⟨*Golgi* vesicles⟩

Golgi apparatus *noun* [Camillo *Golgi*] (1916)
: a cytoplasmic organelle that consists of a stack of smooth membranous saccules and associated vesicles and that is active in the modification and transport of proteins — called also *Golgi complex;* see CELL illustration

Golgi body *noun* (1925)
: GOLGI APPARATUS; *also* **:** DICTYOSOME

go·liard \ˈgōl-yərd, -ˌyärd\ *noun* [Middle English, from Middle French, goliard, glutton, from *gole* throat, from Latin *gula* — more at GLUTTON] (15th century)
: a wandering student of the 12th or 13th century given to the writing of satiric Latin verse and to convivial living and minstrelsy
— go·liar·dic \gōl-ˈyär-dik\ *adjective*

Go·li·ath \gə-ˈlī-əth\ *noun* [Hebrew *Golyath*]
1 : a Philistine champion who in I Samuel 17 is killed by David
2 : GIANT

gol·li·wog *also* **gol·ly·wog** *or* **gol·li·wogg** \ˈgä-lē-ˌwäg\ *noun* [*Golliwogg,* an animated doll in children's fiction by Bertha Upton (died 1912) American writer] (1895)
1 : a grotesque black doll
2 : a person resembling a golliwog

gol·ly \ˈgä-lē\ *interjection* [euphemism for *God*] (1775)
— used as a mild oath or to express surprise

Go·mor·rah \gə-ˈmȯr-ə\ *noun* [from *Gomorrah,* ancient city destroyed by God for its wickedness in Genesis 19] (1907)
: a place notorious for vice and corruption

gon- *or* **gono-** *combining form* [Greek, from *gonos* procreation, seed, from *gignesthai* to be born — more at KIN]
: sexual **:** generative **:** semen **:** seed ⟨*gono*cyte⟩

-gon *noun combining form* [New Latin *-gonum,* from Greek *-gōnon,* from *gōnia* angle; akin to Greek *gony* knee — more at KNEE]
: figure having (so many) angles ⟨deca*gon*⟩

go·nad \ˈgō-ˌnad\ *noun* [New Latin *gonad-, gonas,* from Greek *gonos*] (1880)
: a reproductive gland (as an ovary or testis) that produces gametes
— go·nad·al \gō-ˈna-dᵊl\ *adjective*

go·nad·ec·to·my \ˌgō-nə-ˈdek-tə-mē\ *noun, plural* **-mies** (1915)
: surgical removal of an ovary or testis
— go·nad·ec·to·mized \-ˌmīzd\ *adjective*

go·nad·o·trop·ic \gō-ˌna-də-ˈträ-pik\ *also* **go·nad·o·tro·phic** \-ˈtrō-fik, -ˈträ-\ *adjective* (circa 1923)
: acting on or stimulating the gonads

go·nad·o·tro·pin \-ˈtrō-pən\ *also* **go·nad·o·tro·phin** \-fən\ *noun* (1931)
: a gonadotropic hormone (as follicle-stimulating hormone)

Gond \ˈgänd\ *noun* (1801)
: a member of a Dravidian or pre-Dravidian people of central India

Gondi \ˈgän-dē\ *noun* (1855)

□ USAGE
golf Like *assault* and *fault, golf* went through the earliest phase of its life as a word with no *l* either in spelling or in pronunciation. Though one no longer hears *assault* and *fault* sounded without an \l\, *golf* is still pronounced without it by some. The pronunciation without \l\ has been proscribed by some critics as "old-fashioned." The forms with an audible \l\ are more common, but no one should be penalized for using the older form without \l\.

: the Dravidian language of the Gonds

gon·do·la \'gän-də-lə (*usual for sense 1*), gän-'dō-\ *noun* [Italian dialect (Venetian), probably from Middle Greek *kontoura* small vessel] (1549)
1 : a long narrow flat-bottomed boat with a high prow and stern used on the canals of Venice
2 : a heavy flat-bottomed boat used on New England rivers and on the Ohio and Mississippi rivers
3 : a railroad car with no top, a flat bottom, and fixed sides that is used chiefly for hauling heavy bulk commodities
4 a : an elongated car attached to the underside of an airship **b :** an often spherical airtight enclosure suspended from a balloon for carrying passengers or instruments **c :** an enclosed car suspended from a cable and used for transporting passengers; *especially* **:** one used as a ski lift

gondola 1

gon·do·lier \,gän-də-'lir\ *noun* (1603)
: one who propels a gondola

gone \'gȯn *also* 'gän\ *adjective* [from past participle of *go*] (1598)
1 a : DEAD **b :** LOST, RUINED ⟨lost looks and *gone* faculties —Penelope Gilliatt⟩ **c :** characterized by sinking or dropping ⟨the empty or *gone* feeling in the abdomen so common in elevators —H. G. Armstrong⟩
2 a : INVOLVED, ABSORBED ⟨far *gone* in hysteria⟩ **b :** possessed with a strong attachment or a foolish or unreasoning love or desire **:** INFATUATED — often used with *on* ⟨was real *gone* on that man —Pete Martin⟩ **c :** PREGNANT ⟨she's six months *gone*⟩
3 : PAST ⟨memories of *gone* summers —John Cheever⟩
4 *slang* **:** GREAT ⟨a real *gone* fashion reporter —Inez Robb⟩

G_1 phase \,jē-'wən-\ *noun* [growth] (1966)
: the period in the cell cycle from the end of cell division to the beginning of DNA replication — compare G_2 PHASE, M PHASE, S PHASE

gon·er \'gȯ-nər *also* 'gä-nər\ *noun* (1854)
: one whose case is hopeless

gon·fa·lon \'gän-fə-,län, -lən\ *noun* [Italian *gonfalone*] (1595)
1 : the ensign of certain princes or states (as the medieval republics of Italy)
2 : a flag that hangs from a crosspiece or frame

gong \'gäŋ, 'gȯŋ\ *noun* [Malay & Javanese, of imitative origin] (circa 1590)
1 : a disk-shaped percussion instrument that produces a resounding tone when struck with a usually padded hammer
2 a : a saucer-shaped bell (as in a fire alarm) that is struck by a mechanical hammer **b :** a wire rod wound in a flat spiral for sounding the time or chime or alarm (as in a clock)
3 *British* **:** MEDAL
— gong *intransitive verb*

Gon·go·rism \'gäŋ-gə-,ri-zəm\ *noun* [Spanish *gongorismo*, from Luis de *Góngora* y Argote (died 1627) Spanish poet] (1813)
: a literary style characterized by studied obscurity and by the use of various ornate devices
— gon·go·ris·tic \,gäŋ-gə-'ris-tik\ *adjective*

go·nid·i·um \gō-'ni-dē-əm\ *noun, plural* **-ia** \-dē-ə\ [New Latin, from *gon-* + *-idium*] (1882)
: an asexual reproductive cell or group of cells especially in algae (as volvox)

gon·if, gon·iff \'gä-nəf\ *variant of* GANEF

go·ni·om·e·ter \,gō-nē-'ä-mə-tər\ *noun* [Greek *gōnia* angle] (1766)
1 : an instrument for measuring angles
2 : DIRECTION FINDER
— go·nio·met·ric \-nē-ə-'me-trik\ *adjective*
— go·ni·om·e·try \-nē-'ä-mə-trē\ *noun*

-gonium *combining form* [New Latin, from Greek *gonos*]
1 : germ cell ⟨spermato*gonium*⟩
2 : reproductive structure of a plant or fungus ⟨oo*gonium*⟩

gon·o·coc·cus \,gä-nə-'kä-kəs\ *noun, plural* **-coc·ci** \-'käk-,sī, -(,)sē; -'kä-,kī, -(,)kē\ [New Latin] (1889)
: a pus-producing bacterium (*Neisseria gonorrhoeae*) that causes gonorrhea
— gon·o·coc·cal \-'kä-kəl\ *adjective*

gon·o·cyte \'gä-nə-,sīt\ *noun* [International Scientific Vocabulary] (1900)
: a cell that produces gametes; *especially* **:** GAMETOCYTE

go–no–go \'gō-'nō-,gō\ *adjective* (circa 1945)
1 : being or relating to a required decision to continue or stop a course of action
2 : being or relating to a point at which a go-no-go decision must be made

gon·o·phore \'gä-nə-,fōr, -,fȯr\ *noun* [International Scientific Vocabulary] (1859)
: an attached reproductive zooid of a hydroid colony

gon·o·pore \'gä-nə-,pōr, -,pȯr\ *noun* (1897)
: a genital pore in some invertebrates and especially some insects

gon·or·rhea \,gä-nə-'rē-ə\ *noun* [New Latin, from Late Latin, morbid loss of semen, from Greek *gonorrhoia*, from *gon-* + *-rrhoia* -rrhea] (circa 1526)
: a contagious inflammation of the genital mucous membrane caused by the gonococcus — called also *clap*
— gon·or·rhe·al \-'rē-əl\ *adjective*

-gony *noun combining form* [Latin *-gonia*, from Greek, from *gonos*]
: generation **:** reproduction **:** manner of coming into being ⟨iso*gony*⟩

gon·zo \'gän-(,)zō\ *adjective* [origin unknown] (1971)
: idiosyncratically subjective but engagé ⟨*gonzo* journalism⟩; *also* **:** BIZARRE

goo \'gü\ *noun* [perhaps short for *burgoo*] (1900)
1 : a viscid or sticky substance
2 : sentimental tripe
— goo·ey \-ē\ *adjective*
— goo·ey·ness \-nəs\ *noun*

goo·ber \'gü-bər, 'gu̇-\ *noun* [of Bantu origin; akin to Kimbundu *ŋguba* peanut] (1833)
Southern & Midland **:** PEANUT

[1]good \'gu̇d\ *adjective* **bet·ter** \'be-tər\; **best** \'best\ [Middle English, from Old English *gōd;* akin to Old High German *guot* good, Sanskrit *gadhya* what one clings to] (before 12th century)
1 a (1) **:** of a favorable character or tendency ⟨*good* news⟩ (2) **:** BOUNTIFUL, FERTILE ⟨*good* land⟩ (3) **:** HANDSOME, ATTRACTIVE ⟨*good* looks⟩ **b** (1) **:** SUITABLE, FIT ⟨*good* to eat⟩ (2) **:** free from injury or disease ⟨one *good* arm⟩ (3) **:** not depreciated ⟨bad money drives out *good*⟩ (4) **:** commercially sound ⟨a *good* risk⟩ (5) **:** that can be relied on ⟨*good* for another year⟩ ⟨*good* for a hundred dollars⟩ ⟨always *good* for a laugh⟩ (6) **:** PROFITABLE, ADVANTAGEOUS ⟨made a very *good* deal⟩ **c** (1) **:** AGREEABLE, PLEASANT ⟨had a *good* time⟩ (2) **:** SALUTARY, WHOLESOME ⟨*good* for a cold⟩ (3) **:** AMUSING, CLEVER ⟨a *good* joke⟩ **d** (1) **:** of a noticeably large size or quantity **:** CONSIDERABLE ⟨won by a *good* margin⟩ ⟨a *good* bit of the time⟩ (2) **:** FULL ⟨waited a *good* hour⟩ (3) — used as an intensive ⟨a *good* many of us⟩ **e** (1) **:** WELL-FOUNDED, COGENT ⟨*good* reasons⟩ (2) **:** TRUE ⟨holds *good* for society at large⟩ (3) **:** deserving of respect **:** HONORABLE ⟨in *good* standing⟩ (4) **:** legally valid or effectual ⟨*good* title⟩ **f** (1) **:** ADEQUATE, SATISFACTORY ⟨*good* care⟩ — often used in faint praise ⟨his serve is only *good* —Frank Deford⟩ (2) **:** conforming to a standard ⟨*good* English⟩ (3) **:** CHOICE, DISCRIMINATING ⟨*good* taste⟩ (4) **:** containing less fat and being less tender than higher grades — used of meat and especially of beef
2 a (1) **:** VIRTUOUS, RIGHT, COMMENDABLE ⟨a *good* person⟩ ⟨*good* conduct⟩ (2) **:** KIND, BENEVOLENT ⟨*good* intentions⟩ **b :** UPPER-CLASS ⟨a *good* family⟩ **c :** COMPETENT, SKILLFUL ⟨a *good* doctor⟩ **d** (1) **:** LOYAL ⟨a *good* party man⟩ ⟨a *good* Catholic⟩ (2) **:** CLOSE ⟨a *good* friend⟩ **e :** free from infirmity or sorrow ⟨I feel *good*⟩ ▫
— good·ish \'gu̇-dish\ *adjective*
— as good as : in effect **:** VIRTUALLY ⟨*as good as* dead⟩
— as good as gold 1 : of the highest worth or reliability ⟨his promise is *as good as gold*⟩ **2 :** well-behaved ⟨the child was *as good as gold*⟩
— good and \,gu̇d-ᵊn\ **:** VERY, ENTIRELY ⟨was *good and* mad⟩

[2]good *noun* (before 12th century)
1 a : something that is good **b** (1) **:** something conforming to the moral order of the universe (2) **:** praiseworthy character **:** GOODNESS **c :** a good element or portion
2 a : advancement of prosperity or well-being ⟨the *good* of the community⟩ ⟨it's for your own *good*⟩ **b :** something useful or beneficial ⟨it's no *good* trying⟩
3 a : something that has economic utility or satisfies an economic want **b** *plural* **:** personal property having intrinsic value but usually excluding money, securities, and negotiable instruments **c** *plural* **:** CLOTH **d** *plural* **:** something manufactured or produced for sale **:** WARES, MERCHANDISE ⟨canned *goods*⟩ **e** *plural, British* **:** FREIGHT
4 : good persons — used with *the*
5 *plural* **a :** the qualities required to achieve an end **b :** proof of wrongdoing ⟨didn't have the *goods* on him —T. G. Cooke⟩
— for good *also* **for good and all :** FOREVER, PERMANENTLY
— in good with : in a favored position with
— to the good 1 : for the best **:** BENEFICIAL ⟨efforts to restrict credit were all *to the good* —*Time*⟩ **2 :** in a position of net gain or profit ⟨wound up $10 *to the good*⟩

[3]good *adverb* (13th century)
1 : WELL ⟨he showed me how *good* I was doing —Herbert Gold⟩

▫ USAGE
good An old notion that it is wrong to say "I feel good" in reference to health still occasionally appears in print. The origins of this notion, which goes back to the turn of the century, are obscure, but they seem to combine someone's idea that *good* should be reserved to describe virtue and uncertainty about whether an adverb or an adjective should follow *feel*. Today nearly everyone agrees that both *good* and *well* can be predicate adjectives after *feel*. Both are used to express good health, but *good* may connote good spirits in addition to good health.

2 — used as an intensive ⟨a *good* 200 pounds⟩ ⟨a *good* long time⟩ ◻
good book *noun, often G&B capitalized* (1860)
: BIBLE
good–bye *or* **good–by** \gu̇d-ˈbī, gə(d)-\ *noun* [alteration of *God be with you*] (circa 1580)
1 : a concluding remark or gesture at parting — often used interjectionally
2 : a taking of leave ⟨a tearful *good-bye*⟩ ◆
good deal *noun* (1881)
: a considerable quantity or extent **:** LOT ⟨knows a *good deal* about disease⟩ ⟨a *good deal* faster⟩
good faith *noun* (1893)
: honesty or lawfulness of purpose
good fellow *noun* (13th century)
: an affable companionable person
— good–fel·low·ship \ˈgu̇d-ˈfe-lō-ˌship, -ˈfe-lə-\ *noun*
good–for–noth·ing \ˈgu̇d-fər-ˌnə-thiŋ\ *adjective* (1533)
: of no use or value
— good–for–nothing *noun*
Good Friday *noun* [from its special sanctity] (13th century)
: the Friday before Easter observed in churches as the anniversary of the crucifixion of Christ and in some states of the U.S. as a legal holiday
good–heart·ed \ˈgu̇d-ˈhär-təd\ *adjective* (1552)
: having a kindly generous disposition
— good–heart·ed·ly *adverb*
— good–heart·ed·ness *noun*
good–hu·mored \-ˈ(h)yü-mərd\ *adjective* (1662)
: GOOD-NATURED, CHEERFUL
— good–hu·mored·ly *adverb*
— good–hu·mored·ness *noun*
good life *noun* (1946)
: a life marked by a high standard of living
good–look·ing \ˈgu̇d-ˈlu̇-kiŋ\ *adjective* (1780)
: having a pleasing or attractive appearance
— good–look·er \-ˈlu̇-kər\ *noun*
good·ly \ˈgu̇d-lē\ *adjective* **good·li·er; -est** (before 12th century)
1 : pleasantly attractive
2 : significantly large **:** CONSIDERABLE ⟨a *goodly* number⟩
good·man \ˈgu̇d-mən\ *noun* (13th century)
1 *archaic* **:** the master of a household
2 *archaic* **:** MR.
good–na·tured \-ˈnā-chərd\ *adjective* (1577)
: of a pleasant cheerful cooperative disposition
synonym see AMIABLE
— good–na·tured·ly *adverb*
— good–na·tured·ness *noun*
good–neighbor *adjective* (1936)
: marked by principles of friendship, cooperation, and noninterference in the internal affairs of another country ⟨a *good-neighbor* policy⟩
good·ness \ˈgu̇d-nəs\ *noun* (before 12th century)
1 : the quality or state of being good
2 — used interjectionally or in phrases especially to express mild surprise or shock ⟨oh, my *goodness!*⟩ ⟨*goodness* knows⟩
3 : the nutritious, flavorful, or beneficial part of something
goodness of fit (circa 1939)
: the conformity between an experimental result and theoretical expectation or between data and an approximating curve
good offices *noun plural* (1904)
: services as a mediator
good old boy *or* **good ol' boy** *or* **good ole boy** \gu̇-dōl(d)-ˌbȯi\ *noun* (circa 1967)
: a usually white Southerner who conforms to the social behavior of his peers
Good Samaritan *noun* (1846)
: SAMARITAN 2
good–tem·pered \ˈgu̇d-ˈtem-pərd\ *adjective* (1768)
: not easily vexed
— good–tem·pered·ly *adverb*
— good–tem·pered·ness *noun*
good·wife \ˈgu̇d-ˌwīf\ *noun* (13th century)
1 *archaic* **:** the mistress of a household
2 *archaic* **:** MRS.
good·will \ˌgu̇d-ˈwil\ *noun* (before 12th century)
1 a : a kindly feeling of approval and support **:** benevolent interest or concern **b** (1) **:** the favor or prestige that a business has acquired beyond the mere value of what it sells (2) **:** the value of projected earnings increases of a business especially as part of its purchase price (3) **:** the value of other intangible assets (as tax credits) of a business especially as part of its purchase price
2 a : cheerful consent **b :** willing effort
— good·willed \-ˈwild\ *adjective*
[1]**goody** \ˈgu̇-dē\ *noun* [alteration of *goodwife*] (1559)
archaic **:** a usually married woman of lowly station — used as a title preceding a surname
[2]**goody** *or* **good·ie** *noun, plural* **good·ies** (1756)
1 : something that is particularly attractive, pleasurable, good, or desirable
2 *chiefly British* **:** one that is good; *especially* **:** an opponent of the villain (as in a motion picture)
goody–goody \ˌgu̇-dē-ˈgu̇-dē\ *adjective* (1871)
: affectedly or ingratiatingly good or proper
— goody–goody *noun*
Goody Two–shoes \ˈgu̇-dē-ˈtü-ˌshüz\ *noun, often S capitalized* [from *Goody Two-Shoes,* heroine of a children's story perhaps by Oliver Goldsmith] (1934)
: a person who is goody-goody; *also* **:** a person who is uncommonly good
— goody–two–shoes *adjective*
[1]**goof** \ˈgüf\ *noun* [probably alteration of English dialect *goff* (simpleton)] (1915)
1 : a silly or stupid person
2 : BLUNDER
[2]**goof** (1932)
intransitive verb
1 a : to spend time idly or foolishly — usually used with *off* **b :** to engage in playful activity — usually used with *around*
2 : to make a usually foolish or careless mistake **:** BLUNDER — often used with *up*
transitive verb
: to make a mess of **:** BUNGLE — usually used with *up*
— goof on *slang* **:** to make fun of **:** KID, PUT ON ⟨you're *goofing on* me, right?⟩
goof·ball \ˈgüf-ˌbȯl\ *noun* (1950)
1 *slang* **:** a barbiturate sleeping pill
2 : a goofy person
go off *intransitive verb* (1579)
1 : EXPLODE
2 : to burst forth or break out suddenly or noisily
3 : to go forth, out, or away **:** LEAVE
4 : to undergo decline or deterioration
5 : to follow the expected or desired course **:** PROCEED ⟨the party *went off* well⟩
6 : to make a characteristic noise **:** SOUND
— go off the deep end 1 : to enter recklessly on a course **2 :** to become very much excited
goof–off \ˈgü-ˌfȯf\ *noun* (1953)
: one who evades work or responsibility
goofy \ˈgü-fē\ *adjective* **goof·i·er; -est** (1921)
: being crazy, ridiculous, or mildly ludicrous **:** SILLY
— goof·i·ly \-fə-lē\ *adverb*
— goof·i·ness \-fē-nəs\ *noun*
goo·gly–eyed \ˈgü-glē-ˌīd\ *adjective* [by alteration] (1926)
: GOGGLE-EYED
goo·gol \ˈgü-ˌgȯl\ *noun* [coined by Milton Sirotta (born about 1929) nephew of Edward Kasner (died 1955) American mathematician] (1938)
: the figure 1 followed by 100 zeroes equal to 10^{100} ◆
goo·gol·plex \-ˌpleks\ *noun* (1938)
: the figure 1 followed by a googol of zeroes equal to

$$10^{googol} \text{ or } 10^{10^{100}}$$

[1]**goo–goo** \ˈgü-(ˌ)gü\ *adjective* [perhaps alteration of [2]*goggle*] (1900)
: LOVING, ENTICING — used chiefly in the phrase *goo-goo eyes*
[2]**goo–goo** *noun, plural* **goo–goos** [from *goo*d *go*vernment] (1912)
: a member or advocate of a political reform movement
[1]**gook** \ˈgük\ *noun* [origin unknown] (1935)
: a nonwhite or non-American person; *specifically* **:** ASIAN — usually used disparagingly
[2]**gook** \ˈgu̇k, ˈgük\ *variant of* GUCK
goon \ˈgün\ *noun* [probably short for English dialect *gooney* (simpleton)] (1921)
1 : a stupid person
2 : a man hired to terrorize or eliminate opponents
go on *intransitive verb* (15th century)

◻ USAGE
good Adverbial *good* has been under attack from the schoolroom since the 19th century. Insistence on *well* rather than *good* has resulted in a split in connotation: *well* is standard, neutral, and colorless, while *good* is emotionally charged and emphatic. This makes *good* the adverb of choice in sports ⟨"I'm seeing the ball real *good*" is what you hear —Roger Angell⟩. In such contexts as ⟨listen up. And listen *good* —Alex Karras⟩ ⟨lets fly with his tomatoes before they can flee. He gets Clarence *good* —Charles Dickinson⟩ *good* cannot be adequately replaced by *well.* Adverbial *good* is primarily a spoken form; in writing it occurs in reported and fictional speech and in generally familiar or informal contexts.

◇ WORD HISTORY
good–bye In Spanish *adios* and French *adieu* "farewell, good-bye," we see an explicit wish that the person addressed should be in the care of God (*dios, dieu*). The same sentiment lies at the origin of *good-bye,* which comes from the phrase *God be with you.* Gradually eroded over time, it appears in such versions as *God be wy you* (in the 16th century), *God b'y you* (a century later), and numerous other versions before settling into *good-bye* in the 19th century, the final form buttressed by the example of *good night* and *good day. Good-bye* was further shortened simply to *bye,* at which point reduction could scarcely proceed further. To some speakers, indeed, this meager monosyllable seemed in need of fattening, so they produced the reduplication *bye-bye.*

googol A *googol* is a very big number whose name was coined by a small boy. In the late 1930s American mathematician Edward Kasner (1878–1955) found himself working with numbers as large as 1 followed by a hundred zeroes. While it is possible to write this number, using scientific notation, simply as 10^{100}, Kasner felt that having a name for it would facilitate discussing it. According to his own account, he asked his nine-year-old nephew Milton Sirotta to supply a name, promising that he would indeed use the word. Milton produced *googol,* Kasner kept his promise, and the word was accepted in the mathematical community.

1 a : to continue on or as if on a journey ⟨life *goes on*⟩ ⟨*went on* to greater things⟩ **b :** to keep on **:** CONTINUE ⟨*went on* smoking⟩ **c :** PROCEED ⟨*went on* to win the election⟩
2 : to take place **:** HAPPEN ⟨what's *going on*⟩
3 : to talk especially in an effusive manner ⟨the way people *go on* about their ancestors —Hamilton Basso⟩

goo·ney *or* **goo·ny** \ˈgü-nē\ *noun, plural* **gooneys** *or* **goonies** [probably from English dialect *gooney* (simpleton)] (1895)
: BLACK-FOOTED ALBATROSS; *broadly* **:** ALBATROSS

goop \ˈgüp\ *noun* [probably alteration of *goo*] (circa 1958)
: GOO, GUNK

goo·san·der \gü-ˈsan-dər\ *noun* [origin unknown] (1766)
: the common merganser (*Mergus merganser*) of the northern hemisphere

[1]**goose** \ˈgüs\ *noun, plural* **geese** \ˈgēs\ [Middle English *gos,* from Old English *gōs;* akin to Old High German *gans* goose, Latin *anser,* Greek *chēn*] (before 12th century)
1 a : any of numerous large waterfowl (family Anatidae) that are intermediate between the swans and ducks and have long necks, feathered lores, and reticulate tarsi **b :** a female goose as distinguished from a gander
2 : SIMPLETON, DOLT
3 *plural* **goos·es :** a tailor's smoothing iron with a gooseneck handle
4 *plural* **goos·es :** a poke between the buttocks

[2]**goose** *transitive verb* **goosed; goos·ing** (circa 1880)
1 : to poke between the buttocks with an upward thrust
2 : to increase the activity, speed, power, intensity, or amount of **:** SPUR ⟨an effort to *goose* newsstand sales⟩

goose·ber·ry \ˈgüs-ˌber-ē, ˈgüz-, -b(ə-)rē, *chiefly British* ˈguz-\ *noun* (1573)
1 a : the acid usually prickly fruit of any of several shrubs (genus *Ribes*) of the saxifrage family **b :** a shrub bearing gooseberries
2 : CURRANT 2

gooseberry: leaves and fruit

goose bumps *noun plural* (1933)
: a roughness of the skin produced by erection of its papillae especially from cold, fear, or a sudden feeling of excitement

goose egg *noun* (1866)
: ZERO, NOTHING; *especially* **:** a score of zero in a game or contest

goose·fish \ˈgüs-ˌfish\ *noun* (1807)
: any of a family (Lophiidae) of pediculate fishes with a large flattened head, a fringe of flaps along each side of the lower jaw, head, and body, and a long stalk on the head with a flap of flesh at the tip for attracting prey; *especially* **:** MONKFISH

goose·flesh \ˈgüs-ˌflesh\ *noun* (circa 1810)
: GOOSE BUMPS

goose·foot \-ˌfu̇t\ *noun, plural* **goose·foots** (1548)
: any of a genus (*Chenopodium*) or family (Chenopodiaceae, the goosefoot family) of glabrous herbs with utricular fruit

goose·grass \-ˌgras\ *noun* (1530)
1 : CLEAVERS
2 : YARD GRASS

goose·neck \ˈgüs-ˌnek\ *noun, often attributive* (1688)
1 : something (as a flexible jointed metal pipe) curved like the neck of a goose or U-shaped
2 : a truck trailer (as for transporting livestock) with a projecting front end designed to attach to the bed of a pickup truck
— goose·necked \-ˌnekt\ *adjective*

goose pimples *noun plural* (circa 1889)
: GOOSE BUMPS

goose–step \ˈgüs-ˌstep\ *intransitive verb* (1879)
1 : to march in a goose step
2 : to practice an unthinking conformity

goose step *noun* (1806)
: a straight-legged stiff-kneed step used by troops of some armies when passing in review

goos·ey \ˈgü-sē\ *adjective* **goos·i·er; -est** (1811)
1 : resembling a goose
2 a : affected with goose bumps **:** SCARED **b :** very nervous **c :** reacting strongly when goosed or startled

go out *intransitive verb* (before 12th century)
1 a : to go forth, abroad, or outdoors; *specifically* **:** to leave one's house **b** (1) **:** to take the field as a soldier (2) **:** to participate as a principal in a duel **c :** to travel as or as if a colonist or immigrant **d :** to work away from home
2 a : to come to an end **b :** to give up office **:** RESIGN **c :** to become obsolete or unfashionable **d** (1) **:** to play the last card of one's hand (2) **:** to reach or exceed the total number of points required for game in cards
3 : to take part in social activities
4 : to go on strike
5 : BREAK, COLLAPSE
6 : to become a candidate ⟨*went out* for the football team⟩

go over *intransitive verb* (1645)
1 : to go on a journey
2 : to become converted
3 : to receive approval **:** SUCCEED ⟨his plan *went over* well⟩

[1]**go·pher** \ˈgō-fər\ *noun* [origin unknown] (1791)
1 : a burrowing land tortoise (*Gopherus polyphemus*) of the southern U.S.; *broadly* **:** any of several related land tortoises — called also *gopher tortoise*
2 a : any of a family (Geomyidae) of burrowing rodents of western North America, Central America, and the southern U.S. that are the size of a large rat and have large cheek pouches opening beside the mouth — called also *pocket gopher* **b :** any of several small ground squirrels (genus *Citellus*) of the prairie region of North America closely related to the chipmunks
3 : GOPHER BALL

[2]**gopher** *variant of* GOFER

gopher ball *noun* (circa 1949)
: a pitched baseball hit for a home run

gopher snake *noun* (1837)
1 : INDIGO SNAKE
2 : BULL SNAKE

Gor·di·an knot \ˈgȯr-dē-ən-\ *noun*
1 : an intricate problem; *especially* **:** a problem insoluble in its own terms — usually used in the phrase *cut the Gordian knot*
2 : a knot tied by Gordius, king of Phrygia, held to be capable of being untied only by the future ruler of Asia, and cut by Alexander the Great with his sword

Gor·don setter \ˈgȯr-dən-\ *noun* [Alexander, 4th Duke of *Gordon* (died 1827) Scottish sportsman] (1865)
: any of a breed of large bird dogs that have a long flat black-and-tan coat

Gordon setter

[1]**gore** \ˈgōr, ˈgȯr\ *noun* [Middle English, from Old English *gāra;* akin to Old English *gār* spear, and perhaps to Greek *chaion* shepherd's staff] (before 12th century)
1 : a small usually triangular piece of land
2 a : a tapering or triangular piece (as of cloth in a skirt) **b :** an elastic gusset for providing a snug fit in a shoe

[2]**gore** *transitive verb* **gored; gor·ing** (1548)
1 : to cut into a tapering triangular form
2 : to provide with a gore

[3]**gore** *transitive verb* **gored; gor·ing** [Middle English] (15th century)
: to pierce or wound with something pointed (as a horn or knife) ⟨*gored* by a bull⟩

[4]**gore** *noun* [Middle English, filth, from Old English *gor*] (1563)
: BLOOD; *especially* **:** clotted blood

[1]**gorge** \ˈgȯrj\ *noun* [Middle English, from Middle French, from Late Latin *gurga,* alteration of *gurges,* from Latin, whirlpool — more at VORACIOUS] (14th century)
1 : THROAT — often used with *rise* to indicate revulsion accompanied by a sensation of constriction ⟨my *gorge* rises at the sight of blood⟩
2 a : a hawk's crop **b :** STOMACH, BELLY
3 : the entrance into an outwork (as a bastion) of a fort
4 : a narrow passage through land; *especially* **:** a narrow steep-walled canyon or part of a canyon
5 : a primitive device used instead of a fishhook that consists of an object (as a piece of bone attached in the middle of a line) easy to swallow but difficult to eject
6 : a mass choking a passage ⟨a river dammed by an ice *gorge*⟩
7 : the line on the front of a coat or jacket formed by the crease of the lapel and collar

[2]**gorge** *verb* **gorged; gorg·ing** (14th century)
intransitive verb
: to eat greedily or to repletion
transitive verb
1 a : to stuff to capacity **:** GLUT **b :** to fill completely or to the point of distension ⟨veins *gorged* with blood⟩
2 : to consume greedily
synonym see SATIATE
— gorg·er *noun*

[3]**gorge** *noun* (1854)
: the act or an instance of gorging

gor·geous \ˈgȯr-jəs\ *adjective* [Middle English *gorgayse,* from Middle French *gorgias* elegant, from *gorgias* wimple, from *gorge* gorget] (15th century)
: splendidly or showily brilliant or magnificent
synonym see SPLENDID
— gor·geous·ly *adverb*
— gor·geous·ness *noun*

gor·get \ˈgȯr-jət\ *noun* [Middle English, from Middle French, from *gorge*] (15th century)
1 : a piece of armor protecting the throat — see ARMOR illustration
2 a : an ornamental collar **b :** a part of a wimple covering the throat and shoulders

gor·gon \ˈgȯr-gən\ *noun* [Latin *Gorgon-, Gorgo,* from Greek *Gorgōn*]
1 *capitalized* **:** any of three snake-haired sisters in Greek mythology whose appearance turns the beholder to stone
2 : an ugly or repulsive woman
— Gor·go·ni·an \gȯr-ˈgō-nē-ən\ *adjective*

gor·go·ni·an \gȯr-ˈgō-nē-ən\ *noun* [New Latin *Gorgonia,* a coral genus, from Latin, coral, from *Gorgon-, Gorgo*] (1835)
: any of an order (Gorgonacea) of colonial anthozoans with a usually horny and branching axial skeleton
— gorgonian *adjective*

gor·gon·ize \ˈgȯr-gə-ˌnīz\ *transitive verb* **-ized; -iz·ing** (1609)
: to have a paralyzing or mesmerizing effect on **:** STUPEFY, PETRIFY

Gor·gon·zo·la \ˌgȯr-gən-ˈzō-lə\ *noun* [Italian, from *Gorgonzola,* Italy] (1878)
: a pungent blue cheese of Italian origin

go·ril·la \gə-ˈri-lə\ *noun* [New Latin, from Greek *Gorillai,* plural, a tribe of hairy women

mentioned in an account of a voyage around Africa] (1847)
1 : an anthropoid ape (*Gorilla gorilla*) of western equatorial Africa related to the chimpanzee but less erect and much larger
2 a : an ugly or brutal man **b :** THUG, GOON ◆

gor·man·dise *chiefly British variant of* GORMANDIZE

gor·man·dize \'gȯr-mən-ˌdīz\ *verb* **-dized; -diz·ing** [*gormand,* alteration of *gourmand*] (1548)
intransitive verb
: to eat gluttonously or ravenously
transitive verb
: to eat greedily **:** DEVOUR
— gor·man·diz·er *noun*

gorm·less \'gȯrm-ləs\ *adjective* [alteration of English dialect *gaumless,* from *gaum* attention, understanding (from Middle English *gome,* from Old Norse *gaum, gaumr*) + *-less*] (1883)
chiefly British **:** lacking intelligence **:** STUPID

go–round \'gō-ˌrau̇nd\ *noun* (1891)
: GO-AROUND

gorp \'gȯrp\ *noun* [origin unknown] (1968)
: a snack consisting of high-energy food (as raisins and nuts)

gorse \'gȯrs\ *noun* [Middle English *gorst,* from Old English; akin to Old High German *gersta* barley, Latin *hordeum*] (before 12th century)
: a spiny yellow-flowered European shrub (*Ulex europaeus*) of the legume family; *broadly* **:** any of several related plants (genera *Ulex* and *Genista*)
— gorsy \'gȯr-sē\ *adjective*

gory \'gōr-ē, 'gȯr-\ *adjective* **gor·i·er; -est** (15th century)
1 : covered with gore **:** BLOODSTAINED
2 : BLOODCURDLING, SENSATIONAL ⟨wanted to hear the *gory* details⟩
synonym see BLOODY

gosh \'gäsh, 'gȯsh\ *interjection* [euphemism for *God*] (1757)
— used as a mild oath or to express surprise

gos·hawk \'gäs-ˌhȯk\ *noun* [Middle English *goshawke,* from Old English *gōshafoc,* from *gōs* goose + *hafoc* hawk] (before 12th century)
: any of several long-tailed accipitrine hawks with short rounded wings; *especially* **:** a hawk (*Accipiter gentilis*) of the northern parts of both the Old and the New World that is larger than a crow and has a white stripe above and behind the eye

gos·ling \'gäz-liŋ, 'gȯz-, -lən\ *noun* [Middle English, from *gos* goose] (14th century)
1 : a young goose
2 : a foolish or callow person

go–slow \ˌgō-'slō\ *noun* (1926)
British **:** SLOWDOWN

[1]gos·pel \'gäs-pəl\ *noun* [Middle English, from Old English *gōdspel* (translation of Late Latin *evangelium*), from *gōd* good + *spell* tale — more at SPELL] (before 12th century)
1 a *often capitalized* **:** the message concerning Christ, the kingdom of God, and salvation **b** *capitalized* **:** one of the first four New Testament books telling of the life, death, and resurrection of Jesus Christ; *also* **:** a similar apocryphal book **c :** an interpretation of the Christian message ⟨the social *gospel*⟩
2 *capitalized* **:** a lection from one of the New Testament Gospels
3 : the message or teachings of a religious teacher
4 : something accepted as infallible truth or as a guiding principle ⟨the *gospel* of conservation —R. M. Hodesh⟩
5 : gospel music

[2]gospel *adjective* (before 12th century)
1 a : having a basis in or being in accordance with the gospel **:** EVANGELICAL ⟨ordained to the *gospel* ministry —*Christian Century*⟩ **b :** marked by special or fervid emphasis on the gospel ⟨a *gospel* meeting⟩
2 : of, relating to, or being religious songs of American origin associated with evangelism and popular devotion and marked by simple melody and harmony and elements of folk songs and blues

gos·pel·er *or* **gos·pel·ler** \'gäs-p(ə-)lər\ *noun* (1506)
1 : one who reads or sings the liturgical Gospel
2 : one who preaches or propounds a gospel

gospel side *noun, often G capitalized* [from the custom of reading the Gospel from this side] (1891)
: the left side of an altar or chancel as one faces it

[1]gos·sa·mer \'gä-sə-mər *also* 'gäz-mər, 'gä-zə-\ *noun* [Middle English *gossomer,* from *gos* goose + *somer* summer] (14th century)
1 : a film of cobwebs floating in air in calm clear weather
2 : something light, delicate, or insubstantial ⟨the *gossamer* of youth's dreams —Andrea Parke⟩
— gos·sa·mery \-mə-rē\ *adjective*

[2]gossamer *adjective* (circa 1807)
: extremely light, delicate, or tenuous

gos·san \'gä-sᵊn\ *noun* [Cornish *gossen,* from *gōs* blood] (1776)
: decomposed rock or vein material of reddish or rusty color that results from oxidized pyrites

[1]gos·sip \'gä-səp\ *noun* [Middle English *gossib,* from Old English *godsibb,* from *god* god + *sibb* kinsman, from *sibb* related — more at SIB] (before 12th century)
1 a *dialect British* **:** GODPARENT **b :** COMPANION, CRONY **c :** a person who habitually reveals personal or sensational facts about others
2 a : rumor or report of an intimate nature **b :** a chatty talk **c :** the subject matter of gossip ◆
— gos·sip·ry \-sə-prē\ *noun*

[2]gossip *intransitive verb* (1627)
: to relate gossip
— gos·sip·er *noun*

gos·sip·mon·ger \'gä-səp-ˌməŋ-gər, -ˌmäŋ-\ *noun* (1836)
: a person who starts or spreads gossip

gos·sipy \'gä-sə-pē\ *adjective* (1818)
: characterized by, full of, or given to gossip ⟨a *gossipy* letter⟩ ⟨*gossipy* neighbors⟩

gos·sy·pol \'gä-sə-ˌpȯl, -ˌpōl\ *noun* [International Scientific Vocabulary, ultimately from Latin *gossypion* cotton] (1899)
: a toxic phenolic pigment $C_{30}H_{30}O_8$ in cottonseed

got *past and past participle of* GET

Goth \'gäth\ *noun* [Middle English *Gothes, Gotes* (plural), partly from Old English *Gotan* (plural); partly from Late Latin *Gothi* (plural)] (14th century)
: a member of a Germanic people that overran the Roman Empire in the early centuries of the Christian era

[1]Goth·ic \'gä-thik\ *adjective* (1591)
1 a : of, relating to, or resembling the Goths, their civilization, or their language **b :** TEUTONIC, GERMANIC **c :** MEDIEVAL **d :** UNCOUTH, BARBAROUS
2 a : of, relating to, or having the characteristics of a style of architecture developed in northern France and spreading through western Europe from the middle of the 12th century to the early 16th century that is characterized by the converging of weights and strains at isolated points upon slender vertical piers and counterbalancing buttresses and by pointed arches and vaulting **b :** of or relating to an architectural style reflecting the influence of the medieval Gothic
3 *often not capitalized* **:** of or relating to a style of fiction characterized by the use of desolate or remote settings and macabre, mysterious, or violent incidents
— goth·i·cal·ly \-thi-k(ə-)lē\ *adverb*
— Goth·ic·ness \-thik-nəs\ *noun*

[2]Gothic *noun* (1691)
1 a : BLACK LETTER **b :** SANS SERIF
2 : Gothic art style or decoration; *specifically* **:** the Gothic architectural style
3 : the East Germanic language of the Goths — see INDO-EUROPEAN LANGUAGES table
4 *often not capitalized* **:** a work of fiction in the gothic style

Gothic arch *noun* (1739)
: a pointed arch; *especially* **:** one with a joint instead of a keystone at its apex

Goth·i·cism \'gä-thə-ˌsi-zəm\ *noun* (1710)
1 : barbarous lack of taste or elegance
2 : conformity to or practice of Gothic style
— Goth·i·cist \-sist\ *noun*

goth·i·cize \-ˌsīz\ *transitive verb* **-cized; -ciz·ing** *often capitalized* (1712)
: to make Gothic

Goth·ick *chiefly British variant of* GOTHIC

Gothic Revival *noun* (1869)
: an artistic style or movement of the 18th and 19th centuries inspired by and imitative of the Gothic style especially in architecture

go through *intransitive verb* (1513)
1 : to continue firmly or obstinately to the end ⟨I was *going through* with it if it killed me —A. W. Long⟩
2 a : to receive approval or sanction **:** PASS **b :** to come to a desired or satisfactory conclusion

◇ WORD HISTORY

gorilla In the sixth or early fifth century B.C., the Carthaginian navigator Hanno made a voyage around the west coast of Africa, perhaps sailing as far as Sherbro Island in modern Sierra Leone. An account of his journey was inscribed on tablets that have since disappeared, though a Greek translation survives. Near the furthest point on their voyage Hanno and his men came upon an island "full of wild people, the greater part of whom were females, hairy on their bodies, whom our interpreters called Gorillas. We pursued them, but could not capture the males . . . but we captured three females, who, biting and scratching their captors, refused to go along with them. We killed and skinned them and brought the skins to Carthage." Whether Hanno actually saw a species of anthropoid ape or something else we will never know, but when the gorilla of western equatorial Africa was first described scientifically in 1847, it was given the Latin name *Troglodytes gorilla.* This binomial was superseded by *Gorilla gorilla,* and the vernacular name of the animal in English and other European languages has become simply *gorilla.*

gossip In Old English *sibb* occurs as a noun meaning "kinship" and as an adjective meaning "related by blood or kinship." By the 11th century a compound had been formed from the noun *sibb* prefixed by *god,* the predecessor of Modern English *god.* A *godsibb,* therefore, was a person spiritually related to another, specifically by being a sponsor at baptism. Today we would call such a person a *godmother* or *godfather,* using *god* in the same way. By the 14th century the *d* had begun to disappear, and *godsibb* developed into *gossib* and then, through devoicing of the final consonant, to *gossip,* the form that is still current. The meaning, too, had begun to change, and the sense of *gossip* as a close friend or comrade developed alongside the sense of a godparent. Since close friends naturally talk about family and other friends, it was only a short step to the *gossip* of today, a person no longer necessarily friend, relative, or sponsor, but someone filled with irresistible tidbits of rumor.

go to *intransitive verb* (15th century)
1 *archaic* — used interjectionally as an exhortation ⟨and they said one to another, *go to,* let us make brick —Genesis 11:3 (Authorized Version)⟩
2 *archaic* — used interjectionally to express disapproval or disbelief ⟨*go to, go to;* you have known what you should not —Shakespeare⟩
gotten *past participle of* GET
Göt·ter·däm·mer·ung \ˌgə(r)-tər-ˈde-mə-ˌrüŋ, -ˈdä-\ *noun* [German, literally, twilight of the gods, from *Götter* (plural of *Gott* god) + *Dämmerung* twilight] (1909)
: a collapse (as of a society or regime) marked by catastrophic violence and disorder
gouache \ˈgwäsh\ *noun* [French, from Italian *guazzo,* literally, puddle, probably from Latin *aquatio* watering place, from *aquari* to fetch water, from *aqua* water — more at ISLAND] (1882)
1 : a method of painting with opaque watercolors
2 a : a picture painted by gouache **b :** the pigment used in gouache
Gou·da \ˈgü-də\ *noun* [*Gouda,* Netherlands] (1885)
: a mild cheese of Dutch origin that is similar to Edam but contains more fat
¹gouge \ˈgau̇j\ *noun* [Middle English *gowge,* from Middle French *gouge,* from Late Latin *gulbia,* of Celtic origin; akin to Old Irish *gulban* beak, sting] (14th century)
1 : a chisel with a concavo-convex cross section
2 a : the act of gouging **b :** a groove or cavity scooped out
3 : an excessive or improper exaction **:** EXTORTION
²gouge *transitive verb* **gouged; goug·ing** (1570)
1 : to scoop out with or as if with a gouge
2 a : to force out (an eye) with the thumb **b :** to thrust the thumb into the eye of
3 : to subject to extortion or undue exaction **:** OVERCHARGE
— goug·er *noun*
gou·lash \ˈgü-ˌläsh, -ˌlash\ *noun* [Hungarian *gulyás,* short for *gulyáshús,* literally, herdsman's meat] (1866)
1 : a stew made with meat (as beef), assorted vegetables, and paprika
2 : a round in bridge played with hands produced by a redistribution of previously dealt cards
3 : a mixture of heterogeneous elements **:** JUMBLE
go under *intransitive verb* (1848)
: to be overwhelmed, destroyed, or defeated **:** FAIL
go up *intransitive verb* (15th century)
1 *chiefly British* **:** to attend a university
2 *of an actor* **:** to become confused
— go up in flames : BURN
— go up in smoke : to be destroyed by or as if by burning
gou·ra·mi \gü-ˈrä-mē\ *noun, plural* **-mi** *or* **-mis** *also* **-mies** [Malay dialect (Java), from Javanese *graméh*] (1878)
: any of numerous African and Asian tropical freshwater fishes (order Perciformes): as **a :** a large Asian food fish (*Osphronemus goramy,* family Osphronemidae) **b :** any of various small fishes (families Belontiidae and Helostomatidae) often kept in aquariums
gourd \ˈgōrd, ˈgȯrd, ˈgu̇rd\ *noun* [Middle English *gourde,* from Middle French, from Latin *cucurbita*] (14th century)
1 : any of a family (Cucurbitaceae, the gourd family) of chiefly herbaceous tendril-bearing vines including the cucumber, melon, squash, and pumpkin
2 : the fruit of a gourd **:** PEPO; *especially* **:** any of various hard-rinded inedible fruits of plants of two genera (*Lagenaria* and *Cucurbita*) often used for ornament or for vessels and utensils
— out of one's gourd *also* **off one's gourd :** CRAZY
gourde \ˈgu̇rd\ *noun* [American French] (circa 1858)
— see MONEY table
gour·mand \ˈgu̇r-ˌmänd, -mənd\ *noun* [Middle English, from Middle French *gourmant*] (15th century)
1 : one who is excessively fond of eating and drinking
2 : one who is heartily interested in good food and drink
synonym see EPICURE
— gour·mand·ism \ˈgu̇r-ˌmän-ˌdi-zəm, -mən-\ *noun*
— gour·man·dize \-ˌdīz\ *intransitive verb*
gour·man·dise \ˌgu̇r-män-ˈdēz\ *noun* [French, from Middle French, from *gourmand*] (15th century)
: appreciation of or interest in good food and drink **:** GOURMANDISM
gour·met \ˈgu̇r-ˌmā, gu̇r-ˈ\ *noun* [French, from Middle French, alteration of *gromet* boy servant, vintner's assistant, probably ultimately from Middle English *grom* groom] (1820)
: a connoisseur of food and drink; *broadly* **:** CONNOISSEUR 2 ⟨a film *gourmet*⟩
synonym see EPICURE
— gourmet *adjective*
gout \ˈgau̇t\ *noun* [Middle English *goute,* from Old French, gout, drop, from Latin *gutta* drop] (13th century)
1 : a metabolic disease marked by a painful inflammation of the joints, deposits of urates in and around the joints, and usually an excessive amount of uric acid in the blood
2 : a mass or aggregate especially of something fluid often gushing or bursting forth
— gouty \-ē\ *adjective*
gov·ern \ˈgə-vərn\ *verb* [Middle English, from Old French *governer,* from Latin *gubernare* to steer, govern, from Greek *kybernan*] (14th century)
transitive verb
1 a : to exercise continuous sovereign authority over; *especially* **:** to control and direct the making and administration of policy in **b :** to rule without sovereign power and usually without having the authority to determine basic policy
2 a *archaic* **:** MANIPULATE **b :** to control the speed of (as a machine) especially by automatic means
3 a : to control, direct, or strongly influence the actions and conduct of **b :** to exert a determining or guiding influence in or over ⟨income must *govern* expenditure⟩ **c :** to hold in check **:** RESTRAIN
4 : to require (a word) to be in a certain case
5 : to serve as a precedent or deciding principle for ⟨customs that *govern* human decisions⟩
intransitive verb
1 : to prevail or have decisive influence **:** CONTROL
2 : to exercise authority
— gov·ern·able \-vər-nə-bəl\ *adjective*
gov·er·nance \ˈgə-vər-nən(t)s\ *noun* (14th century)
: GOVERNMENT
gov·ern·ess \ˈgə-vər-nəs\ *noun* (15th century)
1 : a woman who governs
2 : a woman entrusted with the care and supervision of a child especially in a private household
gov·ern·essy \ˈgə-vər-ni-sē\ *adjective* (1872)
: characteristic of or resembling a governess (as in primness)
gov·ern·ment \ˈgə-vər(n)-mənt, -və-mənt; ˈgə-bᵊm-ənt, -vᵊm-\ *noun, often attributive* (14th century)
1 : the act or process of governing; *specifically* **:** authoritative direction or control
2 *obsolete* **:** moral conduct or behavior **:** DISCRETION
3 a : the office, authority, or function of governing **b** *obsolete* **:** the term during which a governing official holds office
4 : the continuous exercise of authority over and the performance of functions for a political unit **:** RULE
5 a : the organization, machinery, or agency through which a political unit exercises authority and performs functions and which is usually classified according to the distribution of power within it **b :** the complex of political institutions, laws, and customs through which the function of governing is carried out
6 : the body of persons that constitutes the governing authority of a political unit or organization: as **a :** the officials comprising the governing body of a political unit and constituting the organization as an active agency **b** *capitalized* **:** the executive branch of the U.S. federal government **c** *capitalized* **:** a small group of persons holding simultaneously the principal political executive offices of a nation or other political unit and being responsible for the direction and supervision of public affairs: (1) **:** such a group in a parliamentary system constituted by the cabinet or by the ministry (2) **:** ADMINISTRATION 4b
7 : POLITICAL SCIENCE
— gov·ern·men·tal \ˌgə-vər(n)-ˈmen-tᵊl\ *adjective*
— gov·ern·men·tal·ize \-tᵊl-ˌīz\ *transitive verb*
— gov·ern·men·tal·ly \-tᵊl-ē\ *adverb*
gov·ern·men·tal·ism \ˌgə-vər(n)-ˈmen-tᵊl-ˌi-zəm\ *noun* (1848)
1 : a theory advocating extension of the sphere and degree of government activity
2 : the tendency toward extension of the role of government
— gov·ern·men·tal·ist \-tᵊl-ist\ *noun*
gov·ern·men·tese \ˌgə-vər-mən-ˈtēz\ *noun* (1944)
: jargon held to be characteristic of government officials
gov·er·nor \ˈgə-vᵊn-ər *also* ˈgə-vər-nər\ *noun* (14th century)
1 : one that governs: as **a :** one that exercises authority especially over an area or group **b :** an official elected or appointed to act as ruler, chief executive, or nominal head of a political unit **c :** COMMANDING OFFICER **d :** the managing director and usually the principal officer of an institution or organization **e :** a member of a group that directs or controls an institution or society
2 : TUTOR
3 a *slang* **:** one looked upon as governing **b :** MISTER, SIR — usually used as a term of address
4 a : an attachment to a machine (as a gasoline engine) for automatic control or limitation of speed **b :** a device giving automatic control (as of pressure or temperature)
— gov·er·nor·ate \-ət, -ˌāt\ *noun*
governor–general *noun, plural* **governors–general** *or* **governor–generals** (1586)
: a governor of high rank; *especially* **:** one who governs a large territory or has deputy governors under him
gov·er·nor·ship \ˈgə-vᵊn-ər-ˌship *also* ˈgə-vər-nər-\ *noun* (1658)
1 : the office of governor
2 : the period of incumbency of a governor
gow·an \ˈgau̇-ən\ *noun* [probably alteration of Middle English *gollan*] (1570)
chiefly Scottish **:** DAISY 1; *broadly* **:** a white or yellow field flower

— **gow·any** \-ə-nē\ *adjective, chiefly Scottish*

gown \'gau̇n\ *noun* [Middle English, from Middle French *goune,* from Late Latin *gunna,* a fur or leather garment] (14th century)
1 a : a loose flowing outer garment formerly worn by men **b :** a distinctive robe worn by a professional or academic person **c :** a woman's dress **d** (1) **:** DRESSING GOWN (2) **:** NIGHTGOWN **e :** a coverall worn in an operating room
2 : the body of students and faculty of a college or university ⟨rivalry between town and *gown*⟩
— **gown** *transitive verb*

gowns·man \'gau̇nz-mən\ *noun* (1627)
: a professional or academic person

goy \'gȯi\ *noun, plural* **goy·im** \'gȯi-əm\ *also* **goys** [Yiddish, from Hebrew *gōy* people, nation] (1841)
: GENTILE 1 — sometimes used disparagingly
— **goy·ish** \'gȯi-ish\ *adjective*

graaf·ian follicle \ˌgrä-fē-ən-, ˌgra-\ *noun, often G capitalized* [Regnier de *Graaf* (died 1673) Dutch anatomist] (1883)
: a liquid-filled cavity in a mammalian ovary containing a mature egg before ovulation

¹grab \'grab\ *verb* **grabbed; grab·bing** [obsolete Dutch or Low German *grabben*] (circa 1581)
transitive verb
1 : to take or seize by or as if by a sudden motion or grasp ⟨*grab* up an ax⟩ ⟨*grabbed* the opportunity⟩ ⟨*grab* attention⟩
2 : to obtain unscrupulously ⟨*grab* public lands⟩
3 : to take hastily ⟨*grab* a bite to eat⟩ ⟨*grab* a cab⟩
4 a : to seize the attention of ⟨the technique of *grabbing* an audience —Pauline Kael⟩ **b :** to impress favorably and deeply
intransitive verb
: to make a grab
synonym see TAKE
— **grab·ber** *noun*

²grab *adjective* (1608)
1 : intended to be grabbed ⟨a *grab* rail⟩
2 : taken at random ⟨*grab* samples of rocks⟩

³grab *noun* (1777)
1 a : something grabbed **b :** a sudden snatch **c :** an unlawful or unscrupulous seizure
2 *chiefly British* **a :** a device for clutching an object **b :** CLAMSHELL
— **up for grabs :** available for anyone to take, win, or settle

grab bag *noun* (1855)
1 : a receptacle (as a bag) containing small articles which are to be drawn (as at a party or fair) without being seen
2 : a miscellaneous collection **:** POTPOURRI

grab·ble \'gra-bəl\ *intransitive verb* **grab·bled; grab·bling** \-b(ə-)liŋ\ [Dutch *grabbelen,* from Middle Dutch, frequentative of *grabben*] (circa 1580)
1 : to search with the hand **:** GROPE
2 : to lie or fall prone **:** SPRAWL
— **grab·bler** \-b(ə-)lər\ *noun*

grab·by \'gra-bē\ *adjective* **grab·bi·er; -est** (1910)
1 : tending to grab **:** GRASPING, GREEDY
2 : having the power to grab the attention ⟨*grabby* previews⟩

gra·ben \'grä-bən\ *noun* [German, ditch, from Old High German *grabo,* from *graban* to dig — more at GRAVE] (1896)
: a depressed segment of the crust of the earth or a celestial body (as the moon) bounded on at least two sides by faults

¹grace \'grās\ *noun* [Middle English, from Old French, from Latin *gratia* favor, charm, thanks, from *gratus* pleasing, grateful; akin to Sanskrit *gṛṇāti* he praises] (12th century)
1 a : unmerited divine assistance given humans for their regeneration or sanctification **b :** a virtue coming from God **c :** a state of sanctification enjoyed through divine grace
2 a : APPROVAL, FAVOR ⟨stayed in his good *graces*⟩ **b** *archaic* **:** MERCY, PARDON **c :** a special favor **:** PRIVILEGE ⟨each in his place, by right, not *grace,* shall rule his heritage —Rudyard Kipling⟩ **d :** disposition to or an act or instance of kindness, courtesy, or clemency **e :** a temporary exemption **:** REPRIEVE
3 a : a charming or attractive trait or characteristic **b :** a pleasingly graceful appearance or effect **:** CHARM **c :** ease and suppleness of movement or bearing
4 — used as a title of address or reference for a duke, a duchess, or an archbishop
5 : a short prayer at a meal asking a blessing or giving thanks
6 *plural, capitalized* **:** three sister goddesses in Greek mythology who are the givers of charm and beauty
7 : a musical trill, turn, or appoggiatura
8 a : sense of propriety or right ⟨had the *grace* not to run for elective office —Calvin Trillin⟩ **b :** the quality or state of being considerate or thoughtful
synonym see MERCY

²grace *transitive verb* **graced; grac·ing** (1585)
1 : to confer dignity or honor on
2 : ADORN, EMBELLISH ⟨graveled walks *graced* with statues —J. A. Michener⟩

grace·ful \'grās-fəl\ *adjective* (circa 1586)
: displaying grace in form or action **:** pleasing or attractive in line, proportion, or movement
— **grace·ful·ly** \-fə-lē\ *adverb*
— **grace·ful·ness** *noun*

grace·less \'grā-sləs\ *adjective* (14th century)
1 : lacking in divine grace **:** IMMORAL, UNREGENERATE
2 a : lacking a sense of propriety **b :** devoid of attractive qualities
3 : artistically inept or unbeautiful
— **grace·less·ly** *adverb*
— **grace·less·ness** *noun*

grace note *noun* (circa 1823)
1 : a musical note added as an ornament; *especially* **:** APPOGGIATURA
2 : a small addition or embellishment

grace period *noun* (1945)
: a period of time beyond a due date during which a financial obligation may be met without penalty or cancellation

grac·ile \'gra-səl, -ˌsīl\ *adjective* [Latin *gracilis*] (1623)
1 : SLENDER, SLIGHT
2 : GRACEFUL
3 : relating to, resembling, or being any of the primitive relatively small slender hominids (genus *Australopithecus*) characterized especially by molars and incisors of similar size that are adapted to an omnivorous diet — compare ROBUST
— **grac·ile·ness** *noun*
— **gra·cil·i·ty** \grə-'si-lə-tē\ *noun*

gra·ci·o·so \ˌgrä-sē-'ō-(ˌ)sō, -(ˌ)zō\ *noun, plural* **-sos** [Spanish, from *gracioso,* adjective, agreeable, amusing, from Latin *gratiosus*] (1749)
: a buffoon in Spanish comedy

gra·cious \'grā-shəs\ *adjective* [Middle English, from Middle French *gracieus,* from Latin *gratiosus* enjoying favor, agreeable, from *gratia*] (14th century)
1 a *obsolete* **:** GODLY **b** *archaic* **:** PLEASING, ACCEPTABLE
2 a : marked by kindness and courtesy **b :** GRACEFUL **c :** marked by tact and delicacy **:** URBANE **d :** characterized by charm, good taste, generosity of spirit, and the tasteful leisure of wealth and good breeding ⟨*gracious* living⟩
3 : MERCIFUL, COMPASSIONATE — used conventionally of royalty and high nobility ☆
— **gra·cious·ly** *adverb*
— **gra·cious·ness** *noun*

grack·le \'gra-kəl\ *noun* [New Latin *Gracula,* genus name, alteration of Latin *graculus* jackdaw] (1772)
1 : any of a genus (*Quiscalus* of the family Icteridae) of large American blackbirds having iridescent black plumage
2 : any of various Asian starlings (as the hill mynahs)

¹grad \'grad\ *noun or adjective* [by shortening] (circa 1871)
: GRADUATE

²grad *noun* [French *grade* degree, from Latin *gradus*] (1898)
: one hundredth of a right angle

gra·da·tion \grā-'dā-shən, grə-\ *noun* (1549)
1 a : a series forming successive stages **b :** a step or place in an ordered scale
2 : an advance by regular degrees
3 : a gradual passing from one tint or shade to another
4 : the act or process of grading
5 : ABLAUT
— **gra·da·tion·al** \-shnəl, -shə-nᵊl\ *adjective*
— **gra·da·tion·al·ly** *adverb*

¹grade \'grād\ *noun* [French, from Latin *gradus* step, degree, from Latin *gradi* to step, go; akin to Lithuanian *gridyti* to go, wander] (1796)
1 a (1) **:** a stage in a process (2) **:** a position in a scale of ranks or qualities **b :** a degree of severity in illness ⟨*grade* III carcinoma⟩ **c :** a class organized for the work of a particular year of a school course **d :** a military or naval rank
2 a : a class of things of the same stage or degree **b :** a mark indicating a degree of accomplishment in school **c :** a standard of food quality
3 a : the degree of inclination of a road or slope; *also* **:** a sloping road **b :** a datum or reference level; *especially* **:** ground level
4 : a domestic animal with one parent purebred and the other of inferior breeding
5 *plural* **:** the elementary school system
— **grade·less** \-ləs\ *adjective*

²grade *verb* **grad·ed; grad·ing** (1659)
transitive verb
1 a : to arrange in grades **:** SORT **b :** to arrange in a scale or series **c :** to assign to a grade or assign a grade to
2 : to level off to a smooth horizontal or sloping surface
intransitive verb
1 a : to form a series **b :** BLEND
2 : to be of a particular grade
— **grad·able** \'grā-də-bəl\ *adjective*

³grade *adjective* (1852)
: being, involving, or yielding domestic animals of improved but not pure stock ⟨*grade* ewes⟩ ⟨*grade* breeding⟩

-grade \ˌgrād\ *adjective combining form* [French, from Latin *-gradus,* from *gradi*]
: walking ⟨planti*grade*⟩

grade crossing *noun* (circa 1890)
: a crossing of highways, railroad tracks, or pedestrian walks or combinations of these on the same level

grade point *noun* (1951)

☆ SYNONYMS
Gracious, cordial, affable, genial, sociable mean markedly pleasant and easy in social intercourse. GRACIOUS implies courtesy and kindly consideration ⟨the *gracious* award winner thanked her colleagues⟩. CORDIAL stresses warmth and heartiness ⟨our host was *cordial* as he greeted us⟩. AFFABLE implies easy approachability and readiness to respond pleasantly to conversation or requests or proposals ⟨though wealthy, she was *affable* to all⟩. GENIAL stresses cheerfulness and even joviality ⟨a *genial* companion with a ready quip⟩. SOCIABLE suggests a genuine liking for the companionship of others ⟨*sociable* people who enjoy entertaining⟩.

: one of the points assigned to each course credit (as in a college) in accordance with the letter grade earned in the course — called also *quality point*

grade point average *noun* (1966)
: the average obtained by dividing the total number of grade points earned by the total number of credits attempted — called also *quality point average*

grad·er \ˈgrā-dər\ *noun* (1832)
1 : one that grades
2 : a machine for leveling earth
3 : a pupil in a school grade ⟨a fifth *grader*⟩

grade school *noun* (1869)
: ELEMENTARY SCHOOL
— grade–school·er \ˈgrād-ˌskü-lər\ *noun*

grade separation *noun* (circa 1949)
: a highway or railroad crossing using an underpass or overpass

grade up *transitive verb* (1903)
: to improve by breeding females to purebred males

gra·di·ent \ˈgrā-dē-ənt\ *noun* [Latin *gradient-, gradiens,* present participle of *gradi*] (1835)
1 a : the rate of regular or graded ascent or descent **:** INCLINATION **b :** a part sloping upward or downward
2 : change in the value of a quantity (as temperature, pressure, or concentration) with change in a given variable and especially per unit distance in a specified direction
3 : the vector sum of the partial derivatives with respect to the three coordinate variables *x, y,* and *z* of a scalar quantity whose value varies from point to point
4 : a graded difference in physiological activity along an axis (as of the body or an embryonic field)
5 : change in response with distance from the stimulus

gra·di·om·e·ter \ˌgrā-dē-ˈä-mə-tər\ *noun* [*gradient* + *-o-* + *-meter*] (1899)
: an instrument for measuring the gradient of a physical quantity (as the earth's magnetic field)

¹grad·u·al \ˈgra-jə-wəl, -jəl, ˈgraj wəl\ *noun, often capitalized* [Middle English, from Medieval Latin *graduale,* from Latin *gradus* step, from its being sung on the steps of the altar] (15th century)
1 : a book containing the choral parts of the Mass
2 : a pair of verses (as from the Psalms) proper after the Epistle in the Mass

²gradual *adjective* [Medieval Latin *gradualis,* from Latin *gradus*] (1692)
1 : proceeding by steps or degrees
2 : moving, changing, or developing by fine or often imperceptible degrees
— grad·u·al·ly *adverb*
— grad·u·al·ness *noun*

grad·u·al·ism \-jə-wə-ˌli-zəm, -jə-ˌli-\ *noun* (1835)
1 : the policy of approaching a desired end by gradual stages
2 : the evolution of new species by gradual accumulation of small genetic changes over long periods of time; *also* **:** a theory or model of evolution emphasizing this — compare PUNCTUATED EQUILIBRIUM
— grad·u·al·ist \-list\ *noun or adjective*

grad·u·and \ˌgra-jə-ˈwand\ *noun* [Medieval Latin *graduandus,* gerundive of *graduare*] (1882)
British **:** one about to graduate **:** a candidate for a degree

¹grad·u·ate \ˈgra-jə-wət, -ˌwāt, ˈgraj-wət\ *noun* (15th century)
1 : a holder of an academic degree or diploma
2 : a graduated cup, cylinder, or flask

²graduate *adjective* (15th century)
1 : holding an academic degree or diploma
2 : of, relating to, or engaged in studies beyond the first or bachelor's degree ⟨*graduate* school⟩ ⟨a *graduate* student⟩

³grad·u·ate \ˈgra-jə-ˌwāt\ *verb* **-at·ed; -at·ing** [Medieval Latin *graduare,* from Latin *gradus* step, degree] (15th century)
transitive verb
1 a : to grant an academic degree or diploma to **b :** to be graduated from
2 a : to mark with degrees of measurement **b :** to divide into grades or intervals
3 : to admit to a particular standing or grade
intransitive verb
1 : to receive an academic degree or diploma
2 : to pass from one stage of experience, proficiency, or prestige to a usually higher one
3 : to change gradually ▪
— grad·u·a·tor \-ˌwā-tər\ *noun*

grad·u·at·ed *adjective* (1861)
of a tax **:** increasing in rate with increase in taxable base **:** PROGRESSIVE ⟨*graduated* income tax⟩

graduated cylinder *noun* (1948)
: a tall narrow container with a volume scale used especially for measuring liquids

grad·u·a·tion \ˌgra-jə-ˈwā-shən\ *noun* (1594)
1 : a mark on an instrument or vessel indicating degrees or quantity; *also* **:** these marks
2 a : the award or acceptance of an academic degree or diploma **b :** COMMENCEMENT
3 : arrangement in degrees or ranks

Graeco- — see GRECO-

graf·fi·to \grə-ˈfē-(ˌ)tō, gra-, grä-\ *noun, plural* **-ti** \-(ˌ)tē\ [Italian, incised inscription, from *graffiare* to scratch, probably from *grafio* stylus, from Latin *graphium*] (1851)
: an inscription or drawing made on some public surface (as a rock or wall); *also* **:** a message or slogan written as or as if as a graffito ▪
— graf·fi·tist \-ˈfē-tist\ *noun*

¹graft \ˈgraft\ *noun* [Middle English *graffe, grafte,* from Middle French *grafe,* from Medieval Latin *graphium,* from Latin, stylus, from Greek *grapheion,* from *graphein* to write — more at CARVE] (14th century)
1 a : a grafted plant **b :** SCION 1 **c :** the point of insertion of a scion upon a stock
2 a : the act of grafting **b :** something grafted; *specifically* **:** living tissue used in grafting

graft 1c: *a* scion, *b* stock

²graft (14th century)
transitive verb
1 a : to cause (a scion) to unite with a stock; *also* **:** to unite (plants or scion and stock) to form a graft **b :** to propagate (a plant) by grafting
2 a : to join or unite as if by grafting **b :** to attach (a chemical unit) to a main molecular chain
3 : to implant (living tissue) surgically
intransitive verb
1 : to become grafted
2 : to perform grafting
— graft·er *noun*

³graft *noun* [English dialect *graft,* verb, to work, perhaps alteration of ¹*grave* (to dig)] (1853)
chiefly British **:** WORK, LABOR

⁴graft [origin unknown] (1859)
transitive verb
: to get (illicit gain) by graft
intransitive verb
: to practice graft

⁵graft *noun* (1865)
: the acquisition of gain (as money) in dishonest or questionable ways; *also* **:** illegal or unfair gain

graft·age \ˈgraf-tij\ *noun* (circa 1895)
: the principles and practice of grafting

graft–versus–host *adjective* (1972)
: relating to or being the bodily condition that results when cells from a tissue or organ transplant mount an immunological attack against the cells or tissues of the host

gra·ham cracker \ˈgram-, ˈgrā-əm-\ *noun* [*graham flour*] (1882)
: a slightly sweet cracker made of whole wheat flour

graham flour *noun* [Sylvester *Graham* (died 1851) American dietary reformer] (1834)
: whole wheat flour

grail \ˈgrā(ə)l\ *noun* [Middle English *graal,* from Middle French, bowl, grail, from Medieval Latin *gradalis*]
1 *capitalized* **:** the cup or platter used according to medieval legend by Christ at the Last Supper and thereafter the object of knightly quests
2 : the object of an extended or difficult quest

¹grain \ˈgrān\ *noun* [Middle English, partly from Middle French *grain* cereal grain, from Latin *granum;* partly from Middle French *graine* seed, kermes, from Latin *grana,* plural of *granum* — more at CORN] (14th century)
1 a (1) *obsolete* **:** a single small hard seed (2) **:** a seed or fruit of a cereal grass **:** CARYOPSIS **b :** the seeds or fruits of various food plants including the cereal grasses and in commercial and statutory usage other plants (as the soybean) **c :** plants producing grain
2 a (1) **:** a small hard particle or crystal (2) **:** any of the particles produced in a photographic material by its development; *also* **:** the size of such grains in the aggregate (3) **:** an individual crystal in a metal **b :** a minute portion or particle **c :** the least amount possible ⟨a *grain* of truth⟩
3 a : kermes or a scarlet dye made from it **b :** cochineal or a brilliant scarlet dye made from it **c :** a fast dye **d** *archaic* **:** COLOR, TINT
4 a : a granulated surface or appearance **b :** the outer or hair side of a skin or hide
5 : a unit of weight based on the weight of a grain of wheat taken as an average of the weight of grains from the middle of the ear — see WEIGHT table
6 a : the stratification of the wood fibers in a piece of wood **b :** a texture due to constituent particles or fibers ⟨the *grain* of a rock⟩ **c :** the direction of threads in cloth
7 : tactile quality
8 a : natural disposition **:** TEMPER ⟨against my *grain*⟩ **b :** a basic or characteristic quality ⟨doctrines . . . very much in the American *grain* —R. W. Noland⟩
— grained \ˈgrānd\ *adjective*

²grain (1530)
transitive verb
1 : INGRAIN

☐ USAGE

graduate In the 19th century the transitive sense (1a) was prescribed; the intransitive ⟨I *graduated* from college⟩ was condemned. The intransitive prevailed nonetheless, and today it is the sense likely to be prescribed and the newer transitive (sense 1b) the one condemned. All three are standard. The intransitive is currently the most common, the new transitive the least common.

graffito The plural *graffiti* is sometimes used with a singular verb as a mass noun ⟨the *graffiti* is being covered with fresh paint —*Springfield (Mass.) Union*⟩ ⟨*graffiti* comes in various styles —S. K. Oberbeck⟩ but this use is not yet as well established as the mass-noun use of *data.* Use of *graffiti* as a singular count noun is still quite rare and is not standard.

2 : to form into grains **:** GRANULATE
3 : to paint in imitation of the grain of wood or stone
4 : to feed with grain
intransitive verb
: to become granular **:** GRANULATE
— grain·er *noun*

grain alcohol *noun* (1889)
: ETHANOL

grain elevator *noun* (1852)
: a building for elevating, storing, discharging, and sometimes processing grain

grain·field \ˈgrān-ˌfēld\ *noun* (circa 1818)
: a field where grain is grown

grain of salt (1647)
: a skeptical attitude

grains of paradise (15th century)
: the pungent seeds of a West African plant (*Aframomum melegueta*) of the ginger family that are used as a spice

grain sorghum *noun* (1920)
: any of several sorghums cultivated primarily for grain — compare SORGO

grainy \ˈgrā-nē\ *adjective* **grain·i·er; -est** (15th century)
1 : resembling or having some characteristic of grain **:** not smooth or fine
2 *of a photograph* **:** appearing to be composed of grain-like particles
— grain·i·ness *noun*

¹gram \ˈgram\ *noun* [obsolete Portuguese (now spelled *grão*), grain, from Latin *granum*] (1702)
: any of several leguminous plants (as a chickpea) grown especially for their seed; *also* **:** their seeds

²gram \ˈgram\ *noun* [French *gramme,* from Late Latin *gramma,* a small weight, from Greek *grammat-, gramma* letter, writing, a small weight, from *graphein* to write — more at CARVE] (1810)
1 : a metric unit of mass equal to 1/1000 kilogram and nearly equal to the mass of one cubic centimeter of water at its maximum density — see METRIC SYSTEM table
2 : the weight of a gram under standard gravity

³gram *noun* [by shortening & alteration] (circa 1934)
: GRANDMOTHER

-gram \ˌgram\ *noun combining form* [Latin *-gramma,* from Greek, from *gramma*]
: drawing **:** writing **:** record ⟨chrono*gram*⟩ ⟨tele*gram*⟩

grama \ˈgra-mə\ *noun* [Spanish, from Latin *gramina,* plural of *gramen* grass] (1828)
: any of several pasture grasses (genus *Bouteloua*) of the western U.S.

gram–atomic weight *noun* (1927)
: the mass of one mole of an element equal in grams to the atomic weight — called also *gram-atom*

gram calorie *noun* (1902)
: CALORIE 1a

gram equivalent *noun* (circa 1897)
: the quantity of an element, group, or compound that has a mass in grams equal to the equivalent weight

gra·mer·cy \grə-ˈmər-sē\ *interjection* [Middle English *grand mercy,* from Middle French *grand merci* great thanks] (14th century)
archaic — used to express gratitude or surprise

gram·i·ci·din \ˌgra-mə-ˈsī-dᵊn\ *noun* [*gram*-positive + *-i-* + *-cide* + *-in*] (1940)
: any of several toxic crystalline polypeptide antibiotics produced by a soil bacterium (*Bacillus brevis*) and used against gram-positive bacteria in local infections

gra·min·e·ous \grə-ˈmi-nē-əs\ *adjective* [Latin *gramineus,* from *gramin-, gramen* grass] (circa 1658)
: of or relating to a grass

gram·i·niv·o·rous \ˌgra-mə-ˈni-v(ə-)rəs\ *adjective* [Latin *gramin-, gramen*] (1739)
: feeding on grass or the seeds of grass ⟨*graminivorous* locusts⟩ ⟨*graminivorous* birds⟩

gram·mar \ˈgra-mər\ *noun* [Middle English *gramere,* from Middle French *gramaire,* modification of Latin *grammatica,* from Greek *grammatikē,* from feminine of *grammatikos* of letters, from *grammat-, gramma* — more at GRAM] (14th century)
1 a : the study of the classes of words, their inflections, and their functions and relations in the sentence **b :** a study of what is to be preferred and what avoided in inflection and syntax
2 a : the characteristic system of inflections and syntax of a language **b :** a system of rules that defines the grammatical structure of a language
3 a : a grammar textbook **b :** speech or writing evaluated according to its conformity to grammatical rules
4 : the principles or rules of an art, science, or technique ⟨a *grammar* of the theater⟩
word history see GLAMOUR
— gram·mar·i·an \grə-ˈmer-ē-ən, -ˈmar-\ *noun*

grammar school *noun* (14th century)
1 a : a secondary school emphasizing Latin and Greek in preparation for college **b :** a British college preparatory school
2 : a school intermediate between primary school and high school
3 : ELEMENTARY SCHOOL

gram·mat·i·cal \grə-ˈma-ti-kəl\ *adjective* (1530)
1 : of or relating to grammar
2 : conforming to the rules of grammar
— gram·mat·i·cal·i·ty \-ˌma-tə-ˈka-lə-tē\ *noun*
— gram·mat·i·cal·ly \-ˈma-ti-k(ə-)lē\ *adverb*
— gram·mat·i·cal·ness \-kəl-nəs\ *noun*

grammatical meaning *noun* (1769)
: the part of meaning that varies from one inflectional form to another (as from *plays* to *played* to *playing*) — compare LEXICAL MEANING

gramme *chiefly British variant of* ²GRAM

gram molecular weight *noun* (circa 1902)
: the mass of one mole of a compound equal in grams to the molecular weight — called also *gram-molecule*

Gram·my \ˈgra-mē\ *service mark*
— used for the annual presentation of a statuette for notable achievement in the recording industry

gram–neg·a·tive \ˈgram-ˈne-gə-tiv\ *adjective* (1907)
: not holding the purple dye when stained by Gram's stain — used chiefly of bacteria

gram·o·phone \ˈgra-mə-ˌfōn\ *noun* [from *Gramophone,* a trademark] (1887)
: PHONOGRAPH

gramp \ˈgramp\ *or* **gramps** \ˈgram(p)s\ *noun, plural* **gramps** [by shortening & alteration] (circa 1900)
: GRANDFATHER

gram–pos·i·tive \ˈgram-ˈpä-zə-tiv, -ˈpäz-tiv\ *adjective* (1907)
: holding the purple dye when stained by Gram's stain — used chiefly of bacteria

gram·pus \ˈgram-pəs\ *noun* [alteration of Middle English *graspey, grapay,* from Middle French *graspeis,* from *gras* fat (from Latin *crassus*) + *peis* fish, from Latin *piscis* — more at CRASS, FISH] (circa 1529)
1 : a dolphin (*Grampus griseus*) of temperate and tropical seas; *also* **:** any of various small cetaceans
2 : the giant whip scorpion (*Mastigoproctus giganteus*) of the southern U.S.

Gram's stain \ˈgramz-\ *or* **Gram stain** \ˈgram-\ *noun* [Hans C. J. *Gram* (died 1938) Danish physician] (1903)
1 : a method for the differential staining of bacteria by treatment with a watery solution of iodine and the iodide of potassium after staining with a triphenylmethane dye (as crystal violet) — called also *Gram's method*
2 : the chemicals used in Gram's stain

gram–vari·able \ˈgram-ˈver-ē-ə-bəl, -ˈvar-\ *adjective* (1956)
: staining irregularly or inconsistently by Gram's stain

grana *plural of* GRANUM

gran·a·dil·la \ˌgran-ə-ˈdi-lə, -ˈdē-(y)ə\ *noun* [Spanish, diminutive of *granada* pomegranate, from Late Latin *granata* — more at GRENADE] (1613)
: the oblong fruit of various passionflowers (especially *Passiflora quadrangularis* of tropical America) used as a dessert; *also* **:** a plant that produces granadillas

gra·na·ry \ˈgrā-nə-rē, ˈgra-\ *noun, plural* **-ries** [Latin *granarium,* from *granum* grain] (1530)
1 a : a storehouse for threshed grain **b :** a region producing grain in abundance
2 : a chief source or storehouse

¹grand \ˈgrand\ *adjective* [Middle French, large, great, grand, from Latin *grandis*] (1584)
1 a : having more importance than others **:** FOREMOST **b :** having higher rank than others bearing the same general designation ⟨the *grand* champion⟩
2 a : INCLUSIVE, COMPREHENSIVE ⟨the *grand* total of all money paid out⟩ **b :** DEFINITIVE, INCONTROVERTIBLE ⟨*grand* example⟩
3 : CHIEF, PRINCIPAL
4 : large and striking in size, scope, extent, or conception ⟨*grand* design⟩
5 a : LAVISH, SUMPTUOUS ⟨a *grand* celebration⟩ **b :** marked by a regal form and dignity **c :** fine or imposing in appearance or impression **d :** LOFTY, SUBLIME ⟨writing in the *grand* style⟩
6 a : pretending to social superiority **:** SUPERCILIOUS **b :** intended to impress ⟨a person of *grand* gestures⟩
7 : very good **:** WONDERFUL ⟨a *grand* time⟩ ☆
— grand·ly \ˈgran-(d)lē\ *adverb*
— grand·ness \ˈgran(d)-nəs\ *noun*

²grand *noun* (1840)
1 : GRAND PIANO
2 *slang* **:** a thousand dollars

gran·dam \ˈgran-ˌdam, -dəm\ *noun* [Middle English *graundam,* from Anglo-French *graund dame,* literally, great lady] (13th century)
1 *or* **gran·dame** \-ˌdām\ **a :** GRANDMOTHER **b :** an old woman
2 *or* **grand·dam** \-ˌdam, -dəm\ **:** a dam's or sire's dam — used of an animal

grand·aunt \ˈgran-ˈdant, -ˈdȧnt\ *noun* (1826)
: the aunt of one's father or mother — called also *great-aunt*

grand·ba·by \ˈgran(d)-ˌbā-bē\ *noun* (1916)
: an infant grandchild

grand·child \-ˌchīld\ *noun* (1587)
: the child of one's son or daughter

☆ SYNONYMS
Grand, magnificent, imposing, stately, majestic, grandiose mean large and impressive. GRAND adds to greatness of size the implications of handsomeness and dignity ⟨a *grand* staircase⟩. MAGNIFICENT implies an impressive largeness proportionate to scale without sacrifice of dignity or good taste ⟨*magnificent* paintings⟩. IMPOSING implies great size and dignity but especially stresses impressiveness ⟨an *imposing* edifice⟩. STATELY may suggest poised dignity, erectness of bearing, handsomeness of proportions, ceremonious deliberation of movement ⟨the *stately* procession⟩. MAJESTIC combines the implications of IMPOSING and STATELY and usually adds a suggestion of solemn grandeur ⟨a *majestic* waterfall⟩. GRANDIOSE implies a size or scope exceeding ordinary experience ⟨*grandiose* hydroelectric projects⟩ but is most commonly applied derogatorily to inflated pretension or absurd exaggeration ⟨*grandiose* schemes⟩.

grand·dad *or* **gran·dad** \ˈgran-ˌdad\ *noun* (1782)
: GRANDFATHER

grand·dad·dy \-ˌda-dē\ *also* **gran·dad·dy** *noun* (1769)
1 : GRANDFATHER
2 : one that is the first, earliest, or most venerable of its kind

grand·daugh·ter \-ˌdȯ-tər\ *noun* (1611)
: the daughter of one's son or daughter

grand duchess *noun* (circa 1757)
1 : a woman who rules a grand duchy in her own right
2 : the wife or widow of a grand duke

grand duchy *noun* (1835)
: the territory or dominion of a grand duke or grand duchess

grand duke *noun* (circa 1693)
1 : the sovereign duke of any of various European states
2 : a male descendant of a Russian czar in the male line

grande dame \ˈgrän-ˈdäm, gräⁿd-dàm\ *noun, plural* **grandes dames** \-ˈdäm(z), -dàm(z)\ *also* **grande dames** *same*\ [French, literally, great lady] (1775)
: a usually elderly woman of great prestige or ability

gran·dee \gran-ˈdē\ *noun* [Spanish *grande,* from *grande,* adjective, large, great, from Latin *grandis*] (1598)
: a man of elevated rank or station; *especially* : a Spanish or Portuguese nobleman of the first rank

gran·deur \ˈgran-jər, -ˌju̇r, -ˌd(y)u̇r, -d(y)ər\ *noun* [Middle English, from Middle French, from *grand*] (1600)
1 : the quality or state of being grand : MAGNIFICENCE ⟨the glory that was Greece and the *grandeur* that was Rome —E. A. Poe⟩
2 : an instance or example of grandeur

¹grand·fa·ther \ˈgran(d)-ˌfä-t͟hər, -ˌfà-\ *noun* (15th century)
: the father of one's father or mother; *also* : ANCESTOR 1a
— **grand·fa·ther·ly** \-lē\ *adjective*

²grandfather *transitive verb* (1972)
: to permit to continue under a grandfather clause

grandfather clause *noun* (1900)
: a clause creating an exemption based on circumstances previously existing; *especially* : a provision in several southern state constitutions designed to enfranchise poor whites and disfranchise Negroes by waiving high voting requirements for descendants of men voting before 1867

grandfather clock *noun* [from the song *My Grandfather's Clock* (1876) by Henry C. Work (died 1884) American songwriter] (1909)
: a tall pendulum clock that stands on the floor — called also *grandfather, grandfather's clock*

grand finale *noun* (1800)
: a climactic finale (as of an opera)

grand fir *noun* (1897)
: a lofty fir tree (*Abies grandis*) of the northwestern Pacific coastal region of North America with cylindrical greenish cones and soft wood

Grand Gui·gnol \ˌgräⁿ-gēn-ˈyȯl, -ˈyōl\ *noun* [*Le Grand Guignol,* small theater in Montmartre, Paris, specializing in such performances] (1908)
: dramatic entertainment featuring the gruesome or horrible ◆
— **Grand Guignol** *adjective*

gran·di·flo·ra \ˌgran-də-ˈflōr-ə, -ˈflȯr-\ *noun* [New Latin, from Latin *grandis* great + *flor-, flos* flower — more at BLOW] (1944)
: a bush rose derived from crosses of floribunda and hybrid tea roses and characterized by production of blooms both singly and in clusters on the same plant

gran·dil·o·quence \gran-ˈdi-lə-kwən(t)s\ *noun* [probably from Middle French, from Latin *grandiloquus* using lofty language, from *grandis* + *loqui* to speak] (1589)
: a lofty, extravagantly colorful, pompous, or bombastic style, manner, or quality especially in language
— **gran·dil·o·quent** \-kwənt\ *adjective*
— **gran·dil·o·quent·ly** *adverb*

gran·di·ose \ˈgran-dē-ˌōs, ˌgran-dē-ˈ\ *adjective* [French, from Italian *grandioso,* from *grande* great, from Latin *grandis*] (1840)
1 : characterized by affectation of grandeur or splendor or by absurd exaggeration
2 : impressive because of uncommon largeness, scope, effect, or grandeur
synonym see GRAND
— **gran·di·ose·ly** *adverb*
— **gran·di·ose·ness** *noun*
— **gran·di·os·i·ty** \ˌgran-dē-ˈä-sə-tē\ *noun*

gran·di·o·so \ˌgrän-dē-ˈō-(ˌ)sō, ˌgran-, -(ˌ)zō\ *adverb or adjective* [Italian] (circa 1859)
: in a broad and noble style — used as a direction in music

grand jury *noun* (15th century)
: a jury that examines accusations against persons charged with crime and if the evidence warrants makes formal charges on which the accused persons are later tried
— **grand juror** *noun*

grand·kid \ˈgran(d)-ˌkid\ *noun* (1927)
: GRANDCHILD

Grand Lama *noun* (1807)
: DALAI LAMA

grand larceny *noun* (1828)
: larceny of property of a value greater than that fixed as constituting petit larceny

grand·ma \ˈgran(d)-ˌmä, -ˌmȯ; ˈgra-ˌmä, -ˌmȯ\ *noun* (1867)
: GRANDMOTHER

grand mal \ˈgrän(d)-ˌmäl, ˈgräⁿ-ˌmäl, -ˌmal; ˈgran(d)-ˌmal\ *noun* [French, literally, great illness] (1897)
: severe epilepsy

grand manner *noun* (1775)
: an elevated or grand style (as in music or literature)

grand march *noun* (1898)
: an opening ceremony at a ball that consists of a march participated in by all the guests

grand marshal *noun* (1951)
: a person honored as the ceremonial marshal of a parade

grand master *noun* (1724)
1 : the chief officer of a principal lodge in various fraternal orders (as Freemasonry)
2 : an expert player (as of chess) who has consistently scored high in international competition

grand·moth·er \ˈgran(d)-ˌmə-t͟hər\ *noun* (15th century)
: the mother of one's father or mother; *also* : a female ancestor
— **grand·moth·er·ly** \-lē\ *adjective*

grand·neph·ew \ˈgran(d)-ˈnef-(ˌ)yü, *chiefly British* -ˈnev-\ *noun* (circa 1639)
: a grandson of one's brother or sister

grand·niece \-ˈnēs\ *noun* (circa 1830)
: a granddaughter of one's brother or sister

grand old man *noun* (1860)
: a venerated practitioner or former practitioner of an art, profession, or sport ⟨the *grand old man* of jazz⟩

grand opera *noun* (1803)
: opera in which the plot is serious or tragic and the entire text is set to music

grand·pa \ˈgran(d)-ˌpä, -ˌpȯ; ˈgram-ˌpä, -ˌpȯ\ *noun* (circa 1889)
: GRANDFATHER

grand·par·ent \ˈgran(d)-ˌpar-ənt, -ˌper-\ *noun* (1830)
: a parent of one's father or mother
— **grand·pa·ren·tal** \ˌgran(d)-pə-ˈren-tᵊl\ *adjective*
— **grand·par·ent·hood** \ˈgran(d)-ˈpar-ent-ˌhu̇d, -ˈper-\ *noun*

grand piano *noun* (1834)
: a piano with horizontal frame and strings — compare UPRIGHT PIANO

grand prix \ˈgräⁿ-ˈprē\ *noun, plural* **grand prix** *also* **grands prix** \-ˈprē(z)\ *often G&P capitalized* [French *Grand Prix de Paris,* an international horse race established 1863, literally, grand prize of Paris] (1863)
1 : the highest level of international equestrian competition; *also* : a contest at this level
2 : one of a series of international formula car races

grand·sire \ˈgran(d)-ˌsīr\ *noun* [Middle English] (14th century)
1 *or* **grand·sir** \ˈgran(t)-sər\ *dialect* : GRANDFATHER
2 *archaic* : FOREFATHER
3 *archaic* : an aged man
4 : a dam's or sire's sire — used of an animal

grand slam *noun* (1814)
1 : the winning of all the tricks in one hand of a card game (as bridge)
2 : a clean sweep or total success; *specifically* : the winning of all the major or specified tournaments on a tour ⟨twice won the tennis *grand slam*⟩
3 : a home run made with the bases loaded
— **grand–slam** *adjective*

grand·son \ˈgran(d)-ˌsən\ *noun* (1586)
: the son of one's son or daughter

¹grand·stand \-ˌstand\ *noun* (1834)
1 : a usually roofed stand for spectators at a racecourse or stadium
2 : AUDIENCE

²grandstand *adjective* (1893)
: done for show or to impress onlookers ⟨a *grandstand* play⟩

³grandstand *intransitive verb* (circa 1917)
: to play or act so as to impress onlookers
— **grand·stand·er** *noun*

grand theft *noun* (circa 1930)
: GRAND LARCENY

grand tour *noun* (1670)
1 : an extended tour of the Continent that was formerly a usual part of the education of young British gentlemen
2 : an extensive and usually educational tour

◇ WORD HISTORY

Grand Guignol Long before cinematic violence became explicit, the Grand-Guignol theater in the Montmartre district of Paris presented to audiences graphic depictions of stabbings, assaults with sulphuric acid, and assorted other mayhem. Founded in 1895 as the Théâtre Salon, it adopted its better-known name and characteristic formula in 1897. On a single night's bill, one-act light comedies alternated with dramas of horror, the latter featuring cleverly arranged (for the time) special effects and plenty of gore. *Guignol* in French denotes both "hand puppet" and, as a proper noun, the leading character in a tradition of children's puppet theater, but aside from the uniformity of repertoire there was nothing puppetlike about the Grand-Guignol. The theater remained a Parisian attraction until 1962, when it finally closed, having outlived an era when stage violence could provide a real shock.

grand touring car *noun* (1970)
: a usually 2-passenger coupe

grand touring car

grand·un·cle \'gran-'dəŋ-kəl\ *noun* (15th century)
: an uncle of one's father or mother

grand unified theory *noun* (1978)
: any of several theories that seek to unite in a single mathematical framework the electromagnetic and weak forces with the strong force or with the strong force and gravity — called also *grand unification theory*

grange \'grānj\ *noun* [Middle English, from Middle French, from Medieval Latin *granica,* from Latin *granum* grain] (14th century)
1 *archaic* **:** GRANARY, BARN
2 : FARM; *especially* **:** a farmhouse with outbuildings
3 *capitalized* **:** one of the lodges of a national fraternal association originally made up of farmers; *also* **:** the association itself

grang·er \'grān-jər\ *noun* (1873)
1 *capitalized* **:** a member of a Grange
2 *chiefly West* **:** FARMER, HOMESTEADER

grang·er·ism \'grān-jə-,ri-zəm\ *noun* (1875)
: the policy or methods of the grangers

gra·ni·ta \grə-'nē-tə\ *noun* [Italian, from feminine of *granito,* past participle of *granire*] (1869)
: a coarse-textured ice confection typically made from fruit

gran·ite \'gra-nət\ *noun* [Italian *granito,* from past participle of *granire* to granulate, from *grano* grain, from Latin *granum*] (1646)
1 : a very hard natural igneous rock formation of visibly crystalline texture formed essentially of quartz and orthoclase or microcline and used especially for building and for monuments
2 : unyielding firmness or endurance ⟨the cold *granite* of Puritan formalism —V. L. Parrington⟩
— gran·ite·like \-,līk\ *adjective*
— gra·nit·ic \gra-'ni-tik\ *adjective*
— gran·it·oid \'gra-nə-,tȯid\ *adjective*

gran·ite·ware \'gra-nət-,war, -,wer\ *noun* (1878)
: ironware with grayish or bluish mottled enamel

gra·niv·o·rous \grə-'niv-rəs, grā-, -'ni-və-\ *adjective* [Latin *granum* grain] (1646)
: feeding on seeds or grain ⟨*granivorous* rodents⟩

gran·ny *or* **gran·nie** \'gra-nē\ *noun, plural* **grannies** [by shortening & alteration] (1663)
1 a : GRANDMOTHER **b :** a fussy person
2 *chiefly Southern & south Midland* **:** MIDWIFE

granny dress *noun* (1909)
: a long loose-fitting dress usually with high neck and long sleeves

granny glasses *noun plural* (1966)
: spectacles with usually small oval, round, or square lenses and metal frames

granny knot *noun* (1853)
: an insecure knot often made instead of a square knot — see KNOT illustration

Granny Smith \-'smith\ *noun* [Maria Ann *Smith* (died 1870) who cultivated it near Sydney, Australia] (1895)
: a tart green apple of Australian origin

grano- *combining form* [German, from *granit,* from Italian *granito*]
: granite **:** granitic ⟨*grano*diorite⟩

grano·di·o·rite \,gra-nō-'dī-ə-,rīt\ *noun* (1893)
: a granular intrusive quartzose igneous rock intermediate between granite and quartz-containing diorite with plagioclase predominant over orthoclase
— grano·di·o·rit·ic \-,dī-ə-'ri-tik\ *adjective*

gra·no·la \grə-'nō-lə\ *noun* [from *Granola,* a trademark] (1970)
: a mixture typically of rolled oats and various added ingredients (as brown sugar, raisins, coconut, and nuts) that is eaten especially for breakfast or as a snack

grano·lith·ic \,gra-nə-'li-thik\ *adjective* (1881)
: relating to or composed of a mixture of crushed granite and cement

grano·phyre \'gran-ə-,fī(-ə)r\ *noun* [International Scientific Vocabulary, from *grano-* + French *-phyre* (as in *porphyre* porphyry)] (1882)
: a porphyritic igneous rock chiefly of feldspar and quartz with granular groundmass
— grano·phyr·ic \,gran-ə-'fir-ik\ *adjective*

¹grant \'grant\ *transitive verb* [Middle English, from Old French *creanter, graanter,* from (assumed) Vulgar Latin *credentare,* from Latin *credent-, credens,* present participle of *credere* to believe — more at CREED] (13th century)
1 a : to consent to carry out for a person **:** allow fulfillment of ⟨*grant* a request⟩ **b :** to permit as a right, privilege, or favor ⟨luggage allowances *granted* to passengers⟩
2 : to bestow or transfer formally ⟨*grant* a scholarship to a student⟩; *specifically* **:** to give the possession or title of by a deed
3 a : to be willing to concede **b :** to assume to be true ⟨*granting* that you are correct, you may find it hard to prove your point⟩ ☆
— grant·able \'gran-tə-bəl\ *adjective*
— grant·er \-tər\ *noun*
— grant·or \'gran-tər, -,tȯr; gran-'tȯr\ *noun*

²grant *noun* (13th century)
1 : the act of granting
2 : something granted; *especially* **:** a gift (as of land or money) for a particular purpose
3 a : a transfer of property by deed or writing **b :** the instrument by which such a transfer is made; *also* **:** the property so transferred
4 : a minor territorial division of Maine, New Hampshire, or Vermont originally granted by the state to an individual or institution

grant·ee \gran-'tē\ *noun* (15th century)
: one to whom a grant is made

grant–in–aid \,gran-tᵊn-'ād\ *noun, plural* **grants–in–aid** \,gran(t)-sᵊn-'ād\ (1851)
1 : a grant or subsidy for public funds paid by a central to a local government in aid of a public undertaking
2 : a grant or subsidy to a school or individual for an educational or artistic project

grants·man \'gran(t)-smən\ *noun* (1966)
: a specialist in grantsmanship

grants·man·ship \-,ship\ *noun* [*grants* + *-manship*] (1961)
: the art of obtaining grants (as for research)
word history see GAMESMANSHIP

granul- *or* **granuli-** *or* **granulo-** *combining form* [Late Latin *granulum*]
: granule ⟨*granulo*cyte⟩

gran·u·lar \'gran-yə-lər\ *adjective* (1794)
: consisting of or appearing to consist of granules **:** GRAINY
— gran·u·lar·i·ty \,gran-yə-'lar-ə-tē\ *noun*

gran·u·late \'gran-yə-,lāt\ *verb* **-lat·ed; -lat·ing** (1666)
transitive verb
: to form or crystallize into grains or granules
intransitive verb
: to form granulations ⟨an open *granulating* wound⟩
— gran·u·la·tor \-,lā-tər\ *noun*

gran·u·la·tion \,gran-yə-'lā-shən\ *noun* (1612)
1 : the act or process of granulating **:** the condition of being granulated
2 : one of the minute red granules of new capillaries formed on the surface of a wound in healing
3 : GRANULE 2

granulation tissue *noun* (1873)
: tissue made up of granulations that temporarily replaces lost tissue in a wound

gran·ule \'gran-(,)yü(ə)l\ *noun* [Late Latin *granulum,* diminutive of Latin *granum* grain] (1652)
1 : a small particle; *especially* **:** one of numerous particles forming a larger unit
2 : any of the small short-lived brilliant spots on the sun's photosphere

gran·u·lite \'gran-yə-,līt\ *noun* (1849)
: a granular metamorphic rock consisting mainly of feldspar and quartz
— gran·u·lit·ic \,gran-yə-'li-tik\ *adjective*

gran·u·lo·cyte \'gran-yə-lō-,sīt\ *noun* [International Scientific Vocabulary] (1906)
: a polymorphonuclear white blood cell with granule-containing cytoplasm
— gran·u·lo·cyt·ic \,gran-yə-lō-'si-tik\ *adjective*

gran·u·lo·cy·to·poi·e·sis \'gran-yə-lō-,sī-tə-pȯi-'ē-səs\ *noun* [New Latin] (1944)
: the formation of blood granulocytes typically in the bone marrow

gran·u·lo·ma \,gran-yə-'lō-mə\ *noun, plural* **-mas** *or* **-ma·ta** \-mə-tə\ (1861)
: a mass or nodule of chronically inflamed tissue with granulations that is usually associated with an infective process
— gran·u·lo·ma·tous \-mə-təs\ *adjective*

granuloma in·gui·na·le \-,iŋ-gwə-'na-lē, -'nä-, -'nā-\ *noun* [New Latin, literally, inguinal granuloma] (1918)
: a sexually transmitted disease characterized by ulceration and formation of granulations on the genitalia and in the groin area and caused by a bacterium (*Calymmatobacterium granulomatis,* synonym *Donovania granulomatis*)

gran·u·lo·sa cell \,gran-yə-'lō-sə-\ *noun* [New Latin *granulosa,* from feminine of *granulosus* granulose] (1936)
: one of the estrogen-secreting cells of the epithelial lining of a graafian follicle or its follicular precursor

gran·u·lose \'gran-yə-,lōs\ *adjective* (1852)
: GRANULAR; *especially* **:** having the surface roughened with granules

gran·u·lo·sis \,gran-yə-'lō-səs\ *noun, plural* **-lo·ses** \-,sēz\ [New Latin] (1949)
: a virus disease of insect larvae distinguished by the presence of minute granular inclusions in infected cells

gra·num \'grā-nəm\ *noun, plural* **gra·na** \-nə\ [New Latin, from Latin, grain — more at CORN] (1894)
: one of the lamellar stacks of chlorophyll-containing thylakoids found in plant chloroplasts

grape \'grāp\ *noun, often attributive* [Middle English, from Old French *crape, grape* hook, grape stalk, bunch of grapes, grape, of Germanic origin; akin to Old High German *krāpfo* hook] (14th century)
1 : a smooth-skinned juicy greenish white to deep red or purple berry eaten dried or fresh as a fruit or fermented to produce wine

☆ SYNONYMS
Grant, concede, vouchsafe, accord, award mean to give as a favor or a right. GRANT implies giving to a claimant or petitioner something that could be withheld ⟨*granted* them a new hearing⟩. CONCEDE implies yielding something reluctantly in response to a rightful or compelling claim ⟨even her critics *concede* she can be charming⟩. VOUCHSAFE implies granting something as a courtesy or an act of gracious condescension ⟨*vouchsafed* the secret to only a few chosen disciples⟩. ACCORD implies giving to another what is due or proper ⟨*accorded* all the honors befitting a head of state⟩. AWARD implies giving what is deserved or merited usually after a careful weighing of pertinent factors ⟨*awarded* the company a huge defense contract⟩.

2 : any of numerous woody vines (genus *Vitis* of the family Vitaceae, the grape family) that usually climb by tendrils, produce clustered fruits that are grapes, and are nearly cosmopolitan in cultivation
3 : GRAPESHOT
— grape·like \-ˌlīk\ *adjective*

grape·fruit \ˈgrāp-ˌfrüt\ *noun* (1814)
1 *plural* **grapefruit** *or* **grapefruits :** a large citrus fruit with a bitter yellow rind and inner skin and a highly flavored somewhat acid juicy pulp
2 : a small roundheaded tree (*Citrus paradisi*) that produces grapefruit

grape hyacinth *noun* (1733)
: any of several small bulbous spring-flowering herbs (genus *Muscari*) of the lily family with racemes of usually blue flowers

grape·shot \ˈgrāp-ˌshät\ *noun* (1747)
: an antipersonnel weapon consisting of a cluster of small iron balls shot from a cannon

grape sugar *noun* (1831)
: DEXTROSE

grape·vine \ˈgrāp-ˌvīn\ *noun* (circa 1736)
1 : GRAPE 2
2 a : an informal person-to-person means of circulating information or gossip ⟨heard about the meeting through the *grapevine*⟩ **b :** a secret source of information

[1]graph \ˈgraf\ *noun* [short for *graphic formula*] (1886)
1 : the collection of all points whose coordinates satisfy a given relation (as a function)
2 : a diagram (as a series of one or more points, lines, line segments, curves, or areas) that represents the variation of a variable in comparison with that of one or more other variables

[2]graph *transitive verb* (1898)
1 : to represent by a graph
2 : to plot on a graph

[3]graph *noun* [probably from *-graph*] (1933)
1 : a written or printed representation of a basic unit of speech (as a phoneme or syllable); *especially* **:** GRAPHEME 2
2 : a single occurrence of a letter of an alphabet in any of its various shapes

-graph \ˌgraf\ *noun combining form* [Middle French *-graphe,* from Latin *-graphum,* from Greek *-graphon,* from neuter of *-graphos* written, from *graphein* to write — more at CARVE]
1 : something written or drawn ⟨mono*graph*⟩
2 [French *-graphe,* from Late Latin *-graphus*] **:** instrument for making or transmitting records or images ⟨chrono*graph*⟩

graph·eme \ˈgra-ˌfēm\ *noun* (1935)
1 : the set of units of a writing system (as letters and letter combinations) that represent a phoneme
2 : a unit (as a letter) of a writing system
— gra·phe·mic \gra-ˈfē-mik\ *adjective*
— gra·phe·mi·cal·ly \-mi-k(ə-)lē\ *adverb*

gra·phe·mics \gra-ˈfē-miks\ *noun plural but singular or plural in construction* (1951)
: the study and analysis of a writing system in terms of graphemes

[1]graph·ic \ˈgra-fik\ *also* **graph·i·cal** \-fi kəl\ *adjective* [Latin *graphicus,* from Greek *graphikos,* from *graphein*] (1637)
1 : formed by writing, drawing, or engraving
2 *usually graphic* **a :** marked by clear lifelike or vividly realistic description **b :** vividly or plainly shown or described
3 a : of or relating to the pictorial arts; *also* **:** PICTORIAL **b :** of, relating to, or involving such reproductive methods as those of engraving, etching, lithography, photography, serigraphy, and woodcut **c :** of or relating to the art of printing **d :** relating or according to graphics
4 *usually graphical* **:** of, relating to, or represented by a graph
5 : of or relating to the written or printed word or the symbols or devices used in writing or printing to represent sound or convey meaning ☆
— graph·i·cal·ly \-fi-k(ə-)lē\ *adverb*
— graph·ic·ness \-fik-nəs\ *noun*

[2]graphic *noun* (1944)
1 a : a product of graphic art **b** *plural* **:** the graphic media
2 a : a graphic representation (as a picture, map, or graph) used especially for illustration **b** *plural but singular or plural in construction* **:** the art or science of drawing a representation of an object on a two-dimensional surface according to mathematical rules of projection
3 *plural but singular or plural in construction* **:** the process whereby a computer displays graphics on a CRT and an operator can manipulate them (as with a light pen)
4 : a printed message superimposed on a television picture

-graphic *or* **-graphical** *adjective combining form* [Late Latin *-graphicus,* from Greek *-graphikos,* from *graphikos*]
: written or transmitted in a (specified) way ⟨stylo*graphic*⟩

graphic arts *noun plural* (1882)
: the fine and applied arts of representation, decoration, and writing or printing on flat surfaces together with the techniques and crafts associated with them

graphic design *noun* (1956)
: the art or profession of using design elements (as typography and images) to convey information or create an effect; *also* **:** a product of this art
— graphic designer *noun*

graphic equalizer *noun* (1969)
: an electronic device for adjusting the frequency response of an audio system by means of a number of slides each of which controls the response for a band centered on a particular frequency

graphic novel *noun* (1978)
: a fictional story for adults that is presented in comic-strip format and published as a book

graphics tablet *noun* (1980)
: a device by which pictorial information is entered into a computer in a manner similar to drawing

graph·ite \ˈgra-ˌfīt\ *noun* [German *Graphit,* from Greek *graphein* to write] (1796)
1 : a soft black lustrous form of carbon that conducts electricity and is used in lead pencils and electrolytic anodes, as a lubricant, and as a moderator in nuclear reactors
2 : a composite material in which carbon fibers are the reinforcing material
— gra·phit·ic \gra-ˈfi-tik\ *adjective*

graph·i·tize \ˈgra-fə-ˌtīz, -ˌfīt-ˌīz\ *transitive verb* **-tized; -tiz·ing** (1899)
: to convert into graphite
— graph·i·tiz·able \-ˌtī-zə-bəl, -ˌfīt-īz-\ *adjective*
— graph·i·ti·za·tion \ˌgra-fə-tə-ˈzā-shən, -ˌfī-\ *noun*

grapho- *combining form* [French, from Middle French, from Greek, from *graphē,* from *graphein* to write]
: writing ⟨*grapho*lect⟩

graph·o·lect \ˈgra-fə-ˌlekt\ *noun* [*grapho-* + *-lect* (as in *dialect*)] (1977)
: a standard written language

gra·phol·o·gist \gra-ˈfä-lə-jist\ *noun* (1885)
: a specialist in graphology

gra·phol·o·gy \-jē\ *noun* [French *graphologie,* from *grapho-* + *-logie* -logy] (1886)
: the study of handwriting especially for the purpose of character analysis
— graph·o·log·i·cal \ˌgra-fə-ˈlä-ji-kəl\ *adjective*

graph paper *noun* (1927)
: paper ruled for drawing graphs

-graphy *noun combining form* [Latin *-graphia,* from Greek, from *graphein*]
1 : writing or representation in a (specified) manner or by a (specified) means or of a (specified) object ⟨steno*graphy*⟩ ⟨photo*graphy*⟩
2 : writing on a (specified) subject or in a (specified) field ⟨hagio*graphy*⟩

grap·nel \ˈgrap-nᵊl\ *noun* [Middle English *grapenel,* from (assumed) Middle French *grapinel,* diminutive of *grapin,* diminutive of *grape* hook — more at GRAPE] (14th century)
: a small anchor with usually four or five flukes used in dragging or grappling operations and for anchoring a dory or skiff — see ANCHOR illustration

grap·pa \ˈgrä-pə\ *noun* [Italian, from Italian dialect, grape stalk, of Germanic origin; akin to Old High German *krāpfo* hook] (circa 1893)
: a dry colorless brandy distilled from fermented grape pomace

[1]grap·ple \ˈgra-pəl\ *noun* [Old French *grappelle,* diminutive of *grape* hook — more at GRAPE] (1601)
1 a : the act or an instance of grappling **b :** a hand-to-hand struggle **c :** a contest for superiority or mastery
2 : a bucket similar to a clamshell but usually having more jaws

[2]grapple *verb* **grap·pled; grap·pling** \ˈgra-p(ə-)liŋ\ (1530)
transitive verb
1 : to seize with or as if with a grapple
2 : to come to grips with **:** WRESTLE
3 : to bind closely
intransitive verb
1 : to make a ship fast with a grappling iron
2 : to come to grips
3 : to use a grapple
— grap·pler \-p(ə-)lər\ *noun*

grap·pling *noun* (1582)
1 : GRAPNEL
2 : GRAPPLING IRON

grappling iron *noun* (1538)
: a hooked iron for anchoring a boat, grappling ships to each other, or recovering sunken objects — called also *grappling hook*

grap·to·lite \ˈgrap-tə-ˌlīt\ *noun* [Greek *graptos* painted (from *graphein* to write, paint) + English *-lite* — more at CARVE] (1841)
: any of numerous extinct fossil colonial Paleozoic animals (group Graptolitoidea) with zooids in cups along a chitinous support

grapy *or* **grap·ey** \ˈgrā-pē\ *adjective* **grap·i·er; -est** (1594)
: of or relating to grapes; *especially, of wine* **:** having the taste or aroma of fresh grapes
— grap·i·ness \-nəs\ *noun*

[1]grasp \ˈgrasp\ *verb* [Middle English *graspen*] (14th century)
intransitive verb
: to make the motion of seizing **:** CLUTCH
transitive verb
1 : to take or seize eagerly
2 : to clasp or embrace especially with the fingers or arms

☆ SYNONYMS
Graphic, vivid, picturesque mean giving a clear visual impression in words. GRAPHIC stresses the evoking of a clear lifelike picture ⟨a *graphic* account of combat⟩. VIVID suggests an impressing on the mind of the vigorous aliveness of something ⟨a *vivid* re-creation of an exciting event⟩. PICTURESQUE suggests the presentation of a striking or effective picture composed of features notable for their distinctness and charm ⟨a *picturesque* account of his travels⟩.

3 : to lay hold of with the mind **:** COMPREHEND
synonym see TAKE
— **grasp·able** \ˈgras-pə-bəl\ *adjective*
— **grasp·er** *noun*
— **grasp at straws :** to reach for or try anything in desperation
— **grasp the nettle :** to act boldly

²grasp *noun* (1561)
1 a : HANDLE **b :** EMBRACE
2 : HOLD, CONTROL
3 a : the reach of the arms **b :** the power of seizing and holding or attaining
4 : mental hold or comprehension especially when broad

grasp·ing *adjective* (1748)
: desiring material possessions urgently and excessively and often to the point of ruthlessness
synonym see COVETOUS
— **grasp·ing·ly** \ˈgras-piŋ-lē\ *adverb*
— **grasp·ing·ness** *noun*

¹grass \ˈgras\ *noun, often attributive* [Middle English *gras,* from Old English *græs;* akin to Old High German *gras* grass, Old English *grōwan* to grow] (before 12th century)
1 : herbage suitable or used for grazing animals
2 : any of a large family (Gramineae) of monocotyledonous mostly herbaceous plants with jointed stems, slender sheathing leaves, and flowers borne in spikelets of bracts
3 : land (as a lawn or a turf racetrack) covered with growing grass ⟨keep off the *grass*⟩ ⟨the horse had never won on *grass*⟩
4 *plural* **:** leaves or plants of grass
5 : a state or place of retirement ⟨put out to *grass*⟩
6 [short for *grasshopper,* rhyming slang for *copper*] *slang British* **:** a police informer
7 : electronic noise on a radarscope that takes the form of vertical lines resembling lawn grass
8 : MARIJUANA
— **grass·less** \-ləs\ *adjective*
— **grass·like** \-ˌlīk\ *adjective*

²grass (circa 1500)
transitive verb
1 : to feed (livestock) on grass sometimes without grain or other concentrates
2 : to cover with grass; *especially* **:** to seed to grass
intransitive verb
1 : to produce grass
2 *slang British* **:** INFORM 2 — often used with *on*

grass carp *noun* (1885)
: an herbivorous cyprinid fish (*Ctenopharyngodon idella*) of Russia and mainland China that has been introduced elsewhere to control aquatic weeds — called also *white amur*

grass cloth *noun* (1857)
: a lustrous plain textile of usually loosely woven fibers

grass court *noun* (1883)
: a tennis court with a grass surface

grass·hop·per \ˈgras-ˌhä-pər\ *noun* (14th century)
1 : any of numerous plant-eating orthopterous insects (Acrididae, Tettigoniidae, and some related families) having the hind legs adapted for leaping and sometimes engaging in migratory flights in which whole regions may be stripped of vegetation
2 : a cocktail made with crème de menthe, crème de cacao, and light cream

grass·land \-ˌland\ *noun* (1682)
1 : farmland occupied chiefly by forage plants and especially grasses
2 a : land on which the natural dominant plant forms are grasses and forbs **b :** an ecological community in which the characteristic plants are grasses

grass·roots \ˈgras-ˌrüts, -ˌrůts\ *also* **grass·root** \-ˌrüt, -ˌrůt\ *adjective* (1907)
1 : BASIC, FUNDAMENTAL ⟨the *grassroots* factor in deciding to buy a house⟩
2 : being, originating, or operating in or at the grass roots ⟨a *grassroots* organization⟩ ⟨*grassroots* political support⟩
3 : not adapted from or added to an existing facility or operation **:** totally new ⟨a *grassroots* refinery⟩

grass roots *noun plural but singular or plural in construction, often attributive* (1901)
1 : the very foundation or source
2 : the basic level of society or of an organization especially as viewed from higher or more centralized positions of power

grass tree *noun* (1802)
: any of a genus (*Xanthorrhoea*) of Australian plants of the lily family with a thick woody trunk bearing a cluster of stiff linear leaves and a terminal spike of small flowers

grass widow *noun* (1528)
1 *chiefly dialect* **a :** a discarded mistress **b :** a woman who has had an illegitimate child
2 a : a woman whose husband is temporarily away from her **b :** a woman divorced or separated from her husband

grass widower *noun* (1862)
1 : a man divorced or separated from his wife
2 : a man whose wife is temporarily away from him

grassy \ˈgra-sē\ *adjective* **grass·i·er; -est** (15th century)
1 a : covered or abounding with grass ⟨*grassy* lawns⟩ **b :** having a flavor or odor of grass ⟨*grassy* tea⟩ ⟨wine with a *grassy* bouquet⟩
2 : resembling grass especially in color

grat *past of* GREET

¹grate \ˈgrāt\ *verb* **grat·ed; grat·ing** [Middle English, from Middle French *grater* to scratch, of Germanic origin; akin to Old High German *krazzōn* to scratch] (14th century)
transitive verb
1 *archaic* **:** ABRADE
2 : to reduce to small particles by rubbing on something rough ⟨*grate* cheese⟩
3 : FRET, IRRITATE
4 a : to gnash or grind noisily **b :** to cause to make a rasping sound **c :** to utter in a harsh voice
intransitive verb
1 : to rub or rasp noisily
2 : to cause irritation **:** JAR ⟨dry, cerebral talk that tends to *grate* on the nerves —Hollis Alpert⟩
— **grat·er** *noun*
— **grat·ing·ly** \ˈgrā-tiŋ-lē\ *adverb*

²grate *noun* [Middle English, from Medieval Latin *crata, grata* hurdle, alteration of Latin *cratis* — more at HURDLE] (14th century)
1 a : a barred frame for cooking over a fire **b :** a frame or bed of iron bars to hold a stove or furnace fire **c :** FIREPLACE
2 : GRATING 2
3 *obsolete* **:** CAGE, PRISON

³grate *transitive verb* **grat·ed; grat·ing** (1547)
: to furnish with a grate

grate·ful \ˈgrāt-fəl\ *adjective* [obsolete *grate* pleasing, thankful, from Latin *gratus* — more at GRACE] (1552)
1 a : appreciative of benefits received **b :** expressing gratitude
2 a : affording pleasure or contentment **:** PLEASING **b :** pleasing by reason of comfort supplied or discomfort alleviated
— **grate·ful·ly** \-fə-lē\ *adverb*
— **grate·ful·ness** *noun*

grat·i·cule \ˈgra-tə-ˌkyü(ə)l\ *noun* [French, from Latin *craticula* fine latticework, diminutive of *cratis* wickerwork, hurdle] (1914)
1 : RETICLE
2 : the network of lines of latitude and longitude upon which a map is drawn

grat·i·fi·ca·tion \ˌgra-tə-fə-ˈkā-shən\ *noun* (1576)
1 : REWARD, RECOMPENSE; *especially* **:** GRATUITY
2 : the act of gratifying **:** the state of being gratified
3 : a source of satisfaction or pleasure

grat·i·fy \ˈgra-tə-ˌfī\ *transitive verb* **-fied; -fy·ing** [Middle French *gratifier,* from Latin *gratificari* to show kindness to, from *gratus* + *-ificari,* passive of *-ificare* -ify] (1539)
1 *archaic* **:** REMUNERATE, REWARD
2 : to be a source of or give pleasure or satisfaction to ⟨it *gratified* him to have his wife wear jewels —Willa Cather⟩
3 : to give in to **:** INDULGE, SATISFY ⟨*gratify* a whim⟩

grat·i·fy·ing *adjective* (circa 1611)
: giving pleasure or satisfaction **:** PLEASING
— **grat·i·fy·ing·ly** \-iŋ-lē\ *adverb*

gra·tin \ˈgra-tᵊn, ˈgrä-\ *noun* [French, from Middle French, from *grater* to scratch] (1806)
: a brown crust formed on food that has been cooked au gratin; *also* **:** a dish so cooked
2 : UPPER CRUST ⟨the *gratin* of London society⟩

gra·ti·né *or* **gra·ti·née** \ˌgra-tᵊn-ˈā, ˌgrä-\ *adjective* [French, from past participle of *gratiner* to cook au gratin, from *gratin*] (1931)
: AU GRATIN

gra·ti·née *or* **gra·ti·nee** \ˌgra-tᵊn-ˈā, ˌgrä-\ *transitive verb, past & past part* **gra·ti·néed** *or* **gra·ti·need** (1974)
: to cook au gratin

grat·ing \ˈgrā-tiŋ\ *noun* (1626)
1 : a wooden or metal lattice used to close or floor any of various openings
2 : a partition, covering, or frame of parallel bars or crossbars
3 : a system of close equidistant and parallel lines or bars ruled on a polished surface to produce spectra by diffraction

gra·tis \ˈgra-təs, ˈgrā-\ *adverb or adjective* [Middle English, from Latin *gratiis, gratis,* from ablative plural of *gratia* favor — more at GRACE] (15th century)
: without charge or recompense **:** FREE

grat·i·tude \ˈgra-tə-ˌtüd, -ˌtyüd\ *noun* [Middle English, from Middle French or Medieval Latin; Middle French, from Medieval Latin *gratitudo,* from Latin *gratus* grateful] (1565)
: the state of being grateful **:** THANKFULNESS

gra·tu·itous \grə-ˈtü-ə-təs, -ˈtyü-\ *adjective* [Latin *gratuitus,* from *gratus*] (1656)
1 a : given unearned or without recompense **b :** not involving a return benefit, compensation, or consideration **c :** costing nothing **:** FREE
2 : not called for by the circumstances **:** UNWARRANTED ⟨*gratuitous* insolence⟩ ⟨a *gratuitous* assumption⟩
— **gra·tu·itous·ly** *adverb*
— **gra·tu·itous·ness** *noun*

gra·tu·ity \grə-ˈtü-ə-tē, -ˈtyü-\ *noun, plural* **-ities** (1540)
: something given voluntarily or beyond obligation usually for some service; *especially* **:** TIP

grat·u·late \ˈgra-chə-ˌlāt\ *transitive verb* [Latin *gratulatus,* past participle of *gratulari* — more at CONGRATULATE] (1584)
archaic **:** CONGRATULATE
— **grat·u·la·tion** \ˌgra-chə-ˈlā-shən\ *noun*
— **grat·u·la·to·ry** \ˈgra-chə-lə-ˌtōr-ē, -ˌtȯr-\ *adjective*

grau·pel \ˈgrau̇-pəl\ *noun* [German] (1889)
: granular snow pellets — called also *soft hail*

Grau·stark \ˈgrau̇-ˌstärk, ˈgrȯ-\ *noun* [*Graustark,* imaginary country in the novel *Graustark* (1901) by George B. McCutcheon (died 1928) American novelist] (1941)
: an imaginary land of high romance; *also* **:** a highly romantic piece of writing
— **Grau·stark·ian** \grau̇-ˈstär-kē-ən, grȯ-\ *adjective*

gra·va·men \grə-ˈvā-mən\ *noun, plural* **-vamens** *or* **-vam·i·na** \-ˈva-mə-nə\ [Late Latin, burden, from Latin *gravare* to burden, from *gravis*] (1647)
: the material or significant part of a grievance or complaint

¹grave \ˈgrāv\ *transitive verb* **graved; graven** \ˈgrā-vən\ *or* **graved; grav·ing** [Middle

English, from Old English *grafan;* akin to Old High German *graban* to dig, Old Church Slavonic pog*reti* to bury] (before 12th century)
1 *archaic* **:** DIG, EXCAVATE
2 a : to carve or shape with a chisel **:** SCULPTURE **b :** to carve or cut (as letters or figures) into a hard surface **:** ENGRAVE
3 : to impress or fix (as a thought) deeply

²grave *noun* [Middle English, from Old English *græf;* akin to Old High German *grab* grave, Old English *grafan* to dig] (before 12th century)
1 : an excavation for burial of a body; *broadly* **:** a burial place
2 : DEATH 1, 4

³grave *transitive verb* **graved; grav·ing** [Middle English *graven*] (15th century)
: to clean and pay with pitch ⟨*grave* a ship's bottom⟩

⁴grave \ˈgrāv, *in sense 5 often* ˈgräv\ *adjective* **grav·er; grav·est** [Middle French, from Latin *gravis* heavy, grave — more at GRIEVE] (1539)
1 a *obsolete* **:** AUTHORITATIVE, WEIGHTY **b :** meriting serious consideration **:** IMPORTANT ⟨*grave* problems⟩ **c :** likely to produce great harm or danger ⟨a *grave* mistake⟩ **d :** significantly serious **:** CONSIDERABLE, GREAT ⟨*grave* importance⟩
2 : having a serious and dignified quality or demeanor ⟨a *grave* and thoughtful look⟩
3 : drab in color **:** SOMBER
4 : low-pitched in sound
5 a *of an accent mark* **:** having the form ` **b :** marked with a grave accent **c :** of the variety indicated by a grave accent
synonym see SERIOUS
— grave·ly *adverb*
— grave·ness *noun*

⁵grave \ˈgrāv, ˈgräv\ *noun* (1609)
: a grave accent ` used to show that a vowel is pronounced with a fall of pitch (as in ancient Greek), that a vowel has a certain quality (as *è* in French), that a final *e* is stressed and close and that a final *o* is stressed and low (as in Italian), that a syllable has a degree of stress between maximum and minimum (as in phonetic transcription), or that the *e* of the English ending *-ed* is to be pronounced (as in "this cursèd day")

⁶gra·ve \ˈgrä-(ˌ)vā\ *adverb or adjective* [Italian, literally, grave, from Latin *gravis*] (1683)
: slowly and solemnly — used as a direction in music

¹grav·el \ˈgra-vəl\ *noun* [Middle English, from Old French *gravele,* diminutive of *grave, greve* pebbly ground, beach] (13th century)
1 *obsolete* **:** SAND
2 a : loose rounded fragments of rock **b :** a stratum or deposit of gravel; *also* **:** a surface covered with gravel ⟨a *gravel* road⟩
3 : a deposit of small calculous concretions in the kidneys and urinary bladder

²gravel *transitive verb* **-eled** *or* **-elled; -el·ing** *or* **-el·ling** \ˈgrav-liŋ, ˈgra-və-\ (1543)
1 : to cover or spread with gravel
2 a : PERPLEX, CONFOUND **b :** IRRITATE, NETTLE

³gravel *adjective* (1939)
: GRAVELLY 2 — used of the human voice

grav·el–blind \ˈgra-vəl-ˌblīnd\ *adjective* [suggested by *sand-blind*] (1596)
: having very weak vision

grave·less \ˈgrāv-ləs\ *adjective* (1606)
1 : not buried ⟨these *graveless* bones⟩
2 : not requiring graves **:** DEATHLESS ⟨the *graveless* home of the blessed⟩

grav·el·ly \ˈgrav-lē, ˈgra-və-\ *adjective* (14th century)
1 : of, containing, or covered with gravel
2 : having a rough or grating sound ⟨a *gravelly* voice⟩

graven image *noun* [*graven,* past participle of ¹*grave*] (14th century)
: an object of worship carved usually from wood or stone **:** IDOL

grav·er \ˈgrā-vər\ *noun* [Middle English] (13th century)
1 : SCULPTOR, ENGRAVER
2 : any of various cutting or shaving tools used in graving or in hand metal-turning

Graves' disease \ˈgrāvz-\ *noun* [Robert J. *Graves* (died 1853) Irish physician] (1868)
: a common form of hyperthyroidism characterized by goiter and often a slight protrusion of the eyeballs

grave·side \ˈgrāv-ˌsīd\ *noun* (1838)
: the area beside a grave ⟨mourners at *graveside*⟩

grave·stone \ˈgrāv-ˌstōn\ *noun* (14th century)
: a burial monument

grave·yard \-ˌyärd\ *noun* (1773)
1 : CEMETERY
2 : something resembling a graveyard ⟨an automobile *graveyard*⟩

graveyard shift *noun* (1908)
: a work shift beginning late at night (as 11 o'clock); *also* **:** the workers on such a shift

gravi- *combining form* [Middle French, from Latin, from *gravis*]
: weight ⟨*gravi*metric⟩

grav·id \ˈgra-vəd\ *adjective* [Latin *gravidus,* from *gravis* heavy] (1597)
1 : PREGNANT
2 : distended with or full of eggs
— gra·vid·i·ty \grə-ˈvi-də-tē\ *noun*

grav·i·da \ˈgra-və-də\ *noun, plural* **-i·das** *or* **-i·dae** \-və-ˌdē\ [Latin, from feminine of *gravidus*] (1926)
: a pregnant woman — often used with a number to indicate the number of pregnancies a woman has had ⟨a 4-*gravida*⟩

gra·vi·me·ter \grə-ˈvi-mə-tər, ˈgra-və-ˌmē-\ *noun* [French *gravimètre,* from *gravi-* + *-mètre* -meter] (1932)
: a sensitive weighing instrument for measuring variations in the gravitational field of the earth or moon

gravi·met·ric \ˌgra-və-ˈme-trik\ *adjective* (1873)
1 : of or relating to measurement by weight
2 : of or relating to variations in the gravitational field determined by means of a gravimeter
— gravi·met·ri·cal·ly \-tri-k(ə-)lē\ *adverb*

gra·vim·e·try \grə-ˈvi-mə-trē\ *noun* (1858)
: the measurement of weight, a gravitational field, or density

graving dock *noun* (1840)
: DRY DOCK

grav·i·tas \ˈgra-və-ˌtäs, -ˌtas\ *noun* [Latin] (1924)
: high seriousness (as in a person's bearing or in the treatment of a subject)

grav·i·tate \ˈgra-və-ˌtāt\ *intransitive verb* **-tat·ed; -tat·ing** (1692)
1 : to move under the influence of gravitation
2 a : to move toward something **b :** to be drawn or attracted especially by natural inclination ⟨youngsters . . . *gravitate* toward a strong leader —Rose Friedman⟩

grav·i·ta·tion \ˌgra-və-ˈtā-shən\ *noun* (circa 1645)
1 : a force manifested by acceleration toward each other of two free material particles or bodies or of radiant-energy quanta **:** GRAVITY 3a(2)
2 : the action or process of gravitating
— grav·i·ta·tion·al \-shnəl, -shə-nᵊl\ *adjective*
— grav·i·ta·tion·al·ly *adverb*
— grav·i·ta·tive \ˈgra-və-ˌtā-tiv\ *adjective*

gravitational lens *noun* (1950)
: a massive celestial object (as a galaxy) that bends and focuses the light of another more distant object (as a quasar) by gravity and that is usually detected by the multiple images it forms of the second object

gravitational wave *noun* (1906)
: a hypothetical wave held to travel at the speed of light and to propagate the gravitational field — called also *gravity wave*

grav·i·ton \ˈgra-və-ˌtän\ *noun* [International Scientific Vocabulary *gravity* + ²*-on*] (1942)
: a hypothetical particle with zero charge and rest mass that is held to be the quantum of the gravitational field

grav·i·ty \ˈgra-və-tē\ *noun, plural* **-ties** *often attributive* [Middle French or Latin; Middle French *gravité,* from Latin *gravitat-, gravitas,* from *gravis*] (1509)
1 a : dignity or sobriety of bearing **b :** IMPORTANCE, SIGNIFICANCE; *especially* **:** SERIOUSNESS **c :** a serious situation or problem
2 : WEIGHT
3 a (1) **:** the gravitational attraction of the mass of the earth, the moon, or a planet for bodies at or near its surface (2) **:** a fundamental physical force that is responsible for interactions which occur because of mass between particles, between aggregations of matter (as stars and planets), and between particles (as photons) and aggregations of matter, that is 10^{39} times weaker than the strong force, and that extends over infinite distances but is dominant over macroscopic distances especially between aggregations of matter — called also *gravitation, gravitational force*; compare ELECTROMAGNETISM 2a, STRONG FORCE, WEAK FORCE **b :** ACCELERATION OF GRAVITY **c :** SPECIFIC GRAVITY

grav·lax *or* **grav·laks** \ˈgräv-ˌläks\ *noun* [Swedish *gravlax* or Norwegian *gravlaks,* from *grav* buried + Swedish *lax,* Norwegian *laks* salmon — more at LOX] (1977)
: salmon usually cured with salt, pepper, dill, and aquavit

gra·vure \grə-ˈvyu̇r, grā-\ *noun* [French, from *graver* to grave, of Germanic origin; akin to Old High German *graban* to dig, engrave — more at GRAVE] (1893)
: PHOTOGRAVURE

gra·vy \ˈgrā-vē\ *noun, plural* **gravies** [Middle English *gravey,* from Middle French *gravé*] (14th century)
1 : a sauce made from the thickened and seasoned juices of cooked meat
2 a : something additional or unexpected that is pleasing or valuable ⟨with expenses now paid, future money is pure *gravy* —K. Crossen⟩ **b :** unearned or illicit gain **:** ⁵GRAFT

gravy train *noun* (circa 1927)
: a much exploited source of easy money; *also* **:** GRAVY 2a

¹gray \ˈgrā\ *adjective* [Middle English, from Old English *grǣg;* akin to Old High German *grīs, grāo* gray] (before 12th century)
1 a : of the color gray **b :** tending toward gray ⟨blue-*gray* eyes⟩ **c :** dull in color
2 : having the hair gray **:** HOARY
3 : clothed in gray
4 a : lacking cheer or brightness in mood, outlook, style, or flavor; *also* **:** DISMAL, GLOOMY ⟨a *gray* day⟩ **b :** prosaically ordinary **:** DULL, UNINTERESTING
5 : having an intermediate and often vaguely defined position, condition, or character ⟨an ethically *gray* area⟩
— gray·ly *adverb*
— gray·ness *noun*

²gray *noun* (13th century)
1 : something (as an animal, garment, cloth, or spot) of a gray color
2 : any of a series of neutral colors ranging between black and white
3 : one who wears a gray uniform: as **a :** a soldier in the Confederate army during the American Civil War **b :** the Confederate army

³gray (14th century)
transitive verb
: to make gray
intransitive verb

1 : to become gray ⟨*graying* hair⟩
2 : AGE; *also* **:** to contain an increasing percentage of older people

gray·beard \'grā-ˌbird\ *noun* (circa 1580)
: an old man

gray birch *noun* (1851)
1 : a small birch (*Betula populifolia*) of northeastern North America that has many lateral branches, grayish white bark, triangular leaves, and soft weak wood and that occurs especially in old fields reverting to woodland
2 : YELLOW BIRCH

gray eminence *noun* [translation of French *Éminence grise,* nickname of Père Joseph (François Joseph du Tremblay) (died 1638) French monk and diplomat who was confidant of Cardinal Richelieu, styled *Éminence rouge* (red eminence); from the colors of their respective habits] (1941)
: a person who exercises power behind the scenes

gray·fish \'grā-ˌfish\ *noun* (1917)
: DOGFISH

gray fox *noun* (circa 1679)
: a fox (*Urocyon cinereoargenteus*) with coarse gray hair and white underparts that occurs from southern Canada to northern South America

gray·ish \'grā-ish\ *adjective* (1562)
1 : somewhat gray
2 *of a color* **:** low in saturation

gray·ling \'grā-liŋ\ *noun, plural* **grayling** *also* **graylings** (15th century)
: any of several freshwater salmonoid fishes (genus *Thymallus*) valued as food and sport fishes

gray market *noun* (1946)
: a market employing irregular but not illegal methods; *especially* **:** a market that legally circumvents authorized channels of distribution to sell goods at prices lower than those intended by the manufacturer

gray matter *noun* (1840)
1 : neural tissue especially of the brain and spinal cord that contains nerve-cell bodies as well as nerve fibers and has a brownish gray color
2 : BRAINS, INTELLECT

gray squirrel *noun* (1674)
: a common light gray to black squirrel (*Sciurus carolinensis*) that is native to eastern North America and has been introduced into England

gray·wacke \'grā-ˌwak, -ˌwa-kə\ *noun* [partial translation of German *Grauwacke*] (1811)
: a coarse usually dark gray sandstone or fine-grained conglomerate composed of firmly cemented fragments (as of quartz or feldspar)

gray whale *noun* (1860)
: a large baleen whale (*Eschrichtius robustus*) of the northern Pacific having short jaws and no dorsal fin

gray wolf *noun* (1814)
: a large usually gray Holarctic wolf (*Canis lupus*) now restricted to northern North America and Asia — called also *timber wolf*

¹graze \'grāz\ *verb* **grazed; graz·ing** [Middle English *grasen,* from Old English *grasian,* from *græs* grass] (before 12th century)
intransitive verb
1 : to feed on growing herbage, attached algae, or phytoplankton
2 : to eat small amounts of various foods several times a day
transitive verb
1 a : to crop and eat in the field **b :** to feed on the herbage of
2 a : to put to graze ⟨*grazed* the cows on the meadow⟩ **b :** to put cattle to graze on
3 : to supply herbage for the grazing of
— graze·able *or* **graz·able** \'grā-zə-bəl\ *adjective*
— graz·er *noun*

²graze *noun* (1692)
1 : an act of grazing
2 : herbage for grazing

³graze *verb* **grazed; graz·ing** [perhaps from ¹*graze*] (1604)
transitive verb
1 : to touch lightly in passing
2 : ABRADE, SCRATCH ⟨*grazed* her knee when she fell⟩
intransitive verb
: to touch or rub against something in passing

⁴graze *noun* (1847)
: a scraping along a surface or an abrasion made by it; *especially* **:** a superficial abrasion of the skin

gra·zier \'grā-zhər\ *noun* (15th century)
1 : a person who grazes cattle; *broadly* **:** RANCHER
2 *Australian* **:** a sheep raiser

grazing *noun* (1517)
: herbage or land for grazing

¹grease \'grēs\ *noun* [Middle English *grese,* from Old French *craisse, graisse,* from (assumed) Vulgar Latin *crassia,* from Latin *crassus* fat] (13th century)
1 a : rendered animal fat **b :** oily matter **c :** a thick lubricant
2 : wool as it comes from the sheep retaining the natural oils or fats
— grease·less \'grēs-ləs\ *adjective*
— grease·proof \'grēs-'prüf\ *adjective*
— in the grease *of wool or fur* **:** in the natural uncleaned condition

²grease \'grēs, 'grēz\ *transitive verb* **greased; greas·ing** (14th century)
1 : to smear or daub with grease
2 : to lubricate with grease
3 : to soil with grease
4 : to hasten the process or progress of; *also* **:** FACILITATE
— grease the hand of *or* **grease the palm of :** BRIBE

grease·ball \'grēs-ˌbȯl\ *noun* (circa 1922)
: a person of Hispanic or Mediterranean descent — usually used disparagingly

grease monkey *noun* (1928)
: MECHANIC

grease·paint \'grēs-ˌpānt\ *noun* (1886)
1 : a melted tallow or grease used in theater makeup
2 : theater makeup

grease pencil *noun* (1944)
: a pencil in which the marking substance is pigment and grease

greaseproof paper *noun* (1900)
British **:** a heavy stiff waxed paper — called also *greaseproof*

greas·er \'grē-zər, -sər\ *noun* [¹*grease*] (1641)
1 : one that greases
2 : a native or inhabitant of Latin America or a Mediterranean land; *especially* **:** MEXICAN — usually used disparagingly
3 : an aggressive swaggering young white male usually of working-class background

grease·wood \'grēs-ˌwu̇d\ *noun* (1838)
: a low stiff shrub (*Sarcobatus vermiculatus*) of the goosefoot family common in alkaline soils in the western U.S.; *also* **:** any of various related or similar shrubs

greasy \'grē-sē, -zē\ *adjective* **greas·i·er; -est** (1514)
1 a : smeared or soiled with grease ⟨*greasy* clothes⟩ **b :** oily in appearance, texture, or manner ⟨his *greasy* smile —Jack London⟩ **c :** SLIPPERY
2 : containing an unusual amount of grease ⟨*greasy* food⟩
— greas·i·ly \-sə-lē, -zə-\ *adverb*
— greas·i·ness \-sē-nəs, -zē-\ *noun*

greasy spoon *noun* (circa 1925)
: a dingy small cheap restaurant

¹great \'grāt, *Southern also* 'gre(ə)t\ *adjective* [Middle English *grete,* from Old English *grēat;* akin to Old High German *grōz* large] (before 12th century)
1 a : notably large in size **:** HUGE **b :** of a kind characterized by relative largeness — used in plant and animal names **c :** ELABORATE, AMPLE ⟨*great* detail⟩
2 a : large in number or measure **:** NUMEROUS ⟨*great* multitudes⟩ **b :** PREDOMINANT ⟨the *great* majority⟩
3 : remarkable in magnitude, degree, or effectiveness ⟨*great* bloodshed⟩
4 : full of emotion ⟨*great* with anger⟩
5 a : EMINENT, DISTINGUISHED ⟨a *great* poet⟩ **b :** chief or preeminent over others — often used in titles ⟨Lord *Great* Chamberlain⟩ **c :** ARISTOCRATIC, GRAND ⟨*great* ladies⟩
6 : long continued ⟨a *great* while⟩
7 : PRINCIPAL, MAIN ⟨a reception in the *great* hall⟩
8 : more remote in a family relationship by a single generation than a specified relative ⟨*great*-grandfather⟩
9 : markedly superior in character or quality; *especially* **:** NOBLE ⟨*great* of soul⟩
10 a : remarkably skilled ⟨*great* at tennis⟩ **b :** marked by enthusiasm **:** KEEN ⟨*great* on science fiction⟩
11 — used as a generalized term of approval ⟨had a *great* time⟩ ⟨it was just *great*⟩
— great·ness *noun*

²great *adverb* (13th century)
: in a great manner **:** SUCCESSFULLY, WELL ⟨things are going *great*⟩

³great *noun, plural* **great** *or* **greats** (13th century)
: an outstandingly superior or skillful person

great ape *noun* (1949)
: any of a family (Pongidae) of primates including the gorilla, orangutan, and chimpanzees — called also *pongid*

great auk *noun* (circa 1828)
: an extinct large flightless auk (*Pinguinus impennis*) formerly abundant along North Atlantic coasts

great–aunt *noun* (1656)
: GRANDAUNT

Great Bear *noun*
: URSA MAJOR

great blue heron *noun* (1835)
: a large slaty-blue American heron (*Ardea herodias*) with a crested head

great circle *noun* (1594)
: a circle formed on the surface of a sphere by the intersection of a plane that passes through the center of the sphere; *specifically* **:** such a circle on the surface of the earth an arc of which connecting two terrestrial points constitutes the shortest distance on the earth's surface between them

great·coat \'grāt-ˌkōt\ *noun* (circa 1685)
: a heavy overcoat

Great Dane *noun* (1774)
: any of a breed of tall massive powerful smooth-coated dogs

great divide *noun* [the *Great Divide,* North American watershed] (1868)
: a significant point of division; *especially* **:** DEATH

great·en \'grā-tᵊn\ *verb* **great·ened; great·en·ing** \'grā-tᵊn-iŋ\ (1614)
transitive verb
: to make greater
intransitive verb
: to become greater

great·er *adjective, often capitalized* [comparative of *great*] (1882)
: consisting of a central city together with adjacent areas that are naturally or administratively connected with it ⟨*Greater* London⟩

greater yellowlegs *noun plural but singular or plural in construction* (circa 1909)
: a common North American marsh and shorebird (*Tringa melanoleuca*) that is largely gray above and white below with black or dark gray flecks and yellow legs — compare LESSER YELLOWLEGS

greatest common divisor *noun* (circa 1924)
: the largest integer or the polynomial of highest degree that is an exact divisor of each of two or more integers or polynomials — called also *greatest common factor*

great group *noun* (1960)
: a group of soils that is characterized by common characteristics usually developed under the influence of environmental factors (as vegetation and climate) active over a considerable geographic range and that comprises one or more families of soil — called also *great soil group*

great·heart·ed \'grāt-ˌhär-təd\ *adjective* (14th century)
1 : characterized by bravery **:** COURAGEOUS
2 : GENEROUS, MAGNANIMOUS
— great·heart·ed·ly *adverb*
— great·heart·ed·ness *noun*

great horned owl *noun* (1812)
: a large American owl (*Bubo virginianus*) with conspicuous ear tufts

great laurel *noun* (1784)
: a large-leaved evergreen rhododendron (*Rhododendron maximum*) of eastern North America with rosy bell-shaped flowers more or less speckled with green — called also *rosebay rhododendron*

great·ly \'grāt-lē\ *adverb* (13th century)
1 : to a great extent or degree **:** very much ⟨contributed *greatly* to improved relations⟩ ⟨not *greatly* bothered⟩
2 : in a great manner **:** NOBLY, MAGNANIMOUSLY ⟨a man may live *greatly* in the law —O. W. Holmes (died 1935)⟩

Great Mogul *noun* (1588)
: the sovereign of the empire founded in India by the Moguls in the 16th century

great–nephew *noun* (1581)
: GRANDNEPHEW

great–niece *noun* (1884)
: GRANDNIECE

great octave *noun* (circa 1854)
: the musical octave that begins on the second C below middle C — see PITCH illustration

great power *noun, often G&P capitalized* (circa 1890)
: one of the nations that figure most decisively in international affairs **:** SUPERPOWER

Great Pyr·e·nees \-'pir-ə-ˌnēz\ *noun, plural* **Great Pyrenees** (1938)
: any of a breed of large heavy-coated white dogs often used to guard livestock

Great Pyrenees

great room *noun* (1700)
: a large room usually combining several functions (as of a dining room, living room, and family room)

Great Russian *noun* (1854)
: RUSSIAN 1b
— Great Russian *adjective*

great seal *noun* (15th century)
: a large seal that constitutes an emblem of sovereignty and is used especially for the authentication of important documents

great skua *noun* (circa 1954)
: a large seabird (*Catharacta skua*) that is related to the jaeger, has dusky plumage and broad rounded wings, breeds chiefly along arctic and antarctic shores, and forages over most cold and temperate seas

great–uncle *noun* (1656)
: GRANDUNCLE

great vowel shift *noun, often G&V&S capitalized* (1909)
: a change in pronunciation of the long vowels of Middle English that began in the 15th century and continued into the 16th century in which the high vowels were diphthongized and the other vowels were raised

great white shark *noun* (circa 1931)
: a large mackerel shark (*Carcharodon carcharias*) of warm seas that is bluish when young but becomes whitish with age and is a man-eater — called also *white shark*; see SHARK illustration

great year *noun* (circa 1741)
: the period of about 25,800 years required for one complete cycle of the equinoxes around the ecliptic

greave \'grēv\ *noun* [Middle English *greve,* from Middle French] (14th century)
: armor for the leg below the knee

grebe \'grēb\ *noun* [French *grèbe*] (1766)
: any of a family (Podicipedidae) of swimming and diving birds closely related to the loons but having lobate toes — compare DABCHICK

Gre·cian \'grē-shən\ *adjective* [Middle English *greciane,* from Latin *Graecia* Greece] (15th century)
: GREEK
— Grecian *noun*
— gre·cian·ize \-shə-ˌnīz\ *transitive verb, often capitalized*

Gre·cism \'grē-ˌsi-zəm\ *noun* (1570)
1 : a Greek idiom
2 : a quality or style imitative of Greek art or culture

gre·cize \-ˌsīz\ *transitive verb* **gre·cized; gre·ciz·ing** *often capitalized* (1692)
: to make Greek or Hellenistic in character

Gre·co- \'gre-kō, 'grē-\ *or* **Grae·co-** \'gre-kō\ *combining form* [Latin *Graeco-,* from *Graecus*]
1 : Greece **:** Greeks ⟨*Greco*phile⟩
2 : Greek and ⟨*Graeco*-Roman⟩

¹gree \'grē\ *noun* [Middle English, from Middle French *grē* step, degree, from Latin *gradus* — more at GRADE] (14th century)
Scottish **:** MASTERY, SUPERIORITY

²gree *verb* **greed; gree·ing** [Middle English *green,* short for *agreen*] (14th century)
dialect **:** AGREE

greed \'grēd\ *noun* [back-formation from *greedy*] (1609)
: excessive or reprehensible acquisitiveness **:** AVARICE

greedy \'grē-dē\ *adjective* **greed·i·er; -est** [Middle English *gredy,* from Old English *grǣdig;* akin to Old High German *grātac* greedy] (before 12th century)
1 : having a strong desire for food or drink
2 : marked by greed
3 : EAGER, KEEN ⟨elated and *greedy* for the future —Frances G. Patton⟩
synonym see COVETOUS
— greed·i·ly \'grē-dᵊl-ē\ *adverb*
— greed·i·ness \'grē-dē-nəs\ *noun*

¹Greek \'grēk\ *noun* [Middle English *Greke,* from Old English *Grēca,* from Latin *Graecus,* from Greek *Graikos*] (before 12th century)
1 a : a native or inhabitant of ancient or modern Greece **b :** a person of Greek descent
2 a : the language used by the Greeks from prehistoric times to the present constituting a branch of Indo-European — see INDO-EUROPEAN LANGUAGES table **b :** ancient Greek as used from the time of the earliest records to the end of the 2d century A.D. — see INDO-EUROPEAN LANGUAGES table **c** *often not capitalized* [translation of Latin *Graecum* (in the medieval phrase *Graecum est; non potest legi* it is Greek; it cannot be read)] **:** something unintelligible ⟨it's *Greek* to me⟩
3 : a member of a Greek-letter fraternity or sorority

²Greek *adjective* (14th century)
1 : of, relating to, or characteristic of Greece, the Greeks, or Greek ⟨*Greek* architecture⟩
2 a : EASTERN ORTHODOX **b :** of or relating to an Eastern church using the Byzantine rite in Greek **c :** of or relating to the established Orthodox church of Greece

Greek Catholic *noun* (1909)
1 : a member of an Eastern church
2 : a member of an Eastern rite of the Roman Catholic Church

Greek cross *noun* (1725)
: a cross having an upright and a transverse shaft equal in length and intersecting at their middles — see CROSS illustration

Greek fire *noun* (circa 1823)
: an incendiary composition used in warfare by the Byzantine Greeks that is said to have burst into flame on wetting

Greek·less \'grēk-ləs\ *adjective* (1891)
: not proficient in Greek

Greek Orthodox *adjective* (circa 1900)
: EASTERN ORTHODOX; *specifically* **:** GREEK 2c

Greek Revival *noun* (1918)
1 : a style of architecture in the first half of the 19th century marked by the use or imitation of Greek orders
2 : a style of decoration (as of furniture) using or imitating the decorative motifs of ancient Greece

¹green \'grēn\ *adjective* [Middle English *grene,* from Old English *grēne;* akin to Old English *grōwan* to grow] (before 12th century)
1 : of the color green
2 a : covered by green growth or foliage ⟨*green* fields⟩ **b** *of winter* **:** MILD, CLEMENT **c :** consisting of green plants and usually edible herbage ⟨a *green* salad⟩
3 : pleasantly alluring
4 : YOUTHFUL, VIGOROUS
5 : not ripened or matured **:** IMMATURE ⟨*green* apples⟩ ⟨tender *green* grasses⟩
6 : FRESH, NEW
7 a : marked by a pale, sickly, or nauseated appearance **b :** ENVIOUS 1 — used especially in the phrase *green with envy*
8 a : not fully processed or treated: as (1) **:** not aged ⟨*green* liquor⟩ (2) **:** not dressed or tanned ⟨*green* hides⟩ (3) **:** freshly sawed **:** UNSEASONED **b :** not in condition for a particular use
9 a : deficient in training, knowledge, or experience **b :** deficient in sophistication and savoir faire **:** NAIVE **c :** not fully qualified for or experienced in a particular function ⟨*green* horse⟩
10 a *often capitalized* **:** relating to or being an environmentalist political movement **b :** concerned with or supporting environmentalism
— green·ish \'grē-nish\ *adjective*
— green·ish·ness *noun*
— green·ly *adverb*
— green·ness \'grē(n)-nəs\ *noun*
— green around the gills : pale or sickly in appearance

²green (before 12th century)
intransitive verb
: to become green
transitive verb
1 : to make green
2 : REJUVENATE, REVITALIZE

³green *noun* (13th century)
1 : a color whose hue is somewhat less yellow than that of growing fresh grass or of the emerald or is that of the part of the spectrum lying between blue and yellow
2 : something of a green color
3 : green vegetation: as **a** *plural* **:** leafy parts of plants for use as decoration **b** *plural* (1) **:** leafy herbs (as spinach, dandelions, Swiss chard) that are cooked as a vegetable **:** POTHERBS (2) **:** GREEN VEGETABLES
4 : a grassy plain or plot: as **a :** a common or park in the center of a town or village **b :** PUTTING GREEN
5 : MONEY; *especially* **:** GREENBACKS
6 *often capitalized* **:** ENVIRONMENTALIST; *especially* **:** a member of an environmentalist political party
— greeny \'grē-nē\ *adjective*

green alga *noun* (1903)

: any of a division (Chlorophyta) of green-colored algae that have chloroplasts and occur especially in freshwater
green·back \'grēn-ˌbak\ *noun* (1862)
: a legal-tender note issued by the U.S. government
green·back·er \-ˌba-kər\ *noun* (1876)
1 *capitalized* **:** a member of a post-Civil War American political party opposing reduction in the amount of paper money in circulation
2 : one who advocates a paper currency backed only by the U.S. government
— green·back·ism \-ˌki-zəm\ *noun*
green bean *noun* (1847)
: a kidney bean that is used as a snap bean when the pods are colored green
green·belt \'grēn-ˌbelt\ *noun* (1932)
: a belt of parkways, parks, or farmlands that encircles a community
Green Beret *noun* [from the beret worn by Special Forces soldiers] (1962)
: a member of the U.S. Army Special Forces
green·bri·er \-ˌbrī(-ə)r\ *noun* (circa 1785)
: any of a genus (*Smilax*) of woody or herbaceous vines of the lily family; *especially* **:** a prickly vine (*S. rotundifolia*) of the eastern U.S. with umbels of small greenish flowers
green·bug \-ˌbəg\ *noun* (1712)
: a green aphid (*Schizaphis graminum*) very destructive to small grains
green card *noun* [from the fact that it was formerly colored green] (1969)
: an identity card attesting the permanent resident status of an alien in the U.S.
— green–carder *noun*
green corn *noun* (1645)
: the young tender ears of Indian corn
green dragon *noun* (circa 1818)
: an American arum (*Arisaema dracontium*) with digitate leaves, slender green spathe, and elongated spadix
green·ery \'grēn-rē, 'grē-nər-ē\ *noun, plural* **-er·ies** (1797)
1 : green foliage or plants
2 : GREEN 3a
green–eyed \'grē-ˌnīd\ *adjective* (1596)
: JEALOUS
green–eyed monster *noun* (1604)
: JEALOUSY
green·finch \'grēn-ˌfinch\ *noun* (15th century)
: a very common European finch (*Carduelis chloris*) having olive-green and yellow plumage
green fingers *noun plural* (1934)
: GREEN THUMB
green flash *noun* (1912)
: a momentary green appearance of the uppermost part of the sun's disk at sunrise or sunset that results from atmospheric refraction
green·fly \'grēn-ˌflī\ *noun* (circa 1750)
British **:** APHID; *especially* **:** GREEN PEACH APHID
green·gage \-ˌgāj\ *noun* [*green* + Sir William *Gage* (died 1820) English botanist] (1724)
: any of several rather small rounded greenish or greenish yellow cultivated plums
green gland *noun* (circa 1890)
: one of a pair of large green glands in some crustaceans (as crayfishes) that have an excretory function and open at the bases of the larger antennae
green·gro·cer \'grēn-ˌgrō-sər, -shər\ *noun* (1723)
chiefly British **:** a retailer of fresh vegetables and fruit
— green·gro·cery \-ˌgrō-sə-rē, -ˌgrōsh-rē\ *noun*
green·head \'grēn-ˌhed\ *noun* (1837)
: any of several green-eyed horseflies (as *Tabanus nigrovittatus*)
green·heart \-ˌhärt\ *noun* (1756)
: a tropical South American evergreen tree (*Ocotea rodiaei* synonym *Nectandra rodiaei*) of the laurel family with a hard greenish wood; *also* **:** its wood
green·horn \-ˌhȯrn\ *noun* [obsolete *greenhorn* (animal with young horns)] (1682)
1 : an inexperienced or unsophisticated person
2 : a newcomer (as to a country) unacquainted with local manners and customs
¹green·house \-ˌhau̇s\ *noun* (1664)
1 : a structure enclosed (as by glass) and used for the cultivation or protection of tender plants
2 : a clear plastic shell (as a canopy) covering a section of an airplane; *also* **:** a compartment (as for a bombardier) enclosed by such a shell
²greenhouse *adjective* (1969)
: of, relating to, or caused by the greenhouse effect ⟨*greenhouse* warming⟩ ⟨*greenhouse* gas⟩
greenhouse effect *noun* (1937)
: warming of the surface and lower atmosphere of a planet (as the earth or Venus) that is caused by conversion of solar radiation into heat in a process involving selective transmission of short wave solar radiation by the atmosphere, its absorption by the planet's surface, and reradiation as infrared which is absorbed and partly reradiated back to the surface by atmospheric gases
green·ing \'grē-niŋ\ *noun* (1664)
: any of several green-skinned apples
green·keep·er \'grēn-ˌkē-pər\ *or* **greens·keep·er** \'grēnz-\ *noun* (circa 1730)
: a person responsible for the care and upkeep of a golf course
green·let \'grēn-lət\ *noun* (1831)
: VIREO
green light *noun* [from the green traffic light which signals permission to proceed] (1937)
: authority or permission to proceed
green·ling \'grēn-liŋ\ *noun* (circa 1898)
: any of several food fishes (family Hexagrammidae) of the rocky coasts of the northern Pacific; *especially* **:** a common food and sport fish (*Hexagrammos decagrammus*)
green·mail \'grēn-ˌmā(ə)l\ *noun* (1983)
: the practice of buying enough of a company's stock to threaten a hostile takeover and reselling it to the company at a price above market value; *also* **:** the money paid for such stock
— greenmail *transitive verb*
— green·mail·er \-ˌmā-lər\ *noun*
green manure *noun* (1842)
: an herbaceous crop (as clover) plowed under while green to enrich the soil
— green–manure *transitive verb*
green mold *noun* (1919)
: a green or green-spored mold (as of the genera *Penicillium* or *Aspergillus*)
green monkey *noun* (1840)
: a long-tailed monkey of any of several African races of a guenon (*Cercopithecus aethiops*) having greenish-appearing hair and often used in medical research
gree·nock·ite \'grē-nə-ˌkīt\ *noun* [Charles M. Cathcart, Lord *Greenock* (died 1859) English soldier] (1844)
: a mineral consisting of native cadmium sulfide occurring in yellow translucent hexagonal crystals or as an earthy incrustation
green onion *noun* (1847)
: a young onion pulled before the bulb has enlarged and used especially in salads
green paper *noun, often G&P capitalized* (1967)
British **:** a government document that proposes and invites discussion on approaches to a problem
green peach aphid *noun* (1922)
: a nearly cosmopolitan yellowish green aphid (*Myzus persicae*) that is frequently a vector of plant virus diseases
green pepper *noun* (1700)
: a sweet pepper before it turns red at maturity
green revolution *noun* (1968)
: the great increase in production of food grains (as rice and wheat) due to the introduction of high-yielding varieties, to the use of pesticides, and to better management techniques
green·room \'grēn-ˌrüm, -ˌru̇m\ *noun* (1701)
: a room in a theater or concert hall where performers can relax before or after appearances
green·sand \-ˌsand\ *noun* (1796)
: a sedimentary deposit that consists largely of dark greenish grains of glauconite often mingled with clay or sand
greens fee \'grēnz-\ *noun* (1909)
: a fee paid for the privilege of playing on a golf course — called also *green fee*
green·shank \'grēn-ˌshaŋk\ *noun* (1766)
: an Old World sandpiper (*Tringa nebularia*) with greenish legs and a slightly upturned bill
green·sick·ness \-ˌsik-nəs\ *noun* (1583)
: CHLOROSIS
— green·sick *adjective*
green snake *noun* (1709)
: either of two bright green harmless largely insectivorous North American colubrid snakes (*Opheodrys vernalis* and *O. aestivus*)
green soap *noun* (circa 1840)
: a soft soap made from vegetable oils and used especially in the treatment of skin diseases
green·stick fracture \'grēn-ˌstik-\ *noun* (circa 1885)
: a bone fracture in a young individual in which the bone is partly broken and partly bent
green·stone \'grēn-ˌstōn\ *noun* (circa 1784)
1 : NEPHRITE
2 : any of numerous usually altered dark green compact rocks (as diorite)
green·stuff \-ˌstəf\ *noun* (1851)
: green vegetation used as foodstuff
green sunfish *noun* (circa 1896)
: a sunfish (*Lepomis cyanellus*) of the Great Lakes region and southwestward to the Rio Grande that is largely greenish above with a blue spot on each scale
green·sward \-ˌswȯrd\ *noun* (1600)
: turf green with growing grass
green tea *noun* (1704)
: tea that is light in color from incomplete fermentation of the leaf before firing
green thumb *noun* (1943)
: an unusual ability to make plants grow
— green–thumbed \'grēn-'thəmd\ *adjective*
green turtle *noun* (1657)
: a large usually herbivorous sea turtle (*Chelonia mydas*) of warm waters with a smooth greenish or olive-colored shell
green vegetable *noun* (1826)
: a vegetable whose foliage or foliage-bearing stalks are the chief edible part
green·way \'grēn-ˌwā\ *noun* (1966)
: a corridor of undeveloped land in or near a city that is designed for recreational use
Green·wich mean time \'gri-nij, 'gre-, -nich\ *noun, often M&T capitalized* [*Greenwich,* England] (1938)
: the mean solar time of the meridian of Greenwich used as the prime basis of standard time throughout the world — called also *Greenwich time*
green·wing \'grēn-ˌwiŋ\ *noun* (circa 1889)
: GREEN-WINGED TEAL
green–winged teal \'grēn-ˌwiŋ(d)-\ *noun* (1792)
: a small dabbling duck (*Anas crecca*) the male of which has a chestnut head with a green eye patch and a metallic green area on the wing speculum
green·wood \'grēn-ˌwu̇d\ *noun* (14th century)
: a forest green with foliage
¹greet \'grēt\ *transitive verb* [Middle English *greten,* from Old English *grētan;* akin to Old English *grǣtan* to weep] (before 12th century)
1 : to address with expression of kind wishes **:** HAIL
2 a : to meet or react to in a specified manner ⟨*greeted* him with boos⟩ **b :** to occur as a response to ⟨apathy *greeted* the plan⟩
3 : to appear to the perception of ⟨a surprising sight *greeted* her eyes⟩
— greet·er *noun*

²greet *intransitive verb* **grat** \'grat\; **grut·ten** \'grə-tᵊn\ [Middle English *greten,* from Old English *grǣtan;* akin to Old Norse *grāta* to weep] (before 12th century)
Scottish **:** WEEP, LAMENT

greet·ing *noun* (before 12th century)
1 : a salutation at meeting
2 : an expression of good wishes **:** REGARDS — usually used in plural ⟨holiday *greetings*⟩

greeting card *noun* (1898)
: a piece of paper or thin paperboard having any of a variety of shapes and formats and bearing a greeting or message of sentiment

greg·a·rine \'gre-gə-ˌrīn\ *noun* [ultimately from Latin *gregarius*] (1867)
: any of a subclass (Gregarinia) of parasitic vermiform sporozoan protozoans that occur especially in insects and other invertebrates
— gregarine *adjective*

gre·gar·i·ous \gri-'gar-ē-əs, -'ger-\ *adjective* [Latin *gregarius* of a flock or herd, from *greg-, grex* flock, herd] (1668)
1 a : tending to associate with others of one's kind **:** SOCIAL **b :** marked by or indicating a liking for companionship **:** SOCIABLE **c :** of or relating to a social group
2 a *of a plant* **:** growing in a cluster or a colony **b :** living in contiguous nests but not forming a true colony — used especially of wasps and bees ◆
— gre·gar·i·ous·ly *adverb*
— gre·gar·i·ous·ness *noun*

¹Gre·go·ri·an \gri-'gōr-ē-ən, -'gȯr-\ *adjective* (1642)
: of or relating to Pope Gregory XIII or the Gregorian calendar

²Gregorian *adjective* (1653)
1 : of or relating to Pope Gregory I
2 : of, relating to, or having the characteristics of Gregorian chant

³Gregorian *adjective* [Saint *Gregory* the Illuminator (died 332), apostle of Armenia] (1955)
: of or relating to the Armenian national church

Gregorian calendar *noun* (circa 1771)
: a calendar in general use introduced in 1582 by Pope Gregory XIII as a revision of the Julian calendar, adopted in Great Britain and the American colonies in 1752, marked by the suppression of 10 days or after 1700 11 days, and having leap years in every year divisible by four with the restriction that centesimal years are leap years only when divisible by 400 — see MONTH table

Gregorian chant *noun* (1751)
: a monodic and rhythmically free liturgical chant of the Roman Catholic Church

greige \'grā(zh)\ *adjective* [French *grège* raw (of silk), from Italian *greggio*] (1926)
: being in an unbleached undyed state as taken from a loom — used of textiles

grei·sen \'grī-zᵊn\ *noun* [German] (1878)
: a crystalline rock consisting of quartz and mica that is common in Cornwall and Saxony

grem·lin \'grem-lən\ *noun* [origin unknown] (1941)
: a cause of error or equipment malfunction (as in aircraft) conceived of as a small mischievous gnome

gre·nade \grə-'nād\ *noun* [Middle French, pomegranate, from Late Latin *granata,* from Latin, feminine of *granatus* seedy, from *granum* grain — more at CORN] (1591)
: a small missile that contains an explosive or a chemical agent (as tear gas, a flame producer, or a smoke producer) and that is thrown by hand or projected (as by a rifle or special launcher)

gren·a·dier \ˌgren-ə-'dir\ *noun* [French, from *grenade* grenade] (1676)
1 a : a soldier who carries and throws grenades **b :** a member of a special regiment or corps formerly armed with grenades
2 : any of various deep-sea fishes (family Macruridae) that are related to the cods and have an elongate tapering body and compressed pointed tail

gren·a·dine \ˌgre-nə-'dēn, 'gre-nə-ˌ\ *noun* [French, from *grenade* coarse silk fabric, pomegranate] (1852)
1 : an open-weave fabric of various fibers
2 : a moderate reddish orange
3 : a syrup flavored with pomegranates and used in mixed drinks

Gren·del \'gren-dᵊl\ *noun* [Old English]
: a monstrous man-eating descendant of Cain slain by Beowulf in the Old English poem *Beowulf*

Gresh·am's law \'gre-shəmz-\ *noun* [Sir Thomas *Gresham*] (1858)
: an observation in economics: when two coins are equal in debt-paying value but unequal in intrinsic value, the one having the lesser intrinsic value tends to remain in circulation and the other to be hoarded or exported as bullion; *broadly* **:** any process by which inferior products or practices drive out superior ones

Gret·na Green \ˌgret-nə-'grēn\ *noun* [*Gretna Green,* village in Scotland] (1813)
: a place where many eloping couples are married

Gre·vy's zebra \(ˌ)grā-'vēz-\ *noun* [Jules *Grévy*] (1891)
: a zebra (*Equus grevyi*) of eastern Africa with narrow stripes and a white belly

grew *past of* GROW

grew·some *variant of* GRUESOME

grey *variant of* GRAY

grey friar *noun, often G&F capitalized* (14th century)
: a Franciscan friar

grey·hound \'grā-ˌhau̇nd\ *noun* [Middle English *grehound,* from Old English *grīghund,* from *grīg-* (akin to Old Norse *grey* bitch) + *hund* hound] (before 12th century)
: any of a breed of tall slender graceful smooth-coated dogs characterized by swiftness and keen sight and used for coursing game and racing; *also* **:** any of several related dogs

greyhound

grey·lag \-ˌlag\ *noun* [perhaps from *gray* + *lag* (last)] (circa 1713)
: the common gray wild goose (*Anser anser*) of Europe — called also *greylag goose*

grib·ble \'gri-bəl\ *noun* [perhaps alteration of ²*grub*] (1838)
: either of two small marine isopods (*Limnoria lignorum* and *L. tripunctata*) that destroy submerged timber

grid \'grid\ *noun* [back-formation from *gridiron*] (1839)
1 : GRATING
2 a (1) **:** a perforated or ridged metal plate used as a conductor in a storage battery (2) **:** an electrode consisting of a mesh or a spiral of fine wire in an electron tube (3) **:** a network of conductors for distribution of electric power; *also* **:** a network of radio or television stations **b :** a network of uniformly spaced horizontal and perpendicular lines (as for locating points on a map); *also* **:** something resembling such a network ⟨a road *grid*⟩ **c :** GRIDIRON 3; *broadly* **:** FOOTBALL
3 : the starting positions of cars on a racecourse
4 : a device in a photocomposer on which are located the characters to be exposed as the text is composed

grid·der \'gri-dər\ *noun* (1928)
: a football player

grid·dle \'gri-dᵊl\ *noun* [Middle English *gredil* gridiron, from Old North French, from Latin *craticulum,* diminutive of *cratis* wickerwork — more at HURDLE] (14th century)
: a flat metal surface or pan on which food is cooked by dry heat

griddle cake *noun* (1783)
: PANCAKE

grid·iron \'grid-ˌī(-ə)rn\ *noun* [Middle English *gredire*] (14th century)
1 : a grate for broiling food
2 : something consisting of or covered with a network
3 : a football field

grid·lock \-ˌläk\ *noun* (1980)
1 : a traffic jam in which a grid of intersecting streets is so completely congested that no vehicular movement is possible
2 : a situation resembling a gridlock
— gridlock *transitive verb*

grief \'grēf\ *noun* [Middle English *gref,* from Middle French, heavy, grave, from (assumed) Vulgar Latin *grevis,* alteration of Latin *gravis*] (15th century)
1 *obsolete* **:** GRIEVANCE 3
2 a : deep and poignant distress caused by or as if by bereavement **b :** a cause of such suffering
3 a : an unfortunate outcome **:** DISASTER — used chiefly in the phrase *come to grief* **b :** MISHAP, MISADVENTURE **c :** TROUBLE, ANNOYANCE ⟨enough *grief* for one day⟩
synonym see SORROW

griev·ance \'grē-vən(t)s\ *noun* (14th century)
1 *obsolete* **:** SUFFERING, DISTRESS
2 : a cause of distress (as an unsatisfactory working condition) felt to afford reason for complaint or resistance
3 : the formal expression of a grievance **:** COMPLAINT
synonym see INJUSTICE

grievance committee *noun* (1927)
: a committee formed by a labor union or by employer and employees jointly to discuss and where possible to eliminate grievances

griev·ant \-vənt\ *noun* (1958)
: one who submits a grievance for arbitration

grieve \'grēv\ *verb* **grieved; griev·ing** [Middle English *greven,* from Old French *grever,* from Latin *gravare* to burden, from *gravis* heavy, grave; akin to Greek *barys* heavy, Sanskrit *guru*] (13th century)
transitive verb
: to cause to suffer **:** DISTRESS
intransitive verb
: to feel grief **:** SORROW
— griev·er *noun*

griev·ous \'grē-vəs\ *adjective* (13th century)
1 : causing or characterized by severe pain, suffering, or sorrow ⟨a *grievous* wound⟩ ⟨a *grievous* loss⟩
2 : OPPRESSIVE, ONEROUS ⟨*grievous* costs of war⟩
3 : SERIOUS, GRAVE ⟨*grievous* fault⟩

◇ WORD HISTORY

gregarious *Gregarious,* which is not usually an unfavorable word, is the improbable linguistic cousin of *egregious,* which is now usually used of something conspicuously bad. Both have as their base *grex,* the Latin word for "herd" or "flock." Whereas *egregious* is etymologically "apart from the herd or flock," *gregarious* literally means "of or relating to a herd or flock." When *gregarious* first appeared in English in the 17th century, with the meaning "inclined to associate with others of one's kind," it was used of animals naturally living and breeding in groups. Not till the next century was the word applied to people with a marked fondness for companionship. See in addition EGREGIOUS.

— **griev·ous·ly** *adverb*
— **griev·ous·ness** *noun*
grif·fin *or* **grif·fon** \'gri-fən\ *noun* [Middle English *griffon,* from Middle French *grifon,* from *grif,* from Latin *gryphus,* from Greek *gryp-, gryps*] (14th century)
: a mythical animal typically having the head, forepart, and wings of an eagle and the body, hind legs, and tail of a lion

griffin

grif·fon \'gri-fən\ *noun* [French, literally, griffin] (1882)
1 : BRUSSELS GRIFFON
2 : WIREHAIRED POINTING GRIFFON
grift \'grift\ *transitive verb* [*grift,* noun, perhaps alteration of *graft*] (1915)
slang **:** to obtain (money) illicitly (as in a confidence game)
— **grift** *noun, slang*
— **grift·er** *noun, slang*
grig \'grig\ *noun* [Middle English *grege*] (1566)
: a lively lighthearted usually small or young person
gri·gri *variant of* GRIS-GRIS
¹grill \'gril\ *transitive verb* (1668)
1 : to broil on a grill; *also* **:** to fry or toast on a griddle
2 a : to torment as if by broiling **b :** to question intensely ⟨the police *grilled* the suspect⟩
— **grill·er** *noun*
²grill *noun* [French *gril,* from Latin *craticulum* — more at GRIDDLE] (1685)
1 : a cooking utensil of parallel bars on which food is exposed to heat (as from charcoal or electricity)
2 : food that is broiled usually on a grill — compare MIXED GRILL
3 : a usually informal restaurant or dining room
gril·lage \'gri-lij\ *noun* [French, from *griller* to supply with grillwork, from *gril*] (1776)
1 : a framework of timber or steel for support in marshy or treacherous soil
2 : a framework for supporting a load (as a column)
grille *or* **grill** \'gril\ *noun* [French *grille,* alteration of Old French *greille,* from Latin *craticula,* diminutive of *cratis* wickerwork — more at HURDLE] (1686)
1 : a grating forming a barrier or screen; *especially* **:** an ornamental one at the front end of an automobile
2 : an opening covered with a grille
grill·room \'gril-ˌrüm, -ˌru̇m\ *noun* (1883)
: GRILL 3
grill·work \'gril-ˌwərk\ *noun* (1896)
: work constituting or resembling a grille
grilse \'grils\ *noun, plural* **grilse** [Middle English *grills*] (15th century)
: a young Atlantic salmon returning to its native river to spawn for the first time after one winter at sea; *broadly* **:** any of various salmon at such a stage of development
grim \'grim\ *adjective* **grim·mer; grim·mest** [Middle English, from Old English *grimm;* akin to Old High German *grimm* fierce, Greek *chremetizein* to neigh] (before 12th century)
1 : fierce in disposition or action **:** SAVAGE
2 a : stern or forbidding in action or appearance ⟨a *grim* taskmaster⟩ **b :** SOMBER, GLOOMY
3 : ghastly, repellent, or sinister in character ⟨a *grim* tale⟩
4 : UNFLINCHING, UNYIELDING ⟨*grim* determination⟩
— **grim·ly** *adverb*
— **grim·ness** *noun*
gri·mace \'gri-məs, gri-'mās\ *noun* [French, from Middle French, alteration of *grimache,* of Germanic origin; akin to Old English *grīma* mask] (1651)
: a facial expression usually of disgust or disapproval
— **grimace** *intransitive verb*
— **gri·mac·er** *noun*
gri·mal·kin \gri-'mȯ(l)-kən, -'mal-\ *noun* [*gray* + *malkin*] (1630)
: a domestic cat; *especially* **:** an old female cat
grime \'grīm\ *noun* [Middle Dutch *grime* soot, mask; akin to Old English *grīma* mask] (14th century)
: soot, smut, or dirt adhering to or embedded in a surface; *broadly* **:** accumulated dirtiness and disorder
— **grime** *transitive verb*
Grimm's law \'grimz-\ *noun* [Jacob *Grimm*] (1838)
: a statement in historical linguistics: Proto-Indo-European voiceless stops became Proto-Germanic voiceless fricatives (as in Greek *pyr, treis, kardia* compared with English *fire, three, heart*), Proto-Indo-European voiced stops became Proto-Germanic voiceless stops (as in Latin *duo, genus* compared with English *two, kin*), and Proto-Indo-European voiced aspirated stops became Proto-Germanic voiced fricatives (as in Sanskrit *nābhi, madhya* "mid" compared with English *navel,* Old Norse *mithr* "mid")
grim reaper *noun, often G&R capitalized* (circa 1927)
: death especially when personified as a man or skeleton with a scythe
grimy \'grī-mē\ *adjective* **grim·i·er; -est** (1612)
: full of or covered with grime **:** DIRTY
— **grim·i·ness** *noun*
grin \'grin\ *intransitive verb* **grinned; grinning** [Middle English *grennen,* from Old English *grennian;* akin to Old High German *grennen* to snarl] (before 12th century)
: to draw back the lips so as to show the teeth especially in amusement or laughter; *broadly* **:** SMILE
— **grin** *noun*
— **grin·ner** *noun*
— **grin·ning·ly** \'gri-niŋ-lē\ *adverb*
grinch \'grinch\ *noun* [from the *Grinch,* character in the children's story *How the Grinch Stole Christmas* (1957) by Dr. Seuss] (1979)
: KILLJOY, SPOILSPORT
¹grind \'grīnd\ *verb* **ground** \'grau̇nd\; **grinding** [Middle English, from Old English *grindan;* akin to Latin *frendere* to crush, grind] (before 12th century)
transitive verb
1 : to reduce to powder or small fragments by friction (as in a mill or with the teeth)
2 : to wear down, polish, or sharpen by friction ⟨*grind* an ax⟩
3 a : OPPRESS, HARASS **b :** to weaken or destroy gradually — usually used with *down* ⟨poverty *ground* her spirit down⟩
4 a : to press together with a rotating motion ⟨*grind* the teeth⟩ **b :** to rub or press harshly ⟨*ground* the cigarette out⟩
5 : to operate or produce by turning a crank ⟨*grind* a hand organ⟩
intransitive verb
1 : to perform the operation of grinding
2 : to become pulverized, polished, or sharpened by friction
3 : to move with difficulty or friction especially so as to make a grating noise ⟨gears *grinding*⟩
4 : DRUDGE; *especially* **:** to study hard ⟨*grind* for an exam⟩
5 : to rotate the hips in an erotic manner (as in a burlesque striptease)
— **grind·ing·ly** \'grīn-diŋ-lē\ *adverb*
²grind *noun* (13th century)
1 a : an act of grinding **b :** the sound of grinding
2 a : dreary, monotonous, or difficult labor, study, or routine **b :** one who works or studies excessively
3 : the result of grinding; *also* **:** material ground to a particular degree of fineness ⟨a drip *grind* of coffee⟩
4 : the act of rotating the hips in an erotic manner
synonym see WORK
grind·er \'grīn-dər\ *noun* (14th century)
1 a : MOLAR **b** *plural* **:** TEETH
2 : one that grinds
3 : a machine or device for grinding
4 : SUBMARINE 2
grind house *noun* (1927)
: an often shabby movie theater having continuous showings especially of pornographic or violent films
grind out *transitive verb* (1868)
: to produce in a mechanical way
grind·stone \'grīn-ˌstōn\ *noun* (13th century)
1 : MILLSTONE 1
2 : a flat circular stone of natural sandstone that revolves on an axle and is used for grinding, shaping, or smoothing
grin·go \'griŋ-(ˌ)gō\ *noun, plural* **gringos** [Spanish, alteration of *griego* Greek, stranger, from Latin *Graecus* Greek] (1849)
: a foreigner in Spain or Latin America especially when of English or American origin; *broadly* **:** a non-Hispanic person — often used disparagingly ◆
gri·ot \'grē-ˌō\ *noun* [French] (1906)
: any of a class of musician-entertainers of western Africa whose performances include tribal histories and genealogies
¹grip \'grip\ *transitive verb* **gripped; gripping** [Middle English *grippen,* from Old English *grippan;* akin to Old English *grīpan*] (before 12th century)
1 : to seize or hold firmly
2 : to hold the interest of strongly ⟨a story that *grips* the reader⟩
— **grip·per** *noun*
— **grip·ping·ly** \'gri-piŋ-lē\ *adverb*
²grip *noun* (before 12th century)
1 a : a strong or tenacious grasp **b :** strength in gripping **c :** manner or style of gripping
2 a : a firm tenacious hold typically giving control, mastery, or understanding **b :** mental grasp
3 : a part or device for gripping
4 : a part by which something is grasped; *especially* **:** HANDLE
5 : SUITCASE
6 : STAGEHAND
¹gripe \'grīp\ *verb* **griped; grip·ing** [Middle English, from Old English *grīpan;* akin to Old High German *grīfan* to grasp, Lithuanian *griebti*] (before 12th century)
transitive verb
1 *archaic* **:** SEIZE, GRASP
2 a : AFFLICT, DISTRESS **b :** IRRITATE, VEX
3 : to cause pinching and spasmodic pain in the bowels of
intransitive verb
1 : to experience gripes

◇ WORD HISTORY
gringo In English the expression "it's all Greek to me" is nearly equivalent to "I haven't a clue as to what it means." Similarly, Spanish *hablar en griego* means literally "to speak Greek" and figuratively "to speak unintelligibly." In the 18th century, *gringo,* usually explained as an alteration of *griego,* appeared in Spanish with the same sense "unintelligible speech," and became common in Latin America in the meaning "foreigner." *Gringo* made its way into English because Mexicans used it freely to refer derogatorily to Anglo-Americans. There is no basis to the story that American soldiers during the Mexican War (1846–48) often sang Robert Burns's "Green Grow the Rashes, O!" and that "green grow" was misinterpreted as *gringo* by the Mexicans.

2 : to complain with grumbling
— **grip·er** *noun*
²gripe *noun* (13th century)
1 *archaic* **:** CLUTCH, GRASP; *broadly* **:** CONTROL, MASTERY
2 : a pinching spasmodic intestinal pain — usually used in plural
3 : GRIEVANCE, COMPLAINT
grip·man \ˈgrip-mən, -ˌman\ *noun* (1886)
: a cable car operator
grippe \ˈgrip\ *noun* [French, literally, seizure] (1776)
: an acute febrile contagious virus disease; *especially* **:** INFLUENZA 1
— **grippy** \ˈgri-pē\ *adjective*
grip·sack \ˈgrip-ˌsak\ *noun* (1877)
: SUITCASE
gri·saille \gri-ˈzī, -ˈzā(ə)l\ *noun* [French, from *gris* gray, from Middle French — more at GRIZZLE] (1848)
: decoration in tones of a single color and especially gray designed to produce a three-dimensional effect
Gri·sel·da \gri-ˈzel-də\ *noun* [Middle English, from Italian]
: a woman of humble origins in medieval legend who endures tests of wifely patience laid on her by her wellborn husband
gris·eo·ful·vin \ˌgri-zē-ō-ˈfu̇l-vən, ˌgri-sē-, -ˈfəl-\ *noun* [New Latin *griseofulvum,* specific epithet of *Penicillium griseofulvum,* mold from which it is obtained] (1939)
: an antibiotic $C_{17}H_{17}ClO_6$ used systematically in treating superficial fungal infections
gri·sette \gri-ˈzet\ *noun* [French, grisette, cheap unbleached cloth, from *gris*] (1723)
1 : a young French working-class woman
2 : a young woman combining part-time prostitution with some other occupation
gris–gris \ˈgrē-(ˌ)grē\ *noun, plural* **gris–gris** \-(ˌ)grēz\ [French] (1698)
: an amulet or incantation used chiefly by people of black African ancestry
gris·ly \ˈgriz-lē\ *adjective* **gris·li·er; -est** [Middle English, from Old English *grislic,* from *gris-* (akin to Old English *āgrīsan* to fear); akin to Old High German *grīsenlīh* terrible] (12th century)
1 : inspiring horror or intense fear ⟨houses that were dark and *grisly* under the blank, cold sky —D. H. Lawrence⟩
2 : inspiring disgust or distaste ⟨a *grisly* account of the fire⟩
synonym see GHASTLY
— **gris·li·ness** *noun*
grist \ˈgrist\ *noun* [Middle English, from Old English *grīst;* akin to Old English *grindan* to grind] (before 12th century)
1 a : grain or a batch of grain for grinding **b :** the product obtained from a grist of grain including the flour or meal and the grain offals
2 : a required or usual amount
3 : matter of interest or value forming the basis of a story or analysis
4 : something turned to advantage or use — used especially in the phrase *grist for one's mill*
gris·tle \ˈgri-səl, ˌzəl\ *noun* [Middle English *gristil,* from Old English *gristle;* akin to Middle Low German *gristel* gristle] (before 12th century)
: CARTILAGE; *broadly* **:** tough cartilaginous, tendinous, or fibrous matter especially in table meats
gris·tly \ˈgri-sə-lē, ˈgris-lē, ˈgriz-\ *adjective* **gris·tli·er; -est** (14th century)
: consisting of or containing gristle ⟨*gristly* steak⟩
— **gris·tli·ness** *noun*
grist·mill \ˈgrist-ˌmil\ *noun* (1602)
: a mill for grinding grain
¹grit \ˈgrit\ *noun* [Middle English *grete,* from Old English *grēot;* akin to Old High German *grioz* sand] (before 12th century)
1 a : SAND, GRAVEL **b :** a hard sharp granule (as of sand); *also* **:** material (as many abrasives) composed of such granules
2 : any of several sandstones
3 a : the structure of a stone that adapts it to grinding **b :** the size of abrasive particles usually expressed as their mesh
4 : firmness of mind or spirit **:** unyielding courage in the face of hardship or danger
5 *capitalized* **:** a Liberal in Canadian politics
²grit *verb* **grit·ted; grit·ting** (1762)
intransitive verb
: to give forth a grating sound
transitive verb
1 : to cause (as one's teeth) to grind or grate
2 : to cover or spread with grit; *especially* **:** to smooth (as marble) with a coarse abrasive
grith \ˈgrith\ *noun* [Middle English, from Old English, from Old Norse, security] (before 12th century)
: peace, security, or sanctuary imposed or guaranteed in early medieval England under various special conditions
grits \ˈgrits\ *noun plural but singular or plural in construction* [perhaps partly from ¹*grit,* partly from dialect *grit* coarse meal, from Old English *grytt;* akin to Old English *grēot*] (1579)
: coarsely ground hulled grain; *especially* **:** ground hominy with the germ removed
grit·ty \ˈgri-tē\ *adjective* **grit·ti·er; -est** (1598)
1 : containing or resembling grit
2 : courageously persistent **:** PLUCKY
3 : having strong qualities of tough uncompromising realism ⟨a *gritty* novel⟩
— **grit·ti·ly** \ˈgri-tᵊl-ē\ *adverb*
— **grit·ti·ness** \ˈgri-tē-nəs\ *noun*
¹griz·zle \ˈgri-zəl\ *noun* [Middle English *grisel,* adjective, gray, from Middle French, from *gris,* of Germanic origin; akin to Old High German *grīs* gray] (1601)
1 *archaic* **:** gray hair
2 a : a roan coat pattern or color **b :** a gray or roan animal
²grizzle *verb* **griz·zled; griz·zling** \ˈgriz-liŋ, ˈgri-zə-\ (1740)
transitive verb
: to make grayish
intransitive verb
1 : GRIPE, GRUMBLE
2 : to become grayish
griz·zled \ˈgri-zəld\ *adjective* (15th century)
: sprinkled or streaked with gray **:** GRAYING ⟨a *grizzled* beard⟩
¹griz·zly \ˈgriz-lē\ *adjective* **griz·zli·er; -est** (1594)
: GRIZZLED
²grizzly *variant of* GRISLY
grizzly bear *noun* (1791)
: a very large brown bear (*Ursus arctos horribilis*) of the uplands of western North America — called also *grizzly*
groan \ˈgrōn\ *verb* [Middle English *gronen,* from Old English *grānian;* akin to Old High German *grīnan* to growl] (before 12th century)
intransitive verb
1 : to utter a deep moan indicative of pain, grief, or annoyance
2 : to make a harsh sound (as of creaking) under sudden or prolonged strain
transitive verb
: to utter or express with groaning
— **groan** *noun*
groan·er \ˈgrō-nər\ *noun* (1795)
1 : one that groans
2 : a stale or corny joke, observation, or story
¹groat \ˈgrōt\ *noun* [Middle English *grotes,* plural, from Old English *grotan,* plural of *grot;* akin to Old English *grēot* grit] (12th century)
1 *usually plural but singular or plural in construction* **:** hulled grain broken into fragments larger than grits
2 : a grain (as of oats) exclusive of the hull
²groat *noun* [Middle English *groot,* from Middle Dutch] (14th century)
: an old British coin worth four pennies
gro·cer \ˈgrō-sər, -shər\ *noun* [Middle English, from Middle French *grossier* wholesaler, from *gros* coarse, wholesale — more at GROSS] (15th century)
: a dealer in staple foodstuffs, meats, produce, and dairy products and usually household supplies
gro·cery \ˈgrōs-rē, ˈgrō-sə-; ˈgrōsh-rē\ *noun, plural* **-cer·ies** (15th century)
1 *plural* **:** commodities sold by a grocer — usually singular in British usage
2 : a grocer's store
grog \ˈgräg\ *noun* [*Old Grog,* nickname of Edward Vernon (died 1757) English admiral responsible for diluting the sailors' rum] (1770)
1 : alcoholic liquor; *especially* **:** liquor (as rum) cut with water and now often served hot with lemon juice and sugar sometimes added
2 : refractory materials (as crushed pottery and firebricks) used in the manufacture of refractory products (as crucibles) to reduce shrinkage in drying and firing ◆
grog·gy \ˈgrä-gē\ *adjective* **grog·gi·er; -est** [*grog*] (1832)
: weak and unsteady on the feet or in action
— **grog·gi·ly** \ˈgrä-gə-lē\ *adverb*
— **grog·gi·ness** \ˈgrä-gē-nəs\ *noun*
gro·gram \ˈgrä-grəm, ˈgrō-\ *noun* [Middle French *gros grain* coarse texture] (1562)
: a coarse loosely woven fabric of silk, silk and mohair, or silk and wool — compare GROSGRAIN
grog·shop \ˈgräg-ˌshäp\ *noun* (1790)
chiefly British **:** a usually low-class barroom
¹groin \ˈgrȯin\ *noun* [alteration of Middle English *grynde,* from Old English, abyss; akin to Old English *grund* ground] (circa 1532)
1 : the fold or depression marking the juncture of the lower abdomen and the inner part of the thigh; *also* **:** the region of this line
2 a : the projecting curved line along which two intersecting vaults meet **b :** a rib that covers this edge
3 : a rigid structure built out from a shore to protect the shore

groin 2a

◇ WORD HISTORY

grog The English admiral Edward Vernon (1684–1757) is reputed to have been in the habit of wearing a grogram cloak and to have thereby won the nickname "Old Grog" among the sailors under his command. Seamen of the Royal Navy in the West Indies had been given by custom a half-pint of rum as a daily ration, but in 1740 Vernon, alarmed at damage to the physical and moral health of his crews, ordered that the rum should be diluted with water. This mixture, not immediately approved, was christened *grog,* after its godfather. *Grog* eventually gained in popularity, however, and the word was sometimes used as a general term for any liquor, even undiluted. A drunkard could be called *groggy,* and now anyone whose mental state seems impaired, whether it be from drugs or from a blow or from fatigue, can be labeled with that word.

from erosion, to trap sand, or to direct a current for scouring a channel

²groin *transitive verb* (circa 1816)
: to build or equip with groins

grom·met \'grä-mət, 'grə-\ *noun* [obsolete French *gormette* curb of a bridle] (1626)
1 : a flexible loop that serves as a fastening, support, or reinforcement
2 : an eyelet of firm material to strengthen or protect an opening or to insulate or protect something passed through it

grom·well \'gräm-ˌwel, -wəl\ *noun* [Middle English *gromil,* from Middle French] (14th century)
: any of a genus (*Lithospermum*) of plants of the borage family having polished white stony nutlets

¹groom \'grüm, 'grům\ *noun* [Middle English *grom*] (14th century)
1 *archaic* **:** MAN, FELLOW
2 a (1) *archaic* **:** MANSERVANT (2) **:** one of several officers of the English royal household **b :** a person responsible for the feeding, exercising, and stabling of horses
3 : BRIDEGROOM

²groom (1809)
transitive verb
1 : to clean and condition (as a horse or dog)
2 : to make neat or attractive ⟨an impeccably *groomed* woman⟩
3 : to get into readiness for a specific objective **:** PREPARE ⟨was being *groomed* as a presidential candidate⟩
intransitive verb
: to groom oneself

groom·er \'grü-mər\ *noun* (circa 1890)
: one who grooms (as dogs)

grooms·man \'grümz-mən, 'grůmz-\ *noun* (1698)
: a male friend who attends a bridegroom at his wedding

¹groove \'grüv\ *noun* [Middle English *groof;* akin to Old English *grafan* to dig — more at GRAVE] (1659)
1 : a long narrow channel or depression
2 a : a fixed routine **:** RUT **b :** a situation suited to one's abilities or interests **:** NICHE
3 : top form ⟨a great talker when he is in the *groove*⟩
4 : the middle of the strike zone in baseball where a pitch is most easily hit ⟨a fastball right in the *groove*⟩
5 : an enjoyable or exciting experience
6 : a pronounced enjoyable rhythm

²groove *verb* **grooved; groov·ing** (1686)
transitive verb
1 a : to make a groove in **b :** to join by a groove
2 : to perfect by repeated practice ⟨*grooved* her golf swing⟩
3 : to throw (a pitch) in the groove
intransitive verb
1 : to become joined or fitted by a groove
2 : to form a groove
3 : to enjoy oneself intensely
4 : to interact harmoniously ⟨contemporary minds and rock *groove* together —Benjamin DeMott⟩
— groov·er *noun*

groovy \'grü-vē\ *adjective* **groov·i·er; -est** (circa 1937)
1 : MARVELOUS, WONDERFUL, EXCELLENT ⟨felt that this poetry was . . . enjoyable, not to mention *groovy* —R. M. Muccigrosso⟩
2 : HIP ⟨a younger and *groovier* audience —Robert MacKenzie⟩

grope \'grōp\ *verb* **groped; grop·ing** [Middle English, from Old English *grāpian;* akin to Old English *grīpan* to seize] (before 12th century)
intransitive verb
1 : to feel about blindly or uncertainly in search ⟨*grope* for the light switch⟩
2 : to look for something blindly or uncertainly ⟨*grope* for the right words⟩
3 : to feel one's way
transitive verb
1 : FEEL UP
2 : to find (as one's way) by groping
— grope *noun*
— grop·er *noun*

gros·beak \'grōs-ˌbēk\ *noun* [part translation of French *grosbec,* from *gros* thick + *bec* beak] (circa 1678)
: any of several finches of Europe or America having large stout conical bills

gro·schen \'grō-shən, 'grȯ-\ *noun, plural* **groschen** [German] (1946)
— see *schilling* at MONEY table

gros·grain \'grō-ˌgrān\ *noun* [French *gros grain* coarse texture] (1869)
: a strong close-woven corded fabric usually of silk or rayon and often with cotton filler — compare GROGRAM

¹gross \'grōs\ *adjective* [Middle English, from Middle French *gros* thick, coarse, from Latin *grossus*] (14th century)
1 a *archaic* **:** immediately obvious **b** (1) **:** glaringly noticeable usually because of inexcusable badness or objectionableness ⟨*gross* error⟩ (2) **:** OUT-AND-OUT, UTTER ⟨*gross* injustice⟩ **c :** visible without the aid of a microscope
2 a : BIG, BULKY; *especially* **:** excessively fat **b :** growing or spreading with excessive luxuriance
3 a : of, relating to, or dealing with general aspects or broad distinctions **b :** consisting of an overall total exclusive of deductions ⟨*gross* income⟩ — compare NET
4 : made up of material or perceptible elements
5 *archaic* **:** not fastidious in taste **:** UNDISCRIMINATING
6 a : coarse in nature or behavior **:** UNREFINED **b :** gravely deficient in civility or decency **:** crudely vulgar ⟨merely *gross,* a scatological rather than a pornographic impropriety —Aldous Huxley⟩ **c** *slang* **:** inspiring disgust or distaste ⟨that sandwich looks *gross*⟩
7 : deficient in knowledge **:** IGNORANT, UNTUTORED
synonym see COARSE, FLAGRANT
— gross·ly *adverb*
— gross·ness *noun*

²gross *noun* (1579)
1 *obsolete* **:** AMOUNT, SUM
2 : overall total exclusive of deductions

³gross *transitive verb* (1884)
: to earn or bring in (an overall total) exclusive of deductions (as for taxes or expenses)
— gross·er *noun*

⁴gross *noun, plural* **gross** [Middle English *groce,* from Middle French *grosse,* from feminine of *gros*] (14th century)
: an aggregate of 12 dozen things ⟨a *gross* of pencils⟩

gross anatomy *noun* (1888)
: a branch of anatomy that deals with the macroscopic structure of tissues and organs

gross national product *noun* (1947)
: the total value of the goods and services produced by the residents of a nation during a specified period (as a year)

gross–out \'grō-ˌsau̇t\ *noun, often attributive* (1973)
: something inspiring disgust or distaste

gross out *transitive verb* (1968)
: to offend, insult, or disgust by something gross

gros·su·lar \'gräs-yə-lər\ *noun* [New Latin *Grossularia,* genus name of the gooseberry] (1819)
: a variety of garnet that is most commonly green and consists of calcium aluminum silicate

gros·su·la·rite \-lə-ˌrīt\ *noun* [German *Grossularit,* from New Latin *Grossularia*] (circa 1847)
: GROSSULAR

gro·szy \'grȯ-shē\ *also* **grosz** *or* **grosze** \'grȯsh\ *noun, plural* **groszy** [Polish] (1916)
— see *zloty* at MONEY table

grot \'grät\ *noun* [Middle French *grotte,* from Italian *grotta*] (1506)
: GROTTO

¹gro·tesque \grō-'tesk\ *noun* [Middle French & Old Italian; Middle French, from Old Italian (*pittura*) *grottesca,* literally, cave painting, feminine of *grottesco* of a cave, from *grotta*] (1561)
1 a : a style of decorative art characterized by fanciful or fantastic human and animal forms often interwoven with foliage or similar figures that may distort the natural into absurdity, ugliness, or caricature **b :** a piece of work in this style
2 : one that is grotesque
3 : SANS SERIF ◆

²grotesque *adjective* (1603)
: of, relating to, or having the characteristics of the grotesque: as **a :** FANCIFUL, BIZARRE **b :** absurdly incongruous **c :** departing markedly from the natural, the expected, or the typical
synonym see FANTASTIC
— gro·tesque·ly *adverb*
— gro·tesque·ness *noun*

gro·tes·que·rie *also* **gro·tes·que·ry** \grō-'tes-kə-rē\ *noun, plural* **-ries** [*grotesque* + French *-erie* -ery] (circa 1666)
1 : something that is grotesque
2 : the quality or state of being grotesque **:** GROTESQUENESS

grot·to \'grä-(ˌ)tō\ *noun, plural* **grottoes** *also* **grottos** [Italian *grotta, grotto,* from Latin *crypta* cavern, crypt] (1617)
1 : CAVE
2 : an artificial recess or structure made to resemble a natural cave

grot·ty \'grä-tē\ *adjective* **grot·ti·er; -est** [origin unknown] (1964)
chiefly British **:** wretchedly shabby **:** of poor quality

grouch \'grau̇ch\ *noun* [probably alteration of *grutch* (grudge)] (circa 1895)
1 a : a fit of bad temper **b :** GRUDGE, COMPLAINT
2 : a habitually irritable or complaining person **:** GRUMBLER
— grouch *intransitive verb*

grouchy \'grau̇-chē\ *adjective* **grouch·i·er; -est** (circa 1895)
: given to grumbling **:** PEEVISH
— grouch·i·ly \-chə-lē\ *adverb*
— grouch·i·ness \-chē-nəs\ *noun*

¹ground \'grau̇nd\ *noun, often attributive* [Middle English, from Old English *grund;* akin to Old High German *grunt* ground] (before 12th century)

◇ WORD HISTORY

grotesque In the middle of the 15th century at Rome, when antiquarian interest in the city's cultural legacy was high, remnants of ancient Roman decorative painting were found in underground chambers of the Domus Aurea, the vast palace of the emperor Nero that had been largely demolished and covered over by newer structures. Inspired by these finds, Renaissance painters and sculptors created a style of decorative art with fantastic combinations of human and animal forms interwoven with strange fruits and flowers. Examples of the style were dubbed after their ancient inspiration *grottesche,* short for *pitture grottesche,* literally, "cave paintings." The adjective *grottesco,* which acquired the sense " having the characteristics of *grottesche*" was borrowed into French and English, with the English word eventually copying the French form. By about the middle of the 17th century *grotesque* in both languages had developed the more extended senses "bizarre" or "absurdly incongruous" that the word still has today.

1 a : the bottom of a body of water **b** *plural* (1) **:** SEDIMENT 1 (2) **:** ground coffee beans after brewing
2 a : a basis for belief, action, or argument ⟨*ground* for complaint⟩ — often used in plural **b** (1) **:** a fundamental logical condition (2) **:** a basic metaphysical cause
3 a : a surrounding area **:** BACKGROUND **b :** material that serves as a substratum
4 a : the surface of the earth **b :** an area used for a particular purpose ⟨parade *ground*⟩ ⟨fishing *grounds*⟩ **c** *plural* **:** the area around and belonging to a house or other building **d :** an area to be won or defended in or as if in battle **e :** an area of knowledge or special interest ⟨covered a lot of *ground* in his lecture⟩
5 a : SOIL, EARTH **b :** a special soil
6 a : an object that makes an electrical connection with the earth **b :** a large conducting body (as the earth) used as a common return for an electric circuit and as an arbitrary zero of potential **c :** electric connection with a ground
7 : a football offense utilizing primarily running plays
— from the ground up 1 : entirely new or afresh **2 :** from top to bottom **:** THOROUGHLY
— into the ground : beyond what is necessary or tolerable **:** to exhaustion ⟨labored an issue *into the ground* —*Newsweek*⟩
— off the ground : in or as if in flight **:** off to a good start ⟨the program never got *off the ground*⟩
— on the ground : at the scene of action
— to ground 1 : into a burrow ⟨the fox went *to ground*⟩ **2 :** into hiding ⟨might need to make a run for it and go *to ground* someplace —Edward Hoagland⟩

²ground (13th century)
transitive verb
1 a : to bring to or place on the ground **b :** to cause to run aground
2 a : to provide a reason or justification for ⟨our fears about technological change may be well *grounded* —L. K. Williams⟩ **b :** to furnish with a foundation of knowledge **:** BASE ⟨an understanding . . . that is *grounded* in fact —Michael Kimmelman⟩
3 : to connect electrically with a ground
4 a : to restrict to the ground ⟨*ground* a pilot⟩ **b :** to prohibit from taking part in some usual activities ⟨*grounded* her for a week⟩
5 : to throw (a football) intentionally to the ground to avoid being tackled for a loss
intransitive verb
1 : to have a ground or basis **:** RELY
2 : to run aground
3 : to hit a grounder

³ground *past and past participle of* GRIND

ground ball *noun* (1857)
: a batted baseball that bounds or rolls along the ground

ground bass *noun* (1699)
: a short bass passage continually repeated below constantly changing melody and harmony

ground beetle *noun* (1848)
: any of a large cosmopolitan family (Carabidae) of soil-inhabiting usually carnivorous often shiny black or metallic beetles commonly having fused elytra

ground·break·er \'grau̇n(d)-ˌbrā-kər\ *noun* (1940)
: one that innovates **:** PIONEER

ground·break·ing \-ˌbrā-kiŋ\ *adjective* (1907)
: markedly innovative ⟨has written a *groundbreaking* work⟩

ground·burst \-ˌbərst\ *noun* (circa 1951)
: the detonation of a nuclear warhead at ground level

ground–cher·ry \'grau̇n(d)-ˈcher-ē\ *noun* (1807)
: any of numerous chiefly New World plants (genus *Physalis*) of the nightshade family with pulpy fruits in papery husks; *also* **:** the fruit of the ground-cherry

ground cloth *noun* (1931)
: a waterproof sheet placed on the ground for protection (as of a sleeping bag) against soil moisture

ground cover *noun* (1900)
1 : the small plants on a forest floor except young trees
2 a : a planting of low plants (as ivy) that covers the ground in place of turf **b :** a plant adapted for use as ground cover

ground crew *noun* (1934)
: the mechanics and technicians who maintain and service an airplane

ground–effect machine *noun* [from the lift provided by compression of air between the vehicle and the ground] (1962)
: HOVERCRAFT

ground·er \'grau̇n-dər\ *noun* (circa 1867)
: GROUND BALL

ground·fish \'grau̇n(d)-ˌfish\ *noun* (1856)
: a bottom fish; *especially* **:** a marine fish (as a cod, haddock, pollack, or flounder) of commercial importance

ground floor *noun* (1601)
1 : the floor of a house most nearly on a level with the ground — compare FIRST FLOOR
2 : a favorable position or privileged opportunity usually obtained by early participants — used especially in the phrase *in on the ground floor*

ground glass *noun* (1848)
: glass with a light-diffusing surface produced by etching or abrading

ground·hog \'grau̇n(d)-ˌhȯg, -ˌhäg\ *noun* (1742)
: WOODCHUCK

Groundhog Day *noun* [from the legend that a groundhog emerging from its burrow returns to hibernate if it sees its shadow on this day] (1871)
: February 2 that traditionally indicates six more weeks of winter if sunny or an early spring if cloudy

ground·ing \'grau̇n-diŋ\ *noun* (1644)
: training or instruction in the fundamentals of a field of knowledge

ground ivy *noun* (14th century)
: a trailing mint (*Glechoma hederacea*) with rounded leaves and blue-purple flowers

ground·less \'grau̇n(d)-ləs\ *adjective* (1620)
: having no ground or foundation ⟨*groundless* fears⟩
— ground·less·ly *adverb*
— ground·less·ness *noun*

ground·ling \'grau̇n(d)-liŋ\ *noun* (1602)
1 a : a spectator who stood in the pit of an Elizabethan theater **b :** a person of unsophisticated taste
2 : one that lives or works on or near the ground

ground loop *noun* (1928)
: a sharp uncontrollable turn made by an aircraft on the ground and usually caused by an unbalanced drag (as from a wingtip touching the ground)

ground·mass \'grau̇n(d)-ˌmas\ *noun* (1879)
: the fine-grained or glassy base of a porphyry in which the larger distinct crystals are embedded

ground meristem *noun* (1938)
: the part of a primary apical meristem remaining after differentiation of protoderm and procambium

ground·nut \'grau̇n(d)-ˌnət\ *noun* (1602)
1 a : any of several plants having edible tuberous roots; *especially* **:** a North American leguminous vine (*Apios americana*) with pinnate leaves and clusters of brownish purple fragrant flowers **b :** the root of a groundnut
2 *chiefly British* **:** PEANUT

ground·out \'grau̇n-ˌdau̇t\ *noun* (1965)
: a play in baseball in which a batter is put out after hitting a grounder to an infielder

ground pine *noun* (1551)
1 : a European bugle (*Ajuga chamaepitys*) with a resinous odor
2 : any of several club mosses (especially *Lycopodium clavatum* and *L. complanatum*) with long creeping stems and erect branches

ground plan *noun* (1731)
1 : a plan of a floor of a building as distinguished from an elevation
2 : a first or basic plan

ground rent *noun* (1667)
: the rent paid by a lessee for the use of land especially for building

ground rule *noun* (1890)
1 : a sports rule adopted to modify play on a particular field, court, or course
2 : a rule of procedure ⟨*ground rules* for selecting a superintendent —*American School Board Journal*⟩

¹ground·sel \'grau̇n(d)-səl\ *noun* [Middle English *groundeswele,* from Old English *grundeswelge,* from *grund* ground + *swelgan* to swallow — more at SWALLOW] (before 12th century)
: any of various senecios (as the nearly cosmopolitan weed *Senecio vulgaris*)

²groundsel *noun* [Middle English *ground sille,* from *ground* + *sille* sill] (15th century)
: a foundation timber

ground·sheet \'grau̇n(d)-ˌshēt\ *noun* (1907)
: GROUND CLOTH

grounds·keep·er \'grau̇n(d)z-ˌkē-pər\ *noun* (1903)
: a person who cares for the grounds of a usually large property (as a sports field)

ground sloth *noun* (1860)
: any of various often huge extinct American edentates related to the recent sloths

grounds·man \'grau̇n(d)z-mən\ *noun* (1886)
chiefly British **:** GROUNDSKEEPER

ground speed *noun* (1917)
: the speed (as of an airplane) with relation to the ground — compare AIRSPEED

ground squirrel *noun* (1688)
: any of various burrowing rodents (especially genus *Spermophilus*) of North America and Eurasia that are related to the squirrels and often live in colonies especially in open areas — called also *spermophile*

ground state *noun* (1926)
: the state of a physical system (as of an atomic nucleus or an atom) having the least energy of all the possible states — called also *ground level*

ground stroke *noun* (1895)
: a stroke made (as in tennis) by hitting a ball that has rebounded from the ground — compare VOLLEY

ground substance *noun* (1882)
: a more or less homogeneous matrix in which the specific differentiated elements of a system are suspended: **a :** the intercellular substance of tissues **b :** CYTOSOL

ground·swell \'grau̇nd-ˌswel\ *noun* (1818)
1 *usually* **ground swell :** a broad deep undulation of the ocean caused by an often distant gale or seismic disturbance
2 : a rapid spontaneous growth (as of political opinion) ⟨a *groundswell* of support⟩

ground·wa·ter \-ˌwȯ-tər, -ˌwä-\ *noun* (circa 1889)
: water within the earth that supplies wells and springs

ground wave *noun* (1925)
: a radio wave that is propagated along the surface of the earth

ground·wood \'grau̇nd-ˌwu̇d\ *noun* [³*ground*] (1885)
: wood ground up and used to make pulp for paper

ground·work \'grau̇nd-ˌwərk\ *noun* (15th century)

\ə\ **abut** \ᵊ\ **kitten** \ər\ **further** \a\ **ash** \ā\ **ace** \ä\ **mop, mar** \au̇\ **out** \ch\ **chin** \e\ **bet** \ē\ **easy** \g\ **go** \i\ **hit** \ī\ **ice** \j\ **job** \ŋ\ **sing** \ō\ **go** \ȯ\ **law** \ȯi\ **boy** \th\ **thin** \th\ **the** \ü\ **loot** \u̇\ **foot** \y\ **yet** \zh\ **vision** *see also* Guide to Pronunciation

: FOUNDATION, BASIS ⟨laid the *groundwork* for a new program⟩; *also* : preparation made beforehand ⟨the *groundwork* was done before the winter tour —Susan Reiter⟩

ground zero *noun* (1946)
1 : the point directly above, below, or at which a nuclear explosion occurs
2 : the center or origin of rapid, intense, or violent activity or change
3 : the very beginning : SQUARE ONE

¹group \'grüp\ *noun, often attributive* [French *groupe,* from Italian *gruppo,* of Germanic origin; akin to Old High German *kropf* craw — more at CROP] (1686)
1 : two or more figures forming a complete unit in a composition
2 a : a number of individuals assembled together or having some unifying relationship **b** : an assemblage of objects regarded as a unit **c** (1) **:** a military unit consisting of a headquarters and attached battalions (2) **:** a unit of the U.S. Air Force higher than a squadron and lower than a wing
3 a : an assemblage of related organisms — often used to avoid taxonomic connotations when the kind or degree of relationship is not clearly defined **b** (1) **:** two or more atoms joined together or sometimes a single atom forming part of a molecule; *especially* **:** FUNCTIONAL GROUP ⟨a methyl *group*⟩ (2) **:** an assemblage of elements forming one of the vertical columns of the periodic table **c :** a stratigraphic division comprising rocks deposited during an era
4 : a mathematical set that is closed under a binary associative operation, contains an identity element, and has an inverse for every element

²group (1718)
transitive verb
1 : to combine in a group
2 : to assign to a group **:** CLASSIFY
intransitive verb
1 : to form a group
2 : to belong to a group
3 : to make groups of closely spaced hits on a target ⟨the gun *grouped* beautifully —R. C. Ruark⟩
— group·able \'grü-pə-bəl\ *adjective*

group captain *noun* (1919)
: a commissioned officer in the British air force who ranks with a colonel in the army

group dynamics *noun plural but singular or plural in construction* (1939)
: the interacting forces within a small human group; *also* **:** the sociological study of these forces

grou·per \'grü-pər\ *noun, plural* **groupers** *also* **grouper** [Portuguese *garoupa*] (1671)
: any of numerous fishes (family Serranidae and especially genera *Epinephelus* and *Mycteroperca*) that are typically large solitary bottom-dwelling fishes of warm seas

group·ie \'grü-pē\ *noun* (1967)
1 : a fan of a rock group who usually follows the group around on concert tours
2 : an admirer of a celebrity who attends as many of his or her public appearances as possible
3 : ENTHUSIAST, AFICIONADO ⟨a political *groupie*⟩ ⟨golf *groupies*⟩

group·ing \'grü-piŋ\ *noun* (1748)
1 : the act or process of combining in groups
2 : a set of objects combined in a group ⟨a furniture *grouping*⟩

group practice *noun* (1942)
: medicine practiced by a group of associated physicians or dentists (as specialists in different fields) working as partners or as partners and employees

group theory *noun* (1898)
: a branch of mathematics concerned with finding all mathematical groups and determining their properties

group therapy *noun* (1943)
: therapy in the presence of a therapist in which several patients discuss and share their personal problems — called also *group psychotherapy*
— group therapist *noun*

group·think \'grüp-ˌthiŋk\ *noun* [¹*group* + *-think* (as in *doublethink*)] (1952)
: a pattern of thought characterized by self-deception, forced manufacture of consent, and conformity to group values and ethics

grou·pus·cule \grü-'pəs-ˌkyül\ *noun* [French, from *groupe* group + *-uscule* (as in *corpuscule* corpuscle)] (1969)
: a small group of political activists

¹grouse \'grau̇s\ *noun, plural* **grouse** *or* **grouses** [origin unknown] (1531)
: any of numerous chiefly ground-dwelling birds (family Tetraonidae) that are usually of reddish brown or other protective color and have feathered legs and that include many important game birds

²grouse *intransitive verb* **groused; grous·ing** [origin unknown] (1887)
: COMPLAIN, GRUMBLE
— grous·er *noun*

³grouse *noun* (1918)
: COMPLAINT

¹grout \'grau̇t\ *noun* [Middle English, coarse meal, from Old English *grūt;* akin to Old English *grēot* grit] (1638)
1 a : thin mortar used for filling spaces (as the joints in masonry); *also* **:** any of various other materials (as a mixture of cement and water or chemicals that solidify) used for a similar purpose **b :** PLASTER
2 : LEES

²grout *transitive verb* (1838)
1 : to fill up or finish with grout
2 : to fix in place by means of grout ⟨*grout* a bolt into a wall⟩
— grout·er *noun*

grove \'grōv\ *noun* [Middle English, from Old English *grāf*] (before 12th century)
1 : a small wood without underbrush ⟨a picnic *grove*⟩
2 : a planting of fruit or nut trees

grov·el \'grä-vᵊl, 'grə-\ *intransitive verb* **-eled** *or* **-elled; -el·ing** *or* **-el·ling** [back-formation from *groveling* prone, from *groveling,* adverb, from Middle English, from *gruf,* adverb, on the face (from Old Norse *ā grūfu*) + *-ling*] (1593)
1 : to creep with the face to the ground **:** CRAWL
2 a : to lie or creep with the body prostrate in token of subservience or abasement **b :** to abase oneself
3 : to give oneself over to what is base or unworthy **:** WALLOW ⟨*groveling* in self-pity⟩
— grov·el·er \-vᵊl-ər\ *noun*
— grov·el·ing·ly \-vᵊl-iŋ-lē\ *adverb*

groves of academe *often A capitalized* (1849)
: the academic world

grow \'grō\ *verb* **grew** \'grü\; **grown** \'grōn\; **grow·ing** [Middle English, from Old English *grōwan;* akin to Old High German *gruowan* to grow] (before 12th century)
intransitive verb
1 a : to spring up and develop to maturity **b :** to be able to grow in some place or situation ⟨trees that *grow* in the tropics⟩ **c :** to assume some relation through or as if through a process of natural growth ⟨ferns *growing* from the rocks⟩
2 a : to increase in size by assimilation of material into the living organism or by accretion of material in a nonbiological process (as crystallization) **b :** INCREASE, EXPAND ⟨*grows* in wisdom⟩
3 : to develop from a parent source ⟨the book *grew* out of a series of lectures⟩
4 a : to pass into a condition **:** BECOME ⟨*grew* pale⟩ **b :** to have an increasing influence ⟨habit *grows* on a person⟩ **c :** to become increasingly acceptable or attractive ⟨didn't like it at first, but it *grew* on him⟩
transitive verb
1 : to cause to grow ⟨*grow* wheat⟩
2 : DEVELOP 5
— grow·er \'grō(-ə)r\ *noun*
— grow·ing·ly \'grō-iŋ-lē\ *adverb*

growing pains *noun plural* (1810)
1 : pains in the legs of growing children having no demonstrable relation to growth
2 : the stresses and strains attending a new project or development

growing point *noun* (1835)
: the undifferentiated end of a plant shoot from which additional shoot tissues differentiate

¹growl \'grau̇(ə)l\ *verb* [Middle English *groulen, grollen*] (14th century)
intransitive verb
1 a : RUMBLE ⟨his stomach *growled*⟩ **b :** to utter a growl ⟨the dog *growled* at the stranger⟩
2 : to complain angrily
transitive verb
: to utter with a growl **:** utter angrily

²growl *noun* (1727)
: a deep guttural inarticulate sound

growl·er \'grau̇-lər\ *noun* (1753)
1 : one that growls
2 : a container (as a can or pitcher) for beer bought by the measure
3 : a small iceberg

growl·ing \'grau̇-liŋ\ *adjective* (1752)
: marked by a growl ⟨a low *growling* voice⟩ ⟨listened to the *growling* thunder⟩
— growl·ing·ly \-liŋ-lē\ *adverb*

growly \'grau̇-lē\ *adjective* **growl·i·er; -est** (1920)
: resembling a growl ⟨a *growly* voice⟩
— growl·i·ness *noun*

grown \'grōn\ *adjective* (1645)
1 : fully grown **:** MATURE ⟨*grown* men and women⟩
2 : covered or surrounded with vegetation ⟨land well *grown* with trees⟩
3 a : cultivated or produced in a specified way or locality — used in combination ⟨shade-*grown* tobacco⟩ **b :** overgrown with — used in combination ⟨a weed-*grown* patio⟩

¹grown–up \'grō-ˌnəp\ *adjective* (1633)
: not childish or immature **:** ADULT

²grown–up *noun* (1813)
: ADULT

growth \'grōth\ *noun* (1557)
1 a (1) **:** a stage in the process of growing **:** SIZE (2) **:** full growth **b :** the process of growing **c :** progressive development **:** EVOLUTION **d :** INCREASE, EXPANSION ⟨the *growth* of the oil industry⟩
2 a : something that grows or has grown **b :** an abnormal proliferation of tissue (as a tumor) **c :** OUTGROWTH **d :** the result of growth **:** PRODUCT
3 : a producing especially by growing ⟨fruits of his own *growth*⟩

growth company *noun* (1959)
: a company that grows at a greater rate than the economy as a whole and that usually directs a relatively high proportion of income back into the business

growth factor *noun* (1926)
: a substance (as a vitamin B_{12} or an interleukin) that promotes growth and especially cellular growth

growth hormone *noun* (1924)
1 : a vertebrate polypeptide hormone that is secreted by the anterior lobe of the pituitary gland and regulates growth — called also *somatotropin*
2 : GROWTH REGULATOR

growth regulator *noun* (1936)
: any of various synthetic or naturally occurring plant substances (as an auxin or gibberellin) that regulate growth

growth ring *noun* (1907)

: a layer of wood (as an annual ring) produced during a single period of growth

growthy \'grō-thē\ *adjective* **growth·i·er; -est** (1884)
of livestock **:** exceptionally fast in growing and gaining weight
— **growth·i·ness** \-nəs\ *noun*

grow up *intransitive verb* (1535)
: to grow toward or arrive at full stature or physical or mental maturity ⟨*growing up* intellectually, socially, and physically⟩

groyne \'grȯin\ *noun* [by alteration] (1582)
: GROIN 3

1grub \'grəb\ *verb* **grubbed; grub·bing** [Middle English *grubben;* akin to Old English *grafan* to dig — more at GRAVE] (14th century)
transitive verb
1 : to clear by digging up roots and stumps
2 : to dig up by or as if by the roots
intransitive verb
1 a : to dig in the ground especially for something that is difficult to find or extract **b :** to search about ⟨*grubbed* in the countryside for food —*Lamp*⟩
2 : TOIL, DRUDGE ⟨folks who *grub* for money —James Street⟩
— **grub·ber** *noun*

2grub *noun* [Middle English *grubbe,* from *grubben*] (15th century)
1 : a soft thick wormlike larva of an insect
2 a : one who does menial work **:** DRUDGE **b :** a slovenly person
3 : FOOD

grub·by \'grə-bē\ *adjective* **grub·bi·er; -est** (1725)
1 : infested with fly maggots
2 a : DIRTY, GRIMY ⟨*grubby* hands⟩ **b :** SLOVENLY, SLOPPY
3 : worthy of contempt **:** BASE ⟨*grubby* political motives⟩
— **grub·bi·ly** \'grə-bə-lē\ *adverb*
— **grub·bi·ness** \'grə-bē-nəs\ *noun*

1grub·stake \'grəb-ˌstāk\ *noun* (1863)
1 : supplies or funds furnished a mining prospector on promise of a share in his discoveries
2 : material assistance (as a loan) provided for launching an enterprise or for a person in difficult circumstances

2grubstake *transitive verb* (1879)
: to provide with a grubstake
— **grub·stak·er** *noun*

Grub Street \'grəb-\ *noun* [*Grub Street,* London, formerly inhabited by literary hacks] (1630)
: the world or category of needy literary hacks

1grudge \'grəj\ *transitive verb* **grudged; grudg·ing** [Middle English *grucchen, grudgen* to grumble, complain, from Middle French *groucier,* of Germanic origin; akin to Middle High German *grogezen* to howl] (14th century)
: to be unwilling to give or admit **:** give or allow reluctantly or resentfully ⟨didn't *grudge* the time⟩
— **grudg·er** *noun*

2grudge *noun* (15th century)
: a feeling of deep-seated resentment or ill will
synonym see MALICE

grudg·ing \'grə-jiŋ\ *adjective* (circa 1533)
1 : UNWILLING, RELUCTANT
2 : done, given, or allowed unwillingly, reluctantly, or sparingly ⟨*grudging* compliance⟩
— **grudg·ing·ly** \-jiŋ-lē\ *adverb*

gru·el \'grü-əl\ *noun* [Middle English *grewel,* from Middle French *gruel,* of Germanic origin; akin to Old English *grūt* grout] (14th century)
1 : a thin porridge
2 *chiefly British* **:** PUNISHMENT

gru·el·ing *or* **gru·el·ling** \'grü-ə-liŋ\ *adjective* [from present participle of obsolete *gruel* (to exhaust)] (1852)
: trying or taxing to the point of exhaustion **:** PUNISHING ⟨a *grueling* race⟩
— **gru·el·ing·ly** \-liŋ-lē\ *adverb*

grue·some \'grü-səm\ *adjective* [alteration of earlier *growsome,* from English dialect *grow, grue* to shiver, from Middle English *gruen,* probably from Middle Dutch *grūwen;* akin to Old High German in*grūēn* to shiver] (1816)
: inspiring horror or repulsion **:** GRISLY
synonym see GHASTLY
— **grue·some·ly** *adverb*
— **grue·some·ness** *noun*

1gruff \'grəf\ *adjective* [Dutch *grof;* akin to Old High German *grob* coarse, *hruf* scurf — more at DANDRUFF] (1706)
1 : rough, brusque, or stern in manner, speech, or aspect ⟨a *gruff* reply⟩
2 : being deep and harsh **:** HOARSE ⟨a *gruff* voice⟩
synonym see BLUFF
— **gruff·ly** *adverb*
— **gruff·ness** *noun*

2gruff *transitive verb* (1706)
: to utter in a gruff voice or manner

grum·ble \'grəm-bəl\ *verb* **grum·bled; grum·bling** \-b(ə-)liŋ\ [probably from Middle French *grommeler,* ultimately from Middle Dutch *grommen;* akin to Old High German *grimm* grim] (circa 1586)
intransitive verb
1 : to mutter in discontent
2 : GROWL, RUMBLE
transitive verb
: to express with grumbling
— **grumble** *noun*
— **grum·bler** \-b(ə-)lər\ *noun*
— **grum·bling·ly** \-b(ə-)liŋ-lē\ *adverb*
— **grum·bly** \-b(ə-)lē\ *adjective*

grum·met \'grə-mət\ *variant of* GROMMET

1grump \'grəmp\ *noun* [obsolete English *grumps* (snubs, slights)] (1844)
1 : a fit of ill humor or sulkiness — usually used in plural
2 : a person given to complaining

2grump (1875)
intransitive verb
1 : SULK
2 : GRUMBLE, COMPLAIN
transitive verb
: to utter in a grumpy manner

grumpy \'grəm-pē\ *adjective* **grump·i·er; -est** (1778)
: moodily cross **:** SURLY
— **grump·i·ly** \-pə-lē\ *adverb*
— **grump·i·ness** \-pē-nəs\ *noun*

grunge \'grənj\ *noun* [back-formation from *grungy*] (1965)
1 : one that is grungy
2 : rock music incorporating elements of punk rock and heavy metal; *also* **:** the untidy working-class fashions typical of fans of grunge

grun·gy \'grən-jē\ *adjective* **grun·gi·er; -est** [origin unknown] (1965)
: shabby or dirty in character or condition

grun·ion \'grən-yən\ *noun* [probably from Spanish *gruñón* grunter] (1917)
: a silversides (*Leuresthes tenuis*) of the California coast notable for the regularity with which it comes inshore to spawn at nearly full moon

1grunt \'grənt\ *verb* [Middle English *grunten,* from Old English *grunnettan,* frequentative of *grunian,* of imitative origin] (before 12th century)
intransitive verb
: to utter a grunt
transitive verb
: to utter with a grunt
— **grunt·er** *noun*

2grunt *noun* (1553)
1 a : the deep short sound characteristic of a hog **b :** a similar sound
2 [from the noise it makes when taken from the water] **:** any of a family (Haemulidae synonym Pomadasyidae) of chiefly tropical marine bony fishes
3 : a dessert made by dropping biscuit dough on top of boiling berries and steaming ⟨blueberry *grunt*⟩

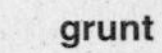
grunt 2

4 : a U.S. army or marine foot soldier especially in the Vietnam war
5 : one who does routine unglamorous work — often used attributively ⟨*grunt* work⟩

grun·tle \'grən-tᵊl\ *transitive verb* **grun·tled; grun·tling** \'grənt-liŋ, 'grən-tᵊl-iŋ\ [back-formation from *disgruntle*] (1926)
: to put in a good humor ⟨were *gruntled* with a good meal and good conversation —W. P. Webb⟩

grutch \'grəch\ *transitive verb* [Middle English *grucchen*] (14th century)
obsolete **:** BEGRUDGE

grutten *past participle of* GREET

Gru·yère \grü-'yer, grē-\ *noun* [*Gruyère,* district in Switzerland] (1802)
1 : a firm cheese with small holes and a nutty flavor that is of Swiss origin
2 : a process cheese made from natural Gruyère

gryph·on *variant of* GRIFFIN

G–string \'jē-ˌstriŋ\ *noun* [origin unknown] (1878)
: a strip of cloth passed between the legs and supported by a waist cord that is worn especially by striptease dancers

G suit *noun* [gravity *suit*] (1944)
: a suit designed to counteract the physiological effects of acceleration on an aviator or astronaut

GT \ˌjē-'tē\ *noun* [Italian *G*ran *T*urismo] (1966)
: GRAND TOURING CAR

GTP \ˌjē-(ˌ)tē-'pē\ *noun* [*g*uanosine *t*ri*p*hosphate] (1961)
: an energy-rich nucleotide analogous to ATP that is composed of guanine, ribose, and three phosphate groups and is necessary for peptide-bond formation during protein synthesis — called also *guanosine triphosphate*

G$_2$ phase \ˌjē-'tü-\ *noun* [growth] (1968)
: the period in the cell cycle from the completion of DNA replication to the beginning of cell division — compare G$_1$ PHASE, M PHASE, S PHASE

gua·ca·mo·le \ˌgwä-kə-'mō-lē\ *noun* [Mexican Spanish, from Nahuatl *āhuacamōlli,* from *āhuacatl* avocado + *mōlli* sauce] (1920)
: pureed or mashed avocado seasoned with condiments
word history see AVOCADO

gua·cha·ro \'gwä-chə-ˌrō\ *noun, plural* **-ros** *or* **-roes** [American Spanish *guácharo*] (1830)
: OILBIRD

guai·ac \'gwī-ˌak, 'gī-\ *noun* [New Latin *Guaiacum*] (circa 1751)
: GUAIACUM 2

guai·a·cum \'gwī-ə-kəm, 'gī-\ *noun* [New Latin, from Spanish *guayaco,* from Taino *guayacan*] (1553)
1 : any of a genus (*Guaiacum*) of the caltrop family of tropical American evergreen trees and shrubs having pinnate leaves, usually blue flowers, and capsular fruit
2 a : the hard greenish brown wood of a guaiacum (especially *Guaiacum officinale*) **b :** a resin with a faint balsamic odor that is obtained from the trunk of two guaiacums (*G. officinale* and *G. sanctum*) and often used as a clinical reagent

guan \'gwän\ *noun* [alteration of earlier *quam,* probably from Cuna (Chibchan language of Panama) *kwama*] (1743)
: any of various large tropical American lowland-forest birds (family Cracidae) that somewhat resemble turkeys

guan

gua·na·co \gwə-'nä-(,)kō\ *noun, plural* **-cos** *also* **-co** [Spanish, from Quechua *wanaku*] (1604)
: a South American mammal (*Lama guanicoe*) of dry open country that has a soft thick fawn-colored coat and is related to the camel but lacks a dorsal hump

gua·neth·i·dine \gwä-'ne-thə-,dēn\ *noun* [*guanidine* + *eth-*] (1959)
: a drug $C_{10}H_{22}N_4$ used especially as the sulfate in treating severe high blood pressure

gua·ni·dine \'gwä-nə-,dēn\ *noun* [International Scientific Vocabulary, from *guanine*] (circa 1864)
: a base CH_5N_3 derived from guanine, found especially in young tissues, and used in organic synthesis and as a parasympathetic stimulant in medicine especially as the hydrochloride salt

gua·nine \'gwä-,nēn\ *noun* [*guano* + *-ine;* from its being found especially in guano] (1850)
: a purine base $C_5H_5N_5O$ that codes genetic information in the polynucleotide chain of DNA or RNA — compare ADENINE, CYTOSINE, THYMINE, URACIL

gua·no \'gwä-(,)nō\ *noun* [Spanish, from Quechua *wanu* fertilizer, dung] (1604)
: a substance composed chiefly of the excrement of seafowl and used as a fertilizer; *also* : a similar substance (as bat excrement or cannery waste) especially when used for fertilizer

gua·no·sine \'gwä-nə-,sēn\ *noun* [*guan-* (as in *guanine*) + ribose + *-ine*] (1909)
: a nucleoside $C_{10}H_{13}N_5O_5$ composed of guanine and ribose

guanosine triphosphate *noun* (circa 1962)
: GTP

guar \'gwär\ *noun* [Hindi *guār*] (1882)
: a drought-tolerant legume (*Cyamopsis tetragonoloba*) cultivated in warm regions as a vegetable, for forage, and for its seeds which produce guar gum

gua·ra·ni \,gwär-ə-'nē\ *noun* [Spanish *guaraní*] (1797)
1 *capitalized* **a** *plural* **guarani** *or* **guaranis** : a member of a Tupi-Guaranian people of Bolivia, Paraguay, and southern Brazil **b** : the language of the Guarani
2 *plural* **guaranies** *also* **guaranis** — see MONEY table

[1]**guar·an·tee** \,gar-ən-'tē, ,gär- *also* 'gar-ən-, *or* 'gär-ən-,\ *noun* [probably alteration of [1]*guaranty*] (1710)
1 : GUARANTOR
2 : GUARANTY 1
3 : an assurance for the fulfillment of a condition: as **a** : an agreement by which one person undertakes to secure another in the possession or enjoyment of something **b** : an assurance of the quality of or of the length of use to be expected from a product offered for sale often with a promise of reimbursement
4 : GUARANTY 4

[2]**guarantee** *transitive verb* **-teed; -tee·ing** (1791)
1 : to undertake to answer for the debt, default, or miscarriage of
2 : to engage for the existence, permanence, or nature of : undertake to do or secure ⟨*guarantee* the winning of three tricks⟩
3 : to give security to
4 : to assert confidently

guar·an·tor \,gar-ən-'tȯr, 'gar-ən-tər, ,gär-, 'gär-\ *noun* (circa 1828)
1 : one that makes or gives a guaranty
2 : one that guarantees

[1]**guar·an·ty** \'gar-ən-tē, 'gär-\ *noun, plural* **-ties** [Middle French *garantie,* from Old French, from *garantir* to guarantee, from *garant* warrant, of Germanic origin; akin to Old High German *werēnto* guarantor — more at WARRANT] (1592)
1 : an undertaking to answer for the payment of a debt or the performance of a duty of another in case of the other's default or miscarriage
2 : GUARANTEE 3
3 : GUARANTOR
4 : something given as security : PLEDGE
5 : the protection of a right afforded by legal provision (as in a constitution)

[2]**guaranty** *transitive verb* **-tied; -ty·ing** (1753)
: GUARANTEE

[1]**guard** \'gärd\ *noun* [Middle English *garde,* from Middle French, from Old French, from *garder* to guard, defend, of Germanic origin; akin to Old High German *wartēn* to watch, take care — more at WARD] (15th century)
1 : one assigned to protect or oversee another: as **a** : a person or a body of persons on sentinel duty **b** *plural* : troops attached to the person of the sovereign **c** *British* : CONDUCTOR
2 a : a defensive state or attitude ⟨asked him out when his *guard* was down⟩ **b** : a defensive position (as in boxing)
3 a : the act or duty of protecting or defending **b** : the state of being protected : PROTECTION
4 : a protective or safety device; *specifically* : a device for protecting a machine part or the operator of a machine
5 *archaic* : PRECAUTION
6 a : a position or player next to the center in a football line **b** : a player stationed in the backcourt in basketball
— **off guard** : in an unprepared or unsuspecting state
— **on guard** : defensively watchful : ALERT

[2]**guard** (1500)
transitive verb
1 : to protect an edge of with an ornamental border
2 a : to protect from danger especially by watchful attention : make secure ⟨police *guarding* our cities⟩ **b** : to stand at the entrance of as if on guard or as a barrier **c** : to tend to carefully : PRESERVE, PROTECT ⟨*guarded* their privacy⟩
3 *archaic* : ESCORT
4 a : to watch over so as to prevent escape, disclosure, or indiscretion **b** : to attempt to prevent (an opponent) from playing effectively or scoring
intransitive verb
: to watch by way of caution or defense : stand guard
synonym see DEFEND
— **guard·er** *noun*

[1]**guar·dant** \'gär-dənt\ *adjective* [Middle French *gardant,* present participle of *garder* to guard, look at] (1572)
: having the head turned toward the spectator — used of a heraldic animal whose body is seen from the side ⟨a lion passant *guardant*⟩

[2]**guardant** *noun* (1591)
obsolete : GUARDIAN

guard cell *noun* (1875)
: one of the two crescent-shaped epidermal cells that border and open and close a plant stoma

guard·ed \'gär-dəd\ *adjective* (1709)
: CAUTIOUS, CIRCUMSPECT
— **guard·ed·ly** *adverb*
— **guard·ed·ness** *noun*

guard hair *noun* (1913)
: one of the long coarse hairs forming a protective coating over the undercoat of a mammal

guard·house \'gärd-,haủs\ *noun* (1592)
1 : a building occupied by a guard or used as a headquarters by soldiers on guard duty
2 : a military jail

guard·ian \'gär-dē-ən\ *noun* (15th century)
1 : one that guards : CUSTODIAN
2 : a superior of a Franciscan monastery
3 : one who has the care of the person or property of another
— **guard·ian·ship** \-,ship\ *noun*

guardian angel *noun* (circa 1631)
: an angel believed to have special care of a particular individual; *broadly* : SAVIOR 1, PROTECTOR

guard of honor (1887)
: HONOR GUARD

guard·rail \'gär-,drāl\ *noun* (1860)
: a railing for guarding against danger or trespass; *especially* : a barrier (as of steel cables) placed along the edge of a highway at dangerous points

guard·room \'gär-,drüm, -,drủm\ *noun* (1762)
1 : a room occupied by a military guard during its term of duty
2 : a room where military prisoners are confined

guards·man \'gärdz-mən\ *noun* (1817)
: a member of a military body called *guard* or *guards*

guar gum *noun* (1950)
: a gum that consists of the ground endosperm of guar seeds and is used especially as a thickening agent and as a sizing material for paper and textiles

Guar·ne·ri·us \gwär-'nir-ē-əs, -'ner-\ *noun* [New Latin, from Italian *Guarneri*] (1866)
: a violin made by one of the Italian Guarneri family in the 17th and 18th centuries

gua·va \'gwä-və\ *noun* [Spanish *guaba, guayaba,* perhaps from Taino] (1604)
1 : any of several tropical American shrubs or small trees (genus *Psidium*) of the myrtle family; *especially* : a shrubby tree (*P. guajava*) widely cultivated for its sweet acid yellow or pink fruit
2 : the globose to pear-shaped fruit of a guava

gua·ya·bera \,gwī-ə-'ber-ə\ *noun* [American Spanish] (1947)
: a usually short-sleeved lightweight sport shirt designed to be worn untucked

gua·yu·le \gwī-'ü-lē, wī-\ *noun* [American Spanish, from Nahuatl *cuauholli* or *huauholli*] (1906)
: a much-branched composite shrub (*Parthenium argentatum*) of Mexico and the southwestern U.S. cultivated as a source of rubber

gu·ber·na·to·ri·al \,gü-bə(r)-nə-'tōr-ē-əl, ,gyü-, ,gü-, -'tȯr-\ *adjective* [Latin *gubernator* governor, steersman, from *gubernare* to govern — more at GOVERN] (1734)
: of or relating to a governor

guck \'gək\ *noun* [perhaps alteration of *goo*] (1949)
: oozy sloppy dirt or debris; *broadly* : GOO, GUNK

[1]**gud·geon** \'gə-jən\ *noun* [Middle English *gudyon,* from Middle French *goujon*] (15th century)
1 : PIVOT 1, JOURNAL
2 : a socket for a rudder pintle

[2]**gudgeon** *noun* [Middle English *gojune,* from Middle French *gouvion, gougon,* from Latin *gobion-, gobio,* alteration of *gobius* — more at GOBY] (15th century)
: a small European freshwater fish (*Gobio gobio*) related to the carps and often used for food or bait

gudgeon pin *noun* (1891)
: WRIST PIN

Gud·run \'gủd-,rün\ *noun* [Old Norse *Guthrūn*]
: the wife of Sigurd and later of Atli in Norse mythology

guel·der rose \'gel-də(r)-\ *noun* [*Guelderland, Gelderland,* Netherlands] (1597)

: a bush of a cultivated variety of the cranberry bush with large globose heads of sterile flowers

Guelf *or* **Guelph** \'gwelf\ *noun* [Italian *Guelfo*] (1579)
: a member of a papal and popular political party in medieval Italy that opposed the authority of the German emperors in Italy — compare GHIBELLINE

gue•non \'gwe-nən; gə-nōⁿ\ *noun* [French] (1838)
: any of various long-tailed chiefly arboreal African monkeys (genera *Cercopithecus* and *Erythrocebus*)

guer•don \'gər-dᵊn\ *noun* [Middle English, from Middle French, from Old French, of Germanic origin; akin to Old High German *widarlōn* reward] (14th century)
: REWARD, RECOMPENSE
— **guerdon** *transitive verb*

gue•ri•don \gā-rē-dōⁿ\ *noun* [French *guéridon*, from *Gueridon*, character in 17th century farces and popular songs] (1853)
: a small usually ornately carved and embellished stand or table

guern•sey \'gərn-zē\ *noun, plural* **guern•seys** *often capitalized* [*Guernsey*, Channel islands] (1834)
: any of a breed of fawn and white dairy cattle that are larger than the jersey and produce rich yellowish milk

guernsey

¹guer•ril•la *or* **gue•ril•la** \gə-'ri-lə, ge-, g(y)i-\ *noun* [Spanish *guerrilla*, from diminutive of *guerra* war, of Germanic origin; akin to Old High German *werra* strife — more at WAR] (1809)
: a person who engages in irregular warfare especially as a member of an independent unit carrying out harassment and sabotage ◇

²guerrilla *adjective* (1811)
: of, relating to, or characteristic of guerrillas especially in being aggressive, radical, or unconventional

guerrilla theater *noun* (1968)
: STREET THEATER

¹guess \'ges\ *verb* [Middle English *gessen*, probably of Scandinavian origin; akin to Old Norse *geta* to get, guess — more at GET] (14th century)
transitive verb
1 : to form an opinion of from little or no evidence
2 : BELIEVE, SUPPOSE ⟨I *guess* you're right⟩
3 : to arrive at a correct conclusion about by conjecture, chance, or intuition ⟨*guess* the answer⟩
intransitive verb
: to make a guess
— **guess•able** \'ge-sə-bəl\ *adjective*
— **guess•er** *noun*

²guess *noun* (14th century)
: CONJECTURE, SURMISE

guess•ti•mate \'ges-tə-mət\ *noun* [blend of *guess* and *estimate*] (1923)
: an estimate usually made without adequate information
— **guess•ti•mate** \-,māt\ *transitive verb*

guess•work \'ges-,wərk\ *noun* (1725)
: work performed or results obtained by guess
: CONJECTURE

¹guest \'gest\ *noun* [Middle English *gest*, from Old Norse *gestr;* akin to Old English *giest* guest, stranger, Latin *hostis* stranger, enemy] (13th century)
1 a : a person entertained in one's house **b :** a person to whom hospitality is extended **c :** a person who pays for the services of an establishment (as a hotel or restaurant)
2 : an organism (as an insect) sharing the dwelling of another; *especially* **:** INQUILINE
3 : a substance that is incorporated in a host substance
4 : a usually prominent person not a regular member of a cast or organization who appears in a program or performance

²guest (14th century)
transitive verb
: to receive as a guest
intransitive verb
: to appear as a guest

guest worker *noun* (1967)
: a foreign laborer working temporarily in an industrialized usually European country

guff \'gəf\ *noun* [probably imitative] (1888)
: NONSENSE, HUMBUG

guf•faw \(,)gə-'fȯ, 'gə-,fȯ\ *noun* [imitative] (1720)
: a loud or boisterous burst of laughter
— **guf•faw** \(,)gə-'fȯ\ *intransitive verb*

gug•gle \'gə-gəl\ *intransitive verb* **gug•gled; gug•gling** \-g(ə-)liŋ\ [imitative] (1611)
: GURGLE
— **guggle** *noun*

guid•able \'gī-də-bəl\ *adjective* (1676)
: capable of being guided

guid•ance \'gī-dᵊn(t)s\ *noun* (1590)
1 : the act or process of guiding
2 a : the direction provided by a guide **b :** advice on vocational or educational problems given to students
3 : the process of controlling the course of a projectile by a built-in mechanism

¹guide \'gīd\ *noun* [Middle English, from Middle French, from Old Provençal *guida*, of Germanic origin; akin to Old English *wītan* to look after, *witan* to know — more at WIT] (14th century)
1 a : one that leads or directs another's way **b :** a person who exhibits and explains points of interest **c :** something that provides a person with guiding information **d :** SIGNPOST 1 **e :** a person who directs another's conduct or course of life
2 a : a device for steadying or directing the motion of something **b :** a ring or loop for holding the line of a fishing rod in position **c :** a sheet or a card with projecting tab for labeling inserted in a card index to facilitate reference
3 : a member of a unit on whom the movements or alignments of a military command are regulated — used especially in commands ⟨*guide* right⟩

²guide *verb* **guid•ed; guid•ing** (14th century)
transitive verb
1 : to act as a guide to **:** direct in a way or course
2 a : to direct, supervise, or influence usually to a particular end **b :** to superintend the training or instruction of
intransitive verb
: to act or work as a guide ☆
— **guid•er** *noun*

guide•book \'gīd-,bu̇k\ *noun* (1814)
: HANDBOOK 1; *especially* **:** a book of information for travelers

guided missile *noun* (1945)
: a missile whose course may be altered during flight (as by a target-seeking radar device)

guide dog *noun* (1932)
: a dog trained to lead the blind

guide•line \'gīd-,līn\ *noun* (1785)
: a line by which one is guided: as **a :** a cord or rope to aid a passer over a difficult point or to permit retracing a course **b :** an indication or outline of policy or conduct

guide•post \-,pōst\ *noun* (1761)
1 : INDICATION, SIGN
2 : GUIDELINE b

guide•way \-,wā\ *noun* (1876)
: a channel or track for controlling the line of motion of something

guide word *noun* (circa 1928)
: either of the terms at the head of a page of an alphabetical reference work (as a dictionary) indicating the alphabetically first and last words on the page

gui•don \'gī-,dän, -dᵊn\ *noun* [Middle French, from *guide*] (1548)
1 : a small flag; *especially* **:** one borne by a military unit as a unit marker
2 : one who carries a guidon

guid•will•ie \gœ̄d-'wi-lē, gid-\ *adjective* [Scots *guidwill* goodwill] (1788)
Scottish **:** CORDIAL, CHEERING

guild \'gild\ *noun* [Middle English *gilde*, from Old Norse *gildi* payment, guild; akin to Old English *gield* tribute, guild — more at GELD] (14th century)
: an association of people with similar interests or pursuits; *especially* **:** a medieval association of merchants or craftsmen
— **guild•ship** \'gil(d)-,ship\ *noun*

guil•der \'gil-dər\ *noun* [Middle English *gylder, gyldren*, modification of Middle Dutch *gulden*] (15th century)
: GULDEN

guild•hall \'gild-,hȯl\ *noun* (14th century)
: a hall where a guild or corporation usually assembles

guilds•man \'gil(d)z-mən\ *noun* (1873)
1 : a guild member
2 : an advocate of guild socialism

guild socialism *noun* (1912)

☆ SYNONYMS
Guide, lead, steer, pilot, engineer mean to direct in a course or show the way to be followed. GUIDE implies intimate knowledge of the way and of all its difficulties and dangers ⟨*guided* the scouts through the cave⟩. LEAD implies showing the way and often keeping those that follow under control and in order ⟨*led* his team to victory⟩. STEER implies an ability to keep to a chosen course and stresses the capacity of maneuvering correctly ⟨*steered* the ship through a narrow channel⟩. PILOT suggests guidance over a dangerous, intricate, or complicated course ⟨*piloted* the bill through the Senate⟩. ENGINEER implies guidance by one who finds ways to avoid or overcome difficulties in achieving an end or carrying out a plan ⟨*engineered* his son's election to the governorship⟩.

◇ WORD HISTORY
guerrilla In Spanish, the word *guerrilla*, a diminutive of *guerra* "war," was used from the early 16th century for a party of skirmishers who engaged the enemy in advance of the main force. After a popular uprising against Napoleonic rule in Spain broke out in 1808, *guerrilla* was applied to bands of Spanish partisans who harrassed French troops. The Duke of Wellington, commander of the British army in Spain, used the word in a dispatch, and it came to be adopted in Britain, though by a curious misunderstanding it was applied to a single partisan rather than to a band. The error may perhaps have been due to Wellington himself, though by 1812 the French borrowing *guérilla* was being used in a parallel way. *Guerrilla* also became an attributive adjective at an early date, appearing in a poem by Sir Walter Scott in 1811; the still common phrase *guerrilla warfare* was in use by at least the 1840s.

: an early 20th century English socialistic theory advocating state ownership of industry with control and management by guilds of workers

guile \ˈgī(ə)l\ *noun* [Middle English, from Old French, probably of Germanic origin; akin to Old English *wigle* divination — more at WITCH] (13th century)
1 : deceitful cunning **:** DUPLICITY
2 *obsolete* **:** STRATAGEM, TRICK
— guile·ful \-fəl\ *adjective*
— guile·ful·ly \-fə-lē\ *adverb*
— guile·ful·ness *noun*

guile·less \ˈgī(ə)l-ləs\ *adjective* (1728)
: INNOCENT, NAIVE
— guile·less·ly *adverb*
— guile·less·ness *noun*

Guil·lain–Bar·ré syndrome \ˌgē-ˌlan-ˌbä-ˈrā-, ˌgē-ya^{n}-\ *noun* [Georges *Guillain* (died 1961) French physician and Jean A. *Barré* (born 1880) French neurologist] (1940)
: a polyneuritis of unknown cause characterized especially by muscle weakness and paralysis

guil·le·met \ˌgē-(y)ə-ˈmā, ˌgi-lə-ˈmet\ *noun* [French, from diminutive of *Guillaume* William (perhaps a printer's name)] (circa 1905)
: either of the marks « or » used as quotation marks in French writing

guil·le·mot \ˈgi-lə-ˌmät\ *noun* [French, from Middle French, diminutive of *Guillaume* William] (1678)
1 *British* **:** a common murre (*Uria aalge*)
2 : any of a genus (*Cepphus*) of narrow-billed auks of northern seas

guil·loche \gi-ˈlōsh, gē-ˈ(y)ōsh\ *noun* [French *guillochis*] (circa 1842)
1 : an architectural ornament formed of two or more interlaced bands with openings containing round devices
2 : a pattern (as on metalwork) made by interlacing curved lines

guil·lo·tine \ˈgi-lə-ˌtēn; ˌgē-(y)ə-ˈ, ˈgē-(y)ə-ˌ\ *noun* [French, from Joseph *Guillotin* (died 1814) French physician] (1793)
1 : a machine for beheading by means of a heavy blade that slides down in vertical guides
2 : a shearing machine or instrument (as a paper cutter) that in action resembles a guillotine
3 *chiefly British* **:** closure by the imposition of a predetermined time limit on the consideration of specific sections of a bill or portions of other legislative business ◆
— guillotine *transitive verb*

guilt \ˈgilt\ *noun* [Middle English, delinquency, guilt, from Old English *gylt* delinquency] (before 12th century)
1 : the fact of having committed a breach of conduct especially violating law and involving a penalty; *broadly* **:** guilty conduct
2 a : the state of one who has committed an offense especially consciously **b :** feelings of culpability especially for imagined offenses or from a sense of inadequacy **:** SELF-REPROACH
3 : a feeling of culpability for offenses

guilt·less \ˈgilt-ləs\ *adjective* (13th century)
: INNOCENT
— guilt·less·ly *adverb*
— guilt·less·ness *noun*

guilty \ˈgil-tē\ *adjective* **guilt·i·er; -est** (before 12th century)
1 : justly chargeable with or responsible for a usually grave breach of conduct or a crime
2 *obsolete* **:** justly liable to or deserving of a penalty
3 a : suggesting or involving guilt ⟨*guilty* looks⟩ **b :** aware of or suffering from guilt ⟨*guilty* consciences⟩
synonym see BLAMEWORTHY
— guilt·i·ly \-tə-lē\ *adverb*
— guilt·i·ness \-tē-nəs\ *noun*

guimpe \ˈgamp, ˈgimp\ *noun* [French, from Old French *guimple,* of Germanic origin; akin to Old English *wimpel* wimple] (1850)
1 : a blouse worn under a jumper or pinafore
2 : a wide cloth used by some nuns to cover the neck and shoulders
3 [by alteration] **:** [1]GIMP

guin·ea \ˈgi-nē\ *noun* [*Guinea,* Africa, supposed source of the gold from which it was made] (1664)
1 : an English gold coin issued from 1663 to 1813 and fixed in 1717 at 21 shillings
2 : a unit of value equal to one pound and one shilling

guinea fowl *noun* (1788)
: an African bird (*Numida meleagris*) related to the pheasants, raised for food in many parts of the world, and marked by a bare neck and head and slaty plumage speckled with white; *broadly* **:** any of several related birds

guinea fowl

guinea grass *noun* (1785)
: a tall African forage grass (*Panicum maximum*) introduced into tropical America and the southern U.S.

guinea hen *noun* (1599)
: a female guinea fowl; *broadly* **:** GUINEA FOWL

guinea pepper *noun* (1597)
: GRAINS OF PARADISE

guinea pig *noun* (1664)
1 : a small stout-bodied short-eared tailless domesticated rodent (*Cavia porcellus*) often kept as a pet and widely used in biological research — called also *cavy*
2 : a subject of research, experimentation, or testing

guinea worm *noun* (1699)
: a slender nematode worm (*Dracunculus medinensis*) of tropical regions parasitic in mammals including humans and having an adult female that infests subcutaneous tissues and may attain a length of several feet

Guin·e·vere \ˈgwi-nə-ˌvir, *British also* ˈgi-\ *noun*
: the wife of King Arthur and mistress of Lancelot

gui·pure \gi-ˈpyu̇r, -ˈpu̇r\ *noun* [French] (1843)
: a heavy large-patterned decorative lace

gui·ro \ˈwē-(ˌ)rō, ˈgwir-(ˌ)ō\ *noun* [American Spanish *güiro,* literally, calabash] (1898)
: a percussion instrument of Latin-American origin made of a serrated gourd and played by scraping a stick along its surface

gui·sard \ˈgī-zərd\ *noun* [obsolete Scots *gyze* to disguise, from Middle English *gyzen* to dress, from *guise, gyze* guise] (1626)
chiefly Scottish **:** MASKER, MUMMER

guise \ˈgīz\ *noun* [Middle English, from Old French, of Germanic origin; akin to Old High German *wīsa* manner — more at WISE] (13th century)
1 : a form or style of dress **:** COSTUME
2 a *obsolete* **:** MANNER, FASHION **b** *archaic* **:** a customary way of speaking or behaving
3 a : external appearance **:** SEMBLANCE **b :** PRETEXT

gui·tar \gə-ˈtär, gi-, *especially Southern & Midland also* ˈgi-ˌtär\ *noun* [French *guitare,* from Spanish *guitarra,* from Arabic *qītār,* from Greek *kithara* cithara] (1621)
: a flat-bodied stringed instrument with a long fretted neck and usually six strings plucked with a pick or with the fingers

guitar: *1* electric, *2* acoustic

— gui·tar·ist \-ist\ *noun*

gui·tar·fish \-ˌfish\ *noun* (circa 1900)
: any of several viviparous rays (family Rhinobatidae) somewhat like a guitar in shape viewed from above

Gu·ja·ra·ti \ˌgü-jə-ˈrä-tē, ˌgu̇-\ *noun* [Hindi *gujarātī,* from *Gujarāt* Gujarat] (1808)
1 *or* **Gu·je·ra·ti** *same*\ **:** the Indo-Aryan language of Gujarat and neighboring regions in northwestern India
2 *or* **Guj·ra·ti** \güj-ˈrä-, gu̇j-ˈrä-\ **:** a member of a people chiefly of Gujarat speaking the Gujarati language

gul \ˈgül\ *noun* [Persian] (1813)
archaic **:** ROSE 1

gu·lag \ˈgü-ˌläg\ *noun, often capitalized* [Russian, from *G*lavnoe *u*pravlenie ispravitel'no-trudovykh *lag*ereĭ chief administration of corrective labor camps] (1974)
: the penal system of the U.S.S.R. consisting of a network of labor camps; *also* **:** LABOR CAMP 1

gu·lar \ˈg(y)ü-lər\ *adjective* [Latin *gula* throat — more at GLUTTON] (1828)
: of, relating to, or situated on the throat

gulch \ˈgəlch\ *noun* [perhaps from English dialect *gulch* to gulp, from Middle English *gulchen*] (1832)
: a deep or precipitous cleft **:** RAVINE; *especially* **:** one occupied by a torrent

gul·den \ˈgül-dən, ˈgu̇l-\ *noun, plural* **gul·dens** *or* **gulden** [Middle English (Scots), from Middle Dutch *gulden florijn* golden florin] (15th century)
— see MONEY table

gules \ˈgyü(ə)lz\ *noun, plural* **gules** [Middle English *goules,* from Middle French] (14th century)
: the heraldic color red

[1]gulf \ˈgəlf\ *noun* [Middle English *goulf,* from Middle French *golfe,* from Italian *golfo,* from Late Latin *colpus,* from Greek *kolpos* bosom, gulf; akin to Old English *hwealf* vault, Old High German *walbo*] (15th century)
1 : a part of an ocean or sea extending into the land
2 : a deep chasm **:** ABYSS
3 : WHIRLPOOL
4 : a wide gap ⟨the *gulf* between generations⟩

[2]gulf *transitive verb* (1807)
: ENGULF

gulf·weed \ˈgəlf-ˌwēd\ *noun* [*Gulf* of Mexico] (1674)
: any of several sargassums; *especially* **:** a branching olive-brown seaweed (*Sargassum bacciferum*) of tropical American seas with numerous berrylike air vesicles

◇ WORD HISTORY
guillotine Joseph-Ignace Guillotin (1738–1814) was a French physician and member of the National Assembly during the early months of the French Revolution. On October 10, 1789, in a proposed egalitarian revision of the penal code, Guillotin advocated that all death sentences be carried out by decapitation—a method formerly reserved to the nobility, common criminals usually being hung—and that the executioner's sword or ax be replaced for humane reasons by a machine. Various devices for beheading were already in existence, and one was perfected by a German mechanic named Schmitt, under the supervision of a French surgeon, Antoine Louis. The new machine, first used in 1792 and made famous during the Reign of Terror of 1793–94, was initially and appropriately called the *Louison* or *Louisette.* But *guillotine,* maliciously suggested by a journalist of royalist sympathies in 1790, was readily accepted, most likely in part because Guillotin was a better-known figure than Louis and was disliked by other politicians for his bluntness of speech.

[1]**gull** \ˈgəl\ *noun* [Middle English, of Celtic origin; akin to Welsh *gwylan* gull] (15th century)
: any of numerous long-winged web-footed aquatic birds (subfamily Larinae of the family Laridae); *especially* **:** a usually gray and white bird (as of the genus *Larus*) differing from a tern in usually larger size, stouter build, thicker somewhat hooked bill, less pointed wings, and short unforked tail

[2]**gull** *transitive verb* [obsolete *gull* gullet, from Middle English *golle,* from Middle French *goule*] (circa 1550)
: to take advantage of (one who is foolish or unwary) **:** DECEIVE
synonym see DUPE

[3]**gull** *noun* (1594)
: a person who is easily deceived or cheated **:** DUPE

Gul·lah \ˈgə-lə\ *noun* (1822)
1 : a member of a group of blacks inhabiting the sea islands and coastal districts of South Carolina, Georgia, and northeastern Florida
2 : an English-based creole spoken by the Gullahs that is marked by vocabulary and grammatical elements from various African languages

gul·let \ˈgə-lət\ *noun* [Middle English *golet,* from Middle French *goulet,* diminutive of *goule* throat, from Latin *gula* — more at GLUTTON] (14th century)
1 : ESOPHAGUS; *broadly* **:** THROAT
2 : an invagination of the protoplasm in various protozoans (as a paramecium) that sometimes functions in the intake of food
3 : the space between the tips of adjacent saw teeth

gull·ible *also* **gull·able** \ˈgə-lə-bəl\ *adjective* (1818)
: easily duped or cheated
— gull·ibil·i·ty \ˌgə-lə-ˈbi-lə-tē\ *noun*
— gull·ibly \ˈgə-lə-blē\ *adverb*

Gul·li·ver \ˈgə-lə vər\ *noun*
: an Englishman in Jonathan Swift's satire *Gulliver's Travels* who makes voyages to the imaginary lands of the Lilliputians, Brobdingnagians, Laputans, and Houyhnhnms

[1]**gul·ly** \ˈgu̇-lē, ˈgə-\ *noun, plural* **gullies** [short for English dialect *gully knife*] (1582)
dialect British **:** a large knife

[2]**gul·ly** *also* **gul·ley** \ˈgə-le\ *noun, plural* **gullies** [obsolete English *gully* (gullet)] (1637)
1 : a trench which was originally worn in the earth by running water and through which water often runs after rains
2 : a small valley or gulch

[3]**gul·ly** \ˈgə-lē\ *verb* **gul·lied; gul·ly·ing** (1754)
transitive verb
: to make gullies in
intransitive verb
: to undergo erosion **:** form gullies

gully erosion *noun* (1928)
: soil erosion produced by running water

gu·los·i·ty \g(y)ü-ˈlä-sə-tē\ *noun* [Middle English *gulosite,* from Late Latin *gulositas,* from Latin *gulosus* gluttonous, from *gula* gullet] (15th century)
: excessive appetite **:** GREEDINESS

gulp \ˈgəlp\ *verb* [Middle English, from a Middle Dutch or Middle Low German word akin to Dutch & Frisian *gulpen* to bubble forth, drink deep; akin to Old English *gielpan* to boast — more at YELP] (14th century)
transitive verb
1 : to swallow hurriedly or greedily or in one swallow
2 : to keep back as if by swallowing ⟨*gulp* down a sob⟩
3 : to take in readily as if by swallowing ⟨*gulp* down knowledge⟩
intransitive verb
: to catch the breath as if in taking a long drink
— gulp *noun*
— gulp·er *noun*

[1]**gum** \ˈgəm\ *noun* [Middle English *gome,* from Old English *gōma* palate; akin to Old High German *guomo* palate, and perhaps to Greek *chaos* abyss] (before 12th century)
: the tissue that surrounds the necks of teeth and covers the alveolar parts of the jaws; *broadly* **:** the alveolar portion of a jaw with its enveloping soft tissues

[2]**gum** *transitive verb* **gummed; gum·ming** (1777)
1 : to enlarge gullets of (a saw)
2 : to chew with the gums

[3]**gum** *noun* [Middle English *gomme,* from Middle French, from Latin *cummi, gummi,* from Greek *kommi,* from Egyptian *qmyt*] (14th century)
1 a : any of numerous colloidal polysaccharide substances of plant origin that are gelatinous when moist but harden on drying and are salts of complex organic acids — compare MUCILAGE 1 **b :** any of various plant exudates (as an oleoresin or gum resin)
2 : a substance or deposit resembling a plant gum (as in sticky or adhesive quality)
3 a : a tree (as a black gum) that yields gum **b** *Australian* **:** EUCALYPTUS
4 : the wood or lumber of a gum; *especially* **:** that of the sweet gum
5 : CHEWING GUM

[4]**gum** *verb* **gummed; gum·ming** (1597)
transitive verb
: to clog, impede, or damage with or as if with gum ⟨*gum* up the works⟩
intransitive verb
1 : to exude or form gum
2 : to become gummy
— gum·mer *noun*

gum arabic *noun* (14th century)
: a water-soluble gum obtained from several acacias (especially *Acacia senegal*) and used especially in the manufacture of inks, adhesives, pharmaceuticals, and confections

[1]**gum·bo** \ˈgəm-(ˌ)bō\ *noun, plural* **gumbos** [American French *gombo,* of Bantu origin; akin to Umbundu *ochinggômbo* okra] (1805)
1 : a soup thickened with okra pods or filé and containing meat or seafoods and usually vegetables
2 : OKRA 1
3 a : any of various fine-grained silty soils especially of the central U.S. that when wet become impervious and soapy or waxy and very sticky **b :** a heavy sticky mud
4 : MIXTURE, MÉLANGE
— gumbo *adjective*

[2]**gumbo** *noun, often capitalized* [American French *gombo,* perhaps from Kongo *nkômbô* runaway slave] (1838)
: CREOLE 4a

gum·boil \ˈgəm-ˌbȯil\ *noun* (1753)
: an abscess in the gum

gum·bo–lim·bo \ˌgəm-bō-ˈlim-(ˌ)bō\ *noun* [origin unknown] (1837)
: a tree (*Bursera simaruba* of the family Burseraceae) of southern Florida and the American tropics that has a smooth coppery bark and supplies a reddish resin used locally in cements and varnishes

gum boot *noun* (1850)
: a rubber boot

gum·drop \ˈgəm-ˌdräp\ *noun* (1860)
: a sugar-coated candy made usually from corn syrup with gelatin or gum arabic

gum·ma \ˈgə-mə\ *noun, plural* **gummas** *also* **gum·ma·ta** \ˈgə-mə-tə\ [New Latin *gummat-, gumma,* from Late Latin, gum, alteration of Latin *gummi* gum] (circa 1722)
: a tumor of gummy or rubbery consistency that is characteristic of the tertiary stage of syphilis
— gum·ma·tous \-mə-təs\ *adjective*

gum·mite \ˈgə-ˌmīt\ *noun* (1868)
: a yellow to reddish brown mixture of hydrous oxides of uranium, thorium, and lead

gum·mo·sis \ˌgə-ˈmō-səs\ *noun* [New Latin] (1882)
: a pathological production of gummy exudate in a plant; *also* **:** a plant disease marked by gummosis

gum·mous \ˈgə-məs\ *adjective* (1669)
: resembling or composed of gum

gum·my \ˈgə-mē\ *adjective* **gum·mi·er; -est** (14th century)
1 : VISCOUS, STICKY
2 a : consisting of or containing gum **b :** covered with gum
— gum·mi·ness *noun*

gump·tion \ˈgəm(p)-shən\ *noun* [origin unknown] (1719)
1 *chiefly dialect* **:** COMMON SENSE, HORSE SENSE
2 : ENTERPRISE, INITIATIVE ⟨lacked the *gumption* to try⟩

gum resin *noun* (1712)
: a product consisting essentially of a mixture of gum and resin usually obtained by making an incision in a plant and allowing the juice which exudes to solidify

[1]**gum·shoe** \ˈgəm-ˌshü\ *noun* (1906)
: DETECTIVE

[2]**gumshoe** *intransitive verb* **gum·shoed; gum·shoe·ing** (1930)
: to engage in detective work

gum tragacanth *noun* (1573)
: TRAGACANTH

gum tree *noun* (1676)
: [3]GUM 3

gum turpentine *noun* (1926)
: TURPENTINE 2a

gum·wood \ˈgəm-ˌwu̇d\ *noun* (1709)
: [3]GUM 4

[1]**gun** \ˈgən\ *noun* [Middle English *gonne, gunne*] (14th century)
1 a : a piece of ordnance usually with high muzzle velocity and comparatively flat trajectory **b :** a portable firearm (as a rifle or handgun) **c :** a device that throws a projectile
2 a : a discharge of a gun especially as a salute or signal **b :** a signal marking a beginning or ending
3 a : HUNTER **b :** GUNMAN
4 : something suggesting a gun in shape or function
5 : THROTTLE
— gunned \ˈgənd\ *adjective*
— under the gun : under pressure or attack

[2]**gun** *verb* **gunned; gun·ning** (1622)
intransitive verb
: to hunt with a gun
transitive verb
1 a : to fire on **b :** SHOOT ⟨*gunned* down by a hit man⟩
2 a : to open up the throttle of so as to increase speed ⟨*gun* the engine⟩ **b :** FIRE 3b ⟨*gunned* the ball to first base⟩
— gun for : to aim at or go after with determination or effort

gun·boat \ˈgən-ˌbōt\ *noun* (1777)
: an armed ship of shallow draft

gunboat diplomacy *noun* (1927)
: diplomacy backed by the use or threat of military force

gun control *noun* (1969)
: regulation of the selling, owning, and use of guns

gun·cot·ton \-ˌkä-tᵊn\ *noun* (1846)
: NITROCELLULOSE; *especially* **:** an explosive highly nitrated product used chiefly in smokeless powder

gun·dog \-ˌdȯg\ *noun* (1744)
: a dog trained to work with hunters by locating and retrieving game

gun·fight \-ˌfīt\ *noun* (1659)
: a hostile encounter in which antagonists fire upon each other
— gun·fight·er \-ˌfī-tər\ *noun*

gun·fire \-ˌfī(-ə)r\ *noun* (1801)
: the firing of guns

gun·flint \-ˌflint\ *noun* (1731)
: a small sharp flint fashioned to ignite the priming in a flintlock

gung ho \ˈgəŋ-ˈhō\ *adjective* [*Gung ho!,* motto (interpreted as meaning "work together") adopted by certain U.S. marines, from Chinese (Beijing) *gōnghé,* short for *Zhōngguó Gōngyè Hézuò Shè* Chinese Industrial Cooperative Society] (1942)
: extremely or overly zealous or enthusiastic ◆

gun·ite \ˈgə-ˌnīt\ *noun* (1914)
: a building material consisting of a mixture of cement, sand, and water that is sprayed onto a mold

gunk \ˈgəŋk\ *noun* [from *Gunk,* trademark for a cleaning solvent] (1943)
: filthy, sticky, or greasy matter
— **gunky** \ˈgəŋ-kē\ *adjective*

gun lap *noun* (circa 1949)
: the final lap of a race in track signaled by the firing of a gun as the leader begins the lap

gun·man \-mən\ *noun* (1624)
1 : a man armed with a gun; *especially* **:** a professional killer
2 : a man noted for speed or skill in handling a gun

gun·met·al \ˈgən-ˌme-tᵊl\ *noun* (1541)
1 : a metal used for guns; *specifically* **:** a bronze formerly much used as a material for cannon
2 : an alloy or metal treated to imitate nearly black tarnished copper-alloy gunmetal
3 : a bluish gray color

gun moll \-ˌmäl, -ˌmȯl\ *noun* [slang *gun* thief, rascal, by shortening & alteration from *gonoph, ganef* thief — more at GANEF] (circa 1908)
slang **:** MOLL 2b

Gun·nar \ˈgü-ˌnär, ˈgü-, -nər\ *noun* [Old Norse *Gunnarr*]
: the king of the Nibelungs and husband of Brynhild in Norse mythology

[1]**gun·nel** \ˈgə-nᵊl\ *variant of* GUNWALE

[2]**gunnel** *noun* [origin unknown] (1740)
: a small slimy elongate north Atlantic blenny (*Pholis gunnellus*); *broadly* **:** any fish of the family (Pholidae) to which the gunnel belongs

gun·ner \ˈgə-nər\ *noun* (14th century)
1 : a soldier or airman who operates or aims a gun
2 : one who hunts with a gun
3 : a warrant officer who supervises ordnance and ordnance stores

gun·nery \ˈgən-rē, ˈgə-nə-\ *noun* (1605)
: the use of guns; *especially* **:** the science of the flight of projectiles and of the effective use of guns

gunnery sergeant *noun* (circa 1961)
: a noncommissioned officer in the marine corps ranking above a staff sergeant and below a master sergeant or first sergeant

gun·ny·sack \ˈgə-nē-ˌsak\ *noun* [*gunny* coarse fabric, of Indo-Aryan origin; akin to Hindi *gon* sack, Punjabi *gūṇī*] (1862)
: a sack made of a coarse heavy fabric (as burlap)

gun·play \ˈgən-ˌplā\ *noun* (1881)
: the shooting of small arms with intent to scare or kill

gun·point \-ˌpȯint\ *noun* (1951)
: the muzzle of a gun
— **at gunpoint :** under a threat of death by being shot

gun·pow·der \-ˌpau̇-dər\ *noun* (15th century)
: an explosive mixture of potassium nitrate, charcoal, and sulfur used in gunnery and blasting; *broadly* **:** any of various powders used in guns as propelling charges

gun room *noun* (1626)
: quarters on a British warship originally used by the gunner and his mates but now by midshipmen and junior officers

gun·run·ner \ˈgən-ˌrə-nər\ *noun* (1899)
: one that traffics in contraband arms and ammunition
— **gun·run·ning** \-ˌrə-niŋ\ *noun*

gun·sel \ˈgən(t)-səl\ *noun* [slang *gunsel* catamite, perhaps modification of Yiddish *gendzl* gosling] (1944)
slang **:** GUNMAN

gun·ship \ˈgən-ˌship\ *noun* (1966)
: a helicopter or cargo aircraft armed with rockets and machine guns

gun·shot \ˈgən-ˌshät\ *noun* (15th century)
1 : shot or a projectile fired from a gun
2 : the range of a gun
3 : the firing of a gun

gun–shy \-ˌshī\ *adjective* (1884)
1 : afraid of loud noise (as that of a gun)
2 : markedly distrustful, afraid, or cautious

gun·sling·er \-ˌsliŋ-ər\ *noun* (1928)
: a person noted for speed and skill in handling and shooting a gun especially in the American West

gun·sling·ing \-ˌsliŋ-iŋ\ *noun* (circa 1944)
: the shooting of a gun especially in a gunfight

gun·smith \-ˌsmith\ *noun* (1588)
: one who designs, makes, or repairs small firearms
— **gun·smith·ing** \-ˌsmi-thiŋ\ *noun*

Gun·ter's chain \ˈgən-tərz-\ *noun* [Edmund *Gunter*] (circa 1679)
: a chain 66 feet (20.1 meters) long that is the unit of length for surveys of U.S. public lands

Gun·ther \ˈgu̇n-tər\ *noun* [German]
: a Burgundian king and husband of Brunhild in Germanic legend

gun·wale *also* **gun·nel** \ˈgə-nᵊl\ *noun* [Middle English *gonne-wale,* from *gonne* gun + [1]*wale;* from its former use as a support for guns] (15th century)
: the upper edge of a ship's or boat's side
— **to the gunwales :** as full as possible

gup·py \ˈgə-pē\ *noun, plural* **guppies** [R.J.L. *Guppy* (died 1916) Trinidadian naturalist] (1925)
: a small live-bearer (*Poecilia reticulata* of the family Poeciliidae) of Barbados, Trinidad, and Venezuela often kept as an aquarium fish

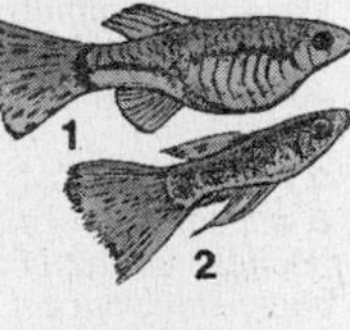

guppy: *1* female, *2* male

gur·gle \ˈgər-gəl\ *intransitive verb* **gur·gled; gur·gling** \-g(ə-)liŋ\ [probably imitative] (1713)
1 : to flow in a broken irregular current ⟨the brook *gurgling* over the rocks⟩
2 : to make a sound like that of a gurgling liquid ⟨the baby *gurgling* in his crib⟩
— **gurgle** *noun*

Gur·kha \ˈgu̇r-kə, ˈgər-\ *noun* [*Ghurka,* member of a Rajput clan who dominated Nepal in the 18th century] (1811)
: a soldier from Nepal in the British or Indian army

gur·nard \ˈgər-nərd\ *noun, plural* **gurnard** *or* **gurnards** [Middle English, from Middle French *gornart,* irregular from *grognier* to grunt, from Latin *grunnire,* of imitative origin] (14th century)
: SEA ROBIN — compare FLYING GURNARD

gur·ney \ˈgər-nē\ *noun, plural* **gurneys** [probably from the name *Gurney*] (1939)
: a wheeled cot or stretcher

gur·ry \ˈgər-ē, ˈgə-rē\ *noun* [origin unknown] (1850)
: fishing offal

gu·ru \ˈgu̇r-(ˌ)ü *also* gə-ˈrü\ *noun, plural* **gurus** [Hindi *gurū,* from Sanskrit *guru,* from *guru,* adjective, heavy, venerable — more at GRIEVE] (1613)
1 : a personal religious teacher and spiritual guide in Hinduism
2 a : a teacher and especially intellectual guide in matters of fundamental concern **b :** one who is an acknowledged leader or chief proponent **c :** a person with knowledge or expertise **:** EXPERT

[1]**gush** \ˈgəsh\ *verb* [Middle English *guschen*] (15th century)
intransitive verb
1 : to issue copiously or violently
2 : to emit a sudden copious flow
3 : to make an effusive display of affection or enthusiasm ⟨an aunt *gushing* over the baby⟩
transitive verb
1 : to emit in a copious free flow
2 : to say or write effusively
— **gush·ing·ly** \ˈgə-shiŋ-lē\ *adverb*

[2]**gush** *noun* (circa 1682)
1 a : a sudden outpouring **b :** something emitted in a gushing forth
2 : an effusive display or outpouring

gush·er \ˈgə-shər\ *noun* (1864)
: one that gushes; *specifically* **:** an oil well with a copious natural flow

gushy \ˈgə-shē\ *adjective* **gush·i·er; -est** (1845)
: marked by effusive sentimentality
— **gush·i·ly** \ˈgə-shə-lē\ *adverb*
— **gush·i·ness** \ˈgə-shē-nəs\ *noun*

gus·set \ˈgə-sət\ *noun* [Middle English, piece of armor covering the joints in a suit of armor, from Middle French *gousset*] (circa 1570)
1 : a usually diamond-shaped or triangular insert in a seam (as of a sleeve, pocketbook, or shoe upper) to provide expansion or reinforcement
2 : a plate or bracket for strengthening an angle in framework (as in a building or bridge)
— **gusset** *transitive verb*

gus·sy up \ˈgə-sē-ˈəp\ *transitive verb* **gussied up; gus·sy·ing up** [origin unknown] (1952)
: DRESS UP, EMBELLISH

[1]**gust** \ˈgəst\ *noun* [Middle English *guste,* from Latin *gustus;* akin to Latin *gustare* to taste — more at CHOOSE] (15th century)
1 *obsolete* **a :** the sensation of taste **b :** INCLINATION, LIKING
2 : keen delight

[2]**gust** *noun* [probably from Old Norse *gustr;* akin to Old High German *gussa* flood, and perhaps to Old English *gēotan* to pour — more at FOUND] (1588)
1 : a sudden brief rush of wind
2 : a sudden outburst **:** SURGE ⟨a *gust* of emotion⟩
— **gust·i·ly** \ˈgəs-tə-lē\ *adverb*

◇ WORD HISTORY

gung ho The Chinese Industrial Cooperative Society was organized in 1938 to develop light industry in areas of China not overrun by the Japanese, and was supported by both Communists and Chiang Kai-shek's Nationalists. The Chinese name of the society, *Zhōngguó Gōngyè Hézuò Shè,* was conventionally shortened to *Gōnghé.* A U.S. Marine Corps lieutenant colonel, Evans Carlson, while an observer with the Communists' 8th Route Army in 1937–38, became an admirer of their tactics. When Carlson formed the Marines' 2nd Raider Battalion at Camp Elliott, California, in February, 1942, he tried to instill the Communists' unity of purpose in his men. He told them that "Gung ho" was the motto of the Chinese Industrial Cooperatives (which Carlson seems to have associated with the Communists) and meant "work together." (In fact, though *gōng* may be translated "work" and *hé* "to join," *Gōnghé* is simply the shortened form of a name and not the Chinese equivalent of "to work together.") Carlson's training program soon lapsed, but *gung ho* spread beyond the 2nd Raider Battalion. At first alluding positively to Marine Corps esprit, it eventually took on the negative sense "excessively zealous" and spread from military to civilian use.

— **gust·i·ness** \-tē-nəs\ *noun*
— **gusty** \-tē\ *adjective*
³gust *intransitive verb* (1813)
: to blow in gusts ⟨winds *gusting* up to 40 mph⟩
gus·ta·tion \ˌgəs-ˈtā-shən\ *noun* [Latin *gustation-, gustatio,* from *gustare*] (1599)
: the act or sensation of tasting
gus·ta·to·ry \ˈgəs-tə-ˌtōr-ē, -ˌtȯr-\ *adjective* (1684)
: relating to or associated with eating or the sense of taste
— **gus·ta·to·ri·ly** \ˌgəs-tə-ˈtōr-ə-lē, -ˈtȯr-\ *adverb*
gus·to \ˈgəs-(ˌ)tō\ *noun, plural* **gustoes** [Italian, from Latin *gustus,* past participle] (1620)
1 a : an individual or special taste ⟨different *gustoes*⟩ **b :** enthusiastic and vigorous enjoyment or appreciation **c :** vitality marked by an abundance of vigor and enthusiasm
2 *archaic* **:** artistic style
¹gut \ˈgət\ *noun* [Middle English, from Old English *guttas,* plural; probably akin to Old English *gēotan* to pour] (before 12th century)
1 a (1) **:** BOWELS, ENTRAILS — usually used in plural (2) **:** the basic visceral or emotional part of a person **b :** the alimentary canal or part of it (as the intestine or stomach) **c :** BELLY, ABDOMEN **d :** CATGUT
2 *plural* **:** the inner essential parts ⟨the *guts* of a car⟩
3 : a narrow passage; *also* **:** a narrow waterway or small creek
4 : the sac of silk taken from a silkworm ready to spin its cocoon and drawn out into a thread for use as a snell
5 *plural* **:** fortitude and stamina in coping with what alarms, repels, or discourages **:** COURAGE, PLUCK
6 : GUT COURSE
²gut *transitive verb* **gut·ted; gut·ting** (14th century)
1 a : EVISCERATE **b :** to extract all the essential passages or portions from
2 a : to destroy the inside of ⟨fire *gutted* the building⟩ **b :** to destroy the essential power or effectiveness of ⟨inflation *gutting* the economy⟩
— **gut it out :** PERSEVERE
³gut *adjective* (1964)
1 : arising from one's inmost self **:** VISCERAL ⟨a *gut* reaction⟩
2 : having strong impact or immediate relevance ⟨*gut* issues⟩
gut·buck·et \ˈgət-ˌbə-kət\ *noun* (1929)
1 : BARRELHOUSE 2
2 : a homemade bass fiddle consisting of a stick attached to an inverted washtub and having a single string
gut course *noun* (1948)
: a course (as in college) that is easily passed
gut·less \ˈgət-ləs\ *adjective* (1900)
1 : lacking courage **:** COWARDLY
2 : lacking significance or vitality
— **gut·less·ness** *noun*
gutsy \ˈgət-sē\ *adjective* **guts·i·er; -est** (circa 1893)
1 : marked by courage, pluck, or determination ⟨a *gutsy* little fighter⟩ ⟨a *gutsy* decision⟩
2 a : expressing or characterized by basic physical senses or passions ⟨*gutsy* macho talk⟩ ⟨*gutsy* country blues⟩ **b :** rough or plain in style **:** not bland or sophisticated ⟨a *gutsy* soup⟩
— **gut·si·ly** \-sə-lē\ *adverb*
— **guts·i·ness** *noun*
gut·ta \ˈgə-tə, ˈgu̇-tə\ *noun, plural* **gut·tae** \ˈgə-ˌtē, ˈgu̇-, -ˌtī\ [Latin, literally, drop] (1563)
: one of a series of ornaments in the Doric entablature that is usually in the form of a frustum of a cone
gut·ta–per·cha \ˌgə-tə-ˈpər-chə\ *noun* [Malay *gĕtah-pĕrcha,* from *gĕtah* sap, latex + *pĕrcha* scrap, rag] (1845)
: a tough plastic substance from the latex of several Malaysian trees (genera *Payena* and *Palaquium*) of the sapodilla family that resembles rubber but contains more resin and is used especially as insulation and in dentistry
gut·ta·tion \ˌgə-ˈtā-shən\ *noun* [Latin *gutta* drop] (circa 1889)
: the exudation of liquid water from the uninjured surface of a plant leaf
¹gut·ter \ˈgə-tər\ *noun* [Middle English *goter,* from Middle French *goutiere,* from *goute* drop, from Latin *gutta*] (14th century)
1 a : a trough along the eaves to catch and carry off rainwater **b :** a low area (as at the edge of a street) to carry off surface water (as to a sewer) **c :** a trough or groove to catch and direct something ⟨the *gutters* of a bowling alley⟩
2 : a white space formed by the adjoining inside margins of two facing pages (as of a book)
3 : the lowest or most vulgar level or condition of human life
²gutter (14th century)
transitive verb
1 : to cut or wear gutters in
2 : to provide with a gutter
intransitive verb
1 a : to flow in rivulets **b** *of a candle* **:** to melt away through a channel out of the side of the cup hollowed out by the burning wick
2 : to incline downward in a draft ⟨the candle flame *guttering*⟩
³gutter *adjective* (15th century)
: of, relating to, or characteristic of the gutter; *especially* **:** marked by extreme vulgarity, cheapness, or indecency ⟨*gutter* politics⟩
gut·ter·ing \-iŋ\ *noun* (1703)
1 : material for gutters
2 : GUTTER 1a
gutter out *intransitive verb* (1875)
1 : to become gradually weaker and then go out ⟨the candle *guttered out*⟩
2 : to end feebly or undramatically ⟨his screen career had slowly *guttered out*⟩
gut·ter·snipe \ˈgə-tər-ˌsnīp\ *noun* (circa 1869)
1 : STREET ARAB
2 : a person of the lowest moral or economic station
— **gut·ter·snip·ish** \-ˌsnī-pish\ *adjective*
gut·tur·al \ˈgə-tə-rəl, ˈgə-trəl\ *adjective* [Middle French, probably from Medieval Latin *gutturalis,* from Latin *guttur* throat] (1594)
1 : articulated in the throat ⟨*guttural* sounds⟩
2 : VELAR
3 : being or marked by utterance that is strange, unpleasant, or disagreeable
— **guttural** *noun*
— **gut·tur·al·ism** \ˈgə-tə-rə-ˌli-zəm, ˈgə-trə-\ *noun*
gut·ty \ˈgə-tē\ *adjective* **gut·ti·er; -est** (1947)
1 : GUTSY 1 ⟨a *gutty* quarterback⟩
2 : having a vigorous challenging quality ⟨*gutty* realism⟩
gut–wrench·ing \ˈgət-ˌren-chiŋ\ *adjective* (1974)
: causing mental or emotional anguish
¹guy \ˈgī\ *noun* [probably from Dutch *gei* brail] (1623)
: a rope, chain, rod, or wire attached to something as a brace or guide — called also *guyline*
²guy *transitive verb* (1712)
: to steady or reinforce with a guy
³guy *noun* [*Guy* Fawkes] (1806)
1 *often capitalized* **:** a grotesque effigy of Guy Fawkes traditionally displayed and burned in England on Guy Fawkes Day
2 *chiefly British* **:** a person of grotesque appearance
3 a : MAN, FELLOW **b :** PERSON — used in plural to refer to the members of a group regardless of sex ⟨saw her and the rest of the *guys*⟩ ◆
⁴guy *transitive verb* (1854)
: to make fun of **:** RIDICULE
Guy Fawkes Day \ˈgī-ˈfȯks-\ *noun* (1825)
: November 5 observed in England in commemoration of the seizure of Guy Fawkes in 1605 for an attempt to blow up the houses of Parliament
guy·ot \ˈgē-(ˌ)ō\ *noun* [Arnold H. *Guyot* (died 1884) American geographer & geologist] (1946)
: a flat-topped seamount
guz·zle \ˈgə-zəl\ *verb* **guz·zled; guz·zling** \ˈgəz-liŋ, ˈgə-zə-\ [origin unknown] (1583)
intransitive verb
: to drink especially liquor greedily, continually, or habitually
transitive verb
: to drink greedily or habitually ⟨*guzzle* beer⟩
— **guz·zler** \ˈgəz-lər, ˈgə-zə-\ *noun*
gwe·duc \ˈgü-ē-ˌdək\ *variant of* GEODUCK
gybe \ˈjīb\ *variant of* JIBE
gym \ˈjim\ *noun* (circa 1871)
1 : GYMNASIUM
2 : PHYSICAL EDUCATION
3 : a usually metal frame supporting an assortment of outdoor play equipment (as a swing, seesaw, and rings)
gym·kha·na \jim-ˈkä-nə, -ˈka-\ *noun* [probably modification of Hindi *gēdkhānā,* literally, ball court] (1877)
: a meet featuring sports contests or athletic skills: as **a :** competitive games on horseback **b :** a timed contest for automobiles featuring a series of events designed to test driving skill
gym·na·si·um *sense 1* jim-ˈnā-zē-əm, -zhəm; *sense 2 usually* gim-ˈnä-zē-əm\ *noun, plural* **-na·si·ums** *or* **-na·sia** \-ˈnā-zē-ə, -ˈnā-zhə; -ˈnä-zē-ə\ [Latin, exercise ground, school, from Greek *gymnasion,* from *gymnazein* to exercise naked, from *gymnos* naked — more at NAKED] (1598)
1 a : a large room used for various indoor sports (as basketball or boxing) and usually equipped with gymnastic apparatus **b :** a building (as on a college campus) containing space and equipment for various indoor sports activities and usually including spectator accommodations, locker and shower rooms, offices, classrooms, and a swimming pool

◇ WORD HISTORY
guy On November 4, 1605, Guy Fawkes was arrested in London and later executed for having planted barrels of gunpowder in the cellars of the houses of Parliament as part of a conspiracy to blow up the buildings. The failure of this conspiracy, now known as the Gunpowder Plot, is still celebrated in England on the evening of November 5, Guy Fawkes Day, when bonfires are traditionally lit and fireworks set off. On the bonfires are burned effigies of Guy Fawkes made from old clothes stuffed with straw and rags. By the early part of the 19th century these effigies had come to be called *guys*. The use of this word was extended to other similar effigies and then to a person of grotesque appearance or dress. In the U.S., where Guy Fawkes' demise has apparently never been celebrated, the word *guy* has had an extraordinary career, passing in the course of the last two centuries from the meaning "person who is an object of derision or pity" to "man, fellow" to a gender-neutral "person." The last sense is heard particularly in the phrase *you guys,* first attested in 1896, which sometimes amounts to little more than a familiar plural form of the second-person pronoun.

2 [German, from Latin, school] **:** a European secondary school that prepares students for the university ◆

gym·nast \'jim-ˌnast, -nəst\ *noun* [Middle French *gymnaste,* from Greek *gymnastēs* trainer, from *gymnazein*] (1594)
: one trained in gymnastics

¹gym·nas·tic \jim-'nas-tik\ *adjective* (1574)
: of or relating to gymnastics **:** ATHLETIC
— **gym·nas·ti·cal·ly** \-ti-k(ə-)lē\ *adverb*

²gymnastic *noun* (1652)
1 *plural but singular in construction* **a :** physical exercises designed to develop strength and coordination **b :** a competitive sport in which individuals perform optional and prescribed acrobatic feats mostly on special apparatus in order to demonstrate strength, balance, and body control
2 : an exercise in intellectual or artistic dexterity ⟨my earlier philosophic study had been an intellectual *gymnastic* —John Dewey⟩ ⟨mental *gymnastics*⟩
3 : a physical feat or contortion ⟨the *gymnastics* necessary for the killer to have swung from the fire escape —E. D. Radin⟩

gym·nos·o·phist \jim-'nä-sə-fist\ *noun* [Latin *gymnosophista,* from Greek *gymnosophistēs,* from *gymnos* + *sophistēs* wise man, sophist] (15th century)
: any of a sect of ascetics in ancient India who went naked and practiced meditation

gym·no·sperm \'jim-nə-ˌspərm\ *noun* [ultimately from Greek *gymnos* + *sperma* seed — more at SPERM] (circa 1838)
: any of a class (Gymnospermae) of woody vascular seed plants (as conifers) that produce naked seeds not enclosed in an ovary and that in some instances have motile spermatozoids
— **gym·no·sper·mous** \ˌjim-nə-'spər-məs\ *adjective*
— **gym·no·sper·my** \'jim-nə-ˌspər-mē\ *noun*

gyn- *or* **gyno-** *combining form* [Greek *gyn-,* from *gynē* woman — more at QUEEN]
: female reproductive organ **:** ovary ⟨*gyno*phore⟩

gy·nae·col·o·gy *chiefly British variant of* GYNECOLOGY

gyn·an·dro·morph \ˌgīn-'an-drə-ˌmȯrf, ˌjin-\ *noun* [International Scientific Vocabulary, from Greek *gynandros* + *-morph*] (circa 1890)
: an abnormal individual exhibiting characters of both sexes in various parts of the body **:** a sexual mosaic
— **gyn·an·dro·mor·phic** \(ˌ)gīn-ˌan-drə-'mȯr-fik, (ˌ)jin-\ *adjective*
— **gyn·an·dro·mor·phism** \-ˌfi-zəm\ *noun*
— **gyn·an·dro·mor·phy** \ˌgīn-'an-drə-ˌmȯr-fē, ˌjin-\ *noun*

gyn·an·drous \ˌgīn-'an-drəs, ˌjin-\ *adjective* [Greek *gynandros* of doubtful sex, from *gynē* woman + *andr-, anēr* man — more at ANDR-] (1807)
: having the androecium and gynoecium united in a column

-gyne *noun combining form* [Greek *gynē*]
: female reproductive organ ⟨tricho*gyne*⟩

gynec- *or* **gyneco-** *also* **gynaec-** *or* **gynaeco-** *combining form* [Greek *gynaik-, gynaiko-,* from *gynaik-, gynē* woman — more at QUEEN]
: woman ⟨*gyneco*logy⟩

gy·ne·coc·ra·cy \ˌgī-ni-'kä-krə-sē, ˌji-\ *noun, plural* **-cies** [Greek *gynaikokratia,* from *gynaik-* + *-kratia* -cracy] (1612)
: political supremacy of women
— **gy·ne·co·crat·ic** \ˌgī-ni-kō-'kra-tik, ˌji-\ *adjective*

gy·ne·coid \'gī-ni-ˌkȯid, 'ji-\ *adjective* (1907)
: typical or characteristic of the human female ⟨*gynecoid* pelvis⟩

gy·ne·col·o·gy \ˌgī-nə-'kä-lə-jē, ˌji-\ *noun* [International Scientific Vocabulary] (circa 1847)
: a branch of medicine that deals with the diseases and routine physical care of the reproductive system of women
— **gy·ne·co·log·ic** \ˌgī-ni-kə-'lä-jik, ˌji-\ *or* **gy·ne·co·log·i·cal** \-ji-kəl\ *adjective*
— **gy·ne·col·o·gist** \ˌgī-nə-'kä-lə-jist, ˌji-\ *noun*

gy·neco·mas·tia \ˌgī-nə-kō-'mas-tē-ə\ *noun* [New Latin, from *gynec-* + Greek *mastos* breast + New Latin *-ia*] (1881)
: excessive development of the breast in the male

gy·noe·ci·um \ji-'nē-shē-əm, gī-, -sē-\ *noun, plural* **-cia** \-shē-ə, -sē-\ [New Latin, alteration of Latin *gynaeceum* women's apartments, from Greek *gynaikeion,* from *gynaik-, gynē*] (1832)
: the aggregate of carpels or pistils in a flower

gy·no·gen·e·sis \ˌgī-nə-'je-nə-səs\ *noun* [New Latin] (1925)
: development in which the embryo contains only maternal chromosomes due to activation of an egg by a sperm that degenerates without fusing with the egg nucleus
— **gy·no·ge·net·ic** \-jə-'ne-tik\ *adjective*

gy·no·phore \'gī-nə-ˌfōr, 'ji-, -ˌfȯr\ *noun* (1821)
: a prolongation of the receptacle (as in a caper flower) with the gynoecium at its apex

-gynous *adjective combining form* [New Latin *-gynus,* from Greek *-gynos,* from *gynē* woman — more at QUEEN]
1 : of, relating to, or having (such or so many) wives ⟨mono*gynous*⟩
2 : situated (in a specified place) in relation to a female organ of a plant ⟨hypo*gynous*⟩

-gyny *noun combining form*
1 : existence of or condition of having (such or so many) wives ⟨poly*gyny*⟩
2 : condition of being situated (in a specified place) in relation to a female organ of a plant ⟨epi*gyny*⟩

¹gyp \'jip\ *noun* [probably short for *gypsy*] (1750)
1 *British* **:** a college servant
2 a : CHEAT, SWINDLER **b :** FRAUD, SWINDLE

²gyp *verb* **gypped; gyp·ping** (1880)
: CHEAT

gyp·se·ous \'jip-sē-əs\ *adjective* (1661)
: resembling, containing, or consisting of gypsum ⟨*gypseous* clay loam⟩

gyp·sif·er·ous \jip-'si-f(ə-)rəs\ *adjective* (circa 1847)
: containing gypsum

gyp·soph·i·la \jip-'sä-fə-lə\ *noun* [New Latin, from Latin *gypsum* + New Latin *-phila* -phil] (1771)
: any of a large genus (*Gypsophila*) of Old World herbs of the pink family having small delicate usually paniculate flowers

gyp·sum \'jip-səm\ *noun* [Latin, from Greek *gypsos*] (14th century)
1 : a widely distributed mineral consisting of hydrous calcium sulfate that is used especially as a soil amendment and in making plaster of paris
2 : PLASTERBOARD

gyp·sy \'jip-sē\ *intransitive verb* **gyp·sied; gyp·sy·ing** (circa 1627)
: to live or roam like a Gypsy

Gyp·sy \'jip-sē\ *noun, plural* **Gypsies** [by shortening & alteration from *Egyptian*] (1537)
1 : a member of a traditionally itinerant people who originated in northern India and now live chiefly in south and southwest Asia, Europe, and North America
2 : ROMANY 2
3 *not capitalized* **:** one that resembles a Gypsy; *especially* **:** WANDERER

gypsy cab *noun* (1964)
: a taxicab licensed only to answer calls; *especially* **:** such a cab that cruises in search of passengers

gypsy moth *noun* (1819)
: an Old World tussock moth (*Lymantria dispar*) that was introduced about 1869 into the U.S. and has a grayish brown mottled hairy caterpillar which is a destructive defoliator of many trees

gyr- *or* **gyro-** *combining form* [probably from Middle French, from Latin, from Greek, from *gyros* rounded]
1 : ring **:** circle **:** spiral ⟨*gyro*magnetic⟩
2 : gyroscope ⟨*gyro*compass⟩

gyr·ase \'jī-ˌrās\ *noun* [*gyr-* + *-ase*] (1976)
: an enzyme that catalyzes the breaking and rejoining of bonds linking adjacent nucleotides in DNA to generate supercoiled DNA helices

¹gy·rate \'jī-ˌrāt\ *adjective* (1830)
: winding or coiled around **:** CONVOLUTED ⟨*gyrate* branches of a tree⟩

²gyrate *intransitive verb* **gy·rat·ed; gy·rat·ing** (1830)
1 : to revolve around a point or axis
2 : to oscillate with or as if with a circular or spiral motion
— **gy·ra·tor** \-ˌrā-tər\ *noun*
— **gy·ra·to·ry** \'jī-rə-ˌtōr-ē, -ˌtȯr-\ *adjective*

gy·ra·tion \jī-'rā-shən\ *noun* (1615)
1 : an act or instance of gyrating
2 : something (as a coil of a shell) that is gyrate
— **gy·ra·tion·al** \-shnəl, -shə-nᵊl\ *adjective*

¹gyre \'jī(-ə)r\ *noun* [Latin *gyrus,* from Greek *gyros*] (1566)
: a circular or spiral motion or form; *especially* **:** a giant circular oceanic surface current
— **gy·ral** \'jī-rəl\ *adjective*

²gyre *intransitive verb* **gyred; gyr·ing** [Late Latin *gyrare,* from Latin *gyrus*] (1593)
: to move in a circle or spiral

gy·rene \jī-'rēn\ *noun* [probably alteration of *marine*] (1944)
slang **:** a U.S. marine

gyr·fal·con \'jər-ˌfal-kən, -ˌfȯl- *also* -ˌfō-kən\ *noun* [Middle English *gerfaucun,* from Old French *girfaucon,* probably from *gir* vulture (from Old High German *gīr*) + *faucon* falcon] (14th century)
: an arctic falcon (*Falco rusticolus*) that occurs in several color forms and is the largest of all falcons

gyrfalcon

◇ WORD HISTORY

gymnasium The ancient Greeks placed a high value on both physical and mental fitness. Each important city in Greece had a public area set aside in which young men would gather to exercise, compete in sports, and receive training in philosophy, music, and literature. The Greeks were accustomed to exercise without clothing, and the word for "exercise ground," *gymnasion,* was a derivative of *gymnazesthai* "to exercise in the nude," itself a derivative of *gymnos* "naked." *Gymnasion* came to mean more generally "school," especially "a philosophic school," through the association of philosophers such as Plato with particular gymnasia. Both the physical and intellectual meanings of *gymnasion* were taken over by the Latin borrowing *gymnasium,* and are reflected in modern European languages. Loans of the Latin word in central and eastern Europe, such as German *Gymnasium,* Polish *gimnazjum,* and Russian *gimnaziya,* denote an elite secondary school that traditionally stressed the classical languages. In English, on the other hand, *gymnasium,* often shortened to *gym,* is associated solely with physical activity—though not exercise in the nude—and a gymnasium is by no means necessarily attached to a school.

[1]**gy·ro** \'jī-(ˌ)rō\ *noun, plural* **gyros** (1910)
1 : GYROCOMPASS
2 : GYROSCOPE

[2]**gy·ro** \'yē-ˌrō, 'zhir-ō\ *noun, plural* **gyros** [New Greek *gyros* turn, from Greek; from the rotation of the meat on a spit] (1971)
: a sandwich especially of lamb and beef, tomato, and onion on pita bread

Gy·ro \'jī-(ˌ)rō\ *noun, plural* **Gyros** [*Gyro* International (association)] (1971)
: a member of a major international service club

gy·ro·com·pass \'jī-rō-ˌkəm-pəs *also* -ˌkäm-\ *noun* (1910)
: a compass consisting of a continuously driven gyroscope whose spinning axis is confined to a horizontal plane so that the earth's rotation causes it to assume a position parallel to the earth's axis and thus point to the true north

gy·ro·fre·quen·cy \-ˌfrē-kwən(t)-sē\ *noun* (1938)
: the frequency with which a charged particle (as an electron) executes spiral gyrations in moving obliquely across a magnetic field

gyro horizon *noun* (1938)
: ARTIFICIAL HORIZON

gy·ro·mag·net·ic \ˌjī-rō-mag-'ne-tik\ *adjective* (1922)
: of or relating to the magnetic properties of a rotating electrical particle

gyromagnetic ratio *noun* (1922)
: the ratio of the magnetic moment of a spinning charged particle to its angular momentum

gy·ron \'jī-rən\ *noun* [Middle French *giron* gore, of Germanic origin; akin to Old High German *gēra* wedge-shaped object, Old English *gār* spear — more at GORE] (1572)
: a heraldic charge of triangular form having one side at the edge of the field and the opposite angle usually at the fess point

gy·ro·plane \'jī-rə-ˌplān\ *noun* [International Scientific Vocabulary] (1907)
: an airplane balanced and supported by the aerodynamic forces acting on rapidly rotating horizontal or slightly inclined airfoils

gy·ro·scope \'jī-rə-ˌskōp, *British also* 'gī-\ *noun* [French] (1856)
: a wheel or disk mounted to spin rapidly about an axis and also free to rotate about one or both of two axes perpendicular to each other and to the axis of spin so that a rotation of one of the two mutually perpendicular axes results from application of torque to the other when the wheel is spinning and so that the entire apparatus offers considerable opposition depending on the angular momentum to any torque that would change the direction of the axis of spin
— gy·ro·scop·ic \ˌjī-rə-'skä-pik\ *adjective*
— gy·ro·scop·i·cal·ly \-pi-k(ə-)lē\ *adverb*

gy·ro·sta·bi·liz·er \ˌjī-rō-'stā-bə-ˌlī-zər\ *noun* (1921)
: a stabilizing device (as for a ship or airplane) that consists of a continuously driven gyro spinning about a vertical axis and pivoted so that its axis of spin may be tipped fore-and-aft in the vertical plane and that serves to oppose sideways motion — called also *gy·ro·stat* \'jī-rə-ˌstat\

gy·rus \'jī-rəs\ *noun, plural* **gy·ri** \'jī-ˌrī\ [New Latin, from Latin, circle — more at GYRE] (circa 1842)
: a convoluted ridge (as a convolution of the brain) between anatomical grooves

gyve \'jīv, 'gīv\ *noun* [Middle English] (13th century)
: FETTER, SHACKLE
— gyve *transitive verb*

H *is the eighth letter of the English alphabet and of most alphabets closely related to that of English. It came through Latin via Etruscan from Greek, which took it from a Phoenician letter* hēth, *standing for a laryngeal aspirate (a consonant with the sound of a very forceful English* h*). In Greek it was called* eta *and was assigned to the long open* e *sound (as in English* café; *see* E*) except in the western alphabet, where it designated the rough breathing (that is, the sound of* h *in* helium*). This aspirated pronunciation passed through Etruscan and Latin to English. Small* h *evolved from a late Roman form of the capital in which the second upright line had been shortened to nearly the height of the horizontal cross-stroke. In a later Roman cursive script, the cross-stroke and shortened right vertical stroke were merged into a single downward-curving line.*

h \'āch\ *noun, plural* **h's** *or* **hs** \'ā-chəz\ *often capitalized, often attributive* (before 12th century)
1 a : the 8th letter of the English alphabet **b :** a graphic representation of this letter **c :** a speech counterpart of orthographic *h*
2 : a graphic device for reproducing the letter *h*
3 : one designated *h* especially as the 8th in order or class
4 : something shaped like the letter H

ha \'hä\ *interjection* [Middle English, from Old English] (before 12th century)
— used especially to express surprise or joy

Ha·ba·cuc \'ha-bə-ˌkək, hə-'ba-kək\ *noun* [Late Latin, from Hebrew *Ḥăbhaqqūq*]
: HABAKKUK

Ha·bak·kuk \'ha-bə-ˌkək, hə-'ba-kək\ *noun* [Hebrew *Ḥăbhaqqūq*]
1 : a Hebrew prophet of 7th century B.C. Judah who prophesied an imminent Chaldean invasion
2 : a prophetic book of canonical Jewish and Christian Scripture — see BIBLE table

ha·ba·ne·ra \ˌ(h)ä-bə-'ner-ə\ *noun* [Spanish *(danza) habanera,* literally, Havanan dance] (1878)
1 : a Cuban dance in slow duple time
2 : the music for the habanera

hab·da·lah \ˌhäv-də-'lä, häv-'dó-lə\ *noun, often capitalized* [Hebrew *habhdālāh* separation] (1733)
: a Jewish ceremony marking the close of a Sabbath or holy day

ha·be·as cor·pus \'hā-bē-əs-'kór-pəs\ *noun* [Middle English, from Medieval Latin, literally, you should have the body (the opening words of the writ)] (15th century)
1 : any of several common-law writs issued to bring a party before a court or judge; *especially* **:** HABEAS CORPUS AD SUBJICIENDUM
2 : the right of a citizen to obtain a writ of habeas corpus as a protection against illegal imprisonment

habeas corpus ad sub·ji·ci·en·dum \-ˌad-səb-ˌyi-kē-'en-dəm\ *noun* [New Latin, literally, you should have the body for submitting] (1768)
: a writ for inquiring into the lawfulness of the restraint of a person who is imprisoned or detained in another's custody

hab·er·dash·er \'ha-bə(r)-ˌda-shər\ *noun* [Middle English *haberdassher,* from modification of Anglo-French *hapertas* petty merchandise] (14th century)
1 *British* **:** a dealer in notions
2 : a dealer in men's clothing and accessories

hab·er·dash·ery \-ˌda-sh(ə-)rē\ *noun, plural* **-er·ies** (1593)
1 : goods sold by a haberdasher
2 : a haberdasher's shop

ha·ber·geon \'ha-bər-jən, hə-'bər-jē-ən, -jən\ *noun* [Middle English *haubergeoun,* from Middle French *haubergeon,* diminutive of *hauberc* hauberk] (14th century)
1 : a medieval jacket of mail shorter than a hauberk
2 : HAUBERK

Ha·ber process \'hä-bər-\ *noun* [Fritz *Haber* (died 1934) German chemist] (1916)
: a catalytic process for synthesizing ammonia from nitrogen and hydrogen

hab·ile \'ha-bəl, -ˌbīl\ *adjective* [French, from Latin *habilis* — more at ABLE] (15th century)
: having general skill **:** ABLE, SKILLFUL

ha·bil·i·ment \hə-'bi-lə-mənt\ *noun* [Middle French *habillement,* from *habiller* to dress, prepare, from Old French *abiller,* from *bille* log — more at BILLET] (15th century)
1 *plural* **:** characteristic apparatus **:** FITTINGS ⟨the *habiliments* of civilization —W. P. Webb⟩
2 a : the dress characteristic of an occupation or occasion — usually used in plural **b :** CLOTHES — usually used in plural

ha·bil·i·tate \hə-'bi-lə-ˌtāt\ *verb* **-tat·ed; -tat·ing** [Late Latin *habilitatus,* past participle of *habilitare,* from Latin *habilitas* ability — more at ABILITY] (1604)
transitive verb
1 : to make fit or capable (as for functioning in society)
2 : CLOTHE, DRESS
intransitive verb
: to qualify oneself
— ha·bil·i·ta·tion \-ˌbi-lə-'tā-shən\ *noun*

[1]hab·it \'ha-bət\ *noun* [Middle English, from Old French, from Latin *habitus* condition, character, from *habēre* to have, hold — more at GIVE] (13th century)
1 *archaic* **:** CLOTHING
2 a : a costume characteristic of a calling, rank, or function ⟨a nun's *habit*⟩ **b :** a costume worn for horseback riding
3 : manner of conducting oneself **:** BEARING
4 : bodily appearance or makeup ⟨a man of fleshy *habit*⟩
5 : the prevailing disposition or character of a person's thoughts and feelings **:** mental makeup
6 : a settled tendency or usual manner of behavior
7 a : a behavior pattern acquired by frequent repetition or physiologic exposure that shows itself in regularity or increased facility of performance **b :** an acquired mode of behavior that has become nearly or completely involuntary **c :** ADDICTION
8 : characteristic mode of growth or occurrence ⟨a grass similar to Indian corn in *habit*⟩
9 *of a crystal* **:** characteristic assemblage of forms at crystallization leading to a usual appearance ☆

[2]habit *transitive verb* (1594)
: CLOTHE, DRESS

hab·it·able \'ha-bə-tə-bəl *also* hə-'bi-tə-\ *adjective* (14th century)
: capable of being lived in **:** suitable for habitation
— hab·it·abil·i·ty \ˌha-bə-tə-'bi-lə-tē\ *noun*
— hab·it·able·ness \'ha-bə-tə-bəl-nes\ *noun*
— hab·it·ably \-blē\ *adverb*

ha·bi·tant *noun* (15th century)
1 \'ha-bə-tənt\ **:** INHABITANT, RESIDENT
2 \ˌ(h)a-bi-'täⁿ\ *also* **ha·bi·tan** \-'täⁿ\ **:** a settler or descendant of a settler of French origin working as a farmer in Canada

hab·i·tat \'ha-bə-ˌtat\ *noun* [Latin, it inhabits, from *habitare*] (1796)
1 a : the place or environment where a plant or animal naturally or normally lives and grows **b :** the typical place of residence of a person or a group **c :** a housing for a controlled physical environment in which people can live under surrounding inhospitable conditions (as under the sea)
2 : the place where something is commonly found

hab·i·ta·tion \ˌha-bə-'tā-shən\ *noun* [Middle English *habitacioun,* from Middle French *habitation,* from Latin *habitation-, habitatio,* from *habitare* to inhabit, frequentative of *habēre*] (14th century)
1 : the act of inhabiting **:** OCCUPANCY
2 : a dwelling place **:** RESIDENCE
3 : SETTLEMENT, COLONY

hab·it–form·ing \'ha-bət-ˌfór-miŋ\ *adjective* (1913)
: inducing the formation of an addiction

ha·bit·u·al \hə-'bi-ch(ə-)wəl, ha-, -'bi-chəl\ *adjective* (1611)
1 : having the nature of a habit **:** being in accordance with habit **:** CUSTOMARY ⟨*habitual* smoking⟩
2 : doing, practicing, or acting in some manner by force of habit ⟨*habitual* drunkard⟩
3 : resorted to on a regular basis ⟨our *habitual* diet⟩
4 : inherent in an individual ⟨*habitual* grace⟩
synonym see USUAL
— ha·bit·u·al·ly *adverb*
— ha·bit·u·al·ness *noun*

ha·bit·u·ate \hə-'bi-chə-ˌwāt, ha-\ *verb* **-at·ed; -at·ing** (15th century)
transitive verb
1 : to make used to something **:** ACCUSTOM
2 : FREQUENT 1
intransitive verb
1 : to cause habituation
2 : to undergo habituation ⟨*habituate* to a stimulus⟩

ha·bit·u·a·tion \-ˌbi-chə-'wā-shən\ *noun* (15th century)
1 : the process of habituating **:** the state of being habituated

☆ SYNONYMS
Habit, practice, usage, custom, wont mean a way of acting fixed through repetition. HABIT implies a doing unconsciously and often compulsively ⟨had a *habit* of tapping his fingers⟩. PRACTICE suggests an act or method followed with regularity and usually through choice ⟨our *practice* is to honor all major credit cards⟩. USAGE suggests a customary action so generally followed that it has become a social norm ⟨western-style dress is now common *usage* in international business⟩. CUSTOM applies to a practice or usage so steadily associated with an individual or group as to have almost the force of unwritten law ⟨the *custom* of wearing black at funerals⟩. WONT usually applies to an habitual manner, method, or practice of an individual or group ⟨as was her *wont,* she slept until noon⟩.

2 a : tolerance to the effects of a drug acquired through continued use **b :** psychological dependence on a drug after a period of use — compare ADDICTION
3 : decrease in responsiveness upon repeated exposure to a stimulus

hab·i·tude \'ha-bə-ˌtüd, ˌtyüd\ *noun* (14th century)
1 *archaic* **:** native or essential character
2 *obsolete* **:** habitual association
3 a : habitual disposition or mode of behavior or procedure **b :** CUSTOM

ha·bi·tué *also* **ha·bi·tue** \hə-'bi-chə-ˌwā, ha-, -ˌbi-chə-'\ *noun* [French, from past participle of *habituer* to frequent, from Late Latin *habituare* to habituate, from Latin *habitus*] (1818)
1 : one who may be regularly found in or at a particular place or kind ofplace ⟨café *habitués*⟩
2 : DEVOTEE ⟨an *habitué* of science-fiction movies⟩

hab·i·tus \'ha-bə-təs\ *noun, plural* **habitus** \-təs, -ˌtüs\ [New Latin, from Latin] (1886)
: HABIT; *specifically* **:** body build and constitution especially as related to predisposition to disease

ha·boob \hə-'büb\ *noun* [Arabic *habūb* violent storm] (1897)
: a violent dust storm or sandstorm especially of Sudan

Habs·burg \'haps-, 'häps-\ *variant of* HAPSBURG

ha·ček \'hä-ˌchek\ *noun* [Czech *háček,* literally, little hook] (1953)
: a diacritic ˇ placed over a letter (as in *č*) to modify it **:** an inverted circumflex

ha·cen·da·do \ˌ(h)ä-s^{ə}n-'dä-(ˌ)dō\ *also* **ha·ci·en·da·do** \ˌhä-sē-en-\ *noun, plural* **-dos** [Spanish, from *hacienda*] (1840)
: the owner or proprietor of a hacienda

^{1}**ha·chure** \ha-'shu̇r\ *noun* [French, from *hacher* to chop up, hash] (1858)
: a short line used for shading and denoting surfaces in relief (as in map drawing) and drawn in the direction of slope

^{2}**hachure** *transitive verb* **ha·chured; ha·chur·ing** (circa 1859)
: to shade with or show by hachures

ha·ci·en·da \ˌ(h)ä-sē-'en-də\ *noun* [Spanish, from Old Spanish *facienda,* from Latin, literally, things to be done, neuter plural of *faciendus,* gerundive of *facere* to do — more at DO] (circa 1772)
1 : a large estate especially in a Spanish-speaking country **:** PLANTATION
2 : the main dwelling of a hacienda

^{1}**hack** \'hak\ *verb* [Middle English *hakken,* from Old English *-haccian;* akin to Old High German *hacchōn* to hack, Old English *hōc* hook] (13th century)
transitive verb
1 a : to cut or sever with repeated irregular or unskillful blows **b :** to cut or shape by or as if by crude or ruthless strokes ⟨*hacking* out new election districts⟩ **c :** ANNOY, VEX — often used with *off*
2 : to clear or make by or as if by cutting away vegetation ⟨*hacked* his way through the brush⟩
3 a : to manage successfully ⟨just couldn't *hack* the new job⟩ **b :** TOLERATE ⟨I can't *hack* all this noise⟩
intransitive verb
1 a : to make chopping strokes or blows ⟨*hacked* at the weeds⟩; *also* **:** to make cuts as if by chopping ⟨*hacking* away at the work force⟩ **b :** to play inexpert golf
2 : to cough in a short dry manner
3 : LOAF — usually used with *around*
4 a : to write computer programs for enjoyment **b :** to gain access to a computer illegally

^{2}**hack** *noun* (14th century)
1 : an implement for hacking
2 : NICK, NOTCH
3 : a short dry cough
4 : a hacking stroke or blow
5 : restriction to quarters as punishment for naval officers — usually used in the phrase *under hack*

^{3}**hack** *noun* [short for *hackney*] (circa 1721)
1 a (1) **:** a horse let out for common hire (2) **:** a horse used in all kinds of work **b :** a horse worn out in service **:** JADE **c :** a light easy saddle horse; *especially* **:** a three-gaited saddle horse **d :** a ride on a horse
2 a : HACKNEY **b** (1) **:** TAXICAB (2) **:** CABDRIVER
3 a : a person who works solely for mercenary reasons **:** HIRELING ⟨party *hacks*⟩ **b :** a writer who works on order; *also* **:** a writer who aims solely for commercial success

^{4}**hack** *adjective* (circa 1734)
1 : working for hire especially with mediocre professional standards
2 : performed by, suited to, or characteristic of a hack ⟨*hack* writing⟩
3 : HACKNEYED, TRITE

^{5}**hack** (circa 1888)
intransitive verb
1 : to ride or drive at an ordinary pace or over the roads especially as distinguished from racing or hunting
2 : to operate a taxicab
transitive verb
: to ride (a horse) at an ordinary pace

^{6}**hack** *noun* [origin unknown] (circa 1914)
slang **:** a guard especially at a prison

hack·a·more \'ha-kə-ˌmōr, -ˌmȯr\ *noun* [by folk etymology from Spanish *jáquima* bridle] (1850)
: a bridle with a loop capable of being tightened about the nose in place of a bit or with a slip noose passed over the lower jaw

hack·ber·ry \'hak-ˌber-ē\ *noun* [alteration of *hagberry* (a cherry resembling the chokecherry)] (1779)
: any of a genus (*Celtis*) of trees and shrubs of the elm family with small often edible berries; *also* **:** its wood

hacker \'ha-kər\ *noun* (14th century)
1 : one that hacks
2 : a person who is inexperienced or unskilled at a particular activity ⟨a tennis *hacker*⟩
3 : an expert at programming and solving problems with a computer
4 : a person who illegally gains access to and sometimes tampers with information in a computer system

hack·ie \'ha-kē\ *noun* (circa 1926)
: CABDRIVER

^{1}**hack·le** \'ha-kəl\ *noun* [Middle English *hakell;* akin to Old High German *hāko* hook — more at HOOK] (15th century)
1 a : one of the long narrow feathers on the neck or saddle of a bird **b :** the neck plumage of the domestic fowl
2 : a comb or board with long metal teeth for dressing flax, hemp, or jute
3 *plural* **a :** erectile hairs along the neck and back especially of a dog **b :** TEMPER, DANDER
4 a : an artificial fishing fly made chiefly of the filaments of a cock's neck feathers **b :** filaments of cock feather projecting from the head of an artificial fly

^{2}**hackle** *transitive verb* **hack·led; hack·ling** \'ha-k(ə-)liŋ\ (1616)
: to comb out with a hackle
— hack·ler \-k(ə-)lər\ *noun*

hack·ly \'ha-k(ə-)lē\ *adjective* (1796)
: having the appearance of something hacked **:** JAGGED

hack·man \'hak-mən\ *noun* (1796)
: CABDRIVER

hack·ma·tack \'hak-mə-ˌtak\ *noun* [earlier *hakmantak,* probably from Western Abenaki (Algonquian language of New Hampshire and Vermont)] (1792)
: TAMARACK

^{1}**hack·ney** \'hak-nē\ *noun, plural* **hack·neys** [Middle English *hakeney*] (14th century)

hackney 1c

1 a : a horse suitable for ordinary riding or driving **b :** a trotting horse used chiefly for driving **c** *often capitalized* **:** any of an English breed of rather compact usually chestnut, bay, or brown high-stepping horses
2 *obsolete* **:** one that works for hire
3 : a carriage or automobile kept for hire

^{2}**hackney** *adjective* (1589)
1 : kept for public hire
2 : HACKNEYED
3 *archaic* **:** done or suitable for doing by a drudge

^{3}**hackney** *transitive verb* **hack·neyed; hack·ney·ing** (1596)
1 a : to make common or frequent use of **b :** to make trite, vulgar, or commonplace
2 *archaic* **:** to make sophisticated or jaded

hackney coach *noun* (1635)
: a coach kept for hire; *especially* **:** a four-wheeled carriage drawn by two horses and having seats for six persons

hack·neyed \'hak-nēd\ *adjective* (1749)
: lacking in freshness or originality
synonym see TRITE

hack·saw \'hak-ˌsȯ\ *noun* (1654)
: a fine-tooth saw with a blade under tension in a frame that is used for cutting hard materials (as metal)

hack·work \-ˌwərk\ *noun* (1851)
: literary, artistic, or professional work done on order usually according to formula and in conformity with commercial standards

had *past and past participle of* HAVE

ha·dal \'hā-d^{ə}l\ *adjective* [French, from *Hadès* Hades] (1959)
: of, relating to, or being the parts of the ocean below 6000 meters

had·dock \'ha-dək\ *noun, plural* **haddock** *also* **haddocks** [Middle English *haddok*] (14th century)
: an important food fish (*Melanogrammus aeglefinus*) that is usually smaller than the related cod and that occurs on both sides of the North Atlantic

Ha·des \'hā-(ˌ)dēz\ *noun* [Greek *Haidēs*]
1 : PLUTO
2 : the underground abode of the dead in Greek mythology
3 : SHEOL
4 *often not capitalized* **:** HELL

ha·dith \hə-'dēth\ *noun, plural* **hadith** *or* **ha·diths** *often capitalized* [Arabic *ḥadīth,* literally, speech, report] (circa 1817)
1 : a narrative record of the sayings or customs of Muhammad and his companions
2 : the collective body of traditions relating to Muhammad and his companions

hadj, hadji *variant of* HAJJ, HAJJI

Had·ley cell \'had-lē-\ *noun* [George *Hadley* (died 1768) English scientific writer] (1969)
: a pattern of atmospheric circulation in which warm air rises near the equator, cools as it travels poleward at high altitude, sinks as cold air, and warms as it travels equatorward; *also* **:** a similar atmospheric circulation pattern on another planet (as Mars)

hadn't \'ha-d^{ə}nt, -d^{ə}n, *dialect also* 'ha-t^{ə}n(t) *or* 'hant\ (1695)
: had not

had·ron \'ha-ˌdrän\ *noun* [International Scientific Vocabulary *hadr-* thick, heavy (from Greek *hadros* thick) + ^{2}*-on*] (1962)

\ə\ **abut** \$^{ə}$\ **kitten** \ər\ **further** \a\ **ash** \ā\ **ace** \ä\ **mop, mar** \au̇\ **out** \ch\ **chin** \e\ **bet** \ē\ **easy** \g\ **go** \i\ **hit** \ī\ **ice** \j\ **job** \ŋ\ **sing** \ō\ **go** \ȯ\ **law** \ȯi\ **boy** \th\ **thin** \t͟h\ **the** \ü\ **loot** \u̇\ **foot** \y\ **yet** \zh\ **vision** *see also* Guide to Pronunciation

: any of the subatomic particles that are made up of quarks and are subject to the strong force
— ha·dron·ic \ha-ˈdrä-nik\ *adjective*

had·ro·saur \ˈha-drə-ˌsȯr\ *noun* [New Latin *Hadrosaurus,* genus name, from Greek *hadros* thick, bulky + *sauros* lizard] (1877)
: any of a family (Hadrosauridae) of medium-sized bipedal herbivorous dinosaurs of the Upper Cretaceous with long flat snouts and an often crested skull

hadst \ˈhadst, hədst, *or* t *for* d\ *archaic past 2d singular of* HAVE

hae \ˈhā\ *chiefly Scottish variant of* HAVE

haem *chiefly British variant of* HEME

haem- *or* **haemo-** *chiefly British variant of* HEM-

haema- *chiefly British variant of* HEMA-

haemat- *or* **haemato-** *chiefly British variant of* HEMAT-

hae·ma·tite *British variant of* HEMATITE

-haemia *chiefly British variant of* -EMIA

haet \ˈhāt\ *noun* [contraction of Scots *hae it* (as in *Deil hae it!* Devil take it!)] (1603)
chiefly Scottish **:** a small quantity **:** WHIT, BIT

haf·fet \ˈha-fət\ *noun* [Middle English (Scots) *halfheid,* from Middle English *half* half + *hed* head] (1513)
Scottish **:** CHEEK, TEMPLE

haf·ni·um \ˈhaf-nē-əm\ *noun* [New Latin, from *Hafnia* (Copenhagen), Denmark] (1923)
: a metallic element that resembles zirconium in its chemical properties and occurs in zirconium minerals and that readily absorbs neutrons — see ELEMENT table

1haft \ˈhaft\ *noun* [Middle English, from Old English *hæft;* akin to Old English *hebban* to lift — more at HEAVE] (before 12th century)
: the handle of a weapon or tool

2haft *transitive verb* (15th century)
: to set in or furnish with a haft

haf·ta·rah *or* **haf·to·rah** \häf-ˈtȯr-ə, ˌhäf-tə-ˈrä\ *noun* [Hebrew *haphṭārāh* conclusion] (1723)
: one of the biblical selections from the Books of the Prophets read after the parashah in the Jewish synagogue service

1hag \ˈhag\ *noun* [Middle English *hagge* demon, old woman] (14th century)
1 : an ugly, slatternly, or evil-looking old woman
2 *archaic* **a :** a female demon **b :** an evil or frightening spirit **:** HOBGOBLIN
3 : WITCH
— hag·gish \ˈha-gish\ *adjective*

2hag *noun* [Scots, break in a moor, from Old Norse *hǫgg* cut, cleft; akin to Old English *hēawan* to hew] (1662)
1 *British* **:** QUAGMIRE, BOG
2 *British* **:** a firm spot in a bog

Ha·gar \ˈhā-ˌgär, -gər\ *noun* [Hebrew *Hāghār*]
: a concubine of Abraham driven into the desert with her son Ishmael because of Sarah's jealousy according to the account in Genesis

hag·fish \ˈhag-ˌfish\ *noun* (1611)
: any of a family (Myxinidae) of marine cyclostomes that are related to the lampreys and in general resemble eels but have a round mouth surrounded by barbels and that feed upon other fishes and invertebrates by boring into their bodies

Hag·ga·dah \hə-ˈgä-də, hä-, -ˈgȯ-\ *noun, plural* **Hag·ga·doth** \-ˈgä-ˌdōt, -ˈgȯ-, -ˌdōth\ [Hebrew *haggādhāh*] (1856)
1 : ancient Jewish lore forming especially the nonlegal part of the Talmud
2 : the prayer book containing the seder ritual
— hag·ga·dic \-ˈgä-dik, -ˈgȯ-\ *adjective, often capitalized*

hag·ga·dist \-ˈgä-dist, -ˈgȯ-\ *noun, often capitalized* (1886)
1 : a haggadic writer
2 : a student of the Haggadah
— hag·ga·dis·tic \ˌha-gə-ˈdis-tik, ˌhä-\ *adjective, often capitalized*

Hag·gai \ˈha-gē-ˌī, ˈha-ˌgī\ *noun* [Hebrew *Ḥaggai*]
1 : a Hebrew prophet who flourished about 500 B.C. and who advocated that the Temple in Jerusalem be rebuilt
2 : a prophetic book of canonical Jewish and Christian Scriptures — see BIBLE table

1hag·gard \ˈha-gərd\ *adjective* [Middle French *hagard*] (1567)
1 *of a hawk* **:** not tamed
2 a : wild in appearance **b :** having a worn or emaciated appearance **:** GAUNT ⟨*haggard* faces looked up sadly from out of the straw —W. M. Thackeray⟩
— hag·gard·ly *adverb*
— hag·gard·ness *noun*

2haggard *noun* (1567)
1 : an adult hawk caught wild
2 *obsolete* **:** an intractable person

hag·gis \ˈha-gəs\ *noun* [Middle English *hageyse*] (15th century)
: a traditionally Scottish dish that consists of the heart, liver, and lungs of a sheep or a calf minced with suet, onions, oatmeal, and seasonings boiled in the stomach of the animal

1hag·gle \ˈha-gəl\ *verb* **hag·gled; hag·gling** \-g(ə-)liŋ\ [frequentative of *hag* (to hew)] (1599)
transitive verb
1 : to cut roughly or clumsily **:** HACK
2 *archaic* **:** to annoy or exhaust with wrangling
intransitive verb
: BARGAIN, WRANGLE
— hag·gler \-g(ə-)lər\ *noun*

2haggle *noun* (1858)
: an act or instance of haggling

hagi- *or* **hagio-** *combining form* [Late Latin, from Greek, from *hagios*]
1 : holy ⟨*hagio*scope⟩
2 : saints ⟨*hagio*graphy⟩

Ha·gi·og·ra·pha \ˌha-gē-ˈä-grə-fə, ˌhā-, -jē-\ *noun plural but singular or plural in construction* [Late Latin, from Late Greek, from *hagio-* + *graphein* to write — more at CARVE] (1583)
: the third part of the Jewish scriptures — see BIBLE table

ha·gi·og·ra·pher \-fər\ *noun* (1849)
: a writer of hagiography

ha·gio·graph·ic \ˌha-gē-ə-ˈgra-fik, ˌhā-, -jē-\ *also* **ha·gio·graph·i·cal** \-fi-kəl\ *adjective* (1819)
1 : of or relating to hagiography
2 : of or relating to the Hagiographa

ha·gi·og·ra·phy \-gē-ˈä-grə-fē, -jē-\ *noun* (1821)
1 : biography of saints or venerated persons
2 : idealizing or idolizing biography

ha·gi·ol·o·gy \-gē-ˈä-lə-jē, -jē-\ *noun* (1807)
1 : literature dealing with venerated persons or writings
2 : a list of venerated figures
— ha·gi·o·log·ic \-gē-ə-ˈlä-jik, -jē-\ *or* **ha·gi·o·log·i·cal** \-ji-kəl\ *adjective*

ha·gio·scope \ˈha-gē-ə-ˌskōp, ˈhā-jē-\ *noun* (circa 1840)
: an opening in the interior walls of a cruciform church so placed as to afford a view of the altar to those in the transept
— ha·gio·scop·ic \ˌha-gē-ə-ˈskä-pik, ˌhā-, -jē-\ *adjective*

hag·ride \ˈhag-ˌrīd\ *transitive verb* **-rode** \-ˌrōd\; **-rid·den** \-ˌri-dᵊn\ (1702)
: HARASS, TORMENT

hah *variant of* HA

1ha–ha \(ˈ)hä-ˈhä\ *interjection* [Middle English, from Old English *ha ha*] (before 12th century)
— used to express amusement or derision

2ha–ha \ˈhä-ˌhä\ *noun* [French *haha*] (1749)
: SUNK FENCE

hahn·ium \ˈhä-nē-əm\ *noun* [New Latin, from Otto *Hahn*] (1970)
: UNNILPENTIUM

Hai·da \ˈhī-də\ *noun, plural* **Haida** *or* **Haidas** (1841)
1 : a member of an American Indian people of the Queen Charlotte islands, British Columbia, and Prince of Wales island, Alaska
2 : the language of the Haida people

haik \ˈhīk\ *noun* [Arabic *ḥā'ik*] (1713)
: a voluminous piece of usually white cloth worn as an outer garment in northern Africa

haik

hai·ku \ˈhī-(ˌ)kü\ *noun, plural* **haiku** [Japanese] (1902)
: an unrhymed verse form of Japanese origin having three lines containing usually 5, 7, and 5 syllables respectively; *also* **:** a poem in this form usually having a seasonal reference — compare TANKA

1hail \ˈhā(ə)l\ *noun* [Middle English, from Old English *hægl;* akin to Old High German *hagal* hail] (before 12th century)
1 : precipitation in the form of small balls or lumps usually consisting of concentric layers of clear ice and compact snow
2 : something that gives the effect of a shower of hail ⟨a *hail* of rifle fire⟩

2hail *intransitive verb* (before 12th century)
1 : to precipitate hail ⟨it was *hailing* hard⟩
2 : to pour down or strike like hail

3hail *interjection* [Middle English, from Old Norse *heill,* from *heill* healthy — more at WHOLE] (13th century)
1 *archaic* — used as a salutation
2 — used to express acclamation ⟨*hail* to the chief —Sir Walter Scott⟩

4hail (13th century)
transitive verb
1 a : SALUTE, GREET **b :** to greet with enthusiastic approval **:** ACCLAIM
2 : to greet or summon by calling ⟨*hail* a taxi⟩
intransitive verb
: to call out; *especially* **:** to call a greeting to a passing ship
— hail·er \hā-lər\ *noun*
— hail from : to be or have been native to or a resident of

5hail *noun* (1500)
1 : an exclamation of greeting or acclamation
2 : a calling to attract attention
3 : hearing distance ⟨stayed within *hail*⟩

hail–fel·low \ˈhāl-ˌfe-(ˌ)lō\ *adjective* (1580)
: HAIL-FELLOW-WELL-MET
— hail–fellow *noun*

hail–fel·low–well–met \-lō-ˌwel-ˈmet, -lə-ˌwel-\ *adjective* [from the archaic salutation "Hail, fellow! Well met!"] (1581)
: heartily informal **:** COMRADELY
— hail–fellow–well–met *noun*

Hail Mary *noun* [translation of Medieval Latin *Ave, Maria,* from the opening words] (15th century)
: a Roman Catholic prayer to the Virgin Mary that consists of salutations and a plea for her intercession

hail·stone \ˈhā(ə)l-ˌstōn\ *noun* (before 12th century)
: a pellet of hail

hail·storm \-ˌstȯrm\ *noun* (15th century)
: a storm accompanied by hail

hair \ˈhar, ˈher\ *noun, often attributive* [Middle English, from Old English *hǣr;* akin to Old High German *hār* hair] (before 12th century)
1 a : a slender threadlike outgrowth of the epidermis of an animal; *especially* **:** one of the usually pigmented filaments that form the characteristic coat of a mammal **b :** the hairy covering of

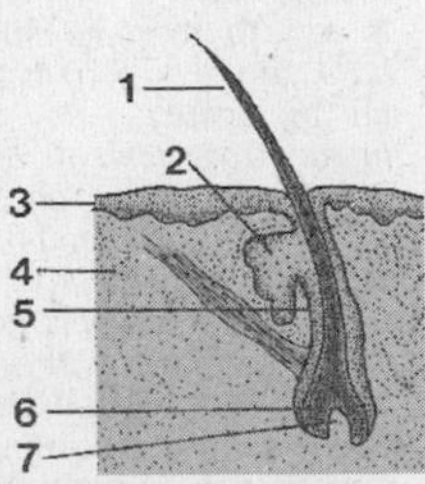

hair 1a: *1* **shaft,** *2* **sebaceous gland,** *3* **epidermis,** *4* **dermis,** *5* **hair follicle,** *6* **bulb,** *7* **papilla**

an animal or a body part; *especially* **:** the coating of hairs on a human head
2 : HAIRCLOTH
3 a : a minute distance or amount ⟨won by a *hair*⟩ **b :** a precise degree ⟨aligned to a *hair*⟩
4 *obsolete* **:** NATURE, CHARACTER
5 : a filamentous structure that resembles hair ⟨leaf *hair*⟩
— hair·less *adjective*
— hair·less·ness *noun*
— hair·like \-ˌlīk\ *adjective*

hair ball *noun* (1712)
: a compact mass of hair formed in the stomach especially of a shedding animal (as a cat) that cleanses its coat by licking

¹hair·breadth \ˈhar-ˌbretth, ˈher-, -ˌbreth, -ˌbredth\ *or* **hairs·breadth** \ˈharz-, ˈherz-\ *noun* (1561)
: a very small distance or margin

²hairbreadth *adjective* (1604)
: very narrow **:** CLOSE ⟨a *hairbreadth* escape⟩

hair·brush \ˈhar-ˌbrəsh, ˈher-\ *noun* (1599)
: a brush for the hair

hair cell *noun* (circa 1890)
: a cell with hairlike processes; *especially* **:** one of the sensory cells in the auditory epithelium of the organ of Corti

hair·cloth \ˈhar-ˌklȯth, ˈher-\ *noun* (1500)
: any of various stiff wiry fabrics especially of horsehair or camel hair used for upholstery or for stiffening in garments

hair·cut \-ˌkət\ *noun* (1899)
: the act, process, or result of cutting and shaping the hair
— hair·cut·ter \-ˌkə-tər\ *noun*
— hair·cut·ting \-ˌkə-tiŋ\ *noun*

hair·do \-ˌdü\ *noun, plural* **hairdos** (1932)
: a way of wearing the hair **:** COIFFURE

hair·dress·er \-ˌdre-sər\ *noun* (1770)
1 : one whose occupation is the dressing or cutting of hair
2 *British* **:** BARBER

hair·dress·ing \-ˌdre-siŋ\ *noun* (1771)
1 a : the action or process of washing, cutting, curling, or arranging the hair **b :** the occupation of a hairdresser
2 : a preparation for grooming and styling the hair

haired \ˈhard, ˈherd\ *adjective* (14th century)
: having hair especially of a specified kind — usually used in combination ⟨dark-*haired*⟩

hair follicle *noun* (1838)
: the tubular epithelial sheath that surrounds the lower part of the hair shaft and encloses at the bottom a vascular papilla supplying the growing basal part of the hair with nourishment

hair·line \-ˈlīn\ *noun* (1846)
1 : a very slender line: as **a :** a tiny line or crack on a surface ⟨a *hairline* bone fracture⟩ **b :** a fine line connecting thicker strokes in a printed letter
2 : HAIRBREADTH
3 a : a textile design consisting of lengthwise or crosswise lines usually one thread wide **b :** a fabric with such a design
4 a : the outline of scalp hair especially on the forehead **b :** the way the hair frames the face
— hairline *adjective*

hair·piece \-ˌpēs\ *noun* (1926)
1 : supplementary hair (as a switch) used in some feminine coiffures
2 : TOUPEE 2

¹hair·pin \-ˌpin\ *noun* (1779)
1 : a pin to hold the hair in place; *specifically* **:** a long U-shaped pin
2 : something shaped like a hairpin; *specifically* **:** a sharp U-shaped turn in a road

²hairpin *adjective* (1887)
: having the shape of a hairpin ⟨a *hairpin* turn⟩; *also* **:** having hairpin turns ⟨a steep *hairpin* road⟩

hair–rais·er \ˈhar-ˌrā-zər, ˈher-\ *noun* (1897)
: THRILLER

hair–rais·ing \-ˌrā-ziŋ\ *adjective* (1900)
: causing terror, excitement, or astonishment
— hair–rais·ing·ly \-ziŋ-lē\ *adverb*

hair seal *noun* (1865)
: any of a family (Phocidae) of seals having a coarse hairy coat, the hind limbs reduced to swimming flippers, and no external ears — called also *true seal*; compare EARED SEAL

hair shirt *noun* (14th century)
1 : a shirt made of rough animal hair worn next to the skin as a penance
2 : one that irritates like a hair shirt

hair·split·ter \ˈhar-ˌspli-tər, ˈher-\ *noun* (1849)
: one that makes excessively fine distinctions in reasoning **:** QUIBBLER
— hair·split·ting \-ˌspli-tiŋ\ *adjective or noun*

hair·spring \-ˌspriŋ\ *noun* (1830)
: a slender spiraled recoil spring that regulates the motion of the balance wheel of a timepiece

hair·streak \-ˌstrēk\ *noun* (1816)
: any of a subfamily (Theclinae of the family Lycaenidae) of small butterflies usually having striped markings on the underside of the wings and thin filamentous projections from the hind wings

hair·style \ˈhar-ˌstīl, ˈher-\ *noun* (1913)
: HAIRDO

hair·styl·ing \-ˌstī-liŋ\ *noun* (1936)
: the work of a hairstylist

hair·styl·ist \-ˌstī-list\ *noun* (1935)
: HAIRDRESSER; *especially* **:** one who does creative styling of coiffures

hair–trigger *adjective* (1834)
1 : immediately responsive to the slightest stimulus ⟨a *hair-trigger* temper⟩
2 : delicately adjusted or easily disrupted

hair trigger *noun* (1806)
: a gun trigger so adjusted as to permit the firearm to be fired by a very slight pressure

hair·worm \ˈhar-ˌwərm, ˈher-\ *noun* (1658)
1 : any of a phylum (Nematomorpha) of elongated worms that have separate sexes, are parasitic in arthropods as larvae, and are free-living in water as adults — called also *horsehair worm*
2 : any of a genus (*Capillaria*) of nematode worms that include serious parasites of the alimentary tract of fowls and tissue and organ parasites of mammals

hairy \ˈhar-ē, ˈher-\ *adjective* **hair·i·er; -est** (14th century)
1 a : covered with hair or hairlike material **b :** having a downy fuzz on the stems and leaves
2 : made of or resembling hair
3 a : tending to cause nervous tension (as from danger) ⟨a *hairy* adventure⟩ **b :** difficult to deal with or comprehend ⟨a *hairy* math problem⟩
— hair·i·ness \ˈhar-ē-nəs, ˈher-\ *noun*

hairy cell leukemia *noun* (1979)
: a lymphocytic leukemia that is usually of B cell origin and is characterized by malignant cells with a ciliated appearance

hairy–chest·ed \-ˌches-təd\ *adjective* (circa 1937)
: characterized by especially exaggerated or stereotypical manliness

hairy vetch *noun* (1901)
: a Eurasian vetch (*Vicia villosa*) extensively cultivated as a cover and early forage crop

hairy woodpecker *noun* (circa 1728)
: a common North American woodpecker (*Picoides villosus*) closely resembling the downy woodpecker but larger with a longer bill

Hai·tian \ˈhā-shən *also* ˈhā-tē-ən\ *noun* (1805)
1 : a native or inhabitant of Haiti
2 : HAITIAN CREOLE
— Haitian *adjective*

Haitian Creole *noun* (circa 1938)
: a French-based creole spoken by Haitians

hajj \ˈhaj\ *noun* [Arabic *ḥajj*] (1673)
: the pilgrimage to Mecca prescribed as a religious duty for Muslims

hajji \ˈha-jē\ *noun* [Arabic *ḥajjī*, from *ḥajj*] (1609)
: one who has made a pilgrimage to Mecca — often used as a title

hake \ˈhāk\ *noun* [Middle English] (14th century)
: any of several marine food fishes (as of the genera *Merluccius* and *Urophycis*) that are related to the common Atlantic cod

¹ha·kim \ˈhä-kəm\ *noun* [Arabic *ḥākim*] (1611)
: an administrator in a Muslim country

²ha·kim \hə-ˈkēm\ *noun* [Arabic *ḥakīm*, literally, wise one] (1638)
: a physician in a Muslim country

hal- *or* **halo-** *combining form* [French, from Greek, from *hals* — more at SALT]
1 : salt ⟨*halo*phyte⟩
2 [International Scientific Vocabulary, from *halogen*] **:** halogen ⟨*hal*ide⟩

ha·la·kah *or* **ha·la·cha** \hä-ˈlä-k̲ə, ˌhä-lə-ˈk̲ä\ *noun, often capitalized* [Hebrew *hălākhāh*, literally, way] (1856)
: the body of Jewish law supplementing the scriptural law and forming especially the legal part of the Talmud
— ha·lak·ic \hə-ˈla-kik, hä-ˈlä-\ *adjective, often capitalized*

ha·la·la \hə-ˈlä-lə\ *noun, plural* **halala** *or* **halalas** [Arabic] (1970)
— see *riyal* at MONEY table

ha·la·tion \hā-ˈlā-shən\ *noun* [*halo* + *-ation*] (1859)
1 : the spreading of light beyond its proper boundaries in a developed photographic image
2 : a bright ring that sometimes surrounds a bright object on a television screen

hal·berd \ˈhal-bərd, ˈhȯl-\ *also* **hal·bert** \-bərt\ *noun* [Middle English, from Middle French *halleburde*, from Middle High German *helmbarte*, from *helm* handle (from Old High German *helmo*) + *barte* ax, from Old High German *barta*; akin to Old High German *bart* beard — more at HELM, BEARD] (15th century)
: a weapon especially of the 15th and 16th centuries consisting typically of a battle-ax and pike mounted on a handle about six feet long

¹hal·cy·on \ˈhal-sē-ən\ *noun* [Middle English *alceon*, from Latin *halcyon*, from Greek *alkyōn*, *halkyōn*] (14th century)
1 : a bird identified with the kingfisher and held in ancient legend to nest at sea about the time of the winter solstice and to calm the waves during incubation
2 : KINGFISHER ◆

◇ WORD HISTORY

halcyon According to Greek myth, Alkyone, the daughter of Aeolus, god of the winds, was so distraught on learning that her husband had been killed in a shipwreck that she threw herself into the sea and was changed into a kingfisher. Henceforth the bird was known to the Greeks as *alkyōn* or *halkyōn* (the latter by association with *hals*, meaning "sea"). Legend had it that the kingfisher built a floating nest on the sea every year before the winter solstice. For two weeks before and after the solstice Aeolus charmed the wind and the waves while the bird incubated its eggs, and *alkyonides hemerai* (literally, "kingfisher days") became a proverbial expression for a time of peace and tranquility. The Romans borrowed the Greek word as *alcyon* or *halcyon*, and the Latin word found its way into Middle English as a poetic name for the kingfisher. Subsequently, *halcyon* has been used as an adjective meaning first "calm, tranquil" and later "happy" or "prosperous." It usually modifies *days* or another noun denoting a period of time.

²halcyon *adjective* (1540)
1 : of or relating to the halcyon or its nesting period
2 a : CALM, PEACEFUL **b :** HAPPY, GOLDEN **c :** PROSPEROUS, AFFLUENT

¹hale \'hā(ə)l\ *adjective* [partly from Middle English (northern) *hale,* from Old English *hāl;* partly from Middle English *hail,* from Old Norse *heill* — more at WHOLE] (before 12th century)
: free from defect, disease, or infirmity **:** SOUND; *also* **:** retaining exceptional health and vigor ⟨a *hale* and hearty old man⟩
synonym see HEALTHY

²hale *transitive verb* **haled; hal·ing** [Middle English *halen,* from Old French *haler* — more at HAUL] (13th century)
1 : HAUL, PULL
2 : to compel to go

ha·ler \'hä-lər, -,ler\ *noun, plural* **ha·le·ru** \'hä-lə-,rü\ [Czech *haléř,* genitive plural *haléřů*] (circa 1934)
— see *koruna* at MONEY table

¹half \'haf, 'håf\ *noun, plural* **halves** \'havz, 'håvz\ [Middle English, from Old English *healf;* akin to Old High German *halb* half] (before 12th century)
1 a : either of two equal parts into which a thing is divisible; *also* **:** a part of a thing approximately equal to the remainder — often used without *of* ⟨*half* the distance⟩ **b :** half an hour — used in designation of time
2 : one of a pair: as **a :** PARTNER **b :** SEMESTER, TERM **c :** either of the two equal periods that together make up the playing time of some games (as football); *also* **:** the midpoint in playing time ⟨the score was tied at the *half*⟩
3 : HALF-DOLLAR
4 : HALFBACK
— by half : by a great deal
— by halves : in part **:** HALFHEARTEDLY
— half again as : one-and-a-half times as ⟨*half again as* many⟩
— in half : into two equal or nearly equal parts

²half *adjective* (before 12th century)
1 a : being one of two equal parts ⟨a *half* share⟩ ⟨a *half* sheet of paper⟩ **b** (1) **:** amounting to approximately half ⟨a *half* mile⟩ ⟨a *half* million⟩ (2) **:** falling short of the full or complete thing **:** PARTIAL ⟨*half* measures⟩ ⟨a *half* smile⟩
2 : extending over or covering only half ⟨a *half* window⟩ ⟨a *half* mask⟩
— half·ness *noun*

³half *adverb* (12th century)
1 a : in an equal part or degree ⟨the crowd was *half* jeering, *half* respectful⟩ **b :** not completely **:** PARTIALLY ⟨*half* persuaded⟩
2 : by any means **:** AT ALL ⟨her singing isn't *half* bad⟩

half–and–half \,haf-ᵊn-'haf, ,håf-ᵊn-'håf\ *noun* (1756)
: something that is approximately half one thing and half another: as **a :** a mixture of two malt beverages (as dark and light beer) **b :** a mixture of cream and whole milk
— half–and–half *adjective or adverb*

half–assed \'haf-'ast, 'håf-'åst\ *adjective* (circa 1932)
1 : lacking significance, adequacy, or completeness — often considered vulgar
2 : lacking intelligence, character, or effectiveness — often considered vulgar
— half–assed *adverb*

half·back \'haf-,bak, 'håf-\ *noun* (1882)
1 : one of the backs stationed near either flank in football
2 : a player stationed immediately behind the forward line (as in field hockey, soccer, or rugby)

half–baked \-'bākt\ *adjective* (1621)
1 a : lacking adequate planning or forethought ⟨a *half-baked* scheme for getting rich⟩ **b :** lacking in judgment, intelligence, or common sense
2 : imperfectly baked **:** UNDERDONE

half·beak \-,bēk\ *noun* (1880)
: any of various narrow-bodied fishes of warm waters that have an elongated lower jaw and are grouped with the flying fishes (family Exocoetidae) or placed in their own family (Hemiramphidae)

half–blood \-'bləd\ *or* **half–blood·ed** \-'blə-dəd\ *adjective* (1605)
: having half blood or being a half blood

half blood *noun* (1553)
1 a : the relation between persons having only one parent in common **b :** a person so related to another
2 : HALF-BREED
3 : GRADE 4

half boot *noun* (1787)
: a boot with a top reaching above the ankle and ending below the knee

half–bound \'haf-,baůnd, 'håf-\ *adjective* (1775)
of a book **:** bound in material of two qualities with the material of better quality on the spine and corners
— half binding *noun*

half–bred \-,bred\ *adjective* (1701)
: having one purebred parent
— half–bred *noun*

half–breed \-,brēd\ *noun* (1760)
: the offspring of parents of different races; *especially* **:** the offspring of an American Indian and a white person — often used disparagingly
— half–breed *adjective*

half brother *noun* (14th century)
: a brother related through one parent only

half–caste \'haf-,kast, 'håf-\ *noun* (1789)
: one of mixed racial descent **:** HALF-BREED
— half–caste *adjective*

half cock *noun* (1745)
1 : the position of the hammer of a firearm when about half retracted and held by the sear so that it cannot be operated by a pull on the trigger
2 *chiefly British* **:** a state of inadequate preparation or mental confusion ⟨go off at *half cock*⟩

half–cocked \'haf-'käkt, 'håf-\ *adjective* (1809)
1 : being at half cock
2 : lacking adequate preparation or forethought ⟨go off *half-cocked*⟩

half–court \'haf-'kōrt, -'kȯrt, 'håf-\ *noun* (1888)
: a dividing line that separates a playing court into equal halves (as in basketball); *also* **:** the area comprising each half

half crown *noun* (1542)
: a British coin worth two shillings and sixpence used as legal tender until 1970

half dime *noun* (1792)
: a silver 5-cent coin struck by the U.S. mint in 1792 and from 1794 to 1873

half disme *noun* (1792)
: a half dime struck in 1792

half–dol·lar \'haf-'dä-lər, 'håf-\ *noun* (1786)
1 : a coin representing one half of a dollar
2 : the sum of 50 cents

half duplex *noun* (1950)
: a mode of communication especially with a computer via telephone line in which information can be sent in only one direction at a time — compare DUPLEX

half eagle *noun* (1786)
: a 5-dollar gold piece issued by the U.S. from 1795 to 1916 and in 1929

half–hardy *adjective* (1824)
of a plant **:** able to withstand a moderately low temperature but injured by severe freezing and surviving the winter in cold climates only if carefully protected

half·heart·ed \'haf-'här-təd, 'håf-\ *adjective* (15th century)
: lacking heart, spirit, or interest
— half·heart·ed·ly *adverb*
— half·heart·ed·ness *noun*

half hitch *noun* (1769)
: a simple knot tied by passing the end of a line around an object, across the main part of the line, and then through the resulting loop — see KNOT illustration

half hour *noun* (15th century)
1 : thirty minutes
2 : the middle point of an hour
— half–hour·ly \'haf-'aů(-ə)r-lē, 'håf-\ *adverb or adjective*

half–knot \'haf-,nät, 'håf-\ *noun* (1933)
: a knot intertwining the ends of two cords and used in tying other knots

half–length \'haf-'leŋ(k)th, 'håf-\ *noun* (1699)
: something (as a portrait) that is or represents only half the complete length

half–life \-,līf\ *noun* (1907)
1 : the time required for half of something to undergo a process: as **a :** the time required for half of the atoms of a radioactive substance to become disintegrated **b :** the time required for half the amount of a substance (as a drug or radioactive tracer) in or introduced into a living system or ecosystem to be eliminated or disintegrated by natural processes
2 : a period of usefulness or popularity preceding decline or obsolescence ⟨slang usually has a short *half-life*⟩

half–light \-,līt\ *noun* (1625)
: dim grayish light

half line *noun* (circa 1914)
: a straight line extending from a point indefinitely in one direction only

¹half–mast \-'mast\ *noun* (1627)
: a point some distance but not necessarily halfway down below the top of a mast or staff or the peak of a gaff

²half–mast *transitive verb* (1891)
: to cause to hang at half-mast ⟨*half-mast* a flag⟩

half–moon \'haf-,mün, 'håf-\ *noun* (15th century)
1 : the moon when half its disk appears illuminated
2 : something shaped like a crescent
3 : the lunule of a fingernail
— half–moon *adjective*

half nelson *noun* (1889)
: a wrestling hold in which one arm is thrust under the corresponding arm of an opponent and the hand placed on the back of the opponent's neck — compare FULL NELSON

half note *noun* (1597)
: a musical note with the time value of ½ of a whole note — see NOTE illustration

half·pen·ny \'hāp-nē, 'hā-pə-, *US also* 'haf-,pe-nē, 'håf-\ *noun* (13th century)
1 *plural* **halfpence** \'hā-pən(t)s, *US also* 'haf-,pen(t)s, 'håf-\ *or* **halfpennies :** a former British coin representing one half of a penny
2 : the sum of half a penny
3 : a small amount
— halfpenny *adjective*

¹half–pint \'haf-,pīnt, 'håf-\ *noun* (1611)
1 : half a pint
2 : a short, small, or inconsequential person

²half–pint *adjective* (circa 1926)
: of less than average size **:** DIMINUTIVE

half plane *noun* (1891)
: the part of a plane on one side of an indefinitely extended straight line drawn in the plane

half rest *noun* (circa 1899)
: a musical rest corresponding in time value to a half note — see REST illustration

half shell *noun* (1860)
: either of the valves of a bivalve
— on the half shell : served in a half shell ⟨oysters *on the half shell*⟩

half sister *noun* (13th century)
: a sister related through one parent only

half–slip \'haf-,slip, 'håf-\ *noun* (circa 1948)
: a topless slip with an elasticized waistband

half–sole *transitive verb* (1795)
: to put half soles on

half sole *noun* (1865)
: a shoe sole extending from the shank forward

half sovereign *noun* (circa 1504)
: a British gold coin worth 10 shillings

half–space \'haf-,spās, 'håf-\ *noun* (1962)
: the part of three-dimensional euclidean space lying on one side of a plane

half–staff \-'staf\ *noun* (1708)
: HALF-MAST

half step *noun* (1904)
1 : a walking step of 15 inches or in double time of 18 inches
2 : a musical interval (as E-F or B-C) equivalent to one twelfth of an octave — called also *semitone*

half–timber *or* **half–tim•bered** \'haf-'tim-bərd, 'håf-\ *adjective* (circa 1876)
of a building **:** constructed of wood framing with spaces filled with masonry
— **half–tim•ber•ing** \-b(ə-)riŋ\ *noun*

half•time \-,tīm\ *noun* (1871)
: an intermission between halves of a game or contest (as in football or basketball)

half–time *adjective* (1961)
: involving or working half the standard hours
— **half–time** *adverb*

half title *noun* (1879)
: the title of a book appearing alone on a right-hand page immediately preceding the title page; *also* **:** the page itself

half•tone \'haf-,tōn, 'håf-\ *noun* (1651)
1 : HALF STEP 2
2 a : any of the shades of gray between the darkest and the lightest parts of a photographic image **b :** a photoengraving made from an image photographed through a screen and then etched so that the details of the image are reproduced in dots
— **halftone** *adjective*

half–track \-,trak\ *noun* (1935)
1 : an endless chain-track drive system that propels a vehicle supported in front by a pair of wheels
2 : a motor vehicle propelled by half-tracks; *specifically* **:** one lightly armored for military use
— **half–track** *or* **half–tracked** \-,trakt\ *adjective*

half–truth \-,trüth\ *noun* (1658)
1 : a statement that is only partially true
2 : a statement that mingles truth and falsehood with deliberate intent to deceive

half volley *noun* (1843)
: a stroke of a ball (as in tennis) at the instant it rebounds from the ground
— **half–volley** *verb*

half•way \'haf-'wā, 'håf-\ *adjective* (1694)
1 : midway between two points
2 : PARTIAL
— **halfway** *adverb*

halfway house *noun* (1694)
1 a : a place to stop midway on a journey **b :** a halfway place in a progression
2 : a residence for formerly institutionalized individuals (as mental patients, drug addicts, or convicts) that is designed to facilitate their readjustment to private life

half–wit \'haf-,wit, 'håf-\ *noun* (1640)
: a foolish or imbecilic person
— **half–wit•ted** \-'wi təd\ *adjective*
half–wit•ted•ness *noun*

half–world \-,wərld\ *noun* (1870)
: DEMIMONDE

hal•i•but \'ha-lə-bət *also* 'hä-\ *noun, plural* **halibut** *also* **halibuts** [Middle English *halybutte,* from *haly, holy* holy + *butte* flatfish, from Middle Dutch or Middle Low German *but;* from its being eaten on holy days] (14th century)
: any of several marine flatfishes (especially *Hippoglossus hippoglossus* of the Atlantic and *H. stenolepis* of the Pacific) that are widely used for food and include some of the largest bony fishes

ha•lide \'ha-,līd, 'hā-\ *noun* (1876)
: a binary compound of a halogen with a more electropositive element or radical

hal•i•dom \'ha-lə-dəm\ *or* **hal•i•dome** \-lə-,dōm\ *noun* [Middle English, from Old English *hāligdōm,* from *hālig* holy + *-dōm* -dom] (before 12th century)
archaic **:** something held sacred

ha•lite \'ha-,līt, 'hā-\ *noun* (1868)
: ROCK SALT

hal•i•to•sis \,ha-lə-'tō-səs\ *noun* [New Latin, from Latin *halitus* breath, from *halare* to breathe — more at EXHALE] (1874)
: a condition of having fetid breath

hall \'hȯl\ *noun* [Middle English *halle,* from Old English *heall;* akin to Latin *cella* small room, *celare* to conceal — more at HELL] (before 12th century)
1 a : the castle or house of a medieval king or noble **b :** the chief living room in such a structure
2 : the manor house of a landed proprietor
3 : a large usually imposing building for public or semipublic purposes
4 a (1) **:** a building used by a college or university for some special purpose (2) **:** DORMITORY **b :** a college or a division of a college at some universities **c** (1) **:** the common dining room of an English college (2) **:** a meal served there
5 a : the entrance room of a building **:** LOBBY **b :** a corridor or passage in a building
6 : a large room for assembly **:** AUDITORIUM
7 : a place used for public entertainment

hal•lah \'kä-lə, 'hä-\ *variant of* CHALLAH

Hall effect \'hȯl-\ *noun* [Edwin H. *Hall* (died 1938) American physicist] (circa 1889)
: a potential difference observed between the edges of a conducting strip carrying a longitudinal current when placed in a magnetic field perpendicular to the plane of the strip

Hal•lel \hä-'lā(ə)l\ *noun* [Hebrew *hallēl* praise] (1702)
: a selection comprising Psalms 113–118 chanted during Jewish feasts (as the Passover)

¹hal•le•lu•jah \,ha-lə-'lü-yə\ *interjection* [Hebrew *hallĕlūyāh* praise (ye) the Lord] (1535)
— used to express praise, joy, or thanks

²hallelujah *noun* (1625)
: a shout or song of praise or thanksgiving

hal•liard *variant of* HALYARD

¹hall•mark \'hȯl-,märk\ *noun* [Goldsmiths' *Hall,* London, England, where gold and silver articles were assayed and stamped] (1721)
1 a : an official mark stamped on gold and silver articles in England to attest their purity **b :** a mark or device placed or stamped on an article of trade to indicate origin, purity, or genuineness
2 : a distinguishing characteristic, trait, or feature ⟨the dramatic flourishes which are the *hallmark* of the trial lawyer —Marion K. Sanders⟩

²hallmark *transitive verb* (1773)
: to stamp with a hallmark

hal•lo \hə-'lō, ha-\ *or* **hal•loo** \-'lü\ *variant of* HOLLO

Hall of Fame (circa 1909)
1 : a structure housing memorials to famous or illustrious individuals usually chosen by a group of electors
2 : a group of individuals in a particular category (as a sport) who have been selected as particularly illustrious
— **Hall of Fam•er** \-'fā-mər\

hal•low \'ha-(,)lō\ *transitive verb* [Middle English *halowen,* from Old English *hālgian,* from *hālig* holy — more at HOLY] (before 12th century)
1 : to make holy or set apart for holy use
2 : to respect greatly **:** VENERATE
synonym see DEVOTE

hal•lowed \'ha-(,)lōd, 'ha-ləd, *in the Lord's Prayer often* 'ha-lə-wəd\ *adjective* (before 12th century)
1 : HOLY, CONSECRATED
2 : SACRED, REVERED ⟨*hallowed* customs⟩

Hal•low•een *also* **Hal•low•e'en** \,ha-lə-'wēn, ,hä-\ *noun* [short for *All Hallow Even* (All Saints' Eve)] (circa 1700)
: October 31 observed especially with dressing up in disguise, trick-or-treating, and displaying jack-o'-lanterns during the evening ■

Hal•low•mas \'ha-lō-,mas, 'ha-lə-, -məs\ *noun* [short for Middle English *Alholowmesse,* from Old English *ealra halgena mæsse,* literally, all saints' mass] (14th century)
: ALL SAINTS' DAY

halls of ivy [from the traditional training of ivy on the walls of older college buildings] (1965)
: UNIVERSITY, COLLEGE

Hall•statt *also* **Hall•stadt** \'hȯl-,stat; 'häl-,shtät, -,stät\ *adjective* [*Hallstatt,* Austria] (1899)
: of or relating to the earlier period of the Iron Age in Europe

hal•lu•ci•nate \hə-'lü-sᵊn-,āt\ *verb* **-nat•ed; -nat•ing** [Latin *hallucinatus,* past participle of *hallucinari, allucinari* to prate, dream, modification of Greek *alyein* to be distressed, to wander] (circa 1834)
transitive verb
1 : to affect with visions or imaginary perceptions
2 : to perceive or experience as an hallucination
intransitive verb
: to have hallucinations
— **hal•lu•ci•na•tor** \-,ā-tər\ *noun*

hal•lu•ci•na•tion \hə-,lü-sᵊn-'ā-shən\ *noun* (1629)
1 a : perception of objects with no reality usually arising from disorder of the nervous system or in response to drugs (as LSD) **b :** the object so perceived
2 : an unfounded or mistaken impression or notion **:** DELUSION

hal•lu•ci•na•to•ry \hə-'lü-sᵊn-ə-,tōr-ē, -'lüs-nə-, -,tȯr-\ *adjective* (1830)
1 : tending to produce hallucination ⟨*hallucinatory* drugs⟩
2 : resembling, involving, or being an hallucination ⟨*hallucinatory* dreams⟩ ⟨an *hallucinatory* figure⟩

hal•lu•ci•no•gen \hə-'lü-sᵊn-ə-jən\ *noun* [*hallucination* + *-o-* + *-gen*] (1954)
: a substance that induces hallucinations
— **hal•lu•ci•no•gen•ic** \-,lü-sᵊn-ə-'je-nik\ *adjective or noun*

hal•lu•ci•no•sis \hə-,lü-sᵊn-'ō-səs\ *noun* [New Latin] (1905)
: a pathological mental state characterized by hallucinations

hal•lux \'ha-ləks\ *noun, plural* **hal•lu•ces** \'ha-lə-,sēz, 'hal-yə-\ [New Latin, from Latin *hallus, hallux*] (1831)

□ USAGE
Halloween Nothing sends chills down the spine of some English speakers like the pronunciation \,hä-lə-'wēn\. The first half of the word, it is argued, comes from the verb *hallow,* whose first syllable is pronounced \'ha\. In earlier centuries, though, *hallow* had a second pronunciation rhyming with *swallow,* and so \,hä-lə-'wēn\ is a natural pronunciation of the name of the October holiday. The disputed variant is common throughout the United States but not in Great Britain. Those who insist that the pronunciation of *Halloween* should be the sum of its parts should consider their own pronunciations of the first vowels of *Christmas* and *Michaelmas.*

: the innermost digit (as the big toe) of a hind or lower limb

hall·way \'hȯl-ˌwā\ *noun* (1876)
1 : an entrance hall
2 : CORRIDOR

hal·ma \'hal-mə\ *noun* [Greek *halma* leap, from *hallesthai* to leap — more at SALLY] (1889)
: a game played on a square board and having rules similar to those of Chinese checkers

¹ha·lo \'hā-(ˌ)lō\ *noun, plural* **halos** *or* **haloes** [Latin *halos,* from Greek *halōs* threshing floor, disk, halo] (1603)
1 : a circle of light appearing to surround the sun or moon and resulting from refraction or reflection of light by ice particles in the atmosphere
2 : something resembling a halo: as **a :** NIMBUS **b :** a region of space surrounding a galaxy that is sparsely populated with luminous objects (as globular clusters) but is believed to contain a great deal of dark matter **c :** a differentiated zone surrounding a central zone or object
3 : the aura of glory, veneration, or sentiment surrounding an idealized person or thing

²halo *transitive verb* (1801)
: to form into or surround with a halo ⟨rainbows *haloed* the waterfalls —Michael Crawford⟩

halo- — see HAL-

hal·o·car·bon \'ha-lə-ˌkär-bən\ *noun* (1950)
: any of various compounds (as fluorocarbon) of carbon and one or more halogens

hal·o·cline \'ha-lə-ˌklīn\ *noun* (1960)
: a usually vertical gradient in salinity (as of the ocean)

halo effect *noun* (circa 1928)
: generalization from the perception of one outstanding personality trait to an overly favorable evaluation of the whole personality

hal·o·gen \'ha-lə-jən\ *noun* [Swedish, from *hal-* + *-gen*] (1842)
: any of the five elements fluorine, chlorine, bromine, iodine, and astatine that form part of group VII A of the periodic table and exist in the free state normally as diatomic molecules
— ha·log·e·nous \ha-'lä-jə-nəs\ *adjective*

ha·lo·ge·nate \'ha-lə-jə-ˌnāt, ha-'lä-jə-\ *transitive verb* **-nat·ed; -nat·ing** (1882)
: to treat or cause to combine with a halogen
— ha·lo·ge·na·tion \ˌha-lə-jə-'nā-shən, haˌlä-jə-\ *noun*

hal·o·ge·ton \ˌha-lə-'jē-ˌtän\ *noun* [New Latin, from *hal-* + Greek *geitōn* neighbor] (1943)
: a coarse annual herb (*Halogeton glomeratus*) of the goosefoot family that is a noxious weed in western American ranges

hal·o·mor·phic \ˌha-lə-'mȯr-fik\ *adjective* (circa 1938)
of a soil **:** developed in the presence of neutral or alkali salts or both

hal·o·per·i·dol \ˌha-lō-'per-ə-ˌdȯl, -ˌdōl\ *noun* [*hal-* + *piperidine* + ¹*-ol*] (1960)
: a depressant $C_{21}H_{23}ClFNO_2$ of the central nervous system used especially as an antipsychotic drug

hal·o·phile \'ha-lə-ˌfīl\ *noun* [International Scientific Vocabulary] (1923)
: an organism that flourishes in a salty environment
— hal·o·phil·ic \ˌha-lə-'fi-lik\ *adjective*

hal·o·phyte \'ha-lə-ˌfīt\ *noun* [International Scientific Vocabulary] (circa 1886)
: a plant (as saltbush or sea lavender) that grows in salty soil and usually has a physiological resemblance to a true xerophyte
— hal·o·phyt·ic \ˌha-lə-'fi-tik\ *adjective*

hal·o·thane \'ha-lə-ˌthān\ *noun* [*halo-* + *ethane*] (1957)
: a nonexplosive inhalational anesthetic $C_2HBrClF_3$

¹halt \'hȯlt\ *adjective* [Middle English, from Old English *healt;* akin to Old High German *halz* lame] (before 12th century)
: LAME

²halt *intransitive verb* (before 12th century)
1 : to walk or proceed lamely **:** LIMP
2 : to stand in perplexity or doubt between alternate courses **:** WAVER
3 : to display weakness or imperfection **:** FALTER

³halt *noun* [German, from Middle High German, from *halt,* imperative of *halten* to hold, from Old High German *haltan* — more at HOLD] (circa 1598)
: STOP

⁴halt (1656)
intransitive verb
1 : to cease marching or journeying
2 : DISCONTINUE, TERMINATE ⟨the project *halted* for lack of funds⟩
transitive verb
1 : to bring to a stop ⟨the strike *halted* subways and buses⟩
2 : to cause the discontinuance of **:** END

¹hal·ter \'hȯl-tər\ *noun* [Middle English, from Old English *hælftre;* akin to Old High German *halftra* halter, Old English *hielfe* helve] (before 12th century)
1 a : a rope or strap for leading or tying an animal **b :** a headstall usually with noseband and throatlatch to which a lead may be attached
2 : a rope for hanging criminals **:** NOOSE; *also* **:** death by hanging
3 : a woman's blouse that leaves the back, arms, and midriff bare and that is typically held in place by straps around the neck and across the back

halter 1b

²halter *transitive verb* **hal·tered; hal·ter·ing** \-t(ə-)riŋ\ (14th century)
1 a : to catch with or as if with a halter; *also* **:** to put a halter on **b :** HANG
2 : to put restraint upon **:** HAMPER

hal·ter·break \'hȯl-tər-ˌbrāk\ *transitive verb* **-broke** \-ˌbrōk\; **-bro·ken** \-ˌbrō-kən\; **-break·ing** (1837)
: to break (as a colt) to a halter

hal·tere \'hȯl-ˌtir, 'hal-\ *noun, plural* **halteres** \'hȯl-ˌtirz, 'hal-; hȯl-'tir-ēz, hal-\ [New Latin *halter,* from Latin, jumping weight, from Greek *haltēr,* from *hallesthai* to leap — more at SALLY] (1823)
: one of a pair of club-shaped organs in a dipteran insect that are the modified second pair of wings and function as sensory flight stabilizers

halt·ing \'hȯl-tiŋ\ *adjective* (1585)
: marked by a lack of sureness or effectiveness ⟨spoke in a *halting* manner⟩
— halt·ing·ly \-tiŋ-lē\ *adverb*

hal·vah *or* **hal·va** \häl-'vä; 'häl-(ˌ)vä, -və\ *noun* [Yiddish *halva,* from Romanian, from Turkish *helva,* from Arabic *ḥalwā* sweetmeat] (1846)
: a flaky confection of crushed sesame seeds in a base of syrup (as of honey)

halve \'hav, 'häv\ *transitive verb* **halved; halv·ing** [Middle English, from *half* half] (13th century)
1 a : to divide into two equal parts **b :** to reduce to one half ⟨*halving* the present cost⟩ **c :** to share equally
2 : to play (as a hole in golf) in the same number of strokes as one's opponent

halv·ers \'ha-vərz, 'hä-\ *noun plural* (1517)
: half shares **:** HALVES

halves *plural of* HALF

hal·yard \'hal-yərd\ *noun* [Middle English *halier,* from *halen* to pull — more at HALE] (14th century)
: a rope or tackle for hoisting and lowering something (as sails)

¹ham \'ham\ *noun* [Middle English *hamme,* from Old English *hamm;* akin to Old High German *hamma* ham, Greek *knēmē* shinbone, Old Irish *cnáim* bone] (before 12th century)
1 a : the hollow of the knee **b :** a buttock with its associated thigh — usually used in plural
2 : a cut of meat consisting of a thigh; *especially* **:** one from a hog
3 [short for *hamfatter,* from "The *Ham-fat* Man," minstrel song] **a :** a showy performer; *especially* **:** an actor performing in an exaggerated theatrical style **b :** a licensed operator of an amateur radio station
4 : a cushion used especially by tailors for pressing curved areas of garments
— ham *adjective*

²ham *verb* **hammed; ham·ming** (1933)
transitive verb
: to execute with exaggerated speech or gestures **:** OVERACT
intransitive verb
: to overplay a part

Ham \'ham\ *noun* [Hebrew]
: a son of Noah held to be the progenitor of the Egyptians, Nubians, and Canaanites

ham·a·dry·ad \ˌha-mə-'drī-əd, -ˌad\ *noun* [Latin *hamadryad-, hamadryas,* from Greek, from *hama* together with + *dryad-, dryas* dryad — more at SAME] (14th century)
1 : WOOD NYMPH
2 : KING COBRA

ham·a·dry·as baboon \ˌha-mə-'drī-əs-\ *noun* [New Latin *hamadryas,* from Latin] (circa 1890)
: a baboon (*Papio hamadryas*) that has a reddish pink muzzle and a large bare patch of pink skin on each buttock and that was venerated by the ancient Egyptians — called also *sacred baboon*

ha·mal *also* **ham·mal** \hə-'mäl\ *noun* [Arabic *ḥammāl* porter] (circa 1760)
: a porter in eastern countries (as Turkey)

Ha·man \'hā-mən\ *noun* [Hebrew *Hāmān*]
: an enemy of the Jews hanged according to the book of Esther for plotting their destruction

ha·man·tasch \'hä-mən-ˌtäsh, 'hō-, -ˌtōsh\ *noun, plural* **-tasch·en** \-ˌtä-shən, -ˌtō-\ [Yiddish *homentash,* from *Homen* Haman + *tash* pocket, bag] (1927)
: a three-cornered pastry with a filling (as of poppy seeds or prunes) traditionally eaten during the Jewish holiday Purim

ha·mar·tia \ˌhä-ˌmär-'tē-ə\ *noun* [Greek, from *hamartanein* to miss the mark, err] (1927)
: TRAGIC FLAW

ha·mate \'hā-ˌmāt\ *noun* [Latin *hamatus* hooked, from *hamus* hook] (1924)
: a bone on the inner side of the second row of the carpus in mammals

ham·burg·er \'ham-ˌbər-gər\ *or* **ham·burg** \-ˌbərg\ *noun* [German *Hamburger* of Hamburg, Germany] (1884)
1 a : ground beef **b :** a patty of ground beef
2 : a sandwich consisting of a patty of hamburger in a split round bun

¹hame \'hām\ *noun* [Middle English] (14th century)
: one of two curved supports attached to the collar of a draft horse to which the traces are fastened

²hame *Scottish variant of* HOME

ham–fist·ed \'ham-ˌfis-təd\ *adjective* (1928)
: HAM-HANDED

ham–hand·ed \-ˌhan-dəd\ *adjective* (1918)
: lacking dexterity or grace **:** HEAVY-HANDED
— ham–hand·ed·ly \-lē\ *adverb*
— ham–hand·ed·ness \-dəd-nəs\ *noun*

Ham·il·to·ni·an \ˌha-məl-'tō-nē-ən\ *noun* [Sir William *Hamilton* (died 1865) Irish mathematician] (1933)
: a function that is used to describe a dynamic system (as the motion of a particle) in terms of components of momentum and coordinates of space and time and that is equal to the total energy of the system when time is not explicitly part of the function — compare LAGRANGIAN

Ham·il·to·ni·an·ism \-nē-ə-ˌni-zəm\ *noun* (1901)

: the political principles and ideas held by or associated with Alexander Hamilton that center around a belief in a strong central government, broad interpretation of the federal constitution, encouragement of an industrial and commercial economy, and a general distrust of the political capacity or wisdom of the common man
— **Hamiltonian** \-'tō-nē-ən\ *adjective*

Ham·ite \'ha-ˌmīt\ *noun* [*Ham*] (1854)
: a member of a Hamitic-speaking people

¹Ham·it·ic \ha-'mi-tik, hə-\ *adjective* (1844)
: of, relating to, or characteristic of the Hamites or one of the Hamitic languages

²Hamitic *noun* (circa 1890)
: HAMITIC LANGUAGES

Hamitic languages *noun plural* (circa 1890)
: any of various groupings of non-Semitic Afro-Asiatic languages (as Berber, Egyptian, and Cushitic) that were formerly thought to comprise a single branch of the Afro-Asiatic family

Ham·i·to–Se·mit·ic \ˌha-mə-(ˌ)tō-sə-'mi-tik, hə-'mi-tō-\ *adjective* (1901)
: of, relating to, or constituting the Afro-Asiatic languages
— **Hamito–Semitic** *noun*

ham·let \'ham-lət\ *noun* [Middle English, from Middle French *hamelet,* diminutive of *ham* village, of Germanic origin; akin to Old English *hām* village, home] (before 12th century)
: a small village

Ham·let \'ham-lət\ *noun*
: a legendary Danish prince and hero of Shakespeare's play *Hamlet*

¹ham·mer \'ha-mər\ *noun* [Middle English *hamer,* from Old English *hamor;* akin to Old High German *hamar* hammer, and perhaps to Old Church Slavonic *kamen-, kamy* stone, Greek *akmē* point, edge — more at EDGE] (before 12th century)
1 a : a hand tool consisting of a solid head set crosswise on a handle and used for pounding **b :** a power tool that often substitutes a metal block or a drill for the hammerhead
2 : something that resembles a hammer in form or action: as **a :** a lever with a striking head for ringing a bell or striking a gong **b** (1) **:** an arm that strikes the cap in a percussion lock to ignite the propelling charge (2) **:** a part of the action of a modern gun that strikes the primer of the cartridge in firing or that strikes the firing pin to ignite the cartridge **c :** MALLEUS **d :** GAVEL **e** (1) **:** a padded mallet in a piano action for striking a string (2) **:** a hand mallet for playing on various percussion instruments (as a xylophone)
3 : a metal sphere thrown for distance in the hammer throw
4 : ACCELERATOR b
— **under the hammer :** for sale at auction

²hammer *verb* **ham·mered; ham·mer·ing** \'ha-mər-iŋ, 'ham-riŋ\ (14th century)
intransitive verb
1 : to strike blows especially repeatedly with or as if with a hammer **:** POUND
2 : to make repeated efforts; *especially* **:** to reiterate an opinion or attitude ⟨the lectures all *hammered* away at the same points⟩
transitive verb
1 a : to beat, drive, or shape with repeated blows of a hammer **b :** to fasten or build with a hammer
2 : to strike or drive with a force suggesting a hammer blow or repeated blows ⟨*hammered* the ball over the fence⟩ ⟨tried to *hammer* me into submission⟩
— **ham·mer·er** \'ha-mər-ər\ *noun*

hammer and sickle *noun* (1921)
: an emblem consisting of a crossed hammer and sickle used especially as a symbol of Soviet Communism

hammer and tongs *adverb* (circa 1780)
: with great force, vigor, or violence ⟨went at it *hammer and tongs*⟩
— **hammer–and–tongs** *adjective*

hammer dulcimer *noun* (1953)
: DULCIMER 1 — called also *hammered dulcimer*

hammered *adjective* (1522)
: having surface indentations produced or appearing to have been produced by hammering ⟨*hammered* copper⟩

ham·mer·head \'ha-mər-ˌhed\ *noun* (1562)
1 : the striking part of a hammer
2 : BLOCKHEAD
3 : any of a family (Sphyrnidae) of active voracious medium-sized sharks that have the eyes at the ends of lateral extensions of the flattened head — see SHARK illustration

ham·mer·less \-ləs\ *adjective* (1875)
: having the hammer concealed ⟨a *hammerless* revolver⟩

ham·mer·lock \-ˌläk\ *noun* (1897)
: a wrestling hold in which an opponent's arm is held bent behind his back; *broadly* **:** a strong hold

hammer mill *noun* (1610)
: a grinder or crusher in which materials are broken up by hammers

hammer out *transitive verb* (circa 1632)
: to produce or bring about as if by repeated blows ⟨*hammered out* an agreement⟩

hammer throw *noun* (1898)
: a field event in which a usually 16-pound metal sphere attached to a flexible handle is thrown for distance

ham·mer·toe \'ha-mər-ˌtō\ *noun* (circa 1885)
: a deformed claw-shaped toe and especially the second that results from permanent angular flexion between one or both phalangeal joints

¹ham·mock \'ha-mək\ *noun* [Spanish *hamaca,* from Taino] (1555)
: a swinging couch or bed usually made of netting or canvas and slung by cords from supports at each end

²hammock *noun* [origin unknown] (1555)
1 : HUMMOCK
2 : a fertile area in the southern U.S. and especially Florida that is usually higher than its surroundings and that is characterized by hardwood vegetation and deep humus-rich soil

ham·my \'ha-mē\ *adjective* **ham·mi·er; -est** (1929)
: marked by exaggerated and usually self-conscious theatricality
— **ham·mi·ly** \'ha-mə-lē\ *adverb*
— **ham·mi·ness** \'ha-mē-nəs\ *noun*

¹ham·per \'ham-pər\ *transitive verb* **ham·pered; ham·per·ing** \-p(ə-)riŋ\ [Middle English] (14th century)
1 a : to restrict the movement of by bonds or obstacles **:** IMPEDE **b :** to interfere with the operation of **:** DISRUPT
2 a : CURB, RESTRAIN **b :** to interfere with **:** ENCUMBER ☆

²hamper *noun* [Middle English *hampere,* alteration of *hanaper,* literally, case to hold goblets, from Middle French *hanapier,* from *hanap* goblet, of Germanic origin; akin to Old English *hnæpp* bowl] (14th century)
: a large basket usually with a cover for packing, storing, or transporting articles (as food or laundry)

Hamp·shire \'ham(p)-ˌshir, -shər\ *noun* [*Hampshire,* England] (1918)
1 : any of a British breed of large hornless black-faced mutton-producing sheep — called also *Hampshire Down*
2 : any of an American breed of black white-belted swine

ham·ster \'ham(p)-stər\ *noun* [German, from Old High German *hamustro,* of Slavic origin; akin to Old Russian *chomĕstorŭ* hamster, of Iranian origin; akin to Avestan *hamaēstar-* oppressor] (1607)
: any of a subfamily (Cricetinae) of small Old World rodents having very large cheek pouches

¹ham·string \'ham-ˌstriŋ\ *noun* (1565)
1 a : either of two groups of tendons at the back of the human knee **b :** HAMSTRING MUSCLE
2 : a large tendon above and behind the hock of a quadruped

²hamstring *transitive verb* **-strung** \-ˌstrəŋ\; **-string·ing** \-ˌstriŋ-iŋ\ (1641)
1 : to make ineffective or powerless **:** CRIPPLE ⟨*hamstrung* by guilt⟩
2 : to cripple by cutting the leg tendons

hamstring muscle *noun* (circa 1888)
: any of three muscles at the back of the thigh that function to flex and rotate the leg and extend the thigh

ham·u·lus \'ham-yə-ləs\ *noun, plural* **-u·li** \-ˌlī, -ˌlē\ [New Latin, from Latin, diminutive of *hamus* hook] (circa 1751)
: a hook or hooked process (as of a bone)

ham·za *or* **ham·zah** \'ham-zə, 'häm-\ *noun* [Arabic *hamzah,* literally, compression] (1938)
: the sign for a glottal stop in Arabic orthography usually represented in English by an apostrophe

Han \'hän\ *noun* [Chinese (Beijing) *Hàn*] (1736)
1 : a Chinese dynasty dated 207B.C.–A.D.220 and marked by centralized control through an appointive bureaucracy, a revival of learning, and the penetration of Buddhism
2 : the Chinese peoples especially as distinguished from non-Chinese (as Mongolian) elements in the population

¹hand \'hand\ *noun, often attributive* [Middle English, from Old English; akin to Old High German *hant* hand] (before 12th century)
1 a (1) **:** the terminal part of the vertebrate forelimb when modified (as in humans) as a grasping organ (2) **:** the forelimb segment (as the terminal section of a bird's wing) of a vertebrate higher than the fishes that corresponds to the hand irrespective of its form or functional specialization **b :** a part serving the function of or resembling a hand; as (1) **:** the hind foot of an ape (2) **:** the chela of a crustacean **c :** something resembling a hand: as (1) **:** an indicator or pointer on a dial ⟨the *hands* of a clock⟩ (2) **:** INDEX 5 (3) **:** a cluster of bananas developed from a single flower group (4) **:** a branched rootstock of ginger (5) **:** a bunch of large leaves (as of tobacco) tied together usually with another leaf
2 a : personal possession — usually used in plural ⟨the documents fell into the *hands* of

☆ SYNONYMS
Hamper, trammel, clog, fetter, shackle, manacle mean to hinder or impede in moving, progressing, or acting. HAMPER may imply the effect of any impeding or restraining influence ⟨*hampered* the investigation by refusing to cooperate⟩. TRAMMEL suggests entangling by or confining within a net ⟨rules that *trammel* the artist's creativity⟩. CLOG usually implies a slowing by something extraneous or encumbering ⟨a court system *clogged* by frivolous suits⟩. FETTER suggests a restraining so severe that freedom to move or progress is almost lost ⟨a nation *fettered* by an antiquated class system⟩. SHACKLE and MANACLE are stronger than FETTER and suggest total loss of freedom ⟨a mind *shackled* by stubborn prejudice⟩ ⟨a people *manacled* by tyranny⟩.

the enemy⟩ **b :** CONTROL, SUPERVISION — usually used in plural ⟨left the matter in her *hands*⟩
3 a : SIDE, DIRECTION ⟨men fighting on either *hand*⟩ **b :** one of two sides or aspects of an issue or argument ⟨on the one *hand* we can appeal for peace, and on the other, declare war⟩
4 : a pledge especially of betrothal or bestowal in marriage
5 a : style of penmanship **:** HANDWRITING ⟨wrote in a fancy *hand*⟩ **b :** SIGNATURE
6 a : SKILL, ABILITY ⟨tried her *hand* at sailing⟩ **b :** an instrumental part ⟨had a *hand* in the victory⟩
7 : a unit of measure equal to 4 inches (10.2 centimeters) used especially for the height of horses
8 a : assistance or aid especially involving physical effort ⟨lend a *hand*⟩ **b :** PARTICIPATION, INTEREST **c :** a round of applause
9 a (1) **:** a player in a card game or board game (2) **:** the cards or pieces held by a player **b :** a single round in a game **c :** the force or solidity of one's position (as in negotiations)
10 a : a person who performs or executes a particular work ⟨two portraits by the same *hand*⟩ **b** (1) **:** a person employed at manual labor or general tasks ⟨a ranch *hand*⟩ (2) **:** WORKER, EMPLOYEE ⟨employed over a hundred *hands*⟩ **c :** a member of a ship's crew ⟨all *hands* on deck⟩ **d :** a person skilled in a particular action or pursuit **e :** a specialist or veteran in a usually designated activity or region ⟨a China *hand*⟩
11 a : HANDIWORK, DOINGS **b :** style of execution **:** WORKMANSHIP ⟨the *hand* of a master⟩ **c :** the feel of or tactile reaction to something (as silk or leather)
12 : a punch made with a specified hand ⟨knocked him out with a good right *hand*⟩
— at hand 1 : near in time or place **:** within reach ⟨use whatever ingredients are *at hand*⟩ **2 :** currently receiving or deserving attention ⟨the business *at hand*⟩
— at the hands of *also* **at the hand of :** by or through the action of
— by hand 1 : with the hands or a handworked implement (as a tool or pen) rather than with a machine **2 :** from one individual directly to another ⟨deliver the document *by hand*⟩
— in hand 1 : in one's possession or control **2 :** in preparation **3 :** under consideration
— on all hands *or* **on every hand :** EVERYWHERE
— on hand 1 : in present possession or readily available **2 :** about to appear **:** PENDING **3 :** in attendance **:** PRESENT
— on one's hands : in one's possession or care ⟨too much time *on my hands*⟩
— out of hand 1 : without delay or deliberation; *also* **:** in a summary or peremptory manner **2 :** done with **:** FINISHED **3 :** out of control **4 :** with the hands ⟨fruit eaten *out of hand*⟩
— to hand 1 : into possession **2 :** within reach

²hand *adverb* (before 12th century)
: with the hands rather than by machine

³hand *transitive verb* (15th century)
1 a *obsolete* **:** to touch or manage with the hands; *also* **:** to deal with **b :** FURL
2 : to lead, guide, or assist with the hand ⟨*hand* a lady into a bus⟩
3 a : to give, pass, or transmit with the hand ⟨*hand* a letter to her⟩ **b :** to present or provide with ⟨*handed* him a surprise⟩
— hand it to : to give credit to **:** concede the excellence of

hand and foot *adverb* (before 12th century)
: TOTALLY, ASSIDUOUSLY

hand ax *noun* (13th century)
1 : a short-handled ax intended for use with one hand
2 : a prehistoric stone tool having one end pointed for cutting and the other end rounded for holding in the hand

hand·bag \ˈhan(d)-ˌbag\ *noun* (1862)
1 : SUITCASE
2 : a bag held in the hand or hung from a shoulder strap and used for carrying small personal articles and money

hand·ball \-ˌbȯl\ *noun* (1886)
1 : a game played in a walled court or against a single wall or board by two or four players who use their hands to strike the ball
2 : a small rubber ball used in handball

hand·bar·row \-ˌbar-(ˌ)ō\ *noun* (15th century)
: a flat rectangular frame with handles at both ends that is carried by two persons

hand·bas·ket \-ˌbas-kət\ *noun* (15th century)
: a small portable basket — usually used in the phrase *to hell in a handbasket* denoting rapid and utter ruination

hand·bell \-ˌbel\ *noun* (before 12th century)
: a small bell with a handle; *especially* **:** one of a set tuned in a scale for musical performance

hand·bill \-ˌbil\ *noun* (1753)
: a small printed sheet to be distributed (as for advertising) by hand

hand·blown \-ˈblōn\ *adjective* (1925)
: made by glassblowing and molded by hand

hand·book \-ˌbu̇k\ *noun* (before 12th century)
1 a : a book capable of being conveniently carried as a ready reference **:** MANUAL **b :** a concise reference book covering a particular subject
2 a : a bookmaker's book of bets **b :** a place where bookmaking is carried on

hand·breadth \-ˌbretth, -ˌbreth, -ˌbredth\ *noun* (before 12th century)
: any of various units of length varying from about 2½ to 4 inches based on the breadth of a hand

hand·car \ˈhan(d)-ˌkär\ *noun* (1850)
: a small four-wheeled railroad car propelled by a hand-operated mechanism or by a small motor

hand·cart \-ˌkärt\ *noun* (1640)
: a cart drawn or pushed by hand

hand cheese *noun* (1890)
: a soft cheese with a sharp pungent odor and flavor that was originally molded by hand

hand·clasp \ˈhan(d)-ˌklasp\ *noun* (1583)
: HANDSHAKE

¹hand·craft \-ˌkraft\ *noun* (before 12th century)
: HANDICRAFT

²handcraft *transitive verb* (1947)
: to fashion by handicraft

hand·crafts·man \-ˌkraf(t)-smən\ *noun* (15th century)
: a person who is skilled in handicraft
— hand·crafts·man·ship \-ship\ *noun*

¹hand·cuff \-ˌkəf\ *noun* (1695)
: a metal fastening that can be locked around a wrist and is usually connected by a chain or bar with another such fastening — usually used in plural

²handcuff *transitive verb* (1720)
1 : to apply handcuffs to **:** MANACLE
2 : to hold in check **:** make ineffective or powerless

hand down *transitive verb* (1692)
1 : to transmit in succession (as from father to son)
2 : to make official formulation of and express (the opinion of a court)

hand·ed \ˈhan-dəd\ *adjective* (15th century)
1 : having a hand or hands especially of a specified kind or number — usually used in combination ⟨a large-*handed* man⟩
2 : using a specified hand or number of hands — used in combination ⟨right-*handed*⟩ ⟨a one-*handed* catch⟩

hand·ed·ness \-nəs\ *noun* (1915)
1 : a tendency to use one hand rather than the other
2 a : the property of an object (as a molecule) of not being identical with its mirror image **b :** either of the two configurations of an object that may exist in forms which are nonidentical mirror images

hand·fast \ˈhan(d)-ˌfast\ *noun* [Middle English, from Old English *handfæst*] (1611)
archaic **:** a contract or covenant especially of betrothal or marriage

hand–feed \ˈhan(d)-ˈfēd\ *transitive verb* **-fed** \-ˈfed\; **-feed·ing** (1805)
: to feed (as animals) by hand

hand·ful \ˈhan(d)-ˌfu̇l\ *noun, plural* **handfuls** \-ˌfu̇lz\ *also* **hands·ful** \ˈhan(d)z-ˌfu̇l\ (before 12th century)
1 : as much or as many as the hand will grasp
2 : a small quantity or number
3 : as much as one can manage

hand glass *noun* (1882)
: a small mirror with a handle

hand·grip \ˈhan(d)-ˌgrip\ *noun* (before 12th century)
1 : a grasping with the hand
2 : HANDLE
3 *plural* **:** hand-to-hand combat

hand·gun \-ˌgən\ *noun* (15th century)
: a firearm (as a revolver or pistol) designed to be held and fired with one hand

hand·held \-ˌheld, -ˈheld\ *adjective* (1923)
: held in the hand; *especially* **:** designed to be operated while being held in the hand ⟨*handheld* computers⟩
— handheld \-ˌheld\ *noun*

hand·hold \ˈhand-ˌhōld\ *noun* (1643)
1 : HOLD, GRIP
2 : something to hold on to (as in mountain climbing)

hand–hold·ing \-ˌhōl-diŋ\ *noun* (1967)
: solicitous attention, support, or instruction (as in the use of new technology)

¹hand·i·cap \ˈhan-di-ˌkap, -dē-\ *noun* [obsolete English *handicap* (a game in which forfeit money was held in a cap), from *hand in cap*] (1754)
1 a : a race or contest in which an artificial advantage is given or disadvantage imposed on a contestant to equalize chances of winning **b :** an advantage given or disadvantage imposed usually in the form of points, strokes, weight to be carried, or distance from the target or goal
2 a : a disadvantage that makes achievement unusually difficult **b :** a physical disability ◆

²handicap *transitive verb* **-capped; -cap·ping** (1852)
1 a : to give a handicap to **b :** to assess the relative winning chances of (contestants) or the likely winner of (a contest)
2 : to put at a disadvantage

◇ WORD HISTORY
handicap *Handicap,* from *hand in cap,* was an old form of barter. Two people who wished to make an exchange asked a third to act as umpire. All three put forfeit money in a cap, into which each of the two barterers inserted a hand. The umpire described the goods to be traded and set the additional amount the owner of the less valuable article should pay the other in order that the exchange might be fair. The barterers withdrew their hands from the cap empty to signify the refusal of the umpire's decision, or full to indicate acceptance. If the hands of both barterers were full, the exchange was made and the umpire pocketed the forfeit money. If both barterers came up empty-handed, there was no exchange but the umpire took the money. Otherwise, each barterer kept his own property, and the one who had accepted the umpire's decision took the forfeit money as well. Later, a horse race arranged in accordance with the rules of handicap was called a *handicap race.* The umpire decided how much extra weight the better horse should carry. The word was eventually extended to other contests, and also came to signify the advantage or disadvantage imposed.

hand•i•capped *adjective* (1915)
: having a physical or mental disability that substantially limits activity especially in relation to employment or education; *also* : of or reserved for handicapped persons ⟨*handicapped* parking spaces⟩
hand•i•cap•per \-ˌka-pər\ *noun* (1754)
1 : a person who assigns handicaps
2 : a person who predicts the winners in a race (as a horse race)
3 : a person who competes with a (specified) handicap (as in golf) — usually used in combination ⟨a 5-*handicapper*⟩
hand•i•craft \ˈhan-di-ˌkraft, -dē-\ *noun* [Middle English *handi-crafte,* alteration of *handcraft*] (13th century)
1 a : manual skill **b :** an occupation requiring skill with the hands
2 : the articles fashioned by those engaged in handicraft
— hand•i•craft•er \-ˌkraf-tər\ *noun*
hand•i•crafts•man \-ˌkraf(t)-smən\ *noun* (1551)
: a person who engages in a handicraft : ARTISAN
Hand•ie–Talk•ie \ˌhan-dē-ˈtȯ-kē\ *trademark*
— used for a small portable radio transmitter-receiver
hand•i•ly \ˈhan-də-lē\ *adverb* (1719)
1 : in a dexterous manner
2 : EASILY ⟨defeated the other candidate *handily*⟩
3 : conveniently nearby
hand in *transitive verb* (1837)
: SUBMIT 2 ⟨*hand in* your homework⟩
hand in glove *or* **hand and glove** *adverb* (1680)
: in extremely close relationship or agreement ⟨working *hand in glove* with the police⟩
hand in hand *adverb* (15th century)
1 : with hands clasped (as in intimacy or affection)
2 : in close association : TOGETHER
hand•i•work \ˈhan-di-ˌwərk, -dē-\ *noun* [Middle English *handiwerk,* from Old English *handgeweorc,* from *hand* + *geweorc,* from *ge-* (collective prefix) + *weorc* work — more at CO-] (before 12th century)
1 a : work done by the hands **b :** work done personally
2 : the product of handiwork
hand•ker•chief \ˈhaŋ-kər-chəf, -(ˌ)chif, -ˌchēf\ *noun, plural* **-chiefs** *also* **-chieves** \-chəfs, -(ˌ)chifs, -ˌchēvz, -ˌchēfs, -chəvz, -(ˌ)chivz\ (1530)
1 : a small usually square piece of cloth used for usually personal purposes (as blowing the nose) or as a clothing accessory
2 : KERCHIEF 1
¹han•dle \ˈhan-dᵊl\ *noun* [Middle English *handel,* from Old English *handle;* akin to Old English *hand*] (before 12th century)
1 : a part that is designed especially to be grasped by the hand
2 : something that resembles a handle
3 a : TITLE 8 **b :** NAME; *also* : NICKNAME
4 : HAND 11c
5 : the total amount of money bet on a race, game, or event
6 : a means of understanding or controlling ⟨can't quite get a *handle* on things⟩
— han•dled \-dᵊld\ *adjective*
— han•dle•less \-dᵊl-(l)əs\ *adjective*
— off the handle : into a state of sudden and violent anger — usually used with *fly*
²handle *verb* **han•dled; han•dling** \ˈhan(d)-liŋ, ˈhan-dᵊl-iŋ\ (before 12th century)
transitive verb
1 a : to try or examine (as by touching, feeling, or moving) with the hand ⟨*handle* silk to judge its weight⟩ **b :** to manage with the hands ⟨*handle* a horse⟩
2 a : to deal with in writing or speaking or in the plastic arts **b :** to have overall responsibility for supervising or directing : MANAGE ⟨a lawyer *handles* all my affairs⟩ **c :** to train and act as second for (a boxer) **d :** to put up with : STAND ⟨can't *handle* the heat⟩
3 : to act on or perform a required function with regard to ⟨*handle* the day's mail⟩
4 : to engage in the buying, selling, or distributing of (a commodity)
intransitive verb
: to act, behave, or feel in a certain way when handled or directed ⟨a car that *handles* well⟩
— han•dle•able \-dᵊl-ə-bəl\ *adjective*
han•dle•bar \ˈhan-dᵊl-ˌbär\ *noun* (1886)
: a straight or bent bar with a handle at each end; *specifically* : one used to steer a bicycle or similar vehicle — usually used in plural
handlebar mustache *noun* (1933)
: a heavy mustache with long sections that curve upward at each end
hand lens *noun* (1930)
: a magnifying glass to be held in the hand
han•dler \ˈhan(d)-lər, ˈhan-dᵊl-ər\ *noun* (14th century)
1 : one that handles something
2 a : a person in immediate physical charge of an animal; *especially* : a person who exhibits dogs at shows or field trials **b :** a person who trains or acts as second for a boxer **c :** a manager of a political figure or campaign
hand•less \ˈhan(d)-ləs\ *adjective* (15th century)
1 : having no hands
2 : inefficient in manual tasks : CLUMSY
han•dling \ˈhan(d)-liŋ, ˈhan-dᵊl-iŋ\ *noun* (before 12th century)
1 a : the action of one that handles something **b :** a process by which something is handled in a commercial transaction; *especially* : the packaging and shipping of an object or material (as to a consumer)
2 : the manner in which something is treated (as in a musical, literary, or art work)
hand•list \ˈhan(d)-ˌlist\ *noun* (1859)
: a list (as of books) for purposes of reference or checking
hand•made \ˈhan(d)-ˈmād\ *adjective* (1613)
: made by hand or by a hand process
hand•maid•en \-ˌmā-dᵊn\ *also* **hand•maid** \-ˌmād\ *noun* (13th century)
1 : a personal maid or female servant
2 : something whose essential function is to serve or assist ⟨criticism is not the enemy of art but rather its *handmaiden* —Gary Michael⟩
hand–me–down \ˈhan(d)-mē-ˌdau̇n\ *adjective* (1827)
1 : put in use by one person or group after being used, discarded, or handed down by another ⟨*hand-me-down* clothes⟩ ⟨*hand-me-down* anecdotes⟩
2 : ready-made and usually cheap and shoddy
— hand–me–down *noun*
hand off (1949)
transitive verb
: to hand (a football) to a nearby teammate on a play
intransitive verb
: to hand off a football
— hand•off \ˈhan-ˌdȯf\ *noun*
hand on *transitive verb* (1865)
: HAND DOWN
hand organ *noun* (1796)
: a barrel organ operated by a hand crank
hand•out \ˈhan-ˌdau̇t\ *noun* (1882)
1 : a portion of food, clothing, or money given to or as if to a beggar
2 : a folder or circular of information for free distribution
3 : a prepared statement released to the news media
hand out *transitive verb* (1877)
1 a : to give without charge **b :** to give freely
2 : ADMINISTER ⟨*handed out* a severe punishment⟩
hand over *transitive verb* (1816)
: to yield control of
— hand•over \ˈhan-ˌdō-vər\ *noun*
hand over fist *adverb* (1825)
: quickly and in large amounts
hand•pick \ˈhan(d)-ˈpik\ *transitive verb* (1831)
1 : to pick by hand as opposed to a machine process
2 : to select personally or for personal ends
hand•press \-ˌpres\ *noun* (1679)
: a hand-operated press
hand•print \-ˌprint\ *noun* (1886)
: an impression of a hand on a surface
hand puppet *noun* (1947)
: PUPPET 1a
hand•rail \ˈhan(d)-ˌrāl\ *noun* (1793)
: a narrow rail for grasping with the hand as a support
hand running *adverb* (1828)
dialect : in unbroken succession
hand•saw \ˈhan(d)-ˌsȯ\ *noun* (14th century)
: a saw designed to be used with one hand
hands•breadth \ˈhan(d)z-ˌbredth, -ˌbretth\ *variant of* HANDBREADTH
hands down \ˈhan(d)z-ˈdau̇n\ *adverb* (1867)
1 : without much effort : EASILY
2 : without question
— hands–down \ˈhan(d)z-ˌdau̇n\ *adjective*
¹hand•sel \ˈhan(t)-səl\ *noun* [Middle English *hansell*] (14th century)
1 : a gift made as a token of good wishes or luck especially at the beginning of a new year
2 : something received first (as in a day of trading) and taken to be a token of good luck
3 a : a first installment : earnest money **b** : EARNEST, FORETASTE
²handsel *transitive verb* **-seled** *or* **-selled; -sel•ing** *or* **-sel•ling** \-s(ə-)liŋ\ (15th century)
1 : to give a handsel to
2 : to inaugurate with a token or gesture of luck or pleasure
3 : to use or do for the first time
hand•set \ˈhan(d)-ˌset\ *noun* (circa 1919)
: a combined telephone transmitter and receiver mounted on a handle
hand•shake \-ˌshāk\ *noun* (1873)
: a clasping usually of right hands by two people (as in greeting or farewell)
hands–off \ˈhan(d)-ˈzȯf\ *adjective* (1902)
: characterized by noninterference ⟨a *hands-off* policy toward the internal affairs of other nations⟩
hand•some \ˈhan(t)-səm\ *adjective* **hand•som•er; -est** [Middle English *handsom* easy to manipulate] (1530)
1 *chiefly dialect* : APPROPRIATE, SUITABLE
2 : moderately large : SIZABLE ⟨a painting that commanded a *handsome* price⟩
3 : marked by skill or cleverness : ADROIT
4 : marked by graciousness or generosity : LIBERAL ⟨*handsome* contributions to charity⟩
5 : having a pleasing and usually impressive or dignified appearance
synonym see BEAUTIFUL
— hand•some•ly *adverb*
— hand•some•ness *noun*
hands–on \ˈhan(d)-ˈzȯn, -ˈzän\ *adjective* (1969)
1 : relating to, being, or providing direct practical experience in the operation or functioning of something ⟨*hands-on* training⟩; *also* : involving or allowing use of or touching with the hands ⟨a *hands-on* museum display⟩
2 : characterized by active personal involvement ⟨a *hands-on* manager⟩
hand•spike \ˈhan(d)-ˌspīk\ *noun* [by folk etymology from Dutch *handspaak,* from *hand* hand + *spaak* pole; akin to Old English *spāca* spoke] (1615)
: a bar used as a lever
hand•spring \-ˌspriŋ\ *noun* (1875)

: an acrobatic feat in which the body turns forward or backward in a full circle from a standing position and lands first on the hands and then on the feet

hand·stand \\-ˌstand\\ *noun* (1899)
: an act of supporting the body on the hands with the trunk and legs balanced in the air

hand–to–hand \\ˈhan-tə-ˈhand, -də-\\ *adjective* (1836)
: involving physical contact or close enough range for physical contact ⟨*hand-to-hand* fighting⟩

hand to hand *adverb* (circa 1533)
: at very close range

hand–to–mouth \\-ˈmau̇th\\ *adjective* (1748)
: having or providing nothing to spare beyond basic necessities ⟨a *hand-to-mouth* existence⟩

hand truck *noun* (1920)
: a small hand-propelled truck; *especially* **:** TRUCK 3b

hand up *transitive verb* (1970)
of a jury **:** to deliver (an indictment) to a judge or higher judicial authority

hand·wheel \\ˈhan(d)-ˌ(h)wēl\\ *noun* (circa 1889)
: a wheel worked by hand

hand·work \\ˈhand-ˌwərk\\ *noun* (before 12th century)
: work done with the hands and not by machines **:** HANDIWORK
— hand·work·er \\-ˌwər-kər\\ *noun*

hand·wo·ven \\-ˈwō-vən\\ *adjective* (1880)
1 : produced on a hand-operated loom
2 : woven by hand ⟨*handwoven* baskets⟩

hand–wring·ing \\-ˌriŋ-iŋ\\ *noun* (1922)
: an overwrought expression of concern or guilt
— hand·wring·er *noun*

hand·write \\-ˌrīt\\ *transitive verb* **-wrote** \\-ˌrōt\\; **-writ·ten** \\-ˌri-tᵊn\\; **-writ·ing** \\-ˌrī-tiŋ\\ [back-formation from *handwriting*] (circa 1853)
: to write by hand

hand·writ·ing \\ˈhand-ˌrī-tiŋ\\ *noun* (15th century)
1 : writing done by hand; *especially* **:** the form of writing peculiar to a particular person
2 : something written by hand
— handwriting on the wall : an omen of one's unpleasant fate

hand·wrought \\ˈhand-ˈrȯt\\ *adjective* (1876)
: fashioned by hand or chiefly by hand processes ⟨*handwrought* silver⟩

handy \\ˈhan-dē\\ *adjective* **hand·i·er; -est** (1650)
1 a : conveniently near **b :** convenient for use **c** *of a ship* **:** easily handled
2 : clever in using the hands especially in a variety of useful ways ⟨*handy* with a hammer as well as with a paintbrush⟩
— hand·i·ness *noun*

handy·man \\-dē-ˌman\\ *noun* (1872)
1 : one who does odd jobs
2 : one competent in a variety of small skills or inventive or ingenious in repair or maintenance work — called also *handyperson*

¹hang \\ˈhaŋ\\ *verb* **hung** \\ˈhəŋ\\ *also* **hanged** \\ˈhaŋd\\; **hang·ing** \\ˈhaŋ-iŋ\\ [partly from Middle English *hon,* from Old English *hōn,* transitive verb; partly from Middle English *hangen,* from Old English *hangian,* intransitive verb & transitive verb; both akin to Old High German *hāhan,* transitive verb, to hang, *hangēn,* intransitive verb — more at CUNCTATION] (before 12th century)
transitive verb
1 a : to fasten to some elevated point without support from below **:** SUSPEND **b :** to suspend by the neck until dead — often *hanged* in the past; often used as a mild oath ⟨I'll be *hanged*⟩ **c :** to fasten so as to allow free motion within given limits upon a point of suspension ⟨*hang* a door⟩ **d :** to adjust the hem of (a skirt) so as to hang evenly and at a proper height
2 : to furnish with hanging decorations (as flags or bunting)
3 : to hold or bear in a suspended or inclined manner ⟨*hung* his head in shame⟩
4 : to apply to a wall ⟨*hang* wallpaper⟩
5 : to display (pictures) in a gallery
6 : to throw (as a curveball) so that it fails to break properly
7 : to make (a turn) especially while driving ⟨*hang* a right⟩ ⟨*hung* a quick U-turn —Tom Clancy⟩
intransitive verb
1 a : to remain suspended or fastened to some point above without support from below **:** DANGLE **b :** to die by hanging — often *hanged* in the past ⟨he *hanged* for his crimes⟩
2 : to remain poised or stationary in the air ⟨clouds *hanging* low overhead⟩
3 : LINGER, PERSIST
4 : to be imminent **:** IMPEND ⟨doom *hung* over the nation⟩
5 : to fall or droop from a usually tense or taut position
6 : DEPEND ⟨election *hangs* on one vote⟩
7 a (1) **:** to take hold for support **:** CLING ⟨she *hung* on his arm⟩ (2) **:** to keep persistent contact ⟨dogs *hung* to the trail⟩ **b :** to be burdensome or oppressive ⟨time *hangs* on his hands⟩
8 : to be uncertain or in suspense ⟨the decision is still *hanging*⟩
9 : to lean, incline, or jut over or downward
10 : to be in a state of rapt attention ⟨*hung* on her every word⟩
11 : to fit or fall from the figure in easy lines ⟨the coat *hangs* loosely⟩
12 *of a thrown ball* **:** to fail to break or drop as intended ■
— hang·able \\ˈhaŋ-ə-bəl\\ *adjective*
— hang fire 1 : to be slow in the explosion of a charge after its primer has been discharged **2 :** DELAY, HESITATE **3 :** to remain unsettled or unresolved
— hang it up : to cease an activity or effort
— hang loose : to remain calm or relaxed
— hang one on 1 : to inflict a blow on **2** *slang* **:** to get very drunk
— hang one's hat : to situate oneself in (as a residence or place of employment)
— hang tough : to remain resolute in the face of adversity **:** HANG IN

²hang *noun* (circa 1797)
1 : the manner in which a thing hangs
2 : DECLIVITY, SLOPE; *also* **:** DROOP
3 : facility with or an understanding of something ⟨can't get the *hang* of this⟩
4 : a hesitation or slackening in motion or in a course
— give a hang *or* **care a hang :** to be the least bit concerned or worried

hang about *intransitive verb* (1849)
British **:** HANG AROUND

¹han·gar \\ˈhaŋ-ər, ˈhaŋ-gər\\ *noun* [French] (1852)
: SHELTER, SHED; *especially* **:** a covered and usually enclosed area for housing and repairing aircraft

²hangar *transitive verb* (1943)
: to place or store in a hangar

hang around (1830)
intransitive verb
1 : to pass time or stay aimlessly **:** loiter idly
2 : to spend one's time in company
transitive verb
: to pass time or stay aimlessly in or at ⟨*hung around* the house all day⟩

hang back *intransitive verb* (1581)
1 : to drag behind others
2 : to be reluctant

¹hang·dog \\ˈhaŋ-ˌdȯg\\ *adjective* (1677)
1 : SAD, DEJECTED
2 : SHEEPISH

²hangdog *noun* (1687)
: a despicable or miserable person

hang·er \\ˈhaŋ-ər\\ *noun* (15th century)
1 : one that hangs or causes to be hung or hanged
2 : something that hangs, overhangs, or is suspended: as **a :** a decorative strip of cloth **b :** a small sword formerly used by seamen **c** *chiefly British* **:** a small wood on steeply sloping land
3 : a device by which or to which something is hung or hangs: as **a :** a strap on a sword belt by which a sword or dagger can be suspended **b :** a loop by which a garment is hung up **c :** a device that fits inside or around a garment for hanging from a hook or rod

hang·er–on \\-ˌȯn, -ˌän\\ *noun, plural* **hangers–on** (1542)
: one that hangs around a person, place, or institution especially for personal gain

hang glider *noun* (1930)
: a kitelike glider from which a harnessed rider hangs while gliding down from a cliff or hill
— hang glid·ing *noun*

hang glider

hang in *intransitive verb* (1966)
: to refuse to be discouraged or intimidated **:** show pluck ⟨*hang in* there⟩

¹hang·ing \\ˈhaŋ-iŋ\\ *adjective* (12th century)
1 : situated or lying on steeply sloping ground
2 a : jutting out **:** OVERHANGING ⟨a *hanging* rock⟩ **b :** supported only by the wall on one side ⟨a *hanging* staircase⟩
3 *archaic* **:** downcast in appearance
4 : adapted for sustaining a hanging object
5 : deserving, likely to cause, or prone to inflict death by hanging

²hanging *noun* (14th century)
1 : an execution by strangling or breaking the neck by a suspended noose
2 : something hung: as **a :** CURTAIN **b :** a covering (as a tapestry) for a wall
3 : a downward slope **:** DECLIVITY

hanging indention *noun* (1904)
: indention of all the lines of a paragraph except the first

hang·man \\ˈhaŋ-mən\\ *noun* (14th century)
: one who hangs a condemned person; *also* **:** a public executioner

hang·nail \\-ˌnāl\\ *noun* [by folk etymology from *agnail* (inflammation about the nail), from Middle English, corn on the foot or toe, from Old English *angnægl,* from *ang-* (akin to *enge* tight, painful) + *nægl* nail — more at ANGER] (1678)
: a bit of skin hanging loose at the side or root of a fingernail ◆

hang off *intransitive verb* (1641)
: HANG BACK

hang on *intransitive verb* (circa 1719)
1 : to keep hold **:** hold onto something
2 : to persist tenaciously ⟨a cold that *hung on* all spring⟩
3 : HOLD ON 2 ⟨*hang on* a second while I look it up⟩
— hang on to : to hold, grip, or keep tenaciously ⟨*hang on to* your money⟩

hang·out \\ˈhaŋ-ˌau̇t\\ *noun* (circa 1893)

☐ USAGE
hang For both transitive and intransitive senses 1b the past and past participle *hung,* as well as *hanged,* is standard. *Hanged* is most appropriate for official executions ⟨he was to be *hanged,* cut down whilst still alive . . . and his bowels torn out —Louis Allen⟩ but *hung* is also used ⟨gave orders that she should be *hung* —Peter Quennell⟩. *Hung* is more appropriate for less formal hangings ⟨by morning I'll be *hung* in effigy —Ronald Reagan⟩.

: a favorite place for spending time; *also* **:** a place frequented for entertainment or for socializing

hang out (14th century)
intransitive verb
1 : to protrude and droop
2 a *slang* **:** LIVE, RESIDE **b :** HANG AROUND
transitive verb
: to display outside as an announcement to the public — used chiefly in the phrase *hang out one's shingle*

hang•over \'haŋ-ˌō-vər\ *noun* (1894)
1 : something (as a surviving custom) that remains from what is past
2 a : disagreeable physical effects following heavy consumption of alcohol or the use of drugs **b :** a letdown following great excitement or excess

hang•tag \'haŋ-ˌtag\ *noun* (1952)
: a tag attached to an article of merchandise giving information about its material and proper care

hang time *noun* (1976)
: the amount of time a kicked football remains in the air; *also* **:** the length of time a leaping athlete is in the air

hang together *intransitive verb* (1551)
1 : to remain united **:** stand by one another
2 : to have unity **:** form a consistent or coherent whole

Hang•town fry \'haŋ-ˌtau̇n-\ *noun, often F capitalized* [*Hangtown,* nickname for Placerville, Calif.] (1949)
: an omelet or scrambled eggs containing oysters

han•gul \'hän-ˌgül\ *noun, often capitalized* [Korean *hangŭl*] (1946)
: the alphabetic script in which Korean is written

hang–up \'haŋ-ˌəp\ *noun* (1959)
: a source of mental or emotional difficulty; *broadly* **:** PROBLEM

hang up (12th century)
transitive verb
1 a : to place on a hook or hanger designed for the purpose ⟨*hang up* your coat⟩ **b :** to replace (a telephone receiver) on the cradle so that the connection is broken
2 : to keep delayed, suspended, or held up
3 : to cause to stick or snag immovably ⟨the ship was *hung up* on a sandbar⟩
intransitive verb
1 : to break a telephone connection
2 : to become stuck or snagged so as to be immovable

ha•ni•wa \'hä-nə-ˌwä\ *noun plural, often capitalized* [Japanese] (1931)
: large hollow baked clay sculptures placed on ancient Japanese burial mounds

hank \'haŋk\ *noun* [Middle English, of Scandinavian origin; akin to Old Norse *hǫnk* hank; akin to Old English *hangian* to hang] (14th century)
1 : COIL, LOOP; *specifically* **:** a coiled or looped bundle (as of yarn or rope) usually containing a definite yardage
2 : any of a series of rings or clips by which a jib or staysail is attached to a stay

han•ker \'haŋ-kər\ *intransitive verb* **han•kered; han•ker•ing** \-k(ə-)riŋ\ [probably from Flemish *hankeren,* frequentative of *hangen* to hang; akin to Old English *hangian*] (1642)
: to have a strong or persistent desire **:** YEARN — often used with *for* or *after*
synonym see LONG
— han•ker•er \-kər-ər\ *noun*

han•kie *or* **han•ky** \'haŋ-kē\ *noun, plural* **hankies** [*hand*kerchief + *-ie*] (1895)
: HANDKERCHIEF

han•ky–pan•ky \ˌhaŋ-kē-'paŋ-kē\ *noun* [origin unknown] (1841)
1 : questionable or underhanded activity
2 : sexual dalliance

¹Han•o•ve•ri•an \ˌha-nə-'vir-ē-ən, -'ver-\ *adjective* [*Hanover,* Germany] (circa 1775)
1 : of, relating to, or supporting the German ducal house of Hanover
2 : of or relating to the British royal house that ruled from 1714 to 1901

²Hanoverian *noun* (1827)
: a member or supporter of the ducal or of the British royal Hanoverian house

Han•sa \'han(t)-sə, 'hän-(ˌ)zä\ *or* **Hanse** \'han(t)s, 'hän-zə\ *noun* [*Hansa* from Medieval Latin, from Middle Low German *hanse; Hanse* from Middle English *Hanze,* from Middle French *hanse,* from Middle Low German] (15th century)
1 : a league originally constituted of merchants of various free German cities dealing abroad in the medieval period and later of the cities themselves and organized to secure greater safety and privileges in trading
2 : a medieval merchant guild or trading association
— Han•se•at•ic \ˌhan(t)-sē-'a-tik\ *noun or adjective*

Han•sard \'han(t)-sərd, 'han-ˌsärd\ *noun* [Luke *Hansard*] (circa 1859)
: the official published report of debates in a Commonwealth parliament

han•sel *variant of* HANDSEL

Han•sen's disease \'han(t)-sənz-\ *noun* [Armauer *Hansen* (died 1912) Norwegian physician] (1938)
: LEPROSY

han•som \'han(t)-səm\ *noun* [Joseph A. *Hansom* (died 1882) English architect] (1847)
: a light 2-wheeled covered carriage with the driver's seat elevated behind — called also *hansom cab*

hansom

hant \'hant\ *dialect variant of* HAUNT

Ha•nuk•kah \'hä-nə-kə, 'kä-\ *noun* [Hebrew *ḥănukkāh* dedication] (1891)
: an 8-day Jewish holiday beginning on the 25th of Kislev and commemorating the rededication of the Temple of Jerusalem after its defilement by Antiochus of Syria

hao \'hau̇\ *noun, plural* **hao** [Vietnamese *hào*] (1948)
: a monetary unit of Vietnam equal to ¹⁄₁₀ dong

hao•le \'hau̇-lē, -(ˌ)lā\ *noun* [Hawaiian] (1834)
: one who is not descended from the aboriginal Polynesian inhabitants of Hawaii; *especially* **:** WHITE — sometimes used disparagingly

¹hap \'hap\ *noun* [Middle English, from Old Norse *happ* good luck; akin to Old English *gehæp* suitable, Old Church Slavonic *kobĭ* lot, fate] (13th century)
1 : HAPPENING 1
2 : CHANCE, FORTUNE

²hap *intransitive verb* **happed; hap•ping** (14th century)
: HAPPEN

³hap *transitive verb* **happed; hap•ping** [Middle English *happen*] (14th century)
dialect **:** CLOTHE, COVER

⁴hap *noun* (1724)
dialect **:** something (as a bed quilt or cloak) that serves as a covering or wrap

ha•pa hao•le \ˌhä-pə-'hau̇-lē, -(ˌ)lā\ *adjective* [Hawaiian, from *hapa* half (from English *half*) + *haole*] (1919)
: of part-white ancestry or origin; *especially* **:** of white and Hawaiian ancestry

ha•pax le•go•me•non \ˌha-ˌpaks-li-'gä-mə-ˌnän, ˌhä-ˌpäks-, -nən\ *noun, plural* **hapax le•go•me•na** \-nə\ [Greek, something said only once] (1882)
: a word or form occurring only once in a document or corpus

ha'•pen•ny \'hāp-nē, 'hā-pə-\ *noun* [by contraction] (circa 1550)
: HALFPENNY

¹hap•haz•ard \(ˌ)hap-'ha-zərd\ *noun* [¹*hap* + *hazard*] (1576)
: CHANCE

²haphazard *adjective* (1671)
: marked by lack of plan, order, or direction
synonym see RANDOM
— haphazard *adverb*
— hap•haz•ard•ly *adverb*
— hap•haz•ard•ness *noun*
— hap•haz•ard•ry \-zər-drē\ *noun*

hapl- *or* **haplo-** *combining form* [New Latin, from Greek, from *haploos,* from *ha-* one (akin to *homos* same) + *-ploos* multiplied by; akin to Latin *-plex* -fold — more at SAME, -FOLD]
1 : single ⟨*haplo*logy⟩
2 : haploid ⟨*hapl*ont⟩

hap•less \'ha-pləs\ *adjective* (14th century)
: having no luck **:** UNFORTUNATE
— hap•less•ly *adverb*
— hap•less•ness *noun*

hap•loid \'ha-ˌplȯid\ *adjective* [International Scientific Vocabulary, from Greek *haploeidēs* single, from *haploos*] (1908)
: having the gametic number of chromosomes or half the number characteristic of somatic cells — compare DIPLOID
— haploid *noun*
— hap•loi•dy \-ˌplȯi-dē\ *noun*

hap•lol•o•gy \ha-'plä-lə-jē\ *noun* (1895)
: contraction of a word by omission of one or more similar sounds or syllables (as in *mineralogy* for **mineralogy* or \'prä-blē\ for *probably*)

hap•lont \'ha-ˌplänt\ *noun* [International Scientific Vocabulary] (1920)
: an organism (as some primitive algae) having a diploid zygote that undergoes meiosis to produce haploid cells — compare DIPLONT
— hap•lon•tic \ha-'plän-tik\ *adjective*

hap•lo•type \'ha-plō-ˌtīp\ *noun* (1969)
: a set of genes that determine different antigens but are closely enough linked to be inherited as a unit; *also* **:** the antigenic phenotype determined by a haplotype

hap•ly \'ha-plē\ *adverb* (14th century)
: by chance, luck, or accident

hap•pen \'ha-pən, -pᵊm\ *intransitive verb* **hap•pened; hap•pen•ing** \'hap-niŋ, 'ha-pə-\ [Middle English, from *hap*] (14th century)
1 : to occur by chance — often used with *it* ⟨it so *happens* I'm going your way⟩
2 : to come into being or occur as an event, process, or result ⟨mistakes will *happen*⟩ ⟨what good things *happened* to you?⟩

◇ WORD HISTORY
hangnail *Hangnail* is altered by folk etymology from *angnail* or *agnail,* which in its earliest sense did not mean what we now call a hangnail. In Old English *angnægl* was used to mean "corn on the foot," with the second element, *nægl,* referring not to a fingernail or toenail but rather to the nail we drive in with a hammer, the head of an iron nail being likened to a hard corn. By the 16th century the association of *-nail* with the body's nails led to a new sense, "a sore or inflammation around a fingernail or toenail." The first element *ang-* or *ag-,* which is related to an Old English adjective *enge,* meaning "tight, painful," had ceased to be interpretable since Old English. In the next century it was altered to *hang-,* and the dominant sense of both *hangnail* and *agnail* came to be "a bit of loose skin at the side or root of a fingernail."

3 : to do, encounter, or attain something by or as if by chance ⟨*happened* to overhear the plotters⟩
4 a : to meet or discover something by chance ⟨*happened* upon a system that worked —Richard Corbin⟩ **b :** to come or go casually **:** make a chance appearance ⟨he *happened* into the room just then⟩
5 : to come especially by way of injury or harm ⟨I promise nothing will *happen* to you⟩

hap·pen·chance \'ha-pən-ˌchan(t)s, 'ha-p^{ə}m-\ *noun* (1876)
: HAPPENSTANCE

hap·pen·ing *noun* (1551)
1 : something that happens **:** OCCURRENCE
2 : an event or series of events designed to evoke a spontaneous reaction to sensory, emotional, or spiritual stimuli
3 : something (as an event) that is particularly interesting, entertaining, or important

hap·pen·stance \'ha-pən-ˌstan(t)s, 'ha-p^{ə}m-\ *noun* [*happen* + circum*stance*] (1897)
: a circumstance especially that is due to chance
— hap·pen·stance *adjective*

hap·pi·ly \'ha-pə-lē\ *adverb* (14th century)
1 a : in a fortunate manner **b :** as it fortunately happens ⟨*happily*, some boyhood pleasures don't change —P. A. Witteman⟩
2 *archaic* **:** by chance
3 : in a happy manner or state ⟨lived *happily* ever after⟩
4 : in an adequate or fitting manner **:** SUCCESSFULLY

hap·pi·ness \'ha-pi-nəs\ *noun* (15th century)
1 *obsolete* **:** good fortune **:** PROSPERITY
2 a : a state of well-being and contentment **:** JOY **b :** a pleasurable or satisfying experience
3 : FELICITY, APTNESS

hap·py \'ha-pē\ *adjective* **hap·pi·er; -est** [Middle English, from *hap*] (14th century)
1 : favored by luck or fortune **:** FORTUNATE
2 : notably fitting, effective, or well adapted **:** FELICITOUS ⟨a *happy* choice⟩
3 a : enjoying or characterized by well-being and contentment **:** JOYOUS **b :** expressing or suggestive of happiness **:** PLEASANT **c :** GLAD, PLEASED **d :** having or marked by an atmosphere of good fellowship **:** FRIENDLY
4 a : characterized by a dazed irresponsible state ⟨a punch-*happy* boxer⟩ **b :** impulsively or obsessively quick to use or do something ⟨trigger-*happy*⟩ **c :** enthusiastic about something to the point of obsession **:** OBSESSED ⟨education-conscious and statistic-*happy* —Helen Rowen⟩
synonym see LUCKY, FIT

hap·py–go–lucky \ˌha-pē-gō-'lə-kē\ *adjective* (1856)
: blithely unconcerned **:** CAREFREE

happy hour *noun* (1961)
: a period of time during which the price of drinks (as at a bar) is reduced or hors d'oeuvres are served free

happy hunting ground *noun* (1837)
1 : the paradise of some American Indian tribes to which the souls of warriors and hunters pass after death to spend a happy hereafter in hunting and feasting
2 : a choice or profitable area of activity or exploitation

happy talk *noun* (1973)
: informal talk among the participants in a television news broadcast; *also* **:** a broadcast format featuring such talk

Haps·burg \'haps-ˌbərg, 'häps-ˌbu̇rg\ *adjective* [*Habsburg,* Aargau, Switzerland] (circa 1895)
: of or relating to the German royal house to which belong the rulers of Austria from 1278 to 1918, the rulers of Spain from 1516 to 1700, and many of the Holy Roman emperors
— Hapsburg *noun*

hap·ten \'hap-ˌten\ *noun* [German, from Greek *haptein* to fasten] (1921)
: a small separable part of an antigen that reacts specifically with an antibody but is incapable of stimulating antibody production except in combination with a carrier protein molecule
— hap·ten·ic \hap-'te-nik\ *adjective*

hap·tic \'hap-tik\ *adjective* [International Scientific Vocabulary, from Greek *haptesthai* to touch] (circa 1890)
1 : relating to or based on the sense of touch
2 : characterized by a predilection for the sense of touch ⟨a *haptic* person⟩

hap·to·glo·bin \'hap-tə-ˌglō-bən\ *noun* [International Scientific Vocabulary, from Greek *haptein* + International Scientific Vocabulary hemo*globin*] (1941)
: any of several carbohydrate-containing serum alpha globulins that can combine with free hemoglobin in the plasma

hara–kiri \ˌhar-i-'kir-ē, -'kar-ē\ *noun* [Japanese *harakiri,* from *hara* belly + *kiri* cutting] (1840)
1 : ritual suicide by disembowelment practiced by the Japanese samurai or formerly decreed by a court in lieu of the death penalty
2 : SUICIDE 1b

[1]ha·rangue \hə-'raŋ\ *noun* [Middle English *arang,* from Middle French *arenge,* from Old Italian *aringa,* from *aringare* to speak in public, from *aringo* public assembly, of Germanic origin; akin to Old High German *hring* ring] (15th century)
1 : a speech addressed to a public assembly
2 : a ranting speech or writing
3 : LECTURE

[2]harangue *verb* **ha·rangued; ha·rangu·ing** (1660)
intransitive verb
: to make a harangue **:** DECLAIM
transitive verb
: to address in a harangue ⟨*haranguing* me . . . on the folly of my ways —Jay Jacobs⟩
— ha·rangu·er *noun*

ha·rass \hə-'ras, 'har-əs\ *transitive verb* [French *harasser,* from Middle French, from *harer* to set a dog on, from Old French *hare,* interjection used to incite dogs, of Germanic origin; akin to Old High German *hier* here — more at HERE] (1617)
1 a : EXHAUST, FATIGUE **b :** to annoy persistently
2 : to worry and impede by repeated raids ⟨*harassed* the enemy⟩ ▢
synonym see WORRY
— ha·rass·er *noun*
— ha·rass·ment \-mənt\ *noun*

[1]har·bin·ger \'här-bən-jər\ *noun* [Middle English *herbergere,* from Middle French, host, from *herberge* hostelry, of Germanic origin; akin to Old High German *heriberga*] (14th century)
1 *archaic* **:** a person sent ahead to provide lodgings
2 a : one that pioneers in or initiates a major change **:** PRECURSOR **b :** one that presages or foreshadows what is to come ◆
synonym see FORERUNNER

[2]harbinger *transitive verb* (1646)
: to be a harbinger of **:** PRESAGE

[1]har·bor \'här-bər\ *noun* [Middle English *herberge, herberwe,* from Old English *herebeorg* military quarters, from *here* army (akin to Old High German *heri*) + *beorg* refuge; akin to Old English *burg* fortified town — more at HARRY, BOROUGH] (12th century)
1 : a place of security and comfort **:** REFUGE
2 : a part of a body of water protected and deep enough to furnish anchorage; *especially* **:** one with port facilities
word history see HARBINGER
— har·bor·ful \-ˌfu̇l\ *noun*
— har·bor·less \-ləs\ *adjective*

[2]harbor *verb* **har·bored; har·bor·ing** \-b(ə-)riŋ\ (12th century)
transitive verb
1 a : to give shelter or refuge to **b :** to be the home or habitat of ⟨the ledges still *harbor* rattlesnakes⟩; *broadly* **:** CONTAIN 2
2 : to hold especially persistently in the mind **:** CHERISH ⟨*harbored* a grudge⟩
intransitive verb
1 : to take shelter in or as if in a harbor
2 : LIVE
— har·bor·er \-bər-ər\ *noun*

har·bor·age \-bə-rij\ *noun* (15th century)
: SHELTER, HARBOR

har·bor·mas·ter \'här-bər-ˌmas-tər\ *noun* (1769)
: an officer who executes the regulations respecting the use of a harbor

harbor seal *noun* (1766)
: a small seal (*Phoca vitulina*) that occurs along oceanic coasts of the northern hemisphere and often ascends rivers

har·bor·side \'här-bər-ˌsīd\ *adjective* (1924)
: located next to a harbor

har·bour *chiefly British variant of* HARBOR

[1]hard \'härd\ *adjective* [Middle English, from Old English *heard;* akin to Old High German *hart* hard, Greek *kratos* strength] (before 12th century)
1 a : not easily penetrated **:** not easily yielding to pressure **b** *of cheese* **:** not capable of being spread **:** very firm
2 a *of liquor* (1) **:** having a harsh or acid taste (2) **:** strongly alcoholic **b :** characterized by the presence of salts (as of calcium or magnesium) that prevent lathering with soap ⟨*hard* water⟩
3 a : of or relating to radiation of relatively high penetrating power ⟨*hard* X rays⟩ **b :** having or producing relatively great photographic contrast ⟨a *hard* negative⟩
4 a : metallic as distinct from paper ⟨*hard* money⟩ **b** *of currency* **:** convertible into gold **:** stable in value **c :** usable as currency ⟨paid in *hard* cash⟩ **d** *of currency* **:** readily acceptable in international trade **e :** being high and firm ⟨*hard* prices⟩
5 a : firmly and closely twisted ⟨*hard* yarns⟩ **b :** having a smooth close napless finish ⟨a *hard* worsted⟩

▢ USAGE
harass Some usage critics insist that *harass* should be pronounced with stress on the penultimate syllable as in *embarrass* and not on the final syllable as in *morass.* Americans prefer final stress in harass by about 2 to 1, whereas the British prefer the other variant in the same ratio. The final stress variant began to take hold in American speech around the beginning of the twentieth century and by mid-century had become a standard pronunciation. Both variants are acceptable on either side of the Atlantic.

◇ WORD HISTORY
harbinger A modern harbinger is simply a precursor or forerunner, but the duties of earlier harbingers were more specific. In late medieval and early modern times, they were people sent before an army or a royal progress to find lodgings for the whole company. Earlier the form of the word was *herbergere,* and the first English *herbergeres* were the hosts themselves, the actual providers of lodgings. The Old French word from which the English was borrowed was itself a loanword from Frankish, a Germanic language of Gaul. Old French *herberge* took from its Frankish source both the literal meaning "army encampment" and the figurative extension "hostelry." Modern English *harbor* is another descendant of the same common prehistoric West Germanic word as the Frankish source of *herberge,* having the meaning "a place of security and comfort" before developing the sense "a sheltered place providing anchorage for ships."

6 a : physically fit ⟨in good *hard* condition⟩ **b :** resistant to stress (as disease) **c :** free of weakness or defects
7 a (1) **:** FIRM, DEFINITE ⟨reached a *hard* agreement⟩ (2) **:** not speculative or conjectural **:** FACTUAL ⟨*hard* evidence⟩ (3) **:** important or informative rather than sensational or entertaining ⟨*hard* news⟩ **b :** CLOSE, SEARCHING ⟨gave a *hard* look⟩ **c :** free from sentimentality or illusion **:** REALISTIC ⟨good *hard* sense⟩ **d :** lacking in responsiveness **:** OBDURATE, UNFEELING ⟨a *hard* heart⟩
8 a (1) **:** difficult to bear or endure ⟨*hard* luck⟩ ⟨*hard* times⟩ (2) **:** OPPRESSIVE, INEQUITABLE ⟨sales taxes are *hard* on the poor⟩ ⟨a *hard* restriction⟩ **b** (1) **:** lacking consideration, compassion, or gentleness **:** CALLOUS ⟨a *hard* greedy landlord⟩ (2) **:** INCORRIGIBLE, TOUGH ⟨a *hard* gang⟩ **c** (1) **:** harsh, severe, or offensive in tendency or effect ⟨said some *hard* things⟩ (2) **:** RESENTFUL ⟨*hard* feelings⟩ (3) **:** STRICT, UNRELENTING ⟨drives a *hard* bargain⟩ **d :** INCLEMENT ⟨*hard* winter⟩ **e** (1) **:** intense in force, manner, or degree ⟨*hard* blows⟩ (2) **:** demanding the exertion of energy **:** calling for stamina and endurance ⟨*hard* work⟩ (3) **:** performing or carrying on with great energy, intensity, or persistence ⟨a *hard* worker⟩ **f :** EXTREME 4a ⟨the *hard* political right⟩
9 a : characterized by sharp or harsh outline, rigid execution, and stiff drawing **b :** sharply defined **:** STARK ⟨*hard* shadows⟩ **c :** lacking in shading, delicacy, or resonance ⟨*hard* singing tones⟩ **d :** sounding as in *arcing* and *geese* respectively — used of *c* and *g* **e :** suggestive of toughness or insensitivity ⟨*hard* eyes⟩
10 a (1) **:** difficult to accomplish or resolve **:** TROUBLESOME ⟨*hard* problems⟩ ⟨the true story was *hard* to come by⟩ (2) **:** difficult to comprehend or explain ⟨a *hard* concept⟩ **b :** having difficulty in doing something ⟨*hard* of hearing⟩ **c :** difficult to magnetize or demagnetize
11 : being at once addictive and gravely detrimental to health ⟨such *hard* drugs as heroin⟩
12 : resistant to biodegradation ⟨*hard* detergents⟩ ⟨*hard* pesticides like DDT⟩
13 : being, schooled in, or using the methods of the natural sciences and especially of the physical sciences ⟨a *hard* scientist⟩ ☆

²hard *adverb* (before 12th century)
1 a : with great or utmost effort or energy **:** STRENUOUSLY ⟨were *hard* at work⟩ ⟨the children played *hard*⟩ **b :** in a violent manner **:** FIERCELY **c :** to the full extent — usually used in nautical directions ⟨steer *hard* aport⟩ **d :** to an immoderate degree ⟨hitting the bottle *hard*⟩ **e :** in a searching, close, or concentrated manner ⟨stared *hard* at me⟩
2 a : in such a manner as to cause hardship, difficulty, or pain **b :** with rancor, bitterness, or grief ⟨took the defeat *hard*⟩
3 : in a firm manner **:** TIGHTLY
4 : to the point of hardness ⟨frozen *hard*⟩
5 : close in time or space ⟨stands *hard* by the river⟩

hard–and–fast \ˌhär-dᵊn-ˈfast\ *adjective* (1867)
: not to be modified or evaded **:** STRICT ⟨a *hard-and-fast* rule⟩

hard–ass \ˈhär-ˌdas\ *noun* (1972)
: a tough, demanding, or uncompromising person — often considered vulgar
— hard–assed \-ˌdast\ *adjective*

hard•back \ˈhärd-ˌbak\ *noun* (1952)
: a book bound in hard covers

hard•ball \-ˌbȯl\ *noun* (circa 1883)
1 : BASEBALL
2 : forceful uncompromising methods employed to gain an end ⟨played political *hardball*⟩

hard–bit•ten \-ˈbi-tᵊn\ *adjective* (1784)
1 : inclined to bite hard
2 : seasoned or steeled by difficult experience **:** TOUGH

hard•board \ˈhärd-ˌbōrd, -ˌbȯrd\ *noun* (1925)
: a very dense fiberboard usually having one smooth face

hard–boil \-ˈbȯi(ə)l\ *transitive verb* [back-formation from *hard-boiled*] (1895)
: to cook (an egg) in the shell until both white and yolk have solidified

hard–boiled \-ˈbȯi(ə)ld\ *adjective* (1886)
1 a : devoid of sentimentality **:** TOUGH ⟨a *hard-boiled* drill sergeant⟩ **b :** of, relating to, or being a detective story featuring a tough unsentimental protagonist and a matter-of-fact attitude towards violence
2 : HARDHEADED, PRACTICAL ⟨handle aid programs on a friendly but *hard-boiled* business basis —*N.Y. Times*⟩

hard•boot \-ˌbüt\ *noun* (1922)
: an especially small-time horseman

hard•bound \-ˌbau̇nd\ *adjective* (1926)
: HARDCOVER

hard candy *noun* (1925)
: a candy made of sugar and corn syrup boiled without crystallizing

hard•case \ˈhärd-ˈkās\ *adjective* (1896)
: HARD-BITTEN, TOUGH

hard case \-ˌkās\ *noun* (1836)
: a tough or hardened person

hard cheese *noun* (1861)
chiefly British **:** tough luck — often used interjectionally

hard cider *noun* (1789)
: fermented apple juice

hard clam *noun* (1846)
: a clam with a thick hard shell; *specifically* **:** QUAHOG

hard coal *noun* (1846)
: ANTHRACITE

hard–coat•ed \ˈhärd-ˈkō-təd\ *adjective* (circa 1898)
of a dog **:** having a crisp harsh-textured coat

hard copy *noun* (1954)
: a copy of textual or graphic information (as from microfilm or computer storage) produced on paper in normal size

hard–core \-ˈkōr, -ˈkȯr\ *adjective* (1940)
1 a : of, relating to, or being part of a hard core ⟨*hard-core* poverty⟩ ⟨the *hard-core* unemployed⟩ **b :** CONFIRMED, DIE-HARD ⟨*hard-core* rock fans⟩ ⟨a *hard-core* liberal⟩
2 *of pornography* **:** containing explicit descriptions of sex acts or scenes of actual sex acts — compare SOFT-CORE
3 : characterized by or being the purest or most basic form of something **:** FUNDAMENTAL ⟨a room gussied up in *hard-core* French provincial style —John Canaday⟩

hard core *noun* (1936)
1 : a central or fundamental and usually enduring group or part: as **a :** a relatively small enduring core of society marked by apparent resistance to change or inability to escape a persistent wretched condition (as poverty or chronic unemployment) **b :** a militant or fiercely loyal faction
2 *usually* **hard•core** \-ˌkōr, -ˌkȯr\ *chiefly British* **:** hard material in pieces (as broken bricks or stone) used as a bottom (as in making roads and in foundations)

hard•cov•er \ˈhärd-ˈkə-vər\ *adjective* (1949)
1 : having rigid boards on the sides covered in cloth or paper ⟨*hardcover* books⟩
2 : of or relating to hardcover books ⟨*hardcover* sales⟩
— hardcover *noun*

hard disk *noun* (1978)
: a rigid metal disk that is sealed against dust and is used as a high-capacity storage device for a microcomputer

hard drive *noun* (1983)
: a disk drive containing one or more hard disks

hard–edge \ˈhärd-ˈej\ *adjective* (1961)
: of or relating to abstract painting characterized by geometric forms with clearly defined boundaries

hard–edged \-ˈejd\ *adjective* (1954)
: having a tough, driving, or sharp quality ⟨*hard-edged* stories about life in the city⟩

hard•en \ˈhär-dᵊn\ *verb* **hard•ened; hard•en•ing** \ˈhärd-niŋ, ˈhär-dᵊn-iŋ\ (13th century)
transitive verb
1 : to make hard or harder
2 : to confirm in disposition, feelings, or action; *especially* **:** to make callous ⟨*hardened* his heart⟩
3 a : INURE, TOUGHEN ⟨*harden* troops⟩ **b :** to inure to unfavorable environmental conditions (as cold) — often used with *off* ⟨*harden* off plants⟩
4 : to protect from blast, heat, or radiation (as by a thick barrier or placement underground)
intransitive verb
1 : to become hard or harder
2 a : to become firm, stable, or settled **b :** to assume an appearance of harshness or severity ⟨her face *hardened* at the word⟩

hard•en•er \ˈhärd-nər, ˈhär-dᵊn-ər\ *noun* (1611)
: one that hardens; *especially* **:** a substance added (as to a paint or varnish) to harden the film

hard•en•ing *noun* (1953)
: SCLEROSIS ⟨*hardening* of the arteries⟩

hard•fist•ed \ˈhärd-ˈfis-təd\ *adjective* (circa 1656)
1 : STINGY, CLOSEFISTED
2 : HARDHANDED 2

hard goods *noun plural* (1934)
: DURABLES

hard•hack \ˈhärd-ˌhak\ *noun* (1814)
: a shrubby American spirea (*Spiraea tomentosa*) with dense terminal panicles of pink or occasionally white flowers and leaves having a hairy and yellow to rust-colored underside

hard•hand•ed \-ˈhan-dəd\ *adjective* (1590)
1 : having hands made hard by labor
2 : STRICT, OPPRESSIVE
— hard•hand•ed•ness *noun*

hard hat *usually* -ˈhat *for 1 and* -ˌhat *for 2 & 3*\ *noun* (1926)
1 a : a protective hat made of rigid material (as metal or fiberglass) and worn especially by construction workers **b :** a construction worker
2 : a conservative who is intolerant of opposing views

hard•head \ˈhärd-ˌhed\ *noun* (15th century)
1 a : a hardheaded person **b :** BLOCKHEAD
2 *plural* **hardheads** *also* **hardhead :** any of several fishes especially with a spiny or bony head; *especially* **:** ATLANTIC CROAKER

hard•head•ed \-ˈhe-dəd\ *adjective* (1583)
1 : STUBBORN, WILLFUL
2 : concerned with or involving practical considerations **:** SOBER, REALISTIC ⟨some *hardheaded* advice⟩ ⟨a *hardheaded* observer of winds and tides⟩
— hard•head•ed•ly *adverb*
— hard•head•ed•ness *noun*

hard–heart•ed \-ˈhär-təd\ *adjective* (13th century)
: lacking in sympathetic understanding **:** UNFEELING, PITILESS

☆ SYNONYMS
Hard, difficult, arduous mean demanding great exertion or effort. HARD implies the opposite of all that is easy ⟨farming is *hard* work⟩. DIFFICULT implies the presence of obstacles to be surmounted or puzzles to be resolved and suggests the need of skill, patience, or courage ⟨the *difficult* ascent of the main face of the mountain⟩. ARDUOUS stresses the need of laborious and persevering exertion ⟨the *arduous* task of rebuilding⟩.

— **hard-heart·ed·ly** *adverb*
— **hard-heart·ed·ness** *noun*
hard–hit·ting \-'hi-tiŋ\ *adjective* (1926)
: strikingly effective in force or result ⟨a *hard-hitting* exposé⟩ ⟨plain *hard-hitting* English⟩
har·di·hood \'här-dē-,hu̇d\ *noun* (1570)
1 a : resolute courage and fortitude **b :** resolute and self-assured audacity often carried to the point of impudent insolence
2 : VIGOR, ROBUSTNESS
synonym see TEMERITY
har·di·ment \-mənt\ *noun* [Middle English, from Middle French, from Old French, from *hardi* bold, hardy] (14th century)
1 *archaic* **:** HARDIHOOD
2 *obsolete* **:** a bold deed
har·ding·grass \'här-diŋ-,gras\ *noun, often capitalized* [R. R. *Harding* (flourished about 1900) Australian botanist] (1917)
: a perennial grass (*Phalaris tuberosa* variety *stenoptera*) widely used as a forage grass
hard knocks *noun plural* (1913)
: rough unsparing treatment (as in use or in life) — often used in the phrase *school of hard knocks*
hard labor *noun* (1841)
: compulsory labor of imprisoned criminals as a part of the prison discipline
hard–line \'härd-'līn\ *adjective* (1962)
: advocating or involving a rigidly uncompromising course of action
— **hard–lin·er** \-'lī-nər\ *noun*
hard lines *noun plural* (1824)
chiefly British **:** hard luck — often used interjectionally
hard·ly \'härd-lē\ *adverb* (before 12th century)
1 : with force **:** VIGOROUSLY
2 : in a severe manner **:** HARSHLY
3 : with difficulty **:** PAINFULLY
4 a — used to emphasize a minimal amount ⟨I *hardly* knew her⟩ ⟨almost new — *hardly* a scratch on it⟩ **b** — used to soften a negative ⟨you can't *hardly* tell who anyone is —G. B. Shaw⟩
5 : certainly not ⟨that news is *hardly* surprising⟩ ▫
hard maple *noun* (1790)
: SUGAR MAPLE
hard·mouthed \'härd-'mau̇thd, -'mau̇tht\ *adjective* (1617)
1 *of a horse* **:** not sensitive to the bit
2 : OBSTINATE, STUBBORN
hard·ness \-nəs\ *noun* (before 12th century)
1 : the quality or state of being hard
2 a : the cohesion of the particles on the surface of a mineral as determined by its capacity to scratch another or be itself scratched — compare MOHS' SCALE **b :** resistance of metal to indentation under a static load or to scratching
hard·nose \'härd-,nōz\ *noun* (circa 1960)
: a hard-nosed person
hard–nosed \'härd-'nōzd\ *adjective* (circa 1927)
1 : being tough, stubborn, or uncompromising
2 : HARDHEADED 2, TOUGH-MINDED
hard–of–hear·ing \,här-də(v)-'hir-iŋ\ *adjective* (1564)
: of or relating to a defective but functional sense of hearing
hard–on \'här-,dȯn, -,dän\ *noun, plural* **hard–ons** (circa 1893)
: an erection of the penis — sometimes considered vulgar
hard palate *noun* (circa 1847)
: the bony anterior part of the palate forming the roof of the mouth
hard·pan \'härd-,pan\ *noun* (1817)
1 : a cemented or compacted and often clayey layer in soil that is impenetrable by roots
2 : a fundamental part **:** BEDROCK
hard pine *noun* (1884)
: a pine (as longleaf pine or pitch pine) that has hard wood and leaves usually in groups of two or three; *also* **:** the wood of a hard pine
hard–pressed \'härd-'prest\ *adjective* (1825)
: HARD PUT; *also* **:** being under financial strain
hard put *adjective* (1893)
: barely able **:** faced with difficulty or perplexity ⟨was *hard put* to find an explanation⟩
hard rock *noun* (1967)
: rock music marked by a heavy regular beat, high amplification, and usually frenzied performances
hard rubber *noun* (1860)
: a firm rubber or rubber product; *especially* **:** a normally black horny substance made by vulcanizing natural rubber with high percentages of sulfur
hard sauce *noun* (1880)
: a creamed mixture of butter and powdered sugar often with added cream and flavoring (as vanilla or rum)
hard·scrab·ble \'härd-,skra-bəl\ *adjective* (1804)
1 a : being or relating to a place of barren or barely arable soil ⟨a *hardscrabble* farm⟩ ⟨*hardscrabble* prairies⟩ **b :** getting a meager living from poor soil ⟨a *hardscrabble* farmer⟩
2 : marked by poverty ⟨a *hardscrabble* cotton town⟩ ⟨a *hardscrabble* childhood⟩
hard sell *noun* (1952)
: aggressive high-pressure salesmanship — compare SOFT SELL
hard–set \'härd-'set\ *adjective* (1813)
: RIGID, FIXED
hard–shell \-,shel\ *or* **hard–shelled** \-,sheld\ *adjective* (1838)
: FUNDAMENTAL 2b, FUNDAMENTALIST ⟨a *hard-shell* preacher⟩ ⟨*hard-shell* Baptists⟩; *also* **:** UNCOMPROMISING, HIDEBOUND ⟨a *hard-shell* conservative⟩
hard–shell clam *noun* (1799)
: QUAHOG — called also *hard-shelled clam*
hard–shell crab *noun* (1902)
: a crab that has not recently shed its shell — called also *hard-shelled crab*
hard·ship \'härd-,ship\ *noun* (13th century)
1 : PRIVATION, SUFFERING
2 : something that causes or entails suffering or privation
hard·stand \-,stand\ *noun* (1944)
: a paved area for parking an airplane
hard·stand·ing \-,stan-diŋ\ *noun* (1944)
chiefly British **:** HARDSTAND; *also* **:** PARKING LOT
hard stone *noun* (1931)
: an opaque usually semiprecious stone that can be shaped or carved (as for jewelery or mosaics)
hard·tack \-,tak\ *noun, plural* **hardtack** *or* **hardtacks** (1836)
1 : a saltless hard biscuit, bread, or cracker
2 : any of several mountain mahoganies (especially *Cercocarpus betuloides*)
hard–times token *noun* (1922)
: any of the tokens issued during the controversy between the Jackson administration and the bank of the U.S.
hard·top \'härd-,täp\ *noun* (1950)
: an automobile or a motorboat having a permanent rigid top; *also* **:** such an automobile styled to resemble a convertible
hard up *adjective* (1821)
1 : short of money
2 : poorly provided ⟨*hard up* for friends⟩
hard·ware \'härd-,war, -,wer\ *noun* (circa 1515)
1 : ware (as fittings, cutlery, tools, utensils, or parts of machines) made of metal
2 : major items of equipment or their components used for a particular purpose ⟨educational *hardware*⟩: as **a :** military equipment **b :** the physical components (as electronic and electrical devices) of a vehicle (as a spacecraft) or an apparatus (as a computer)
hardware cloth *noun* (circa 1914)
: rugged galvanized screening
hard wheat *noun* (1812)
: a wheat with hard flinty kernels that are high in gluten and that yield a flour especially suitable for bread and macaroni
hard·wired \'härd-,wīrd\ *adjective* (1968)
: implemented in the form of permanent electronic circuits; *also* **:** having permanent electrical connections ⟨*hardwired* phone⟩
hard–won \-'wən\ *adjective* (circa 1843)
: gained by great effort
¹hard·wood \'härd-,wu̇d\ *noun* (1568)
1 : the wood of an angiospermous tree as distinguished from that of a coniferous tree
2 : a tree that yields hardwood
²hardwood *adjective* (circa 1817)
1 : having or made of hardwood ⟨*hardwood* floors⟩
2 : consisting of mature woody tissue ⟨*hardwood* cuttings⟩
hard·work·ing \'härd-'wər-kiŋ\ *adjective* (1774)
: INDUSTRIOUS
har·dy \'här-dē\ *adjective* **har·di·er; -est** [Middle English *hardi,* from Old French, from (assumed) Old French *hardir* to make hard, of Germanic origin; akin to Old English *heard* hard] (13th century)
1 : BOLD, BRAVE
2 : AUDACIOUS, BRAZEN
3 a : inured to fatigue or hardships **:** ROBUST **b :** capable of withstanding adverse conditions ⟨*hardy* outdoor furniture⟩ ⟨*hardy* plants⟩ ⟨*hardy* cattle⟩
— **har·di·ly** \'här-dᵊl-ē\ *adverb*
— **har·di·ness** \'här-dē-nəs\ *noun*
Har·dy–Wein·berg law \,här-dē-'wīn-,bərg-\ *noun* [G. H. *Hardy* (died 1947) English mathematician and W. *Weinberg,* 20th century German scientist] (1950)
: a fundamental principle of population genetics: population gene frequencies and genotype frequencies remain constant from generation to generation if mating is random and if mutation, selection, immigration, and emigration do not occur — called also *Hardy-Weinberg principle*
¹hare \'har, 'her\ *noun, plural* **hare** *or* **hares** [Middle English, from Old English *hara;* akin to Old High German *haso* hare, Sanskrit *śaśa,* Old English *hasu* gray] (before 12th century)
: any of various swift long-eared lagomorph mammals (family Leporidae and especially genus *Lepus*) that are usually solitary or sometimes live in pairs and have the young open-eyed and furred at birth — compare RABBIT 1A
²hare *intransitive verb* **hared; har·ing** (1719)
: to go swiftly **:** ³TEAR
hare and hounds *noun* (circa 1845)
: a game in which some of the players leave a trail and others try to follow the trail to find and catch them
hare·bell \'har-,bel, 'her-\ *noun* (1765)
: a slender blue-flowered herb (*Campanula rotundifolia*) with linear leaves on the stem
hare·brained \-'brānd\ *adjective* (1534)
: FOOLISH 1, 2a
Ha·re Krish·na \'här-ē-'krish-nə, ,har-\ *noun, plural* **Hare Krishna** *or* **Hare Krishnas**

☐ USAGE
hardly *Hardly* in sense 5 is used sometimes with *not* for emphasis ⟨just another day at the office? Not *hardly*⟩. In sense 4b with a negative verb (as *can't, wouldn't, didn't*) it does not make a double negative but softens the negative. In "you can't hardly find a red one," the sense is that you can find a red one, but only with difficulty; in "you can't find a red one," the sense is that red ones are simply not available. Use of *hardly* with a negative verb is a speech form; it is most commonly heard in Southern and Midland speech areas. In other speech areas and in all discursive prose, *hardly* is normally used with a positive ⟨you can *hardly* find a red one⟩.

[from *Hare Krishna,* phrase in a chant, from Hindi *hare Kṛṣṇa* O Krishna!] (1969)
: a member of a religious group dedicated to the worship of the Hindu god Krishna

hare·lip \'har-'lip, 'her-\ *noun* (1567)
: a congenital deformity characterized by a cleft upper lip resulting from failure of the embryonic parts of the lip to unite

har·em \'har-əm, 'her-\ *noun* [Arabic *ḥarīm,* literally, something forbidden & *ḥaram,* literally, sanctuary] (1623)
1 a : a usually secluded house or part of a house allotted to women in a Muslim household **b :** the wives, concubines, female relatives, and servants occupying a harem
2 : a group of women associated with one man
3 : a group of females associated with one male — used of polygamous animals

harem pants *noun* [from their oriental appearance] (1952)
: women's loose trousers that fit closely at the ankle

har·i·cot \'(h)ar-i-ˌkō\ *noun* [French] (1653)
: the ripe seed or the unripe pod of any of several beans (genus *Phaseolus* and especially *P. vulgaris*)

ha·ri·jan \ˌhär-i-'jän\ *noun, often capitalized* [Sanskrit *harijana* one belonging to the god Vishnu, from *Hari* Vishnu + *jana* person] (1931)
: a member of the outcaste group in India **:** UNTOUCHABLE

hari–kari \ˌhar-i-'kar-ē, -'kir-\ *variant of* HARA-KIRI

hark \'härk\ *intransitive verb* [Middle English *herkien;* akin to Old High German *hōrechen* to listen, Old English *hīeran* to hear] (14th century)
: to pay close attention **:** LISTEN

hark back *intransitive verb* (1829)
1 : to turn back to an earlier topic or circumstance
2 : to go back to something as an origin or source

harken *intransitive verb* (before 12th century)
1 : HEARKEN
2 : HARK BACK — usually used with *back*

har·le·quin \'här-li-k(w)ən\ *noun* [ultimately from Italian *arlecchino,* from Middle French *Helquin,* a demon] (1590)
1 a *capitalized* **:** a character in comedy and pantomime with a shaved head, masked face, variegated tights, and wooden sword **b :** BUFFOON
2 a : a variegated pattern (as of a textile) **b :** a combination of patches on a solid ground of contrasting color (as in the coats of some dogs)

Harlequin

har·le·quin·ade \ˌhär-li-k(w)ə-'nād\ *noun* (1790)
: a play or pantomime in which Harlequin has a leading role

har·lot \'här-lət\ *noun* [Middle English, from Old French *harlot* rogue] (15th century)
: PROSTITUTE

har·lot·ry \-lə-trē\ *noun, plural* **-ries** (14th century)
1 : sexual profligacy **:** PROSTITUTION
2 : an unprincipled or immoral woman ⟨he sups tonight with a *harlotry* —Shakespeare⟩

[1]harm \'härm\ *noun* [Middle English, from Old English *hearm;* akin to Old High German *harm* injury, Old Church Slavonic *sramŭ* shame] (before 12th century)
1 : physical or mental damage **:** INJURY
2 : MISCHIEF, HURT

[2]harm *transitive verb* (before 12th century)
: to cause harm to
synonym see INJURE
— harm·er *noun*

har·mat·tan \ˌhär-mə-'tan, här-'ma-tᵊn\ *noun, often capitalized* [Twi *haramata*] (1671)
: a dust-laden wind on the Atlantic coast of Africa in some seasons

harm·ful \'härm-fəl\ *adjective* (14th century)
: of a kind likely to be damaging **:** INJURIOUS
— harm·ful·ly \-fə-lē\ *adverb*
— harm·ful·ness *noun*

harm·less \'härm-ləs\ *adjective* (14th century)
1 : free from harm, liability, or loss
2 : lacking capacity or intent to injure **:** INNOCUOUS
— harm·less·ly *adverb*
— harm·less·ness *noun*

[1]har·mon·ic \här-'mä-nik\ *adjective* (1570)
1 : MUSICAL
2 : of or relating to musical harmony or a harmonic
3 : pleasing to the ear **:** HARMONIOUS
4 : of an integrated nature **:** CONGRUOUS
— har·mon·i·cal·ly \-ni-k(ə-)lē\ *adverb*

[2]harmonic *noun* (1777)
1 a : OVERTONE; *especially* **:** one whose vibration frequency is an integral multiple of that of the fundamental **b :** a flutelike tone produced on a stringed instrument by touching a vibrating string at a nodal point
2 : a component frequency of a complex wave (as of electromagnetic energy) that is an integral multiple of the fundamental frequency

har·mon·i·ca \här-'mä-ni-kə\ *noun* [Italian *armonica,* feminine of *armonico* harmonious] (1762)
1 : GLASS HARMONICA
2 : a small rectangular wind instrument with free reeds recessed in air slots from which tones are sounded by exhaling and inhaling
— har·mon·i·cist \-nə-sist\ *noun*

harmonic analysis *noun* (1867)
: the expression of a periodic function as a sum of sines and cosines and specifically by a Fourier series

harmonic mean *noun* (1856)
: the reciprocal of the arithmetic mean of the reciprocals of a finite set of numbers

harmonic motion *noun* (1867)
: a periodic motion (as of a sounding violin string or swinging pendulum) that has a single frequency or amplitude or a periodic motion that is composed of two or more such simple periodic motions

harmonic progression *noun* (1856)
: a sequence of numbers whose reciprocals form an arithmetic progression

harmonic series *noun* (1866)
: a series of the form

$$1 + \frac{1}{2^\alpha} + \frac{1}{3^\alpha} + \frac{1}{4^\alpha} \cdot \cdot \cdot$$

which diverges for $0 \leq \alpha \leq 1$ and converges for $\alpha > 1$

har·mo·ni·ous \här-'mō-nē-əs\ *adjective* (1530)
1 : musically concordant
2 : having the parts agreeably related **:** CONGRUOUS ⟨blended into a *harmonious* whole⟩
3 : marked by accord in sentiment or action
— har·mo·ni·ous·ly *adverb*
— har·mo·ni·ous·ness *noun*

har·mo·nise *British variant of* HARMONIZE

har·mo·ni·um \här-'mō-nē-əm\ *noun* [French, from Middle French *harmonie, armonie*] (1847)
: REED ORGAN

har·mo·nize \'här-mə-ˌnīz\ *verb* **-nized; -niz·ing** (15th century)
intransitive verb
1 : to play or sing in harmony
2 : to be in harmony
transitive verb
1 : to bring into consonance or accord
2 : to provide or accompany with harmony
— har·mo·ni·za·tion \ˌhär-mə-nə-'zā-shən\ *noun*
— har·mo·niz·er \'här-mə-ˌnī-zər\ *noun*

har·mo·ny \'här-mə-nē\ *noun, plural* **-nies** [Middle English *armony,* from Middle French *armonie,* from Latin *harmonia,* from Greek, joint, harmony, from *harmos* joint — more at ARM] (14th century)
1 *archaic* **:** tuneful sound **:** MELODY
2 a : the combination of simultaneous musical notes in a chord **b :** the structure of music with respect to the composition and progression of chords **c :** the science of the structure, relation, and progression of chords
3 a : pleasing or congruent arrangement of parts ⟨a painting exhibiting *harmony* of color and line⟩ **b :** CORRESPONDENCE, ACCORD ⟨lives in *harmony* with her neighbors⟩ **c :** internal calm **:** TRANQUILLITY
4 a : an interweaving of different accounts into a single narrative **b :** a systematic arrangement of parallel literary passages (as of the Gospels) for the purpose of showing agreement or harmony

harm's way *noun* (circa 1661)
: a dangerous place or situation ⟨was placed in *harm's way*⟩ ⟨got them out of *harm's way*⟩

[1]har·ness \'här-nəs\ *noun* [Middle English *herneis* baggage, gear, from Old French] (14th century)
1 a : the gear other than a yoke of a draft animal **b :** GEAR, EQUIPMENT; *especially* **:** military equipment for a horse or man
2 a : occupational surroundings or routine ⟨get back into *harness* after a vacation⟩ **b :** close association ⟨ability to work in *harness* with others —R. P. Brooks⟩
3 a : something that resembles a harness (as in holding or fastening something) ⟨a parachute *harness*⟩ **b :** prefabricated wiring with insulation and terminals ready to be attached
4 : a part of a loom which holds and controls the heddles

[2]harness *transitive verb* (14th century)
1 a : to put a harness on **b :** to attach by means of a harness
2 : to tie together **:** YOKE
3 : UTILIZE ⟨*harness* the computer's potential⟩

harness horse *noun* (1861)
: a horse for racing or working in harness

harness racing *noun* (1901)
: the sport of racing standardbred horses harnessed to 2-wheeled sulkies

[1]harp \'härp\ *noun* [Middle English, from Old English *hearpe;* akin to Old High German *harpha* harp] (before 12th century)
1 : a plucked stringed instrument consisting of a resonator, an arched or angled neck that may be supported by a post, and strings of graded length that are perpendicular to the soundboard
2 : something that resembles a harp
3 : HARMONICA 2
— harp·ist \'här-pist\ *noun*

harp 1

[2]harp *intransitive verb* (before 12th century)
1 : to play on a harp
2 : to dwell on or recur to a subject tiresomely or monotonously — usually used with *on*

harp·er \'här-pər\ *noun* (before 12th century)
1 : a harp player
2 : one that harps

har·poon \här-'pün\ *noun* [probably from Dutch *harpoen,* from Old French *harpon* brooch, from *harper* to grapple] (1625)
: a barbed spear or javelin used especially in hunting large fish or whales
— harpoon *transitive verb*
— har·poon·er *noun*

\ə\ abut \ᵊ\ kitten \ər\ further \a\ ash \ā\ ace \ä\ mop, mar \au̇\ out \ch\ chin \e\ bet \ē\ easy \g\ go \i\ hit \ī\ ice \j\ job \ŋ\ sing \ō\ go \ȯ\ law \ȯi\ boy \th\ thin \th\ the \ü\ loot \u̇\ foot \y\ yet \zh\ vision *see also* Guide to Pronunciation

harp seal *noun* [from the shape of its markings] (1766)
: a dark-faced seal (*Phoca groenlandicus*) of the North Atlantic that is a variable light gray with the male usually having a dark crescent on its back and sides

harp·si·chord \'härp-si-ˌkȯrd\ *noun* [modification of Italian *arpicordo,* from *arpa* harp + *corda* string] (1611)
: a stringed instrument resembling a grand piano but usually having two keyboards and two or more strings for each note and producing tones by the plucking of strings with plectra
— **harp·si·chord·ist** \-ˌkȯr-dist\ *noun*

har·py \'här-pē\ *noun, plural* **harpies** [Latin *Harpyia,* from Greek] (1513)
1 *capitalized* **:** a foul malign creature in Greek mythology that is part woman and part bird
2 a : a predatory person **:** LEECH **b :** a shrewish woman

harpy eagle *noun* (1830)
: a large powerful crested eagle (*Harpia harpyja*) of Central and South America

har·que·bus \'här-kwi-(ˌ)bəs, -kə-bəs\ *noun* [Middle French *harquebuse, arquebuse,* modification of Middle Dutch *hakebusse,* from *hake* hook + *busse* tube, box, gun, from Late Latin *buxis* box] (1532)
: a matchlock gun invented in the 15th century which was portable but heavy and was usually fired from a support
— **har·que·bus·ier** \ˌhär-kwi-(ˌ)bə-'sir, -kə-bə-\ *noun*

har·ri·dan \'har-ə-dᵊn\ *noun* [perhaps modification of French *haridelle* old horse, gaunt woman] (circa 1700)
: SHREW 2

har·ried \'har-ēd\ *adjective* (circa 1915)
: beset by problems **:** HARASSED

¹har·ri·er \'har-ē-ər\ *noun* [irregular from ¹*hare*] (1542)
1 : any of a breed of hunting dogs resembling a small foxhound and originally bred for hunting rabbits
2 : a runner on a cross-country team

²harrier *noun* [alteration of *harrower,* from ¹*harrow*] (1556)
: any of a genus (*Circus*) of slender hawks having long angled wings and long legs and feeding chiefly on small mammals, reptiles, and insects

³harrier *noun* (1596)
: one that harries

¹har·row \'har-(ˌ)ō\ *transitive verb* [Middle English *harwen,* from Old English *hergian*] (before 12th century)
archaic **:** PILLAGE, PLUNDER

²harrow *noun* [Middle English *harwe,* probably of Scandinavian origin; akin to Old Norse *hervi* harrow] (14th century)
: a cultivating implement set with spikes, spring teeth, or disks and used primarily for pulverizing and smoothing the soil

³harrow *transitive verb* (14th century)
1 : to cultivate with a harrow
2 : TORMENT, VEX
— **har·row·er** \'har-ə-wər\ *noun*

har·rumph \hə-'rəm(p)f\ *intransitive verb* [imitative] (1942)
1 : to clear the throat in a pompous way
2 : to comment disapprovingly

har·ry \'har-ē\ *transitive verb* **har·ried; har·ry·ing** [Middle English *harien,* from Old English *hergian;* akin to Old High German *heriōn* to lay waste, *heri* army, Greek *koiranos* ruler] (before 12th century)
1 : to make a pillaging or destructive raid on **:** ASSAULT
2 : to force to move along by harassing ⟨*harrying* the terrified horses down out of the mountains —R. A. Sokolov⟩
3 : to torment by or as if by constant attack
synonym see WORRY

harsh \'härsh\ *adjective* [Middle English *harsk,* of Scandinavian origin; akin to Norwegian *harsk* harsh] (14th century)
1 : having a coarse uneven surface that is rough or unpleasant to the touch
2 a : causing a disagreeable or painful sensory reaction **:** IRRITATING **b :** physically discomforting **:** PAINFUL
3 : unduly exacting **:** SEVERE
4 : lacking in aesthetic appeal or refinement **:** CRUDE
synonym see ROUGH
— **harsh·ly** *adverb*
— **harsh·ness** *noun*

harsh·en \'här-shən\ *verb* **harsh·ened; harsh·en·ing** \-sh(ə-)niŋ\ (1824)
transitive verb
: to make (as a voice) harsh
intransitive verb
: to become harsh ⟨saw the grain of his skin *harshening* over face bones —Elizabeth Bowen⟩

hart \'härt\ *noun* [Middle English *hert,* from Old English *heort;* akin to Latin *cervus* hart, Greek *keras* horn — more at HORN] (before 12th century)
chiefly British **:** the male of the red deer especially when over five years old **:** STAG — compare HIND

harte·beest \'här-tə-ˌbēst\ *noun* [obsolete Afrikaans (now *hartbees*), from Dutch, from *hart* deer + *beest* beast] (1786)
: either of two large African antelopes (*Alcelaphus buselaphus* and *Sigmoceros lichtensteini*) with long faces and short annulate divergent horns; *also* **:** a smaller antelope (*Damaliscus hunteri* synonym *Beatragus hunteri*) of eastern Africa having a horizontal white line between the eyes

harts·horn \'härts-ˌhȯrn\ *noun* [from the earlier use of hart's horns as the chief source of ammonia] (1685)
: a preparation of ammonia used as smelling salts

har·um–scar·um \ˌhar-əm-'skar-əm, ˌher-əm-'sker-\ *adjective* [perhaps from archaic *hare* (to harass) + *scare*] (1751)
: RECKLESS, IRRESPONSIBLE
— **harum–scarum** *adverb*

ha·rus·pex \hə-'rəs-ˌpeks, 'har-əs-\ *noun, plural* **ha·rus·pi·ces** \hə-'rəs-pə-ˌsēz\ [Latin, from *haru-* (akin to *chordē* gut, cord) + *-spex,* from *specere* to look — more at YARN, SPY] (1584)
: a diviner in ancient Rome basing his predictions on inspection of the entrails of sacrificial animals

ha·rus·pi·ca·tion \hə-ˌrəs-pə-'kā-shən\ *noun* (1871)
chiefly British **:** an act or instance of foretelling something

¹har·vest \'här-vəst\ *noun, often attributive* [Middle English *hervest,* from Old English *hærfest;* akin to Latin *carpere* to pluck, gather, Greek *karpos* fruit] (before 12th century)
1 : the season for gathering in agricultural crops
2 : the act or process of gathering in a crop
3 a : a mature crop (as of grain or fruit) **:** YIELD **b :** the quantity of a natural product gathered in a single season ⟨deer *harvest*⟩ ⟨ice *harvest*⟩
4 : an accumulated store or productive result ⟨a fantastic revenue *harvest* from lower tax rates —Peter Passell⟩

²harvest (15th century)
transitive verb
1 a : to gather in (a crop) **:** REAP **b :** to gather, catch, hunt, or kill (as salmon, oysters, or deer) for human use, sport, or population control **c :** to remove or extract (as living cells, tissues, or organs) from culture or a living or recently deceased body especially for transplanting
2 a : to accumulate a store of ⟨has now *harvested* this new generation's scholarly labors —M. J. Wiener⟩ **b :** to win by achievement ⟨the team *harvested* several awards⟩
intransitive verb
: to gather in a crop especially for food
— **har·vest·able** \-və-stə-bəl\ *adjective*
— **har·vest·er** *noun*

harvest fly *noun* (circa 1753)
: CICADA

harvest home *noun* (1596)
1 : the gathering or the time of harvest
2 : a feast at the close of harvest
3 : a song sung by the reapers at the close of the harvest

har·vest·man \'här-vəs(t)-mən\ *noun* (1830)
: an arachnid (order Phalangida) that superficially resembles a true spider but has a small rounded body and very long slender legs — called also *daddy longlegs*

harvest mite *noun* (1873)
: CHIGGER 2

harvest moon *noun* (1706)
: the full moon nearest the time of the September equinox

har·vest·time \'här-vəs(t)-ˌtīm\ *noun* (14th century)
: the time during which an annual crop (as wheat) is harvested

has *present 3d singular of* HAVE

has–been \'haz-ˌbin, *chiefly British* -ˌbēn\ *noun* (1606)
: one that has passed the peak of effectiveness or popularity

ha·sen·pfef·fer \'hä-zᵊn-ˌ(p)fe-fər, 'hä-sᵊn-\ *noun* [German, from *Hase* hare + *Pfeffer* pepper] (1892)
: a highly seasoned stew made of marinated rabbit meat

¹hash \'hash\ *transitive verb* [French *hacher,* from Old French *hachier,* from *hache* battle-ax, of Germanic origin; akin to Old High German *hāppa* sickle; akin to Greek *koptein* to cut — more at CAPON] (1590)
1 a : to chop (as meat and potatoes) into small pieces **b :** CONFUSE, MUDDLE
2 : to talk about **:** REVIEW — often used with *over* or *out*

²hash *noun* (circa 1663)
1 : chopped food; *specifically* **:** chopped meat mixed with potatoes and browned
2 : a restatement of something that is already known
3 a : HODGEPODGE, JUMBLE **b :** a confused muddle ⟨made a *hash* of the whole project⟩

³hash *noun* (1959)
: HASHISH

hash browns *noun plural* (1951)
: boiled potatoes that have been diced, mixed with chopped onions and shortening, and fried usually until they form a browned cake — called also *hash brown potatoes, hashed brown potatoes, hashed browns*

Hash·em·ite *or* **Hash·im·ite** \'ha-shə-ˌmīt\ *noun* [*Hashim,* great-grandfather of Muhammad] (1697)
: a member of an Arab family having common ancestry with Muhammad and founding dynasties in countries of the eastern Mediterranean

hash house *noun* (1869)
: an inexpensive eating place

hash·ish \'ha-ˌshēsh, ha-'shēsh\ *noun* [Arabic *ḥashīsh*] (1598)
: the concentrated resin from the flowering tops of the female hemp plant (*Cannabis sativa*) that is smoked, chewed, or drunk for its intoxicating effect — called also *charas*; compare BHANG, MARIJUANA

hash mark *noun* (1907)
1 : SERVICE STRIPE
2 : INBOUNDS LINE

Ha·sid \'ha-səd, 'kä-\ *noun, plural* **Ha·si·dim** \'ha-sə-dəm, kä-'sē-\ [Hebrew *ḥāsīdh* pious] (1812)
1 : a member of a Jewish sect of the second century B.C. opposed to Hellenism and devoted to the strict observance of the ritual law

2 *also* **Has•sid :** a member of a Jewish mystical sect founded in Poland about 1750 in opposition to rationalism and ritual laxity
— **Ha•sid•ic** *also* **Has•sid•ic** \ha-'si-dik, hä-, kä-\ *adjective*

Ha•si•dism \'ha-sə-,di-zəm, 'hä-, 'kä-\ *noun* (1893)
1 : the practices and beliefs of the Hasidim
2 : the Hasidic movement

Has•mo•nae•an *or* **Has•mo•ne•an** \,haz-mə-'nē-ən\ *noun* [Late Latin *Asmonaeus* Hasmon, ancestor of the Maccabees, from Greek *Asamōnaios*] (1620)
: a member of the Maccabees
— **Hasmonaean** *or* **Hasmonean** *adjective*

hasn't \'ha-zᵊnt, -zᵊn\ (1746)
: has not

hasp \'hasp\ *noun* [Middle English, alteration from Old English *hæpse;* akin to Middle High German *haspe* hasp] (before 12th century)
: any of several devices for fastening; *especially* **:** a fastener especially for a door or lid consisting of a hinged metal strap that fits over a staple and is secured by a pin or padlock

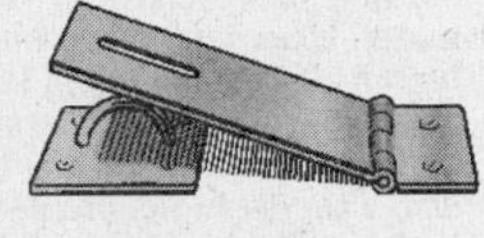
hasp

— **hasp** *transitive verb*

¹has•sle \'ha-səl\ *noun* (1945)
1 : a heated often protracted argument **:** WRANGLE
2 : a violent skirmish **:** FIGHT
3 a : a state of confusion **:** TURMOIL **b :** an annoying or troublesome concern

²hassle *verb* **has•sled; has•sling** \-s(ə-)liŋ\ [perhaps blend of *harass* and *hustle*] (1951)
intransitive verb
: ARGUE, FIGHT ⟨*hassled* with the umpire⟩
transitive verb
: to annoy persistently or acutely **:** HARASS ⟨he gets *hassled* in the street because he dresses funny —William Kloman⟩

has•sock \'ha-sək\ *noun* [Middle English, sedge, from Old English *hassuc*] (before 12th century)
1 : TUSSOCK
2 a : a cushion for kneeling ⟨a church *hassock*⟩ **b :** a padded cushion or low stool that serves as a seat or leg rest

hast \'hast, (h)əst\ *archaic present 2d singular of* HAVE

has•tate \'has-,tāt\ *adjective* [New Latin *hastatus,* from Latin *hasta* spear — more at YARD] (1788)
1 : triangular with sharp basal lobes spreading away from the base of the petiole ⟨*hastate* leaves⟩ — see LEAF illustration
2 : shaped like a spear or the head of a spear ⟨a *hastate* spot of a bird⟩

¹haste \'hāst\ *noun* [Middle English, from Middle French, of Germanic origin; akin to Old English *hǣst* violence] (14th century)
1 : rapidity of motion **:** SWIFTNESS
2 : rash or headlong action **:** PRECIPITATENESS ⟨the beauty of speed uncontaminated by *haste* —*Harper's*⟩
3 : undue eagerness to act ☆

²haste *verb* **hast•ed; hast•ing** (14th century)
transitive verb
archaic **:** to urge on **:** HASTEN
intransitive verb
: to move or act swiftly

has•ten \'hā-sᵊn\ *verb* **has•tened; has•ten•ing** \'hās-niŋ, 'hā-sᵊn-iŋ\ (1568)
intransitive verb
: to move or act quickly
transitive verb
1 : to urge on ⟨*hastened* her to the door —A. J. Cronin⟩
2 : ACCELERATE ⟨*hasten* the coming of a new order —D. W. Brogan⟩
— **has•ten•er** \'hās-nər, 'hā-sᵊn-ər\ *noun*

hast•i•ly \'hā-stə-lē\ *adverb* (14th century)
: in haste **:** HURRIEDLY

hasty \'hā-stē\ *adjective* **hast•i•er; -est** (14th century)
1 a *archaic* **:** rapid in action or movement **:** SPEEDY **b :** done or made in a hurry **c :** fast and typically superficial ⟨made a *hasty* examination of the wound⟩
2 : EAGER, IMPATIENT
3 : PRECIPITATE, RASH
4 : prone to anger **:** IRRITABLE
synonym see FAST
— **hast•i•ness** *noun*

hasty pudding *noun* (1599)
1 *British* **:** a porridge of oatmeal or flour boiled in water
2 *New England* **a :** cornmeal mush **b :** INDIAN PUDDING

¹hat \'hat\ *noun* [Middle English, from Old English *hæt;* akin to Old High German *huot* head covering — more at HOOD] (before 12th century)
1 : a covering for the head usually having a shaped crown and brim
2 a : a distinctive head covering worn as a symbol of office **b :** an office, position, or role assumed by or as if by the wearing of a special hat
— **hat•less** \-ləs\ *adjective*

²hat *verb* **hat•ted; hat•ting** (15th century)
transitive verb
: to furnish or provide with a hat
intransitive verb
: to make or supply hats

hat•band \'hat-,band\ *noun* (15th century)
: a band (as of fabric, leather, or cord) around the crown of a hat just above the brim

hat•box \-,bäks\ *noun* (1794)
1 : a box for holding or storing a hat
2 : a usually round piece of luggage designed especially for carrying hats

¹hatch \'hach\ *noun* [Middle English *hache,* from Old English *hæc;* akin to Middle Dutch *hecke* trapdoor] (before 12th century)
1 : a small door or opening (as in an airplane or spaceship) ⟨an escape *hatch*⟩
2 a : an opening in the deck of a ship or in the floor or roof of a building **b :** the covering for such an opening **c :** HATCHWAY **d :** COMPARTMENT
3 : FLOODGATE

²hatch *verb* [Middle English *hacchen;* akin to Middle High German *hecken* to mate] (13th century)
intransitive verb
1 : to produce young by incubation
2 a : to emerge from an egg, chrysalis, or pupa **b :** to give forth young or imagoes
3 : to incubate eggs **:** BROOD
transitive verb
1 a : to produce (young) from an egg by applying natural or artificial heat **b :** INCUBATE 1
2 : to bring into being **:** ORIGINATE; *especially* **:** to concoct in secret
— **hatch•abil•i•ty** \,ha-chə-'bi-lə-tē\ *noun*
— **hatch•able** \'ha-chə-bəl\ *adjective*
— **hatch•er** *noun*

³hatch *noun* (1601)
1 : an act or instance of hatching
2 : a brood of hatched young

⁴hatch *transitive verb* [Middle English *hachen,* from Middle French *hacher* to inlay, chop up, from Old French *hachier* — more at HASH] (15th century)
1 : to inlay with narrow bands of distinguishable material ⟨a silver handle *hatched* with gold⟩
2 : to mark (as a drawing or engraving) with fine closely spaced lines

⁵hatch *noun* (1658)
: LINE; *especially* **:** one used to give the effect of shading

hatch•back \'hach-,bak\ *noun* (1970)
: an automobile the back of which consists of a hatch that opens upward; *also* **:** the back itself

hat•check \'hat-,chek\ *adjective* (1917)
: employed in checking hats and articles of outdoor clothing ⟨a *hatcheck* girl⟩

hatch•ery \'ha-chə-rē\ *noun, plural* **-er•ies** (1880)
: a place for hatching eggs (as of poultry or fish)

hatch•et \'ha-chət\ *noun* [Middle English *hachet,* from Middle French *hachette,* diminutive of *hache* battle-ax — more at HASH] (14th century)
1 : a short-handled ax often with a hammerhead to be used with one hand
2 : TOMAHAWK

hatchet face *noun* (circa 1666)
: a thin sharp face
— **hatch•et–faced** \'ha-chət-,fāst\ *adjective*

hatchet job *noun* (1944)
: a forceful or malicious verbal attack

hatchet man *noun* (1880)
1 : one hired for murder, coercion, or attack
2 a : a writer specializing in invective **b :** a person hired to perform underhanded or unscrupulous tasks (as ruin reputations)

hatchet work *noun* (1944)
: the work of a hatchet man

hatch•ing *noun* (1662)
: the engraving or drawing of fine lines in close proximity especially to give an effect of shading; *also* **:** the pattern so made

hatch•ling \'hach-liŋ\ *noun* (1899)
: a recently hatched animal

hatch•ment \'hach-mənt\ *noun* [perhaps alteration of *achievement*] (1548)
: a panel on which a coat of arms of a deceased person is temporarily displayed

hatch•way \'hach-,wā\ *noun* (1626)
: a passage giving access usually by a ladder or stairs to an enclosed space (as a cellar); *also* **:** HATCH 2a

¹hate \'hāt\ *noun, often attributive* [Middle English, from Old English *hete;* akin to Old High German *haz* hate, Greek *kēdos* care] (before 12th century)
1 a : intense hostility and aversion usually deriving from fear, anger, or sense of injury **b :** extreme dislike or antipathy **:** LOATHING ⟨had a great *hate* of hard work⟩
2 : an object of hatred ⟨a generation whose finest *hate* had been big business —F. L. Paxson⟩

²hate *verb* **hat•ed; hat•ing** (before 12th century)
transitive verb
1 : to feel extreme enmity toward ⟨*hates* his country's enemies⟩

☆ SYNONYMS
Haste, hurry, speed, expedition, dispatch mean quickness in movement or action. HASTE applies to personal action and implies urgency and precipitancy and often rashness ⟨marry in *haste*⟩. HURRY often has a strong suggestion of agitated bustle or confusion ⟨in the *hurry* of departure she forgot her toothbrush⟩. SPEED suggests swift efficiency in movement or action ⟨exercises to increase your reading *speed*⟩. EXPEDITION and DISPATCH both imply speed and efficiency in handling affairs but EXPEDITION stresses ease or efficiency of performance and DISPATCH stresses promptness in concluding matters ⟨the case came to trial with *expedition*⟩ ⟨paid bills with *dispatch*⟩.

2 : to have a strong aversion to **:** find very distasteful ⟨*hated* to have to meet strangers⟩ ⟨*hate* hypocrisy⟩
intransitive verb
: to express or feel extreme enmity or active hostility ☆
— **hat·er** *noun*
— **hate one's guts :** to hate someone with great intensity
hate crime *noun* (1989)
: any of various crimes (as assault or defacement of property) when motivated by hostility to the victim as a member of a group (as one based on color, creed, gender, or sexual orientation)
hate·ful \ˈhāt-fəl\ *adjective* (14th century)
1 : full of hate **:** MALICIOUS
2 : deserving of or arousing hate
— **hate·ful·ly** \-fə-lē\ *adverb*
— **hate·ful·ness** *noun*
hath \ˈhath, (h)əth\ *archaic present 3d singular of* HAVE
hatha yo·ga \ˈhə-tə-ˈyō-gə, ˈhä-\ *noun* [Sanskrit *haṭha* force + *yoga* yoga] (1890)
: a system of physical exercises for the control and perfection of the body that constitutes one of the four chief Hindu disciplines
hat in hand *adverb* (1851)
: in an attitude of respectful humility ⟨have to go *hat in hand* to apologize⟩
hat·mak·er \ˈhat-ˌmā-kər\ *noun* (15th century)
: one who makes hats
ha·tred \ˈhā-trəd\ *noun* [Middle English, from *hate* + Old English *rǣden* condition — more at KINDRED] (12th century)
1 : HATE
2 : prejudiced hostility or animosity ⟨old racial prejudices and national *hatreds* —Peter Thomson⟩
hat·ter \ˈha-tər\ *noun* (14th century)
: one that makes, sells, or cleans and repairs hats
hat trick *noun* [probably from the former practice of rewarding the feat with the gift of a hat] (1877)
1 : the retiring of three batsmen with three consecutive balls by a bowler in cricket
2 : the scoring of three goals in one game (as of hockey or soccer) by a single player
3 : a succession of three victories, successes, or related accomplishments ⟨scored a *hat trick* when her three best steers corralled top honors —*People*⟩
hau·berk \ˈhȯ-(ˌ)bərk\ *noun* [Middle English, from Old French *hauberc,* of Germanic origin; akin to Old English *healsbeorg* neck armor] (14th century)
: a tunic of chain mail worn as defensive armor from the 12th to the 14th century

1 hauberk

haugh \ˈhȯ(k)\ *noun* [Middle English (Scots) *halch,* from Old English *healh* corner of land; akin to Old English *holh* hole] (before 12th century)
Scottish **:** a low-lying meadow by the side of a river
haugh·ty \ˈhȯ-tē, ˈhä-\ *adjective* **haugh·ti·er; -est** [obsolete *haught,* from Middle English *haute,* from Middle French *haut,* literally, high, from Latin *altus* — more at OLD] (15th century)
: blatantly and disdainfully proud
synonym see PROUD
— **haugh·ti·ly** \ˈhȯ-tᵊl-ē, ˈhä-\ *adverb*
— **haugh·ti·ness** \ˈhȯ-tē-nəs, ˈhä-\ *noun*
¹haul \ˈhȯl\ *verb* [Middle English *halen* to pull, from Old French *haler,* of Germanic origin; akin to Middle Dutch *halen* to pull; akin to Old English *geholian* to obtain] (13th century)
transitive verb
1 a : to exert traction on **:** DRAW ⟨*haul* a wagon⟩ **b :** to obtain or move by or as if by hauling ⟨was *hauled* to parties night after night by his wife⟩ **c :** to transport in a vehicle **:** CART
2 : to change the course of (a ship) especially so as to sail closer to the wind
3 : to bring before an authority for interrogation or judgment **:** HALE ⟨*haul* traffic violators into court⟩
intransitive verb
1 : to exert traction **:** PULL
2 : to move along **:** PROCEED
3 : to furnish transportation
4 *of the wind* **:** SHIFT
— **haul ass :** to move quickly — often considered vulgar
²haul *noun* (1670)
1 a : the act or process of hauling **:** PULL **b :** a device for hauling
2 a : the result of an effort to obtain, collect, or win ⟨the burglar's *haul*⟩ **b :** the fish taken in a single draft of a net
3 a : transportation by hauling **b :** the length or course of a transportation route ⟨a long *haul*⟩ **c :** a quantity transported **:** LOAD
haul·age \ˈhȯ-lij\ *noun* (1826)
1 : the act or process of hauling
2 : a charge made for hauling
haul·er \ˈhȯ-lər\ *noun* (1674)
: one that hauls: as **a :** a commercial establishment or worker whose business is hauling **b :** an automotive vehicle for hauling goods or material
haul·ier \ˈhȯl-yər\ *British variant of* HAULER
haulm \ˈhȯm\ *noun* [Middle English *halm,* from Old English *healm;* akin to Old High German *halm* stem, Latin *culmus* stalk, Greek *kalamos* reed] (before 12th century)
: the stems or tops of crop plants (as peas or potatoes) especially after the crop has been gathered
haul off *intransitive verb* (1870)
: to get ready — used with *and* and a following verb describing a usually sudden and violent act ⟨I *hauled off* and hit him⟩
haunch \ˈhȯnch, ˈhänch\ *noun* [Middle English *haunche,* from Old French *hanche,* of Germanic origin; akin to Middle Dutch *hanke* haunch] (13th century)
1 a : HIP 1a **b :** HINDQUARTER 2 — usually used in plural
2 : HINDQUARTER 1
3 : either side of an arch between the springing and the crown
— on one's haunches : in a squatting position
¹haunt \ˈhȯnt, ˈhänt\ *verb* [Middle English, from Old French *hanter,* probably from Old Norse *heimta* to lead home, pull, claim, from *heimr* home] (14th century)
transitive verb
1 a : to visit often **:** FREQUENT **b :** to continually seek the company of
2 a : to have a disquieting or harmful effect on **:** TROUBLE ⟨problems we ignore now will come back to *haunt* us⟩ **b :** to recur constantly and spontaneously to ⟨the tune *haunted* her⟩ **c :** to reappear continually in ⟨a sense of tension that *haunts* his writing⟩
3 : to visit or inhabit as a ghost
intransitive verb
1 : to stay around or persist **:** LINGER
2 : to appear habitually as a ghost
— **haunt·er** *noun*
— **haunt·ing·ly** \ˈhȯn-tiŋ-lē, ˈhän-\ *adverb*
²haunt \ˈhȯnt, ˈhänt, *2 is usually* ˈhant\ *noun* (14th century)
1 : a place habitually frequented
2 *chiefly dialect* **:** GHOST
Hau·sa \ˈhau̇-sə, -zə\ *noun, plural* **Hausa** *or* **Hausas** (1853)
1 : the Chadic language of the Hausa people widely used in western Africa as a trade language
2 : a member of a black people of northern Nigeria and southern Niger
haus·frau \ˈhau̇s-ˌfrau̇\ *noun* [German, from *Haus* house + *Frau* woman, wife] (1798)
: HOUSEWIFE
haus·tel·lum \hȯ-ˈste-ləm\ *noun, plural* **-la** \-lə\ [New Latin, diminutive of Latin *haustrum* scoop on a waterwheel, from *haurire* to drink, draw — more at EXHAUST] (1816)
: a proboscis (as of an insect) adapted to suck blood or juices of plants
haus·to·ri·al \hȯ-ˈstōr-ē-əl, -ˈstȯr-\ *adjective* (1894)
: of, relating to, or having a haustorium
haus·to·ri·um \-ē-əm\ *noun, plural* **-ria** \-ē-ə\ [New Latin, from Latin *haurire*] (1875)
: a food-absorbing outgrowth of a plant organ (as a hypha or stem)
haut·bois *or* **haut·boy** \ˈ(h)ō-ˌbȯi\ *noun, plural* **hautbois** \-ˌbȯiz\ *or* **haut·boys** [Middle French *hautbois,* from *haut* high + *bois* wood] (1575)
: OBOE
haute \ˈōt\ *also* **haut** \ˈōt, ˈō\ *adjective* [French] (1787)
: FASHIONABLE, HIGH-CLASS ⟨*haute* interior decorators⟩ ⟨a store filled with *haute* kitsch⟩
haute cou·ture \ˌōt-kü-ˈtu̇r\ *noun* [French, literally, high sewing] (1908)
: the houses or designers that create exclusive and often trend-setting fashions for women; *also* **:** the fashions created
haute cui·sine \-kwi-ˈzēn\ *noun* [French, literally, high cooking] (1928)
: artful or elaborate cuisine; *especially* **:** traditionally elaborate French cuisine
haute école \-ā-ˈkȯl, -ˈkəl\ *noun* [French, literally, high school] (1858)
: a highly stylized form of classical riding **:** advanced dressage
hau·teur \hȯ-ˈtər, (h)ō-\ *noun* [French, from *haut* high — more at HAUGHTY] (circa 1628)
: ARROGANCE, HAUGHTINESS
haut monde \ō-ˈmänd, ō-mōⁿd\ *also* **haute monde** \ōt-\ *noun* [French, literally, high world] (1864)
: high society
Ha·vana \hə-ˈva-nə\ *noun* [probably from Spanish *habano,* from *habano* of Havana, from La *Habana* (Havana), Cuba] (1826)
1 : a cigar made from Cuban tobacco
2 : a tobacco originally grown in Cuba
Ha·var·ti \hə-ˈvär-tē\ *noun* [*Havarti,* locale in Denmark] (1957)
: a semisoft Danish cheese having a porous texture and usually a mild flavor
havdalah *variant of* HABDALAH
¹have \ˈhav, (h)əv, v; *in "have to" meaning "must" usually* ˈhaf\ *verb* **had** \ˈhad, (h)əd, d\; **hav·ing** \ˈha-viŋ\; **has** \ˈhaz, (h)əz, z, s; *in "has to" meaning "must" usually* ˈhas\ [Middle English, from Old English *habban;* akin to Old High German *habēn* to have, and perhaps to *hevan* to lift — more at HEAVE] (before 12th century)
transitive verb
1 a : to hold or maintain as a possession, privilege, or entitlement ⟨they *have* a new car⟩ ⟨I *have* my rights⟩ **b :** to hold in one's use, service, regard, or at one's disposal ⟨the group will *have* enough tickets for everyone⟩ ⟨we don't *have* time to stay⟩ **c :** to hold, include, or contain as a part or whole ⟨the car *has* power brakes⟩ ⟨April *has* 30 days⟩

☆ SYNONYMS
Hate, detest, abhor, abominate, loathe mean to feel strong aversion or intense dislike for. HATE implies an emotional aversion often coupled with enmity or malice ⟨*hated* the enemy with a passion⟩. DETEST suggests violent antipathy ⟨*detests* cowards⟩. ABHOR implies a deep often shuddering repugnance ⟨a crime *abhorred* by all⟩. ABOMINATE suggests strong detestation and often moral condemnation ⟨every society *abominates* incest⟩. LOATHE implies utter disgust and intolerance ⟨*loathed* self-appointed moral guardians⟩.

2 : to feel obligation in regard to — usually used with an infinitive with *to* ⟨we *have* things to do⟩ ⟨*have* a deadline to meet⟩
3 : to stand in a certain relationship to ⟨*has* three fine children⟩ ⟨we will *have* the wind at our backs⟩
4 a : to acquire or get possession of **:** OBTAIN ⟨these shoes are the best to be *had*⟩ **b :** RECEIVE ⟨*had* news⟩ **c :** ACCEPT; *specifically* **:** to accept in marriage **d :** to copulate with
5 a : to be marked or characterized by (a quality, attribute, or faculty) ⟨both *have* red hair⟩ ⟨*has* a way with words⟩ **b :** EXHIBIT, SHOW ⟨*had* the gall to refuse⟩ **c :** USE, EXERCISE ⟨*have* mercy on us⟩
6 a : to experience especially by submitting to, undergoing, or suffering ⟨I *have* a cold⟩ **b :** to make the effort to perform (an action) or engage in (an activity) ⟨*have* a look at that cut⟩ **c :** to entertain in the mind ⟨*have* an opinion⟩
7 a : to cause or command to do something — used with the infinitive without *to* ⟨*have* the children stay⟩ **b :** to cause to be in a certain place or state ⟨*has* people around at all times⟩
8 : ALLOW ⟨we'll *have* no more of that⟩
9 : to be competent in ⟨*has* only a little French⟩
10 a : to hold in a position of disadvantage or certain defeat ⟨we *have* him now⟩ **b :** to take advantage of **:** TRICK, FOOL ⟨been *had* by a partner⟩
11 : BEGET, BEAR ⟨*have* a baby⟩
12 : to partake of ⟨*have* dinner⟩ ⟨*have* a smoke⟩
13 : BRIBE, SUBORN ⟨can be *had* for a price⟩
verbal auxiliary
1 — used with the past participle to form the present perfect, past perfect, or future perfect ⟨*has* gone home⟩ ⟨*had* already eaten⟩ ⟨will *have* finished dinner by then⟩
2 : to be compelled, obliged, or required — used with an infinitive with *to* or *to* alone ⟨we *had* to go⟩ ⟨do what you *have* to⟩ ⟨it *has* to be said⟩
— had better *or* **had best :** would be wise to
— have at : to go at or deal with **:** ATTACK
— have coming : to deserve or merit what one gets, benefits by, or suffers ⟨he *had* that *coming*⟩
— have done : FINISH, STOP
— have done with : to bring to an end **:** have no further concern with ⟨let us *have done with* name-calling⟩
— have had it 1 : to have had or have done all one is going to be allowed to **2 :** to have experienced, endured, or suffered all one can
— have it : ASSERT, CLAIM ⟨rumor *has it* that he was drunk⟩
— have it in for : to intend to do harm to
— have it out : to settle a matter of contention by discussion or a fight
— have none of : to refuse to have anything to do with
— have one's eye on 1 a : to look at **b :** to watch constantly and attentively **2 :** to have as an objective
— have to do with 1 : to deal with ⟨the story *has to do with* real people —Alice M. Jordan⟩ **2 :** to have a specified relationship with or effect on ⟨the size of the brain *has* nothing *to do with* intelligence —Ruth Benedict⟩

²have \'hav\ *noun* (1836)
: one that is well-endowed especially in material wealth

have·lock \'hav-,läk, -lək\ *noun* [Sir Henry *Havelock*] (1861)
: a covering attached to a cap to protect the neck from the sun or bad weather

ha·ven \'hā-vən\ *noun* [Middle English, from Old English *hæfen;* akin to Middle High German *habene* harbor] (before 12th century)
1 : HARBOR, PORT
2 : a place of safety **:** ASYLUM
3 : a place offering favorable opportunities or conditions ⟨a tourist's *haven*⟩
— haven *transitive verb*

have–not \'hav-,nät, -'nät\ *noun* (1836)
: one that is poor especially in material wealth

haven't \'ha-vənt, 'ha-bᵊm(t)\ (1777)
: have not

have on *transitive verb* (before 12th century)
1 : WEAR ⟨*has on* a new suit⟩
2 *chiefly British* **:** to trick or deceive intentionally **:** PUT ON 5
3 : to have plans for ⟨what do you *have on* for tomorrow⟩

ha·ver \'hā-vər\ *intransitive verb* [origin unknown] (1866)
chiefly British **:** to hem and haw

hav·er·sack \'ha-vər-,sak\ *noun* [French *havresac,* from German *Habersack* bag for oats, from *Haber* oats + *Sack* bag] (1749)
: a bag similar to a knapsack but worn over one shoulder

ha·ver·sian canal \hə-'vər-zhən-\ *noun, often H capitalized* [Clopton *Havers* (died 1702) English physician & anatomist] (1842)
: any of the small canals through which the blood vessels ramify in bone

haversian system *noun, often H capitalized* (circa 1846)
: a haversian canal with the concentrically arranged laminae of bone that surround it

¹hav·oc \'ha-vək, -vik\ *noun* [Middle English *havok,* from Anglo-French, modification of Old French *havot* plunder] (15th century)
1 : wide and general destruction **:** DEVASTATION
2 : great confusion and disorder ⟨children can create *havoc* in a house⟩

²havoc *transitive verb* **hav·ocked; hav·ock·ing** (1577)
: to lay waste **:** DESTROY

¹haw \'hȯ\ *noun* [Middle English *hawe,* from Old English *haga* — more at HEDGE] (before 12th century)
1 : a hawthorn berry
2 : HAWTHORN

²haw *noun* [origin unknown] (15th century)
: NICTITATING MEMBRANE; *especially* **:** an inflamed nictitating membrane of a domesticated mammal

³haw *intransitive verb* [imitative] (1632)
1 : to utter the sound represented by *haw* ⟨hemmed and *hawed* before answering⟩
2 : EQUIVOCATE ⟨the administration hemmed and *hawed* over the students' demands⟩

⁴haw *interjection* (1679)
— often used to indicate a vocalized pause in speaking

⁵haw \'hȯ\ *verb* [origin unknown] (1777)
verb imperative
— used as a direction to turn to the left; compare GEE
intransitive verb
: to turn to the near or left side

Hawaii–Aleutian time *noun* (1983)
: the time of the 10th time zone west of Greenwich that includes the Hawaiian islands and the Aleutians west of the Fox group

Ha·wai·ian \hə-'wä-yən, -'wī-(y)ən, -'wȯ-yən\ *noun* (circa 1864)
1 : a native or resident of Hawaii; *especially* **:** one of Polynesian ancestry
2 : the Polynesian language of the Hawaiians
— Hawaiian *adjective*

Hawaiian goose *noun* (circa 1909)
: NENE

Hawaiian guitar *noun* (1928)
: a usually electric stringed instrument having a long fretted neck and six to eight steel strings that are plucked while being pressed with a movable steel bar for a glissando effect

Hawaiian shirt *noun* (1962)
: a usually short-sleeved sport shirt with a colorful pattern

haw·finch \'hȯ-,finch\ *noun* [¹*haw*] (circa 1674)
: an Old World finch (*Coccothraustes coccothraustes*) with a large heavy bill and short thick neck and the male marked with black, white, and brown

¹hawk \'hȯk\ *noun* [Middle English *hauk,* from Old English *hafoc;* akin to Old High German *habuh* hawk, Russian *kobets* a falcon] (before 12th century)
1 : any of numerous diurnal birds of prey belonging to a suborder (Falcones of the order Falconiformes) and including all the smaller members of this group; *especially* **:** ACCIPITER — compare OWL
2 : a small board or metal sheet with a handle on the underside used to hold mortar
3 : one who takes a militant attitude and advocates immediate vigorous action; *especially* **:** a supporter of a war or warlike policy — compare DOVE
— hawk·ish \'hȯ-kish\ *adjective*
— hawk·ish·ly *adverb*
— hawk·ish·ness *noun*

²hawk (14th century)
intransitive verb
1 : to hunt birds by means of a trained hawk
2 : to soar and strike like a hawk
transitive verb
: to hunt on the wing like a hawk

³hawk *verb* [imitative] (1581)
transitive verb
: to raise by trying to clear the throat ⟨*hawk* up phlegm⟩
intransitive verb
: to utter a harsh guttural sound in or as if in hawking

⁴hawk *noun* (1604)
: an audible effort to force up phlegm from the throat

⁵hawk *transitive verb* [back-formation from ²*hawker*] (1713)
: to offer for sale by calling out in the street ⟨*hawking* newspapers⟩; *broadly* **:** SELL

¹hawk·er \'hȯ-kər\ *noun* [Middle English, from Old English *hafocere,* from *hafoc*] (before 12th century)
: FALCONER

²hawker *noun* [by folk etymology from Low German *höker,* from Middle Low German *hōker,* from *hōken* to peddle] (1512)
: one that hawks wares

Hawk·eye \'hȯ-,kī\ *noun* (1823)
: a native or resident of Iowa — used as a nickname

hawk·eyed \'hȯ-,kīd\ *adjective* (circa 1818)
: having keen sight

hawk·moth \'hȯk-,mȯth\ *noun* (1785)
: any of a family (Sphingidae) of stout-bodied moths with a long proboscis, long narrow more or less pointed forewings, and small hind wings — called also *sphinx*

hawks·bill \'hȯks-,bil\ *noun* (1712)
: a small brown or brown and yellow sea turtle (*Eretmochelys imbricata*) of warm waters that has a carapace of overlapping plates — compare TORTOISESHELL 1

hawksbill

hawk·shaw \'hȯk-,shȯ\ *noun* [from *Hawkshaw,* detective in the play *The Ticket of Leave Man* (1863) by Tom Taylor] (1888)
: DETECTIVE

hawk·weed \'hȯk-,wēd\ *noun* (1562)

: any of a genus (*Hieracium*) of composite plants usually having flower heads with red or orange rays — compare ORANGE HAWKWEED

hawse \'hȯz\ *noun* [alteration of Middle English *halse,* from Old Norse *hals* neck, hawse — more at COLLAR] (14th century)
1 a : the part of a ship's bow that contains the hawseholes **b :** HAWSEHOLE
2 : the distance between a ship's bow and her anchor

hawse•hole \-ˌhōl\ *noun* (1664)
: a hole in the bow of a ship through which a cable passes

haw•ser \'hȯ-zər\ *noun* [Middle English, from Anglo-French *hauceour,* from Middle French *haucier* to hoist, from (assumed) Vulgar Latin *altiare,* from Latin *altus* high — more at OLD] (13th century)
: a large rope for towing, mooring, or securing a ship

haw•ser–laid \'hȯ-zər-ˌlād\ *adjective* (1769)
: composed of three ropes laid together right-handed with each containing three strands twisted together

haw•thorn \'hȯ-ˌthȯrn\ *noun* [Middle English *hawethorn,* from Old English *hagathorn,* from *haga* hawthorn + *thorn* — more at HEDGE] (before 12th century)
: any of a genus (*Crataegus*) of spring-flowering spiny shrubs of the rose family with glossy and often lobed leaves, white or pink fragrant flowers, and small red fruits

Haw•thorne effect \'hȯ-ˌthȯrn-\ *noun* [from the *Hawthorne* Works of the Western Electric Co., Cicero, Ill., where its existence was established by experiment] (1962)
: the stimulation to output or accomplishment that results from the mere fact of being under observation; *also* **:** such an increase in output or accomplishment

[1]hay \'hā\ *noun* [Middle English *hey,* from Old English *hīeg;* akin to Old High German *hewi* hay, Old English *hēawan* to hew] (before 12th century)
1 : herbage and especially grass mowed and cured for fodder
2 : REWARD
3 *slang* **:** BED
4 : a small sum of money ⟨a saving of . . . $14 million is not *hay* —H. C. Schonberg⟩

[2]hay (1556)
intransitive verb
: to cut, cure, and store hay
transitive verb
: to feed with hay

hay•cock \'hā-ˌkäk\ *noun* (13th century)
: a somewhat rounded conical pile of hay

hay fever *noun* (1829)
: an acute allergic rhinitis and conjunctivitis; *especially* **:** POLLINOSIS

hay•lage \'hā-lij\ *noun* [*hay* + si*lage*] (circa 1958)
: a stored forage that is essentially a grass silage wilted to 35 to 50 percent moisture

hay•loft \'hā-ˌlȯft\ *noun* (1573)
: a loft especially for storing hay

hay•mak•er \-ˌmā-kər\ *noun* (1912)
: a powerful blow

hay•mow \-ˌmau̇\ *noun* (15th century)
: a mow especially of or for hay

hay•rack \-ˌrak\ *noun* (1825)
1 : a feeding rack that holds hay for livestock
2 : a frame mounted on the running gear of a wagon and used especially in hauling hay or straw; *also* **:** a wagon equipped with a hayrack

hay•rick \-ˌrik\ *noun* (15th century)
: a relatively large sometimes thatched outdoor pile of hay **:** HAYSTACK

hay•ride \-ˌrīd\ *noun* (1896)
: a pleasure ride usually at night by a group in a wagon, sleigh, or open truck partly filled with straw or hay

hay–scent•ed fern \'hā-ˌsen-təd-\ *noun* (1915)
: a common fern (*Dennstaedtia punctilobula*) of eastern North America with fragrant finely divided pale green fronds

hay•seed \'hā-ˌsēd\ *noun, plural* **hayseed** *or* **hayseeds** (1577)
1 a : seed shattered from hay **b :** clinging bits of straw or chaff from hay
2 *plural hayseeds* **:** BUMPKIN, YOKEL

hay•stack \-ˌstak\ *noun* (15th century)
1 : a stack of hay
2 : a vertical standing wave in turbulent river waters

hay•wire \-ˌwīr\ *adverb or adjective* [from the use of baling wire for makeshift repairs] (1929)
1 : being out of order or having gone wrong ⟨the radio went *haywire*⟩
2 : emotionally or mentally upset or out of control **:** CRAZY ⟨is going *haywire* with grief⟩

ha•zan \k̲ə-'zän, 'k̲ä-zᵊn\ *noun, plural* **ha•za•nim** \k̲ə-'zä-nəm\ [Late Hebrew *ḥazzān*] (1650)
1 : an official of a Jewish synagogue or community of the period when the Talmud was compiled
2 : CANTOR 2

[1]haz•ard \'ha-zərd\ *noun* [Middle English, from Middle French *hasard,* from Arabic *az-zahr* the die] (14th century)
1 : a game of chance like craps played with two dice
2 : a source of danger
3 a : CHANCE, RISK **b :** a chance event **:** ACCIDENT
4 *obsolete* **:** STAKE 3a
5 : a golf-course obstacle ◇
— at hazard : at stake

[2]hazard *transitive verb* (1530)
: VENTURE, RISK ⟨*hazard* a guess as to the outcome⟩

haz•ard•ous \'ha-zər-dəs\ *adjective* (1585)
1 : depending on hazard or chance
2 : involving or exposing one to risk (as of loss or harm) ⟨a *hazardous* occupation⟩ ⟨disposing of *hazardous* waste⟩
synonym see DANGEROUS
— haz•ard•ous•ly *adverb*
— haz•ard•ous•ness *noun*

[1]haze \'hāz\ *noun* [probably back-formation from *hazy*] (1706)
1 a : fine dust, smoke, or light vapor causing lack of transparency of the air **b :** a cloudy appearance in a transparent liquid or solid; *also* **:** a dullness of finish (as on furniture)
2 : something suggesting atmospheric haze; *especially* **:** vagueness of mind or mental perception

[2]haze *verb* **hazed; haz•ing** (1801)
transitive verb
: to make hazy, dull, or cloudy
intransitive verb
: to become hazy or cloudy

[3]haze *transitive verb* **hazed; haz•ing** [origin unknown] (1840)
1 a : to harass by exacting unnecessary or disagreeable work **b :** to harass by banter, ridicule, or criticism
2 : to haze by way of initiation ⟨*haze* the fraternity pledges⟩
3 *West* **:** to drive (as cattle or horses) from horseback
— haz•er *noun*

[1]ha•zel \'hā-zəl\ *noun* [Middle English *hasel,* from Old English *hæsel;* akin to Old High German *hasal* hazel, Latin *corulus*] (before 12th century)
1 : any of a genus (*Corylus* and especially the American *C. americana* and the European *C. avellana*) of shrubs or small trees of the birch family bearing nuts enclosed in a leafy involucre
2 : a light brown to strong yellowish brown

[2]hazel *adjective* (14th century)
1 : consisting of hazels or of the wood of the hazel
2 : of the color hazel

hazel hen *noun* (1661)
: a European woodland grouse (*Bonasa bonasia*) related to the ruffed grouse — called also *hazel grouse*

ha•zel•nut \'hā-zəl-ˌnət\ *noun* (before 12th century)
: the nut of a hazel

haz•ing \'hā-ziŋ\ *noun* (circa 1855)
: the action of hazing; *especially* **:** an initiation process involving harassment

hazy \'hā-zē\ *adjective* **haz•i•er; -est** [origin unknown] (1625)
1 : obscured or made dim or cloudy by or as if by haze
2 : VAGUE, INDEFINITE ⟨has only a *hazy* recollection⟩; *also* **:** UNCERTAIN ⟨I'm *hazy* on that point⟩
— haz•i•ly \-zə-lē\ *adverb*
— haz•i•ness \-zē-nəs\ *noun*

H–bomb \'āch-ˌbäm\ *noun* (1950)
: HYDROGEN BOMB

HDL \ˌāch-(ˌ)dē-'el\ *noun* [*h*igh-*d*ensity *l*ipoprotein] (circa 1965)
: a cholesterol-poor protein-rich lipoprotein of blood plasma correlated with reduced risk of atherosclerosis — compare LDL

[1]he \'hē, ē\ *pronoun* [Middle English, from Old English *hē;* akin to Old English *hēo* she, *hit* it, Old High German *hē* he, Latin *cis, citra* on this side, Greek *ekeinos* that person] (before 12th century)
1 : that male one who is neither speaker nor hearer ⟨*he* is my father⟩ — compare HIM, HIS, IT, SHE, THEY
2 — used in a generic sense or when the sex of the person is unspecified ⟨*he* that hath ears to hear, let him hear —Matthew 11:15 (Authorized Version)⟩ ⟨one should do the best *he* can⟩ ■

[2]he \'hē\ *noun* (before 12th century)
1 : a male person or animal
2 : one that is strongly masculine or has strong masculine appeal — usually used in combination ⟨that's what I call *he*-literature —Sinclair Lewis⟩

[3]he \'hā\ *noun* [Hebrew *hē'*] (1639)
: the 5th letter of the Hebrew alphabet — see ALPHABET table

[1]head \'hed\ *noun* [Middle English *hed,* from Old English *hēafod;* akin to Old High German *houbit* head, Latin *caput*] (before 12th century)
1 : the upper or anterior division of the animal body that contains the brain, the chief sense organs, and the mouth
2 a : the seat of the intellect **:** MIND ⟨two *heads* are better than one⟩ **b :** a person with respect to mental qualities ⟨let wiser *heads* prevail⟩ **c :** natural aptitude or talent ⟨a good *head* for figures⟩ **d :** mental or emotional control **:** POISE ⟨a level *head*⟩ **e :** HEADACHE
3 : the obverse of a coin — usually used in plural ⟨*heads,* I win⟩

◇ WORD HISTORY

hazard *Hazard* was once only a game played with dice. The French archbishop William of Tyre, writing his Latin history of the Crusades late in the 12th century, explained the origin of this hazard and its name (*hasard* in Old French). The game was invented, he said, to pass the time during the siege of a castle in Palestine called *Hazard* or *Azart.* Unfortunately for the credibility of William's theory, the name of the castle seems actually to have been *'Ain Zarba,* and the name of the game was certainly never that. The French word did not originate in the time of the Crusades, however, and most likely comes from post-classical Arabic *az-zahr* "the die (one of the dice)." Old French *hasard* was borrowed into Middle English, and within a few centuries what had been a venture on the outcome of a throw of the dice could be any venture or risk.

4 a : PERSON, INDIVIDUAL ⟨count *heads*⟩ **b** *plural* **head :** one of a number (as of domestic animals)
5 a : the end that is upper or higher or opposite the foot ⟨the *head* of the table⟩ ⟨*head* of a sail⟩ **b :** the source of a stream **c :** either end of something (as a drum) whose two ends need not be distinguished
6 : DIRECTOR, LEADER: as **a :** HEADMASTER **b :** one in charge of a division or department in an office or institution ⟨the *head* of the English department⟩
7 a : CAPITULUM 2 **b :** the foliaged part of a plant especially when consisting of a compact mass of leaves or close fructification
8 a : the leading element of a military column or a procession **b :** HEADWAY
9 a : the uppermost extremity or projecting part of an object **:** TOP **b :** the striking part of a weapon, tool, or implement **c :** the rounded proximal end of a long bone (as the humerus) **d :** the end of a muscle nearest the origin — compare ORIGIN **e :** the oval part of a printed musical note
10 a : a body of water kept in reserve at a height; *also* **:** the containing bank, dam, or wall **b :** a mass of water in motion
11 a : the difference in elevation between two points in a body of fluid **b :** the resulting pressure of the fluid at the lower point expressible as this height; *broadly* **:** pressure of a fluid
12 a : the bow and adjacent parts of a ship **b :** a ship's toilet; *broadly* **:** TOILET
13 : the approximate length of the head of a horse ⟨won by a *head*⟩
14 : the place of leadership, honor, or command ⟨at the *head* of her class⟩
15 a (1) **:** a word or series of words often in larger letters placed at the beginning of a passage or at the top of a page in order to introduce or categorize (2) **:** a separate part or topic **b :** a portion of a page or sheet that is above the first line of printing
16 : the foam or scum that rises on a fermenting or effervescing liquid (as beer)
17 a : the part of a boil, pimple, or abscess at which it is likely to break **b :** culminating point of action **:** CRISIS ⟨events came to a *head*⟩
18 a : a part or attachment of a machine or machine tool containing a device (as a cutter or drill); *also* **:** the part of an apparatus that performs the chief function or a particular function **b :** an electromagnet used as a transducer in magnetic recording for recording on, reading, or erasing a magnetic medium (as tape)
19 : an immediate constituent of a construction that can have the same grammatical function as the whole (as *man* in "an old man", "a very old man", or "the man in the street")
20 a : one who uses a drug **b :** DEVOTEE ⟨chili *heads*⟩
21 : FELLATIO, CUNNILINGUS — usually used with *give;* often considered vulgar
— by the head : drawing the greater depth of water forward
— off one's head : CRAZY, DISTRACTED
— out of one's head : DELIRIOUS
— over one's head 1 : beyond one's comprehension or competence ⟨the most awful intellectual detail, all of it *over my head* —E. B. White⟩ **2 :** so as to pass over one's superior standing or authority ⟨went *over my head* to complain⟩
[2]head *adjective* (before 12th century)
1 : of, relating to, or intended for the head
2 : PRINCIPAL, CHIEF ⟨*head* cook⟩
3 : situated at the head
4 : coming from in front ⟨*head* sea⟩
[3]head (14th century)
transitive verb
1 : BEHEAD
2 a : to put a head on **:** fit a head to ⟨*head* an arrow⟩ **b :** to form the head or top of ⟨tower *headed* by a spire⟩
3 : to act as leader or head to ⟨*head* a revolt⟩
4 a : to get in front of so as to hinder, stop, or turn back **b :** to take a lead over (as a racehorse) **:** SURPASS **c :** to pass (a stream) by going round above the source
5 a : to put something at the head of (as a list) **b :** to stand as the first or leading member of ⟨*heads* the list of heroes⟩
6 : to set the course of ⟨*head* a ship northward⟩
7 : to drive (as a soccer ball) with the head
intransitive verb
1 : to form a head ⟨this cabbage *heads* early⟩
2 : to point or proceed in a certain direction ⟨the fleet was *heading* out⟩
3 : to have a source **:** ORIGINATE
head•ache \'he-ˌdāk\ *noun* (before 12th century)
1 : pain in the head
2 : a vexatious or baffling situation or problem
— head•achy \-ˌdā-kē\ *adjective*
head and shoulders *adverb* (circa 1864)
: beyond comparison **:** by far ⟨*head and shoulders* above the competition⟩
head•band \'hed-ˌband\ *noun* (1535)
1 : a band worn on or around the head
2 : a narrow strip of cloth sewn or glued by hand to a book at the extreme ends of the backbone
head•board \-ˌbōrd, -ˌbȯrd\ *noun* (1730)
: a board forming the head (as of a bed)
head•cheese \-ˌchēz\ *noun* (1841)
: a jellied loaf or sausage made from edible parts of the head, feet, and sometimes the tongue and heart especially of a pig
head cold *noun* (1937)
: a common cold centered in the nasal passages and adjacent mucous tissues
head•dress \'hed-ˌdres\ *noun* (1703)
: an often elaborate covering for the head

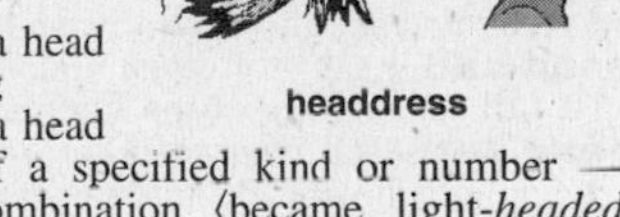
headdress

head•ed \'he-dəd\ *adjective* (13th century)
1 : having a head or a heading
2 : having a head or heads of a specified kind or number — used in combination ⟨became light-*headed* from the fever⟩ ⟨a round*headed* screw⟩
head•er \'he-dər\ *noun* (15th century)
1 : one that removes heads; *especially* **:** a grain-harvesting machine that cuts off the grain heads and elevates them to a wagon
2 a : a brick or stone laid in a wall with its end toward the face of the wall **b :** a beam fitted at one side of an opening to support free ends of floor joists, studs, or rafters **c :** a horizontal structural or finish piece over an opening **:** LINTEL **d :** a conduit (as an exhaust pipe for a many-cylindered engine) into which a number of smaller conduits open **e :** a mounting plate through which electrical terminals pass from a sealed device (as a transistor)
3 : a fall or dive headfirst
4 : a shot or pass in soccer made by heading the ball
5 : HEAD 15a(1)
head•first \'hed-'fərst\ *adverb* (circa 1828)
: with the head foremost **:** HEADLONG ⟨dove *headfirst* into the waves⟩
— headfirst *adjective*
head•fore•most \-'fōr-ˌmōst, -'fȯr-\ *adverb* (1697)
: HEADFIRST
head•gate \'hed-ˌgāt\ *noun* (1832)
: a gate for controlling the water flowing into a channel (as an irrigation ditch)
head•gear \-ˌgir\ *noun* (15th century)
1 : a covering or protective device for the head
2 : a harness for a horse's head
head•hunt•er \-ˌhən-tər\ *noun* (1853)
1 : one that engages in head-hunting
2 : a recruiter of personnel especially at the executive level
head–hunt•ing \-ˌhən-tiŋ\ *noun* (1853)
1 : the act or custom of seeking out, decapitating, and preserving the heads of enemies as trophies
2 : a seeking to deprive usually political enemies of position or influence
head•ing \'he-diŋ\ *noun* (1676)
1 a : something that forms or serves as a head; *especially* **:** an inscription, headline, or title standing at the top or beginning (as of a letter or chapter) **b :** the address and date at the beginning of a letter showing its place and time of origin
2 : the compass direction in which the longitudinal axis of a ship or aircraft points; *broadly* **:** DIRECTION
3 : DRIFT 6
head•lamp \-ˌlamp\ *noun* (1885)
: HEADLIGHT
head•land \'hed-lənd, -ˌland\ *noun* (before 12th century)
1 : unplowed land at the ends of furrows or near a fence

□ USAGE
he Since English has no third-person singular pronoun of common gender, the problem of how to refer to a singular noun antecedent that can apply to either sex—such as *person*—has no perfect solution. From the earliest days of the language two approaches have been generally adopted: the use of *he* and the use of *they.* By the 18th century, grammarians had begun to be troubled by the use of the plural pronoun to refer to a singular antecedent. Such use violated number agreement. On the other hand, the use of *he* violated gender agreement. Some writers avoided the problem by using two pronouns ⟨the reader's heart (if *he* or *she* have any) —Henry Fielding⟩. The grammarians then hit upon the idea that the masculine pronoun included the feminine, and the generic *he* was created. Grammarians and usage commentators of the next two centuries have in general insisted on the generic *he* and have disparaged both *he or she* and *they.* But in the later 20th century generic *he* has met greater resistance, and more writers have sought alternative solutions, no single one of which works well in all cases. The consistent use of *she* can readily seem laughable in sentences like "show me a person who isn't concerned about her weight and I'll show you a man." The alternation of masculine and feminine pronouns can produce sentences like "Have the patient lie down no matter how she feels. Have him do this even though she acts all right and insists you leave him alone." So *he or she* has come in for increased use, but the paired pronouns are hard to use consistently. Some writers have started off with *he or she* and failed to follow through, lapsing into *he* or *him* in subsequent sentences. Many handbooks suggest recasting sentences in the plural so that the plural pronouns can be used without violation of number agreement. But here the common disapproval of *they* can cause writers to produce sentences in which a plural noun antecedent is followed by *he or she.* Those for whom the old solution of using the plural pronoun, now quite widely adopted, will not work and those who dislike generic *he* will have to be especially alert in using *he or she* or another approach. See in addition ANYBODY, EVERYBODY, NOBODY, THEY.

2 : a point of usually high land jutting out into a body of water **:** PROMONTORY

head·less \-ləs\ *adjective* (before 12th century)
1 a : having no head **b :** having the head cut off **:** BEHEADED
2 : having no chief
3 : lacking good sense or prudence **:** FOOLISH
— **head·less·ness** *noun*

head·light \-ˌlīt\ *noun* (1861)
1 : a light with a reflector and special lens mounted on the front of a vehicle to illuminate the road ahead; *also* **:** the beam cast by a headlight
2 : a light worn on the forehead (as of a miner or physician)

[1]head·line \-ˌlīn\ *noun* (1824)
1 : words set at the head of a passage or page to introduce or categorize
2 a : a head of a newspaper story or article usually printed in large type and giving the gist of the story or article that follows **b** *plural* **:** front-page news ⟨the scandal made *headlines*⟩

[2]headline *transitive verb* (1891)
1 : to provide with a headline
2 : to publicize highly
3 : to be engaged as a leading performer in (a show)

head·lin·er \ˈhed-ˌlī-nər\ *noun* (1896)
1 : the principal performer in a show **:** STAR; *broadly* **:** PERSONALITY 4b
2 : fabric covering the inside of the roof of an automobile

head linesman *noun* (circa 1949)
: a football linesman

head·lock \ˈhed-ˌläk\ *noun* (1905)
: a hold in which a wrestler encircles his opponent's head with one arm

[1]head·long \-ˈlȯŋ\ *adverb* [Middle English *hedlong,* alteration of *hedling,* from *hed* head] (14th century)
1 : HEADFIRST
2 : without deliberation **:** RECKLESSLY
3 : without pause or delay

[2]head·long \-ˌlȯŋ\ *adjective* (circa 1550)
1 *archaic* **:** STEEP, PRECIPITOUS
2 : lacking in calmness or restraint **:** PRECIPITATE ⟨a *headlong* torrent of emotion⟩
3 : plunging headforemost
synonym see PRECIPITATE

head louse *noun* (1547)
: one of a variety (*Pediculus humanus capitis*) of the common louse that lives on the human scalp

head·man *noun* (before 12th century)
1 a \ˈhed-ˈman\ **:** FOREMAN, OVERSEER **b** \-ˈman, -ˌman\ **:** a lesser chief of a primitive community
2 \-mən\ **:** HEADSMAN

head·mas·ter \ˈhed-ˌmas-tər, -ˈmas-\ *noun* (1576)
: a man heading the staff of a private school **:** PRINCIPAL
— **head·mas·ter·ship** \-ˌship\ *noun*

head·mis·tress \-ˌmis-trəs, -ˈmis-\ *noun* (1872)
: a woman heading the staff of a private school

head·most \ˈhed-ˌmōst\ *adjective* (1628)
: most advanced **:** LEADING

head·note \-ˌnōt\ *noun* (1855)
1 : a prefixed note of comment or explanation
2 : a note prefixed to the report of a decided legal case

head off *transitive verb* (1841)
: to turn back or turn aside **:** BLOCK, PREVENT ⟨*head* them *off* at the pass⟩ ⟨attempts to *head off* the imminent crisis⟩

[1]head–on \ˈhed-ˈȯn, -ˈän\ *adverb* (1840)
1 : with the head or front making the initial contact ⟨the cars collided *head-on*⟩
2 : in direct opposition, confrontation, or contradiction ⟨met the problem *head-on*⟩

[2]head–on *adjective* (1903)
1 : having the front facing in the direction of initial contact or line of sight ⟨a *head-on* collision⟩
2 : FRONTAL

head over heels *adverb* (1771)
1 a : in or as if in a somersault **:** HELTER-SKELTER **b :** UPSIDE DOWN
2 : very much **:** DEEPLY ⟨*head over heels* in love⟩

head·phone \ˈhed-ˌfōn\ *noun* (1914)
: an earphone held over the ear by a band worn on the head — usually used in plural

head·piece \-ˌpēs\ *noun* (1535)
1 a : a protective or defensive covering for the head **b :** an ornamental, ceremonial, or traditional covering for the head
2 : BRAINS, INTELLIGENCE
3 : an ornament especially at the beginning of a chapter

head·pin \-ˌpin\ *noun* (1927)
: a bowling pin that stands foremost in the arrangement of pins

head·quar·ter \ˈhed-ˌkwȯ(r)-tər, -ˌkȯr-, (ˈ)hed-ˈ\ (1903)
transitive verb
: to place in headquarters
intransitive verb
: to make one's headquarters

head·quar·ters \-tərz\ *noun plural but singular or plural in construction* (1647)
1 : a place from which a commander performs the functions of command
2 : the administrative center of an enterprise

head·rest \-ˌrest\ *noun* (1853)
1 : a support for the head
2 : HEAD RESTRAINT

head restraint *noun* (1967)
: a resilient pad at the top of the back of an automobile seat especially for preventing whiplash injury

head rhyme *noun* (circa 1943)
: ALLITERATION

head·room \ˈhed-ˌrüm, -ˌru̇m\ *noun* (1851)
: vertical space in which to stand, sit, or move

head·sail \-ˌsāl, -səl\ *noun* (1627)
: a sail set forward of the foremast

head·set \-ˌset\ *noun* (1921)
1 : an attachment for holding an earphone and transmitter at one's head
2 : a pair of headphones

head·ship \-ˌship\ *noun* (1582)
: the position, office, or dignity of a head

head shop *noun* (1968)
: a shop specializing in articles (as hashish pipes and roach clips) of interest to drug users

head·shrink·er \-ˌshriŋ-kər, *especially Southern* -ˌsriŋ-\ *noun* (1950)
: SHRINK 3

heads·man \ˈhedz-mən\ *noun* (1601)
: one that beheads **:** EXECUTIONER

head·space \ˈhed-ˌspās\ *noun* (1936)
: the volume above a liquid or solid in a closed container

head·spring \ˈhed-ˌspriŋ\ *noun* (14th century)
: FOUNTAINHEAD, SOURCE

head·stall \-ˌstȯl\ *noun* (14th century)
: a part of a bridle or halter that encircles the head

head·stand \-ˌstand\ *noun* (circa 1934)
: the gymnastic feat of standing on one's head usually with support from the hands

head start *noun* (1886)
1 : an advantage granted or achieved at the beginning of a race, a chase, or a competition ⟨a 10-minute *head start*⟩
2 : a favorable or promising beginning

head·stock \ˈhed-ˌstäk\ *noun* (1731)
: a bearing or pedestal for a revolving or moving part; *specifically* **:** a part of a lathe that holds the revolving spindle and its attachments

head·stone \-ˌstōn\ *noun* (1775)
: a memorial stone at the head of a grave

head·stream \-ˌstrēm\ *noun* (14th century)
: a stream that is the source of a river

head·strong \-ˌstrȯŋ\ *adjective* (14th century)
1 : not easily restrained **:** impatient of control, advice, or suggestions
2 : directed by ungovernable will ⟨violent *headstrong* actions⟩
synonym see UNRULY

heads–up \ˈhed-ˈzəp\ *adjective* (1947)
: ALERT, RESOURCEFUL ⟨*heads-up* football⟩

heads up *interjection* (circa 1941)
— used as a warning to look out for danger especially overhead or to clear a passageway

head–to–head *adverb or adjective* (circa 1728)
: in a direct confrontation or encounter usually between individuals

head·wait·er \ˈhed-ˈwā-tər\ *noun* (1805)
: the head of the dining-room staff of a restaurant or hotel

head·wa·ter \-ˌwȯ-tər, -ˌwä-\ *noun* (1802)
: the source of a stream — usually used in plural

head·way \-ˌwā\ *noun* (1748)
1 a : motion or rate of motion in a forward direction **b :** ADVANCE, PROGRESS
2 : headroom (as under an arch) sufficient to allow passage
3 : the time interval between two vehicles traveling in the same direction on the same route

head·word \ˈhed-ˌwərd\ *noun* (circa 1823)
1 : a word or term placed at the beginning (as of a chapter or an entry in an encyclopedia)
2 : HEAD 19

head·work \-ˌwərk\ *noun* (1837)
: mental labor; *especially* **:** clever thinking

heady \ˈhe-dē\ *adjective* **head·i·er; -est** (14th century)
1 a : WILLFUL, RASH ⟨*heady* opinions⟩ **b :** VIOLENT, IMPETUOUS
2 a : tending to intoxicate or make giddy or elated ⟨*heady* wine⟩ ⟨being in such distinguished company was a *heady* experience⟩ **b :** GIDDY, EXHILARATED ⟨*heady* with his success⟩ **c :** RICH ⟨a *heady* sauce⟩ ⟨a *heady* variety⟩ **d :** IMPRESSIVE ⟨a man of *heady* accomplishments⟩
3 a : marked by or showing good judgment **:** SHREWD, INTELLIGENT **b :** intellectually stimulating or demanding
— **head·i·ly** \ˈhe-dᵊl-ē\ *adverb*
— **head·i·ness** \ˈhe-dē-nəs\ *noun*

heal \ˈhē(ə)l\ *verb* [Middle English *helen,* from Old English *hǣlan;* akin to Old High German *heilen* to heal, Old English *hāl* whole — more at WHOLE] (before 12th century)
transitive verb
1 a : to make sound or whole ⟨*heal* a wound⟩ **b :** to restore to health
2 a : to cause (an undesirable condition) to be overcome **:** MEND ⟨the troubles . . . had not been forgotten, but they had been *healed* —William Power⟩ **b :** to patch up (a breach or division) ⟨*heal* a breach between friends⟩
3 : to restore to original purity or integrity ⟨*healed* of sin⟩
intransitive verb
: to return to a sound state

heal·er \ˈhē-lər\ *noun* (12th century)
1 : one that heals
2 : a Christian Science practitioner

health \ˈhelth *also* ˈheltth\ *noun, often attributive* [Middle English *helthe,* from Old English *hǣlth,* from *hāl*] (before 12th century)
1 a : the condition of being sound in body, mind, or spirit; *especially* **:** freedom from physical disease or pain **b :** the general condition of the body ⟨in poor *health*⟩ ⟨enjoys good *health*⟩
2 a : flourishing condition **:** WELL-BEING ⟨defending the *health* of the beloved oceans —Peter Wilkinson⟩ **b :** general condition or state ⟨poor economic *health*⟩
3 : a toast to someone's health or prosperity

health food *noun* (1882)
: a food promoted as highly conducive to health

health·ful \'helth-fəl *also* 'heltth-\ *adjective* (14th century)
1 : beneficial to health of body or mind
2 : HEALTHY ⟨he felt incapable of looking into the girl's pretty, *healthful* face —Saul Bellow⟩ ☆
usage see HEALTHY
— health·ful·ness *noun*

health insurance *noun* (1901)
: insurance against loss through illness of the insured; *especially* **:** insurance providing compensation for medical expenses

health maintenance organization *noun* (1973)
: HMO

health spa *noun* (1960)
: a commercial establishment (as a resort) providing facilities devoted to health and fitness

healthy \'hel-thē *also* 'helt-\ *adjective* **health·i·er; -est** (1552)
1 : enjoying health and vigor of body, mind, or spirit **:** WELL
2 : evincing health ⟨a *healthy* complexion⟩
3 : conducive to health ⟨walk three miles every day . . . a beastly bore, but *healthy* —G. S. Patton⟩
4 a : PROSPEROUS, FLOURISHING **b :** not small or feeble **:** CONSIDERABLE ☆ □
— health·i·ly \-thə-lē\ *adverb*
— health·i·ness \-thē-nəs\ *noun*

¹heap \'hēp\ *noun* [Middle English *heep,* from Old English *hēap;* akin to Old High German *houf* heap] (before 12th century)
1 : a collection of things thrown one on another **:** PILE
2 : a great number or large quantity **:** LOT

²heap *transitive verb* (before 12th century)
1 a : to throw or lay in a heap **:** pile or collect in great quantity ⟨his sole object was to *heap* up riches⟩ **b :** to form or round into a heap ⟨*heaped* the dirt into a mound⟩ **c :** to form a heap on **:** load heavily ⟨*heap* the plates with food⟩
2 : to accord or bestow lavishly or in large quantities ⟨*heaped* honors upon them⟩

hear \'hir\ *verb* **heard** \'hərd\; **hear·ing** \'hir-iŋ\ [Middle English *heren,* from Old English *hīeran;* akin to Old High German *hōren* to hear, and probably to Latin *cavēre* to be on guard, Greek *akouein* to hear] (before 12th century)
transitive verb
1 : to perceive or apprehend by the ear
2 : to gain knowledge of by hearing
3 a : to listen to with attention **:** HEED **b :** ATTEND ⟨*hear* mass⟩
4 a : to give a legal hearing to **b :** to take testimony from ⟨*hear* witnesses⟩
intransitive verb
1 : to have the capacity of apprehending sound
2 a : to gain information **:** LEARN **b :** to receive communication ⟨*heard* from her recently⟩
3 : to entertain the idea — used in the negative ⟨wouldn't *hear* of it⟩
4 — often used in the expression *Hear! Hear!* to express approval (as during a speech)
— hear·er \'hir-ər\ *noun*

hear·ing *noun* (13th century)
1 a : the process, function, or power of perceiving sound; *specifically* **:** the special sense by which noises and tones are received as stimuli **b :** EARSHOT
2 a : opportunity to be heard, to present one's side of a case, or to be generally known or appreciated **b** (1) **:** a listening to arguments (2) **:** a preliminary examination in criminal procedure **c :** a session (as of a legislative committee) in which testimony is taken from witnesses
3 *chiefly dialect* **:** a piece of news

hearing aid *noun* (1922)
: an electronic device usually worn by a person for amplifying sound before it reaches the receptor organs

hearing dog *noun* (1952)
: a dog trained to alert its deaf or hearing-impaired owner to sounds (as of a doorbell, alarm, or telephone) — called also *hearing ear dog*

hear·ken \'här-kən\ *verb* **hear·kened; hear·ken·ing** \'härk-niŋ, 'här-kə-\ [Middle English *herknen,* from Old English *heorcnian;* akin to Old High German *hōrechen* to listen, Old English *hīeran* to hear] (before 12th century)
intransitive verb
1 : LISTEN
2 : to give respectful attention
transitive verb
archaic **:** to give heed to **:** HEAR

hearken back *intransitive verb* (1933)
: HARK BACK

hear·say \'hir-ˌsā\ *noun* (circa 1532)
: RUMOR

hearsay evidence *noun* (1753)
: evidence based not on a witness's personal knowledge but on another's statement not made under oath

¹hearse \'hərs\ *noun* [Middle English *herse,* from Middle French *herce* harrow, frame for holding candles, from Latin *hirpic-, hirpex* harrow] (14th century)
1 a : an elaborate framework erected over a coffin or tomb to which memorial verses or epitaphs are attached **b :** a triangular candelabrum for 15 candles used especially at Tenebrae
2 a *archaic* **:** COFFIN **b** *obsolete* **:** BIER 2
3 : a vehicle for conveying the dead to the grave ◇

²hearse *transitive verb* **hearsed; hears·ing** (1592)
1 a *archaic* **:** to place on or in a hearse **b :** to convey in a hearse
2 : BURY

¹heart \'härt\ *noun* [Middle English *hert,* from Old English *heorte;* akin to Old High German *herza* heart, Latin *cord-, cor,* Greek *kardia*] (before 12th century)
1 a : a hollow muscular organ of vertebrate animals that by its rhythmic contraction acts as a force pump maintaining the circulation of the blood **b :** a structure in an invertebrate animal functionally analogous to the vertebrate heart **c :** BREAST, BOSOM **d :** something resembling a heart in shape; *specifically* **:** a stylized representation of a heart
2 a : a playing card marked with a stylized figure of a red heart **b** *plural* **:** the suit comprising cards marked with hearts **c** *plural but singular or plural in construction* **:** a game in which the object is to avoid taking tricks containing hearts

1 2 3 4 5 6

heart 1a: ***1*** **aorta,** ***2*** **pulmonary artery,** ***3*** **left atrium,** ***4*** **left ventricle,** ***5*** **right ventricle,** ***6*** **right atrium**

☆ SYNONYMS
Healthful, wholesome, salubrious, salutary mean favorable to the health of mind or body. HEALTHFUL implies a positive contribution to a healthy condition ⟨a *healthful* diet⟩. WHOLESOME applies to what benefits, builds up, or sustains physically, mentally, or spiritually ⟨*wholesome* foods⟩ ⟨the movie is *wholesome* family entertainment⟩. SALUBRIOUS applies chiefly to the helpful effects of climate or air ⟨cool and *salubrious* weather⟩. SALUTARY describes something corrective or beneficially effective, even though it may in itself be unpleasant ⟨a *salutary* warning that increased production⟩.

☆ SYNONYMS
Healthy, sound, wholesome, robust, hale, well mean enjoying or indicative of good health. HEALTHY implies full strength and vigor as well as freedom from signs of disease ⟨a *healthy* family⟩. SOUND emphasizes the absence of disease, weakness, or malfunction ⟨a *sound* heart⟩. WHOLESOME implies appearance and behavior indicating soundness and balance ⟨a face with a *wholesome* glow⟩. ROBUST implies the opposite of all that is delicate or sickly ⟨a lively, *robust* little boy⟩. HALE applies particularly to robustness in old age ⟨still *hale* at the age of eighty⟩. WELL implies nothing more than just freedom from disease or illness ⟨in her declining years she has never been a *well* person⟩.

□ USAGE
healthy Since 1881 usage writers have been trying to draw a distinction between *healthy* and *healthful. Healthy,* they say, should be used in the sense "enjoying health, evincing health" and *healthful* in the sense "conducive to health." And each word is used in the prescribed way part of the time. The problem is that both words are used in the latter sense ⟨on the high, *healthful* promontory on the right bank —David Levering Lewis⟩ ⟨on horseback, attending to my farm or other concerns, which I find *healthful* to my body —Thomas Jefferson⟩ ⟨prescribed the very *healthful* exercise of woodcutting —T. Coraghessan Boyle⟩ ⟨having been born & brought up in a mountainous & *healthy* country —Thomas Jefferson⟩ ⟨bastion of right thinking, vegetarianism and self-improvement, . . . the single *healthiest* spot on the planet —T. Coraghessan Boyle⟩ ⟨more genteel or *healthier* living habits —J. M. Richards⟩. The commentators would like to eliminate this use of *healthy* from the language, but as you can see from the second Thomas Jefferson quotation, this use had long been established when its first critic was writing. In fact, it first appeared in the middle of the 16th century. The distinction is simply a fabrication that one may observe or ignore and still be in good company.

◇ WORD HISTORY
hearse In medieval French the word *herce,* which literally meant "harrow," was applied to other things that were thought to resemble the toothed cultivating implement, such as a portcullis and a frame like a chandelier for holding candles. The latter sense when borrowed into Middle English as *herce* or *herse* was applied specifically to a framework for candles, statues or other objects that was placed over the coffin or tomb of a distinguished person. From this starting point the English word diverged further and further in meaning from its French source; its reference was transferred from the framework to the coffin itself or the stand on which the coffin was placed, and then to the vehicle that transports a coffin, which is the ordinary sense of the word today. Completely obscured for people today is the relation of *hearse* to *rehearse,* which comes from a medieval French verb meaning "to harrow again."

3 a : PERSONALITY, DISPOSITION ⟨a cold *heart*⟩ **b** *obsolete* **:** INTELLECT
4 : the emotional or moral as distinguished from the intellectual nature: as **a :** generous disposition **:** COMPASSION ⟨a leader with *heart*⟩ **b :** LOVE, AFFECTIONS ⟨won her *heart*⟩ **c :** COURAGE, ARDOR ⟨never lost *heart*⟩
5 : one's innermost character, feelings, or inclinations ⟨knew it in his *heart*⟩ ⟨a man after my own *heart*⟩
6 a : the central or innermost part **:** CENTER **b :** the essential or most vital part of something **c :** the younger central compact part of a leafy rosette (as a head of lettuce)
— at heart : in essence **:** BASICALLY, ESSENTIALLY
— by heart : by rote or from memory
— to heart : with deep concern

²heart *transitive verb* (before 12th century)
1 *archaic* **:** HEARTEN
2 *archaic* **:** to fix in the heart

heart·ache \'härt-ˌāk\ *noun* (1602)
: anguish of mind **:** SORROW

heart attack *noun* (1928)
: an acute episode of heart disease (as myocardial infarction) due to insufficient blood supply to the heart muscle especially when caused by a coronary thrombosis or a coronary occlusion

heart·beat \'härt-ˌbēt\ *noun* (1850)
1 : one complete pulsation of the heart
2 : the vital center or driving impulse

heart block *noun* (1903)
: incoordination of the heartbeat in which the atria and ventricles beat independently and which is marked by decreased cardiac output

heart·break \'härt-ˌbrāk\ *noun* (14th century)
: crushing grief

heart·break·er \-ˌbrā-kər\ *noun* (1863)
: one that causes heartbreak

heart·break·ing \-ˌbrā-kiŋ\ *adjective* (1591)
1 a : causing intense sorrow or distress **b :** extremely trying or difficult
2 : producing an intense emotional reaction or response ⟨*heartbreaking* beauty⟩
— heart·break·ing·ly \-kiŋ-lē\ *adverb*

heart·bro·ken \-ˌbrō-kən\ *adjective* (circa 1586)
: overcome by sorrow

heart·burn \-ˌbərn\ *noun* (1597)
: a burning discomfort behind the lower part of the sternum usually related to spasm of the lower end of the esophagus or of the upper part of the stomach

heart·burn·ing \-ˌbər-niŋ\ *noun* (1513)
: intense or rancorous jealousy or resentment

heart disease *noun* (1864)
: an abnormal organic condition of the heart or of the heart and circulation

heart·ed \'här-təd\ *adjective* (13th century)
1 : having a heart especially of a specified kind — usually used in combination ⟨a faint-*hearted* leader⟩ ⟨a light*hearted* wanderer⟩
2 : seated in the heart

heart·en \'här-tᵊn\ *transitive verb* **heart·ened; heart·en·ing** \'härt-niŋ, 'har-tᵊn-iŋ\ (1526)
: to give heart to **:** CHEER
synonym see ENCOURAGE
— heart·en·ing·ly \-niŋ-lē, -tᵊn-iŋ-\ *adverb*

heart failure *noun* (1894)
1 : a condition in which the heart is unable to pump blood at an adequate rate or in adequate volume
2 : cessation of heartbeat **:** DEATH

heart·felt \'härt-ˌfelt\ *adjective* (1734)
: deeply felt **:** EARNEST
synonym see SINCERE

heart–free \'härt-ˌfrē\ *adjective* (1748)
: not in love

hearth \'härth\ *noun* [Middle English *herth,* from Old English *heorth;* akin to Old High German *herd* hearth, and probably to Sanskrit *kūḍayāti* he scorches] (before 12th century)
1 a : a brick, stone, or cement area in front of a fireplace **b :** the floor of a fireplace; *also* **:** FIREPLACE **c :** the lowest section of a furnace; *especially* **:** the section of a furnace on which the ore or metal is exposed to the flame or heat
2 : HOME
3 : a vital or creative center ⟨the central *hearth* of occidental civilization —A. L. Kroeber⟩

hearth·stone \-ˌstōn\ *noun* (14th century)
1 : stone forming a hearth
2 : HOME

heart·i·ly \'här-tᵊl-ē\ *adverb* (14th century)
1 : in a hearty manner
2 a : with all sincerity **:** WHOLEHEARTEDLY **b :** with zest or gusto
3 : WHOLLY, THOROUGHLY ⟨*heartily* sick of all this talk⟩

heart·land \'härt-ˌland\ *noun* (1904)
: a central area: as **a :** a central land area (as northern Eurasia) having strategic advantages **b :** a central geographical region especially of the U.S. in which mainstream or traditional values predominate

heart·less \-ləs\ *adjective* (14th century)
1 *archaic* **:** SPIRITLESS
2 : lacking feeling **:** CRUEL
— heart·less·ly *adverb*
— heart·less·ness *noun*

heart–lung machine *noun* (1953)
: a mechanical pump that maintains circulation during heart surgery by shunting blood away from the heart, oxygenating it, and returning it to the body

heart·rend·ing \'härt-ˌren-diŋ\ *adjective* (1594)
: HEARTBREAKING 1a
— heart·rend·ing·ly \-diŋ-lē\ *adverb*

hearts·ease \'härts-ˌēz\ *noun* (15th century)
1 : peace of mind **:** TRANQUILLITY
2 : any of various violas; *especially* **:** JOHNNY-JUMP-UP

heart·sick \'härt-ˌsik\ *adjective* (1526)
: very despondent **:** DEPRESSED
— heart·sick·ness *noun*

heart·some \'hert-səm\ *adjective* (1596)
chiefly Scottish **:** giving spirit or vigor **:** ANIMATING, ENLIVENING
— heart·some·ly *adverb, chiefly Scottish*

heart·sore \'härt-ˌsōr, -ˌsȯr\ *adjective* (1591)
: HEARTSICK

heart·string \-ˌstriŋ\ *noun* (15th century)
1 *obsolete* **:** a nerve once believed to sustain the heart
2 : the deepest emotions or affections ⟨pulled at his *heartstrings*⟩

heart·throb \-ˌthräb\ *noun* (1839)
1 : the throb of a heart
2 a : sentimental emotion **:** PASSION **b :** SWEETHEART

¹heart–to–heart \'härt-tə-'härt\ *adjective* (1867)
: SINCERE, FRANK ⟨a *heart-to-heart* talk⟩

²heart–to–heart *noun* (1910)
: a frank, serious, and often intimate conversation

heart·warm·ing \'härt-ˌwȯr-miŋ\ *adjective* (1899)
: inspiring sympathetic feeling **:** CHEERING
— heart–warm·er *noun*

heart–whole \-ˌhōl\ *adjective* (1600)
1 : HEART-FREE
2 : SINCERE, GENUINE

heart·wood \-ˌwu̇d\ *noun* (1810)
: the older harder nonliving central wood of trees that is usually darker, denser, less permeable, and more durable than the surrounding sapwood

heart·worm \-ˌwərm\ *noun* (1888)
: a filarial worm (*Dirofilaria immitis*) that is a parasite especially in the right heart of dogs and is transmitted by mosquitoes; *also* **:** infestation or disease caused by the heartworm

¹hearty \'här-tē\ *adjective* **heart·i·er; -est** (14th century)
1 a : giving unqualified support ⟨a *hearty* endorsement⟩ **b :** enthusiastically or exuberantly cordial **:** JOVIAL **c :** expressed unrestrainedly
2 a : exhibiting vigorous good health **b** (1) **:** having a good appetite (2) **:** abundant, rich, or flavorful enough to satisfy the appetite
3 : VIGOROUS, VEHEMENT ⟨a *hearty* pull⟩
synonym see SINCERE
— heart·i·ness *noun*

²hearty *noun, plural* **heart·ies** (1803)
: a hearty fellow; *also* **:** SAILOR

¹heat \'hēt\ *verb* [Middle English *heten,* from Old English *hǣtan;* akin to Old English *hāt* hot] (before 12th century)
intransitive verb
1 : to become warm or hot
2 : to start to spoil from heat
transitive verb
1 : to make warm or hot
2 : EXCITE
— heat·able \'hē-tə-bəl\ *adjective*

²heat *noun* [Middle English *hete,* from Old English *hǣte, hǣtu;* akin to Old English *hāt* hot] (before 12th century)
1 a (1) **:** a condition of being hot **:** WARMTH (2) **:** a marked or notable degree of hotness **b :** pathological excessive bodily temperature **c :** a hot place or situation **d** (1) **:** a period of heat (2) **:** a single complete operation of heating; *also* **:** the quantity of material so heated **e** (1) **:** added energy that causes substances to rise in temperature, fuse, evaporate, expand, or undergo any of various other related changes, that flows to a body by contact with or radiation from bodies at higher temperatures, and that can be produced in a body (as by compression) (2) **:** the energy associated with the random motions of the molecules, atoms, or smaller structural units of which matter is composed **f :** appearance, condition, or color of a body as indicating its temperature
2 a : intensity of feeling or reaction **:** PASSION **b :** the height or stress of an action or condition ⟨in the *heat* of battle⟩ **c :** sexual excitement especially in a female mammal; *specifically* **:** ESTRUS
3 : a single continuous effort: as **a :** a single round of a contest (as a race) having two or more rounds for each contestant **b :** one of several preliminary contests held to eliminate less competent contenders
4 : pungency of flavor
5 a *slang* (1) **:** the intensification of law-enforcement activity or investigation (2) **:** POLICE **b :** PRESSURE, COERCION **c :** ABUSE, CRITICISM ⟨took a lot of *heat*⟩
6 : SMOKE 8
— heat·less \'hēt-ləs\ *adjective*
— heat·proof \-'prüf\ *adjective*

heat cramps *noun plural* (1938)
: a condition that is marked by sudden development of cramps in skeletal muscles and that results from prolonged work in high temperatures accompanied by profuse perspiration with loss of sodium chloride from the body

heat·ed \'hē-təd\ *adjective* (1886)
: marked by anger ⟨a *heated* argument⟩
— heat·ed·ly *adverb*

heat engine *noun* (circa 1895)
: a mechanism (as an internal-combustion engine) for converting heat energy into mechanical or electrical energy

heat·er \'hē-tər\ *noun* (15th century)
1 : one that heats; *especially* **:** a device that imparts heat or holds something to be heated
2 : FASTBALL

heat exchanger *noun* (1902)
: a device (as an automobile radiator) for transferring heat from one fluid to another without allowing them to mix

heat exhaustion *noun* (1939)
: a condition marked by weakness, nausea, dizziness, and profuse sweating that results from physical exertion in a hot environment — called also *heat prostration;* compare HEATSTROKE

heath \'hēth\ *noun* [Middle English *heth,* from

Old English *hǣth;* akin to Old High German *heida* heather, Old Welsh *coit* forest] (before 12th century)
1 a : a tract of wasteland **b :** an extensive area of rather level open uncultivated land usually with poor coarse soil, inferior drainage, and a surface rich in peat or peaty humus
2 a : any of a family (Ericaceae, the heath family) of shrubby dicotyledonous and often evergreen plants that thrive on open barren usually acid and ill-drained soil; *especially* **:** an evergreen subshrub of either of two genera (*Erica* and *Calluna*) with whorls of needlelike leaves and clusters of small flowers **b :** any of various plants that resemble true heaths
— heath·less \-ləs\ *adjective*
— heath·like \-ˌlīk\ *adjective*
— heathy \ˈhē-thē\ *adjective*

¹hea·then \ˈhē-thən\ *adjective* [Middle English *hethen,* from Old English *hǣthen;* akin to Old High German *heidan* heathen, and probably to Old English *hǣth* heath] (before 12th century)
1 : of or relating to heathens, their religions, or their customs
2 : STRANGE, UNCIVILIZED

²heathen *noun, plural* **heathens** *or* **heathen** (before 12th century)
1 : an unconverted member of a people or nation that does not acknowledge the God of the Bible
2 : an uncivilized or irreligious person
— hea·then·dom \-dəm\ *noun*
— hea·then·ism \-thə-ˌni-zəm\ *noun*
— hea·then·ize \-thə-ˌnīz\ *transitive verb*

hea·then·ish \ˈhē-thə-nish\ *adjective* (1593)
: resembling or characteristic of heathens **:** BARBAROUS
— hea·then·ish·ly *adverb*

¹heath·er \ˈhe-thər\ *noun* [Middle English (northern) *hather*] (14th century)
: HEATH 2a; *especially* **:** a common Eurasian heath (*Calluna vulgaris*) of northern and alpine regions that has small crowded sessile leaves and racemes of tiny usually purplish pink flowers and is naturalized in the northeastern U.S.

²heather *adjective* (1615)
: HEATHERY

heath·ery \ˈheth-rē, ˈhe-thə-\ *adjective* (1535)
1 : of, relating to, or resembling heather
2 : having flecks of various colors ⟨a soft *heathery* tweed⟩

heath hen *noun* (1644)
: a now extinct grouse (*Tympanuchus cupido cupido*) of the northeastern U.S. — compare PRAIRIE CHICKEN

heath·land \ˈhēth-ˌland\ *noun* (1819)
: HEATH 1

heat lightning *noun* (1834)
: vivid and extensive flashes of electric light without thunder seen near the horizon especially at the close of a hot day and ascribed to far-off lightning reflected by high clouds

heat prostration *noun* (1938)
: HEAT EXHAUSTION

heat pump *noun* (1894)
: an apparatus for heating or cooling a building by transferring heat by mechanical means from or to a reservoir (as the ground, water, or air) outside the building

heat rash *noun* (1887)
: PRICKLY HEAT

heat shield *noun* (1962)
: a barrier of ablative material to protect a space capsule from heat on its entry into an atmosphere

heat sink *noun* (1936)
: a substance or device for the absorption or dissipation of unwanted heat (as from a process or an electronic device)

heat·stroke \ˈhēt-ˌstrōk\ *noun* (1874)
: a condition marked especially by cessation of sweating, extremely high body temperature, and collapse that results from prolonged exposure to high temperature — compare HEAT EXHAUSTION

heat–treat \ˈhēt-ˌtrēt\ *transitive verb* (1907)
: to subject to heat; *especially* **:** to treat (as metals) by heating and cooling in a way that will produce desired properties
— heat treater *noun*
— heat treatment *noun*

heat wave *noun* (1893)
: a period of unusually hot weather

¹heave \ˈhēv\ *verb* **heaved** *or* **hove** \ˈhōv\; **heav·ing** [Middle English *heven,* from Old English *hebban;* akin to Old High German *hevan* to lift, Latin *capere* to take] (before 12th century)
transitive verb
1 *obsolete* **:** ELEVATE
2 : LIFT, RAISE
3 : THROW, CAST
4 a : to cause to swell or rise **b :** to displace (as a rock stratum) especially by a fault
5 : to utter with obvious effort or with a deep breath
6 : HAUL, DRAW
intransitive verb
1 : LABOR, STRUGGLE
2 : RETCH
3 a : to rise and fall rhythmically **b :** PANT
4 a : PULL, PUSH **b :** to move a ship in a specified direction or manner **c** *past usually* **hove :** to move in an indicated way — used of a ship
5 : to rise or become thrown or raised up
synonym see LIFT
— heav·er *noun*
— heave to : to bring a ship to a stop

²heave *noun* (circa 1571)
1 a : an effort to heave or raise **b :** HURL, CAST
2 : an upward motion **:** RISING; *especially* **:** a rhythmical rising
3 : horizontal displacement especially by the faulting of a rock
4 *plural but singular or plural in construction* **:** chronic pulmonary emphysema of the horse resulting in difficult expiration, heaving of the flanks, and a persistent cough

heave–ho \ˈhēv-ˈhō\ *noun* [from *heave ho!,* interjection used when heaving on a rope] (1947)
: DISMISSAL ⟨gave him the old *heave-ho*⟩

heav·en \ˈhe-vən\ *noun* [Middle English *heven,* from Old English *heofon;* akin to Old High German *himil* heaven] (before 12th century)
1 : the expanse of space that seems to be over the earth like a dome **:** FIRMAMENT — usually used in plural
2 a *often capitalized* **:** the dwelling place of the Deity and the joyful abode of the blessed dead **b :** a spiritual state of everlasting communion with God
3 *capitalized* **:** GOD 1
4 : a place or condition of utmost happiness
5 *Christian Science* **:** a state of thought in which sin is absent and the harmony of divine Mind is manifest

heav·en·ly \-lē\ *adjective* (before 12th century)
1 : of or relating to heaven or the heavens **:** CELESTIAL ⟨the *heavenly* choirs⟩ ⟨use a telescope to study the *heavenly* bodies⟩
2 a : suggesting the blessed state of heaven **:** BEATIFIC ⟨*heavenly* peace⟩ **b :** DELIGHTFUL
— heav·en·li·ness *noun*

heav·en–sent \-ˌsent\ *adjective* (circa 1649)
: PROVIDENTIAL

heav·en·ward \-wərd\ *adverb or adjective* (13th century)
: toward heaven

heav·en·wards \-wərdz\ *adverb* (1650)
: HEAVENWARD

heavier–than–air *adjective* (1903)
: of greater weight than the air displaced

heavi·ly \ˈhe-və-lē\ *adverb* (before 12th century)
1 : to a great degree **:** SEVERELY
2 : slowly and laboriously **:** DULLY
3 *archaic* **:** with sorrow **:** GRIEVOUSLY
4 : in a heavy manner

Heav·i·side layer \ˈhe-vē-ˌsīd-\ *noun* [Oliver *Heaviside*] (1912)
: IONOSPHERE

¹heavy \ˈhe-vē\ *adjective* **heavi·er; -est** [Middle English *hevy,* from Old English *hefig;* akin to Old High German *hebīc* heavy, Old English *hebban* to lift — more at HEAVE] (before 12th century)
1 a : having great weight; *also* **:** characterized by mass or weight ⟨how *heavy* is it?⟩ **b :** having a high specific gravity **:** having great weight in proportion to bulk **c** (1) *of an isotope* **:** having or being atoms of greater than normal mass for that element (2) *of a compound* **:** containing heavy isotopes
2 : hard to bear; *specifically* **:** GRIEVOUS, AFFLICTIVE ⟨a *heavy* sorrow⟩
3 : of weighty import **:** SERIOUS
4 : DEEP, PROFOUND
5 a : borne down by something oppressive **:** BURDENED **b :** PREGNANT; *especially* **:** approaching parturition
6 a : slow or dull from loss of vitality or resiliency **:** SLUGGISH **b :** lacking sparkle or vivacity **:** DRAB **c :** lacking mirth or gaiety **:** DOLEFUL **d :** characterized by declining prices
7 : dulled with weariness **:** DROWSY
8 : greater in quantity or quality than the average of its kind or class: as **a :** of unusually large size or amount ⟨a *heavy* turnout⟩ ⟨*heavy* traffic⟩ **b :** of great force ⟨*heavy* seas⟩ **c :** threatening to rain or snow **d** (1) **:** impeding motion (2) **:** full of clay and inclined to hold water **e :** coming as if from a depth **:** LOUD ⟨*heavy* breathing⟩ **f :** THICK, COARSE **g :** OPPRESSIVE ⟨*heavy* odor⟩ **h :** STEEP, ACUTE **i :** LABORIOUS, DIFFICULT ⟨*heavy* going⟩ **j :** IMMODERATE ⟨a *heavy* smoker⟩ **k :** more powerful than usual for its kind ⟨*heavy* cavalry⟩ ⟨*heavy* cruiser⟩ **l :** of large capacity or output
9 a : very rich and hard to digest ⟨*heavy* desserts⟩ **b :** not properly raised or leavened ⟨*heavy* bread⟩
10 : producing goods (as coal, steel, or chemicals) used in the production of other goods ⟨*heavy* industry⟩
11 a : having stress ⟨*heavy* rhythm⟩ — used especially of syllables in accentual verse **b :** being the strongest degree of stress in speech
12 : relating to theatrical parts of a grave or somber nature
13 : LONG 9 ⟨*heavy* on ideas⟩
14 : IMPORTANT, PROMINENT ⟨a *heavy* politician⟩ ☆
— heavi·ness *noun*

²heavy *adverb* (before 12th century)
: in a heavy manner **:** HEAVILY

☆ SYNONYMS
Heavy, weighty, ponderous, cumbrous, cumbersome mean having great weight. HEAVY implies that something has greater density or thickness than the average of its kind or class ⟨a *heavy* child for his age⟩. WEIGHTY suggests having actual and not just relative weight ⟨a load of *weighty* boxes⟩. PONDEROUS implies having great weight because of size and massiveness with resulting great inertia ⟨*ponderous* elephants in a circus parade⟩. CUMBROUS and CUMBERSOME imply heaviness and bulkiness that make for difficulty in grasping, moving, carrying, or manipulating ⟨wrestled with the *cumbrous* furniture⟩ ⟨early cameras were *cumbersome* and inconvenient⟩.

³heavy *noun, plural* **heav·ies** (1897)
1 : HEAVYWEIGHT 2
2 a : a theatrical role of a dignified or somber character; *also* **:** an actor playing such a role **b :** VILLAIN 4 **c :** someone or something influential, serious, or important

heavy chain *noun* (1964)
: either of the two larger of the four polypeptide chains comprising antibodies — compare LIGHT CHAIN

heavy cream *noun* (1930)
: a cream that is markedly thick; *especially* **:** cream that by law contains not less than 36 percent butterfat

heavy–du·ty \'he-vē-'dü-tē, -'dyü-\ *adjective* (1914)
1 : able or designed to withstand unusual strain
2 : INTENSIVE ⟨*heavy-duty* bargaining⟩

heavy–foot·ed \-'fu̇-təd\ *adjective* (1625)
: heavy and slow in movement

heavy–hand·ed \-'han-dəd\ *adjective* (1647)
1 : CLUMSY
2 : OPPRESSIVE, HARSH
— **heavy–hand·ed·ly** *adverb*
— **heavy–hand·ed·ness** *noun*

heavy·heart·ed \-'här-təd\ *adjective* (14th century)
: DESPONDENT, SADDENED
— **heavy·heart·ed·ly** *adverb*
— **heavy·heart·ed·ness** *noun*

heavy hitter *noun* (1976)
: BIG SHOT, HEAVY

heavy hydrogen *noun* (1933)
: DEUTERIUM

heavy metal *noun* (1974)
: energetic and highly amplified electronic rock music having a hard beat

heavy·set \ˌhe-vē-'set\ *adjective* (1922)
: stocky and compact and sometimes tending to stoutness in build

heavy water *noun* (1933)
1 : the compound D_2O composed of deuterium and oxygen — called also *deuterium oxide*
2 : water enriched in deuterium

heavy·weight \'he-vē-ˌwāt\ *noun* (1857)
1 : one that is above average in weight
2 : one in the usually heaviest class of contestants: as **a :** a boxer in an unlimited weight division — compare LIGHT HEAVYWEIGHT **b :** a weight lifter weighing more than 198 pounds
3 : BIG SHOT, HEAVY

heb·do·mad \'heb-də-ˌmad\ *noun* [Latin *hebdomad-, hebdomas,* from Greek, from *hebdomos* seventh, from *hepta* seven — more at SEVEN] (1545)
1 : a group of seven
2 : a period of seven days **:** WEEK

heb·dom·a·dal \heb-'dä-mə-dᵊl\ *adjective* (1711)
: WEEKLY
— **heb·dom·a·dal·ly** \-dᵊl-ē\ *adverb*

hebe \'hēb\ *noun, often capitalized* [short for *Hebrew*] (1932)
: JEW — usually taken to be offensive

He·be \'hē-bē\ *noun* [Latin, from Greek *Hēbē*]
: the Greek goddess of youth and a cupbearer to the gods

he·be·phre·nia \ˌhē-bə-'frē-nē-ə, -'fre-nē-\ *noun* [New Latin, irregular from Greek *hēbētēs* young adult (from *hēbē* youth) + English *-phrenia;* from the childish behavior which is often found with it] (1883)
: a form of schizophrenia characterized especially by incoherence, delusions lacking an underlying theme, and affect that is flat, inappropriate, or silly
— **he·be·phre·nic** \-'fre-nik, -'frē-nik\ *adjective or noun*

heb·e·tate \'he-bə-ˌtāt\ *transitive verb* **-tat·ed; -tat·ing** [Latin *hebetatus,* past participle of *hebetare,* from *hebet-, hebes* dull] (1574)
: to make dull or obtuse
— **heb·e·ta·tion** \ˌhe-bə-'tā-shən\ *noun*

heb·e·tude \'he-bə-ˌtüd, -ˌtyüd\ *noun* [Late Latin *hebetudo,* from *hebēre* to be dull; akin to Latin *hebes* dull] (circa 1621)
: LETHARGY, DULLNESS
— **heb·e·tu·di·nous** \ˌhe-bə-'tü-dᵊn-əs, -'tyü-\ *adjective*

He·bra·ic \hi-'brā-ik\ *adjective* [Middle English *Ebrayke,* from Late Latin *Hebraicus,* from Greek *Hebraikos,* from *Hebraios*] (14th century)
: of, relating to, or characteristic of the Hebrews or their language or culture
— **He·bra·i·cal·ly** \-'brā-ə-k(ə-)lē\ *adverb*

He·bra·ism \'hē-(ˌ)brā-ˌi-zəm\ *noun* (1570)
1 : a characteristic feature of Hebrew occurring in another language
2 : the thought, spirit, or practice characteristic of the Hebrews
3 : a moral theory or emphasis attributed to the Hebrews

He·bra·ist \-ˌbrā-ist\ *noun* (circa 1755)
: a specialist in Hebrew and Hebraic studies

He·bra·is·tic \ˌhē-brā-'is-tik\ *adjective* (1690)
1 : marked by Hebraisms
2 : HEBRAIC

he·bra·ize \'hē-brā-ˌīz\ *verb* **-ized; -iz·ing** *often capitalized* (1645)
intransitive verb
: to use Hebraisms
transitive verb
: to make Hebraic in character or form
— **he·bra·iza·tion** \ˌhē-ˌbrā-ə-'zā-shən\ *noun, often capitalized*

He·brew \'hē-(ˌ)brü\ *noun* [Middle English *Ebreu,* from Old French, from Late Latin *Hebraeus,* from Latin, adjective, from Greek *Hebraios,* from Aramaic *'Ebrai*] (13th century)
1 a : the Semitic language of the ancient Hebrews **b :** any of various later forms of this language
2 : a member of or descendant from one of a group of northern Semitic peoples including the Israelites; *especially* **:** ISRAELITE
— **Hebrew** *adjective*

He·brews \'hē-(ˌ)brüz\ *noun plural but singular in construction*
: a theological treatise addressed to early Christians and included as a book in the New Testament — see BIBLE table

Hec·ate \'he-kə-tē, 'he-kət\ *noun* [Latin, from Greek *Hekatē*]
: a Greek goddess associated especially with the underworld, night, and witchcraft

hec·a·tomb \'he-kə-ˌtōm\ *noun* [Latin *hecatombe,* from Greek *hekatombē,* from *hekaton* hundred + *-bē;* akin to Greek *bous* cow — more at HUNDRED, COW] (circa 1592)
1 : an ancient Greek and Roman sacrifice of 100 oxen or cattle
2 : the sacrifice or slaughter of many victims

heck \'hek\ *noun* [euphemism] (1887)
: HELL ⟨a *heck* of a lot of money⟩

heck·le \'he-kəl\ *transitive verb* **heck·led; heck·ling** \-k(ə-)liŋ\ [Middle English *hekelen* to dress flax, scratch, from *heckele* hackle; akin to Old High German *hāko* hook — more at HOOK] (circa 1825)
: to harass and try to disconcert with questions, challenges, or gibes **:** BADGER
synonym see BAIT
— **heck·ler** \-k(ə-)lər\ *noun*

hect- *or* **hecto-** *combining form* [French, irregular from Greek *hekaton*]
: hundred ⟨*hect*are⟩

hect·are \'hek-ˌtar, -ˌter, -ˌtär\ *noun* [French, from *hect-* + *are* ²are] (1810)
— see METRIC SYSTEM table

hec·tic \'hek-tik\ *adjective* [Middle English *etyk,* from Middle French *etique,* from Late Latin *hecticus,* from Greek *hektikos* habitual, consumptive, from *echein* to have — more at SCHEME] (14th century)
1 : of, relating to, or being a fluctuating but persistent fever (as in tuberculosis)
2 : having a hectic fever
3 : RED, FLUSHED
4 : characterized by activity, excitement, or confusion ⟨the *hectic* days before Christmas⟩
— **hec·ti·cal·ly** \-ti-k(ə-)lē\ *adverb*

hec·to·gram \'hek-tə-ˌgram\ *noun* [French *hectogramme,* from *hect-* + *gramme* gram] (1810)
— see METRIC SYSTEM table

hec·to·graph \-ˌgraf\ *noun* [German *Hektograph,* from *hekto-* hect- + *-graph* -graph] (1880)
: a machine for making copies of a writing or drawing produced on a gelatin surface
— **hectograph** *transitive verb*

hec·to·li·ter \'hek-tə-ˌlē-tər\ *noun* [French *hectolitre,* from *hect-* + *litre* liter] (1810)
— see METRIC SYSTEM table

hec·to·me·ter \'hek-tə-ˌmē-tər, hek-'tä-mə-tər\ *noun* [French *hectomètre,* from *hect-* + *mètre* meter] (1810)
— see METRIC SYSTEM table

¹hec·tor \'hek-tər\ *noun* [Latin, from Greek *Hektōr*]
1 *capitalized* **:** a son of Priam, husband of Andromache, and Trojan champion slain by Achilles
2 : BULLY, BRAGGART ◆

²hector *verb* **hec·tored; hec·tor·ing** \-t(ə-)riŋ\ (1660)
intransitive verb
: to play the bully **:** SWAGGER
transitive verb
: to intimidate or harass by bluster or personal pressure
synonym see BAIT
— **hec·tor·ing·ly** \-t(ə-)riŋ-lē\ *adverb*

Hec·u·ba \'he-kyə-bə\ *noun* [Latin, from Greek *Hekabē*]
: the wife of Priam in Homer's *Iliad*

he'd \'hēd, ēd\ (circa 1600)
: he had **:** he would

hed·dle \'he-dᵊl\ *noun* [probably alteration of Middle English *helde,* from Old English *hefeld;* akin to Old Norse *hafald* heddle, Old English *hebban* to lift — more at HEAVE] (1513)
: one of the sets of parallel cords or wires that with their mounting compose the harness used to guide warp threads in a loom

he·der \'k̲ā-dər, 'k̲e-\ *noun* [Yiddish *kheyder,* from Hebrew *ḥedher* room] (1882)
: an elementary Jewish school in which children are taught to read the Torah and other books in Hebrew

¹hedge \'hej\ *noun* [Middle English *hegge,* from Old English *hecg;* akin to Old English *haga* hedge, hawthorn] (before 12th century)
1 a : a fence or boundary formed by a dense row of shrubs or low trees **b :** BARRIER, LIMIT
2 : a means of protection or defense (as against financial loss)
3 : a calculatedly noncommittal or evasive statement

²hedge *verb* **hedged; hedg·ing** (14th century)

◇ WORD HISTORY
hector In Homer's account of the Trojan War, Hector, the great Trojan hero, is portrayed as the ideal warrior: fearless combatant, loving son and husband, steadfast friend. After his death at the hands of Achilles, he is buried with the highest honors. From the 14th century the name *Hector* has been used allusively in English literature for a valiant warrior. The word apparently acquired its pejorative sense in the latter half of the 17th century. At that time there appeared on the streets of London rowdy toughs who were known as *Hectors.* While they may have perceived themselves as gallant young blades, to the general populace they were blustering, swaggering bullies who brandished their swords, intimidated passersby, and engaged in vandalism. Thus *Hector* deteriorated into a synonym for a braggart and a bully. The conversion of *hector* into a verb soon followed.

transitive verb
1 : to enclose or protect with or as if with a hedge **:** ENCIRCLE
2 : to hem in or obstruct with or as if with a barrier **:** HINDER ⟨*hedged* about by special regulations and statutes —Sandi Rosenbloom⟩
3 : to protect oneself from losing or failing by a counterbalancing action ⟨*hedge* a bet⟩
intransitive verb
1 : to plant, form, or trim a hedge
2 : to evade the risk of commitment especially by leaving open a way of retreat **:** TRIM
3 : to protect oneself financially: as **a :** to buy or sell commodity futures as a protection against loss due to price fluctuation **b :** to minimize the risk of a bet
— hedg·er *noun*
— hedg·ing·ly \ˈhe-jiŋ-lē\ *adverb*

[3]hedge *adjective* (14th century)
1 : of, relating to, or designed for a hedge
2 : born, living, or made near or as if near hedges **:** ROADSIDE
3 : INFERIOR 3

hedge fund *noun* (1967)
: an investing group usually in the form of a limited partnership that employs speculative techniques in the hope of obtaining large capital gains

hedge·hog \ˈhej-ˌhȯg, -ˌhäg\ *noun* (15th century)
1 a : any of a subfamily (Erinaceinae) of Old World nocturnal insectivores that have both hair and spines which they present outwardly by rolling themselves up when threatened **b :** any of several spiny mammals (as a porcupine)
2 a : a military defensive obstacle (as of barbed wire) **b :** a well-fortified military stronghold

hedgehog 1a

hedge·hop \-ˌhäp\ *intransitive verb* [back-formation from *hedgehopper*] (1926)
: to fly an airplane close to the ground and rise over obstacles as they appear
— hedge·hop·per *noun*

hedge·pig \-ˌpig\ *noun* (1605)
: HEDGEHOG

hedge·row \-ˌrō\ *noun* (before 12th century)
: a row of shrubs or trees enclosing or separating fields

he·don·ic \hi-ˈdä-nik\ *adjective* (1656)
1 : of, relating to, or characterized by pleasure
2 : of, relating to, or characterized by hedonism
— he·don·i·cal·ly \-ni-k(ə-)lē\ *adverb*

he·do·nism \ˈhē-dᵊn-ˌi-zəm\ *noun* [Greek *hēdonē* pleasure; akin to Greek *hēdys* sweet — more at SWEET] (1856)
1 : the doctrine that pleasure or happiness is the sole or chief good in life
2 : a way of life based on or suggesting the principles of hedonism
— he·do·nist \-dᵊn-ist\ *noun*
— he·do·nis·tic \ˌhē-dᵊn-ˈis-tik\ *adjective*
— he·do·nis·ti·cal·ly \-ti-k(ə-)lē\ *adverb*

-he·dral \ˈhē-drəl\ *adjective combining form* [New Latin *-hedron*]
: having (such) a surface or (such or so many) surfaces ⟨di*hedral*⟩

-he·dron \ˈhē-drən\ *noun combining form, plural* **-hedrons** *or* **-he·dra** \-drə\ [New Latin, from Greek *-edron,* from *hedra* seat — more at SIT]
: crystal or geometrical figure having a (specified) form or number of surfaces ⟨penta*hedron*⟩ ⟨trapezo*hedron*⟩

hee·bie–jee·bies \ˌhē-bē-ˈjē-bēz\ *noun plural* [coined by Billy DeBeck (died 1942) American cartoonist] (1923)
: JITTERS, CREEPS

[1]heed \ˈhēd\ *verb* [Middle English, from Old English *hēdan;* akin to Old High German *huota* guard, Old English *hōd* hood] (before 12th century)
intransitive verb
: to pay attention
transitive verb
: to give consideration or attention to **:** MIND ⟨*heed* what he says⟩ ⟨*heed* the call⟩

[2]heed *noun* (14th century)
: ATTENTION, NOTICE

heed·ful \ˈhēd-fəl\ *adjective* (1548)
: taking heed **:** ATTENTIVE ⟨*heedful* of what they were doing⟩
— heed·ful·ly \-fə-lē\ *adverb*
— heed·ful·ness *noun*

heed·less \-ləs\ *adjective* (1579)
: not taking heed **:** INCONSIDERATE, THOUGHTLESS ⟨*heedless* follies of unbridled youth —John DeBruyn⟩
— heed·less·ly *adverb*
— heed·less·ness *noun*

hee–haw \ˈhē-ˌhȯ, -ˈhȯ\ *noun* [imitative] (1815)
1 : the bray of a donkey
2 : a loud rude laugh **:** GUFFAW
— hee–haw *intransitive verb*

[1]heel \ˈhē(ə)l\ *noun* [Middle English, from Old English *hēla;* akin to Old Norse *hæll* heel, Old English *hōh* — more at HOCK] (before 12th century)
1 a : the back of the human foot below the ankle and behind the arch **b :** the part of the hind limb of other vertebrates that is homologous with the human heel
2 : an anatomical structure suggestive of the human heel; *especially* **:** the part of the palm of the hand nearest the wrist
3 : one of the crusty ends of a loaf of bread
4 a : the part (as of a shoe) that covers the human heel **b :** a solid attachment of a shoe or boot forming the back of the sole under the heel of the foot
5 : a rear, low, or bottom part: as **a :** the after end of a ship's keel or the lower end of a mast **b :** the base of a tuber or cutting of a plant used for propagation **c :** the base of a ladder
6 : a contemptible person
— heel·less \ˈhē(ə)l-ləs\ *adjective*
— by the heels : in a tight grip
— down at heel *or* **down at the heel :** in or into a run-down or shabby condition
— on the heels of : immediately following
— to heel 1 : close behind **2 :** into agreement or line
— under heel : under control or subjection

[2]heel (1605)
transitive verb
1 a : to furnish with a heel **b :** to supply especially with money
2 a : to exert pressure on, propel, or strike with the heel ⟨*heeled* her horse⟩ **b :** to urge (as a lagging animal) by following closely or by nipping at the heels ⟨dogs *heeling* cattle⟩
intransitive verb
: to move along at someone's heels

[3]heel *verb* [alteration of Middle English *heelden,* from Old English *hieldan;* akin to Old High German *hald* inclined, Lithuanian *šalis* side, region] (1575)
intransitive verb
: to lean to one side **:** TIP; *especially, of a boat or ship* **:** to lean temporarily (as from the action of wind or waves) — compare LIST
transitive verb
: to cause (a boat) to heel

[4]heel *noun* (1760)
: a tilt (as of a boat) to one side; *also* **:** the extent of such a tilt

heel–and–toe \ˌhē-lən-ˈtō\ *adjective* (1827)
: marked by a stride in which the heel of one foot touches the ground before the toe of the other foot leaves it ⟨*heel-and-toe* walking⟩

heel·ball \ˈhē(ə)l-ˌbȯl\ *noun* (1822)
: a composition of wax and lampblack used by shoemakers for polishing and by antiquarians for making rubbings of inscriptions

heel·er \ˈhē-lər\ *noun* (1665)
1 a : one that heels **b :** AUSTRALIAN CATTLE DOG
2 a : a henchman of a local political boss **b :** a worker for a local party organization; *especially* **:** WARD HEELER

heel fly *noun* (1878)
: CATTLE GRUB

heel·piece \ˈhē(ə)l-ˌpēs\ *noun* (1709)
: a piece designed for or forming the heel (as of a shoe)

heel·tap \-ˌtap\ *noun* (1780)
: a small quantity of alcoholic beverage remaining (as in a glass after drinking)

[1]heft \ˈheft\ *noun* [from *heave,* after such pairs as *weave : weft*] (15th century)
1 a : WEIGHT, HEAVINESS **b :** IMPORTANCE, INFLUENCE
2 *archaic* **:** the greater part of something **:** BULK

[2]heft *transitive verb* (1661)
1 : to heave up **:** HOIST
2 : to test the weight of by lifting ⟨*hefting* the rod . . . to get the feel of it —*Consumer Reports*⟩

hefty \ˈhef-tē\ *adjective* **heft·i·er; -est** (1867)
1 : quite heavy
2 a : marked by bigness, bulk, and usually strength ⟨a *hefty* football player⟩ **b :** POWERFUL, MIGHTY **c :** impressively large **:** SUBSTANTIAL ⟨*hefty* portions⟩
— heft·i·ly \-tə-lē\ *adverb*
— heft·i·ness \-tē-nəs\ *noun*

he·gari \hi-ˈgar-ē, -ˈgar-ə, -ˈger-; ˈhī-ˌgir\ *noun* [Arabic dialect (Sudan) *hegiri*] (1919)
: any of several Sudanese grain sorghums having chalky white seeds including one grown in the southwestern U.S.

[1]He·ge·li·an \hā-ˈgā-lē-ən, hi-\ *adjective* (1838)
: of, relating to, or characteristic of Hegel, his philosophy, or his dialectic method

[2]Hegelian *noun* (1843)
: a follower of Hegel **:** an adherent of Hegelianism

He·ge·li·an·ism \-lē-ə-ˌni-zəm\ *noun* (1846)
: the philosophy of Hegel that places ultimate reality in ideas rather than in things and that uses dialectic to comprehend an absolute idea behind phenomena

he·ge·mo·ny \hi-ˈje-mə-nē, -ˈge-; ˈhe-jə-ˌmō-nē\ *noun* [Greek *hēgemonia,* from *hēgemōn* leader, from *hēgeisthai* to lead — more at SEEK] (1567)
: preponderant influence or authority over others **:** DOMINATION
— heg·e·mon·ic \ˌhe-jə-ˈmä-nik, ˌhe-gə-\ *adjective*

he·gi·ra *also* **he·ji·ra** \hi-ˈjī-rə, ˈhe-jə-rə\ *noun* [the *Hegira,* flight of Muhammad from Mecca in A.D. 622, from Medieval Latin, from Arabic *hijrah,* literally, flight] (1753)
: a journey especially when undertaken to escape from a dangerous or undesirable situation **:** EXODUS

Hei·del·berg man \ˈhī-dᵊl-ˌbərg-, -ˌberg-\ *noun* [*Heidelberg,* Germany] (1920)
: an early Pleistocene man known from a massive fossilized jaw with distinctly human dentition and now classified with the pithecanthropines

heif·er \ˈhe-fər\ *noun* [Middle English *hayfare,* from Old English *hēahfore*] (before 12th century)
: a young cow; *especially* **:** one that has not had a calf

\ə\ **abut** \ᵊ\ **kitten** \ər\ **further** \a\ **ash** \ā\ **ace** \ä\ **mop, mar** \au̇\ **out** \ch\ **chin** \e\ **bet** \ē\ **easy** \g\ **go** \i\ **hit** \ī\ **ice** \j\ **job** \ŋ\ **sing** \ō\ **go** \ȯ\ **law** \ȯi\ **boy** \th\ **thin** \t͟h\ **the** \ü\ **loot** \u̇\ **foot** \y\ **yet** \zh\ **vision** *see also* Guide to Pronunciation

heigh–ho \ˈhī-ˈhō, ˈhā-\ *interjection* (circa 1553)
— used typically to express boredom, weariness, or sadness or sometimes as a cry of encouragement

height \ˈhīt, ÷ˈhītth\ *noun* [Middle English *heighthe,* from Old English *hīehthu;* akin to Old High German *hōhida* height, Old English *hēah* high] (before 12th century)
1 a : the highest part **:** SUMMIT **b :** highest or most advanced point **:** ZENITH ⟨at the *height* of his powers⟩
2 a : the distance from the bottom to the top of something standing upright **b :** the extent of elevation above a level
3 : the condition of being tall or high
4 a : an extent of land rising to a considerable degree above the surrounding country **b :** a high point or position
5 *obsolete* **:** an advanced social rank ☆ ◆

height·en \ˈhī-tᵊn\ *verb* **height·ened; height·en·ing** \ˈhīt-niŋ, ˈhī-tᵊn-iŋ\ (1523)
transitive verb
1 a : to increase the amount or degree of **:** AUGMENT **b :** to make brighter or more intense **:** DEEPEN **c :** to bring out more strongly **:** point up **d :** to make more acute **:** SHARPEN
2 a : to raise high or higher **:** ELEVATE **b :** to raise above the ordinary or trite
3 *obsolete* **:** ELATE
intransitive verb
1 *archaic* **:** GROW, RISE
2 a : to become great or greater in amount, degree, or extent **b :** to become brighter or more intense

height to paper (1771)
: the height of printing type standardized at 0.9186 inch (2.333 centimeters) in English-speaking countries

Heim·lich maneuver \ˈhīm-lik-\ *noun* [Henry J. *Heimlich* (born 1920) American surgeon] (1974)
: the manual application of sudden upward pressure on the upper abdomen of a choking victim to force a foreign object from the windpipe

hei·nie \ˈhī-nē\ *noun* [alteration of ^{2}*hinder*] (1940)
slang **:** BUTTOCKS

hei·nous \ˈhā-nəs\ *adjective* [Middle English, from Middle French *haineus,* from *haine* hate, from *hair* to hate, of Germanic origin; akin to Old High German *haz* hate — more at HATE] (14th century)
: hatefully or shockingly evil **:** ABOMINABLE
— **hei·nous·ly** *adverb*
— **hei·nous·ness** *noun*

^{1}**heir** \ˈar, ˈer\ *noun* [Middle English, from Old French, from Latin *hered-, heres;* akin to Greek *chēros* bereaved] (13th century)
1 : one who inherits or is entitled to inherit property
2 : one who inherits or is entitled to succeed to a hereditary rank, title, or office ⟨*heir* to the throne⟩
3 : one who receives or is entitled to receive some endowment or quality from a parent or predecessor
— **heir·less** \-ləs\ *adjective*
— **heir·ship** \-ˌship\ *noun*

^{2}**heir** *transitive verb* (14th century)
chiefly dialect **:** INHERIT

heir apparent *noun, plural* **heirs apparent** (14th century)
1 : an heir whose right to an inheritance is indefeasible in law if he survives the legal ancestor
2 : HEIR PRESUMPTIVE
3 : one whose succession especially to a position or role appears certain under existing circumstances

heir at law (1729)
: an heir in whom an intestate's real property is vested by operation of law

heir·ess \ˈar-əs, ˈer-\ *noun* (1607)
: a female heir; *especially* **:** a female heir to great wealth

heir·loom \ˈar-ˌlüm, ˈer-\ *noun* [Middle English *heirlome,* from *heir* + *lome* implement — more at LOOM] (15th century)
1 : a piece of property that descends to the heir as an inseparable part of an inheritance of real property
2 : something of special value handed on from one generation to another

heir presumptive *noun, plural* **heirs presumptive** (circa 1737)
: an heir whose legal right to an inheritance may be defeated (as by the birth of a nearer relative)

Hei·sen·berg uncertainty principle \ˈhī-zᵊn-ˌbərg-, -ˌbərk-\ *noun* [Werner *Heisenberg*] (1939)
: UNCERTAINTY PRINCIPLE — called also *Heisenberg's uncertainty principle*

^{1}**heist** \ˈhīst\ *transitive verb* [variant of ^{1}*hoist*] (1865)
1 *chiefly dialect* **:** HOIST
2 *slang* **a :** to commit armed robbery on **b :** STEAL 1a

^{2}**heist** *noun* (1930)
slang **:** armed robbery **:** HOLDUP; *also* **:** THEFT

Hel \ˈhel\ *noun* [Old Norse]
: the Norse goddess of the dead and queen of the underworld

hela cell \ˈhe-lə-\ *noun, often H&1stL capitalized* [Henrietta Lacks (died 1951) patient from whom the cells were taken] (1953)
: a cell of a continuously cultured strain isolated from a human uterine cervical carcinoma in 1951 and used in biomedical research especially to culture viruses

held *past and past participle of* HOLD

hel·den·te·nor \ˈhel-dən-ˌtā-ˌnȯr, -ˌnōr, -ˌte-nər\ *noun, often capitalized* [German, from *Held* hero + *Tenor* tenor] (circa 1903)
: a tenor with a powerful dramatic voice well suited to heroic (as Wagnerian) roles

Helen of Troy \ˌhe-lə-nəv-ˈtrȯi\
: the wife of Menelaus whose abduction by Paris brings about the Trojan War

^{1}**heli-** *or* **helio-** *combining form* [Latin, from Greek *hēli-, hēlio-,* from *hēlios* — more at SOLAR]
: sun ⟨*helio*centric⟩

^{2}**heli-** *combining form* [by shortening]
: helicopter ⟨*heli*port⟩

he·li·a·cal \hi-ˈlī-ə-kəl\ *adjective* [Late Latin *heliacus,* from Greek *hēliakos,* from *hēlios*] (1545)
: relating to or near the sun — used especially of the last setting of a star before and its first rising after invisibility due to conjunction with the sun
— **he·li·a·cal·ly** \-k(ə-)lē\ *adverb*

helic- *or* **helico-** *combining form* [Greek *helik-, heliko-,* from *helik-, helix* spiral — more at HELIX]
: helix **:** spiral ⟨*helic*al⟩

he·li·cal \ˈhe-li-kəl, ˈhē-\ *adjective* (1591)
: of, relating to, or having the form of a helix; *broadly* **:** SPIRAL 1a
— **he·li·cal·ly** \-k(ə-)lē\ *adverb*

he·li·coid \ˈhe-lə-ˌkȯid, ˈhē-\ *or* **he·li·coi·dal** \ˌhe-lə-ˈkȯi-dᵊl, ˌhē-\ *adjective* (1704)
1 : forming or arranged in a spiral
2 : having the form of a flat coil or flattened spiral ⟨*helicoid* snail shell⟩

hel·i·con \ˈhe-lə-ˌkän, -i-kən\ *noun* [probably from Greek *helik-, helix* + English *-on* (as in *bombardon*); from its tube's forming a spiral encircling the player's body] (1875)
: a large circular tuba similar to a sousaphone but lacking an adjustable bell

^{1}**he·li·cop·ter** \ˈhe-lə-ˌkäp-tər, ˈhē-\ *noun* [French *hélicoptère,* from Greek *heliko-* + *pteron* wing — more at FEATHER] (1887)
: an aircraft whose lift is derived from the aerodynamic forces acting on one or more powered rotors turning about substantially vertical axes

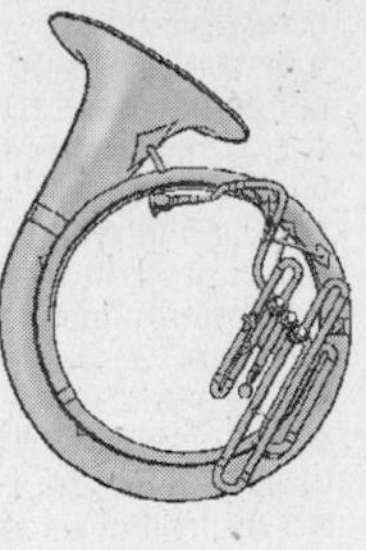
helicon

^{2}**helicopter** (1952)
intransitive verb
: to travel by helicopter
transitive verb
: to transport by helicopter

he·lio·cen·tric \ˌhē-lē-ō-ˈsen-trik\ *adjective* (1685)
1 : referred to or measured from the sun's center or appearing as if seen from it
2 : having or relating to the sun as center — compare GEOCENTRIC

he·lio·graph \-ˌgraf\ *noun* [International Scientific Vocabulary] (1877)
: an apparatus for telegraphing by means of the sun's rays flashed from a mirror
— **heliograph** *transitive verb*

he·lio·graph·ic \ˌhē-lē-ə-ˈgra-fik\ *adjective* (1706)
: measured on the sun's disk ⟨*heliographic* latitude⟩

he·li·ol·a·try \ˌhē-lē-ˈä-lə-trē\ *noun* (circa 1828)
: sun worship
— **he·li·ol·a·trous** \-trəs\ *adjective*

he·li·om·e·ter \ˌhē-lē-ˈä-mə-tər\ *noun* [French *héliomètre,* from *hélio-* ^{1}heli- + *-mètre* -meter] (1753)
: a visual telescope that has a divided objective designed for measuring the apparent diameter of the sun but also used for measuring angles between celestial bodies or between points on the moon
— **he·lio·met·ric** \ˌhē-lē-ō-ˈme-trik\ *adjective*
— **he·lio·met·ri·cal·ly** \-tri-k(ə-)lē\ *adverb*

He·li·os \ˈhē-lē-əs, -(ˌ)ōs\ *noun* [Greek *Hēlios*]

☆ SYNONYMS

Height, altitude, elevation mean vertical distance either between the top and bottom of something or between a base and something above it. HEIGHT refers to something measured vertically whether high or low ⟨a wall two meters in *height*⟩. ALTITUDE and ELEVATION apply to height as measured by angular measurement or atmospheric pressure; ALTITUDE is preferable when referring to vertical distance above the surface of the earth or above sea level; ELEVATION is used especially in reference to vertical height on land ⟨fly at an *altitude* of 10,000 meters⟩ ⟨Denver is a city with a high *elevation*⟩.

◇ WORD HISTORY

height The word *height* stands in peculiar contrast to *width, breadth,* and *length* in having a final \t\ sound rather than \th\. Only a couple of centuries ago, however, there was a *heighth* or *highth* that competed in popularity with *height* and was common enough in literary English to be used by Milton in *Paradise Lost.* The form with final *t* arose in northern dialects of Middle English while the *gh* sound in *heighth* was still pronounced, having approximately the sound of \k\ in the German name *Bach* or the Scots pronunciation of *loch.* But because few other English words ended in the consonant cluster \kth\, the second sound lost its fricative quality and became a simple \t\, remaining so even after the \k\ sound had been lost. In dialects of southern England the cluster was preserved until the \k\ sound was lost, but by the 19th century *heighth,* despite its correspondence to the other nouns of dimension, had been largely ousted by *height.*

: the god of the sun in Greek mythology — compare SOL

he·lio·stat \'hē-lē-ə-,stat\ *noun* [New Latin *heliostata,* from [1]*heli-* + Greek *-statēs* -stat] (1747)
: an instrument consisting of a mirror mounted on an axis moved by clockwork by which a sunbeam is steadily reflected in one direction

he·lio·trope \'hē-lē-ə-,trōp, 'hēl-yə-, *British also* 'hel-yə-\ *noun* [Latin *heliotropium,* from Greek *hēliotropion,* from *hēlio-* [1]heli- + *tropos* turn; from its flowers' turning toward the sun — more at TROPE] (circa 1626)
1 : any of a genus (*Heliotropium*) of herbs or shrubs of the borage family — compare GARDEN HELIOTROPE
2 : BLOODSTONE
3 : a variable color averaging a moderate to reddish purple

heliotrope 1

he·li·ot·ro·pism \,hē-lē-'ä-trə-,pi-zəm\ *noun* (circa 1854)
: phototropism in which sunlight is the orienting stimulus
— **he·lio·tro·pic** \-lē-ə-'trō-pik, -'trä-\ *adjective*

he·lio·zo·an \,hē-lē-ə-'zō-ən\ *noun* [New Latin *Heliozoa,* from [1]*heli-* + *-zoa*] (circa 1889)
: any of a class (Heliozoa) of free-living holozoic usually freshwater rhizopod protozoans that reproduce by binary fission or budding

he·li·pad \'he-lə-,pad, 'hē-\ *noun* (1960)
: HELIPORT

he·li·port \-,pōrt, -,pȯrt\ *noun* (1948)
: a landing and takeoff place for a helicopter

he·li–ski·ing \-,skē-iŋ\ *noun* (1976)
: downhill skiing on remote mountains reached by helicopter

he·li·um \'hē-lē-əm, 'hēl-yəm\ *noun* [New Latin, from Greek *hēlios*] (1872)
: a light colorless nonflammable gaseous element found especially in natural gases and used chiefly for inflating airships and balloons, for filling incandescent lamps, and for cryogenic research — see ELEMENT table

he·lix \'hē-liks\ *noun, plural* **he·li·ces** \'he-lə-,sēz, 'hē-\ *also* **he·lix·es** \'hē-lik-səz\ [Latin, from Greek; akin to Greek *eilyein* to roll, wrap — more at VOLUBLE] (1563)
1 : something spiral in form: as **a :** an ornamental volute **b :** a coil formed by winding wire around a uniform tube
2 : the incurved rim of the external ear
3 : a curve traced on a cylinder or cone by the rotation of a point crossing its right sections at a constant oblique angle; *broadly* **:** SPIRAL 1b

hell \'hel\ *noun* [Middle English, from Old English; akin to Old English *helan* to conceal, Old High German *helan,* Latin *celare,* Greek *kalyptein*] (before 12th century)
1 a (1) **:** a nether world in which the dead continue to exist **:** HADES (2) **:** the nether realm of the devil and the demons in which the damned suffer everlasting punishment — often used in curses ⟨go to *hell*⟩ or as a generalized term of abuse ⟨the *hell* with it⟩ **b** *Christian Science* **:** ERROR 2b, SIN
2 a : a place or state of misery, torment, or wickedness ⟨war is *hell* —W. T. Sherman⟩ **b :** a place or state of turmoil or destruction ⟨all *hell* broke loose⟩ **c :** a severe scolding ⟨got *hell* for coming in late⟩ **d :** unrestrained fun or sportiveness ⟨the kids were full of *hell*⟩ — often used in the phrase *for the hell of it* especially to suggest action on impulse or without a serious motive ⟨decided to go for the *hell* of it⟩
3 *archaic* **:** a tailor's receptacle
4 — used as an interjection ⟨*hell,* I don't know!⟩ or as an intensive ⟨hurts like *hell*⟩ ⟨funny as *hell*⟩; often used in the phrase *hell of a* ⟨it was one *hell* of a good fight⟩ or *hell out of* ⟨scared the *hell* out of him⟩ or with *the* or *in* ⟨moved way the *hell* up north⟩ ⟨what in *hell* is wrong, now?⟩
— **from hell :** being the worst or most dreadful of its kind
— **hell on :** very hard on or destructive to ⟨the constant traveling is *hell on* your digestive system⟩
— **hell or high water :** difficulties of whatever kind or size ⟨will stand by her convictions come *hell or high water*⟩
— **hell to pay :** dire consequences ⟨if he's late there'll be *hell to pay*⟩

he'll \'hē(ə)l, 'hil, ēl, il\ (1588)
: he will **:** he shall

hel·la·cious \,he-'lā-shəs\ *adjective* [*hell* + *-acious* (as in *audacious*)] (1943)
1 : exceptionally powerful or violent
2 : remarkably good
3 : extremely difficult
4 : extraordinarily large
— **hel·la·cious·ly** *adverb*

hell·ben·der \'hel-,ben-dər\ *noun* (1812)
: a large aquatic usually gray salamander (*Cryptobranchus alleganiensis*) of the Ohio valley region

hellbender

hell–bent \-,bent\ *adjective* (1835)
: stubbornly and often recklessly determined or intent ⟨*hell-bent* on winning⟩
— **hell–bent** *adverb*

hell·broth \-,brȯth\ *noun* (1605)
: a brew for working black magic

hell·cat \-,kat\ *noun* (circa 1605)
1 : WITCH 2
2 : a violently temperamental person; *especially* **:** an ill-tempered woman

hel·le·bore \'he-lə-,bōr, -,bȯr\ *noun* [Middle English *elebre,* from Latin *elleborus, helleborus,* from Greek *helleboros*] (15th century)
1 : any of a genus (*Helleborus*) of poisonous herbs of the buttercup family having showy flowers with petaloid sepals; *also* **:** the dried rhizome or an extract or powder of this formerly used in medicine
2 : a poisonous herb (genus *Veratrum*) of the lily family; *also* **:** the dried rhizome of a hellebore (*Veratrum album* or *V. viride*) or a powder or extract of this containing alkaloids used as a cardiac and respiratory depressant and as an insecticide

Hel·lene \'he-,lēn\ *noun* [Greek *Hellēn*] (1662)
: GREEK 1a

[1]**Hel·len·ic** \he-'le-nik, hə-\ *adjective* (1644)
: of or relating to Greece, its people, or its language; *specifically* **:** of or relating to ancient Greek history, culture, or art before the Hellenistic period

[2]**Hellenic** *noun* (1847)
: GREEK 2a

Hel·le·nism \'he-lə-,ni-zəm\ *noun* (1609)
1 : GRECISM 1
2 : devotion to or imitation of ancient Greek thought, customs, or styles
3 : Greek civilization especially as modified in the Hellenistic period by oriental influences
4 : a body of humanistic and classical ideals associated with ancient Greece and including reason, the pursuit of knowledge and the arts, moderation, civic responsibility, and bodily development

Hel·le·nist \-nist\ *noun* (1613)
1 : a person living in Hellenistic times who was Greek in language, outlook, and way of life but was not Greek in ancestry; *especially* **:** a hellenized Jew
2 : a specialist in the language or culture of ancient Greece

Hel·le·nis·tic \,he-lə-'nis-tik\ *adjective* (circa 1706)
1 : of or relating to Greek history, culture, or art after Alexander the Great
2 : of or relating to the Hellenists
— **Hel·le·nis·ti·cal·ly** \-ti-k(ə-)lē\ *adverb*

hel·le·nize \'he-lə-,nīz\ *verb* **-nized; -niz·ing** *often capitalized* (1613)
intransitive verb
: to become Greek or Hellenistic
transitive verb
: to make Greek or Hellenistic in form or culture
— **hel·le·ni·za·tion** \,he-lə-nə-'zā-shən\ *noun, often capitalized*

hell·er \'he-lər\ *noun* (circa 1895)
: HELLION

hel·leri \'he-lə-,rī, -(,)rē\ *noun, plural* **-ler·ies** [New Latin (specific epithet of *Xiphophorus helleri*), from C. *Heller,* 20th century tropical fish collector] (1931)
1 : SWORDTAIL
2 : any of various brightly colored aquarium fishes developed by hybridization of swordtails and platys

hell·fire \'hel-,fīr\ *noun* (before 12th century)
: the eternal fire of hell that tortures sinners
— **hellfire** *adjective*

[1]**hell–for–leather** *adverb* (1889)
: in a hell-for-leather manner **:** at full speed ⟨rode *hell-for-leather* down the trail⟩

[2]**hell–for–leather** *adjective* (1920)
: marked by determined recklessness, great speed, or lack of restraint ⟨a cocky, *hell-for-leather* fighting man —H. H. Martin⟩

hell·gram·mite \'hel-grə-,mīt\ *noun* [origin unknown] (1866)
: a carnivorous aquatic North American insect larva that is the young form of a dobsonfly (especially *Corydalis cornutus*) and is used for fish bait

hell·hole \'hel-,hōl\ *noun* (1866)
: a place of extreme misery or squalor

hell·hound \-,hau̇nd\ *noun* (before 12th century)
1 : a dog represented in mythology as a guardian of the underworld
2 : a fiendish person

hel·lion \'hel-yən\ *noun* [probably alteration (influenced by *hell*) of *hallion* (scamp)] (1787)
: a troublesome or mischievous person

hell·ish \'he-lish\ *adjective* (circa 1530)
: of, resembling, or befitting hell; *broadly* **:** TERRIBLE
— **hell·ish·ly** *adverb*
— **hell·ish·ness** *noun*

hel·lo \hə-'lō, he-\ *noun, plural* **hellos** [alteration of *hollo*] (1889)
: an expression or gesture of greeting — used interjectionally in greeting, in answering the telephone, or to express surprise

hell–rais·er \'hel-,rā-zər\ *noun* (1914)
: one given to wild, boisterous, or intemperate behavior
— **hell–rais·ing** \-,rā-ziŋ\ *noun or adjective*

[1]**helm** \'helm\ *noun* [Middle English, from Old English] (before 12th century)
: HELMET 1

[2]**helm** *transitive verb* (before 12th century)
: to cover or furnish with a helmet

[3]**helm** *noun* [Middle English *helme,* from Old English *helma;* akin to Old High German *helmo* tiller] (before 12th century)
1 a : a lever or wheel controlling the rudder of a ship for steering; *broadly* **:** the entire apparatus for steering a ship **b :** position of the helm with respect to the amidships position ⟨turn the *helm* hard alee⟩
2 : a position of control **:** HEAD ⟨a new dean is at the *helm* of the medical school⟩

\ə\ **abut** \ᵊ\ **kitten** \ər\ **further** \a\ **ash** \ā\ **ace** \ä\ **mop, mar** \au̇\ **out** \ch\ **chin** \e\ **bet** \ē\ **easy** \g\ **go** \i\ **hit** \ī\ **ice** \j\ **job** \ŋ\ **sing** \ō\ **go** \ȯ\ **law** \ȯi\ **boy** \th\ **thin** \t͟h\ **the** \ü\ **loot** \u̇\ **foot** \y\ **yet** \zh\ **vision** *see also* Guide to Pronunciation

⁴helm *transitive verb* (1603)
: to direct with or as if with a helm **:** STEER

hel·met \'hel-mət\ *noun* [Middle French, diminutive of *helme* helmet, of Germanic origin; akin to Old English *helm* helmet, Old High German *helan* to conceal — more at HELL] (15th century)
1 : a covering or enclosing headpiece of ancient or medieval armor — see ARMOR illustration
2 : any of various protective head coverings usually made of a hard material to resist impact
3 : something resembling a helmet
— hel·met·ed \-mə-təd\ *adjective*
— hel·met·like \-mət-,līk\ *adjective*

hel·minth \'hel-,min(t)th\ *noun* [Greek *helminth-, helmis*] (1852)
: a parasitic worm (as a tapeworm, liver fluke, ascarid, or leech); *especially* **:** an intestinal worm
— hel·min·thic \hel-'min(t)-thik\ *adjective*

hel·min·thi·a·sis \,hel-mən-'thī-ə-səs\ *noun, plural* **-a·ses** \-ə-,sēz\ [New Latin] (circa 1811)
: infestation with or disease caused by parasitic worms

hel·min·thol·o·gy \-'thä-lə-jē\ *noun* (1819)
: a branch of zoology concerned with helminths; *especially* **:** the study of parasitic worms

helms·man \'helmz-mən\ *noun* (1627)
: the person at the helm **:** STEERSMAN
— helms·man·ship \-,ship\ *noun*

he·lo \'hē-(,)lō\ *noun, plural* **helos** [by shortening & alteration] (1968)
: HELICOPTER

hel·ot \'he-lət\ *noun* [Latin *Helotes,* plural, from Greek *Heilōtes*] (1579)
1 *capitalized* **:** a member of a class of serfs in ancient Sparta
2 : SERF, SLAVE
— hel·ot·ry \'he-lə-trē\ *noun*

hel·ot·ism \'he-lə-,ti-zəm\ *noun* (circa 1900)
: the physiological relation existing in a lichen where a fungus appears to control an alga

¹help \'help; *Southern often* 'hep *also* 'heəp\ *verb* [Middle English, from Old English *helpan;* akin to Old High German *helfan* to help, and perhaps to Lithuanian *šelpti*] (before 12th century)
transitive verb
1 : to give assistance or support to ⟨*help* a child with homework⟩
2 a : to make more pleasant or bearable **:** IMPROVE, RELIEVE ⟨bright curtains will *help* the room⟩ ⟨took an aspirin to *help* her headache⟩ **b** *archaic* **:** RESCUE, SAVE
3 a : to be of use to **:** BENEFIT **b :** to further the advancement of **:** PROMOTE
4 a : to change for the better **b :** to refrain from **:** AVOID **c :** to keep from occurring **:** PREVENT ⟨they couldn't *help* the accident⟩ **d :** to restrain (oneself) from doing something ⟨knew they shouldn't go but couldn't *help* themselves⟩
5 : to serve with food or drink especially at a meal ⟨told the guests to *help* themselves⟩
6 : to appropriate something for (oneself) ⟨*helped* himself to the car keys⟩
intransitive verb
1 : give assistance or support — often used with *out*
2 : to be of use or benefit
synonym see IMPROVE
— so help me : upon my word **:** believe it or not

²help *noun* (before 12th century)
1 : AID, ASSISTANCE
2 : a source of aid ⟨printed *helps* to the memory —C. S. Braden⟩
3 : REMEDY, RELIEF ⟨there was no *help* for it⟩
4 a : one who serves or assists another (as in housework) **:** HELPER **b :** EMPLOYEE ⟨*help* wanted⟩ — often used collectively ⟨the hired *help*⟩

help·er *noun* (13th century)
: one that helps; *especially* **:** a relatively unskilled worker who assists a skilled worker usually by manual labor

helper T cell *noun* (1976)
: a T cell that participates in an immune response by recognizing a foreign antigen and secreting lymphokines to activate T cell and B cell proliferation, that usually carries CD4 molecular markers on its cell surface, and that is reduced to 20 percent or less of normal numbers in AIDS — called also *helper cell, T-helper cell*

help·ful \'help-fəl; *Southern often* 'hep- *also* 'heəp-\ *adjective* (14th century)
: of service or assistance **:** USEFUL
— help·ful·ly \-fə-lē\ *adverb*
— help·ful·ness *noun*

help·ing *noun* (1883)
: a portion of food **:** SERVING

helping hand *noun* (15th century)
: HAND 8a

help·ing verb *noun* (1711)
: an auxiliary verb

help·less \'help-ləs; *Southern often* 'hep-ləs *also* 'heəp-\ *adjective* (before 12th century)
1 : lacking protection or support **:** DEFENSELESS
2 a : marked by an inability to act or react ⟨the crowd looked on in *helpless* horror —*Current Biography*⟩ **b :** not able to be controlled or restrained ⟨*helpless* laughter⟩
— help·less·ly *adverb*
— help·less·ness *noun*

help·mate \'help-,māt; *Southern often* 'hep- *also* 'heəp-\ *noun* [by folk etymology from *helpmeet*] (circa 1714)
: one who is a companion and helper; *especially* **:** WIFE

help·meet \-,mēt\ *noun* [²*help* + *meet,* adjective] (1673)
: HELPMATE

¹hel·ter–skel·ter \,hel-tər-'skel-tər\ *adverb* [perhaps from Middle English *skelten* to come, go] (1593)
1 : in undue haste, confusion, or disorder ⟨ran *helter-skelter*, getting in each other's way —F. V. W. Mason⟩
2 : in a haphazard manner

²helter–skelter *noun* (1713)
1 : a disorderly confusion **:** TURMOIL
2 *British* **:** a spiral slide around a tower at an amusement park

³helter–skelter *adjective* (1785)
1 : confusedly hurried **:** PRECIPITATE
2 : marked by a lack of order or plan **:** HAPHAZARD ⟨the *helter-skelter* arrangement of the papers, all mussed and frayed —Jean Stafford⟩

helve \'helv\ *noun* [Middle English, from Old English *hielfe;* probably akin to Old English *helma* helm] (before 12th century)
: a handle of a tool or weapon **:** HAFT

Hel·ve·tii \hel-'vē-shē-,ī\ *noun plural* [Latin] (1889)
: an early Celtic people in the area of western Switzerland at the time of Julius Caesar

¹hem \'hem\ *noun* [Middle English, from Old English; akin to Middle High German *hemmen* to hem in, Armenian *kamel* to press] (before 12th century)
1 : a border of a cloth article doubled back and stitched down
2 : RIM, MARGIN ⟨bright green *hem* of reeds about the ponds —R. M. Lockley⟩

²hem *verb* **hemmed; hem·ming** (14th century)
transitive verb
1 a : to finish with a hem **b :** BORDER, EDGE
2 : to surround in a restrictive manner **:** CONFINE — usually used with *in* ⟨*hemmed* in by enemy troops⟩
intransitive verb
: to make a hem in sewing
— hem·mer *noun*

³hem \'hem\ *intransitive verb* **hemmed; hem·ming** (15th century)
1 : to utter the sound represented by *hem* ⟨*hemmed* and hawed before answering⟩
2 : EQUIVOCATE ⟨the administration *hemmed* and hawed over the students' demands⟩

⁴hem *usually read as* 'hem\ *interjection* [imitative]
— often used to indicate a vocalized pause in speaking

hem- *or* **hemo-** *combining form* [Middle French *hemo-,* from Latin *haem-, haemo-,* from Greek *haim-, haimo-,* from *haima*]
: blood ⟨*hem*agglutination⟩ ⟨*hemo*flagellate⟩

hema- *combining form* [New Latin, from Greek *haima*]
: HEM- ⟨*hema*cytometer⟩

he·ma·cy·tom·e·ter \,hē-mə-sī-'tä-mə-tər\ *noun* (1877)
: an instrument for counting blood cells

hem·ag·glu·ti·na·tion \,hē-mə-,glü-tᵊn-'ā-shən\ *noun* (1907)
: agglutination of red blood cells
— hem·ag·glu·ti·nate \-'glü-tᵊn-,āt\ *transitive verb*

hem·ag·glu·ti·nin \-'glü-tᵊn-ən\ *noun* [International Scientific Vocabulary] (circa 1903)
: an agglutinin (as an antibody or viral capsid protein) that causes hemagglutination

he–man \'hē-,man\ *noun* (1832)
: a strong virile man

hem·an·gi·o·ma \,hē-,man-jē-'ō-mə\ *noun* [New Latin, from *hem-* + *angioma*] (circa 1890)
: a usually benign tumor made up of blood vessels that typically occurs as a purplish or reddish slightly elevated area of skin

hemat- *or* **hemato-** *combining form* [Latin *haemat-, haemato-,* from Greek *haimat-, haimato-,* from *haimat-, haima*]
: HEM- ⟨*hemato*genous⟩

he·ma·tin \'hē-mə-tən\ *noun* (1845)
: a brownish black or bluish black derivative $C_{34}H_{33}N_4O_5Fe$ of oxidized heme; *also* **:** any of several similar compounds

he·ma·tin·ic \,hē-mə-'ti-nik\ *noun* (1855)
: an agent that tends to stimulate blood cell formation or to increase the hemoglobin in the blood
— hematinic *adjective*

he·ma·tite \'hē-mə-,tīt\ *noun* (1540)
: a mineral constituting an important iron ore and occurring in crystals or in a red earthy form
— he·ma·ti·tic \,hē-mə-'ti-tik\ *adjective*

he·mat·o·crit \hi-'ma-tə-krət, -,krit\ *noun* [International Scientific Vocabulary *hemat-* + Greek *kritēs* judge, from *krinein* to judge — more at CERTAIN] (circa 1903)
1 : an instrument for determining usually by centrifugation the relative amounts of plasma and corpuscles in blood
2 : the ratio of the volume of packed red blood cells to the volume of whole blood as determined by a hematocrit

he·ma·tog·e·nous \,hē-mə-'tä-jə-nəs\ *adjective* (1886)
1 : producing blood
2 : involving, spread by, or arising in the blood ⟨*hematogenous* spread of infection⟩

he·ma·to·log·ic \,hē-mə-tᵊl-'ä-jik\ *also* **he·ma·to·log·i·cal** \-ji-kəl\ *adjective* (1854)
: of or relating to blood or to hematology

he·ma·tol·o·gy \,hē-mə-'tä-lə-jē\ *noun* (circa 1811)
: a medical science that deals with the blood and blood-forming organs
— he·ma·tol·o·gist \-jist\ *noun*

he·ma·to·ma \-'tō-mə\ *noun, plural* **-mas** *or* **-ma·ta** \-mə-tə\ [New Latin] (circa 1849)
: a mass of usually clotted blood that forms in a tissue, organ, or body space as a result of a broken blood vessel

he·ma·toph·a·gous \-'tä-fə-gəs\ *adjective* [International Scientific Vocabulary] (circa 1854)
: feeding on blood

he·ma·to·poi·e·sis \hi-,ma-tə-pȯi-'ē-səs, ,hē-mə-tō-\ *noun* [New Latin] (circa 1854)

: the formation of blood or of blood cells in the living body
— **he·ma·to·poi·et·ic** \-'e-tik\ *adjective*

he·ma·to·por·phy·rin \-'pȯr-fə-rən\ *noun* [International Scientific Vocabulary] (1885)
: any of several isomeric porphyrins $C_{34}H_{38}O_6N_4$ that are hydrated derivatives of protoporphyrins; *especially* : the deep red crystalline pigment obtained by treating hematin or heme with acid

he·ma·tox·y·lin \ˌhē-mə-'täk-sə-lən\ *noun* [International Scientific Vocabulary, from New Latin *Haematoxylon,* genus of plants] (circa 1847)
: a crystalline phenolic compound $C_{16}H_{14}O_6$ found in logwood and used chiefly as a biological stain

he·ma·tu·ria \-'tu̇r-ē-ə, -'tyu̇r-\ *noun* [New Latin] (circa 1811)
: the presence of blood or blood cells in the urine

heme \'hēm\ *noun* [International Scientific Vocabulary, from *hematin*] (1925)
: the deep red iron-containing prosthetic group $C_{34}H_{32}N_4O_4Fe$ of hemoglobin and myoglobin

hem·el·y·tron \he-'me-lə-ˌträn\ *noun, plural* **-tra** \-trə\ [New Latin, from *hemi-* + *elytron*] (circa 1889)
: one of the basally thickened anterior wings of various insects (as true bugs)

hem·ero·cal·lis \ˌhe-mə-rō-'ka-ləs\ *noun* [New Latin, from Greek *hēmerokalles,* from *hēmera* day + *kallos* beauty] (1625)
: DAYLILY

hem·er·y·thrin \hē-'mer-ə-thrən\ *noun* [*hem-* + *erythr-* + [1]*-in*] (1903)
: an iron-containing respiratory pigment in the blood of various invertebrates (as some annelids)

hemi- *prefix* [Middle English, from Latin, from Greek *hēmi-* — more at SEMI-]
: half ⟨*hemi*hedral⟩

-hemia — see -EMIA

hemi·ac·e·tal \ˌhe-mē-'a-sə-ˌtal\ *noun* (1893)
: any of a class of compounds characterized by the grouping C(OH)(OR) where R is an alkyl group and usually formed as intermediates in the preparation of acetals from aldehydes or ketones

he·mic \'hē-mik\ *adjective* (1857)
: of, relating to, or produced by the blood or the circulation of blood ⟨a *hemic* murmur⟩

hemi·cel·lu·lose \ˌhe-mi-'sel-yə-ˌlōs, -ˌlōz\ *noun* [International Scientific Vocabulary] (1891)
: any of various plant polysaccharides less complex than cellulose and easily hydrolyzable to simple sugars and other products

hemi·chor·date \-'kȯr-dət, -'kȯr-ˌdāt\ *noun* [New Latin *Hemichordata,* from *hemi-* + *Chordata* chordates] (1885)
: any of a phylum (Hemichordata) of wormlike marine animals (as an acorn worm) that have in the proboscis an outgrowth of the pharyngeal wall which superficially resembles the notochord of chordates

hemi·cy·cle \'he-mi-ˌsī-kəl\ *noun* [French *hémicycle,* from Latin *hemicyclium,* from Greek *hēmikyklion,* from *hēmi-* + *kyklos* circle — more at CYCLE] (13th century)
: a curved or semicircular structure or arrangement

hemi·demi·semi·qua·ver \ˌhe-mi-ˌde-mi-'se-mi-ˌkwā-vər\ *noun* (1853)
: SIXTY-FOURTH NOTE

hemi·he·dral \ˌhe-mi-'hē-drəl\ *adjective* [*hemi-* + *-hedron*] (1837)
of a crystal : having half the faces required by complete symmetry — compare HOLOHEDRAL, TETARTOHEDRAL

hemi·hy·drate \-'hī-ˌdrāt\ *noun* (circa 1901)
: a hydrate (as plaster of paris) containing half a mole of water to one mole of the compound forming the hydrate
— **hemi·hy·drat·ed** \-ˌdrā-təd\ *adjective*

hemi·me·tab·o·lous \ˌhe-mi-mə-'ta-bə-ləs\ *adjective* (1870)
: characterized by incomplete metamorphosis ⟨*hemimetabolous* insects⟩ — compare HOLOMETABOLOUS

hemi·mor·phic \ˌhe-mi-'mȯr-fik\ *adjective* [International Scientific Vocabulary] (circa 1859)
of a crystal : having different crystalline forms at each end of a crystallographic axis
— **hemi·mor·phism** \-ˌfi-zəm\ *noun*

he·min \'hē-mən\ *noun* [International Scientific Vocabulary] (circa 1857)
: a red-brown to blue-black crystalline salt $C_{34}H_{32}N_4O_4FeCl$ derived from oxidized heme but usually obtained in a characteristic crystalline form from hemoglobin

hemi·o·la \ˌhe-mē-'ō-lə\ *noun* [Late Latin *hemiolia,* from Greek *hēmiolia* ratio of one and a half to one, from *hēmi-* + *holos* whole — more at SAFE] (circa 1934)
: a musical rhythmic alteration in which six equal notes may be heard as two groups of three or three groups of two

hemi·ple·gia \ˌhe-mi-'plē-j(ē-)ə\ *noun* [New Latin, from Middle Greek *hēmiplēgia* paralysis, from Greek *hēmi-* + *-plēgia* -plegia] (1600)
: total or partial paralysis of one side of the body that results from disease of or injury to the motor centers of the brain
— **hemi·ple·gic** \-jik\ *adjective or noun*

he·mip·ter·an \hi-'mip-tə-rən\ *noun* [ultimately from Greek *hēmi-* + *pteron* wing — more at FEATHER] (circa 1864)
: any of a large order (Hemiptera) of hemimetabolous insects (as the true bugs) that have hemelytra and mouthparts adapted to piercing and sucking
— **he·mip·ter·ous** \-rəs\ *adjective*

hemi·sphere \'he-mə-ˌsfir\ *noun* [Middle English *hemispere,* from Latin *hemisphaerium,* from Greek *hēmisphairion,* from *hēmi-* + *sphairion,* diminutive of *sphaira* sphere] (14th century)
1 a : a half of the celestial sphere divided into two halves by the horizon, the celestial equator, or the ecliptic **b :** half of a spherical or roughly spherical body (as a planet); *specifically* : the northern or southern half of the earth divided by the equator or the eastern or western half divided by a meridian **c :** the inhabitants of a terrestrial hemisphere
2 : REALM, PROVINCE
3 : one of two half spheres formed by a plane through the sphere's center
4 : a map or projection of a celestial or terrestrial hemisphere
5 : CEREBRAL HEMISPHERE
— **hemi·spher·ic** \ˌhe-mə-'sfir-ik, -'sfer-\ *or* **hemi·spher·i·cal** \-'sfir-i-kəl, -'sfer-\ *adjective*

hemi·stich \'he-mi-ˌstik\ *noun* [Latin *hemistichium,* from Greek *hēmistichion,* from *hēmi-* + *stichos* line, verse; akin to Greek *steichein* to go — more at STAIR] (1575)
: half a poetic line of verse usually divided by a caesura

hemi·zy·gous \ˌhe-mi-'zī-gəs\ *adjective* (circa 1921)
: having or characterized by one or more genes (as in a genetic deficiency or in an X chromosome paired with a Y chromosome) that have no allelic counterparts

hem·line \'hem-ˌlīn\ *noun* (1923)
: the line formed by the lower edge of a dress, skirt, or coat

hem·lock \'hem-ˌläk\ *noun* [Middle English *hemlok,* from Old English *hemlic*] (before 12th century)
1 a : any of several poisonous herbs (as a poison hemlock or a water hemlock) of the carrot family having finely cut leaves and small white flowers **b :** a drug or lethal drink prepared from the poison hemlock (*Conium maculatum*)
2 : any of a genus (*Tsuga*) of evergreen coniferous trees of the pine family; *also* : the soft light splintery wood of a hemlock

hemo- — see HEM-

he·mo·chro·ma·to·sis \ˌhē-mə-ˌkrō-mə-'tō-səs\ *noun* [New Latin, from *hem-* + *chromat-* + *-osis*] (1899)
: a metabolic disorder involving the deposition of iron-containing pigments in the tissues and characterized by bronzing of the skin, diabetes, and weakness

he·mo·coel \'hē-mə-ˌsēl\ *noun* (1839)
: a body cavity (as in arthropods or some mollusks) that contains blood or hemolymph and functions as part of the circulatory system

he·mo·cy·a·nin \ˌhē-mō-'sī-ə-nən\ *noun* [International Scientific Vocabulary *hem-* + *cyan-* + [1]*-in*] (1885)
: a colorless copper-containing respiratory pigment in the circulatory fluid of various arthropods and mollusks

he·mo·cyte \'hē-mə-ˌsīt\ *noun* [International Scientific Vocabulary] (circa 1903)
: a blood cell especially of an invertebrate animal

he·mo·cy·tom·e·ter \ˌhē-mə-sī-'tä-mə-tər\ *noun* [International Scientific Vocabulary] (1877)
: HEMACYTOMETER

he·mo·di·al·y·sis \ˌhē-mō-dī-'a-lə-səs\ *noun* (1947)
: the process of removing blood from an artery (as of a kidney patient), purifying it by dialysis, adding vital substances, and returning it to a vein

he·mo·di·lu·tion \-dī-'lü-shən, -də-\ *noun* (1939)
: decreased concentration of cells and solids in the blood resulting from gain of fluid

he·mo·dy·nam·ic \-dī-'na-mik, -də-\ *adjective* (1907)
1 : of, relating to, or involving hemodynamics
2 : relating to or functioning in the mechanics of blood circulation
— **he·mo·dy·nam·i·cal·ly** \-mi-k(ə-)lē\ *adverb*

he·mo·dy·nam·ics \-miks\ *noun plural but singular or plural in construction* (circa 1857)
1 : a branch of physiology that deals with the circulation of the blood
2 : the forces or mechanisms involved in circulation

he·mo·fla·gel·late \ˌhē-mō-'fla-jə-lət, -ˌlāt; -flə-'je-lət\ *noun* (1909)
: a flagellate (as a trypanosome) that is a blood parasite

he·mo·glo·bin \'hē-mə-ˌglō-bən\ *noun* [International Scientific Vocabulary, short for earlier *hematoglobulin*] (1869)
1 : an iron-containing respiratory pigment of vertebrate red blood cells that consists of a globin composed of four subunits each of which is linked to a heme molecule, that functions in oxygen transport to the tissues after conversion to oxygenated form in the gills or lungs, and that assists in carbon dioxide transport back to the gills or lungs after surrender of its oxygen
2 : any of numerous iron-containing respiratory pigments of invertebrates and some plants (as yeasts)

he·mo·glo·bin·op·a·thy \ˌhē-mə-ˌglō-bə-'nä-pə-thē\ *noun, plural* **-thies** (1957)
: a blood disorder (as sickle-cell anemia) caused by a genetically determined change in the molecular structure of hemoglobin

hemoglobin S *noun* (1954)
: an abnormal hemoglobin that occurs in the red blood cells in sickle-cell anemia and sickle-cell trait

\ə\ **abut** \ᵊ\ **kitten** \ər\ **further** \a\ **ash** \ā\ **ace** \ä\ **mop, mar** \au̇\ **out** \ch\ **chin** \e\ **bet** \ē\ **easy** \g\ **go** \i\ **hit** \ī\ **ice** \j\ **job** \ŋ\ **sing** \ō\ **go** \ȯ\ **law** \ȯi\ **boy** \th\ **thin** \t͟h\ **the** \ü\ **loot** \u̇\ **foot** \y\ **yet** \zh\ **vision** *see also* Guide to Pronunciation

he•mo•glo•bin•uria \ˌhē-mə-ˌglō-bə-ˈnu̇r-ē-ə, -ˈnyu̇r-\ *noun* [New Latin] (1866)
: the presence of free hemoglobin in the urine
— **he•mo•glo•bin•uric** \-ˈnu̇r-ik, -ˈnyu̇r-\ *adjective*

he•mo•lymph \ˈhē-mə-ˌlim(p)f\ *noun* (1885)
: the circulatory fluid of various invertebrate animals that is functionally comparable to the blood and lymph of vertebrates

he•mo•ly•sin \ˌhē-mə-ˈlī-sᵊn\ *noun* [International Scientific Vocabulary] (1900)
: a substance that causes the dissolution of red blood cells

he•mo•ly•sis \hi-ˈmä-lə-səs, ˌhē-mə-ˈlī-səs\ *noun* [New Latin] (1890)
: lysis of red blood cells with liberation of hemoglobin
— **he•mo•lyt•ic** \ˌhē-mə-ˈli-tik\ *adjective*

hemolytic anemia *noun* (1938)
: anemia caused by excessive destruction (as in chemical poisoning, infection, or sickle-cell anemia) of red blood cells

hemolytic disease of the newborn (1948)
: ERYTHROBLASTOSIS FETALIS

he•mo•lyze \ˈhē-mə-ˌlīz\ *verb* **-lyzed; -lyz•ing** [irregular from *hemolysis*] (1902)
transitive verb
: to cause hemolysis of
intransitive verb
: to undergo hemolysis

he•mo•phil•ia \ˌhē-mə-ˈfi-lē-ə\ *noun* [New Latin] (1872)
: a sex-linked hereditary blood defect that occurs almost exclusively in males and is characterized by delayed clotting of the blood and consequent difficulty in controlling hemorrhage even after minor injuries

¹he•mo•phil•i•ac \-ˈfi-lē-ˌak\ *adjective* (1896)
: of, resembling, or affected with hemophilia

²hemophiliac *noun* (1897)
: one affected with hemophilia — called also *bleeder*

he•mo•phil•ic \-ˈfi-lik\ *noun or adjective* (1864)
: HEMOPHILIAC

he•mo•poi•e•sis \ˌhē-mə-pȯi-ˈē-səs\ *noun* [New Latin] (circa 1900)
: HEMATOPOIESIS
— **he•mo•poi•et•ic** \-ˈe-tik\ *adjective*

he•mo•pro•tein \-ˈprō-ˌtēn, -ˈprō-tē-ən\ *noun* (1948)
: a conjugated protein (as hemoglobin or cytochrome) whose prosthetic group is a porphyrin combined with iron

he•mop•ty•sis \hi-ˈmäp-tə-səs\ *noun* [New Latin, from *hem-* + Greek *ptysis* act of spitting, from *ptyein* to spit — more at SPEW] (1646)
: expectoration of blood from some part of the respiratory tract

¹hem•or•rhage \ˈhem-rij, ˈhe-mə-\ *noun* [Latin *haemorrhagia,* from Greek *haimorrhagia,* from *haimo-* hem- + *-rrhagia*] (1671)
: a copious discharge of blood from the blood vessels
— **hem•or•rhag•ic** \ˌhe-mə-ˈra-jik\ *adjective*

²hemorrhage *verb* **-rhaged; -rhag•ing** (1928)
intransitive verb
: to undergo heavy or uncontrollable bleeding
transitive verb
: to lose rapidly and uncontrollably ⟨*hemorrhage* money⟩

hemorrhagic fever *noun* (1948)
: any of a diverse group of arthropod-borne virus diseases characterized by a sudden onset, fever, aching, bleeding in the internal organs, petechiae, and shock

hem•or•rhoid \ˈhem-ˌrȯid, ˈhe-mə-\ *noun* [Middle English *emeroides,* plural, from Middle French *hemorrhoides,* from Latin *haemorrhoidae,* from Greek *haimorrhoides,* from *haimorrhoos* flowing with blood, from *haimo-* hem- + *rhein* to flow — more at STREAM] (14th century)
: a mass of dilated veins in swollen tissue at the margin of the anus or nearby within the rectum — usually used in plural; called also *piles*

¹hem•or•rhoid•al \ˌhem-ˈrȯi-dᵊl, ˌhe-mə-\ *noun* (15th century)
: a hemorrhoidal part (as an artery or vein)

²hemorrhoidal *adjective* (1651)
1 : of, relating to, or involving hemorrhoids
2 : RECTAL

he•mo•sid•er•in \ˌhē-mō-ˈsi-də-rən\ *noun* [International Scientific Vocabulary *hem-* + *sider-* + ¹*-in*] (circa 1885)
: a yellowish brown granular pigment that is formed in some phagocytic cells by the breakdown of hemoglobin and is probably essentially a denatured form of ferritin

he•mo•sta•sis \ˌhē-mə-ˈstā-səs\ *noun* [New Latin, from Greek *haimostasis* styptic, from *haimo-* hem- + *-stasis*] (1843)
: arrest of bleeding

he•mo•stat \ˈhē-mə-ˌstat\ *noun* (circa 1900)
: HEMOSTATIC; *especially* **:** an instrument for compressing a bleeding vessel

¹he•mo•stat•ic \ˌhē-mə-ˈsta-tik\ *noun* (circa 1706)
: a hemostatic agent

²hemostatic *adjective* (1834)
1 : of or caused by hemostasis
2 : serving to check bleeding

hemp \ˈhemp\ *noun* [Middle English, from Old English *hænep;* akin to Old High German *hanaf* hemp, Greek *kannabis*] (before 12th century)
1 a : a tall widely cultivated Asian herb (*Cannabis sativa*) of the mulberry family with tough bast fiber used especially for cordage **b :** the fiber of hemp **c :** a psychoactive drug (as marijuana or hashish) from hemp
2 : a fiber (as jute) from a plant other than the true hemp; *also* **:** a plant yielding such fiber

hemp•en \ˈhem-pən\ *adjective* (14th century)
: composed of hemp

hemp nettle *noun* (1801)
: any of a genus (*Galeopsis*) of coarse Old World herbs of the mint family; *especially* **:** a bristly Eurasian herb (*G. tetrahit*) naturalized in the U.S. as a weed

¹hem•stitch \ˈhem-ˌstich\ *transitive verb* (1839)
: to decorate (as a border) with hemstitch
— **hem•stitch•er** *noun*

²hemstitch *noun* (1853)
1 : decorative needlework similar to drawnwork that is used especially on or next to the stitching line of hems
2 : a stitch used in hemstitching

hemstitch

hen \ˈhen\ *noun* [Middle English, from Old English *henn;* akin to Old English *hana* rooster — more at CHANT] (before 12th century)
1 a : a female chicken especially over a year old; *broadly* **:** a female bird **b :** the female of various mostly aquatic animals (as lobsters or fish)
2 : WOMAN; *especially* **:** a fussy middle-aged woman

hen and chickens *noun* (1884)
: any of several plants having offsets, runners, or flowers that send out shoots; *especially* **:** HOUSELEEK

hen•bane \ˈhen-ˌbān\ *noun* (14th century)
: a poisonous fetid Old World herb (*Hyoscyamus niger*) of the nightshade family having sticky hairy dentate leaves and yellowish brown flowers and yielding hyoscyamine and scopolamine

hen•bit \ˈhen-ˌbit\ *noun* (1597)
: a Eurasian herb (*Lamium amplexicaule*) of the mint family that has scalloped reniform leaves and purplish flowers and is naturalized in North America

hence \ˈhen(t)s\ *adverb* [Middle English *hennes, henne,* from Old English *heonan;* akin to Old High German *hinnan* away, Old English *hēr* here] (13th century)
1 : from this place **:** AWAY
2 a *archaic* **:** HENCEFORTH **b :** from this time
3 : because of a preceding fact or premise **:** THEREFORE
4 : from this source or origin
— **from hence** *archaic* **:** from this place **:** from this time

hence•forth \ˈhen(t)s-ˌfōrth, -ˌfȯrth, hen(t)s-ˈ\ *adverb* (14th century)
: from this point on

hence•for•ward \hen(t)s-ˈfȯr-wərd\ *adverb* (14th century)
: HENCEFORTH

hench•man \ˈhench-mən\ *noun* [Middle English *henshman, hengestman* groom, from *hengest* stallion (from Old English) + *man;* akin to Old High German *hengist* gelding] (15th century)
1 *obsolete* **:** a squire or page to a person of high rank
2 a : a trusted follower **:** a right-hand man **b :** a political follower whose support is chiefly for personal advantage **c :** a member of a gang

hen•deca•syl•lab•ic \(ˌ)hen-ˌde-kə-sə-ˈla-bik\ *adjective* [Latin *hendecasyllabus,* from Greek *hendeka* eleven (from *hen-, heis* one + *deka* ten) + *syllabē* syllable — more at SAME, TEN] (circa 1751)
: consisting of 11 syllables or composed of verses of 11 syllables
— **hendecasyllabic** *noun*
— **hen•deca•syl•la•ble** \hen-ˈde-kə-ˌsi-lə-bəl, (ˌ)hen-ˌde-kə-ˈ\ *noun*

hen•di•a•dys \hen-ˈdī-ə-dəs\ *noun* [Late Latin *hendiadys, hendiadyoin,* modification of Greek *hen dia dyoin,* literally, one through two] (circa 1577)
: the expression of an idea by the use of usually two independent words connected by *and* (as *nice and warm*) instead of the usual combination of independent word and its modifier (as *nicely warm*)

hen•e•quen \ˈhe-ni-kən, ˌhe-ni-ˈken\ *noun* [Spanish *henequén*] (1880)
: a strong yellowish or reddish hard fiber obtained from the leaves of a tropical American agave chiefly in Yucatán and used especially for binder twine; *also* **:** a plant (*Agave fourcroydes*) that yields henequen

hen•house \ˈhen-ˌhau̇s\ *noun* (circa 1513)
: a house or shelter for fowl

Hen•le's loop \ˈhen-lēz-\ *noun* (circa 1890)
: LOOP OF HENLE

¹hen•na \ˈhe-nə\ *noun* [Arabic *ḥinnā'*] (1600)
1 : a reddish brown dye obtained from leaves of the henna plant and used especially on hair
2 : an Old World tropical shrub or small tree (*Lawsonia inermis*) of the loosestrife family with small opposite leaves and axillary panicles of fragrant white flowers

²henna *transitive verb* (1919)
: to dye (as hair) with henna

hen•nery \ˈhe-nə-rē\ *noun, plural* **-ner•ies** (1850)
: a poultry farm; *also* **:** an enclosure for poultry

heno•the•ism \ˈhe-nə-(ˌ)thē-ˌi-zəm\ *noun* [German *Henotheismus,* from Greek *hen-, heis* one + *theos* god — more at SAME] (1860)
: the worship of one god without denying the existence of other gods
— **heno•the•ist** \-ˌthē-ist\ *noun*
— **heno•the•is•tic** \ˌhe-nə-thē-ˈis-tik\ *adjective*

hen party *noun* (circa 1885)
: a party for women only

hen•peck \ˈhen-ˌpek\ *transitive verb* (1688)
: to subject (one's husband) to persistent nagging and domination

hen•ry \'hen-rē\ *noun, plural* **henrys** *or* **henries** [Joseph *Henry*] (circa 1890)
: the practical meter-kilogram-second unit of inductance equal to the self-inductance of a circuit or the mutual inductance of two circuits in which the variation of one ampere per second results in an induced electromotive force of one volt

hent \'hent\ *transitive verb* [Middle English, from Old English *hentan* — more at HUNT] *archaic* (before 12th century)
: SEIZE

hen track *noun* (1907)
: an illegible or scarcely legible mark intended as handwriting — called also *hen scratch*

¹hep \'hep, 'həp, 'hət\ *interjection* [origin unknown] (1862)
— used to mark a marching cadence

²hep \'hep\ *adjective* [origin unknown] (1904)
: ⁴HIP

hep•a•rin \'he-pə-rən\ *noun* [International Scientific Vocabulary, from Greek *hēpar* liver] (1918)
: a mucopolysaccharide sulfuric acid ester that is found especially in liver, that prolongs the clotting time of blood, and that is used medically
— hep•a•rin•ized \-rə-ˌnīzd\ *adjective*

hepat- *or* **hepato-** *combining form* [Latin, from Greek *hēpat-, hēpato-,* from *hēpat-, hēpar;* akin to Latin *jecur* liver]
1 : liver ⟨*hepat*ectomy⟩ ⟨*hepato*toxic⟩
2 : hepatic and ⟨*hepato*cellular⟩

hep•a•tec•to•my \ˌhe-pə-'tek-tə-mē\ *noun, plural* **-mies** (circa 1890)
: excision of the liver or of part of the liver
— hep•a•tec•to•mized \-ˌmīzd\ *adjective*

¹he•pat•ic \hi-'pa-tik\ *adjective* [Latin *hepaticus,* from Greek *hēpatikos,* from *hēpat-, hēpar*] (1599)
: of, relating to, affecting, associated with, supplying, or draining the liver ⟨a *hepatic* complaint⟩ ⟨*hepatic* arteries⟩

²hepatic *noun* (1900)
: LIVERWORT

he•pat•i•ca \hi-'pa-ti-kə\ *noun* [New Latin, from Medieval Latin, liverwort, from Latin, feminine of *hepaticus*] (1578)
: any of a genus (*Hepatica*) of herbs of the buttercup family with lobed leaves and delicate flowers

hep•a•ti•tis \ˌhe-pə-'tī-təs\ *noun, plural* **-tit•i•des** \-'ti-tə-ˌdēz\ [New Latin] (circa 1751)
1 : inflammation of the liver
2 : a disease or condition (as hepatitis A or hepatitis B) marked by inflammation of the liver

hepatitis A *noun* (1972)
: an acute usually benign hepatitis caused by an RNA virus that does not persist in the blood serum and is transmitted especially in food and water contaminated with infected fecal matter — called also *infectious hepatitis*

hepatitis B *noun* (1972)
: a sometimes fatal hepatitis caused by a double-stranded DNA virus that tends to persist in the blood serum and is transmitted especially by contact with infected blood (as by transfusion) or blood products — called also *serum hepatitis*

he•pa•to•cel•lu•lar \ˌhe-pə-tō-'sel-yə-lər, hiˌpa-tə-'sel-\ *adjective* (1940)
: of or involving hepatocytes ⟨*hepatocellular* carcinoma⟩

he•pa•to•cyte \hi-'pa-tə-ˌsīt, 'he-pə-tə-\ *noun* (1965)
: an epithelial parenchymatous cell of the liver

hep•a•to•ma \ˌhe-pə-'tō-mə\ *noun, plural* **-mas** *or* **-ma•ta** \-mə-tə\ [New Latin] (circa 1923)
: a usually malignant tumor of the liver

he•pa•to•meg•a•ly \ˌhe-pə-tō-'me-gə-lē, hiˌpa-tə-'me-\ *noun, plural* **-lies** (circa 1901)
: enlargement of the liver

he•pa•to•pan•cre•as \-'paŋ-krē-əs, -'pan-\ *noun* (1884)
: a glandular structure (as of a crustacean) that combines the digestive functions of the vertebrate liver and pancreas

hep•a•to•tox•ic \-'täk-sik\ *adjective* (1926)
: relating to or causing injury to the liver ⟨*hepatotoxic* drugs⟩

hep•a•to•tox•ic•i•ty \-täk-'si-sə-tē\ *noun* (1952)
1 : a state of toxic damage to the liver
2 : a tendency or capacity to cause hepatotoxicity

hep•cat \'hep-ˌkat\ *noun* (1938)
: HIPSTER

He•phaes•tus \hi-'fes-təs, -'fēs-\ *noun* [Latin, from Greek *Hēphaistos*]
: the Greek god of fire and metalworking — compare VULCAN

hepped up *adjective* (1947)
: ENTHUSIASTIC

Hep•ple•white \'he-pəl-ˌhwīt, -ˌwīt\ *adjective* [George *Hepplewhite*] (1897)
: of, relating to, or imitating a style of furniture originating in late 18th century England

hepta- *or* **hept-** *combining form* [Greek, from *hepta* — more at SEVEN]
1 : seven ⟨*hepta*meter⟩
2 : containing seven atoms, groups, or equivalents ⟨*hept*ane⟩

hep•ta•chlor \'hep-tə-ˌklōr, -ˌklȯr\ *noun* [*hepta-* + *chlor*ine] (1949)
: a cyclodiene chlorinated hydrocarbon pesticide $C_{10}H_5Cl_7$ that causes liver disease in animals and is a suspected human carcinogen

hep•tad \'hep-ˌtad\ *noun* [Greek *heptad-, heptas,* from *hepta*] (1660)
: a group of seven

hep•ta•gon \'hep-tə-ˌgän\ *noun* [Greek *heptagōnos* heptagonal, from *hepta* + *gōnia* angle — more at -GON] (1570)
: a polygon of seven angles and seven sides
— hep•tag•o•nal \hep-'ta-gə-nᵊl\ *adjective*

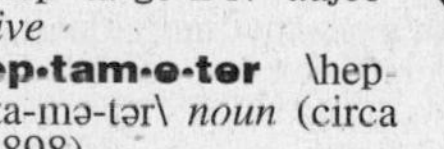

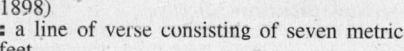

heptagon

hep•tam•e•ter \hep-'ta-mə-tər\ *noun* (circa 1898)
: a line of verse consisting of seven metrical feet

hep•tane \'hep-ˌtān\ *noun* (1877)
: any of several isomeric alkanes C_7H_{16}; *especially* **:** the liquid normal isomer occurring in petroleum and used especially as a solvent and in determining octane numbers

hep•tar•chy \'hep-ˌtär-kē\ *noun* (1576)
: a hypothetical confederacy of seven Anglo-Saxon kingdoms of the 7th and 8th centuries

Hep•ta•teuch \'hep-tə-ˌtük, -ˌtyük\ *noun* [Late Latin *heptateuchos,* from Greek, from *hepta* + *teuchos* book — more at PENTATEUCH] (1678)
: the first seven books of the canonical Jewish and Christian Scriptures

hep•tose \'hep-ˌtōs, -ˌtōz\ *noun* (1890)
: any of various monosaccharides $C_7H_{14}O_7$ containing seven carbon atoms in a molecule

¹her \(h)ər, 'hər\ *adjective* [Middle English *hire,* from Old English *hiere,* genitive of *hēo* she — more at HE] (before 12th century)
: of or relating to her or herself especially as possessor, agent, or object of an action ⟨*her* house⟩ ⟨*her* research⟩ ⟨*her* rescue⟩ — compare ¹SHE

²her *pronoun objective case of* SHE

He•ra \'hir-ə, 'hē-rə, 'her-ə\ *noun* [Greek *Hēra, Hērē*]
: the sister and consort of Zeus — compare JUNO

Her•a•cles \'her-ə-ˌklēz\ *noun* [Greek *Hēraklēs*]
: HERCULES

¹her•ald \'her-əld\ *noun* [Middle English, from Middle French *hiraut,* from an (assumed) Germanic compound whose first component is akin to Old High German *heri* army, and whose second is akin to Old High German *waltan* to rule — more at HARRY, WIELD] (14th century)
1 a : an official at a tournament of arms with duties including the making of announcements and the marshaling of combatants **b :** an officer with the status of ambassador acting as official messenger between leaders especially in war **c** (1) **:** OFFICER OF ARMS (2) **:** an officer of arms ranking above a pursuivant and below a king of arms
2 : an official crier or messenger
3 a : one that precedes or foreshadows **b :** one that conveys news or proclaims **:** ANNOUNCER ⟨it was the lark, the *herald* of the morn —Shakespeare⟩ **c :** one who actively promotes or advocates **:** EXPONENT
synonym see FORERUNNER

²herald *transitive verb* (14th century)
1 : to give notice of **:** ANNOUNCE
2 a : to greet especially with enthusiasm **:** HAIL **b :** PUBLICIZE
3 : to signal the approach of **:** FORESHADOW

he•ral•dic \he-'ral-dik, hə-\ *adjective* (1772)
: of or relating to heralds or heraldry
— he•ral•di•cal•ly \-di-k(ə-)lē\ *adverb*

her•ald•ry \'her-əl-drē\ *noun, plural* **-ries** (1572)
1 : the practice of devising, blazoning, and granting armorial insignia and of tracing and recording genealogies
2 : an armorial ensign; *broadly* **:** INSIGNIA
3 : PAGEANTRY

herb \'ərb, *US also & British usually* 'hərb\ *noun, often attributive* [Middle English *herbe,* from Old French, from Latin *herba*] (14th century)
1 : a seed-producing annual, biennial, or perennial that does not develop persistent woody tissue but dies down at the end of a growing season
2 : a plant or plant part valued for its medicinal, savory, or aromatic qualities
3 *slang* **:** MARIJUANA 2
— herb•like \'(h)ər-ˌblīk\ *adjective*
— herby \'(h)ər-bē\ *adjective*

her•ba•ceous \ˌ(h)ər-'bā-shəs\ *adjective* (1646)
1 a : of, relating to, or having the characteristics of an herb **b** *of a stem* **:** having little or no woody tissue and persisting usually for a single growing season
2 : having the texture, color, or appearance of a leaf

herb•age \'(h)ər-bij\ *noun* (14th century)
1 : herbaceous vegetation (as grass) especially when used for grazing
2 : the succulent parts of herbaceous plants

¹herb•al \'(h)ər-bəl\ *noun* (1516)
1 : a book about plants especially with reference to their medicinal properties
2 *archaic* **:** HERBARIUM 1

²herbal *adjective* (1612)
: of, relating to, or made of herbs

herb•al•ist \'(h)ər-bə-list\ *noun* (1589)
1 : one who practices healing by the use of herbs
2 : one who collects or grows herbs

her•bar•i•um \ˌ(h)ər-'bar-ē-əm, -'ber-\ *noun, plural* **-ia** \-ē-ə\ (1776)
1 : a collection of dried plant specimens usually mounted and systematically arranged for reference
2 : a place that houses an herbarium

herb doctor *noun* (1828)
: HERBALIST 1

herbed \'(h)ərbd\ *adjective* (1950)
: seasoned with herbs

her•bi•cide \'(h)ər-bə-ˌsīd\ *noun* [Latin *herba* + International Scientific Vocabulary *-cide*] (1899)

: an agent used to destroy or inhibit plant growth
— **her·bi·cid·al** \ˌ(h)ər-bə-ˈsī-dᵊl\ *adjective*
— **her·bi·cid·al·ly** \-dᵊl-ē\ *adverb*
her·bi·vore \ˈ(h)ər-bə-ˌvōr, -ˌvȯr\ *noun* [New Latin *Herbivora,* group of mammals, from neuter plural of *herbivorus*] (1854)
: a plant-eating animal
her·biv·o·rous \ˌ(h)ər-ˈbiv-rəs, -ˈbi-və-\ *adjective* [New Latin *herbivorus,* from Latin *herba* grass + *-vorus* -vorous] (1661)
: feeding on plants
— **her·biv·o·ry** \-ˈbi-və-rē\ *noun*
herb Rob·ert \ˈ(h)ərb-ˈrä-bərt\ *noun* [Medieval Latin *herba Roberti,* probably from *Robertus* (Saint Robert) (died 1067) French ecclesiastic] (13th century)
: a sticky low geranium (*Geranium robertianum*) with small reddish purple flowers
Her·cu·le·an \ˌhər-kyə-ˈlē-ən, ˌhər-ˈkyü-lē-\ *adjective* (1593)
1 : of, relating to, or characteristic of Hercules
2 *often not capitalized* **:** of extraordinary power, extent, intensity, or difficulty
Her·cu·les \ˈhər-kyə-ˌlēz\ *noun* [Latin, from Greek *Hēraklēs*]
1 : a mythical Greek hero renowned for his great strength and especially for performing 12 labors imposed on him by Hera
2 [Latin (genitive *Herculis*)] **:** a northern constellation between Corona Borealis and Lyra
Her·cu·les'–club \ˈhər-kyə-ˌlēz-ˌkləb\ *noun* (1847)
1 : a small prickly eastern U.S. tree (*Aralia spinosa*) of the ginseng family — called also *angelica tree*
2 : a small prickly southern U.S. tree (*Zanthoxylum clava-herculis*) of the rue family
¹herd \ˈhərd\ *noun* [Middle English, from Old English *heord;* akin to Old High German *herta* herd, Middle Welsh *cordd* troop, Lithuanian *kerdžius* shepherd] (before 12th century)
1 a : a number of animals of one kind kept together under human control **b :** a congregation of gregarious wild animals
2 a (1) **:** a group of people usually having a common bond (2) **:** a large assemblage of like things **b :** the undistinguished masses **:** CROWD ⟨isolate the individual prophets from the *herd* —Norman Cousins⟩
— **herd·like** \-ˌlīk\ *adjective*
²herd (13th century)
transitive verb
1 a : to gather, lead, or drive as if in a herd ⟨seventy-five boys and girls were *herded* by six or eight teachers —W. A. White⟩ **b :** to keep or move (animals) together
2 : to place in a group
intransitive verb
1 : to assemble or move in a herd
2 : to place oneself in a group **:** ASSOCIATE
herd·er \ˈhər-dər\ *noun* (1635)
: one that herds; *specifically* **:** HERDSMAN 1
herds·man \ˈhərdz-mən\ *noun* (1603)
1 : a manager, breeder, or tender of livestock
2 *capitalized* **:** BOÖTES
¹here \ˈhir\ *adverb* [Middle English, from Old English *hēr;* akin to Old High German *hier* here, Old English *hē* he] (before 12th century)
1 a : in or at this place ⟨turn *here*⟩ — often used interjectionally especially in answering a roll call **b :** NOW ⟨*here* it's morning already⟩ **c :** in an arbitrary location ⟨a book *here,* a paper there⟩
2 : at or in this point, particular, or case ⟨*here* we agree⟩
3 : in the present life or state
4 : HITHER ⟨come *here*⟩
5 — used interjectionally in rebuke or encouragement
— **here goes** — used interjectionally to express resolution or resignation especially at the beginning of a difficult or unpleasant undertaking
— **neither here nor there :** having no interest or relevance **:** of no consequence ⟨comfort is *neither here nor there* to a real sailor⟩
²here *adjective* (15th century)
1 — used for emphasis especially after a demonstrative pronoun or after a noun modified by a demonstrative adjective ⟨this book *here*⟩
2 *nonstandard* — used for emphasis after a demonstrative adjective but before the noun modified ⟨this *here* book⟩
³here *noun* (1605)
: this place
here·abouts \ˈhir-ə-ˌbau̇ts\ *or* **here·about** \-ˌbau̇t\ *adverb* (13th century)
: in this vicinity
¹here·af·ter \hir-ˈaf-tər\ *adverb* (before 12th century)
1 : after this in sequence or in time
2 : in some future time or state
²hereafter *noun, often capitalized* (1546)
1 : FUTURE
2 : an existence beyond earthly life
³hereafter *adjective* (1591)
archaic **:** FUTURE
here and now *noun* (1829)
: the present time — used with *the* ⟨man's obligation is in the *here and now* —W. H. Whyte⟩
here and there *adverb* (14th century)
1 : in one place and another
2 : from time to time
here·away \ˈhir-ə-ˌwā\ *or* **here·aways** \-ˌwāz\ *adverb* (14th century)
dialect **:** HEREABOUTS
here·by \hir-ˈbī, ˈhir-ˌ\ *adverb* (13th century)
: by this means
her·e·dit·a·ment \ˌher-ə-ˈdi-tə-mənt\ *noun* [Medieval Latin *hereditamentum,* from Late Latin *hereditare* to inherit, from Latin *hered-, heres*] (15th century)
: heritable property
he·red·i·tar·i·an \hə-ˌre-də-ˈter-ē-ən\ *noun* (1881)
: an advocate of the theory that individual differences in human beings can be accounted for primarily on the basis of genetics
— **hereditarian** *adjective*
he·red·i·tary \hə-ˈre-də-ˌter-ē\ *adjective* [Latin *hereditarius,* from *hereditas*] (15th century)
1 a : genetically transmitted or transmittable from parent to offspring **b :** characteristic of or fostered by one's predecessors
2 a : received or passing by inheritance or required to pass by inheritance or by reason of birth **b :** having title or possession through inheritance or by reason of birth
3 : of a kind established by tradition ⟨*hereditary* enemies⟩
4 : of or relating to inheritance or heredity
synonym see INNATE
— **he·red·i·tar·i·ly** \-ˌre-də-ˈter-ə-lē\ *adverb*
he·red·i·ty \hə-ˈre-də-tē\ *noun* [Middle French *heredité,* from Latin *hereditat-, hereditas,* from *hered-, heres* heir — more at HEIR] (circa 1540)
1 a : INHERITANCE **b :** TRADITION
2 a : the sum of the qualities and potentialities genetically derived from one's ancestors **b :** the transmission of such qualities from ancestor to descendant through the genes
Her·e·ford \ˈhər-fərd *sometimes* ˈher-ə-\ *noun* [*Hereford* former county in England] (1805)
: any of a breed of hardy red-coated beef cattle of English origin with white faces and markings
here·in \hir-ˈin\ *adverb* (before 12th century)
: in this
here·in·above \(ˌ)hir-ˌin-ə-ˈbəv\ *adverb* (circa 1812)
: at a prior point in this writing or document
here·in·af·ter \ˌhir-ə-ˈnaf-tər\ *adverb* (1590)
: in the following part of this writing or document
here·in·be·fore \(ˌ)hir-ˌin-bi-ˈfōr, -ˈfȯr\ *adverb* (1687)
: in the preceding part of this writing or document
here·in·be·low \-bi-ˈlō\ *adverb* (1946)
: at a subsequent point in this writing or document
here·of \hir-ˈəv, -ˈäv\ *adverb* (before 12th century)
: of this
here·on \-ˈȯn, -ˈän\ *adverb* (12th century)
: on this
He·re·ro \hə-ˈrer-(ˌ)ō, ˈher-ə-ˌrō\ *noun, plural* **Herero** *or* **Hereros** (1880)
: a member of a Bantu people of central Namibia
he·re·si·arch \hə-ˈrē-zē-ˌärk, ˈher-ə-sē-\ *noun* [Late Latin *haeresiarcha,* from Late Greek *hairesiarchēs,* from *hairesis* + Greek *-archēs* -arch] (1624)
: an originator or chief advocate of a heresy
her·e·sy \ˈher-ə-sē\ *noun, plural* **-sies** [Middle English *heresie,* from Old French, from Late Latin *haeresis,* from Late Greek *hairesis,* from Greek, action of taking, choice, sect, from *hairein* to take] (13th century)
1 a : adherence to a religious opinion contrary to church dogma **b :** denial of a revealed truth by a baptized member of the Roman Catholic Church **c :** an opinion or doctrine contrary to church dogma
2 a : dissent or deviation from a dominant theory, opinion, or practice **b :** an opinion, doctrine, or practice contrary to the truth or to generally accepted beliefs or standards
her·e·tic \ˈher-ə-ˌtik\ *noun* (14th century)
1 : a dissenter from established church dogma; *especially* **:** a baptized member of the Roman Catholic Church who disavows a revealed truth
2 : one who dissents from an accepted belief or doctrine **:** NONCONFORMIST
he·ret·i·cal \hə-ˈre-ti-kəl\ *also* **he·re·tic** \ˈher-ə-ˌtik\ *adjective* (15th century)
1 : of, relating to, or characterized by heresy
2 : of, relating to, or characterized by departure from accepted beliefs or standards **:** UNORTHODOX
— **he·ret·i·cal·ly** \hə-ˈre-ti-k(ə-)lē\ *adverb*
here·to \hir-ˈtü\ *adverb* (12th century)
: to this writing or document
here·to·fore \ˈhir-tə-ˌfōr, -ˌfȯr, ˌhir-tə-ˈ\ *adverb* (13th century)
: up to this time **:** HITHERTO
here·un·der \hir-ˈən-dər\ *adverb* (15th century)
: under or in accordance with this writing or document
here·un·to \hir-ˈən-(ˌ)tü, ˌhir-(ˌ)ən-ˈtü\ *adverb* (1509)
: to this
here·up·on \ˈhir-ə-ˌpȯn, -ˌpän, ˌhir-ə-ˈ\ *adverb* (12th century)
: on this **:** immediately after this
here·with \hir-ˈwith, -ˈwith\ *adverb* (before 12th century)
1 : with this communication **:** enclosed in this
2 : HEREBY
He·rez \hə-ˈrez\ *or* **He·riz** \-ˈriz\ *noun* [*Herez, Heriz,* town in Iran] (circa 1922)
: a Persian rug characterized by a large central geometric medallion and by angular floral designs
her·i·ot \ˈher-ē-ət\ *noun* [Middle English, from Old English *heregeatwe,* plural, military equipment, from *here* army (akin to Old High German *heri* army) + *geatwe* equipment — more at HARRY] (before 12th century)
: a feudal duty or tribute due under English law to a lord on the death of a tenant
her·i·ta·bil·i·ty \ˌher-ə-tə-ˈbi-lə-tē\ *noun* (1832)
1 : the quality or state of being heritable
2 : the proportion of observed variation in a particular trait (as intelligence) that can be attributed to inherited genetic factors in contrast to environmental ones

her·i·ta·ble \'her-ə-tə-bəl\ *adjective* (14th century)
1 : capable of being inherited or of passing by inheritance
2 : HEREDITARY

her·i·tage \'her-ə-tij\ *noun* [Middle English, from Middle French, from *heriter* to inherit, from Late Latin *hereditare,* from Latin *hered-, heres* heir — more at HEIR] (13th century)
1 : property that descends to an heir
2 a : something transmitted by or acquired from a predecessor **:** LEGACY, INHERITANCE **b :** TRADITION
3 : something possessed as a result of one's natural situation or birth **:** BIRTHRIGHT ⟨the nation's *heritage* of tolerance⟩

her·i·tor \'her-ə-tər\ *noun* (15th century)
: one that inherits **:** INHERITOR

herky–jerky \'hər-kē-'jər-kē\ *adjective* [reduplication of *jerky*] (1957)
: characterized by sudden, irregular, or unpredictable movement or style

herm \'hərm\ *noun* [Latin *hermes,* from Greek *hermēs* statue of Hermes, herm, from *Hermēs*] (circa 1580)
: a statue in the form of a square stone pillar surmounted by a bust or head especially of Hermes

her·ma \'hər-mə\ *noun, plural* **her·mae** \-ˌmē, -ˌmī\ *or* **her·mai** \-ˌmī\ (1638)
: HERM

her·maph·ro·dite \(ˌ)hər-'ma-frə-ˌdīt\ *noun* [Middle English *hermofrodite,* from Latin *hermaphroditus,* from Greek *hermaphroditos,* from *Hermaphroditos*] (14th century)
1 : an animal or plant having both male and female reproductive organs
2 : something that is a combination of diverse elements
— **hermaphrodite** *adjective*
— **her·maph·ro·dit·ic** \(ˌ)hər-ˌma-frə-'di-tik\ *adjective*
— **her·maph·ro·dit·ism** \-'ma-frə-ˌdī-ˌti-zəm\ *noun*

Her·maph·ro·di·tus \(ˌ)hər-ˌma-frə-'dī-təs\ *noun* [Latin, from Greek *Hermaphroditos,* from *Hermēs* + *Aphroditē* Aphrodite]
: a son of Hermes and Aphrodite who becomes joined in one body with a nymph while bathing

her·ma·typ·ic \ˌhər-mə-'ti-pik\ *adjective* [Greek *herma* prop, reef + *typtein* to strike, coin + English *-ic* — more at TYPE] (1950)
: building reefs ⟨*hermatypic* corals⟩

her·me·neu·ti·cal \ˌhər-mə-'nü-ti-kəl, -'nyü-\ *or* **her·me·neu·tic** \-tik\ *adjective* [Greek *hermēneutikos,* from *hermēneuein* to interpret, from *hermēneus* interpreter] (1678)
: of or relating to hermeneutics **:** INTERPRETATIVE
— **her·me·neu·ti·cal·ly** \-ti-k(ə-)lē\ *adverb*

her·me·neu·tics \-tiks\ *noun plural but singular or plural in construction* (1737)
: the study of the methodological principles of interpretation (as of the Bible)

Her·mes \'hər-(ˌ)mēz\ *noun* [Latin, from Greek *Hermēs*]
: a Greek god of commerce, eloquence, invention, travel, and theft who serves as herald and messenger of the other gods — compare MERCURY

Hermes Tris·me·gis·tus \-ˌtris-mə-'jis-təs\ *noun* [Medieval Latin, from Greek *Hermēs trismegistos,* literally, Hermes thrice greatest]
: a legendary author of works embodying magical, astrological, and alchemical doctrines

her·met·ic \(ˌ)hər-'me-tik\ *also* **her·met·i·cal** \-ti-kəl\ *adjective* [New Latin *hermeticus,* from *Hermet-, Hermes Trismegistus*] (1605)
1 *often capitalized* **a :** of or relating to the Gnostic writings or teachings arising in the first three centuries A.D. and attributed to Hermes Trismegistus **b :** relating to or characterized by occultism or abstruseness **:** RECONDITE
2 [from the belief that Hermes Trismegistus invented a magic seal to keep vessels airtight] **a :** AIRTIGHT ⟨*hermetic* seal⟩ **b :** impervious to external influence ⟨trapped inside the *hermetic* military machine —Jack Newfield⟩ **c :** RECLUSE, SOLITARY ⟨leads a *hermetic* life⟩ ◆
— **her·met·i·cal·ly** \-ti-k(ə-)lē\ *adverb*

her·met·i·cism \-'me-tə-ˌsi-zəm\ *noun, often capitalized* (1897)
: HERMETISM

her·me·tism \'hər-mə-ˌti-zəm\ *noun, often capitalized* (1897)
1 a : a system of ideas based on hermetic teachings **b :** adherence to or practice of hermetic doctrine
2 : the practice of being hermetically mysterious ⟨it is not . . . willful *hermetism,* if the message of their art is veiled and indirect —R. J. Goldwater⟩
— **her·me·tist** \-mə-tist\ *noun*

her·mit \'hər-mət\ *noun* [Middle English *eremite,* from Old French, from Late Latin *eremita,* from Late Greek *erēmitēs,* from Greek, adjective, living in the desert, from *erēmia* desert, from *erēmos* desolate] (12th century)
1 a : one that retires from society and lives in solitude especially for religious reasons **:** RECLUSE **b** *obsolete* **:** BEADSMAN
2 : a spiced molasses cookie
— **her·mit·ism** \'hər-mə-ˌti-zəm\ *noun*

her·mit·age \'hər-mə-tij\ *noun* (14th century)
1 a : the habitation of a hermit **b :** a secluded residence or private retreat **:** HIDEAWAY **c :** MONASTERY
2 : the life or condition of a hermit

Her·mi·tage \ˌ(h)er-mi-'täzh\ *noun* [Tain-l'*Ermitage,* commune in France] (1680)
: a red or white Rhone valley wine

hermit crab *noun* (1735)
: any of numerous chiefly marine decapod crustaceans (especially families Diogenidae, Paguridae, and Parapaguridae) having soft asymmetrical abdomens and occupying the empty shells of gastropods

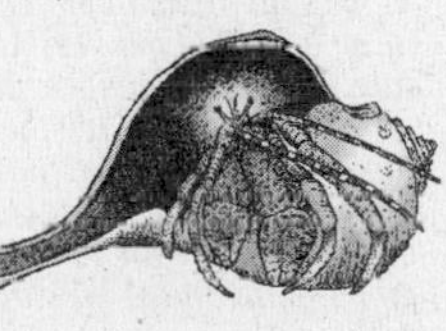
hermit crab

Her·mi·tian matrix \er-'mē-shən-, ˌhər-'mi-shən-\ *noun* [Charles *Hermite* (died 1901) French mathematician] (1935)
: a square matrix having the property that each pair of elements in the *i*th row and *j*th column and in the *j*th row and *i*th column are conjugate complex numbers

hern \'hern, 'hərn\ *dialect variant of* HERON

her·nia \'hər-nē-ə\ *noun, plural* **-ni·as** *or* **-ni·ae** \-nē-ˌē, -nē-ˌī\ [Latin — more at YARN] (14th century)
: a protrusion of an organ or part through connective tissue or through a wall of the cavity in which it is normally enclosed — called also *rupture*
— **her·ni·al** \-nē-əl\ *adjective*

her·ni·ate \'hər-nē-ˌāt\ *intransitive verb* **-at·ed; -at·ing** (circa 1922)
: to protrude through an abnormal body opening **:** RUPTURE
— **her·ni·a·tion** \ˌhər-nē-'ā-shən\ *noun*

he·ro \'hir-(ˌ)ō, 'hē-(ˌ)rō\ *noun, plural* **heroes** [Latin *heros,* from Greek *hērōs*] (14th century)
1 a : a mythological or legendary figure often of divine descent endowed with great strength or ability **b :** an illustrious warrior **c :** a man admired for his achievements and noble qualities **d :** one that shows great courage
2 a : the principal male character in a literary or dramatic work **b :** the central figure in an event, period, or movement
3 *plural usually* **heros :** SUBMARINE 2
4 : an object of extreme admiration and devotion **:** IDOL

Hero *noun* [Latin, from Greek *Hērō*]
: a legendary priestess of Aphrodite loved by Leander

¹he·ro·ic \hi-'rō-ik *also* her-'ō- *or* hē-'rō-\ *also* **he·ro·ical** \-i-kəl\ *adjective* (1549)
1 : of, relating to, or resembling heroes especially of antiquity
2 a : exhibiting or marked by courage and daring **b :** supremely noble or self-sacrificing
3 a : of impressive size, power, extent, or effect **:** POTENT ⟨*heroic* doses⟩ ⟨a *heroic* voice⟩ **b** (1) **:** of great intensity **:** EXTREME, DRASTIC ⟨*heroic* effort⟩ (2) **:** of a kind that is likely only to be undertaken to save a life ⟨*heroic* surgery⟩
4 : of, relating to, or constituting drama written during the Restoration in heroic couplets and concerned with a conflict between love and honor
— **he·ro·ical·ly** \-i-k(ə-)lē\ *adverb*

²heroic *noun* (1596)
1 : a heroic verse or poem
2 *plural* **a :** flamboyantly heroic language or action **b :** heroic action or behavior **c :** determined effort especially in the face of difficulty

heroic couplet *noun* (1889)
: a rhyming couplet in iambic pentameter

he·roi·com·ic \hi-ˌrō-i-'kä-mik\ *or* **he·roi·com·i·cal** \-'kä-mi-kəl\ *adjective* [French *héroïcomique,* from *héroïque* heroic + *comique* comic] (1756)
: comic by being ludicrously noble, bold, or elevated

heroic poem *noun* (1693)
: an epic or a poem in epic style

heroic stanza *noun* (circa 1922)
: a rhymed quatrain in heroic verse with a rhyme scheme of *abab* — called also *heroic quatrain*

heroic verse *noun* (1586)
1 : dactylic hexameter especially of epic verse of classical times — called also *heroic meter*
2 : the iambic pentameter used especially in English epic poetry during the 17th and 18th centuries — called also *heroic line, heroic meter*

her·o·in \'her-ə-wən\ *noun* [from *Heroin,* a trademark] (1898)
: a strongly physiologically addictive narcotic $C_{21}H_{23}NO_5$ that is made by acetylation of but is more potent than morphine and that is prohibited for medical use in the U.S. but is used illicitly for its euphoric effects
— **her·o·in·ism** \-wə-ˌni-zəm\ *noun*

◇ WORD HISTORY
hermetic Thoth was an ancient Egyptian god of wisdom whose cult in the Nile valley endured for millennia. The Greeks who settled in Egypt after its conquest by Alexander the Great and created a blend of Egyptian and Hellenistic culture identified their god Hermes with Thoth, who was then renamed in Greek *Hermēs Trismegistos* "thrice-greatest Hermes," a rough translation of one of Thoth's Egyptian titles. Thoth-Hermes was the reputed author of a set of treatises, the *Hermetica,* containing revelation on mystical subjects as well as astrological and alchemical speculation, nearly all of which was Greek rather than Egyptian in origin. In medieval alchemical tradition Hermes Trismegistus, as his name was latinized, was the inventor of a seal that kept vessels airtight. From this tradition come the most common meanings of *hermetic* in English, "airtight" and by extension "impervious to external influence."

her·o·ine \'her-ə-wən, 'hir-\ *noun* [Latin *heroina,* from Greek *hērōinē,* feminine of *hērōs*] (1609)
1 a : a mythological or legendary woman having the qualities of a hero **b :** a woman admired and emulated for her achievements and qualities
2 a : the principal female character in a literary or dramatic work **b :** the central female figure in an event or period

her·o·ism \'her-ə-,wi-zəm *also* 'hir-\ *noun* (1717)
1 : heroic conduct especially as exhibited in fulfilling a high purpose or attaining a noble end
2 : the qualities of a hero

he·ro·ize \'hē-(,)rō-,īz, 'hir-(,)ō-; 'her-ə-,wīz\ *transitive verb* **-ized; -iz·ing** (1738)
: to make heroic

her·on \'her-ən\ *noun, plural* **herons** *also* **heron** [Middle English *heiroun,* from Middle French *hairon,* of Germanic origin; akin to Old High German *heigaro* heron] (14th century)
: any of various long-necked wading birds (family Ardeidae) with a long tapering bill, large wings, and soft plumage

her·on·ry \-ən-rē\ *noun, plural* **-ries** (1616)
: a heron rookery

hero–worship *transitive verb* (1884)
: to feel or express hero worship for
— hero–worshiper *noun*

hero worship *noun* (1774)
1 : veneration of a hero
2 : foolish or excessive adulation for an individual

her·pes \'hər-(,)pēz\ *noun* [Latin, from Greek *herpēs,* from *herpein* to creep — more at SERPENT] (14th century)
: any of several inflammatory virus diseases of the skin characterized by clusters of vesicles; *especially* **:** HERPES SIMPLEX
— her·pet·ic \(,)hər-'pe-tik\ *adjective*

her·pes sim·plex \-'sim-,pleks\ *noun* [New Latin, literally, simple herpes] (1907)
: either of two virus diseases marked in one case by groups of watery blisters on the skin or mucous membranes (as of the mouth and lips) above the waist and in the other by such blisters on the genitals

her·pes·vi·rus \-'vī-rəs\ *noun* (1925)
: any of a group of DNA-containing viruses that replicate in cell nuclei and produce herpes

herpes zos·ter \-'zäs-tər\ *noun* [New Latin, literally, girdle herpes] (1807)
: an acute viral inflammation of the sensory ganglia of spinal and cranial nerves associated with a vesicular eruption and neuralgic pains and caused by reactivation of the poxvirus causing chicken pox — called also *shingles*

her·pe·tol·o·gy \,hər-pə-'tä-lə-jē\ *noun* [Greek *herpeton* quadruped, reptile, from neuter of *herpetos* crawling, from *herpein*] (1824)
: a branch of zoology dealing with reptiles and amphibians
— her·pe·to·log·i·cal \-tə-'lä-ji-kəl\ *adjective*
— her·pe·tol·o·gist \,hər-pə-'tä-lə-jist\ *noun*

Herr \(,)her\ *noun, plural* **Her·ren** \,her-ən, (,)hern\ [German] (1653)
— used among German-speaking people as a title equivalent to *Mr.*

her·ren·volk \'her-ən-,fōk, -,fȯlk\ *noun, often capitalized* [German] (1940)
: MASTER RACE

her·ring \'her-iŋ\ *noun, plural* **herring** *or* **herrings** [Middle English *hering,* from Old English *hǣring;* akin to Old High German *hārinc* herring] (before 12th century)
1 : either of two clupeid food fishes (genus *Clupeus*): **a :** one (*C. harengus*) that is abundant in the temperate and colder parts of the North Atlantic and that in the adult state is preserved by smoking or salting and in the young state is extensively canned and sold as sardines **b :** one (*C. pallasi* synonym *C. h. pallasi*) of the North Pacific harvested especially for its roe
2 : CLUPEID

¹her·ring·bone \'her-iŋ-,bōn\ *noun, often attributive* (1659)
1 : a pattern made up of rows of parallel lines which in any two adjacent rows slope in opposite directions
2 a : a twilled fabric with a herringbone pattern; *also* **:** a suit made of this fabric **b :** a herringbone arrangement (as of materials or parts)
3 : a method in skiing of ascending a slope by herringboning

herringbone 1

²herringbone (1787)
transitive verb
1 : to produce a herringbone pattern on
2 : to arrange in a herringbone pattern
intransitive verb
1 : to produce a herringbone pattern
2 : to ascend a slope by toeing out on skis and placing the weight on the inner side

herring gull *noun* (1857)
: a common large gull (*Larus argentatus*) of the northern hemisphere that as an adult is largely white with a gray mantle, dark wing tips, pink feet, and yellow bill

hers \'hərz\ *pronoun, singular or plural in construction*
: that which belongs to her — used without a following noun as a pronoun equivalent in meaning to the adjective *her*

her·self \(h)ər-'self, *Southern also* -'sef\ *pronoun* (before 12th century)
1 : that identical female one — compare ¹SHE; used reflexively, for emphasis, in absolute constructions, or in place of *her* especially when joined to another object ⟨she considers *herself* lucky⟩ ⟨she *herself* did it⟩ ⟨*herself* an orphan, she understood the situation⟩ ⟨accepted the award for her colleagues and *herself*⟩
2 : her normal, healthy, or sane condition or self
3 *chiefly Irish & Scottish* **:** a woman of consequence; *especially* **:** the mistress of the house

her·sto·ry \'hər-st(ə-)rē\ *noun, plural* **-ries** [blend of *her* and *history*] (1971)
: HISTORY; *specifically* **:** history considered or presented from a feminist viewpoint or with special attention to the experience of women

hertz \'hərts, 'herts\ *noun, plural* **hertz** [Heinrich R. *Hertz*] (circa 1928)
: a unit of frequency equal to one cycle per second — abbreviation *Hz*

he's \'hēz, ēz\ (1588)
: he is **:** he has

he/she \'hē-'shē; 'hē-ər-; 'hē-'slash-\ *pronoun* (1974)
: he or she — used in writing as a pronoun of common gender

Hesh·van \'kesh-vən\ *noun* [Hebrew *Ḥeshwān*] (circa 1769)
: the 2d month of the civil year or the 8th month of the ecclesiastical year in the Jewish calendar — see MONTH table

hes·i·tance \'he-zə-tən(t)s\ *noun* (1601)
: HESITANCY

hes·i·tan·cy \-tən(t)-sē\ *noun, plural* **-cies** (1617)
1 : the quality or state of being hesitant: as **a :** INDECISION **b :** RELUCTANCE ⟨took that drastic step only with the greatest *hesitancy*⟩
2 : HESITATION 1

hes·i·tant \'he-zə-tənt\ *adjective* (1647)
: tending to hesitate
synonym see DISINCLINED
— hes·i·tant·ly *adverb*

hes·i·tate \'he-zə-,tāt\ *verb* **-tat·ed; -tat·ing** [Latin *haesitatus,* past participle of *haesitare* to stick fast, hesitate, frequentative of *haerēre* to stick] (circa 1623)
intransitive verb
1 : to hold back in doubt or indecision
2 : to delay momentarily **:** PAUSE
3 : STAMMER
transitive verb
: to hold back from in doubt or uncertainty ⟨wouldn't *hesitate* to commit herself⟩ ☆
— hes·i·tat·er *noun*
— hes·i·tat·ing·ly \-,tā-tiŋ-lē\ *adverb*

hes·i·ta·tion \,he-zə-'tā-shən\ *noun* (14th century)
1 : an act or instance of hesitating
2 : a pausing or faltering in speech

Hes·pe·ri·an \he-'spir-ē-ən\ *adjective* [Latin *Hesperia,* the west, from Greek, from feminine of *hesperios* of the evening, western, from *hesperos* evening — more at WEST] (15th century)
: WESTERN, OCCIDENTAL

Hes·per·i·des \he-'sper-ə-,dēz\ *noun plural* [Latin, from Greek]
1 : the nymphs in classical mythology who guard with the aid of a dragon a garden in which golden apples grow
2 : a legendary garden at the western extremity of the world producing golden apples

hes·per·i·din \he-'sper-ə-dᵊn\ *noun* [New Latin *hesperidium* orange, from Latin *Hesperides*] (1838)
: a crystalline glycoside $C_{28}H_{34}O_{15}$ found in most citrus fruits and especially in orange peel

hes·per·id·i·um \,hes-pə-'ri-dē-əm\ *noun, plural* **-id·ia** \-dē-ə\ [New Latin] (circa 1866)
: a berry (as an orange or lime) having a leathery rind

Hes·per·us \'hes-p(ə-)rəs\ *noun* [Middle English, from Latin, from Greek *Hesperos*] (14th century)
: EVENING STAR 1

hes·sian \'he-shən\ *noun* (1729)
1 *capitalized* **a :** a native of Hesse **b :** a German mercenary serving in the British forces during the American Revolution; *broadly* **:** a mercenary soldier
2 *chiefly British* **:** BURLAP

Hessian boot *noun* (1809)
: a high boot that extends to just below the knee and is commonly ornamented with a tassel and that was introduced into England by the Hessians early in the 19th century

Hessian fly *noun* (1786)
: a small dipteran fly (*Mayetiola destructor*) that is destructive to wheat in the U.S.

hess·ite \'he-,sīt\ *noun* [German *Hessit,* from Henry *Hess* (died 1850) Swiss chemist] (1849)
: a mineral consisting of a lead-gray sectile silver telluride

hes·so·nite \'he-sᵊn-,īt\ *variant of* ESSONITE

hest \'hest\ *noun* [Middle English *hest, hes,* from Old English *hǣs;* akin to Old English *hātan* to command — more at HIGHT] (before 12th century)
archaic **:** COMMAND, PRECEPT

Hes·tia \'hes-tē-ə; 'hes-chə, 'hesh-\ *noun* [Greek]
: the Greek goddess of the hearth and chief goddess of domestic activity — compare VESTA

☆ SYNONYMS
Hesitate, waver, vacillate, falter mean to show irresolution or uncertainty. HESITATE implies a pause before deciding or acting or choosing ⟨*hesitated* before answering the question⟩. WAVER implies hesitation after seeming to decide and so connotes weakness or a retreat ⟨*wavered* in his support of the rebels⟩. VACILLATE implies prolonged hesitation from inability to reach a firm decision ⟨*vacillated* until events were out of control⟩. FALTER implies a wavering or stumbling and often connotes nervousness, lack of courage, or outright fear ⟨never once *faltered* during her testimony⟩.

he·tae·ra \hi-ˈtir-ə\ *or* **he·tai·ra** \-ˈtī-rə\ *noun, plural* **he·tae·rae** \-ˈtir-(ˌ)ē\ *or* **he·taeras** *or* **hetairas** *or* **he·tai·rai** \-ˈtī-ˌrī\ [Greek *hetaira,* literally, companion, feminine of *hetairos*] (1820)
1 : one of a class of highly cultivated courtesans in ancient Greece
2 : DEMIMONDAINE

heter- *or* **hetero-** *combining form* [Middle French or Late Latin; Middle French, from Late Latin, from Greek, from *heteros;* akin to Greek *heis* one — more at SAME]
1 : other than usual **:** other **:** different ⟨*hetero*phyllous⟩
2 : containing atoms of different kinds ⟨*het*erocyclic⟩

het·ero \ˈhe-tə-ˌrō\ *noun, plural* **-er·os** (1933)
: HETEROSEXUAL
— hetero *adjective*

het·ero·at·om \ˈhe-tə-rō-ˌa-təm\ *noun* (1900)
: an atom other than carbon in the ring of a heterocyclic compound

het·ero·aux·in \ˌhe-tə-rō-ˈȯk-sən\ *noun* (1935)
: INDOLEACETIC ACID

het·ero·cer·cal \-ˈsər-kəl\ *adjective* (1838)
1 *of a fish tail fin* **:** having the upper lobe larger than the lower with the vertebral column extending into the upper lobe
2 : having or relating to a heterocercal tail fin

het·ero·chro·ma·tin \-ˈkrō-mə-tən\ *noun* [German] (1932)
: densely staining chromatin that appears as nodules in or along chromosomes and contains relatively few genes
— het·ero·chro·mat·ic \-krə-ˈma-tik\ *adjective*

¹het·ero·clite \ˈhe-tə-rə-ˌklīt\ *noun* (1580)
1 : a word irregular in inflection; *especially* **:** a noun irregular in declension
2 : one that deviates from common rules or forms

²heteroclite *adjective* [Middle French or Late Latin; Middle French, from Late Latin *heteroclitus,* from Greek *heteroklitos,* from *heter-* + *klinein* to lean, inflect — more at LEAN] (1598)
: deviating from common forms or rules

het·ero·cy·clic \ˌhe-tə-rō-ˈsī-klik, -ˈsi-\ *adjective* [International Scientific Vocabulary] (1899)
: relating to, characterized by, or being a ring composed of atoms of more than one kind
— het·ero·cy·cle \ˈhe-tə-rō-ˌsī-kəl\ *noun*
— heterocyclic *noun*

het·ero·cyst \ˈhe-tə-rō-ˌsist\ *noun* (1872)
: a large transparent thick-walled cell that is found in the filaments of some blue-green algae and is the site of nitrogen fixation
— het·ero·cys·tous \ˌhe-tə-rō-ˈsis-təs\ *adjective*

het·ero·dox \ˈhe-tə-rə-ˌdäks, ˈhe-trə-\ *adjective* [Late Latin *heterodoxus,* from Greek *heterodoxos,* from *heter-* + *doxa* opinion — more at DOXOLOGY] (circa 1650)
1 : contrary to or different from an acknowledged standard, a traditional form, or an established religion **:** UNORTHODOX, UNCONVENTIONAL ⟨a *heterodox* book⟩ ⟨*heterodox* ideas⟩
2 : holding unorthodox opinions or doctrines

het·ero·doxy \-ˌdäk-sē\ *noun, plural* **-dox·ies** (1659)
1 : the quality or state of being heterodox
2 : a heterodox opinion or doctrine

het·ero·du·plex \ˌhe-tə-rō-ˈdü-ˌpleks, -ˈdyü-\ *noun* (1962)
: a nucleic-acid molecule composed of two chains with each derived from a different parent molecule
— heteroduplex *adjective*

¹het·ero·dyne \ˈhe-tə-rə-ˌdīn, ˈhe-trə-\ *adjective* [*heter-* + *-dyne,* modification of Greek *dynamis* power — more at DYNAMIC] (1908)
: of or relating to the production of an electrical beat between two radio frequencies of which one usually is that of a received signal-carrying current and the other that of an uninterrupted current introduced into the apparatus; *also* **:** of or relating to the production of a beat between two optical frequencies

²heterodyne *transitive verb* **-dyned; -dyn·ing** (1923)
: to combine (as a radio frequency) with a different frequency so that a beat is produced

het·er·oe·cious \ˌhe-tə-ˈrē-shəs\ *adjective* [*heter-* + Greek *oikia* house — more at VICINITY] (1882)
: passing through the different stages in the life cycle on alternate and often unrelated hosts ⟨*heteroecious* insects⟩
— het·er·oe·cism \-ˈrē-ˌsi-zəm\ *noun*

het·ero·ga·mete \ˌhe-tə-rō-ˈga-ˌmēt *also* -gə-ˈmēt\ *noun* [International Scientific Vocabulary] (1897)
: either of a pair of gametes that differ in form, size, or behavior and occur typically as large nonmotile oogametes and small motile sperms

het·ero·ga·met·ic \-gə-ˈme-tik\ *adjective* (1910)
: forming two kinds of gametes of which one produces male offspring and the other female offspring ⟨the human male is *heterogametic*⟩
— het·ero·gam·e·ty \-ˈgam-ə-tē\ *noun*

het·er·og·a·mous \ˌhe-tə-ˈrä-gə-məs\ *adjective* (1839)
: having or characterized by fusion of unlike gametes — compare ANISOGAMOUS, ISOGAMOUS

het·er·og·a·my \-mē\ *noun* (circa 1894)
1 : sexual reproduction involving fusion of unlike gametes often differing in size, structure, and physiology
2 : the condition of reproducing by heterogamy

het·ero·ge·ne·ity \ˌhe-tə-rō-jə-ˈnē-ə-tē, ˌhe-trō-\ *noun* (1641)
: the quality or state of being heterogeneous

het·ero·ge·neous \ˌhe-tə-rə-ˈjē-nē-əs, ˌhe-trə-, -nyəs\ *adjective* [Medieval Latin *heterogeneus,* from Greek *heterogenēs,* from *heter-* + *genos* kind — more at KIN] (1630)
: consisting of dissimilar or diverse ingredients or constituents **:** MIXED
— het·ero·ge·neous·ly *adverb*
— het·ero·ge·neous·ness *noun*

het·er·og·e·nous \ˌhe-tə-ˈrä-jə-nəs\ *adjective* (1695)
: HETEROGENEOUS

het·er·og·e·ny \-nē\ *noun* (1838)
: a heterogenous collection or group

het·er·og·o·ny \ˌhe-tə-ˈrä-gə-nē\ *noun* (circa 1887)
1 : ALTERNATION OF GENERATIONS; *especially* **:** alternation of a dioecious with a parthenogenetic generation
2 : ALLOMETRY
— het·ero·gon·ic \ˌhe-tə-rə-ˈgä-nik\ *adjective*

het·ero·graft \ˈhe-tə-rō-ˌgraft\ *noun* (1923)
: a graft of tissue taken from a donor of one species and grafted into a recipient of another species — called also *xenograft*; compare HOMOGRAFT

het·ero·kary·on \ˌhe-tə-rō-ˈkar-ē-ˌän, -ən\ *noun* [New Latin, from *heter-* + *karyon,* caryon nucleus, from Greek *karyon* nut, kernel] (1941)
: a cell (as in the mycelium of a fungus) that contains two or more genetically unlike nuclei

het·ero·kary·o·sis \ˌhe-tə-rō-ˌkar-ē-ˈō-səs\ *noun* [New Latin] (1916)
: the condition of having cells that are heterokaryons
— het·ero·kary·ot·ic \-ē-ˈä-tik\ *adjective*

het·er·ol·o·gous \-ˈrä-lə-gəs\ *adjective* [*heter-* + *-logous* (as in *homologous*)] (1893)
: derived from a different species ⟨*heterologous* DNAs⟩ ⟨*heterologous* transplants⟩
— het·er·ol·o·gous·ly *adverb*

het·ero·ly·sis \ˌhe-tə-ˈrä-lə-səs, -ə-rə-ˈlī-səs\ *noun* [New Latin] (1938)
: decomposition of a compound into two oppositely charged particles or ions
— het·ero·lyt·ic \-ə-rə-ˈli-tik\ *adjective*

het·ero·mor·phic \ˌhe-tə-rə-ˈmȯr-fik\ *adjective* [International Scientific Vocabulary] (circa 1859)
1 : deviating from the usual form
2 : exhibiting diversity of form or forms ⟨*heteromorphic* pairs of chromosomes⟩
— het·ero·mor·phism \-ˌfi-zəm\ *noun*

het·er·on·o·mous \ˌhe-tə-ˈrä-nə-məs\ *adjective* (circa 1871)
: subject to external controls and impositions

het·er·on·o·my \-mē\ *noun* [*heter-* + *-nomy* (as in *autonomy*)] (1798)
: subjection to something else; *especially* **:** a lack of moral freedom or self-determination

het·ero·nym \ˈhe-tə-rə-ˌnim\ *noun* (circa 1889)
: one of two or more homographs (as a *bass* voice and *bass,* a fish) that differ in pronunciation and meaning

het·ero·phile \ˈhe-tə-rə-ˌfīl\ *or* **het·ero·phil** \-ˌfil\ *adjective* (1920)
: relating to or being any of a group of antigens in organisms of different species that induce the formation of antibodies that will cross-react with the other antigens of the group; *also* **:** being or relating to any of these antibodies

het·er·oph·o·ny \ˌhe-tə-ˈrä-fə-nē\ *noun, plural* **-nies** [Greek *heterophōnia* diversity of note, from *heter-* + *-phōnia* -phony] (1919)
: independent variation on a single melody by two or more voices

het·ero·phyl·lous \ˌhe-tə-rō-ˈfi-ləs\ *adjective* (circa 1828)
: having the foliage leaves of more than one form on the same plant or stem
— het·ero·phyl·ly \ˈhe-tə-rō-ˌfi-lē\ *noun*

het·ero·ploid \ˈhe-tə-rə-ˌplȯid\ *adjective* [International Scientific Vocabulary] (1926)
: having a chromosome number that deviates from and is not an integral multiple of the number characteristic of a given species
— heteroploid *noun*
— het·ero·ploi·dy \-ˌplȯi-dē\ *noun*

het·er·op·ter·ous \ˌhe-tə ˈräp-tə-rəs\ *adjective* [ultimately from Greek *heter-* + *pteron* wing — more at FEATHER] (1895)
: of or relating to an insect order or suborder (Heteroptera) comprising the true bugs

¹het·ero·sex·u·al \ˌhe-tə-rō-ˈsek-sh(ə-)wəl, -ˈsek-shəl\ *adjective* [International Scientific Vocabulary] (1892)
1 a : of, relating to, or characterized by a tendency to direct sexual desire toward the opposite sex **b :** of, relating to, or involving sexual intercourse between individuals of opposite sex
2 : of or relating to different sexes
— het·ero·sex·u·al·i·ty \-ˌsek-shə-ˈwa-lə-tē\ *noun*
— het·ero·sex·u·al·ly \-ˈsek-sh(ə-)wə-lē, -ˈsek-shə-lē\ *adverb*

²heterosexual *noun* (1920)
: a heterosexual person

het·er·o·sis \ˌhe-tə-ˈrō-səs\ *noun* [New Latin] (1911)
: the marked vigor or capacity for growth often exhibited by crossbred animals or plants — called also *hybrid vigor*
— het·er·ot·ic \-ˈrä-tik\ *adjective*

het·ero·spo·ry \ˈhe-tə-rə-ˌspōr-ē, -ˌspȯr-; ˌhe-tə-ˈräs-pə-rē\ *noun* (1898)
: the production of microspores and megaspores (as in seed plants and some ferns)
— het·ero·spo·rous \ˌhe-tə-rə-ˈspōr-əs, -ˈspȯr-; -ˈräs-pə-rəs\ *adjective*

het·ero·thal·lic \ˌhe-tə-rō-ˈtha-lik\ *adjective* [*heter-* + *thallus* + *-ic*] (1904)
1 : having two or more morphologically similar haploid phases or types of which individuals from the same type are mutually sterile but individuals from different types are cross-fertile ⟨*heterothallic* fungi⟩ ⟨*heterothallic* spores⟩
2 : DIOECIOUS
— het·ero·thal·lism \-ˈtha-ˌli-zəm\ *noun*

het·ero·top·ic \-rə-ˈtä-pik\ *adjective* [*heter-* + Greek *topos* place] (1878)
: occurring in an abnormal place ⟨*heterotopic* bone formation⟩ ⟨*heterotopic* liver transplantation⟩

het·ero·troph \ˈhe-tə-rə-ˌtrōf, -ˌträf\ *noun* (circa 1900)
: a heterotrophic individual

het·ero·tro·phic \ˌhe-tə-rə-ˈtrō-fik\ *adjective* (1893)
: requiring complex organic compounds of nitrogen and carbon for metabolic synthesis
— het·ero·tro·phi·cal·ly \-fi-k(ə-)lē\ *adverb*
— het·ero·tro·phy \ˌhe-tə-ˈrä-trə-fē, ˈhe-tə-rə-ˌtrō-\ *noun*

het·ero·typ·ic \ˌhe-tə-rō-ˈti-pik\ *adjective* (1889)
: different in kind, arrangement, or form

het·ero·zy·go·sis \-(ˌ)zī-ˈgō-səs\ *noun* [New Latin] (1902)
: HETEROZYGOSITY

het·ero·zy·gos·i·ty \-ˈgä-sə-tē\ *noun* (1912)
: the state of being heterozygous

het·ero·zy·gote \-ˈzī-ˌgōt\ *noun* (1902)
: a heterozygous individual

het·ero·zy·gous \-gəs\ *adjective* (1902)
: having the two alleles at corresponding loci on homologous chromosomes different for one or more loci

heth \ˈkāt, ˈkāth, ˈket, ˈketh\ *noun* [Hebrew *ḥēth*] (1823)
: the 8th letter of the Hebrew alphabet — see ALPHABET table

het·man \ˈhet-mən\ *noun, plural* **hetmans** [Ukrainian *het'man*] (1710)
: a cossack leader

het up \ˈhet-ˈəp\ *adjective* [*het,* dialect past of *heat*] (1909)
: highly excited **:** UPSET

heu·land·ite \ˈhyü-lən-ˌdīt\ *noun* [Henry *Heuland,* 19th century English mineral collector] (1822)
: a zeolite consisting of a hydrous aluminosilicate of sodium and calcium

[1]heu·ris·tic \hyu̇-ˈris-tik\ *adjective* [German *heuristisch,* from New Latin *heuristicus,* from Greek *heuriskein* to discover; akin to Old Irish *fo-fúair* he found] (1821)
: involving or serving as an aid to learning, discovery, or problem-solving by experimental and especially trial-and-error methods ⟨*heuristic* techniques⟩ ⟨a *heuristic* assumption⟩; *also* **:** of or relating to exploratory problem-solving techniques that utilize self-educating techniques (as the evaluation of feedback) to improve performance ⟨a *heuristic* computer program⟩
— heu·ris·ti·cal·ly \-ti-k(ə-)lē\ *adverb*

[2]heuristic *noun* (1860)
1 : the study or practice of heuristic procedure
2 : heuristic argument
3 : a heuristic method or procedure

hew \ˈhyü\ *verb* **hewed; hewed** *or* **hewn** \ˈhyün\; **hew·ing** [Middle English, from Old English *hēawan;* akin to Old High German *houwan* to hew, Lithuanian *kauti* to forge, Latin *cudere* to beat] (before 12th century)
transitive verb
1 : to cut with blows of a heavy cutting instrument
2 : to fell by blows of an ax ⟨*hew* a tree⟩
3 : to give form or shape to with or as if with heavy cutting blows ⟨*hewed* their farms from the wilderness —J. T. Shotwell⟩
intransitive verb
1 : to make cutting blows
2 : CONFORM, ADHERE — often used in the phrase *hew to the line* ⟨no pressure . . . on newspapers to *hew* to the official line —*N.Y. Times Magazine*⟩
— hew·er *noun*

[1]hex \ˈheks\ *verb* [Pennsylvania German *hexe,* from German *hexen,* from *Hexe* witch, from Old High German *hagzissa;* akin to Middle English *hagge* hag] (1830)
intransitive verb
: to practice witchcraft
transitive verb
1 : to put a hex on
2 : to affect as if by an evil spell **:** JINX ⟨giving in to an unscientific fear of *hexing* the whole project —Daniel Lang⟩
— hex·er *noun*

[2]hex *noun* (1856)
1 : a person who practices witchcraft **:** WITCH
2 : SPELL, JINX

[3]hex *adjective* (1924)
: HEXAGONAL ⟨a bolt with a *hex* head⟩

[4]hex *adjective or noun* (1970)
: HEXADECIMAL

hexa- *or* **hex-** *combining form* [Greek, from *hex* six — more at SIX]
1 : six ⟨*hexa*ploid⟩
2 : containing six atoms, groups, or equivalents ⟨*hex*ane⟩

hexa·chlo·ro·eth·ane \ˌhek-sə-ˌklōr-ō-ˈe-ˌthān, -ˌklȯr-\ *or* **hexa·chlor·eth·ane** \-ˌklōr-ˈe-, -ˌklȯr-\ *noun* [International Scientific Vocabulary] (1898)
: a toxic crystalline compound C_2Cl_6 used especially in smoke bombs and in the control of liver flukes in ruminants

hexa·chlo·ro·phene \-ˈklōr-ə-ˌfēn, -ˈklȯr-\ *noun* [*hexa-* + *chlor-* + *phenol*] (1948)
: a powdered phenolic bacteria-inhibiting agent $C_{13}Cl_6H_6O_2$

hexa·chord \ˈhek-sə-ˌkȯrd\ *noun* [*hexa-* + Greek *chordē* string — more at YARN] (1730)
: a diatonic series of six tones having a semitone between the third and fourth tones

hexa·dec·i·mal \ˌhek-sə-ˈdes-məl, -ˈde-sə-\ *adjective* (1954)
: of, relating to, or being a number system with a base of 16
— hexadecimal *noun*

hexa·gon \ˈhek-sə-ˌgän\ *noun* [Greek *hexagōnon,* neuter of *hexagōnos* hexagonal, from *hexa-* + *gōnia* angle — more at -GON] (1570)
: a polygon of six angles and six sides

hex·ag·o·nal \hek-ˈsa-gə-nᵊl\ *adjective* (1571)
1 : having six angles and six sides
2 : having a hexagon as section or base
3 : relating to or being a crystal system characterized by three equal lateral axes intersecting at angles of 60 degrees and a vertical axis of variable length at right angles
— hex·ag·o·nal·ly \-nᵊl-ē\ *adverb*

hexa·gram \ˈhek-sə-ˌgram\ *noun* [International Scientific Vocabulary] (1871)
: a plane figure that has the shape of a 6-pointed star, that consists of two intersecting congruent equilateral triangles having the same point as center and their sides parallel, and that can be formed by constructing external equilateral triangles on the sides of a regular hexagon — compare SOLOMON'S SEAL 2

hexagram

hexa·he·dron \ˌhek-sə-ˈhē-drən\ *noun, plural* **-drons** *also* **-dra** \-drə\ [Late Latin, from Greek *hexaedron,* from neuter of *hexaedros* of six surfaces, from *hexa-* + *hedra* seat — more at SIT] (1571)
: a polyhedron of six faces (as a cube)

hexa·hy·drate \-ˈhī-ˌdrāt\ *noun* (1908)
: a chemical compound with six molecules of water

hex·am·e·ter \hek-ˈsa-mə-tər\ *noun* [Latin, from Greek *hexametron,* from neuter of *hexametros* having six measures, from *hexa-* + *metron* measure — more at MEASURE] (1546)
: a line of verse consisting of six metrical feet

hexa·me·tho·ni·um \ˌhek-sə-mə-ˈthō-nē-əm\ *noun* [*hexa-* + *meth-* + *-onium*] (1949)
: either of two compounds $C_{12}H_{30}Br_2N_2$ or $C_{12}H_{30}Cl_2N_2$ used as ganglionic blocking agents in the treatment of hypertension

hexa·meth·y·lene·tet·ra·mine \ˌhek-sə-ˈme-thə-ˌlēn-ˈte-trə-ˌmēn\ *noun* [International Scientific Vocabulary *hexa-* + *methylene* + *tetra-* + *amine*] (1888)
: a crystalline compound $C_6H_{12}N_4$ used especially as an accelerator in vulcanizing rubber and as a urinary antiseptic — compare METHENAMINE

hex·ane \ˈhek-ˌsān\ *noun* [International Scientific Vocabulary] (1877)
: any of several isomeric volatile liquid alkanes C_6H_{14} found in petroleum

hex·a·no·ic acid \ˌhek-sə-ˈnō-ik-\ *noun* [International Scientific Vocabulary *hexane* + *-oic*] (1926)
: CAPROIC ACID

hexa·ploid \ˈhek-sə-ˌplȯid\ *adjective* [International Scientific Vocabulary] (1912)
: having or being six times the monoploid chromosome number
— hexaploid *noun*
— hexa·ploi·dy \-ˌplȯi-dē\ *noun*

[1]hexa·pod \ˈhek-sə-ˌpäd\ *noun* [Greek *hexapod-, hexapous* having six feet, from *hexa-* + *pod-, pous* foot — more at FOOT] (1668)
: INSECT 1b

[2]hexapod *adjective* (circa 1847)
1 : six-footed
2 : of or relating to insects

Hexa·teuch \ˈhek-sə-ˌtük, -ˌtyük\ *noun* [*hexa-* + Greek *teuchos* book — more at PENTATEUCH] (1878)
: the first six books of the Bible

hex·e·rei \ˌhek-sə-ˈrī\ *noun* [Pennsylvania German, from German, from *Hexe* witch] (1898)
: WITCHCRAFT

hexo·bar·bi·tal \ˌhek-sə-ˈbär-bə-ˌtȯl\ *noun* [*hexo-* (from *hexa-*) + *barbital*] (1941)
: a barbiturate $C_{12}H_{16}N_2O_3$ used as a sedative and hypnotic and in the form of its soluble sodium salt as an intravenous anesthetic of short duration

hexo·ki·nase \ˌhek-sə-ˈkī-ˌnās, -ˌnāz\ *noun* [*hexose* + *kinase*] (1930)
: any of a group of enzymes that accelerate the phosphorylation of hexoses (as in the formation of glucose-6-phosphate from glucose and ATP) in carbohydrate metabolism

hex·os·a·min·i·dase \ˌhek-ˌsä-sə-ˈmi-nə-ˌdās, -ˌdāz\ *noun* [*hexose* + *amino* + *-ide* + *-ase*] (1969)
: either of two hydrolytic enzymes that catalyze the splitting off of a hexose from a ganglioside and are deficient in some metabolic diseases (as a variant of Tay-Sachs disease)

hex·o·san \ˈhek-sə-ˌsan\ *noun* (1894)
: a polysaccharide yielding only hexoses on hydrolysis

hex·ose \ˈhek-ˌsōs, -ˌsōz\ *noun* [International Scientific Vocabulary] (1892)
: a monosaccharide (as glucose) containing six carbon atoms in the molecule

hex·yl \ˈhek-səl\ *noun* [International Scientific Vocabulary] (1869)
: an alkyl radical C_6H_{13}— derived from a hexane

hex·yl·res·or·cin·ol \ˌhek-səl-rə-ˈzȯr-sᵊn-ˌȯl, -ˌōl\ *noun* (1924)
: a crystalline phenol $C_{12}H_{18}O_2$ used as an antiseptic and anthelmintic

hey \ˈhā\ *interjection* [Middle English] (13th century)
— used especially to call attention or to express interrogation, surprise, or exultation

[1]hey·day \ˈhā-ˌdā\ *interjection* [irregular from *hey*] (1672)
archaic — used to express elation or wonder

[2]heyday *noun* (1590)

1 *archaic* **:** high spirits
2 **:** the period of one's greatest strength, vigor, or prosperity

hey presto \(ˈ)hā-ˈpres-(ˌ)tō\ *interjection* (1731)
British **:** suddenly as if by magic

Hez·e·ki·ah \ˌhe-zə-ˈkī-ə\ *noun* [Hebrew *Ḥizqīyāh*]
: a king of Judah under whom the kingdom underwent a ruinous Assyrian invasion at the end of the 8th century B.C.

hi \ˈhī(-ē)\ *interjection* [Middle English *hy*] (15th century)
— used especially as a greeting

hi·a·tal \hī-ˈā-tᵊl\ *adjective* (1909)
: of, relating to, or involving a hiatus

hiatal hernia *noun* (circa 1944)
: a hernia in which an anatomical part (as the stomach) protrudes through the esophageal hiatus of the diaphragm — called also *hiatus hernia*

hi·a·tus \hī-ˈā-təs\ *noun* [Latin, from *hiare* to yawn — more at YAWN] (1563)
1 a : a break in or as if in a material object **:** GAP ⟨the *hiatus* between the theory and the practice of the party —J. G. Colton⟩ **b :** a gap or passage in an anatomical part or organ
2 a : an interruption in time or continuity **:** BREAK **b :** the occurrence of two vowel sounds without pause or intervening consonantal sound

Hi·a·wa·tha \ˌhī-ə-ˈwȯ-thə, ˌhē-ə-, -ˈwä-\ *noun*
: the Indian hero of Longfellow's poem *The Song of Hiawatha*

hi·ba·chi \hi-ˈbä-chē\ *noun* [Japanese] (1863)
: a charcoal brazier

hi·ber·nac·u·lum \ˌhī-bər-ˈna-kyə-ləm\ *noun, plural* **-la** \-lə\ [New Latin, from Latin, winter residence, from *hibernare*] (1789)
: a shelter occupied during the winter by a dormant animal (as an insect or reptile)

hi·ber·nal \hī-ˈbər nᵊl\ *adjective* (1646)
: of, relating to, or occurring in winter

hi·ber·nate \ˈhī-bər-ˌnāt\ *intransitive verb* **-nat·ed; -nat·ing** [Latin *hibernatus*, past participle of *hibernare* to pass the winter, from *hibernus* of winter; akin to Latin *hiems* winter, Greek *cheimōn*] (circa 1802)
1 : to pass the winter in a torpid or resting state
2 : to be or become inactive or dormant
— hi·ber·na·tion \ˌhī-bər-ˈnā-shən\ *noun*
— hi·ber·na·tor \ˈhī-bər-ˌnā-tər\ *noun*

¹Hi·ber·ni·an \hī-ˈbər-nē-ən\ *adjective* [Latin *Hibernia* Ireland] (1632)
: of, relating to, or characteristic of Ireland or the Irish

²Hibernian *noun* (1709)
: a native or inhabitant of Ireland

Hi·ber·no- *combining form* [Latin *Hibernia*]
1 : Irish and ⟨*Hiberno*-British⟩
2 : Irish ⟨*Hiberno*-English⟩

Hiberno–English *noun* (1985)
: the English language spoken in Ireland

hi·bis·cus \hī-ˈbis-kəs, hə-\ *noun* [New Latin, from Latin, marshmallow] (1706)
: any of a large genus (*Hibiscus*) of herbs, shrubs, or small trees of the mallow family with large showy flowers and usually dentate leaves

¹hic·cup *also* **hic·cough** \ˈhi-(ˌ)kəp\ *noun* [imitative] (circa 1580)
1 : a spasmodic inhalation with closure of the glottis accompanied by a peculiar sound
2 : an attack of hiccuping — usually used in plural but singular or plural in construction

²hiccup *also* **hiccough** *intransitive verb* **hic·cuped** *also* **hic·cupped; hic·cup·ing** *also* **hic·cup·ping** (circa 1580)
: to make a hiccup; *also* **:** to be affected with hiccups

hic ja·cet \ˈhik-ˈjā-sət, ˈhēk-ˈyä-kət\ *noun* [Latin, literally, here lies] (1654)
: EPITAPH

¹hick \ˈhik\ *noun* [*Hick*, nickname for *Richard*] (circa 1690)
: an unsophisticated provincial person
— hick·ish \ˈhi-kish\ *adjective*

²hick *adjective* (1920)
: UNSOPHISTICATED, PROVINCIAL ⟨a *hick* town⟩

¹hick·ey \ˈhi-kē\ *noun, plural* **hickeys** [origin unknown] (1913)
: DEVICE, GADGET

²hickey *noun, plural* **hickeys** [origin unknown] (circa 1915)
1 a : PIMPLE **b :** a temporary red mark on the skin (as one produced by biting and sucking)
2 *plural also* **hick·ies :** a small imperfection in printing

hick·o·ry \ˈhi-k(ə-)rē\ *noun, plural* **-ries** [short for obsolete *pokahickory*, from Virginia Algonquian *pawcohiccora* food prepared from pounded nuts] (1670)
1 a : any of a genus (*Carya*) of North American hardwood trees of the walnut family that often have sweet edible nuts **b :** the usually tough wood of a hickory
2 : a switch or cane (as of hickory wood) used especially for punishing a child
— hickory *adjective*

hid \ˈhid\ *adjective* (12th century)
: HIDDEN

hi·dal·go \hi-ˈdal-(ˌ)gō, ē-ˈt͟häl-\ *noun, plural* **-gos** *often capitalized* [Spanish, from Old Spanish *fijo dalgo*, literally, son of something] (1594)
: a member of the lower nobility of Spain

Hi·dat·sa \hi-ˈdät-sə\ *noun, plural* **Hidatsa** *also* **Hidatsas** (1873)
1 : a member of an American Indian people of the Missouri River valley in North Dakota
2 : the Siouan language of the Hidatsa

hid·den \ˈhi-dᵊn\ *adjective* (13th century)
1 : being out of sight or not readily apparent **:** CONCEALED
2 : OBSCURE, UNEXPLAINED, UNDISCLOSED
— hid·den·ness \-nəs\ *noun*

hidden agenda *noun* (1971)
: an ulterior motive

hid·den·ite \ˈhi-dᵊn-ˌīt\ *noun* [William E. *Hidden* (died 1918) American mineralogist] (1881)
: a transparent yellow to green spodumene valued as a gem

hidden tax *noun* (1936)
1 : a tax that is ultimately paid by someone other than the person on whom it is levied
2 : an economic inequity that reduces one's real income or buying power

¹hide \ˈhīd\ *noun* [Middle English, from Old English *hīgid, hīd*] (before 12th century)
: any of various old English units of land area; *especially* **:** a unit of 120 acres

²hide *verb* **hid** \ˈhid\; **hid·den** \ˈhi-dᵊn\ *or* **hid; hid·ing** \ˈhī-diŋ\ [Middle English *hiden*, from Old English *hȳdan;* akin to Greek *keuthein* to conceal] (before 12th century)
transitive verb
1 a : to put out of sight **:** SECRETE **b :** to conceal for shelter or protection **:** SHIELD
2 : to keep secret
3 : to screen from or as if from view **:** OBSCURE
4 : to turn (the eyes or face) away in shame or anger
intransitive verb
1 : to remain out of sight — often used with *out*
2 : to seek protection or evade responsibility ☆
— hid·er \ˈhī-dər\ *noun*

³hide *noun* (14th century)
chiefly British **:** BLIND 2

⁴hide *noun* [Middle English, from Old English *hȳd;* akin to Old High German *hūt* hide, Latin *cutis* skin, Greek *kytos* hollow vessel] (before 12th century)
1 : the skin of an animal whether raw or dressed — used especially of large heavy skins
2 : the life or physical well-being of a person ⟨betrayed his friend to save his own *hide*⟩
— hide or hair *or* **hide nor hair :** a vestige or trace of someone or something ⟨a wife he hadn't seen *hide or hair* of in over 20 years —H. L. Davis⟩

⁵hide *transitive verb* **hid·ed; hid·ing** (circa 1825)
: to give a beating to **:** FLOG

hide–and–seek \ˌhī-dᵊn-ˈsēk\ *noun* (circa 1727)
: a children's game in which one player does not look while others hide and then goes to find them

hide·away \ˈhī-də-ˌwā\ *noun* (1926)
: RETREAT, HIDEOUT

hide·bound \ˈhīd-ˌbau̇nd\ *adjective* (1559)
1 *of a domestic animal* **:** having a dry skin lacking in pliancy and adhering closely to the underlying flesh
2 : having an inflexible or ultraconservative character

hid·eous \ˈhi-dē-əs\ *adjective* [alteration of Middle English *hidous*, from Middle French, from *hisde, hide* terror] (14th century)
1 : offensive to the senses and especially to sight **:** exceedingly ugly
2 : morally offensive **:** SHOCKING
— hid·eos·i·ty \ˌhi-dē-ˈä-sə-tē\ *noun*
— hid·eous·ly *adverb*
— hid·eous·ness *noun*

hide·out \ˈhī-ˌdau̇t\ *noun* (1904)
: a place of refuge, retreat, or concealment

hid·ey–hole *or* **hidy–hole** \ˈhī-dē-ˌhōl\ *noun* [alteration of earlier *hiding-hole*] (1817)
: HIDEAWAY

hie \ˈhī\ *verb* **hied; hy·ing** *or* **hie·ing** [Middle English, from Old English *hīgian* to strive, hasten] (12th century)
intransitive verb
: to go quickly **:** HASTEN
transitive verb
: to cause (oneself) to go quickly

hi·er·arch \ˈhī-(ə-)ˌrärk\ *noun* [Middle English *ierarchis*, plural, from Middle French or Medieval Latin; Middle French *hierarche*, from Medieval Latin *hierarcha*, from Greek *hierarchēs*, from *hieros* sacred + *-archēs* -arch] (15th century)
1 : a religious leader in a position of authority
2 : a person high in a hierarchy
— hi·er·ar·chal \ˌhī-(ə-)ˈrär-kəl\ *adjective*

hi·er·ar·chi·cal \ˌhī-(ə-)ˈrär-ki-kəl *also* hir-ˈär-\ *or* **hi·er·ar·chic** \-kik\ *adjective* (1561)
: of, relating to, or arranged in a hierarchy
— hi·er·ar·chi·cal·ly \-ki-k(ə-)lē\ *adverb*

hi·er·ar·chize \ˈhī-(ə-)ˌrär-ˌkīz\ *transitive verb* **-chized; -chiz·ing** (1884)
: to arrange in a hierarchy

hi·er·ar·chy \ˈhī-(ə-)ˌrär-kē *also* ˈhi(-ə)r-ˌär-\ *noun, plural* **-chies** (14th century)
1 : a division of angels
2 a : a ruling body of clergy organized into orders or ranks each subordinate to the one above it; *especially* **:** the bishops of a province

☆ SYNONYMS
Hide, conceal, screen, secrete, bury mean to withhold or withdraw from sight. HIDE may or may not suggest intent ⟨*hide* in the closet⟩ ⟨a house *hidden* in the woods⟩. CONCEAL usually does imply intent and often specifically implies a refusal to divulge ⟨*concealed* the weapon⟩. SCREEN implies an interposing of something that prevents discovery ⟨a house *screened* by trees⟩. SECRETE suggests a depositing in a place unknown to others ⟨*secreted* the amulet inside his shirt⟩. BURY implies covering up so as to hide completely ⟨*buried* the treasure⟩.

or nation **b :** church government by a hierarchy
3 : a body of persons in authority
4 : the classification of a group of people according to ability or to economic, social, or professional standing; *also* **:** the group so classified
5 : a graded or ranked series ⟨Christian *hierarchy* of values⟩ ⟨a machine's *hierarchy* of responses⟩

hi·er·at·ic \ˌhī-(ə-)ˈra-tik\ *adjective* [Latin *hieraticus* sacerdotal, from Greek *hieratikos,* from *hierasthai* to perform priestly functions, from *hieros* sacred; probably akin to Sanskrit *iṣara* vigorous] (1669)
1 : constituting or belonging to a cursive form of ancient Egyptian writing simpler than the hieroglyphic
2 : SACERDOTAL
3 : highly stylized or formal ⟨*hieratic* poses⟩
— hi·er·at·i·cal·ly \-ti-k(ə-)lē\ *adverb*

hi·ero·dule \ˈhī-(ə-)rō-ˌdü(ə)l, hī-ˈer-ə-, -ˌdyü(ə)l\ *noun* [Late Latin *hierodulus,* from Greek *hierodoulos,* from *hieron* temple + *doulos* slave] (1835)
: a slave or prostitute in the service of a temple (as in ancient Greece)

hi·ero·glyph \ˈhī-(ə-)rə-ˌglif\ *noun* [French *hiéroglyphe,* from Middle French, back-formation from *hieroglyphique*] (1598)
1 : a character used in a system of hieroglyphic writing
2 : something that resembles a hieroglyph

¹hi·ero·glyph·ic \ˌhī-(ə-)rə-ˈgli-fik\ *also* **hi·ero·glyph·i·cal** \-fi-kəl\ *adjective* [Middle French *hieroglyphique,* from Late Latin *hieroglyphicus,* from Greek *hieroglyphikos,* from *hieros* + *glyphein* to carve — more at CLEAVE] (1585)
1 : written in, constituting, or belonging to a system of writing mainly in pictorial characters
2 : inscribed with hieroglyphic
3 : resembling hieroglyphic in difficulty of decipherment
— hi·ero·glyph·i·cal·ly \-fi-k(ə-)lē\ *adverb*

²hieroglyphic *noun* (1586)
1 : HIEROGLYPH
2 : a system of hieroglyphic writing; *specifically* **:** the picture script of the ancient Egyptian priesthood — often used in plural but singular or plural in construction

hieroglyphic 2

3 : something that resembles a hieroglyph especially in difficulty of decipherment

hi·ero·phant \ˈhī-(ə-)rə-ˌfant, hī-ˈer-ə-fənt\ *noun* [Late Latin *hierophanta,* from Greek *hierophantēs,* from *hieros* + *phainein* to show — more at FANCY] (1677)
1 : a priest in ancient Greece; *specifically* **:** the chief priest of the Eleusinian mysteries
2 a : EXPOSITOR **b :** ADVOCATE
— hi·ero·phan·tic \ˌhī-(ə-)rə-ˈfan-tik, (ˌ)hī-ˌer-ə-\ *adjective*

hi·fa·lu·tin *variant of* HIGHFALUTIN

hi–fi \ˈhī-ˈfī\ *noun* (1950)
1 : HIGH FIDELITY
2 : equipment for reproduction of sound with high fidelity

hig·gle \ˈhi-gəl\ *intransitive verb* **hig·gled; hig·gling** \-g(ə-)liŋ\ [probably alteration of *haggle*] (1633)
: HAGGLE
— hig·gler \-g(ə-)lər\ *noun*

hig·gle·dy–pig·gle·dy \ˌhi-gəl-dē-ˈpi-gəl-dē\ *adverb* [origin unknown] (circa 1598)
: in a confused, disordered, or random manner ⟨tiny hovels piled *higgledy-piggledy* against each other —Edward Behr⟩
— higgledy–piggledy *adjective*

¹high \ˈhī\ *adjective* [Middle English, from Old English *hēah;* akin to Old High German *hōh* high, Lithuanian *kaukaras* hill] (before 12th century)
1 a : having large extension upward **:** taller than average, usual, or expected ⟨a *high* wall⟩ **b :** having a specified elevation **:** TALL ⟨six feet *high*⟩ — often used in combinations ⟨sky-*high*⟩ ⟨waist-*high*⟩
2 a (1) **:** advanced toward the acme or culmination ⟨*high* summer⟩ (2) **:** advanced toward the most active or culminating period ⟨on the Riviera during *high* season⟩ (3) **:** constituting the late, most fully developed, or most creative stage or period ⟨*high* Gothic⟩ (4) **:** advanced in complexity, development, or elaboration ⟨the *higher* apes⟩ ⟨*higher* mathematics⟩ **b :** verging on lateness — usually used in the phrase *high time* **c :** long past **:** REMOTE ⟨*high* antiquity⟩
3 : elevated in pitch ⟨a *high* note⟩
4 : relatively far from the equator ⟨*high* latitude⟩
5 : rich in quality **:** LUXURIOUS ⟨*high* living⟩
6 : slightly tainted ⟨*high* game⟩; *also* **:** MALODOROUS ⟨smelled rather *high*⟩
7 : exalted in character **:** NOBLE ⟨*high* purposes⟩
8 : of greater degree, amount, cost, value, or content than average, usual, or expected ⟨*high* prices⟩ ⟨food *high* in iron⟩ ⟨the *high* bid⟩
9 : of relatively great importance: as **a :** foremost in rank, dignity, or standing ⟨*high* officials⟩ **b :** SERIOUS, GRAVE ⟨*high* crimes⟩ **c :** observed with the utmost solemnity ⟨*high* religious observances⟩ **d :** CRITICAL, CLIMACTIC ⟨the *high* point of the novel⟩ **e :** intellectually or artistically of the first order ⟨*high* culture⟩ **f :** marked by sublime, heroic, or stirring events or subject matter ⟨*high* tragedy⟩ ⟨*high* adventure⟩
10 : FORCIBLE, STRONG ⟨*high* winds⟩
11 : stressing matters of doctrine and ceremony; *specifically* **:** HIGH CHURCH
12 a : filled with or expressing great joy or excitement ⟨*high* spirits⟩ **b :** INTOXICATED; *also* **:** excited or stupefied by or as if by a drug
13 : articulated with some part of the tongue close to the palate ⟨a *high* vowel⟩ ☆
— high on : enthusiastically in favor or support of

²high *adverb* (before 12th century)
1 : at or to a high place, altitude, level, or degree ⟨climbed *higher*⟩ ⟨passions ran *high*⟩
2 : WELL, LUXURIOUSLY — often used in the phrases *high off the hog* and *high on the hog*

³high *noun* (13th century)
1 : an elevated place or region: as **a :** HILL, KNOLL **b :** the space overhead **:** SKY — usually used with *on* **c :** HEAVEN — usually used with *on*
2 : a region of high barometric pressure — called also *anticyclone*
3 a : a high point or level **:** HEIGHT ⟨sales reached a new *high*⟩ **b :** the transmission gear of a vehicle (as an automobile) giving the highest speed of travel
4 a : an excited, euphoric, or stupefied state produced by or as if by a drug **b :** a state of elation or high spirits

high altar *noun* (before 12th century)
: the principal altar in a church

high analysis *adjective* (1949)
of a fertilizer **:** containing more than 20 percent of total plant nutrients

high and dry *adjective* (1822)
1 : being out of reach of the current or tide or out of the water
2 : being in a helpless or abandoned position

high and low *adverb* (14th century)
: EVERYWHERE

high–and–mighty *adjective* (1654)
: characterized by arrogance **:** IMPERIOUS

¹high·ball \ˈhī-ˌbȯl\ *noun* (1897)
1 : a railroad signal for a train to proceed at full speed
2 : an iced drink containing liquor (as whiskey) and water or a carbonated beverage and served in a tall glass

²highball *intransitive verb* (1912)
: to go at full or high speed ⟨a *highballing* express train⟩

high beam *noun* (1939)
: the long-range focus of a vehicle headlight

high·bind·er \ˈhī-ˌbīn-dər\ *noun* [the *Highbinders,* gang of ruffians in New York City (about 1806)] (1876)
1 : a professional killer operating in the Chinese quarter of an American city
2 : a corrupt politician

high blood pressure *noun* (1916)
: HYPERTENSION

high·born \ˈhī-ˈbȯrn\ *adjective* (13th century)
: of noble birth

high·boy \-ˌbȯi\ *noun* (1891)
: a tall chest of drawers with a legged base

highboy

high·bred \-ˈbred\ *adjective* (1674)
: coming from superior stock

high·brow \-ˌbrau̇\ *noun* (circa 1903)
: a person who possesses or has pretensions to superior learning or culture
— highbrow *adjective*
— high·browed \-ˌbrau̇d\ *adjective*
— high·brow·ism \-ˌbrau̇-ˌi-zəm\ *noun*

high·bush \-ˈbu̇sh\ *adjective* (1805)
: forming a notably tall or erect bush; *also* **:** borne on a highbush plant

highbush blueberry *noun* (1913)
: a variable moisture-loving North American shrub (*Vaccinium corymbosum*) that is the source of most cultivated blueberries; *also* **:** its fruit

high chair *noun* (1848)
: a child's chair with long legs, a footrest, and usually a feeding tray

High Church *adjective* (1687)
: favoring especially in Anglican worship the sacerdotal, liturgical, ceremonial, and traditional elements in worship

High Churchman *noun* (1687)
: an Anglican who adheres to High Church elements in worship

high–class \ˈhī-ˈklas\ *adjective* (1864)
: of superior quality or status

high comedy *noun* (1895)
: comedy employing subtle characterizations and witty dialogue — compare LOW COMEDY

high command *noun* (1917)
1 : the supreme headquarters of a military force
2 : the highest leaders in an organization

high commissioner *noun* (1881)
: a principal or a high-ranking commissioner; *especially* **:** an ambassadorial representative of the government of one country stationed in another

high–con·cept \ˈhī-ˈkän-ˌsept\ *adjective* (1985)

☆ SYNONYMS
High, tall, lofty mean above the average in height. HIGH implies marked extension upward and is applied chiefly to things which rise from a base or foundation or are placed at a conspicuous height above a lower level ⟨a *high* hill⟩ ⟨a *high* ceiling⟩. TALL applies to what grows or rises high by comparison with others of its kind and usually implies relative narrowness ⟨a *tall* thin man⟩. LOFTY suggests great or imposing altitude ⟨*lofty* mountain peaks⟩.

: having or exploiting elements (as fast action, glamour, or suspense) that appeal to a wide audience ⟨*high-concept* movies⟩

high–count \'hī-'kau̇nt\ *adjective* (1926)
: having a large number of warp and weft yarns to the square inch ⟨*high-count* percale sheeting⟩

high court *noun* (14th century)
: SUPREME COURT

high–density lipoprotein *noun* (1960)
: HDL

high–end \'hī-'end\ *adjective* (1980)
: UPSCALE ⟨*high-end* boutiques⟩

high–energy *adjective* (1934)
1 a : having such speed and kinetic energy as to exhibit relativistic departure from classical laws of motion — used especially of elementary particles whose velocity has been imparted by an accelerator **b :** of or relating to high-energy particles ⟨a *high-energy* reaction⟩
2 : yielding a relatively large amount of energy when undergoing hydrolysis ⟨*high-energy* phosphate bonds in ATP⟩

high–energy physics *noun* (1964)
: PARTICLE PHYSICS

higher criticism *noun* (1836)
: study of biblical writings to determine their literary history and the purpose and meaning of the authors — compare LOWER CRITICISM
— **higher critic** *noun*

higher education *noun* (1866)
: education beyond the secondary level; *especially* : education provided by a college or university

higher law *noun* (1844)
: a principle of divine or moral law that is considered to be superior to constitutions and enacted legislation

higher learning *noun* (1926)
: education, learning, or scholarship on the collegiate or university level

high·er–up \,hī-ə-'rəp, 'hī-ə-,\ *noun* (1911)
: a superior officer or official

high explosive *noun* (1877)
: an explosive (as TNT) that generates gas with extreme rapidity and has a shattering effect

high·fa·lu·tin \,hī-fə-'lü-tᵊn\ *adjective* [perhaps from ²*high* + alteration of *fluting*, present participle of *flute*] (1839)
1 : PRETENTIOUS
2 : expressed in or marked by the use of high-flown bombastic language : POMPOUS

high fashion *noun* (1945)
1 : HIGH STYLE
2 : HAUTE COUTURE

high fidelity *noun* (1934)
: the reproduction of an effect (as sound or an image) with a high degree of faithfulness to the original
— **high–fidelity** *adjective*

high five *noun* (1981)
: a slapping of upraised right hands by two people (as in celebration)
— **high–five** *verb*

high·fli·er *or* **high·fly·er** \'hī-'flī(-ə)r\ *noun* (circa 1961)
1 : a stock whose price rises much more rapidly than the market average
2 : a company whose stock is a highflier
3 : an ambitiously competitive person with high aspirations

high–flown \'hī-'flōn\ *adjective* (1647)
1 : exceedingly or excessively high or favorable
2 : having an excessively embellished or inflated character : PRETENTIOUS ⟨inflated rhetoric and *high-flown* vocabulary —James Yaffe⟩

high–fly·ing \-'flī-iŋ\ *adjective* (1581)
1 : marked by extravagance, pretension, or excessive ambition
2 : rising to considerable height

high frequency *noun* (1892)
: a radio frequency between very high frequency and medium frequency — see RADIO FREQUENCY table

high gear *noun* (1896)
1 : HIGH 3b
2 : a state of intense or maximum activity — usually used with *in*

High German *noun* (1673)
1 : German as natively used in southern and central Germany
2 : GERMAN 3b

high–grade *adjective* (1878)
1 : of superior grade or quality ⟨*high-grade* bonds⟩
2 : being near the upper or most favorable extreme of a specified range

high ground *noun* (1944)
: a position of advantage or preeminence

high–hand·ed \-'han-dəd\ *adjective* (1631)
: having or showing no regard for the rights, concerns, or feelings of others : ARBITRARY, OVERBEARING
— **high–hand·ed·ly** *adverb*
— **high–hand·ed·ness** *noun*

high–hat \'hī-'hat\ *adjective* (1924)
: SUPERCILIOUS, SNOBBISH
— **high–hat** *transitive verb*

high hat *noun* (1889)
1 : ¹BEAVER 2
2 *or* **hi–hat :** a pair of cymbals operated by a foot pedal

high heels *noun plural* (1671)
: shoes with high heels

High Holiday *noun* (1946)
: either of two important Jewish holidays: **a** : ROSH HASHANAH **b :** YOM KIPPUR

high horse *noun* (1721)
: an arrogant and unyielding mood or attitude

high·jack \'hī-,jak\ *variant of* HIJACK

high jinks *noun plural* (1825)
: boisterous or rambunctious carryings-on : carefree antics or horseplay

high jump *noun* (1895)
: a jump for height over a horizontal bar in a track-and-field contest
— **high jumper** *noun*

¹high·land \'hī-lənd\ *noun* (before 12th century)
: elevated or mountainous land

²highland *adjective* (15th century)
1 : of or relating to a highland
2 *capitalized* **:** of or relating to the Highlands of Scotland

high·land·er \-lən-dər\ *noun* (1610)
1 : an inhabitant of a highland
2 *capitalized* **:** an inhabitant of the Highlands of Scotland

Highland fling *noun* (1804)
: a lively Scottish folk dance

high–lev·el \'hī-'le-vəl\ *adjective* (1876)
1 : occurring, done, or placed at a high level
2 : being of high importance or rank ⟨*high-level* diplomats⟩
3 : of, relating to, or being a computer programming language (as BASIC or Pascal) which is similar to a natural language (as English) and in which each statement is translated by a compiler usually into several machine language instructions

¹high·light \'hī-,līt\ *noun* (circa 1889)
1 : the lightest spot or area (as in a painting) : any of several spots in a modeled drawing or painting that receives the greatest amount of illumination
2 : something (as an event or detail) that is of major significance or special interest

²highlight *transitive verb* **-light·ed; -light·ing** (1927)
1 : to throw a strong light on
2 a : to center attention on **b :** to constitute a highlight of

high–low–jack \,hī-,lō-'jak\ *noun* (1818)
: an all-fours game in which scores are made by winning the highest trump, the lowest trump, the jack of trumps, and either the ten of trumps or the most points

high·ly \'hī-lē\ *adverb* (before 12th century)
1 : in or to a high place, level, or rank
2 : in or to a high degree or amount
3 : with approval : FAVORABLY

high mass *noun, often H&M capitalized* (12th century)
: a mass marked by the singing of prescribed parts by the celebrant and the choir or congregation

high–mind·ed \'hī-'mīn-dəd\ *adjective* (1556)
: marked by elevated principles and feelings; *also* : PRETENTIOUS ⟨too *high-minded* to read any fiction —Alfred Kazin⟩
— **high–mind·ed·ly** *adverb*
— **high–mind·ed·ness** *noun*

high–muck–a–muck \,hī-'mə-ki-,mək\ *or* **high–muck·e·ty–muck** \,hī-'mə-kə-tē-,mək\ *noun* [by folk etymology from Chinook Jargon *hayo makamak* plenty to eat] (1856)
: an important and often arrogant person ◆

high·ness \'hī-nəs\ *noun* (before 12th century)
1 : the quality or state of being high
2 — used as a title for a person of exalted rank (as a king or prince)

high noon *noun* (1523)
1 : precisely noon
2 : the most advanced, flourishing, or creative stage or period ⟨the *high noon* of her career⟩

high–octane *adjective* (1932)
1 : having a high octane number and hence good antiknock properties ⟨*high-octane* gasoline⟩
2 : HIGH-POWERED

high–pitched \'hī-'picht\ *adjective* (1748)
1 : having a high pitch ⟨a *high-pitched* voice⟩
2 : marked by or exhibiting strong feeling : AGITATED ⟨a *high-pitched*, almost frantic campaign —Geoffrey Rice⟩

high place *noun* (14th century)
: a temple or altar used by the ancient Semites and built usually on a hill or elevation

high polymer *noun* (1942)
: a substance (as polystyrene) consisting of molecules that are large multiples of units of low molecular weight

high–pow·ered \'hī-'pau̇(-ə)rd\ *also* **high–pow·er** \-'pau̇(-ə)r\ *adjective* (1893)
1 : having great drive, energy, or capacity : DYNAMIC ⟨a *high-powered* executive⟩

◇ WORD HISTORY

high-muck-a-muck Chinook Jargon was a pidgin spoken in the Pacific Northwest during the 19th century. Its vocabulary was drawn largely from lower Chinook, once spoken at the mouth of the Columbia River, and Nootka, spoken by an Indian people living on western Vancouver Island. A small number of words from Chinook Jargon have found their way into American and Canadian English, of which the most curious is the phrase *hayo makamak*, literally "plenty of food," both elements of which are Nootka in origin (*hayo*, literally, "ten" and *ma·ho·maq-* "eat"). Somehow this phrase was transplanted from the Northwest to central California, where it surfaced in 1856 as a colloquialism meaning not "plenty of food," but "big shot." Presumably the first syllable was transformed by folk-etymology into *high*, though how *makamak* came to be interpreted as "person" is unclear. Indians of local importance in Northwest Coast cultures would give potlatches at which *hayo makamak* would be distributed, but there is no evidence that the phrase was transferred from food to person in Chinook Jargon. At any rate, once established, further variants of *high-muck-a-muck* have proliferated, such as *high-muckety-muck*, *high-monkey-monk*, and—by a second round of folk-etymology—*high mogul*.

2 : having or conferring great influence ⟨a *high-powered* job⟩

¹high–pressure *adjective* (1824)
1 a : having or involving a high or comparatively high pressure especially greatly exceeding that of the atmosphere **b :** having a high barometric pressure
2 a : using or involving aggressive and insistent sales techniques **b :** imposing or involving severe strain or tension ⟨*high-pressure* occupations⟩

²high–pressure *transitive verb* (1926)
: to sell or influence by high-pressure tactics

high priest *noun* (14th century)
1 : a chief priest especially of the ancient Jewish Levitical priesthood traditionally traced from Aaron
2 : a priest of the Melchizedek priesthood in the Mormon Church
3 : the head of a movement or chief exponent of a doctrine or an art
— high priesthood *noun*

high priestess *noun* (1645)
: a chief priestess

high relief *noun* (circa 1828)
: sculptural relief in which at least half of the circumference of the modeled form projects

high–rise \ˈhī-ˈrīz\ *adjective* (1954)
1 : being multistory and equipped with elevators ⟨*high-rise* apartments⟩
2 : of, relating to, or characterized by high-rise buildings
— high–rise \ˈhī-ˌ\ *noun*

high road *noun* (1709)
1 : HIGHWAY
2 : the easiest course
3 : an ethical course

high roller *noun* (1881)
1 : a person who spends freely in luxurious living
2 : a person who gambles recklessly or for high stakes
— high–rolling *adjective*

¹high school *noun* (1824)
: a school especially in the U.S. usually including grades 9–12 or 10–12
— high school·er \-ˈskü-lər\ *noun*

²high school *noun* [translation of French *haute école*] (1884)
: a system of advanced exercises in horsemanship

high sea *noun* (before 12th century)
: the open part of a sea or ocean especially outside territorial waters — usually used in plural

high sign *noun* (1902)
: a gesture used as a signal (as of approval or warning) — usually used in the phrase *give the high sign*

high–sound·ing \ˈhī-ˈsau̇n-diŋ\ *adjective* (1784)
: POMPOUS, IMPOSING

high–speed \ˈhī-ˈspēd\ *adjective* (1873)
1 : operated or adapted for operation at high speed
2 : relating to the production of short-exposure photographs of rapidly moving objects or events of short duration

high–spir·it·ed \-ˈspir-ə-təd\ *adjective* (circa 1631)
: characterized by a bold or energetic spirit
— high–spir·it·ed·ly *adverb*
— high–spir·it·ed·ness *noun*

high·spot \ˈhī-ˌspät\ *noun* (1910)
: HIGHLIGHT 2

high–stick·ing \-ˌsti-kiŋ\ *noun* (1947)
: the act of carrying the blade of the stick at an illegal height in ice hockey

high street *noun* (before 12th century)
British **:** a main or principal street

high–strung \ˈhī-ˈstrəŋ\ *adjective* (1748)
: having an extremely nervous or sensitive temperament

high style *noun* (1939)
: the newest style in fashion or design usually adopted by a limited number of people

hight \ˈhīt\ *adjective* [Middle English, past participle (earlier past) of *hoten* to command, call, be called, from Old English *hātan;* akin to Old High German *heizzan* to command, call] (15th century)
archaic **:** being called **:** NAMED

high table *noun* (1711)
: an elevated table in the dining room of a British college for use by the master and fellows and distinguished guests

high·tail \ˈhī-ˌtāl\ *intransitive verb* (1925)
: to move at full speed or rapidly often in making a retreat — usually used with *it*

high tea *noun* (1831)
: a fairly substantial late afternoon or early evening meal at which tea is served

high tech \-ˈtek\ *noun* (1973)
1 : HIGH TECHNOLOGY
2 : a style of interior design featuring industrial products, materials, or designs
— high–tech *adjective*

high technology *noun* (1968)
: scientific technology involving the production or use of advanced or sophisticated devices especially in the fields of electronics and computers

high–tension *adjective* (1905)
: having or using a high voltage

high–test *adjective* (1923)
: meeting a high standard; *also* **:** HIGH-OCTANE

high tide *noun* (1745)
1 : the tide when the water is at its greatest elevation
2 : culminating point **:** CLIMAX

high–toned \ˈhī-ˈtōnd\ *adjective* (1807)
1 : high in social, moral, or intellectual quality
2 : PRETENTIOUS, POMPOUS

high–top \ˈhī-ˌtäp\ *adjective* (1966)
: extending up over the ankle ⟨*high-top* sneakers⟩
— high–tops *noun plural*

high treason *noun* (15th century)
: TREASON 2

high–volt·age \ˈhī-ˈvōl-tij\ *adjective* (1948)
: marked by great energy **:** ELECTRIC, DYNAMIC ⟨a *high-voltage* performance⟩

high–water *adjective* (1856)
: unusually short ⟨*high-water* pants⟩

high water *noun* (15th century)
: a high stage of the water in a river or lake; *also* **:** HIGH TIDE

high–water mark *noun* (1814)
: highest point **:** PEAK

high·way \ˈhī-ˌwā\ *noun* (before 12th century)
: a public way; *especially* **:** a main direct road

high·way·man \-mən\ *noun* (1649)
: a person who robs travelers on a road

highway robbery *noun* (1778)
1 : robbery committed on or near a public highway usually against travelers
2 : excessive profit or advantage derived from a business transaction

high–wire \ˈhī-ˌwī(-ə)r\ *adjective* (1956)
1 : involving great risk ⟨a financial *high-wire* act⟩
2 : DARING ⟨*high-wire* prose⟩

high–wrought \ˈhī-ˈrȯt\ *adjective* (1604)
: extremely agitated

high yellow *noun* (1923)
: a black person of light complexion — called also *high yal·ler* \-ˈya-lər\; often taken to be offensive

hi–hat *variant of* HIGH HAT 2

hi·jack \ˈhī-ˌjak\ *transitive verb* [origin unknown] (1923)
1 a : to steal by stopping a vehicle on the highway **b :** to commandeer (a flying airplane) especially by coercing the pilot at gunpoint **c :** to stop and steal from (a vehicle in transit) **d :** KIDNAP
2 a : to steal or rob as if by hijacking **b :** to subject to extortion or swindling
— hijack *noun*
— hi·jack·er *noun*

hi·jinks \ˈhī-ˌjiŋks\ *variant of* HIGH JINKS

¹hike \ˈhīk\ *verb* **hiked; hik·ing** [perhaps akin to ¹*hitch*] (1804)
intransitive verb
1 a : to go on a hike **b :** to travel by any means
2 : to rise up; *especially* **:** to work upward out of place ⟨skirt had *hiked* up in back⟩
transitive verb
1 a : to move, pull, or raise with a sudden motion ⟨*hiked* himself onto the top bunk⟩ **b :** SNAP 6b **c :** to raise in amount sharply or suddenly ⟨*hike* rents⟩
2 : to take on a hike
— hik·er *noun*

²hike *noun* (1865)
1 : a long walk especially for pleasure or exercise
2 : an increase especially in quantity or amount ⟨a new wage *hike*⟩
3 : SNAP 11

hi·lar \ˈhī-lər\ *adjective* (circa 1859)
: of, relating to, or located near a hilum

hi·lar·i·ous \hi-ˈlar-ē-əs, hī-, -ˈler-\ *adjective* [irregular from Latin *hilarus, hilaris* cheerful, from Greek *hilaros*] (circa 1840)
: marked by or affording hilarity
— hi·lar·i·ous·ly *adverb*
— hi·lar·i·ous·ness *noun*

hi·lar·i·ty \-ə-tē\ *noun* (15th century)
: high spirits that may be carried to the point of boisterous conviviality or merriment

Hil·bert space \ˈhil-bərt-\ *noun* [David *Hilbert*] (1939)
: a vector space for which a scalar product is defined and in which every Cauchy sequence composed of elements in the space converges to a limit in the space

hil·ding \ˈhil-diŋ\ *noun* [*hilding,* adjective (base)] (1592)
archaic **:** a base contemptible person

¹hill \ˈhil\ *noun* [Middle English, from Old English *hyll;* akin to Latin *collis* hill, *culmen* top] (before 12th century)
1 : a usually rounded natural elevation of land lower than a mountain
2 : an artificial heap or mound (as of earth)
3 : several seeds or plants planted in a group rather than a row
4 : SLOPE, INCLINE

²hill *transitive verb* (1581)
1 : to form into a heap
2 : to draw earth around the roots or base of
— hill·er *noun*

hill·bil·ly \ˈhil-ˌbi-lē\ *noun, plural* **-lies** [¹*hill* + *Billy,* nickname for *William*] (1900)
: a person from a backwoods area

hillbilly music *noun* (1943)
: COUNTRY MUSIC

hill climb *noun* (1905)
: a road race for automobiles or motorcycles in which competitors are individually timed up a hill

hill·crest \ˈhil-ˌkrest\ *noun* (circa 1898)
: the top line of a hill

hill mynah *noun* (circa 1890)
: a largely black Asian mynah (*Gracula religiosa*) often tamed and taught to pronounce words

hill·ock \ˈhi-lək\ *noun* (14th century)
: a small hill
— hill·ocky \-lə-kē\ *adjective*

Hill reaction \ˈhil-\ *noun* [Robert *Hill* (died 1991) British biochemist] (1950)
: the light-dependent transfer of electrons by chloroplasts in photosynthesis that results in the cleavage of water molecules and liberation of oxygen

hill·side \-ˌsīd\ *noun* (14th century)
: a part of a hill between the top and the foot

hill·top \ˈhil-ˌtäp\ *noun* (15th century)
: the highest part of a hill

hilly \ˈhi-lē\ *adjective* **hill·i·er; -est** (14th century)
: abounding in hills

hilt \'hilt\ *noun* [Middle English, from Old English; akin to Old High German *helza* hilt] (before 12th century)
: a handle especially of a sword or dagger
— to the hilt : to the very limit **:** COMPLETELY

hi·lum \'hī-ləm\ *noun, plural* **hi·la** \-lə\ [New Latin, from Latin, trifle] (circa 1753)
1 : a scar on a seed (as a bean) marking the point of attachment of the ovule
2 : a notch in or opening from a bodily part suggesting the hilum of a bean

him \im, 'him\ *pronoun objective case of* HE

Hi·ma·la·yan \ˌhi-mə-'lā-ən, hi-'mäl-yən, -'mä-lē-ən\ *noun* [*Himalaya* Mountains] (1949)
: any of a breed of domestic cats developed by crossing the Persian and the Siamese and having the stocky build and long thick coat of the former and the blue eyes and coat patterns of the latter

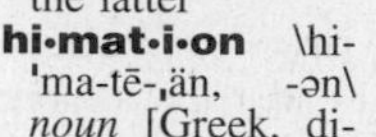
Himalayan

hi·mat·i·on \hi-'ma-tē-ˌän, -ən\ *noun* [Greek, diminutive of *heimat-, heima* garment; akin to Greek *hennynai* to clothe — more at WEAR] (1850)
: a rectangular cloth draped over the left shoulder and about the body and worn as a garment in ancient Greece

him/her \'him-'hər, 'hi-mər-, 'him-'slash-\ *pronoun objective case of* HE/SHE

him·self \(h)im-'self, *Southern also* -'sef\ *pronoun* (before 12th century)
1 a : that identical male one — compare ¹HE; used reflexively, for emphasis, in absolute constructions, and in place of *him* especially when joined to another object ⟨considers *himself* lucky⟩ ⟨he *himself* did it⟩ ⟨*himself* unhappy, he understood the situation⟩ ⟨a gift to his wife and *himself*⟩ **b** — used reflexively when the sex of the antecedent is unspecified ⟨everyone must fend for *himself*⟩
2 : his normal, healthy, or sane condition or self
3 *chiefly Irish & Scottish* **:** a man of consequence; *especially* **:** the master of the house

hin \'hin\ *noun* [Hebrew *hīn,* from Egyptian *hnw*] (14th century)
: an ancient Hebrew unit of liquid measure equal to about 1.5 gallons (5.7 liters)

Hi·na·ya·na \ˌhi-nə-'yä-nə, ˌhē-\ *noun* [Sanskrit *hīnayāna* lesser vehicle] (1868)
: THERAVADA
— Hi·na·ya·nist \-'yä-nist\ *noun*
— Hi·na·ya·nis·tic \-yä-'nis-tik\ *adjective*

¹hind \'hīnd\ *noun, plural* **hinds** *also* **hind** [Middle English, from Old English; akin to Old High German *hinta* hind, Greek *kemas* young deer] (before 12th century)
1 : the female of the red deer — compare HART
2 : any of various spotted groupers (especially genus *Epinephelus*)

²hind *noun* [Middle English *hine* servant, farmhand, from Old English *hīna,* genitive of *hīwan,* plural, members of a household; akin to Old High German *hīwo* spouse, Latin *civis* fellow citizen, and probably to Old Irish *cóim* dear, Lithuanian *šeima* family] (1520)
1 : a British farm assistant
2 *archaic* **:** RUSTIC

³hind *adjective* [Middle English, probably back-formation from Old English *hinder,* adverb, behind; akin to Old High German *hintar,* preposition, behind] (14th century)
: of or forming the part that follows or is behind **:** REAR

hind·brain \'hīn(d)-ˌbrān\ *noun* (1888)
1 : the posterior of the three primary divisions of the developing vertebrate brain or the corresponding part of the adult brain that includes the cerebellum, the medulla oblongata, and in mammals the pons and that controls autonomic functions and equilibrium — called also *rhombencephalon;* compare METENCEPHALON, MYELENCEPHALON
2 : the posterior segment of the brain of an invertebrate

¹hin·der \'hin-dər\ *verb* **hin·dered; hin·der·ing** \-d(ə-)riŋ\ [Middle English *hindren,* from Old English *hindrian;* akin to Old English *hinder* behind] (before 12th century)
transitive verb
1 : to make slow or difficult the progress of **:** HAMPER
2 : to hold back **:** CHECK
intransitive verb
: to delay, impede, or prevent action ☆
— hin·der·er \-dər-ər\ *noun*

²hind·er \'hīn-dər\ *adjective* [Middle English, from Old English *hinder,* adverb] (13th century)
: situated behind or in the rear **:** POSTERIOR

hind·gut \'hīn(d)-ˌgət\ *noun* (1878)
: the posterior part of the alimentary canal

Hin·di \'hin-(ˌ)dē\ *noun* [Hindi *hindī,* from *Hind* India, from Persian] (1801)
1 : a literary and official language of northern India
2 : a complex of Indo-Aryan languages and dialects of northern India for which Hindi is the usual literary language
— Hindi *adjective*

hind·most \'hīn(d)-ˌmōst\ *adjective* (14th century)
: farthest to the rear **:** LAST

hind·quar·ter \-ˌkwȯ(r)-tər, -ˌkȯ(r)-\ *noun* (1881)
1 : one side of the back half of the carcass of a quadruped including a leg and usually one or more ribs
2 *plural* **:** the hind pair of legs of a quadruped; *broadly* **:** all the structures of a quadruped that lie posterior to the attachment of the hind legs to the trunk

hin·drance \'hin-drən(t)s\ *noun* (1526)
1 : the state of being hindered
2 : IMPEDIMENT
3 : the action of hindering

hind·sight \'hīn(d)-ˌsīt\ *noun* (1866)
: perception of the nature of an event after it has happened

¹Hin·du *also* **Hin·doo** \'hin-(ˌ)dü\ *noun* [Persian *Hindū* inhabitant of India, from *Hind* India] (1662)
1 : an adherent of Hinduism
2 : a native or inhabitant of India

²Hindu *also* **Hindoo** *adjective* (1698)
: of, relating to, or characteristic of the Hindus or Hinduism

Hindu–Arabic *adjective* (1925)
: relating to, being, or composed of Arabic numerals ⟨*Hindu-Arabic* numeration system⟩

Hindu calendar *noun* (circa 1909)
: a lunar calendar usually dating from 3101 B.C. and used especially in India

Hin·du·ism \'hin-(ˌ)dü-ˌi-zəm\ *noun* (1809)
: the dominant religion of India that emphasizes dharma with its resulting ritual and social observances and often mystical contemplation and ascetic practices

¹Hin·du·stani *also* **Hin·do·stani** \ˌhin-dü-'sta-nē, -'stä-nē\ *noun* [Hindi *Hindūstānī,* from Persian *Hindūstān* India] (1808)
: a group of Indo-Aryan dialects of northern India of which literary Hindi and Urdu are considered diverse written forms

²Hindustani *also* **Hindostani** *adjective* (1800)
: of or relating to Hindustan or its people or Hindustani

hind wing *noun* (1899)
: either of the posterior wings of a 4-winged insect

¹hinge \'hinj\ *noun* [Middle English *heng;* akin to Middle Dutch *henge* hook, Old English *hangian* to hang] (14th century)
1 a : a jointed or flexible device on which a door, lid, or other swinging part turns **b :** a flexible ligamentous joint **c :** a small piece of thin gummed paper used in fastening a postage stamp in an album
2 : a determining factor **:** TURNING POINT

²hinge *verb* **hinged; hing·ing** (1719)
intransitive verb
: to be contingent on a single consideration or point — used with *on* or *upon*
transitive verb
: to attach by or furnish with hinges

hinge joint *noun* (1802)
: a joint between bones (as at the elbow) that permits motion in only one plane

hin·ny \'hi-nē\ *noun, plural* **hinnies** [Latin *hinnus,* from Greek *innos*] (1688)
: a hybrid between a stallion and a female donkey — compare MULE

¹hint \'hint\ *noun* [probably alteration of obsolete *hent* act of seizing, from *hent* verb] (1604)
1 *archaic* **:** OPPORTUNITY, TURN
2 a : a statement conveying by implication what it is preferred not to say explicitly **b :** an indirect or summary suggestion ⟨helpful *hints*⟩
3 : a slight indication of the existence, approach, or nature of something **:** CLUE
4 : a very small amount **:** SUGGESTION

²hint (1648)
transitive verb
: to convey indirectly and by allusion rather than explicitly ⟨a suspicion that she scarcely dared to *hint*⟩
intransitive verb
: to give a hint — usually used with *at*
synonym see SUGGEST
— hint·er *noun*

hin·ter·land \'hin-tər-ˌland, -lənd\ *noun* [German, from *hinter* hinder + *Land*] (1890)
1 : a region lying inland from a coast
2 a : a region remote from urban areas **b :** a region lying beyond major metropolitan or cultural centers

¹hip \'hip\ *noun* [Middle English *hipe,* from Old English *hēope;* akin to Old High German *hiafo* hip] (before 12th century)
: ROSE HIP

²hip *noun* [Middle English, from Old English *hype;* akin to Old High German *huf* hip] (before 12th century)
1 a : the laterally projecting region of each side of the lower or posterior part of the mammalian trunk formed by the lateral parts of the pelvis and upper part of the femur together with the fleshy parts covering them **b :** HIP JOINT
2 : the external angle formed by the meeting of two sloping sides of a roof that have their wall plates running in different directions

☆ SYNONYMS
Hinder, impede, obstruct, block mean to interfere with the activity or progress of. HINDER stresses causing harmful or annoying delay or interference with progress ⟨rain *hindered* the climb⟩. IMPEDE implies making forward progress difficult by clogging, hampering, or fettering ⟨tight clothing that *impedes* movement⟩. OBSTRUCT implies interfering with something in motion or in progress by the sometimes intentional placing of obstacles in the way ⟨the view was *obstructed* by billboards⟩. BLOCK implies complete obstruction to passage or progress ⟨a landslide *blocked* the road⟩.

³hip *interjection* [origin unknown] (1827)
— used in a cheer ⟨*hip hip* hooray⟩

⁴hip *adjective* **hip·per; hip·pest** [alteration of ²*hep*] (1904)
: characterized by a keen informed awareness of or involvement in the newest developments or styles

⁵hip *transitive verb* **hipped; hip·ping** (circa 1932)
: to make aware **:** TELL, INFORM

⁶hip *noun* (1952)
: HIPNESS

hip and thigh *adverb* (1560)
: in a fierce or ruthless manner **:** UNSPARINGLY

hip·bone \'hip-'bōn, -ˌbōn\ *noun* (12th century)
: INNOMINATE BONE

hip boot *noun* (1893)
: a waterproof boot reaching to the hips

hip–hop \'hip-ˌhäp\ *noun, often attributive* [perhaps from ⁴*hip* + ¹*hop*] (1983)
: a subculture especially of inner-city youths whose amusements include rap music, graffiti, and break dancing; *also* **:** an element or art form prevalent within this subculture

hip joint *noun* (1794)
: the articulation between the femur and the innominate bone

hip·line \'hip-ˌlīn\ *noun* (1907)
1 : an arbitrary line encircling the fullest part of the hips
2 : body circumference at the hips

hip·ness \'hip-nəs\ *noun* (1946)
: the quality or state of being hip

¹hipped \'hipt\ *adjective* (1508)
1 : having hips especially of a specified kind — often used in combination ⟨a broad-*hipped* person⟩
2 : constructed with hips ⟨a *hipped* roof⟩

²hipped *adjective* [*hip* (hypochondria)] (circa 1710)
: DEPRESSED

³hipped *adjective* [⁵*hip*] (1920)
: extremely absorbed or interested ⟨*hipped* on astrology⟩

hip·pie *or* **hip·py** \'hi-pē\ *noun, plural* **hip·pies** [⁴*hip* + *-ie*] (1965)
: a usually young person who rejects the mores of established society (as by dressing unconventionally or favoring communal living) and advocates a nonviolent ethic; *broadly* **:** a long-haired unconventionally dressed young person
— hip·pie·dom \-pē-dəm\ *noun*
— hip·pie·ness *or* **hip·pi·ness** \-pē-nəs\ *noun*

hip·po \'hi-(ˌ)pō\ *noun, plural* **hippos** (1872)
: HIPPOPOTAMUS

hip·po·cam·pus \ˌhi-pə-'kam-pəs\ *noun, plural* **-pi** \-ˌpī, -(ˌ)pē\ [New Latin, from Greek *hippokampos* sea horse, from *hippos* horse + *kampos* sea monster — more at EQUINE] (1706)
: a curved elongated ridge that extends over the floor of the descending horn of each lateral ventricle of the brain and consists of gray matter covered on the ventricular surface with white matter
— hip·po·cam·pal \-pəl\ *adjective*

hip·po·cras \'hi-pə-ˌkras\ *noun* [Middle English *ypocras,* from *Ypocras* Hippocrates, to whom its invention was ascribed] (14th century)
: a mulled wine popular in medieval Europe

Hip·po·crat·ic \ˌhi-pə-'kra-tik\ *adjective* (circa 1620)
: of or relating to Hippocrates or to the school of medicine that took his name

Hippocratic oath *noun* (1747)
: an oath embodying a code of medical ethics usually taken by those about to begin medical practice

Hip·po·crene \'hi-pə-ˌkrēn, ˌhi-pə-'krē-nē\ *noun* [Latin, from Greek *Hippokrēnē*] (1605)
: a fountain on Mount Helicon sacred to the Muses and believed to be a source of poetic inspiration

hip·po·drome \'hi-pə-ˌdrōm\ *noun* [Middle French, from Latin *hippodromos,* from Greek, from *hippos* + *dromos* racecourse — more at DROMEDARY] (1585)
1 : an oval stadium for horse and chariot races in ancient Greece
2 : an arena for equestrian performances

hip·po·griff \-ˌgrif\ *noun* [French *hippogriffe,* from Italian *ippogrifo,* from *ippo-* (from Greek *hippos* horse) + *grifo* griffin, from Latin *gryphus*] (circa 1656)
: a legendary animal having the foreparts of a griffin and the body of a horse

Hip·pol·y·ta \hi-'pä-lə-tə\ *noun* [Latin, from Greek *Hippolytē*]
: a queen of the Amazons given in marriage to Theseus by Hercules

Hip·pol·y·tus \-ə-təs\ *noun* [Latin, from Greek *Hippolytos*]
: a son of Theseus falsely accused of amorous advances by his stepmother and killed by his father through the agency of Poseidon

Hip·pom·e·nes \hi-'pä-mə-nēz\ *noun* [Latin, from Greek *Hippomenēs*]
: the successful suitor of Atalanta in Greek mythology

hip·po·pot·a·mus \ˌhi-pə-'pä-tə-məs\ *noun, plural* **-mus·es** *or* **-mi** \-ˌmī, -(ˌ)mē\ [Latin, from Greek *hippopotamos,* alteration of *hippos potamios,* literally, riverine horse] (1563)
: a very large herbivorous 4-toed chiefly aquatic artiodactyl mammal (*Hippopotamus amphibius*) of sub-Saharan Africa with an extremely large head and mouth, bare and very thick grayish skin, and short legs; *also* **:** a smaller closely related mammal (*Choeropsis liberiensis*) of western Africa ◆

hippopotamus

hip·py \'hi-pē\ *adjective* (1919)
: having large hips

hip roof *noun* (circa 1741)
: a roof having sloping ends and sloping sides — see ROOF illustration

hip–shoot·ing \'hip-ˌshü-tiŋ\ *noun* (1951)
: action or reaction that is quick and often reckless
— hip shooter *noun*

hip·ster \'hip-stər\ *noun* [⁴*hip*] (circa 1941)
: a person who is unusually aware of and interested in new and unconventional patterns (as in jazz or the use of stimulants)

hip·ster·ism \-stə-ˌri-zəm\ *noun* (1958)
1 : HIPNESS
2 : the way of life characteristic of hipsters

hi·ra·ga·na \ˌhir-ə-'gä-nə\ *noun* [Japanese, from *hira-* ordinary + *kana* syllabary] (1859)
: the cursive script that is one of two sets of symbols of Japanese syllabic writing — compare KATAKANA

¹hire \'hī(-ə)r\ *noun* [Middle English, from Old English *hȳr;* akin to Old Saxon *hūria* hire] (before 12th century)
1 a : payment for the temporary use of something **b :** payment for labor or personal services **:** WAGES
2 a : the act or an instance of hiring **b :** the state of being hired **:** EMPLOYMENT
3 *British* **:** RENTAL — often used attributively
4 : one who is hired ⟨starting wage for the new *hires*⟩
— for hire *also* **on hire :** available for use or service in return for payment

²hire *verb* **hired; hir·ing** (before 12th century)
transitive verb
1 a : to engage the personal services of for a set sum ⟨*hire* a crew⟩ **b :** to engage the temporary use of for a fixed sum ⟨*hire* a hall⟩
2 : to grant the personal services of or temporary use of for a fixed sum ⟨*hire* themselves out⟩
3 : to get done for pay ⟨*hire* the mowing done⟩
intransitive verb
: to take employment ⟨*hire* out as a guide during the tourist season⟩ ☆
— hir·er *noun*

hired gun *noun* (1971)
: an expert hired to do a specific job; *especially* **:** HIRELING — often used disparagingly

hire·ling \'hī(-ə)r-liŋ\ *noun* (before 12th century)
: a person who serves for hire especially for purely mercenary motives

hire purchase *noun* (1895)
chiefly British **:** purchase on the installment plan

hiring hall *noun* (1934)
: a union-operated placement office where registered applicants are referred in rotation to jobs

hir·sute \'hər-ˌsüt, 'hir-, ˌhər-', hir-'\ *adjective* [Latin *hirsutus;* akin to Latin *horrēre* to bristle — more at HORROR] (1621)
: HAIRY 1; *especially* **:** covered with coarse stiff hairs ⟨*hirsute* leaf⟩
— hir·sute·ness *noun*

hir·sut·ism \'hər-sə-ˌti-zəm, 'hir-\ *noun* (1927)
: excessive growth of hair of normal or abnormal distribution

hi·ru·din \hi-'rü-dᵊn, 'hir-(y)ə-\ *noun* [from *Hirudin,* a trademark] (1905)
: an anticoagulant extracted from the buccal glands of a leech

¹his \(h)iz, ˌhiz\ *adjective* [Middle English, from Old English, genitive of *hē* he] (before 12th century)
: of or relating to him or himself especially as possessor, agent, or object of an action ⟨*his* house⟩ ⟨*his* writings⟩ ⟨*his* confirmation⟩ — compare ¹HE

☆ SYNONYMS
Hire, let, lease, rent, charter mean to engage or grant for use at a price. HIRE and LET, strictly speaking, are complementary terms, HIRE implying the act of engaging or taking for use and LET the granting of use ⟨we *hired* a car for the summer⟩ ⟨decided to *let* the cottage to a young couple⟩. LEASE strictly implies a letting under the terms of a contract but is often applied to hiring on a lease ⟨the diplomat *leased* an apartment for a year⟩. RENT stresses the payment of money for the full use of property and may imply either hiring or letting ⟨instead of buying a house, they decided to *rent*⟩ ⟨will not *rent* to families with children⟩. CHARTER applies to the hiring or letting of a vehicle usually for exclusive use ⟨*charter* a bus to go to the game⟩.

◇ WORD HISTORY
hippopotamus The bulky African mammal that spends its daytime hours submerged up to its eyes in a river owes its English name to the ancient Greeks. The historian Herodotus referred to this animal, which he may have seen in Egypt, as *ho hippos ho potamios* "the riverine horse." This epithet was reduced by later Greek writers to *hippopotamos* (which by normal rules of Greek compounding should mean "horse river"). Despite its name, the hippopotamus, which may weigh up to five tons, is more closely related to hogs than horses. Greek *hippos* "horse" was also applied to a species of marine fish that bore even less resemblance to a horse than a hippopotamus does—paralleling to some degree English *sea horse* (which resembles a horse) and *redhorse* (which does not).

²his \'hiz\ *pronoun, singular or plural in construction* (before 12th century)
: that which belongs to him — used without a following noun as a pronoun equivalent in meaning to the adjective *his*

his/her \'hiz-'hər, 'hi-zər-'hər, 'hiz-'slash-\ *adjective* (1952)
: his or her — used in writing as an adjective of common gender

His·pan·ic \hi-'spa-nik\ *adjective* [Latin *hispanicus,* from *Hispania* Iberian Peninsula, Spain] (circa 1889)
: of, relating to, or being a person of Latin American descent living in the U.S.; *especially* **:** one of Cuban, Mexican, or Puerto Rican origin
— **Hispanic** *noun*
— **His·pan·i·cism** \-'spa-nə-,si-zəm\ *noun*
— **His·pan·i·cist** \-sist\ *noun*
— **His·pan·i·cize** \-,sīz\ *transitive verb*

his·pa·ni·dad \,is-,pa-ni-'thä(th)\ *noun* (1941)
: HISPANISM 1

his·pa·nism \'his-pə-,ni-zəm\ *noun, often capitalized* (1940)
1 : a movement to reassert the cultural unity of Spain and Latin America
2 : a characteristic feature of Spanish occurring in another language

His·pa·nist \-nist\ *noun* (1786)
: a scholar specially informed in Spanish or Portuguese language, literature, linguistics, or civilization

His·pa·no \hi-'spa-(,)nō, 'his-pə-,nō\ *noun, plural* **-nos** [American Spanish *hispano,* probably short for *hispanoamericano,* literally, Spanish-American] (1946)
: a native or resident of the southwestern U.S. descended from Spaniards settled there before annexation; *also* **:** MEXICAN 1

his·pid \'his-pəd\ *adjective* [Latin *hispidus;* akin to Latin *horrēre*] (1646)
: rough or covered with bristles, stiff hairs, or minute spines ⟨*hispid* leaf⟩

hiss \'his\ *verb* [Middle English, of imitative origin] (14th century)
intransitive verb
: to make a sharp sibilant sound ⟨the crowd *hissed* in disapproval⟩ ⟨*hissing* steam⟩
transitive verb
1 : to express disapproval of by hissing
2 : to utter or whisper angrily or threateningly and with a hiss
— **hiss** *noun*
— **hiss·er** *noun*

his·self \(h)i-'self, -'sef\ *pronoun* (12th century)
chiefly dialect **:** HIMSELF 1

hissy \'hi-sē\ *noun* [perhaps by shortening & alteration from *hysterical*] (circa 1934)
chiefly Southern & southern Midland **:** TANTRUM — called also *hissy fit*

¹hist \s *often prolonged and usually with* p *preceding and* t *following; often read as* 'hist\ *interjection* [origin unknown] (1592)
— used to attract attention

²hist \'hīst\ *dialect variant of* HOIST

hist- *or* **histo-** *combining form* [French, from Greek *histos* mast, loom beam, web, from *histanai* to cause to stand — more at STAND]
: tissue ⟨*histo*physiology⟩

his·ta·mi·nase \hi-'sta-mə-,nās, 'his-tə-mə-, -,nāz\ *noun* [International Scientific Vocabulary] (1930)
: a widely occurring flavoprotein enzyme that oxidizes histamine and various diamines

his·ta·mine \'his-tə-,mēn, -mən\ *noun* [International Scientific Vocabulary *hist-* + *amine*] (circa 1913)
: a compound $C_5H_9N_3$ especially of mammalian tissues that causes dilatation of capillaries, contraction of smooth muscle, and stimulation of gastric acid secretion, that is released during allergic reactions, and that is formed by decarboxylation of histidine

his·ta·min·er·gic \,his-tə-mə-'nər-jik\ *adjective* [International Scientific Vocabulary] (1936)
of autonomic nerve fibers **:** liberating or activated by histamine ⟨*histaminergic* receptors⟩

his·ti·dine \'his-tə-,dēn\ *noun* [International Scientific Vocabulary *hist-* + *-idine*] (1896)
: a crystalline essential amino acid $C_6H_9N_3O_2$ formed by the hydrolysis of most proteins

his·tio·cyte \'his-tē-ə-,sīt\ *noun* [Greek *histion* web (diminutive of *histos*) + International Scientific Vocabulary *-cyte*] (1924)
: a nonmotile macrophage of extravascular tissues and especially connective tissue
— **his·tio·cyt·ic** \,his-tē-ə-'si-tik\ *adjective*

his·to·chem·is·try \,his-tō-'ke-mə-strē\ *noun* [International Scientific Vocabulary] (circa 1860)
: a science that combines the techniques of biochemistry and histology in the study of the chemical constitution of cells and tissues
— **his·to·chem·i·cal** \-'ke-mi-kəl\ *adjective*
— **his·to·chem·i·cal·ly** \-k(ə-)lē\ *adverb*

his·to·com·pat·i·bil·i·ty \'his-(,)tō-kəm-,pa-tə-'bi-lə-tē\ *noun* (1948)
: a state of mutual tolerance that allows some tissues to be grafted effectively to others

his·to·gen·e·sis \,his-tə-'je-nə-səs\ *noun* [New Latin] (circa 1854)
: the formation and differentiation of tissues
— **his·to·ge·net·ic** \-jə-'ne-tik\ *adjective*

his·to·gram \'his-tə-,gram\ *noun* [Greek *histos* mast, web + English *-gram*] (1891)
: a representation of a frequency distribution by means of rectangles whose widths represent class intervals and whose areas are proportional to the corresponding frequencies

his·tol·o·gy \his-'tä-lə-jē\ *noun, plural* **-gies** [French *histologie,* from *hist-* + *-logie* -logy] (circa 1847)
1 : a branch of anatomy that deals with the minute structure of animal and plant tissues as discernible with the microscope
2 : a treatise on histology
3 : tissue structure or organization
— **his·to·log·i·cal** \,his-tə-'lä-ji-kəl\ *or* **his·to·log·ic** \-'lä-jik\ *adjective*
— **his·to·log·i·cal·ly** \-ji-k(ə-)lē\ *adverb*
— **his·tol·o·gist** \his-'tä-lə-jist\ *noun*

his·tol·y·sis \his-'tä-lə-səs\ *noun* [New Latin] (circa 1857)
: the breakdown of bodily tissues

his·tone \'his-,tōn\ *noun* [German *Histon*] (1885)
: any of various simple water-soluble proteins that are rich in the basic amino acids lysine and arginine and are complexed with DNA in the nucleosomes of eukaryotic chromatin

his·to·pa·thol·o·gy \,his-tō-pə-'thä-lə-jē, -pa-\ *noun* [International Scientific Vocabulary] (1896)
1 : a branch of pathology concerned with the tissue changes characteristic of disease
2 : the tissue changes that affect a part or accompany a disease
— **his·to·path·o·log·ic** \-,pa-thə-'lä-jik\ *or* **his·to·path·o·log·i·cal** \-ji-kəl\ *adjective*
— **his·to·path·o·log·i·cal·ly** \-ji-k(ə-)lē\ *adverb*
— **his·to·pa·thol·o·gist** \-pə-'thä-lə-jist, -pa-\ *noun*

his·to·phys·i·ol·o·gy \-,fi-zē-'ä-lə-jē\ *noun* (circa 1886)
1 : a branch of physiology concerned with the function and activities of tissues
2 : structural and functional tissue organization
— **his·to·phys·i·o·log·i·cal** \-ē-ə-'lä-ji-kəl\ *or* **his·to·phys·i·o·log·ic** \-jik\ *adjective*

his·to·plas·mo·sis \,his-tə-plaz-'mō-səs\ *noun* [New Latin, from *Histoplasma,* genus of fungi] (1907)
: a respiratory disease with symptoms like those of influenza that is caused by a fungus (*Histoplasma capsulatum*) and is marked by benign involvement of lymph nodes of the trachea and bronchi or by severe progressive generalized involvement of the lymph nodes and the reticuloendothelial system

his·to·ri·an \hi-'stōr-ē-ən, -'stȯr-, -'stär-\ *noun* (15th century)
1 : a student or writer of history; *especially* **:** one that produces a scholarly synthesis
2 : a writer or compiler of a chronicle

his·tor·ic \hi-'stȯr-ik, -'stär-\ *adjective* (1607)
: HISTORICAL: as **a :** famous or important in history ⟨*historic* battlefields⟩ **b :** having great and lasting importance ⟨an *historic* occasion⟩ **c :** known or established in the past ⟨*historic* interest rates⟩ **d :** dating from or preserved from a past time or culture ⟨*historic* buildings⟩ ⟨*historic* artifacts⟩

his·tor·i·cal \-i-kəl\ *adjective* (15th century)
1 a : of, relating to, or having the character of history **b :** based on history **c :** used in the past and reproduced in historical presentations
2 : famous in history **:** HISTORIC a
3 a : SECONDARY 1c **b :** DIACHRONIC ⟨*historical* grammar⟩
— **his·tor·i·cal·ness** \-i-kəl-nəs\ *noun*

his·tor·i·cal·ly \-i-k(ə-)lē\ *adverb* (1550)
1 : in accordance with or with respect to history ⟨an *historically* accurate account⟩
2 : in the past ⟨*historically,* stagnant cities seldom have recovered —Jane Jacobs⟩

historical materialism *noun* (1925)
: the Marxist theory of history and society that holds that ideas and social institutions develop only as the superstructure of a material economic base — compare DIALECTICAL MATERIALISM

historical present *noun* (1867)
: the present tense used in relating past events

his·tor·i·cism \hi-'stȯr-ə-,si-zəm, -'stär-\ *noun* (1895)
: a theory, doctrine, or style that emphasizes the importance of history: as **a :** a theory in which history is seen as a standard of value or as a determinant of events **b :** a style (as in architecture) characterized by the use of traditional forms and elements
— **his·tor·i·cist** \-sist\ *adjective or noun*

his·to·ric·i·ty \,his-tə-'ri-sə-tē\ *noun* (1880)
: historical actuality

his·tor·i·cize \hi-'stȯr-ə-,sīz, -'stär-\ *verb* **-cized; -ciz·ing** (1846)
transitive verb
: to make historical
intransitive verb
: to use historical material

his·to·ri·og·ra·pher \hi-,stōr-ē-'ä-grə-fər, -,stȯr-\ *noun* [Middle French *historiographeur,* from Late Latin *historiographus,* from Greek *historiographos,* from *historia* + *graphein* to write — more at CARVE] (15th century)
: HISTORIAN

his·to·ri·og·ra·phy \-fē\ *noun* (1569)
1 a : the writing of history; *especially* **:** the writing of history based on the critical examination of sources, the selection of particulars from the authentic materials, and the synthesis of particulars into a narrative that will stand the test of critical methods **b :** the principles, theory, and history of historical writing ⟨a course in *historiography*⟩
2 : the product of historical writing **:** a body of historical literature
— **his·to·rio·graph·i·cal** \-ē-ə-'gra-fi-kəl\ *also* **his·to·rio·graph·ic** \-fik\ *adjective*
— **his·to·rio·graph·i·cal·ly** \-ē-ə-'gra-fi-k(ə-)lē\ *adverb*

his·to·ry \'his-t(ə-)rē\ *noun, plural* **-ries** [Latin *historia,* from Greek, inquiry, history, from *histōr, istōr* knowing, learned; akin to Greek *eidenai* to know — more at WIT] (14th century)
1 : TALE, STORY
2 a : a chronological record of significant events (as affecting a nation or institution) often including an explanation of their causes **b :** a treatise presenting systematically related natural phenomena **c :** an account of a patient's medical background **d :** an established record ⟨a prisoner with a *history* of violence⟩
3 : a branch of knowledge that records and explains past events ⟨medieval *history*⟩
4 a : events that form the subject matter of a history **b :** events of the past **c :** one that is finished or done for ⟨the winning streak was *history*⟩ ⟨you're *history*⟩ **d :** previous treatment, handling, or experience (as of a metal)
his·tri·on·ic \,his-trē-'ä-nik\ *adjective* [Late Latin *histrionicus,* from Latin *histrion-, histrio* actor] (1648)
1 : deliberately affected **:** THEATRICAL
2 : of or relating to actors, acting, or the theater
synonym see DRAMATIC
— his·tri·on·i·cal·ly \-ni-k(ə-)lē\ *adverb*
his·tri·on·ics \-niks\ *noun plural but singular or plural in construction* (1864)
1 : theatrical performances
2 : deliberate display of emotion for effect
¹hit \'hit\ *verb* **hit; hit·ting** [Middle English, from Old English *hittan,* from Old Norse *hitta* to meet with, hit] (before 12th century)
transitive verb
1 a : to reach with or as if with a blow **b :** to come in contact with ⟨the ball *hit* the window⟩
2 a : to cause to come into contact **b :** to deliver (as a blow) by action **c :** to apply forcefully or suddenly ⟨*hit* the brakes⟩
3 : to affect especially detrimentally ⟨farmers *hit* by drought⟩
4 : to make a request of ⟨*hit* his friend for 10 dollars⟩ — often used with *up*
5 : to discover or meet especially by chance
6 a : to accord with **:** SUIT **b :** REACH, ATTAIN ⟨prices *hit* a new high⟩ **c :** to arrive or appear at, in, or on ⟨*hit* town⟩ ⟨the best time to *hit* the stores⟩ **d** *of fish* **:** to bite at or on **e :** to reflect accurately ⟨*hit* the right note⟩ **f :** to reach or strike (as a target) especially for a score in a game or contest ⟨couldn't seem to *hit* the basket⟩ **g :** BAT 2b
7 : to indulge in excessively ⟨*hit* the bottle⟩
intransitive verb
1 a : to strike a blow **b :** to arrive with a forceful effect like that of a blow ⟨the storm *hit*⟩
2 a : to come into contact with something **b :** ATTACK **c** *of a fish* **:** STRIKE 11b **d :** BAT 1
3 : to succeed in attaining or coming up with something — often used with *on* or *upon* ⟨*hit* on a solution⟩
4 *obsolete* **:** to be in agreement **:** SUIT
5 *of an internal-combustion engine* **:** to fire the charge in the cylinders
— hit·ter *noun*
— hit it big : to achieve great success
— hit it off : to get along well **:** become friends ⟨they *hit it off* immediately⟩
— hit on : to make especially sexual overtures to
— hit one's stride : to reach one's best speed or highest potential
— hit the books : to study especially with intensity
— hit the fan : to have a major usually undesirable impact
— hit the hay *or* **hit the sack :** to go to bed
— hit the high points *or* **hit the high spots :** to touch on or at the most important points or places
— hit the jackpot : to become notably and unexpectedly successful
— hit the nail on the head : to be exactly right
— hit the road : LEAVE, TRAVEL; *also* **:** to set out
— hit the roof *or* **hit the ceiling :** to give vent to a burst of anger or angry protest
— hit the spot : to give complete or special satisfaction — used especially of food or drink
²hit *noun* (15th century)
1 : an act or instance of hitting or being hit ⟨more *hits* than misses⟩ ⟨took a financial *hit*⟩
2 a : a stroke of luck **b :** a great success
3 : a telling or critical remark
4 : BASE HIT
5 : a quantity of a narcotic drug ingested at one time
6 : a premeditated murder committed especially by a member of a crime syndicate
— hit·less \'hit-ləs\ *adjective*
hit–and–miss \,hi-tᵊn-'mis\ *adjective* (1897)
: sometimes successful and sometimes not **:** not reliably good or successful
¹hit–and–run \-'rən\ *adjective* (1899)
1 : being or relating to a hit-and-run in baseball
2 : being or involving a motor-vehicle driver who does not stop after being involved in an accident
3 : involving or intended for quick specific action or results
²hit–and–run *noun* (1904)
: a baseball play calling for a runner on first to begin running as a pitch is delivered and for the batter to attempt to hit the pitch
³hit–and–run *intransitive verb* (1966)
: to execute a hit-and-run play in baseball
¹hitch \'hich\ *verb* [Middle English *hytchen*] (14th century)
transitive verb
1 : to move by jerks or with a tug
2 a : to catch or fasten by or as if by a hook or knot ⟨*hitched* his horse to the fence post⟩ **b** (1) **:** to connect (a vehicle or implement) with a source of motive power ⟨*hitch* a rake to a tractor⟩ (2) **:** to attach (a source of motive power) to a vehicle or instrument ⟨*hitch* the horses to the wagon⟩ **c :** to join in marriage ⟨got *hitched*⟩
3 : HITCHHIKE
intransitive verb
1 : to move with halts and jerks **:** HOBBLE
2 a : to become entangled, made fast, or linked **b :** to become joined in marriage
3 : HITCHHIKE
— hitch·er *noun*
²hitch *noun* (1664)
1 : LIMP
2 : a sudden movement or pull **:** JERK ⟨gave his trousers a *hitch*⟩
3 a : a sudden halt **:** STOPPAGE **b :** a usually unforeseen difficulty or obstacle ⟨the plan went off without a *hitch*⟩
4 : the act or fact of catching hold
5 : a connection between a vehicle or implement and a detachable source of power (as a tractor or horse)
6 : a delimited period especially of military service
7 : any of various knots used to form a temporary noose in a line or to secure a line temporarily to an object
8 : LIFT 5b
hitch·hike \'hich-,hīk\ (1926)
intransitive verb
1 : to travel by securing free rides from passing vehicles
2 : to be carried or transported by chance or unintentionally ⟨destructive insects *hitchhiking* on ships⟩
transitive verb
: to solicit and obtain (a free ride) especially in a passing vehicle
— hitch·hik·er *noun*
hitch up *intransitive verb* (1817)
: to hitch a draft animal or team to a vehicle
hi–tech *variant of* HIGH-TECH
¹hith·er \'hi-t͟hər\ *adverb* [Middle English *hider, hither,* from Old English *hider;* akin to Gothic *hidre* hither, Latin *citra* on this side — more at HE] (before 12th century)
: to this place
²hither *adjective* (14th century)
: being on the near or adjacent side
hith·er·most \-,mōst\ *adjective* (1563)
: nearest on this side
hith·er·to \-,tü, ,hi-t͟hər-'tü\ *adverb* (13th century)
: up to this or that time
hith·er·ward \'hi-t͟hə(r)-wərd\ *adverb* (before 12th century)
: HITHER
Hit·ler·ism \'hit-lə-,ri-zəm\ *noun* (1930)
: the principles and policies associated with Hitler
— Hit·ler·ite \-,rīt\ *noun or adjective*
hit list *noun* (1972)
: a list of persons or programs to be opposed or eliminated; *broadly* **:** a list of those targeted for special attention or treatment
hit man *noun* (1968)
1 : a professional assassin who works for a crime syndicate
2 : HATCHET MAN
hit–or–miss \,hit-ər-'mis\ *adjective* (1848)
: marked by a lack of care, forethought, system, or plan; *also* **:** HIT-AND-MISS
hit or miss *adverb* (1606)
: in a hit-or-miss manner **:** HAPHAZARDLY
hit parade *noun* (1929)
: a group or listing of the most popular or noteworthy items of a particular kind (as popular songs)
Hit·tite \'hi-,tīt\ *noun* [Hebrew *Ḥittī,* from Hittite *ḫatti*] (1608)
1 : a member of a conquering people in Asia Minor and Syria with an empire in the 2d millennium B.C.
2 : the extinct Indo-European language of the Hittites — see INDO-EUROPEAN LANGUAGES table
— Hittite *adjective*
HIV \,āch-,ī-'vē\ *noun* (1986)
: any of a group of retroviruses and especially HIV-1 that infect and destroy helper T cells of the immune system causing the marked reduction in their numbers that is diagnostic of AIDS — called also *AIDS virus, human immunodeficiency virus*
¹hive \'hīv\ *noun* [Middle English, from Old English *hȳf;* perhaps akin to Old Norse *hūfr* ship's hull, Latin *cūpa* tub, Sanskrit *kūpa* cave] (before 12th century)
1 : a container for housing honeybees
2 : a colony of bees
3 : a place swarming with activity
— hive·less \-ləs\ *adjective*
²hive *verb* **hived; hiv·ing** (14th century)
intransitive verb
1 *of bees* **:** to enter and take possession of a hive
2 : to reside in close association
transitive verb
1 : to collect into a hive
2 : to store up in or as if in a hive
hive off (circa 1856)
intransitive verb
chiefly British **:** to break away from or as if from a group **:** become separate
transitive verb
chiefly British **:** to make separate: as **a :** to remove from a group ⟨*hive off* the youngest campers into another room⟩ **b :** to assign (as assets or responsibilities) to another **c :** SPIN OFF
hives \'hīvz\ *noun plural but singular or plural in construction* [origin unknown] (circa 1500)
: URTICARIA
HIV–1 \-'wən\ *noun* (1986)
: a retrovirus that is the most common HIV

hiz·zon·er \ˌhi-ˈzä-nər\ *noun, often capitalized* [alteration of *his honor*] (circa 1924)
— used as a title for a man holding the office of mayor

HLA \ˌāch-(ˌ)el-ˈā\ *noun* [*h*uman *l*eukocyte *a*ntigen] (1968)
1 : the major histocompatibility complex in humans
2 : a genetic locus, gene, or antigen of HLA — often used with one or more letters to designate a locus or with letters and a number to designate an allele at the locus or the antigen corresponding to the locus and allele

HMO \ˌāch-(ˌ)em-ˈō\ *noun* (1972)
: an organization that provides comprehensive health care to voluntarily enrolled individuals and families in a particular geographic area by member physicians with limited referral to outside specialists and that is financed by fixed periodic payments determined in advance — called also *health maintenance organization*

ho \ˈhō\ *interjection* [Middle English] (15th century)
— used especially to attract attention to something specified ⟨land *ho*⟩

hoa·gie *also* **hoa·gy** \ˈhō-gē\ *noun, plural* **hoagies** [origin unknown] (1955)
: SUBMARINE 2

[1]hoar \ˈhōr, ˈhȯr\ *adjective* [Middle English *hor,* from Old English *hār;* akin to Old High German *hēr* hoary] (before 12th century)
: HOARY

[2]hoar *noun* [Middle English *hor,* hoariness, from *hor,* adjective] (1567)
: FROST 1c

[1]hoard \ˈhōrd, ˈhȯrd\ *noun* [Middle English *hord,* from Old English; akin to Gothic *huzd* treasure, Old English *hȳdan* to hide] (before 12th century)
: a supply or fund stored up and often hidden away

[2]hoard (before 12th century)
transitive verb
1 : to lay up a hoard of
2 : to keep (as one's thoughts) to oneself
intransitive verb
: to lay up a hoard
— hoard·er *noun*

hoard·ing \ˈhōr-diŋ, ˈhȯr-\ *noun* [*hourd, hoard* (hoarding)] (circa 1823)
1 : a temporary board fence put about a building being erected or repaired — called also *hoard*
2 *British* **:** BILLBOARD

hoar·frost \ˈhōr-ˌfrȯst, ˈhȯr-\ *noun* (14th century)
: FROST 1c

hoarse \ˈhōrs, ˈhȯrs\ *adjective* **hoars·er; hoars·est** [Middle English *hos, hors,* probably from (assumed) Old Norse *hārs;* akin to Old English *hās* hoarse, Old High German *heis*] (before 12th century)
1 : rough or harsh in sound **:** GRATING ⟨a *hoarse* voice⟩
2 : having a hoarse voice ⟨shouted himself *hoarse*⟩
— hoarse·ly *adverb*
— hoarse·ness *noun*

hoars·en \ˈhōr-sᵊn, ˈhȯr-\ *verb* **hoars·ened; hoars·en·ing** \ˈhōrs-niŋ, ˈhȯrs-, ˈhōr-sᵊ-, ˈhȯr-\ (1748)
transitive verb
: to make hoarse
intransitive verb
: to become hoarse

hoary \ˈhōr-ē, ˈhȯr-\ *adjective* **hoar·i·er; -est** (1530)
1 : gray or white with or as if with age
2 : extremely old **:** ANCIENT ⟨*hoary* legends⟩
— hoar·i·ness *noun*

hoa·tzin \wä(t)-ˈsēn\ *noun* [Nahuatl *huāctzin* the laughing falcon (*Herpetotheres cachinnans*)] (1661)
: a crested large South American bird (*Opisthocomos hoazin* synonym *O. hoatzin*) with blue facial skin, red eyes, brown plumage marked with white above, and claws on the first and second digits of the wing when young

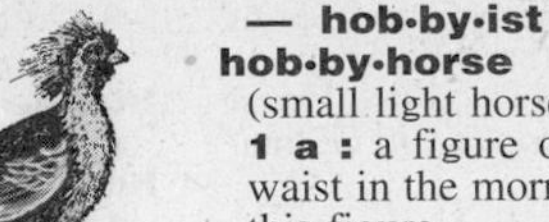
hoatzin

[1]hoax \ˈhōks\ *transitive verb* [probably contraction of *hocus*] (circa 1796)
: to trick into believing or accepting as genuine something false and often preposterous
synonym see DUPE
— hoax·er *noun*

[2]hoax *noun* (1808)
1 : an act intended to trick or dupe **:** IMPOSTURE
2 : something accepted or established by fraud or fabrication

[1]hob \ˈhäb\ *noun* [Middle English *hobbe,* from *Hobbe,* nickname for Robert] (15th century)
1 *dialect English* **:** HOBGOBLIN, ELF
2 : MISCHIEF, TROUBLE — used with *play* and *raise* ⟨always raising *hob*⟩

[2]hob *noun* [origin unknown] (1511)
1 : a projection at the back or side of a fireplace on which something may be kept warm
2 : a cutting tool used for cutting the teeth of worm wheels or gears

[3]hob *transitive verb* **hobbed; hob·bing** (1799)
1 : to cut with a hob
2 : to furnish with hobnails

Hobbes·ian \ˈhäb-zē-ən\ *adjective* (1776)
: of or relating to the English philosopher Thomas Hobbes or Hobbism

Hob·bism \ˈhä-ˌbi-zəm\ *noun* (1691)
: the philosophical system of Thomas Hobbes; *especially* **:** the Hobbesian theory that people have a fundamental right to self-preservation and to pursue selfish aims but will relinquish these rights to an absolute monarch in the interest of common safety and happiness
— Hob·bist \ˈhä-bist\ *noun or adjective*

hob·bit \ˈhä-bət\ *noun* [coined by J.R.R. Tolkien] (1937)
: a member of a fictitious peaceful and genial race of small humanlike creatures that dwell underground

[1]hob·ble \ˈhä-bəl\ *verb* **hob·bled; hob·bling** \-b(ə-)liŋ\ [Middle English *hoblen;* akin to Middle Dutch *hobbelen* to turn, roll] (14th century)
intransitive verb
: to move along unsteadily or with difficulty; *especially* **:** to limp along
transitive verb
1 : to cause to limp **:** make lame **:** CRIPPLE
2 [probably alteration of *hopple* (to hobble)] **a :** to fasten together the legs of (as a horse) to prevent straying **:** FETTER **b :** to place under handicap **:** HAMPER, IMPEDE
— hob·bler \-b(ə-)lər\ *noun*

[2]hobble *noun* (1726)
1 : a hobbling movement
2 *archaic* **:** an awkward situation
3 : something used to hobble an animal

hob·ble·bush \ˈhä-bəl-ˌbu̇sh\ *noun* (circa 1818)
: a white-flowered shrubby viburnum (*Viburnum alnifolium*) of northeastern North America having serrate rounded leaves and red berries

hob·ble·de·hoy \ˈhä-bəl-di-ˌhȯi\ *noun* [origin unknown] (1540)
: an awkward gawky youth

hobble skirt *noun* (1911)
: a skirt constricted at the bottom

[1]hob·by \ˈhä-bē\ *noun, plural* **hobbies** [Middle English *hoby,* from Middle French *hobé*] (15th century)
: a small Old World falcon (*Falco subbuteo*) formerly trained to catch small birds (as larks)

[2]hobby *noun, plural* **hobbies** [short for *hobbyhorse*] (1816)
: a pursuit outside one's regular occupation engaged in especially for relaxation
— hob·by·ist \-bē-ist\ *noun*

hob·by·horse \ˈhä-bē-ˌhȯrs\ *noun* [*hobby* (small light horse)] (1557)
1 a : a figure of a horse fastened about the waist in the morris dance **b :** a dancer wearing this figure
2 *obsolete* **:** BUFFOON
3 a : a stick having an imitation horse's head at one end that a child pretends to ride **b :** ROCKING HORSE **c :** a toy horse suspended by springs from a frame
4 a : a topic to which one constantly reverts **b :** [2]HOBBY

hob·gob·lin \ˈhäb-ˌgäb-lən\ *noun* (1530)
1 : a mischievous goblin
2 : BOGEY 2, BUGABOO

hob·nail \-ˌnāl\ *noun* [[2]*hob*] (1592)
: a short large-headed nail for studding shoe soles
— hob·nailed \-ˌnāld\ *adjective*

hob·nob \-ˌnäb\ *intransitive verb* **hob·nobbed; hob·nob·bing** [from the obsolete phrase *drink hobnob* (to drink alternately to one another)] (1763)
1 *archaic* **:** to drink sociably
2 : to associate familiarly ◇
— hob·nob·ber *noun*

[1]ho·bo \ˈhō-(ˌ)bō\ *noun, plural* **hoboes** *also* **hobos** [origin unknown] (1889)
1 : a migratory worker
2 : a homeless and usually penniless person who wanders from place to place

[2]hobo *intransitive verb* (1906)
: to live or travel in the manner of a hobo

Hob·son's choice \ˈhäb-sənz-\ *noun* [Thomas *Hobson* (died 1631) English liveryman, who required every customer to take the horse nearest the door] (1649)
: an apparently free choice when there is no real alternative

[1]hock \ˈhäk\ *noun* [Middle English *hoch, hough,* from Old English *hōh* heel; akin to Old Norse *hāsin* hock] (1540)
1 a : the tarsal joint or region in the hind limb of a digitigrade quadruped (as the horse) corresponding to the human ankle but elevated and bending backward — see HORSE illustration **b :** a joint of a fowl's leg that corresponds to the hock of a quadruped

◇ WORD HISTORY

hobnob *Hobnob,* which first appears in Shakespeare in this form, meant in various contexts "one way or another," "hit or miss," or "give and take." This linguistic Siamese twin appears originally to have been a phrase, because from its earliest appearance as *habbe or nhabbe* in 1542, *hob* is commonly separated from *nob* by an intervening word, as *hob or nob, hob a nob,* or *hob and nob.* The form *habnab,* attested earlier and probably older than *hobnob,* may descend from a joining of early Middle English *habbe,* a subjunctive form of *haven* "to have" and *nabbe,* the negative form of *habbe* (from Old English *nabban,* a contraction of *ne* "not" and *habban* "to have"). The original meaning of *habbe (or) nabbe* would thus have been "may he have or may he not have." In the 18th century, people drinking to each other's health began to use *hob and nob* and its variants—first with the verb *drink* and then as a verb itself—to suggest alternation, first one person and then another lifting a glass. To *hob-and-nob* or to *hobnob* with someone began in the 19th century to refer to simple social intercourse, with no implication of drinking.

2 : a small cut of meat from either the front or hind leg just above the foot ⟨ham *hocks*⟩

[2]hock *noun, often capitalized* [modification of German *Hochheimer,* from *Hochheim,* Germany] (circa 1625)
chiefly British **:** RHINE WINE 1

[3]hock *transitive verb* (1878)
: PAWN
— hock·er *noun*

[4]hock *noun* [Dutch *hok* pen, prison] (1883)
1 a : [2]PAWN 2 ⟨got his watch out of *hock*⟩ **b :** DEBT 3 ⟨in *hock* to the bank⟩
2 : PRISON

hock·ey \'hä-kē\ *noun* [perhaps from Middle French *hoquet* shepherd's crook, diminutive of *hoc* hook, of Germanic origin; akin to Old English *hōc* hook] (1527)
1 : FIELD HOCKEY
2 : ICE HOCKEY

hock·shop \'häk-ˌshäp\ *noun* (1871)
: PAWNSHOP

ho·cus \'hō-kəs\ *transitive verb* **ho·cussed** *or* **ho·cused; ho·cus·sing** *or* **ho·cus·ing** [obsolete *hocus,* noun, short for *hocus-pocus*] (1675)
1 : to perpetrate a trick or hoax on **:** DECEIVE
2 : to befuddle often with drugged liquor; *also* **:** DOPE, DRUG ⟨*hocussed* the favorite before the race⟩

[1]ho·cus–po·cus \ˌhō-kəs-'pō-kəs\ *noun* [probably from *hocus pocus,* imitation Latin phrase used by jugglers] (1647)
1 : SLEIGHT OF HAND
2 : nonsense or sham used especially to cloak deception

[2]hocus–pocus *transitive verb* **-cussed** *or* **-cused; -cus·sing** *or* **-cus·ing** (1774)
: to play tricks on

hod \'häd\ *noun* [probably from Middle Dutch *hodde;* akin to Middle High German *hotte* cradle] (1573)
1 : a tray or trough that has a pole handle and that is borne on the shoulder for carrying loads (as of mortar or brick)
2 : a coal scuttle

hod 1

hod carrier *noun* (1771)
: a laborer employed in carrying supplies to bricklayers, stonemasons, cement finishers, or plasterers on the job

hodge·podge \'häj-ˌpäj\ *noun* [alteration of *hotchpotch*] (15th century)
: a heterogeneous mixture **:** JUMBLE

Hodg·kin's disease \'häj-kinz-\ *noun* [Thomas *Hodgkin* (died 1866) English physician] (1865)
: a neoplastic disease that is characterized by progressive enlargement of lymph nodes, spleen, and liver and by progressive anemia

ho·do·scope \'hä-də-ˌskōp, 'hō-\ *noun* [Greek *hodos* road, path + English *-scope*] (circa 1933)
: an instrument for tracing the paths of ionizing particles by means of ion counters in close array

[1]hoe \'hō\ *noun* [Middle English *howe,* from Middle French *houe,* of Germanic origin; akin to Old High German *houwa* mattock, *houwan* to hew — more at HEW] (14th century)
1 : any of various implements for tilling, mixing, or raking; *especially* **:** an implement with a thin flat blade on a long handle used especially for cultivating, weeding, or loosening the earth around plants
2 : BACKHOE

[2]hoe *verb* **hoed; hoe·ing** (15th century)
intransitive verb
: to use a hoe **:** work with a hoe
transitive verb
1 : to weed, cultivate, or thin (a crop) with a hoe
2 : to remove (weeds) by hoeing
3 : to dress or cultivate (land) by hoeing
— ho·er \'hō(-ə)r\ *noun*

hoe·cake \'hō-ˌkāk\ *noun* (1745)
: a small cake made of cornmeal

hoe·down \-ˌdau̇n\ *noun* (1841)
1 : SQUARE DANCE
2 : a gathering featuring hoedowns

[1]hog \'hȯg, 'häg\ *noun, plural* **hogs** *also* **hog** [Middle English *hogge,* from Old English *hogg*] (14th century)
1 : a domestic swine especially when weighing more than 120 pounds; *broadly* **:** any of various wild and domestic swine
2 *usually* **hogg** *British* **:** a young unshorn sheep; *also* **:** wool from such a sheep
3 a : a selfish, gluttonous, or filthy person **b :** one that uses something to excess ⟨old cars that are gas *hogs*⟩

[2]hog *verb* **hogged; hog·ging** (1769)
transitive verb
1 : to cut (a horse's mane) short **:** ROACH
2 : to cause to arch
3 : to take in excess of one's due ⟨*hog* the credit⟩
4 : to tear up or shred (as waste wood) into bits by machine
intransitive verb
: to become curved upward in the middle — used of a ship's bottom or keel

ho·gan \'hō-ˌgän\ *noun* [Navajo *hooghan*] (1871)
: a Navajo Indian dwelling usually made of logs and mud with a door traditionally facing east

hogan

hog·back \'hȯg-ˌbak, 'häg-\ *noun* (1840)
: a ridge of land formed by the outcropping edges of tilted strata; *broadly* **:** a ridge with a sharp summit and steeply sloping sides

hog cholera *noun* (1859)
: a highly infectious often fatal virus disease of swine characterized by fever, loss of appetite, weakness, erythematous lesions especially in light-skinned animals, and severe leukopenia

hog·fish \'hȯg-ˌfish, 'häg-\ *noun* (1734)
: a large West Indian and Florida wrasse (*Lachnolaimus maximus*) often used for food

hog·get \'hä-gət, 'hȯ-\ *noun* [Middle English, from [1]*hog* + *-et*] (15th century)
chiefly British **:** HOG 2

hog·gish \'hȯ-gish, 'hä-\ *adjective* (15th century)
: grossly selfish, gluttonous, or filthy
— hog·gish·ly *adverb*
— hog·gish·ness *noun*

hog heaven *noun* (1945)
: an extremely satisfying state or situation

Hog·ma·nay \ˌhäg-mə-'nā, 'häg-mə-ˌ\ *noun* [origin unknown] (circa 1680)
1 *Scottish* **:** the eve of New Year's Day
2 *Scottish* **:** a gift solicited or given at Hogmanay

hog·nose snake \'hȯg-ˌnōz-, 'häg-\ *noun* (1736)
: any of a genus (*Heterodon*) of rather small harmless stout-bodied North American colubrid snakes with keeled scales and an upturned snout that seldom bite but hiss wildly and often play dead when disturbed — called also *hog-nosed snake, puff adder*

hog score *noun* [*hog* (curling stone that fails to reach the score)] (1685)
: a line which is marked across a curling rink seven yards from the tee and beyond which a stone must pass or be removed from the ice — called also *hog line*

hogs·head \'hȯgz-ˌhed, 'hägz-\ *noun* (14th century)
1 : a large cask or barrel
2 : any of various units of capacity; *especially* **:** a U.S. unit equal to 63 gallons (238 liters)

hog sucker *noun* (1877)
: a North American sucker (*Hypentelium nigricans*) that is brassy olive marked with brown and is sometimes used for food

hog–tie \'hȯg-ˌtī, 'häg-\ *transitive verb* (1894)
1 : to tie together the feet of
2 : to make helpless

hog·wash \-ˌwȯsh, -ˌwäsh\ *noun* (15th century)
1 : SWILL 2a, SLOP
2 : NONSENSE, BALDERDASH

hog–wild \-'wī(ə)ld\ *adjective* (1904)
: lacking in restraint of judgment or temper ⟨*hog-wild* enthusiasm⟩ ⟨would go *hog-wild* if unconfined by constitutional limitations —Leo Egan⟩

[1]Ho·hen·stau·fen \'hō-ən-ˌshtau̇-fən, -ˌstau̇-\ *noun* (circa 1895)
: a member of the Hohenstaufen family; *especially* **:** a Hohenstaufen monarch

[2]Hohenstaufen *adjective* (1921)
: of or relating to a princely German family that reigned over the Holy Roman Empire from 1138–1254 and over Sicily from 1194–1266

[1]Ho·hen·zol·lern \ˌhō-ən-'zä-lərn, -'zȯ-\ *noun* (circa 1895)
: a member of the Hohenzollern family; *especially* **:** a Hohenzollern monarch

[2]Hohenzollern *adjective* (1924)
: of or relating to a princely German family that reigned in Prussia from 1701–1918 and in Germany from 1871–1918

Ho·ho·kam \ˌhō-hō-'käm\ *noun, plural* **Hohokam** [O'odham (Uto-Aztecan language of southern Arizona) *huhugam,* literally, those who have gone] (1884)
: a member of a prehistoric desert culture of the southwestern U.S. centering in the Gila Valley of Arizona and characterized especially by irrigated agriculture
— Hohokam *adjective*

ho–hum \'hō-'həm\ *adjective* (1969)
1 : ROUTINE, DULL ⟨a *ho-hum* existence⟩
2 : BORED, INDIFFERENT ⟨a *ho-hum* reaction⟩

ho hum *interjection* [imitative] (1924)
— used to express weariness, boredom, or disdain

hoick \'hȯik\ *transitive verb* [probably alteration of [1]*hike*] (1898)
: to move or pull abruptly **:** YANK ⟨was *hoicked* out of my job —Vincent Sheean⟩

hoi pol·loi \ˌhȯi-pə-'lȯi\ *noun plural* [Greek, the many] (1837)
: the general populace **:** MASSES ■

hoise \'hȯiz\ *transitive verb* **hoised** \'hȯizd\ *or* **hoist** \'hȯist\; **hois·ing** \'hȯi-ziŋ\ [alteration of *hysse* to hoist, perhaps from Low German *hissen*] (1509)
: HOIST
— hoist with one's own petard : victimized or hurt by one's own scheme

[1]hoist \'hȯist, *chiefly dialect* 'hīst\ *verb* [alteration of *hoise*] (15th century)
transitive verb
1 : LIFT, RAISE; *especially* **:** to raise into position by or as if by means of tackle
2 : DRINK 1 ⟨*hoist* a few beers⟩

□ USAGE
hoi polloi Since *hoi polloi* is a transliteration of the Greek for "the many," some 20th century critics assert that the phrase should not be preceded by *the.* They find *the* to be redundant—an opinion that fails to recognize that *hoi* means nothing at all in English. Nonetheless, the opinion has influenced the omission of *the* in the usage of some writers. But most writers use *the,* which is normal English grammar.

intransitive verb
: to become hoisted : RISE
synonym see LIFT
— **hoist·er** *noun*

[2]hoist *noun* (1654)
1 : an act of hoisting **:** LIFT
2 : an apparatus for hoisting
3 : the height of a flag when viewed flying

[1]hoi·ty–toi·ty \ˌhȯi-tē-ˈtȯi-tē, ˌhī-tē-ˈtī-tē\ *noun* [rhyming compound from English dialect *hoit* to play the fool] (1668)
: thoughtless giddy behavior

[2]hoity–toity *adjective* (1690)
1 : thoughtlessly silly or frivolous **:** FLIGHTY
2 : marked by an air of assumed importance **:** HIGHFALUTIN

hoke \ˈhōk\ *transitive verb* **hoked; hok·ing** [*hokum*] (1925)
: to give an impressive but false value or quality to **:** FAKE — usually used with *up* ⟨used parts of B-grade movies to *hoke* up a film —Robert Sherrill⟩

hok·ey \ˈhō-kē\ *adjective* **hok·i·er; -est** (1927)
1 : [1]CORNY 3 ⟨the usual *hokey* melodrama⟩
2 : obviously contrived **:** PHONY ⟨the plots are tricky but not *hokey* —Cleveland Amory⟩
— **hok·ey·ness** *or* **hok·i·ness** \ˈhō-kē-nəs\ *noun*
— **hok·i·ly** \ˈhō-kə-lē\ *adverb*

ho·key·po·key \ˌhō-kē-ˈpō-kē\ *noun* (circa 1878)
1 : HOCUS-POCUS 2
2 : ice cream sold by street vendors

hok·ku \ˈhō-(ˌ)kü\ *noun, plural* **hokku** [Japanese] (1898)
: HAIKU

ho·kum \ˈhō-kəm\ *noun* [probably blend of *hocus-pocus* and *bunkum*] (1917)
1 : a device used (as by showmen) to evoke a desired audience response
2 : pretentious nonsense **:** BUNKUM

hol- *or* **holo-** *combining form* [Middle English, from Old French, from Latin, from Greek, from *holos* whole — more at SAFE]
1 : complete **:** total ⟨*holo*hedral⟩
2 : completely **:** totally ⟨*hol*andric⟩

hol·an·dric \hō-ˈlan-drik, hä-\ *adjective* [International Scientific Vocabulary, from *hol-* + *andr-* + *-ic*] (1930)
: transmitted by a gene in the nonhomologous portion of the Y chromosome

Hol·arc·tic \hō-ˈlärk-tik, hä-, -ˈlär-tik\ *adjective* (1883)
: of, relating to, or being the biogeographic region including the northern parts of the Old and the New Worlds and comprising the Nearctic and Palearctic regions or subregions

[1]hold \ˈhōld\ *verb* **held** \ˈheld\; **hold·ing** [Middle English, from Old English *healdan;* akin to Old High German *haltan* to hold, and perhaps to Latin *celer* rapid, Greek *klonos* agitation] (before 12th century)
transitive verb
1 a : to have possession or ownership of or have at one's disposal ⟨*holds* property worth millions⟩ ⟨the bank *holds* the title to the car⟩ **b :** to have as a privilege or position of responsibility ⟨*hold* a professorship⟩ **c :** to have as a mark of distinction ⟨*holds* the record for the 100-yard dash⟩ ⟨*holds* a PhD⟩
2 : to keep under restraint ⟨*hold* price increases to a minimum⟩: as **a :** to prevent free expression of ⟨*hold* your temper⟩ **b :** to prevent from some action ⟨ordered the troops to *hold* fire⟩ ⟨the only restraining motive which may *hold* the hand of a tyrant —Thomas Jefferson⟩ **c :** to keep back from use ⟨ask them to *hold* a room for us⟩ ⟨I'll have a hot dog, and *hold* the mustard⟩ **d :** to delay temporarily the handling of ⟨please *hold* all my calls⟩
3 : to make liable or accountable or bound to an obligation ⟨I'll *hold* you to your promise⟩
4 a : to have or maintain in the grasp ⟨*hold* my hand⟩ ⟨this is how you *hold* the racket⟩; *also* **:** AIM, POINT ⟨*held* a gun on them⟩ **b :** to support in a particular position or keep from falling or moving ⟨*hold* me up so I can see⟩ ⟨*hold* the ladder steady⟩ ⟨a clamp *holds* the whole thing together⟩ ⟨*hold* your head up⟩ **c :** to bear the pressure of **:** SUPPORT ⟨can the roof *hold* all of that weight⟩
5 : to prevent from leaving or getting away ⟨*hold* the train⟩: as **a :** to avoid emitting or letting out ⟨how long can you *hold* your breath⟩ **b :** to restrain as or as if a captive ⟨the suspect was *held* without bail⟩ ⟨*held* them at gunpoint⟩; *also* **:** to have strong appeal to ⟨the book *held* my interest throughout⟩
6 a : to enclose and keep in a container or within bounds **:** CONTAIN ⟨the jug *holds* one gallon⟩ ⟨this corral will not *hold* all of the horses⟩ **b :** to be able to consume easily or without undue effect ⟨can't *hold* any more pie⟩; *especially* **:** to be able to drink (alcoholic beverages) without becoming noticeably drunk ⟨can't *hold* your liquor⟩ **c :** ACCOMMODATE ⟨the restaurant *holds* 400 diners⟩ **d :** to have as a principal or essential feature or attribute ⟨the book *holds* a number of surprises⟩; *also* **:** to have in store ⟨no one knows what the future *holds*⟩
7 a : to have in the mind or express as a judgment, opinion, or belief ⟨I *hold* the view that this is wrong⟩ ⟨*hold* a grudge⟩ ⟨*holding* that it is nobody's business but his —Jack Olsen⟩ — often used with *against* ⟨in America they *hold* everything you say against you —Paul McCartney⟩ **b :** to think of in a particular way **:** REGARD ⟨were *held* in high esteem⟩
8 a : to assemble for and carry on the activity of ⟨*held* a convention⟩ **b :** to cause to be carried on **:** CONDUCT ⟨will *hold* a seminar⟩ **c :** to produce or sponsor especially as a public exhibition ⟨will *hold* an art show⟩
9 a : to maintain occupation, control, or defense of ⟨the troops *held* the ridge⟩; *also* **:** to resist the offensive efforts or advance of ⟨*held* the opposing team to just two points⟩ **b :** to maintain (a certain condition, situation, or course of action) without change ⟨*hold* a course due east⟩
10 : to cover (a part of the body) especially for protection ⟨had to *hold* their ears because of the cold⟩
intransitive verb
1 a : to maintain position **:** refuse to give ground ⟨the defensive line is *holding*⟩ **b :** to continue in the same way or to the same degree **:** LAST ⟨hopes the weather will *hold*⟩ — often used with *up*
2 : to derive right or title — often used with *of* or *from*
3 : to be or remain valid **:** APPLY ⟨the rule *holds* in most cases⟩ — often used in the phrase *hold true*
4 : to maintain a grasp on something **:** remain fastened to something ⟨the anchor *held* in the rough sea⟩
5 : to go ahead as one has been going ⟨*held* south for several miles⟩
6 : to bear or carry oneself ⟨asked him to *hold* still⟩
7 : to forbear an intended or threatened action **:** HALT, PAUSE — often used as a command
8 : to stop counting during a countdown
9 *slang* **:** to have illicit drug material in one's possession
synonym see CONTAIN
— **hold a candle to :** to qualify for comparison with
— **hold court :** to be the center of attention among friends or admirers
— **hold forth :** to speak at length **:** EXPATIATE
— **hold hands :** to engage one's hand with another's especially as an expression of affection
— **hold one's horses :** to slow down or stop for a moment — usually used in the imperative
— **hold one's own :** to do well in the face of difficulty or opposition
— **hold one's tongue :** to keep silent
— **hold sway :** to have a dominant influence **:** RULE
— **hold the bag 1 :** to be left empty-handed **2 :** to bear alone a responsibility that should have been shared by others
— **hold the fort 1 :** to maintain a firm position **2 :** to take care of usual affairs ⟨is *holding the fort* until the manager returns⟩
— **hold the line :** to maintain the current position or situation ⟨*hold the line* on prices⟩
— **hold to :** to give firm assent to **:** adhere to strongly ⟨*holds to* his promise⟩
— **hold to account :** to hold responsible
— **hold water :** to stand up under criticism or analysis
— **hold with :** to agree with or approve of

[2]hold *noun* (14th century)
1 : STRONGHOLD 1
2 a : CONFINEMENT, CUSTODY **b :** PRISON
3 a (1) **:** the act or the manner of holding or grasping **:** GRIP ⟨released his *hold* on the handle⟩ (2) **:** a manner of grasping an opponent in wrestling **b :** a nonphysical bond that attaches, restrains, or constrains or by which something is affected, controlled, or dominated ⟨has lost its *hold* on the broad public —Oscar Cargill⟩ **c :** full comprehension ⟨get *hold* of exactly what is happening —J. P. Lyford⟩ **d :** full or immediate control **:** POSSESSION ⟨get *hold* of yourself⟩ ⟨wants to get *hold* of a road map⟩ **e :** TOUCH 14 — used with *of* ⟨tried to get *hold* of me⟩
4 : something that may be grasped as a support
5 a : FERMATA **b :** the time between the onset and the release of a vocal articulation
6 : a sudden motionless posture at the end of a dance
7 a : an order or indication that something is to be reserved or delayed **b :** a delay in a countdown (as in launching a spacecraft)
— **on hold 1 :** into a state of interruption during a telephone call when one party switches to another line without totally disconnecting the other party **2 :** into a state or period of indefinite suspension ⟨put our plans *on hold*⟩

[3]hold *noun* [alteration of *hole*] (1591)
1 : the interior of a ship below decks; *especially* **:** the cargo deck of a ship
2 : the cargo compartment of a plane

hold·all \ˈhōl-ˌdȯl\ *noun* (1851)
chiefly British **:** an often cloth traveling case or bag

hold·back \ˈhōl(d)-ˌbak\ *noun* (1581)
1 : something that retains or restrains
2 a : the act of holding back **b :** something held back

hold back (1535)
transitive verb
1 : to hinder the progress or achievement of **:** RESTRAIN
2 : to refrain from revealing or parting with
intransitive verb
1 : to keep oneself in check
2 : to refrain from revealing or parting with something

hold down \ˈhōl-ˌdau̇n\ *noun* (1888)
1 : something used to fasten an object in place
2 a : an act of holding down **b :** LIMIT ⟨agreed to wage-rate *hold-downs*⟩

hold down *transitive verb* (1533)
1 : to keep within limits ⟨*hold* the noise *down*⟩
2 : to assume the responsibility for ⟨*holding down* two jobs⟩

hold·en \ˈhōl-dən\ *archaic past participle of* HOLD

hold·er \ˈhōl-dər\ *noun* (14th century)

\ə\ **abut** \ᵊ\ **kitten** \ər\ **further** \a\ **ash** \ā\ **ace** \ä\ **mop, mar** \au̇\ **out** \ch\ **chin** \e\ **bet** \ē\ **easy** \g\ **go** \i\ **hit** \ī\ **ice** \j\ **job** \ŋ\ **sing** \ō\ **go** \ȯ\ **law** \ȯi\ **boy** \th\ **thin** \t͟h\ **the** \ü\ **loot** \u̇\ **foot** \y\ **yet** \zh\ **vision** *see also* Guide to Pronunciation

1 : a person that holds: as **a** (1) **:** OWNER (2) **:** TENANT **b :** a person in possession of and legally entitled to receive payment of a bill, note, or check
2 : a device that holds ⟨cigarette *holder*⟩
holder in due course (1882)
: one other than the original recipient who holds a legally effective negotiable instrument (as a promissory note) and who has a right to collect from and no responsibility toward the issuer
hold·fast \'hōl(d)-ˌfast\ *noun* (1566)
1 : something to which something else may be firmly secured
2 a : a part by which a plant clings to a flat surface **b :** an organ by which a parasitic animal attaches itself to its host
¹hold·ing \'hōl-diŋ\ *noun* (15th century)
1 a : land held especially by a vassal or tenant **b :** property (as land or securities) owned — usually used in plural
2 : a ruling of a court especially on an issue of law raised in a case — compare DICTUM
3 : something that holds
²holding *adjective* (1568)
1 : having the effect of holding back or delaying something ⟨the [war] represented a *holding* action against the spread of world Communism —Sidney Offit⟩
2 : intended for usually temporary storage or retention ⟨a *holding* tank⟩
holding company *noun* (1906)
: a company whose primary business is holding a controlling interest in the securities of other companies — compare INVESTMENT COMPANY
holding pattern *noun* (circa 1952)
1 : the usually oval course flown (as over an airport) by aircraft awaiting clearance especially to land
2 : a state of waiting or suspended activity or progress
hold off (15th century)
transitive verb
1 : to block from an objective **:** DELAY
2 : to defer action on **:** POSTPONE
3 : to fight to a standoff **:** WITHSTAND
intransitive verb
: to defer or temporarily stop doing something
hold on *intransitive verb* (13th century)
1 a : to maintain a condition or position **:** PERSIST **b :** to maintain a grasp on something **:** HANG ON
2 : to await something (as a telephone connection) desired or requested; *broadly* **:** WAIT
— hold on to : to maintain possession of or adherence to
hold·out \'hōl-ˌdau̇t\ *noun* (1945)
: one that holds out (as in negotiations); *also* **:** an instance of holding out
hold out (1585)
intransitive verb
1 : to remain unsubdued or operative **:** continue to cope or function
2 : to refuse to go along with others in a concerted action or to come to an agreement ⟨*holding out* for a shorter workweek⟩
transitive verb
1 : to present as something realizable **:** PROFFER
2 : to represent to be
— hold out on : to withhold something (as information) from
hold·over \'hōl-ˌdō-vər\ *noun* (1893)
: one that is held over
hold over (1647)
intransitive verb
: to continue (as in office) for a prolonged period
transitive verb
1 a : POSTPONE, DEFER **b :** to retain in a condition or position from an earlier period
2 : to prolong the engagement of ⟨the film was *held over* another week⟩
hold·up \'hōl-ˌdəp\ *noun* (1837)
1 : DELAY
2 : a robbery carried out at gunpoint
hold up (1851)
transitive verb
1 : to rob at gunpoint
2 : DELAY, IMPEDE
3 : to call attention to **:** single out ⟨his work was *held up* to ridicule⟩ ⟨*hold* this *up* as perfection —*Times Literary Supplement*⟩
intransitive verb
: to continue in the same condition without failing or losing effectiveness or force ⟨you seem to be *holding up* under the strain⟩
¹hole \'hōl\ *noun* [Middle English, from Old English *hol* (from neuter of *hol,* adjective, hollow) & *holh;* akin to Old High German *hol,* adjective, hollow; perhaps akin to Old English *helan* to conceal — more at HELL] (before 12th century)
1 a : an opening through something **:** PERFORATION ⟨have a *hole* in my coat⟩ **b :** an area where something is missing **:** GAP: as (1) **:** a serious discrepancy **:** FLAW, WEAKNESS ⟨there are *holes* in your logic⟩ (2) **:** an opening in a defensive formation; *especially* **:** the area of a baseball field between the positions of shortstop and third baseman (3) **:** a defect in a crystal (as of a semiconductor) that is due to an electron's having left its normal position in one of the crystal bonds and that is equivalent in many respects to a positively charged particle
2 : a hollowed-out place: as **a :** a cave, pit, or well in the ground **b :** BURROW **c :** an unusually deep place in a body of water
3 : a wretched or dreary place
4 a : a shallow cylindrical hole in the putting green of a golf course into which the ball is played **b :** a part of the golf course from tee to putting green ⟨just beginning play on the third *hole*⟩; *also* **:** the play on such a hole as a unit of scoring ⟨won the *hole* by two strokes⟩
5 a : an awkward position or circumstance **:** FIX ⟨got the rebels out of a *hole* at the battle —Kenneth Roberts⟩ **b :** a position of owing or losing money ⟨$10 million in the *hole*⟩ ⟨raising money to get out of the *hole*⟩
— in the hole 1 : having a score below zero **2 :** at a disadvantage
²hole *verb* **holed; hol·ing** (before 12th century)
transitive verb
1 : to make a hole in
2 : to drive or hit into a hole
intransitive verb
: to make a hole in something
hole-and-corner *adjective* (1835)
1 : being or carried on in a place away from public view **:** CLANDESTINE
2 : INSIGNIFICANT
hole card *noun* (1908)
: a card in stud poker that is properly dealt facedown and that the holder need not expose before the showdown
hole in one (1925)
: ACE 4
hole-in-the-wall *noun, plural* **holes-in-the-wall** (1856)
: a small and often unpretentious out-of-the-way place (as a restaurant)
hole out *intransitive verb* (1857)
: to play one's ball into the hole in golf
hole up (1875)
intransitive verb
: to take refuge or shelter in or as if in a hole or cave
transitive verb
: to place in or as if in a refuge or hiding place
hol·ey \'hō-lē\ *adjective* (13th century)
: having holes
¹hol·i·day \'hä-lə-ˌdā, *British usually* 'hä-lə-dē\ *noun* [Middle English, from Old English *hāligdæg,* from *hālig* holy + *dæg* day] (before 12th century)
1 : HOLY DAY
2 : a day on which one is exempt from work; *specifically* **:** a day marked by a general suspension of work in commemoration of an event
3 *chiefly British* **:** a period of relaxation **:** VACATION — often used in the phrase *on holiday;* often used in plural
²holiday *intransitive verb* (1869)
: to take or spend a holiday especially in travel or at a resort
— hol·i·day·er *noun*
hol·i·day·mak·er \'hä-lə-dē-ˌmā-kər, 'hä-lə-ˌdā-\ *noun* (1836)
chiefly British **:** VACATIONER
hol·i·days \-lə-ˌdāz, *British usually* -lə-dēz\ *adverb* (circa 1961)
: on holidays repeatedly **:** on any holiday
ho·li·er-than-thou \ˌhō-lē-ər-t͟hən-'t͟hau̇\ *adjective* (1859)
: marked by an air of superior piety or morality
¹ho·li·ness \'hō-lē-nəs\ *noun* (before 12th century)
1 : the quality or state of being holy — used as a title for various high religious dignitaries ⟨his *holiness* the pope⟩
2 : SANCTIFICATION 2
²holiness *adjective, often capitalized* (1888)
: emphasizing the doctrine of the second blessing; *specifically* **:** of or relating to a perfectionist movement arising in U.S. Protestantism in the late 19th century
ho·lism \'hō-ˌli-zəm\ *noun* [*hol-* + *-ism*] (1926)
1 : a theory that the universe and especially living nature is correctly seen in terms of interacting wholes (as of living organisms) that are more than the mere sum of elementary particles
2 : a holistic study or method of treatment
— ho·list \'hō-list\ *noun*
ho·lis·tic \hō-'lis-tik\ *adjective* (1926)
1 : of or relating to holism
2 : relating to or concerned with wholes or with complete systems rather than with the analysis of, treatment of, or dissection into parts ⟨*holistic* medicine attempts to treat both the mind and the body⟩ ⟨*holistic* ecology views man and the environment as a single system⟩
— ho·lis·ti·cal·ly \-ti-k(ə-)lē\ *adverb*
hol·land \'hä-lənd\ *noun, often capitalized* [Middle English *holand,* from *Holand,* county in the Netherlands, from Middle Dutch *Holland*] (14th century)
: a cotton or linen fabric in plain weave usually heavily sized or glazed and used for window shades, bookbinding, and clothing
hol·lan·daise \ˌhä-lən-'dāz\ *noun* [French *sauce hollandaise,* literally, Dutch sauce] (1907)
: a rich sauce made basically of butter, egg yolks, and lemon juice or vinegar
Hol·lands \'hä-lən(d)z\ *noun* [Dutch *hollandsch,* from *hollandsch genever* Dutch gin] (1788)
: gin made in the Netherlands — called also *Holland gin*
¹hol·ler \'hä-lər\ *verb* **hol·lered; hol·ler·ing** \'häl-riŋ, 'hä-lə-\ [alteration of *hollo*] (1699)
intransitive verb
1 : to cry out (as to attract attention or in pain) **:** SHOUT
2 : GRIPE, COMPLAIN
transitive verb
: to call out (a word or phrase)
²holler *noun* (1825)
1 : SHOUT, CRY
2 : COMPLAINT
3 : a freely improvised work song of black Americans ⟨field *hollers*⟩
³holler *chiefly dialect variant of* HOLLOW
Hol·ler·ith card \'hä-lə-rith-\ *noun* [Herman *Hollerith* (died 1929) American engineer] (1946)
: PUNCHED CARD

Hollerith code *noun* (1962)
: a system for encoding alphanumeric information on punched cards

[1]**hol·lo** *also* **hol·loa** \ˈhä-(ˌ)lō\ *or* **hol·la** \ˈhä-lə\ (14th century)
intransitive verb
: to cry hollo **:** HOLLER
transitive verb
1 : to call or cry hollo to
2 : to utter loudly **:** HOLLER

[2]**hollo** *also* **holloa** *or* **holla** *noun, plural* **hollos** *also* **holloas** *or* **hollas** (15th century)
: an exclamation or call of hollo

[3]**hol·lo** \hä-ˈlō, hə-; ˈhä-(ˌ)\ *also* **hol·loa** \hä-ˈlō, hə-\ *or* **hol·la** \hə-ˈlä, ˈhä-(ˌ)\ *interjection* [origin unknown] (1588)
1 — used to attract attention (as when a fox is spied during a fox hunt)
2 — used as a call of encouragement or jubilation

[1]**hollow** *noun* (before 12th century)
1 : an unfilled space **:** CAVITY, HOLE
2 : a depressed or low part of a surface; *especially* **:** a small valley or basin

[2]**hol·low** \ˈhä-(ˌ)lō\ *adjective* **hol·low·er** \ˈhä-lə-wər\; **hol·low·est** \-lə-wəst\ [Middle English *holw, holh,* from *holh* hole, den, from Old English *holh* hole, hollow — more at HOLE] (13th century)
1 : having an indentation or inward curve **:** CONCAVE, SUNKEN
2 : having a cavity within ⟨*hollow* tree⟩
3 : lacking in real value, sincerity, or substance **:** FALSE, MEANINGLESS ⟨*hollow* promises⟩ ⟨a victory over a weakling is *hollow* and without triumph —Ernest Beaglehole⟩
4 : reverberating like a sound made in or by beating on a large empty enclosure **:** MUFFLED
synonym see VAIN
— **hol·low·ly** \ˈhä-lō-lē, -lə-lē\ *adverb*
— **hol·low·ness** *noun*

[3]**hollow** (15th century)
transitive verb
1 : to make hollow
2 : to form by a hollowing action — usually used with *out* ⟨rain barrels *hollowed* out from trees —Robert Shaplen⟩
intransitive verb
: to become hollow

[4]**hollow** *adverb* (1601)
1 : so as to have a hollow sound
2 : COMPLETELY, THOROUGHLY ⟨an ongoing story that has the old cowboy-and-Indians genre beat *hollow* —Barbara Bannon⟩ — often used with *all*

hol·low·ware *or* **hol·lo·ware** \ˈhä-(ˌ)lō-ˌwar, -ˌwer\ *noun* (1682)
: vessels (as bowls, cups, or vases) usually of pottery, glass, or metal that have a significant depth and volume — compare FLATWARE

hol·ly \ˈhä-lē\ *noun, plural* **hollies** [Middle English *holin, holly,* from Old English *holen;* akin to Old High German *hulis* holly, Middle Irish *cuilenn*] (before 12th century)
1 : any of a genus (*Ilex* of the family Aquifoliaceae, the holly family) of trees and shrubs; *especially* **:** either of two (*I. opaca* of the eastern U.S. and *I. aquifolium* of Eurasia) with spiny-margined evergreen leaves and usually red berries often used for Christmas decorations
2 : the foliage or branches of the holly

hol·ly·hock \ˈhä-lē-ˌhäk, -ˌhȯk\ *noun* [Middle English *holihoc* marshmallow, from *holi* holy + *hoc* mallow, from Old English] (1548)
: a tall widely cultivated perennial Chinese herb (*Alcea rosea* synonym *Althaea rosea*) of the mallow family with large coarse rounded leaves and tall spikes of showy flowers

hollyhock

[1]**Hol·ly·wood** \ˈhä-lē-ˌwu̇d\ *noun* [*Hollywood,* district of Los Angeles, Calif.] (1923)
: the American motion-picture industry
— **Hol·ly·wood·ish** \-ˌwu̇-dish\ *adjective*

[2]**Hollywood** *adjective* (1935)
: of or characteristic of people in the American motion-picture industry ⟨*Hollywood* lifestyle⟩

Hollywood bed *noun* (1947)
: a mattress on a box spring supported by low legs and often having an upholstered headboard

holm \ˈhō(l)m\ *noun* [Middle English, from Old English, from Old Norse *hōlmr;* akin to Old English *hyll* hill] (before 12th century)
British **:** a small inland or inshore island; *also* **:** BOTTOMS

Holmes·ian \ˈhōm-zē-ən *also* ˈhōlm-\ *adjective* [Sherlock *Holmes,* detective in stories by Sir Arthur Conan Doyle] (1929)
: of, characteristic of, or suggestive of the detective Sherlock Holmes

hol·mi·um \ˈhō(l)-mē-əm\ *noun* [New Latin, from *Holmia* Stockholm, Sweden] (1879)
: a metallic element of the rare-earth group that occurs with yttrium and forms highly magnetic compounds — see ELEMENT table

holm oak *noun* (1597)
: a southern European evergreen oak (*Quercus ilex*)

holo- — see HOL-

ho·lo·blas·tic \ˌhō-lə-ˈblas-tik, ˌhä-\ *adjective* [International Scientific Vocabulary] (1872)
: characterized by cleavage planes that divide the whole egg into distinct and separate though coherent blastomeres — compare MEROBLASTIC

ho·lo·caust \ˈhō-lə-ˌkȯst, ˈhä- *also* -ˌkäst *or* ˈhȯ-lə-ˌkȯst\ *noun* [Middle English, from Old French *holocauste,* from Late Latin *holocaustum,* from Greek *holokauston,* from neuter of *holokaustos* burnt whole, from *hol-* + *kaustos* burnt, from *kaiein* to burn — more at CAUSTIC] (13th century)
1 : a sacrifice consumed by fire
2 : a thorough destruction involving extensive loss of life especially through fire ⟨a nuclear *holocaust*⟩
3 a *often capitalized* **:** the mass slaughter of European civilians and especially Jews by the Nazis during World War II — usually used with *the* **b :** a mass slaughter of people; *especially* **:** GENOCIDE ◇

Ho·lo·cene \ˈhō-lə-ˌsēn, ˈhä-\ *adjective* [International Scientific Vocabulary] (1897)
: of, relating to, or being the present or post-Pleistocene geologic epoch — see GEOLOGIC TIME table
— **Holocene** *noun*

ho·lo·crine \-krən, -ˌkrīn, -ˌkrēn\ *adjective* [International Scientific Vocabulary *hol-* + Greek *krinein* to separate — more at CERTAIN] (circa 1905)
: producing or being a secretion resulting from lysis of secretory cells ⟨*holocrine* gland⟩

ho·lo·en·zyme \ˌhō-lō-ˈen-ˌzīm\ *noun* [International Scientific Vocabulary] (1943)
: a catalytically active enzyme consisting of an apoenzyme combined with its cofactor

Ho·lo·fer·nes \ˌhä-lə-ˈfər-(ˌ)nēz, ˌhō-\ *noun* [Late Latin, from Greek *Holophernēs*]
: a general of Nebuchadnezzar's who led an Assyrian army against Israel and was beheaded in his sleep by Judith

ho·lo·gram \ˈhō-lə-ˌgram, ˈhä-\ *noun* (1949)
: a three-dimensional image reproduced from a pattern of interference produced by a split coherent beam of radiation (as a laser); *also* **:** the pattern of interference itself

ho·lo·graph \ˈhō-lə-ˌgraf, ˈhä-\ *noun* [Late Latin *holographus,* from Late Greek *holographos,* from Greek *hol-* + *graphein* to write — more at CARVE] (circa 1623)
: a document wholly in the handwriting of its author; *also* **:** the handwriting itself ⟨a letter in the president's *holograph*⟩
— **holograph** *or* **ho·lo·graph·ic** \ˌhō-lə-ˈgra-fik, ˌhä-\ *adjective*

ho·log·ra·phy \hō-ˈlä-grə-fē\ *noun* (1964)
: the art or process of making or using a hologram
— **ho·lo·graph** \ˈhō-lə-ˌgraf, ˈhä-\ *transitive verb*
— **ho·log·ra·pher** \hō-ˈlä-grə-fər\ *noun*
— **ho·lo·graph·ic** \ˌhō-lə-ˈgra-fik, ˌhä-\ *adjective*
— **ho·lo·graph·i·cal·ly** \-fi-k(ə-)lē\ *adverb*

ho·lo·he·dral \ˌhō-lə-ˈhē-drəl, ˌhä-\ *adjective* (1837)
of a crystal **:** having all the faces required by complete symmetry — compare HEMIHEDRAL, TETARTOHEDRAL

ho·lo·me·tab·o·lous \ˌhō-lō-mə-ˈta-bə-ləs, ˌhä-\ *adjective* (1870)
: characterized by complete metamorphosis ⟨*holometabolous* insects⟩ — compare HEMIMETABOLOUS
— **ho·lo·me·tab·o·lism** \-ˌli-zəm\ *noun*

ho·lo·phras·tic \ˌhō-lə-ˈfras-tik, ˌhä-\ *adjective* [International Scientific Vocabulary *hol-* + *-phrastic* (from Greek *phrazein* to point out, declare)] (1860)
: expressing a complex of ideas in a single word or in a fixed phrase

ho·lo·phyt·ic \-ˈfi-tik\ *adjective* (1885)
: obtaining food after the manner of a green plant by photosynthetic activity

ho·lo·thu·ri·an \-ˈthu̇r-ē-ən, -ˈthyu̇r-\ *noun* [ultimately from Greek *holothourion* water polyp] (circa 1842)
: SEA CUCUMBER
— **holothurian** *adjective*

ho·lo·type \ˈhō-lə-ˌtīp, ˈhä-\ *noun* (1897)
1 : the single specimen designated by an author as the type of a species or lesser taxon at the time of establishing the group
2 : the type of a species or lesser taxon designated at a date later than that of establishing a group or by another person than the author of the taxon
— **ho·lo·typ·ic** \ˌhō-lə-ˈti-pik, ˌhä-\ *adjective*

ho·lo·zo·ic \ˌhō-lə-ˈzō-ik, ˌhä-\ *adjective* (1885)
: characterized by food procurement after the manner of most animals by the ingestion of

◇ WORD HISTORY
holocaust The ancient Greek word *holokauston* meant literally "something completely burnt up." In the Septuagint, the Greek translation of Jewish scripture dating from the early centuries B.C., it renders the Hebrew word *kālīl* "sacrifice consumed wholly on the altar." Borrowed from Greek through Latin, the English word *holocaust* by the 18th century had begun to lose as its primary sense "burnt offering" and to refer more generally to a destructive fire or any massive catastrophe, without implying sacrifice. It was the "catastrophe" sense of *holocaust* that Jewish historians writing in English adopted in the late 1950's when they used the word as a proper noun to refer to a single event, the Nazi slaughter of European Jews during World War II, in part to render Modern Hebrew *hurban* (literally, "destruction") or *shoa* (literally, "disaster, catastrophe"). *Holocaust* has in recent decades been applied to the mass murder of civilians at other times in history, and on occasion is virtually synonymous with *genocide.*

complex organic matter **:** HETEROTROPHIC ⟨*holozoic* nutrition⟩

holp \ˈhō(l)p\ *chiefly dialect past of* HELP

hol·pen \ˈhō(l)-pən\ *chiefly dialect past participle of* HELP

hols \ˈhälz\ *noun plural* [short for *holidays*] (1905)
British **:** VACATION 2

Hol·stein \ˈhōl-ˌstēn, -ˌstīn\ *noun* [short for *Holstein-Friesian*] (1865)
: any of a breed of large usually black-and-white dairy cattle originally from northern Holland and Friesland that produce large quantities of comparatively low-fat milk

Hol·stein–Frie·sian \-ˈfrē-zhən\ *noun* [*Holstein,* Germany, its later locality + *Friesian* (variant of *Frisian*)] (1889)
: HOLSTEIN

hol·ster \ˈhōl-stər\ *noun* [Dutch; akin to Old English *heolstor* cover, *helan* to conceal — more at HELL] (1663)
: a leather or fabric case for carrying a firearm on the person (as on the hip or chest), on a saddle, or in a vehicle

holt \ˈhōlt\ *noun* [Middle English, from Old English; akin to Old High German *holz* wood, Greek *klados* twig] (before 12th century)
archaic **:** a small woods **:** COPPICE

ho·lus–bo·lus \ˌhō-ləs-ˈbō-ləs\ *adverb* [probably reduplication of *bolus*] (1857)
: all at once

ho·ly \ˈhō-lē\ *adjective* **ho·li·er; -est** [Middle English, from Old English *hālig;* akin to Old English *hāl* whole — more at WHOLE] (before 12th century)
1 : exalted or worthy of complete devotion as one perfect in goodness and righteousness
2 : DIVINE ⟨for the Lord our God is *holy* —Psalms 99:9 (Authorized Version)⟩
3 : devoted entirely to the deity or the work of the deity ⟨a *holy* temple⟩ ⟨*holy* prophets⟩
4 a : having a divine quality ⟨*holy* love⟩ **b :** venerated as or as if sacred ⟨*holy* scripture⟩ ⟨a *holy* relic⟩ ⟨lampooning the *holy* conventions of playwriting —William Zinsser⟩
5 — used as an intensive ⟨this is a *holy* mess⟩ ⟨he was a *holy* terror when he drank —Thomas Wolfe⟩; often used in combination as a mild oath ⟨*holy* smoke⟩
— ho·li·ly \-lə-lē\ *adverb*

holy city *noun* (14th century)
: a city that is the center of religious worship and traditions

Holy Communion *noun* (1548)
: COMMUNION 2a

holy day *noun* (before 12th century)
: a day set aside for special religious observance

holy day of obligation (1909)
: a feast on which Roman Catholics are duty-bound to attend mass

Holy Father *noun* (15th century)
: POPE 1

Holy Ghost *noun* (before 12th century)
: the third person of the Trinity **:** HOLY SPIRIT

Holy Grail *noun* (1590)
: GRAIL

Holy Joe \ˈhō-lē-ˈjō\ *noun* (circa 1874)
slang **:** PARSON, CHAPLAIN

Holy Office *noun* (circa 1741)
: a congregation of the curia charged with protecting faith and morals

holy of holies [translation of Late Latin *sanctum sanctorum,* translation of Hebrew *qōdhesh haq-qŏdhāshīm*] (1641)
: the innermost and most sacred chamber of the Jewish tabernacle and temple

holy oil *noun* (14th century)
: olive oil blessed by a bishop for use in a sacrament or sacramental

holy order *noun, often H&O capitalized* (14th century)
1 a : MAJOR ORDER — usually used in plural **b :** one of the orders of the ministry in the Anglican or Episcopal church
2 : the rite or sacrament of ordination — usually used in plural

Holy Roller *noun* (1842)
: a member of one of the Protestant sects whose worship meetings are characterized by spontaneous expressions of emotional excitement — often taken to be offensive

Holy Roman Empire *noun* (1728)
: an empire consisting primarily of a loose confederation of German and Italian territories under the suzerainty of an emperor and existing from the 9th or 10th century to 1806

Holy Saturday *noun* (14th century)
: the Saturday before Easter

Holy See *noun* (1765)
: the see of the pope

Holy Spirit *noun* (14th century)
: the third person of the Christian Trinity

ho·ly·stone \ˈhō-lē-ˌstōn\ *noun* (circa 1823)
: a soft sandstone used to scrub a ship's decks
— holystone *transitive verb*

Holy Synod *noun* (1768)
: the governing body of a national Eastern church

Holy Thursday *noun* (13th century)
1 : ASCENSION DAY
2 : MAUNDY THURSDAY

holy war *noun* (1691)
: a war waged by religious partisans to propagate or defend their faith

holy water *noun* (before 12th century)
: water blessed by a priest and used as a purifying sacramental

Holy Week *noun* (1710)
: the week before Easter during which the last days of Christ's life are commemorated

holy writ *noun, often H&W capitalized* (before 12th century)
1 : BIBLE 1
2 : a writing or utterance having unquestionable authority ⟨its financial precepts were not necessarily *Holy Writ* —Herbert Stein⟩

Holy Year *noun* (1900)
: a Roman Catholic jubilee year

hom- *or* **homo-** *combining form* [Latin, from Greek, from *homos* — more at SAME]
1 : one and the same **:** similar **:** alike ⟨*homo*graph⟩ ⟨*homo*sporous⟩
2 : homosexual ⟨*homo*phobia⟩

hom·age \ˈä-mij, ˈhä-\ *noun* [Middle English, from Old French *hommage,* from *homme* man, vassal, from Latin *homin-, homo* human being; akin to Old English *guma* human being, Latin *humus* earth — more at HUMBLE] (14th century)
1 a : a feudal ceremony by which a man acknowledges himself the vassal of a lord **b :** the relationship between a feudal lord and his vassal **c :** an act done or payment made in meeting the obligations of vassalage
2 a : expression of high regard **:** RESPECT — often used with *pay* **b :** something that shows respect or attests to the worth or influence of another **:** TRIBUTE ⟨his long life filled with international *homages* to his unique musical talent —*People*⟩
synonym see HONOR
usage see HUMOR

hom·ag·er \ˈä-mi-jər, ˈhä-\ *noun* (15th century)
: VASSAL

hom·bre \ˈäm-brā, ˈəm-, ˈōm-, -ˌbrē\ *noun* [Spanish, man, from Latin *homin-, homo*] (1846)
: GUY, FELLOW

hom·burg \ˈhäm-ˌbərg\ *noun* [*Bad Homburg,* Germany] (1894)
: a man's felt hat with a stiff curled brim and a high crown creased lengthwise

homburg

[1]home \ˈhōm\ *noun* [Middle English *hom,* from Old English *hām* village, home; akin to Old High German *heim* home] (before 12th century)
1 a : one's place of residence **:** DOMICILE **b :** HOUSE
2 : the social unit formed by a family living together
3 a : a familiar or usual setting **:** congenial environment; *also* **:** the focus of one's domestic attention ⟨*home* is where the heart is⟩ **b :** HABITAT
4 a : a place of origin ⟨salmon returning to their *home* to spawn⟩; *also* **:** one's own country ⟨having troubles at *home* and abroad⟩ **b :** HEADQUARTERS 2 ⟨*home* of the dance company⟩
5 : an establishment providing residence and care for people with special needs ⟨*homes* for the elderly⟩ ⟨a *home* for unwed mothers⟩
6 : the objective in various games; *especially* **:** HOME PLATE
— at home 1 : relaxed and comfortable **:** at ease ⟨felt completely *at home* on the stage⟩ **2 :** in harmony with the surroundings **3 :** on familiar ground **:** KNOWLEDGEABLE ⟨teachers *at home* in their subject fields⟩

[2]home *adverb* (before 12th century)
1 : to, from, or at one's home ⟨go *home*⟩ ⟨left *home* at the age of 12⟩
2 a : to a final, closed, or ultimate position ⟨drive a nail *home*⟩ **b :** to or at an ultimate objective (as a goal or finish line)
3 : to a vital sensitive core ⟨the truth struck *home*⟩
— home free : out of jeopardy **:** in a comfortable position with respect to some objective

[3]home *adjective* (1552)
1 : of, relating to, or being a home, place of origin, or base of operations ⟨*home* office⟩ ⟨checkers in position on their *home* squares⟩
2 : prepared, done, or designed for use in a home ⟨*home* remedies⟩ ⟨*home* cooking⟩ ⟨a *home* videotape system⟩
3 : operating or occurring in a home area ⟨the *home* team⟩ ⟨*home* games⟩

[4]home *verb* **homed; hom·ing** (1765)
intransitive verb
1 : to go or return home
2 *of an animal* **:** to return accurately to one's home or natal area from a distance
3 : to proceed to or toward a source of radiated energy used as a guide ⟨missiles *home* in on radar⟩
4 : to proceed or direct attention toward an objective ⟨science is *homing* in on the mysterious human process —Sam Glucksberg⟩
transitive verb
: to send to or provide with a home

home·body \ˈhōm-ˌbä-dē\ *noun* (1821)
: one whose life centers in the home

[1]home·bound \ˈhōm-ˌbau̇nd\ *adjective* [*home* + [1]*bound*] (circa 1625)
: going homeward **:** bound for home ⟨*homebound* travelers⟩

[2]homebound *adjective* [*home* + [5]*bound*] (1882)
: confined to the home

home·boy \ˈhōm-ˌbȯi\ *noun* (1927)
1 : a boy or man from one's neighborhood, hometown, or region
2 : a fellow member of a youth gang

home·bred \ˈhōm-ˈbred\ *adjective* (1587)
: produced at home **:** INDIGENOUS

home brew *noun* (1853)
: an alcoholic beverage (as beer) made at home

home·built \ˈhōm-ˈbilt\ *adjective* (1676)
: HOMEMADE 1

home·com·ing \ˈhōm-ˌkə-miŋ\ *noun* (14th century)
1 : a return home
2 : the return of a group of people usually on a special occasion to a place formerly frequented or regarded as home; *especially* **:** an annual celebration for alumni at a college or university

home computer *noun* (1976)

: a small inexpensive microcomputer
home economics *noun plural but singular or plural in construction* (1899)
: the theory and practice of homemaking — called also *home ec* \-ˈek\
— **home economist** *noun*
home fries *noun plural* (1951)
: potatoes that have usually been parboiled, sliced, and then fried — called also *home fried potatoes*
home front *noun* (1919)
: the sphere of civilian activity in war
home•grown \ˈhōm-ˈgrōn\ *adjective* (1827)
1 : grown or produced at home or in a particular local area ⟨*homegrown* vegetables⟩ ⟨*homegrown* films⟩
2 : native to or characteristic of a particular area ⟨the festival will feature *homegrown* artists⟩
home•land \-ˌland *also* -lənd\ *noun* (1670)
1 : native land **:** FATHERLAND
2 : a state or area set aside to be a state for a people of a particular national, cultural, or racial origin; *especially* **:** BANTUSTAN
home•less \-ləs\ *adjective* (1615)
: having no home or permanent place of residence
— **home•less•ness** *noun*
home•like \ˈhōm-ˌlīk\ *adjective* (1817)
: characteristic of a home
home•ly \ˈhōm-lē\ *adjective* **home•li•er; -est** (14th century)
1 : suggestive or characteristic of a home
2 : being something familiar with which one is at home ⟨satisfy themselves with houses, furniture, books and clothes that were worn and *homely* and friendly to the touch —Brendan Gill⟩
3 a : unaffectedly natural **:** SIMPLE **b :** not elaborate or complex ⟨*homely* virtues⟩
4 : plain or unattractive in appearance
— **home•li•ness** *noun*
home•made \ˈhō(m)-ˈmād\ *adjective* (circa 1659)
1 : made in the home, on the premises, or by one's own efforts
2 : of domestic manufacture
home•mak•er \ˈhōm-ˌmā-kər\ *noun* (1876)
: one who manages a household especially as a wife and mother
— **home•mak•ing** \-kiŋ\ *noun or adjective*
homeo- *or* **homoe-** *or* **homoeo-** *also* **homoio-** *combining form* [Latin & Greek; Latin *homoeo-*, from Greek *homoi-, homoio-*, from *homoios*, from *homos* same — more at SAME]
: like **:** similar ⟨*homeo*stasis⟩ ⟨*homeo*thermic⟩
2 : homeotic ⟨*homeo*box⟩
ho•meo•box \ˈhō-mē-ə-ˌbäks\ *noun* (1984)
: a short usually highly conserved DNA sequence in various eukaryotic genes (as many homeotic genes) that codes for a peptide which may be a DNA-binding protein
ho•meo•mor•phism \ˌhō-mē-ə-ˈmȯr-ˌfi-zəm\ *noun* [International Scientific Vocabulary] (1854)
: a function that is a one-to-one mapping between sets such that both the function and its inverse are continuous and that in topology exists for geometric figures which can be transformed one into the other by an elastic deformation
— **ho•meo•mor•phic** \-ˈmȯr-fik\ *adjective*
ho•meo•path•ic \ˌhō-mē-ə-ˈpa-thik\ *adjective* (1830)
1 : of or relating to homeopathy
2 : of a diluted or insipid nature ⟨a *homeopathic* abolitionist —W. A. White⟩
— **ho•meo•path•i•cal•ly** \-thi-k(ə-)lē\ *adverb*
ho•me•op•a•thy \ˌhō-mē-ˈä-pə-thē, ˌhä-\ *noun* [German *Homöopathie*, from *homöo-* homeo- + *-pathie* -pathy] (1826)
: a system of medical practice that treats a disease especially by the administration of minute doses of a remedy that would in healthy persons produce symptoms similar to those of the disease
— **ho•meo•path** \ˈhō-mē-ə-ˌpath\ *noun*
ho•meo•sta•sis \ˌhō-mē-ō-ˈstā-səs\ *noun* [New Latin] (1926)
: a relatively stable state of equilibrium or a tendency toward such a state between the different but interdependent elements or groups of elements of an organism, population, or group
— **ho•meo•stat•ic** \-ˈsta-tik\ *adjective*
ho•meo•ther•mic \-ˈthər-mik\ *adjective* (1870)
: WARM-BLOODED 1
— **ho•meo•therm** \ˈhō-mē-ō-ˌthərm\ *noun*
— **ho•meo•ther•my** \-ˌthər-mē\ *noun*
ho•me•o•tic \ˌhō-mē-ˈä-tik, ˌhä-\ *adjective* [from *homeosis, homoeosis* a shift in structural development, from Greek *homoiōsis* assimilation, resemblance, from *homoioun* to make like, from *homoios*] (circa 1903)
: relating to or being a gene producing a usually major shift in structural development
home plate *noun* (1875)
: a 5-sided rubber slab at one corner of a baseball diamond at which a batter stands when batting and which must be touched by a base runner in order to score
home•port \ˈhōm-ˌpōrt, -ˌpȯrt\ *transitive verb* (1957)
: to provide with or assign to a home port
home port *noun* (circa 1891)
: the port from which a ship hails or from which it is documented
¹ho•mer \ˈhō-mər\ *noun* [Hebrew *ḥōmer*] (1535)
: an ancient Hebrew unit of capacity equal to about 10½ or later 11½ bushels or 100 gallons (378 liters)
²hom•er \ˈhō-mər\ *noun* [¹*home*] (1868)
1 : HOME RUN
2 : HOMING PIGEON
³hom•er *intransitive verb* (1940)
: to hit a home run
home range *noun* (1884)
: the area to which an animal usually confines its daily activities
Ho•mer•ic \hō-ˈmer-ik\ *adjective* (circa 1771)
1 : of, relating to, or characteristic of the Greek poet Homer, his age, or his writings
2 : of epic proportions **:** HEROIC ⟨*Homeric* feats of reporting —Stanley Walker⟩
— **Ho•mer•i•cal•ly** \-i-k(ə-)lē\ *adverb*
home•room \ˈhōm-ˌrüm, -ˌru̇m\ *noun* (1915)
: a classroom where pupils report especially at the beginning of each school day
home rule *noun* (1860)
: self-government or limited autonomy in internal affairs by a dependent political unit (as a territory or municipality)
home run *noun* (1856)
: a hit in baseball that enables the batter to make a complete circuit of the bases and score a run
home•school \ˈhōm-ˌskül\ (1986)
intransitive verb
: to teach school subjects to one's children at home
transitive verb
: to teach (one's children) at home
— **home•school•er** \-ˌskü-lər\ *noun*
home screen *noun* (1968)
: TELEVISION 2
home•sick \ˈhōm-ˌsik\ *adjective* (1756)
: longing for home and family while absent from them
— **home•sick•ness** *noun*
home•site \-ˌsīt\ *noun* (1911)
: a location of or suitable for a home
¹home•spun \-ˌspən\ *adjective* (1591)
1 a : spun or made at home **b :** made of homespun
2 : SIMPLE, HOMELY ⟨*homespun* philosophy⟩
²homespun *noun* (1607)
: a loosely woven usually woolen or linen fabric originally made from homespun yarn
home stand *noun* (1965)
: a series of baseball games played at a team's home field
home•stay \ˈhōm-ˌstā\ *noun* (1956)
: a period during which a visitor in a foreign country lives with a local family
¹home•stead \ˈhōm-ˌsted, -stid\ *noun* (before 12th century)
1 a : the home and adjoining land occupied by a family **b :** an ancestral home **c :** HOUSE
2 : a tract of land acquired from U.S. public lands by filing a record and living on and cultivating the tract
²home•stead \-ˌsted\ (1872)
transitive verb
: to acquire or occupy as a homestead
intransitive verb
: to acquire or settle on land under a homestead law
— **home•stead•er** \-ˌste-dər\ *noun*
homestead law *noun* (1850)
1 : a law exempting a homestead from attachment or sale under execution for general debts
2 : any of several legislative acts authorizing the sale of public lands in homesteads
home•stretch \ˈhōm-ˈstrech\ *noun* (1841)
1 : the part of a racecourse between the last turn and the winning post
2 : a final stage (as of a project)
home•town \-ˈtau̇n\ *noun, often attributive* (1912)
: the city or town where one was born or grew up; *also* **:** the place of one's principal residence
home truth *noun* (1711)
1 : an unpleasant fact that jars the sensibilities
2 : a statement of undisputed fact
¹home•ward \ˈhōm-wərd\ *or* **home•wards** \-wərdz\ *adverb* (before 12th century)
: toward home ⟨look *homeward*, angel —John Milton⟩
²homeward *adjective* (1566)
: being or going in the direction of home
home•work \ˈhōm-ˌwərk\ *noun* (circa 1683)
1 : piecework done at home for pay
2 : an assignment given to a student to be completed outside the regular class period
3 : preparatory reading or research (as for a discussion or a debate)
hom•ey \ˈhō-mē\ *adjective* **hom•i•er; -est** (1856)
: HOMELIKE ⟨a restaurant with a *homey* atmosphere⟩
— **hom•ey•ness** *or* **hom•i•ness** *noun*
ho•mi•cid•al \ˌhä-mə-ˈsī-dᵊl, ˌhō-\ *adjective* (1725)
: of, relating to, or tending toward homicide
— **ho•mi•cid•al•ly** \-dᵊl-ē\ *adverb*
ho•mi•cide \ˈhä-mə-ˌsīd, ˈhō-\ *noun* [in sense 1, from Middle English, from Middle French, from Latin *homicida*, from *homo* human being + *-cida* -cide; in sense 2, from Middle English, from Middle French, from Latin *homicidium*, from *homo* + *-cidium* -cide] (14th century)
1 : a person who kills another
2 : a killing of one human being by another
hom•i•let•ic \ˌhä-mə-ˈle-tik\ *or* **hom•i•let•i•cal** \-ti-kəl\ *adjective* [Late Latin *homileticus*, from Greek *homilētikos* of conversation, from *homilein*] (1644)
1 : of, relating to, or resembling a homily
2 : of or relating to homiletics; *also* **:** PREACHY
hom•i•let•ics \-tiks\ *noun plural but singular in construction* (1830)
: the art of preaching
hom•i•ly \ˈhä-mə-lē\ *noun, plural* **-lies** [Middle English *omelie*, from Middle French, from Late Latin *homilia*, from Late Greek, from Greek, conversation, discourse, from *homilein* to consort with, address, from *homilos* crowd,

assembly; akin to Greek *homos* same — more at SAME] (14th century)
1 : a usually short sermon
2 : a lecture or discourse on or of a moral theme
3 : an inspirational catchphrase; *also* **:** PLATITUDE

homing pigeon *noun* (1886)
: a racing pigeon trained to return home

hom·i·nid \'hä-mə-nəd, -ˌnid\ *noun* [New Latin *Hominidae,* from *Homin-, Homo* + *-idae*] (circa 1889)
: any of a family (Hominidae) of erect bipedal primate mammals comprising recent humans together with extinct ancestral and related forms
— hominid *adjective*

hom·i·ni·za·tion \ˌhä-mə-nə-'zā-shən\ *noun* [Latin *homin-, homo* + English *-ization*] (1952)
: the evolutionary development of human characteristics that differentiate hominids from their primate ancestors

hom·i·noid \'hä-mə-ˌnȯid\ *noun* [New Latin *Hominoidea,* from *Homin-, Homo* + *-oidea,* suffix of higher taxa, from Latin *-oïdes* ²-oid] (1949)
: any of a superfamily (Hominoidea) of primates including recent hominids, gibbons, and pongids together with extinct ancestral and related forms (as of the genera *Proconsul* and *Dryopithecus*)
— hominoid *adjective*

hom·i·ny \'hä-mə-nē\ *noun* [Virginia Algonquian *-homen,* literally, that treated (in the way specified)] (1629)
: kernels of corn that have been soaked in a caustic solution (as of lye) and then washed to remove the hulls

hominy grits *noun plural but singular or plural in construction* (1879)
: GRITS

¹ho·mo \'hō-(ˌ)mō\ *noun, plural* **homos** [New Latin *Homin-, Homo,* from Latin, human being — more at HOMAGE] (1596)
: any of a genus (*Homo*) of primate mammals that includes modern humans (*H. sapiens*) and several extinct related species

²homo *noun, plural* **homos** [by shortening] (1929)
: HOMOSEXUAL — often used disparagingly

homo- — see HOM-

ho·mo·cer·cal \ˌhō-mə-'sər-kəl, ˌhä-\ *adjective* (1838)
1 *of a fish tail fin* **:** having the upper and lower lobes approximately symmetrical and the vertebral column ending at or near the middle of the base
2 : having or relating to a homocercal tail fin

homoe- — see HOMEO-

ho·mo·erot·ic \ˌhō-mō-i-'rä-tik\ *adjective* (1916)
: HOMOSEXUAL
— ho·mo·erot·i·cism \-'rä-tə-ˌsi-zəm\ *noun*

ho·mo·ga·met·ic \ˌhō-mō-gə-'me-tik, ˌhä-\ *adjective* (1910)
: forming gametes which all have the same type of sex chromosome

ho·mog·a·my \hō-'mä-gə-mē\ *noun* [German *Homogamie,* from *hom-* + *-gamie* -gamy] (1897)
: the mating of like with like
— ho·mog·a·mous \-məs\ *adjective*

ho·mog·e·nate \hō-'mä-jə-ˌnāt, hə-\ *noun* (1941)
: a product of homogenizing

ho·mo·ge·ne·i·ty \ˌhō-mə-jə-'nē-ə-tē, -'nā- *also* ÷-'nī-; *especially British* ˌhä-\ *noun* (1625)
1 : the quality or state of being homogeneous
2 : the state of having identical distribution functions or values ⟨*homogeneity* of variances⟩

ho·mo·ge·neous \-'jē-nē-əs, -nyəs\ *adjective* [Medieval Latin *homogeneus, homogenus,* from Greek *homogenēs,* from *hom-* + *genos* kind — more at KIN] (1641)
1 : of the same or a similar kind or nature
2 : of uniform structure or composition throughout ⟨a culturally *homogeneous* neighborhood⟩
3 : having the property that if each variable is replaced by a constant times that variable the constant can be factored out **:** having each term of the same degree if all variables are considered ⟨a *homogeneous* equation⟩
— ho·mo·ge·neous·ly *adverb*
— ho·mo·ge·neous·ness *noun*

ho·mog·e·ni·sa·tion, ho·mog·e·nise *British variant of* HOMOGENIZATION, HOMOGENIZE

ho·mog·e·ni·za·tion \hō-ˌmä-jə-nə-'zā-shən, hə-\ *noun* (1908)
1 : the act or process of homogenizing
2 : the quality or state of being homogenized

ho·mog·e·nize \hō-'mä-jə-ˌnīz, hə-\ *verb* **-nized; -niz·ing** (1886)
transitive verb
1 a : to blend (diverse elements) into a uniform mixture **b :** to make homogeneous
2 a : to reduce to small particles of uniform size and distribute evenly usually in a liquid **b :** to reduce the particles of so that they are uniformly small and evenly distributed; *specifically* **:** to break up the fat globules of (milk) into very fine particles
intransitive verb
: to become homogenized
— ho·mog·e·niz·er *noun*

ho·mog·e·nous \-nəs\ *adjective* (1919)
1 : HOMOPLASTIC 2
2 : HOMOGENEOUS

ho·mo·graft \'hō-mə-ˌgraft, 'hä-\ *noun* (1923)
: a graft of tissue taken from a donor of the same species as the recipient — compare HETEROGRAFT

ho·mo·graph \'hä-mə-ˌgraf, 'hō-\ *noun* (1873)
: one of two or more words spelled alike but different in meaning or derivation or pronunciation (as the *bow* of a ship, a *bow* and arrow)
— ho·mo·graph·ic \ˌhä-mə-'gra-fik, ˌhō-\ *adjective*

homoio- — see HOMEO-

ho·moio·therm, ho·moio·ther·mic *variant of* HOMEOTHERM, HOMEOTHERMIC

ho·moi·ou·si·an \hō-ˌmȯi-'ü-zē-ən, hä-, -'ü-sē-\ *noun* [Late Greek *homoiousios* of like substance, from Greek *homoi-* homeo- + *ousia* essence, substance, from *ont-, ōn,* present participle of *einai* to be — more at IS] (1732)
: an adherent of an ecclesiastical party of the 4th century holding that the Son is essentially like the Father but not of the same substance

ho·mol·o·gate \hō-'mä-lə-ˌgāt, hə-\ *transitive verb* **-gat·ed; -gat·ing** [Medieval Latin *homologatus,* past participle of *homologare* to agree, from Greek *homologein,* from *homologos*] (1593)
: SANCTION, ALLOW; *especially* **:** to approve or confirm officially
— ho·mol·o·ga·tion \-ˌmä-lə-'gā-shən\ *noun*

ho·mo·log·i·cal \ˌhō-mə-'lä-ji-kəl, ˌhä-\ *adjective* (circa 1847)
1 : HOMOLOGOUS
2 : of or relating to topological homology theory ⟨*homological* algebra⟩
— ho·mo·log·i·cal·ly \-ji-k(ə-)lē\ *adverb*

ho·mol·o·gize \hō-'mä-lə-ˌjīz, hə-\ *transitive verb* **-gized; -giz·ing** (1811)
1 : to make homologous
2 : to demonstrate the homology of
— ho·mol·o·giz·er *noun*

ho·mol·o·gous \hō-'mä-lə-gəs, hə-\ *adjective* [Greek *homologos* agreeing, from *hom-* + *legein* to say — more at LEGEND] (1660)
1 a : having the same relative position, value, or structure: as (1) **:** exhibiting biological homology (2) **:** having the same or allelic genes with genetic loci usually arranged in the same order ⟨*homologous* chromosomes⟩ **b :** belonging to or consisting of a chemical series whose successive members have a regular difference in composition especially of one methylene group
2 : derived from or developed in response to organisms of the same species

ho·mo·logue *or* **ho·mo·log** \'hō-mə-ˌlȯg, 'hä-, -ˌläg\ *noun* (1848)
: something (as a chemical compound or a chromosome) homologous

ho·mol·o·gy \hō-'mä-lə-jē, hə-\ *noun, plural* **-gies** (circa 1656)
1 : a similarity often attributable to common origin
2 a : likeness in structure between parts of different organisms due to evolutionary differentiation from the same or a corresponding part of a remote ancestor — compare ANALOGY **b :** correspondence in structure between different parts of the same individual
3 : similarity of nucleotide or amino-acid sequence in nucleic acids, peptides, or proteins
4 : a branch of the theory of topology concerned with partitioning space into geometric components (as points, lines, and triangles) and with the study of the number and interrelationships of these components especially by the use of group theory — called also *homology theory*; compare COHOMOLOGY

ho·mo·lyt·ic \ˌhō-mə-'li-tik, ˌhä-\ *adjective* (1941)
of a chemical compound **:** decomposing into two uncharged atoms or radicals
— ho·mol·y·sis \hō-'mä-lə-səs\ *noun*

ho·mo·mor·phism \ˌhō-mə-'mȯr-ˌfi-zəm, ˌhä-\ *noun* [International Scientific Vocabulary] (1935)
: a mapping of a mathematical set (as a group, ring, or vector space) into or onto another set or itself in such a way that the result obtained by applying the operations to elements of the first set is mapped onto the result obtained by applying the corresponding operations to their respective images in the second set
— ho·mo·mor·phic \-fik\ *adjective*

ho·mo·nu·cle·ar \ˌhō-mə-'nü-klē-ər, ˌhä-, -'nyü-, ÷-kyə-lər\ *adjective* (1930)
: of or relating to a molecule composed of identical nuclei

hom·onym \'hä-mə-ˌnim, 'hō-\ *noun* [Latin *homonymum,* from Greek *homōnymon,* from neuter of *homōnymos*] (1697)
1 a : HOMOPHONE **b :** HOMOGRAPH **c :** one of two or more words spelled and pronounced alike but different in meaning (as the noun *quail* and the verb *quail*)
2 : NAMESAKE
3 : a taxonomic designation rejected as invalid because the identical term has been used to designate another group of the same rank — compare SYNONYM
— hom·onym·ic \ˌhä-mə-'ni-mik, ˌhō-\ *adjective*

hom·on·y·mous \hō-'mä-nə-məs\ *adjective* [Latin *homonymus* having the same name, from Greek *homōnymos,* from *hom-* + *onyma, onoma* name — more at NAME] (1621)
1 : AMBIGUOUS
2 : having the same designation
3 : of, relating to, or being homonyms **:** HOMONYMIC
— hom·on·y·mous·ly *adverb*

hom·on·y·my \-mē\ *noun* (1597)
: the quality or state of being homonymous

ho·mo·ou·si·an \hō-ˌmō-'ü-zē-ən, hä-, -'ü-sē-\ *noun* [Late Greek *homoousios* of the same substance, from Greek *hom-* + *ousia* substance — more at HOMOIOUSIAN] (1565)
: an adherent of an ecclesiastical party of the 4th century holding to the doctrine of the Nicene Creed that the Son is of the same substance with the Father

ho·mo·phile \'hō-mə-ˌfīl\ *adjective* [*hom-* + ²*-phil*] (1960)
: GAY 4b

ho·mo·phobe \-ˌfōb\ *noun* (1975)
: a person characterized by homophobia

ho·mo·pho·bia \ˌhō-mə-ˈfō-bē-ə\ *noun* (1969)
: irrational fear of, aversion to, or discrimination against homosexuality or homosexuals
— **ho·mo·pho·bic** \-ˈfō-bik\ *adjective*

ho·mo·phone \ˈhä-mə-ˌfōn, ˈhō-\ *noun* [International Scientific Vocabulary] (1843)
1 : one of two or more words pronounced alike but different in meaning or derivation or spelling (as the words *to, too,* and *two*)
2 : a character or group of characters pronounced the same as another character or group
— **ho·moph·o·nous** \hō-ˈmä-fə-nəs\ *adjective*

ho·mo·pho·nic \ˌhä-mə-ˈfä-nik, ˌhō-, -ˈfō-\ *adjective* [Greek *homophōnos* being in unison, from *hom-* + *phōnē* sound — more at BAN] (circa 1879)
1 : CHORDAL
2 : of or relating to homophones
— **ho·moph·o·ny** \hō-ˈmä-fə-nē\ *noun*

ho·mo·plas·tic \ˌhō-mə-ˈplas-tik, ˌhä-\ *adjective* (1870)
1 : of or relating to homoplasy
2 : of, relating to, or derived from another individual of the same species ⟨*homoplastic* grafts⟩

ho·mo·pla·sy \ˈhō-mə-ˌplā-sē, ˈhä-, -ˌpla-; hō-ˈmä-plə-sē\ *noun* (1870)
: correspondence between parts or organs acquired as the result of parallel evolution or convergence

ho·mo·po·lar \ˌhō-mə-ˈpō-lər, ˌhä-\ *adjective* (1896)
1 *of a motor or generator* **:** using or producing direct current without the use of commutators
2 : of or relating to a union of atoms of like polarity **:** NONIONIC

ho·mo·pol·y·mer \-ˈpä-lə-mər\ *noun* (1946)
: a polymer (as polyethylene) consisting of identical monomer units
— **ho·mo·pol·y·mer·ic** \-ˌpä-lə-ˈmer-ik\ *adjective*

ho·mop·ter·an \hō-ˈmäp-tə-rən\ *noun* [ultimately from Greek *hom-* + *pteron* wing — more at FEATHER] (circa 1842)
: any of a large order or suborder (Homoptera) of insects (as cicadas, aphids, and scale insects) that have sucking mouthparts
— **homopteran** *adjective*
— **ho·mop·ter·ous** \-rəs\ *adjective*

homos *plural of* HOMO

Ho·mo sa·pi·ens \ˌhō-(ˌ)mō-ˈsā-pē-ˌenz, -ənz, *especially British* -ˈsa-pē-ənz\ *noun* [New Latin, species name, from *Homo,* genus name + *sapiens,* specific epithet, from Latin, wise, intelligent — more at HOMO, SAPIENT] (1802)
: HUMANKIND

ho·mo·sce·das·tic·i·ty \ˌhō-mō-si-ˌdas-ˈti-sə-tē, ˌhä-\ *noun* [*hom-* + Greek *skedastikos* able to disperse, from *skedannynai* to disperse] (1905)
: the property of having equal statistical variances
— **ho·mo·sce·das·tic** \-ˈdas-tik\ *adjective*

[1]**ho·mo·sex·u·al** \ˌhō-mə-ˈsek-sh(ə-)wəl, -ˈsek-shəl\ *adjective* (1892)
1 : of, relating to, or characterized by a tendency to direct sexual desire toward another of the same sex
2 : of, relating to, or involving sexual intercourse between persons of the same sex
— **ho·mo·sex·u·al·ly** *adverb*

[2]**homosexual** *noun* (1902)
: a homosexual person and especially a male

ho·mo·sex·u·al·i·ty \ˌhō-mə-ˌsek-shə-ˈwa-lə-tē\ *noun* (1892)
1 : the quality or state of being homosexual
2 : erotic activity with another of the same sex

ho·mo·spo·rous \ˌhō-mə-ˈspōr-əs, ˌhä-, -ˈspȯr-; hō-ˈmäs-pə-rəs\ *adjective* (1887)
: producing asexual spores of one kind only

ho·mo·spo·ry \ˈhō-mə-ˌspōr-ē, ˈhä-, -ˌspȯr-; hō-ˈmäs-pə-rē\ *noun* (1903)
: the production by various plants (as the club mosses and horsetails) of asexual spores of only one kind

ho·mo·thal·lic \ˌhō-mō-ˈtha-lik\ *adjective* [*hom-* + Greek *thallein* to sprout, grow — more at THALLUS] (1904)
1 : having a haploid phase that produces two kinds of gametes capable of fusing to form a zygote — used especially of algae and fungi
2 : MONOECIOUS
— **ho·mo·thal·lism** \-ˈtha-ˌli-zəm\ *noun*

ho·mo·trans·plant \ˌhō-mō-ˈtran(t)s-ˌplant, ˌhä-\ *noun* (1927)
: HOMOGRAFT
— **ho·mo·trans·plan·ta·tion** \-ˌtran(t)s-ˌplan-ˈtā-shən\ *noun*

ho·mo·zy·go·sis \ˌhō-mə-zi-ˈgō-səs, ˌhä-\ *noun* [New Latin] (1905)
: HOMOZYGOSITY

ho·mo·zy·gos·i·ty \-ˈgä-sə-tē\ *noun* (1916)
: the state of being homozygous

ho·mo·zy·gote \-ˈzī-ˌgōt\ *noun* [International Scientific Vocabulary] (1902)
: a homozygous individual

ho·mo·zy·gous \-ˈzī-gəs\ *adjective* (1902)
: having the two genes at corresponding loci on homologous chromosomes identical for one or more loci
— **ho·mo·zy·gous·ly** *adverb*

ho·mun·cu·lus \hō-ˈməŋ-kyə-ləs\ *noun, plural* **-li** \-ˌlī, -ˌlē\ [Latin, diminutive of *homin-, homo* human being — more at HOMAGE] (1656)
1 : a little man **:** MANIKIN
2 : a miniature adult that in the theory of preformation is held to inhabit the germ cell and to produce a mature individual merely by an increase in size

homy *variant of* HOMEY

hon \ˈhən\ *noun* (circa 1906)
: HONEY 2a

hon·cho \ˈhän-(ˌ)chō\ *noun, plural* **honchos** [Japanese *hanchō* squad leader, from *han* squad + *chō* head, chief] (1947)
: BOSS, BIG SHOT; *also* **:** HOTSHOT ◆

[1]**hone** \ˈhōn\ *noun* [Middle English, from Old English *hān* stone; akin to Old Norse *hein* whetstone, Latin *cot-, cos,* Sanskrit *śiśāti* he whets] (14th century)
: WHETSTONE

[2]**hone** *transitive verb* **honed; hon·ing** (1826)
1 : to sharpen or smooth with a whetstone
2 : to make more acute, intense, or effective **:** WHET ⟨helped her *hone* her comic timing —Patricia Bosworth⟩
— **hon·er** *noun*

[3]**hone** *intransitive verb* **honed; hon·ing** [Middle French *hoigner* to grumble] (1600)
1 *dialect* **:** YEARN — often used with *for* or *after*
2 *dialect* **:** GRUMBLE, MOAN

hone in *intransitive verb* [alteration of *home in*] (1965)
: to move toward or focus attention on an objective ⟨looking back for the ball *honing in* —George Plimpton⟩ ⟨a missile *honing in* on its target —Bob Greene⟩ ⟨*hones in* on the plights and victories of the common man —Lisa Russell⟩ □

[1]**hon·est** \ˈä-nəst\ *adjective* [Middle English, from Middle French *honeste,* from Latin *honestus* honorable, from *honos, honor* honor] (14th century)
1 a : free from fraud or deception **:** LEGITIMATE, TRUTHFUL ⟨an *honest* plea⟩ **b :** GENUINE, REAL ⟨making *honest* stops at stop signs —*Christian Science Monitor*⟩ **c :** HUMBLE, PLAIN ⟨good *honest* food⟩
2 a : REPUTABLE, RESPECTABLE ⟨*honest* decent people⟩ **b** *chiefly British* **:** GOOD, WORTHY
3 : CREDITABLE, PRAISEWORTHY ⟨an *honest* day's work⟩
4 a : marked by integrity **b :** marked by free, forthright, and sincere expression **:** FRANK ⟨an *honest* appraisal⟩ **c :** INNOCENT, SIMPLE
synonym see UPRIGHT

[2]**honest** *adverb* (1596)
1 : in an honest manner **:** HONESTLY ⟨I have ever found thee *honest* true —Shakespeare⟩
2 : with all sincerity ⟨I didn't do it, *honest*⟩

honest broker *noun* (circa 1884)
: a neutral mediator

hon·est·ly \ˈä-nəst-lē\ *adverb* (14th century)
1 : in an honest manner: as **a :** without cheating ⟨counted the ballots *honestly*⟩ **b :** REALLY, GENUINELY ⟨was *honestly* scared⟩ **c :** without frills ⟨food *honestly* prepared⟩
2 : to be honest **:** to tell the truth ⟨*honestly,* I don't know⟩

hon·es·ty \ˈä-nəs-tē\ *noun, plural* **-ties** (14th century)
1 *obsolete* **:** CHASTITY
2 a : fairness and straightforwardness of conduct **b :** adherence to the facts **:** SINCERITY
3 : any of a genus (*Lunaria*) of European herbs of the mustard family with toothed leaves and flat disk-shaped siliques ☆

☆ SYNONYMS
Honesty, honor, integrity, probity mean uprightness of character or action. HONESTY implies a refusal to lie, steal, or deceive in any way. HONOR suggests an active or anxious regard for the standards of one's profession, calling, or position. INTEGRITY implies trustworthiness and incorruptibility to a degree that one is incapable of being false to a trust, responsibility, or pledge. PROBITY implies tried and proven honesty or integrity.

□ USAGE
hone in The few commentators who have noticed *hone in* consider it to be a mistake for *home in.* It may have arisen from *home in* by the weakening of the \m\ sound to \n\ or may perhaps simply be due to the influence of *hone.* Even though it seems to have established itself in American English (and its mention in a British usage book suggests it is used in British English too), your use of it especially in writing is likely to be called a mistake. *Home in* or in figurative use *zero in* is an easy alternative.

◇ WORD HISTORY
honcho *Honcho* is a relic of the large U.S. presence in Japan during the years following World War II, when the Japanese word *hanchō* "leader of the squad, section, or group" entered English. We are uncertain of the exact route by which this word found its way into American military argot in the mid-1950s, though it is known that the Japanese applied *hanchō* to British or Australian officers in charge of work parties in prisoner-of-war camps. Judging by the scarcity of print evidence, *honcho* was not well established in general colloquial English before the 1960s. By that time it had become part of American political jargon—an area where it continues to flourish in the 1990s. *Honcho* has occasionally been used as a verb meaning "to supervise," and in 1993 the cartoonist Garry Trudeau produced the derivative *honchette* to describe a female big shot—surely a sign of the naturalization of *honcho* in English.

¹hon·ey \'hə-nē\ *noun, plural* **honeys** [Middle English *hony,* from Old English *hunig;* akin to Old High German *honag* honey, Latin *canicae* bran] (before 12th century)
1 a : a sweet viscid material elaborated out of the nectar of flowers in the honey sac of various bees **b :** a sweet fluid resembling honey that is collected or elaborated by various insects
2 a : a loved one **:** SWEETHEART, DEAR **b :** a superlative example
3 : the quality or state of being sweet **:** SWEETNESS

²honey *verb* **hon·eyed** *also* **hon·ied** \'hə-nēd\; **hon·ey·ing** (14th century)
transitive verb
1 : to sweeten with or as if with honey
2 : to speak ingratiatingly to **:** FLATTER
intransitive verb
: to use blandishments or cajolery

³honey *adjective* (14th century)
1 : of, relating to, or resembling honey
2 : much loved **:** DEAR

hon·ey·bee \'hə-nē-ˌbē\ *noun* (15th century)
: a honey-producing bee (*Apis* and related genera); *especially* **:** a European bee (*A. mellifera*) introduced worldwide and kept in hives for the honey it produces

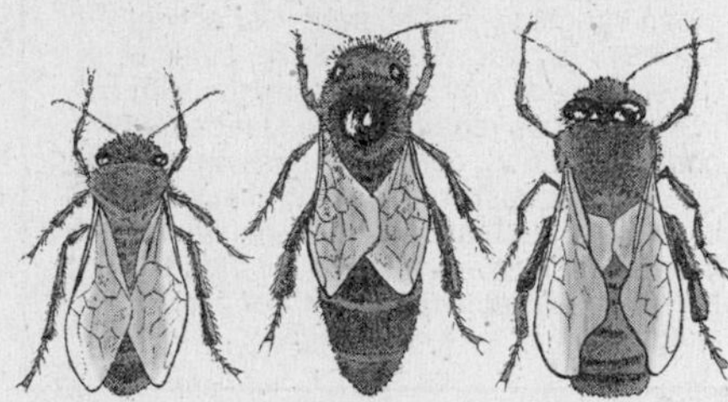

honeybees, left to right: worker, queen, drone

¹hon·ey·comb \-ˌkōm\ *noun* (before 12th century)
1 : a mass of hexagonal wax cells built by honeybees in their nest to contain their brood and stores of honey
2 : something that resembles a honeycomb in structure or appearance; *especially* **:** a strong lightweight cellular structural material

²honeycomb (1774)
transitive verb
1 a : to cause to be full of cavities like a honeycomb **b :** to make into a checkered pattern **:** FRET
2 a : to penetrate into every part **:** FILL **b :** SUBVERT, WEAKEN
intransitive verb
: to become pitted, checked, or cellular

hon·ey·creep·er \'hə-nē-ˌkrē-pər\ *noun* (1872)
1 : any of numerous small bright-colored oscine birds (especially genera *Cyanerpes* and *Chlorophanes* of the family Coerebidae) of tropical America
2 : any of a family (Drepanididae) of oscine birds that are found only in Hawaii

hon·ey·dew \-ˌdü, -ˌdyü\ *noun* (1577)
: a saccharine deposit secreted on the leaves of plants usually by aphids or scale insects or sometimes by a fungus

honeydew melon *noun* (1916)
: a pale smooth-skinned winter melon with sweet greenish flesh

hon·ey·eat·er \-ˌē-tər\ *noun* (1822)
: any of a family (Meliphagidae) of oscine birds chiefly of the South Pacific that have a long protrusible tongue adapted for extracting nectar and small insects from flowers

hon·ey·guide \-ˌgīd\ *noun* (1777)
: any of a family (Indicatoridae) of small plainly colored nonpasserine birds that inhabit Africa, the Himalayas, and the East Indies and that include some which lead people or animals to the nests of bees

honey locust *noun* (1743)
: a tall usually spiny North American leguminous tree (*Gleditsia triacanthos*) with very hard wood and long twisted pods containing a sweet edible pulp and seeds that resemble beans

hon·ey·moon \'hə-nē-ˌmün\ *noun* [from the idea that the first month of marriage is the sweetest] (1546)
1 : a period of harmony immediately following marriage
2 : a period of unusual harmony especially following the establishment of a new relationship
3 : a trip or vacation taken by a newly married couple
— honeymoon *intransitive verb*
— hon·ey·moon·er *noun*

honey sac *noun* (circa 1909)
: a distension of the esophagus of a bee in which honey is elaborated — called also *honey stomach*

hon·ey·suck·le \'hə-nē-ˌsə-kəl\ *noun* [Middle English *honysoukel* clover, alteration of *honysouke,* from Old English *hunisūce,* from *hunig* honey + *sūcan* to suck] (1548)
: any of a genus (*Lonicera* of the family Caprifoliaceae, the honeysuckle family) of shrubs with opposite leaves and often showy flowers rich in nectar; *broadly* **:** any of various plants (as a columbine or azalea) with tubular flowers rich in nectar

honeysuckle

hong \'häŋ, 'hȯŋ\ *noun* [Chinese (Guangdong) *hòhng,* literally, row] (1726)
: a commercial establishment or house of foreign trade in China

¹honk \'häŋk, 'hȯŋk\ *verb* [imitative] (circa 1835)
intransitive verb
1 : to make the characteristic cry of a goose
2 : to make a sound resembling the cry of a goose
transitive verb
: to cause (as a horn) to honk
— honk·er *noun*

²honk *noun* (1854)
: the characteristic cry of a goose; *also* **:** a similar sound

hon·ky *or* **hon·kie** *also* **hon·key** \'hȯŋ-kē, 'häŋ-\ *noun, plural* **honkies** *also* **honkeys** [probably alteration of *Hunky*] (1967)
: a white person — usually used disparagingly

¹hon·ky–tonk \'häŋ-kē-ˌtäŋk, 'hȯŋ-kē-ˌtȯŋk\ *noun* [origin unknown] (circa 1909)
1 : a usually tawdry nightclub or dance hall; *especially* **:** one that features country music
2 : a district marked by places of cheap entertainment

²honky–tonk *adjective* (circa 1920)
1 : of, used in, or being a form of ragtime piano playing performed typically on an upright piano
2 : marked by or characteristic of honky-tonks

¹hon·or \'ä-nər\ *noun* [Middle English, from Old French *honor,* from Latin *honos, honor*] (13th century)
1 a : good name or public esteem **:** REPUTATION **b :** a showing of usually merited respect **:** RECOGNITION ⟨pay *honor* to our founder⟩
2 : PRIVILEGE
3 : a person of superior standing — now used especially as a title for a holder of high office ⟨if Your *Honor* please⟩
4 : one whose worth brings respect or fame **:** CREDIT ⟨an *honor* to the profession⟩
5 : the center point of the upper half of an armorial escutcheon
6 : an evidence or symbol of distinction: as **a :** an exalted title or rank **b** (1) **:** BADGE, DECORATION (2) **:** a ceremonial rite or observance ⟨buried with full military *honors*⟩ **c :** an award in a contest or field of competition **d** *archaic* **:** a gesture of deference **:** BOW **e** *plural* (1) **:** an academic distinction conferred on a superior student (2) **:** a course of study for superior students supplementing or replacing a regular course
7 : CHASTITY, PURITY ⟨fought fiercely for her *honor* and her life —Barton Black⟩
8 a : a keen sense of ethical conduct **:** INTEGRITY **b :** one's word given as a guarantee of performance
9 *plural* **:** social courtesies or civilities extended by a host ⟨did the *honors* at the table⟩
10 a (1) **:** an ace, king, queen, jack, or ten especially of the trump suit in bridge (2) **:** the scoring value of honors held in bridge — usually used in plural **b :** the privilege of playing first from the tee in golf ☆

²honor *transitive verb* **hon·ored; hon·or·ing** \'ä-nə-riŋ, 'än-riŋ\ (13th century)
1 a : to regard or treat with honor or respect **b :** to confer honor on
2 a : to live up to or fulfill the terms of ⟨*honor* a commitment⟩ **b :** to accept as payment ⟨*honor* a credit card⟩
3 : to salute with a bow in square dancing
— hon·or·ee \ˌä-nə-'rē\ *noun*
— hon·or·er \'ä-nər-ər\ *noun*

hon·or·able \'ä-nər-(ə-)bəl, 'än-rə-\ *adjective* (14th century)
1 : deserving of honor
2 a : of great renown **:** ILLUSTRIOUS **b :** entitled to honor — used as a title for the children of certain British noblemen and for various government officials
3 : performed or accompanied with marks of honor or respect
4 a : attesting to creditable conduct **b :** consistent with an untarnished reputation ⟨an *honorable* withdrawal⟩
5 : characterized by integrity **:** guided by a high sense of honor and duty
synonym see UPRIGHT
— hon·or·abil·i·ty \ˌän-rə-'bi-lə-tē, ˌä-nə-\ *noun*
— hon·or·able·ness \'än-rə-bəl-nəs, 'ä-nə-\ *noun*
— hon·or·ably \-blē\ *adverb*

honorable mention *noun* (1866)
: a distinction conferred (as in a contest or exhibition) on works or persons of exceptional merit but not deserving of top honors

hon·o·rar·i·um \ˌä-nə-'rer-ē-əm\ *noun, plural* **-ia** \-ē-ə\ *also* **-i·ums** [Latin, from neuter of *honorarius*] (1658)
: a payment for a service (as making a speech) on which custom or propriety forbids a price to be set

hon·or·ary \'ä-nə-ˌrer-ē\ *adjective* [Latin *honorarius,* from *honor*] (1614)
1 a : having or conferring distinction **b :** COMMEMORATIVE
2 : dependent on honor for fulfillment

☆ SYNONYMS
Honor, homage, reverence, deference mean respect and esteem shown to another. HONOR may apply to the recognition of one's right to great respect or to any expression of such recognition ⟨the nomination is an *honor*⟩. HOMAGE adds the implication of accompanying praise ⟨paying *homage* to Shakespeare⟩. REVERENCE implies profound respect mingled with love, devotion, or awe ⟨great *reverence* for my father⟩. DEFERENCE implies a yielding or submitting to another's judgment or preference out of respect or reverence ⟨showed no *deference* to their elders⟩. See in addition HONESTY.

3 a : conferred or elected in recognition of achievement or service without the usual prerequisites or obligations ⟨an *honorary* degree⟩ ⟨an *honorary* member⟩ **b :** UNPAID, VOLUNTARY ⟨an *honorary* chairman⟩
— **hon·or·ari·ly** \ˌä-nə-ˈrer-ə-lē\ *adverb*
— **honorary** *noun*

honor guard *noun* (1925)
: a guard assigned to a ceremonial duty (as to accompany a casket at a military funeral)

hon·or·if·ic \ˌä-nə-ˈri-fik\ *adjective* (1650)
1 : conferring or conveying honor ⟨*honorific* titles⟩
2 : belonging to or constituting a class of grammatical forms used in speaking to or about a social superior
— **honorific** *noun*
— **hon·or·if·i·cal·ly** \-fi-k(ə-)lē\ *adverb*

honor roll *noun* (1909)
: a roster of names of persons deserving honor; *especially* **:** a list of students achieving academic distinction

honor society *noun* (1927)
: a society for the recognition of scholarly achievement especially of undergraduates

honor system *noun* (1904)
: a system (as at a college or prison) whereby persons are trusted to abide by the regulations (as for a code of conduct) without supervision or surveillance

hon·our, hon·our·able *chiefly British variant of* HONOR, HONORABLE

[1]**hooch** \ˈhüch\ *noun* [short for *hoochinoo,* a distilled liquor made by the Hoochinoo (Hutsnuwu) Indians, a Tlingit tribe] (1897)
slang **:** alcoholic liquor especially when inferior or illicitly made or obtained ◇

[2]**hooch** *or* **hootch** \ˈhüch\ *noun* [modification of Japanese *uchi* house] (1960)
slang **:** a usually thatched hut; *broadly* **:** DWELLING

[1]**hood** \ˈhu̇d\ *noun* [Middle English, from Old English *hōd;* akin to Old High German *huot* head covering, *huota* guard] (before 12th century)
1 a (1) **:** a flexible covering for the head and neck (2) **:** a protective covering for the head and face **b :** a covering for a hawk's head and eyes **c :** a covering for a horse's head; *also* **:** BLINDER
2 a : an ornamental scarf worn over an academic gown that indicates by its color the wearer's college or university **b :** a color marking or crest on the head of an animal or an expansion of the head that suggests a hood
3 a : something resembling a hood in form or use **b :** a cover for parts of mechanisms; *specifically* **:** the movable metal covering over the engine of an automobile **c** *chiefly British* **:** a top cover over the passenger section of a vehicle usually designed to be folded back **d :** an enclosure or canopy provided with a draft for carrying off disagreeable or noxious fumes, sprays, smokes, or dusts **e :** a covering for an opening (as a companion hatch) on a boat
— **hood** *transitive verb*
— **hood·like** \-ˌlīk\ *adjective*

[2]**hood** \ˈhu̇d, ˈhüd\ *noun* (1930)
: HOODLUM
— **hood·y** \ˈhu̇-dē, ˈhü-\ *adjective*

[3]**hood** \ˈhu̇d\ *noun* [by shortening] (1967)
: NEIGHBORHOOD 4

-hood \ˌhu̇d\ *noun suffix* [Middle English *-hod,* from Old English *-hād;* akin to Old High German *-heit* state, Gothic *haidus* way, manner]
1 : state **:** condition **:** quality **:** character ⟨widower*hood*⟩ ⟨hardi*hood*⟩
2 : time **:** period ⟨child*hood*⟩
3 : instance of a (specified) state or quality ⟨false*hood*⟩
4 : individuals sharing a (specified) state or character ⟨brother*hood*⟩

hood·ed \ˈhu̇-dəd\ *adjective* (15th century)
1 : having a hood
2 : shaped like a hood ⟨*hooded* spathes⟩
3 a : having the head conspicuously different in color from the rest of the body ⟨*hooded* bird⟩ **b :** having a crest on the head that suggests a hood ⟨*hooded* seals⟩ **c :** having the skin at each side of the neck capable of expansion by movements of the ribs ⟨*hooded* cobra⟩
4 : half-closed ⟨*hooded* eyes⟩
— **hood·ed·ness** *noun*

hood·lum \ˈhüd-ləm, ˈhu̇d-\ *noun* [perhaps from German dialect (Swabia) *hudelum* disorderly] (1871)
1 : THUG; *especially* **:** one who commits acts of violence
2 : a young ruffian
— **hood·lum·ish** \-lə-mish\ *adjective*
— **hood·lum·ism** \-ˌmi-zəm\ *noun*

hood·man–blind \ˌhu̇d-mən-ˈblīnd\ *noun* (1565)
archaic **:** BLINDMAN'S BUFF

[1]**hoo·doo** \ˈhü-(ˌ)dü\ *noun, plural* **hoodoos** [of African origin; akin to Hausa *hu*^{3}*'du*^{3}*'ba*^{1} to arouse resentment] (1875)
1 : a body of practices of sympathetic magic traditional especially among blacks in the southern U.S.
2 : a natural column of rock in western North America often in fantastic form
3 : something that brings bad luck
— **hoo·doo·ism** \-ˌi-zəm\ *noun*

[2]**hoodoo** *transitive verb* (1886)
: to cast a spell on; *broadly* **:** to be a source of misfortune to

hood·wink \ˈhu̇d-ˌwiŋk\ *transitive verb* [[1]*hood* + *wink*] (1562)
1 *archaic* **:** BLINDFOLD
2 *obsolete* **:** HIDE
3 : to deceive by false appearance **:** DUPE
— **hood·wink·er** *noun*

hoo·ey \ˈhü-ē\ *noun* [origin unknown] (1924)
: NONSENSE

[1]**hoof** \ˈhu̇f, ˈhüf\ *noun, plural* **hooves** \ˈhüvz, ˈhu̇vz\ *also* **hoofs** [Middle English, from Old English *hōf;* akin to Old High German *huof* hoof, Sanskrit *śapha*] (before 12th century)
1 : a curved covering of horn that protects the front of or encloses the ends of the digits of an ungulate mammal and that corresponds to a nail or claw
2 : a hoofed foot especially of a horse
— **on the hoof** *of a meat animal* **:** before butchering **:** LIVING ⟨90¢ a pound *on the hoof*⟩

hoof 2: ***1, 2, 3, 4*** **parts of wall (*1* toe, *2* side walls, *3* quarters, *4* buttresses), *5* bulbs, *6* sole, *7* white line, *8* frog**

[2]**hoof** (1641)
transitive verb
1 : WALK ⟨*hoofed* it to the lecture hall⟩
2 : KICK, TRAMPLE
intransitive verb
: to move on the feet; *especially* **:** DANCE

hoof–and–mouth disease *noun* (1884)
: FOOT-AND-MOUTH DISEASE

hoof·beat \ˈhu̇f-ˌbēt, ˈhüf-\ *noun* (1847)
: the sound of a hoof striking a hard surface (as the ground)

hoofed \ˈhu̇ft, ˈhüft, ˈhüvd, ˈhu̇vd\ *or* **hooved** \ˈhüvd, ˈhu̇vd\ *adjective* (1513)
: furnished with hooves **:** UNGULATE

hoof·er \ˈhu̇-fər, ˈhü-\ *noun* (circa 1918)
: a professional dancer

hoof·print \ˈhu̇f-ˌprint, ˈhüf-\ *noun* (1804)
: an impression or hollow made by a hoof

hoo–ha \ˈhü-ˌhä\ *noun* [probably from Yiddish *hu-ha* uproar, exclamation of surprise] (1931)
: UPROAR

[1]**hook** \ˈhu̇k\ *noun* [Middle English, from Old English *hōc;* akin to Middle Dutch *hoec* fishhook, corner, Lithuanian *kengė* hook] (before 12th century)
1 a : a curved or bent device for catching, holding, or pulling **b :** something intended to attract and ensnare
2 : something curved or bent like a hook; *especially, plural* **:** FINGERS
3 : a flight or course of a ball that deviates from straight in a direction opposite to the dominant hand of the player propelling it; *also* **:** a ball following such a course — compare SLICE
4 : a short blow delivered with a circular motion by a boxer while the elbow remains bent and rigid
5 : HOOK SHOT
6 : BUTTONHOOK
7 : quick or summary removal — used with *get* or *give* ⟨the pitcher got the *hook* after giving up three runs⟩
8 : a device especially in music or writing that catches the attention
9 : a selling point or marketing scheme
— **by hook or by crook :** by any means
— **off the hook 1 :** out of trouble **2 :** free of responsibility or accountability
— **on one's own hook :** by oneself **:** INDEPENDENTLY

[2]**hook** (13th century)
transitive verb
1 : to form into a hook **:** CROOK
2 a : to seize or make fast by or as if by a hook **b :** to connect by or as if by a hook — often used with *up*
3 : STEAL, PILFER
4 : to make (as a rug) by drawing loops of yarn, thread, or cloth through a coarse fabric with a hook
5 : to hit or throw (a ball) so that a hook results
intransitive verb
1 : to form a hook **:** CURVE
2 : to become hooked
3 : to work as a prostitute

hoo·kah \ˈhu̇-kə, ˈhü-\ *noun* [Arabic *ḥuqqah* bottle of a water pipe] (1763)
: WATER PIPE 2

hook and eye *noun* (circa 1626)

◇ WORD HISTORY

hooch The Tlingit Indians of the Alaskan panhandle, like indigenous peoples elsewhere in the world confronted with drastic cultural disruption, suffered from epidemic alcoholism in the 19th century. After Alaska was purchased from Russia in 1867, contact with U.S. troops, as well as an influx of American merchants ready to sell whiskey, greatly exacerbated the Indian drinking problem, to the point that the importation of distilled spirits into Alaska was banned in 1870. The Tlingit resourcefully responded by distilling their own liquor. Some Anglo-American, whose identity remains uncertain, introduced the Indians to a crude style of pot still, with mash made from potatoes, berries, and a crucial imported ingredient, molasses. The foul-tasting outcome was called by whites *hoochinoo* or *hoochenoo,* after the Hoochinoo or Hutsnuwu, a Tlingit tribe of Admiralty, Chichagof, and Baranof Islands. (The Tlingit name of the tribe, *xucnu·wú,* means literally "grizzly bear fort.") It is not clear that the Hutsnuwu drank or made more alcohol than their neighbors, but nonetheless *hoochinoo* became a common word for "illicit liquor" in Alaska, western Canada, and the western U.S. In the 1890s it was clipped to *hooch,* which in the Prohibition Era spread throughout American English.

: a 2-part fastening device (as on a garment or a door) consisting of a metal hook that catches over a bar or into a loop

hook and ladder truck *noun* (1865)
: a piece of mobile fire apparatus carrying ladders and usually other fire-fighting and rescue equipment — called also *hook and ladder, ladder truck*

hook check *noun* (circa 1939)
: an act or instance of attempting to knock the puck away from an opponent in ice hockey by hooking it with the stick

hooked \ˈhu̇kt, *1 is also* ˈhu̇-kəd\ *adjective* (before 12th century)
1 : having the form of a hook
2 : provided with a hook
3 : made by hooking ⟨a *hooked* rug⟩
4 a : addicted to narcotics **b :** fascinated by or devoted to something ⟨*hooked* on skiing⟩

1hook·er \ˈhu̇-kər\ *noun* (1567)
1 : one that hooks
2 : DRINK ⟨a *hooker* of Scotch⟩
3 : PROSTITUTE

2hooker *noun* [Dutch *hoeker,* alteration of Middle Dutch *hoecboot,* from *hoec* fishhook + *boot* boat] (1801)
: a one-masted fishing boat used on the English and Irish coasts; *also* **:** a small clumsy boat

Hooke's law \ˈhu̇ks-\ *noun* [Robert *Hooke*] (1853)
: a statement in physics: the stress within an elastic solid is proportional to the strain responsible for it

hook·let \ˈhu̇k-lət\ *noun* (circa 1839)
: a small hook

hook, line and sinker *adverb* [from analogy with a well-hooked fish] (1838)
: without hesitation or reservation **:** COMPLETELY ⟨fell for the story *hook, line and sinker*⟩

hook shot *noun* (circa 1932)
: a shot in basketball made usually while standing sideways to the basket by swinging the ball up in an arc with the far hand

hook·up \ˈhu̇k-ˌəp\ *noun* (1903)
1 : a state of cooperation or alliance
2 : an assemblage (as of circuits) used for a specific purpose (as radio transmission); *also* **:** the plan of such an assemblage
3 : an arrangement of mechanical parts; *also* **:** CONNECTION ⟨a campsite with electric, water, and sewer *hookups*⟩

hook up *intransitive verb* (1925)
: to become associated especially in a working or social relationship

hook·worm \ˈhu̇k-ˌwərm\ *noun* (1902)
1 : any of several parasitic nematode worms (family Ancylostomatidae) that have strong buccal hooks or plates for attaching to the host's intestinal lining and that include serious bloodsucking pests
2 : ANCYLOSTOMIASIS

hooky *or* **hook·ey** \ˈhu̇-kē\ *noun, plural* **hook·ies** *or* **hookeys** [probably from slang *hook, hook it* (to make off)] (circa 1848)
: TRUANT — used chiefly in the phrase *play hooky*

hoo·li·gan \ˈhü-li-gən\ *noun* [perhaps from Patrick *Hooligan* (flourished 1898) Irish hoodlum in Southwark, London] (1898)
: RUFFIAN, HOODLUM
— hoo·li·gan·ism \-gə-ˌni-zəm\ *noun*

1hoop \ˈhüp *also* ˈhu̇p\ *noun, often attributive* [Middle English, from Old English *hōp;* akin to Middle Dutch *hoep* ring, hoop] (12th century)
1 : a circular strip used especially for holding together the staves of containers or as a plaything
2 a : a circular figure or object **:** RING **b :** the rim of a basketball goal; *broadly* **:** the entire goal
3 : a circle or series of circles of flexible material used to expand a woman's skirt
4 : BASKETBALL — usually used in plural
— hoop·like \-ˌlīk\ *adjective*

2hoop *transitive verb* (15th century)
: to bind or fasten with or as if with a hoop
— hoop·er *noun*

hoop·la \ˈhü-ˌplä, ˈhu̇-\ *noun* [French *houp-là,* interjection] (1877)
: excited commotion **:** TO-DO; *also* **:** BALLYHOO

hoo·poe \ˈhü-(ˌ)pü, -(ˌ)pō\ *noun* [alteration of obsolete *hoop,* from Middle French *huppe,* from Latin *upupa,* of imitative origin] (1668)
: a crested Old World nonpasserine bird (*Upupa epops* of the family Upupidae) having a slender decurved bill and barred black-and-white wings and tail

hoop·skirt \ˈhüp-ˈskərt *also* ˈhu̇p-\ *noun* (1857)
: a skirt stiffened with or as if with hoops

hoo·rah \hu̇-ˈrä, -ˈrȯ\, **hoo·ray** \-ˈrā\ *variant of* HURRAH

hoose·gow \ˈhüs-ˌgau̇\ *noun* [Spanish *juzgado* panel of judges, courtroom, from past participle of *juzgar* to judge, from Latin *judicare* — more at JUDGE] (1909)
: JAIL

Hoo·sier \ˈhü-zhər\ *noun* [perhaps alteration of English dialect *hoozer* anything large of its kind] (1826)
: a native or resident of Indiana — used as a nickname
— Hoosier *adjective*

1hoot \ˈhüt\ *verb* [Middle English *houten,* of imitative origin] (13th century)
intransitive verb
1 : to shout or laugh usually derisively
2 : to make the natural throat noise of an owl or a similar cry
3 : to make a loud clamorous mechanical sound
transitive verb
1 : to assail or drive out by hooting ⟨*hooted* down the speaker⟩
2 : to express or utter with hoots ⟨*hooted* their disapproval⟩
— hoot·er *noun*

2hoot *noun* (15th century)
1 : a sound of hooting; *especially* **:** the cry of an owl
2 : a minimum amount or degree **:** the least bit ⟨don't give a *hoot*⟩
3 : something or someone amusing ⟨the play is a real *hoot*⟩
— hooty \ˈhü-tē\ *adjective*

3hoot \ˈhüt\ *or* **hoots** \ˈhüts\ *interjection* [origin unknown] (1540)
chiefly Scottish — used to express impatience, dissatisfaction, or objection

hoo·te·nan·ny \ˈhü-tᵊn-ˌa-nē\ *noun, plural* **-nies** [origin unknown] (1925)
1 *chiefly dialect* **:** GADGET
2 : a gathering at which folksingers entertain often with the audience joining in

Hoo·ver·ville \ˈhü-vər-ˌvil\ *noun* [Herbert *Hoover* + *-ville*] (1933)
: a shantytown of temporary dwellings during the depression years in the U.S.; *broadly* **:** any similar area of temporary dwellings

1hop \ˈhäp\ *verb* **hopped; hop·ping** [Middle English *hoppen,* from Old English *hoppian;* probably akin to Old English *hype* hip] (before 12th century)
intransitive verb
1 : to move by a quick springy leap or in a series of leaps; *also* **:** to move as if by hopping ⟨*hop* in the car⟩
2 : to make a quick trip especially by air
transitive verb
1 : to jump over ⟨*hop* a fence⟩
2 : to ride on ⟨*hopped* a flight⟩; *also* **:** to ride surreptitiously and without authorization ⟨*hop* a freight train⟩

2hop *noun* (1508)
1 a : a short brisk leap especially on one leg **b :** BOUNCE, REBOUND ⟨shortstop scooped it up on the first *hop*⟩
2 : DANCE 3
3 a : a flight in an aircraft **b :** a short trip

3hop *noun* [Middle English *hoppe,* from Middle Dutch; akin to Old High German *hopfo* hop] (15th century)
1 *plural* **:** the ripe dried pistillate catkins of a hop used especially to impart a bitter flavor to malt liquors
2 : a twining vine (*Humulus lupulus*) of the mulberry family with 3-lobed or 5-lobed leaves and inconspicuous flowers of which the pistillate ones are in glandular cone-shaped catkins

4hop *transitive verb* **hopped; hop·ping** (1572)
: to impregnate with hops

1hope \ˈhōp\ *verb* **hoped; hop·ing** [Middle English, from Old English *hopian;* akin to Middle High German *hoffen* to hope] (before 12th century)
intransitive verb
1 : to cherish a desire with anticipation ⟨*hopes* for a promotion⟩
2 *archaic* **:** TRUST
transitive verb
1 : to desire with expectation of obtainment
2 : to expect with confidence **:** TRUST
synonym see EXPECT
— hop·er *noun*
— hope against hope : to hope without any basis for expecting fulfillment

2hope *noun* (before 12th century)
1 *archaic* **:** TRUST, RELIANCE
2 a : desire accompanied by expectation of or belief in fulfillment ⟨came in *hopes* of seeing you⟩; *also* **:** expectation of fulfillment or success ⟨no *hope* of a cure⟩ **b :** someone or something on which hopes are centered ⟨our only *hope* for victory⟩ **c :** something hoped for

hope chest *noun* (1911)
: a young woman's accumulation of clothes and domestic furnishings (as silver and linen) kept in anticipation of her marriage; *also* **:** a chest for such an accumulation

1hope·ful \ˈhōp-fəl\ *adjective* (1568)
1 : having qualities which inspire hope ⟨*hopeful* signs of economic recovery⟩
2 : full of hope **:** inclined to hope
— hope·ful·ness *noun*

2hopeful *noun* (1720)
: ASPIRANT ⟨Olympic *hopefuls*⟩

hope·ful·ly \ˈhōp-fə-lē\ *adverb* (circa 1639)
1 : in a hopeful manner
2 : it is hoped **:** I hope **:** we hope ▫

hope·less \ˈhō-pləs\ *adjective* (1534)
1 a : having no expectation of good or success **:** DESPAIRING **b :** not susceptible to remedy or cure **c :** incapable of redemption or improvement
2 a : giving no ground for hope **:** DESPERATE **b :** incapable of solution, management, or accomplishment **:** IMPOSSIBLE
synonym see DESPONDENT
— hope·less·ness *noun*

hope·less·ly \-lē\ *adverb* (1616)

▫ USAGE
hopefully In the early 1960s the second sense of *hopefully,* which had been in sporadic use since around 1932, underwent a surge of popular use. A surge of popular criticism followed in reaction, but the criticism took no account of the grammar of adverbs. *Hopefully* in its second sense is a member of a class of adverbs known as disjuncts. Disjuncts serve as a means by which the author or speaker can comment directly to the reader or hearer usually on the content of the sentence to which they are attached. Many other adverbs (as *interestingly, frankly, clearly, luckily, unfortunately*) are similarly used; most are so ordinary as to excite no comment or interest whatsoever. The second sense of *hopefully* is entirely standard.

: in a hopeless manner — used especially as an intensifier ⟨the formerly despised and *hopelessly* middle-class game of golf —Ejner Jensen⟩

hop·head \'häp-ˌhed\ *noun* (1911)
slang **:** a drug addict

hop hornbeam *noun* (1785)
: a chiefly eastern U.S. tree (*Ostrya virginiana*) of the birch family with fruiting clusters resembling hops

Ho·pi \'hō-(ˌ)pē\ *noun, plural* **Hopi** *or* **Hopis** [Hopi *hópi,* literally, good, peaceful] (1877)
1 : a member of an American Indian people of northeastern Arizona
2 : the Uto-Aztecan language of the Hopi people

hop·lite \'häp-ˌlīt\ *noun* [Greek *hoplitēs,* from *hoplon* tool, weapon, from *hepein* to care for, work at — more at SEPULCHRE] (circa 1741)
: a heavily armed infantry soldier of ancient Greece

hop–o'–my–thumb \ˌhä-pə-mə-'thəm\ *noun* [earlier *hop on my thumb,* imperative issued to one supposedly small enough to be held in the hand] (1530)
: a very small person

hopped–up \'häpt-'əp, -ˌəp\ *adjective* (circa 1924)
1 a : being under the influence of a narcotic **b :** full of enthusiasm or excitement; *also* **:** overly excited **c :** more exciting or attractive than normal or usual
2 : having more than usual power **:** being souped up

hop·per \'hä-pər\ *noun* (13th century)
1 a : one that hops **b :** a leaping insect; *specifically* **:** an immature hopping form of an insect (as a grasshopper or locust)
2 [from the shaking motion of hoppers used to feed grain into a mill] **a :** a usually funnel-shaped receptacle for delivering material (as grain or coal); *also* **:** any of various other receptacles for the temporary storage of material **b :** a freight car with a floor sloping to one or more hinged doors for discharging bulk materials — called also *hopper car* **c :** a box in which a bill to be considered by a legislative body is dropped **d :** a tank holding liquid and having a device for releasing its contents through a pipe

¹hop·ping \'hä-piŋ\ *adverb* (1675)
: EXTREMELY, VIOLENTLY — used in the phrase *hopping mad*

²hopping *adjective* (1785)
1 : intensely active **:** BUSY ⟨they kept us *hopping*⟩
2 : extremely angry

³hopping *noun* (1879)
: a going from one place to another of the same kind — usually used in combination ⟨gallery-*hopping*⟩

hopping John \ˌhä-pən-'jän, -piŋ-\ *or* **hoppin' John** \ˌhä-pən-\ *noun, often H capitalized* (1838)
: a dish made essentially of cowpeas, rice, and salt pork or bacon

hop·py \'hä-pē\ *adjective* **hop·pi·er; -est** (circa 1889)
: having the taste or aroma of hops — used especially of ale or beer

hop·sack \'häp-ˌsak\ *also* **hop·sack·ing** \-ˌsa-kiŋ\ *noun* [Middle English *hopsak* sack for hops, from *hoppe* ³hop + *sak* sack] (1888)
: a rough-surfaced loosely woven clothing fabric

¹hop·scotch \'häp-ˌskäch\ *noun* [¹*hop* + *scotch* (line)] (1801)
: a child's game in which a player tosses an object (as a stone) into areas of a figure outlined on the ground and hops through the figure and back to regain the object

²hopscotch *intransitive verb* (1918)
: to move as if by hopping ⟨*hopscotched* across Europe⟩

hop, skip, and jump *noun* (1760)
: a short distance

hop, step, and jump *noun* (circa 1719)
: TRIPLE JUMP

ho·ra *also* **ho·rah** \'hōr-ə, 'hȯr-ə\ *noun* [New Hebrew *hōrāh,* from Romanian *horă*] (1878)
: a circle dance of Romania and Israel

Ho·rae \'hōr-ˌē, 'hȯr-, -ˌī\ *noun plural* [Latin, from Greek *Hōrai*]
: the Greek goddesses of the seasons

ho·ra·ry \'hōr-ər-ē, 'hȯr-, 'här-\ *adjective* [Medieval Latin *horarius,* from Latin *hora* hour — more at HOUR] (1632)
: of or relating to an hour; *also* **:** HOURLY

Ho·ra·tio Al·ger \hə-'rā-shō-'al-jər\ *adjective* (1925)
: of, relating to, or resembling the fiction of Horatio Alger in which success is achieved through self-reliance and hard work

Ho·ra·tius \hə-'rā-sh(ē-)əs\ *noun* [Latin]
: a hero in Roman legend noted for his defense of a bridge over the Tiber against the Etruscans

horde \'hōrd, 'hȯrd\ *noun* [Middle French, German, & Polish; Middle French & German, from Polish *horda,* from Ukrainian dialect *gorda,* alteration of Ukrainian *orda,* from Old Russian, from Turkic *orda, ordu* khan's residence] (1555)
1 a : a political subdivision of central Asian nomads **b :** a people or tribe of nomadic life
2 : a teeming crowd or throng **:** SWARM ◇
synonym see CROWD

hore·hound \'hōr-ˌhau̇nd, 'hȯr-\ *noun* [Middle English *horhoune,* from Old English *hārhūne,* from *hār* hoary + *hūne* horehound — more at HOAR] (before 12th century)
1 a : a bitter mint (*Marrubium vulgare*) with downy leaves **b :** an extract or confection made from this plant
2 : any of several mints resembling the horehound

ho·ri·zon \hə-'rī-zᵊn\ *noun* [Middle English *orizon,* from Late Latin *horizont-, horizon,* from Greek *horizont-, horizōn,* from present participle of *horizein* to bound, define, from *horos* boundary; perhaps akin to Latin *urvum* curved part of a plow] (14th century)
1 a : the apparent junction of earth and sky **b :** the great circle on the celestial sphere formed by the intersection of a plane tangent to the earth's surface at an observer's position with the celestial sphere **c :** range of perception or experience **d :** something that might be attained ⟨new *horizons*⟩
2 a : the geological deposit of a particular time usually identified by distinctive fossils **b :** any of the reasonably distinct layers of soil or its underlying material in a vertical section of land **c :** a cultural area or level of development indicated by separated groups of artifacts
— ho·ri·zon·al \-'rī-zᵊn-əl\ *adjective*

ho·ri·zon·less \-ləs\ *adjective* (circa 1839)
1 a : having no horizon **b :** ENDLESS 1
2 : HOPELESS

hor·i·zon·tal \ˌhȯr-ə-'zän-tᵊl, ˌhär-\ *adjective* (1555)
1 a : of, relating to, or situated near the horizon **b :** parallel to, in the plane of, or operating in a plane parallel to the horizon or to a base line **:** LEVEL ⟨*horizontal* distance⟩ ⟨*horizontal* engine⟩
2 : relating to, directed toward, or consisting of individuals or entities of similar status or on the same level ⟨*horizontal* mergers⟩ ⟨*horizontal* hostility⟩
— horizontal *noun*
— hor·i·zon·tal·i·ty \-ˌzän-'ta-lə-tē\ *noun*
— hor·i·zon·tal·ly \-'zän-tᵊl-ē\ *adverb*

horizontal bar *noun* (1827)
1 : a steel bar supported in a horizontal position approximately eight feet above the floor and used for swinging feats in gymnastics
2 : an event in gymnastics competition in which the horizontal bar is used

hor·mo·go·ni·um \ˌhȯr-mə-'gō-nē-əm\ *noun, plural* **-nia** \-nē-ə\ [New Latin, from Greek *hormos* chain, necklace + New Latin *-gonium* — more at SERIES] (1880)
: a portion of a filament in many blue-green algae that becomes detached as a reproductive body

hor·mon·al \hȯr-'mō-nᵊl\ *adjective* (1926)
: of, relating to, or effected by hormones
— hor·mon·al·ly \-nᵊl-ē\ *adverb*

hor·mone \'hȯr-ˌmōn\ *noun* [Greek *hormōn,* present participle of *horman* to stir up, from *hormē* impulse, assault; akin to Greek *ornynai* to rouse — more at RISE] (1905)
1 : a product of living cells that circulates in body fluids or sap and produces a specific effect on the activity of cells remote from its point of origin; *especially* **:** one exerting a stimulatory effect on a cellular activity
2 : a synthetic substance that acts like a hormone
— hor·mone·like \-ˌlīk\ *adjective*

horn \'hȯrn\ *noun* [Middle English, from Old English; akin to Old High German *horn,* Latin *cornu,* Greek *keras*] (before 12th century)
1 a : one of the usually paired bony processes that arise from the head of many ungulates and that are found in some extinct mammals and reptiles: as (1) **:** one of the permanent paired hollow sheaths of keratin usually present in both sexes of cattle and their relatives that function chiefly for defense and arise from a bony core anchored to the skull — see COW illustration (2) **:** ANTLER (3) **:** a permanent solid horn of keratin that is attached to the nasal bone of a rhinoceros (4) **:** one of a pair of permanent bone protuberances from the skull of a giraffe or okapi that are covered with hairy skin **b :** a part like an animal's horn attributed especially to the devil **c :** a natural projection or excrescence from an animal resembling or suggestive of a horn **d** (1) **:** the tough fibrous material consisting chiefly of keratin that covers or forms the horns of cattle and related animals, hooves, or other horny parts (as claws or nails) (2) **:** a manufactured product (as a plastic) resembling horn **c :** a hollow horn used to hold something

◇ WORD HISTORY

horde In a 1985 novel, a young boy is described as owning "a horde of teddy bears." That *horde* can now be linked with something as innocuous as teddy bears demonstrates how far the word has strayed from its original application to divisions of the Mongols and Turks, whose forays beyond Central Asia in the Middle Ages terrified the world. The source of *horde* is ultimately Turkic *ordu* or *ordō,* meaning "royal camp" or "royal residence" in the earliest inscriptions. The word was borrowed from Turkic into Mongolian as *ordu* and after the division of Genghis Khan's empire following his death was associated with the field armies and encampments, such as the Golden Horde, that governed each part of the empire. The Turkic and Mongolian word was borrowed into Old Russian as *orda,* and, with the addition of a murmured \h\ in some forms of Ukrainian, was borrowed into Polish as *horda* and hence into Western European languages. The indistinct European image of a Central Asian army as a nomadic swarm led to the eventual generalization of the word *horde* to mean a large number of almost anything—even teddy bears.

2 : something resembling or suggestive of a horn: as **a :** one of the curved ends of a crescent **b :** a sharp mountain peak **c :** a body of land or water shaped like a horn **d :** a beak-shaped part of an anvil **e :** a high pommel of a saddle **f :** CORNU
3 a : an animal's horn used as a wind instrument **b :** a brass wind instrument: as (1) **:** HUNTING HORN (2) **:** FRENCH HORN **c :** a wind instrument used in a jazz band; *especially* **:** TRUMPET **d :** a usually electrical device that makes a noise like that of a horn
4 : a source of strength
5 : one of the equally disadvantageous alternatives presented by a dilemma
6 *slang* **:** TELEPHONE
— horn *adjective*
— horned \ˈhȯrnd *also* ˈhȯr-nəd\ *adjective*
— horned·ness \ˈhȯr-nəd-nəs, ˈhȯrn(d)-nəs\ *noun*
— horn·less \ˈhȯrn-ləs\ *adjective*
— horn·less·ness *noun*
— horn·like \-ˌlīk\ *adjective*

horn·beam \ˈhȯrn-ˌbēm\ *noun* (14th century) **:** any of a genus (*Carpinus*) of trees of the birch family having smooth gray bark and hard white wood

horn·bill \-ˌbil\ *noun* (1773) **:** any of a family (Bucerotidae) of large nonpasserine Old World birds having enormous bills

horn·blende \-ˌblend\ *noun* [German] (1770) **:** a mineral that is the common dark variety of aluminous amphibole; *broadly* **:** AMPHIBOLE 2
— horn·blend·ic \hȯrn-ˈblen-dik\ *adjective*

horn·book \ˈhȯrn-ˌbu̇k\ *noun* (circa 1595)
1 : a child's primer consisting of a sheet of parchment or paper protected by a sheet of transparent horn
2 : a rudimentary treatise

horned lizard *noun* (1806) **:** HORNED TOAD

horned owl *noun* (14th century) **:** any of various owls having conspicuous tufts of feathers on the head

horned pout *noun* (1837) **:** a bullhead (genus *Ameiurus*); *especially* **:** a common bullhead (*A. nebulosus*) of the eastern U.S. that has been introduced into streams of the Pacific coast

horned toad \ˈhȯrnd-; ˈhȯr-nəd-, -nət-\ *noun* (1806) **:** any of several small harmless insectivorous lizards (genus *Phrynosoma*) of the western U.S. and Mexico having hornlike spines

horned viper *noun* (1767) **:** CERASTES

hor·net \ˈhȯr-nət\ *noun* [Middle English *hernet,* from Old English *hyrnet;* akin to Old High German *hornaz* hornet, Latin *crabro*] (before 12th century) **:** any of the larger social wasps (family Vespidae) — compare YELLOW JACKET

hornet's nest *noun* (circa 1740)
1 : a troublesome or hazardous situation
2 : an angry reaction ⟨must have known that his frank comments . . . would stir up a *hornet's nest* —*U.S. Investor*⟩

horn·fels \ˈhȯrn-ˌfelz\ *noun* [German, from *Horn* horn + *Fels* cliff, rock] (1854) **:** a fine-grained silicate rock produced by metamorphism especially of slate

horn fly *noun* (1708) **:** a small black European fly (*Haematobia irritans*) that has been introduced into North America where it is a blood-sucking pest of cattle

horn in *intransitive verb* (1911) **:** to participate without invitation or consent **:** INTRUDE

horn·ist \ˈhȯr-nist\ *noun* (1836) **:** one who plays a French horn

horn–mad \ˈhȯrn-ˈmad\ *adjective* (1579) **:** furiously enraged

horn of plenty (circa 1586) **:** CORNUCOPIA

horn·pipe \ˈhȯrn-ˌpīp\ *noun* (15th century)
1 : a single-reed wind instrument consisting of a wooden or bone pipe with finger holes, a bell, and mouthpiece usually of horn
2 : a lively folk dance of the British Isles originally accompanied by hornpipe playing

horn–rims \-ˌrimz\ *noun plural* (1927) **:** glasses with horn rims

horn·stone \ˈhȯrn-ˌstōn\ *noun* (1728) **:** a mineral that is a variety of quartz much like flint but more brittle

horn·swog·gle \-ˌswä-gəl\ *transitive verb* **-swog·gled; -swog·gling** \-g(ə-)liŋ\ [origin unknown] (circa 1829) *slang* **:** BAMBOOZLE, HOAX

horn·tail \-ˌtāl\ *noun* (1884) **:** any of various hymenopterous insects (family Siricidae) related to the typical sawflies but having larvae that burrow in woody plants and on the females a stout hornlike ovipositor for depositing the egg

horn·worm \-ˌwərm\ *noun* (1676) **:** a hawkmoth caterpillar having a hornlike tail process — compare TOMATO HORNWORM

horn·wort \-ˌwərt, -ˌwȯrt\ *noun* (circa 1805) **:** any of a genus (*Ceratophyllum*) of rootless thin-stemmed aquatic herbs that have flowers with a sepaloid perianth and a single carpel

horny \ˈhȯr-nē\ *adjective* **horn·i·er; -est** (14th century)
1 a : of or made of horn **b :** HARD, CALLOUS ⟨*horny*-handed⟩ **c :** compact and homogeneous with a dull luster — used of a mineral
2 : having horns
3 [*horn* erect penis + [1]-y] **a :** desiring sexual gratification **b :** excited sexually
— horn·i·ness *noun*

hor·o·loge \ˈhȯr-ə-ˌlōj, ˈhär-\ *noun* [Middle English, from Middle French, from Latin *horologium,* from Greek *hōrologion,* from *hōra* hour + *legein* to gather — more at YEAR, LEGEND] (14th century) **:** a timekeeping device

hor·o·log·i·cal \ˌhȯr-ə-ˈlä-ji-kəl\ *adjective* (15th century) **:** of or relating to a horologe or horology

ho·rol·o·gist \hə-ˈrä-lə-jist\ *noun* (1798)
1 : a person skilled in the practice or theory of horology
2 : a maker of clocks or watches

ho·rol·o·gy \-jē\ *noun* [Greek *hōra* + English *-logy*] (1819)
1 : the science of measuring time
2 : the art of making instruments for indicating time

horo·scope \ˈhȯr-ə-ˌskōp, ˈhär-\ *noun* [Middle English *oruscope,* from Middle French *horoscope,* from Latin *horoscopus,* from Greek *hōroskopos,* from *hōra* + *skopos* watcher; akin to Greek *skopein* to look at — more at SPY] (14th century)
1 : a diagram of the relative positions of planets and signs of the zodiac at a specific time (as at one's birth) for use by astrologers in inferring individual character and personality traits and in foretelling events of a person's life
2 : an astrological forecast

hor·ren·dous \hȯ-ˈren-dəs, hä-, hə-\ *adjective* [Latin *horrendus,* from gerundive of *horrēre*] (1659) **:** perfectly horrid **:** DREADFUL ⟨the tax rate was *horrendous*⟩
— hor·ren·dous·ly *adverb*

hor·rent \ˈhȯr-ənt, ˈhär-\ *adjective* [Latin *horrent-, horrens,* present participle of *horrēre*] (1667)
1 *archaic* **:** covered with bristling points **:** BRISTLED
2 *archaic* **:** standing up like bristles **:** BRISTLING

hor·ri·ble \ˈhȯr-ə-bəl, ˈhär-\ *adjective* [Middle English *orrible, horrible,* from Middle French, from Latin *horribilis,* from *horrēre*] (14th century)
1 : marked by or conducive to horror
2 : extremely unpleasant or disagreeable
— horrible *noun*
— hor·ri·ble·ness *noun*
— hor·ri·bly \-blē\ *adverb*

hor·rid \ˈhȯr-əd, ˈhär-\ *adjective* [Latin *horridus,* from *horrēre*] (1590)
1 *archaic* **:** ROUGH, BRISTLING
2 : innately offensive or repulsive: **a :** inspiring horror **:** SHOCKING **b :** inspiring disgust or loathing **:** NASTY
— hor·rid·ly *adverb*
— hor·rid·ness *noun*

hor·rif·ic \hȯ-ˈri-fik, hä-\ *adjective* (1653) **:** having the power to horrify ⟨a *horrific* account of the tragedy⟩
— hor·rif·i·cal·ly \-fi-k(ə-)lē\ *adverb*

hor·ri·fy \ˈhȯr-ə-ˌfī, ˈhär-\ *transitive verb* **-fied; -fy·ing** (1791)
1 : to cause to feel horror
2 : to fill with distaste **:** SHOCK
synonym see DISMAY
— hor·ri·fy·ing·ly \-ˌfī-iŋ-lē\ *adverb*

[1]**hor·ror** \ˈhȯr-ər, ˈhär-\ *noun* [Middle English *horrour,* from Middle French *horror,* from Latin, action of bristling, from *horrēre* to bristle, shiver; akin to Sanskrit *harṣate* he is excited] (14th century)
1 a : painful and intense fear, dread, or dismay ⟨astonishment giving place to *horror* on the faces of the people about me —H. G. Wells⟩ **b :** intense aversion or repugnance
2 a : the quality of inspiring horror **:** repulsive, horrible, or dismal quality or character ⟨contemplating the *horror* of their lives —Liam O'Flaherty⟩ **b :** something that inspires horror
3 *plural* **:** a state of extreme depression or apprehension

[2]**horror** *adjective* (1797) **:** calculated to inspire feelings of dread or horror **:** BLOODCURDLING ⟨a *horror* story⟩

hor·ror–struck \-ˌstrək\ *adjective* (1814) **:** struck with horror ⟨stood *horror-struck* as they watched . . . their own city destroyed —*Nashville Tennessean*⟩

hors de com·bat \ˌȯr-də-kōⁿ-ˈbä\ *adverb or adjective* [French] (1757) **:** out of combat **:** DISABLED

hors d'oeuvre \ȯr-ˈdərv\ *noun, plural* **hors d'oeuvres** *also* **hors d'oeuvre** \-ˈdərv(z)\ [French *hors-d'œuvre,* literally, outside of the work] (1714) **:** any of various savory foods usually served as appetizers

[1]**horse** \ˈhȯrs\ *noun, plural* **hors·es** *also* **horse** [Middle English *hors,* from Old English *hors;* akin to Old High German *hros* horse] (before 12th century)

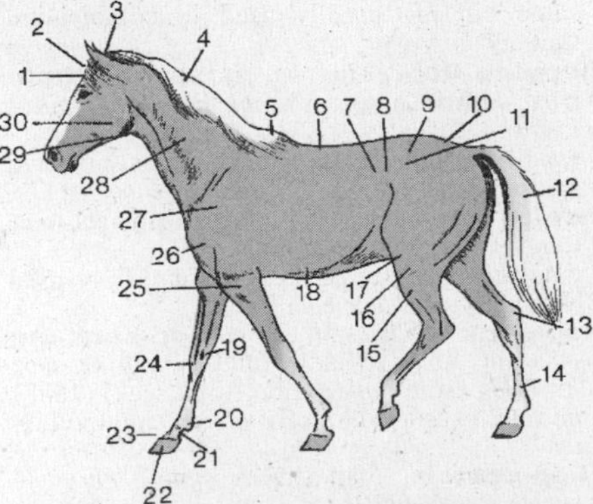

horse 1a(1): *1* **forehead,** *2* **forelock,** *3* **poll,** *4* **mane,** *5* **withers,** *6* **back,** *7* **flank,** *8* **loin,** *9* **haunch,** *10* **croup,** *11* **point of hip,** *12* **tail,** *13* **hock,** *14* **cannon,** *15* **gaskin,** *16* **thigh,** *17* **stifle,** *18* **barrel,** *19* **chestnut,** *20* **fetlock,** *21* **pastern,** *22* **hoof,** *23* **coronet,** *24* **knee,** *25* **forearm,** *26* **chest,** *27* **shoulder,** *28* **neck,** *29* **throatlatch,** *30* **cheek**

1 a (1) **:** a large solid-hoofed herbivorous mammal (*Equus caballus,* family Equidae, the horse family) domesticated since a prehistoric period and used as a beast of burden, a draft animal, or for riding (2) **:** RACEHORSE ⟨play the *horses*⟩ **b :** a male horse; *especially* **:** STALLION **c :** a recent or extinct animal (as a zebra, ass, or onager) of the horse family
2 a : JACKSTAY **b :** a frame usually with legs used for supporting something (as planks or staging) **c** (1) **:** POMMEL HORSE (2) **:** VAULTING HORSE
3 *horse plural* **:** CAVALRY
4 : a mass of the same geological character as the wall rock occurring within a vein
5 : HORSEPOWER
6 *slang* **:** HEROIN
7 : an athlete whose performance is consistently strong and reliable ⟨a team with the *horses* to win the pennant⟩
— horse·less \'hȯr-sləs\ *adjective*
— horse·like \'hȯrs-ˌlīk\ *adjective*
— from the horse's mouth : from the original source

[2]horse *verb* **horsed; hors·ing** (before 12th century)
transitive verb
1 : to provide with a horse
2 : to move by brute force
intransitive verb
of a mare **:** to be in heat

[3]horse *adjective* (15th century)
1 a : of or relating to a horse **b :** hauled or powered by a horse ⟨a *horse* barge⟩
2 : large or coarse of its kind
3 : mounted on horses ⟨*horse* guards⟩

horse–and–buggy *adjective* (circa 1926)
1 : of or relating to the era before the advent of certain socially revolutionizing inventions (as the automobile)
2 : clinging to outdated attitudes or ideas **:** OLD-FASHIONED

horse around *intransitive verb* (circa 1928)
: to engage in horseplay ⟨*horse around* together, joking and laughing and pushing each other —D. K. Shipler⟩; *also* **:** FOOL AROUND 1

[1]horse·back \'hȯrs-ˌbak\ *noun* (14th century)
: the back of a horse

[2]horseback *adverb* (1727)
: on horseback

[3]horseback *adjective* (1879)
: given without thorough consideration ⟨a *horseback* opinion⟩

horse·bean \'hȯrs-ˌbēn\ *noun* (1684)
1 : BROAD BEAN
2 : JERUSALEM THORN

horse·car \-ˌkär\ *noun* (1833)
1 : a streetcar drawn by horses
2 : a car fitted for transporting horses

horse chestnut *noun* (1597)
1 : a large Asian tree (*Aesculus hippocastanum* of the family Hippocastanaceae, the horse-chestnut family) that has palmate leaves and erect conical clusters of showy flowers and is widely cultivated as an ornamental and shade tree and naturalized as an escape; *also* **:** BUCKEYE
2 : the large glossy brown seed of a horse chestnut

horse coper *noun* (1614)
British **:** COPER

horse·feath·ers \'hȯrs-ˌfe-t͟hərz\ *noun plural* (1928)
slang **:** NONSENSE, BALDERDASH

horse·flesh \-ˌflesh\ *noun* (15th century)
: horses considered especially with reference to riding, driving, or racing

horse·fly \-ˌflī\ *noun* (14th century)
: any of a family (Tabanidae) of swift usually large two-winged flies with bloodsucking females

horse gentian *noun* (1837)
: FEVERWORT

horse·hair \'hȯrs-ˌhar, -ˌher\ *noun* (14th century)
1 : the hair of a horse especially from the mane or tail
2 : cloth made from horsehair

horsehair worm *noun* (circa 1753)
: HAIRWORM 1 — called also *horsehair snake*

horse·hide \'hȯrs-ˌhīd\ *noun* (14th century)
1 : the dressed or raw hide of a horse
2 : the ball used in the game of baseball

horse latitudes *noun plural* (1777)
: either of two belts or regions in the neighborhood of 30° N and 30° S latitude characterized by high pressure, calms, and light variable winds

horse·laugh \'hȯrs-ˌlaf, -ˌlàf\ *noun* (1713)
: a loud boisterous laugh **:** GUFFAW

horseless carriage *noun* (1895)
: AUTOMOBILE

horse mackerel *noun* (circa 1705)
1 : any of various large carangid food fishes; *especially* **:** JACK MACKEREL
2 : BLUEFIN TUNA

horse·man \'hȯrs-mən\ *noun* (14th century)
1 : a rider or driver of horses; *especially* **:** one whose skill is exceptional
2 : a person skilled in caring for or managing horses
3 : a person who breeds or raises horses
— horse·man·ship \-ˌship\ *noun*

horse·mint \'hȯrs-ˌmint\ *noun* (before 12th century)
: any of various coarse mints; *especially* **:** MONARDA

horse nettle *noun* (circa 1818)
: a coarse prickly weed (*Solanum carolinense*) of the nightshade family with bright yellow fruit resembling berries

horse opera *noun* (1927)
: WESTERN 2

horse·play \'hȯrs-ˌplā\ *noun* (1589)
: rough or boisterous play

horse·play·er \-ər\ *noun* (1947)
: one who habitually bets on horse races

horse·pow·er \'hȯrs-ˌpau̇(-ə)r\ *noun* (1806)
1 : the power that a horse exerts in pulling
2 : a unit of power equal in the U.S. to 746 watts and nearly equivalent to the English gravitational unit of the same name that equals 550 foot-pounds of work per second
3 : effective power ⟨intellectual *horsepower*⟩ ⟨computing *horsepower*⟩

horse race *noun* (1954)
: a close contest (as in politics)

horse·rad·ish \'hȯrs-ˌra-dish, -ˌre-\ *noun* (1597)
1 : a tall coarse white-flowered herb (*Armoracia lapathifolia*) of the mustard family
2 : a condiment made from ground-up horseradish root

horse's ass \'hȯr-səz-\ *noun* (circa 1942)
: a stupid or incompetent person **:** BLOCKHEAD — often considered vulgar

horse sense *noun* (1832)
: COMMON SENSE

horse·shit \'hȯrs-ˌshit, 'hȯrsh-\ *noun* (1946)
: NONSENSE, BUNK — usually considered vulgar

horse·shoe \'hȯrs-ˌshü, 'hȯrsh-\ *noun* (14th century)
1 : a usually U-shaped band of iron fitted and nailed to the rim of a horse's hoof to protect it
2 : something (as a valley) shaped like a horseshoe
3 *plural* **:** a game like quoits played with horseshoes or with horseshoe-shaped pieces of metal
— horseshoe *transitive verb*
— horse·sho·er \-ˌshü-ər\ *noun*

horseshoe arch *noun* (circa 1816)
: an arch having an intrados that widens above the springing before narrowing to a rounded or pointed crown — see ARCH illustration

horseshoe crab *noun* (1797)
: any of several closely related marine arthropods (order Xiphosura and class Merostomata) with a broad crescentic cephalothorax — called also *king crab*

horse show *noun* (1856)
: an exhibition of horses that usually includes competition in riding, driving, and jumping

horse·tail \'hȯrs-ˌtāl\ *noun* (15th century)
: EQUISETUM

horse trade *noun* (1846)
: negotiation accompanied by shrewd bargaining and reciprocal concessions ⟨a political *horse trade*⟩
— horse–trade *intransitive verb*
— horse trader *noun*

horse·weed \'hȯrs-ˌwēd\ *noun* (1790)
1 : a common North American fleabane (*Conyza canadensis* synonym *Erigeron canadensis*) with linear leaves and small discoid heads of yellowish flowers
2 : a coarse annual ragweed (*Ambrosia trifida*)

horse·whip \'hȯrs-ˌwip, 'hȯrs-ˌhwip\ *transitive verb* (1768)
: to flog with or as if with a whip made to be used on a horse
— horse·whip·per *noun*

horse·wom·an \'hȯrs-ˌwu̇m-ən\ *noun* (circa 1578)
1 : a woman rider or driver of horses; *especially* **:** one whose skill is exceptional
2 : a woman skilled in caring for or managing horses
3 : a woman who breeds or raises horses

hors·ey *or* **horsy** \'hȯr-sē\ *adjective* **hors·i·er; -est** (1591)
1 : of, relating to, or resembling a horse
2 : having to do with horses or horse racing
3 : characteristic of the manners, dress, or tastes of horsemen or horsewomen
— hors·i·ly \-sə-lē\ *adverb*
— hors·i·ness \-sē-nəs\ *noun*

horst \'hȯrst\ *noun* [German, literally, thicket] (1893)
: a block of the earth's crust separated by faults from adjacent relatively depressed blocks

hor·ta·tive \'hȯr-tə-tiv\ *adjective* [Late Latin *hortativus,* from Latin *hortatus,* past participle of *hortari* to urge — more at YEARN] (1623)
: giving exhortation **:** ADVISORY
— hor·ta·tive·ly *adverb*

hor·ta·to·ry \'hȯr-tə-ˌtōr-ē, -ˌtȯr-\ *adjective* (1586)
: HORTATIVE, EXHORTATORY

hor·ti·cul·ture \'hȯr-tə-ˌkəl-chər\ *noun* [Latin *hortus* garden + English *-i-* + *culture* — more at YARD] (1678)
: the science and art of growing fruits, vegetables, flowers, or ornamental plants
— hor·ti·cul·tur·al \ˌhȯr-tə-'kəl-chə-rəl\ *adjective*
— hor·ti·cul·tur·al·ly \-rə-lē\ *adverb*
— hor·ti·cul·tur·ist \-rist\ *noun*

Ho·rus \'hōr-əs, 'hȯr-\ *noun* [Late Latin, from Greek *Hōros,* from Egyptian *Ḥr*]
: the Egyptian god of light and the son of Osiris and Isis

ho·san·na *also* **ho·san·nah** \hō-'za-nə *also* -'zä-\ *interjection* [Middle English *osanna,* from Old English, from Late Latin, from Greek *hōsanna,* from Hebrew *hōshīʽāh-nnā* pray, save (us)!] (before 12th century)
— used as a cry of acclamation and adoration
— hosanna *noun*

HO scale \(')ā-'chō-\ *noun* [*h*alf + *O (gaug*
(1939)
: a scale of 3.5 millimeters to one foo
especially for model toys (as automo
trains)

[1]hose \'hōz\ *noun, plural* **hose**
[Middle English, from Old E
stocking, husk; akin to Old High
leg covering] (before 12th cen

1 *plural hose* **a** (1) **:** a cloth leg covering that sometimes covers the foot (2) **:** STOCKING, SOCK **b** (1) **:** a close-fitting garment covering the legs and waist that is usually attached to a doublet by points (2) **:** short breeches reaching to the knee
2 : a flexible tube for conveying fluids (as from a faucet or hydrant)

²hose *transitive verb* **hosed; hos·ing** (1889)
1 : to spray, water, or wash with a hose — often used with *down* ⟨*hose* down a stable floor⟩
2 *slang* **a :** to deprive of something due or expected **:** TRICK, CHEAT **b :** to shoot with automatic weapons fire

Ho·sea \hō-ˈzā-ə, -ˈzē-\ *noun* [Hebrew *Hōshēaʻ*]
1 : a Hebrew prophet of the 8th century B.C.
2 : a prophetic book of canonical Jewish and Christian Scripture — see BIBLE table

ho·sel \ˈhō-zəl\ *noun* [diminutive of ¹*hose*] (1899)
: a socket in the head of a golf club into which the shaft is inserted

hose·pipe \ˈhōz-ˌpīp\ *noun* (1835)
chiefly British **:** HOSE 2

ho·siery \ˈhōzh-rē, ˈhōz-; ˈhō-zhə-rē, -zə-\ *noun* (1796)
1 : HOSE 1a
2 *chiefly British* **:** KNITWEAR

hos·pice \ˈhäs-pəs\ *noun* [French, from Latin *hospitium,* from *hospit-, hospes* host — more at HOST] (1818)
1 : a lodging for travelers, young persons, or the underprivileged especially when maintained by a religious order
2 : a facility or program designed to provide a caring environment for supplying the physical and emotional needs of the terminally ill

hos·pi·ta·ble \hä-ˈspi-tə-bəl, ˈhäs-(ˌ)pi-\ *adjective* (circa 1570)
1 a : given to generous and cordial reception of guests **b :** promising or suggesting generous and cordial welcome **c :** offering a pleasant or sustaining environment
2 : readily receptive **:** OPEN ⟨*hospitable* to new ideas⟩
— hos·pi·ta·bly \-blē\ *adverb*

hos·pi·tal \ˈhäs-(ˌ)pi-tᵊl\ *noun, often attributive* [Middle English, from Middle French, from Medieval Latin *hospitale* hospice, guest house, from neuter of Latin *hospitalis* of a guest, from *hospit-, hospes*] (14th century)
1 : a charitable institution for the needy, aged, infirm, or young
2 : an institution where the sick or injured are given medical or surgical care — usually used in British English without an article after a preposition
3 : a repair shop for specified small objects ⟨clock *hospital*⟩

hos·pi·tal·ise *British variant of* HOSPITALIZE

hos·pi·tal·i·ty \ˌhäs-pə-ˈta-lə-tē\ *noun, plural* **-ties** (14th century)
: hospitable treatment, reception, or disposition

[illegible]**spitality suite** *noun* (1963)
[illegible] or suite especially in a hotel set aside [illegible] socializing especially for busi-[illegible]

[illegible]-ˌpi-tᵊl-ˌīz\ *transitive verb* [illegible] 1899)
[illegible] as a patient

[illegible]**n** \ˌhäs-ˌpi-tᵊl-ə-ˈzā-[illegible]

¹h[illegible]**pi·tal·er** \-tᵊl-ər\ [illegible]*italier,* from Mid-[illegible]al Latin *hospi-*[illegible]*itale*] (14th cen-[illegible]
[illegible] order estab-[illegible]ntury
[illegible] from Old [illegible]om Latin, [illegible]4th cen-[illegible]

1 : ARMY
2 : a very large number **:** MULTITUDE

²host *intransitive verb* (15th century)
: to assemble in a host usually for a hostile purpose

³host *noun* [Middle English *hoste* host, guest, from Old French, from Latin *hospit-, hospes,* probably from *hostis*] (14th century)
1 a : one that receives or entertains guests socially, commercially, or officially **b :** one that provides facilities for an event or function ⟨our college served as *host* for the basketball tournament⟩
2 a : a living animal or plant affording subsistence or lodgment to a parasite **b :** the larger, stronger, or dominant member of a commensal or symbiotic pair **c :** an individual into which a tissue, part, or embryo is transplanted from another
3 : a mineral or rock that is older than the minerals or rocks in it; *also* **:** a substance that contains a usually small amount of another substance incorporated in its structure
4 : a radio or television emcee
5 : a computer that controls communications in a network that administers a database

⁴host *transitive verb* (15th century)
1 : to serve as host to, at, or for ⟨*host* friends⟩ ⟨*host* a dinner⟩
2 : EMCEE ⟨*hosted* a series of TV programs⟩

⁵host *noun, often capitalized* [Middle English *hoste, oste,* from Middle French *hoiste,* from Late Latin & Latin; Late Latin *hostia* Eucharist, from Latin, sacrifice] (14th century)
: the eucharistic bread

hos·ta \ˈhō-stə, ˈhä-\ *noun* [New Latin, from Nicolaus *Host* (died 1834) Austrian botanist] (1828)
: PLANTAIN LILY

hos·tage \ˈhäs-tij\ *noun* [Middle English, from Old French, from *hoste*] (13th century)
1 a : a person held by one party in a conflict as a pledge that promises will be kept or terms met by the other party **b :** a person taken by force to secure the taker's demands
2 : one that is involuntarily controlled by an outside influence

¹hos·tel \ˈhäs-tᵊl\ *noun* [Middle English, from Old French, from Medieval Latin *hospitale* hospice] (14th century)
1 : INN
2 a *chiefly British* **:** a supervised institutional residence **b :** a supervised lodging for usually young travelers — called also *youth hostel*

²hostel *intransitive verb* **-teled** *or* **-telled; -tel·ing** *or* **-tel·ling** (14th century)
: to stay at hostels overnight in the course of traveling

hos·tel·er *or* **hos·tel·ler** \ˈhäs-tə-lər\ *noun* (14th century)
1 : one that lodges guests or strangers
2 : a traveler who stops at hostels overnight

hos·tel·ry \ˈhäs-tᵊl-rē\ *noun, plural* **-ries** (14th century)
: INN, HOTEL

¹host·ess \ˈhōs-təs\ *noun* (14th century)
1 : a woman who entertains socially
2 a : a woman in charge of a public dining room who seats diners **b :** a female employee on a public conveyance (as an airplane) who manages the provisioning of food and attends passengers **c :** a woman who acts as a dancing partner or companion to male patrons in a dance hall or bar

²hostess (1927)
intransitive verb
: to act as hostess
transitive verb
: to serve as hostess to

hos·tile \ˈhäs-tᵊl, -ˌtīl\ *adjective* [Middle French or Latin; Middle French, from Latin *hostilis,* from *hostis*] (1580)
1 a : of or relating to an enemy ⟨*hostile* fire⟩ **b :** marked by malevolence ⟨a *hostile* act⟩ **c :** openly opposed or resisting ⟨a *hostile* critic⟩ ⟨*hostile* to new ideas⟩ **d :** not hospitable ⟨a *hostile* environment⟩
2 a : of or relating to the opposing party in a legal controversy ⟨a *hostile* witness⟩ **b :** adverse to the interests of a property owner or corporation management ⟨a *hostile* takeover⟩
— hostile *noun*
— hos·tile·ly \-tᵊl-(l)ē, -ˌtīl-lē\ *adverb*

hos·til·i·ty \hä-ˈsti-lə-tē\ *noun, plural* **-ties** (15th century)
1 a : deep-seated usually mutual ill will **b** (1) **:** hostile action (2) *plural* **:** overt acts of warfare **:** WAR
2 : conflict, opposition, or resistance in thought or principle
synonym see ENMITY

hos·tler \ˈhäs-lər, ˈäs-\ *noun* [Middle English, innkeeper, hostler, from *hostel*] (14th century)
1 : one who takes care of horses or mules
2 : one who moves locomotives in and out of a roundhouse; *also* **:** one who services locomotives

host·ly \ˈhōst-lē\ *adjective* (1893)
: of or appropriate to a host

¹hot \ˈhät\ *adjective* **hot·ter; hot·test** [Middle English, from Old English *hāt;* akin to Old High German *heiz* hot, Lithuanian *kaisti* to get hot] (before 12th century)
1 a : having a relatively high temperature **b :** capable of giving a sensation of heat or of burning, searing, or scalding **c :** having heat in a degree exceeding normal body heat
2 a : VIOLENT, STORMY ⟨a *hot* temper⟩ ⟨a *hot* battle⟩; *also* **:** ANGRY ⟨got *hot* about the remark⟩ **b :** sexually excited or receptive; *also* **:** SEXY **c :** EAGER, ZEALOUS ⟨*hot* for reform⟩ **d** *of jazz* **:** emotionally exciting and marked by strong rhythms and free melodic improvisations
3 : having or causing the sensation of an uncomfortable degree of body heat ⟨*hot* and tired⟩ ⟨it's *hot* in here⟩
4 a : newly made **:** FRESH ⟨a *hot* scent⟩ ⟨*hot* off the press⟩ **b :** close to something sought ⟨*hot* on the trail⟩
5 a : suggestive of heat or of burning or glowing objects ⟨*hot* colors⟩ **b :** PUNGENT, PEPPERY
6 a : of intense and immediate interest ⟨some *hot* gossip⟩ **b :** unusually lucky or favorable ⟨on a *hot* streak⟩ **c :** temporarily capable of unusual performance (as in a sport) **d :** currently popular or in demand ⟨a *hot* commodity⟩ **e :** very good ⟨a *hot* idea⟩ ⟨not feeling too *hot*⟩ **f :** ABSURD, UNBELIEVABLE ⟨wants to fight the champ? that's a *hot* one⟩
7 a : electrically energized especially with high voltage **b :** RADIOACTIVE; *also* **:** dealing with radioactive material **c** *of an atom or molecule* **:** being in an excited state
8 a : recently and illegally obtained ⟨*hot* jewels⟩ **b :** wanted by the police; *also* **:** unsafe for a fugitive
9 : FAST ⟨a *hot* new fighter plane⟩ ⟨a *hot* lap around the track⟩
— hot·ness *noun*
— hot·tish \ˈhä-tish\ *adjective*
— hot under the collar : extremely exasperated or angry

²hot *adverb* (before 12th century)
1 : HOTLY
2 : FAST, QUICKLY

³hot *noun* (13th century)
1 : HEAT 1d(1) ⟨the *hot* of the day⟩
2 : one that is hot (as a hot meal or a horse just after a workout)
3 *plural* **:** strong sexual desire — used with *the*

⁴hot *transitive verb* **hot·ted; hot·ting** (1561)
chiefly Southern, southern Midland, & British **:** HEAT, WARM — usually used with *up*

hot air *noun* (1873)
: empty talk

hot·bed \ˈhät-ˌbed\ *noun* (1626)

1 : a bed of soil enclosed in glass, heated especially by fermenting manure, and used for forcing or for raising seedlings
2 : an environment that favors rapid growth or development ⟨a *hotbed* of activity⟩

hot·blood \-ˌbləd\ *noun* (1798)
1 : one that is hot-blooded; *especially* **:** one having strong passions or a quick temper
2 : THOROUGHBRED 1

hot–blood·ed \-ˈblə-dəd\ *adjective* (1598)
1 : easily excited **:** PASSIONATE
2 *of a horse* **:** having Arab or Thoroughbred ancestors
— hot–blood·ed·ness *noun*

hot·box \-ˌbäks\ *noun* (1848)
: a journal bearing (as of a railroad car) overheated by friction

hot button *noun* (1975)
: an emotional and usually controversial issue or concern that triggers immediate intense reaction
— hot–button *adjective*

hot·cake \-ˌkāk\ *noun* (1683)
: PANCAKE
— like hotcakes : at a rapid rate ⟨selling *like hotcakes*⟩

hotch \ˈhäch\ *intransitive verb* [Middle English, probably from Middle French *hocher* to shake, from Old French *hochier*] (15th century)
1 *Scottish* **:** WIGGLE, FIDGET
2 *chiefly Scottish* **:** SWARM

hotch·pot \ˈhäch-ˌpät\ *noun* [Anglo-French *hochepot,* from Old French, hotchpotch] (1552)
: the combining of properties into a common lot to ensure equality of division among heirs

hotch·potch \ˈhäch-ˌpäch\ *noun* [Middle English *hochepot,* from Middle French, from Old French, from *hochier* to shake + *pot*] (1583)
1 a : a thick soup or stew of vegetables, potatoes, and usually meat **b :** HODGEPODGE
2 : HOTCHPOT

hot comb *noun* (1970)
: a metal comb usually electrically heated for straightening or styling the hair
— hot–comb *transitive verb*

hot corner *noun* (1903)
: the fielding position of the third baseman in baseball

hot·dog \ˈhät-ˌdȯg\ *intransitive verb* [²*hot dog*] (1962)
: to perform in a conspicuous or often ostentatious manner; *especially* **:** to perform fancy stunts and maneuvers (as while surfing or skiing)
— hot·dog·ger \-ˌdȯ-gər\ *noun*

¹hot dog \ˈhät-ˌdȯg\ *noun* (circa 1900)
1 : FRANKFURTER; *especially* **:** a frankfurter heated and served in a long split roll
2 [perhaps from ²*hot dog*] **:** one that hotdogs; *also* **:** SHOW-OFF

²hot dog \ˈhät-ˈdȯg\ *interjection* (circa 1906)
— used to express approval or gratification

ho·tel \hō-ˈtel, ˈhō-ˌ\ *noun* [French *hôtel,* from Old French *hostel* hostel] (1765)
: an establishment that provides lodging and usually meals, entertainment, and various personal services for the public **:** INN
— ho·tel·dom \-dəm\ *noun*

Hotel (1952)
— a communications code word for the letter *h*

ho·te·lier \hō-ˈtel-yər; ˌō-tᵊl-ˈyā, ˌȯ-\ *noun* [French *hôtelier,* from Old French *hostelier,* from *hostel*] (1905)
: a proprietor or manager of a hotel

ho·tel·man \hō-ˈtel-ˌman, -mən\ *noun* (1920)
: one who is engaged in the hotel business especially in a supervisory or managerial capacity

hot flash *noun* (1910)
: a sudden brief flushing and sensation of heat caused by dilation of skin capillaries usually associated with menopausal endocrine imbalance — called also *hot flush*

¹hot·foot \ˈhät-ˌfu̇t\ *adverb* (14th century)
: in haste

²hotfoot *intransitive verb* (1896)
: to go hotfoot **:** HURRY — usually used with *it*

³hotfoot *noun, plural* **hotfoots** (1934)
: a practical joke in which a match is surreptitiously inserted between the upper and the sole of a victim's shoe and lighted

hot·head \ˈhät-ˌhed\ *noun* (1660)
: a hotheaded person

hot·head·ed \-ˈhe-dəd\ *adjective* (1641)
: FIERY, IMPETUOUS
— hot·head·ed·ly *adverb*
— hot·head·ed·ness *noun*

¹hot·house \-ˌhau̇s\ *noun* (1511)
1 *obsolete* **:** BORDELLO
2 : a greenhouse maintained at a high temperature especially for the culture of tropical plants
3 : HOTBED 2

²hothouse *adjective* (1838)
1 : grown in a hothouse
2 : suggestive of growth and development in a hothouse ⟨a *hothouse* existence⟩; *also* **:** suggesting a hothouse ⟨a *hothouse* atmosphere⟩

hot line *noun* (1955)
1 : a direct telephone line in constant operational readiness so as to facilitate immediate communication
2 : a usually toll-free telephone service available to the public for some specific purpose ⟨a consumer *hot line*⟩

hot·ly \ˈhät-lē\ *adverb* (15th century)
: in a hot manner ⟨a *hotly* contested series⟩

hot–melt \ˈhät-ˌmelt\ *noun* (1939)
: a fast-drying nonvolatile adhesive applied hot in the molten state

hot metal *noun* (1960)
: composition in which the type is cast from molten metal

hot money *noun* (1936)
: investment funds intended for the highest short-term rate of return

hot pepper *noun* (1945)
: any of various small and usually thin-walled capsicum fruits of marked pungency; *also* **:** a plant bearing hot peppers

hot plate *noun* (1845)
1 : a heated iron plate for cooking
2 : a simple portable appliance for heating or for cooking in limited spaces

hot pot *noun* (1851)
1 : a stew of meat and vegetables
2 : FIREPOT 2

hot potato *noun* (1846)
: a controversial question or issue that involves unpleasant or dangerous consequences for anyone dealing with it

hot pursuit *noun* (circa 1922)
: close continuous pursuit of a fleeing suspected lawbreaker or hostile military force especially across territorial lines

hot rod *noun* (1945)
: an automobile rebuilt or modified for high speed and fast acceleration
— hot–rod \ˈhät-ˈräd\ *verb*
— hot–rod·der \ˈhät-ˌrä-dər\ *noun*

hot seat *noun* (1925)
1 *slang* **:** ELECTRIC CHAIR
2 : a position of uneasiness, embarrassment, or anxiety

hot·shot \ˈhät-ˌshät\ *noun, often attributive* (circa 1925)
1 : a fast freight
2 : a person who is conspicuously talented or successful

hot spot *noun* (1929)
1 : a place of more than usual interest, activity, or action
2 : a place in the upper mantle of the earth at which hot magma from the lower mantle upwells to melt through the crust usually in the interior of a tectonic plate to form a volcanic feature; *also* **:** a place in the crust overlying a hot spot

hot spring *noun* (1669)
: a spring whose water issues at a temperature higher than that of its surroundings

hot stuff *noun* (1889)
: someone or something unusually good

Hot·ten·tot \ˈhä-tᵊn-ˌtät\ *noun* [Afrikaans] (1677)
1 : a member of any of a group of Khoisan-speaking pastoral peoples of southern Africa
2 : the group of Khoisan languages spoken by the Hottentots

hot ticket *noun* (1972)
: someone or something very popular **:** RAGE

hot tub *noun* (1975)
: a large usually wooden tub of hot water in which bathers soak and usually socialize

hot up (1878)
intransitive verb
chiefly British **:** to increase in intensity, pace, or excitement ⟨air raids began to *hot up* about the beginning of February —George Orwell⟩
transitive verb
chiefly British **:** to make livelier, speedier, or more intense

hot war *noun* (1947)
: a conflict involving actual fighting — compare COLD WAR

hot water *noun* (1537)
: TROUBLE 4, DIFFICULTY ⟨was in *hot water* with the authorities⟩

hot–wire \ˈhät-ˌwī(-ə)r\ *transitive verb* (1954)
: to start (as an automobile) by short-circuiting the ignition system

¹hound \ˈhau̇nd\ *noun* [Middle English, from Old English *hund;* akin to Old High German *hunt* dog, Latin *canis,* Greek *kyōn*] (before 12th century)
1 a : DOG **b :** a dog of any of various hunting breeds typically having large drooping ears and a deep voice and following their prey by scent
2 : a mean or despicable person
3 : DOGFISH
4 : a person who pursues like a hound; *especially* **:** one who avidly seeks or collects something ⟨autograph *hounds*⟩

²hound *transitive verb* (1528)
1 : to pursue with or as if with hounds
2 : to drive or affect by persistent harassing
synonym see BAIT
— hound·er *noun*

hounds \ˈhau̇n(d)z\ *noun plural* [Middle English *houne,* from Old Norse *hūnn* knob at the top of a masthead] (15th century)
: the framing at the masthead of a ship that supports the heel of the topmast and the upper parts of the lower rigging

hound's–tongue \ˈhau̇n(d)z-ˌtəŋ\ *noun* (before 12th century)
: any of various coarse plants (genus *Cynoglossum,* especially *C. officinale*) of the borage family having tongue-shaped leaves and reddish flowers

hounds·tooth check *or* **hound's–tooth check** \ˈhau̇n(d)z-ˌtüth-\ *noun* (1937)
: a usually small broken-check textile pattern

houndstooth check

hour \ˈau̇(-ə)r\ *noun* [Middle English, from Old French *heure,* from Late Latin & Latin; Late Latin *hora* canonical hour, from Latin, hour of the day, from Greek *hōra* — more at YEAR] (13th century)
1 : a time or office for daily liturgical devotion; *especially* **:** CANONICAL HOUR
2 : the 24th part of a day **:** 60 minutes

3 a : the time of day reckoned in two 12-hour periods **b** *plural* **:** the time reckoned in one 24-hour period from midnight to midnight using a 4-digit number of which the first two digits indicate the hour and the last two digits indicate the minute ⟨in the military 4:30 p.m. is called 1630 *hours*⟩
4 a : a customary or particular time ⟨lunch *hour*⟩ ⟨in our *hour* of need⟩ **b** *plural* **:** time of going to bed ⟨keeps late *hours*⟩; *also* **:** time of working ⟨banker's *hours*⟩
5 : an angular unit of right ascension equal to 15 degrees measured along the celestial equator
6 : the work done or distance traveled at normal rate in an hour ⟨the city was two *hours* away⟩
7 a : a class session **b :** CREDIT HOUR, SEMESTER HOUR
— after hours : after the regular quitting or closing time

hour angle *noun* (circa 1837)
: the angle between the celestial meridian of an observer and the hour circle of a celestial object measured westward from the meridian

hour circle *noun* (1690)
: a circle on the celestial sphere that passes through both celestial poles

¹hour·glass \'au̇(-ə)r-ˌglas\ *noun* (circa 1515)
: an instrument for measuring time consisting of a glass vessel having two compartments from the uppermost of which a quantity of sand, water, or mercury runs in an hour into the lower one

hourglass

²hourglass *adjective* (circa 1834)
: shaped like an hourglass ⟨an *hourglass* figure⟩

hour hand *noun* (1669)
: the short hand that marks the hours on the face of a watch or clock

hou·ri \'hu̇r-ē, 'hü-rē\ *noun* [French, from Persian *hūri,* from Arabic *ḥūrīyah*] (1737)
1 : one of the beautiful maidens that in Muslim belief live with the blessed in paradise
2 : a voluptuously beautiful young woman

hour–long \'au̇(-ə)r-'lȯŋ\ *adjective* (1803)
: lasting an hour

¹hour·ly \'au̇(-ə)r-lē\ *adverb* (15th century)
: at or during every hour; *also* **:** FREQUENTLY, CONTINUALLY

²hourly *adjective* (circa 1530)
1 a : occurring hour by hour ⟨*hourly* bus service⟩ **b :** FREQUENT, CONTINUAL ⟨in *hourly* expectation of the rain's stopping⟩
2 : computed in terms of an hour ⟨an *hourly* wage⟩
3 : paid by the hour ⟨*hourly* workers⟩

¹house \'hau̇s\ *noun, plural* **hous·es** \'hau̇-zəz *also* -səz\ *often attributive* [Middle English *hous,* from Old English *hūs;* akin to Old High German *hūs* house] (before 12th century)
1 : a building that serves as living quarters for one or a few families **:** HOME
2 a (1) **:** a shelter or refuge (as a nest or den) of a wild animal (2) **:** a natural covering (as a test or shell) that encloses and protects an animal or a colony of zooids **b :** a building in which something is housed ⟨a carriage *house*⟩
3 a : one of the 12 equal sectors in which the celestial sphere is divided in astrology **b :** a zodiacal sign that is the seat of a planet's greatest influence
4 a : HOUSEHOLD **b :** a family including ancestors, descendants, and kindred ⟨the *house* of Tudor⟩
5 a : a residence for a religious community or for students **b :** the community or students in residence
6 a : a legislative, deliberative, or consultative assembly; *especially* **:** one constituting a division of a bicameral body **b :** the building or chamber where such an assembly meets **c :** a quorum of such an assembly
7 a : a place of business or entertainment **b** (1) **:** a business organization ⟨a publishing *house*⟩ (2) **:** a gambling establishment **c :** the audience in a theater or concert hall ⟨a full *house* on opening night⟩
8 : the circular area 12 feet in diameter surrounding the tee and within which a curling stone must rest in order to count
9 : a type of dance music mixed by a disc jockey that features overdubbing with a heavy repetitive drumbeat and repeated electronic melody lines
— house·ful \'hau̇s-ˌfu̇l\ *noun*
— house·less \'hau̇-sləs\ *adjective*
— house·less·ness *noun*
— on the house : without charge **:** FREE

²house \'hau̇z\ *verb* **housed; hous·ing** (before 12th century)
transitive verb
1 a : to provide with living quarters or shelter **b :** to store in a house
2 : to encase, enclose, or shelter as if by putting in a house
3 : to serve as shelter for **:** CONTAIN
intransitive verb
: to take shelter **:** LODGE

house arrest *noun* (1936)
: confinement often under guard to one's house or quarters instead of in prison

house·boat \'hau̇s-ˌbōt\ *noun* (1790)
: a boat fitted for use as a dwelling; *especially* **:** a pleasure craft with a broad beam, a usually shallow draft, and a large superstructure resembling a house
— house·boat·er \-ˌbō-tər\ *noun*

house·bound \'hau̇s-ˌbau̇nd\ *adjective* (1878)
: confined to the house

house·boy \-ˌbȯi\ *noun* (circa 1898)
: HOUSEMAN

house·break \-ˌbrāk\ *transitive verb* **-broke** \-ˌbrōk\; **-bro·ken** \-ˌbrō-kən\; **-break·ing** [back-formation from *housebroken*] (1944)
1 : to make housebroken
2 a : to teach acceptable social manners to **b :** TAME, SUBDUE

house·break·ing \'hau̇s-ˌbrā-kiŋ\ *noun* (1617)
: an act of breaking open and entering the dwelling house of another with a felonious purpose
— house·break·er \-ˌbrā-kər\ *noun*

house·bro·ken \-ˌbrō-kən\ *adjective* (1900)
1 : trained to excretory habits acceptable in indoor living — used of a household pet
2 : made tractable or polite

house call *noun* (1960)
: a visit (as by a doctor or a repair person) to a home to provide a requested service

house·carl \-ˌkär(-ə)l\ *noun* [Old English *hūscarl,* from Old Norse *hūskarl,* from *hūs* house + *karl* man; akin to Old English *ceorl* churl] (before 12th century)
: a member of the bodyguard of a Danish or early English king or noble

house cat *noun* (1607)
: CAT 1a

house·clean \'hau̇s-ˌklēn\ *verb* [back-formation from *housecleaning*] (1863)
intransitive verb
1 : to clean a house and its furniture
2 : to get rid of unwanted or undesirable items or people
transitive verb
1 : to clean the surfaces and furnishings of
2 : to improve or reform by ridding of undesirable people or practices
— house·clean·ing *noun*

house·coat \'hau̇s-ˌkōt\ *noun* (1913)
: a woman's often long-skirted informal garment for wear around the house

house cricket *noun* (1774)
: a widely distributed cricket (*Acheta domesticus*) usually living in or about dwellings

house detective *noun* (1898)
: a person who is employed (as by a hotel) to prevent disorderly or improper conduct of patrons

house·dress \'hau̇s-ˌdres\ *noun* (1897)
: a dress with simple lines that is suitable for housework and is made usually of a washable fabric

house·fa·ther \-ˌfä-t͟hər, -ˌfȧ-\ *noun* (1901)
: a man in charge of a dormitory, hall, or hostel

house finch *noun* (1869)
: a small finch (*Carpodacus mexicanus*) that has a male with a red head, breast, and rump and that is native to Mexico and the western U.S. and has been introduced in the eastern U.S.

house·fly \'hau̇s-ˌflī\ *noun* (15th century)
: a cosmopolitan dipteran fly (*Musca domestica*) that is often about human habitations and may act as a mechanical vector of diseases (as typhoid fever); *also* **:** any of various flies of similar appearance or habitat

house·front \-ˌfrənt\ *noun* (1838)
: the facade of a house

house girl *noun* (1835)
: HOUSEMAID

house·guest \'hau̇s-ˌgest\ *noun* (1917)
: GUEST 1a

¹house·hold \'hau̇s-ˌhōld, 'hau̇-ˌsōld\ *noun* (14th century)
: those who dwell under the same roof and compose a family; *also* **:** a social unit comprised of those living together in the same dwelling

²household *adjective* (14th century)
1 : of or relating to a household **:** DOMESTIC ⟨cooking and other *household* arts⟩
2 : FAMILIAR, COMMON ⟨a *household* name⟩

house·hold·er \'hau̇s-ˌhōl-dər, 'hau̇-ˌsōl-\ *noun* (14th century)
: a person who occupies a house or tenement alone or as the head of a household

household troops *noun plural* (1711)
: troops appointed to attend and guard a sovereign or the residence of a sovereign

house·hus·band \'hau̇s-ˌhəz-bənd\ *noun* (1955)
: a husband who does housekeeping usually while his wife earns the family income

house·keep \'hau̇s-ˌkēp\ *intransitive verb* **-kept** \-ˌkept\; **-keep·ing** [back-formation from *housekeeper*] (1842)
: to perform the routine duties (as cooking and cleaning) of managing a house

house·keep·er \-ˌkē-pər\ *noun* (1607)
1 : a woman employed to keep house
2 : HOUSEWIFE 1

house·keep·ing \-piŋ\ *noun* (1550)
1 : the management of a house and home affairs
2 : the care and management of property and the provision of equipment and services (as for an industrial organization)
3 : the routine tasks that must be done in order for a system to function or to function efficiently

¹hou·sel \'hau̇-zəl\ *noun* [Middle English, from Old English *hūsel* sacrifice, Eucharist; akin to Gothic *hunsl* sacrifice] (before 12th century)
archaic **:** the Eucharist or the act of administering or receiving it

²housel *transitive verb* (before 12th century)
archaic **:** to administer communion to

house·leek \'hau̇s-ˌlēk\ *noun* (14th century)
: a pink-flowered thick-leaved European plant (*Sempervivum tectorum*) of the orpine family that tends to form clusters of rosettes and is often grown in rock gardens; *broadly* **:** SEMPERVIVUM

house·lights \'hau̇s-ˌlīts\ *noun plural* (1920)
: the lights that illuminate the auditorium of a theater

house·maid \'hau̇s-ˌmād\ *noun* (circa 1694)
: a female servant employed to do housework

housemaid's knee *noun* [so called from its occurrence among women who work a great deal on their knees] (1831)
: a swelling over the knee due to an enlargement of the bursa in the front of the patella

house•man \'haủs-mən, -ˌman\ *noun* (1920)
: a person who performs general work about a house or hotel

house•mas•ter \-ˌmas-tər\ *noun* (1884)
: a master in charge of a house in a boy's boarding school

house•mate \'haủs-ˌmāt\ *noun* (circa 1810)
: a person who lives in the same house with another

house•moth•er \'haủs-ˌmə-t͟hər\ *noun* (1882)
: a woman acting as hostess, chaperon, and often housekeeper in a group residence

house mouse *noun* (1835)
: a common nearly cosmopolitan grayish-brown mouse (*Mus musculus*) that usually lives and breeds about buildings, may act as a vector of diseases, and is an important laboratory animal

house of assembly (1653)
: a legislative body or the lower house of a legislature (as in various British colonies, protectorates, and countries of the Commonwealth)

House of Burgesses (1658)
: the colonial representative assembly of Virginia

house of cards (1903)
: a structure, situation, or institution that is insubstantial, shaky, or in constant danger of collapse

House of Commons (1621)
: the lower house of the British and Canadian parliaments

house of correction (circa 1576)
: an institution where persons who have committed a minor offense and are considered capable of reformation are confined

house of delegates (1783)
: HOUSE 6a; *especially* **:** the lower house of the state legislature in Maryland, Virginia, and West Virginia

House of Lords (1818)
: the upper house of the British Parliament composed of the lords temporal and spiritual

house of representatives (1716)
: the lower house of a legislative body (as the U.S. Congress)

house of studies (1929)
: an educational institution serving scholars of a religious order — called also *house of study*

house organ *noun* (1907)
: a periodical distributed by a business concern among its employees, sales personnel, or customers

house•paint•er \'haủs-ˌpān-tər\ *noun* (1689)
: one whose business or occupation is painting houses

house•par•ent \-ˌpar-ənt, -ˌper-\ *noun*
: an adult in charge of a dormitory, hall, hostel, or group residence

house party *noun* (1876)
: a party lasting over one or more nights at a residence (as a home or fraternity house)

house•per•son \'haủs-ˌpər-sᵊn\ *noun* (1974)
: a person who does housekeeping

house•plant \'haủs-ˌplant\ *noun* (1871)
: a plant grown or kept indoors

house–proud \'haủs-ˌpraủd\ *adjective* (1849)
chiefly British **:** proud of one's house or housekeeping

hous•er \'haủ-zər\ *noun* (1940)
: one that promotes or administers housing projects

house–rais•ing \'haủs-ˌrā-ziŋ\ *noun* (1704)
: the joint erection of a house or its framework by a gathering of neighbors

house•room \-ˌrüm, -ˌrủm\ *noun* (1582)
: space for accommodation in or as if in a house ⟨given *houseroom* by a family all too eager to have a celebrity in their midst —Walter Kerr⟩

house rule *noun* (1947)
: a rule (as in a game) that applies only among a certain group or in a certain place

house seat *noun* (1948)
: a theater seat reserved by the management for a special guest

house sitter *noun* (1971)
: a person who occupies a dwelling to provide security and maintenance while the tenant is away
— house–sit \'haủ(s)-ˌsit\ *intransitive verb*
— house–sit•ting \-ˌsi-tiŋ\ *noun*

house sparrow *noun* (1674)
: a sparrow (*Passer domesticus*) native to Eurasia that has been introduced worldwide and is found especially in urban and agricultural areas — called also *English sparrow*

house–to–house \ˌhaủs-tə-'haủs\ *adjective* (1859)
: going or done by going from one building to the next ⟨*house-to-house* fighting⟩

house•top \'haủs-ˌtäp\ *noun* (1526)
: ROOF; *especially* **:** the level surface of a flat roof
— from the housetops : for all to hear **:** OPENLY ⟨shouting their grievances *from the housetops*⟩

house trailer *noun* (1937)
: MOBILE HOME

house–train \'haủs-ˌtrān\ *transitive verb* (1924)
chiefly British **:** HOUSEBREAK

house•wares \'haủs-ˌwarz, -ˌwerz\ *noun plural* (1921)
: furnishings for a house; *especially* **:** small articles of household equipment (as cooking utensils or small appliances)

house•warm•ing \'haủs-ˌwȯr-miŋ\ *noun* (1577)
: a party to celebrate the taking possession of a house or premises

house•wife \'haủs-ˌwīf; *especially 2 & in early poetry* 'hə-zəf *or* -səf\ *noun, plural* **house•wives** \'haủs-ˌwīvz *also* 'haủz-ˌwīvz; 'hə-zəfs, -zəvz, -səfs, -səvz\ (13th century)
1 : a married woman in charge of a household
2 : a pocket-size container for small articles (as thread)
— house•wife•li•ness \-lē-nəs\ *noun*
— house•wife•ly \-lē\ *adjective*
— house•wif•ery \-ˌwī-f(ə-)rē; *British* -(ˌ)wi-f(ə-)rē *also* 'hə-zə-frē\ *noun*
— house•wif•ey \'haủs-ˌwī-fē\ *adjective*

house•work \'haủs-ˌwərk\ *noun* (1841)
: the work of housekeeping

¹hous•ing \'haủ-ziŋ\ *noun* (14th century)
1 a : SHELTER, LODGING **b :** dwellings provided for people
2 a : a niche for a sculpture **b :** the space taken out of a structural member (as a timber) to admit the insertion of part of another
3 : something that covers or protects: as **a :** a case or enclosure (as for a mechanical part or an instrument) **b :** a casing (as an enclosed bearing) in which a shaft revolves **c :** a support (as a frame) for mechanical parts

²housing *noun* [Middle English, from *house* housing (from Middle French *houce*, of Germanic origin) + *-ing*; akin to Middle High German *hulft* covering] (15th century)
: CAPARISON 1

housing development *noun* (1951)
: a group of individual dwellings or apartment houses typically of similar design that are usually built and sold or leased by one management

housing estate *noun* (1920)
British **:** HOUSING DEVELOPMENT

housing project *noun* (circa 1937)
: a publicly supported and administered housing development planned usually for low-income families

Hou•yhn•hnm \'hwi-nəm, hü-'i-nəm\ *noun*
: a member of a race of horses endowed with reason in Swift's *Gulliver's Travels*

hove *past and past participle of* HEAVE

hov•el \'hə-vəl, 'hä-\ *noun* [Middle English] (15th century)
1 : an open shed or shelter
2 : TABERNACLE
3 : a small, wretched, and often dirty house **:** HUT

hov•er \'hə-vər, 'hä-\ *intransitive verb* **hov•ered; hov•er•ing** \-v(ə-)riŋ\ [Middle English *hoveren*, frequentative of *hoven* to hover] (15th century)
1 a : to hang fluttering in the air or on the wing **b :** to remain suspended over a place or object
2 a : to move to and fro near a place **:** fluctuate around a given point ⟨unemployment *hovered* around 10%⟩ **b :** to be in a state of uncertainty, irresolution, or suspense
— hover *noun*
— hov•er•er \-vər-ər\ *noun*

hov•er•craft \-vər-ˌkraft\ *noun* (1959)
: a vehicle that is supported above the surface of land or water by a cushion of air produced by downwardly directed fans

¹how \'haủ\ *adverb* [Middle English, from Old English *hū*; akin to Old High German *hwuo* how, Old English *hwā* who — more at WHO] (before 12th century)
1 a : in what manner or way **b :** for what reason **:** WHY **c :** with what meaning **:** to what effect **d :** by what name or title ⟨*how* art thou called —Shakespeare⟩
2 : to what degree or extent
3 : in what state or condition ⟨*how* are you⟩
4 : at what price ⟨*how* a score of ewes now —Shakespeare⟩ ▫
— how about : what do you say to or think of ⟨*how about* it, are you going?⟩
— how come : how does it happen that **:** WHY

²how *conjunction* (before 12th century)
1 a : the way or manner in which ⟨remember *how* they fought⟩; *also* **:** the state or condition in which **b :** THAT ⟨told them *how* he had a situation —Charles Dickens⟩
2 : HOWEVER, AS ⟨a reader can shift his attention *how* he likes —William Empson⟩

³how \'haủ\ *noun* (1533)
1 : a question about manner or method
2 : MANNER, METHOD

¹how•be•it \haủ-'bē-ət\ *conjunction* (14th century)
: ALTHOUGH

²howbeit *adverb* (15th century)
: NEVERTHELESS

▫ USAGE
how *How come* is a familiar phrase of obscure origin that first came to attention as an Americanism in the 19th century. It seems to scare writers on usage: most of them treat it as a highly informal or even slangy speech form and warn against its use in writing. But it is precisely in writing that the phrase surprises us—it is gaining improved status in print. It now appears commonly in the general prose of good writers ⟨if the car was going that fast, *how come* it wasn't more seriously damaged? —Calvin Trillin⟩ ⟨*how come* I never hear those sounds in the daytime? —Andy Rooney⟩ ⟨that's *how come* she knows about food —Harold Brodkey⟩ ⟨*The Tempest* begins with exactly the same *dramatis personae* as the *Mabinogi* . . . . *How come?* —Robert Graves⟩. This sort of use began about the end of World War II. The phrase is still not used in highly elevated prose.

how·dah \'hau̇-də\ *noun* [Hindi *hauda,* from Arabic *hawdaj*] (1774)
: a seat or covered pavilion on the back of an elephant or camel

howdah

how·dy \'hau̇-dē\ *interjection* [alteration of *how do ye*] (1712)
— used to express greeting
— **howdy** *verb*

howe \'hau̇, 'hō\ *noun* [Middle English (northern) *holl* hollow place, from Old English *hol,* from *hol,* adjective, hollow — more at HOLE] (before 12th century)
Scottish **:** HOLLOW, VALLEY

[1]**how·ev·er** \hau̇-'e-vər\ *conjunction* (14th century)
1 : in whatever manner or way that ⟨will help *however* I can⟩
2 *archaic* **:** ALTHOUGH

[2]**however** *adverb* (14th century)
1 a : in whatever manner or way ⟨shall serve you, sir, truly, *however* else —Shakespeare⟩ **b :** to whatever degree or extent ⟨has done this for *however* many thousands of years —Emma Hawkridge⟩
2 : in spite of that **:** on the other hand ⟨still seems possible, *however,* that conditions will improve⟩ ⟨would like to go; *however,* I think I'd better not⟩
3 : how in the world ⟨*however* did you manage to do it⟩

howff *or* **howf** \'hau̇f, 'hōf\ *noun* [Dutch *hof* enclosure; akin to Old English *hof* enclosure, and perhaps to *hufil* hill] (1711)
Scottish **:** HAUNT, RESORT

how·it·zer \'hau̇-ət-sər\ *noun* [Dutch *houwitser,* ultimately from Czech *houfnice* ballista] (1695)
: a short cannon used to fire projectiles at medium muzzle velocities and with relatively high trajectories

howl \'hau̇(ə)l\ *verb* [Middle English *houlen;* akin to Middle High German *hiulen* to howl] (14th century)
intransitive verb
1 : to emit a loud sustained doleful sound characteristic of members of the dog family
2 : to cry out loudly and without restraint under strong impulse (as pain, grief, or amusement)
3 : to go on a spree or rampage
transitive verb
1 : to utter with unrestrained outcry
2 : to drown out or cause to fail by adverse outcry — used especially with *down* ⟨*howled* down the speaker⟩
— **howl** *noun*

howl·er \'hau̇-lər\ *noun* (1800)
1 a : one that howls **b :** HOWLER MONKEY
2 : a humorous and ridiculous blunder

howler monkey *noun* (1932)
: any of a genus (*Alouatta*) of South and Central American monkeys that have a long prehensile tail and enlargement of the hyoid and laryngeal apparatus enabling them to make loud howling noises

howler monkey

howl·ing \'hau̇-liŋ\ *adjective* (1599)
1 : producing or marked by a sound resembling a sustained howl ⟨a *howling* storm⟩
2 : DESOLATE, WILD ⟨a *howling* wilderness⟩
3 : very great **:** PRONOUNCED ⟨a *howling* success⟩
— **howl·ing·ly** *adverb*

how·so·ev·er \ˌhau̇-sə-'we-vər\ *adverb* (14th century)
1 : in whatever manner
2 : to whatever degree or extent

[1]**how–to** \'hau̇-'tü\ *adjective* (1926)
: giving practical instruction and advice (as on a craft) ⟨*how-to* books on all sorts of hobbies —Harry Milt⟩

[2]**how–to** *noun* (1954)
: a practical method or instruction ⟨the *how-tos* of balancing a checkbook⟩

[1]**hoy** \'hȯi\ *interjection* [Middle English] (14th century)
— used in attracting attention or in driving animals

[2]**hoy** *noun* [Middle English, from Middle Dutch *hoei*] (15th century)
1 : a small usually sloop-rigged coasting ship
2 : a heavy barge for bulky cargo

hoya \'hȯi-(y)ə\ *noun* [New Latin, from Thomas *Hoy* (died 1821) English gardener] (1851)
: any of a genus (*Hoya*) of climbing Asian and Australian evergreen shrubs of the milkweed family

hoy·den \'hȯi-dᵊn\ *noun* [perhaps from obsolete Dutch *heiden* country lout, from Middle Dutch, heathen; akin to Old English *hǣthen* heathen] (1676)
: a girl or woman of saucy, boisterous, or carefree behavior
— **hoy·den·ish** \-ish\ *adjective*

hoyle \'hȯi(ə)l\ *noun, often capitalized* [Edmond *Hoyle* (died 1769) English writer on games] (1906)
: an encyclopedia of the rules of indoor games and especially card games

Hsia \shē-'ä\ *noun* [Chinese (Beijing) *Xià*] (circa 1909)
: the legendary first dynasty of Chinese history traditionally dated from about 2200–1766 B.C.

HTLV \ˌāch-(ˌ)tē-(ˌ)el-'vē\ *noun* [*h*uman *T*-cell *l*ymphotropic *v*irus] (1980)
: any of several retroviruses — often used with a number or Roman numeral to indicate type

HTLV–III \-(ˌ)vē-'thrē\ *noun* (1984)
: HIV-1

hua·ra·che \wə-'rä-chē, hə-\ *noun* [Mexican Spanish, from Tarascan *kʷaráči*] (1892)
: a low-heeled sandal having an upper made of interwoven leather strips

hub \'həb\ *noun* [probably alteration of [2]*hob*] (1649)
1 : the central part of a circular object (as a wheel or propeller)
2 a : a center of activity **:** FOCAL POINT **b :** an airport or city through which an airline routes most of its traffic
3 : a steel punch from which a working die for a coin or medal is made

hub–and–spoke *adjective* (1980)
: being or relating to a system of routing air traffic in which a major airport serves as a central point for coordinating flights to and from other airports

Hub·bard squash \'hə-bərd-\ *noun* [probably from the name *Hubbard*] (1868)
: any of various often large variably green winter squashes — called also *Hubbard*

hub·ble–bub·ble \'hə-bəl-ˌbə-bəl\ *noun* [reduplication of *bubble*] (1634)
1 : WATER PIPE 2
2 : a flurry of sound or activity **:** COMMOTION

hub·bub \'hə-ˌbəb\ *noun* [perhaps of Irish origin; akin to Scottish Gaelic *ub ub,* interjection of contempt] (1555)
1 : NOISE, UPROAR
2 : CONFUSION, TURMOIL

hub·by \'hə-bē\ *noun, plural* **hubbies** [by alteration] (1688)
: HUSBAND

hub·cap \'həb-ˌkap\ *noun* (1903)
: a removable usually metal cap over the end of an axle; *especially* **:** one used on the wheel of a motor vehicle

hu·bris \'hyü-brəs\ *noun* [Greek *hybris*] (1884)
: exaggerated pride or self-confidence
— **hu·bris·tic** \hyü-'bris-tik\ *adjective*

huck \'hək\ *noun* (1851)
: HUCKABACK

huck·a·back \'hə-kə-ˌbak\ *noun* [origin unknown] (1690)
: an absorbent durable fabric of cotton, linen, or both used chiefly for towels

huck·le·ber·ry \'hə-kəl-ˌber-ē\ *noun* [perhaps alteration of *hurtleberry* (huckleberry)] (1670)
1 : any of a genus (*Gaylussacia*) of American shrubs of the heath family; *also* **:** the edible dark blue to black usually acid berry (especially of *G. baccata*) with 10 nutlets
2 : BLUEBERRY

[1]**huck·ster** \'hək-stər\ *noun* [Middle English *hukster,* from Middle Dutch *hokester,* from *hoeken* to peddle] (13th century)
1 : HAWKER, PEDDLER
2 : one who produces promotional material for commercial clients especially for radio or television
— **huck·ster·ism** \-stə-ˌri-zəm\ *noun*

[2]**huckster** *verb* **huck·stered; huck·ster·ing** \-st(ə-)riŋ\ (1592)
intransitive verb
: HAGGLE
transitive verb
1 : to deal in or bargain over
2 : to promote by showmanship

[1]**hud·dle** \'hə-dᵊl\ *verb* **hud·dled; hud·dling** \'həd-liŋ, 'hə-dᵊl-iŋ\ [probably from or akin to Middle English *hoderen* to huddle] (1579)
transitive verb
1 *British* **:** to arrange carelessly or hurriedly
2 a : to crowd together **b :** to draw (oneself) together **:** CROUCH
3 : to wrap closely in (as clothes)
intransitive verb
1 a : to gather in a close-packed group **b :** to curl up **:** CROUCH
2 a : to hold a consultation **b :** to gather in a huddle in football
— **hud·dler** \'həd-lər, 'hə-dᵊl-ər\ *noun*

[2]**huddle** *noun* (1586)
1 : a close-packed group **:** BUNCH ⟨*huddles* of children⟩ ⟨a *huddle* of cottages⟩
2 a : MEETING, CONFERENCE ⟨secret *huddles* were held by five leading Republicans —*Newsweek*⟩ **b :** a brief gathering of football players away from the line of scrimmage to receive instructions (as from the quarterback) for the next down

Hu·di·bras·tic \ˌhyü-də-'bras-tik\ *adjective* [irregular from *Hudibras,* satirical poem by Samuel Butler (died 1680)] (1712)
1 : written in humorous octosyllabic couplets
2 : MOCK-HEROIC
— **Hudibrastic** *noun*

hue \'hyü\ *noun* [Middle English *hewe,* from Old English *hīw;* akin to Old Norse *hȳ* plant down, Gothic *hiwi* form] (before 12th century)
1 : COMPLEXION, ASPECT ⟨political parties of every *hue* —Louis Wasserman⟩
2 a : COLOR **b :** gradation of color **c :** the attribute of colors that permits them to be classed as red, yellow, green, blue, or an intermediate between any contiguous pair of these colors — compare BRIGHTNESS 2, LIGHTNESS 2, SATURATION 4

hue and cry *noun* [*hue* (outcry)] (15th century)
1 a : a loud outcry formerly used in the pursuit of one who is suspected of a crime **b :** the pursuit of a suspect or a written proclamation for the capture of a suspect
2 : a clamor of alarm or protest
3 : HUBBUB

hued \'hyüd\ *adjective* (before 12th century)
: COLORED — usually used in combination ⟨green-*hued*⟩

[1]**huff** \'həf\ *verb* [imitative] (1583)
intransitive verb
1 : to emit puffs (as of breath or steam)

2 a : to make empty threats **:** BLUSTER **b :** to react or behave indignantly ⟨*huffed* off in anger⟩
transitive verb
1 : to puff up **:** INFLATE
2 *archaic* **:** to treat with contempt **:** BULLY
3 : to make angry
4 : to utter with indignation or scorn
²huff *noun* (1599)
: a usually peevish and transitory spell of anger or resentment ⟨quit in a *huff*⟩
synonym see OFFENSE
huff•ish \'hə-fish\ *adjective* (circa 1755)
: ARROGANT, SULKY
huffy \'hə-fē\ *adjective* **huff•i•er; -est** (1677)
1 : HAUGHTY, ARROGANT
2 a : roused to indignation **:** IRRITATED **b :** easily offended **:** TOUCHY
— huff•i•ly \'hə-fə-lē\ *adverb*
— huff•i•ness \-fē-nəs\ *noun*
hug \'həg\ *transitive verb* **hugged; hug•ging** [perhaps of Scandinavian origin; akin to Old Norse *hugga* to soothe] (1567)
1 : to press tightly especially in the arms
2 a : CONGRATULATE **b :** to hold fast **:** CHERISH ⟨*hugged* his miseries like a sulky child —John Buchan⟩
3 : to stay close to ⟨the road *hugs* the river⟩
— hug *noun*
— hug•ga•ble \'hə-gə-bəl\ *adjective*
huge \'hyüj, 'yüj\ *adjective* **hug•er; hug•est** [Middle English, from Old French *ahuge*] (12th century)
: very large or extensive: as **a :** of great size or area **b :** great in scale or degree ⟨a *huge* deficit⟩ **c :** great in scope or character ⟨a man of *huge* talent⟩
synonym see ENORMOUS
— huge•ly *adverb*
— huge•ness *noun*
huge•ous \'hyü-jəs, 'yü-\ *adjective* (1519)
: HUGE
— huge•ous•ly *adverb*
¹hug•ger–mug•ger \'hə-gər-,mə-gər\ *noun* [origin unknown] (1529)
1 : SECRECY
2 : CONFUSION, MUDDLE
²hugger–mugger *adjective* (1692)
1 : SECRET
2 : of a confused or disorderly nature **:** JUMBLED
— hugger–mugger *adverb*
hug–me–tight \'həg-mē-,tīt\ *noun* (1860)
: a woman's short usually knitted sleeveless close-fitting jacket
Hu•gue•not \'hyü-gə-,nät\ *noun* [Middle French, alteration of Middle French dialect *eyguenot,* adherent of a Swiss political movement, from German dialect *Eidgnosse* confederate] (1565)
: a member of the French Reformed communion especially of the 16th and 17th centuries
— Hu•gue•not•ic \,hyü-gə-'nä-tik\ *adjective*
— Hu•gue•not•ism \'hyü-gə-,nä-,ti-zəm\ *noun*
huh *a grunt articulated as a syllabic* m *or* n *with a voiceless onset, or as the syllable* 'hə *or* 'hə", *often ending in a glottal stop, and uttered with a range of intonations; often read as* 'hə\ *interjection* [imitative of a grunt] (1608)
— used to express surprise, disbelief, or confusion, or as an inquiry inviting affirmative reply
Hui•chol \wē-'chōl\ *noun, plural* **Huichol** *or* **Hui•cho•les** [Mexican Spanish] (1900)
1 : a member of an American Indian people of the mountains between Zacatecas and Nayarit, Mexico
2 : the Uto-Aztecan language of the Huichol people
— Huichol *adjective*
hui•sa•che \wē-'sä-chē\ *noun* [Mexican Spanish, from Nahuatl *huixachi,* from *huitzli* thorn + *ixachi* a great amount, many] (1838)
: a widely cultivated thorny shrubby acacia (*Acacia farnesiana*) of the southern U.S. and tropical America
hu•la \'hü-lə\ *also* **hu•la–hu•la** \,hü-lə-'hü-lə\ *noun* [Hawaiian] (1825)
: a sinuous Polynesian dance characterized by rhythmic movement of the hips and mimetic gestures with the hands and often accompanied by chants and rhythmic drumming
hula hoop *noun* (1958)
: a plastic hoop that is twirled around the body
¹hulk \'həlk\ *noun* [Middle English *hulke,* from Old English *hulc,* probably from Medieval Latin *holcas,* from Greek *holkas,* from *helkein* to pull — more at SULCUS] (before 12th century)
1 a : a heavy clumsy ship **b** (1) **:** the body of an old ship unfit for service (2) **:** a ship used as a prison — usually used in plural ⟨every prisoner sent to the *hulks* —Kenneth Roberts⟩ **c :** an abandoned wreck or shell (as of a building or automobile)
2 : one that is bulky or unwieldy ⟨a big *hulk* of a man⟩
²hulk *intransitive verb* (circa 1825)
1 *dialect English* **:** to move ponderously
2 : to appear impressively large or massive **:** LOOM
hulk•ing \'həl-kiŋ\ *adjective* (1698)
: PONDEROUS, MASSIVE
¹hull \'həl\ *noun* [Middle English, from Old English *hulu;* akin to Old High German *hala* hull, Old English *helan* to conceal — more at HELL] (before 12th century)
1 a : the outer covering of a fruit or seed **b :** the persistent calyx or involucre that subtends some fruits
2 a : the frame or body of a ship or boat exclusive of masts, yards, sails, and rigging **b :** the main body of a usually large or heavy craft or vehicle (as an airship or tank)
3 : COVERING, CASING
— hull–less \'həl-ləs\ *adjective*
²hull *transitive verb* (14th century)
: to remove the hulls of **:** SHUCK
— hull•er *noun*
hul•la•ba•loo \'hə-lə-bə-,lü\ *noun, plural* **-loos** [perhaps from *hallo* + Scots *balloo,* interjection used to hush children] (1762)
: DIN; *also* **:** UPROAR
hull down *adverb or adjective* (1775)
of a ship **:** at such a distance that only the superstructure is visible
hulled corn *noun* (1788)
: whole grain corn from which the hulls have been removed by soaking or boiling in lye water
hul•lo \(,)hə-'lō\ *chiefly British variant of* HELLO
¹hum \'həm\ *verb* **hummed; hum•ming** [Middle English *hummen;* akin to Middle High German *hummen* to hum, Middle Dutch *hommel* bumblebee] (14th century)
intransitive verb
1 a : to utter a sound like that of the speech sound \m\ prolonged **b :** to make the natural noise of an insect in motion or a similar sound **:** DRONE **c :** to give forth a low continuous blend of sound
2 a : to be busily active ⟨the museum *hummed* with visitors⟩ **b :** to run smoothly ⟨the business started to *hum*⟩
transitive verb
1 : to sing with the lips closed and without articulation
2 : to affect or express by humming ⟨*hummed* his displeasure⟩
— hum *noun*
— hum•ma•ble \'hə-mə-bəl\ *adjective*
²hum *chiefly British variant of* HEM
¹hu•man \'hyü-mən, 'yü-\ *adjective* [Middle English *humain,* from Middle French, from Latin *humanus;* akin to Latin *homo* human being — more at HOMAGE] (14th century)
1 : of, relating to, or characteristic of humans
2 : consisting of humans
3 a : having human form or attributes **b :** susceptible to or representative of the sympathies and frailties of human nature ⟨such an inconsistency is very *human* —P. E. More⟩
usage see HUMOR
— hu•man•ness \-mən-nəs\ *noun*
²human *noun* (circa 1533)
: a bipedal primate mammal (*Homo sapiens*) **:** MAN; *broadly* **:** any living or extinct member of the family (Hominidae) to which the primate belongs
— hu•man•like \-mən-,līk\ *adjective*
human being *noun* (1800)
: HUMAN
hu•mane \hyü-'mān, yü-\ *adjective* [Middle English *humain*] (circa 1500)
1 : marked by compassion, sympathy, or consideration for humans or animals
2 : characterized by or tending to broad humanistic culture **:** HUMANISTIC ⟨*humane* studies⟩
— hu•mane•ly *adverb*
— hu•mane•ness \-'mān-nəs\ *noun*
human ecology *noun* (1907)
1 : a branch of sociology dealing especially with the spatial and temporal interrelationships between humans and their economic, social, and political organization
2 : the ecology of human communities and populations especially as concerned with preservation of environmental quality (as of air or water) through proper application of conservation and civil engineering practices
human engineering *noun* (1920)
1 : management of humans and their affairs especially in industry
2 : ERGONOMICS
human immunodeficiency virus *noun* (1986)
: HIV
hu•man•ism \'hyü-mə-,ni-zəm, 'yü-\ *noun* (1832)
1 a : devotion to the humanities **:** literary culture **b :** the revival of classical letters, individualistic and critical spirit, and emphasis on secular concerns characteristic of the Renaissance
2 : HUMANITARIANISM
3 : a doctrine, attitude, or way of life centered on human interests or values; *especially* **:** a philosophy that usually rejects supernaturalism and stresses an individual's dignity and worth and capacity for self-realization through reason
— hu•man•ist \-nist\ *noun or adjective*
— hu•man•is•tic \,hyü-mə-'nis-tik, ,yü-\ *adjective*
— hu•man•is•ti•cal•ly \-ti-k(ə-)lē\ *adverb*
hu•man•i•tar•i•an \hyü-,ma-nə-'ter-ē-ən, yü-\ *noun* (1844)
: a person promoting human welfare and social reform **:** PHILANTHROPIST
— humanitarian *adjective*
— hu•man•i•tar•i•an•ism \-ē-ə-,ni-zəm\ *noun*
hu•man•i•ty \hyü-'ma-nə-tē, yü-\ *noun, plural* **-ties** (14th century)
1 : the quality or state of being humane
2 a : the quality or state of being human **b** *plural* **:** human attributes or qualities ⟨his work has the ripeness of the 18th century, and its rough *humanities* —Pamela H. Johnson⟩
3 *plural* **:** the branches of learning (as philosophy or languages) that investigate human constructs and concerns as opposed to natural processes (as in physics or chemistry)
4 : MANKIND
hu•man•ize \'hyü-mə-,nīz, 'yü-\ *transitive verb* **-ized; -iz•ing** (1603)

1 a : to represent as human **:** attribute human qualities to **b :** to adapt to human nature or use
2 : to make humane
— hu·man·iza·tion \ˌhyü-mə-nə-ˈzā-shən, ˌyü-\ *noun*
— hu·man·iz·er *noun*

hu·man·kind \ˈhyü-mən-ˌkīnd, ˈyü-\ *noun singular but singular or plural in construction* (circa 1645)
: the human race

hu·man·ly \ˈhyü-mən-lē, ˈyü-\ *adverb* (15th century)
1 a : with regard to human needs and emotions ⟨provide *humanly* for those who are not needed in the economy —E. F. Bacon⟩ **b :** with regard to or in keeping with human proneness to error or weakness ⟨*humanly* inaccurate⟩
2 a : from a human viewpoint ⟨*humanly* speaking, the process works . . . like this —Elizabeth Janeway⟩ **b :** within the range of human capacity ⟨did everything *humanly* possible⟩ **c :** by humans ⟨*humanly* made⟩

human nature *noun* (1668)
: the nature of humans; *especially* **:** the fundamental dispositions and traits of humans

hu·man·oid \ˈhyü-mə-ˌnȯid, ˈyü-\ *adjective* (1918)
: having human form or characteristics ⟨*humanoid* dentition⟩ ⟨*humanoid* robots⟩
— humanoid *noun*

human relations *noun plural but usually singular in construction* (1946)
1 : a study of human problems arising from organizational and interpersonal relations (as in industry)
2 : a course, study, or program designed to develop better interpersonal and intergroup adjustments

human resources *noun plural* (1975)
: PERSONNEL 1a, 2 ⟨director of *human resources*⟩

human rights *noun plural* (1791)
: rights (as freedom from unlawful imprisonment, torture, and execution) regarded as belonging fundamentally to all persons

hu·mate \ˈhyü-ˌmāt, ˈyü-\ *noun* (1844)
: a salt or ester of a humic acid

¹hum·ble \ˈhəm-bəl *also chiefly Southern* ˈəm-\ *adjective* **hum·bler** \-b(ə-)lər\; **hum·blest** \-b(ə-)ləst\ [Middle English, from Old French, from Latin *humilis* low, humble, from *humus* earth; akin to Greek *chthōn* earth, *chamai* on the ground] (13th century)
1 : not proud or haughty **:** not arrogant or assertive
2 : reflecting, expressing, or offered in a spirit of deference or submission ⟨a *humble* apology⟩
3 a : ranking low in a hierarchy or scale **:** INSIGNIFICANT, UNPRETENTIOUS **b :** not costly or luxurious ⟨a *humble* contraption⟩
usage see HUMOR
— hum·ble·ness \-bəl-nəs\ *noun*
— hum·bly \-blē\ *adverb*

²humble *transitive verb* **hum·bled; hum·bling** \-b(ə-)liŋ\ (14th century)
1 : to make humble in spirit or manner
2 : to destroy the power, independence, or prestige of
— hum·bler \-b(ə-)lər\ *noun*
— hum·bling·ly \-b(ə-)liŋ-lē\ *adverb*

hum·ble–bee \ˈhəm-bəl-ˌbē\ *noun* [Middle English *humbylbee,* from *humbyl-* (akin to Middle Dutch *hommel* bumblebee) + *bee* — more at HUM] (15th century)
: BUMBLEBEE

humble pie *noun* (1830)
: a figurative serving of humiliation usually in the form of a forced submission, apology, or retraction — often used in the phrase *eat humble pie* ◆

¹hum·bug \ˈhəm-ˌbəg\ *noun* [origin unknown] (1751)
1 a : something designed to deceive and mislead **b :** a willfully false, deceptive, or insincere person
2 : an attitude or spirit of pretense and deception
3 : NONSENSE, DRIVEL
4 *British* **:** a hard usually mint-flavored candy
synonym see IMPOSTURE
— hum·bug·gery \-ˌbə-g(ə-)rē\ *noun*

²humbug *verb* **hum·bugged; hum·bug·ging** (1751)
transitive verb
: DECEIVE, HOAX
intransitive verb
: to engage in a hoax or deception

hum·ding·er \ˈhəm-ˈdiŋ-ər\ *noun* [probably alteration of *hummer*] (circa 1904)
: a striking or extraordinary person or thing

hum·drum \ˈhəm-ˌdrəm\ *adjective* [reduplication of *hum*] (1553)
: MONOTONOUS, DULL
— humdrum *noun*

hu·mec·tant \hyü-ˈmek-tənt\ *noun* [Latin *humectant-, humectans,* present participle of *humectare* to moisten, from *humectus* moist, from *humēre* to be moist — more at HUMOR] (circa 1857)
: a substance that promotes retention of moisture
— humectant *adjective*

hu·mer·al \ˈhyü-mə-rəl\ *adjective* (1615)
1 : of, relating to, or situated in the region of the humerus or shoulder
2 : of, relating to, or being a body part analogous to the humerus or shoulder
— humeral *noun*

humeral veil *noun* (1853)
: an oblong vestment worn around the shoulders and over the hands by a priest holding a sacred vessel

hu·mer·us \ˈhyü-mə-rəs\ *noun, plural* **hu·meri** \-ˌrī, -ˌrē\ [Middle English, from Latin *humerus, umerus* upper arm, shoulder; akin to Gothic *ams* shoulder, Greek *ōmos*] (15th century)
: the long bone of the upper arm or forelimb extending from the shoulder to the elbow

hu·mic \ˈhyü-mik, ˈyü-\ *adjective* (1844)
: of, relating to, or derived at least in part from humus

humic acid *noun* (1844)
: any of various organic acids obtained from humus

hu·mid \ˈhyü-məd, ˈyü-\ *adjective* [French or Latin; French *humide,* from Latin *humidus,* from *humēre*] (15th century)
: containing or characterized by perceptible moisture especially to the point of being oppressive
synonym see WET
— hu·mid·ly *adverb*

hu·mid·i·fi·er \hyü-ˈmid-ə-ˌfī(-ə)r, yü-\ *noun* (1884)
: a device for supplying or maintaining humidity

hu·mid·i·fy \-ˌfī\ *transitive verb* **-fied; -fy·ing** (1885)
: to make humid
— hu·mid·i·fi·ca·tion \-ˌmi-də-fə-ˈkā-shən\ *noun*

hu·mid·i·stat \hyü-ˈmi-də-ˌstat, yü-\ *noun* (circa 1904)
: an instrument for regulating or maintaining the degree of humidity

hu·mid·i·ty \hyü-ˈmi-də-tē, yü-\ *noun, plural* **-ties** (15th century)
: a moderate degree of wetness especially of the atmosphere — compare RELATIVE HUMIDITY

hu·mi·dor \ˈhyü-mə-ˌdȯr, ˈyü-\ *noun* [*humid* + *-or* (as in *cuspidor*)] (1903)
: a case or enclosure (as for storing cigars) in which the air is kept properly humidified

hu·mi·fi·ca·tion \ˌhyü-mə-fə-ˈkā-shən, ˌyü-\ *noun* (1897)
: formation of or conversion into humus

hu·mi·fied \ˈhyü-mə-ˌfīd, ˈyü-\ *adjective* (1906)
: converted into humus

hu·mil·i·ate \hyü-ˈmi-lē-ˌāt, yü-\ *transitive verb* **-at·ed; -at·ing** [Late Latin *humiliatus,* past participle of *humiliare,* from Latin *humilis* low — more at HUMBLE] (circa 1534)
: to reduce to a lower position in one's own eyes or others' eyes **:** MORTIFY
— hu·mil·i·a·tion \-ˌmi-lē-ˈā-shən\ *noun*

hu·mil·i·at·ing \hyü-ˈmi-lē-ˌā-tiŋ, yü-\ *adjective* (1757)
: extremely destructive to one's self-respect or dignity **:** HUMBLING
— hu·mil·i·at·ing·ly \-tiŋ-lē\ *adverb*

hu·mil·i·ty \hyü-ˈmi-lə-tē, yü-\ *noun* (14th century)
: the quality or state of being humble

hum·mer \ˈhə-mər\ *noun* (1605)
1 : one that hums
2 : HUMMINGBIRD
3 : HUMDINGER
4 : FASTBALL

hum·ming·bird \ˈhə-miŋ-ˌbərd\ *noun* (1637)
: any of a family (Trochilidae) of tiny brightly colored nonpasserine New World birds related to the swifts and like them having narrow wings with long primaries, a slender bill, and a very extensile tongue

hum·mock \ˈhə-mək\ *noun* [alteration of ²*hammock*] (1555)
1 : a rounded knoll or hillock
2 : a ridge of ice
3 : ²HAMMOCK 2
— hummock *verb*
— hum·mocky \-mə-kē\ *adjective*

hum·mus \ˈhə-məs, ˈhu̇-\ *noun* [Arabic *ḥummuṣ* chickpeas] (1950)
: a paste of pureed chickpeas usually mixed with sesame oil or sesame paste and eaten as a dip or sandwich spread

hu·mon·gous \hyü-ˈməŋ-gəs, yü-, -ˈmäŋ-\ *also* **hu·mun·gous** \-ˈməŋ-gəs\ *adjective* [perhaps alteration of *huge* + *monstrous*] (circa 1967)
: extremely large **:** HUGE

¹hu·mor \ˈhyü-mər, ˈyü-\ *noun* [Middle English *humour,* from Middle French *humeur,* from Medieval Latin & Latin; Medieval Latin *humor,* from Latin *humor, umor* moisture; akin to Old Norse *vǫkr* damp, Latin *humēre* to be moist, and perhaps to Greek *hygros* wet] (14th century)

◇ WORD HISTORY

humble pie The phrase *humble pie* appears to be a simple culinary metaphor for humility, though beneath its surface is a pun on an actual but now seldom-eaten dish. *Umble pie* is a pastry stuffed and baked with *umbles,* the edible viscera of an animal, typically a game animal such as a deer. The word *umbles* is altered from earlier *numbles* (the initial *n* apparently being dropped due to the proximity of the following nasal consonant). *Numbles* was a Middle English borrowing from Old French *nombles,* ultimately descended from Latin *lumbuli,* diminutive of *lumbi* (singular *lumbus*) "loins." (The word *loin* itself is from a French descendant of another derivative of *lumbus;* related words going back directly to Latin are *lumbar* and *lumbago.*) The loss of initial \h\ in British regional speech would have contributed to the association of *umble pie* with the adjective *humble,* though, curiously, the metaphorical *humble pie* is first noted in East Anglia, an area where the \h\ has not traditionally been dropped, and the spelling *humble pie* in reference to the literal pastry already appears in the 17th century—suggesting that the form with \h\ may have originated through something other than punning.

1 a : a normal functioning bodily semifluid or fluid (as the blood or lymph) **b :** a secretion (as a hormone) that is an excitant of activity **2 a** *in medieval physiology* **:** a fluid or juice of an animal or plant; *specifically* **:** one of the four fluids entering into the constitution of the body and determining by their relative proportions a person's health and temperament **b :** characteristic or habitual disposition or bent **:** TEMPERAMENT ⟨of cheerful *humor*⟩ **c :** an often temporary state of mind imposed especially by circumstances ⟨was in no *humor* to listen⟩ **d :** a sudden, unpredictable, or unreasoning inclination **:** WHIM ⟨the uncertain *humors* of nature⟩
3 a : that quality which appeals to a sense of the ludicrous or absurdly incongruous **b :** the mental faculty of discovering, expressing, or appreciating the ludicrous or absurdly incongruous **c :** something that is or is designed to be comical or amusing ▫ ◇
synonym see WIT
— out of humor : out of sorts

²humor *transitive verb* **hu•mored; hu•mor•ing** \'hyüm-riŋ, 'yüm-, 'hyü-mə-, 'yü-\ (1588)
1 : to soothe or content by indulgence
2 : to adapt oneself to
synonym see INDULGE

hu•mor•al \'hyü-mə-rəl, 'yü-\ *adjective* (15th century)
1 : of, relating to, proceeding from, or involving a bodily humor (as a hormone)
2 : relating to or being the part of immunity or the immune response that involves antibodies secreted by B cells and circulating in bodily fluids

hu•mor•esque \,hyü-mə-'resk, ,yü-\ *noun* [German *Humoreske,* from *Humor,* from Medieval Latin] (1889)
: a typically whimsical or fanciful musical composition

hu•mor•ist \'hyü-mə-rist, 'yü-\ *noun* (1589)
1 *archaic* **:** a person subject to whims
2 : a person specializing in or noted for humor

hu•mor•is•tic \,hyü-mə-'ris-tik, ,yü-\ *adjective* (1818)
: HUMOROUS

hu•mor•less \'hyü-mər-ləs\ *adjective* (circa 1847)
1 : lacking a sense of humor
2 : lacking humorous characteristics
— hu•mor•less•ly \-lē\ *adverb*
— hu•mor•less•ness *noun*

hu•mor•ous \'hyüm-rəs, 'yüm-, 'hyü-mə-, 'yü-\ *adjective* (15th century)
1 *obsolete* **:** HUMID
2 a : full of or characterized by humor **:** JOCULAR **b :** indicating or expressive of a sense of humor
synonym see WITTY
— hu•mor•ous•ly *adverb*
— hu•mor•ous•ness *noun*

hu•mour *chiefly British variant of* HUMOR

¹hump \'həmp\ *noun* [akin to Middle Low German *hump* bump] (1709)
1 : a rounded protuberance: as **a :** HUMPBACK 1 **b :** a fleshy protuberance on the back of an animal (as a camel, bison, or whale) **c** (1) **:** MOUND, HUMMOCK (2) **:** MOUNTAIN, RANGE ⟨the Himalayan *hump*⟩
2 *British* **:** a fit of depression or sulking
3 : a difficult, trying, or critical phase or obstacle — often used in the phrase *over the hump*
— humped *adjective*

²hump (1835)
transitive verb
1 : to copulate with — often considered vulgar
2 : to exert (oneself) vigorously
3 : to make humpbacked **:** HUNCH
4 *chiefly British* **:** to put or carry on the back **:** LUG; *also* **:** TRANSPORT
intransitive verb
1 : to exert oneself **:** HUSTLE
2 : to move swiftly **:** RACE

hump•back \-,bak, *for 1 also* -'bak\ *noun* (1697)
1 : a humped or crooked back; *also* **:** KYPHOSIS
2 : HUNCHBACK 1
3 : HUMPBACK WHALE
4 : PINK SALMON

hump•backed \-'bakt\ *adjective* (1681)
1 : having a humped back
2 : convexly curved ⟨a *humpbacked* bridge⟩

humpback whale *noun* (1725)
: a large baleen whale (*Megaptera novaeangliae*) that is black above and white below and has very long flippers, and fleshy tubercles along the snout

humpback whale

¹humph \'həm(p)f\ (1814)
intransitive verb
: to utter a humph
transitive verb
: to utter (as a remark) in a tone suggestive of a humph

²humph *a snort articulated as a syllabic* m *or* n *with a voiceless onset and ending in a nasal* h *or a glottal stop; often read as* 'həm(p)f\ *interjection* [imitative of a grunt] (1815)
— used to express doubt or contempt

hump•ty–dump•ty \,həm(p)-tē-'dəm(p)-tē\ *noun, plural* **-dumpties** *often H&D capitalized* [*Humpty-Dumpty,* egg-shaped nursery-rhyme character who fell from a wall and broke into bits] (1883)
: something that once broken is impossible or almost impossible to put back together

humpy \'həm-pē\ *adjective* **hump•i•er; -est** (1708)
1 : full of humps
2 : covered with humps

hu•mus \'hyü-məs, 'yü-\ *noun* [New Latin, from Latin, earth — more at HUMBLE] (1796)
: a brown or black complex variable material resulting from partial decomposition of plant or animal matter and forming the organic portion of soil

hum•vee \,həm-'vē, 'həm-,\ *noun* [from HMMWV (*H*igh *M*obility *M*ultipurpose *W*heeled *V*ehicle)] (1982) *often capitalized*
: a diesel-powered multipurpose U.S. military vehicle that replaced the jeep

Hun \'hən\ *noun* [Middle English, from Old English *Hunas,* plural, from Late Latin *Hunni,* plural] (before 12th century)
1 : a member of a nomadic central Asian people gaining control of a large part of central and eastern Europe under Attila about A.D. 450
2 a *often not capitalized* **:** a person who is wantonly destructive **:** VANDAL **b :** GERMAN; *especially* **:** a German soldier — usually used disparagingly

Hu•nan \'hü-'nän\ *or* **Hu•na•nese** \,hü-nə-'nēz, -'nēs\ *adjective* [*Hunan,* China] (1970)
: of, relating to, or being a hot and spicy style of Chinese cooking

¹hunch \'hənch\ *verb* [origin unknown] (1581)
transitive verb
1 : JOSTLE, SHOVE
2 : to thrust or bend over into a humped or crooked position
intransitive verb
1 : to thrust oneself forward
2 a : to assume a bent or crooked posture **b :** to draw oneself into a ball **:** curl up **c :** HUDDLE, SQUAT

²hunch *noun* (1630)
1 : an act or instance of hunching **:** PUSH
2 a : a thick piece **:** LUMP **b :** HUMP
3 : a strong intuitive feeling concerning especially a future event or result

hunch•back \'hənch-,bak\ *noun* (1712)
1 : a person with a humpback
2 : HUMPBACK 1
— hunch•backed \-'bakt\ *adjective*

hun•dred \'hən-drəd, -dərd\ *noun, plural* **hundreds** *or* **hundred** [Middle English, from Old English, from *hund* hundred + *-red* (akin to Gothic *rathjo* account, number); akin to Latin *centum* hundred, Greek he*katon,* Old English *tien* ten — more at TEN, REASON] (before 12th century)
1 — see NUMBER table
2 *hundreds plural* **a :** the numbers 100 to 999 **b :** a great number ⟨*hundreds* of times⟩
3 : a 100-dollar bill
4 : a subdivision of some English and American counties
— hundred *adj*
— hun•dred•fold \-,fōld\ *adjective or adverb*
— hun•dredth \-drədth, -drətth\ *adjective or noun*

hun•dred–per•cent•er \-pər-'sen-tər\ *noun* [*hundred-percent (American)*] (1921)
: a thoroughgoing nationalist
— hun•dred–per•cent•ism \-'sen-,ti-zəm\ *noun*

hundreds place *noun* (1937)
: the place three to the left of the decimal point in a number expressed in the Arabic system of notation

hun•dred•weight \'hən-drəd-,wāt, -dərd-,wāt\ *noun, plural* **hundredweight** *or* **hundredweights** (1577)

□ USAGE

humor Some speakers insist that in *humor* and similar words (as *human, humble, homage*) the *h* must be sounded. In the beginning, however, it was not so: these words were first brought into English without an *h* in the pronunciation or, in some cases, in the spelling. In time the literate classes adopted a spelling pronunciation of *humor,* sounding the *h* contrary to established usage. The *h*-less pronunciation came to be seen as a mark of poor education, and so even today critics advocate an initial \h\ in *humor.* Those who do not pronounce the \h\, however, have centuries of usage to vindicate their pronunciation.

◇ WORD HISTORY

humor A theory of physiology first articulated by the ancient Greeks and popular throughout the Middle Ages and Renaissance held that the composition of the human body included four fluids: yellow bile or choler, black bile or melancholy, blood, and phlegm. The proper proportion of these fluids determined both the physical and mental well-being of an individual. In Middle English these fluids were referred to as *humours,* ultimately from Latin *umor,* meaning "moisture" or "bodily fluid." If in a natural combination, a *humour* was *goode* and stimulated health and good spirits, but if unnatural or corrupted, a *humour* was *ille* and led to sickness and ill spirits. In the course of the 16th and 17th centuries, *good humour* and *ill humour* lost their original physiological senses and began to refer strictly to mental states, whether fixed disposition or passing mood, and we continue to use *humor* in this way. The use of *humors* to mean "whims" led to other common modern applications of *humor,* for a person's faculty for appreciating the ludicrous (as in *sense of humor*) and correspondingly, the quality that appeals to this faculty (as in *to see the humor in a situation*).

1 : a unit of weight equal to 100 pounds — called also *short hundredweight;* see WEIGHT table
2 *British* **:** a unit of weight equal to 112 pounds — called also *long hundredweight*

[1]hung *past and past participle of* HANG

[2]hung *adjective* (1848)
: unable to reach a decision or verdict ⟨a *hung* jury⟩; *also, British* **:** not having a political party with an overall majority ⟨a *hung* parliament⟩

Hun·gar·i·an \ˌhəŋ-ˈger-ē-ən, -ˈgar-\ *noun* (1553)
1 a : a native or inhabitant of Hungary **:** MAGYAR **b :** a person of Hungarian descent
2 : the Finno-Ugric language of the Hungarians
— Hungarian *adjective*

[1]hun·ger \ˈhəŋ-gər\ *noun* [Middle English, from Old English *hungor;* akin to Old High German *hungar* hunger, Lithuanian *kanka* torture] (before 12th century)
1 a : a craving or urgent need for food or a specific nutrient **b :** an uneasy sensation occasioned by the lack of food **c :** a weakened condition brought about by prolonged lack of food
2 : a strong desire **:** CRAVING
— from hunger : very bad or inept ⟨the jokes were *from hunger* —Mordecai Richler⟩

[2]hunger *verb* **hun·gered; hun·ger·ing** \-g(ə-)riŋ\ (before 12th century)
intransitive verb
1 : to feel or suffer hunger
2 : to have an eager desire
transitive verb
: to make hungry
synonym see LONG

hunger strike *noun* (1889)
: refusal (as by a prisoner) to eat enough to sustain life
— hunger striker *noun*

hung·over \ˈhəŋ-ˈō-vər\ *adjective* (1941)
: suffering from a hangover

hun·gry \ˈhəŋ-grē\ *adjective* **hun·gri·er; -est** [Middle English, from Old English *hungrig;* akin to Old English *hungor*] (before 12th century)
1 a : feeling hunger **b :** characterized by or characteristic of hunger or appetite
2 a : EAGER, AVID ⟨*hungry* for affection⟩ **b :** strongly motivated (as by ambition)
3 : not rich or fertile **:** BARREN
— hun·gri·ly \-grə-lē\ *adverb*
— hun·gri·ness \-grē-nəs\ *noun*

hung up *adjective* (1948)
1 : delayed or detained for a time
2 : anxiously nervous
3 : being much involved with: as **a :** being infatuated **b :** ENTHUSIASTIC **c :** PREOCCUPIED

hunk \ˈhəŋk\ *noun* [Flemish *hunke*] (circa 1813)
1 : a large lump, piece, or portion ⟨a *hunk* of bread⟩
2 : an attractive well-built man

hun·ker \ˈhəŋ-kər\ *intransitive verb* **hun·kered; hun·ker·ing** \-k(ə-)riŋ\ [perhaps of Scandinavian origin; akin to Old Norse *hūka* to squat; akin to Middle Low German *hōken* to squat, peddle — more at HAWKER] (1720)
1 : CROUCH, SQUAT — usually used with *down*
2 : to settle in or dig in for a sustained period — used with *down* ⟨*hunker* down for a good long wait —*New Yorker*⟩

hun·kers \ˈhəŋ-kərz\ *noun plural* (1756)
: HAUNCHES

hunks \ˈhəŋ(k)s\ *noun plural but singular in construction* [origin unknown] (1602)
: a surly ill-natured person; *especially* **:** MISER

hunky \ˈhəŋ-kē\ *adjective* **hunk·i·er; -est** (1911)
: muscular and usually attractive ⟨*hunky* men⟩

Hun·ky *also* **Hun·kie** \ˈhəŋ-kē\ *noun, plural* **Hunkies** [alteration of *Hungarian*] (circa 1896)
: a person of central or east European birth or descent — usually used disparagingly

hun·ky–do·ry \ˌhəŋ-kē-ˈdōr-ē, -ˈdȯr-\ *adjective* [obsolete English dialect *hunk* (home base) + *-dory* (origin unknown)] (1866)
: quite satisfactory **:** FINE

Hun·nish \ˈhə-nish\ *adjective* (1820)
: relating to or resembling the Huns

[1]hunt \ˈhənt\ *verb* [Middle English, from Old English *huntian;* akin to Old English *hentan* to seize] (before 12th century)
transitive verb
1 a : to pursue for food or in sport ⟨*hunt* buffalo⟩ **b :** to manage in the search for game ⟨*hunts* a pack of dogs⟩
2 a : to pursue with intent to capture ⟨*hunted* the escapees⟩ **b :** to search out **:** SEEK
3 : to drive or chase especially by harrying ⟨members . . . were *hunted* from their homes —J. T. Adams⟩
4 : to traverse in search of prey ⟨*hunts* the woods⟩
intransitive verb
1 : to take part in a hunt
2 : to attempt to find something
3 : to oscillate alternately to each side (as of a neutral point) or to run alternately faster and slower — used especially of a device or machine

[2]hunt *noun* (14th century)
1 : the act, the practice, or an instance of hunting
2 : a group of mounted hunters and their hunting dogs

hunt–and–peck \ˌhənt-^{ə}n-ˈpek\ *noun* (1939)
: a method of typing in which one looks at the keyboard and types using usually the forefingers

hunt·er \ˈhən-tər\ *noun* (13th century)
1 a : a person who hunts game **b :** a dog used or trained for hunting **c :** a horse used or adapted for use in hunting with hounds; *especially* **:** a fast strong horse trained for cross-country work and jumping
2 : one that searches for something
3 : a pocket watch with a hinged protective cover

hunt·er–gath·er·er \ˈhən-tər-ˈga-t͟hər-ər\ *noun* (1977)
: a member of a culture in which food is obtained by hunting, fishing, and foraging rather than by agriculture or animal husbandry

hunter green *noun* (circa 1930)
: a dark yellowish green

hunt·ing *noun* (before 12th century)
1 : the act of one that hunts; *specifically* **:** the pursuit of game
2 : the process of hunting
3 a : a periodic variation in speed of a synchronous electrical machine **b :** a self-induced and undesirable oscillation of a variable above and below the desired value in an automatic control system **c :** a continuous attempt by an automatically controlled system to find a desired equilibrium condition

hunting horn *noun* (1694)
: a signal horn used in the chase; *specifically* **:** a coiled circular horn with a flared bell and a cup-shaped mouthpiece

Hun·ting·ton's chorea \ˈhən-tiŋ-tənz-\ *noun* [George *Huntington* (died 1916) American physician] (1889)
: hereditary chorea usually developing in adult life and progressing to dementia — called also *Huntington's disease*

hunt·ress \ˈhən-trəs\ *noun* (14th century)
: a woman who hunts game; *also* **:** a female animal that hunts prey

hunts·man \ˈhən(t)-smən\ *noun* (1567)
1 : HUNTER 1a
2 : a person who manages a hunt and looks after the hounds

hup \ˈhəp, ˈhət\ *interjection* [probably alteration of [1]*hep*] (1951)
— used to mark a marching cadence

hur·dies \ˈhər-dēz\ *noun plural* [origin unknown] (14th century)
dialect British **:** RUMP

[1]hur·dle \ˈhər-d^{ə}l\ *noun* [Middle English *hurdel,* from Old English *hyrdel;* akin to Old High German *hurt* hurdle, Latin *cratis* wickerwork, hurdle] (before 12th century)
1 a : a portable panel usually of wattled withes and stakes used especially for enclosing land or livestock **b :** a frame or sled formerly used in England for dragging traitors to execution
2 a : an artificial barrier over which racers must leap **b** *plural* **:** any of various track events in which a series of hurdles must be surmounted
3 : BARRIER, OBSTACLE

hurdle 2a

[2]hurdle *transitive verb* **hur·dled; hur·dling** \ˈhərd-liŋ, ˈhər-d^{ə}l-iŋ\ (1896)
1 : to leap over especially while running (as in a sporting competition)
2 : OVERCOME, SURMOUNT
— hur·dler \ˈhərd-lər, ˈhər-d^{ə}l-ər\ *noun*

hur·dy–gur·dy \ˌhər-dē-ˈgər-dē, ˈhər-dē-ˌ\ *noun, plural* **-gurdies** [probably imitative] (1749)
1 : a stringed instrument in which sound is produced by the friction of a rosined wheel turned by a crank against the strings and the pitches are varied by keys
2 : any of various mechanical musical instruments (as the barrel organ)

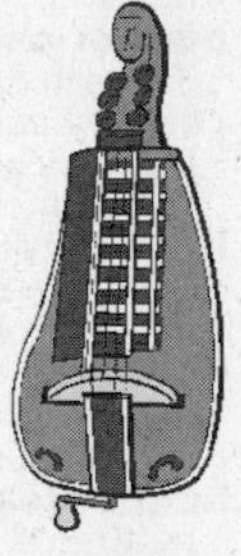
hurdy-gurdy 1

hurl \ˈhər(-ə)l\ *verb* **hurled; hurl·ing** \ˈhər-liŋ\ [Middle English] (13th century)
intransitive verb
1 : RUSH, HURTLE
2 : PITCH 5a, b
transitive verb
1 : to send or thrust with great vigor ⟨the forces that were to be *hurled* against the Turks —N. T. Gilroy⟩
2 : to throw down with violence
3 a : to throw forcefully **:** FLING ⟨*hurled* the manuscript into the fire⟩ ⟨*hurled* myself over the fence⟩ **b :** PITCH 2a
4 : to utter with vehemence ⟨*hurled* insults at the police⟩
synonym see THROW
— hurl *noun*
— hurl·er \ˈhər-lər\ *noun*

hurl·ing \ˈhər-liŋ\ *noun* (circa 1600)
: an Irish game resembling field hockey played between two teams of 15 players each

hur·ly \ˈhər-lē\ *noun* [probably short for *hurly-burly*] (1594)
: UPROAR, TUMULT

hur·ly–bur·ly \ˌhər-lē-ˈbər-lē\ *noun* [probably alteration & reduplication of *hurling,* gerund of *hurl*] (1539)
: UPROAR, TUMULT
— hurly–burly *adjective*

Hu·ron \ˈhyu̇r-ən, -ˌän\ *noun, plural* **Hurons** *or* **Huron** [French, literally, boor] (1658)
: a member of a confederacy of American Indian peoples formerly occupying the country between Georgian Bay and Lake Ontario

[1]hur·rah \hu̇-ˈrȯ, -ˈrä, ˈhü-ˌ\ *noun* (1686)
1 a : EXCITEMENT, FANFARE **b :** CHEER 7
2 : FUSS

[2]hur·rah \hu̇-ˈrȯ, -ˈrä\ *also* **hur·ray** \hu̇-ˈrā\ *interjection* [perhaps from German *hurra*](1716)
— used to express joy, approbation, or encouragement

Hur·ri·an \'hu̇r-ē-ən\ *noun* (1911)
1 : a member of an ancient non-Semitic people of northern Mesopotamia, Syria, and eastern Asia Minor about 1500 B.C.
2 : the language of the Hurrian people

[1]hur·ri·cane \'hər-ə-ˌkān, -i-kən, 'hə-rə-, 'hə-ri-\ *noun* [Spanish *huracán,* from Taino *hurakán*] (1555)
1 : a tropical cyclone with winds of 74 miles (118 kilometers) per hour or greater that occurs especially in the western Atlantic, that is usually accompanied by rain, thunder, and lightning, and that sometimes moves into temperate latitudes — see BEAUFORT SCALE table
2 : something resembling a hurricane especially in its turmoil

[2]hurricane *adjective* (1894)
: having or being a glass chimney providing protection from wind ⟨a *hurricane* lamp⟩

hur·ried \'hər-ēd, 'hə-rēd\ *adjective* (1667)
1 : going or working at speed
2 : done in a hurry **:** HASTY
— **hur·ried·ly** \'hər-əd-lē, 'hə-rəd-\ *adverb*
— **hur·ried·ness** \'hər-ēd-nəs, 'hə-rēd-\ *noun*

[1]hur·ry \'hər-ē, 'hə-rē\ *verb* **hur·ried; hur·ry·ing** [perhaps from Middle English *horyen*] (1592)
transitive verb
1 a : to carry or cause to go with haste ⟨*hurry* them to the hospital⟩ **b :** to impel to rash or precipitate action
2 a : to impel to greater speed **:** PROD ⟨used spurs to *hurry* the horse⟩ **b :** EXPEDITE **c :** to perform with undue haste ⟨*hurry* a minuet⟩
intransitive verb
: to move or act with haste ⟨please *hurry* up⟩
— **hur·ri·er** *noun*

[2]hurry *noun* (1600)
1 : disturbed or disorderly activity **:** COMMOTION
2 a : agitated and often bustling or disorderly haste **b :** a state of eagerness or urgency **:** RUSH
synonym see HASTE
— **in a hurry :** without delay **:** as rapidly as possible ⟨the police got there *in a hurry*⟩

hur·ry–scur·ry *or* **hur·ry–skur·ry** \ˌhər-ē-'skər-ē, ˌhə-rē-'skə-rē\ *noun* [reduplication of [2]*hurry*] (1754)
: a confused rush **:** TURMOIL
— **hurry–scurry** *adjective or adverb*

hur·ry–up \'hər-ē-ˌəp, 'hə-rē-\ *adjective* (1902)
: speeded up **:** completed in a hurry ⟨a *hurry-up* dinner⟩

[1]hurt \'hərt\ *verb* **hurt; hurt·ing** [Middle English, probably from Old French *hurter* to collide with, probably of Germanic origin; akin to Old Norse *hrūtr* male sheep] (13th century)
transitive verb
1 a : to inflict with physical pain **:** WOUND **b :** to do substantial or material harm to **:** DAMAGE ⟨the dry summer has *hurt* the land⟩
2 a : to cause emotional pain or anguish to **:** OFFEND **b :** to be detrimental to **:** HAMPER ⟨charges of graft *hurt* my chances of being elected⟩
intransitive verb
1 a : to suffer pain or grief **b :** to be in need — usually used with *for* ⟨*hurting* for money⟩
2 : to cause damage or distress ⟨hit where it *hurts*⟩
synonym see INJURE
— **hurt** *adjective*
— **hurt·er** *noun*

[2]hurt *noun* (13th century)
1 : a cause of injury or damage **:** BLOW
2 a : a bodily injury or wound **b :** mental distress or anguish **:** SUFFERING
3 : WRONG, HARM

hurt·ful \'hərt-fəl\ *adjective* (1526)
: causing injury, detriment, or suffering **:** DAMAGING
— **hurt·ful·ly** \-fə-lē\ *adverb*
— **hurt·ful·ness** *noun*

hur·tle \'hər-tᵊl\ *verb* **hur·tled; hur·tling** \'hərt-liŋ, 'hər-tᵊl-iŋ\ [Middle English *hurtlen* to collide, frequentative of *hurten* to cause to strike, hurt] (14th century)
intransitive verb
: to move rapidly or forcefully
transitive verb
: HURL, FLING
— **hurtle** *noun*

hurt·less \'hərt-ləs\ *adjective* (1549)
: causing no pain or injury **:** HARMLESS

[1]hus·band \'həz-bənd\ *noun* [Middle English *husbonde,* from Old English *hūsbonda* master of a house, from Old Norse *hūsbōndi,* from *hūs* house + *bōndi* householder; akin to Old Norse *būa* to inhabit; akin to Old English *būan* to dwell — more at BOWER] (13th century)
1 : a male partner in a marriage
2 *British* **:** MANAGER, STEWARD
3 : a frugal manager
— **hus·band·ly** *adjective*

[2]husband *transitive verb* (15th century)
1 a : to manage prudently and economically **b :** to use sparingly **:** CONSERVE
2 *archaic* **:** to find a husband for **:** MATE
— **hus·band·er** *noun*

hus·band·man \'həz-bən(d)-mən\ *noun* (14th century)
1 : one that plows and cultivates land **:** FARMER
2 : a specialist in a branch of farm husbandry

hus·band·ry \'həz-bən-drē\ *noun* (14th century)
1 *archaic* **:** the care of a household
2 : the control or judicious use of resources **:** CONSERVATION
3 a : the cultivation or production of plants and animals **:** AGRICULTURE **b :** the scientific control and management of a branch of farming and especially of domestic animals

[1]hush \'həsh\ *verb* [back-formation from *husht* (hushed), from Middle English *hussht,* from *huissht,* interjection used to enjoin silence] (1546)
transitive verb
1 : CALM, QUIET ⟨*hushed* the children as they entered the library⟩
2 : to put at rest **:** MOLLIFY
3 : to keep from public knowledge **:** SUPPRESS ⟨*hush* the story up⟩
intransitive verb
: to become quiet

[2]hush *adjective* (1602)
1 *archaic* **:** SILENT, STILL
2 : intended to prevent the dissemination of certain information ⟨*hush* money⟩

[3]hush *noun* (1689)
: a silence or calm especially following noise **:** QUIET

hush–hush \'həsh-ˌhəsh\ *adjective* (1916)
: SECRET, CONFIDENTIAL

hush puppy *noun* [from its occasional use as food for dogs] (circa 1918)
chiefly Southern & southern Midland **:** cornmeal dough shaped into small balls and fried in deep fat — usually used in plural

[1]husk \'həsk\ *noun* [Middle English] (14th century)
1 a : a usually dry or membranous outer covering (as a pod or hull or one composed of bracts) of various seeds and fruits; *also* **:** one of the constituent parts **b :** a carob pod
2 a : an outer layer **:** SHELL **b :** an emptied shell **:** REMNANT **c :** a supporting framework

[2]husk *transitive verb* (1562)
: to strip the husk from
— **husk·er** *noun*

husk·ing *noun* (1692)
: CORNHUSKING — called also *husking bee*

husk–tomato *noun* (1895)
: GROUND-CHERRY

[1]husky \'həs-kē\ *adjective* **husk·i·er; -est** (1552)
: resembling, containing, or full of husks

[2]hus·ky \'həs-kē\ *adjective* **hus·ki·er; -est** [probably from *husk* (huskiness), from obsolete *husk* (to have a dry cough)] (circa 1722)
: hoarse with or as if with emotion
— **hus·ki·ly** \-kə-lē\ *adverb*
— **hus·ki·ness** \-kē-nəs\ *noun*

[3]hus·ky *noun, plural* **huskies** [shortening of *Huskemaw, Uskemaw* Eskimo, from Cree *aškime·w;* akin to Montagnais (Algonquian language of eastern Canada) *aiachkime8* Micmac, Eskimo — more at ESKIMO] (1852)
1 : a heavy-coated working dog of the New World arctic region
2 : SIBERIAN HUSKY

[4]hus·ky *noun, plural* **huskies** (1864)
: one that is husky

[5]hus·ky *adjective* **hus·ki·er; -est** [probably from [1]*husk*] (1869)
1 : BURLY, ROBUST
2 : LARGE

hus·sar \(ˌ)hə-'zär, -'sär\ *noun* [Hungarian *huszár* hussar, (obsolete) highway robber, from Serbo-Croatian *husar* pirate, from Medieval Latin *cursarius* — more at CORSAIR] (1532)
: a member of any of various European units originally modeled on the Hungarian light cavalry of the 15th century

Huss·ite \'hə-ˌsīt, 'hu̇-\ *noun* [New Latin *Hussita,* from John *Huss*] (1532)
: a member of the Bohemian religious and nationalist movement originating with John Huss
— **Hussite** *adjective*
— **Huss·it·ism** \-ˌsī-ˌti-zəm\ *noun*

hus·sy \'hə-sē, -zē\ *noun, plural* **hussies** [alteration of Middle English *huswif* housewife, from *hus* house + *wif* wife, woman] (1505)
1 : a lewd or brazen woman
2 : a saucy or mischievous girl

hus·tings \'həs-tiŋz\ *noun plural but singular or plural in construction* [Middle English, from Old English *hūsting,* from Old Norse *hūsthing,* from *hūs* house + *thing* assembly] (before 12th century)
1 a : a local court formerly held in various English municipalities and still held infrequently in London **b :** a local court in some cities in Virginia
2 a : a raised platform used until 1872 for the nomination of candidates for the British Parliament and for election speeches **b :** an election platform **:** STUMP **c :** the proceedings or locale of an election campaign ◆

◇ WORD HISTORY

hustings Hustings, now the haunt of politicians, were originally assemblies. Late Old English *hūsting,* the ancestor of modern *husting,* was a loan from Old Norse *hūsthing,* a compound of *hūs* "house" and *thing* "assembly"; the modifying *hūs* presumably signified that the assembly took place in a building rather than outdoors. After the Norman Conquest the *husting,* originally a sort of royal council, became a court hearing civil cases, especially one in London that met at the Guildhall, London's original seat of municipal government. *Hustings* was also apparently applied to a dais at one end of the Guildhall's Gothic meeting room, where the Court of Hustings met and where the lord mayor, sheriffs, and other officials were elected. At this point the history of *hustings* becomes murky, but it appears that about 1700, at a time when English electoral politics were taking something like their modern form, *hustings* was applied to any platform where a parliamentary seat was contested. In Hanoverian England this was typically a substantial structure erected in a large open space, where nomination speeches were made and the results of polling announced. Today in both Britain and the U.S. the hustings are usually the figurative rather than literal site of electioneering.

hus·tle \'hə-səl\ *verb* **hus·tled; hus·tling** \'hə-s(ə-)liŋ\ [Dutch *husselen* to shake, from Middle Dutch *hutselen,* frequentative of *hutsen*] (1751)
transitive verb
1 a : JOSTLE, SHOVE **b :** to convey forcibly or hurriedly **c :** to urge forward precipitately
2 a : to obtain by energetic activity ⟨*hustle* up new customers⟩ **b :** to sell something to or obtain something from by energetic and especially underhanded activity ⟨*hustling* the suckers⟩ **c :** to sell or promote energetically and aggressively ⟨*hustling* a new product⟩ **d :** to lure less skillful players into competing against oneself at (a gambling game) ⟨*hustle* pool⟩
intransitive verb
1 : SHOVE, PRESS
2 : HASTEN, HURRY
3 a : to make strenuous efforts to obtain especially money or business **b :** to obtain money by fraud or deception **c :** to engage in prostitution
4 : to play a game or sport in an alert aggressive manner
— **hustle** *noun*
— **hus·tler** \'hə-slər\ *noun*

[1]**hut** \'hət\ *noun* [Middle French *hutte,* of Germanic origin; akin to Old High German *hutta* hut; probably akin to Old English *hȳd* skin, hide] (1658)
1 : an often small and temporary dwelling of simple construction **:** SHACK
2 : a simple shelter from the elements
— **hut** *verb*

[2]**hut** \'hət, 'həp\ *interjection* [probably alteration of [1]*hep*] (1948)
— used to mark a marching cadence

hutch \'həch\ *noun* [Middle English *huche,* from Old French] (13th century)
1 a : a chest or compartment for storage **b :** a cupboard usually surmounted by open shelves
2 : a pen or coop for an animal
3 : SHACK, SHANTY

hut·ment \'hət-mənt\ *noun* (1889)
1 : a collection of huts **:** ENCAMPMENT
2 : HUT

Hut·ter·ite \'hə-tə-ˌrīt, 'hü-\ *noun* [Jakob *Hutter* (died 1536) Moravian Anabaptist] (1910)
: a member of a Mennonite sect of northwestern U.S. and Canada living communally and holding property in common
— **Hut·te·ri·an** \ˌhə-'tir-ē-ən, hü-\ *adjective*

Hu·tu \'hü-(ˌ)tü\ *noun, plural* **Hutu** *or* **Hutus** (1952)
: a member of a Bantu-speaking people of Rwanda and Burundi

hutz·pah *or* **hutz·pa** *variant of* CHUTZPAH

huz·zah *or* **huz·za** \(ˌ)hə-'zä\ *noun* [origin unknown] (1573)
: an expression or shout of acclaim — often used interjectionally to express joy or approbation

hy·a·cinth \'hī-ə-(ˌ)sin(t)th, -sən(t)th\ *noun* [Latin *hyacinthus,* a precious stone, a flowering plant, from Greek *hyakinthos*] (1553)
1 a : a precious stone of the ancients sometimes held to be the sapphire **b :** a gem zircon or essonite
2 a : a plant of the ancients held to be a lily, iris, larkspur, or gladiolus **b :** a bulbous herb (*Hyacinthus orientalis*) of the lily family that is native to the Mediterranean region but is widely grown for its dense spikes of fragrant flowers — compare GRAPE HYACINTH, WATER HYACINTH
3 : a light violet to moderate purple
— **hy·a·cin·thine** \ˌhī-ə-'sin(t)-thən\ *adjective*

hyacinth 2b

Hy·a·cin·thus \ˌhī-ə-'sin(t)-thəs\ *noun* [Latin, from Greek *Hyakinthos*]
: a youth loved and accidentally killed by Apollo who memorializes him with a hyacinth growing from the youth's blood

Hy·a·des \'hī-ə-ˌdēz\ *noun plural* [Latin, from Greek] (14th century)
: a V-shaped cluster of stars in the head of the constellation Taurus held by the ancients to indicate rainy weather when they rise with the sun

hy·ae·na *variant of* HYENA

hyal- *or* **hyalo-** *combining form* [Late Latin, glass, from Greek, from *hyalos*]
: glass **:** glassy **:** hyaline ⟨*hyal*uronic acid⟩

[1]**hy·a·line** \'hī-ə-lən, -ˌlīn\ *adjective* [Late Latin *hyalinus,* from Greek *hyalinos,* from *hyalos*] (circa 1661)
: transparent or nearly so and usually homogeneous

[2]**hy·a·line** \'hī-ə-lən, -ˌlīn, *in sense 2* -lən *or* -ˌlēn\ *noun* (1667)
1 : something (as the clear atmosphere) that is transparent
2 *or* **hy·a·lin** \-lən\ **:** any of several translucent nitrogenous substances related to chitin, found especially around cells, and readily stained by eosin

hyaline cartilage *noun* (1855)
: translucent bluish white cartilage with the cells embedded in an apparently homogeneous matrix present in joints and respiratory passages and forming most of the fetal skeleton

hy·a·lite \'hī-ə-ˌlīt\ *noun* [German *Hyalit,* from Greek *hyalos*] (1794)
: a colorless opal that is clear as glass or sometimes translucent or whitish

hy·a·loid \-ˌlȯid\ *adjective* [Greek *hyaloeidēs,* from *hyalos*] (1835)
: GLASSY, TRANSPARENT

hy·a·lo·plasm \hī-'a-lə-ˌpla-zəm, 'hī-ə-lō-\ *noun* [probably from German *Hyaloplasma,* from *hyal-* + *-plasma* -plasm] (1886)
: CYTOSOL

hy·al·uron·ic acid \ˌhīl-yu̇-'rä-nik-, ˌhī-əl-yu̇-\ *noun* [International Scientific Vocabulary] (1934)
: a viscous mucopolysaccharide acid that occurs especially in the vitreous humor, the umbilical cord, and synovial fluid and as a cementing substance in the subcutaneous tissue

hy·al·uron·i·dase \-'rä-nə-ˌdās, -ˌdāz\ *noun* [International Scientific Vocabulary, from *hyaluronic* (*acid*) + *-idase* (as in *glucosidase*)] (1940)
: a mucolytic enzyme that facilitates the spread of fluids through tissues by lowering the viscosity of hyaluronic acid

hy·brid \'hī-brəd\ *noun* [Latin *hybrida*] (1601)
1 : an offspring of two animals or plants of different races, breeds, varieties, species, or genera
2 : a person whose background is a blend of two diverse cultures or traditions
3 a : something heterogeneous in origin or composition **:** COMPOSITE ⟨artificial *hybrids* of DNA and RNA⟩ ⟨a *hybrid* of medieval and Renaissance styles⟩ **b :** something (as a power plant, vehicle, or electronic circuit) that has two different types of components performing essentially the same function
— **hybrid** *adjective*
— **hy·brid·ism** \-brə-ˌdi-zəm\ *noun*
— **hy·brid·i·ty** \hī-'bri-də-tē\ *noun*

hybrid computer *noun* (1968)
: a computer system consisting of a combination of analog and digital computer systems

hy·brid·ize \'hī-brə-ˌdīz\ *verb* **-ized; -iz·ing** (1845)
transitive verb
: to cause to produce hybrids **:** INTERBREED
intransitive verb
: to produce hybrids
— **hy·brid·iza·tion** \ˌhī-brə-də-'zā-shən\ *noun*
— **hy·brid·iz·er** *noun*

hy·brid·oma \ˌhī-brə-'dō-mə\ *noun* (1978)
: a hybrid cell produced by the fusion of an antibody-producing lymphocyte with a tumor cell and used to culture continuously a specific monoclonal antibody

hybrid tea rose *noun* (1926)
: any of numerous moderately hardy cultivated hybrid bush roses grown especially for their strongly recurrent bloom of large usually scentless flowers — called also *hybrid tea*

hybrid vigor *noun* (1918)
: HETEROSIS

hy·bris \'hī-brəs, 'hē-\ *variant of* HUBRIS

hy·da·thode \'hī-də-ˌthōd\ *noun* [International Scientific Vocabulary, from Greek *hydat-, hydōr* water + *hodos* road] (1895)
: a specialized pore on the leaves of higher plants that functions in the exudation of water

hy·da·tid \'hī-də-təd, -ˌtid\ *noun* [Greek *hydatid-, hydatis* watery cyst, from *hydat-, hydōr*] (1683)
: the larval cyst of a tapeworm (genus *Echinococcus*) occurring as a fluid-filled sac containing daughter cysts in which scolices develop

hydr- *or* **hydro-** *combining form* [Middle English *ydr-, ydro-,* from Old French, from Latin *hydr-, hydro-,* from Greek, from *hydōr* — more at WATER]
1 a : water ⟨*hydr*ous⟩ ⟨*hydro*electric⟩ **b :** liquid ⟨*hydro*kinetic⟩
2 : hydrogen **:** containing or combined with hydrogen ⟨*hydro*carbon⟩ ⟨*hydr*oxyl⟩
3 : hydroid ⟨*hydro*medusa⟩

Hy·dra \'hī-drə\ *noun* [Middle English *Ydra,* from Latin *Hydra,* from Greek]
1 : a many-headed serpent or monster in Greek mythology that was slain by Hercules and each head of which when cut off was replaced by two others
2 *not capitalized* **:** a multifarious evil not to be overcome by a single effort
3 [Latin (genitive *Hydrae*), from Greek] **:** a southern constellation of great length that lies south of Cancer, Sextans, Corvus, and Virgo and is represented on old maps by a serpent
4 *not capitalized* [New Latin, from Latin, Hydra] **:** any of numerous small tubular freshwater hydrozoan polyps (*Hydra* and related genera) having at one end a mouth surrounded by tentacles

hy·dra–head·ed \ˌhī-drə-'he-dəd\ *adjective* (1599)
: having many centers or branches ⟨a *hydra-headed* organization⟩

hy·dral·azine \hī-'dra-lə-ˌzēn\ *noun* [*hydr-* + phth*alic* (acid) + *azine*] (1952)
: a vasodilator $C_8H_8N_4$ used in the treatment of hypertension

hy·dran·gea \hī-'drān-jə\ *noun* [New Latin, from *hydr-* + Greek *angeion* vessel — more at ANGI-] (circa 1753)
: any of a genus (*Hydrangea*) of shrubs and one woody vine of the saxifrage family with opposite leaves and showy clusters of usually sterile white or tinted flowers

hy·drant \'hī-drənt\ *noun* (1806)
1 : a discharge pipe with a valve and spout at which water may be drawn from a water main (as for fighting fires) — called also *fireplug*
2 : FAUCET

hy·dranth \'hī-ˌdran(t)th\ *noun* [International Scientific Vocabulary *hydr-* + Greek *anthos* flower — more at ANTHOLOGY] (1874)
: one of the nutritive zooids of a hydroid colony

hy·drase \'hī-ˌdrās, -ˌdrāz\ *noun* (1943)
: an enzyme that promotes the addition or removal of water to or from its substrate

[1]**hy·drate** \'hī-ˌdrāt\ *noun* (1802)
: a compound formed by the union of water with some other substance

[2]**hydrate** *verb* **hy·drat·ed; hy·drat·ing** (1850)
transitive verb
: to cause to take up or combine with water or the elements of water

intransitive verb
: to become a hydrate
— **hy·dra·tion** \hī-ˈdrā-shən\ *noun*
— **hy·dra·tor** \ˈhī-ˌdrā-tər\ *noun*

hy·drau·lic \hī-ˈdrȯ-lik\ *adjective* [Latin *hydraulicus,* from Greek *hydraulikos,* from *hydraulis* hydraulic organ, from *hydr-* + *aulos* reed instrument — more at ALVEOLUS] (1661)
1 : operated, moved, or effected by means of water
2 a : of or relating to hydraulics ⟨*hydraulic* engineer⟩ **b :** of or relating to water or other liquid in motion ⟨*hydraulic* erosion⟩
3 : operated by the resistance offered or the pressure transmitted when a quantity of liquid (as water or oil) is forced through a comparatively small orifice or through a tube ⟨*hydraulic* brakes⟩
4 : hardening or setting under water ⟨*hydraulic* cement⟩
— **hy·drau·li·cal·ly** \-li-k(ə-)lē\ *adverb*

hy·drau·lics \hī-ˈdrȯ-liks\ *noun plural but singular in construction* (1671)
: a branch of science that deals with practical applications (as the transmission of energy or the effects of flow) of liquid (as water) in motion

hy·dra·zide \ˈhī-drə-ˌzīd\ *noun* (1888)
: any of a class of compounds resulting from the replacement of hydrogen by an acid group in hydrazine or in one of its derivatives

hy·dra·zine \ˈhī-drə-ˌzēn\ *noun* [International Scientific Vocabulary] (1887)
: a colorless fuming corrosive strongly reducing liquid base N_2H_4 used especially in fuels for rocket and jet engines; *also* **:** an organic base derived from this compound

hy·dra·zo·ic acid \ˌhī-drə-ˈzō-ik-\ *noun* [*hydr-* + *azo-* + [1]*-ic*] (1894)
: a colorless volatile poisonous explosive liquid HN_3 that has a foul odor and yields explosive salts of heavy metals

hy·dric \ˈhī-drik\ *adjective* (1926)
: characterized by, relating to, or requiring an abundance of moisture ⟨a *hydric* habitat⟩ ⟨a *hydric* plant⟩ — compare MESIC, XERIC

-hydric *adjective suffix*
: containing acid hydrogen ⟨mono*hydric*⟩

hy·dride \ˈhī-ˌdrīd\ *noun* (1869)
: a compound of hydrogen with a more electropositive element or group

hy·dri·od·ic acid \ˌhī-drē-ˈä-dik-\ *noun* [International Scientific Vocabulary] (1819)
: an aqueous solution of hydrogen iodide HI that is a strong acid resembling hydrochloric acid chemically and that is also a strong reducing agent

[1]hy·dro \ˈhī-(ˌ)drō\ *noun, plural* **hydros** [short for *hydropathic establishment*] (1882)
British **:** an establishment offering hydropathic treatment (as for weight loss) **:** HEALTH SPA

[2]hydro *noun, often attributive* [short for *hydropower*] (1916)
: hydroelectric power

hy·dro·bi·ol·o·gy \ˌhī-drō-bī-ˈä-lə-jē\ *noun* (1926)
: the biology of bodies or units of water; *especially* **:** LIMNOLOGY
— **hy·dro·bi·o·log·i·cal** \-ˌbī-ə-ˈlä-ji-kəl\ *adjective*
— **hy·dro·bi·ol·o·gist** \-bī-ˈä-lə-jist\ *noun*

hy·dro·bro·mic acid \ˌhī-drə-ˈbrō-mik-\ *noun* [International Scientific Vocabulary] (1836)
: an aqueous solution of hydrogen bromide HBr that is a strong acid resembling hydrochloric acid chemically, that is a weak reducing agent, and that is used especially for making bromides

hy·dro·car·bon \ˈhī-drō-ˌkär-bən\ *noun* (1826)
: an organic compound (as acetylene or butane) containing only carbon and hydrogen and often occurring in petroleum, natural gas, coal, and bitumens

hy·dro·cele \ˈhī-drə-ˌsēl\ *noun* [Latin, from Greek *hydrokēlē,* from *hydr-* + *kēlē* tumor — more at -CELE] (1597)
: an accumulation of serous fluid in a sacculated cavity (as the scrotum)

hy·dro·ce·phal·ic \ˌhī-drō-sə-ˈfa-lik\ *adjective* (1815)
: relating to, characterized by, or affected with hydrocephalus
— **hydrocephalic** *noun*

hy·dro·ceph·a·lus \-ˈse-fə-ləs\ *also* **hy·dro·ceph·a·ly** \-lē\ *noun* [New Latin *hydrocephalus,* from Late Latin, hydrocephalic, adjective, from Greek *hydrokephalos,* from *hydr-* + *kephalē* head — more at CEPHALIC] (1670)
: an abnormal increase in the amount of cerebrospinal fluid within the cranial cavity that is accompanied by expansion of the cerebral ventricles, enlargement of the skull and especially the forehead, and atrophy of the brain

hy·dro·chlo·ric acid \ˌhī-drə-ˈklōr-ik-, -ˈklȯr-\ *noun* [International Scientific Vocabulary] (circa 1828)
: an aqueous solution of hydrogen chloride HCl that is a strong corrosive irritating acid, is normally present in dilute form in gastric juice, and is widely used in industry and in the laboratory

hy·dro·chlo·ride \-ˈklōr-ˌīd, -ˈklȯr-\ *noun* (1826)
: a chemical complex composed of an organic base (as an alkaloid) in association with hydrogen chloride

hy·dro·chlo·ro·thi·a·zide \-ˌklōr-ə-ˈthī-ə-ˌzīd, -ˌklȯr-\ *noun* [*hydr-* + *chlor-* + *thiaz*ine + *-ide*] (1958)
: a diuretic and antihypertensive drug $C_7H_8ClN_3O_4S_2$

hy·dro·col·loid \ˌhī-drə-ˈkä-ˌlȯid\ *noun* (1916)
: a substance that yields a gel with water
— **hy·dro·col·loi·dal** \-kə-ˈlȯi-dᵊl, -kä-\ *adjective*

hy·dro·cor·ti·sone \-ˈkȯr-tə-ˌsōn, -ˌzōn\ *noun* (1951)
: a glucocorticoid $C_{21}H_{30}O_5$ of the adrenal cortex that is a derivative of cortisone and is used in the treatment of rheumatoid arthritis — called also *cortisol*

hy·dro·crack·ing \ˈhī-drə-ˌkra-kiŋ\ *noun* (1940)
: the cracking of hydrocarbons in the presence of hydrogen
— **hy·dro·crack** \-ˌkrak\ *transitive verb*
— **hy·dro·crack·er** *noun*

hy·dro·cy·an·ic acid \ˌhī-drō-sī-ˈa-nik-\ *noun* [International Scientific Vocabulary] (1819)
: an aqueous solution of hydrogen cyanide HCN that is a poisonous weak acid and is used chiefly in fumigating and in organic synthesis

hy·dro·dy·nam·ic \-dī-ˈna-mik\ *also* **hy·dro·dy·nam·i·cal** \-mi-kəl\ *adjective* [New Latin *hydrodynamicus,* from *hydr-* + *dynamicus* dynamic] (circa 1828)
: of, relating to, or involving principles of hydrodynamics
— **hy·dro·dy·nam·i·cal·ly** \-mi k(ə-)lē\ *adverb*

hy·dro·dy·nam·ics \-miks\ *noun plural but singular in construction* (1779)
: a branch of physics that deals with the motion of fluids and the forces acting on solid bodies immersed in fluids and in motion relative to them — compare HYDROSTATICS
— **hy·dro·dy·nam·i·cist** \-ˈna-mə-sist\ *noun*

hy·dro·elec·tric \ˌhī-drō-i-ˈlek-trik\ *adjective* [International Scientific Vocabulary] (1884)
: of or relating to production of electricity by waterpower ⟨constructed a *hydroelectric* power plant at the dam site⟩
— **hy·dro·elec·tri·cal·ly** \-tri-k(ə-)lē\ *adverb*
— **hy·dro·elec·tric·i·ty** \-ˌlek-ˈtri-sə-tē, -ˈtris-tē\ *noun*

hy·dro·flu·or·ic acid \ˌhī-drō-ˈflȯr-ik, -ˈflu̇r-\ *noun* [International Scientific Vocabulary] (circa 1872)
: an aqueous solution of hydrogen fluoride HF that is a weak poisonous acid, that resembles hydrochloric acid chemically but attacks silica and silicates, and that is used especially in finishing and etching glass

hy·dro·foil \ˈhī-drə-ˌfȯil\ *noun* (1919)
1 : a body similar to an airfoil but designed for action in or on water
2 : a motorboat that has metal plates or fins attached by struts fore and aft for lifting the hull clear of the water as speed is attained

hydrofoil 2

hy·dro·gen \ˈhī-drə-jən, -dər-\ *noun* [French *hydrogène,* from *hydr-* + *-gène* -gen; from the fact that water is generated by its combustion] (1791)
: a nonmetallic element that is the simplest and lightest of the elements, is normally a colorless odorless highly flammable diatomic gas, and is used especially in synthesis — compare DEUTERIUM, TRITIUM; see ELEMENT table
— **hy·drog·e·nous** \hī-ˈdrä-jə-nəs\ *adjective*

hy·drog·e·nase \hī-ˈdrä-jə-ˌnās, -ˌnāz\ *noun* (1900)
: an enzyme of various microorganisms that promotes the formation and utilization of gaseous hydrogen

hy·dro·ge·nate \hī-ˈdrä-jə-ˌnāt, ˈhī-drə-\ *transitive verb* **-nat·ed; -nat·ing** (1809)
: to combine or treat with or expose to hydrogen; *especially* **:** to add hydrogen to the molecule of (an unsaturated organic compound)
— **hy·dro·ge·na·tion** \hī-ˌdrä-jə-ˈnā-shən, ˌhī-drə-\ *noun*

hydrogen bomb *noun* (1947)
: a bomb whose violent explosive power is due to the sudden release of atomic energy resulting from the fusion of light nuclei (as of hydrogen atoms) at very high temperature and pressure to form helium nuclei

hydrogen bond *noun* (1923)
: an electrostatic attraction between a hydrogen atom in one polar molecule (as of water) and a small electronegative atom (as of oxygen, nitrogen, or fluorine) in usually another molecule of the same or a different polar substance
— **hydrogen bonding** *noun*

hydrogen bromide *noun* (1885)
: a colorless irritating gas HBr that fumes in moist air and yields hydrobromic acid when dissolved in water

hydrogen chloride *noun* (1869)
: a colorless pungent poisonous gas HCl that fumes in moist air and yields hydrochloric acid when dissolved in water

hydrogen cyanide *noun* (1882)
1 : a poisonous usually gaseous compound HCN that has the odor of bitter almonds
2 : HYDROCYANIC ACID

hydrogen fluoride *noun* (1885)
: a colorless corrosive fuming usually gaseous compound HF that yields hydrofluoric acid when dissolved in water

hydrogen iodide *noun* (1885)
: an acrid colorless gas HI that fumes in moist air and yields hydriodic acid when dissolved in water

hydrogen ion *noun* (1896)
1 : the cation H^+ of acids consisting of a hydrogen atom whose electron has been transferred to the anion of the acid
2 : HYDRONIUM

hydrogen peroxide *noun* (1872)
: an unstable compound H_2O_2 used especially as an oxidizing and bleaching agent, an antiseptic, and a propellant

hydrogen sulfide *noun* (1873)
: a flammable poisonous gas H_2S that has an odor suggestive of rotten eggs and is found especially in many mineral waters and in putrefying matter

hy·dro·graph·ic \ˌhī-drə-ˈgra-fik\ *adjective* [French *hydrographique,* from Middle French, from *hydr-* + *-graphique* -graphic] (1665)
1 : of or relating to the characteristic features (as flow or depth) of bodies of water
2 : relating to the charting of bodies of water
— **hy·drog·ra·pher** \hī-ˈdrä-grə-fər\ *noun*
— **hy·drog·ra·phy** \-grə-fē\ *noun*

¹hy·droid \ˈhī-ˌdrȯid\ *adjective* [ultimately from New Latin *Hydra*] (circa 1864)
: of or relating to a hydrozoan; *especially* **:** resembling a typical hydra

²hydroid *noun* (1865)
: HYDROZOAN; *especially* **:** a hydrozoan polyp as distinguished from a medusa

hy·dro·ki·net·ic \ˌhī-drō-kə-ˈne-tik, -(ˌ)kī-\ *adjective* (1876)
: of or relating to the motions of fluids or the forces which produce or affect such motions — compare HYDROSTATIC

hy·dro·lase \ˈhī-drə-ˌlās, -ˌlāz\ *noun* [International Scientific Vocabulary, from New Latin *hydrolysis* + International Scientific Vocabulary *-ase*] (1910)
: a hydrolytic enzyme

hydrologic cycle *noun* (1936)
: the sequence of conditions through which water passes from vapor in the atmosphere through precipitation upon land or water surfaces and ultimately back into the atmosphere as a result of evaporation and transpiration — called also *hydrological cycle*

hy·drol·o·gy \hī-ˈdrä-lə-jē\ *noun* [New Latin *hydrologia,* from Latin *hydr-* + *-logia* -logy] (1762)
: a science dealing with the properties, distribution, and circulation of water on and below the earth's surface and in the atmosphere
— **hy·dro·log·ic** \ˌhī-drə-ˈlä-jik\ *or* **hy·dro·log·i·cal** \-ji-kəl\ *adjective*
— **hy·dro·log·i·cal·ly** \-ji-k(ə-)lē\ *adverb*
— **hy·drol·o·gist** \hī-ˈdrä-lə-jist\ *noun*

hy·dro·ly·sate \hī-ˈdrä-lə-ˌsāt\ *also* **hy·dro·ly·zate** \-ˌzāt\ *noun* (1915)
: a product of hydrolysis

hy·dro·ly·sis \hī-ˈdrä-lə-səs\ *noun* [New Latin] (1880)
: a chemical process of decomposition involving the splitting of a bond and the addition of the hydrogen cation and the hydroxide anion of water
— **hy·dro·lyt·ic** \ˌhī-drə-ˈli-tik\ *adjective*
— **hy·dro·lyt·i·cal·ly** \-ti-k(ə-)lē\ *adverb*

hy·dro·lyze \ˈhī-drə-ˌlīz\ *verb* **-lyzed; -lyz·ing** [International Scientific Vocabulary, from New Latin *hydrolysis*] (1880)
transitive verb
: to subject to hydrolysis
intransitive verb
: to undergo hydrolysis
— **hy·dro·lyz·able** \-ˌlī-zə-bəl\ *adjective*

hy·dro·mag·net·ic \ˌhī-drō-mag-ˈne-tik\ *adjective* (1943)
: MAGNETOHYDRODYNAMIC

hy·dro·man·cy \ˈhī-drə-ˌman(t)-sē\ *noun* [Middle English *ydromancie,* from Middle French, from Latin *hydromantia,* from *hydr-* + *-mantia* -mancy] (14th century)
: divination by the appearance or motion of liquids (as water)

hy·dro·me·chan·i·cal \ˌhī-drō-mi-ˈka-ni-kəl\ *adjective* (1825)
: relating to a branch of mechanics that deals with the equilibrium and motion of fluids and of solid bodies immersed in them
— **hy·dro·me·chan·ics** \-ˈka-niks\ *noun plural but singular in construction*

hy·dro·me·du·sa \ˌhī-drō-mi-ˈdü-sə, -ˈdyü-, -zə\ *noun, plural* **-sae** \-ˌsē, -ˌzē\ [New Latin] (circa 1889)
: a medusa (as of the orders Anthomedusae and Leptomedusae) produced as a bud from a hydroid

hy·dro·met·al·lur·gy \ˌhī-drō-ˈme-tᵊl-ˌər-jē\ *noun* [International Scientific Vocabulary] (circa 1859)
: the treatment of ores by wet processes (as leaching)
— **hy·dro·met·al·lur·gi·cal** \-ˌme-tᵊl-ˈər-ji-kəl\ *adjective*
— **hy·dro·met·al·lur·gist** \-ˈme-tᵊl-ˌər-jist\ *noun*

hy·dro·me·te·or \ˌhī-drō-ˈmē-tē-ər, -tē-ˌȯr\ *noun* [International Scientific Vocabulary] (1857)
: a product (as fog, rain, or hail) formed by the condensation of atmospheric water vapor

hy·dro·me·te·o·rol·o·gy \-ˌmē-tē-ə-ˈrä-lə-jē\ *noun* (circa 1859)
: a branch of meteorology that deals with water in the atmosphere especially as precipitation
— **hy·dro·me·te·o·ro·log·i·cal** \-tē-ˌȯr-ə-ˈlä-ji-kəl, -ˌär-ə-, -ə-rə-\ *adjective*
— **hy·dro·me·te·o·rol·o·gist** \-tē-ə-ˈrä-lə-jist\ *noun*

hy·drom·e·ter \hī-ˈdrä-mə-tər\ *noun* (1675)
: an instrument for determining the specific gravity of a liquid (as battery acid or an alcohol solution) and hence its strength
— **hy·dro·met·ric** \ˌhī-drə-ˈme-trik\ *adjective*

hy·dro·mor·phic \ˌhī-drə-ˈmȯr-fik\ *adjective* (1938)
of a soil **:** developed in the presence of an excess of moisture which tends to suppress aerobic factors in soil-building

hy·dron·ic \hī-ˈdrä-nik\ *adjective* [*hydr-* + *-onic* (as in *electronic*)] (1946)
: of, relating to, or being a system of heating or cooling that involves transfer of heat by a circulating fluid (as water or vapor) in a closed system of pipes
— **hy·dron·i·cal·ly** \-ni-k(ə-)lē\ *adverb*

hy·dro·ni·um \hī-ˈdrō-nē-əm\ *noun* [International Scientific Vocabulary *hydr-* + *-onium*] (1908)
: a hydrated hydrogen ion H_3O^+

hy·drop·a·thy \hī-ˈdrä-pə-thē\ *noun* [International Scientific Vocabulary] (1843)
: the empirical use of water in the treatment of disease — compare HYDROTHERAPY
— **hy·dro·path·ic** \ˌhī-drə-ˈpa-thik\ *adjective*

hy·dro·per·ox·ide \ˌhī-drō-pə-ˈräk-ˌsīd\ *noun* (1937)
: a compound containing an O_2H group

hy·dro·phane \ˈhī-drə-ˌfān\ *noun* (1784)
: a semitranslucent opal that becomes translucent or transparent on immersion in water

hy·dro·phil·ic \ˌhī-drə-ˈfi-lik\ *adjective* [New Latin *hydrophilus,* from Greek *hydr-* + *-philos* -philous] (1916)
: of, relating to, or having a strong affinity for water ⟨*hydrophilic* proteins⟩
— **hy·dro·phi·lic·i·ty** \ˌhī-drə-fi-ˈli-sə-tē\ *noun*

hy·dro·pho·bia \ˌhī-drə-ˈfō-bē-ə\ *noun* [Late Latin, from Greek, from *hydr-* + *-phobia* -phobia] (1547)
1 : RABIES
2 : a morbid dread of water

hy·dro·pho·bic \-ˈfō-bik\ *adjective* (1807)
1 : of, relating to, or suffering from hydrophobia
2 : lacking affinity for water
— **hy·dro·pho·bic·i·ty** \-fō-ˈbi-sə-tē\ *noun*

hy·dro·phone \ˈhī-drə-ˌfōn\ *noun* (1860)
: an instrument for listening to sound transmitted through water

hy·dro·phyte \-ˌfīt\ *noun* [International Scientific Vocabulary] (1832)
1 : a perennial vascular aquatic plant having its overwintering buds under water
2 : a plant growing in water or in soil too waterlogged for most plants to survive
— **hy·dro·phyt·ic** \ˌhī-drə-ˈfi-tik\ *adjective*

¹hy·dro·plane \ˈhī-drə-ˌplān\ *noun* (1904)
1 : a powerboat designed for racing that skims the surface of the water
2 : SEAPLANE

²hydroplane *intransitive verb* (1962)
: to skim on water; *especially, of a vehicle* **:** to skid on a wet surface (as pavement) because a film of water on the surface causes the tires to lose contact with it

hy·dro·pon·ics \ˌhī-drə-ˈpä-niks\ *noun plural but singular in construction* [*hydr-* + *-ponics* (as in *geoponics* agriculture)] (1937)
: the growing of plants in nutrient solutions with or without an inert medium to provide mechanical support
— **hy·dro·pon·ic** \-nik\ *adjective*
— **hy·dro·pon·i·cal·ly** \-ni-k(ə-)lē\ *adverb*

hy·dro·pow·er \ˈhī-drə-ˌpau̇(-ə)r\ *noun* (1933)
: hydroelectric power

hy·dro·qui·none \ˌhī-drō-kwi-ˈnōn, -ˈkwi-ˌnōn\ *noun* [International Scientific Vocabulary] (circa 1872)
: a white crystalline strongly reducing phenol $C_6H_6O_2$ used especially as a photographic developer and as an antioxidant and stabilizer

hy·dro·sere \ˈhī-drə-ˌsir\ *noun* (1920)
: an ecological sere originating in an aquatic habitat

hy·dro·sol \ˈhī-drə-ˌsäl, -ˌsȯl\ *noun* [*hydr-* + *-sol* (from *solution*)] (1864)
: a sol in which the liquid is water
— **hy·dro·sol·ic** \ˌhī-drə-ˈsäl-ik\ *adjective*

hy·dro·space \-ˌspās\ *noun* (1963)
: the regions beneath the surface of the ocean

hy·dro·sphere \-ˌsfir\ *noun* [International Scientific Vocabulary] (1887)
: the aqueous vapor of the atmosphere; *broadly* **:** the aqueous envelope of the earth including bodies of water and aqueous vapor in the atmosphere
— **hy·dro·spher·ic** \ˌhī-drə-ˈsfir-ik, -ˈsfer-\ *adjective*

hy·dro·stat·ic \ˌhī-drə-ˈsta-tik\ *adjective* [probably from New Latin *hydrostaticus,* from *hydr-* + *staticus* static] (1666)
: of or relating to fluids at rest or to the pressures they exert or transmit — compare HYDROKINETIC
— **hy·dro·stat·i·cal·ly** \-ti-k(ə-)lē\ *adverb*

hy·dro·stat·ics \-tiks\ *noun plural but singular in construction* (1660)
: a branch of physics that deals with the characteristics of fluids at rest and especially with the pressure in a fluid or exerted by a fluid on an immersed body — compare HYDRODYNAMICS

hy·dro·ther·a·py \ˌhī-drə-ˈther-ə-pē\ *noun* [International Scientific Vocabulary] (1876)
: the use of water especially externally in the treatment of disease or disability

hy·dro·ther·mal \ˌhī-drə-ˈthər-məl\ *adjective* [International Scientific Vocabulary] (1849)
: of or relating to hot water — used especially of the formation of minerals by hot solutions rising from a cooling magma
— **hy·dro·ther·mal·ly** \-mə-lē\ *adverb*

hy·dro·tho·rax \-ˈthōr-ˌaks, -ˈthȯr-\ *noun* [New Latin] (1793)
: an excess of serous fluid in the pleural cavity; *especially* **:** an effusion resulting from failing circulation (as in heart disease or from lung infection)

hy·drot·ro·pism \hī-ˈdrä-trə-ˌpi-zəm\ *noun* [International Scientific Vocabulary] (circa 1882)
: a tropism (as in plant roots) in which water or water vapor is the orienting factor

— **hy·dro·tro·pic** \ˌhī-drə-ˈtrō-pik, -ˈträ-\ *adjective*

hy·drous \ˈhī-drəs\ *adjective* (1826)
: containing water usually in chemical association (as in hydrates)

hy·drox·ide \hī-ˈdräk-ˌsīd\ *noun* [International Scientific Vocabulary] (1851)
1 : the univalent anion OH^- consisting of one atom of hydrogen and one of oxygen — called also *hydroxide ion*
2 : an ionic compound of hydroxide with an element or group

hy·droxy \hī-ˈdräk-sē\ *adjective* [International Scientific Vocabulary, from *hydroxyl*] (1812)
: being or containing hydroxyl; *especially* **:** containing hydroxyl especially in place of hydrogen — usually used in combination ⟨*hydroxy*acetic acid⟩

hy·droxy·ap·a·tite \hī-ˌdräk-sē-ˈa-pə-ˌtīt\ *or* **hy·drox·yl·ap·a·tite** \-sə-ˈla-pə-ˌtīt\ *noun* (1912)
: a complex phosphate of calcium $Ca_5(PO_4)_3OH$ that occurs as a mineral and is the chief structural element of vertebrate bone

hy·droxy·bu·tyr·ic acid \hī-ˌdräk-sē-byü-ˈtir-ik-\ *noun* (1879)
: a hydroxy derivative $C_4H_8O_3$ of butyric acid that is excreted in increased quantities in the urine in diabetes

hy·drox·yl \hī-ˈdräk-səl\ *noun* [*hydr-* + *ox-* + *-yl*] (1869)
1 : the chemical group or ion OH that consists of one atom of hydrogen and one of oxygen and is neutral or positively charged
2 : HYDROXIDE 1
— **hy·drox·yl·ic** \ˌhī-ˌdräk-ˈsi-lik\ *adjective*

hy·drox·yl·amine \hī-ˈdräk-sə-lə-ˌmēn, ˌhī-ˌdräk-ˈsi-lə-ˌmēn\ *noun* [International Scientific Vocabulary] (1869)
: a colorless odorless nitrogenous base NH_3O that resembles ammonia in its reactions but is less basic and that is used especially as a reducing agent

hy·drox·y·lase \hī-ˈdräk-sə-ˌlās, -ˌlāz\ *noun* (1953)
: any of a group of enzymes that catalyze oxidation reactions in which one of the two atoms of molecular oxygen is incorporated into the substrate and the other is used to oxidize NADH or NADPH

hy·drox·yl·ate \hī-ˈdräk-sə-ˌlāt\ *transitive verb* **-at·ed; -at·ing** (circa 1909)
: to introduce hydroxyl into
— **hy·drox·yl·ation** \-ˌdräk-sə-ˈlā-shən\ *noun*

hy·droxy·pro·line \hī-ˌdräk-sē-ˈprō-ˌlēn\ *noun* (1905)
: an amino acid $C_5H_9NO_3$ that occurs naturally as a constituent of collagen

hy·droxy·tryp·ta·mine \-ˈtrip-tə-ˌmēn\ *noun* (1949)
: SEROTONIN

hy·droxy·urea \-yu̇-ˈrē-ə\ *noun* (1949)
: a compound $CH_4N_2O_2$ used to treat some forms of leukemia

hy·droxy·zine \hī-ˈdräk-sə-ˌzēn\ *noun* [*hydroxy-* + pipera*zine*] (1956)
: a compound $C_{21}H_{27}ClN_2O_2$ used as an antihistaminic and tranquilizer

hy·dro·zo·an \ˌhī-drə-ˈzō-ən\ *noun* [ultimately from Greek *hydr-* + *zōion* animal — more at ZO-] (1877)
: any of a class (Hydrozoa) of coelenterates that includes simple and compound polyps and jellyfishes having no stomodeum or gastric tentacles
— **hydrozoan** *adjective*

hy·e·na \hī-ˈē-nə\ *noun, plural* **hyenas** *also* **hyena** [Middle English *hyene,* from Latin *hyaena,* from Greek *hyaina,* from *hys* hog — more at SOW] (14th century)
: any of several large strong nocturnal carnivorous Old World mammals (family Hyaenidae) that usually feed as scavengers
— **hy·e·nic** \-ˈē-nik, -ˈe-nik\ *adjective*

Hy·ge·ia \hī-ˈjē-ə\ *noun* [Latin, from Greek *Hygieia*]
: the goddess of health in Greek mythology

hy·giene \ˈhī-ˌjēn *also* hī-ˈ\ *noun* [French *hygiène* & New Latin *hygieina,* from Greek, neuter plural of *hygieinos* healthful, from *hygiēs* healthy; akin to Sanskrit *su* well and to Latin *vivus* living — more at QUICK] (1671)
1 : a science of the establishment and maintenance of health
2 : conditions or practices (as of cleanliness) conducive to health
— **hy·gien·ic** \ˌhī-ˈjē-nik, -ˈje- *also* -jē-ˈe-nik\ *adjective*
— **hy·gien·i·cal·ly** \-ni-k(ə-)lē\ *adverb*
— **hy·gien·ist** \hī-ˈjē-nist, ˈhī-ˌjē-, hī-ˈje-\ *noun*

hy·gien·ics \ˌhī-ˈjē-niks, -ˈje- *also* -jē-ˈe-niks\ *noun plural but singular in construction* (1855)
: HYGIENE 1

hygr- *also* **hygro-** *combining form* [Greek, from *hygros* wet — more at HUMOR]
: humidity **:** moisture ⟨*hygro*phyte⟩

hy·gro·graph \ˈhī-grə-ˌgraf\ *noun* [International Scientific Vocabulary] (circa 1864)
: an instrument for recording automatically variations in atmospheric humidity

hy·grom·e·ter \hī-ˈgrä-mə-tər\ *noun* [probably from French *hygromètre,* from *hygr-* + *-mètre* -meter] (1670)
: any of several instruments for measuring the humidity of the atmosphere
— **hy·gro·met·ric** \ˌhī-grə-ˈme-trik\ *adjective*

hy·groph·i·lous \hī-ˈgrä-fə-ləs\ *adjective* (1863)
: living or growing in moist places

hy·gro·phyte \ˈhī-grə-ˌfīt\ *noun* [International Scientific Vocabulary] (1903)
: HYDROPHYTE
— **hy·gro·phyt·ic** \ˌhī-grə-ˈfi-tik\ *adjective*

hy·gro·scop·ic \ˌhī-grə-ˈskä-pik\ *adjective* [*hygroscope,* an instrument showing changes in humidity + [1]*-ic;* from the use of such materials in the hygroscope] (1790)
1 : readily taking up and retaining moisture
2 : taken up and retained under some conditions of humidity and temperature ⟨*hygroscopic* water in clay⟩
— **hy·gro·scop·ic·i·ty** \-(ˌ)skä-ˈpi-sə-tē\ *noun*

hying *present participle of* HIE

Hyk·sos \ˈhik-ˌsäs, -ˌsōs\ *adjective* [Greek *Hyksōs,* dynasty ruling Egypt, perhaps from Egyptian *ḥqꜣ* ruler + *ḫꜣst* foreign land] (1602)
: of or relating to a Semite dynasty that ruled Egypt from about the 18th to the 16th century B.C.

hy·la \ˈhī-lə\ *noun* [New Latin, from Latin *Hylas,* a companion of Hercules] (1842)
: any of a genus (*Hyla*) of tree frogs

hy·lo·zo·ism \ˌhī-lə-ˈzō-ˌi-zəm\ *noun* [Greek *hylē* matter, literally, wood + *zōos* alive, living; akin to Greek *zōē* life — more at QUICK] (1678)
: a doctrine held especially by early Greek philosophers that all matter has life
— **hy·lo·zo·ist** \-ˈzō-ist\ *noun*
— **hy·lo·zo·is·tic** \-zō-ˈis-tik\ *adjective*

hy·men \ˈhī-mən\ *noun* [Late Latin, from Greek *hymēn* membrane] (1615)
: a fold of mucous membrane partly closing the orifice of the vagina
— **hy·men·al** \-mə-nᵊl\ *adjective*

Hymen *noun* [Latin, from Greek *Hymēn*]
: the Greek god of marriage

[1]**hy·me·ne·al** \ˌhī-mə-ˈnē-əl\ *adjective* [Latin *hymenaeus* wedding song, wedding, from Greek *hymenaios,* from *Hymēn*] (1602)
: NUPTIAL
— **hy·me·ne·al·ly** \-ˈnē-ə-lē\ *adverb*

[2]**hymeneal** *noun* (1655)
1 *plural, archaic* **:** NUPTIALS
2 *archaic* **:** a wedding hymn

hy·me·ni·um \hī-ˈmē-nē-əm\ *noun, plural* **-nia** \-nē-ə\ *or* **-niums** [New Latin, from Greek *hymēn*] (1830)
: a spore-bearing layer in fungi consisting of a group of asci or basidia often interspersed with sterile structures

hy·me·nop·ter·an \ˌhī-mə-ˈnäp-tə-rən\ *noun* [New Latin *Hymenoptera,* from Greek, neuter plural of *hymenopteros* membrane-winged, from *hymēn* + *pteron* wing — more at FEATHER] (circa 1842)
: any of an order (Hymenoptera) of highly specialized insects with complete metamorphosis that include the bees, wasps, ants, ichneumon flies, sawflies, gall wasps, and related forms, often associate in large colonies with complex social organization, and have usually four membranous wings and the abdomen generally borne on a slender pedicel
— **hymenopteran** *adjective*
— **hy·me·nop·ter·ous** \-rəs\ *adjective*

hy·me·nop·ter·on \-tə-ˌrän, -rən\ *noun, plural* **-tera** \-rə\ *also* **-terons** [New Latin, from Greek, neuter of *hymenopteros*] (1877)
: HYMENOPTERAN

[1]**hymn** \ˈhim\ *noun* [Middle English *ymne,* from Old English *ymen,* from Latin *hymnus* song of praise, from Greek *hymnos*] (before 12th century)
1 a : a song of praise to God **b :** a metrical composition adapted for singing in a religious service
2 : a song of praise or joy
3 : something resembling a hymn **:** PAEAN

[2]**hymn** *verb* **hymned** \ˈhimd\; **hymn·ing** \ˈhi-miŋ\ (1667)
transitive verb
: to praise or worship in or as if in hymns
intransitive verb
: to sing a hymn

hym·nal \ˈhim-nəl\ *noun* [Middle English *hymnale,* from Medieval Latin, from Latin *hymnus*] (15th century)
: a collection of church hymns

hym·na·ry \ˈhim-nə-rē\ *noun, plural* **-ries** (1888)
: HYMNAL

hymn·book \ˈhim-ˌbu̇k\ *noun* (before 12th century)
: HYMNAL

hym·no·dy \ˈhim-nə-dē\ *noun* [Late Latin *hymnodia,* from Greek *hymnōidia,* from *hymnos* + *aeidein* to sing — more at ODE] (1711)
1 : hymn singing
2 : hymn writing
3 : the hymns of a time, place, or church

hym·nol·o·gy \him-ˈnä-lə-jē\ *noun* [Greek *hymnologia* singing of hymns, from *hymnos* + *-logia* -logy] (circa 1638)
1 : HYMNODY
2 : the study of hymns

hy·oid \ˈhī-ˌȯid\ *adjective* [New Latin *hyoides* hyoid bone] (1842)
: of or relating to the hyoid bone

hyoid bone *noun* [New Latin *hyoides,* from Greek *hyoeidēs* shaped like the letter upsilon (Υ, υ), being the hyoid bone, from *y, hy* upsilon] (circa 1811)
: a bone or complex of bones situated at the base of the tongue and supporting the tongue and its muscles

hy·o·scine \ˈhī-ə-ˌsēn\ *noun* [International Scientific Vocabulary *hyos*cyamine + *-ine*] (1872)
: SCOPOLAMINE; *especially* **:** the levorotatory form of scopolamine

hy·o·scy·a·mine \ˌhī-ə-ˈsī-ə-ˌmēn\ *noun* [German *Hyoscyamin,* from New Latin *Hyoscyamus,* genus of herbs, from Latin, henbane, from Greek *hyoskyamos,* literally, swine's

bean, from *hyos* (genitive of *hys* swine) + *kyamos* bean — more at SOW] (1858)
: a poisonous crystalline alkaloid $C_{17}H_{23}NO_3$ of which atropine is a racemic mixture; *especially* **:** its levorotatory form found especially in the plants belladonna and henbane and used similarly to atropine

hyp- — see HYPO-

hyp·abys·sal \ˌhi-pə-ˈbi-səl, ˌhī-\ *adjective* [International Scientific Vocabulary] (1895)
: of or relating to a fine-grained igneous rock usually formed at a moderate distance below the surface
— **hyp·abys·sal·ly** \-sə-lē\ *adverb*

hy·pae·thral \hī-ˈpē-thrəl\ *adjective* [Latin *hypaethrus* exposed to the open air, from Greek *hypaithros,* from *hypo-* + *aithēr* ether, air — more at ETHER] (1794)
1 : having a roofless central space ⟨*hypaethral* temple⟩
2 : open to the sky

hy·pal·la·ge \hī-ˈpa-lə-jē, hi-\ *noun* [Latin, from Greek *hypallagē,* literally, interchange, from *hypallassein* to interchange, from *hypo-* + *allassein* to change, from *allos* other — more at ELSE] (1586)
: an interchange of two elements in a phrase or sentence from a more logical to a less logical relationship (as in "a mind is a terrible thing to waste" for "to waste a mind is a terrible thing")

hy·pan·thi·um \hī-ˈpan(t)-thē-əm\ *noun, plural* **-thia** \-thē-ə\ [New Latin, from *hypo-* + *anth-* + *-ium*] (circa 1855)
: an enlargement of the floral receptacle bearing on its rim the stamens, petals, and sepals and often enlarging and surrounding the fruits (as in the rose hip)

¹hype \ˈhīp\ *noun* [by shortening & alteration] (1924)
1 *slang* **:** a narcotics addict
2 *slang* **:** HYPODERMIC

²hype *transitive verb* **hyped; hyping** (1938)
1 : STIMULATE, ENLIVEN — usually used with *up*
2 : INCREASE
— **hyped–up** \ˈhīp-ˈdəp\ *adjective*

³hype *transitive verb* **hyped; hyp·ing** [origin unknown] (circa 1931)
1 : PUT ON, DECEIVE
2 : to promote or publicize extravagantly

⁴hype *noun* (1955)
1 : DECEPTION, PUT-ON
2 : PUBLICITY; *especially* **:** promotional publicity of an extravagant or contrived kind

hy·per \ˈhī-pər\ *adjective* [short for *hyperactive*] (1971)
1 : HIGH-STRUNG, EXCITABLE; *also* **:** highly excited
2 : extremely active

hyper- *prefix* [Middle English *iper-,* from Latin *hyper-,* from Greek, from *hyper* — more at OVER]
1 : above **:** beyond **:** SUPER- ⟨*hyper*market⟩
2 a : excessively ⟨*hyper*sensitive⟩ **b :** excessive ⟨*hyper*emia⟩
3 : that is or exists in a space of more than three dimensions ⟨*hyper*space⟩

hy·per·acu·ity
hy·per·acute
hy·per·aes·thet·ic
hy·per·ag·gres·sive
hy·per·alert
hy·per·arid
hy·per·arous·al
hy·per·aware
hy·per·aware·ness
hy·per·ca·tab·o·lism
hy·per·cau·tious
hy·per·charged
hy·per·civ·i·lized
hy·per·co·ag·u·la·bil·i·ty
hy·per·co·ag·u·la·ble
hy·per·com·pet·i·tive
hy·per·con·cen·tra·tion
hy·per·con·scious
hy·per·con·scious·ness
hy·per·de·vel·op·ment
hy·per·ef·fi·cient
hy·per·emo·tion·al
hy·per·emo·tion·al·i·ty
hy·per·en·dem·ic
hy·per·en·er·get·ic
hy·per·ex·cit·abil·i·ty
hy·per·ex·cit·able
hy·per·ex·cit·ed
hy·per·ex·cite·ment
hy·per·ex·cre·tion
hy·per·fas·tid·i·ous
hy·per·func·tion
hy·per·func·tion·al
hy·per·func·tion·ing
hy·per·im·mune
hy·per·im·mu·ni·za·tion
hy·per·im·mu·nize
hy·per·in·flat·ed
hy·per·in·ner·va·tion
hy·per·in·tel·lec·tu·al
hy·per·in·tel·li·gent
hy·per·in·tense
hy·per·in·vo·lu·tion
hy·per·ma·nia
hy·per·man·ic
hy·per·mas·cu·line
hy·per·met·a·bol·ic
hy·per·me·tab·o·lism
hy·per·mo·bil·i·ty
hy·per·mod·ern
hy·per·mod·ern·ist
hy·per·mu·ta·bil·i·ty
hy·per·mu·ta·ble
hy·per·na·tion·al·is·tic
hy·per·phys·i·cal
hy·per·pig·men·ta·tion
hy·per·pig·ment·ed
hy·per·pro·duc·er
hy·per·pro·duc·tion
hy·per·pure
hy·per·ra·tio·nal
hy·per·ra·tio·nal·i·ty
hy·per·re·ac·tive
hy·per·re·ac·tiv·i·ty
hy·per·re·ac·tor
hy·per·re·spon·sive
hy·per·ro·man·tic
hy·per·sa·line
hy·per·sa·lin·i·ty
hy·per·sal·i·va·tion
hy·per·se·cre·tion
hy·per·sen·si·ti·za·tion
hy·per·sen·si·tize
hy·per·som·no·lence
hy·per·stat·ic
hy·per·stim·u·late
hy·per·stim·u·la·tion
hy·per·sus·cep·ti·bil·i·ty
hy·per·sus·cep·ti·ble
hy·per·tense
hy·per·typ·i·cal
hy·per·vig·i·lance
hy·per·vig·i·lant
hy·per·vir·u·lent
hy·per·vis·cos·i·ty

hy·per·acid·i·ty \ˌhī-pər-ə-ˈsi-də-tē\ *noun* (circa 1890)
: the condition of containing more than the normal amount of acid
— **hy·per·ac·id** \ˌhī-pər-ˈa-səd\ *adjective*

hy·per·ac·tive \ˌhī-pər-ˈak-tiv\ *adjective* (1867)
: affected with or exhibiting hyperactivity; *broadly* **:** more active than is usual or desirable
— **hyperactive** *noun*

hy·per·ac·tiv·i·ty \-ˌak-ˈti-və-tē\ *noun* (1888)
: the state or condition of being excessively or pathologically active; *especially* **:** ATTENTION DEFICIT DISORDER

hy·per·aes·the·sia *variant of* HYPERESTHESIA

hy·per·al·i·men·ta·tion \ˌhī-pər-ˌa-lə-mən-ˈtā-shən\ *noun* (1967)
: the administration of nutrients by intravenous feeding especially to patients who cannot ingest food through the alimentary tract

hy·per·bar·ic \ˌhī-pər-ˈbar-ik\ *adjective* [*hyper-* + *bar-* + ¹*-ic*] (1962)
: of, relating to, or utilizing greater than normal pressure (as of oxygen) ⟨*hyperbaric* chamber⟩ ⟨*hyperbaric* medicine⟩
— **hy·per·bar·i·cal·ly** \-i-k(ə-)lē\ *adverb*

hy·per·bo·la \hī-ˈpər-bə-lə\ *noun, plural* **-las** *or* **-lae** \-(ˌ)lē\ [New Latin, from Greek *hyperbolē*] (1668)
: a plane curve generated by a point so moving that the difference of the distances from two fixed points is a constant **:** a curve formed by the intersection of a double right circular cone with a plane that cuts both halves of the cone

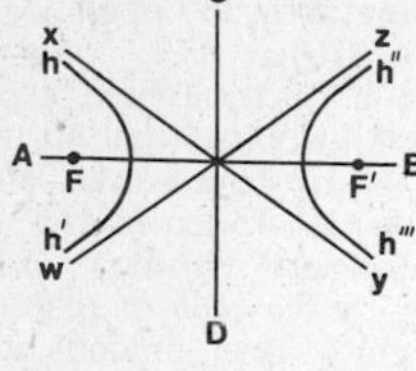

hyperbola: *AB, CD* axes; *F, F'* foci; *xy, zw* asymptotes; *h, h', h'', h'''* hyperbola

hy·per·bo·le \hī-ˈpər-bə-(ˌ)lē\ *noun* [Latin, from Greek *hyperbolē* excess, hyperbole, hyperbola, from *hyperballein* to exceed, from *hyper-* + *ballein* to throw — more at DEVIL] (15th century)
: extravagant exaggeration (as "mile-high ice-cream cones")
— **hy·per·bo·list** \-list\ *noun*

¹hy·per·bol·ic \ˌhī-pər-ˈbä-lik\ *also* **hy·per·bol·i·cal** \-li-kəl\ *adjective* (15th century)
: of, relating to, or marked by hyperbole
— **hy·per·bol·i·cal·ly** \-li-k(ə-)lē\ *adverb*

²hyperbolic *adjective* (1646)
1 : of, relating to, or being analogous to a hyperbola
2 : of, relating to, or being a space in which more than one line parallel to a given line passes through a point ⟨*hyperbolic* geometry⟩

hyperbolic function *noun* (circa 1890)
: any of a set of six functions analogous to the trigonometric functions but related to the hyperbola in a way similar to that in which the trigonometric functions are related to a circle

hyperbolic paraboloid *noun* (1842)
: a saddle-shaped quadric surface whose sections by planes parallel to one coordinate plane are hyperbolas while those sections by planes parallel to the other two are parabolas if proper orientation of the coordinate axes is assumed

hy·per·bo·lize \hī-ˈpər-bə-ˌlīz\ *verb* **-lized; -liz·ing** (1599)
intransitive verb
: to indulge in hyperbole
transitive verb
: to exaggerate to a hyperbolic degree

hy·per·bo·loid \-ˌlȯid\ *noun* (1743)
: a quadric surface whose sections by planes parallel to one coordinate plane are ellipses while those sections by planes parallel to the other two are hyperbolas if proper orientation of the axes is assumed
— **hy·per·bo·loi·dal** \(ˌ)hī-ˌpər-bə-ˈlȯi-dᵊl\ *adjective*

¹hy·per·bo·re·an \ˌhī-pər-ˈbōr-ē-ən, -ˈbȯr-; -(ˌ)pər-bə-ˈrē-ən\ *noun* [Latin *Hyperborei* (plural), from Greek *Hyperboreoi,* from *hyper-* + *Boreas*] (15th century)
1 *often capitalized* **:** a member of a people held by the ancient Greeks to live beyond the north wind in a region of perpetual sunshine
2 : an inhabitant of a cool northern climate

²hyperborean *adjective* (1591)
1 : of or relating to an extreme northern region **:** FROZEN
2 : of or relating to any of the arctic peoples

hy·per·cal·ce·mia \ˌhī-pər-ˌkal-ˈsē-mē-ə\ *noun* [New Latin] (1925)
: an excess of calcium in the blood
— **hy·per·cal·ce·mic** \-ˈsē-mik\ *adjective*

hy·per·cap·nia \-ˈkap-nē-ə\ *noun* [New Latin, from *hyper-* + Greek *kapnos* smoke; probably akin to Lithuanian *Kvapas* breath] (1908)
: the presence of excessive amounts of carbon dioxide in the blood
— **hy·per·cap·nic** \-nik\ *adjective*

hy·per·cat·a·lex·is \-ˌka-tə-ˈlek-səs\ *noun, plural* **-lex·es** \-ˈlek-ˌsēz\ [New Latin] (circa 1890)
: the occurrence of an additional syllable after the final complete foot or dipody in a line of verse
— **hy·per·cat·a·lec·tic** \-ˈlek-tik\ *adjective*

hy·per·charge \ˈhī-pər-ˌchärj\ *noun* [short for *hyperonic charge,* from *hyperon*] (1956)
: a quantum characteristic of a group of subatomic particles governed by the strong force that is related to strangeness and is represented by a number equal to twice the average value of the electric charge of the group

hy·per·cho·les·ter·ol·emia \ˌhī-pər-kə-ˌles-tə-rə-ˈlē-mē-ə\ *noun* [New Latin] (circa 1894)
: the presence of excess cholesterol in the blood
— **hy·per·cho·les·ter·ol·emic** \-mik\ *adjective*

hy·per·com·plex \ˌhī-pər-ˈkäm-ˌpleks\ *adjective* (circa 1889)
: of, relating to, or being the most general form of number that extends the complex number to an expression of the same type (as a quaternion) involving a finite number of units or components in which addition is by

components and multiplication does not have all of the properties of real or complex numbers

hy·per·cor·rect \ˌhī-pər-kə-ˈrekt\ *adjective* (1922)
: of, relating to, or characterized by the production of a nonstandard linguistic form or construction on the basis of a false analogy (as "badly" in "my eyes have gone badly" and "widely" in "open widely")
— **hy·per·cor·rec·tion** \-ˈrek-shən\ *noun*
— **hy·per·cor·rect·ly** \-ˈrek-(t)lē\ *adverb*
— **hy·per·cor·rect·ness** \-ˈrek(t)-nəs\ *noun*

hy·per·crit·ic \ˌhī-pər-ˈkri-tik\ *noun* [New Latin *hypercriticus,* from *hyper-* + Latin *criticus* critic] (1633)
: a carping or unduly censorious critic
— **hy·per·crit·i·cism** \-ˈkri-tə-ˌsi-zəm\ *noun*

hy·per·crit·i·cal \-ˈkri-ti-kəl\ *adjective* (1605)
: meticulously or excessively critical
synonym see CRITICAL
— **hy·per·crit·i·cal·ly** \-k(ə-)lē\ *adverb*

hy·per·cube \ˈhī-pər-ˌkyüb\ *noun* (1909)
1 : a geometric figure (as a tesseract) in Euclidean space of *n* dimensions that is analogous to a cube in three dimensions
2 : a computer architecture in which each processor is connected to *n* others based on analogy to a hypercube of *n* dimensions

hy·per·emia \ˌhī-pər-ˈē-mē-ə\ *noun* [New Latin] (circa 1839)
: excess of blood in a body part **:** CONGESTION
— **hy·per·emic** \-mik\ *adjective*

hy·per·es·the·sia \ˌhī-pər-es-ˈthē-zh(ē-)ə\ *noun* [New Latin, from *hyper-* + *-esthesia* (as in *anesthesia*)] (circa 1852)
: unusual or pathological sensitivity of the skin or of a particular sense
— **hy·per·es·thet·ic** \-ˈthe-tik\ *adjective*

hy·per·eu·tec·tic \ˌhī-pər-yu̇-ˈtek-tik\ *adjective* (1902)
: HYPEREUTECTOID

hy·per·eu·tec·toid \-ˌtȯid\ *adjective* (1908)
: containing the minor component in excess of that contained in the eutectoid

hy·per·ex·tend \ˌhī-pər-ik-ˈstend\ *transitive verb* (1883)
: to extend so that the angle between bones of a joint is greater than normal ⟨a *hyperextended* elbow⟩
— **hy·per·ex·ten·sion** \-ˈsten(t)-shən\ *noun*

hy·per·fine \ˈhī-pər-ˌfīn\ *adjective* (1926)
: being or relating to a fine-structure multiplet occurring in an atomic spectrum that is due to interaction between electrons and nuclear spin

hy·per·fo·cal distance \ˌhī-pər-ˈfō-kəl-\ *noun* [International Scientific Vocabulary] (1905)
: the nearest distance upon which a photographic lens may be focused to produce satisfactory definition at infinity

hy·per·ga·my \hī-ˈpər-gə-mē\ *noun, plural* **-mies** (1883)
: marriage into an equal or higher caste or social group

hy·per·geo·met·ric distribution \ˌhī-pər-ˌjē-ə-ˈme-trik-\ *noun* (1950)
: a probability function that gives the probability of obtaining exactly *x* elements of one kind and $n - x$ elements of another if *n* elements are chosen at random without replacement from a finite population containing *N* elements of which *M* are of the first kind and $N - M$ are of the second kind and has the form

$$f(x) = \frac{\binom{M}{x}\binom{N-M}{n-x}}{\binom{N}{n}} \text{ where } \binom{M}{x} = \frac{M!}{x!(M-x)!}$$

hy·per·gly·ce·mia \ˌhī-pər-glī-ˈsē-mē-ə\ *noun* [New Latin] (1894)
: excess of sugar in the blood
— **hy·per·gly·ce·mic** \-mik\ *adjective*

hy·per·gol·ic \ˌhī-pər-ˈgä-lik\ *adjective* [German *Hypergol* a hypergolic fuel, probably from *hyper-* + *erg-* + *-ol* (hydrocarbon)] (1947)
1 : igniting upon contact of components without external aid (as a spark)
2 : of, relating to, or using hypergolic fuel
— **hy·per·gol·i·cal·ly** \-li-k(ə-)lē\ *adverb*

hy·per·hi·dro·sis \ˌhī-pər-hī-ˈdrō-səs\ *noun* [New Latin *hidrosis* perspiration, from Greek *hidrōsis,* from *hidroun* to sweat, from *hidrōs* sweat — more at SWEAT] (circa 1860)
: generalized or local excessive sweating

hy·per·in·fla·tion \ˌhī-pər-in-ˈflā-shən\ *noun* (1930)
: inflation growing at a very high rate in a very short time; *also* **:** a period of hyperinflation
— **hy·per·in·fla·tion·ary** \-shə-ˌner-ē\ *adjective*

hy·per·in·su·lin·ism \ˌhī-pə-ˈrin(t)-s(ə-)lə-ˌni-zəm\ *noun* [International Scientific Vocabulary] (1924)
: the presence of excess insulin in the body resulting in hypoglycemia

Hy·pe·ri·on \hī-ˈpir-ē-ən\ *noun* [Latin, from Greek *Hyperiōn*]
: a Titan and the father of Aurora, Selene, and Helios

hy·per·ir·ri·ta·bil·i·ty \ˌhī-pər-ˌir-ə-tə-ˈbi-lə-tē\ *noun* (1913)
: abnormally great or uninhibited response to stimuli
— **hy·per·ir·ri·ta·ble** \-ˈir-ə-tə-bəl\ *adjective*

hy·per·ker·a·to·sis \ˌhī-pər-ˌker-ə-ˈtō-səs\ *noun, plural* **-to·ses** \-ˈtō-ˌsēz\ [New Latin] (1841)
: hypertrophy of the corneous layer of the skin
— **hy·per·ker·a·tot·ic** \-ˈtä-tik\ *adjective*

hy·per·ki·ne·sia \-kə-ˈnē-zh(ē-)ə, -kī-\ *noun* [New Latin, from *hyperkinesis*] (circa 1848)
: HYPERKINESIS

hy·per·ki·ne·sis \-ˈnē-səs\ *noun* [New Latin] (circa 1855)
1 : abnormally increased and sometimes uncontrollable activity or muscular movements
2 : a condition especially of childhood characterized by hyperactivity

hy·per·ki·net·ic \-ˈne-tik\ *adjective* (1888)
: of, relating to, or affected with hyperkinesis or hyperactivity ⟨the *hyperkinetic* child⟩; *also* **:** characterized by fast-paced or frenetic activity

hy·per·li·pe·mia \ˌhī-pər-lī-ˈpē-mē-ə\ *noun* [New Latin] (circa 1894)
: the presence of excess fat or lipids in the blood
— **hy·per·li·pe·mic** \-mik\ *adjective*

hy·per·lip·id·emia \-ˌli-pə-ˈdē-mē-ə\ *noun* [New Latin] (1961)
: HYPERLIPEMIA

hy·per·mar·ket \ˈhī-pər-ˌmär-kət\ *noun* (1970)
: a very large store that carries products found in a supermarket as well as merchandise commonly found in department stores

hy·per·me·ter \hī-ˈpər-mə-tər\ *noun* [Late Latin *hypermetrus* hypercatalectic, from Greek *hypermetros* beyond measure, beyond the meter, from *hyper-* + *metron* measure, meter] (circa 1656)
1 : a verse marked by hypercatalexis
2 : a period comprising more than two or three cola
— **hy·per·met·ric** \ˌhī-pər-ˈme-trik\ *or* **hy·per·met·ri·cal** \-tri-kəl\ *adjective*

hy·per·me·tro·pia \ˌhī-pər-mi-ˈtrō-pē-ə\ *noun* [New Latin, from Greek *hypermetros* + New Latin *-opia*] (1868)
: HYPEROPIA
— **hy·per·me·tro·pic** \-ˈtrō-pik, -ˈträ-\ *adjective*

hy·perm·ne·sia \ˌhī-(ˌ)pərm-ˈnē-zh(ē-)ə\ *noun* [New Latin, from *hyper-* + *-mnesia* (as in *amnesia*)] (1882)
: abnormally vivid or complete memory or recall of the past
— **hy·perm·ne·sic** \-ˈnē-zik, -sik\ *adjective*

hy·per·on \ˈhī-pə-ˌrän\ *noun* [probably from *hyper-* + [2]*-on*] (1953)
: a fundamental particle of the baryon group that is greater in mass than a proton or neutron

hy·per·opia \ˌhī-pə-ˈrō-pē-ə\ *noun* [New Latin] (1884)
: a condition in which visual images come to a focus behind the retina of the eye and vision is better for distant than for near objects — called also *farsightedness*
— **hy·per·opic** \-ˈrō-pik, -ˈrä-\ *adjective*

hy·per·os·to·sis \ˌhī-pər-ˌäs-ˈtō-səs\ *noun, plural* **-to·ses** \-ˈtō-ˌsēz\ [New Latin] (circa 1836)
: excessive growth or thickening of bone tissue
— **hy·per·os·tot·ic** \-ˈtä-tik\ *adjective*

hy·per·par·a·site \ˌhī-pər-ˈpar-ə-ˌsīt\ *noun* (circa 1889)
: a parasite that is parasitic upon another parasite
— **hy·per·par·a·sit·ic** \-ˌpar-ə-ˈsi-tik\ *adjective*
— **hy·per·par·a·sit·ism** \-ˈpar-ə-ˌsī-ˌti-zəm, -ˈpar-ə-sə-\ *noun*

hy·per·para·thy·roid·ism \-ˌpar-ə-ˈthī-ˌrȯi-ˌdi-zəm\ *noun* (1917)
: the presence of excess parathyroid hormone in the body resulting in disturbance of calcium metabolism with increase in serum calcium and decrease in inorganic phosphorus, loss of calcium from bone, and renal damage with frequent kidney-stone formation

hy·per·pha·gia \-ˈfā-j(ē-)ə\ *noun* [New Latin] (1941)
: abnormally increased appetite for consumption of food frequently associated with injury to the hypothalamus
— **hy·per·phag·ic** \-ˈfa-jik\ *adjective*

hy·per·pi·tu·ita·rism \-pə-ˈtü-ə-tə-ˌri-zəm, -ˈtü-ə-ˌtri-, -ˈtyü-\ *noun* [International Scientific Vocabulary] (1909)
: excessive production of growth hormones by the pituitary gland
— **hy·per·pi·tu·itary** \-ˈtü-ə-ˌter-ē, -ˈtyü-\ *adjective*

hy·per·plane \ˈhī-pər-ˌplān\ *noun* (1903)
: a figure in hyperspace corresponding to a plane in ordinary space

hy·per·pla·sia \ˌhī-pər-ˈplā-zh(ē-)ə\ *noun* [New Latin] (1861)
: an abnormal or unusual increase in the elements composing a part (as cells composing a tissue)
— **hy·per·plas·tic** \-ˈplas-tik\ *adjective*

hy·per·ploid \ˈhī-pər-ˌplȯid\ *adjective* [International Scientific Vocabulary] (1930)
: having a chromosome number slightly greater than an exact multiple of the monoploid number
— **hyperploid** *noun*
— **hy·per·ploi·dy** \-ˌplȯi-dē\ *noun*

hy·per·pnea \ˌhī-pər-ˈnē-ə, -ˌpərp-ˈnē-\ *noun* [New Latin] (circa 1860)
: abnormally rapid or deep breathing
— **hy·per·pne·ic** \-ˈnē-ik\ *adjective*

hy·per·po·lar·ize \ˌhī-pər-ˈpō-lə-ˌrīz\ (1950) *transitive verb*
: to produce an increase in potential difference across (a biological membrane) ⟨a *hyperpolarized* nerve cell⟩

intransitive verb
: to undergo or produce an increase in potential difference across something
— **hy·per·po·lar·iza·tion** \-ˌpō-lə-rə-ˈzā-shən\ *noun*

hy·per·py·rex·ia \-pī-ˈrek-sē-ə\ *noun* [New Latin] (1875)
: exceptionally high fever (as in a particular disease)

hy·per·re·al·ism \ˌhī-pər-ˈrē-ə-ˌli-zəm, -ˈri-ə-\ *noun* (1937)
: realism in painting characterized by depiction of real life in an unusual or striking manner — compare PHOTO-REALISM
— **hy·per·re·al·ist** \-list\ *adjective*
— **hy·per·re·al·is·tic** \-ˌrē-ə-ˈlis-tik, -ˌri-\ *adjective*

hy·per·sen·si·tive \ˌhī-pər-ˈsen(t)-s(ə-)tiv\ *adjective* (1871)
1 : excessively or abnormally sensitive
2 : abnormally susceptible physiologically to a specific agent (as a drug or antigen)
— **hy·per·sen·si·tive·ness** *noun*
— **hy·per·sen·si·tiv·i·ty** \-ˌsen(t)-sə-ˈti-və-tē\ *noun*

hy·per·sex·u·al \-ˈsek-sh(ə-)wəl, -ˈsek-shəl\ *adjective* (1942)
: exhibiting unusual or excessive concern with or indulgence in sexual activity
— **hy·per·sex·u·al·i·ty** \-ˌsek-shə-ˈwa-lə-tē\ *noun*

hy·per·son·ic \-ˈsä-nik\ *adjective* [International Scientific Vocabulary] (1946)
1 : of or relating to speed five or more times that of sound in air — compare SONIC
2 : moving, capable of moving, or utilizing air currents that move at hypersonic speed ⟨*hypersonic* wind tunnel⟩
— **hy·per·son·i·cal·ly** \-ni-k(ə-)lē\ *adverb*

hy·per·space \ˈhī-pər-ˌspās\ *noun* (1867)
1 : space of more than three dimensions
2 : a fictional space held to support extraordinary events (as travel faster than the speed of light)

hy·per·sthene \ˈhī-pərs-ˌthēn\ *noun* [French *hypersthène,* from Greek *hyper-* + *sthenos* strength] (circa 1808)
: an orthorhombic grayish or greenish black or dark brown pyroxene
— **hy·per·sthe·nic** \ˌhī-pərs-ˈthe-nik, -ˈthē-\ *adjective*

hy·per·sur·face \ˈhī-pər-ˌsər-fəs\ *noun* (circa 1909)
: a figure that is the analogue in hyperspace of a surface in three-dimensional space

hy·per·ten·sion \ˌhī-pər-ˈten(t)-shən\ *noun* [International Scientific Vocabulary] (1893)
: abnormally high blood pressure and especially arterial blood pressure; *also* **:** the systemic condition accompanying high blood pressure
— **hy·per·ten·sive** \ˌhī-pər-ˈten(t)-siv\ *adjective or noun*

hy·per·text \ˈhī-pər-ˌtekst\ *noun* (1986)
: a database format in which information related to that on a display can be accessed directly from the display

hy·per·ther·mia \ˌhī-pər-ˈthər-mē-ə\ *noun* [New Latin, from *hyper-* + *therm-* + *-ia*] (1887)
: exceptionally high fever especially when induced artificially for therapeutic purposes
— **hy·per·ther·mic** \-mik\ *adjective*

hy·per·thy·roid \-ˈthī-ˌrȯid\ *adjective* [back-formation from *hyperthyroidism*] (1916)
: of, relating to, or affected with hyperthyroidism

hy·per·thy·roid·ism \-ˌrȯi-ˌdi-zəm\ *noun* [International Scientific Vocabulary] (circa 1900)
: excessive functional activity of the thyroid gland; *also* **:** the resulting condition marked especially by increased metabolic rate, enlargement of the thyroid gland, rapid heart rate, and high blood pressure

hy·per·to·nia \ˌhī-pər-ˈtō-nē-ə\ *noun* (circa 1842)
: HYPERTONICITY

hy·per·ton·ic \-ˈtä-nik\ *adjective* [International Scientific Vocabulary] (1855)
1 : exhibiting excessive tone or tension ⟨a *hypertonic* baby⟩ ⟨a *hypertonic* bladder⟩
2 : having a higher osmotic pressure than a surrounding medium or a fluid under comparison

hy·per·to·nic·i·ty \-tə-ˈni-sə-tē\ *noun* (1886)
: the quality or state of being hypertonic

¹hy·per·tro·phy \hī-ˈpər-trə-fē\ *noun, plural* **-phies** [probably from New Latin *hypertrophia,* from *hyper-* + *-trophia* -trophy] (1834)
1 : excessive development of an organ or part; *specifically* **:** increase in bulk (as by thickening of muscle fibers) without multiplication of parts
2 : exaggerated growth or complexity
— **hy·per·tro·phic** \ˌhī-pər-ˈtrō-fik\ *adjective*

²hypertrophy *intransitive verb* **-phied; -phy·ing** (1883)
: to undergo hypertrophy

hy·per·ur·ban·ism \ˌhī-pər-ˈər-bə-ˌni-zəm\ *noun* (1925)
: use of hypercorrect forms in language; *also* **:** such a form

hy·per·uri·ce·mia \ˌhī-pər-ˌyu̇r-ə-ˈsē-mē-ə\ *noun* [New Latin] (circa 1894)
: excess uric acid in the blood

hy·per·ve·loc·i·ty \-və-ˈlä-sə-tē, -ˈläs-tē\ *noun* (1949)
: a high or relatively high velocity; *especially* **:** one greater than 10,000 feet (3048 meters) per second

hy·per·ven·ti·late \ˌhī-pər-ˈven-tᵊl-ˌāt\ *intransitive verb* (1931)
: to breathe rapidly and deeply **:** undergo hyperventilation

hy·per·ven·ti·la·tion \-ˌven-tᵊl-ˈā-shən\ *noun* (1928)
: excessive rate and depth of respiration leading to abnormal loss of carbon dioxide from the blood

hy·per·vi·ta·min·osis \-ˌvī-tə-mə-ˈnō-səs\ *noun, plural* **-oses** \-ˈnō-ˌsēz\ [New Latin] (1928)
: an abnormal state resulting from excessive intake of one or more vitamins

hy·pha \ˈhī-fə\ *noun, plural* **hy·phae** \-(ˌ)fē\ [New Latin, from Greek *hyphē* web; akin to Greek *hyphos* web — more at WEAVE] (1866)
: one of the threads that make up the mycelium of a fungus, increase by apical growth, and are coenocytic or transversely septate
— **hy·phal** \-fəl\ *adjective*

¹hy·phen \ˈhī-fən\ *noun* [Late Latin & Greek; Late Latin, from Greek, from *hyph' hen* under one, from *hypo* under + *hen,* neuter of *heis* one — more at UP, SAME] (circa 1620)
: a punctuation mark - used especially to divide or to compound words, word elements, or numbers
— **hy·phen·less** \-ləs\ *adjective*

²hyphen *transitive verb* (1814)
: HYPHENATE

¹hy·phen·ate \ˈhī-fə-ˌnāt\ *transitive verb* **-ated; -at·ing** (circa 1889)
: to connect (as two words) or divide (as a word at the end of a line of print) with a hyphen
— **hy·phen·ation** \ˌhī-fə-ˈnā-shən\ *noun*

²hyphenate *noun* [from the hyphens in the titles of such people, as *producer-director*] (1974)
: a person who performs more than one function (as on a filmmaking project)

hyphenated *adjective* [from the use of hyphenated words (as German-American) to designate foreign-born citizens of the U.S.] (circa 1893)
: of, relating to, or being an individual or unit of mixed or diverse background or composition ⟨*hyphenated* Americans⟩

hypn- *or* **hypno-** *combining form* [French, from Late Latin, from Greek, from *hypnos* — more at SOMNOLENT]
1 : sleep ⟨*hypno*pompic⟩
2 : hypnotism ⟨*hypno*therapy⟩

hyp·na·go·gic *also* **hyp·no·go·gic** \ˌhip-nə-ˈgä-jik, -ˈgō-\ *adjective* [French *hypnagogique,* from Greek *hypn-* + *-agōgos* leading, inducing, from *agein* to lead — more at AGENT] (1886)
: of, relating to, or associated with the drowsiness preceding sleep

hyp·noid \ˈhip-ˌnȯid\ *or* **hyp·noi·dal** \hip-ˈnȯi-dᵊl\ *adjective* (1898)
: of or relating to sleep or hypnosis

hyp·no·pom·pic \ˌhip-nə-ˈpäm-pik\ *adjective* [*hypn-* + Greek *pompē* act of sending — more at POMP] (circa 1901)
: associated with the semiconsciousness preceding waking ⟨*hypnopompic* illusions⟩

Hyp·nos \ˈhip-nəs, -ˌnōs\ *noun* [Greek]
: the Greek god of sleep

hyp·no·sis \hip-ˈnō-səs\ *noun, plural* **-noses** \-ˌsēz\ [New Latin] (1876)
1 : a state that resembles sleep but is induced by a person whose suggestions are readily accepted by the subject
2 : any of various conditions that resemble sleep
3 : HYPNOTISM 1

hyp·no·ther·a·py \ˌhip-nō-ˈther-ə-pē\ *noun* (1897)
1 : the treatment of disease by hypnotism
2 : psychotherapy that facilitates suggestion, reeducation, or analysis by means of hypnosis
— **hyp·no·ther·a·pist** \-pist\ *noun*

¹hyp·not·ic \hip-ˈnä-tik\ *adjective* [French or Late Latin; French *hypnotique,* from Late Latin *hypnoticus,* from Greek *hypnōtikos,* from *hypnoun* to put to sleep, from *hypnos*] (1625)
1 : tending to produce sleep **:** SOPORIFIC
2 a : of or relating to hypnosis or hypnotism **b :** readily holding the attention ⟨a *hypnotic* personality⟩ ⟨a simple *hypnotic* beat⟩
— **hyp·not·i·cal·ly** \-ti-k(ə-)lē\ *adverb*

²hypnotic *noun* (1681)
1 : a sleep-inducing agent **:** SOPORIFIC
2 : one that is or can be hypnotized

hyp·no·tism \ˈhip-nə-ˌti-zəm\ *noun* (1842)
1 : the study or act of inducing hypnosis — compare MESMERISM
2 : HYPNOSIS 1
— **hyp·no·tist** \-tist\ *noun*

hyp·no·tize \-ˌtīz\ *transitive verb* **-tized; -tiz·ing** (1843)
1 : to induce hypnosis in
2 : to dazzle or overcome by or as if by suggestion ⟨a voice that *hypnotizes* its hearers⟩ ⟨drivers *hypnotized* by speed⟩
— **hyp·no·tiz·abil·i·ty** \ˌhip-nə-ˌtī-zə-ˈbi-lə-tē\ *noun*
— **hyp·no·tiz·able** \ˈhip-nə-ˌtī-zə-bəl\ *adjective*

¹hy·po \ˈhī-(ˌ)pō\ *noun, plural* **hypos** (1711)
: HYPOCHONDRIA

²hypo *noun, plural* **hypos** [short for *hyposulfite* thiosulfate] (1855)
: SODIUM THIOSULFATE

³hypo *noun, plural* **hypos** (1925)
1 : HYPODERMIC SYRINGE
2 : HYPODERMIC INJECTION
3 : STIMULUS

⁴hypo *transitive verb* (1942)
: STIMULATE ⟨do everything possible to *hypo* the economy —Clem Morgello⟩

hypo- *or* **hyp-** *prefix* [Middle English *ypo-,* from Old French, from Late Latin *hypo-, hyp-,* from Greek, from *hypo* — more at UP]
1 : under **:** beneath **:** down ⟨*hypo*blast⟩ ⟨*hypo*dermic⟩
2 : less than normal or normally ⟨*hyp*esthesia⟩ ⟨*hypo*tension⟩
3 : in a lower state of oxidation **:** in a low and usually the lowest position in a series of compounds ⟨*hypo*chlorous acid⟩ ⟨*hypo*xanthine⟩

hy·po·al·ler·gen·ic \ˌhī-pō-ˌa-lər-ˈje-nik\ *adjective* (1940)

: having little likelihood of causing an allergic response ⟨*hypoallergenic* cosmetics⟩ ⟨*hypoallergenic* foods⟩

hy·po·blast \ˈhī-pə-ˌblast\ *noun* (1875)
: the endoderm of an embryo

hy·po·cal·ce·mia \ˌhī-pō-ˌkal-ˈsē-mē-ə\ *noun* [New Latin] (1925)
: a deficiency of calcium in the blood
— **hy·po·cal·ce·mic** \-mik\ *adjective*

hy·po·caust \ˈhī-pə-ˌkȯst\ *noun* [Latin *hypocaustum,* from Greek *hypokauston,* from *hypokaiein* to light a fire under, from *hypo-* + *kaiein* to burn] (1678)
: an ancient Roman central heating system with underground furnace and tile flues to distribute the heat

hy·po·cen·ter \ˈhī-pə-ˌsen-tər\ *noun* (1905)
1 : the focus of an earthquake — compare EPICENTER 1
2 : the point on the earth's surface directly below the center of a nuclear bomb explosion
— **hy·po·cen·tral** \ˌhī-pə-ˈsen-trəl\ *adjective*

hy·po·chlo·rite \ˌhī-pə-ˈklōr-ˌīt, -ˈklȯr-\ *noun* (circa 1849)
: a salt or ester of hypochlorous acid

hy·po·chlo·rous acid \ˌhī-pə-ˈklōr-əs-, -ˈklȯr-\ *noun* [International Scientific Vocabulary] (1841)
: an unstable strongly oxidizing but weak acid HClO obtained in solution along with hydrochloric acid by reaction of chlorine with water and used especially in the form of salts as an oxidizing agent, bleaching agent, disinfectant, and chlorinating agent

hy·po·chon·dria \ˌhī-pə-ˈkän-drē-ə\ *noun* [New Latin, from Late Latin, plural, upper abdomen (formerly regarded as the seat of hypochondria), from Greek, literally, the parts under the cartilage (of the breastbone), from *hypo-* + *chondros* cartilage] (1668)
: extreme depression of mind or spirits often centered on imaginary physical ailments; *specifically* **:** HYPOCHONDRIASIS ◆

¹hy·po·chon·dri·ac \-drē-ˌak\ *adjective* [French *hypochondriaque,* from Greek *hypochondriakos,* from *hypochondria*] (1599)
1 : HYPOCHONDRIACAL
2 : of, relating to, or being the two regions of the abdomen lying on either side of the epigastric region and above the lumbar regions

²hypochondriac *noun* (1639)
: one affected by hypochondria

hy·po·chon·dri·a·cal \-kən-ˈdrī-ə-kəl, -ˌkän-\ *adjective* (1621)
: affected or produced by hypochondria
— **hy·po·chon·dri·a·cal·ly** \-k(ə-)lē\ *adverb*

hy·po·chon·dri·a·sis \-ˈdrī-ə-səs\ *noun, plural* **-a·ses** \-ˌsēz\ [New Latin] (1766)
: morbid concern about one's health especially when accompanied by delusions of physical disease

hy·po·chro·mic anemia \ˌhī-pə-ˈkrō-mik-\ *noun* (1934)
: an anemia marked by deficient hemoglobin and usually microcytic red blood cells

hy·po·co·rism \hī-ˈpä-kə-ˌri-zəm; ˌhī-pə-ˈkōr-ˌi-, -ˈkȯr-\ *noun* [Late Latin *hypocorisma,* from Greek *hypokorisma,* from *hypokorizesthai* to call by pet names, from *hypo-* + *korizesthai* to caress, from *koros* boy, *korē* girl] (1850)
1 : a pet name
2 : the use of pet names
— **hy·po·co·ris·tic** \ˌhī-pə-kə-ˈris-tik\ *or* **hy·po·co·ris·ti·cal** \-ti-kəl\ *adjective*
— **hy·po·co·ris·ti·cal·ly** \-ti-k(ə-)lē\ *adverb*

hy·po·cot·yl \ˈhī-pə-ˌkä-tᵊl\ *noun* [International Scientific Vocabulary *hypo-* + *cotyl*edon] (1880)
: the part of the axis of a plant embryo or seedling below the cotyledon — see PLUMULE illustration

hy·poc·ri·sy \hi-ˈpä-krə-sē *also* hī-\ *noun, plural* **-sies** [Middle English *ypocrisie,* from Old French, from Late Latin *hypocrisis,* from Greek *hypokrisis* act of playing a part on the stage, hypocrisy, from *hypokrinesthai* to answer, act on the stage, from *hypo-* + *krinein* to decide — more at CERTAIN] (13th century)
1 : a feigning to be what one is not or to believe what one does not; *especially* **:** the false assumption of an appearance of virtue or religion
2 : an act or instance of hypocrisy

hyp·o·crite \ˈhi-pə-ˌkrit\ *noun* [Middle English *ypocrite,* from Old French, from Late Latin *hypocrita,* from Greek *hypokritēs* actor, hypocrite, from *hypokrinesthai*] (13th century)
: a person who puts on a false appearance of virtue or religion
— **hypocrite** *adjective*

hyp·o·crit·i·cal \ˌhi-pə-ˈkri-ti-kəl\ *adjective* (1561)
: characterized by hypocrisy; *also* **:** being a hypocrite
— **hyp·o·crit·i·cal·ly** \-k(ə-)lē\ *adverb*

hy·po·cy·cloid \ˌhī-pō-ˈsī-ˌklȯid\ *noun* (1843)
: a curve traced by a point on the circumference of a circle rolling internally on the circumference of a fixed circle

hy·po·der·mal \ˌhī-pə-ˈdər-məl\ *adjective* (1854)
1 : of or relating to a hypodermis
2 : lying beneath an outer skin or epidermis

¹hy·po·der·mic \-mik\ *adjective* [International Scientific Vocabulary] (1863)
1 : adapted for use in or administered by injection beneath the skin
2 : of or relating to the parts beneath the skin
3 : resembling a hypodermic injection in effect **:** STIMULATING
— **hy·po·der·mi·cal·ly** \-mi-k(ə-)lē\ *adverb*

²hypodermic *noun* (circa 1889)
1 : HYPODERMIC INJECTION
2 : HYPODERMIC SYRINGE

hypodermic injection *noun* (1868)
: an injection made into the subcutaneous tissues

hypodermic needle *noun* (circa 1909)
1 : NEEDLE 1c(1)
2 : a hypodermic syringe complete with needle

hypodermic syringe *noun* (1893)
: a small syringe used with a hollow needle for injection of material into or beneath the skin

hy·po·der·mis \ˌhī-pə-ˈdər-məs\ *noun* [New Latin] (circa 1866)
1 : the tissue immediately beneath the epidermis of a plant especially when modified to serve as a supporting and protecting layer
2 : the cellular layer that underlies and secretes the chitinous cuticle (as of an arthropod)
3 : SUPERFICIAL FASCIA

hy·po·dip·loid \ˌhī-pō-ˈdi-ˌplȯid\ *adjective* (1962)
: having slightly fewer than the diploid number of chromosomes
— **hy·po·dip·loi·dy** \-ˌplȯi-dē\ *noun*

hy·po·eu·tec·toid \ˌhī-pō-yu̇-ˈtek-ˌtȯid\ *adjective* (1911)
: containing less of the minor component than is contained in the eutectoid

hy·po·gas·tric \ˌhī-pə-ˈgas-trik\ *adjective* [French *hypogastrique,* from *hypogastre* hypogastric region, from Greek *hypogastrion,* from *hypo-* + *gastr-, gastēr* belly — more at GASTRIC] (1656)
: of or relating to the lower median region of the abdomen

hy·po·ge·al \ˌhī-pə-ˈjē-əl\ *or* **hy·po·ge·an** \-ˈjē-ən\ *or* **hy·po·ge·ous** \-ˈjē-əs\ *adjective* [Late Latin *hypogeus* subterranean, from Greek *hypogaios,* from *hypo-* + *gaia* earth] (1686)
1 : growing or living below the surface of the ground
2 *of a cotyledon* **:** remaining below the ground while the epicotyl elongates

hy·po·gene \ˈhī-pə-ˌjēn\ *adjective* [*hypo-* + Greek *-genēs* born, produced — more at -GEN] (1831)
: formed, crystallized, or lying at depths below the earth's surface **:** PLUTONIC — used of various rocks

hy·po·ge·um \ˌhī-pə-ˈjē-əm\ *noun, plural* **-gea** \-ˈjē-ə\ [Latin, from Greek *hypogaion,* from neuter of *hypogaios*] (1706)
: the subterranean part of an ancient building; *also* **:** an ancient underground burial chamber

hy·po·glos·sal \ˌhī-pə-ˈglä-səl\ *adjective* (1831)
: of or relating to the hypoglossal nerves

hypoglossal nerve *noun* (1848)
: either of the 12th and final pair of cranial nerves which are motor nerves arising from the medulla oblongata and supplying muscles of the tongue in higher vertebrates — called also *hypoglossal*

hy·po·gly·ce·mia \ˌhī-pō-glī-ˈsē-mē-ə\ *noun* [New Latin] (circa 1894)
: abnormal decrease of sugar in the blood
— **hy·po·gly·ce·mic** \-mik\ *adjective or noun*

hy·pog·y·nous \hī-ˈpä-jə-nəs\ *adjective* (1821)
1 *of a floral organ* **:** inserted upon the receptacle or axis below the gynoecium and free from it
2 : having hypogynous floral organs
— **hy·pog·y·ny** \-nē\ *noun*

hy·po·ka·le·mia \ˌhī-pō-ˌkā-ˈlē-mē-ə\ *noun* [New Latin, from *hypo-* + *kal*ium potassium (from *kali* alkali, from Arabic *qily* saltwort) + *-emia*] (1949)
: a deficiency of potassium in the blood
— **hy·po·ka·le·mic** \-mik\ *adjective*

hy·po·lim·ni·on \ˌhī-pō-ˈlim-nē-ˌän, -nē-ən\ *noun, plural* **-nia** \-nē-ə\ [New Latin, from *hypo-* + Greek *limnion,* diminutive of *limnē* lake — more at LIMNETIC] (1928)
: the part of a lake below the thermocline made up of water that is stagnant and of essentially uniform temperature except during the period of overturn

◇ WORD HISTORY

hypochondria A hypochondriac's ailment is deemed to be mental rather than physical, but etymologically hypochondria has nothing to do with the head. The ancient Greek noun *chondros* meant "cartilage," and the adjective *hypochondrios* meant "located under the cartilage of the breastbone and ribs." Formed from this adjective was the plural noun *hypochondria,* designating the area of the abdomen below the rib cartilage and above the navel. In Greek medical doctrine the spleen and kidneys, located in this part of the abdomen, secreted black bile, or melancholy, an excess of which caused low spirits, and the ancient physician Galen characterized depression as *hypochondriakon nosēma* "the ailment of the upper abdominal organs." The English adjective *hypochondriac,* borrowed from Greek through French, was used in the 17th century to mean "suffering from depression," and later in the century the noun *hypochondria* came into use, either as a Latinism created by removing the final letter from *hypochondriac,* or by construing the Greek and Latin neuter plural noun *hypochondria* as feminine singular. The narrowing of *hypochondria* to morbid concern about one's health—in more formal medical terminology *hypochondriasis*—dates from the late 18th and 19th centuries.

hy·po·mag·ne·se·mia \ˌhī-pə-ˌmag-nə-ˈsē-mē-ə\ *noun* [New Latin, from *hypo-* + *magnesium* + *-emia*] (1933)
: deficiency of magnesium in the blood especially of cattle

hy·po·ma·nia \ˌhī-pə-ˈmā-nē-ə, -nyə\ *noun* [New Latin] (1882)
: a mild mania especially when part of a manic-depressive cycle
— **hy·po·man·ic** \-ˈma-nik\ *adjective*

hy·po·morph \ˈhī-pə-ˌmȯrf\ *noun* (1926)
: a mutant gene having a similar but weaker effect than the corresponding wild-type gene
— **hy·po·mor·phic** \ˌhī-pə-ˈmȯr-fik\ *adjective*

hy·po·para·thy·roid·ism \ˈhī-pō-ˌpar-ə-ˈthī-ˌrȯi-ˌdi-zəm\ *noun* (1910)
: deficiency of parathyroid hormone in the body; *also* **:** the resultant abnormal state marked by low serum calcium and a tendency to chronic tetany

hy·po·phar·ynx \-ˈfar-iŋ(k)s, -ˈfer-\ *noun* [New Latin] (1826)
1 : an appendage or thickened fold on the floor of the mouth of many insects that resembles a tongue
2 : the laryngeal part of the pharynx extending from the hyoid bone to the lower margin of the cricoid cartilage

hy·po·phy·se·al *also* **hy·po·phy·si·al** \(ˌ)hī-ˌpä-fə-ˈsē-əl, ˌhī-pə-, -ˈzē-; ˌhī-pə-ˈfi-zē-əl\ *adjective* [irregular from New Latin *hypophysis*] (1882)
: of or relating to the hypophysis

hy·poph·y·sec·to·mize \(ˌ)hī-ˌpä-fə-ˈsek-tə-ˌmīz\ *transitive verb* **-mized; -miz·ing** (1910)
: to remove the pituitary gland from

hy·poph·y·sec·to·my \-mē\ *noun, plural* **-mies** (1909)
: surgical removal of the pituitary gland

hy·poph·y·sis \hī-ˈpä-fə-səs\ *noun, plural* **-y·ses** \-ˌsēz\ [New Latin, from Greek, attachment underneath, from *hypophyein* to grow beneath, from *hypo-* + *phyein* to grow, produce — more at BE] (1825)
: PITUITARY GLAND

hy·po·pi·tu·ita·rism \ˌhī-pō-pə-ˈtü-ə-tə-ˌri-zəm, -ˈtü-ə-ˌtri-, -ˈtyü-\ *noun* [International Scientific Vocabulary] (1909)
: deficient production of growth hormones by the pituitary gland
— **hy·po·pi·tu·itary** \-ˈtü-ə-ˌter-ē, -ˈtyü-\ *adjective*

hy·po·pla·sia \-ˈplā-zh(ē-)ə\ *noun* [New Latin] (1889)
: a condition of arrested development in which an organ or part remains below the normal size or in an immature state
— **hy·po·plas·tic** \-ˈplas-tik\ *adjective*

hy·po·ploid \ˈhī-pō-ˌplȯid\ *adjective* (1930)
: having a chromosome number slightly less than an exact multiple of the monoploid number
— **hypoploid** *noun*

hy·po·sen·si·ti·za·tion \ˌhī-pō-ˌsen(t)-sə-tə-ˈzā-shən, -ˌsen(t)-stə-ˈzā-\ *noun* (1922)
: the state or process of being reduced in sensitivity especially to an allergen **:** DESENSITIZATION
— **hy·po·sen·si·tize** \-ˈsen(t)-sə-ˌtīz\ *transitive verb*

hy·po·spa·di·as \ˌhī-pə-ˈspā-dē-əs\ *noun* [New Latin, from Greek, man with hypospadias, from *hypo-* + *-spadias*, from *-spad-*, *-spas* something torn, from *span* to tear, pluck off] (circa 1855)
: an abnormality of the penis in which the urethra opens on the undersurface

hy·pos·ta·sis \hī-ˈpäs-tə-səs\ *noun, plural* **-ta·ses** \-ˌsēz\ [Late Latin, substance, sediment, from Greek, support, foundation, substance, sediment, from *hyphistasthai* to stand under, support, from *hypo-* + *histasthai* to be standing — more at STAND] (1590)
1 a : something that settles at the bottom of a fluid **b :** the settling of blood in the dependent parts of an organ or body
2 : PERSON 3
3 a : the substance or essential nature of an individual **b :** something that is hypostatized
4 [New Latin, from Late Latin] **:** failure of a gene to produce its usual effect when coupled with another gene that is epistatic toward it
— **hy·po·stat·ic** \ˌhī-pə-ˈsta-tik\ *adjective*
— **hy·po·stat·i·cal·ly** \-ti-k(ə-)lē\ *adverb*

hy·pos·ta·tize \hī-ˈpäs-tə-ˌtīz\ *transitive verb* **-tized; -tiz·ing** [Greek *hypostatos* substantially existing, from *hyphistasthai*] (1829)
: to attribute real identity to (a concept)
— **hy·pos·ta·ti·za·tion** \-ˌpäs-tə-tə-ˈzā-shən\ *noun*

hy·po·stome \ˈhī-pə-ˌstōm\ *noun* [International Scientific Vocabulary *hypo-* + *-stome* (from Greek *stoma* mouth) — more at STOMACH] (circa 1862)
: any of several structures associated with the mouth: as **a :** the manubrium of a hydrozoan **b :** a rodlike organ that arises at the base of the beak in various mites and ticks

hy·po·style \ˈhī-pə-ˌstīl\ *adjective* [Greek *hypostylos,* from *hypo-* + *stylos* pillar — more at STEER] (1831)
: having the roof resting on rows of columns
— **hypostyle** *noun*

hy·po·tac·tic \ˌhī-pə-ˈtak-tik\ *adjective* [Greek *hypotaktikos,* from *hypotassein*] (1896)
: of or relating to hypotaxis

hy·po·tax·is \-ˈtak-səs\ *noun* [New Latin, from Greek, subjection, from *hypotassein* to arrange under, from *hypo-* + *tassein* to arrange] (1883)
: syntactic subordination (as by a conjunction)

hy·po·ten·sion \ˌhī-pō-ˈten(t)-shən\ *noun* [International Scientific Vocabulary] (1893)
: abnormally low blood pressure

[1]hy·po·ten·sive \ˌhī-pō-ˈten(t)-siv\ *adjective* (1904)
1 : characterized by or due to hypotension
2 : causing low blood pressure or a lowering of blood pressure ⟨*hypotensive* drugs⟩

[2]hypotensive *noun* (1941)
: a person with hypotension

hy·pot·e·nuse \hī-ˈpä-tᵊn-ˌüs, -ˌyüz\ *also* **hy·poth·e·nuse** \-ˈpä-thən-\ *noun* [Latin *hypotenusa,* from Greek *hypoteinousa,* from feminine of *hypoteinōn,* present participle of *hypoteinein* to subtend, from *hypo-* + *teinein* to stretch — more at THIN] (1594)
1 : the side of a right-angled triangle that is opposite the right angle
2 : the length of a hypotenuse

ac **hypotenuse**

hy·po·tha·lam·ic \ˌhī-pō-thə-ˈla-mik\ *adjective* (1899)
: of or relating to the hypothalamus

hy·po·thal·a·mus \-ˈtha-lə-məs\ *noun* [New Latin] (1896)
: a basal part of the diencephalon that lies beneath the thalamus on each side, forms the floor of the third ventricle, and includes vital autonomic regulatory centers

[1]hy·poth·e·cate \hi-ˈpä-thə-ˌkāt, hī-\ *transitive verb* **-cat·ed; -cat·ing** [Medieval Latin *hypothecare* to pledge, from Late Latin *hypotheca* pledge, from Greek *hypothēkē,* from *hypotithenai* to put under, deposit as a pledge] (1681)
: to pledge as security without delivery of title or possession
— **hy·poth·e·ca·tion** \-ˌpä-thə-ˈkā-shən\ *noun*
— **hy·poth·e·ca·tor** \-ˈpä-thə-ˌkā-tər\ *noun*

[2]hy·poth·e·cate \hī-ˈpä-thə-ˌkāt\ *transitive verb* **-cat·ed; -cat·ing** [Greek *hypothēkē* suggestion, from *hypotithenai*] (1906)
: HYPOTHESIZE

hy·po·ther·mal \ˌhī-pō-ˈthər-məl\ *adjective* (1944)
: of or relating to a hydrothermal metalliferous ore vein deposited at high temperature

hy·po·ther·mia \-ˈthər-mē-ə\ *noun* [New Latin] (circa 1886)
: subnormal temperature of the body
— **hy·po·ther·mic** \-mik\ *adjective*

hy·poth·e·sis \hī-ˈpä-thə-səs\ *noun, plural* **-e·ses** \-ˌsēz\ [Greek, from *hypotithenai* to put under, suppose, from *hypo-* + *tithenai* to put — more at DO] (circa 1656)
1 a : an assumption or concession made for the sake of argument **b :** an interpretation of a practical situation or condition taken as the ground for action
2 : a tentative assumption made in order to draw out and test its logical or empirical consequences
3 : the antecedent clause of a conditional statement ☆

hy·poth·e·size \-ˌsīz\ *verb* **-sized; -siz·ing** (1738)
intransitive verb
: to make a hypothesis
transitive verb
: to adopt as a hypothesis

hy·po·thet·i·cal \ˌhī-pə-ˈthe-ti-kəl\ *adjective* (1588)
: being or involving a hypothesis **:** CONJECTURAL ⟨*hypothetical* arguments⟩ ⟨a *hypothetical* situation⟩
— **hy·po·thet·i·cal·ly** \-ti-k(ə-)lē\ *adverb*

hy·po·thet·i·co–de·duc·tive \ˌhī-pə-ˈthe-ti-ˌkō-di-ˈdək-tiv\ *adjective* (1912)
: relating to, being, or making use of the method of proposing hypotheses and testing their acceptability or falsity by determining whether their logical consequences are consistent with observed data

hy·po·thy·roid \ˌhī-pō-ˈthī-ˌrȯid\ *adjective* (1909)
: of, relating to, or affected with hypothyroidism

hy·po·thy·roid·ism \-ˌrȯi-ˌdi-zəm\ *noun* [International Scientific Vocabulary] (1905)
: deficient activity of the thyroid gland; *also* **:** a resultant bodily condition characterized by lowered metabolic rate and general loss of vigor

hy·po·to·nia \ˌhī-pə-ˈtō-nē-ə, -pō-\ *noun* [New Latin] (circa 1886)
: the state of having hypotonic muscle tone

hy·po·ton·ic \ˌhī-pə-ˈtä-nik, -pō-\ *adjective* [International Scientific Vocabulary] (1895)
1 : having deficient tone or tension ⟨*hypotonic* children⟩
2 : having a lower osmotic pressure than a surrounding medium or a fluid under comparison ⟨*hypotonic* organisms⟩
— **hy·po·to·nic·i·ty** \-tə-ˈni-sə-tē\ *noun*

hy·po·xan·thine \ˌhī-pō-ˈzan-ˌthēn\ *noun* [International Scientific Vocabulary] (circa 1857)
: a purine base $C_5H_4N_4O$ found in plant and animal tissues that yields xanthine on oxidation

hyp·ox·emia \ˌhi-ˌpäk-ˈsē-mē-ə, ˌhī-\ *noun* [New Latin] (circa 1886)
: deficient oxygenation of the blood
— **hyp·ox·emic** \-mik\ *adjective*

hyp·ox·ia \hi-ˈpäk-sē-ə, hī-\ *noun* [New Latin] (1941)

☆ SYNONYMS
Hypothesis, theory, law mean a formula derived by inference from scientific data that explains a principle operating in nature. HYPOTHESIS implies insufficient evidence to provide more than a tentative explanation ⟨a *hypothesis* explaining the extinction of the dinosaurs⟩. THEORY implies a greater range of evidence and greater likelihood of truth ⟨the *theory* of evolution⟩. LAW implies a statement of order and relation in nature that has been found to be invariable under the same conditions ⟨the *law* of gravitation⟩.

: a deficiency of oxygen reaching the tissues of the body
— **hyp•ox•ic** \-sik\ *adjective*

hyp•som•e•ter \hip-'sä-mə-tər\ *noun* [International Scientific Vocabulary, from Greek *hypsos* height (akin to Old English *ūp* up) + International Scientific Vocabulary *-meter* — more at UP] (1927)
: any of various instruments for determining the height of trees by triangulation

hyp•so•met•ric \ˌhip-sə-'me-trik\ *adjective* [Greek *hypsos*] (1845)
: of, relating to, or indicating elevation (as on a map) ⟨*hypsometric* curve⟩

hy•rax \'hī-ˌraks\ *noun, plural* **hy•rax•es** \-ˌrak-səz\ *also* **hy•ra•ces** \-rə-ˌsēz\ [Greek *hyrak-, hyrax* shrew] (1832)
: any of a family (Procaviidae) of small ungulate mammals of Africa and the Middle East characterized by thickset body with short legs and ears and rudimentary tail, feet with soft pads and broad nails, and teeth of which the molars resemble those of the rhinoceros and the incisors those of rodents — called also *coney, dassie*

hyrax

hy•son \'hī-sᵊn\ *noun* [Chinese (Guangdong) *héichēun,* literally, bright spring] (1740)
: a Chinese green tea made from thinly rolled and twisted leaves

hys•sop \'hi-səp\ *noun* [Middle English *ysop,* from Old English *ysope,* from Latin *hyssopus,* from Greek *hyssōpos,* of Semitic origin; akin to Hebrew *ēzōbh* hyssop] (before 12th century)
1 : a plant used in purificatory sprinkling rites by the ancient Hebrews
2 : a European mint (*Hyssopus officinalis*) that has highly aromatic and pungent leaves and is sometimes used as a potherb

hyster- *or* **hystero-** *combining form* [French or Latin; French *hystér-,* from Latin *hyster-,* from Greek, from *hystera*]
: womb ⟨*hyster*ectomy⟩

hys•ter•ec•to•my \ˌhis-tə-'rek-tə-mē\ *noun, plural* **-mies** (circa 1886)
: surgical removal of the uterus
— **hys•ter•ec•to•mized** \-tə-ˌmīzd\ *adjective*

hys•ter•e•sis \ˌhis-tə-'rē-səs\ *noun, plural* **-e•ses** \-ˌsēz\ [New Latin, from Greek *hysterēsis* shortcoming, from *hysterein* to be late, fall short, from *hysteros* later — more at OUT] (1881)
: a retardation of an effect when the forces acting upon a body are changed (as if from viscosity or internal friction); *especially* **:** a lagging in the values of resulting magnetization in a magnetic material (as iron) due to a changing magnetizing force
— **hys•ter•et•ic** \-'re-tik\ *adjective*

hys•te•ria \his-'ter-ē-ə, -'tir-\ *noun* [New Latin, from English *hysteric,* adjective, from Latin *hystericus,* from Greek *hysterikos,* from *hystera* womb; from the Greek notion that hysteria was peculiar to women and caused by disturbances of the uterus] (1801)
1 : a psychoneurosis marked by emotional excitability and disturbances of the psychic, sensory, vasomotor, and visceral functions
2 : behavior exhibiting overwhelming or unmanageable fear or emotional excess ⟨political *hysteria*⟩ ◆
— **hys•ter•ic** \-'ter-ik\ *noun*
— **hys•ter•i•cal** \-'ter-i-kəl\ *also* **hysteric** *adjective*
— **hys•ter•i•cal•ly** \-i-k(ə-)lē\ *adverb*

hys•ter•ics \-'ter-iks\ *noun plural but singular or plural in construction* (1721)
: a fit of uncontrollable laughter or crying

hys•ter•oid \'his-tə-ˌrȯid\ *adjective* (circa 1855)
: resembling or tending toward hysteria

hys•ter•on prot•er•on \ˌhis-tə-ˌrän-'prä-tə-ˌrän, -tə-rən-'prä-tə-rən, -'prō-\ *noun* [Late Latin, from Greek, literally, (the) later earlier, (the) latter first] (1565)
: a figure of speech consisting of the reversal of a natural or rational order (as in "then came the thunder and the lightning")

hys•ter•ot•o•my \ˌhis-tə-'rä-tə-mē\ *noun, plural* **-mies** [New Latin *hysterotomia,* from *hyster-* + *-tomia* -tomy] (1801)
: surgical incision of the uterus; *especially* **:** CESAREAN SECTION

◇ WORD HISTORY
hysteria Since the earliest days of civilization physicians have attempted to treat psychological disorders that manifest themselves physically, through tics and tremors, seizures, and a variety of other symptoms, including what sufferers experience as numbness and partial paralysis. Following ancient Egyptian medical lore, Greek physicians from Hippocrates to Galen believed that these ailments were unique to women and resulted from disturbances in the uterus. The collective labeling of such symptoms in Greek as *hysterika pathē* "sufferings of the womb," rendered in Latin as *hysterica passio,* gave rise to the term *hysteria* in English around the end of the 18th century, with counterparts in other European languages, such as French *hystérie* and German *Hysterie.* The boundaries of what constitutes hysteria, its diverse expression in different cultures, and the difficulties in separating it from genuine neurological disease are all issues that still preoccupy psychiatrists. The word itself has gradually been slipping from the usage of medical professionals. Most of the phenomena traditionally called *hysteria* are now put under the rubric *conversion reaction,* based on Freud's German term *Konversionshysterie,* which embodies his hypothesis that psychic anxiety is "converted" into bodily expression.

I is the ninth letter of the English alphabet and of most alphabets closely related to it. The character can be traced back through Latin, Etruscan and Greek (iota) to Phoenician yōdh, and probably ultimately to Egyptian. In Phoenician i represented a fricative (a consonant pronounced with air friction) articulated like the y in English yet, but it also had the value of a vowel in certain positions, and only the vowel pronunciation was taken over by the Greeks. In Latin and Old English, i had both vowel and consonant y sounds. Several hundred years later, i was used in English to spell both the vowel sound and the consonant sound j (as in modern jewel). By the 17th century, English used the J letter form (see J) exclusively for the consonant, and I exclusively for the vowel. Small i developed through a reduction in the size of the capital letter and the addition of an accent mark (later a dot) above the vertical stroke to distinguish it from adjacent letters (as m, n, and u) having identically formed strokes.

i \ˈī\ *noun, plural* **i's** *or* **is** \ˈīz\ *often capitalized, often attributive* (before 12th century)
1 a : the 9th letter of the English alphabet **b :** a graphic representation of this letter **c :** a speech counterpart of orthographic *i*
2 : ONE — see NUMBER table
3 : a graphic device for reproducing the letter *i*
4 : one designated *i* especially as the 9th in order or class
5 : something shaped like the letter I
6 : a unit vector parallel to the x-axis
7 [abbreviation for *incomplete*] **a :** a grade rating a student's work as incomplete **b :** one graded or rated with an I
8 : I FORMATION

¹I \ˈī, ə\ *pronoun* [Middle English, from Old English *ic;* akin to Old High German *ih* I, Latin *ego,* Greek *egō*] (before 12th century)
: the one who is speaking or writing ⟨*I* feel fine⟩ — compare ME, MINE, MY, WE
usage see ME

²I \ˈī\ *noun, plural* **I's** *or* **Is** \ˈīz\ (1539)
: someone aware of possessing a personal individuality **:** SELF

-i- [Middle English, from Old French, from Latin, thematic vowel of most nouns and adjectives in combination]
— used as a connective vowel to join word elements especially of Latin origin ⟨matr*i*lineal⟩ ⟨rat*i*cide⟩

¹-ia *noun suffix* [New Latin, from Latin & Greek, suffix forming feminine nouns]
1 : pathological condition ⟨hyster*ia*⟩
2 : genus of plants or animals ⟨Fuchs*ia*⟩
3 : territory **:** world **:** society ⟨suburb*ia*⟩

²-ia *noun plural suffix* [New Latin, from Latin (neuter plural of *-ius,* adjective ending) & Greek, neuter plural of *-ios,* adjective ending]
1 : higher taxon (as class or order) consisting of (such plants or animals) ⟨Saur*ia*⟩
2 : things derived from or relating to (something specified) ⟨militar*ia*⟩

³-ia *plural of* -IUM

Ia·go \ē-ˈä-(ˌ)gō\ *noun*
: the villain of Shakespeare's tragedy *Othello*

-ial *adjective suffix* [Middle English, from Middle French, from Latin *-ialis,* from *-i-* + *-alis* -al]
: ¹-AL ⟨manor*ial*⟩

iamb \ˈī-ˌam(b)\ *or* **iam·bus** \ī-ˈam-bəs\ *noun, plural* **iambs** \ˈī-ˌamz\ *or* **iam·bus·es** [Latin *iambus,* from Greek *iambos*] (1586)
: a metrical foot consisting of one short syllable followed by one long syllable or of one unstressed syllable followed by one stressed syllable (as in *above*)
— iam·bic \ī-ˈam-bik\ *adjective or noun*

-ian — see -AN

-iana — see -ANA

-iasis *noun suffix, plural* **-iases** [New Latin, from Latin, from Greek, suffix of action, from denominative verbs in *-ian, -iazein*]
: disease having characteristics of or produced by (something specified) ⟨onchocerc*iasis*⟩ ⟨ancylostom*iasis*⟩

-iatric *also* **-iatrical** *adjective combining form* [New Latin *-iatria*]
: of or relating to (such) medical treatment or healing ⟨ped*iatric*⟩

-iatrics *noun plural combining form but singular or plural in construction*
: medical treatment ⟨ped*iatrics*⟩

iat·ro·gen·ic \(ˌ)ī-ˌa-trə-ˈje-nik\ *adjective* [Greek *iatros* physician + English *-genic*] (1924)
: induced inadvertently by a physician or surgeon or by medical treatment or diagnostic procedures ⟨an *iatrogenic* rash⟩
— iat·ro·gen·i·cal·ly \-ˈje-ni-k(ə-)lē\ *adverb*

-iatry *noun combining form* [French *-iatrie,* from New Latin *-iatria,* from Greek *iatreia* art of healing, from *iatros* healer, from *iasthai* to heal]
: medical treatment **:** healing ⟨pod*iatry*⟩

I band \ˈī-\ *noun* (1948)
: an isotropic band of a striated muscle fiber

I beam *noun* (circa 1889)
: an iron or steel beam that is I-shaped in cross section

¹Ibe·ri·an \ī-ˈbir-ē-ən\ *noun* [*Iberia,* ancient region of the Caucasus] (1601)
: a member of one or more peoples anciently inhabiting the Caucasus in Asia between the Black and Caspian seas
— Iberian *adjective*

²Iberian *noun* [*Iberia,* peninsula in Europe] (1611)
1 a : a member of one or more peoples anciently inhabiting parts of the peninsula comprising Spain and Portugal **b :** a native or inhabitant of Spain or Portugal or the Basque region
2 : one or more of the languages of the ancient Iberians
— Iberian *adjective*

ibex \ˈī-ˌbeks\ *noun, plural* **ibex** *or* **ibex·es** [Latin] (1607)
: any of several wild goats (genus *Capra,* especially *C. ibex*) living chiefly in high mountain areas of the Old World and having large recurved horns transversely ridged in front

ibi·dem \ˈi-bə-ˌdem; i-ˈbī-dəm, -ˈbē-\ *adverb* [Latin] (circa 1771)
: in the same place

-ibility — see -ABILITY

ibis \ˈī-bəs\ *noun, plural* **ibis** *or* **ibis·es** [Middle English, from Latin, from Greek, from Egyptian *hb*] (14th century)
: any of various wading birds (family Threskiornithidae) related to the herons but distinguished by a long slender downwardly curved bill

ibis

Ibi·zan hound \i-ˈbē-zən-\ *noun* [*Ibiza* Island] (1960)
: any of a breed of slender agile medium-sized hunting dogs developed in the Balearic Islands with a short and smooth or a wire-haired coat

Ibizan hound

-ible — see -ABLE

Ibo \ˈē-(ˌ)bō\ *noun, plural* **Ibo** *or* **Ibos** (1732)
1 : a member of a people of the area around the lower Niger in Africa
2 : the language of the Ibo people

Ib·sen·ism \ˈib-sə-ˌni-zəm, ˈip-\ *noun* (1890)
1 : dramatic invention or construction characteristic of Ibsen
2 : championship of Ibsen's plays and ideas
— Ib·sen·ite \-ˌnīt\ *noun or adjective*

ibu·pro·fen \ˌī-byu̇-ˈprō-fən *also* ī-ˈbyü-prə-fən\ *noun* [*is-* + *bu*tyl + *pro*pionic acid + *-fen* (alteration of *phenyl*)] (1969)
: a nonsteroidal anti-inflammatory drug $C_{13}H_{18}O_2$ used to relieve pain and fever

¹IC \ˌī-ˈsē\ *noun* (1947)
: IMMEDIATE CONSTITUENT

²IC *noun* (1966)
: INTEGRATED CIRCUIT

¹-ic *adjective suffix* [Middle English, from Old French & Latin; Old French *-ique,* from Latin *-icus* — more at -Y]
1 : having the character or form of **:** being ⟨panoram*ic*⟩ **:** consisting of ⟨run*ic*⟩
2 a : of or relating to ⟨alderman*ic*⟩ **b :** related to, derived from, or containing ⟨alcohol*ic*⟩
3 : in the manner of **:** like that of **:** characteristic of ⟨Byron*ic*⟩
4 : associated or dealing with ⟨Ved*ic*⟩ **:** utilizing ⟨electron*ic*⟩
5 : characterized by **:** exhibiting ⟨nostalg*ic*⟩ **:** affected with ⟨allerg*ic*⟩
6 : caused by ⟨amoeb*ic*⟩
7 : tending to produce ⟨analges*ic*⟩
8 : having a valence relatively higher than in compounds or ions named with an adjective ending in *-ous* ⟨ferr*ic* iron⟩

²-ic *noun suffix*
: one having the character or nature of **:** one belonging to or associated with **:** one exhibiting or affected by **:** one that produces

-ical *adjective suffix* [Middle English, from Late Latin *-icalis* (as in *clericalis* clerical, *radicalis* radical)]
: -IC ⟨symmetr*ical*⟩ ⟨geolog*ical*⟩ — sometimes differing from *-ic* in that adjectives formed with *-ical* have a wider or more transferred semantic range than corresponding adjectives in *-ic*

Ic·a·rus \ˈi-kə-rəs\ *noun* [Latin, from Greek *Ikaros*]
: the son of Daedalus who to escape imprisonment flies by means of artificial wings but falls into the sea and drowns when the wax of his wings melts as he flies too near the sun

ICBM \ˌī-ˌsē-(ˌ)bē-ˈem\ *noun, plural* **ICBM's** *or* **ICBMs** \-ˈemz\ (1955)
: an intercontinental ballistic missile

¹ice \'īs\ *noun, often attributive* [Middle English *is,* from Old English *īs;* akin to Old High German *īs* ice, Avestan *isu-* icy] (before 12th century)
1 a : frozen water **b :** a sheet or stretch of ice
2 : a substance resembling ice; *especially* **:** the solid state of a substance usually found as a gas or liquid ⟨ammonia *ice* in the rings of Saturn⟩
3 : a state of coldness (as from formality or reserve)
4 a : a frozen dessert containing a flavoring (as fruit juice); *especially* **:** one containing no milk or cream **b** *British* **:** a serving of ice cream
5 *slang* **:** DIAMONDS; *broadly* **:** JEWELRY
6 : an undercover premium paid to a theater employee for choice theater tickets
— ice·less \'ī-sləs\ *adjective*
— on ice 1 : with every likelihood of being won or accomplished **2 :** in reserve or safekeeping

²ice *verb* **iced; ic·ing** (15th century)
transitive verb
1 a : to coat with or convert into ice **b :** to chill with ice **c :** to supply with ice
2 : to cover with or as if with icing
3 : to put on ice
4 : SECURE 1b ⟨made two free throws . . . to *ice* the win —Jack McCallum⟩
5 : to shoot (an ice hockey puck) the length of the rink and beyond the opponents' goal line
6 *slang* **:** KILL 1a
intransitive verb
1 : to become ice-cold
2 a : to become covered with ice — often used with *up* or *over* **b :** to have ice form inside ⟨the carburetor *iced* up⟩

ice age *noun* (1873)
1 : a time of widespread glaciation
2 *I&A capitalized* **:** the Pleistocene glacial epoch

ice ax *noun* (1820)
: a combination pick and adze with a spiked handle that is used in mountain climbing

ice ax

ice bag *noun* (1883)
: a waterproof bag to hold ice for local application of cold to the body

ice·berg \'īs-,bərg\ *noun* [probably part translation of Danish or Norwegian *isberg,* from *is* ice + *berg* mountain] (1820)
1 : a large floating mass of ice detached from a glacier
2 : an emotionally cold person
3 : ICEBERG LETTUCE

iceberg lettuce *noun* (1893)
: any of various crisp light green lettuces that when mature have the leaves arranged in a compact head

ice·blink \-,bliŋk\ *noun* (1817)
: a glare in the sky over an ice field

ice·boat \-,bōt\ *noun* (circa 1819)
: a skeleton boat or frame on runners propelled on ice usually by sails

ice·boat·ing \-,bō-tiŋ\ *noun* (1885)
: the sport of sailing in iceboats
— ice·boat·er \-,bō-tər\ *noun*

ice·bound \-,bau̇nd\ *adjective* (1641)
: surrounded, obstructed, or covered by ice

ice·box \-,bäks\ *noun* (1846)
: REFRIGERATOR

ice·break·er \-,brā-kər\ *noun* (1875)
1 : a ship equipped to make and maintain a channel through ice
2 : something that breaks the ice on a project or occasion; *especially* **:** MIXER 1c

ice cap *noun* (circa 1860)
1 : an ice bag shaped to the head
2 : a cover of perennial ice and snow; *specifically* **:** a glacier forming on an extensive area of relatively level land and flowing outward from its center

ice–cold \'īs-'kōld\ *adjective* (before 12th century)
: extremely cold

ice–cream *adjective* (1890)
: of a color similar to that of vanilla ice cream

ice cream \,īs-'krēm, 'īs-,\ *noun* (1744)
: a sweet flavored frozen food containing cream or butterfat and usually eggs

ice–cream chair *noun* [from its use in ice cream parlors] (1949)
: a small armless chair with a circular seat for use at a table (as at a café)

ice–cream cone *noun* (1909)
: a thin crisp edible cone for holding ice cream; *also* **:** one filled with ice cream

ice dancing *noun* (1925)
: a sport in which ice-skating pairs perform to music routines similar to ballroom dances

ice·fall \'īs-,fȯl\ *noun* (1817)
1 : a frozen waterfall
2 : the mass of usually jagged blocks into which a glacier may break when it moves down a steep declivity

ice field *noun* (1694)
1 : an extensive sheet of sea ice
2 : ICE CAP 2

ice floe *noun* (1819)
: a usually large flat free mass of floating sea ice

ice fog *noun* (1856)
: a fog composed of ice particles

ice hockey *noun* (1883)
: a game played on an ice rink by two teams of six players on skates whose object is to drive a puck into the opponents' goal with a hockey stick

ice·house \'īs-,hau̇s, 'ī-,sau̇s\ *noun* (1687)
: a building in which ice is made or stored

¹Ice·lan·dic \ī-'slan-dik\ *adjective* (1674)
: of, relating to, or characteristic of Iceland, the Icelanders, or Icelandic

²Icelandic *noun* (circa 1824)
: the North Germanic language of Iceland

Ice·land moss \'ī-slən(d)-, -,slan(d)-\ *noun* (1805)
: a lichen (*Cetraria islandica*) of mountainous and arctic regions sometimes used in medicine or as food

Iceland poppy *noun* (1884)
: a poppy (*Papaver nudicaule*) of holarctic regions often cultivated for its usually single showy flowers

Iceland spar *noun* (1771)
: a doubly refracting transparent calcite

ice·man \'īs-,man\ *noun* (1833)
1 : one who sells or delivers ice
2 : a man skilled in traveling on ice

ice milk *noun* (1947)
: a sweetened frozen food made of skim milk

Ice·ni \ī-'sē-,nī\ *noun plural* [Latin] (circa 1891)
: an ancient British people that under their queen Boudicca revolted against the Romans in A.D. 60
— Ice·ni·an \-'sē-nē-ən\ *or* **Ice·nic** \-'sē-nik, -'se-\ *adjective*

ice–out \'īs-,au̇t\ *noun* (1951)
: the disappearance of ice from the surface of a body of water (as a lake) as a result of thawing

ice pack *noun* (1853)
: an expanse of pack ice

ice pick *noun* (circa 1877)
: a hand tool ending in a spike for chipping ice

ice plant *noun* (1753)
: an Old World annual herb (*Mesembryanthemum crystallinum*) of the carpetweed family that has fleshy foliage covered with glistening papillate dots or vesicles and is widely naturalized in warm regions; *broadly* **:** FIG MARIGOLD

ice point *noun* (1903)
: the freezing point of water of 0° Celsius or 273.15 Kelvin at standard atmospheric pressure

ice sheet *noun* (1873)
: ICE CAP 2

ice show *noun* (1948)
: an entertainment consisting of various exhibitions by ice-skaters usually with musical accompaniment

ice–skate \'ī(s)-,skāt\ *intransitive verb* (circa 1948)
: to skate on ice
— ice–skat·er \-,skā-tər\ *noun*

ice skate *noun* (1897)
: a shoe with a metal runner attached for ice-skating

ice storm *noun* (1876)
: a storm in which falling rain freezes on contact

ice water *noun* (1722)
: chilled or iced water especially served as a beverage

ich·neu·mon \ik-'nü-mən, -'nyü-\ *noun* [Middle English, from Latin, from Greek *ichneumōn,* literally, tracker, from *ichneuein* to track, from *ichnos* footprint] (15th century)
1 : a mongoose (*Herpestes ichneumon*) of Africa, southern Europe, and southwestern Asia
2 : ICHNEUMON FLY

ichneumon fly *noun* (1713)
: any of a large superfamily (Ichneumonoidea) of hymenopterous insects whose larvae are usually internal parasites of other insect larvae and especially of caterpillars

ichor \'ī-,kȯr, -kər\ *noun* [Greek *ichōr*] (15th century)
1 : a thin watery or blood-tinged discharge
2 : an ethereal fluid taking the place of blood in the veins of the ancient Greek gods
— ichor·ous \-kə-rəs\ *adjective*

ichthy- *or* **ichthyo-** *combining form* [Latin, from Greek, from *ichthys;* akin to Armenian *jukn* fish, Lithuanian *žuvis*]
: fish ⟨*ichthy*ology⟩

ich·thyo·fau·na \,ik-thē-ō-'fȯ-nə, -'fä-\ *noun* [New Latin] (1883)
: the fish life of a region
— ich·thyo·fau·nal \-'fȯ-nᵊl, -'fä-\ *adjective*

ich·thy·ol·o·gy \,ik-thē-'ä-lə-jē\ *noun* (1646)
: a branch of zoology that deals with fishes
— ich·thy·o·log·i·cal \-thē-ə-'lä-ji-kəl\ *adjective*
— ich·thy·o·log·i·cal·ly \-k(ə-)lē\ *adverb*
— ich·thy·ol·o·gist \-thē-'ä-lə-jist\ *noun*

ich·thy·oph·a·gous \,ik-thē-'ä-fə-gəs\ *adjective* [Greek *ichthyophagos,* from *ichthy-* + *-phagos* -phagous] (circa 1828)
: eating or subsisting on fish

ich·thyo·saur \'ik-thē-ə-,sȯr\ *noun* [ultimately from Greek *ichthy-* + *sauros* lizard] (1830)
: any of an order (Ichthyosauria) of extinct marine reptiles of the Mesozoic specialized for aquatic life by a streamlined body with a long snout, limbs reduced to small fins for steering, and a large lunate caudal fin
— ich·thyo·sau·ri·an \,ik-thē-ə-'sȯr-ē-ən\ *adjective or noun*

-ician *noun suffix* [Middle English, from Old French *-icien,* from Latin *-ica* (as in *rhetorica* rhetoric) + Old French *-ien* -ian]
: specialist **:** practitioner ⟨beaut*ician*⟩

ici·cle \'ī-,si-kəl\ *noun* [Middle English *isikel,* from *is* ice + *ikel* icicle, from Old English *gicel;* akin to Old High German *ihilla* icicle, Middle Irish *aig* ice] (14th century)
1 : a pendent mass of ice formed by the freezing of dripping water
2 : an emotionally cold person

3 : a long narrow strip (as of foil) used to decorate a Christmas tree ◇

[1]ic·ing \ˈī-siŋ\ *noun* (1769)
1 : a sweet flavored usually creamy mixture used to coat baked goods (as cupcakes) — called also *frosting*
2 : something that adds to the interest, value, or appeal of an item or event — often used in the phrase *icing on the cake*

[2]icing *noun* (1948)
: an act by an ice-hockey player of shooting a puck from within the defensive zone or defensive half of the rink beyond the opponents' goal line but not into the goal

ick·er \ˈi-kər\ *noun* [(assumed) Middle English (Scots dialect), from Old English *ēar, eher* — more at EAR] (1513)
Scottish **:** a head of grain

icky \ˈi-kē\ *adjective* **ick·i·er; -est** [perhaps baby talk alteration of *sticky*] (1929)
: offensive to the senses or sensibilities **:** DISTASTEFUL ⟨put off by her *icky* triteness —Renata Adler⟩
— ick·i·ness \-nəs\ *noun*

icon \ˈī-ˌkän\ *noun* [Latin, from Greek *eikōn,* from *eikenai* to resemble] (1572)
1 : a usually pictorial representation **:** IMAGE
2 [Late Greek *eikōn,* from Greek] **:** a conventional religious image typically painted on a small wooden panel and used in the devotions of Eastern Christians
3 : an object of uncritical devotion **:** IDOL
4 : EMBLEM, SYMBOL ⟨the house became an *icon* of 1860's residential architecture —Paul Goldberger⟩
5 a : a sign (as a word or graphic symbol) whose form suggests its meaning **b :** a graphic symbol on a computer display screen that suggests the purpose of an available function
— icon·ic \ī-ˈkä-nik\ *adjective*
— icon·i·cal·ly \-ni-k(ə-)lē\ *adverb*

icon- *or* **icono-** *combining form* [Greek *eikon-, eikono-,* from *eikon-, eikōn*]
: image ⟨*icono*latry⟩

ico·nic·i·ty \ˌī-kə-ˈni-sə-tē\ *noun* (1946)
: correspondence between form and meaning ⟨the *iconicity* of the Roman numeral III⟩

icon·o·clasm \ī-ˈkä-nə-ˌkla-zəm\ *noun* (1797)
: the doctrine, practice, or attitude of an iconoclast

icon·o·clast \-ˌklast\ *noun* [Medieval Latin *iconoclastes,* from Middle Greek *eikonoklastēs,* literally, image destroyer, from Greek *eikono-* + *klan* to break — more at CLAST] (1641)
1 : one who destroys religious images or opposes their veneration
2 : one who attacks settled beliefs or institutions
— icon·o·clas·tic \(ˌ)ī-ˌkä-nə-ˈklas-tik\ *adjective*
— icon·o·clas·ti·cal·ly \-ti-k(ə-)lē\ *adverb*

ico·nog·ra·pher \ˌī-kə-ˈnä-grə-fər\ *noun* (1888)
1 : a maker of figures or drawings especially of a conventional type
2 : a student of iconography

icon·o·graph·ic \(ˌ)ī-ˌkä-nə-ˈgra-fik\ *or* **icon·o·graph·i·cal** \-fi-kəl\ *adjective* (circa 1855)
1 : of or relating to iconography
2 : representing something by pictures or diagrams
— icon·o·graph·i·cal·ly \-fi-k(ə-)lē\ *adverb*

ico·nog·ra·phy \ˌī-kə-ˈnä-grə-fē\ *noun, plural* **-phies** [Medieval Latin *iconographia,* from Greek *eikonographia* sketch, description, from *eikonographein* to describe, from *eikon-* + *graphein* to write — more at CARVE] (1678)
1 : pictorial material relating to or illustrating a subject
2 : the traditional or conventional images or symbols associated with a subject and especially a religious or legendary subject
3 : the imagery or symbolism of a work of art, an artist, or a body of art
4 : ICONOLOGY

ico·nol·a·try \-ˈnä-lə-trē\ *noun* (1624)
: the worship of images or icons

ico·nol·o·gy \-ˈnä-lə-jē\ *noun* [French *iconologie,* from *icono-* icon- + *-logie* -logy] (circa 1736)
: the study of icons or artistic symbolism
— icon·o·log·i·cal \(ˌ)ī-ˌkä-nᵊl-ˈä-ji-kəl\ *adjective*

icon·o·scope \ī-ˈkä-nə-ˌskōp\ *noun* [from *Iconoscope,* a trademark] (1932)
: a camera tube containing an electron gun and a photoemissive mosaic screen of which each cell produces a charge proportional to the varying light intensity of the image focused on the screen

ico·nos·ta·sis \ˌī-kə-ˈnäs-tə-səs, (ˌ)ī-ˈkä-nə-ˌstä-səs\ *noun, plural* **-ta·ses** \-ˌsēz\ [modification of Middle Greek *eikonostasion,* from Late Greek, shrine, from Greek *eikono-* + *-stasion* (from *histanai* to stand) — more at STAND] (1833)
: a screen or partition with doors and tiers of icons that separates the bema from the nave in Eastern churches

ico·sa·he·dral \(ˌ)ī-ˌkō-sə-ˈhē-drəl, -ˌkä-\ *adjective* (circa 1828)
: of or having the form of an icosahedron

ico·sa·he·dron \-drən\ *noun, plural* **-drons** *or* **-dra** \-drə\ [Greek *eikosaedron,* from *eikosi* twenty + *-edron* -hedron — more at VIGESIMAL] (1570)
: a polyhedron having 20 faces

-ics *noun plural suffix but singular or plural in construction* [[1]*-ic* + [2]*-s;* translation of Greek *-ika,* from neuter plural of *-ikos* -ic]
1 : study **:** knowledge **:** skill **:** practice ⟨linguis*tics*⟩ ⟨electron*ics*⟩
2 : characteristic actions or activities ⟨acrobat*ics*⟩
3 : characteristic qualities, operations, or phenomena ⟨mechan*ics*⟩

ic·ter·ic \ik-ˈter-ik\ *adjective* (circa 1600)
: of, relating to, or affected with jaundice

ic·ter·us \ˈik-tə-rəs\ *noun* [New Latin, from Greek *ikteros*] (circa 1706)
: JAUNDICE

ic·tus \ˈik-təs\ *noun* [Latin, literally, blow, from *icere* to strike] (1752)
: the recurring stress or beat in a rhythmic or metrical series of sounds

icy \ˈī-sē\ *adjective* **ic·i·er; -est** (before 12th century)
1 a : covered with, abounding in, or consisting of ice **b :** intensely cold
2 : characterized by coldness **:** FRIGID ⟨an *icy* stare⟩; *also* **:** STEELY ⟨*icy* nerves⟩
— ic·i·ly \-sə-lē\ *adverb*
— ic·i·ness \-sē-nəs\ *noun*

id \ˈid\ *noun* [New Latin, from Latin, it] (1924)
: the one of the three divisions of the psyche in psychoanalytic theory that is completely unconscious and is the source of psychic energy derived from instinctual needs and drives — compare EGO, SUPEREGO

[1]-id *noun suffix* [in sense 1, from Latin *-ides,* masculine patronymic suffix, from Greek *-idēs;* in sense 2, from Italian *-ide,* from Latin *-id-, -is,* feminine patronymic suffix, from Greek]
1 : one belonging to a (specified) dynastic line ⟨Abbas*id*⟩
2 : meteor associated with or radiating from a (specified) constellation or comet ⟨Perse*id*⟩

[2]-id *noun suffix* [probably from Latin *-id-, -is,* feminine patronymic suffix, from Greek]
: body **:** particle ⟨chromat*id*⟩

I'd \ˈīd\ (circa 1592)
: I would **:** I had **:** I should

-idae *noun plural suffix* [New Latin, from Latin, from Greek *-idai,* plural of *-idēs*]
: members of the family of — in names of zoological families ⟨Fel*idae*⟩

Ida·ho \ˈī-də-ˌhō\ *noun, plural* **Idahos** *or* **Idahoes** (1934)
: an elongated baking potato grown especially in the state of Idaho

ID card \ˈī-ˈdē-\ *noun* (circa 1945)
: a card bearing identifying data (as age or organizational membership) about the individual whose name appears thereon — called also *identification card, identity card*

-ide *also* **-id** *noun suffix* [German & French; German *-id,* from French *-ide* (as in *oxide*)]
1 : binary chemical compound — added to the contracted name of the nonmetallic or more electronegative element ⟨hydrogen sul*fide*⟩ or group ⟨cyan*ide*⟩
2 : chemical compound derived from or related to another (usually specified) compound ⟨anhydr*ide*⟩ ⟨glucos*ide*⟩

idea \ī-ˈdē-ə *also* ˈī-(ˌ)dē-ə *or* ˈī-dē\ *noun* [Middle English, from Latin, from Greek, from *idein* to see — more at WIT] (14th century)
1 a : a transcendent entity that is a real pattern of which existing things are imperfect representations **b :** a standard of perfection **:** IDEAL **c :** a plan for action **:** DESIGN
2 *archaic* **:** a visible representation of a conception **:** a replica of a pattern
3 a *obsolete* **:** an image recalled by memory **b :** an indefinite or unformed conception **c :** an entity (as a thought, concept, sensation, or image) actually or potentially present to consciousness
4 : a formulated thought or opinion
5 : whatever is known or supposed about something ⟨a child's *idea* of time⟩
6 : the central meaning or chief end of a particular action or situation
7 *Christian Science* **:** an image in Mind ☆

☆ SYNONYMS
Idea, concept, conception, thought, notion, impression mean what exists in the mind as a representation (as of something comprehended) or as a formulation (as of a plan). IDEA may apply to a mental image or formulation of something seen or known or imagined, to a pure abstraction, or to something assumed or vaguely sensed ⟨innovative *ideas*⟩ ⟨my *idea* of paradise⟩. CONCEPT may apply to the idea formed by consideration of instances of a species or genus or, more broadly, to any idea of what a thing ought to be ⟨a society with no *concept* of private property⟩. CONCEPTION is often interchangeable with CONCEPT; it may stress the process of imagining or formulating rather than the result ⟨our changing *conception* of what constitutes art⟩. THOUGHT is likely to suggest the result of reflecting, reasoning, or meditating rather than of imagining ⟨commit your *thoughts* to paper⟩. NOTION suggests an idea not much resolved by analysis or reflection and may suggest the capricious or accidental ⟨you have the oddest *notions*⟩. IMPRESSION applies to an idea or notion resulting immediately from some stimulation of the senses ⟨the first *impression* is of soaring height⟩.

◇ WORD HISTORY
icicle The Old English noun *gicel* is found as early as the first half of the 8th century translating Latin *stiria* "icicle." *Gicel* became Middle English *ikyl* or *ikel* and later Modern English *ickle,* which was still in use around the middle of this century in Derbyshire and Yorkshire. The word for *ice* in Old English was *is,* and in a manuscript of about the year 1000 we find *stiria* translated with seeming redundancy as *ises gicel,* that is, "icicle of ice." Some three hundred years later in Middle English appeared the compound that we know today as *icicle,* which means precisely what *ises gicel* meant a thousand years ago.

— **idea·less** \ī-ˈdē-ə-ləs\ *adjective*

[1]**ide·al** \ī-ˈdē(-ə)l, ˈī-ˌ\ *adjective* [Middle English *ydeall,* from Late Latin *idealis,* from Latin *idea*] (15th century)
1 : existing as an archetypal idea
2 a : existing as a mental image or in fancy or imagination only; *broadly* **:** lacking practicality **b :** relating to or constituting mental images, ideas, or conceptions
3 a : of, relating to, or embodying an ideal **b :** conforming exactly to an ideal, law, or standard **:** PERFECT ⟨an *ideal* gas⟩ — compare REAL 2b(3)
4 : of or relating to philosophical idealism

[2]**ideal** *noun* (15th century)
1 : a standard of perfection, beauty, or excellence
2 : one regarded as exemplifying an ideal and often taken as a model for imitation
3 : an ultimate object or aim of endeavor **:** GOAL
4 : a subset of a mathematical ring that is closed under addition and subtraction and contains the products of any given element of the subset with each element of the ring
synonym see MODEL
— **ide·al·less** \ī-ˈdē(-ə)l-ləs\ *adjective*

ide·al·ise *British variant of* IDEALIZE

ide·al·ism \ī-ˈdē-(ə-)ˌliz-əm, ˈī-(ˌ)dē-\ *noun* (1796)
1 a (1) **:** a theory that ultimate reality lies in a realm transcending phenomena (2) **:** a theory that the essential nature of reality lies in consciousness or reason **b** (1) **:** a theory that only the perceptible is real (2) **:** a theory that only mental states or entities are knowable
2 a : the practice of forming ideals or living under their influence **b :** something that is idealized
3 : literary or artistic theory or practice that affirms the preeminent value of imagination as compared with faithful copying of nature — compare REALISM

[1]**ide·al·ist** \-(ə-)list\ *noun* (1701)
1 a : an adherent of a philosophical theory of idealism **b :** an artist or author who advocates or practices idealism in art or writing
2 : one guided by ideals; *especially* **:** one that places ideals before practical considerations

[2]**idealist** *adjective* (1875)
: IDEALISTIC

ide·al·is·tic \(ˌ)ī-ˌdē-(ə-)ˈlis-tik, ˌī-dē-\ *adjective* (1829)
: of or relating to idealists or idealism
— **ide·al·is·ti·cal·ly** \-ti-k(ə-)lē\ *adverb*

ide·al·i·ty \ˌī-dē-ˈa-lə-tē\ *noun, plural* **-ties** (1817)
1 a : the quality or state of being ideal **b :** existence only in idea
2 : something imaginary or idealized

ide·al·ize \ī-ˈdē-(ə-)ˌlīz\ *verb* **-ized; -iz·ing** (1786)
intransitive verb
1 : to form ideals
2 : to work idealistically
transitive verb
1 a : to give an ideal form or value to **b :** to attribute ideal characteristics to
2 : to treat idealistically
— **ide·al·i·za·tion** \-ˌdē-(ə-)lə-ˈzā-shən\ *noun*
— **ide·al·iz·er** \-ˈdē-(ə-)ˌlī-zər\ *noun*

ide·al·ly \ī-ˈdē-ə-lē, -ˈdē(-ə)l-lē\ *adverb* (1598)
1 : in idea or imagination **:** MENTALLY
2 : in relation to an exemplar
3 a : conformably to or in respect to an ideal **:** PERFECTLY **b :** for best results ⟨*ideally,* the counselor should vary his techniques for each applicant —T. M. Martinez⟩ **c :** in accordance with an ideal or typical standard **:** CLASSICALLY

ideal point *noun* (1879)
: a point added to the plane or to space to eliminate special cases; *specifically* **:** the point at infinity added in projective geometry as the assumed intersection of two parallel lines

ide·ate \ˈī-dē-ˌāt\ *verb* **-at·ed; -at·ing** (1610)
transitive verb
: to form an idea or conception of
intransitive verb
: to form an idea

ide·a·tion \ˌī-dē-ˈā-shən\ *noun* (1818)
: the forming of ideas (as of things not present to the senses)

ide·a·tion·al \-shnəl, -shə-n^{ə}l\ *adjective* (1853)
: of, relating to, or produced by ideation; *broadly* **:** of or relating to ideas
— **ide·a·tion·al·ly** *adverb*

idée fixe \(ˌ)ē-ˌdā-ˈfēks\ *noun, plural* **idées fixes** *same*\ [French, literally, fixed idea] (1836)
: an idea that dominates one's mind especially for a prolonged period **:** OBSESSION

idem \ˈī-ˌdem, ˈē-, ˈi-\ *pronoun* [Middle English, from Latin, same — more at IDENTITY] (14th century)
: something previously mentioned **:** SAME

idem·po·tent \ˈī-dəm-ˌpō-t^{ə}nt\ *adjective* [Latin *idem* same + *potent-, potens* having power — more at POTENT] (1870)
: relating to or being a mathematical quantity which when applied to itself under a given binary operation (as multiplication) equals itself; *also* **:** relating to or being an operation under which a mathematical quantity is idempotent
— **idempotent** *noun*

iden·tic \ī-ˈden-tik, ə-\ *adjective* (1649)
: IDENTICAL: as **a :** constituting a diplomatic action or expression in which two or more governments follow precisely the same course or employ an identical form **b :** constituting an action or expression in which a government follows precisely the same course or employs identical forms with reference to two or more other governments

iden·ti·cal \ī-ˈden-ti-kəl, ə-\ *adjective* [probably from Medieval Latin *identicus,* from Late Latin *identitas*] (1599)
1 : being the same **:** SELFSAME ⟨the *identical* place we stopped before⟩
2 : having such close resemblance as to be essentially the same ⟨*identical* hats⟩ — often used with *to* or *with*
3 a : having the same cause or origin ⟨*identical* infections⟩ **b :** MONOZYGOTIC
synonym see SAME
— **iden·ti·cal·ly** \-k(ə-)lē\ *adverb*
— **iden·ti·cal·ness** \-kəl-nəs\ *noun*

iden·ti·fi·ca·tion \ī-ˌden-tə-fə-ˈkā-shən, ə-\ *noun* (1644)
1 a : an act of identifying **:** the state of being identified **b :** evidence of identity
2 a : psychological orientation of the self in regard to something (as a person or group) with a resulting feeling of close emotional association **b :** a largely unconscious process whereby an individual models thoughts, feelings, and actions after those attributed to an object that has been incorporated as a mental image

identification card *noun* (1908)
: ID CARD

iden·ti·fi·er \ī-ˈden-tə-ˌfī(-ə)r, ə-\ *noun* (1889)
: one that identifies

iden·ti·fy \ī-ˈden-tə-ˌfī, ə-\ *verb* **-fied; -fy·ing** (1644)
transitive verb
1 a : to cause to be or become identical **b :** to conceive as united (as in spirit, outlook, or principle) ⟨groups that are *identified* with conservation⟩
2 a : to establish the identity of **b :** to determine the taxonomic position of (a biological specimen)
intransitive verb
1 : to be or become the same
2 : to practice psychological identification ⟨*identify* with the hero of a novel⟩
— **iden·ti·fi·able** \-ˌden-tə-ˈfī-ə-bəl\ *adjective*
— **iden·ti·fi·ably** \-blē\ *adverb*

iden·ti·ty \ī-ˈden-tə-tē, ə-, -ˈde-nə-\ *noun, plural* **-ties** [Middle French *identité,* from Late Latin *identitat-, identitas,* probably from Latin *identidem* repeatedly, contraction of *idem et idem,* literally, same and same] (1570)
1 a : sameness of essential or generic character in different instances **b :** sameness in all that constitutes the objective reality of a thing **:** ONENESS
2 a : the distinguishing character or personality of an individual **:** INDIVIDUALITY **b :** the relation established by psychological identification
3 : the condition of being the same with something described or asserted ⟨establish the *identity* of stolen goods⟩
4 : an equation that is satisfied for all values of the symbols
5 : IDENTITY ELEMENT

identity card *noun* (1900)
: ID CARD

identity crisis *noun* (1954)
1 : personal psychosocial conflict especially in adolescence that involves confusion about one's social role and often a sense of loss of continuity to one's personality
2 : a state of confusion in an institution or organization regarding its nature or direction

identity element *noun* (1902)
: an element (as 0 in the set of all integers under addition or 1 in the set of positive integers under multiplication) that leaves any element of the set to which it belongs unchanged when combined with it by a specified operation

identity matrix *noun* (circa 1929)
: a square matrix that has numeral 1's along the principal diagonal and 0's elsewhere

ideo- *combining form* [French *idéo-,* from Greek *idea*]
: idea ⟨*ideo*gram⟩

ideo·gram \ˈi-dē-ə-ˌgram, ˈī-\ *noun* (circa 1840)
1 : a picture or symbol used in a system of writing to represent a thing or an idea but not a particular word or phrase for it; *especially* **:** one that represents not the object pictured but some thing or idea that the object pictured is supposed to suggest
2 : LOGOGRAM
— **ideo·gram·ic** *or* **ideo·gram·mic** \ˌi-dē-ə-ˈgra-mik, ˌī-\ *adjective*
— **ideo·gram·mat·ic** \-dē-ō-grə-ˈma-tik\ *adjective*

ideo·graph \ˈi-dē-ə-ˌgraf, ˈī-\ *noun* (1835)
: IDEOGRAM
— **ideo·graph·ic** \ˌi-dē-ə-ˈgra-fik, ˌī-\ *adjective*
— **ideo·graph·i·cal·ly** \-fi-k(ə-)lē\ *adverb*

ide·og·ra·phy \ˌi-dē-ˈä-grə-fē, ˌī-\ *noun* (1836)
1 : the use of ideograms
2 : the representation of ideas by graphic symbols

ideo·log·i·cal \ˌī-dē-ə-ˈlä-ji-kəl, ˌi-\ *also* **ideo·log·ic** \-ˈlä-jik\ *adjective* (1797)
1 : relating to or concerned with ideas
2 : of, relating to, or based on ideology
— **ideo·log·i·cal·ly** \-ˈlä-ji-k(ə-)lē\ *adverb*

ide·ol·o·gize \ˌī-dē-ˈä-lə-ˌjīz, ˌi-\ *transitive verb* **-gized; -giz·ing** (1860)
: to give an ideological character or interpretation to; *especially* **:** to change or interpret in relation to a sociopolitical ideology often seen as biased or limited

ideo·logue *also* **idea·logue** \ˈī-dē-ə-ˌlȯg, -ˌläg\ *noun* [French *idéologue,* back-formation from *idéologie*] (1815)
1 : an impractical idealist **:** THEORIST
2 : an often blindly partisan advocate or adherent of a particular ideology

ide·ol·o·gy \ˌī-dē-ˈä-lə-jē, ˌi-\ *also* **ide·al·o·gy** \-ˈä-lə-jē, -ˈa-\ *noun, plural* **-gies** [French *idéologie,* from *idéo-* ideo- + *-logie* -logy] (1813)
1 : visionary theorizing
2 a : a systematic body of concepts especially about human life or culture **b :** a manner or the content of thinking characteristic of an individual, group, or culture **c :** the integrated assertions, theories and aims that constitute a sociopolitical program
— **ide·ol·o·gist** \-jist\ *noun*

ideo·mo·tor \ˌī-dē-ə-ˈmō-tər, ˌi-\ *adjective* [International Scientific Vocabulary] (1867)
: not reflex but motivated by an idea ⟨*ideomotor* muscular activity⟩

ides \ˈīdz\ *noun plural but singular or plural in construction* [Middle English, from Middle French, from Latin *idus*] (14th century)
: the 15th day of March, May, July, or October or the 13th day of any other month in the ancient Roman calendar; *broadly* **:** this day and the seven days preceding it

-idin *or* **-idine** *noun suffix* [International Scientific Vocabulary *-ide* + *-in, -ine*]
: chemical compound related in origin or structure to another compound ⟨tolu*idine*⟩ ⟨guan*idine*⟩

idio- *combining form* [Greek, from *idios* — more at IDIOT]
: one's own **:** personal **:** separate **:** distinct ⟨*idio*blast⟩

id·i·o·blast \ˈi-dē-ə-ˌblast\ *noun* [International Scientific Vocabulary] (1822)
: a plant cell (as a sclereid) that differs markedly from neighboring cells
— **id·io·blas·tic** \ˌi-dē-ə-ˈblas-tik\ *adjective*

id·i·o·cy \ˈi-dē-ə-sē\ *noun, plural* **-cies** (circa 1529)
1 : extreme mental retardation commonly due to incomplete or abnormal development of the brain
2 : something notably stupid or foolish

id·i·o·graph·ic \ˌi-dē-ə-ˈgra-fik\ *adjective* [International Scientific Vocabulary] (circa 1890)
: relating to or dealing with something concrete, individual, or unique

id·i·o·lect \ˈi-dē-ə-ˌlekt\ *noun* [*idio-* + *-lect* (as in *dialect*)] (1948)
: the language or speech pattern of one individual at a particular period of life
— **id·io·lec·tal** \ˌi-dē-ə-ˈlek-t^{ə}l\ *adjective*

id·i·om \ˈi-dē-əm\ *noun* [Middle French & Late Latin; Middle French *idiome,* from Late Latin *idioma* individual peculiarity of language, from Greek *idiōmat-, idiōma,* from *idiousthai* to appropriate, from *idios*] (1588)
1 a : the language peculiar to a people or to a district, community, or class **:** DIALECT **b :** the syntactical, grammatical, or structural form peculiar to a language
2 : an expression in the usage of a language that is peculiar to itself either grammatically (as *no, it wasn't me*) or in having a meaning that cannot be derived from the conjoined meanings of its elements (as *Monday week* for "the Monday a week after next Monday")
3 : a style or form of artistic expression that is characteristic of an individual, a period or movement, or a medium or instrument ⟨the modern jazz *idiom*⟩; *broadly* **:** MANNER, STYLE ⟨a new culinary *idiom*⟩

id·i·o·mat·ic \ˌi-dē-ə-ˈma-tik\ *adjective* (1712)
1 : of, relating to, or conforming to idiom
2 : peculiar to a particular group, individual, or style
— **id·i·o·mat·i·cal·ly** \-ti-k(ə-)lē\ *adverb*
— **id·i·o·mat·ic·ness** \-tik-nəs\ *noun*

id·i·o·mor·phic \ˌi-dē-ə-ˈmȯr-fik\ *adjective* [Greek *idiomorphos,* from *idio-* + *-morphos* -morphous] (1887)
: having the proper form or shape — used of minerals whose crystalline growth has not been interfered with

id·i·o·path·ic \ˌi-dē-ə-ˈpa-thik\ *adjective* (1669)
1 : arising spontaneously or from an obscure or unknown cause **:** PRIMARY
2 : peculiar to the individual
— **id·i·o·path·i·cal·ly** \-ˈpa-thi-k(ə-)lē\ *adverb*

id·i·o·syn·cra·sy \ˌi-dē-ə-ˈsiŋ-krə-sē\ *noun, plural* **-sies** [Greek *idiosynkrasia,* from *idio-* + *synkerannynai* to blend, from *syn-* + *kerannynai* to mingle, mix — more at CRATER] (1604)
1 a : a peculiarity of constitution or temperament **:** an individualizing characteristic or quality **b :** individual hypersensitiveness (as to a drug or food)
2 : characteristic peculiarity (as of temperament); *broadly* **:** ECCENTRICITY
— **id·i·o·syn·crat·ic** \ˌi-dē-ō-(ˌ)sin-ˈkra-tik\ *adjective*
— **id·i·o·syn·crat·i·cal·ly** \-ˈkra-ti-k(ə-)lē\ *adverb*

id·i·ot \ˈi-dē-ət\ *noun* [Middle English, from Latin *idiota* ignorant person, from Greek *idiōtēs* one in a private station, layman, ignorant person, from *idios* one's own, private; akin to Latin *suus* one's own — more at SUICIDE] (14th century)
1 : a person affected with idiocy; *especially* **:** a feebleminded person having a mental age not exceeding three years and requiring complete custodial care
2 : a foolish or stupid person ◆
— **idiot** *adjective*

idiot box *noun* (circa 1955)
: TELEVISION

id·i·ot·ic \ˌi-dē-ˈä-tik\ *also* **id·i·ot·i·cal** \-ˈä-ti-kəl\ *adjective* (1713)
1 : characterized by idiocy
2 : showing complete lack of thought or common sense **:** FOOLISH
— **id·i·ot·i·cal·ly** \-ti-k(ə-)lē\ *adverb*

[1]**id·i·o·tism** \ˈi-dē-ə-ˌti-zəm\ *noun* [Middle French *idiotisme,* from Latin *idiotismus* common speech, from Greek *idiōtismos,* from *idiōtēs*] (1588)
1 *obsolete* **:** IDIOM 1
2 : IDIOM 2

[2]**id·i·ot·ism** \ˈi-dē-ə-(ˌ)ti-zəm\ *noun* [*idiot* + *-ism*] (circa 1611)
archaic **:** IDIOCY 2

idiot light *noun* (1966)
: a colored light on an instrument panel (as of an automobile) designed to give a warning (as of low oil pressure)

idiot sa·vant \ˈē-ˌdyō-sä-ˈvä^{n}, *or same as* IDIOT *and* SAVANT *for respective singular and plural forms*\ *noun, plural* **idiots savants** \-ˌdyō-sä-ˈvä^{n}(z)\ *or* **idiot savants** \-ˈvä^{n}(z)\ [French, literally, learned idiot] (1927)
1 : a mentally defective person who exhibits exceptional skill or brilliance in some limited field
2 : a person who is highly knowledgeable about one subject but knows little about anything else

-idium *noun suffix, plural* **-idiums** *or* **-idia** [New Latin, from Greek *-idion,* diminutive suffix]
: small one ⟨anther*idium*⟩

[1]**idle** \ˈī-d^{ə}l\ *adjective* **idler** \ˈīd-lər, ˈī-d^{ə}l-ər\; **idlest** \ˈīd-ləst, ˈī-d^{ə}l-əst\ [Middle English *idel,* from Old English *īdel;* akin to Old High German *ītal* worthless] (before 12th century)
1 : lacking worth or basis **:** VAIN ⟨*idle* chatter⟩ ⟨*idle* pleasure⟩
2 : not occupied or employed: as **a :** having no employment **:** INACTIVE ⟨*idle* workers⟩ **b :** not turned to normal or appropriate use ⟨*idle* funds⟩ ⟨*idle* farmland⟩ **c :** not scheduled to compete ⟨the team will be *idle* tomorrow⟩
3 a : SHIFTLESS, LAZY **b :** having no evident lawful means of support
synonym see VAIN, INACTIVE
— **idle·ness** \ˈī-d^{ə}l-nəs\ *noun*
— **idly** \ˈīd-lē, ˈī-d^{ə}l-ē\ *adverb*

[2]**idle** *verb* **idled; idling** \ˈīd-liŋ, ˈī-d^{ə}l-iŋ\ (1592)
intransitive verb
1 a : to spend time in idleness **b :** to move idly
2 : to run at low power and often disconnected usually so that power is not used for useful work ⟨the engine is *idling*⟩
transitive verb
1 : to pass in idleness
2 : to make idle ⟨workers *idled* by a strike⟩
3 : to cause to idle
— **idler** \ˈīd-lər, ˈī-d^{ə}l-ər\ *noun*

idler pulley *noun* (circa 1890)
: a guide or tightening pulley for a belt or chain

idler wheel *noun* (1805)
1 : a wheel, gear, or roller used to transfer motion or to guide or support something
2 : IDLER PULLEY

idler wheel 1

idlesse \ˈīd-ləs, īd-ˈles\ *noun* [Middle English, from *idle* + *-esse* (as in *richesse* wealth) — more at RICHES] (15th century)
: the quality or state of being idle **:** IDLENESS

ido·crase \ˈī-də-ˌkrās, ˈi-, -ˌkrāz\ *noun* [French, from Greek *eidos* form + *krasis* mixture, from *kerannynai* to mix — more at CRATER] (1804)
: a mineral that is a complex silicate of calcium, magnesium, iron, and aluminum

idol \ˈī-d^{ə}l\ [Middle English, from Old French *idole,* from Late Latin *idolum,* from Greek *eidōlon* image, idol; akin to Greek *eidos* form — more at IDYLL] (13th century)
1 : a representation or symbol of an object of worship; *broadly* **:** a false god
2 a : a likeness of something **b** *obsolete* **:** PRETENDER, IMPOSTOR
3 : a form or appearance visible but without substance ⟨an enchanted phantom, a lifeless *idol* —P. B. Shelley⟩
4 : an object of extreme devotion ⟨a movie *idol*⟩; *also* **:** IDEAL 2
5 : a false conception **:** FALLACY

idol·a·ter *or* **idol·a·tor** \ī-ˈdä-lə-tər\ *noun* [Middle English *idolatrer,* from Middle French *idolatre,* from Late Latin *idololatres,* from Greek *eidōlolatrēs,* from *eidōlon* + *-latrēs* -later] (14th century)
1 : a worshiper of idols
2 : a person that admires intensely and often blindly one that is not usually a subject of worship

idol·a·trous \ī-ˈdä-lə-trəs\ *adjective* (1543)
1 : of or relating to idolatry
2 : having the character of idolatry ⟨the religion of *idolatrous* nationalism —Aldous Huxley⟩
3 : given to idolatry
— **idol·a·trous·ly** *adverb*

◇ WORD HISTORY
idiot To be called "one's own man" is usually to receive a compliment. To be called an *idiot* is a blatant insult. Surprisingly, there is an etymological equivalence in the two characterizations. In Greek *idiōtēs,* a derivative of *idios* "one's own," meant "a private person," that is, one who did not hold public office. Through a series of natural extensions, *idiōtēs* came to mean "a common man" and then "an ignorant person." The imputation of ignorance is more benign than one might suspect, for it implied a lack of book learning, not innate stupidity. The word became *idiota* in Latin, from whence it was borrowed into Middle English in the then-standard sense of "an ignorant person." By the 16th century, however, the meaning of *idiot* had deteriorated to the point where it no longer meant a lack of education but rather of basic intelligence.

— **idol·a·trous·ness** *noun*

idol·a·try \-trē\ *noun, plural* **-tries** (13th century)
1 : the worship of a physical object as a god
2 : immoderate attachment or devotion to something

idol·ize \ˈī-dᵊl-ˌīz\ *verb* **-ized; -iz·ing** (1598)
transitive verb
: to worship as a god; *broadly* **:** to love or admire to excess ⟨the common people whom he so *idolized* —*Times Literary Supplement*⟩
intransitive verb
: to practice idolatry
— **idol·i·za·tion** \ˌī-dᵊl-ə-ˈzā-shən\ *noun*
— **idol·iz·er** \ˈī-dᵊl-ˌī-zər\ *noun*

idyll *also* **idyl** \ˈī-dᵊl, *British usually* ˈi-(ˌ)dil\ *noun* [Latin *idyllium,* from Greek *eidyllion,* from diminutive of *eidos* form; akin to Greek *idein* to see — more at WIT] (1586)
1 a : a simple descriptive work in poetry or prose that deals with rustic life or pastoral scenes or suggests a mood of peace and contentment **b :** a narrative poem (as Tennyson's *Idylls of the King*) treating an epic, romantic, or tragic theme
2 a : a lighthearted carefree episode that is a fit subject for an idyll **b :** a romantic interlude

idyl·lic \ī-ˈdi-lik, *chiefly British* i-\ *adjective* (1856)
1 : of, relating to, or being an idyll
2 : pleasing or picturesque in natural simplicity
— **idyl·li·cal·ly** \-ˈdi-li-k(ə-)lē\ *adverb*

-ie *also* **-y** *noun suffix* [Middle English]
1 : little one **:** dear little one ⟨bird*ie*⟩ ⟨sonn*y*⟩
2 : one belonging to **:** one having to do with ⟨town*ie*⟩ **:** one who is ⟨preem*ie*⟩
3 : one of (such) a kind or quality ⟨cut*ie*⟩ ⟨tough*ie*⟩

-ier — see -ER

^{1}**if** \ˈif, əf\ *conjunction* [Middle English, from Old English *gif;* akin to Old High German *ibu* if] (before 12th century)
1 a : in the event that **b :** allowing that **c :** on the assumption that **d :** on condition that
2 : WHETHER ⟨asked *if* the mail had come⟩ ⟨I doubt *if* I'll pass the course⟩
3 — used as a function word to introduce an exclamation expressing a wish ⟨*if* it would only rain⟩
4 : even though ⟨an interesting *if* untenable argument⟩
— **if anything :** on the contrary even **:** perhaps even ⟨*if anything,* you ought to apologize⟩

^{2}**if** \ˈif\ *noun* (1513)
1 : CONDITION, STIPULATION ⟨the question . . . depends on too many *ifs* to allow an answer —*Encounter*⟩
2 : SUPPOSITION

-iferous *adjective combining form* [Middle English, from Latin *-ifer,* from *-i-* + *-fer* -ferous]
: -FEROUS

iff \ˈif-ᵊn(d)-ˈōn-lē-ˌif; ˈif, *sometimes read with a prolonged* f\ *conjunction* [alteration of ^{1}*if*] (1955)
: if and only if ⟨two figures are congruent *iff* one can be placed over the other so that they coincide⟩

if·fy \ˈi-fē\ *adjective* [^{1}*if*] (1937)
: abounding in contingencies or unknown qualities or conditions **:** UNCERTAIN ⟨an *iffy* proposition⟩
— **if·fi·ness** *noun*

-ification *noun suffix* [Middle English *-ificacioun,* from Middle French & Latin; Middle French *-ification,* from Latin *-ification-, -ificatio,* from *-i-* + *-ficatio* -fication]
: -FICATION ⟨desert*ification*⟩

-iform *adjective combining form* [Middle French & Latin; Middle French *-iforme,* from Latin *-iformis,* from *-i-* + *-formis* -form]
: -FORM ⟨patell*iform*⟩

I formation *noun* (1951)
: an offensive football formation in which the running backs line up in a line directly behind the quarterback — compare T FORMATION

-ify *verb suffix* [Middle English *-ifien,* from Old French *-ifier,* from Latin *-ificare,* from *-i-* + *-ficare* -fy]
: -FY

IgA \ˌī-(ˌ)jē-ˈā\ *noun* [*i*mmuno*g*lobulin] (1969)
: a class of immunoglobulins found in external bodily secretions (as saliva, tears, and sweat)

Ig·bo \ˈig-(ˌ)bō\ *variant of* IBO

IgE \ˌī-(ˌ)jē-ˈē\ *noun* (1969)
: a class of immunoglobulins that function especially in allergic reactions

IgG \-ˈjē\ *noun* (1969)
: a class of immunoglobulins that includes the most common antibodies circulating in the blood

ig·loo \ˈi-(ˌ)glü\ *noun, plural* **igloos** [Inuit *iglu* house] (1856)
1 : an Eskimo house usually made of sod, wood, or stone when permanent or of blocks of snow or ice in the shape of a dome when built for temporary purposes

igloo 1

2 : a building or structure shaped like a dome

IgM \ˌī-(ˌ)jē-ˈem\ *noun* (1969)
: a class of immunoglobulins that includes antibodies that appear early in the immune response

ig·ne·ous \ˈig-nē-əs\ *adjective* [Latin *igneus,* from *ignis* fire; akin to Sanskrit *agni* fire] (1664)
1 : of, relating to, or resembling fire **:** FIERY
2 a : relating to, resulting from, or suggestive of the intrusion or extrusion of magma or volcanic activity **b :** formed by solidification of magma ⟨*igneous* rock⟩

ig·nes·cent \ig-ˈne-sᵊnt\ *adjective* [Latin *ignescent-, ignescens,* present participle of *ignescere* to catch fire, from *ignis*] (circa 1828)
: VOLATILE

ig·nim·brite \ˈig-nəm-ˌbrīt\ *noun* [German *Ignimbrit,* from Latin *ignis* + *imbr-* (from *imber* rain) + German *-it* ^{1}-ite — more at IMBRICATE] (1932)
: a hard rock formed by solidification of chiefly fine deposits of volcanic ash

ig·nis fat·u·us \ˈig-nəs-ˈfa-chə-wəs, -ˈfach-wəs\ *noun, plural* **ig·nes fat·ui** \-ˌnēz-ˈfa-chə-ˌwī\ [Medieval Latin, literally, foolish fire] (circa 1563)
1 : a light that sometimes appears in the night over marshy ground and is often attributable to the combustion of gas from decomposed organic matter
2 : a deceptive goal or hope

ig·nite \ig-ˈnīt\ *verb* **ig·nit·ed; ig·nit·ing** [Latin *ignitus,* past participle of *ignire* to ignite, from *ignis*] (1666)
transitive verb
1 : to subject to fire or intense heat; *especially* **:** to render luminous by heat
2 a : to set afire; *also* **:** KINDLE **b :** to cause (a fuel) to burn
3 : to heat up **:** EXCITE ⟨oppression that *ignited* the hatred of the people⟩
intransitive verb
1 : to catch fire
2 : to begin to glow
— **ig·nit·abil·i·ty** \ig-ˌnī-tə-ˈbi-lə-tē\ *noun*
— **ig·nit·able** *also* **ig·nit·ible** \-ˈnī-tə-bəl\ *adjective*
— **ig·nit·er** *also* **ig·ni·tor** \-ˈnī-tər\ *noun*

ig·ni·tion \ig-ˈni-shən\ *noun* (1612)
1 : the act or action of igniting: as **a :** the starting of a fire **b :** the heating of a plasma to a temperature high enough to sustain nuclear fusion
2 : the process or means (as an electric spark) of igniting a fuel mixture

ig·ni·tron \ig-ˈnī-ˌträn\ *noun* [Latin *ignis* fire + English *-tron*] (1933)
: a mercury-containing rectifier tube in which the arc is struck again at the beginning of each cycle by a special electrode separately energized by an auxiliary circuit

ig·no·ble \ig-ˈnō-bəl\ *adjective* [Middle English, from Middle French, from Latin *ignobilis,* from *in-* + Old Latin *gnobilis* noble] (15th century)
1 : of low birth or common origin **:** PLEBEIAN
2 : characterized by baseness, lowness, or meanness
synonym see MEAN
— **ig·no·bil·i·ty** \ˌig-nō-ˈbi-lə-tē\ *noun*
— **ig·no·ble·ness** \ig-ˈnō-bəl-nəs\ *noun*
— **ig·no·bly** \-blē *also* -bə-lē\ *adverb*

ig·no·min·i·ous \ˌig-nə-ˈmi-nē-əs\ *adjective* (15th century)
1 : marked with or characterized by disgrace or shame **:** DISHONORABLE
2 : deserving of shame or infamy **:** DESPICABLE
3 : HUMILIATING, DEGRADING ⟨suffered an *ignominious* defeat⟩
— **ig·no·min·i·ous·ly** *adverb*
— **ig·no·min·i·ous·ness** *noun*

ig·no·mi·ny \ˈig-nə-ˌmi-nē, -mə-nē *also* ig-ˈnä-mə-nē\ *noun, plural* **-nies** [Middle French or Latin; Middle French *ignominie,* from Latin *ignominia,* from *ig-* (as in *ignorare* to be ignorant of, ignore) + *nomin-, nomen* name, repute — more at NAME] (1540)
1 : deep personal humiliation and disgrace
2 : disgraceful or dishonorable conduct, quality, or action
synonym see DISGRACE

ig·no·ra·mus \ˌig-nə-ˈrā-məs *also* -ˈra-\ *noun, plural* **-mus·es** *also* **-mi** \-mē\ [*Ignoramus,* ignorant lawyer in *Ignoramus* (1615), play by George Ruggle, from Latin, literally, we are ignorant of] (circa 1616)
: an utterly ignorant person **:** DUNCE ◆

ig·no·rance \ˈig-n(ə-)rən(t)s\ *noun* (13th century)
: the state or fact of being ignorant

ig·no·rant \ˈig-n(ə-)rənt\ *adjective* (14th century)
1 a : destitute of knowledge or education ⟨an *ignorant* society⟩; *also* **:** lacking knowledge or comprehension of the thing specified ⟨parents

◇ WORD HISTORY

ignoramus *Ignoramus* was a Latin satire written by a Cambridge University tutor named George Ruggle (1575–1622) and first performed in 1615. The title character, whose name in Latin means literally "we do not know," was Ruggle's send-up of unlearned but crafty lawyers. He was inspired by a real attorney named Francis Brakin, who had won a court dispute between the town of Cambridge and the university and thus aroused the ire of local scholars. The success of the play, which was performed twice before King James I in 1615, led to the generic use of *ignoramus* in the 17th century for any woefully ignorant person. The character's name may have been suggested by the legal term *ignoramus,* written on bills of indictment when the evidence presented seemed to be insufficient to justify prosecution. In these cases *ignoramus* meant "we take no notice of (i.e., we do not recognize) the indictment." Such a reference would have been familiar to Ruggle's audience and appropriate to his satire of the legal profession.

ignorant of modern mathematics⟩ **b :** resulting from or showing lack of knowledge or intelligence ⟨*ignorant* errors⟩
2 : UNAWARE, UNINFORMED ☆
— ig·no·rant·ly *adverb*
— ig·no·rant·ness *noun*

ig·no·ra·tio elen·chi \ˌig-nə-ˈrä-tē-ˌō-i-ˈleŋ-ˌkē\ *noun* [Latin, literally, ignorance of proof] (1588)
: a fallacy in logic of supposing a point proved or disproved by an argument proving or disproving something not at issue

ig·nore \ig-ˈnōr, -ˈnȯr\ *transitive verb* **ig·nored; ig·nor·ing** [obsolete *ignore* to be ignorant of, from French *ignorer,* from Latin *ignorare,* from *ignarus* ignorant, unknown, from *in-* + *gnoscere, noscere* to know — more at KNOW] (1801)
1 : to refuse to take notice of
2 : to reject (a bill of indictment) as ungrounded
synonym see NEGLECT
— ig·nor·able \-ˈnōr-ə-bəl, -ˈnȯr-\ *adjective*
— ig·nor·er *noun*

Igo·rot \ˌē-gə-ˈrōt\ *noun, plural* **Igorot** *or* **Igorots** (1821)
1 : a member of any of several related peoples of northwestern Luzon, Philippines
2 : any of the Austronesian languages of the Igorot

Igraine \i-ˈgrān\ *noun*
: the wife of Uther and mother of King Arthur

igua·na \i-ˈgwä-nə\ *noun* [Spanish, from Arawak & Carib *iwana*] (1555)
: any of various large herbivorous typically dark-colored tropical American lizards (family Iguanidae) that have a serrated dorsal crest and are important as human food in their native habitat; *broadly* **:** any of various large lizards

iguana

iguan·odon \i-ˈgwä-nə-ˌdän\ *noun* [New Latin *Iguanodont-, Iguanodon,* from Spanish *iguana* + Greek *odōn* tooth, from *odont-, odous* — more at TOOTH] (1830)
: any of a genus (*Iguanodon*) of very large herbivorous dinosaurs of the early Cretaceous

IHS \ˌī-ˌāch-ˈes\ [Late Latin, part transliteration of Greek ΙΗΣ, abbreviation for ΙΗΣΟΥΣ *Iēsous* Jesus]
— used as a Christian symbol and monogram for *Jesus*

ikat \ˈē-ˌkät\ *noun* [Malay, tying] (1927)
: a fabric in which the yarns have been tie-dyed before weaving

ike·ba·na \ˌi-kā-ˈbä-nə, ˌi-ki-, ˌē-\ *noun* [Japanese, from *ikeru* to keep alive, arrange + *hana* flower] (1901)
: the Japanese art of flower arranging that emphasizes form and balance

ikon *variant of* ICON

il- — see IN-

ilang–ilang *variant of* YLANG-YLANG

¹-ile *adjective suffix* [Middle English, from Middle French, from Latin *-ilis*]
: tending to or capable of ⟨contract*ile*⟩

²-ile *noun suffix* [*-ile* (as in *quartile,* noun)]
: segment of a (specified) size in a frequency distribution ⟨dec*ile*⟩

il·e·i·tis \ˌi-lē-ˈī-təs\ *noun* [New Latin] (circa 1855)
: inflammation of the ileum

il·e·um \ˈi-lē-əm\ *noun, plural* **il·ea** \-lē-ə\ [New Latin, alteration of Latin *ilia,* plural, groin, viscera] (1682)
: the last division of the small intestine extending between the jejunum and large intestine
— il·e·al \-lē-əl\ *adjective*

il·e·us \ˈi-lē-əs\ *noun* [Latin, from Greek *eileos,* from *eilyein* to roll — more at VOLUBLE] (1693)
: mechanical or functional obstruction of the bowel

ilex \ˈī-ˌleks\ *noun* [Middle English, from Latin] (14th century)
1 : HOLM OAK
2 : HOLLY 1

il·i·ac \ˈi-lē-ˌak\ *also* **il·i·al** \ˈi-lē-əl\ *adjective* [Late Latin *iliacus,* from Latin *ilium*] (1541)
: of, relating to, or located near the ilium

Il·i·ad \ˈi-lē-əd, -ˌad\ *noun* [*Iliad,* ancient Greek epic poem attributed to Homer, from Latin *Iliad-, Ilias,* from Greek, from *Ilion* Troy] (1603)
1 a : a series of miseries or disastrous events **b :** a series of exploits regarded as suitable for an epic
2 : a long narrative; *especially* **:** an epic in the Homeric tradition
— Il·i·ad·ic \ˌi-lē-ˈa-dik\ *adjective*

il·i·um \ˈi-lē-əm\ *noun, plural* **il·ia** \-lē-ə\ [New Latin, alteration of Latin *ilia*] (1706)
: the dorsal, upper, and largest one of the three bones composing either lateral half of the pelvis

¹ilk \ˈilk\ *pronoun* [Middle English, from Old English *ilca,* from a prehistoric compound whose constituents are akin respectively to Gothic *is* he (akin to Latin *is* he, that) and Old English *gelīc* like — more at ITERATE, LIKE] (before 12th century)
chiefly Scottish **:** SAME — used with *that* especially in the names of landed families

²ilk *noun* (1790)
: SORT, KIND ⟨the rejection of these books or others of like *ilk* —Kathleen Molz⟩

³ilk *pronoun* [Middle English, adjective & pronoun, from Old English *ylc, ǣlc* — more at EACH] (before 12th century)
chiefly Scottish **:** EACH

il·ka \ˈil-kə\ *adjective* [Middle English, from *ilk* + *a* (indefinite article)] (13th century)
chiefly Scottish **:** EACH, EVERY

¹ill \ˈil\ *adjective* **worse** \ˈwərs\; **worst** \ˈwərst\ [Middle English, from Old Norse *illr*] (12th century)
1 a *chiefly Scottish* **:** IMMORAL, VICIOUS **b :** resulting from, accompanied by, or indicative of an evil or malevolent intention ⟨*ill* deeds⟩ **c :** attributing evil or an objectionable quality ⟨held an *ill* opinion of his neighbors⟩
2 a : causing suffering or distress ⟨*ill* weather⟩ **b** *comparative also* **ill·er** (1) **:** not normal or sound ⟨*ill* health⟩ (2) **:** not in good health; *also* **:** NAUSEATED
3 a : not suited to circumstances or not to one's advantage **:** UNLUCKY ⟨an *ill* omen⟩ **b :** involving difficulty **:** HARD ⟨an *ill* man to please⟩
4 a : not meeting an accepted standard ⟨*ill* manners⟩ **b** *archaic* **:** notably unskillful or inefficient
5 : UNFRIENDLY, HOSTILE ⟨*ill* feeling⟩

²ill *adverb* **worse; worst** (13th century)
1 a : with displeasure or hostility **b :** in a harsh manner **c :** so as to reflect unfavorably ⟨spoke *ill* of the neighbors⟩
2 : in a reprehensible manner
3 : HARDLY, SCARCELY ⟨can *ill* afford such extravagances⟩
4 a : in an unfortunate manner **:** BADLY, UNLUCKILY ⟨*ill* fares the land . . . where wealth accumulates, and men decay —Oliver Goldsmith⟩ **b :** in a faulty, inefficient, insufficient, or unpleasant manner — often used in combination ⟨the methods used may be *ill*-adapted to the aims in view —R. M. Hutchins⟩

³ill *noun* (13th century)
1 : the reverse of good **:** EVIL
2 a : MISFORTUNE, DISTRESS **b** (1) **:** AILMENT, SICKNESS (2) **:** something that disturbs or afflicts **:** TROUBLE ⟨economic and social *ills*⟩
3 : something that reflects unfavorably ⟨spoke no *ill* of him⟩

I'll \ˈī(ə)l\ (1567)
: I will **:** I shall

ill–ad·vised \ˌi-ləd-ˈvīzd\ *adjective* (circa 1592)
: resulting from or showing lack of wise and sufficient counsel or deliberation ⟨an *ill-advised* decision⟩
— ill–ad·vis·ed·ly \-ˈvī-zəd-lē\ *adverb*

ill at ease *adjective* (14th century)
: not feeling easy **:** UNCOMFORTABLE

il·la·tion \i-ˈlā-shən\ *noun* [Late Latin *illation-, illatio,* from Latin, action of bringing in, from *inferre* (past participle *illatus*) to bring in, from *in-* + *ferre* to carry — more at TOLERATE, BEAR] (1533)
1 : the action of inferring **:** INFERENCE
2 : a conclusion inferred

¹il·la·tive \ˈi-lə-tiv, i-ˈlā-\ *noun* (1591)
1 : a word (as *therefore*) or phrase (as *as a consequence*) introducing an inference
2 : ILLATION 2

²illative *adjective* (1611)
: INFERENTIAL
— il·la·tive·ly *adverb*

il·laud·able \(ˌ)i(l)-ˈlȯ-də-bəl\ *adjective* [Latin *illaudabilis,* from *in-* + *laudabilis* laudable] (1589)
: deserving no praise
— il·laud·ably \-blē\ *adverb*

ill–be·ing \ˈil-ˈbē-iŋ\ *noun* (1840)
: a condition of being deficient in health, happiness, or prosperity

ill–bod·ing \-ˈbō-diŋ\ *adjective* (1591)
: boding evil **:** INAUSPICIOUS

ill–bred \-ˈbred\ *adjective* (1604)
: badly brought up or showing bad upbringing **:** IMPOLITE

¹il·le·gal \(ˌ)i(l)-ˈlē-gəl\ *adjective* [Middle French or Medieval Latin; Middle French *illegal,* from Medieval Latin *illegalis,* from Latin *in-* + *legalis* legal] (1538)
: not according to or authorized by law **:** UNLAWFUL, ILLICIT; *also* **:** not sanctioned by official rules (as of a game)
— il·le·gal·i·ty \ˌi-li-ˈga-lə-tē\ *noun*
— il·le·gal·ly \(ˌ)i(l)-ˈlē-gə-lē\ *adverb*

²illegal *noun* (1939)
: an illegal immigrant

il·le·gal·ize \(ˌ)i(l)-ˈlē-gə-ˌlīz\ *transitive verb* (circa 1818)
: to make or declare illegal
— il·le·gal·i·za·tion \-ˌlē-gə-lə-ˈzā-shən\ *noun*

il·leg·i·ble \(ˌ)i(l)-ˈle-jə-bəl\ *adjective* (1640)
: not legible **:** UNDECIPHERABLE ⟨*illegible* writing⟩
— il·leg·i·bil·i·ty \-ˌle-jə-ˈbi-lə-tē\ *noun*
— il·leg·i·bly \-ˈle-jə-blē\ *adverb*

il·le·git·i·ma·cy \ˌi-li-ˈji-tə-mə-sē\ *noun* (1680)
1 : the quality or state of being illegitimate
2 : BASTARDY 2

il·le·git·i·mate \-ˈji-tə-mət\ *adjective* (1536)
1 : not recognized as lawful offspring; *specifically* **:** born of parents not married to each other
2 : not rightly deduced or inferred **:** ILLOGICAL

☆ SYNONYMS
Ignorant, illiterate, unlettered, untutored, unlearned mean not having knowledge. IGNORANT may imply a general condition or it may apply to lack of knowledge or awareness of a particular thing ⟨an *ignorant* fool⟩ ⟨*ignorant* of nuclear physics⟩. ILLITERATE applies to either an absolute or a relative inability to read and write ⟨much of the population is still *illiterate*⟩. UNLETTERED implies ignorance of the knowledge gained by reading ⟨an allusion meaningless to the *unlettered*⟩. UNTUTORED may imply lack of schooling in the arts and ways of civilization ⟨strange monuments built by an *untutored* people⟩. UNLEARNED suggests ignorance of advanced subjects ⟨poetry not for academics but for the *unlearned* masses⟩.

3 : departing from the regular **:** ERRATIC
4 a : not sanctioned by law **:** ILLEGAL **b :** not authorized by good usage **c** *of a taxon* **:** published but not in accordance with the rules of the relevant international code
— il·le·git·i·mate·ly *adverb*
ill–fat·ed \'il-'fā-təd\ *adjective* (1710)
1 : having or destined to a hapless fate **:** UNFORTUNATE ⟨an *ill-fated* expedition⟩
2 : that causes or marks the beginning of misfortune
ill–fa·vored \-'fā-vərd\ *adjective* (circa 1530)
1 : unattractive in physical appearance; *especially* **:** having an ugly face
2 : OFFENSIVE, OBJECTIONABLE
ill–got·ten \-'gä-tᵊn\ *adjective* (1552)
: acquired by illicit or improper means ⟨*ill-gotten* gains⟩
ill–hu·mored \'il-'hyü-mərd, -'yü-\ *adjective* (1687)
: SURLY, IRRITABLE
— ill–hu·mored·ly *adverb*
il·lib·er·al \(ˌ)i(l)-'li-b(ə-)rəl\ *adjective* [Middle French or Latin; Middle French, from Latin *illiberalis* ignoble, stingy, from Latin *in-* + *liberalis* liberal] (1535)
: not liberal: as **a** *archaic* (1) **:** lacking a liberal education (2) **:** lacking culture and refinement **b :** not requiring the background of a liberal arts education ⟨*illiberal* occupations⟩ **c** *archaic* **:** not generous **:** STINGY **d :** not broad-minded **:** BIGOTED **e :** opposed to liberalism
— il·lib·er·al·i·ty \-ˌli-bə-'ra-lə-tē\ *noun*
— il·lib·er·al·ly \-'li-b(ə-)rə-lē\ *adverb*
— il·lib·er·al·ness \-b(ə-)rəl-nəs\ *noun*
il·lib·er·al·ism \-b(ə-)rə-ˌli-zəm\ *noun* (1839)
: opposition to or lack of liberalism
il·lic·it \(ˌ)i(l)-'li-sət\ *adjective* [Latin *illicitus,* from *in-* + *licitus* lawful — more at LICIT] (1506)
: not permitted **:** UNLAWFUL
— il·lic·it·ly *adverb*
il·lim·it·able \(ˌ)i(l)-'li-mə-tə-bəl\ *adjective* (1596)
: incapable of being limited or bounded **:** MEASURELESS ⟨the *illimitable* reaches of space and time⟩
— il·lim·it·abil·i·ty \-ˌli-mə-tə-'bi-lə-tē\ *noun*
— il·lim·it·able·ness \-'li-mə-tə-bəl-nəs\ *noun*
— il·lim·it·ably \-blē\ *adverb*
Il·li·nois \ˌi-lə-'nȯi *also* -'nȯiz\ *noun, plural* **Illinois** [French, of Algonquian origin] (1703)
1 *plural* **:** a confederacy of American Indian peoples of Illinois, Iowa, and Wisconsin
2 : a member of any of the Illinois peoples
il·liq·uid \(ˌ)i(l)-'lik-wəd\ *adjective* (1913)
1 : not being cash or readily convertible into cash ⟨*illiquid* holdings⟩
2 : deficient in liquid assets ⟨an *illiquid* bank⟩
— il·li·quid·i·ty \-ˌli-'kwi-də-tē\ *noun*
il·lite \'i-ˌlīt\ *noun* [*Ill*inois, state of U.S. + ¹*-ite*] (1937)
: any of a group of clay minerals having essentially the crystal structure of muscovite
— il·lit·ic \i-'li-tik\ *adjective*
il·lit·er·a·cy \(ˌ)i(l)-'li-t(ə-)rə-sē\ *noun, plural* **-cies** (1660)
1 : the quality or state of being illiterate; *especially* **:** inability to read or write
2 : a mistake or crudity (as in speaking) typical of one who is illiterate
il·lit·er·ate \(ˌ)i(l)-'li-t(ə-)rət\ *adjective* [Middle English, from Latin *illiteratus,* from *in-* + *litteratus* literate] (15th century)
1 : having little or no education; *especially* **:** unable to read or write
2 a : showing or marked by a lack of familiarity with language and literature **b :** violating approved patterns of speaking or writing
3 : showing or marked by a lack of acquaintance with the fundamentals of a particular field of knowledge
synonym see IGNORANT
— illiterate *noun*
— il·lit·er·ate·ly *adverb*
— il·lit·er·ate·ness *noun*
ill–man·nered \'il-'ma-nərd\ *adjective* (15th century)
: having bad manners **:** RUDE
ill–na·tured \-'nā-chərd\ *adjective* (1605)
1 : MALEVOLENT, SPITEFUL
2 : having a bad disposition **:** CROSS, SURLY
— ill–na·tured·ly *adverb*
ill·ness \'il-nəs\ *noun* (circa 1500)
1 *obsolete* **a :** WICKEDNESS **b :** UNPLEASANTNESS
2 a : an unhealthy condition of body or mind **b :** SICKNESS 2
il·lo·cu·tion·ary \ˌi-lə-'kyü-shə-ˌner-ē, ˌi(l)-lō-\ *adjective* [²*in-* + *locution*] (1955)
: relating to or being the communicative effect (as commanding or requesting) of an utterance ⟨"There's a snake under you" may have the *illocutionary* force of a warning⟩
il·log·ic \(ˌ)i(l)-'lä-jik\ *noun* [back-formation from *illogical*] (1856)
: the quality or state of being illogical **:** ILLOGICALITY
il·log·i·cal \-ji-kəl\ *adjective* (1588)
1 : not observing the principles of logic
2 : devoid of logic **:** SENSELESS
— il·log·i·cal·i·ty \-ˌlä-jə-'ka-lə-tē\ *noun*
— il·log·i·cal·ly \-'lä-ji-k(ə-)lē\ *adverb*
— il·log·i·cal·ness \-kəl-nəs\ *noun*
ill–sort·ed \'il-'sȯr-təd\ *adjective* (1691)
1 : not well matched ⟨he and his wife were an *ill-sorted* pair —Lord Byron⟩
2 *Scottish* **:** much displeased
ill–starred \-'stärd\ *adjective* (1604)
: ILL-FATED, UNLUCKY ⟨an *ill-starred* venture⟩
ill–tem·pered \-'tem-pərd\ *adjective* (1601)
: ILL-NATURED, QUARRELSOME
— ill–tem·pered·ly \-lē\ *adverb*
ill–treat \-'trēt\ *transitive verb* (1689)
: to treat cruelly or improperly **:** MALTREAT
— ill–treat·ment \-mənt\ *noun*
il·lume \i-'lüm\ *transitive verb* **il·lumed; il·lum·ing** (1602)
: ILLUMINATE
il·lu·mi·nance \i-'lü-mə-nən(t)s\ *noun* (circa 1938)
: ILLUMINATION 2
il·lu·mi·nant \-nənt\ *noun* (1644)
: an illuminating device or substance
¹il·lu·mi·nate \i-'lü-mə-nət\ *adjective* (15th century)
1 *archaic* **:** brightened with light
2 *archaic* **:** intellectually or spiritually enlightened
²il·lu·mi·nate \-ˌnāt\ *transitive verb* **-nat·ed; -nat·ing** [Middle English, from Latin *illuminatus,* past participle of *illuminare,* from *in-* + *luminare* to light up, from *lumin-, lumen* light — more at LUMINARY] (15th century)
1 a : to enlighten spiritually or intellectually **b** (1) **:** to supply or brighten with light (2) **:** to make luminous or shining **c** *archaic* **:** to set alight **d :** to subject to radiation
2 : to make clear **:** ELUCIDATE
3 : to make illustrious or resplendent
4 : to decorate (as a manuscript) with gold or silver or brilliant colors or with often elaborate designs or miniature pictures
— il·lu·mi·nat·ing·ly \-ˌnā-tiŋ-lē\ *adverb*
— il·lu·mi·na·tor \-ˌnā-tər\ *noun*
³il·lu·mi·nate \-nət\ *noun* (1600)
archaic **:** one having or claiming unusual enlightenment
il·lu·mi·na·ti \i-ˌlü-mə-'nä-tē\ *noun plural* [Italian & New Latin; Italian, from New Latin, from Latin, plural of *illuminatus*] (1599)
1 *capitalized* **:** any of various groups claiming special religious enlightenment
2 : persons who are or who claim to be unusually enlightened
il·lu·mi·na·tion \i-ˌlü-mə-'nā-shən\ *noun* (14th century)
1 : the action of illuminating or state of being illuminated: as **a :** spiritual or intellectual enlightenment **b** (1) **:** a lighting up (2) **:** decorative lighting or lighting effects **c :** decoration by the art of illuminating
2 : the luminous flux per unit area on an intercepting surface at any given point
3 : one of the decorative features used in the art of illuminating or in decorative lighting
il·lu·mi·na·tive \i-'lü-mə-ˌnā-tiv\ *adjective* (1644)
: of, relating to, or producing illumination **:** ILLUMINATING
il·lu·mine \i-'lü-mən\ *transitive verb* **-mined; -min·ing** (14th century)
: ILLUMINATE
— il·lu·min·able \-mə-nə-bəl\ *adjective*
il·lu·mi·nism \-mə-ˌni-zəm\ *noun* (1798)
1 : belief in or claim to a personal enlightenment not accessible to mankind in general
2 *capitalized* **:** beliefs or claims viewed as forming doctrine or principles of Illuminati
— il·lu·mi·nist \-nist\ *noun*
ill–us·age \'il-'yü-sij, -zij\ *noun* (1593)
: harsh, unkind, or abusive treatment
ill–use \-'yüz\ *transitive verb* (1841)
: to use badly **:** MALTREAT, ABUSE
il·lu·sion \i-'lü-zhən\ *noun* [Middle English, from Middle French, from Late Latin *illusion-, illusio,* from Latin, action of mocking, from *illudere* to mock at, from *in-* + *ludere* to play, mock — more at LUDICROUS] (14th century)
1 a *obsolete* **:** the action of deceiving **b** (1) **:** the state or fact of being intellectually deceived or misled **:** MISAPPREHENSION (2) **:** an instance of such deception

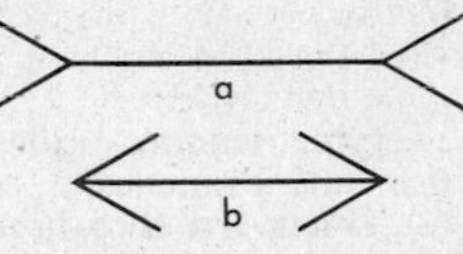

illusion 2a(1): ***a*** **and** ***b*** **are equal in length**

2 a (1) **:** a misleading image presented to the vision (2) **:** something that deceives or misleads intellectually **b** (1) **:** perception of something objectively existing in such a way as to cause misinterpretation of its actual nature (2) **:** HALLUCINATION 1 (3) **:** a pattern capable of reversible perspective
3 : a fine plain transparent bobbinet or tulle usually made of silk and used for veils, trimmings, and dresses
— il·lu·sion·al \-'lüzh-nəl, -'lü-zhə-nᵊl\ *adjective*
il·lu·sion·ary \i-'lü-zhə-ˌner-ē\ *adjective* (1886)
: ILLUSORY
il·lu·sion·ism \i-'lü-zhə-ˌni-zəm\ *noun* (1911)
: the use of artistic techniques (as perspective or shading) to create the illusion of reality especially in a work of art
il·lu·sion·ist \i-'lüzh-nist, -'lü-zhə-\ *noun* (1850)
: one who produces illusory effects: as **a :** one (as an artist) whose work is marked by illusionism **b :** a sleight-of-hand performer or a magician
— il·lu·sion·is·tic \-ˌlü-zhə-'nis-tik\ *adjective*
— il·lu·sion·is·ti·cal·ly \-ti-k(ə-)lē\ *adverb*
il·lu·sive \i-'lü-siv, -ziv\ *adjective* (1606)
: ILLUSORY
— il·lu·sive·ly *adverb*
— il·lu·sive·ness *noun*
il·lu·so·ry \i-'lüs-rē, -'lüz-; -'lü-sə-, -zə-\ *adjective* (circa 1631)

: based on or producing illusion : DECEPTIVE ⟨*illusory* hopes⟩
synonym see APPARENT
— **il·lu·so·ri·ly** \-rə-lē\ *adverb*
— **il·lu·so·ri·ness** \-rē-nəs\ *noun*

il·lus·trate \ˈi-ləs-ˌtrāt *also* i-ˈləs-\ *verb* **-trat·ed; -trat·ing** [Latin *illustratus,* past participle of *illustrare,* from *in-* + *lustrare* to purify, make bright — more at LUSTER] (1526)
transitive verb
1 *obsolete* **a :** ENLIGHTEN **b :** to light up
2 a *archaic* **:** to make illustrious **b** *obsolete* (1) **:** to make bright (2) **:** ADORN
3 a : to make clear **:** CLARIFY **b :** to make clear by giving or by serving as an example or instance **c :** to provide with visual features intended to explain or decorate ⟨*illustrate* a book⟩
4 : to show clearly **:** DEMONSTRATE
intransitive verb
: to give an example or instance
— **il·lus·tra·tor** \ˈi-ləs-ˌtrā-tər *also* i-ˈləs-\ *noun*

il·lus·tra·tion \ˌi-ləs-ˈtrā-shən *also* i-ˌləs-\ *noun* (14th century)
1 a : the action of illustrating **:** the condition of being illustrated **b** *archaic* **:** the action of making illustrious or honored or distinguished
2 : something that serves to illustrate: as **a :** an example or instance that helps make something clear **b :** a picture or diagram that helps make something clear or attractive
synonym see INSTANCE
— **il·lus·tra·tion·al** \-shnəl, -shə-nᵊl\ *adjective*

il·lus·tra·tive \i-ˈləs-trə-tiv *also* ˈi-lə-ˌstrā-\ *adjective* (1643)
: serving, tending, or designed to illustrate ⟨*illustrative* examples⟩
— **il·lus·tra·tive·ly** *adverb*

il·lus·tri·ous \i-ˈləs-trē-əs\ *adjective* [Latin *illustris,* probably from *illustrare*] (1588)
1 : notably or brilliantly outstanding because of dignity or achievements or actions **:** EMINENT
2 *archaic* **a :** shining brightly with light **b :** clearly evident
synonym see FAMOUS
— **il·lus·tri·ous·ly** *adverb*
— **il·lus·tri·ous·ness** *noun*

il·lu·vi·al \i-ˈlü-vē-əl\ *adjective* (1924)
: of, relating to, or marked by illuviation or illuviated materials or areas

il·lu·vi·a·tion \i-ˌlü-vē-ˈā-shən\ *noun* [²*in-* + *-luviation* (as in *eluviation*)] (1928)
: accumulation of dissolved or suspended soil materials in one area or horizon as a result of eluviation from another
— **il·lu·vi·at·ed** \i-ˈlü-vē-ˌā-təd\ *adjective*

ill will *noun* (14th century)
: unfriendly feeling
synonym see MALICE

ill–wish·er \ˈil-ˌwi-shər, -ˈwi-\ *noun* (1607)
: one that wishes ill to another

il·ly \ˈi(l)-lē\ *adverb* (15th century)
: not wisely or well **:** BADLY, ILL ⟨his *illy* concealed pride —Della Lutes⟩

Il·lyr·i·an \i-ˈlir-ē-ən\ *noun* (1584)
1 : a native or inhabitant of ancient Illyria
2 : the poorly attested Indo-European languages of the Illyrians — see INDO-EUROPEAN LANGUAGES table
— **Illyrian** *adjective*

il·men·ite \ˈil-mə-ˌnīt\ *noun* [German *Ilmenit,* from *Ilmen* range, Ural Mts., Russia] (circa 1827)
: a usually massive iron-black mineral composed of iron, titanium, and oxygen that is a titanium ore

Ilo·ca·no *or* **Ilo·ka·no** \ˌē-lə-ˈkä-(ˌ)nō, ˌi-\ *noun, plural* **Ilocano** *or* **Ilocanos** *or* **Ilokano** *or* **Ilokanos** (1898)
1 : a member of a major people of northern Luzon in the Philippines
2 : the Austronesian language of the Ilocano people

im- — see IN-

I'm \ˈīm\ (circa 1594)
: I am

¹im·age \ˈi-mij\ *noun* [Middle English, from Old French, short for *imagene,* from Latin *imagin-, imago;* perhaps akin to Latin *imitari* to imitate] (13th century)
1 : a reproduction or imitation of the form of a person or thing; *especially* **:** an imitation in solid form **:** STATUE
2 a : the optical counterpart of an object produced by an optical device (as a lens or mirror) or an electronic device **b :** a likeness of an object produced on a photographic material
3 a : exact likeness **:** SEMBLANCE ⟨God created man in his own *image* —Genesis 1:27 (Revised Standard Version)⟩ **b :** a person strikingly like another person ⟨she is the *image* of her mother⟩
4 a : a tangible or visible representation **:** INCARNATION ⟨the *image* of filial devotion⟩ **b** *archaic* **:** an illusory form **:** APPARITION
5 a (1) **:** a mental picture of something not actually present **:** IMPRESSION (2) **:** a mental conception held in common by members of a group and symbolic of a basic attitude and orientation ⟨a disorderly courtroom can seriously tarnish a community's *image* of justice —Herbert Brownell⟩ **b :** IDEA, CONCEPT
6 : a vivid or graphic representation or description
7 : FIGURE OF SPEECH
8 : a popular conception (as of a person, institution, or nation) projected especially through the mass media ⟨promoting a corporate *image* of brotherly love and concern —R. C. Buck⟩
9 : a set of values given by a mathematical function (as a homomorphism) that corresponds to a particular subset of the domain

²image *transitive verb* **im·aged; im·ag·ing** (14th century)
1 : to call up a mental picture of **:** IMAGINE
2 : to describe or portray in language especially in a vivid manner
3 a : to create a representation of; *also* **:** to form an image of **b :** to represent symbolically
4 a : REFLECT, MIRROR **b :** to make appear **:** PROJECT
— **im·ag·er** \ˈi-mi-jər\ *noun*

image orthicon *noun* (1945)
: a highly sensitive television image tube

im·ag·ery \ˈi-mij-rē, -mi-jə-\ *noun, plural* **-er·ies** (14th century)
1 a : the product of image makers **:** IMAGES; *also* **:** the art of making images **b :** pictures produced by an imaging system
2 : figurative language
3 : mental images; *especially* **:** the products of imagination

image tube *noun* (1936)
: an electron tube in which incident electromagnetic radiation (as light or infrared) produces a visible image on its fluorescent screen duplicating the original pattern of radiation — called also *image converter*

imag·in·able \i-ˈmaj-nə-bəl, -ˈma-jə-\ *adjective* (14th century)
: capable of being imagined **:** CONCEIVABLE
— **imag·in·able·ness** *noun*
— **imag·in·ably** \-blē\ *adverb*

¹imag·i·nal \i-ˈma-jə-nᵊl\ *adjective* [*imagine* + ¹*-al*] (1647)
: of or relating to imagination, images, or imagery

²ima·gi·nal \i-ˈma-jə-nᵊl, -ˈmā-; -ˈmä-gə-\ *adjective* [New Latin *imagin-, imago*] (1877)
: of or relating to the insect imago

imag·i·nary \i-ˈma-jə-ˌner-ē\ *adjective* (14th century)
1 a : existing only in imagination **:** lacking factual reality **b :** formed or characterized imaginatively or arbitrarily ⟨his canvases, chiefly *imaginary,* somber landscapes —*Current Biography*⟩
2 : containing or relating to the imaginary unit ⟨*imaginary* roots⟩ ☆
— **imag·i·nari·ly** \i-ˌma-jə-ˈner-ə-lē\ *adverb*
— **imag·i·nari·ness** \-ˈma-jə-ˌner-ē-nəs\ *noun*

imaginary number *noun* (circa 1911)
: a complex number (as $2 + 3i$) in which the coefficient of the imaginary unit is not zero — called also *imaginary;* compare PURE IMAGINARY

imaginary part *noun* (circa 1929)
: the part of a complex number (as $3i$ in $2 + 3i$) that has the imaginary unit as a factor

imaginary unit *noun* (circa 1911)
: the positive square root of minus 1 denoted by i or $+\sqrt{-1}$

imag·i·na·tion \i-ˌma-jə-ˈnā-shən\ *noun* [Middle English, from Middle French, from Latin *imagination-, imaginatio,* from *imaginari*] (14th century)
1 : the act or power of forming a mental image of something not present to the senses or never before wholly perceived in reality
2 a : creative ability **b :** ability to confront and deal with a problem **:** RESOURCEFULNESS **c :** the thinking or active mind **:** INTEREST ⟨stories that fired the *imagination*⟩
3 a : a creation of the mind; *especially* **:** an idealized or poetic creation **b :** fanciful or empty assumption

imag·i·na·tive \i-ˈmaj-nə-tiv; -ˈma-jə-ˌnā-, -nə-\ *adjective* (14th century)
1 a : of, relating to, or characterized by imagination **b :** devoid of truth **:** FALSE
2 : given to imagining **:** having a lively imagination
3 : of or relating to images; *especially* **:** showing a command of imagery
— **imag·i·na·tive·ly** *adverb*
— **imag·i·na·tive·ness** *noun*

imag·ine \i-ˈma-jən\ *verb* **imag·ined; imag·in·ing** \-ˈmaj-niŋ, -ˈma-jə-\ [Middle English, from Middle French *imaginer,* from Latin *imaginari,* from *imagin-, imago* image] (14th century)
transitive verb
1 : to form a mental image of (something not present)
2 *archaic* **:** PLAN, SCHEME
3 : SUPPOSE, GUESS ⟨I *imagine* it will rain⟩
4 : to form a notion of without sufficient basis **:** FANCY ⟨*imagines* himself to be a charming conversationalist⟩
intransitive verb
1 : to use the imagination
2 : BELIEVE 3
synonym see THINK

im·ag·ing \ˈi-mi-jiŋ\ *noun* (1967)
: the action or process of producing an image especially by means other than visible light ⟨acoustic *imaging*⟩ — compare MAGNETIC RESONANCE IMAGING

im·ag·ism \ˈi-mi-ˌji-zəm\ *noun, often capitalized* (1912)

☆ SYNONYMS
Imaginary, fanciful, visionary, fantastic, chimerical, quixotic mean unreal or unbelievable. IMAGINARY applies to something which is fictitious and purely the product of one's imagination ⟨an *imaginary* desert isle⟩. FANCIFUL suggests the free play of the imagination ⟨a teller of *fanciful* stories⟩. VISIONARY stresses impracticality or incapability of realization ⟨*visionary* schemes⟩. FANTASTIC implies incredibility or strangeness beyond belief ⟨a *fantastic* world inhabited by monsters⟩. CHIMERICAL combines the implication of VISIONARY and FANTASTIC ⟨*chimerical* dreams of future progress⟩. QUIXOTIC implies a devotion to romantic or chivalrous ideals unrestrained by ordinary prudence and common sense ⟨a *quixotic* crusade⟩.

: a 20th century movement in poetry advocating free verse and the expression of ideas and emotions through clear precise images
— **im·ag·ist** \-mi-jist\ *noun or adjective, often capitalized*
— **im·ag·is·tic** \ˌi-mi-ˈjis-tik\ *adjective*
— **im·ag·is·ti·cal·ly** \-ti-k(ə-)lē\ *adverb*

ima·go \i-ˈmä-(ˌ)gō, -ˈmā-\ *noun, plural* **ima·goes** *or* **ima·gi·nes** \-ˈmā-gə-ˌnēz, -ˈmä-; -ˈmā-jə-, -ˈma-\ [New Latin, from Latin, image] (circa 1797)
1 : an insect in its final, adult, sexually mature, and typically winged state
2 : an idealized mental image of another person or the self
word history see LARVA

imam \i-ˈmäm, ē-ˈ, -ˈmam\ *noun, often capitalized* [Arabic *imām*] (1613)
1 : the prayer leader of a mosque
2 : a Muslim leader of the line of Ali held by Shiites to be the divinely appointed, sinless, infallible successors of Muhammad
3 : any of various rulers that claim descent from Muhammad and exercise spiritual and temporal leadership over a Muslim region

imam·ate \-ˈmä-ˌmāt, -ˈma-\ *noun, often capitalized* (circa 1741)
1 : the office of an imam
2 : the region or country ruled by an imam

ima·ret \i-ˈmär-ət\ *noun* [Turkish] (1613)
: an inn or hospice in Turkey

Ima·ri \i-ˈmär-ē\ *noun* [*Imari,* Japan] (1875)
: a multicolored Japanese porcelain usually characterized by elaborate floral designs
— **Imari** *adjective*

im·bal·ance \(ˌ)im-ˈba-lən(t)s\ *noun* (circa 1890)
: lack of balance **:** the state of being out of equilibrium or out of proportion ⟨a vitamin *imbalance*⟩ ⟨racial *imbalance* in schools⟩
— **im·bal·anced** \-lən(t)st\ *adjective*

im·be·cile \ˈim-bə-səl, -ˌsil\ *noun* [French *imbécile,* noun, from adjective, weak, weak-minded, from Latin *imbecillus*] (1802)
1 : a mentally deficient person; *especially* **:** a feebleminded person having a mental age of three to seven years and requiring supervision in the performance of routine daily tasks of self-care
2 : FOOL, IDIOT
— **imbecile** *or* **im·be·cil·ic** \ˌim-bə-ˈsi-lik\ *adjective*

im·be·cil·i·ty \ˌim-bə-ˈsi-lə-tē\ *noun, plural* **-ties** (1533)
1 : the quality or state of being imbecile or an imbecile
2 a : utter foolishness; *also* **:** FUTILITY **b :** something that is foolish or nonsensical

im·bed *variant of* EMBED

im·bibe \im-ˈbīb\ *verb* **im·bibed; im·bib·ing** [in sense 1, from Middle English *enbiben,* from Middle French *embiber,* from Latin *imbibere* to drink in, conceive, from *in-* + *bibere* to drink; in other senses, from Latin *imbibere* — more at POTABLE] (14th century)
transitive verb
1 *archaic* **:** SOAK, STEEP
2 a : to receive into the mind and retain ⟨*imbibe* moral principles⟩ **b :** to assimilate or take into solution
3 a : DRINK **b :** to take in or up ⟨a sponge *imbibes* moisture⟩
intransitive verb
1 : DRINK 2
2 a : to take in liquid **b :** to absorb or assimilate moisture, gas, light, or heat
— **im·bib·er** *noun*

im·bi·bi·tion \ˌim-bə-ˈbi-shən\ *noun* (15th century)
: the act or action of imbibing; *especially* **:** the taking up of fluid by a colloidal system resulting in swelling
— **im·bi·bi·tion·al** \-ˈbish-nəl, -ˈbi-shə-nᵊl\ *adjective*

im·bit·ter *variant of* EMBITTER

im·bo·som *variant of* EMBOSOM

¹im·bri·cate \ˈim-bri-kət\ *adjective* [Late Latin *imbricatus,* past participle of *imbricare* to cover with pantiles, from Latin *imbric-, imbrex* pantile, from *imbr-, imber* rain; akin to Greek *ombros* rain] (circa 1610)
: lying lapped over each other in regular order ⟨*imbricate* scales⟩

²im·bri·cate \ˈim-brə-ˌkāt\ *transitive verb* **-cat·ed; -cat·ing** (1784)
: OVERLAP; *especially* **:** to overlap like roof tiles

im·bri·ca·tion \ˌim-brə-ˈkā-shən\ *noun* (1713)
1 : an overlapping of edges (as of tiles or scales)
2 : a decoration or pattern showing imbrication

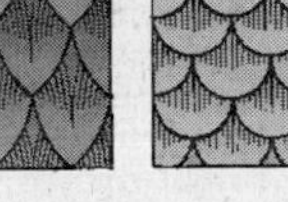
imbrication

im·bro·glio \im-ˈbrōl-(ˌ)yō\ *noun, plural* **-glios** [Italian, from *imbrogliare* to entangle, from Middle French *embrouiller* — more at EMBROIL] (1750)
1 : a confused mass
2 a : an intricate or complicated situation (as in a drama or novel) **b :** an acutely painful or embarrassing misunderstanding **c :** a violently confused or bitterly complicated altercation **:** EMBROILMENT

im·brown *variant of* EMBROWN

im·brue \im-ˈbrü\ *transitive verb* **im·brued; im·bru·ing** [Middle English *enbrewen,* probably from Middle French *abrevrer, embevrer* to soak, drench, ultimately from Latin *bibere* to drink — more at POTABLE] (15th century)
: STAIN

im·brute \im-ˈbrüt\ *verb* **im·brut·ed; im·brut·ing** (1634)
intransitive verb
: to sink to the level of a brute
transitive verb
: to degrade to the level of a brute

im·bue \-ˈbyü\ *transitive verb* **im·bued; im·bu·ing** [Latin *imbuere*] (1555)
1 : to permeate or influence as if by dyeing ⟨the spirit that *imbues* the new constitution⟩
2 : to tinge or dye deeply
synonym see INFUSE

im·id·az·ole \ˌi-mə-ˈda-ˌzōl\ *noun* [International Scientific Vocabulary] (1892)
: a white crystalline heterocyclic base $C_3H_4N_2$ that is an antimetabolite related to histidine; *broadly* **:** any of various derivatives of this

im·ide \ˈi-ˌmīd\ *noun* [International Scientific Vocabulary, alteration of *amide*] (1857)
: a compound containing the NH group that is derived from ammonia by replacement of two hydrogen atoms by a metal or an equivalent of acid groups — compare AMIDE
— **im·id·ic** \i-ˈmi-dik\ *adjective*

im·i·do \ˈi-mə-ˌdō\ *adjective* (1881)
: relating to or containing the NH group or its substituted form NR united to one or two acid groups

im·ine \ˈi-ˌmēn\ *noun* [International Scientific Vocabulary, alteration of *amine*] (1883)
: a compound containing the NH group or its substituted form NR that is derived from ammonia by replacement of two hydrogen atoms by a hydrocarbon group or other nonacid organic group

im·i·no \ˈi-mə-ˌnō\ *adjective* (1903)
: relating to or containing the NH group or its substituted form NR united to a group other than an acid group

imip·ra·mine \i-ˈmi-prə-ˌmēn\ *noun* [*imide* + *propyl* + *amine*] (1958)
: a tricyclic antidepressant drug $C_{19}H_{24}N_2$

im·i·ta·ble \ˈi-mə-tə-bəl\ *adjective* (1550)
: capable or worthy of being imitated or copied

im·i·tate \ˈi-mə-ˌtāt\ *transitive verb* **-tat·ed; -tat·ing** [Latin *imitatus,* past participle of *imitari* — more at IMAGE] (1534)
1 : to follow as a pattern, model, or example
2 : to be or appear like **:** RESEMBLE
3 : to produce a copy of **:** REPRODUCE
4 : MIMIC, COUNTERFEIT ⟨can *imitate* his father's booming voice⟩
synonym see COPY
— **im·i·ta·tor** \-ˌtā-tər\ *noun*

¹im·i·ta·tion \ˌi-mə-ˈtā-shən\ *noun* (14th century)
1 : an act or instance of imitating
2 : something produced as a copy **:** COUNTERFEIT
3 : a literary work designed to reproduce the style of another author
4 : the repetition by one voice of a melody, phrase, or motive stated earlier in the composition by a different voice
5 : the quality of an object in possessing some of the nature or attributes of a transcendent idea
6 : the assumption of behavior observed in other individuals

²imitation *adjective* (1840)
: resembling something else that is usually genuine and of better quality **:** not real ⟨*imitation* leather⟩

im·i·ta·tive \ˈi-mə-ˌtā-tiv, *especially British* -tə-tiv\ *adjective* (1584)
1 a : marked by imitation ⟨acting is an *imitative* art⟩ **b :** reproducing or representing a natural sound **:** ONOMATOPOEIC ⟨"hiss" is an *imitative* word⟩ **c :** exhibiting mimicry
2 : inclined to imitate
3 : imitating something superior **:** COUNTERFEIT
— **im·i·ta·tive·ly** *adverb*
— **im·i·ta·tive·ness** *noun*

im·mac·u·la·cy \i-ˈma-kyə-lə-sē\ *noun* (1799)
: the quality or state of being immaculate

im·mac·u·late \i-ˈma-kyə-lət\ *adjective* [Middle English *immaculat,* from Latin *immaculatus,* from *in-* + *maculatus* stained — more at MACULATE] (15th century)
1 : having no stain or blemish **:** PURE
2 : containing no flaw or error
3 a : spotlessly clean **b :** having no colored spots or marks ⟨petals *immaculate*⟩
— **im·mac·u·late·ly** *adverb*

Immaculate Conception *noun* (1687)
1 : the conception of the Virgin Mary in which as decreed in Roman Catholic dogma her soul was preserved free from original sin by divine grace
2 : December 8 observed as a Roman Catholic feast in commemoration of the Immaculate Conception

im·mane \i-ˈmān\ *adjective* [Latin *immanis,* from *in-* + *manus* good — more at MATURE] (1602)
archaic **:** HUGE; *also* **:** monstrous in character

im·ma·nence \ˈi-mə-nən(t)s\ *noun* (1816)
: the quality or state of being immanent **:** INHERENCE

im·ma·nen·cy \-nən(t)-sē\ *noun* (1659)
: IMMANENCE

im·ma·nent \-nənt\ *adjective* [Late Latin *immanent-, immanens,* present participle of *immanēre* to remain in place, from Latin *in-* + *manēre* to remain — more at MANSION] (1535)
: remaining or operating within a domain of reality or realm of discourse **:** INHERENT; *specifically* **:** having existence or effect only within the mind or consciousness — compare TRANSCENDENT
— **im·ma·nent·ly** *adverb*

im·ma·nent·ism \-nən-ˌti-zəm\ *noun* (1907)
: any of several theories according to which God or an abstract mind or spirit pervades the world
— **im·ma·nent·ist** \-nən-tist, -ˌnen-\ *noun or adjective*

\ə\ **abut** \ᵊ\ **kitten** \ər\ **further** \a\ **ash** \ā\ **ace**
\ä\ **mop, mar** \au̇\ **out** \ch\ **chin** \e\ **bet** \ē\ **easy**
\g\ **go** \i\ **hit** \ī\ **ice** \j\ **job** \ŋ\ **sing** \ō\ **go**
\ȯ\ **law** \ȯi\ **boy** \th\ **thin** \t͟h\ **the** \ü\ **loot** \u̇\ **foot**
\y\ **yet** \zh\ **vision** *see also* Guide to Pronunciation

— **im·ma·nent·is·tic** \ˌi-mə-nən-ˈtis-tik\ *adjective*

Im·man·u·el \i-ˈman-yə-wəl, -yəl\ *noun* [Middle English *Emanuel,* from Late Latin *Emmanuel,* from Greek *Emmanouēl,* from Hebrew *ʽimmānū'ēl,* literally, with us is God] (15th century)
: MESSIAH 1

im·ma·te·ri·al \ˌi-mə-ˈtir-ē-əl\ *adjective* [Middle English *immateriel,* from Middle French, from Late Latin *immaterialis,* from Latin *in-* + Late Latin *materialis* material] (14th century)
1 : not consisting of matter **:** INCORPOREAL
2 : of no substantial consequence **:** UNIMPORTANT

im·ma·te·ri·al·ism \-ē-ə-ˌli-zəm\ *noun* (1713)
: a philosophical theory that material things have no reality except as mental perceptions
— **im·ma·te·ri·al·ist** \-list\ *noun*

im·ma·te·ri·al·i·ty \ˌi-mə-ˌtir-ē-ˈa-lə-tē\ *noun, plural* **-ties** (1570)
1 : the quality or state of being immaterial
2 : something immaterial

im·ma·te·ri·al·ize \-ˈtir-ē-ə-ˌlīz\ *transitive verb* (1661)
: to make immaterial or incorporeal

im·ma·ture \ˌi-mə-ˈtu̇r, -ˈtyu̇r, -ˈchu̇r\ *adjective* [Latin *immaturus,* from *in-* + *maturus* mature] (1548)
1 *archaic* **:** PREMATURE
2 a : lacking complete growth, differentiation, or development ⟨a thin *immature* soil⟩ **b :** having the potential capacity to attain a definitive form or state **:** CRUDE, UNFINISHED ⟨a vigorous but *immature* school of art⟩ **c :** exhibiting less than an expected degree of maturity ⟨emotionally *immature* adults⟩
— **immature** *noun*
— **im·ma·ture·ly** *adverb*
— **im·ma·tu·ri·ty** \-ˈtu̇r-ə-tē, -ˈtyu̇r-, -ˈchu̇r-\ *noun*

im·mea·sur·able \(ˌ)i(m)-ˈme-zh-rə-bəl, -ˈmāzh-; -ˈme-zhə-rə-, -ˈmā-, -zhər-bəl\ *adjective* (14th century)
: incapable of being measured; *broadly* **:** indefinitely extensive
— **im·mea·sur·able·ness** *noun*
— **im·mea·sur·ably** \-blē\ *adverb*

im·me·di·a·cy \i-ˈmē-dē-ə-sē, *British often* -ˈmē-jə-sē\ *noun, plural* **-cies** (1605)
1 : the quality or state of being immediate
2 : something that is immediate — usually used in plural

im·me·di·ate \i-ˈmē-dē-ət, *British often* -ˈmē-jit\ *adjective* [Middle English, from Late Latin *immediatus,* from Latin *in-* + Late Latin *mediatus* intermediate — more at MEDIATE] (15th century)
1 a : acting or being without the intervention of another object, cause, or agency **:** DIRECT ⟨the *immediate* cause of death⟩ **b :** present to the mind independently of other states or factors ⟨*immediate* awareness⟩ **c :** involving or derived from a single premise ⟨an *immediate* inference⟩
2 : being next in line or relation ⟨the *immediate* family⟩
3 a : existing without intervening space or substance ⟨brought into *immediate* contact⟩ **b :** being near at hand ⟨the *immediate* neighborhood⟩
4 a : occurring, acting, or accomplished without loss or interval of time **:** INSTANT ⟨an *immediate* need⟩ **b** (1) **:** near to or related to the present ⟨the *immediate* past⟩ (2) **:** of or relating to the here and now **:** CURRENT ⟨too busy with *immediate* concerns to worry about the future⟩
5 : directly touching or concerning a person or thing ⟨the child's *immediate* world is the classroom⟩

immediate constituent *noun* (1933)
: any of the meaningful constituents directly forming a larger linguistic construction (as a phrase or sentence)

¹im·me·di·ate·ly \i-ˈmē-dē-ət-lē *also* -ˈmē-dit-, *British often* -ˈmē-jət-\ *adverb* (15th century)
1 : in direct connection or relation **:** DIRECTLY ⟨the parties *immediately* involved in the case⟩ ⟨the house *immediately* beyond this one⟩
2 : without interval of time **:** STRAIGHTWAY

²immediately *conjunction* (1839)
chiefly British **:** AS SOON AS

im·me·di·ate·ness *noun* (1633)
: IMMEDIACY 1

im·med·i·ca·ble \(ˌ)i(m)-ˈme-di-kə-bəl\ *adjective* [Latin *immedicabilis,* from *in-* + *medicabilis* medicable] (1533)
: INCURABLE ⟨wounds *immedicable* —John Milton⟩
— **im·med·i·ca·bly** \-blē\ *adverb*

Im·mel·mann \ˈi-məl-mən\ *noun* [Max *Immelmann*] (1917)
: a maneuver in which an airplane reverses direction by executing half of a loop upwards followed by half of a roll — called also *Immelmann turn*

im·me·mo·ri·al \ˌi-mə-ˈmōr-ē-əl, -ˈmȯr-\ *adjective* [probably from French *immémorial,* from Middle French, from Medieval Latin *immemorialis* lacking memory, from Latin *in-* + *memorialis* memorial] (1602)
: extending or existing since beyond the reach of memory, record, or tradition ⟨existing from time *immemorial*⟩
— **im·me·mo·ri·al·ly** \-ē-ə-lē\ *adverb*

im·mense \i-ˈmen(t)s\ *adjective* [Middle English, from Middle French, from Latin *immensus* immeasurable, from *in-* + *mensus,* past participle of *metiri* to measure — more at MEASURE] (15th century)
1 : marked by greatness especially in size or degree; *especially* **:** transcending ordinary means of measurement ⟨the *immense* and boundless universe⟩
2 : supremely good **:** EXCELLENT
synonym see ENORMOUS
— **im·mense·ly** *adverb*
— **im·mense·ness** *noun*

im·men·si·ty \i-ˈmen(t)-sə-tē\ *noun, plural* **-ties** (15th century)
1 : the quality or state of being immense
2 : something immense

im·men·su·ra·ble \(ˌ)i(m)-ˈmen(t)s-rə-bəl, -ˈmen(t)sh-; -ˈmen(t)-sə-, -shə-\ *adjective* [Middle English, from Late Latin *immensurabilis,* from Latin *in-* + Late Latin *mensurabilis* measurable — more at MENSURABLE] (15th century)
: IMMEASURABLE

im·merge \i-ˈmərj\ *intransitive verb* **im·merged; im·merg·ing** [Latin *immergere*] (1706)
: to plunge into or immerse oneself in something

im·merse \i-ˈmərs\ *transitive verb* **im·mersed; im·mers·ing** [Middle English, from Latin *immersus,* past participle of *immergere,* from *in-* + *mergere* to merge] (15th century)
1 : to plunge into something that surrounds or covers; *especially* **:** to plunge or dip into a fluid
2 : ENGROSS, ABSORB ⟨completely *immersed* in his work⟩
3 : to baptize by immersion

im·mers·ible \i-ˈmər-sə-bəl\ *adjective* (circa 1846)
: capable of being totally submerged in water without damage to the heating element ⟨an *immersible* electric frying pan⟩

im·mer·sion \i-ˈmər-zhən, -shən\ *noun* (15th century)
: an act of immersing **:** a state of being immersed; *specifically* **:** baptism by complete submersion of the person in water

immersion heater *noun* (1914)
: a usually electric unit that heats the liquid in which it is immersed

im·mesh \i-ˈmesh\ *variant of* ENMESH

im·me·thod·i·cal \ˌi-mə-ˈthä-di-kəl\ *adjective* (1605)
: not methodical
— **im·me·thod·i·cal·ly** \-k(ə-)lē\ *adverb*

im·mi·grant \ˈi-mi-grənt\ *noun* (1789)
: one that immigrates: as **a :** a person who comes to a country to take up permanent residence **b :** a plant or animal that becomes established in an area where it was previously unknown
— **immigrant** *adjective*

im·mi·grate \ˈi-mə-ˌgrāt\ *verb* **-grat·ed; -grat·ing** [Latin *immigratus,* past participle of *immigrare* to remove, go in, from *in-* + *migrare* to migrate] (circa 1623)
intransitive verb
: to enter and usually become established; *especially* **:** to come into a country of which one is not a native for permanent residence
transitive verb
: to bring in or send as immigrants
— **im·mi·gra·tion** \ˌi-mə-ˈgrā-shən\ *noun*
— **im·mi·gra·tion·al** \-shnəl, -shə-nᵊl\ *adjective*

im·mi·nence \ˈi-mə-nən(t)s\ *noun* (1606)
1 : something imminent; *especially* **:** impending evil or danger
2 : the quality or state of being imminent

im·mi·nen·cy \-nən-sē\ *noun* (1665)
: IMMINENCE 2

im·mi·nent \ˈi-mə-nənt\ *adjective* [Latin *imminent-, imminens,* present participle of *imminēre* to project, threaten, from *in-* + *-minēre* (akin to Latin *mont-, mons* mountain) — more at MOUNT] (1528)
: ready to take place; *especially* **:** hanging threateningly over one's head ⟨was in *imminent* danger of being run over⟩
— **im·mi·nent·ly** *adverb*

im·min·gle \i-ˈmiŋ-gəl\ *verb* (1606)
: BLEND, INTERMINGLE

im·mis·ci·ble \(ˌ)i(m)-ˈmi-sə-bəl\ *adjective* (1671)
: incapable of mixing or attaining homogeneity
— **im·mis·ci·bil·i·ty** \-ˌmi-sə-ˈbi-lə-tē\ *noun*

im·mit·i·ga·ble \(ˌ)i(m)-ˈmi-ti-gə-bəl\ *adjective* [Late Latin *immitigabilis,* from Latin *in-* + *mitigare* to mitigate] (1576)
: not capable of being mitigated
— **im·mit·i·ga·bly** \-blē\ *adverb*

im·mit·tance \(ˌ)i(m)-ˈmi-tᵊn(t)s\ *noun* [*impedance* + *admittance*] (circa 1948)
: electrical admittance or impedance

im·mix \i-ˈmiks\ *transitive verb* [back-formation from *immixed* mixed in, from Middle English *immixte,* from Latin *immixtus,* past participle of *immiscēre,* from *in-* + *miscēre* to mix — more at MIX] (15th century)
: to mix intimately **:** COMMINGLE
— **im·mix·ture** \-ˈmiks-chər\ *noun*

im·mo·bile \(ˌ)i(m)-ˈmō-bəl, -ˌbīl *also* -ˌbēl\ *adjective* [Middle English *in-mobill,* from Latin *immobilis,* from *in-* + *mobilis* mobile] (14th century)
1 : incapable of being moved **:** FIXED
2 : not moving **:** MOTIONLESS ⟨keep the patient *immobile*⟩
— **im·mo·bil·i·ty** \ˌi-(ˌ)mō-ˈbi-lə-tē\ *noun*

im·mo·bi·lism \i-ˈmō-bə-ˌli-zəm\ *noun* (1949)
: a policy of extreme conservatism and opposition to change

im·mo·bi·lize \i-ˈmō-bə-ˌlīz\ *transitive verb* (1871)
: to make immobile: as **a :** to prevent freedom of movement or effective use of ⟨the planes were *immobilized* by bad weather⟩ **b :** to reduce or eliminate motion of (the body or a part) by mechanical means or by strict bed rest **c :** to withhold (money or capital) from circulation
— **im·mo·bi·li·za·tion** \-ˌmō-bə-lə-ˈzā-shən\ *noun*
— **im·mo·bi·liz·er** \-ˈmō-bə-ˌlī-zər\ *noun*

im·mod·er·a·cy \(ˌ)i(m)-ˈmä-d(ə-)rə-sē\ *noun* (1682)
: lack of moderation

im·mod·er·ate \-d(ə-)rət\ *adjective* [Middle English *immoderat,* from Latin *immoderatus,* from *in-* + *moderatus,* past participle of *moderare* to moderate] (14th century)
: exceeding just, usual, or suitable bounds ⟨*immoderate* pride⟩ ⟨an *immoderate* appetite⟩
synonym see EXCESSIVE
— **im·mod·er·ate·ly** *adverb*
— **im·mod·er·ate·ness** *noun*
— **im·mod·er·a·tion** \(ˌ)i-ˌmä-də-ˈrā-shən\ *noun*

im·mod·est \(ˌ)i(m)-ˈmä-dəst\ *adjective* [Latin *immodestus,* from *in-* + *modestus* modest] (circa 1570)
: not modest; *specifically* : not conforming to the sexual mores of a particular time or place
— **im·mod·est·ly** *adverb*
— **im·mod·es·ty** \-də-stē\ *noun*

im·mo·late \ˈi-mə-ˌlāt\ *transitive verb* **-lat·ed; -lat·ing** [Latin *immolatus,* past participle of *immolare,* from *in-* + *mola* spelt grits; from the custom of sprinkling victims with sacrificial meal; akin to Latin *molere* to grind — more at MEAL] (1548)
1 : to offer in sacrifice; *especially* : to kill as a sacrificial victim
2 : to kill or destroy often by fire
— **im·mo·la·tor** \-ˌlā-tər\ *noun*

im·mo·la·tion \ˌi-mə-ˈlā-shən\ *noun* (15th century)
1 : the act of immolating : the state of being immolated
2 : something that is immolated

im·mor·al \(ˌ)i(m)-ˈmȯr-əl, -ˈmär-\ *adjective* (1660)
: not moral; *broadly* : conflicting with generally or traditionally held moral principles
— **im·mor·al·ly** \-ə-lē\ *adverb*

im·mor·al·ist \-ə-list\ *noun* (1697)
: an advocate of immorality
— **im·mor·al·ism** \-ˌli-zəm\ *noun*

im·mo·ral·i·ty \ˌi (ˌ)mȯ-ˈra-lə-tē, ˌi-mə-\ *noun* (circa 1566)
1 : the quality or state of being immoral; *especially* : UNCHASTITY
2 : an immoral act or practice

[1]**im·mor·tal** \(ˌ)i-ˈmȯr-tᵊl\ *adjective* [Middle English, from Latin *immortalis,* from *in-* + *mortalis* mortal] (14th century)
1 : exempt from death ⟨the *immortal* gods⟩
2 : exempt from oblivion : IMPERISHABLE ⟨*immortal* fame⟩
3 : connected with or relating to immortality
— **im·mor·tal·ly** \-tᵊl-ē\ *adverb*

[2]**immortal** *noun* (1684)
1 a : one exempt from death **b** *plural, often capitalized* : the gods of the Greek and Roman pantheon
2 a : a person whose fame is lasting **b** *capitalized* : any of the 40 members of the Académie Française

im·mor·tal·ise *British variant of* IMMORTALIZE

im·mor·tal·i·ty \ˌi-ˌmȯr-ˈta-lə-tē\ *noun* (14th century)
: the quality or state of being immortal: **a :** unending existence **b :** lasting fame

im·mor·tal·ize \i-ˈmȯr-tᵊl-ˌīz\ *transitive verb* **-ized; -iz·ing** (circa 1566)
: to make immortal
— **im·mor·tal·i·za·tion** \-ˌmȯr-tᵊl-ə-ˈzā-shən\ *noun*
— **im·mor·tal·iz·er** \-ˈmȯr-tᵊl-ˌī-zər\ *noun*

im·mor·telle \ˌi-ˌmȯr-ˈtel\ *noun* [French, from feminine of *immortel* immortal, from Latin *immortalis*] (1832)
: EVERLASTING 3

im·mo·tile \(ˌ)i(m)-ˈmō-tᵊl, -ˌtīl\ *adjective* (1872)
: lacking motility

[1]**im·mov·able** \(ˌ)i(m)-ˈmü-və-bəl\ *adjective* (14th century)
1 : incapable of being moved; *broadly* : not moving or not intended to be moved
2 a : STEADFAST, UNYIELDING **b :** not capable of being moved emotionally
— **im·mov·abil·i·ty** \-ˌmü-və-ˈbi-lə-tē\ *noun*
— **im·mov·able·ness** \-ˈmü-və-bəl-nəs\ *noun*
— **im·mov·ably** \-blē\ *adverb*

[2]**immovable** *noun* (1588)
1 : one that cannot be moved
2 *plural* : real property as opposed to movable property

im·mune \i-ˈmyün\ *adjective* [Middle English, from Latin *immunis,* from *in-* + *munia* services, obligations; akin to Latin *munus* service — more at MEAN] (15th century)
1 a : FREE, EXEMPT ⟨*immune* from further taxation⟩ **b :** marked by protection ⟨some criminal leaders are *immune* from arrest⟩
2 : not susceptible or responsive ⟨*immune* to all pleas⟩; *especially* : having a high degree of resistance to a disease ⟨*immune* to diphtheria⟩
3 a : having or producing antibodies or lymphocytes capable of reacting with a specific antigen ⟨an *immune* serum⟩ **b :** produced by, involved in, or concerned with immunity or an immune response ⟨*immune* agglutinins⟩ ⟨*immune* globulins⟩
— **immune** *noun*

immune response *noun* (1953)
: a bodily response to an antigen that occurs when lymphocytes identify the antigenic molecule as foreign and induce the formation of antibodies and lymphocytes capable of reacting with it and rendering it harmless — called also *immune reaction*

immune system *noun* (circa 1919)
: the bodily system that protects the body from foreign substances, cells, and tissues by producing the immune response and that includes especially the thymus, spleen, lymph nodes, special deposits of lymphoid tissue (as in the gastrointestinal tract and bone marrow), lymphocytes including the B cells and T cells, and antibodies

im·mu·ni·ty \i-ˈmyü-nə tē\ *noun, plural* **-ties** (14th century)
: the quality or state of being immune; *especially* : a condition of being able to resist a particular disease especially through preventing development of a pathogenic microorganism or by counteracting the effects of its products

im·mu·nize \ˈi-myə-ˌnīz\ *transitive verb* **-nized; -niz·ing** (1892)
: to make immune
— **im·mu·ni·za·tion** \ˌi-myə-nə-ˈzā-shən *also* i-ˌmyü-nə-\ *noun*

immuno- *combining form* [International Scientific Vocabulary, from *immune*]
1 : physiological immunity ⟨*immuno*logy⟩
2 : immunologic ⟨*immuno*chemistry⟩ : immunologically ⟨*immuno*competent⟩ : immunology and ⟨*immuno*genetics⟩

im·mu·no·as·say \ˌi-myə-nō-ˈa-ˌsā, i-ˌmyü-nō-, -a-ˈsā\ *noun* (1959)
: the identification of a substance (as a protein) based on its capacity to act as an antigen
— **im·mu·no·as·say·able** \-a-ˈsā-ə-bəl\ *adjective*

im·mu·no·blot \ˈi-myə-nə-ˌblät, i-ˈmyü-(ˌ)nō-\ *noun* (1982)
: a blot in which a radioactively labeled antibody is used as the molecular probe
— **im·mu·no·blot·ting** \-ˌblä-tiŋ\ *noun*

im·mu·no·chem·is·try \ˌi-myə-nō-ˈke-mə-strē, i-ˌmyü-nō-\ *noun* [International Scientific Vocabulary] (1907)
: a branch of chemistry that deals with the chemical aspects of immunology
— **im·mu·no·chem·i·cal** \-ˈke-mi-kəl\ *adjective*
— **im·mu·no·chem·i·cal·ly** \-k(ə-)lē\ *adverb*
— **im·mu·no·chem·ist** \-ˈke-mist\ *noun*

im·mu·no·com·pe·tence \-ˈkäm-pə-tən(t)s\ *noun* (1967)
: the capacity for a normal immune response
— **im·mu·no·com·pe·tent** \-tənt\ *adjective*

im·mu·no·com·pro·mised \-ˈkäm-prə-ˌmīzd\ *adjective* (1979)
: having the immune system impaired or weakened (as by drugs or illness) ⟨*immunocompromised* patients⟩

im·mu·no·cy·to·chem·is·try \-ˌsī-tō-ˈke-mə-strē\ *noun* (1960)
: the application of biochemistry to cellular immunology
— **im·mu·no·cy·to·chem·i·cal** \-ˈkə-mi-kəl\ *adjective*
— **im·mu·no·cy·to·chem·i·cal·ly** \-ˈke-mi-k(ə-)lē\ *adverb*

im·mu·no·de·fi·cien·cy \-di-ˈfi-shən(t)-sē\ *noun* (1969)
: inability to produce a normal complement of antibodies or immunologically sensitized T cells especially in response to specific antigens
— **im·mu·no·de·fi·cient** \-shənt\ *adjective*

im·mu·no·di·ag·no·sis \-ˌdī-ig-ˈnō-səs, -əg-\ *noun* (1972)
: diagnosis (as of cancer) by immunological methods
— **im·mu·no·di·ag·nos·tic** \-ˈnäs-tik\ *adjective*

im·mu·no·dif·fu·sion \-di-ˈfyü-zhən\ *noun* (1959)
: any of several techniques for obtaining a precipitate between an antibody and its specific antigen by suspending one in a gel and letting the other migrate through it from a well or by letting both antibody and antigen migrate through the gel from separate wells to form an area of precipitation

im·mu·no·elec·tro·pho·re·sis \-ə-ˌlek-trə-fə-ˈrē-səs\ *noun, plural* **-re·ses** \-ˌsēz\ (1958)
: electrophoretic separation of proteins followed by identification by the formation of precipitates through specific immunologic reactions
— **im·mu·no·elec·tro·pho·ret·ic** \-ˈre-tik\ *adjective*
— **im·mu·no·elec·tro·pho·ret·i·cal·ly** \-ti-k(ə-)lē\ *adverb*

im·mu·no·flu·o·res·cence \-(ˌ)flȯ-ˈres-ᵊn(t)s, -(ˌ)flō-, -(ˌ)flü(-ə)-\ *noun* (1960)
: the labeling of antibodies or antigens with fluorescent dyes especially for the purpose of demonstrating the presence of a particular antigen or antibody in a tissue preparation or smear
— **im·mu·no·flu·o·res·cent** \-ᵊnt\ *adjective*

im·mu·no·gen \i-ˈmyü-nə-jən, -ˌjen\ *noun* [from *Immunogen,* a trademark] (1959)
: an antigen that produces an immune response

im·mu·no·ge·net·ics \ˌi-myə-nō-jə-ˈne-tiks, i-ˌmyü-nō-\ *noun plural but singular in construction* (1936)
: a branch of immunology concerned with the interrelations of heredity, disease, and the immune system and its components (as antibodies)
— **im·mu·no·ge·net·ic** \-tik\ *adjective*
— **im·mu·no·ge·net·i·cal·ly** \-ti-k(ə-)lē\ *adverb*
— **im·mu·no·ge·net·i·cist** \-jə-ˈne-tə-sist\ *noun*

im·mu·no·gen·ic \ˌi-myə-nō-ˈje-nik, i-ˌmyü-nō-\ *adjective* (circa 1923)
: relating to or producing an immune response ⟨*immunogenic* substances⟩

— **im·mu·no·gen·e·sis** \-ˈje-nə-səs\ *noun*
— **im·mu·no·ge·nic·i·ty** \-jə-ˈni-sə-tē\ *noun*

im·mu·no·glob·u·lin \-ˈglä-byə-lən\ *noun* (1953)
: ANTIBODY

im·mu·no·he·ma·tol·o·gy \-ˌhē-mə-ˈtä-lə-jē\ *noun* (1950)
: a branch of immunology that deals with the immunologic properties of blood
— **im·mu·no·he·ma·to·log·ic** \-ˌhē-mə-t^{ə}l-ˈä-jik\ *or* **im·mu·no·he·ma·to·log·i·cal** \-ˈä-ji-kəl\ *adjective*
— **im·mu·no·he·ma·tol·o·gist** \-ˌhē-mə-ˈtä-lə-jist\ *noun*

im·mu·no·his·to·chem·i·cal \-ˌhis-tō-ˈke-mi-kəl\ *adjective* (1960)
: of or relating to the application of histochemical and immunologic methods to chemical analysis of living cells and tissues
— **im·mu·no·his·to·chem·is·try** \-ˈke-mə-strē\ *noun*

im·mu·nol·o·gy \ˌi-myə-ˈnä-lə-jē\ *noun* [International Scientific Vocabulary] (1910)
: a science that deals with the immune system and the cell-mediated and humoral aspects of immunity and immune responses
— **im·mu·no·log·ic** \-nə-ˈlä-jik\ *or* **im·mu·no·log·i·cal** \-ˈlä-ji-kəl\ *adjective*
— **im·mu·no·log·i·cal·ly** \-ji-k(ə-)lē\ *adverb*
— **im·mu·nol·o·gist** \ˌi-myə-ˈnä-lə-jist\ *noun*

im·mu·no·mod·u·la·tor \ˌi-myə-nō-ˈmä-jə-ˌlā-tər, i-ˌmyü-nō-\ *noun* (1977)
: a substance that affects the functioning of the immune system
— **im·mu·no·mod·u·la·to·ry** \-ˈmä-jə-lə-ˌtōr-ē, -ˌtȯr-\ *adjective*

im·mu·no·pa·thol·o·gy \-pə-ˈthä-lə-jē, -pa-\ *noun* (1959)
: a branch of medicine that deals with immune responses associated with disease
— **im·mu·no·path·o·log·ic** \-ˌpa-thə-ˈlä-jik\ *or* **im·mu·no·path·o·log·i·cal** \-ji-kəl\ *adjective*
— **im·mu·no·pa·thol·o·gist** \-pə-ˈthä-lə-jist, -pa-\ *noun*

im·mu·no·pre·cip·i·ta·tion \-pri-ˌsi-pə-ˈtā-shən\ *noun* (1966)
: precipitation of a complex of an antibody and its specific antigen
— **im·mu·no·pre·cip·i·tate** \-ˈsi-pə-tət, -pə-ˌtāt\ *noun*
— **im·mu·no·pre·cip·i·tate** \-pə-ˌtāt\ *transitive verb*

im·mu·no·re·ac·tive \-rē-ˈak-tiv\ *adjective* (1966)
: reacting to particular antigens or haptens ⟨*immunoreactive* lymphocytes⟩
— **im·mu·no·re·ac·tiv·i·ty** \-(ˌ)rē-ˌak-ˈti-və-tē\ *noun*

im·mu·no·reg·u·la·to·ry \-ˈre-gyə-lə-ˌtōr-ē, -ˌtȯr-\ *adjective* (1971)
: of or relating to the regulation of the immune system ⟨*immunoregulatory* T cells⟩
— **im·mu·no·reg·u·la·tion** \-ˌre-gyə-ˈlā-shən, -gə-\ *noun*

im·mu·no·sor·bent \-ˈsȯr-bənt\ *adjective* (1966)
: relating to or using a substrate consisting of a specific antibody or antigen chemically combined with an insoluble substance (as cellulose) to selectively remove the corresponding specific antigen or antibody from solution
— **immunosorbent** *noun*

im·mu·no·sup·pres·sion \-sə-ˈpre-shən\ *noun* (1963)
: suppression (as by drugs) of natural immune responses
— **im·mu·no·sup·press** \-sə-ˈpres\ *transitive verb*
— **im·mu·no·sup·pres·sant** \-ˈpre-s^{ə}nt\ *noun or adjective*
— **im·mu·no·sup·pres·sive** \-ˈpre-siv\ *adjective*

im·mu·no·ther·a·py \-ˈther-ə-pē\ *noun* [International Scientific Vocabulary] (circa 1911)
: treatment of or prophylaxis against disease by attempting to produce active or passive immunity
— **im·mu·no·ther·a·peu·tic** \-ˌther-ə-ˈpyü-tik\ *adjective*

im·mure \i-ˈmyu̇r\ *transitive verb* **im·mured; im·mur·ing** [Medieval Latin *immurare,* from Latin *in-* + *murus* wall — more at MUNITION] (1583)
1 a : to enclose within or as if within walls **b :** IMPRISON
2 : to build into a wall; *especially* **:** to entomb in a wall
— **im·mure·ment** \-ˈmyu̇r-mənt\ *noun*

im·mu·ta·ble \(ˌ)i(m)-ˈmyü-tə-bəl\ *adjective* [Middle English, from Latin *immutabilis,* from *in-* + *mutabilis* mutable] (15th century)
: not capable of or susceptible to change
— **im·mu·ta·bil·i·ty** \-ˌmyü-tə-ˈbi-lə-tē\ *noun*
— **im·mu·ta·ble·ness** \-ˈmyü-tə-bəl-nəs\ *noun*
— **im·mu·ta·bly** \-blē\ *adverb*

[1]**imp** \ˈimp\ *noun* [Middle English *impe,* from Old English *impa,* from *impian* to imp] (before 12th century)
1 *obsolete* **:** SHOOT, BUD; *also* **:** GRAFT
2 a : a small demon **:** FIEND **b :** a mischievous child **:** URCHIN ◆

[2]**imp** *transitive verb* [Middle English, from Old English *impian* to graft, from (assumed) Vulgar Latin *imputare,* from Late Latin *impotus* grafted shoot, from Greek *emphytos* implanted, from *emphyein* to implant, from *em-* [2]en- + *phyein* to bring forth — more at BE] (15th century)
1 : to graft or repair (a wing, tail, or feather) with a feather to improve a falcon's flying capacity
2 : to equip with wings

[1]**im·pact** \im-ˈpakt\ *verb* [Latin *impactus,* past participle of *impingere* to push against — more at IMPINGE] (1601)
transitive verb
1 a : to fix firmly by or as if by packing or wedging **b :** to press together
2 a : to have an impact on **:** impinge on **b :** to strike forcefully; *also* **:** to cause to strike forcefully
intransitive verb
1 : to have an impact
2 : to impinge or make contact especially forcefully
— **im·pac·tive** \im-ˈpak-tiv\ *adjective*
— **im·pac·tor** *also* **im·pact·er** \-tər\ *noun*

[2]**im·pact** \ˈim-ˌpakt\ *noun* (1781)
1 a : an impinging or striking especially of one body against another **b :** a forceful contact, collision, or onset; *also* **:** the impetus communicated in or as if in a collision
2 : the force of impression of one thing on another **:** a significant or major effect ⟨the *impact* of science on our society⟩ ⟨an environmental *impact* study⟩

impact crater *noun* (1895)
: CRATER 1b

im·pact·ed \im-ˈpak-təd\ *adjective* (circa 1616)
1 a : packed or wedged in **b :** deeply entrenched **:** not easily changed or removed
2 *of a tooth* **:** wedged between the jawbone and another tooth
3 : of, relating to, or being an area (as a school district) providing tax-supported services to a population having a large proportion of federal employees and especially those living or working on tax-exempt federal property ⟨aid to education in *impacted* areas⟩

im·pac·tion \im-ˈpak-shən\ *noun* (1739)
: the act of becoming or the state of being impacted; *especially* **:** lodgment of something (as feces) in a body passage or cavity

impact printer *noun* (1968)
: a printing device in which a printing element directly strikes a surface (as in a typewriter)

im·paint \im-ˈpānt\ *transitive verb* (1596)
obsolete **:** PAINT, DEPICT

im·pair \im-ˈpar, -ˈper\ *transitive verb* [Middle English *empeiren,* from Middle French *empeirer,* from (assumed) Vulgar Latin *impejorare,* from Latin *in-* + Late Latin *pejorare* to make worse — more at PEJORATIVE] (14th century)
: to damage or make worse by or as if by diminishing in some material respect ⟨his health was *impaired* by overwork⟩ ⟨the strike seriously *impaired* community services⟩
synonym see INJURE
— **im·pair·er** *noun*
— **im·pair·ment** \-mənt\ *noun*

im·paired \-ˈpard, -ˈperd\ *adjective* (1605)
: being in a less than perfect or whole condition: as **a :** handicapped or functionally defective — often used in combination ⟨hearing-*impaired*⟩ **b** *chiefly Canadian* **:** DRUNK ⟨driving while *impaired*⟩

im·pa·la \im-ˈpa-lə, -ˈpä-\ *noun, plural* **impalas** *or* **impala** [Zulu] (1875)
: a large brownish African antelope (*Aepyceros melampus*) that in the male has slender lyrate horns

impala

im·pale \im-ˈpā(ə)l\ *transitive verb* **im·paled; im·pal·ing** [Middle French & Medieval Latin; Middle French *empaler,* from Medieval Latin *impalare,* from Latin *in-* + *palus* stake — more at POLE] (1605)
1 : to join (coats of arms) on a heraldic shield divided vertically by a pale
2 a : to pierce with or as if with something pointed; *especially* **:** to torture or kill by fixing on a sharp stake **b :** to fix in an inescapable or helpless position
— **im·pale·ment** \-mənt\ *noun*
— **im·pal·er** \-ˈpā-lər\ *noun*

im·pal·pa·ble \(ˌ)im-ˈpal-pə-bəl\ *adjective* (1509)
1 a : incapable of being felt by touch **:** INTANGIBLE ⟨the *impalpable* aura of power that emanated from him —Osbert Sitwell⟩ **b :** so finely divided that no grains or grit can be felt ⟨rock worn to an *impalpable* powder⟩
2 : not readily discerned by the mind
— **im·pal·pa·bil·i·ty** \-ˌpal-pə-ˈbi-lə-tē\ *noun*
— **im·pal·pa·bly** \-ˈpal-pə-blē\ *adverb*

im·pan·el \im-ˈpa-n^{ə}l\ *transitive verb* (15th century)
: to enroll in or on a panel ⟨*impanel* a jury⟩

◇ WORD HISTORY

imp *Imp,* which now usually refers to either a small demon or a mischievous child, has its origin in the unlikely field of horticulture. Old English *impa* or *impe* meant "graft" or "young shoot of a plant" and was a derivative of the verb *impian* "to graft." By later Middle English *impe* could mean metaphorically "offspring" or "child," a sense that survived at least into the 19th century. The frequency of phrases such as *imp of the devil* (i.e., "offspring of the devil") and *imp of hell* led to the use of *imp* without qualification to denote a demon, the evil parentage of the offspring being understood. By the 18th century the demonic sense had, as had the words *demon* and *devil* themselves, given rise to a less forceful usage: *imp* could be applied to nothing more than a naughty child.

im·par·a·dise \im-ˈpar-ə-ˌdīs, -ˌdīz\ *transitive verb* **-dised; -dis·ing** (1592)
: ENRAPTURE

im·par·i·ty \(ˌ)im-ˈpar-ə-tē\ *noun, plural* **-ties** [Late Latin *imparitas,* from Latin *impar* unequal, from *in-* + *par* equal] (1563)
: INEQUALITY, DISPARITY

im·part \im-ˈpärt\ *transitive verb* [Middle French & Latin; Middle French *impartir,* from Latin *impartire,* from *in-* + *partire* to divide, part] (15th century)
1 : to give, convey, or grant from or as if from a store ⟨her experience *imparted* authority to her words⟩ ⟨the flavor *imparted* by herbs⟩
2 : to communicate the knowledge of **:** DISCLOSE ⟨*imparted* my scheme to no one⟩
— **im·par·ta·tion** \ˌim-ˌpär-ˈtā-shən\ *noun*
— **im·part·ment** \im-ˈpärt-mənt\ *noun*

im·par·tial \(ˌ)im-ˈpär-shəl\ *adjective* (1587)
: not partial or biased **:** treating or affecting all equally
synonym see FAIR
— **im·par·tial·i·ty** \-ˌpär-shē-ˈa-lə-tē, -ˌpär-ˈsha-\ *noun*
— **im·par·tial·ly** \-ˈpär-sh(ə-)lē\ *adverb*

im·par·ti·ble \(ˌ)im-ˈpär-tə-bəl\ *adjective* [Late Latin *impartibilis,* from Latin *in-* + Late Latin *partibilis* divisible, from Latin *partire*] (14th century)
: not partible **:** not subject to partition
— **im·par·ti·bly** \-blē\ *adverb*

im·pass·able \(ˌ)im-ˈpa-sə-bəl\ *adjective* (1568)
: incapable of being passed, traveled, crossed, or surmounted
— **im·pass·abil·i·ty** \-ˌpa-sə-ˈbi-lə-tē\ *noun*
— **im·pass·able·ness** \-ˈpa-sə-bəl-nəs\ *noun*
— **im·pass·ably** \-blē\ *adverb*

im·passe \ˈim-ˌpas, im-ˈ\ *noun* [French, from *in-* + *passer* to pass] (1851)
1 a : a predicament affording no obvious escape **b :** DEADLOCK
2 : an impassable road or way **:** CUL-DE-SAC

im·pas·si·ble \(ˌ)im-ˈpa-sə-bəl\ *adjective* [Middle English, from Middle French or Late Latin; Middle French, from Late Latin *impassibilis,* from Latin *in-* + Late Latin *passibilis* passible] (14th century)
1 a : incapable of suffering or of experiencing pain **b :** inaccessible to injury
2 : incapable of feeling **:** IMPASSIVE
— **im·pas·si·bil·i·ty** \-ˌpa-sə-ˈbi-lə-tē\ *noun*
— **im·pas·si·bly** \-ˈpa-sə-blē\ *adverb*

im·pas·sion \im-ˈpa-shən\ *transitive verb* **im·pas·sioned; im·pas·sion·ing** \-sh(ə-)niŋ\ [probably from Italian *impassionare,* from *in-* (from Latin) + *passione* passion, from Late Latin *passio-, passio*] (1591)
: to arouse the feelings or passions of

impassioned *adjective* (1603)
: filled with passion or zeal **:** showing great warmth or intensity of feeling ☆

im·pas·sive \(ˌ)im-ˈpa-siv\ *adjective* (1605)
1 a *archaic* **:** unsusceptible to pain **b :** unsusceptible to physical feeling **:** INSENSIBLE **c :** unsusceptible to or destitute of emotion **:** APATHETIC
2 : giving no sign of feeling or emotion **:** EXPRESSIONLESS ☆
— **im·pas·sive·ly** *adverb*
— **im·pas·sive·ness** *noun*
— **im·pas·siv·i·ty** \ˌim-ˌpa-ˈsi-və-tē\ *noun*

im·paste \im-ˈpāst\ *transitive verb* [Italian *impastare,* from *in-* (from Latin) + *pasta* paste, from Late Latin] (1576)
obsolete **:** to make into a paste or crust

im·pas·to \im-ˈpas-(ˌ)tō, -ˈpäs-\ *noun, plural* **-tos** [Italian, from *impastare*] (1784)
1 : the thick application of a pigment to a canvas or panel in painting; *also* **:** the body of pigment so applied
2 : raised decoration on ceramic ware usually of slip or enamel
— **im·pas·toed** \-(ˌ)tōd\ *adjective*

im·pa·tience \(ˌ)im-ˈpā-shən(t)s\ *noun* (13th century)
: the quality or state of being impatient

im·pa·tiens \im-ˈpā-shənz, -shən(t)s\ *noun* [New Latin, from Latin, impatient] (1885)
: any of a widely distributed genus (*Impatiens,* family Balsaminaceae) of annual herbs with irregular spurred or saccate flowers and dehiscent capsules — compare TOUCH-ME-NOT

im·pa·tient \(ˌ)im-ˈpā-shənt\ *adjective* [Middle English *impacient,* from Middle French, from Latin *impatient-, impatiens,* from *in-* + *patient-, patiens* patient] (14th century)
1 a : not patient **:** restless or short of temper especially under irritation, delay, or opposition **b :** INTOLERANT ⟨*impatient* of delay⟩
2 : prompted or marked by impatience ⟨an *impatient* reply⟩
3 : eagerly desirous **:** ANXIOUS ⟨*impatient* to get home⟩
— **im·pa·tient·ly** *adverb*

im·pawn \im-ˈpȯn, -ˈpän\ *transitive verb* (1596)
archaic **:** to put in pawn **:** PLEDGE

[1]im·peach \im-ˈpēch\ *transitive verb* [Middle English *empechen,* from Middle French *empeechier* to hinder, from Late Latin *impedicare* to fetter, from Latin *in-* + *pedica* fetter, from *ped-, pes* foot — more at FOOT] (14th century)
1 a : to bring an accusation against **b :** to charge with a crime or misdemeanor; *specifically* **:** to charge (a public official) before a competent tribunal with misconduct in office
2 : to cast doubt on; *especially* **:** to challenge the credibility or validity of ⟨*impeach* the testimony of a witness⟩
3 : to remove from office especially for misconduct ■
— **im·peach·able** \-ˈpē-chə-bəl\ *adjective*
— **im·peach·ment** \-ˈpēch-mənt\ *noun*

[2]impeach *noun* (1590)
obsolete **:** CHARGE, IMPEACHMENT

im·pearl \im-ˈpər(-ə)l\ *transitive verb* [probably from Middle French *emperler,* from *en-* + *perle* pearl] (15th century)
: to form into pearls; *also* **:** to form of or adorn with pearls

im·pec·ca·ble \(ˌ)im-ˈpek-ə-bəl\ *adjective* [Latin *impeccabilis,* from *in-* + *peccare* to sin] (1531)
1 : not capable of sinning or liable to sin
2 : free from fault or blame **:** FLAWLESS ⟨spoke *impeccable* French⟩
— **im·pec·ca·bil·i·ty** \-ˌpe-kə-ˈbi-lə-tē\ *noun*
— **im·pec·ca·bly** \-ˈpe-kə-blē\ *adverb*

im·pe·cu·nious \ˌim-pi-ˈkyü-nyəs, -nē-əs\ *adjective* [[1]*in-* + obsolete English *pecunious* rich, from Middle English, from Latin *pecuniosus,* from *pecunia* money — more at FEE] (1596)
: having very little or no money usually habitually **:** PENNILESS
— **im·pe·cu·ni·os·i·ty** \-ˌkyü-nē-ˈä-sə-tē\ *noun*
— **im·pe·cu·nious·ly** *adverb*
— **im·pe·cu·nious·ness** *noun*

im·ped·ance \im-ˈpē-dᵊn(t)s\ *noun* (1886)
: something that impedes **:** HINDRANCE; *as* **a :** the apparent opposition in an electrical circuit to the flow of an alternating current that is analogous to the actual electrical resistance to a direct current and that is the ratio of effective electromotive force to the effective current **b :** the ratio of the pressure to the volume displacement at a given surface in a sound-transmitting medium

im·pede \im-ˈpēd\ *transitive verb* **im·ped·ed; im·ped·ing** [Latin *impedire,* from *in-* + *ped-, pes* foot — more at FOOT] (1605)
: to interfere with or slow the progress of
synonym see HINDER
— **im·ped·er** *noun*

im·ped·i·ment \im-ˈpe-də-mənt\ *noun* (14th century)
1 : something that impedes; *especially* **:** an organic obstruction to speech
2 : a bar or hindrance (as lack of sufficient age) to a lawful marriage

im·ped·i·men·ta \(ˌ)im-ˌpe-də-ˈmen-tə\ *noun plural* [Latin, plural of *impedimentum* impediment, from *impedire*] (1600)
1 : APPURTENANCES, EQUIPMENT
2 : things that impede

im·pel \im-ˈpel\ *transitive verb* **im·pelled; im·pel·ling** [Latin *impellere,* from *in-* + *pellere* to drive — more at FELT] (15th century)
1 : to urge or drive forward or on by or as if by the exertion of strong moral pressure **:** FORCE
2 : to impart motion to **:** PROPEL
synonym see MOVE

☆ SYNONYMS

Impassioned, passionate, ardent, fervent, fervid, perfervid mean showing intense feeling. IMPASSIONED implies warmth and intensity without violence and suggests fluent verbal expression ⟨an *impassioned* plea for justice⟩. PASSIONATE implies great vehemence and often violence and wasteful diffusion of emotion ⟨a *passionate* denunciation⟩. ARDENT implies an intense degree of zeal, devotion, or enthusiasm ⟨an *ardent* supporter of human rights⟩. FERVENT stresses sincerity and steadiness of emotional warmth or zeal ⟨*fervent* good wishes⟩. FERVID suggests warmly and spontaneously and often feverishly expressed emotion ⟨*fervid* love letters⟩. PERFERVID implies the expression of exaggerated or overwrought feelings ⟨*perfervid* expressions of patriotism⟩.

Impassive, stoic, phlegmatic, apathetic, stolid mean unresponsive to something that might normally excite interest or emotion. IMPASSIVE stresses the absence of any external sign of emotion in action or facial expression ⟨met the news with an *impassive* look⟩. STOIC implies an apparent indifference to pleasure or especially to pain often as a matter of principle or self-discipline ⟨was resolutely *stoic* even in adversity⟩. PHLEGMATIC implies a temperament or constitution hard to arouse ⟨a *phlegmatic* man unmoved by tears⟩. APATHETIC may imply a puzzling or deplorable indifference or inertness ⟨charitable appeals met an *apathetic* response⟩. STOLID implies an habitual absence of interest, responsiveness, or curiosity ⟨*stolid* workers wedded to routine⟩.

□ USAGE

impeach Back in 1970 a newspaperman took an informal poll of people who worked for the paper and discovered that 10 out of 13 believed *impeach* meant "to remove from office" rather than "to charge with misconduct in office." This is a meaning many people know—our evidence goes back to before World War I—but seldom see in print. (It is sense 3 here.) But the strong presence of this meaning in people's minds can make many uses of *impeach* ambiguous ⟨(who, by the way, was later *impeached* for corruption!) —Julian Huxley⟩. Was he removed or merely charged with corruption? A writer must reckon with the sense "remove from office" when writing for the general public and take care to make the context clear.

im·pel·ler *also* **im·pel·lor** \im-ˈpe-lər\ *noun* (1685)
1 : one that impels
2 : ROTOR; *also* **:** a blade of a rotor
im·pend \im-ˈpend\ *intransitive verb* [Latin *impendēre,* from *in-* + *pendēre* to hang — more at PENDANT] (1599)
1 a : to hover threateningly **:** MENACE **b :** to be about to occur
2 *archaic* **:** to hang suspended
im·pen·dent \im-ˈpen-dənt\ *adjective* (1590)
: being near at hand **:** APPROACHING
im·pen·e·tra·bil·i·ty \(ˌ)im-ˌpe-nə-trə-ˈbi-lə-tē\ *noun* (1665)
1 : the inability of two portions of matter to occupy the same space at the same time
2 : the quality or state of being impenetrable
im·pen·e·tra·ble \(ˌ)im-ˈpe-nə-trə-bəl\ *adjective* [Middle English *impenetrabel,* from Middle French *impenetrable,* from Latin *impenetrabilis,* from *in-* + *penetrabilis* penetrable] (15th century)
1 a : incapable of being penetrated or pierced **b :** inaccessible to knowledge, reason, or sympathy **:** IMPERVIOUS
2 : incapable of being comprehended **:** INSCRUTABLE
— **im·pen·e·tra·bly** \-blē\ *adverb*
im·pen·i·tence \(ˌ)im-ˈpe-nə-tən(t)s\ *noun* (1624)
archaic **:** the quality or state of being impenitent
im·pen·i·tent \-tənt\ *adjective* [Middle English, from Late Latin *impaenitent-, impaenitens,* from Latin *in-* + *paenitent-, paenitens* penitent] (15th century)
: not penitent
— **im·pen·i·tent·ly** *adverb*
¹im·per·a·tive \im-ˈper-ə-tiv\ *adjective* [Middle English *imperatyf,* from Late Latin *imperativus,* from Latin *imperatus,* past participle of *imperare* to command — more at EMPEROR] (15th century)
1 a : of, relating to, or constituting the grammatical mood that expresses the will to influence the behavior of another **b :** expressive of a command, entreaty, or exhortation **c :** having power to restrain, control, and direct
2 : not to be avoided or evaded **:** NECESSARY ⟨an *imperative* duty⟩
synonym see MASTERFUL
— **im·per·a·tive·ly** *adverb*
— **im·per·a·tive·ness** *noun*
²imperative *noun* (1530)
1 : the imperative mood or a verb form or verbal phrase expressing it
2 : something that is imperative: as **a :** COMMAND, ORDER **b :** RULE, GUIDE **c :** an obligatory act or duty **d :** an imperative judgment or proposition
im·pe·ra·tor \ˌim-pə-ˈrä-tər, -ˌtȯr\ *noun* [Latin — more at EMPEROR] (circa 1580)
: a commander in chief or emperor of the ancient Romans
— **im·per·a·to·ri·al** \(ˌ)im-ˌper-ə-ˈtōr-ē-əl, -ˈtȯr-\ *adjective*
im·per·ceiv·able \ˌim-pər-ˈsē-və-bəl\ *adjective* (circa 1617)
archaic **:** IMPERCEPTIBLE
im·per·cep·ti·ble \ˌim-pər-ˈsep-tə-bəl\ *adjective* [Middle French, from Medieval Latin *imperceptibilis,* from Latin *in-* + Late Latin *perceptibilis* perceptible] (15th century)
: not perceptible by a sense or by the mind **:** extremely slight, gradual, or subtle
— **im·per·cep·ti·bly** \-ˈsep-tə-blē\ *adverb*
im·per·cep·tive \ˌim-pər-ˈsep-tiv\ *adjective* (1661)
: not perceptive
— **im·per·cep·tive·ness** *noun*
im·per·cip·i·ence \-ˈsi-pē-ən(t)s\ *noun* (1891)
: the quality or state of being imperceptive
— **im·per·cip·i·ent** \-ənt\ *adjective*
¹im·per·fect \(ˌ)im-ˈpər-fikt\ *adjective* [alteration of Middle English *imperfit,* from Middle French *imparfait,* from Latin *imperfectus,* from *in-* + *perfectus* perfect] (1570)
1 : not perfect: as **a :** DEFECTIVE **b** *of a flower* **:** having stamens or pistils but not both **c :** lacking or not involving sexual reproduction ⟨the *imperfect* stage of a fungus⟩
2 : of, relating to, or constituting a verb tense used to designate a continuing state or an incomplete action especially in the past
3 : not enforceable at law
— **im·per·fect·ly** \-fik(t)-lē\ *adverb*
— **im·per·fect·ness** \-fik(t)-nəs\ *noun*
²imperfect *noun* (1871)
: an imperfect tense; *also* **:** the verb form expressing it
imperfect fungus *noun* (circa 1895)
: any of various fungi (order Fungi Imperfecti synonym Deuteromycetes) of which only the conidial stage is known
im·per·fec·tion \ˌim-pər-ˈfek-shən\ *noun* (14th century)
: the quality or state of being imperfect; *also* **:** FAULT, BLEMISH
im·per·fec·tive \ˌim-pər-ˈfek-tiv\ *adjective* (1887)
of a verb form or aspect **:** expressing action as incomplete or without reference to completion or as reiterated — compare PERFECTIVE
— **imperfective** *noun*
im·per·fo·rate \(ˌ)im-ˈpər-f(ə-)rət, -fə-ˌrāt\ *adjective* (1673)
1 : having no opening or aperture; *specifically* **:** lacking the usual or normal opening
2 *of a stamp or a sheet of stamps* **:** lacking perforations or roulettes
¹im·pe·ri·al \im-ˈpir-ē-əl\ *adjective* [Middle English, from Middle French, from Late Latin *imperialis,* from Latin *imperium* command, empire] (14th century)
1 a : of, relating to, befitting, or suggestive of an empire or an emperor **b** (1) **:** of or relating to the United Kingdom as distinguished from the constituent parts (2) **:** of or relating to the Commonwealth and British Empire
2 a : SOVEREIGN **b :** REGAL, IMPERIOUS
3 : of superior or unusual size or excellence
4 : belonging to the official British series of weights and measures — see WEIGHT table
— **im·pe·ri·al·ly** \-ə-lē\ *adverb*
²imperial *noun* (circa 1524)
1 *capitalized* **:** an adherent or soldier of the Holy Roman emperor
2 : EMPEROR
3 [French *impériale;* from the beard worn by Napoléon III] **:** a pointed beard growing below the lower lip
4 : something of unusual size or excellence
im·pe·ri·al·ism \im-ˈpir-ē-ə-ˌli-zəm\ *noun* (1851)
1 : imperial government, authority, or system
2 : the policy, practice, or advocacy of extending the power and dominion of a nation especially by direct territorial acquisitions or by gaining indirect control over the political or economic life of other areas; *broadly* **:** the extension or imposition of power, authority, or influence ⟨union *imperialism*⟩
— **im·pe·ri·al·ist** \-list\ *noun or adjective*
— **im·pe·ri·al·is·tic** \-ˌpir-ē-ə-ˈlis-tik\ *adjective*
— **im·pe·ri·al·is·ti·cal·ly** \-ti-k(ə-)lē\ *adverb*
imperial moth *noun* (circa 1904)
: a large American saturniid moth (*Eacles imperialis*) marked with yellow and lilac or purplish brown

imperial moth

im·per·il \im-ˈper-əl\ *transitive verb* **-iled** *or* **-illed; -il·ing** *or* **-il·ling** (15th century)
: to bring into peril **:** ENDANGER
— **im·per·il·ment** \-mənt\ *noun*
im·pe·ri·ous \im-ˈpir-ē-əs\ *adjective* [Latin *imperiosus,* from *imperium*] (1540)
1 : befitting or characteristic of one of eminent rank or attainments **:** COMMANDING, DOMINANT ⟨an *imperious* manner⟩ **b :** marked by arrogant assurance **:** DOMINEERING
2 : intensely compelling **:** URGENT ⟨the *imperious* problems of the new age —J. F. Kennedy⟩
synonym see MASTERFUL
— **im·pe·ri·ous·ly** *adverb*
— **im·pe·ri·ous·ness** *noun*
im·per·ish·able \(ˌ)im-ˈper-i-shə-bəl\ *adjective* (1648)
1 : not perishable or subject to decay
2 : enduring or occurring forever ⟨*imperishable* fame⟩
— **im·per·ish·abil·i·ty** \-ˌper-i-shə-ˈbi-lə-tē\ *noun*
— **imperishable** *noun*
— **im·per·ish·able·ness** \-ˈper-i-shə-bəl-nəs\ *noun*
— **im·per·ish·ably** \-blē\ *adverb*
im·pe·ri·um \im-ˈpir-ē-əm\ *noun* [Latin — more at EMPIRE] (1651)
1 a : supreme power or absolute dominion **:** CONTROL **b :** EMPIRE
2 : the right to command or to employ the force of the state **:** SOVEREIGNTY
im·per·ma·nence \(ˌ)im-ˈpərm-nən(t)s, -ˈpər-mə-\ *noun* (1796)
: the quality or state of being impermanent
im·per·ma·nen·cy \-nən(t)-sē\ *noun* (1648)
: IMPERMANENCE
im·per·ma·nent \-nənt\ *adjective* (1653)
: not permanent **:** TRANSIENT
— **im·per·ma·nent·ly** *adverb*
im·per·me·able \(ˌ)im-ˈpər-mē-ə-bəl\ *adjective* [Late Latin *impermeabilis,* from Latin *in-* + Late Latin *permeabilis* permeable] (1697)
: not permitting passage (as of a fluid) through its substance; *broadly* **:** IMPERVIOUS
— **im·per·me·abil·i·ty** \-ˌpər-mē-ə-ˈbi-lə-tē\ *noun*
im·per·mis·si·ble \ˌim-pər-ˈmi-sə-bəl\ *adjective* (1858)
: not permissible
— **im·per·mis·si·bil·i·ty** \-ˌmi-sə-ˈbi-lə-tē\ *noun*
— **im·per·mis·si·bly** \-ˈmi-sə-blē\ *adverb*
im·per·son·al \(ˌ)im-ˈpərs-nəl, -ˈpər-sᵊn-əl\ *adjective* [Middle English, from Late Latin *impersonalis,* from Latin *in-* + Late Latin *personalis* personal] (15th century)
1 a : denoting the verbal action of an unspecified agent and hence used with no expressed subject (as *methinks*) or with a merely formal subject (as *rained* in *it rained*) **b** *of a pronoun* **:** INDEFINITE
2 a : having no personal reference or connection ⟨*impersonal* criticism⟩ **b :** not engaging the human personality or emotions ⟨the machine as compared with the hand tool is an *impersonal* agency —John Dewey⟩ **c :** not existing as a person **:** not having human qualities or characteristics
— **im·per·son·al·i·ty** \-ˌpər-sᵊn-ˈa-lə-tē\ *noun*
— **im·per·son·al·ly** \-ˈpərs-nə-lē, -ˈpər-sᵊn-ə-lē\ *adverb*
im·per·son·al·ize \(ˌ)im-ˈpərs-nə-ˌlīz, -ˈpər-sᵊn-ə-\ *transitive verb* (circa 1899)
: to make impersonal
— **im·per·son·al·i·za·tion** \-ˌpərs-nə-lə-ˈzā-shən, -ˌpər-sᵊn-ə-\ *noun*
im·per·son·ate \im-ˈpər-sᵊn-ˌāt\ *transitive verb* **-at·ed; -at·ing** (1715)
: to assume or act the character of **:** PERSONATE
— **im·per·son·a·tion** \-ˌpər-sᵊn-ˈā-shən\ *noun*
— **im·per·son·a·tor** \-ˈpər-sᵊn-ˌā-tər\ *noun*
im·per·ti·nence \(ˌ)im-ˈpər-tᵊn-ən(t)s, -ˈpərt-nən(t)s\ *noun* (1603)
1 : the quality or state of being impertinent: as **a :** IRRELEVANCE, INAPPROPRIATENESS **b :** INCIVILITY, INSOLENCE

2 **:** an instance of impertinence

im·per·ti·nen·cy \-ən(t)-sē, -nən(t)-\ *noun, plural* **-cies** (1589)
: IMPERTINENCE

im·per·ti·nent \(ˌ)im-ˈpər-tᵊn-ənt, -ˈpərt-nənt\ *adjective* [Middle English, from Middle French, from Late Latin *impertinent-, impertinens,* from Latin *in-* + *pertinent-, pertinens,* present participle of *pertinēre* to pertain] (14th century)
1 : not pertinent **:** IRRELEVANT
2 a : not restrained within due or proper bounds especially of propriety or good taste ⟨*impertinent* curiosity⟩ **b :** given to or characterized by insolent rudeness ⟨an *impertinent* answer⟩ ☆
— **im·per·ti·nent·ly** *adverb*

im·per·turb·able \ˌim-pər-ˈtər-bə-bəl\ *adjective* [Middle English, from Late Latin *imperturbabilis,* from Latin *in-* + *perturbare* to perturb] (15th century)
: marked by extreme calm, impassivity, and steadiness **:** SERENE
synonym see COOL
— **im·per·turb·abil·i·ty** \-ˌtər-bə-ˈbi-lə-tē\ *noun*
— **im·per·turb·ably** \-ˈtər-bə-blē\ *adverb*

im·per·vi·ous \(ˌ)im-ˈpər-vē-əs\ *adjective* [Latin *impervius,* from *in-* + *pervius* pervious] (1650)
1 a : not allowing entrance or passage **:** IMPENETRABLE ⟨a coat *impervious* to rain⟩ **b :** not capable of being damaged or harmed ⟨a carpet *impervious* to rough treatment⟩
2 : not capable of being affected or disturbed ⟨*impervious* to criticism⟩
— **im·per·vi·ous·ly** *adverb*
— **im·per·vi·ous·ness** *noun*

im·pe·tig·i·nous \ˌim-pə-ˈti-jə-nəs\ *adjective* (1620)
: of, relating to, or resembling impetigo ⟨*impetiginous* lesions⟩

im·pe·ti·go \ˌim-pə-ˈtē-(ˌ)gō, ˈtī-\ *noun* [Latin, from *impetere* to attack — more at IMPETUS] (14th century)
: an acute contagious staphylococcal or streptococcal skin disease characterized by vesicles, pustules, and yellowish crusts

im·pe·trate \ˈim-pə-ˌtrāt\ *transitive verb* **-trat·ed; -trat·ing** [Latin *impetratus,* past participle of *impetrare,* from *in-* + *patrare* to accomplish — more at PERPETRATE] (circa 1534)
1 : to obtain by request or entreaty
2 : to ask for **:** ENTREAT
— **im·pe·tra·tion** \ˌim-pə-ˈtrā-shən\ *noun*

im·pet·u·os·i·ty \im-ˌpe-chə-ˈwä-sə-tē\ *noun, plural* **-ties** (15th century)
1 : the quality or state of being impetuous
2 : an impetuous action or impulse

im·pet·u·ous \im-ˈpech-wəs, -ˈpe-chə-\ *adjective* [Middle English, from Middle French *impetueux,* from Late Latin *impetuosus,* from Latin *impetus*] (14th century)
1 : marked by impulsive vehemence or passion ⟨an *impetuous* temperament⟩
2 : marked by force and violence of movement or action ⟨an *impetuous* wind⟩
synonym see PRECIPITATE
— **im·pet·u·ous·ly** *adverb*
— **im·pet·u·ous·ness** *noun*

im·pe·tus \ˈim-pə-təs\ *noun* [Latin, assault, impetus, from *impetere* to attack, from *in-* + *petere* to go to, seek — more at FEATHER] (1641)
1 a (1) **:** a driving force **:** IMPULSE (2) **:** INCENTIVE, STIMULUS **b :** stimulation or encouragement resulting in increased activity
2 : the property possessed by a moving body in virtue of its mass and its motion — used of bodies moving suddenly or violently to indicate the origin and intensity of the motion

im·pi·e·ty \(ˌ)im-ˈpī-ə-tē\ *noun, plural* **-ties** (14th century)
1 : the quality or state of being impious **:** IRREVERENCE
2 : an impious act

im·pinge \im-ˈpinj\ *intransitive verb* **im·pinged; im·ping·ing** [Latin *impingere,* from *in-* + *pangere* to fasten, drive in — more at PACT] (1605)
1 : to strike or dash especially with a sharp collision ⟨I heard the rain *impinge* upon the earth —James Joyce⟩
2 : to have an effect **:** make an impression ⟨waiting for the germ of a new idea to *impinge* upon my mind —Phyllis Bentley⟩
3 : ENCROACH, INFRINGE ⟨*impinge* on other people's rights⟩
— **im·pinge·ment** \-ˈpinj-mənt\ *noun*

im·pi·ous \ˈim-pē-əs, (ˌ)im-ˈpī-\ *adjective* [Latin *impius,* from *in-* + *pius* pious] (1542)
: not pious **:** lacking in reverence or proper respect (as for God or one's parents) **:** IRREVERENT
— **im·pi·ous·ly** *adverb*

imp·ish \ˈim-pish\ *adjective* (1652)
: of, relating to, or befitting an imp; *especially* **:** MISCHIEVOUS
— **imp·ish·ly** *adverb*
— **imp·ish·ness** *noun*

im·pla·ca·ble \(ˌ)im-ˈpla-kə-bəl, -ˈplā-\ *adjective* [Middle French or Latin; Middle French, from Latin *implacabilis,* from *in-* + *placabilis* placable] (15th century)
: not placable **:** not capable of being appeased, significantly changed, or mitigated ⟨an *implacable* enemy⟩
— **im·pla·ca·bil·i·ty** \-ˌpla-kə-ˈbi-lə-tē, -ˌplā-\ *noun*
— **im·pla·ca·bly** \-ˈpla-kə-blē, -ˈplā-\ *adverb*

[1]**im·plant** \im-ˈplant\ *transitive verb* (15th century)
1 a : to fix or set securely or deeply ⟨a ruby *implanted* in the idol's forehead⟩ **b :** to set permanently in the consciousness or habit patterns **:** INCULCATE
2 : to insert in a living site (as for growth, slow release, or formation of an organic union) ⟨subcutaneously *implanted* hormone pellets⟩ ☆
— **im·plant·able** \-ˈplan-tə-bəl\ *adjective*
— **im·plant·er** \im-ˈplan-tər\ *noun*

[2]**im·plant** \ˈim-ˌplant\ *noun* (1890)
: something (as a graft or pellet) implanted in tissue

im·plan·ta·tion \ˌim-ˌplan-ˈtā-shən\ *noun* (1578)
1 a : the act or process of implanting something **b :** the state resulting from being implanted
2 *in placental mammals* **:** the process of attachment of the embryo to the maternal uterine wall

im·plau·si·ble \(ˌ)im-ˈplȯ-zə-bəl\ *adjective* (circa 1677)
: not plausible **:** provoking disbelief
— **im·plau·si·bil·i·ty** \-ˌplȯ-zə-ˈbi-lə-tē\ *noun*
— **im·plau·si·bly** \-ˈplȯ-zə-blē\ *adverb*

im·plead \im-ˈplēd\ *transitive verb* [Middle English *empleden,* from Middle French *emplaider,* from Old French *emplaidier,* from *en-* + *plaidier* to plead] (14th century)
: to sue or prosecute at law

[1]**im·ple·ment** \ˈim-plə-mənt\ *noun* [Middle English, from Late Latin *implementum* action of filling up, from Latin *implēre* to fill up, from *in-* + *plēre* to fill — more at FULL] (15th century)
1 : an article serving to equip ⟨the *implements* of religious worship⟩
2 : a device used in the performance of a task
3 : one that serves as an instrument or tool ⟨the partnership agreement does not seem to be a very potent *implement* —H. B. Hoffman⟩ ☆

[2]**im·ple·ment** \-ˌment\ *transitive verb* (1806)
1 : CARRY OUT, ACCOMPLISH; *especially* **:** to give practical effect to and ensure of actual fulfillment by concrete measures
2 : to provide instruments or means of expression for
— **im·ple·men·ta·tion** \ˌim-plə-mən-ˈtā-shən, -ˌmen-\ *noun*
— **im·ple·men·ter** *or* **im·ple·men·tor** \ˈim-plə-ˌmən-tər\ *noun*

im·pli·cate \ˈim-plə-ˌkāt\ *transitive verb* **-cat·ed; -cat·ing** [Middle English, to convey by implication, from Medieval Latin *implicatus,* past participle of *implicare,* from Latin, to entwine, involve — more at EMPLOY] (15th century)
1 : to involve as a consequence, corollary, or natural inference **:** IMPLY
2 *archaic* **:** to fold or twist together **:** ENTWINE
3 a : to bring into intimate or incriminating connection **b :** to involve in the nature or operation of something

im·pli·ca·tion \ˌim-plə-ˈkā-shən\ *noun* (15th century)
1 a : the act of implicating **:** the state of being implicated **b :** close connection; *especially* **:** an incriminating involvement

☆ SYNONYMS

Impertinent, officious, meddlesome, intrusive, obtrusive mean given to thrusting oneself into the affairs of others. IMPERTINENT implies exceeding the bounds of propriety in showing interest or curiosity or in offering advice ⟨resented their *impertinent* interference⟩. OFFICIOUS implies the offering of services or attentions that are unwelcome or annoying ⟨*officious* friends made the job harder⟩. MEDDLESOME stresses an annoying and usually prying interference in others' affairs ⟨a *meddlesome* landlord⟩. INTRUSIVE implies a tactless or otherwise objectionable thrusting into others' affairs ⟨tried to be helpful without being *intrusive*⟩. OBTRUSIVE stresses improper or offensive conspicuousness of interfering actions ⟨expressed an *obtrusive* concern for his safety⟩.

Implant, inculcate, instill, inseminate, infix mean to introduce into the mind. IMPLANT implies teaching that makes for permanence of what is taught ⟨*implanted* a love of reading in her students⟩. INCULCATE implies persistent or repeated efforts to impress on the mind ⟨tried to *inculcate* in him high moral standards⟩. INSTILL stresses gradual, gentle imparting of knowledge over a long period of time ⟨*instill* traditional values in your children⟩. INSEMINATE applies to a sowing of ideas in many minds so that they spread through a class or nation ⟨*inseminated* an unquestioning faith in technology⟩. INFIX stresses firmly inculcating a habit of thought ⟨*infixed* a chronic cynicism⟩.

Implement, tool, instrument, appliance, utensil mean a relatively simple device for performing work. IMPLEMENT may apply to anything necessary to perform a task ⟨crude stone *implements*⟩ ⟨farm *implements*⟩. TOOL suggests an implement adapted to facilitate a definite kind or stage of work and suggests the need of skill more strongly than IMPLEMENT ⟨a carpenter's *tools*⟩. INSTRUMENT suggests a device capable of delicate or precise work ⟨the dentist's *instruments*⟩. APPLIANCE refers to a tool or instrument utilizing a power source and suggests portability or temporary attachment ⟨household *appliances*⟩. UTENSIL applies to a device used in domestic work or some routine unskilled activity ⟨kitchen *utensils*⟩.

2 a : the act of implying **:** the state of being implied **b** (1) **:** a logical relation between two propositions that fails to hold only if the first is true and the second is false (2) **:** a logical relationship between two propositions in which if the first is true the second is true (3) **:** a statement exhibiting a relation of implication
3 : something implied: as **a :** SUGGESTION **b :** a possible significance ⟨the book has political *implications*⟩
— im·pli·ca·tive \ˈim-plə-ˌkā-tiv, im-ˈpli-kə-\ *adjective*
— im·pli·ca·tive·ly *adverb*
— im·pli·ca·tive·ness *noun*

im·plic·it \im-ˈpli-sət\ *adjective* [Latin *implicitus,* past participle of *implicare*] (1599)
1 a : capable of being understood from something else though unexpressed **:** IMPLIED ⟨an *implicit* assumption⟩ **b :** involved in the nature or essence of something though not revealed, expressed, or developed **:** POTENTIAL ⟨a sculptor may see different figures *implicit* in a block of stone —John Dewey⟩ **c** *of a mathematical function* **:** defined by an expression in which the dependent variable and the one or more independent variables are not separated on opposite sides of an equation — compare EXPLICIT 4
2 : being without doubt or reserve **:** UNQUESTIONING
— im·plic·it·ly *adverb*
— im·plic·it·ness *noun*

implicit differentiation *noun* (circa 1889)
: the process of finding the derivative of a dependent variable in an implicit function by differentiating each term separately, by expressing the derivative of the dependent variable as a symbol, and by solving the resulting expression for the symbol

im·plode \im-ˈplōd\ *verb* **im·plod·ed; im·plod·ing** [²*in-* + *-plode* (as in *explode*)] (1881)
intransitive verb
1 a : to burst inward ⟨a blow causing a vacuum tube to *implode*⟩ **b :** to undergo violent compression ⟨massive stars which *implode*⟩
2 : to collapse inward as if from external pressure; *also* **:** to become greatly reduced as if from collapsing
transitive verb
: to cause to implode

im·plore \im-ˈplōr, -ˈplȯr\ *transitive verb* **im·plored; im·plor·ing** [Middle French or Latin; Middle French *implorer,* from Latin *implorare,* from *in-* + *plorare* to cry out] (circa 1540)
1 : to call upon in supplication **:** BESEECH
2 : to call or pray for earnestly **:** ENTREAT
synonym see BEG
— im·plor·ing·ly *adverb*

im·plo·sion \im-ˈplō-zhən\ *noun* [²*in-* + *-plosion* (as in *explosion*)] (1877)
1 : the inrush of air in forming a suction stop
2 : the action of imploding
3 : the act or action of bringing to or as if to a center; *also* **:** INTEGRATION ⟨this *implosion* of cultures makes realistic for the first time the age-old vision of a world culture —Kenneth Keniston⟩
— im·plo·sive \-ˈplō-siv, -ziv\ *adjective or noun*

im·ply \im-ˈplī\ *transitive verb* **im·plied; im·ply·ing** [Middle English *emplien,* from Middle French *emplier,* from Latin *implicare*] (14th century)
1 *obsolete* **:** ENFOLD, ENTWINE
2 : to involve or indicate by inference, association, or necessary consequence rather than by direct statement ⟨rights *imply* obligations⟩
3 : to contain potentially
4 : to express indirectly ⟨his silence *implied* consent⟩
synonym see SUGGEST
usage see INFER

im·po·lite \ˌim-pə-ˈlīt\ *adjective* [Latin *impolitus,* from *in-* + *politus* polite] (1739)
: not polite **:** RUDE
— im·po·lite·ly *adverb*
— im·po·lite·ness *noun*

im·pol·i·tic \(ˌ)im-ˈpä-lə-ˌtik\ *adjective* (circa 1600)
: not politic **:** RASH
— im·po·lit·i·cal \-pə-ˈli-ti-kəl\ *adjective*
— im·po·lit·i·cal·ly \-ˈli-ti-k(ə-)lē\ *adverb*
— im·pol·i·tic·ly \-ˈpä-lə-ˌti-klē\ *adverb*

im·pon·der·a·ble \(ˌ)im-ˈpän-d(ə-)rə-bəl\ *adjective* [Medieval Latin *imponderabilis,* from Latin *in-* + Late Latin *ponderabilis* ponderable] (1794)
: not ponderable **:** incapable of being weighed or evaluated with exactness
— im·pon·der·a·bil·i·ty \-ˌpän-d(ə-)rə-ˈbi-lə-tē\ *noun*
— imponderable *noun*
— im·pon·der·a·bly \-ˈpän-d(ə-)rə-blē\ *adverb*

im·pone \im-ˈpōn\ *transitive verb* **im·poned; im·pon·ing** [Latin *imponere* to put upon, from *in-* + *ponere* to put — more at POSITION] (circa 1623)
obsolete **:** WAGER, BET

¹im·port \im-ˈpōrt, -ˈpȯrt, ˈim-ˌ\ *verb* [Middle English, from Latin *importare* to bring into, from *in-* + *portare* to carry — more at FARE] (15th century)
transitive verb
1 a : to bear or convey as meaning or portent **:** SIGNIFY **b** *archaic* **:** EXPRESS, STATE **c :** IMPLY
2 : to bring from a foreign or external source; *especially* **:** to bring (as merchandise) into a place or country from another country
3 *archaic* **:** to be of importance to **:** CONCERN
intransitive verb
: to be of consequence **:** MATTER
— im·port·able \im-ˈpōr-tə-bəl, -ˈpȯr-, ˈim-ˌ\ *adjective*
— im·port·er *noun*

²im·port \ˈim-ˌpōrt, -ˌpȯrt\ *noun* (circa 1568)
1 : IMPORTANCE; *especially* **:** relative importance ⟨it is hard to determine the *import* of this decision⟩
2 : PURPORT, SIGNIFICATION
3 : something that is imported
4 : IMPORTATION

im·por·tance \im-ˈpȯr-tᵊn(t)s, *especially Southern & New England* -tən(t)s, -dən(t)s\ *noun* (1508)
1 a : the quality or state of being important **:** CONSEQUENCE **b :** an important aspect or bearing **:** SIGNIFICANCE
2 *obsolete* **:** IMPORT, MEANING
3 *obsolete* **:** IMPORTUNITY
4 *obsolete* **:** a weighty matter ☆

im·por·tan·cy \-tᵊn(t)-sē, -tən(t)-, -dən(t)-\ *noun* (1540)
archaic **:** IMPORTANCE

im·por·tant \im-ˈpȯr-tᵊnt, *especially Southern & New England* -tənt, -dənt\ *adjective* [Middle English *importante,* from Medieval Latin *important-, importans,* present participle of *importare* to signify, from Latin, to bring into] (15th century)
1 : marked by or indicative of significant worth or consequence **:** valuable in content or relationship
2 *obsolete* **:** IMPORTUNATE, URGENT
3 : giving evidence of a feeling of self-importance
usage see IMPORTANTLY

im·por·tant·ly \-lē\ *adverb* (1647)
1 : in an important way ⟨contributed *importantly* to the language of the field —Ernst Mayr⟩ ⟨cleared his throat *importantly* and waited —E. K. Gann⟩ ⟨the real story is *importantly* different —Alexander Woollcott⟩
2 : it is important that ⟨more *importantly* he stands a chance of having his publication barred —H. L. Mencken⟩ ■

im·por·ta·tion \ˌim-ˌpōr-ˈtā-shən, -ˌpȯr-, -pər-\ *noun* (1601)
1 : the act or practice of importing
2 : something imported

imported cabbageworm *noun* (1892)
: a small cosmopolitan white butterfly (*Pieris rapae*) or its larva which is a pest of cruciferous plants and especially cabbage

imported fire ant *noun* (circa 1949)
: either of two mound-building South American fire ants (*Solenopsis richteri* and *S. invicta*) introduced into the southeastern U.S. that are agricultural pests and can produce stings requiring medical attention

im·por·tu·nate \im-ˈpȯr-chə-nət, -tyu̇-nət\ *adjective* (1529)
1 : troublesomely urgent **:** overly persistent in request or demand
2 : TROUBLESOME
— im·por·tu·nate·ly *adverb*
— im·por·tu·nate·ness *noun*

¹im·por·tune \ˌim-pər-ˈtün, -ˈtyün; im-ˈpȯr-ˌ, -chən\ *adjective* [Middle English, from Middle French & Latin; Middle French *importun,* from Latin *importunus,* from *in-* + *-portunus* (as in *opportunus* fit) — more at OPPORTUNE] (15th century)
: IMPORTUNATE
— im·por·tune·ly *adverb*

²importune *verb* **-tuned; -tun·ing** (1530)
transitive verb
1 a : to press or urge with troublesome persistence **b** *archaic* **:** to request or beg for urgently
2 : ANNOY, TROUBLE
intransitive verb
: to beg, urge, or solicit persistently or troublesomely
synonym see BEG
— im·por·tun·er *noun*

im·por·tu·ni·ty \ˌim-pər-ˈtü-nə-tē, -ˈtyü-\ *noun, plural* **-ties** (15th century)
1 : the quality or state of being importunate
2 : an importunate request or demand

☆ SYNONYMS
Importance, consequence, moment, weight, significance mean a quality or aspect having great worth or significance. IMPORTANCE implies a value judgment of the superior worth or influence of something or someone ⟨a region with no cities of *importance*⟩. CONSEQUENCE generally implies importance because of probable or possible effects ⟨the style you choose is of little *consequence*⟩. MOMENT implies conspicuous or self-evident consequence ⟨a decision of great *moment*⟩. WEIGHT implies a judgment of the immediate relative importance of something ⟨the argument carried no *weight* with the judge⟩. SIGNIFICANCE implies a quality or character that should mark a thing as important but that is not self-evident and may or may not be recognized ⟨the treaty's *significance*⟩.

□ USAGE
importantly A number of commentators have objected to *importantly* as a sentence modifier (sense 2) and have recommended *important* instead. Actually both the adverb and the adjective are in reputable standard use in this function. *Important* is always used with *more* or *most* ⟨had bronze weapons and composite bows; more *important,* they utilized the horse and war chariot —Harry A. Gailey, Jr.⟩ ⟨second and most *important,* the book contains no important woman character —F. Scott Fitzgerald⟩. *Importantly* is somewhat more flexible in not requiring *more* or *most* ⟨sticks and, just as *importantly,* unsticks easily —Phoebe Hawkins⟩ ⟨*importantly,* the leaven in the mixture is quality —George O'Brien⟩.

im·pose \im-ˈpōz\ *verb* **im·posed; im·pos·ing** [Middle French *imposer,* from Latin *imponere,* literally, to put upon (perfect indicative *imposui*), from *in-* + *ponere* to put — more at POSITION] (1581)
transitive verb
1 a : to establish or apply by authority ⟨*impose* a tax⟩ ⟨*impose* new restrictions⟩ ⟨*impose* penalties⟩ **b :** to establish or bring about as if by force ⟨those limits *imposed* by our own inadequacies —C. H. Plimpton⟩
2 a : PLACE, SET **b :** to arrange (as pages) in the proper order for printing
3 : PASS OFF ⟨*impose* fake antiques on the public⟩
4 : to force into the company or on the attention of another ⟨*impose* oneself on others⟩
intransitive verb
: to take unwarranted advantage of something ⟨*imposed* on his good nature⟩
— im·pos·er *noun*

im·pos·ing \im-ˈpō-ziŋ\ *adjective* (1786)
: impressive in size, bearing, dignity, or grandeur
synonym see GRAND
— im·pos·ing·ly \-ziŋ-lē\ *adverb*

im·po·si·tion \ˌim-pə-ˈzi-shən\ *noun* (14th century)
1 : something imposed: as **a :** LEVY, TAX **b :** an excessive or uncalled-for requirement or burden
2 : the act of imposing
3 : DECEPTION
4 : the order of arrangement of imposed pages

im·pos·si·bil·i·ty \(ˌ)im-ˌpä-sə-ˈbi-lə-tē\ *noun* (14th century)
1 : the quality or state of being impossible
2 : something impossible

im·pos·si·ble \(ˌ)im-ˈpä-sə-bəl\ *adjective* [Middle English, from Middle French & Latin; Middle French, from Latin *impossibilis,* from *in-* + *possibilis* possible] (14th century)
1 a : incapable of being or of occurring **b :** felt to be incapable of being done, attained, or fulfilled **:** insuperably difficult
2 a : extremely undesirable **:** UNACCEPTABLE **b :** extremely awkward or difficult to deal with
— im·pos·si·ble·ness *noun*

im·pos·si·bly \-blē\ *adverb* (circa 1580)
1 : not possibly
2 : to an improbable degree **:** UNBELIEVABLY ⟨an *impossibly* green lawn⟩

¹im·post \ˈim-ˌpōst\ *noun* [Middle French, from Medieval Latin *impositum,* from Latin, neuter of *impositus,* past participle of *imponere*] (1568)
: something imposed or levied **:** TAX

²impost *noun* [French *imposte,* ultimately from Latin *impositus*] (1664)
: a block, capital, or molding from which an arch springs — see ARCH illustration

im·pos·tor *or* **im·pos·ter** \im-ˈpäs-tər\ *noun* [Late Latin *impostor,* from Latin *imponere*] (1624)
: one that assumes false identity or title for the purpose of deception

im·pos·tume \im-ˈpäs-ˌchüm\ *or* **im·pos·thume** \-ˌthüm, -ˌthyüm\ *noun* [Middle English *emposteme,* ultimately from Greek *apostēma,* from *aphistanai* to remove, from *apo-* + *histanai* to cause to stand — more at STAND] (14th century)
archaic **:** ABSCESS

im·pos·ture \im-ˈpäs-chər\ *noun* [Late Latin *impostura,* from Latin *impositus, impostus,* past participle of *imponere*] (1537)
1 : the act or practice of deceiving by means of an assumed character or name
2 : an instance of imposture ☆

im·po·tence \ˈim-pə-tən(t)s\ *noun* (15th century)
: the quality or state of being impotent

im·po·ten·cy \-tən(t)-sē\ *noun* (15th century)
: IMPOTENCE

im·po·tent \ˈim-pə-tənt\ *adjective* [Middle English, from Middle French & Latin; Middle French, from Latin *impotent-, impotens,* from *in-* + *potent-, potens* potent] (14th century)
1 a : not potent **:** lacking in power, strength, or vigor **:** HELPLESS **b :** unable to copulate; *broadly* **:** STERILE — usually used of males
2 *obsolete* **:** incapable of self-restraint **:** UNGOVERNABLE
— impotent *noun*
— im·po·tent·ly *adverb*

im·pound \im-ˈpau̇nd\ *transitive verb* (15th century)
1 a : to shut up in or as if in a pound **:** CONFINE **b :** to seize and hold in the custody of the law **c :** to take possession of ⟨she was dismissed and her manuscript *impounded* —Jonathan Weiner⟩
2 : to collect and confine (water) in or as if in a reservoir

im·pound·ment \-ˈpau̇n(d)-mənt\ *noun* (circa 1665)
1 : the act of impounding **:** the state of being impounded
2 : a body of water formed by impounding

im·pov·er·ish \im-ˈpäv-rish, -ˈpä-və-\ *transitive verb* [Middle English *enpoverisen,* from Middle French *empovriss-,* stem of *empovrir,* from *en-* + *povre* poor — more at POOR] (15th century)
1 : to make poor
2 : to deprive of strength, richness, or fertility by depleting or draining of something essential
synonym see DEPLETE
— im·pov·er·ish·er *noun*
— im·pov·er·ish·ment \-mənt\ *noun*

impoverished *adjective* (1950)
of a fauna or flora **:** represented by few species or individuals

im·prac·ti·ca·ble \(ˌ)im-ˈprak-ti-kə-bəl\ *adjective* (1653)
1 : IMPASSABLE ⟨an *impracticable* road⟩
2 : not practicable **:** incapable of being performed or accomplished by the means employed or at command
— im·prac·ti·ca·bil·i·ty \-ˌprak-ti-kə-ˈbi-lə-tē\ *noun*
— im·prac·ti·ca·bly \-ˈprak-ti-kə-blē\ *adverb*

im·prac·ti·cal \(ˌ)im-ˈprak-ti-kəl\ *adjective* (1865)
: not practical: as **a :** not wise to put into or keep in practice or effect **b :** incapable of dealing sensibly or prudently with practical matters **c :** IMPRACTICABLE **d :** IDEALISTIC
— im·prac·ti·cal·i·ty \-ˌprak-ti-ˈka-lə-tē\ *noun*
— im·prac·ti·cal·ly \-ˈprak-ti-k(ə-)lē\ *adverb*

im·pre·cate \ˈim-pri-ˌkāt\ *verb* **-cat·ed; -cat·ing** [Latin *imprecatus,* past participle of *imprecari,* from *in-* + *precari* to pray — more at PRAY] (1613)
transitive verb
: to invoke evil on **:** CURSE
intransitive verb
: to utter curses

im·pre·ca·tion \ˌim-pri-ˈkā-shən\ *noun* (15th century)
1 : CURSE
2 : the act of imprecating
— im·pre·ca·to·ry \ˈim-pri-kə-ˌtōr-ē, im-ˈpre-kə-, -ˌtȯr-\ *adjective*

im·pre·cise \ˌim-pri-ˈsīs\ *adjective* (1805)
: not precise **:** INEXACT, VAGUE
— im·pre·cise·ly *adverb*
— im·pre·cise·ness *noun*
— im·pre·ci·sion \-ˈsi-zhən\ *noun*

im·preg·na·ble \im-ˈpreg-nə-bəl\ *adjective* [Middle English *imprenable,* from Middle French, from *in-* + *prenable* vulnerable to capture, from *prendre* to take — more at PRIZE] (15th century)
1 : incapable of being taken by assault **:** UNCONQUERABLE
2 : UNASSAILABLE; *also* **:** IMPENETRABLE
— im·preg·na·bil·i·ty \(ˌ)im-ˌpreg-nə-ˈbi-lə-tē\ *noun*
— im·preg·na·ble·ness \im-ˈpreg-nə-bəl-nəs\ *noun*
— im·preg·na·bly \-blē\ *adverb*

im·preg·nant \im-ˈpreg-nənt\ *noun* (1926)
: a substance used for impregnating another substance

¹im·preg·nate \im-ˈpreg-ˌnāt, ˈim-ˌ\ *transitive verb* **-nat·ed; -nat·ing** [Late Latin *impraegnatus,* past participle of *impraegnare,* from Latin *in-* + *praegnas* pregnant] (1605)
1 a : to cause to be filled, imbued, permeated, or saturated **b :** to permeate thoroughly
2 : to make pregnant **:** FERTILIZE
synonym see SOAK
— im·preg·na·tion \(ˌ)im-ˌpreg-ˈnā-shən\ *noun*
— im·preg·na·tor \im-ˈpreg-ˌnā-tər, ˈim-ˌ\ *noun*

²im·preg·nate \im-ˈpreg-nət\ *adjective* (1646)
: being filled or saturated

im·pre·sa \im-ˈprā-zə, -sə\ *noun* [Italian, literally, undertaking] (1588)
: a device with a motto used in the 16th and 17th centuries; *broadly* **:** EMBLEM

im·pre·sa·rio \ˌim-prə-ˈsär-ē-ˌō, -ˈsar-, -ˈzär-\ *noun, plural* **-ri·os** [Italian, from *impresa* undertaking, from *imprendere* to undertake, from (assumed) Vulgar Latin *imprehendere* — more at EMPRISE] (1746)
1 : the promoter, manager, or conductor of an opera or concert company
2 : a person who puts on or sponsors an entertainment (as a television show or sports event)
3 : MANAGER, DIRECTOR

¹im·press \im-ˈpres\ *verb* [Middle English, from Latin *impressus,* past participle of *imprimere,* from *in-* + *premere* to press — more at PRESS] (14th century)
transitive verb
1 a : to apply with pressure so as to imprint **b :** to produce (as a mark) by pressure **c :** to mark by or as if by pressure or stamping
2 a : to produce a vivid impression of **b :** to affect especially forcibly or deeply **:** INFLUENCE
3 : TRANSFER, TRANSMIT
intransitive verb
: to produce an impression
synonym see AFFECT
— im·press·ibil·i·ty \-ˌpre-sə-ˈbi-lə-tē\ *noun*
— im·press·ible \-ˈpre-sə-bəl\ *adjective*

²im·press \ˈim-ˌpres *also* im-ˈ\ *noun* (1590)

☆ SYNONYMS
Imposture, fraud, sham, fake, humbug, counterfeit mean a thing made to seem other than it is. IMPOSTURE applies to any situation in which a spurious object or performance is passed off as genuine ⟨their claim of environmental concern is an *imposture*⟩. FRAUD usually implies a deliberate perversion of the truth ⟨the diary was exposed as a *fraud*⟩. SHAM applies to fraudulent imitation of a real thing or action ⟨condemned the election as a *sham*⟩. FAKE implies an imitation of or substitution for the genuine but does not necessarily imply dishonesty ⟨these jewels are *fakes;* the real ones are in the vault⟩. HUMBUG suggests elaborate pretense usually so flagrant as to be transparent ⟨creating publicity by foisting *humbugs* on a gullible public⟩. COUNTERFEIT applies especially to the close imitation of something valuable ⟨20-dollar bills that were *counterfeits*⟩.

1 : a characteristic or distinctive mark **:** STAMP ⟨the *impress* of a fresh and vital intelligence is stamped . . . in his work —Lytton Strachey⟩
2 : IMPRESSION, EFFECT ⟨have an *impress* on history⟩
3 : the act of impressing
4 a : a mark made by pressure **:** IMPRINT **b :** an image of something formed by or as if by pressure; *especially* **:** SEAL **c :** a product of pressure or influence

³im•press \im-ˈpres\ *transitive verb* [²*in-* + ³*press*] (1596)
1 : to levy or take by force for public service; *especially* **:** to force into naval service
2 a : to procure or enlist by forcible persuasion **b :** FORCE ⟨*impressed* him into a white coat for the Christmas festivities —Nancy Hale⟩

⁴im•press \ˈim-ˌpres *also* im-ˈ\ *noun* (1602)
: IMPRESSMENT

im•pres•sion \im-ˈpre-shən\ *noun* (14th century)
1 a : a characteristic, trait, or feature resulting from some influence ⟨the *impression* on behavior produced by the social milieu⟩ **b :** an effect of alteration or improvement ⟨the settlement left little *impression* on the wilderness⟩ **c :** a telling image impressed on the senses or the mind
2 : the effect produced by impressing: as **a :** a stamp, form, or figure resulting from physical contact **b :** an imprint of the teeth and adjacent portions of the jaw for use in dentistry **c :** an especially marked and often favorable influence or effect on feeling, sense, or mind
3 : the act of impressing: as **a :** an affecting by stamping or pressing **b :** a communicating of a mold, trait, or character by an external force or influence
4 a : the amount of pressure with which an inked printing surface deposits its ink on the paper **b :** one instance of the meeting of a printing surface and the material being printed; *also* **:** a single print or copy so made **c :** all the copies (as of a book) printed in one continuous operation from a single makeready
5 : an often indistinct or imprecise notion or remembrance
6 a : the first coat of color in painting **b :** a coat of paint for ornament or preservation
7 : an imitation or representation of salient features in an artistic or theatrical medium; *especially* **:** an imitation in caricature of a noted personality as a form of theatrical entertainment
synonym see IDEA

im•pres•sion•able \im-ˈpre-sh(ə-)nə-bəl\ *adjective* (1836)
: capable of being easily impressed
— im•pres•sion•abil•i•ty \-ˌpre-sh(ə-)nə-ˈbi-lə-tē\ *noun*

im•pres•sion•ism \im-ˈpre-shə-ˌni-zəm\ *noun* (1882)
1 *often capitalized* **:** a theory or practice in painting especially among French painters of about 1870 of depicting the natural appearances of objects by means of dabs or strokes of primary unmixed colors in order to simulate actual reflected light
2 a : the depiction (as in literature) of scene, emotion, or character by details intended to achieve a vividness or effectiveness more by evoking subjective and sensory impressions than by recreating an objective reality **b :** a style of musical composition designed to create subtle moods and impressions

im•pres•sion•ist \im-ˈpre-sh(ə-)nist\ *noun* (1876)
1 *often capitalized* **:** one (as a painter) who practices or adheres to the theories of impressionism
2 : an entertainer who does impressions

im•pres•sion•is•tic \(ˌ)im-ˌpre-shə-ˈnis-tik\ *adjective* (1886)
1 *or* **im•pres•sion•ist** \im-ˈpre-sh(ə-)nist\ *often capitalized* **:** of, relating to, or constituting impressionism
2 : based on or involving impression as distinct from expertise or fact ⟨intuitions and *impressionistic* anecdotal accounts —Sidney Hook⟩
— im•pres•sion•is•ti•cal•ly \(ˌ)im-ˌpre-shə-ˈnis-ti-k(ə-)lē\ *adverb*

im•pres•sive \im-ˈpre-siv\ *adjective* (1598)
: making or tending to make a marked impression **:** having the power to excite attention, awe, or admiration
synonym see MOVING
— im•pres•sive•ly *adverb*
— im•pres•sive•ness *noun*

im•press•ment \im-ˈpres-mənt\ *noun* (1787)
: the act of seizing for public use or of impressing into public service

im•pres•sure \im-ˈpre-shər\ *noun* (1600)
archaic **:** a mark made by pressure **:** IMPRESSION

im•prest \ˈim-ˌprest\ *noun* [obsolete *imprest* to lend, probably from Italian *imprestare*] (1568)
: a loan or advance of money

im•pri•ma•tur \ˌim-prə-ˈmä-ˌtu̇r, im-ˈpri-mə-ˌtu̇r, -ˌtyu̇r\ *noun* [New Latin, let it be printed, from *imprimere* to print, from Latin, to imprint, impress — more at IMPRESS] (1640)
1 a : a license to print or publish especially by Roman Catholic episcopal authority **b :** approval of a publication under circumstances of official censorship
2 a : SANCTION, APPROVAL **b :** IMPRINT **c :** a mark of approval or distinction

im•pri•mis \im-ˈprī-məs, -ˈprē-\ *adverb* [Middle English *imprimis,* from Latin *in primis* among the first (things)] (15th century)
: in the first place — used to introduce a list of items or considerations

¹im•print \im-ˈprint, ˈim-ˌ\ (14th century)
transitive verb
1 : to mark by or as if by pressure **:** IMPRESS
2 a : to fix indelibly or permanently (as on the memory) **b :** to subject to or induce by imprinting ⟨an *imprinted* preference⟩
intransitive verb
: to undergo imprinting
— im•print•er \-ˈprin-tər, -ˌprin-\ *noun*

²im•print \ˈim-ˌprint\ *noun* [Middle English *enpreent,* from Middle French *empreinte,* from feminine of *empreint,* past participle of *empreindre* to imprint, from Latin *imprimere*] (15th century)
: something imprinted or printed: as **a :** a mark or depression made by pressure ⟨the fossil *imprint* of a dinosaur's foot⟩ **b :** an identifying name (as of a publisher) placed conspicuously on a product; *also* **:** the name under which a publisher issues books **c :** an indelible distinguishing effect or influence

im•print•ing \ˈim-ˌprin-tiŋ, im-ˈ\ *noun* (1937)
: a rapid learning process that takes place early in the life of a social animal (as a greylag goose) and establishes a behavior pattern (as recognition of and attraction to its own kind or a substitute)

im•pris•on \im-ˈpri-zᵊn\ *transitive verb* [Middle English, from Old French *emprisoner,* from *en-* + *prison* prison] (14th century)
: to put in or as if in prison **:** CONFINE
— im•pris•on•ment \-mənt\ *noun*

im•prob•a•ble \(ˌ)im-ˈprä-bə-bəl, -ˈpräb-bəl\ *adjective* [Middle French or Latin; Middle French, from Latin *improbabilis,* from *in-* + *probabilis* probable] (1598)
: unlikely to be true or to occur; *also* **:** unlikely but real or true
— im•prob•a•bil•i•ty \-ˌprä-bə-ˈbi-lə-tē\ *noun*
— im•prob•a•bly \-ˈprä-bə-blē, -ˈpräb-blē\ *adverb*

¹im•promp•tu \im-ˈpräm(p)-(ˌ)tü, -(ˌ)tyü\ *noun* [French, from *impromptu* extemporaneously, from Latin *in promptu* in readiness] (1683)
1 : something that is impromptu
2 : a musical composition suggesting improvisation

²impromptu *adjective* (1764)
1 : made, done, or formed on or as if on the spur of the moment **:** IMPROVISED
2 : composed or uttered without previous preparation **:** EXTEMPORANEOUS
— impromptu *adverb*

im•prop•er \(ˌ)im-ˈprä-pər\ *adjective* [Middle English, from Middle French *impropre,* from Latin *improprius,* from *in-* + *proprius* proper] (15th century)
: not proper: as **a :** not in accord with fact, truth, or right procedure **:** INCORRECT ⟨*improper* inference⟩ **b :** not regularly or normally formed or not properly so called **c :** not suited to the circumstances, design, or end ⟨*improper* medicine⟩ **d :** not in accord with propriety, modesty, good manners, or good taste
synonym see INDECOROUS
— im•prop•er•ly *adverb*
— im•prop•er•ness *noun*

improper fraction *noun* (1542)
: a fraction whose numerator is equal to, larger than, or of equal or higher degree than the denominator

improper integral *noun* (circa 1942)
: a definite integral whose region of integration is unbounded or includes a point at which the integrand is undefined or tends to infinity

im•pro•pri•e•ty \ˌim-p(r)ə-ˈprī-ə-tē\ *noun, plural* **-ties** [French or Late Latin; French *impropriété,* from Late Latin *improprietat-, improprietas,* from Latin *improprius*] (1607)
1 : an improper or indecorous act or remark; *especially* **:** an unacceptable use of a word or of language
2 : the quality or state of being improper

im•prove \im-ˈprüv\ *verb* **im•proved; im•prov•ing** [Anglo-French *emprouer* to invest profitably, from Old French *en-* + *prou* advantage, from Late Latin *prode* — more at PROUD] (circa 1529)
transitive verb
1 *archaic* **:** EMPLOY, USE
2 a : to enhance in value or quality **:** make better **b :** to increase the value of (land or property) by making it more useful for humans (as by cultivation or the erection of buildings) **c :** to grade and drain (a road) and apply surfacing material other than pavement
3 : to use to good purpose
intransitive verb
1 : to advance or make progress in what is desirable
2 : to make useful additions or amendments ☆
— im•prov•abil•i•ty \(ˌ)im-ˌprü-və-ˈbi-lə-tē\ *noun*
— im•prov•able \-ˈprü-və-bəl\ *adjective*
— im•prov•er *noun*

im•prove•ment \im-ˈprüv-mənt\ *noun* (circa 1550)
1 : the act or process of improving
2 a : the state of being improved; *especially* **:** enhanced value or excellence **b :** an instance of such improvement **:** something that enhances value or excellence

☆ SYNONYMS
Improve, better, help, ameliorate mean to make more acceptable or to bring nearer a standard. IMPROVE and BETTER are general and interchangeable and apply to what can be made better whether it is good or bad ⟨measures to further *improve* the quality of medical care⟩ ⟨immigrants hoping to *better* their lot⟩. HELP implies a bettering that still leaves room for improvement ⟨a coat of paint would *help* that house⟩. AMELIORATE implies making more tolerable or acceptable conditions that are hard to endure ⟨tried to *ameliorate* the lives of people in the tenements⟩.

im·prov·i·dence \(ˌ)im-ˈprä-və-dən(t)s, -ˌden(t)s\ *noun* (15th century)
: the quality or state of being improvident

im·prov·i·dent \-dənt, -ˌdent\ *adjective* [Late Latin *improvident-, improvidens,* from Latin *in-* + *provident-, providens* provident] (1624)
: not provident **:** not foreseeing and providing for the future
— **im·prov·i·dent·ly** *adverb*

im·pro·vi·sa·tion \(ˌ)im-ˌprä-və-ˈzā-shən, ˌim-prə-və- *also* ˌim-prə-(ˌ)vī-\ *noun* (1786)
1 : the act or art of improvising
2 : something (as a musical or dramatic composition) improvised
— **im·pro·vi·sa·tion·al** \-shnəl, -shə-nᵊl\ *adjective*
— **im·pro·vi·sa·tion·al·ly** *adverb*

im·prov·i·sa·tor \im-ˈprä-və-ˌzā-tər\ *noun* (1795)
: one that improvises
— **im·prov·i·sa·to·ri·al** \(ˌ)im-ˌprä-və-zə-ˈtōr-ē-əl, -ˈtȯr-\ *adjective*
— **im·pro·vi·sa·to·ry** \im-ˈprä-və-zə-ˌtōr-ē, ˌim-prə-ˈvī-zə-, -ˌtȯr-\ *adjective*

im·pro·vi·sa·to·re \(ˌ)im-ˌprä-və-zə-ˈtōr-ē, ˌim-prə-ˌvē-zə-, -ˈtȯr-\ *noun, plural* **-to·ri** \-ˈtōr-ē, -ˈtȯr-\ *or* **-tores** [Italian *improvvisatore,* from *improvvisare*] (1765)
: one that improvises (as verse) usually extemporaneously

im·pro·vise \ˈim-prə-ˌvīz *also* ˌim-prə-ˈ\ *verb* **-vised; -vis·ing** [French *improviser,* from Italian *improvvisare,* from *improvviso* sudden, from Latin *improvisus,* literally, unforeseen, from *in-* + *provisus,* past participle of *providēre* to see ahead — more at PROVIDE] (1826)
transitive verb
1 : to compose, recite, play, or sing extemporaneously
2 : to make, invent, or arrange offhand
3 : to fabricate out of what is conveniently on hand
intransitive verb
: to improvise something
— **im·pro·vis·er** *or* **im·pro·vi·sor** \-ˌvī-zər, -ˈvī-\ *noun*

im·pru·dence \(ˌ)im-ˈprü-dᵊn(t)s\ *noun* (15th century)
1 : the quality or state of being imprudent
2 : an imprudent act

im·pru·dent \-dᵊnt\ *adjective* [Middle English, from Latin *imprudent-, imprudens,* from *in-* + *prudent-, prudens* prudent] (14th century)
: not prudent **:** lacking discretion
— **im·pru·dent·ly** *adverb*

im·pu·dence \ˈim-pyə-dən(t)s\ *noun* (14th century)
: the quality or state of being impudent

im·pu·dent \-dənt\ *adjective* [Middle English, from Latin *impudent-, impudens,* from *in-* + *pudent-, pudens,* present participle of *pudēre* to feel shame] (14th century)
1 *obsolete* **:** lacking modesty
2 : marked by contemptuous or cocky boldness or disregard of others **:** INSOLENT
— **im·pu·dent·ly** *adverb*

im·pu·dic·i·ty \ˌim-pyü-ˈdi-sə-tē\ *noun* [Middle French *impudicité,* from Latin *impudicus* immodest, from *in-* + *pudicus* modest, from *pudēre*] (1528)
: lack of modesty **:** SHAMELESSNESS

im·pugn \im-ˈpyün\ *transitive verb* [Middle English, to assail, from Middle French *impugner,* from Latin *inpugnare,* from *in-* + *pugnare* to fight — more at PUNGENT] (14th century)
1 : to assail by words or arguments **:** oppose or attack as false or lacking integrity
2 *obsolete* **a :** ASSAIL **b :** RESIST
— **im·pugn·able** \-ˈpyü-nə-bəl\ *adjective*
— **im·pugn·er** \-nər\ *noun*

im·puis·sance \(ˌ)im-ˈpwi-sᵊn(t)s, (ˌ)im-ˈpyü-ə-sən(t)s; ˌim-pyü-ˈi-sᵊn(t)s\ *noun* [Middle English, from Middle French, from *in-* + *puissance* puissance, power] (15th century)
: WEAKNESS, POWERLESSNESS

im·puis·sant \-sᵊnt, -sənt\ *adjective* [French] (1629)
: WEAK, POWERLESS

[1]im·pulse \ˈim-ˌpəls, im-ˈ\ *transitive verb* **impulsed; im·puls·ing** (1611)
: to give an impulse to

[2]im·pulse \ˈim-ˌpəls\ *noun* [Latin *impulsus,* from *impellere* to impel] (1647)
1 a : INSPIRATION, MOTIVATION **b :** a force so communicated as to produce motion suddenly **c :** INCENTIVE
2 a : the act of driving onward with sudden force **:** IMPULSION **b :** motion produced by such an impulsion **:** IMPETUS **c :** a wave of excitation transmitted through tissues and especially nerve fibers and muscles that results in physiological activity or inhibition
3 a : a sudden spontaneous inclination or incitement to some usually unpremeditated action **b :** a propensity or natural tendency usually other than rational
4 a : the product of the average value of a force and the time during which it acts **:** the change in momentum produced by the force **b :** PULSE 4a
synonym see MOTIVE

im·pul·sion \im-ˈpəl-shən\ *noun* (15th century)
1 a : the act of impelling **:** the state of being impelled **b :** an impelling force **c :** an onward tendency derived from an impulsion
2 : IMPULSE 3a
3 : COMPULSION 2

im·pul·sive \im-ˈpəl-siv\ *adjective* (15th century)
1 : having the power of or actually driving or impelling
2 : actuated by or prone to act on impulse
3 : acting momentarily
synonym see SPONTANEOUS
— **im·pul·sive·ly** *adverb*
— **im·pul·sive·ness** *noun*
— **im·pul·siv·i·ty** \-ˌpəl-ˈsi-və-tē\ *noun*

im·pu·ni·ty \im-ˈpyü-nə-tē\ *noun* [Middle French or Latin; Middle French *impunité,* from Latin *impunitat-, impunitas,* from *impune* without punishment, from *in-* + *poena* punishment — more at PAIN] (1532)
: exemption or freedom from punishment, harm, or loss

im·pure \(ˌ)im-ˈpyu̇r\ *adjective* [Middle English, from Middle French & Latin; Middle French, from Latin *impurus,* from *in-* + *purus* pure] (15th century)
: not pure: as **a :** LEWD, UNCHASTE **b :** containing something unclean **:** FOUL ⟨*impure* water⟩ **c :** ritually unclean **d :** mixed or impregnated with an extraneous and usually unwanted substance ⟨an *impure* chemical⟩
— **im·pure·ly** *adverb*
— **im·pure·ness** *noun*

im·pu·ri·ty \(ˌ)im-ˈpyu̇r-ə-tē\ *noun, plural* **-ties** (15th century)
1 : something that is impure or makes something else impure
2 : the quality or state of being impure

im·pu·ta·tion \ˌim-pyə-ˈtā-shən\ *noun* (1581)
1 : the act of imputing: as **a :** ATTRIBUTION, ASCRIPTION **b :** ACCUSATION **c :** INSINUATION
2 : something imputed
— **im·pu·ta·tive** \im-ˈpyü-tə-tiv\ *adjective*
— **im·pu·ta·tive·ly** *adverb*

im·pute \im-ˈpyüt\ *transitive verb* **im·put·ed; im·put·ing** [Middle English *inputen,* from Latin *imputare,* from *in-* + *putare* to consider] (14th century)
1 : to lay the responsibility or blame for often falsely or unjustly
2 : to credit to a person or a cause **:** ATTRIBUTE ⟨our vices as well as our virtues have been *imputed* to bodily derangement —B. N. Cardozo⟩
synonym see ASCRIBE
— **im·put·abil·i·ty** \-ˌpyü-tə-ˈbi-lə-tē\ *noun*
— **im·put·able** \-ˈpyü-tə-bəl\ *adjective*

[1]in \ˈin, ən, ᵊn\ *preposition* [Middle English, from Old English; akin to Old High German *in* in, Latin *in,* Greek *en*] (before 12th century)
1 a — used as a function word to indicate inclusion, location, or position within limits ⟨*in* the lake⟩ ⟨wounded *in* the leg⟩ ⟨*in* the summer⟩ **b :** INTO 1 ⟨went *in* the house⟩
2 — used as a function word to indicate means, medium, or instrumentality ⟨written *in* pencil⟩ ⟨bound *in* leather⟩
3 a — used as a function word to indicate limitation, qualification, or circumstance ⟨alike *in* some respects⟩ ⟨left *in* a hurry⟩ **b :** INTO 2a ⟨broke *in* pieces⟩
4 — used as a function word to indicate purpose ⟨said *in* reply⟩
5 — used as a function word to indicate the larger member of a ratio ⟨one *in* six is eligible⟩

[2]in \ˈin\ *adverb* (before 12th century)
1 a (1) **:** to or toward the inside especially of a house or other building ⟨come *in*⟩ (2) **:** to or toward some destination or particular place ⟨flew *in* on the first plane⟩ (3) **:** at close quarters **:** NEAR ⟨play close *in*⟩ **b :** so as to incorporate ⟨mix *in* the flour⟩ — often used in combination ⟨built-*in* bookcases⟩ **c :** to or at an appropriate place ⟨fit a piece *in*⟩
2 a : within a particular place; *especially* **:** within the customary place of residence or business ⟨the doctor is *in*⟩ **b :** in the position of participant, insider, or officeholder — often used with *on* ⟨*in* on the joke⟩ **c** (1) **:** on good terms (2) **:** in a specified relation ⟨*in* bad with the boss⟩ (3) **:** in a position of assured or definitive success **d :** in vogue or season **e** *of an oil well* **:** in production **f :** in one's presence, possession, or control ⟨after the crops are *in*⟩ **g :** from a condition of indistinguishability to one of clarity ⟨fade *in*⟩
— **in for :** certain to experience ⟨*in for* a rude awakening⟩

[3]in \ˈin\ *adjective* (1599)
1 a : that is located inside or within ⟨the *in* part⟩ **b :** that is in position, operation, or power ⟨the *in* party⟩ **c :** INSIDE 2
2 : that is directed or bound inward **:** INCOMING ⟨the *in* train⟩
3 a : extremely fashionable ⟨the *in* thing to do⟩ **b :** keenly aware of and responsive to what is new and smart ⟨the *in* crowd⟩

[4]in \ˈin\ *noun* (1764)
1 : one who is in office or power or on the inside ⟨a matter of *ins* versus outs⟩
2 : INFLUENCE, PULL ⟨enjoyed some sort of *in* with the commandant —Henriette Roosenburg⟩

[1]in- *or* **il-** *or* **im-** *or* **ir-** *prefix* [Middle English, from Middle French, from Latin; akin to Old English *un-*]
: not **:** NON-, UN- — usually *il-* before *l* ⟨*il*logical⟩, *im-* before *b, m,* or *p* ⟨*im*balance⟩ ⟨*im*moral⟩ ⟨*im*practical⟩, *ir-* before *r* ⟨*ir*reducible⟩, and *in-* before other sounds ⟨*in*conclusive⟩

[2]in- *or* **il-** *or* **im-** *or* **ir-** *prefix* [Middle English, from Middle French, from Latin, from *in* in, into]
1 : in **:** within **:** into **:** toward **:** on — usually *il-* before *l* ⟨*il*luviation⟩, *im-* before *b, m,* or *p* ⟨*im*mingle⟩, *ir-* before *r* ⟨*ir*radiance⟩, and *in-* before other sounds ⟨*in*filtrate⟩
2 : [1]EN- ⟨*im*brute⟩ ⟨*im*peril⟩ ⟨*in*spirit⟩

[1]-in *noun suffix* [French *-ine,* from Latin *-īna,* feminine of *-īnus* of or belonging to — more at -EN]
1 a : neutral chemical compound ⟨insul*in*⟩ **b :** enzyme ⟨pancreat*in*⟩ **c :** antibiotic ⟨penicil*lin*⟩
2 : [2]-INE 1a, b ⟨epinephr*in*⟩

3 : pharmaceutical product ⟨niacin⟩
²-in *combining form* [sit-*in*]
: organized public protest by means of or in favor of **:** demonstration ⟨teach-*in*⟩ ⟨love-*in*⟩
in·abil·i·ty \ˌi-nə-ˈbi-lə-tē\ *noun* [Middle English *inabilite,* from Middle French *inhabilité,* from *in-* + *habilité* ability] (15th century)
: lack of sufficient power, resources, or capacity ⟨his *inability* to do math⟩
in ab·sen·tia \ˌin-ab-ˈsen(t)-sh(ē-)ə\ *adverb* [Latin] (1886)
: in absence ⟨gave him the award *in absentia*⟩
in·ac·ces·si·ble \ˌi-nik-ˈse-sə-bəl, (ˌ)i-ˌnak-\ *adjective* [Middle English, from Middle French or Late Latin; Middle French, from Late Latin *inaccessibilis,* from Latin *in-* + Late Latin *accessibilis* accessible] (15th century)
: not accessible
— in·ac·ces·si·bil·i·ty \-ˌse-sə-ˈbi-lə-tē\ *noun*
— in·ac·ces·si·bly \-ˈse-sə-blē\ *adverb*
in·ac·cu·ra·cy \(ˌ)i-ˈna-kyə-rə-sē, -k(ə-)rə-sē\ *noun, plural* **-cies** (circa 1755)
1 : the quality or state of being inaccurate
2 : MISTAKE, ERROR
in·ac·cu·rate \-ˈa-kyə-rət, -k(ə-)rət\ *adjective* (1738)
: not accurate **:** FAULTY
— in·ac·cu·rate·ly \-kyə-rət-lē, -k(ə-)rət-, -kyərt-\ *adverb*
in·ac·tion \(ˌ)i-ˈnak-shən\ *noun* (1707)
: lack of action or activity **:** IDLENESS
in·ac·ti·vate \(ˌ)i-ˈnak-tə-ˌvāt\ *transitive verb* (1906)
: to make inactive
— in·ac·ti·va·tion \-ˌnak-tə-ˈvā-shən\ *noun*
in·ac·tive \(ˌ)i-ˈnak-tiv\ *adjective* (1664)
: not active: as **a** (1) **:** SEDENTARY (2) **:** INDOLENT, SLUGGISH **b** (1) **:** being out of use (2) **:** relating to or being members of the armed forces who are not performing or available for military duties (3) *of a disease* **:** QUIESCENT **c** (1) **:** chemically inert (2) **:** optically neutral in polarized light **d :** biologically inert especially because of the loss of some quality (as infectivity or antigenicity) ☆
— in·ac·tive·ly *adverb*
— in·ac·tiv·i·ty \-ˌnak-ˈti-və-tē\ *noun*
in·ad·e·qua·cy \(ˌ)i-ˈna-di-kwə-sē\ *noun, plural* **-cies** (1787)
1 : the quality or state of being inadequate
2 : INSUFFICIENCY, DEFICIENCY
in·ad·e·quate \-kwət\ *adjective* (1671)
: not adequate **:** INSUFFICIENT; *also* **:** not capable
— in·ad·e·quate·ly *adverb*
— in·ad·e·quate·ness *noun*
in·ad·mis·si·ble \ˌi-nəd-ˈmi-sə-bəl\ *adjective* (1776)
: not admissible
— in·ad·mis·si·bil·i·ty \-ˌmi-sə-ˈbi-lə-tē\ *noun*
— in·ad·mis·si·bly \-ˈmi-sə-blē\ *adverb*
in·ad·ver·tence \ˌi-nəd-ˈvər-tᵊn(t)s\ *noun* [Middle English, from Medieval Latin *inadvertentia,* from Latin *in-* + *advertent-, advertens,* present participle of *advertere* to advert] (15th century)
1 : the fact or action of being inadvertent
2 : a result of inattention **:** OVERSIGHT
in·ad·ver·ten·cy \-tᵊn(t)-sē\ *noun, plural* **-cies** (1592)
: INADVERTENCE
in·ad·ver·tent \-tᵊnt\ *adjective* [back-formation from *inadvertence*] (1653)
1 : not focusing the mind on a matter **:** INATTENTIVE
2 : UNINTENTIONAL
— in·ad·ver·tent·ly *adverb*
in·ad·vis·able \ˌi-nəd-ˈvī-zə-bəl\ *adjective* (1870)
: not advisable
— in·ad·vis·abil·i·ty \-ˌvī-zə-ˈbi-lə-tē\ *noun*

-inae *noun plural suffix* [New Latin *-īnae,* from Latin, feminine plural of *-īnus*]
: members of the subfamily of — in all names of zoological subfamilies in recent classifications ⟨Fel*inae*⟩
in·alien·able \(ˌ)i-ˈnāl-yə-nə-bəl, -ˈnā-lē-ə-nə-\ *adjective* [probably from French *inaliénable,* from *in-* + *aliénable* alienable] (circa 1645)
: incapable of being alienated, surrendered, or transferred ⟨*inalienable* rights⟩
— in·alien·abil·i·ty \-ˌnāl-yə-nə-ˈbi-lə-tē, -ˌnā-lē-ə-nə-\ *noun*
— in·alien·ably \-ˈnāl-yə-nə-blē, -ˈnā-lē-ə-nə-\ *adverb*
in·al·ter·able \(ˌ)i-ˈnȯl-t(ə-)rə-bəl\ *adjective* (1541)
: not alterable **:** UNALTERABLE
— in·al·ter·abil·i·ty \-ˌnȯl-t(ə-)rə-ˈbi-lə-tē\ *noun*
— in·al·ter·able·ness \-ˈnȯl-t(ə-)rə-bəl-nəs\ *noun*
— in·al·ter·ably \-blē\ *adverb*
in·amo·ra·ta \i-ˌna-mə-ˈrä-tə\ *noun* [Italian *innamorata,* from feminine of *innamorato,* past participle of *innamorare* to inspire with love, from *in-* (from Latin) + *amore* love, from Latin *amor* — more at AMOROUS] (1651)
: a woman with whom one is in love or has intimate relations
in–and–in \ˌin-ən(d)-ˈin\ *adverb or adjective* (1765)
: in repeated generations of the same or closely related stock ⟨families . . . of one blood through mating or marrying *in-and-in* —F. H. Giddings⟩ ⟨*in-and-in* breeding⟩
¹inane \i-ˈnān\ *adjective* **inan·er; -est** [Latin *inanis*] (1662)
1 : EMPTY, INSUBSTANTIAL
2 : lacking significance, meaning, or point **:** SILLY
synonym see INSIPID
— inane·ly *adverb*
— inane·ness \-ˈnān-nəs\ *noun*
²inane *noun* (1677)
: void or empty space ⟨a voyage into the limitless *inane* —V. G. Childe⟩
in·an·i·mate \(ˌ)i-ˈna-nə-mət\ *adjective* [Middle English, from Late Latin *inanimatus,* from Latin *in-* + *animatus,* past participle of *animare* to animate] (15th century)
1 : not animate: **a :** not endowed with life or spirit **b :** lacking consciousness or power of motion
2 : not animated or lively **:** DULL
— in·an·i·mate·ly *adverb*
— in·an·i·mate·ness *noun*
in·a·ni·tion \ˌi-nə-ˈni-shən\ *noun* (14th century)
: the quality or state of being empty: **a :** the loss of vitality that results from lack of food and water **b :** the absence or loss of social, moral, or intellectual vitality or vigor
inan·i·ty \i-ˈna-nə-tē\ *noun, plural* **-ties** (1603)
1 : the quality or state of being inane: as **a :** lack of substance **:** EMPTINESS **b :** vapid, pointless, or fatuous character **:** SHALLOWNESS
2 : something that is inane
in·ap·par·ent \ˌi-nə-ˈpar-ənt, -ˈper-\ *adjective* (1626)
: not apparent
— in·ap·par·ent·ly *adverb*
in·ap·peas·able \ˌi-nə-ˈpē-zə-bəl\ *adjective* (1803)
: UNAPPEASABLE
in·ap·pe·tence \(ˌ)i-ˈna-pə-tən(t)s\ *noun* (circa 1691)
: loss or lack of appetite
in·ap·pli·ca·ble \(ˌ)i-ˈna-pli-kə-bəl *also* ˌi-nə-ˈpli-kə-\ *adjective* (1656)
: not applicable **:** IRRELEVANT
— in·ap·pli·ca·bil·i·ty \-ˌna-pli-kə-ˈbi-lə-tē *also* ˌi-nə-ˌpli-kə-\ *noun*
— in·ap·pli·ca·bly \(ˌ)i-ˈna-pli-kə-blē *also* ˌi-nə-ˈpli-kə-\ *adverb*

in·ap·po·site \(ˌ)i-ˈna-pə-zət\ *adjective* (1661)
: not apposite
— in·ap·po·site·ly *adverb*
— in·ap·po·site·ness *noun*
in·ap·pre·cia·ble \ˌi-nə-ˈprē-shə-bəl, -ˈpri-sh(ē-)ə-bəl\ *adjective* [probably from French *inappréciable,* from Middle French *inappreciable,* from *in-* + *appreciable*] (1802)
: too small to be perceived
— in·ap·pre·cia·bly \-blē\ *adverb*
in·ap·pre·cia·tive \ˌi-nə-ˈprē-shə-tiv, -ˈpri- *also* -ˈprē-shē-ˌā-\ *adjective* (1869)
: not appreciative
— in·ap·pre·cia·tive·ly *adverb*
— in·ap·pre·cia·tive·ness *noun*
in·ap·proach·able \ˌi-nə-ˈprō-chə-bəl\ *adjective* (circa 1828)
: not approachable **:** INACCESSIBLE
in·ap·pro·pri·ate \ˌi-nə-ˈprō-prē-ət\ *adjective* (1804)
: not appropriate **:** UNSUITABLE
— in·ap·pro·pri·ate·ly *adverb*
— in·ap·pro·pri·ate·ness *noun*
in·apt \(ˌ)i-ˈnapt\ *adjective* (circa 1670)
: not apt: **a :** not suitable **b :** INEPT
— in·apt·ly \-ˈnap(t)-lē\ *adverb*
— in·apt·ness \-nəs\ *noun*
in·ap·ti·tude \(ˌ)i-ˈnap-tə-ˌtüd, -ˌtyüd\ *noun* (1620)
: lack of aptitude
in·ar·gu·able \(ˌ)i-ˈnär-gyə-wə-bəl\ *adjective* (circa 1875)
: not arguable
— in·ar·gu·ably \-blē\ *adverb*
in·ar·tic·u·la·cy \ˌi-(ˌ)när-ˈti-kyə-lə-sē\ *noun* (1921)
: the quality or state of being inarticulate
¹in·ar·tic·u·late \-kyə-lət\ *adjective* [Late Latin *inarticulatus,* from Latin *in-* + *articulatus,* past participle of *articulare* to utter distinctly — more at ARTICULATE] (1603)
1 a *of a sound* **:** uttered or formed without the definite articulations of intelligible speech **b** (1) **:** incapable of speech especially under stress of emotion **:** MUTE (2) **:** incapable of being expressed by speech ⟨*inarticulate* fear⟩ (3) **:** not voiced or expressed **:** UNSPOKEN ⟨society functions on many *inarticulate* premises⟩
2 : incapable of giving coherent, clear, or effective expression to one's ideas or feelings
3 [New Latin *inarticulatus,* from Latin *in-* + New Latin *articulatus* articulate] **:** relating to, characteristic of, or being an inarticulate or its shell
— in·ar·tic·u·late·ly *adverb*
— in·ar·tic·u·late·ness *noun*
²inarticulate *noun* (1952)
: any of a class (Inarticulata) of brachiopods lacking a hinge connecting the two shell valves
in·ar·tis·tic \ˌi-när-ˈtis-tik\ *adjective* (1859)
1 : not conforming to the principles of art

☆ SYNONYMS
Inactive, idle, inert, passive, supine mean not engaged in work or activity. INACTIVE applies to anyone or anything not in action or in operation or at work ⟨on *inactive* status as an astronaut⟩ ⟨*inactive* accounts⟩. IDLE applies to persons that are not busy or occupied or to their powers or their implements ⟨workers were *idle* in the fields⟩. INERT as applied to things implies powerlessness to move or to affect other things; as applied to persons it suggests an inherent or habitual indisposition to activity ⟨*inert* ingredients in drugs⟩ ⟨an *inert* citizenry⟩. PASSIVE implies immobility or lack of normally expected response to an external force or influence and often suggests deliberate submissiveness or self-control ⟨*passive* resistance⟩. SUPINE applies only to persons and commonly implies abjectness or indolence ⟨a *supine* willingness to play the fool⟩.

2 : not appreciative of art
— **in·ar·tis·ti·cal·ly** \-ti-k(ə-)lē\ *adverb*
in·as·much as \ˌi-nəz-ˈmə-chəz, -ˈməch-ˌaz\ *conjunction* (14th century)
1 : in the degree that **:** INSOFAR AS
2 : in view of the fact that **:** SINCE
in·at·ten·tion \ˌi-nə-ˈten(t)-shən\ *noun* (circa 1670)
: failure to pay attention
in·at·ten·tive \-ˈten-tiv\ *adjective* (1692)
: not attentive
— **in·at·ten·tive·ly** *adverb*
— **in·at·ten·tive·ness** *noun*
in·au·di·ble \(ˌ)i-ˈnȯ-də-bəl\ *adjective* [Late Latin *inaudibilis,* from Latin *in-* + Late Latin *audibilis* audible] (1601)
: not audible
— **in·au·di·bil·i·ty** \(ˌ)i-ˌnȯ-də-ˈbi-lə-tē\ *noun*
— **in·au·di·bly** \(ˌ)i-ˈnȯ-də-blē\ *adverb*
[1]**in·au·gu·ral** \i-ˈnȯ-gyə-rəl, -g(ə-)rəl\ *adjective* (1689)
1 : of or relating to an inauguration
2 : marking a beginning **:** first in a projected series
[2]**inaugural** *noun* (1832)
1 : an inaugural address
2 : INAUGURATION
in·au·gu·rate \i-ˈnȯ-gyə-ˌrāt, -gə-ˌrāt\ *transitive verb* **-rat·ed; -rat·ing** [Latin *inauguratus,* past participle of *inaugurare,* literally, to practice augury, from *in-* + *augurare* to augur; from the rites connected with augury] (1606)
1 : to induct into an office with suitable ceremonies
2 a : to dedicate ceremoniously **:** observe formally the beginning of **b :** to bring about the beginning of
synonym see BEGIN
— **in·au·gu·ra·tor** \-ˌrā-tər\ *noun*
in·au·gu·ra·tion \-ˌnȯ-gyə-ˈrā-shən, -gə-\ *noun* (1569)
: an act of inaugurating; *especially* **:** a ceremonial induction into office
Inauguration Day *noun* (1829)
: January 20 following a presidential election on which the president of the U.S. is inaugurated
in·aus·pi·cious \ˌi-ˌnȯ-ˈspi-shəs\ *adjective* (1592)
: not auspicious
— **in·aus·pi·cious·ly** *adverb*
— **in·aus·pi·cious·ness** *noun*
in·au·then·tic \ˌi-ˌnȯ-ˈthen-tik\ *adjective* (1860)
: not authentic
— **in·au·then·tic·i·ty** \ˌi-ˌnȯ-ˌthen-ˈti-sə-tē, -thən-\ *noun*
in–be·tween \ˌin-bi-ˈtwēn\ *adjective or noun* (1815)
: INTERMEDIATE
in between *adverb or preposition* (1892)
: BETWEEN
[1]**in·board** \ˈin-ˌbōrd, -ˌbȯrd\ *adverb* (1830)
1 : inside the line of a ship's bulwarks or hull
2 : toward the center line of a vehicle or craft (as a ship or aircraft)
[2]**inboard** *adjective* (1847)
1 : located inboard ⟨an *inboard* engine⟩ ⟨an *inboard* spoiler⟩
2 *of a boat* **:** having an inboard engine
[3]**inboard** *noun* (1950)
: a boat with an inboard motor
in·born \ˈin-ˈbȯrn\ *adjective* (1513)
1 : present from or as if from birth
2 : HEREDITARY, INHERITED
synonym see INNATE
in·bound \ˈin-ˌbau̇nd\ *adjective* (1894)
: inward bound
in·bounds \ˈin-ˈbau̇n(d)z\ *adjective* (1968)
: involving putting a basketball in play by passing it onto the court from out of bounds
inbounds line *noun* (circa 1961)
: either of two broken lines running the length of a football field at right angles to the yard lines
in·breathe \ˈin-ˌbrēth\ *transitive verb* (14th century)
: to breathe (something) in **:** INHALE
in·bred \ˈin-ˈbred\ *adjective* (circa 1592)
1 : rooted and ingrained in one's nature as deeply as if implanted by heredity ⟨an *inbred* love of freedom⟩
2 [from past participle of *inbreed*] **:** subjected to or produced by inbreeding
synonym see INNATE
— **in·bred** \ˈin-ˌbred\ *noun*
in·breed \ˈin-ˈbrēd\ *verb* **-bred** \-ˈbred\; **-breed·ing** (1599)
transitive verb
: to subject to inbreeding
intransitive verb
: to engage in inbreeding
in·breed·ing \ˈin-ˌbrē-diŋ\ *noun* (circa 1842)
1 : the interbreeding of closely related individuals especially to preserve and fix desirable characters of and to eliminate unfavorable characters from a stock
2 : confinement to a narrow range or a local or limited field of choice
in·built \ˈin-ˈbilt\ *adjective* (1923)
chiefly British **:** BUILT-IN
In·ca \ˈiŋ-kə\ *noun* [Spanish, from Quechua *inka* ruler of the Inca empire] (1594)
1 a : a member of the Quechuan peoples of Peru maintaining an empire until the Spanish conquest **b :** a king or noble of the Inca empire
2 : a member of any people under Inca influence
— **In·ca·ic** \in-ˈkā-ik\ *adjective*
— **In·can** \ˈiŋ-kən\ *adjective*
in·cal·cu·la·ble \(ˌ)in-ˈkal-kyə-lə-bəl\ *adjective* (1795)
: not capable of being calculated: as **a :** very great **b :** not predictable **:** UNCERTAIN
— **in·cal·cu·la·bil·i·ty** \-ˌkal-kyə-lə-ˈbi-lə-tē\ *noun*
— **in·cal·cu·la·bly** \-ˈkal-kyə-lə-blē\ *adverb*
in·ca·les·cence \ˌin-kə-ˈle-sᵊn(t)s, ˌiŋ-\ *noun* [Latin *incalescere* to become warm, from *in-* + *calescere* to become warm, inchoative of *calēre* to be warm — more at LEE] (1646)
: a growing warm or ardent
— **in·ca·les·cent** \-sᵊnt\ *adjective*
in camera *adverb* [New Latin, literally, in a chamber] (1882)
: in private **:** SECRETLY
in·can·desce \ˌin-kən-ˈdes *also* -(ˌ)kan-\ *intransitive verb* **-desced; -desc·ing** [Latin *incandescere*] (1874)
: to be or become incandescent
in·can·des·cence \ˌin-kən-ˈde-sᵊn(t)s *also* -(ˌ)kan-\ *noun* (1656)
: the quality or state of being incandescent; *especially* **:** emission by a hot body of radiation that makes it visible
in·can·des·cent \-sᵊnt\ *adjective* [probably from French, from Latin *incandescent-, incandescens,* present participle of *incandescere* to become hot, from *in-* + *candescere* to become hot, from *candēre* to glow — more at CANDID] (1794)
1 a : white, glowing, or luminous with intense heat **b :** strikingly bright, radiant, or clear **c :** marked by brilliance especially of expression ⟨*incandescent* wit⟩ **d :** characterized by glowing zeal **:** ARDENT ⟨*incandescent* affection⟩
2 a : of, relating to, or being light produced by incandescence **b :** producing light by incandescence
— **in·can·des·cent·ly** *adverb*
incandescent lamp *noun* (1881)
: an electric lamp in which a filament gives off light when heated to incandescence by an electric current — called also *incandescent, incandescent bulb, incandescent lightbulb, lightbulb*
in·cant \in-ˈkant\ *intransitive verb* [Latin *incantare*] (1945)
: RECITE, UTTER
in·can·ta·tion \ˌin-ˌkan-ˈtā-shən\ *noun* [Middle English *incantacioun,* from Middle French *incantation,* from Late Latin *incantation-, incantatio,* from Latin *incantare* to enchant — more at ENCHANT] (14th century)
: a use of spells or verbal charms spoken or sung as a part of a ritual of magic; *also* **:** a written or recited formula of words designed to produce a particular effect
— **in·can·ta·tion·al** \-shnəl, -shə-nᵊl\ *adjective*
— **in·can·ta·to·ry** \in-ˈkan-tə-ˌtōr-ē, -ˌtȯr-\ *adjective*

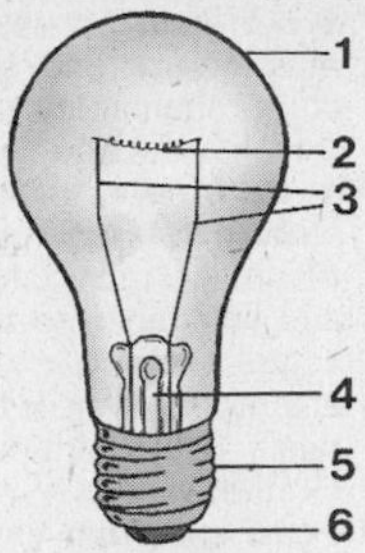

incandescent lamp: *1* **bulb containing gas,** *2* **filament,** *3* **connecting and supporting wires,** *4* **exhaust tube,** *5* **screw base,** *6* **base contact**

in·ca·pa·ble \(ˌ)in-ˈkā-pə-bəl\ *adjective* [Middle French, from *in-* + *capable* capable] (1594)
1 : lacking capacity, ability, or qualification for the purpose or end in view: as **a** *archaic* **:** not able to take in, hold, or keep **b** *archaic* **:** not receptive **c :** not being in a state or of a kind to admit **:** INSUSCEPTIBLE **d :** not able or fit for the doing or performance **:** INCOMPETENT
2 : lacking legal qualification or power (as by reason of mental incompetence) **:** DISQUALIFIED
— **in·ca·pa·bil·i·ty** \-ˌkā-pə-ˈbi-lə-tē\ *noun*
— **in·ca·pa·ble·ness** \-ˈkā-pə-bəl-nəs\ *noun*
— **in·ca·pa·bly** \-blē\ *adverb*
in·ca·pac·i·tate \ˌin-kə-ˈpa-sə-ˌtāt\ *transitive verb* **-tat·ed; -tat·ing** (1657)
1 : to make legally incapable or ineligible
2 : to deprive of capacity or natural power **:** DISABLE
— **in·ca·pac·i·ta·tion** \-ˌpa-sə-ˈtā-shən\ *noun*
in·ca·pac·i·ty \ˌin-kə-ˈpa-sə-tē, -ˈpas-tē\ *noun, plural* **-ties** [French *incapacité,* from Middle French, from *in-* + *capacité* capacity] (1611)
: the quality or state of being incapable; *especially* **:** lack of physical or intellectual power or of natural or legal qualifications
in·car·cer·ate \in-ˈkär-sə-ˌrāt\ *transitive verb* **-at·ed; -at·ing** [Latin *incarceratus,* past participle of *incarcerare,* from *in-* + *carcer* prison] (1560)
1 : to put in prison
2 : to subject to confinement
— **in·car·cer·a·tion** \(ˌ)in-ˌkär-sə-ˈrā-shən\ *noun*
in·car·di·na·tion \(ˌ)in-ˌkär-dᵊn-ˈā-shən\ *noun* [Late Latin *incardination-, incardinatio,* from *incardinare* to ordain as chief priest, from Latin *in-* [2]in- + Late Latin *cardinalis* principal — more at CARDINAL] (1897)
: the formal acceptance by a diocese of a clergyman from another diocese
[1]**in·car·na·dine** \in-ˈkär-nə-ˌdīn, -ˌdēn, -dən\ *adjective* [Middle French *incarnadin,* from Old Italian *incarnadino,* from *incarnato* flesh-colored, from Late Latin *incarnatus*] (1591)
1 : having the pinkish color of flesh
2 : RED; *especially* **:** BLOODRED
[2]**incarnadine** *transitive verb* **-dined; -din·ing** (1605)
: to make incarnadine **:** REDDEN
[1]**in·car·nate** \in-ˈkär-nət, -ˌnāt\ *adjective* [Middle English *incarnat,* from Late Latin *incarnatus,* past participle of *incarnare* to incarnate, from Latin *in-* + *carn-, caro* flesh — more at CARNAL] (14th century)

1 a : invested with bodily and especially human nature and form **b :** made manifest or comprehensible **:** EMBODIED ⟨a fiend *incarnate*⟩
2 : INCARNADINE ⟨*incarnate* clover⟩

²in·car·nate \in-ˈkär-ˌnāt, ˈin-ˌ\ *transitive verb* **-nat·ed; -nat·ing** (1533)
: to make incarnate: as **a :** to give bodily form and substance to **b** (1) **:** to give a concrete or actual form to **:** ACTUALIZE (2) **:** to constitute an embodiment or type of ⟨no one culture *incarnates* every important human value —Denis Goulet⟩

in·car·na·tion \ˌin-(ˌ)kär-ˈnā-shən\ *noun* (14th century)
1 a (1) **:** the embodiment of a deity or spirit in some earthly form (2) *capitalized* **:** the union of divinity with humanity in Jesus Christ **b :** a concrete or actual form of a quality or concept; *especially* **:** a person showing a trait or typical character to a marked degree ⟨she is the *incarnation* of goodness⟩
2 : the act of incarnating **:** the state of being incarnate
3 : a particular physical form or state **:** VERSION ⟨in another *incarnation* he might be a first vice-president —Walter Teller⟩ ⟨TV and movie *incarnations* of the story⟩

in·case *variant of* ENCASE

in case *conjunction* (14th century)
1 : IF ⟨*in case* we are surprised, keep by me —Washington Irving⟩
2 : as a precaution against the event that ⟨carries a gun *in case* he is attacked⟩

in·cau·tion \(ˌ)in-ˈkȯ-shən\ *noun* (circa 1720)
: lack of caution **:** HEEDLESSNESS

in·cau·tious \-shəs\ *adjective* (circa 1703)
: lacking in caution **:** CARELESS
— in·cau·tious·ly *adverb*
— in·cau·tious·ness *noun*

in·cen·di·a·rism \in-ˈsen-dē-ə-ˌri-zəm\ *noun* (circa 1710)
: incendiary action or behavior

¹in·cen·di·ary \in-ˈsen-dē-ˌer-ē; -ˈsen-də-rē, -dyə-\ *noun, plural* **-ar·ies** [Middle English, from Latin *incendiarius,* from *incendium* conflagration, from *incendere*] (15th century)
1 a : a person who deliberately sets fire to a building or other property **b :** an incendiary agent (as a bomb)
2 : a person who excites factions, quarrels, or sedition **:** AGITATOR

²incendiary *adjective* (15th century)
1 : of, relating to, or involving a deliberate burning of property
2 : tending to excite or inflame **:** INFLAMMATORY ⟨*incendiary* speeches⟩
3 a : igniting combustible materials spontaneously **b :** of, relating to, or being a weapon (as a bomb) designed to start fires

¹in·cense \ˈin-ˌsen(t)s\ *noun* [Middle English *encens,* from Old French, from Late Latin *incensum,* from Latin, neuter of *incensus,* past participle of *incendere* to set on fire, from *in-* + *-cendere* to burn; akin to Latin *candēre* to glow — more at CANDID] (13th century)
1 : material used to produce a fragrant odor when burned
2 : the perfume exhaled from some spices and gums when burned; *broadly* **:** a pleasing scent
3 : pleasing attention **:** FLATTERY

²incense *transitive verb* **in·censed; in·cens·ing** (13th century)
1 : to apply or offer incense to
2 : to perfume with incense

³in·cense \in-ˈsen(t)s\ *transitive verb* **incensed; in·cens·ing** [Middle English *encensen,* from Middle French *incenser,* from Latin *incensus*] (15th century)
1 *archaic* **:** to cause (a passion or emotion) to become aroused
2 : to arouse the extreme anger or indignation of

in·cen·ter \ˈin-ˌsen-tər\ *noun* [*in*scribe + ¹*center*] (circa 1890)
: the single point in which the three bisectors of the interior angles of a triangle intersect and which is the center of the inscribed circle

in·cen·tive \in-ˈsen-tiv\ *noun* [Middle English, from Late Latin *incentivum,* from neuter of *incentivus* stimulating, from Latin, setting the tune, from *incentus,* past participle of *incinere* to set the tune, from *in-* + *canere* to sing — more at CHANT] (15th century)
: something that incites or has a tendency to incite to determination or action
synonym see MOTIVE
— incentive *adjective*

in·cep·tion \in-ˈsep-shən\ *noun* [Middle English *incepcion,* from Latin *inception-, inceptio,* from *incipere* to begin, from *in-* + *capere* to take] (15th century)
: an act, process, or instance of beginning **:** COMMENCEMENT
synonym see ORIGIN

¹in·cep·tive \in-ˈsep-tiv\ *noun* (1612)
: an inchoative verb

²inceptive *adjective* (1656)
1 : INCHOATIVE 2
2 : of or relating to a beginning
— in·cep·tive·ly *adverb*

in·cer·ti·tude \(ˌ)in-ˈsər-tə-ˌtüd, -ˌtyüd\ *noun* [Middle English, from Middle French, from Late Latin *incertitudo,* from Latin *in-* + Late Latin *certitudo* certitude] (15th century)
: UNCERTAINTY: **a :** absence of assurance or confidence **:** DOUBT **b :** the quality or state of being unstable or insecure

in·ces·san·cy \(ˌ)in-ˈse-sᵊn(t)-sē\ *noun* (1615)
: the quality or state of being incessant

in·ces·sant \(ˌ)in-ˈse-sᵊnt\ *adjective* [Middle English *incessaunt,* from Late Latin *incessant-, incessans,* from Latin *in-* + *cessant-, cessans,* present participle of *cessare* to delay — more at CEASE] (15th century)
: continuing or following without interruption **:** UNCEASING
synonym see CONTINUAL
— in·ces·sant·ly *adverb*

in·cest \ˈin-ˌsest\ *noun* [Middle English, from Latin *incestus* sexual impurity, from *incestus* impure, from *in-* + *castus* pure — more at CASTE] (13th century)
: sexual intercourse between persons so closely related that they are forbidden by law to marry; *also* **:** the statutory crime of such a relationship

in·ces·tu·ous \in-ˈses-chə-wəs, -ˈsesh-\ *adjective* (1532)
1 : constituting or involving incest
2 : guilty of incest
— in·ces·tu·ous·ly *adverb*
— in·ces·tu·ous·ness *noun*

¹inch \ˈinch\ *noun* [Middle English, from Old English *ynce,* from Latin *uncia* — more at OUNCE] (before 12th century)
1 : a unit of length equal to 1⁄36 yard — see WEIGHT table
2 : a small amount, distance, or degree ⟨is like cutting a dog's tail off by *inches* —Milton Friedman⟩
3 *plural* **:** STATURE, HEIGHT
4 a : a fall (as of rain or snow) sufficient to cover a surface or to fill a gauge to the depth of one inch **b :** a degree of atmospheric or other pressure sufficient to balance the weight of a column of liquid (as mercury) one inch high in a barometer or manometer
— every inch : to the utmost degree ⟨looks *every inch* a winner⟩
— within an inch of : almost to the point of

²inch (1599)
intransitive verb
: to move by small degrees **:** progress slowly ⟨the long line of people *inching* up the stairs⟩
transitive verb
: to cause to move slowly ⟨sooner or later they begin *inching* prices back up —*Forbes*⟩

³inch *noun* [Middle English (Scots dialect), from Scottish Gaelic *innis*] (15th century)
chiefly Scottish **:** ISLAND

inched \ˈincht\ *adjective* (1605)
: measuring a specified number of inches

-inch·er \ˈin-chər\ *noun combining form*
: one that has a dimension of a specified number of inches ⟨a four-*incher*⟩

inch·meal \ˈinch-ˌmēl, -ˈmē(ə)l\ *adverb* [¹*inch* + *-meal* (as in *piecemeal*)] (1530)
: LITTLE BY LITTLE, GRADUALLY

in·cho·ate \in-ˈkō-ət, ˈin-kə-ˌwāt\ *adjective* [Latin *inchoatus,* past participle of *inchoare* to start work on, perhaps from *in-* + *cohum* part of a yoke to which the beam of a plow is fitted] (1534)
: being only partly in existence or operation; *especially* **:** imperfectly formed or formulated **:** FORMLESS ⟨misty, *inchoate* suspicions that all is not well with the nation —J. M. Perry⟩
— in·cho·ate·ly *adverb*
— in·cho·ate·ness *noun*

in·cho·a·tive \in-ˈkō-ə-tiv\ *adjective* (circa 1631)
1 : INITIAL, FORMATIVE ⟨the *inchoative* stages⟩
2 : denoting the beginning of an action, state, or occurrence — used of verbs
— inchoative *noun*
— in·cho·a·tive·ly *adverb*

inch·worm \ˈinch-ˌwərm\ *noun* (circa 1861)
: LOOPER 1

in·ci·dence \ˈin(t)-sə-dən(t)s, -ˌden(t)s\ *noun* (1626)
1 a : ANGLE OF INCIDENCE **b :** the arrival of something (as a projectile or a ray of light) at a surface
2 a : an act or the fact or manner of falling upon or affecting **:** OCCURRENCE **b :** rate of occurrence or influence ⟨a high *incidence* of crime⟩

¹in·ci·dent \ˈin(t)-sə-dənt, -ˌdent\ *noun* [Middle English, from Middle French, from Medieval Latin *incident-, incidens,* from Latin, present participle of *incidere* to fall into, from *in-* + *cadere* to fall — more at CHANCE] (15th century)
1 : something dependent on or subordinate to something else of greater or principal importance
2 a : an occurrence of an action or situation that is a separate unit of experience **:** HAPPENING **b :** an accompanying minor occurrence or condition **:** CONCOMITANT
3 : an action likely to lead to grave consequences especially in diplomatic matters ⟨a serious border *incident*⟩
synonym see OCCURRENCE

²incident *adjective* (15th century)
1 : occurring or likely to occur especially as a minor consequence or accompaniment ⟨the confusion *incident* to moving day⟩
2 : dependent on or relating to another thing in law
3 : falling or striking on something ⟨*incident* light rays⟩

¹in·ci·den·tal \ˌin(t)-sə-ˈden-tᵊl\ *adjective* (1616)
1 : being likely to ensue as a chance or minor consequence ⟨social obligations *incidental* to the job⟩
2 : occurring merely by chance or without intention or calculation

²incidental *noun* (1707)
1 *plural* **:** minor items (as of expense) that are not particularized
2 : something that is incidental

in·ci·den·tal·ly \-ˈden-tᵊl-ē, *especially for 2* -ˈdent-lē\ *adverb* (1665)
1 : in an incidental manner **:** not intentionally ⟨the arrant nonsense of some of his statements is *incidentally* hilarious —John Lahr⟩
2 : by way of interjection or digression **:** by the way ⟨fortunate in having a good teacher . . . —still living, *incidentally* —John Fischer⟩

incidental music *noun* (1864)
: descriptive music played during a play to project a mood or to accompany stage action

in·cin·er·ate \in-'si-nə-ˌrāt\ *transitive verb* **-at·ed; -at·ing** [Medieval Latin *incineratus,* past participle of *incinerare,* from Latin *in-* + *ciner-, cinis* ashes; akin to Greek *konis* dust, ashes] (1555)
: to cause to burn to ashes
— in·cin·er·a·tion \-ˌsi-nə-'rā-shən\ *noun*

in·cin·er·a·tor \in-'si-nə-ˌrā-tər\ *noun* (1883)
: one that incinerates; *especially* **:** a furnace or a container for incinerating waste materials

in·cip·i·ence \in-'si-pē-ən(t)s\ *noun* (circa 1864)
: INCIPIENCY

in·cip·i·en·cy \-ən(t)-sē\ *noun* (1817)
: the state or fact of being incipient **:** BEGINNING

in·cip·i·ent \-ənt\ *adjective* [Latin *incipient-, incipiens,* present participle of *incipere* to begin — more at INCEPTION] (1669)
: beginning to come into being or to become apparent ⟨an *incipient* solar system⟩ ⟨evidence of *incipient* racial tension⟩
— in·cip·i·ent·ly *adverb*

in·ci·pit \'in(t)-sə-pət, 'iŋ-kə-ˌpit; in-'si-pət, -'ki-\ *noun* [Latin, it begins, from *incipere*] (1897)
: the first part **:** BEGINNING; *specifically* **:** the opening words of a text of a medieval manuscript or early printed book

in·ci·sal \in-'sī-zəl, -səl\ *adjective* (1903)
: relating to, involving, or being the cutting edge or surface of a tooth (as an incisor)

in·cise \in-'sīz, -'sīs\ *transitive verb* **in·cised; in·cis·ing** [Middle French or Latin; Middle French *inciser,* from Latin *incisus,* past participle of *incidere,* from *in-* + *caedere* to cut] (1567)
1 : to cut into
2 a : to carve figures, letters, or devices into **:** ENGRAVE **b :** to carve (as an inscription) into a surface

incised *adjective* (15th century)
1 : cut in **:** ENGRAVED; *especially* **:** decorated with incised figures
2 : having a margin that is deeply and sharply notched ⟨an *incised* leaf⟩

in·ci·sion \in-'si-zhən\ *noun* (14th century)
1 a : CUT, GASH; *specifically* **:** a wound made especially in surgery by incising the body **b :** a marginal notch (as in a leaf)
2 : an act of incising something
3 : the quality or state of being incisive

in·ci·sive \in-'sī-siv\ *adjective* (circa 1850)
: impressively direct and decisive (as in manner or presentation)
— in·ci·sive·ly *adverb*
— in·ci·sive·ness *noun*

in·ci·sor \in-'sī-zər\ *noun* (1666)
: a front tooth typically adapted for cutting; *especially* **:** one of the cutting teeth in mammals located between the canines when canines are present — see TOOTH illustration

in·ci·ta·tion \ˌin-ˌsī-'tā-shən, ˌin(t)-sə-\ *noun* (15th century)
1 : an act of inciting **:** STIMULATION
2 : something that incites to action **:** INCENTIVE

in·cite \in-'sīt\ *transitive verb* **in·cit·ed; in·cit·ing** [Middle French *inciter,* from Latin *incitare,* from *in-* + *citare* to put in motion — more at CITE] (15th century)
: to move to action **:** stir up **:** spur on **:** urge on ☆
— in·cit·ant \-'sī-t^{ə}nt\ *noun*
— in·cite·ment \-'sīt-mənt\ *noun*
— in·cit·er *noun*

in·ci·vil·i·ty \ˌin(t)-sə-'vi-lə-tē\ *noun* [Middle French *incivilité,* from Late Latin *incivilitat-, incivilitas,* from *incivilis,* from Latin *in-* + *civilis* civil] (1584)
1 : the quality or state of being uncivil
2 : a rude or discourteous act

in·clem·en·cy \(ˌ)in-'kle-mən(t)-sē\ *noun* (1559)
: the quality or state of being inclement

in·clem·ent \(ˌ)in-'kle-mənt, 'in-klə-\ *adjective* [Latin *inclement-, inclemens,* from *in-* + *clement-, clemens* clement] (1621)
: lacking mildness: as **a** *archaic* **:** severe in temper or action **:** UNMERCIFUL **b :** physically severe **:** STORMY ⟨*inclement* weather⟩
— in·clem·ent·ly *adverb*

in·clin·able \in-'klī-nə-bəl\ *adjective* (15th century)
: having a tendency or inclination; *also* **:** disposed to favor or think well of

in·cli·na·tion \ˌin-klə-'nā-shən, ˌiŋ-\ *noun* (14th century)
1 a *obsolete* **:** natural disposition **:** CHARACTER **b :** a particular disposition of mind or character **:** PROPENSITY; *especially* **:** LIKING ⟨had little *inclination* for housekeeping⟩
2 : an act or the action of bending or inclining: as **a :** BOW, NOD **b :** a tilting of something
3 a : a deviation from the true vertical or horizontal **:** SLANT; *also* **:** the degree of such deviation **b :** an inclined surface **:** SLOPE **c** (1) **:** the angle determined by two lines or planes (2) **:** the angle made by a line with the x-axis measured counterclockwise from the positive direction of that axis
4 : a tendency to a particular aspect, state, character, or action ⟨the clutch has an *inclination* to slip⟩
— in·cli·na·tion·al \-shnəl, -shə-n^{ə}l\ *adjective*

[1]in·cline \in-'klīn\ *verb* **in·clined; in·clin·ing** [Middle English, from Middle French *incliner,* from Latin *inclinare,* from *in-* + *clinare* to lean — more at LEAN] (14th century)
intransitive verb
1 : to bend the head or body forward **:** BOW
2 : to lean, tend, or become drawn toward an opinion or course of conduct
3 : to deviate from a line, direction, or course; *specifically* **:** to deviate from the vertical or horizontal
transitive verb
1 : to cause to stoop or bow **:** BEND
2 : to have influence on **:** PERSUADE ⟨his love of books *inclined* him toward a literary career⟩
3 : to give a bend or slant to ☆
— in·clin·er *noun*

[2]in·cline \'in-ˌklīn\ *noun* (1846)
: an inclined plane **:** GRADE, SLOPE

in·clined \in-'klīnd, *2 also* 'in-ˌ\ *adjective* (14th century)
1 : having inclination, disposition, or tendency
2 a : having a leaning or slope **b :** making an angle with a line or plane

inclined plane *noun* (1710)
: a plane surface that makes an oblique angle with the plane of the horizon

in·clin·ing \in-'klī-niŋ\ *noun* (14th century)
1 : INCLINATION
2 *archaic* **:** PARTY, FOLLOWING

in·cli·nom·e·ter \ˌin-klə-'nä-mə-tər, ˌiŋ-; ˌin-ˌklī-\ *noun* (1852)
: an instrument for indicating the inclination to the horizontal of an axis (as of an airplane)

in·clip \in-'klip\ *transitive verb* (1608)
archaic **:** CLASP, ENCLOSE

inclose, inclosure *variant of* ENCLOSE, ENCLOSURE

in·clude \in-'klüd\ *transitive verb* **in·clud·ed; in·clud·ing** [Middle English, from Latin *includere,* from *in-* + *claudere* to close — more at CLOSE] (15th century)
1 : to shut up **:** ENCLOSE
2 : to take in or comprise as a part of a whole
3 : to contain between or within ⟨two sides and the *included* angle⟩ ☆
— in·clud·able *or* **in·clud·ible** \-'klü-də-bəl\ *adjective*

in·clu·sion \in-'klü-zhən\ *noun* [Latin *inclusion-, inclusio,* from *includere*] (1600)
1 : the act of including **:** the state of being included
2 : something that is included: as **a :** a gaseous, liquid, or solid foreign body enclosed in a mass (as of a mineral) **b :** a passive usually temporary product of cell activity (as a starch grain) within the cytoplasm or nucleus
3 : a relation between two classes that exists when all members of the first are also members of the second — compare MEMBERSHIP 3

inclusion body *noun* (circa 1919)
: an inclusion, abnormal structure, or foreign cell within a cell; *specifically* **:** an intracellular body that is characteristic of some virus diseases and that is the site of virus multiplication

in·clu·sive \in-'klü-siv, -ziv\ *adjective* (15th century)
1 : comprehending stated limits or extremes ⟨from Monday to Friday *inclusive*⟩
2 a : broad in orientation or scope **b :** covering or intended to cover all items, costs, or services
— in·clu·sive·ly *adverb*
— in·clu·sive·ness *noun*

inclusive disjunction *noun* (1942)
: a complex sentence in logic that is true when either or both of its constituent propositions are true — see TRUTH TABLE table

inclusive of *preposition* (1709)
: including or taking into account ⟨the cost of building *inclusive of* materials⟩

in·co·erc·ible \ˌin-kō-'ər-sə-bəl\ *adjective* (1710)

☆ SYNONYMS

Incite, instigate, abet, foment mean to spur to action. INCITE stresses a stirring up and urging on, and may or may not imply initiating ⟨*inciting* a riot⟩. INSTIGATE definitely implies responsibility for initiating another's action and often connotes underhandedness or evil intention ⟨*instigated* a conspiracy⟩. ABET implies both assisting and encouraging ⟨aiding and *abetting* the enemy⟩. FOMENT implies persistence in goading ⟨*fomenting* rebellion⟩.

Incline, bias, dispose, predispose mean to influence one to have or take an attitude toward something. INCLINE implies a tendency to favor one of two or more actions or conclusions ⟨I *incline* to agree⟩. BIAS suggests a settled and predictable leaning in one direction and connotes unfair prejudice ⟨the experience *biased* him against foreigners⟩. DISPOSE suggests an affecting of one's mood or temper so as to incline one toward something ⟨her nature *disposes* her to trust others⟩. PREDISPOSE implies the operation of a disposing influence well in advance of the opportunity to manifest itself ⟨does fictional violence *predispose* them to accept real violence?⟩.

Include, comprehend, embrace, involve mean to contain within as part of the whole. INCLUDE suggests the containment of something as a constituent, component, or subordinate part of a larger whole ⟨the price of dinner *includes* dessert⟩. COMPREHEND implies that something comes within the scope of a statement or definition ⟨his system *comprehends* all history⟩. EMBRACE implies a gathering of separate items within a whole ⟨her faith *embraces* both Christian and non-Christian beliefs⟩. INVOLVE suggests inclusion by virtue of the nature of the whole, whether by being its natural or inevitable consequence ⟨the new job *involves* a lot of detail⟩.

: incapable of being controlled, checked, or confined

in·cog·i·tant \in-ˈkä-jə-tənt\ *adjective* [Latin *incogitant-, incogitans,* from *in-* + *cogitant-, cogitans,* present participle of *cogitare* to cogitate] (1628)
: THOUGHTLESS, INCONSIDERATE

in·cog·ni·ta \ˌin-ˌkäg-ˈnē-tə *also* in-ˈkäg-nə-tə\ *adverb or adjective* [Italian, feminine of *incognito*] (1638)
: INCOGNITO — used only of a woman
— **incognita** *noun*

[1]**in·cog·ni·to** \ˌin-ˌkäg-ˈnē-(ˌ)tō *also* in-ˈkäg-nə-ˌtō\ *adverb or adjective* [Italian, from Latin *incognitus* unknown, from *in-* + *cognitus,* past participle of *cognoscere* to know — more at COGNITION] (1635)
: with one's identity concealed

[2]**incognito** *noun, plural* **-tos** (1638)
1 : one appearing or living incognito
2 : the state or assumed identity of one living or traveling incognito or incognita

in·cog·ni·zant \(ˌ)in-ˈkäg-nə-zənt\ *adjective* (1837)
: lacking awareness or consciousness
— **in·cog·ni·zance** \-zən(t)s\ *noun*

in·co·her·ence \ˌin-kō-ˈhir-ən(t)s, -ˈher-\ *noun* (1611)
1 : the quality or state of being incoherent
2 : something that is incoherent

in·co·her·ent \-ənt\ *adjective* (1626)
1 : lacking coherence: as **a :** lacking cohesion **:** LOOSE **b :** lacking orderly continuity, arrangement, or relevance **:** INCONSISTENT
2 : lacking clarity or intelligibility **:** INCOMPREHENSIBLE
— **in·co·her·ent·ly** *adverb*

in·com·bus·ti·ble \ˌin-kəm-ˈbəs-tə-bəl\ *adjective* [Middle English, probably from Middle French, from *in-* + *combustible* combustible] (15th century)
: not combustible **:** incapable of being burned
— **in·com·bus·ti·bil·i·ty** \-ˌbəs-tə-ˈbi-lə-tē\ *noun*
— **incombustible** *noun*

in·come \ˈin-ˌkəm *also* ˈin-kəm *or* ˈiŋ-kəm\ *noun* (14th century)
1 : a coming in **:** ENTRANCE, INFLUX ⟨fluctuations in the nutrient *income* of a body of water⟩
2 : a gain or recurrent benefit usually measured in money that derives from capital or labor; *also* **:** the amount of such gain received in a period of time ⟨has an *income* of $20,000 a year⟩

income account *noun* (1869)
: a financial statement of a business showing the details of revenues, costs, expenses, losses, and profits for a given period — called also *income statement*

income bond *noun* (circa 1864)
: a bond that pays interest at a rate based on the issuer's earnings

income tax *noun* (1799)
: a tax on the net income of an individual or a business

[1]**in·com·ing** \ˈin-ˌkə-miŋ\ *noun* (14th century)
1 : the act of coming in **:** ARRIVAL
2 : INCOME — usually used in plural

[2]**incoming** *adjective* (1753)
1 : coming in **:** ARRIVING ⟨an *incoming* ship⟩ ⟨*incoming* mail⟩
2 : taking a new place or position especially as part of a succession ⟨the *incoming* president⟩
3 : just starting or beginning ⟨the *incoming* year⟩

in·com·men·su·ra·ble \ˌin-kə-ˈmen(t)s-rə-bəl, -ˈmen(t)sh-; -ˈmen(t)-sə-, -shə-\ *adjective* (1570)
: not commensurable; *broadly* **:** lacking a basis of comparison in respect to a quality normally subject to comparison
— **in·com·men·su·ra·bil·i·ty** \-ˌmen(t)s-rə-ˈbi-lə-tē, -ˌmen(t)sh-; -ˌmen(t)-sə-, -shə-\ *noun*
— **incommensurable** *noun*
— **in·com·men·su·ra·bly** \-ˈmen(t)s-rə-blē, -ˈmen(t)sh-; -ˈmen(t)-sə-, -shə-\ *adverb*

in·com·men·su·rate \-ˈmen(t)s-rət, -ˈmen(t)sh-; -ˈmen(t)-sə-, -shə-\ *adjective* (1650)
: not commensurate: as **a :** INCOMMENSURABLE **b :** INADEQUATE **c :** DISPROPORTIONATE

in·com·mode \ˌin-kə-ˈmōd\ *transitive verb* **-mod·ed; -mod·ing** [Middle French *incommoder,* from Latin *incommodare,* from *incommodus* inconvenient, from *in-* + *commodus* convenient — more at COMMODE] (1598)
: to give inconvenience or distress to **:** DISTURB

in·com·mo·di·ous \ˌin-kə-ˈmō-dē-əs\ *adjective* (1551)
: not commodious **:** INCONVENIENT
— **in·com·mo·di·ous·ly** *adverb*
— **in·com·mo·di·ous·ness** *noun*

in·com·mod·i·ty \-ˈmä-də-tē\ *noun* (15th century)
: a source of inconvenience **:** DISADVANTAGE

in·com·mu·ni·ca·ble \ˌin-kə-ˈmyü-ni-kə-bəl\ *adjective* [Middle French or Late Latin; Middle French, from Late Latin *incommunicabilis,* from Latin *in-* + Late Latin *communicabilis* communicable] (1568)
: not communicable: as **a :** incapable of being communicated or imparted **b :** UNCOMMUNICATIVE
— **in·com·mu·ni·ca·bil·i·ty** \-ˌmyü-ni-kə-ˈbi-lə-tē\ *noun*
— **in·com·mu·ni·ca·bly** \-ˈmyü-ni-kə-blē\ *adverb*

in·com·mu·ni·ca·do \-ˌmyü-nə-ˈkä-(ˌ)dō\ *adverb or adjective* [Spanish *incomunicado,* from past participle of *incomunicar* to deprive of communication, from *in-* (from Latin) + *comunicar* to communicate, from Latin *communicare*] (1844)
: without means of communication; *also* **:** in solitary confinement

in·com·mu·ni·ca·tive \-ˈmyü-nə-ˌkā-tiv, -ni-kə-tiv\ *adjective* (1670)
: UNCOMMUNICATIVE

in·com·mut·able \ˌin-kə-ˈmyü-tə-bəl\ *adjective* [Middle English, from Latin *incommutabilis,* from *in-* + *commutabilis* commutable] (15th century)
: not commutable: as **a :** not interchangeable **b :** UNCHANGEABLE
— **in·com·mut·ably** \-blē\ *adverb*

in·com·pa·ra·ble \(ˌ)in-ˈkäm-p(ə-)rə-bəl, ÷ˌin-kəm-ˈpar-ə-\ *adjective* [Middle English, from Middle French, from Latin *incomparabilis,* from *in-* + *comparabilis* comparable] (15th century)
1 : eminent beyond comparison **:** MATCHLESS
2 : not suitable for comparison
— **in·com·pa·ra·bil·i·ty** \(ˌ)in-ˌkäm-p(ə-)rə-ˈbi-lə-tē, ÷ˌin-kəm-ˌpar-ə-\ *noun*
— **in·com·pa·ra·bly** \(ˌ)in-ˈkäm-p(ə-)rə-blē, ÷ˌin-kəm-ˈpar-ə-\ *adverb*

in·com·pat·i·bil·i·ty \ˌin-kəm-ˌpa-tə-ˈbi-lə-tē\ *noun, plural* **-ties** (1611)
1 a : the quality or state of being incompatible **b :** lack of interfertility between two plants
2 *plural* **:** mutually antagonistic things or qualities

in·com·pat·i·ble \ˌin-kəm-ˈpa-tə-bəl\ *adjective* [Middle English, from Middle French & Medieval Latin; Middle French, from Medieval Latin *incompatibilis,* from Latin *in-* + Medieval Latin *compatibilis* compatible] (15th century)
1 : incapable of being held by one person at one time — used of offices that make conflicting demands on the holder
2 : not compatible: as **a :** incapable of association or harmonious coexistence ⟨*incompatible* colors⟩ **b :** unsuitable for use together because of undesirable chemical or physiological effects ⟨*incompatible* drugs⟩ **c :** not both true ⟨*incompatible* propositions⟩ **d :** incapable of blending into a stable homogeneous mixture
— **incompatible** *noun*
— **in·com·pat·i·bly** \-blē\ *adverb*

in·com·pe·tence \(ˌ)in-ˈkäm-pə-tən(t)s\ *noun* (1663)
: the state or fact of being incompetent

in·com·pe·ten·cy \-tən(t)-sē\ *noun* (1611)
: INCOMPETENCE

in·com·pe·tent \(ˌ)in-ˈkäm-pə-tənt\ *adjective* [Middle French *incompétent,* from *in-* + *compétent* competent] (1597)
1 : not legally qualified
2 : inadequate to or unsuitable for a particular purpose
3 a : lacking the qualities needed for effective action **b :** unable to function properly ⟨*incompetent* heart valves⟩
— **incompetent** *noun*
— **in·com·pe·tent·ly** *adverb*

in·com·plete \ˌin-kəm-ˈplēt\ *adjective* [Middle English *incompleet,* from Late Latin *incompletus,* from Latin *in-* + *completus* complete] (14th century)
1 : not complete **:** UNFINISHED: as **a :** lacking a part; *especially* **:** lacking one or more sets of floral organs **b** *of insect metamorphosis* **:** characterized by the absence of a pupal stage between the immature stages and the adult of an insect in which the young usually resemble the adult — compare COMPLETE 5
2 *of a football pass* **:** not legally caught
— **in·com·plete·ly** *adverb*
— **in·com·plete·ness** *noun*

in·com·pli·ant \ˌin-kəm-ˈplī-ənt\ *adjective* (1647)
: not compliant or pliable

in·com·pre·hen·si·ble \(ˌ)in-ˌkäm-pri-ˈhen(t)-sə-bəl\ *adjective* [Middle English, from Latin *incomprehensibilis,* from *in-* + *comprehensibilis* comprehensible] (14th century)
1 *archaic* **:** having or subject to no limits
2 : impossible to comprehend **:** UNINTELLIGIBLE
— **in·com·pre·hen·si·bil·i·ty** \-ˌhen(t)-sə-ˈbi-lə-tē\ *noun*
— **in·com·pre·hen·si·ble·ness** *noun*
— **in·com·pre·hen·si·bly** \-ˈhen(t)-sə-blē\ *adverb*

in·com·pre·hen·sion \-ˈhen(t)-shən\ *noun* (1605)
: lack of comprehension or understanding

in·com·press·ible \ˌin-kəm-ˈpre-sə-bəl\ *adjective* (circa 1736)
: incapable of or resistant to compression

in·com·put·able \ˌin-kəm-ˈpyü-tə-bəl\ *adjective* (1606)
: not computable **:** very great
— **in·com·put·ably** \-blē\ *adverb*

in·con·ceiv·able \ˌin-kən-ˈsē-və-bəl\ *adjective* (circa 1631)
: not conceivable: as **a :** impossible to comprehend **b :** UNBELIEVABLE
— **in·con·ceiv·abil·i·ty** \-ˌsē-və-ˈbi-lə-tē\ *noun*
— **in·con·ceiv·able·ness** \-ˈsē-və-bəl-nəs\ *noun*
— **in·con·ceiv·ably** \-blē\ *adverb*

in·con·cin·ni·ty \ˌin-kən-ˈsi-nə-tē\ *noun* [Latin *inconcinnitas,* from *in-* + *concinnitas* concinnity] (circa 1616)
: lack of suitability or congruity **:** INELEGANCE

in·con·clu·sive \ˌin-kən-ˈklü-siv, -ziv\ *adjective* (1707)
: leading to no conclusion or definite result
— **in·con·clu·sive·ly** *adverb*
— **in·con·clu·sive·ness** *noun*

in·con·dite \in-ˈkän-dət, -ˌdīt\ *adjective* [Latin *inconditus,* from *in-* + *conditus,* past participle of *condere* to put together, from *com-* + *-dere* to put — more at DO] (1539)
: badly put together **:** CRUDE

in·con·for·mi·ty \ˌin-kən-ˈfȯr-mə-tē\ *noun* (1594)
: NONCONFORMITY

in·con·gru·ence \ˌin-kən-ˈgrü-ən(t)s, (ˌ)in-ˈkäŋ-grə-wən(t)s\ *noun* (1610)
: INCONGRUITY

in·con·gru·ent \-ənt, -wənt\ *adjective* [Middle English, from Latin *incongruent-, incongruens,* from *in-* + *congruent-, congruens* congruent] (15th century)
: not congruent ⟨*incongruent* triangles⟩
— **in·con·gru·ent·ly** *adverb*

in·con·gru·i·ty \ˌin-kən-ˈgrü-ə-tē, -ˌkän-\ *noun, plural* **-ties** (circa 1532)
1 : the quality or state of being incongruous
2 : something that is incongruous

in·con·gru·ous \(ˌ)in-ˈkäŋ-grə-wəs\ *adjective* [Late Latin *incongruus,* from Latin *in-* + *congruus* congruous] (1611)
: lacking congruity: as **a :** not harmonious **:** INCOMPATIBLE ⟨*incongruous* colors⟩ **b :** not conforming **:** DISAGREEING ⟨conduct *incongruous* with principle⟩ **c :** inconsistent within itself ⟨an *incongruous* story⟩ **d :** lacking propriety **:** UNSUITABLE ⟨*incongruous* manners⟩
— **in·con·gru·ous·ly** *adverb*
— **in·con·gru·ous·ness** *noun*

in·con·scient \(ˌ)in-ˈkän(t)-shənt\ *adjective* [probably from French, from *in-* + *conscient* mindful, from Latin *conscient-, consciens,* present participle of *conscire* to be conscious — more at CONSCIENCE] (1885)
: UNCONSCIOUS, MINDLESS

in·con·sec·u·tive \ˌin-kən-ˈse-kyə-tiv, -kə-tiv\ *adjective* (1831)
: not consecutive

in·con·se·quence \(ˌ)in-ˈkän(t)-sə-ˌkwen(t)s, -si-kwən(t)s\ *noun* (1588)
: the quality or state of being inconsequent

in·con·se·quent \-ˌkwent, -kwənt\ *adjective* [Late Latin *inconsequent-, inconsequens,* from Latin *in-* + *consequent-, consequens* consequent] (1579)
1 a : lacking reasonable sequence **:** ILLOGICAL **b :** INCONSECUTIVE
2 : INCONSEQUENTIAL 2
3 : IRRELEVANT
— **in·con·se·quent·ly** *adverb*

in·con·se·quen·tial \(ˌ)in-ˌkän(t)-sə-ˈkwen(t)-shəl\ *adjective* (1621)
1 a : ILLOGICAL **b :** IRRELEVANT
2 : of no significance **:** UNIMPORTANT
— **in·con·se·quen·ti·al·i·ty** \-ˌkwen(t)-she-ˈa-lə-tē\ *noun*
— **in·con·se·quen·tial·ly** \-ˈkwen(t)-shle, -ˈkwen(t)-shə-\ *adverb*

in·con·sid·er·able \ˌin-kən-ˈsi-dər-(ə-)bəl, -ˈsi-drə-bəl\ *adjective* [French, from *in-* + *considerable* considerable, from Medieval Latin *considerabilis*] (1637)
: not considerable **:** TRIVIAL
— **in·con·sid·er·able·ness** *noun*
— **in·con·sid·er·ably** \-blē\ *adverb*

in·con·sid·er·ate \ˌin-kən-ˈsi-d(ə-)rət\ *adjective* [Middle English *inconsyderatt,* from Latin *inconsideratus,* from *in-* + *consideratus* considerate] (15th century)
1 a : HEEDLESS, THOUGHTLESS **b :** careless of the rights or feelings of others
2 : not adequately considered **:** ILL-ADVISED
— **in·con·sid·er·ate·ly** *adverb*
— **in·con·sid·er·ate·ness** *noun*
— **in·con·sid·er·ation** \-ˌsi-də-ˈrā-shən\ *noun*

in·con·sis·tence \ˌin-kən-ˈsis-tən(t)s\ *noun* (1643)
: INCONSISTENCY

in·con·sis·ten·cy \ˌin-kən-ˈsis-tən(t)-sē\ *noun* (1647)
1 : an instance of being inconsistent
2 : the quality or state of being inconsistent

in·con·sis·tent \-tənt\ *adjective* (1646)
: lacking consistency: as **a :** not compatible with another fact or claim ⟨*inconsistent* statements⟩ **b :** containing incompatible elements ⟨an *inconsistent* argument⟩ **c :** incoherent or illogical in thought or actions **:** CHANGEABLE **d :** not satisfiable by the same set of values for the unknowns ⟨*inconsistent* equations⟩ ⟨*inconsistent* inequalities⟩
— **in·con·sis·tent·ly** *adverb*

in·con·sol·able \ˌin-kən-ˈsō-lə-bəl\ *adjective* [Latin *inconsolabilis,* from *in-* + *consolabilis* consolable] (1596)
: incapable of being consoled **:** DISCONSOLATE
— **in·con·sol·able·ness** *noun*
— **in·con·sol·ably** \-blē\ *adverb*

in·con·so·nance \(ˌ)in-ˈkän(t)-s(ə-)nən(t)s\ *noun* (circa 1811)
: lack of consonance or harmony **:** DISAGREEMENT

in·con·so·nant \-s(ə-)nənt\ *adjective* (1658)
: not consonant **:** DISCORDANT

in·con·spic·u·ous \ˌin-kən-ˈspi-kyə-wəs\ *adjective* [Latin *inconspicuus,* from *in-* + *conspicuus* conspicuous] (1648)
: not readily noticeable
— **in·con·spic·u·ous·ly** *adverb*
— **in·con·spic·u·ous·ness** *noun*

in·con·stan·cy \(ˌ)in-ˈkän(t)-stən(t)-sē\ *noun* (1526)
: the quality or state of being inconstant

in·con·stant \-stənt\ *adjective* [Middle English, from Middle French, from Latin *inconstant-, inconstans,* from *in-* + *constant-, constans* constant] (15th century)
: likely to change frequently without apparent or cogent reason ☆
— **in·con·stant·ly** *adverb*

in·con·sum·able \ˌin-kən-ˈsü-mə-bəl\ *adjective* (1646)
: not capable of being consumed
— **in·con·sum·ably** \-blē\ *adverb*

in·con·test·able \ˌin-kən-ˈtes-tə-bəl\ *adjective* [French, from *in-* + *contestable,* from *contester* to contest] (1673)
: not contestable **:** INDISPUTABLE
— **in·con·test·abil·i·ty** \-ˌtes-tə-ˈbi-lə-tē\ *noun*
— **in·con·test·ably** \-ˈtes-tə-blē\ *adverb*

in·con·ti·nence \(ˌ)in-ˈkän-tᵊn-ən(t)s\ *noun* (14th century)
: the quality or state of being incontinent: as **a :** failure to restrain sexual appetite **:** UNCHASTITY **b :** inability of the body to control the evacuative functions

in·con·ti·nen·cy \-ən(t)-sē\ *noun* (15th century)
: INCONTINENCE

¹in·con·ti·nent \(ˌ)in-ˈkän-tᵊn-ənt\ *adjective* [Middle English, from Middle French or Latin; Middle French, from Latin *incontinent-, incontinens,* from *in-* + *continent-, continens* continent] (14th century)
: not continent: as **a** (1) **:** lacking self-restraint (2) **:** not being under control **b :** unable to retain a bodily discharge (as urine) voluntarily

²incontinent *adverb* [Middle English, from Middle French, from Late Latin *in continenti*] (15th century)
: ¹INCONTINENTLY

¹in·con·ti·nent·ly *adverb* (15th century)
: without delay **:** IMMEDIATELY

²incontinently *adverb* (circa 1552)
: in an incontinent or unrestrained manner: as **a :** without moral restraint **:** LEWDLY **b :** without due or reasonable consideration

in·con·trol·la·ble \ˌin-kən-ˈtrō-lə-bəl\ *adjective* (1599)
: UNCONTROLLABLE

in·con·tro·vert·ible \(ˌ)in-ˌkän-trə-ˈvər-tə-bəl\ *adjective* (1646)
: not open to question **:** INDISPUTABLE ⟨*incontrovertible* evidence⟩
— **in·con·tro·vert·ibly** \-blē\ *adverb*

¹in·con·ve·nience \ˌin-kən-ˈvē-nyən(t)s\ *noun* [Middle English, harm, damage, from Late Latin *inconvenientia,* from Latin *inconvenient-, inconveniens*] (1547)
1 : something that is inconvenient
2 : the quality or state of being inconvenient

²inconvenience *transitive verb* **-nienced; -nienc·ing** (circa 1656)
: to subject to inconvenience **:** put to trouble

in·con·ve·nien·cy \ˌin-kən-ˈvē-nyən(t)-sē\ *noun, plural* **-cies** (circa 1552)
: INCONVENIENCE

in·con·ve·nient \ˌin-kən-ˈvē-nyənt\ *adjective* [Middle English, from Middle French, from Latin *inconvenient-, inconveniens,* from *in-* + *convenient-, conveniens* convenient] (1651)
: not convenient especially in giving trouble or annoyance **:** INOPPORTUNE
— **in·con·ve·nient·ly** *adverb*

in·con·vert·ible \-ˈvər-tə-bəl\ *adjective* [probably from Late Latin *inconvertibilis,* from Latin *in-* + *convertibilis* convertible] (1646)
: not convertible: as **a** *of paper money* **:** not exchangeable for coin **b** *of a currency* **:** not exchangeable for a foreign currency
— **in·con·vert·ibil·i·ty** \-ˌvər-tə-ˈbi-lə-tē\ *noun*
— **in·con·vert·ibly** \-ˈvər-tə-blē\ *adverb*

in·con·vinc·ible \ˌin-kən-ˈvin(t)-sə-bəl\ *adjective* (1674)
: incapable of being convinced

in·co·or·di·na·tion \-(ˌ)kō-ˌȯr-dᵊn-ˈā-shən\ *noun* (1876)
: lack of coordination especially of muscular movements resulting from loss of voluntary control

¹in·cor·po·rate \in-ˈkȯr-pə-ˌrāt\ *verb* **-rat·ed; -rat·ing** [Middle English, from Late Latin *incorporatus,* past participle of *incorporare,* from Latin *in-* + *corpor-, corpus* body — more at MIDRIFF] (14th century)
transitive verb
1 a : to unite or work into something already existent so as to form an indistinguishable whole **b :** to blend or combine thoroughly
2 a : to form into a legal corporation **b :** to admit to membership in a corporate body
3 : to give material form to **:** EMBODY
intransitive verb
1 : to unite in or as one body
2 : to form or become a corporation
— **in·cor·po·ra·ble** \-p(ə-)rə-bəl\ *adjective*
— **in·cor·po·ra·tion** \-ˌkȯr-pə-ˈrā-shən\ *noun*
— **in·cor·po·ra·tive** \-ˈkȯr-pə-ˌrā-tiv, -p(ə-)rə-tiv\ *adjective*
— **in·cor·po·ra·tor** \-pə-ˌrā-tər\ *noun*

²in·cor·po·rate \in-ˈkȯr-p(ə-)rət\ *adjective* (14th century)
: INCORPORATED

in·cor·po·rat·ed \-pə-ˌrā-təd\ *adjective* (1599)
1 : united in one body
2 : formed into a legal corporation

in·cor·po·re·al \ˌin-(ˌ)kȯr-ˈpȯr-ē-əl, -ˈpȯr-\ *adjective* [Middle English *incorporealle,* from Latin *incorporeus,* from *in-* + *corporeus* corporeal] (15th century)

☆ SYNONYMS
Inconstant, fickle, capricious, mercurial, unstable mean lacking firmness or steadiness (as in purpose or devotion). INCONSTANT implies an incapacity for steadiness and an inherent tendency to change ⟨an *inconstant* friend⟩. FICKLE suggests unreliability because of perverse changeability and incapacity for steadfastness ⟨performers discover how *fickle* fans can be⟩. CAPRICIOUS suggests motivation by sudden whim or fancy and stresses unpredictability ⟨an utterly *capricious* critic⟩. MERCURIAL implies a rapid changeability in mood ⟨made anxious by her boss's *mercurial* temperament⟩. UNSTABLE implies an incapacity for remaining in a fixed position or steady course and applies especially to a lack of emotional balance ⟨too *unstable* to hold a job⟩.

1 : not corporeal **:** having no material body or form
2 : of, relating to, or constituting a right that is based on property (as bonds or patents) which has no intrinsic value
— in·cor·po·re·al·ly \-ə-lē\ *adverb*

in·cor·po·re·ity \(,)in-,kȯr-pə-'rē-ə-tē\ *noun* (1601)
: the quality or state of being incorporeal **:** IMMATERIALITY

in·cor·rect \,in-kə-'rekt\ *adjective* [Middle English, from Middle French or Latin; Middle French, from Latin *incorrectus,* from *in-* + *correctus* correct] (15th century)
1 *obsolete* **:** not corrected or chastened
2 a : INACCURATE, FAULTY **b :** not true **:** WRONG
3 : UNBECOMING, IMPROPER
— in·cor·rect·ly \-'rek(t)-lē\ *adverb*
— in·cor·rect·ness \-nəs\ *noun*

in·cor·ri·gi·ble \(,)in-'kȯr-ə-jə-bəl, -'kär-\ *adjective* [Middle English, from Late Latin *incorrigibilis,* from Latin *in-* + *corrigere* to correct — more at CORRECT] (14th century)
: incapable of being corrected or amended: as **a** (1) **:** not reformable **:** DEPRAVED (2) **:** DELINQUENT **b :** not manageable **:** UNRULY **c :** UNALTERABLE, INVETERATE
— in·cor·ri·gi·bil·i·ty \-,kȯr-ə-jə-'bi-lə-tē, -,kär-\ *noun*
— incorrigible *noun*
— in·cor·ri·gi·ble·ness \-'kȯr-ə-jə-bəl-nəs, -'kär-\ *noun*
— in·cor·ri·gi·bly \-blē\ *adverb*

in·cor·rupt \,in-kə-'rəpt\ *also* **in·cor·rupt·ed** \-'rəp-təd\ *adjective* [Middle English, from Latin *incorruptus,* from *in-* + *corruptus* corrupt] (14th century)
: free from corruption: as **a** *obsolete* **:** not affected with decay **b :** not defiled or depraved **:** UPRIGHT **c :** free from error
— in·cor·rupt·ly \-'rəp(t)-lē\ *adverb*
— in·cor·rupt·ness \-nəs\ *noun*

in·cor·rupt·ible \,in-kə-'rəp-tə-bəl\ *adjective* (14th century)
: incapable of corruption: as **a :** not subject to decay or dissolution **b :** incapable of being bribed or morally corrupted
— in·cor·rupt·ibil·i·ty \-,rəp-tə-'bi-lə-tē\ *noun*
— incorruptible *noun*
— in·cor·rupt·ibly \-'rəp-tə-blē\ *adverb*

in·cor·rup·tion \,in-kə-'rəp-shən\ *noun* (14th century)
archaic **:** the quality or state of being free from physical decay

¹in·crease \in-'krēs, 'in-,\ *verb* **in·creased; in·creas·ing** [Middle English *encresen,* from Middle French *encreistre,* from Latin *increscere,* from *in-* + *crescere* to grow — more at CRESCENT] (14th century)
intransitive verb
1 : to become progressively greater (as in size, amount, number, or intensity)
2 : to multiply by the production of young
transitive verb
1 : to make greater **:** AUGMENT
2 *obsolete* **:** ENRICH ☆
— in·creas·able \-'krē-sə-bəl, -,krē-\ *adjective*
— in·creas·er *noun*

²in·crease \'in-,krēs, in-'\ *noun* (14th century)
1 : the act or process of increasing: as **a :** addition or enlargement in size, extent, or quantity **b** *obsolete* **:** PROPAGATION
2 : something that is added to an original stock or amount by augmentation or growth (as offspring, produce, profit)

in·creas·ing·ly \in-'krē-siŋ-lē, 'in-,krē-\ *adverb* (14th century)
: to an increasing degree

in·cre·ate \,in-krē-'āt, in-'krē-ət\ *adjective* [Middle English *increat,* from Late Latin *increatus,* from Latin *in-* + *creatus,* past participle of *creare* to create] (15th century)
: UNCREATED

in·cred·i·ble \(,)in-'kre-də-bəl\ *adjective* [Middle English, from Latin *incredibilis,* from *in-* + *credibilis* credible] (15th century)
: too extraordinary and improbable to be believed; *also* **:** hard to believe
— in·cred·i·bil·i·ty \-,kre-də-'bi-lə-tē\ *noun*
— in·cred·i·ble·ness \-'kre-də-bəl-nəs\ *noun*

in·cred·i·bly \-blē\ *adverb* (circa 1500)
1 : in an incredible manner
2 : EXTREMELY ⟨*incredibly* difficult⟩

in·cre·du·li·ty \,in-kri-'dü-lə-tē, -'dyü-\ *noun* (15th century)
: the quality or state of being incredulous **:** DISBELIEF

in·cred·u·lous \(,)in-'kre-jə-ləs, -dyə-ləs\ *adjective* [Latin *incredulus,* from *in-* + *credulus* credulous] (1579)
1 : unwilling to admit or accept what is offered as true **:** not credulous **:** SKEPTICAL
2 : INCREDIBLE
3 : expressing incredulity ◻
— in·cred·u·lous·ly *adverb*

in·cre·ment \'iŋ-krə-mənt, 'in-\ *noun* [Middle English, from Latin *incrementum,* from *increscere* to increase] (15th century)
1 : the action or process of increasing especially in quantity or value **:** ENLARGEMENT
2 a : something gained or added **b :** one of a series of regular consecutive additions **c :** a minute increase in quantity
3 : the amount or degree by which something changes; *especially* **:** the amount of positive or negative change in the value of one or more of a set of variables
— in·cre·men·tal \,iŋ-krə-'men-tᵊl, ,in-\ *adjective*
— in·cre·men·tal·ly \-tᵊl-ē\ *adverb*

in·cre·men·tal·ism \,iŋ-krə-'men-tᵊl-,i-zəm\ *noun* (1966)
: a policy or advocacy of a policy of political or social change by degrees **:** GRADUALISM
— in·cre·men·tal·ist \-tᵊl-ist\ *noun*

incremental repetition *noun* (1918)
: repetition in each stanza (as of a ballad) of part of the preceding stanza usually with a slight change in wording for dramatic effect

in·cres·cent \in-'kre-sᵊnt\ *adjective* [Latin *increscent-, increscens,* present participle of *increscere*] (circa 1658)
: becoming gradually greater **:** WAXING ⟨the *increscent* moon⟩

in·crim·i·nate \in-'kri-mə-,nāt\ *transitive verb* **-nat·ed; -nat·ing** [Late Latin *incriminatus,* past participle of *incriminare,* from Latin *in-* + *crimin-, crimen* crime] (circa 1736)
: to charge with or show evidence or proof of involvement in a crime or fault
— in·crim·i·na·tion \-,kri-mə-'nā-shən\ *noun*
— in·crim·i·na·to·ry \-'krim-nə-,tōr-ē, -'kri-mə-, -,tȯr-\ *adjective*

incrust *variant of* ENCRUST

in·crus·ta·tion \,in-,krəs-'tā-shən\ *noun* [Latin *incrustation-, incrustatio,* from *incrustare* to encrust] (1644)
1 a : a crust or hard coating **b :** a growth or accumulation (as of habits, opinions, or customs) resembling a crust
2 : the act of encrusting **;** the state of being encrusted
3 a : OVERLAY a **b :** INLAY

in·cu·bate \'iŋ-kyə-,bāt, 'in-\ *verb* **-bat·ed; -bat·ing** [Latin *incubatus,* past participle of *incubare,* from *in-* + *cubare* to lie] (circa 1721)
transitive verb
1 : to sit on (eggs) so as to hatch by the warmth of the body; *also* **:** to maintain (as an embryo or a chemically active system) under conditions favorable for hatching, development, or reaction
2 : to cause (as an idea) to develop
intransitive verb
1 : to sit on eggs
2 : to undergo incubation
— in·cu·ba·tive \-,bā-tiv\ *adjective*
— in·cu·ba·to·ry \-kyə-bə-,tōr-ē, -,tȯr-; -,bā-tə-rē\ *adjective*

in·cu·ba·tion \,iŋ-kyə-'bā-shən, ,in-\ *noun* (1646)
1 : the act or process of incubating
2 : INCUBATION PERIOD

incubation period *noun* (1879)
: the period between the infection of an individual by a pathogen and the manifestation of the disease it causes

in·cu·ba·tor \'iŋ-kyə-,bā-tər, 'in-\ *noun* (1857)
: one that incubates: as **a :** an apparatus by which eggs are hatched artificially **b :** an apparatus with a chamber used to provide controlled environmental conditions especially for the cultivation of microorganisms or the care and protection of premature or sick babies

in·cu·bus \'iŋ-kyə-bəs, 'in-\ *noun, plural* **-bi** \-,bī, -,bē\ *also* **-bus·es** [Middle English, from Late Latin, from Latin *incubare*] (13th century)
1 : an evil spirit that lies on persons in their sleep; *especially* **:** one that has sexual intercourse with women while they are sleeping — compare SUCCUBUS
2 : NIGHTMARE 2
3 : one that oppresses or burdens like a nightmare

in·cul·cate \in-'kəl-,kāt, 'in-(,)\ *transitive verb* **-cat·ed; -cat·ing** [Latin *inculcatus,* past participle of *inculcare,* literally, to tread on, from *in-* + *calcare* to trample, from *calc-, calx* heel] (1550)
: to teach and impress by frequent repetitions or admonitions
synonym see IMPLANT
— in·cul·ca·tion \,in-(,)kəl-'kā-shən\ *noun*
— in·cul·ca·tor \in-'kəl-,kā-tər, 'in-(,)\ *noun*

in·cul·pa·ble \(,)in-'kəl-pə-bəl\ *adjective* (15th century)
: free from guilt **:** BLAMELESS

in·cul·pate \in-'kəl-,pāt, 'in-(,)\ *transitive verb* **-pat·ed; -pat·ing** [Late Latin *inculpatus,* from Latin *in-* + *culpatus,* past participle of *culpare* to blame, from *culpa* guilt] (1799)
: INCRIMINATE
— in·cul·pa·tion \,in-(,)kəl-'pā-shən\ *noun*
— in·cul·pa·to·ry \in-'kəl-pə-,tōr-ē, -,tȯr-\ *adjective*

in·cult \in-'kəlt\ *adjective* [Latin *incultus,* from *in-* + *cultus,* past participle of *colere* to cultivate — more at WHEEL] (1599)

☆ SYNONYMS
Increase, enlarge, augment, multiply mean to make or become greater. INCREASE used intransitively implies progressive growth in size, amount, or intensity ⟨his waistline *increased* with age⟩; used transitively it may imply simple not necessarily progressive addition ⟨*increased* her landholdings⟩. ENLARGE implies expansion or extension that makes greater in size or capacity ⟨*enlarged* the kitchen⟩. AUGMENT implies addition to what is already well grown or well developed ⟨the inheritance *augmented* his fortune⟩. MULTIPLY implies increase in number by natural generation or by indefinite repetition of a process ⟨with each attempt the problems *multiplied*⟩.

◻ USAGE
incredulous Sense 2 has been revived in this century after a couple of centuries of disuse. Although it is a sense with good literary precedent—among others Shakespeare used it—many people think it is a result of confusion with *incredible,* which is still the usual word in this sense.

: COARSE, UNCULTURED

in•cum•ben•cy \in-ˈkəm-bən(t)-sē\ *noun, plural* **-cies** (circa 1608)
1 : something that is incumbent **:** DUTY
2 : the quality or state of being incumbent
3 : the sphere of action or period of office of an incumbent

¹in•cum•bent \in-ˈkəm-bənt\ *noun* [Middle English, from Latin *incumbent-, incumbens,* present participle of *incumbere* to lie down on, from *in-* + *-cumbere* to lie down; akin to *cubare* to lie] (15th century)
1 : the holder of an office or ecclesiastical benefice
2 : OCCUPANT

²incumbent *adjective* (1567)
1 : imposed as a duty **:** OBLIGATORY
2 : having the status of an incumbent; *especially* **:** occupying a specified office
3 : lying or resting on something else
4 : bent over so as to rest on or touch an underlying surface

incumber *variant of* ENCUMBER

in•cu•na•ble \in-ˈkyü-nə-bəl\ *noun* [French, from New Latin *incunabulum*] (1886)
: INCUNABULUM

in•cu•nab•u•lum \ˌin-kyə-ˈna-byə-ləm, ˌiŋ-\ *noun, plural* **-la** \-lə\ [New Latin, from Latin *incunabula,* plural, bands holding the baby in a cradle, from *in-* + *cunae* cradle] (circa 1859)
1 : a book printed before 1501
2 : a work of art or of industry of an early period

in•cur \in-ˈkər\ *transitive verb* **in•curred; in•cur•ring** [Middle English *incurren,* from Latin *incurrere,* literally, to run into, from *in-* + *currere* to run — more at CAR] (15th century)
: to become liable or subject to **:** bring down upon oneself ⟨*incur* expenses⟩

in•cur•able \(ˌ)in-ˈkyu̇r-ə-bəl\ *adjective* [Middle English, from Middle French or Late Latin; Middle French, from Late Latin *incurabilis,* from Latin *in-* + *curabilis* curable] (14th century)
: not curable ⟨an *incurable* disease⟩; *broadly* **:** not likely to be changed or corrected ⟨*incurable* optimism⟩
— **incurable** *noun*
— **in•cur•ably** \-blē\ *adverb*

in•cu•ri•ous \(ˌ)in-ˈkyu̇r-ē-əs\ *adjective* [Latin *incuriosus,* from *in-* + *curiosus* curious] (circa 1618)
: lacking a normal or usual curiosity **:** UNINTERESTED ⟨a blank *incurious* stare⟩
synonym see INDIFFERENT
— **in•cu•ri•os•i•ty** \-ˌkyu̇r-ē-ˈä-sə-tē\ *noun*
— **in•cu•ri•ous•ly** \-ˈkyu̇r-ē-əs-lē\ *adverb*
— **in•cu•ri•ous•ness** *noun*

in•cur•rence \in-ˈkər-ən(t)s, -ˈkə-rən(t)s\ *noun* (circa 1656)
: the act or process of incurring

in•cur•rent \-ənt, -rənt\ *adjective* [Latin *incurrent-, incurrens,* present participle of *incurrere*] (circa 1856)
: giving passage to a current that flows inward

in•cur•sion \in-ˈkər-zhən\ *noun* [Middle English, from Middle French or Latin; Middle French, from Latin *incursion-, incursio,* from *incurrere*] (15th century)
1 : a hostile entrance into a territory **:** RAID
2 : an entering in or into (as an activity or undertaking) ⟨his only *incursion* into the arts⟩

in•cur•vate \ˈin-ˌkər-ˌvāt, (ˌ)in-ˈkər-\ *transitive verb* **-vat•ed; -vat•ing** (1578)
: to cause to curve inward **:** BEND
— **in•cur•vate** \ˈin-ˌkər-ˌvāt, (ˌ)in-ˈkər-vət\ *adjective*
— **in•cur•va•tion** \ˌin-ˌkər-ˈvā-shən\ *noun*
— **in•cur•va•ture** \(ˌ)in-ˈkər-və-ˌchu̇r, -chər, -ˌtyu̇r, -ˌtu̇r\ *noun*

in•curve \(ˌ)in-ˈkərv, ˈin-ˌ\ *transitive verb* [Latin *incurvare,* from *in-* + *curvare* to curve, from *curvus* curved — more at CURVE] (1610)
: to bend so as to curve inward

in•cus \ˈiŋ-kəs\ *noun, plural* **in•cu•des** \iŋ-ˈkyü-(ˌ)dēz, ˈiŋ-kyə-ˌdēz\ [New Latin, from Latin, anvil, from *incudere*] (1669)
: the middle bone of a chain of three small bones in the ear of a mammal — called also *anvil;* see EAR illustration

in•cuse \in-ˈkyüz, -ˈkyüs\ *adjective* [Latin *incusus,* past participle of *incudere* to stamp, strike, from *in-* + *cudere* to beat — more at HEW] (1818)
: formed by stamping or punching in — used chiefly of old coins or features of their design

Ind \ˈind, ˈīnd\ *noun* (13th century)
1 *archaic* **:** India
2 *obsolete* **:** Indies

ind- *or* **indo-** *combining form* [International Scientific Vocabulary, from Latin *indicum* — more at INDIGO]
1 : indigo ⟨*ind*oxyl⟩
2 : resembling indigo (as in color) ⟨*indo*phenol⟩

in•da•ba \in-ˈdä-bə\ *noun* [Zulu *indaba* affair] (1827)
chiefly South African **:** CONFERENCE, PARLEY

in•da•gate \ˈin-də-ˌgāt\ *transitive verb* **-gat•ed; -gat•ing** [Latin *indagatus,* past participle of *indagare,* from *indago* ring of hunters encircling game, act of searching, from Old Latin *indu* in + Latin *agere* to drive — more at END-, AGENT] (circa 1623)
: to search into **:** INVESTIGATE
— **in•da•ga•tion** \ˌin-də-ˈgā-shən\ *noun*
— **in•da•ga•tor** \ˈin-də-ˌgā-tər\ *noun*

in•debt•ed \in-ˈde-təd\ *adjective* [Middle English *indetted,* from Old French *endeté,* past participle of *endeter* to involve in debt, from *en-* + *dete* debt] (13th century)
1 : owing gratitude or recognition to another **:** BEHOLDEN
2 : owing money

in•debt•ed•ness *noun* (1647)
1 : the condition of being indebted
2 : something (as an amount of money) that is owed

in•de•cen•cy \(ˌ)in-ˈdē-sᵊn(t)-sē\ *noun* (1589)
1 : the quality or state of being indecent
2 : something (as a word or action) that is indecent

in•de•cent \-sᵊnt\ *adjective* [Middle French or Latin; Middle French *indécent,* from Latin *indecent-, indecens,* from *in-* + *decent-, decens* decent] (circa 1587)
: not decent; *especially* **:** grossly unseemly or offensive to manners or morals
— **in•de•cent•ly** *adverb*

indecent assault *noun* (1861)
: an immoral act or series of acts exclusive of rape committed against another person without consent

indecent exposure *noun* (1851)
: intentional exposure of part of one's body (as the genitals) in a place where such exposure is likely to be an offense against the generally accepted standards of decency

in•de•ci•pher•able \ˌin-di-ˈsī-f(ə-)rə-bəl\ *adjective* (1802)
: incapable of being deciphered

in•de•ci•sion \ˌin-di-ˈsi-zhən\ *noun* [French *indécision,* from *indécis* undecided, from Late Latin *indecisus,* from Latin *in-* + *decisus,* past participle of *decidere* to decide] (circa 1763)
: a wavering between two or more possible courses of action **:** IRRESOLUTION

in•de•ci•sive \ˌin-di-ˈsī-siv\ *adjective* (1726)
1 : not decisive **:** INCONCLUSIVE
2 : marked by or prone to indecision **:** IRRESOLUTE
3 : not clearly marked out **:** INDEFINITE
— **in•de•ci•sive•ly** *adverb*
— **in•de•ci•sive•ness** *noun*

in•de•clin•able \ˌin-di-ˈklī-nə-bəl\ *adjective* [Middle English, from Middle French, from Late Latin *indeclinabilis,* from Latin *in-* + Late Latin *declinabilis* capable of being inflected, from Latin *declinare* to inflect — more at DECLINE] (14th century)
: having no grammatical inflections

in•de•com•pos•able \ˌin-ˌdē-kəm-ˈpō-zə-bəl\ *adjective* (1807)
: not capable of being separated into component parts or elements

in•de•co•rous \(ˌ)in-ˈde-k(ə-)rəs; ˌin-di-ˈkōr-əs, -ˈkȯr-\ *adjective* [Latin *indecorus,* from *in-* + *decorus* decorous] (1682)
: not decorous **:** conflicting with accepted standards of good conduct or good taste ☆
— **in•de•co•rous•ly** *adverb*
— **in•de•co•rous•ness** *noun*

in•de•co•rum \ˌin-di-ˈkōr-əm, -ˈkȯr-\ *noun* [Latin, neuter of *indecorus*] (1575)
1 : something that is indecorous
2 : lack of decorum **:** IMPROPRIETY

in•deed \in-ˈdēd\ *adverb* (14th century)
1 : without any question **:** TRULY, UNDENIABLY — often used interjectionally to express irony or disbelief or surprise
2 : in reality
3 : all things considered **:** as a matter of fact

in•de•fat•i•ga•ble \ˌin-di-ˈfa-ti-gə-bəl\ *adjective* [Middle French, from Latin *indefatigabilis,* from *in-* + *defatigare* to fatigue, from *de-* + *fatigare* to fatigue] (1611)
: incapable of being fatigued **:** UNTIRING
— **in•de•fa•ti•ga•bil•i•ty** \-ˌfa-ti-gə-ˈbi-lə-tē\ *noun*
— **in•de•fat•i•ga•ble•ness** \-ˈfa-ti-gə-bəl-nəs\ *noun*
— **in•de•fat•i•ga•bly** \-blē\ *adverb*

in•de•fea•si•ble \-ˈfē-zə-bəl\ *adjective* (1548)
: not capable of being annulled or voided or undone ⟨an *indefeasible* right⟩
— **in•de•fea•si•bil•i•ty** \-ˌfē-zə-ˈbi-lə-tē\ *noun*
— **in•de•fea•si•bly** \-ˈfē-zə-blē\ *adverb*

in•de•fec•ti•ble \-ˈfek-tə-bəl\ *adjective* (1659)
1 : not subject to failure or decay **:** LASTING
2 : free of faults **:** FLAWLESS
— **in•de•fec•ti•bil•i•ty** \-ˌfek-tə-ˈbi-lə-tē\ *noun*
— **in•de•fec•ti•bly** \-ˈfek-tə-blē\ *adverb*

in•de•fen•si•ble \-ˈfen(t)-sə-bəl\ *adjective* (1529)
1 a : incapable of being maintained as right or valid **:** UNTENABLE **b :** incapable of being justified or excused **:** INEXCUSABLE
2 : incapable of being protected against physical attack
— **in•de•fen•si•bil•i•ty** \-ˌfen(t)-sə-ˈbi-lə-tē\ *noun*
— **in•de•fen•si•bly** \-ˈfen(t)-sə-blē\ *adverb*

in•de•fin•able \-ˈfī-nə-bəl\ *adjective* (1810)
: incapable of being precisely described or analyzed

☆ SYNONYMS
Indecorous, improper, unseemly, unbecoming, indelicate mean not conforming to what is accepted as right, fitting, or in good taste. INDECOROUS suggests a violation of accepted standards of good manners ⟨*indecorous* behavior⟩. IMPROPER applies to a broader range of transgressions of rules not only of social behavior but of ethical practice or logical procedure or prescribed method ⟨*improper* use of campaign contributions⟩. UNSEEMLY adds a suggestion of special inappropriateness to a situation or an offensiveness to good taste ⟨remarried with *unseemly* haste⟩. UNBECOMING suggests behavior or language that does not suit one's character or status ⟨conduct *unbecoming* to an officer⟩. INDELICATE implies a lack of modesty or of tact or of refined perception of feeling ⟨*indelicate* expressions for bodily functions⟩.

— **in·de·fin·abil·i·ty** \-ˌfī-nə-ˈbi-lə-tē\ *noun*
— **indefinable** *noun*
— **in·de·fin·able·ness** \-ˈfī-nə-bəl-nəs\ *noun*
— **in·de·fin·ably** \-blē\ *adverb*

in·def·i·nite \(ˌ)in-ˈdef-nət, -ˈde-fə-\ *adjective* [Latin *indefinitus,* from *in-* + *definitus* definite] (1530)
: not definite: as **a :** typically designating an unidentified, generic, or unfamiliar person or thing ⟨the *indefinite* articles *a* and *an*⟩ ⟨*indefinite* pronouns⟩ **b :** not precise **:** VAGUE **c :** having no exact limits **d** *of floral organs* **:** numerous and difficult to ascertain in number
— **indefinite** *noun*
— **in·def·i·nite·ly** *adverb*
— **in·def·i·nite·ness** *noun*

indefinite integral *noun* (circa 1877)
: any function whose derivative is a given function

in·de·his·cent \ˌin-di-ˈhi-s^{ə}nt\ *adjective* (1832)
: remaining closed at maturity ⟨*indehiscent* fruits⟩
— **in·de·his·cence** \-s^{ə}n(t)s\ *noun*

in·del·i·ble \in-ˈde-lə-bəl\ *adjective* [Middle English *indelyble,* from Medieval Latin *indelibilis,* alteration of Latin *indelebilis,* from *in-* + *delēre* to delete] (15th century)
1 a : that cannot be removed, washed away, or erased **b :** making marks that cannot easily be removed ⟨an *indelible* pencil⟩
2 a : LASTING ⟨*indelible* memories⟩ **b :** UNFORGETTABLE, MEMORABLE ⟨an *indelible* performance⟩
— **in·del·i·bil·i·ty** \(ˌ)in-ˌde-lə-ˈbi-lə-tē\ *noun*
— **in·del·i·bly** \in-ˈde-lə-blē\ *adverb*

in·del·i·ca·cy \(ˌ)in-ˈde-li-kə-sē\ *noun* (1712)
1 : the quality or state of being indelicate
2 : something that is indelicate

in·del·i·cate \-li-kət\ *adjective* (1742)
: not delicate: **a** (1) **:** lacking in or offending against propriety **:** IMPROPER (2) **:** verging on the indecent **:** COARSE **b :** marked by a lack of feeling for the sensibilities of others **:** TACTLESS
synonym see INDECOROUS
— **in·del·i·cate·ly** *adverb*
— **in·del·i·cate·ness** *noun*

in·dem·ni·fi·ca·tion \in-ˌdem-nə-fə-ˈkā-shən\ *noun* (1732)
1 a : the action of indemnifying **b :** the condition of being indemnified
2 : INDEMNITY 2b

in·dem·ni·fy \in-ˈdem-nə-ˌfī\ *transitive verb* **-fied; -fy·ing** [Latin *indemnis* unharmed, from *in-* + *damnum* damage] (circa 1611)
1 : to secure against hurt, loss, or damage
2 : to make compensation to for incurred hurt, loss, or damage
synonym see PAY
— **in·dem·ni·fi·er** \-ˌfī(-ə)r\ *noun*

in·dem·ni·ty \in-ˈdem-nə-tē\ *noun, plural* **-ties** (15th century)
1 a : security against hurt, loss, or damage **b :** exemption from incurred penalties or liabilities
2 a : INDEMNIFICATION 1 **b :** something that indemnifies

in·de·mon·stra·ble \ˌin-di-ˈmän(t)-strə-bəl, (ˌ)in-ˈdem-ən-strə-\ *adjective* (1570)
: incapable of being demonstrated **:** not subject to proof
— **in·de·mon·stra·bly** \-blē\ *adverb*

[1]in·dent \in-ˈdent\ *verb* [Middle English, from Middle French *endenter,* from Old French, from *en-* + *dent* tooth, from Latin *dent-, dens* — more at TOOTH] (14th century)
transitive verb
1 a : to divide (a document) so as to produce sections with irregular edges that can be matched for authentication **b :** to draw up (as a deed) in two or more exactly corresponding copies
2 : to notch the edge of **:** make jagged
3 : INDENTURE
4 : to set (as a line of a paragraph) in from the margin
5 *chiefly British* **:** to order by an indent
intransitive verb
1 *obsolete* **:** to make a formal or express agreement
2 : to form an indentation
3 *chiefly British* **:** to make out an indent for something
— **in·dent·er** *noun*
— **indent on 1** *chiefly British* **:** to make a requisition on **2** *chiefly British* **:** to draw on

[2]in·dent \in-ˈdent, ˈin-ˌ\ *noun* (15th century)
1 a : INDENTURE 1 **b :** a certificate issued by the U.S. at the close of the American Revolution for the principal or interest on the public debt
2 *chiefly British* **a :** an official requisition **b :** a purchase order for goods especially when sent from a foreign country
3 : INDENTION

[3]in·dent \in-ˈdent\ *transitive verb* [Middle English *endenten,* from *en-* + *denten* to dent] (15th century)
1 : to force inward so as to form a depression
2 : to form a dent in
— **in·dent·er** *noun*

[4]in·dent \in-ˈdent, ˈin-ˌ\ *noun* (1596)
: INDENTATION

in·den·ta·tion \ˌin-ˌden-ˈtā-shən\ *noun* (circa 1728)
1 a : an angular cut in an edge **:** NOTCH **b :** a recess in a surface
2 : the action of indenting **:** the condition of being indented
3 : DENT
4 : INDENTION 2b

in·den·tion \in-ˈden(t)-shən\ *noun* (1763)
1 *archaic* **:** INDENTATION 1
2 a : the action of indenting **:** the condition of being indented **b :** the blank space produced by indenting

[1]in·den·ture \in-ˈden(t)-shər\ *noun* [Middle English *endenture,* from Middle French, from *endenter*] (14th century)
1 a (1) **:** a document or a section of a document that is indented (2) **:** a formal or official document usually executed in two or more copies (3) **:** a contract binding one person to work for another for a given period of time — usually used in plural **b :** a formal certificate (as an inventory or voucher) prepared for purposes of control **c :** a document stating the terms under which a security (as a bond) is issued
2 : INDENTATION 1
3 [[3]*indent*] **:** DENT

[2]indenture *transitive verb* **in·den·tured; in·den·tur·ing** \-ˈden(t)-shriŋ, -ˈden(t)-shə-riŋ\ (1676)
: to bind (as an apprentice) by or as if by indentures

indentured servant *noun* (1723)
: a person who signs and is bound by indentures to work for another for a specified time especially in return for payment of travel expenses and maintenance

in·de·pen·dence \ˌin-də-ˈpen-dən(t)s\ *noun* (1640)
1 : the quality or state of being independent
2 *archaic* **:** COMPETENCE 1

Independence Day *noun* (1791)
: a civil holiday for the celebration of the anniversary of the beginnings of national independence; *specifically* **:** July 4 observed as a legal holiday in the U.S. in commemoration of the adoption of the Declaration of Independence in 1776

in·de·pen·den·cy \ˌin-də-ˈpen-dən(t)-sē\ *noun* (circa 1611)
1 : INDEPENDENCE 1
2 *capitalized* **:** the Independent polity or movement
3 : an independent political unit

[1]in·de·pen·dent \ˌin-də-ˈpen-dənt\ *adjective* (1611)
1 : not dependent: as **a** (1) **:** not subject to control by others **:** SELF-GOVERNING (2) **:** not affiliated with a larger controlling unit **b** (1) **:** not requiring or relying on something else **:** not contingent ⟨an *independent* conclusion⟩ (2) **:** not looking to others for one's opinions or for guidance in conduct (3) **:** not bound by or committed to a political party **c** (1) **:** not requiring or relying on others (as for care or livelihood) ⟨*independent* of her parents⟩ (2) **:** being enough to free one from the necessity of working for a living ⟨a man of *independent* means⟩ **d :** showing a desire for freedom ⟨an *independent* manner⟩ **e** (1) **:** not determined by or capable of being deduced or derived from or expressed in terms of members (as axioms or equations) of the set under consideration; *especially* **:** having linear independence ⟨an *independent* set of vectors⟩ (2) **:** having the property that the joint probability (as of events or samples) or the joint probability density function (as of random variables) equals the product of the probabilities or probability density functions of separate occurrence
2 *capitalized* **:** of or relating to the Independents
3 a : MAIN 5 ⟨an *independent* clause⟩ **b :** neither deducible from nor incompatible with another statement ⟨*independent* postulates⟩
synonym see FREE
— **in·de·pen·dent·ly** *adverb*

[2]independent *noun* (1644)
1 *capitalized* **:** a sectarian of an English religious movement for congregational autonomy originating in the late 16th century, giving rise to Congregationalists, Baptists, and Friends, and forming one of the major political groupings of the period of Cromwell
2 : one that is independent; *especially, often capitalized* **:** one that is not bound by or definitively committed to a political party

independent assortment *noun* (circa 1948)
: formation of random combinations of chromosomes in meiosis and of genes on different pairs of homologous chromosomes by the passage according to the laws of probability of one of each diploid pair of homologous chromosomes into each gamete independently of each other pair

independent variable *noun* (1852)
: a mathematical variable whose value is specified first and determines the value of one or more other values in an expression or function

in–depth \(ˈ)in-ˈdepth\ *adjective* (1965)
: COMPREHENSIVE, THOROUGH ⟨an *in-depth* study⟩

in·de·scrib·able \ˌin-di-ˈskrī-bə-bəl\ *adjective* (1794)
1 : that cannot be described ⟨an *indescribable* sensation⟩
2 : surpassing description ⟨*indescribable* joy⟩
— **in·de·scrib·able·ness** *noun*
— **in·de·scrib·ably** \-blē\ *adverb*

in·de·struc·ti·ble \-ˈstrək-tə-bəl\ *adjective* [probably from Late Latin *indestructibilis,* from Latin *in-* + *destructus,* past participle of *destruere* to tear down — more at DESTROY] (1667)
: incapable of being destroyed, ruined, or rendered ineffective
— **in·de·struc·ti·bil·i·ty** \-ˌstrək-tə-ˈbi-lə-tē\ *noun*
— **in·de·struc·ti·ble·ness** \-ˈstrək-tə-bəl-nəs\ *noun*
— **in·de·struc·ti·bly** \-blē\ *adverb*

in·de·ter·min·able \ˌin-di-ˈtərm-nə-bəl, -ˈtər-mə-\ *adjective* (15th century)
1 : incapable of being definitely decided or settled
2 : incapable of being definitely fixed or ascertained
— **in·de·ter·min·ably** \-blē\ *adverb*

in·de·ter·mi·na·cy \-nə-sē\ *noun* (1649)

: the quality or state of being indeterminate

indeterminacy principle *noun* (circa 1928)
: UNCERTAINTY PRINCIPLE

in·de·ter·mi·nate \ˌin-di-ˈtərm-nət, -ˈtər-mə-\ *adjective* [Middle English *indeterminat,* from Late Latin *indeterminatus,* from Latin *in-* + *determinatus,* past participle of *determinare* to determine] (14th century)
1 a : not definitely or precisely determined or fixed **:** VAGUE **b :** not known in advance **c :** not leading to a definite end or result
2 : having an infinite number of solutions ⟨a system of *indeterminate* equations⟩
3 : being one of the seven undefined mathematical expressions

$$\frac{0}{0}, \frac{\infty}{\infty}, \infty \cdot 0, 1^{\infty}, 0^{0}, \infty^{0}, \infty - \infty$$

4 : characterized by sequential flowering from the lateral or basal buds to the central or uppermost buds; *also* **:** characterized by growth in which the main stem continues to elongate indefinitely without being limited by a terminal inflorescence — compare DETERMINATE 4
— **in·de·ter·mi·nate·ly** *adverb*
— **in·de·ter·mi·nate·ness** *noun*
— **in·de·ter·mi·na·tion** \-ˌtər-mə-ˈnā-shən\ *noun*

in·de·ter·min·ism \-ˈtər-mə-ˌni-zəm\ *noun* (1874)
1 a : a theory that the will is free and that deliberate choice and actions are not determined by or predictable from antecedent causes **b :** a theory that holds that not every event has a cause
2 : the quality or state of being indeterminate; *especially* **:** UNPREDICTABILITY
— **in·de·ter·min·ist** \-ˈtərm-nist, -ˈtər-mə-\ *noun*
— **in·de·ter·min·is·tic** \-ˌtər-mə-ˈnis-tik\ *adjective*

[1]**in·dex** \ˈin-ˌdeks\ *noun, plural* **in·dex·es** *or* **in·di·ces** \-də-ˌsēz\ [Latin *indic-, index,* from *indicare* to indicate] (1571)
1 a : a device (as the pointer on a scale or the gnomon of a sundial) that serves to indicate a value or quantity **b :** something (as a physical feature or a mode of expression) that leads one to a particular fact or conclusion **:** INDICATION
2 : a list (as of bibliographical information or citations to a body of literature) arranged usually in alphabetical order of some specified datum (as author, subject, or keyword): as **a :** a list of items (as topics or names) treated in a printed work that gives for each item the page number where it may be found **b :** THUMB INDEX **c :** a bibliographical analysis of groups of publications that is usually published periodically
3 : a list of restricted or prohibited material; *specifically, capitalized* **:** a list of books the reading of which is prohibited or restricted for Roman Catholics by the church authorities
4 *plural usually* **indices :** a number or symbol or expression (as an exponent) associated with another to indicate a mathematical operation to be performed or to indicate use or position in an arrangement
5 : a character ☞ used to direct attention to a note or paragraph — called also *fist*
6 a : a number (as a ratio) derived from a series of observations and used as an indicator or measure; *specifically* **:** INDEX NUMBER **b :** the ratio of one dimension of a thing (as an anatomical structure) to another dimension

[2]**index** (1720)
transitive verb
1 a : to provide with an index **b :** to list in an index ⟨all persons and places mentioned are carefully *indexed*⟩
2 : to serve as an index of
3 : to regulate (as wages, prices, or interest rates) by indexation
intransitive verb
: to index something
— **in·dex·er** *noun*

in·dex·ation \ˌin-dek-ˈsā-shən\ *noun* (1960)
: a system of economic control in which certain variables (as wages and interest) are tied to a cost-of-living index so that both rise or fall at the same rate and the detrimental effect of inflation is theoretically eliminated

index finger *noun* (1849)
: FOREFINGER

index fossil *noun* (1900)
: a fossil usually with a narrow time range and wide spatial distribution that is used in the identification of related geologic formations

[1]**in·dex·i·cal** \(ˌ)in-ˈdek-si-kəl\ *adjective* (circa 1828)
1 : of or relating to an index
2 a : varying in reference with the individual speaker ⟨the *indexical* words *I, here, now*⟩ **b :** associated with or identifying an individual speaker ⟨*indexical* features of speech⟩

[2]**indexical** *noun* (1971)
: an indexical word, sign, or feature

indexing *noun* (1957)
: INDEXATION

index number *noun* (circa 1896)
: a number used to indicate change in magnitude (as of cost or price) as compared with the magnitude at some specified time usually taken as 100

index of refraction (1829)
: the ratio of the speed of radiation (as light) in one medium (as a vacuum) to that in another medium

indi- — see IND-

In·dia \ˈin-dē-ə\ (circa 1952)
— a communications code word for the letter *i*

india ink *noun, often 1st I capitalized* (1665)
1 : a solid black pigment (as specially prepared lampblack) used in drawing and lettering
2 : a fluid ink consisting usually of a fine suspension of india ink in a liquid

In·dia·man \ˈin-dē-ə-mən\ *noun* (1709)
: a merchant ship formerly used in trade with India; *especially* **:** a large sailing ship used in this trade

In·di·an \ˈin-dē-ən, *dialect* -jən *or* -din\ *noun* (14th century)
1 a : a native or inhabitant of India or of the East Indies **b :** a person of Indian descent
2 a [from the belief held by Columbus that the lands he discovered were part of Asia] **:** AMERICAN INDIAN **b :** one of the native languages of American Indians
— **Indian** *adjective*
— **In·di·an·ness** *noun*

Indian agent *noun* (1807)
: an official representative of the U.S. federal government to American Indian tribes especially on reservations

Indian club *noun* (1857)
: a usually wooden club shaped like a large bottle or tenpin that is swung for gymnastic exercise

Indian corn *noun* (1617)
1 : a tall widely cultivated American cereal grass (*Zea mays*) bearing seeds on elongated ears
2 : the ears of Indian corn; *also* **:** its edible seeds

Indian elephant *noun* (1607)
: ELEPHANT 1b

Indian file *noun* (1758)
: SINGLE FILE

Indian giver *noun* (circa 1848)
: a person who gives something to another and then takes it back or expects an equivalent in return
— **Indian giving** *noun*

Indian hemp *noun* (1619)
1 : an American dogbane (*Apocynum cannabinum*) with milky juice, tough fibrous bark, and an emetic and cathartic root
2 : HEMP 1

In·di·an·ism \ˈin-dē-ə-ˌni-zəm\ *noun* (1651)
1 : the qualities or culture distinctive of American Indians
2 : policy designed to further the interests or culture of American Indians
— **In·di·an·ist** \-nist\ *adjective or noun*

In·di·an·ize \ˈin-dē-ə-ˌnīz\ *transitive verb* **-ized; -iz·ing** (1702)
1 : to cause to acquire or conform to the characteristics, culture, or usage of American Indians or of India
2 : to bring (as a region) under the cultural or political influence or control of India
— **In·di·an·i·za·tion** \ˌin-dē-ə-(ˌ)nī-ˈzā-shən\ *noun*

Indian licorice *noun* (circa 1890)
: ROSARY PEA 1

Indian meal *noun* (1609)
: CORNMEAL

Indian paintbrush *noun* (circa 1892)
1 : any of a genus (*Castilleja*) of herbaceous plants of the snapdragon family that have brightly colored bracts — called also *painted cup*
2 : ORANGE HAWKWEED

Indian paintbrush

Indian pipe *noun* (circa 1818)
: a waxy white saprophytic herb (*Monotropa uniflora* of the family Monotropaceae, the Indian-pipe family) of Asia and North America having leaves reduced to scales

Indian pipe

Indian pudding *noun* (1722)
: a baked pudding made chiefly of cornmeal, milk, and molasses

Indian red *noun* (circa 1753)
1 : any of various usually dark red pigments consisting chiefly of iron oxide
2 : a strong or moderate reddish brown

Indian sign *noun* (1910)
: HEX, SPELL

Indian summer *noun* (1778)
1 : a period of warm or mild weather in late autumn or early winter
2 : a happy or flourishing period occurring toward the end of something ⟨life in the *Indian summer* of Czarist Russia —John Davenport⟩

Indian tobacco *noun* (circa 1618)
1 : an American wild lobelia (*Lobelia inflata*) with small blue or white flowers
2 : a wild tobacco (*Nicotiana bigelovii*) found in dry valleys from southern California to southern Oregon

In·di·an–wres·tle \ˈin-dē-ən-ˌre-səl, -ˌra-\ *intransitive verb* [back-formation from *Indian wrestling*] (1938)
: to engage in Indian wrestling

Indian wrestling *noun* (1913)
1 : wrestling in which two people lie side by side on their backs in reversed position locking their near arms and raising and locking the corresponding legs and attempt to force each other's leg down and turn the other wrestler facedown
2 : wrestling in which two people stand face to face gripping usually their right hands and setting the outsides of the corresponding feet together and attempt to force each other off balance
3 : ARM WRESTLING

India paper *noun* (1768)

\ə\ abut \ᵊ\ kitten \ər\ further \a\ ash \ā\ ace \ä\ mop, mar \au̇\ out \ch\ chin \e\ bet \ē\ easy \g\ go \i\ hit \ī\ ice \j\ job \ŋ\ sing \ō\ go \ȯ\ law \ȯi\ boy \th\ thin \t͟h\ the \ü\ loot \u̇\ foot \y\ yet \zh\ vision *see also* Guide to Pronunciation

1 : a thin absorbent paper used especially for proving inked intaglio surfaces (as steel engravings)
2 : a thin tough opaque printing paper
india rubber *noun, often I capitalized* (1790)
: [1]RUBBER 2a
In·dic \'in-dik\ *adjective* (1877)
1 : of or relating to the subcontinent of India **:** INDIAN
2 : of, relating to, or constituting the Indo-Aryan branch of the Indo-European languages
— Indic *noun*
in·di·can \'in-də-,kan\ *noun* [Latin *indicum* indigo — more at INDIGO] (1859)
1 : an indigo-forming substance $C_8H_7NO_4S$ found as a salt in urine and other animal fluids; *also* **:** its potassium salt $C_8H_6KNO_4S$
2 : a glucoside $C_{14}H_{17}NO_6$ occurring especially in the indigo plant and being a source of natural indigo
in·di·cant \'in-di-kənt\ *noun* (1623)
: something that serves to indicate
in·di·cate \'in-də-,kāt\ *transitive verb* **-cat·ed; -cat·ing** [Latin *indicatus,* past participle of *indicare,* from *in-* + *dicare* to proclaim, dedicate — more at DICTION] (circa 1609)
1 a : to point out or point to **b :** to be a sign, symptom, or index of ⟨the high fever *indicates* a serious condition⟩ **c :** to demonstrate or suggest the necessity or advisability of ⟨*indicated* the need for a new school⟩
2 : to state or express briefly ⟨*indicated* a desire to cooperate⟩
in·di·ca·tion \,in-də-'kā-shən\ *noun* (15th century)
1 a : something that serves to indicate **b :** something that is indicated as advisable or necessary
2 : the action of indicating
— in·di·ca·tion·al \-shnəl, -shə-nᵊl\ *adjective*
[1]**in·dic·a·tive** \in-'di-kə-tiv\ *adjective* (15th century)
1 : of, relating to, or constituting a verb form or set of verb forms that represents the denoted act or state as an objective fact ⟨the *indicative* mood⟩
2 : serving to indicate ⟨actions *indicative* of fear⟩
— in·dic·a·tive·ly *adverb*
[2]**indicative** *noun* (1530)
1 : the indicative mood of a language
2 : a form in the indicative mood
in·di·ca·tor \'in-də-,kā-tər\ *noun* (1666)
1 : one that indicates: as **a :** an index hand (as on a dial) **:** POINTER **b** (1) **:** GAUGE 2b, DIAL 4a (2) **:** an instrument for automatically making a diagram that indicates the pressure in and volume of the working fluid of an engine throughout the cycle
2 a : a substance (as litmus) used to show visually (as by change of color) the condition of a solution with respect to the presence of a particular material (as a free acid or alkali) **b :** TRACER 4b
3 : an organism or ecological community so strictly associated with particular environmental conditions that its presence is indicative of the existence of these conditions
4 : any of a group of statistical values (as level of employment) that taken together give an indication of the health of the economy
— in·dic·a·to·ry \in-'di-kə-,tōr-ē, -,tȯr-\ *adjective*
indices *plural of* INDEX
in·di·cia \in-'di-sh(ē-)ə\ *noun plural* [Latin, plural of *indicium* sign, from *indicare*] (circa 1626)
1 : distinctive marks **:** INDICATIONS
2 : postal markings often imprinted on mail or on labels to be affixed to mail
in·dict \in-'dīt\ *transitive verb* [alteration of earlier *indite,* from Middle English *inditen,* from Anglo-French *enditer,* from Old French, to write down — more at INDITE] (circa 1626)
1 : to charge with a fault or offense **:** CRITICIZE, ACCUSE
2 : to charge with a crime by the finding or presentment of a jury (as a grand jury) in due form of law
— in·dict·er *or* **in·dict·or** \-'dī-tər\ *noun*
in·dict·able \-'dī-tə-bəl\ *adjective* (circa 1706)
1 : subject to being indicted **:** liable to indictment
2 : making one liable to indictment ⟨an *indictable* offense⟩
in·dic·tion \in-'dik-shən\ *noun* [Middle English *indiccioun,* from Late Latin *indiction-, indictio,* from Latin, proclamation, from *indicere* to proclaim, from *in-* + *dicere* to say — more at DICTION] (14th century)
: a 15-year cycle used as a chronological unit in several ancient and medieval systems
in·dict·ment \in-'dīt-mənt\ *noun* (14th century)
1 a : the action or the legal process of indicting **b :** the state of being indicted
2 : a formal written statement framed by a prosecuting authority and found by a jury (as a grand jury) charging a person with an offense
3 : an expression of strong disapproval ⟨an *indictment* of contemporary morality⟩
in·die \'in-dē\ *noun* [by shortening & alteration from *independent*] (1928)
: something (as an unaffiliated record or motion-picture production company) independent
— indie *adjective*
in·dif·fer·ence \in-'di-fərn(t)s, -f(ə-)rən(t)s\ *noun* (15th century)
1 : the quality, state, or fact of being indifferent
2 a *archaic* **:** lack of difference or distinction between two or more things **b :** absence of compulsion to or toward one thing or another
in·dif·fer·en·cy \-fərn(t)-sē, -f(ə-)rən(t)-sē\ *noun* (15th century)
archaic **:** INDIFFERENCE
in·dif·fer·ent \in-'di-fərnt, -f(ə-)rənt\ *adjective* [Middle English, from Middle French or Latin; Middle French, regarded as neither good nor bad, from Latin *indifferent-, indifferens,* from *in-* + *different-, differens,* present participle of *differre* to be different — more at DIFFER] (14th century)
1 : marked by impartiality **:** UNBIASED
2 a : that does not matter one way or the other **b :** of no importance or value one way or the other
3 a : marked by no special liking for or dislike of something ⟨*indifferent* about which task he was given⟩ **b :** marked by a lack of interest, enthusiasm, or concern for something **:** APATHETIC ⟨*indifferent* to suffering and poverty⟩
4 : being neither excessive nor inadequate **:** MODERATE ⟨hills of *indifferent* size⟩
5 a : being neither good nor bad **:** MEDIOCRE ⟨does *indifferent* work⟩ **b :** being neither right nor wrong
6 : characterized by lack of active quality **:** NEUTRAL
7 a : not differentiated **b :** capable of development in more than one direction; *especially* **:** not yet embryologically determined ☆
— in·dif·fer·ent·ly *adverb*
in·dif·fer·ent·ism \-fərn-,ti-zəm, -f(ə-)rən-\ *noun* (1827)
: INDIFFERENCE; *specifically* **:** belief that all religions are equally valid
— in·dif·fer·ent·ist \-fərn-tist, -f(ə-)rən-\ *noun*
in·di·gence \'in-di-jən(t)s\ *noun* (14th century)
: a level of poverty in which real hardship and deprivation are suffered and comforts of life are wholly lacking
synonym see POVERTY
in·di·gene \'in-də-,jēn\ *also* **in·di·gen** \-di-jən, -də-,jen\ *noun* [Latin *indigena*] (1598)
: NATIVE
in·dig·e·nize \in-'di-jə-,nīz\ *transitive verb* **-nized; -niz·ing** (1951)
: to cause to have indigenous characteristics
— in·dig·e·ni·za·tion \-,di-jə-nə-'zā-shən, -,nī-'\ *noun*
in·dig·e·nous \in-'di-jə-nəs\ *adjective* [Late Latin *indigenus,* from Latin *indigena,* noun, native, from Old Latin *indu, endo* in, within + Latin *gignere* to beget — more at END-, KIN] (1646)
1 : having originated in and being produced, growing, living, or occurring naturally in a particular region or environment
2 : INNATE, INBORN
synonym see NATIVE
— in·dig·e·nous·ly *adverb*
— in·dig·e·nous·ness *noun*
in·di·gent \'in-di-jənt\ *adjective* [Middle English, from Middle French, from Latin *indigent-, indigens,* present participle of *indigēre* to need, from Old Latin *indu* + Latin *egēre* to need; perhaps akin to Old High German *ekrōdi* thin] (15th century)
1 : suffering from indigence **:** IMPOVERISHED
2 a *archaic* **:** DEFICIENT **b** *archaic* **:** totally lacking in something specified
— indigent *noun*
in·di·gest·ed \,in-(,)dī-'jes-təd, -də-\ *adjective* (1587)
: not carefully thought out or arranged **:** FORMLESS
in·di·gest·ible \-'jes-tə-bəl\ *adjective* [Middle English, from Late Latin *indigestibilis,* from Latin *in-* + Late Latin *digestibilis* digestible] (15th century)
: not digestible **:** not easily digested
— in·di·gest·ibil·i·ty \-,jes-tə-'bi-lə-tē\ *noun*
— indigestible *noun*
in·di·ges·tion \-'jes-chən, -'jesh-\ *noun* (14th century)
1 : inability to digest or difficulty in digesting something
2 : a case or attack of indigestion
in·dign \in-'dīn\ *adjective* [Middle English *indigne,* from Middle French, from Latin *indignus*] (14th century)
1 *archaic* **:** UNWORTHY, UNDESERVING
2 *obsolete* **:** UNBECOMING, DISGRACEFUL
in·dig·nant \in-'dig-nənt\ *adjective* [Latin *indignant-, indignans,* present participle of *indignari* to be indignant, from *indignus* unworthy, from *in-* + *dignus* worthy — more at DECENT] (1590)
: filled with or marked by indignation ⟨became *indignant* at the accusation⟩
— in·dig·nant·ly *adverb*
in·dig·na·tion \,in-dig-'nā-shən\ *noun* (14th century)

☆ SYNONYMS
Indifferent, unconcerned, incurious, aloof, detached, disinterested mean not showing or feeling interest. INDIFFERENT implies neutrality of attitude from lack of inclination, preference, or prejudice ⟨*indifferent* to the dictates of fashion⟩. UNCONCERNED suggests a lack of sensitivity or regard for others' needs or troubles ⟨*unconcerned* about the homeless⟩. INCURIOUS implies an inability to take a normal interest due to dullness of mind or to self-centeredness ⟨*incurious* about the world⟩. ALOOF suggests a cool reserve arising from a sense of superiority or disdain for inferiors or from shyness ⟨*aloof* from his coworkers⟩. DETACHED implies an objective attitude achieved through absence of prejudice or selfishness ⟨observed family gatherings with *detached* amusement⟩. DISINTERESTED implies a circumstantial freedom from concern for personal or especially financial advantage that enables one to judge or advise without bias ⟨judged by a panel of *disinterested* observers⟩.

: anger aroused by something unjust, unworthy, or mean
synonym see ANGER
in·dig·ni·ty \in-'dig-nə-tē\ *noun, plural* **-ties** [Latin *indignitat-, indignitas,* from *indignus*] (1584)
1 a : an act that offends against a person's dignity or self-respect **:** INSULT **b :** humiliating treatment
2 *obsolete* **:** lack or loss of dignity or honor
in·di·go \'in-di-,gō\ *noun, plural* **-gos** *or* **-goes** [Italian dialect, from Latin *indicum,* from Greek *indikon,* from neuter of *indikos* Indic, from *Indos* India] (1555)
1 a : a blue vat dye obtained from plants (as indigo plants) **b :** the principal coloring matter $C_{16}H_{10}N_2O_2$ of natural indigo usually synthesized as a blue powder with a coppery luster
2 : INDIGO PLANT
3 : a deep reddish blue
indigo bunting *noun* (1783)
: a common small American finch (*Passerina cyanea*) of which the male is largely indigo-blue in spring and summer
indigo plant *noun* (1757)
: a plant that yields indigo; *especially* **:** any of a genus (*Indigofera*) of leguminous herbs
indigo snake *noun* (circa 1885)
: a large blue-black or brownish colubrid snake (*Drymarchon corais*) of the southeastern U.S. and Texas to Argentina — called also *gopher snake*
in·di·go·tin \in-'di-gə-tən, ,in-di-'gō-tᵊn\ *noun* [French *indigotine,* irregular from *indigo* indigo] (1838)
: INDIGO 1b
in·di·rect \,in-də-'rekt, -(,)dī-\ *adjective* [Middle English, from Medieval Latin *indirectus,* from Latin *in-* + *directus* direct — more at DRESS] (14th century)
: not direct: as **a** (1) **:** deviating from a direct line or course **:** ROUNDABOUT (2) **:** not going straight to the point ⟨an *indirect* accusation⟩ (3) **:** being or involving proof of a proposition or theorem by demonstration that its negation leads to an absurdity or contradiction **b :** not straightforward and open **:** DECEITFUL **c :** not directly aimed at or achieved ⟨*indirect* consequences⟩ **d :** stating what a real or supposed original speaker said with changes in wording that conform the statement grammatically to the sentence in which it is included ⟨*indirect* discourse⟩ ⟨an *indirect* question⟩ **e :** not effected by the action of the people or the electorate ⟨*indirect* government representation⟩
— **in·di·rect·ly** \-'rek(t)-lē\ *adverb*
— **in·di·rect·ness** \-nəs\ *noun*
indirect cost *noun* (circa 1909)
: a cost that is not identifiable with a specific product, function, or activity
indirect evidence *noun* (1824)
: evidence that establishes immediately collateral facts from which the main fact may be inferred **:** CIRCUMSTANTIAL EVIDENCE
in·di·rec·tion \,in-də-'rek-shən, -(,)dī-\ *noun* (1590)
1 a : indirect action or procedure **b :** lack of direction **:** AIMLESSNESS
2 a : lack of straightforwardness and openness **:** DECEITFULNESS **b :** something (as an act or statement) marked by lack of straightforwardness ⟨hated diplomatic *indirections —Review of Reviews*⟩
indirect lighting *noun* (1922)
: lighting in which the light emitted by a source is diffusely reflected (as by the ceiling)
indirect object *noun* (1879)
: a grammatical object representing the secondary goal of the action of its verb (as *her* in "I gave her the book")
in·dis·cern·ible \,in-di-'sər-nə-bəl, -'zər-\ *adjective* (1635)
: incapable of being discerned **:** not recognizable as distinct
in·dis·ci·plin·able \(,)in-,di-sə-'pli-nə-bəl; -'di-sə-plə-\ *adjective* (1600)
: not subject to or capable of being disciplined
in·dis·ci·pline \(,)in-'di-sə-plən\ *noun* (1783)
: lack of discipline
— **in·dis·ci·plined** \-plənd, -(,)plind\ *adjective*
in·dis·cov·er·able \,in-dis-'kəv-rə-bəl, -'kə-və-\ *adjective* (1640)
: not discoverable
in·dis·creet \,in-di-'skrēt\ *adjective* [Middle English *indiscrete,* from Middle French & Late Latin; Middle French *indiscret,* from Late Latin *indiscretus,* from Latin, indistinguishable, from *in-* + *discretus,* past participle of *discernere* to separate — more at DISCERN] (15th century)
: not discreet **:** IMPRUDENT
— **in·dis·creet·ly** *adverb*
— **in·dis·creet·ness** *noun*
in·dis·cre·tion \,in-dis-'kre-shən\ *noun* (14th century)
1 : lack of discretion **:** IMPRUDENCE ⟨dietary *indiscretion*⟩
2 a : something (as an act or remark) marked by lack of discretion **b :** an act at variance with the accepted morality of a society ⟨resigned because of financial *indiscretions*⟩
in·dis·crim·i·nate \,in-dis-'krim-nət, -'kri-mə-\ *adjective* (circa 1598)
1 a : not marked by careful distinction **:** deficient in discrimination and discernment ⟨*indiscriminate* reading habits⟩ ⟨*indiscriminate* mass destruction⟩ **b :** HAPHAZARD, RANDOM ⟨*indiscriminate* application of a law⟩
2 a : PROMISCUOUS, UNRESTRAINED ⟨*indiscriminate* sexual behavior⟩ **b :** HETEROGENEOUS, MOTLEY ⟨an *indiscriminate* collection⟩
— **in·dis·crim·i·nate·ly** *adverb*
— **in·dis·crim·i·nate·ness** *noun*
in·dis·crim·i·nat·ing \-'kri-mə-,nā-tiŋ\ *adjective* (circa 1767)
: not discriminating
— **in·dis·crim·i·nat·ing·ly** \-tiŋ-lē\ *adverb*
in·dis·crim·i·na·tion \-,kri-mə-'nā-shən\ *noun* (1649)
: lack of discrimination
in·dis·pens·able \,in-di-'spen(t)-sə-bəl\ *adjective* (1653)
1 : not subject to being set aside or neglected ⟨an *indispensable* obligation⟩
2 : absolutely necessary **:** ESSENTIAL ⟨an *indispensable* member of the staff⟩
— **in·dis·pens·abil·i·ty** \-,spen(t)-sə-'bi-lə-tē\ *noun*
— **indispensable** *noun*
— **in·dis·pens·able·ness** \-'spen(t)-sə-bəl-nəs\ *noun*
— **in·dis·pens·ably** \-blē\ *adverb*
in·dis·pose \,in-di-'spōz\ *transitive verb* **-posed; -pos·ing** [probably back-formation from *indisposed*] (1657)
1 a : to make unfit **:** DISQUALIFY **b :** to make averse **:** DISINCLINE
2 *archaic* **:** to cause to be in poor physical health
in·dis·posed \-'spōzd\ *adjective* (15th century)
1 : slightly ill
2 : AVERSE
in·dis·po·si·tion \(,)in-,dis-pə-'zi-shən\ *noun* (15th century)
: the condition of being indisposed: **a :** DISINCLINATION **b :** a usually slight illness
in·dis·put·able \,in-di-'spyü-tə-bəl, (,)in-'dis-pyə-\ *adjective* [Late Latin *indisputabilis,* from Latin *in-* + *disputabilis* disputable] (1551)
: not disputable **:** UNQUESTIONABLE ⟨*indisputable* proof⟩
— **in·dis·put·able·ness** *noun*
— **in·dis·put·ably** \-blē\ *adverb*
in·dis·so·cia·ble \,in-di-'sō-sh(ē-)ə-bəl, -sē-ə-\ *adjective* (1855)
: not dissociated **:** INSEPARABLE
— **in·dis·so·cia·bly** \-blē\ *adverb*
in·dis·sol·u·ble \,in-di-'säl-yə-bəl\ *adjective* (1542)
: not dissoluble; *especially* **:** incapable of being annulled, undone, or broken **:** PERMANENT ⟨an *indissoluble* contract⟩
— **in·dis·sol·u·bil·i·ty** \-,säl-yə-'bi-lə-tē\ *noun*
— **in·dis·sol·u·ble·ness** \-'säl-yə-bəl-nəs\ *noun*
— **in·dis·sol·u·bly** \-blē\ *adverb*
in·dis·tinct \,in-di-'stiŋ(k)t\ *adjective* [Latin *indistinctus,* from *in-* + *distinctus* distinct] (1526)
: not distinct: as **a :** not sharply outlined or separable **:** BLURRED ⟨*indistinct* figures in the fog⟩ **b :** FAINT, DIM ⟨an *indistinct* light in the distance⟩ **c :** not clearly recognizable or understandable **:** UNCERTAIN
— **in·dis·tinct·ly** \-'stiŋ(k)t-lē, -'stiŋ-klē\ *adverb*
— **in·dis·tinct·ness** \-'stiŋ(k)t-nəs, -'stiŋk-nəs\ *noun*
in·dis·tinc·tive \-'stiŋ(k)-tiv\ *adjective* (1846)
: lacking distinctive qualities
in·dis·tin·guish·able \,in-di-'stiŋ-gwi-shə-bəl, -'stiŋ-wi-\ *adjective* (1606)
: not distinguishable: as **a :** indeterminate in shape or structure **b :** not clearly recognizable or understandable **c :** lacking identifying or individualizing qualities
— **in·dis·tin·guish·abil·i·ty** \-,stiŋ-gwi-shə-'bi-lə-tē, -,stiŋ-wi-\ *noun*
— **in·dis·tin·guish·able·ness** \-'stiŋ-gwi-shə-bəl-nəs, -'stiŋ-wi-\ *noun*
— **in·dis·tin·guish·ably** \-blē\ *adverb*
in·dite \in-'dīt\ *transitive verb* **in·dit·ed; in·dit·ing** [Middle English *enditen,* from Old French *enditer* to write down, proclaim, from (assumed) Vulgar Latin *indictare* to proclaim, frequentative of Latin *indicere* to proclaim, from *in-* + *dicere* to say — more at DICTION] (14th century)
1 a : MAKE UP, COMPOSE ⟨*indite* a poem⟩ **b :** to give literary or formal expression to **c :** to put down in writing ⟨*indite* a message⟩
2 *obsolete* **:** DICTATE
— **in·dit·er** *noun*
in·di·um \'in-dē-əm\ *noun* [International Scientific Vocabulary *ind-* + New Latin *-ium*] (1864)
: a malleable fusible silvery metallic element that is chiefly trivalent, occurs especially in sphalerite ores, and is used as a plating for bearings, in alloys having a low melting point, and in the making of transistors — see ELEMENT table
indium antimonide *noun* (1957)
: a synthetic compound InSb of indium and antimony that is a semiconducting and photosensitive material and is used especially in infrared photodetectors
¹in·di·vid·u·al \,in-də-'vij-wəl, -'vi-jə-wəl, -'vi-jəl\ *adjective* [Medieval Latin *individualis,* from Latin *individuus* indivisible, from *in-* + *dividuus* divided, from *dividere* to divide] (15th century)
1 *obsolete* **:** INSEPARABLE
2 a : of, relating to, or distinctively associated with an individual ⟨an *individual* effort⟩ **b :** being an individual or existing as an indivisible whole **c :** intended for one person ⟨an *individual* serving⟩
3 : existing as a distinct entity **:** SEPARATE
4 : having marked individuality ⟨an *individual* style⟩
synonym see SPECIAL, CHARACTERISTIC
— **in·di·vid·u·al·ly** *adverb*
²individual *noun* (1605)
1 a : a particular being or thing as distinguished from a class, species, or collection: as

(1) **:** a single human being as contrasted with a social group or institution ⟨a teacher who works with *individuals*⟩ (2) **:** a single organism as distinguished from a group **b :** a particular person ⟨are you the *individual* I spoke with on the telephone?⟩
2 : an indivisible entity
3 : the reference of a name or variable of the lowest logical type in a calculus

in·di·vid·u·al·ise *British variant of* INDIVIDUALIZE

in·di·vid·u·al·ism \ˌin-də-ˈvij-wə-ˌli-zəm, -ˈvi-jə-wə-, -ˈvi-jə-ˌli-\ *noun* (1827)
1 a (1) **:** a doctrine that the interests of the individual are or ought to be ethically paramount; *also* **:** conduct guided by such a doctrine (2) **:** the conception that all values, rights, and duties originate in individuals **b :** a theory maintaining the political and economic independence of the individual and stressing individual initiative, action, and interests; *also* **:** conduct or practice guided by such a theory
2 a : INDIVIDUALITY **b :** an individual peculiarity **:** IDIOSYNCRASY

in·di·vid·u·al·ist \-list\ *noun* (1840)
1 : one that pursues a markedly independent course in thought or action
2 : one that advocates or practices individualism
— individualist *or* **in·di·vid·u·al·is·tic** \-ˌvij-wə-ˈlis-tik, -ˌvi-jə-wə-, -ˌvi-jə-ˈlis-\ *adjective*
— in·di·vid·u·al·is·ti·cal·ly \-ˈlis-ti-k(ə-)lē\ *adverb*

in·di·vid·u·al·i·ty \-ˌvi-jə-ˈwa-lə-tē\ *noun, plural* **-ties** (1614)
1 a : total character peculiar to and distinguishing an individual from others **b :** PERSONALITY
2 *archaic* **:** the quality or state of being indivisible
3 : separate or distinct existence
4 : INDIVIDUAL, PERSON

in·di·vid·u·al·ize \-ˈvij-wə-ˌlīz, -ˈvi-jə-wə-, -ˈvi-jə-ˌlīz\ *transitive verb* **-ized; -iz·ing** (1637)
1 : to make individual in character
2 : to treat or notice individually **:** PARTICULARIZE
3 : to adapt to the needs or special circumstances of an individual ⟨*individualize* teaching according to student ability⟩
— in·di·vid·u·al·i·za·tion \-ˌvij-wə-lə-ˈzā-shən, -ˌvi-jə-wə-, -ˌvi-jə-lə-\ *noun*

individual medley *noun* (circa 1949)
: a swimming race in which each contestant swims each part of the course with a different stroke

individual retirement account *noun* (1974)
: IRA

in·di·vid·u·ate \ˌin-də-ˈvi-jə-ˌwāt\ *transitive verb* **-at·ed; -at·ing** (1614)
1 : to give individuality to
2 : to form into a distinct entity

in·di·vid·u·a·tion \-ˌvi-jə-ˈwā-shən\ *noun* (1628)
1 : the act or process of individuating: as **a** (1) **:** the development of the individual from the universal (2) **:** the determination of the individual in the general **b :** the process by which individuals in society become differentiated from one another **c :** regional differentiation along a primary embryonic axis
2 : the state of being individuated; *specifically* **:** INDIVIDUALITY

in·di·vis·i·ble \ˌin-də-ˈvi-zə-bəl\ *adjective* [Middle English, from Late Latin *indivisibilis,* from Latin *in-* + Late Latin *divisibilis* divisible] (14th century)
: not divisible
— in·di·vis·i·bil·i·ty \-ˌvi-zə-ˈbi-lə-tē\ *noun*
— indivisible *noun*
— in·di·vis·i·bly \-ˈvi-zə-blē\ *adverb*

indo- — see IND-

Indo- *combining form* [Greek, from *Indos* India]
1 : India or the East Indies ⟨*Indo*-Pakistani⟩
2 : Indo-European ⟨*Indo*-Hittite⟩

In·do–Ar·y·an \ˌin-dō-ˈar-ē-ən, -ˈer-; -ˈär-yən\ *noun* (1881)
1 : a member of one of the peoples of the Indian subcontinent speaking an Indo-European language
2 : one of the early Indo-European invaders of southern Asia
3 : a branch of the Indo-European language family that includes Hindi, Bengali, Punjabi, and other languages spoken primarily in India, Pakistan, Bangladesh, and Sri Lanka — see INDO-EUROPEAN LANGUAGES table
— Indo–Aryan *adjective*

In·do–Chi·nese \-ˌchī-ˈnēz, -ˈnēs\ *noun* (circa 1934)
1 : a native or inhabitant of Indochina
2 : SINO-TIBETAN
— Indo–Chinese *adjective*

in·doc·ile \(ˌ)in-ˈdä-səl *also* -ˌsīl, *especially British* -ˈdō-ˌsīl\ *adjective* [Middle French, from Latin *indocilis,* from *in-* + *docilis* docile] (1603)
: unwilling or indisposed to be taught or disciplined **:** INTRACTABLE
— in·do·cil·i·ty \ˌin-dä-ˈsi-lə-tē, -dō-\ *noun*

in·doc·tri·nate \in-ˈdäk-trə-ˌnāt\ *transitive verb* **-nat·ed; -nat·ing** [probably from Middle English *endoctrinen,* from Middle French *endoctriner,* from Old French, from *en-* + *doctrine* doctrine] (1626)
1 : to instruct especially in fundamentals or rudiments **:** TEACH
2 : to imbue with a usually partisan or sectarian opinion, point of view, or principle
— in·doc·tri·na·tion \(ˌ)in-ˌdäk-trə-ˈnā-shən\ *noun*
— in·doc·tri·na·tor \in-ˈdäk-trə-ˌnā-tər\ *noun*

¹In·do–Eu·ro·pe·an \ˌin-dō-ˌyu̇r-ə-ˈpē-ən\ *adjective* (1814)
: of, relating to, or constituting the Indo-European languages

²Indo–European *noun* (1864)
1 : INDO-EUROPEAN LANGUAGES
2 a : an unrecorded prehistoric language from which the Indo-European languages are descended **b :** a member of the people speaking this language

In·do–Eu·ro·pe·an·ist \-ˌyu̇r-ə-ˈpē-ə-nist\ *noun* (1926)
: a specialist in Indo-European linguistics

Indo–European languages *noun plural* (1864)
: a family of languages comprising those spoken in most of Europe and in the parts of the world colonized by Europeans since 1500 and also in Persia, the subcontinent of India, and some other parts of Asia

In·do–Ger·man·ic \ˌin-dō-jər-ˈma-nik\ *adjective* (1835)
: INDO-EUROPEAN
— Indo–Germanic *noun*

In·do–Hit·tite \-ˈhi-ˌtīt\ *noun* (1930)
1 : a hypothetical parent language of Indo-European and Anatolian
2 : a language family including Indo-European and Anatolian
— Indo–Hittite *adjective*

In·do–Ira·ni·an \-ir-ˈā-nē-ən\ *adjective* (1876)
: of, relating to, or constituting a subfamily of the Indo-European languages that consists of the Indo-Aryan and the Iranian branches — see INDO-EUROPEAN LANGUAGES table
— Indo–Iranian *noun*

in·dole \ˈin-ˌdōl\ *noun* [International Scientific Vocabulary *ind-* + *-ole*] (1869)
: a crystalline compound C_8H_7N that is a decomposition product of proteins containing tryptophan, that can be made synthetically, and that is used in perfumes; *also* **:** a derivative of indole

in·dole·ace·tic acid \ˌin-(ˌ)dō-lə-ˈsē-tik-\ *noun* (1937)
: a crystalline plant hormone $C_{10}H_9NO_2$ that is a naturally occurring auxin promoting growth and rooting of plants — called also *heteroauxin*

in·dole·bu·tyr·ic acid \ˌin-(ˌ)dōl-byü-ˈtir-ik-\ *noun* (1936)
: a crystalline acid $C_{12}H_{13}NO_2$ similar to indoleacetic acid in its effects on plants

in·do·lence \ˈin-də-lən(t)s\ *noun* (1710)
: inclination to laziness **:** SLOTH

in·do·lent \-lənt\ *adjective* [Late Latin *indolent-, indolens* insensitive to pain, from Latin *in-* + *dolent-, dolens,* present participle of *dolēre* to feel pain] (1663)
1 a : causing little or no pain **b :** slow to develop or heal
2 a : averse to activity, effort, or movement **:** habitually lazy **b :** conducing to or encouraging laziness ⟨*indolent* heat⟩ **c :** exhibiting indolence ⟨an *indolent* sigh⟩
synonym see LAZY
— in·do·lent·ly *adverb*

In·dol·o·gy \(ˌ)in-ˈdä-lə-jē\ *noun* (1888)
: the study of India and its people
— In·dol·o·gist \-jist\ *noun*

in·do·meth·a·cin \ˌin-dō-ˈme-thə-sən\ *noun* [*ind*ole + *meth-* + *ac*etic acid + ¹*-in*] (1963)
: a nonsteroidal drug $C_{19}H_{16}ClNO_4$ with anti-inflammatory, analgesic, and antipyretic properties used especially in treating arthritis

in·dom·i·ta·ble \in-ˈdä-mə-tə-bəl\ *adjective* [Late Latin *indomitabilis,* from Latin *in-* + *domitare* to tame — more at DAUNT] (1830)
: incapable of being subdued **:** UNCONQUERABLE ⟨*indomitable* courage⟩
— in·dom·i·ta·bil·i·ty \(ˌ)in-ˌdä-mə-tə-ˈbi-lə-tē\ *noun*
— in·dom·i·ta·ble·ness \in-ˈdä-mə-tə-bəl-nəs\ *noun*
— in·dom·i·ta·bly \-blē\ *adverb*

In·do·ne·sian \ˌin-də-ˈnē-zhən, -shən\ *noun* (1850)
1 : a native or inhabitant of the Malay Archipelago
2 a : a native or inhabitant of the Republic of Indonesia **b :** the language based on Malay that is the national language of the Republic of Indonesia
— Indonesian *adjective*

in·door \ˈin-ˈdōr, -ˈdȯr\ *adjective* (1711)
1 : of or relating to the interior of a building
2 : living, located, or carried on within a building ⟨an *indoor* sport⟩

in·doors \ˈin-ˈdōrz, (ˌ)in-, -ˈdȯrz\ *adverb* (1832)
: in or into a building

in·do·phe·nol \ˌin-dō-ˈfē-ˌnōl, -ˌnȯl, ˌin-(ˌ)dō-fi-ˈ\ *noun* [International Scientific Vocabulary] (circa 1881)
: any of various blue or green dyes

in·dorse, in·dorse·ment *variant of* ENDORSE, ENDORSEMENT

in·dox·yl \in-ˈdäk-səl\ *noun* [International Scientific Vocabulary *ind-* + hydr*oxyl*] (circa 1886)
: a crystalline compound C_8H_7NO found in plants and animals or synthesized as a step in indigo manufacture

in·draft \ˈin-ˌdraft, -ˌdráft\ *noun* (1594)
1 : an inward flow or current (as of air or water)
2 : a drawing or pulling in

in·drawn \ˈin-ˌdrȯn\ *adjective* (1751)
1 : ALOOF, RESERVED
2 : drawn in

in·dri \ˈin-drē\ *noun, plural* **indris** [French, from Malagasy *indry* look!] (1839)
: a large black-and-white Madagascan lemur (*Indri indri*) that is about two feet long with a rudimentary tail

in·du·bi·ta·ble \(ˌ)in-ˈdü-bə-tə-bəl, -ˈdyü-\ *adjective* [Middle English *indubitabyll,* from Latin *indubitabilis,* from *in-* + *dubitabilis* dubitable] (15th century)

INDO-EUROPEAN LANGUAGES

BRANCH	GROUP		LANGUAGES AND MAJOR DIALECTS[1] ANCIENT	MEDIEVAL	MODERN	PROVENIENCE
Anatolian			*Hittite, Lydian, Lycian* *Luwian, Palaic* *Hieroglyphic Luwian*			ancient Asia Minor
INDO-IRANIAN	Kafiri (Nuristani)				languages of eastern Afghanistan	Afghanistan
	Indo-Aryan	Sanskritic	Sanskrit, Pali *Prakrits*	*Prakrits*		India
					Shina, Khowar	Upper Indus valley
					Kashmiri	Kashmir
					Lahnda	western Punjab
					Sindhi	Sind
					Panjabi	Punjab
					Rajasthani	Rajasthan
					Gujarati	Gujarat
					Marathi	western India
					Konkani	western India
					Oriya	Orissa
					Bengali	Bangladesh, West Bengal
					Assamese	Assam
					Bhojpuri	Bihar, Uttar Pradesh
					Hindi	northern India
					Urdu	Pakistan, India
					Nepali	Nepal
					Sinhalese	Sri Lanka
					Romany	uncertain
	Iranian	West	*Old Persian*	*Pahlavi*		Persia
				Persian	Persian	Iran
					Kurdish	Iran, Iraq, Turkey
					Baluchi	Pakistan
					Tajiki	Tajikistan
		East	Avestan			ancient Persia
				Sogdian, Khotanese		central Asia
					Pashto	Afghanistan, Pakistan
					Ossetic	Caucasus
Armenian				Armenian	Armenian	Asia Minor, Caucasus
Greek or Hellenic			Greek	Greek	Greek	Greece, the eastern Mediterranean
BALTO-SLAVIC	Baltic			*Old Prussian*		East Prussia
					Lithuanian	Lithuania
					Latvian	Latvia
	Slavic	South		Old Church Slavonic		
					Slovene	Slovenia
					Serbo-Croatian	Croatia, Bosnia and Herzegovina, Serbia, Montenegro
					Macedonian	Macedonia
					Bulgarian	Bulgaria
		West			Czech, Slovak	Czechoslovakia
					Polish, Kashubian	Poland
					Wendish, *Polabian*	Germany
		East		*Old Russian*	Russian	Russia
					Ukrainian	Ukraine
					Belorussian	Belarus
Albanian					Albanian	Albania, Serbia, Macedonia
Scantily recorded and of uncertain affinities within Indo-European			*Ligurian, Messapian,* *Illyrian, Thracian,* *Phrygian*			ancient Italy Balkans Asia Minor
Tocharian				*Tocharian A* *Tocharian B*		central Asia
GERMANIC	East			*Gothic*		eastern Europe
	North			*Old Norse*	Icelandic	Iceland
					Faeroese	Faeroe Islands
					Norwegian	Norway
					Swedish	Sweden
					Danish	Denmark
	West			*Old High German*	German	Germany, Switzerland, Austria
				Middle High German		
					Yiddish	Germany, eastern Europe
				Old Saxon	Low German	northern Germany
				Middle Low German		
				Middle Dutch	Dutch	Netherlands
					Afrikaans	South Africa
					Flemish	Belgium
				Old Frisian	Frisian	Netherlands, Germany
				Old English	English	England
				Middle English		
ITALIC			*Venetic, Oscan, Umbrian,* *Faliscan*			ancient Italy
			Latin			
					Portuguese	Portugal
					Spanish	Spain
					Judeo-Spanish	Mediterranean lands
					Catalan	Spain (Catalonia)
				Old Provençal	Provençal	southern France
				Old French	French	France, Belgium, Switzerland
				Middle French		
					Italian	Italy, Switzerland
					Rhaeto-Romance	Switzerland, Italy
					Sardinian	Sardinia
					Dalmation	Adriatic coast
					Romanian	Romania, Balkans
CELTIC	Continental		*Gaulish*			Gaul
	Goidelic			*Old Irish*	Irish	Ireland
				Middle Irish		
					Scottish Gaelic	Scotland
					Manx	Isle of Man
	Brythonic			*Old Welsh*	Welsh	Wales
				Middle Welsh		
				Old Cornish	*Cornish*	Cornwall
				Old Breton	Breton	Brittany

[1]Italics denote dead languages. Languages listed in roman type in the ancient or medieval column survive only in some special use, as in literary composition or liturgy.

: too evident to be doubted : UNQUESTIONABLE
— **in·du·bi·ta·bil·i·ty** \-ˌdü-bə-tə-ˈbi-lə-tē, -ˌdyü-\ *noun*
— **in·du·bi·ta·ble·ness** \-ˈdü-bə-tə-bəl-nəs, -ˈdyü-\ *noun*
— **in·du·bi·ta·bly** \-blē\ *adverb*

in·duce \in-ˈdüs, -ˈdyüs\ *transitive verb* **in·duced; in·duc·ing** [Middle English, from Latin *inducere,* from *in-* + *ducere* to lead — more at TOW] (14th century)
1 a : to move by persuasion or influence **b :** to call forth or bring about by influence or stimulation
2 a : EFFECT, CAUSE **b :** to cause the formation of **c :** to produce (as an electric current) by induction
3 : to determine by induction; *specifically* **:** to infer from particulars

in·duce·ment \in-ˈdüs-mənt, -ˈdyüs-\ *noun* (1594)
1 : a motive or consideration that leads one to action or to additional or more effective actions
2 : the act or process of inducing
3 : matter presented by way of introduction or background to explain the principal allegations of a legal cause, plea, or defense
synonym see MOTIVE

in·duc·er \-ˈdü-sər, -ˈdyü-\ *noun* (1554)
: one that induces; *especially* **:** a substance that is capable of activating a structural gene by combining with and inactivating a genetic repressor

in·duc·ible \in-ˈdü-sə-bəl, -ˈdyü-\ *adjective* (circa 1677)
: capable of being induced; *especially* **:** formed by a cell in response to the presence of its substrate ⟨*inducible* enzymes⟩ — compare CONSTITUTIVE
— **in·duc·ibil·i·ty** \-ˌdü-sə-ˈbi-lə-tē, -ˌdyü-\ *noun*

in·duct \in-ˈdəkt\ *transitive verb* [Middle English, from Medieval Latin *inductus,* past participle of *inducere,* from Latin] (14th century)
1 : to put in formal possession (as of a benefice or office) **:** INSTALL ⟨was *inducted* as president of the college⟩
2 a : to admit as a member ⟨*inducted* into a scholastic society⟩ **b :** INTRODUCE, INITIATE **c :** to enroll for military training or service (as under a selective service act)
3 : LEAD, CONDUCT

in·duc·tance \in-ˈdək-tən(t)s\ *noun* (1886)
1 a : a property of an electric circuit by which an electromotive force is induced in it by a variation of current either in the circuit itself or in a neighboring circuit **b :** the measure of this property that is equal to the ratio of the induced electromotive force to the rate of change of the inducing current
2 : a circuit or a device possessing inductance

in·duct·ee \(ˌ)in-ˌdək-ˈtē, in-ˈdək-ˌ\ *noun* (1940)
: one who is inducted

in·duc·tion \in-ˈdək-shən\ *noun* (14th century)
1 a : the act or process of inducting (as into office) **b :** an initial experience **:** INITIATION **c :** the formality by which a civilian is inducted into military service
2 a (1) **:** inference of a generalized conclusion from particular instances — compare DEDUCTION 2a (2) **:** a conclusion arrived at by induction **b :** mathematical demonstration of the validity of a law concerning all the positive integers by proving that it holds for the integer 1 and that if it holds for all the integers preceding a given integer it must hold for the next following integer
3 : a preface, prologue, or introductory scene especially of an early English play
4 a : the act of bringing forward or adducing (as facts or particulars) **b :** the act of causing or bringing on or about **c :** the process by which an electrical conductor becomes electrified when near a charged body, by which a magnetizable body becomes magnetized when in a magnetic field or in the magnetic flux set up by a magnetomotive force, or by which an electromotive force is produced in a circuit by varying the magnetic field linked with the circuit **d :** the inspiration of the fuel-air charge from the carburetor into the combustion chamber of an internal-combustion engine **e :** the sum of the processes by which the fate of embryonic cells is determined and morphogenetic differentiation brought about

induction coil *noun* (1837)
: an apparatus for obtaining intermittent high voltage consisting of a primary coil through which the direct current flows, an interrupter, and a secondary coil of a larger number of turns in which the high voltage is induced

induction heating *noun* (1919)
: heating of material by means of an electric current that is caused to flow through the material or its container by electromagnetic induction

induction motor *noun* (1897)
: an alternating-current motor in which torque is produced by the reaction between a varying magnetic field generated in the stator and the current induced in the coils of the rotor

in·duc·tive \in-ˈdək-tiv\ *adjective* (15th century)
1 : leading on **:** INDUCING
2 : of, relating to, or employing mathematical or logical induction ⟨*inductive* reasoning⟩
3 : of or relating to inductance or electrical induction
4 : INTRODUCTORY
5 : involving the action of an embryological inductor **:** tending to produce induction
— **in·duc·tive·ly** *adverb*

in·duc·tor \in-ˈdək-tər\ *noun* (1652)
1 : one that inducts
2 a : a part of an electrical apparatus that acts upon another or is itself acted upon by induction **b :** REACTOR 2
3 : ORGANIZER 2

in·due *variant of* ENDUE

in·dulge \in-ˈdəlj\ *verb* **in·dulged; in·dulg·ing** [Latin *indulgēre* to be complaisant] (circa 1623)
transitive verb
1 a : to give free rein to **b :** to take unrestrained pleasure in **:** GRATIFY
2 a : to yield to the desire of **:** HUMOR **b :** to treat with excessive leniency, generosity, or consideration
intransitive verb
: to indulge oneself ☆
— **in·dulg·er** *noun*

in·dul·gence \in-ˈdəl-jən(t)s\ *noun* (14th century)
1 : remission of part or all of the temporal and especially purgatorial punishment that according to Roman Catholicism is due for sins whose eternal punishment has been remitted and whose guilt has been pardoned (as through the sacrament of reconciliation)
2 : the act of indulging **:** the state of being indulgent
3 a : an indulgent act **b :** an extension of time for payment or performance granted as a favor
4 a : the act of indulging in something; *especially* **:** SELF-INDULGENCE **b :** something indulged in ⟨walk off gastronomic *indulgences* —Barbara L. Michaels⟩

in·dul·gent \in-ˈdəl-jənt\ *adjective* [Latin *indulgent-, indulgens,* present participle of *indulgēre*] (1509)
: indulging or characterized by indulgence; *especially* **:** LENIENT
— **in·dul·gent·ly** *adverb*

in·dult \ˈin-ˌdəlt, in-ˈ\ *noun* [Middle English (Scots), from Medieval Latin *indultum,* from Late Latin, grant, from Latin, neuter of *indultus,* past participle of *indulgēre*] (15th century)
: a special often temporary dispensation granted in the Roman Catholic Church

¹in·du·rate \ˈin-də-rət, -dyə-; in-ˈdu̇r-ət, -ˈdyu̇r-\ *adjective* (14th century)
: physically or morally hardened

²in·du·rate \ˈin-də-ˌrāt, -dyə-\ *verb* **-rat·ed; -rat·ing** [Latin *induratus,* past participle of *indurare,* from *in-* + *durare* to harden, from *durus* hard — more at DURING] (1538)
transitive verb
1 : to make unfeeling, stubborn, or obdurate
2 : to make hardy **:** INURE
3 : to make hard ⟨great heat *indurates* clay⟩
4 : to establish firmly **:** CONFIRM
intransitive verb
1 : to grow hard **:** HARDEN
2 : to become established

in·du·rat·ed \-ˌrā-təd\ *adjective* (1604)
: having become firm or hard especially by increase of fibrous elements ⟨*indurated* tissue⟩

in·du·ra·tion \ˌin-də-ˈrā-shən, -dyə-\ *noun* (14th century)
: the process of or condition produced by growing hard; *specifically* **:** sclerosis especially when associated with inflammation
— **in·du·ra·tive** \ˈin-də-ˌrā-tiv, -dyə-; in-ˈdu̇r-ə-, -ˈdyu̇r-\ *adjective*

in·du·si·um \in-ˈdü-zē-əm, -ˈdyü-, -zhē-\ *noun, plural* **-sia** \-zē-ə, -zhē-\ [New Latin, from Latin, tunic] (1807)
: an investing outgrowth or membrane: as **a :** an outgrowth of a fern frond that invests the sori **b :** the annulus of a fungus especially when large and full

¹in·dus·tri·al \in-ˈdəs-trē-əl\ *adjective* (15th century)
1 : of or relating to industry
2 : derived from human industry ⟨*industrial* wealth⟩
3 : engaged in industry ⟨the *industrial* classes⟩
4 : used in or developed for use in industry ⟨*industrial* diamonds⟩; *also* **:** HEAVY-DUTY ⟨an *industrial* zipper⟩
5 : characterized by highly developed industries ⟨an *industrial* nation⟩
— **in·dus·tri·al·ly** \-trē-ə-lē\ *adverb*

²industrial *noun* (1865)
1 a : one that is employed in industry **b :** a company engaged in industrial production or service
2 : a stock or bond issued by an industrial corporation or enterprise

industrial action *noun* (circa 1931)
British **:** JOB ACTION

industrial archaeology *noun* (1951)
: the study of the buildings, machinery, and equipment of the industrial revolution
— **industrial archaeologist** *noun*

industrial arts *noun plural but singular in construction* (circa 1925)
: a subject taught in elementary and secondary schools that aims at developing manual skill and familiarity with tools and machines

industrial engineering *noun* (circa 1924)

☆ SYNONYMS
Indulge, pamper, humor, spoil, baby, mollycoddle mean to show undue favor to a person's desires and feelings. INDULGE implies excessive compliance and weakness in gratifying another's or one's own desires ⟨*indulged* myself with food at the slightest excuse⟩. PAMPER implies inordinate gratification of desire for luxury and comfort with consequent enervating effect ⟨*pampered* by the amenities of modern living⟩. HUMOR stresses a yielding to a person's moods or whims ⟨*humored* him by letting him tell the story⟩. SPOIL stresses the injurious effects on character by indulging or pampering ⟨foolish parents *spoil* their children⟩. BABY suggests excessive care, attention, or solicitude ⟨*babying* students by grading too easily⟩. MOLLYCODDLE suggests an excessive degree of care and attention to another's health or welfare ⟨refused to *mollycoddle* her malingering son⟩.

: engineering that deals with the design, improvement, and installation of integrated systems (as of people, materials, and energy) in industry
— **industrial engineer** *noun*
in·dus·tri·al·ise *British variant of* INDUSTRIALIZE
in·dus·tri·al·ism \in-ˈdəs-trē-ə-ˌli-zəm\ *noun* (1831)
: social organization in which industries and especially large-scale industries are dominant
in·dus·tri·al·ist \-list\ *noun* (1864)
: one owning or engaged in the management of an industry **:** MANUFACTURER
in·dus·tri·al·ize \in-ˈdəs-trē-ə-ˌlīz\ *verb* **-ized; -iz·ing** (1882)
transitive verb
: to make industrial ⟨*industrialize* an agricultural region⟩
intransitive verb
: to become industrial
— **in·dus·tri·al·i·za·tion** \-ˌdəs-trē-ə-lə-ˈzā-shən\ *noun*
industrial melanism *noun* (1943)
: genetically determined melanism as a population phenomenon especially in moths in which the proportion of dark individuals tends to increase due to differential predation especially by birds which more easily find and eat lighter-colored individuals in habitats darkened by industrial pollution
industrial psychology *noun* (1917)
: the application of the findings and methods of experimental, clinical, and social psychology to industrial concerns
— **industrial psychologist** *noun*
industrial relations *noun plural* (1904)
: the dealings or relationships of a usually large business or industrial enterprise with its own workers, with labor in general, with governmental agencies, or with the public
industrial revolution *noun* (1848)
: a rapid major change in an economy (as in England in the late 18th century) marked by the general introduction of power-driven machinery or by an important change in the prevailing types and methods of use of such machines
industrial school *noun* (1853)
: a school specializing in the teaching of industrial arts; *specifically* **:** one for juvenile delinquents
industrial sociology *noun* (1948)
: sociological analysis directed at institutions and social relationships within and largely controlled or affected by industry
in·dus·tri·al–strength \in-ˈdəs-trē-əl-ˈstreŋ(k)th, -ˈstren(t)th\ *adjective* (1976)
: of industrial quality; *also* **:** marked by more than usual power or durability ⟨*industrial-strength* boots⟩ ⟨an *industrial-strength* voice⟩
industrial union *noun* (1902)
: a labor union open to workers in an industry irrespective of their occupation or craft — compare CRAFT UNION
in·dus·tri·ous \in-ˈdəs-trē-əs\ *adjective* (15th century)
1 *obsolete* **:** SKILLFUL, INGENIOUS
2 : persistently active **:** ZEALOUS
3 : constantly, regularly, or habitually occupied **:** DILIGENT
synonym see BUSY
— **in·dus·tri·ous·ly** *adverb*
— **in·dus·tri·ous·ness** *noun*
in·dus·try \ˈin-(ˌ)dəs-trē\ *noun, plural* **-tries** [Middle English *industrie* skill, employment involving skill, from Middle French, from Latin *industria* diligence, from *industrius* diligent, from Old Latin *indostruus,* perhaps from *indu* in + *-struus* (akin to Latin *struere* to build) — more at END-, STREW] (15th century)
1 : diligence in an employment or pursuit; *especially* **:** steady or habitual effort
2 a : systematic labor especially for some useful purpose or the creation of something of value **b :** a department or branch of a craft, art, business, or manufacture; *especially* **:** one that employs a large personnel and capital especially in manufacturing **c :** a distinct group of productive or profit-making enterprises ⟨the banking *industry*⟩ **d :** manufacturing activity as a whole ⟨the nation's *industry*⟩
3 : work devoted to the study of a particular subject or author ⟨the Shakespeare *industry*⟩
synonym see BUSINESS
in·dwell \in-ˈdwel, ˈin-ˌ\ (14th century)
intransitive verb
: to exist as an inner activating spirit, force, or principle
transitive verb
: to exist within as an activating spirit, force, or principle
— **in·dwell·er** *noun*
in·dwell·ing \ˈin-ˌdwe-liŋ\ *adjective* (1646)
1 : being an inner activating or guiding force
2 : left within a bodily organ or passage especially to promote drainage — used of an implanted tube (as a catheter)
¹-ine *adjective suffix*
1 [Middle English *-in, -ine,* from Middle French & Latin; Middle French *-in,* from Latin *-īnus* — more at -EN] **:** of or relating to ⟨estuar*ine*⟩
2 [Middle English *-in, -ine,* from Middle French & Latin; Middle French *-in,* from Latin *-ĭnus,* from Greek *-inos* — more at -EN] **:** made of **:** like ⟨opal*ine*⟩
²-ine *noun suffix* [Middle English *-ine, -in,* from Middle French & Latin; Middle French *-ine,* from Latin *-īna,* from feminine of *-īnus,* adjective suffix]
1 : chemical substance: as **a :** halogen element ⟨chlor*ine*⟩ **b :** basic or base-containing carbon compound that contains nitrogen ⟨quin*ine*⟩ ⟨cyst*ine*⟩ **c :** mixture of compounds (as of hydrocarbons) ⟨gasol*ine*⟩ **d :** hydride ⟨ars*ine*⟩
2 : -IN 1a
3 : commercial product or material ⟨glass*ine*⟩
in·ebri·ant \i-ˈnē-brē-ənt\ *noun* (1819)
: INTOXICANT
¹in·ebri·ate \i-ˈnē-brē-ˌāt\ *transitive verb* **-at·ed; -at·ing** [Latin *inebriatus,* past participle of *inebriare,* from *in-* + *ebriare* to intoxicate, from *ebrius* drunk] (15th century)
1 : to exhilarate or stupefy as if by liquor
2 : to make drunk **:** INTOXICATE
— **in·ebri·a·tion** \-ˌnē-brē-ˈā-shən\ *noun*
²in·ebri·ate \i-ˈnē-brē-ət, -ˌāt\ *adjective* (15th century)
1 : affected by alcohol **:** DRUNK
2 : addicted to excessive drinking
³in·ebri·ate \-ət\ *noun* (circa 1796)
: one who is drunk; *especially* **:** DRUNKARD
in·ebri·at·ed \-brē-ˌā-təd\ *adjective* (1609)
: exhilarated or confused by or as if by alcohol **:** INTOXICATED
in·ebri·ety \ˌi-ni-ˈbrī-ə-tē\ *noun* [probably blend of *inebriation* and *ebriety* drunkenness] (1801)
: the state of being inebriated **:** DRUNKENNESS
in·ed·i·ble \(ˌ)i-ˈne-də-bəl\ *adjective* (circa 1834)
: not fit to be eaten
in·ed·u·ca·ble \(ˌ)i-ˈne-jə-kə-bəl\ *adjective* (1884)
: incapable of being educated
— **in·ed·u·ca·bil·i·ty** \-ˌne-jə-kə-ˈbi-lə-tē\ *noun*
in·ef·fa·ble \(ˌ)i-ˈne-fə-bəl\ *adjective* [Middle English, from Middle French, from Latin *ineffabilis,* from *in-* + *effabilis* capable of being expressed, from *effari* to speak out, from *ex-* + *fari* to speak — more at BAN] (14th century)
1 a : incapable of being expressed in words **:** INDESCRIBABLE ⟨*ineffable* joy⟩ **b :** UNSPEAKABLE ⟨*ineffable* disgust⟩
2 : not to be uttered **:** TABOO ⟨the *ineffable* name of Jehovah⟩
— **in·ef·fa·bil·i·ty** \-ˌne-fə-ˈbi-lə-tē\ *noun*
— **in·ef·fa·ble·ness** \-ˈne-fə-bəl-nəs\ *noun*
— **in·ef·fa·bly** \-blē\ *adverb*
in·ef·face·able \ˌi-nə-ˈfā-sə-bəl\ *adjective* [probably from French *ineffaçable,* from Middle French, from *in-* + *effaçable* effaceable] (1804)
: not effaceable **:** INERADICABLE
— **in·ef·face·abil·i·ty** \-ˌfā-sə-ˈbi-lə-tē\ *noun*
— **in·ef·face·ably** \-ˈfā-sə-blē\ *adverb*
in·ef·fec·tive \ˌi-nə-ˈfek-tiv\ *adjective* (1649)
1 : not producing an intended effect **:** INEFFECTUAL ⟨*ineffective* lighting⟩
2 : not capable of performing efficiently or as expected **:** INCAPABLE ⟨an *ineffective* executive⟩
— **in·ef·fec·tive·ly** *adverb*
— **in·ef·fec·tive·ness** *noun*
in·ef·fec·tu·al \ˌi-nə-ˈfek-chə(-wə)l, -ˈfeksh-wəl\ *adjective* (15th century)
1 : not producing the proper or intended effect **:** FUTILE
2 : INEFFECTIVE 2
— **in·ef·fec·tu·al·i·ty** \-ˌfek-chə-ˈwa-lə-tē\ *noun*
— **in·ef·fec·tu·al·ly** \-ˈfek-chə(-wə)-lē, -ˈfek-shwə-\ *adverb*
— **in·ef·fec·tu·al·ness** *noun*
in·ef·fi·ca·cious \(ˌ)i-ˌne-fə-ˈkā-shəs\ *adjective* (1658)
: lacking the power to produce a desired effect **:** INEFFECTIVE
— **in·ef·fi·ca·cious·ly** *adverb*
— **in·ef·fi·ca·cious·ness** *noun*
in·ef·fi·ca·cy \(ˌ)i-ˈne-fi-kə-sē\ *noun* [Late Latin *inefficacia,* from Latin *inefficac-, inefficax* inefficacious, from *in-* + *efficac-, efficax* efficacious] (circa 1615)
: lack of power to produce a desired effect
in·ef·fi·cien·cy \ˌi-nə-ˈfi-shən(t)-sē\ *noun, plural* **-cies** (1749)
1 : the quality or state of being inefficient
2 : something that is inefficient
in·ef·fi·cient \-ˈfi-shənt\ *adjective* (1750)
: not efficient: as **a :** not producing the effect intended or desired **b :** wasteful of time or energy ⟨*inefficient* operating procedures⟩ **c :** INCAPABLE, INCOMPETENT ⟨an *inefficient* worker⟩
— **inefficient** *noun*
— **in·ef·fi·cient·ly** *adverb*
in·egal·i·tar·i·an \ˌi-ni-ˌga-lə-ˈter-ē-ən\ *adjective* (1940)
: marked by disparity in social and economic standing
in·elas·tic \ˌi-nə-ˈlas-tik\ *adjective* (1748)
: not elastic: as **a :** slow to react or respond to changing conditions **b :** INFLEXIBLE, UNYIELDING
— **in·elas·tic·i·ty** \ˌi-ni-ˌlas-ˈti-s(ə-)tē, (ˌ)i-ˌnē-ˌlas-\ *noun*
inelastic collision *noun* (1937)
: a collision in which part of the kinetic energy of the colliding particles changes into another form of energy (as radiation)
inelastic scattering *noun* (1938)
: a scattering of particles as the result of inelastic collision in which the total kinetic energy of the colliding particles changes
in·el·e·gance \(ˌ)i-ˈne-li-gən(t)s\ *noun* (1726)
: lack of elegance
in·el·e·gant \-gənt\ *adjective* [Middle French, from Latin *inelegant-, inelegans,* from *in-* + *elegant-, elegans* elegant] (circa 1570)
: lacking in refinement, grace, or good taste
— **in·el·e·gant·ly** *adverb*
in·el·i·gi·ble \(ˌ)i-ˈne-lə-jə-bəl\ *adjective* [French *inéligible,* from *in-* + *éligible* eligible] (1770)
: not eligible: as **a :** not qualified for an office or position **b :** not permitted under football rules to catch a forward pass
— **in·el·i·gi·bil·i·ty** \-ˌne-lə-jə-ˈbi-lə-tē\ *noun*

— **ineligible** *noun*
in·el·o·quent \(ˌ)i-ˈne-lə-kwənt\ *adjective* (circa 1530)
: not eloquent
— **in·el·o·quent·ly** *adverb*
in·eluc·ta·ble \ˌi-ni-ˈlək-tə-bəl\ *adjective* [Latin *ineluctabilis,* from *in-* + *eluctari* to struggle clear of, from *ex-* + *luctari* to struggle, wrestle; akin to Latin *luxus* dislocated — more at LOCK] (circa 1623)
: not to be avoided, changed, or resisted **:** INEVITABLE
— **in·eluc·ta·bil·i·ty** \-ˌlək-tə-ˈbi-lə-tē\ *noun*
— **in·eluc·ta·bly** \-ˈlək-tə-blē\ *adverb*
in·elud·ible \ˌi-ni-ˈlü-də-bəl\ *adjective* (1662)
: INESCAPABLE
in·enar·ra·ble \ˌi-ni-ˈnar-ə-bəl\ *adjective* [Middle English, from Middle French, from Latin *inenarrabilis,* from *in-* + *enarrare* to explain in detail, from *e-* + *narrare* to narrate] (15th century)
: incapable of being narrated **:** INDESCRIBABLE
in·ept \i-ˈnept\ *adjective* [Middle French *inepte,* from Latin *ineptus,* from *in-* + *aptus* apt] (1542)
1 : lacking in fitness or aptitude **:** UNFIT
2 : lacking sense or reason **:** FOOLISH
3 : not suitable to the time, place, or occasion **:** inappropriate often to an absurd degree
4 : generally incompetent **:** BUNGLING
synonym see AWKWARD
— **in·ept·ly** \-ˈnep(t)-lē\ *adverb*
— **in·ept·ness** \-nəs\ *noun*
in·ep·ti·tude \(ˌ)i-ˈnep-tə-ˌtüd, -ˌtyüd\ *noun* [Latin *ineptitudo,* from *ineptus*] (1615)
: the quality or state of being inept; *especially* **:** INCOMPETENCE
in·equal·i·ty \ˌi-ni-ˈkwä-lə-tē\ *noun* [Middle English *inequalite,* from Middle French *inequalité,* from Latin *inaequalitat-, inaequalitas,* from *inaequalis* unequal, from *in-* + *aequalis* equal] (15th century)
1 : the quality of being unequal or uneven: as **a :** lack of evenness **b :** social disparity **c :** disparity of distribution or opportunity **d :** the condition of being variable **:** CHANGEABLENESS
2 : an instance of being unequal
3 : a formal statement of inequality between two quantities usually separated by a sign of inequality (as <, >, or ≠ signifying respectively *is less than, is greater than,* or *is not equal to*)
in·eq·ui·ta·ble \(ˌ)i-ˈne-kwə-tə-bəl\ *adjective* (1667)
: not equitable **:** UNFAIR
— **in·eq·ui·ta·bly** \-blē\ *adverb*
in·eq·ui·ty \(ˌ)i-ˈne-kwə-tē\ *noun* (1556)
1 : INJUSTICE, UNFAIRNESS
2 : an instance of injustice or unfairness
in·equi·valve \(ˌ)i-ˈnē-kwə-ˌvalv\ *also* **in·equi·valved** \-ˌvalvd\ *adjective* (1776)
: having the valves unequal in size and form — used of a bivalve mollusk or shell
in·erad·i·ca·ble \ˌi-ni-ˈra-di-kə-bəl\ *adjective* (1818)
: incapable of being eradicated
— **in·erad·i·ca·bil·i·ty** \-ˌra-di-kə-ˈbi-lə-tē\ *noun*
— **in·erad·i·ca·bly** \-ˈra-di-kə-blē\ *adverb*
in·er·ran·cy \(ˌ)i-ˈner-ən(t)-sē\ *noun* (circa 1834)
: exemption from error **:** INFALLIBILITY ⟨the question of biblical *inerrancy*⟩
in·er·rant \-ənt\ *adjective* [Latin *inerrant-, inerrans,* from *in-* + *errant-, errans,* present participle of *errare* to err] (1837)
: free from error
in·ert \i-ˈnərt\ *adjective* [Latin *inert-, iners* unskilled, idle, from *in-* + *art-, ars* skill — more at ARM] (1647)
1 : lacking the power to move
2 : very slow to move or act **:** SLUGGISH
3 : deficient in active properties; *especially* **:** lacking a usual or anticipated chemical or biological action
synonym see INACTIVE
— **inert** *noun*
— **in·ert·ly** *adverb*
— **in·ert·ness** *noun*
inert gas *noun* (1898)
: NOBLE GAS
in·er·tia \i-ˈnər-shə, -shē-ə\ *noun* [New Latin, from Latin, lack of skill, from *inert-, iners*] (1713)
1 a : a property of matter by which it remains at rest or in uniform motion in the same straight line unless acted upon by some external force **b :** an analogous property of other physical quantities (as electricity)
2 : indisposition to motion, exertion, or change **:** INERTNESS
— **in·er·tial** \-shəl\ *adjective*
— **in·er·tial·ly** \-ˈnər-sh(ə-)lē\ *adverb*
inertial guidance *noun* (circa 1948)
: guidance (as of an aircraft or spacecraft) by means of self-contained automatically controlling devices that respond to inertial forces — called also *inertial navigation*
in·es·cap·able \ˌi-nə-ˈskā-pə-bəl\ *adjective* (1792)
: incapable of being avoided, ignored, or denied **:** INEVITABLE
— **in·es·cap·ably** \-blē\ *adverb*
in·es·sen·tial \ˌi-nə-ˈsen(t)-shəl\ *adjective* (1677)
1 : having no essence
2 : not essential **:** UNESSENTIAL
— **inessential** *noun*
in·es·ti·ma·ble \(ˌ)i-ˈnes-tə-mə-bəl\ *adjective* [Middle English, from Middle French, from Latin *inaestimabilis,* from *in-* + *aestimabilis* estimable] (14th century)
1 : incapable of being estimated or computed ⟨storms caused *inestimable* damage⟩
2 : too valuable or excellent to be measured or appreciated ⟨has performed an *inestimable* service for his country⟩
— **in·es·ti·ma·bly** \-blē\ *adverb*
in·ev·i·ta·ble \i-ˈne-və-tə-bəl\ *adjective* [Middle English, from Latin *inevitabilis,* from *in-* + *evitabilis* evitable] (14th century)
: incapable of being avoided or evaded
— **in·ev·i·ta·bil·i·ty** \-ˌne-və-tə-ˈbi-lə-tē\ *noun*
— **in·ev·i·ta·ble·ness** \-ˈne-və-tə-bəl-nəs\ *noun*
in·ev·i·ta·bly \-blē\ *adverb* (15th century)
1 : in an inevitable way
2 : as is to be expected ⟨*inevitably,* it rained⟩
in·ex·act \ˌi-nig-ˈzakt\ *adjective* [French, from *in-* + *exact* exact] (circa 1828)
1 : not precisely correct or true **:** INACCURATE ⟨an *inexact* translation⟩
2 : not rigorous and careful ⟨an *inexact* thinker⟩
— **in·ex·act·ly** \-ˈzak(t)-lē\ *adverb*
— **in·ex·act·ness** \-nəs\ *noun*
in·ex·ac·ti·tude \(ˌ)i-ˌnig-ˈzak-tə-ˌtüd, -ˌtyüd\ *noun* [French, from *inexact*] (1782)
1 : lack of exactitude or precision
2 : an instance of inexactness
in ex·cel·sis \ˌin-ik-ˈsel-səs *also* -ˈchel-\ *adverb* [Late Latin, on high] (1602)
: in the highest degree
in·ex·cus·able \ˌi-nik-ˈskyü-zə-bəl\ *adjective* [Middle English, from Latin *inexcusabilis,* from *in-* + *excusabilis* excusable] (15th century)
: being without excuse or justification
— **in·ex·cus·able·ness** *noun*
— **in·ex·cus·ably** \-blē\ *adverb*
in·ex·haust·ible \ˌi-nig-ˈzȯ-stə-bəl\ *adjective* (1601)
: not exhaustible: as **a :** incapable of being used up ⟨*inexhaustible* riches⟩ **b :** incapable of being wearied or worn out ⟨an *inexhaustible* hiker⟩
— **in·ex·haust·ibil·i·ty** \-ˌzȯ-stə-ˈbi-lə-tē\ *noun*
— **in·ex·haust·ible·ness** \-ˈzȯ-stə-bəl-nəs\ *noun*
— **in·ex·haust·ibly** \-blē\ *adverb*
in·ex·is·tence \ˌi-nig-ˈzis-tən(t)s\ *noun* (circa 1623)
: absence of existence **:** NONEXISTENCE
in·ex·is·tent \-tənt\ *adjective* [Late Latin *inexsistent-, inexsistens,* from Latin *in-* + *exsistent-, exsistens,* present participle of *exsistere* to exist] (1646)
: not having existence **:** NONEXISTENT
in·ex·o·ra·ble \(ˌ)i-ˈneks-rə-bəl, -ˈnek-sə-, -ˈneg-zə-rə-\ *adjective* [Latin *inexorabilis,* from *in-* + *exorabilis* pliant, from *exorare* to prevail upon, from *ex-* + *orare* to speak — more at ORATION] (1553)
: not to be persuaded or moved by entreaty **:** RELENTLESS
— **in·ex·o·ra·bil·i·ty** \(ˌ)i-ˌneks-rə-ˈbi-lə-tē, -ˌnek-sə-, -ˌneg-zə-\ *noun*
— **in·ex·o·ra·ble·ness** \-ˈneks-rə-bəl-nəs, -ˈnek-sə-, -ˈneg-zə-\ *noun*
— **in·ex·o·ra·bly** \-blē\ *adverb*
in·ex·pe·di·ence \ˌi-nik-ˈspē-dē-ən(t)s\ *noun* (1608)
: INEXPEDIENCY
in·ex·pe·di·en·cy \-ən(t)-sē\ *noun* (1641)
: the quality or fact of being inexpedient
in·ex·pe·di·ent \-ənt\ *adjective* (1608)
: not expedient **:** INADVISABLE
— **in·ex·pe·di·ent·ly** *adverb*
in·ex·pen·sive \ˌi-nik-ˈspen(t)-siv\ *adjective* (circa 1846)
: reasonable in price **:** CHEAP
— **in·ex·pen·sive·ly** *adverb*
— **in·ex·pen·sive·ness** *noun*
in·ex·pe·ri·ence \ˌi-nik-ˈspir-ē-ən(t)s\ *noun* [Middle French, from Late Latin *inexperientia,* from Latin *in-* + *experientia* experience] (1598)
1 : lack of practical experience
2 : lack of knowledge of the ways of the world
— **in·ex·pe·ri·enced** \-ən(t)st\ *adjective*
in·ex·pert \(ˌ)i-ˈnek-ˌspərt, ˌi-nik-ˈ\ *adjective* [Middle English, from Middle French, from Latin *inexpertus,* from *in-* + *expertus* expert] (15th century)
: not expert **:** UNSKILLED
— **in·ex·pert** \-ˈnek-ˌspərt\ *noun*
— **in·ex·pert·ly** \-ˈnek-ˌspərt-lē, ˌi-nik-ˈ\ *adverb*
— **in·ex·pert·ness** *noun*
in·ex·pi·a·ble \(ˌ)i-ˈnek-spē-ə-bəl\ *adjective* [Middle English *inexpyable,* from Latin *inexpiabilis,* from *in-* + *expiare* to expiate] (15th century)
1 : not capable of being atoned for
2 *obsolete* **:** IMPLACABLE, UNAPPEASABLE
— **in·ex·pi·a·bly** \-blē\ *adverb*
in·ex·plain·able \ˌi-nik-ˈsplā-nə-bəl\ *adjective* (circa 1923)
: INEXPLICABLE
in·ex·pli·ca·ble \ˌi-nik-ˈspli-kə-bəl, (ˌ)i-ˈnek-(ˌ)spli-\ *adjective* [Middle English, from Middle French, from Latin *inexplicabilis,* from *in-* + *explicabilis* explicable] (15th century)
: incapable of being explained, interpreted, or accounted for
— **in·ex·pli·ca·bil·i·ty** \ˌi-nik-ˌspli-kə-ˈbi-lə-tē, (ˌ)i-ˌnek-(ˌ)spli-\ *noun*
— **in·ex·pli·ca·ble·ness** \ˌi-nik-ˈspli-kə-bəl-nəs, (ˌ)i-ˈnek-(ˌ)spli-\ *noun*
— **in·ex·pli·ca·bly** \-blē\ *adverb*
in·ex·plic·it \ˌi-nik-ˈspli-sət\ *adjective* (circa 1812)
: not explicit
in·ex·press·ible \-ˈspre-sə-bəl\ *adjective* (1625)
: not capable of being expressed **:** INDESCRIBABLE
— **in·ex·press·ibil·i·ty** \-ˌspre-sə-ˈbi-lə-tē\ *noun*

— **in·ex·press·ible·ness** \-ˈspre-sə-bəl-nəs\ *noun*
— **in·ex·press·ibly** \-blē\ *adverb*
in·ex·pres·sive \-ˈspre-siv\ *adjective* (1652)
1 *archaic* **:** INEXPRESSIBLE
2 : lacking expression or meaning ⟨an *inexpressive* face⟩
— **in·ex·pres·sive·ly** *adverb*
— **in·ex·pres·sive·ness** *noun*
in·ex·pug·na·ble \ˌi-nik-ˈspəg-nə-bəl, -ˈspyü-nə-\ *adjective* [Middle English *in-expungnabull,* from Middle French, from Latin *inexpugnabilis,* from *in-* + *expugnare* to take by storm, from *ex-* + *pugnare* to fight — more at PUNGENT] (15th century)
1 : incapable of being subdued or overthrown **:** IMPREGNABLE ⟨an *inexpugnable* position⟩
2 : STABLE, FIXED ⟨*inexpugnable* hatred⟩
— **in·ex·pug·na·ble·ness** *noun*
— **in·ex·pug·na·bly** \-blē\ *adverb*
in·ex·pung·ible \ˌi-nik-ˈspən-jə-bəl\ *adjective* [*in-* + *expunge*] (1888)
: incapable of being obliterated
in ex·ten·so \ˌin-ik-ˈsten(t)-(ˌ)sō\ *adverb* [Medieval Latin] (1826)
: at full length ⟨the passage was quoted *in extenso*⟩
in·ex·tin·guish·able \ˌi-nik-ˈstiŋ-gwi-shə-bəl, -ˈstiŋ-wi-\ *adjective* (15th century)
: not extinguishable **:** UNQUENCHABLE ⟨an *inextinguishable* flame⟩ ⟨an *inextinguishable* longing⟩
— **in·ex·tin·guish·ably** \-blē\ *adverb*
in ex·tre·mis \ˌin-ik-ˈstrē-məs, -ˈstrā-\ *adverb* [Latin] (circa 1530)
: in extreme circumstances; *especially* **:** at the point of death
in·ex·tri·ca·ble \ˌi-nik-ˈstri-kə-bəl, (ˌ)i-ˈnek-(ˌ)stri-\ *adjective* [Middle English, from Middle French or Latin; Middle French, from Latin *inextricabilis,* from *in-* + *extricabilis* extricable] (15th century)
1 : forming a maze or tangle from which it is impossible to get free
2 a : incapable of being disentangled or untied ⟨an *inextricable* knot⟩ **b :** not capable of being solved
— **in·ex·tri·ca·bil·i·ty** \ˌi-nik-ˌstri-kə-ˈbi-lə-tē, (ˌ)i-ˌnek-(ˌ)stri-\ *noun*
— **in·ex·tri·ca·bly** \ˌi-nik-ˈstri-kə-ble, (ˌ)i-ˈnek-(ˌ)stri-\ *adverb*
in·fal·li·ble \(ˌ)in-ˈfa-lə-bəl\ *adjective* [Middle English, from Medieval Latin *infallibilis,* from Latin *in-* + Late Latin *fallibilis* fallible] (15th century)
1 : incapable of error **:** UNERRING ⟨an *infallible* memory⟩
2 : not liable to mislead, deceive, or disappoint **:** CERTAIN ⟨an *infallible* remedy⟩
3 : incapable of error in defining doctrines touching faith or morals
— **in·fal·li·bil·i·ty** \-ˌfa-lə-ˈbi-lə-tē\ *noun*
— **in·fal·li·bly** \-ˈfa-lə-blē\ *adverb*
in·fall·ing \ˈin-ˌfo-liŋ\ *adjective* (1964)
: moving under the influence of gravity toward a celestial object (as a black hole)
— **in·fall** \-ˌfȯl\ *noun*
in·fa·mous \ˈin-fə-məs\ *adjective* [Middle English, from Latin *infamis,* from *in-* + *fama* fame] (14th century)
1 : having a reputation of the worst kind
2 : causing or bringing infamy **:** DISGRACEFUL
3 : convicted of an offense bringing infamy
— **in·fa·mous·ly** *adverb*
in·fa·my \-mē\ *noun, plural* **-mies** (15th century)
1 : evil reputation brought about by something grossly criminal, shocking, or brutal
2 a : an extreme and publicly known criminal or evil act **b :** the state of being infamous
synonym see DISGRACE
in·fan·cy \ˈin-fən(t)-sē\ *noun, plural* **-cies** (14th century)
1 : early childhood
2 : a beginning or early period of existence
3 : the legal status of an infant

[1]**in·fant** \ˈin-fənt\ *noun* [Middle English *enfaunt,* from Middle French *enfant,* from Latin *infant-, infans,* from *infant-, infans,* adjective, incapable of speech, young, from *in-* + *fant-, fans,* present participle of *fari* to speak — more at BAN] (14th century)
1 : a child in the first period of life
2 : a person who is not of full age **:** MINOR ◇
[2]**infant** *adjective* (circa 1586)
1 : intended for young children
2 : being in an early stage of development
3 : of, relating to, or being in infancy
in·fan·ta \in-ˈfan-tə, -ˈfän-\ *noun* [Spanish & Portuguese, feminine of *infante*] (1593)
: a daughter of a Spanish or Portuguese monarch
in·fan·te \in-ˈfan-tē, -ˈfän-(ˌ)tā\ *noun* [Spanish & Portuguese, literally, infant, from Latin *infant-, infans*] (1555)
: a younger son of a Spanish or Portuguese monarch
in·fan·ti·cide \in-ˈfan-tə-ˌsīd\ *noun* [Late Latin *infanticidium,* from Latin *infant-, infans* + *-i-* + *-cidium* -cide] (circa 1656)
1 : the killing of an infant
2 [Late Latin *infanticida,* from Latin *infant-, infans* + *-i-* + *-cida* -cide] **:** one who kills an infant
— **in·fan·ti·ci·dal** \-ˌfan-tə-ˈsī-dᵊl\ *adjective*
in·fan·tile \ˈin-fən-ˌtīl, -tᵊl, -ˌtēl, -(ˌ)til\ *adjective* (1696)
1 : of or relating to infants or infancy
2 : suitable to or characteristic of an infant; *especially* **:** very immature
— **in·fan·til·i·ty** \ˌin-fən-ˈti-lə-tē\ *noun*
infantile paralysis *noun* (1843)
: POLIOMYELITIS
in·fan·til·ism \ˈin-fən-ˌtī-ˌli-zəm, -tə-ˌli-; in-ˈfan-tᵊl-ˌi-\ *noun* (1895)
1 : retention of childish physical, mental, or emotional qualities in adult life; *especially* **:** failure to attain sexual maturity
2 : an act or expression that indicates lack of maturity
in·fan·til·ize \ˈin-fən-ˌtī-ˌlīz, -fən-tᵊl-ˌīz; in-ˈfan-tᵊl-ˌīz\ *transitive verb* **-ized; -iz·ing** (1943)
1 : to make or keep infantile
2 : to treat as if infantile
— **in·fan·til·i·za·tion** \ˌin-fən-ˌtī-lə-ˈzā-shən, -fən-tᵊl-ə-; in-ˌfan-tᵊl-ə-\ *noun*
in·fan·tine \ˈin-fən-ˌtīn, -ˌtēn\ *adjective* (1603)
: INFANTILE, CHILDISH
in·fan·try \ˈin-fən-trē\ *noun, plural* **-tries** [Middle French & Old Italian; Middle French *infanterie,* from Old Italian *infanteria,* from *infante* boy, foot soldier, from Latin *infant-, infans*] (1579)
1 a : soldiers trained, armed, and equipped to fight on foot **b :** a branch of an army composed of these soldiers
2 : an infantry regiment or division ◇
in·fan·try·man \-trē-mən\ *noun* (1883)
: an infantry soldier
infant school *noun* (1824)
British **:** a school for children aged five to seven or eight
in·farct \ˈin-ˌfärkt, in-ˈ\ *noun* [Latin *infarctus,* past participle of *infarcire* to stuff, from *in-* + *farcire* to stuff] (1873)
: an area of necrosis in a tissue or organ resulting from obstruction of the local circulation by a thrombus or embolus
— **in·farct·ed** \in-ˈfärk-təd\ *adjective*
— **in·farc·tion** \in-ˈfärk-shən\ *noun*
in·fare \ˈin-ˌfar, -ˌfer\ *noun* [Middle English, entrance, from Old English *infær,* from *in* + *fær* way, from *faran* to go — more at FARE] (1595)
chiefly dialect **:** a reception for a newly married couple
[1]**in·fat·u·ate** \in-ˈfa-chə-wət\ *adjective* (15th century)
: being in an infatuated state or condition

[2]**in·fat·u·ate** \-ˌwāt\ *transitive verb* **-at·ed; -at·ing** [Latin *infatuatus,* past participle of *infatuare,* from *in-* + *fatuus* fatuous] (1533)
1 : to cause to be foolish **:** deprive of sound judgment
2 : to inspire with a foolish or extravagant love or admiration
— **in·fat·u·a·tion** \-ˌfa-chə-ˈwā-shən\ *noun*
in·fau·na \ˈin-ˌfȯ-nə, -ˌfä-\ *noun* [New Latin, from [2]*in-* + *fauna*] (1914)
: benthic fauna living in the substrate and especially in a soft sea bottom — compare EPIFAUNA
— **in·fau·nal** \-ˌfȯ-nᵊl, -ˌfä-\ *adjective*
in·fea·si·ble \(ˌ)in-ˈfē-zə-bəl\ *adjective* (1533)
: not feasible **:** IMPRACTICABLE
— **in·fea·si·bil·i·ty** \-ˌfē-zə-ˈbi-lə-tē\ *noun*
in·fect \in-ˈfekt\ *transitive verb* [Middle English, from Latin *infectus,* past participle of *inficere,* from *in-* + *facere* to make, do — more at DO] (14th century)
1 : to contaminate with a disease-producing substance or agent (as bacteria)
2 a : to communicate a pathogen or a disease to **b** *of a pathogenic organism* **:** to invade (an individual or organ) usually by penetration **c** *of a computer virus* **:** to become transmitted and copied to (as a computer)

◇ WORD HISTORY

infant To the frustrated parent of a sick, bawling infant incapable of saying just where it hurts, the etymology of *infant* must seem especially apt. In Latin the adjective *infans* literally meant "not speaking, incapable of speech." The noun *infans* designated a very young child who had not yet learned to talk. Later, however, the scope of *infans* was broadened to include any child, no matter how talkative. When the word was adopted into the Romance languages, the broader usage was carried over also. Like French, from which it had directly borrowed the word, English originally used *infant* for any child. Over time, English reverted to the earlier Latin sense, restricting *infant* to a child still young enough to be called a baby.

infantry In the 20th century U.S. Army, the infantry's training film epithet is "queen of battle," and the combat infantryman's badge is a coveted mark of honor, though colloquialisms such as *dogface* and *grunt* express another viewpoint about the status of the ordinary foot soldier. The history of the word *infantry,* moreover, reveals a rather lowly origin for this branch of arms. The Italian word *fante,* descended from Latin *infans* "infant, child," originally meant "child," later "youth, boy," and then "servant." In the early 14th century, *fante* also took on the sense "foot soldier." In the High Middle Ages he was typically a lowly figure, who was often no more than a servant in a horseman's retinue and of little value on the battlefield. But in the 15th and 16th centuries the situation changed, as foot soldiers equipped first with longbows and crossbows and then with pikes and matchlock guns achieved dominance over mounted troops in cumbersome armor. The *fanteria,* that is, the *fanti* or foot soldiers collectively, became a significant branch of arms, and the Italian word, in the more Latinate form *infanteria,* was borrowed into French as *infanterie* in the 1500s and later in the same century into English as *infantry.*

3 a : CONTAMINATE, CORRUPT ⟨the inflated writing that *infects* such stories⟩ **b :** to work upon or seize upon so as to induce sympathy, belief, or support ⟨trying to *infect* their salespeople with their enthusiasm⟩
— **in·fec·tor** \-ˈfek-tər\ *noun*

in·fec·tion \in-ˈfek-shən\ *noun* (14th century)
1 : the act or result of affecting injuriously
2 : an infective agent or material contaminated with an infective agent
3 a : the state produced by the establishment of an infective agent in or on a suitable host **b :** a disease resulting from infection
4 : an act or process of infecting; *also* **:** the establishment of a pathogen in its host after invasion
5 : the communication of emotions or qualities through example or contact

in·fec·tious \-shəs\ *adjective* (1542)
1 a : capable of causing infection **b :** communicable by infection — compare CONTAGIOUS
2 : that corrupts or contaminates
3 : spreading or capable of spreading rapidly to others ⟨their enthusiasm was *infectious*⟩ ⟨an *infectious* grin⟩
— **in·fec·tious·ly** *adverb*
— **in·fec·tious·ness** *noun*

infectious hepatitis *noun* (circa 1941)
: HEPATITIS A

infectious mononucleosis *noun* (1920)
: an acute infectious disease associated with Epstein-Barr virus and characterized by fever, swelling of lymph nodes, and lymphocytosis

in·fec·tive \in-ˈfek-tiv\ *adjective* (14th century)
1 : producing or capable of producing infection
2 : affecting others **:** INFECTIOUS
— **in·fec·tiv·i·ty** \(ˌ)in-ˌfek-ˈti-və-tē\ *noun*

in·fe·lic·i·tous \ˌin-fi-ˈli-sə-təs\ *adjective* (1835)
: not appropriate in application or expression
— **in·fe·lic·i·tous·ly** *adverb*

in·fe·lic·i·ty \-sə-tē\ *noun, plural* **-ties** [Middle English *infelicite* unhappiness, from Latin *infelicitas*, from *infelic-, infelix* unhappy, from *in-* + *felic-, felix* fruitful — more at FEMININE] (1617)
1 : the quality or state of being infelicitous
2 : something that is infelicitous

in·fer \in-ˈfər\ *verb* **in·ferred; in·fer·ring** [Middle French or Latin; Middle French *inferer*, from Latin *inferre*, literally, to carry or bring into, from *in-* + *ferre* to carry — more at BEAR] (1528)
transitive verb
1 : to derive as a conclusion from facts or premises ⟨we see smoke and *infer* fire —L. A. White⟩ — compare IMPLY
2 : GUESS, SURMISE ⟨your letter . . . allows me to *infer* that you are as well as ever —O. W. Holmes (died 1935)⟩
3 a : to involve as a normal outcome of thought **b :** to point out **:** INDICATE ⟨this doth *infer* the zeal I had to see him —Shakespeare⟩
4 : SUGGEST, HINT ⟨another survey . . . *infers* that two-thirds of all present computer installations are not paying for themselves —H. R. Chellman⟩
intransitive verb
: to draw inferences ⟨men . . . have observed, *inferred*, and reasoned . . . to all kinds of results —John Dewey⟩ ☆ ▪
— **in·fer·able** *also* **in·fer·ri·ble** \in-ˈfər-ə-bəl\ *adjective*
— **in·fer·rer** \-ˈfər-ər\ *noun*

in·fer·ence \ˈin-f(ə-)rən(t)s, -fərn(t)s\ *noun* (1594)
1 : the act or process of inferring: as **a :** the act of passing from one proposition, statement, or judgment considered as true to another whose truth is believed to follow from that of the former **b :** the act of passing from statistical sample data to generalizations (as of the value of population parameters) usually with calculated degrees of certainty
2 : something that is inferred; *especially* **:** a proposition arrived at by inference
3 : the premises and conclusion of a process of inferring

in·fer·en·tial \ˌin-fə-ˈren(t)-shəl\ *adjective* [Medieval Latin *inferentia*, from Latin *inferent-, inferens*, present participle of *inferre*] (1657)
1 : relating to, involving, or resembling inference
2 : deduced or deducible by inference

in·fer·en·tial·ly \-ˈren(t)-sh(ə-)lē\ *adverb* (1691)
: by way of inference **:** through inference

in·fe·ri·or \in-ˈfir-ē-ər\ *adjective* [Middle English, from Latin, comparative of *inferus* lower — more at UNDER] (15th century)
1 : situated lower down **:** LOWER
2 a : of low or lower degree or rank **b :** of poor quality **:** MEDIOCRE
3 : of little or less importance, value, or merit ⟨always felt *inferior* to his older brother⟩
4 a : situated below another and especially another similar superior part of an upright body **b :** situated in a relatively low posterior or ventral position in a quadrupedal body **c** (1) **:** situated below another plant part or organ (2) **:** ABAXIAL
5 : relating to or being a subscript
— **inferior** *noun*
— **in·fe·ri·or·i·ty** \(ˌ)in-ˌfir-ē-ˈȯr-ə-tē, -ˈär-\ *noun*
— **in·fe·ri·or·ly** \in-ˈfir-ē-ər-lē\ *adverb*

inferior conjunction *noun* (1833)
: a conjunction of an inferior planet with the sun in which the planet is aligned between the earth and the sun

inferiority complex *noun* (1922)
1 : an acute sense of personal inferiority often resulting either in timidity or through overcompensation in exaggerated aggressiveness
2 : a collective sense of cultural, regional, or national inferiority

inferior planet *noun* (1658)
: either of the planets Mercury and Venus whose orbits lie within that of the earth

in·fer·nal \in-ˈfər-nᵊl\ *adjective* [Middle English, from Middle French, from Late Latin *infernalis*, from *infernus* hell, from Latin, lower, from *inferus*] (14th century)
1 : of or relating to a nether world of the dead
2 a : of or relating to hell **b :** HELLISH, DIABOLICAL
3 : DAMNABLE ⟨an *infernal* nuisance⟩
— **in·fer·nal·ly** \-nᵊl-ē\ *adverb*

infernal machine *noun* (1810)
: a machine or apparatus maliciously designed to explode and destroy life or property; *especially* **:** a concealed or disguised bomb

in·fer·no \in-ˈfər-(ˌ)nō\ *noun, plural* **-nos** [Italian, hell, from Late Latin *infernus*] (1834)
: a place or a state that resembles or suggests hell ⟨the *inferno* of war⟩; *also* **:** intense heat ⟨the roaring *inferno* of the blast furnace⟩

in·fer·tile \(ˈ)in-ˈfər-tᵊl\ *adjective* [Middle French, from Late Latin *infertilis*, from Latin *in-* + *fertilis* fertile] (1597)
: not fertile or productive ⟨*infertile* eggs⟩ ⟨*infertile* fields⟩
— **in·fer·til·i·ty** \ˌin-(ˌ)fər-ˈti-lə-tē\ *noun*

in·fest \in-ˈfest\ *transitive verb* [French *infester*, from Latin *infestare*, from *infestus* hostile] (1602)
1 : to spread or swarm in or over in a troublesome manner ⟨a slum *infested* with crime⟩ ⟨shark-*infested* waters⟩
2 : to live in or on as a parasite
— **in·fes·tant** \-ˈfes-tənt\ *noun*
— **in·fes·ta·tion** \ˌin-ˌfes-ˈtā-shən\ *noun*
— **in·fest·er** \in-ˈfes-tər\ *noun*

in·fi·del \ˈin-fə-dᵊl, -fə-ˌdel\ *noun* [Middle English *infidele*, from Middle French, from Late Latin *infidelis* unbelieving, from Latin, unfaithful, from *in-* + *fidelis* faithful — more at FIDELITY] (15th century)
1 : one who is not a Christian or who opposes Christianity
2 a : an unbeliever with respect to a particular religion **b :** one who acknowledges no religious belief
3 : a disbeliever in something specified or understood
— **infidel** *adjective*

in·fi·del·i·ty \ˌin-fə-ˈde-lə-tē, -(ˌ)fī-\ *noun, plural* **-ties** (15th century)
1 : lack of belief in a religion
2 a : unfaithfulness to a moral obligation **:** DISLOYALTY **b :** marital unfaithfulness or an instance of it

in·field \ˈin-ˌfēld\ *noun* (1606)
1 : a field near a farmhouse
2 a : the area of a baseball field enclosed by the three bases and home plate **b :** the defensive positions comprising first base, second base, shortstop, and third base; *also* **:** the players who play these positions
3 : the area enclosed by a racetrack or running track

in·field·er \-ˌfēl-dər\ *noun* (1867)
: a baseball player who plays in the infield

infield hit *noun* (1912)
: a base hit on a ball that does not leave the infield

infield out *noun* (1926)
: a ground ball on which the batter is put out by an infielder

in·fight·ing \ˈin-ˌfī-tiŋ\ *noun* (1816)

☆ SYNONYMS

Infer, deduce, conclude, judge, gather mean to arrive at a mental conclusion. INFER implies arriving at a conclusion by reasoning from evidence; if the evidence is slight, the term comes close to *surmise* ⟨from that remark, I *inferred* that they knew each other⟩. DEDUCE often adds to INFER the special implication of drawing a particular inference from a generalization ⟨denied we could *deduce* anything important from human mortality⟩. CONCLUDE implies arriving at a necessary inference at the end of a chain of reasoning ⟨*concluded* that only the accused could be guilty⟩. JUDGE stresses a weighing of the evidence on which a conclusion is based ⟨*judge* people by their actions⟩. GATHER suggests an intuitive forming of a conclusion from implications ⟨*gathered* their desire to be alone without a word⟩.

□ USAGE

infer Sir Thomas More is the first writer known to have used both *infer* and *imply* in their approved senses (1528). He is also the first to have used *infer* in a sense close in meaning to *imply* (1533). Both of these uses of *infer* coexisted without comment until some time around the end of World War I. Since then, senses 3 and 4 of *infer* have been frequently condemned as an undesirable blurring of a useful distinction. The actual blurring has been done by the commentators. Sense 3, descended from More's use of 1533, does not occur with a personal subject. When objections arose, they were to a use with a personal subject (now sense 4). Since dictionaries did not recognize this use specifically, the objectors assumed that sense 3 was the one they found illogical, even though it had been in respectable use for four centuries. The actual usage condemned was a spoken one never used in logical discourse. At present sense 4 is found in print chiefly in letters to the editor and other informal prose, not in serious intellectual writing. The controversy over sense 4 has apparently reduced the frequency of use of sense 3.

1 : fighting or boxing at close quarters
2 : rough-and-tumble fighting
3 : prolonged and often bitter dissension or rivalry among members of a group or organization ⟨bureaucratic *infighting*⟩
— **in·fight** \ˈin-ˌfīt\ *intransitive verb*
— **in·fight·er** \-ˌfī-tər\ *noun*

in·fil·trate \in-ˈfil-ˌtrāt, ˈin-(ˌ)\ *verb* **-trat·ed; -trat·ing** (1758)
transitive verb
1 : to cause (as a liquid) to permeate something by penetrating its pores or interstices
2 : to pass into or through (a substance) by filtering or permeating
3 : to pass (troops) singly or in small groups through gaps in the enemy line
4 : to enter or become established in gradually or unobtrusively usually for subversive purposes ⟨the intelligence staff had been *infiltrated* by spies⟩
intransitive verb
: to enter, permeate, or pass through a substance or area by filtering or by insinuating gradually
— **infiltrate** *noun*
— **in·fil·tra·tion** \ˌin-(ˌ)fil-ˈtrā-shən\ *noun*
— **in·fil·tra·tive** \ˈin-(ˌ)fil-ˌtrā-tiv, in-ˈfil-trə-\ *adjective*
— **in·fil·tra·tor** \in-ˈfil-ˌtrā-tər, ˈin-(ˌ)\ *noun*

[1]**in·fi·nite** \ˈin-fə-nət\ *adjective* [Middle English *infinit,* from Middle French or Latin; Middle French, from Latin *infinitus,* from *in-* + *finitus* finite] (14th century)
1 : extending indefinitely **:** ENDLESS ⟨*infinite* space⟩
2 : immeasurably or inconceivably great or extensive **:** INEXHAUSTIBLE ⟨*infinite* patience⟩
3 : subject to no limitation or external determination
4 a : extending beyond, lying beyond, or being greater than any preassigned finite value however large ⟨*infinite* number of positive numbers⟩ **b :** extending to infinity ⟨*infinite* plane surface⟩ **c :** characterized by an infinite number of elements or terms ⟨an *infinite* set⟩ ⟨an *infinite* series⟩
— **in·fi·nite·ly** *adverb*
— **in·fi·nite·ness** *noun*

[2]**infinite** *noun* (1535)
: something that is infinite (as in extent, duration, or number)

[1]**in·fin·i·tes·i·mal** \(ˌ)in-ˌfi-nə-ˈte-sə-məl, -zə-məl\ *noun* [New Latin *infinitesimus* infinite in rank, from Latin *infinitus*] (1706)
: an infinitesimal quantity or variable

[2]**infinitesimal** *adjective* (1710)
1 : taking on values arbitrarily close to but greater than zero
2 : immeasurably or incalculably small
— **in·fin·i·tes·i·mal·ly** \-mə-lē\ *adverb*

infinitesimal calculus *noun* (1801)
: CALCULUS 1b

in·fin·i·ti·val \(ˌ)in-ˌfi-nə-ˈtī-vəl\ *adjective* (1869)
: relating to the infinitive

[1]**in·fin·i·tive** \in-ˈfi-nə-tiv\ *adjective* [Middle English *infinityf,* from Late Latin *infinitivus,* from Latin *infinitus*] (15th century)
: formed with the infinitive
— **in·fin·i·tive·ly** *adverb*

[2]**infinitive** *noun* (1530)
: a verb form normally identical in English with the first person singular that performs some functions of a noun and at the same time displays some characteristics of a verb and that is used with *to* (as in "I asked him *to go*") except with auxiliary and various other verbs (as in "no one saw him *leave*")

in·fin·i·tude \in-ˈfi-nə-ˌtüd, -ˌtyüd\ *noun* (1641)
1 : the quality or state of being infinite **:** INFINITENESS
2 : something that is infinite especially in extent
3 : an infinite number or quantity

in·fin·i·ty \in-ˈfi-nə-tē\ *noun, plural* **-ties** (14th century)
1 a : the quality of being infinite **b :** unlimited extent of time, space, or quantity **:** BOUNDLESSNESS
2 : an indefinitely great number or amount ⟨an *infinity* of stars⟩
3 a : the limit of the value of a function or variable when it tends to become numerically larger than any preassigned finite number **b :** a part of a geometric magnitude that lies beyond any part whose distance from a given reference position is finite ⟨do parallel lines ever meet if they extend to *infinity*⟩ **c :** a transfinite number (as aleph-null)
4 : a distance so great that the rays of light from a point source at that distance may be regarded as parallel

in·firm \in-ˈfərm\ *adjective* [Middle English, from Latin *infirmus,* from *in-* + *firmus* firm] (14th century)
1 : of poor or deteriorated vitality; *especially* **:** feeble from age
2 : weak of mind, will, or character **:** IRRESOLUTE, VACILLATING
3 : not solid or stable **:** INSECURE
synonym see WEAK
— **in·firm·ly** *adverb*

in·fir·ma·ry \in-ˈfərm-rē, -ˈfər-mə-\ *noun, plural* **-ries** (15th century)
: a place where the infirm or sick are lodged for care and treatment

in·fir·mi·ty \in-ˈfər-mə-tē\ *noun, plural* **-ties** (14th century)
1 a : the quality or state of being infirm **b :** the condition of being feeble **:** FRAILTY
2 : DISEASE, MALADY
3 : a personal failing **:** FOIBLE ⟨one of the besetting *infirmities* of living creatures is egotism —A. J. Toynbee⟩

[1]**in·fix** \ˈin-ˌfiks, in-ˈ\ *transitive verb* [Latin *infixus,* past participle of *infigere,* from *in-* + *figere* to fasten — more at FIX] (1502)
1 : to fasten or fix by piercing or thrusting in
2 : to impress firmly in the consciousness or disposition
3 : to insert (as a sound or letter) as an infix
synonym see IMPLANT
— **in·fix·ation** \ˌin-(ˌ)fik-ˈsā-shən\ *noun*

[2]**in·fix** \ˈin-ˌfiks\ *noun* (1881)
: a derivational or inflectional affix appearing in the body of a word (as Sanskrit *-n-* in *vindami* "I know" as contrasted with *vid* "to know")

[3]**in·fix** *same as* [2]\ *adjective* (1971)
: characterized by placement of a binary operator between the operands ⟨*a* + *b* is expressed in *infix* notation⟩ — compare POSTFIX, PREFIX

in fla·gran·te de·lic·to \ˌin-flə-ˌgrän-tē-di-ˈlik-(ˌ)tō, -ˌgran-\ *adverb* (1772)
: FLAGRANTE DELICTO

in·flame \in-ˈflām\ *verb* **in·flamed; in·flam·ing** [Middle English *enflamen,* from Middle French *enflamer,* from Latin *inflammare,* from *in-* + *flamma* flame] (14th century)
transitive verb
1 a : to excite to excessive or uncontrollable action or feeling **b :** to make more heated or violent **:** INTENSIFY ⟨insults served only to *inflame* the feud⟩
2 : to set on fire **:** KINDLE
3 : to cause to redden or grow hot from anger or excitement
4 : to cause inflammation in (bodily tissue)
intransitive verb
1 : to burst into flame
2 : to become excited or angered
3 : to become affected with inflammation
— **in·flam·er** *noun*

in·flam·ma·ble \in-ˈfla-mə-bəl\ *adjective* [French, from Medieval Latin *inflammabilis,* from Latin *inflammare*] (1605)
1 : FLAMMABLE
2 : easily inflamed, excited, or angered **:** IRASCIBLE
— **in·flam·ma·bil·i·ty** \-ˌfla-mə-ˈbi-lə-tē\ *noun*
— **inflammable** *noun*
— **in·flam·ma·ble·ness** \-ˈfla-mə-bəl-nəs\ *noun*
— **in·flam·ma·bly** \-blē\ *adverb*

in·flam·ma·tion \ˌin-flə-ˈmā-shən\ *noun* (15th century)
1 : a local response to cellular injury that is marked by capillary dilatation, leukocytic infiltration, redness, heat, and pain and that serves as a mechanism initiating the elimination of noxious agents and of damaged tissue
2 : the act of inflaming **:** the state of being inflamed

in·flam·ma·to·ry \in-ˈfla-mə-ˌtōr-ē, -ˌtȯr-\ *adjective* (circa 1711)
1 : tending to excite anger, disorder, or tumult **:** SEDITIOUS
2 : tending to inflame or excite the senses
3 : accompanied by or tending to cause inflammation
— **in·flam·ma·to·ri·ly** \-ˌfla-mə-ˈtōr-ə-lē, -ˈtȯr-\ *adverb*

in·flat·able \in-ˈflā-tə-bəl\ *adjective* (1878)
: capable of being inflated ⟨an *inflatable* boat⟩
— **inflatable** *noun*

in·flate \in-ˈflāt\ *verb* **in·flat·ed; in·flat·ing** [Middle English, from Latin *inflatus,* past participle of *inflare,* from *in-* + *flare* to blow — more at BLOW] (15th century)
transitive verb
1 : to swell or distend with air or gas
2 : to puff up **:** ELATE
3 : to expand or increase abnormally or imprudently
intransitive verb
: to become inflated
synonym see EXPAND
— **in·fla·tor** *or* **in·flat·er** \-ˈflā-tər\ *noun*

inflated *adjective* (1652)
1 : elaborated or heightened by artificial or empty means ⟨an *inflated* style of writing⟩
2 : distended with air or gas
3 : expanded to an abnormal or unjustifiable volume or level ⟨*inflated* prices⟩
4 : being hollow and enlarged or distended

in·fla·tion \in-ˈflā-shən\ *noun* (14th century)
1 : an act of inflating **:** a state of being inflated: as **a :** DISTENSION **b :** a hypothetical extremely brief period of very rapid expansion of the universe immediately following the big bang **c :** empty pretentiousness **:** POMPOSITY
2 : an increase in the volume of money and credit relative to available goods and services resulting in a continuing rise in the general price level

in·fla·tion·ary \-shə-ˌner-ē\ *adjective* (1920)
: of, characterized by, or productive of inflation

inflationary spiral *noun* (1931)
: a continuous rise in prices that is sustained by the tendency of wage increases and cost increases to react on each other

in·fla·tion·ism \in-ˈflā-shə-ˌni-zəm\ *noun* (1919)
: the policy of economic inflation
— **in·fla·tion·ist** \-sh(ə-)nist\ *noun or adjective*

in·flect \in-ˈflekt\ *verb* [Middle English, from Latin *inflectere,* from *in-* + *flectere* to bend] (15th century)
transitive verb
1 : to turn from a direct line or course **:** CURVE
2 : to vary (a word) by inflection **:** DECLINE, CONJUGATE
3 : to change or vary the pitch of (as the voice)
intransitive verb
: to become modified by inflection
— **in·flect·able** \-ˈflek-tə-bəl\ *adjective*
— **in·flec·tive** \-ˈflek-tiv\ *adjective*

in·flec·tion \in-ˈflek-shən\ *noun* (1531)
1 : the act or result of curving or bending **:** BEND
2 : change in pitch or loudness of the voice
3 a : the change of form that words undergo to mark such distinctions as those of case, gender, number, tense, person, mood, or voice **b :** a form, suffix, or element involved in such variation **c :** ACCIDENCE
4 a : change in curvature of an arc or curve from concave to convex or conversely **b :** INFLECTION POINT

in·flec·tion·al \-shnəl, -shə-nᵊl\ *adjective* (1832)
: of, relating to, or characterized by inflection ⟨an *inflectional* suffix⟩
— in·flec·tion·al·ly *adverb*

inflection point *noun* (circa 1721)
: a point on a curve that separates an arc concave upward from one concave downward and vice versa

in·flexed \ˈin-ˌflekst\ *adjective* [Latin *inflexus,* past participle of *inflectere*] (1661)
: bent or turned abruptly inward or downward or toward the axis ⟨*inflexed* petals⟩

in·flex·i·ble \(ˌ)in-ˈflek-sə-bəl\ *adjective* [Middle English, from Latin *inflexibilis,* from *in-* + *flexibilis* flexible] (14th century)
1 : rigidly firm in will or purpose **:** UNYIELDING
2 : not readily bent **:** lacking or deficient in suppleness
3 : incapable of change **:** UNALTERABLE ☆
— in·flex·i·bil·i·ty \-ˌflek-sə-ˈbi-lə-tē\ *noun*
— in·flex·i·ble·ness \-ˈflek-sə-bəl-nəs\ *noun*
— in·flex·i·bly \-blē\ *adverb*

in·flex·ion *chiefly British variant of* INFLECTION

in·flict \in-ˈflikt\ *transitive verb* [Latin *inflictus,* past participle of *infligere,* from *in-* + *fligere* to strike — more at PROFLIGATE] (1566)
1 : AFFLICT
2 a : to give by or as if by striking ⟨*inflict* pain⟩ **b :** to cause (something unpleasant) to be endured ⟨*inflict* my annual message upon the church —Mark Twain⟩
— in·flict·er *or* **in·flic·tor** \-ˈflik-tər\ *noun*
— in·flic·tive \-tiv\ *adjective*

in·flic·tion \in-ˈflik-shən\ *noun* (1534)
1 : the act of inflicting
2 : something (as punishment or suffering) that is inflicted

in–flight \ˈin-ˈflīt, (ˌ)in-\ *adjective* (1944)
: made, carried out, or provided for use or enjoyment while in flight ⟨*in-flight* movies⟩

in·flo·res·cence \ˌin-flə-ˈre-sᵊn(t)s\ *noun* [New Latin *inflorescentia,* from Late Latin *inflorescent-, inflorescens,* present participle of *inflorescere* to begin to bloom, from Latin *in-* + *florescere* to begin to bloom — more at FLORESCENCE] (1760)
1 a (1) **:** the mode of development and arrangement of flowers on an axis (2) **:** a floral axis with its appendages; *also* **:** a flower cluster **b :** a cluster of reproductive organs on a moss usually subtended by a bract
2 : the budding and unfolding of blossoms **:** FLOWERING

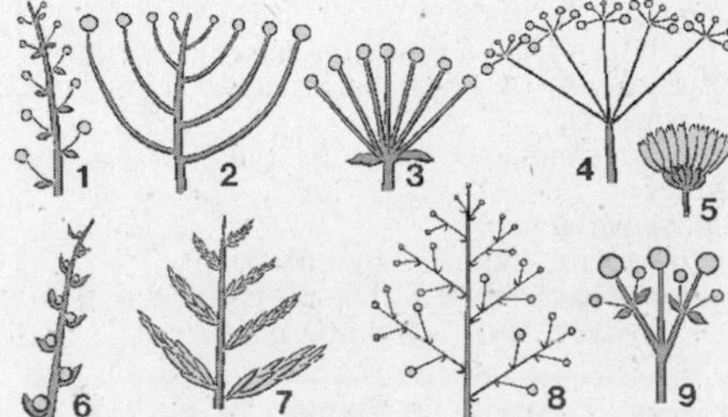

inflorescence 1a(1): ***1*** **raceme,** ***2*** **corymb,** ***3*** **umbel,** ***4*** **compound umbel,** ***5*** **capitulum,** ***6*** **spike,** ***7*** **compound spike,** ***8*** **panicle,** ***9*** **cyme**

in·flow \ˈin-ˌflō\ *noun* (1839)
: a flowing in ⟨the *inflow* of air⟩ ⟨an *inflow* of funds⟩

¹in·flu·ence \ˈin-ˌflü-ən(t)s, *especially Southern* in-ˈ\ *noun* [Middle English, from Middle French, from Medieval Latin *influentia,* from Latin *influent-, influens,* present participle of *influere* to flow in, from *in-* + *fluere* to flow — more at FLUID] (14th century)
1 a : an ethereal fluid held to flow from the stars and to affect the actions of humans **b :** an emanation of occult power held to derive from stars
2 : an emanation of spiritual or moral force
3 a : the act or power of producing an effect without apparent exertion of force or direct exercise of command **b :** corrupt interference with authority for personal gain
4 : the power or capacity of causing an effect in indirect or intangible ways **:** SWAY
5 : one that exerts influence ☆
— under the influence : affected by alcohol **:** DRUNK ⟨was arrested for driving *under the influence*⟩

²influence *transitive verb* **-enced; -enc·ing** (1658)
1 : to affect or alter by indirect or intangible means **:** SWAY
2 : to have an effect on the condition or development of **:** MODIFY
synonym see AFFECT
— in·flu·ence·able \-ən(t)-sə-bəl\ *adjective*

¹in·flu·ent \ˈin-ˌflü-ənt, in-ˈ\ *adjective* (15th century)
: flowing in

²influent *noun* (1859)
1 : something that flows in: as **a :** a tributary stream **b :** fluid input into a reservoir or process
2 : a factor (as a particular animal) modifying the balance and stability of an ecological community

¹in·flu·en·tial \ˌin-(ˌ)flü-ˈen(t)-shəl\ *adjective* (1570)
: exerting or possessing influence
— in·flu·en·tial·ly \-ˈen(t)-sh(ə-)lē\ *adverb*

²influential *noun* (1831)
: one who has great influence

in·flu·en·za \ˌin-(ˌ)flü-ˈen-zə\ *noun* [Italian, literally, influence, from Medieval Latin *influentia;* from the belief that epidemics were due to the influence of the stars] (1743)
1 : an acute highly contagious virus disease caused by various strains of a myxovirus (family Orthomyxoviridae) and characterized by sudden onset, fever, prostration, severe aches and pains, and progressive inflammation of the respiratory mucous membrane; *broadly* **:** a human respiratory infection of undetermined cause
2 : any of numerous febrile usually virus diseases of domestic animals marked by respiratory symptoms, inflammation of mucous membranes, and often systemic involvement ◆
— in·flu·en·zal \-zəl\ *adjective*

in·flux \ˈin-ˌfləks\ *noun* [Medieval Latin *influxus,* from Latin *influere*] (1626)
: a coming in ⟨an *influx* of tourists⟩

in·fo \ˈin-(ˌ)fō\ *noun* (1913)
: INFORMATION

in·fold \in-ˈfōld\ (15th century)
transitive verb
: ENFOLD, ENVELOP
intransitive verb
: to fold inward or toward one another

in·fo·mer·cial \ˈin-(ˌ)fō-ˌmər-shəl, -fə-\ *noun* [*information* + ²*commercial*] (1981)
: a television program that is an extended advertisement often including a discussion or demonstration

in·form \in-ˈfȯrm\ *verb* [Middle English, from Middle French *enformer,* from Latin *informare,* from *in-* + *forma* form] (14th century)
transitive verb
1 *obsolete* **:** to give material form to
2 a : to give character or essence to ⟨the principles which *inform* modern teaching⟩ **b :** to be the characteristic quality of **:** ANIMATE ⟨the compassion that *informs* her work⟩
3 *obsolete* **:** GUIDE, DIRECT
4 *obsolete* **:** to make known
5 : to communicate knowledge to ⟨*inform* a prisoner of his rights⟩
intransitive verb
1 : to impart information or knowledge
2 : to give information (as of another's wrongdoing) to an authority ☆

☆ SYNONYMS

Inflexible, obdurate, adamant mean unwilling to alter a predetermined course or purpose. INFLEXIBLE implies rigid adherence or even slavish conformity to principle ⟨*inflexible* in their demands⟩. OBDURATE stresses hardness of heart and insensitivity to appeals for mercy or the influence of divine grace ⟨*obdurate* in his refusal to grant clemency⟩. ADAMANT implies utter immovability in the face of all temptation or entreaty ⟨*adamant* that the work should continue⟩. See in addition STIFF.

Influence, authority, prestige, weight, credit mean power exerted over the minds or behavior of others. INFLUENCE may apply to a force exercised and received consciously or unconsciously ⟨used her *influence* to get the bill passed⟩. AUTHORITY implies the power of winning devotion or allegiance or of compelling acceptance and belief ⟨his opinions lacked *authority*⟩. PRESTIGE implies the ascendancy given by conspicuous excellence or reputation for superiority ⟨the *prestige* of the newspaper⟩. WEIGHT implies measurable or decisive influence in determining acts or choices ⟨their wishes obviously carried much *weight*⟩. CREDIT suggests influence that arises from the confidence of others ⟨his *credit* with the press⟩.

◇ WORD HISTORY

influenza The Italian word *influenza* was, like English *influence,* originally an astrological term. The effect that the stars and planets were assumed to have on humans was ascribed in a sometimes literal, sometimes metaphorical way to the "inflow" of some ethereal liquid from the heavens. Also like the English word, *influenza* came to denote any sort of effect produced without seeming exertion by its source, but the Italian word in the Middle Ages was also applied more narrowly to outbreaks of disease supposedly brought about by unusual conjunctions of the planets. In the 17th and 18th centuries, the scope of *influenza* narrowed still further, to refer specifically to the infectious disease we now call by this name. The word first appeared in English through reports of an influenza outbreak in Italy during the winter of 1743, and it reappeared in reports of the major European epidemics of 1761–62 and 1802–03, when the disease also reached Britain. In 1933, after the great influenza pandemic of 1918–19 had killed half a million people in the U.S., American and British researchers identified the myxovirus that causes the disease.

in·for·mal \(ˌ)in-ˈfȯr-məl\ *adjective* (1585)
1 : marked by the absence of formality or ceremony ⟨an *informal* meeting⟩ ⟨an *informal* group⟩
2 : characteristic of or appropriate to ordinary, casual, or familiar use ⟨*informal* English⟩ ⟨*informal* clothes⟩
— **in·for·mal·i·ty** \-(ˌ)fȯr-ˈma-lə-tē, -fər-\ *noun*
— **in·for·mal·ly** \-ˈfȯr-mə-lē\ *adverb*

in·for·mant \in-ˈfȯr-mənt\ *noun* (1693)
: a person who gives information: as **a :** INFORMER **b :** one who supplies cultural or linguistic data in response to interrogation by an investigator

in for·ma pau·pe·ris \ˌin-ˈfȯr-mə-ˈpȯ-pə-rəs, -ˈpau̇-\ *adjective or adverb* [Latin, in the form of a pauper] (1592)
: as a poor person

in·for·mat·ics \ˌin-fər-ˈma-tiks\ *noun plural but singular in construction* [International Scientific Vocabulary *informat*ion + *-ics*] (circa 1967)
chiefly British **:** INFORMATION SCIENCE

in·for·ma·tion \ˌin-fər-ˈmā-shən\ *noun* (14th century)
1 : the communication or reception of knowledge or intelligence
2 a (1) **:** knowledge obtained from investigation, study, or instruction (2) **:** INTELLIGENCE, NEWS (3) **:** FACTS, DATA **b :** the attribute inherent in and communicated by one of two or more alternative sequences or arrangements of something (as nucleotides in DNA or binary digits in a computer program) that produce specific effects **c** (1) **:** a signal or character (as in a communication system or computer) representing data (2) **:** something (as a message, experimental data, or a picture) which justifies change in a construct (as a plan or theory) that represents physical or mental experience or another construct **d :** a quantitative measure of the content of information; *specifically* **:** a numerical quantity that measures the uncertainty in the outcome of an experiment to be performed
3 : the act of informing against a person
4 : a formal accusation of a crime made by a prosecuting officer as distinguished from an indictment presented by a grand jury
— **in·for·ma·tion·al** \-shnəl, -shə-nᵊl\ *adjective*
— **in·for·ma·tion·al·ly** *adverb*

information retrieval *noun* (1950)
: the techniques of storing and recovering and often disseminating recorded data especially through the use of a computerized system

information science *noun* (1960)
: the collection, classification, storage, retrieval, and dissemination of recorded knowledge treated both as a pure and as an applied science

information theory *noun* (1950)
: a theory that deals statistically with information, with the measurement of its content in terms of its distinguishing essential characteristics or by the number of alternatives from which it makes a choice possible, and with the efficiency of processes of communication between humans and machines

in·for·ma·tive \in-ˈfȯr-mə-tiv\ *adjective* (1655)
: imparting knowledge **:** INSTRUCTIVE
— **in·for·ma·tive·ly** *adverb*
— **in·for·ma·tive·ness** *noun*

in·for·ma·to·ry \-mə-ˌtōr-ē, -ˌtȯr-\ *adjective* (circa 1879)
: conveying information
— **in·for·ma·to·ri·ly** \in-ˌfȯr-mə-ˈtōr-ə-lē, -ˈtȯr-\ *adverb*

in·formed \in-ˈfȯrmd\ *adjective* (15th century)
1 a : having information ⟨*informed* sources⟩ ⟨*informed* observers⟩ **b :** based on possession of information ⟨an *informed* opinion⟩
2 : EDUCATED, KNOWLEDGEABLE ⟨what the *informed* person should know⟩
— **in·formed·ly** \-ˈfȯrmd-lē, -ˈfȯr-məd-lē\ *adverb*

informed consent *noun* (circa 1967)
: consent to surgery by a patient or to participation in a medical experiment by a subject after achieving an understanding of what is involved

in·form·er \in-ˈfȯr-mər\ *noun* (14th century)
1 : one that imparts knowledge or news
2 : one that informs against another; *specifically* **:** one who makes a practice especially for a financial reward of informing against others for violations of penal laws

in·fo·tain·ment \ˌin-(ˌ)fō-ˈtān-mənt\ *noun* [*in*formation + enter*tainment*] (1982)
: a television program that presents information (as news) in a manner intended to be entertaining

in·fra \ˈin-frə *also* -ˌfrä\ *adverb* [Latin] (circa 1740)
: later in this writing **:** BELOW ⟨for additional examples see *infra*⟩

infra- *prefix* [Latin *infra* — more at UNDER]
1 : below ⟨*infra*human⟩ ⟨*infra*sonic⟩
2 : within ⟨*infra*specific⟩
3 : below in a scale or series ⟨*infra*red⟩

in·frac·tion \in-ˈfrak-shən\ *noun* [Middle English, from Medieval Latin *infraction-, infractio,* from Latin, subduing, from *infringere* to break — more at INFRINGE] (15th century)
: the act of infringing **:** VIOLATION
— **in·fract** \in-ˈfrakt\ *transitive verb*

in·fra dig \ˈin-frə-ˈdig, -(ˌ)frä-\ *adjective* [short for Latin *infra dignitatem*] (1824)
: being beneath one's dignity **:** UNDIGNIFIED ⟨while his work . . . was financially profitable, it was just a bit *infra dig* —John McCarten⟩

in·fra·hu·man \ˌin-frə-ˈhyü-mən, -(ˌ)frä-, -ˈyü-\ *adjective* (1847)
: less or lower than human ⟨*infrahuman* primates⟩
— **infrahuman** *noun*

in·fran·gi·ble \(ˌ)in-ˈfran-jə-bəl\ *adjective* [Middle French, from Late Latin *infrangibilis,* from Latin *in-* + *frangere* to break — more at BREAK] (1597)
1 : not capable of being broken or separated into parts
2 : not to be infringed or violated
— **in·fran·gi·bil·i·ty** \-ˌfran-jə-ˈbi-lə-tē\ *noun*
— **in·fran·gi·bly** \-ˈfran-jə-blē\ *adverb*

in·fra·red \ˌin-frə-ˈred, -(ˌ)frä-, -fə-\ *adjective* (1881)
1 : situated outside the visible spectrum at its red end — used of radiation having a wavelength between about 700 nanometers and 1 millimeter
2 : relating to, producing, or employing infrared radiation ⟨*infrared* therapy⟩
3 : sensitive to infrared radiation ⟨*infrared* photographic film⟩
— **infrared** *noun*

in·fra·son·ic \ˌin-frə-ˈsä-nik, -(ˌ)frä-\ *adjective* (1927)
1 : having or relating to a frequency below the audibility range of the human ear
2 : utilizing or produced by infrasonic waves or vibrations

in·fra·spe·cif·ic \-spi-ˈsi-fik\ *adjective* (1939)
: included within a species ⟨*infraspecific* variability⟩

in·fra·struc·ture \ˈin-frə-ˌstrək-chər, -(ˌ)frä-\ *noun* (1927)
1 : the underlying foundation or basic framework (as of a system or organization)
2 : the permanent installations required for military purposes
3 : the system of public works of a country, state, or region; *also* **:** the resources (as personnel, buildings, or equipment) required for an activity

in·fre·quence \(ˌ)in-ˈfrē-kwən(t)s\ *noun* (1644)
: INFREQUENCY

in·fre·quen·cy \-kwən(t)-sē\ *noun* (1677)
: rarity of occurrence

in·fre·quent \(ˌ)in-ˈfrē-kwənt\ *adjective* [Latin *infrequent-, infrequens,* from *in-* + *frequent-, frequens* frequent] (1531)
1 : seldom happening or occurring **:** RARE
2 : placed or occurring at wide intervals in space or time ⟨a slope scattered with *infrequent* pines⟩ ⟨*infrequent* visits⟩ ☆
— **in·fre·quent·ly** *adverb*

in·fringe \in-ˈfrinj\ *verb* **in·fringed; in·fring·ing** [Medieval Latin *infringere,* from Latin, to break, crush, from *in-* + *frangere* to break — more at BREAK] (1533)
transitive verb
1 : to encroach upon in a way that violates law or the rights of another ⟨*infringe* a patent⟩
2 *obsolete* **:** DEFEAT, FRUSTRATE
intransitive verb
: ENCROACH — used with *on* or *upon* ⟨*infringe* on our rights⟩
synonym see TRESPASS
— **in·fring·er** *noun*

in·fringe·ment \in-ˈfrinj-mənt\ *noun* (1628)
1 : the act of infringing **:** VIOLATION
2 : an encroachment or trespass on a right or privilege

in·fun·dib·u·lar \ˌin-(ˌ)fən-ˈdi-byə-lər\ *adjective* (1795)
1 : INFUNDIBULIFORM
2 : of, relating to, or having an infundibulum

in·fun·dib·u·li·form \-lə-ˌfȯrm\ *adjective* [New Latin *infundibulum* + English *-iform*] (1752)
: having the form of a funnel or cone

in·fun·dib·u·lum \ˌin-(ˌ)fən-ˈdi-byə-ləm\ *noun, plural* **-la** \-lə\ [New Latin, from Latin, funnel — more at FUNNEL] (1543)
: any of various funnel-shaped organs or parts: as **a :** the hollow conical process of gray matter by which the pituitary gland is continuous with the brain **b :** the calyx of a kidney **c :** the abdominal opening of a fallopian tube

¹in·fu·ri·ate \in-ˈfyu̇r-ē-ˌāt\ *transitive verb* **-at·ed; -at·ing** [Medieval Latin *infuriatus,*

☆ SYNONYMS
Inform, acquaint, apprise, notify mean to make one aware of something. INFORM implies the imparting of knowledge especially of facts or occurrences ⟨*informed* us of the crisis⟩. ACQUAINT lays stress on introducing to or familiarizing with ⟨*acquaint* yourself with the keyboard⟩. APPRISE implies communicating something of special interest or importance ⟨keep us *apprised* of the situation⟩. NOTIFY implies sending notice of something requiring attention or demanding action ⟨*notified* the witness when to appear⟩.

Infrequent, uncommon, scarce, rare, sporadic mean not common or abundant. INFREQUENT implies occurrence at wide intervals in space or time ⟨*infrequent* family visits⟩. UNCOMMON suggests a frequency below normal expectation ⟨smallpox is now *uncommon* in many countries⟩. SCARCE implies falling short of a standard or required abundance ⟨jobs were *scarce* during the Depression⟩. RARE suggests extreme scarcity or infrequency and often implies consequent high value ⟨*rare* first editions⟩. SPORADIC implies occurrence in scattered instances or isolated outbursts ⟨*sporadic* cases of influenza⟩.

past participle of *infuriare,* from Latin *in-* + *furia* fury] (1667)
: to make furious
— **in·fu·ri·at·ing·ly** \-ˌā-tiŋ-lē\ *adverb*
— **in·fu·ri·a·tion** \-ˌfyu̇r-ē-ˈā-shən\ *noun*

2in·fu·ri·ate \in-ˈfyu̇r-ē-ət\ *adjective* (1667)
: furiously angry

in·fuse \in-ˈfyüz\ *transitive verb* **in·fused; in·fus·ing** [Middle English, to pour in, from Middle French & Latin; Middle French *infuser,* from Latin *infusus,* past participle of *infundere* to pour in, from *in-* + *fundere* to pour — more at FOUND] (1526)
1 a : to cause to be permeated with something (as a principle or quality) that alters usually for the better ⟨*infuse* the team with confidence⟩ **b :** INTRODUCE, INSINUATE ⟨a new spirit was *infused* into American art —*American Guide Series: New York*⟩
2 : INSPIRE, ANIMATE ⟨the sense of purpose that *infuses* scientific research⟩
3 : to steep in liquid (as water) without boiling so as to extract the soluble constituents or principles ☆
— **in·fus·er** *noun*

in·fus·ible \(ˌ)in-ˈfyü-zə-bəl\ *adjective* (1555)
: incapable of being fused **:** very difficult to fuse
— **in·fus·ibil·i·ty** \-ˌfyü-zə-ˈbi-lə-tē\ *noun*
— **in·fus·ible·ness** \-ˈfyü-zə-bəl-nəs\ *noun*

in·fu·sion \in-ˈfyü-zhən\ *noun* (15th century)
1 : the act or process of infusing
2 : a product obtained by infusing
3 : the continuous slow introduction of a solution especially into a vein

in·fu·so·ri·al earth \ˌin-fyü-ˈzōr-ē-əl-, -ˈsōr-, -ˈzȯr-, -ˈsȯr-\ *noun* (1882)
: KIESELGUHR

in·fu·so·ri·an \-ē-ən\ *noun* [ultimately from Latin *infusus*] (1859)
: any of a heterogeneous group of minute organisms found especially in decomposing infusions of organic matter; *especially* **:** a ciliated protozoan
— **infusorian** *adjective*

1-ing \iŋ *also* ēŋ; *in some dialects & in other dialects informally* in, ən *also* ēn; *after certain consonants* ^{ə}n, ^{ə}m, ^{ə}ŋ\ *noun suffix* [Middle English, from Old English *-ung, -ing,* suffix forming nouns from verbs; akin to Old High German *-ung,* suffix forming nouns from verbs]
1 : action or process ⟨runn*ing*⟩ ⟨sleep*ing*⟩ **:** instance of an action or process ⟨a meet*ing*⟩
2 a : product or result of an action or process ⟨an engrav*ing*⟩ — often in plural ⟨earn*ings*⟩ **b :** something used in an action or process ⟨a bed cover*ing*⟩ ⟨the lin*ing* of a coat⟩
3 : action or process connected with (a specified thing) ⟨boat*ing*⟩
4 : something connected with, consisting of, or used in making (a specified thing) ⟨scaffold*ing*⟩ ⟨shirt*ing*⟩
5 : something related to (a specified concept) ⟨off*ing*⟩

2-ing *noun suffix* [Middle English, from Old English *-ing, -ung;* akin to Old High German *-ing* one of a (specified) kind]
: one of a (specified) kind ⟨sweet*ing*⟩

3-ing *verb suffix or adjective suffix* [Middle English, probably from 1*-ing*]
— used to form the present participle ⟨sail*ing*⟩ and sometimes to form an adjective resembling a present participle but not derived from a verb ⟨swashbuckl*ing*⟩ ■

in·gath·er \ˈin-ˌga-t͟hər, -ˌge-\ (1557)
transitive verb
: to gather in
intransitive verb
: ASSEMBLE
— **in·gath·er·ing** \-ˌgath-riŋ, -ˌget͟h-; -ˌga-t͟hə-, -ˌge-\ *noun*

in·ge·nious \in-ˈjēn-yəs\ *adjective* [Middle English *ingenyous,* from Middle French *ingenieux,* from Latin *ingeniosus,* from *ingenium* natural capacity — more at ENGINE] (15th century)
1 *obsolete* **:** showing or calling for intelligence, aptitude, or discernment
2 : marked by especial aptitude at discovering, inventing, or contriving
3 : marked by originality, resourcefulness, and cleverness in conception or execution ⟨an *ingenious* contraption⟩
synonym see CLEVER
— **in·ge·nious·ly** *adverb*
— **in·ge·nious·ness** *noun*

in·ge·nue *or* **in·gé·nue** \ˈan-jə-ˌnü, ˈän-; ˈa^n-zhə-, ˈä^n-\ *noun* [French *ingénue,* feminine of *ingénu* ingenuous, from Latin *ingenuus*] (1848)
1 : a naive girl or young woman
2 : the stage role of an ingenue; *also* **:** an actress playing such a role

in·ge·nu·i·ty \ˌin-jə-ˈnü-ə-tē, -ˈnyü-\ *noun, plural* **-ties** (circa 1592)
1 *obsolete* **:** CANDOR, INGENUOUSNESS
2 a : skill or cleverness in devising or combining **:** INVENTIVENESS **b :** cleverness or aptness of design or contrivance
3 : an ingenious device or contrivance

1in·gen·u·ous \in-ˈjen-yə-wəs\ *adjective* [by alteration] (1588)
obsolete **:** INGENIOUS

2ingenuous *adjective* [Latin *ingenuus* native, free born, from *in-* + *gignere* to beget — more at KIN] (1588)
1 *obsolete* **:** NOBLE, HONORABLE
2 a : showing innocent or childlike simplicity and candidness **b :** lacking craft or subtlety
synonym see NATURAL
— **in·gen·u·ous·ly** *adverb*
— **in·gen·u·ous·ness** *noun*

in·gest \in-ˈjest\ *transitive verb* [Latin *ingestus,* past participle of *ingerere* to carry in, from *in-* + *gerere* to bear] (1620)
: to take in for or as if for digestion
— **in·gest·ible** \-ˈjes-tə-bəl\ *adjective*
— **in·ges·tion** \-ˈjes-chən, -ˈjesh-\ *noun*
— **in·ges·tive** \-ˈjes-tiv\ *adjective*

in·ges·ta \in-ˈjes-tə\ *noun plural* [New Latin, from Latin, neuter plural of *ingestus*] (1727)
: material taken into the body by way of the digestive tract

in·gle \ˈiŋ-gəl, ˈiŋ-əl\ *noun* [Scottish Gaelic *aingeal*] (1508)
1 : a fire in a fireplace
2 : FIREPLACE
3 : CORNER, ANGLE

in·gle·nook \-ˌnu̇k\ *noun* (1772)
: a nook by a large open fireplace; *also* **:** a bench or settle occupying this nook

in·glo·ri·ous \(ˌ)in-ˈglōr-ē-əs, -ˈglȯr-\ *adjective* [Latin *inglorius,* from *in-* + *gloria* glory] (1573)
1 : SHAMEFUL, IGNOMINIOUS
2 : not glorious **:** lacking fame or honor
— **in·glo·ri·ous·ly** *adverb*
— **in·glo·ri·ous·ness** *noun*

in·got \ˈiŋ-gət\ *noun* [Middle English, perhaps modification of Middle French *lingot* ingot of metal, incorrectly divided as *l'ingot,* as if from *le* the, from Latin *ille* that] (14th century)
1 : a mold in which metal is cast
2 : a mass of metal cast into a convenient shape for storage or transportation to be later processed

ingot iron *noun* (1877)
: iron containing only small proportions of impurities (as less than 0.05 percent carbon)

1in·grain \(ˌ)in-ˈgrān\ *transitive verb* (circa 1641)
: to work indelibly into the natural texture or mental or moral constitution
synonym see INFUSE

2in·grain \ˈin-ˌgrān\ *adjective* (1766)
1 a : made of fiber that is dyed before being spun into yarn **b :** made of yarn that is dyed before being woven or knitted
2 : thoroughly worked in **:** INNATE

3in·grain \ˈin-ˌgrān\ *noun* (circa 1890)
: innate quality or character

in·grained \ˈin-ˌgrānd, (ˌ)in-ˈ\ *adjective* (1599)
1 : worked into the grain or fiber
2 : forming a part of the essence or inmost being **:** DEEP-SEATED ⟨*ingrained* prejudice⟩
— **in·grained·ly** \ˈin-ˌgrā-nəd-lē, ˈin-ˌgrānd-lē, (ˌ)in-ˈ\ *adverb*

in·grate \ˈin-ˌgrāt\ *noun* [Latin *ingratus* ungrateful, from *in-* + *gratus* grateful — more at GRACE] (1622)
: an ungrateful person

in·gra·ti·ate \in-ˈgrā-shē-ˌāt\ *transitive verb* **-at·ed; -at·ing** [2*in-* + Latin *gratia* grace] (1622)
: to gain favor or favorable acceptance for by deliberate effort — usually used with *with* ⟨*ingratiate* themselves with the community leaders —William Attwood⟩
— **in·gra·ti·a·tion** \-ˌgrā-shē-ˈā-shən\ *noun*
— **in·gra·tia·to·ry** \-ˈgrā-sh(ē-)ə-ˌtōr-ē, -ˌtȯr-\ *adjective*

ingratiating *adjective* (1655)
1 : capable of winning favor **:** PLEASING ⟨an *ingratiating* smile⟩
2 : intended or adopted in order to gain favor **:** FLATTERING
— **in·gra·ti·at·ing·ly** \-ˈgrā-shē-ˌā-tiŋ-lē\ *adverb*

in·grat·i·tude \(ˌ)in-ˈgra-tə-ˌtüd, -ˌtyüd\ *noun* [Middle English, from Middle French, from Medieval Latin *ingratitudo,* from Latin *in-* + Late Latin *gratitudo* gratitude] (14th century)
: forgetfulness of or poor return for kindness received **:** UNGRATEFULNESS

☆ SYNONYMS
Infuse, suffuse, imbue, ingrain, inoculate, leaven mean to introduce one thing into another so as to affect it throughout. INFUSE implies a pouring in of something that gives new life or significance ⟨new members *infused* enthusiasm into the club⟩. SUFFUSE implies a spreading through of something that gives an unusual color or quality ⟨a room *suffused* with light⟩. IMBUE implies the introduction of a quality that fills and permeates the whole being ⟨*imbue* students with intellectual curiosity⟩. INGRAIN, used only in the passive or past participle, suggests the deep implanting of a quality or trait ⟨clung to *ingrained* habits⟩. INOCULATE implies an imbuing or implanting with a germinal idea and often suggests surreptitiousness or subtlety ⟨an electorate *inoculated* with dangerous ideas⟩. LEAVEN implies introducing something that enlivens, tempers, or markedly alters the total quality ⟨a serious play *leavened* with comic moments⟩.

□ USAGE
-ing Though the pronunciation of *-ing* with the consonant \n\, misleadingly referred to as "dropping the *g,*" is often deprecated, this pronunciation is frequently heard. It is not known for certain why the Middle English present participle ending *-ende* was replaced by *-ing.* Analogy with the earlier noun suffix *-ing* probably had something to do with it. In early Modern English, present participles were regularly formed with *-ing* pronounced \iŋg\ (as can still be heard in a few dialects) and later \iŋ\. Evidence also shows that some speakers used \in\ and by the 18th century this pronunciation became widespread. Though teachers (with some success) campaigned against it, \in\ remained a feature of the speech of many of the best speakers in Britain and the U.S. well into this century. It has by now lost its respectability, at least when attention is drawn to it, but throughout the U.S. it persists largely unnoticed and in some dialects it predominates over \iN\.

in•gre•di•ent \in-'grē-dē-ənt\ *noun* [Middle English, from Latin *ingredient-, ingrediens,* present participle of *ingredi* to go into, from *in-* + *gradi* to go — more at GRADE] (15th century)
: something that enters into a compound or is a component part of any combination or mixture **:** CONSTITUENT
synonym see ELEMENT
— **ingredient** *adjective*

in•gress \'in-,gres\ *noun* [Middle English, from Latin *ingressus,* from *ingredi*] (15th century)
1 : the act of entering **:** ENTRANCE
2 : the power or liberty of entrance or access
— **in•gres•sion** \in-'gre-shən\ *noun*

in•gres•sive \in-'gre-siv\ *adjective* (1649)
1 : of, relating to, or involving ingress; *especially* **:** produced by ingress of air into the vocal tract ⟨*ingressive* sounds⟩
2 : INCHOATIVE 2
— **ingressive** *noun*
— **in•gres•sive•ness** *noun*

in–group \'in-,grüp\ *noun* (1907)
1 : a group with which one feels a sense of solidarity or community of interests — compare OUT-GROUP
2 : CLIQUE

in•grow•ing \'in-,grō-iŋ\ *adjective* (1869)
: growing or tending inward

in•grown \-,grōn\ *adjective* (1670)
1 : grown in; *specifically* **:** having the free tip or edge embedded in the flesh ⟨an *ingrown* toenail⟩
2 : having the direction of growth or activity or interest inward rather than outward **:** WITHDRAWN
— **in•grown•ness** \(,)in-'grōn-nəs\ *noun*

in•growth \'in-,grōth\ *noun* (1870)
1 : a growing inward (as to fill a void)
2 : something that grows in or into a space

in•gui•nal \'iŋ-gwə-nᵊl\ *adjective* [Middle English *inguynale,* from Latin *inguinalis,* from *inguin-, inguen* groin — more at ADEN-] (15th century)
: of, relating to, or situated in the region of the groin or in either of the lowest lateral regions of the abdomen ⟨*inguinal* hernia⟩

in•gur•gi•tate \in-'gər-jə-,tāt\ *transitive verb* **-tat•ed; -tat•ing** [Latin *ingurgitatus,* past participle of *ingurgitare,* from *in-* + *gurgit-, gurges* whirlpool — more at VORACIOUS] (circa 1570)
: to swallow greedily or in large quantities **:** GUZZLE
— **in•gur•gi•ta•tion** \(,)in-,gər-jə-'tā-shən\ *noun*

in•hab•it \in-'ha-bət\ *verb* [Middle English *enhabiten,* from Middle French & Latin; Middle French *enhabiter,* from Latin *inhabitare,* from *in-* + *habitare* to dwell, frequentative of *habēre* to have — more at GIVE] (14th century)
transitive verb
1 : to occupy as a place of settled residence or habitat **:** live in ⟨*inhabit* a small house⟩
2 : to be present in or occupy in any manner or form ⟨the human beings who *inhabit* this tale —Al Newman⟩
intransitive verb
archaic **:** to have residence in a place **:** DWELL
— **in•hab•it•able** \-bə-tə-bəl\ *adjective*
— **in•hab•it•er** *noun*

in•hab•i•tan•cy \in-'ha-bə-tən(t)-sē\ *noun* (1681)
: INHABITATION

in•hab•i•tant \in-'ha-bə-tənt\ *noun* (15th century)
: one that occupies a particular place regularly, routinely, or for a period of time ⟨*inhabitants* of large cities⟩ ⟨the tapeworm is an *inhabitant* of the intestine⟩

in•hab•i•ta•tion \in-,ha-bə-'tā-shən\ *noun* (15th century)
: the act of inhabiting **:** the state of being inhabited

inhabited *adjective* (15th century)
: having inhabitants

in•hal•ant \in-'hā-lənt\ *noun* (circa 1834)
: something (as an allergen or medication) that is inhaled
— **inhalant** *adjective*

in•ha•la•tion \,in-hə-'lā-shən, ,i-nᵊl-'ā-\ *noun* (circa 1623)
1 : the act or an instance of inhaling
2 : material (as medication) to be taken in by inhaling
— **in•ha•la•tion•al** \-shnəl, -shə-nᵊl\ *adjective*

in•ha•la•tor \'in-hə-,lā-tər, 'i-nᵊl-,ā-\ *noun* (1925)
: a device providing a mixture of oxygen and carbon dioxide for breathing that is used especially in conjunction with artificial respiration

in•hale \in-'hā(ə)l\ *verb* **in•haled; in•hal•ing** [*in-* + ex*hale*] (1725)
transitive verb
1 : to draw in by breathing
2 : to take in eagerly or greedily ⟨*inhaled* about four meals at once —Ring Lardner⟩
intransitive verb
: to breathe in
— **in•hale** \in-', -'in-,\ *noun*

in•hal•er \in-'hā-lər\ *noun* (1778)
1 : a device by means of which medicinal material is inhaled
2 : one that inhales

in•har•mon•ic \,in-(,)här-'mä-nik\ *adjective* (circa 1828)
: not harmonic

in•har•mo•ni•ous \-'mō-nē-əs\ *adjective* (1711)
1 : not harmonious **:** DISCORDANT
2 : not fitting or congenial **:** CONFLICTING
— **in•har•mo•ni•ous•ly** *adverb*
— **in•har•mo•ni•ous•ness** *noun*

in•har•mo•ny \(,)in-'här-mə-nē\ *noun* (1799)
: DISCORD

in•here \in-'hir\ *intransitive verb* **in•hered; in•her•ing** [Latin *inhaerēre,* from *in-* + *haerēre* to adhere] (15th century)
: to be inherent

in•her•ence \in-'hir-ən(t)s, -'her-\ *noun* (1577)
: the quality, state, or fact of inhering

in•her•ent \-ənt\ *adjective* [Latin *inhaerent-, inhaerens,* present participle of *inhaerēre*] (1581)
: involved in the constitution or essential character of something **:** belonging by nature or habit **:** INTRINSIC
— **in•her•ent•ly** *adverb*

in•her•it \in-'her-ət\ *verb* [Middle English *enheriten* to make one an heir, inherit, from Middle French *enheriter* to make one an heir, from Late Latin *inhereditare,* from Latin *in-* + *hereditas* inheritance — more at HEREDITY] (14th century)
transitive verb
1 : to come into possession of or receive especially as a right or divine portion ⟨and every one who has left houses or brothers or sisters . . . for my name's sake, will receive a hundredfold, and *inherit* eternal life —Matthew 19:29 (Revised Standard Version)⟩
2 a : to receive from an ancestor as a right or title descendible by law at the ancestor's death **b :** to receive as a devise or legacy
3 : to receive from ancestors by genetic transmission ⟨*inherit* a strong constitution⟩
4 : to have in turn or receive as if from an ancestor ⟨*inherited* the problem from his predecessor⟩
intransitive verb
: to take or hold a possession or rights by inheritance
— **in•her•i•tor** \-ə-tər\ *noun*
— **in•her•i•tress** \-ə-trəs\ *or* **in•her•i•trix** \-ə-(,)triks\ *noun*

in•her•it•able \in-'her-ə-tə-bəl\ *adjective* (15th century)
1 : capable of being inherited **:** TRANSMISSIBLE
2 : capable of taking by inheritance
— **in•her•it•abil•i•ty** \-,her-ə-tə-'bi-lə-tē\ *noun*
— **in•her•it•able•ness** \-'her-ə-tə-bəl-nəs\ *noun*

in•her•i•tance \in-'her-ə-tən(t)s\ *noun* (14th century)
1 a : the act of inheriting property **b :** the reception of genetic qualities by transmission from parent to offspring **c :** the acquisition of a possession, condition, or trait from past generations
2 : something that is or may be inherited
3 a : TRADITION **b :** a valuable possession that is a common heritage from nature
4 *obsolete* **:** POSSESSION

inheritance tax *noun* (1841)
1 : an excise in the form of a percentage of the value of the property received that is levied on the privilege of an heir to inherit property
2 : DEATH TAX; *especially* **:** ESTATE TAX

in•hib•in \in-'hi-bən\ *noun* [Latin *inhibēre* to inhibit + English [1]*-in*] (1932)
: a human hormone that is secreted by Sertoli cells in the male and granulosa cells in the female and that inhibits the secretion of follicle-stimulating hormone

in•hib•it \in-'hi-bət\ *verb* [Middle English, from Latin *inhibitus,* past participle of *inhibēre,* from *in-* [2]in- + *habēre* to have — more at HABIT] (15th century)
transitive verb
1 : to prohibit from doing something
2 a : to hold in check **:** RESTRAIN **b :** to discourage from free or spontaneous activity especially through the operation of inner psychological impediments or of social controls
intransitive verb
: to cause inhibition
synonym see FORBID
— **in•hib•i•tive** \-bə-tiv\ *adjective*
— **in•hib•i•to•ry** \-bə-,tōr-ē, -,tȯr-\ *adjective*

in•hi•bi•tion \,in-hə-'bi-shən, ,i-nə-\ *noun* (14th century)
1 a : the act of inhibiting **:** the state of being inhibited **b :** something that forbids, debars, or restricts
2 : an inner impediment to free activity, expression, or functioning: as **a :** a psychic activity imposing restraint upon another activity **b :** a restraining of the function of a bodily organ or an agent (as an enzyme)

in•hib•i•tor \in-'hi-bə-tər\ *noun* (circa 1611)
: one that inhibits; *especially* **:** an agent that slows or interferes with a chemical action

in•hold•ing \'in-,hōl-diŋ\ *noun* (1947)
: privately owned land inside the boundary of a national park

in•ho•mo•ge•ne•i•ty \(,)in-,hō-mə-jə-'nē-ə-tē, -'nā- *also* ÷-'nī-; *especially British* -,hä-mə-\ *noun, plural* **-ties** (1899)
1 : the condition of not being homogeneous
2 : a part that is not homogeneous with the larger uniform mass in which it occurs; *especially* **:** a localized collection of matter in the universe
— **in•ho•mo•ge•neous** \-'jē-nē-əs, -nyəs\ *adjective*

in•hos•pi•ta•ble \,in-(,)hä-'spi-tə-bəl, (,)in-'häs-(,)pi-\ *adjective* (circa 1570)
1 : not showing hospitality **:** not friendly or receptive
2 : providing no shelter or sustenance
— **in•hos•pi•ta•ble•ness** *noun*
— **in•hos•pi•ta•bly** \-blē\ *adverb*

in•hos•pi•tal•i•ty \(,)in-,häs-pə-'ta-lə-tē\ *noun* (circa 1576)
: the quality or state of being inhospitable

in–house \'in-,hau̇s, -'hau̇s\ *adjective* (circa 1956)

: existing, originating, or carried on within a group or organization or its facilities **:** not outside ⟨an *in-house* publication⟩ ⟨a company's *in-house* staff⟩
— **in–house** *adverb*

in·hu·man \(ˌ)in-ˈhyü-mən, -ˈyü-\ *adjective* [Middle English *inhumayne,* from Middle French & Latin; Middle French *inhumain,* from Latin *inhumanus,* from *in-* + *humanus* human] (15th century)
1 a : lacking pity, kindness, or mercy **:** SAVAGE ⟨an *inhuman* tyrant⟩ **b :** COLD, IMPERSONAL ⟨his usual quiet, almost *inhuman* courtesy —F. Tennyson Jesse⟩ **c :** not worthy of or conforming to the needs of human beings ⟨*inhuman* living conditions⟩
2 : of or suggesting a nonhuman class of beings
— **in·hu·man·ly** *adverb*
— **in·hu·man·ness** \-mən-nəs\ *noun*

in·hu·mane \ˌin-(ˌ)hyü-ˈmān, -(ˌ)yü-\ *adjective* [Middle French *inhumain* & Latin *inhumanus*] (1599)
: not humane **:** INHUMAN 1
— **in·hu·mane·ly** *adverb*

in·hu·man·i·ty \-ˈma-nə-tē\ *noun, plural* **-ties** (15th century)
1 a : the quality or state of being cruel or barbarous **b :** a cruel or barbarous act
2 : absence of warmth or geniality **:** IMPERSONALITY

in·hume \in-ˈhyüm\ *transitive verb* **in·humed; in·hum·ing** [probably from French *inhumer,* from Medieval Latin *inhumare,* from Latin *in-* + *humus* earth — more at HUMBLE] (1604)
: BURY, INTER
— **in·hu·ma·tion** \ˌin-hyü-ˈmā-shən\ *noun*

in·im·i·cal \i-ˈni-mi-kəl\ *adjective* [Late Latin *inimicalis,* from Latin *inimicus* enemy — more at ENEMY] (1573)
1 : being adverse often by reason of hostility or malevolence
2 a : having the disposition of an enemy **:** HOSTILE **b :** reflecting or indicating hostility **:** UNFRIENDLY
— **in·im·i·cal·ly** \-mi-k(ə-)lē\ *adverb*

in·im·i·ta·ble \(ˌ)i-ˈni-mə-tə-bəl\ *adjective* [Middle English, from Middle French or Latin; Middle French, from Latin *inimitabilis,* from *in-* + *imitabilis* imitable] (15th century)
: not capable of being imitated **:** MATCHLESS
— **in·im·i·ta·ble·ness** *noun*
— **in·im·i·ta·bly** \-blē\ *adverb*

in·i·on \ˈi-nē-ˌän, -ən\ *noun* [New Latin, from Greek, back of the head, diminutive of *in-, is* sinew, tendon] (circa 1811)
: the external occipital protuberance of the skull

in·iq·ui·tous \i-ˈni-kwə-təs\ *adjective* (1726)
: characterized by iniquity
synonym see VICIOUS
— **in·iq·ui·tous·ly** *adverb*
— **in·iq·ui·tous·ness** *noun*

in·iq·ui·ty \-kwə-tē\ *noun, plural* **-ties** [Middle English *iniquite,* from Middle French *iniquité,* from Latin *iniquitat-, iniquitas,* from *iniquus* uneven, from *in-* + *aequus* equal] (14th century)
1 : gross injustice **:** WICKEDNESS
2 : a wicked act or thing **:** SIN

¹ini·tial \i-ˈni-shəl\ *adjective* [Middle French & Latin; Middle French, from Latin *initialis,* from *initium* beginning, from *inire* to go into, from *in-* + *ire* to go — more at ISSUE] (1526)
1 : of or relating to the beginning **:** INCIPIENT
2 : placed at the beginning **:** FIRST
— **ini·tial·ly** \i-ˈni-sh(ə-)lē\ *adverb*
— **ini·tial·ness** \i-ˈni-shəl-nəs\ *noun*

²initial *noun* (1627)
1 a : the first letter of a name **b** *plural* **:** the first letter of each word in a full name ⟨found that their *initials* were identical⟩
2 : a large letter beginning a text or a division or paragraph
3 : ANLAGE, PRECURSOR; *specifically* **:** a meristematic cell

³initial *transitive verb* **ini·tialed** *or* **ini·tialled; ini·tial·ing** *or* **ini·tial·ling** \i-ˈni-sh(ə-)liŋ\ (circa 1864)
1 : to affix an initial to
2 : to authenticate or give preliminary approval to by affixing the initials of an authorizing representative

ini·tial·ism \i-ˈni-shə-ˌli-zəm\ *noun* (1899)
: an acronym formed from initial letters

ini·tial·ize \-ˌlīz\ *transitive verb* **-ized; -iz·ing** (1957)
: to set (as a computer program counter) to a starting position, value, or configuration
— **ini·tial·i·za·tion** \i-ˌni-sh(ə-)lə-ˈzā-shən\ *noun*

initial rhyme *noun* (1838)
: ALLITERATION

initial side *noun* (1957)
: a stationary straight line that contains a point about which another straight line is rotated to form an angle — compare TERMINAL SIDE

¹ini·ti·ate \i-ˈni-shē-ˌāt\ *transitive verb* **-at·ed; -at·ing** [Late Latin *initiatus,* past participle of *initiare,* from Latin, to induct, from *initium*] (circa 1573)
1 : to cause or facilitate the beginning of **:** set going ⟨*initiate* a program of reform⟩ ⟨enzymes that *initiate* fermentation⟩
2 : to induct into membership by or as if by special rites
3 : to instruct in the rudiments or principles of something **:** INTRODUCE
synonym see BEGIN
— **ini·ti·a·tor** \-ˌā-tər\ *noun*

²ini·tiate \i-ˈni-sh(ē-)ət\ *adjective* (1605)
1 *obsolete* **:** relating to an initiate
2 a : initiated or properly admitted (as to membership or an office) **b :** instructed in some secret knowledge

³ini·tiate \i-ˈni-sh(ē-)ət\ *noun* (1811)
1 : a person who is undergoing or has undergone an initiation
2 : a person who is instructed or adept in some special field

ini·ti·a·tion \i-ˌni-shē-ˈā-shən\ *noun* (1583)
1 a : the act or an instance of initiating **b :** the process of being initiated **c :** the rites, ceremonies, ordeals, or instructions with which one is made a member of a sect or society or is invested with a particular function or status
2 : the condition of being initiated into some experience or sphere of activity **:** KNOWLEDGEABLENESS

¹ini·tia·tive \i-ˈni-shə-tiv *also* -shē-ə-tiv\ *adjective* (1795)
: of or relating to initiation **:** INTRODUCTORY, PRELIMINARY

²initiative *noun* (1793)
1 : an introductory step ⟨took the *initiative* in attempting to settle the issue⟩
2 : energy or aptitude displayed in initiation of action **:** ENTERPRISE ⟨showed great *initiative*⟩
3 a : the right to initiate legislative action **b :** a procedure enabling a specified number of voters by petition to propose a law and secure its submission to the electorate or to the legislature for approval — compare REFERENDUM 1
— **on one's own initiative :** at one's own discretion **:** independently of outside influence or control

ini·tia·to·ry \i-ˈni-sh(ē-)ə-ˌtōr-ē, -ˌtȯr-\ *adjective* (circa 1615)
1 : constituting a beginning
2 : tending or serving to initiate

in·ject \in-ˈjekt\ *transitive verb* [Latin *injectus,* past participle of *inicere,* from *in-* + *jacere* to throw — more at JET] (1601)
1 a : to introduce into something forcefully ⟨*inject* fuel into an engine⟩ **b :** to force a fluid into (as for medical purposes)
2 : to introduce as an element or factor in or into some situation or subject ⟨condemning any attempt to *inject* religious bigotry into the campaign —*Current Biography*⟩
— **in·ject·able** \-ˈjek-tə-bəl\ *adjective or noun*
— **in·jec·tor** \-ˈjek-tər\ *noun*

in·jec·tant \-ˈjek-tənt\ *noun* (1950)
: a substance that is injected into something

in·jec·tion \in-ˈjek-shən\ *noun* (15th century)
1 a : an act or instance of injecting **b :** the placing of an artificial satellite or a spacecraft into an orbit or on a trajectory; *also* **:** the time or place at which injection occurs
2 : something (as a medication) that is injected
3 : a mathematical function that is a one-to-one mapping — compare BIJECTION, SURJECTION

injection molding *noun* (1932)
: a method of forming articles (as of plastic) by heating the molding material until it can flow and injecting it into a mold
— **injection–molded** *adjective*

in·jec·tive \in-ˈjek-tiv\ *adjective* (1952)
: being a one-to-one mathematical function

in–joke \ˈin-ˌjōk, -ˈjōk\ *noun* (1964)
: a joke for or about a select group of people

in·ju·di·cious \ˌin-jü-ˈdi-shəs\ *adjective* (1649)
: not judicious **:** INDISCREET, UNWISE
— **in·ju·di·cious·ly** *adverb*
— **in·ju·di·cious·ness** *noun*

in·junc·tion \in-ˈjəŋ(k)-shən\ *noun* [Middle English *injunccion,* from Middle French & Late Latin; Middle French *injonction,* from Late Latin *injunction-, injunctio,* from Latin *injungere* to enjoin — more at ENJOIN] (15th century)
1 : the act or an instance of enjoining **:** ORDER, ADMONITION
2 : a writ granted by a court of equity whereby one is required to do or to refrain from doing a specified act
— **in·junc·tive** \-ˈjəŋ(k)-tiv\ *adjective*

in·jure \ˈin-jər\ *transitive verb* **in·jured; in·jur·ing** \ˈinj-riŋ, ˈin-jə-\ [Middle English *enjuren,* from Middle French *enjurier,* from Late Latin *injuriare,* from Latin *injuria* injury] (15th century)
1 a : to do an injustice to **:** WRONG **b :** to harm, impair, or tarnish the standing of **c :** to give pain to ⟨*injure* a person's pride⟩
2 a : to inflict bodily hurt on **b :** to impair the soundness of **c :** to inflict material damage or loss on ☆
— **in·jur·er** \ˈin-jər-ər\ *noun*

in·ju·ri·ous \in-ˈju̇r-ē-əs\ *adjective* (15th century)
1 : inflicting or tending to inflict injury **:** DETRIMENTAL ⟨*injurious* to health⟩
2 : ABUSIVE, DEFAMATORY ⟨speak not *injurious* words —George Washington⟩
— **in·ju·ri·ous·ly** *adverb*
— **in·ju·ri·ous·ness** *noun*

☆ SYNONYMS
Injure, harm, hurt, damage, impair, mar mean to affect injuriously. INJURE implies the inflicting of anything detrimental to one's looks, comfort, health, or success ⟨badly *injured* in an accident⟩. HARM often stresses the inflicting of pain, suffering, or loss ⟨careful not to *harm* the animals⟩. HURT implies inflicting a wound to the body or to the feelings ⟨*hurt* by their callous remarks⟩. DAMAGE suggests injury that lowers value or impairs usefulness ⟨a table *damaged* in shipping⟩. IMPAIR suggests a making less complete or efficient by deterioration or diminution ⟨years of smoking had *impaired* his health⟩. MAR applies to injury that spoils perfection (as of a surface) or causes disfigurement ⟨the text is *marred* by many typos⟩.

in·ju·ry \'inj-rē, 'in-jə-\ *noun, plural* **-ries** [Middle English *injurie,* from Latin *injuria,* from *injurus* injurious, from *in-* + *jur-, jus* right — more at JUST] (14th century)
1 a : an act that damages or hurts **:** WRONG **b :** violation of another's rights for which the law allows an action to recover damages
2 : hurt, damage, or loss sustained
synonym see INJUSTICE

in·jus·tice \(,)in-'jəs-təs\ *noun* [Middle English, from Middle French, from Latin *injustitia,* from *injustus* unjust, from *in-* + *justus* just] (14th century)
1 : absence of justice **:** violation of right or of the rights of another **:** UNFAIRNESS
2 : an unjust act **:** WRONG ☆

[1]ink \'iŋk\ *noun, often attributive* [Middle English *enke,* from Old French, from Late Latin *encaustum,* from neuter of Latin *encaustus* burned in, from Greek *enkaustos,* verbal of *enkaiein* to burn in — more at ENCAUSTIC] (13th century)
1 : a colored usually liquid material for writing and printing
2 : the black protective secretion of a cephalopod
3 *slang* **:** PUBLICITY 2d
— ink·i·ness \'iŋ-kē-nəs\ *noun*
— inky \'iŋ-kē\ *adjective*

[2]ink *transitive verb* (1562)
1 : to put ink on ⟨*ink* a pen⟩; *also* **:** to draw or write on in ink
2 a : SIGN 2a ⟨*inked* a new contract⟩ **b :** SIGN 4

ink·ber·ry \'iŋk-,ber-ē\ *noun* [from the use of the berries for making ink] (1765)
1 a : a holly (*Ilex glabra*) of eastern North America with evergreen oblong leathery leaves and small usually black berries **b :** POKEWEED
2 : the fruit of an inkberry

ink·blot test \'iŋk-,blät-\ *noun* (1928)
: any of several psychological tests (as a Rorschach test) based on the interpretation of irregular figures (as blots of ink)

[1]ink·horn \'iŋk-,hȯrn\ *noun* (14th century)
: a small portable bottle (as of horn) for holding ink

[2]inkhorn *adjective* (1543)
: ostentatiously learned **:** PEDANTIC ⟨*inkhorn* terms⟩

in–kind \'in-'kīnd\ *adjective* (1973)
: consisting of something (as goods or commodities) other than money ⟨*in-kind* relief for the poor⟩

ink–jet \'iŋk-'jet\ *adjective* (1976)
: of, relating to, or being a dot matrix printer in which electrically charged droplets of ink are projected onto the paper

in·kle \'iŋ-kəl\ *noun* [origin unknown] (1541)
: a colored linen tape or braid woven on a very narrow loom and used for trimming; *also* **:** the thread used

in·kling \'iŋ-kliŋ\ *noun* [Middle English *yngkiling* whisper, mention, probably from *inclen* to hint at; akin to Old English *inca* suspicion] (1513)
1 : a slight indication or suggestion **:** HINT, CLUE ⟨there was no path—no *inkling* even of a track —*New Yorker*⟩
2 : a slight knowledge or vague notion ⟨had not the faintest *inkling* of what it was all about —H. W. Carter⟩

ink·stand \'iŋk-,stand\ *noun* (1773)
: INKWELL; *also* **:** a stand with fittings for holding ink and pens

ink·stone \'iŋk-,stōn\ *noun* (circa 1889)
: a stone used in Chinese art and calligraphy on which dry ink and water are mixed

ink·well \'iŋk-,wel\ *noun* (circa 1875)
: a container (as in a desk) for ink

inky cap *noun* (1923)
: a mushroom (genus *Coprinus,* especially *C. atramentarius*) whose pileus deliquesces into an inky fluid after the spores have matured — called also *ink cap*

in·laid \'in-'lād\ *adjective* (1598)
1 a : set into a surface in a decorative design ⟨tables with *inlaid* marble⟩ **b :** decorated with a design or material set into a surface ⟨a table with an *inlaid* top⟩
2 *of linoleum* **:** having a design that goes all the way through to the backing

inky cap

[1]in·land \'in-,land, -lənd\ *adjective* (1546)
1 *chiefly British* **:** not foreign **:** DOMESTIC
2 : of or relating to the interior of a country

[2]inland *noun* (1573)
: the interior part of a country

[3]inland *adverb* (1600)
: into or toward the interior

in·land·er \'in-,lan-dər, -lən-\ *noun* (1610)
: one who lives inland

in–law \'in-,lȯ\ *noun* [mother-*in-law,* etc.] (1894)
: a relative by marriage

[1]in·lay \(,)in-'lā, 'in-,\ *transitive verb* **in·laid** \-'lād, -,lād\; **in·lay·ing** (1596)
1 a : to set into a surface or ground material **b :** to adorn with insertions **c :** to insert (as a color plate) into a mat or other reinforcement
2 : to rub, beat, or fuse (as wire) into an incision in metal, wood, or stone
— in·lay·er *noun*

[2]in·lay \'in-,lā\ *noun* (1667)
1 : inlaid work or a decorative inlaid pattern
2 : a tooth filling shaped to fit a cavity and then cemented into place

in·let \'in-,let, -lət\ *noun* [from its letting water in] (circa 1576)
1 a : a bay or recess in the shore of a sea, lake, or river; *also* **:** CREEK **b :** a narrow water passage between peninsulas or through a barrier island leading to a bay or lagoon
2 : a way of entering; *especially* **:** an opening for intake

in·li·er \'in-,lī(-ə)r\ *noun* [[3]*in* + out*lier*] (circa 1859)
1 : a mass of rock whose outcrop is surrounded by rock of younger age
2 : a distinct area or formation completely surrounded by another; *also* **:** ENCLAVE

in–line \'in-'līn, ,in-\ *adjective or adverb* (1929)
: having the parts or units arranged in a straight line; *also* **:** being so arranged

in–line engine \(,)in-'līn-, 'in-,\ *noun* (1929)
: an internal combustion engine in which the cylinders are arranged in one or more straight lines

in–line skate *noun* (1989)
: a roller skate whose four wheels are set in-line for greater speed and maneuverability
— in–line skating *noun*

[1]in lo·co pa·ren·tis \in-'lō-kō-pə-'ren-təs\ *adverb* [Latin] (1828)
: in the place of a parent

[2]in loco parentis *noun* (1968)
: regulation or supervision by an administrative body (as at a university) acting in loco parentis

in·ly \'in-lē\ *adverb* (before 12th century)
1 : INWARDLY
2 : in a manner suggesting great depth of knowledge or understanding **:** THOROUGHLY

in·mate \'in-,māt\ *noun* (1589)
: any of a group occupying a single place of residence; *especially* **:** a person confined (as in a prison or hospital)

in me·di·as res \in-'me-dē-əs-'rās, -'mē-dē-əs-'rēz\ *adverb* [Latin, literally, into the midst of things] (1786)
: in or into the middle of a narrative or plot

in me·mo·ri·am \,in-mə-'mōr-ē-əm, -'mȯr-\ *preposition* [Latin] (1850)
: in memory of — used especially in epitaphs

in–mi·grant \'in-,mī-grənt\ *noun* (1942)
: one that in-migrates

in–mi·grate \'in-,mī-,grāt\ *intransitive verb* (1942)
: to move into or come to live in a region or community especially as part of a large-scale and continuing movement of population — compare OUT-MIGRATE
— in–mi·gra·tion \,in-mī-'grā-shən\ *noun*

in·most \'in-,mōst\ *adjective* [Middle English, from Old English *innemest,* superlative of *inne,* adverb, in, within, from *in,* adverb] (before 12th century)
: deepest within **:** farthest from the outside

[1]inn \'in\ *noun* [Middle English, from Old English; akin to Old Norse *inni* dwelling, inn, Old English *in,* adverb] (before 12th century)
1 a : an establishment for the lodging and entertaining of travelers **b :** TAVERN
2 : a residence formerly provided for British students in London and especially for students of law

[2]inn *intransitive verb* (14th century)
: to put up at an inn

in·nards \'i-nərdz\ *noun plural* [alteration of *inwards*] (circa 1825)
1 : the internal organs of a human being or animal; *especially* **:** VISCERA
2 : the internal parts especially of a structure or mechanism

in·nate \i-'nāt, 'i-,\ *adjective* [Middle English *innat,* from Latin *innatus,* past participle of *innasci* to be born in, from *in-* + *nasci* to be born — more at NATION] (15th century)
1 : existing in, belonging to, or determined by factors present in an individual from birth **:** NATIVE, INBORN ⟨*innate* behavior⟩
2 : belonging to the essential nature of something **:** INHERENT
3 : originating in or derived from the mind or the constitution of the intellect rather than from experience ☆
— in·nate·ly *adverb*
— in·nate·ness *noun*

☆ SYNONYMS

Injustice, injury, wrong, grievance mean an act that inflicts undeserved hurt. INJUSTICE applies to any act that involves unfairness to another or violation of his rights ⟨the *injustices* suffered by the lower classes⟩. INJURY applies in law specifically to an injustice for which one may sue to recover compensation ⟨libel constitutes a legal *injury*⟩. WRONG applies also in law to any act punishable according to the criminal code; it may apply more generally to any flagrant injustice ⟨determined to right society's *wrongs*⟩. GRIEVANCE applies to any circumstance or condition that constitutes an injustice to the sufferer and gives just ground for complaint ⟨investigating employee *grievances*⟩.

Innate, inborn, inbred, congenital, hereditary mean not acquired after birth. INNATE applies to qualities or characteristics that are part of one's inner essential nature ⟨an *innate* sense of fair play⟩. INBORN suggests a quality or tendency either actually present at birth or so marked and deep-seated as to seem so ⟨her *inborn* love of nature⟩. INBRED suggests something either acquired from parents by heredity or so deeply rooted and ingrained as to seem acquired in that way ⟨*inbred* political loyalties⟩. CONGENITAL and HEREDITARY refer to what is acquired before or at birth, the former to things acquired during fetal development and the latter to things transmitted from one's ancestors ⟨a *congenital* heart murmur⟩ ⟨eye color is *hereditary*⟩.

\ə\ **abut** \ᵊ\ **kitten** \ər\ **further** \a\ **ash** \ā\ **ace** \ä\ **mop, mar** \au̇\ **out** \ch\ **chin** \e\ **bet** \ē\ **easy** \g\ **go** \i\ **hit** \ī\ **ice** \j\ **job** \ŋ\ **sing** \ō\ **go** \ȯ\ **law** \ȯi\ **boy** \th\ **thin** \t͟h\ **the** \ü\ **loot** \u̇\ **foot** \y\ **yet** \zh\ **vision** *see also* Guide to Pronunciation

in·ner \ˈi-nər\ *adjective* [Middle English, from Old English *innera,* comparative of *inne* within] (before 12th century)
1 a : situated farther in ⟨the *inner* bark⟩ **b :** being near a center especially of influence ⟨the *inner* circles of political power⟩
2 : of or relating to the mind or spirit ⟨the *inner* life⟩
— inner *noun*
— in·ner·ly *adverb*

inner city *noun* (1961)
: the usually older, poorer, and more densely populated central section of a city
— inner–city *adjective*

in·ner–di·rect·ed \ˈi-nər-də-ˈrek-təd, -(ˌ)dī-\ *adjective* (1950)
: directed in thought and action by one's own scale of values as opposed to external norms

inner ear *noun* (circa 1923)
: the essential organ of hearing and equilibrium that is located in the temporal bone, is innervated by the auditory nerve, and includes the vestibule, the semicircular canals, and the cochlea

inner light *noun, often I&L capitalized* (1856)
: a divine presence held (as in Quaker doctrine) to enlighten and guide the soul

¹in·ner·most \ˈi-nər-ˌmōst\ *adjective* (14th century)
: farthest inward **:** INMOST

²innermost *noun* (14th century)
: the inmost part

inner planet *noun* (1951)
: any of the planets Mercury, Venus, Earth, and Mars whose orbits are within the asteroid belt

inner product *noun* (circa 1911)
: SCALAR PRODUCT

in·ner·sole \ˈi-nər-ˈsōl\ *noun* (circa 1892)
: INSOLE

inner space *noun* (1958)
1 : space at or near the earth's surface and especially under the sea
2 : one's inner self

in·ner·spring \ˈi-nər-ˈspriŋ\ *adjective* (1928)
: having coil springs inside a padded casing ⟨*innerspring* mattress⟩

inner tube *noun* (1895)
: an airtight rubber tube inside a pneumatic tire to hold air under pressure

in·ner·vate \i-ˈnər-ˌvāt, ˈi-(ˌ)nər-\ *transitive verb* **-vat·ed; -vat·ing** (1870)
: to supply with nerves
— in·ner·va·tion \ˌi-(ˌ)nər-ˈvā-shən, i-ˌnər-\ *noun*

in·ning \ˈi-niŋ\ *noun* [²in + ¹-ing] (1735)
1 a *plural but singular or plural in construction* **:** a division of a cricket match **b :** a division of a baseball game consisting of a turn at bat for each team; *also* **:** a baseball team's turn at bat ending with the third out **c :** a player's turn (as in horseshoes, pool, or croquet)
2 : a chance or opportunity for action or accomplishment — usually used in plural but singular or plural in construction ⟨that momentous *innings* which was to project him into world politics —*Times Literary Supplement*⟩

inn·keep·er \ˈin-ˌkē-pər\ *noun* (15th century)
1 : a proprietor of an inn
2 : HOTELMAN

in·no·cence \ˈi-nə-sən(t)s\ *noun* (14th century)
1 a : freedom from guilt or sin through being unacquainted with evil **:** BLAMELESSNESS **b :** CHASTITY **c :** freedom from legal guilt of a particular crime or offense **d** (1) **:** freedom from guile or cunning **:** SIMPLICITY (2) **:** lack of worldly experience or sophistication **e :** lack of knowledge **:** IGNORANCE ⟨written in entire *innocence* of the Italian language —E. R. Bentley⟩
2 : one that is innocent
3 : BLUET

in·no·cen·cy \-sən(t)-sē\ *noun, plural* **-cies** (14th century)
: INNOCENCE; *also* **:** an innocent action or quality

in·no·cent \ˈi-nə-sənt\ *adjective* [Middle English, from Middle French, from Latin *innocent-, innocens,* from *in-* + *nocent-, nocens* wicked, from present participle of *nocēre* to harm — more at NOXIOUS] (14th century)
1 a : free from guilt or sin especially through lack of knowledge of evil **:** BLAMELESS ⟨an *innocent* child⟩ **b :** harmless in effect or intention ⟨searching for a hidden motive in even the most *innocent* conversation —Leonard Wibberley⟩; *also* **:** CANDID ⟨gave me an *innocent* gaze⟩ **c :** free from legal guilt or fault; *also* **:** LAWFUL ⟨a wholly *innocent* transaction⟩
2 a : lacking or reflecting a lack of sophistication, guile, or self-consciousness **:** ARTLESS, INGENUOUS **b :** IGNORANT ⟨almost entirely *innocent* of Latin —C. L. Wrenn⟩; *also* **:** UNAWARE ⟨perfectly *innocent* of the confusion he had created —B. R. Haydon⟩
3 : lacking or deprived of something ⟨her face *innocent* of cosmetics —Marcia Davenport⟩
— innocent *noun*
— in·no·cent·ly *adverb*

in·noc·u·ous \i-ˈnä-kyə-wəs\ *adjective* [Latin *innocuus,* from *in-* + *nocēre*] (1598)
1 : producing no injury **:** HARMLESS
2 : not likely to give offense or to arouse strong feelings or hostility **:** INOFFENSIVE, INSIPID
— in·noc·u·ous·ly *adverb*
— in·noc·u·ous·ness *noun*

in·nom·i·nate \i-ˈnä-mə-nət\ *adjective* [Late Latin *innominatus,* from Latin *in-* + *nominatus,* past participle of *nominare* to nominate] (1638)
: having no name **:** UNNAMED; *also* **:** ANONYMOUS

innominate artery *noun* (1870)
: BRACHIOCEPHALIC ARTERY

innominate bone *noun* (1866)
: the large flaring bone that constitutes a lateral half of the pelvis in mammals and is composed of the ilium, ischium, and pubis which are fused into one bone in the adult

innominate vein *noun* (1876)
: BRACHIOCEPHALIC VEIN

in·no·vate \ˈi-nə-ˌvāt\ *verb* **-vat·ed; -vat·ing** [Latin *innovatus,* past participle of *innovare,* from *in-* + *novus* new — more at NEW] (1548)
transitive verb
1 : to introduce as or as if new
2 *archaic* **:** to effect a change in ⟨the dictates of my father were . . . not to be altered, *innovated,* or even discussed —Sir Walter Scott⟩
intransitive verb
: to make changes **:** do something in a new way
— in·no·va·tor \-ˌvā-tər\ *noun*
— in·no·va·to·ry \ˈi-nə-və-ˌtōr-ē, -ˌtȯr-; ˈi-nə-ˌvā-tə-rē\ *adjective*

in·no·va·tion \ˌi-nə-ˈvā-shən\ *noun* (15th century)
1 : the introduction of something new
2 : a new idea, method, or device **:** NOVELTY
— in·no·va·tion·al \-shnəl, -shə-nᵊl\ *adjective*

in·no·va·tive \ˈi-nə-ˌvā-tiv\ *adjective* (1608)
: characterized by, tending to, or introducing innovations
— in·no·va·tive·ly *adverb*
— in·no·va·tive·ness *noun*

Inns of Court (15th century)
1 : the four sets of buildings in London belonging to four societies of students and practitioners of the law
2 : the four societies that alone admit to practice at the English bar

in·nu·en·do \ˌin-yə-ˈwen-(ˌ)dō\ *noun, plural* **-dos** *or* **-does** [Latin, by hinting, from *innuere* to hint, from *in-* + *nuere* to nod — more at NUMEN] (1678)
1 a : an oblique allusion **:** HINT, INSINUATION; *especially* **:** a veiled or equivocal reflection on character or reputation **b :** the use of such allusions ⟨resorting to *innuendo*⟩
2 : a parenthetical explanation introduced into the text of a legal document ◆

In·nu·it *variant of* INUIT

in·nu·mer·a·ble \i-ˈnüm-rə-bəl, -ˈnyüm-; -ˈn(y)ü-mə-\ *adjective* [Middle English, from Latin *innumerabilis,* from *in-* + *numerabilis* numerable] (14th century)
: too many to be numbered **:** COUNTLESS; *also* **:** very many
— in·nu·mer·a·bly \-blē\ *adverb*

in·nu·mer·ate \-rət\ *adjective* (1959)
: marked by an ignorance of mathematics and the scientific approach
— in·nu·mer·a·cy \-rə-sē\ *noun*
— innumerate *noun*

in·nu·mer·ous \-rəs\ *adjective* [Latin *innumerus,* from *in-* + *numerus* number] (1531)
: INNUMERABLE

in·ob·ser·vance \ˌi-nəb-ˈzər-vən(t)s\ *noun* [French & Latin; French, from Latin *inobservantia,* from *in-* + *observantia* observance] (1611)
1 : lack of attention **:** HEEDLESSNESS
2 : failure to fulfill **:** NONOBSERVANCE
— in·ob·ser·vant \-vənt\ *adjective*

in·oc·u·lant \i-ˈnä-kyə-lənt\ *noun* (1898)
: INOCULUM

in·oc·u·late \i-ˈnä-kyə-ˌlāt\ *transitive verb* **-lat·ed; -lat·ing** [Middle English, to insert a bud in a plant, from Latin *inoculatus,* past participle of *inoculare,* from *in-* + *oculus* eye, bud — more at EYE] (1721)
1 a : to introduce a microorganism into ⟨*inoculate* mice with anthrax⟩ ⟨beans *inoculated* with nitrogen-fixing bacteria⟩ **b :** to introduce (as a microorganism) into a suitable situation for growth **c :** to introduce immunologically active material (as an antibody or antigen) into especially in order to treat or prevent a disease ⟨*inoculate* children against diphtheria⟩
2 : to introduce something into the mind of
3 : to protect as if by inoculation ◆

◇ WORD HISTORY

innuendo The verb *innuere* in classical Latin meant "to nod, beckon, or make a sign to" a person, and in Medieval Latin more generally "to hint" or "to insinuate." The ablative case of the gerund was *innuendo,* which meant literally "by hinting." In medieval legal documents *innuendo* introduced inserted remarks, meaning in effect "to wit" or "that is to say," and the word was adopted with the same function into English legal usage. By the late 17th century *innuendo* was used as a noun referring to the insertion itself, and more broadly to any indirect suggestion or veiled allusion. The notion of the invidious possibilities of such a remark came to predominate so that today *innuendo* typically refers to an insinuation that is at best catty, at worst defamatory.

inoculate In both Latin and English, the word for "eye" (*oculus* in Latin) has been used metaphorically to denote something that looks like or is suggestive of a person's organ of sight. The circular markings on a peacock's tail and the undeveloped buds on a potato are common examples of this use of *eye.* In horticulture, as the Romans learned, the *oculus* or bud from one plant can be grafted onto another plant for propagation. The Latin verb *inoculare* was derived from *oculus* to refer to this procedure. On the basis of the past participle, *inoculatus,* this verb was borrowed into English in the 15th century with the same meaning. In the 18th century medical researchers discovered that introducing a small amount of an infective agent into a person made that person immune to a normal attack of the disease. By analogy with the implanting of a bud, the inserting of an infective agent into the body was also referred to by the verb *inoculate.*

synonym see INFUSE
— **in·oc·u·la·tive** \-ˌlā-tiv\ *adjective*
— **in·oc·u·la·tor** \-ˌlā-tər\ *noun*

in·oc·u·la·tion \i-ˌnä-kyə-ˈlā-shən\ *noun* (1714)
1 : the act or process or an instance of inoculating; *especially* **:** the introduction of a pathogen or antigen into a living organism to stimulate the production of antibodies
2 : INOCULUM

in·oc·u·lum \i-ˈnä-kyə-ləm\ *noun, plural* **-la** \-lə\ [New Latin, from Latin *inoculare*] (1902)
: material used for inoculation

in·of·fen·sive \ˌi-nə-ˈfen(t)-siv\ *adjective* (1646)
1 : causing no harm or injury
2 a : giving no provocation **:** PEACEABLE **b :** not objectionable to the senses
— **in·of·fen·sive·ly** *adverb*
— **in·of·fen·sive·ness** *noun*

in·op·er·a·ble \(ˌ)i-ˈnä-p(ə-)rə-bəl\ *adjective* [probably from French *inopérable*] (1886)
1 : not treatable or remediable by surgery ⟨*inoperable* cancer⟩
2 : INOPERATIVE

in·op·er·a·tive \-ˈnä-p(ə-)rə-tiv, -ˈnä-pə-ˌrā-\ *adjective* (circa 1631)
: not operative: as **a :** not functioning ⟨an *inoperative* clock⟩ **b :** having no effect or force ⟨an *inoperative* law⟩
— **in·op·er·a·tive·ness** *noun*

in·oper·cu·late \ˌi-nō-ˈpər-kyə-lət\ *adjective* (circa 1836)
: having no operculum
— **inoperculate** *noun*

in·op·por·tune \(ˌ)i-ˌnä-pər-ˈtün, -ˈtyün\ *adjective* [Latin *inopportunus,* from *in-* + *opportunus* opportune] (circa 1507)
: INCONVENIENT, UNSEASONABLE
— **in·op·por·tune·ly** *adverb*
— **in·op·por·tune·ness** \-ˈt(y)ün-nəs\ *noun*

in order that *conjunction* (1711)
: THAT 2a(1)

in·or·di·nate \i-ˈnȯr-dᵊn-ət, -ˈnȯrd-nət\ *adjective* [Middle English *inordinat,* from Latin *inordinatus,* from *in-* + *ordinatus,* past participle of *ordinare* to arrange — more at ORDAIN] (14th century)
1 *archaic* **:** DISORDERLY, UNREGULATED
2 : exceeding reasonable limits **:** IMMODERATE
synonym see EXCESSIVE
— **in·or·di·nate·ly** *adverb*
— **in·or·di·nate·ness** *noun*

in·or·gan·ic \ˌi-(ˌ)nȯr-ˈga-nik\ *adjective* (1794)
1 a (1) **:** being or composed of matter other than plant or animal **:** MINERAL (2) **:** forming or belonging to the inanimate world **b :** of, relating to, or dealt with by a branch of chemistry concerned with substances not usually classed as organic
2 : not arising from natural growth **:** ARTIFICIAL; *also* **:** lacking structure, character, or vitality ⟨dull *inorganic* things, without individuality or prestige —John Buchan⟩
— **in·or·gan·i·cal·ly** \-ni-k(ə-)lē\ *adverb*

in·os·cu·late \i-ˈnäs-kyə-ˌlāt\ *verb* **-lat·ed; -lat·ing** [^{2}*in-* + *osculate*] (1671)
: JOIN, UNITE
— **in·os·cu·la·tion** \(ˌ)i-ˌnäs-kyə-ˈlā-shən\ *noun*

ino·si·tol \i-ˈnō-sə-ˌtȯl, ī-ˈnō-, -ˌtōl\ *noun* [International Scientific Vocabulary, from *inosite* inositol, from Greek *in-, is* sinew + International Scientific Vocabulary ^{2}*-ose* + ^{1}*-ite*] (1891)
: any of several crystalline stereoisomeric cyclic alcohols $C_6H_{12}O_6$; *especially* **:** MYOINOSITOL

ino·tro·pic \ˌē-nə-ˈtrō-pik, ˌī-nə-, -ˈträ-\ *adjective* [International Scientific Vocabulary *ino-* (from Greek *in-, is* sinew) + *-tropic*] (1903)
: relating to or influencing the force of muscular contractions

in·pa·tient \ˈin-ˌpā-shənt\ *noun* (1760)
: a hospital patient who receives lodging and food as well as treatment — compare OUTPATIENT

in per·so·nam \ˌin-pər-ˈsō-ˌnam, -ˌnäm\ *adverb or adjective* [Late Latin, against a person] (circa 1860)
: against a person for the purpose of imposing a liability or obligation — used especially of legal actions, judgments, or jurisdiction; compare IN REM

in pet·to \in-ˈpe-(ˌ)tō\ *adverb or adjective* [Italian, literally, in the breast] (circa 1674)
1 : in private **:** SECRETLY
2 : in miniature

in·pour·ing \ˈin-ˌpōr-iŋ, -ˌpȯr-\ *noun* (1721)
: INRUSH

in–print \ˈin-ˈprint\ *adjective* (1950)
: being in print

in–pro·cess \(ˌ)in-ˈprä-ˌses, -ˈprō-, -səs\ *adjective* (1925)
: of, relating to, or being goods in manufacture as distinguished from raw materials or from finished products

in pro·pria per·so·na \in-ˈprō-prē-ə-pər-ˈsō-nə\ *adverb* [Medieval Latin] (1654)
: in one's own person or character **:** PERSONALLY; *especially* **:** without the assistance of an attorney

^{1}**in·put** \ˈin-ˌpu̇t\ *noun* (1753)
1 : something that is put in: as **a :** an amount put in ⟨increased *input* of fertilizer increases crop yield⟩ **b :** power or energy put into a machine or system for storage, conversion in kind, or conversion of characteristics usually with the intent of sizable recovery in the form of output **c :** a component of production (as land, labor, or raw materials) **d :** information fed into a data processing system or computer **e :** ADVICE, OPINION, COMMENT
2 : the means by which or the point at which an input (as of energy, material, or data) is made
3 : the act or process of putting in

^{2}**input** *transitive verb* **in·put·ted** *or* **input; in·put·ting** (1946)
: to enter (as data) into a computer or data processing system

in·quest \ˈin-ˌkwest\ *noun* [Middle English, from Old French *enqueste,* from (assumed) Vulgar Latin *inquaesta,* feminine of *inquaestus,* past participle of *inquaerere* to inquire] (13th century)
1 a : a judicial or official inquiry or examination especially before a jury ⟨a coroner's *inquest*⟩ **b :** a body of people (as a jury) assembled to hold such an inquiry **c :** the finding of the jury upon such inquiry or the document recording it
2 : INQUIRY, INVESTIGATION

in·qui·etude \(ˌ)in-ˈkwī-ə-ˌtüd, -ˌtyüd\ *noun* [Middle English, from Middle French or Late Latin; Middle French, from Late Latin *inquietudo,* from Latin *inquietus* disturbed, from *in-* + *quietus* quiet] (15th century)
: disturbed state **:** DISQUIETUDE

in·qui·line \ˈin-kwə-ˌlīn, ˈiŋ-, -lən\ *noun* [Latin *inquilinus* tenant, lodger, from *in-* + *colere* to cultivate, dwell — more at WHEEL] (1879)
: an animal that lives habitually in the nest or abode of some other species

in·quire \in-ˈkwīr\ *verb* **in·quired; in·quir·ing** [Middle English *enquiren,* from Old French *enquerre,* from (assumed) Vulgar Latin *inquaerere,* alteration of Latin *inquirere,* from *in-* + *quaerere* to seek] (13th century)
transitive verb
1 : to ask about ⟨some kindred spirit shall *inquire* thy fate —Thomas Gray⟩
2 : to search into **:** INVESTIGATE
intransitive verb
1 : to put a question **:** seek for information by questioning ⟨*inquired* about the horses⟩
2 : to make investigation or inquiry — often used with *into*
synonym see ASK
— **in·quir·er** *noun*
— **in·quir·ing·ly** \-ˈkwī-riŋ-lē\ *adverb*
— **inquire after :** to ask about the health of

in·qui·ry \in-ˈkwīr-ē, ˈin-ˌ; ˈin-kwə-rē, ˈiŋ-; ˈin-ˌkwir-ē\ *noun, plural* **-ries** (15th century)
1 : examination into facts or principles **:** RESEARCH
2 : a request for information
3 : a systematic investigation often of a matter of public interest

in·qui·si·tion \ˌin-kwə-ˈzi-shən, ˌiŋ-\ *noun* [Middle English *inquisicioun,* from Middle French *inquisition,* from Latin *inquisition-, inquisitio,* from *inquirere*] (14th century)
1 : the act of inquiring **:** EXAMINATION
2 : a judicial or official inquiry or examination usually before a jury; *also* **:** the finding of the jury
3 a *capitalized* **:** a former Roman Catholic tribunal for the discovery and punishment of heresy **b :** an investigation conducted with little regard for individual rights **c :** a severe questioning
— **in·qui·si·tion·al** \-ˈzi-sh(ə-)nᵊl\ *adjective*

in·quis·i·tive \in-ˈkwi-zə-tiv\ *adjective* (14th century)
1 : given to examination or investigation
2 : inclined to ask questions; *especially* **:** inordinately or improperly curious about the affairs of others
synonym see CURIOUS
— **in·quis·i·tive·ly** *adverb*
— **in·quis·i·tive·ness** *noun*

in·quis·i·tor \in-ˈkwi-zə-tər\ *noun* (1504)
: one who inquires or makes inquisition; *especially* **:** one who is unduly harsh, severe, or hostile in making an inquiry
— **in·quis·i·to·ri·al** \-ˌkwi-zə-ˈtōr-ē-əl, -ˈtȯr-\ *adjective*
— **in·quis·i·to·ri·al·ly** \-ē-ə-lē\ *adverb*

in re \in-ˈrā, -ˈrē\ *preposition* [Latin] (1877)
: in the matter of **:** CONCERNING, RE — often used in the title or name of a law case

in rem \in-ˈrem\ *adverb or adjective* [Late Latin] (circa 1860)
: against a thing (as a right, status, or property) — used especially of legal actions, judgments, or jurisdiction; compare IN PERSONAM

in–residence *adjective* (1845)
: being officially associated with an organization in a specified capacity — usually used in combination ⟨writer-*in-residence* at the university⟩

in·ro \ˈin-(ˌ)rō\ *noun, plural* **inro** [Japanese *inrō*] (1617)
: a small compartmented and usually ornamented container hung from an obi to hold small objects (as medicines and perfumes)

in·road \ˈin-ˌrōd\ *noun* (1548)
1 : a sudden hostile incursion **:** RAID
2 : an advance or penetration often at the expense of someone or something — usually used in plural

in·rush \ˈin-ˌrəsh\ *noun* (1817)
: a crowding or flooding in

in·sa·lu·bri·ous \ˌin-sə-ˈlü-brē-əs\ *adjective* [Latin *insalubris,* from *in-* + *salubris* healthful — more at SAFE] (1615)
: not conducive to health **:** UNWHOLESOME ⟨an *insalubrious* climate⟩
— **in·sa·lu·bri·ty** \-brə-tē\ *noun*

ins and outs \ˌin-zən(d)-ˈau̇ts\ *noun plural* (circa 1670)
1 : characteristic peculiarities or technicalities
2 : RAMIFICATIONS

in·sane \(ˌ)in-ˈsān\ *adjective* [Latin *insanus,* from *in-* + *sanus* sane] (circa 1550)
1 : mentally disordered **:** exhibiting insanity
2 : used by, typical of, or intended for insane persons ⟨an *insane* asylum⟩

\ə\ **abut** \ᵊ\ **kitten** \ər\ **further** \a\ **ash** \ā\ **ace** \ä\ **mop, mar** \au̇\ **out** \ch\ **chin** \e\ **bet** \ē\ **easy** \g\ **go** \i\ **hit** \ī\ **ice** \j\ **job** \ŋ\ **sing** \ō\ **go** \ȯ\ **law** \ȯi\ **boy** \th\ **thin** \t͟h\ **the** \ü\ **loot** \u̇\ **foot** \y\ **yet** \zh\ **vision** *see also* Guide to Pronunciation

3 : ABSURD ⟨an *insane* scheme for making money⟩
— **in·sane·ly** *adverb*
— **in·sane·ness** \-'sān-nəs\ *noun*

in·san·i·tary \(,)in-'sa-nə-,ter-ē\ *adjective* (1874)
: unclean enough to endanger health **:** CONTAMINATED
— **in·san·i·ta·tion** \in-,sa-nə-'tā-shən\ *noun*

in·san·i·ty \in-'sa-nə-tē\ *noun, plural* **-ties** (1590)
1 a : a deranged state of the mind usually occurring as a specific disorder (as schizophrenia) and usually excluding such states as mental retardation, psychoneurosis, and various character disorders **b :** a mental disorder
2 : such unsoundness of mind or lack of understanding as prevents one from having the mental capacity required by law to enter into a particular relationship, status, or transaction or as removes one from criminal or civil responsibility
3 a : extreme folly or unreasonableness **b :** something utterly foolish or unreasonable

in·sa·tia·ble \(,)in-'sā-shə-bəl\ *adjective* [Middle English *insaciable,* from Middle French, from Latin *insatiabilis,* from *in-* + *satiare* to satisfy — more at SATIATE] (15th century)
: incapable of being satisfied **:** QUENCHLESS ⟨had an *insatiable* desire for wealth⟩
— **in·sa·tia·bil·i·ty** \(,)in-,sā-shə-'bi-lə-tē\ *noun*
— **in·sa·tia·ble·ness** \(,)in-'sā-shə-bəl-nəs\ *noun*
— **in·sa·tia·bly** \-blē\ *adverb*

in·sa·tiate \(,)in-'sā-sh(ē-)ət\ *adjective* (15th century)
: INSATIABLE
— **in·sa·tiate·ly** *adverb*
— **in·sa·tiate·ness** *noun*

in·scribe \in-'skrīb\ *transitive verb* [Middle English, from Latin *inscribere,* from *in-* + *scribere* to write — more at SCRIBE] (15th century)
1 a : to write, engrave, or print as a lasting record **b :** to enter on a list **:** ENROLL
2 a : to write, engrave, or print characters upon **b :** to autograph or address (a book) as a gift
3 : to dedicate to someone
4 : to draw within a figure so as to touch in as many places as possible ⟨a regular polygon *inscribed* in a circle⟩
5 *British* **:** to register the name of the holder of (a security)
— **in·scrib·er** *noun*

in·scrip·tion \in-'skrip-shən\ *noun* [Middle English *inscripcioun,* from Latin *inscription-, inscriptio,* from *inscribere*] (14th century)
1 a : something that is inscribed; *also* **:** SUPERSCRIPTION **b :** EPIGRAPH 2 **c :** the wording on a coin, medal, seal, or currency note
2 : the dedication of a book or work of art
3 a : the act of inscribing **b :** the entering of a name on or as if on a list **:** ENROLLMENT
4 *British* **a :** the act of inscribing securities **b** *plural* **:** inscribed securities
— **in·scrip·tion·al** \-shnəl, -shə-nᵊl\ *adjective*

in·scrip·tive \in-'skrip-tiv\ *adjective* (1740)
: relating to or constituting an inscription
— **in·scrip·tive·ly** *adverb*

in·scroll \in-'skrōl\ *transitive verb* (1596)
archaic **:** to write on a scroll **:** RECORD

in·scru·ta·ble \in-'skrü-tə-bəl\ *adjective* [Middle English, from Late Latin *inscrutabilis,* from Latin *in-* + *scrutari* to search — more at SCRUTINY] (15th century)
: not readily investigated, interpreted, or understood **:** MYSTERIOUS
— **in·scru·ta·bil·i·ty** \-,skrü-tə-'bi-lə-tē\ *noun*
— **in·scru·ta·ble·ness** \-'skrü-tə-bəl-nəs\ *noun*
— **in·scru·ta·bly** \-blē\ *adverb*

in·sculp \in-'skəlp\ *transitive verb* [Middle English, from Latin *insculpere,* from *in-* + *scalpere* to scratch, carve — more at SHELF] (15th century)
archaic **:** ENGRAVE, SCULPTURE

in·seam \'in-,sēm\ *noun* (1886)
: the seam on the inside of the leg of a pair of pants; *also* **:** the length of this seam

in·sect \'in-,sekt\ *noun* [Latin *insectum,* from neuter of *insectus,* past participle of *insecare* to cut into, from *in-* + *secare* to cut — more at SAW] (1601)
1 a : any of numerous small invertebrate animals (as spiders or centipedes) that are more or less obviously segmented **b :** any of a class (Insecta) of arthropods (as bugs or bees) with well-defined head, thorax, and abdomen, only three pairs of legs, and typically one or two pairs of wings
2 : a trivial or contemptible person ◆
— **insect** *adjective*

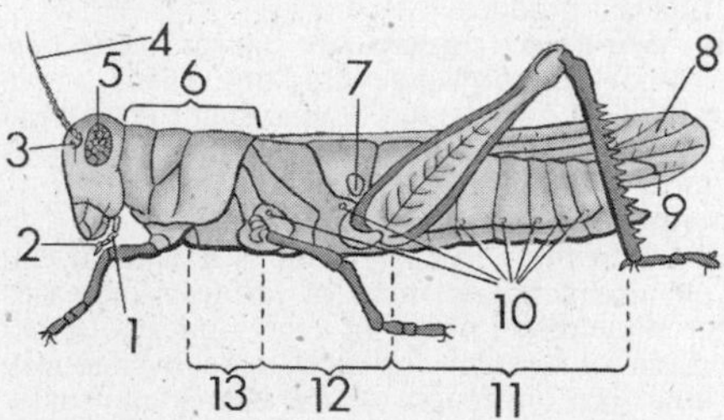

insect 1b: *1* **labial palpus,** *2* **maxillary palpus,** *3* **simple eye,** *4* **antenna,** *5* **compound eye,** *6* **prothorax,** *7* **tympanum,** *8* **wing,** *9* **ovipositor,** *10* **spiracles,** *11* **abdomen,** *12* **metathorax,** *13* **mesothorax**

in·sec·ta·ry \'in-,sek-tə-rē, in-'\ *noun, plural* **-ries** (1888)
: a place for the keeping or rearing of living insects

in·sec·ti·cid·al \(,)in-,sek-tə-'sī-dᵊl\ *adjective* (1857)
1 : destroying or controlling insects
2 : of or relating to an insecticide
— **in·sec·ti·cid·al·ly** \-dᵊl-ē\ *adverb*

in·sec·ti·cide \in-'sek-tə-,sīd\ *noun* [International Scientific Vocabulary] (1865)
: an agent that destroys insects

in·sec·tile \(,)in-'sek-tᵊl, -,tīl, -(,)til\ *adjective* (circa 1626)
: being or suggestive of an insect

in·sec·ti·vore \in-'sek-tə-,vōr, -,vȯr\ *noun* [New Latin *Insectivora,* from Latin *insectum* + *-vorus* -vorous] (1840)
1 : any of an order (Insectivora) of mammals comprising forms (as moles, shrews, and hedgehogs) that are mostly small, insectivorous, and nocturnal
2 : an insectivorous plant or animal

in·sec·tiv·o·rous \,in-,sek-'tiv-rəs, -'ti-və-\ *adjective* (1661)
: depending on insects as food

in·se·cure \,in-si-'kyu̇r\ *adjective* [Medieval Latin *insecurus,* from Latin *in-* + *securus* secure] (1649)
1 : not confident or sure **:** UNCERTAIN ⟨feeling somewhat *insecure* of his reception⟩
2 : not adequately guarded or sustained **:** UNSAFE ⟨an *insecure* investment⟩
3 : not firmly fastened or fixed **:** SHAKY ⟨the hinge is loose and *insecure*⟩
4 a : not highly stable or well-adjusted ⟨an *insecure* marriage⟩ **b :** deficient in assurance **:** beset by fear and anxiety ⟨always felt *insecure* in a group of strangers⟩
— **in·se·cure·ly** *adverb*
— **in·se·cure·ness** *noun*
— **in·se·cu·ri·ty** \-'kyu̇r-ə-tē\ *noun*

in·sel·berg \'in(t)-səl-,bərg, 'in-zəl-, -,berg\ *noun, plural* **-bergs** *also* **-ber·ge** \-,bər-gə, -,ber-\ [German, from *Insel* island + *Berg* mountain] (1913)
: an isolated mountain

in·sem·i·nate \in-'se-mə-,nāt\ *transitive verb* **-nat·ed; -nat·ing** [Latin *inseminatus,* past participle of *inseminare,* from *in-* + *semin-, semen* seed — more at SEMEN] (circa 1623)
1 : SOW
2 : to introduce semen into the genital tract of (a female)
synonym see IMPLANT
— **in·sem·i·na·tion** \-,se-mə-'nā-shən\ *noun*

in·sem·i·na·tor \-'se-mə-,nā-tər\ *noun* (1944)
: one that inseminates cattle artificially

in·sen·sate \(,)in-'sen-,sāt, -sət\ *adjective* [Late Latin *insensatus,* from Latin *in-* + Late Latin *sensatus* having sense, from Latin *sensus* sense] (15th century)
1 : lacking sense or understanding; *also* **:** FOOLISH
2 : lacking animate awareness or sensation
3 : lacking humane feeling **:** BRUTAL
— **in·sen·sate·ly** *adverb*

in·sen·si·ble \(,)in-'sen(t)-sə-bəl\ *adjective* [Middle English, from Middle French & Latin; Middle French, from Latin *insensibilis,* from *in-* + *sensibilis* sensible] (14th century)
1 : IMPERCEPTIBLE ⟨dampened by an *insensible* dew⟩; *broadly* **:** SLIGHT, GRADUAL ⟨*insensible* motion⟩
2 : incapable or bereft of feeling or sensation: as **a :** not endowed with life or spirit **:** INSENTIENT ⟨*insensible* earth⟩ **b :** UNCONSCIOUS ⟨knocked *insensible* by a sudden blow⟩ **c :** lacking sensory perception or ability to react ⟨*insensible* to pain⟩
3 a : lacking emotional response **:** APATHETIC, INDIFFERENT ⟨*insensible* to fear⟩ **b :** UNAWARE ⟨*insensible* of their danger⟩
4 *archaic* **:** STUPID, SENSELESS
5 : not intelligible **:** MEANINGLESS
6 : lacking delicacy or refinement
— **in·sen·si·bil·i·ty** \(,)in-,sen(t)-sə-'bi-lə-tē\ *noun*
— **in·sen·si·ble·ness** \(,)in-'sen(t)-sə-bəl-nəs\ *noun*
— **in·sen·si·bly** \-blē\ *adverb*

in·sen·si·tive \(,)in-'sen(t)-s(ə-)tiv\ *adjective* (1834)
1 a : not responsive or susceptible ⟨*insensitive* to the demands of the public⟩ **b :** lacking feeling or tact ⟨so *insensitive* as to laugh at someone in pain⟩
2 : not physically or chemically sensitive
— **in·sen·si·tive·ly** *adverb*
— **in·sen·si·tive·ness** *noun*
— **in·sen·si·tiv·i·ty** \(,)in-,sen(t)-sə-'ti-və-tē\ *noun*

in·sen·tient \(,)in-'sen(t)-sh(ē-)ənt\ *adjective* (1764)
: lacking perception, consciousness, or animation
— **in·sen·tience** \-sh(ē-)ən(t)s\ *noun*

in·sep·a·ra·ble \(,)in-'se-p(ə-)rə-bəl\ *adjective* [Middle English, from Latin *inseparabilis,* from *in-* + *separabilis* separable] (14th century)

◇ WORD HISTORY

insect The distinct components into which insects' bodies are divided—head, thorax, and abdomen—inspired the Greek name used for them by Aristotle: *entomon,* the "notched" or "segmented" animal. (*Entomon* is a noun derived from the verb *entemnein* "to cut up" or "to cut into.") The Roman naturalist Pliny used *insectum,* a literal translation of Greek *entomon,* as his name for the invertebrate, and this Latin word has provided the modern scientific name for the class comprising insects as well as the ordinary English word for one such creature. The name of the discipline dealing with insects, *entomology,* reverts to a Greek base, however, like the names of most other branches of zoology.

: incapable of being separated or disjoined
— **in·sep·a·ra·bil·i·ty** \(,)in-,se-p(ə-)rə-'bi-lə-tē\ *noun*
— **inseparable** *noun*
— **in·sep·a·ra·ble·ness** \(,)in-'se-p(ə-)rə-bəl-nəs\ *noun*
— **in·sep·a·ra·bly** \-blē\ *adverb*

¹in·sert \in-'sərt\ *verb* [Latin *insertus,* past participle of *inserere,* from *in-* + *serere* to join — more at SERIES] (1529)
transitive verb
1 : to put or thrust in ⟨*insert* the key in the lock⟩ ⟨*insert* a spacecraft into orbit⟩
2 : to put or introduce into the body of something **:** INTERPOLATE ⟨*insert* a change in a manuscript⟩
3 : to set in and make fast; *especially* **:** to insert by sewing between two cut edges
4 : to place into action (as in a game) ⟨*insert* a new pitcher⟩
intransitive verb
of a muscle **:** to be in attachment to the part to be moved
synonym see INTRODUCE
— **in·sert·er** *noun*

²in·sert \'in-,sərt\ *noun* (circa 1889)
: something that is inserted or is for insertion; *especially* **:** written or printed material inserted (as between the leaves of a book)

in·ser·tion \in-'sər-shən\ *noun* (1539)
1 : something that is inserted: as **a :** the part of a muscle that inserts **b :** the mode or place of attachment of an organ or part **c :** embroidery or needlework inserted as ornament between two pieces of fabric **d :** a section of genetic material that is inserted into an existing gene sequence
2 a : the act or process of inserting **b :** the mutational process producing a genetic insertion
— **in·ser·tion·al** \-shnəl, -shə-nᵊl\ *adjective*

in–ser·vice \'in-'sər-vəs\ *adjective* (1928)
1 : going on or continuing while one is fully employed ⟨*in-service* teacher education workshops⟩
2 : of, relating to, or being one that is fully employed ⟨*in-service* police officers⟩

¹in·set \'in-,set\ *noun* (1559)
1 a : a place where something flows in **:** CHANNEL **b :** a setting or flowing in
2 : something that is inset: as **a :** a small graphic representation (as a map or picture) set within a larger one **b :** a piece of cloth set into a garment (as for decoration) **c :** a part or section of a utensil that fits into an outer part

²in·set \'in-,set, in-'\ *transitive verb* **inset** *or* **in·set·ted; in·set·ting** (1658)
1 : SET IN; *especially* **:** to insert within something else in such a way as to be visible
2 : to provide with an inset

¹in·shore \'in-'shōr, -'shȯr\ *adjective* (1701)
1 : situated, living, or carried on near shore
2 : moving toward shore ⟨an *inshore* current⟩

²inshore *adverb* (1748)
: to or toward shore ⟨boats driven *inshore* by the storm⟩

¹in·side \(,)in-'sīd, 'in-,\ *noun* (14th century)
1 a : an interior or internal part or place **:** the part within **b :** inward nature, thoughts, or feeling **c :** VISCERA, ENTRAILS — usually used in plural
2 : an inner side or surface
3 a : a position of power, trust, or familiarity ⟨only someone on the *inside* could have told⟩ **b :** confidential information ⟨has the *inside* on what happened at the convention⟩
4 : the area nearest a specified or implied point of reference: as **a :** the side of home plate nearest the batter **b :** the middle portion of a playing area **c :** the area near or underneath the basket in basketball

²inside *adverb* (15th century)
1 : on the inner side
2 : in or into the interior
3 : to or on the inside
4 : in prison

³inside *adjective* (1611)
1 : of, relating to, or being on or near the inside ⟨an *inside* pitch⟩
2 a : relating or known to a select group ⟨*inside* information⟩ **b :** BEHIND-THE-SCENES

⁴inside *preposition* (1791)
1 a : in or into or as if in or into the interior of **b :** on the inner side of
2 : WITHIN ⟨*inside* an hour⟩

inside address *noun* (circa 1941)
: ADDRESS 5c

inside of *preposition* (1839)
: INSIDE

inside out *adverb* (circa 1600)
1 : in such a manner that the inner surface becomes the outer ⟨turned the shirt *inside out*⟩
2 : to a thorough degree ⟨knows the subject *inside out*⟩
3 : in or into a state of disarray often involving drastic reorganization ⟨turned the business *inside out*⟩

in·sid·er \(,)in-'sī-dər, 'in-,\ *noun* (1848)
: a person recognized or accepted as a member of a group, category, or organization: as **a :** a person who is in a position of power or has access to confidential information **b :** one (as an officer or director) who is in a position to have special knowledge of the affairs of or to influence the decisions of a company

insider trading *noun* (1966)
: the illegal use of insider information for profit in financial trading

inside track *noun* (1857)
: an advantageous competitive position ⟨the owner's son has the *inside track* for the job⟩

in·sid·i·ous \in-'si-dē-əs\ *adjective* [Latin *insidiosus,* from *insidiae* ambush, from *insidēre* to sit in, sit on, from *in-* + *sedēre* to sit — more at SIT] (1545)
1 a : awaiting a chance to entrap **:** TREACHEROUS **b :** harmful but enticing **:** SEDUCTIVE ⟨*insidious* drugs⟩
2 a : having a gradual and cumulative effect **:** SUBTLE ⟨the *insidious* pressures of modern life⟩ **b** *of a disease* **:** developing so gradually as to be well established before becoming apparent
— **in·sid·i·ous·ly** *adverb*
— **in·sid·i·ous·ness** *noun*

in·sight \'in-,sīt\ *noun* (13th century)
1 : the power or act of seeing into a situation **:** PENETRATION
2 : the act or result of apprehending the inner nature of things or of seeing intuitively
synonym see DISCERNMENT

in·sight·ful \'in-,sīt-fəl, in-'\ *adjective* (1907)
: exhibiting or characterized by insight
— **in·sight·ful·ly** *adverb*

in·sig·nia \in-'sig-nē-ə\ *or* **in·sig·ne** \-(,)nē\ *noun, plural* **-nia** *or* **-ni·as** [Latin *insignia,* plural of *insigne* mark, badge, from neuter of *insignis* marked, distinguished, from *in-* + *signum* mark — more at SIGN] (1648)
1 : a badge of authority or honor
2 : a distinguishing mark or sign

in·sig·nif·i·cance \,in(t)-sig-'ni-fi-kən(t)s\ *noun* (1699)
: the quality or state of being insignificant

in·sig·nif·i·can·cy \-kən(t)-sē\ *noun, plural* **-cies** (1651)
1 : INSIGNIFICANCE
2 : an insignificant thing or person

in·sig·nif·i·cant \-kənt\ *adjective* (1651)
: not significant: as **a :** lacking meaning or import **:** INCONSEQUENTIAL **b :** not worth considering **:** UNIMPORTANT **c :** lacking weight, position, or influence **:** CONTEMPTIBLE **d :** small in size, quantity, or number
— **in·sig·nif·i·cant·ly** *adverb*

in·sin·cere \,in-sin-'sir, -sən-\ *adjective* [Latin *insincerus,* from *in-* + *sincerus* sincere] (1634)
: not sincere **:** HYPOCRITICAL
— **in·sin·cere·ly** *adverb*
— **in·sin·cer·i·ty** \-'ser-ə-tē *also* -'sir-\ *noun*

in·sin·u·ate \in-'sin-yə-,wāt\ *verb* **-at·ed; -at·ing** [Latin *insinuatus,* past participle of *insinuare,* from *in-* + *sinuare* to bend, curve, from *sinus* curve] (1529)
transitive verb
1 a : to introduce (as an idea) gradually or in a subtle, indirect, or covert way ⟨*insinuate* doubts into a trusting mind⟩ **b :** to impart or communicate with artful or oblique reference
2 : to introduce (as oneself) by stealthy, smooth, or artful means
intransitive verb
1 *archaic* **:** to enter gently, slowly, or imperceptibly **:** CREEP
2 *archaic* **:** to ingratiate oneself
synonym see INTRODUCE, SUGGEST
— **in·sin·u·a·tive** \-,wā-tiv\ *adjective*
— **in·sin·u·a·tor** \-,wā-tər\ *noun*

insinuating *adjective* (1591)
1 : winning favor and confidence by imperceptible degrees **:** INGRATIATING
2 : tending gradually to cause doubt, distrust, or change of outlook often in a slyly subtle manner ⟨*insinuating* remarks⟩
— **in·sin·u·at·ing·ly** *adverb*

in·sin·u·a·tion \(,)in-,sin-yə-'wā-shən\ *noun* (1526)
1 : the act or process of insinuating
2 : something that is insinuated; *especially* **:** a sly, subtle, and usually derogatory utterance

in·sip·id \in-'si-pəd\ *adjective* [French & Late Latin; French *insipide,* from Late Latin *insipidus,* from Latin *in-* + *sapidus* savory, from *sapere* to taste — more at SAGE] (1609)
1 : lacking taste or savor **:** TASTELESS
2 : lacking in qualities that interest, stimulate, or challenge **:** DULL, FLAT ☆
— **in·si·pid·i·ty** \,in-sə-'pi-də-tē\ *noun*
— **in·sip·id·ly** \in-'si-pəd-lē\ *adverb*

in·sist \in-'sist\ *verb* [Middle French or Latin; Middle French *insister,* from Latin *insistere* to stand upon, persist, from *in-* + *sistere* to take a stand; akin to Latin *stare* to stand — more at STAND] (1586)
intransitive verb
1 : to be emphatic, firm, or resolute about something intended, demanded, or required ⟨they *insist* on going⟩
2 *archaic* **:** PERSIST
transitive verb
: to maintain in a persistent or positive manner ⟨*insisted* that the story was true⟩

in·sis·tence \in-'sis-tən(t)s\ *noun* (15th century)
1 : the act or an instance of insisting
2 : the quality or state of being insistent **:** URGENCY

in·sis·ten·cy \-tən(t)-sē\ *noun, plural* **-cies** (1859)
: INSISTENCE

☆ SYNONYMS
Insipid, vapid, flat, jejune, banal, inane mean devoid of qualities that make for spirit and character. INSIPID implies a lack of sufficient taste or savor to please or interest ⟨an *insipid* romance with platitudes on every page⟩. VAPID suggests a lack of liveliness, force, or spirit ⟨an exciting story given a *vapid* treatment⟩. FLAT applies to things that have lost their sparkle or zest ⟨although well-regarded in its day, the novel now seems *flat*⟩. JEJUNE suggests a lack of rewarding or satisfying substance ⟨a *jejune* and gassy speech⟩. BANAL stresses the complete absence of freshness, novelty, or immediacy ⟨a *banal* tale of unrequited love⟩. INANE implies a lack of any significant or convincing quality ⟨an *inane* interpretation of the play⟩.

in·sis·tent \in-ˈsis-tənt\ *adjective* [Latin *insistent-, insistens,* present participle of *insistere*] (1868)
1 : disposed to insist **:** PERSISTENT
2 : compelling attention ⟨the *insistent* pounding of the waves⟩
— in·sis·tent·ly *adverb*

in si·tu \(ˌ)in-ˈsī-(ˌ)tü, -ˈsi-, -(ˌ)tyü *also* -ˈsē-, -(ˌ)chü\ *adverb or adjective* [Latin, in position] (1740)
: in the natural or original position or place

in·so·bri·ety \ˌin-sə-ˈbrī-ə-tē, -sō-\ *noun* (circa 1611)
: lack of sobriety or moderation; *especially* **:** intemperance in drinking

in·so·cia·ble \(ˌ)in-ˈsō-shə-bəl\ *adjective* [Latin *insociabilis,* from *in-* + *sociabilis* sociable] (1588)
: not sociable
— in·so·cia·bil·i·ty \-(ˌ)sō-shə-ˈbi-lə-tē\ *noun*
— in·so·cia·bly \-ˈsō-shə-blē\ *adverb*

in·so·far \ˌin-sə-ˈfär\ *adverb* (1596)
: to such extent or degree

insofar as *conjunction* (15th century)
: to the extent or degree that

in·so·la·tion \ˌin-(ˌ)sō-ˈlā-shən\ *noun* [French or Latin; French, from Latin *insolation-, insolatio,* from *insolare* to expose to the sun, from *in-* + *sol* sun — more at SOLAR] (1617)
1 : exposure to the sun's rays
2 : SUNSTROKE
3 a : solar radiation that has been received **b :** the rate of delivery of direct solar radiation per unit of horizontal surface; *broadly* **:** that relating to total solar radiation

in·sole \ˈin-ˌsōl\ *noun* (circa 1861)
1 : an inside sole of a shoe
2 : a loose thin strip placed inside a shoe for warmth or comfort

in·so·lence \ˈin(t)-s(ə-)lən(t)s\ *noun* (14th century)
1 : the quality or state of being insolent
2 : an instance of insolent conduct or treatment

in·so·lent \-s(ə-)lənt\ *adjective* [Middle English, from Latin *insolent-, insolens* unaccustomed, overbearing, from *in-* + *solens,* present participle of *solēre* to be accustomed; perhaps akin to Latin *sodalis* comrade — more at SIB] (14th century)
1 : insultingly contemptuous in speech or conduct **:** OVERBEARING
2 : exhibiting boldness or effrontery **:** IMPUDENT
synonym see PROUD
— insolent *noun*
— in·so·lent·ly *adverb*

in·sol·u·bi·lize \(ˌ)in-ˈsäl-yə-bə-ˌlīz\ *transitive verb* (1897)
: to make insoluble
— in·sol·u·bi·li·za·tion \-ˌsäl-yə-bə-lə-ˈzā-shən\ *noun*

in·sol·u·ble \(ˌ)in-ˈsäl-yə-bəl\ *adjective* [Middle English *insolible,* from Latin *insolubilis,* from *in-* + *solvere* to free, dissolve — more at SOLVE] (14th century)
: not soluble: as **a** *archaic* **:** INDISSOLUBLE **b :** having or admitting of no solution or explanation **c :** incapable of being dissolved in a liquid; *also* **:** soluble only with difficulty or to a slight degree
— in·sol·u·bil·i·ty \-ˌsäl-yə-ˈbi-lə-tē\ *noun*
— insoluble *noun*
— in·sol·u·ble·ness \-ˈsäl-yə-bəl-nəs\ *noun*
— in·sol·u·bly \-blē\ *adverb*

in·solv·able \(ˌ)in-ˈsäl-və-bəl, -ˈsȯl-\ *adjective* (1693)
: admitting no solution ⟨an apparently *insolvable* problem⟩
— in·solv·ably \-blē\ *adverb*

in·sol·vent \(ˌ)in-ˈsäl-vənt, -ˈsȯl-\ *adjective* (1591)
1 a (1) **:** unable to pay debts as they fall due in the usual course of business (2) **:** having liabilities in excess of a reasonable market value of assets held **b :** insufficient to pay all debts ⟨an *insolvent* estate⟩ **c :** not up to a normal standard or complement **:** IMPOVERISHED
2 : relating to or for the relief of insolvents
— in·sol·ven·cy \-vən(t)-sē\ *noun*
— insolvent *noun*

in·som·nia \in-ˈsäm-nē-ə\ *noun* [Latin, from *insomnis* sleepless, from *in-* + *somnus* sleep — more at SOMNOLENT] (circa 1623)
: prolonged and usually abnormal inability to obtain adequate sleep
— in·som·ni·ac \-nē-ˌak\ *adjective or noun*

in·so·much as \ˌin(t)-sō-ˈməch-\ *conjunction* (14th century)
: INASMUCH AS

insomuch that *conjunction* (14th century)
: SO 1

in·sou·ci·ance \in-ˈsü-sē-ən(t)s, a^{n}-süs-yä^{n}s\ *noun* [French, from *in-* + *soucier* to trouble, disturb, from Latin *sollicitare* — more at SOLICIT] (1799)
: lighthearted unconcern **:** NONCHALANCE
— in·sou·ci·ant \in-ˈsü-sē-ənt, a^{n}-süs-yä^{n}\ *adjective*
— in·sou·ci·ant·ly \in-ˈsü-sē-ənt-lē\ *adverb*

in·soul *variant of* ENSOUL

in·span \in-ˈspan, ˈin-ˌ\ *verb* [Afrikaans, from Dutch *inspannen*] (circa 1827)
chiefly South African **:** YOKE, HARNESS

in·spect \in-ˈspekt\ *verb* [Latin *inspectus,* past participle of *inspicere,* from *in-* + *specere* to look — more at SPY] (circa 1623)
transitive verb
1 : to view closely in critical appraisal **:** look over
2 : to examine officially ⟨*inspects* the barracks every Friday⟩
intransitive verb
: to make an inspection
synonym see SCRUTINIZE
— in·spec·tive \-ˈspek-tiv\ *adjective*

in·spec·tion \in-ˈspek-shən\ *noun* (14th century)
1 a : the act of inspecting **b :** recognition of a familiar pattern leading to immediate solution of a mathematical problem ⟨solve an equation by *inspection*⟩
2 : a checking or testing of an individual against established standards

inspection arms *noun* [from the command *inspection arms!*] (circa 1884)
: a position in the manual of arms in which the rifle is held at port arms with the chamber open for inspection; *also* **:** a command to assume this position

in·spec·tor \in-ˈspek-tər\ *noun* (1602)
1 : a person employed to inspect something
2 a : a police officer who is in charge of usually several precincts and ranks below a superintendent or deputy superintendent **b :** a person appointed to oversee a polling place
— in·spec·tor·ship \-ˌship\ *noun*

in·spec·tor·ate \in-ˈspek-t(ə-)rət\ *noun* (1762)
1 : the office, position, work, or district of an inspector
2 : a body of inspectors

inspector general *noun* (1702)
: a person who heads an inspectorate or a system of inspection (as of an army)

in·sphere *variant of* ENSPHERE

in·spi·ra·tion \ˌin(t)-spə-ˈrā-shən, -(ˌ)spi-\ *noun* (14th century)
1 a : a divine influence or action on a person believed to qualify him or her to receive and communicate sacred revelation **b :** the action or power of moving the intellect or emotions **c :** the act of influencing or suggesting opinions
2 : the act of drawing in; *specifically* **:** the drawing of air into the lungs
3 a : the quality or state of being inspired **b :** something that is inspired ⟨a scheme that was pure *inspiration*⟩
4 : an inspiring agent or influence
— in·spi·ra·tion·al \-shnəl, -shə-nᵊl\ *adjective*
— in·spi·ra·tion·al·ly *adverb*

in·spi·ra·tor \ˈin(t)-spə-ˌrā-tər, -(ˌ)spi-\ *noun* (1624)
: one that inspires ⟨teachers who are *inspirators* of the young⟩

in·spi·ra·to·ry \in-ˈspī-rə-ˌtōr-ē, ˈin(t)-sp(ə-)rə-, -ˌtȯr-\ *adjective* (1773)
: of, relating to, used for, or associated with inspiration

in·spire \in-ˈspīr\ *verb* **in·spired; in·spir·ing** [Middle English, from Middle French & Latin; Middle French *inspirer,* from Latin *inspirare,* from *in-* + *spirare* to breathe] (14th century)
transitive verb
1 a : to influence, move, or guide by divine or supernatural inspiration **b :** to exert an animating, enlivening, or exalting influence on ⟨was particularly *inspired* by the Romanticists⟩ **c :** to spur on **:** IMPEL, MOTIVATE ⟨threats don't necessarily *inspire* people to work⟩ **d :** AFFECT ⟨seeing the old room again *inspired* him with nostalgia⟩
2 a *archaic* **:** to breathe or blow into or upon **b** *archaic* **:** to infuse (as life) by breathing
3 a : to communicate to an agent supernaturally **b :** to draw forth or bring out ⟨thoughts *inspired* by a visit to the cathedral⟩
4 : INHALE 1
5 a : BRING ABOUT, OCCASION ⟨the book was *inspired* by his travels in the Far East⟩ **b :** INCITE
6 : to spread (rumor) by indirect means or through the agency of another
intransitive verb
: INHALE
— in·spir·er *noun*

inspired *adjective* (15th century)
: outstanding or brilliant in a way or to a degree suggestive of divine inspiration ⟨gave an *inspired* performance⟩

inspiring *adjective* (1717)
: having an animating or exalting effect

in·spir·it \in-ˈspir-ət\ *transitive verb* (15th century)
: to fill with spirit
synonym see ENCOURAGE
— in·spir·it·ing·ly \-ə-tiŋ-lē\ *adverb*

in·spis·sate \in-ˈspi-ˌsāt, ˈin(t)-spə-ˌsāt\ *transitive verb* **-sat·ed; -sat·ing** [Late Latin *inspissatus,* past participle of *inspissare,* from Latin *in-* + *spissus* slow, dense; akin to Greek *spidnos* compact, Lithuanian *spisti* to form a swarm] (1626)
: to make thick or thicker
— in·spis·sa·tion \ˌin(t)-spə-ˈsā-shən, (ˌ)in-ˌspi-ˈsā-\ *noun*
— in·spis·sa·tor \in-ˈspi-ˌsā-tər, ˈin(t)-spə-ˌsā-\ *noun*

in·spis·sat·ed \in-ˈspi-ˌsā-təd, ˈin(t)-spə-ˌsā-\ *adjective* (1603)
: thickened in consistency; *broadly* **:** made or having become thick, heavy, or intense

in·sta·bil·i·ty \ˌin(t)-stə-ˈbi-lə-tē\ *noun* (15th century)
: the quality or state of being unstable; *especially* **:** lack of emotional or mental stability

in·sta·ble \(ˌ)in-ˈstā-bəl\ *adjective* [Middle French or Latin; Middle French, from Latin *instabilis,* from *in-* + *stabilis* stable] (15th century)
: UNSTABLE

in·stall *also* **in·stal** \in-ˈstȯl\ *transitive verb* **in·stalled; in·stall·ing** [Middle English, from Middle French *installer,* from Medieval Latin *installare,* from Latin *in-* + Medieval Latin *stallum* stall, from Old High German *stal*] (15th century)

1 a : to place in an office or dignity by seating in a stall or official seat **b :** to induct into an office, rank, or order ⟨*installed* the new president⟩
2 : to establish in an indicated place, condition, or status ⟨*installing* herself in front of the fireplace⟩
3 : to set up for use or service ⟨had an exhaust fan *installed* in the kitchen⟩
— in·stall·er *noun*

in·stal·la·tion \ˌin(t)-stə-ˈlā-shən\ *noun* (15th century)
1 : the act of installing **:** the state of being installed
2 : something that is installed for use
3 : a military camp, fort, or base

¹in·stall·ment *also* **in·stal·ment** \in-ˈstȯl-mənt\ *noun* (1589)
: INSTALLATION 1

²installment *also* **instalment** *noun* [alteration of earlier *estallment* payment by installment, derivative of Old French *estaler* to place, fix, from *estal* place, of Germanic origin; akin to Old High German *stal* place, stall] (1776)
1 : one of the parts into which a debt is divided when payment is made at intervals
2 a : one of several parts (as of a publication) presented at intervals **b :** one part of a serial story
— installment *adjective*

installment plan *noun* (1876)
: a system of paying for goods by installments

¹in·stance \ˈin(t)-stən(t)s\ *noun* (14th century)
1 a *archaic* **:** urgent or earnest solicitation **b :** INSTIGATION, REQUEST ⟨am writing to you at the *instance* of my client⟩ **c** *obsolete* **:** an impelling cause or motive
2 a *archaic* **:** EXCEPTION **b :** an individual illustrative of a category or brought forward in support or disproof of a generalization **c** *obsolete* **:** TOKEN, SIGN
3 : the institution and prosecution of a lawsuit **:** SUIT
4 : a step, stage, or situation viewed as part of a process or series of events ⟨prefers, in this *instance*, to remain anonymous —*Times Literary Supplement*⟩ ☆
— for in·stance \fə-ˈrin(t)-stən(t)s, ˈfrin(t)-\ **:** as an instance or example ⟨older people like my grandmother, *for instance*⟩

²instance *transitive verb* **in·stanced; in·stanc·ing** (1601)
1 : to illustrate or demonstrate by an instance
2 : to mention as a case or example **:** CITE

in·stan·cy \ˈin(t)-stən(t)-sē\ *noun, plural* **-cies** (1515)
1 : URGENCY, INSISTENCE
2 : nearness of approach **:** IMMINENCE
3 : immediacy of occurrence or action **:** INSTANTANEOUSNESS

¹in·stant \ˈin(t)-stənt\ *noun* [Middle English, from Medieval Latin *instant-, instans,* from *instant-, instans,* adjective, instant, from Latin] (14th century)
1 : an infinitesimal space of time; *especially* **:** a point in time separating two states ⟨at the *instant* of death⟩
2 : the present or current month

²instant *adjective* [Middle English, from Middle French or Latin; Middle French, from Latin *instant-, instans,* from present participle of *instare* to stand upon, urge, from *in-* + *stare* to stand — more at STAND] (15th century)
1 : IMPORTUNATE, URGENT
2 a : PRESENT, CURRENT ⟨previous felonies not related to the *instant* crime⟩ **b :** of or occurring in the present month
3 : IMMEDIATE, DIRECT ⟨the play was an *instant* success⟩
4 a (1) **:** premixed or precooked for easy final preparation ⟨*instant* pudding⟩ (2) **:** appearing in or as if in ready-to-use form ⟨*instant* culture⟩ **b :** immediately soluble in water ⟨*instant* coffee⟩
5 : produced or occurring with or as if with extreme rapidity and ease
— in·stant·ness *noun*

in·stan·ta·neous \ˌin(t)-stən-ˈtā-nē-əs, -nyəs\ *adjective* [Medieval Latin *instantaneus,* from *instant-, instans,* noun] (1651)
1 : done, occurring, or acting without any perceptible duration of time ⟨death was *instantaneous*⟩
2 : done without any delay being purposely introduced ⟨took *instantaneous* action to correct the abuse⟩
3 : occurring or present at a particular instant ⟨*instantaneous* velocity⟩
— in·stan·ta·ne·ity \ˌin-ˌstan-tᵊn-ˈē-ə-tē, ˌin(t)-stən-tə-ˈnē-\ *noun*
— in·stan·ta·neous·ly \ˌin(t)-stən-ˈtā-nē-əs-lē, -nyəs-lē\ *adverb*
— in·stan·ta·neous·ness *noun*

in·stan·ter \in-ˈstan-tər\ *adverb* [Medieval Latin, from *instant-, instans*] (1688)
: at once

in·stan·ti·ate \in-ˈstan(t)-shē-ˌāt\ *transitive verb* **-at·ed; -at·ing** (1949)
: to represent (an abstraction) by a concrete instance ⟨heroes *instantiate* ideals —W. J. Bennett⟩
— in·stan·ti·a·tion \-ˌstan(t)-shē-ˈā-shən\ *noun*

¹in·stant·ly \ˈin(t)-stənt-lē\ *adverb* (15th century)
1 : with importunity **:** URGENTLY
2 : without the least delay **:** IMMEDIATELY

²instantly *conjunction* (1793)
: as soon as ⟨he ran across the grass *instantly* he perceived his mother —W. M. Thackeray⟩

instant replay *noun* (1966)
: a videotape recording of an action (as a play in football) that can be played back (as in slow motion) immediately after the action has been completed; *also* **:** the playing of such a recording

in·star \ˈin-ˌstär\ *noun* [New Latin, from Latin, equivalent] (1895)
: a stage in the life of an arthropod (as an insect) between two successive molts; *also* **:** an individual in a specified instar

in·state \in-ˈstāt\ *transitive verb* (1603)
1 *obsolete* **a :** INVEST, ENDOW **b :** BESTOW, CONFER
2 : to set or establish in a rank or office **:** INSTALL

in sta·tu quo \in-ˌstā-(ˌ)tü-ˈkwō, -ˌsta-; -ˌsta-(ˌ)chü-\ *adverb* [New Latin, literally, in the state in which] (1602)
: in the former or same state

in·stau·ra·tion \ˌin-ˌstȯ-ˈrā-shən, ˌin(t)-stə-\ *noun* [Latin *instauration-, instauratio,* from *instaurare* to renew, restore — more at STORE] (circa 1603)
1 : restoration after decay, lapse, or dilapidation
2 : an act of instituting or establishing something

in·stead \in-ˈsted\ *adverb* (1667)
1 : as a substitute or equivalent ⟨was going to write but called *instead*⟩
2 : as an alternative to something expressed or implied **:** RATHER ⟨longed *instead* for a quiet country life⟩

in·stead of \in-ˈste-də(v), -ˈsti-\ *preposition* [Middle English *in sted of*] (13th century)
: in place of **:** as a substitute for or alternative to

in·step \ˈin-ˌstep\ *noun* (15th century)
1 : the arched middle portion of the human foot in front of the ankle joint; *especially* **:** its upper surface
2 : the part of a shoe or stocking that fits over the instep

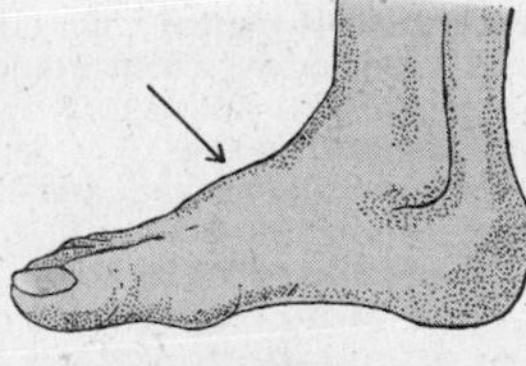
instep 1

in·sti·gate \ˈin(t)-stə-ˌgāt\ *transitive verb* **-gat·ed; -gat·ing** [Latin *instigatus,* past participle of *instigare* — more at STICK] (1542)
: to goad or urge forward **:** PROVOKE
synonym see INCITE
— in·sti·ga·tion \ˌin(t)-stə-ˈgā-shən\ *noun*
— in·sti·ga·tive \ˈin(t)-stə-ˌgā-tiv\ *adjective*
— in·sti·ga·tor \-ˌgā-tər\ *noun*

in·still *also* **in·stil** \in-ˈstil\ *transitive verb* **in·stilled; in·still·ing** [Middle English, from Middle French & Latin; Middle French *instiller,* from Latin *instillare,* from *in-* + *stillare* to drip, from *stilla* drop] (15th century)
1 : to cause to enter drop by drop ⟨*instill* medication into the infected eye⟩
2 : to impart gradually ⟨*instilling* a love of learning in children⟩
synonym see IMPLANT
— in·stil·la·tion \ˌin(t)-stə-ˈlā-shən, -(ˌ)sti-\ *noun*
— in·still·er \in-ˈsti-lər\ *noun*
— in·still·ment \-mənt\ *noun*

¹in·stinct \ˈin-ˌstiŋ(k)t\ *noun* [Middle English, from Latin *instinctus* impulse, from *instinguere* to incite; akin to Latin *instigare* to instigate] (15th century)
1 : a natural or inherent aptitude, impulse, or capacity ⟨had an *instinct* for the right word⟩
2 a : a largely inheritable and unalterable tendency of an organism to make a complex and specific response to environmental stimuli without involving reason **b :** behavior that is mediated by reactions below the conscious level
— in·stinc·tu·al \in-ˈstiŋ(k)-chə-wəl, -chəl, -shwəl\ *adjective*
— in·stinc·tu·al·ly *adverb*

²in·stinct \in-ˈstiŋ(k)t, ˈin-ˌ\ *adjective* (1667)
1 *obsolete* **:** impelled by an inner or animating or exciting agency
2 : profoundly imbued **:** INFUSED ⟨my mood, *instinct* with romance —S. J. Perelman⟩

in·stinc·tive \in-ˈstiŋ(k)-tiv\ *adjective* (15th century)
1 : of, relating to, or being instinct
2 : prompted by natural instinct or propensity **:** arising spontaneously ⟨an *instinctive* fear of innovation —V. L. Parrington⟩
synonym see SPONTANEOUS
— in·stinc·tive·ly *adverb*

¹in·sti·tute \ˈin(t)-stə-ˌtüt, -ˌtyüt\ *transitive verb* **-tut·ed; -tut·ing** [Middle English, from

☆ SYNONYMS
Instance, case, illustration, example, sample, specimen mean something that exhibits distinguishing characteristics in its category. INSTANCE applies to any individual person, act, or thing that may be offered to illustrate or explain ⟨an *instance* of history repeating itself⟩. CASE is used to direct attention to a real or assumed occurrence or situation that is to be considered, studied, or dealt with ⟨a *case* of mistaken identity⟩. ILLUSTRATION applies to an instance offered as a means of clarifying or illuminating a general statement ⟨a telling *illustration* of Murphy's Law⟩. EXAMPLE applies to a typical, representative, or illustrative instance or case ⟨a typical *example* of bureaucratic waste⟩. SAMPLE implies a part or unit taken at random from a larger whole and so presumed to be typical of its qualities ⟨show us a *sample* of your work⟩. SPECIMEN applies to any example or sample whether representative or merely existent and available ⟨one of the finest *specimens* of the jeweler's art⟩.

Latin *institutus,* past participle of *instituere,* from *in-* + *statuere* to set up — more at STATUTE] (14th century)
1 : to establish in a position or office
2 a : to originate and get established **:** ORGANIZE ⟨*instituted* reading clinics⟩ **b :** to set going **:** INAUGURATE ⟨*instituting* an investigation of the charges⟩
— **in·sti·tut·er** *or* **in·sti·tu·tor** \-ˌtü-tər, -ˌtyü-\ *noun*

²institute *noun* (1546)
: something that is instituted: as **a** (1) **:** an elementary principle recognized as authoritative (2) *plural* **:** a collection of such principles and precepts; *especially* **:** a legal compendium **b :** an organization for the promotion of a cause **:** ASSOCIATION ⟨a research *institute*⟩ ⟨an *institute* for the blind⟩ **c :** an educational institution and especially one devoted to technical fields **d :** a usually brief intensive course of instruction on selected topics relating to a particular field ⟨an urban studies *institute*⟩

in·sti·tu·tion \ˌin(t)-stə-ˈtü-shən, -ˈtyü-\ *noun* (14th century)
1 : an act of instituting **:** ESTABLISHMENT
2 a : a significant practice, relationship, or organization in a society or culture ⟨the *institution* of marriage⟩; *also* **:** something or someone firmly associated with a place or thing ⟨she has become an *institution* in the theater⟩ **b :** an established organization or corporation (as a college or university) especially of a public character; *also* **:** ASYLUM 4
— **in·sti·tu·tion·al** \-shnəl, -shə-nᵊl\ *adjective*
— **in·sti·tu·tion·al·ly** *adverb*

in·sti·tu·tion·al·ise *British variant of* INSTITUTIONALIZE

in·sti·tu·tion·al·ism \-shnə-ˌli-zəm, -shə-nᵊl-ˌi-zəm\ *noun* (1862)
1 : emphasis on organization (as in religion) at the expense of other factors
2 : public institutional care of handicapped, delinquent, or dependent persons
3 : an economic school of thought that emphasizes the role of social institutions in influencing economic behavior
— **in·sti·tu·tion·al·ist** \-shnə-list, -shə-nᵊl-ist\ *noun*

in·sti·tu·tion·al·ize \-shnə-ˌlīz, -shə-nᵊl-ˌīz\ *transitive verb* **-ized; -iz·ing** (1865)
1 : to make into an institution **:** give character of an institution to ⟨*institutionalized* housing⟩; *especially* **:** to incorporate into a structured and often highly formalized system ⟨*institutionalized* values⟩
2 : to put in the care of an institution ⟨*institutionalize* alcoholics⟩
— **in·sti·tu·tion·al·i·za·tion** \-ˌtü-shnə-lə-ˈzā-shən, -shə-nᵊl-ə-ˈzā-\ *noun*

in–store \ˈin-ˈstōr, -ˈstȯr\ *adjective* (1961)
: relating to or being an operation or activity located or taking place inside a store ⟨*in-store* consumer survey⟩

in·struct \in-ˈstrəkt\ *transitive verb* [Middle English, from Latin *instructus,* past participle of *instruere,* from *in-* + *struere* to build — more at STRUCTURE] (15th century)
1 : to give knowledge to **:** TEACH, TRAIN
2 : to provide with authoritative information or advice ⟨the judge *instructed* the jury⟩
3 : to give an order or command to **:** DIRECT
synonym see TEACH, COMMAND

in·struc·tion \in-ˈstrək-shən\ *noun* (15th century)
1 a : PRECEPT ⟨prevailing cultural *instructions*⟩ **b :** a direction calling for compliance **:** ORDER — usually used in plural ⟨had *instructions* not to admit strangers⟩ **c** *plural* **:** an outline or manual of technical procedure **:** DIRECTIONS **d :** a code that tells a computer to perform a particular operation
2 : the action, practice, or profession of teaching
— **in·struc·tion·al** \-shnəl, -shə-nᵊl\ *adjective*

in·struc·tive \in-ˈstrək-tiv\ *adjective* (1611)
: carrying a lesson **:** ENLIGHTENING
— **in·struc·tive·ly** *adverb*
— **in·struc·tive·ness** *noun*

in·struc·tor \in-ˈstrək-tər\ *noun* (15th century)
: one that instructs **:** TEACHER; *especially* **:** a college teacher below professorial rank
— **in·struc·tor·ship** \-ˌship\ *noun*

in·struc·tress \-ˈstrək-trəs\ *noun* (1630)
: a woman who is an instructor

¹in·stru·ment \ˈin(t)-strə-mənt\ *noun* [Middle English, from Latin *instrumentum,* from *instruere* to arrange, instruct] (14th century)
1 : a device used to produce music
2 a : a means whereby something is achieved, performed, or furthered **b :** one used by another as a means or aid **:** DUPE, TOOL
3 : IMPLEMENT; *especially* **:** one designed for precision work
4 : a formal legal document (as a deed, bond, or agreement)
5 a : a measuring device for determining the present value of a quantity under observation **b :** an electrical or mechanical device used in navigating an airplane; *especially* **:** such a device used as the sole means of navigating
synonym see IMPLEMENT

²in·stru·ment \-ˌment\ *transitive verb* (1752)
1 : to address a legal instrument to
2 : to score for musical performance **:** ORCHESTRATE
3 : to equip with instruments especially for measuring and recording data ⟨an *instrumented* spacecraft⟩

in·stru·men·tal \ˌin(t)-strə-ˈmen-tᵊl\ *adjective* (14th century)
1 a : serving as a means, agent, or tool ⟨was *instrumental* in organizing the strike⟩ **b :** of, relating to, or done with an instrument or tool
2 : relating to, composed for, or performed on a musical instrument
3 : of, relating to, or being a grammatical case or form expressing means or agency
4 : of or relating to instrumentalism
5 : OPERANT 3 ⟨*instrumental* learning⟩ ⟨*instrumental* conditioning⟩
— **instrumental** *noun*
— **in·stru·men·tal·ly** \-tᵊl-ē\ *adverb*

in·stru·men·tal·ism \-ˌi-zəm\ *noun* (1909)
: a doctrine that ideas are instruments of action and that their usefulness determines their truth

in·stru·men·tal·ist \-ist\ *noun* (1823)
1 : a player on a musical instrument
2 : an exponent of instrumentalism
— **instrumentalist** *adjective*

in·stru·men·tal·i·ty \ˌin(t)-strə-mən-ˈta-lə-tē, -ˌmen-\ *noun, plural* **-ties** (1651)
1 : the quality or state of being instrumental
2 : MEANS, AGENCY

in·stru·men·ta·tion \ˌin(t)-strə-mən-ˈtā-shən, -ˌmen-\ *noun* (1845)
1 : the arrangement or composition of music for instruments especially for a band or orchestra
2 : the use or application of instruments (as for observation, measurement, or control)
3 : instruments for a particular purpose; *also* **:** a selection or arrangement of instruments

instrument flying *noun* (1928)
: navigation of an airplane by instruments only

instrument landing *noun* (1938)
: a landing made with limited visibility by means of instruments and by ground radio direction

instrument panel *noun* (1922)
: a panel on which instruments are mounted; *especially* **:** DASHBOARD 2

in·sub·or·di·nate \ˌin(t)-sə-ˈbȯr-dᵊn-ət, -ˈbȯrd-nət\ *adjective* (circa 1828)
: disobedient to authority
— **insubordinate** *noun*
— **in·sub·or·di·nate·ly** *adverb*
— **in·sub·or·di·na·tion** \-ˌbȯr-dᵊn-ˈā-shən\ *noun*

in·sub·stan·tial \ˌin(t)-səb-ˈstan(t)-shəl\ *adjective* [probably from French *insubstantiel,* from Late Latin *insubstantialis,* from Latin *in-* + Late Latin *substantialis* substantial] (1607)
: not substantial: as **a :** lacking substance or material nature **b :** lacking firmness or solidity **:** FLIMSY
— **in·sub·stan·ti·al·i·ty** \-ˌstan(t)-shē-ˈa-lə-tē\ *noun*

in·suf·fer·able \(ˌ)in-ˈsə-f(ə-)rə-bəl\ *adjective* (15th century)
: not to be endured **:** INTOLERABLE ⟨an *insufferable* bore⟩
— **in·suf·fer·able·ness** *noun*
— **in·suf·fer·ably** \-blē\ *adverb*

in·suf·fi·cien·cy \ˌin(t)-sə-ˈfi-shən(t)-sē\ *noun, plural* **-cies** (1526)
1 : the quality or state of being insufficient: as **a :** lack of mental or moral fitness **:** INCOMPETENCE ⟨the *insufficiency* of this person for public office⟩ **b :** lack of adequate supply ⟨*insufficiency* of provisions⟩ **c :** lack of physical power or capacity; *specifically* **:** inability of an organ or body part to function normally
2 : something that is insufficient or falls short of expectations

in·suf·fi·cient \ˌin(t)-sə-ˈfi-shənt\ *adjective* [Middle English, from Middle French, from Late Latin *insufficient-, insufficiens,* from Latin *in-* + *sufficient-, sufficiens* sufficient] (14th century)
: not sufficient **:** INADEQUATE; *especially* **:** lacking adequate power, capacity, or competence
— **in·suf·fi·cient·ly** *adverb*

in·suf·fla·tion \ˌin(t)-sə-ˈflā-shən, in-ˌsə-ˈflā-\ *noun* [Middle English *insufflacion,* from Middle French *insufflation,* from Late Latin *insufflation-, insufflatio,* from *insufflare* to blow upon, from Latin *in-* + *sufflare* to inflate, from *sub-* + *flare* to blow — more at BLOW] (15th century)
: an act or the action of blowing on, into, or in: as **a :** a Christian ceremonial rite of exorcism performed by breathing on a person **b :** the act of blowing something (as a gas, powder, or vapor) into a body cavity
— **in·suf·flate** \ˈin(t)-sə-ˌflāt, in-ˈsə-ˌflāt\ *transitive verb*
— **in·suf·fla·tor** \-ˌflā-tər\ *noun*

in·su·lant \ˈin(t)-sə-lənt\ *noun* (circa 1929)
chiefly British **:** INSULATION 2

in·su·lar \ˈin(t)-sü-lər, -syü-, ˈin-shə-lər\ *adjective* [Late Latin *insularis,* from Latin *insula* island] (1611)
1 a : of, relating to, or constituting an island **b :** dwelling or situated on an island ⟨*insular* residents⟩
2 : characteristic of an isolated people; *especially* **:** being, having, or reflecting a narrow provincial viewpoint
3 : of or relating to an island of cells or tissue
— **in·su·lar·ism** \-lə-ˌri-zəm\ *noun*
— **in·su·lar·i·ty** \ˌin(t)-sü-ˈlar-ə-tē, -syü-, ˌin-shə-ˈlar-\ *noun*
— **in·su·lar·ly** \ˈin(t)-sü-lər-lē, -syü-, ˈin-shə-\ *adverb*

in·su·late \ˈin(t)-sə-ˌlāt\ *transitive verb* **-lat·ed; -lat·ing** [Latin *insula*] (circa 1741)
: to place in a detached situation **:** ISOLATE; *especially* **:** to separate from conducting bodies by means of nonconductors so as to prevent transfer of electricity, heat, or sound

in·su·la·tion \ˌin(t)-sə-ˈlā-shən\ *noun* (1798)
1 a : the action of insulating **b :** the state of being insulated
2 : material used in insulating

in·su·la·tor \ˈin(t)-sə-ˌlā-tər\ *noun* (1801)
: one that insulates: as **a :** a material that is a poor conductor (as of electricity or heat) — compare SEMICONDUCTOR **b :** a device made of an electrical insulating material and used for separating or supporting conductors

in·su·lin \ˈin(t)-s(ə-)lən\ *noun* [New Latin *insula* islet (of Langerhans), from Latin, island] (1914)

: a protein pancreatic hormone secreted by the islets of Langerhans that is essential especially for the metabolism of carbohydrates and is used in the treatment and control of diabetes mellitus

insulin shock *noun* (1925)
: hypoglycemia associated with the presence of excessive insulin in the system and characterized by progressive development of coma

1in·sult \in-ˈsəlt\ *verb* [Middle French or Latin; Middle French *insulter,* from Latin *insultare,* literally, to spring upon, from *in-* + *saltare* to leap — more at SALTATION] (1540)
intransitive verb
archaic **:** to behave with pride or arrogance **:** VAUNT
transitive verb
: to treat with insolence, indignity, or contempt **:** AFFRONT; *also* **:** to affect offensively or damagingly ⟨doggerel that *insults* the reader's intelligence⟩
synonym see OFFEND
— **in·sult·er** *noun*
— **in·sult·ing·ly** \in-ˈsəl-tiŋ-lē\ *adverb*

2in·sult \ˈin-ˌsəlt\ *noun* (1671)
1 : a gross indignity
2 : injury to the body or one of its parts; *also* **:** something that causes or has a potential for causing such insult ⟨pollution and other environmental *insults*⟩

in·su·per·a·ble \(ˌ)in-ˈsü-p(ə-)rə-bəl\ *adjective* [Middle English, from Middle French & Latin; Middle French, from Latin *insuperabilis,* from *in-* + *superare* to surmount, from *super* over — more at OVER] (14th century)
: incapable of being surmounted, overcome, passed over, or solved ⟨*insuperable* difficulties⟩
— **in·su·per·a·bly** \-blē\ *adverb*

in·sup·port·able \ˌin(t)-sə-ˈpōr-tə-bəl, -ˈpȯr-\ *adjective* [Middle French or Late Latin; Middle French, from Late Latin *insupportabilis,* from Latin *in-* + *supportare* to support] (circa 1530)
: not supportable: **a :** more than can be endured ⟨*insupportable* pain⟩ **b :** impossible to justify ⟨*insupportable* charges⟩
— **in·sup·port·ably** \-blē\ *adverb*

in·sup·press·ible \ˌin(t)-sə-ˈpre-sə-bəl\ *adjective* (1610)
: IRREPRESSIBLE

in·sur·able \in-ˈshu̇r-ə-bəl\ *adjective* (1810)
: that may be insured
— **in·sur·abil·i·ty** \-ˌshu̇r-ə-ˈbi-lə-tē\ *noun*

1in·sur·ance \in-ˈshu̇r-ən(t)s *also* ˈin-ˌ\ *noun* (1651)
1 a : the business of insuring persons or property **b :** coverage by contract whereby one party undertakes to indemnify or guarantee another against loss by a specified contingency or peril **c :** the sum for which something is insured
2 : a means of guaranteeing protection or safety ⟨the contract is your *insurance* against price changes⟩

2insurance *adjective* (1954)
: being a score that adds to a team's lead and makes it impossible for the opposing team to tie the game with its next score ⟨*insurance* run⟩

in·sure \in-ˈshu̇r\ *verb* **in·sured; in·sur·ing** [Middle English, to assure, probably alteration of *assuren*] (1635)
transitive verb
1 : to provide or obtain insurance on or for
2 : to make certain especially by taking necessary measures and precautions
intransitive verb
: to contract to give or take insurance
synonym see ENSURE

insured *noun* (1681)
: a person whose life or property is insured

in·sur·er \in-ˈshu̇r-ər\ *noun* (1654)
: one that insures; *specifically* **:** an insurance underwriter

in·sur·gence \in-ˈsər-jən(t)s\ *noun* (1847)
: an act or the action of being insurgent **:** INSURRECTION

in·sur·gen·cy \-jən(t)-sē\ *noun, plural* **-cies** (1803)
1 : the quality or state of being insurgent; *specifically* **:** a condition of revolt against a government that is less than an organized revolution and that is not recognized as belligerency
2 : INSURGENCE

1in·sur·gent \-jənt\ *noun* [Latin *insurgent-, insurgens,* present participle of *insurgere* to rise up, from *in-* + *surgere* to rise — more at SURGE] (1765)
1 : a person who revolts against civil authority or an established government; *especially* **:** a rebel not recognized as a belligerent
2 : one who acts contrary to the policies and decisions of one's own political party

2insurgent *adjective* (1814)
: rising in opposition to civil authority or established leadership **:** REBELLIOUS
— **in·sur·gent·ly** *adverb*

in·sur·mount·able \ˌin(t)-sər-ˈmau̇n-tə-bəl\ *adjective* (1690)
: incapable of being surmounted **:** INSUPERABLE ⟨*insurmountable* problems⟩
— **in·sur·mount·ably** \-blē\ *adverb*

in·sur·rec·tion \ˌin(t)-sə-ˈrek-shən\ *noun* [Middle English, from Middle French, from Late Latin *insurrection-, insurrectio,* from *insurgere*] (15th century)
: an act or instance of revolting against civil authority or an established government
synonym see REBELLION
— **in·sur·rec·tion·al** \-shnəl, -shə-nᵊl\ *adjective*
— **in·sur·rec·tion·ary** \-shə-ˌner-ē\ *adjective or noun*
— **in·sur·rec·tion·ist** \-sh(ə-)nist\ *noun*

in·sus·cep·ti·ble \ˌin(t)-sə-ˈsep-tə-bəl\ *adjective* (1603)
: not susceptible ⟨*insusceptible* to flattery⟩
— **in·sus·cep·ti·bil·i·ty** \-ˌsep-tə-ˈbi-lə-tē\ *noun*
— **in·sus·cep·ti·bly** \ˌin(t)-sə-ˈsep-tə-blē\ *adverb*

in·tact \in-ˈtakt\ *adjective* [Middle English *intacte,* from Latin *intactus,* from *in-* + *tactus,* past participle of *tangere* to touch — more at TANGENT] (15th century)
1 : untouched especially by anything that harms or diminishes **:** ENTIRE, UNINJURED
2 *of a living body or its parts* **:** having no relevant component removed or destroyed: **a :** physically virginal **b :** not castrated
synonym see PERFECT
— **in·tact·ness** \-ˈtak(t)-nəs\ *noun*

in·ta·glio \in-ˈtal-(ˌ)yō, -ˈtäl-; -ˈta-glē-ˌō, -ˈtä-\ *noun, plural* **-glios** [Italian, from *intagliare* to engrave, cut, from Medieval Latin *intaliare,* from Latin *in-* + Late Latin *taliare* to cut — more at TAILOR] (1644)
1 a : an engraving or incised figure in stone or other hard material depressed below the surface so that an impression from the design yields an image in relief **b :** the art or process of executing intaglios **c :** printing (as in die stamping and gravure) done from a plate in which the image is sunk below the surface
2 : something (as a gem) carved in intaglio

intaglio 1a

in·take \ˈin-ˌtāk\ *noun* (15th century)
1 : an opening through which fluid enters an enclosure
2 a : a taking in **b** (1) **:** the amount taken in (2) **:** something (as energy) taken in **:** INPUT

1in·tan·gi·ble \(ˌ)in-ˈtan-jə-bəl\ *adjective* [French or Medieval Latin; French, from Medieval Latin *intangibilis,* from Latin *in-* + Late Latin *tangibilis* tangible] (1640)
: not tangible **:** IMPALPABLE
— **in·tan·gi·bil·i·ty** \-ˌtan-jə-ˈbi-lə-tē\ *noun*
— **in·tan·gi·ble·ness** \-ˈtan-jə-bəl-nəs\ *noun*
— **in·tan·gi·bly** \-blē\ *adverb*

2intangible *noun* (1914)
: something intangible; *specifically* **:** an asset (as goodwill) that is not corporeal

in·tar·sia \in-ˈtär-sē-ə\ *noun* [German, modification of Italian *intarsio*] (1867)
1 : a mosaic usually of wood fitted into a support; *also* **:** the art or process of making such a mosaic
2 : a colored design knitted on both sides of a fabric (as in a sweater)

in·te·ger \ˈin-ti-jər\ *noun* [Latin, adjective, whole, entire — more at ENTIRE] (1571)
1 : any of the natural numbers, the negatives of these numbers, or zero
2 : a complete entity

in·te·gra·ble \ˈin-ti-grə-bəl\ *adjective* (circa 1741)
: capable of being integrated ⟨*integrable* functions⟩
— **in·te·gra·bil·i·ty** \ˌin-ti-grə-ˈbi-lə-tē\ *noun*

1in·te·gral \ˈin-ti-grəl (*usually so in mathematics*); in-ˈte-grəl *also* -ˈtē- *also* ÷ˈin-trə-gəl\ *adjective* (1551)
1 a : essential to completeness **:** CONSTITUENT ⟨an *integral* part of the curriculum⟩ **b** (1) **:** being, containing, or relating to one or more mathematical integers (2) **:** relating to or concerned with mathematical integrals or integration **c :** formed as a unit with another part ⟨a seat with *integral* headrest⟩
2 : composed of integral parts
3 : lacking nothing essential **:** ENTIRE
— **in·te·gral·i·ty** \ˌin-tə-ˈgra-lə-tē\ *noun*
— **in·te·gral·ly** \ˈin-ti-grə-lē; in-ˈte-grə- *also* -ˈtē-\ *adverb*

2integral *noun* (circa 1741)
: the result of a mathematical integration — compare DEFINITE INTEGRAL, INDEFINITE INTEGRAL

integral calculus *noun* (circa 1741)
: a branch of mathematics concerned with the theory and applications (as in the determination of lengths, areas, and volumes and in the solution of differential equations) of integrals and integration

integral domain *noun* (1937)
: a mathematical ring in which multiplication is commutative, which has a multiplicative identity element, and which contains no pair of nonzero elements whose product is zero ⟨the integers under the operations of addition and multiplication form an *integral domain*⟩

in·te·grand \ˈin-tə-ˌgrand\ *noun* [Latin *integrandus,* gerundive of *integrare*] (1897)
: a mathematical expression to be integrated

in·te·grate \ˈin-tə-ˌgrāt\ *verb* **-grat·ed; -grat·ing** [Latin *integratus,* past participle of *integrare,* from *integr-, integer*] (1638)
transitive verb
1 : to form, coordinate, or blend into a functioning or unified whole **:** UNITE
2 : to find the integral of (as a function or equation)
3 a : to unite with something else **b :** to incorporate into a larger unit
4 a : to end the segregation of and bring into equal membership in society or an organization **b :** DESEGREGATE ⟨*integrate* school districts⟩
intransitive verb
: to become integrated

integrated *adjective* (1922)
1 : marked by the unified control of all aspects of industrial production from raw materials

through distribution of finished products ⟨*integrated* companies⟩ ⟨*integrated* production⟩
2 : characterized by integration and especially racial integration ⟨an *integrated* society⟩ ⟨*integrated* schools⟩

integrated circuit *noun* (1962)
: a tiny complex of electronic components and their connections that is produced in or on a small slice of material (as silicon)
— **integrated circuitry** *noun*

in·te·gra·tion \ˌin-tə-ˈgrā-shən\ *noun* (1620)
1 : the act or process or an instance of integrating: as **a :** incorporation as equals into society or an organization of individuals of different groups (as races) **b :** coordination of mental processes into a normal effective personality or with the individual's environment
2 a : the operation of finding a function whose differential is known **b :** the operation of solving a differential equation

in·te·gra·tion·ist \-sh(ə-)nist\ *noun* (1951)
: a person who believes in, advocates, or practices social integration
— **integrationist** *adjective*

in·te·gra·tive \ˈin-tə-ˌgrā-tiv\ *adjective* (1862)
: serving to integrate or favoring integration **:** directed toward integration

in·te·gra·tor \-ˌgrā-tər\ *noun* (1876)
: one that integrates; *especially* **:** a device or computer unit that totalizes variable quantities in a manner comparable to mathematical integration

in·teg·ri·ty \in-ˈte-grə-tē\ *noun* [Middle English *integrite,* from Middle French & Latin; Middle French *intégrité,* from Latin *integritat-, integritas,* from *integr-, integer* entire] (14th century)
1 : firm adherence to a code of especially moral or artistic values **:** INCORRUPTIBILITY
2 : an unimpaired condition **:** SOUNDNESS
3 : the quality or state of being complete or undivided **:** COMPLETENESS
synonym see HONESTY

in·teg·u·ment \in-ˈte-gyə-mənt\ *noun* [Latin *integumentum,* from *integere* to cover, from *in-* + *tegere* to cover — more at THATCH] (circa 1611)
: something that covers or encloses; *especially* **:** an enveloping layer (as a skin, membrane, or husk) of an organism or one of its parts
— **in·teg·u·men·ta·ry** \-ˈmen-t(ə-)rē\ *adjective*

in·tel·lect \ˈin-tᵊl-ˌekt\ *noun* [Middle English, from Middle French or Latin; Middle French, from Latin *intellectus,* from *intellegere* to understand — more at INTELLIGENT] (14th century)
1 a : the power of knowing as distinguished from the power to feel and to will **:** the capacity for knowledge **b :** the capacity for rational or intelligent thought especially when highly developed
2 : a person with great intellectual powers

in·tel·lec·tion \ˌin-tᵊl-ˈek-shən\ *noun* (1579)
1 : an act of the intellect **:** THOUGHT
2 : exercise of the intellect **:** REASONING

in·tel·lec·tive \-ˈek-tiv\ *adjective* (15th century)
: having, relating to, or belonging to the intellect **:** RATIONAL
— **in·tel·lec·tive·ly** *adverb*

¹in·tel·lec·tu·al \ˌin-tᵊl-ˈek-chə-wəl, -chəl, -shwəl\ *adjective* (14th century)
1 a : of or relating to the intellect or its use **b :** developed or chiefly guided by the intellect rather than by emotion or experience **:** RATIONAL **c :** requiring use of the intellect
2 a : given to study, reflection, and speculation **b :** engaged in activity requiring the creative use of the intellect
— **in·tel·lec·tu·al·i·ty** \-ˌek-chə-ˈwa-lə-tē\ *noun*
— **in·tel·lec·tu·al·ly** \-ˈek-chə-wə-lē, -chə-lē, -shwə-lē\ *adverb*
— **in·tel·lec·tu·al·ness** \-ˈek-chə-wəl-nəs, -chəl-, -shwəl-\ *noun*

²intellectual *noun* (1615)
1 *plural, archaic* **:** intellectual powers
2 : an intellectual person

in·tel·lec·tu·al·ism \ˌin-tᵊl-ˈek-chə-wə-ˌli-zəm, -chə-ˌli-, -shwə-ˌli-\ *noun* (1838)
: devotion to the exercise of intellect or to intellectual pursuits
— **in·tel·lec·tu·al·ist** \-list\ *noun or adjective*
— **in·tel·lec·tu·al·is·tic** \-ˌek-chə-wə-ˈlis-tik, -chə-ˈlis-, -shwə-ˈlis-\ *adjective*

in·tel·lec·tu·al·ize \ˌin-tᵊl-ˈek-chə-wə-ˌlīz, -chə-ˌlīz, -shwə-ˌlīz\ *transitive verb* **-ized; -iz·ing** (circa 1819)
: to give rational form or content to
— **in·tel·lec·tu·al·i·za·tion** \-ˌek-chə-wə-lə-ˈzā-shən, -chə-lə-, -shwə-lə-\ *noun*
— **in·tel·lec·tu·al·iz·er** \-ˈek-chə-wə-ˌlī-zər, -chə-ˌlī-, -shwə-ˌlī-\ *noun*

in·tel·li·gence \in-ˈte-lə-jən(t)s\ *noun* [Middle English, from Middle French, from Latin *intelligentia,* from *intelligent-, intelligens* intelligent] (14th century)
1 a (1) **:** the ability to learn or understand or to deal with new or trying situations **:** REASON; *also* **:** the skilled use of reason (2) **:** the ability to apply knowledge to manipulate one's environment or to think abstractly as measured by objective criteria (as tests) **b** *Christian Science* **:** the basic eternal quality of divine Mind **c :** mental acuteness **:** SHREWDNESS
2 a : an intelligent entity; *especially* **:** ANGEL **b :** intelligent minds or mind ⟨cosmic *intelligence*⟩
3 : the act of understanding **:** COMPREHENSION
4 a : INFORMATION, NEWS **b :** information concerning an enemy or possible enemy or an area; *also* **:** an agency engaged in obtaining such information
5 : the ability to perform computer functions

intelligence quotient *noun* (1916)
: a number used to express the apparent relative intelligence of a person that is the ratio multiplied by 100 of the mental age as reported on a standardized test to the chronological age

in·tel·li·genc·er \in-ˈte-lə-jən(t)-sər; -ˈte-lə-ˌjen(t)-, -ˌte-lə-ˈ\ *noun* (1581)
1 : a secret agent **:** SPY
2 : a bringer of news **:** REPORTER

intelligence test *noun* (1914)
: a test designed to determine the relative mental capacity of a person

in·tel·li·gent \in-ˈte-lə-jənt\ *adjective* [Latin *intelligent-, intelligens,* present participle of *intelligere, intellegere* to understand, from *inter-* + *legere* to gather, select — more at LEGEND] (1509)
1 a : having or indicating a high or satisfactory degree of intelligence and mental capacity **b :** revealing or reflecting good judgment or sound thought **:** SKILLFUL
2 a : possessing intelligence **b :** guided or directed by intellect **:** RATIONAL
3 a : guided or controlled by a computer; *especially* **:** using a built-in microprocessor for automatic operation, for processing of data, or for achieving greater versatility — compare DUMB 7 **b :** able to produce printed material from digital signals ⟨an *intelligent* copier⟩ ☆
— **in·tel·li·gen·tial** \-ˌte-lə-ˈjen(t)-shəl\ *adjective*
— **in·tel·li·gent·ly** \-ˈte-lə-jənt-lē\ *adverb*

in·tel·li·gent·sia \in-ˌte-lə-ˈjen(t)-sē-ə, -ˈgen(t)-\ *noun* [Russian *intelligentsiya,* from Latin *intelligentia* intelligence] (1907)
: intellectuals who form an artistic, social, or political vanguard or elite

in·tel·li·gi·ble \in-ˈte-lə-jə-bəl\ *adjective* [Middle English, from Latin *intelligibilis,* from *intelligere*] (14th century)
1 : apprehensible by the intellect only
2 : capable of being understood or comprehended
— **in·tel·li·gi·bil·i·ty** \-ˌte-lə-jə-ˈbi-lə-tē\ *noun*
— **in·tel·li·gi·ble·ness** \-ˈte-lə-jə-bəl-nəs\ *noun*
— **in·tel·li·gi·bly** \-blē\ *adverb*

in·tem·per·ance \(ˌ)in-ˈtem-p(ə-)rən(t)s\ *noun* (15th century)
: lack of moderation; *especially* **:** habitual or excessive drinking of intoxicants

in·tem·per·ate \-p(ə-)rət\ *adjective* [Middle English *intemperat,* from Latin *intemperatus,* from *in-* + *temperatus,* past participle of *temperare* to temper] (14th century)
: not temperate ⟨*intemperate* criticism⟩; *especially* **:** given to excessive use of intoxicating liquors
— **in·tem·per·ate·ly** *adverb*
— **in·tem·per·ate·ness** *noun*

in·tend \in-ˈtend\ *verb* [Middle English *entenden, intenden,* from Middle French *entendre* to purpose, from Latin *intendere* to stretch out, direct, aim at, from *in-* + *tendere* to stretch — more at THIN] (14th century)
transitive verb
1 : to direct the mind on
2 *archaic* **:** to proceed on (a course)
3 a : SIGNIFY, MEAN **b :** to refer to
4 a : to have in mind as a purpose or goal **:** PLAN **b :** to design for a specified use or future
intransitive verb
archaic **:** SET OUT, START
— **in·tend·er** *noun*

in·ten·dance \in-ˈten-dən(t)s\ *noun* (1739)
1 : MANAGEMENT, SUPERINTENDENCE
2 : an administrative department

in·ten·dant \-dənt\ *noun* [French, from Middle French, from Latin *intendent-, intendens,* present participle of *intendere* to intend, attend] (1652)
: an administrative official (as a governor) especially under the French, Spanish, or Portuguese monarchies

¹intended *adjective* (15th century)
1 : expected to be such in the future ⟨an *intended* career⟩ ⟨his *intended* bride⟩
2 : INTENTIONAL
— **in·tend·ed·ly** *adverb*

²intended *noun* (1767)
: the person to whom another is engaged

intending *adjective* (1788)
: PROSPECTIVE, ASPIRING ⟨an *intending* teacher⟩

in·tend·ment \in-ˈten(d)-mənt\ *noun* (14th century)
: the true meaning or intention especially of a law

in·ten·er·ate \in-ˈte-nə-ˌrāt\ *transitive verb* **-at·ed; -at·ing** [²*in-* + Latin *tener* soft, tender — more at TENDER] (1595)
: to make tender **:** SOFTEN
— **in·ten·er·a·tion** \-ˌte-nə-ˈrā-shən\ *noun*

in·tense \in-ˈten(t)s\ *adjective* [Middle English, from Middle French, from Latin *intensus,* from past participle of *intendere* to stretch out] (15th century)
1 a : existing in an extreme degree ⟨the excitement was *intense*⟩ ⟨*intense* pain⟩ **b :** having or showing a characteristic in extreme degree ⟨*intense* colors⟩

☆ SYNONYMS
Intelligent, clever, alert, quick-witted mean mentally keen or quick. INTELLIGENT stresses success in coping with new situations and solving problems ⟨an *intelligent* person could assemble it fast⟩. CLEVER implies native ability or aptness and sometimes suggests a lack of more substantial qualities ⟨*clever* with words⟩. ALERT stresses quickness in perceiving and understanding ⟨*alert* to new technology⟩. QUICK-WITTED implies promptness in finding answers in debate or in devising expedients in moments of danger or challenge ⟨no match for his *quick-witted* opponent⟩.

2 : marked by or expressive of great zeal, energy, determination, or concentration ⟨*intense* effort⟩
3 a : exhibiting strong feeling or earnestness of purpose ⟨an *intense* student⟩ **b :** deeply felt
— **in·tense·ly** *adverb*
— **in·tense·ness** *noun*

in·ten·si·fi·er \in-ˈten(t)-sə-ˌfī(-ə)r\ *noun* (1835)
: one that intensifies; *especially* **:** INTENSIVE

in·ten·si·fy \in-ˈten(t)-sə-ˌfī\ *verb* **-fied; -fy·ing** (1817)
transitive verb
1 : to make intense or more intensive **:** STRENGTHEN
2 a : to increase the density and contrast of (a photographic image) by chemical treatment **b :** to make more acute **:** SHARPEN
intransitive verb
: to become intense or more intensive **:** grow stronger or more acute
— **in·ten·si·fi·ca·tion** \-ˌten(t)s-fə-ˈkā-shən, -ˌten(t)-sə-\ *noun*

in·ten·sion \in-ˈten(t)-shən\ *noun* (1604)
1 : INTENSITY
2 : CONNOTATION 3
— **in·ten·sion·al** \-ˈtench-nəl, -ˈten(t)-shə-nᵊl\ *adjective*
— **in·ten·sion·al·i·ty** \-ˌten(t)-shə-ˈna-lə-tē\ *noun*
— **in·ten·sion·al·ly** \-ˈtench-nə-lē, -ˈten(t)-shə-nᵊl-ē\ *adverb*

in·ten·si·ty \in-ˈten(t)-sə-tē\ *noun, plural* **-ties** (1665)
1 : the quality or state of being intense; *especially* **:** extreme degree of strength, force, energy, or feeling
2 : the magnitude of a quantity (as force or energy) per unit (as of area, charge, mass, or time)
3 : SATURATION 4a

¹in·ten·sive \in-ˈten(t)-siv\ *adjective* (15th century)
: of, relating to, or marked by intensity or intensification: as **a :** highly concentrated ⟨*intensive* study⟩ **b :** tending to strengthen or increase; *especially* **:** tending to give force or emphasis ⟨*intensive* adverb⟩ **c :** constituting or relating to a method designed to increase productivity by the expenditure of more capital and labor rather than by increase in scope ⟨*intensive* farming⟩
— **in·ten·sive·ly** *adverb*
— **in·ten·sive·ness** *noun*

²intensive *noun* (1813)
: an intensive linguistic element

intensive care *noun* (1963)
: special medical equipment and services for taking care of seriously ill patients ⟨heart patients in *intensive care*⟩ ⟨an *intensive care* unit⟩

¹in·tent \in-ˈtent\ *noun* [Middle English *entent,* from Old French, from Late Latin *intentus,* from Latin, act of stretching out, from *intendere*] (13th century)
1 a : the act or fact of intending **:** PURPOSE; *especially* **:** the design or purpose to commit a wrongful or criminal act ⟨admitted wounding him with *intent*⟩ **b :** the state of mind with which an act is done **:** VOLITION
2 : a usually clearly formulated or planned intention **:** AIM
3 a : MEANING, SIGNIFICANCE **b :** CONNOTATION 3
synonym see INTENTION

²intent *adjective* [Latin *intentus,* from past participle of *intendere*] (14th century)
1 : directed with strained or eager attention **:** CONCENTRATED
2 : having the mind, attention, or will concentrated on something or some end or purpose ⟨*intent* on their work⟩
— **in·tent·ly** *adverb*
— **in·tent·ness** *noun*

in·ten·tion \in-ˈten(t)-shən\ *noun* (14th century)
1 : a determination to act in a certain way **:** RESOLVE
2 : IMPORT, SIGNIFICANCE
3 a : what one intends to do or bring about **b :** the object for which a prayer, mass, or pious act is offered
4 : a process or manner of healing of incised wounds
5 : CONCEPT; *especially* **:** a concept considered as the product of attention directed to an object of knowledge
6 *plural* **:** purpose with respect to marriage ☆

in·ten·tion·al \in-ˈtench-nəl, -ˈten(t)-shə-nᵊl\ *adjective* (circa 1727)
1 : done by intention or design **:** INTENDED ⟨*intentional* damage⟩
2 a : of or relating to epistemological intention **b :** having external reference
synonym see VOLUNTARY
— **in·ten·tion·al·i·ty** \-ˌten(t)-shə-ˈna-lə-tē\ *noun*
— **in·ten·tion·al·ly** \in-ˈtench-nə-lē, -ˈten(t)-shə-nᵊl-ē\ *adverb*

in·ter \in-ˈtər\ *transitive verb* **in·terred; in·ter·ring** [Middle English *enteren,* from Middle French *enterrer,* from (assumed) Vulgar Latin *interrare,* from *in-* + Latin *terra* earth — more at TERRACE] (14th century)
: to deposit (a dead body) in the earth or in a tomb

inter- *prefix* [Middle English *inter-, enter-,* from Middle French & Latin; Middle French *inter-, entre-,* from Latin *inter-,* from *inter;* akin to Old High German *untar* among, Greek *enteron* intestine, Old English *in* in]
1 : between **:** among **:** in the midst ⟨*inter*crop⟩ ⟨*inter*penetrate⟩ ⟨*inter*stellar⟩
2 : reciprocal ⟨*inter*relation⟩ **:** reciprocally ⟨*inter*marry⟩
3 : located between ⟨*inter*face⟩
4 : carried on between ⟨*inter*national⟩
5 : occurring between ⟨*inter*borough⟩ **:** intervening ⟨*inter*glacial⟩
6 : shared by, involving, or derived from two or more ⟨*inter*faith⟩
7 : between the limits of **:** within ⟨*inter*tropical⟩
8 : existing between ⟨*inter*communal⟩ ⟨*inter*company⟩

in·ter–Af·ri·can
in·ter·agen·cy
in·ter·al·le·lic
in·ter–Amer·i·can
in·ter·an·i·ma·tion
in·ter·an·nu·al
in·ter·as·so·ci·a·tion
in·ter·atom·ic
in·ter·avail·abil·i·ty
in·ter·bank
in·ter·ba·sin
in·ter·bed
in·ter·be·hav·ior
in·ter·be·hav·ior·al
in·ter·bor·ough
in·ter·branch
in·ter·cal·i·bra·tion
in·ter·cam·pus
in·ter·caste
in·ter·cell
in·ter·cel·lu·lar
in·ter·chain
in·ter·chan·nel
in·ter·chro·mo·som·al
in·ter·church
in·ter·city
in·ter·clan
in·ter·class
in·ter·club
in·ter·clus·ter
in·ter·coast·al
in·ter·co·lo·nial
in·ter·com·mu·nal
in·ter·com·mu·ni·ty
in·ter·com·pa·ny
in·ter·com·pare
in·ter·com·par·i·son
in·ter·com·pre·hen·si·bil·i·ty
in·ter·cor·po·rate
in·ter·cor·re·late
in·ter·cor·re·la·tion
in·ter·cor·ti·cal
in·ter·coun·try
in·ter·coun·ty
in·ter·cou·ple
in·ter·cra·ter
in·ter·crys·tal·line
in·ter·cul·tur·al
in·ter·cul·tur·al·ly
in·ter·cul·ture
in·ter·deal·er
in·ter·de·nom·i·na·tion·al
in·ter·de·part·men·tal
in·ter·de·part·men·tal·ly
in·ter·de·pend
in·ter·de·pen·dence
in·ter·de·pen·den·cy
in·ter·de·pen·dent
in·ter·de·pen·dent·ly
in·ter·di·a·lec·tal
in·ter·dis·trict
in·ter·di·vi·sion·al
in·ter·do·min·ion
in·ter·elec·trode
in·ter·elec·tron
in·ter·elec·tron·ic
in·ter·ep·i·dem·ic
in·ter·eth·nic
in·ter·fac·ul·ty
in·ter·fa·mil·ial
in·ter·fam·i·ly
in·ter·fi·ber
in·ter·firm
in·ter·flow
in·ter·flu·vi·al
in·ter·fold
in·ter·fra·ter·ni·ty
in·ter·gang
in·ter·gen·er·a·tion
in·ter·gen·er·a·tion·al
in·ter·ge·ner·ic
in·ter·graft
in·ter·gran·u·lar
in·ter·group
in·ter·hemi·spher·ic
in·ter·in·di·vid·u·al
in·ter·in·dus·try
in·ter·in·flu·ence
in·ter·in·sti·tu·tion·al
in·ter·in·volve
in·ter·ion·ic
in·ter·is·land
in·ter·ju·ris·dic·tion·al
in·ter·la·cus·trine
in·ter·lam·i·nar
in·ter·lay
in·ter·lay·er
in·ter·lend
in·ter·li·brary
in·ter·lin·er
in·ter·lob·u·lar
in·ter·lo·cal
in·ter·male
in·ter·mar·gin·al
in·ter·mem·brane
in·ter·men·stru·al
in·ter·min·is·te·ri·al
in·ter·mi·tot·ic
in·ter·mo·lec·u·lar
in·ter·mo·lec·u·lar·ly
in·ter·moun·tain
in·ter·nu·cle·ar
in·ter·nu·cle·on
in·ter·nu·cle·on·ic
in·ter·nu·cle·o·tide
in·ter·ob·serv·er
in·ter·ocean
in·ter·oce·an·ic
in·ter·of·fice
in·ter·op·er·a·tive
in·ter·or·bit·al
in·ter·or·gan
in·ter·or·ga·ni·za·tion·al
in·ter·pan·dem·ic
in·ter·par·ish
in·ter·pa·ro·chi·al
in·ter·par·ox·ys·mal
in·ter·par·ti·cle
in·ter·par·ty
in·ter·per·cep·tu·al
in·ter·per·me·ate
in·ter·pha·lan·ge·al
in·ter·plan·e·tary
in·ter·plu·vi·al
in·ter·point
in·ter·pop·u·la·tion
in·ter·pop·u·la·tion·al
in·ter·pro·fes·sion·al
in·ter·pro·vin·cial
in·ter·psy·chic
in·ter·re·gion·al
in·ter·re·li·gious
in·ter·re·nal
in·ter·row
in·ter·school
in·ter·sec·tion·al
in·ter·seg·ment
in·ter·seg·men·tal
in·ter·sen·so·ry
in·ter·so·ci·etal
in·ter·so·ci·ety
in·ter·stage
in·ter·sta·tion
in·ter·stim·u·la·tion
in·ter·stim·u·lus
in·ter·strain
in·ter·strand
in·ter·strat·i·fi·ca·tion
in·ter·strat·i·fy
in·ter·sub·sti·tut·abil·i·ty
in·ter·sub·sti·tut·able
in·ter·sys·tem
in·ter·term

☆ SYNONYMS
Intention, intent, purpose, design, aim, end, object, objective, goal mean what one intends to accomplish or attain. INTENTION implies little more than what one has in mind to do or bring about ⟨announced his *intention* to marry⟩. INTENT suggests clearer formulation or greater deliberateness ⟨the clear *intent* of the statute⟩. PURPOSE suggests a more settled determination ⟨being successful was her *purpose* in life⟩. DESIGN implies a more carefully calculated plan ⟨the order of events came by accident, not *design*⟩. AIM adds to these implications of effort directed toward attaining or accomplishing ⟨her *aim* was to raise film to an art form⟩. END stresses the intended effect of action often in distinction or contrast to the action or means as such ⟨willing to use any means to achieve his *end*⟩. OBJECT may equal END but more often applies to a more individually determined wish or need ⟨his constant *object* was the achievement of pleasure⟩. OBJECTIVE implies something tangible and immediately attainable ⟨their *objective* is to seize the oil fields⟩. GOAL suggests something attained only by prolonged effort and hardship ⟨worked years to reach her *goals*⟩.

in·ter·ter·mi·nal
in·ter·ter·ri·to·ri·al
in·ter·trans·lat·able
in·ter·tri·al
in·ter·trib·al
in·ter·troop
in·ter·union
in·ter·unit
in·ter·uni·ver·si·ty
in·ter·ur·ban
in·ter·val·ley
in·ter·ven·tric·u·lar
in·ter·ver·te·bral
in·ter·vil·lage
in·ter·vis·i·bil·i·ty
in·ter·vis·i·ble
in·ter·vis·i·ta·tion
in·ter·war
in·ter·work
in·ter·work·ing
in·ter·zon·al
in·ter·zone

in·ter·a·bang *variant of* INTERROBANG

in·ter·act \ˌin-tə-ˈrakt\ *intransitive verb* (1839)
: to act upon one another

in·ter·ac·tant \-ˈrak-tənt\ *noun* (1949)
: one that interacts

in·ter·ac·tion \ˌin-tə-ˈrak-shən\ *noun* (1832)
: mutual or reciprocal action or influence
— in·ter·ac·tion·al \-shnəl, -shə-nᵊl\ *adjective*

in·ter·ac·tive \-ˈrak-tiv\ *adjective* (1832)
1 : mutually or reciprocally active
2 : of, relating to, or being a two-way electronic communication system (as a telephone, cable television, or a computer) that involves a user's orders (as for information or merchandise) or responses (as to a poll)
— in·ter·ac·tive·ly *adverb*

in·ter alia \ˌin-tər-ˈā-lē-ə, -ˈä-\ *adverb* [Latin] (1665)
: among other things

in·ter ali·os \-lē-ˌōs\ *adverb* [Latin] (circa 1670)
: among other persons

in·ter·al·lied \ˌin-tər-ˈa-ˌlīd, -ə-ˈlīd\ *adjective* (1917)
: relating to, composed of, or involving allies

in·ter·breed \ˌin-tər-ˈbrēd\ *verb* **-bred** \-ˈbred\; **-breed·ing** (1859)
intransitive verb
: to breed together: as **a :** CROSSBREED **b :** to breed within a closed population
transitive verb
: to cause to breed together

in·ter·ca·la·ry \in-ˈtər-kə-ˌler-ē, ˌin-tər-ˈka-lə-rē\ *adjective* [Latin *intercalarius,* from *intercalare*] (1614)
1 a : inserted in a calendar ⟨an *intercalary* day⟩ **b** *of a year* **:** containing an intercalary period (as a day or month)
2 : inserted between other things or parts **:** INTERPOLATED

in·ter·ca·late \in-ˈtər-kə-ˌlāt\ *transitive verb* **-lat·ed; -lat·ing** [Latin *intercalatus,* past participle of *intercalare,* from *inter-* + *calare* to proclaim, call — more at LOW] (1603)
1 : to insert (as a day) in a calendar
2 : to insert between or among existing elements or layers
synonym see INTRODUCE
— in·ter·ca·la·tion \-ˌtər-kə-ˈlā-shən\ *noun*

in·ter·cede \ˌin-tər-ˈsēd\ *intransitive verb* **-ced·ed; -ced·ing** [Latin *intercedere,* from *inter-* + *cedere* to go] (1597)
: to intervene between parties with a view to reconciling differences **:** MEDIATE
synonym see INTERPOSE
— in·ter·ced·er *noun*

in·ter·cen·sal \ˌin-tər-ˈsen(t)-səl\ *adjective* (1887)
: occurring between censuses ⟨*intercensal* estimates⟩ ⟨*intercensal* period⟩

¹in·ter·cept \ˌin-tər-ˈsept\ *transitive verb* [Middle English, from Latin *interceptus,* past participle of *intercipere,* from *inter-* + *capere* to take, seize — more at HEAVE] (15th century)
1 *obsolete* **:** PREVENT, HINDER
2 a : to stop, seize, or interrupt in progress or course or before arrival **b :** to receive (a communication or signal directed elsewhere) usually secretly
3 *obsolete* **:** to interrupt communication or connection with
4 : to include (part of a curve, surface, or solid) between two points, curves, or surfaces ⟨the part of a circumference *intercepted* between two radii⟩
5 a : to gain possession of (an opponent's pass) **b :** to intercept a pass thrown by (an opponent)

²in·ter·cept \ˈin-tər-ˌsept\ *noun* (1821)
1 : the distance from the origin to a point where a graph crosses a coordinate axis
2 : INTERCEPTION; *especially* **:** the interception of a missile by an interceptor or of a target by a missile
3 : a message, code, or signal that is intercepted (as by monitoring radio communications)

in·ter·cep·tion \ˌin-tər-ˈsep-shən\ *noun* (15th century)
1 a : the action of intercepting **b :** the state of being intercepted
2 : something that is intercepted; *especially* **:** an intercepted forward pass

in·ter·cep·tor *also* **in·ter·cept·er** \ˌin-tər-ˈsep-tər\ *noun* (1598)
: one that intercepts; *specifically* **:** a light high-speed fast-climbing fighter plane or missile designed for defense against raiding bombers or missiles

in·ter·ces·sion \ˌin-tər-ˈse-shən\ *noun* [Middle English, from Middle French or Latin; Middle French, from Latin *intercession-, intercessio,* from *intercedere*] (15th century)
1 : the act of interceding
2 : prayer, petition, or entreaty in favor of another
— in·ter·ces·sion·al \-ˈsesh-nəl, -ˈse-shə-nᵊl\ *adjective*
— in·ter·ces·sor \-ˈse-sər\ *noun*
— in·ter·ces·so·ry \-ˈses-rē, -ˈse-sə-rē\ *adjective*

¹in·ter·change \ˌin-tər-ˈchānj\ *verb* [Middle English *entrechaungen,* from Middle French *entrechangier,* from Old French, from *entre-* inter- + *changier* to change] (14th century)
transitive verb
1 : to put each of (two things) in the place of the other
2 : EXCHANGE
intransitive verb
: to change places mutually
— in·ter·chang·er *noun*

²in·ter·change \ˈin-tər-ˌchānj\ *noun* (15th century)
1 : the act, process, or an instance of interchanging **:** EXCHANGE
2 : a junction of two or more highways by a system of separate levels that permit traffic to pass from one to another without the crossing of traffic streams

in·ter·change·able \ˌin-tər-ˈchān-jə-bəl\ *adjective* (14th century)
: capable of being interchanged; *especially* **:** permitting mutual substitution ⟨*interchangeable* parts⟩
— in·ter·change·abil·i·ty \-ˌchān-jə-ˈbi-lə-tē\ *noun*
— in·ter·change·able·ness \-ˈchān-jə-bəl-nəs\ *noun*
— in·ter·change·ably \-blē\ *adverb*

in·ter·col·le·giate \ˌin-tər-kə-ˈlē-jət, -jē-ət\ *adjective* (circa 1874)
: existing, carried on, or participating in activities between colleges ⟨*intercollegiate* athletics⟩

in·ter·co·lum·ni·a·tion \ˌin-tər-kə-ˌləm-nē-ˈā-shən\ *noun* [Latin *intercolumnium* space between two columns, from *inter-* + *columna* column] (1624)
1 : the clear space between the columns of a series
2 : the system of spacing of the columns of a colonnade

in·ter·com \ˈin-tər-ˌkäm\ *noun* [short for *intercommunication system*] (1940)
: a two-way communication system with a microphone and loudspeaker at each station for localized use

in·ter·com·mu·ni·cate \ˌin-tər-kə-ˈmyü-nə-ˌkāt\ *intransitive verb* (1586)
1 : to exchange communication with one another
2 : to afford passage from one to another
— in·ter·com·mu·ni·ca·tion \-ˌmyü-nə-ˈkā-shən\ *noun*

intercommunication system *noun* (1911)
: INTERCOM

in·ter·com·mu·nion \ˌin-tər-kə-ˈmyü-nyən\ *noun* (1921)
: interdenominational participation in communion

in·ter·con·nect \ˌin-tər-kə-ˈnekt\ (1865)
transitive verb
: to connect with one another
intransitive verb
: to be or become mutually connected
— in·ter·con·nec·tion \-ˈnek-shən\ *noun*

interconnected *adjective* (1865)
1 : mutually joined or related ⟨*interconnected* highways⟩ ⟨*interconnected* political issues⟩
2 : having internal connections between the parts or elements
— in·ter·con·nec·ted·ness *noun*

in·ter·con·ti·nen·tal \ˌin-tər-ˌkän-tᵊn-ˈen-tᵊl\ *adjective* (circa 1855)
1 : extending among continents or carried on between continents
2 : capable of traveling between continents ⟨*intercontinental* ballistic missile⟩

in·ter·con·ver·sion \ˌin-tər-kən-ˈvər-zhən, -shən\ *noun* (1865)
: mutual conversion ⟨*interconversion* of chemical compounds⟩
— in·ter·con·vert \-ˈvərt\ *transitive verb*
— in·ter·con·vert·ibil·i·ty \-ˌvər-tə-ˈbi-lə-tē\ *noun*
— in·ter·con·vert·ible \-ˈvər-tə-bəl\ *adjective*

in·ter·cool·er \ˌin-tər-ˈkü-lər\ *noun* (1899)
: a device for cooling a fluid (as air) between successive heat-generating processes

in·ter·cos·tal \ˌin-tər-ˈkäs-tᵊl\ *adjective* [New Latin *intercostalis,* from Latin *inter-* + *costa* rib — more at COAST] (1597)
: situated or extending between the ribs ⟨*intercostal* spaces⟩ ⟨*intercostal* muscles⟩
— intercostal *noun*

in·ter·course \ˈin-tər-ˌkōrs, -ˌkȯrs\ *noun* [Middle English *intercurse,* probably from Middle French *entrecours,* from Medieval Latin *intercursus,* from Latin, act of running between, from *intercurrere* to run between, from *inter-* + *currere* to run — more at CAR] (15th century)
1 : connection or dealings between persons or groups
2 : exchange especially of thoughts or feelings **:** COMMUNION
3 : physical sexual contact between individuals that involves the genitalia of at least one person ⟨heterosexual *intercourse*⟩ ⟨anal *intercourse*⟩ ⟨oral *intercourse*⟩; *especially* **:** SEXUAL INTERCOURSE 1

in·ter·crop \ˌin-tər-ˈkräp, ˈin-tər-ˌ\ (1898)
transitive verb
: to grow a crop in between (another)
intransitive verb
: to grow two or more crops simultaneously (as in alternate rows) on the same plot
— in·ter·crop \ˈin-tər-ˌkräp\ *noun*

¹in·ter·cross \ˌin-tər-ˈkrȯs\ *verb* (1711)
: CROSS

²in·ter·cross \ˈin-tər-ˌkrȯs\ *noun* (1859)
: an instance or a product of crossbreeding

in·ter·cur·rent \ˌin-tər-ˈkər-ənt, -ˈkə-rənt\ *adjective* [Latin *intercurrent-, intercurrens,* present participle of *intercurrere*] (1611)
: occurring during and modifying the course of another disease ⟨an *intercurrent* infection⟩

in·ter·cut \ˌin-tər-ˈkət\ (1938)
transitive verb
1 : to insert (a contrasting camera shot) into a take by cutting

2 : to insert a contrasting camera shot into (a take) by cutting

intransitive verb

: to alternate contrasting camera shots by cutting

in·ter·den·tal \\ˌin-tər-ˈden-tᵊl\\ *adjective* (circa 1874)

1 : situated or intended for use between the teeth

2 : formed with the tip of the tongue between the upper and lower front teeth

— in·ter·den·tal·ly \\-tᵊl-ē\\ *adverb*

[1]in·ter·dict \\ˈin-tər-ˌdikt\\ *noun* [Middle English, alteration of *entredit,* from Old French, from Latin *interdictum* prohibition, from neuter of *interdictus,* past participle of *interdicere* to interpose, forbid, from *inter-* + *dicere* to say — more at DICTION] (15th century)

1 : a Roman Catholic ecclesiastical censure withdrawing most sacraments and Christian burial from a person or district

2 : a prohibitory decree **:** PROHIBITION

[2]in·ter·dict \\ˌin-tər-ˈdikt\\ *transitive verb* (15th century)

1 : to lay under or prohibit by an interdict

2 : to forbid in a usually formal or authoritative manner

3 : to destroy, damage, or cut off (as an enemy line of supply) by firepower to stop or hamper an enemy

synonym see FORBID

— in·ter·dic·tion \\-ˈdik-shən\\ *noun*

— in·ter·dic·tive \\-ˈdik-tiv\\ *adjective*

— in·ter·dic·tor \\-tər\\ *noun*

— in·ter·dic·to·ry \\-t(ə-)rē\\ *adjective*

in·ter·dif·fu·sion \\-di-ˈfyü-zhən\\ *noun* (circa 1872)

: the process of diffusing and mixing freely so as to approach a homogeneous mixture

— in·ter·dif·fuse \\-ˈfyüz\\ *intransitive verb*

in·ter·dig·i·tate \\-ˈdi-jə-ˌtāt\\ *intransitive verb* **-tat·ed; -tat·ing** [*inter-* + Latin *digitus* finger — more at TOE] (circa 1849)

: to become interlocked like the fingers of folded hands

— in·ter·dig·i·ta·tion \\-ˌdi-jə-ˈtā-shən\\ *noun*

in·ter·dis·ci·plin·ary \\-ˈdi-sə-plə-ˌner-ē\\ *adjective* (1926)

: involving two or more academic, scientific, or artistic disciplines

[1]in·ter·est \\ˈin-t(ə-)rəst; ˈin-tə-ˌrest, -ˌtrest; ˈin-tərst\\ *noun* [Middle English, probably alteration of earlier *interesse,* from Anglo-French & Medieval Latin; Anglo-French, from Medieval Latin, from Latin, to be between, make a difference, concern, from *inter-* + *esse* to be — more at IS] (15th century)

1 a (1) **:** right, title, or legal share in something (2) **:** participation in advantage and responsibility **b :** BUSINESS, COMPANY

2 a : a charge for borrowed money generally a percentage of the amount borrowed **b :** an excess above what is due or expected ⟨returned the insults with *interest*⟩

3 : ADVANTAGE, BENEFIT; *also* **:** SELF-INTEREST

4 : SPECIAL INTEREST

5 a : a feeling that accompanies or causes special attention to an object or class of objects **:** CONCERN **b :** something that arouses such attention **c :** a quality in a thing arousing interest

[2]interest *transitive verb* (1608)

1 : to induce or persuade to participate or engage

2 : to engage the attention or arouse the interest of

in·ter·est·ed *adjective* (1602)

1 : having the attention engaged ⟨*interested* listeners⟩

2 : being affected or involved ⟨*interested* parties⟩

— in·ter·est·ed·ly *adverb*

interest group *noun* (1908)

: a group of persons having a common identifying interest that often provides a basis for action

in·ter·est·ing \\ˈin-t(ə-)rəs-tiŋ; ˈin-tə-ˌres-, ˈin-ˌtres-; ˈin-tərs-\\ *adjective* (1768)

: holding the attention **:** arousing interest

— in·ter·est·ing·ness *noun*

in·ter·est·ing·ly \\-lē\\ *adverb* (1811)

1 : in an interesting manner

2 : as a matter of interest

[1]in·ter·face \\ˈin-tər-ˌfās\\ *noun* (1882)

1 : a surface forming a common boundary of two bodies, spaces, or phases ⟨an oil-water *interface*⟩

2 a : the place at which independent and often unrelated systems meet and act on or communicate with each other ⟨the man-machine *interface*⟩ **b :** the means by which interaction or communication is achieved at an interface

— in·ter·fa·cial \\ˌin-tər-ˈfā-shəl\\ *adjective*

[2]interface (1962)

transitive verb

1 : to connect by means of an interface ⟨*interface* a machine with a computer⟩

2 : to serve as an interface for

intransitive verb

1 : to become interfaced

2 : to interact or coordinate harmoniously

in·ter·fac·ing \\-ˌfā-siŋ\\ *noun* (1942)

: fabric sewn between the facing and the outside of a garment (as in a collar or cuff) for stiffening and shape retention

in·ter·faith \\ˌin-tər-ˈfāth\\ *adjective* (1932)

: involving persons of different religious faiths

in·ter·fere \\ˌin-tə(r)-ˈfir\\ *intransitive verb* **-fered; -fer·ing** [Middle English *enterferen,* from Middle French *(s')entreferir* to strike one another, from Old French, from *entre-* inter- + *ferir* to strike, from Latin *ferire* — more at BORE] (15th century)

1 : to interpose in a way that hinders or impedes **:** come into collision or be in opposition

2 : to strike one foot against the opposite foot or ankle in walking or running — used especially of horses

3 : to enter into or take a part in the concerns of others

4 : to act reciprocally so as to augment, diminish, or otherwise affect one another — used of waves

synonym see INTERPOSE

— in·ter·fer·er *noun*

in·ter·fer·ence \\-ˈfir-ən(t)s\\ *noun* (1783)

1 a : the act or process of interfering **b :** something that interferes **:** OBSTRUCTION

2 : the mutual effect on meeting of two wave trains (as of light or sound) that constitutes alternating areas of increased and decreased amplitude (as light and dark lines or louder and softer sound)

3 a : the legal blocking of an opponent in football to make way for the ballcarrier **b :** the illegal hindering of an opponent in sports

4 : partial or complete inhibition or sometimes facilitation of other genetic crossovers in the vicinity of a chromosomal locus where a preceding crossover has occurred

5 a : confusion of a received radio signal due to the presence of noise (as atmospherics) or signals from two or more transmitters on a single frequency **b :** something that produces such confusion

6 : the disturbing effect of new learning on the performance of previously learned behavior with which it is inconsistent

— in·ter·fer·en·tial \\-fə-ˈren(t)-shəl, -ˌfir-ˈen(t)-\\ *adjective*

in·ter·fer·o·gram \\ˌin-tə(r)-ˈfir-ə-ˌgram\\ *noun* (1921)

: a photographic record made by an apparatus for recording optical interference phenomena

in·ter·fer·om·e·ter \\ˌin-tə(r)-fə-ˈrä-mə-tər, -ˌfi-ˈrä-\\ *noun* [International Scientific Vocabulary] (1897)

: an instrument that utilizes the interference of waves (as of light) for precise determinations (as of distance or wavelength)

— in·ter·fer·o·met·ric \\-ˌfir-ə-ˈme-trik\\ *adjective*

— in·ter·fer·o·met·ri·cal·ly \\-tri-k(ə-)lē\\ *adverb*

— in·ter·fer·om·e·try \\-fə-ˈrä-mə-trē, -ˌfi-ˈrä-\\ *noun*

in·ter·fer·on \\ˌin-tə(r)-ˈfir-ˌän\\ *noun* [*interfere* + [2]*-on*] (1957)

: any of a group of heat-stable soluble basic antiviral glycoproteins of low molecular weight that are produced usually by cells exposed to the action of a virus, sometimes to the action of another intracellular parasite (as a bacterium), or experimentally to the action of some chemicals

in·ter·fer·tile \\ˌin-tər-ˈfər-tᵊl\\ *adjective* (1899)

: capable of interbreeding

— in·ter·fer·til·i·ty \\-(ˌ)fər-ˈti-lə-tē\\ *noun*

in·ter·file \\ˌin-tər-ˈfī(ə)l\\ *transitive verb* (1950)

: to arrange in or add to a file **:** FILE

in·ter·fluve \\ˈin-tər-ˌflüv\\ *noun* [*inter-* + Latin *fluvius* river — more at FLUVIAL] (1895)

: the area between adjacent streams flowing in the same direction

in·ter·fuse \\ˌin-tər-ˈfyüz\\ *verb* [Latin *interfusus,* past participle of *interfundere* to pour between, from *inter-* + *fundere* to pour — more at FOUND] (1593)

transitive verb

1 : to combine by fusing **:** BLEND

2 : to add as if by fusing **:** INFUSE

intransitive verb

: BLEND, FUSE

— in·ter·fu·sion \\-ˈfyü-zhən\\ *noun*

in·ter·ga·lac·tic \\ˌin-tər-gə-ˈlak-tik\\ *adjective* (1928)

1 : situated in or relating to the spaces between galaxies

2 : of, relating to, or occurring in outer space ⟨*intergalactic* battles⟩

in·ter·gla·cial \\-ˈglā-shəl\\ *noun* (1867)

: a warm period between glacial epochs

— interglacial *adjective*

in·ter·gov·ern·men·tal \\-ˌgə-vər(n)-ˈmen-tᵊl\\ *adjective* (1927)

: existing or occurring between two or more governments or levels of government

in·ter·gra·da·tion \\-grā-ˈdā-shən, -grə-\\ *noun* (1874)

: the condition of an individual or population that intergrades

— in·ter·gra·da·tion·al \\-shnəl, -shə-nᵊl\\ *adjective*

[1]in·ter·grade \\ˌin-tər-ˈgrād\\ *intransitive verb* (1874)

: to merge gradually one with another through a continuous series of intermediate forms

[2]in·ter·grade \\ˈin-tər-ˌgrād\\ *noun* (1888)

: an intermediate form

in·ter·growth \\ˈin-tər-ˌgrōth\\ *noun* (1844)

: a growing between or together; *also* **:** the product of such growth

[1]in·ter·im \\ˈin-tə-rəm\\ *noun* [Latin, adverb, meanwhile, from *inter* between — more at INTER-] (circa 1580)

: an intervening time **:** INTERVAL

[2]interim *adjective* (1604)

: done, made, appointed, or occurring for an interim

[1]in·te·ri·or \\in-ˈtir-ē-ər\\ *adjective* [Middle French & Latin; Middle French, from Latin, comparative of (assumed) Old Latin *interus* inward, on the inside; akin to Latin *inter*] (15th century)

1 : lying, occurring, or functioning within the limiting boundaries **:** INNER ⟨an *interior* point of a triangle⟩

2 : belonging to mental or spiritual life ⟨a simple *interior* piety⟩
3 : belonging to the inner constitution or concealed nature of something ⟨*interior* meaning of a poem⟩
4 : lying away or remote from the border or shore
— **in·te·ri·or·ly** \in-ˈtir-ē-ər-lē\ *adverb*

²interior *noun* (1596)
1 : the inner or spiritual nature **:** CHARACTER
2 : the interior part (as of a country or island)
3 : the internal or inner part of a thing **:** INSIDE
4 : the internal affairs of a state or nation
5 : a representation (as in a play or movie) of the interior of a building

interior angle *noun* (1756)
1 : the inner of the two angles formed where two sides of a polygon come together
2 : any of the four angles formed in the area between a pair of parallel lines when a third line cuts them

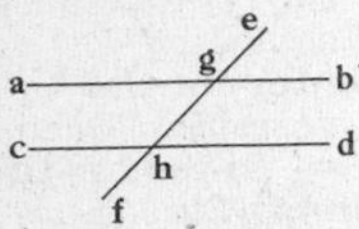

agh, bgh, ghd, ghc
interior angle 2

interior decoration *noun* (1807)
: INTERIOR DESIGN

interior decorator *noun* (1867)
: INTERIOR DESIGNER, DECORATOR

interior design *noun* (1927)
: the art or practice of planning and supervising the design and execution of architectural interiors and their furnishings

interior designer *noun* (1938)
: one who specializes in interior design

in·te·ri·or·ise *British variant of* INTERIORIZE

in·te·ri·or·i·ty \(ˌ)in-ˌtir-ē-ˈȯr-ə-tē, -ˈär-\ *noun* (1701)
: interior quality or character; *also* **:** inner life or substance ⟨characters that lack *interiority*⟩

in·te·ri·or·ize \in-ˈtir-ē-ə-ˌrīz\ *transitive verb* **-ized; -iz·ing** (1906)
: to make interior; *especially* **:** to make a part of one's own inner being or mental structure
— **in·te·ri·or·i·za·tion** \-ˌtir-ē-ə-rə-ˈzā-shən\ *noun*

interior monologue *noun* (1922)
: a usually extended representation in monologue of a fictional character's thought and feeling

in·ter·ject \ˌin-tər-ˈjekt\ *transitive verb* [Latin *interjectus,* past participle of *intericere,* from *inter-* + *jacere* to throw — more at JET] (1588)
: to throw in between or among other things **:** INTERPOLATE ⟨*interject* a remark⟩
synonym see INTRODUCE
— **in·ter·jec·tor** \-ˈjek-tər\ *noun*
— **in·ter·jec·to·ry** \-t(ə-)rē\ *adjective*

in·ter·jec·tion \ˌint-ər-ˈjek-shən\ *noun* (15th century)
1 a : the act of uttering exclamations **:** EJACULATION **b :** the act of putting in between **:** INTERPOSITION
2 : an ejaculatory utterance usually lacking grammatical connection: as **a :** a word or phrase used in exclamation (as *Heavens! Dear me!*) **b :** a cry or inarticulate utterance (as *Alas! ouch! phooey! ugh!*) expressing an emotion
3 : something that is interjected or that interrupts

in·ter·jec·tion·al \-shnəl, -shə-nᵊl\ *adjective* (1761)
1 : of, relating to, or constituting an interjection **:** EJACULATORY
2 : thrown in between other words **:** PARENTHETICAL
— **in·ter·jec·tion·al·ly** *adverb*

in·ter·lace \ˌin-tər-ˈlās\ *verb* [Middle English *entrelacen,* from Middle French *entrelacer,* from Old French *entrelacier,* from *entre-* inter- + *lacier* to lace] (14th century)
transitive verb
1 : to unite by or as if by lacing together **:** INTERWEAVE
2 : to vary by alternation or intermixture **:** INTERSPERSE ⟨narrative *interlaced* with anecdotes⟩
intransitive verb
: to cross one another as if woven together **:** INTERTWINE
— **in·ter·lace·ment** \-mənt\ *noun*

in·ter·lard \ˌin-tər-ˈlärd\ *transitive verb* [Middle French *entrelarder,* from Old French, from *entre* inter- + *larder* to lard, from *lard,* noun] (circa 1587)
: to vary by intermixture **:** INTERSPERSE, INTERLACE

in·ter·leave \ˌin-tər-ˈlēv\ *transitive verb* **-leaved; -leav·ing** (1668)
: to arrange in or as if in alternate layers

in·ter·leu·kin \ˌin-tər-ˈlü-kən\ *noun* [*inter-* + *leuk-* + ¹*-in*] (1979)
: any of several compounds of low molecular weight that are produced by lymphocytes, macrophages, and monocytes and that function especially in regulation of the immune system and especially cell-mediated immunity

in·ter·leu·kin–1 \-ˈwən\ *noun* (1979)
: an interleukin produced especially by monocytes and macrophages that regulates immune responses by activating lymphocytes and mediates other biological processes (as the onset of fever) usually associated with infection and inflammation

in·ter·leu·kin–2 \-ˈtü\ *noun* (1979)
: an interleukin produced by antigen-stimulated helper T cells in the presence of interleukin-1 that induces proliferation of immune cells (as T cells and B cells) and is used experimentally especially in treating certain cancers

¹in·ter·line \ˌin-tər-ˈlīn\ *transitive verb* [Middle English *enterlinen,* from Medieval Latin *interlineare,* from Latin *inter-* + *linea* line] (15th century)
: to insert between lines already written or printed
— **in·ter·lin·e·a·tion** \-ˌli-nē-ˈā-shən\ *noun*

²interline *transitive verb* [Middle English, from *inter-* + *linen* to line] (15th century)
: to provide (a garment) with an interlining

³interline *adjective* (1897)
: relating to, involving, or carried by two or more transportation lines

¹in·ter·lin·ear \ˌin-tər-ˈli-nē-ər\ *adjective* [Middle English *interliniare,* from Medieval Latin *interlinearis,* from Latin *inter-* + *linea* line] (15th century)
1 : inserted between lines already written or printed
2 : written or printed in different languages or texts in alternate lines
— **in·ter·lin·ear·ly** *adverb*

²interlinear *noun* (1850)
: a book having interlinear matter; *especially* **:** a book in a foreign language with interlinear translation

in·ter·lin·ing \ˈin-tər-ˌlī-niŋ\ *noun* (1881)
: a lining (as of a coat) sewn between the ordinary lining and the outside fabric

in·ter·link \ˌin-tər-ˈliŋk\ *transitive verb* (1587)
: to link together
— **in·ter·link** \ˈin-tər-ˌliŋk\ *noun*

¹in·ter·lock \ˌin-tər-ˈläk\ (1632)
intransitive verb
: to become locked together or interconnected
transitive verb
1 : to lock together **:** UNITE
2 : to connect so that the motion or operation of any part is constrained by another

²in·ter·lock \ˈin-tər-ˌläk\ *noun* (1874)
1 : the quality, state, sense, or an instance of being interlocked
2 : an arrangement in which the operation of one part or mechanism automatically brings about or prevents the operation of another ⟨a safety *interlock*⟩
3 a : a stretchable fabric made on a circular knitting machine and consisting of two ribbed fabrics joined by interlocking **b :** a garment made of interlock

in·ter·loc·u·tor \ˌin-tər-ˈlä-kyə-tər\ *noun* [Latin *interloqui* to speak between, issue an interlocutory decree, from *inter-* + *loqui* to speak] (1514)
1 : one who takes part in dialogue or conversation
2 : a man in the middle of the line in a minstrel show who questions the end men and acts as leader

in·ter·loc·u·to·ry \-kyə-ˌtȯr-ē, -ˌtȯr-\ *adjective* (15th century)
: pronounced during the progress of a legal action and having only provisional force ⟨*interlocutory* decree⟩

in·ter·lope \ˌin-tər-ˈlōp, ˈin-tər-ˌ\ *intransitive verb* **-loped; -lop·ing** [probably back-formation from *interloper,* from *inter-* + *-loper* (akin to Middle Dutch *lōpen* to run, Old English *hlēapan* to leap) — more at LEAP] (1615)
1 : to encroach on the rights (as in trade) of others
2 : INTRUDE, INTERFERE

in·ter·lop·er \ˌin-tər-ˈlō-pər, ˈin-tər-ˌ\ *noun* (circa 1590)
: one that interlopes: as **a :** an illegal or unlicensed trader **b :** one that intrudes in a place or sphere of activity

in·ter·lude \ˈin-tər-ˌlüd\ *noun* [Middle English *enterlude,* from Medieval Latin *interludium,* from Latin *inter-* + *ludus* play — more at LUDICROUS] (14th century)
1 : a usually short simple play or dramatic entertainment
2 : an intervening or interruptive period, space, or event **:** INTERVAL
3 : a musical composition inserted between the parts of a longer composition, a drama, or a religious service

in·ter·lu·nar \ˌin-tər-ˈlü-nər\ *also* **in·ter·lu·na·ry** \-nə-rē\ *adjective* [probably from Middle French *interlunaire,* from Latin *interlunium* interlunar period, from *inter-* + *luna* moon — more at LUNAR] (1598)
: relating to the interval between old and new moon when the moon is invisible

in·ter·mar·riage \ˌin-tər-ˈmar-ij\ *noun* (1579)
1 : ENDOGAMY
2 : marriage between members of different groups

in·ter·mar·ry \-ˈmar-ē\ *intransitive verb* (1574)
1 a : to marry each other **b :** to marry within a group
2 : to become connected by intermarriage

in·ter·med·dle \ˌin-tər-ˈme-dᵊl\ *intransitive verb* [Middle English *entermedlen,* from Middle French *entremedler,* from Old French, from *entre-* inter- + *medler* to mix — more at MEDDLE] (15th century)
: to meddle impertinently and officiously and usually so as to interfere
— **in·ter·med·dler** \-ˈmed-lər, -ˈme-dᵊl-ər\ *noun*

in·ter·me·di·a·cy \ˌin-tər-ˈmē-dē-ə-sē\ *noun* (1713)
1 : the act or action of intermediating
2 : the quality or state of being intermediate

¹in·ter·me·di·ary \ˌin-tər-ˈmē-dē-ˌer-ē\ *adjective* (1788)
1 : INTERMEDIATE
2 : acting as a mediator ⟨an *intermediary* agent⟩ ⟨an *intermediary* particle⟩

²intermediary *noun, plural* **-ar·ies** (1791)
1 a : MEDIATOR, GO-BETWEEN **b :** MEDIUM, MEANS
2 : an intermediate form, product, or stage

¹in·ter·me·di·ate \ˌin-tər-ˈmē-dē-ət\ *adjective* [Medieval Latin *intermediatus,* from Latin *intermedius,* from *inter-* + *medius* mid, middle — more at MID] (15th century)

1 : being or occurring at the middle place, stage, or degree or between extremes
2 : of or relating to an intermediate school ⟨an *intermediate* curriculum⟩
— **in·ter·me·di·ate·ly** *adverb*
— **in·ter·me·di·ate·ness** *noun*

²intermediate *noun* (1650)
1 : one that is intermediate
2 : MEDIATOR, GO-BETWEEN
3 a : a chemical compound synthesized from simpler compounds and usually intended to be used in later syntheses of more complex products **b :** a usually short-lived chemical species formed in a reaction as an intermediate step between the starting material and the final product
4 : an automobile larger than a compact but smaller than a full-sized automobile

³in·ter·me·di·ate \-dē-ˌāt\ *intransitive verb* [Medieval Latin *intermediatus,* past participle of *intermediare,* from Latin *inter-* + Late Latin *mediare* to mediate] (1610)
1 : INTERVENE, INTERPOSE
2 : to act as an intermediate

intermediate host *noun* (1878)
1 : a host which is normally used by a parasite in the course of its life cycle and in which it may multiply asexually but not sexually — compare DEFINITIVE HOST
2 a : RESERVOIR 3 **b :** VECTOR

intermediate school *noun* (1842)
1 : JUNIOR HIGH SCHOOL
2 : a school usually comprising grades 4 to 6

intermediate vector boson *noun* (1968)
: any of three particles that mediate the weak force — called also *intermediate boson*; compare W PARTICLE, Z PARTICLE

in·ter·me·di·a·tion \ˌin-tər-ˌmē-dē-ˈā-shən\ *noun* (1602)
: the act of coming between **:** INTERVENTION, MEDIATION

in·ter·me·din \ˌin-tər-ˈmē dᵊn\ *noun* (1932)
: MELANOCYTE-STIMULATING HORMONE

in·ter·ment \in-ˈtər-mənt\ *noun* (14th century)
: the act or ceremony of interring

in·ter·mesh \ˌin-tər-ˈmesh\ (circa 1903)
intransitive verb
: INTERLOCK
transitive verb
: to mesh together **:** INTERLOCK

in·ter·me·tal·lic \ˌin-tər mə-ˈta-lik\ *adjective* (1900)
: composed of two or more metals or of a metal and a nonmetal; *especially* **:** being an alloy having a characteristic crystal structure and usually a definite composition ⟨*intermetallic* compound⟩
— **intermetallic** *noun*

in·ter·mez·zo \ˌin-tər-ˈmet-(ˌ)sō, -ˈmed-(ˌ)zō\ *noun, plural* **-zi** \-(ˌ)sē, -(ˌ)zē\ *or* **-zos** [Italian, ultimately from Latin *intermedius* intermediate] (1771)
1 : a short light entr'acte
2 a : a movement coming between the major sections of an extended musical work (as an opera) **b :** a short independent instrumental composition
3 : a usually brief interlude or diversion

in·ter·mi·na·ble \(ˌ)in-ˈtərm-nə-bəl, -ˈtər-mə-\ *adjective* [Middle English, from Late Latin *interminabilis,* from Latin *in-* + *terminare* to terminate] (15th century)
: having or seeming to have no end; *especially* **:** wearisomely protracted ⟨an *interminable* sermon⟩
— **in·ter·mi·na·ble·ness** *noun*
— **in·ter·mi·na·bly** \-blē\ *adverb*

in·ter·min·gle \ˌin-tər-ˈmiŋ-gəl\ *verb* (15th century)
: INTERMIX

in·ter·mis·sion \ˌin-tər-ˈmi-shən\ *noun* [Middle English *intermyssyown,* from Latin *intermission-, intermissio,* from *intermittere*] (15th century)
1 : the act of intermitting **:** the state of being intermitted
2 : an interval between the parts of an entertainment (as the acts of a play)
— **in·ter·mis·sion·less** *adjective*

in·ter·mit \-ˈmit\ *verb* **-mit·ted; -mit·ting** [Latin *intermittere,* from *inter-* + *mittere* to send] (circa 1542)
transitive verb
: to cause to cease for a time or at intervals **:** DISCONTINUE
intransitive verb
: to be intermittent
— **in·ter·mit·ter** *noun*

in·ter·mit·tence \-ˈmi-tən(t)s\ *noun* (1796)
: the quality or state of being intermittent

in·ter·mit·ten·cy \-ˈmi-tən(t)-sē\ *noun* (1662)
: INTERMITTENCE

in·ter·mit·tent \-ˈmi-tᵊnt\ *adjective* [Latin *intermittent-, intermittens,* present participle of *intermittere*] (1601)
: coming and going at intervals **:** not continuous ⟨*intermittent* rain⟩; *also* **:** OCCASIONAL ⟨*intermittent* trips abroad⟩
— **in·ter·mit·tent·ly** *adverb*

in·ter·mix \ˌin-tər-ˈmiks\ *verb* [back-formation from obsolete *intermixt* intermingled, from Latin *intermixtus,* past participle of *intermiscēre* to intermix, from *inter-* + *miscēre* to mix — more at MIX] (1542)
transitive verb
: to mix together
intransitive verb
: to become mixed together
— **in·ter·mix·ture** \-ˈmiks-chər\ *noun*

in·ter·mod·al \-ˈmō-dəl\ *adjective* (1963)
1 : being or involving transportation by more than one form of carrier during a single journey
2 : used for intermodal transport

in·ter·mod·u·la·tion \-ˌmä-jə-ˈlā-shən\ *noun* [International Scientific Vocabulary] (1931)
: the production in an electrical device of currents having frequencies equal to the sums and differences of frequencies supplied to the device or of their harmonics

in·ter·mon·tane \ˌin-tər-ˈmän-ˌtān\ *or* **in·ter·mont** \ˈin-tər-ˌmänt\ *adjective* [Latin *mont-, mons* mount] (1807)
: situated between mountains ⟨an *intermontane* basin⟩

¹in·tern *or* **in·terne** \in-ˈtərn, ˈin-ˌ\ *adjective* [Middle French *interne,* from Latin *internus*] (circa 1500)
archaic **:** INTERNAL

²in·tern \ˈin-ˌtərn, in-ˈ\ *transitive verb* (1866)
: to confine or impound especially during a war ⟨*intern* enemy aliens⟩
— **in·tern·ee** \(ˌ)in-ˌtər-ˈnē\ *noun*
— **in·tern·ment** \in-ˈtərn-mənt, ˈin-ˌ\ *noun*

³in·tern *also* **in·terne** \ˈin-ˌtərn\ *noun* [French *interne,* from *interne,* adjective] (circa 1879)
: an advanced student or graduate usually in a professional field (as medicine or teaching) gaining supervised practical experience (as in a hospital or classroom)
— **in·tern·ship** \-ˌship\ *noun*

⁴in·tern \ˈin-ˌtərn\ *intransitive verb* (circa 1928)
: to act as an intern

in·ter·nal \in-ˈtər-nᵊl, ˈin-ˌ\ *adjective* [Middle English *internalle,* from Latin *internus;* akin to Latin *inter* between] (15th century)
1 : existing or situated within the limits or surface of something: as **a** (1) **:** situated near the inside of the body (2) **:** situated on the side toward the median plane of the body **b :** of, relating to, or occurring within the confines of an organized structure (as a club, company, or state) ⟨*internal* affairs⟩
2 : relating or belonging to or existing within the mind
3 : INTRINSIC, INHERENT ⟨*internal* evidence of forgery in a document⟩
4 : present or arising within an organism or one of its parts ⟨*internal* stimulus⟩
5 : applied or intended for application through the stomach by being swallowed ⟨an *internal* remedy⟩
— **in·ter·nal·i·ty** \ˌin-ˌtər-ˈna-lə-tē\ *noun*
— **in·ter·nal·ly** \in-ˈtər-nᵊl-ē\ *adverb*

internal combustion engine *noun* (1884)
: a heat engine in which the combustion that generates the heat takes place inside the engine proper instead of in a furnace

in·ter·nal·ise *British variant of* INTERNALIZE

in·ter·nal·ize \in-ˈtər-nᵊl-ˌīz\ *transitive verb* **-ized; -iz·ing** (1884)
: to give a subjective character to; *specifically* **:** to incorporate (as values or patterns of culture) within the self as conscious or subconscious guiding principles through learning or socialization
— **in·ter·nal·i·za·tion** \-ˌtər-nᵊl-ə-ˈzā-shən\ *noun*

internal medicine *noun* (circa 1904)
: a branch of medicine that deals with the diagnosis and treatment of diseases not requiring surgery

internal respiration *noun* (circa 1890)
: an exchange of gases between the cells of the body and the blood by way of the fluid bathing the cells — compare EXTERNAL RESPIRATION

internal rhyme *noun* (1903)
: rhyme between a word within a line and another either at the end of the same line or within another line

internal secretion *noun* (1895)
: HORMONE

¹in·ter·na·tion·al \ˌin-tər-ˈnash-nəl, -ˈna-shə-nᵊl\ *adjective* (1780)
1 : of, relating to, or affecting two or more nations ⟨*international* trade⟩
2 : of, relating to, or constituting a group or association having members in two or more nations ⟨*international* movement⟩
3 : active, known, or reaching beyond national boundaries ⟨an *international* reputation⟩
— **in·ter·na·tion·al·i·ty** \-ˌna-shə-ˈna-lə-tē\ *noun*
— **in·ter·na·tion·al·ly** \-ˈnash-nə-lē, -ˈna-shə-nᵊl-ē\ *adverb*

²international *noun* (1870)
: one that is international; *especially* **:** an organization of international scope

international date line *noun* (circa 1909)
: an arbitrary line approximately along the 180th meridian designated as the place where each calendar day begins

in·ter·na·tion·al·ise *British variant of* INTERNATIONALIZE

in·ter·na·tion·al·ism \-ˈnash-nə-ˌli-zəm, -ˈna-shə-nᵊl-ˌi-zəm\ *noun* (1851)
1 : international character, principles, interests, or outlook
2 a : a policy of cooperation among nations **b :** an attitude or belief favoring such a policy
— **in·ter·na·tion·al·ist** \-list, -ist\ *noun or adjective*

in·ter·na·tion·al·ize \ˌin-tər-ˈnash-nə-ˌlīz, -ˈna-shə-nᵊl-ˌīz\ *transitive verb* (circa 1864)
: to make international; *also* **:** to place under international control ⟨a proposal to *internationalize* the city⟩
— **in·ter·na·tion·al·i·za·tion** \-ˌnash-nə-lə-ˈzā-shən, -ˌna-shə-nᵊl-ə-\ *noun*

international law *noun* (circa 1828)
: a body of rules that control or affect the rights of nations in their relations with each other

International Phonetic Alphabet *noun* (1898)
: IPA

international pitch *noun* (1904)

: a tuning standard of 440 vibrations per second for A above middle C

international relations *noun plural but singular in construction* (1951)
: a branch of political science concerned with relations between nations and primarily with foreign policies

International Scientific Vocabulary *noun* (circa 1959)
: a part of the vocabulary of the sciences and other specialized studies that consists of words or other linguistic forms current in two or more languages and differing from New Latin in being adapted to the structure of the individual languages in which they appear

International Style *noun* (1932)
1 : a style in European art of the 14th and early 15th centuries marked by sinuous line, rich color, and decorative surface detail
2 : a style in architecture developed in the 1920s that uses modern materials (as steel, glass, and reinforced concrete), expresses structure directly, and eliminates nonstructural ornament

International System of Units *noun* (1961)
: a system of units based on the metric system and developed and refined by international convention especially for scientific work

international unit *noun* (1922)
: a quantity of a biologic (as a vitamin) that produces a particular biological effect agreed upon as an international standard

in·ter·ne·cine \ˌin-tər-ˈne-ˌsēn, -ˈnē-sᵊn, -ˈnē-ˌsīn, -nə-ˈsēn; in-ˈtər-nə-ˌsēn\ *adjective* [Latin *internecinus,* from *internecare* to destroy, kill, from *inter-* + *necare* to kill, from *nec-, nex* violent death — more at NOXIOUS] (1663)
1 : marked by slaughter **:** DEADLY; *especially* **:** mutually destructive
2 : of, relating to, or involving conflict within a group ⟨bitter *internecine* feuds⟩ ◆

in·ter·neu·ron \ˌin-tər-ˈnu̇r-ˌän, -ˈnyu̇r-\ *noun* (1939)
: an internuncial neuron
— in·ter·neu·ro·nal \-ˈnu̇r-ə-nᵊl, -ˈnyu̇r-; -nu̇-ˈrō-, -nyu̇-ˈ\ *adjective*

in·ter·nist \ˈin-ˌtər-nist\ *noun* (1904)
: a specialist in internal medicine

in·ter·node \ˈin-tər-ˌnōd\ *noun* [Latin *internodium,* from *inter-* + *nodus* knot] (1667)
: an interval or part between two nodes (as of a stem)
— in·ter·nod·al \ˌin-tər-ˈnō-dᵊl\ *adjective*

in·ter·nun·ci·al \ˌin-tər-ˈnən(t)-sē-əl, -ˈnu̇n(t)-\ *adjective* (1845)
1 : of or relating to an internuncio
2 : serving to link sensory and motor neurons

in·ter·nun·cio \ˌin-tər-ˈnən(t)-sē-ˌō, -ˈnu̇n(t)-\ *noun* [Italian *internunzio,* from Latin *internuntius, internuncius,* from *inter-* + *nuntius, nuncius* messenger] (1641)
1 : a messenger between two parties **:** GO-BETWEEN
2 : a papal legate of lower rank than a nuncio

in·ter·o·cep·tive \ˌin-tə-rō-ˈsep-tiv\ *adjective* [*inter*ior + *-o-* + *-ceptive* (as in *receptive*)] (circa 1921)
: of, relating to, or being stimuli arising within the body and especially in the viscera

in·ter·o·cep·tor \-tər\ *noun* (1906)
: a sensory receptor excited by interoceptive stimuli

in·ter·op·er·a·bil·i·ty \ˌin-tə-ˌrä-p(ə-)rə-ˈbi-lə-tē\ *noun* (1977)
: ability of a system (as a weapons system) to use the parts or equipment of another system
— in·ter·op·er·a·ble \-ˈrä-p(ə-)rə-bəl\ *adjective*

in·ter·pel·late \ˌin-tər-ˈpe-ˌlāt, -pə-ˈlāt\ *transitive verb* **-lat·ed; -lat·ing** [Latin *interpellatus,* past participle of *interpellare* to interrupt, from *inter-* + *-pellare* (from *pellere* to drive) — more at FELT] (1874)
: to question (as a foreign minister) formally concerning an official action or policy or personal conduct
— in·ter·pel·la·tion \-pə-ˈlā-shən\ *noun*
— in·ter·pel·la·tor \-ˈpe-ˌlā-tər, -pə-ˈlā-\ *noun*

in·ter·pen·e·trate \ˌin-tər-ˈpe-nə-ˌtrāt\ (circa 1810)
intransitive verb
: to penetrate mutually
transitive verb
: to penetrate between, within, or throughout **:** PERMEATE
— in·ter·pen·e·tra·tion \-ˌpe-nə-ˈtrā-shən\ *noun*

in·ter·per·son·al \-ˈpərs-nəl, -ˈpər-sᵊn-əl\ *adjective* (1842)
: being, relating to, or involving relations between persons
— in·ter·per·son·al·ly *adverb*

in·ter·phase \ˈin-tər-ˌfāz\ *noun* (1913)
: the interval between the end of one mitotic or meiotic division and the beginning of another

in·ter·plant \ˌin-tər-ˈplant\ *transitive verb* (1911)
: to plant a crop between (plants of another kind); *also* **:** to set out young trees among (existing growth)

in·ter·play \ˈin-tər-ˌplā\ *noun* (1862)
: INTERACTION ⟨the *interplay* of opposing forces⟩
— in·ter·play \ˌin-tər-ˈ, ˈin-tər-ˌ\ *intransitive verb*

in·ter·plead \ˌin-tər-ˈplēd\ *intransitive verb* [Anglo-French *enterpleder,* from *enter-* inter- + *pleder* to plead, from Old French *plaidier* — more at PLEAD] (1567)
: to go to trial with each other in order to determine a right on which the action of a third party depends

¹in·ter·plead·er \-ˈplē-dər\ *noun* [Anglo-French *enterpleder,* from *enterpleder,* verb] (1567)
: a proceeding to enable a person to compel parties making the same claim against him to litigate the matter between themselves

²interpleader *noun* (circa 1846)
: one that interpleads

in·ter·po·late \in-ˈtər-pə-ˌlāt\ *verb* **-lat·ed; -lat·ing** [Latin *interpolatus,* past participle of *interpolare* to refurbish, alter, interpolate, from *inter-* + *-polare* (from *polire* to polish)] (1612)
transitive verb
1 a : to alter or corrupt (as a text) by inserting new or foreign matter **b :** to insert (words) into a text or into a conversation
2 : to insert between other things or parts **:** INTERCALATE
3 : to estimate values of (a function) between two known values
intransitive verb
: to make insertions (as of estimated values)
synonym see INTRODUCE
— in·ter·po·la·tion \-ˌtər-pə-ˈlā-shən\ *noun*
— in·ter·po·la·tive \-ˈtər-pə-ˌlā-tiv\ *adjective*
— in·ter·po·la·tor \-ˌlā-tər\ *noun*

in·ter·pose \ˌin-tər-ˈpōz\ *verb* **-posed; -pos·ing** [Middle French *interposer,* from Latin *interponere* (perfect indicative *interposui*), from *inter-* + *ponere* to put — more at POSITION] (1582)
transitive verb
1 a : to place in an intervening position **b :** to put (oneself) between **:** INTRUDE
2 : to put forth by way of interference or intervention
3 : to introduce or throw in between the parts of a conversation or argument
intransitive verb
1 : to be or come between
2 : to step in between parties at variance **:** INTERVENE
3 : INTERRUPT ☆
— in·ter·pos·er *noun*

in·ter·po·si·tion \-pə-ˈzi-shən\ *noun* (14th century)
1 a : the act of interposing **b :** the action of a state whereby its sovereignty is placed between its citizens and the federal government
2 : something interposed

in·ter·pret \in-ˈtər-prət, -pət\ *verb* [Middle English, from Middle French & Latin; Middle French *interpreter,* from Latin *interpretari,* from *interpret-, interpres* agent, negotiator, interpreter] (14th century)
transitive verb
1 : to explain or tell the meaning of **:** present in understandable terms
2 : to conceive in the light of individual belief, judgment, or circumstance **:** CONSTRUE
3 : to represent by means of art **:** bring to realization by performance or direction ⟨*interprets* a role⟩
intransitive verb
: to act as an interpreter between speakers of different languages
synonym see EXPLAIN
— in·ter·pret·abil·i·ty \-ˌtər-prə-tə-ˈbi-lə-tē, -pə-tə-\ *noun*
— in·ter·pret·able \-ˈtər-prə-tə-bəl, -pə-tə-\ *adjective*

in·ter·pre·ta·tion \in-ˌtər-prə-ˈtā-shən, -pə-\ *noun* (14th century)
1 : the act or the result of interpreting **:** EXPLANATION
2 : a particular adaptation or version of a work, method, or style
3 : a teaching technique that combines factual with stimulating explanatory information ⟨natural history *interpretation* program⟩

☆ SYNONYMS
Interpose, interfere, intervene, mediate, intercede mean to come or go between. INTERPOSE often implies no more than this ⟨*interposed* herself between him and the door⟩. INTERFERE implies hindering ⟨noise *interfered* with my concentration⟩. INTERVENE may imply an occurring in space or time between two things or a stepping in to stop a conflict ⟨quarreled until the manager *intervened*⟩. MEDIATE implies intervening between hostile factions ⟨*mediated* between the parties⟩. INTERCEDE implies acting in behalf of an offender in begging mercy or forgiveness ⟨*interceded* on our behalf⟩. See in addition INTRODUCE.

◇ WORD HISTORY
internecine In Latin the verb *necare* meant "to kill" and the verb *internecare* "to kill without exception, massacre." In this instance the familiar Latin prefix *inter-* did not mean "between" but served to signify the completion of an action. The derivative adjective *internecinus* thus had the meaning of "fought to the death, devastating." In 1755, when Samuel Johnson published his great dictionary, it became clear that he had apparently interpreted the *inter-* of *internecine* as meaning "between" and so had defined the word as "endeavouring mutual destruction." In 1828, when Noah Webster defined *internecine,* he gave only the original sense of "deadly, destructive"; the Johnsonian sense of "mutually destructive" was not added until later revisions. Ironically, the invented Johnsonian sense won favor among the learned and superseded the original meaning entirely. In the 20th century the meaning of *internecine* was extended to include bitter—but bloodless—conflict within a group, and this has become the dominant sense.

— **in·ter·pre·ta·tion·al** \-shnəl, -shə-nᵊl\ *adjective*

— **in·ter·pre·ta·tive** \in-ˈtər-prə-ˌtā-tiv *also* -prə-tə-tiv\ *adjective*

— **in·ter·pre·ta·tive·ly** *adverb*

— **in·ter·pre·tive** \-ˈtər-prə-tiv, -pə-\ *adjective*

— **in·ter·pre·tive·ly** *adverb*

in·ter·pret·er \in-ˈtər-prə-tər, -pə-\ *noun* (14th century)
1 : one that interprets: as **a :** one who translates orally for parties conversing in different languages **b :** one who explains or expounds
2 a : a machine that prints on punched cards the symbols recorded in them by perforations **b :** a computer program that translates an instruction into machine language and executes it before going to the next instruction

in·ter·prox·i·mal \ˌin-tər-ˈpräk-sə-məl\ *adjective* (1897)
: situated or used in the areas between adjoining teeth ⟨*interproximal* space⟩

in·ter·pu·pil·lary \ˌin-tər-ˈpyü-pə-ˌler-ē\ *adjective* (circa 1904)
: extending between the pupils of the eyes; *also* **:** extending between the centers of a pair of spectacle lenses ⟨*interpupillary* distance⟩

in·ter·ra·cial \-ˈrā-shəl\ *adjective* (1888)
: of, involving, or designed for members of different races

— **in·ter·ra·cial·ly** \-shə-lē\ *adverb*

interred *past and past participle of* INTER

in·ter·reg·num \ˌin-tə-ˈreg-nəm\ *noun, plural* **-nums** *or* **-na** \-nə\ [Latin, from *inter-* + *regnum* reign — more at REIGN] (1590)
1 : the time during which a throne is vacant between two successive reigns or regimes
2 : a period during which the normal functions of government or control are suspended
3 : a lapse or pause in a continuous series

in·ter·re·late \ˌin-tə(r)-ri-ˈlāt\ (1888)
transitive verb
: to bring into mutual relation
intransitive verb
: to have mutual relationship

— **in·ter·re·la·tion** \-ˈlā-shən\ *noun*

— **in·ter·re·la·tion·ship** \-ˌship\ *noun*

in·ter·re·lat·ed \-ˈlā-təd\ *adjective* (1827)
: having a mutual or reciprocal relation

— **in·ter·re·lat·ed·ly** *adverb*

— **in·ter·re·lat·ed·ness** *noun*

interring *present participle of* INTER

in·ter·ro·bang \in-ˈter-ə-ˌbaŋ\ *noun* [*interrogation* (point) + *bang* (printers' slang for *exclamation point*)] (1967)
: a punctuation mark ‽ designed for use especially at the end of an exclamatory rhetorical question

in·ter·ro·gate \in-ˈter-ə-ˌgāt\ *transitive verb* **-gat·ed; -gat·ing** [Latin *interrogatus,* past participle of *interrogare,* from *inter-* + *rogare* to ask — more at RIGHT] (15th century)
1 : to question formally and systematically
2 : to give or send out a signal to (as a transponder or computer) for triggering an appropriate response
synonym see ASK

— **in·ter·ro·ga·tee** \-ˌter-ə-(ˌ)gā-ˈtē\ *noun*

— **in·ter·ro·ga·tion** \-ˌter-ə-ˈgā-shən\ *noun*

— **in·ter·ro·ga·tion·al** \-shnəl, -shə-nᵊl\ *adjective*

interrogation point *noun* (circa 1864)
: QUESTION MARK

[1]**in·ter·rog·a·tive** \ˌin-tə-ˈrä-gə-tiv\ *adjective* (15th century)
1 a : used in a question **b :** having the form or force of a question
2 : INQUISITIVE, QUESTIONING

— **in·ter·rog·a·tive·ly** *adverb*

[2]**interrogative** *noun* (1522)
1 : a word (as *who, what, which*) or a particle (as Latin *-ne*) used in asking questions
2 : QUESTION 1a

in·ter·ro·ga·tor \in-ˈter-ə-ˌgā-tər\ *noun* (1751)
1 : one that interrogates
2 : a radio transmitter and receiver for sending out a signal that triggers a transponder and for receiving and displaying the reply

[1]**in·ter·rog·a·to·ry** \ˌin-tə-ˈrä-gə-ˌtōr-ē, -ˌtȯr-\ *noun, plural* **-ries** (1533)
: a formal question or inquiry; *especially* **:** a written question required to be answered under direction of a court

[2]**interrogatory** *adjective* (1576)
: INTERROGATIVE

in·ter·ro·gee \in-ˌter-ə-ˈgē\ *noun* (1919)
: one who is interrogated

[1]**in·ter·rupt** \ˌin-tə-ˈrəpt\ *verb* [Middle English, from Latin *interruptus,* past participle of *interrumpere,* from *inter-* + *rumpere* to break — more at REAVE] (15th century)
transitive verb
1 : to stop or hinder by breaking in
2 : to break the uniformity or continuity of
intransitive verb
: to break in upon an action; *especially* **:** to break in with questions or remarks while another is speaking

— **in·ter·rupt·ible** \-ˈrəp-tə-bəl\ *adjective*

— **in·ter·rup·tion** \-ˈrəp-shən\ *noun*

— **in·ter·rup·tive** \-ˈrəp-tiv\ *adverb*

[2]**in·ter·rupt** \ˌin-tə-ˈrəpt, ˈin-tə-ˌ\ *noun* (1957)
: a feature of a computer that permits the execution of a program to be interrupted in order to perform a special set of operations; *also* **:** the interruption itself

in·ter·rupt·er *also* **in·ter·rup·tor** \ˌin-tə-ˈrəp-tər\ *noun* (circa 1512)
: one that interrupts; *especially* **:** a device for interrupting an electric current usually automatically

in·ter·scho·las·tic \ˌin-tər-skə-ˈlas-tik\ *adjective* (1879)
: existing or carried on between schools ⟨*interscholastic* athletics⟩

in·ter se \ˌin-tər-ˈsā, -ˈsē\ *adverb or adjective* [Latin] (1845)
: among or between themselves

in·ter·sect \ˌin-tər-ˈsekt\ *verb* [Latin *intersectus,* past participle of *intersecare,* from *inter-* + *secare* to cut — more at SAW] (1615)
transitive verb
: to pierce or divide by passing through or across **:** CROSS
intransitive verb
1 : to meet and cross at a point
2 : to share a common area **:** OVERLAP

in·ter·sec·tion \ˌin-tər-ˈsek-shən, *especially in sense 2* ˈin-tər-ˌ\ *noun* (1559)
1 : the act or process of intersecting
2 : a place or area where two or more things (as streets) intersect
3 a : the set of elements common to two or more sets; *especially* **:** the set of points common to two geometric configurations **b :** the operation of finding the intersection of two or more sets

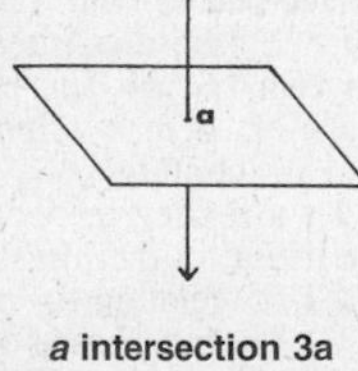

a **intersection 3a**

in·ter·ser·vice \ˌin-tər-ˈsər-vəs\ *adjective* (1946)
: existing between or relating to two or more of the armed services ⟨*interservice* rivalry⟩

in·ter·ses·sion \ˈin-tər-ˌse-shən\ *noun* (1932)
: a period between two academic sessions or terms sometimes utilized for brief concentrated courses

in·ter·sex \ˈin-tər-ˌseks\ *noun* [International Scientific Vocabulary] (1910)
: an intersexual individual

in·ter·sex·u·al \ˌin-tər-ˈsek-shə-wəl, -shwəl, -shəl\ *adjective* [International Scientific Vocabulary] (circa 1866)
1 : existing between sexes ⟨*intersexual* hostility⟩
2 : intermediate in sexual characters between a typical male and a typical female

— **in·ter·sex·u·al·i·ty** \-ˌsek-shə-ˈwa-lə-tē\ *noun*

— **in·ter·sex·u·al·ly** \-ˈsek-shə-wə-lē, -shwə-lē, -shə-lē\ *adverb*

[1]**in·ter·space** \ˈin-tər-ˌspās\ *noun* (15th century)
: an intervening space **:** INTERVAL

[2]**in·ter·space** \ˌin-tər-ˈspās\ *transitive verb* (1685)
: to occupy or fill the space between

in·ter·spe·cif·ic \ˌin-tər-spi-ˈsi-fik\ *also* **in·ter·spe·cies** \-ˈspē-(ˌ)shēz, -(ˌ)sēz\ *adjective* (1889)
: existing, occurring, or arising between species ⟨*interspecific* hybrid⟩

in·ter·sperse \ˌin-tər-ˈspərs\ *transitive verb* **-spersed; -spers·ing** [Latin *interspersus* interspersed, from *inter-* + *sparsus,* past participle of *spargere* to scatter — more at SPARK] (1566)
1 : to place something at intervals in or among
2 : to insert at intervals among other things ⟨*interspersing* drawings throughout the text⟩

— **in·ter·sper·sion** \-ˈspər-zhən, -shən\ *noun*

in·ter·sta·di·al \ˌin-tər-ˈstā-dē-əl\ *noun* [International Scientific Vocabulary *inter-* + New Latin *stadium* stage, phase, from Latin — more at STADIUM] (1914)
: a subdivision within a glacial stage marking a temporary retreat of the ice

[1]**in·ter·state** \ˌin-tər-ˈstāt\ *adjective* (1844)
: of, connecting, or existing between two or more states especially of the U.S. ⟨*interstate* commerce⟩

[2]**in·ter·state** \ˈin-tər-ˌstāt\ *noun* (1968)
: any of a system of expressways connecting most major U.S. cities — called also *interstate highway*

in·ter·stel·lar \-ˈste-lər\ *adjective* (1626)
: located, taking place, or traveling among the stars especially of the Milky Way galaxy

in·ter·ster·ile \-ˈster-əl, *chiefly British* -ˌīl\ *adjective* (1916)
: incapable of producing offspring by interbreeding

— **in·ter·ste·ril·i·ty** \-stə-ˈri-lə-tē\ *noun*

in·ter·stice \in-ˈtər-stəs\ *noun, plural* **-stic·es** \-stə-ˌsēz, -stə-səz\ [Middle English, from Latin *interstitium,* from *inter-* + *stit-, -stes* standing (as in *superstes* standing over) — more at SUPERSTITION] (15th century)
1 a : a space that intervenes between things; *especially* **:** one between closely spaced things **b :** a gap or break in something generally continuous ⟨the *interstices* of society⟩ ⟨passages of genuine literary merit in the *interstices* of the ludicrous . . . plots —Joyce Carol Oates⟩
2 : a short space of time between events

in·ter·sti·tial \ˌin-tər-ˈsti-shəl\ *adjective* (1646)
1 : relating to or situated in the interstices
2 a : situated within but not restricted to or characteristic of a particular organ or tissue — used especially of fibrous tissue **b :** affecting the interstitial tissues of an organ or part
3 : being or relating to a crystalline compound in which usually small atoms or ions of a nonmetal occupy holes between the larger metal atoms or ions in the crystal lattice

— **in·ter·sti·tial·ly** \-shə-lē\ *adverb*

in·ter·sub·jec·tive \ˌin-tər-səb-ˈjek-tiv\ *adjective* (1899)
1 : involving or occurring between separate conscious minds ⟨*intersubjective* communication⟩
2 : accessible to or capable of being established for two or more subjects **:** OBJECTIVE

— **in·ter·sub·jec·tive·ly** *adverb*

— **in·ter·sub·jec·tiv·i·ty** \-(ˌ)səb-ˌjek-ˈti-və-tē\ *noun*

in·ter·tes·ta·men·tal \ˌtes-tə-ˈmen-tᵊl\ *adjective* (1929)
: of, relating to, or forming the period of two centuries between the composition of the last book of the Old Testament and the first book of the New Testament

in·ter·tid·al \-ˈtī-dᵊl\ *adjective* (1883)
: of, relating to, or being the part of the littoral zone above low-tide mark
— **in·ter·tid·al·ly** \-dᵊl-ē\ *adverb*

in·ter·tie \ˈin-tər-ˌtī\ *noun* (1951)
: an interconnection permitting passage of current between two or more electric utility systems

in·ter·till \ˌin-tər-ˈtil\ *transitive verb* (1912)
: to cultivate between the rows of (a crop)
— **in·ter·till·age** \-ˈti-lij\ *noun*

in·ter·trop·i·cal \-ˈträ-pi-kəl\ *adjective* (1794)
1 : situated between or within the tropics
2 : relating to regions within the tropics **:** TROPICAL

in·ter·twine \-ˈtwīn\ (1641)
transitive verb
: to unite by twining one with another
intransitive verb
: to twine about one another; *also* **:** to become mutually involved
— **in·ter·twine·ment** \-mənt\ *noun*

in·ter·twist \-ˈtwist\ *verb* (circa 1659)
: INTERTWINE
— **in·ter·twist** \ˈin-tər-ˌtwist\ *noun*

in·ter·val \ˈin-tər-vəl\ *noun* [Middle English *intervalle,* from Middle French, from Latin *intervallum* space between ramparts, interval, from *inter-* + *vallum* rampart — more at WALL] (14th century)
1 a : a space of time between events or states **b** *British* **:** INTERMISSION
2 a : a space between objects, units, points, or states **b :** difference in pitch between tones
3 : a set of real numbers between two numbers either including or excluding one or both of them
4 : one of a series of fast-paced runs interspersed with jogging for training (as of a runner)
— **in·ter·val·lic** \ˌin-tər-ˈva-lik\ *adjective*

in·ter·vale \ˈin-tər-vəl, -ˌvāl\ *noun* [obsolete *intervale* interval] (1647)
chiefly New England **:** BOTTOMLAND

in·ter·val·om·e·ter \ˌin-tər-və-ˈlä-mə-tər\ *noun* (1933)
: a device that operates a control (as for a camera shutter) at regular intervals

in·ter·vene \ˌin-tər-ˈvēn\ *intransitive verb* **-vened; -ven·ing** [Latin *intervenire* to come between, from *inter-* + *venire* to come — more at COME] (1587)
1 : to occur, fall, or come between points of time or events
2 : to enter or appear as an irrelevant or extraneous feature or circumstance
3 : to come in or between by way of hindrance or modification ⟨*intervene* to stop a fight⟩
4 : to occur or lie between two things
5 a : to become a third party to a legal proceeding begun by others for the protection of an alleged interest **b :** to interfere usually by force or threat of force in another nation's internal affairs especially to compel or prevent an action or to maintain or alter a condition
synonym see INTERPOSE
— **in·ter·ven·tion** \-ˈven(t)-shən\ *noun*

in·ter·ve·nor \-ˈvē-nər, -ˌnȯr\ *or* **in·ter·ven·er** \-ˈvē-nər\ *noun* (1621)
: one who intervenes; *especially* **:** one who intervenes as a third party in a legal proceeding

in·ter·ven·tion·ism \-ˈven(t)-shə-ˌni-zəm\ *noun* (1923)
: the theory or practice of intervening; *specifically* **:** governmental interference in economic affairs at home or in political affairs of another country
— **in·ter·ven·tion·ist** \-ˈvench-nist, -ˈven(t)-shə-nist\ *noun or adjective*

in·ter·ver·te·bral disk \ˌin-tər-ˈvər-tə-brəl-, -(ˌ)vər-ˈtē-\ *noun* (circa 1860)
: any of the tough elastic disks that are interposed between the centra of adjoining vertebrae and that consist of an outer fibrous ring enclosing an inner pulpy nucleus

in·ter·view \ˈin-tər-ˌvyü\ *noun* [Middle French *entrevue,* from (*s'*)*entrevoir* to see one another, meet, from *entre-* inter- + *voir* to see — more at VIEW] (1514)
1 : a formal consultation usually to evaluate qualifications (as of a prospective student or employee)
2 a : a meeting at which information is obtained (as by a reporter, television commentator, or pollster) from a person **b :** a report or reproduction of information so obtained
3 : INTERVIEWEE
— **interview** *verb*
— **in·ter·view·er** *noun*

in·ter·view·ee \ˌin-tər-(ˌ)vyü-ˈē\ *noun* (1884)
: one who is interviewed

in·ter vi·vos \ˈin-tər-ˈvē-ˌvōs, -ˈvī-\ *adverb or adjective* [Late Latin] (1837)
: between living persons ⟨transaction *inter vivos*⟩; *especially* **:** from one living person to another ⟨*inter vivos* gifts⟩ ⟨property transferred *inter vivos*⟩

in·ter·vo·cal·ic \ˌin-tər-vō-ˈka-lik\ *adjective* (1887)
: immediately preceded and immediately followed by a vowel
— **in·ter·vo·cal·i·cal·ly** \-li-k(ə-)lē\ *adverb*

in·ter·weave \ˌin-tər-ˈwēv\ *verb* **-wove** \-ˈwōv\ *also* **-weaved; -wo·ven** \-ˈwō-vən\ *also* **-weaved; -weav·ing** (1598)
transitive verb
1 : to weave together
2 : to mix or blend together ⟨*interweaving* his own insights . . . with letters and memoirs —Phoebe Adams⟩
intransitive verb
: INTERTWINE, INTERMINGLE
— **in·ter·weave** \ˈin-tər-ˌwēv\ *noun*
— **in·ter·wo·ven** \ˌin-tər-ˈwō-vən\ *adjective*

in·tes·ta·cy \in-ˈtes-tə-sē\ *noun* (1767)
: the quality or state of being or dying intestate

¹in·tes·tate \in-ˈtes-ˌtāt, -tət\ *adjective* [Middle English, from Latin *intestatus,* from *in-* + *testatus* testate] (14th century)
1 : having made no valid will ⟨died *intestate*⟩
2 : not disposed of by will ⟨an *intestate* estate⟩

²intestate *noun* (1658)
: one who dies intestate

in·tes·ti·nal \in-ˈtes-tə-nᵊl, -ˈtes(t)-nəl, -ˈtes-ᵊn-əl, *British often* ˌin-(ˌ)tes-ˈtī-nᵊl\ *adjective* (15th century)
1 : affecting or occurring in the intestine; *also* **:** living in the intestine
2 : of, relating to, or being the intestine
— **in·tes·ti·nal·ly** *adverb*

intestinal fortitude *noun* [euphemism for *guts*] (circa 1937)
: COURAGE, STAMINA

¹in·tes·tine \in-ˈtes-tən\ *adjective* [Middle English, from Middle French or Latin; Middle French *intestin,* from Latin *intestinus,* from *intus* within — more at ENT-] (15th century)
: INTERNAL; *specifically* **:** of or relating to the internal affairs of a state or country ⟨*intestine* war⟩

²intestine *noun* [Middle English, from Middle French *intestin,* from Latin *intestinum,* from neuter of *intestinus*] (15th century)
: the tubular part of the alimentary canal that extends from the stomach to the anus

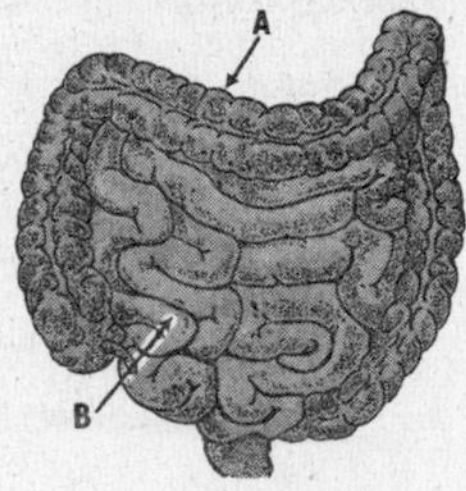

intestine: *A* large intestine, *B* small intestine

in·ti \ˈin-tē\ *noun* [Quechua, sun] (1985)
: the basic monetary unit of Peru from 1985 to 1990

in·ti·ma \ˈin-tə-mə\ *noun, plural* **-mae** \-ˌmē, -ˌmī\ *or* **-mas** [New Latin, from Latin, feminine of *intimus*] (1873)
: the innermost coat of an organ (as a blood vessel) consisting usually of an endothelial layer backed by connective tissue and elastic tissue
— **in·ti·mal** \-məl\ *adjective*

in·ti·ma·cy \ˈin-tə-mə-sē\ *noun, plural* **-cies** (1641)
1 : the state of being intimate **:** FAMILIARITY
2 : something of a personal or private nature

¹in·ti·mate \ˈin-tə-ˌmāt\ *transitive verb* **-mat·ed; -mat·ing** [Late Latin *intimatus,* past participle of *intimare* to put in, announce, from Latin *intimus* innermost, superlative of (assumed) Old Latin *interus* inward — more at INTERIOR] (1522)
1 : to make known especially publicly or formally **:** ANNOUNCE
2 : to communicate delicately and indirectly **:** HINT
synonym see SUGGEST
— **in·ti·mat·er** *noun*
— **in·ti·ma·tion** \ˌin-tə-ˈmā-shən\ *noun*

²in·ti·mate \ˈin-tə-mət\ *adjective* [alteration of obsolete *intime,* from Latin *intimus*] (1632)
1 a : INTRINSIC, ESSENTIAL **b :** belonging to or characterizing one's deepest nature
2 : marked by very close association, contact, or familiarity ⟨*intimate* knowledge of the law⟩
3 a : marked by a warm friendship developing through long association **b :** suggesting informal warmth or privacy ⟨*intimate* clubs⟩
4 : of a very personal or private nature
— **in·ti·mate·ly** *adverb*
— **in·ti·mate·ness** *noun*

³in·ti·mate \ˈin-tə-mət\ *noun* (1659)
: an intimate friend or confidant

in·tim·i·date \in-ˈti-mə-ˌdāt\ *transitive verb* **-dat·ed; -dat·ing** [Medieval Latin *intimidatus,* past participle of *intimidare,* from Latin *in-* + *timidus* timid] (1646)
: to make timid or fearful **:** FRIGHTEN; *especially* **:** to compel or deter by or as if by threats ☆
— **in·tim·i·dat·ing·ly** \-ˌdā-tiŋ-lē\ *adverb*
— **in·tim·i·da·tion** \-ˌti-mə-ˈdā-shən\ *noun*
— **in·tim·i·da·tor** \-ˈti-mə-ˌdā-tər\ *noun*

in·tim·i·da·to·ry \-ˈti-mə-də-ˌtōr-ē, -ˌtȯr-\ *adjective* (circa 1846)
: tending to intimidate

in·tinc·tion \in-ˈtiŋ(k)-shən\ *noun* [Late Latin *intinction-, intinctio* baptism, from Latin *intingere* to dip in, from *in-* + *tingere* to dip, moisten. — more at TINGE] (1872)
: the administration of the sacrament of Communion by dipping the bread in the wine and giving both together to the communicant

☆ SYNONYMS
Intimidate, cow, bulldoze, bully, browbeat mean to frighten into submission. INTIMIDATE implies inducing fear or a sense of inferiority into another ⟨*intimidated* by so many other bright freshmen⟩. COW implies reduction to a state where the spirit is broken or all courage is lost ⟨not at all *cowed* by the odds against making it in show business⟩. BULLDOZE implies an intimidating or an overcoming of resistance usually by urgings, demands, or threats ⟨*bulldozed* the city council into approving the plan⟩. BULLY implies intimidation through threats, insults, or aggressive behavior ⟨*bullied* into giving up their lunch money⟩. BROWBEAT implies a cowing through arrogant, scornful, contemptuous, or insolent treatment ⟨*browbeat* the witness into a contradiction⟩.

in·tine \'in-ˌtēn\ *noun* [probably from German, from Latin *intus* within + New Latin *in-* fibrous tissue, from Greek *in-, is* tendon] (1835)
: the inner mostly cellulose wall of a spore (as a pollen grain)

in·tit·ule \in-'ti-(ˌ)chü(ə)l\ *transitive verb* **-uled; -ul·ing** [Middle French *intituler,* from Late Latin *intitulare,* from Latin *in-* + *titulus* title] (15th century)
British **:** to furnish (as a legislative act) with a title or designation

in·to \'in-(ˌ)tü, -tə\ *preposition* [Middle English, from Old English *intō,* from [2]*in* + *tō* to] (before 12th century)
1 — used as a function word to indicate entry, introduction, insertion, superposition, or inclusion ⟨came *into* the house⟩ ⟨enter *into* an alliance⟩
2 a : to the state, condition, or form of ⟨got *into* trouble⟩ **b :** to the occupation, action, or possession of ⟨go *into* farming⟩ **c :** involved with or interested in ⟨*into* hard drugs⟩ ⟨*into* Latin epigrammatists⟩
3 — used as a function word to indicate a period of time or an extent of space part of which is passed or occupied ⟨far *into* the night⟩
4 : in the direction of ⟨looking *into* the sun⟩
5 : to a position of contact with **:** AGAINST ⟨ran *into* a wall⟩
6 — used as a function word to indicate the dividend in division ⟨dividing 3 *into* 6 gives 2⟩

in·tol·er·a·ble \(ˌ)in-'täl-rə-bəl, -'tä-lə-rə-, -'tä-lər-bəl\ *adjective* [Middle English, from Latin *intolerabilis,* from *in-* + *tolerabilis* tolerable] (15th century)
1 : not tolerable **:** UNBEARABLE ⟨*intolerable* pain⟩
2 : EXCESSIVE
— in·tol·er·a·bil·i·ty \-ˌtäl-rə-'bi-lə-tē, -ˌtä-lə-rə-, -ˌtä-lər-\ *noun*
— in·tol·er·a·ble·ness \-'täl-rə-bəl-nəs, -'tä-lə-rə-, -'tä-lər-bəl-\ *noun*
— in·tol·er·a·bly \-blē\ *adverb*

in·tol·er·ance \(ˌ)in-'täl-rən(t)s, -'tä-lə-\ *noun* (1765)
: the quality or state of being intolerant; *especially* **:** exceptional sensitivity (as to a drug)

in·tol·er·ant \-rənt\ *adjective* (circa 1735)
1 : unable or unwilling to endure ⟨a plant *intolerant* of direct sunlight⟩ ⟨*intolerant* of criticism⟩
2 a : unwilling to grant equal freedom of expression especially in religious matters **b :** unwilling to grant or share social, political, or professional rights **:** BIGOTED
— in·tol·er·ant·ly *adverb*
— in·tol·er·ant·ness *noun*

in·to·nate \'in-tə-ˌnāt, -(ˌ)tō-\ *transitive verb* **-nat·ed; -nat·ing** (1795)
: INTONE, UTTER

in·to·na·tion \ˌin-tə-'nā-shən, -(ˌ)tō-\ *noun* (1620)
1 : the act of intoning and especially of chanting
2 : something that is intoned; *specifically* **:** the opening tones of a Gregorian chant
3 : the ability to play or sing notes in tune
4 : manner of utterance; *specifically* **:** the rise and fall in pitch of the voice in speech
— in·to·na·tion·al \-shnəl, -shə-n^{ə}l\ *adjective*

in·tone \in-'tōn\ *verb* **in·toned; in·ton·ing** [Middle English *entonen,* from Middle French *entoner,* from Medieval Latin *intonare,* from Latin *in-* + *tonus* tone] (15th century)
transitive verb
: to utter in musical or prolonged tones **:** recite in singing tones or in a monotone
intransitive verb
: to utter something in singing tones or in monotone
— in·ton·er *noun*

in to·to \in-'tō-(ˌ)tō\ *adverb* [Latin, on the whole] (1796)
: TOTALLY, ENTIRELY

in·tox·i·cant \in-'täk-si-kənt\ *noun* (1863)
: something that intoxicates; *especially* **:** an alcoholic drink
— intoxicant *adjective*

[1]in·tox·i·cate \-si-kət\ *adjective* (15th century)
archaic **:** INTOXICATED

[2]in·tox·i·cate \-sə-ˌkāt\ *transitive verb* **-cat·ed; -cat·ing** [Middle English, from Medieval Latin *intoxicatus,* past participle of *intoxicare,* from Latin *in-* + *toxicum* poison — more at TOXIC] (15th century)
1 : POISON
2 a : to excite or stupefy by alcohol or a drug especially to the point where physical and mental control is markedly diminished **b :** to excite or elate to the point of enthusiasm or frenzy

in·tox·i·cat·ed \-sə-ˌkā-təd\ *adjective* (1576)
: affected by or as if by alcohol **:** DRUNK
— in·tox·i·cat·ed·ly \-ˌkā-təd-lē\ *adverb*

in·tox·i·ca·tion \in-ˌtäk-sə-'kā-shən\ *noun* (15th century)
1 : an abnormal state that is essentially a poisoning ⟨intestinal *intoxication*⟩
2 a : the condition of being drunk **:** INEBRIATION **b :** a strong excitement or elation

in·tra- \'in-trə, -(ˌ)trä\ *prefix* [Late Latin, from Latin *intra,* from (assumed) Old Latin *interus,* adjective, inward — more at INTERIOR]
1 a : within ⟨*intra*galactic⟩ **b :** during ⟨*intra*day⟩ **c :** between layers of ⟨*intra*dermal⟩
2 : INTRO- ⟨an *intra*muscular injection⟩

in·tra–ar·te·ri·al \-är-'tir-ē-əl\ *adjective* (1897)
: situated or occurring within, administered into, or involving entry by way of an artery
— in·tra–ar·te·ri·al·ly \-ē-ə-lē\ *adverb*

in·tra–ar·tic·u·lar \-är-'ti-kyə-lər\ *adjective* (circa 1888)
: situated within, occurring within, or administered by entry into a joint

in·tra·car·di·ac \-'kär-dē-ˌak\ *also* **in·tra·car·di·al** \-dē-əl\ *adjective* (1876)
: situated or occurring within or introduced or involving entry into the heart ⟨*intracardiac* surgery⟩ ⟨an *intracardiac* catheter⟩
— in·tra·car·di·al·ly \-dē-ə-lē\ *adverb*

in·tra·cel·lu·lar \-'sel-yə-lər\ *adjective* (1876)
: existing, occurring, or functioning within a cell ⟨*intracellular* parasites⟩
— in·tra·cel·lu·lar·ly *adverb*

in·tra·ce·re·bral \-sə-'rē-brəl, -'ser-ə-\ *adjective* (1881)
: situated in, introduced into, or made into the cerebrum ⟨*intracerebral* injections⟩ ⟨*intracerebral* bleeding⟩
— in·tra·ce·re·bral·ly \-brə-lē\ *adverb*

in·tra·com·pa·ny \-'kəmp-nē, -'kəm-pə-\ *adjective* (1926)
: occurring within or taking place between branches or employees of a company ⟨*intracompany* income⟩

in·tra·cra·ni·al \-'krā-nē-əl\ *adjective* (circa 1849)
: existing or occurring within the cranium; *also* **:** affecting or involving intracranial structures
— in·tra·cra·ni·al·ly \-nē-ə-lē\ *adverb*

in·trac·ta·ble \(ˌ)in-'trak-tə-bəl\ *adjective* [Latin *intractabilis,* from *in-* + *tractabilis* tractable] (1531)
1 : not easily governed, managed, or directed **:** OBSTINATE
2 : not easily manipulated or wrought ⟨*intractable* metal⟩
3 : not easily relieved or cured ⟨*intractable* pain⟩
synonym see UNRULY
— in·trac·ta·bil·i·ty \(ˌ)in-ˌtrak-tə-'bi-lə-tē\ *noun*
— in·trac·ta·bly \-blē\ *adverb*

in·tra·cu·ta·ne·ous \ˌin-trə-kyu̇-'tā-nē-əs, -(ˌ)trä-\ *adjective* (1885)
: INTRADERMAL
— in·tra·cu·ta·ne·ous·ly *adverb*

in·tra·day \'in-trə-ˌdā, -(ˌ)trä-\ *adjective* (1950)
: occurring in the course of a single day ⟨the market showed wide *intraday* fluctuations⟩

in·tra·der·mal \ˌin-trə-'dər-məl, -(ˌ)trä-\ *adjective* (circa 1900)
: situated, occurring, or done within or between the layers of the skin; *also* **:** administered by entering the skin ⟨*intradermal* injections⟩
— in·tra·der·mal·ly \-mə-lē\ *adverb*

intradermal test *noun* (1916)
: a test for immunity or hypersensitivity made by injecting a minute amount of diluted antigen into the skin

in·tra·dos \'in-trə-ˌdäs, -ˌdō; in-'trā-ˌdäs\ *noun, plural* **-dos** \-ˌdōz, -ˌdäs\ *or* **-dos·es** \-ˌdä-səz\ [French, from Latin *intra* within + French *dos* back — more at DOSSIER] (1772)
: the interior curve of an arch — see ARCH illustration

in·tra·ga·lac·tic \ˌin-trə-gə-'lak-tik, -(ˌ)trä-\ *adjective* (1964)
: situated or occurring within the confines of a single galaxy

in·tra·gen·ic \-'je-nik\ *adjective* (1937)
: being or occurring within a gene ⟨*intragenic* recombination⟩ ⟨*intragenic* mutation⟩

in·tra·mo·lec·u·lar \-mə-'le-kyə-lər\ *adjective* [International Scientific Vocabulary] (1884)
: existing or acting within the molecule; *also* **:** formed by reaction between different parts of the same molecule
— in·tra·mo·lec·u·lar·ly *adverb*

in·tra·mu·ral \-'myu̇r-əl\ *adjective* (1846)
1 a : being or occurring within the limits usually of a community, organization, or institution **b :** competed only within the student body ⟨*intramural* sports⟩
2 : situated or occurring within the substance of the walls of an organ
— in·tra·mu·ral·ly \-ə-lē\ *adverb*

in·tra·mus·cu·lar \-'məs-kyə-lər\ *adjective* [International Scientific Vocabulary] (1874)
: situated in, occurring in, or administered by entering a muscle
— in·tra·mus·cu·lar·ly *adverb*

in·tra·na·sal \-'nā-zəl\ *adjective* (1886)
: lying within or administered by way of the nasal structures
— in·tra·na·sal·ly \-'nāz-lē, -'nā-zə-\ *adverb*

in·tran·si·geance \in-'tran(t)-sə-jən(t)s, -'tran-zə-\ *noun* [French] (1899)
: INTRANSIGENCE
— in·tran·si·geant \-jənt\ *adjective or noun*
— in·tran·si·geant·ly *adverb*

in·tran·si·gence \-jən(t)s\ *noun* (1882)
: the quality or state of being intransigent

in·tran·si·gent \-jənt\ *adjective* [Spanish *intransigente,* from *in-* + *transigente,* present participle of *transigir* to compromise, from Latin *transigere* to come to an agreement — more at TRANSACT] (circa 1879)
1 a : refusing to compromise or to abandon an extreme position or attitude **:** UNCOMPROMISING **b :** IRRECONCILABLE
2 : characteristic of an intransigent person
— intransigent *noun*
— in·tran·si·gent·ly *adverb*

in·tran·si·tive \(ˌ)in-'tran(t)-sə-tiv, -'tran-zə-; -'tran(t)s-tiv\ *adjective* [Late Latin *intransitivus,* from Latin *in-* + Late Latin *transitivus* transitive] (1612)
: not transitive; *especially* **:** characterized by not having or containing a direct object ⟨an *intransitive* verb⟩
— in·tran·si·tive·ly *adverb*

— **in·tran·si·tive·ness** *noun*
— **in·tran·si·tiv·i·ty** \(ˌ)in-ˌtran(t)-sə-ˈti-və-tē, -ˌtran-zə-\ *noun*
in·tra·oc·u·lar \ˌin-trə-ˈä-kyə-lər, -(ˌ)trä-\ *adjective* [International Scientific Vocabulary] (1826)
: implanted in, occurring in, or administered by entering the eyeball
— **in·tra·oc·u·lar·ly** *adverb*
in·tra·per·i·to·ne·al \ˌin-trə-ˌper-ə-tᵊn-ˈē-əl\ *adjective* (circa 1836)
: existing within or administered by entering the peritoneum
— **in·tra·per·i·to·ne·al·ly** \-ˈē-ə-lē\ *adverb*
in·tra·per·son·al \-ˈpərs-nəl, -ˈpər-sᵊn-əl\ *adjective* (1909)
: occurring within the individual mind or self ⟨*intrapersonal* concerns of the aged⟩
in·tra·plate \ˈin-trə-ˌplāt, -(ˌ)trä-\ *adjective* (1973)
: relating to or occurring within the interior of a tectonic plate ⟨an *intraplate* earthquake⟩
in·tra·pop·u·la·tion \ˈin trə-ˌpä-pyə-ˈlā-shən, -(ˌ)trä-\ *adjective* (1959)
: occurring within or taking place between members of a population
in·tra·pre·neur \ˌin-trə-prə-ˈnər, -ˈnu̇r, -ˈnyu̇r\ *noun* [*intra-* + entre*preneur*] (1982)
: a corporate executive who develops new enterprises within the corporation
— **in·tra·pre·neur·ial** \-ˈnər-ē-əl, -ˈn(y)u̇r-\ *adjective*
in·tra·psy·chic \ˌin-trə-ˈsī-kik, -(ˌ)trä-\ *adjective* (1917)
: being or occurring within the psyche, mind, or personality
— **in·tra·psy·chi·cal·ly** \-ki-k(ə-)lē\ *adverb*
in·tra·spe·cies \-ˈspē-(ˌ)shēz, -(ˌ)sēz\ *adjective* (1927)
: INTRASPECIFIC
in·tra·spe·cif·ic \-spi-ˈsi-fik\ *adjective* (1919)
: occurring within a species or involving members of one species
in·tra·state \-ˈstāt\ *adjective* (1903)
: existing or occurring within a state
in·tra·the·cal \-ˈthē-kəl\ *adjective* (1887)
: introduced into or occurring in the space under the arachnoid membrane of the brain or spinal cord
— **in·tra·the·cal·ly** \-k(ə-)lē\ *adverb*
in·tra·tho·rac·ic \-thə-ˈra-sik\ *adjective* [International Scientific Vocabulary] (1862)
: situated or occurring within the thorax ⟨*intrathoracic* pressure⟩
— **in·tra·tho·rac·i·cal·ly** *adverb*
in·tra·uter·ine \-ˈyü-tə-rən, -ˌrīn\ *adjective* [International Scientific Vocabulary] (circa 1836)
: situated, used, or occurring within the uterus; *also* **:** involving the part of development that takes place in the uterus
intrauterine device *noun* (1964)
: a device inserted and left in the uterus to prevent effective conception — called also *intrauterine contraceptive device, IUD*
in·tra·vas·cu·lar \ˌin-trə-ˈvas-kyə-lər, -(ˌ)trä-\ *adjective* (1876)
: situated in, occurring in, or administered by entry into a blood vessel ⟨*intravascular* thrombosis⟩ ⟨an *intravascular* injection⟩
— **in·tra·vas·cu·lar·ly** *adverb*
in·tra·ve·nous \ˌin-trə-ˈvē-nəs\ *adjective* [International Scientific Vocabulary] (circa 1849)
: situated, performed, or occurring within or entering by way of a vein; *also* **:** used in intravenous procedures
— **in·tra·ve·nous·ly** *adverb*
in·tra·ven·tric·u·lar \ˌin-trə-ven-ˈtri-kyə-lər, -(ˌ)trä-\ *adjective* (1882)
: situated within, occurring within, or administered into a ventricle
— **in·tra·ven·tric·u·lar·ly** *adverb*
in·tra·vi·tal \-ˈvī-tᵊl\ *adjective* [International Scientific Vocabulary] (circa 1890)
1 : performed upon or found in a living subject
2 : having or utilizing the property of staining cells without killing them — compare SUPRAVITAL
— **in·tra·vi·tal·ly** \-tᵊl-ē\ *adverb*
in·tra·vi·tam \-ˈvī-ˌtam, -ˈwē-ˌtäm\ *adjective* [New Latin *intra vitam* during life] (1881)
: INTRAVITAL
in·tra·zon·al \ˌin-trə-ˈzō-nᵊl, -(ˌ)trä-\ *adjective* (1927)
: of, relating to, or being a soil or a major soil group marked by relatively well-developed characteristics that are determined primarily by essentially local factors (as the parent material) rather than climate and vegetation — compare AZONAL, ZONAL
intreat *archaic variant of* ENTREAT
intrench *variant of* ENTRENCH
in·trep·id \in-ˈtre-pəd\ *adjective* [Latin *intrepidus,* from *in-* + *trepidus* alarmed — more at TREPIDATION] (1697)
: characterized by resolute fearlessness, fortitude, and endurance
— **in·tre·pid·i·ty** \ˌin-trə-ˈpi-də-tē\ *noun*
— **in·trep·id·ly** \in-ˈtre-pəd-lē\ *adverb*
— **in·trep·id·ness** *noun*
in·tri·ca·cy \ˈin-tri-kə-sē\ *noun, plural* **-cies** (1602)
1 : the quality or state of being intricate
2 : something intricate ⟨the *intricacies* of a plot⟩
in·tri·cate \ˈin-tri-kət\ *adjective* [Middle English, from Latin *intricatus,* past participle of *intricare* to entangle, from *in-* + *tricae* trifles, complications] (15th century)
1 : having many complexly interrelating parts or elements **:** COMPLICATED
2 : difficult to resolve or analyze
synonym see COMPLEX
— **in·tri·cate·ly** *adverb*
— **in·tri·cate·ness** *noun*
in·tri·gant *or* **in·tri·guant** \ˌin-trē-ˈgänt, ˌan-, -ˈgäⁿ\ *noun* [French *intrigant,* from Italian *intrigante,* present participle of *intrigare*] (1781)
: one that intrigues
¹in·trigue \in-ˈtrēg\ *verb* **in·trigued; in·trigu·ing** [French *intriguer,* from Italian *intrigare,* from Latin *intricare*] (1612)
transitive verb
1 : CHEAT, TRICK
2 : to accomplish by intrigue ⟨*intrigued* myself into the club⟩
3 *obsolete* **:** ENTANGLE
4 : to arouse the interest, desire, or curiosity of ⟨*intrigued* by the tale⟩
intransitive verb
: to carry on an intrigue; *especially* **:** PLOT, SCHEME
— **in·trigu·er** *noun*
²in·trigue \ˈin-ˌtrēg, in-ˈ\ *noun* (1647)
1 a : a secret scheme **:** MACHINATION **b :** the practice of engaging in intrigues
2 : a clandestine love affair
synonym see PLOT
in·trigu·ing \in-ˈtrē-giŋ\ *adjective* (1909)
: engaging the interest to a marked degree **:** FASCINATING
— **in·trigu·ing·ly** \-giŋ-lē\ *adverb*
in·trin·sic \in-ˈtrin-zik, -ˈtrin(t)-sik\ *adjective* [Middle French *intrinsèque* internal, from Late Latin *intrinsecus,* from Latin, adverb, inwardly; akin to Latin *intra* within — more at INTRA-] (1642)
1 a : belonging to the essential nature or constitution of a thing ⟨the *intrinsic* worth of a gem⟩ ⟨the *intrinsic* brightness of a star⟩ **b :** being or relating to a semiconductor in which the concentration of charge carriers is characteristic of the material itself instead of the content of any impurities it contains
2 a : originating or due to causes within a body, organ, or part ⟨an *intrinsic* metabolic disease⟩ **b :** originating and included wholly within an organ or part ⟨*intrinsic* muscles⟩ — compare EXTRINSIC 1b
— **in·trin·si·cal·ly** \-zi-k(ə-)lē, -si-\ *adverb*
in·trin·si·cal \-zi-kəl, -si-\ *adjective* (circa 1548)
archaic **:** INTRINSIC
intrinsic factor *noun* (1930)
: a substance produced by normal gastrointestinal mucosa that facilitates absorption of vitamin B_{12}
in·tro \ˈin-(ˌ)trō\ *noun, plural* **intros** (circa 1899)
: INTRODUCTION
intro- *prefix* [Middle English, from Middle French, from Latin, from *intro* inside, to the inside, from (assumed) Old Latin *interus,* adjective, inward]
1 : in **:** into ⟨*intro*jection⟩
2 : inward **:** within ⟨*intro*vert⟩ — compare EXTRO-
in·tro·duce \ˌin-trə-ˈdüs, -ˈdyüs\ *transitive verb* **-duced; -duc·ing** [Middle English, from Latin *introducere,* from *intro-* + *ducere* to lead — more at TOW] (15th century)
1 : to lead or bring in especially for the first time
2 a : to bring into play **b :** to bring into practice or use **:** INSTITUTE
3 : to lead to or make known by a formal act, announcement, or recommendation: as **a :** to cause to be acquainted **b :** to present formally at court or into society **c :** to present or announce formally or officially or by an official reading **d :** to make preliminary explanatory or laudatory remarks about **e :** to bring (as an actor or singer) before the public for the first time
4 : PLACE, INSERT
5 : to bring to a knowledge of something ☆
— **in·tro·duc·er** *noun*
in·tro·duc·tion \ˌin-trə-ˈdək-shən\ *noun* [Middle English *introduccioun* act of introducing, from Middle French *introduction,* from Latin *introduction-, introductio,* from *introducere*] (14th century)
1 : something that introduces: as **a** (1) **:** a part of a book or treatise preliminary to the main portion (2) **:** a preliminary treatise or course of study **b :** a short introductory musical passage
2 : the act or process of introducing **:** the state of being introduced
3 : a putting in **:** INSERTION
4 : something introduced; *specifically* **:** a new or exotic plant or animal
in·tro·duc·to·ry \ˌin-trə-ˈdək-t(ə-)rē\ *adjective* (1605)
: of, relating to, or being a first step that sets something going or in proper perspective ⟨an *introductory* course in calculus⟩

☆ SYNONYMS
Introduce, insert, insinuate, interpolate, intercalate, interpose, interject mean to put between or among others. INTRODUCE is a general term for bringing or placing a thing or person into a group or body already in existence ⟨*introduced* a new topic into the conversation⟩. INSERT implies putting into a fixed or open space between or among ⟨*inserted* a clause in the contract⟩. INSINUATE implies introducing gradually or by gentle pressure ⟨*insinuated* himself into the group⟩. INTERPOLATE applies to the inserting of something extraneous or spurious ⟨*interpolated* her own comments into the report⟩. INTERCALATE suggests an intrusive inserting of something in an existing series or sequence ⟨new chapters *intercalated* with the old⟩. INTERPOSE suggests inserting an obstruction or cause of delay ⟨*interpose* barriers to communication⟩. INTERJECT implies an abrupt or forced introduction ⟨*interjected* a question⟩.

— **in·tro·duc·to·ri·ly** \-t(ə-)rə-lē\ *adverb*

in·tro·gres·sion \,in-trə-'gre-shən\ *noun* [*intro-* + *-gression* (as in *regression*)] (1938)
: the entry or introduction of a gene from one gene complex into another
— **in·tro·gres·sant** \-'gre-s^{ə}nt\ *adjective or noun*
— **in·tro·gres·sive** \-'gre-siv\ *adjective*

in·troit \'in-,trȯit, -,trō-ət, in-'\ *noun* [Middle English, from Middle French *introite,* from Medieval Latin *introitus,* from Latin, entrance, from *introire* to go in, from *intro-* + *ire* to go — more at ISSUE] (15th century)
1 *often capitalized* **:** the first part of the traditional proper of the Mass consisting of an antiphon, verse from a psalm, and the Gloria Patri
2 : a piece of music sung or played at the beginning of a worship service

in·tro·ject \,in-trə-'jekt\ *transitive verb* [back-formation from *introjection,* from International Scientific Vocabulary *intro-* + pro*jection*] (1925)
: to incorporate (attitudes or ideas) into one's personality unconsciously
— **in·tro·jec·tion** \-'jek-shən\ *noun*

in·tro·mis·sion \,in-trə-'mi-shən\ *noun* [French, from Middle French, from Medieval Latin *intromission-, intromissio,* from Latin *intromittere*] (1601)
: the act or process of intromitting; *especially* **:** the insertion or period of insertion of the penis in the vagina in copulation

in·tro·mit \-'mit\ *transitive verb* **-mit·ted; -mit·ting** [Latin *intromittere,* from *intro-* + *mittere* to send] (circa 1588)
: to send or put in **:** INSERT
— **in·tro·mit·tent** \-'mi-t^{ə}nt\ *adjective*
— **in·tro·mit·ter** \-tər\ *noun*

in·tron \'in-,trän\ *noun* [*inter*vening sequence + ^{2}*-on*] (1978)
: a polynucleotide sequence in a nucleic acid that does not code information for protein synthesis and is removed before translation of messenger RNA — compare EXON

in·trorse \'in-,trȯrs\ *adjective* [Latin *introrsus,* adverb, inward, from *intro-* + *versus* toward, from past participle of *vertere* to turn — more at WORTH] (1842)
: facing inward or toward the axis of growth ⟨an *introrse* anther⟩

in·tro·spec·tion \,in-trə-'spek-shən\ *noun* [Latin *introspectus,* past participle of *introspicere* to look inside, from *intro-* + *specere* to look — more at SPY] (circa 1677)
: a reflective looking inward **:** an examination of one's own thoughts and feelings
— **in·tro·spect** \-'spekt\ *verb*
— **in·tro·spec·tion·al** \-'spek-shnəl, -shə-n^{ə}l\ *adjective*
— **in·tro·spec·tive** \-'spek-tiv\ *adjective*
— **in·tro·spec·tive·ly** *adverb*
— **in·tro·spec·tive·ness** *noun*

in·tro·spec·tion·ism \-shə-,ni-zəm\ *noun* (1922)
: a doctrine that psychology must be based essentially on data derived from introspection — compare BEHAVIORISM
— **in·tro·spec·tion·ist** \-sh(ə-)nist\ *or* **in·tro·spec·tion·is·tic** \-,spek-shə-'nis-tik\ *adjective*
— **introspectionist** *noun*

in·tro·ver·sion \,in-trə-'vər-zhən, -shən\ *noun* [*intro-* + *-version* (as in *diversion*)] (1654)
1 : the act of introverting **:** the state of being introverted
2 : the state or tendency toward being wholly or predominantly concerned with and interested in one's own mental life
— **in·tro·ver·sive** \-'vər-siv, -ziv\ *adjective*
— **in·tro·ver·sive·ly** *adverb*

^{1}**in·tro·vert** \'in-trə-,vərt\ *transitive verb* [*intro-* + *-vert* (as in *divert*)] (1669)
: to turn inward or in upon itself: as **a :** to concentrate or direct upon oneself **b :** to produce psychological introversion in

^{2}**introvert** *noun* (1883)
1 : something (as the retractile proboscis of some worms) that is or can be drawn in especially by invagination
2 : one whose personality is characterized by introversion; *broadly* **:** a reserved or shy person

in·trude \in-'trüd\ *verb* **in·trud·ed; in·trud·ing** [Middle English, from Latin *intrudere* to thrust in, from *in-* + *trudere* to thrust — more at THREAT] (15th century)
intransitive verb
1 : to thrust oneself in without invitation, permission, or welcome
2 : to enter as a geological intrusion
transitive verb
1 : to thrust or force in or upon someone or something especially without permission, welcome, or fitness
2 : to cause to enter as if by force
— **in·trud·er** *noun*

in·tru·sion \in-'trü-zhən\ *noun* [Middle English, from Middle French, from Medieval Latin *intrusion-, intrusio,* from Latin *intrudere*] (15th century)
1 : the act of intruding or the state of being intruded; *especially* **:** the act of wrongfully entering upon, seizing, or taking possession of the property of another
2 : the forcible entry of molten rock or magma into or between other rock formations; *also* **:** the intruded magma

in·tru·sive \in-'trü-siv, -ziv\ *adjective* (15th century)
1 a : characterized by intrusion **b :** intruding where one is not welcome or invited
2 a : projecting inward ⟨an *intrusive* arm of the sea⟩ **b** (1) *of a rock* **:** having been forced while in a plastic state into cavities or between layers (2) **:** PLUTONIC
3 : having nothing that corresponds to a sound or letter in orthography or etymon ⟨*intrusive* \t\ in \'mints\ for *mince*⟩
synonym see IMPERTINENT
— **intrusive** *noun*
— **in·tru·sive·ly** *adverb*
— **in·tru·sive·ness** *noun*

intrust *variant of* ENTRUST

in·tu·ba·tion \,in-(,)tü-'bā-shən, -(,)tyü-, -tə-\ *noun* (1887)
: the introduction of a tube into a hollow organ (as the trachea)
— **in·tu·bate** \'in-(,)tü-,bāt, -(,)tyü-, -tə-\ *transitive verb*

in·tu·it \in-'tü-ət, -'tyü-\ *transitive verb* (1855)
: to apprehend by intuition
— **in·tu·it·able** \-ə-tə-bəl\ *adjective*

in·tu·i·tion \,in-tü̇-'wi-shən, -tyü̇-\ *noun* [Middle English *intuycyon,* from Late Latin *intuition-, intuitio* act of contemplating, from Latin *intuēri* to look at, contemplate, from *in-* + *tuēri* to look at] (15th century)
1 : quick and ready insight
2 a : immediate apprehension or cognition **b :** knowledge or conviction gained by intuition **c :** the power or faculty of attaining to direct knowledge or cognition without evident rational thought and inference
— **in·tu·i·tion·al** \-'wish-nəl, -'wi-shə-n^{ə}l\ *adjective*

in·tu·i·tion·ism \,in-tü̇-'wi-shə-,ni-zəm\ *noun* (1847)
1 a : a doctrine that objects of perception are intuitively known to be real **b :** a doctrine that there are basic truths intuitively known
2 : a doctrine that right or wrong or fundamental principles about what is right and wrong can be intuited
3 : a philosophical thesis that human beings have a direct intuitive understanding of mathematics and that rejects the principle that every mathematical statement must be true or false
— **in·tu·i·tion·ist** \-sh(ə-)nist\ *adjective or noun*

in·tu·i·tive \in-'tü-ə-tiv, -'tyü-\ *adjective* (1621)
1 a : known or perceived by intuition **:** directly apprehended ⟨had an *intuitive* awareness of his sister's feelings⟩ **b :** knowable by intuition
2 : knowing or perceiving by intuition
3 : possessing or given to intuition or insight ⟨an *intuitive* mind⟩
— **in·tu·i·tive·ly** *adverb*
— **in·tu·i·tive·ness** *noun*

in·tu·mes·cence \,in-tü̇-'me-s^{ə}n(t)s, -tyü̇-\ *noun* [French, from Latin *intumescere* to swell up, from *in-* + *tumescere* to swell — more at TUMESCENCE] (circa 1656)
: a swollen or enlarged part of a plant or animal; *also* **:** the process of swelling up or enlarging

in·tu·mes·cent \-s^{ə}nt\ *adjective* (1953)
of paint **:** swelling and charring when exposed to flame

in·tus·sus·cept \,in-tə-sə-'sept\ *verb* [back-formation from *intussusception*] (1802)
transitive verb
: to take in by or cause to undergo intussusception; *especially* **:** INVAGINATE
intransitive verb
: to undergo intussusception

in·tus·sus·cep·tion \-'sep-shən\ *noun* [Latin *intus* within + *susception-, susceptio* action of undertaking, from *suscipere* to take up — more at SUSCEPTIBLE] (1707)
: a drawing in of something from without: as **a :** INVAGINATION; *especially* **:** the slipping of a length of intestine into an adjacent portion usually producing obstruction **b :** the assimilation of new material and its dispersal among preexistent matter
— **in·tus·sus·cep·tive** \-'sep-tiv\ *adjective*

In·u·it \'i-nü̇-wət, -nyü̇-\ *noun* [Inuit *inuit,* plural of *inuk* person] (1765)
1 *plural* **Inuit** *or* **Inuits a** (1) **:** the Eskimo people of North America and Greenland (2) **:** the Eskimo people of Canada **b :** a member of such people
2 a : ESKIMO 2 **b :** the group of Eskimo dialects spoken from northwestern Canada to Greenland

Inuk·ti·tut \i-'nük-tə-,tüt\ *noun* (1974)
: the group of Eskimo dialects spoken by the Inuit of central and eastern arctic Canada

in·u·lin \'in-yə-lən\ *noun* [probably from German *Inulin,* from Latin *inula* elecampane] (1813)
: a tasteless white polysaccharide found especially dissolved in the sap of the roots and rhizomes of composite plants

in·unc·tion \i-'nəŋ(k)-shən\ *noun* [Middle English, from Latin *inunction-, inunctio,* from *inunguere* to anoint — more at ANOINT] (15th century)
: an act of applying oil or ointment **:** ANOINTING

in·un·date \'i-(,)nən-,dāt\ *transitive verb* **-dat·ed; -dat·ing** [Latin *inundatus,* past participle of *inundare,* from *in-* + *unda* wave — more at WATER] (1590)
1 : to cover with a flood **:** OVERFLOW
2 : OVERWHELM
— **in·un·da·tion** \,i-(,)nən-'dā-shən\ *noun*
— **in·un·da·tor** \'i-(,)nən-,dā-tər\ *noun*
— **in·un·da·to·ry** \i-'nən-də-,tōr-ē, -,tȯr-\ *adjective*

Inu·pi·at \i-'nü-pē-,ät, -'nyü-\ *also* **Inu·pi·aq** \-,äk\ *noun* [Inupiat *in^{y}upiaq,* plural *in^{y}upiat,* literally, real person] (1967)
1 *plural* **Inupiat** *or* **Inupiats** *also* **Inupiaq** *or* **Inupiaqs :** a member of the Eskimo people of northern Alaska
2 : the language of the Inupiat people

in·ure \i-'nu̇r, -'nyu̇r\ *verb* **in·ured; in·ur·ing** [Middle English *enuren,* from *en-* + *ure,* noun,

use, custom, from Middle French *uevre* work, practice, from Latin *opera* work — more at OPERA] (15th century)
transitive verb
: to accustom to accept something undesirable
intransitive verb
: to become of advantage
— **in·ure·ment** \-mənt\ *noun*

in·urn \i-ˈnərn\ *transitive verb* (1602)
1 : ENTOMB
2 : to place (as cremated remains) in an urn

in utero \in-ˈyü-tə-ˌrō\ *adverb or adjective* [Latin] (1713)
: in the uterus **:** before birth ⟨a disease acquired *in utero*⟩ ⟨an *in utero* diagnosis⟩

in·utile \(ˌ)in-ˈyü-tᵊl, -ˌtīl\ *adjective* [Middle English, from Middle French, from Latin *inutilis,* from *in-* + *utilis* useful — more at UTILITY] (15th century)
: USELESS, UNUSABLE
— **in·util·i·ty** \ˌin-yü-ˈti-lə-tē\ *noun*

in vac·uo \in-ˈva-kyə-ˌwō\ *adverb* [New Latin] (1660)
: in a vacuum

in·vade \in-ˈvād\ *transitive verb* **in·vad·ed; in·vad·ing** [Middle English, from Latin *invadere,* from *in-* + *vadere* to go — more at WADE] (15th century)
1 : to enter for conquest or plunder
2 : to encroach upon **:** INFRINGE
3 a : to spread over or into as if invading **:** PERMEATE ⟨doubts *invade* his mind⟩ **b :** to affect injuriously and progressively ⟨gangrene *invades* healthy tissue⟩
synonym see TRESPASS
— **in·vad·er** *noun*

in·vag·i·nate \in-ˈva-jə-ˌnāt\ *verb* **-nat·ed; -nat·ing** [Medieval Latin *invaginatus,* past participle of *invaginare,* from Latin *in-* + *vagina* sheath] (circa 1656)
transitive verb
1 : ENCLOSE, SHEATHE
2 : to fold in so that an outer becomes an inner surface
intransitive verb
: to undergo invagination

in·vag·i·na·tion \-ˌva-jə-ˈnā-shən\ *noun* (circa 1658)
1 : an act or process of invaginating; *especially* **:** the formation of a gastrula by an infolding of part of the wall of the blastula
2 : an invaginated part

¹in·val·id \(ˌ)in-ˈva-ləd\ *adjective* [Latin *invalidus* weak, from *in-* + *validus* strong — more at VALID] (1542)
: not valid: **a :** being without foundation or force in fact, truth, or law **b :** logically inconsequent
— **in·val·id·ly** \-lē\ *adverb*

²in·va·lid \ˈin-və-ləd, *British usually* -ˌlēd\ *adjective* [Latin & French; French *invalide,* from Latin *invalidus*] (1642)
1 : suffering from disease or disability **:** SICKLY
2 : of, relating to, or suited to one that is sick

³invalid *same as* ²\ *noun* (1709)
: one that is sickly or disabled

⁴in·va·lid \ˈin-və-ləd, -ˌlid, *British usually* -ˌlēd *or* ˌin-və-ˈlēd\ *transitive verb* (1787)
1 : to remove from active duty by reason of sickness or disability
2 : to make sickly or disabled

in·val·i·date \(ˌ)in-ˈva-lə-ˌdāt\ *transitive verb* (1649)
: to make invalid; *especially* **:** to weaken or destroy the cogency of
synonym see NULLIFY
— **in·val·i·da·tion** \-ˌva-lə-ˈdā-shən\ *noun*
— **in·val·i·da·tor** \in-ˈva-lə-ˌdā-tər\ *noun*

in·va·lid·ism \ˈin-və-lə-ˌdi-zəm\ *noun* (1794)
: a chronic condition of being an invalid

in·val·id·i·ty \ˌin-və-ˈli-də-tē, -va-\ *noun, plural* **-ties** (circa 1550)
1 : lack of validity or cogency
2 : incapacitating bodily disability; *also* **:** INVALIDISM

in·valu·able \(ˌ)in-ˈval-yə-bəl, -yə-wə-bəl\ *adjective* [¹*in-* + *value,* verb + *-able*] (1576)
: valuable beyond estimation **:** PRICELESS
— **in·valu·able·ness** *noun*
— **in·valu·ably** \-blē\ *adverb*

in·var \ˈin-ˌvär\ *noun, often capitalized* [from *Invar,* a trademark] (1902)
: an iron-nickel alloy that expands little on heating

in·vari·able \(ˌ)in-ˈver-ē-ə-bəl, -ˈvar-\ *adjective* (15th century)
: not changing or capable of change **:** CONSTANT
— **in·vari·abil·i·ty** \-ˌver-ē-ə-ˈbi-lə-tē, -ˌvar-\ *noun*
— **invariable** *noun*
— **in·vari·ably** \-blē\ *adverb*

in·vari·ance \(ˌ)in-ˈver-ē-ən(t)s, -ˈvar-\ *noun* (1878)
: the quality or state of being invariant

in·vari·ant \-ənt\ *adjective* (1874)
: CONSTANT, UNCHANGING; *specifically* **:** unchanged by specified mathematical or physical operations or transformations ⟨*invariant* factor⟩
— **invariant** *noun*

in·va·sion \in-ˈvā-zhən\ *noun* [Middle English *invasioune,* from Middle French *invasion,* from Late Latin *invasion-, invasio,* from Latin *invadere* to invade] (15th century)
1 : an act of invading; *especially* **:** incursion of an army for conquest or plunder
2 : the incoming or spread of something usually hurtful

in·va·sive \-siv, -ziv\ *adjective* (1598)
1 : of, relating to, or characterized by military aggression
2 : tending to spread; *especially* **:** tending to invade healthy tissue ⟨*invasive* cancer cells⟩
3 : tending to infringe
4 : involving entry into the living body (as by incision or by insertion of an instrument) ⟨*invasive* diagnostic techniques⟩
— **in·va·sive·ness** *noun*

¹in·vec·tive \in-ˈvek-tiv\ *adjective* [Middle English *invectif,* from Middle French, from Latin *invectivus,* from *invectus,* past participle of *invehere*] (15th century)
: of, relating to, or characterized by insult or abuse
— **in·vec·tive·ly** *adverb*
— **in·vec·tive·ness** *noun*

²invective *noun* (1523)
1 : an abusive expression or speech
2 : insulting or abusive language **:** VITUPERATION
synonym see ABUSE

in·veigh \in-ˈvā\ *intransitive verb* [Latin *invehi* to attack, inveigh, passive of *invehere* to carry in, from *in-* + *vehere* to carry — more at WAY] (1529)
: to protest or complain bitterly or vehemently **:** RAIL
— **in·veigh·er** *noun*

in·vei·gle \in-ˈvā-gəl *sometimes* -ˈvē-\ *transitive verb* **in·vei·gled; in·vei·gling** \-g(ə-)liŋ\ [modification of Middle French *aveugler* to blind, hoodwink, from Old French *avogler,* from *avogle* blind, from Medieval Latin *ab oculis,* literally, lacking eyes] (1539)
1 : to win over by wiles **:** ENTICE
2 : to acquire by ingenuity or flattery ◆
synonym see LURE
— **in·vei·gle·ment** \-gəl-mənt\ *noun*
— **in·vei·gler** \-g(ə-)lər\ *noun*

in·vent \in-ˈvent\ *transitive verb* [Middle English, from Latin *inventus,* past participle of *invenire* to come upon, find, from *in-* + *venire* to come — more at COME] (15th century)
1 *archaic* **:** FIND, DISCOVER
2 : to devise by thinking **:** FABRICATE
3 : to produce (as something useful) for the first time through the use of the imagination or of ingenious thinking and experiment
— **in·ven·tor** \-ˈven-tər\ *noun*
— **in·ven·tress** \-ˈven-trəs\ *noun*

in·ven·tion \in-ˈven(t)-shən\ *noun* (14th century)
1 : DISCOVERY, FINDING
2 : productive imagination **:** INVENTIVENESS
3 a : something invented: as (1) **:** a product of the imagination; *especially* **:** a false conception (2) **:** a device, contrivance, or process originated after study and experiment **b :** a short keyboard composition featuring two or three part counterpoint
4 : the act or process of inventing

in·ven·tive \in-ˈven-tiv\ *adjective* (15th century)
1 : adept or prolific at producing inventions **:** CREATIVE
2 : characterized by invention
— **in·ven·tive·ly** *adverb*
— **in·ven·tive·ness** *noun*

¹in·ven·to·ry \ˈin-vən-ˌtōr-ē, -ˌtȯr-\ *noun, plural* **-ries** (1523)
1 a : an itemized list of current assets: as (1) **:** a catalog of the property of an individual or estate (2) **:** a list of goods on hand **b :** a survey of natural resources **c :** a list of traits, preferences, attitudes, interests, or abilities used to evaluate personal characteristics or skills
2 : the quantity of goods or materials on hand **:** STOCK
3 : the act or process of taking an inventory
— **in·ven·to·ri·al** \ˌin-vən-ˈtōr-ē-əl\ *adjective*
— **in·ven·to·ri·al·ly** \-ē-ə-lē\ *adverb*

²inventory *transitive verb* **-ried; -ry·ing** (1602)
: to make an inventory of **:** CATALOG

in·ver·ness \ˌin-vər-ˈnes\ *noun* [*Inverness,* Scotland] (1863)
: a loose belted coat having a cape with a close-fitting round collar

inverness

¹in·verse \(ˌ)in-ˈvərs, ˈin-ˌ\ *adjective* [Latin *inversus,* from past participle of *invertere*] (15th century)
1 : opposite in order, nature, or effect
2 : being an inverse function ⟨*inverse* sine⟩

²inverse *noun* (circa 1681)
1 : something of a contrary nature or quality **:** OPPOSITE, REVERSE
2 : a proposition or theorem formed by contradicting both the subject and predicate or both the hypothesis and conclusion

◇ WORD HISTORY
inveigle To inveigle is not to blind, yet when we permit ourselves to be inveigled we are, in a sense, blinded by our tempter's wiles. The ancestor of our word *inveigle* is probably a phrase meaning "blind" in the Latin spoken around the time of the Roman Empire's fall. Literally, Latin *orbus ab oculis* would have meant "deprived of eyes," and in a text of the 5th or 6th century A.D. *ab oculis* occurs by itself with the meaning "blind." This phrase is the most likely source of French *aveugle* "blind," from which was derived *aveugler* "to blind." Although *aveugler* was used both literally and figuratively, only the figurative sense was borrowed by English in the late 15th century. English *inveigle* originally meant "to blind or delude in judgment." This sense is now obsolete, but the present meaning, "to win over by wiles," is not far removed. The substitution of *in-* for *a-*, taken to be a prefix, was probably based on Anglo-French usage, in which forms such as *aveogler* and *enveegler* were employed indiscriminately.

of a given proposition or theorem ⟨the *inverse* of "if A then B" is "if not-A then not-B"⟩ — compare CONTRAPOSITIVE
3 a : INVERSE FUNCTION; *also* **:** an operation (as subtraction) that undoes the effect of another operation **b :** a set element that is related to another element in such a way that the result of applying a given binary operation to them is an identity element of the set

inverse function *noun* (1816)
: a function that is derived from a given function by interchanging the two variables — compare LOGARITHMIC FUNCTION

in·verse·ly \ˈin-ˌvərs-lē, (ˌ)in-ˈ\ *adverb* (1660)
1 : in an inverse order or manner
2 : in the manner of inverse variation ⟨varies *inversely*⟩

inversely proportional *adjective* (circa 1890)
: related by inverse variation — compare DIRECTLY PROPORTIONAL

inverse square law *noun* (1921)
: a statement in physics: a given physical quantity (as illumination) varies with the distance from the source inversely as the square of the distance

inverse variation *noun* (1936)
1 : mathematical relationship between two variables which can be expressed by an equation in which the product of two variables is equal to a constant
2 : an equation or function expressing inverse variation — compare DIRECT VARIATION

in·ver·sion \in-ˈvər-zhən, -shən\ *noun* (1586)
1 : a reversal of position, order, form, or relationship: as **a** (1) **:** a change in normal word order; *especially* **:** the placement of a verb before its subject (2) **:** the process or result of changing or reversing the relative positions of the notes of a musical interval, chord, or phrase **b :** the condition of being turned inward or inside out **c :** a breaking off of a chromosome section and its subsequent reattachment in inverted position; *also* **:** a chromosomal section that has undergone this process
2 : the act or process of inverting
3 a : a change in the order of the terms of a mathematical proportion effected by inverting each ratio **b :** the operation of forming the inverse of a magnitude, a function, an operation, or an element
4 a : the conversion of dextrorotatory sucrose into a levorotatory mixture of glucose and fructose **b :** a change from one stereochemical figuration at a chiral center in a usually organic molecule to the opposite configuration that is brought about by a reaction in which a substitution of one group is made for a different group
5 : HOMOSEXUALITY
6 : an increase of temperature with height through a layer of air

in·ver·sive \-ˈvər-siv, -ziv\ *adjective* (1875)
: marked by inversion

[1]in·vert \in-ˈvərt\ *transitive verb* [Latin *invertere,* from *in-* + *vertere* to turn — more at WORTH] (1533)
1 a : to reverse in position, order, or relationship **b :** to subject to inversion
2 a : to turn inside out or upside down **b :** to turn inward
3 : to find the mathematical reciprocal of ⟨to divide using fractions, *invert* the divisor and multiply⟩
synonym see REVERSE

[2]in·vert \ˈin-ˌvərt\ *noun* (1838)
: one characterized by inversion; *especially* **:** HOMOSEXUAL

in·ver·tase \in-ˈvər-ˌtās, -ˌtāz; ˈin-vər-ˌ\ *noun* [International Scientific Vocabulary] (1887)
: an enzyme that catalyzes the hydrolysis of sucrose

in·ver·te·brate \(ˌ)in-ˈvər-tə-brət, -ˌbrāt\ *adjective* [New Latin *invertebratus,* from Latin *in-* + New Latin *vertebratus* vertebrate] (1838)
1 : lacking a spinal column; *also* **:** of or relating to invertebrate animals
2 : lacking in strength or vitality **:** WEAK
— invertebrate *noun*

inverted comma *noun* (1789)
chiefly British **:** QUOTATION MARK

in·vert·er \in-ˈvər-tər\ *noun* (1611)
1 : one that inverts
2 : a device for converting direct current into alternating current

in·vert·ible \in-ˈvər-tə-bəl\ *adjective* (1881)
: capable of being inverted or subjected to inversion ⟨an *invertible* matrix⟩

invert sugar *noun* (1880)
: a mixture of dextrose and levulose found in fruits or produced artificially by the inversion of sucrose

[1]in·vest \in-ˈvest\ *transitive verb* [Latin *investire* to clothe, surround, from *in-* + *vestis* garment — more at WEAR] (circa 1534)
1 [Medieval Latin *investire,* from Latin, to clothe] **a :** to array in the symbols of office or honor **b :** to furnish with power or authority **c :** to grant someone control or authority over **:** VEST
2 : to cover completely **:** ENVELOP
3 : CLOTHE, ADORN
4 [Middle French *investir,* from Old Italian *investire,* from Latin, to surround] **:** to surround with troops or ships so as to prevent escape or entry
5 : to endow with a quality **:** INFUSE

[2]invest *verb* [Italian *investire* to clothe, invest money, from Latin, to clothe] (1613)
transitive verb
1 : to commit (money) in order to earn a financial return
2 : to make use of for future benefits or advantages
intransitive verb
: to make an investment
— in·vest·able \-ˈves-tə-bəl\ *adjective*
— in·ves·tor \-tər\ *noun*

in·ves·ti·gate \in-ˈves-tə-ˌgāt\ *verb* **-gat·ed; -gat·ing** [Latin *investigatus,* past participle of *investigare* to track, investigate, from *in-* + *vestigium* footprint, track] (circa 1510)
transitive verb
: to observe or study by close examination and systematic inquiry
intransitive verb
: to make a systematic examination; *especially* **:** to conduct an official inquiry
— in·ves·ti·ga·tion \-ˌves-tə-ˈgā-shən\ *noun*
— in·ves·ti·ga·tion·al \-shnəl, -shə-nᵊl\ *adjective*
— in·ves·ti·ga·tive \-ˈves-tə-ˌgā-tiv\ *adjective*
— in·ves·ti·ga·tor \-ˌgā-tər\ *noun*
— in·ves·ti·ga·to·ry \-ˈves-ti-gə-ˌtōr-ē, -ˌtȯr-\ *adjective*

in·ves·ti·ture \in-ˈves-tə-ˌchu̇r, -chər, -ˌtyu̇r, -ˌtu̇r\ *noun* [Middle English, from Medieval Latin *investitura,* from *investitus,* past participle of *investire*] (14th century)
1 : the act of establishing in office or ratifying
2 : something that covers or adorns

[1]in·vest·ment \in-ˈves(t)-mənt\ *noun* [[1]*invest*] (1597)
1 a *archaic* **:** VESTMENT **b :** an outer layer **:** ENVELOPE
2 : INVESTITURE 1
3 : BLOCKADE, SIEGE

[2]investment *noun* [[2]*invest*] (1615)
: the outlay of money usually for income or profit **:** capital outlay; *also* **:** the sum invested or the property purchased

investment company *noun* (circa 1917)
: a company whose primary business is holding securities of other companies purely for investment purposes — compare HOLDING COMPANY

in·vet·er·a·cy \in-ˈve-t(ə-)rə-sē\ *noun* [*inveterate* + *-cy*] (circa 1719)
: the quality or state of being obstinate or persistent **:** TENACITY

in·vet·er·ate \in-ˈve-t(ə-)rət\ *adjective* [Middle English, from Latin *inveteratus,* from past participle of *inveterare* to age (transitive verb), from *in-* + *veter-, vetus* old — more at WETHER] (14th century)
1 : firmly established by long persistence ⟨the *inveterate* tendency to overlook the obvious⟩
2 : confirmed in a habit **:** HABITUAL ⟨an *inveterate* liar⟩ ☆
— in·vet·er·ate·ly *adverb*

in·vi·a·ble \(ˌ)in-ˈvī-ə-bəl\ *adjective* [International Scientific Vocabulary] (1918)
: incapable of surviving especially because of a deleterious genetic constitution
— in·vi·a·bil·i·ty \-ˌvī-ə-ˈbi-lə-tē\ *noun*

in·vid·i·ous \in-ˈvi-dē-əs\ *adjective* [Latin *invidiosus* envious, invidious, from *invidia* envy — more at ENVY] (1606)
1 : tending to cause discontent, animosity, or envy ⟨the *invidious* task of arbitration⟩
2 : ENVIOUS
3 a : of an unpleasant or objectionable nature **:** OBNOXIOUS ⟨*invidious* remarks⟩ **b :** of a kind to cause harm or resentment ⟨an *invidious* comparison⟩
synonym see REPUGNANT
— in·vid·i·ous·ly *adverb*
— in·vid·i·ous·ness *noun*

in·vig·i·late \in ˈvi-jə-ˌlāt\ *verb* **-lat·ed; -lat·ing** [Latin *invigilatus,* past participle of *invigilare* to stay awake, be watchful, from *in-* + *vigilare* to stay awake — more at VIGILANT] (1553)
intransitive verb
: to keep watch; *especially, British* **:** to supervise students at an examination
transitive verb
: SUPERVISE, MONITOR
— in·vig·i·la·tion \-ˌvi-jə-ˈlā-shən\ *noun*
— in·vig·i·la·tor \-ˈvi-jə-ˌlā-tər\ *noun*

in·vig·o·rate \in-ˈvi-gə-ˌrāt\ *transitive verb* **-rat·ed; -rat·ing** [probably from *in-* + *vigor*] (1646)
: to give life and energy to **:** ANIMATE; *also* **:** STIMULATE 1
— in·vig·o·rat·ing·ly \-ˌrā-tiŋ-lē\ *adverb*
— in·vig·o·ra·tion \-ˌvi-gə-ˈrā-shən\ *noun*
— in·vig·o·ra·tor \-ˈvi-gə-ˌrā-tər\ *noun*

in·vin·ci·ble \(ˌ)in-ˈvin(t)-sə-bəl\ *adjective* [Middle English, from Middle French, from Late Latin *invincibilis,* from Latin *in-* + *vincere* to conquer — more at VICTOR] (15th century)
: incapable of being conquered, overcome, or subdued
— in·vin·ci·bil·i·ty \-ˌvin(t)-sə-ˈbi-lə-tē\ *noun*
— in·vin·ci·ble·ness \-ˈvin(t)-sə-bəl-nəs\ *noun*
— in·vin·ci·bly \-blē\ *adverb*

☆ SYNONYMS
Inveterate, confirmed, chronic mean firmly established. INVETERATE applies to a habit, attitude or feeling of such long existence as to be practically ineradicable or unalterable ⟨an *inveterate* smoker⟩. CONFIRMED implies a growing stronger and firmer with time so as to resist change or reform ⟨a *confirmed* bachelor⟩. CHRONIC suggests something that is persistent or endlessly recurrent and troublesome ⟨a *chronic* complainer⟩.

in·vi·o·la·ble \(,)in-ˈvī-ə-lə-bəl\ *adjective* [Middle English, from Middle French or Latin; Middle French, from Latin *inviolabilis,* from *in-* + *violare* to violate] (15th century)
1 : secure from violation or profanation
2 : secure from assault or trespass **:** UNASSAILABLE
— **in·vi·o·la·bil·i·ty** \-ˌvī-ə-lə-ˈbi-lə-tē\ *noun*
— **in·vi·o·la·ble·ness** \-ˈvī-ə-lə-bəl-nəs\ *noun*
— **in·vi·o·la·bly** \-blē\ *adverb*

in·vi·o·la·cy \(,)in-ˈvī-ə-lə-sē\ *noun* (circa 1846)
: the quality or state of being inviolate

in·vi·o·late \-ˈvī-ə-lət\ *adjective* (15th century)
: not violated or profaned; *especially* **:** PURE
— **in·vi·o·late·ly** *adverb*
— **in·vi·o·late·ness** *noun*

in·vis·cid \(,)in-ˈvi-səd\ *adjective* (circa 1889)
1 : having zero viscosity
2 : of or relating to an inviscid fluid ⟨*inviscid* flow⟩

in·vis·i·ble \(,)in-ˈvi-zə-bəl\ *adjective* [Middle English, from Middle French, from Latin *invisibilis,* from *in-* + *visibilis* visible] (14th century)
1 a : incapable by nature of being seen **b :** inaccessible to view **:** HIDDEN
2 : IMPERCEPTIBLE, INCONSPICUOUS
3 a : not appearing in published financial statements **b :** not reflected in statistics
— **in·vis·i·bil·i·ty** \-ˌvi-zə-ˈbi-lə-tē\ *noun*
— **invisible** *noun*
— **in·vis·i·ble·ness** \-ˈvi-zə-bəl-nəs\ *noun*
— **in·vis·i·bly** \-blē\ *adverb*

in·vi·ta·tion \ˌin-və-ˈtā-shən\ *noun* (15th century)
1 a : the act of inviting **b :** an often formal request to be present or participate
2 : INCENTIVE, INDUCEMENT

in·vi·ta·tion·al \-shnəl, -shə-nᵊl\ *adjective* (1922)
1 : prepared or entered in response to a request ⟨an *invitational* article⟩
2 : limited to invited participants ⟨an *invitational* tournament⟩
— **invitational** *noun*

¹in·vi·ta·to·ry \in-ˈvī-tə-ˌtōr-ē, -ˌtȯr-\ *adjective* (14th century)
: containing an invitation

²invitatory *noun, plural* **-ries** (14th century)
: an invitatory psalm or antiphon

¹in·vite \in-ˈvīt\ *transitive verb* **in·vit·ed; in·vit·ing** [Middle French or Latin; Middle French *inviter,* from Latin *invitare*] (1533)
1 a : to offer an incentive or inducement to **:** ENTICE **b :** to increase the likelihood of
2 a : to request the presence or participation of **b :** to request formally **c :** to urge politely **:** WELCOME
— **in·vit·er** *noun*

²in·vite \ˈin-ˌvīt\ *noun* (1659)
: INVITATION 1

in·vi·tee \ˌin-və-ˈtē, -ˌvī-\ *noun* (1803)
: an invited person

inviting \in-ˈvī-tiŋ\ *adjective* (1604)
: ATTRACTIVE, TEMPTING
— **in·vit·ing·ly** \-tiŋ-lē\ *adverb*

in vi·tro \in-ˈvē-(ˌ)trō, -ˈvi-\ *adverb or adjective* [New Latin, literally, in glass] (circa 1894)
: outside the living body and in an artificial environment

in vi·vo \in-ˈvē-(ˌ)vō\ *adverb or adjective* [New Latin, literally, in the living] (1901)
: in the living body of a plant or animal

in·vo·cate \ˈin-və-ˌkāt\ *transitive verb* (1526)
archaic **:** INVOKE

in·vo·ca·tion \ˌin-və-ˈkā-shən\ *noun* [Middle English *invocacioun,* from Middle French *invocation,* from Latin *invocation-, invocatio,* from *invocare*] (14th century)
1 a : the act or process of petitioning for help or support; *specifically, often capitalized* **:** a prayer of entreaty (as at the beginning of a service of worship) **b :** a calling upon for authority or justification
2 : a formula for conjuring **:** INCANTATION
3 : an act of legal or moral implementation **:** ENFORCEMENT
— **in·vo·ca·tion·al** \-shnəl, -shə-nᵊl\ *adjective*
— **in·voc·a·to·ry** \in-ˈvä-kə-ˌtōr-ē, -ˌtȯr-\ *adjective*

¹in·voice \ˈin-ˌvȯis\ *noun* [modification of Middle French *envois,* plural of *envoi* message — more at ENVOI] (1560)
1 : an itemized list of goods shipped usually specifying the price and the terms of sale **:** BILL
2 : a consignment of merchandise

²invoice *transitive verb* **in·voiced; in·voic·ing** (1698)
: to send an invoice for or to

in·voke \in-ˈvōk\ *transitive verb* **in·voked; in·vok·ing** [Middle English, from Middle French *invoquer,* from Latin *invocare,* from *in-* + *vocare* to call, from *voc-, vox* voice — more at VOICE] (15th century)
1 a : to petition for help or support **b :** to appeal to or cite as authority
2 : to call forth by incantation **:** CONJURE
3 : to make an earnest request for **:** SOLICIT
4 : to put into effect or operation **:** IMPLEMENT
5 : BRING ABOUT, CAUSE
— **in·vok·er** *noun*

in·vo·lu·cre \ˈin-və-ˌlü-kər\ *noun* [French, from New Latin *involucrum*] (1794)
: one or more whorls of bracts situated below and close to a flower, flower cluster, or fruit
— **in·vo·lu·cral** \ˌin-və-ˈlü-krəl\ *adjective*
— **in·vo·lu·crate** \-krət, -ˌkrāt\ *adjective*

in·vo·lu·crum \ˌin-və-ˈlü-krəm\ *noun, plural* **-cra** \-krə\ [New Latin, sheath, involucre, from Latin, wrapper, from *involvere* to wrap] (circa 1677)
: a surrounding envelope or sheath

in·vol·un·tary \(,)in-ˈvä-lən-ˌter-ē\ *adjective* [Middle English *involuntari,* from Late Latin *involuntarius,* from Latin *in-* + *voluntarius* voluntary] (15th century)
1 : done contrary to or without choice
2 : COMPULSORY
3 : not subject to control of the will **:** REFLEX
— **in·vol·un·tari·ly** \-ˌvä-lən-ˈter-ə-lē\ *adverb*
— **in·vol·un·tari·ness** \-ˈvä-lən-ˌter-ē-nəs\ *noun*

involuntary manslaughter *noun* (circa 1879)
: manslaughter resulting from the failure to perform a legal duty expressly required to safeguard human life, from the commission of an unlawful act not constituting a felony, or from the commission of a lawful act in a negligent or improper manner

involuntary muscle *noun* (1840)
: muscle governing reflex functions and not under direct voluntary control; *especially* **:** SMOOTH MUSCLE

¹in·vo·lute \ˈin-və-ˌlüt\ *adjective* [Latin *involutus* concealed, from past participle of *involvere*] (1661)
1 a : curled spirally **b** (1) **:** curled or curved inward (2) **:** having the edges rolled over the upper surface toward the midrib ⟨an *involute* leaf⟩ **c :** having the form of an involute ⟨a gear with *involute* teeth⟩
2 : INVOLVED, INTRICATE

²involute *noun* (circa 1796)
: a curve traced by a point on a thread kept taut as it is unwound from another curve ⟨*involute* of a circle⟩

³in·vo·lute \ˌin-və-ˈlüt\ *intransitive verb* **-lut·ed; -lut·ing** (1816)
1 : to become involute
2 a : to return to a former condition **b :** to become cleared up **:** DISAPPEAR

in·vo·lu·tion \ˌin-və-ˈlü-shən\ *noun* [Latin *involution-, involutio,* from *involvere*] (circa 1611)
1 a (1) **:** the act or an instance of enfolding or entangling **:** INVOLVEMENT (2) **:** an involved grammatical construction usually characterized by the insertion of clauses between the subject and predicate **b :** COMPLEXITY, INTRICACY
2 : EXPONENTIATION
3 a : an inward curvature or penetration **b :** the formation of a gastrula by ingrowth of cells formed at the dorsal lip
4 : a shrinking or return to a former size
5 : the regressive alterations of a body or its parts characteristic of the aging process; *specifically* **:** decline marked by a decrease of bodily vigor and in women by menopause
— **in·vo·lu·tion·al** \-shnəl, -shə-nᵊl\ *adjective*

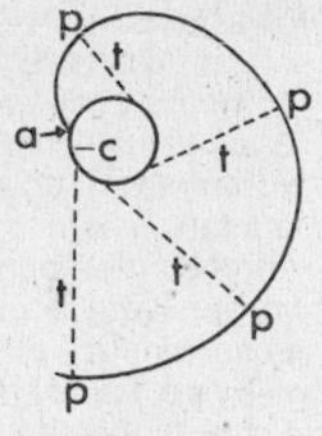

involute *a, p, p, p* traced by any point *p* of the thread *t* unwinding from curve *c*

in·volve \in-ˈvälv, -ˈvȯlv *also* -ˈväv *or* -ˈvȯv\ *transitive verb* **in·volved; in·volv·ing** [Middle English, to roll up, wrap, from Latin *involvere,* from *in-* + *volvere* to roll — more at VOLUBLE] (14th century)
1 *archaic* **:** to enfold or envelop so as to encumber
2 a : to engage as a participant ⟨workers *involved* in building a house⟩ **b :** to oblige to take part ⟨right of Congress to *involve* the nation in war⟩ **c :** to occupy (as oneself) absorbingly; *especially* **:** to commit (as oneself) emotionally ⟨was *involved* with a married man⟩
3 : to surround as if with a wrapping **:** ENVELOP
4 a *archaic* **:** to wind, coil, or wreathe about **b :** to relate closely **:** CONNECT
5 a : to have within or as part of itself **:** INCLUDE **b :** to require as a necessary accompaniment **:** ENTAIL **c :** to have an effect on
synonym see INCLUDE
— **in·volve·ment** \-ˈvälv-mənt, -ˈvȯlv-\ *noun*
— **in·volv·er** *noun*

in·volved \-ˈvälvd, -ˈvȯlvd *also* -ˈvävd *or* -ˈvȯvd\ *adjective* (15th century)
1 : INVOLUTE, TWISTED
2 a : marked by extreme and often needless or excessive complexity **b :** difficult to deal with because of complexity or disorder
3 : being affected or implicated
synonym see COMPLEX
— **in·volv·ed·ly** \-ˈväl-vəd-lē, -ˈvȯl- *also* -ˈvä-vəd- *or* -ˈvȯ-\ *adverb*

in·vul·ner·a·ble \(,)in-ˈvəl-n(ə-)rə-bəl, -nər-bəl\ *adjective* [Latin *invulnerabilis,* from *in-* + *vulnerare* to wound — more at VULNERABLE] (1595)
1 : incapable of being wounded, injured, or harmed
2 : immune to or proof against attack **:** IMPREGNABLE
— **in·vul·ner·a·bil·i·ty** \-ˌvəl-n(ə-)rə-ˈbi-lə-tē\ *noun*
— **in·vul·ner·a·ble·ness** \-ˈvəl-n(ə-)rə-bəl-nəs, -nər-bəl-\ *noun*
— **in·vul·ner·a·bly** \-blē\ *adverb*

¹in·ward \ˈin-wərd\ *adjective* [Middle English, from Old English *inweard* (akin to Old High German *inwert*), from *in* + *-weard* ward] (before 12th century)
1 : situated on the inside **:** INNER
2 a : of or relating to the mind or spirit ⟨*inward* peace⟩ **b :** absorbed in one's own mental or spiritual life **:** INTROSPECTIVE
3 : marked by close acquaintance **:** FAMILIAR
4 : directed toward the interior

²inward *or* **in·wards** \-wərdz\ *adverb* (before 12th century)

1 : toward the inside, center, or interior
2 : toward the inner being

³inward *noun* (before 12th century)
1 : something that is inward
2 in·wards \ˈin-ərdz, -wərdz\ *plural* **:** INNARDS

Inward Light *noun* (1706)
: INNER LIGHT

in·ward·ly \ˈin-wərd-lē\ *adverb* (before 12th century)
1 : in the innermost being **:** MENTALLY, SPIRITUALLY
2 a : beneath the surface **:** INTERNALLY ⟨bled *inwardly*⟩ **b :** to oneself **:** PRIVATELY ⟨cursed *inwardly*⟩

in·ward·ness *noun* (14th century)
1 : internal quality or substance
2 : close acquaintance **:** FAMILIARITY
3 : fundamental nature **:** ESSENCE
4 : absorption in one's own mental or spiritual life

in·weave \(ˌ)in-ˈwēv\ *transitive verb* **-wove** \-ˈwōv\ *also* **-weaved; -wo·ven** \-ˈwō-vən\ *also* **-weaved; -weav·ing** (15th century)
: INTERWEAVE, INTERLACE

in–wrought \(ˌ)in-ˈrȯt, ˈin-ˌ\ *adjective* (1637)
1 : having decoration worked in **:** ORNAMENTED; *especially* **:** decorated with embroidery
2 : worked in especially as decoration

in–your–face *adjective* (1982)
: characterized by or being bold and often arrogant aggressiveness ⟨*in-your-face* basketball⟩

Io \ˈī-(ˌ)ō\ *noun* [Latin, from Greek *Iō*]
: a maiden loved by Zeus and changed by him into a heifer so that she might escape the jealous rage of Hera

iod- *or* **iodo-** *combining form* [French *iode*]
: iodine ⟨*iod*ize⟩ ⟨*iodo*form⟩

io·date \ˈī-ə-ˌdāt, -dət\ *noun* [French, from *iode*] (1826)
: a salt of iodic acid

iod·ic acid \ī-ˈä-dik-\ *noun* [French *iodique,* from *iode*] (1826)
: a crystalline oxidizing solid HIO_3 formed by oxidation of iodine

io·dide \ˈī-ə-ˌdīd\ *noun* [International Scientific Vocabulary] (1822)
: a salt of hydriodic acid; *also* **:** the univalent anion I^- of such a salt

io·din·ate \ˈī-ə-də-ˌnāt\ *transitive verb* **-at·ed; -at·ing** (1908)
: to treat or cause to combine with iodine or a compound of iodine
— io·din·ation \ˌī-ə-də-ˈnā-shən\ *noun*

io·dine \ˈī-ə-ˌdīn, -dᵊn, -ˌdēn\ *noun, often attributive* [French *iode,* from Greek *ioeidēs* violet colored, from *ion* violet] (1814)
1 : a nonmetallic halogen element obtained usually as heavy shining blackish gray crystals and used especially in medicine, photography, and analysis — see ELEMENT table
2 : a tincture of iodine used especially as a topical antiseptic

io·dise *British variant of* IODIZE

io·dize \ˈī-ə-ˌdīz\ *transitive verb* **io·dized; io·diz·ing** (1841)
: to treat with iodine or an iodide ⟨*iodized* salt⟩

io·do·form \ī-ˈō-də-ˌfȯrm, -ˈä-\ *noun* [International Scientific Vocabulary *iod-* + *-form* (as in *chloroform*)] (1838)
: a yellow crystalline volatile compound CHI_3 with a penetrating persistent odor that is used as an antiseptic dressing

io·do·phor \-ˌfȯr\ *noun* [*iod-* + Greek *-phoros* carrier — more at -PHORE] (1952)
: a complex of iodine and a surface-active agent that releases iodine gradually and serves as a disinfectant

io·dop·sin \ˌī-ə-ˈdäp-sən\ *noun* [*iod-* (from Greek *ioeidēs* violet colored) + Greek *opsis* sight, vision + English *-in* — more at OPTIC] (1938)
: a photosensitive violet pigment in the retinal cones that is similar to rhodopsin but more labile, is formed from vitamin A, and is important in daylight vision

io moth \ˈī-(ˌ)ō-\ *noun* [Latin *Io*] (1870)
: a chiefly North American saturniid moth (*Automeris io*) having a large circular eyelike spot on the upper surface of each hind wing and a larva with stinging spines

ion \ˈī-ən, ˈī-ˌän\ *noun* [Greek, neuter of *iōn,* present participle of *ienai* to go — more at ISSUE] (circa 1834)
1 : an atom or group of atoms that carries a positive or negative electric charge as a result of having lost or gained one or more electrons
2 : a charged subatomic particle (as a free electron)

-ion *noun suffix* [Middle English *-ioun, -ion,* from Old French *-ion,* from Latin *-ion-, -io*]
1 a : act or process ⟨validat*ion*⟩ **b :** result of an act or process ⟨regulat*ion*⟩
2 : state or condition ⟨hydrat*ion*⟩

ion engine *noun* (1958)
: a reaction engine deriving thrust from the ejection of a stream of ionized particles

ion exchange *noun* (1923)
: a reversible interchange of one kind of ion present on an insoluble solid with another of like charge present in a solution surrounding the solid with the reaction being used especially for softening or demineralizing water, the purification of chemicals, or the separation of substances
— ion exchanger *noun*

ion exchange resin *noun* (1943)
: an insoluble material of high molecular weight that contains groups which can be exchanged with ions in a solution with which it is in contact

Io·ni·an \ī-ˈō-nē-ən\ *noun* [*Ionia,* Asia Minor] (1550)
1 : a member of any of the Greek peoples who settled on the islands of the Aegean Sea and the western shore of Asia Minor toward the end of the second millennium B.C.
2 : a native or inhabitant of Ionia
— Ionian *adjective*

ion·ic \ī-ˈä-nik\ *adjective* [International Scientific Vocabulary] (1890)
1 : of, relating to, existing as, or characterized by ions ⟨*ionic* gases⟩ ⟨the *ionic* charge⟩
2 : based on or functioning by means of ions ⟨*ionic* conduction⟩ ⟨an *ionic* lattice⟩
— ion·ic·i·ty \ˌī-ə-ˈni-sə-tē\ *noun*

¹Ion·ic \ī-ˈä-nik\ *adjective* [Latin & Middle French; Middle French *ionique,* from Latin *ionicus,* from Greek *iōnikos,* from *Iōnia* Ionia] (1585)
1 : of or relating to the ancient Greek architectural order distinguished especially by fluted columns on bases and scroll volutes in its capitals — see ORDER illustration
2 : of or relating to Ionia or the Ionians

²Ionic *noun* (1668)
: a dialect of ancient Greek spoken especially in Ionia and the Cyclades

ionic bond *noun* (1939)
: a chemical bond formed between oppositely charged species because of their mutual electrostatic attraction

ion·ise *British variant of* IONIZE

io·ni·um \ī-ˈō-nē-əm\ *noun* [*ion;* from its ionizing action] (1907)
: a natural radioactive isotope of thorium having a mass number of 230

ionization chamber *noun* (1904)
: a partially evacuated tube provided with electrodes so that its conductivity due to the ionization of the residual gas reveals the presence of ionizing radiation

ion·ize \ˈī-ə-ˌnīz\ *verb* **ion·ized; ion·iz·ing** [International Scientific Vocabulary] (1898)
transitive verb
: to convert wholly or partly into ions
intransitive verb
: to become ionized
— ion·iz·able \-ˌnī-zə-bəl\ *adjective*
— ion·i·za·tion \ˌī-ə-nə-ˈzā-shən\ *noun*
— ion·iz·er \ˈī-ə-ˌnī-zər\ *noun*

ion·o·phore \ī-ˈä-nə-ˌfōr, -ˌfȯr\ *noun* (circa 1955)
: a compound that facilitates transmission of an ion (as of calcium) across a lipid barrier (as in a cell membrane) by combining with the ion or by increasing the permeability of the barrier to it

ion·o·sphere \ī-ˈä-nə-ˌsfir\ *noun* (1926)
: the part of the earth's atmosphere in which ionization of atmospheric gases affects the propagation of radio waves, which extends from about 30 miles (50 kilometers) to the exosphere, which is divided into regions of one or more layers whose altitudes and degrees of ionization vary with time of day, season, and solar cycle, and which is contiguous with the upper portion of the mesosphere and the thermosphere; *also* **:** a comparable region of charged particles surrounding another celestial body (as Venus)
— ion·o·spher·ic \ī-ˌä-nə-ˈsfir-ik, -ˈsfer-\ *adjective*
— ion·o·spher·i·cal·ly \-i-k(ə-)lē\ *adverb*

ion·to·pho·re·sis \(ˌ)ī-ˌän-tə-fə-ˈrē-səs\ *noun, plural* **-re·ses** \-ˌsēz\ [New Latin, from *ionto-* ion (from Greek *iont-, iōn,* present participle of *ienai*) + *-phoresis*] (1909)
: the introduction of an ionized substance (as a drug) through intact skin by the application of a direct electric current
— ion·to·pho·ret·ic \-ˈre-tik\ *adjective*
— ion·to·pho·ret·i·cal·ly \-ti-k(ə-)lē\ *adverb*

io·ta \ī-ˈō-tə\ *noun* [Latin, from Greek *iōta,* of Semitic origin; akin to Hebrew *yōdh* yod] (1607)
1 : the 9th letter of the Greek alphabet — see ALPHABET table
2 : an infinitesimal amount **:** JOT

IOU \ˌī-(ˌ)ō-ˈyü\ *noun* [from the pronunciation of *I owe you*] (1795)
1 : a paper that has on it the letters IOU, a stated sum, and a signature and that is given as an acknowledgment of debt
2 : DEBT, OBLIGATION

-ious *adjective suffix* [Middle English, partly from Old French *-ious, -ieux,* from Latin *-iosus,* from *-i-,* penultimate vowel of some noun stems + *-osus* -ous; partly from Latin *-ius,* adjective suffix]
: -OUS ⟨edac*ious*⟩

IPA \ˌī-(ˌ)pē-ˈā\ *noun* [*I*nternational *P*honetic *A*lphabet] (1933)
: an alphabet designed to represent each human speech sound with a unique symbol

ip·e·cac \ˈi-pi-ˌkak\ *or* **ipe·ca·cu·a·nha** \ˌi-pi-ˌka-kyə-ˈwä-nə, -nyə\ *noun* [Portuguese *ipecacuanha,* from Tupi *ipekaaguéne*] (1682)
1 : the dried rhizome and roots of either of two tropical American plants (*Cephaelis acuminata* and *C. ipecacuanha*) of the madder family used especially as a source of emetine; *also* **:** either of these plants
2 : an emetic and expectorant drug that contains emetine and is prepared from ipecac especially as a syrup for use in treating accidental poisoning

Iph·i·ge·nia \ˌi-fə-jə-ˈnī-ə\ *noun* [Latin, from Greek *Iphigeneia*]
: a daughter of Agamemnon nearly sacrificed by him to Artemis but saved by her and made a priestess

ipro·ni·a·zid \ˌī-prə-ˈnī-ə-zəd\ *noun* [blend of *isoniazid* and *propyl*] (1952)
: a derivative $C_9H_{13}N_3O$ of isoniazid that is used as a monoamine oxidase inhibitor and was formerly used in treating tuberculosis

ip·se dix·it \ˈip-sē-ˈdik-sət\ *noun* [Latin, he himself said it] (15th century)

: an assertion made but not proved : DICTUM

ip·si·lat·er·al \ˌip-si-ˈla-t(ə-)rəl\ *adjective* [International Scientific Vocabulary, from Latin *ipse* self, himself + *later-, latus* side] (1907)
: situated or appearing on or affecting the same side of the body
— **ip·si·lat·er·al·ly** \-t(ə-)rə-lē\ *adverb*

ip·sis·si·ma ver·ba \ip-ˈsi-sə-mə-ˈvər-bə\ *noun plural* [New Latin, literally, the selfsame words] (1807)
: the exact language used by someone quoted

ip·so fac·to \ˈip-(ˌ)sō-ˈfak-(ˌ)tō\ *adverb* [New Latin, literally, by the fact itself] (1548)
: by that very fact or act : as an inevitable result

IQ \ˌī-ˈkyü\ *noun* (1920)
: INTELLIGENCE QUOTIENT

ir- — see IN-

IRA \ˌī-(ˌ)är-ˈā; ˈī-rə\ *noun, plural* **IRAs** [individual retirement account] (1974)
: a savings account in which a person may deposit up to a stipulated amount each year with the deposits deductible from taxable income and both deposits and interest taxable after the person's retirement

Ira·ni·an \i-ˈrā-nē-ən, -ˈra-, -ˈrä-; ī-ˈ\ *noun* (1789)
1 : a native or inhabitant of Iran
2 : a branch of the Indo-European family of languages that includes Persian — see INDO-EUROPEAN LANGUAGES table
— **Iranian** *adjective*

iras·ci·ble \i-ˈra-sə-bəl\ *adjective* [Middle French, from Late Latin *irascibilis,* from Latin *irasci* to become angry, be angry, from *ira*] (circa 1530)
: marked by hot temper and easily provoked anger
— **iras·ci·bil·i·ty** \-ˌra-sə-ˈbi-lə-tē\ *noun*
— **iras·ci·ble·ness** \i-ˈra-sə-bəl-nəs\ *noun*
— **iras·ci·bly** \-blē\ *adverb*

irate \ī-ˈrāt, ˈī-ˌ, i-ˈrāt\ *adjective* (1838)
1 : roused to ire ⟨an *irate* taxpayer⟩
2 : arising from anger ⟨*irate* words⟩
— **irate·ly** *adverb*
— **irate·ness** *noun*

ire \ˈīr\ *noun* [Middle English, from Middle French, from Latin *ira;* perhaps akin to Greek *oistros* gadfly, frenzy] (14th century)
: intense and usually openly displayed anger
synonym see ANGER
— **ire** *transitive verb*
— **ire·ful** \-fəl\ *adjective*

ire·nic \ī-ˈre-nik, -ˈrē-\ *adjective* [Greek *eirēnikos,* from *eirēnē* peace] (circa 1864)
: favoring, conducive to, or operating toward peace, moderation, or conciliation
— **ire·ni·cal·ly** \-ˈre-ni-k(ə-)lē, -ˈrē-\ *adverb*

irid- *or* **irido-** *combining form*
1 [Latin *irid-, iris*] : rainbow ⟨*irid*escent⟩
2 [New Latin *irid-, iris*] : iris of the eye ⟨*irido*logy⟩

ir·i·des·cence \ˌir-ə-ˈde-sᵊn(t)s\ *noun* (1804)
1 : a lustrous rainbowlike play of color caused by differential refraction of light waves (as from an oil slick, soap bubble, or fish scales) that tends to change as the angle of view changes
2 : a lustrous or attractive quality or effect ⟨the agglomeration of power which . . . gives them a certain *iridescence* of glamor and superiority —Margaret Landon⟩

ir·i·des·cent \-sᵊnt\ *adjective* (1791)
: having or exhibiting iridescence
— **ir·i·des·cent·ly** *adverb*

irid·ic \i-ˈri-dik, ī-ˈri-dik\ *adjective* (circa 1890)
: of or relating to the iris of the eye

irid·i·um \i-ˈri-dē-əm\ *noun* [New Latin, from Latin *irid-, iris;* from the colors produced by its dissolving in hydrochloric acid] (1804)
: a silver-white hard brittle very heavy metallic element — see ELEMENT table

ir·i·dol·o·gy \ˌī-rə-ˈdä-lə-jē\ *noun, plural* **-gies** (circa 1923)
: the study of the iris of the eye for indications of bodily health and disease
— **ir·i·dol·o·gist** \-jist\ *noun*

ir·id·os·mine \ˌir-ə-ˈdäz-ˌmēn\ *noun* [German, from New Latin *iridium* + *osmium*] (1827)
: a mineral that is a native iridium osmium alloy usually containing some rhodium and platinum

¹iris \ˈī-rəs\ *noun, plural* **iris·es** *or* **iri·des** \ˈī-rə-ˌdēz, ˈir-ə-\ [Middle English, from Latin *irid-, iris* rainbow, iris plant, from Greek, rainbow, iris plant, iris of the eye — more at WIRE] (15th century)
1 : RAINBOW
2 [New Latin *irid-, iris,* from Greek] **a** : the opaque contractile diaphragm perforated by the pupil and forming the colored portion of the eye — see EYE illustration **b** : IRIS DIAPHRAGM; *also* : a similar device with a circular opening that can be varied in size
3 *or plural* **iris** [New Latin *Irid-, Iris,* genus name, from Latin] : any of a large genus (*Iris* of the family Iridaceae, the iris family) of perennial herbaceous plants with linear usually basal leaves and large showy flowers

²iris *transitive verb* (1816)
: to make iridescent

Iris \ˈī-rəs\ *noun* [Latin, from Greek]
: the Greek goddess of the rainbow and a messenger of the gods

iris diaphragm *noun* (1867)
: an adjustable diaphragm of thin opaque plates that can be turned by a ring so as to change the diameter of a central opening usually to regulate the aperture of a lens

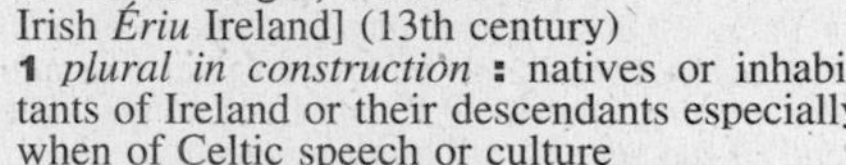
iris diaphragm

Irish \ˈī-rish\ *noun* [Middle English, from (assumed) Old English *Īrisc,* from *Īras* Irishmen, of Celtic origin; akin to Old Irish *Ériu* Ireland] (13th century)
1 *plural in construction* : natives or inhabitants of Ireland or their descendants especially when of Celtic speech or culture
2 **a** : the Celtic language of Ireland especially as used since the later medieval period **b** : English spoken by the Irish
3 : IRISH WHISKEY
— **Irish** *adjective*

Irish bull *noun* (1802)
: an incongruous statement (as "it was hereditary in his family to have no children")

Irish coffee *noun* (1950)
: hot sugared coffee and Irish whiskey topped with whipped cream

Irish confetti *noun* (1922)
: a rock or brick used as a missile

Irish Gaelic *noun* (1891)
: IRISH 2a

Irish·ism \ˈī-ri-ˌshi-zəm\ *noun* (1734)
1 : a word, phrase, or expression characteristic of the Irish
2 : IRISH BULL

Irish·ly \ˈī-rish-lē\ *adverb* (1571)
: in a manner characteristic of the Irish

Irish mail *noun* (1908)
: a 3- or 4-wheeled toy vehicle activated by a hand lever

Irish·man \ˈī-rish-mən\ *noun* (13th century)
1 : a native or inhabitant of Ireland
2 : one that is of Irish descent

Irish moss *noun* (1845)
1 : the dried and bleached plants of a red alga (especially *Chondrus crispus*) used as an agent for thickening or emulsifying or as a demulcent
2 : a red alga that is a source of Irish moss — called also *carrageen*

Irish·ness \ˈī-rish-nəs\ *noun* (1804)
: the fact or quality of being Irish

Irish potato *noun* (1664)
: POTATO 2b

Irish·ry \ˈī-rish-rē\ *noun, plural* **-ries** (14th century)
1 : IRISH 1
2 **a** : Irish quality or character **b** : an Irish peculiarity or trait

Irish setter *noun* (1866)
: any of a breed of bird dogs generally comparable to English setters but with a mahogany-red coat

Irish stew *noun* (1814)
: a stew having as its principal ingredients meat (as lamb), potatoes, and onions in a thick gravy

Irish terrier *noun* (1857)
: any of a breed of active medium-sized terriers developed in Ireland and characterized by a dense close usually reddish wiry coat

Irish water spaniel *noun* (1885)
: any of a breed of large retrievers characterized by a topknot, a heavy curly coat which is dark liver in color, and a short-haired tail

Irish whiskey *noun* (1798)
: whiskey made in Ireland chiefly of barley

Irish wolfhound *noun* (1880)
: any of a breed of very tall heavily built hounds having a rough wiry coat

Irish wolfhound

Irish·wom·an \ˈī-rish-ˌwu̇-mən\ *noun* (15th century)
: a woman born in Ireland or of Irish descent

iri·tis \ī-ˈrī-təs\ *noun* [New Latin, irregular from *iris*] (1818)
: inflammation of the iris of the eye

¹irk \ˈərk\ *transitive verb* [Middle English] (15th century)
: to make weary, irritated, or bored
synonym see ANNOY

²irk *noun* (circa 1570)
1 : the fact of being annoying
2 : a source of annoyance

irk·some \ˈərk-səm\ *adjective* (15th century)
: tending to irk : TEDIOUS ⟨an *irksome* task⟩
— **irk·some·ly** *adverb*
— **irk·some·ness** *noun*

iro·ko \ə-ˈrō-(ˌ)kō\ *noun* [Yoruba *ìrokò*] (1890)
: a large tropical western African tree (*Chlorophora excelsa*) of the mulberry family having strong streaky insect-resistant wood which is often used as a teak substitute; *also* : this wood

¹iron \ˈī(-ə)rn\ *noun* [Middle English, from Old English *īsern, īren;* akin to Old High German *īsarn* iron] (before 12th century)
1 : a heavy malleable ductile magnetic silver-white metallic element that readily rusts in moist air, occurs native in meteorites and combined in most igneous rocks, is the most used of metals, and is vital to biological processes — see ELEMENT table
2 : something made of iron: as **a** *plural* : shackles for the hands or legs **b** : a heated metal implement used for branding or cauterizing **c** : a household device usually with a flat metal base that is heated to smooth, finish, or press (as cloth) **d** : STIRRUP — usually used in plural **e** : any of a series of numbered golf clubs having relatively thin metal heads — compare WOOD
3 : great strength, hardness, or determination
usage see ASK
— **iron in the fire 1** : a matter requiring close attention **2** : a prospective course of action

²iron *adjective* (before 12th century)
1 : of, relating to, or made of iron
2 : resembling iron
3 **a** : strong and healthy : ROBUST **b** : INFLEXIBLE, UNRELENTING ⟨*iron* determination⟩ **c** : holding or binding fast ⟨an *iron* grip⟩
— **iron·ness** \ˈī(-ə)rn-nəs\ *noun*

³iron (15th century)
transitive verb
1 : to furnish or cover with iron

2 : to shackle with irons
3 a : to smooth with or as if with a heated iron ⟨*iron* a shirt⟩ **b :** to remove (as wrinkles) by ironing
intransitive verb
: to smooth or press cloth or clothing with a heated iron

Iron Age *noun* (1879)
: the period of human culture characterized by the smelting of iron and its use in industry beginning somewhat before 1000 B.C. in western Asia and Egypt

iron·bark \'ī(-ə)rn-ˌbärk\ *noun* (1799)
: any of several Australian eucalypti having hard gray bark and heavy hard durable wood used especially in heavy construction; *also* **:** this wood

iron·bound \-'bau̇nd\ *adjective* (14th century)
: bound with or as if with iron: as **a :** HARSH, RUGGED ⟨*ironbound* coast⟩ **b :** STERN, RIGOROUS ⟨*ironbound* traditions⟩

¹iron·clad \-'klad\ *adjective* (circa 1847)
1 : sheathed in iron armor — used especially of naval vessels
2 : so firm or secure as to be unbreakable: as **a :** BINDING ⟨an *ironclad* oath⟩ **b :** having no obvious weakness ⟨an *ironclad* case against the defendant⟩

²iron·clad \-ˌklad\ *noun* (1862)
: an armored naval vessel especially of the mid to late nineteenth century

iron curtain *noun* (1819)
1 : an impenetrable barrier ⟨the *iron curtain* between the ego and the unconscious —C. J. Rolo⟩
2 : a political, military, and ideological barrier that cuts off and isolates an area; *specifically, often capitalized* **:** one isolating an area under Soviet control

iron·er \'ī(-ə)r-nər\ *noun* (1773)
: one that irons; *specifically* **:** MANGLE

iron·fisted \'ī(-ə)rn-'fis-təd\ *adjective* (1599)
1 : STINGY, MISERLY
2 : being both harsh and ruthless ⟨*ironfisted* methods⟩

iron gray *noun* (before 12th century)
: a slightly greenish dark gray

iron hand *noun* (1703)
: stern or rigorous control ⟨ruled with an *iron hand*⟩
— **iron·hand·ed** \'ī(-ə)rn-'han-dəd\ *adjective*

iron·heart·ed \'ī(-ə)rn-'här-təd\ *adjective* (1600)
: CRUEL, HARD-HEARTED

iron horse *noun* (1840)
: LOCOMOTIVE 1; *especially* **:** a steam locomotive

iron·ic \ˌī-'rä-nik *also* i-'rä-\ *or* **iron·i·cal** \-ni-kəl\ *adjective* (1576)
1 : relating to, containing, or constituting irony
2 : given to irony
synonym see SARCASTIC
— **iron·i·cal·ness** \-ni-kəl-nəs\ *noun*

iron·i·cal·ly \-ni-k(ə-)lē\ *adverb* (1576)
1 : in an ironic manner
2 : it is ironic, curious, or surprising

iron·ing \'ī(-ə)r-niŋ\ *noun* (circa 1710)
1 : the action or process of smoothing or pressing with or as if with a heated iron
2 : clothes ironed or to be ironed

ironing board *noun* (1843)
: a flat padded cloth-covered surface on which clothes are ironed

iro·nist \'ī-rə-nist\ *noun* (1727)
: one who uses irony especially in the development of a literary work or theme

iro·nize \-ˌnīz\ *verb* **-nized; -niz·ing** (1602)
transitive verb
: to make ironic in appearance or effect
intransitive verb
: to use irony **:** speak or behave ironically

iron lung *noun* (1932)
: a device for artificial respiration in which rhythmic alternations in the air pressure in a chamber surrounding a patient's chest force air into and out of the lungs

iron maiden *noun* (circa 1895)
: a supposed medieval torture device consisting of a hollow iron statue or coffin in the shape of a woman that is lined with spikes which impale the enclosed victim

iron man *noun* (1914)
: a man of unusual physical endurance

iron·mas·ter \'ī(-ə)rn-ˌmas-tər\ *noun* (1674)
: a manufacturer of iron

iron·mon·ger \-ˌməŋ-gər, -ˌmäŋ-\ *noun* (14th century)
British **:** a dealer in iron and hardware

iron·mon·gery \-g(ə-)rē\ *noun* (1711)
: something made of metal; *especially, British* **:** HARDWARE 1

iron out *transitive verb* (1870)
1 : to make smooth or flat by or as if by pressing
2 : to resolve or work out a solution to ⟨*ironed out* their differences⟩

iron oxide *noun* (1885)
: any of several oxides of iron: as **a :** FERRIC OXIDE **b :** FERROUS OXIDE

iron pyrites *noun* (circa 1817)
: PYRITE — called also *iron pyrite*

iron ration *noun* (1876)
: an emergency ration

iron·side \-ˌsīd\ *noun* (13th century)
: a man of great strength or bravery

iron·stone \'ī(-ə)rn-ˌstōn\ *noun* (1522)
1 : a hard sedimentary rock rich in iron; *especially* **:** a siderite in a coal region
2 : IRONSTONE CHINA

ironstone china *noun* (1825)
: a hard heavy durable white pottery developed in England early in the 19th century

iron sulfide *noun* (1885)
: any of several sulfides of iron

iron·ware \'ī(-ə)rn-ˌwar, -ˌwer\ *noun* (15th century)
: articles made of iron

iron·weed \-ˌwēd\ *noun* (1819)
: any of several mostly weedy American composite plants (genus *Vernonia*) usually having alternate leaves and perfect red, purple, or white tubular flowers in terminal cymose heads

iron·wood \-ˌwu̇d\ *noun* (1657)
1 : any of numerous trees and shrubs (as a hornbeam or hop hornbeam) with exceptionally tough or hard wood
2 : the wood of an ironwood

iron·work \-ˌwərk\ *noun* (15th century)
1 : work in iron; *also* **:** something made of iron
2 *plural but singular or plural in construction* **:** a mill or building where iron or steel is smelted or heavy iron or steel products are made
— **iron·work·er** \-ˌwər-kər\ *noun*

iro·ny \'ī-rə-nē *also* 'ī(-ə)r-nē\ *noun, plural* **-nies** [Latin *ironia,* from Greek *eirōnia,* from *eirōn* dissembler] (1502)
1 : a pretense of ignorance and of willingness to learn from another assumed in order to make the other's false conceptions conspicuous by adroit questioning — called also *Socratic irony*
2 a : the use of words to express something other than and especially the opposite of the literal meaning **b :** a usually humorous or sardonic literary style or form characterized by irony **c :** an ironic expression or utterance
3 a (1) **:** incongruity between the actual result of a sequence of events and the normal or expected result (2) **:** an event or result marked by such incongruity **b :** incongruity between a situation developed in a drama and the accompanying words or actions that is understood by the audience but not by the characters in the play — called also *dramatic irony, tragic irony*
synonym see WIT

Ir·o·quoi·an \ˌir-ə-'kwȯi-ən\ *noun* (1697)
1 : a member of any of the peoples constituting the Iroquois
2 : an American Indian language family of eastern North America including Cayuga, Cherokee, Mohawk, Onondaga, Oneida, Seneca, and Tuscarora
— **Iroquoian** *adjective*

Ir·o·quois \'ir-ə-ˌkwȯi *also* -ˌkwä\ *noun, plural* **Iroquois** \-ˌkwȯi(z), -ˌkwä(z)\ [French, probably of Algonquian origin] (1666)
1 *plural* **:** an American Indian confederacy originally of New York consisting of the Cayuga, Mohawk, Oneida, Onondaga, and Seneca and later including the Tuscarora
2 : a member of any of the Iroquois peoples

ir·ra·di·ance \i-'rā-dē-ən(t)s\ *noun* (1667)
1 : RADIANCE 1
2 : the density of radiation incident on a given surface usually expressed in watts per square centimeter or square meter

ir·ra·di·ate \i-'rā-dē-ˌāt\ *verb* **-at·ed; -at·ing** [Latin *irradiatus,* past participle of *irradiare,* from *in-* + *radius* ray] (1603)
transitive verb
1 a : to cast rays of light upon **:** ILLUMINATE **b :** to enlighten intellectually or spiritually **c :** to affect or treat by radiant energy (as heat); *specifically* **:** to treat by exposure to radiation
2 : to emit like rays of light **:** RADIATE ⟨*irradiating* strength and comfort⟩
intransitive verb
archaic **:** to emit rays **:** SHINE
— **ir·ra·di·a·tive** \-ˌā-tiv\ *adjective*
— **ir·ra·di·a·tor** \-ˌā-tər\ *noun*

ir·ra·di·a·tion \i-ˌrā-dē-'ā-shən\ *noun* (1901)
1 : exposure to radiation (as X rays or alpha particles)
2 : application of radiation (as X rays or ultraviolet light) especially for therapeutic purposes

ir·rad·i·ca·ble \i-'ra-di-kə-bəl, ˌi(r)-\ *adjective* [Medieval Latin *irradicabilis,* from Latin *in-* + *radic-, radix* root — more at ROOT] (1728)
: impossible to eradicate **:** DEEP-ROOTED
— **ir·rad·i·ca·bly** \-blē\ *adverb*

¹ir·ra·tio·nal \i-'ra-sh(ə-)nəl, ˌi(r)-\ *adjective* [Middle English, from Latin *irrationalis,* from *in-* + *rationalis* rational] (14th century)
: not rational: as **a** (1) **:** not endowed with reason or understanding (2) **:** lacking usual or normal mental clarity or coherence **b :** not governed by or according to reason ⟨*irrational* fears⟩ **c** *Greek & Latin prosody* (1) *of a syllable* **:** having a quantity other than that required by the meter (2) *of a foot* **:** containing such a syllable **d** (1) **:** being an irrational number ⟨an *irrational* root of an equation⟩ (2) **:** having a numerical value that is an irrational number ⟨a length that is *irrational*⟩
— **ir·ra·tio·nal·i·ty** \-ˌra-shə-'na-lə-tē\ *noun*
— **ir·ra·tio·nal·ly** \-'rash(ə-)nə-lē\ *adverb*

²irrational *noun* (1646)
1 : an irrational being
2 : IRRATIONAL NUMBER

ir·ra·tio·nal·ism \-'ra-sh(ə-)nə-ˌli-zəm\ *noun* (1811)
1 : a system emphasizing intuition, instinct, feeling, or faith rather than reason or holding that the universe is governed by irrational forces
2 : the quality or state of being irrational
— **ir·ra·tio·nal·ist** \-list\ *noun or adjective*
— **ir·ra·tio·nal·is·tic** \-ˌra-sh(ə-)nə-'lis-tik\ *adjective*

irrational number *noun* (1551)
: a number that can be expressed as an infinite decimal with no set of consecutive digits repeating itself indefinitely and that cannot be expressed as the quotient of two integers

ir·re·al \i-ˈrē(-ə)l, ˌi(r)-, -ˈri(-ə)l\ *adjective* (1944)
: not real

ir·re·al·i·ty \ˌir-ē-ˈa-lə-tē\ *noun* (1803)
: UNREALITY

ir·re·claim·able \ˌir-i-ˈklā-mə-bəl\ *adjective* (1662)
: incapable of being reclaimed
— **ir·re·claim·ably** \-blē\ *adverb*

¹ir·rec·on·cil·able \i-ˌre-kən-ˈsī-lə-bəl, -ˈre-kən-ˌ, ˌi(r)-\ *adjective* (1599)
: impossible to reconcile
— **ir·rec·on·cil·abil·i·ty** \-ˌre-kən-ˌsī-lə-ˈbi-lə-tē\ *noun*
— **ir·rec·on·cil·able·ness** \-ˌre-kən-ˈsī-lə-bəl-nəs, -ˈre-kən-ˌ\ *noun*
— **ir·rec·on·cil·ably** \-blē\ *adverb*

²irreconcilable *noun* (1748)
: one that is irreconcilable; *especially* **:** a member of a group (as a political party) opposing compromise or collaboration

ir·re·cov·er·able \ˌir-i-ˈkəv-rə-bəl, -ˈkə-və-\ *adjective* (15th century)
: not capable of being recovered or rectified : IRREPARABLE
— **ir·re·cov·er·able·ness** *noun*
— **ir·re·cov·er·ably** \-blē\ *adverb*

ir·re·cu·sa·ble \ˌir-i-ˈkyü-zə-bəl\ *adjective* [Late Latin *irrecusabilis,* from Latin *in-* + *recusare* to reject, refuse — more at RECUSANT] (1776)
: not subject to exception or rejection
— **ir·re·cu·sa·bly** \-blē\ *adverb*

ir·re·deem·able \ˌir-i-ˈdē-mə-bəl\ *adjective* (1609)
1 : not redeemable: as **a :** not terminable by payment of the principal ⟨*irredeemable* bond⟩ **b :** INCONVERTIBLE a
2 : being beyond remedy : HOPELESS ⟨*irredeemable* mistakes⟩
— **ir·re·deem·ably** \-blē\ *adverb*

ir·re·den·ta \ˌir-i-ˈden-tə\ *noun* [Italian *Italia irredenta,* literally, unredeemed Italy, Italian-speaking territory not incorporated in Italy] (1914)
: a territory historically or ethnically related to one political unit but under the political control of another

ir·re·den·tism \-ˈden-ˌti-zəm\ *noun* (1883)
: a political principle or policy directed toward the incorporation of irredentas within the boundaries of their historically or ethnically related political unit
— **ir·re·den·tist** \-ˈden-tist\ *noun or adjective*

ir·re·duc·ible \ˌir-i-ˈdü-sə-bəl, -ˈdyü-\ *adjective* (1633)
: impossible to transform into or restore to a desired or simpler condition ⟨an *irreducible* matrix⟩; *specifically* **:** incapable of being factored into polynomials of lower degree with coefficients in some given field (as the rational numbers) or integral domain (as the integers) ⟨an *irreducible* equation⟩
— **ir·re·duc·ibil·i·ty** \-ˌdü-sə-ˈbi-lə-tē, -ˌdyü-\ *noun*
— **ir·re·duc·ibly** \-ˈdü-sə-blē, -ˈdyü-\ *adverb*

ir·re·flex·ive \ˌir-i-ˈflek-siv\ *adjective* (circa 1890)
: being a relation for which the reflexive property does not hold for any element of a given set

ir·re·form·able \ˌir-i-ˈfȯr-mə-bəl\ *adjective* (1609)
1 : incapable of being reformed : INCORRIGIBLE
2 : not subject to revision or alteration ⟨*irreformable* dogma⟩
— **ir·re·form·abil·i·ty** \-ˌfȯr-mə-ˈbi-lə-tē\ *noun*

ir·re·fra·ga·ble \i-ˈre-frə-gə-bəl, ˌi(r)-; ˌir-i-ˈfra-gə-\ *adjective* [Late Latin *irrefragabilis,* from Latin *in-* + *refragari* to oppose, from *re-* + *-fragari* (as in *suffragari* to vote for); akin to Latin *suffragium* suffrage] (1533)
1 : impossible to refute ⟨*irrefragable* arguments⟩
2 : impossible to break or alter ⟨*irrefragable* rules⟩
— **ir·re·fra·ga·bil·i·ty** \i-ˌre-frə-gə-ˈbi-lə-tē, ˌi(r)-; ˌir-i-ˌfra-gə-\ *noun*
— **ir·ref·ra·ga·bly** \i-ˈre-frə-gə-blē, ˌi(r)-; ˌir-i-ˈfra-gə-\ *adverb*

ir·re·fut·able \ˌir-i-ˈfyü-tə-bəl; i-ˈre-fyə-tə-, ˌi(r)-\ *adjective* [Late Latin *irrefutabilis,* from Latin *in-* + *refutare* to refute] (1620)
: impossible to refute : INCONTROVERTIBLE ⟨*irrefutable* proof⟩
— **ir·re·fut·abil·i·ty** \ˌir-i-ˌfyü-tə-ˈbi-lə-tē; i-ˌre-fyə-tə-, ˌi(r)-\ *noun*
— **ir·re·fut·ably** \ˌir-i-ˈfyü-tə-blē; i-ˈre-fyə-tə-, ˌi(r)-\ *adverb*

ir·re·gard·less \ˌir-i-ˈgärd-ləs\ *adverb* [probably blend of *irrespective* and *regardless*] (circa 1912)
nonstandard : REGARDLESS ■

¹ir·reg·u·lar \i-ˈre-gyə-lər, ˌi(r)-\ *adjective* [Middle English *irreguler,* from Middle French, from Late Latin *irregularis* not in accordance with rule, from Latin *in-* + *regularis* regular] (14th century)
1 a : not being or acting in accord with laws, rules, or established custom ⟨*irregular* conduct⟩ **b :** not conforming to the usual pattern of inflection ⟨*irregular* verbs⟩; *specifically* : STRONG 16 **c :** not following a usual or prescribed procedure; *especially, British* **:** celebrated without either proclamation of the banns or publication of intention to marry ⟨*irregular* marriage⟩
2 : not belonging to or a part of a regular organized group; *specifically* **:** not belonging to a regular army but raised for a special purpose ⟨*irregular* troops⟩
3 a : lacking perfect symmetry or evenness ⟨an *irregular* coastline⟩ **b :** ZYGOMORPHIC ⟨*irregular* flowers⟩
4 : lacking continuity or regularity especially of occurrence or activity ⟨*irregular* employment⟩ ☆
— **ir·reg·u·lar·ly** *adverb*

²irregular *noun* (15th century)
: one that is irregular: as **a :** a soldier who is not a member of a regular military force **b** *plural* **:** merchandise that has minor defects or that falls next below the manufacturer's standard for firsts

ir·reg·u·lar·i·ty \i-ˌre-gyə-ˈla-rə-tē, ˌi(r)-\ *noun, plural* **-ties** [Middle English *irregularite,* from Middle French *irregularité,* from Medieval Latin *irregularitat-, irregularitas,* from Late Latin *irregularis*] (14th century)
1 : something that is irregular (as improper or dishonest conduct) ⟨alleged *irregularities* in the city government⟩
2 : the quality or state of being irregular
3 : CONSTIPATION

ir·rel·a·tive \-ˈre-lə-tiv\ *adjective* (1640)
: not relative: **a :** not related **b :** IRRELEVANT
— **ir·rel·a·tive·ly** *adverb*

ir·rel·e·vance \-ˈre-lə-vən(t)s\ *noun* (1847)
1 : the quality or state of being irrelevant
2 : something irrelevant

ir·rel·e·van·cy \-vən(t)-sē\ *noun, plural* **-cies** (1592)
: IRRELEVANCE

ir·rel·e·vant \-vənt\ *adjective* (1786)
: not relevant : INAPPLICABLE ⟨that statement is *irrelevant* to your argument⟩
— **ir·rel·e·vant·ly** *adverb*

ir·re·li·gion \ˌir-i-ˈli-jən\ *noun* [Middle French or Late Latin; Middle French, from Late Latin *irreligion-, irreligio,* from Latin *in-* + *religion-, religio* religion] (1598)
: the quality or state of being irreligious
— **ir·re·li·gion·ist** \-ˈlij-nist, -ˈli-jə-\ *noun*

ir·re·li·gious \-ˈli-jəs\ *adjective* (15th century)
1 : neglectful of religion : lacking religious emotions, doctrines, or practices ⟨so *irreligious* that they exploit popular religion for professional purposes —G. B. Shaw⟩
2 : indicating lack of religion
— **ir·re·li·gious·ly** *adverb*

ir·re·me·able \i-ˈrē-mē-ə-bəl, ˌi(r)-\ *adjective* [Latin *irremeabilis,* from *in-* + *remeare* to go back, from *re-* + *meare* to go — more at PERMEATE] (1569)
: offering no possibility of return

ir·re·me·di·a·ble \ˌir-i-ˈmē-dē-ə-bəl\ *adjective* [Middle English, from Latin *irremediabilis,* from *in-* + *remediabilis* remediable] (15th century)
: not remediable; *specifically* : INCURABLE
— **ir·re·me·di·a·ble·ness** *noun*
— **ir·re·me·di·a·bly** \-blē\ *adverb*

ir·re·mov·able \-ˈmü-və-bəl\ *adjective* (1598)
: not removable
— **ir·re·mov·abil·i·ty** \-ˌmü-və-ˈbi-lə-tē\ *noun*
— **ir·re·mov·ably** \-ˈmü-və-blē\ *adverb*

ir·rep·a·ra·ble \i-ˈre-p(ə-)rə-bəl, ˌi(r)- *also* ÷ˌir-(r)ə-ˈpar-ə-bəl\ *adjective* [Middle English, from Middle French, from Latin *irreparabilis,* from *in-* + *reparabilis* reparable] (15th century)
: not reparable : IRREMEDIABLE ⟨*irreparable* damage⟩
— **ir·rep·a·ra·ble·ness** *noun*
— **ir·rep·a·ra·bly** \-blē\ *adverb*

ir·re·peal·able \ˌir-i-ˈpē-lə-bəl\ *adjective* (1633)
: not repealable
— **ir·re·peal·abil·i·ty** \-ˌpē-lə-ˈbi-lə-tē\ *noun*

ir·re·place·able \-ˈplā-sə-bəl\ *adjective* (1807)
: not replaceable
— **ir·re·place·abil·i·ty** \-ˌplā-sə-ˈbi-lə-tē\ *noun*
— **ir·re·place·able·ness** \-ˈplā-sə-bəl-nəs\ *noun*
— **ir·re·place·ably** \-blē\ *adverb*

ir·re·press·ible \-ˈpre-sə-bəl\ *adjective* (1811)
: impossible to repress, restrain, or control ⟨*irrepressible* curiosity⟩
— **ir·re·press·ibil·i·ty** \-ˌpre-sə-ˈbi-lə-tē\ *noun*
— **ir·re·press·ibly** \-ˈpre-sə-blē\ *adverb*

ir·re·proach·able \-ˈprō-chə-bəl\ *adjective* (1634)

☆ SYNONYMS
Irregular, anomalous, unnatural mean not conforming to rule, law, or custom. IRREGULAR implies not conforming to a law or regulation imposed for the sake of uniformity in method, practice, or conduct ⟨concerned about his *irregular* behavior⟩. ANOMALOUS implies not conforming to what might be expected because of the class or type to which it belongs or the laws that govern its existence ⟨her drive made her an *anomalous figure in a sleepy organization*⟩. UNNATURAL suggests what is contrary to nature or to principles or standards felt to be essential to the well-being of civilized society ⟨prisoners treated with *unnatural* cruelty⟩.

□ USAGE
irregardless *Irregardless* originated in dialectal American speech in the early 20th century. Its fairly widespread use in speech called it to the attention of usage commentators as early as 1927. The most frequently repeated remark about it is that "there is no such word." There is such a word, however. It is still used primarily in speech, although it can be found from time to time in edited prose. Its reputation has not risen over the years, and it is still a long way from general acceptance. Use *regardless* instead.

: not reproachable **:** BLAMELESS, IMPECCABLE ⟨*irreproachable* conduct⟩
— **ir·re·proach·abil·i·ty** \-ˌprō-chə-ˈbi-lə-tē\ *noun*
— **ir·re·proach·able·ness** \-ˈprō-chə-bəl-nəs\ *noun*
— **ir·re·proach·ably** \-blē\ *adverb*

ir·re·pro·duc·ible \i-ˌrē-prə-ˈdü-sə-bəl, ˌi(r)-, -ˈdyü-\ *adjective* (1868)
: not reproducible
— **ir·re·pro·duc·ibil·i·ty** \-ˌdü-sə-ˈbi-lə-tē, -ˌdyü-\ *noun*

ir·re·sist·ible \ˌir-i-ˈzis-tə-bəl\ *adjective* (1597)
: impossible to resist ⟨an *irresistible* attraction⟩
— **ir·re·sist·ibil·i·ty** \-ˌzis-tə-ˈbi-lə-tē\ *noun*
— **ir·re·sist·ible·ness** \-ˈzis-tə-bəl-nəs\ *noun*
— **ir·re·sist·ibly** \-blē\ *adverb*

ir·re·sol·u·ble \-ˈzäl-yə-bəl\ *adjective* [Latin *irresolubilis,* from *in-* + *resolvere* to resolve] (1666)
1 *archaic* **:** INDISSOLUBLE
2 : having or admitting of no solution or explanation

ir·res·o·lute \i-ˈre-zə-ˌlüt, ˌi(r)-, -lət\ *adjective* (1579)
: uncertain how to act or proceed **:** VACILLATING
— **ir·res·o·lute·ly** \-ˌlüt-lē, -lət-; -ˌre-zə-ˈlüt-\ *adverb*
— **ir·res·o·lute·ness** \-ˌlüt-nəs, -lət-, -ˈlüt-\ *noun*
— **ir·res·o·lu·tion** \-ˌre-zə-ˈlü-shən\ *noun*

ir·re·solv·able \ˌir-i-ˈzäl-və-bəl, -ˈzȯl-\ *adjective* (1660)
: incapable of being resolved; *also* **:** not analyzable

ir·re·spec·tive of \ˌir-i-ˈspek-tiv-\ *preposition* (1839)
: REGARDLESS OF ⟨free public schools open to all *irrespective of* race, color, or creed —J. B. Conant⟩

[1]ir·re·spon·si·ble \ˌir-i-ˈspän(t)-sə-bəl\ *adjective* (1648)
: not responsible: as **a :** not answerable to higher authority ⟨an *irresponsible* dictatorship⟩ **b :** said or done with no sense of responsibility ⟨*irresponsible* accusations⟩ **c :** lacking a sense of responsibility **d :** unable especially mentally or financially to bear responsibility
— **ir·re·spon·si·bil·i·ty** \ˌir-i-ˌspän(t)-sə-ˈbi-lə-tē\ *noun*
— **ir·re·spon·si·ble·ness** \ˌir-i-ˈspän(t)-sə-bəl-nəs\ *noun*
— **ir·re·spon·si·bly** \-blē\ *adverb*

[2]irresponsible *noun* (1894)
: one that is irresponsible

ir·re·spon·sive \ˌir-i-ˈspän(t)-siv\ *adjective* (circa 1846)
: not responsive; *especially* **:** not able, ready, or inclined to respond
— **ir·re·spon·sive·ness** *noun*

ir·re·triev·able \ˌir-i-ˈtrē-və-bəl\ *adjective* (1702)
: not retrievable **:** impossible to regain or recover
— **ir·re·triev·abil·i·ty** \-ˌtrē-və-ˈbi-lə-tē\ *noun*
— **ir·re·triev·ably** \ˌir-i-ˈtrē-və-blē\ *adverb*

ir·rev·er·ence \i-ˈrev-rənts, ˌi(r)-, -ˈre-və-; -ˈre-vərn(t)s\ *noun* (14th century)
1 : lack of reverence
2 : an irreverent act or utterance

ir·rev·er·ent \-rənt; -vərnt\ *adjective* [Middle English, from Latin *irreverent-, irreverens,* from *in-* + *reverent-, reverens* reverent] (15th century)
: lacking proper respect or seriousness; *also* **:** SATIRIC
— **ir·rev·er·ent·ly** *adverb*

ir·re·vers·ible \ˌir-i-ˈvər-sə-bəl\ *adjective* (1630)
: not reversible
— **ir·re·vers·ibil·i·ty** \-ˌvər-sə-ˈbi-lə-tē\ *noun*
— **ir·re·vers·ibly** \-ˈvər-sə-blē\ *adverb*

ir·rev·o·ca·ble \i-ˈre-və-kə-bəl, ˌi(r)- *sometimes* ˌir-(r)ə-ˈvō-kə-\ *adjective* [Middle English, from Latin *irrevocabilis,* from *in-* + *revocabilis* revocable] (14th century)
: not possible to revoke **:** UNALTERABLE ⟨an *irrevocable* decision⟩
— **ir·rev·o·ca·bil·i·ty** \-ˌre-və-kə-ˈbi-lə-tē, ˌir-(r)ə-ˌvō-kə-\ *noun*
— **ir·rev·o·ca·ble·ness** \-ˈre-və-kə-bəl-nəs, ˌir-(r)ə-ˈvō-kə-\ *noun*
— **ir·rev·o·ca·bly** \-blē\ *adverb*

ir·ri·den·ta *variant of* IRREDENTA

ir·ri·gate \ˈir-ə-ˌgāt\ *verb* **-gat·ed; -gat·ing** [Latin *irrigatus,* past participle of *irrigare,* from *in-* + *rigare* to water; perhaps akin to Old High German *regan* rain — more at RAIN] (1615)
transitive verb
1 : WET, MOISTEN: as **a :** to supply (as land) with water by artificial means **b :** to flush (a body part) with a stream of liquid (as in removing a foreign body or medicating)
2 : to refresh as if by watering
intransitive verb
: to practice irrigation
— **ir·ri·ga·tion** \ˌir-ə-ˈgā-shən\ *noun*
— **ir·ri·ga·tor** \ˈir-ə-ˌgā-tər\ *noun*

ir·ri·ta·bil·i·ty \ˌir-ə-tə-ˈbi-lə-tē\ *noun, plural* **-ties** (1755)
1 : the property of protoplasm and of living organisms that permits them to react to stimuli
2 : the quality or state of being irritable: as **a :** quick excitability to annoyance, impatience, or anger **:** PETULANCE **b :** abnormal or excessive excitability of an organ or part of the body

ir·ri·ta·ble \ˈir-ə-tə-bəl\ *adjective* (1662)
: capable of being irritated: as **a :** easily exasperated or excited **b :** responsive to stimuli
— **ir·ri·ta·ble·ness** *noun*
— **ir·ri·ta·bly** \-blē\ *adverb*

[1]ir·ri·tant \ˈir-ə-tənt\ *adjective* (1636)
: causing irritation; *specifically* **:** tending to produce physical irritation

[2]irritant *noun* (1802)
: something that irritates or excites

ir·ri·tate \ˈir-ə-ˌtāt\ *verb* **-tat·ed; -tat·ing** [Latin *irritatus,* past participle of *irritare*] (1598)
transitive verb
1 : to provoke impatience, anger, or displeasure in **:** ANNOY
2 : to induce irritability in or of
intransitive verb
: to cause or induce displeasure or irritation ☆
— **ir·ri·tat·ing·ly** \-ˌtā-tiŋ-lē\ *adverb*

irritated *adjective* (1595)
: subjected to irritation; *especially* **:** roughened, reddened, or inflamed by an irritant ⟨*irritated* eyes⟩

ir·ri·ta·tion \ˌir-ə-ˈtā-shən\ *noun* (15th century)
1 a : the act of irritating **b :** something that irritates **c :** the state of being irritated
2 : a condition of irritability, soreness, roughness, or inflammation of a bodily part

ir·ri·ta·tive \ˈir-ə-ˌtā-tiv\ *adjective* (1644)
1 : serving to excite **:** IRRITATING
2 : accompanied with or produced by irritation ⟨*irritative* coughing⟩

ir·ro·ta·tion·al \ˌir-(r)ō-ˈtā-shnəl, -shə-n^{ə}l\ *adjective* (1875)
1 : not rotating or involving rotation ⟨an *irrotational* electric field⟩
2 : free of vortices ⟨*irrotational* flow⟩

ir·rupt \i-ˈrəpt\ *intransitive verb* [Latin *irruptus,* past participle of *irrumpere,* from *in-* + *rumpere* to break — more at REAVE] (1886)
1 : to rush in forcibly or violently
2 *of a natural population* **:** to undergo a sudden upsurge in numbers especially when natural ecological balances and checks are disturbed
3 : ERUPT 1c ⟨the crowd *irrupted* in a fervor of patriotism —*Time*⟩
— **ir·rup·tion** \-ˈrəp-shən\ *noun*

ir·rup·tive \-ˈrəp-tiv\ *adjective* (1593)
: irrupting or tending to irrupt
— **ir·rup·tive·ly** *adverb*

is [Middle English, from Old English; akin to Old High German *ist* is (from *sīn* to be), Latin *est* (from *esse* to be), Greek *esti* (from *einai* to be)] *present 3d singular of* BE, *dialect present 1st & 2d singular of* BE, *substandard present plural of* BE

is- *or* **iso-** *combining form* [Late Latin, from Greek, from *isos* equal]
1 : equal **:** homogeneous **:** uniform ⟨*is*entropic⟩
2 : isomeric ⟨*iso*cyanate⟩
3 : for or from different individuals of the same species ⟨*iso*agglutinin⟩

Isaac \ˈī-zik, -zək\ *noun* [Late Latin, from Hebrew *Yiṣḥāq*]
: the son of Abraham and father of Jacob according to the account in Genesis

Isa·iah \ī-ˈzā-ə, *chiefly British* -ˈzī-\ *noun* [Hebrew *Yĕsha'yāhū*]
1 : a major Hebrew prophet in Judah about 740 to 701 B.C.
2 : a prophetic book of canonical Jewish and Christian Scripture — see BIBLE table

Isa·ias \-əs\ *noun* [Late Latin, from Greek *Ēsaias,* from Hebrew *Yĕsha'yāhū*]
: ISAIAH

is·al·lo·bar \(ˌ)ī-ˈsa-lə-ˌbär\ *noun* [International Scientific Vocabulary *is-* + *all-* + *-bar* (as in *isobar*)] (1909)
: an imaginary line or a line on a chart connecting the places of equal change of atmospheric pressure within a specified time
— **is·al·lo·bar·ic** \ˌī-ˌsa-lə-ˈbär-ik, -ˈbar-\ *adjective*

is·chae·mia *chiefly British variant of* ISCHEMIA

is·che·mia \is-ˈkē-mē-ə\ *noun* [New Latin *ischaemia,* from *ischaemus* styptic, from Greek *ischaimos,* from *ischein* to restrain (akin to Greek *echein* to hold) + *haima* blood — more at SCHEME] (circa 1860)
: localized tissue anemia due to obstruction of the inflow of arterial blood
— **is·che·mic** \-mik\ *adjective*

is·chi·um \ˈis-kē-əm\ *noun, plural* **is·chia** \-ə\ [Latin, hip joint, from Greek *ischion*] (1646)
: the dorsal and posterior of the three principal bones composing either half of the pelvis
— **is·chi·al** \-əl\ *adjective*

-ise *verb suffix, chiefly British*
: -IZE

☆ SYNONYMS
Irritate, exasperate, nettle, provoke, rile, peeve mean to excite a feeling of anger or annoyance. IRRITATE implies an often gradual arousing of angry feelings that may range from mere impatience to rage ⟨constant nagging that *irritated* me greatly⟩. EXASPERATE suggests galling annoyance and the arousing of extreme impatience ⟨his *exasperating* habit of putting off needed decisions⟩. NETTLE suggests a sharp but passing annoyance or stinging ⟨your pompous attitude *nettled* several people⟩. PROVOKE implies an arousing of strong annoyance that may excite to action ⟨remarks made solely to *provoke* her⟩. RILE implies inducing an angry or resentful agitation ⟨the new work schedules *riled* the employees⟩. PEEVE suggests arousing fretful often petty or querulous irritation ⟨a toddler *peeved* at being refused a cookie⟩.

is·en·tro·pic \ˌī-sᵊn-ˈtrō-pik, -ˈträ-\ *adjective* (1873)
: of or relating to equal or constant entropy; *especially* **:** taking place without change of entropy
— **is·en·tro·pi·cal·ly** \-ˈtrō-pi-k(ə-)lē, -ˈträ-\ *adverb*

Iseult \is-ˈült, iz-\ *noun* [Old French *Isolt, Iseut*]
: ISOLDE

-ish *adjective suffix* [Middle English, from Old English *-isc;* akin to Old High German *-isc,* -ish, Greek *-iskos,* diminutive suffix]
1 : of, relating to, or being — chiefly in adjectives indicating nationality or ethnic group ⟨Finn*ish*⟩
2 a : characteristic of ⟨boy*ish*⟩ ⟨Pollyanna*ish*⟩ **b :** inclined or liable to ⟨book*ish*⟩ ⟨qualm*ish*⟩
3 a : having a touch or trace of ⟨purpl*ish*⟩ **:** somewhat ⟨dark*ish*⟩ **b :** having the approximate age of ⟨forty*ish*⟩

Ish·ma·el \ˈish-(ˌ)mā-əl, -mē-\ *noun* [Hebrew *Yishmāʽēl*]
1 : the outcast son of Abraham and Hagar according to the account in Genesis
2 : a social outcast

Ish·ma·el·ite \-ə-ˌlīt\ *noun* (14th century)
1 : a descendant of Ishmael
2 : ISHMAEL 2
— **Ish·ma·el·it·ish** \-ˌlī-tish\ *adjective*
— **Ish·ma·el·it·ism** \-ˌlī-ˌti-zəm\ *noun*

isin·glass \ˈī-zᵊn-ˌglas, ˈī-ziŋ-\ *noun* [probably by folk etymology from obsolete Dutch *huizenblas,* from Middle Dutch *huusblase,* from *huus* sturgeon + *blase* bladder] (1545)
1 : a semitransparent whitish very pure gelatin prepared from the air bladders of fishes (as sturgeons) and used especially as a clarifying agent and in jellies and glue
2 : MICA

Isis \ˈī-səs\ *noun* [Latin *Isid-, Isis,* from Greek, from Egyptian *ʼst*]
: an Egyptian nature goddess and wife and sister of Osiris

Is·lam \is-ˈläm, iz-, -ˈlam, ˈis-ˌ, ˈiz-ˌ\ *noun* [Arabic *islām* submission (to the will of God)] (1817)
1 : the religious faith of Muslims including belief in Allah as the sole deity and in Muhammad as his prophet
2 a : the civilization erected upon Islamic faith **b :** the group of modern nations in which Islam is the dominant religion
— **Is·lam·ic** \is-ˈlä-mik, iz-, -ˈla-\ *adjective*
— **Is·lam·ics** \-miks\ *noun plural but singular or plural in construction*

Islamic calendar *noun* (1974)
: a lunar calendar reckoned from the Hegira in A.D. 622 and organized in cycles of 30 years — see MONTH table

Islamic era *noun* (1975)
: the era used in Muslim countries for numbering Islamic calendar years since the Hegira

Is·lam·ism \is-ˈlä-ˌmi-zəm, iz-, -ˈla-; ˈiz-lə-\ *noun* (1747)
: the faith, doctrine, or cause of Islam
— **Is·lam·ist** \-mist\ *noun*

Is·lam·ize \ˈiz-lə-ˌmīz; is-ˈlä-ˌmīz, iz-, -ˈla-\ *transitive verb* **-ized; -iz·ing** (circa 1846)
: to make Islamic; *especially* **:** to convert to Islam
— **Is·lam·i·za·tion** \ˌiz-lə-mə-ˈzā-shən; is-ˌlä-mə-, iz-, -ˌla-\ *noun*

¹is·land \ˈī-lənd\ *noun* [alteration (influenced by Old French *isle*) of earlier *iland,* from Middle English, from Old English *īgland* (akin to Old Norse *eyland*), from *īg* island (akin to Old English *ēa* river, Latin *aqua* water) + *land* land] (before 12th century)
1 : a tract of land surrounded by water and smaller than a continent
2 : something resembling an island especially in its isolated or surrounded position: as **a :** a usually raised area within a thoroughfare, parking lot, or driveway used especially to separate or direct traffic **b :** a superstructure on the deck of a ship (as an aircraft carrier) **c :** a kitchen counter that is approachable from all sides
3 : an isolated group or area; *especially* **:** an isolated ethnological group

²island *transitive verb* (1661)
1 a : to make into or as if into an island **b :** to dot with or as if with islands
2 : ISOLATE

is·land·er \ˈī-lən-dər\ *noun* (circa 1550)
: a native or inhabitant of an island

is·land–hop \ˈī-lənd-ˌhäp\ *intransitive verb* (1944)
: to travel from island to island in a chain

island universe *noun* (1867)
: a galaxy other than the Milky Way

¹isle \ˈī(ə)l\ *noun* [Middle English, from Old French, from Latin *insula*] (13th century)
: ISLAND; *especially* **:** ISLET

²isle *transitive verb* **isled; isl·ing** (circa 1576)
1 : to make an isle of
2 : to place on or as if on an isle

is·let \ˈī-lət\ *noun* (1538)
: a little island

islet of Lang·er·hans \-ˈläŋ-ər-ˌhänz, -ˌhän(t)s\ [Paul *Langerhans* (died 1888) German physician] (1896)
: any of the groups of small slightly granular endocrine cells that form anastomosing trabeculae among the tubules and alveoli of the pancreas and secrete insulin and glucagon — called also *islet*

ism \ˈi-zəm\ *noun* [*-ism*] (1680)
: a distinctive doctrine, cause, or theory

-ism *noun suffix* [Middle English *-isme,* from Middle French & Latin; Middle French, partly from Latin *-isma* (from Greek) & partly from Latin *-ismus,* from Greek *-ismos;* Greek *-isma* & *-ismos,* from verbs in *-izein* -ize]
1 a : act **:** practice **:** process ⟨critic*ism*⟩ ⟨plagiar*ism*⟩ **b :** manner of action or behavior characteristic of a (specified) person or thing ⟨animal*ism*⟩ **c :** prejudice or discrimination on the basis of a (specified) attribute ⟨rac*ism*⟩ ⟨sex*ism*⟩
2 a : state **:** condition **:** property ⟨barbarian*ism*⟩ **b :** abnormal state or condition resulting from excess of a (specified) thing ⟨alcohol*ism*⟩ or marked by resemblance to (such) a person or thing ⟨mongol*ism*⟩
3 a : doctrine **:** theory **:** cult ⟨Buddh*ism*⟩ **b :** adherence to a system or a class of principles ⟨stoic*ism*⟩
4 : characteristic or peculiar feature or trait ⟨colloquial*ism*⟩

isn't \ˈi-zᵊnt, -ᵊn, *dialect also* ˈi-dᵊn(t) *or* ˈin(t)\ (1710)
: is not

iso- — see IS-

iso·ag·glu·ti·nin \ˌī-(ˌ)sō-ə-ˈglü-tᵊn-ən\ *noun* (circa 1903)
: an agglutinin from one individual that is specific for a cell type (as of red blood cells) possessed by some individuals of the same species

iso·al·lox·a·zine \-ə-ˈläk-sə-ˌzēn\ *noun* [*is-* + *all*antoic + *ox*alic + *azine*] (1936)
: a yellow solid $C_{10}H_6N_4O_2$ that is the precursor of various flavins (as riboflavin)

iso·an·ti·body \-ˈan-ti-ˌbä-dē\ *noun* (1919)
: ALLOANTIBODY

iso·an·ti·gen \-ˈan-ti-jən\ *noun* [International Scientific Vocabulary] (1936)
: ALLOANTIGEN
— **iso·an·ti·gen·ic** \ˌī-(ˌ)sō-ˌan-ti-ˈje-nik\ *adjective*

iso·bar \ˈī-sə-ˌbär\ *noun* [International Scientific Vocabulary *is-* + *-bar* (from Greek *baros* weight); akin to Greek *barys* heavy — more at GRIEVE] (circa 1864)
1 : an imaginary line or a line on a map or chart connecting or marking places on the surface of the earth where the height of the barometer reduced to sea level is the same either at a given time or for a certain period
2 : one of two or more atoms or elements having the same atomic weights or mass numbers but different atomic numbers

iso·bar·ic \ˌī-sə-ˈbär-ik, -ˈbar-\ *adjective* [International Scientific Vocabulary] (1878)
1 : of or relating to an isobar
2 : characterized by constant or equal pressure ⟨an *isobaric* process⟩

iso·bu·tane \ˌī-sō-ˈbyü-ˌtān\ *noun* [International Scientific Vocabulary] (1876)
: a gaseous branched-chain hydrocarbon $(CH_3)_3CH$ isomeric with normal butane that is used especially as a fuel

iso·bu·tyl·ene \-ˈbyü-tᵊl-ˌēn\ *noun* [International Scientific Vocabulary] (1872)
: a gaseous butylene C_4H_8 used especially in making butyl rubber and gasoline components

iso·ca·lo·ric \-kə-ˈlȯr-ik, -ˈlōr-, -ˈlär-; -ˈka-lə-rik\ *adjective* (1922)
: having similar caloric values ⟨*isocaloric* diets⟩

iso·car·box·az·id \-ˌkär-ˈbäk-sə-zəd\ *noun* [*is-* + *carb-* + *ox-* + hydr*azide*] (1959)
: an antidepressant drug $C_{12}H_{13}N_3O_2$

iso·chro·mo·some \ˌī-sə-ˈkrō-mə-ˌsōm, -ˌzōm\ *noun* (1939)
: a chromosome produced by transverse splitting of the centromere so that both arms are derived from the dyad on one side of the centromere of the parental chromosome and each arm has identical genes arranged in the same order counting away from the centromere

iso·chron \ˈī-sə-ˌkrän\ *or* **iso·chrone** \-ˌkrōn\ *noun* [International Scientific Vocabulary *is-* + *-chron* (from Greek *chronos* time)] (1881)
: an imaginary line or a line on a chart connecting points at which an event occurs simultaneously or which represents the same time or time difference

iso·chro·nal \ī-ˈsä-krə-nᵊl, ˌī-sə-ˈkrō-\ *adjective* [Greek *isochronos,* from *is-* + *chronos* time] (1680)
: uniform in time **:** having equal duration **:** recurring at regular intervals
— **iso·chro·nal·ly** \-nᵊl-ē\ *adverb*
— **iso·chro·nism** \ī-ˈsä-krə-ˌni-zəm, ˌī-sə-ˈkrō-\ *noun*

iso·chro·nous \ī-ˈsä-krə-nəs, ˌī-sə-ˈkrō-\ *adjective* [Greek *isochronos*] (1706)
: ISOCHRONAL
— **iso·chro·nous·ly** *adverb*

iso·cit·ric acid \ˌī-sə-ˈsi-trik-\ *noun* (1869)
: a crystalline isomer of citric acid that occurs especially as an intermediate stage in the Krebs cycle

iso·cy·a·nate \ˌī-sō-ˈsī-ə-ˌnāt, -nət\ *noun* [International Scientific Vocabulary] (1872)
: an ester of isomeric cyanic acid used especially in plastics and adhesives

iso·cy·clic \-ˈsī-klik, -ˈsi-\ *adjective* [International Scientific Vocabulary] (1900)
: having or being a ring composed of atoms of only one element; *especially* **:** CARBOCYCLIC

iso·di·a·met·ric \-ˌdī-ə-ˈme-trik\ *adjective* [International Scientific Vocabulary] (circa 1879)
: having equal diameters

iso·dose \ˈī-sə-ˌdōs\ *adjective* [International Scientific Vocabulary] (1922)
: of or relating to points or zones in a medium that receive equal doses of radiation

iso·elec·tric \ˌī-sō-i-ˈlek-trik\ *adjective* [International Scientific Vocabulary] (1877)
1 : having or representing zero difference of electric potential
2 : being the pH at which the electrolyte will not migrate in an electrical field ⟨the *isoelectric* point of a protein⟩

isoelectric focusing *noun* (1971)
: an electrophoretic technique for separating proteins by causing them to migrate under the

influence of an electric field through a medium (as a gel) having a pH gradient to locations with pH values corresponding to their isoelectric points

iso·elec·tron·ic \-i-ˌlek-ˈträ-nik\ *adjective* [International Scientific Vocabulary] (1926)
: having the same number of electrons or valence electrons
— **iso·elec·tron·i·cal·ly** \-ni-k(ə-)lē\ *adverb*

iso·en·zyme \-ˈen-ˌzīm\ *noun* (1960)
: any of two or more chemically distinct but functionally similar enzymes
— **iso·en·zy·mat·ic** \-ˌen-zə-ˈma-tik, -zī-\ *adjective*
— **iso·en·zy·mic** \-en-ˈzī-mik\ *adjective*

iso·ga·mete \-ˈga-ˌmēt *also* -gə-ˈmēt\ *noun* [International Scientific Vocabulary] (1891)
: a gamete indistinguishable in form or size or behavior from another gamete with which it can unite to form a zygote
— **iso·ga·met·ic** \-gə-ˈme-tik\ *adjective*

isog·a·mous \ī-ˈsä-gə-məs\ *adjective* [probably from (assumed) New Latin *isogamus,* from *is-* + *-gamus* -gamous] (1887)
: having or involving isogametes — compare HETEROGAMOUS
— **isog·a·my** \-mē\ *noun*

iso·ge·ne·ic \ˌī-sō-jə-ˈnē-ik, -ˈnā-\ *adjective* [alteration of *isogenic*] (1963)
: SYNGENEIC ⟨an *isogeneic* graft⟩

iso·gen·ic \-ˈje-nik\ *adjective* [*is-* + *gene* + [1]*-ic*] (circa 1931)
: characterized by essentially identical genes ⟨identical twins are *isogenic*⟩

iso·gloss \ˈī-sə-ˌgläs, -ˌglȯs\ *noun* [International Scientific Vocabulary *is-* + Greek *glōssa* language — more at GLOSS] (1925)
1 : a boundary line between places or regions that differ in a particular linguistic feature
2 : a line on a map representing an isogloss
— **iso·gloss·al** \ˌī-sə-ˈglä-səl, -ˈglȯ-\ *adjective*
— **iso·gloss·ic** \-ˈglä-sik, -ˈglȯ-\ *adjective*

iso·gon·ic \ˌī-sə-ˈgä-nik\ *adjective* [*isogony,* from *is-* + *-gony*] (1924)
: exhibiting equivalent relative growth of parts such that size relations remain constant
— **isog·o·ny** \ī-ˈsä-gə-nē\ *noun*

isogonic line *noun* (circa 1859)
: an imaginary line or a line on a map joining points on the earth's surface at which the magnetic declination is the same — called also *isogonal*

iso·graft \ˈī-sə-ˌgraft\ *noun* (1958)
: a homograft between genetically identical or nearly identical individuals
— **isograft** *transitive verb*

iso·gram \ˈī-sə-ˌgram\ *noun* (1889)
: ISOLINE

iso·hy·et \ˌī-sō-ˈhī-ət\ *noun* [International Scientific Vocabulary *is-* + Greek *hyetos* rain, from *hyein* to rain; probably akin to Tocharian B *swese* rain] (1899)
: a line on a map or chart connecting areas of equal rainfall
— **iso·hy·et·al** \-ə-təl\ *adjective*

iso·la·ble \ˈī-sə-lə-bəl *also* ˈi-\ *also* **iso·lat·able** \ˌī-sə-ˈlā-tə-bəl *also* ˌi-\ *adjective* (circa 1855)
: capable of being isolated

[1]**iso·late** \ˈī-sə-ˌlāt *also* ˈi-\ *transitive verb* **-lat·ed; -lat·ing** [back-formation from *isolated* set apart, from French *isolé,* from Italian *isolato,* from *isola* island, from Latin *insula*] (1807)
1 : to set apart from others; *also* **:** QUARANTINE
2 : to select from among others; *especially* **:** to separate from another substance so as to obtain pure or in a free state
3 : INSULATE
— **iso·la·tor** \-ˌlā-tər\ *noun*

[2]**iso·late** \-lət, -ˌlāt\ *adjective* (1819)
: being alone **:** SOLITARY, ISOLATED

[3]**iso·late** \-lət, -ˌlāt\ *noun* (1890)
1 : a product of isolating **:** an individual, population, or kind obtained by or resulting from selection or separation
2 : an individual socially withdrawn or removed from society

isolated *adjective* (1763)
1 : occurring alone or once **:** UNIQUE
2 : SPORADIC

iso·la·tion \ˌī-sə-ˈlā-shən *also* ˌi-\ *noun* (1833)
: the action of isolating **:** the condition of being isolated
synonym see SOLITUDE

iso·la·tion·ism \-shə-ˌni-zəm\ *noun* (1922)
: a policy of national isolation by abstention from alliances and other international political and economic relations
— **iso·la·tion·ist** \-sh(ə-)nist\ *noun or adjective*

Isol·de \i-ˈzōl-də, -ˈsōl-\ *noun* [German, from Old French *Isolt, Iseut*]
1 : an Irish princess married to King Mark of Cornwall and loved by Tristram
2 : the daughter of the King of Brittany and wife of Tristram

iso·leu·cine \ˌī-sō-ˈlü-ˌsēn\ *noun* [International Scientific Vocabulary] (1903)
: a crystalline essential amino acid $C_6H_{13}NO_2$ isomeric with leucine

iso·line \ˈī-(ˌ)sō-ˌlīn\ *noun* (1944)
: a line on a map or chart along which there is a constant value (as of temperature or rainfall)

iso·mer \ˈī-sə-mər\ *noun* [International Scientific Vocabulary, back-formation from *isomeric*] (1866)
1 : one of two or more compounds, radicals, or ions that contain the same number of atoms of the same elements but differ in structural arrangement and properties
2 : a nuclide isomeric with one or more others

isom·er·ase \ī-ˈsä-mə-ˌrās, -ˌrāz\ *noun* (1927)
: an enzyme that catalyzes the conversion of its substrate to an isomeric form

iso·mer·ic \ˌī-sə-ˈmer-ik\ *adjective* [German *isomerisch,* from Greek *isomerēs* equally divided, from *is-* + *meros* part — more at MERIT] (1838)
: of, relating to, or exhibiting isomerism

isom·er·ism \ī-ˈsä-mə-ˌri-zəm\ *noun* (1839)
1 : the relationship of two or more chemical species that are isomers
2 : the relation of two or more nuclides with the same mass numbers and atomic numbers but different energy states and rates of radioactive decay

isom·er·ize \ī-ˈsä-mə-ˌrīz\ *verb* **-ized; -iz·ing** (1891)
intransitive verb
: to become changed into an isomeric form
transitive verb
: to cause to isomerize
— **isom·er·i·za·tion** \-ˌsä-mə-rə-ˈzā-shən\ *noun*

iso·met·ric \ˌī-sə-ˈme-trik\ *adjective* (circa 1855)
1 : of, relating to, or characterized by equality of measure; *especially* **:** relating to or being a crystallographic system characterized by three equal axes at right angles
2 : of, relating to, involving, or being muscular contraction (as in isometrics) against resistance, without significant shortening of muscle fibers, and with marked increase in muscle tone — compare ISOTONIC
— **iso·met·ri·cal·ly** \-tri-k(ə-)lē\ *adverb*

isometric line *noun* (circa 1911)
1 : a line representing changes of pressure or temperature under conditions of constant volume
2 : a line (as a contour line) drawn on a map and indicating a true constant value throughout its extent

iso·met·rics \ˌī-sə-ˈme-triks\ *noun plural but singular or plural in construction* (1962)
: exercise or a system of exercises in which opposing muscles are so contracted that there is little shortening but great increase in tone of muscle fibers involved

isom·e·try \ī-ˈsä-mə-trē\ *noun, plural* **-tries** (1941)
: a mapping of a metric space onto another or onto itself so that the distance between any two points in the original space is the same as the distance between their images in the second space ⟨rotation and translation are *isometries* of the plane⟩

iso·mor·phic \ˌī-sə-ˈmȯr-fik\ *adjective* (1862)
1 a : being of identical or similar form, shape, or structure ⟨*isomorphic* crystals⟩ **b :** having sporophytic and gametophytic generations alike in size and shape
2 : related by an isomorphism ⟨*isomorphic* mathematical rings⟩
— **iso·mor·phi·cal·ly** \-fi-k(ə-)lē\ *adverb*

iso·mor·phism \ˌī-sə-ˈmȯr-ˌfi-zəm\ *noun* [International Scientific Vocabulary] (circa 1828)
1 : the quality or state of being isomorphic: as **a :** similarity in organisms of different ancestry resulting from convergence **b :** similarity of crystalline form between chemical compounds
2 : a one-to-one correspondence between two mathematical sets; *especially* **:** a homomorphism that is one-to-one — compare ENDOMORPHISM

iso·mor·phous \ˌī-sə-ˈmȯr-fəs\ *adjective* (circa 1828)
: ISOMORPHIC 1a
— **iso·morph** \ˈī-sə-ˌmȯrf\ *noun*

iso·ni·a·zid \ˌī-sə-ˈnī-ə-zəd\ *noun* [*is-* + *nicotinic acid* + *hydrazide*] (1952)
: a crystalline compound $C_6H_7N_3O$ used in treating tuberculosis

iso·oc·tane \ˌī-sō-ˈäk-ˌtān\ *noun* [International Scientific Vocabulary] (1909)
: an octane of branched-chain structure or a mixture of such octanes; *especially* **:** a flammable liquid octane used in determining the octane number of fuels

iso·pach \ˈī-sə-ˌpak\ *noun* [*is-* + *-pach* (from Greek *pachys* thick) — more at PACHYDERM] (circa 1918)
: an isoline that connects points of equal thickness of a geological stratum formation or group of formations

iso·phote \ˈī-sə-ˌfōt\ *noun* [International Scientific Vocabulary *is-* + *-phote* (from Greek *phōt-, phōs* light) — more at FANCY] (circa 1909)
: a curve on a chart joining points of equal light intensity from a given source
— **iso·phot·al** \ˌī-sə-ˈfō-təl\ *adjective*

iso·pi·es·tic \ˌī-sō-pē-ˈes-tik, -pī-\ *adjective* [*is-* + Greek *piestos,* verbal of *piezein* to press — more at PIEZO-] (1873)
: of, relating to, or marked by equal pressure

iso·pleth \ˈī-sə-ˌpleth\ *noun* [International Scientific Vocabulary *is-* + Greek *plēthos* quantity; akin to Greek *plēthein* to be full — more at FULL] (1908)
1 : an isoline on a graph showing the occurrence or frequency of a phenomenon as a function of two variables
2 : a line on a map connecting points at which a given variable has a specified constant value
— **iso·pleth·ic** \ˌī-sə-ˈple-thik\ *adjective*

iso·pod \ˈī-sə-ˌpäd\ *noun* [ultimately from Greek *is-* + *pod-, pous* foot — more at FOOT] (circa 1835)
: any of a large order (Isopoda) of small sessile-eyed crustaceans with the body composed of seven free thoracic segments each bearing a pair of similar legs
— **isopod** *adjective*

iso·pren·a·line \ˌī-sə-ˈpre-nᵊl-ən\ *noun* [probably from *isopropyl* + *adrenaline*] (1951)

: ISOPROTERENOL

iso·prene \ˈī-sə-ˌprēn\ *noun* [probably from *is-* + *propyl* + *-ene*] (1860)
: a flammable liquid unsaturated hydrocarbon C_5H_8 used especially in synthetic rubber

iso·pren·oid \ˌī-sə-ˈprē-ˌnȯid\ *adjective* (1940)
: relating to, containing, or being a branched-chain grouping characteristic of isoprene

iso·pro·pyl \ˌī-sə-ˈprō-pəl\ *noun, often attributive* [International Scientific Vocabulary] (1866)
: the alkyl radical isomeric with normal propyl — often used in combination

isopropyl alcohol *noun* (1872)
: a volatile flammable alcohol C_3H_8O used especially as a solvent and rubbing alcohol

iso·pro·ter·e·nol \ˌī-sə-prō-ˈter-ə-ˌnȯl, -ˌnōl\ *noun* [*isopropyl* + *arterenol* norepinephrine, from *Arterenol,* a trademark] (1957)
: a drug $C_{11}H_{17}NO_3$ used in the treatment of asthma

iso·pyc·nic \ˌī-sō-ˈpik-nik\ *adjective* [*is-* + Greek *pyknos* dense] (circa 1890)
1 : of, relating to, or marked by equal or constant density
2 : being or produced by a technique (as centrifugation) in which the components of a mixture are separated on the basis of differences in density

isos·ce·les \ī-ˈsäs-ˌlēz, -ˈsä-sə-\ *adjective* [Late Latin *isosceles* having two equal sides, from Greek *isoskelēs,* from *is-* + *skelos* leg; perhaps akin to Old English *sceol* wry] (1551)
1 *of a triangle* **:** having two equal sides — see TRIANGLE illustration
2 *of a trapezoid* **:** having the two nonparallel sides equal

is·os·mot·ic \ˌī-ˌsäz-ˈmä-tik, -ˌsäs-\ *adjective* [International Scientific Vocabulary] (1895)
: of, relating to, or exhibiting equal osmotic pressure ⟨*isosmotic* solutions⟩
— **is·os·mot·i·cal·ly** \-ˈmä-ti-k(ə-)lē\ *adverb*

iso·spin \ˈī-sə-ˌspin\ *noun* (1961)
: a quantum characteristic of a group of closely related subatomic particles (as a proton and a neutron) handled mathematically like ordinary spin with the possible orientations in a hypothetical space specifying the number of particles of differing electric charge comprising the group — called also *isotopic spin*

isos·ta·sy \ī-ˈsäs-tə-sē\ *noun* [International Scientific Vocabulary *is-* + Greek *-stasia* condition of standing, from *histanai* to cause to stand — more at STAND] (1889)
1 : general equilibrium in the earth's crust maintained by a yielding flow of rock material beneath the surface under gravitative stress
2 : the quality or state of being subjected to equal pressure from every side
— **iso·stat·ic** \ˌī-sə-ˈsta-tik\ *adjective*
— **iso·stat·i·cal·ly** \-ˈsta-ti-k(ə-)lē\ *adverb*

iso·tach \ˈī-sə-ˌtak\ *noun* [International Scientific Vocabulary *is-* + *-tach* (from Greek *tachys* quick)] (1947)
: a line on a map or chart connecting points of equal wind speed

iso·tac·tic \ˌī-sə-ˈtak-tik\ *adjective* (1955)
: having or relating to a stereochemical regularity of structure in the repeating units of a polymer — compare ATACTIC

iso·therm \ˈī-sə-ˌthərm\ *noun* [French *isotherme,* adjective] (1859)
1 : a line on a map or chart of the earth's surface connecting points having the same temperature at a given time or the same mean temperature for a given period
2 : a line on a chart representing changes of volume or pressure under conditions of constant temperature

iso·ther·mal \ˌī-sə-ˈthər-məl\ *adjective* [French *isotherme,* from *is-* + *-therme* (from Greek *thermos* hot) — more at THERM] (1826)
1 : of, relating to, or marked by equality of temperature
2 : of, relating to, or marked by changes of volume or pressure under conditions of constant temperature
— **iso·ther·mal·ly** \-mə-lē\ *adverb*

iso·ton·ic \ˌī-sə-ˈtä-nik\ *adjective* [International Scientific Vocabulary] (1891)
1 : of, relating to, or being muscular contraction in the absence of significant resistance, with marked shortening of muscle fibers, and without great increase in muscle tone — compare ISOMETRIC
2 : ISOSMOTIC — used of solutions
— **iso·ton·i·cal·ly** \-ni-k(ə-)lē\ *adverb*
— **iso·to·nic·i·ty** \-tō-ˈni-sə-tē\ *noun*

iso·tope \ˈī-sə-ˌtōp\ *noun* [*is-* + Greek *topos* place] (1913)
1 : any of two or more species of atoms of a chemical element with the same atomic number and nearly identical chemical behavior but with differing atomic mass or mass number and different physical properties
2 : NUCLIDE
— **iso·to·pic** \ˌī-sə-ˈtä-pik, -ˈtō-\ *adjective*
— **iso·to·pi·cal·ly** \-ˈtä-pi-k(ə-)lē, -ˈtō-\ *adverb*

isotopic spin *noun* (1937)
: ISOSPIN

iso·tro·pic \ˌī-sə-ˈtrō-pik, -ˈträ-\ *adjective* [International Scientific Vocabulary] (circa 1860)
: exhibiting properties (as velocity of light transmission) with the same values when measured along axes in all directions ⟨an *isotropic* crystal⟩
— **isot·ro·py** \ī-ˈsä-trə-pē\ *noun*

iso·zyme \ˈī-sə-ˌzīm\ *noun* (1959)
: ISOENZYME
— **iso·zy·mic** \ˌī-sə-ˈzī-mik\ *adjective*

Is·ra·el \ˈiz-rē-əl, -(ˌ)rā- *also* ˈis- *or* ˈiz-rəl\ *noun* [Middle English, from Old English, from Late Latin, from Greek *Israēl,* from Hebrew *Yiśrāʼēl*]
1 : JACOB 2
2 : the Jewish people
3 : a people chosen by God
— **Israel** *adjective*

¹Is·rae·li \iz-ˈrā-lē *also* ˌiz-rə-ˈā-lē\ *adjective* [New Hebrew *yiśrĕʼēlī,* from Hebrew, Israelite, noun & adjective, from *Yiśrāʼēl*] (1948)
: of or relating to the people or the republic of Israel

²Israeli *noun, plural* **Israelis** *also* **Israeli** (1948)
: a native or inhabitant of the republic of Israel

¹Is·ra·el·ite \ˈiz-r(ē-)ə-ˌlīt\ *noun* [Middle English, from Late Latin *Israelita,* from Greek *Israēlitēs,* from *Israēl*] (14th century)
: a descendant of the Hebrew patriarch Jacob; *specifically* **:** a native or inhabitant of the ancient northern kingdom of Israel

²Israelite *adjective* (1851)
: of or relating to Israel or to the Israelites

Is·sa·char \ˈi-sə-ˌkär\ *noun* [Late Latin, from Greek, from Hebrew *Yiśśākhār*]
: a son of Jacob and the traditional eponymous ancestor of one of the tribes of Israel

is·sei \(ˌ)ē-ˈsā, ˈē-ˌsā\ *noun, plural* **issei** *often capitalized* [Japanese, literally, first generation] (1937)
: a Japanese immigrant especially to the U.S.

is·su·able \ˈi-shü-ə-bəl\ *adjective* (circa 1570)
1 : open to contest, debate, or litigation
2 : authorized for issue ⟨bonds *issuable* under the merger terms⟩
3 : possible as a result or consequence
— **is·su·ably** \-blē\ *adverb*

is·su·ance \ˈi-shə-wən(t)s\ *noun* (1863)
: ISSUE 2, 9a

is·su·ant \-wənt\ *adjective* (1610)
1 *of a heraldic animal* **:** rising with only the upper part visible
2 *archaic* **:** coming forth **:** EMERGING

¹is·sue \ˈi-(ˌ)shü, *chiefly Southern* ˈi-shə, *chiefly British* ˈis-(ˌ)yü\ *noun* [Middle English, exit, proceeds, from Middle French, from Old French, from *issir* to come out, go out, from Latin *exire* to go out, from *ex-* + *ire* to go; akin to Gothic *iddja* he went, Greek *ienai* to go, Sanskrit *eti* he goes] (14th century)
1 *plural* **:** proceeds from a source of revenue (as an estate)
2 : the action of going, coming, or flowing out **:** EGRESS, EMERGENCE
3 : a means or place of going out **:** EXIT, OUTLET
4 : OFFSPRING, PROGENY ⟨died without *issue*⟩
5 a : a final outcome that usually constitutes a solution (as of a problem) or resolution (as of a difficulty) **b** *obsolete* **:** a final conclusion or decision about something arrived at after consideration **c** *archaic* **:** TERMINATION, END ⟨hope that his enterprise would have a prosperous *issue* —T. B. Macaulay⟩
6 a : a matter that is in dispute between two or more parties **b :** the point at which an unsettled matter is ready for a decision ⟨brought the matter to an *issue*⟩
7 : a discharge (as of blood) from the body
8 a : something coming forth from a specified source ⟨*issues* of a disordered imagination⟩ **b** *obsolete* **:** DEED
9 a : the act of publishing or officially giving out or making available ⟨the next *issue* of commemorative stamps⟩ ⟨*issue* of supplies by the quartermaster⟩ **b :** the thing or the whole quantity of things given out at one time ⟨read the latest *issue*⟩
— **is·sue·less** \ˈi-shü-ləs\ *adjective*
— **at issue 1 :** in a state of controversy **:** in disagreement **2** *also* **in issue :** under discussion or in dispute

²issue *verb* **is·sued; is·su·ing** (14th century)
intransitive verb
1 a : to go, come, or flow out **b :** to come forth **:** EMERGE **c :** to come to an issue of law or fact in pleading
2 : ACCRUE ⟨profits *issuing* from the sale of the stock⟩
3 : to descend from a specified parent or ancestor
4 : to be a consequence or final outcome **:** EMANATE, RESULT
5 : to appear or become available through being officially put forth or distributed
6 : EVENTUATE, TERMINATE
transitive verb
1 : to cause to come forth **:** DISCHARGE, EMIT
2 a : to put forth or distribute usually officially ⟨government *issued* a new airmail stamp⟩ ⟨*issue* orders⟩ **b :** to send out for sale or circulation **:** PUBLISH **c** *British* **:** PROVIDE 2b, SUPPLY
synonym see SPRING
— **is·su·er** *noun*

¹-ist *noun suffix* [Middle English *-iste,* from Old French & Latin; Old French *-iste,* from Latin *-ista, -istes,* from Greek *-istēs,* from verbs in *-izein* -ize]
1 a : one that performs a (specified) action ⟨cycl*ist*⟩ **:** one that makes or produces a (specified) thing ⟨novel*ist*⟩ **b :** one that plays a (specified) musical instrument ⟨harp*ist*⟩ **c :** one that operates a (specified) mechanical instrument or contrivance ⟨automobil*ist*⟩
2 : one that specializes in a (specified) art or science or skill ⟨geolog*ist*⟩ ⟨ventriloqu*ist*⟩
3 : one that adheres to or advocates a (specified) doctrine or system or code of behavior ⟨social*ist*⟩ ⟨royal*ist*⟩ ⟨hedon*ist*⟩ or that of a (specified) individual ⟨Calvin*ist*⟩ ⟨Darwin*ist*⟩

²-ist *adjective suffix*
: of, relating to, or characteristic of ⟨elit*ist*⟩

¹isth·mi·an \ˈis-mē-ən\ *noun* (1601)
1 : a native or inhabitant of an isthmus
2 *capitalized* **:** a native or inhabitant of the Isthmus of Panama

²isthmian *adjective* (1603)
: of, relating to, or situated in or near an isthmus: as **a** *often capitalized* **:** of or relating to the Isthmus of Corinth in Greece or the games

held there in ancient times **b** *often capitalized* **:** of or relating to the Isthmus of Panama connecting the North American and South American continents

isth·mic \'is-mik\ *adjective* (1585)
: of or relating to an anatomical isthmus

isth·mus \'is-məs\ *noun* [Latin, from Greek *isthmos*] (1555)
1 : a narrow strip of land connecting two larger land areas
2 : a narrow anatomical part or passage connecting two larger structures or cavities

is·tle \'ist-lē\ *noun* [American Spanish *ixtle,* from Nahuatl *īchtli*] (1883)
: a strong fiber (as for cordage or basketwork) from various tropical American plants (as of the genus *Agave*)

[1]**it** \'it, ət\ *pronoun* [Middle English, from Old English *hit* — more at HE] (before 12th century)
1 : that one — used as subject or direct object or indirect object of a verb or object of a preposition usually in reference to a lifeless thing ⟨took a quick look at the house and noticed *it* was very old⟩, a plant ⟨there is a rosebush near the fence and *it* is now blooming⟩, a person or animal whose sex is unknown or disregarded ⟨don't know who *it* is⟩, a group of individuals or things, or an abstract entity ⟨beauty is everywhere and *it* is a source of joy⟩; compare HE, ITS, SHE, THEY
2 — used as subject of an impersonal verb that expresses a condition or action without reference to an agent ⟨*it* is raining⟩
3 a — used as anticipatory subject or object of a verb ⟨*it* is necessary to repeat the whole thing⟩; often used to shift emphasis to a part of a statement other than the subject ⟨*it* was in this city that the treaty was signed⟩ **b** — used with many verbs as a direct object with little or no meaning ⟨footed *it* back to camp⟩
4 — used to refer to an explicit or implicit state of affairs or circumstances ⟨how is *it* going⟩
5 : a crucial or climactic point ⟨this is *it*⟩

[2]**it** \'it\ *noun* (1842)
: the player in a game who performs the principal action of the game (as trying to find others in hide-and-seek)

it·a·con·ic acid \ˌi-tə-'kä-nik-\ *noun* [International Scientific Vocabulary, anagram of *aconitic acid,* $C_3H_3(COOH)_3$, from *aconite*] (circa 1872)
: a crystalline dicarboxylic acid $C_5H_6O_4$ obtained usually by fermentation of sugars with molds (genus *Aspergillus*) and used as a monomer for polymers and polyesters

[1]**Ital·ian** \ə-'tal-yən, i- *also* ˌī-\ *noun* (14th century)
1 a : a native or inhabitant of Italy **b :** a person of Italian descent
2 : the Romance language of the Italians

[2]**Italian** *adjective* (15th century)
: of, relating to, or characteristic of Italy, the Italians, or Italian

ital·ian·ate \-yə-ˌnāt\ *transitive verb* **-at·ed; -at·ing** *often capitalized* (1553)
: ITALIANIZE

Ital·ian·ate \-nət, -ˌnāt\ *adjective* (1572)
: Italian in quality or characteristics

Italian dressing *noun* (circa 1902)
: a salad dressing flavored especially with garlic and oregano

Italian greyhound *noun* (1743)
: any of a breed of toy dogs resembling the standard greyhound in miniature

Italian greyhound

ital·ian·ise *often capitalized, British variant of* ITALIANIZE

Ital·ian·ism \ə-'tal-yə-ˌni-zəm, i- *also* ī-\ *noun* (1594)
1 a : a quality characteristic of Italy or the Italian people **b :** a characteristic feature of Italian occurring in another language
2 a : specialized interest in or emulation of Italian qualities or achievements **b :** promotion or love of Italian policies or ideals

ital·ian·ize \ə-'tal-yə-ˌnīz, i- *also* ī-\ *verb* **-ized; -iz·ing** *often capitalized* (circa 1611)
intransitive verb
: to act Italian; *specifically* **:** to follow the style or technique of recognized Italian painters
transitive verb
: to make Italian (as in appearance or behavior)
— **Ital·ian·i·za·tion** \-ˌtal-yə-nə-'zā-shən\ *noun*

Italian sandwich *noun* (circa 1953)
: SUBMARINE 2

Italian sonnet *noun* (1879)
: a sonnet consisting of an octave rhyming *abba abba* and a sestet rhyming in any of various patterns (as *cde cde* or *cdc dcd*) — called also *Petrarchan sonnet*

[1]**ital·ic** \ə-'ta-lik, i-, ī-\ *adjective* (1612)
1 a : of or relating to a type style with characters that slant upward to the right (as in "*these words are italic*") — compare ROMAN **b :** of or relating to a style of slanted cursive handwriting developed in the 15th and 16th centuries
2 *capitalized* **:** of or relating to ancient Italy, its peoples, or their Indo-European languages

[2]**italic** *noun* (1676)
1 : an italic character or type
2 *capitalized* **:** a branch of the Indo-European language family that includes Latin, Oscan, and Umbrian — see INDO-EUROPEAN LANGUAGES table

ital·i·cise *British variant of* ITALICIZE

ital·i·cize \ə-'ta-lə-ˌsīz, i-, ī-\ *transitive verb* **-cized; -ciz·ing** (1795)
1 : to print in italics or underscore with a single line
2 : EMPHASIZE ⟨the microphone *italicizes* every curdled top note —P. G. Davis⟩
— **ital·i·ci·za·tion** \-ˌta-lə-sə-'zā-shən\ *noun*

Ita·lo- \i-'ta-lō *also* 'i-tᵊl-ō\ *combining form*
1 : Italian ⟨*Italo*phile⟩
2 : Italian and ⟨*Italo*-Austrian⟩

Ita·lo·phile \i-'ta-lə-ˌfīl\ *adjective* (circa 1902)
: friendly to or favoring what is Italian
— **Italophile** *noun*

[1]**itch** \'ich\ *verb* [Middle English *icchen,* from Old English *giccan;* akin to Old High German *juechen* to itch] (before 12th century)
intransitive verb
1 a : to have an itch ⟨her arm *itched*⟩ **b :** to produce an itchy sensation ⟨long underwear that *itches*⟩
2 : to have a restless desire or hankering for something ⟨were *itching* to go outside⟩
transitive verb
1 : to cause to itch
2 : VEX, IRRITATE

[2]**itch** *noun* (before 12th century)
1 a : an uneasy irritating sensation in the upper surface of the skin usually held to result from mild stimulation of pain receptors **b :** a skin disorder accompanied by such a sensation; *especially* **:** a contagious eruption caused by a mite (*Sarcoptes scabiei*) that burrows in the skin and causes intense itching
2 a : a restless usually constant often compulsive desire ⟨an *itch* to travel⟩ **b :** LUST, PRURIENCE
— **itch·i·ness** \'i-chē-nəs\ *noun*
— **itchy** \'i-chē\ *adjective*

it'd \'i-təd, ˌid\ (1859)
: it had **:** it would

[1]**-ite** *noun suffix* [Middle English, from Old French & Latin; Old French, from Latin *-ita, -ites,* from Greek *-itēs*]
1 a : native **:** resident ⟨Brooklyn*ite*⟩ **b :** descendant ⟨Ephraim*ite*⟩ **c :** adherent **:** follower ⟨Jacob*ite*⟩ ⟨Pusey*ite*⟩
2 a (1) **:** product ⟨metabol*ite*⟩ (2) **:** commercially manufactured product ⟨ebon*ite*⟩ **b :** -ITOL ⟨mann*ite*⟩
3 [New Latin *-ites,* from Latin] **:** fossil ⟨ammon*ite*⟩
4 : mineral ⟨erythr*ite*⟩ **:** rock ⟨anorthos*ite*⟩
5 [French, from Latin *-ita, -ites*] **:** segment or constituent part of a body or of a bodily part ⟨som*ite*⟩ ⟨dendr*ite*⟩

[2]**-ite** *noun suffix* [French, alteration of *-ate* -ate, from New Latin *-atum*]
: salt or ester of an acid with a name ending in *-ous* ⟨nitr*ite*⟩

[1]**item** \'ī-ˌtem, 'ī-təm\ *adverb* [Middle English, from Latin, from *ita* thus] (14th century)
: and in addition **:** ALSO — used to introduce each article in a list or enumeration

[2]**item** \'ī-təm\ *noun* (1571)
1 *obsolete* **:** WARNING, HINT
2 : a distinct part in an enumeration, account, or series **:** ARTICLE
3 : an object of attention, concern, or interest
4 : a separate piece of news or information
5 : a couple in a romantic or sexual relationship ☆

[3]**item** \'ī-təm\ *transitive verb* (1601)
1 *archaic* **:** COMPUTE, RECKON
2 *archaic* **:** to set down the particular details of

item·ise *British variant of* ITEMIZE

item·i·za·tion \ˌī-tə-mə-'zā-shən\ *noun* (1894)
: the act of itemizing; *also* **:** an itemized list

item·ize \'ī-tə-ˌmīz\ *transitive verb* **-ized; -iz·ing** (1857)
: to set down in detail or by particulars **:** LIST ⟨*itemized* all expenses⟩

it·er·ance \'i-tə-rən(t)s\ *noun* (1604)
: REPETITION

it·er·ant \-rənt\ *adjective* (1626)
: marked by repetition, reiteration, or recurrence ⟨*iterant* echoes⟩

it·er·ate \'i-tə-ˌrāt\ *transitive verb* **-at·ed; -at·ing** [Latin *iteratus,* past participle of *iterare,* from *iterum* again; akin to Latin *is* he, that, *ita* thus, Sanskrit *itara* the other, *iti* thus] (1533)
: to say or do again or again and again **:** REITERATE

it·er·a·tion \ˌi-tə-'rā-shən\ *noun* (15th century)
1 : the action or a process of iterating or repeating: as **a :** a procedure in which repetition of a sequence of operations yields results successively closer to a desired result **b :** the repetition of a sequence of computer instructions a specified number of times or until a condition is met — compare RECURSION
2 : one execution of a sequence of operations or instructions in an iteration

it·er·a·tive \'i-tə-ˌrā-tiv, -rə-\ *adjective* (15th century)
: involving repetition: as **a :** expressing repetition of a verbal action **b :** relating to or being iteration of an operation or procedure
— **it·er·a·tive·ly** *adverb*

☆ SYNONYMS
Item, detail, particular mean one of the distinct parts of a whole. ITEM applies to each thing specified separately in a list or in a group of things that might be listed or enumerated ⟨every *item* on the list⟩. DETAIL applies to one of the small component parts of a larger whole such as a task, building, painting, narration, or process ⟨leave the *details* to others⟩. PARTICULAR stresses the smallness, singleness, and especially the concreteness of a detail or item ⟨a description that included few *particulars*⟩.

ithy·phal·lic \ˌi-thi-ˈfa-lik\ *adjective* [Late Latin *ithyphallicus,* from Greek *ithyphallikos,* from *ithyphallos* erect phallus, from *ithys* straight + *phallos* phallus] (1795)
1 : of or relating to the phallus carried in procession in ancient festivals of Bacchus
2 a : having an erect penis — usually used of figures in an art representation **b :** OBSCENE, LEWD

itin·er·an·cy \ī-ˈti-nə-rən(t)-sē\ *noun* (1789)
1 : a system (as in the Methodist Church) of rotating ministers who itinerate
2 a : the act of itinerating **b :** the state of being itinerant

itin·er·ant \-rənt\ *adjective* [Late Latin *itinerant-, itinerans,* present participle of *itinerari* to journey, from Latin *itiner-, iter* journey, way; akin to Hittite *itar* way, Latin *ire* to go — more at ISSUE] (circa 1576)
: traveling from place to place; *especially* **:** covering a circuit ⟨*itinerant* preacher⟩
— **itinerant** *noun*
— **itin·er·ant·ly** *adverb*

itin·er·ary \ī-ˈti-nə-ˌrer-ē, ə-, *chiefly British* ÷-ˈti-nə-rē\ *noun, plural* **-ar·ies** (15th century)
1 : the route of a journey or tour or the proposed outline of one
2 : a travel diary
3 : a traveler's guidebook
— **itinerary** *adjective*

itin·er·ate \ī-ˈti-nə-ˌrāt, ə-\ *intransitive verb* **-at·ed; -at·ing** (1775)
: to travel a preaching or judicial circuit
— **itin·er·a·tion** \-ˌti-nə-ˈrā-shən\ *noun*

-itious *adjective suffix* [Latin *-icius, -itius*]
: of, relating to, or having the characteristics of ⟨excrement*itious*⟩

-itis *noun suffix, plural* **-itises** *also* **-itides** *or* **-ites** [New Latin, from Latin & Greek; Latin, from Greek, from feminine of *-itēs* -ite]
1 : disease or inflammation ⟨bronch*itis*⟩
2 *plural usually -itises* **:** condition likened to a disease — chiefly in nonce formations ⟨television*itis*⟩

it'll \ˈi-tᵊl\ (1824)
: it will **:** it shall

-itol *noun suffix* [International Scientific Vocabulary 1*-ite* + 1*-ol*]
: polyhydroxy alcohol usually related to a sugar ⟨mann*itol*⟩

its \ˈits, əts\ *adjective* (1598)
: of or relating to it or itself especially as possessor, agent, or object of an action ⟨going to *its* kennel⟩ ⟨a child proud of *its* first drawings⟩ ⟨*its* final enactment into law⟩

it's \ˈits, əts\ (circa 1555)
: it is **:** it has

it·self \it-ˈself, ət-, *Southern also* -ˈsef\ *pronoun* (before 12th century)
1 : that identical one — compare IT 1; used reflexively ⟨watched the cat giving *itself* a bath⟩, for emphasis ⟨the letter *itself* was missing⟩, or in absolute constructions ⟨*itself* a splendid specimen of classic art, it has been exhibited throughout the world⟩
2 : its normal, healthy, or sane condition
— **in itself :** in its own nature **:** INTRINSICALLY ⟨was not *in itself* bad⟩

it·ty–bit·ty \ˈi-tē-ˈbi-tē\ *or* **it·sy–bit·sy** \ˈit-sē-ˈbit-sē\ *adjective* [probably from baby talk for *little bit*] (1938)
: extremely small **:** TINY

-ity *noun suffix, plural* **-ities** [Middle English *-ite,* from Old French or Latin; Old French *-ité,* from Latin *-itat-, -itas,* from *-i-* (stem vowel of adjectives) + *-tat-, -tas* -ity; akin to Greek *-tēt-, -tēs* -ity]
: quality **:** state **:** degree ⟨alkalin*ity*⟩ ⟨theatrical*ity*⟩

IUD \ˌī-(ˌ)yü-ˈdē\ *noun* (1965)
: INTRAUTERINE DEVICE

-ium *noun suffix*
1 [New Latin, from Latin, ending of some neuter nouns] **a :** a chemical element ⟨sod*ium*⟩ **b :** cation ⟨tetrazol*ium*⟩
2 *plural* **-iums** *or* **-ia** [New Latin, from Latin, from Greek *-ion*] **:** small one **:** mass — especially in botanical terms ⟨pollin*ium*⟩

IV \ˌī-ˈvē\ *noun, plural* **IVs** [*intravenous*] (circa 1955)
: an apparatus used to administer an intravenous injection or feeding; *also* **:** such an injection or feeding

-ive *adjective suffix* [Middle English *-if, -ive,* from Middle French & Latin; Middle French *-if,* from Latin *-ivus*]
: that performs or tends toward an (indicated) action ⟨amus*ive*⟩

I've \ˈīv\ (1586)
: I have

iver·mec·tin \ˌī-vər-ˈmek-tən\ *noun* [perhaps from *di-* + New Latin a*vermitilis* (specific epithet of *Streptomyces avermitilis,* bacterium from which it is derived) + English *-ect-* (of unknown origin) + 1*-in*] (1981)
: a drug mixture of two structurally similar semisynthetic lactones that is used in veterinary medicine as an anthelmintic, acaricide, and insecticide and in human medicine to treat onchocerciasis

ivied \ˈī-vēd\ *adjective* (circa 1771)
1 : overgrown with ivy
2 : ACADEMIC

ivo·ry \ˈīv-rē, ˈī-və-rē\ *noun, plural* **-ries** [Middle English *ivorie,* from Old French *ivoire,* from Latin *eboreus* of ivory, from *ebor-, ebur* ivory, from Egyptian *ꜣb, ꜣbw* elephant, ivory] (13th century)
1 a : the hard creamy-white modified dentine that composes the tusks of a tusked mammal (as an elephant, walrus, or narwhal) **b :** a tusk that yields ivory
2 : a variable color averaging a pale yellow
3 *slang* **:** TOOTH
4 : something (as a piano key) made of ivory or of a similar substance
— **ivory** *adjective*

ivo·ry·bill \-ˌbil\ *noun* (1787)
: IVORY-BILLED WOODPECKER

ivo·ry–billed woodpecker \-ˈbild-\ *noun* (1811)
: a very large black-and-white woodpecker (*Campephilus principalis*) of the southeastern U.S. and Cuba that has a showy red crest in the male and is presumed extinct in the U.S.

ivory-billed woodpecker

ivory black *noun* (1634)
: a fine black pigment made by calcining ivory

ivory nut *noun* (circa 1847)
: the nutlike seed of a South American palm (*Phytelephas macrocarpa*) containing a very hard endosperm used for carving and turning — compare VEGETABLE IVORY

ivory tower *noun* [translation of French *tour d'ivoire*] (1911)
1 : an impractical often escapist attitude marked by aloof lack of concern with or interest in practical matters or urgent problems
2 : a secluded place that affords the means of treating practical issues with an impractical often escapist attitude; *especially* **:** a place of learning
— **ivory–tower** *adjective*
— **ivo·ry–tow·er·ish** \-ˈtau̇(-ə)rish\ *adjective*

ivo·ry–tow·ered \-ˈtau̇(-ə)rd\ *adjective* (1937)
: divorced from reality and practical matters ⟨an *ivory-towered* recluse⟩

1ivy \ˈī-vē\ *noun, plural* **ivies** [Middle English, from Old English *īfig;* akin to Old High German *ebah* ivy] (before 12th century)
1 : a widely cultivated ornamental climbing or prostrate or sometimes shrubby chiefly Eurasian vine (*Hedera helix*) of the ginseng family with evergreen leaves, small yellowish flowers, and black berries
2 : POISON IVY
3 *often capitalized* **:** an Ivy League college

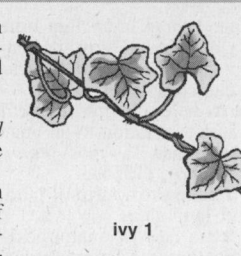

ivy 1

2ivy *adjective* [from the prevalence of ivy-covered buildings on the campuses of older U.S. colleges] (1933)
1 : ACADEMIC
2 : IVY LEAGUE

Ivy League *adjective* (1939)
1 : of, relating to, or characteristic of a group of long-established eastern U.S. colleges widely regarded as high in scholastic and social prestige
2 : of, relating to, or characteristic of the students of Ivy League colleges

Ivy Leaguer *noun* (1943)
: a student at or a graduate of an Ivy League college

iwis \ē-ˈwis, ī-\ *adverb* [Middle English, from Old English *gewis* certain; akin to Old High German *giwisso* certainly, Old English *witan* to know — more at WIT] (12th century)
archaic **:** SURELY

Ix·i·on \ik-ˈsī-ən\ *noun* [Latin, from Greek *Ixiōn*]
: a Thessalian king bound by Zeus to a burning wheel in Tartarus for attempting to seduce Hera

ix·o·did \ˈik-sə-ˌdid, ik-ˈsō-dəd\ *adjective* [ultimately from Greek *ixōdēs* sticky, from *ixos* birdlime] (circa 1909)
: of, relating to, or being a typical tick (family Ixodidae)
— **ixodid** *noun*

Iyar \ˈē-ˌyär\ *noun* [Hebrew *Iyyār*] (1737)
: the 8th month of the civil year or the 2d month of the ecclesiastical year in the Jewish calendar — see MONTH table

-i·za·tion \ə-ˈzā-shən *also* (ˌ)ī-ˈzā-\ *noun suffix*
: action, process, or result of making ⟨social*ization*⟩

-ize *verb suffix* [Middle English *-isen,* from Old French *-iser,* from Late Latin *-izare,* from Greek *-izein*]
1 a (1) **:** cause to be or conform to or resemble ⟨system*ize*⟩ ⟨American*ize*⟩ **:** cause to be formed into ⟨union*ize*⟩ (2) **:** subject to a (specified) action ⟨plagiar*ize*⟩ (3) **:** impregnate or treat or combine with ⟨alumin*ize*⟩ **b :** treat like ⟨idol*ize*⟩ **c :** treat according to the method of ⟨bowdler*ize*⟩
2 a : become **:** become like ⟨crystall*ize*⟩ **b :** be productive in or of ⟨hypothes*ize*⟩ **:** engage in a (specified) activity ⟨philosoph*ize*⟩ **c :** adopt or spread the manner of activity or the teaching of ⟨Platon*ize*⟩ ▫

iz·zard \ˈi-zərd\ *noun* [alteration of earlier *ezod, ezed,* probably from Middle French *et zede* and Z] (circa 1726)
chiefly dialect **:** the letter *z*

▫ USAGE
-ize The suffix *-ize* has been productive in English since the time of Thomas Nashe (1567–1601), who claimed credit for introducing it into English to remedy the surplus of monosyllabic words. Almost any noun or adjective can be made into a verb by adding *-ize* ⟨hospital*ize*⟩ ⟨familiar*ize*⟩; many technical terms are coined this way ⟨oxid*ize*⟩ as well as verbs of ethnic derivation ⟨American*ize*⟩ and verbs derived from proper names ⟨bowdler*ize*⟩ ⟨mesmer*ize*⟩. Nashe noted in 1591 that his coinages in *-ize* were being complained about, and to this day new words in *-ize* ⟨final*ize*⟩ ⟨priorit*ize*⟩ are sure to draw critical fire.

J is the tenth letter of the English alphabet. In Latin i *was used for either a vowel sound or for a consonant sound like the* y *in English* yet. *The letter form* j *was developed as a lengthened form of* i *during the Middle Ages and was used in certain positions, such as at the beginning of a word as an equivalent for the shorter form. In those positions it had exactly the same value as* i. *Because the initial* i *was usually consonantal, the* j *that so often took its place gradually became differentiated from* i *in sound as well as form. It was not, however, until the 17th century that the distinction of* j *as a consonant and* i *as a vowel was fully established and the capital form* J *introduced. In English, what is now the regular sound of* j, *as in* jet, *dates from the 11th century. It came with the Normans from the continent, where it was the consonant sound of* i *and* j *in French.*

j \ˈjā\ *noun, plural* **j's** *or* **js** \ˈjāz\ *often capitalized, often attributive* (15th century)
1 a : the 10th letter of the English alphabet **b :** a speech counterpart of orthographic *j*
2 a : a graphic device for reproducing the letter *j* **b :** a unit vector parallel to the y-axis
3 : one designated *j* especially as the 10th in order or class
4 : something shaped like the letter J

[1]jab \ˈjab\ *verb* **jabbed; jab·bing** [alteration of *job* to strike] (1827)
transitive verb
1 a : to pierce with or as if with a sharp object **:** STAB **b :** to poke quickly or abruptly **:** THRUST
2 : to strike with a short straight blow
intransitive verb
1 : to make quick or abrupt thrusts with a sharp object
2 : to strike a person with a short straight blow

[2]jab *noun* (1872)
: an act of jabbing; *especially* **:** a short straight boxing punch delivered with the leading hand

[1]jab·ber \ˈja-bər\ *verb* **jab·bered; jab·ber·ing** \ˈja-b(ə-)riŋ\ [Middle English *jaberen,* of imitative origin] (15th century)
intransitive verb
: to talk rapidly, indistinctly, or unintelligibly
transitive verb
: to speak rapidly or indistinctly
— jab·ber·er \ˈja-bər-ər\ *noun*

[2]jabber *noun* (1727)
: GIBBERISH, CHATTER

jab·ber·wocky \ˈja-bər-ˌwä-kē\ *noun* [*Jabberwocky,* nonsense poem by Lewis Carroll] (1908)
: meaningless speech or writing ◆

jab·i·ru \ˌzha-bə-ˈrü\ *noun* [Portuguese, from Tupi & Guarani *jabirú*] (1774)
: any of several large tropical storks

jab·o·ran·di \ˌzha-bə-ˌran-ˈdē, -ˈran-dē\ *noun* [Portuguese, from Tupi *yaborandí*] (circa 1875)
: the dried leaves of two South American shrubs (*Pilocarpus jaborandi* and *P. microphyllus*) of the rue family that are a source of pilocarpine

ja·bot \zha-ˈbō, ˈja-ˌbō\ *noun* [French] (1823)
1 : a fall of lace or cloth attached to the front of a neckband and worn especially by men in the 18th century
2 : a pleated frill of cloth or lace attached down the center front of a woman's blouse or dress

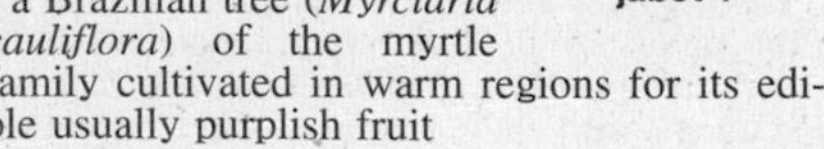
jabot 1

ja·bo·ti·ca·ba \zhə-ˌbü-ti-ˈkä-bə\ *noun* [Portuguese, from Tupi] (1824)
: a Brazilian tree (*Myrciaria cauliflora*) of the myrtle family cultivated in warm regions for its edible usually purplish fruit

ja·cal \hə-ˈkäl\ *noun, plural* **ja·ca·les** \-ˈkä-(ˌ)lās\ *also* **ja·cals** [Mexican Spanish, from Nahuatl *xahcalli*] (1838)
: a hut in Mexico and southwestern U.S. with a thatched roof and walls made of upright poles or sticks covered and chinked with mud or clay

jac·a·mar \ˈzha-kə-ˌmär\ *noun* [French, from Tupi *jacamá-ciri*] (1825)
: any of a family (Galbulidae) of usually iridescent green or bronze insectivorous birds of Central and South American tropical forests

ja·ca·na \jə-ˈkä-nə, ˌzha-sᵊn-ˈäⁿ\ *noun* [Portuguese *jaçanã,* from Tupi & Guarani] (circa 1753)
: any of a family (Jacanidae) of long-legged and long-toed tropical wading birds that frequent coastal freshwater marshes and ponds

jac·a·ran·da \ˌja-kə-ˈran-də\ *noun* [New Latin, from Portuguese *jacarandá* a tree of this genus, from Tupi] (circa 1753)
: any of a genus (*Jacaranda*) of pinnate-leaved tropical American trees of the trumpet-creeper family with showy usually blue flowers in panicles

ja·cinth \ˈjā-sᵊn(t)th, ˈja-\ *noun* [Middle English *iacinct,* from Old French *jacinthe,* from Latin *hyacinthus,* a flowering plant, a gem] (13th century)
1 : HYACINTH
2 : a gem more nearly orange in color than a hyacinth

[1]jack \ˈjak\ *noun* [Middle English *Jacke,* familiar term of address to a social inferior, nickname for *Johan* John] (1548)
1 a : MAN — usually used as an intensive in such phrases as *every man jack* **b** *often capitalized* **:** SAILOR **c** (1) **:** SERVANT, LABORER (2) **:** LUMBERJACK
2 : any of various usually mechanical devices: as **a :** a device for turning a spit **b :** a usually portable mechanism or device for exerting pressure or lifting a heavy body a short distance
3 : something that supports or holds in position: as **a :** an iron bar at a topgallant masthead to support a royal mast and spread the royal shrouds **b :** a wooden brace fastened behind a scenic unit in a stage set to prop it up
4 a : any of several fishes; *especially* **:** any of various carangids **b :** a male donkey **c :** JACKRABBIT **d :** any of several birds (as a jackdaw)
5 a : a small white target ball in lawn bowling **b :** a small national flag flown by a ship **c** (1) *plural but singular in construction* **:** a game played with a set of small objects that are tossed, caught, and moved in various figures (2) **:** a small 6-pointed metal object used in the game of jacks
6 a : a playing card carrying the figure of a soldier or servant and ranking usually below the queen **b :** JACKPOT 1a(2)
7 *slang* **:** MONEY
8 : a female fitting in an electric circuit used with a plug to make a connection with another circuit
9 a : APPLEJACK **b :** BRANDY
10 : JACKKNIFE 2
11 : MONTEREY JACK

[2]jack (circa 1841)
intransitive verb
: to hunt or fish at night with a jacklight
transitive verb
1 : to hunt or fish for at night with a jacklight
2 a : to move or lift by or as if by a jack **b :** to raise the level of — usually used with *up* ⟨*jack* up the price⟩ **c :** to take to task
— jack·er *noun*

jack·al \ˈja-kəl *also* -ˌkȯl\ *noun* [Turkish *çakal,* from Persian *shagāl,* of Indo-Aryan origin; akin to Sanskrit *sr̥gāla* jackal] (1603)
1 : any of several small omnivorous canids (as *Canis aureus*) of Africa and Asia having large ears, long legs, and bushy tails
2 a : a person who performs routine or menial tasks for another **b :** a person who serves or collaborates with another especially in the commission of base acts

Jack–a–Lent \ˈja-kə-ˌlent\ *noun* [[1]*jack* + *a* (of) + *Lent*] (1598)
1 : a small stuffed puppet set up to be pelted for fun in Lent
2 : a simple or insignificant person

jack·a·napes \ˈja-kə-ˌnāps\ *noun* [Middle English *Jack Napis,* nickname for William de la Pole (died 1450) duke of Suffolk] (1526)
1 : MONKEY, APE
2 a : an impudent or conceited fellow **b :** a saucy or mischievous child

jack·ass \ˈjak-ˌas\ *noun* (1727)
1 : DONKEY; *especially* **:** a male donkey
2 : a stupid person **:** FOOL

jack·ass·ery \ˈjak-ˌa-sə-rē, -ˌas-rē\ *noun, plural* **-er·ies** (1833)
: foolish action or behavior

jack bean *noun* (1885)
: a bushy annual tropical American legume (genus *Canavalia*); *especially* **:** a plant (*C. ensiformis*) grown especially for forage

jack·boot \ˈjak-ˌbüt\ *noun* (1686)
1 : a heavy military boot made of glossy black leather extending above the knee and worn especially during the 17th and 18th centuries
2 : the spirit or policy of militarism or totalitarianism
3 : a laceless military boot reaching to the calf

jack·boot·ed \-ˌbü-təd\ *adjective* (1846)
1 : wearing jackboots
2 : ruthlessly and violently oppressive ⟨*jackbooted* force⟩

◇ WORD HISTORY

jabberwocky Lewis Carroll's poem "Jabberwocky," published in *Through the Looking-Glass,* the second of his Alice-in-Wonderland books, is a wonderful example of nonsense verse. In the opening lines "Twas brillig and the slithy toves/Did gyre and gimble in the wabe," the words appear to be English, and we can tell which are nouns, adjectives, and verbs, but as evocative as they are they mean nothing. Lewis Carroll simply created them for the purposes of the poem. The word *jabberwocky,* a mock derivative of *jabberwock,* the fantastic monster in the poem, has become a generic term for words (or sometimes acts) that seem as though they should make sense but do not. In one bound *jabberwocky* has leapt from nonsense to metalanguage (language we use to talk about language).

\ə\ **abut** \ᵊ\ **kitten** \ər\ **further** \a\ **ash** \ā\ **ace** \ä\ **mop, mar** \au̇\ **out** \ch\ **chin** \e\ **bet** \ē\ **easy** \g\ **go** \i\ **hit** \ī\ **ice** \j\ **job** \ŋ\ **sing** \ō\ **go** \ȯ\ **law** \ȯi\ **boy** \th\ **thin** \t͟h\ **the** \ü\ **loot** \u̇\ **foot** \y\ **yet** \zh\ **vision** *see also* Guide to Pronunciation

jack cheese *noun, often J capitalized* [alteration of David *Jacks,* 19th century California landowner] (1912)
: MONTEREY JACK

jack crevalle *noun* (1948)
: a carangid fish (*Caranx hippos*) that is an important food fish along the west coast of Florida

jack·daw \ˈjak-ˌdȯ\ *noun* (1543)
1 : a common black and gray bird (*Corvus monedula*) of Eurasia and northern Africa that is related to but smaller than the common crow
2 : GRACKLE 1

¹jack·et \ˈja-kət\ *noun* [Middle English *jaket,* from Middle French *jaquet,* diminutive of *jaque* short jacket, from *jacque* peasant, from the name *Jacques* James] (15th century)
1 a : a garment for the upper body usually having a front opening, collar, lapels, sleeves, and pockets **b :** something worn or fastened around the body but not for use as clothing
2 a (1) **:** the natural covering of an animal (2) **:** the fur or wool of a mammal **b :** the skin of a potato
3 : an outer covering or casing: as **a** (1) **:** a thermally nonconducting cover (2) **:** a covering that encloses an intermediate space through which a temperature-controlling fluid circulates (3) **:** a tough cold-worked metal casing that forms the outer shell of a built-up bullet **b** (1) **:** a wrapper or open envelope for a document (2) **:** an envelope for enclosing registered mail during delivery from one post office to another **c** (1) **:** a detachable protective cover for a book (2) **:** a paper or paperboard envelope for a phonograph record
— jack·et·less \-ləs\ *adjective*

²jacket *transitive verb* (1856)
: to put a jacket on **:** enclose in or with a jacket

Jack Frost *noun* (1826)
: frost or frosty weather personified

jack·fruit \ˈjak-ˌfrüt\ *noun* [Portuguese *jaca* jackfruit, from Malayalam *cakka*] (1830)
: a large tropical Asian tree (*Artocarpus heterophyllus*) related to the breadfruit that yields a fine-grained yellow wood and immense fruits which contain an edible pulp and nutritious seeds; *also* **:** its fruit

jack·ham·mer \ˈjak-ˌha-mər\ *noun* (1916)
1 : a pneumatically operated percussive rock-drilling tool usually held in the hands
2 : a device in which a tool (as a chisel for breaking up pavements) is driven percussively by compressed air
— jackhammer *verb*

jack–in–the–box \ˈja-kən-t͟hə-ˌbäks\ *noun, plural* **jack–in–the–box·es** *or* **jacks–in–the–box** (1702)
: a toy consisting of a small box out of which a figure (as of a clown's head) springs when the lid is raised

jack–in–the–pul·pit \ˌja-kən-t͟hə-ˈpu̇l-ˌpit, -pət *also* -ˈpəl-\ *noun, plural* **jack–in–the–pulpits** *or* **jacks–in–the–pulpit** (1837)
: an American spring-flowering woodland herb (*Arisaema triphyllum* synonym *A. atrorubens*) of the arum family having an upright club-shaped spadix arched over by a green and purple spathe

jack-in-the-pulpit

¹jack·knife \ˈjak-ˌnīf\ *noun* (1711)
1 : a large strong pocketknife
2 : a dive executed headfirst in which the diver bends from the waist and touches the ankles while holding the knees unbent and then straightens out

²jackknife (1806)
transitive verb
1 : to cut with a jackknife
2 : to cause to double up like a jackknife
intransitive verb
1 : to double up like a jackknife
2 : to turn and form an angle of 90 degrees or less with each other — used especially of a tractor-trailer combination

jack·leg \ˈjak-ˌleg, -ˌlāg\ *adjective* [¹*jack* + *-leg* (as in *blackleg*)] (1850)
1 a : lacking skill or training **:** AMATEUR ⟨a *jackleg* carpenter⟩ **b :** characterized by unscrupulousness, dishonesty, or lack of professional standards ⟨a *jackleg* lawyer⟩
2 : designed as a temporary expedient **:** MAKESHIFT
— jackleg *noun*

jack·light \-ˌlīt\ *noun* (circa 1841)
: a light used especially in hunting or fishing at night

jack mackerel *noun* (1882)
: a California carangid food fish (*Trachurus symmetricus*) that is iridescent green or bluish above and silvery below

jack–of–all–trades \ˌja-kə-ˌvȯl-ˈtrādz\ *noun, plural* **jacks–of–all–trades** (1618)
: a person who can do passable work at various tasks **:** a handy versatile person

jack off *verb* [probably alteration of *jerk off*] (1959)
: MASTURBATE — usually considered vulgar

jack–o'–lan·tern \ˈja-kə-ˌlan-tərn\ *noun* (1750)
1 : IGNIS FATUUS
2 : a lantern made of a pumpkin cut to look like a human face
3 : a large orangish gill fungus (*Omphalotus olearius* synonym *Clitocybe illudens*) that is poisonous and luminescent

jack pine *noun* (1883)
: a slender pine (*Pinus banksiana*) of northern North America that has two stout needles in each fascicle and wood used especially for pulpwood

jack plane *noun* (circa 1816)
: a medium-sized general-purpose plane used in carpentry and joinery

jack·pot \ˈjak-ˌpät\ *noun* (1881)
1 a (1) **:** a hand or game of draw poker in which a pair of jacks or better is required to open (2) **:** a large pot (as in poker) formed by the accumulation of stakes from previous play **b** (1) **:** a combination on a slot machine that wins a top prize or all the coins available for paying out (2) **:** the sum so won **c :** a large fund of money or other reward formed by the accumulation of unwon prizes
2 : an impressive often unexpected success or reward
3 *chiefly West* **:** a tight spot **:** JAM

jack·rab·bit \-ˌra-bət\ *noun* [¹*jack* (jackass) + *rabbit;* from its long ears] (1863)
: any of several large hares (genus *Lepus*) of western North America having very long ears and long hind legs

Jack Russell terrier \ˈjak-ˈrə-səl-\ *noun* [*Jack* (John) *Russell* (died 1883) English clergyman & dog breeder] (1961)
: any of a breed of small terriers having a usually white coat with brown, black, or tricolor markings — called also *Jack Russell*

jack salmon *noun* (1871)
1 : WALLEYE 3
2 : GRILSE

jack·screw \ˈjak-ˌskrü\ *noun* (1769)
: SCREW JACK

jack·smelt \-ˌsmelt\ *noun* (1949)
: a large silversides (*Atherinopsis californiensis*) of the Pacific coast of North America that is the chief commercial smelt of the California markets

Jack·son Day \ˈjak-sən-\ *noun* [Andrew *Jackson;* from his defense of New Orleans] (1885)
: January 8 celebrated as a legal holiday in Louisiana to commemorate the successful defense of New Orleans in 1815

jack stand *noun* (1968)
: a stand whose height may be adjusted and fixed by a pin and which is used to support an automobile that has been raised by a jack

jack·stay \-ˌstā\ *noun* (circa 1840)
1 : an iron rod, wooden bar, or wire rope along a yard of a ship to which the sails are fastened
2 : a support of wood, iron, or rope running up a mast on which the parrel of a yard travels

jack·straw \ˈjak-ˌstrȯ\ *noun* (1801)
1 *plural but singular in construction* **:** a game in which a set of straws or thin strips is let fall in a heap with each player in turn trying to remove one at a time without disturbing the rest
2 : one of the pieces used in the game jackstraws

jack–tar \-ˈtär\ *noun, often capitalized* (1781)
: SAILOR

jack–up \ˈjak-ˌəp\ *noun* (1965)
: a drilling rig used in offshore drilling whose drilling platform is a barge from which legs are lowered to the bottom when over the drill site and which is raised above the water and supported on the legs to conduct drilling operations

Ja·cob \ˈjā-kəb\ *noun* [Late Latin, from Greek *Iacōb,* from Hebrew *Ya'ăqōbh*]
1 : a son of Isaac and Rebekah, the twin brother of Esau, and heir of God's promise of blessing to Abraham
2 : the ancient Hebrew nation

Jac·o·be·an \ˌja-kə-ˈbē-ən\ *adjective* [New Latin *Jacobaeus,* from *Jacobus* James] (1844)
: of, relating to, or characteristic of James I of England or his age
— Jacobean *noun*

jacobean lily *noun, often J capitalized* [Late Latin *Jacobus* (Saint James)] (circa 1774)
: a Mexican bulbous herb (*Sprekelia formosissima*) of the amaryllis family cultivated for its bright red solitary flower

Ja·co·bi·an \jə-ˈkō-bē-ən, yä-\ *noun* [K. G. J. *Jacobi* (died 1851) German mathematician] (1881)
: a determinant defined for a finite number of functions of the same number of variables in which each row consists of the first partial derivatives of the same function with respect to each of the variables

Jac·o·bin \ˈja-kə-bən\ *noun* [Middle English, from Middle French, from Medieval Latin *Jacobinus,* from Late Latin *Jacobus* (Saint James); from the location of the first Dominican convent in the street of Saint James, Paris] (14th century)
1 : DOMINICAN
2 [French, from *Jacobin* Dominican; from the group's founding in the Dominican convent in Paris] **:** a member of an extremist or radical political group; *especially* **:** a member of such a group advocating egalitarian democracy and engaging in terrorist activities during the French Revolution of 1789
— Jac·o·bin·ic \ˌja-kə-ˈbi-nik\ *or* **Jac·o·bin·i·cal** \-ni-kəl\ *adjective*
— Jac·o·bin·ism \ˈja-kə-bə-ˌni-zəm\ *noun*

¹Jac·o·bite \ˈja-kə-ˌbīt\ *noun* [Middle English, from Medieval Latin *Jacobita,* from *Jacobus* Baradaeus (Jacob Baradai) (died 578) Syrian monk] (15th century)
: a member of any of various Monophysite Eastern churches; *especially* **:** a member of the Monophysite Syrian church

²Jacobite *noun* [*Jacobus* (James II)] (1689)
: a partisan of James II of England or of the Stuarts after the revolution of 1688
— Jac·o·bit·i·cal \ˌja-kə-ˈbi-ti-kəl\ *adjective*
— Jac·o·bit·ism \ˈja-kə-ˌbī-ˌti-zəm\ *noun*

Ja·cob's ladder \ˈjā-kəbz-\ *noun* [from the ladder seen in a dream by Jacob in Genesis 28:12] (1733)
1 : any of a genus (*Polemonium*) of herbs of the phlox family that have pinnate leaves, an herbaceous calyx, a bell-shaped corolla with stamens with their filaments bent downward,

and a several-seeded capsule; *especially* **:** a perennial (*P. caeruleum*) of European origin with bright blue or occasionally white flowers
2 : a marine ladder of rope or chain with wooden or iron rungs

Ja·cob·son's organ \'jā-kəb-sənz-\ *noun* [Ludvig L. *Jacobson* (died 1843) Danish anatomist] (1885)
: either of a pair of small pits that are situated one on either side of the nasal septum (as in higher vertebrates) or in the buccal cavity and that are rudimentary in humans but developed in reptiles, amphibians, and some mammals as chemoreceptors

Ja·co·bus \jə-'kō-bəs\ *noun* [*Jacobus* (James I), during whose reign unites were coined] (1612)
: UNITE

jac·o·net \'ja-kə-,net\ *noun* [modification of Urdu *jagannāthī*] (1769)
: a lightweight cotton cloth used for clothing and bandages

jac·quard \'ja-,kärd\ *noun, often capitalized, often attributive* [Joseph *Jacquard*] (1841)
1 a : the control mechanism of a Jacquard loom **b :** JACQUARD LOOM
2 : a fabric of intricate variegated weave or pattern

Jacquard loom *noun* (1851)
: a loom designed to weave fabrics of intricate design whose control mechanism makes use of cards with holes punched in them

jac·que·rie \,zhä-kə-'rē, ,zha-\ *noun, often capitalized* [Middle French, from the French peasant revolt in 1358, from *jacque* peasant — more at JACKET] (1523)
: a peasants' revolt

jac·ti·ta·tion \,jak-tə-'tā-shən\ *noun* [Late Latin *jactitation-, jactitatio,* from *jactitare,* frequentative of Latin *jactare* to throw — more at JET] (1665)
: a tossing to and fro or jerking and twitching of the body

Ja·cuz·zi \jə-'kü-zē\ *trademark*
— used for a whirlpool bath and a recreational bathing tub or pool

¹jade \'jād\ *noun* [Middle English] (14th century)
1 : a broken-down, vicious, or worthless horse
2 a : a disreputable woman **b :** a flirtatious girl

²jade *verb* **jad·ed; jad·ing** (1598)
transitive verb
1 a : to wear out by overwork or abuse **b :** to tire or dull through repetition or excess
2 *obsolete* **:** to make ridiculous
intransitive verb
: to become weary or dulled
synonym see TIRE

³jade *noun* [French, from obsolete Spanish (*piedra de la*) *ijada,* literally, loin stone, ultimately from Latin *ilia,* plural, flanks; from the belief that jade cures renal colic] (1657)
1 : either of two tough compact typically green gemstones that take a high polish: **a :** JADEITE **b :** NEPHRITE
2 : a sculpture or artifact of jade
3 : JADE GREEN ◆

jaded *adjective* (1600)
1 : fatigued by overwork **:** EXHAUSTED
2 : dulled by experience or by surfeit
— **jad·ed·ly** *adverb*
— **jad·ed·ness** *noun*

jade green *noun* (1892)
: a light bluish green

jade·ite \'jā-,dīt\ *noun* [French] (1865)
: a monoclinic mineral that is a silicate of sodium and aluminum and is a jade
— **ja·dit·ic** \jā-'di-tik\ *adjective*

jade plant *noun* (1944)
: any of several succulent plants (genus *Crassula*) of the orpine family cultivated as foliage plants

jae·ger \'yā-gər\ *noun* [German *Jäger*] (1809)
1 a : HUNTER, HUNTSMAN **b :** one attending a person of rank or wealth and wearing hunter's costume
2 : any of several large dark-colored birds (genus *Stercorarius* of the family Stercorariidae) of northern seas that are strong fliers and that tend to harass weaker birds until they drop or disgorge their prey

¹jag \'jag\ *verb* **jagged; jag·ging** [Middle English *jaggen*] (15th century)
transitive verb
1 *chiefly dialect* **:** PRICK, STAB
2 : to cut indentations into; *also* **:** to form teeth on (a saw) by cutting indentations
intransitive verb
1 : PRICK, THRUST
2 : to move in jerks
— **jag·ger** *noun*

²jag *noun* (1578)
: a sharp projecting part **:** BARB

³jag *noun* [origin unknown] (1597)
1 : a small load
2 a : a state or feeling of exhilaration or intoxication usually induced by liquor **b :** SPREE ⟨a crying *jag*⟩

jag·ged \'ja-gəd\ *adjective* (1577)
1 : having a sharply uneven edge or surface
2 : having a harsh, rough, or irregular quality
— **jag·ged·ly** *adverb*
— **jag·ged·ness** *noun*

jag·gery \'ja-gə-rē\ *noun* [Portuguese *jágara,* probably from Malayalam *chakkara* sugar] (1598)
: an unrefined brown sugar made from palm sap

jag·gy \'ja-gē\ *adjective* (1717)
: JAGGED, NOTCHED

jag·uar \'ja-,gwär, -gyə-,wär, -gwər, *especially British* 'ja-gyə-wər\ *noun* [Spanish *yaguar* & Portuguese *jaguar,* from Guarani *yaguara* & Tupi *jaguara*] (1604)
: a large cat (*Panthera onca* synonym *Felis onca*) chiefly of Central and South America that is larger and stockier than the leopard and is brownish yellow or buff with black spots

jaguar

jag·ua·run·di \,zha-gwə-'rən-dē\ *also* **jag·ua·ron·di** \-rän-\ *noun* [American Spanish, from Guarani *yaguarundi*] (circa 1885)
: a slender long-tailed short-legged black, gray, or reddish wildcat (*Felis yagouaroundi*) chiefly of Central and South America

Jah \'jä\ *noun* [Hebrew *Yāh*] (1975)
: the Supreme Being of Rastafarianism

Jah·veh \'yä-(,)vā\ *variant of* YAHWEH

jai alai \'hī-,lī, ,hī-ə-'lī\ *noun* [Spanish, from Basque, from *jai* festival + *alai* merry] (1903)
: a court game somewhat like handball played usually by two or four players with a ball and a long curved wicker basket strapped to the wrist

¹jail \'jā(ə)l\ *noun* [Middle English *jaiole,* from Old French, from Late Latin *caveola,* diminutive of Latin *cavea* cage — more at CAGE] (13th century)
1 : a place of confinement for persons held in lawful custody; *specifically* **:** such a place under the jurisdiction of a local government (as a county) for the confinement of persons awaiting trial or those convicted of minor crimes — compare PRISON
2 : confinement in a jail

²jail *transitive verb* (1604)
: to confine in or as if in a jail

jail·bait \'jā(ə)l-,bāt\ *noun* (1930)
: a girl under the age of consent with whom sexual intercourse is unlawful and constitutes statutory rape

jail·bird \-,bərd\ *noun* (1603)
: a person confined in jail; *specifically* **:** an habitual criminal

jail·break \-,brāk\ *noun* (1910)
: a forcible escape from jail

jail·er *or* **jail·or** \'jā-lər\ *noun* (13th century)
1 : a keeper of a jail
2 : one that restricts another's liberty as if by imprisonment

jail·house \'jā(ə)l-,haùs\ *noun* (1812)
: JAIL 1

jailhouse lawyer *noun* (1969)
: a prison inmate self-taught in the law who tries to gain release through legal maneuvers or who advises fellow inmates on their legal problems

Jain \'jīn\ *or* **Jai·na** \'jī-nə\ *noun* [Hindi *Jain,* from Sanskrit *Jaina*] (1805)
: an adherent of Jainism

Jain·ism \'jī-,ni-zəm\ *noun* (1858)
: a religion of India originating in the 6th century B.C. and teaching liberation of the soul by right knowledge, right faith, and right conduct

jake \'jāk\ *adjective* [origin unknown] (1914)
slang **:** ALL RIGHT, FINE

jake leg \'jāk-,leg, -,lāg\ *noun* [*jake* grain alcohol flavored with an alcoholic extract of ginger] (1932)
: a paralysis caused by drinking improperly distilled or contaminated liquor

jakes \'jāks\ *noun plural but singular or plural in construction* [perhaps from French *Jacques* James] (1538)
: PRIVY 1

jal·ap \'ja-ləp, 'jä-\ *noun* [French & Spanish; French *jalap,* from Spanish *jalapa,* from *Jalapa,* Mexico] (1644)
1 a : the dried tuberous root of a Mexican plant (*Ipomoea purga* synonym *Exogonium purga*) of the morning-glory family; *also* **:** a powdered purgative drug prepared from it that contains resinous glycosides **b :** the root or derived drug of plants related to the one supplying jalap
2 : a plant yielding jalap

ja·la·pe·ño *also* **ja·la·pe·no** \,hä-lə-'pā-(,)nyō, ,ha-, -'pā-(,)nō, -'pē-(,)nō\ *noun, plural* **-ños** *also* **-nos** [Mexican Spanish, from *jalapeño,* adjective, of Jalapa] (1939)
: a small plump dark green Mexican hot pepper — called also *jalapeño pepper*

ja·lopy \jə-'lä-pē\ *noun, plural* **ja·lop·ies** [origin unknown] (1928)
: a dilapidated old vehicle (as an automobile)

jal·ou·sie \'ja-lə-sē\ *noun* [French, literally, jealousy, from Old French *jelous* jealous] (1766)

◇ WORD HISTORY

jade Gemstones were once believed to have magical and medicinal properties. Jade was supposed to be especially effective in combating kidney disorders, and it owes its name to this belief. The green gemstone was known as *piedra de la ijada* or "loin stone" to the 16th century Spanish who brought it home from the New World. Not only in Spain but throughout western Europe jade became popular both as an ornament and as a cure or preventive for internal disorders. English *jade* was borrowed from French, in which *jade* had developed from earlier *ejade,* through construing *l'ejade,* the form of the word when preceded by the definite article, as *le jade; ejade,* in turn, was borrowed from Spanish *ijada,* taken to be the name of the stone.

1 : a blind with adjustable horizontal slats for admitting light and air while excluding direct sun and rain
2 : a window made of adjustable glass louvers that control ventilation

[1]jam \'jam\ *verb* **jammed; jam·ming** [origin unknown] (1706)
intransitive verb
1 a : to become blocked or wedged **b :** to become unworkable through the jamming of a movable part
2 : to force one's way into a restricted space
3 : to take part in a jam session
transitive verb
1 a : to press into a close or tight position ⟨*jam* his hat on⟩ **b** (1) **:** to cause to become wedged so as to be unworkable ⟨*jam* the typewriter keys⟩ (2) **:** to make unworkable by jamming **c :** to block passage of **:** OBSTRUCT **d :** to fill often to excess **:** PACK ⟨the crowd *jammed* the theater⟩
2 : to push forcibly; *especially* **:** to apply (brakes) suddenly and forcibly — used with *on*
3 : CRUSH, BRUISE
4 a : to make unintelligible by sending out interfering signals or messages **b :** to make (as a radar apparatus) ineffective by jamming signals or by causing reflection of radar waves
5 : to pitch inside to (a batter)
— jam·mer \'ja-mər\ *noun*

[2]jam *noun* (1805)
1 a : an act or instance of jamming **b :** a crowded mass that impedes or blocks ⟨a traffic *jam*⟩
2 a : the quality or state of being jammed **b :** the pressure or congestion of a crowd **:** CRUSH
3 : a difficult state of affairs **:** FIX ⟨got into a *jam*⟩
4 : JAM SESSION
5 : DUNK SHOT

[3]jam *noun* [probably from [1]*jam*] (circa 1736)
: a food made by boiling fruit and sugar to a thick consistency
— jam·my \'ja-mē\ *adjective*

Jamaica rum *noun* (circa 1900)
: a heavy-bodied rum made by slow fermentation and marked by a pungent bouquet

jamb \'jam\ *noun* [Middle English *jambe,* from Middle French, literally, leg, from Late Latin *gamba* — more at GAMBIT] (14th century)
1 : an upright piece or surface forming the side of an opening (as for a door, window, or fireplace)
2 : a projecting columnar part or mass

jam·ba·laya \ˌjəm-bə-'lī-ə\ *noun* [Louisiana French, from Provençal *jambalaia*] (1872)
1 : rice cooked usually with ham, sausage, chicken, shrimp, or oysters and seasoned with herbs
2 : a mixture of diverse elements ⟨curious *jambalayas* of competing elements —Neil Hickey⟩

jam·beau \'jam-(ˌ)bō\ *noun, plural* **jam·beaux** \-(ˌ)bōz\ [Middle English, from (assumed) Anglo-French, from Middle French *jambe*] (14th century)
: a piece of medieval armor for the leg below the knee — see ARMOR illustration

jam·bo·ree \ˌjam-bə-'rē\ *noun* [origin unknown] (1864)
1 : a noisy or unrestrained carouse
2 a : a large festive gathering **b :** a national or international camping assembly of Boy Scouts
3 : a long mixed program of entertainment

James \'jāmz\ *noun* [Middle English, from Middle French, from (assumed) Vulgar Latin *Jacomus,* alteration of Late Latin *Jacobus, Jacob* Jacob]
1 : an apostle, son of Zebedee, and brother of the apostle John according to the Gospel accounts
2 : an apostle and son of Alphaeus according to the Gospel accounts — called also *James the Less*
3 : a brother of Jesus traditionally held to be the author of the New Testament Epistle of James
4 : a moral lecture addressed to early Christians and included as a book in the New Testament — see BIBLE table

jam·mies \'ja-mēz\ *noun plural* [by shortening & alteration] (1973)
: PAJAMAS 2

jam–pack \'jam-'pak\ *transitive verb* (1924)
: to pack tightly or to excess

Jams \'jamz\ *trademark*
— used for knee-length loose-fitting swim trunks

jam session *noun* [[2]*jam*] (1933)
: an often impromptu performance by a group especially of jazz musicians that is characterized by improvisation

Jam·shid *or* **Jam·shyd** \jam-'shēd\ *noun* [Persian *Jamshīd*]
: an early legendary king of Persia who reigned for 700 years

jam–up \'jam-ˌəp\ *noun* (1941)
: JAM 1

Jane Doe \'jān-'dō\ *noun* (1936)
: a female party to legal proceedings whose true name is unknown

Jane·ite \'jā-ˌnīt\ *noun* (1896)
: an enthusiastic admirer of Jane Austen's writings

[1]jan·gle \'jaŋ-gəl\ *verb* **jan·gled; jan·gling** \-g(ə-)liŋ\ [Middle English, from Middle French *jangler,* of Germanic origin; akin to Middle Dutch *jangelen* to grumble] (14th century)
intransitive verb
1 : to talk idly
2 : to quarrel verbally
3 : to make a harsh or discordant often ringing sound ⟨keys *jangling* in my pocket⟩
transitive verb
1 : to utter or sound in a discordant, babbling, or chattering way
2 a : to cause to sound harshly or inharmoniously **b :** to excite to tense irritation ⟨*jangled* nerves⟩
— jan·gler \-g(ə-)lər\ *noun*

[2]jangle *noun* (14th century)
1 : idle talk
2 : noisy quarreling
3 : a discordant often ringing sound **:** DISCORD

jan·gly \'jaŋ-g(ə-)lē\ *adjective* (1892)
: marked by jangling **:** having a jangling quality ⟨*jangly* earrings⟩ ⟨*jangly* guitar music⟩

jan·is·sary *or* **jan·i·zary** \'ja-nə-ˌser-ē, -ˌzer-\ *noun, plural* **-sar·ies** *or* **-zar·ies** [Italian *gianizzero,* from Turkish *yeniçeri,* from *yeni* new + *çeri* soldier] (1529)
1 *often capitalized* **:** a soldier of an elite corps of Turkish troops organized in the 14th century and abolished in 1826
2 : a member of a group of loyal or subservient troops, officials, or supporters

jan·i·tor \'ja-nə-tər\ *noun* [Latin, from *janus* arch, gate] (1629)
1 : DOORKEEPER
2 : one who keeps the premises of a building (as an apartment or office) clean, tends the heating system, and makes minor repairs
— jan·i·to·ri·al \ˌja-nə-'tōr-ē-əl, -'tȯr-\ *adjective*

Jan·sen·ism \'jan(t)-sə-ˌni-zəm\ *noun* [French *jansénisme,* from Cornelis *Jansen*] (circa 1656)
1 : a system of doctrine based on moral determinism, defended by various reformist factions among 17th and 18th century western European Roman Catholic clergy, religious, and scholars, and condemned as heretical by papal authority
2 : a puritanical attitude (as toward sex)
— Jan·sen·ist \-nist\ *noun*
— Jan·sen·is·tic \ˌjan(t)-sə-'nis-tik\ *adjective*

Jan·u·ary \'jan-yə-ˌwer-ē\ *noun, plural* **-ar·ies** *or* **-ar·ys** [Middle English *Januarie,* from Latin *Januarius,* 1st month of the ancient Roman year, from *Janus*] (14th century)
: the 1st month of the Gregorian calendar

Ja·nus \'jā-nəs\ *noun* [Latin]
: a Roman god that is identified with doors, gates, and all beginnings and that is represented artistically with two opposite faces

Janus–faced \-ˌfāst\ *adjective* (1682)
: having two contrasting aspects; *especially* **:** DUPLICITOUS, TWO-FACED

Janus green *noun* [probably from *Janus,* a trademark] (1898)
: a basic azine dye used especially as a biological stain (as for mitochondria)

Jap \'jap\ *noun or adjective* (1886)
: JAPANESE — usually used disparagingly

JAP \'jap\ *noun* (circa 1973)
: JEWISH AMERICAN PRINCESS — usually used disparagingly

[1]ja·pan \jə-'pan\ *adjective* (1673)
: of, relating to, or originating in Japan **:** of a kind or style characteristic of Japanese workmanship

[2]japan *noun* (1688)
1 a : any of several varnishes yielding a hard brilliant finish **b :** a hard dark coating containing asphalt and a drier that is used especially on metal and fixed by heating — called also *japan black*
2 : work (as lacquer ware) finished and decorated in the Japanese manner

[3]japan *transitive verb* **ja·panned; ja·pan·ning** (1688)
1 : to cover with or as if with a coat of japan
2 : to give a high gloss to
— ja·pan·ner *noun*

Jap·a·nese \ˌja-pə-'nēz, -'nēs\ *noun, plural* **Japanese** (1604)
1 a : a native or inhabitant of Japan **b :** a person of Japanese descent
2 : the language of the Japanese
— Japanese *adjective*

Japanese an·drom·e·da \-ˌan-'drä-mə-də\ *noun* [New Latin *Andromeda,* genus of plants, from Latin *Andromeda,* Ethiopian princess, from Greek *Andromedē*] (1948)
: a shrubby evergreen Asian heath (*Pieris japonica*) with glossy leaves and drooping clusters of usually whitish flowers

Japanese beetle *noun* (1900)
: a small metallic green and brown scarab beetle (*Popillia japonica*) that has been introduced into America from Japan and as a grub feeds on the roots of grasses and decaying vegetation and as an adult eats foliage and fruits

Japanese cedar *noun* (circa 1880)
: a large evergreen tree (*Cryptomeria japonica*) grown especially in China and Japan for its valuable soft wood

Japanese iris *noun* (1883)
: any of various beardless garden irises (especially *Iris kaempferi*) with very large showy flowers

Japanese lacquer *noun* (1876)
: LACQUER 1b

Japanese maple *noun* (1898)
: a maple (*Acer palmatum*) of Japan, China, and Korea with purple flowers and usually deeply parted leaves that is widely cultivated as a shrub or small tree

Japanese millet *noun* (1900)
: a coarse annual grass (*Echinochloa frumentacea*) cultivated especially in Asia for its edible seeds

Japanese plum *noun* (1893)
: a tree (*Prunus salicina*) native to China and cultivated in Japan that bears Japanese plums; *also* **:** the large usually yellow to light red fruit of a Japanese plum

Japanese quail *noun* (1963)
: a quail (*Coturnix japonica* synonym *C. coturnix japonica*) from China and Japan that is often used in laboratory research

Japanese quince *noun* (1900)
: a hardy Chinese ornamental shrub (*Chaenomeles speciosa* synonym *C. lagenaria*) of the rose family with scarlet, pink, or white flowers

Japanese spurge *noun* (1924)
: a low Japanese herb or subshrub (*Pachysandra terminalis*) of the box family often used as a ground cover

Jap·a·nize \'ja-pə-ˌnīz\ *transitive verb* **-nized; -niz·ing** (1890)
1 : to make Japanese
2 : to bring (as an area or industry) under the influence of Japan
— Jap·a·ni·za·tion \ˌja-pə-nə-'zā-shən\ *noun*

Japan wax *noun* (1859)
: a yellowish fat obtained from the berries of several sumacs (as *Rhus verniciflua* and *R. succedanea*) and used chiefly in polishes

¹jape \'jāp\ *verb* **japed; jap·ing** [Middle English] (14th century)
intransitive verb
: to say or do something jokingly or mockingly
transitive verb
: to make mocking fun of
— jap·er \'jā-pər\ *noun*
— jap·ery \'jā-p(ə-) rē\ *noun*

²jape *noun* (14th century)
: something designed to arouse amusement or laughter: as **a :** an amusing literary or dramatic production **b :** GIBE

Ja·pheth \'jā-fəth\ *noun* [Latin *Japheth* or Greek *Iapheth,* from Hebrew *Yepheth*]
: a son of Noah held to be the progenitor of the Medes and Greeks

ja·po·nais·e·rie \zhä-pō-ne-zə-'rē, -nez-'rē\ *noun, often capitalized* [French, from *japonais* Japanese] (1896)
: a style in art reflecting Japanese qualities or motifs; *also* **:** an object or decoration in this style

ja·pon·i·ca \jə-'pä-ni-kə\ *noun* [New Latin, from feminine of *Japonicus* Japanese, from *Japonia* Japan] (1819)
: JAPANESE QUINCE

¹jar \'jär\ *verb* **jarred; jar·ring** [probably of imitative origin] (1526)
intransitive verb
1 a : to make a harsh or discordant sound **b :** to have a harshly disagreeable or disconcerting effect **c :** to be out of harmony; *specifically* **:** BICKER
2 : to undergo severe vibration
transitive verb
: to cause to jar: as **a :** to affect disagreeably **:** UNSETTLE **b :** to make unstable **:** SHAKE
— jar·ring·ly \'jär-iŋ-lē\ *adverb*

²jar *noun* (1537)
1 a : a state or manifestation of discord or conflict **b :** a harsh grating sound
2 a : a sudden or unexpected shake **b :** an unsettling shock **c :** an unpleasant break or conflict in rhythm, flow, or transition

³jar *noun* [Middle French *jarre,* from Old Provençal *jarra,* from Arabic *jarrah* earthen water vessel] (1592)
1 : a widemouthed container made typically of earthenware or glass
2 : as much as a jar will hold
— jar·ful \-ˌfu̇l\ *noun*

⁴jar *noun* [alteration of earlier *char* turn, from Middle English — more at CHARE] (1674)
archaic **:** the position of being ajar — usually used in the phrase *on the jar*

jar·di·niere *or* **jar·di·nière** \ˌjär-dᵊn-'ir, ˌzhär-dᵊn-'yer, -'er\ *noun* [French *jardinière,* literally, female gardener] (1841)
1 a : an ornamental stand for plants or flowers **b :** a large usually ceramic flowerpot holder
2 : a garnish for meat consisting of several cooked vegetables cut into pieces

¹jar·gon \'jär-gən, -ˌgän\ *noun* [Middle English, from Middle French] (14th century)
1 a : confused unintelligible language **b :** a strange, outlandish, or barbarous language or dialect **c :** a hybrid language or dialect simplified in vocabulary and grammar and used for communication between peoples of different speech
2 : the technical terminology or characteristic idiom of a special activity or group
3 : obscure and often pretentious language marked by circumlocutions and long words
— jar·gon·ish \-gə-nish\ *adjective*
— jar·gon·is·tic \ˌjär-gə-'nis-tik\ *adjective*

²jargon *intransitive verb* (14th century)
1 : TWITTER, WARBLE
2 : JARGONIZE

jar·gon·ize \'jär-gə-ˌnīz\ *verb* **-ized; -iz·ing** (1803)
intransitive verb
: to speak or write jargon
transitive verb
1 : to express in jargon
2 : to make into jargon

jar·goon \jär-'gün\ *or* **jar·gon** \-'gän\ *noun* [French *jargon* — more at ZIRCON] (1769)
: a colorless, pale yellow, or smoky zircon

jar·head \'jär-ˌhed\ *noun* (circa 1960)
slang **:** MARINE 2

jarl \'yär(-ə)l\ *noun* [Old Norse — more at EARL] (1820)
: a Scandinavian noble ranking immediately below the king

jar·rah \'jar-ə\ *noun* [Nyungar (Australian aboriginal language of southwest Western Australia) *jarily*] (circa 1866)
: a eucalyptus (*Eucalyptus marginata*) of western Australia with rough bark and alternate leaves; *also* **:** its wood

jas·mine \'jaz mən\ *noun* [French *jasmin,* from Arabic *yāsamīn,* from Persian] (1562)
1 a : any of numerous often climbing shrubs (genus *Jasminum*) of the olive family that usually have extremely fragrant flowers; *especially* **:** a tall-climbing semievergreen Asian shrub (*J. officinale*) with fragrant white flowers from which oil is extracted for use in perfumes **b :** any of numerous plants having sweet-scented flowers; *especially* **:** YELLOW JESSAMINE
2 : a light yellow

jasmine 1a

Ja·son \'jā-sᵊn\ *noun* [Latin *Iason,* from Greek *Iasōn*]
: a legendary Greek hero distinguished for his successful quest of the Golden Fleece

jas·per \'jas-pər\ *noun* [Middle English *jaspre,* from Middle French, from Latin *jaspis,* from Greek *iaspis,* of Semitic origin; akin to Hebrew *yāshĕpheh* jasper] (14th century)
1 : an opaque cryptocrystalline quartz of any of several colors; *especially* **:** green chalcedony
2 : colored stoneware with raised white decoration
3 : a blackish green
— jas·pery \-pə-rē\ *adjective*

jas·per·ware \'jas-pər-ˌwar, -wer\ *noun* (1863)
: JASPER 2

jas·sid \'ja-səd\ *noun* [ultimately from Greek *Iasos,* town in Asia Minor] (1892)
: any of numerous small leafhoppers that include many economically significant pests of cultivated plants; *broadly* **:** LEAFHOPPER

Jat \'jät\ *noun* [Hindi *Jāṭ*] (1622)
: a member of an Indo-Aryan people of the Punjab and Uttar Pradesh

jaunce \'jȯn(t)s, 'jän(t)s\ *intransitive verb* [origin unknown] (1593)
archaic **:** PRANCE

jaun·dice \'jȯn-dəs, 'jän-\ *noun* [Middle English *jaundis,* from Middle French *jaunisse,* from *jaune* yellow, from Latin *galbinus* greenish yellow] (14th century)
1 : yellowish pigmentation of the skin, tissues, and body fluids caused by the deposition of bile pigments
2 : a disease or abnormal condition characterized by jaundice
3 : a state or attitude characterized by satiety, distaste, or hostility

jaun·diced \-dəst\ *adjective* (1640)
1 : affected with or as if with jaundice
2 : exhibiting or influenced by envy, distaste, or hostility ⟨a *jaundiced* eye⟩

¹jaunt \'jȯnt, 'jänt\ *intransitive verb* [origin unknown] (1575)
1 *archaic* **:** to trudge about
2 : to make a usually short journey for pleasure

²jaunt *noun* (1592)
1 *archaic* **:** a tiring trip
2 : an excursion undertaken especially for pleasure

jaunting car *noun* (1801)
: a light 2-wheeled open horse-drawn vehicle used especially in Ireland with lengthwise seats placed face-to-face or back to back

jaun·ty \'jȯn-tē, 'jän-\ *adjective* **jaun·ti·er; -est** [modification of French *gentil*] (1662)
1 *archaic* **a :** GENTEEL **b :** STYLISH
2 : sprightly in manner or appearance **:** LIVELY
— jaun·ti·ly \'jȯn-tᵊl-ē, 'jän-\ *adverb*
— jaun·ti·ness \'jȯn-tē-nəs, 'jän-\ *noun*

ja·va \'ja-və, 'jä-, -vē\ *noun, often capitalized* [*Java,* island of Indonesia] (1850)
: COFFEE

Ja·va man \'jä-və-, 'ja-\ *noun* (1911)
: a Pleistocene hominid known from fragmentary skeletons found in Trinil and Djetis, Java and now classified with the pithecanthropines

Ja·va·nese \ˌja-və-'nēz, ˌjä-, -'nēs\ *noun, plural* **Javanese** [*Java* + *-nese* (as in *Japanese*)] (1704)
1 : a member of an Indonesian people inhabiting the island of Java
2 : an Austronesian language of the Javanese people
— Javanese *adjective*

jav·e·lin \'jav-lən, 'ja-və-\ *noun* [Middle English *chafeveleyn,* from Middle French *javeline,* alteration of *javelot,* of Celtic origin; akin to Old Irish *gabul* forked stick] (15th century)
1 : a light spear thrown as a weapon of war or in hunting
2 : a slender usually metal shaft at least 260 centimeters long that is thrown for distance in a field event

ja·ve·li·na \ˌhä-və-'lē-nə\ *noun* [American Spanish *jabalina,* from Spanish, feminine of *jabalí* wild boar, from Arabic *jabalīy*] (1822)
: PECCARY

Ja·velle water \zha-'vel, zhə-\ *noun* [*Javel,* former village in France] (1890)
: an aqueous solution of sodium hypochlorite used as a disinfectant or a bleaching agent

¹jaw \'jȯ\ *noun* [Middle English] (14th century)
1 a : either of two complex cartilaginous or bony structures in most vertebrates that border the mouth, support the soft parts enclosing it, usually bear teeth on their oral margin, and are an upper that is more or less firmly fused with the skull and a lower that is hinged, movable, and articulated with the temporal bone of either side **b :** the parts constituting the walls of the mouth and serving to open and close it —

usually used in plural **c :** any of various organs of invertebrates that perform the function of the vertebrate jaws
2 : something resembling the jaw of an animal: as **a :** one of the sides of a narrow pass or channel **b :** either of two or more opposable parts that open and close for holding or crushing something between them
3 a : a space lying between or as if between open jaws ⟨escaped from out of the *jaws* of the whale⟩ **b :** a position or situation in which one is threatened ⟨rode into the *jaws* of danger⟩
4 : a friendly chat

[2]jaw (1748)
intransitive verb
: to talk especially abusively, indignantly, or long-windedly
transitive verb
: to talk to in a scolding or boring manner

jaw·bone \'jȯ-ˌbōn\ *noun* (15th century)
: JAW 1a; *especially* **:** MANDIBLE

jaw·bon·ing \-ˌbō-niŋ\ *noun* (1969)
: the use of public appeals (as by a president) to influence the actions especially of business and labor leaders; *broadly* **:** the use of spoken persuasion
— jaw·bone \-ˌbōn\ *transitive verb*

jaw·break·er \-ˌbrā-kər\ *noun* (1839)
1 : a word difficult to pronounce
2 : a round hard candy

jawed \'jȯd\ *adjective* (circa 1529)
: having jaws ⟨*jawed* fishes⟩ — usually used in combination ⟨square-*jawed*⟩ ⟨a 3-*jawed* chuck⟩

jaw·less fish \'jȯ-ləs-\ *noun* (circa 1941)
: any of the taxonomic group (Agnatha) of primitive vertebrates without jaws including cyclostomes and extinct related forms — compare BONY FISH, CARTILAGINOUS FISH

jaw·line \'jȯ-ˌlīn\ *noun* (1924)
: the outline of the lower jaw

[1]jay \'jā\ *noun* [Middle English, from Middle French *jai,* from Late Latin *gaius*] (14th century)
1 a : a predominantly fawn-colored Old World bird (*Garrulus glandarius*) of the crow family with a black-and-white crest and wings marked with black, white, and blue **b :** any of various usually crested and largely blue chiefly New World birds that are related to the common Old World jay and have roving habits and harsh voices
2 a : an impertinent chatterer **b :** DANDY 1 **c :** GREENHORN
3 : a moderate blue

[2]jay *noun* (circa 1889)
1 : the letter *j*
2 : JOINT 4

jay·bird \'jā-ˌbərd\ *noun* (1661)
: [1]JAY 1, 2

Jay·cee \ˌjā-'sē\ *noun* [from the initials of *J*unior *C*itizens, former name of the organization] (1938)
: a member of a major national and international civic organization

jay·gee \'jā-'jē\ *noun* [*j*unior *g*rade] (1943)
: LIEUTENANT JUNIOR GRADE

jay·hawk·er \'jā-ˌhȯ-kər\ *noun* (1858)
1 a *often capitalized* **:** a member of a band of antislavery guerrillas in Kansas and Missouri before and during the Civil War **b :** BANDIT
2 *capitalized* **:** a native or resident of Kansas — used as a nickname

jay·vee \ˌjā-'vē\ *noun* [*j*unior *v*arsity] (1937)
1 : JUNIOR VARSITY
2 : a member of a junior varsity team

jay·walk \'jā-ˌwȯk\ *intransitive verb* (1919)
: to cross a street carelessly or in an illegal manner so as to be endangered by traffic
— jay·walk·er *noun*

[1]jazz \'jaz\ *noun, often attributive* [origin unknown] (1913)
1 a : American music developed especially from ragtime and blues and characterized by propulsive syncopated rhythms, polyphonic ensemble playing, varying degrees of improvisation, and often deliberate distortions of pitch and timbre **b :** popular dance music influenced by jazz and played in a loud rhythmic manner
2 : empty talk **:** HUMBUG ⟨spouted all the scientific *jazz* —Pete Martin⟩
3 : similar but unspecified things **:** STUFF ⟨that wind, and the waves, and all that *jazz* —John Updike⟩
— jazz·like \-ˌlīk\ *adjective*

[2]jazz (1917)
transitive verb
1 a : ENLIVEN — usually used with *up* **b :** ACCELERATE
2 : to play in the manner of jazz
intransitive verb
1 : to go here and there **:** GAD
2 : to dance to or play jazz

jazz·man \'jaz-ˌman, -mən\ *noun* (1926)
: a jazz musician

jazz–rock \-'räk\ *noun* (1968)
: a blend of jazz and rock music

jazzy \'ja-zē\ *adjective* **jazz·i·er; -est** (1919)
1 : having the characteristics of jazz
2 : marked by unrestraint, animation, or flashiness
— jazz·i·ly \'ja-zə-lē\ *adverb*
— jazz·i·ness \'ja-zē-nəs\ *noun*

J–bar lift \'jā-ˌbär-\ *noun* (1954)
: a ski lift having a series of J-shaped bars each of which pulls one skier

jeal·ous \'je-ləs\ *adjective* [Middle English *jelous,* from Old French, from (assumed) Vulgar Latin *zelosus,* from Late Latin *zelus* zeal — more at ZEAL] (13th century)
1 a : intolerant of rivalry or unfaithfulness **b :** disposed to suspect rivalry or unfaithfulness
2 : hostile toward a rival or one believed to enjoy an advantage
3 : vigilant in guarding a possession ⟨new colonies were *jealous* of their new independence —Scott Buchanan⟩
— jeal·ous·ly *adverb*
— jeal·ous·ness *noun*

jeal·ou·sy \'je-lə-sē\ *noun, plural* **-sies** (13th century)
1 : a jealous disposition, attitude, or feeling
2 : zealous vigilance

jean \'jēn\ *noun* [short for *jean fustian,* from Middle English *Gene* Genoa, Italy + *fustian*] (1577)
1 : a durable twilled cotton cloth used especially for sportswear and work clothes
2 : pants that are usually close-fitting and made especially of jean or denim — usually used in plural

[1]jeep \'jēp\ *noun* [probably from *g. p.* (abbreviation of *general purpose*)] (1940)
: a small general-purpose motor vehicle with 80-inch wheelbase, 1/4-ton capacity, and four-wheel drive used by the U.S. army in World War II; *also* **:** a similar but larger and more powerful U.S. army vehicle ◆

[2]jeep *intransitive verb* (1942)
: to travel by jeep

Jeep *trademark*
— used for a civilian automotive vehicle

jee·pers \'jē-pərz\ *also* **jee·pers cree·pers** \-'krē-pərz\ *interjection* [*jeepers,* euphemism for *Jesus; jeepers creepers,* euphemism for *Jesus Christ*] (1927)
— used as a mild oath

jeep·ney \'jēp-nē\ *noun* [*jeep* + *jitney*] (circa 1949)
: a Philippine jitney bus converted from a jeep

[1]jeer \'jir\ *verb* [origin unknown] (1561)
intransitive verb
: to speak or cry out with derision or mockery
transitive verb
: to deride with jeers **:** TAUNT
synonym see SCOFF
— jeer·er *noun*
— jeer·ing·ly \-iŋ-lē\ *adverb*

[2]jeer *noun* (1625)
: a jeering remark or sound **:** TAUNT

jeez \'jēz\ *interjection* [euphemism for *Jesus*] (1923)
— used as a mild oath or introductory expletive (as to express surprise)

je·fe \'hā-(ˌ)fā, 'he-; 'he-fē\ *noun* [Spanish, from French *chef,* from Middle French *chief* — more at CHIEF] (1903)
: CHIEF, LEADER

Jef·fer·son Da·vis's Birthday \'je-fər-sən-'dā-və-səz-\ *noun* (1929)
: the first Monday in June observed as a legal holiday in many Southern states

Jef·fer·son Day \'je-fər-sən-\ *noun* (1936)
: April 13 observed as a holiday in Alabama in commemoration of Thomas Jefferson's birthday

Jef·frey pine \'je-frē-\ *noun* [John *Jeffrey* (died 1854) Scottish botanical explorer] (1858)
: a pine (*Pinus jeffreyi*) of the western U.S. having long needles in groups of three

je·had *variant of* JIHAD

Je·hosh·a·phat \ji-'hä-sə-ˌfat, -shə-\ *noun* [Hebrew *Yĕhōshāphāth*]
: a king of Judah who brought Judah into an alliance with the northern kingdom of Israel in the 9th century B.C.

Je·ho·vah \ji-'hō-və\ *noun* [New Latin, reading (as *Yĕhōwāh*) of Hebrew *yhwh* Yahweh with the vowel points of *'adhōnāy* my lord] (1530)
: GOD 1

Jehovah's Witness *noun* (1932)
: a member of a group that witness by distributing literature and by personal evangelism to beliefs in the theocratic rule of God, the sinfulness of organized religions and governments, and an imminent millennium

je·hu \'jē-(ˌ)hyü, -(ˌ)hü\ *noun* [Hebrew *Yēhū*]
1 *capitalized* **:** a king of Israel in the 9th century B.C. who according to the account in II Kings had Jezebel killed in accordance with Elijah's prophecy
2 : a driver of a coach or cab

je·ju·nal \ji-'jü-nᵊl\ *adjective* (circa 1887)
: of or relating to the jejunum

je·june \ji-'jün\ *adjective* [Latin *jejunus* empty of food, hungry, meager] (1646)
1 : lacking nutritive value ⟨*jejune* diets⟩
2 : devoid of significance or interest **:** DULL ⟨*jejune* lectures⟩

◇ WORD HISTORY

jeep A common etymology of *jeep* claims that the word was vocalized from *g.p.* (*gener*al *purpose*), stenciled on military vehicles. Whatever merit this hypothesis has, *jeep* in its early days saw a wide range of applications. Perhaps the progenitor of all of them was a character named Eugene the Jeep that first appeared in March 1936 in Elzie C. Segar's comic strip *Thimble Theater,* featuring Popeye. In a short story of 1938, *jeep* means "nerd," a direction that seems to have been continued by the sense "trainee," U.S. Army slang during World War II. In 1936 *jeep* was applied to an oil company's recording device, and in 1940 to an agricultural tractor converted to military use. In the same year, a wire-service article defined *jeep* as Army vernacular for "any very small plane, helicopter, or gadget." To the Air Corps, *jeep* was a small trainer, and to the Navy an escort carrier. The now familiar use of *jeep* for a 1/4-ton military vehicle was first noted in February 1941, but in the early years of the war, some soldiers applied *jeep* to a now forgotten 3/4-ton command car, while the 1/4-ton vehicle was known as a *peep.*

3 : JUVENILE, PUERILE ⟨*jejune* reflections on life and art⟩ ◻ ◇
synonym see INSIPID
— je·june·ly *adverb*
— je·june·ness \-ˈjün-nəs\ *noun*

je·ju·num \ji-ˈjü-nəm\ *noun* [Middle English, from Medieval Latin, from neuter of Latin *jejunus*] (14th century)
: the section of the small intestine that comprises the first two fifths beyond the duodenum and that is larger, thicker-walled, and more vascular and has more circular folds than the ileum

Je·kyll and Hyde \ˈje-kə-lən-ˈhīd *also* ˈjē- *or* ˈjā-\ *noun* [Dr. *Jekyll* & Mr. *Hyde,* representing the two-sided personality of the protagonist in *The Strange Case of Dr. Jekyll and Mr. Hyde* (1886) by R. L. Stevenson] (circa 1922)
: one having a two-sided personality one side of which is good and the other evil

jell \ˈjel\ *verb* [back-formation from *jelly*] (1869)
intransitive verb
1 : to come to the consistency of jelly **:** CONGEAL, SET
2 : to take shape and achieve distinctness **:** become cohesive
transitive verb
: to cause to jell

jellied gasoline *noun* (1944)
: NAPALM

Jell–O \ˈje-(ˌ)lō\ *trademark*
— used for a gelatin dessert usually with the flavor and color of fruit

[1]jel·ly \ˈje-lē\ *noun, plural* **jellies** [Middle English *gelly,* from Middle French *gelee,* from feminine of *gelé,* past participle of *geler* to freeze, congeal, from Latin *gelare* — more at COLD] (14th century)
1 : a soft somewhat elastic food product made usually with gelatin or pectin; *especially* **:** a fruit product made by boiling sugar and the juice of fruit
2 : a substance resembling jelly in consistency
3 : a state of fear or irresolution
4 : a shapeless structureless mass **:** PULP
— jel·ly·like \-ˌlīk\ *adjective*

[2]jelly *verb* **jel·lied; jel·ly·ing** (1590)
intransitive verb
1 : JELL
2 : to make jelly
transitive verb
: to bring to the consistency of jelly

jelly bean *noun* (1905)
: a sugar-glazed bean-shaped candy

jel·ly·fish \ˈje-lē-ˌfish\ *noun* (1841)
1 a : a free-swimming marine coelenterate that is the sexually reproducing form of a hydrozoan or scyphozoan and has a nearly transparent saucer-shaped body and extensile marginal tentacles studded with stinging cells **b :** SIPHONOPHORE **c :** CTENOPHORE
2 : a person lacking backbone or firmness

jelly roll *noun* (1895)
: a thin sheet of sponge cake spread with jelly and rolled up

jel·u·tong \ˈje-lə-ˌtȯŋ\ *noun* [Malay *jĕlutong*] (circa 1836)
1 : any of several trees (genus *Dyera*) of the dogbane family
2 : the resinous rubbery latex of a jelutong (especially *Dyera costulata*) used especially as a chicle substitute

jem·my \ˈje-mē\ *noun, plural* **jemmies** [from the name *Jemmy*] (circa 1811)
British **:** JIMMY

je ne sais quoi \zhə-nə-ˌsā-ˈkwä\ *noun* [French, literally, I know not what] (circa 1656)
: something that cannot be adequately described or expressed

jen·net \ˈhe-nā, ˈje-nət\ *noun* [Middle English *genett,* from Middle French *genet,* from Catalan, Zenete (member of a Berber people), horse] (15th century)
1 : a small Spanish horse
2 a : a female donkey **b :** HINNY

jen·ny \ˈje-nē\ *noun, plural* **jennies** [from the name *Jenny*] (1600)
1 a : a female bird ⟨*jenny* wren⟩ **b :** a female donkey
2 : SPINNING JENNY

je·on \(ˌ)jā-ˈȯn\ *noun, plural* **jeon** [Korean *chŏn*] (circa 1969)
: the chon of South Korea

jeop·ard \ˈje-pərd\ *transitive verb* [Middle English, back-formation from *jeopardie*] (14th century)
: JEOPARDIZE

jeop·ar·dise *British variant of* JEOPARDIZE

jeop·ar·dize \ˈje-pər-ˌdīz\ *transitive verb* **-dized; -diz·ing** (1582)
: to expose to danger or risk **:** IMPERIL

jeop·ar·dy \ˈje-pər-dē\ *noun* [Middle English *jeopardie,* from Anglo-French *juparti,* from Old French *jeu parti* alternative, literally, divided game] (14th century)
1 : exposure to or imminence of death, loss, or injury **:** DANGER
2 : the danger that an accused person is subjected to when on trial for a criminal offense ◇

je·quir·i·ty bean \jə-ˈkwir-ə-tē-\ *noun* [French *jékwirity*] (circa 1889)
1 : the poisonous scarlet and black seed of the rosary pea often used for beads
2 : ROSARY PEA 1

jer·boa \jər-ˈbō-ə, jer-\ *noun* [New Latin, from Arabic *yarbūʿ*] (1662)
: any of several social nocturnal Old World jumping rodents (family Dipodidae) with long hind legs and long tail

jer·e·mi·ad \ˌjer-ə-ˈmī-əd, -ˌad\ *noun* [French *jérémiade,* from *Jérémie* Jeremiah, from Late Latin *Jeremias*] (1780)
: a prolonged lamentation or complaint; *also* **:** a cautionary or angry harangue ◇

Jer·e·mi·ah \-ˈmī-ə\ *noun* [Late Latin *Jeremias,* from Greek *Hieremias,* from Hebrew *Yirmĕyāh*]
1 : a major Hebrew prophet of the 6th and 7th centuries B.C.
2 : one who is pessimistic about the present and foresees a calamitous future
3 : a prophetic book of canonical Jewish and Christian Scripture — see BIBLE table

Jer·e·mi·as \-ˈmī-əs\ *noun* [Late Latin]
: JEREMIAH

[1]jerk \ˈjərk\ *noun* [probably alteration of *yerk*] (1575)
1 : a single quick motion of short duration
2 a : jolting, bouncing, or thrusting motions **b :** a tendency to produce spasmodic motions
3 a : an involuntary spasmodic muscular movement due to reflex action **b** *plural* **:** involuntary twitchings due to nervous excitement
4 : an annoyingly stupid or foolish person
5 : the pushing of a weight from shoulder height to a position overhead in weight lifting

[2]jerk (1589)
transitive verb
1 : to give a quick suddenly arrested push, pull, or twist to
2 : to propel or move with or as if with a quick suddenly arrested motion
3 : to mix and serve (as sodas) behind a soda fountain
intransitive verb
1 : to make a sudden spasmodic motion
2 : to move in short abrupt motions or with frequent jolts
— jerk·er *noun*

[3]jerk *transitive verb* [back-formation from [1]*jerky*] (1707)
: to preserve (meat) in long sun-dried slices

jerk around *transitive verb* (1941)
: to treat badly especially by being underhanded or inconsistent

◻ USAGE

jejune The propriety of the third sense of *jejune* is still a matter for discussion by commentators on the language. The meaning may have been invented by H. L. Mencken ⟨I changed the meaning of *jejune* eight or ten years ago. That is to say, I added a new special meaning, to wit, that of youthful feebleness —H. L. Mencken (letter, 1922)⟩. Or Mencken may have taken a bit more credit than he deserved ⟨we pass in the world for sects and schools, for erudition and piety, and we are all the time *jejune* babes —Ralph Waldo Emerson (1841)⟩. No matter who used it first, the meaning is established.

◇ WORD HISTORY

jejune In Latin the adjective *jejunus* meant "having eaten nothing, fasting." When the neuter form *jejunum* was borrowed into English, it was used to denote a section of the small intestine, so named because it was said to be found empty at the time of death. In the early 17th century the adjective *jejune* began appearing in English in the sense "fasting," later extended to mean "undernourished." Over time figurative use of the "undernourished" sense came to prevail over its literal applications. Once this dominance was well established, the meaning of *jejune* evolved into "insipid, vapid." In a semantic development of the 20th century, *jejune* also took on the sense "juvenile" or "puerile," perhaps as a result of confusion with *jeune,* an unrelated French word meaning "young."

jeopardy In French *jeu parti* literally means "divided game." In medieval France the basic requirement for a *jeu parti* was the involvement of alternative possibilities or opposed viewpoints. For example, a *jeu parti* could take the form of a poetic dialogue presenting opposing points of view, or it could be a situation in a game such as chess in which the merits of alternative plays is uncertain. It was in this latter sense that the word was borrowed into Middle English. *Jeopardy* soon came to be applied to any situation that seems to provide equal opportunities for success or failure. Soon afterwards, the word came to refer to a state in which the possibility of injury, pain, or loss is greater than even.

jeremiad Jeremiah (*ca* 650–*ca* 570 B.C.) was a Biblical prophet who was intimately involved in the political and religious events of his time, including the fall of Jerusalem to the Babylonians and the subsequent exile of many Judaeans to Babylon. Jeremiah's messages were condemnations of his people for their wicked ways, worship of false gods, and social injustices. He rebuked his Judaean leaders and countrymen for not readily submitting to the yoke of the Babylonians. (He claimed that God had willed it.) Even while exiled in Egypt, he continued his chastisement. When Jeremiah was not lamenting the plight of his people, he was lamenting his own, a life filled with the hardships of a prophet bearing an unpopular message. According to tradition, his exasperated companions finally stoned him to death. *Jeremias,* the prophet's name in Late Latin, was gallicized as *Jérémie,* from which the French derived *jérémiade,* the immediate source of English *jeremiad.*

jer·kin \ˈjər-kən\ *noun* [origin unknown] (1519)
: a close-fitting hip-length usually sleeveless jacket

jerk off *verb* (circa 1904)
: MASTURBATE — usually considered vulgar

jerk·wa·ter \ˈjərk-ˌwȯ-tər, -ˌwä-\ *adjective* [from *jerkwater* rural train] (1888)
1 : remote and unimportant ⟨*jerkwater* towns⟩
2 : TRIVIAL

[1]**jer·ky** \ˈjər-kē\ *noun* [Spanish *charqui,* from Quechua *ch'arki*] (1850)
: jerked meat

[2]**jerky** \ˈjər-kē\ *adjective* **jerk·i·er; -est** (1858)
1 a : moving along with or marked by fits and starts **b :** characterized by abrupt transitions
2 : INANE, FOOLISH
— jerk·i·ly \-kə-lē\ *adverb*
— jerk·i·ness \-kē-nəs\ *noun*

jer·o·bo·am \ˌjer-ə-ˈbō-əm\ *noun* [*Jeroboam* I (died about 912 B.C.) king of the northern kingdom of Israel] (1816)
: an oversize wine bottle holding about three liters

jer·ri·can *or* **jerry can** \ˈjer-ē-ˌkan\ *noun* [*Jerry + can;* from its German design] (1943)
: a narrow flat-sided liquid container holding five gallons (about 19 liters)

Jer·ry \ˈjer-ē\ *noun, plural* **Jerries** [by shortening & alteration] (1898)
chiefly British **:** GERMAN

jer·ry–build \ˈjer-ē-ˌbild\ *transitive verb* **-built** \-ˌbilt\; **-build·ing** [back-formation from *jerry-built*] (1885)
: to build cheaply and flimsily
— jer·ry–build·er *noun*

jer·ry–built *adjective* [origin unknown] (1869)
1 : built cheaply and unsubstantially
2 : carelessly or hastily put together

jer·sey \ˈjər-zē\ *noun, plural* **jerseys** [*Jersey,* one of the Channel Islands] (1587)
1 : a plain weft-knitted fabric made of wool, cotton, nylon, rayon, or silk and used especially for clothing
2 : any of various close-fitting usually circular-knitted garments especially for the upper body
3 *often capitalized* **:** any of a breed of small short-horned predominantly yellowish brown or fawn dairy cattle noted for their rich milk

Jersey pine *noun* (1743)
: VIRGINIA PINE

Je·ru·sa·lem artichoke \jə-ˈrü-s(ə-)ləm-, -ˈrüz-ləm-, -ˈrü-zə-\ *noun* [*Jerusalem* by folk etymology from Italian *girasole* girasole] (1641)
: a perennial American sunflower (*Helianthus tuberosus*) widely cultivated for its tubers that are used as a vegetable and as a livestock feed

Jerusalem cherry *noun* [*Jerusalem,* Palestine] (1788)
: either of two plants (*Solanum pseudocapsicum* and *S. capsicastrum*) of the nightshade family cultivated as ornamental houseplants for their orange to red berries

Jerusalem cricket *noun* (1947)
: a large-headed burrowing nocturnal orthopteran insect (*Stenopelmatus fuscus*) of the southwestern U.S.

Jerusalem cricket

Jerusalem thorn *noun* (1866)
: a tropical American leguminous spiny shrub or shrubby tree (*Parkinsonia aculeata*) with pinnate leaves and showy racemose yellow flowers that is widely cultivated

jess \ˈjes\ *noun* [Middle English *ges,* from Middle French *gies,* from plural of *jet* throw, from *jeter* to throw — more at JET] (14th century)
: a short strap secured on the leg of a hawk and usually provided with a ring for attaching a leash
— jessed \ˈjest\ *adjective*

jes·sa·mine \ˈjes-mən, ˈje-sə-\ *variant of* JASMINE

Jes·se \ˈje-sē\ *noun* [Hebrew *Yishay*]
: the father of David, king of Israel, according to the account in I Samuel

jest \ˈjest\ *noun* [Middle English *geste* idle tale, story in verse, from Old French, from Latin *gesta* deeds, from neuter plural of *gestus,* past participle of *gerere* to bear, wage] (circa 1548)
1 : an utterance (as a jeer or quip) intended to be taken as mockery or humor
2 a : PRANK **b :** a ludicrous circumstance or incident
3 a : a frivolous mood or manner ⟨spoken in *jest*⟩ **b :** gaiety and merriment
4 : LAUGHINGSTOCK
synonym see FUN
— jest *verb*

jest·er \ˈjes-tər\ *noun* (14th century)
1 : FOOL 2a
2 : one given to jests

Je·su·it \ˈje-zü-ət, -zhü- *also* -zyü-\ *noun* [New Latin *Jesuita,* from Late Latin *Jesus*] (1559)
1 : a member of the Roman Catholic Society of Jesus founded by Saint Ignatius Loyola in 1534 and devoted to missionary and educational work
2 : one given to intrigue or equivocation
— je·su·it·ic \ˌje-zü-ˈ(w)i-tik, -zhü-, -zyü-\ *or* **je·su·it·i·cal** \-ti-kəl\ *adjective, often capitalized*
— je·su·it·i·cal·ly \-ti-k(ə-)lē\ *adverb, often capitalized*
— je·su·it·ism \ˈje-zü-ə-ˌti-zəm, -zhü-, -zyü-\ *or* **je·su·it·ry** \-ə-trē\ *noun, often capitalized*

Je·sus \ˈjē-zəs, -zəz *also* -ˌzəs *and* -ˌzəz\ *noun* [Late Latin, from Greek *Iēsous,* from Hebrew *Yēshūaʿ*]
1 : the Jewish religious teacher whose life, death, and resurrection as reported by the Evangelists are the basis of the Christian message of salvation — called also *Jesus Christ*
2 *Christian Science* **:** the highest human corporeal concept of the divine idea rebuking and destroying error and bringing to light man's immortality

[1]**jet** \ˈjet\ *noun* [Middle English, from Middle French *jaiet,* from Latin *gagates,* from Greek *gagatēs,* from *Gagas,* town and river in Asia Minor] (14th century)
1 : a compact velvet-black coal that takes a good polish and is often used for jewelry
2 : an intense black

[2]**jet** *adjective* (1716)
: of the color jet

[3]**jet** *verb* **jet·ted; jet·ting** [French *jeter,* literally, to throw, from Middle French, from Latin *jactare* to throw, frequentative of *jacere* to throw; akin to Greek *hienai* to send] (1692)
intransitive verb
: to spout forth **:** GUSH
transitive verb
: to emit in a stream **:** SPOUT

[4]**jet** *noun* (circa 1696)
1 a (1) **:** a usually forceful stream of fluid (as water or gas) discharged from a narrow opening or a nozzle (2) **:** a narrow stream of material (as plasma) emanating or appearing to emanate from a celestial object (as a radio galaxy) **b :** a nozzle for a jet of fluid
2 : something issuing as if in a jet ⟨talk poured from her in a brilliant *jet —Time*⟩
3 a : JET ENGINE **b :** an airplane powered by one or more jet engines
4 : a long narrow current of high-speed winds (as a jet stream)
— jet·like \-ˌlīk\ *adjective*

[5]**jet** *intransitive verb* **jet·ted; jet·ting** (1949)
1 : to travel by jet airplane
2 : to move or progress by or as if by jet propulsion

jet·bead \ˈjet-ˌbēd\ *noun* (circa 1930)
: a shrub (*Rhodotypos scandens*) that has black shining fruit and is used as an ornamental

jet–black \-ˈblak\ *adjective* (15th century)
: black as jet

je·té \zhə-ˈtā\ *noun* [French, from past participle of *jeter*] (1830)
: a springing jump in ballet made from one foot to the other in any direction

jet engine *noun* (1943)
: an engine that produces motion as a result of the rearward discharge of a jet of fluid; *specifically* **:** an airplane engine that uses atmospheric oxygen to burn fuel and produces a rearward discharge of heated air and exhaust gases — see AIRPLANE illustration

jet lag *noun* (1969)
: a condition that is characterized by various psychological and physiological effects (as fatigue and irritability), occurs following long flight through several time zones, and probably results from disruption of circadian rhythms in the human body
— jet–lagged \ˈjet-ˌlagd\ *adjective*

jet·lin·er \ˈjet-ˌlī-nər\ *noun* (1949)
: a jet-propelled airliner

jet·port \ˈjet-ˌpōrt, -ˌpȯrt\ *noun* (1961)
: an airport designed to handle jet airplanes

jet–pro·pelled \ˈjet-prə-ˈpeld\ *adjective* (1877)
1 : moving by jet propulsion
2 : suggestive of the speed and force of a jet airplane

jet propulsion *noun* (1867)
: propulsion of a body produced by the forwardly directed forces of the reaction resulting from the rearward discharge of a jet of fluid; *especially* **:** propulsion of an airplane by jet engines

jet·sam \ˈjet-səm\ *noun* [alteration of *jettison*] (1570)
1 : the part of a ship, its equipment, or its cargo that is cast overboard to lighten the load in time of distress and that sinks or is washed ashore
2 : FLOTSAM 2

jet set *noun* (1951)
: an international social group of wealthy individuals who frequent fashionable resorts
— jet–set·ter \-ˌse-tər\ *noun*
— jet–set·ting \-ˌse-tiŋ\ *adjective*

jet stream *noun* (1947)
: a long narrow meandering current of high-speed winds near the tropopause blowing from a generally westerly direction and often exceeding a speed of 250 miles (402 kilometers) per hour

[1]**jet·ti·son** \ˈje-tə-sən, -zən\ *noun* [Middle English *jetteson,* from Anglo-French *getteson,* from Old French *getaison* action of throwing, from Latin *jactation-, jactatio,* from *jactare* — more at JET] (15th century)
: a voluntary sacrifice of cargo to lighten a ship's load in time of distress

[2]**jettison** *transitive verb* (1848)
1 : to make jettison of
2 : to get rid of as superfluous or encumbering **:** DISCARD
3 : to drop from an airplane or spacecraft in flight
— jet·ti·son·able \-sə-nə-bəl, -zə-\ *adjective*

[1]**jet·ty** \ˈje-tē\ *noun, plural* **jetties** [Middle English *jette,* from Middle French *jetee,* from feminine of *jeté,* past participle of *jeter* to throw — more at JET] (15th century)
1 a : a structure extended into a sea, lake, or river to influence the current or tide or to protect a harbor **b :** a protecting frame of a pier
2 : a landing wharf

[2]**jetty** *intransitive verb* **jet·tied; jet·ty·ing** (1598)
: PROJECT, JUT

[3]jetty *adjective* (1586)
: black as jet

jeu d'es·prit \zhœ̄-des-prē\ *noun, plural* **jeux d'esprit** *same*\ [French, literally, play of the mind] (1712)
: a witty comment or composition

jeu·nesse do·rée \zhœ-nes-dó-rā\ *noun* [French, gilded youth] (1836)
: young people of wealth and fashion

Jew \'jü\ *noun* [Middle English, from Old French *gyu,* from Latin *Judaeus,* from Greek *Ioudaios,* from Hebrew *Yĕhūdhī,* from *Yĕhūdhāh* Judah, Jewish kingdom] (13th century)
1 a : a member of the tribe of Judah **b :** ISRAELITE
2 : a member of a nation existing in Palestine from the 6th century B.C. to the 1st century A.D.
3 : a person belonging to a continuation through descent or conversion of the ancient Jewish people
4 : one whose religion is Judaism

[1]jew·el \'jü-əl, 'jül *also* 'jü(ə)l\ *noun, often attributive* [Middle English *juel,* from Old French, probably diminutive of *jeu* game, play, from Latin *jocus* game, joke — more at JOKE] (13th century)
1 : an ornament of precious metal often set with stones or decorated with enamel and worn as an accessory of dress
2 : one that is highly esteemed
3 : a precious stone **:** GEM
4 : a bearing for a pivot (as in a watch) made of crystal, glass, or a gem
— jew·el·like \-,līk\ *adjective*

[2]jewel *transitive verb* **-eled** *or* **-elled; -el·ing** *or* **-el·ling** (1601)
1 : to adorn or equip with jewels
2 : to give beauty to as if with jewels **:** EMBELLISH

jewel box *noun* (1831)
1 : a small box or case designed to hold jewelry
2 : something small and exquisite
3 : a clear plastic case for a compact disc

jew·el·er *or* **jew·el·ler** \'jü-ə-lər, 'jü-lər *also* 'jü(-ə)-lər\ *noun* (14th century)
1 : one who makes or repairs jewelry
2 : one who deals in jewelry, precious stones, watches, and usually silverware and china

jew·el·lery *chiefly British variant of* JEWELRY

jew·el·ry \'jü-əl-rē, 'jül-rē, 'jü(-ə)l-\ *noun* (14th century)
: JEWELS; *especially* **:** objects of precious metal often set with gems and worn for personal adornment

jew·el·weed \-,wēd\ *noun* (1818)
: TOUCH-ME-NOT

Jew·ess \'jü-əs\ *noun* (14th century)
: a Jewish girl or woman — sometimes taken to be offensive

jew·fish \'jü-,fish\ *noun* (1679)
: any of various large groupers (especially *Epinephelus itajara*) that are usually dusky green, brown, or blackish, thickheaded, and rough-scaled

Jew·ish \'jü-ish\ *adjective* (circa 1546)
: of, relating to, or characteristic of the Jews; *also* **:** being a Jew
— Jew·ish·ly *adverb*
— Jew·ish·ness *noun*

Jewish American Princess *noun* (1979)
: a stereotypical well-to-do or spoiled American Jewish girl or woman — called also *Jewish Princess;* often used disparagingly

Jewish calendar *noun* (circa 1888)
: a calendar in use among Jewish peoples that is reckoned from the year 3761 B.C. and dates in its present form from about A.D. 360 — see MONTH table

Jew·ry \'jü(-ə)r-ē, 'jü-rē\ *noun* (14th century)
1 *plural* **Jewries :** a community of Jews
2 : the Jewish people

JEWISH YEARS 5754–5773

JEWISH YEAR		A.D.
5754	begins	September 16, 1993
5755	begins	September 6, 1994
5756	begins	September 25, 1995
5757	begins	September 14, 1996
5758	begins	October 2, 1997
5759	begins	September 21, 1998
5760	begins	September 11, 1999
5761	begins	September 30, 2000
5762	begins	September 18, 2001
5763	begins	September 7, 2002
5764	begins	September 27, 2003
5765	begins	September 16, 2004
5766	begins	October 4, 2005
5767	begins	September 23, 2006
5768	begins	September 13, 2007
5769	begins	September 30, 2008
5770	begins	September 19, 2009
5771	begins	September 9, 2010
5772	begins	September 29, 2011
5773	begins	September 17, 2012

Jew's harp *or* **Jews' harp** \'jüz-,härp, 'jüs-\ *noun* (1595)
: a small lyre-shaped instrument that when held between the teeth gives tones from a metal tongue struck by the finger

Jew's harp

Jez·e·bel \'je-zə-,bel\ *noun* [Hebrew *Īzebhel*]
1 : the Phoenician wife of Ahab who according to the account in I and II Kings pressed the cult of Baal on the Israelite kingdom but was finally killed in accordance with Elijah's prophecy
2 *often not capitalized* **:** an impudent, shameless, or morally unrestrained woman

JHVH *variant of* YHWH

jiao \jē-'au̇\ *noun* [Chinese (Beijing) *jiăo*] (1949)
: a monetary unit of the People's Republic of China equal to 1/10 yuan

[1]jib \'jib\ *noun* [origin unknown] (1661)
: a triangular sail set on a stay extending usually from the head of the foremast to the bowsprit or the jibboom; *also* **:** the small triangular headsail on a sloop — see SAIL illustration

[2]jib *noun* [probably by shortening & alteration from *gibbet*] (1764)
1 : the projecting arm of a crane
2 : a derrick boom

[3]jib *intransitive verb* **jibbed; jib·bing** [probably from *jib* to shift from one side of a ship to the other, perhaps from [1]*jib*] (1811)
: to refuse to proceed further **:** BALK
— jib·ber *noun*

jib·boom \'ji(b)-'büm\ *noun* [[1]*jib* + *boom*] (1748)
: a spar that forms an extension of the bowsprit

[1]jibe \'jīb\ *variant of* GIBE

[2]jibe *verb* **jibed; jib·ing** [perhaps modification of Dutch *gijben*] (1693)
intransitive verb
1 : to shift suddenly and forcibly from one side to the other — used of a fore-and-aft sail
2 : to change a vessel's course when sailing with the wind so that as the stern passes through the eye of the wind the boom swings to the opposite side
transitive verb
: to cause to jibe

[3]jibe *intransitive verb* **jibed; jib·ing** [origin unknown] (1813)
: to be in accord **:** AGREE

ji·ca·ma \'hē-kə-mə\ *noun* [Mexican Spanish *jícama,* from Nahuatl *xīcamatl*] (circa 1909)
: an edible starchy tuberous root of a leguminous tropical American vine (*Pachyrhizus erosus*)

Ji·ca·ril·la \,hē-kə-'rē-yə\ *noun, plural* **Jicarilla** *or* **Jicarillas** [American Spanish *apaches de la xicarilla,* literally, gourd-cup Apaches, from *Cerro de la Xicarilla,* literally, gourd-cup peak, unidentified mountain in Jicarilla territory] (1850)
1 : a member of an Apache people originally of southeastern Colorado, northern New Mexico, and adjacent areas and now living chiefly in northern New Mexico
2 : the language of the Jicarilla people

jiff \'jif\ *noun* (1797)
: JIFFY

jif·fy \'ji-fē\ *noun, plural* **jiffies** [origin unknown] (1779)
: MOMENT, INSTANT ⟨ready in a *jiffy*⟩

[1]jig \'jig\ *noun* [perhaps from Middle French *giguer* to frolic, from *gigue* fiddle, of Germanic origin; akin to Old High German *gīga* fiddle; akin to Old Norse *geiga* to turn aside] (circa 1560)
1 a : any of several lively springy dances in triple rhythm **b :** music to which a jig may be danced
2 : TRICK, GAME — used chiefly in the phrase *the jig is up*
3 a : any of several fishing devices that are jerked up and down or drawn through the water **b :** a device used to maintain mechanically the correct positional relationship between a piece of work and the tool or between parts of work during assembly **c :** a device in which crushed ore is concentrated or coal is cleaned by agitating in water
— in jig time : in a short time **:** QUICKLY

[2]jig *verb* **jigged; jig·ging** (1604)
intransitive verb
1 a : to move with rapid jerky motions **b :** to dance a jig
2 : to fish with a jig
transitive verb
1 : to dance in the rapid lively manner of a jig
2 a : to give a rapid jerky motion to **b :** to separate (a mineral or ore from waste) with a jig
3 : to catch (a fish) with a jig
4 : to machine by means of a jig-controlled tool operation

[3]jig *noun* [short for *jigaboo* black person] (1927)
: BLACK 4 — usually taken to be offensive

[1]jig·ger \'ji-gər\ *noun* (1675)
1 : one that jigs or operates a jig
2 : any of several sails
3 : JIG 3a
4 a (1) **:** a mechanical device usually with a jerky reciprocating motion (2) **:** a mold or a machine incorporating a revolving mold on which ceramic items (as plates) are formed **b :** GADGET, DOODAD
5 : a measure used in mixing drinks that usually holds 1 to 2 ounces (30 to 60 milliliters)

[2]jigger *noun* [perhaps from Wolof *jiga* insect] (1781)
: CHIGGER

[3]jigger *verb* [frequentative of [2]*jig*] (1867)
intransitive verb
: to jerk up and down
transitive verb
: to alter or rearrange especially by manipulating ⟨*jigger* an election district⟩

jig·gery–pok·ery \'ji-gər-ē-'pō-kər-ē\ *noun* [probably alteration of Scots *joukery-pawkery,* from *jouk* to dodge, cheat + *pawk* trick, wile] (circa 1892)
: underhanded manipulation or dealings **:** TRICKERY

\ə\ **abut** \ᵊ\ **kitten** \ər\ **further** \a\ **ash** \ā\ **ace** \ä\ **mop, mar** \au̇\ **out** \ch\ **chin** \e\ **bet** \ē\ **easy** \g\ **go** \i\ **hit** \ī\ **ice** \j\ **job** \ŋ\ **sing** \ō\ **go** \ó\ **law** \ói\ **boy** \th\ **thin** \t͟h\ **the** \ü\ **loot** \u̇\ **foot** \y\ **yet** \zh\ **vision** *see also* Guide to Pronunciation

jig·gle \ˈji-gəl\ *verb* **jig·gled; jig·gling** \-g(ə-)liŋ\ [frequentative of [2]*jig*] (1836)
transitive verb
: to cause to move with quick little jerks or oscillating motions
intransitive verb
: to move from or as if from being jiggled
— jiggle *noun*
— jig·gly \-g(ə-)lē\ *adjective*

[1]**jig·saw** \ˈjig-ˌsȯ\ *noun* (1873)
1 : a machine saw with a narrow vertically reciprocating blade for cutting curved and irregular lines or ornamental patterns in openwork
2 : SCROLL SAW 1
3 : JIGSAW PUZZLE

[2]**jigsaw** *transitive verb* (1873)
1 : to cut or form by or as if by a jigsaw
2 : to arrange or place in an intricate or interlocking way

[3]**jigsaw** *adjective* (1884)
: suggesting a jigsaw puzzle or its separate pieces

jigsaw puzzle *noun* (1919)
: a puzzle consisting of small irregularly cut pieces that are to be fitted together to form a picture; *also* **:** something suggesting a jigsaw puzzle

ji·had \ji-ˈhäd, *chiefly British* -ˈhad\ *noun* [Arabic *jihād*] (1869)
1 : a holy war waged on behalf of Islam as a religious duty
2 : a crusade for a principle or belief

jil·lion \ˈjil-yən\ *noun* [*j* + *-illion* (as in *million*)] (circa 1942)
: an indeterminately large number
— jillion *adjective*

[1]**jilt** \ˈjilt\ *transitive verb* (1673)
: to drop (a lover) capriciously or unfeelingly
— jilt·er *noun*

[2]**jilt** *noun* [alteration of *jillet* flirtatious girl] (circa 1674)
: one who jilts a lover

jim crow \ˈjim-ˈkrō\ *noun, often J&C capitalized* [*Jim Crow,* stereotype black man in a 19th century song-and-dance act] (1838)
1 : BLACK 4 — usually taken to be offensive
2 : ethnic discrimination especially against blacks by legal enforcement or traditional sanctions
— jim crow *adjective, often J&C capitalized*
— jim crow·ism \-ˌi-zəm\ *noun, often J&C capitalized*

jim–dan·dy \ˈjim-ˈdan-dē\ *noun* [from the name *Jim*] (1887)
: something excellent of its kind

jim·jams \ˈjim-ˌjamz\ *noun plural* [perhaps alteration of *delirium tremens*] (1852)
: JITTERS

jim·mies \ˈji-mēz\ *noun plural* [origin unknown] (circa 1947)
: tiny rod-shaped bits of usually chocolate-flavored candy often sprinkled on ice cream

[1]**jim·my** \ˈji-mē\ *noun, plural* **jimmies** [from the name *Jimmy*] (1848)
: a short crowbar

[2]**jimmy** *transitive verb* **jim·mied; jim·my·ing** (1893)
: to force open with or as if with a jimmy ⟨the burglar *jimmied* a window⟩

jim·son·weed \ˈjim(p)-sən-ˌwēd\ *noun, often capitalized* [*Jamestown,* Va.] (1832)
: a poisonous tall coarse annual weed (*Datura stramonium*) of the nightshade family with rank-smelling foliage and large white or violet trumpet-shaped flowers succeeded by globose prickly fruits

[1]**jin·gle** \ˈjiŋ-gəl\ *verb* **jin·gled; jin·gling** \-g(ə-)liŋ\ [Middle English *ginglen,* of imitative origin] (14th century)
intransitive verb
1 : to make a light clinking or tinkling sound
2 : to rhyme or sound in a catchy repetitious manner
transitive verb
: to cause to jingle
— jin·gler \-g(ə-)lər\ *noun*

[2]**jingle** *noun* (1599)
1 a : a light clinking or tinkling sound **b :** a catchy repetition of sounds in a poem
2 a : something that jingles **b :** a short verse or song marked by catchy repetition
— jin·gly \-g(ə-)lē\ *adjective*

[1]**jin·go** \ˈjiŋ-(ˌ)gō\ *interjection* [probably euphemism for *Jesus*] (1694)
— used as a mild oath usually in the phrase *by jingo*

[2]**jingo** *noun, plural* **jingoes** [from the fact that the phrase *by jingo* appeared in the refrain of a chauvinistic song] (1878)
: one characterized by jingoism
— jin·go·ish \-ish\ *adjective*

jin·go·ism \ˈjiŋ-(ˌ)gō-ˌi-zəm\ *noun* (1878)
: extreme chauvinism or nationalism marked especially by a belligerent foreign policy ◇
— jin·go·ist \-ist\ *noun or adjective*
— jin·go·is·tic \ˌjiŋ-gō-ˈis-tik\ *adjective*
— jin·go·is·ti·cal·ly \-ti-k(ə-)lē\ *adverb*

[1]**jink** \ˈjiŋk\ *intransitive verb* [origin unknown] (1785)
: to move quickly or unexpectedly with sudden turns and shifts (as in dodging)

[2]**jink** *noun* (1786)
1 : a quick evasive turn **:** SLIP
2 *plural* **:** PRANKS, FROLICS; *especially* **:** HIGH JINKS

jin·ni \ˈjē-nē, ˈji-, jə-ˈnē\ *or* **jinn** \ˈjin\ *noun, plural* **jinn** *or* **jinns** [Arabic *jinnīy* demon] (1684)
1 : one of a class of spirits that according to Muslim demonology inhabit the earth, assume various forms, and exercise supernatural power
2 : GENIE 2

jin·rick·sha *or* **jin·rik·i·sha** \jin-ˈrik-ˌshȯ\ *noun* [Japanese] (1874)
: RICKSHA

[1]**jinx** \ˈjiŋ(k)s\ *noun* [perhaps alteration of *jynx* wryneck; from the use of wrynecks in witchcraft] (1911)
: one that brings bad luck; *also* **:** the state or spell of bad luck brought on by a jinx

[2]**jinx** *transitive verb* (1917)
: to foredoom to failure or misfortune **:** bring bad luck to

ji·pi·ja·pa \ˌhē-pē-ˈhä-pə\ *noun* [Spanish, from *Jipijapa,* Ecuador] (1858)
1 : a Central and South American plant (*Carludovica palmata* of the family Cyclanthaceae) resembling a palm
2 : PANAMA

jit·ney \ˈjit-nē\ *noun, plural* **jitneys** [origin unknown] (1903)
1 *slang* **:** NICKEL 2a(1)
2 [from the original 5 cent fare] **:** BUS 1a; *especially* **:** a small bus that carries passengers over a regular route on a flexible schedule

[1]**jit·ter** \ˈji-tər\ *noun* [origin unknown] (1929)
1 *plural* **:** a sense of panic or extreme nervousness ⟨had a bad case of the *jitters* before his performance⟩
2 : the state of mind or the movement of one that jitters
3 : irregular random movement (as of a pointer or an image on a television screen); *also* **:** vibratory motion

[2]**jitter** *intransitive verb* (1931)
1 : to be nervous or act in a nervous way
2 : to make continuous fast repetitive movements

[1]**jit·ter·bug** \ˈji-tər-ˌbəg\ *noun* (1938)
1 : a jazz variation of the two-step in which couples swing, balance, and twirl in standardized patterns and often with vigorous acrobatics
2 : one who dances the jitterbug

[2]**jitterbug** *intransitive verb* (1939)
1 : to dance the jitterbug
2 : to move around or back and forth with quick often jerky movements especially to confuse or disconcert an opponent in sports

jit·tery \ˈji-tə-rē\ *adjective* (1931)
1 : suffering from the jitters
2 : marked by jittering movements
— jit·ter·i·ness \-nəs\ *noun*

jiu·jit·su, jiu·jut·su *variant of* JUJITSU

[1]**jive** \ˈjīv\ *noun* [origin unknown] (1928)
1 : swing music or the dancing performed to it
2 a : glib, deceptive, or foolish talk **b :** the jargon of hipsters **c :** a special jargon of difficult or slang terms
— jivey \ˈjī-vē\ *adjective*

[2]**jive** *verb* **jived; jiv·ing** (1928)
transitive verb
1 : TEASE 3, CAJOLE
2 : SWING 5
intransitive verb
1 : KID
2 : to dance to or play jive

[3]**jive** *adjective* (1953)
slang **:** PHONY

jo \ˈjō\ *noun, plural* **joes** [alteration of *joy*] (circa 1529)
chiefly Scottish **:** SWEETHEART, DEAR

[1]**job** \ˈjäb\ *noun* [perhaps from obsolete English *job* lump] (circa 1627)
1 a : a piece of work; *especially* **:** a small miscellaneous piece of work undertaken on order at a stated rate **b :** the object or material on which work is being done **c :** something produced by or as if by work ⟨do a better *job* next time⟩ **d :** an example of a usually specified type **:** ITEM ⟨this *job* is round-necked and sleeveless —Lois Long⟩
2 a : something done for private advantage ⟨suspected the whole incident was a put-up *job*⟩ **b :** a criminal enterprise; *specifically* **:** ROBBERY **c :** a damaging or destructive bit of work ⟨did a *job* on him⟩
3 a (1) **:** something that has to be done **:** TASK (2) **:** an undertaking requiring unusual exertion ⟨it was a real *job* to talk over that noise⟩ **b :** a specific duty, role, or function **c :** a regular remunerative position **d** *chiefly British* **:** state of affairs — used with *bad* or *good* ⟨it was a good *job* you didn't hit the old man —E. L. Thomas⟩
synonym see TASK
— on the job : at work

[2]**job** *verb* **jobbed; job·bing** (1694)
intransitive verb
1 : to do odd or occasional pieces of work for hire
2 : to carry on public business for private gain
3 : to carry on the business of a middleman or wholesaler
transitive verb

◇ WORD HISTORY

jingoism The word *jingoism* grew out of an issue that bedeviled 19th century European diplomats, the Eastern Question. As the Ottoman Empire, "the sick man of Europe," gradually collapsed, the other major European powers came into constant conflict on when and how to carve it up among themselves. Britain's policy was to support Turkey in order to prevent any other single power from overwhelming it, and pursuit of this aim led to war with Russia in 1854–55 and a near-war in 1878, when Russian troops, in support of their Slavic coreligionists fighting Ottoman rule in the Balkans, reached the outskirts of Constantinople. With the dispatch of the British fleet to Constantinople, anti-Russian sentiment mounted and the words of a music-hall song expressed the patriotic fervor of some Britons: "We don't want to fight, yet by jingo if we do, we've got the ships, we've got the men, we've got the money too." The song made *jingo* a catchword for a militant patriot. War never came, and the Eastern Question was temporarily settled by an international conference in Berlin, but *jingoism* has remained in English as a byword for excessive patriotism.

1 : to buy and sell (as stock) for profit **:** SPECULATE
2 : to hire or let by the job or for a period of service
3 : to get, deal with, or effect by jobbery
4 : to do or cause to be done by separate portions or lots **:** SUBCONTRACT
5 : to penalize or deprive unfairly

3job *adjective* (1710)
1 *British* **:** that is for hire for a given service or period
2 : used in, engaged in, or done as job work ⟨a *job* shop⟩
3 : of or relating to a job or to employment ⟨a guarantee of *job* security⟩

Job \'jōb\ *noun* [Latin, from Greek *Iōb,* from Hebrew *Iyyōbh*]
1 : the hero of the book of Job who endures afflictions with fortitude and faith
2 : a narrative and poetic book of canonical Jewish and Christian Scripture — see BIBLE table

job action *noun* (circa 1968)
: a temporary action (as a slowdown) by workers as a protest and means of forcing compliance with demands

job·ber \'jä-bər\ *noun* (1670)
: one that jobs: as **a** (1) **:** WHOLESALER; *specifically* **:** a wholesaler who operates on a small scale or who sells only to retailers and institutions (2) **:** STOCKJOBBER **a b :** a person who works by the job

job·bery \'jä-b(ə-)rē\ *noun* (1832)
: the act or practice of jobbing; *especially* **:** corruption in public office

job·hold·er \'jäb-,hōl-dər\ *noun* (1904)
: a person having a regular job

job–hop·ping \-,hä-piŋ\ *noun* (circa 1952)
: the practice of moving from job to job
— job–hop·per \-,hä-pər\ *noun*

job·less \'jäb-ləs\ *adjective* (1919)
1 : having no job
2 : of or relating to those having no job
— job·less·ness *noun*

job lot *noun* (1851)
1 : a miscellaneous collection of goods for sale as a lot usually to a retailer
2 : a miscellaneous and usually inferior collection or group

Job's comforter \'jōbz-\ *noun* [from the tone of the speeches made to Job by his friends] (1738)
: a person who discourages or depresses while seemingly giving comfort and consolation

Job's tears *noun plural* (1597)
1 : hard pearly white seeds often used as beads
2 *singular in construction* **:** an Asian grass (*Coix lacryma-jobi*) whose seeds are Job's tears

Job's tears 2

Jo·cas·ta \jō-'kas-tə\ *noun* [Latin, from Greek *Iokastē*]
: a queen of Thebes who marries Oedipus not knowing that he is her son

1jock \'jäk\ *noun* (1826)
1 : JOCKEY
2 : DISC JOCKEY

2jock *noun* [*jockstrap*] (1922)
1 : ATHLETIC SUPPORTER
2 : ATHLETE; *especially* **:** a college athlete

1jock·ey \'jä-kē\ *noun, plural* **jockeys** [*Jockey,* Scots nickname for *John*] (1670)
1 : a person who rides or drives a horse especially as a professional in a race
2 : a person who operates or works with a specified vehicle, device, or object **:** OPERATOR ⟨an accountant, a pencil *jockey* — with almost no association with the out-of-doors —James Selder⟩

2jockey *verb* **jock·eyed; jock·ey·ing** (1708)
transitive verb
1 : to deal shrewdly or fraudulently with
2 a : to ride or drive (a horse) as a jockey **b :** DRIVE, OPERATE
3 a : to maneuver or manipulate by adroit or devious means ⟨was *jockeyed* out of the job⟩ **b :** to change the position of by a series of movements ⟨*jockey* a truck into position⟩
intransitive verb
1 : to act as a jockey
2 : to maneuver for advantage — often used in the phrase *jockey for position*

jockey club *noun* (1775)
: an association for the promotion and regulation of horse racing

jock itch *noun* [2*jock*] (1950)
: ringworm of the crotch **:** TINEA CRURIS

jock·strap \'jäk-,strap\ *noun* [English slang *jock* penis + English *strap*] (1886)
: ATHLETIC SUPPORTER

jo·cose \jō-'kōs, jə-\ *adjective* [Latin *jocosus,* from *jocus* joke] (1673)
1 : given to joking **:** MERRY
2 : characterized by joking **:** HUMOROUS
synonym see WITTY
— jo·cose·ly *adverb*
— jo·cose·ness *noun*
— jo·cos·i·ty \jō-'kä-sə-tē, jə-\ *noun*

joc·u·lar \'jä-kyə-lər\ *adjective* [Latin *jocularis,* from *joculus,* diminutive of *jocus*] (1626)
1 : given to jesting **:** habitually jolly or jocund
2 : characterized by jesting **:** PLAYFUL
synonym see WITTY
— joc·u·lar·i·ty \,jä-kyə-'lar-ə-tē\ *noun*
— joc·u·lar·ly \'jä-kyə-lər-lē\ *adverb*

jo·cund \'jä-kənd *also* 'jō-(,)kənd\ *adjective* [Middle English, from Late Latin *jocundus,* alteration of Latin *jucundus,* from *juvare* to help] (14th century)
: marked by or suggestive of high spirits and lively mirthfulness ⟨a poet could not but be gay, in such a *jocund* company —William Wordsworth⟩
synonym see MERRY
— jo·cun·di·ty \jō-'kən-də-tē, jä-\ *noun*
— jo·cund·ly \'jä-kənd-lē, 'jō-(,)\ *adverb*

jodh·pur \'jäd-(,)pər\ *noun* [*Jodhpur,* India] (1899)
1 *plural* **:** riding breeches cut full through the hips and close-fitting from knee to ankle
2 : an ankle-high boot fastened with a strap that is buckled at the side — called also *jodhpur boot*

joe \'jō\ *noun, often capitalized* [from *Joe,* nickname for *Joseph*] (1846)
: FELLOW, GUY ⟨an average *joe*⟩

Jo·el \'jō(-ə)l\ *noun* [Latin, from Greek *Iōēl,* from Hebrew *Yō'ēl*]
1 : the traditionally assumed author of the book of Joel
2 : a narrative and apocalyptic book of canonical Jewish and Christian Scripture — see BIBLE table

joe–pye weed \'jō-'pī-\ *noun* [origin unknown] (circa 1818)
: any of several tall American perennial composite herbs (especially *Eupatorium maculatum* and *E. purpureum*) with whorled leaves and corymbose heads of typically purple tubular flowers

jo·ey \'jō-ē\ *noun* [origin unknown] (1839)
Australian **:** a baby animal; *especially* **:** a baby kangaroo

1jog \'jäg, 'jȯg\ *verb* **jogged; jog·ging** [probably alteration of *shog*] (1548)
transitive verb
1 : to give a slight shake or push to **:** NUDGE
2 : to rouse to alertness ⟨*jogged* his memory⟩
3 : to cause (as a horse) to go at a jog
4 : to align the edges of (piled sheets of paper) by hitting or shaking against a flat surface
intransitive verb
1 : to move up and down or about with a short heavy motion ⟨his . . . holster *jogging* against his hip —Thomas Williams⟩
2 a : to run or ride at a slow trot **b :** to go at a slow, leisurely, or monotonous pace **:** TRUDGE

2jog *noun* (1635)
1 : a slight shake **:** PUSH
2 a : a movement, pace, or instance of jogging (as for exercise) **b :** a horse's slow measured trot

3jog *noun* [probably alteration of 2*jag*] (1715)
1 a : a projecting or retreating part (as of a line or surface) **b :** the space in the angle of a jog
2 : a brief abrupt change in direction

4jog *intransitive verb* **jogged; jog·ging** (1953)
: to make a jog ⟨the road *jogs* to the right⟩

jog·ger \'jä-gər, 'jȯ-\ *noun* (circa 1700)
1 : one that jogs
2 : a device for jogging piled sheets of paper

1jog·gle \'jä-gəl\ *verb* **jog·gled; jog·gling** \-g(ə-)liŋ\ [frequentative of 1*jog*] (1513)
transitive verb
: to shake slightly
intransitive verb
: to move shakily or jerkily
— jog·gler \-g(ə-)lər\ *noun*

2joggle *noun* (circa 1727)
: 2JOG 2a

3joggle *noun* [diminutive of 3*jog*] (1793)
1 : a notch or tooth in a joining surface (as of a piece of building material) to prevent slipping
2 : a dowel for joining two adjacent blocks of masonry

4joggle *transitive verb* **jog·gled; jog·gling** \'jä-g(ə-)liŋ\ (1820)
: to join by means of a joggle so as to prevent sliding apart

jog trot *noun* (1796)
1 : 2JOG 2b
2 : a routine habit or course of action

Jo·han·nine \jō-'ha-,nīn, -nən\ *adjective* [Late Latin *Johannes* John] (1861)
: of, relating to, or characteristic of the apostle John or the New Testament books ascribed to him

Jo·han·nis·berg Riesling \yō-'hä-nəs-,berg-\ *noun* [*Johannisberg,* village in Germany] (1976)
: a Riesling produced in the U.S. (as in California)

john \'jän\ *noun* [from the name *John*] (1856)
1 : TOILET
2 : a prostitute's client

John \'jän\ *noun* [Late Latin *Johannes,* from Greek *Iōannēs,* from Hebrew *Yōḥānān*]
1 : a Jewish prophet who according to Gospel accounts foretold Jesus's messianic ministry and baptized him — called also *John the Baptist*
2 : an apostle who according to various Christian traditions wrote the fourth Gospel, the three Johannine Epistles, and the Book of Revelation
3 : the fourth Gospel in the New Testament — see BIBLE table
4 : any of three short didactic letters addressed to early Christians and included in the New Testament — see BIBLE table

John Barleycorn *noun* (circa 1620)
: alcoholic liquor personified

john·boat \'jän-,bōt\ *noun* [from the name *John*] (1905)
: a narrow flat-bottomed square-ended boat usually propelled by a pole or paddle and used on inland waterways

John Bull \-'bu̇l\ *noun* [*John Bull,* character typifying the English nation in *The History of John Bull* (1712) by John Arbuthnot] (1778)
1 : the English nation personified **:** the English people
2 : a typical Englishman
— John Bull·ish \-'bu̇-lish\ *adjective*

— **John Bull·ish·ness** *noun*
— **John Bull·ism** \-ˌli-zəm\ *noun*
John Doe \-ˈdō\ *noun* (1768)
1 : a party to legal proceedings whose true name is unknown
2 : an average man
John Do·ry \-ˈdōr-ē, -ˈdȯr-\ *noun, plural* **John Dories** [earlier *dory,* from Middle English *dorre,* from Middle French *doree,* literally, gilded one] (1754)
: a common yellow to olive food fish (*Zeus faber*) of Europe and Africa with an oval compressed body, long dorsal spines, and a dark spot on each side; *also* **:** a closely related and possibly identical fish (*Z. capensis*) widely distributed in southern seas
Joh·ne's disease \ˈyō-nəz-\ *noun* [Heinrich A. *Johne* (died 1910) German bacteriologist] (1907)
: a chronic often fatal contagious enteritis of ruminants and especially of cattle that is caused by a bacterium (*Mycobacterium paratuberculosis*) and is characterized by persistent diarrhea and gradual emaciation
John Han·cock \jän-ˈhan-ˌkäk\ *noun* [*John Hancock;* from the prominence of his signature on the Declaration of Independence] (1903)
: an autograph signature
John Hen·ry \-ˈhen-rē\ *noun* [from the name *John Henry,* from confusion with *John Hancock*] (1914)
: an autograph signature
John Mark *noun*
: MARK 1a
john·ny \ˈjä-nē\ *noun, plural* **johnnies** [from the name *Johnny*] (1673)
1 *often capitalized* **:** FELLOW, GUY
2 : a short-sleeved collarless gown that is open in the back and is worn by persons (as hospital patients) undergoing medical examination or treatment
john·ny·cake \ˈjä-nē-ˌkāk\ *noun* [probably from the name *Johnny*] (1739)
: a bread made with cornmeal
John·ny–come–late·ly \ˈjä-nē-(ˌ)kəm-ˈlāt-lē\ *noun, plural* **Johnny–come–latelies** *or* **Johnnies–come–lately** (1839)
1 : a late or recent arrival **:** NEWCOMER
2 : UPSTART ⟨established families tend to hold themselves above the *Johnny-come-latelies* —William Zeckendorf (died 1976)⟩
John·ny–jump–up \ˌjä-nē-ˈjəmp-ˌəp\ *noun* (1842)
1 : a common and long-cultivated European viola (*Viola tricolor*) which has short-spurred flowers usually blue or purple mixed with white and yellow and from which most of the garden pansies are derived; *broadly* **:** any of various small-flowered cultivated pansies
2 : any of various American violets
John·ny–on–the–spot \ˈjä-nē-ˌȯn-t͟hə-ˈspät, -ˌän-\ *noun* (1896)
: a person who is on hand and ready to perform a service or respond to an emergency
Johnny Reb \-ˈreb\ *noun* [from the name *Johnny* + *reb* rebel] (1865)
: a Confederate soldier
John·son·ese \ˌjän(t)-sə-ˈnēz, -ˈnēs\ *noun* [Samuel *Johnson*] (1843)
: a literary style characterized by balanced phraseology and Latinate diction
john·son·grass \ˈjän(t)-sən-ˌgras\ *noun, often capitalized* [William *Johnston* (died 1859) American agriculturist] (1884)
: a tall perennial sorghum (*Sorghum halepense*) originally of the Mediterranean region that is widely used for forage in warm areas and often becomes naturalized
joie de vi·vre \ˌzhwä-də-ˈvēvrᵊ\ *noun* [French, literally, joy of living] (1889)
: keen or buoyant enjoyment of life
¹join \ˈjȯin\ *verb* [Middle English, from Old French *joindre,* from Latin *jungere* — more at YOKE] (13th century)
transitive verb
1 a : to put or bring together so as to form a unit ⟨*join* two blocks of wood with glue⟩ **b :** to connect (as points) by a line **c :** ADJOIN
2 : to put or bring into close association or relationship ⟨*joined* in marriage⟩
3 : to engage in (battle)
4 a : to come into the company of ⟨*joined* us for lunch⟩ **b :** to associate oneself with ⟨*joined* the church⟩
intransitive verb
1 a : to come together so as to be connected ⟨nouns *join* to form compounds⟩ **b :** ADJOIN ⟨the two estates *join*⟩
2 : to come into close association or relationship: as **a :** to form an alliance **b :** to become a member of a group **c :** to take part in a collective activity ⟨*join* in singing⟩ ☆
— **join·able** \ˈjȯi-nə-bəl\ *adjective*
²join *noun* (1825)
1 : JOINT
2 : UNION 2d
join·der \ˈjȯin-dər\ *noun* [French *joindre* to join, from Old French] (1601)
1 : CONJUNCTION 1
2 a (1) **:** a joining of parties as plaintiffs or defendants in a suit (2) **:** a joining of causes of action or defense **b :** acceptance of an issue tendered
join·er \ˈjȯi-nər\ *noun* (14th century)
: one that joins: as **a :** a person whose occupation is to construct articles by joining pieces of wood **b :** a gregarious or civic-minded person who joins many organizations
join·ery \ˈjȯi-nə-rē, ˈjȯin-rē\ *noun* (1678)
1 : the art or trade of a joiner
2 : work done by a joiner
join·ing \ˈjȯi-niŋ\ *noun* (14th century)
1 : the act or an instance of joining one thing to another **:** JUNCTURE
2 a : the place or manner of being joined together **b :** something that joins two things together
¹joint \ˈjȯint\ *noun* [Middle English *jointe,* from Old French, from *joindre*] (13th century)
1 a (1) **:** the point of contact between elements of an animal skeleton with the parts that surround and support it (2) **:** NODE 5b **b :** a part or space included between two articulations, knots, or nodes **c :** a large piece of meat for roasting
2 a : a place where two things or parts are joined **b :** a space between the adjacent surfaces of two bodies joined and held together (as by cement or mortar) **c :** a fracture or crack in rock not accompanied by dislocation **d :** the flexing part of a cover along either backbone edge of a book **e :** the junction of two or more members of a framed structure **f :** a union formed by two abutting rails in a track including the elements (as bars and bolts) necessary to hold the abutting rails together **g :** an area at which two ends, surfaces, or edges are attached
3 a : a shabby or disreputable place of entertainment **b :** PLACE, ESTABLISHMENT **c** *slang* **:** PRISON 2
4 : a marijuana cigarette
— **joint·ed** \ˈjȯin-təd\ *adjective*
— **joint·ed·ly** *adverb*
— **joint·ed·ness** *noun*
— **out of joint 1 a** *of a bone* **:** having the head slipped from its socket **b :** at variance **2 a :** DISORDERED 2a **b :** being out of humor **:** DISSATISFIED
²joint *adjective* [Middle English, from Middle French, from past participle of *joindre,* from Old French] (14th century)
1 : UNITED, COMBINED ⟨the *joint* influences of culture and climate⟩
2 : common to two or more: as **a** (1) **:** involving the united activity of two or more ⟨a *joint* effort⟩ (2) **:** constituting an activity, operation, or organization in which elements of more than one armed service participate ⟨*joint* maneuvers⟩ (3) **:** constituting an action or expression of two or more governments ⟨*joint* peace talks⟩ **b :** shared by or affecting two or more ⟨a *joint* fine⟩
3 : united, joined, or sharing with another (as in a right or status) ⟨*joint* heirs⟩
4 : being a function of or involving two or more variables and especially random variables
— **joint·ly** *adverb*
³joint *verb* [¹*joint*] (1530)
transitive verb
1 : to separate the joints of (as meat)
2 a : to unite by a joint **:** fit together **b :** to provide with a joint **:** ARTICULATE **c :** to prepare (as a board) for joining by planing the edge
intransitive verb
1 : to fit as if by joints ⟨the stones *joint* neatly⟩
2 : to form joints as a stage in growth — used especially of small grains
Joint Chiefs of Staff (1946)
: a military advisory group composed of the chiefs of staff of the army and air force, the chief of naval operations, and sometimes the commandant of the marine corps
joint·er \ˈjȯin-tər\ *noun* (1678)
: one that joints; *especially* **:** any of various tools used in making joints
joint grass *noun* (1835)
: a coarse creeping grass (*Paspalum distichum*) with jointed stems that is used for fodder and for erosion control
joint resolution *noun* (1838)
: a resolution passed by both houses of a legislative body that has the force of law when signed by or passed over the veto of the executive
join·tress \ˈjȯin-trəs\ *noun* (1602)
: a woman having a legal jointure
joint–stock company *noun* (1776)
: a company or association consisting of individuals organized to conduct a business for gain and having a joint stock of capital represented by shares owned individually by the members and transferable without the consent of the group
join·ture \ˈjȯin-chər\ *noun* (14th century)
1 a : an act of joining **:** the state of being joined **b :** JOINT
2 a : an estate settled on a wife to be taken by her in lieu of dower **b :** a settlement on the wife of a freehold estate for her lifetime
joint·worm \ˈjȯint-ˌwərm\ *noun* (1851)
: the larva of any of several small chalcid wasps (genus *Harmolita*) that attacks the stems of grain and causes swellings like galls at or just above the first joint
joist \ˈjȯist\ *noun* [Middle English *joiste,* from Middle French *giste,* from (assumed) Vulgar

☆ SYNONYMS
Join, combine, unite, connect, link, associate, relate mean to bring or come together into some manner of union. JOIN implies a bringing into contact or conjunction of any degree of closeness ⟨*joined* forces in an effort to win⟩. COMBINE implies some merging or mingling with corresponding loss of identity of each unit ⟨*combined* jazz and rock to create a new music⟩. UNITE implies somewhat greater loss of separate identity ⟨the colonies *united* to form a republic⟩. CONNECT suggests a loose or external attachment with little or no loss of identity ⟨a mutual defense treaty *connected* the two nations⟩. LINK may imply strong connection or inseparability of elements still retaining identity ⟨a name forever *linked* with liberty⟩. ASSOCIATE stresses the mere fact of frequent occurrence or existence together in space or in logical relation ⟨opera is popularly *associated* with high society⟩. RELATE suggests the existence of a real or presumed logical connection ⟨*related* what he observed to what he already knew⟩.

Latin *jacitum,* from Latin *jacēre* to lie — more at ADJACENT] (15th century)
: any of the small timbers or metal beams ranged parallel from wall to wall in a structure to support a floor or ceiling

joists

jo·jo·ba \hə-ˈhō-bə\ *noun* [Mexican Spanish] (1923)
: a shrub or small tree (*Simmondsia chinensis* synonym *S. californica*) of the box family of southwestern North America with edible seeds that yield a valuable liquid wax

[1]joke \ˈjōk\ *noun* [Latin *jocus;* perhaps akin to Old High German *gehan* to say, Sanskrit *yācati* he asks] (1670)
1 a : something said or done to provoke laughter; *especially* **:** a brief oral narrative with a climactic humorous twist **b** (1) **:** the humorous or ridiculous element in something (2) **:** an instance of jesting **:** KIDDING ⟨can't take a *joke*⟩ **c :** PRACTICAL JOKE **d :** LAUGHINGSTOCK
2 : something not to be taken seriously **:** a trifling matter ⟨consider his skiing a *joke* —Harold Callender⟩ — often used in negative construction ⟨it is no *joke* to be lost in the desert⟩

[2]joke *verb* **joked; jok·ing** (1670)
intransitive verb
: to make jokes **:** JEST
transitive verb
: to make the object of a joke **:** KID
— jok·ing·ly \ˈjō-kiŋ-lē\ *adverb*

jok·er \ˈjō-kər\ *noun* (1729)
1 a : a person given to joking **:** WAG **b :** FELLOW, GUY; *especially* **:** an insignificant, obnoxious, or incompetent person ⟨a shame to let a *joker* like this win —Harold Robbins⟩
2 : a playing card added to a pack as a wild card or as the highest-ranking card
3 a (1) **:** an ambiguous or apparently immaterial clause inserted in a legislative bill to make it inoperative or uncertain in some respect (2) **:** an unsuspected, misleading, or misunderstood clause, phrase, or word in a document that nullifies or greatly alters it **b :** something (as an expedient or stratagem) held in reserve to gain an end or escape from a predicament **c :** an unsuspected or not readily apparent fact, factor, or condition that thwarts or nullifies a seeming advantage

jok·ey *also* **joky** \ˈjō-kē\ *adjective* **jok·i·er; -est** (circa 1825)
1 : given to joking
2 : HUMOROUS, COMICAL
3 : amusingly ridiculous **:** LAUGHABLE
— jok·i·ly \-kə-lē\ *adverb*
— jok·i·ness \-kē-nəs\ *noun*

jol·li·fi·ca·tion \ˌjä-li-fə-ˈkā-shən\ *noun* (1809)
: FESTIVITY, MERRYMAKING

jol·li·ty \ˈjä-lə-tē\ *noun, plural* **-ties** (14th century)
1 : the quality or state of being jolly **:** MERRIMENT
2 *British* **:** a festive gathering

[1]jol·ly \ˈjä-lē\ *adjective* **jol·li·er; -est** [Middle English *joli,* from Old French] (14th century)
1 a (1) **:** full of high spirits **:** JOYOUS (2) **:** given to conviviality **:** JOVIAL **b :** expressing, suggesting, or inspiring gaiety **:** CHEERFUL
2 : extremely pleasant or agreeable **:** SPLENDID
synonym see MERRY

[2]jolly *adverb* (1549)
: VERY ⟨would . . . do as they were *jolly* well told —John Stockbridge⟩

[3]jolly *verb* **jol·lied; jol·ly·ing** (1610)
intransitive verb
: to engage in good-natured banter
transitive verb
: to put or try to put in good humor especially to gain an end

[4]jolly *noun, plural* **jollies** (1905)
1 *chiefly British* **:** a good time **:** JOLLIFICATION
2 *plural* **:** KICKS ⟨get their *jollies* by reenacting famous murders —H. F. Waters⟩

jol·ly boat \ˈjä-lē-\ *noun* [origin unknown] (circa 1741)
: a ship's boat of medium size used for general-purpose work

Jol·ly Rog·er \ˈjä-lē-ˈrä-jər\ *noun* [probably from [1]*jolly* + the name *Roger*] (circa 1785)
: a black flag with a white skull and crossbones formerly used by pirates as their ensign

[1]jolt \ˈjōlt\ *verb* [probably blend of obsolete *joll* to strike and *jot* to bump] (1596)
transitive verb
1 : to cause to move with a sudden jerky motion
2 : to give a knock or blow to; *specifically* **:** to jar with a quick or hard blow
3 a : to disturb the composure of ⟨crudely *jolted* out of that mood —Virginia Woolf⟩ **b :** to interfere with roughly, abruptly, and disconcertingly ⟨determination to pursue his own course was *jolted* badly —F. L. Paxson⟩
intransitive verb
: to move with a sudden jerky motion
— jolt·er *noun*

[2]jolt *noun* (1599)
1 : an abrupt sharp jerky blow or movement
2 a (1) **:** a sudden feeling of shock, surprise, or disappointment (2) **:** an event or development causing such a feeling ⟨the defeat was quite a *jolt*⟩ **b :** a serious check or reverse ⟨a severe financial *jolt*⟩
3 : a small potent or bracing portion ⟨a *jolt* of horseradish⟩
— jolty \ˈjōl-tē\ *adjective*

jolt–wag·on \ˈjōlt-ˌwa-gən\ *noun* (1886)
Midland **:** a farm wagon

Jo·nah \ˈjō-nə, *3 is also* -nər\ *noun* [Hebrew *Yōnāh*]
1 : an Israelite prophet who according to the account in the book of Johah resisted a divine call to preach repentance to the people of Nineveh, was swallowed and vomited by a great fish, and eventually carried out his mission
2 : a narrative book of canonical Jewish and Christian Scripture — see BIBLE table
3 : one believed to bring bad luck

Jo·nas \ˈjō-nəs\ *noun* [Late Latin, from Hebrew *Yōnāh*]
: JONAH

Jon·a·than \ˈjä-nə-thən\ *noun* [Hebrew *Yōnāthān*]
1 : a son of Saul and friend of David according to the account in I Samuel
2 : AMERICAN; *especially* **:** a New Englander
3 : any of a variety of red-skinned apple

jon·gleur \zhō^{n}-ˈglər\ *noun* [French, from Old French *jogleour* — more at JUGGLER] (1779)
: an itinerant medieval entertainer proficient in juggling, acrobatics, music, and recitation

jon·quil \ˈjän-kwəl, ˈjäŋ-\ *noun* [French *jonquille,* from Spanish *junquillo,* diminutive of *junco* reed, from Latin *juncus*] (1629)
: a Mediterranean perennial bulbous herb (*Narcissus jonquilla*) of the amaryllis family with long linear leaves that is widely cultivated for its yellow or white fragrant short-tubed clustered flowers — compare DAFFODIL

Jor·dan almond \ˈjȯr-d^{ə}n-\ *noun* [Middle English *jardin almande,* from Middle French *jardin* garden + Middle English *almande* almond] (1615)
: a large Spanish almond especially when salted or coated with sugar of various colors

Jor·dan curve \zhȯr-ˈdä^{n}-, ˈjȯr-d^{ə}n-\ *noun* [Camille *Jordan* (died 1922) French mathematician] (1900)
: SIMPLE CLOSED CURVE

Jordan curve theorem *noun* (1947)
: a fundamental theorem of topology: every simple closed curve divides the plane into two regions for which it is the common boundary

jo·rum \ˈjȯr-əm, ˈjȯr-\ *noun* [perhaps from *Joram* in the Bible who "brought with him vessels of silver" (2 Samuel 8:10—Authorized Version)] (1730)
: a large drinking vessel or its contents

jo·seph \ˈjō-zəf *also* -səf\ *noun* [Latin, from Greek *Iōsēph,* from Hebrew *Yōsēph*]
1 *capitalized* **a :** a son of Jacob who according to the account in Genesis rose to high political office in Egypt after being sold into slavery by his brothers **b :** the husband of Mary the mother of Jesus according to the Gospel accounts
2 : a long cloak worn especially by women in the 18th century

Jo·seph·ite \-zə-ˌfīt, -sə-\ *noun* (1890)
: a member of Saint Joseph's Society of the Sacred Heart founded in 1871 in Baltimore, Md. and devoted to missionary work among black Americans

Joseph of Ar·i·ma·thea \-ˌar-ə-mə-ˈthē-ə\
: a rich councillor of the Sanhedrin who according to the Gospel accounts placed the body of Jesus in his own tomb and according to medieval legend took the Holy Grail to England

Jo·seph·son junction \ˈjō-zəf-sən- *also* -səf-sən-\ *noun* [Brian D. *Josephson*] (1965)
: an electronic fast-switching device that consists of two layers of superconducting metal separated by a thin layer of insulator through which low current flows but increased current causes the insulator to block the flow

[1]josh \ˈjäsh\ *verb* [origin unknown] (1852)
transitive verb
: to tease good-naturedly **:** KID
intransitive verb
: to engage in banter **:** JOKE
— josh·er *noun*

[2]josh *noun* (1878)
: a good-humored joke **:** JEST

Josh·ua \ˈjä-sh(ə-)wə\ *noun* [Hebrew *Yĕhōshūaʻ*]
1 : the divinely commissioned successor of Moses and military leader of the Israelites during the conquest of Canaan according to the account in the book of Joshua
2 : a mainly narrative book of canonical Jewish and Christian Scripture — see BIBLE table

Joshua tree *noun* (1884)
: a tall branched arborescent yucca (*Yucca brevifolia*) of the southwestern U.S. that has clustered greenish white flowers

Joshua tree

joss \ˈjäs, ˈjȯs\ *noun* [Chinese Pidgin English, from Portuguese *deus* god, from Latin — more at DEITY] (1711)
: a Chinese idol or cult image

joss house *noun* (1771)
: a Chinese temple or shrine

joss stick *noun* (1845)
: a slender stick of incense burned in front of a joss

[1]jos·tle \ˈjä-səl\ *verb* **jos·tled; jos·tling** \-s(ə-)liŋ\ [alteration of *justle,* frequentative of [1]*joust*] (1546)
intransitive verb
1 a : to come in contact or into collision **b :** to make one's way by pushing and shoving **c :** to exist in close proximity
2 : to vie in gaining an objective **:** CONTEND
transitive verb
1 a : to come in contact or into collision with **b :** to force by pushing **:** ELBOW **c :** to stir up **:** AGITATE **d :** to exist in close proximity with
2 : to vie with in attaining an objective

[2]jostle *noun* (1611)
1 : a jostling encounter or experience

\ə\ **abut** \$^{ə}$\ **kitten** \ər\ **further** \a\ **ash** \ā\ **ace** \ä\ **mop, mar** \au̇\ **out** \ch\ **chin** \e\ **bet** \ē\ **easy** \g\ **go** \i\ **hit** \ī\ **ice** \j\ **job** \ŋ\ **sing** \ō\ **go** \ȯ\ **law** \ȯi\ **boy** \th\ **thin** \t͟h\ **the** \ü\ **loot** \u̇\ **foot** \y\ **yet** \zh\ **vision** *see also* Guide to Pronunciation

2 : the state of being crowded or jostled together
Jos·ue \'jä-shü-(,)ē\ *noun* [Late Latin, from Hebrew *Yĕhōshūa'*]
: JOSHUA
¹jot \'jät\ *noun* [Latin *iota, jota* iota] (1500)
: the least bit **:** IOTA ◆
²jot *transitive verb* **jot·ted; jot·ting** (1721)
: to write briefly or hurriedly **:** set down in the form of a note ⟨*jot* this down⟩
jotting *noun* (1814)
: a brief note **:** MEMORANDUM
Jo·tun *also* **Jo·tunn** \'yō-tᵊn, -,tün\ *noun* [Old Norse *jǫtunn*]
: a member of a race of giants in Norse mythology
Jo·tun·heim *also* **Jo·tunn·heim** \-,hīm, -,hām\ *noun* [Old Norse *Jǫtunheimar*]
: the home of the Jotuns in Norse mythology
jou·al \zhü-'al, -'äl, -'ål\ *noun* [Canadian French, rendering of a nonstandard pronunciation of French *cheval* horse] (1962)
: spoken Canadian French; *especially* **:** the local forms of the spoken French of Quebec that differ the most from prescribed forms
joule \'jü(ə)l *also* ÷'jaü(ə)l\ *noun* [James P. *Joule*] (1882)
: a unit of work or energy equal to the work done by a force of one newton acting through a distance of one meter
¹jounce \'jaün(t)s\ *verb* **jounced; jounc·ing** [Middle English] (15th century)
intransitive verb
: to move in an up-and-down manner **:** BOUNCE
transitive verb
: to cause to jounce
²jounce *noun* (circa 1787)
: JOLT
jouncy \'jaün(t)-sē\ *adjective* **jounc·i·er; -est** (1943)
: marked by a jouncing motion or effect
jour·nal \'jər-nᵊl\ *noun* [Middle English, service book containing the day hours, from Middle French, from *journal,* adjective, daily, from Latin *diurnalis,* from *diurnus* of the day, from *dies* day — more at DEITY] (15th century)
1 a : a record of current transactions; *especially* **:** a book of original entry in double-entry bookkeeping **b :** an account of day-to-day events **c :** a record of experiences, ideas, or reflections kept regularly for private use **d :** a record of transactions kept by a deliberative or legislative body **e :** LOG 3, 4
2 a : a daily newspaper **b :** a periodical dealing especially with matters of current interest
3 : the part of a rotating shaft, axle, roll, or spindle that turns in a bearing
journal box *noun* (circa 1859)
: a metal housing to support and protect a journal bearing
jour·nal·ese \,jər-nᵊl-'ēz, -'ēs\ *noun* (1882)
: a style of writing held to be characteristic of newspapers
jour·nal·ism \'jər-nᵊl-,i-zəm\ *noun* (1833)
1 a : the collection and editing of news for presentation through the media **b :** the public press **c :** an academic study concerned with the collection and editing of news or the management of a news medium
2 a : writing designed for publication in a newspaper or magazine **b :** writing characterized by a direct presentation of facts or description of events without an attempt at interpretation **c :** writing designed to appeal to current popular taste or public interest
jour·nal·ist \-nᵊl-ist\ *noun* (1693)
1 a : a person engaged in journalism; *especially* **:** a writer or editor for a news medium **b :** a writer who aims at a mass audience
2 : a person who keeps a journal
jour·nal·is·tic \,jər-nᵊl-'is-tik\ *adjective* (1829)
: of, relating to, or characteristic of journalism or journalists
— jour·nal·is·ti·cal·ly \-ti-k(ə-)lē\ *adverb*
jour·nal·ize \'jər-nᵊl-,īz\ *verb* **-ized; -iz·ing** (1766)
transitive verb
: to record in a journal
intransitive verb
1 : to keep a journal in accounting
2 : to keep a personal journal
— jour·nal·iz·er *noun*
¹jour·ney \'jər-nē\ *noun, plural* **journeys** [Middle English, from Old French *journee* day's journey, from *jour* day, from Late Latin *diurnum,* from Latin, neuter of *diurnus*] (13th century)
1 : travel or passage from one place to another **:** TRIP
2 *chiefly dialect* **:** a day's travel
3 : something suggesting travel or passage from one place to another ⟨the *journey* from youth to maturity⟩ ⟨a *journey* through time⟩
²journey *verb* **jour·neyed; jour·ney·ing** (14th century)
intransitive verb
: to go on a journey **:** TRAVEL
transitive verb
: to travel over or through
— jour·ney·er *noun*
jour·ney·man \-mən\ *noun* [Middle English, from *journey* journey, a day's labor + *man*] (15th century)
1 : a worker who has learned a trade and works for another person usually by the day
2 : an experienced reliable worker or performer especially as distinguished from one who is brilliant or colorful ⟨a good *journeyman* trumpeter —*New Yorker*⟩ ⟨a *journeyman* outfielder⟩
jour·ney·work \-,wərk\ *noun* (1601)
1 : work done by a journeyman
2 : HACKWORK
¹joust \'jaüst *sometimes* 'jəst *or* 'jüst\ *intransitive verb* [Middle English, from Middle French *juster* to unite, joust, from (assumed) Vulgar Latin *juxtare,* from Latin *juxta* near; akin to Latin *jungere* to join — more at YOKE] (14th century)
1 a : to fight on horseback as a knight or man-at-arms **b :** to engage in combat with lances on horseback
2 : to engage in personal combat or competition
— joust·er *noun*
²joust *noun* (14th century)
1 a : a combat on horseback between two knights with lances especially as part of a tournament **b** *plural* **:** TOURNAMENT
2 : a personal combat or competition **:** STRUGGLE
Jove \'jōv\ *noun* [Latin *Jov-, Juppiter*]
: JUPITER — often used interjectionally to express surprise or agreement especially in the phrase *by Jove*
jo·vial \'jō-vē-əl, -vyəl\ *adjective* (1592)
1 *capitalized* **:** of or relating to Jove
2 : markedly good-humored especially as evidenced by jollity and conviviality
synonym see MERRY
— jo·vi·al·i·ty \,jō-vē-'a-lə-tē\ *noun*
— jo·vial·ly \'jō-vē-ə-lē, -vyə-\ *adverb*
Jo·vi·an \'jō-vē-ən\ *adjective* (1530)
: of, relating to, or characteristic of the god or planet Jupiter
jow \'jaü\ *noun* [Middle English *jollen* to knock] (1515)
chiefly Scottish **:** STROKE, TOLL
jo·war \jə-'wär\ *noun* [Hindi *joār, juwār,* from Sanskrit *yavākāra* barley-shaped, from *yava* barley + *karoti* he makes] (1800)
: DURRA
¹jowl \'jaü(ə)l *sometimes* 'jōl\ *noun* [alteration of Middle English *choll*] (15th century)
: a cut of fish consisting of the head and usually adjacent parts
²jowl *noun* [alteration of Middle English *cholle,* probably from Old English *ceole* throat — more at GLUTTON] (1591)
: usually slack flesh (as a dewlap, wattle, or the pendulous part of a double chin) associated with the cheeks, lower jaw, or throat
³jowl *noun* [alteration of Middle English *chavel,* from Old English *ceafl;* akin to Middle High German *kivel* jaw, Avestan *zafar-* mouth] (1598)
1 a : JAW; *especially* **:** MANDIBLE **b :** one of the lateral halves of the mandible
2 a : CHEEK 1 **b :** the cheek meat of a hog ⟨a dinner of boiled *jowls*⟩
jowly \'jaü-lē *sometimes* 'jō-\ *adjective* **jowl·i·er; -est** (circa 1873)
: having marked jowls **:** having full or saggy flesh about the lower cheeks and jaw area ⟨elderly man with a disillusioned *jowly* face —John Dos Passos⟩
¹joy \'jöi\ *noun* [Middle English, from Old French *joie,* from Latin *gaudia,* plural of *gaudium,* from *gaudēre* to rejoice; probably akin to Greek *gēthein* to rejoice] (13th century)
1 a : the emotion evoked by well-being, success, or good fortune or by the prospect of possessing what one desires **:** DELIGHT **b :** the expression or exhibition of such emotion **:** GAIETY
2 : a state of happiness or felicity **:** BLISS
3 : a source or cause of delight
— joy·less \-ləs\ *adjective*
— joy·less·ly *adverb*
— joy·less·ness *noun*
²joy (14th century)
intransitive verb
: to experience great pleasure or delight **:** REJOICE
transitive verb
1 *archaic* **:** GLADDEN
2 *archaic* **:** ENJOY
joy·ance \'jöi-ən(t)s\ *noun* (circa 1586)
archaic **:** DELIGHT, ENJOYMENT
joy·ful \'jöi-fəl\ *adjective* (13th century)
: experiencing, causing, or showing joy **:** HAPPY
— joy·ful·ly \-fə-lē\ *adverb*
— joy·ful·ness *noun*
joy·ous \'jöi-əs\ *adjective* (14th century)
: JOYFUL
— joy·ous·ly *adverb*
— joy·ous·ness *noun*
joy·pop \'jöi-,päp\ *intransitive verb* (1953)
: to use habit-forming drugs occasionally or irregularly without becoming addicted
— joy·pop·per *noun*
joy·ride \'jöi-,rīd\ *noun* (1909)
1 : a ride taken for pleasure (as in a car or aircraft); *especially* **:** an automobile ride marked by reckless driving (as in a stolen car)
2 : conduct or action resembling a joyride especially in disregard of cost or consequences
— joyride *intransitive verb*
— joy·rid·er \-,rī-dər\ *noun*
— joy·rid·ing *noun*

◇ WORD HISTORY

jot "Till heaven and earth pass, one jot or one tittle shall in no wise pass from the law, till all be fulfilled." These are the words, in the King James Version, of Christ's assurance (Matthew 5:18) that He had not "come to destroy the law or the prophets." Not the smallest letter, not even a single stroke of a letter, we are told, will be lost. The word *jot* first appeared in English in William Tyndale's translation of Matthew's Gospel, in 1526. Tyndale simply put in an anglicized form the Latin word *jota* (or *iota*), itself simply a transliteration of the Greek name of the ninth letter in the Greek alphabet. The Gospel writer may have had in mind *yōdh,* the smallest letter in the Hebrew alphabet. The transfer across language boundaries was easily made because the Greek equivalent, *iōta,* was also the smallest letter in its alphabet. A *jot* is now simply "a very small part."

joy·stick \-ˌstik\ *noun* [perhaps from English slang *joystick* penis] (1910)
1 : a lever in an airplane that operates the elevators by a fore-and-aft motion and the ailerons by a side-to-side motion
2 : a control for any of various devices (as a computer display) that resembles an airplane's joystick especially in being capable of motion in two or more directions

J/psi particle \ˌjā-ˈsī-, -ˈpsī-\ *noun* (1977)
: an unstable neutral fundamental particle of the meson group that has a mass about 6000 times the mass of an electron — called also *J particle, psi particle*

ju·ba \ˈjü-bə\ *noun* [origin unknown] (1834)
: a dance of Southern plantation blacks accompanied by complexly rhythmic hand clapping and slapping of the knees and thighs

Ju·bal \ˈjü-bəl\ *noun* [Hebrew *Yūbhāl*]
: a descendant of Cain who according to the account in Genesis is the father of those who play the harp and organ

ju·bi·lance \ˈjü-bə-lən(t)s\ *noun* (1864)
: JUBILATION 1

ju·bi·lant \ˈjü-bə-lənt\ *adjective* (1667)
: EXULTANT
— ju·bi·lant·ly *adverb*

ju·bi·lar·i·an \ˌjü-bə-ˈler-ē-ən, -ˈlar-\ *noun* (1782)
: one celebrating a jubilee

ju·bi·late \ˈjü-bə-ˌlāt\ *intransitive verb* **-lat·ed; -lat·ing** [Latin *jubilatus,* past participle of *jubilare;* akin to Middle High German *jū* (exclamation of joy), Greek *iygē* shout] (circa 1641)
: REJOICE

Ju·bi·la·te \ˌyü-bə-ˈlä-ˌtā, ˌjü-\ *noun* [Latin, 2d person plural imperative of *jubilare*] (1549)
1 a : the 100th Psalm in the Authorized Version **b** *not capitalized* **:** a joyous song or outburst
2 : the third Sunday after Easter

ju·bi·la·tion \ˌjü-bə-ˈlā-shən\ *noun* (14th century)
1 : an act of rejoicing **:** the state of being jubilant
2 : an expression of great joy

¹ju·bi·lee \ˈjü-bə-(ˌ)lē, ˌjü-bə-ˈlē\ *noun* [Middle English, from Middle French & Late Latin; Middle French *jubilé,* from Late Latin *jubilaeus,* modification of Late Greek *iōbēlaios,* from Hebrew *yōbhēl* ram's horn, jubilee] (14th century)
1 *often capitalized* **:** a year of emancipation and restoration provided by ancient Hebrew law to be kept every 50 years by the emancipation of Hebrew slaves, restoration of alienated lands to their former owners, and omission of all cultivation of the land
2 a : a special anniversary; *especially* **:** a 50th anniversary **b :** a celebration of such an anniversary
3 a : a period of time proclaimed by the Roman Catholic pope ordinarily every 25 years as a time of special solemnity **b :** a special plenary indulgence granted during a year of jubilee to Roman Catholics who perform certain specified works of repentance and piety
4 a : JUBILATION **b :** a season of celebration
5 : an Afro-American religious song usually referring to a time of future happiness ◆

²jubilee *adjective, often capitalized* (1951)
: FLAMBÉ ⟨cherries *jubilee*⟩

Ju·dah \ˈjü-də\ *noun* [Hebrew *Yĕhūdhāh*]
: a son of Jacob and the traditional eponymous ancestor of one of the tribes of Israel

Ju·da·ic \jü-ˈdā-ik\ *also* **Ju·da·ical** \-ˈdā-ə-kəl\ *adjective* [Middle English *Judeical,* from Latin *judaicus,* from Greek *ioudaikos,* from *Ioudaios* Jew — more at JEW] (15th century)
: of, relating to, or characteristic of Jews or Judaism

Ju·da·ica \-ˈdā-ə-kə\ *noun plural* [Latin, neuter plural of *Judaicus*] (1923)
: literary or historical materials relating to Jews or Judaism

Ju·da·ism \ˈjü-də-ˌi-zəm, ˈjü-dē-, ˈjü-(ˌ)dā-, *British also* ˈjü-ˌdi-zəm\ *noun* (14th century)
1 : a religion developed among the ancient Hebrews and characterized by belief in one transcendent God who has revealed himself to Abraham, Moses, and the Hebrew prophets and by a religious life in accordance with Scriptures and rabbinic traditions
2 : conformity to Jewish rites, ceremonies, and practices
3 : the cultural, social, and religious beliefs and practices of the Jews
4 : the whole body of Jews **:** the Jewish people

Ju·da·ist \ˈjü-də-ist, ˈjü-dē-, jü-ˈdā-\ *noun* (circa 1846)
: one that believes in or practices Judaism
— Ju·da·is·tic \ˌjü-də-ˈis-tik, ˌjü-dē-, ˌjü-(ˌ)dā-\ *adjective*

Ju·da·ize \ˈjü-də-ˌīz, ˈjü-dē-, ˈjü-(ˌ)dā-\ *verb* **-ized; -iz·ing** (1582)
intransitive verb
: to adopt the customs, beliefs, or character of a Jew
transitive verb
: to make Jewish
— Ju·da·i·za·tion \ˌjü-də-ə-ˈzā-shən, ˌjü-dē-ə-, ˌjü-(ˌ)dā-ə-\ *noun*
— Ju·da·iz·er \ˈjü-də-ˌīz-ər, ˈjü-dē-, ˈjü-(ˌ)dā-\ *noun*

Ju·das \ˈjü-dəs\ *noun* [Late Latin, from Greek *Ioudas,* from Hebrew *Yĕhūdhāh*]
1 a : the apostle who in the Gospel accounts betrayed Jesus **b :** a son of James and one of the twelve apostles
2 : TRAITOR; *especially* **:** one who betrays under the guise of friendship
3 *not capitalized* **:** PEEPHOLE — called also *judas hole, judas window*

Judas Is·car·i·ot \-is-ˈkar-ē-ət\ *noun* [Late Latin *Judas Iscariotes,* from Greek *Ioudas Iskariōtēs*]
: JUDAS 1a

Judas tree *noun* [from the belief that Judas Iscariot hanged himself from a tree of this kind] (1668)
: any of a genus (*Cercis*) of leguminous trees and shrubs (as a redbud) often cultivated for their showy flowers; *especially* **:** a Eurasian tree (*C. siliquastrum*) with purplish rosy flowers

¹jud·der \ˈjə-dər\ *intransitive verb* [probably alteration of *shudder*] (1931)
chiefly British **:** to vibrate with intensity ⟨the engine stalled and kept *juddering* —Roy Spicer⟩

²judder *noun* (1935)
chiefly British **:** the action or sound of juddering

Jude \ˈjüd\ *noun* [Late Latin *Judas*]
1 : the author of the New Testament Epistle of Jude
2 : a short hortatory epistle addressed to early Christians and included as a book in the New Testament — see BIBLE table

Ju·deo–Chris·tian \jü-ˌdā-ō-ˈkris-chən, -ˈkrish- *also* ˌjü-dē-ō- *or* jü-ˌdē-ō-\ *adjective* [Latin *Judaeus* Jew — more at JEW] (1899)
: having historical roots in both Judaism and Christianity

Ju·deo–Span·ish \-ˈspa-nish\ *noun* (1851)
: the Romance language of Sephardic Jews especially in the Balkans and Asia Minor

¹judge \ˈjəj\ *verb* **judged; judg·ing** [Middle English *juggen,* from Old French *jugier,* from Latin *judicare,* from *judic-, judex* judge, from *jus* right, law + *dicere* to decide, say — more at JUST, DICTION] (13th century)
transitive verb
1 : to form an opinion about through careful weighing of evidence and testing of premises
2 : to sit in judgment on **:** TRY
3 : to determine or pronounce after inquiry and deliberation
4 : GOVERN, RULE — used of a Hebrew tribal leader
5 : to form an estimate or evaluation of; *especially* **:** to form a negative opinion about ⟨shouldn't *judge* him because of his accent⟩
6 : to hold as an opinion **:** GUESS, THINK ⟨I *judge* she knew what she was doing⟩
intransitive verb
1 : to form an opinion
2 : to decide as a judge
synonym see INFER
— judg·er *noun*

²judge *noun* [Middle English *juge,* from Middle French, from Latin *judex*] (14th century)
: one who judges: as **a :** a public official authorized to decide questions brought before a court **b** *often capitalized* **:** a tribal hero exercising leadership among the Hebrews after the death of Joshua **c :** one appointed to decide in a contest or competition **:** UMPIRE **d :** one who gives an authoritative opinion **e :** CRITIC
— judge·ship \-ˌship\ *noun*

judge advocate *noun* (1748)
1 : an officer assigned to the judge advocate general's corps or department
2 : a staff officer serving as legal adviser to a military commander

judge advocate general *noun* (1862)
: the senior legal officer and chief legal adviser in the army, air force, or navy

Judg·es \ˈjə-jəz\ *noun*
: a narrative and historical book of Jewish and Christian Scripture — see BIBLE table

judg·mat·ic \ˌjəj-ˈma-tik\ *or* **judg·mat·i·cal** \-ti-kəl\ *adjective* [probably irregular from *judgment*] (1826)
: JUDICIOUS
— judg·mat·i·cal·ly \-ti-k(ə-)lē\ *adverb*

judg·ment *or* **judge·ment** \ˈjəj-mənt\ *noun* (13th century)
1 a : a formal utterance of an authoritative opinion **b :** an opinion so pronounced
2 a : a formal decision given by a court **b** (1) **:** an obligation (as a debt) created by the decree of a court (2) **:** a certificate evidencing such a decree
3 a *capitalized* **:** the final judging of mankind by God **b :** a divine sentence or decision; *specifically* **:** a calamity held to be sent by God
4 a : the process of forming an opinion or evaluation by discerning and comparing **b :** an opinion or estimate so formed
5 a : the capacity for judging **:** DISCERNMENT **b :** the exercise of this capacity
6 : a proposition stating something believed or asserted
synonym see SENSE

◇ WORD HISTORY
jubilee Ancient Hebrew law established every 50th year as a year of emancipation and restoration. All Hebrew slaves were freed; lands were restored to their former owners; fields were left uncultivated. This year took its name, *yōbhēl,* from the ram's horn trumpets sounded to proclaim its advent. Since neither the Greeks nor the Romans held comparable celebrations, when the Hebrew scriptures were translated into Greek and later into Latin, the Hebrew name for this joyful time was retained. The Greeks simply put *yōbhēl* into Greek form as *iōbēlaios.* Under the influence of the Latin verb *jubilare,* which meant "to let out joyful shouts," *iōbēlaios* became *jubilaeus* in Late Latin. Since its introduction into English, in John Wycliffe's translation of the Bible, *jubilee* has broadened in meaning considerably. It is now used for any 50th anniversary and even for other special anniversaries.

judg•men•tal \ˌjəj-ˈmen-tᵊl\ *adjective* (1909)
1 : of, relating to, or involving judgment
2 : characterized by a tendency to judge harshly
— judg•men•tal•ly \-ē\

judgment day *noun* (1591)
1 *J&D capitalized* **:** the day of God's judgment of mankind at the end of the world according to various theologies
2 : a day of final judgment

ju•di•ca•to•ry \ˈjü-di-kə-ˌtōr-ē, -ˌtȯr-\ *noun, plural* **-ries** (circa 1575)
1 : JUDICIARY 1a
2 : JUDICATURE 2

ju•di•ca•ture \ˈjü-di-kə-ˌchu̇r, -chər, -ˌtyu̇r, -ˌtu̇r\ *noun* [Middle French, from Medieval Latin *judicatura,* from Latin *judicatus,* past participle of *judicare*] (circa 1530)
1 : the action of judging **:** the administration of justice
2 : a court of justice
3 : JUDICIARY 1

ju•di•cial \ju̇-ˈdi-shəl\ *adjective* [Middle English, from Latin *judicialis,* from *judicium* judgment, from *judex*] (14th century)
1 a : of or relating to a judgment, the function of judging, the administration of justice, or the judiciary ⟨*judicial* processes⟩ **b :** belonging to the branch of government that is charged with trying all cases that involve the government and with the administration of justice within its jurisdiction — compare EXECUTIVE, LEGISLATIVE
2 : ordered or enforced by a court ⟨*judicial* decisions⟩
3 : of, characterized by, or expressing judgment **:** CRITICAL 1b
4 : arising from a judgment of God
5 : belonging or appropriate to a judge or the judiciary
— ju•di•cial•ly \-ˈdi-sh(ə-)lē\ *adverb*

judicial review *noun* (circa 1924)
1 : REVIEW 5
2 : a constitutional doctrine that gives to a court system the power to annul legislative or executive acts which the judges declare to be unconstitutional

ju•di•cia•ry \ju̇-ˈdi-shē-ˌer-ē, -ˈdi-shə-rē\ *noun* [*judiciary,* adjective, from Latin *judiciarius* judicial, from *judicium*] (1787)
1 a : a system of courts of law **b :** the judges of these courts
2 : a branch of government in which judicial power is vested
— judiciary *adjective*

ju•di•cious \ju̇-ˈdi-shəs\ *adjective* (1598)
: having, exercising, or characterized by sound judgment **:** DISCREET
synonym see WISE
— ju•di•cious•ly *adverb*
— ju•di•cious•ness *noun*

Ju•dith \ˈjü-dəth\ *noun* [Late Latin, from Greek *Ioudith,* from Hebrew *Yĕhūdhīth*]
1 : the Jewish heroine who saves the city of Bethulia in the book of Judith
2 : a book of Scripture included in the Roman Catholic canon of the Old Testament and in the Protestant Apocrypha — see BIBLE table

ju•do \ˈjü-(ˌ)dō\ *noun* [Japanese *jūdō,* from *jū* weakness, gentleness + *dō* art] (1889)
: a sport developed from jujitsu that emphasizes the use of quick movement and leverage to throw an opponent
— judo•ist \-ˌō-ist\ *noun*

[1]jug \ˈjəg\ *noun* [perhaps from *Jug,* nickname for *Joan*] (1538)
1 a *chiefly British* **:** a small pitcher **b** (1) **:** a large deep usually earthenware or glass container with a narrow mouth and a handle (2) **:** the contents of such a container **:** JUGFUL
2 : JAIL, PRISON

[2]jug *transitive verb* **jugged; jug•ging** (1747)
1 : to stew (as a hare) in an earthenware vessel
2 : JAIL, IMPRISON

ju•gate \ˈjü-ˌgāt, -gət\ *adjective* [New Latin *jugum* yoke] (circa 1887)
1 : having parts arranged in pairs **:** PAIRED
2 : having a jugum

jug band *noun* (circa 1933)
: a band that uses primitive or improvised instruments (as jugs, washboards, and kazoos) to play blues, jazz, and folk music

jug•ful \ˈjəg-ˌfu̇l\ *noun* (1831)
1 : as much as a jug will hold
2 : a great deal — used in the phrase *not by a jugful*

jug•ger•naut \ˈjə-gər-ˌnȯt, -ˌnät\ *noun* [Hindi *Jagannāth,* literally, lord of the world, title of Vishnu] (1841)
1 *chiefly British* **:** a large heavy truck
2 : a massive inexorable force, campaign, movement, or object that crushes whatever is in its path ⟨an advertising *juggernaut*⟩ ⟨a political *juggernaut*⟩ ◆

[1]jug•gle \ˈjə-gəl\ *verb* **jug•gled; jug•gling** \-g(ə-)liŋ\ [Middle English *jogelen,* from Middle French *jogler* to joke, from Latin *joculari,* from *joculus,* diminutive of *jocus* joke] (15th century)
intransitive verb
1 : to perform the tricks of a juggler
2 : to engage in manipulation especially in order to achieve a desired end
transitive verb
1 a : to practice deceit or trickery on **:** BEGUILE **b :** to manipulate or rearrange especially in order to achieve a desired end ⟨*juggle* an account to hide a loss⟩
2 a : to toss in the manner of a juggler **b :** to hold or balance precariously
3 : to handle or deal with usually several things (as obligations) at one time so as to satisfy often competing requirements ⟨*juggle* the responsibilities of family life and full-time job —Jane S. Gould⟩

[2]juggle *noun* (1664)
: an act or instance of juggling: **a :** a trick of magic **b :** a show of manual dexterity **c :** an act of manipulation especially to achieve a desired end

jug•gler \ˈjə-g(ə-)lər\ *noun* [Middle English *jogelour,* from Old English *geogelere* jester, from Old French *jogleour,* from Latin *joculator,* from *joculari*] (14th century)
1 a : one who performs tricks or acts of magic or deftness **b :** one skilled in keeping several objects in motion in the air at the same time by alternately tossing and catching them
2 : one who manipulates especially in order to achieve a desired end

jug•glery \ˈjə-glə-rē\ *noun* (14th century)
1 : the art or practice of a juggler
2 : manipulation or trickery especially to achieve a desired end

[1]jug•u•lar \ˈjə-gyə-lər *also* ˈjü- *also* -g(ə-)lər\ *adjective* [Late Latin *jugularis,* from Latin *jugulum* collarbone, throat, from *jugum* yoke] (1597)
1 : of or relating to the throat or neck
2 : of or relating to the jugular vein

[2]jugular *noun* (1615)
1 : JUGULAR VEIN
2 : the most vital or vulnerable part of something ⟨showed an instinct for the *jugular* in competition⟩

jugular vein *noun* (1597)
: any of several veins of each side of the neck that return blood from the head

ju•gum \ˈjü-gəm\ *noun, plural* **ju•ga** \-gə\ *or* **jugums** [New Latin, from Latin, yoke — more at YOKE] (1893)
: the most posterior and basal region of an insect's wing modified in some lepidopterans into a lobe that couples the fore and hind wings during flight

jug wine *noun* (1972)
: table wine sold in large bottles

[1]juice \ˈjüs\ *noun* [Middle English *jus,* from Old French, broth, juice, from Latin; akin to Old Norse *ostr* cheese, Greek *zymē* leaven, Sanskrit *yūṣa* broth] (14th century)
1 : the extractable fluid contents of cells or tissues
2 a *plural* **:** the natural fluids of an animal body **b :** the liquid or moisture contained in something
3 a : the inherent quality of a thing **:** ESSENCE **b :** STRENGTH, VIGOR, VITALITY
4 : a medium (as electricity or gasoline) that supplies power
5 *slang* **:** LIQUOR
6 *slang* **:** exorbitant interest exacted of a borrower under the threat of violence
7 *slang* **:** INFLUENCE, CLOUT
8 : a motivating, inspiring, or enabling force or factor ⟨creative *juices*⟩
— juice•less \ˈjüs-ləs\ *adjective*

[2]juice *transitive verb* **juiced; juic•ing** (1603)
1 : to add juice to
2 : to extract the juice of

juiced \ˈjüst\ *adjective* (1592)
1 : containing juice — usually used in combination ⟨precious-*juiced* flowers —Shakespeare⟩
2 *slang* **:** DRUNK 1a

juice•head \ˈjüs-ˌhed\ *noun* (1955)
slang **:** ALCOHOLIC

juic•er \ˈjü-sər\ *noun* (1938)
1 : an appliance for extracting juice from fruit or vegetables
2 *slang* **:** a heavy or habitual drinker

juice up *transitive verb* (1955)
: to give life, energy, or spirit to

juicy \ˈjü-sē\ *adjective* **juic•i•er; -est** (15th century)
1 : having much juice **:** SUCCULENT
2 : rewarding or profitable especially financially **:** FAT ⟨*juicy* contract⟩ ⟨a *juicy* dramatic role⟩
3 a : rich in interest **:** COLORFUL ⟨*juicy* details⟩ **b :** SENSATIONAL, RACY ⟨a *juicy* scandal⟩ **c :** full of vitality **:** LUSTY
— juic•i•ly \-sə-lē\ *adverb*
— juic•i•ness \-sē-nəs\ *noun*

ju•jit•su *or* **ju•jut•su** \jü-ˈjit-(ˌ)sü\ *noun* [Japanese *jūjutsu,* from *jū* weakness, gentleness + *jutsu* art, skill] (1875)
: an art of weaponless fighting employing holds, throws, and paralyzing blows to subdue or disable an opponent

ju•ju \ˈjü-(ˌ)jü\ *noun* [of West African origin; akin to Hausa *jùju* fetish] (1894)
1 : a fetish, charm, or amulet of West African peoples
2 : the magic attributed to or associated with jujus

◇ WORD HISTORY

juggernaut *Jagannātha* (meaning "world-lord" in Sanskrit) is a form of the Hindu deity Krishna worshipped mainly in northeastern India. The god's chief annual festival involves a procession in which his image is mounted on an enormous wooden chariot, about 35 feet square and 45 feet high, which is wheeled from his temple at Puri, in the state of Orissa, to his nearby country house. Accounts by western travelers dating back to the Middle Ages claimed that some of Jagannātha's devotees would allow themselves to be crushed beneath the wheels of the chariot in sacrifice to their god. British officials in the 19th century determined that these reports were greatly exaggerated; occasionally people with painful illnesses or suicidal tendencies would throw themselves in front of the chariot, but most deaths during the celebration resulted accidentally, through the crush of thousands of pilgrims striving to touch the god's vehicle. Despite the spuriousness of the belief, the English form *juggernaut* began to be used in the 19th century to denote a massive, inexorable force or object that crushes everything in its path.

ju·jube \'jü-ˌjüb, *especially for 2* 'jü-jü-ˌbē\ *noun* [Middle English, from Medieval Latin *jujuba,* alteration of Latin *zizyphum,* from Greek *zizyphon*] (14th century)
1 a : an edible drupaceous fruit of any of several trees (genus *Ziziphus*) of the buckthorn family; *especially* **:** one of an Asian tree (*Z. jujuba*) **b :** a tree producing this fruit
2 : a fruit-flavored gumdrop or lozenge

juke \'jük\ *verb* **juked; juk·ing** [probably alteration of English dialect *jouk* to cheat, deceive] (1967)
transitive verb
: to fake out of position (as in football)
intransitive verb
: to juke someone

juke·box \'jük-ˌbäks\ *noun* [*juke* brothel, akin to Gullah *juke* disorderly, of West African origin; akin to Bambara *dzugu* wicked] (1939)
: a coin-operated phonograph or compact-disc player that automatically plays recordings selected from its list

juke joint *noun* (1937)
: a small inexpensive establishment for eating, drinking, or dancing to the music of a jukebox

ju·lep \'jü-ləp\ *noun* [Middle English, from Middle French, from Arabic *julāb,* from Persian *gulāb,* from *gul* rose + *āb* water] (14th century)
1 : a drink consisting of sweet syrup, flavoring, and water
2 : a drink consisting of a liquor (as bourbon or brandy) and sugar poured over crushed ice and garnished with mint

Ju·lian calendar \'jül-yən-\ *noun* [Latin *julianus,* from Gaius *Julius* Caesar] (circa 1771)
: a calendar introduced in Rome in 46 B.C. establishing the 12-month year of 365 days with each fourth year having 366 days and the months each having 31 or 30 days except for February which has 28 or in leap years 29 days — compare GREGORIAN CALENDAR

[1]**ju·li·enne** \ˌjü-lē-'en, ˌzhü-\ *noun* [French, short for *potage à la julienne,* probably from *Julienne* woman's name] (1841)
1 : a consommé containing julienned vegetables
2 a : food (as meat or vegetables) that has been julienned **b :** a preparation or garnish of julienned food ⟨a *julienne* of leeks⟩
— julienne *adjective*

[2]**julienne** *transitive verb* **-enned; -en·ning** (circa 1930)
: to slice into thin strips the size of matchsticks ⟨wash and *julienne* the carrots⟩

Ju·liet \'jül-yət; ˌjü-lē-'et, 'jü-lē-ˌ\ *noun*
: the heroine of Shakespeare's tragedy *Romeo and Juliet* who dies for love of Romeo

Ju·li·ett \ˌjü-lē-'et\ [probably irregular from *Juliet*] (1952)
— a communications code word for the letter *j*

Ju·ly \jü-'lī\ *noun* [Middle English *Julie,* from Old English *Julius,* from Latin, from Gaius *Julius* Caesar] (13th century)
: the 7th month of the Gregorian calendar

Ju·ma·da \jü-'ma-də\ *noun* [Arabic *Jumāda*] (1771)
: either of two months of the Islamic year: **a :** the 5th month **b :** the 6th month — see MONTH table

[1]**jum·ble** \'jəm-bəl\ *verb* **jum·bled; jum·bling** \-b(ə-)liŋ\ [perhaps imitative] (circa 1529)
intransitive verb
: to move in a confused or disordered manner
transitive verb
: to mix into a confused or disordered mass — often used with *up*

[2]**jumble** *noun* (1661)
1 a : a mass of things mingled together without order or plan **:** HODGEPODGE **b :** a state of confusion
2 *British* **:** articles for a rummage sale

[3]**jumble** *noun* [origin unknown] (1615)
: a small thin usually ring-shaped sugared cookie or cake

jumble sale *noun* (1898)
British **:** RUMMAGE SALE

jum·bo \'jəm-(ˌ)bō\ *noun, plural* **jumbos** [*Jumbo,* a huge elephant exhibited by P. T. Barnum] (1883)
: a very large specimen of its kind
— jumbo *adjective*

[1]**jump** \'jəmp\ *verb* [probably akin to Low German *gumpen* to jump] (1530)
intransitive verb
1 a : to spring into the air **:** LEAP; *especially* **:** to spring free from the ground or other base by the muscular action of feet and legs **b :** to move suddenly or involuntarily **:** START **c :** to move over a position occupied by an opponent's piece in a board game often thereby capturing the piece **d :** to undergo a vertical or lateral displacement owing to improper alignment of the film on a projector mechanism **e :** to start out or forward **:** BEGIN — usually used with *off* ⟨*jump* off to a big lead⟩ **f :** to move energetically **:** HUSTLE **g :** to go from one sequence of instructions in a computer program to another ⟨*jump* to a subroutine⟩
2 : COINCIDE, AGREE
3 a : to move haphazardly or irregularly **:** shift abruptly ⟨*jumped* from job to job⟩ **b :** to change or abandon employment especially in violation of contract **c :** to rise suddenly in rank or status **d :** to undergo a sudden sharp change in value ⟨prices *jumped*⟩ **e :** to make a jump in bridge **f :** to make a hurried judgment ⟨*jump* to conclusions⟩ **g :** to show eagerness ⟨*jumped* at the chance⟩ **h :** to enter eagerly ⟨*jump* on the bandwagon⟩
4 : to make a sudden physical or verbal attack ⟨*jumped* on him for his criticism⟩
5 : to bustle with activity ⟨the bar was *jumping* with young people⟩
transitive verb
1 a : to leap over ⟨*jump* a hurdle⟩ **b :** to move over (a piece) in a board game **c :** to act, move, or begin before (as a signal) ⟨*jump* the green light⟩ **d :** to leap aboard ⟨*jump* a freight⟩
2 *obsolete* **:** RISK, HAZARD
3 a : to escape from **:** AVOID **b :** to leave hastily or in violation of contract ⟨*jump* town without paying their bills — Hamilton Basso⟩ **c :** to depart from (a normal course) ⟨*jump* the track⟩
4 a : to make a sudden physical or verbal attack on **b :** to occupy illegally ⟨*jump* a mining claim⟩
5 a (1) **:** to cause to leap (2) **:** to cause (game) to break cover **:** START, FLUSH **b :** to elevate in rank or status **c :** to raise (a bridge partner's bid) by more than one rank **d :** to increase suddenly and sharply
— jump bail : to abscond after being released from prison on bail
— jump ship 1 : to leave the company of a ship without authority **2 :** to desert a cause or party especially abruptly **:** DEFECT
— jump the gun 1 : to start in a race before the starting signal **2 :** to act, move, or begin something before the proper time
— jump the queue *British* **:** to advance directly to or as if to the head of a line

[2]**jump** *adverb* (1539)
obsolete **:** EXACTLY, PAT

[3]**jump** *noun* (circa 1552)
1 a (1) **:** an act of jumping **:** LEAP (2) **:** any of several sports competitions featuring a leap, spring, or bound (3) **:** a space cleared or covered by a leap (4) **:** an obstacle to be jumped over or from **b :** a sudden involuntary movement **:** START **c :** a move made in a board game by jumping **d :** a transfer from one sequence of instructions in a computer program to a different sequence ⟨conditional *jump*⟩
2 *obsolete* **:** VENTURE
3 a (1) **:** a sharp sudden increase (2) **:** a bid in bridge of more tricks than are necessary to overcall the preceding bid — compare SHIFT **b :** an abrupt change or transition **c** (1) **:** a quick short journey (2) **:** one in a series of moves from one place to another
4 : an advantage at the start ⟨desirous of getting the *jump* on the competition —Elmer Davis⟩

jump ball *noun* (1924)
: a method of putting a basketball into play by tossing it into the air between two opponents who jump up and attempt to tap the ball to a teammate

jump boot *noun* (1942)
: a boot worn especially by paratroopers

jump cut *noun* (1948)
: a sudden often jarring cut from one shot or scene to another without intervening devices (as fade-outs or dissolves); *broadly* **:** an abrupt transition (as in a narrative)
— jump–cut \'jəmp-ˌkət\ *verb*

[1]**jump·er** \'jəm-pər\ *noun* (1611)
1 : a person who jumps
2 a : any of various devices operating with a jumping motion **b :** any of several sleds **c :** a short wire used to close a break or cut out part of a circuit
3 : any of several jumping animals; *especially* **:** a saddle horse trained to jump obstacles
4 : JUMP SHOT

[2]**jum·per** \'jəm-pər\ *noun* [probably from English dialect *jump* jumper] (1853)
1 : a loose blouse or jacket worn by workmen
2 : a sleeveless one-piece dress worn usually with a blouse
3 : a child's coverall — usually used in plural
4 *chiefly British* **:** SWEATER 2

jumper cables *noun plural* (circa 1926)
: a pair of heavy-duty electrical cables with alligator clamps used to make a connection for jump-starting a vehicle

jump hook *noun* (1982)
: a hook shot in which the player jumps before releasing the ball

jumping bean *noun* (circa 1889)
: a seed of any of several Mexican shrubs (genera *Sebastiania* and *Sapium*) of the spurge family that tumbles about because of the movements of the larva of a small moth (*Carpocapsa saltitans*) inside it

jumping jack *noun* (1883)
1 : a toy figure of a man jointed and made to jump or dance by means of strings or a sliding stick
2 : a conditioning exercise performed from a standing position by jumping to a position with legs spread and hands touching overhead and then to the original position — called also *side-straddle hop*

jumping mouse *noun* (1826)
: any of several small hibernating North American rodents (family Zapodidae) with long hind legs and tail and no cheek pouches

jumping–off place \ˌjəm-piŋ-'ȯf-\ *noun* (1826)
1 : a remote or isolated place
2 : a place or point from which an enterprise, investigation, or discussion is launched — called also *jumping-off point*

jumping plant louse *noun* (circa 1901)
: any of numerous plant lice (family Psyllidae) with the femurs thickened and adapted for leaping

jumping spider *noun* (1813)
: any of a family (Salticidae) of small spiders that stalk and leap upon their prey

jump jet *noun* (1964)
: a military jet aircraft with vertical takeoff and landing capability

jump–off \'jəm-ˌpȯf\ *noun* (1917)
1 : the start of a race or an attack

2 : a jumping competition to break a tie at the end of regular competition (as in a horse show)

jump pass *noun* (circa 1948)
: a pass made by a player (as in football or basketball) while jumping

jump rope *noun* (1834)
: a rope used for exercises and children's games that involve jumping over the usually twirling rope each time it reaches its lowest point; *also* **:** a game played with a jump rope

jump seat *noun* (circa 1864)
1 : a movable carriage seat
2 : a folding seat between the front and rear seats of a passenger automobile

jump shot *noun* (1948)
: a shot in basketball made by jumping into the air and releasing the ball with one or both hands at the peak of the jump

jump–start \ˈjəmp-ˈstärt\ *transitive verb* (1973)
1 : to start (an engine or vehicle) by temporary connection to an external power source (as another vehicle's battery)
2 a : to get off to a speedy start ⟨advertising can *jump-start* a political campaign⟩ **b :** to impart fresh or renewed energy to **:** ENERGIZE ⟨a plan to *jump-start* the stagnant economy⟩
— jump start *noun*

jump•suit \ˈjəmp-ˌsüt\ *noun* (1944)
1 : a coverall worn by parachutists for jumping
2 : a one-piece garment consisting of a blouse or shirt with attached trousers or shorts

jumpy \ˈjəm-pē\ *adjective* **jump•i•er; -est** (1869)
1 : characterized by jumps or sudden variations
2 : NERVOUS, JITTERY
— jump•i•ness *noun*

jun \ˈjən\ *noun, plural* **jun** [Korean *chŏn*] (1966)
: the chon of North Korea

jun•co \ˈjəŋ-(ˌ)kō\ *noun, plural* **juncos** *or* **juncoes** [New Latin, from Spanish, reed — more at JONQUIL] (1887)
: any of a genus (*Junco*) of small widely distributed American finches usually having a pink bill, ashy gray head and back, and conspicuous white lateral tail feathers

junc•tion \ˈjəŋ(k)-shən\ *noun* [Latin *junction-, junctio,* from *jungere* to join — more at YOKE] (1711)
1 : an act of joining **:** the state of being joined
2 a : a place or point of meeting **b :** an intersection of roads especially where one terminates **c :** a point (as in a thermocouple) at which dissimilar metals make contact **d :** an interface in a semiconductor device between regions with different electrical characteristics
3 : something that joins
— junc•tion•al \-shnəl, -shə-nᵊl\ *adjective*

junc•tur•al \ˈjəŋ(k)-chə-rəl, ˈjəŋ(k)-shrəl\ *adjective* (1942)
: of or relating to phonetic juncture

junc•ture \ˈjəŋ(k)-chər\ *noun* (14th century)
1 a : JOINT, CONNECTION **b :** the manner of transition or mode of relationship between two consecutive sounds in speech
2 : an instance of joining **:** UNION, JUNCTION
3 : a point of time; *especially* **:** one made critical by a concurrence of circumstances ☆

June \ˈjün\ *noun* [Middle English, from Middle French & Latin; Middle French *Juin,* from Latin *Junius*] (14th century)
: the 6th month of the Gregorian calendar

June•ber•ry \ˈjün-ˌber-ē\ *noun* (circa 1810)
: SERVICEBERRY

june bug *noun, often J capitalized* (1829)
: any of numerous rather large leaf-eating scarab beetles (subfamily Melolonthinae) that fly chiefly in late spring and have larvae that are white grubs which live in soil and feed chiefly on the roots of grasses and other plants — called also *june beetle*

Jung•ian \ˈyu̇ŋ-ē-ən\ *adjective* (circa 1930)
: of, relating to, or characteristic of C. G. Jung or his psychological doctrines
— Jungian *noun*

jun•gle \ˈjəŋ-gəl\ *noun, often attributive* [Hindi *jaṅgal* forest, from Sanskrit *jaṅgala* desert region] (1776)
1 a : an impenetrable thicket or tangled mass of tropical vegetation **b :** a tract overgrown with thickets or masses of vegetation
2 : a hobo camp
3 a (1) **:** a confused or disordered mass of objects **:** JUMBLE (2) **:** something that baffles or frustrates by its tangled or complex character **:** MAZE ⟨the *jungle* of housing laws —Bernard Taper⟩ **b :** a place of ruthless struggle for survival ⟨the city is a *jungle* where no one is safe after dark —Stuart Chase⟩
— jun•gle•like \-gə(l)-ˌlīk\ *adjective*
— jun•gly \-g(ə-)lē\ *adjective*

jun•gled \-gəld\ *adjective* (1842)
: abounding in jungle ⟨a *jungled* island⟩

jungle fowl *noun* (circa 1825)
: any of several Asian wild birds (genus *Gallus*) related to the pheasants; *especially* **:** a bird (*G. gallus*) of southeastern Asia from which domestic fowls have probably descended

jungle gym *noun* [from *Junglegym,* a trademark] (1923)
: a structure of vertical and horizontal bars for use by children at play

¹ju•nior \ˈjün-yər\ *adjective* [Latin, comparative of *juvenis* young — more at YOUNG] (13th century)
1 a : less advanced in age **:** YOUNGER — used chiefly to distinguish a son with the same given name as his father **b** (1) **:** YOUTHFUL (2) **:** designed for young people and especially adolescents **c :** of more recent date and therefore inferior or subordinate ⟨a *junior* lien⟩
2 a : lower in standing or rank ⟨*junior* partners⟩ **b :** duplicating or suggesting on a smaller scale something typically large or powerful ⟨a *junior* gale⟩
3 : of or relating to juniors or the class of juniors at an educational institution ⟨the *junior* prom⟩

²junior *noun* [Latin, noun & adjective] (1526)
1 a (1) **:** a person who is younger than another ⟨a man six years my *junior*⟩ (2) **:** a male child **:** SON (3) **:** a young person **b :** a clothing size for women and girls with slight figures
2 a : a person holding a lower position in a hierarchy of ranks **b :** a student in the next-to-the-last year before graduating from an educational institution
3 *capitalized* **:** a member of a program of the Girl Scouts for girls in the third through sixth grades in school

ju•nior•ate \ˈjün-yə-ˌrāt, -rət\ *noun* (1845)
1 : a course of high school or college study for candidates for the priesthood, brotherhood, or sisterhood; *specifically* **:** one preparatory to the course in philosophy
2 : a seminary for juniorate training

junior college *noun* (1899)
: an educational institution that offers two years of studies corresponding to those in the first two years of a four-year college and that often offers technical, vocational, and liberal studies to the adults of a community

junior high school *noun* (1909)
: a school usually including grades 7 to 9 — called also *junior high*

Junior Leaguer *noun* (1938)
: a member of a league of young women organized for volunteer service to civic and social organizations

junior miss *noun* (1927)
1 : an adolescent girl
2 : JUNIOR 1b

junior varsity *noun* (1949)
: a team composed of members lacking the experience or qualification required for the varsity

ju•ni•per \ˈjü-nə-pər\ *noun* [Middle English *junipere,* from Latin *juniperus*] (14th century)
1 : any of numerous shrubs or trees (genus *Juniperus*) of the cypress family with leaves resembling needles or scales and female cones usually resembling berries
2 : any of several coniferous trees resembling true junipers

juniper oil *noun* (circa 1889)
: an acrid essential oil obtained from the fruit of a common juniper (*Juniperus communis*) and used especially in gin and liqueurs

juniper tar *noun* (circa 1930)
: a tarry liquid used locally in treating skin diseases and obtained by distillation from the wood of a Eurasian juniper (*Juniperus oxycedrus*) — called also *cade oil, juniper tar oil*

¹junk \ˈjəŋk\ *noun* [Middle English *jonke*] (14th century)
1 : pieces of old cable or cordage used especially to make gaskets, mats, swabs, or oakum
2 a (1) **:** old iron, glass, paper, or other waste that may be used again in some form (2) **:** secondhand, worn, or discarded articles (3) **:** CLUTTER 1b **b :** something of poor quality **:** TRASH **c :** something of little meaning, worth, or significance
3 *slang* **:** NARCOTICS; *especially* **:** HEROIN
4 : JUNK BOND
5 : baseball pitches other than fastballs (as curveballs or change-ups)
— junky \ˈjəŋ-kē\ *adjective*

²junk *transitive verb* (1916)
: to get rid of as worthless **:** SCRAP
synonym see DISCARD

³junk *noun* [Portuguese *junco,* from Javanese *joṅ*] (1613)
: any of various ships of Chinese waters with bluff lines, a high poop and overhanging stem, little or no keel, high pole masts, and a deep rudder

junk

junk art *noun* (1962)
: three-dimensional art made from discarded material (as metal, mortar, glass, or wood)
— junk artist *noun*

junk bond *noun* (1976)
: a high-risk bond that offers a high yield and is often issued to finance a takeover of a company

junk•er \ˈjəŋ-kər\ *noun* [¹*junk* + ²*-er*] (1944)
: something (as an automobile) of such age and condition as to be ready for scrapping

☆ SYNONYMS

Juncture, exigency, emergency, contingency, pinch, straits, crisis mean a critical or crucial time or state of affairs. JUNCTURE stresses the significant concurrence or convergence of events ⟨an important *juncture* in our country's history⟩. EXIGENCY stresses the pressure of restrictions or urgency of demands created by a special situation ⟨provide for *exigencies*⟩. EMERGENCY applies to a sudden unforeseen situation requiring prompt action to avoid disaster ⟨the presence of mind needed to deal with *emergencies*⟩. CONTINGENCY implies an emergency or exigency that is regarded as possible but uncertain of occurrence ⟨*contingency* plans⟩. PINCH implies urgency or pressure for action to a less intense degree than EXIGENCY or EMERGENCY ⟨come through in a *pinch*⟩. STRAITS applies to a troublesome situation from which escape is extremely difficult ⟨in dire *straits*⟩. CRISIS applies to a juncture whose outcome will make a decisive difference ⟨a *crisis* of confidence⟩.

Jun·ker \'yůŋ-kər\ *noun* [German, from Old High German *junchērro,* literally, young lord] (1554)
: a member of the Prussian landed aristocracy
— Jun·ker·dom \-kər-dəm\ *noun*
— Jun·ker·ism \-kə-ˌri-zəm\ *noun*

¹jun·ket \'jəŋ-kət\ *noun* [Middle English *ioncate,* ultimately from (assumed) Vulgar Latin *juncata,* from Latin *juncus* rush] (15th century)
1 : a dessert of sweetened flavored milk set with rennet
2 a : a festive social affair **b :** TRIP, JOURNEY; *especially* **:** a trip made by an official at public expense ◆

²junket *intransitive verb* (1555)
1 : FEAST, BANQUET
2 : to go on a junket
— jun·ke·teer \ˌjəŋ-kə-'tir\ *or* **jun·ket·er** \'jəŋ-kə-tər\ *noun*

junk food *noun* (1971)
1 : food that is high in calories but low in nutritional content
2 : something that is appealing or enjoyable but of little or no real value ⟨the ultimate in *junk food* for young minds —Cleveland Amory⟩

junk·ie *also* **junky** \'jəŋ-kē\ *noun, plural* **junk·ies** (1923)
1 *slang* **a :** a narcotics peddler or addict **b :** a person who derives inordinate pleasure from or who is dependent on something ⟨sugar *junkie*⟩ ⟨television news *junkie*⟩
2 : a junk dealer

junk mail *noun* (1954)
: third-class mail (as advertising circulars) that is often addressed to "occupant" or "resident"

junk sculpture *noun* (1965)
: JUNK ART

junk·yard \'jəŋk-ˌyärd\ *noun* (1880)
: a yard used to store sometimes resalable junk

Ju·no \'jü-(ˌ)nō\ *noun*
: the wife of Jupiter, queen of heaven, and goddess of light, birth, women, and marriage — compare HERA

Ju·no·esque \ˌjü-(ˌ)nō-'esk\ *adjective* (1888)
: marked by stately beauty

jun·ta \'hůn-tə, 'jən-, 'hən-\ *noun* [Spanish, from feminine of *junto* joined, from Latin *junctus,* past participle of *jungere* to join — more at YOKE] (1622)
1 : a council or committee for political or governmental purposes; *especially* **:** a group of persons controlling a government especially after a revolutionary seizure of power
2 : JUNTO

jun·to \'jən-(ˌ)tō\ *noun, plural* **juntos** [probably alteration of *junta*] (1623)
: a group of persons joined for a common purpose

Ju·pi·ter \'jü-pə-tər\ *noun* [Latin]
1 : the chief Roman god, husband of Juno, and god of light, of the sky and weather, and of the state and its welfare and its laws — compare ZEUS
2 : the largest of the planets and fifth in order from the sun — see PLANET table

ju·ral \'jůr-əl\ *adjective* [Latin *jur-, jus* law] (1635)
1 : of or relating to law
2 : of or relating to rights or obligations
— ju·ral·ly \-ə-lē\ *adverb*

Ju·ras·sic \jů-'ra-sik\ *adjective* [French *jurassique,* from *Jura* mountain range] (1831)
: of, relating to, or being the period of the Mesozoic era between the Triassic and the Cretaceous or the corresponding system of rocks marked by the presence of dinosaurs and the first appearance of birds — see GEOLOGIC TIME table
— Jurassic *noun*

ju·rat \'jůr-ˌat\ *noun* [short for Latin *juratum* (*est*) it has been sworn, 3d singular perfect passive of *jurare* to swear — more at JURY] (1796)
: a certificate added to an affidavit stating when, before whom, and where it was made

ju·rel \hü-'rel\ *noun* [Spanish] (1772)
: any of several food fishes (family Carangidae) of warm seas

ju·rid·i·cal \jů-'ri-di-kəl\ *also* **ju·rid·ic** \-dik\ *adjective* [Latin *juridicus,* from *jur-, jus* + *dicere* to say — more at DICTION] (1502)
1 : of or relating to the administration of justice or the office of a judge
2 : of or relating to law or jurisprudence **:** LEGAL
— ju·rid·i·cal·ly \-di-k(ə-)lē\ *adverb*

ju·ris·con·sult \ˌjůr-əs-'kän-ˌsəlt, -kən-'\ *noun* [Latin *jurisconsultus,* from *juris* (genitive of *jus*) + *consultus,* past participle of *consulere* to consult] (1605)
: JURIST; *especially* **:** one learned in international and public law

ju·ris·dic·tion \ˌjůr-əs-'dik-shən\ *noun* [Middle English *jurisdiccioun,* from Middle French & Latin; Middle French *juridiction,* from Latin *jurisdiction-, jurisdictio,* from *juris* + *diction-, dictio* act of saying — more at DICTION] (14th century)
1 : the power, right, or authority to interpret and apply the law
2 a : the authority of a sovereign power to govern or legislate **b :** the power or right to exercise authority **:** CONTROL
3 : the limits or territory within which authority may be exercised
synonym see POWER
— ju·ris·dic·tion·al \-shnəl, -shə-nᵊl\ *adjective*
— ju·ris·dic·tion·al·ly *adverb*

Ju·ris Doctor \'jůr-əs-\ *noun* [Latin, doctor of law] (1969)
: a degree equivalent to bachelor of laws

ju·ris·pru·dence \ˌjůr-əs-'prü-dᵊn(t)s\ *noun* (circa 1656)
1 a : a system or body of law **b :** the course of court decisions
2 : the science or philosophy of law
3 : a department of law ⟨medical *jurisprudence*⟩
— ju·ris·pru·den·tial \-prü-'den(t)-shəl\ *adjective*
— ju·ris·pru·den·tial·ly \-'den(t)-sh(ə-)lē\ *adverb*

ju·ris·pru·dent \-'prü-dᵊnt\ *noun* [Late Latin *jurisprudent-, jurisprudens,* from Latin *juris* + *prudent-, prudens* skilled, prudent] (1628)
: JURIST

ju·rist \'jůr-ist\ *noun* [Middle French *juriste,* from Medieval Latin *jurista,* from Latin *jur-, jus*] (15th century)
: one having a thorough knowledge of law; *especially* **:** JUDGE

ju·ris·tic \jů-'ris-tik\ *adjective* (1831)
1 : of or relating to a jurist or jurisprudence
2 : of, relating to, or recognized in law
— ju·ris·ti·cal·ly \-ti-k(ə-)lē\ *adverb*

ju·ror \'jůr-ər, 'jůr-ˌȯr\ *noun* (14th century)
1 a : a member of a jury **b :** a person summoned to serve on a jury
2 : a person who takes an oath (as of allegiance)

¹ju·ry \'jůr-ē\ *noun, plural* **juries** [Middle English *jure,* from Anglo-French *juree,* from Old French *jurer* to swear, from Latin *jurare,* from *jur-, jus*] (13th century)
1 : a body of persons sworn to give a verdict on some matter submitted to them; *especially* **:** a body of persons legally selected and sworn to inquire into any matter of fact and to give their verdict according to the evidence
2 : a committee for judging and awarding prizes at a contest or exhibition
3 : one (as the public or test results) that will decide — used especially in the phrase *the jury is still out*

²jury *adjective* [origin unknown] (1616)
: improvised for temporary use especially in an emergency **:** MAKESHIFT ⟨a *jury* mast⟩ ⟨a *jury* rig⟩

³jury *transitive verb* **jur·ied; jury·ing** (1947)
: to select material as appropriate for exhibition in (as an art show) — used chiefly as a participle ⟨a *juried* show⟩

ju·ry–rig \'jůr-ē-ˌrig, -'rig\ *transitive verb* [²*jury*] (1788)
: to erect, construct, or arrange in a makeshift fashion

jus gen·ti·um \'yüs-'gen-tē-əm\ *noun* [Latin, law of nations] (1548)
: INTERNATIONAL LAW

jus san·gui·nis \-'säŋ-gwə-nəs\ *noun* [Latin, right of blood] (1902)
: a rule that a child's citizenship is determined by its parents' citizenship

jus·sive \'jə-siv\ *noun* [Latin *jussus,* past participle of *jubēre* to order; akin to Greek *hysminē* battle] (1846)
: a word, form, case, or mood expressing command
— jussive *adjective*

jus so·li \'yüs-'sō-ˌlē\ *noun* [Latin, right of the soil] (1902)
: a rule that the citizenship of a child is determined by the place of its birth

¹just \'jəst, 'jüst\ *variant of* JOUST

²just \'jəst\ *adjective* [Middle English, from Middle French & Latin; Middle French *juste,* from Latin *justus,* from *jus* right, law; akin to Sanskrit *yos* welfare] (14th century)
1 a : having a basis in or conforming to fact or reason **:** REASONABLE ⟨a *just* but not a generous decision⟩ **b** *archaic* **:** faithful to an original **c :** conforming to a standard of correctness **:** PROPER ⟨*just* proportions⟩
2 a (1) **:** acting or being in conformity with what is morally upright or good **:** RIGHTEOUS ⟨a *just* war⟩ (2) **:** being what is merited **:** DESERVED ⟨a *just* punishment⟩ **b :** legally correct **:** LAWFUL ⟨*just* title to an estate⟩
synonym see FAIR, UPRIGHT
— just·ly *adverb*
— just·ness \'jəs(t)-nəs\ *noun*

³just \'jəst, 'jist, 'jest *also without* t\ *adverb* (15th century)
1 a : EXACTLY, PRECISELY ⟨*just* right⟩ **b :** very recently ⟨the bell *just* rang⟩
2 a : by a very small margin **:** BARELY ⟨*just* too late⟩ **b :** IMMEDIATELY, DIRECTLY ⟨*just* west of here⟩

◇ WORD HISTORY

junket *Juncus* is the Latin word for "rush." The stems of that marsh plant have long been the source of plaiting for mats and baskets. French inherited *juncus* as *jonc,* and derivatives of *jonc* in the medieval French of Normandy came to denote rush baskets, as well as soft, unripened cheeses made in such baskets. One such derivative is presumed to be the source of Middle English *ioncate* (with the suffix perhaps influenced by Medieval Latin *juncata*), a name for a sweetened dish of curds and cream; in the form *junket* this word is still applied to desserts made from sweetened milk or cream. Back in the 16th century, however, we find that *junket* embarked on a long semantic journey away from desserts. Its meaning was extended from any sweet dish to a feast or banquet that presumably featured such dishes, and then to any banquet, a usage that lasted at least into the 19th century. In the U.S. *junket* became a picnic or pleasure trip, and by the later 19th century the word had entered the political vocabulary as a derogatory expression for a politician's trip at taxpayers' expense.

3 a : ONLY, SIMPLY ⟨*just* last year⟩ ⟨*just* be yourself⟩ **b :** QUITE, VERY ⟨*just* wonderful⟩
4 : PERHAPS, POSSIBLY ⟨it *just* might work⟩
— just about : ALMOST ⟨the work is *just about* done⟩

just–folks \'jəs(t)-'fōks\ *adjective* (1952)
: marked by the absence of formality or sophistication **:** UNPRETENTIOUS ⟨her *just-folks* manner⟩

jus•tice \'jəs-təs\ *noun* [Middle English, from Old English & Old French; Old English *justice,* from Old French *justice,* from Latin *justitia,* from *justus*] (12th century)
1 a : the maintenance or administration of what is just especially by the impartial adjustment of conflicting claims or the assignment of merited rewards or punishments **b :** JUDGE **c :** the administration of law; *especially* **:** the establishment or determination of rights according to the rules of law or equity
2 a : the quality of being just, impartial, or fair **b** (1) **:** the principle or ideal of just dealing or right action (2) **:** conformity to this principle or ideal **:** RIGHTEOUSNESS **c :** the quality of conforming to law
3 : conformity to truth, fact, or reason **:** CORRECTNESS

justice of the peace (15th century)
: a local magistrate empowered chiefly to administer summary justice in minor cases, to commit for trial, and to administer oaths and perform marriages

jus•ti•cia•ble \,jə-'sti-sh(ē-)ə-bəl\ *adjective* (15th century)
1 : liable to trial in a court of justice ⟨a *justiciable* offense⟩
2 : capable of being decided by legal principles or by a court of justice
— jus•ti•cia•bil•i•ty \,jə-,sti-sh(ē-)ə-'bi-lə-tē\ *noun*

jus•ti•ci•ar \,jə-'sti-shē-ər, -,är\ *noun* [Middle English, from Medieval Latin *justitiarius,* from Latin *justitia*] (circa 1580)
: the chief political and judicial officer of the Norman and later kings of England until the 13th century

jus•ti•fi•able \'jəs-tə-,fī-ə-bəl\ *adjective* (1561)
: capable of being justified **:** EXCUSABLE ⟨*justifiable* family pride —*Current Biography*⟩
— jus•ti•fi•abil•i•ty \,jəs-tə-,fī-ə-'bi-lə-tē\ *noun*
— jus•ti•fi•ably \'jəs-tə-,fī-ə-blē\ *adverb*

jus•ti•fi•ca•tion \,jəs-tə-fə-'kā-shən\ *noun* (14th century)
1 : the act, process, or state of being justified by God
2 a : the act or an instance of justifying **:** VINDICATION **b :** something that justifies
3 : the process or result of justifying lines of text

jus•ti•fi•ca•tive \'jəs-tə-fə-,kā-tiv\ *adjective* (1611)
: JUSTIFICATORY

jus•ti•fi•ca•to•ry \,jəs-'ti-fi-kə-,tōr-ē, -,tȯr-; 'jəs-tə-fə-,kā-tə-rē\ *adjective* (1579)
: tending or serving to justify **:** VINDICATORY

jus•ti•fy \'jəs-tə-,fī\ *verb* **-fied; -fy•ing** [Middle English *justifien,* from Middle French or Late Latin; Middle French *justifier,* from Late Latin *justificare,* from Latin *justus*] (14th century)
transitive verb
1 a : to prove or show to be just, right, or reasonable **b** (1) **:** to show to have had a sufficient legal reason (2) **:** to qualify (oneself) as a surety by taking oath to the ownership of sufficient property
2 a *archaic* **:** to administer justice to **b** *archaic* **:** ABSOLVE **c :** to judge, regard, or treat as righteous and worthy of salvation
3 a : to space (as lines of text) so that the lines come out even at the margin **b :** to make even by justifying ⟨*justified* margins⟩
intransitive verb
1 a : to show a sufficient lawful reason for an act done **b :** to qualify as bail or surety
2 : to justify lines of text
synonym see MAINTAIN
— jus•ti•fi•er \-,fī(-ə)r\ *noun*

just–in–time *noun, often attributive* (1981)
: a manufacturing strategy wherein parts are produced or delivered only as needed

¹jut \'jət\ *verb* **jut•ted; jut•ting** [perhaps short for ²*jutty*] (circa 1573)
intransitive verb
: to extend out, up, or forward **:** PROJECT ⟨mountains *jutting* into the sky⟩ ⟨a *jutting* jaw⟩
transitive verb
: to cause to project

²jut *noun* (1786)
: something that juts **:** PROJECTION

jute \'jüt\ *noun* [Bengali *jhuṭo*] (1746)
: the glossy fiber of either of two East Indian plants (*Corchorus olitorius* and *C. capsularis*) of the linden family used chiefly for sacking, burlap, and twine; *also* **:** a plant producing jute

Jute \'jüt\ *noun* [Middle English, from Medieval Latin *Jutae* Jutes, of Germanic origin; akin to Old English *Eotenas* Jutes] (14th century)
: a member of a Germanic people invading England from the Continent and settling in Kent in the 5th century
— Jut•ish \'jü-tish\ *adjective*

¹jut•ty \'jə-tē\ *noun, plural* **jutties** [Middle English] (15th century)
1 *archaic* **:** JETTY
2 : a projecting part of a building

²jutty *transitive verb* **jut•tied; jut•ty•ing** (1599)
obsolete **:** to project beyond

ju•ve•nes•cence \,jü-və-'ne-sᵊn(t)s\ *noun* (1800)
: the state of being youthful or of growing young
— ju•ve•nes•cent \-sᵊnt\ *adjective*

¹ju•ve•nile \'jü-və-,nīl, -nᵊl\ *adjective* [French or Latin; French *juvénile,* from Latin *juvenilis,* from *juvenis* young person — more at YOUNG] (1625)
1 a : physiologically immature or undeveloped **:** YOUNG **b :** derived from sources within the earth and coming to the surface for the first time — used especially of water and gas
2 : of, relating to, characteristic of, or suitable for children or young people ⟨*juvenile* books⟩
3 : reflecting psychological or intellectual immaturity **:** CHILDISH

²juvenile *noun* (1733)
1 a : a young person **:** YOUTH **b :** a book for children or young people
2 : a young individual resembling an adult of its kind except in size and reproductive activity: as **a :** a fledged bird not yet in adult plumage **b :** a 2-year-old racehorse
3 : an actor or actress who plays youthful parts

juvenile court *noun* (1899)
: a court that has special jurisdiction over delinquent and dependent children usually up to the age of 18

juvenile delinquency *noun* (1816)
1 : conduct by a juvenile characterized by antisocial behavior that is beyond parental control and therefore subject to legal action
2 : a violation of the law committed by a juvenile and not punishable by death or life imprisonment
— juvenile delinquent *noun*

juvenile hormone *noun* (1940)
: an insect hormone that is secreted by the corpora allata, inhibits maturation to the imago, and plays a role in reproduction

juvenile officer *noun* (1954)
: a police officer charged with the detection, prosecution, and care of juvenile delinquents

ju•ve•nil•ia \'jü-və-'ni-lē-ə\ *noun plural* [Latin, neuter plural of *juvenilis*] (1622)
1 : compositions produced in the artist's or author's youth
2 : artistic or literary compositions suited to or designed for the young

ju•ve•nil•i•ty \,jü-və-'ni-lə-tē\ *noun, plural* **-ties** (circa 1623)
1 : the quality or state of being juvenile **:** YOUTHFULNESS
2 a : immaturity of thought or conduct **b :** an instance of being juvenile

jux•ta•pose \'jək-stə-,pōz\ *transitive verb* **-posed; -pos•ing** [probably back-formation from *juxtaposition*] (1851)
: to place side by side ⟨*juxtapose* unexpected combinations of colors, shapes and ideas —J. F. T. Bugental⟩

juxtaposed *adjective* (1855)
: placed side by side **:** being in juxtaposition
synonym see ADJACENT

jux•ta•po•si•tion \,jək-stə-pə-'zi-shən\ *noun* [Latin *juxta* near + English *position* — more at JOUST] (1654)
: the act or an instance of placing two or more things side by side; *also* **:** the state of being so placed
— jux•ta•po•si•tion•al \-'zish-nəl, -'zi-shə-nᵊl\ *adjective*

K

K is the eleventh letter of the English alphabet. It came through Latin, via Etruscan, from Greek, where it was called kappa; *Greek took it from the Phoenician* kaph. *In all these languages* k *represented practically the same sound as in English, but in Latin, the Romance languages, and Old English it was largely replaced by* c. *Eventually, however, the phonetic ambiguity of that letter, which stood for both the* g *and* k *sounds in Latin and presented other problems of interpretation in Romance and Old English, led to the revival of the old letter for the sound* k. *This is true in English especially before* e *and* i *(as in* kine, *plural for* cow*), before* n *(as in* knife, *from Old English* cnīf*), and more recently in loan words from many sources (such as* kugel *from German,* karaoke *from Japanese, and* Koran *from Arabic). In many words (for example,* knead, knight, *and* know*)* k *has become silent before* n, *though once it was pronounced. It is frequent also after short vowels in the combination* ck, *which is pronounced like simple* k, *as in* black. *The small form of* k *developed through a downsizing of the capital form.*

---

k \'kā\ *noun, plural* **k's** *or* **ks** \'kāz\ *often capitalized, often attributive* (before 12th century)
1 a : the 11th letter of the English alphabet **b :** a graphic representation of this letter **c :** a speech counterpart of orthographic *k*
2 : a graphic device for reproducing the letter *k*
3 : one designated *k* especially as the 11th in order or class
4 : something shaped like the letter K
5 : a unit vector parallel to the z-axis
6 [*kilo-*] **:** THOUSAND ⟨a salary of $24*K*⟩
7 [*kilo-*] **:** a unit of computer storage capacity equal to 1024 bytes ⟨a computer memory of 64*K*⟩
8 *capitalized* **:** STRIKEOUT

Kaa•ba \'kä-bə\ *noun* [Arabic *ka'bah,* literally, cubic building] (1734)
: a small stone building in the court of the Great Mosque at Mecca that contains a sacred black stone and is the goal of Islamic pilgrimage and the point toward which Muslims turn in praying

ka•ba•la *or* **kab•ba•la** *or* **kab•ba•lah** *variant of* CABALA

ka•bob \kə-'bäb *also* 'kä-,\ *noun* [Persian, Hindi, Arabic & Turkish; Persian & Hindi *kabāb,* from Arabic, from Turkish *kebap*] (1673)
: cubes of meat (as lamb or beef) marinated and cooked with vegetables (as onions, tomatoes, and green peppers) usually on a skewer

Ka•bu•ki \kə-'bü-kē, 'kä-bü-(,)kē\ *noun* [Japanese] (1899)
: traditional Japanese popular drama with singing and dancing performed in a highly stylized manner

Ka•byle \kə-'bī(ə)l\ *noun* [Arabic *qabā'il,* plural of *qabīlah* tribe] (1738)
1 : a member of a Berber people living in the mountainous coastal area east of Algiers
2 : the Berber language of the Kabyles

ka•chi•na \kə-'chē-nə\ *noun* [Hopi *qacína*] (1888)
1 : one of the deified ancestral spirits believed among the Hopi and other Pueblo Indians to visit the pueblos at intervals
2 : one of the elaborately masked kachina impersonators that dance at agricultural ceremonies
3 : a doll representing a kachina

kad•dish \'kä-dish\ *noun, often capitalized* [Aramaic *qaddīsh* holy] (1613)
: a Jewish prayer recited in the daily ritual of the synagogue and by mourners at public services after the death of a close relative

kaf•fee•klatsch \'kò-fē-,klach, 'kä-fē-, -,kläch, -,kləch\ *noun, often capitalized* [German, from *Kaffee* coffee + *Klatsch* gossip] (1888)
: an informal social gathering for coffee and conversation

Kaf•fir *or* **Kaf•ir** \'ka-fər\ *noun* [Arabic *kāfir* infidel] (1778)
1 *archaic* **:** a member of a group of southern African Bantu-speaking peoples
2 *often not capitalized, chiefly South African* **:** a black African — usually used disparagingly

kaf•fi•yeh \kə-'fē-ə\ *noun* [Arabic *kūfīyah, kaffīyah,* from al-*Kufah* Al Kufa, town in Iraq] (circa 1817)
: an Arab headdress consisting of a square of cloth folded to form a triangle and held on by a cord

kaf•ir \'ka-fər\ *noun* (circa 1785)
: a grain sorghum (*Sorghum caffrorum*) with stout short-jointed somewhat juicy stalks and erect heads

Kaf•ir \'ka-fər\ *noun* [Arabic *kāfir*] (1759)
: a member of a group of peoples of the Hindu Kush in northeastern Afghanistan

Kaf•iri \'ka-fə-rē\ *noun* (1901)
: the group of languages spoken by the Kafirs that constitutes a distinct branch of Indo-Iranian — see INDO-EUROPEAN LANGUAGES table

Kaf•ka•esque \,käf-kə-'esk, ,kaf-\ *adjective* (1946)
: of, relating to, or suggestive of Franz Kafka or his writings; *especially* **:** having a nightmarishly complex, bizarre, or illogical quality ⟨*Kafkaesque* bureaucratic delays⟩

kaf•tan *variant of* CAFTAN

ka•hu•na \kə-'hü-nə\ *noun* [Hawaiian] (1886)
: a Hawaiian witch doctor

kai•nite \'kī-,nīt, 'kā-\ *also* **kai•nit** \kī-'nēt\ *noun* [German *Kainit,* from Greek *kainos* new — more at RECENT] (1868)
: a natural salt $KMg(SO_4)Cl \cdot 3H_2O$ consisting of a hydrous sulfate and chloride of magnesium and potassium that is used as a fertilizer and as a source of potassium and magnesium compounds

kai•ser \'kī-zər\ *noun* [Middle English, from Old Norse *keisari;* akin to Old High German *keisur* emperor; both from a prehistoric Germanic word borrowed from Latin *Caesar,* cognomen of the Emperor Augustus] (13th century)
: EMPEROR; *especially* **:** the ruler of Germany from 1871 to 1918 ◆
— **kai•ser•dom** \-zər-dəm\ *noun*
— **kai•ser•ism** \-zə-,ri-zəm\ *noun*

kai•se•rin \'kī-zə-rən\ *noun* [German, feminine of *Kaiser*] (circa 1888)
: the wife of a kaiser

kaiser roll *noun* (circa 1898)
: a crusty roll often used for sandwiches

ka•ka \'kä-kə\ *noun* [Maori] (circa 1774)
: an olive brown New Zealand parrot (*Nestor meridionalis*) with gray and red markings

ka•ka•po \,kä-kə-'pō\ *noun, plural* **-pos** [Maori] (1843)
: a large chiefly nocturnal burrowing New Zealand parrot (*Strigops habroptilus*) that has green and brown barred plumage and well-developed wings with little power of flight

ka•ke•mo•no \,kä-ki-'mō-(,)nō\ *noun, plural* **-nos** [Japanese] (1890)
: a vertical Japanese ornamental pictorial or calligraphic scroll

ka•ki•e•mon \,kä-kē-'ā-,män\ *noun, often capitalized* [Sakaida *Kakiemon* (flourished 1650) Japanese potter] (1890)
: a Japanese porcelain decorated with enamel

ka•la–azar \,kä-lə-ə-'zär, ,ka-\ *noun* [Hindi *kālā-āzār* black disease, from Hindi *kālā* black + Persian *āzār* disease] (1882)
: a severe infectious disease chiefly of tropical areas that is marked by fever, progressive anemia, leukopenia, and enlargement of the spleen and liver and caused by a flagellate (*Leishmania donovani*) transmitted by the bite of sand flies

ka•lan•choe \,ka-lən-'kō-ē *also* kə-'laŋ-kə-(,)wē *or* 'ka-lən-,chō\ *noun* [New Latin] (1830)
: any of a genus (*Kalanchoe*) of chiefly African tropical succulent herbs or shrubs of the orpine family often cultivated as ornamentals — compare BRYOPHYLLUM

kale \'kā(ə)l\ *noun* [Scots, from Middle English (northern) *cal,* from Old English *cāl* — more at COLE] (14th century)
1 a : COLE **b :** a hardy cabbage (*Brassica oleracea acephala*) with curled often finely incised leaves that do not form a dense head
2 *slang* **:** MONEY

◇ WORD HISTORY

kaiser Augustus, the man whom history regards as the first Roman emperor, was born Gaius Octavius; but when he was adopted by his great uncle Julius Caesar in the latter's will, he took the new name Gaius Julius Caesar Octavianus. Subsequent Roman emperors preserved *Caesar* as part of their full name, and the word was used generically in Latin with the meaning "emperor." Several early Germanic languages borrowed the word in this sense. Old English *cāsere* passed into Old Scots and northern dialects of Middle English as *casere* or *kasere,* but this word was displaced in Middle English and early Modern English by forms such as *cayser* and *keyser,* borrowed from Old Norse *keisari.* Although the latter words were occasionally resurrected by later authors as deliberate archaisms, Modern English *kaiser* is borrowed from Modern German *Kaiser,* used to denote the Holy Roman Emperor and (after the demise of the Holy Roman Empire in 1806) his successor, the emperor of Austria. *Kaiser* is most closely associated in English with the three emperors of Germany's Second Reich (1871–1918). Since the abdication of the Austrian and German emperors at the end of World War I, *kaiser* has become purely a historical term.

ka•lei•do•scope \kə-ˈlī-də-ˌskōp\ *noun* [Greek *kalos* beautiful + *eidos* form + English *-scope* — more at IDYLL] (1817)
1 : an instrument containing loose bits of colored material (as glass or plastic) between two flat plates and two plane mirrors so placed that changes of position of the bits of material are reflected in an endless variety of patterns
2 : something resembling a kaleidoscope: as **a :** a variegated changing pattern or scene ⟨the lake a *kaleidoscope* of changing colors — Robert Gibbings⟩ **b :** a succession of changing phases or actions ⟨a . . . *kaleidoscope* of shifting values, information, fashions —Frank McLaughlin⟩
— ka•lei•do•scop•ic \-ˌlī-də-ˈskä-pik\ *adjective*
— ka•lei•do•scop•i•cal•ly \-pi-k(ə-)lē\ *adverb*

ka•lends *variant of* CALENDS

kal•li•din \ˈka-lə-dən\ *noun* [German, from *Kalli*krein + Pepti*d* peptide + *-in*] (1950)
: either of two vasodilator kinins formed from blood plasma globulin by the action of kallikrein: **a :** BRADYKININ **b :** one with a terminal lysine amino-acid residue added to bradykinin

kal•li•kre•in \ˌka-lə-ˈkrē-ən, kə-ˈli-krē-ən\ *noun* [German, from Greek *kallikreas* sweetbread, pancreas, from *kalli-* beautiful (from *kallos* beauty) + *kreas* flesh — more at RAW] (1930)
: a hypotensive protease that liberates kinins from blood plasma proteins and is used therapeutically for vasodilation

Kal•muck *or* **Kal•muk** \ˈkal-ˌmək, kal-ˈ\ *or* **Kal•myk** \kal-ˈmik\ *noun* [Russian *kalmyk,* from Volga Tatar *kalmık*] (1613)
1 : a member of a Buddhist Mongol people originally of Dzungaria living mainly northwest of the Caspian Sea in Russia
2 : the Mongolian language of the Kalmucks

Ka•ma \ˈkä-mə\ *noun* [Sanskrit *Kāma,* from *kāma* love]
: the Hindu god of love

ka•ma•ai•na \ˌkä-mə-ˈī-nə\ *noun* [Hawaiian *kamaʻāina,* from *kama* child + *ʻāina* land] (1903)
: one who has lived in Hawaii for a long time

kame \ˈkām\ *noun* [Scots, kame, comb, from Middle English (northern) *camb* comb, from Old English] (1795)
: a short ridge, hill, or mound of stratified drift deposited by glacial meltwater

Ka•me•ha•me•ha Day \kə-ˌmā-ə-ˈmā-(ˌ)hä-\ *noun* (1925)
: June 11 observed as a holiday in Hawaii in commemoration of the birthday of Kamehameha I

¹ka•mi•ka•ze \ˌkä-mi-ˈkä-zē\ *noun* [Japanese, literally, divine wind] (1945)
1 : a member of a Japanese air attack corps in World War II assigned to make a suicidal crash on a target (as a ship)
2 : an airplane containing explosives to be flown in a suicide crash on a target ◆

²kamikaze *adjective* (1945)
1 : of, relating to, being, or resembling a kamikaze
2 : having or showing reckless disregard for safety or personal welfare

kam•pong \ˈkäm-ˌpȯŋ, ˈkam-\ *noun* [Malay] (1844)
: a hamlet or village in a Malay-speaking country

ka•na \ˈkä-nə\ *noun, plural* **kana** *often attributive* [Japanese] (1727)
: a Japanese system of syllabic writing having characters that can be used exclusively for writing foreign words or in combination with kanji (as for indicating pronunciations or grammatical inflections); *also* **:** a single character belonging to the kana system — compare HIRAGANA, KATAKANA

Ka•nak \kə-ˈnäk, -ˈnak\ *noun* [French *canaque,* probably from English *Kanaka* South Sea islander, from Hawaiian, person] (1910)
: a native Melanesian inhabitant of New Caledonia

kana•my•cin \ˌka-nə-ˈmī-sᵊn\ *noun* [New Latin *kanamyceticus,* specific epithet of *Streptomyces kanamyceticus*] (1957)
: a broad-spectrum antibiotic from a Japanese soil streptomyces (*Streptomyces kanamyceticus*)

Kan•a•rese \ˌka-nə-ˈrēz, -ˈrēs\ *noun, plural* **Kanarese** [*Kanara,* India] (1847)
1 : KANNADA
2 : a member of a Kannada-speaking people of Karnataka, southern India

kan•ban \ˈkän-ˌbän\ *noun, often attributive* [Japanese, sign, placard; from the cards used on assembly lines to signal that parts are needed] (1982)
: JUST-IN-TIME

kan•ga•roo \ˌkaŋ-gə-ˈrü\ *noun, plural* **-roos** [Guugu Yimidhirr (Australian aboriginal language of northern Queensland) *gaŋurru*] (1770)
: any of various herbivorous leaping marsupial mammals (family Macropodidae) of Australia, New Guinea, and adjacent islands with a small head, large ears, long powerful hind legs, a long thick tail used as a support and in balancing, and rather small forelegs not used in progression ◆

kangaroo

kangaroo court *noun* (1853)
1 : a mock court in which the principles of law and justice are disregarded or perverted
2 : a court characterized by irresponsible, unauthorized, or irregular status or procedures
3 : judgment or punishment given outside of legal procedure

kangaroo rat *noun* (1867)
: any of a genus (*Dipodomys*) of pouched nocturnal burrowing rodents of arid parts of the western U.S. and Mexico

kangaroo rat

kan•ji \ˈkän-(ˌ)jē\ *noun, plural* **kanji** *often attributive* [Japanese] (1920)
: a Japanese system of writing that utilizes characters borrowed or adapted from Chinese writing; *also* **:** a single character belonging to the kanji system — compare KANA

Kan•na•da \ˈkä-nə-də *also* ˈka-\ *noun* [Kannada *kannaḍa*] (1856)
: the major Dravidian language of Karnataka, southern India

kan•te•le \ˈkän-tə-lə\ *noun* [Finnish] (circa 1903)
: a traditional Finnish zither originally having five strings but now having as many as thirty

ka•o•lin \ˈkā-ə-lən\ *noun* [French *kaolin,* from *Gaoling* hill in China] (circa 1741)
: a fine usually white clay that is used in ceramics and refractories, as a filler or extender, and in medicine especially as an adsorbent in the treatment of diarrhea

ka•o•lin•ite \-lə-ˌnīt\ *noun* (1867)
: a mineral consisting of a hydrous silicate of aluminum that constitutes the principal mineral in kaolin
— ka•o•lin•it•ic \ˌkā-ə-lə-ˈni-tik\ *adjective*

ka•on \ˈkā-ˌän\ *noun* [International Scientific Vocabulary *ka* kay (from *K-meson,* its earlier name) + ²*-on*] (1958)
: an unstable meson that occurs in both charged and neutral forms and is about 970 times more massive than an electron

ka•pell•mei•ster \kə-ˈpel-ˌmī-stər, kä-\ *noun, often capitalized* [German, from *Kapelle* choir + *Meister* master] (1838)
: the director of a choir or orchestra

kaph \ˈkäf, ˈkȯf\ *noun* [Hebrew, literally, palm of the hand] (circa 1823)
: the 11th letter of the Hebrew alphabet — see ALPHABET table

ka•pok \ˈkā-ˌpäk\ *noun* [Malay] (circa 1750)
: a mass of silky fibers that clothe the seeds of the ceiba tree and are used especially as a filling for mattresses, life preservers, and sleeping bags and as insulation

Ka•po•si's sarcoma \ˈka-pə-sēz- *also* -shēz-, kə-ˈpō-sēz-\ *noun* [Moritz *Kaposi* (died 1902) Hungarian dermatologist] (1916)
: a neoplastic disease associated especially with AIDS, affecting especially the skin and mucous membranes, and characterized usually by the formation of pink to reddish-brown or bluish plaques, macules, papules, or nodules

kap•pa \ˈka-pə\ *noun* [Middle English, from Greek, of Semitic origin; akin to Hebrew *kaph*] (15th century)
: the 10th letter of the Greek alphabet — see ALPHABET table

ka•put *also* **ka•putt** \kə-ˈpu̇t, kä-, -ˈpüt\ *adjective* [German *kaputt,* from French *capot* not having made a trick at piquet] (1895)
1 : utterly finished, defeated, or destroyed

◇ WORD HISTORY

kamikaze In 1274 and 1281 Kublai Khan, the Mongol emperor of China, sent out great fleets to conquer Japan. Providential storms dispersed the fleets on both occasions and by saving the country reinforced the Japanese belief that the Shinto gods would forever protect Japan from invasion. The storms became known in Japanese history as *kamikaze,* literally, "divine wind." In the fall of 1944, as Japanese military fortunes in World War II declined dramatically, pilots were recruited who were willing to give up their lives for their country by crashing specially prepared planes—in effect flying bombs—into American ships. These pilots were members of several units called collectively *Kamikaze Tokubetsu Kōgekitai* (or *Tokkōtai*) ("Kamikaze Special Attack Force") after the storms that had saved Japan some seven centuries earlier. The kamikaze's drastic methods made a greater impression on Americans than did his patriotic goal. *Kamikaze,* originally borrowed into English as a name for the suicidal pilot, is now an adjective meaning "suicidal" or "reckless of personal safety or welfare."

kangaroo *Kangaroo* is the oldest and probably the best-known word borrowed from an Australian Aboriginal language, though for many years its origin was in doubt. During a voyage along the Australian coast in 1770, the English explorer James Cook was forced to beach and repair his ship near the Endeavor River in northern Queensland. Cook's party recorded a number of words from the native Australians in a language now called Guugu Yimidhirr. Among these words was one pronounced /gaŋurru/, recorded as *kangaroo,* and under this name the Australian marsupial became known when Cook returned to Europe. Controversy began when in 1820 a British naval officer revisited the Endeavor River area but failed to elicit the word again. This led to the notion that Cook's party had been mistaken, and that *kangaroo,* if it was a native Australian word, must have meant something quite different from "kangaroo." Not until 1972 was the mystery cleared up, when the linguist John Haviland began the first thorough study of Guugu Yimidhirr. He again elicited the word, more than 200 years after Cook's expedition, as the name for a large, dark kangaroo, probably the male of the species *Macropus giganteus.*

2 : unable to function **:** USELESS ⟨my battery went *kaput* —Henry James Jr.⟩
3 : hopelessly outmoded ◆

kar·a·bi·ner *variant of* CARABINER

Kara·ism \'kar-ə-ˌi-zəm\ *noun* [Late Hebrew *qĕrāīm* Karaites, from Hebrew *qārā* to read] (circa 1883)
: a Jewish doctrine originating in Baghdad in the 8th century that rejects rabbinism and talmudism and bases its tenets on Scripture alone
— Kara·ite \-ˌīt\ *noun*

kar·a·kul \'kar-ə-kəl\ *noun* [*Karakul,* village in Uzbekistan] (1853)
1 *often capitalized* **:** any of a breed of hardy fat-tailed sheep from Uzbekistan with a narrow body and coarse wiry fur
2 : the tightly curled glossy coat of the newborn lamb of a karakul valued as fur

karakul 1

kar·a·o·ke \ˌkar-ē-'ō-kē, kə-'rō-kē, ˌkä-rä-'ō-(ˌ)kā\ *noun* [Japanese, from *kara* empty + *ōke,* short for *ōkesutora* orchestra] (1982)
: a device that plays instrumental accompaniments for a selection of songs to which the user sings along and that records the user's singing with the music

kar·at \'kar-ət\ *noun* [Middle English, probably from Middle French *carat,* from Medieval Latin *carratus* unit of weight for precious stones — more at CARAT] (15th century)
: a unit of fineness for gold equal to 1/24 part of pure gold in an alloy

ka·ra·te \kə-'rä-tē\ *noun* [Japanese, from *kara* empty + *te* hand] (1955)
: an Oriental art of self-defense in which an attacker is disabled by crippling kicks and punches
— ka·ra·te·ist \-tē-ist\ *noun*

ka·ra·ya gum \kə-'rī-ə-\ *noun* [Hindi *karāyal* resin] (1916)
: any of several vegetable gums similar to tragacanth and often used as substitutes for it that are obtained from tropical Asian trees (genera *Sterculia* of the family Sterculiaceae and *Cochlospermum* of the family Bixaceae); *especially* **:** one derived from an Indian tree (*S. urens*)

Ka·re·lian \kə-'rē-lē-ən, -'rēl-yən\ *noun* (1855)
1 : a native or inhabitant of Karelia
2 : the Finno-Ugric language of the Karelians
— Karelian *adjective*

Ka·ren \kə-'ren\ *noun, plural* **Karen** *or* **Karens** (1833)
1 : a member of a group of peoples of eastern and southern Myanmar
2 a : a group of languages spoken by the Karen peoples **b :** a language of this group

kar·ma \'kär-mə *also* 'kər-\ *noun* [Sanskrit *karma* fate, work] (1827)
1 *often capitalized* **:** the force generated by a person's actions held in Hinduism and Buddhism to perpetuate transmigration and in its ethical consequences to determine the nature of the person's next existence
2 : VIBRATION 4
— kar·mic \-mik\ *adjective*

ka·roo *or* **kar·roo** \kə-'rü\ *noun, plural* **karoos** *or* **karroos** [Afrikaans *karo*] (1789)
: a dry tableland of southern Africa

ka·ross \kə-'räs\ *noun* [Afrikaans *karos*] (1785)
: a simple garment or rug of skins used especially by native tribesmen of southern Africa

karst \'kärst\ *noun* [German] (1902)
: an irregular limestone region with sinks, underground streams, and caverns
— karst·ic \'kär-stik\ *adjective*

kart \'kärt\ *noun* [probably from *GoKart,* a trademark] (1959)
: a miniature motorcar used especially in racing
— kart·ing *noun*

kary- *or* **karyo-** *also* **cary-** *or* **caryo-** *combining form* [New Latin, from Greek *karyon* nut — more at CAREEN]
1 : nucleus of a cell ⟨*karyo*kinesis⟩
2 : nut **:** kernel ⟨*cary*opsis⟩

kar·y·og·a·my \ˌkar-ē-'ä-gə-mē\ *noun, plural* **-mies** (1891)
: the fusion of cell nuclei (as in fertilization)

kar·yo·ki·ne·sis \ˌkar-ē-ō-kə-'nē-səs, -kī-\ *noun* [New Latin] (1882)
1 : the nuclear phenomena characteristic of mitosis
2 : the whole process of mitosis
— kar·yo·ki·net·ic \-'ne-tik\ *adjective*

kar·y·ol·o·gy \ˌkar-ē-'ä-lə-jē\ *noun* [International Scientific Vocabulary] (1895)
1 : the minute cytological characteristics of the cell nucleus especially with regard to the chromosomes
2 : a branch of cytology concerned with the karyology of cell nuclei
— kar·y·o·log·i·cal \-ē-ə-'lä-ji-kəl\ *also* **kar·y·o·log·ic** \-jik\ *adjective*

kar·yo·lymph \'kar-ē-ō-ˌlim(p)f\ *noun* [International Scientific Vocabulary] (1899)
: NUCLEAR SAP

kar·yo·some \'kar-ē-ə-ˌsōm\ *noun* [International Scientific Vocabulary] (1889)
: a mass of chromatin in a cell nucleus that resembles a nucleolus

¹kar·yo·type \'kar-ē-ə-ˌtīp\ *noun* [International Scientific Vocabulary] (1929)
: the chromosomal characteristics of a cell; *also* **:** the chromosomes themselves or a representation of them
— kar·yo·typ·ic \ˌkar-ē-ə-'ti-pik\ *adjective*
— kar·yo·typ·i·cal·ly \-pi-k(ə-)lē\ *adverb*

²karyotype *transitive verb*
: to determine the karyotype of

Kas·bah *variant of* CASBAH

ka·sha \'kä-shə, 'ka-\ *noun* [Russian] (1808)
1 : a porridge made usually from buckwheat groats
2 : kasha grain before cooking

Ka·shan \kə-'shän\ *noun* [*Kashan,* Iran] (1920)
: an Oriental rug with floral motifs in soft colors

Kash·mir *variant of* CASHMERE

Kash·miri \kash-'mir-ē, kazh-\ *noun, plural* **Kashmiris** *or* **Kashmiri** (1880)
1 : an Indo-Aryan language spoken in Kashmir
2 : a native or inhabitant of Kashmir

kash·ruth *or* **kash·rut** \kä-'shrüt, -'shrüth\ *noun* [Hebrew *kashrūth,* literally, fitness] (1907)
1 : the Jewish dietary laws
2 : the state of being kosher

Ka·shu·bi·an \kə-'shü-bē-ən\ *noun* [*Kashube* a member of a Slavic people] (1919)
: a Slavic language spoken in the vicinity of Gdansk

kat *variant of* KHAT

ka·ta \'kä-(ˌ)tä\ *noun, plural* **kata** *or* **katas** [Japanese, literally, model, pattern] (1945)
: a set combination of positions and movements (as in karate) performed as an exercise

kat·a·bat·ic \ˌka-tə-'ba-tik\ *adjective* [Greek *katabatos* descending, verbal of *katabainein* to go down, from *kata-* cata- + *bainein* to go — more at COME] (1918)
: relating to or being a wind produced by the flow of cold dense air down a slope (as of a mountain or glacier) in an area subject to radiational cooling

ka·ta·ka·na \ˌkä-tə-'kä-nə\ *noun* [Japanese, from *kata* part + *kana* kana] (1727)
: the form of Japanese syllabic writing used especially for scientific terms, official documents, and words adopted from other languages — compare HIRAGANA

ka·tchi·na *or* **ka·tci·na** *variant of* KACHINA

Ka·tha·re·vu·sa \ˌkä-thə-'re-və-ˌsä\ *noun* [New Greek *kathareuousa,* from Greek, feminine of *kathareuōn,* present participle of *kathareuein* to be pure, from *katharos* pure] (1936)
: modern Greek conforming to classic Greek usage

ka·thar·sis *variant of* CATHARSIS

ka·ty·did \'kā-tē-ˌdid\ *noun* [imitative] (1784)
: any of several large green American long-horned grasshoppers usually having stridulating organs on the forewings of the males that produce a loud shrill sound

kat·zen·jam·mer \'kat-sən-ˌja-mər\ *noun* [German, from *Katze* cat + *Jammer* distress] (1849)
1 : HANGOVER
2 : DISTRESS 2
3 : a discordant clamor

kau·ri \'kau̇(-ə)r-ē\ *noun* [Maori *kawri*] (1823)
1 : any of various trees (genus *Agathis*) of the araucaria family; *especially* **:** a tall timber tree (*A. australis*) of New Zealand having fine white straight-grained wood
2 : a light-colored to brown resin from the kauri tree found as a fossil in the ground or collected from living trees and used especially in varnishes and linoleum — called also *kauri gum, kauri resin*

ka·va \'kä-və\ *noun* [Tongan & Marquesan, literally, bitter] (1810)
1 : an Australasian shrubby pepper (*Piper methysticum*) from whose crushed root an intoxicating beverage is made
2 : the beverage made from kava

kay \'kā\ *noun* (14th century)
: the letter *k*

Kay \'kā\ *noun*
: a boastful overbearing knight of the Round Table who is foster brother and seneschal of King Arthur

kay·ak \'kī-ˌak\ *noun* [Inuit *qayaq*] (1757)
1 : an Eskimo canoe made of a frame covered with skins except for a small opening in the center and propelled by a double-bladed paddle
2 : a portable boat styled like an Eskimo kayak
— kay·ak·er \-ˌa-kər\ *noun*
— kay·ak·ing \-kiŋ\ *noun*

¹kayo \(ˌ)kā-'ō, 'kā-(ˌ)ō\ *noun* [pronunciation of *KO,* abbreviation] (1920)
: KNOCKOUT

²kayo *transitive verb* **kay·oed; kayo·ing** (1923)
: KNOCK OUT

Ka·zak \kə-'zak, -'zäk\ *noun* [*Kazak* (Kazakh), town in Azerbaijan] (1900)
: an Oriental rug in bold colors with geometric designs or stylized plant and animal forms

◇ WORD HISTORY

kaput To win all of the tricks in the card game piquet is *faire capot* "to make capot" in French, while *être capot* "to be capot" is to have lost all of the tricks in a game. In German *capot* was borrowed as *kaput,* and from the sense of having lost a game *kaput* developed the meanings "finished" and "broken." *Kaput* began to appear in English writing in the late 19th century, generally in contexts where a German word would be appropriate or else printed in italics or glossed to show that it was not yet a fully anglicized word. However, during and after World War II *kaput* gained greater currency in English in the senses "utterly finished," "useless," or "hopelessly outmoded" and is now completely established in our language.

Ka·zakh \kə-ˈzak, -ˈzäk\ *noun* [Russian *kazakh,* from Kazakh *kazak*] (1832)
1 : a member of a Turkic people of Kazakhstan and other countries of central Asia
2 : the language of the Kazakhs

ka·zoo \kə-ˈzü\ *noun, plural* **kazoos** [imitative] (1884)
: an instrument that imparts a buzzing quality to the human voice and that usually consists of a small metal or plastic tube with a side hole covered by a thin membrane

K–band \ˈkā-\ *noun* (circa 1948)
: a segment of the radio spectrum that lies between 10.9 GHz and 36.0 GHz and spans the upper superhigh-frequency and lower extremely-high-frequency bands and that is used especially for police radars, satellite communication, and astronomical observation

kea \ˈkē-ə\ *noun* [Maori] (1862)
: a large mostly dull green New Zealand parrot (*Nestor notabilis*) that is normally insectivorous but sometimes destroys sheep by slashing the back to feed on the kidney fat

ke·bab *or* **ke·bob** *variant of* KABOB

keb·buck *or* **keb·bock** \ˈke-bək\ *noun* [Middle English (Scots dialect) *cabok,* from Scottish Gaelic *ceapag*] (15th century)
dialect British **:** a whole wheel or ball of cheese

¹kedge \ˈkej\ *transitive verb* **kedged; kedg·ing** [Middle English *caggen*] (1627)
: to move (a ship) by means of a line attached to a kedge dropped at the distance and in the direction desired

²kedge *noun* (circa 1769)
: a small anchor used especially in kedging

ked·ge·ree \ˈke-jə-rē\ *noun* [Hindi *khicṛī,* from Sanskrit *khiccā*] (1662)
1 : an Indian dish of seasoned rice, beans, lentils, and sometimes smoked fish
2 : cooked or smoked fish, rice, hard-boiled eggs, and seasoning heated in cream

¹keek \ˈkēk\ *intransitive verb* [Middle English *kiken*] (14th century)
chiefly Scottish **:** PEEP, LOOK

²keek *noun* (1721)
chiefly Scottish **:** PEEP, LOOK

¹keel \ˈkē(ə)l\ *verb* [Middle English *kelen,* from Old English *cēlan,* from *cōl* cool] (before 12th century)
chiefly dialect **:** COOL

²keel *noun* [Middle English *kele,* from Middle Dutch *kiel;* akin to Old English *cēol* ship] (14th century)
: a flat-bottomed barge used especially on the Tyne to carry coal

³keel *noun* [Middle English *kele,* from Old Norse *kjǫlr;* akin to Old English *ceole* throat, beak of a ship — more at GLUTTON] (14th century)
1 a : the chief structural member of a boat or ship that extends longitudinally along the center of its bottom and that often projects from the bottom; *also* **:** this projection **b :** SHIP
2 : a projection suggesting a keel; *especially* **:** CARINA
— keeled \ˈkē(ə)ld\ *adjective*
— keel·less \ˈkē(ə)l-ləs\ *adjective*

⁴keel *intransitive verb* (1832)
1 : to fall in or as if in a faint — usually used with *over*
2 : to heel or lean precariously

⁵keel *noun* [Middle English (Scots dialect) *keyle*] (15th century)
chiefly dialect **:** RED OCHER

keel·boat \ˈkē(ə)l-ˌbōt\ *noun* (1695)
: a shallow covered keeled riverboat that is usually rowed, poled, or towed and that is used for freight

keel·haul \-ˌhȯl\ *transitive verb* [Dutch *kielhalen,* from *kiel* keel + *halen* to haul] (1666)
1 : to haul under the keel of a ship as punishment or torture
2 : to rebuke severely

keel·son \ˈkel-sən, ˈkē(ə)l-\ *noun* [Middle English *kelswayn,* probably of Scandinavian origin; akin to Swedish *kölsvin* keelson] (1598)
: a longitudinal structure running above and fastened to the keel of a ship in order to stiffen and strengthen its framework

¹keen \ˈkēn\ *adjective* [Middle English *kene* brave, sharp, from Old English *cēne* brave; akin to Old High German *kuoni* brave] (13th century)
1 a : having a fine edge or point **:** SHARP ⟨a *keen* sword⟩ **b :** affecting one as if by cutting ⟨*keen* sarcasm⟩ **c :** pungent to the sense ⟨a *keen* scent⟩
2 a (1) **:** showing a quick and ardent responsiveness **:** ENTHUSIASTIC ⟨a *keen* swimmer⟩ (2) **:** EAGER ⟨was *keen* to begin⟩ **b** *of emotion or feeling* **:** INTENSE ⟨the *keen* delight in the chase —F. W. Maitland⟩
3 a : intellectually alert **:** having or characteristic of a quick penetrating mind ⟨a *keen* student⟩ ⟨a *keen* awareness of the problem⟩; *also* **:** shrewdly astute ⟨*keen* bargainers⟩ **b :** sharply contested ⟨*keen* debate⟩ **c :** extremely sensitive in perception ⟨*keen* eyes⟩
4 : WONDERFUL, EXCELLENT
synonym see SHARP, EAGER
— keen·ly *adverb*
— keen·ness \ˈkēn-nəs\ *noun*

²keen *noun* (1830)
: a lamentation for the dead uttered in a loud wailing voice or sometimes in a wordless cry

³keen *verb* [Irish *caoinim* I lament] (1845)
intransitive verb
1 a : to lament with a keen **b :** to make a sound suggestive of a keen
2 : to lament, mourn, or complain loudly
transitive verb
: to utter by keening
— keen·er *noun*

¹keep \ˈkēp\ *verb* **kept** \ˈkept\; **keep·ing** [Middle English *kepen,* from Old English *cēpan;* perhaps akin to Old High German *chapfēn* to look] (before 12th century)
transitive verb
1 : to take notice of by appropriate conduct **:** FULFILL: as **a :** to be faithful to ⟨*keep* a promise⟩ **b :** to act fittingly in relation to ⟨*keep* the Sabbath⟩ **c :** to conform to in habits or conduct ⟨*keep* late hours⟩ **d :** to stay in accord with (a beat) ⟨*keep* time⟩
2 : PRESERVE, MAINTAIN: as **a :** to watch over and defend ⟨*keep* us from harm⟩ **b** (1) **:** to take care of **:** TEND ⟨*keep* a garden⟩ (2) **:** SUPPORT (3) **:** to maintain in a good, fitting, or orderly condition — usually used with *up* **c :** to continue to maintain ⟨*keep* watch⟩ **d** (1) **:** to cause to remain in a given place, situation, or condition ⟨*keep* him waiting⟩ (2) **:** to preserve (food) in an unspoiled condition **e :** to have or maintain in one's service or at one's disposal ⟨*keep* a mistress⟩ — often used with *on* ⟨*kept* the cook on⟩; *also* **:** to lodge or feed for pay ⟨*keep* boarders⟩ **f** (1) **:** to maintain a record in ⟨*keep* a diary⟩ (2) **:** to enter in a book ⟨*keep* records⟩ **g :** to have customarily in stock for sale
3 a : to restrain from departure or removal **:** DETAIN ⟨*keep* children in after school⟩ **b :** HOLD BACK, RESTRAIN ⟨*keep* them from going⟩ ⟨*kept* him back with difficulty⟩ **c :** SAVE, RESERVE ⟨*keep* some for later⟩ ⟨*kept* some out for a friend⟩ **d :** to refrain from revealing ⟨*keep* a secret⟩
4 a : to retain in one's possession or power ⟨*kept* the money we found⟩ **b :** to refrain from granting, giving, or allowing ⟨*kept* the news back⟩ **c :** to have in control ⟨*keep* your temper⟩
5 : to confine oneself to ⟨*keep* my room⟩
6 a : to stay or continue in ⟨*keep* the path⟩ ⟨*keep* your seat⟩ **b :** to stay or remain on or in usually against opposition **:** HOLD ⟨*kept* her ground⟩
7 : CONDUCT, MANAGE ⟨*keep* a tearoom⟩
intransitive verb
1 *chiefly British* **:** LIVE, LODGE
2 a : to maintain a course, direction, or progress ⟨*keep* to the right⟩ **b :** to continue usually without interruption ⟨*keep* talking⟩ ⟨*keep* quiet⟩ ⟨*keep* on smiling⟩ **c :** to persist in a practice ⟨*kept* bothering them⟩ ⟨*kept* on smoking in spite of warnings⟩
3 : STAY, REMAIN ⟨*keep* out of the way⟩ ⟨*keep* off the grass⟩: as **a :** to stay even — usually used with *up* ⟨*keep* up with the Joneses⟩ **b :** to remain in good condition ⟨meat will *keep* in the freezer⟩ **c :** to remain secret ⟨the secret would *keep*⟩ **d :** to call for no immediate action ⟨the matter will *keep* until morning⟩
4 : ABSTAIN, REFRAIN ⟨can't *keep* from talking⟩
5 : to be in session ⟨school will *keep* through the winter —W. M. Thayer⟩
6 *of a quarterback* **:** to retain possession of a football especially after faking a handoff ☆☆
— keep an eye on : WATCH
— keep at : to persist in doing or concerning oneself with
— keep company : to go together as frequent companions or in courtship
— keep house : to manage a household
— keep one's distance *or* **keep at a distance :** to stay aloof **:** maintain a reserved attitude
— keep one's eyes open *or* **keep one's eyes peeled :** to be on the alert **:** be watchful
— keep one's hand in : to keep in practice
— keep pace : to stay even; *also* **:** KEEP UP
— keep step : to keep in step
— keep to 1 a : to stay in **b :** to limit oneself to **2 :** to abide by
— keep to oneself 1 : to keep secret ⟨*kept* the facts *to myself*⟩ **2 :** to remain solitary or apart from other people

²keep *noun* (1579)
1 a *archaic* **:** CUSTODY, CHARGE **b :** MAINTENANCE
2 : one that keeps or protects: as **a :** FORTRESS, CASTLE; *specifically* **:** the strongest and securest part of a medieval castle **b :** one whose job is to keep or tend **c :** PRISON, JAIL
3 : the means or provisions by which one is kept ⟨earned his *keep*⟩
4 : KEEPER 4

☆ SYNONYMS

Keep, observe, celebrate, commemorate mean to notice or honor a day, occasion, or deed. KEEP stresses the idea of not neglecting or violating ⟨*kept* the Sabbath by refraining from work⟩. OBSERVE suggests marking the occasion by ceremonious performance ⟨not all holidays are *observed* nationally⟩. CELEBRATE suggests acknowledging an occasion by festivity ⟨traditionally *celebrates* Thanksgiving with a huge dinner⟩. COMMEMORATE suggests that an occasion is marked by observances that remind one of the origin and significance of the day ⟨*commemorate* Memorial Day with the laying of wreaths⟩.

Keep, retain, detain, withhold, reserve mean to hold in one's possession or under one's control. KEEP may suggest a holding securely in one's possession, custody, or control ⟨*keep* this while I'm gone⟩. RETAIN implies continued keeping, especially against threatened seizure or forced loss ⟨managed to *retain* their dignity even in poverty⟩. DETAIN suggests a delay in letting go ⟨*detained* them for questioning⟩. WITHHOLD implies restraint in letting go or a refusal to let go ⟨*withheld* information from the authorities⟩. RESERVE suggests a keeping in store for future use ⟨*reserve* some of your energy for the last mile⟩.

— **for keeps 1 a :** with the provision that one keep what one has won ⟨played marbles *for keeps*⟩ **b :** with deadly seriousness **2 :** for an indefinitely long time **:** PERMANENTLY **3 :** with the result of ending the matter

keep back *intransitive verb* (1837)
: to refrain from approaching or advancing near something ⟨police asked the spectators to *keep back*⟩

keep down *transitive verb* (1581)
1 : to keep in control ⟨*keep* expenses *down*⟩
2 : to prevent from growing, advancing, or succeeding

keep•er \ˈkē-pər\ *noun* (14th century)
1 : one that keeps: as **a :** PROTECTOR **b :** GAMEKEEPER **c :** WARDEN **d :** CUSTODIAN **e** *chiefly British* **:** CURATOR
2 : any of various devices for keeping something in position
3 a : one fit or suitable for keeping; *especially* **:** a fish large enough to be legally caught and kept **b :** a domestic animal considered with respect to how easy it is to care for ⟨an easy *keeper*⟩
4 : an offensive football play in which the quarterback runs with the ball

keep•ing \ˈkē-piŋ\ *noun* (14th century)
1 : the act of one that keeps: as **a :** CUSTODY, MAINTENANCE **b :** OBSERVANCE **c :** a reserving or preserving for future use
2 a : the means by which something is kept **:** SUPPORT, PROVISION **b :** the state of being kept or the condition in which something is kept ⟨the house is in good *keeping*⟩
3 : CONFORMITY ⟨in *keeping* with good taste⟩ ⟨out of *keeping* with the decor⟩

keep•sake \ˈkēp-ˌsāk\ *noun* [[1]*keep* + [1]*-sake* (as in *namesake*)] (1790)
: something kept or given to be kept as a memento

keep up (15th century)
transitive verb
: to persist or persevere in ⟨*kept up* the good work⟩; *also* **:** MAINTAIN, SUSTAIN ⟨*keep* standards *up*⟩
intransitive verb
1 : to keep adequately informed or up-to-date ⟨*keep up* on international affairs⟩
2 : to continue without interruption ⟨rain *kept up* all night⟩
3 : to maintain contact or relations with ⟨*keep up* with old friends⟩

kees•hond \ˈkās-ˌhȯnt\ *noun, plural* **kees•hon•den** \-ˌhȯn-dən\ [Dutch, probably from *Kees,* nickname for *Cornelis* Cornelius + *hond* dog, from Middle Dutch; akin to Old English *hund* hound] (1926)
: any of a Dutch breed of compact medium-sized dogs that have a dense heavy grayish coat and a foxy head

kef \ˈkēf, ˈkef, ˈkāf\ *noun* [Arabic *kayf* pleasure] (1808)
1 : a state of dreamy tranquillity
2 : a smoking material (as marijuana) that produces kef

kef•fi•yeh *variant of* KAFFIYEH

ke•fir \ke-ˈfir; ˈkē-fər, ˈke-\ *noun* [Russian] (1884)
: a beverage of fermented cow's milk

keg \ˈkeg, *dialect* ˈkag, ˈkāg\ *noun* [Middle English *kag,* of Scandinavian origin; akin to Old Norse *kaggi* keg] (circa 1632)
1 : a small cask or barrel having a capacity of 30 gallons (136.4 liters) or less
2 : the contents of a keg

keg•ler \ˈke-glər, ˈkā-\ *noun* [German, from *kegeln* to bowl, from *Kegel* bowling pin, from Old High German *kegil* stake, peg] (1932)
: [1]BOWLER

keg•ling \ˈke-gliŋ, ˈkā-\ *noun* (1938)
: BOWLING

keis•ter \ˈkēs-tər, ˈkīs-\ *or* **kees•ter** \ˈkēs-\ *noun* [English slang *keister* satchel] (1931)
slang **:** BUTTOCKS

ke•lim *variant of* KILIM

kel•ly green \ˈke-lē-\ *noun, often K capitalized* [from the common Irish name *Kelly;* from the association of Ireland with the color green] (circa 1935)
: a strong yellowish green

ke•loid \ˈkē-ˌlȯid\ *noun* [French *kéloïde,* from Greek *chēlē* claw] (1854)
: a thick scar resulting from excessive growth of fibrous tissue
— **keloid** *adjective*
— **ke•loi•dal** \kē-ˈlȯi-dᵊl\ *adjective*

kelp \ˈkelp\ *noun* [Middle English *culp*] (14th century)
1 a : any of various large brown seaweeds (order Laminariales) **b :** a mass of large seaweeds
2 : the ashes of seaweed used especially as a source of iodine

kelp bass *noun* (circa 1936)
: a mottled sea bass (*Paralabrax clathratus*) that occurs along the Pacific coast of the U.S. and is an important sport fish

[1]kel•pie \ˈkel-pē\ *noun* [perhaps from Scottish Gaelic *cailpeach, colpach* heifer, colt] (1747)
: a water sprite of Scottish folklore that delights in or brings about the drowning of wayfarers

[2]kelpie *noun* [*Kelpie,* name of a dog of this breed] (1903)
: any of a breed of energetic working dogs developed in Australia from British sheepdogs

Kelt \ˈkelt\, **Kelt•ic** \ˈkel-tik\ *variant of* CELT, CELTIC

kel•vin \ˈkel-vən\ *noun* (1968)
: the base unit of temperature in the International System of Units that is equal to 1/273.16 of the Kelvin scale temperature of the triple point of water

Kelvin *adjective* [William Thomson, Lord *Kelvin*] (1908)
: relating to, conforming to, or having a thermometric scale on which the unit of measurement equals the Celsius degree and according to which absolute zero is 0 K, the equivalent of −273.15°C

kemp \ˈkemp\ *noun* [Middle English *kempe* coarse hair, probably of Scandinavian origin; akin to Old Norse *kampr* mustache; akin to Old English *cenep* mustache] (1641)
: a coarse fiber especially of wool that is usually short, wavy, and white, has little affinity for dye, and is used in mixed wools

Kemp's ridley \ˈkemps-\ *noun* [Richard M. *Kemp* (flourished 1873) American amateur naturalist] (1979)
: RIDLEY a

kempt \ˈkem(p)t\ *adjective* [back-formation from *unkempt*] (1929)
: neatly kept **:** TRIM

[1]ken \ˈken\ *verb* **kenned; ken•ning** [Middle English *kennen,* from Old English *cennan* to make known & Old Norse *kenna* to perceive; both akin to Old English *can* know — more at CAN] (13th century)
transitive verb
1 *archaic* **:** SEE
2 *chiefly dialect* **:** RECOGNIZE
3 *chiefly Scottish* **:** KNOW
intransitive verb
chiefly Scottish **:** KNOW

[2]ken *noun* (1590)
1 a : the range of vision **b :** SIGHT, VIEW ⟨'tis double death to drown in *ken* of shore —Shakespeare⟩
2 : the range of perception, understanding, or knowledge ⟨abstract words that are beyond the *ken* of young children —Lois M. Rettie⟩

ke•naf \kə-ˈnaf\ *noun* [Persian] (1891)
: an African hibiscus (*Hibiscus cannabinus*) widely cultivated for its fiber; *also* **:** the fiber used especially for cordage

Ken•dal green \ˈken-dᵊl-\ *noun* [*Kendal,* England] (1514)
: a green woolen cloth resembling homespun or tweed

ken•do \ˈken-(ˌ)dō\ *noun* [Japanese *kendō,* from *ken* sword + *dō* art] (1921)
: a Japanese sport of fencing with bamboo swords

[1]ken•nel \ˈke-nᵊl\ *noun* [Middle English *kenel,* ultimately from (assumed) Vulgar Latin *canile,* from Latin *canis* dog — more at HOUND] (14th century)
1 a : a shelter for a dog or cat **b :** an establishment for the breeding or boarding of dogs or cats
2 : a pack of dogs

[2]kennel *verb* **-neled** *or* **-nelled; -nel•ing** *or* **-nel•ling** (1552)
intransitive verb
: to take shelter in or as if in a kennel
transitive verb
: to put or keep in or as if in a kennel

[3]kennel *noun* [alteration of *cannel* gutter] (15th century)
: a gutter in a street

[1]ken•ning \ˈke-niŋ\ *noun* [Middle English, sight, view, from gerund of *kennen*] (1786)
chiefly Scottish **:** a perceptible but small amount

[2]kenning *noun* [Old Norse, from *kenna*] (1883)
: a metaphorical compound word or phrase (as *swan-road* for *ocean*) used especially in Old English and Old Norse poetry

ke•no \ˈkē-(ˌ)nō\ *noun* [French *quine,* set of five winning numbers in a lottery + English *-o* (as in *lotto*)] (1814)
: a game resembling bingo

ken•speck•le \ˈken-ˌspe-kəl\ *adjective* [probably of Scandinavian origin; akin to Norwegian *kjennspak* quick to recognize] (1616)
chiefly Scottish **:** CONSPICUOUS

kent•ledge \ˈkent-lij\ *noun* [origin unknown] (1607)
: pig iron or scrap metal used as ballast

Ken•tucky bluegrass \kən-ˈtə-kē-\ *noun* [*Kentucky,* state of U.S.] (1849)
: a valuable Old World pasture and meadow grass (*Poa pratensis*) naturalized in America — called also *bluegrass*

Kentucky coffee tree *noun* (1785)
: a tall North American leguminous tree (*Gymnocladus dioica*) with bipinnate leaves and large woody brown pods whose seeds have been used as a substitute for coffee

Kentucky rifle *noun* (1832)
: a muzzle-loading long-barreled flintlock rifle developed in the 18th century in Pennsylvania and used extensively on the American frontier

Ke•ogh plan \ˈkē-(ˌ)ō-\ *noun* [Eugene James *Keogh* (died 1989) American politician] (1974)
: an individual retirement account for the self-employed

ke•pi \ˈkā-pē, ˈke-\ *also* **ké•pi** \ˈkā-\ *noun* [French *képi,* from German dialect (Switzerland) *käppi* cap] (1861)
: a military cap with a round flat top sloping toward the front and a visor

kept *past and past participle of* KEEP

kerat- *or* **kerato-** *combining form* [International Scientific Vocabulary, from Greek *kerato-, keras* horn — more at HORN]
: cornea ⟨*kerat*itis⟩

ker•a•tin \ˈker-ə-tᵊn\ *noun* [International Scientific Vocabulary] (circa 1849)
: any of various sulfur-containing fibrous proteins that form the chemical basis of horny epidermal tissues (as hair and nails)
— **ke•ra•ti•nous** \kə-ˈra-tᵊn-əs\ *adjective*

ke•ra•ti•ni•za•tion \ˌker-ə-tə-nə-ˈzā-shən, kə-ˌra-tᵊn-ə-\ *noun* (circa 1887)
: conversion into keratin or keratinous tissue
— **ke•ra•ti•nize** \ˈker-ə-tə-ˌnīz, kə-ˈra-tᵊn-ˌīz\ *verb*

ke•ra•ti•no•phil•ic \ˌker-ə-tə-nə-ˈfi-lik, kə-ˌra-tᵊn-ə-\ *adjective* (1946)

\ə\ **abut** \ᵊ\ **kitten** \ər\ **further** \a\ **ash** \ā\ **ace** \ä\ **mop, mar** \au̇\ **out** \ch\ **chin** \e\ **bet** \ē\ **easy** \g\ **go** \i\ **hit** \ī\ **ice** \j\ **job** \ŋ\ **sing** \ō\ **go** \ȯ\ **law** \ȯi\ **boy** \th\ **thin** \t͟h\ **the** \ü\ **loot** \u̇\ **foot** \y\ **yet** \zh\ **vision** *see also* Guide to Pronunciation

: exhibiting affinity for keratin (as in hair, skin, feathers, or horns) — used chiefly of fungi capable of growing on such materials

ker·a·ti·tis \ˌker-ə-ˈtī-təs\ *noun, plural* **-tit·i·des** \-ˈti-tə-ˌdēz\ [New Latin] (1858)
: inflammation of the cornea of the eye

ker·a·to·con·junc·ti·vi·tis \ˈker-ə-(ˌ)tō-kən-ˌjəŋ(k)-tə-ˈvī-təs\ *noun* [New Latin] (1887)
: combined inflammation of the cornea and conjunctiva

ker·a·to·plas·ty \ˈker-ə-tō-ˌplas-tē\ *noun, plural* **-ties** (circa 1857)
: plastic surgery on the cornea; *especially* : corneal grafting

ker·a·to·sis \ˌker-ə-ˈtō-səs\ *noun, plural* **-to·ses** \-ˌsēz\ [New Latin] (1885)
: an area of skin marked by overgrowth of horny tissue
— **ker·a·tot·ic** \-ˈtä-tik\ *adjective*

kerb \ˈkərb\ *noun* (1805)
British : CURB 5

ker·chief \ˈkər-chəf, -ˌchēf\ *noun, plural* **ker·chiefs** \-chəfs, -ˌchēfs\ *also* **ker·chieves** \-ˌchēvz\ [Middle English *courchef,* from Old French *cuevrechief,* from *covrir* to cover + *chief* head — more at CHIEF] (13th century)
1 : a square of cloth used as a head covering or worn as a scarf around the neck
2 : HANDKERCHIEF 1
— **ker·chiefed** \-chəft, -ˌchēft\ *adjective*

kerf \ˈkərf\ *noun* [Middle English, action of cutting, from Old English *cyrf;* akin to Old English *ceorfan* to carve — more at CARVE] (1523)
1 : a slit or notch made by a saw or cutting torch
2 : the width of cut made by a saw or cutting torch

ker·fuf·fle \kər-ˈfə-fəl\ *noun* [alteration of *carfuffle,* from Scots *car-* (probably from Scottish Gaelic *cearr* wrong, awkward) + *fuffle* to become disheveled] (1946)
chiefly British : DISTURBANCE, FUSS

Ker·man \kər-ˈmän, ker-\ *variant of* KIRMAN

ker·mes \ˈkər-(ˌ)mēz\ *noun* [French *kermès,* from Arabic *qirmiz*] (1603)
: the dried bodies of the females of various scale insects (genus *Kermes*) that are found on a Mediterranean oak (*Quercus coccifera*) and constitute a red dyestuff; *also* : the dye

ker·mis \ˈkər-məs\ *or* **ker·mess** \-məs, -ˌmes\ *or* **ker·messe** \-məs, -ˌmes\ *noun* [Dutch *kermis,* from Middle Dutch *kercmisse,* from *kerc, kerke* church + *misse* mass, church festival] (1577)
1 : an outdoor festival of the Low Countries
2 : a fair held usually for charitable purposes

[1]**kern** *or* **kerne** \ˈkərn, ˈkern\ *noun* [Middle English *kerne,* from Middle Irish *cethern* band of soldiers] (15th century)
1 : a light-armed foot soldier of medieval Ireland or Scotland
2 : YOKEL

[2]**kern** \ˈkərn\ *noun* [French *carne* corner, from French dialect, from Latin *cardin-, cardo* hinge] (1683)
: a part of a typeset letter that projects beyond its side bearings

ker·nel \ˈkər-nᵊl\ *noun* [Middle English, from Old English *cyrnel,* diminutive of *corn*] (before 12th century)
1 *chiefly dialect* : a fruit seed
2 : the inner softer part of a seed, fruit stone, or nut
3 : a whole seed of a cereal ⟨a *kernel* of corn⟩
4 : a central or essential part : GERM ⟨like many stereotypes . . . this one too contains some *kernels* of truth —S. M. Lyman⟩
5 : a subset of the elements of one set (as a group) that a function (as a homomorphism) maps onto an identity element of another set

kern·ite \ˈkər-ˌnīt\ *noun* [Kern Co., Calif.] (1927)
: a mineral that consists of a hydrous borate of sodium and is an important source of borax

ker·o·gen \ˈker-ə-jən\ *noun* [Greek *kēros* wax + English *-gen* — more at CERUMEN] (1906)
: bituminous material occurring in shale and yielding oil when heated

ker·o·sene *or* **ker·o·sine** \ˈker-ə-ˌsēn, ˌker-ə-ˈ, ˈkar-, ˌkar-\ *noun* [Greek *kēros* + English *-ene* (as in *camphene*)] (1854)
: a flammable hydrocarbon oil usually obtained by distillation of petroleum and used for a fuel and as a solvent and thinner

ker·ria \ˈker-ē-ə\ *noun* [New Latin, from William *Kerr* (died 1814) English gardener] (1823)
: a shrub (*Kerria japonica*) of the rose family that is native to China and Japan and has solitary yellow and often double flowers

ker·ry \ˈker-ē\ *noun, plural* **kerries** *often capitalized* [County *Kerry,* Ireland] (1829)
: any of an Irish breed of small hardy black dairy cattle

Kerry blue terrier *noun* (1922)
: any of an Irish breed of medium-sized terriers with a long squarish head, deep chest, and silky bluish coat

ker·sey \ˈkər-zē\ *noun, plural* **kerseys** [Middle English, from *Kersey,* England] (14th century)
1 a : a coarse ribbed woolen cloth for hose and work clothes **b** : a heavy wool or wool and cotton fabric used especially for uniforms and coats
2 : a garment of kersey

ker·sey·mere \ˈkər-zē-ˌmir\ *noun* [alteration of *cassimere*] (1798)
: a fine woolen fabric with a close nap made in fancy twill weaves

ke·ryg·ma \kə-ˈrig-mə\ *noun* [Greek *kērygma,* from *kēryssein* to proclaim, from *kēryx* herald — more at CADUCEUS] (1889)
: the apostolic proclamation of salvation through Jesus Christ
— **ker·yg·mat·ic** \ˌker-ig-ˈma-tik\ *adjective*

kes·trel \ˈkes-trəl\ *noun* [Middle English *castrel,* from Middle French *crecerelle,* from *crecelle* rattle; from its cry] (15th century)
: any of various small chiefly Old World falcons (genus *Falco*) that usually hover in the air while searching for prey: as **a** : a common Eurasian falcon (*F. tinnunculus*) **b** : an American falcon (*F. sparverius*) having a reddish brown back and tail and bluish gray wings

ket- *or* **keto-** *combining form* [International Scientific Vocabulary]
: ketone ⟨*ketosis*⟩

ketch \ˈkech\ *noun* [alteration of *catch,* from Middle English *cache*] (circa 1649)
: a fore-and-aft rigged vessel similar to a yawl but with a larger mizzen and with the mizzenmast stepped farther forward

ketch

ketch·up \ˈke-chəp, ˈka-\ *noun* [Malay *kĕchap* fish sauce] (circa 1690)
: a seasoned pureed condiment usually made from tomatoes

ke·tene \ˈkē-ˌtēn\ *noun* [International Scientific Vocabulary] (1907)
: a colorless poisonous gas C_2H_2O of penetrating odor used especially as an acetylating agent; *also* : any of various derivatives of this compound

ke·to \ˈkē-(ˌ)tō\ *adjective* [*ket-*] (1891)
: of or relating to a ketone; *also* : containing a ketone group

ke·to·gen·e·sis \ˌkē-tō-ˈje-nə-səs\ *noun* [New Latin] (1915)
: the production of ketone bodies (as in diabetes)
— **ke·to·gen·ic** \-ˈje-nik\ *adjective*

ke·to·glu·tar·ic acid \-glü-ˈtar-ik-\ *noun* (1908)
: either of two crystalline keto derivatives $C_5H_6O_5$ of glutaric acid; *especially* : the alpha keto isomer formed in various metabolic processes (as the Krebs cycle)

ke·tone \ˈkē-ˌtōn\ *noun* [German *Keton,* alteration of *Aceton* acetone] (1851)
: an organic compound (as acetone) with a carbonyl group attached to two carbon atoms
— **ke·ton·ic** \kē-ˈtä-nik\ *adjective*

ketone body *noun* (1915)
: any of the three compounds acetoacetic acid, acetone, and beta-hydroxybutyric acid which are normal intermediates in lipid metabolism and accumulate in the blood and urine in abnormal amounts in conditions of impaired metabolism (as diabetes mellitus)

ke·tose \ˈkē-ˌtōs, -ˌtōz\ *noun* [International Scientific Vocabulary] (1902)
: a sugar (as fructose) containing one ketone group per molecule

ke·to·sis \kē-ˈtō-səs\ *noun* [New Latin] (1917)
: an abnormal increase of ketone bodies in the body
— **ke·tot·ic** \-ˈtä-tik\ *adjective*

ke·to·ste·roid \ˌkē-tō-ˈstir-ˌȯid *also* -ˈster-\ *noun* [International Scientific Vocabulary] (1939)
: a steroid (as cortisone or estrone) containing a ketone group

ket·tle \ˈke-tᵊl\ *noun* [Middle English *ketel,* from Old Norse *ketill* (akin to Old English *cietel* kettle), both from a prehistoric Germanic word borrowed from Latin *catillus,* diminutive of *catinus* bowl] (13th century)
1 : a metallic vessel usually used for boiling liquids; *especially* : TEAKETTLE
2 : KETTLEDRUM
3 a : POTHOLE 1b **b** : a steep-sided hollow without surface drainage especially in a deposit of glacial drift

ket·tle·drum \-ˌdrəm\ *noun* (1602)
: a percussion instrument that consists of a hollow brass, copper, or fiberglass hemisphere with a calfskin or plastic head whose tension can be changed to vary the pitch

kettledrum

kettle of fish (1742)
1 : a bad state of affairs : MESS
2 : something to be considered or reckoned with : MATTER ⟨books and discs . . . were two very different *kettles of fish* —Roland Gelatt⟩

Kew·pie \ˈkyü-pē\ *trademark*
— used for a small chubby doll with a topknot of hair

[1]**key** \ˈkē\ *noun* [Middle English, from Old English *cǣg;* akin to Old Frisian *kēi* key] (before 12th century)
1 a : a usually metal instrument by which the bolt of a lock is turned **b** : any of various devices having the form or function of such a key
2 a : a means of gaining or preventing entrance, possession, or control **b** : an instrumental or deciding factor
3 a : something that gives an explanation or identification or provides a solution ⟨the *key* to a riddle⟩ **b** : a list of words or phrases giving an explanation of symbols or abbreviations **c** : an aid to interpretation or identification : CLUE **d** : an arrangement of the salient characters of a group of plants or animals or of taxa designed to facilitate identification **e** : a map legend
4 a (1) : COTTER PIN (2) : COTTER **b** : a keystone in an arch **c** : a small piece of wood or metal used as a wedge or for preventing motion between parts

5 a : one of the levers of a keyboard musical instrument that actuates the mechanism and produces the tones **b :** a lever that controls a vent in the side of a woodwind instrument or a valve in a brass instrument **c :** a part to be depressed by a finger that serves as one unit of a keyboard
6 : SAMARA
7 : a system of tones and harmonies generated from a hierarchical scale of seven tones based on a tonic ⟨the *key* of G major⟩
8 a : characteristic style or tone **b :** the tone or pitch of a voice **c :** the predominant tone of a photograph with respect to its lightness or darkness
9 : a decoration or charm resembling a key
10 : a small switch for opening or closing an electric circuit ⟨a telegraph *key*⟩
11 : the set of instructions governing the encipherment and decipherment of messages
12 : a free-throw area in basketball
— keyed \ˈkēd\ *adjective*
— key·less \ˈkē-ləs\ *adjective*

²key (14th century)
transitive verb
1 : to lock with or as if with a key **:** FASTEN: as **a :** to secure (as a pulley on a shaft) by a key **b :** to finish off (an arch) by inserting a keystone
2 : to regulate the musical pitch of
3 : to bring into harmony or conformity **:** make appropriate **:** ATTUNE ⟨remarks *keyed* to a situation⟩
4 : to identify (a biological specimen) by a key
5 : to provide with identifying or explanatory cross-references ⟨instructions *keyed* to accompanying drawings —John Gartner⟩
6 : to make nervous, tense, or excited — usually used with *up* ⟨was *keyed* up over her impending operation⟩
7 : KEYBOARD — often used with *in*
intransitive verb
1 : to use a key
2 : to observe the position or movement of an opposing player in football in order to anticipate the play — usually used with *on*
3 : KEYBOARD

³key *adjective* (1913)
: IMPORTANT, FUNDAMENTAL ⟨*key* issues⟩

⁴key *noun* [Spanish *cayo,* from Taino] (1697)
: a low island or reef; *specifically* **:** any of the coral islets off the southern coast of Florida

⁵key *noun* [by shortening & alteration from *kilo*] (1968)
slang **:** a kilogram especially of marijuana or heroin

¹key·board \ˈkē-ˌbōrd, -ˌbȯrd\ *noun* (1819)
1 : a bank of keys on a musical instrument (as a piano) that usually consists of seven white and five raised black keys to the octave
2 : an assemblage of systematically arranged keys by which a machine or device is operated
3 : a board on which keys for locks are hung

²keyboard (1961)
transitive verb
: to capture or set (as data or text) by means of a keyboard
intransitive verb
: to operate a machine (as for typesetting) by means of a keyboard
— key·board·er *noun*

key·board·ist \ˈkē-ˌbōr-dist, -ˌbȯr-\ *noun* (1973)
: a person who plays a keyboard musical instrument

key·but·ton \ˈkē-ˌbə-tᵊn\ *noun* (circa 1920)
: KEY 5c

key club *noun* [from the key to the premises provided to each member] (1962)
: an informal private club serving liquor and providing entertainment

key deer *noun, often K capitalized* (1950)
: a very small rare white-tailed deer (*Odocoileus virginianus clavium*) native to the Florida Keys

key grip \ˌkē-\ *noun* (1979)
: the technician in charge of moving and setting up camera tracks and scenery in a motion-picture or television production

¹key·hole \ˈkē-ˌhōl\ *noun* (circa 1592)
1 : a hole for receiving a key
2 : KEY 12

²keyhole *adjective* (1937)
1 : revealingly intimate ⟨a *keyhole* report⟩
2 : intent on revealing intimate details ⟨*keyhole* columnists⟩

keyhole saw *noun* (1777)
: a narrow pointed fine-toothed handsaw used for cutting curves of short radius

key light *noun* (circa 1937)
: the main light illuminating a photographic subject

key lime *noun, often K capitalized* (1929)
: a lime grown especially in the Florida Keys and adjacent areas

key lime pie *noun, often K capitalized* (1954)
: a usually meringue-topped lime-custard pie traditionally made from key limes

Keynes·ian·ism \ˈkān-zē-ə-ˌni-zəm\ *noun* (1946)
: the economic theories and programs ascribed to John M. Keynes and his followers; *specifically* **:** the advocacy of monetary and fiscal programs by government to increase employment and spending
— Keynes·ian \ˈkān-zē-ən\ *noun or adjective*

¹key·note \ˈkē-ˌnōt\ *noun* (1776)
1 : the first and harmonically fundamental tone of a scale
2 : the fundamental or central fact, idea, or mood ⟨sadness is the *keynote* of this little collection —*Books Abroad*⟩

²keynote *transitive verb* (1914)
1 : to set the keynote of
2 : to deliver the keynote address at
— key·not·er *noun*

keynote address *noun* (circa 1908)
: an address designed to present the issues of primary interest to an assembly (as a political convention) and often to arouse unity and enthusiasm — called also *keynote speech*

keynote speaker *noun* (1950)
: one who delivers a keynote address

key·pad \ˈkē-ˌpad\ *noun* (1975)
: a small often handheld keyboard

¹key·punch \ˈkē-ˌpənch\ *noun* (1918)
: a machine with a keyboard used to cut holes or notches in punched cards

²keypunch *transitive verb* (1959)
: to enter (data) on punched cards with a keypunch
— key·punch·er *noun*

key signature *noun* (1875)
: the sharps or flats placed after a clef in music to indicate the key

key·stone \ˈkē-ˌstōn\ *noun* (circa 1637)
1 : the wedge-shaped piece at the crown of an arch that locks the other pieces in place — see ARCH illustration
2 : something on which associated things depend for support ⟨determination, a *keystone* of the puritan ethic —L. S. Lewis⟩

¹key·stroke \-ˌstrōk\ *noun* (circa 1910)
: the act or an instance of depressing a key on a keyboard

²keystroke *transitive verb* (1966)
: KEYBOARD

key·way \-ˌwā\ *noun* (circa 1864)
1 : a groove or channel for a key
2 : the aperture for the key in a lock having a flat metal key

key word *noun* (1859)
: a word that is a key: as **a :** a word exemplifying the meaning or value of a letter or symbol **b** *usually* **key·word** \-ˌwərd\ **:** a significant word from a title or document used as an index to content

khad·dar \ˈkä-dər\ *or* **kha·di** \ˈkä-dē\ *noun* [Hindi *khādar, khādī*] (circa 1885)
: homespun cotton cloth of India

kha·ki \ˈka-kē, ˈkä-, *Canadian often* ˈkär-\ *noun* [Hindi *khākī* dust-colored, from *khāk* dust, from Persian] (1857)
1 : a light yellowish brown
2 a : a khaki-colored cloth made usually of cotton or wool and used especially for military uniforms **b :** a garment of this cloth; *especially* **:** a military uniform — usually used in plural ◆
— khaki *adjective*

Khal·kha \ˈkal-kə, ˈkal-kə\ *noun* (1873)
1 : a member of a Mongol people of Outer Mongolia
2 : the language of the Khalkha people used as the official language of the Mongolian People's Republic

kham·sin \kam-ˈsēn\ *noun* [Arabic *rīḥ al-khamsīn* the wind of the fifty (days between Easter and Pentecost)] (1685)
: a hot southerly Egyptian wind

¹khan \ˈkän *also* ˈkan\ *noun* [Middle English *caan,* from Middle French, of Turkic origin; akin to Turkish *han* prince] (15th century)
1 : a medieval sovereign of China and ruler over the Turkish, Tatar, and Mongol tribes
2 : a local chieftain or man of rank in some countries of central Asia

²khan *noun* [Arabic *khān*] (1614)
: a caravansary or rest house in some Asian countries

khan·ate \ˈkä-ˌnāt *also* ˈka-\ *noun* (1799)
: the state or jurisdiction of a khan

khap·ra beetle \ˈka-prə-, ˈkä-\ *noun* [Hindi *khaprā,* literally, destroyer] (1928)
: a dermestid beetle (*Trogoderma granarium*) that is native to the Indian subcontinent and is now a serious pest of stored grain in most parts of the world

khat \ˈkät\ *noun* [Arabic *qāt*] (1858)
: a shrub (*Catha edulis*) cultivated in the Middle East and Africa for its leaves and buds that are the source of an habituating stimulant when chewed or used as a tea

khe·dive \kə ˈdēv\ *noun* [French *khédive,* from Turkish *hidiv*] (1867)
: a ruler of Egypt from 1867 to 1914 governing as a viceroy of the sultan of Turkey
— khe·div·ial \-ˈdē vē-əl\ *or* **khe·div·al** \-ˈdē-vəl\ *adjective*

◇ WORD HISTORY
khaki *Khaki,* like *bungalow, cot,* and *shampoo,* is a by-product of British rule in India, though for the average user of English these words have lost all association with the Raj. The Hindi word *khākī* "dust-colored, brown" was applied to the color of a uniform cloth used by units of Britain's Indian Army in the mid-19th century. Dull brown field uniforms were most likely originally adopted for their ease of maintenance on dirty trails, though as the range of a rifle increased and battlefield tactics changed in the later 19th century, it was doubtless realized that drab clothing was valuable as camouflage. By the time of the South African War (1899–1902), all colonial units of the British Army were dressed in the color, and *khaki* became linked with support for the army and a militaristic foreign policy. (The vote of 1900, when patriotic Britons returned to Parliament the same Conservatives who had started the war with the Boers, went down in history as the "Khaki Election.") Contemporary retailers in the U.S., however, to judge by their advertising, associate khaki clothing with fashionable casualness and a summery look rather than with military uniforms or British India.

Khmer \kə-ˈmer\ *noun, plural* **Khmer** *or* **Khmers** (1876)
1 : a member of an aboriginal people of Cambodia
2 : the Mon-Khmer language of the Khmer people that is the official language of Cambodia
— Khmer *adjective*

Khoi·khoi \ˈkȯi-ˌkȯi\ *noun, plural* **Khoikhoi** (1791)
: HOTTENTOT

Khoi·san \ˈkȯi-ˌsän, -ˈsän\ *noun* [*Khoi*khoi + *San*] (1930)
1 : a group of African peoples speaking Khoisan languages
2 : a family of African languages comprising principally Hottentot and the Bushman languages

khoum \ˈküm, ˈk̲üm\ *noun* [modification of Arabic *khums,* literally, one fifth] (1973)
— see *ouguiya* at MONEY table

Kho·war \ˈkō-ˌwär\ *noun* (1882)
: an Indo-Aryan language of northwest Pakistan

ki·ang \kē-ˈäŋ\ *noun* [Tibetan *rkyaṅ*] (1869)
: an Asian wild ass (*Equus hemionus kiang* synonym *E. kiang*) usually with reddish back and sides and white underparts, muzzle, and legs

kiaugh \ˈkyäk̲\ *noun* [origin unknown] (1786)
Scottish **:** TROUBLE, ANXIETY

kib·be *or* **kib·beh** *or* **kib·bi** \ˈki-bē\ *noun* [Arabic *kubbah*] (1937)
: a Near Eastern dish of ground lamb and bulgur that is eaten cooked or raw

¹kib·ble \ˈki-bəl\ *transitive verb* **kib·bled; kib·bling** \-b(ə-)liŋ\ [origin unknown] (circa 1790)
: to grind coarsely ⟨*kibbled* dog biscuit⟩ ⟨*kibbled* grain⟩

²kibble *noun* (1942)
: coarsely ground meal or grain (as for animal feed)

kib·butz \ki-ˈbu̇ts, -ˈbüts\ *noun, plural* **kib·but·zim** \-ˌbu̇t-ˈsēm, -ˌbüt-\ [New Hebrew *qibbūṣ*] (1944)
: a communal farm or settlement in Israel

kib·butz·nik \-ˈbu̇t-snik, -ˈbüt-\ *noun* [Yiddish *kibutsnik,* from *kibuts* kibbutz (from New Hebrew *qibbūṣ*) + *-nik,* agent suffix] (1947)
: a member of a kibbutz

kibe \ˈkīb\ *noun* [Middle English] (14th century)
: an ulcerated chilblain especially on the heel

ki·bitz *also* **kib·bitz** \ˈki-bəts, kə-ˈbits\ *verb* [Yiddish *kibetsn*] (1927)
intransitive verb
: to act as a kibitzer
transitive verb
: to observe as a kibitzer; *especially* **:** to be a kibitzer at ⟨*kibitz* a card game⟩

ki·bitz·er *also* **kib·bitz·er** \ˈki-bət-sər, kə-ˈbit-\ *noun* (1922)
: one who looks on and often offers unwanted advice or comment especially at a card game; *broadly* **:** one who offers opinions

ki·bosh \ˈkī-ˌbäsh, kī-ˈ; ki-ˈbäsh\ *noun* [origin unknown] (1836)
: something that serves as a check or stop ⟨put the *kibosh* on that⟩ ◆
— kibosh *transitive verb*

¹kick \ˈkik\ *verb* [Middle English *kiken*] (14th century)
intransitive verb
1 a : to strike out with the foot or feet **b :** to make a kick in football
2 a : to show opposition **:** RESIST, REBEL **b :** to protest strenuously or urgently **:** express grave discontent; *broadly* **:** COMPLAIN
3 : to function with vitality and energy ⟨alive and *kicking*⟩
4 *of a firearm* **:** to recoil when fired
5 : to go from one place to another as circumstance or whim dictates
6 : to run at a faster speed during the last part of a race
transitive verb
1 a : to strike, thrust, or hit with the foot **b :** to strike suddenly and forcefully as if with the foot **c :** to remove from a position or status ⟨*kicked* him off the team⟩
2 : to score by kicking a ball
3 : to heap reproaches upon (oneself) ⟨*kicked* themselves for not going⟩
4 : to free oneself of (as a drug habit)
— kick·able \ˈki-kə-bəl\ *adjective*
— kick ass : to kick butt — often considered vulgar
— kick butt : to use forceful or coercive measures in order to achieve a purpose; *also* **:** to succeed or win overwhelmingly
— kick over the traces : to cast off restraint, authority, or control
— kick the bucket : DIE
— kick up one's heels 1 : to show sudden delight **2 :** to have a lively time
— kick upstairs : to promote to a higher but less desirable position

²kick *noun* (1530)
1 a : a blow or sudden forceful thrust with the foot; *specifically* **:** a sudden propelling of a ball with the foot **b :** the power to kick **c :** a rhythmic motion of the legs used in swimming **d :** a burst of speed in racing
2 : a sudden forceful jolt or thrust suggesting a kick; *especially* **:** the recoil of a gun
3 : POCKET, WALLET
4 a : a feeling or expression of opposition or objection ⟨a *kick* against the administration⟩ **b :** the grounds for objection
5 a : an effect suggestive of a kick ⟨chili with a *kick*⟩ **b :** a stimulating or pleasurable effect or experience **c :** pursuit of an absorbing or obsessive new interest ⟨a skiing *kick*⟩
6 : a sudden and striking surprise, revelation, or turn of events

kick around (1839)
intransitive verb
1 : to wander or pass time aimlessly
2 : to lie about mostly unnoticed or forgotten
transitive verb
1 : to treat in an inconsiderate or high-handed fashion
2 : to consider, examine, or discuss from various angles

kick–ass \ˈkik-ˌas\ *adjective* (1970)
: strikingly tough, aggressive, or uncompromising — often considered vulgar

kick·back \ˈkik-ˌbak\ *noun* (1920)
1 : a sharp violent reaction
2 : a return of a part of a sum received often because of confidential agreement or coercion ⟨every city contract had been let with a ten percent *kickback* to city officials —D. K. Shipler⟩

kick·board \-ˌbōrd, -ˌbȯrd\ *noun* (1949)
: a buoyant rectangular board held by a swimmer while developing kicking techniques

kick·box·ing \-ˌbäk-siŋ\ *noun* (1971)
: boxing in which boxers are permitted to kick with bare feet as in karate
— kick·box·er \-sər\ *noun*

kick·er *noun* (circa 1580)
1 : one that kicks or kicks something
2 : KICK 6

kick in (1908)
transitive verb
: CONTRIBUTE
intransitive verb
1 *slang* **:** DIE
2 : to make a contribution

kick·off \ˈkik-ˌȯf\ *noun* (1857)
1 : a kick that puts the ball into play in a football or soccer game
2 : COMMENCEMENT 1 ⟨the campaign *kickoff*⟩

kick off (1857)
intransitive verb
1 : to start or resume play in football by a placekick
2 : to begin proceedings
3 *slang* **:** DIE
transitive verb
: to mark the beginning of

kick out *transitive verb* (1697)
: to dismiss or eject forcefully or summarily

kick over (1951)
intransitive verb
: to begin to fire — used of an internal combustion engine
transitive verb
: TURN OVER 1b

kick pleat *noun* (1926)
: a short inverted pleat (as at the bottom of a skirt) used to give breadth

kick·shaw \ˈkik-ˌshȯ\ *noun* [by folk etymology from French *quelque chose* something] (1597)
1 : a fancy dish **:** DELICACY
2 : TRINKET, GEWGAW

kick·stand \ˈkik-ˌstand\ *noun* [from its being put in position by a kick] (1947)
: a swiveling metal bar or rod for holding up a 2-wheeled vehicle (as a bicycle) when not in use

kick turn *noun* (1910)
: a standing half turn in skiing made by swinging one ski high with a jerk and planting it in the desired direction and then lifting the other ski into a parallel position

kick·up \ˈkik-ˌəp\ *noun* (circa 1793)
: a noisy quarrel **:** ROW

kick up (1756)
transitive verb
1 : to stir up **:** PROVOKE ⟨*kick up* a fuss⟩
2 : to cause to rise upward ⟨clouds of dust *kicked up* by passing cars⟩
intransitive verb
: to give evidence of disorder

kicky \ˈki-kē\ *adjective* (1966)
: providing a kick or thrill **:** EXCITING; *also* **:** excitingly fashionable

¹kid \ˈkid\ *noun* [Middle English *kide,* of Scandinavian origin; akin to Old Norse *kith* kid] (13th century)
1 a : a young goat **b :** a young individual of various animals related to the goat
2 a : the flesh, fur, or skin of a kid **b :** something made of kid
3 : a young person; *especially* **:** CHILD — often used as a generalized reference to one especially younger or less experienced ⟨the *kid* on the pro golf tour⟩ ⟨poor *kid*⟩
— kid·dish \ˈki-dish\ *adjective*

²kid *intransitive verb* **kid·ded; kid·ding** (15th century)

◇ WORD HISTORY

kibosh The word *kibosh* is first recorded in 1836 in an early sketch by Charles Dickens (spelled *kye-bosk,* with a perhaps erroneous final *k*). Nine years later *kye-bosh* resurfaces in a dictionary of British underworld argot defined as "eighteen pence"; the connection between this and the more familiar "check" or "stop" sense is unclear. For over a century this peculiar colloquialism has taxed the ingenuity of etymologists, with minimal results: beyond the fact that it was prominent enough in lower-class London speech to attract Dickens's attention in the 1830s we know nothing for certain. Claims were once made that the word was Yiddish, though no remotely plausible Yiddish source has ever been produced. Another hypothesis is that its source is Irish *caidhp bháis,* literally, "coif (or cap) of death," variously explained as headgear a judge put on when pronouncing a death sentence, or—somewhat more plausibly—as a covering pulled over the face of a dead person when a coffin was closed. But evidence for any metaphorical use of this phrase in Irish appears to be lacking, and *kibosh* is not recorded in English as spoken in Ireland until decades after Dickens must have heard it in London.

: to bring forth young — used of a goat or an antelope

[3]kid *verb* **kid·ded; kid·ding** [probably from [1]*kid*] (1902)
transitive verb
1 : to deceive as a joke ⟨it's the truth; I wouldn't *kid* you⟩
2 : to make fun of
intransitive verb
: to engage in good-humored fooling or horseplay — often used with *around*
— kid·der *noun*
— kid·ding·ly \'ki-diŋ-lē\ *adverb*

Kid·der·min·ster \'ki-dər-ˌmin(t)-stər\ *noun* [*Kidderminster,* England] (1836)
: an ingrain carpet — called also *Kidderminster carpet*

kid·die *or* **kid·dy** \'ki-dē\ *noun, plural* **kiddies** *often attributive* [[1]*kid*] (1889)
: a small child

kid·dush \'ki-dəsh, -dish; ki-'düsh\ *noun* [Late Hebrew *qiddūsh* sanctification] (1753)
: a ceremonial blessing pronounced over wine or bread in a Jewish home or synagogue on a holy day (as the Sabbath)

kid–glove \'kid-'gləv\ *adjective* (1888)
: marked by extreme care or deference ⟨*kid-glove* treatment⟩

kid glove *noun* (1832)
: a dress glove made of kid leather
— kid–gloved \'kid-'gləvd\ *adjective*
— with kid gloves : with special consideration

kid leather *noun* (1687)
1 : a soft pliable leather made from kidskin
2 : a glove leather made from lambskin or goatskin

kid·nap \'kid-ˌnap\ *transitive verb* **-napped** *or* **-naped** \-ˌnapt\; **-nap·ping** *or* **-nap·ing** [probably back-formation from *kidnapper,* from *kid* + obsolete *napper* thief] (1682)
: to seize and detain or carry away by unlawful force or fraud and often with a demand for ransom
— kid·nap·pee *or* **kid·nap·ee** \ˌkid-ˌna-'pē\ *noun*
— kid·nap·per *or* **kid·nap·er** *noun*

kid·ney \'kid-nē\ *noun, plural* **kidneys** [Middle English] (14th century)
1 a : one of a pair of vertebrate organs situated in the body cavity near the spinal column that excrete waste products of metabolism, in humans are bean-shaped organs about 4½ inches (11½ centimeters) long lying behind the peritoneum in a mass of fatty tissue, and consist chiefly of nephrons by which urine is secreted, collected, and discharged into a main cavity whence it is conveyed by the ureter to the bladder **b :** any of various excretory organs of invertebrate animals
2 : the kidney of an animal eaten as food
3 : sort or kind especially with regard to temperament ⟨a nice helpful guy, of a different *kidney* entirely from the . . . Secret Police —Paula Lecler⟩

kidney bean *noun* (1548)
1 : an edible and nutritious seed of any cultivated bean of the common species (*Phaseolus vulgaris*); *especially* **:** a large dark red bean seed
2 : a plant bearing kidney beans

kidney stone *noun* (1946)
: a calculus in the kidney

kid·skin \'kid-ˌskin\ *noun* (14th century)
: the skin of a young or sometimes a mature goat; *also* **:** KID LEATHER

kid stuff *noun* (1929)
1 : something befitting or appropriate only to children
2 : something extremely simple or easy

kiel·ba·sa \kēl-'bä-sə, kil- *also* ki-'bä-sə\ *noun, plural* **-basas** *also* **-ba·sy** \-'bä-sē\ [Polish *kiełbasa*] (circa 1939)
: a smoked sausage of Polish origin

kie·sel·guhr \'kē-zəl-ˌgu̇r\ *noun* [German *Kieselgur*] (1875)
: loose or porous diatomite

kie·ser·ite \'kē-zə-ˌrīt\ *noun* [German *Kieserit,* from Dietrich *Kieser* (died 1862) German physician] (1862)
: a mineral that is a white hydrous magnesium sulfate

kif \'kif, 'kēf\ *variant of* KEF

kike \'kīk\ *noun* [origin unknown] (1904)
: JEW — usually taken to be offensive

Ki·ku·yu \kē-'kü-(ˌ)yü\ *noun, plural* **Kikuyu** *or* **Kikuyus** (1894)
1 : a member of a Bantu-speaking people of Kenya
2 : the Bantu language of the Kikuyu people

kil·der·kin \'kil-dər-kən\ *noun* [Middle English, from Middle Dutch *kindekijn,* from Medieval Latin *quintale* quintal] (14th century)
1 : an English unit of capacity equal to ½ barrel
2 : CASK

ki·lim \kē-'lēm\ *noun* [Turkish, from Persian *kilīm*] (1881)
: a pileless handwoven reversible rug or covering made in Turkey, Kurdistan, the Caucasus, Iran, and western Turkestan

[1]kill \'kil\ *verb* [Middle English, perhaps from (assumed) Old English *cyllan;* akin to Old English *cwellan* to kill — more at QUELL] (14th century)
transitive verb
1 a : to deprive of life **b** (1) **:** to slaughter (as a hog) for food (2) **:** to convert a food animal into (a kind of meat) by slaughtering
2 a : to put an end to ⟨*kill* competition⟩ **b :** DEFEAT, VETO ⟨*killed* the amendment⟩ **c :** to mark for omission; *also* **:** DELETE
3 a : to destroy the vital or essential quality of ⟨*killed* the pain with drugs⟩ **b :** to cause to stop ⟨*kill* the motor⟩ **c :** to check the flow of current through
4 : to make a markedly favorable impression on ⟨she *killed* the audience⟩
5 : to get through uneventfully ⟨*kill* time⟩; *also* **:** to get through (the time of a penalty) without being scored on ⟨*kill* a penalty⟩
6 a : to cause extreme pain to **b :** to tire almost to the point of collapse
7 : to hit (a shot) so hard in various games that a return is impossible
8 : to consume (as a drink) totally
intransitive verb
1 : to deprive of life
2 : to make a markedly favorable impression ⟨was dressed to *kill*⟩ ☆

[2]kill *noun* (1814)
1 : an act or instance of killing
2 : something killed: as **a** (1) **:** an animal shot in a hunt (2) **:** animals killed in a hunt, season, or particular period of time **b :** an enemy unit (as an airplane or ship) destroyed by military action **c :** a return shot in any of various games (as badminton, handball, or table tennis) that is too hard for an opponent to handle

[3]kill *noun, often capitalized* [Dutch *kil*] (1669)
: CHANNEL, CREEK — used chiefly in place names in Delaware, Pennsylvania, and New York

kill·deer \'kil-ˌdir\ *noun, plural* **killdeers** *or* **killdeer** [imitative] (1731)
: an American plover (*Charadrius vociferus*) characterized by two black breast bands and a plaintive penetrating cry

killdeer

[1]kill·er \'ki-lər\ *noun* (15th century)
1 : one that kills
2 : KILLER WHALE
3 a : one that has a forceful, violent, or striking impact **b :** one that is extremely difficult to deal with

[2]killer *adjective* (1951)
1 : strikingly impressive or effective ⟨a *killer* smile⟩ ⟨a *killer* résumé⟩
2 : extremely difficult to deal with ⟨a *killer* fastball⟩; *also* **:** causing death or devastation ⟨a *killer* tornado⟩

killer bee *noun* (1976)
: AFRICANIZED BEE

killer cell *noun* (1972)
: a T cell that functions in cell-mediated immunity by destroying a cell (as a tumor cell) having a specific antigenic molecule on its surface — called also *killer T cell*

killer instinct *noun* (1931)
: an aggressive tenacious urge for domination in a struggle to attain a set goal

killer whale *noun* (1884)
: a small gregarious whale (*Orcinus orca*) that is black with a white ventral side and white oval-shaped patches behind the eyes and attains a length of 20 to 30 feet (6.1 to 9.1 meters) — called also *orca*

kil·li·fish \'ki-li-ˌfish\ *noun* [*killie* killifish (perhaps from [3]*kill*) + *fish*] (1836)
1 : any of a family (Cyprinodontidae) of numerous small oviparous fishes much used as bait and in mosquito control
2 : any of a family (Poeciliidae) of livebearers

[1]kill·ing \'ki-liŋ\ *noun* (15th century)
1 : the act of one that kills
2 : KILL 2a
3 : a sudden notable gain or profit

[2]killing *adjective* (15th century)
1 : that kills or relates to killing
2 : highly amusing
3 : extremely difficult to deal with ⟨the suspense is *killing*⟩; *also* **:** calling for great strength, stamina, or endurance ⟨a *killing* schedule⟩
— kill·ing·ly \-lē\ *adverb*

kill·joy \'kil-ˌjȯi\ *noun* (1776)
: one who spoils the pleasure of others

kill off *transitive verb* (1607)
: to destroy in large numbers or totally

kiln \'kiln, 'kil\ *noun* [Middle English *kilne,* from Old English *cyln,* from Latin *culina* kitchen, from *coquere* to cook — more at COOK] (before 12th century)
: an oven, furnace, or heated enclosure used for processing a substance by burning, firing, or drying
— kiln *transitive verb*

ki·lo \'kē-(ˌ)lō *also* 'ki-\ *noun, plural* **kilos** (1870)
: KILOGRAM

Kilo (1952)
— a communications code word for the letter *k*

kilo- *combining form* [French, modification of Greek *chilioi*]
: thousand ⟨*kilo*ton⟩

☆ SYNONYMS
Kill, slay, murder, assassinate, dispatch, execute mean to deprive of life. KILL merely states the fact of death caused by an agency in any manner ⟨*killed* in an accident⟩ ⟨frost *killed* the plants⟩. SLAY is a chiefly literary term implying deliberateness and violence but not necessarily motive ⟨*slew* thousands of the Philistines⟩. MURDER specifically implies stealth and motive and premeditation and therefore full moral responsibility ⟨convicted of *murdering* a rival⟩. ASSASSINATE applies to deliberate killing openly or secretly often for political motives ⟨terrorists *assassinated* the Senator⟩. DISPATCH stresses quickness and directness in putting to death ⟨*dispatched* the sentry with one bullet⟩. EXECUTE stresses putting to death as a legal penalty ⟨*executed* by lethal gas⟩.

\ə\ abut \ᵊ\ kitten \ər\ further \a\ ash \ā\ ace \ä\ mop, mar \au̇\ out \ch\ chin \e\ bet \ē\ easy \g\ go \i\ hit \ī\ ice \j\ job \ŋ\ sing \ō\ go \ȯ\ law \ȯi\ boy \th\ thin \t͟h\ the \ü\ loot \u̇\ foot \y\ yet \zh\ vision *see also* Guide to Pronunciation

ki·lo·bar \ˈkē-lə-ˌbär, ˈki-lə-\ *noun* [International Scientific Vocabulary] (1926)
: a unit of pressure equal to 1000 bars

ki·lo·base \-ˌbās\ *noun* (1975)
: a unit of measure of the length of a nucleic-acid chain that equals one thousand base pairs

ki·lo·bit \-ˌbit\ *noun* [International Scientific Vocabulary] (1961)
1 : 1000 bits
2 : 1024 bits

ki·lo·byte \-ˌbīt\ *noun* [from the fact that 1024 (2^{10}) is the power of 2 closest to 1000] (1970)
: 1024 bytes

ki·lo·cal·o·rie \-ˌka-lə-rē, -ˈkal-rē\ *noun* [International Scientific Vocabulary] (1894)
: CALORIE 1b

kilo·cy·cle \ˈki-lə-ˌsī-kəl\ *noun* [International Scientific Vocabulary] (1921)
: 1000 cycles; *especially* **:** KILOHERTZ

ki·lo·gauss \ˈkē-lə-ˌgau̇s, ˈki-lə-\ *noun* [International Scientific Vocabulary] (1895)
: 1000 gauss

ki·lo·gram \-ˌgram\ *noun* [French *kilogramme,* from *kilo-* + *gramme* gram] (1797)
1 : the base unit of mass in the International System of Units that is equal to the mass of a prototype agreed upon by international convention and that is nearly equal to the mass of 1000 cubic centimeters of water at the temperature of its maximum density — see METRIC SYSTEM table
2 : a unit of force equal to the weight of a kilogram mass under a gravitational attraction equal to that of the earth

kilogram–meter *noun* (1886)
: the meter-kilogram-second gravitational unit of work and energy equal to the work done by a kilogram force acting through a distance of one meter in the direction of the force **:** about 7.235 foot-pounds

ki·lo·hertz \ˈki-lə-ˌhərts, ˈkē-lə-, -ˌherts\ *noun* [International Scientific Vocabulary] (1929)
: 1000 hertz

ki·lo·joule \-ˌjü(ə)l\ *noun* [International Scientific Vocabulary] (circa 1889)
: 1000 joules

kilo·li·ter \ˈki-lə-ˌlē-tər\ *noun* [French *kilolitre,* from *kilo-* + *litre* liter] (1810)
— see METRIC SYSTEM table

ki·lo·me·ter \÷kə-ˈlä-mə-tər, ÷ki-; ˈki-lə-ˌmē-tər\ *noun* [French *kilomètre,* from *kilo-* + *mètre* meter] (1810)
— see METRIC SYSTEM table ▪

ki·lo·par·sec \ˈki-lə-ˌpär-ˌsek, ˈkē-lə-\ *noun* (1922)
: 1000 parsecs

ki·lo·pas·cal \ˌki-lə-ˌpas-ˈkal\ *noun* (1978)
: 1000 pascals

ki·lo·rad \ˈki-lə-ˌrad\ *noun* [International Scientific Vocabulary] (1965)
: 1000 rads

ki·lo·ton \-ˌtən *also* -ˌtän\ *noun* (1950)
1 : 1000 tons
2 : an explosive force equivalent to that of 1000 tons of TNT

ki·lo·volt \-ˌvōlt\ *noun* [International Scientific Vocabulary] (circa 1898)
: a unit of potential difference equal to 1000 volts

kilo·watt \ˈki-lə-ˌwät\ *noun* [International Scientific Vocabulary] (1884)
: 1000 watts

kilowatt–hour *noun* (1892)
: a unit of work or energy equal to that expended by one kilowatt in one hour or to 3.6 million joules

[1]kilt \ˈkilt\ *verb* [Middle English, of Scandinavian origin; akin to Old Norse *kjalta* lap, fold of a gathered skirt] (14th century)
transitive verb
1 *chiefly dialect* **:** to tuck up (as a skirt)
2 : to equip with a kilt
intransitive verb
: to move nimbly

[2]kilt *noun* (circa 1730)
1 : a knee-length pleated skirt usually of tartan worn by men in Scotland and by Scottish regiments in the British armies
2 : a garment that resembles a Scottish kilt

kil·ter \ˈkil-tər\ *noun* [origin unknown] (1628)
: proper or usual state or condition **:** ORDER ⟨out of *kilter*⟩

kilt·ie \ˈkil-tē\ *noun* (1842)
1 *or* **kilty :** one who wears a kilt
2 : a shoe with a long slashed tongue that folds over the instep; *also* **:** such a tongue

kim·ber·lite \ˈkim-bər-ˌlīt\ *noun* [*Kimberley,* South Africa + [1]*-ite*] (1887)
: an agglomerate biotite-peridotite that occurs in pipes especially in southern Africa and that often contains diamonds

Kim·bun·du \kim-ˈbu̇n-(ˌ)dü\ *noun* (circa 1895)
: a Bantu language of northern Angola

kim·chi *also* **kim·chee** \ˈkim-chē\ *noun* [Korean *kimch'i*] (1898)
: a vegetable pickle seasoned with garlic, red pepper, and ginger that is the national dish of Korea

ki·mo·no \kə-ˈmō-(ˌ)nō, -nə\ *noun, plural* **-nos** [Japanese, clothes, from *ki* wearing + *mono* thing] (1886)
1 : a long robe with wide sleeves traditionally worn with a broad sash as an outer garment by the Japanese
2 : a loose dressing gown or jacket
— ki·mo·noed \-(ˌ)nōd, -nəd\ *adjective*

kimono 1

[1]kin \ˈkin\ *noun* [Middle English, from Old English *cynn;* akin to Old High German *chunni* race, Latin *genus* birth, race, kind, Greek *genos,* Latin *gignere* to beget, Greek *gignesthai* to be born] (before 12th century)
1 : a group of persons of common ancestry **:** CLAN
2 a : one's relatives **:** KINDRED **b :** KINSMAN ⟨he wasn't any *kin* to you —Jean Stafford⟩
3 *archaic* **:** KINSHIP

[2]kin *adjective* (1597)
: KINDRED, RELATED

-kin \kən\ *also* **-kins** \kənz\ *noun suffix* [Middle English, from Middle Dutch *-kin;* akin to Old High German *-chīn,* diminutive suffix]
: little ⟨cat*kin*⟩ ⟨baby*kins*⟩

ki·na \ˈkē-nə\ *noun* [New Guinea Pidgin, literally, a kind of shell] (1975)
— see MONEY table

ki·nase \ˈkī-ˌnās, -ˌnāz\ *noun* [International Scientific Vocabulary, from *kinetic*] (1947)
: an enzyme that catalyzes the transfer of phosphate groups from a high-energy phosphate-containing molecule (as ATP) to a substrate

[1]kind \ˈkīnd\ *noun* [Middle English *kinde,* from Old English *cynd;* akin to Old English *cynn* kin] (before 12th century)
1 a *archaic* **:** NATURE **b** *archaic* **:** FAMILY, LINEAGE
2 *archaic* **:** MANNER
3 : fundamental nature or quality **:** ESSENCE
4 a : a group united by common traits or interests **:** CATEGORY **b :** a specific or recognized variety ⟨what *kind* of car do you drive⟩ **c :** a doubtful or barely admissible member of a category ⟨a *kind* of gray⟩
5 a : goods or commodities as distinguished from money ⟨payment in *kind*⟩ **b :** the equivalent of what has been offered or received
synonym see TYPE
— all kinds of 1 : MANY ⟨likes *all kinds of* sports⟩ **2 :** plenty of ⟨has *all kinds of* time⟩

[2]kind *adjective* (14th century)
1 *chiefly dialect* **:** AFFECTIONATE, LOVING
2 a : of a sympathetic or helpful nature **b :** of a forbearing nature **:** GENTLE **c :** arising from or characterized by sympathy or forbearance ⟨a *kind* act⟩
3 : of a kind to give pleasure or relief

kin·der·gar·ten \ˈkin-də(r)-ˌgär-tᵊn, -dᵊn\ *noun* [German, from *Kinder* children + *Garten* garden] (1852)
: a school or class for children usually from four to six years old

kin·der·gart·ner \-ˌgärt-nər, -ˌgärd-\ *also* **kin·der·gar·ten·er** \-ˌgär-tᵊn-ər, -dᵊn-\ *noun* (1889)
1 : a teacher at a kindergarten
2 : a child attending or of an age to attend kindergarten

kind·heart·ed \ˌkīnd-ˈhär-təd\ *adjective* (1535)
: marked by a sympathetic nature
— kind·heart·ed·ly *adverb*
— kind·heart·ed·ness *noun*

[1]kin·dle \ˈkin-dᵊl\ *verb* **kin·dled; kin·dling** \ˈkin(d)-liŋ, ˈkin-dᵊl-iŋ\ [Middle English, probably modification of Old Norse *kynda;* akin to Old High German *cuntesal* fire] (13th century)
transitive verb
1 : to start (a fire) burning **:** LIGHT
2 a : to stir up **:** AROUSE ⟨*kindle* interest⟩ **b :** to bring into being **:** START
3 : to cause to glow **:** ILLUMINATE
intransitive verb
1 : to catch fire
2 a : to flare up **b :** to become animated
3 : to become illuminated
— kin·dler \ˈkin(d)-lər, ˈkin-dᵊl-ər\ *noun*

[2]kindle *verb* **kin·dled; kin·dling** [Middle English, from *kindle* young animal, probably from *kinde,* noun, kind] (13th century)
transitive verb
: BEAR — used especially of a rabbit
intransitive verb
: to bring forth young — used especially of a rabbit

kind·less \ˈkīn(d)-ləs\ *adjective* (13th century)
1 *obsolete* **:** INHUMAN
2 : DISAGREEABLE, UNCONGENIAL
— kind·less·ly *adverb*

kind·li·ness \ˈkīn(d)-lē-nəs\ *noun* (15th century)
1 : the quality or state of being kindly
2 : a kindly deed

kin·dling \ˈkin(d)-liŋ, ˈkin-lən\ *noun* (1513)
: easily combustible material for starting a fire

[1]kind·ly \ˈkīn(d)-lē\ *adjective* **kind·li·er; -est** [Middle English, from Old English *cyndelīc,* from *cynd*] (before 12th century)
1 a *obsolete* **:** NATURAL **b** *archaic* **:** LAWFUL
2 : of an agreeable or beneficial nature **:** PLEASANT ⟨*kindly* climate⟩

□ USAGE

kilometer In North American speech *kilometer* is most often pronounced with primary stress on the second syllable. This pronunciation is also heard frequently in British speech. Those who object to second syllable stress say that the first syllable should be stressed in accord with the stress patterns of *centimeter, millimeter,* etc. However, the pronunciation of *kilometer* does not parallel that of other metric compounds. From 1828 to 1841 Noah Webster indicated only second syllable stress, and his successor added a first syllable stress variant in the first Merriam-Webster dictionary of 1847. Thus, both pronunciations are venerable. Most scientists use second syllable stress, although first syllable stress seems to occur with a higher rate of frequency among scientists than among nonscientists.

3 : of a sympathetic or generous nature

[2]**kindly** *adverb* (before 12th century)
1 a : in the normal way **:** NATURALLY ⟨old wounds which had healed *kindly* —*American Mercury*⟩ **b :** READILY ⟨did not take *kindly* to suggestions⟩
2 a : in a kind manner **:** SYMPATHETICALLY **b :** as a gesture of goodwill ⟨would take it *kindly* if you would put in a good word⟩ **c :** in a gracious manner **:** COURTEOUSLY ⟨they *kindly* invited us along⟩ **d :** as a matter of courtesy **:** PLEASE ⟨would you *kindly* order me a cab⟩
3 *chiefly Southern* **:** SOMEWHAT, KIND OF ⟨it's *kindly* embarrassing —Walter Davis⟩

kind•ness \ˈkīn(d)-nəs\ *noun* (13th century)
1 : a kind deed **:** FAVOR
2 a : the quality or state of being kind **b** *archaic* **:** AFFECTION

kind of *adverb* (1775)
1 : to a moderate degree **:** SOMEWHAT ⟨it's *kind of* late to begin⟩
2 : in a way that approximates **:** MORE OR LESS ⟨*kind of* sneaked up on us⟩

[1]**kin•dred** \ˈkin-drəd\ *noun* [Middle English, from *kin* + Old English *rǣden* condition, from *rǣdan* to advise, read] (12th century)
1 a : a group of related individuals **b :** one's relatives
2 : family relationship **:** KINSHIP

[2]**kindred** *adjective* (14th century)
1 : of a similar nature or character **:** LIKE
2 : of the same ancestry

kine \ˈkīn\ *archaic plural of* COW

kin•e•ma \ˈki-nə-mə\ *British variant of* CINEMA

ki•ne•mat•ics \ˌki-nə-ˈma-tiks *also* ˌkī-\ *noun plural but singular in construction* [French *cinématique,* from Greek *kinēmat-, kinēma* motion, from *kinein* to move] (1840)
: a branch of dynamics that deals with aspects of motion apart from considerations of mass and force
— ki•ne•mat•ic \-tik\ *or* **ki•ne•mat•i•cal** \-ti-kəl\ *adjective*
— ki•ne•mat•i•cal•ly \-ti-k(ə-)lē\ *adverb*

[1]**ki•ne•scope** \ˈki-nə-ˌskōp *also* ˈkī-\ *noun* [from *Kinescope,* a trademark] (1930)
1 : PICTURE TUBE
2 : a motion picture made from an image on a picture tube

[2]**kinescope** *transitive verb* **-scoped; -scop•ing** (1949)
: to make a kinescope of

ki•ne•sics \kə-ˈnē-siks, kī-, -ziks\ *noun plural but singular in construction* [Greek *kinēsis* motion + English *-ics*] (1952)
: a systematic study of the relationship between nonlinguistic body motions (as blushes, shrugs, or eye movement) and communication

ki•ne•si•ol•o•gy \kə-ˌnē-sē-ˈä-lə-jē, kī-, -zē-\ *noun* [Greek *kinēsis*] (1894)
: the study of the principles of mechanics and anatomy in relation to human movement

ki•ne•sis \kə-ˈnē-səs, kī-\ *noun, plural* **ki•ne•ses** \-ˌsēz\ [New Latin, from Greek *kinēsis*] (1905)
: a movement that lacks directional orientation and depends upon the intensity of stimulation

-kinesis *noun combining form, plural* **-kineses** [New Latin, from Greek *kinēsis,* from *kinein* to move; akin to Latin *ciēre* to move]
1 : division ⟨karyo*kinesis*⟩
2 : production of motion ⟨tele*kinesis*⟩

kin•es•the•sia \ˌki-nəs-ˈthē-zh(ē-)ə, ˌkī-\ *or* **kin•es•the•sis** \-ˈthē-səs\ *noun, plural* **-the•sias** *or* **-the•ses** \-ˌsēz\ [New Latin, from Greek *kinein* + *aisthēsis* perception — more at ANESTHESIA] (1880)
: a sense mediated by end organs located in muscles, tendons, and joints and stimulated by bodily movements and tensions; *also* **:** sensory experience derived from this sense
— kin•es•thet•ic \-ˈthe-tik\ *adjective*
— kin•es•thet•i•cal•ly \-ti-k(ə-)lē\ *adverb*

kinet- *or* **kineto-** *combining form* [Greek *kinētos* moving]
: movement **:** motion ⟨*kineto*some⟩

ki•net•ic \kə-ˈne-tik *also* kī-\ *adjective* [Greek *kinētikos,* from *kinētos,* from *kinein*] (1864)
1 : of or relating to the motion of material bodies and the forces and energy associated therewith
2 a : ACTIVE, LIVELY **b :** DYNAMIC, ENERGIZING
3 : of or relating to kinetic art
— ki•net•i•cal•ly \-ti-k(ə-)lē\ *adverb*

kinetic art *noun* (1961)
: art (as sculpture or assemblage) having mechanical parts which can be set in motion
— kinetic artist *noun*

kinetic energy *noun* (1870)
: energy associated with motion

ki•net•i•cist \kə-ˈne-tə-sist *also* kī-\ *noun* (1960)
1 : a specialist in kinetics
2 : a person who works in kinetic art **:** KINETIC ARTIST

ki•net•ics \kə-ˈne-tiks *also* kī-\ *noun plural but singular or plural in construction* (circa 1859)
1 a : a branch of science that deals with the effects of forces upon the motions of material bodies or with changes in a physical or chemical system **b :** the rate of change in such a system
2 : the mechanism by which a physical or chemical change is effected

kinetic theory *noun* (1864)
: either of two theories in physics based on the fact that the minute particles of a substance are in vigorous motion: **a :** a theory that the particles of a gas move in straight lines with high average velocity, continually encounter one another and thus change their individual velocities and directions, and cause pressure by their impact against the walls of a container — called also *kinetic theory of gases* **b :** a theory that the temperature of a substance increases with an increase in either the average kinetic energy of the particles or the average potential energy of separation (as in fusion) of the particles or in both when heat is added — called also *kinetic theory of heat*

ki•ne•tin \ˈkī-nə-tən\ *noun* (1955)
: a cytokinin $C_{10}H_9N_5O$ that increases mitosis and callus formation

ki•net•o•chore \kə-ˈne-tə-ˌkōr, kī-, -ˌkȯr\ *noun* [*kinet-* + Greek *chōros* place] (1934)
: CENTROMERE

ki•net•o•plast \kə-ˈne-tə-ˌplast, kī-\ *noun* [International Scientific Vocabulary] (1925)
: an extranuclear cell organelle within a mitochondrion especially of trypanosomes that contains DNA

ki•net•o•scope \kə-ˈne-tə-ˌskōp, kī-\ *noun* [from *Kinetoscope,* a trademark] (1894)
: a device for viewing through a magnifying lens a sequence of pictures on an endless band of film moved continuously over a light source and a rapidly rotating shutter that creates an illusion of motion

ki•net•o•some \-ˌsōm\ *noun* (1912)
: BASAL BODY

kin•folk \ˈkin-ˌfōk\ *or* **kinfolks** *noun plural* (1873)
: RELATIVES

king \ˈkiŋ\ *noun* [Middle English, from Old English *cyning;* akin to Old High German *kuning* king, Old English *cynn* kin] (before 12th century)
1 a : a male monarch of a major territorial unit; *especially* **:** one whose position is hereditary and who rules for life **b :** a paramount chief
2 *capitalized* **:** GOD, CHRIST
3 : one that holds a preeminent position; *especially* **:** a chief among competitors
4 : the principal piece of each color in chess having the power to move ordinarily one square in any direction and to capture opposing pieces but being obliged never to enter or remain in check
5 : a playing card marked with a stylized figure of a king
6 : a checker that has been crowned

king•bird \-ˌbərd\ *noun* (1778)
: any of various American tyrant flycatchers (genus *Tyrannus*)

king•bolt \-ˌbōlt\ *noun* (1825)
: a vertical bolt by which the forward axle and wheels of a vehicle or the trucks of a railroad car are connected with the other parts

King Charles spaniel \-ˈchär(-ə)lz-\ *noun* [*Charles* II of England] (1833)
: an English toy spaniel having a black and tan coat

king cobra *noun* (1894)
: a large cobra (*Ophiophagus hannah* synonym *Naja hannah*) of southeastern Asia and the Philippines that may attain a length of 18 feet (5.5 meters)

king cobra

king crab *noun* (1698)
1 : HORSESHOE CRAB
2 : any of several very large crabs; *especially* **:** one (*Paralithoides camtschaticus*) of the North Pacific caught commercially for food

king•craft \ˈkiŋ-ˌkraft\ *noun* (1643)
: the art of governing as a king

king•cup \-ˌkəp\ *noun* (1538)
: any of several plants of the buttercup family; *especially* **:** MARSH MARIGOLD

king•dom \ˈkiŋ-dəm\ *noun* (before 12th century)
1 *archaic* **:** KINGSHIP
2 : a politically organized community or major territorial unit having a monarchical form of government headed by a king or queen
3 *often capitalized* **a :** the eternal kingship of God **b :** the realm in which God's will is fulfilled
4 a : a realm or region in which something is dominant **b :** an area or sphere in which one holds a preeminent position
5 a : one of the three primary divisions into which natural objects are commonly classified — compare ANIMAL KINGDOM, MINERAL KINGDOM, PLANT KINGDOM **b :** a major category (as Plantae or Protista) in biological taxonomy that ranks above the phylum and is the highest and most encompassing group

kingdom come *noun* [from the phrase "Thy *kingdom come*" (Matthew 6:10)] (1785)
: the next world **:** HEAVEN

king•fish \ˈkiŋ-ˌfish\ *noun* (1750)
1 : any of several marine croakers (family Sciaenidae) **a :** any of three fishes (*Menticirrhus americanus, M. littoralis,* and *M. saxatilis*) of shallow coastal waters of the Atlantic Ocean **b :** a small silvery food and sport fish (*Genyonemus lineatus*) of inshore waters especially of California — called also *white croaker*
2 : KING MACKEREL
3 : an undisputed master in an area or group

king•fish•er \-ˌfi-shər\ *noun* (15th century)
: any of numerous nonpasserine birds (family Alcedinidae) that are usually crested and bright-colored with a short tail and a long stout sharp bill

King James Version \-ˈjāmz-\ *noun* [*James* I of England] (circa 1889)
: AUTHORIZED VERSION

king•let \ˈkiŋ-lət\ *noun* (1603)

1 : a weak or petty king
2 : any of several small birds (genus *Regulus*) that are related to the gnatcatchers

king·ly \'kiŋ-lē\ *adjective* **king·li·er; -est** (14th century)
1 : having royal rank
2 : of, relating to, or befitting a king
3 : MONARCHICAL
— king·li·ness *noun*
— kingly *adverb*

king mackerel *noun* (circa 1930)
: a mackerel (*Scomberomorus cavalla*) that is noted especially as a fighting sport fish

king·mak·er \'kiŋ-,mā-kər\ *noun* (1599)
: one having great influence over the choice of candidates for political office

king of arms (15th century)
: an officer of arms of the highest rank

king penguin *noun* (1885)
: a large penguin (*Aptenodytes patagonica*) chiefly of subantarctic regions

king·pin \'kiŋ-,pin\ *noun* (1801)
1 : any of several bowling pins: as **a :** HEADPIN **b :** the pin that stands in the middle of a triangular arrangement of bowling pins
2 : the chief person in a group or undertaking
3 a : KINGBOLT **b :** a pin connecting the two parts of a knuckle joint (as in an automobile steering linkage)

king post *noun* (1776)
: a vertical member connecting the apex of a triangular truss (as of a roof) with the base

Kings \'kiŋz\ *noun plural but singular in construction*
1 : either of two narrative and historical books of canonical Jewish and Christian Scripture — see BIBLE table
2 : any of four narrative and historical books in the former Roman Catholic canon of the Old Testament

king salmon *noun* (1881)
: CHINOOK SALMON

King's Bench *noun* (14th century)
: a division in the English superior courts system that hears civil and criminal cases

King's Counsel *noun* (1689)
: a barrister selected to serve as counsel to the British crown

King's English *noun* (1553)
: standard, pure, or correct English speech or usage

king's evil *noun, often K&E capitalized* [from the former belief that it could be healed by a king's touch] (14th century)
: SCROFULA

king·ship \'kiŋ-,ship\ *noun* (14th century)
1 : the position, office, or dignity of a king
2 : the personality of a king
3 : government by a king

king·side \-,sīd\ *noun* (1941)
: the side of a chessboard containing the file on which the king sits at the beginning of the game

king–size \-,sīz\ *or* **king–sized** \-,sīzd\ *adjective* (1942)
1 : longer than the regular or standard size ⟨a *king-size* cigarette⟩
2 : unusually large
3 a : having dimensions of approximately 76 inches by 80 inches (about 1.9 by 2.0 meters) — used of a bed; compare FULL-SIZE, QUEEN-SIZE, TWIN-SIZE **b :** of a size that fits a king-size bed ⟨*king-size* sheets⟩

king snake *noun* (1709)
: any of numerous brightly marked colubrid snakes (genus *Lampropeltis*) chiefly of North and Central America

king's ransom *noun* (circa 1590)
: a very large sum

king·wood \-,wu̇d\ *noun* (circa 1851)
: the wood of any of several tropical American leguminous trees (especially genus *Dalbergia*); *especially* **:** the wood of a Brazilian tree (*D. cearensis*) used especially for furniture

ki·nin \'kī-nən\ *noun* [Greek *kinein* to move, stimulate + English [1]*-in* — more at -KINESIS] (1954)
1 : any of various polypeptide hormones that are formed locally in the tissues and cause dilation of blood vessels and contraction of smooth muscle
2 : CYTOKININ

[1]kink \'kiŋk\ *noun* [Dutch; akin to Middle Low German *kinke* kink] (1678)
1 : a short tight twist or curl caused by a doubling or winding of something upon itself
2 a : a mental or physical peculiarity **:** ECCENTRICITY, QUIRK **b :** WHIM
3 : a clever unusual way of doing something
4 : a cramp in some part of the body
5 : an imperfection likely to cause difficulties in the operation of something

[2]kink (1697)
intransitive verb
: to form a kink
transitive verb
: to make a kink in

kin·ka·jou \'kiŋ-kə-,jü\ *noun* [French, alteration of *quincajou* wolverine, of Algonquian origin; akin to Ojibwa *kwi·nkwaʔa·ke·* wolverine] (1796)
: a nocturnal arboreal omnivorous mammal (*Potos flavus*) found from Mexico to South America that is related to the raccoon and has a long prehensile tail, large eyes, and yellowish brown fur

kinky \'kiŋ-kē\ *adjective* **kink·i·er; -est** (1844)
1 : closely twisted or curled
2 : relating to, having, or appealing to unconventional tastes especially in sex; *also* **:** sexually deviant
3 : OUTLANDISH, FAR-OUT
— kink·i·ly \'kiŋ-kə-lē\ *adverb*
— kink·i·ness \'kiŋ-kē-nəs\ *noun*

kin·ni·kin·nick \,ki-ni-kə-'nik, 'ki-ni-kə-,\ *noun* [of Algonquian origin; akin to Massachuset *kinukkinuk* mixture] (1799)
: a mixture of dried leaves and bark and sometimes tobacco smoked by the Indians and pioneers especially in the Ohio valley; *also* **:** a plant (as a sumac or dogwood) used in it

-kins — see -KIN

kins·folk \'kinz-,fōk\ *noun plural* (15th century)
: RELATIVES

kin·ship \'kin-,ship\ *noun* (1833)
: the quality or state of being kin **:** RELATIONSHIP

kins·man \'kinz-mən\ *noun* (12th century)
: RELATIVE; *specifically* **:** a male relative

kins·wom·an \-,wu̇-mən\ *noun* (14th century)
: a female relative

ki·osk \'kē-,äsk\ *noun* [Turkish *köşk,* from Persian *kūshk* portico] (1625)
1 : an open summerhouse or pavilion
2 : a small structure with one or more open sides that is used to vend merchandise (as newspapers) or services (as film developing)

Ki·o·wa \'kī-ə-,wȯ, -,wä, -,wā\ *noun, plural* **Kiowa** *or* **Kiowas** (1808)
1 : a member of an American Indian people of what are now Colorado, Kansas, New Mexico, Oklahoma, and Texas
2 : the language of the Kiowa people

[1]kip \'kip\ *noun* [obsolete Dutch; akin to Middle Low German *kip* bundle of hides] (circa 1525)
: a bundle of undressed hides of young or small animals; *also* **:** one of the hides

[2]kip *noun* [perhaps from Danish *kippe* cheap tavern] (1879)
1 : BED ⟨ready for the *kip* after this screwball day —K. M. Dodson⟩
2 *chiefly British* **:** SLEEP, NAP ⟨roused the . . . family from their *kip* —Sylvia Margolis⟩

[3]kip *intransitive verb* **kipped; kip·ping** (circa 1889)
British **:** SLEEP — sometimes used with *down* ⟨*kip* down on a spare bed —Alice Glenday⟩

[4]kip *noun* [*ki*lo- + *p*ound] (1914)
: a unit of weight equal to 1000 pounds (4448 newtons) used to express deadweight load

[5]kip \'kip, 'gip\ *noun, plural* **kip** *or* **kips** [Lao *kì:p,* literally, ingot] (1955)
— see MONEY table

[1]kip·per \'ki-pər\ *noun* [Middle English *kypre,* from Old English *cypera;* akin to Old English *coper* copper] (before 12th century)
1 : a male salmon or sea trout during or after the spawning season
2 : a kippered herring or salmon

[2]kipper *transitive verb* **kip·pered; kip·per·ing** \-p(ə-)riŋ\ (1773)
: to cure (split dressed fish) by salting and smoking
— kip·per·er \-pər-ər\ *noun*

Kir·ghiz \kir-'gēz\ *noun, plural* **Kirghiz** *or* **Kir·ghiz·es** [Kirghiz *kırğız*] (1600)
1 : a member of a Turkic people of Kyrgyzstan and adjacent areas of central Asia
2 : the language of the Kirghiz

kirk \'kirk, 'kərk\ *noun* [Middle English (northern dialect), from Old Norse *kirkja,* from Old English *cirice* — more at CHURCH] (12th century)
1 *chiefly Scottish* **:** CHURCH
2 *capitalized* **:** the national church of Scotland as distinguished from the Church of England or the Episcopal Church in Scotland

Kir·li·an photography \'kir-lē-ən-\ *noun* [Semyon D. & Valentina K. *Kirlian* (flourished 1939) Soviet inventors] (1972)
: a process in which an image is obtained by application of a high-frequency electric field to an object so that it radiates a characteristic pattern of luminescence that is recorded on photographic film
— Kirlian photograph *noun*

Kir·man \kər-'män, kir-\ *noun* [*Kirman,* province in Iran] (1876)
: a Persian carpet or rug characterized by elaborate fluid designs and soft colors

kir·mess \'kər-məs\ *variant of* KERMIS

kirsch \'kirsh\ *noun* [German, short for *Kirschwasser,* from *Kirsche* cherry + *Wasser* water] (1869)
: a dry colorless brandy distilled from the fermented juice of the black morello cherry

Kirt·land's warbler \'kərt-lən(d)z-\ *noun* [Jared P. *Kirtland* (died 1877) American naturalist] (1858)
: a rare warbler (*Dendroica kirtlandii*) of northeastern North America that breeds in Michigan and winters in the Bahamas

kir·tle \'kər-tᵊl\ *noun* [Middle English *kirtel,* from Old English *cyrtel,* from (assumed) Old English *curt* short, from Latin *curtus* shortened — more at SHEAR] (before 12th century)
1 : a tunic or coat worn by men especially in the Middle Ages
2 : a long gown or dress worn by women

kish·ke *also* **kish·ka** \'kish-kə\ *noun* [Yiddish *kishke* gut, sausage, of Slavic origin; akin to Polish *kiszka* gut, sausage] (circa 1936)
: beef or fowl casing stuffed (as with meat, flour, and spices) and cooked

Kis·lev \'kis-ləf\ *noun* [Hebrew *Kislēw*] (14th century)
: the 3d month of the civil year or the 9th month of the ecclesiastical year in the Jewish calendar — see MONTH table

kis·met \'kiz-,met, -mət\ *noun, often capitalized* [Turkish, from Arabic *qismah* portion, lot] (1834)
: FATE 1, 2a

[1]kiss \'kis\ *verb* [Middle English, from Old English *cyssan;* akin to Old High German *kussen* to kiss] (before 12th century)
transitive verb
1 : to touch with the lips especially as a mark of affection or greeting

2 : to touch gently or lightly ⟨wind gently *kissing* the trees⟩
intransitive verb
1 : to salute or caress one another with the lips
2 : to come in gentle contact
— **kiss·able** \ˈki-sə-bəl\ *adjective*
— **kiss ass :** to act obsequiously especially to gain favor — usually considered vulgar
— **kiss good–bye 1 :** LEAVE **2 :** to resign oneself to the loss of

[2]kiss *noun* (before 12th century)
1 : a caress with the lips
2 : a gentle touch or contact
3 a : a small drop cookie made of meringue **b :** a bite-size piece of candy often wrapped in paper or foil
4 : an expression of affection ⟨sent him *kisses* in her letter⟩

kiss–and–tell \ˈkis-ᵊn(d)-ˈtel\ *adjective* (1949)
: telling details of private matters ⟨*kiss-and-tell* autobiographies⟩

kiss·er \ˈki-sər\ *noun* (1537)
1 : one that kisses
2 *slang* **a :** MOUTH **b :** FACE

kissing bug *noun* (1899)
: CONENOSE

kissing cousin *noun* (1941)
1 : a person and especially a relative whom one knows well enough to kiss more or less formally upon meeting
2 : one that is closely related in kind to something else

kissing disease *noun* [from the belief that it is frequently transmitted by kissing] (1962)
: INFECTIOUS MONONUCLEOSIS

kiss of death [from the kiss with which Judas betrayed Jesus (Mark 14:44–46)] (1943)
: something (as an act or association) ultimately causing ruin

kiss off *transitive verb* (circa 1935)
: to dismiss lightly ⟨*kisses* the other performers *off* as mere amateurs⟩
— **kiss–off** \ˈki-ˌsȯf\ *noun*

kiss of life (1961)
chiefly British **:** artificial respiration by the mouth-to-mouth method

kiss of peace (circa 1898)
: a ceremonial kiss, embrace, or handclasp used in Christian liturgies and especially the Eucharist as a sign of fraternal unity

kist \ˈkist\ *noun* [Middle English *kiste,* from Old Norse *kista,* ultimately from Latin *cista* — more at CHEST] (14th century)
chiefly Scottish & South African **:** CHEST 1b

[1]kit \ˈkit\ *noun* [Middle English] (14th century)
1 *dialect British* **:** a wooden tub
2 a (1) **:** a collection of articles usually for personal use ⟨a travel *kit*⟩ (2) **:** a set of tools or implements ⟨a carpenter's *kit*⟩ (3) **:** a set of parts to be assembled or worked up ⟨model-airplane *kit*⟩ (4) **:** a packaged collection of related material ⟨convention *kit*⟩ (5) *chiefly British* **:** GEAR ⟨run over to my billet and get some overnight *kit* —Lionel Shapiro⟩ **b :** a container for any of such sets or collections
3 : a group of persons or things — usually used in the phrase *the whole kit and caboodle*

[2]kit *transitive verb* **kit·ted; kit·ting** (1919)
chiefly British **:** EQUIP, OUTFIT — often used with *up* or *out*

[3]kit *noun* [origin unknown] (1519)
: a small narrow violin

[4]kit *noun* (1562)
1 : KITTEN
2 : a young or undersized fur-bearing animal; *also* **:** its pelt

kit bag *noun* [[1]*kit*] (1893)
1 : KNAPSACK
2 : a suitcase usually with sides that fasten at the top or open to the full width of the bag

kitch·en \ˈki-chən\ *noun* [Middle English *kichene,* from Old English *cycene,* from Late Latin *coquina,* from Latin *coquere* to cook — more at COOK] (before 12th century)
1 : a place (as a room) with cooking facilities
2 : the personnel that prepares, cooks, and serves food
3 : CUISINE

kitchen cabinet *noun* (1832)
1 : an informal group of advisers to one in a position of power (as the head of a government)
2 : a cupboard with drawers and shelves for use in a kitchen

kitch·en·ette \ˌki-chə-ˈnet\ *noun* (1903)
: a small kitchen or an alcove containing cooking facilities

kitchen garden *noun* (1580)
: a garden in which plants (as vegetables or herbs) for use in the kitchen are cultivated

kitchen midden *noun* (1863)
: a refuse heap; *specifically* **:** a mound marking the site of a primitive human habitation

kitchen police *noun* (circa 1917)
1 : KP
2 : the work of KPs

kitch·en–sink \-ˈsiŋk\ *adjective* (1941)
1 *chiefly British* **:** portraying or emphasizing the squalid aspects of modern life ⟨the *kitchen-sink* realism of contemporary British drama —*Current Biography*⟩
2 : being or made up of a hodgepodge of disparate elements or ingredients

kitch·en·ware \-ˌwar, -ˌwer\ *noun* (1722)
: utensils and appliances for use in a kitchen

[1]kite \ˈkīt\ *noun* [Middle English, from Old English *cȳta;* akin to Middle High German *kūze* owl] (before 12th century)
1 : any of various usually small hawks (family Accipitridae) with long narrow wings and often a notched or forked tail
2 : a person who preys on others
3 : a light frame covered usually with paper or cloth, often provided with a balancing tail, and designed to be flown in the air at the end of a long string
4 : a check drawn against uncollected funds in a bank account or fraudulently raised before cashing
5 : a light sail used in a light breeze usually in addition to the regular working sails; *especially* **:** SPINNAKER
— **kite·like** *adjective*

[2]kite *verb* **kit·ed; kit·ing** (1839)
transitive verb
1 : to use (a bad check) to get credit or money
2 : to cause to soar ⟨*kited* the prices they charged wealthy clients⟩
intransitive verb
1 a : to go in a rapid, carefree, or flighty manner **b :** to rise rapidly **:** SOAR ⟨the prices of necessities continue to *kite*⟩
2 : to get money or credit by a kite

kit fox *noun* [[4]*kit*] (1805)
1 a : SWIFT FOX **b :** a fox (*Vulpes macrotis*) of the southwestern U.S. and Mexico with exceptionally large ears and a black tip on the tail
2 : the fur or pelt of a kit fox

kith \ˈkith\ *noun* [Middle English, from Old English *cȳthth;* akin to *cūth* known — more at UNCOUTH] (before 12th century)
: familiar friends, neighbors, or relatives ⟨*kith* and kin⟩

kith·a·ra \ˈki-thə-rə\ *noun* [Middle English *cithara,* from Latin, from Greek *kithara*] (14th century)
: an ancient Greek stringed instrument similar to but larger than the lyre and having a box-shaped resonator

kithe \ˈkīth\ *verb* **kithed; kith·ing** [Middle English, from Old English *cȳthan,* from *cūth*] (before 12th century)
transitive verb
chiefly Scottish **:** to make known
intransitive verb
chiefly Scottish **:** to become known

kitsch \ˈkich\ *noun* [German, kitsch, trash] (1925)
: something that appeals to popular or low-brow taste and is often of poor quality
— **kitsch** *adjective*
— **kitschy** \ˈki-chē\ *adjective*

[1]kit·ten \ˈki-tᵊn\ *noun* [Middle English *kitoun,* from (assumed) Old North French *caton,* diminutive of *cat,* from Late Latin *cattus*] (14th century)
: a young cat; *also* **:** an immature individual of various other small mammals

[2]kitten *intransitive verb* **kit·tened; kit·ten·ing** \ˈkit-niŋ, ˈki-tᵊn-iŋ\ (15th century)
: to give birth to kittens

kit·ten·ish \ˈkit-nish, ˈki-tᵊn-ish\ *adjective* (1754)
: resembling a kitten; *especially* **:** coyly playful
— **kit·ten·ish·ly** *adverb*
— **kit·ten·ish·ness** *noun*

kit·ti·wake \ˈki-tē-ˌwāk\ *noun* [imitative] (1661)
: either of two cliff-nesting gulls (*Rissa tridactyla* and *R. brevirostris*) that winter on the open ocean

[1]kit·tle \ˈki-tᵊl\ *transitive verb* **kit·tled; kit·tling** \ˈkit-liŋ, ˈki-tᵊl-iŋ\ [Middle English (northern dialect) *kytyllen*] (before 12th century)
1 *chiefly Scottish* **:** TICKLE
2 *chiefly Scottish* **:** PERPLEX

[2]kittle *adjective* (1568)
chiefly Scottish **:** TICKLISH 2, TOUCHY

[1]kit·ty \ˈki-tē\ *noun, plural* **kitties** (1719)
: CAT 1a; *especially* **:** KITTEN

[2]kitty *noun, plural* **kitties** [[1]*kit*] (circa 1887)
1 : a fund in a poker game made up of contributions from each pot
2 : a sum of money or collection of goods often made up of small contributions **:** POOL

kit·ty–cor·ner *or* **kit·ty–cor·nered** *variant of* CATERCORNER

ki·va \ˈkē-və\ *noun* [Hopi *kíva*] (1871)
: a Pueblo Indian ceremonial structure that is usually round and partly underground

Ki·wa·ni·an \kə-ˈwä-nē-ən\ *noun* [*Kiwanis* (*Club*)] (1921)
: a member of a major national and international service club

ki·wi \ˈkē-(ˌ)wē\ *noun* [Maori] (1835)
1 : any of a small genus (*Apteryx*) of flightless New Zealand birds with rudimentary wings, stout legs, a long bill, and grayish brown hairlike plumage
2 *capitalized* **:** a native or resident of New Zealand — used as a nickname
3 : KIWIFRUIT

kiwi 1

ki·wi·fruit \-ˌfrüt\ *noun* (1966)
: the fruit of a Chinese gooseberry

Klam·ath weed \ˈkla-məth-\ *noun* [*Klamath* (*River*)] (1922)
: a European yellow-flowered perennial Saint-John's-wort (*Hypericum perforatum*) that is naturalized in North America especially in rangelands

Klan \ˈklan\ *noun* [(*Ku Klux*) *Klan*] (1867)
: an organization of Ku Kluxers; *also* **:** a subordinate unit of such an organization
— **Klan·ism** \-ˌi-zəm\ *noun*
— **Klans·man** \ˈklanz-mən\ *noun*

klatch *or* **klatsch** \ˈklach, ˈkläch\ *noun* [German *Klatsch* gossip] (1941)
: a gathering characterized usually by informal conversation

klav·ern \ˈkla-vərn\ *noun, often capitalized* [blend of *Klan* and *cavern*] (circa 1924)
: a local unit of the Klan

Klax·on \'klak-sən\ *trademark*
— used for an electrically operated horn or warning signal

kleb·si·el·la \ˌkleb-zē-'e-lə\ *noun* [New Latin, from Edwin *Klebs* (died 1913) German pathologist] (1928)
: any of a genus (*Klebsiella*) of nonmotile gram-negative frequently encapsulated bacterial rods

Klee·nex \'klē-ˌneks\ *trademark*
— used for a cleansing tissue

Klein bottle \'klīn-\ *noun* [Felix *Klein* (died 1925) German mathematician] (1941)
: a one-sided surface that is formed by passing the narrow end of a tapered tube through the side of the tube and flaring this end out to join the other end

klepht \'kleft\ *noun, often capitalized* [New Greek *klephtēs*, literally, robber, from Greek *kleptēs*, from *kleptein*] (1820)
: a Greek belonging to any of several independent guerrilla communities formed after the Turkish conquest of Greece
— kleph·tic \'klef-tik\ *adjective, often capitalized*

klepto- *combining form* [Greek, from *kleptein* to steal; akin to Gothic *hlifan* to steal, Latin *clepere*]
: stealing **:** theft ⟨*klepto*mania⟩

klep·to·ma·nia \ˌklep-tə-'mā-nē-ə, -nyə\ *noun* [New Latin] (1830)
: a persistent neurotic impulse to steal especially without economic motive

klep·to·ma·ni·ac \-nē-ˌak\ *noun* (1861)
: a person evidencing kleptomania

klez·mer \'klez-mər\ *noun, plural* **klez·mo·rim** \(ˌ)klez-'mȯr-əm\ [Yiddish, from Hebrew *kĕlēy zemer* musical instruments] (1949)
1 : a Jewish instrumentalist especially of traditional eastern European music
2 : the music played by klezmorim

klieg light *or* **kleig light** \'klēg-\ *noun* [John H. *Kliegl* (died 1959) & Anton T. *Kliegl* (died 1919) German-born American lighting experts] (1919)
: a carbon arc lamp used especially in making motion pictures

Kline·fel·ter's syndrome \'klīn-ˌfel-tərz-\ *noun* [Harry F. *Klinefelter* (born 1912) American physician] (1950)
: an abnormal condition in a male characterized by usually two X and one Y chromosomes, infertility, and smallness of the testicles

klis·ter \'klis-tər\ *noun* [Norwegian, literally, paste, from Middle Low German *klīster*] (1936)
: a soft wax used on skis

kloof \'klüf\ *noun* [Afrikaans] (1731)
South African **:** a deep glen **:** RAVINE

kludge \'klüj\ *or* **kluge** \'klüj, 'klü-jē\ *noun* [origin unknown] (1962)
: a system and especially a computer system made up of poorly matched components

klutz \'kləts\ *noun* [Yiddish *klots*, literally, wooden beam, from Middle High German *kloz* lumpy mass — more at CLOUT] (1960)
: a clumsy person
— klutz·i·ness \'klət-sē-nəs\ *noun*
— klutzy \'klət-sē\ *adjective*

kly·stron \'klī-ˌsträn\ *noun* [from *Klystron*, a trademark] (1939)
: an electron tube in which bunching of electrons is produced by electric fields and which is used for the generation and amplification of ultrahigh-frequency current

K–me·son \'kā-\ *noun* (1951)
: KAON

knack \'nak\ *noun* [Middle English *knak*] (14th century)
1 a : a clever trick or stratagem **b :** a clever way of doing something
2 : a special ready capacity that is hard to analyze or teach
3 *archaic* **:** an ingenious device; *broadly* **:** TOY, KNICKKNACK
synonym see GIFT

knack·er \'na-kər\ *noun* [probably from English dialect, saddlemaker] (1812)
1 *British* **:** a buyer of worn-out domestic animals or their carcasses for use especially as animal food or fertilizer
2 *British* **:** a buyer of old structures for their constituent materials

knack·ered \'na-kərd\ *adjective* [English slang *knacker* to kill, tire, perhaps from *knacker*, noun] (1886)
British **:** TIRED, EXHAUSTED

knack·wurst *variant of* KNOCKWURST

[1]knap \'nap\ *noun* [Middle English, from Old English *cnæp;* akin to Old English *cnotta* knot] (before 12th century)
1 *chiefly dialect* **:** a crest of a hill **:** SUMMIT
2 *chiefly dialect* **:** a small hill

[2]knap *transitive verb* **knapped; knap·ping** [Middle English *knappen*, of imitative origin] (15th century)
1 *dialect British* **:** [2]RAP 1
2 : to break with a quick blow; *especially* **:** to shape (as flints) by breaking off pieces
3 *dialect British* **:** SNAP, CROP
4 *dialect British* **:** CHATTER
— knap·per *noun*

knap·sack \'nap-ˌsak\ *noun* [Low German *knappsack* or Dutch *knapzak*, from Low German & Dutch *knappen* to make a snapping noise, eat + Low German *sack* or Dutch *zak* sack] (1603)
: a bag (as of canvas or nylon) strapped on the back and used for carrying supplies or personal belongings
— knap·sacked \-ˌsakt\ *adjective*

knap·weed \-ˌwēd\ *noun* [Middle English *knopwed*, from *knop* knop + *wed* weed] (15th century)
: any of various weedy centaureas; *especially* **:** a widely naturalized European perennial (*Centaurea nigra*) with tough wiry stems and knobby heads of purple flowers

knave \'nāv\ *noun* [Middle English, from Old English *cnafa;* akin to Old High German *knabo* boy] (before 12th century)
1 *archaic* **a :** a boy servant **b :** a male servant **c :** a man of humble birth or position
2 : a tricky deceitful fellow
3 : JACK 6a

knav·ery \'nā-və-rē, 'nāv-rē\ *noun, plural* **-er·ies** (1528)
1 a : RASCALITY **b :** a roguish or mischievous act
2 *obsolete* **:** roguish mischief

knav·ish \'nā-vish\ *adjective* (14th century)
: of, relating to, or characteristic of a knave; *especially* **:** DISHONEST
— knav·ish·ly *adverb*

knead \'nēd\ *transitive verb* [Middle English *kneden*, from Old English *cnedan;* akin to Old High German *knetan* to knead] (before 12th century)
1 a : to work and press into a mass with or as if with the hands ⟨*kneading* dough⟩ **b :** to manipulate or massage with a kneading motion ⟨*kneaded* sore neck muscles⟩
2 : to form or shape by or as if by kneading
— knead·able \'nē-də-bəl\ *adjective*
— knead·er *noun*

[1]knee \'nē\ *noun, often attributive* [Middle English, from Old English *cnēow;* akin to Old High German *kneo* knee, Latin *genu*, Greek *gony*] (before 12th century)
1 a : a joint in the middle part of the human leg that is the articulation between the femur, tibia, and patella; *also* **:** the part of the leg that includes this joint **b** (1) **:** the joint in the hind leg of a four-footed vertebrate that corresponds to the human knee (2) **:** the carpal joint of the foreleg of a four-footed vertebrate **c :** the tarsal joint of a bird **d :** the joint between the femur and tibia of an insect
2 : something resembling the human knee: as **a :** a piece of timber naturally or artificially bent for use in supporting structures coming together at an angle (as the deck beams of a ship) **b :** a rounded or conical process rising from the roots of various swamp-growing trees ⟨cypress *knee*⟩
3 : the part of a garment covering the knee
4 : a blow with the bent knee
— kneed \'nēd\ *adjective*
— to one's knees : into a state of submission or defeat

[2]knee *transitive verb* **kneed; knee·ing** (before 12th century)
1 *archaic* **:** to bend the knee to
2 : to strike with the knee

knee breeches *noun plural* (1833)
: BREECH 1a

knee·cap \'nē-ˌkap\ *noun* (1869)
: PATELLA

knee·cap·ping \'nē-ˌka-piŋ\ *noun* (1974)
: the terroristic act or practice of maiming a person's knees (as by gunshot)
— kneecap *transitive verb*

knee–deep \-'dēp\ *adjective* (15th century)
1 a : sunk to the knees ⟨*knee-deep* in mud⟩ **b :** deeply engaged or occupied ⟨*knee-deep* in work⟩
2 : KNEE-HIGH

knee–high \-'hī\ *adjective* (1743)
: rising or reaching upward to the knees ⟨*knee-high* stockings⟩
— knee–high \-ˌhī\ *noun*

knee·hole \-ˌhōl\ *noun* (1893)
: an open space (as under a desk) for the knees

knee–jerk \'nē-ˌjərk, -'jərk\ *adjective* (1951)
: readily predictable **:** AUTOMATIC ⟨*knee-jerk* reactions⟩; *also* **:** reacting in a readily predictable way ⟨*knee-jerk* liberals⟩

knee jerk *noun* (1876)
: an involuntary forward kick produced by a light blow on the tendon below the patella

kneel \'nē(ə)l\ *intransitive verb* **knelt** \'nelt\ *or* **kneeled; kneel·ing** [Middle English *knelen*, from Old English *cnēowlian;* akin to Old English *cnēow* knee] (before 12th century)
: to bend the knee **:** fall or rest on the knees

kneel·er \'nē-lər\ *noun* (14th century)
1 : one that kneels
2 : something (as a cushion or board) to kneel on

knee·pan \'nē-ˌpan\ *noun* (15th century)
: PATELLA

knee–slap·per \-ˌsla-pər\ *noun* (1966)
: an extremely funny joke, line, or story

knee·sock \-ˌsäk\ *noun* (1964)
: a knee-high sock

[1]knell \'nel\ *verb* [Middle English, from Old English *cnyllan;* akin to Middle High German *erknellen* to toll] (before 12th century)
transitive verb
: to summon or announce by or as if by a knell
intransitive verb
1 : to ring especially for a death, funeral, or disaster **:** TOLL
2 : to sound in an ominous manner or with an ominous effect

[2]knell *noun* (before 12th century)
1 : a stroke or sound of a bell especially when rung slowly (as for a death, funeral, or disaster)
2 : an indication of the end or the failure of something ⟨sounded the death *knell* for our hopes⟩

knew *past of* KNOW

knick·er·bock·er \'ni-kə(r)-ˌbä-kər\ *noun* [Diedrich *Knickerbocker*, fictitious author of *History of New York* (1809) by Washington Irving] (1848)
1 *capitalized* **:** a descendant of the early Dutch settlers of New York; *broadly* **:** a native or resident of the city or state of New York — used as a nickname

2 *plural* **:** KNICKERS ◆
knick·ers \ˈni-kərz\ *noun plural* [short for *knickerbockers*] (1881)
1 : loose-fitting short pants gathered at the knee
2 *chiefly British* **:** UNDERPANTS
word history see KNICKERBOCKER
knick·knack \ˈnik-ˌnak\ *noun* [reduplication of *knack*] (1682)
: a small trivial article usually intended for ornament
[1]**knife** \ˈnīf\ *noun, plural* **knives** \ˈnīvz\ *often attributive* [Middle English *knif,* from Old English *cnīf,* perhaps from Old Norse *knīfr*; akin to Middle Low German *knīf* knife] (before 12th century)
1 a : a cutting instrument consisting of a sharp blade fastened to a handle **b :** a weapon resembling a knife
2 : a sharp cutting blade or tool in a machine
3 : SURGERY 4 — usually used in the phrase *under the knife*
— knife·like \ˈnīf-ˌlīk\ *adjective*
[2]**knife** *verb* **knifed; knif·ing** (1865)
transitive verb
1 : to use a knife on; *specifically* **:** to stab, slash, or wound with a knife
2 : to cut, mark, or spread with a knife
3 : to try to defeat by underhanded means
4 : to move like a knife in ⟨birds *knifing* the autumn sky⟩
intransitive verb
: to cut a way with or as if with a knife blade ⟨the cruiser *knifed* through the heavy seas⟩
knife–edge \ˈnīf-ˌej\ *noun* (1818)
1 : a sharp wedge of steel or other hard material used as a fulcrum for a lever beam in a precision instrument
2 : a sharp narrow knifelike edge
knife·point \-ˌpȯint\ *noun* (circa 1911)
: the point of a knife
— at knifepoint : under a threat of being knifed
[1]**knight** \ˈnīt\ *noun* [Middle English, from Old English *cniht* man-at-arms, boy, servant; akin to Old High German *kneht* youth, military follower] (before 12th century)
1 a (1) **:** a mounted man-at-arms serving a feudal superior; *especially* **:** a man ceremonially inducted into special military rank usually after completing service as page and squire (2) **:** a man honored by a sovereign for merit and in Great Britain ranking below a baronet (3) **:** a person of antiquity equal to a knight in rank **b :** a man devoted to the service of a lady as her attendant or champion **c :** a member of an order or society
2 : either of two pieces of the same color in a set of chessmen having the power to make an L-shaped move of two squares in one row and one square in a perpendicular row over squares that may be occupied
[2]**knight** *transitive verb* (13th century)
: to make a knight of
knight–er·rant \ˈnīt-ˈer-ənt\ *noun, plural* **knights–errant** (14th century)
: a knight traveling in search of adventures in which to exhibit military skill, prowess, and generosity
knight–er·rant·ry \-ˈer-ən-trē\ *noun, plural* **knight–errantries** (1654)
1 : the practice or actions of a knight-errant
2 : quixotic conduct
knight·hood \ˈnīt-ˌhu̇d\ *noun* (13th century)
1 : the rank, dignity, or profession of a knight
2 : the qualities befitting a knight **:** CHIVALRY
3 : knights as a class or body
knight·ly \ˈnīt-lē\ *adjective* (14th century)
1 : of, relating to, or characteristic of a knight
2 : made up of knights
— knight·li·ness *noun*
— knightly *adverb*
Knight of Co·lum·bus \-kə-ˈləm-bəs\ *noun, plural* **Knights of Columbus** [Christopher *Columbus*] (1882)
: a member of a benevolent and fraternal society of Roman Catholic men
Knight of Pyth·i·as \-ˈpi-thē-əs\ *noun, plural* **Knights of Pythias** (1869)
: a member of a secret benevolent and fraternal order
Knight of the Mac·ca·bees \-ˈma-kə-ˌbēz\ *noun, plural* **Knights of the Maccabees** (1922)
: a member of a secret benevolent society
Knight Templar *noun, plural* **Knights Templars** *or* **Knights Templar** (1610)
1 : TEMPLAR 1
2 : a member of an order of Freemasonry conferring three orders in the York rite
knish \kə-ˈnish\ *noun* [Yiddish, from Polish *knysz*] (1916)
: a small round or square of dough stuffed with a filling (as potato) and baked or fried
[1]**knit** \ˈnit\ *verb* **knit** *or* **knit·ted; knit·ting** [Middle English *knitten,* from Old English *cnyttan*; akin to Old English *cnotta* knot] (before 12th century)
transitive verb
1 *chiefly dialect* **:** to tie together
2 a : to link firmly or closely ⟨*knitted* my hands⟩ **b :** to cause to grow together ⟨time and rest will *knit* a fractured bone⟩ **c :** to contract into wrinkles ⟨*knitted* her brow⟩
3 : to form by interlacing yarn or thread in a series of connected loops with needles
intransitive verb
1 : to make knitted fabrics or objects
2 a : to become compact **b :** to grow together **c :** to become drawn together
— knit·ter *noun*
[2]**knit** *noun* (1596)
: KNIT STITCH; *also* **:** a knit fabric
knit stitch *noun* (circa 1885)
: a basic knitting stitch usually made with the yarn at the back of the work by inserting the right needle into the front part of a loop on the left needle from the left side, catching the yarn with the point of the right needle, and bringing it through the first loop to form a new loop — compare PURL STITCH
knit·ting *noun* (15th century)
1 : the action or method of one that knits
2 : work done or being done by one that knits
knit·wear \ˈnit-ˌwar, -ˌwer\ *noun* (1926)
: knitted clothing
knob \ˈnäb\ *noun* [Middle English *knobbe*; akin to Middle Low German *knubbe* knob] (before 12th century)
1 a : a rounded protuberance **:** LUMP **b :** a small rounded ornament or handle
2 : a rounded usually isolated hill or mountain
— knobbed \ˈnäbd\ *adjective*
— knob·by \ˈnä-bē\ *adjective*
knob·bly \ˈnä-b(ə-)lē\ *adjective* (1859)
: having very small knobs ⟨a *knobbly* mattress⟩
knob·ker·rie \ˈnäb-ˌker-ē\ *noun* [Afrikaans *knopkierie,* from *knop* knob + *kierie* club] (1844)
: a short wooden club with a knob at one end used as a missile or in close attack especially by Zulus of southern Africa
[1]**knock** \ˈnäk\ *verb* [Middle English *knoken,* from Old English *cnocian*; akin to Middle High German *knochen* to press] (before 12th century)
intransitive verb
1 : to strike something with a sharp blow
2 : to collide with something
3 a : BUSTLE ⟨heard them *knocking* around in the kitchen⟩ **b :** WANDER ⟨*knocked* about Europe all summer⟩
4 a : to make a pounding noise **b :** to have engine knock
5 : to find fault
transitive verb
1 a (1) **:** to strike sharply (2) **:** to drive, force, or make by or as if by so striking **b :** to set forcibly in motion with a blow
2 : to cause to collide
3 : to find fault with ⟨never *knocks* a friend⟩
— knock cold : KNOCK OUT 2a(1)
— knock dead : to move strongly especially to admiration or applause ⟨a comedian who really *knocks* them *dead*⟩
— knock for a loop 1 a : OVERCOME ⟨*knocked* my opponent *for a loop*⟩ **b :** DEMOLISH ⟨*knocked* our idea *for a loop*⟩ **2 :** DUMBFOUND, AMAZE ⟨the news *knocked* them *for a loop*⟩
— knock one's socks off : to overwhelm or amaze one ⟨a performance that will *knock your socks off*⟩
— knock together : to make or assemble especially hurriedly or in a makeshift way ⟨*knocked together* my own bookcase⟩
[2]**knock** *noun* (14th century)
1 a : a sharp blow **:** RAP, HIT ⟨a loud *knock* on the door⟩ **b** (1) **:** a severe misfortune or hardship (2) **:** SETBACK, REVERSAL
2 a : a pounding noise **b :** a sharp repetitive metallic noise caused by abnormal ignition in an automobile engine
3 : a harsh and often petty criticism ⟨likes praise but can't stand the *knocks*⟩
knock·about \ˈnä-kə-ˌbau̇t\ *adjective* (1880)
1 : suitable for rough use ⟨*knockabout* clothing⟩
2 a : being noisy and rough **:** BOISTEROUS ⟨*knockabout* games⟩ **b :** characterized by boisterous antics and often extravagant burlesque ⟨*knockabout* comedy⟩
3 *of a sailing vessel* **:** having a simplified rig marked by absence of bowsprit and topmast ⟨a *knockabout* sloop⟩
— knockabout *noun*
knock back *transitive verb* (circa 1931)
: DRINK, SWALLOW; *specifically* **:** to toss down (an alcoholic beverage)
[1]**knock·down** \ˈnäk-ˌdau̇n\ *adjective* (1690)
1 : having such force as to strike down or overwhelm ⟨a bewildering assortment of *knockdown* arguments —J. W. Krutch⟩
2 : that can easily be assembled or disassembled ⟨a *knockdown* table⟩
3 *chiefly British* **:** extremely low **:** REDUCED ⟨*knockdown* prices⟩
[2]**knockdown** *noun* (1809)
1 : the action of knocking down

◇ WORD HISTORY

knickerbocker In 1809 Washington Irving published *A History of New York,* a supposed work of the fictitious historian Diedrich Knickerbocker, as a genial satire of the early Dutch settlers of the New Netherlands. Irving most likely chose the name *Knickerbocker* for his fictional alter ego in humorous allusion to his friend Herman Knickerbocker, who was of Dutch descent. *Knickerbocker's History of New York* (as the work came to be known) had such an enduring popularity that by the middle of the 19th century New Yorkers of Dutch descent had taken to calling themselves "genuine Knickerbockers." *Knickerbocker* then went on to become a nickname for any native or resident of New York city or state. The use of *knickerbockers* as a term for knee breeches stems from an edition of Irving's work published in the 1850s with illustrations by English caricaturist George Cruikshank. Cruikshank depicted Diedrich Knickerbocker and his fellow Dutch burghers as wearing loose-fitting breeches gathered just below the knee. A coincidental revival of the fashion for knee breeches at that time prompted the adoption of *knickerbockers* for the garment. The shortened form *knickers* soon followed.

2 : something (as a blow) that strikes down or overwhelms
3 : something (as a piece of furniture) that can be easily assembled or disassembled
knock down \-ˈdau̇n\ *transitive verb* (15th century)
1 : to strike to the ground with or as if with a sharp blow **:** FELL
2 : to dispose of (an item) to a bidder at an auction sale
3 : to take apart **:** DISASSEMBLE
4 : to receive as income or salary **:** EARN ⟨positions where they were able to *knock down* good money —*Infantry Journal*⟩
5 : REDUCE
knock–down–drag–out *or* **knock–down–and–drag–out** *adjective* (1834)
: marked by extreme violence or bitterness and by the showing of no mercy ⟨*knock-down-drag-out* political debates⟩
— knock–down–drag–out *noun*
knock•er \ˈnä-kər\ *noun* (14th century)
1 : one that knocks: as **a :** a metal ring, bar, or hammer hinged to a door for use in knocking **b :** a persistently pessimistic critic
2 : BREAST — usually used in plural; often considered vulgar
knock–knee \ˈnäk-ˈnē, -ˌnē\ *noun* (1879)
: a condition in which the legs curve inward at the knees
— knock–kneed \-ˈnēd\ *adjective*
knock•off \ˈnäk-ˌȯf\ *noun* (1966)
: a copy that sells for less than the original; *broadly* **:** a copy or imitation of someone or something popular
knock off (1649)
intransitive verb
: to stop doing something
transitive verb
1 : to do hurriedly or routinely ⟨*knocked off* one painting after another⟩
2 : DISCONTINUE, STOP ⟨*knocked off* work at five⟩
3 : DEDUCT ⟨*knocked off* a little to make the price more attractive⟩
4 a : KILL ⟨*knocked off* two men . . . on mercenary grounds —Lewis Baker⟩ **b :** OVERCOME, DEFEAT ⟨*knocked off* each center of rebellion⟩
5 : ROB ⟨*knocked off* a couple of banks⟩
6 : to make a knockoff of **:** COPY, IMITATE ⟨*knocks off* popular dress designs⟩
knock•out \ˈnäk-ˌau̇t\ *noun* (1887)
1 a : the act of knocking out **:** the condition of being knocked out **b** (1) **:** the termination of a boxing match when one boxer has been knocked down and is unable to rise and resume boxing within a specified time (2) **:** TECHNICAL KNOCKOUT **c :** a blow that knocks out an opponent
2 : something sensationally striking, appealing, or attractive
— knockout *adjective*
knock out *transitive verb* (1856)
1 : to produce roughly or hastily
2 a (1) **:** to defeat (a boxing opponent) by a knockout (2) **:** to make unconscious ⟨the drug *knocked* him *out*⟩ **b :** to make inoperative or useless ⟨electricity was *knocked out* by the storm⟩ **c :** to get rid of **:** ELIMINATE ⟨*knocked out* illegal gambling⟩
3 : to tire out **:** EXHAUST ⟨*knocked* themselves *out* with work⟩
4 : to cause (an opposing pitcher) to be removed from a baseball game by a batting rally
knockout drops *noun plural* (1895)
: drops of a solution of a drug (as chloral hydrate) put into a drink to produce unconsciousness or stupefaction
knock over *transitive verb* (circa 1814)
1 a (1) **:** to strike to the ground **:** FELL (2) **:** OVERWHELM ⟨was *knocked over* by the news⟩ **b :** ELIMINATE ⟨*knocked over* every difficulty⟩
2 a : STEAL; *especially* **:** HIJACK ⟨*knocks over* a truckload of merchandise —J. B. Martin⟩ **b :** ROB ⟨*knocking over* a bank⟩
knock up *transitive verb* (1663)
1 *British* **:** ROUSE, SUMMON
2 : to make pregnant — sometimes considered vulgar
knock•wurst \ˈnäk-(ˌ)wərst *also* -ˌvu̇(r)st, *sometimes* -ˌvu̇sht\ *noun* [German *Knackwurst,* from *knacken* to crackle (of imitative origin) + *Wurst* wurst] (circa 1929)
: a short thick heavily seasoned sausage
¹knoll \ˈnōl\ *noun* [Middle English *knol,* from Old English *cnoll;* akin to Old Norse *knollr* mountaintop] (before 12th century)
: a small round hill **:** MOUND
²knoll *verb* [Middle English, probably alteration of *knellen* to knell] (15th century)
archaic **:** KNELL
knop \ˈnäp\ *noun* [Middle English, from Old English *-cnoppa* knob] (before 12th century)
: a usually ornamental knob
— knopped \ˈnäpt\ *adjective*
¹knot \ˈnät\ *noun* [Middle English, from Old English *cnotta;* akin to Old High German *knoto* knot] (before 12th century)
1 a : an interlacement of the parts of one or more flexible bodies forming a lump or knob **b :** the lump or knob so formed **c :** a tight constriction or the sense of constriction ⟨my stomach was all in *knots*⟩
2 : something hard to solve **:** PROBLEM ⟨a matter full of legal *knots*⟩
3 : a bond of union; *especially* **:** the marriage bond
4 a : a protuberant lump or swelling in tissue ⟨a *knot* in a gland⟩ **b :** the base of a woody branch enclosed in the stem from which it arises; *also* **:** its section in lumber
5 : a cluster of persons or things **:** GROUP
6 : an ornamental bow of ribbon **:** COCKADE
7 a : a division of the log's line serving to measure a ship's speed **b** (1) **:** one nautical mile per hour (2) **:** one nautical mile — not used technically ■

knot 1b: *1* **Blackwall hitch,** *2* **carrick bend,** *3* **clove hitch,** *4* **cat's-paw,** *5* **figure eight,** *6* **granny knot,** *7* **bowline,** *8* **overhand knot,** *9* **fisherman's bend,** *10* **half hitch,** *11* **square knot,** *12* **slip knot,** *13* **stevedore knot,** *14* **true lover's knot,** *15* **surgeon's knot,** *16* **Turk's head,** *17* **sheet bend,** *18* **timber hitch,** *19* **seizing,** *20* **rolling hitch,** *21* **sheepshank**

²knot *verb* **knot•ted; knot•ting** (1547)
transitive verb
1 : to tie in or with a knot **:** form knots in
2 : to unite closely or intricately **:** ENTANGLE
3 : TIE 4b ⟨*knotted* the score⟩
intransitive verb
: to form knots
— knot•ter *noun*
³knot *noun, plural* **knots** *or* **knot** [Middle English *knott*] (15th century)
: either of two sandpipers (*Calidris canutus* and *C. tenuirostris*) that breed in the Arctic and winter in temperate or warm parts of the New and Old World
knot garden *noun* (1519)
: an elaborately designed garden especially of flowers or herbs
knot•grass \ˈnät-ˌgras\ *noun* (1538)
1 : a cosmopolitan weed (*Polygonum aviculare*) of the buckwheat family with jointed stems, prominent sheathing stipules, and minute flowers; *broadly* **:** any of several congeneric plants
2 : any of several grasses with markedly jointed stems; *especially* **:** JOINT GRASS
knot•hole \-ˌhōl\ *noun* (1726)
: a hole in a board or tree trunk where a knot or branch has come out
knot•ted \ˈnä-təd\ *adjective* (12th century)
1 : tied in or with a knot
2 : full of knots **:** GNARLED
3 : KNOTTY
4 : ornamented with knots or knobs
knot•ty \ˈnä-tē\ *adjective* **knot•ti•er; -est** (13th century)
: marked by or full of knots; *especially* **:** so full of difficulties and complications as to be likely to defy solution ⟨a *knotty* problem⟩
synonym see COMPLEX
— knot•ti•ness *noun*
knotty pine *noun* (circa 1898)
: pine wood that has a decorative distribution of knots and is used especially for interior finish
knot•weed \ˈnät-ˌwēd\ *noun* (1884)
: any of several herbs (genus *Polygonum*) of the buckwheat family with leaves and bracts jointed and having a very short petiole; *broadly* **:** POLYGONUM
knout \ˈnau̇t, *sometimes* ˈnüt\ *noun* [Russian *knut,* of Scandinavian origin; akin to Old Norse *knūtr* knot; akin to Old English *cnotta*] (1716)
: a whip used for flogging
— knout *transitive verb*
¹know \ˈnō\ *verb* **knew** \ˈnü *also* ˈnyü\; **known** \ˈnōn\; **know•ing** [Middle English, from Old English *cnāwan;* akin to Old High German bi*chnāan* to recognize, Latin *gnoscere, noscere* to come to know, Greek *gignōskein*] (before 12th century)
transitive verb
1 a (1) **:** to perceive directly **:** have direct cognition of (2) **:** to have understanding of ⟨importance of *knowing* oneself⟩ (3) **:** to recognize the nature of **:** DISCERN **b** (1) **:** to recognize as being the same as something previously known (2) **:** to be acquainted or familiar with (3) **:** to have experience of
2 a : to be aware of the truth or factuality of **:** be convinced or certain of **b :** to have a practical understanding of ⟨*knows* how to write⟩
3 *archaic* **:** to have sexual intercourse with
intransitive verb
1 : to have knowledge
2 : to be or become cognizant — sometimes used interjectionally with *you* especially as a filler in informal speech
— know•able \ˈnō-ə-bəl\ *adjective*
— know•er \ˈnō(-ə)r\ *noun*
— know from : to have knowledge of ⟨didn't *know from* sibling rivalry —Penny Marshall⟩
²know *noun* (1592)

□ USAGE
knot It is a favorite notion of those who claim a knowledge of things nautical that the knot is a unit of speed only. One who chances to say or write *knots per hour* or *knots an hour* will be quickly corrected. And in truth it is a bit of a jump in meaning from a ship's speed measured by the number of knots tied at fixed intervals in a line run out in a half minute to a nautical mile. But this is no recent jump—it took place in the 18th century, and the earliest attestations are from a couple of famous English sailing men, including this one ⟨the strong tide, though even here it ran five *knots* an hour —Captain James Cook⟩. So criticism of *knots an hour* can elicit the polite reply that what was good enough for Captain Cook is good enough for a mere modern landlubber.

: KNOWLEDGE
— **in the know :** in possession of exclusive knowledge or information; *broadly* **:** WELL-INFORMED

know–all \'nō-,ȯl\ *noun* (circa 1864)
chiefly British **:** KNOW-IT-ALL

know–how \'nō-,hau̇\ *noun* (1838)
: knowledge of how to do something smoothly and efficiently **:** EXPERTISE

[1]**know·ing** \'nō-iŋ\ *noun* (14th century)
: ACQUAINTANCE, COGNIZANCE

[2]**knowing** *adjective* (14th century)
1 : having or reflecting knowledge, information, or intelligence
2 a : shrewdly and keenly alert **:** ASTUTE **b :** indicating possession of exclusive inside knowledge or information ⟨a *knowing* smile⟩
3 : COGNITIVE
4 : DELIBERATE ⟨*knowing* interference in the affairs of another⟩
— **know·ing·ly** *adverb*
— **know·ing·ness** *noun*

know–it–all \'nō-ət-,ȯl\ *noun* (1895)
: one who claims to know everything; *also* **:** one who disdains advice
— **know–it–all** *adjective*

knowl·edge \'nä-lij\ *noun* [Middle English *knowlege,* from *knowlechen* to acknowledge, irregular from *knowen*] (14th century)
1 *obsolete* **:** COGNIZANCE
2 a (1) **:** the fact or condition of knowing something with familiarity gained through experience or association (2) **:** acquaintance with or understanding of a science, art, or technique **b** (1) **:** the fact or condition of being aware of something (2) **:** the range of one's information or understanding ⟨answered to the best of my *knowledge*⟩ **c :** the circumstance or condition of apprehending truth or fact through reasoning **:** COGNITION **d :** the fact or condition of having information or of being learned ⟨a man of unusual *knowledge*⟩
3 *archaic* **:** SEXUAL INTERCOURSE
4 a : the sum of what is known **:** the body of truth, information, and principles acquired by mankind **b** *archaic* **:** a branch of learning ☆

knowl·edge·able \'nä-lij-ə bəl\ *adjective* (1829)
: having or exhibiting knowledge or intelligence
— **knowl·edge·abil·i·ty** \,nä-li-jə-'bi-lə-tē\ *noun*
— **knowl·edge·able·ness** *noun*
— **knowl·edge·ably** \-blē\ *adverb*

knowledge engineering *noun* (1980)
: a branch of artificial intelligence that emphasizes the development and use of expert systems
— **knowledge engineer** *noun*

known \'nōn\ *adjective* (13th century)
: familiar **:** generally recognized ⟨a *known* authority on art⟩

know–noth·ing \'nō-,nə-thiŋ\ *noun* (1827)
1 a : IGNORAMUS **b :** AGNOSTIC
2 *K&N capitalized* **:** a member of a 19th century secret American political organization hostile to the political influence of recent immigrants and Roman Catholics

know–noth·ing·ism \-thiŋ-,i-zəm\ *noun* (1854)
1 *K&N capitalized* **:** the principles and policies of the Know Nothings
2 : the condition of knowing nothing or desiring to know nothing or the conviction that nothing can be known with certainty especially in religion or morality
3 *often K&N capitalized* **:** a mid-twentieth century political attitude characterized by anti-intellectualism, exaggerated patriotism, and fear of foreign subversive influences

knub·by *variant of* NUBBY

[1]**knuck·le** \'nə-kəl\ *noun* [Middle English *knokel;* akin to Middle High German *knöchel* knuckle] (14th century)
1 a : the rounded prominence formed by the ends of the two adjacent bones at a joint — used especially of those at the joints of the fingers **b :** the joint of a knuckle
2 : a cut of meat consisting of the tarsal or carpal joint with the adjoining flesh
3 : something resembling a knuckle: as **a** (1) **:** one of the joining parts of a hinge through which a pin or rivet passes (2) **:** KNUCKLE JOINT **b :** the meeting of two surfaces at a sharp angle (as in a roof) **c :** a pivotal point
4 *plural* **:** a set of metal finger rings or guards attached to a transverse piece and worn over the front of the doubled fist for use as a weapon — called also *brass knuckles*
— **knuck·led** *adjective*

[2]**knuckle** *verb* **knuck·led; knuck·ling** \'nə-k(ə-)liŋ\ (1740)
intransitive verb
: to place the knuckles on the ground in shooting a marble
transitive verb
: to press or rub with the knuckles

knuck·le·ball \'nə-kəl-,bȯl\ *noun* (1910)
: a baseball pitch in which the ball is gripped with the knuckles or the tips of the fingers pressed against the top and thrown with little speed or spin
— **knuck·le·ball·er** \-,bȯ-lər\ *noun*

knuck·le·bone \'nək-əl-,bōn\ *noun* (1577)
1 : a bone (as a metatarsus or metacarpus of a sheep) used in games and formerly in divination
2 *plural but singular in construction* **:** a game played with knucklebones or jacks

knuckle down *intransitive verb* (circa 1864)
: to apply oneself earnestly

knuck·le–dust·er \'nə-kəl-,dəs-tər\ *noun* (1858)
: KNUCKLE 4

knuck·le·head \-,hed\ *noun* (1942)
: DUMBBELL 2
— **knuck·le·head·ed** \-,hed-əd\ *adjective*

knuckle joint *noun* (circa 1864)
: a hinge joint in which a projection with an eye on one piece enters a jaw between two corresponding projections with eyes on another piece and is retained by a pin or rivet

knuckle joint

knuck·ler \'nə-k(ə-)lər\ *noun* (1928)
: KNUCKLEBALL

knuckle under *intransitive verb* (1869)
: GIVE IN, SUBMIT

knur \'nər\ *noun* [Middle English *knorre;* akin to Middle High German *knorre* burl] (14th century)
: a hard excrescence (as on a tree trunk) **:** GNARL

knurl \'nər(-ə)l\ *noun* [probably blend of *knur* and *gnarl*] (1608)
1 : a small protuberance, excrescence, or knob
2 : one of a series of small ridges or beads on a metal surface to aid in gripping
— **knurled** \'nər(-ə)ld\ *adjective*
— **knurly** \'nər-lē\ *adjective*

[1]**KO** \(,)kā-'ō, 'kā-(,)ō\ *noun* [*k*nock *o*ut] (1911)
: KNOCKOUT

[2]**KO** *transitive verb* **KO'd** \kā-'ōd, 'kā-(,)ōd\; **KO'·ing** \-'ō-iŋ, (,)ō-\ (1926)
: to knock out (as in boxing)

koa \'kō-ə\ *noun* [Hawaiian] (1850)
1 : a Hawaiian timber tree (*Acacia koa*) with crescent-shaped leaves and white flowers borne in small round heads
2 : the fine-grained red wood of the koa used especially for furniture

ko·ala \kə-'wä-lə, kō-'ä-\ *noun* [Dharuk (Australian aboriginal language of the Port Jackson area) *gulawaŋ*] (1803)
: an Australian arboreal marsupial (*Phascolarctos cinereus*) that has a broad head, large hairy ears, dense gray fur, and sharp claws and feeds on eucalyptus leaves — called also *koala bear*

koala

ko·an \'kō-,än\ *noun* [Japanese *kōan,* from *kō* public + *an* proposition] (1945)
: a paradox to be meditated upon that is used to train Zen Buddhist monks to abandon ultimate dependence on reason and to force them into gaining sudden intuitive enlightenment

ko·bo \'kō-(,)bō\ *noun, plural* **kobo** [alteration of [1]*copper*] (1972)
— see *naira* at MONEY table

ko·bold \'kō-,bȯld\ *noun* [German — more at COBALT] (1830)
1 : a gnome that in German folklore inhabits underground places
2 : an often mischievous domestic spirit of German folklore

Ko·di·ak bear \'kō-dē-,ak-\ *noun* [*Kodiak* Island, Alaska] (1899)
: a large brown bear of the southern coast of Alaska and adjacent islands

kohl \'kōl\ *noun* [Arabic *kuḥl*] (1799)
: a preparation used especially in Arabia and Egypt to darken the edges of the eyelids

kohl·ra·bi \kōl-'rä-bē *also* -'ra-\ *noun, plural* **-bies** [German, from Italian *cavolo rapa,* from *cavolo* cabbage + *rapa* turnip] (1807)
: any of a race of cabbages (*Brassica oleracea gongylodes*) having a greatly enlarged, fleshy, turnip-shaped edible stem

koi \'kȯi\ *noun, plural* **koi** [Japanese] (1727)
: a carp (*Cyprinus carpio*) bred especially in Japan for large size and a variety of colors and often stocked in ornamental ponds

koi·ne *also* **koi·né** \kȯi-'nā, 'kȯi-,; kē-'nē\ *noun* [Greek *koinē,* from feminine of *koinos* common] (1909)
1 *capitalized* **:** the Greek language commonly spoken and written in eastern Mediterranean countries in the Hellenistic and Roman periods
2 : a dialect or language of a region that has become the common or standard language of a larger area

ko·kan·ee \kō-'ka-nē\ *noun* [perhaps from Shuswap (Salishan language of British Columbia) *kəknǽxʷ*] (1875)
: a small landlocked sockeye salmon — called also *kokanee salmon*

kok–sa·ghyz *or* **kok–sa·gyz** \,kōk-sə-'gēz, ,käk-, -'giz\ *noun* [Russian *kok-sagyz*] (1932)

☆ SYNONYMS
Knowledge, learning, erudition, scholarship mean what is or can be known by an individual or by mankind. KNOWLEDGE applies to facts or ideas acquired by study, investigation, observation, or experience ⟨rich in the *knowledge* of human nature⟩. LEARNING applies to knowledge acquired especially through formal, often advanced, schooling ⟨a book that demonstrates vast *learning*⟩. ERUDITION strongly implies the acquiring of profound, recondite, or bookish learning ⟨an *erudition* unusual even in a scholar⟩. SCHOLARSHIP implies the possession of learning characteristic of the advanced scholar in a specialized field of study or investigation ⟨a work of first-rate literary *scholarship*⟩.

: a perennial Asian dandelion (*Taraxacum kok-saghyz*) cultivated for its fleshy roots that have a high rubber content

ko·la *variant of* COLA

ko·la nut \'kō-lə-\ *noun* [*kola,* perhaps modification of Malinke *kolo* kola nut] (1868)
: the bitter caffeine-containing chestnut-sized seed of a kola tree used especially as a masticatory and in beverages

kola tree *noun* (1937)
: an African tree (genus *Cola,* especially *C. nitida* and *C. acuminata* of the family Sterculiaceae) cultivated in various tropical areas for its kola nuts

ko·lin·sky \kə-'lin(t)-skē\ *noun, plural* **-skies** [origin unknown] (1851)
1 : any of several Asian minks (especially *Mustela siberica*)
2 : the fur or pelt of a kolinsky

kol·khoz \käl-'kȯz, -'kȯs\ *noun, plural* **kol·kho·zy** \-'kȯ-zē\ *or* **kol·khoz·es** \-'kȯ-zəz\ [Russian, from *kol*lektivnoe *khoz*yaĭstvo collective farm] (1921)
: a collective farm of the U.S.S.R.

kol·khoz·nik \käl-'kȯz-nik\ *noun, plural* **-ni·ki** \-ni-kē\ *or* **-niks** [Russian, from *kolkhoz* + *-nik,* agent suffix] (1944)
: a member of a kolkhoz

Kol Ni·dre \kōl-'ni-(,)drā, kȯl-, -drə; -ni-'drā\ *noun* [Aramaic *kol nidhrē* all the vows; from the opening phrase of the prayer] (1881)
: a formula for the annulment of private vows chanted in the synagogue on the eve of Yom Kippur

ko·lo \'kō-(,)lō\ *noun, plural* **kolos** [Serbo-Croatian, literally, circle, wheel; akin to Greek *kyklos* circle — more at WHEEL] (1910)
: a central European folk dance in which dancers form a circle and progress slowly to right or left while one or more dancers perform elaborate steps in the center

ko·mat·ik \kō-'ma-tik\ *noun* [Inuit *qamutik*] (circa 1824)
: an Eskimo sledge with wooden runners and crossbars lashed with rawhide

Ko·mo·do dragon \kə-'mō-dō-\ *noun* [*Komodo* Island, Indonesia] (1927)
: an Indonesian monitor lizard (*Varanus komodoensis*) that is the largest of all known lizards and may attain a length of 10 feet (3 meters)

ko·mon·dor \'kä-mən-,dȯr, 'kō-\ *noun, plural* **-dors** *or* **-dor·ok** \-,dȯr-ək\ [Hungarian] (1931)
: any of a breed of large powerful shaggy-coated white dogs of Hungarian origin that are used to guard sheep

Kom·so·mol \'käm-sə-,mȯl, -,mōl\ *noun* [Russian, from *Kom*munisticheskiĭ *So*yuz *Mol*odezhi Communist Union of Youth] (1925)
: a Russian Communist youth organization

Kon·go \'käŋ-(,)gō\ *noun, plural* **Kongo** *or* **Kongos** (circa 1902)
1 : a member of a Bantu people of the lower Congo river
2 : the Bantu language of the Kongo people

Kon·ka·ni \'käŋ-kə-(,)nē\ *noun* [Marathi *Koṅkaṇī*] (1873)
: an Indo-Aryan language of the west coast of India

koo·doo *variant of* KUDU

kook \'kük\ *noun* [by shortening & alteration from *cuckoo*] (1960)
: one whose ideas or actions are eccentric, fantastic, or insane **:** SCREWBALL

kook·a·bur·ra \'ku̇-kə-,bər-ə, -,bə-rə\ *noun* [Wiradhuri (Australian aboriginal language of central New South Wales) *gugubarra*] (1834)
: a brownish kingfisher (*Dacelo novaeguineae* synonym *D. gigas*) of Australia that is about the size of a crow and has a call resembling loud laughter — called also *laughing jackass*

kooky *also* **kook·ie** \'kü-kē\ *adjective* **kook·i·er; -est** (1959)
: having the characteristics of a kook **:** CRAZY, OFFBEAT
— kook·i·ness *noun*

ko·peck *or* **ko·pek** \'kō-,pek\ *noun* [Russian *kopeĭka*] (1698)
— see *ruble* at MONEY table

koph *variant of* QOPH

kop·je *or* **kop·pie** \'kä-pē\ *noun* [Afrikaans *koppie*] (1848)
: a small hill especially on the African veld

kor *variant of* COR

Ko·ran \kə-'ran, -'rän; 'kōr-,an, 'kȯr-\ *noun* [Arabic *qur'ān*] (circa 1615)
: the book composed of sacred writings accepted by Muslims as revelations made to Muhammad by Allah through the angel Gabriel
— Ko·ran·ic \kə-'ra-nik\ *adjective*

Ko·rat \kō-'rät, kȯ-\ *noun* [*Korat* province, Thailand] (1967)
: any of a breed of shorthaired domestic cats originating in Thailand and having a heart-shaped face, a silver-blue coat, and green eyes

ko·re \'kȯ-(,)rā, 'kō-\ *noun, plural* **ko·rai** \-,rī, -(,)rā\ [Greek *korē* girl; akin to Greek *koros* boy — more at CRESCENT] (1920)
: an ancient Greek statue of a clothed young woman standing with feet together

Ko·re·an \kə-'rē-ən, *especially Southern* (,)kō-\ *noun* (1600)
1 : a native or inhabitant of Korea
2 : the language of the Korean people
— Korean *adjective*

ko·ru·na \'kȯr-ə-,nä, 'kär-\ *noun, plural* **ko·ru·ny** \-ə-nē\ *or* **korunas** *or* **ko·run** \'kȯr-ən, 'kär-\ [Czech, literally, crown, from Latin *corona* — more at CROWN] (1930)
— see MONEY table

¹ko·sher \'kō-shər\ *adjective* [Yiddish, from Hebrew *kāshēr* fit, proper] (1851)
1 a : sanctioned by Jewish law; *especially* **:** ritually fit for use ⟨*kosher* meat⟩ **b :** selling or serving food ritually fit according to Jewish law ⟨a *kosher* restaurant⟩
2 : being proper, acceptable, or satisfactory ⟨found things going on that were not *kosher* —Homer Bigart⟩

²kosher *transitive verb* **ko·shered; ko·sher·ing** \-sh(ə-)riŋ\ (1871)
: to make kosher

³kosher *noun* (1886)
: the observance of kosher practices — used in the phrase *keep kosher*

ko·to \'kō-(,)tō\ *noun* [Japanese] (1795)
: a long Japanese zither having 13 silk strings

kou·miss \kü-'mis, 'kü-məs\ *noun* [Russian *kumys,* of Turkic origin; akin to Turkish *kımız* koumiss] (1598)
: a beverage of fermented mare's milk made originally by the nomadic peoples of central Asia

kou·prey \'kü-,prā\ *noun* [Khmer *ko:prey*] (1940)
: a short-haired ox (*Bos sauveli*) of forests of Cambodia, Thailand, and Vietnam having a large dewlap

kou·ros \'kü-,rȯs\ *noun, plural* **kou·roi** \-(,)rȯi\ [Greek *kouros, koros* boy — more at CRESCENT] (1920)
: an ancient Greek statue of a nude male youth standing with the left leg forward and arms at the sides

¹kow·tow \(,)kau̇-'tau̇, 'kau̇-,\ *noun* [Chinese (Beijing) *kòutóu,* from *kòu* to knock + *tóu* head] (1804)
: an act of kowtowing

²kowtow *intransitive verb* (1826)
1 : to show obsequious deference **:** FAWN
2 : to kneel and touch the forehead to the ground in token of homage, worship, or deep respect

KP \,kā-'pē\ *noun* [*k*itchen *p*olice] (1918)
1 : an enlisted man detailed to assist the cooks in a military mess
2 : the work of KPs

¹kraal \'krȯl, 'kräl\ *noun* [Afrikaans, from Portuguese *curral* pen for cattle, enclosure, from (assumed) Vulgar Latin *currale* enclosure for vehicles — more at CORRAL] (1731)
1 a : a village of southern African natives **b :** the native village community
2 : an enclosure for animals especially in southern Africa

²kraal *transitive verb* (1827)
: to pen in a kraal

kraft \'kräft, 'kraft\ *noun, often attributive* [German, literally, strength, from Old High German — more at CRAFT] (1906)
: a strong paper or paperboard made from wood pulp produced from wood chips boiled in an alkaline solution containing sodium sulfate

krait \'krīt\ *noun* [Hindi *karait*] (1874)
: any of a genus (*Bungarus*) of brightly banded extremely venomous nocturnal elapid snakes of Pakistan, India, southeastern Asia, and adjacent islands

kra·ken \'krä-kən\ *noun* [Norwegian dialect] (1755)
: a fabulous Scandinavian sea monster

kra·ter \'krā-tər, krä-'ter\ *noun* [Greek *kratēr* — more at CRATER] (circa 1736)
: a jar or vase of classical antiquity having a large round body and a wide mouth and used for mixing wine and water

K ration \'kā-\ *noun* [A. B. *K*eys (born 1904) American physiologist] (circa 1940)
: a lightweight packaged ration of emergency foods developed for the U.S. armed forces in World War II

kraut \'krau̇t\ *noun* [German, cabbage, from Old High German *krūt*] (1855)
1 : SAUERKRAUT
2 *often capitalized* **:** GERMAN — usually used disparagingly

Krebs cycle \'krebz-\ *noun* [H. A. *Krebs*] (1941)
: a sequence of reactions in the living organism in which oxidation of acetic acid or acetyl equivalent provides energy for storage in phosphate bonds (as in ATP) — called also *citric acid cycle, tricarboxylic acid cycle*

krem·lin \'krem-lən\ *noun* [obsolete German *Kremelien* the citadel of Moscow, ultimately from Old Russian *kremlĭ*] (1662)
1 : the citadel of a Russian city
2 *capitalized* [the *Kremlin,* citadel of Moscow and seat of government of Russia and formerly of the U.S.S.R.] **:** the Russian government

krem·lin·ol·o·gy \,krem-lə-'nä-lə-jē\ *noun, often capitalized* (1958)
: the study of the policies and practices of the Soviet government
— krem·lin·ol·o·gist \-jist\ *noun, often capitalized*

krep·lach \'krep-lək, -,läk\ *noun* [Yiddish *kreplekh,* plural of *krepl* filled dumpling] (circa 1892)
: square or triangular dumplings filled with ground meat or cheese, boiled or fried, and usually served in soup

kreu·zer \'krȯit-sər\ *noun* [German, from *Kreuz* cross; from its markings] (1547)
: a small coin formerly used in Austria and Germany

krill \'kril\ *noun* [Norwegian *kril* fry of fish] (1907)
: planktonic crustaceans and larvae (order Euphausiacea) that constitute the principal food of baleen whales

krim·mer \'kri-mər\ *noun* [German, from *Krim* Crimea] (1834)
: a gray fur made from the pelts of young lambs of the Crimean Peninsula region

Krio \'krē-ō\ *noun* [Krio, speaker of Krio, Krio language, perhaps from Yoruba *Kiriyó* Christian, ultimately from Portuguese *crioulo* Creole] (1955)
: an English-based creole spoken in Sierra Leone

kris \'krēs\ *noun* [Malay *kĕris*] (circa 1580)
: a Malay or Indonesian dagger with a ridged serpentine blade

Krish·na \'krish-nə, 'krēsh-\ *noun* [Sanskrit *Kṛṣṇa*] (1864)

: a deity or deified hero of later Hinduism worshiped as an incarnation of Vishnu

Krish·na·ism \-ˌi-zəm\ *noun* (1885)
: a widespread form of Hindu religion characterized by the worship of Krishna

Kriss Krin·gle \'kris-'kriŋ-gəl\ *noun* [modification of German *Christkindl* Christ child, Christmas gift, diminutive of *Christkind* Christ child] (1830)
: SANTA CLAUS

[1]kro·na \'krō-nə\ *noun, plural* **kro·nor** \-ˌnȯr, -nər\ [Swedish, literally, crown] (1875)
— see MONEY table

[2]kro·na \'krō-nə\ *noun, plural* **kro·nur** \-nər\ [Icelandic *krōna,* literally, crown] (1886)
— see MONEY table

[1]kro·ne \'krō-nə\ *noun, plural* **kro·ner** \-nər\ [Danish, literally, crown] (1885)
— see MONEY table

[2]kro·ne \'krō-nə\ *noun, plural* **kro·nen** \-nən\ [German, literally, crown] (1895)
1 : the basic monetary unit of Austria from 1892 to 1925
2 : a coin representing one krone

Kro·neck·er delta \'krō-ˌne-kər-\ *noun* [Leopold *Kronecker* (died 1891) German mathematician] (1926)
: a function of two variables that is 1 when the variables have the same value and is 0 when they have different values

Kru·ger·rand \'krü-gə(r)-ˌrand, -ˌränd, -ˌränt\ *noun* [S.J.P. *Kruger* + *rand*] (1967)
: a one-ounce gold coin of the Republic of South Africa

krumm·holz \'krùm-ˌhōlts\ *noun, plural* **krummholz** [German, from *krumm* crooked + *Holz* wood] (1903)
: stunted forest characteristic of timberline

krumm·horn *also* **krum·horn** \'krəm-ˌhȯrn\ *noun* [German *Krummhorn,* from *krumm* curved + *Horn* horn] (circa 1696)
: a Renaissance double-reed woodwind instrument consisting of a curved boxwood tube and having a pierced cap covering the reed

kryp·ton \'krip-ˌtän\ *noun* [Greek, neuter of *kryptos* hidden — more at CRYPT] (1898)
: a colorless relatively inert gaseous element found in air at about one part per million and used especially in electric lamps — see ELEMENT table

Ksha·tri·ya \'ksha-trē-ə, 'cha-\ *noun* [Sanskrit *kṣatriya,* from *kṣatra* dominion — more at CHECK] (1794)
: a Hindu of an upper caste traditionally assigned to governing and military occupations

Ku·che·an \kü-'chē-ən\ *noun* [*Kuche, Kucha,* Sinkiang, China] (circa 1934)
: TOCHARIAN B

ku·chen \'kü-kən, -k̲ən\ *noun, plural* **kuchen** [German, cake, from Old High German *kuocho* — more at CAKE] (1854)
: any of various coffee cakes made from sweet yeast dough

ku·do \'kü-(ˌ)dō, 'kyü-\ *noun, plural* **kudos** [back-formation from *kudos* (taken as a plural)] (1926)
1 : AWARD, HONOR ⟨a score of honorary degrees and . . . other *kudos* —*Time*⟩
2 : COMPLIMENT, PRAISE ⟨to all three should go some kind of special *kudo* for refusing to succumb —Al Hine⟩ ■

ku·dos \'kü-ˌdäs, 'kyü-, -ˌdōs\ *noun* [Greek *kydos*] (1831)
1 : fame and renown resulting from an act or achievement **:** PRESTIGE
2 : praise given for achievement

ku·du \'kü-(ˌ)dü\ *noun, plural* **kudu** *or* **kudus** [Afrikaans *koedoe*] (1777)
: a large grayish brown African antelope (*Tragelaphus strepsiceros*) with large annulated spirally twisted horns; *also* **:** a related antelope (*T. imberbis*)

kud·zu \'kùd-(ˌ)zü, 'kəd-\ *noun* [Japanese *kuzu*] (1876)
: an Asian leguminous vine (*Pueraria lobata* synonym *P. thunbergiana*) that is used for forage and erosion control and that is often a serious weed in the southeastern U.S.

kudu

ku·gel \'kü-gəl\ *noun* [Yiddish *kugl,* from Middle High German *kugel* ball — more at CUDGEL] (1846)
: a baked pudding (as of potatoes or noodles) usually served as a side dish

Ku Klux·er \'kü-'klək-sər *also* 'kyü- *or* 'klü-\ *noun* (1880)
: a member of the Ku Klux Klan
— **Ku Klux·ism** \-'klək-ˌsi-zəm\ *noun*

Ku Klux Klan \'kü-'klək-'sklan *also* 'kyü- *or* 'klü-\ *noun* (1867)
1 : a post-Civil War secret society advocating white supremacy
2 : a 20th century secret fraternal group held to confine its membership to American-born white Christians

ku·lak \'kü-ˌlak, -ˌläk, kü-'\ *noun* [Russian, literally, fist] (1877)
1 : a prosperous or wealthy peasant farmer in 19th century Russia
2 : a farmer characterized by Communists as having excessive wealth

kul·tur \kùl-'tùr\ *noun, often capitalized* [German, from Latin *cultura* culture] (1914)
1 : CULTURE 5
2 : culture emphasizing practical efficiency and individual subordination to the state
3 : German culture held to be superior especially by militant Nazi and Hohenzollern expansionists

Kul·tur·kampf \-ˌkäm(p)f\ *noun* [German, from *Kultur* + *Kampf* conflict] (1879)
: conflict between civil government and religious authorities especially over control of education and church appointments

ku·miss *variant of* KOUMISS

küm·mel \'ki-məl\ *noun* [German, literally, caraway seed, from Old High German *kumīn* cumin] (1864)
: a colorless liqueur flavored principally with caraway seeds

kum·quat \'kəm-ˌkwät\ *noun* [Chinese (Guangdong) *gām-gwāt,* from *gām* gold + *gwāt* citrus fruit] (1699)
: any of several small yellow to orange citrus fruits with sweet spongy rind and somewhat acid pulp that are used chiefly for preserves; *also* **:** a tree or shrub (genus *Fortunella*) of the rue family that bears kumquats

kun·da·li·ni \ˌkùn-də-'lē-nē, ˌkən-\ *noun, often capitalized* [Sanskrit *kuṇḍalinī,* from feminine of *kuṇḍalin* circular, coiled, from *kuṇḍala* ring] (1905)
: the yogic life force that is held to lie coiled at the base of the spine until it is aroused and sent to the head to trigger enlightenment

Kung \'kùŋ, 'küŋ; *in !Kung the initial sound is a type of click*\ *noun, plural* **Kung** (1926)
1 : a member of a people of southern Africa — usually preceded in writing by !
2 : the Khoisan language of the !Kung people — usually preceded in writing by !

kung fu \ˌkəŋ-'fü, ˌkùŋ-\ *noun* [Chinese (Beijing) *gōngfu* skill, art] (1966)
: any of various Chinese arts of self-defense like karate

kunz·ite \'kùn(t)-ˌsīt\ *noun* [G. F. *Kunz* (died 1932) American gem expert] (1903)
: a spodumene that occurs in pinkish lilac crystals and is used as a gem

Kurd \'kùrd, 'kərd\ *noun* (1595)
: a member of a pastoral and agricultural people who inhabit a plateau region in adjoining parts of Turkey, Iran, Iraq, Syria, Armenia, and Azerbaijan
— **Kurd·ish** \'kùr-dish, 'kər-\ *adjective*

Kurdish *noun* (1813)
: the Iranian language of the Kurds

Kur·di·stan \ˌkùr-də-'stan, ˌkər-\ *noun* [*Kurdistan,* Asia] (1904)
: an Oriental rug woven by the Kurds and noted for fine colors

kur·gan \kùr-'gän, -'gan\ *noun* [Russian, of Turkic origin; akin to Turkish *kurgan* fortress, castle] (1889)
: a burial mound of eastern Europe or Siberia

kur·ra·jong \'kər-ə-ˌjòŋ, 'kə-rə-, -ˌjäŋ\ *noun* [Dharuk (Australian aboriginal language of the Port Jackson area) *garajuŋ*] (1823)
: any of several Australian trees or shrubs (family Sterculiaceae) having strong bast fiber used by Australian aborigines; *especially* **:** a widely planted shelter and forage tree (*Brachychiton populneum*)

kur·to·sis \(ˌ)kər-'tō-səs\ *noun* [Greek *kyrtōsis* convexity, from *kyrtos* convex — more at CURVE] (1905)
: the peakedness or flatness of the graph of a frequency distribution especially with respect to the concentration of values near the mean as compared with the normal distribution

ku·ru \'kùr-(ˌ)ü\ *noun* [Fore (language of eastern highland Papua New Guinea)] (1957)
: a rare progressive fatal encephalopathy that is caused by a slow virus, resembles Creutzfeldt-Jakob disease, and occurs among tribesmen in eastern New Guinea

ku·rus \kə-'rüsh\ *noun, plural* **kurus** [Turkish *kuruş*] (1882)
— see *lira* at MONEY table

Ku·te·nai \'kü-tᵊn-ˌā, -ˌē\ *noun, plural* **Kutenai** *or* **Kutenais** (1801)
1 : a member of an American Indian people of the Rocky Mountains in both the U.S. and Canada
2 : the language of the Kutenai people

kvass \'kväs, 'kfäs\ *noun* [Russian *kvas*] (circa 1553)
: a slightly alcoholic beverage of eastern Europe made from fermented mixed cereals and often flavored

[1]kvetch \'kvech, 'kfech\ *intransitive verb* [Yiddish *kvetshn,* literally, to squeeze, pinch, from Middle High German *quetschen*] (circa 1952)
: to complain habitually **:** GRIPE

[2]kvetch *noun* (1964)
1 : an habitual complainer
2 : COMPLAINT 1
— **kvetchy** \'kve-chē, 'kfe-\ *adjective*

Kwa \'kwä\ *noun* (1857)
: a branch of the Niger-Congo language family that is spoken along the African coast and its hinterland from the Ivory Coast to southwestern Nigeria

kwa·cha \'kwä-chə\ *noun, plural* **kwacha** [Bemba or Chichewa (Bantu language of Malawi), literally, it dawns] (1966)

□ USAGE
kudo Some commentators hold that since *kudos* is a singular word it cannot be used as a plural and that the word *kudo* is impossible. But *kudo* does exist; it is simply one of the most recent words created by back-formation from another word misunderstood as a plural. *Kudos* was introduced into English in the 19th century; it was used in contexts where a reader unfamiliar with Greek could not be sure whether it was singular or plural. By the 1920s it began to appear as a plural, and about 25 years later *kudo* began to appear. It may have begun as a misunderstanding, but then so did *cherry* and *pea.*

\ə\ **abut** \ᵊ\ kitten \ər\ **further** \a\ **ash** \ā\ **ace** \ä\ **mop, mar** \aù\ **out** \ch\ **chin** \e\ **bet** \ē\ **easy** \g\ **go** \i\ **hit** \ī\ **ice** \j\ **job** \ŋ\ **sing** \ō\ **go** \ȯ\ **law** \ȯi\ **boy** \th\ **thin** \t̲h̲\ **the** \ü\ **loot** \ù\ **foot** \y\ **yet** \zh\ **vision** *see also* Guide to Pronunciation

— see MONEY table

Kwa·ki·utl \ˌkwä-kē-ˈyü-tᵊl, ˌkwä-ˈkyü-\ *noun, plural* **Kwakiutl** (1848)
1 : a member of an American Indian people of the Canadian Pacific coast
2 : the language of the Kwakiutl people

kwan·za \ˈkwän-zə\ *noun, plural* **kwanzas** *or* **kwanza** [*Kwanza* (Cuanza), river in Angola] (1978)
— see MONEY table

Kwan·za *or* **Kwan·zaa** \ˈkwän-zə\ *noun* [Swahili *kwanza* first] (1972)
: an African-American festival held in late December

kwash·i·or·kor \ˌkwä-shē-ˈȯr-kər, -ȯr-ˈkȯr\ *noun* [Ga (Kwa language of coastal Ghana) *kwàṣìɔkɔ́* influence a child is said to be under when a second child comes] (1935)
: severe malnutrition in infants and children that is caused by a diet high in carbohydrate and low in protein

KWIC \ˈkwik\ *noun* [*k*ey*w*ord *i*n *c*ontext] (1959)
: a computer-generated index alphabetized on a keyword that appears within a brief context

ky·ack \ˈkī-ˌak\ *noun* [origin unknown] (1901)
: a packsack to be swung on either side of a packsaddle

ky·a·nite \ˈkī-ə-ˌnīt\ *noun* [German *Zyanit,* from Greek *kyanos* dark blue enamel, lapis lazuli] (1794)
: an aluminum silicate Al_2SiO_5 that occurs usually in blue thin-bladed triclinic crystals and crystalline aggregates and is sometimes used as a gemstone

kyat \ˈchät\ *noun* [Burmese *cʸat*] (1952)
— see MONEY table

ky·bosh *chiefly British variant of* KIBOSH

ky·mo·gram \ˈkī-mə-ˌgram\ *noun* [International Scientific Vocabulary] (1923)
: a record made by a kymograph

ky·mo·graph \-ˌgraf\ *noun* [Greek *kyma* wave + International Scientific Vocabulary *-graph* — more at CYME] (1872)
: a device which graphically records motion or pressure (as of blood)
— **ky·mo·graph·ic** \ˌkī-mə-ˈgra-fik\ *adjective*
— **ky·mog·ra·phy** \kī-ˈmä-grə-fē\ *noun*

Kym·ric *variant of* CYMRIC

ky·pho·sis \kī-ˈfō-səs\ *noun* [New Latin, from Greek *kyphōsis,* from *kyphos* humpbacked] (1847)
: abnormal backward curvature of the spine
— **ky·phot·ic** \-ˈfä-tik\ *adjective*

ky·rie \ˈkir-ē-ˌā\ *noun, often capitalized* [Middle English, from Medieval Latin, from Late Latin *kyrie eleison,* transliteration of Greek *kyrie eleēson* Lord, have mercy] (14th century)
: a short liturgical prayer that begins with or consists of the words "Lord, have mercy"

ky·rie elei·son \ˈkir-ē-ˌā-ə-ˈlā-(ə-)ˌsän, -(ə-)sən *also* ˈkir-ē-ə-ˈlā-\ *noun, often K&E capitalized* (13th century)
: KYRIE

kyte \ˈkīt\ *noun* [probably from Low German *küt* bowel] (circa 1540)
chiefly Scottish **:** BELLY 1, 2

kythe *variant of* KITHE

L is the twelfth letter of the English alphabet. It came to Latin from Etruscan, and there from Greek, where it was called lambda; *Greek, in turn, took its letter from Phoenician* lāmadh. *Throughout these migrations* L *has not varied greatly either in form or in value. In English it normally represents a lateral continuant, a sound produced by the free flow of air through the sides of the mouth when the center of the tongue touches the palate.* L *is also described as a liquid consonant, that is, one produced without air friction and capable of being prolonged like a vowel. In unaccented syllables the sound often forms a syllable by itself without a vowel sound, as in* medal, little. *It is silent in some words, especially after* a *or* o *and before* k, m, *or* f, *as in* talk, yolk, calm, *and* half. *The small* l *came into being when scribes replaced the horizontal stroke at the foot of the capital letter with a curved hook.*

l \'el\ *noun, plural* **l's** *or* **ls** \'elz\ *often capitalized, often attributive* (before 12th century)
1 a : the 12th letter of the English alphabet **b :** a graphic representation of this letter **c :** a speech counterpart of orthographic *l*
2 : fifty — see NUMBER table
3 : a graphic device for reproducing the letter *l*
4 : one designated *l* especially as the 12th in order or class
5 : something shaped like the letter L; *specifically* **:** [2]ELL 1
6 : [2]EL

l- *prefix* [International Scientific Vocabulary, from *lev-*]
1 \ˌlē-(ˌ)vō, ˌel, 'el\ **:** levorotatory ⟨*l*-tartaric acid⟩
2 \ˌel, 'el\ **:** having a similar configuration at a selected carbon atom to the configuration of levorotatory glyceraldehyde — usually printed as a small capital ⟨L-fructose⟩

[1]**la** \'lȯ, 'lä\ *interjection* [Middle English (northern dialect), from Old English *lā*] (before 12th century)
chiefly dialect — used for emphasis or expressing surprise

[2]**la** \'lä\ *noun* [Middle English, from Medieval Latin, from the syllable sung to this note in a medieval hymn to Saint John the Baptist] (14th century)
: the 6th tone of the diatonic scale in solmization

laa•ger \'lä-gər\ *noun* [obsolete Afrikaans *lager* (now *laer*), from German *Lager,* from Old High German *legar* couch — more at LAIR] (1850)
1 *South African* **:** CAMP; *especially* **:** an encampment protected by a circle of wagons or armored vehicles
2 : a defensive position, policy, or attitude
— laager *intransitive verb*

laa•ri \'lä-(ˌ)rē\ *noun, plural* **laari** [probably from Divehi (Indo-Aryan language of the Maldive Islands), from Persian *lārī* piece of silver wire used as currency] (1983)
— see *rufiyaa* at MONEY table

lab \'lab\ *noun* (circa 1895)
: LABORATORY

Lab \'lab\ *noun* (1957)
: LABRADOR RETRIEVER

la•ba•no•ta•tion \ˌlā-bə-nō-'tā-shən, ˌla-; lə-ˌbä-(ˌ)nō-\ *noun* [Rudolf *Laban* (died 1958) Hungarian dance theorist + English *notation*] (1952)
: a method of recording bodily movement (as in dance) on a staff by means of symbols (as of direction) that can be aligned with musical accompaniment

lab•a•rum \'la-bə-rəm\ *noun* [Late Latin] (1606)
: an imperial standard of the later Roman emperors resembling the vexillum; *especially* **:** the standard bearing the Chi-Rho adopted by Constantine after his conversion to Christianity

lab•da•num \'lab-də-nəm\ *noun* [Medieval Latin *lapdanum*] (14th century)
: a soft dark fragrant bitter oleoresin derived from various rockroses (genus *Cistus*) and used in making perfumes

[1]**la•bel** \'lā-bəl\ *noun* [Middle English, from Middle French] (14th century)
1 *archaic* **:** BAND, FILLET; *specifically* **:** one attached to a document to hold an appended seal
2 : a heraldic charge that consists of a narrow horizontal band with usually three pendants
3 a : a slip (as of paper or cloth) inscribed and affixed to something for identification or description **b :** written or printed matter accompanying an article to furnish identification or other information **c :** a descriptive or identifying word or phrase: as (1) **:** EPITHET (2) **:** a word or phrase used with a dictionary definition to provide additional information **d :** a usually radioactive isotope used in labeling
4 : an adhesive stamp (as for postage or revenue)
5 a (1) **:** a brand of commercial recordings issued under a usually trademarked name (2) **:** a recording so issued (3) **:** a company issuing such recordings **b :** the brand name of a retail store selling clothing, a clothing manufacturer, or a fashion designer

[2]**label** *transitive verb* **la•beled** *or* **la•belled; la•bel•ing** *or* **la•bel•ling** \'lā-b(ə-)liŋ\ (1601)
1 a : to affix a label to **b :** to describe or designate with or as if with a label
2 a : to distinguish (an element or atom) by using an isotope distinctive in some manner (as in mass or radioactivity) **b :** to distinguish (as a compound or cell) by introducing a traceable constituent (as a dye or labeled atom)
— la•bel•able \'lā-bə-lə-bəl\ *adjective*
— la•bel•er \'lā-b(ə-)lər\ *noun*

la•bel•lum \lə-'be-ləm\ *noun, plural* **la•bel•la** \-lə\ [New Latin, from Latin, diminutive of *labrum* lip — more at LIP] (1830)
1 : the median and usually most morphologically distinct member of the corolla of an orchid
2 : a terminal part of the labium or labrum of various insects

labia *plural of* LABIUM

[1]**la•bi•al** \'lā-bē-əl\ *adjective* [Medieval Latin *labialis,* from Latin *labium* lip] (1594)
1 : uttered with the participation of one or both lips ⟨the *labial* sounds \f\, \p\, and \ü\⟩
2 : of, relating to, or situated near the lips or labia
— la•bi•al•ly \-ə-lē\ *adverb*

[2]**labial** *noun* (1668)
: a labial consonant

la•bi•al•ize \'lā-bē-ə-ˌlīz\ *transitive verb* **-ized; -iz•ing** (1867)
: to make labial **:** ROUND 1b(2)
— la•bi•al•i•za•tion \ˌlā-bē-ə-lə-'zā-shən, -byə-lə-\ *noun*

la•bia ma•jo•ra \'lā-bē-ə-mə-'jōr-ə, -'jȯr-\ *noun plural* [New Latin, literally, larger lips] (1838)
: the outer fatty folds of the vulva bounding the vestibule

labia mi•no•ra \-mə-'nōr-ə, -'nȯr-\ *noun plural* [New Latin, literally, smaller lips] (1838)
: the inner highly vascular largely connective-tissue folds of the vulva bounding the vestibule

[1]**la•bi•ate** \'lā-bē-ət, -bē-ˌāt\ *adjective* [New Latin *labiatus,* from Latin *labium*] (1706)
1 : having the limb of a tubular corolla or calyx divided into two unequal parts projecting one over the other like lips ⟨mints and the snapdragon are *labiate*⟩
2 : of or relating to the mint family

[2]**labiate** *noun* (1845)
: a plant of the mint family

la•bile \'lā-ˌbīl, -bəl\ *adjective* [French, from Middle French, prone to err, from Late Latin *labilis,* from Latin *labi* to slip — more at SLEEP] (1603)
1 : readily or continually undergoing chemical, physical, or biological change or breakdown **:** UNSTABLE ⟨a *labile* mineral⟩
2 : readily open to change
— la•bil•i•ty \lā-'bi-lə-tē\ *noun*

labio- *combining form* [Latin *labium*]
: labial and ⟨*labio*dental⟩

la•bio•den•tal \ˌlā-bē-ō-'den-tᵊl\ *adjective* (1669)
: uttered with the participation of the lip and teeth ⟨the *labiodental* sounds \f\ and \v\⟩
— labiodental *noun*

la•bio•ve•lar \-'vē-lər\ *adjective* [International Scientific Vocabulary] (1894)
: both labial and velar ⟨the *labiovelar* sound \w\⟩
— labiovelar *noun*

la•bi•um \'lā-bē-əm\ *noun, plural* **la•bia** \-ə\ [New Latin, from Latin, lip — more at LIP] (1634)
1 : any of the folds at the margin of the vulva — compare LABIA MAJORA, LABIA MINORA
2 : the lower lip of a labiate corolla
3 a : a lower mouthpart of an insect that is formed by the second pair of maxillae united in the middle line **b :** a liplike part of various invertebrates

[1]**la•bor** \'lā-bər\ *noun* [Middle English, from Middle French, from Latin *labor;* perhaps akin to Latin *labare* to totter, *labi* to slip — more at SLEEP] (14th century)
1 a : expenditure of physical or mental effort especially when difficult or compulsory **b** (1) **:** human activity that provides the goods or services in an economy (2) **:** the services performed by workers for wages as distinguished from those rendered by entrepreneurs for profits **c :** the physical activities involved in giving birth; *also* **:** the period of such labor
2 : an act or process requiring labor **:** TASK
3 : a product of labor
4 a : an economic group comprising those who do manual labor or work for wages **b** (1) **:** workers employed in an establishment (2) **:** workers available for employment **c :** the organizations or officials representing groups of workers
5 *usually* **Labour :** the Labour party of the United Kingdom or of another nation of the Commonwealth
synonym see WORK

\ə\ **abut** \ᵊ\ **kitten** \ər\ **further** \a\ **ash** \ā\ **ace** \ä\ **mop, mar** \au̇\ **out** \ch\ **chin** \e\ **bet** \ē\ **easy** \g\ **go** \i\ **hit** \ī\ **ice** \j\ **job** \ŋ\ **sing** \ō\ **go** \ȯ\ **law** \ȯi\ **boy** \th\ **thin** \t͟h\ **the** \ü\ **loot** \u̇\ **foot** \y\ **yet** \zh\ **vision** *see also* Guide to Pronunciation

²labor *verb* **la·bored; la·bor·ing** \-b(ə-)riŋ\ (14th century)
intransitive verb
1 : to exert one's powers of body or mind especially with painful or strenuous effort **:** WORK
2 : to move with great effort ⟨the truck *labored* up the hill⟩
3 : to be in the labor of giving birth
4 : to suffer from some disadvantage or distress ⟨*labor* under a delusion⟩
5 *of a ship* **:** to pitch or roll heavily
transitive verb
1 *archaic* **a :** to spend labor on or produce by labor **b :** to strive to effect or achieve
2 : to treat or work out in often laborious detail ⟨*labor* the obvious⟩
3 : DISTRESS, BURDEN
4 : to cause to labor

³labor *adjective* (1640)
1 : of or relating to labor
2 *capitalized* **:** of, relating to, or constituting a political party held to represent the interests of workers or made up largely of organized labor groups

lab·o·ra·to·ry \ˈla-b(ə-)rə-ˌtōr-ē, -ˌtȯr- *sometimes* ˈla-bə(r)-, *or* lə-ˈbȯr-ə-ˌ, *British usually* lə-ˈbär-ə-t(ə-)rē\ *noun, plural* **-ries** *often attributive* [Medieval Latin *laboratorium,* from Latin *laborare* to labor, from *labor*] (1605)
1 a : a place equipped for experimental study in a science or for testing and analysis; *broadly* **:** a place providing opportunity for experimentation, observation, or practice in a field of study **b :** a place like a laboratory for testing, experimentation, or practice ⟨the *laboratory* of the mind⟩
2 : an academic period set aside for laboratory work

labor camp *noun* (1900)
1 : a penal colony where forced labor is performed
2 : a camp for migratory laborers

Labor Day *noun* (1886)
: a day set aside for special recognition of working people: as **a :** the first Monday in September observed in the U.S. and Canada as a legal holiday **b :** May 1 in many countries

la·bored \ˈlā-bərd\ *adjective* (1608)
: produced or performed with labor ⟨*labored* breathing⟩; *also* **:** lacking ease of expression ⟨a *labored* speech⟩

la·bor·er \-bər-ər\ *noun* (14th century)
: one that labors; *specifically* **:** a person who does unskilled physical work for wages

labor force *noun* (1911)
: WORK FORCE

la·bor–in·ten·sive \ˈlā-bər-in-ˌten(t)-siv\ *adjective* (1953)
: having high labor costs per unit of output; *especially* **:** requiring greater expenditure on labor than in capital

la·bo·ri·ous \lə-ˈbōr-ē-əs, -ˈbȯr-\ *adjective* (14th century)
1 : devoted to labor **:** INDUSTRIOUS
2 : involving or characterized by hard or toilsome effort **:** LABORED
— la·bo·ri·ous·ly *adverb*
— la·bo·ri·ous·ness *noun*

la·bor·ite \ˈlā-bə-ˌrīt\ *noun* (1889)
1 : a member of a group favoring the interests of labor
2 *capitalized* **a :** a member of a political party devoted chiefly to the interests of labor **b** *usually* **La·bour·ite :** a member of the British Labour party

la·bor·sav·ing \ˈlā-bər-ˌsā-viŋ\ *adjective* (circa 1779)
: adapted to replace or decrease human and especially manual labor

labor union *noun* (1866)
: an organization of workers formed for the purpose of advancing its members' interests in respect to wages, benefits, and working conditions

la·bour *chiefly British variant of* LABOR

lab·ra·dor·ite \ˈla-brə-ˌdȯr-ˌīt\ *noun* [*Labrador* Peninsula, Canada] (1814)
: a triclinic feldspar showing a play of several colors

Lab·ra·dor retriever \ˈla-brə-ˌdȯr-\ *noun* [*Labrador,* Newfoundland] (1910)
: any of a breed of compact, strongly built retrievers largely developed in England from stock originating in Newfoundland and having a short dense black, yellow, or chocolate coat — called also *Lab, Labrador*

Labrador tea *noun* (1767)
: a low-growing ericaceous evergreen shrub (*Ledum groenlandicum*) chiefly of northern North America with white or creamy bell-shaped flowers and leaves sometimes used in making tea; *also* **:** a related Rocky Mountain shrub (*L. glandulosum*)

la·bret \ˈlā-brət\ *noun* [Latin *labrum*] (1857)
: an ornament worn in a perforation of the lip

la·brum \ˈlā-brəm\ *noun* [New Latin, from Latin, lip, edge — more at LIP] (1826)
: an upper or anterior mouthpart of an arthropod consisting of a single median piece in front of or above the mandibles

la·bur·num \lə-ˈbər-nəm\ *noun* [New Latin, from Latin] (1567)
: any of a small genus (*Laburnum* and especially *L. anagyroides*) of poisonous leguminous shrubs and trees of Eurasia with pendulous racemes of bright yellow flowers

lab·y·rinth \ˈla-bə-ˌrin(t)th, -rən(t)th\ *noun* [Middle English *laborintus,* from Latin *labyrinthus,* from Greek *labyrinthos*] (14th century)
1 a : a place constructed of or full of intricate passageways and blind alleys **b :** a maze (as in a garden) formed by paths separated by high hedges
2 : something extremely complex or tortuous in structure, arrangement, or character **:** INTRICACY, PERPLEXITY ⟨a *labyrinth* of swamps and channels⟩ ⟨guided them through the *labyrinths* of city life —Paul Blanshard⟩
3 : a tortuous anatomical structure; *especially* **:** the internal ear or its bony or membranous part

lab·y·rin·thi·an \ˌla-bə-ˈrin(t)-thē-ən\ *adjective* (1588)
: LABYRINTHINE

lab·y·rin·thine \-ˈrin(t)-thən; -ˈrin-ˌthīn, -ˌthēn\ *adjective* (1632)
1 : of, relating to, or resembling a labyrinth **:** INTRICATE, INVOLVED
2 : of, relating to, affecting, or originating in the internal ear ⟨*labyrinthine* lesions⟩

lab·y·rin·tho·dont \-ˈrin(t)-thə-ˌdänt\ *noun* [New Latin *Labyrinthodontia,* from Greek *labyrinthos* + *odont-, odous* tooth — more at TOOTH] (circa 1852)
: any of a superorder (Labyrinthodontia) of extinct amphibians of the Late Paleozoic and Early Mesozoic typically having bodies resembling salamanders or crocodiles and considered to be the earliest tetrapod vertebrates
— labyrinthodont *adjective*

lac \ˈlak\ *noun* [Persian *lak* & Hindi *lākh,* from Sanskrit *lākṣā*] (1598)
: a resinous substance secreted by a scale insect (*Laccifer lacca*) and used chiefly in the form of shellac

lac·co·lith \ˈla-kə-ˌlith\ *noun* [Greek *lakkos* pond, reservoir + English *-lith* — more at LAKE] (1879)
: a mass of igneous rock that is intruded between sedimentary beds and produces a domical bulging of the overlying strata
— lac·co·lith·ic \ˌla-kə-ˈli-thik\ *adjective*

¹lace \ˈlās\ *verb* **laced; lac·ing** [Middle English, from Old French *lacier,* from Latin *laqueare* to ensnare, from *laqueus*] (13th century)
transitive verb
1 : to draw together the edges of by or as if by a lace passed through eyelets
2 : to draw or pass (as a lace) through something (as eyelets)
3 : to confine or compress by tightening laces of a garment
4 a : to adorn with or as if with lace **b :** to mark with streaks of color
5 : BEAT, LASH
6 a : to add a dash of liquor to **b :** to add to something to impart pungency, savor, or zest ⟨a sauce *laced* with garlic⟩ ⟨conversation *laced* with sarcasm⟩ **c :** to adulterate with a substance ⟨*laced* a guard's coffee with a sedative⟩
intransitive verb
: to admit of being tied or fastened with a lace
— lac·er *noun*

²lace *noun* [Middle English, from Middle French *laz,* from Latin *laqueus* snare] (14th century)
1 : a cord or string used for drawing together two edges (as of a garment or a shoe)
2 : an ornamental braid for trimming coats or uniforms
3 : an openwork usually figured fabric made of thread or yarn and used for trimmings, household coverings, and entire garments
— laced \ˈlāst\ *adjective*
— lace·less \ˈlās-ləs\ *adjective*
— lace·like \ˈlās-ˌlīk\ *adjective*

lace–curtain *adjective* (1934)
: copying middle-class attributes **:** aspiring to middle-class standing

¹lac·er·ate \ˈla-sə-ˌrāt\ *transitive verb* **-at·ed; -at·ing** [Middle English, from Latin *laceratus,* past participle of *lacerare* to tear; akin to Greek *lakis* tear] (15th century)
1 : to tear or rend roughly
2 : to cause sharp mental or emotional pain to **:** DISTRESS
— lac·er·a·tive \-ˌrā-tiv\ *adjective*

²lac·er·ate \-rət, -ˌrāt\ *or* **lac·er·at·ed** \-ˌrā-təd\ *adjective* (1542)
1 a : torn jaggedly **:** MANGLED **b :** extremely harrowed or distracted
2 : having the edges deeply and irregularly cut ⟨a *lacerate* petal⟩

lac·er·a·tion \ˌla-sə-ˈrā-shən\ *noun* (1597)
1 : the act of lacerating
2 : a torn and ragged wound

lace·wing \ˈlās-ˌwiŋ\ *noun* (1854)
: any of various neuropterous insects (as genera *Chrysopa* and *Hemerobius*) having delicate lacelike wing venation, usually long antennae, and often brilliant eyes — called also *lacewing fly*

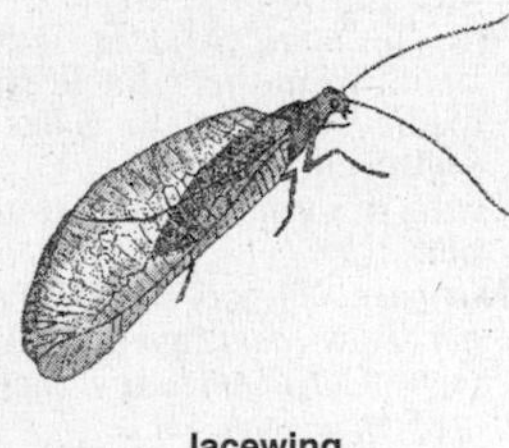
lacewing

lace·work \ˈlās-ˌwərk\ *noun* (1849)
: objects or patterns consisting of or resembling lace

lac·ey *variant of* LACY

la·ches \ˈla-chəz, ˈlā-\ *noun, plural* **laches** [Middle English *lachesse,* from Middle French *laschesse,* from Old French *lasche* lax, ultimately from Latin *laxare* to loosen — more at LEASE] (14th century)
: negligence in the observance of duty or opportunity; *specifically* **:** undue delay in asserting a legal right or privilege

lach·ry·mal *or* **lac·ri·mal** \ˈla-krə-məl\ *adjective* [Middle French or Medieval Latin; Middle French *lacrymal,* from Medieval Latin *lacrimalis,* from Latin *lacrima* tear, from Old Latin *dacrima,* probably from Greek *dakry* — more at TEAR] (15th century)
1 *usually* *lacrimal* **:** of, relating to, or being glands that produce tears
2 : of, relating to, or marked by tears

lach·ry·mose \-ˌmōs\ *adjective* [Latin *lacrimosus,* from *lacrima*] (circa 1727)
1 : given to tears or weeping **:** TEARFUL
2 : tending to cause tears **:** MOURNFUL

— **lach·ry·mose·ly** *adverb*
— **lach·ry·mos·i·ty** \ˌla-krə-ˈmä-sə-tē\ *noun*
lac·ing \ˈlā-siŋ\ *noun* (14th century)
1 : the action of one that laces
2 : something that laces **:** LACE
3 : a contrasting marginal band of color (as on a feather)
4 a : a dash of liquor in a food or beverage **b :** a trace or sprinkling that adds spice or flavor
5 : a decisive defeat
la·cin·i·ate \lə-ˈsi-nē-ət, -ˌāt\ *adjective* [Latin *lacinia* flap; akin to Latin *lacerare*] (circa 1760)
: bordered with a fringe; *especially* **:** cut into deep irregular usually pointed lobes ⟨*laciniate* petals⟩
— **la·cin·i·a·tion** \-ˌsi-nē-ˈā-shən\ *noun*
[1]**lack** \ˈlak\ (13th century)
intransitive verb
1 : to be deficient or missing ⟨time is *lacking* for a full explanation⟩
2 : to be short or have need of something ⟨he will not *lack* for advisers⟩
transitive verb
: to stand in need of **:** suffer from the absence or deficiency of ⟨*lack* the necessities of life⟩
[2]**lack** *noun* [Middle English *lak;* akin to Middle Dutch *lak* lack, Old Norse *lakr* defective] (14th century)
1 : the fact or state of being wanting or deficient
2 : something that is lacking or is needed
lack·a·dai·si·cal \ˌla-kə-ˈdā-zi-kəl\ *adjective* [irregular from *lackaday* + *-ical*] (1768)
: lacking life, spirit, or zest **:** LANGUID
— **lack·a·dai·si·cal·ly** \-k(ə-)lē\ *adverb*
lack·a·day \ˈla-kə-ˌdā\ *interjection* [by alteration & shortening from *alack the day*] (1695) *archaic* — used to express regret or deprecation
[1]**lack·ey** \ˈla-kē\ *noun, plural* **lackeys** [Middle French *laquais*] (1523)
1 a : FOOTMAN 2, SERVANT **b :** someone who does menial tasks or runs errands for another
2 : a servile follower **:** TOADY
[2]**lackey** *verb* **lack·eyed; lack·ey·ing** (1568)
intransitive verb
obsolete **:** to act as a lackey **:** TOADY
transitive verb
: to wait upon or serve obsequiously
lack·lus·ter \ˈlak-ˌləs-tər\ *adjective* (1600)
: lacking in sheen, brilliance, or vitality **:** DULL, MEDIOCRE
— **lackluster** *noun*
la·con·ic \lə-ˈkä-nik\ *adjective* [Latin *laconicus* Spartan, from Greek *lakōnikos;* from the Spartan reputation for terseness of speech] (1589)
: using or involving the use of a minimum of words **:** concise to the point of seeming rude or mysterious
synonym see CONCISE
— **la·con·i·cal·ly** \-ni-k(ə-)lē\ *adverb*
lac·o·nism \ˈla-kə-ˌni-zəm\ *noun* (1570)
1 : brevity or terseness of expression or style
2 : a laconic expression
[1]**lac·quer** \ˈla-kər\ *noun* [Portuguese *lacré* sealing wax, from *laca* lac, from Arabic *lakk,* from Persian *lak* — more at LAC] (1592)
1 a : a spirit varnish (as shellac) **b :** any of various durable natural varnishes; *especially* **:** a varnish obtained from an Asian sumac (*Rhus verniciflua*) — called also *Japanese lacquer*
2 : any of various clear or colored synthetic organic coatings that typically dry to form a film by evaporation of the solvent; *especially* **:** a solution of a cellulose derivative (as nitrocellulose)
[2]**lacquer** *transitive verb* **lac·quered; lac·quer·ing** \-k(ə-)riŋ\ (1688)
1 : to coat with or as if with lacquer
2 : to give a smooth finish or appearance to **:** make glossy
— **lac·quer·er** \-kər-ər\ *noun*
lac·quer·ware \-ˌwar, -ˌwer\ *noun* (1697)
: a decorative article usually made of wood and coated with lacquer; *also* **:** such articles or ware collectively
lac·quer·work \-ˌwərk\ *noun* (circa 1901)
: LACQUERWARE
lac·ri·ma·tion \ˌla-krə-ˈmā-shən\ *noun* (1572)
: the secretion of tears especially when abnormal or excessive
lac·ri·ma·tor *or* **lach·ry·ma·tor** \ˈla-krə-ˌmā-tər\ *noun* [Latin *lacrimare* to weep, from *lacrima* tear — more at LACHRYMAL] (1918)
: a tear-producing substance (as tear gas)
la·crosse \lə-ˈkrȯs\ *noun* [Canadian French *la crosse,* literally, the crosier] (1718)
: a goal game in which players use a long-handled stick that has a triangular head with a mesh pouch for catching, carrying, and throwing the ball
lact- *or* **lacti-** *or* **lacto-** *combining form* [French & Latin; French, from Latin, from *lact-, lac* — more at GALAXY]
1 : milk ⟨*lacto*globulin⟩
2 a : lactic acid ⟨*lact*ate⟩ **b :** lactose ⟨*lact*ase⟩
lact·al·bu·min \ˌlak-ˌtal-ˈbyü-mən\ *noun* [International Scientific Vocabulary] (circa 1857)
: an albumin that is obtained from whey and is similar to serum albumin
lac·tase \ˈlak-ˌtās, -ˌtāz\ *noun* [International Scientific Vocabulary] (1891)
: an enzyme that hydrolyzes beta-galactosides (as lactose) and occurs especially in the intestines of young mammals and in yeasts
[1]**lac·tate** \ˈlak-ˌtāt\ *noun* (circa 1794)
: a salt or ester of lactic acid
[2]**lactate** *intransitive verb* **lac·tat·ed; lac·tat·ing** [Latin *lactatus,* past participle of *lactare,* from *lact-, lac*] (circa 1889)
: to secrete milk
— **lac·ta·tion** \lak-ˈtā-shən\ *noun*
— **lac·ta·tion·al** \-shnəl, -shə-nᵊl\ *adjective*
[1]**lac·te·al** \ˈlak-tē-əl\ *adjective* [Latin *lacteus* of milk, from *lact-, lac*] (circa 1658)
1 : relating to, consisting of, producing, or resembling milk
2 a : conveying or containing a milky fluid **b :** of or relating to the lacteals
[2]**lacteal** *noun* (1680)
: any of the lymphatic vessels arising from the villi of the small intestine and conveying chyle to the thoracic duct
lac·tic \ˈlak-tik\ *adjective* (1790)
1 a : of or relating to milk **b :** obtained from sour milk or whey
2 : involving the production of lactic acid
lactic acid *noun* (1790)
: a hygroscopic organic acid $C_3H_6O_3$ present normally in tissue, produced in carbohydrate matter usually by bacterial fermentation, and used especially in food and medicine and in industry
lac·tif·er·ous \lak-ˈti-f(ə-)rəs\ *adjective* [French or Late Latin; French *lactifère,* from Late Latin *lactifer,* from Latin *lact-, lac* + *-fer*] (circa 1674)
1 : yielding a milky juice ⟨*lactiferous* plants⟩
2 : secreting or conveying milk
lac·to·ba·cil·lus \ˌlak-tō-bə-ˈsi-ləs\ *noun* [New Latin] (1924)
: any of a genus (*Lactobacillus*) of lactic-acid-forming bacteria
lac·to·gen·ic \ˌlak-tə-ˈje-nik\ *adjective* (1933)
: inducing lactation ⟨*lactogenic* hormones⟩
lac·to·glob·u·lin \-ˈglä-byə-lən\ *noun* (1885)
: a crystalline protein fraction that is obtained from the whey of milk
lac·tone \ˈlak-ˌtōn\ *noun* [International Scientific Vocabulary] (1880)
: any of various cyclic esters formed from hydroxy acids
— **lac·ton·ic** \lak-ˈtä-nik\ *adjective*
lac·tose \ˈlak-ˌtōs, -ˌtōz\ *noun* [International Scientific Vocabulary] (1858)
: a disaccharide sugar $C_{12}H_{22}O_{11}$ that is present in milk and yields glucose and galactose upon hydrolysis and yields especially lactic acid upon fermentation
la·cu·na \lə-ˈkü-nə, -ˈkyü-\ *noun, plural* **la·cu·nae** \-ˈkyü-(ˌ)nē, -ˈkü-ˌnī\ *or* **la·cu·nas** \-ˈkü-nəz, -ˈkyü-\ [Latin, pool, pit, gap — more at LAGOON] (1652)
1 : a blank space or a missing part **:** GAP
2 : a small cavity, pit, or discontinuity in an anatomical structure
— **la·cu·nar** \-ˈkü-nər, -ˈkyü-\ *also* **la·cu·nate** \lə-ˈkü-nət, -ˈkyü-, -ˌnāt; ˈla-kyə-ˌnāt\ *adjective*
la·cus·trine \lə-ˈkəs-trən\ *adjective* [French or Italian *lacustre,* from Latin *lacus* lake] (1830)
: of, relating to, formed in, living in, or growing in lakes ⟨*lacustrine* deposits⟩ ⟨*lacustrine* faunas⟩
lacy \ˈlā-sē\ *adjective* **lac·i·er; -est** (1804)
: resembling or consisting of lace
lad \ˈlad\ *noun* [Middle English *ladde*] (14th century)
1 : a male person of any age between early boyhood and maturity **:** BOY, YOUTH
2 : FELLOW, CHAP
lad·a·num \ˈla-dᵊn-əm, ˈlad-nəm\ *variant of* LABDANUM
lad·der \ˈla-dər\ *noun, often attributive* [Middle English, from Old English *hlǣder;* akin to Old High German *leitara* ladder, Old English *hlinian* to lean — more at LEAN] (before 12th century)
1 : a structure for climbing up or down that consists essentially of two long sidepieces joined at intervals by crosspieces on which one may step
2 : something that resembles or suggests a ladder in form or use; *especially* **:** RUN 11a
3 : a series of usually ascending steps or stages **:** SCALE ⟨climbing up the corporate *ladder*⟩
— **lad·der·like** \-ˌlīk\ *adjective*
lad·der–back \-ˌbak\ *adjective* (1908)
of furniture **:** having a back consisting of two upright posts connected by horizontal slats
ladder truck *noun* (1889)
: HOOK AND LADDER TRUCK
lad·die \ˈla-dē\ *noun* (1546)
: a young lad
lade \ˈlād\ *verb* **lad·ed; laded** *or* **lad·en** \ˈlā-dᵊn\; **lad·ing** [Middle English, from Old English *hladan;* akin to Old High German *hladan* to load, Old Church Slavonic *klasti* to place] (before 12th century)
transitive verb
1 a : to put a load or burden on or in **:** LOAD **b :** to put or place as a load especially for shipment **:** SHIP **c :** to load heavily or oppressively
2 : DIP, LADLE
intransitive verb
1 : to take on cargo **:** LOAD
2 : to take up or convey a liquid by dipping
[1]**lad·en** \ˈlā-dᵊn\ *transitive verb* **lad·ened; lad·en·ing** \ˈlād-niŋ, ˈlā-dᵊn-iŋ\ (1514)
: LADE
[2]**laden** *adjective* (before 12th century)
: carrying a load or burden
la–di–da \ˌlä-dē-ˈdä\ *also* **la–de–da** *adjective* [perhaps alteration of *lardy-dardy* foppish] (1895)
: affectedly refined in manners or tastes **:** PRETENTIOUS, ELEGANT
la·dies \ˈlā-dēz\ *noun plural but singular or plural in construction* (1918)
chiefly British **:** LADIES' ROOM
ladies' man *also* **lady's man** *noun* (1784)
: a man who shows a marked fondness for the company of women or is especially attentive to women
ladies' room *noun* (1870)
: a room equipped with lavatories and toilets for the use of women

ladies' tresses *noun plural but singular or plural in construction* (1548)
: any of a widely distributed genus (*Spiranthes*) of terrestrial orchids with slender often twisted spikes of white irregular flowers

La·din \lə-ˈdēn\ *noun* [Rhaeto-Romance, from Latin *Latinum* Latin] (1877)
1 a : a Rhaeto-Romance dialect of Alto Adige in northern Italy **b :** the Rhaeto-Romance dialects of the Engadine Valley in Switzerland
2 : one speaking Ladin as a mother tongue

lad·ing \ˈlā-diŋ\ *noun* (1500)
1 a : LOADING 1 **b :** an act of bailing, dipping, or ladling
2 : CARGO, FREIGHT

la·di·no \lə-ˈdē-(ˌ)nō\ *noun, plural* **-nos** [Spanish, literally, Latin, from Latin *latinus*] (1877)
1 *often capitalized* [American Spanish] **:** a westernized Spanish-speaking Latin American; *especially* **:** MESTIZO
2 *capitalized* [Judeo-Spanish, from Old Spanish] **:** JUDEO-SPANISH

la·di·no clover \lə-ˈdī-(ˌ)nō-, -nə-\ *noun* [perhaps irregular from *Lodi,* Italy + Italian *-ino,* adjective suffix] (1924)
: a large nutritious rapidly growing clover that is a variety of white clover and is widely planted especially for forage — called also *ladino*

[1]la·dle \ˈlā-dᵊl\ *noun* [Middle English *ladel,* from Old English *hlædel,* from *hladan*] (before 12th century)
1 : a deep-bowled long-handled spoon used especially for dipping up and conveying liquids
2 : something resembling a ladle in form or function

[2]ladle *transitive verb* **la·dled; la·dling** \ˈlād-liŋ, ˈlā-dᵊl-iŋ\ (1525)
: to take up and convey in or as if in a ladle

la dol·ce vi·ta \(ˌ)lä-ˈdōl-(ˌ)chā-ˈvē-(ˌ)tä\ *variant of* DOLCE VITA

la·dy \ˈlā-dē\ *noun, plural* **ladies** *often attributive* [Middle English, from Old English *hlǣfdige,* from *hlāf* bread + *-dige* (akin to *dǣge* kneader of bread) — more at LOAF, DAIRY] (before 12th century)
1 a : a woman having proprietary rights or authority especially as a feudal superior **b :** a woman receiving the homage or devotion of a knight or lover
2 *capitalized* **:** VIRGIN MARY — usually used with *Our*
3 a : a woman of superior social position **b :** a woman of refinement and gentle manners **c :** WOMAN, FEMALE — often used in a courteous reference ⟨show the *lady* to a seat⟩ or usually in the plural in address ⟨*ladies* and gentlemen⟩
4 a : WIFE **b :** GIRLFRIEND, MISTRESS
5 a : any of various titled women in Great Britain — used as the customary title of (1) a marchioness, countess, viscountess, or baroness or (2) the wife of a knight, baronet, member of the peerage, or one having the courtesy title of *lord* and used as a courtesy title for the daughter of a duke, marquess, or earl **b :** a female member of an order of knighthood — compare DAME ◆

la·dy·bug \ˈlā-dē-ˌbəg\ *noun* [Our *Lady,* the Virgin Mary] (1699)
: any of numerous small nearly hemispherical often brightly colored beetles (family Coccinellidae) of temperate and tropical regions that usually feed both as larvae and adults on other insects — called also *lady beetle, ladybird, ladybird beetle*

ladybug

lady chapel *noun, often L&C capitalized* (15th century)
: a chapel dedicated to the Virgin Mary

Lady Day *noun* (13th century)
: ANNUNCIATION 1

la·dy·fin·ger \ˈlā-dē-ˌfiŋ-gər\ *noun* (1820)
: a small finger-shaped sponge cake

la·dy·fish \-ˌfish\ *noun* (1712)
1 : BONEFISH 1
2 : a large silvery food and sport fish (*Elops saurus*) that resembles a herring but is related to the tarpon

la·dy–in–wait·ing \ˈlā-dē-in-ˈwā-tiŋ\ *noun, plural* **ladies–in–waiting** (1862)
: a lady of a queen's or a princess's household appointed to wait on her

la·dy–kill·er \ˈlā-dē-ˌki-lər\ *noun* (circa 1810)
: a man who is extremely attractive to women

la·dy·kin \ˈlā-dē-kən\ *noun* (1853)
: a little lady

la·dy·like \-ˌlīk\ *adjective* (1586)
1 : becoming or suitable to a lady
2 : resembling a lady in appearance or manners **:** WELL-BRED
3 a : feeling or showing too much concern about elegance or propriety ⟨*ladylike* embarrassment at not being the wife of a real doctor —Lewis Vogler⟩ **b :** lacking in strength, force, or virility

la·dy·love \ˈlā-dē-ˌləv, ˌlā-dē-ˈ\ *noun* (1733)
: SWEETHEART, MISTRESS

lady of the house (1816)
: the chief female in a household

Lady of the Lake (15th century)
: VIVIAN

la·dy·ship \ˈlā-dē-ˌship\ *noun* (13th century)
: the condition of being a lady **:** rank of lady — used as a title for a woman having the rank of lady ⟨her *Ladyship* is not at home⟩ ⟨if your *Ladyship* please⟩

lady's slipper *noun* (1597)
: any of several North American temperate-zone orchids (as of the genus *Cypripedium*) having flowers whose shape suggests a slipper — called also *lady slipper*

lady's slipper

la·dy's–smock \-ˌsmäk\ *noun* (1588)
: CUCKOOFLOWER 1

lady's thumb *noun* (1837)
: a widely distributed weedy annual herb (*Polygonum persicaria*) that has large lanceolate leaves often with a blackish blotch suggesting a thumbprint

La·er·tes \lā-ˈər-tēz\ *noun* [Latin, from Greek *Laertēs*]
1 : the father of Odysseus in Greek mythology
2 : the son of Polonius and brother of Ophelia in Shakespeare's *Hamlet*

Lae·ta·re Sunday \lā-ˈtär-ē-, -ˈtar-\ *noun* [Latin *laetare,* singular imperative of *laetari* to rejoice] (circa 1870)
: the fourth Sunday in Lent

la·e·trile \ˈlā-ə-(ˌ)tril, -trəl\ *noun, often capitalized* [*lae*vorotatory (levorotatory) + ni*trile*] (1953)
: a drug derived especially from apricot pits that contains amygdalin and has been used in the treatment of cancer although of unproved effectiveness

[1]lag \ˈlag\ *noun* [probably of Scandinavian origin; akin to Norwegian dialect *lagga* to go slowly] (1514)
1 : one that lags or is last
2 a : the act or the condition of lagging **b :** comparative slowness or retardation **c** (1) **:** an amount of lagging or the time during which lagging continues (2) **:** a space of time especially between related events or phenomena **:** INTERVAL
3 : the action of lagging for opening shot (as in marbles or billiards)

[2]lag *verb* **lagged; lag·ging** (1530)
intransitive verb
1 a : to stay or fall behind **:** LINGER, LOITER **b :** to move, function, or develop with comparative slowness **c :** to become retarded in attaining maximum value
2 : to slacken or weaken gradually **:** FLAG
3 : to toss or roll a marble toward a line or a cue ball toward the head cushion to determine order of play
transitive verb
1 : to lag behind ⟨current that *lags* the voltage⟩
2 : to pitch or shoot (as a coin or marble) at a mark
synonym see DELAY
— lag·ger *noun*

[3]lag *adjective* (1552)
: LAST, HINDMOST

[4]lag *noun* [probably of Scandinavian origin; akin to Old Norse *lǫgg* rim of a barrel] (1672)
1 : a barrel stave
2 : a stave, slat, or strip (as of wood or asbestos) forming part of a covering for a cylindrical object

[5]lag *transitive verb* **lagged; lag·ging** (1870)
: to cover or provide with lags

[6]lag *transitive verb* **lagged; lag·ging** [origin unknown] (circa 1812)
1 *slang chiefly British* **:** to transport or jail for crime
2 *slang chiefly British* **:** ARREST

[7]lag *noun* (circa 1812)
1 *slang chiefly British* **a :** a person transported for crime **b :** CONVICT **c :** an ex-convict
2 *slang chiefly British* **:** a jail sentence **:** STRETCH

lag·an \ˈla-gən\ *also* **lag·end** \-gənd\ *noun* [Middle French *lagan* or Medieval Latin *laganum* debris washed up from the sea] (1641)
: goods thrown into the sea with a buoy attached so that they may be found again

Lag b'Omer \ˈläg-ˈbō-mər, ˌläg-bə-ˈō-\ *noun* [Hebrew, 33d in Omer] (circa 1904)
: a Jewish holiday falling on the 33d day of the Omer and commemorating the heroism of Bar Kokhba and Akiba ben Joseph

la·ger \ˈlä-gər\ *noun* [German *Lagerbier* beer made for storage, from *Lager* storehouse + *Bier* beer] (circa 1853)
: a light beer brewed by slow fermentation and matured under refrigeration

[1]lag·gard \ˈla-gərd\ *adjective* (1702)
: lagging or tending to lag **:** DILATORY
— lag·gard·ly *adverb or adjective*
— lag·gard·ness *noun*

◇ WORD HISTORY

lady *Hlǣfdige,* the Old English ancestor of the word *lady,* contains as an initial element *hlāf* "bread, loaf," as does *hlāford,* the ancestor of *lord.* The second element of the compound, *-dige,* is akin to *dǣge* "(female) bread maker," ultimately a derivative of *dāg* "dough"; the word *dǣge* developed in Middle English into *deye* "dairymaid" (dairy and bread-making duties apparently overlapping), which survives in Modern English packed inside the word *dairy.* We might have expected a *hlǣfdige* in Anglo-Saxon times to have been domestic help, but by a semantic shift that has never been adequately explained, this personage was not a servant but rather the female head of a medieval household and the counterpart of a lord. Although more exalted uses of *lady,* into which *hlǣfdige* had developed by later Middle English, survive in traditional collocations such as *lady of the house* and *Our Lady* (the Virgin Mary), in Modern English the word has broadened in sense to "woman of superior social position" and "woman of refinement and gentle manners," becoming the counterpart of *gentleman* rather than *lord.* See in addition LORD.

[2]laggard *noun* (1808)
: one that lags or lingers
lag·ging \ˈla-giŋ\ *noun* (1794)
: a lag or material used for making lags: as **a** : material for thermal insulation especially around a cylindrical object **b** : planking used especially for preventing cave-ins in earthwork or for supporting an arch during construction
la·gniappe \ˈlan-ˌyap, lan-ˈ\ *noun* [American French, from American Spanish *la ñapa* the lagniappe] (1849)
: a small gift given a customer by a merchant at the time of a purchase; *broadly* : something given or obtained gratuitously or by way of good measure
lago·morph \ˈla-gə-ˌmȯrf\ *noun* [ultimately from Greek *lagōs* hare + *morphē* form] (1882)
: any of an order (Lagomorpha) of gnawing herbivorous mammals having two pairs of incisors in the upper jaw one behind the other and comprising the rabbits, hares, and pikas
la·goon \lə-ˈgün\ *noun* [French & Italian; French *lagune,* from Italian *laguna,* from Latin *lacuna* pit, pool, from *lacus* lake] (1673)
1 : a shallow sound, channel, or pond near or communicating with a larger body of water
2 : a shallow artificial pool or pond (as for the processing of sewage or storage of a liquid)
— **la·goon·al** \-ˈgü-n^{ə}l\ *adjective*
La·grang·ian \lə-ˈgrän-jē-ən, -ˈgrä^{n}-zhē-\ *noun* [Joseph-Louis *Lagrange*] (1938)
: a function that describes the state of a dynamic system in terms of position coordinates and their time derivatives and that is equal to the difference between the potential energy and kinetic energy — compare HAMILTONIAN
la·har \ˈlä-ˌhär\ *noun* [Javanese] (1929)
: a mudflow composed of volcanic debris and water
lah–de–dah, lah–dee–dah, lah–di–dah *variant of* LA-DI-DA
Lahn·da \ˈlän-də\ *noun* (1901)
: an Indo-Aryan language of eastern Pakistan
la·ical \ˈlā-ə-kəl\ *or* **la·ic** \ˈlā-ik\ *adjective* [Late Latin *laicus,* from Late Greek *laïkos,* from Greek, of the people, from *laos* people] (circa 1587)
: of or relating to the laity : SECULAR
— **laic** *noun*
— **la·ical·ly** \ˈlā-ə-k(ə-)lē\ *adverb*
la·i·cism \ˈlā-ə-ˌsi-zəm\ *noun* (circa 1909)
: a political system characterized by the exclusion of ecclesiastical control and influence
la·i·cize \ˈlā-ə-ˌsīz\ *transitive verb* **la·i·cized; la·i·ciz·ing** (1870)
1 : to reduce to lay status
2 : to put under the direction of or open to the laity
— **la·i·ci·za·tion** \ˌlā-ə-sə-ˈzā-shən\ *noun*
laid *past and past participle of* LAY
laid–back \ˈlād-ˈbak, ˌlād-\ *adjective* (1969)
: having a relaxed style or character ⟨*laid-back* music⟩
— **laid–back·ness** \-nəs\ *noun*
laid paper \ˈlād-\ *noun* (1839)
: paper watermarked with fine lines running across the grain — compare WOVE PAPER
laigh \ˈlāḵ\ *Scottish variant of* LOW
lain *past participle of* LIE
[1]lair \ˈlar, ˈler\ *noun* [Middle English, from Old English *leger;* akin to Old High German *legar* bed, Old English *licgan* to lie — more at LIE] (before 12th century)
1 *dialect British* : a resting or sleeping place : BED
2 a : the resting or living place of a wild animal : DEN **b** : a refuge or place for hiding
[2]lair *verb* [Scots *lair* mire] (circa 1560)
transitive verb
chiefly Scottish : to cause to sink in mire
intransitive verb
chiefly Scottish : WALLOW
laird \ˈlard, ˈlerd\ *noun* [Middle English (northern dialect) *lord, lard* lord] (14th century)
chiefly Scottish : a landed proprietor
— **laird·ly** \-lē\ *adjective*
lais·ser–faire *chiefly British variant of* LAISSEZ-FAIRE
lais·ser–pas·ser *chiefly British variant of* LAISSEZ-PASSER
lais·sez–faire \ˌle-ˌsā-ˈfar, ˌlā-, -ˌzā-, -ˈfer\ *noun* [French *laissez faire,* imperative of *laisser faire* to let (people) do (as they choose)] (1825)
1 : a doctrine opposing governmental interference in economic affairs beyond the minimum necessary for the maintenance of peace and property rights
2 : a philosophy or practice characterized by a usually deliberate abstention from direction or interference especially with individual freedom of choice and action
— **laissez–faire** *adjective*
lais·sez–pas·ser \-ˌpa-ˈsā\ *noun* [French, from *laissez passer* let (someone) pass] (1914)
: PERMIT, PASS
lai·tance \ˈlā-t^{ə}n(t)s\ *noun* [French, from *lait* milk, from Latin *lact-, lac* — more at GALAXY] (circa 1902)
: an accumulation of fine particles on the surface of fresh concrete due to an upward movement of water (as when excessive mixing water is used)
la·ity \ˈlā-ə-tē\ *noun* [[5]*lay*] (15th century)
1 : the people of a religious faith as distinguished from its clergy
2 : the mass of the people as distinguished from those of a particular profession or those specially skilled
La·ius \ˈlā-əs, ˈlī-əs\ *noun* [Latin, from Greek *Laïos*]
: a king of Thebes slain by his son Oedipus in fulfillment of an oracle
[1]lake \ˈlāk\ *noun, often attributive* [Middle English, from Old French *lac* lake, from Latin *lacus;* akin to Old English *lagu* sea, Greek *lakkos* pond] (12th century)
: a considerable inland body of standing water; *also* : a pool of other liquid (as lava, oil, or pitch)
— **lake·like** \-ˌlīk\ *adjective*
[2]lake *noun* [French *laque* lac, from Old Provençal *laca,* from Arabic *lakk* — more at LACQUER] (1598)
1 a : a purplish red pigment prepared from lac or cochineal **b** : any of numerous usually bright translucent organic pigments composed essentially of a soluble dye absorbed on or combined with an inorganic carrier
2 : CARMINE 2
— **laky** \ˈlā-kē\ *adjective*
[3]lake *verb* **laked; lak·ing** (1903)
transitive verb
: to cause (blood) to undergo a physiological change in which the hemoglobin becomes dissolved in the plasma
intransitive verb
of blood : to undergo the process by which hemoglobin becomes dissolved in the plasma
lake dwelling *noun* (1863)
: a dwelling built on piles in a lake; *specifically* : one built in prehistoric times
— **lake dweller** *noun*
lake·front \ˈlāk-ˌfrənt\ *noun* (1880)
: an area fronting on a lake
lake herring *noun* (1842)
: a cisco (*Coregonus artedii*) found from Lake Memphremagog to Lake Superior and northward and important as a commercial food fish
Lake·land terrier \ˈlāk-lənd-, -ˌland-\ *noun* [*Lakeland,* England] (1928)
: any of an English breed of rather small harsh-coated terriers
lak·er \ˈlā-kər\ *noun* (1823)
: one associated with a lake; *especially* : a fish living in or taken from a lake
lake·shore \ˈlāk-ˌshōr, -ˌshȯr\ *noun* (1798)
: the shore of a lake; *also* : LAKEFRONT
lake·side \-ˌsīd\ *noun* (1560)
: LAKEFRONT
lake trout *noun* (1668)
: any of various trout and salmon found in lakes; *especially* : MACKINAW TROUT
lakh \ˈläk, ˈlak\ *noun* [Hindi *lākh*] (1599)
1 : one hundred thousand ⟨50 *lakhs* of rupees⟩
2 : a great number
— **lakh** *adjective*
La·ko·ta \lə-ˈkō-tə\ *noun, plural* **Lakota** *also* **Lakotas** (1927)
1 : a member of a western division of the Dakota peoples
2 : a dialect of Dakota
-lalia *noun combining form* [New Latin, from Greek *lalia* chatter, from *lalein* to chat]
: speech disorder (of a specified type) ⟨echo*lalia*⟩
lal·lan \ˈla-lən\ *or* **lal·land** \-lən(d)\ *Scottish variant of* LOWLAND
Lal·lans \ˈla-lənz\ *noun* (1785)
: Scots as spoken and written in the lowlands of Scotland
Lal·ly \ˈlä-lē\ *trademark*
— used for a concrete-filled cylindrical steel structural column
lal·ly·gag *variant of* LOLLYGAG
[1]lam \ˈlam\ *verb* **lammed; lam·ming** [of Scandinavian origin; akin to Old Norse *lemja* to thrash; akin to Old English *lama* lame] (1595)
transitive verb
: to beat soundly : THRASH
intransitive verb
1 : STRIKE, THRASH
2 : to flee hastily : SCRAM
[2]lam *noun* (1897)
: sudden or hurried flight especially from the law ⟨on the *lam*⟩
la·ma \ˈlä-mə\ *noun* [Tibetan *blama*] (1654)
: a Lamaist monk
La·ma·ism \ˈlä-mə-ˌi-zəm\ *noun* (1817)
: the Mahayana Buddhism of Tibet and Mongolia marked by tantric and shamanistic ritual and a dominant monastic hierarchy headed by the Dalai Lama
— **La·ma·ist** \-mə-ist\ *noun or adjective*
— **La·ma·is·tic** \ˌlä-mə-ˈis-tik\ *adjective*
La·marck·ism \lə-ˈmär-ˌki-zəm\ *noun* [J.-B. de Monet de *Lamarck*] (1884)
: a theory of organic evolution asserting that environmental changes cause structural changes in animals and plants that are transmitted to offspring
— **La·marck·ian** \-kē-ən\ *adjective*
la·ma·sery \ˈlä-mə-ˌser-ē\ *noun, plural* **-ser·ies** [French *lamaserie,* from *lama* + Persian *sarāī* palace] (1849)
: a monastery of lamas
La·maze \lə-ˈmäz\ *adjective* [Fernand *Lamaze* (died 1957) French obstetrician] (1965)
: relating to or being a method of childbirth that involves psychological and physical preparation by the mother in order to suppress pain and facilitate delivery without drugs
[1]lamb \ˈlam\ *noun* [Middle English, from Old English; akin to Old High German *lamb* lamb] (before 12th century)
1 a : a young sheep; *especially* : one that is less than one year old or without permanent teeth **b** : the young of various animals (as the smaller antelopes) other than sheep
2 a : a gentle or weak person **b** : DEAR, PET **c** : a person easily cheated or deceived especially in trading securities

3 a : the flesh of a lamb used as food **b :** LAMBSKIN
— **lamb·like** \-ˌlīk\ *adjective*
— **lamby** \ˈla-mē\ *adjective*

[2]lamb (1611)
intransitive verb
: to bring forth a lamb
transitive verb
1 : to bring forth (a lamb)
2 : to tend (ewes) at lambing time
— **lamb·er** \ˈla-mər\ *noun*

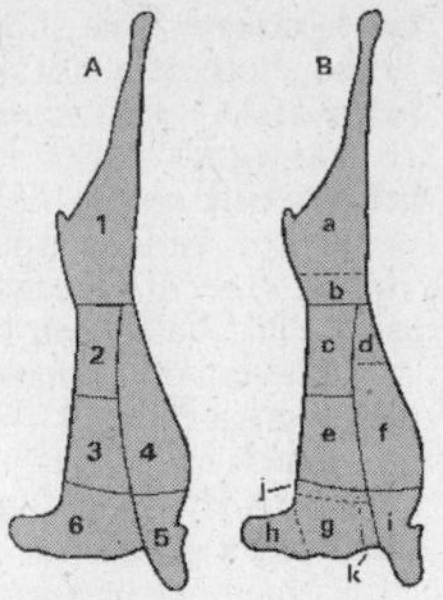

lamb 3a: *A* **wholesale cuts:** *1* **leg,** *2* **loin,** *3* **rack,** *4* **breast,** *5* **shank,** *6* **shoulder;** *B* **retail cuts:** *a* **leg,** *b* **sirloin chops and roast,** *c* **loin chops, rolled loin roast,** *d* **patties and chopped roast,** *e* **rib chops, crown roast,** *f* **riblets, stew, and stuffed or rolled breast,** *g* **shoulder roast, shoulder chops,** *h* **neck slices,** *i* **shanks,** *j* **blade chops,** *k* **arm chops**

lam·baste *or* **lam·bast** \(ˌ)lam-ˈbāst, -ˈbast; ˈlam-ˌ\ *transitive verb* [probably from [1]*lam* + *baste*] (1637)
1 : to assault violently **:** BEAT, WHIP
2 : to attack verbally **:** CENSURE

lamb·da \ˈlam-də\ *noun* [Middle English, from Greek, of Semitic origin; akin to Hebrew *lāmedh* lamed] (15th century)
1 : the 11th letter of the Greek alphabet — see ALPHABET table
2 : an uncharged unstable elementary particle that has a mass 2183 times that of an electron and that decays typically into a nucleon and a pion

lam·ben·cy \ˈlam-bən(t)-sē\ *noun, plural* **-cies** (1817)
: the quality, state, or an instance of being lambent

lam·bent \ˈlam-bənt\ *adjective* [Latin *lambent-, lambens,* present participle of *lambere* to lick — more at LAP] (1647)
1 : playing lightly on or over a surface **:** FLICKERING
2 : softly bright or radiant
3 : marked by lightness or brilliance especially of expression
— **lam·bent·ly** *adverb*

lam·bert \ˈlam-bərt\ *noun* [Johann H. *Lambert* (died 1777) German physicist & philosopher] (1915)
: the centimeter-gram-second unit of brightness equal to the brightness of a perfectly diffusing surface that radiates or reflects one lumen per square centimeter

lamb·kill \ˈlam-ˌkil\ *noun* (1790)
: SHEEP LAUREL

lam·bre·quin \ˈlam-bər-kən, -bri-kən\ *noun* [French] (circa 1725)
1 : a scarf used to cover a knight's helmet
2 : a short decorative drapery for a shelf edge or for the top of a window casing **:** VALANCE

Lam·brus·co \lam-ˈbrü-(ˌ)skō, -ˈbru̇-\ *noun* [Italian, from Latin *labruscum* fruit of the wild grape *Vitis labrusca*] (1943)
: a fruity and fizzy red Italian table wine

lamb's ears *noun plural but usually singular in construction* (1930)
: a widely cultivated perennial herb (*Stachys byzantina* synonym *S. olympica*) of the mint family that occurs in southwestern Asia and has leaves covered with densely matted hairs

lamb·skin \ˈlam-ˌskin\ *noun* (14th century)
: a lamb's skin or a small fine-grade sheepskin or the leather made from either; *specifically* **:** such a skin dressed with the wool on and used especially for winter clothing

lamb's lettuce *noun* (1597)
: CORN SALAD

lamb's–quar·ter \ˈlamz-ˌkwȯ(r)-tər, -ˌkȯ(r)-\ *noun* (1773)
1 : a goosefoot (*Chenopodium album*) having glaucous foliage that is sometimes used as greens — usually used in plural but singular or plural in construction
2 : any of several oraches — usually used in plural but singular or plural in construction

[1]lame \ˈlām\ *adjective* **lam·er; lam·est** [Middle English, from Old English *lama;* akin to Old High German *lam* lame, Lithuanian *limti* to break down] (before 12th century)
1 a : having a body part and especially a limb so disabled as to impair freedom of movement **b :** marked by stiffness and soreness ⟨a *lame* shoulder⟩
2 : lacking needful or desirable substance **:** WEAK, INEFFECTUAL ⟨a *lame* excuse⟩
3 *slang* **:** not being in the know **:** SQUARE
— **lame·ly** *adverb*
— **lame·ness** *noun*

[2]lame *transitive verb* **lamed; lam·ing** (14th century)
1 : to make lame **:** CRIPPLE
2 : to make weak or ineffective **:** DISABLE

[3]lame *noun* (1959)
slang **:** a person who is not in the know **:** SQUARE

[4]lame \ˈlām, ˈlam\ *noun* [Middle French, from Latin *lamina*] (1586)
1 : a thin plate especially of metal **:** LAMINA
2 *plural* **:** small overlapping steel plates joined to slide on one another (as in medieval armor)

la·mé \lä-ˈmā, la-\ *noun* [French] (1922)
: a brocaded clothing fabric made from any of various fibers combined with tinsel filling threads

lame·brain \ˈlām-ˌbrān\ *noun* (1944)
: a dull-witted person **:** DOLT
— **lamebrain** *or* **lame·brained** \-ˈbrānd\ *adjective*

la·med \ˈlä-ˌmed\ *noun* [Hebrew *lāmedh,* literally, ox goad] (1665)
: the 12th letter of the Hebrew alphabet — see ALPHABET table

lame duck *noun* (1761)
1 : one that is weak or that falls behind in ability or achievement; *especially, chiefly British* **:** an ailing company
2 : an elected official or group continuing to hold office during the period between the election and the inauguration of a successor
3 : one whose position or term of office will soon end
— **lame–duck** \ˈlām-ˈdək\ *adjective*

lamell- *or* **lamelli-** *combining form* [New Latin, from *lamella*]
: lamella ⟨*lamelli*form⟩

la·mel·la \lə-ˈme-lə\ *noun, plural* **la·mel·lae** \-ˈme-(ˌ)lē, -ˌlī\ *also* **lamellas** [New Latin, from Latin, diminutive of *lamina* thin plate] (1678)
: a thin flat scale, membrane, or layer: as **a :** one of the thin plates composing the gills of a bivalve mollusk **b :** a gill of a mushroom

la·mel·lar \lə-ˈme-lər\ *adjective* (1794)
1 : composed of or arranged in lamellae
2 : LAMELLIFORM

la·mel·late \lə-ˈme-lət, ˈla-mə-ˌlāt\ *adjective* (1826)
1 : composed of or furnished with lamellae
2 : LAMELLIFORM
— **la·mel·late·ly** *adverb*

la·mel·li·branch \lə-ˈme-lə-ˌbraŋk\ *noun, plural* **-branchs** [New Latin *Lamellibranchia,* from *lamell-* + Latin *branchia* gill] (1855)
: any of a class (Lamellibranchia) of bivalve mollusks (as clams, oysters, and mussels) that have the body bilaterally symmetrical, compressed, and enclosed within the mantle and that build up a shell whose right and left parts are connected by a hinge over the animal's back
— **lamellibranch** *adjective*

la·mel·li·corn \lə-ˈme-lə-ˌkȯrn\ *adjective* [ultimately from New Latin *lamella* + Latin *cornu* horn — more at HORN] (circa 1842)
: of, relating to, or belonging to a superfamily (Scarabaeoidea synonym Lamellicornia) of beetles (as a scarab or stag beetle) characterized by 5-jointed tarsi and club-tipped antennae often angled like an elbow
— **lamellicorn** *noun*

la·mel·li·form \-ˌfȯrm\ *adjective* (1819)
: having the form of a thin plate

[1]la·ment \lə-ˈment\ *verb* [Middle English *lementen,* from Middle French & Latin; Middle French *lamenter,* from Latin *lamentari,* from *lamentum,* noun, lament] (15th century)
intransitive verb
: to mourn aloud **:** WAIL
transitive verb
1 : to express sorrow, mourning, or regret for often demonstratively **:** MOURN
2 : to regret strongly
synonym see DEPLORE

[2]lament *noun* (1591)
1 : a crying out in grief **:** WAILING
2 : DIRGE, ELEGY
3 : COMPLAINT

la·men·ta·ble \ˈla-mən-tə-bəl, ÷lə-ˈmen-\ *adjective* (15th century)
1 : that is to be regretted or lamented **:** DEPLORABLE
2 : expressing grief **:** MOURNFUL
— **la·men·ta·ble·ness** *noun*
— **la·men·ta·bly** \-blē\ *adverb*

lam·en·ta·tion \ˌla-mən-ˈtā-shən\ *noun* (14th century)
: an act or instance of lamenting

Lam·en·ta·tions \-shənz\ *noun plural but singular in construction*
: a poetic book on the fall of Jerusalem in canonical Jewish and Christian Scripture — see BIBLE table

la·ment·ed \lə-ˈmen-təd\ *adjective* (1611)
: mourned for
— **la·ment·ed·ly** *adverb*

la·mia \ˈlā-mē-ə\ *noun* [Middle English, from Latin, from Greek, devouring monster; akin to Greek *lamyros* gluttonous] (14th century)
: a female demon **:** VAMPIRE

lamin- *combining form*
: lamina ⟨*lamin*ar⟩

lam·i·na \ˈla-mə-nə\ *noun, plural* **-nae** \-ˌnē, -ˌnī\ *or* **-nas** [Latin] (circa 1656)
1 : a thin plate or scale **:** LAYER
2 : the expanded part of a foliage leaf
3 : one of the narrow thin parallel plates of soft vascular sensitive tissue that cover the flesh within the wall of a hoof

lam·i·nal \ˈla-mə-nᵊl\ *adjective* (1825)
1 : LAMINAR
2 : produced with the blade of the tongue (as \sh\, \zh\, \ch\, \j\, or \y\) — compare APICAL

lamina pro·pria \-ˈprō-prē-ə\ *noun, plural* **laminae pro·pri·ae** \-prē-ˌē, -ˌī\ [New Latin, literally, proper lamina] (1937)
: a highly vascular layer of connective tissue under the basement membrane lining a layer of epithelium

lam·i·nar \ˈla-mə-nər\ *adjective* (1811)
: arranged in, consisting of, or resembling laminae

laminar flow *noun* (1935)
: streamline flow in a fluid near a solid boundary — compare TURBULENT FLOW

lam·i·nar·ia \ˌla-mə-ˈner-ē-ə, -ˈnar-\ *noun* [New Latin, from Latin *lamina*] (1848)
: any of a genus (*Laminaria*) of large chiefly perennial kelps with an unbranched cylindrical or flattened stipe and a smooth or convoluted blade; *broadly* **:** any of various related kelps (order Laminariales)
— **lam·i·nar·i·an** \-ē-ən\ *adjective or noun*

lam·i·nar·in \ˌlam-ə-ˈner-ən, -ˈnar-\ *noun* [International Scientific Vocabulary *laminar-* (from New Latin *Laminaria*) + [1]*-in*] (circa 1931)
: a polysaccharide found in various brown algae that yields only glucose on hydrolysis

[1]lam·i·nate \ˈla-mə-ˌnāt\ *verb* **-nat·ed; -nat·ing** (1665)

transitive verb
1 : to roll or compress into a thin plate
2 : to separate into laminae
3 a : to make (as a windshield) by uniting superposed layers of one or more materials **b :** to unite (layers of material) by an adhesive or other means
intransitive verb
: to divide into laminae
— lam·i·na·tor \-ˌnā-tər\ *noun*

[2]lam·i·nate \-nət, -ˌnāt\ *adjective* (1668)
1 : consisting of laminae
2 : bearing or covered with laminae

[3]lam·i·nate \-nət, -ˌnāt\ *noun* (1939)
: a product made by laminating

lam·i·nat·ed \-ˌnā-təd\ *adjective* (1665)
1 : LAMINATE 1
2 a : composed of layers of firmly united material **b :** made by bonding or impregnating superposed layers (as of paper, wood, or fabric) with resin and compressing under heat

lam·i·na·tion \ˌla-mə-ˈnā-shən\ *noun* (circa 1676)
1 : the process of laminating
2 : the state of being laminated
3 : a laminated structure
4 : LAMINA

lam·i·ni·tis \ˌla-mə-ˈnī-təs\ *noun* [New Latin] (1843)
: inflammation of the laminae especially in the hoof of a horse — called also *founder*

Lam·mas \ˈla-məs\ *noun* [Middle English *Lammasse,* from Old English *hlāfmæsse,* from *hlāf* loaf, bread + *mæsse* mass; from the fact that formerly loaves from the first ripe grain were consecrated on this day] (before 12th century)
1 : August 1 originally celebrated in England as a harvest festival — called also *Lammas Day*
2 : the time of the year around Lammas Day

Lam·mas·tide \-ˌtīd\ *noun* (14th century)
: LAMMAS 2

lam·mer·gei·er *or* **lam·mer·gey·er** \ˈla-mər-ˌgī(-ə)r\ *noun* [German *Lämmergeier,* from *Lämmer* lambs + *Geier* vulture] (1817)
: a large Old World vulture (*Gypaetus barbatus*) that occurs in mountain regions and in flight resembles a huge falcon

lamp \ˈlamp\ *noun* [Middle English, from Old French *lampe,* from Latin *lampas,* from Greek, from *lampein* to shine; akin to Hittite *lap-* to burn] (13th century)
1 a : any of various devices for producing light or sometimes heat: as (1) **:** a vessel with a wick for burning an inflammable liquid (as oil) to produce light (2) **:** a glass bulb or tube that emits light produced by electricity (as an incandescent lamp or fluorescent lamp) **b :** a decorative appliance housing a lamp that is usually covered by a shade
2 : a celestial body
3 : a source of intellectual or spiritual illumination
4 : EYE 1a — usually used in plural

lamp·black \-ˌblak\ *noun* (1598)
: a finely powdered black soot deposited in incomplete combustion of carbonaceous materials and used chiefly as a pigment (as in paints, enamels, and printing inks)

lamp·brush chromosome \ˈlamp-ˌbrəsh-\ *noun* (1911)
: a greatly enlarged diplotene chromosome that has apparently filamentous granular loops extending from the chromomeres and is characteristic of some animal oocytes

lamp·light \ˈlamp-ˌlīt\ *noun* (14th century)
: the light of a lamp

lamp·light·er \-ˌlī-tər\ *noun* (1750)
: one that lights a lamp

[1]lam·poon \lam-ˈpün\ *noun* [French *lampon*] (1645)
: SATIRE 1; *specifically* **:** a harsh satire usually directed against an individual

[2]lampoon *transitive verb* (circa 1657)
: to make the subject of a lampoon **:** RIDICULE
— lam·poon·er *noun*
— lam·poon·ery \-ˈpü-nə-rē, -ˈpün-rē\ *noun*

lamp·post \ˈlam(p)-ˌpōst\ *noun* (1790)
: a post supporting a usually outdoor lamp or lantern

lam·prey \ˈlam-prē, -ˌprā\ *noun, plural* **lam·preys** [Middle English, from Old French *lampreie,* from Medieval Latin *lampreda*] (14th century)
: any of a family (Petromyzontidae) of eel-shaped freshwater or anadromous jawless fishes that include those cyclostomes having well-developed eyes and a large disk-shaped suctorial mouth armed with horny teeth — called also *lamprey eel*

lamp·shell \ˈlamp-ˌshel\ *noun* [from the resemblance of the shell and its protruding peduncle to an ancient oil lamp with the wick protruding] (1854)
: BRACHIOPOD

lam·ster \ˈlam(p)-stər\ *also* **lam·is·ter** \ˈla-mə-stər\ *noun* [[2]lam + -ster] (1904)
: a fugitive especially from the law

LAN \ˈlan, ˌel-(ˌ)ā-ˈen\ *noun* (1982)
: LOCAL AREA NETWORK

la·nai \lə-ˈnī, lä-\ *noun* [Hawaiian *lānai*] (1823)
: PORCH, VERANDA

Lan·ca·shire \ˈlaŋ-kə-ˌshir, -shər\ *noun* [*Lancashire,* England] (1896)
: a moist crumbly white English cheese that is used especially in cooking

Lan·cas·tri·an \lan-ˈkas-trē-ən, laŋ-\ *adjective* [John of Gaunt, duke of *Lancaster*] (1612)
: of or relating to the English royal house that ruled from 1399 to 1461

[1]lance \ˈlan(t)s\ *noun* [Middle English, from Old French, from Latin *lancea*] (14th century)
1 : a steel-tipped spear carried by mounted knights or light cavalry
2 : any of various sharp objects suggestive of a lance: as **a :** LANCET **b :** a spear used for killing whales or fish
3 : LANCER 1b

[2]lance *verb* **lanced; lanc·ing** [Middle English *launcen,* from Middle French *lancer,* from Late Latin *lanceare,* from Latin *lancea*] (14th century)
transitive verb
1 a : to pierce with or as if with a lance **b :** to open with or as if with a lancet ⟨*lance* a boil⟩
2 : to throw forward **:** HURL
intransitive verb
: to move forward quickly

lance corporal *noun* [*lance* (as in obsolete *lancepesade* lance corporal, from Middle French *lancepessade*)] (1786)
: an enlisted man in the marine corps ranking above a private first class and below a corporal

lance·let \ˈlan(t)-slət\ *noun* (circa 1836)
: any of a subphylum (Cephalochordata) of small translucent marine primitive chordate animals that are fishlike in appearance and usually live partially buried on the ocean floor — called also *amphioxus*

Lan·ce·lot \ˈlan(t)-sə-ˌlät, ˈlän(t)-, -s(ə-)lət\ *noun*
: a knight of the Round Table and lover of Queen Guinevere

lan·ce·o·late \ˈlan(t)-sē-ə-ˌlāt\ *adjective* [Late Latin *lanceolatus,* from Latin *lanceola,* diminutive of *lancea*] (circa 1760)
: shaped like a lance head; *specifically* **:** tapering to a point at the apex and sometimes at the base ⟨*lanceolate* leaves⟩ ⟨*lanceolate* prisms⟩ — see LEAF illustration

lanc·er \ˈlan(t)-sər\ *noun* (1590)
1 a : one who carries a lance **b :** a member of a military unit formerly composed of light cavalry armed with lances
2 *plural but singular in construction* **a :** a set of five quadrilles each in a different meter **b :** the music for such dances

lan·cet \ˈlan(t)-sət\ *noun* (15th century)
1 : a sharp-pointed and commonly 2-edged surgical instrument used to make small incisions
2 a : LANCET WINDOW **b :** LANCET ARCH

lancet arch *noun* (circa 1823)
: an acutely pointed arch — see ARCH illustration

lan·cet·ed \ˈlan(t)-sə-təd\ *adjective* (1855)
: having a lancet arch or lancet windows

lancet window *noun* (1781)
: a high narrow window with an acutely pointed head and without tracery

lance·wood \ˈlan(t)s-ˌwu̇d\ *noun* (1697)
: a tough elastic wood used especially for shafts, fishing rods, and bows; *also* **:** a tree (especially *Oxandra lanceolata*) yielding this wood

lan·ci·nat·ing \ˈlan(t)-sə-ˌnā-tiŋ\ *adjective* [*lancinate* to pierce, from Latin *lancinatus,* past participle of *lancinare;* akin to Latin *lacerare* to rend — more at LACERATE] (1762)
: characterized by piercing or stabbing sensations ⟨*lancinating* pain⟩

[1]land \ˈland\ *noun, often attributive* [Middle English, from Old English; akin to Old High German *lant* land, Middle Irish *lann*] (before 12th century)
1 a : the solid part of the surface of the earth; *also* **:** a corresponding part of a celestial body (as the moon) **b :** ground or soil of a specified situation, nature, or quality ⟨dry *land*⟩ **c :** the surface of the earth and its natural resources
2 : a portion of the earth's solid surface distinguishable by boundaries or ownership ⟨bought *land* in the country⟩: as **a :** COUNTRY ⟨campaigned across the *land*⟩ **b :** a rural area characterized by farming or ranching; *also* **:** farming or ranching as a way of life ⟨wanted to move back to the *land*⟩
3 : REALM, DOMAIN ⟨in the *land* of dreams⟩ — sometimes used in combination ⟨movie*land*⟩
4 : the people of a country ⟨the *land* rose in rebellion⟩
5 : an area of a partly machined surface that is left without machining
— land·less \ˈland-ləs\ *adjective*
— land·less·ness \-nəs\ *noun*

[2]land (13th century)
transitive verb
1 : to set or put on shore from a ship **:** DISEMBARK
2 a : to set down after conveying **b :** to cause to reach or come to rest in a particular place ⟨never *landed* a punch⟩ **c :** to bring to a specified condition ⟨his carelessness *landed* him in trouble⟩ **d :** to bring (as an airplane) to a landing
3 a : to catch and bring in (as a fish) **b :** GAIN, SECURE ⟨*land* a job⟩
intransitive verb
1 a : to go ashore from a ship **:** DISEMBARK **b** *of a ship or boat* **:** to touch at a place on shore
2 a : to come to the end of a course or to a stage in a journey **:** ARRIVE ⟨took the wrong subway and *landed* on the other side of town⟩ **b :** to come to be in a condition or situation ⟨*landed* in jail⟩ **c :** to strike or meet a surface (as after a fall) ⟨*landed* on my head⟩ **d :** to alight on a surface

lan·dau \ˈlan-ˌdau̇, -ˌdȯ\ *noun* [*Landau,* Bavaria, Germany] (1743)
: a four-wheel carriage with a top divided into two sections that can be let down, thrown back, or removed and with a raised seat outside for the driver

landau

\ə\ **abut** \ᵊ\ kitten \ər\ **fur**ther \a\ **a**sh \ā\ **ace** \ä\ m**o**p, m**a**r \au̇\ **ou**t \ch\ **ch**in \e\ b**e**t \ē\ **ea**sy \g\ **g**o \i\ h**i**t \ī\ **i**ce \j\ **j**ob \ŋ\ si**ng** \ō\ **go** \ȯ\ l**aw** \ȯi\ b**oy** \th\ **th**in \th̲\ **th**e \ü\ l**oo**t \u̇\ f**oo**t \y\ **y**et \zh\ vi**si**on *see also* Guide to Pronunciation

lan·dau·let \ˌlan-dᵊl-ˈet\ *noun* (1794)
: a small landau

land bank *noun* (1696)
: a bank that provides financing for land development and for farm mortgages

land·ed \ˈlan-dəd\ *adjective* (15th century)
1 : having an estate in land ⟨*landed* proprietors⟩
2 : consisting in or derived from land or real estate ⟨*landed* wealth⟩

land·er \ˈlan-dər\ *noun* (1859)
: one that lands; *especially* **:** a space vehicle that is designed to land on a celestial body (as the moon or a planet)

land·fall \ˈlan(d)-ˌfȯl\ *noun* (1627)
1 : a sighting or making of land after a voyage or flight
2 : the land first sighted on a voyage or flight

land·fill \-ˌfil\ *noun* (1942)
1 : a system of trash and garbage disposal in which the waste is buried between layers of earth to build up low-lying land — called also *sanitary landfill*
2 : an area built up by landfill

land·form \-ˌfȯrm\ *noun* (1893)
: a natural feature of a land surface

land·grab \-ˌgrab\ *noun* (1860)
: a usually swift acquisition of land often by fraud or force
— **land–grab·ber** \-ˌgra-bər\ *noun*

land grant *noun* (1862)
: a grant of land made by the government especially for roads, railroads, or agricultural colleges

land·hold·er \ˈland-ˌhōl-dər\ *noun* (15th century)
: a holder or owner of land

land·hold·ing \-ˌhōl-diŋ\ *noun* (circa 1890)
1 : the state or fact of holding or owning land
2 : property in land
— **landholding** *adjective*

land·ing *noun* (15th century)
1 : an act or process of one that lands; *especially* **:** a going or bringing to a surface (as land or shore) after a voyage or flight
2 : a place for discharging and taking on passengers and cargo
3 : a level part of a staircase (as at the end of a flight of stairs)

landing craft *noun* (1940)
: any of numerous naval craft designed for conveying troops and equipment from a transport to a beach in an amphibious assault

landing field *noun* (circa 1920)
: a field where aircraft may land and take off

landing gear *noun* (1911)
: the part that supports the weight of an airplane or spacecraft when in contact with the land or water

landing strip *noun* (1930)
: AIRSTRIP

land·la·dy \ˈland-ˌlā-dē\ *noun* (circa 1536)
: a woman who is a landlord

land·line \-ˌlīn\ *noun* (1865)
: a line of communication (as by telephone cable) on land

land·locked \-ˌläkt\ *adjective* (1622)
1 : enclosed or nearly enclosed by land ⟨a *landlocked* country⟩
2 : confined to fresh water by some barrier ⟨*landlocked* salmon⟩

land·lord \-ˌlȯrd\ *noun* (before 12th century)
1 : the owner of property (as land, houses, or apartments) that is leased or rented to another
2 : the master of an inn or lodging house **:** INNKEEPER

land·lord·ism \-ˌlȯr-ˌdi-zəm\ *noun* (1844)
: an economic system or practice by which ownership of land is vested in one who leases it to cultivators

land·lub·ber \-ˌlə-bər\ *noun* (circa 1700)
: LANDSMAN 2 ⟨clumsy *landlubbers* learning to sail⟩
— **land·lub·ber·li·ness** \-bər-lē-nəs\ *noun*
— **land·lub·ber·ly** \-bər-lē\ *adjective*
— **land·lub·bing** \-biŋ\ *adjective*

land·mark \-ˌmärk\ *noun* (before 12th century)
1 : an object (as a stone or tree) that marks the boundary of land
2 a : a conspicuous object on land that marks a locality **b :** an anatomical structure used as a point of orientation in locating other structures
3 : an event or development that marks a turning point or a stage
4 : a structure (as a building) of unusual historical and usually aesthetic interest; *especially* **:** one that is officially designated and set aside for preservation

land·mass \-ˌmas\ *noun* (1856)
: a large area of land ⟨continental *landmasses*⟩

land mine *noun* (1890)
1 : a mine placed on or just below the surface of the ground and designed to be exploded by the weight of vehicles or troops passing over it
2 : BOOBY TRAP 1

land office *noun* (1681)
: a government office in which entries upon and sales of public land are registered

land–office business *noun* (1839)
: extensive and rapid business ⟨money changers . . . did a *land-office business* on payday —F. J. Haskin⟩

land·own·er \ˈland-ˌō-nər\ *noun* (circa 1733)
: an owner of land
— **land·own·er·ship** \-ˌship\ *noun*
— **land·own·ing** \-ˌō-niŋ\ *adjective or noun*

land–poor \ˈlan(d)-ˌpu̇r\ *adjective* (1873)
: owning so much unprofitable or encumbered land as to lack funds to develop the land or pay the charges due on it

Land·ra·ce \ˈlän(d)-ˌrä-sə\ *noun* [Danish, from *land* + *race*] (1935)
: a swine of any of several breeds locally developed in northern Europe

land reform *noun* (1846)
: measures designed to effect a more equitable distribution of agricultural land especially by governmental action; *also* **:** the resulting redistribution

¹land·scape \ˈlan(d)-ˌskāp\ *noun, often attributive* [Dutch *landschap,* from *land* + *-schap* -ship] (1598)
1 a : a picture representing a view of natural inland scenery **b :** the art of depicting such scenery
2 a : the landforms of a region in the aggregate **b :** a portion of territory that can be viewed at one time from one place **c :** a particular area of activity **:** SCENE ⟨political *landscape*⟩
3 *obsolete* **:** VISTA, PROSPECT ◆

²landscape *verb* **land·scaped; land·scap·ing** (1914)
transitive verb
: to modify or ornament (a natural landscape) by altering the plant cover
intransitive verb
: to engage in landscape gardening
— **land·scap·er** *noun*

landscape architect *noun* (1863)
: one who develops land for human use and enjoyment through effective placement of structures, vehicular and pedestrian ways, and plantings
— **landscape architecture** *noun*

landscape gardener *noun* (circa 1763)
: one who is engaged in the development and decorative planting of gardens and grounds
— **landscape gardening** *noun*

land·scap·ist \ˈlan(d)-ˌskā-pist\ *noun* (1843)
: a painter of landscapes

¹land·slide \ˈlan(d)-ˌslīd\ *noun* (1838)
1 : the usually rapid downward movement of a mass of rock, earth, or artificial fill on a slope; *also* **:** the mass that moves down
2 a : a great majority of votes for one side **b :** an overwhelming victory

²landslide *intransitive verb* **-slid** \-ˌslid\; **-slid·ing** \-ˌslī-diŋ\ (1926)
1 : to produce a landslide
2 : to win an election by a heavy majority

land·slip \-ˌslip\ *noun* (1679)
: LANDSLIDE 1

Lands·mål *or* **Lands·maal** \ˈlän(t)s-ˌmȯl\ *noun* [Norwegian, from *land* country + *mål* speech] (1886)
: NYNORSK

lands·man \ˈlan(d)z-mən\ *noun* (1598)
1 : a fellow countryman
2 : one who lives on the land; *especially* **:** one who knows little or nothing of the sea or seamanship

land·ward \ˈland-wərd\ *adverb or adjective* (15th century)
: to or toward the land

land yacht *noun* (1928)
: a 3-wheel wind-driven recreation vehicle consisting usually of a bare-frame structure and a single sail and used especially on areas of firmly packed sand

¹lane \ˈlān\ *noun* [Middle English, from Old English *lanu;* akin to Middle Dutch *lane* lane] (before 12th century)
1 : a narrow passageway between fences or hedges
2 : a relatively narrow way or track: as **a :** an ocean route used by or prescribed for ships **b :** a strip of roadway for a single line of vehicles **c :** AIR LANE **d :** any of several parallel courses on a track or swimming pool in which a competitor must stay during a race **e :** an unmarked lengthwise division of a playing area which defines the playing zone of a particular player **f :** a narrow hardwood surface having pins at one end and a shallow channel along each side that is used in bowling **g :** FREE THROW LANE

²lane *Scottish variant of* LONE

lane·way \ˈlān-ˌwā\ *noun* (1882)
British **:** LANE

lang·bein·ite \ˈlaŋ-ˌbī-ˌnīt\ *noun* [German *Langbeinit,* from A. *Langbein,* 19th century German chemist] (circa 1897)
: a mineral that is a double sulfate of potassium and magnesium used in the fertilizer industry

lang·lauf \ˈläŋ-ˌlau̇f\ *noun* [German, from *lang* long + *Lauf* race] (1927)
: cross-country running or racing on skis
— **lang·lauf·er** \-ˌlau̇-fər\ *noun*

lang·ley \ˈlaŋ-lē\ *noun, plural* **langleys** [Samuel P. *Langley*] (1947)
: a unit of solar radiation equivalent to one gram calorie per square centimeter of irradiated surface

Lan·go·bard \ˈlaŋ-gə-ˌbärd\ *noun* [Latin *Langobardus*] (1788)
: LOMBARD 1a
— **Lan·go·bar·dic** \ˌlaŋ-gə-ˈbär-dik\ *adjective*

lan·gouste \läⁿ-ˈgüst\ *noun* [French, grasshopper, lobster, from Old French *languste,*

◇ WORD HISTORY

landscape The word *landscape* is appropriately enough a loan from the language of the Dutch, who pioneered in making landscape a distinct genre of painting in the 17th century. Traditionally views of natural scenery in a painting were no more than background to the portrait or episode of mythology or history that occupied the foreground. But the figures got smaller and of less significance, and the natural scene became the true subject of the painting in the work of artists such as Jacob van Ruisdael and Meindert Hobbema. In the 18th century the meaning of *landscape* was extended from a painted view of natural scenery to the landforms themselves, a usage that still remains common. The element *-scape* has detached itself from *landscape* and become something like a suffix, forming words such as *seascape, streetscape,* and *moonscape.*

from Old Provençal *langosta,* from (assumed) Vulgar Latin *lacusta,* alteration of Latin *locusta*] (1832)
: SPINY LOBSTER

lan•gous•tine \ˌlaŋ-gə-ˈstēn\ *also* **lan•gos•ti•no** \ˌlaŋ-gə-ˈstē-(ˌ)nō\ *noun, plural* **-tines** *also* **-ti•nos** [*langoustine,* from French, diminutive of *langouste; langostino,* from Spanish, diminutive of *langosta* spiny lobster, locust, from (assumed) Vulgar Latin *lacusta*] (1915)
: any of several small lobsters (genera *Nephropsis* and *Metanephrops* of the family Nephropidae) widely used for food

^{1}**lang syne** \(ˌ)laŋ-ˈzīn, -ˈsīn\ *adverb* [Middle English (Scots), from *lang* long + *syne* since] (15th century)
chiefly Scottish **:** at a distant time in the past

^{2}**lang syne** *noun* (1694)
chiefly Scottish **:** times past ⟨should auld acquaintance be forgot, and days o' auld *lang syne* —Robert Burns⟩

lan•guage \ˈlaŋ-gwij, -wij\ *noun* [Middle English, from Old French, from *langue* tongue, language, from Latin *lingua* — more at TONGUE] (14th century)
1 a : the words, their pronunciation, and the methods of combining them used and understood by a community **b** (1) **:** audible, articulate, meaningful sound as produced by the action of the vocal organs (2) **:** a systematic means of communicating ideas or feelings by the use of conventionalized signs, sounds, gestures, or marks having understood meanings (3) **:** the suggestion by objects, actions, or conditions of associated ideas or feelings ⟨*language* in their very gesture —Shakespeare⟩ (4) **:** the means by which animals communicate (5) **:** a formal system of signs and symbols (as FORTRAN or a calculus in logic) including rules for the formation and transformation of admissible expressions (6) **:** MACHINE LANGUAGE 1
2 a : form or manner of verbal expression; *specifically* **:** STYLE **b :** the vocabulary and phraseology belonging to an art or a department of knowledge **c :** PROFANITY
3 : the study of language especially as a school subject

language arts *noun plural* (1948)
: the subjects (as reading, spelling, literature, and composition) that aim at developing the student's comprehension and capacity for use of written and oral language

langue \läⁿg\ *noun* [French, literally, language] (1924)
: language viewed abstractly as a system of forms and conventions used for communication in a community; *also* **:** COMPETENCE 3 — compare PAROLE

langue d'oc \läŋ-ˈdȯk, läⁿg-dȯk\ *noun* [French, from Old French, literally, language of *oc;* from the Provençal use of the word *oc* for "yes"] (1703)
: PROVENÇAL 2

langue d'oïl \läŋ-ˈdȯi(ə)l, -ˈdȯi; läⁿg-dȯ-ēl, -dȯi\ *noun* [French, from Old French, literally, language of *oïl;* from the French use of the word *oïl* for "yes"] (1703)
: FRENCH 1

lan•guet \ˈlaŋ-gwət, laŋ-ˈgwet\ *noun* [Middle English, from Middle French *languete,* diminutive of *langue*] (15th century)
: something resembling the tongue in form or function

lan•guid \ˈlaŋ-gwəd\ *adjective* [Middle French *languide,* from Latin *languidus,* from *languēre* to languish — more at SLACK] (1597)
1 : drooping or flagging from or as if from exhaustion **:** WEAK
2 : sluggish in character or disposition **:** LISTLESS
3 : lacking force or quickness of movement **:** SLOW
— lan•guid•ly *adverb*
— lan•guid•ness *noun*

HOME LANGUAGES WITH OVER FORTY MILLION SPEAKERS[1]

LANGUAGE	MILLIONS
Mandarin Chinese	865
English	334
Spanish	283
Arabic	197
Bengali	181
Hindi and Urdu	172
Portuguese	161
Russian	156
Japanese	125
German	104
Wu Chinese	94
Panjabi	76
Javanese	76
Telugu	72
Cantonese	70
Korean	70
Marathi	68
Italian	68
Tamil	65
French	65
Vietnamese	61
Awadhi[2]	61
Bhojpuri	58
Southern Min Chinese[3]	55
Turkish	52
Ukrainian	50
Thai and Lao	47
Polish	42
Gujarati	41
Persian	40

[1]Compiled by William W. Gage, using information supplied in *Ethnologue: Languages of the World,* 11th ed. (Dallas: Summer Institute of Linguistics, 1988). Home language used here means the language usually spoken at home.

[2]Indo-Aryan language of eastern and central Uttar Pradesh.

[3]Group of Chinese dialects especially of Xiamen, Shantou, and Taiwan.

lan•guish \ˈlaŋ-gwish\ *intransitive verb* [Middle English, from Middle French *languiss-,* stem of *languir,* from (assumed) Vulgar Latin *languire,* from Latin *languēre*] (14th century)
1 a : to be or become feeble, weak, or enervated **b :** to be or live in a state of depression or decreasing vitality
2 a : to become dispirited **:** PINE ⟨*languishing* in prison⟩ **b :** to suffer neglect ⟨the bill *languished* in the Senate for eight months⟩
3 : to assume an expression of grief or emotion appealing for sympathy
— lan•guish•er *noun*
— lan•guish•ing•ly \-gwi-shiŋ-lē\ *adverb*
— lan•guish•ment \-gwish-mənt\ *noun*

lan•guor \ˈlaŋ-gər *also* -ər\ *noun* [Middle English, from Middle French, from Latin, from *languēre*] (14th century)
1 : weakness or weariness of body or mind
2 : listless indolence or inertia
synonym see LETHARGY

lan•guor•ous \ˈlaŋ-gə-rəs, -grəs *also* -ə-rəs\ *adjective* (15th century)
1 : producing or tending to produce languor ⟨a *languorous* climate⟩
2 : full of or characterized by languor
— lan•guor•ous•ly *adverb*

lan•gur \läŋ-ˈgu̇r\ *noun* [Hindi *lãgūr*] (1825)
: any of several slender long-tailed Asian monkeys (subfamily Colobinae)

langur

lank \ˈlaŋk\ *adjective* [Middle English, from Old English *hlanc;* akin to Old High German *hlanca* loin] (before 12th century)
1 : not well filled out **:** SLENDER, THIN ⟨*lank* cattle⟩
2 : insufficient in quantity, degree, or extent ⟨*lank* grass⟩
3 : hanging straight and limp without spring or curl ⟨*lank* hair⟩
synonym see LEAN
— lank•ly *adverb*
— lank•ness *noun*

lanky \ˈlaŋ-kē\ *adjective* **lank•i•er; -est** (1818)
: ungracefully tall and thin
synonym see LEAN
— lank•i•ly \-kə-lē\ *adverb*
— lank•i•ness \-kē-nəs\ *noun*

lan•ner \ˈla-nər\ *noun* [Middle English *laner,* from Middle French *lanier*] (14th century)
: a falcon (*Falco biarmicus*) of southern Europe, southwestern Asia, and Africa; *specifically* **:** a female lanner

lan•ner•et \ˌla-nə-ˈret\ *noun* (15th century)
: a male lanner

lan•o•lin \ˈla-nᵊl-ən\ *noun* [Latin *lana* wool + International Scientific Vocabulary ^{3}*-ol* + ^{1}*-in*] (1885)
: wool grease especially when refined for use in ointments and cosmetics

lan•ta•na \lan-ˈtä-nə\ *noun* [New Latin, from Italian dialect, viburnum] (1791)
: any of a genus (*Lantana*) of tropical shrubs of the vervain family with showy heads of small bright flowers

lan•tern \ˈlan-tərn\ *noun, often attributive* [Middle English *lanterne,* from Old French, from Latin *lanterna,* from Greek *lamptēr,* from *lampein* to shine — more at LAMP] (13th century)
1 : a usually portable protective case for a light with transparent openings — compare CHINESE LANTERN
2 a *obsolete* **:** LIGHTHOUSE **b :** the chamber in a lighthouse containing the light **c :** a structure with glazed or open sides above an opening in a roof for light or ventilation **d :** a small tower or cupola or one stage of a cupola
3 : PROJECTOR 2b

lantern fish *noun* (circa 1753)
: any of a family (Myctophidae) of small deep-sea bony fishes that have a large mouth, large eyes, and usually numerous photophores

lantern fly *noun* (circa 1753)
: any of several large brightly marked homopterous insects (family Fulgoridae) having the front of the head prolonged into a hollow structure

lantern jaw *noun* (1711)
: an undershot jaw
— lan•tern–jawed \ˈlan-tərn-ˌjȯd\ *adjective*

lantern pinion *noun* (circa 1859)
: a gear pinion having cylindrical bars instead of teeth

lan•tha•nide \ˈlan(t)-thə-ˌnīd\ *noun* [International Scientific Vocabulary] (1926)
: any in a series of elements of increasing atomic numbers beginning with lanthanum (57) or cerium (58) and ending with lutetium (71) — see PERIODIC TABLE table

lan•tha•num \-nəm\ *noun* [New Latin, from Greek *lanthanein* to escape notice — more at LATENT] (1841)
: a white soft malleable metallic element that occurs in rare-earth minerals — see ELEMENT table

lant•horn \ˈlan-tərn\ *noun* (1587)
chiefly British **:** LANTERN

la•nu•gi•nous \lə-ˈnü-jə-nəs, -ˈnyü-\ *adjective* [Latin *lanuginosus,* from *lanugin-, lanugo*] (1575)

\ə\ **abut** \ᵊ\ **kitten** \ər\ **further** \a\ **ash** \ā\ **ace** \ä\ **mop, mar** \au̇\ **out** \ch\ **chin** \e\ **bet** \ē\ **easy** \g\ **go** \i\ **hit** \ī\ **ice** \j\ **job** \ŋ\ **sing** \ō\ **go** \ȯ\ **law** \ȯi\ **boy** \th\ **thin** \t͟h\ **the** \ü\ **loot** \u̇\ **foot** \y\ **yet** \zh\ **vision** *see also* Guide to Pronunciation

: covered with down or fine soft hair : DOWNY

la·nu·go \lə-ˈnü-(ˌ)gō, -ˈnyü-\ *noun* [Latin, down, from *lana* wool — more at WOOL] (15th century)
: a dense cottony or downy growth; *specifically* : the soft woolly hair that covers the fetus of some mammals

lan·yard \ˈlan-yərd\ *noun* [Middle English *lanyer* thong, strap, from Middle French *laniere*] (1626)
1 : a piece of rope or line for fastening something in a ship; *especially* : one of the pieces passing through deadeyes to extend shrouds or stays
2 a : a cord or strap to hold something (as a knife or a whistle) and usually worn around the neck **b :** a cord worn as a symbol of a military citation
3 : a strong line used to activate a system (as in firing a cannon or sounding a whistle)

Lao \ˈlau̇\ *noun, plural* **Lao** *or* **Laos** \ˈlau̇z\ (1808)
1 : a member of a Buddhist people living in Laos and adjacent parts of northeastern Thailand
2 : the Thai language of the Lao people
— Lao *adjective*

La·oc·o·ön \lā-ˈä-kə-ˌwän\ *noun* [Latin, from Greek *Laokoōn*]
: a Trojan priest killed with his sons by two sea serpents after warning the Trojans against the wooden horse

La·od·i·ce·an \lā-ˌä-də-ˈsē-ən, ˌlā-ō-də-\ *adjective* [from the reproach to the church of the Laodiceans in Revelation 3:15–16] (1633)
: lukewarm or indifferent in religion or politics
— Laodicean *noun*

Lao·tian \lā-ˈō-shən, ˈlau̇-shən\ *noun* [probably from French *laotien,* adjective & noun, from *Lao*] (1847)
: LAO
— Laotian *adjective*

[1]**lap** \ˈlap\ *noun* [Middle English *lappe,* from Old English *læppa;* akin to Old High German *lappa* flap] (before 12th century)
1 a : a loose overlapping or hanging panel or flap especially of a garment **b** *archaic* **:** the skirt of a coat or dress
2 a : the clothing that lies on the knees, thighs, and lower part of the trunk when one sits **b :** the front part of the lower trunk and thighs of a seated person
3 : responsible custody **:** CONTROL ⟨going to drop the whole thing in your *lap* —Hamilton Basso⟩
— lap·ful \ˈlap-ˌfu̇l\ *noun*
— the lap of luxury : an environment of great ease, comfort, and wealth ⟨was reared in *the lap of luxury*⟩

[2]**lap** *verb* **lapped; lap·ping** (14th century)
transitive verb
1 a : to fold over or around something **:** WIND **b :** to envelop entirely **:** SWATHE
2 : to fold over especially into layers
3 : to hold protectively in or as if in the lap **:** CUDDLE
4 a : to place over and cover a part of **:** OVERLAP ⟨*lap* shingles on a roof⟩ **b :** to join (as two boards) by a lap joint
5 a : to dress, smooth, or polish (as a metal surface) to a high degree of refinement or accuracy **b :** to shape or fit by working two surfaces together with or without abrasives until a very close fit is produced
6 a : to overtake and thereby lead or increase the lead over (another contestant) by a full circuit of a racecourse **b :** to complete the circuit of (a racecourse)
intransitive verb
1 : FOLD, WIND
2 a : to project beyond or spread over something **b :** to lie partly over or alongside of something or of one another **:** OVERLAP
3 : to traverse a course
— lap·per *noun*

[3]**lap** *noun* (1800)
1 a : the amount by which one object overlaps or projects beyond another **b :** the part of an object that overlaps another
2 : a smoothing and polishing tool usually consisting of a piece of wood, leather, felt, or soft metal in a special shape used with or without an embedded abrasive
3 : a doubling or layering of a flexible substance (as fibers or paper)
4 a : the act or an instance of traversing a course (as a racing track or swimming pool); *also* **:** the distance covered **b :** one segment of a larger unit (as a journey) **c :** one complete turn (as of a rope around a drum)

[4]**lap** *verb* **lapped; lap·ping** [Middle English, from Old English *lapian;* akin to Old High German *laffan* to lick, Latin *lambere,* Greek *laphyssein* to devour] (before 12th century)
intransitive verb
1 : to take in food or drink with the tongue
2 a : to make a gentle intermittent splashing sound **b :** to move in little waves **:** WASH
transitive verb
1 a : to take in (food or drink) with the tongue **b :** to take in or absorb eagerly or quickly — used with *up* ⟨the crowd *lapped* up every word he said⟩
2 : to flow or splash against in little waves
— lap·per *noun*

[5]**lap** *noun* (14th century)
1 a : an act or instance of lapping **b :** the amount that can be carried to the mouth by one lick or scoop of the tongue
2 : a thin or weak beverage or food
3 : a gentle splashing sound

laparo- *combining form* [Greek *lapara* flank, from *laparos* slack]
: abdominal wall ⟨*laparo*tomy⟩

lap·a·ro·scope \ˈla-p(ə-)rə-ˌskōp\ *noun* [International Scientific Vocabulary] (circa 1923)
: a fiberoptic instrument inserted through an incision in the abdominal wall and used to examine visually the interior of the peritoneal cavity

lap·a·ros·co·py \ˌla-pə-ˈräs-kə-pē\ *noun, plural* **-pies** (1916)
1 : visual examination of the abdomen by means of a laparoscope
2 : an operation involving laparoscopy; *especially* **:** one for sterilization of the female or for removal of ova that involves use of a laparoscope to guide surgical procedures within the abdomen
— lap·a·ro·scop·ic \-rə-ˈskä-pik\ *adjective*
— lap·a·ros·co·pist \ˌla-pə-ˈräs-kə-pist\ *noun*

lap·a·rot·o·my \ˌla-pə-ˈrä-tə-mē\ *noun, plural* **-mies** (1878)
: surgical section of the abdominal wall

lap belt *noun* (1952)
: a seat belt that fastens across the lap

lap·board \ˈlap-ˌbōrd, -ˌbȯrd\ *noun* (1804)
: a board used on the lap as a table or desk

lap·dog \-ˌdȯg\ *noun* (1645)
1 : a small dog that may be held in the lap
2 : a servile dependent or follower

la·pel \lə-ˈpel\ *noun* [diminutive of [1]*lap*] (1789)
: the part of a garment that is turned back; *specifically* **:** the fold of the front of a coat that is usually a continuation of the collar
— la·pelled *or* **la·peled** *adjective*

lap·i·dar·i·an \ˌla-pə-ˈder-ē-ən\ *adjective* (1683)
: LAPIDARY 1

[1]**lap·i·dary** \ˈla-pə-ˌder-ē\ *noun, plural* **-dar·ies** (14th century)
1 : a cutter, polisher, or engraver of precious stones usually other than diamonds
2 : the art of cutting gems

[2]**lapidary** *adjective* [Latin *lapidarius* of stone, from *lapid-, lapis* stone] (1724)
1 : having the elegance and precision associated with inscriptions on monumental stone ⟨the *lapidary* phrasing . . . and subtle condensations of emotions . . . reward attentive reading —G. A. Cardwell⟩
2 a : sculptured in or engraved on stone **b :** of or relating to precious stones or the art of cutting them

la·pil·lus \lə-ˈpi-ləs\ *noun, plural* **-li** \-ˌlī, -(ˌ)lē\ [Latin, diminutive of *lapis*] (1747)
: a small stony or glassy fragment of lava ejected in a volcanic eruption

lap·in \ˈla-pən\ *noun* [French] (1905)
1 : RABBIT; *specifically* **:** a castrated male rabbit
2 : rabbit fur usually sheared and dyed

la·pis la·zu·li \ˌlap-əs-ˈla-zə-lē, -ˈla-zhə-\ *noun* [Middle English, from Medieval Latin, from Latin *lapis* + Medieval Latin *lazuli,* genitive of *lazulum* lapis lazuli, from Arabic *lāzaward* — more at AZURE] (15th century)
: a semiprecious stone that is usually rich azure blue and is essentially a complex silicate often with spangles of iron pyrites — called also *lapis*

lap joint *noun* (1823)
: a joint made by overlapping two ends or edges and fastening them together
— lap–joint·ed \ˈlap-ˌjȯin-təd\ *adjective*

La·place transform \lə-ˈpläs-, -ˈplas-\ [Pierre Simon, Marquis de *Laplace*] (1942)
: a transformation of a function $f(x)$ into the function

$$g(t) = \int_0^\infty e^{-xt} f(x)dx$$

that is useful especially in reducing the solution of an ordinary linear differential equation with constant coefficients to the solution of a polynomial equation

Lapp \ˈlap\ *noun* [Swedish] (1859)
1 : a member of a people of northern Scandinavia, Finland, and the Kola Peninsula of northern Russia who are traditionally fishermen, nomadic herders of caribou, and hunters of sea mammals
2 : any or all of the closely related Finno-Ugric languages of the Lapps
— Lapp·ish *adjective or noun*

lap·pet \ˈla-pət\ *noun* (1573)
1 : a fold or flap on a garment or headdress
2 : a flat overlapping or hanging piece

lap robe *noun* (circa 1866)
: a covering (as a blanket) for the legs, lap, and feet especially of a passenger in a car or carriage

Lap·sang souchong \ˈläp-ˌsäŋ-, ˈlap-ˌsaŋ-\ *noun* [origin unknown] (circa 1878)
: a souchong tea having a pronounced smoky flavor and aroma

[1]**lapse** \ˈlaps\ *noun* [Latin *lapsus,* from *labi* to slip — more at SLEEP] (1526)
1 a : a slight error typically due to forgetfulness or inattention ⟨a *lapse* in table manners⟩ **b :** a temporary deviation or fall especially from a higher to a lower state ⟨a *lapse* from grace⟩
2 : a becoming less **:** DECLINE
3 a (1) **:** the termination of a right or privilege through neglect to exercise it within some limit of time (2) **:** termination of coverage for nonpayment of premiums **b :** INTERRUPTION, DISCONTINUANCE ⟨returned to college after a *lapse* of several years⟩
4 : an abandonment of religious faith **:** APOSTASY
5 : a passage of time; *also* **:** INTERVAL
synonym see ERROR

[2]**lapse** *verb* **lapsed; laps·ing** (1611)
intransitive verb
1 a : to fall from an attained and usually high level (as of morals or manners) to one much lower; *also* **:** to depart from an accepted pattern or standard **b :** to sink or slip gradually **:** SUBSIDE ⟨*lapsed* into unconsciousness⟩
2 : to go out of existence **:** CEASE ⟨after a few polite exchanges, the conversation *lapsed*⟩

3 : to pass from one proprietor to another or from an original owner by omission or negligence ⟨allowed the insurance policy to *lapse*⟩
4 a : to glide along **:** PASS ⟨time *lapses*⟩
transitive verb
: to let slip **:** FORFEIT ⟨all of those who have *lapsed* their membership —*AAUP Bulletin*⟩
— **laps·er** *noun*

lapsed *adjective* (1638)
of a person **:** having given up or allowed the lapse of a former position, relationship, or commitment ⟨a *lapsed* Catholic⟩

lapse rate *noun* (1918)
: the adiabatic rate of decrease of atmospheric temperature with increasing altitude

lap·strake \ˈlap-ˌstrāk\ *adjective* (1771)
: CLINKER-BUILT

lap·top \ˈlap-ˌtäp\ *adjective* (1984)
: of a size and design that makes operation and use on one's lap convenient ⟨a *laptop* personal computer⟩ — compare DESKTOP
— **laptop** *noun*

La·pu·tan \lə-ˈpyü-tᵊn\ *noun*
: an inhabitant of a flying island in Swift's *Gulliver's Travels* characterized by a neglect of useful occupations and a devotion to visionary projects
— **Laputan** *adjective*

lap·wing \ˈlap-ˌwiŋ\ *noun* [Middle English, by folk etymology from Old English *hlēapewince;* akin to Old English *hlēapan* to leap and to Old English *wincian* to wink] (14th century)
: a crested Old World plover (*Vanellus vanellus*) noted for its slow irregular flapping flight and shrill wailing cry; *also* **:** any of several related plovers

Lar \ˈlär\ *noun, plural* **Lar·es** \ˈlar-(ˌ)ēz, ˈler-\ [Latin — more at LARVA]
: a tutelary god or spirit associated with Vesta and the Penates as a guardian of the household by the ancient Romans

lar·board \ˈlär-bərd\ *noun* [Middle English *ladeborde*] (14th century)
: ⁵PORT
word history see STARBOARD
— **larboard** *adjective*

lar·ce·ner \ˈlärs-nər, ˈlär-sᵊn-ər\ *noun* (circa 1635)
: LARCENIST

lar·ce·nist \-nist, -sᵊn-ist\ *noun* (1803)
: a person who commits larceny

lar·ce·nous \-nəs, -sᵊn-əs\ *adjective* (1742)
1 : having the character of or constituting larceny
2 : committing larceny
— **lar·ce·nous·ly** *adverb*

lar·ce·ny \-nē, -sᵊn-ē\ *noun, plural* **-nies** [Middle English, from Middle French *larcin* theft, from Latin *latrocinium* robbery, from *latron-, latro* mercenary soldier, probably from (assumed) Greek *latrōn,* from Greek *latron* pay] (15th century)
: the unlawful taking of personal property with intent to deprive the rightful owner of it permanently

larch \ˈlärch\ *noun* [probably from German *Lärche,* from Middle High German *lerche,* from Latin *laric-, larix*] (1548)
: any of a genus (*Larix*) of trees of the pine family with short fascicled deciduous leaves; *also* **:** the wood of a larch

larch

¹lard \ˈlärd\ *transitive verb* (14th century)
1 a : to dress (meat) for cooking by inserting or covering with something (as strips of fat) **b :** to cover or soil with grease
2 : to decorate or intersperse with something ⟨the book is *larded* with anecdotes⟩

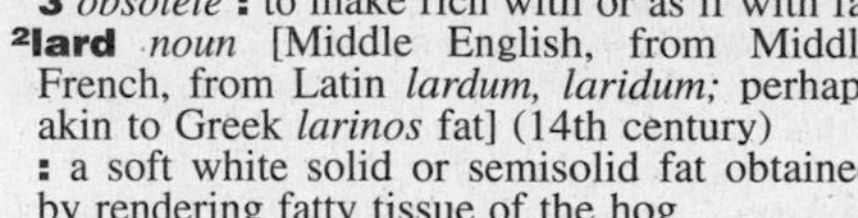

3 *obsolete* **:** to make rich with or as if with fat

²lard *noun* [Middle English, from Middle French, from Latin *lardum, laridum;* perhaps akin to Greek *larinos* fat] (14th century)
: a soft white solid or semisolid fat obtained by rendering fatty tissue of the hog
— **lardy** \ˈlär-dē\ *adjective*

lar·der \ˈlär-dər\ *noun* [Middle English, from Middle French *lardier,* from *lard*] (14th century)
1 : a place where food is stored **:** PANTRY
2 : a supply of food

lar·doon \lär-ˈdün\ *or* **lar·don** \ˈlär-ˌdän\ *noun* [Middle English, from Middle French *lardon* piece of fat pork, from *lard*] (14th century)
: a strip (as of salt pork) with which meat is larded

lares and penates *see* LAR, PENATES\ *noun plural* (1775)
1 : household gods
2 : personal or household effects

¹large \ˈlärj\ *adjective* **larg·er; larg·est** [Middle English, from Old French, from Latin *largus*] (12th century)
1 *obsolete* **:** LAVISH
2 *obsolete* **a :** AMPLE, ABUNDANT **b :** EXTENSIVE, BROAD
3 : having more than usual power, capacity, or scope **:** COMPREHENSIVE ⟨take the *large* view⟩ ⟨will take a *larger* role in the negotiations⟩
4 a : exceeding most other things of like kind especially in quantity or size **:** BIG **b :** dealing in great numbers or quantities ⟨a *large* and highly profitable business⟩
5 *obsolete* **a** *of language or expression* **:** COARSE, VULGAR **b :** lax in conduct **:** LOOSE
6 *of a wind* **:** FAVORABLE
7 : EXTRAVAGANT, BOASTFUL ⟨*large* talk⟩
— **large·ness** *noun*
— **larg·ish** \ˈlär-jish\ *adjective*

²large *adverb* (14th century)
1 *obsolete* **:** in abundance **:** AMPLY, LIBERALLY
2 : with the wind abaft the beam

³large *noun* (14th century)
obsolete **:** LIBERALITY, GENEROSITY
— **at large 1 a :** free of restraint or confinement ⟨the escaped prisoner is still *at large*⟩ **b :** without a specific subject or assignment ⟨critic *at large*⟩ **2 :** at length **3 :** in a general way **4 :** as a whole ⟨society *at large*⟩ **5 :** as the political representative of or to a whole area rather than of one of its subdivisions — used in combination with a preceding noun ⟨a congressman-*at-large*⟩
— **in the large :** on a large scale **:** in general

large calorie *noun* (circa 1909)
: CALORIE 1b

large·heart·ed \ˌlärj-ˈhär-təd\ *adjective* (1645)
: having a generous disposition **:** SYMPATHETIC
— **large·heart·ed·ness** *noun*

large intestine *noun* (1823)
: the more terminal division of the vertebrate intestine that is wider and shorter than the small intestine, typically divided into cecum, colon, and rectum, and concerned especially with the resorption of water and the formation of feces

large·ly \ˈlärj-lē\ *adverb* (13th century)
: in a large manner; *especially* **:** to a large extent **:** MOSTLY, PRIMARILY

large–mind·ed \ˌlärj-ˈmīn-dəd\ *adjective* (1725)
: generous or comprehensive in outlook, range, or capacity
— **large–mind·ed·ly** *adverb*
— **large–mind·ed·ness** *noun*

large·mouth bass \ˈlärj-ˌmau̇th-\ *noun* (1941)
: a large black bass (*Micropterus salmoides*) that is blackish green above and lighter below and has the maxillary bones of the upper jaw extending behind the eyes — called also *largemouth, largemouth black bass*

large–print \-ˈprint\ *adjective* (1968)
: being set in a large size of type (as 14 point or larger) especially for use by the partially sighted ⟨*large-print* books⟩

larger–than–life *adjective* (1950)
: of the sort legends are made of ⟨*larger-than-life* heroes⟩

large–scale integration *noun* (1966)
: the process of placing a large number of circuits on a small chip

lar·gesse *or* **lar·gess** \lär-ˈzhes, lär-ˈjes *also* ˈlär-ˌjes\ *noun* [Middle English *largesse,* from Old French, from *large*] (13th century)
1 : liberal giving (as of money) to or as if to an inferior; *also* **:** something so given
2 : GENEROSITY

¹lar·ghet·to \lär-ˈge-(ˌ)tō\ *noun, plural* **-tos** (circa 1724)
: a movement played larghetto

²larghetto *adverb or adjective* [Italian, somewhat slow, from *largo*] (circa 1801)
: slower than andante but not so slow as largo — used as a direction in music

¹lar·go \ˈlär-(ˌ)gō\ *adverb or adjective* [Italian, slow, broad, from Latin *largus* abundant] (1683)
: at a very slow tempo — used as a direction in music

²largo *noun, plural* **largos** (circa 1724)
: a largo movement

lar·i·at \ˈlar-ē-ət, ˈler-\ *noun* [American Spanish *la reata* the lasso, from Spanish *la* the (feminine of *el,* from Latin *ille* that) + American Spanish *reata* lasso, from Spanish *reatar* to tie again, from *re-* + *atar* to tie, from Latin *aptare* to fit — more at ADAPT] (1832)
: a long light rope (as of hemp or leather) used with a running noose to catch livestock or with or without the noose to tether grazing animals **:** LASSO

¹lark \ˈlärk\ *noun* [Middle English *laveroc, laverke,* from Old English *lāwerce;* akin to Old High German *lērihha* lark] (before 12th century)
: any of a family (Alaudidae) of chiefly Old World songbirds that are usually brownish in color; *especially* **:** SKYLARK — compare MEADOWLARK

²lark *noun* [³*lark*] (circa 1811)
: something done solely for fun or adventure ⟨ran for office on a *lark*⟩

³lark *intransitive verb* [probably alteration of *lake* to frolic] (1813)
: to engage in harmless fun or mischief — often used with *about*
— **lark·er** *noun*

lark·spur \ˈlärk-ˌspər\ *noun* (1578)
1 : DELPHINIUM
2 : any of the delphiniums that are annuals, have the upper two petals of the corolla united and the bottom two missing, are now often placed in a separate genus (*Consolida*), and include several widely cultivated forms

larky \ˈlär-kē\ *adjective* **lark·i·er; -est** (1841)
1 : given to or ready for larking **:** SPORTIVE
2 : resulting from a lark
— **lark·i·ness** \-nəs\ *noun*

lar·ri·gan \ˈlar-i-gən\ *noun* [origin unknown] (1886)
: an oil-tanned moccasin with a leg often reaching the knee

lar·ri·kin \ˈlar-i-kən\ *noun* [origin unknown] (1868)
chiefly Australian **:** HOODLUM, ROWDY
— **larrikin** *adjective*

¹lar·rup \ˈlar-əp\ *noun* [origin unknown] (circa 1820)
dialect **:** ⁵BLOW

²larrup (circa 1823)
transitive verb
1 *dialect* **:** to flog soundly **:** WHIP

\ə\ **abut** \ᵊ\ kitten \ər\ **further** \a\ ash \ā\ ace \ä\ mop, mar \au̇\ **out** \ch\ **chin** \e\ bet \ē\ **easy** \g\ go \i\ hit \ī\ ice \j\ job \ŋ\ sing \ō\ go \ȯ\ law \ȯi\ boy \th\ **thin** \t͟h\ **the** \ü\ loot \u̇\ foot \y\ yet \zh\ vision *see also* Guide to Pronunciation

2 *dialect* **:** to defeat decisively **:** TROUNCE
intransitive verb
dialect **:** to move indolently or clumsily

la·rum \'lär-əm, 'lar-\ *noun* [short for *alarum*] (15th century)
archaic **:** ALARM

lar·va \'lär-və\ *noun, plural* **lar·vae** \-(,)vē, -,vī\ *also* **larvas** [New Latin, from Latin, specter, mask; akin to Latin *lar* Lar] (1768)
1 : the immature, wingless, and often wormlike feeding form that hatches from the egg of many insects, alters chiefly in size while passing through several molts, and is finally transformed into a pupa or chrysalis from which the adult emerges
2 : the early form of an animal (as a frog or sea urchin) that at birth or hatching is fundamentally unlike its parent and must metamorphose before assuming the adult characters ◇
— lar·val \-vəl\ *adjective*

lar·vi·cide \'lär-və-,sīd\ *noun* (circa 1888)
: an agent for killing larval pests
— lar·vi·cid·al \,lär-və-'sī-dᵊl\ *adjective*

laryng- *or* **laryngo-** *combining form* [New Latin, from Greek, from *laryng-, larynx*]
: larynx ⟨*laryng*itis⟩

¹la·ryn·geal \lə-'rin-jəl *also* -jē-əl; ,lar-ən-'jē-əl\ *adjective* (1795)
1 : of, relating to, or used on the larynx
2 : produced by or with constriction of the larynx ⟨*laryngeal* articulation of sounds⟩

²laryngeal *noun* (circa 1902)
1 : an anatomical part (as a nerve or artery) that supplies or is associated with the larynx
2 a : a laryngeal sound **b :** any of a set of several conjectured phonemes reconstructed for Proto-Indo-European chiefly on indirect evidence

lar·yn·gec·to·mee \,lar-ən-,jek-tə-'mē\ *noun* (1956)
: a person who has undergone laryngectomy

lar·yn·gec·to·my \-'jek-tə-mē\ *noun, plural* **-mies** (circa 1888)
: surgical removal of all or part of the larynx
— lar·yn·gec·to·mized \-tə-,mīzd\ *adjective*

lar·yn·gi·tis \,lar-ən-'jī-təs\ *noun* [New Latin] (circa 1834)
: inflammation of the larynx
— lar·yn·git·ic \-'ji-tik\ *adjective*

lar·yn·gol·o·gy \,lar-ən-'gä-lə-jē\ *noun* [International Scientific Vocabulary] (circa 1842)
: a branch of medicine dealing with diseases of the larynx and nasopharynx

la·ryn·go·scope \lə-'riŋ-gə-,skōp, -'rin-jə-\ *noun* [International Scientific Vocabulary] (1860)
: an instrument for examining the interior of the larynx
— lar·yn·gos·co·py \,lar-ən-'gäs-kə-pē\ *noun*

lar·ynx \'lar-iŋ(k)s\ *noun, plural* **la·ryn·ges** \lə-'rin-(,)jēz\ *or* **lar·ynx·es** [New Latin *laryng-, larynx,* from Greek] (1578)
: the modified upper part of the trachea of air-breathing vertebrates that in humans, most other mammals, and a few lower forms contains the vocal cords

la·sa·gna \lə-'zän-yə\ *noun* [Italian *lasagna,* from (assumed) Vulgar Latin *lasania* cooking pot, its contents, from Latin *lasanum* chamber pot, from Greek *lasanon*] (1846)
1 *also* **la·sa·gne** \-yə\ **:** pasta in the form of broad often ruffled ribbons
2 : a baked dish consisting chiefly of layers of boiled lasagna, cheese, and a seasoned sauce of tomatoes and usually meat

las·car \'las-kər\ *noun* [Hindi *lashkar* army] (1615)
: an Indian sailor, army servant, or artilleryman

las·civ·i·ous \lə-'si-vē-əs\ *adjective* [Late Latin *lasciviosus,* from Latin *lascivia* wantonness, from *lascivus* wanton — more at LUST] (15th century)
: LEWD, LUSTFUL
— las·civ·i·ous·ly *adverb*
— las·civ·i·ous·ness *noun*

lase \'lāz\ *intransitive verb* **lased; las·ing** [back-formation from *laser*] (1962)
: to emit coherent light

la·ser \'lā-zər\ *noun, often attributive* [*l*ight *a*mplification by *s*timulated *e*mission of *r*adiation] (1960)
: a device that utilizes the natural oscillations of atoms or molecules between energy levels for generating coherent electromagnetic radiation usually in the ultraviolet, visible, or infrared regions of the spectrum

laser disc *noun* (1980)
: OPTICAL DISK; *especially* **:** one on which programs are recorded for playback on a television set

laser printer *noun* (1979)
: a high-resolution printer for computer output that xerographically prints an image formed by a laser

¹lash \'lash\ *verb* [Middle English] (14th century)
intransitive verb
1 : to move violently or suddenly **:** DASH
2 : to thrash or beat violently ⟨rain *lashed* at the windowpanes⟩
3 : to make a verbal attack or retort — usually used with *out*
transitive verb
1 a : to whip or fling about violently ⟨the big cat *lashed* its tail about threateningly⟩ **b :** to strike or beat with or as if with a whip ⟨waves *lashed* the shore⟩
2 a : to assail with stinging words **b :** DRIVE, WHIP ⟨*lashed* them to fury with his speech⟩
— lash·er *noun*

²lash *noun* (14th century)
1 a (1) **:** a stroke with or as if with a whip (2) **:** the flexible part of a whip; *also* **:** WHIP **b :** punishment by whipping
2 : a beating, whipping, or driving force
3 : a stinging rebuke
4 : EYELASH
5 : the clearance or play between adjacent movable mechanical parts

³lash *transitive verb* [Middle English *lasschyn* to lace, from Middle French *lacier* — more at LACE] (1624)
: to bind with or as if with a line
— lash·er *noun*

lash·ing *noun* (1669)
: something used for binding, wrapping, or fastening

lash·ings \'la-shiŋz, -shənz\ *also* **lash·ins** \-shənz\ *noun plural* [from gerund of ¹*lash*] (1829)
chiefly British **:** a great plenty **:** ABUNDANCE ⟨piles of bread and butter and *lashings* of tea —Molly Weir⟩

lash–up \'lash-,əp\ *noun* [³*lash*] (1898)
1 : something hastily put together or improvised
2 : OUTFIT 3

L–as·par·a·gi·nase \'el-as-'par-ə-jə-,nās, -,nāz\ *noun* (1962)
: an enzyme that breaks down the physiologically commoner form of asparagine, is obtained especially from bacteria, and is used especially to treat leukemia

lass \'las\ *noun* [Middle English *las*] (14th century)
1 : a young woman **:** GIRL
2 : SWEETHEART

Las·sa fever \'la-sə-\ *noun* [*Lassa,* village in Nigeria] (1970)
: a virus disease especially of Africa that is characterized by a high fever, headaches, mouth ulcers, muscle aches, small hemorrhages under the skin, heart and kidney failure, and a high mortality rate

lass·ie \'la-sē\ *noun* (1725)
: LASS 1

las·si·tude \'la-sə-,tüd, -,tyüd\ *noun* [Middle French, from Latin *lassitudo,* from *lassus* weary; probably akin to Old English *læt* late — more at LATE] (15th century)
1 : a condition of weariness or debility **:** FATIGUE
2 : a condition of listlessness **:** LANGUOR
synonym see LETHARGY

¹las·so \'la-(,)sō, la-'sü\ *transitive verb* (1807)
: to capture with or as if with a lasso **:** ROPE
— las·so·er *noun*

²lasso *noun, plural* **lassos** *or* **lassoes** [Spanish *lazo,* from Latin *laqueus* snare] (1808)
: a rope or long thong of leather with a noose used especially for catching horses and cattle **:** LARIAT

¹last \'last\ *verb* [Middle English, from Old English *lǣstan* to last, follow; akin to Old English *lāst* footprint] (before 12th century)
intransitive verb
1 : to continue in time
2 a : to remain fresh or unimpaired **:** ENDURE **b :** to manage to continue (as in a course of action) **c :** to continue to live
transitive verb
1 : to continue in existence or action as long as or longer than — often used with *out* ⟨couldn't *last* out the training program⟩
2 : to be enough for the needs of ⟨the supplies will *last* them a week⟩
synonym see CONTINUE
word history see LEARN
— last·er *noun*

²last *noun* [Middle English, from Old English *lǣste,* from *lāst* footprint; akin to Old High German *leist* shoemaker's last, Latin *lira* furrow — more at LEARN] (before 12th century)
: a form (as of metal or plastic) which is shaped like the human foot and over which a shoe is shaped or repaired
word history see LEARN

³last *transitive verb* (circa 1859)
: to shape with a last
— last·er *noun*

⁴last *adverb* [Middle English, from Old English *latost,* superlative of *læt* late] (before 12th century)
1 : after all others **:** at the end ⟨came *last* and left first⟩
2 : most lately ⟨saw him *last* in Rome⟩
3 : in conclusion ⟨*last,* let's consider the social aspect⟩

⁵last *adjective* (13th century)
1 a : following all the rest ⟨he was the *last* one out⟩ **b :** being the only remaining ⟨our *last* dollar⟩
2 a : belonging to the final stage (as of life) ⟨his *last* hours on earth⟩ **b :** administered to the seriously sick or dying ⟨the *last* rites of the church⟩
3 a : next before the present **:** most recent ⟨*last* week⟩ ⟨his *last* book was a failure⟩ **b :** most up-to-date **:** LATEST ⟨it's the *last* thing in fashion⟩

◇ WORD HISTORY

larva The designations *larva, pupa,* and *imago* for the three post-egg stages of an insect's life cycle are usually credited to Carolus Linnaeus, the founder of modern biological taxonomy. Biologists of the 18th century saw the adult as the only genuine form of the insect, and hence called it the *imago,* the proper "image" or representation of the creature as it actually is. The preceding stages were considered somehow unreal, or at best misleading, and so were named with the Latin words *pupa* "doll" and *larva,* which in classical Latin was a sort of ghostlike demon or a mask representing a demon. Although to modern biology larval stages such as caterpillars are as real and taxonomically valid as adults, the traditional terminology remains with us.

4 a : lowest in rank or standing; *also* **:** WORST **b :** farthest from a specified quality, attitude, or likelihood ⟨would be the *last* person to fall for flattery⟩
5 a : CONCLUSIVE ⟨there is no *last* answer to the problem⟩ **b :** highest in degree **:** SUPREME, ULTIMATE **c :** DISTINCT, SEPARATE — used as an intensive ⟨ate every *last* piece of food⟩ ☆
— last·ly *adverb*

⁶last *noun* (13th century)
: something that is last
— at last *or* **at long last :** at the end of a period of time **:** FINALLY ⟨*at last* you've come home⟩

last–ditch \'las(t)-ˌdich, -'dich\ *adjective* (1937)
1 : fought or conducted from the last ditch **:** waged with desperation or unyielding defiance ⟨*last-ditch* resistance⟩
2 : made as a final effort especially to avert disaster ⟨a *last-ditch* attempt to raise the money⟩

last ditch *noun* (circa 1715)
: a place of final defense or resort

last–gasp \'las(t)-'gasp\ *adjective* (1921)
: done or coming at the very end
— last gasp *noun*

last hurrah *noun* [from *The Last Hurrah* (1956) by Edwin O'Connor (died 1968) American novelist] (1966)
: a final often valedictory effort, production, or appearance ⟨his unsuccessful Senate run was his *last hurrah* —R. W. Daly⟩

last–in first–out *adjective* (1940)
: of, relating to, or being a method of inventory accounting that values stock on hand according to costs at the time of acquisition and not according to the cost of replacement

¹last·ing *adjective* (12th century)
: existing or continuing a long while **:** ENDURING ☆
— last·ing·ly \'las-tiŋ-lē\ *adverb*
— last·ing·ness *noun*

²lasting *noun* (15th century)
1 *archaic* **:** long life
2 : a sturdy cotton or worsted cloth used especially in shoes and luggage

Last Judgment *noun* (14th century)
: the judgment of mankind before God at the end of the world

last minute *noun* (1920)
: the moment just before some climactic, decisive, or disastrous event

last name *noun* (1897)
: SURNAME 2

last rites *noun* (1922)
: EXTREME UNCTION

last straw *noun* [from the fable of the last straw that broke the camel's back when added to its burden] (1848)
: the last of a series (as of events or indignities) that brings one beyond the point of endurance

Last Supper *noun* (14th century)
: the supper eaten by Jesus and his disciples on the night of his betrayal

Last Things *noun plural* [translation of Medieval Latin *Novissima*] (1522)
: events (as the resurrection and divine judgment of all humankind) marking the end of the world **:** eschatological happenings

last word *noun* (1563)
1 : the final remark in a verbal exchange
2 a : the power of final decision **b :** a definitive statement or treatment ⟨this study will surely be the *last word* on the subject for many years⟩
3 : the most advanced, up-to-date, or fashionable exemplar of its kind ⟨the *last word* in sports cars⟩

lat \'lat\ *noun* [Latvian *lats,* from *Latvija* Latvia] (1923)
— see MONEY table

lat·a·kia \ˌla-tə-'kē-ə\ *noun* [*Latakia,* seaport in Syria] (1833)
: a highly aromatic Turkish smoking tobacco

¹latch \'lach\ *intransitive verb* [Middle English *lachen,* from Old English *læccan;* perhaps akin to Greek *lambanein* to take, seize] (13th century)
1 : to lay hold with or as if with the hands or arms — used with *on* or *onto*
2 : to associate oneself intimately and often artfully — used with *on* or *onto* ⟨*latched* onto a rich widow⟩

²latch *noun* (13th century)
: any of various devices in which mating mechanical parts engage to fasten but usually not to lock something: **a :** a fastener (as for a door) consisting essentially of a pivoted bar that falls into a notch **b :** a fastener (as for a door) in which a spring slides a bolt into a hole; *also* **:** NIGHT LATCH

³latch *transitive verb* (15th century)
: to make fast with or as if with a latch

latch·et \'la-chət\ *noun* [Middle English *lachet,* from Middle French, shoestring, from *laz* snare, from Latin *laqueus*] (15th century)
: a narrow leather strap, thong, or lace that fastens a shoe or sandal on the foot

latch·key \'lach-ˌkē\ *noun* (1825)
: a key to an outside and especially a front door

latchkey child *noun* (1944)
: a young child of working parents who must spend part of the day unsupervised (as at home) — called also *latchkey kid*

latch·string \-ˌstriŋ\ *noun* (1791)
: a string on a latch that may be left hanging outside the door to permit the raising of the latch from the outside or drawn inside to prevent intrusion

¹late \'lāt\ *adjective* **lat·er; lat·est** [Middle English, late, slow, from Old English *læt;* akin to Old High German *laz* slow, Old English *lǣtan* to let] (before 12th century)
1 a (1) **:** coming or remaining after the due, usual, or proper time ⟨a *late* spring⟩ (2) **:** of, relating to, or imposed because of tardiness **b :** of or relating to an advanced stage in point of time or development ⟨the *late* Middle Ages⟩; *especially* **:** far advanced toward the close of the day or night ⟨*late* hours⟩
2 a : living comparatively recently — used of persons ⟨the *late* John Doe⟩ and often with reference to a specific relationship or status ⟨his *late* wife⟩ **b :** being something or holding some position or relationship recently but not now ⟨the *late* belligerents⟩ **c :** made, appearing, or happening just previous to the present time especially as the most recent of a succession ⟨our *late* quarrel⟩ ■
synonym see DEAD
— late·ness *noun*

²late *adverb* **lat·er; lat·est** (before 12th century)
1 a : after the usual or proper time ⟨got to work *late*⟩ **b :** at or to an advanced point of time
2 : not long ago **:** RECENTLY ⟨a man *late* of Chicago⟩
— of late : in the period shortly or immediately preceding **:** RECENTLY ⟨has been sick *of late*⟩

late blight *noun* (1900)
: a disease of solanaceous plants (as the potato and tomato) that is caused by a fungus (*Phytophthora infestans*) and is characterized by decay of stems, leaves, and in the potato also of tubers

late·com·er \'lāt-ˌkə-mər\ *noun* (1892)
: one that arrives late; *also* **:** a recent arrival

lat·ed \'lā-təd\ *adjective* (circa 1592)
: BELATED

¹la·teen \lə-'tēn\ *adjective* [French (*voile*) *latine,* literally, Latin (Mediterranean) sail] (circa 1741)
: being or relating to a rig used especially on the north coast of Africa and characterized by a triangular sail extended by a long spar slung to a low mast

²lateen *noun* (circa 1775)
1 *also* **la·teen·er** \-'tē-nər\ **:** a lateen-rigged ship
2 : a lateen sail

Late Greek *noun* (circa 1889)
: the Greek language as used in the 3d to 6th centuries

Late Hebrew *noun* (1951)
: the Hebrew language used by writers from about the 2d century B.C. to the early Middle Ages

Late Latin *noun* (1888)
: the Latin language used by writers in the 3d to 6th centuries

☆ SYNONYMS

Last, final, terminal, eventual, ultimate mean following all others (as in time, order, or importance). LAST applies to something that comes at the end of a series but does not always imply that the series is completed or stopped ⟨*last* page of a book⟩ ⟨*last* news we had of him⟩. FINAL applies to that which definitely closes a series, process, or progress ⟨*final* day of school⟩. TERMINAL may indicate a limit of extension, growth, or development ⟨*terminal* phase of a disease⟩. EVENTUAL applies to something that is bound to follow sooner or later as the final effect of causes already operating ⟨*eventual* defeat of the enemy⟩. ULTIMATE implies the last degree or stage of a long process beyond which further progress or change is impossible ⟨the *ultimate* collapse of the system⟩.

Lasting, permanent, durable, stable mean enduring for so long as to seem fixed or established. LASTING implies a capacity to continue indefinitely ⟨a book that left a *lasting* impression on me⟩. PERMANENT adds usually the implication of being designed or planned to stand or continue indefinitely ⟨*permanent* living arrangements⟩. DURABLE implies power to resist destructive agencies ⟨*durable* fabrics⟩. STABLE implies lastingness because of resistance to being overturned or displaced ⟨a *stable* government⟩.

□ USAGE

late The definition of sense 2a above contains the words "comparatively recently." One of the hardest things for the user of this sense of *late* to decide is how long the word applies. The commentators—mostly newspaper editors—who have given the subject some thought have reached no consensus; their advice runs from "within the preceding decade" to "half a century." This certainly gives you some latitude, and our evidence shows that writers have in general taken advantage of that latitude. Two strictures are placed on the use of *late* by various critics. The first is that *late* should not be used of famous people everyone knows are dead; a couple of instances will show how little this is observed ⟨the *late* Henry Ford —*N.Y. Times*⟩ ⟨the *late* President John F. Kennedy —*Current Biography*⟩. The other is that *late* should not be used if the person was alive at the time referred to; this too is little observed ⟨the Senate Select Intelligence Committee, then . . . headed by the *late* Senator Frank Church —David Burnham⟩ ⟨my *late* sister Lucy at one time decided to be on the safe side —George Bernard Shaw⟩.

late·ly \'lāt-lē\ *adverb* (15th century)
: of late : RECENTLY ⟨has been friendlier *lately*⟩
lat·en \'lā-tᵊn\ *verb* **lat·ened; lat·en·ing** \'lāt-niŋ, 'lā-tᵊn-iŋ\ (1880)
intransitive verb
: to grow late
transitive verb
: to cause to grow late
la·ten·cy \'lā-tᵊn(t)-sē\ *noun, plural* **-cies** (circa 1638)
1 : the quality or state of being latent : DORMANCY
2 : something latent
3 : a stage of personality development that extends from about the age of five to the beginning of puberty and during which sexual urges often appear to lie dormant
4 : LATENT PERIOD 2
latency period *noun* (1910)
1 : LATENCY 3
2 : LATENT PERIOD
La Tène \lä-'ten, -'tän\ *adjective* [*La Tène,* shallows of the Lake of Neuchâtel, Switzerland] (1901)
: of or relating to the later period of the Iron Age in Europe assumed to date from 500 B.C. to A.D. 1
la·ten·si·fi·ca·tion \lā-,ten(t)-sə-fə-'kā-shən, lə-\ *noun* [blend of [1]*latent* and *intensification*] (1940)
: intensification of a latent photographic image by chemical treatment or exposure to light of low intensity
[1]**la·tent** \'lā-tᵊnt\ *adjective* [Middle English, from Latin *latent-, latens,* from present participle of *latēre* to lie hidden; akin to Greek *lanthanein* to escape notice] (15th century)
: present and capable of becoming though not now visible, obvious, or active ⟨a *latent* infection⟩ ☆
— **la·tent·ly** *adverb*
[2]**latent** *noun* (1923)
: a fingerprint (as at the scene of a crime) that is scarcely visible but can be developed for study
latent heat *noun* (circa 1757)
: heat given off or absorbed in a process (as fusion or vaporization) other than a change of temperature
latent period *noun* (1837)
1 : the incubation period of a disease
2 : the interval between stimulation and response
latent root *noun* (1883)
: an eigenvalue of a matrix
lat·er \'lā-tər\ *adverb* (13th century)
: at some time subsequent to a given time : SUBSEQUENTLY, AFTERWARD ⟨four months *later*⟩ ⟨they *later* regretted their decision⟩ — often used with *on* ⟨experience that will be useful *later on*⟩
-later *noun combining form* [Middle English *-latrer,* from Middle French *-latre,* from Late Latin *-latres,* from Greek *-latrēs;* akin to Greek *latron* pay]
: worshiper ⟨biblio*later*⟩
lat·er·ad \'la-tə-,rad\ *adverb* [Latin *later-, latus*] (1814)
: toward the side
[1]**lat·er·al** \'la-tə-rəl *also* 'la-trəl\ *adjective* [Middle English *laterale,* from Latin *lateralis,* from *later-, latus* side] (15th century)
1 : of or relating to the side
2 : situated on, directed toward, or coming from the side
3 : extending from side to side ⟨*lateral* axis of an airplane⟩
4 : produced with passage of breath around the side of a constriction formed with the tongue ⟨\l\ is *lateral*⟩
— **lat·er·al·ly** *adverb*
[2]**lateral** *noun* (1851)
1 : a branch from the main part (as in an irrigation or electrical system)
2 : a pass in football thrown parallel to the line of scrimmage or in a direction away from the opponent's goal
3 : a lateral speech sound
[3]**lateral** *intransitive verb* (1944)
: to throw a lateral
lateral bud *noun* (circa 1892)
: a bud that develops in the axil between a petiole and a stem
lat·er·al·i·za·tion \,la-tə-rə-lə-'zā-shən, ,la-trə-\ *noun* (circa 1899)
: localization of function or activity (as of verbal processes in the brain) on one side of the body in preference to the other
— **lat·er·al·ize** \'la-tə-rə-,līz, 'la-trə-\ *transitive verb*
lateral line *noun* (1870)
: a canal along the side of a fish containing pores that open into tubes supplied with sense organs sensitive to low vibrations; *also* : one of these tubes or sense organs
lat·er·ite \'la-tə-,rīt\ *noun* [Latin *later* brick] (1807)
: a residual product of rock decay that is red in color and has a high content in the oxides of iron and hydroxide of aluminum
— **lat·er·it·ic** \,la-tə-'ri-tik\ *adjective*
lat·er·i·za·tion \,la-tə-rə-'zā-shən\ *noun* (circa 1882)
: the process of conversion of rock to laterite
[1]**lat·est** \'lā-təst\ *adjective* (1588)
1 *archaic* : LAST
2 : most recent ⟨the *latest* news⟩ ⟨the *latest* style⟩
[2]**latest** *noun* (1884)
1 : the latest acceptable time — usually used in the phrase *at the latest*
2 : something that is the most recent or currently fashionable ⟨the *latest* in diving techniques⟩
late·wood \'lāt-,wu̇d\ *noun* (circa 1933)
: SUMMERWOOD
la·tex \'lā-,teks\ *noun, plural* **la·ti·ces** \'lā-tə-,sēz, 'la-\ *or* **la·tex·es** [New Latin *latic-, latex,* from Latin, fluid] (1835)
1 : a milky usually white fluid that is produced by cells of various seed plants (as of the milkweed, spurge, and poppy families) and is the source of rubber, gutta-percha, chicle, and balata
2 : a water emulsion of a synthetic rubber or plastic obtained by polymerization and used especially in coatings (as paint) and adhesives
lath \'lath *also* 'lath\ *noun, plural* **laths** *or* **lath** [Middle English, from (assumed) Old English *lætth-;* akin to Old High German *latta* lath, Welsh *llath* yard] (13th century)
1 : a thin narrow strip of wood nailed to rafters, joists, or studding as a groundwork for slates, tiles, or plaster
2 : a building material in sheets used as a base for plaster
3 : a quantity of laths
— **lath** *transitive verb*
[1]**lathe** \'lāth\ *noun* [probably from Middle English *lath* supporting stand] (circa 1611)
: a machine in which work is rotated about a horizontal axis and shaped by a fixed tool
[2]**lathe** *transitive verb* **lathed; lath·ing** (circa 1903)
: to cut or shape with a lathe
[1]**lath·er** \'la-thər\ *noun* [(assumed) Middle English, from Old English *lēathor;* akin to Latin *lavere* to wash — more at LYE] (before 12th century)
1 a : a foam or froth formed when a detergent (as soap) is agitated in water **b :** foam or froth from profuse sweating (as on a horse)
2 : an agitated or overwrought state : DITHER
— **lath·ery** \-th(ə-)rē\ *adjective*
[2]**lather** *verb* **lath·ered; lath·er·ing** \-th(ə-)riŋ\ (before 12th century)
transitive verb
1 : to spread lather over
2 : to beat severely : FLOG
intransitive verb
: to form a lather or a froth like lather
— **lath·er·er** \-thər-ər\ *noun*
lath·y·rism \'la-thə-,ri-zəm\ *noun* [New Latin *Lathyrus,* from Greek *lathyros,* a type of pea] (circa 1888)
: a diseased condition of humans, domestic animals, and especially horses that results from poisoning by a substance found in some legumes (genus *Lathyrus* and especially *L. sativus*) and is characterized especially by spastic paralysis of the hind or lower limbs
lath·y·rit·ic \,la-thə-'ri-tik\ *adjective* (1960)
: of, relating to, affected with, or characteristic of lathyrism ⟨*lathyritic* rats⟩ ⟨*lathyritic* cartilage⟩
latices *plural of* LATEX
la·tic·i·fer \lā-'ti-sə-fər\ *noun* [International Scientific Vocabulary *latici-* (from New Latin *latic-, latex*) + *-fer*] (circa 1928)
: a plant cell or vessel that contains latex
la·ti·fun·dio \,lä-tə-'fün-dē-,ō\ *noun, plural* **-di·os** [Spanish, from Latin *latifundium*] (circa 1902)
: a latifundium in Spain or Latin America
lat·i·fun·di·um \,la-tə-'fən-dē-əm\ *noun, plural* **-dia** \-dē-ə\ [Latin, from *latus* wide + *fundus* piece of landed property, foundation, bottom — more at BOTTOM] (1630)
: a great landed estate with primitive agriculture and labor often in a state of partial servitude
lat·i·go \'la-ti-,gō\ *noun, plural* **-gos** *also* **-goes** [Spanish *látigo*] (1873)
chiefly West : a long strap on a saddletree of a western saddle to adjust the cinch
[1]**Lat·in** \'la-tᵊn\ *adjective* [Middle English, from Old English, from Latin *Latinus,* from *Latium,* ancient country of Italy] (before 12th century)
1 a : of, relating to, or composed in Latin **b** : ROMANCE
2 : of or relating to Latium or the Latins
3 : of or relating to the part of the Catholic Church that until recently used a Latin rite and forms the patriarchate of the pope
4 : of or relating to the peoples or countries using Romance languages; *specifically* : of or relating to the peoples or countries of Latin America
[2]**Latin** *noun* (before 12th century)
1 : the Italic language of ancient Latium and of Rome and until modern times the dominant language of school, church, and state in western Europe — see INDO-EUROPEAN LANGUAGES table
2 : a member of the people of ancient Latium
3 : a Catholic of the Latin rite
4 : a member of one of the Latin peoples; *specifically* : a native or inhabitant of Latin America
5 : LATIN ALPHABET
Latin alphabet *noun* (1823)
: an alphabet that was used for writing Latin and that has been modified for writing many modern languages
Latin Americanist *noun* (1972)
: a specialist in Latin American civilization
Lat·in·ate \'la-tᵊn-,āt\ *adjective* (1904)

☆ SYNONYMS
Latent, dormant, quiescent, potential mean not now showing signs of activity or existence. LATENT applies to a power or quality that has not yet come forth but may emerge and develop ⟨a *latent* desire for success⟩. DORMANT suggests the inactivity of something (as a feeling or power) as though sleeping ⟨their passion had lain *dormant*⟩. QUIESCENT suggests a usually temporary cessation of activity ⟨the disease was *quiescent*⟩. POTENTIAL applies to what does not yet have existence or effect but is likely soon to have ⟨a *potential* disaster⟩.

: of, relating to, resembling, or derived from Latin

Latin cross *noun* (1797)
: a figure of a cross having a long upright shaft and a shorter crossbar traversing it above the middle — see CROSS illustration

Lat·in·ism \ˈla-tᵊn-ˌi-zəm\ *noun* (circa 1570)
1 : a characteristic feature of Latin occurring in another language
2 : Latin quality or character

Lat·in·ist \ˈla-tᵊn-ist, ˈlat-nist\ *noun* (15th century)
: a specialist in the Latin language or Roman culture

la·tin·i·ty \la-ˈti-nə-tē, lə-\ *noun, often capitalized* (1619)
1 : a manner of speaking or writing Latin
2 : LATINISM 2

lat·in·ize \ˈla-tᵊn-ˌīz\ *verb* **-ized; -iz·ing** *often capitalized* (1589)
transitive verb
1 a *obsolete* **:** to translate into Latin **b :** to give a Latin form to **c :** to introduce Latinisms into **d :** ROMANIZE 2
2 : to make Latin or Italian in doctrine, ideas, or traits; *specifically* **:** to cause to resemble the Roman Catholic Church
intransitive verb
1 : to use Latinisms
2 : to exhibit the influence of the Romans or of the Roman Catholic Church
— lat·in·i·za·tion \ˌla-tᵊn-ə-ˈzā-shən, ˌlat-nə-\ *noun*

La·ti·no \lə-ˈtē-(ˌ)nō\ *noun, plural* **-nos** [American Spanish, probably short for *latinoamericano* Latin American] (1946)
1 : a native or inhabitant of Latin America
2 : a person of Latin-American origin living in the U.S.
— Latino *adjective*

Latin square *noun* (1890)
: a square array which contains *n* different elements with each element occurring *n* times but with no element occurring twice in the same column or row and which is used especially in the statistical design of experiments (as in agriculture)

lat·ish \ˈlā-tish\ *adjective* (1611)
: somewhat late

lat·i·tude \ˈla-tə-ˌtüd, -ˌtyüd\ *noun* [Middle English, from Latin *latitudin-, latitudo,* from *latus* wide; akin to Old Church Slavonic *postĭlati* to spread] (14th century)
1 *archaic* **:** extent or distance from side to side **:** WIDTH
2 : angular distance from some specified circle or plane of reference: as **a :** angular distance north or south from the earth's equator measured through 90 degrees **b :** angular distance of a celestial body from the ecliptic **c :** a region or locality as marked by its latitude

latitude 2a: hemisphere marked with parallels of latitude

3 a *archaic* **:** SCOPE, RANGE **b :** the range of exposures within which a film or plate will produce a negative or positive of satisfactory quality
4 : freedom of action or choice
— lat·i·tu·di·nal \ˌla-tə-ˈtüd-nəl, -ˈtyüd-; -ˈtü-dᵊn-əl, -ˈtyü-\ *adjective*
— lat·i·tu·di·nal·ly *adverb*

lat·i·tu·di·nar·i·an \ˌla-tə-ˌtü-dᵊn-ˈer-ē-ən, -ˌtyü-\ *noun* (1662)
: a person who is broad and liberal in standards of religious belief and conduct
— latitudinarian *adjective*
— lat·i·tu·di·nar·i·an·ism \-ē-ə-ˌni-zəm\ *noun*

lat·ke \ˈlät-kə\ *noun* [Yiddish, pancake, from Ukrainian *oladka*] (1927)
: POTATO PANCAKE

lat·o·sol \ˈla-tə-ˌsȯl\ *noun* [irregular from Latin *later* brick + English *-sol* (as in *podsol,* variant of *podzol*)] (1949)
: a leached red and yellow tropical soil
— lat·o·sol·ic \ˌla-tə-ˈsȯ-lik\ *adjective*

la·trine \lə-ˈtrēn\ *noun* [French, from Latin *latrina,* contraction of *lavatrina,* from *lavare* to wash — more at LYE] (1642)
1 : a receptacle (as a pit in the earth) for use as a toilet
2 : TOILET

-latry *noun combining form* [Middle English *-latrie,* from Old French, from Late Latin *-latria,* from Greek, from *latreia;* akin to Greek *latron* pay]
: worship ⟨helio*latry*⟩

lat·te \ˈlä-(ˌ)tā\ *noun* (1991)
: CAFFE LATTE

lat·ten \ˈla-tᵊn\ *noun* [Middle English *laton,* from Middle French] (14th century)
: a yellow alloy identical to or resembling brass typically hammered into thin sheets and formerly much used for church utensils

lat·ter \ˈla-tər\ *adjective* [Middle English, from Old English *lætra,* comparative of *læt* late] (before 12th century)
1 a : belonging to a subsequent time or period **:** more recent ⟨the *latter* stages of growth⟩ **b :** of or relating to the end ⟨in their *latter* days⟩ **c :** RECENT, PRESENT ⟨affected by *latter* calamities⟩
2 : of, relating to, or being the second of two groups or things or the last of several groups or things referred to ⟨of ham and beef the *latter* meat is cheaper today⟩ ⟨of ham and beef the *latter* is cheaper today⟩

lat·ter–day \ˈla-tər-ˌdā\ *adjective* (1850)
1 : of present or recent times ⟨*latter-day* prophets⟩
2 : of a later or subsequent time

Latter–day Saint *noun, often D capitalized* (1834)
: a member of any of several religious bodies tracing their origin to Joseph Smith in 1830 and accepting the Book of Mormon as divine revelation **:** MORMON

lat·ter·ly *adverb* (1734)
1 : LATER
2 : of late **:** RECENTLY

lat·tice \ˈla-təs\ *noun* [Middle English *latis,* from Middle French *lattis*] (14th century)
1 a : a framework or structure of crossed wood or metal strips **b :** a window, door, or gate having a lattice **c :** a network or design resembling a lattice
2 : a regular geometrical arrangement of points or objects over an area or in space; *specifically* **:** the arrangement of atoms in a crystal
3 : a mathematical set that has some elements ordered and that is such that for any two elements there exists a greatest element in the subset of all elements less than or equal to both and a least element in the subset of all elements greater than or equal to both
— lattice *transitive verb*
— lat·ticed \-tist\ *adjective*

lattice girder *noun* (1852)
: a girder with top and bottom flanges connected by a latticework web

lat·tice·work \ˈla-təs-ˌwərk\ *noun* (15th century)
: a lattice or work made of lattices

la·tus rec·tum \ˈla-təs-ˈrek-təm\ *noun* [New Latin, literally, straight side] (1702)
: a chord of a conic section (as an ellipse) that passes through a focus and is parallel to the directrix

Lat·vi·an \ˈlat-vē-ən\ *noun* (1924)
1 : the Baltic language of the Latvian people
2 : a native or inhabitant of Latvia
— Latvian *adjective*

lau·an \ˈlü-ˌän, lü-ˈ; lau̇-ˈän\ *noun* [Tagalog *lawaan*] (1894)
: any of various Philippine timbers (as of the genera *Shorea* and *Parashorea*) that are light yellow to reddish brown or brown and include some which enter commerce as Philippine mahogany

[1]laud \ˈlȯd\ *noun* [Middle English *laudes* (plural), from Medieval Latin, from Latin, plural of *laud-, laus* praise] (14th century)
1 *plural but singular or plural in construction, often capitalized* **:** an office of solemn praise to God forming with matins the first of the canonical hours
2 : PRAISE, ACCLAIM

[2]laud *transitive verb* [Latin *laudare,* from *laud-, laus*] (14th century)
: PRAISE, EXTOL

laud·able \ˈlȯ-də-bəl\ *adjective* (15th century)
: worthy of praise **:** COMMENDABLE
— laud·able·ness \ˈlȯ-də-bəl-nəs\ *noun*
— laud·ably \-blē\ *adverb*

lau·da·num \ˈlȯd-nəm, ˈlȯ-dᵊn-əm\ *noun* [New Latin] (circa 1603)
1 : any of various formerly used preparations of opium
2 : a tincture of opium

lau·da·tion \lȯ-ˈdā-shən\ *noun* (15th century)
: the act of praising **:** EULOGY

lau·da·tive \ˈlȯ-də-tiv\ *adjective* (15th century)
: LAUDATORY

lau·da·to·ry \ˈlȯ-də-ˌtōr-ē, -ˌtȯr-\ *adjective* (1555)
: of, relating to, or expressing praise

[1]laugh \ˈlaf, ˈläf\ *verb* [Middle English, from Old English *hliehhan;* akin to Old High German *lachēn* to laugh] (before 12th century)
intransitive verb
1 a : to show mirth, joy, or scorn with a smile and chuckle or explosive sound **b :** to find amusement or pleasure in something ⟨*laughed* at his own clumsiness⟩ **c :** to become amused or derisive ⟨a very skeptical public *laughed* at our early efforts —Graenum Berger⟩
2 a : to produce the sound or appearance of laughter ⟨a *laughing* brook⟩ **b :** to be of a kind that inspires joy
transitive verb
1 : to influence or move by laughter ⟨*laughed* the bad singer off the stage⟩
2 : to utter with a laugh
— laugh·ing·ly \ˈla-fiŋ-lē, ˈlä-\ *adverb*

[2]laugh *noun* (1690)
1 : the act of laughing
2 a : a cause for derision or merriment **:** JOKE **b :** an expression of scorn or mockery **:** JEER
3 *plural* **:** DIVERSION, SPORT ⟨play baseball just for *laughs*⟩

laugh·able \ˈla-fə-bəl, ˈlä-\ *adjective* (1596)
: of a kind to provoke laughter or sometimes derision **:** amusingly ridiculous ☆
— laugh·able·ness *noun*
— laugh·ably \-blē\ *adverb*

laugh·er \ˈla-fər, ˈlä-\ *noun* (15th century)
1 : one that laughs

> ☆ SYNONYMS
> **Laughable, ludicrous, ridiculous, comic, comical** mean provoking laughter or mirth. LAUGHABLE applies to anything occasioning laughter ⟨*laughable* attempts at waterskiing⟩. LUDICROUS suggests absurdity or preposterousness that excites both laughter and scorn ⟨a spy thriller with a *ludicrous* plot⟩. RIDICULOUS suggests extreme absurdity, foolishness, or contemptibility ⟨a *ridiculous* portrayal of wartime combat⟩. COMIC applies especially to that which arouses thoughtful amusement ⟨a *comic* character⟩. COMICAL applies to that which arouses spontaneous hilarity ⟨you looked *comical*⟩.

2 : something (as a game) that is easily won or handled

laughing gas *noun* (circa 1842)
: NITROUS OXIDE

laughing gull *noun* (1789)
: an American gull (*Larus atricilla*) having a black head in breeding plumage and black wing tips blending into the gray upper side of the wings

laughing jackass *noun* (1798)
: KOOKABURRA

laughing matter *noun* (circa 1583)
: something not to be taken seriously — usually used in the phrase *no laughing matter*

laugh·ing·stock \'la-fiŋ-ˌstäk, 'lȧ-\ *noun* (1533)
: an object of ridicule

laugh off *transitive verb* (1715)
: to minimize by treating as amusingly or absurdly trivial

laugh·ter \'laf-tər, 'lȧf-\ *noun* [Middle English, from Old English *hleahtor;* akin to Old English *hliehhan*] (before 12th century)
1 : a sound of or as if of laughing
2 *archaic* **:** a cause of merriment

laugh track *noun* (1962)
: laughter to accompany dialogue or action (as of a television program)

launce \'lȯn(t)s, 'län(t)s\ *noun* [probably from [1]*lance*] (1623)
: SAND LANCE

[1]launch \'lȯnch, 'länch\ *verb* [Middle English, from Old North French *lancher,* from Late Latin *lanceare* to wield a lance — more at LANCE] (14th century)
transitive verb
1 a : to throw forward **:** HURL **b :** to release, catapult, or send off (a self-propelled object) ⟨*launch* a rocket⟩
2 a : to set (a boat or ship) afloat **b :** to give (a person) a start ⟨*launched* in a new career⟩ **c** (1) **:** to originate or set in motion **:** INITIATE, INTRODUCE (2) **:** to get off to a good start
intransitive verb
1 a : to spring forward **:** TAKE OFF **b :** to throw oneself energetically **:** PLUNGE
2 a *archaic* **:** to slide down the ways **b :** to make a start

[2]launch *noun* (1749)
: an act or instance of launching

[3]launch *noun* [Spanish or Portuguese; Spanish *lancha,* from Portuguese] (1697)
1 : a large boat that operates from a ship
2 : a small motorboat that is open or that has the forepart of the hull covered

launch·er \'lȯn-chər, 'län-\ *noun* (1911)
: one that launches: as **a :** a device for firing grenades **b :** a device for launching a missile **c :** LAUNCH VEHICLE

launching pad *noun* (1951)
1 : LAUNCHPAD
2 : SPRINGBOARD 2

launch·pad \'lȯnch-ˌpad, 'länch-\ *noun* (1958)
: a nonflammable platform from which a rocket, launch vehicle, or guided missile can be launched

launch vehicle *noun* (circa 1960)
: a rocket used to launch a satellite or spacecraft

launch window *noun* (1962)
: WINDOW 8

[1]laun·der \'lȯn-dər, 'län-\ *verb* **laun·dered; laun·der·ing** \-d(ə-)riŋ\ [Middle English *launder,* noun] (1664)
transitive verb
1 : to wash (as clothes) in water
2 : to make ready for use by washing and ironing ⟨a freshly *laundered* shirt⟩
3 : to transfer (as illegally obtained money or investments) through an outside party to conceal the true source
intransitive verb
: to wash or wash and iron clothing or household linens
— laun·der·er \-dər-ər\ *noun*

[2]launder *noun* [Middle English, launderer, from Middle French *lavandier,* from Medieval Latin *lavandarius,* from Latin *lavandus,* gerundive of *lavare* to wash — more at LYE] (1667)
: TROUGH; *especially* **:** a box conduit conveying particulate material suspended in water in ore dressing

laun·der·ette \ˌlȯn-də-'ret, ˌlän-\ *also* **laun·drette** \-'dret\ *noun* [from *Launderette,* a service mark] (circa 1948)
: a self-service laundry

laun·dress \'lȯn-drəs, 'län-\ *noun* (1550)
: a woman who is a laundry worker

Laun·dro·mat \'lȯn-drə-ˌmat, 'län-\ *service mark*
— used for a self-service laundry

laun·dry \'lȯn-drē, 'län-\ *noun, plural* **laun·dries** (14th century)
1 a : a room for doing the family wash **b :** a commercial laundering establishment
2 : clothes or linens that have been or are to be laundered

laundry list *noun* (1958)
: a usually long list of items ⟨the *laundry list* of new consumer-protection bills —N. C. Miller⟩

laun·dry·man \-mən\ *noun* (1708)
: a man who is a laundry worker

Laun·fal \'lȯn-fəl, 'län-\ *noun*
: a knight of the Round Table in late Arthurian legend

lau·ra \'läv-rə\ *noun* [Late Greek, from Greek, lane] (circa 1752)
: a monastery of an Eastern church

[1]lau·re·ate \'lȯr-ē-ət, 'lär-\ *noun* [Middle English, crowned with laurel as a distinction, from Latin *laureatus,* from *laurea* laurel wreath, from feminine of *laureus* of laurel, from *laurus*] (circa 1529)
: the recipient of honor or recognition for achievement in an art or science; *specifically* **:** POET LAUREATE
— laureate *adjective*
— lau·re·ate·ship \-ˌship\ *noun*

[2]lau·re·ate \'lȯr-ē-ˌāt, 'lär-\ *transitive verb* **-at·ed; -at·ing** (circa 1610)
1 : to crown with or as if with a laurel wreath for excellence or achievement
2 : to appoint to the office of poet laureate
— lau·re·ation \ˌlȯr-ē-'ā-shən, ˌlär-\ *noun*

[1]lau·rel \'lȯr-əl, 'lär-\ *noun* [Middle English *lorel,* from Old French *lorier,* from *lor* laurel, from Latin *laurus*] (14th century)
1 : any of a genus (*Laurus* of the family Lauraceae, the laurel family) of evergreen trees that have alternate entire leaves, small tetramerous flowers surrounded by bracts, and fruits that are ovoid berries; *specifically* **:** a tree (*L. nobilis*) of southern Europe with foliage used by the ancient Greeks to crown victors in the Pythian games

laurel 1

2 : a tree or shrub that resembles the true laurel; *especially* **:** MOUNTAIN LAUREL
3 : a crown of laurel **:** HONOR — usually used in plural

[2]laurel *transitive verb* **-reled** *or* **-relled; -rel·ing** *or* **-rel·ling** (1631)
: to deck or crown with laurel

lau·ric acid \'lȯr-ik-, 'lär-\ *noun* [International Scientific Vocabulary, from Latin *laurus*] (1873)
: a crystalline fatty acid $C_{12}H_{24}O_2$ found especially in coconut oil and used in making soaps, esters, and lauryl alcohol

lau·ryl alcohol \'lȯr-əl-, 'lär-\ *noun* (1922)
: a compound $C_{12}H_{26}O$; *also* **:** a liquid mixture of this and other alcohols used especially in making detergents

la·va \'lä-və, 'la-\ *noun* [Italian, ultimately from Latin *labes* fall; akin to Latin *labi* to slide — more at SLEEP] (1759)
: molten rock that issues from a volcano or from a fissure in the earth's surface; *also* **:** such rock that has cooled and hardened
— la·va·like \-ˌlīk\ *adjective*

la·va·bo \lə-'vä-(ˌ)bō\ *noun, plural* **-bos** [Latin, I shall wash, from *lavare*] (circa 1858)
1 *often capitalized* **:** a ceremony at Mass in which the celebrant washes his hands after offering the oblations and says Psalm 25:6–12 (Douay Version)
2 a : a washbasin and a tank with a spigot that are fastened to a wall **b :** this combination used as a planter

la·vage \lə-'väzh, *British usually* 'la-vij\ *noun* [French, from Middle French, from *laver* to wash, from Latin *lavare*] (circa 1895)
: WASHING; *especially* **:** the therapeutic washing out of an organ

la·va·la·va \ˌlä-və-'lä-və\ *noun* [Samoan, clothing] (circa 1891)
: a rectangular cloth of cotton print worn like a kilt or skirt in Polynesia and especially in Samoa

la·va·liere *or* **la·val·liere** \ˌlä-və-'lir, ˌla-\ *noun* [French *lavallière* necktie with a large knot] (1906)
: a pendant on a fine chain that is worn as a necklace

la·va·lier microphone \ˌlä-və-'lir-, ˌla-\ *or* **lavaliere microphone** *same*\ *noun* (circa 1962)
: a small microphone hung around the neck of the user

la·va·tion \lā-'vā-shən\ *noun* [Latin *lavation-, lavatio,* from *lavare*] (15th century)
: the act or an instance of washing or cleansing

lav·a·to·ry \'la-və-ˌtȯr-ē, -ˌtȯr-, *British* -və-t(ə-)rē\ *noun, plural* **-ries** [Middle English *lavatorie,* from Medieval Latin *lavatorium,* from Latin *lavare* to wash — more at LYE] (14th century)
1 : a vessel (as a basin) for washing; *especially* **:** a fixed bowl or basin with running water and drainpipe for washing
2 : a room with conveniences for washing and usually with one or more toilets
3 : TOILET 3b
— lavatory *adjective*

[1]lave \'lāv\ *noun* [Middle English (northern dialect), from Old English *lāf;* akin to Old English *belīfan* to remain — more at LEAVE] (before 12th century)
chiefly dialect **:** something that is left **:** RESIDUE

[2]lave *verb* **laved; lav·ing** [Middle English, from Old English *lafian,* from Latin *lavare*] (before 12th century)
transitive verb
1 a : WASH, BATHE **b :** to flow along or against
2 : POUR
intransitive verb
archaic **:** to wash oneself **:** BATHE

[1]lav·en·der \'la-vən-dər\ *noun* [Middle English *lavendre,* from Anglo-French, from Medieval Latin *lavandula*] (13th century)
1 a : a Mediterranean mint (*Lavandula angustifolia* synonym *L. officinalis*) widely cultivated for its narrow aromatic leaves and spikes of lilac-purple flowers which are dried and used in sachets **b :** any of several plants congeneric with true lavender and used similarly but often considered inferior
2 : a pale purple

[2]lavender *transitive verb* **lav·en·dered; lav·en·der·ing** \-d(ə-)riŋ\ (1820)
: to sprinkle or perfume with lavender

[1]la·ver \'lā-vər\ *noun* [Middle English *lavour,* from Middle French *lavoir,* from Medieval Latin *lavatorium*] (1535)

: a large basin used for ceremonial ablutions in the ancient Jewish Tabernacle and Temple worship

²la·ver \ˈlā-vər, ˈlä-\ *noun* [New Latin, from Latin, a water plant] (1611)
: any of several common red algae (genus *Porphyra*) with fronds used for stewing or pickling

La·vin·ia \lə-ˈvi-nē-ə\ *noun* [Latin]
: a daughter of King Latinus in Virgil's *Aeneid* who is betrothed to Turnus but marries Aeneas

¹lav·ish \ˈla-vish\ *adjective* [Middle English *lavas* abundance, from Middle French *lavasse* downpour of rain, from *laver* to wash — more at LAVAGE] (15th century)
1 : expending or bestowing profusely **:** PRODIGAL
2 a : expended or produced in abundance **b :** marked by profusion or excess
synonym see PROFUSE
— lav·ish·ly *adverb*
— lav·ish·ness *noun*

²lavish *transitive verb* (1542)
: to expend or bestow with profusion **:** SQUANDER

lav·rock \ˈlāv-rək, ˈlav-\ *or* **la·ver·ock** \ˈla-və-\ *noun* [Middle English *laverok,* from Old English *lāwerce*] (14th century)
chiefly Scottish **:** LARK

¹law \ˈlȯ\ *noun* [Middle English, from Old English *lagu,* of Scandinavian origin; akin to Old Norse *lǫg* law; akin to Old English *licgan* to lie — more at LIE] (before 12th century)
1 a (1) **:** a binding custom or practice of a community **:** a rule of conduct or action prescribed or formally recognized as binding or enforced by a controlling authority (2) **:** the whole body of such customs, practices, or rules (3) **:** COMMON LAW **b** (1) **:** the control brought about by the existence or enforcement of such law (2) **:** the action of laws considered as a means of redressing wrongs; *also* **:** LITIGATION (3) **:** the agency of or an agent of established law **c :** a rule or order that it is advisable or obligatory to observe **d :** something compatible with or enforceable by established law **e :** CONTROL, AUTHORITY
2 a *often capitalized* **:** the revelation of the will of God set forth in the Old Testament **b** *capitalized* **:** the first part of the Jewish scriptures **:** PENTATEUCH, TORAH — see BIBLE table
3 : a rule of construction or procedure ⟨the *laws* of poetry⟩
4 : the whole body of laws relating to one subject
5 a : the legal profession **b :** law as a department of knowledge **:** JURISPRUDENCE **c :** legal knowledge
6 a : a statement of an order or relation of phenomena that so far as is known is invariable under the given conditions **b :** a general relation proved or assumed to hold between mathematical or logical expressions ☆ ◆
— at law : under or within the provisions of the law ⟨enforceable *at law*⟩

²law (circa 1550)
intransitive verb
: LITIGATE
transitive verb
chiefly dialect **:** to sue or prosecute at law

law–abid·ing \ˈlȯ-ə-ˌbī-diŋ\ *adjective* (1834)
: abiding by or obedient to the law
— law–abid·ing·ness *noun*

law·break·er \ˈlȯ-ˌbrā-kər\ *noun* (15th century)
: one who violates the law
— law·break·ing \-kiŋ\ *adjective or noun*

law·ful \ˈlȯ-fəl\ *adjective* (14th century)
1 a : being in harmony with the law ⟨a *lawful* judgment⟩ **b :** constituted, authorized, or established by law **:** RIGHTFUL ⟨*lawful* institutions⟩
2 : LAW-ABIDING ⟨*lawful* citizens⟩ ☆
— law·ful·ly \-f(ə-)lē\ *adverb*
— law·ful·ness \-fəl-nəs\ *noun*

law·giv·er \ˈlȯ-ˌgi-vər\ *noun* (14th century)
1 : one who gives a code of laws to a people
2 : LEGISLATOR

law·less \ˈlȯ-ləs\ *adjective* (12th century)
1 : not regulated by or based on law
2 a : not restrained or controlled by law **:** UNRULY **b :** ILLEGAL
— law·less·ly *adverb*
— law·less·ness *noun*

law·mak·er \ˈlȯ-ˌmā-kər\ *noun* (14th century)
: one that makes laws **:** LEGISLATOR
— law·mak·ing \-kiŋ\ *noun*

law·man \ˈlȯ-mən\ *noun* (1944)
: a law-enforcement officer (as a sheriff or marshal)

law merchant *noun, plural* **laws merchant** (15th century)
: the legal rules formerly applied to cases arising in commercial transactions

¹lawn \ˈlȯn, ˈlän\ *noun* [Middle English *launde,* from Middle French *lande* heath, of Celtic origin; akin to Middle Irish *lann* land — more at LAND] (14th century)
1 *archaic* **:** an open space between woods **:** GLADE
2 : ground (as around a house or in a garden or park) that is covered with grass and is kept mowed
3 : a relatively even layer of bacteria covering the surface of a culture medium
— lawn *or* **lawny** \ˈlȯ-nē, ˈlä-\ *adjective*

²lawn *noun* [Middle English, from *Laon,* France] (15th century)
: a fine sheer linen or cotton fabric of plain weave that is thinner than cambric
— lawny *adjective*

lawn bowling *noun* (circa 1929)
: a bowling game played on a green with wooden balls which are rolled at a jack

lawn mower *noun* (1869)
: a machine for cutting grass on lawns

lawn tennis *noun* (1874)
: TENNIS 2

law of averages (circa 1929)
: the commonsense observation that probability influences everyday life so that over the long term the possible outcomes of a repeated event occur with specific frequencies

law of definite proportions (circa 1909)
: a statement in chemistry: every definite compound always contains the same elements in the same proportions by weight

law of dominance (1942)
: MENDEL'S LAW 3

law of independent assortment (1943)
: MENDEL'S LAW 2

law of large numbers (1911)
: a theorem in mathematical statistics: the probability that the absolute value of the difference between the mean of a population sample and the mean of the population from which it is drawn is greater than an arbitrarily small amount approaches zero as the size of the sample approaches infinity

Law of Moses (14th century)
: PENTATEUCH, TORAH

law of nations (circa 1548)
: INTERNATIONAL LAW

law of parsimony (1837)
: OCCAM'S RAZOR

law of segregation (circa 1909)
: MENDEL'S LAW 1

law of war (1947)
: the code that governs or one of the rules that govern the rights and duties of belligerents in international war

law·ren·ci·um \lȯ-ˈren(t)-sē-əm\ *noun* [New Latin, from Ernest O. *Lawrence*] (1961)
: a short-lived radioactive element that is produced artificially from californium — see ELEMENT table

law·suit \ˈlȯ-ˌsüt\ *noun* (1624)
: a suit in law **:** a case before a court

law·yer \ˈlȯ-yər, ˈlȯi-ər\ *noun* (14th century)
: one whose profession is to conduct lawsuits for clients or to advise as to legal rights and obligations in other matters
— law·yer·like \-ˌlīk\ *adjective*
— law·yer·ly \-lē\ *adjective*

law·yer·ing \ˈlȯ-yə-riŋ, ˈlȯi-ə-\ *noun* (1676)
: the profession or work of a lawyer

☆ SYNONYMS

Law, rule, regulation, precept, statute, ordinance, canon mean a principle governing action or procedure. LAW implies imposition by a sovereign authority and the obligation of obedience on the part of all subject to that authority ⟨obey the *law*⟩. RULE applies to more restricted or specific situations ⟨the *rules* of the game⟩. REGULATION implies prescription by authority in order to control an organization or system ⟨*regulations* affecting nuclear power plants⟩. PRECEPT commonly suggests something advisory and not obligatory communicated typically through teaching ⟨the *precepts* of effective writing⟩. STATUTE implies a law enacted by a legislative body ⟨a *statute* requiring the use of seat belts⟩. ORDINANCE applies to an order governing some detail of procedure or conduct enforced by a limited authority such as a municipality ⟨a city *ordinance*⟩. CANON suggests in nonreligious use a principle or rule of behavior or procedure commonly accepted as a valid guide ⟨the *canons* of good taste⟩. See in addition HYPOTHESIS.

Lawful, legal, legitimate, licit mean being in accordance with law. LAWFUL may apply to conformity with law of any sort (as natural, divine, common, or canon) ⟨the *lawful* sovereign⟩. LEGAL applies to what is sanctioned by law or in conformity with the law, especially as it is written or administered by the courts ⟨*legal* residents of the state⟩. LEGITIMATE may apply to a legal right or status but also, in extended use, to a right or status supported by tradition, custom, or accepted standards ⟨a perfectly *legitimate* question about finances⟩. LICIT applies to a strict conformity to the provisions of the law and applies especially to what is regulated by law ⟨the *licit* use of the drug by hospitals⟩.

◇ WORD HISTORY

law The word *law* is part of the English language's abundant heritage from medieval Scandinavia. It was borrowed into late Old English from an unrecorded Old Norse noun *lagu* (whose existence is inferred from the liguistically later Old Icelandic form *log*). *Lagu* is the plural of *lag* "something laid, layer, stratum." The image of the law as a sort of edifice built up of past decisions and decrees is fairly natural. Other languages have resorted to similar metaphors, using derivatives of verbs meaning "to put or place" or "to erect" in order to express the notion of a law. Hence, for example, we have Greek *thesmos* "law, ordinance" (from *tithenai* "to place"), Latin *statutum* "ordinance, statute" (from *statuere* "to set upright, erect"), and German *Gesetz* "law" (from *setzen* "to put, set"). Old English did not lack an earlier word for "law"; this was *æ* or *æw,* which also meant "custom"—appropriately enough, since law and custom in preliterate societies typically overlap. This word may be akin to Latin *jus* "what is sanctioned, law," the ultimate source of English *jury, perjury, injury,* and a host of other words.

lax \'laks\ *adjective* [Middle English, from Latin *laxus* loose — more at SLACK] (14th century)
1 a *of the bowels* **:** LOOSE, OPEN **b :** having loose bowels
2 : deficient in firmness **:** not stringent ⟨*lax* control⟩ ⟨a *lax* foreman⟩
3 a : not tense, firm, or rigid **:** SLACK ⟨a *lax* rope⟩ **b :** having an open or loose texture **c :** having the constituents spread apart ⟨a *lax* flower cluster⟩
4 : articulated with the muscles involved in a relatively relaxed state (as the vowel \i\ in contrast with the vowel \ē\)
synonym see NEGLIGENT
— lax·a·tion \lak-'sā-shən\ *noun*
— lax·ly \'laks-lē\ *adverb*
— lax·ness *noun*

¹lax·a·tive \'lak-sə-tiv\ *adjective* [Middle English *laxatif,* from Medieval Latin *laxativus,* from Latin *laxatus,* past participle of *laxare* to loosen, from *laxus*] (14th century)
: having a tendency to loosen or relax; *specifically* **:** relieving constipation

²laxative *noun* (14th century)
: a usually mild laxative drug

lax·ity \'lak-sə-tē\ *noun* (1528)
: the quality or state of being lax

¹lay \'lā\ *verb* **laid** \'lād\; **lay·ing** [Middle English *leyen,* from Old English *lecgan;* akin to Old English *licgan* to lie — more at LIE] (before 12th century)
transitive verb
1 : to beat or strike down with force
2 a : to put or set down **b :** to place for rest or sleep; *especially* **:** BURY
3 : to bring forth and deposit (an egg)
4 : CALM, ALLAY ⟨*lay* the dust⟩
5 : BET, WAGER
6 : to press down giving a smooth and even surface
7 a : to dispose or spread over or on a surface ⟨*lay* track⟩ ⟨*lay* plaster⟩ **b :** to set in order or position ⟨*lay* a table for dinner⟩ ⟨*lay* brick⟩ **c :** to put (strands) in place and twist to form a rope, hawser, or cable; *also* **:** to make by so doing ⟨*lay* up rope⟩
8 a : to impose as a duty, burden, or punishment ⟨*lay* a tax⟩ **b :** to put as a burden of reproach ⟨*laid* the blame on her⟩ **c :** to advance as an accusation **:** IMPUTE ⟨the disaster was *laid* to faulty inspection⟩
9 : to place (something immaterial) on something ⟨*lay* stress on grammar⟩
10 : PREPARE, CONTRIVE ⟨a well-*laid* plan⟩
11 a : to bring against or into contact with something **:** APPLY ⟨*laid* the watch to his ear⟩ **b :** to prepare or position for action or operation ⟨*lay* a fire in the fireplace⟩; *also* **:** to adjust (a gun) to the proper direction and elevation
12 : to bring to a specified condition ⟨*lay* waste the land⟩
13 a : ASSERT, ALLEGE ⟨*lay* claim to an estate⟩ **b :** to submit for examination and judgment ⟨*laid* her case before the commission⟩
14 : to copulate with — sometimes considered vulgar
intransitive verb
1 : to produce and deposit eggs
2 *nonstandard* **:** ¹LIE
3 : WAGER, BET
4 *dialect* **:** PLAN, PREPARE
5 a : to apply oneself vigorously ⟨*laid* to his oars⟩ **b :** to proceed to a specified place or position on a ship ⟨*lay* aloft⟩ ▢
— lay on the table 1 : to remove (a parliamentary motion) from consideration indefinitely **2** *British* **:** to put (as legislation) on the agenda

²lay *noun* (1590)
1 : COVERT, LAIR
2 : something (as a layer) that lies or is laid
3 a : line of action **:** PLAN **b :** line of work **:** OCCUPATION
4 a : terms of sale or employment **:** PRICE **b :** share of profit (as on a whaling voyage) paid in lieu of wages
5 a : the amount of advance of any point in a rope strand for one turn **b :** the nature of a fiber rope as determined by the amount of twist, the angle of the strands, and the angle of the threads in the strands
6 : the way in which a thing lies or is laid in relation to something else ⟨the *lay* of the land⟩
7 : the state of one that lays eggs ⟨hens coming into *lay*⟩
8 a : a partner in sexual intercourse — usually considered vulgar **b :** SEXUAL INTERCOURSE — usually considered vulgar

³lay *past of* LIE

⁴lay *noun* [Middle English, from Old French *lai*] (13th century)
1 : a simple narrative poem **:** BALLAD
2 : MELODY, SONG

⁵lay *adjective* [Middle English, from Middle French *lai,* from Late Latin *laicus,* from Greek *laikos* of the people, from *laos* people] (15th century)
1 : of or relating to the laity **:** not ecclesiastical
2 : of or relating to members of a religious house occupied with domestic or manual work ⟨a *lay* brother⟩
3 : not of a particular profession ⟨the *lay* public⟩; *also* **:** lacking extensive knowledge of a particular subject

lay·about \'lā-ə-,bau̇t\ *noun* (1932)
: a lazy shiftless person **:** IDLER

lay·away \'lā-ə-,wā\ *noun* (1944)
: a purchasing agreement by which a retailer agrees to hold merchandise secured by a deposit until the price is paid in full by the customer

lay away *transitive verb* (circa 1928)
: to put aside for future use or delivery

lay–by \'lā-,bī\ *noun* (1939)
1 *British* **:** TURNOUT 4b
2 : the final operation (as a last cultivating) in the growing of a field crop

lay by *transitive verb* (15th century)
1 : to lay aside **:** DISCARD
2 : to store for future use **:** SAVE
3 : to cultivate (as corn) for the last time

lay day *noun* (1845)
1 : one of the days allowed by the charter for loading or unloading a vessel
2 : a day of delay in port

lay down (13th century)
transitive verb
1 : to give up **:** SURRENDER ⟨*lay down* your arms⟩
2 a : ESTABLISH, PRESCRIBE ⟨*lay down* a scale for a map⟩ **b :** to assert or command dogmatically ⟨*lay down* the law⟩
3 : STORE, PRESERVE
4 a : to direct toward a target ⟨*lay down* a barrage⟩ **b :** to hit along the ground ⟨*laid down* a sacrifice bunt⟩
intransitive verb
nonstandard **:** to lie down

¹lay·er \'lā-ər, 'le(-ə)r\ *noun* (13th century)
1 : one that lays (as a worker who lays brick or a hen that lays eggs)
2 a : one thickness, course, or fold laid or lying over or under another **b :** STRATUM **c :** HORIZON 2
3 a : a branch or shoot of a plant that roots while still attached to the parent plant **b :** a plant developed by layering
— lay·ered \'lā-ərd, 'le(-ə)rd\ *adjective*

²layer (1832)
transitive verb
1 : to propagate (a plant) by means of layers
2 a : to place as a layer **b :** to place a layer on top of **c :** to form or arrange in layers
intransitive verb
1 a : to separate into layers **b :** to form out of superimposed layers
2 *of a plant* **:** to form roots where a stem comes in contact with the ground

lay·er·age \'lā-ə-rij, 'le-ə-\ *noun* (1902)
: the practice, art, or process of rooting plants by layering

lay·ette \lā-'et\ *noun* [French, from Middle French, diminutive of *laye* box, from Middle Dutch *lade;* akin to Old English *hladan* to load — more at LADE] (1839)
: a complete outfit of clothing and equipment for a newborn infant

lay figure *noun* [obsolete English *layman* lay figure, from Dutch *leeman*] (1795)
1 : a jointed model of the human body used by artists to show the disposition of drapery
2 : a person likened to a dummy or puppet

lay in *transitive verb* (1579)
: LAY BY, SAVE

laying on of hands (15th century)
: the act of laying hands on a person's head to confer a spiritual blessing (as in Christian ordination, confirmation, or faith healing)

lay·man \'lā-mən\ *noun* (15th century)
1 : a person who is not a member of the clergy
2 : a person who does not belong to a particular profession or who is not expert in some field

lay·off \'lā-,ȯf\ *noun* (1889)
1 : a period of inactivity or idleness
2 : the act of laying off an employee or a workforce; *also* **:** SHUTDOWN

lay off (1748)
transitive verb
1 : to mark or measure off
2 : to cease to employ (a worker) often temporarily
3 *of a bookie* **:** to place all or part of (an accepted bet) with another bookie to reduce the risk
4 a : to leave undisturbed **b :** AVOID, QUIT
intransitive verb
1 : to stop doing or taking something
2 : to leave one alone ⟨wish you'd just *lay off*⟩

lay on (1600)
transitive verb
1 a : to apply by or as if by spreading on a surface ⟨*laying* it *on* thick⟩ **b :** PROVIDE, ARRANGE ⟨food *laid on* in abundance⟩ **c :** HAND OUT ⟨*laid on* awards⟩
2 *chiefly British* **:** HIRE
intransitive verb
: ATTACK, BEAT

lay·out \'lā-,au̇t\ *noun* (1852)
1 : the plan or design or arrangement of something that is laid out: as **a :** DUMMY 5b **b :** final arrangement of matter to be reproduced especially by printing
2 : the act or process of planning or laying out in detail

▢ USAGE
lay LAY has been used intransitively in the sense of "lie" since the 14th century. The practice was unremarked until around 1770; attempts to correct it have been a fixture of schoolbooks ever since. Generations of teachers and critics have succeeded in taming most literary and learned writing, but intransitive *lay* persists in familiar speech and is a bit more common in general prose than one might suspect. Much of the problem lies in the confusing similarity of the principal parts of the two words. Another influence may be a folk belief that *lie* is for people and *lay* is for things. Some commentators are ready to abandon the distinction, suggesting that *lay* is on the rise socially. But if it does rise to respectability, it is sure to do so slowly: many people have invested effort in learning to keep *lie* and *lay* distinct. Remember that even though many people do use *lay* for *lie,* others will judge you unfavorably if you do. See in addition SET.

3 a : something that is laid out ⟨a model train *layout*⟩ **b :** land or structures or rooms used for a particular purpose ⟨a cattle-ranching *lay-out*⟩; *also* **:** PLACE **c :** a set or outfit especially of tools

lay out *transitive verb* (15th century)
1 : DISPLAY, EXHIBIT
2 : SPEND
3 a : to prepare (a corpse) for viewing **b :** to knock flat or unconscious
4 : to plan in detail ⟨*lay out* a campaign⟩
5 : ARRANGE, DESIGN
6 : to mark (work) for drilling, machining, or filing

lay·over \'lā-,ō-vər\ *noun* (1873)
: STOPOVER

lay over (1838)
transitive verb
: POSTPONE
intransitive verb
: to make a stopover

lay·peo·ple \'lā-,pē-pəl\ *noun plural* (15th century)
: LAYPERSONS

lay·per·son \-,pər-sᵊn\ *noun* (1972)
: a member of the laity

lay reader *noun* (1751)
: an Anglican or Roman Catholic layman authorized to conduct parts of the church services not requiring a clergyman

lay to (1796)
intransitive verb
: LIE TO
transitive verb
: to bring (a ship) into the wind and hold stationary

lay·up \'lā-,əp\ *noun* (1925)
1 : the action of laying up or the condition of being laid up
2 : a shot in basketball made from near the basket usually by playing the ball off the backboard

lay up *transitive verb* (14th century)
1 : to store up **:** LAY BY
2 : to disable or confine with illness or injury
3 : to take out of active service

lay·wom·an \'lā-,wu̇-mən\ *noun* (1529)
: a woman who is a member of the laity

la·zar \'la-zər, 'lā-\ *noun* [Middle English, from Medieval Latin *lazarus,* from Late Latin *Lazarus*] (14th century)
: a person afflicted with a repulsive disease; *specifically* **:** LEPER

laz·a·ret·to \,la-zə-'re-(,)tō\ *or* **laz·a·ret** \-'ret\ *also* **laz·a·rette** \-'ret\ *noun, plural* **-rettos** *or* **-rets** *also* **-rettes** [Italian *lazzaretto,* alteration of *Nazaretto,* quarantine station in Venice, from *Santa Maria di Nazareth,* church on the island where it was located] (1549)
1 *usually lazaretto* **:** an institution (as a hospital) for those with contagious diseases
2 : a building or a ship used for detention in quarantine
3 *usually lazaret* **:** a space in a ship between decks used as a storeroom

La·za·rist \'la-zə-rist, lə 'zär-ist\ *noun* [Collège of Saint *Lazare,* Paris, former home of the congregation] (1747)
: VINCENTIAN

Laz·a·rus \'laz-rəs, 'la-zə-\ *noun* [Late Latin, from Greek *Lazaros,* from Hebrew *El'āzār*]
1 : a brother of Mary and Martha raised by Jesus from the dead according to the account in John 11
2 : the diseased beggar in the parable of the rich man and the beggar found in Luke 16

laze \'lāz\ *verb* **lazed; laz·ing** [back-formation from *lazy*] (circa 1592)
intransitive verb
: to act or lie lazily **:** IDLE
transitive verb
: to pass (time) in idleness or relaxation
— laze *noun*

la·zu·lite \'la-zyü-,līt, -zhə-\ *noun* [German *Lazulith,* from Medieval Latin *lazulum* lapis lazuli] (1807)
: an often crystalline azure-blue mineral that is a hydrous phosphate of aluminum, iron, and magnesium

¹la·zy \'lā-zē\ *adjective* **la·zi·er; -est** [perhaps from Middle Low German *lasich* feeble; akin to Middle High German er*leswen* to become weak] (1549)
1 a : disinclined to activity or exertion **:** not energetic or vigorous **b :** encouraging inactivity or indolence ⟨a *lazy* summer day⟩
2 : moving slowly **:** SLUGGISH
3 : DROOPY, LAX ⟨a rabbit with *lazy* ears⟩
4 : placed on its side ⟨*lazy* E livestock brand⟩
5 : not rigorous or strict ⟨*lazy* scholarship⟩ ☆
— la·zi·ly \-zə-lē\ *adverb*
— la·zi·ness \-zē-nəs\ *noun*
— la·zy·ish \-zē-ish\ *adjective*

²lazy *intransitive verb* **la·zied; la·zy·ing** (1612)
: to move or lie lazily **:** LAZE

la·zy·bones \'lā-zē-,bōnz\ *noun plural but singular or plural in construction* (1592)
: a lazy person

lazy eye *noun* (1939)
: AMBLYOPIA; *also* **:** an eye affected with amblyopia

lazy Su·san \-'sü-zᵊn\ *noun* (1917)
: a revolving tray used for serving food, condiments, or relishes

lazy tongs *noun plural* (1836)
: a series of jointed and pivoted bars capable of great extension used to pick up or handle something at a distance

lazy tongs

LCD \,el-(,)sē-'dē\ *noun* [*l*iquid *c*rystal *d*isplay] (1973)
: a constantly operating display (as of the time in a digital watch) that consists of segments of a liquid crystal whose reflectivity varies according to the voltage applied to them

LD_{50} \,el-,dē-'fif-tē\ *noun* [*l*ethal *d*ose] (1942)
: the amount of a toxic agent (as a poison, virus, or radiation) that is sufficient to kill 50% of a population of animals usually within a certain time

LDL \,el-(,)dē-'el\ *noun* [*l*ow-*d*ensity *l*ipoprotein] (1976)
: a cholesterol-rich protein-poor lipoprotein of blood plasma correlated with increased risk of atherosclerosis — compare HDL

L–do·pa \'el-'dō-pə\ *noun* (1939)
: the levorotatory form of dopa that is obtained especially from broad beans or prepared synthetically, stimulates the production of dopamine in the brain, and is used in treating Parkinson's disease

lea *or* **ley** \'lē, 'lā\ *noun* [Middle English *leye,* from Old English *lēah;* akin to Old High German *lōh* thicket, Latin *lucus* grove, *lux* light — more at LIGHT] (before 12th century)
1 : GRASSLAND, PASTURE
2 *usually ley* **:** arable land used temporarily for hay or grazing

¹leach \'lēch\ *variant of* LEECH

²leach *verb* [*leach* vessel through which water is passed to extract lye] (1796)
transitive verb
1 : to dissolve out by the action of a percolating liquid ⟨*leach* out alkali from ashes⟩
2 : to subject to the action of percolating liquid (as water) in order to separate the soluble components
3 a : to remove (nutritive or harmful elements) from soil by percolation **b :** to draw out or remove as if by percolation ⟨all meaning has been *leached* from my life⟩
intransitive verb
: to pass out or through by percolation
— leach·abil·i·ty \,lē-chə-'bi-lə-tē\ *noun*
— leach·able \'lē-chə-bəl\ *adjective*
— leach·er *noun*

leach·ate \'lē-,chāt\ *noun* (1934)
: a solution or product obtained by leaching

¹lead \'lēd\ *verb* **led** \'led\; **lead·ing** [Middle English *leden,* from Old English *lǣdan;* akin to Old High German *leiten* to lead, Old English *līthan* to go] (before 12th century)
transitive verb
1 a : to guide on a way especially by going in advance **b :** to direct on a course or in a direction **c :** to serve as a channel for ⟨a pipe *leads* water to the house⟩
2 : to go through **:** LIVE ⟨*lead* a quiet life⟩
3 a (1) **:** to direct the operations, activity, or performance of ⟨*lead* an orchestra⟩ (2) **:** to have charge of ⟨*lead* a campaign⟩ (3) **:** to suggest to (a witness) the answer desired by asking leading questions **b** (1) **:** to go at the head of ⟨*lead* a parade⟩ (2) **:** to be first in or among ⟨*lead* the league⟩ (3) **:** to have a margin over ⟨*led* his opponent⟩
4 : to bring to some conclusion or condition ⟨*led* to believe otherwise⟩
5 : to begin play with ⟨*lead* trumps⟩
6 a : to aim in front of (a moving object) ⟨*lead* a duck⟩ **b :** to pass a ball or puck just in front of (a moving teammate)
intransitive verb
1 a : to guide someone or something along a way **b :** to lie, run, or open in a specified place or direction ⟨path *leads* uphill⟩ **c :** to guide a dance partner through the steps of a dance
2 a : to be first **b** (1) **:** BEGIN, OPEN (2) **:** to play the first card of a trick, round, or game
3 : to tend toward or have a result ⟨study *leading* to a degree⟩
4 : to direct the first of a series of blows at an opponent in boxing ☐
synonym see GUIDE
— lead one down the garden path *also* **lead one up the garden path**
: HOODWINK, DECEIVE

²lead *noun* (15th century)
1 a (1) **:** LEADERSHIP (2) **:** EXAMPLE, PRECEDENT **b** (1) **:** position at the front **:** VANGUARD (2) **:** INITIATIVE (3) **:** the act or privilege of leading in cards; *also* **:** the card or suit led **c :** a margin or measure of advantage or superiority or position in advance
2 : one that leads: as **a :** LODE 1 **b :** a channel of water especially through a field of ice **c :** INDICATION, CLUE **d :** a principal role in a dramatic production; *also* **:** one who plays such a role **e :** LEASH 1 **f** (1) **:** an introductory section of a news story (2) **:** a news story of chief importance

☆ SYNONYMS
Lazy, indolent, slothful mean not easily aroused to activity. LAZY suggests a disinclination to work or to take trouble ⟨convenience foods for *lazy* cooks⟩. INDOLENT suggests a love of ease and a settled dislike of movement or activity ⟨the heat made us all *indolent*⟩. SLOTHFUL implies a temperamental inability to act promptly or speedily when action or speed is called for ⟨fired for being *slothful* about filling orders⟩.

☐ USAGE
lead The past tense and past participle of *lead* is spelled *led.* Many people—perhaps influenced by the pronunciation of the metal *lead* or by the past tense of the verb *read*—tend to spell it *lead.* This is a common mistake in casual writing and is to be avoided.

\ə\ **abut** \ᵊ\ **kitten** \ər\ **further** \a\ **ash** \ā\ **ace** \ä\ **mop, mar** \au̇\ **out** \ch\ **chin** \e\ **bet** \ē\ **easy** \g\ **go** \i\ **hit** \ī\ **ice** \j\ **job** \ŋ\ **sing** \ō\ **go** \ȯ\ **law** \ȯi\ **boy** \th\ **thin** \t͟h\ **the** \ü\ **loot** \u̇\ **foot** \y\ **yet** \zh\ **vision** *see also* Guide to Pronunciation

3 : an insulated electrical conductor connected to an electrical device
4 : the course of a rope from end to end
5 : the amount of axial advance of a point accompanying a complete turn of a thread (as of a screw or worm)
6 : a position taken by a base runner off a base toward the next
7 : the first punch of a series or an exchange of punches in boxing
— lead·less \-ləs\ *adjective*

[3]lead *adjective* (1828)
: acting or serving as a lead or leader ⟨a *lead* article⟩

[4]lead \ˈled\ *noun, often attributive* [Middle English *leed,* from Old English *lēad;* akin to Middle High German *lōt* lead] (before 12th century)
1 : a heavy soft malleable ductile plastic but inelastic bluish white metallic element found mostly in combination and used especially in pipes, cable sheaths, batteries, solder, and shields against radioactivity — see ELEMENT table
2 a : a plummet for sounding at sea **b** *plural, British* **:** a usually flat lead roof **c** *plural* **:** lead framing for panes in windows **d :** a thin strip of metal used to separate lines of type in printing
3 a : a thin stick of marking substance in or for a pencil **b :** WHITE LEAD
4 : BULLETS, PROJECTILES
5 : TETRAETHYL LEAD
— lead·less \-ləs\ *adjective*

[5]lead \ˈled\ *transitive verb* (14th century)
1 : to cover, line, or weight with lead
2 : to fix (window glass) in position with leads
3 : to put space between the lines of (typeset matter)
4 : to treat or mix with lead or a lead compound ⟨*leaded* gasoline⟩

lead acetate *noun* (1885)
: an acetate of lead; *especially* **:** a poisonous soluble salt $PbC_4H_6O_4{\cdot}3H_2O$

lead arsenate *noun* (circa 1903)
: an arsenate of lead: as **a :** an acid salt $PbHAsO_4$ used especially as an insecticide **b :** a neutral salt $Pb_3(AsO_4)_2$ used especially as an insecticide

lead azide *noun* (1918)
: a crystalline explosive compound $Pb(N_3)_2$ used as a detonating agent

lead carbonate *noun* (1873)
: a carbonate of lead; *especially* **:** a poisonous basic salt $Pb_3(OH)_2(CO_3)_2$ used especially as a white pigment

lead chromate *noun* (circa 1885)
: a chromate of lead; *especially* **:** CHROME YELLOW

lead dioxide *noun* (1885)
: a poisonous compound PbO_2 used especially as an oxidizing agent and as an electrode in batteries

lead·en \ˈle-dᵊn\ *adjective* (before 12th century)
1 a : made of lead **b :** of the color of lead **:** dull gray
2 a : oppressively heavy **b :** SLUGGISH **c :** lacking spirit or animation
— lead·en·ly *adverb*
— lead·en·ness \-dᵊn-(n)əs\ *noun*

lead·er \ˈlē-dər\ *noun* (14th century)
1 : something that leads: as **a :** a primary or terminal shoot of a plant **b :** TENDON, SINEW **c** *plural* **:** dots or hyphens (as in an index) used to lead the eye horizontally **:** ELLIPSIS 2 **d** *chiefly British* **:** a newspaper editorial **e** (1) **:** something for guiding fish into a trap (2) **:** a short length of material for attaching the end of a fishing line to a lure or hook **f :** LOSS LEADER **g :** something that ranks first **h :** a blank section at the beginning or end of a reel of film or recorded tape
2 : a person who leads: as **a :** GUIDE, CONDUCTOR **b** (1) **:** a person who directs a military force or unit (2) **:** a person who has commanding authority or influence **c** (1) **:** the principal officer of a British political party (2) **:** a party member chosen to manage party activities in a legislative body (3) **:** such a party member presiding over the whole legislative body when the party constitutes a majority **d** (1) **:** CONDUCTOR c (2) **:** a first or principal performer of a group
3 : a horse placed in advance of the other horses of a team
— lead·er·less \-ləs\ *adjective*

leader of the opposition (1771)
: the principal member of the opposition party in a British legislative body who is given the status of a salaried government official and an important role in organizing the business of the house

lead·er·ship \ˈlē-dər-ˌship\ *noun* (1821)
1 : the office or position of a leader
2 : capacity to lead
3 : the act or an instance of leading
4 : LEADERS

lead glass *noun* (1856)
: glass containing a high proportion of lead oxide and having extraordinary clarity and brilliance

lead–in \ˈlēd-ˌin\ *noun* (1913)
: something that leads into something else ⟨a *lead-in* to the commercial⟩
— lead–in *adjective*

lead·ing \ˈlē-diŋ\ *adjective* (1597)
1 : coming or ranking first **:** FOREMOST
2 : exercising leadership
3 : providing direction or guidance ⟨a *leading* question⟩
4 : given most prominent display ⟨the *leading* story⟩

leading edge \ˈlē-diŋ-\ *noun* (1877)
1 : the foremost edge of an airfoil
2 : the forward part of something that moves or seems to move

leading lady *noun* (1874)
: an actress who plays the leading female role (as in a play or movie)

leading man *noun* (1827)
: an actor who plays the leading male role (as in a play or movie)

leading tone *noun* (circa 1889)
: the seventh tone of a diatonic scale — called also *leading note*

lead line \ˈled-\ *noun* (15th century)
: SOUNDING LINE

lead·man \ˈlēd-ˌman, -mən\ *noun* (1939)
: a worker in charge of other workers

lead monoxide *noun* (circa 1909)
: a yellow to brownish red poisonous compound PbO used in rubber manufacture and glassmaking

lead·off \ˈlēd-ˌȯf\ *noun* (circa 1886)
1 : a beginning or leading action
2 : one that leads off
— leadoff *adjective*

lead off \ˈlēd-\ (1817)
transitive verb
1 : to make a start on **:** OPEN
2 : to bat first for a baseball team in (an inning)
intransitive verb
: BEGIN; *also* **:** to come on or perform first

lead on *transitive verb* (1598)
: to entice or induce to proceed in a course especially when unwise or mistaken

lead oxide *noun* (circa 1926)
: any of several oxides of lead; *especially* **:** LEAD MONOXIDE

lead pencil \ˈled-\ *noun* (1688)
: a pencil using graphite as the marking material

lead–pipe \ˈled-ˈpīp\ *adjective* (1898)
: CERTAIN, GUARANTEED ⟨a *lead-pipe* cinch⟩

lead·plant \ˈled-ˌplant\ *noun* (circa 1833)
: a leguminous shrub (*Amorpha canescens*) of the western U.S. that has hoary pinnate leaves and bears dull-colored racemose flowers

lead poisoning *noun* (circa 1842)
: chronic intoxication that is produced by the absorption of lead into the system and is characterized by severe colicky pains, a dark line along the gums, and local muscular paralysis

lead·screw \ˈlēd-ˌskrü\ *noun* (circa 1889)
: a threaded rod on which a mechanism travels and can be positioned precisely

leads·man \ˈledz-mən\ *noun* (1857)
: a man who uses a sounding lead to determine depth of water

lead sulfide *noun* (circa 1898)
: an insoluble black compound PbS that occurs naturally as galena and is used in photoconductive cells

lead time \ˈlēd-\ *noun* (1944)
: the time between the beginning of a process or project and the appearance of its results

lead–up \ˈlēd-ˌəp\ *noun* (1942)
: something that leads up to or prepares the way for something else

lead up \ˈlēd-\ *intransitive verb* (1861)
1 : to prepare the way
2 : to make a gradual or indirect approach to a topic

lead·work \ˈled-ˌwərk\ *noun* (1641)
: articles made of or work done in lead

leady \ˈle-dē\ *adjective* **lead·i·er; -est** (14th century)
: containing or resembling lead

[1]leaf \ˈlēf\ *noun, plural* **leaves** \ˈlēvz\ *also* **leafs** \ˈlēfs\ *often attributive* [Middle English *leef,* from Old English *lēaf;* akin to Old High German *loub* leaf] (before 12th century)
1 a (1) **:** a lateral outgrowth from a plant stem that is typically a flattened expanded variably shaped greenish organ, constitutes a unit of the foliage, and functions primarily in food manufacture by photosynthesis (2) **:** a modified leaf primarily engaged in functions other than food manufacture **b** (1) **:** FOLIAGE ⟨trees in full *leaf*⟩ (2) **:** the leaves of a plant as an article of commerce
2 : something suggestive of a leaf: as **a :** a part of a book or folded sheet containing a page on each side **b** (1) **:** a part (as of window shutters, folding doors, or gates) that slides or is hinged (2) **:** the movable parts of a table top **c** (1) **:** a thin sheet or plate of any substance **:** LAMINA (2) **:** metal (as gold or silver) in sheets usually thinner than foil (3) **:** one of the plates of a leaf spring
— leaf·less \ˈlēf-ləs\ *adjective*
— leaf·like \ˈlēf-ˌlīk\ *adjective*

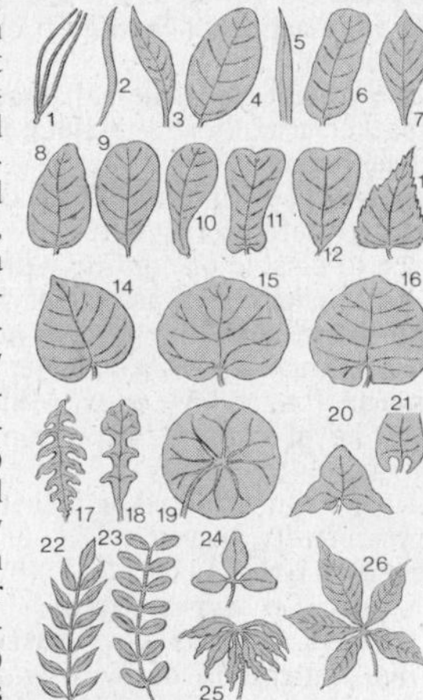

forms of leaf 1a(1): ***1*** **needle-shaped,** ***2*** **linear,** ***3*** **lanceolate,** ***4*** **elliptical,** ***5*** **ensiform,** ***6*** **oblong,** ***7*** **oblanceolate with acuminate tip,** ***8*** **ovate with acute tip,** ***9*** **obovate,** ***10*** **spatulate,** ***11*** **fiddle-shaped,** ***12*** **cuneate,** ***13*** **deltoid,** ***14*** **cordate,** ***15*** **reniform,** ***16*** **orbiculate,** ***17*** **runcinate,** ***18*** **lyrate,** ***19*** **peltate,** ***20*** **hastate,** ***21*** **sagittate,** ***22*** **odd-pinnate,** ***23*** **abruptly pinnate,** ***24*** **trifoliolate,** ***25, 26*** **palmate**

[2]leaf (1611)
intransitive verb
1 : to shoot out or produce leaves
2 : to turn over pages ⟨*leaf* through a book⟩

transitive verb
: to turn over the pages of

leaf·age \'lē-fij\ *noun* (1599)
1 : FOLIAGE 2
2 : the representation of leafage (as in architecture)

leaf bud *noun* (1664)
: a bud that develops into a leafy shoot and does not produce flowers

leaf butterfly *noun* (1882)
: any of a genus (*Kallima*) of nymphalid butterflies of southern Asia and the East Indies that mimic leaves

leaf curl *noun* (1899)
: a plant disease caused by a fungus (genus *Taphrina*) or virus and characterized by curling of leaves; *especially* **:** PEACH LEAF CURL

leafed \'lēft\ *adjective* (1552)
: LEAVED

leaf fat *noun* (circa 1725)
: the fat that lines the abdominal cavity and encloses the kidneys; *especially* **:** that of a hog used in the manufacture of lard

leaf·hop·per \'lēf-ˌhä-pər\ *noun* (1852)
: any of a family (Cicadellidae) of small leaping homopterous insects that suck the juices of plants

leaf lard *noun* (circa 1847)
: high-quality lard made from leaf fat

[1]**leaf·let** \'lēf-lət\ *noun* (1787)
1 a : one of the divisions of a compound leaf **b :** a small or young foliage leaf
2 : a leaflike organ or part
3 : a usually folded printed sheet intended for free distribution

[2]**leaflet** *verb* **-let·ed** *or* **-let·ted; -let·ing** *or* **-let·ting** (1962)
intransitive verb
: to hand out leaflets
transitive verb
: to hand out leaflets to
— leaf·le·teer \ˌlēf-lə-'tir\ *noun*

leaf miner *noun* (1830)
: any of various small insects (as moths or dipteran flies) that in the larval stages burrow in and eat the parenchyma of leaves

leaf mold *noun* (1845)
1 : a compost or layer composed chiefly of decayed vegetable matter
2 : a mold or mildew that affects foliage

leaf roll *noun* (1916)
: a virus disease of the potato that is transmitted by aphids and is characterized by an upward rolling of the leaf margins, smaller tubers, and netlike necrotic areas in the phloem

leaf roller *noun* (1830)
: any of various lepidopterans whose larvae make a nest by rolling up plant leaves

leaf rust *noun* (1865)
: a rust disease of plants and especially of wheat that affects primarily the leaves

leaf scar *noun* (1835)
: the mark left on a stem after a leaf falls

leaf scorch *noun* (circa 1909)
: any of various plant diseases characterized by a burned or scorched appearance of the foliage

leaf spot *noun* (circa 1895)
: any of various plant diseases characterized by discolored often circular spots on the leaves

leaf spring *noun* (circa 1893)
: a spring made of superposed strips or leaves

leaf·stalk \'lēf-ˌstȯk\ *noun* (circa 1776)
: PETIOLE

leaf trace *noun* [[3]*trace*] (1875)
: a trace associated with a leaf

leafy \'lē-fē\ *adjective* **leaf·i·er; -est** (15th century)
1 a : furnished with or abounding in leaves ⟨*leafy* woodlands⟩ **b :** having broad-bladed leaves ⟨mosses, grasses, and *leafy* plants⟩ **c :** consisting chiefly of leaves ⟨*leafy* vegetables⟩
2 : resembling a leaf; *specifically* **:** LAMINATE

leafy liverwort *noun* (1922)
: any of an order (Jungermanniales) of usually epiphytic liverworts with a leafy gametophyte that has one ventral and two dorsal rows of leaves on the stem

leafy spurge *noun* (circa 1889)
: a tall perennial European herb (*Euphorbia esula*) that is naturalized in the northern U.S. and Canada

[1]**league** \'lēg\ *noun* [Middle English *leuge, lege,* from Late Latin *leuga*] (14th century)
1 : any of various units of distance from about 2.4 to 4.6 statute miles (3.9 to 7.4 kilometers)
2 : a square league

[2]**league** *noun* [Middle English (Scots) *ligg,* from Middle French *ligue,* from Old Italian *liga,* from *ligare* to bind, from Latin — more at LIGATURE] (15th century)
1 a : an association of nations or other political entities for a common purpose **b** (1) **:** an association of persons or groups united by common interests or goals (2) **:** a group of sports teams that regularly play one another **c :** an informal alliance
2 : CLASS, CATEGORY

[3]**league** *verb* **leagued; leagu·ing** (1604)
transitive verb
: to unite in a league
intransitive verb
: to form a league

[1]**lea·guer** \'lē-gər\ *noun* [Dutch *leger;* akin to Old High German *legar* bed — more at LAIR] (1577)
1 : a military camp
2 : SIEGE

[2]**leaguer** *transitive verb* (circa 1720)
archaic **:** BESIEGE, BELEAGUER

[3]**leagu·er** \'lē-gər\ *noun* [[2]*league*] (1591)
: a member of a league

[1]**leak** \'lēk\ *verb* [Middle English *leken,* from Old Norse *leka;* akin to Old English *leccan* to moisten, Middle Irish *legaid* it melts] (14th century)
intransitive verb
1 a : to enter or escape through an opening usually by a fault or mistake ⟨fumes *leak* in⟩ **b :** to let a substance or light in or out through an opening
2 a : to become known despite efforts at concealment **b :** to be the source of an information leak
transitive verb
1 : to permit to enter or escape through or as if through a leak
2 : to give out (information) surreptitiously ⟨*leaked* the story to the press⟩
— leak·er \'lē-kər\ *noun*

[2]**leak** *noun* (15th century)
1 a : a crack or hole that usually by mistake admits or lets escape **b :** something that permits the admission or escape of something else usually with prejudicial effect
2 : the act, process, or an instance of leaking
3 : an act of urinating — used especially in the phrase *take a leak; sometimes considered vulgar*
— leak·proof \'lēk-ˌprüf\ *adjective*

leak·age \'lē-kij\ *noun* (15th century)
1 a : the act or process or an instance of leaking **b :** loss of electricity especially due to faulty insulation
2 : something or the amount that leaks

leaky \'lē-kē\ *adjective* **leak·i·er; -est** (15th century)
: permitting fluid to leak in or out
— leak·i·ly \-kə-lē\ *adverb*
— leak·i·ness \-kē-nəs\ *noun*

leal \'lē(ə)l\ *adjective* [Middle English *leel,* from Middle French *leial, leel* — more at LOYAL] (14th century)
chiefly Scottish **:** LOYAL, TRUE
— leal·ly \'lē-ə(l)-lē, 'lēl-lē\ *adverb*

[1]**lean** \'lēn\ *verb* **leaned** \'lēnd, *chiefly British* 'lent\; **lean·ing** \'lē-niŋ\ [Middle English *lenen,* from Old English *hleonian;* akin to Old High German *hlinēn* to lean, Greek *klinein,* Latin *clinare*] (before 12th century)
intransitive verb
1 a : to incline, deviate, or bend from a vertical position **b :** to cast one's weight to one side for support
2 : to rely for support or inspiration
3 : to incline in opinion, taste, or desire
transitive verb
: to cause to lean **:** INCLINE
— lean on : to apply pressure to

[2]**lean** *noun* (1776)
: the act or an instance of leaning **:** INCLINATION

[3]**lean** *adjective* [Middle English *lene,* from Old English *hlǣne*] (before 12th century)
1 a : lacking or deficient in flesh **b :** containing little or no fat
2 : lacking richness, sufficiency, or productiveness
3 : deficient in an essential or important quality or ingredient: as **a** *of ore* **:** containing little valuable mineral **b :** low in combustible component — used especially of fuel mixtures
4 : characterized by economy (as of style, expression, or operation) ☆
— lean·ly *adverb*
— lean·ness \'lēn-nəs\ *noun*

[4]**lean** *transitive verb* (before 12th century)
: to make lean

[5]**lean** *noun* (15th century)
: the part of meat that consists principally of lean muscle

Le·an·der \lē-'an-dər\ *noun* [Latin, from Greek *Leandros*]
: a youth in Greek mythology who swims the Hellespont nightly to visit Hero and who ultimately drowns in one of the crossings

lean·ing \'lē-niŋ\ *noun* (15th century)
: a definite but not decisive attraction or tendency — often used in plural ☆

leant \'lent\ *chiefly British past of* LEAN

[1]**lean–to** \'lēn-ˌtü\ *noun, plural* **lean–tos** \-ˌtüz\ (15th century)

☆ SYNONYMS

Lean, spare, lank, lanky, gaunt, rawboned, scrawny, skinny mean thin because of an absence of excess flesh. LEAN stresses lack of fat and of curving contours ⟨a *lean* racehorse⟩. SPARE suggests leanness from abstemious living or constant exercise ⟨the gymnast's *spare* figure⟩. LANK implies tallness as well as leanness ⟨the *lank* legs of the heron⟩. LANKY suggests awkwardness and loose-jointedness as well as thinness ⟨a *lanky* youth, all arms and legs⟩. GAUNT implies marked thinness or emaciation as from overwork or suffering ⟨a prisoner's *gaunt* face⟩. RAWBONED suggests a large ungainly build without implying undernourishment ⟨a *rawboned* farmer⟩. SCRAWNY and SKINNY imply an extreme leanness that suggests deficient strength and vitality ⟨a *scrawny* chicken⟩ ⟨*skinny* street urchins⟩.

Leaning, propensity, proclivity, penchant mean a strong instinct or liking for something. LEANING suggests a liking or attraction not strong enough to be decisive or uncontrollable ⟨a student with artistic *leanings*⟩. PROPENSITY implies a deeply ingrained and usually irresistible inclination ⟨a *propensity* to offer advice⟩. PROCLIVITY suggests a strong natural proneness usually to something objectionable or evil ⟨a *proclivity* for violence⟩. PENCHANT implies a strongly marked taste in the person or an irresistible attraction in the object ⟨a *penchant* for taking risks⟩.

\ə\ **abut** \ᵊ\ **kitten** \ər\ **further** \a\ **ash** \ā\ **ace** \ä\ **mop, mar** \au̇\ **out** \ch\ **chin** \e\ **bet** \ē\ **easy** \g\ **go** \i\ **hit** \ī\ **ice** \j\ **job** \ŋ\ **sing** \ō\ **go** \ȯ\ **law** \ȯi\ **boy** \th\ **thin** \t͟h\ **the** \ü\ **loot** \u̇\ **foot** \y\ **yet** \zh\ **vision** *see also* Guide to Pronunciation

1 : a wing or extension of a building having a lean-to roof
2 : a rough shed or shelter with a lean-to roof
[2]lean–to *adjective* (1649)
: having only one slope or pitch ⟨*lean-to* roof⟩ — see ROOF illustration
[1]leap \'lēp\ *verb* **leaped** *or* **leapt** \'lēpt *also* 'lept\; **leap·ing** \'lē-piŋ\ [Middle English *lepen,* from Old English *hlēapan;* akin to Old High German *hlouffan* to run] (before 12th century)
intransitive verb
1 : to spring free from or as if from the ground **:** JUMP ⟨*leap* over a fence⟩ ⟨a fish *leaps* out of the water⟩
2 a : to pass abruptly from one state or topic to another **b :** to act precipitately ⟨*leaped* at the chance⟩
transitive verb
: to pass over by leaping
— leap·er \'lē-pər\ *noun*
[2]leap *noun* (before 12th century)
1 a : an act of leaping **:** SPRING, BOUND **b** (1) **:** a place leaped over or from (2) **:** the distance covered by a leap
2 a : a sudden passage or transition ⟨a great *leap* forward⟩ **b :** a choice made in an area of ultimate concern ⟨a *leap* of faith⟩
— by leaps and bounds : with extraordinary rapidity
[1]leap·frog \'lēp-ˌfrȯg, -ˌfräg\ *noun* (1599)
: a game in which one player bends down and is vaulted over by another player
[2]leapfrog *verb* **leap·frogged; leap·frog·ging** (1872)
intransitive verb
: to leap or progress in or as if in leapfrog
transitive verb
1 : to go ahead of (each other) in turn; *specifically* **:** to advance (two military units) by keeping one unit in action while moving the other unit past it to a position farther in front
2 : to evade by or as if by a bypass
leap second *noun* (1971)
: an intercalary second added to Coordinated Universal Time to compensate for the slowing of the earth's rotation and keep Coordinated Universal Time in synchrony with solar time
leap year *noun* (14th century)
1 : a year in the Gregorian calendar containing 366 days with February 29 as the extra day
2 : an intercalary year in any calendar
Lear \'lir\ *noun*
: a legendary king of Britain and hero of Shakespeare's tragedy *King Lear*
learn \'lərn\ *verb* **learned** \'lərnd, 'lərnt\; **learn·ing** [Middle English *lernen,* from Old English *leornian;* akin to Old High German *lernēn* to learn, Old English *last* footprint, Latin *lira* furrow, track] (before 12th century)
transitive verb
1 a (1) **:** to gain knowledge or understanding of or skill in by study, instruction, or experience ⟨*learn* a trade⟩ (2) **:** MEMORIZE ⟨*learn* the lines of a play⟩ **b :** to come to be able ⟨*learn* to dance⟩ **c :** to come to realize ⟨*learned* that honesty paid⟩
2 a *nonstandard* **:** TEACH **b** *obsolete* **:** to inform of something
3 : to come to know **:** HEAR ⟨we just *learned* that he was ill⟩
intransitive verb
: to acquire knowledge or skill or a behavioral tendency ◻ ◆
synonym see DISCOVER
— learn·able \'lər-nə-bəl\ *adjective*
— learn·er *noun*
learned *adjective* (14th century)
1 \'lər-nəd\ **:** characterized by or associated with learning **:** ERUDITE
2 \'lərnd, 'lərnt\ **:** acquired by learning ⟨*learned* behavior⟩
— learn·ed·ly \'lər-nəd-lē\ *adverb*
— learn·ed·ness \-nəd-nəs\ *noun*
learn·ing *noun* (before 12th century)
1 : the act or experience of one that learns
2 : knowledge or skill acquired by instruction or study
3 : modification of a behavioral tendency by experience (as exposure to conditioning)
synonym see KNOWLEDGE
learning curve *noun* (1922)
: a curve plotting performance against practice; *especially* **:** one graphing decline in unit costs with cumulative output
learning disabled *adjective* (1973)
: having difficulty in learning a basic scholastic skill because of a disorder (as dyslexia or attention deficit disorder) that interferes with the learning process
— learning disability *noun*
learnt \'lərnt\ *chiefly British past and past participle of* LEARN
leary *variant of* LEERY
[1]lease \'lēs\ *noun* [Middle English *les,* from Anglo-French, from *lesser*] (14th century)
1 : a contract by which one conveys real estate, equipment, or facilities for a specified term and for a specified rent; *also* **:** the act of such conveyance or the term for which it is made
2 : a piece of land or property that is leased
3 : a continuance or opportunity for continuance ⟨a new *lease* on life⟩
[2]lease *transitive verb* **leased; leas·ing** [Anglo-French *lesser,* from Old French *laissier* to let go, from Latin *laxare* to loosen, from *laxus* slack — more at SLACK] (circa 1570)
1 : to grant by lease
2 : to hold under a lease
synonym see HIRE
— leas·able \'lē-sə-bəl\ *adjective*
lease·back \'lēs-ˌbak\ *noun* (1947)
: the sale of property with the understanding that the seller can lease it from the new owner
lease·hold \'lēs-ˌhōld\ *noun* (1720)
1 : a tenure by lease
2 : property held by lease
— lease·hold·er *noun*
leash \'lēsh\ *noun* [Middle English *lees, leshe,* from Middle French *laisse,* from Old French *laissier*] (14th century)
1 a : a line for leading or restraining an animal **b :** something that restrains
2 a : a set of three animals (as greyhounds, foxes, bucks, or hares) **b :** a set of three
— leash *transitive verb*
leash law *noun* (1966)
: an ordinance requiring dogs to be restrained when not confined to their owner's property
leas·ing \'lē-siŋ, -ziŋ\ *noun* [Middle English *lesing,* from Old English *lēasung,* from *lēasian* to lie, from *lēas* false] (before 12th century)
archaic **:** the act of lying; *also* **:** LIE, FALSEHOOD
[1]least \'lēst\ *adjective superlative of* [1]LITTLE [Middle English *leest,* from Old English *lǣst,* superlative of *lǣssa* less] (before 12th century)
1 : lowest in importance or position
2 a : smallest in size or degree **b :** being a member of a kind distinguished by diminutive size ⟨*least* bittern⟩ **c :** smallest possible **:** SLIGHTEST
[2]least *noun* (12th century)
: one that is least
— at least 1 : at the minimum **2 :** in any case
[3]least *adverb superlative of* [2]LITTLE (13th century)
: in the smallest or lowest degree
— least of all : especially not ⟨no one, *least of all* the children, paid attention⟩
least common denominator *noun* (1875)
: the least common multiple of two or more denominators
least common multiple *noun* (1823)
1 : the smallest common multiple of two or more numbers
2 : the common multiple of lowest degree of two or more polynomials
least squares *noun plural* (1825)
: a method of fitting a curve to a set of points representing statistical data in such a way that the sum of the squares of the distances of the points from the curve is a minimum
least·ways \'lēst-ˌwāz\ *adverb* (14th century)
dialect **:** at least
least·wise \-ˌwīz\ *adverb* (15th century)
: at least
[1]leath·er \'le-t͟hər\ *noun* [Middle English *lether,* from Old English *lether-;* akin to Old High German *leder* leather, Old Irish *lethar*] (13th century)
1 : animal skin dressed for use
2 : the flap of the ear of a dog — see DOG illustration
3 : something wholly or partly made of leather
— leather *adjective*
— leath·er·like \-ˌlīk\ *adjective*
[2]leather *transitive verb* **leath·ered; leath·er·ing** \'le-t͟hə-riŋ, 'let͟h-riŋ\ (13th century)
1 : to cover with leather
2 : to beat with a strap **:** THRASH
leath·er·back \'le-t͟hər-ˌbak\ *noun* (circa 1855)
: the largest existing sea turtle (*Dermochelys coriacea*) distinguished by its flexible carapace composed of a mosaic of small bones embedded in a thick leathery skin

leatherback

leath·er·ette \ˌle-t͟hə-'ret\ *noun, often attributive* [from *Leatherette,* a trademark] (circa 1879)
: simulated leather
leath·er·leaf \'le-t͟hər-ˌlēf\ *noun* (circa 1818)

◻ USAGE
learn *Learn* in the sense of "teach" dates from the 13th century and was standard until at least the early 19th ⟨made them drunk with true Hollands—and then *learned* them the art of making bargains —Washington Irving⟩. But by Mark Twain's time it was receding to a speech form associated chiefly with the less educated ⟨never done nothing for three months but set in his back yard and *learn* that frog to jump —Mark Twain⟩. The present-day status of *learn* has not risen. This use persists in speech, but in writing it appears mainly in the representation of such speech or its deliberate imitation for effect.

◇ WORD HISTORY
learn The verb *learn,* in Old English *leornian,* is paralleled by other words in the old and modern West Germanic languages, such as Old Saxon *līnon* and Modern German *lernen.* All these represent a derivative, formed with a suffixal *-n-,* of a verb meaning "to teach." In Old English this is *lǣran,* which as *lear* "to learn, ascertain" still survives in Scots. (A noun derivative of the "teach" verb is *lore,* continuing Old English *lār.*) The oldest Germanic cognate of Old English *lǣran* is Gothic *laisjan,* a causative verb formed from *lais* "I have learned, know how." A number of decades ago etymologists advanced the hypothesis that *lais* was a derivative of a noun meaning "track" or "furrow": learning how to do something is like following a track. Old English *lāst* "track, footprint, sole of the foot" is most likely a suffixed form of this "track" word, and the ultimate source of both the verb *last* "to continue in time" (Old English *lǣstan* "to follow, carry out") and the noun *last* "form for making a shoe" (Old English *lǣste*).

: a north temperate ericaceous bog shrub (*Chamaedaphne calyculata*) with evergreen coriaceous leaves and small white cylindrical flowers

leath·ern \ˈle-th̲ərn\ *adjective* (before 12th century)
: made of, consisting of, or resembling leather

leath·er·neck \-th̲ər-ˌnek\ *noun* [from the leather collar formerly part of the uniform] (circa 1914)
: a member of the U.S. Marine Corps

leath·er·wood \ˈle-th̲ər-ˌwu̇d\ *noun* (1743)
1 : a small tree (*Dirca palustris*) of the mezereon family with pliant stems and yellow flowers
2 : a small tree (*Cyrilla racemiflora*) of the southeastern U.S. related to the titi

leath·ery \ˈle-th̲ə-rē, ˈleth-rē\ *adjective* (circa 1552)
: resembling leather in appearance or consistency

[1]leave \ˈlēv\ *verb* **left** \ˈleft\; **leav·ing** [Middle English *leven,* from Old English *lǣfan;* akin to Old High German *verleiben* to leave, Old English *belīfan* to be left over, and perhaps to Lithuanian *lipti* to adhere, Greek *lipos* grease, fat] (before 12th century)
transitive verb
1 a (1) **:** BEQUEATH, DEVISE ⟨*left* a fortune to his son⟩ (2) **:** to have remaining after one's death ⟨*leaves* a widow and two children⟩ **b :** to cause to remain as a trace or aftereffect ⟨oil *leaves* a stain⟩ ⟨the wound *left* an ugly scar⟩
2 a : to cause or allow to be or remain in a specified condition ⟨*leave* the door open⟩ ⟨his manner *left* me cold⟩ **b :** to fail to include or take along ⟨*left* the notes at home⟩ ⟨the movie *leaves* a lot out⟩ **c :** to have as a remainder ⟨4 from 7 *leaves* 3⟩ **d :** to permit to be or remain subject to another's action or control ⟨just *leave* everything to me⟩ **e :** LET **f :** to cause or allow to be or remain available ⟨*leave* room for expansion⟩ ⟨wanted to *leave* himself an out⟩
3 a : to go away from **:** DEPART ⟨*leave* the room⟩ **b :** DESERT, ABANDON ⟨*left* his wife⟩ **c :** to terminate association with **:** withdraw from ⟨*left* school before graduation⟩
4 : to put, deposit, or deliver before or in the process of departing ⟨someone *left* a package for you⟩
intransitive verb
: SET OUT, DEPART ▫
— leav·er *noun*
— leave alone : to refrain from bothering or using

[2]leave *noun* [Middle English *leve,* from Old English *lēaf;* akin to Middle High German *loube* permission, Old English *alȳfan* to allow — more at BELIEVE] (before 12th century)
1 a : permission to do something **b :** authorized especially extended absence from duty or employment
2 : an act of leaving **:** DEPARTURE

[3]leave *intransitive verb* **leaved; leav·ing** [Middle English *leven,* from *leef* leaf] (14th century)
: LEAF

leaved *adjective* (13th century)
: having leaves — usually used in combination ⟨palmate-*leaved*⟩ ⟨a four-*leaved* clover⟩

[1]leav·en \ˈle-vən\ *noun* [Middle English *levain,* from Middle French, from (assumed) Vulgar Latin *levamen,* from Latin *levare* to raise — more at LEVER] (14th century)
1 a : a substance (as yeast) used to produce fermentation in dough or a liquid; *especially* **:** SOURDOUGH **b :** a material (as baking powder) used to produce a gas that lightens dough or batter
2 : something that modifies or lightens

[2]leaven *transitive verb* **leav·ened; leav·en·ing** \ˈlev-niŋ, ˈle-və-\ (15th century)
1 : to raise (as bread) with a leaven
2 : to mingle or permeate with some modifying, alleviating, or vivifying element
synonym see INFUSE

leavening *noun* (circa 1626)
: a leavening agent **:** LEAVEN

leave of absence (1771)
1 : permission to be absent from duty or employment
2 : LEAVE 1b

leave off *verb* (14th century)
: STOP, CEASE

leaves *plural of* LEAF

leave–tak·ing \ˈlēv-ˌtā-kiŋ\ *noun* (14th century)
: DEPARTURE, FAREWELL

leav·ings \ˈlē-viŋz\ *noun plural* (14th century)
: REMNANTS, RESIDUE

le·bens·raum \ˈlā-bənz-ˌrau̇m, -bən(t)s-\ *noun, often capitalized* [German, from *Leben* living, life + *Raum* space] (1905)
1 : territory believed especially by Nazis to be necessary for national existence or economic self-sufficiency
2 : space required for life, growth, or activity

[1]lech \ˈlech\ *noun* (circa 1830)
1 : LETCH, LUST
2 : LECHER

[2]lech *intransitive verb* (1911)
: LUST

lech·er \ˈle-chər\ *noun* [Middle English *lechour,* from Old French *lecheor,* from *lechier* to lick, live in debauchery, of Germanic origin; akin to Old High German *leckōn* to lick — more at LICK] (13th century)
: a man who engages in lechery

lech·er·ous \ˈle-chə-rəs, ˈlech-rəs\ *adjective* (14th century)
: given to or suggestive of lechery
— lech·er·ous·ly *adverb*
— lech·er·ous·ness *noun*

lech·ery \-rē\ *noun* (13th century)
: inordinate indulgence in sexual activity **:** LASCIVIOUSNESS

le·chwe \ˈlēch-wē\ *noun, plural* **lechwe** *or* **lechwes** [probably from Sesotho *lets'a*] (1857)
: an antelope (*Kobus leche*) that inhabits wetlands of southern Africa; *also* **:** a related antelope (*K. megaceros*) of the Nile Valley in Sudan and Ethiopia

lec·i·thin \ˈle-sə-thən\ *noun* [International Scientific Vocabulary, from Greek *lekithos* yolk of an egg] (1861)
: any of several waxy hygroscopic phospholipids that are widely distributed in animals and plants, form colloidal solutions in water, and have emulsifying, wetting, and antioxidant properties; *also* **:** a mixture of or substance rich in lecithins

lec·i·thin·ase \-thə-ˌnās, -ˌnāz\ *noun* (1910)
: PHOSPHOLIPASE

lec·tern \ˈlek-tərn\ *noun* [Middle English *lettorne,* from Middle French *letrun,* from Medieval Latin *lectorinum,* from Latin *lector* reader, from *legere* to read — more at LEGEND] (14th century)
: a stand used to support a book in a convenient position for a standing reader; *especially* **:** one from which scripture lessons are read in a church service
usage see PODIUM

lec·tion \ˈlek-shən\ *noun* [Late Latin *lection-, lectio,* from Latin, act of reading — more at LESSON] (1608)
1 : a liturgical lesson for a particular day
2 [New Latin *lection-, lectio,* from Latin] **:** a variant reading of a text

lec·tion·ary \ˈlek-shə-ˌner-ē\ *noun, plural* **-ar·ies** (1780)
: a book or list of lections for the church year

lec·tor \ˈlek-tər, -ˌtȯr\ *noun* [Middle English, from Late Latin, reader of the lessons in a church service, from Latin, reader, from *legere*] (14th century)
: one who assists at a worship service chiefly by reading a lesson

lec·to·type \ˈlek-tə-ˌtīp\ *noun* [Greek *lektos* chosen (from *legein* to gather, choose) + English *type* — more at LEGEND] (circa 1905)
: a specimen chosen as the type of a species or subspecies if the author of the name fails to designate a type

[1]lec·ture \ˈlek-chər, -shər\ *noun* [Middle English, act of reading, from Late Latin *lectura,* from Latin *lectus,* past participle of *legere*] (15th century)
1 : a discourse given before an audience or class especially for instruction
2 : a formal reproof
— lec·ture·ship \-ˌship\ *noun*

[2]lecture *verb* **lec·tured; lec·tur·ing** \ˈlek-chə-riŋ, ˈlek-shriŋ\ (circa 1590)
intransitive verb
: to deliver a lecture or a course of lectures
transitive verb
1 : to deliver a lecture to
2 : to reprove formally
— lec·tur·er \-chər-ər, -shrər\ *noun*

led *past and past participle of* LEAD

LED \ˌel-(ˌ)ē-ˈdē\ *noun* [*l*ight-*e*mitting *d*iode] (1968)
: a semiconductor diode that emits light when a voltage is applied to it and that is used in an electronic display (as for a digital watch)

Le·da \ˈlē-də\ *noun* [Latin, from Greek *Lēda*]
: the mother of Clytemnestra and Castor by her husband Tyndareus and of Helen and Pollux by Zeus who comes to her in the form of a swan

le·der·ho·sen \ˈlā-dər-ˌhō-zᵊn\ *noun plural* [German, from Middle High German *lederhose,* from *leder* leather + *hose* trousers] (1936)
: leather shorts often with suspenders worn especially in Bavaria

ledge \ˈlej\ *noun* [Middle English *legge* bar of a gate] (1535)
1 : a raised or projecting edge or molding intended to protect or check ⟨a window *ledge*⟩
2 : an underwater ridge or reef especially near the shore
3 a : a narrow flat surface or shelf; *especially* **:** one that projects from a wall of rock **b :** rock that is solid or continuous enough to form ledges **:** BEDROCK ⟨the field was full of *ledge*⟩
4 : LODE, VEIN
— ledgy \ˈle-jē\ *adjective*

led·ger \ˈle-jər\ *noun* [Middle English *legger,* probably from *leyen, leggen* to lay] (1588)
1 : a book containing accounts to which debits and credits are posted from books of original entry
2 : a horizontal timber secured to the uprights of scaffolding to support the putlog

ledger line *noun* (1700)
: a short line added above or below a musical staff to extend its range

[1]lee \ˈlē\ *noun* [Middle English, from Old English *hlēo;* perhaps akin to Old High German *lāo* lukewarm, Latin *calēre* to be warm] (before 12th century)
1 : protecting shelter
2 : the side (as of a ship) that is sheltered from the wind

[2]lee *adjective* (15th century)

▫ USAGE
leave *Leave* (sense 2e) with the infinitive but without *to* ⟨*leave* it be⟩ is a mostly spoken idiom used in writing especially for humorous effect. It is not often criticized in British English, but American commentators, adhering to an opinion first expressed in 1881, still dislike it.

1 : of or relating to the side sheltered from the wind — compare WEATHER
2 : facing in the direction of motion of an overriding glacier — used especially of a hillside

lee•board \'lē-ˌbōrd, -ˌbȯrd\ *noun* (1691)
: either of the wood or metal planes attached outside the hull of a sailboat to prevent leeway

¹leech \'lēch\ *noun* [Middle English *leche,* from Old English *lǣce;* akin to Old High German *lāhhi* physician] (before 12th century)
1 *archaic* **:** PHYSICIAN, SURGEON
2 [from its former use by physicians for bleeding patients] **:** any of numerous carnivorous or bloodsucking usually freshwater annelid worms (class Hirudinea) that have typically a flattened lanceolate segmented body with a sucker at each end
3 : a hanger-on who seeks advantage or gain
synonym see PARASITE
— leech•like \-ˌlīk\ *adjective*

²leech (1828)
transitive verb
1 : to bleed by the use of leeches
2 : to drain the substance of **:** EXHAUST
intransitive verb
: to attach oneself to a person as a leech

³leech *noun* [Middle English *leche,* from Middle Low German *līk* boltrope — more at LIGATURE] (15th century)
1 : either vertical edge of a square sail
2 : the after edge of a fore-and-aft sail

leek \'lēk\ *noun* [Middle English, from Old English *lēac;* akin to Old High German *louh* leek] (before 12th century)
: a biennial garden herb (*Allium ampeloprasum porrum*) of the lily family grown for its mildly pungent succulent linear leaves and especially for its thick cylindrical stalk

leek

¹leer \'lir\ *intransitive verb* [probably from obsolete *leer* cheek] (1530)
: to cast a sidelong glance; *especially* **:** to give a leer
— leer•ing•ly \-iŋ-lē\ *adverb*

²leer *noun* (1598)
: a lascivious, knowing, or wanton look

leery \'lir-ē\ *adjective* (1718)
: SUSPICIOUS, WARY — often used with *of*

lees \'lēz\ *noun plural* [Middle English *lie,* from Middle French, from Medieval Latin *lia*] (14th century)
: the sediment of a liquor (as wine) during fermentation and aging **:** DREGS

Lee's Birthday \'lēz-\ *noun* [Robert E. *Lee*] (1910)
: January 19 or the third Monday in January observed as a legal holiday in many southern states

lee shore *noun* (circa 1580)
: a shore lying off a ship's leeward side and constituting a severe danger in storm

¹lee•ward \'lē-wərd, *especially nautical* 'lü-ərd\ *noun* (1549)
: the lee side

²leeward *adjective* (1666)
: being in or facing the direction toward which the wind is blowing; *also* **:** being the side opposite the windward

lee•way \'lē-ˌwā\ *noun* (1669)
1 a : off-course lateral movement of a ship when under way **b :** the angle between the heading and the track of an airplane
2 : an allowable margin of freedom or variation **:** TOLERANCE

¹left \'left\ *adjective* [Middle English, from Old English, weak; akin to Middle Low German *lucht* left; from the left hand's being the weaker in most individuals] (13th century)
1 a : of, relating to, situated on, or being the side of the body in which the heart is mostly located **b :** located nearer to the left hand than to the right **c** (1) **:** located on the left of an observer facing in the same direction as the object specified ⟨stage *left*⟩ (2) **:** located on the left when facing downstream ⟨the *left* bank of a river⟩
2 *often capitalized* **:** of, adhering to, or constituted by the left especially in politics
— left *adverb*

²left *noun* (13th century)
1 a : the left hand **b :** the location or direction of the left side **c :** the part on the left side
2 a : LEFT FIELD **b :** a blow struck with the left fist
3 *often capitalized* **a :** the part of a legislative chamber located to the left of the presiding officer **b :** the members of a continental European legislative body occupying the left as a result of holding more radical political views than other members
4 *capitalized* **a :** those professing views usually characterized by desire to reform or overthrow the established order especially in politics and usually advocating change in the name of the greater freedom or well-being of the common man **b :** a radical as distinguished from a conservative position

³left *past and past participle of* LEAVE

left-bank *adjective, often L&B capitalized* (1929)
: of, relating to, situated in, or characteristic of the bohemian district of Paris on the left bank of the Seine River

left field *noun* (1857)
1 : the position of the player defending left field
2 : the part of the baseball outfield to the left looking out from the plate
3 : a state or position far from the mainstream (as of prevailing opinion) ⟨they were really out in *left field* with that idea⟩ **:** a source of the unexpected or illogical ⟨that question came out of *left field*⟩
— left fielder *noun*

left-hand \'left-'hand, 'lef-'tand\ *adjective* (1598)
1 : situated on the left
2 : LEFT-HANDED

left-hand•ed \-'han-dəd, -'tan-\ *adjective* (14th century)
1 : using the left hand habitually or more easily than the right; *also* **:** swinging from left to right ⟨a *left-handed* batter⟩
2 : relating to, designed for, or done with the left hand
3 : MORGANATIC
4 a : CLUMSY, AWKWARD **b :** INSINCERE, BACKHANDED, DUBIOUS ⟨a *left-handed* compliment⟩
5 a : having a direction contrary to that of the hands of a watch viewed from in front **:** COUNTERCLOCKWISE **b :** having a spiral structure or form that ascends or advances to the left ⟨a *left-handed* rope⟩
— left-handed *adverb*
— left-hand•ed•ly *adverb*
— left-hand•ed•ness *noun*

left-hand•er \-'han-dər, -'tan-\ *noun* (1881)
: a left-handed person

left•ish \'lef-tish\ *adjective* (1934)
: showing leftist tendencies

left•ism \'lef-ˌti-zəm\ *noun* (1920)
1 : the principles and views of the Left; *also* **:** the movement embodying these principles
2 : advocacy of or adherence to the doctrines of the Left
— left•ist \-tist\ *noun or adjective*

¹left•over \'left-ˌō-vər\ *noun* (1891)
1 : something that remains unused or unconsumed; *especially* **:** leftover food served at a later meal — usually used in plural
2 : an anachronistic survival **:** VESTIGE

²left•over *adjective* (1897)
: not consumed or used ⟨*leftover* food⟩ ⟨*leftover* space⟩

left shoulder arms *noun* [from the command *left shoulder arms!*] (circa 1918)
: a position in the manual of arms in which the butt of the rifle is held in the left hand with the barrel resting on the left shoulder; *also* **:** a command to assume this position

left•ward \'left-wərd\ *adjective or adverb* (15th century)
: being toward or on the left

left wing *noun* (1884)
1 : the leftist division of a group (as a political party)
2 : LEFT 4a
— left-wing *adjective*
— left-wing•er \'left-'wiŋ-ər\ *noun*

lefty \'lef-tē\ *noun, plural* **left•ies** (1886)
1 : LEFT-HANDER
2 : an advocate of leftism
— lefty *adjective*

¹leg \'leg *also* 'lāg\ *noun* [Middle English, from Old Norse *leggr*] (14th century)
1 : a limb of an animal used especially for supporting the body and for walking: as **a** (1) **:** one of the paired vertebrate limbs that in bipeds extend from the top of the thigh to the foot (2) **:** the part of such a limb between the knee and foot **b :** the back half of a hindquarter of a meat animal **c :** one of the rather generalized segmental appendages of an arthropod used in walking and crawling
2 a : a pole or bar serving as a support or prop ⟨the *legs* of a tripod⟩ **b :** a branch of a forked or jointed object ⟨the *legs* of a compass⟩
3 a : the part of an article of clothing that covers the leg **b :** the part of the upper (as of a boot) that extends above the ankle
4 : OBEISANCE, BOW — used chiefly in the phrase *to make a leg*
5 : a side of a right triangle that is not the hypotenuse; *also* **:** a side of an isosceles triangle that is not the base
6 a : the course and distance sailed by a boat on a single tack **b :** a portion of a trip **:** STAGE **c :** one section of a relay race **d :** one of several events or games necessary to be won to decide a competition ⟨won the first two *legs* of horse racing's Triple Crown⟩
7 : a branch or part of an object or system
— leg•less \-ləs\ *adjective*
— a leg to stand on : SUPPORT; *especially* **:** a basis for one's position in a controversy
— on one's last legs : at or near the end of one's resources **:** on the verge of failure, exhaustion, or ruin

²leg *intransitive verb* **legged; leg•ging** (1601)
: to use the legs in walking; *especially* **:** RUN

leg•a•cy \'le-gə-sē\ *noun, plural* **-cies** [Middle English *legacie* office of a legate, bequest, from Middle French or Medieval Latin; Middle French, office of a legate, from Medieval Latin *legatia,* from Latin *legatus*] (15th century)
1 : a gift by will especially of money or other personal property **:** BEQUEST
2 : something transmitted by or received from an ancestor or predecessor or from the past ⟨the *legacy* of the ancient philosophers⟩

¹le•gal \'lē-gəl\ *adjective* [Middle French, from Latin *legalis,* from *leg-, lex* law] (circa 1500)
1 : of or relating to law
2 a : deriving authority from or founded on law **:** DE JURE **b :** having a formal status derived from law often without a basis in actual fact **:** TITULAR ⟨a corporation is a *legal* but not a real person⟩ **c :** established by law; *especially* **:** STATUTORY
3 : conforming to or permitted by law or established rules
4 : recognized or made effective by a court of law as distinguished from a court of equity
5 : of, relating to, or having the characteristics of the profession of law or of one of its members
6 : created by the constructions of the law ⟨a *legal* fiction⟩
synonym see LAWFUL
— le•gal•ly \-gə-lē\ *adverb*

²legal *noun* (1526)
: one that conforms to rules or the law
legal age *noun* (circa 1904)
: the age at which a person enters into full adult legal rights and responsibilities (as of making contracts or wills)
legal aid *noun* (1890)
: aid provided by an organization established especially to serve the legal needs of the poor
legal eagle *noun* (1942)
: LAWYER
le·gal·ese \ˌlē-gə-ˈlēz, -ˈlēs\ *noun* (1914)
: the specialized language of the legal profession ⟨replaced *legalese* with plain talk —Steve Weinberg⟩
legal holiday *noun* (1867)
: a holiday established by legal authority and marked by restrictions on work and transaction of official business
le·gal·ise *British variant of* LEGALIZE
le·gal·ism \ˈlē-gə-ˌli-zəm\ *noun* (1838)
1 : strict, literal, or excessive conformity to the law or to a religious or moral code ⟨the institutionalized *legalism* that restricts free choice⟩
2 : a legal term or rule
le·gal·ist \-list\ *noun* (1646)
1 : an advocate or adherent of moral legalism
2 : one that views things from a legal standpoint; *especially* **:** one that places primary emphasis on legal principles or on the formal structure of governmental institutions
— le·gal·is·tic \ˌlē-gə-ˈlis-tik\ *adjective*
— le·gal·is·ti·cal·ly \-ti-k(ə-)lē\ *adverb*
le·gal·i·ty \li-ˈga-lə-tē\ *noun, plural* **-ties** (15th century)
1 : attachment to or observance of law
2 : the quality or state of being legal **:** LAWFULNESS
3 *plural* **:** obligations imposed by law
le·gal·ize \ˈlē-gə-ˌlīz\ *transitive verb* **-ized; -iz·ing** (circa 1716)
: to make legal; *especially* **:** to give legal validity or sanction to
— le·gal·i·za·tion \ˌlē-gə-lə-ˈzā-shən\ *noun*
— le·gal·iz·er \ˈlē-gə-ˌlī-zər\ *noun*
legal pad *noun* (1967)
: a writing tablet of ruled yellow paper that is usually 8.5 by 14 inches (22 by 36 centimeters)
legal reserve *noun* (circa 1902)
: the minimum amount of bank deposits or life insurance company assets required by law to be kept as reserves
legal tender *noun* (1739)
: money that is legally valid for the payment of debts and that must be accepted for that purpose when offered
¹leg·ate \ˈle-gət\ *noun* [Middle English, from Middle French & Latin; Middle French *legat,* from Latin *legatus* deputy, emissary, from past participle of *legare* to depute, send as emissary, bequeath, from *leg-, lex*] (12th century)
: a usually official emissary
— leg·ate·ship \-ˌship\ *noun*
²le·gate \li-ˈgāt\ *transitive verb* **le·gat·ed; le·gat·ing** [*legatus,* past participle of *legare* to bequeath] (15th century)
: BEQUEATH 1
— le·ga·tor \-ˈgā-tər\ *noun*
leg·a·tee \ˌle-gə-ˈtē\ *noun* (circa 1688)
: one to whom a legacy is bequeathed or a devise is given
leg·a·tine \ˈle-gə-ˌtēn, -ˌtīn\ *adjective* (1611)
: of, headed by, or enacted under the authority of a legate
le·ga·tion \li-ˈgā-shən\ *noun* (14th century)
1 : the sending forth of a legate
2 : a body of deputies sent on a mission; *specifically* **:** a diplomatic mission in a foreign country headed by a minister
3 : the official residence and office of a diplomatic minister in a foreign country
¹le·ga·to \li-ˈgä-(ˌ)tō\ *adverb or adjective* [Italian, literally, tied] (circa 1811)
: in a manner that is smooth and connected (as between successive tones) — used especially as a direction in music
²legato *noun* (1885)
: a smooth and connected manner of performance (as of music); *also* **:** a passage of music so performed
leg·end \ˈle-jənd\ *noun* [Middle English *legende,* from Middle French & Medieval Latin; Middle French *legende,* from Medieval Latin *legenda,* from Latin, feminine of *legendus,* gerundive of *legere* to gather, select, read; akin to Greek *legein* to gather, say, *logos* speech, word, reason] (14th century)
1 a : a story coming down from the past; *especially* **:** one popularly regarded as historical although not verifiable **b :** a body of such stories ⟨a place in the *legend* of the frontier⟩ **c :** a popular myth of recent origin **d :** a person or thing that inspires legends **e :** the subject of a legend ⟨its violence was *legend* even in its own time —William Broyles Jr.⟩
2 a : an inscription or title on an object (as a coin) **b :** CAPTION 2b **c :** an explanatory list of the symbols on a map or chart ◆
leg·end·ary \ˈle-jən-ˌder-ē\ *adjective* (circa 1587)
1 : of, relating to, or characteristic of legend or a legend
2 : WELL-KNOWN, FAMOUS
synonym see FICTITIOUS
— leg·en·dari·ly \ˌle-jən-ˈder-ə-lē\ *adverb*
leg·end·ry \ˈle-jən-drē\ *noun* (1849)
: a body of legends
leg·er \ˈle-jər\ *variant of* LEDGER
leg·er·de·main \ˌle-jər-də-ˈmān\ *noun* [Middle English, from Middle French *leger de main* light of hand] (15th century)
1 : SLEIGHT OF HAND
2 : a display of skill or adroitness
le·ger·i·ty \lə-ˈjer-ə-tē, le-\ *noun* [Middle French *legereté,* from Old French, lightness, from *leger* light, from (assumed) Vulgar Latin *leviarius,* from Latin *levis* — more at LIGHT] (1561)
: alert facile quickness of mind or body
leger line *noun* (circa 1775)
: LEDGER LINE
leges *plural of* LEX
leg·ged \ˈle-gəd *also* ˈlā-, *British usually* ˈlegd\ *adjective* (15th century)
: having a leg or legs especially of a specified kind or number — often used in combination ⟨a four-*legged* animal⟩
leg·ging *or* **leg·gin** \ˈle-gən *also* ˈlā-, -giŋ\ *noun* (1751)
: a covering (as of leather or cloth) for the leg — usually used in plural; *also* **:** TIGHTS
leg·gy \ˈle-gē *also* ˈlā-\ *adjective* **leg·gi·er; -est** (1787)
1 : having disproportionately long legs
2 : having long and attractive legs
3 : SPINDLY — used of a plant
— leg·gi·ness \-nəs\ *noun*
leg·hold trap \ˈleg-ˌhōld-\ *noun* (1973)
: a jawed usually steel trap that is used to hold a wild mammal and operates by springing closed and clamping onto the leg of the animal that steps on it
leg·horn \ˈleg-ˌhȯrn; ˈle-ˌgȯrn, -gərn\ *noun* [*Leghorn,* Italy] (1740)
1 a : a fine plaited straw made from an Italian wheat **b :** a hat of this straw
2 : any of a Mediterranean breed of small hardy fowls noted for their large production of white eggs
leg·i·ble \ˈle-jə-bəl\ *adjective* [Middle English, from Late Latin *legibilis,* from Latin *legere* to read] (14th century)
1 : capable of being read or deciphered **:** PLAIN
2 : capable of being discovered or understood
— leg·i·bil·i·ty \ˌle-jə-ˈbi-lə-tē\ *noun*
— leg·i·bly \ˈle-jə-blē\ *adverb*
¹le·gion \ˈlē-jən\ *noun* [Middle English, from Old French, from Latin *legion-, legio,* from *legere* to gather — more at LEGEND] (13th century)
1 : the principal unit of the Roman army comprising 3000 to 6000 foot soldiers with cavalry
2 : a large military force; *especially* **:** ARMY 1a
3 : a very large number **:** MULTITUDE
4 : a national association of ex-servicemen
²legion *adjective* (1678)
: MANY, NUMEROUS ⟨the problems are *legion*⟩
¹le·gion·ary \ˈlē-jə-ˌner-ē\ *adjective* [Latin *legionarius,* from *legion-, legio*] (15th century)
: of, relating to, or constituting a legion
²legionary *noun, plural* **-ar·ies** (1598)
: LEGIONNAIRE
le·gion·naire \ˌlē-jə-ˈnar, -ˈner\ *noun* [French *légionnaire,* from Latin *legionarius*] (1818)
: a member of a legion
Legionnaires' disease \-ˈnarz-, -ˈnerz-\ *also* **Legionnaire's disease** *noun* [from its first recognized occurrence at an American Legion convention in 1976] (1976)
: a lobar pneumonia caused by a bacterium (*Legionella pneumophila*)
Legion of Honor (1827)
: a French order conferred as a reward for civil or military merit
Legion of Merit (1943)
: a U.S. military decoration awarded for exceptionally meritorious conduct in the performance of outstanding services
leg·is·late \ˈle-jəs-ˌlāt\ *verb* **-lat·ed; -lat·ing** [back-formation from *legislator*] (1805)
intransitive verb
: to perform the function of legislation; *specifically* **:** to make or enact laws
transitive verb
: to cause, create, provide, or bring about by or as if by legislation
leg·is·la·tion \ˌle-jəs-ˈlā-shən\ *noun* (circa 1655)
1 : the action of legislating; *specifically* **:** the exercise of the power and function of making rules (as laws) that have the force of authority by virtue of their promulgation by an official organ of a state or other organization
2 : the enactments of a legislator or a legislative body
3 : a matter of business for or under consideration by a legislative body
¹leg·is·la·tive \ˈle-jəs-ˌlā-tiv, -lə-\ *adjective* (circa 1641)
1 a : having the power or performing the function of legislating **b :** belonging to the branch of government that is charged with

◇ WORD HISTORY
legend The literal meaning of the Latin verb *legere* was "to gather," but it acquired extended senses such as "to pick out (sights or sounds)," "to make one's way over" and finally "to read." In Medieval Latin *legenda,* the gerundive of *legere,* meaning literally "a thing to be read," was used in specific reference to the story of the life of a saint. One of the most popular books of the later Middle Ages was the *Legenda Aurea* ("Golden Legend") of Jacobus de Voragine, a collection of saints' lives that incorporated a generous measure of fanciful material along with fact. When the word was borrowed, via French, into Middle English as *legende,* it continued to be used with specific reference to a saint's story. By the 17th century, however, *legend* had come to be applied to any story handed down from earlier times that, like the traditional saints' lives, was at best unverifiable.

such powers as making laws, levying and collecting taxes, and making financial appropriations — compare EXECUTIVE, JUDICIAL
2 a : of or relating to a legislature ⟨*legislative* committees⟩ **b :** composed of members of a legislature ⟨*legislative* caucus⟩ **c :** created by a legislature especially as distinguished from an executive or judicial body **d :** designed to assist a legislature or its members ⟨a *legislative* research agency⟩
3 : of, concerned with, or created by legislation
— leg·is·la·tive·ly *adverb*

²legislative *noun* (1642)
: the body or department exercising the power and function of legislating **:** LEGISLATURE

legislative assembly *noun, often L&A capitalized* (1836)
1 : a bicameral legislature (as in an American state)
2 : the lower house of a bicameral legislature
3 : a unicameral legislature; *especially* **:** one in a Canadian province

legislative council *noun, often L&C capitalized* (1787)
1 : a permanent committee chosen from both houses that meets between sessions of a state legislature to study state problems and plan a legislative program
2 : a unicameral legislature (as in a British colony)
3 : the upper house of a British bicameral legislature

leg·is·la·tor \'le-jəs-ˌlā-ˌtȯr, -ˌlā-tər *also* ˌle-jəs-'lā-ˌtȯr\ *noun* [Latin *legis lator,* literally, proposer of a law, from *legis* (genitive of *lex* law) + *lator* proposer, from *ferre* (past participle *latus*) to carry, propose — more at TOLERATE, BEAR] (1603)
: one that makes laws especially for a political unit; *especially* **:** a member of a legislative body
— leg·is·la·to·ri·al \ˌle-jəs-lə-'tōr-ē-əl, -'tȯr-\ *adjective*
— leg·is·la·tor·ship \'le-jəs-ˌlā-tər-ˌship\ *noun*

leg·is·la·ture \'le-jəs-ˌlā-chər *also* ˌle-jəs-', *British often* 'le-jəs-lə-\ *noun* (circa 1676)
: a body of persons having the power to legislate; *specifically* **:** an organized body having the authority to make laws for a political unit

le·gist \'lē-jist\ *noun* [Middle English, from Middle French *legiste,* from Medieval Latin *legista,* from Latin *leg-, lex*] (15th century)
: a specialist in law; *especially* **:** one learned in Roman or civil law

le·git \li-'jit\ *adjective* (1908)
slang **:** LEGITIMATE

le·git·i·ma·cy \li-'ji-tə-mə-sē\ *noun* (1691)
: the quality or state of being legitimate

¹le·git·i·mate \li-'ji-tə-mət\ *adjective* [Middle English *legitimat,* from Medieval Latin *legitimatus,* past participle of *legitimare* to legitimate, from Latin *legitimus* legitimate, from *leg-, lex* law] (15th century)
1 a : lawfully begotten; *specifically* **:** born in wedlock **b :** having full filial rights and obligations by birth ⟨a *legitimate* child⟩
2 : being exactly as purposed **:** neither spurious nor false ⟨*legitimate* grievance⟩ ⟨a *legitimate* practitioner⟩
3 a : accordant with law or with established legal forms and requirements ⟨a *legitimate* government⟩ **b :** ruling by or based on the strict principle of hereditary right ⟨a *legitimate* king⟩
4 : conforming to recognized principles or accepted rules and standards ⟨*legitimate* advertising expenditure⟩ ⟨*legitimate* inference⟩
5 : relating to plays acted by professional actors but not including revues, burlesque, or some forms of musical comedy ⟨the *legitimate* theater⟩
synonym see LAWFUL
— le·git·i·mate·ly *adverb*

²le·git·i·mate \-ˌmāt\ *transitive verb* **-mat·ed; -mat·ing** (1531)
: to make legitimate: **a** (1) **:** to give legal status or authorization to (2) **:** to show or affirm to be justified **b :** to put (a bastard) in the state of a legitimate child before the law by legal means
— le·git·i·ma·tion \-ˌji-tə-'mā-shən\ *noun*
— le·git·i·mat·or \-'ji-tə-ˌmā-tər\ *noun*

le·git·i·ma·tize \li-'ji-tə-mə-ˌtīz\ *transitive verb* **-tized; -tiz·ing** (1791)
: LEGITIMATE

le·git·i·mise *British variant of* LEGITIMIZE

le·git·i·mism \li-'ji-tə-ˌmi-zəm\ *noun, often capitalized* (1877)
: adherence to the principles of political legitimacy or to a person claiming legitimacy
— le·git·i·mist \-mist\ *noun, often capitalized*
— legitimist *adjective*

le·git·i·mize \-ˌmīz\ *transitive verb* **-mized; -miz·ing** (1848)
: LEGITIMATE
— le·git·i·mi·za·tion \-ˌji-tə-mə-'zā-shən\ *noun*
— le·git·i·miz·er \-'ji-tə-ˌmī-zər\ *noun*

leg·man \'leg-ˌman *also* 'lāg-\ *noun* (1923)
1 : a reporter assigned usually to gather information
2 : an assistant who performs various subordinate tasks (as gathering information or running errands)

leg–of–mut·ton *or* **leg–o'–mut·ton** \ˌle-gə(v)-'mə-tᵊn *also* ˌlāg-\ *adjective* (1840)
: having the approximately triangular shape or outline of a leg of mutton ⟨*leg-of-mutton* sleeve⟩ ⟨*leg-of-mutton* sail⟩

leg out *transitive verb* (1965)
: to make (as a base hit) by fast running

leg–pull \'leg-ˌpu̇l *also* 'lāg-\ *noun* [from the phrase *to pull one's leg*] (1915)
: a humorous deception or hoax

leg·room \-ˌrüm, -ˌru̇m\ *noun* (1926)
: space in which to extend the legs while seated

le·gume \'le-ˌgyüm, li-'gyüm\ *noun* [French *légume,* from Latin *legumin-, legumen* leguminous plant, from *legere* to gather — more at LEGEND] (1676)
1 a : the fruit or seed of leguminous plants (as peas or beans) used for food **b :** a vegetable used for food
2 : any of a large family (Leguminosae synonym Fabaceae) of dicotyledonous herbs, shrubs, and trees having fruits that are legumes (sense 3) or loments, bearing nodules on the roots that contain nitrogen-fixing bacteria, and including important food and forage plants (as peas, beans, or clovers)
3 : a dry dehiscent one-celled fruit developed from a simple superior ovary and usually dehiscing into two valves with the seeds attached to the ventral suture **:** POD

le·gu·mi·nous \li-'gyü-mə-nəs, le-\ *adjective* (15th century)
1 : of, relating to, or consisting of plants that are legumes
2 : resembling a legume

leg up *noun* (1837)
1 : a helping hand **:** BOOST
2 : HEAD START

leg warmer *noun* (1974)
: a usually knitted covering for the leg

leg·work \'leg-ˌwərk *also* 'lāg-\ *noun* (1891)
: active physical work (as in gathering information) that forms the basis of more creative or mentally exacting work (as writing a book)

le·hua \lā-'hü-ə\ *noun* [Hawaiian] (1888)
: a common very showy chiefly Polynesian tree (*Metrosideros collinus*) of the myrtle family having bright red flowers and a hard wood; *also* **:** its flower

¹lei \'lā, 'lā-ˌē\ *noun* [Hawaiian] (1843)
: a wreath or necklace usually of flowers or leaves

²lei \'lā\ *plural of* LEU

Leices·ter \'les-tər\ *noun* [*Leicester,* county in England] (1798)
1 : an individual of either of two English breeds of white-faced long-wool sheep raised especially for mutton
2 : a hard usually orange-colored cheese similar to cheddar

leish·man·ia \lēsh-'ma-nē-ə\ *noun* [New Latin, from Sir W. B. *Leishman* (died 1926) British medical officer] (1914)
: any of a genus (*Leishmania*) of flagellate protozoans that are parasitic in the tissues of vertebrates; *broadly* **:** an organism resembling the leishmanias that is included in the family (Trypanosomatidae) to which they belong
— leish·man·ial \-nē-əl\ *adjective*

leish·man·i·a·sis \ˌlēsh-mə-'nī-ə-səs\ *noun* [New Latin] (1912)
: infection with or disease caused by leishmanias

leis·ter \'lēs-tər\ *noun* [of Scandinavian origin; akin to Old Norse *ljōstr* leister] (circa 1534)
: a spear armed with three or more barbed prongs for catching fish

lei·sure \'lē-zhər, 'le-, 'lā-\ *noun* [Middle English *leiser,* from Middle French *leisir,* from *leisir* to be permitted, from Latin *licēre*] (14th century)
1 : freedom provided by the cessation of activities; *especially* **:** time free from work or duties
2 : EASE, LEISURELINESS
— leisure *adjective*
— at leisure *or* **at one's leisure :** in one's leisure time **:** at one's convenience ⟨read the book *at her leisure*⟩

lei·sured \-zhərd\ *adjective* (1631)
: having leisure **:** LEISURELY

¹lei·sure·ly \-zhər-lē\ *adverb* (15th century)
: without haste **:** DELIBERATELY

²leisurely *adjective* (1604)
: characterized by leisure **:** UNHURRIED
— lei·sure·li·ness *noun*

leisure suit *noun* (1975)
: a suit consisting of a shirt jacket and matching trousers for informal wear

leit·mo·tiv *or* **leit·mo·tif** \'līt-mō-ˌtēf\ *noun* [German *Leitmotiv,* from *leiten* to lead + *Motiv* motive] (circa 1876)
1 : an associated melodic phrase or figure that accompanies the reappearance of an idea, person, or situation especially in a Wagnerian music drama
2 : a dominant recurring theme

¹lek \'lek\ *noun* [Swedish, short for *lekställe* mating ground, from *lek* mating, sport + *ställe* place] (1871)
: an assembly area where animals (as the prairie chicken) carry on display and courtship behavior

²lek *noun, plural* **leks** *or* **le·ke** \'le-kə\ *also* **lek** *or* **le·ku** \'le-(ˌ)kü\ [Albanian] (1927)
— see MONEY table

lek·var \'lek-ˌvär\ *noun* [Hungarian *lekvár* jam] (circa 1958)
: a prune butter used as a pastry filling

le·man \'le-mən, 'lē-\ *noun* [Middle English *lefman, leman,* from *lef* lief] (13th century)
archaic **:** SWEETHEART, LOVER; *especially* **:** MISTRESS

¹lem·ma \'le-mə\ *noun, plural* **lemmas** *or* **lem·ma·ta** \-mə-tə\ [Latin, from Greek *lēmma* thing taken, assumption, from *lambanein* to take — more at LATCH] (1570)
1 : an auxiliary proposition used in the demonstration of another proposition
2 : the argument or theme of a composition prefixed as a title or introduction; *also* **:** the heading or theme of a comment or note on a text
3 : a glossed word or phrase

²lemma *noun* [Greek, husk, from *lepein* to peel — more at LEPER] (1906)
: the lower of the two bracts enclosing the flower in the spikelet of grasses

lem•ming \'le-miŋ\ *noun* [Norwegian] (1713)
: any of various small short-tailed furry-footed rodents (as genera *Lemmus* and *Dicrostonyx*) of circumpolar distribution that are notable for the recurrent mass migrations of a European form (*L. lemmus*) which often continue into the sea where vast numbers are drowned
— **lem•ming•like** \-ˌlīk\ *adjective*

lem•nis•cate \lem-'nis-kət\ *noun* [New Latin *lemniscata,* from feminine of Latin *lemniscatus* with hanging ribbons, from *lemniscus*] (circa 1781)
: a figure-eight shaped curve whose equation in polar coordinates is $\rho^2=a^2 \cos 2\theta$ or $\rho^2=a^2 \sin 2\theta$

lem•nis•cus \lem-'nis-kəs\ *noun, plural* **-nis•ci** \-'nis-ˌkī, -ˌkē; -'ni-ˌsī\ [New Latin, from Latin, ribbon, from Greek *lēmniskos*] (circa 1905)
: a band of fibers and especially nerve fibers
— **lem•nis•cal** \-kəl\ *adjective*

[1]lem•on \'le-mən\ *noun* [Middle English *lymon,* from Middle French *limon,* from Medieval Latin *limon-, limo,* from Arabic *laymūn*] (15th century)
1 a : an acid fruit that is botanically a many-seeded pale yellow oblong berry and is produced by a small thorny tree (*Citrus limon*) **b :** a tree that bears lemons
2 : one (as an automobile) that is unsatisfactory or defective
— **lem•ony** \'le-mə-nē\ *adjective*

lemon 1: branch with fruit and flowers

[2]lemon *adjective* (1598)
1 : of the color lemon yellow
2 a : containing lemon **b :** having the flavor or scent of lemon

lem•on•ade \ˌle-mə-'nād\ *noun* (1604)
: a beverage of sweetened lemon juice mixed with water

lemon balm *noun* (circa 1888)
: a bushy perennial Old World mint (*Melissa officinalis*) often cultivated for its fragrant lemon-flavored leaves

lem•on•grass \'le-mən-ˌgras\ *noun* (1801)
: a grass (*Cymbopogon citratus*) of robust habit that grows in tropical regions, is used as an herb, and is the source of an essential oil with an odor of lemon or verbena

lemon law *noun* (1982)
: a law offering car buyers relief (as by repair, replacement, or refund) for defects detected during a specified period after purchase

lemon shark *noun* (1942)
: a medium-sized requiem shark (*Negaprion brevirostris*) of the warm Atlantic that is yellowish brown to gray above with yellow or greenish sides

lemon sole *noun* (1876)
: any of several flatfishes and especially flounders: as **a :** a bottom-dwelling flounder (*Microstomus kitt*) of the northeastern Atlantic that is an important food fish **b :** WINTER FLOUNDER

lemon yellow *noun* (1807)
: a brilliant greenish yellow color

lem•pi•ra \lem-'pir-ə\ *noun* [American Spanish, from *Lempira,* 16th century Indian chief] (circa 1934)
— see MONEY table

le•mur \'lē-mər\ *noun* [New Latin, from Latin *lemures,* plural, ghosts] (1795)
: any of various arboreal chiefly nocturnal mammals that were formerly widespread but are now largely confined to Madagascar, are related to the monkeys but are usually regarded as constituting a distinct superfamily (Lemuroidea), and usually have a muzzle like a fox, large eyes, very soft woolly fur, and a long furry tail ◆

le•mu•res \'le-mə-ˌrās, 'lem-yə-ˌrēz\ *noun plural* [Latin] (1555)
: spirits of the unburied dead exorcised from homes in early Roman religious rites

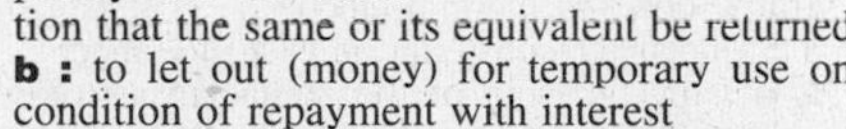

lemur

lend \'lend\ *verb* **lent** \'lent\; **lend•ing** [Middle English *lenen, lenden,* from Old English *lǣnan,* from *lǣn* loan — more at LOAN] (before 12th century)
transitive verb
1 a : to give for temporary use on condition that the same or its equivalent be returned **b :** to let out (money) for temporary use on condition of repayment with interest
2 a : to give the assistance or support of **:** AFFORD, FURNISH ⟨a dispassionate and scholarly manner which *lends* great force to his criticisms —*Times Literary Supplement*⟩ **b :** to adapt or apply (oneself) readily **:** ACCOMMODATE ⟨a topic that *lends* itself admirably to class discussion⟩
intransitive verb
: to make a loan
usage see LOAN
— **lend•able** \'len-də-bəl\ *adjective*
— **lend•er** *noun*

lending library *noun* (1708)
: a library from which materials are lent; *especially* **:** RENTAL LIBRARY

lend–lease \'lend-'lēs\ *noun* [U.S. *Lend-Lease* Act (1941)] (1941)
: the transfer of goods and services to an ally to aid in a common cause with payment made by a return of the original items or their use in the cause or by a similar transfer of other goods and services
— **lend–lease** *transitive verb*

length \'leŋ(k)th, 'len(t)th\ *noun, plural* **lengths** \'leŋ(k)ths, 'len(t)ths, 'leŋ(k)s\ [Middle English *lengthe,* from Old English *lengthu,* from *lang* long] (before 12th century)
1 a : the longer or longest dimension of an object **b :** a measured distance or dimension ⟨10 feet in *length*⟩ — see METRIC SYSTEM table, WEIGHT table **c :** the quality or state of being long
2 a : duration or extent in time **b :** relative duration or stress of a sound
3 a : distance or extent in space **b :** the length of something taken as a unit of measure ⟨his horse led by a *length*⟩
4 : the degree to which something (as a course of action or a line of thought) is carried — often used in plural ⟨went to great *lengths* to learn the truth⟩
5 a : a long expanse or stretch **b :** a piece constituting or usable as part of a whole or of a connected series **:** SECTION ⟨a *length* of pipe⟩
6 : a vertical dimension of an article of clothing
— **at length 1 :** FULLY, COMPREHENSIVELY **2 :** at last **:** FINALLY

length•en \'leŋ(k)-thən, 'len(t)-\ *verb* **length•ened; length•en•ing** \'leŋ(k)th-niŋ, 'len(t)th-; 'leŋ(k)-thə-, 'len(t)-\ (14th century)
transitive verb
: to make longer
intransitive verb
: to grow longer
synonym see EXTEND
— **length•en•er** \'leŋ(k)th-nər, 'len(t)th-; 'leŋ(k)-thə-, 'len(t)-\ *noun*

length•ways \'leŋ(k)th-ˌwāz, 'len(t)th-\ *adverb* (1599)
: LENGTHWISE

length•wise \-ˌwīz\ *adverb* (circa 1580)
: in the direction of the length **:** LONGITUDINALLY
— **lengthwise** *adjective*

lengthy \'leŋ(k)-thē, 'len(t)-\ *adjective* **length•i•er; -est** (1689)
1 : protracted excessively **:** OVERLONG
2 : EXTENDED, LONG
— **length•i•ly** \-thə-lē\ *adverb*
— **length•i•ness** \-thē-nəs\ *noun*

le•nience \'lē-nyən(t)s, -nē-ən(t)s\ *noun* (1796)
: LENIENCY

le•nien•cy \'lē-nē-ən(t)-sē, -nyən(t)-sē\ *noun, plural* **-cies** (1780)
1 : the quality or state of being lenient
2 : a lenient disposition or practice
synonym see MERCY

le•nient \'lē-nē-ənt, -nyənt\ *adjective* [Latin *lenient-, leniens,* present participle of *lenire* to soften, soothe, from *lenis* soft, mild; probably akin to Lithuanian *lėnas* tranquil — more at LET] (1652)
1 : exerting a soothing or easing influence **:** relieving pain or stress
2 : of mild and tolerant disposition; *especially* **:** INDULGENT
— **le•nient•ly** *adverb*

Leni–Len•a•pe *or* **Len•ni–Len•a•pe** \ˌle-nē-'le-nə-pē, -lə-'nä-pē\ *noun* [Delaware (Unami dialects) *lə̀ni-lənáˑpˑe*] (circa 1782)
: DELAWARE 1

Le•nin•ism \'le-nə-ˌni-zəm\ *noun* (1918)
: the political, economic, and social principles and policies advocated by Lenin; *especially* **:** the theory and practice of communism developed by or associated with Lenin
— **Le•nin•ist** \-nist\ *noun or adjective*
— **Le•nin•ite** \-ˌnīt\ *noun or adjective*

le•nis \'lē-nəs, 'lā-\ *adjective* [New Latin, from Latin, mild, smooth] (circa 1897)
: produced with an articulation that is lax in relation to another speech sound ⟨\t\ in *gutter* is *lenis,* \t\ in *toe* is fortis⟩

len•i•tion \lə-'ni-shən\ *noun* [Latin *lenire*] (1912)
: the change from fortis to lenis articulation

len•i•tive \'le-nə-tiv\ *adjective* [Middle English *lenitif,* from Middle French, from Medieval Latin *lenitivus,* from Latin *lenitus,* past participle of *lenire*] (15th century)
: alleviating pain or harshness **:** SOOTHING
— **lenitive** *noun*
— **len•i•tive•ly** *adverb*

len•i•ty \'le-nə-tē\ *noun* (1548)
: the quality or state of being lenient **:** CLEMENCY

le•no \'lē-(ˌ)nō\ *noun* [perhaps from French *linon* linen fabric, lawn, from Middle French *lin* flax, linen, from Latin *linum* flax] (1821)

◇ WORD HISTORY
lemur In ancient Roman belief the *lemures* were the spirits of dead people who returned at night to disturb the living. They were placated by a ritual performed by the male head of a household on midnight of the *Lemuria,* three days in May during which temples were closed and marriages forbidden. In the late 17th and 18th centuries visitors to the island of Madagascar encountered some strange-looking creatures. Arboreal and nocturnal, these big-eyed, monkeylike mammals presented a strikingly eerie appearance to their first European observers. Specimens of the ring-tailed lemur (*Lemur catta*) were fittingly dubbed in scientific Latin *lemures* by the Swedish biologist Linnaeus, after the Roman ghosts. This word has been borrowed into English as *lemur.*

1 : an open weave in which pairs of warp yarns cross one another and thereby lock the filling yarn in position
2 : a fabric made with a leno weave

¹lens *also* **lense** \'lenz\ *noun* [New Latin *lent-, lens,* from Latin, lentil; from its shape] (1693)
1 a : a piece of transparent material (as glass) that has two opposite regular surfaces either both curved or one curved and the other plane and that is used either singly or combined in an optical instrument for forming an image by focusing rays of light **b :** a combination of two or more simple lenses **c :** a piece of glass or plastic used (as in safety goggles or sunglasses) to protect the eye
2 : a device for directing or focusing radiation other than light (as sound waves, radio microwaves, or electrons)
3 : something shaped like a double-convex optical lens ⟨*lens* of sandstone⟩
4 : a highly transparent biconvex lens-shaped or nearly spherical body in the eye that focuses light rays (as upon the retina) — see EYE illustration
5 : something that facilitates and influences perception, comprehension, or evaluation ⟨the author's own *lens* seems blurred by bias —Seymour Topping⟩
— **lensed** \'lenzd\ *adjective*
— **lens·less** \'lenz-ləs\ *adjective*

²lens *transitive verb* (1942)
: to make a motion picture of **:** FILM

lens·man \-mən, -ˌman\ *noun* (1938)
: PHOTOGRAPHER

Lent \'lent\ *noun* [Middle English *lente* springtime, Lent, from Old English *lencten;* akin to Old High German *lenzin* spring] (13th century)
: the 40 weekdays from Ash Wednesday to Easter observed by the Roman Catholic, Eastern, and some Protestant churches as a period of penitence and fasting

len·ta·men·te \ˌlen-tə-'men-(ˌ)tā\ *adverb or adjective* [Italian, from *lento* slow] (1724)
: LENTO

len·tan·do \len-'tän-(ˌ)dō\ *adverb or adjective* [Italian] (circa 1847)
: becoming slower — used as a direction in music

Lent·en \'len-tᵊn\ *adjective* (before 12th century)
: of, relating to, or suitable for Lent; *especially* **:** MEAGER ⟨*Lenten* fare⟩

len·tic \'len-tik\ *adjective* [Latin *lentus* sluggish] (circa 1938)
: of, relating to, or living in still waters (as lakes, ponds, or swamps) — compare LOTIC

len·ti·cel \'len-tə-ˌsel\ *noun* [New Latin *lenticella,* diminutive of Latin *lent-, lens* lentil] (circa 1864)
: a loose aggregation of cells which penetrates the surface (as of a stem) of a woody plant and through which gases are exchanged between the atmosphere and the underlying tissues

len·tic·u·lar \len-'ti-kyə-lər\ *adjective* [Middle English, from Latin *lenticularis* lentil-shaped, from *lenticula* lentil] (15th century)
1 : having the shape of a double-convex lens
2 : of or relating to a lens
3 : provided with or utilizing lenticules ⟨a *lenticular* screen⟩

len·ti·cule \'len-tə-ˌkyü(ə)l\ *noun* [Latin *lenticula*] (1942)
1 : any of the minute lenses on the base side of a film used in stereoscopic or color photography
2 : any of the tiny corrugations or grooves molded or embossed into the surface of a projection screen

len·til \'len-tᵊl\ *noun* [Middle English, from Old French *lentille,* from Latin *lenticula,* diminutive of *lent-, lens*] (13th century)
1 : a widely cultivated Eurasian annual leguminous plant (*Lens culinaris*) with flattened edible seeds and leafy stalks used as fodder
2 : the seed of the lentil

len·tis·si·mo \len-'ti-sə-ˌmō\ *adverb or adjective* [Italian, superlative of *lento*] (circa 1903)
: at a very slow tempo — used as a direction in music

len·ti·vi·rus \ˌlen-tə-'vī-rəs\ *noun* [New Latin, from Latin *lentus* slow + New Latin *virus*] (1982)
: any of a group of retroviruses that cause slowly progressive often fatal animal diseases

len·to \'len-(ˌ)tō\ *adverb or adjective* [Italian, from *lento,* adjective, slow, from Latin *lentus* pliant, sluggish, slow — more at LITHE] (circa 1724)
: at a slow tempo — used especially as a direction in music

Leo \'lē-(ˌ)ō\ *noun* [Latin (genitive *Leonis*), literally, lion — more at LION]
1 : a northern constellation east of Cancer
2 a : the 5th sign of the zodiac in astrology — see ZODIAC table **b :** one born under this sign
— **Le·o·nine** \'lē-ə-ˌnīn\ *adjective*

le·one \lē-'ōn\ *noun, plural* **leones** *or* **leone** [*Sierra Leone*] (1964)
— see MONEY table

Le·o·nid \'lē-ə-nid\ *noun, plural* **Leonids** *or* **Le·on·i·des** \lē-'ä-nə-ˌdēz\ [Latin *Leon-, Leo;* from their appearing to radiate from a point in Leo] (1876)
: any of the meteors in a meteor shower occurring every year about November 14

le·o·nine \'lē-ə-ˌnīn\ *adjective* [Middle English, from Latin *leoninus,* from *leon-, leo*] (14th century)
: of, relating to, suggestive of, or resembling a lion

leop·ard \'le-pərd\ *noun* [Middle English, from Old French *leupart,* from Late Latin *leopardus,* from Greek *leopardos,* from *leōn* lion + *pardos* leopard] (13th century)
1 : a large strong cat (*Panthera pardus*) of southern Asia and Africa that is adept at climbing and is usually tawny or buff with black spots arranged in rosettes — called also *panther*
2 : a heraldic representation of a lion passant guardant
— **leop·ard·ess** \-pər-dəs\ *noun*

leopard 1

leopard frog *noun* (1839)
: a common North American frog (*Rana pipiens*) that is bright green or brown with large black white-margined blotches on the back; *also* **:** a similar frog (*R. sphenocephala*) of the southeastern U.S.

leopard seal *noun* (1893)
: a spotted slate gray seal (*Hydrurga leptonyx*) of the southern Atlantic to the southern Pacific that often feeds on other seals

le·o·tard \'lē-ə-ˌtärd\ *noun* [Jules *Léotard,* (died 1870) French aerial gymnast] (1886)
: a close-fitting one-piece garment worn especially by dancers, acrobats, and aerialists; *also* **:** TIGHTS — often used in plural
— **le·o·tard·ed** \-ˌtär-dəd\ *adjective*

Lep·cha \'lep-chə\ *noun, plural* **Lepcha** *or* **Lepchas** (1819)
1 : a member of a people of Sikkim, India
2 : the Tibeto-Burman language of the Lepcha people

lep·er \'le-pər\ *noun* [Middle English, from *lepre* leprosy, from Middle French, from Late Latin *lepra,* from Greek, from *lepein* to peel; perhaps akin to Lithuanian *lopas* piece, scrap] (14th century)
1 : a person affected with leprosy
2 : a person shunned for moral or social reasons

lepid- *or* **lepido-** *combining form* [New Latin, from Greek, from *lepid-, lepis* scale, from *lepein*]
: flake **:** scale ⟨*Lepido*ptera⟩

le·pid·o·lite \li-'pi-dᵊl-ˌīt\ *noun* [German *Lepidolith,* from *lepid-* + *-lith*] (circa 1796)
: a variable mineral that consists of a mica containing lithium and is used especially in glazes and enamels

lep·i·dop·tera \ˌle-pə-'däp-tə-rə\ *noun plural* (circa 1773)
: insects that are lepidopterans

lep·i·dop·ter·an \-rən\ *noun* [New Latin *Lepidoptera,* from *lepid-* + Greek *pteron* wing — more at FEATHER] (circa 1901)
: any of a large order (Lepidoptera) of insects comprising the butterflies, moths, and skippers that as adults have four broad or lanceolate wings usually covered with minute overlapping and often brightly colored scales and that as larvae are caterpillars
— **lepidopteran** *adjective*
— **lep·i·dop·ter·ous** \-tə-rəs\ *adjective*

lep·i·dop·ter·ist \-tə-rist\ *noun* (1826)
: a specialist in lepidopterology

lep·i·dop·ter·ol·o·gy \-ˌdäp-tə-'rä-lə-jē\ *noun* (1898)
: a branch of entomology concerned with lepidopterans
— **lep·i·dop·ter·o·log·i·cal** \-tə-rə-'lä-ji-kəl\ *adjective*
— **lep·i·dop·ter·ol·o·gist** \-tə-'rä-lə-jist\ *noun*

lep·i·dote \'le-pə-ˌdōt\ *noun* [Greek *lepidōtos* scaly, from *lepid-, lepis*] (circa 1836)
: a rhododendron covered with scurf or scurfy scales

lep·re·chaun \'lep-rə-ˌkän, -ˌkȯn\ *noun* [Irish *leipreachán*] (1604)
: a mischievous elf of Irish folklore usually believed to reveal the hiding place of treasure if caught
— **lep·re·chaun·ish** \-ˌkä-nish, -ˌkȯ-\ *adjective*

le·pro·ma·tous \lə-'prä-mə-təs, -'prō-\ *adjective* [New Latin *lepromat-, leproma* leprous lesion, from Late Latin *lepra*] (1898)
: characterized by, exhibiting, or being leprosy with infective superficial granulomatous nodules

lep·ro·sar·i·um \ˌle-prə-'ser-ē-əm\ *noun, plural* **-i·ums** *or* **-ia** \-ē-ə\ [Medieval Latin, from Late Latin *leprosus*] (circa 1846)
: a hospital for leprosy patients

lep·ro·sy \'le-prə-sē\ *noun* [Middle English *lepruse,* from *leprous*] (15th century)
1 : a chronic disease caused by a bacillus (*Mycobacterium leprae*) and characterized by the formation of nodules or of macules that enlarge and spread accompanied by loss of sensation with eventual paralysis, wasting of muscle, and production of deformities and mutilations
2 : a morally or spiritually harmful influence
— **lep·rot·ic** \le-'prä-tik\ *adjective*

lep·rous \'le-prəs\ *adjective* [Middle English, from Late Latin *leprosus* leprous, from *lepra* leprosy — more at LEPER] (13th century)
1 a : infected with leprosy **b :** of, relating to, or resembling leprosy or a leper
2 : SCALY, SCURFY
— **lep·rous·ly** *adverb*

-lepsy *noun combining form* [Middle French *-lepsie,* from Late Latin *-lepsia,* from Greek *-lēpsia,* from *lēpsis,* from *lambanein* to take, seize — more at LATCH]
: taking **:** seizure ⟨narco*lepsy*⟩

lep·to·ceph·a·lus \ˌlep-tə-'se-fə-ləs\ *noun, plural* **-li** \-ˌlī, -ˌlē\ [New Latin, from Greek *leptos* + *kephalē* head — more at CEPHALIC] (1769)
: a long thin small-headed transparent pelagic first larva of various eels

¹lep·ton \lep-'tän\ *noun, plural* **lep·ta** \-'tä\ [New Greek, from Greek, a small coin, from neuter of *leptos* peeled, slender, small, from *lepein* to peel — more at LEPER] (circa 1741) — see *drachma* at MONEY table

²lep·ton \ˈlep-ˌtän\ *noun* [Greek *leptos* + English ²*-on*] (circa 1929)
: any of a family of particles (as electrons, muons, and neutrinos) that have spin quantum number ½ and that experience no strong forces
— lep·ton·ic \lep-ˈtä-nik\ *adjective*

lep·to·some \ˈlep-tə-ˌsōm\ *adjective* [German *Leptosom,* from Greek *leptos* + *sōma* body] (1931)
: ECTOMORPHIC
— leptosome *noun*

lep·to·spire \-ˌspīr\ *noun* [New Latin *Leptospira,* from Greek *leptos* + Latin *spira* coil — more at SPIRE] (1952)
: any of a genus (*Leptospira*) of slender aerobic spirochetes that are free-living or parasitic in mammals
— lep·to·spir·al \ˌlep-tə-ˈspī-rəl\ *adjective*

lep·to·spi·ro·sis \ˌlep-tə-spī-ˈrō-səs\ *noun, plural* **-ro·ses** \-ˌsēz\ [New Latin] (circa 1926)
: any of several diseases of humans and domestic animals that are caused by infection with leptospires

lep·to·tene \ˈlep-tə-ˌtēn\ *noun* [International Scientific Vocabulary] (1912)
: a stage of meiotic prophase immediately preceding synapsis in which the chromosomes appear as fine discrete threads
— leptotene *adjective*

¹les·bi·an \ˈlez-bē-ən\ *adjective, often capitalized* (1591)
1 : of or relating to Lesbos
2 [from the reputed homosexual band associated with Sappho of Lesbos] **:** of or relating to homosexuality between females

²lesbian *noun, often capitalized* (circa 1890)
: a female homosexual

les·bi·an·ism \ˈlez-bē-ə-ˌni-zəm\ *noun* (1870)
: female homosexuality

lèse–ma·jes·té *or* **lese maj·es·ty** \ˌlāz-ˈma-jə-stē, ˌlez-, ˌlēz-\ *noun* [Middle French *lese majesté,* from Latin *laesa majestas,* literally, injured majesty] (1536)
1 a : a crime (as treason) committed against a sovereign power **b :** an offense violating the dignity of a ruler as the representative of a sovereign power
2 : a detraction from or affront to dignity or importance

le·sion \ˈlē-zhən\ *noun* [Middle English, from Middle French, from Latin *laesion-, laesio,* from *laedere* to injure] (15th century)
1 : INJURY, HARM
2 : an abnormal change in structure of an organ or part due to injury or disease; *especially* **:** one that is circumscribed and well defined
— le·sioned \-zhənd\ *adjective*

les·pe·de·za \ˌles-pə-ˈdē-zə\ *noun* [New Latin, irregular from V. M. de *Zespedes* (flourished 1785) Spanish governor of East Florida] (1891)
: any of a genus (*Lespedeza*) of herbaceous or shrubby leguminous plants including some widely used for forage, soil improvement, and especially hay

¹less \ˈles\ *adjective comparative of* ¹LITTLE [Middle English, partly from Old English *lǣs,* adverb & noun; partly from *lǣssa,* adjective; akin to Old Frisian *lēs* less] (before 12th century)
1 : constituting a more limited number ⟨*less* than three⟩
2 : of lower rank, degree, or importance ⟨no *less* a person than the president himself⟩
3 a : of reduced size, extent, or degree **b :** more limited in quantity ⟨in *less* time⟩ □

²less *adverb comparative of* ²LITTLE (before 12th century)
: to a lesser extent or degree
— less and less : to a progressively smaller size or extent
— less than : by no means **:** not at all ⟨*less than* honest in his replies⟩

³less *noun, plural* **less** (before 12th century)
1 : a smaller portion or quantity
2 : something of less importance

⁴less *preposition* (15th century)
: diminished by **:** MINUS

-less *adjective suffix* [Middle English *-les, -lesse,* from Old English *-lēas,* from *lēas* devoid, false; akin to Old High German *lōs* loose, Old English *losian* to get lost — more at LOSE]
1 : destitute of **:** not having ⟨wit*less*⟩ ⟨child*less*⟩
2 : unable to be acted on or to act (in a specified way) ⟨daunt*less*⟩ ⟨fade*less*⟩

les·see \le-ˈsē\ *noun* [Middle English, from Anglo-French, from *lessé,* past participle of *lesser* to lease — more at LEASE] (15th century)
: one that holds real or personal property under a lease

less·en \ˈle-sᵊn\ *verb* **less·ened; less·en·ing** \ˈles-niŋ, ˈle-sᵊn-iŋ\ (13th century)
intransitive verb
: to shrink in size, number, or degree **:** DECREASE
transitive verb
1 : to reduce in size, extent, or degree
2 a *archaic* **:** to represent as of little value **b :** to lower in status or dignity **:** DEGRADE
synonym see DECREASE

¹less·er \ˈle-sər\ *adjective, comparative of* ¹LITTLE (13th century)
: of less size, quality, degree, or significance **:** of lower status

²lesser *adverb* (1539)
: LESS ⟨*lesser*-known⟩

lesser celandine *noun* (circa 1890)
: a yellow-flowered Eurasian perennial herb (*Ranunculus ficaria*) of the buttercup family naturalized in North America

lesser cornstalk borer *noun* (circa 1925)
: a pyralid moth (*Elasmopalpus lignosellus*) having slender greenish larvae that burrow in the stalk especially of Indian corn near ground level

lesser peach tree borer *noun* (circa 1924)
: a moth (*Synanthedon pictipes* family Aegeriidae) whose larva is a borer in the forks and crotches of stone-fruit trees and especially the peach

lesser yellowlegs *noun plural but singular or plural in construction* (circa 1903)
: a common American marsh and shore bird (*Tringa flavipes*) that closely resembles the greater yellowlegs in color and markings but is smaller with a shorter more slender bill

¹les·son \ˈle-sᵊn\ *noun* [Middle English, from Old French *leçon,* from Late Latin *lection-, lectio,* from Latin, act of reading, from *legere* to read — more at LEGEND] (13th century)
1 : a passage from sacred writings read in a service of worship
2 a : a piece of instruction **b :** a reading or exercise to be studied by a pupil **c :** a division of a course of instruction
3 a : something learned by study or experience ⟨his years of travel had taught him valuable *lessons*⟩ **b :** an instructive example ⟨the *lessons* of history⟩ **c :** REPRIMAND

²lesson *transitive verb* **les·soned; les·son·ing** \ˈle-sə-niŋ, ˈles-niŋ\ (1555)
1 : to give a lesson to **:** INSTRUCT
2 : LECTURE, REBUKE

les·sor \ˈle-ˌsȯr, le-ˈsȯr\ *noun* [Middle English *lessour,* from Anglo-French, from *lesser* to lease] (14th century)
: one that conveys property by lease

lest \ˈlest\ *conjunction* [Middle English *les the, leste,* from Old English *thȳ lǣs the,* from *thȳ* (instrumental of *thæt* that) + *lǣs* + *the,* relative particle] (before 12th century)
: for fear that — used after an expression denoting fear or apprehension ⟨worried *lest* she should be late⟩ ⟨hesitant to speak out *lest* he be fired⟩

¹let \ˈlet\ *transitive verb* **let·ted; letted** *or* **let; let·ting** [Middle English *letten,* from Old English *lettan* to delay, hinder; akin to Old High German *lezzen* to delay, hurt, Old English *lǣt* late] (before 12th century)
archaic **:** HINDER, PREVENT

²let *noun* (12th century)
1 : something that impedes **:** OBSTRUCTION
2 : a shot or point in racket games that does not count and must be replayed

³let *verb* **let; let·ting** [Middle English *leten,* from Old English *lǣtan;* akin to Old High German *lāzzan* to permit, and perhaps to Lithuanian *lėnas* tranquil] (before 12th century)
transitive verb
1 : to cause to **:** MAKE ⟨*let* me know⟩
2 a : to offer or grant for rent or lease ⟨*let* rooms⟩ **b :** to assign especially after bids ⟨*let* a contract⟩
3 a : to give opportunity to or fail to prevent ⟨live and *let* live⟩ ⟨a break in the clouds *let* us see the summit⟩ ⟨*let* the opportunity slip⟩ **b** — used in the imperative to introduce a request or proposal ⟨*let* us pray⟩ **c** — used as an auxiliary to express a warning ⟨*let* him try⟩
4 : to free from or as if from confinement ⟨*let* out a scream⟩ ⟨*let* blood⟩
5 : to permit to enter, pass, or leave ⟨*let* them through⟩ ⟨*let* them off with a warning⟩
6 : to make an adjustment to ⟨*let* out the waist⟩
intransitive verb
1 : to become rented or leased
2 : to become awarded to a contractor
synonym see HIRE
— let alone : to leave undisturbed; *also* **:** to leave to oneself
— let fly : to hurl an object
— let go : to dismiss from employment ⟨the firm *let* him *go* at the end of the month⟩
— let it all hang out : to reveal one's true feelings **:** act without dissimulation
— let one have it : to subject to vigorous assault
— let rip 1 : to utter or release without restraint ⟨*let* 'er *rip*⟩ **2 :** to do or utter something without restraint ⟨*let rip* at the press⟩
— let the cat out of the bag : to give away a secret

-let *noun suffix* [Middle English, from Middle French *-elet,* from *-el,* diminutive suffix (from Latin *-ellus*) + *-et*]
1 : small one ⟨book*let*⟩
2 : article worn on ⟨wrist*let*⟩

let alone *conjunction* (1812)
: to say nothing of **:** not to mention — used to add an example of narrower range by way

□ USAGE
less The traditional view is that *less* applies to matters of degree, value, or amount and modifies collective nouns, mass nouns, or nouns denoting an abstract whole while *fewer* applies to matters of number and modifies plural nouns. *Less* has been used to modify plural nouns since the days of King Alfred and the usage, though roundly decried, appears to be increasing. *Less* is more likely than *fewer* to modify plural nouns when distances, sums of money, and a few fixed phrases are involved ⟨*less* than 100 miles⟩ ⟨an investment of *less* than $2000⟩ ⟨in 25 words or *less*⟩ and as likely as *fewer* to modify periods of time ⟨in *less* (or *fewer*) than four hours⟩.

of contrast especially in negative contexts ⟨believed that he would never walk again *let alone* play golf —*Sports Illustrated*⟩ ⟨great to read but bloody to speak, *let alone* sing —Robertson Davies⟩

letch \'lech\ *noun* [back-formation from *letcher,* alteration of *lecher*] (circa 1796)
: CRAVING; *specifically* **:** sexual desire

let·down \'let-ˌdau̇n\ *noun* (1768)
1 a : DISCOURAGEMENT, DISAPPOINTMENT **b :** a slackening of effort **:** RELAXATION
2 : the descent of an aircraft or spacecraft to the point at which a landing approach is begun

let down *transitive verb* (12th century)
1 : to allow to descend gradually
2 a : to fail to support ⟨felt her parents had *let* her *down*⟩ **b :** to fall short of the expectations of ⟨the plot *lets* you *down* at the end⟩

¹le·thal \'lē-thəl\ *adjective* [Latin *letalis, lethalis,* from *letum* death] (1604)
1 a : of, relating to, or causing death ⟨death by *lethal* injection⟩ **b :** capable of causing death ⟨*lethal* chemicals⟩
2 : gravely damaging or destructive **:** DEVASTATING ⟨a *lethal* attack on his reputation⟩
synonym see DEADLY
— le·thal·i·ty \lē-'tha-lə-tē\ *noun*
— le·thal·ly \'lē-thə-lē\ *adverb*

²lethal *noun* (1917)
1 : an abnormality of genetic origin causing the death of the organism possessing it
2 : LETHAL GENE

lethal gene *noun* (1939)
: a gene that in some (as homozygous) conditions may prevent development or cause the death of an organism or its germ cells — called also *lethal factor, lethal mutant, lethal mutation*

le·thar·gic \lə-'thär-jik, le-\ *adjective* (14th century)
1 : of, relating to, or characterized by lethargy **:** SLUGGISH
2 : INDIFFERENT, APATHETIC
— le·thar·gi·cal·ly \-ji-k(ə-)lē\ *adverb*

leth·ar·gy \'le-thər-jē\ *noun* [Middle English *litargie,* from Medieval Latin *litargia,* from Late Latin *lethargia,* from Greek *lēthargia,* from *lēthargos* forgetful, lethargic, irregular from *lēthē*] (14th century)
1 : abnormal drowsiness
2 : the quality or state of being lazy, sluggish, or indifferent ☆

le·the \'lē-thē\ *noun* [Latin, from Greek *Lēthē,* from *lēthē* forgetfulness; akin to Greek *lanthanein* to escape notice, *lanthanesthai* to forget — more at LATENT]
1 *capitalized* **:** a river in Hades whose waters cause drinkers to forget their past
2 : OBLIVION, FORGETFULNESS
— le·the·an \'lē-thē-ən, li-'thē-\ *adjective, often capitalized*

let on *intransitive verb* (1725)
1 : to make acknowledgment **:** ADMIT ⟨knows more than he *lets on*⟩
2 : to reveal a secret ⟨nobody *let on* about the surprise party⟩
3 : PRETEND ⟨*let on* to being a stranger⟩

let out *intransitive verb* (1888)
: to conclude a session or performance ⟨school *let out* in June⟩

let's \'lets, *in rapid speech* 'les\ (1573)
: let us

Lett \'let\ *noun* [German *Lette,* ultimately from Latvian *latvis*] (1589)
: LATVIAN 2

¹let·ter \'le-tər\ *noun* [Middle English, from Old French *lettre,* from Latin *littera* letter of the alphabet, *litterae,* plural, epistle, literature] (13th century)
1 : a symbol usually written or printed representing a speech sound and constituting a unit of an alphabet
2 a : a direct or personal written or printed message addressed to a person or organization **b :** a written communication containing a grant — usually used in plural
3 *plural but singular or plural in construction* **a :** LITERATURE, BELLES LETTRES **b :** LEARNING
4 : the strict or outward sense or significance ⟨the *letter* of the law⟩
5 a : a single piece of type **b :** a style of type
6 : the initial of a school awarded to a student for achievement usually in athletics

²letter (1668)
transitive verb
1 : to set down in letters **:** PRINT
2 : to mark with letters
intransitive verb
: to win an athletic letter
— let·ter·er \-tər-ər\ *noun*

³let·ter \'le-tər\ *noun* (1552)
: one that rents or leases

let·ter·boxed \'le-tər-ˌbäkst\ *adjective* [perhaps from the resemblance of the picture on the TV screen or the bands above and below the picture to slots in a mailbox] (1989)
of a video recording **:** formatted so as to display the full rectangular frame of a wide-screen motion picture
— let·ter·box·ing \-ˌbäk-siŋ\ *noun*

letter carrier *noun* (circa 1552)
: a person who delivers mail

let·tered \'le-tərd\ *adjective* (14th century)
1 a : LEARNED, EDUCATED **b :** of, relating to, or characterized by learning **:** CULTURED
2 : inscribed with or as if with letters

let·ter·form \'le-tər-ˌfȯrm\ *noun* (1908)
: the shape of a letter of an alphabet especially from the standpoint of design or development

let·ter·head \-ˌhed\ *noun* (circa 1887)
1 : a sheet of stationery printed or engraved usually with the name and address of an organization
2 : the heading at the top of a letterhead

let·ter·ing \'le-tə-riŋ\ *noun* (1811)
: letters used in an inscription

let·ter·man \'le-tər-mən\ *noun* (1926)
: an athlete who has earned a letter in a school sport

letter missive *noun, plural* **letters missive** [Middle English, from Middle French *lettre missive* letter intended to be sent] (15th century)
: a letter from a superior authority conveying a command, recommendation, permission, or invitation

letter of credence (14th century)
: a formal document attesting to the power of a diplomatic agent to act for the issuing government — called also *letters of credence*

letter of credit (1645)
1 : a letter addressed by a banker to a correspondent certifying that a person named therein is entitled to draw on the writer's credit up to a certain sum
2 : a letter addressed by a banker to a person to whom credit is given authorizing drafts on the issuing bank or on a bank in the person's country up to a certain sum and guaranteeing to accept the drafts if duly made

letter of intent (circa 1942)
: a written statement of the intention to enter into a formal agreement

let·ter–per·fect \'le-tər-'pər-fikt\ *adjective* (1845)
: correct to the smallest detail; *especially* **:** VERBATIM

let·ter·press \'le-tər-ˌpres\ *noun* (circa 1765)
1 : the process of printing from an inked raised surface especially when the paper is impressed directly upon the surface
2 *chiefly British* **:** text (as of a book) distinct from pictorial illustrations

letters close \-'klōs\ *noun plural* (1903)
: letters issued by a government or sovereign to a private person in a private matter

letter sheet *noun* (1851)
: a sheet of stationery that can be folded and sealed with the message inside to form its own envelope

letters of administration (15th century)
: a letter evidencing the right of an administrator to administer the goods or estate of a deceased person

letters of marque \-'märk\ (15th century)
: written authority granted to a private person by a government to seize the subjects of a foreign state or their goods; *specifically* **:** a license granted to a private person to fit out an armed ship to plunder the enemy

let·ter·spac·ing \'le-tər-ˌspā-siŋ\ *noun* (1917)
: insertion of space between the letters of a word

letters patent *noun plural* (14th century)
: a writing (as from a sovereign) that confers on a designated person a grant in a form open for public inspection

¹Lett·ish \'le-tish\ *adjective* (1831)
: of or relating to the Latvians or their language

²Lettish *noun* (1841)
: LATVIAN 1

let·tre de ca·chet \'le-trə-də-ˌka-'shā\ *noun, plural* **lettres de cachet** \-trə(z)-\ [French, literally, letter with a seal] (1718)
: a letter bearing an official seal and usually authorizing imprisonment without trial of a named person

let·tuce \'le-təs\ *noun* [Middle English *letuse,* from Middle French *laitues,* plural of *laitue,* from Latin *lactuca,* from *lact-, lac* milk; from its milky juice — more at GALAXY] (14th century)
: any of a genus (*Lactuca*) of composite plants; *especially* **:** a common garden vegetable (*L. sativa*) whose succulent leaves are used especially in salads ◆

let·up \'let-ˌəp\ *noun* (1837)
: a lessening of effort or intensity

let up *intransitive verb* (1787)
1 a : to diminish or slow down **:** SLACKEN **b :** CEASE, STOP
2 : to become less severe — used with *on*

leu \'leu̇\ *noun, plural* **lei** \'lā\ [Romanian, literally, lion, from Latin *leo* — more at LION] (1879)
— see MONEY table

leuc- *or* **leuco-** *chiefly British variant of* LEUK-

☆ SYNONYMS
Lethargy, languor, lassitude, stupor, torpor mean physical or mental inertness. LETHARGY implies such drowsiness or aversion to activity as is induced by disease, injury, or drugs ⟨months of *lethargy* followed my accident⟩. LANGUOR suggests inertia induced by an enervating climate or illness or love ⟨*languor* induced by a tropical vacation⟩. LASSITUDE stresses listlessness or indifference resulting from fatigue or poor health ⟨a depression marked by *lassitude*⟩. STUPOR implies a deadening of the mind and senses by shock, narcotics, or intoxicants ⟨lapsed into an alcoholic *stupor*⟩. TORPOR implies a state of suspended animation as of hibernating animals but may suggest merely extreme sluggishness ⟨a once alert mind now in a *torpor*⟩.

◇ WORD HISTORY
lettuce Many types of lettuce have a milky white juice, and it is to this property that the vegetable owes its name. The earliest known Middle English form, *letuse,* is apparently borrowed from Old French *laitues,* the plural of *laitue.* The Old French word is descended in turn from *lactuca,* the Latin word that is still used as the scientific name for the vegetable. *Lactuca* is a derivative of *lac,* the Latin word for "milk."

leu·cine \ˈlü-ˌsēn\ *noun* [French, from *leuc-* leuk-] (1826)
: a white crystalline essential amino acid $C_6H_{13}NO_2$ obtained by the hydrolysis of most dietary proteins

leu·cite \ˈlü-ˌsīt\ *noun* [German *Leuzit,* from *leuz-* leuk-] (1799)
: a white or gray mineral consisting of a silicate of potassium and aluminum and occurring in igneous rocks
— **leu·cit·ic** \lü-ˈsi-tik\ *adjective*

leu·co·ci·din \ˌlü-kə-ˈsī-dᵊn\ *noun* [International Scientific Vocabulary *leuc-* + *-cide* + [1]*-in*] (1894)
: a bacterial substance that destroys white blood cells

leu·co·plast \ˈlü-kə-ˌplast\ *noun* [International Scientific Vocabulary] (1886)
: a colorless plastid especially in the cytoplasm of interior plant tissues that is potentially capable of developing into a chloroplast

leuk- *or* **leuko-** *combining form* [New Latin *leuc-, leuco-,* from Greek *leuk-, leuko-,* from *leukos* — more at LIGHT]
1 : white **:** colorless **:** weakly colored ⟨*leuko*cyte⟩ ⟨*leuk*orrhea⟩
2 : leukocyte ⟨*leuk*emia⟩
3 : white matter of the brain ⟨*leuko*tomy⟩

leu·kae·mia *chiefly British variant of* LEUKEMIA

leu·kae·mo·gen·e·sis *chiefly British variant of* LEUKEMOGENESIS

leu·ke·mia \lü-ˈkē-mē-ə\ *noun* [New Latin] (circa 1855)
: an acute or chronic disease in humans and other warm-blooded animals characterized by an abnormal increase in the number of white blood cells in the tissues and often in the blood
— **leu·ke·mic** \-mik\ *adjective or noun*

leu·ke·mo·gen·e·sis \lü-ˌkē-mə-ˈje-nə-səs\ *noun* (1942)
: induction or production of leukemia
— **leu·ke·mo·gen·ic** \-ˈje-nik\ *adjective*

leu·ke·moid \-ˌmȯid\ *adjective* (1926)
: resembling leukemia but not involving the same changes in the blood-forming organs

leu·ko·cyte \ˈlü-kə-ˌsīt\ *noun* [International Scientific Vocabulary] (1870)
: WHITE BLOOD CELL
— **leu·ko·cyt·ic** \ˌlü-kə-ˈsi-tik\ *adjective*

leu·ko·cy·to·sis \ˌlü-kə-sī-ˈtō-səs, -kə-sə-\ *noun* [New Latin] (1866)
: an increase in the number of white blood cells in the circulating blood

leu·ko·dys·tro·phy \ˌlü-kō-ˈdis-trə-fē\ *noun, plural* **-phies** (1962)
: any of several genetically determined diseases characterized by degeneration of the white matter of the brain

leu·ko·pe·nia \ˌlü-kə-ˈpē-nē-ə\ *noun* [New Latin] (1898)
: a condition in which the number of white blood cells circulating in the blood is abnormally low
— **leu·ko·pe·nic** \-nik\ *adjective*

leu·ko·pla·kia \ˌlü-kō-ˈplā-kē-ə\ *noun* [New Latin, from *leuk-* + Greek *plak-, plax* flat surface — more at FLUKE] (circa 1888)
: an abnormal condition in which thickened white patches of epithelium occur on the mucous membranes (as of the mouth or vulva); *also* **:** a lesion or lesioned area of leukoplakia
— **leu·ko·pla·kic** \-ˈplā-kik\ *adjective*

leu·ko·poi·e·sis \-pȯi-ˈē-səs\ *noun* [New Latin] (circa 1913)
: the formation of white blood cells
— **leu·ko·poi·et·ic** \-ˈe-tik\ *adjective*

leu·kor·rhea \ˌlü-kə-ˈrē-ə\ *noun* [New Latin] (circa 1797)
: a whitish viscid discharge from the vagina resulting from inflammation or congestion of the mucous membrane
— **leu·kor·rhe·al** \-ˈrē-əl\ *adjective*

leu·ko·sis \lü-ˈkō-səs\ *noun, plural* **-ko·ses** \-ˌsēz\ [New Latin] (1922)
: LEUKEMIA; *especially* **:** any of various leukemic diseases of poultry

leu·kot·o·my \lü-ˈkä-tə-mē\ *noun, plural* **-mies** (1937)
: LOBOTOMY

leu·ko·tri·ene \ˌlü-kō-ˈtrī-ˌēn\ *noun* (1980)
: any of a group of eicosanoids that participate in allergic responses

lev \ˈlef\ *noun, plural* **le·va** \ˈle-və\ [Bulgarian, literally, lion] (circa 1900)
— see MONEY table

lev- *or* **levo-** *combining form* [French *lévo-,* from Latin *laevus* left; akin to Greek *laios* left]
1 : levorotatory ⟨*lev*ulose⟩
2 : to the left ⟨*levo*rotatory⟩

Le·val·loi·si·an \ˌle-və-ˈlȯi-zē-ən, lə-ˌval-ˈwä-zē-\ *adjective* [*Levallois*-Perret, suburb of Paris, France] (1932)
: of or relating to a Middle Paleolithic culture characterized by a technique of manufacturing tools by striking flakes from a flint nodule

le·vant \lə-ˈvant\ *intransitive verb* [perhaps from Spanish *levantar* to break camp, ultimately from Latin *levare*] (1797)
chiefly British **:** to run away from a debt

le·vant·er \lə-ˈvan-tər\ *noun* (1668)
1 *capitalized* **:** a native or inhabitant of the Levant
2 : a strong easterly Mediterranean wind

Le·vant storax \lə-ˈvant-\ *noun* (1937)
: STORAX 1a

le·va·tor \li-ˈvā-tər\ *noun, plural* **lev·a·to·res** \ˌle-və-ˈtōr-(ˌ)ēz\ *or* **le·va·tors** \li-ˈvā-tərz\ [New Latin, from Latin *levare* to raise — more at LEVER] (1615)
: a muscle that serves to raise a body part — compare DEPRESSOR

[1]**le·vee** \ˈle-vē; lə-ˈvē, -ˈvā\ *noun* [French *lever,* from Middle French, act of arising, from (*se*) *lever* to rise] (1672)
1 : a reception held by a person of distinction on rising from bed
2 : an afternoon assembly at which the British sovereign or his or her representative receives only men
3 : a reception usually in honor of a particular person

[2]**lev·ee** \ˈle-vē\ *noun* [French *levée,* from Old French, act of raising, from *lever* to raise — more at LEVER] (circa 1720)
1 a : an embankment for preventing flooding **b :** a river landing place **:** PIER
2 : a continuous dike or ridge (as of earth) for confining the irrigation areas of land to be flooded

[3]**lev·ee** \ˈle-vē\ *transitive verb* **lev·eed; lev·ee·ing** (1832)
: to provide with a levee

[1]**lev·el** \ˈle-vəl\ *noun* [Middle English, plumb line, from Middle French *livel,* from (assumed) Vulgar Latin *libellum,* alteration of Latin *libella,* from diminutive of *libra* weight, balance] (14th century)
1 : a device for establishing a horizontal line or plane by means of a bubble in a liquid that shows adjustment to the horizontal by movement to the center of a slightly bowed glass tube
2 : a measurement of the difference of altitude of two points by means of a level
3 : horizontal condition; *especially* **:** equilibrium of a fluid marked by a horizontal surface of even altitude ⟨water seeks its own *level*⟩
4 a : an approximately horizontal line or surface taken as an index of altitude **b :** a practically horizontal surface or area (as of land)
5 : a position in a scale or rank ⟨funded at the national *level*⟩
6 a : a line or surface that cuts perpendicularly all plumb lines that it meets and hence would everywhere coincide with a surface of still water **b :** the plane of the horizon or a line in it
7 : a horizontal passage in a mine intended for regular working and transportation
8 : a concentration of a constituent especially of a body fluid (as blood)
9 : the magnitude of a quantity considered in relation to an arbitrary reference value; *broadly* **:** MAGNITUDE, INTENSITY ⟨a high *level* of hostility⟩
— **on the level :** BONA FIDE, HONEST

[2]**level** *verb* **-eled** *or* **-elled; -el·ing** *or* **-el·ling** \ˈle-və-liŋ, ˈlev-liŋ\ (15th century)
transitive verb
1 : to make (a line or surface) horizontal **:** make flat or level ⟨*level* a field⟩ ⟨*level* off a house lot⟩
2 a : to bring to a horizontal aiming position **b :** AIM, DIRECT ⟨*leveled* a charge of fraud⟩
3 : to bring to a common level or plane **:** EQUALIZE ⟨love *levels* all ranks —W. S. Gilbert⟩
4 a : to lay level with or as if with the ground **:** RAZE **b :** to knock down ⟨*leveled* him with one punch⟩
5 : to make (as color) even or uniform
6 : to find the heights of different points in (a piece of land) especially with a surveyor's level
intransitive verb
1 : to attain or come to a level ⟨the plane *leveled* off at 10,000 feet⟩
2 : to aim a gun or other weapon horizontally
3 : to bring persons or things to a level
4 : to deal frankly and openly

[3]**level** *adjective* (15th century)
1 a : having no part higher than another **:** conforming to the curvature of the liquid parts of the earth's surface **b :** parallel with the plane of the horizon **:** HORIZONTAL
2 a : even or unvarying in height **b :** equal in advantage, progression, or standing **c :** proceeding monotonously or uneventfully **d** (1) **:** STEADY, UNWAVERING ⟨gave him a *level* look⟩ (2) **:** CALM, UNEXCITED ⟨spoke in *level* tones⟩
3 : REASONABLE, BALANCED ⟨arrive at a justly proportional and *level* judgment on this affair —Sir Winston Churchill⟩
4 : distributed evenly ⟨*level* stress⟩
5 : being a surface perpendicular to all lines of force in a field of force **:** EQUIPOTENTIAL
6 : suited to a particular rank or plane of ability or achievement ⟨top-*level* thinking⟩
7 : of or relating to the spreading out of a cost or charge in even payments over a period of time ☆
— **lev·el·ly** \ˈle-və(l)-lē\ *adverb*
— **lev·el·ness** \-vəl-nəs\ *noun*
— **level best :** very best

level crossing *noun* (circa 1841)
British **:** GRADE CROSSING

lev·el·er *or* **lev·el·ler** \ˈle-və-lər, ˈlev-lər\ *noun* (1598)

☆ SYNONYMS
Level, flat, plane, even, smooth mean having a surface without bends, curves, or irregularities. LEVEL applies to a horizontal surface that lies on a line parallel with the horizon ⟨the vast prairies are nearly *level*⟩. FLAT applies to a surface devoid of noticeable curvatures, prominences, or depressions ⟨the work surface must be *flat*⟩. PLANE applies to any real or imaginary flat surface in which a straight line between any two points on it lies wholly within that surface ⟨the *plane* sides of a crystal⟩. EVEN applies to a surface that is noticeably flat or level or to a line that is observably straight ⟨trim the hedge so it is *even*⟩. SMOOTH applies especially to a polished surface free of irregularities ⟨a *smooth* skating rink⟩.

1 : one that levels
2 a *capitalized* **:** one of a group of radicals arising during the English Civil War and advocating equality before the law and religious toleration **b :** one favoring the removal of political, social, or economic inequalities **c :** something that tends to reduce or eliminate differences among individuals

lev·el·head·ed \ˌle-vəl-ˈhe-dəd\ *adjective* (1879)
: having sound judgment **:** SENSIBLE
— lev·el·head·ed·ness *noun*

leveling rod *noun* (circa 1889)
: a graduated rod used in measuring the vertical distance between a point on the ground and the line of sight of a surveyor's level

level off *intransitive verb* (1919)
: to approach or reach a steady rate, volume, or amount **:** STABILIZE ⟨expect prices to *level off*⟩

level of significance (1925)
: the probability of rejecting the null hypothesis in a statistical test when it is true — called also *significance level*

[1]**le·ver** \ˈle-vər, ˈlē-\ *noun* [Middle English, from Middle French *levier,* from *lever* to raise, from Latin *levare,* from *levis* light in weight — more at LIGHT] (14th century)
1 a : a bar used for prying or dislodging something **b :** an inducing or compelling force **:** TOOL ⟨use food as a political *lever* —*Time*⟩
2 a : a rigid piece that transmits and modifies force or motion when forces are applied at two points and it turns about a third; *specifically* **:** a rigid bar used to exert a pressure or sustain a weight at one point of its length by the application of a force at a second and turning at a third on a fulcrum **b :** a projecting piece by which a mechanism is operated or adjusted

[2]**lever** *transitive verb* **le·vered; le·ver·ing** \ˈle-və-riŋ, ˈlē-; ˈlev-riŋ, ˈlēv-\ (1876)
1 : to pry, raise, or move with or as if with a lever
2 : to operate (a device) in the manner of a lever

[1]**le·ver·age** \ˈle-və-rij, ˈlē-; ˈlev-rij, ˈlēv-\ *noun* (1830)
1 : the action of a lever or the mechanical advantage gained by it
2 : POWER, EFFECTIVENESS ⟨trying to gain more political *leverage*⟩
3 : the use of credit to enhance one's speculative capacity

[2]**leverage** *transitive verb* **-aged; -ag·ing** (1937)
: to provide (as a corporation) or supplement (as money) with leverage; *also* **:** to enhance as if by supplying with financial leverage

le·ver·aged \ˈle-və-rijd, ˈlē-; ˈlev-rijd, ˈlēv-\ *adjective* (1953)
1 : having a high proportion of debt relative to equity
2 *of the purchase of a company* **:** made with borrowed money that is secured by the assets of the company bought ⟨a *leveraged* buyout⟩

lev·er·et \ˈle-və-rət, ˈlev-rət\ *noun* [Middle English, from (assumed) Middle French *levret,* from Middle French *levre* hare, from Latin *lepor-, lepus*] (15th century)
: a hare in its first year

Le·vi \ˈlē-ˌvī\ *noun* [Late Latin, from Hebrew *Lēwī*]
: a son of Jacob and the traditional eponymous ancestor of the priestly tribe of Levi

levi·able \ˈle-vē-ə-bəl\ *adjective* (15th century)
: capable of being levied or levied upon

le·vi·a·than \li-ˈvī-ə-thən\ *noun* [Middle English, from Late Latin, from Hebrew *liwyāthān*]
1 a *often capitalized* **:** a sea monster defeated by Yahweh in various scriptural accounts **b :** a large sea animal
2 *capitalized* **:** the political state; *especially* **:** a totalitarian state having a vast bureaucracy
3 : something large or formidable
— leviathan *adjective*

levi·er \ˈle-vē-ər\ *noun* (15th century)
: one that levies

lev·i·gate \ˈle-və-ˌgāt\ *transitive verb* **-gat·ed; -gat·ing** [Latin *levigatus,* past participle of *levigare* to make smooth, from *levis* smooth (akin to Greek *leios* smooth and perhaps to Latin *linere* to smear) + *-igare* (akin to *agere* to drive) — more at LIME, AGENT] (1612)
1 : POLISH, SMOOTH
2 a : to grind to a fine smooth powder while in moist condition **b :** to separate (fine powder) from coarser material by suspending in a liquid
— lev·i·ga·tion \ˌle-və-ˈgā-shən\ *noun*

lev·in \ˈle-vən\ *noun* [Middle English *levene*] (13th century)
archaic **:** LIGHTNING

le·vi·rate \ˈle-və-rət, ˈlē-, -ˌrāt\ *noun* [Latin *levir* husband's brother; akin to Old English *tācor* husband's brother, Greek *daēr*] (1725)
: the sometimes compulsory marriage of a widow to a brother of her deceased husband
— le·vi·rat·ic \ˌle-və-ˈra-tik, ˌlē-\ *adjective*

Le·vi's \ˈlē-ˌvīz\ *trademark*
— used especially for blue denim jeans

lev·i·tate \ˈle-və-ˌtāt\ *verb* **-tat·ed; -tat·ing** [*levity*] (1673)
intransitive verb
: to rise or float in the air especially in seeming defiance of gravitation
transitive verb
: to cause to levitate

lev·i·ta·tion \ˌle-və-ˈtā-shən\ *noun* (1668)
: the act or process of levitating; *especially* **:** the rising or lifting of a person or thing by means held to be supernatural
— lev·i·ta·tion·al \-shnəl, -shə-nᵊl\ *adjective*

Le·vite \ˈlē-ˌvīt\ *noun* (14th century)
: a member of the priestly Hebrew tribe of Levi; *specifically* **:** a Levite of non-Aaronic descent assigned to lesser ceremonial offices under the Levitical priests of the family of Aaron

Le·vit·i·cal \li-ˈvi-ti-kəl\ *adjective* [Late Latin *Leviticus*] (1535)
: of or relating to the Levites or to Leviticus

Le·vit·i·cus \-kəs\ *noun* [Late Latin, literally, of the Levites]
: the third book of canonical Jewish and Christian Scripture consisting mainly of priestly legislation — see BIBLE table

lev·i·ty \ˈle-və-tē\ *noun* [Latin *levitat-, levitas,* from *levis* light in weight — more at LIGHT] (1564)
1 : excessive or unseemly frivolity
2 : lack of steadiness **:** CHANGEABLENESS

le·vo \ˈlē-(ˌ)vō\ *adjective* (1938)
: LEVOROTATORY

levo- — see LEV-

levo·do·pa \ˌle-və-ˈdō-pə\ *noun* (1969)
: L-DOPA

le·vo·ro·ta·to·ry \-ˈrō-tə-ˌtōr-ē, -ˌtȯr-\ *or* **le·vo·ro·ta·ry** \-ˈrō-tə-rē\ *adjective* (1873)
: turning toward the left or counterclockwise; *especially* **:** rotating the plane of polarization of light to the left — compare DEXTROROTATORY

lev·u·lose \ˈlev-yə-ˌlōs, -ˌlōz\ *noun* [International Scientific Vocabulary, irregular from *lev-* + [2]*-ose*] (1871)
: FRUCTOSE 2

[1]**levy** \ˈle-vē\ *noun, plural* **lev·ies** [Middle English, from Old French *levee,* act of raising — more at LEVEE] (13th century)
1 a : the imposition or collection of an assessment **b :** an amount levied
2 a : the enlistment or conscription of men for military service **b :** troops raised by levy

[2]**levy** *verb* **lev·ied; levy·ing** (14th century)
transitive verb
1 a : to impose or collect by legal authority ⟨*levy* a tax⟩ **b :** to require by authority
2 : to enlist or conscript for military service
3 : to carry on (war) **:** WAGE
intransitive verb
: to seize property

lewd \ˈlüd\ *adjective* [Middle English *lewed* vulgar, from Old English *lǣwede* laical, ignorant] (14th century)
1 *obsolete* **:** EVIL, WICKED
2 a : sexually unchaste or licentious **b :** OBSCENE, VULGAR
— lewd·ly *adverb*
— lewd·ness *noun*

lew·is \ˈlü-əs\ *noun* [probably from the name *Lewis*] (1743)
: an iron dovetailed tenon that is made in sections, can be fitted into a dovetail mortise, and is used in hoisting large stones

Lew·is acid \ˈlü-əs-\ *noun* [Gilbert N. *Lewis* (died 1946) American chemist] (1944)
: a substance that is capable of accepting an unshared pair of electrons from a base to form a covalent bond

lew·is·ite \ˈlü-ə-ˌsīt\ *noun* [Winford L. *Lewis* (died 1943) American chemist] (1919)
: a colorless or brown vesicant liquid $C_2H_2AsCl_3$ developed as a poison gas for war use

lex \ˈleks\ *noun, plural* **le·ges** \ˈlā-(ˌ)gās\ [Latin *leg-, lex*] (circa 1775)
: LAW

lex·eme \ˈlek-ˌsēm\ *noun* [Greek *lexis* word, speech + English *-eme*] (1940)
: a meaningful linguistic unit that is an item in the vocabulary of a language
— lex·em·ic \lek-ˈsē-mik\ *adjective*

lex·i·cal \ˈlek-si-kəl\ *adjective* (1836)
1 : of or relating to words or the vocabulary of a language as distinguished from its grammar and construction
2 : of or relating to a lexicon or to lexicography
— lex·i·cal·i·ty \ˌlek-sə-ˈka-lə-tē\ *noun*
— lex·i·cal·ly \ˈlek-si-k(ə-)lē\ *adverb*

lex·i·cal·i·sa·tion *British variant of* LEXICALIZATION

lex·i·cal·i·za·tion \ˌlek-si-kə-lə-ˈzā-shən\ *noun* (1949)
1 : the realization of a meaning in a single word or morpheme rather than in a grammatical construction
2 : the treatment of a formerly freely composed, grammatically regular, and semantically transparent phrase or inflected form as a formally or semantically idiomatic expression
— lex·i·cal·ize \ˈlek-si-kə-ˌlīz\ *transitive verb*

lexical meaning *noun* (1933)
: the meaning of the base (as the word *play*) in a paradigm (as *plays, played, playing*) — compare GRAMMATICAL MEANING

lex·i·cog·ra·pher \ˌlek-sə-ˈkä-grə-fər\ *noun* [Late Greek *lexikographos,* from *lexikon* + Greek *-graphos* writer, from *graphein* to write] (1658)
: an author or editor of a dictionary

lex·i·cog·ra·phy \ˌlek-sə-ˈkä-grə-fē\ *noun* (1680)
1 : the editing or making of a dictionary
2 : the principles and practices of dictionary making
— lex·i·co·graph·i·cal \ˌlek-si-kō-ˈgra-fi-kəl\ *or* **lex·i·co·graph·ic** \-fik\ *adjective*
— lex·i·co·graph·i·cal·ly \-fi-k(ə-)lē\ *adverb*

lex·i·col·o·gy \ˌlek-sə-ˈkä-lə-jē\ *noun* [French *lexicologie,* from *lexico-* (from Late Greek *lexiko-,* from *lexikon*) + *-logie* -logy] (circa 1828)
: a branch of linguistics concerned with the signification and application of words
— lex·i·col·o·gist \-jist\ *noun*

lex·i·con \ˈlek-sə-ˌkän *also* -kən\ *noun, plural* **lex·i·ca** \-kə\ *or* **lexicons** [Late Greek *lexikon,* from neuter of *lexikos* of words, from Greek *lexis* word, speech, from *legein* to say — more at LEGEND] (1603)

1 : a book containing an alphabetical arrangement of the words in a language and their definitions **:** DICTIONARY
2 a : the vocabulary of a language, an individual speaker or group of speakers, or a subject **b :** the total stock of morphemes in a language
3 : REPERTOIRE, INVENTORY

lex·is \'lek-səs\ *noun, plural* **lex·es** \-ˌsēz\ [Greek, speech, word] (1960)
: LEXICON 2a

ley *variant of* LEA

Ley·den jar \'lī-dᵊn-\ *noun* [*Leiden, Leyden,* Netherlands] (1825)
: an electrical capacitor consisting of a glass jar coated inside and outside with metal foil and having the inner coating connected to a conducting rod passed through the insulating stopper

L–form \'el-ˌfȯrm\ *noun* [*L*ister Institute, London, where it was first isolated] (1948)
: a variant bacterium formed especially under stressful conditions and usually lacking a cell wall

Lha·sa ap·so \'lä-sə-'äp-(ˌ)sō, 'la-sə-'ap-\ *noun, plural* **Lhasa apsos** *often A capitalized* [*Lhasa,* Tibet + Tibetan *apso*] (1935)
: any of a Tibetan breed of small dogs that have a dense coat of long hard straight hair, a heavy fall over the eyes, heavy whiskers and beard, and a well-feathered tail curled over the back — called also *Lhasa*

Lhasa apso

li·a·bil·i·ty \ˌlī ə 'bi-lə-tē\ *noun, plural* **-ties** (circa 1809)
1 a : the quality or state of being liable **b :** PROBABILITY
2 : something for which one is liable; *especially* **:** pecuniary obligation **:** DEBT — usually used in plural
3 : one that acts as a disadvantage **:** DRAWBACK

li·a·ble \'lī-ə-bəl, *especially in sense 2 often* 'lī-bəl\ *adjective* [Middle English *lyable,* from (assumed) Anglo-French, from Old French *lier* to bind, from Latin *ligare* — more at LIGATURE] (15th century)
1 a : obligated according to law or equity **:** RESPONSIBLE **b :** subject to appropriation or attachment
2 a : being in a position to incur — used with *to* ⟨*liable* to a fine⟩ **b :** exposed or subject to some usually adverse contingency or action ⟨watch out or you're *liable* to fall⟩ ☆ ☐

li·aise \lē-'āz\ *intransitive verb* **li·aised; li·ais·ing** [back-formation from *liaison*] (1928)
chiefly British
1 : to establish liaison
2 : to act as a liaison officer

li·ai·son \'lē-ə-ˌzän, lē-'ā-, ÷'lā-ə-\ *noun* [French, from Middle French, from *lier,* from Old French] (1759)
1 : a binding or thickening agent used in cooking
2 a : a close bond or connection **:** INTERRELATIONSHIP **b :** an illicit sexual relationship **:** AFFAIR 3a
3 a : communication for establishing and maintaining mutual understanding and cooperation (as between parts of an armed force) **b :** one that establishes and maintains liaison
4 : the pronunciation of an otherwise absent consonant sound at the end of the first of two consecutive words the second of which begins with a vowel sound and follows without pause

li·a·na \lē-'ä-nə, -'a-\ *also* **li·ane** \-'än, -'an\ *noun* [French *liane*] (1796)
: any of various usually woody vines especially of tropical rain forests that root in the ground

li·ar \'lī(-ə)r\ *noun* [Middle English, from Old English *lēogere,* from *lēogan* to lie — more at LIE] (before 12th century)
: one that tells lies

Li·as \'lī-əs\ *adjective* [*Lias,* division of the European Jurassic, from French, from English, a limestone rock] (1813)
chiefly British **:** LIASSIC

Li·as·sic \lī-'a-sik\ *adjective* [modification of French *liasique,* from *Lias*] (1833)
: of, relating to, or being a subdivision of the European Jurassic

lib \'lib\ *noun* (1970)
: LIBERATION 2

li·ba·tion \lī-'bā-shən\ *noun* [Latin *libation-, libatio,* from *libare* to pour as an offering; akin to Greek *leibein* to pour] (14th century)
1 a : an act of pouring a liquid as a sacrifice (as to a deity) **b :** a liquid (as wine) used in a libation
2 a : an act or instance of drinking often ceremoniously **b :** BEVERAGE; *especially* **:** a drink containing alcohol
— **li·ba·tion·ary** \-shə-ˌner-ē\ *adjective*

lib·ber \'li-bər\ *noun* [*lib*] (1971)
: a person who supports a liberation movement especially for women — often used disparagingly

li·bec·cio \li-'be-chē-ˌō, -'be-chō\ *or* **li·bec·chio** \-'be-kē-ˌō\ *noun* [Italian *libeccio*] (1667)
: a southwest wind in Italy

[1]li·bel \'lī-bəl\ *noun* [Middle English, written declaration, from Middle French, from Latin *libellus,* diminutive of *liber* book] (14th century)
1 a : a written statement in which a plaintiff in certain courts sets forth the cause of action or the relief sought **b** *archaic* **:** a handbill especially attacking or defaming someone
2 a : a written or oral defamatory statement or representation that conveys an unjustly unfavorable impression **b** (1) **:** a statement or representation published without just cause and tending to expose another to public contempt (2) **:** defamation of a person by written or representational means (3) **:** the publication of blasphemous, treasonable, seditious, or obscene writings or pictures (4) **:** the act, tort, or crime of publishing such a libel ◇

[2]libel *verb* **-beled** *or* **-belled; -bel·ing** *or* **-bel·ling** \-b(ə-)liŋ\ (1570)
intransitive verb
: to make libelous statements
transitive verb
: to make or publish a libel against
— **li·bel·er** \-b(ə-)lər\ *noun*
— **li·bel·ist** \-bə-list\ *noun*

li·bel·ant *or* **li·bel·lant** \'lī-bə-lənt\ *noun* (1726)
: one that institutes a suit by a libel

li·bel·ee *or* **li·bel·lee** \ˌlī-bə-'lē\ *noun* (circa 1856)
: one against whom a libel has been filed in a court

li·bel·ous *or* **li·bel·lous** \'lī-b(ə-)ləs\ *adjective* (1619)
: constituting or including a libel **:** DEFAMATORY ⟨a *libelous* statement⟩

Li·be·ra \'lē-bə-ˌrä, 'lē-brə\ *noun* [Latin, literally, deliver, imperative of *liberare* to liberate; from the first word of the responsory] (circa 1903)
: a Roman Catholic funeral responsory

[1]lib·er·al \'li-b(ə-)rəl\ *adjective* [Middle English, from Middle French, from Latin *liberalis* suitable for a freeman, generous, from *liber* free; perhaps akin to Old English *lēodan* to grow, Greek *eleutheros* free] (14th century)
1 a : of, relating to, or based on the liberal arts ⟨*liberal* education⟩ **b** *archaic* **:** of or befitting a man of free birth
2 a : marked by generosity **:** OPENHANDED ⟨a *liberal* giver⟩ **b :** given or provided in a generous and openhanded way ⟨a *liberal* meal⟩ **c :** AMPLE, FULL
3 *obsolete* **:** lacking moral restraint **:** LICENTIOUS
4 : not literal or strict **:** LOOSE ⟨a *liberal* translation⟩
5 : BROAD-MINDED; *especially* **:** not bound by authoritarianism, orthodoxy, or traditional forms
6 a : of, favoring, or based upon the principles of liberalism **b** *capitalized* **:** of or constituting a political party advocating or associated with the principles of political liberalism; *especially* **:** of or constituting a political party in the United Kingdom associated with ideals of individual especially economic freedom, greater individual participation in government, and

☆ SYNONYMS
Liable, open, exposed, subject, prone, susceptible, sensitive mean being by nature or through circumstances likely to experience something adverse. LIABLE implies a possibility or probability of incurring something because of position, nature, or particular situation ⟨*liable* to get lost⟩. OPEN stresses a lack of barriers preventing incurrence ⟨a claim *open* to question⟩. EXPOSED suggests lack of protection or powers of resistance against something actually present or threatening ⟨*exposed* to infection⟩. SUBJECT implies an openness for any reason to something that must be suffered or undergone ⟨all reports are *subject* to review⟩. PRONE stresses natural tendency or propensity to incur something ⟨*prone* to procrastination⟩. SUSCEPTIBLE implies conditions existing in one's nature or individual constitution that make incurrence probable ⟨very *susceptible* to flattery⟩. SENSITIVE implies a readiness to respond to or be influenced by forces or stimuli ⟨unduly *sensitive* to criticism⟩. See in addition RESPONSIBLE.

☐ USAGE
liable Both *liable* and *apt* when followed by an infinitive are used nearly interchangeably with *likely.* Although conflicting advice has been given over the years, most current commentators accept *apt* when so used. They generally recommend limiting *liable* to situations having an undesirable outcome, and our evidence shows that in edited writing it is more often so used than not.

◇ WORD HISTORY
libel The ancient Romans used the Latin noun *liber* to refer to the inner bark of a tree. Because this material was used to write on before the introduction of papyrus, *liber* came to be applied to a written document, especially a lengthy one; after papyrus became common, a *liber* was usually an entire papyrus scroll, hence a "book." *Libellus,* the diminutive of *liber,* was borrowed into French as *libelle,* which was in turn borrowed into 14th century English as *libel.* Both French and English originally used the word in the literal sense "little book," but in the 16th century, handbills and leaflets, being likened to little books, were also called libels in English. Because they were a popular means of spreading gossip and defamatory statements about famous people, the meaning of *libel* was extended to such defamatory statements as well as to the crime of writing them.

constitutional, political, and administrative reforms designed to secure these objectives ☆
— **lib·er·al·ly** \-b(ə-)rə-lē\ *adverb*
— **lib·er·al·ness** *noun*

²liberal *noun* (1820)
: a person who is liberal: as **a :** one who is open-minded or not strict in the observance of orthodox, traditional, or established forms or ways **b** *capitalized* **:** a member or supporter of a liberal political party **c :** an advocate or adherent of liberalism especially in individual rights

liberal arts *noun plural* (14th century)
1 : the medieval studies comprising the trivium and quadrivium
2 : the studies (as language, philosophy, history, literature, abstract science) in a college or university intended to provide chiefly general knowledge and to develop the general intellectual capacities (as reason and judgment) as opposed to professional or vocational skills

lib·er·al·ism \'li-b(ə-)rə-ˌli-zəm\ *noun* (1819)
1 : the quality or state of being liberal
2 a *often capitalized* **:** a movement in modern Protestantism emphasizing intellectual liberty and the spiritual and ethical content of Christianity **b :** a theory in economics emphasizing individual freedom from restraint and usually based on free competition, the self-regulating market, and the gold standard **c :** a political philosophy based on belief in progress, the essential goodness of the human race, and the autonomy of the individual and standing for the protection of political and civil liberties **d** *capitalized* **:** the principles and policies of a Liberal party
— **lib·er·al·ist** \-b(ə-)rə-list\ *noun or adjective*
— **lib·er·al·is·tic** \ˌli-b(ə-)rə-'lis-tik\ *adjective*

lib·er·al·ise *British variant of* LIBERALIZE

lib·er·al·i·ty \ˌli-bə-'ra-lə-tē\ *noun, plural* **-ties** (14th century)
: the quality or state of being liberal; *also* **:** an instance of this

lib·er·al·ize \'li-b(ə-)rə-ˌlīz\ *verb* **-ized; -iz·ing** (1774)
transitive verb
: to make liberal or more liberal
intransitive verb
: to become liberal or more liberal
— **lib·er·al·iza·tion** \ˌli-b(ə-)rə-lə-'zā-shən\ *noun*
— **lib·er·al·iz·er** \'li-b(ə-)rə-ˌlī-zər\ *noun*

lib·er·ate \'li-bə-ˌrāt\ *transitive verb* **-at·ed; -at·ing** [Latin *liberatus,* past participle of *liberare,* from *liber*] (circa 1623)
1 : to set at liberty **:** FREE; *specifically* **:** to free (as a country) from domination by a foreign power
2 : to free from combination ⟨*liberate* the gas by adding acid⟩
3 : to take or take over illegally or unjustly ⟨material *liberated* from a nearby construction site —Thorne Dreyer⟩
synonym see FREE
— **lib·er·a·tor** \-ˌā-tər\ *noun*

liberated *adjective* (1946)
: freed from or opposed to traditional social and sexual attitudes or roles ⟨a *liberated* woman⟩ ⟨a *liberated* marriage⟩

lib·er·a·tion \ˌli-bə-'rā-shən\ *noun* (15th century)
1 : the act of liberating **:** the state of being liberated
2 : a movement seeking equal rights and status for a group ⟨women's *liberation*⟩
— **lib·er·a·tion·ist** \-sh(ə-)nist\ *noun*

liberation theology *noun* (1972)
: a religious movement especially among Roman Catholic clergy in Latin America that combines political philosophy usually of a Marxist orientation with a theology of salvation as liberation from injustice
— **liberation theologian** *noun*

lib·er·tar·i·an \ˌli-bər-'ter-ē-ən\ *noun* (1789)
1 : an advocate of the doctrine of free will
2 a : a person who upholds the principles of absolute and unrestricted liberty especially of thought and action **b** *capitalized* **:** a member of a political party advocating libertarian principles
— **libertarian** *adjective*
— **lib·er·tar·i·an·ism** \-ē-ə-ˌni-zəm\ *noun*

lib·er·tin·age \'li-bər-ˌtē-nij\ *noun* (1611)
: LIBERTINISM

lib·er·tine \'li-bər-ˌtēn\ *noun* [Middle English *libertyn* freedman, from Latin *libertinus,* from *libertinus,* adjective, of a freedman, from *libertus* freedman, from *liber*] (1577)
1 : a freethinker especially in religious matters — usually used disparagingly
2 : a person who is unrestrained by convention or morality; *specifically* **:** one leading a dissolute life
— **libertine** *adjective*

lib·er·tin·ism \'li-bər-ˌtē-ˌni-zəm, -tə-\ *noun* (1611)
: the quality or state of being libertine **:** the behavior of a libertine

lib·er·ty \'li-bər-tē\ *noun, plural* **-ties** [Middle English, from Middle French *liberté,* from Latin *libertat-, libertas,* from *liber* free — more at LIBERAL] (14th century)
1 : the quality or state of being free: **a :** the power to do as one pleases **b :** freedom from physical restraint **c :** freedom from arbitrary or despotic control **d :** the positive enjoyment of various social, political, or economic rights and privileges **e :** the power of choice
2 a : a right or immunity enjoyed by prescription or by grant **:** PRIVILEGE **b :** permission especially to go freely within specified limits
3 : an action going beyond normal limits: as **a :** a breach of etiquette or propriety **:** FAMILIARITY **b :** RISK, CHANCE ⟨took foolish *liberties* with his health⟩ **c :** a violation of rules or a deviation from standard practice **d :** a distortion of fact
4 : a short authorized absence from naval duty usually for less than 48 hours
synonym see FREEDOM
— **at liberty 1 :** FREE **2 :** at leisure **:** UNOCCUPIED

liberty cap *noun* (1803)
: a close-fitting conical cap used as a symbol of liberty by the French revolutionists and in the U.S. before 1800

liberty pole *noun* (1770)
: a tall flagstaff surmounted by a liberty cap or the flag of a republic and set up as a symbol of liberty

li·bid·i·nal \lə-'bi-dᵊn-əl, -'bid-nᵊl\ *adjective* (1922)
: of or relating to the libido
— **li·bid·i·nal·ly** *adverb*

li·bid·i·nous \-dᵊn-əs, -'bid-nəs\ *adjective* [Middle English, from Middle French *libidineus,* from Latin *libidinosus,* from *libidin-, libido*] (15th century)
1 : having or marked by lustful desires **:** LASCIVIOUS
2 : LIBIDINAL
— **li·bid·i·nous·ly** *adverb*
— **li·bid·i·nous·ness** *noun*

li·bi·do \lə-'bē-(ˌ)dō *also* 'li-bə-ˌdō\ *noun, plural* **-dos** [New Latin *libidin-, libido,* from Latin, desire, lust, from *libēre* to please — more at LOVE] (1909)
1 : emotional or psychic energy that in psychoanalytic theory is derived from primitive biological urges and that is usually goal-directed
2 : sexual drive

li·bra *for 1 & 2a* 'lē-brə, *sometimes* 'lī-brə, *for 2b* 'lē-brə *or* 'lēv-rə\ *noun* [Middle English, from Latin (genitive *Librae*), literally, scales, pound]
1 *capitalized* **a :** a southern zodiacal constellation between Virgo and Scorpio represented by a pair of scales **b** (1) **:** the 7th sign of the zodiac in astrology — see ZODIAC table (2) **:** one born under the sign of Libra
2 a *plural* **li·brae** \'lī-ˌbrē, 'lē-ˌbrī\ [Latin] **:** an ancient Roman unit of weight equal to 327.45 grams **b** [Spanish & Portuguese, from Latin] **:** any of various Spanish, Portuguese, Colombian, or Venezuelan units of weight

Li·bran \'lē-brən, 'lī-\ *noun* (1911)
: LIBRA 1b(2)

li·brar·i·an \lī-'brer-ē-ən\ *noun* (1703)
: a specialist in the care or management of a library
— **li·brar·i·an·ship** \-ˌship\ *noun*

li·brary \'lī-ˌbrer-ē; *British usually & US sometimes* -brər-ē; *US sometimes* -brē, ÷-ˌber-ē\ *noun, plural* **-brar·ies** [Middle English, from Medieval Latin *librarium,* from Latin, neuter of *librarius* of books, from *libr-, liber* inner bark, rind, book] (14th century)
1 a : a place in which literary, musical, artistic, or reference materials (as books, manuscripts, recordings, or films) are kept for use but not for sale **b :** a collection of such materials
2 a : a collection resembling or suggesting a library ⟨a *library* of computer programs⟩ ⟨wine *library*⟩ **b :** MORGUE 2
3 a : a series of related books issued by a publisher **b :** a collection of publications on the same subject
4 : a collection of sequences of DNA and especially recombinant DNA that are maintained in a suitable cellular environment and that represent the genetic material of a particular organism or tissue ■

library paste *noun* (1953)
: a thick white adhesive made from starch

library science *noun* (circa 1904)
: the study or the principles and practices of library care and administration

li·bra·tion \lī-'brā-shən\ *noun* [Latin *libration-, libratio,* from *librare* to balance, from *libra* scales] (1669)
: an oscillation in the apparent aspect of a secondary body (as a planet or a satellite) as seen from the primary object around which it revolves
— **li·bra·tion·al** \-shnəl, -shə-nᵊl\ *adjective*
— **li·bra·to·ry** \'lī-brə-ˌtōr-ē, -ˌtȯr-\ *adjective*

☆ SYNONYMS
Liberal, generous, bountiful, munificent mean giving or given freely and unstintingly. LIBERAL suggests openhandedness in the giver and largeness in the thing or amount given ⟨a teacher *liberal* with her praise⟩. GENEROUS stresses warmhearted readiness to give more than size or importance of the gift ⟨a *generous* offer of help⟩. BOUNTIFUL suggests lavish, unremitting giving or providing ⟨children spoiled by *bountiful* presents⟩. MUNIFICENT suggests a scale of giving appropriate to lords or princes ⟨a *munificent* foundation grant⟩.

□ USAGE
library While the pronunciation \'lī-ˌbrer-ē\ is the most frequent variant in the U.S., the other variants are not uncommon. The contraction \'lī-brē\ and the dissimilated form \'lī-ˌber-ē\ result from the relative difficulty of repeating \r\ in successive syllables; our files contain citations for these variants from educated speakers, including college presidents and professors, as well as with somewhat greater frequency from less educated speakers.

libration point *noun* (1962)
: any of five positions in the plane of a celestial system consisting of one massive body orbiting another at which the gravitational influences of the two bodies are approximately equal

li·bret·tist \lə-ˈbre-tist\ *noun* (1862)
: the writer of a libretto

li·bret·to \lə-ˈbre-(ˌ)tō\ *noun, plural* **-tos** *or* **-ti** \-(ˌ)tē\ [Italian, diminutive of *libro* book, from Latin *libr-, liber*] (1742)
1 : the text of a work (as an opera) for the musical theater
2 : the book containing a libretto

li·bri·form \ˈlī-brə-ˌfȯrm\ *adjective* [Latin *libr-, liber* + International Scientific Vocabulary *-iform*] (1877)
: resembling phloem fibers

Lib·ri·um \ˈli-brē-əm\ *trademark*
— used for a preparation of chlordiazepoxide

Lib·y·an \ˈli-bē-ən\ *noun* (15th century)
1 : a native or inhabitant of Libya
2 : a language of ancient North Africa probably ancestral to Berber dialects
— Libyan *adjective*

lice *plural of* LOUSE

[1]**li·cense** *or* **li·cence** \ˈlī-sᵊn(t)s\ *noun* [Middle English, from Middle French *licence,* from Latin *licentia,* from *licent-, licens,* present participle of *licēre* to be permitted] (14th century)
1 a : permission to act **b :** freedom of action
2 a : a permission granted by competent authority to engage in a business or occupation or in an activity otherwise unlawful **b :** a document, plate, or tag evidencing a license granted
3 a : freedom that allows or is used with irresponsibility **b :** disregard for standards of personal conduct **:** LICENTIOUSNESS
4 : deviation from fact, form, or rule by an artist or writer for the sake of the effect gained
synonym see FREEDOM

[2]**license** *also* **licence** *transitive verb* **li·censed; li·cens·ing** (15th century)
1 a : to issue a license to **b :** to permit or authorize especially by formal license
2 : to give permission or consent to
— li·cens·able \-sᵊn(t)-sə-bəl\ *adjective*
— li·cens·er \-sər\ *or* **li·cen·sor** \-sər, ˌlī-sᵊn-ˈsȯr\ *noun*

licensed practical nurse *noun* (1951)
: a person who has undergone training and obtained a license (as from a state) conferring authorization to provide routine care for the sick

li·cens·ee \ˌlī-sᵊn(t)-ˈsē\ *noun* (circa 1864)
: one that is licensed

license plate *noun* (1926)
: a plate or tag (as of metal) attesting that a license has been secured and usually bearing a registration number

li·cen·sure \ˈlī-sᵊn-shər, -ˌshu̇r\ *noun* (circa 1846)
: the granting of licenses especially to practice a profession; *also* **:** the state of being licensed

licente *plural of* SENTE

li·cen·ti·ate \lī-ˈsen(t)-shē-ət, *especially in sense 2* li-\ *noun* [Medieval Latin *licentiatus,* from past participle of *licentiare* to allow, from Latin *licentia*] (1555)
1 : one who has a license granted especially by a university to practice a profession
2 : an academic degree ranking below that of doctor given by some European universities

li·cen·tious \lī-ˈsen(t)-shəs\ *adjective* [Latin *licentiosus,* from *licentia*] (1535)
1 : lacking legal or moral restraints; *especially* **:** disregarding sexual restraints
2 : marked by disregard for strict rules of correctness
— li·cen·tious·ly *adverb*
— li·cen·tious·ness *noun*

li·chee *variant of* LITCHI

li·chen \ˈlī-kən, *British also* ˈli-chən\ *noun* [Latin, from Greek *leichēn, lichēn,* from *leichein* to lick] (circa 1657)
1 : any of several skin diseases characterized by a papular eruption
2 : any of numerous complex thallophytic plants made up of an alga and a fungus growing in symbiotic association on a solid surface (as a rock)
— li·chened \-kənd\ *adjective*
— li·chen·ous \-kə-nəs\ *adjective*

li·chen·ol·o·gy \ˌlī-kə-ˈnä-lə-jē\ *noun* [International Scientific Vocabulary] (1855)
: the study of lichens
— li·chen·o·log·i·cal \ˌlī-kə-nə-ˈlä-ji-kəl\ *adjective*
— li·chen·ol·o·gist \ˌlī-kə-ˈnä-lə-jist\ *noun*

lich–gate *variant of* LYCH-GATE

licht \ˈlikt\ *Scottish variant of* LIGHT

lic·it \ˈli-sət\ *adjective* [Middle English, from Middle French *licite,* from Latin *licitus,* from past participle of *licēre* to be permitted — more at LICENSE] (15th century)
: conforming to the requirements of the law **:** not forbidden by law **:** PERMISSIBLE
synonym see LAWFUL
— lic·it·ly *adverb*

[1]**lick** \ˈlik\ *verb* [Middle English, from Old English *liccian;* akin to Old High German *leckōn* to lick, Latin *lingere,* Greek *leichein*] (before 12th century)
transitive verb
1 a (1) **:** to draw the tongue over ⟨*lick* a stamp⟩ (2) **:** to flicker over like a tongue **b :** to take into the mouth with the tongue **:** LAP
2 a : to strike repeatedly **:** THRASH **b :** to get the better of **:** OVERCOME, DEFEAT ⟨has *licked* every problem⟩
intransitive verb
1 : to lap with or as if with the tongue
2 : to dart like a tongue ⟨flames *licking* out of windows⟩
— lick into shape : to put into proper form or condition
— lick one's wounds : to recover from defeat or disappointment

[2]**lick** *noun* (1603)
1 a : an act or instance of licking **b :** a small amount **:** BIT ⟨couldn't swim a *lick*⟩ **c :** a hasty careless effort
2 a : a sharp hit **:** BLOW **b :** a directed effort **:** CRACK — usually used in plural; usually used in the phrase *get in one's licks*
3 a : a natural salt deposit (as a salt spring) that animals lick **b :** a block of often medicated saline preparation given to livestock to lick
4 : a musical figure; *specifically* **:** an interpolated and usually improvised figure or flourish
5 : a critical thrust **:** DIG, BARB
— lick and a promise : a perfunctory performance of a task

lick·er·ish \ˈli-k(ə-)rish\ *adjective* [alteration of *lickerous,* from Middle English *likerous,* from (assumed) Old North French, from Old North French *leckeur* lecher] (14th century)
1 : GREEDY, DESIROUS
2 *obsolete* **:** tempting to the appetite
3 : LECHEROUS
— lick·er·ish·ly *adverb*
— lick·er·ish·ness *noun*

lick·e·ty–split \ˌli-kə-tē-ˈsplit\ *adverb* [probably irregular from [1]*lick* + *split*] (circa 1859)
: at great speed

lick·ing *noun* (1756)
1 : a sound thrashing **:** DRUBBING
2 : DEFEAT

lick·spit·tle \ˈlik-ˌspi-tᵊl\ *noun* (1825)
: a fawning subordinate **:** TOADY

lic·o·rice \ˈli-k(ə-)rish, -k(ə-)rəs\ *noun* [Middle English *licorice,* from Old French, from Late Latin *liquiritia,* alteration of Latin *glycyrrhiza,* from Greek *glykyrrhiza,* from *glykys* sweet + *rhiza* root — more at DULCET, ROOT] (13th century)
1 a : the dried root of a European leguminous plant (*Glycyrrhiza glabra*) with pinnate leaves and spikes of blue flowers; *also* **:** an extract of this used especially in medicine, liquors, and confectionery **b :** a candy flavored with licorice
2 : a plant yielding licorice

lic·tor \ˈlik-tər\ *noun* [Middle English *littour,* from Latin *lictor*] (14th century)
: an ancient Roman officer who bore the fasces as the insignia of his office and whose duties included accompanying the chief magistrates in public appearances

[1]**lid** \ˈlid\ *noun* [Middle English, from Old English *hlid;* akin to Old High German *hlit* cover, and probably to Old English *hlinian* to lean — more at LEAN] (before 12th century)
1 : a movable cover for the opening of a hollow container (as a vessel or box)
2 : EYELID
3 : the operculum in mosses
4 *slang* **:** HAT
5 : something that confines, limits, or suppresses **:** CHECK, RESTRAINT
6 : an ounce of marijuana

[2]**lid** *transitive verb* **lid·ded; lid·ding** (13th century)
: to cover or supply with a lid

li·dar \ˈlī-ˌdär\ *noun* [*l*ight + ra*dar*] (1963)
: a device that is similar in operation to radar but emits pulsed laser light instead of microwaves

lidded *adjective* (before 12th century)
1 : having or covered with a lid ⟨a *lidded* tureen⟩
2 : having lids especially of a specified kind — usually used in combination ⟨heavy-*lidded* eyes⟩

lid·less \ˈlid-ləs\ *adjective* (14th century)
1 : having no lid
2 *archaic* **:** WATCHFUL

li·do \ˈlē-(ˌ)dō\ *noun, plural* **lidos** [*Lido,* Italy] (1860)
: a fashionable beach resort

li·do·caine \ˈlī-də-ˌkān\ *noun* [acetani*lid* + *-o-* + *-caine*] (circa 1949)
: a crystalline compound $C_{14}H_{22}N_2O$ that is used in the form of its hydrochloride as a local anesthetic and as an antiarrhythmic agent

[1]**lie** \ˈlī\ *intransitive verb* **lay** \ˈlā\; **lain** \ˈlān\; **ly·ing** \ˈlī-iŋ\ [Middle English, from Old English *licgan;* akin to Old High German *ligen* to lie, Latin *lectus* bed, Greek *lechos*] (before 12th century)
1 a : to be or to stay at rest in a horizontal position **:** be prostrate **:** REST, RECLINE ⟨*lie* motionless⟩ ⟨*lie* asleep⟩ **b :** to assume a horizontal position — often used with *down* **c** *archaic* **:** to reside temporarily **:** stay for the night **:** LODGE **d :** to have sexual intercourse — used with *with* **e :** to remain inactive (as in concealment) ⟨*lie* in wait⟩
2 : to be in a helpless or defenseless state ⟨the town *lay* at the mercy of the invaders⟩
3 *of an inanimate thing* **:** to be or remain in a flat or horizontal position upon a broad support ⟨books *lying* on the table⟩
4 : to have direction **:** EXTEND ⟨the route *lay* to the west⟩
5 a : to occupy a certain relative place or position ⟨hills *lie* behind us⟩ **b :** to have a place in relation to something else ⟨the real reason *lies* deeper⟩ **c :** to have an effect through mere presence, weight, or relative position ⟨remorse *lay* heavily on him⟩ **d :** to be sustainable or admissible
6 : to remain at anchor or becalmed
7 a : to have place **:** EXIST ⟨the choice *lay* between fighting or surrendering⟩ **b :** CONSIST, BELONG ⟨the success of the book *lies* in its direct style⟩ ⟨responsibility *lay* with the adults⟩
8 : REMAIN; *especially* **:** to remain unused, unsought, or uncared for
usage see LAY

— **li·er** \'lī(-ə)r\ *noun*
— **lie low 1 :** to lie prostrate, defeated, or disgraced **2 :** to stay in hiding **:** strive to avoid notice **3 :** to bide one's time **:** remain secretly ready for action

²lie *noun* (1697)
1 *chiefly British* **:** LAY 6
2 : the position or situation in which something lies ⟨a golf ball in a difficult *lie*⟩
3 : the haunt of an animal (as a fish) **:** COVERT
4 *British* **:** an act or instance of lying or resting

³lie *verb* **lied; ly·ing** \'lī-iŋ\ [Middle English, from Old English *lēogan;* akin to Old High German *liogan* to lie, Old Church Slavonic *lŭgati*] (before 12th century)
intransitive verb
1 : to make an untrue statement with intent to deceive
2 : to create a false or misleading impression
transitive verb
: to bring about by telling lies ⟨*lied* his way out of trouble⟩ ☆

⁴lie *noun* [Middle English *lige, lie,* from Old English *lyge;* akin to Old High German *lugī,* Old English *lēogan* to lie] (before 12th century)
1 a : an assertion of something known or believed by the speaker to be untrue with intent to deceive **b :** an untrue or inaccurate statement that may or may not be believed true by the speaker
2 : something that misleads or deceives
3 : a charge of lying

lieb·frau·milch \'lēp-ˌfrau̇-ˌmilk, 'lēb-, -ˌmilk, -ˌmilsh\ *noun* [German, alteration of *Liebfrauenmilch,* from *Liebfrauenstift,* religious foundation in Worms, Germany + *Milch* milk] (1833)
: a fruity white Rhine wine

lie by *intransitive verb* (1613)
: to remain inactive **:** REST

lied \'lēt\ *noun, plural* **lie·der** \'lē-dər\ [German, song, from Old High German *liod*] (1852)
: a German art song especially of the 19th century

Lie·der·kranz \'lē-dər-ˌkran(t)s, -ˌkrän(t)s\ *trademark*
— used for a pungent surface-ripened cheese

lie detector *noun* (1909)
: an instrument for detecting physiological evidence of the tension that accompanies lying

lie down *intransitive verb* (1888)
1 : to submit meekly or abjectly to defeat, disappointment, or insult ⟨won't take that criticism *lying down*⟩
2 : to fail to perform or to neglect one's part deliberately ⟨*lying down* on the job⟩

¹lief \'lēf, 'lēv\ *adjective* [Middle English *lief, lef,* from Old English *lēof;* akin to Old English *lufu* love] (before 12th century)
1 *archaic* **:** DEAR, BELOVED
2 *archaic* **:** WILLING, GLAD

²lief \'lēv, 'lēf\ *adverb* (13th century)
: SOON, GLADLY ⟨I'd as *lief* go as not⟩

¹liege \'lēj\ *adjective* [Middle English, from Old French, from Late Latin *laeticus,* from *laetus* serf, of Germanic origin; akin to Old Frisian *let* serf] (14th century)
1 a : having the right to feudal allegiance or service ⟨his *liege* lord⟩ **b :** obligated to render feudal allegiance and service
2 : FAITHFUL, LOYAL

²liege *noun* (14th century)
1 a : a vassal bound to feudal service and allegiance **b :** a loyal subject
2 : a feudal superior to whom allegiance and service are due

liege man *noun* (14th century)
1 : VASSAL
2 : a devoted follower

lie–in \'lī-ˌin\ *noun* (1963)
: an act of lying down (as in a public place) in organized protest or as a means of forcing compliance with demands

lien \'lēn, 'lē-ən\ *noun* [Middle French, tie, band, from Latin *ligamen,* from *ligare* to bind — more at LIGATURE] (1531)
1 : a charge upon real or personal property for the satisfaction of some debt or duty ordinarily arising by operation of law
2 : the security interest created by a mortgage

lie off *intransitive verb* (1573)
1 : to hold back in the early part of a race
2 : to keep a little away from the shore or another ship
3 : to cease work for a time

lie over *intransitive verb* (circa 1847)
: to await disposal or attention at a later time

li·erne \lē-'ərn, -'ern\ *noun* [French, from Middle French, probably from *lier* to bind, from Latin *ligare*] (1842)
: a rib in Gothic vaulting that passes from one intersection of the principal ribs to another

lie to *intransitive verb* (1711)
of a ship **:** to stay stationary with head to windward

lieu \'lü\ *noun* [Middle English *liue,* from Old French *lieu,* from Latin *locus* — more at STALL] (14th century)
archaic **:** PLACE, STEAD
— **in lieu :** INSTEAD
— **in lieu of :** in the place of **:** instead of

lie up *intransitive verb* (1699)
1 : to go into or remain in a dock
2 : to stay in bed or at rest

lieu·ten·an·cy \lü-'te-nən(t)-sē, *British* le(f)-'ten(t)-\ *noun* (15th century)
: the office, rank, or commission of a lieutenant

lieu·ten·ant \-'te-nənt\ *noun* [Middle English, from Middle French, from *lieu* + *tenant* holding, from *tenir* to hold, from Latin *tenēre* — more at THIN] (14th century)
1 a : an official empowered to act for a higher official **b :** an aide or representative of another in the performance of duty **:** ASSISTANT
2 a (1) **:** FIRST LIEUTENANT (2) **:** SECOND LIEUTENANT **b :** a commissioned officer in the navy or coast guard ranking above a lieutenant junior grade and below a lieutenant commander **c :** a fire or police department officer ranking below a captain

lieutenant colonel *noun* (1598)
: a commissioned officer in the army, air force, or marine corps ranking above a major and below a colonel

lieutenant commander *noun* (1839)
: a commissioned officer in the navy or coast guard ranking above a lieutenant and below a commander

lieutenant general *noun* (1589)
: a commissioned officer in the army, air force, or marine corps who ranks above a major general and whose insignia is three stars

lieutenant governor *noun* (1595)
: a deputy or subordinate governor: as **a :** an elected official serving as deputy to the governor of an American state **b :** the formal head of the government of a Canadian province appointed by the federal government as the representative of the crown
— **lieutenant governorship** *noun*

lieutenant junior grade *noun, plural* **lieutenants junior grade** (circa 1909)
: a commissioned officer in the navy or coast guard ranking above an ensign and below a lieutenant

¹life \'līf\ *noun, plural* **lives** \'līvz\ [Middle English *lif,* from Old English *līf;* akin to Old English *libban* to live — more at LIVE] (before 12th century)
1 a : the quality that distinguishes a vital and functional being from a dead body **b :** a principle or force that is considered to underlie the distinctive quality of animate beings — compare VITALISM 1 **c :** an organismic state characterized by capacity for metabolism, growth, reaction to stimuli, and reproduction
2 a : the sequence of physical and mental experiences that make up the existence of an individual **b :** one or more aspects of the process of living ⟨sex *life* of the frog⟩
3 : BIOGRAPHY 1
4 : spiritual existence transcending physical death
5 a : the period from birth to death **b :** a specific phase of earthly existence ⟨adult *life*⟩ **c :** the period from an event until death ⟨a judge appointed for *life*⟩ **d :** a sentence of imprisonment for the remainder of a convict's life
6 : a way or manner of living
7 : LIVELIHOOD
8 : a vital or living being; *specifically* **:** PERSON ⟨many *lives* were lost in the disaster⟩
9 : an animating and shaping force or principle
10 : SPIRIT, ANIMATION ⟨there was no *life* in her dancing⟩
11 : the form or pattern of something existing in reality ⟨painted from *life*⟩
12 : the period of duration, usefulness, or popularity of something ⟨the expected *life* of flashlight batteries⟩
13 : the period of existence (as of a subatomic particle) — compare HALF-LIFE
14 : a property (as resilience or elasticity) of an inanimate substance or object resembling the animate quality of a living being
15 : living beings (as of a particular kind or environment) ⟨forest *life*⟩
16 a : human activities **b :** animate activity and movement ⟨stirrings of *life*⟩ **c :** the activities of a given sphere, area, or time ⟨the political *life* of the country⟩
17 : one providing interest and vigor ⟨*life* of the party⟩
18 : an opportunity for continued viability ⟨gave the patient a new *life*⟩
19 *capitalized, Christian Science* **:** GOD 1b
20 : something resembling animate life ⟨a grant saved the project's *life*⟩

²life *adjective* (13th century)
1 : of or relating to animate being
2 : LIFELONG ⟨a *life* member⟩
3 : using a living model ⟨a *life* class⟩
4 : of, relating to, or provided by life insurance ⟨a *life* policy⟩

life–and–death *also* **life–or–death** *adjective* (1822)
: involving or culminating in life or death **:** vitally important as if involving life or death

life belt *noun* (circa 1858)
1 *chiefly British* **:** a life preserver in the form of a buoyant belt
2 : SAFETY BELT

life·blood \'līf-'bləd, -ˌbləd\ *noun* (1579)
1 : blood regarded as the seat of vitality
2 : a vital or life-giving force ⟨freedom of inquiry is the *lifeblood* of a university⟩

life·boat \-ˌbōt\ *noun* (1801)
: a sturdy buoyant boat (as one carried by a ship) for use in an emergency and especially in saving lives at sea

life buoy *noun* (1801)
: a ring-shaped life preserver

☆ SYNONYMS
Lie, prevaricate, equivocate, palter, fib mean to tell an untruth. LIE is the blunt term, imputing dishonesty ⟨*lied* about where he had been⟩. PREVARICATE softens the bluntness of LIE by implying quibbling or confusing the issue ⟨during the hearings the witness did his best to *prevaricate*⟩. EQUIVOCATE implies using words having more than one sense so as to seem to say one thing but intend another ⟨*equivocated* endlessly in an attempt to mislead her inquisitors⟩. PALTER implies making unreliable statements of fact or intention or insincere promises ⟨a swindler *paltering* with his investors⟩. FIB applies to a telling of a trivial untruth ⟨*fibbed* about the price of the new suit⟩.

life–care \'līf-ˌkar, -ˌker\ *adjective* (1980)
: of, relating to, or being a residential complex for elderly people that provides an apartment, personal and social services, and health care for life

life cycle *noun* (1873)
1 : the series of stages in form and functional activity through which an organism passes between successive recurrences of a specified primary stage
2 : LIFE HISTORY 2
3 : a series of stages through which something (as an individual, culture, or manufactured product) passes during its lifetime

life expectancy *noun* (1935)
: the average life span of an individual

life force *noun* (1896)
: ÉLAN VITAL

life–form \'līf-ˈfȯrm, -ˌfȯrm\ *noun* (1899)
: the body form that characterizes a kind of organism (as a species) at maturity; *also* **:** a kind of organism

life•ful \'līf-fəl\ *adjective* (13th century)
archaic **:** full of or giving vitality

life–giv•ing \-ˌgi-viŋ\ *adjective* (1596)
: giving or having power to give life and spirit : INVIGORATING

life•guard \-ˌgärd\ *noun* (1896)
: a usually expert swimmer employed (as at a beach or a pool) to safeguard other swimmers
— **lifeguard** *intransitive verb*

life history *noun* (1870)
1 : the history of an individual or thing
2 : a history of the changes through which an organism passes in its development from the primary stage to its natural death

life insurance *noun* (1809)
: insurance providing for payment of a stipulated sum to a designated beneficiary upon death of the insured

life jacket *noun* (1883)
: a life preserver in the form of a buoyant vest

life•less \'līf-ləs\ *adjective* (before 12th century)
: having no life: **a :** DEAD **b :** INANIMATE **c :** lacking qualities expressive of life and vigor : INSIPID **d :** destitute of living beings
— **life•less•ly** *adverb*
— **life•less•ness** *noun*

life•like \'līf-ˌlīk\ *adjective* (14th century)
: accurately representing or imitating real life ⟨a *lifelike* portrait⟩
— **life•like•ness** *noun*

life•line \'līf-ˌlīn\ *noun* (1700)
1 : a line (as a rope) used for saving or preserving life: as **a :** a line along the outer edge of the deck of a boat or ship **b :** a line used to keep contact with a person (as a diver or astronaut) in a dangerous or potentially dangerous situation
2 : something regarded as indispensable for the maintaining or protection of life

life list *noun* (1960)
: a record kept of all birds sighted and identified by a birder

life•long \'līf-ˌlȯŋ\ *adjective* (1855)
1 : lasting or continuing through life
2 : LONG-STANDING

life•man•ship \'līf-mən-ˌship\ *noun* (1949)
: the skill or practice of achieving superiority or an appearance of superiority over others (as in conversation) by perplexing and demoralizing them

life net *noun* (circa 1904)
: a strong net or sheet (as of canvas) used (as by firefighters) to catch a person jumping from a burning building

life of Ri•ley *also* **life of Reil•ly** \-'rī-lē\ [from the name *Riley* or *Reilly*] (1923)
: a carefree comfortable way of living

life peer *noun* (1869)
: a British peer whose title is not hereditary
— **life peerage** *noun*

life preserver *noun* (1804)
1 : a device (as a life jacket or life buoy) designed to save a person from drowning by providing buoyancy in water
2 *chiefly British* **:** BLACKJACK 3

lif•er \'lī-fər\ *noun* (1827)
1 : a person sentenced to imprisonment for life
2 : a person who makes a career of one of the armed forces
3 : a person who has made a lifelong commitment (as to a way of life)

life raft *noun* (1819)
: a raft usually made of wood or an inflatable material and designed for use by people forced into the water

life ring *noun* (circa 1909)
: LIFE BUOY

life•sav•er \'līf-ˌsā-vər\ *noun* (1887)
1 : one trained to save lives of drowning persons
2 : one that is at once timely and effective in time of distress or need

¹life•sav•ing \-viŋ\ *adjective* (1858)
: designed for or used in saving lives ⟨*lifesaving* drugs⟩

²lifesaving *noun* (1919)
: the skill or practice of saving or protecting the lives especially of drowning persons

life science *noun* (1941)
: a branch of science (as biology, medicine, anthropology, or sociology) that deals with living organisms and life processes — usually used in plural
— **life scientist** *noun*

life–size \'līf-'sīz\ *or* **life–sized** \-'sīzd\ *adjective* (1841)
: of natural size **:** of the size of the original ⟨a *life-size* statue⟩

life span *noun* (1918)
1 : the duration of existence of an individual
2 : the average length of life of a kind of organism or of a material object especially in a particular environment or under specified circumstances

life•style \'līf-'stī(ə)l, -ˌstī(ə)l\ *noun* (1939)
: the typical way of life of an individual, group, or culture

life–support \-ˌsə-pōrt, -pȯrt\ *adjective* (1965)
: providing support necessary to sustain life; *especially* **:** of or relating to a system providing such support ⟨*life-support* equipment⟩

life support *noun* (1975)
: medical life-support equipment ⟨the patient was placed on *life support*⟩

life–support system *noun* (1959)
: an artificial or natural system that provides all or some of the items (as oxygen, food, water, control of temperature and pressure, disposition of carbon dioxide and body wastes) necessary for maintaining life or health

life table *noun* (circa 1859)
: MORTALITY TABLE

¹life•time \'līf-ˌtīm\ *noun* (13th century)
1 a : the duration of the existence of a living being or a thing (as a star or a subatomic particle) **b :** LIFE 12
2 : an amount accumulated or experienced in a lifetime ⟨a *lifetime* of regrets⟩

²lifetime *adjective* (1904)
1 : LIFELONG
2 : measured or achieved over the span of a career ⟨a baseball player's *lifetime* batting average⟩

life vest *noun* (1939)
: LIFE JACKET

life•way \-ˌwā\ *noun* (1948)
: LIFE 6

life•work \-'wərk\ *noun* (1871)
: the entire or principal work of one's lifetime; *also* **:** a work extending over a lifetime

life zone *noun* (1901)
: a region characterized by specific plants and animals

¹lift \'lift\ *noun* [Middle English, from Old English *lyft*] (before 12th century)
chiefly Scottish **:** HEAVENS, SKY

²lift *verb* [Middle English, from Old Norse *lypta;* akin to Old English *lyft* air — more at LOFT] (14th century)
transitive verb
1 a : to raise from a lower to a higher position : ELEVATE **b :** to raise in rank or condition **c :** to raise in rate or amount
2 : to put an end to (a blockade or siege) by withdrawing or causing the withdrawal of investing forces
3 : REVOKE, RESCIND ⟨*lift* an embargo⟩
4 a : STEAL ⟨had her purse *lifted*⟩ **b :** PLAGIARIZE **c :** to take out of normal setting ⟨*lift* a word out of context⟩
5 : to take up (as a root crop or transplants) from the ground
6 : to pay off (an obligation) ⟨*lift* a mortgage⟩
7 : to move from one place to another (as by aircraft) **:** TRANSPORT
8 : to take up (a fingerprint) from a surface
intransitive verb
1 a : ASCEND, RISE ⟨the rocket *lifted* off⟩ **b :** to appear elevated (as above surrounding objects)
2 *of inclement weather* **:** to dissipate and clear
☆
— **lift•able** \'lif-tə-bəl\ *adjective*
— **lift•er** *noun*

³lift *noun* (14th century)
1 : the amount that may be lifted at one time : LOAD
2 a : the action or an instance of lifting **b :** the action or an instance of rising **c :** elevated carriage (as of a body part) **d :** the lifting up (as of a dancer) usually by a partner
3 : a device (as a handle or latch) for lifting
4 : an act of stealing **:** THEFT
5 a : ASSISTANCE, HELP **b :** a ride especially along one's way
6 : a layer in the heel of a shoe
7 : a rise or advance in position or condition
8 : a slight rise or elevation
9 : the distance or extent to which something rises
10 : an apparatus or machine used for hoisting: as **a :** a set of pumps used in a mine **b** *chiefly British* **:** ELEVATOR 1b **c :** an apparatus for raising an automobile (as for repair) **d :** SKI LIFT
11 a : an elevating influence **b :** an elevation of the spirit
12 a : the component of the total aerodynamic force acting on an airplane or airfoil that is perpendicular to the relative wind and that for an airplane constitutes the upward force that

☆ SYNONYMS
Lift, raise, rear, elevate, hoist, heave, boost mean to move from a lower to a higher place or position. LIFT usually implies exerting effort to overcome resistance of weight ⟨*lift* the chair while I vacuum⟩. RAISE carries a stronger implication of bringing up to the vertical or to a high position ⟨scouts *raising* a flagpole⟩. REAR may add an element of suddenness to RAISE ⟨suddenly *reared* itself up on its hind legs⟩. ELEVATE may replace LIFT or RAISE especially when exalting or enhancing is implied ⟨*elevated* the taste of the public⟩. HOIST implies lifting something heavy especially by mechanical means ⟨*hoisted* the cargo on board⟩. HEAVE implies lifting and throwing with great effort or strain ⟨*heaved* the heavy crate inside⟩. BOOST suggests assisting to climb or advance by a push ⟨*boosted* his brother over the fence⟩.

opposes the pull of gravity **b :** an updraft that can be used to increase altitude (as of a sailplane)
13 : an organized movement of people, equipment, or supplies by some form of transportation; *especially* **:** AIRLIFT

lift·gate \ˈlift-ˌgāt\ *noun* (1953)
: a rear panel (as on a station wagon) that opens upward

lift·man \-ˌman\ *noun* (1883)
British **:** an elevator operator

lift·off \-ˌȯf\ *noun* (circa 1956)
: a vertical takeoff by an aircraft or a rocket vehicle or missile

lift truck *noun* (1926)
: a small truck for lifting and transporting loads

lig·a·ment \ˈli-gə-mənt\ *noun* [Middle English, from Medieval Latin & Latin; Medieval Latin *ligamentum,* from Latin, band, tie, from *ligare*] (14th century)
1 : a tough band of tissue connecting the articular extremities of bones or supporting an organ in place
2 : a connecting or unifying bond ⟨the law of nations, the great *ligament* of mankind —Edmund Burke⟩
— lig·a·men·tous \-ˈmen-təs\ *adjective*

li·gand \ˈli-gənd, ˈlī-\ *noun* [Latin *ligandus,* gerundive of *ligare*] (1949)
: a group, ion, or molecule coordinated to a central atom or molecule in a complex

li·gase \ˈlī-ˌgās, -ˌgāz\ *noun* [International Scientific Vocabulary *lig-* (from Latin *ligare*) + *-ase*] (1961)
: SYNTHETASE

li·gate \ˈlī-ˌgāt, lī-ˈ\ *transitive verb* **li·gat·ed; li·gat·ing** [Latin *ligatus*] (1599)
: to tie with a ligature

li·ga·tion \lī-ˈgā-shən\ *noun* (1597)
1 : an act of ligating
2 : something that binds **:** LIGATURE

lig·a·ture \ˈli-gə-ˌchu̇r, -chər, -ˌtu̇r, -ˌtyu̇r\ *noun* [Middle English, from Middle French, from Late Latin *ligatura,* from Latin *ligatus,* past participle of *ligare* to bind, tie; akin to Middle Low German *līk* band, boltrope, Albanian *lidh* I tie] (14th century)
1 a : something that is used to bind; *specifically* **:** a filament (as a thread) used in surgery **b :** something that unites or connects **:** BOND
2 : the action of binding or tying
3 : a compound note in mensural notation indicating a group of musical notes to be sung to one syllable
4 : a printed or written character (as *æ* or *ff*) consisting of two or more letters or characters joined together

[1]light \ˈlīt\ *noun* [Middle English, from Old English *lēoht;* akin to Old High German *lioht* light, Latin *luc-, lux* light, *lucēre* to shine, Greek *leukos* white] (before 12th century)
1 a : something that makes vision possible **b :** the sensation aroused by stimulation of the visual receptors **c :** an electromagnetic radiation in the wavelength range including infrared, visible, ultraviolet, and X rays and traveling in a vacuum with a speed of about 186,281 miles (300,000 kilometers) per second; *specifically* **:** the part of this range that is visible to the human eye
2 a : DAYLIGHT **b :** DAWN
3 : a source of light: as **a :** a celestial body **b :** CANDLE **c :** an electric light
4 *archaic* **:** SIGHT 4a
5 a : spiritual illumination **b :** INNER LIGHT **c :** ENLIGHTENMENT **d :** TRUTH
6 a : public knowledge ⟨facts brought to *light*⟩ **b :** a particular aspect or appearance presented to view ⟨saw the matter in a different *light*⟩
7 : a particular illumination
8 : something that enlightens or informs ⟨shed some *light* on the problem⟩
9 : a medium (as a window) through which light is admitted
10 *plural* **:** a set of principles, standards, or opinions ⟨worship according to one's *lights* —Adrienne Koch⟩
11 : a noteworthy person in a particular place or field
12 : a particular expression of the eye
13 a : LIGHTHOUSE, BEACON **b :** TRAFFIC LIGHT
14 : the representation of light in art
15 : a flame for lighting something
— in the light of 1 : from the point of view of **2** *or* **in light of :** in view of

[2]light *adjective* (before 12th century)
1 : having light **:** BRIGHT ⟨a *light* airy room⟩
2 a : not dark, intense, or swarthy in color or coloring **:** PALE **b** *of colors* **:** medium in saturation and high in lightness ⟨*light* blue⟩
3 *of coffee* **:** served with extra milk or cream

[3]light *verb* **lit** \ˈlit\ *or* **light·ed; light·ing** (before 12th century)
intransitive verb
1 : to become light **:** BRIGHTEN — usually used with *up* ⟨her face *lit* up⟩
2 : to take fire
3 : to ignite something (as a cigarette) — often used with *up*
transitive verb
1 : to set fire to
2 a : to conduct with a light **:** GUIDE **b :** ILLUMINATE ⟨rockets *light* up the sky⟩ **c :** ANIMATE, BRIGHTEN ⟨a smile *lit* up her face⟩

[4]light *adjective* [Middle English, from Old English *lēoht;* akin to Old High German *līhti* light, Latin *levis,* Greek *elachys* small] (before 12th century)
1 a : having little weight **:** not heavy **b :** designed to carry a comparatively small load ⟨a *light* truck⟩ **c :** having relatively little weight in proportion to bulk ⟨aluminum is a *light* metal⟩ **d :** containing less than the legal, standard, or usual weight ⟨a *light* coin⟩
2 a : of little importance **:** TRIVIAL **b :** not abundant ⟨*light* rain⟩ ⟨a *light* lunch⟩
3 a : easily disturbed ⟨a *light* sleeper⟩ **b :** exerting a minimum of force or pressure **:** GENTLE ⟨a *light* touch⟩ **c :** resulting from a very slight pressure **:** FAINT ⟨*light* print⟩
4 a : easily endurable ⟨a *light* illness⟩ **b :** requiring little effort ⟨*light* work⟩
5 : capable of moving swiftly or nimbly ⟨*light* on his feet⟩
6 a : FRIVOLOUS 1a ⟨*light* conduct⟩ **b :** lacking in stability **:** CHANGEABLE ⟨*light* opinions⟩ **c :** sexually promiscuous
7 : free from care **:** CHEERFUL
8 : less powerful but usually more mobile than usual for its kind ⟨*light* cavalry⟩
9 a : made with a lower calorie content or with less of some ingredient (as salt, fat, or alcohol) than usual ⟨*light* beer⟩ ⟨*light* salad dressing⟩ **b :** having a relatively mild flavor
10 a : easily digested ⟨a *light* soup⟩ **b :** well leavened ⟨a *light* crust⟩
11 : coarse and sandy or easily pulverized ⟨*light* soil⟩
12 : DIZZY, GIDDY ⟨felt *light* in the head⟩
13 : intended chiefly to entertain ⟨*light* verse⟩ ⟨*light* comedy⟩
14 a : carrying little or no cargo ⟨the ship returned *light*⟩ **b :** producing goods for direct consumption by the consumer ⟨*light* industry⟩
15 : not bearing a stress or accent ⟨a *light* syllable⟩
16 : having a clear soft quality ⟨a *light* voice⟩
17 : being in debt to the pot in a poker game ⟨three chips *light*⟩
18 : SHORT 5d ⟨*light* on experience⟩
synonym see EASY
— light·ish \ˈlī-tish\ *adjective*

[5]light *adverb* (before 12th century)
1 : LIGHTLY
2 : with little baggage ⟨travel *light*⟩

[6]light *intransitive verb* **lit** \ˈlit\ *or* **light·ed; light·ing** [Middle English, from Old English *līhtan;* akin to Old English *lēoht* light in weight] (before 12th century)
1 : DISMOUNT
2 : SETTLE, ALIGHT ⟨a bird *lit* on the lawn⟩
3 : to fall unexpectedly — usually used with *on* or *upon*
4 : to arrive by chance **:** HAPPEN — usually used with *on* or *upon* ⟨*lit* upon a solution⟩
— light into : to attack forcefully ⟨I *lit into* that food until I'd finished off the heel of the loaf —Helen Eustis⟩

light adaptation *noun* (1900)
: the process including contraction of the pupil and decrease in visual purple by which the eye adapts to conditions of increased illumination

light–adapt·ed \ˈlīt-ə-ˌdap-təd\ *adjective* (1900)
: adjusted for vision in bright light **:** having undergone light adaptation

light air *noun* (circa 1881)
: wind having a speed of 1 to 3 miles (1.6 to 4.8 kilometers) per hour — see BEAUFORT SCALE table

light bread \ˈlīt-ˌbred\ *noun* [[2]light] (1821)
chiefly Southern & Midland **:** bread in loaves made from white flour leavened with yeast

light breeze *noun* (circa 1881)
: wind having a speed of 4 to 7 miles (6.4 to 11 kilometers) per hour — see BEAUFORT SCALE table

light·bulb \ˈlīt-ˌbəlb\ *noun* (1884)
: INCANDESCENT LAMP

light chain *noun* (1964)
: either of the two smaller of the four polypeptide chains comprising antibodies — compare HEAVY CHAIN

light curve *noun* (1890)
: a graph showing the variation in brightness of a celestial object (as a variable star) over a period of time

light–emitting diode *noun* (1970)
: LED

[1]light·en \ˈlī-tᵊn\ *verb* **light·ened; light·en·ing** \ˈlīt-niŋ, ˈlī-tᵊn-iŋ\ [Middle English *lightenen,* from *light*] (14th century)
transitive verb
1 : to make light or clear **:** ILLUMINATE
2 *archaic* **:** ENLIGHTEN
3 : to make (as a color) lighter
intransitive verb
1 a *archaic* **:** to shine brightly **b :** to grow lighter **:** BRIGHTEN
2 : to give out flashes of lightning
— light·en·er \ˈlīt-nər, ˈlī-tᵊn-ər\ *noun*

[2]lighten *verb* **light·ened; light·en·ing** \ˈlīt-niŋ, ˈlī-tᵊn-iŋ\ (14th century)
transitive verb
1 a : to relieve of a burden in whole or in part ⟨the news *lightened* his mind⟩ **b :** to reduce in weight or quantity **:** LESSEN ⟨*lighten* her duties⟩ **c :** to make less wearisome **:** ALLEVIATE ⟨*lighten* our sorrow⟩
2 : CHEER, GLADDEN
intransitive verb
1 : to become lighter or less burdensome
2 : to become more cheerful
synonym see RELIEVE
— light·en·er \ˈlīt-nər, ˈlī-tᵊn-ər\ *noun*

lighten up *intransitive verb* (circa 1968)
: to take things less seriously

[1]ligh·ter \ˈlī-tər\ *noun* [Middle English, from (assumed) Middle Dutch *lichter,* from Middle Dutch *lichten* to unload; akin to Old English *lēoht* light in weight] (14th century)
: a large usually flat-bottomed barge used especially in unloading or loading ships

[2]lighter *transitive verb* (1840)
: to convey by a lighter

[3]light·er \ˈlī-tər\ *noun* (1553)
1 : one that lights or sets a fire
2 : a device for lighting a fire; *especially* **:** a mechanical or electrical device used for lighting cigarettes, cigars, or pipes

ligh·ter·age \ˈlī-tə-rij\ *noun* (15th century)
1 : the loading, unloading, or transportation of goods by means of a lighter
2 : a price paid for lightering

lighter–than–air *adjective* (1903)
: of less weight than the air displaced

light·face \'līt-,fās\ *noun* (circa 1871)
: a typeface having comparatively light thin lines; *also* : printing in lightface
— **light·faced** \-,fāst\ *adjective*

light·fast \-,fast\ *adjective* (1950)
: resistant to light and especially to sunlight; *especially* : colorfast to light
— **light·fast·ness** \-,fas(t)-nəs\ *noun*

light–fin·gered \-,fiŋ-gərd\ *adjective* (1547)
1 : showing adroitness in stealing or a tendency to steal especially by picking pockets or shoplifting
2 : having a light and dexterous touch : NIMBLE
— **light–fin·gered·ness** *noun*

light–foot·ed \-,fu̇-təd\ *also* **light–foot** \-,fu̇t\ *adjective* (15th century)
1 : having a light and springy step
2 : moving gracefully and nimbly

light guide *noun* (1951)
: an optical fiber used especially for telecommunication

light–hand·ed \'līt-,han-dəd\ *adjective* (15th century)
: having a light or delicate touch : FACILE
— **light–hand·ed·ness** *noun*

light–head·ed \-,he-dəd\ *adjective* (1537)
1 : mentally disoriented : DIZZY
2 : lacking in maturity or seriousness : FRIVOLOUS
— **light–head·ed·ly** *adverb*
— **light–head·ed·ness** *noun*

light·heart·ed \-,här-təd\ *adjective* (15th century)
1 : free from care, anxiety, or seriousness : GAY
2 : cheerfully optimistic and hopeful : EASYGOING
— **light·heart·ed·ly** *adverb*
— **light·heart·ed·ness** *noun*

light heavyweight *noun* (1903)
: a boxer in a weight division having a maximum limit of 175 pounds for professionals and 178 pounds for amateurs — compare HEAVYWEIGHT, MIDDLEWEIGHT

light·house \'līt-,hau̇s\ *noun* (1622)
1 : a structure (as a tower) with a powerful light that gives a continuous or intermittent signal to navigators
2 : BEACON 3

light housekeeping *noun* (1904)
1 : domestic work restricted to the less laborious duties
2 : housekeeping in quarters with limited facilities for cooking

light·ing \'lī-tiŋ\ *noun* (before 12th century)
1 a : ILLUMINATION **b :** IGNITION
2 : an artificial supply of light or the apparatus providing it

light·less \'līt-ləs\ *adjective* (before 12th century)
1 : receiving no light : DARK
2 : giving no light

light·ly \'līt-lē\ *adverb* (before 12th century)
: in a light manner: as **a :** with little weight or force : GENTLY **b :** with indifference or carelessness : UNCONCERNEDLY ⟨the problem should not be passed over *lightly* —Shelly Halpern⟩ **c :** with little difficulty : EASILY **d :** GAILY, CHEERFULLY **e :** in an agile manner : NIMBLY, SWIFTLY **f :** in a small degree or amount ⟨*lightly* salted food⟩

light meter *noun* (1921)
: a small and often portable device for measuring illumination; *especially* : EXPOSURE METER

light–mind·ed \'līt-,mīn-dəd\ *adjective* (1611)
: lacking in seriousness : FRIVOLOUS
— **light–mind·ed·ly** *adverb*
— **light–mind·ed·ness** *noun*

[1]light·ness \-nəs\ *noun* (before 12th century)
1 : the quality or state of being illuminated : ILLUMINATION
2 : the attribute of object colors by which the object appears to reflect or transmit more or less of the incident light — compare BRIGHTNESS 2, HUE 2c, SATURATION 4

[2]lightness *noun* (12th century)
1 : the quality or state of being light especially in weight
2 : lack of seriousness and stability of character often accompanied by casual heedlessness
3 a : the quality or state of being nimble **b :** an ease and gaiety of style or manner
4 : a lack of weightiness or force : DELICACY

[1]light·ning \'līt-niŋ\ *noun* [Middle English, from gerund of *lightenen* to lighten] (13th century)
1 : the flashing of light produced by a discharge of atmospheric electricity; *also* : the discharge itself
2 : a sudden stroke of fortune

[2]lightning *adjective* (1640)
: having or moving with or as if with the speed and suddenness of lightning ⟨a *lightning* assault⟩

[3]lightning *intransitive verb* **light·ninged; lightning** (1903)
: to discharge a flash of lightning

lightning arrester *noun* (1860)
: a device for protecting an electrical apparatus from damage by lightning

lightning bug *noun* (1778)
: FIREFLY

lightning rod *noun* (1789)
1 : a grounded metallic rod set up on a structure (as a building) to protect it from lightning
2 : one that serves to divert attack from another or as a frequent target of criticism

light–o'–love \,līt-ə-'ləv\ *also* **light–of–love** \-əv-'\ *noun, plural* **light–o'–loves** *also* **lights–of–love** (1589)
1 : PROSTITUTE
2 : LOVER, PARAMOUR

light opera *noun* (1882)
: OPERETTA

light out *intransitive verb* [[6]light] (1866)
1 : to leave in a hurry ⟨*lit out* for home at once⟩
2 : SET OFF

light pen *noun* (1958)
: a pen-shaped device for direct interaction with a computer through a cathode ray tube display

light pipe *noun* (1950)
: an optical fiber or a solid transparent plastic rod for transmitting light lengthwise

light·plane \'līt-'plān\ *noun* (1923)
: a small and comparatively lightweight airplane; *especially* : a privately owned passenger airplane

light pollution *noun* (1971)
: artificial skylight (as from city lights) that interferes with astronomical observations

light·proof \'līt-,prüf\ *adjective* (1923)
: impenetrable by light

light quantum *noun* (1925)
: PHOTON; *especially* : one of luminous radiation

light–rail \'līt-,rā(ə)l\ *noun, often attributive* (1975)
: a means of urban railway transportation using trolley cars

light reaction *noun* (circa 1929)
: the phase of photosynthesis that requires the presence of light and that involves photophosphorylation

lights \'līts\ *noun plural* [Middle English *lightes,* from *light* light in weight] (12th century)
: the lungs especially of a slaughtered animal

light·ship \'līt-,ship\ *noun* (1837)
: a ship equipped with a brilliant light and moored at a place dangerous to navigation

light show *noun* (1966)
: a kaleidoscopic display of colored lights, slides, and film loops

[1]light·some \'līt-səm\ *adjective* (14th century)
1 : free from care : LIGHTHEARTED
2 : AIRY, NIMBLE
— **light·some·ly** *adverb*
— **light·some·ness** *noun*

[2]lightsome *adjective* (15th century)
1 : well lighted : BRIGHT
2 : giving light

lights–out \'līts-'au̇t\ *noun* (1868)
1 : a command or signal for putting out lights
2 : a prescribed bedtime for persons living under discipline

light table *noun* (circa 1948)
: a device that projects even light through a flat translucent surface over which films or tracings may be spread out and viewed

light·tight \'līt-,tīt\ *adjective* (1884)
: LIGHTPROOF

light trap *noun* (1906)
1 : a device that allows movement of a sliding part or passage of a person (as into a darkroom) but excludes light
2 : a device for collecting or destroying insects that consists of a bright light in association with a trapping or killing medium

light water *noun* (1933)
: WATER 1a — compare HEAVY WATER

[1]light·weight \'līt-,wāt\ *noun* (1773)
1 : one of less than average weight; *specifically* : a boxer in a weight division having a maximum limit of 135 pounds for professionals and 132 pounds for amateurs
2 : one of little consequence or ability ⟨shows up its author as a *lightweight* —C. J. Rolo⟩

[2]lightweight *adjective* (1809)
1 : lacking in earnestness, ability, or profundity : INCONSEQUENTIAL
2 : having less than average weight
3 : of, relating to, or characteristic of a lightweight ⟨the *lightweight* championship⟩

light·wood \'līt-,wu̇d, 'lī-təd\ *noun* (1685)
chiefly Southern : wood used for kindling; *especially* : coniferous wood abounding in pitch

light–year \'līt-,yir\ *noun* (1888)
1 : a unit of length in astronomy equal to the distance that light travels in one year in a vacuum or about 5.88 trillion miles (9.46 trillion kilometers)
2 : an extremely large measure of comparison (as of distance, time, or quality) ⟨seems like *light-years* ago⟩ ⟨has *light-years* more talent⟩ ⟨two minutes and yet *light-years* away from the crowded village —Suzanne Patterson⟩ ■

lign- *or* **ligni-** *or* **ligno-** *combining form* [Latin *lign-, ligni-,* from *lignum*]
: wood ⟨*lign*in⟩ ⟨*ligno*cellulose⟩

lig·ne·ous \'lig-nē-əs\ *adjective* [Latin *ligneus,* from *lignum* wood, probably from *legere* to gather — more at LEGEND] (1626)
: of or resembling wood

lig·ni·fy \'lig-nə-,fī\ *verb* **-fied; -fy·ing** [French *lignifier,* from Latin *lignum*] (circa 1828)
transitive verb
: to convert into wood or woody tissue

□ USAGE

light–year Although some commentators insist that *light-year* is a measure of distance only—literal or figurative—and not a measure of time, there is evidence that in general (though not scientific) prose it is fairly commonly used for time ⟨cramming a . . . *light-year's* thought into three calendar months —Flann O'Brien⟩ ⟨events and circumstances that had seemed *light-years* gone —Brent Staples⟩ ⟨it seems more like a thousand days, and a *light-year* since Father died —James Clavell⟩ ⟨a thousand *light-years* ago, I thought —Dick Francis⟩. This extension of meaning should cause no surprise: how many people think of *year* as denoting distance?

intransitive verb
: to become wood or woody
— **lig·ni·fi·ca·tion** \ˌlig-nə-fə-ˈkā-shən\ *noun*

lig·nin \ˈlig-nən\ *noun* (1822)
: an amorphous polymer related to cellulose that provides rigidity and together with cellulose forms the woody cell walls of plants and the cementing material between them

lig·nite \ˈlig-ˌnīt\ *noun* [French, from Latin *lignum*] (circa 1808)
: a usually brownish black coal intermediate between peat and bituminous coal; *especially* **:** one in which the texture of the original wood is distinct — called also *brown coal*
— **lig·nit·ic** \lig-ˈni-tik\ *adjective*

lig·no·cel·lu·lose \ˌlig-nō-ˈsel-yə-ˌlōs, -ˌlōz\ *noun* [International Scientific Vocabulary] (1900)
: any of several closely related substances constituting the essential part of woody cell walls of plants and consisting of cellulose intimately associated with lignin
— **lig·no·cel·lu·los·ic** \-ˌsel-yə-ˈlō-sik, -zik\ *adjective*

lig·no·sul·fo·nate \-ˈsəl-fə-ˌnāt\ *noun* (1908)
: any of various compounds produced from the spent sulfite liquor in the pulping of softwood in papermaking and used especially for binders and dispersing agents

lig·num vi·tae \ˌlig-nəm-ˈvī-tē\ *noun, plural* **lignum vitaes** [New Latin, literally, wood of life] (1594)
1 : the very hard heavy wood of any of several tropical American guaiacums
2 : a tree yielding lignum vitae

lig·u·la \ˈli-gyə-lə\ *noun, plural* **-lae** \-ˌlē, -ˌlī\ *also* **-las** [New Latin] (circa 1760)
1 : LIGULE
2 : the distal lobed part of the labium of an insect

lig·u·late \ˈli-gyə-lət, -ˌlāt\ *adjective* (1760)
1 : furnished with ligules, ligulae, or ligulate corollas
2 [Latin *ligula*] **:** shaped like a strap ⟨*ligulate* corolla of a ray flower⟩

lig·ule \ˈli-(ˌ)gyü(ə)l\ *noun* [New Latin *ligula,* from Latin, small tongue, strap, from *lingere* to lick — more at LICK] (circa 1847)
: a scalelike projection especially on a plant: as **a :** a thin appendage of a foliage leaf and especially of the sheath of a blade of grass **b :** a ligulate corolla of a ray floret in a composite head

lig·ure \ˈli-ˌgyu̇r, -gyər\ *noun* [Middle English *lygire,* from Late Latin *ligurius,* from Greek *ligyrion*] (13th century)
: a traditional precious stone that is probably the jacinth

lik·able *or* **like·able** \ˈlī-kə-bəl\ *adjective* (1730)
: having qualities that bring about a favorable regard **:** PLEASANT, AGREEABLE
— **lik·abil·i·ty** \ˌlī-kə-ˈbi-lə-tē\ *noun*
— **lik·able·ness** *noun*

[1]like \ˈlīk\ *verb* **liked; lik·ing** [Middle English, from Old English *līcian;* akin to Old English *gelīc* alike] (before 12th century)
transitive verb
1 *chiefly dialect* **:** to be suitable or agreeable to ⟨I like onions but they don't *like* me⟩
2 a : to feel attraction toward or take pleasure in **:** ENJOY ⟨*likes* baseball⟩ **b :** to feel toward **:** REGARD ⟨how would you *like* a change⟩
3 : to wish to have **:** WANT ⟨would *like* a drink⟩
4 : to do well in ⟨this plant *likes* dry soil⟩ ⟨my car does not *like* cold weather⟩
intransitive verb
1 *dialect* **:** APPROVE
2 : to feel inclined **:** CHOOSE, PREFER ⟨leave any time you *like*⟩

[2]like *noun* (1851)
1 : LIKING, PREFERENCE
2 : something that one likes

[3]like *adjective* [Middle English, alteration of *ilich,* from Old English *gelīc* like, alike, from *ge-,* associative prefix + *līc* body; akin to Old High German *gilīh* like, alike, Lithuanian *lygus* like — more at CO-] (13th century)
1 a : the same or nearly the same (as in appearance, character, or quantity) ⟨suits of *like* design⟩ — formerly used with *as, unto, of* ⟨it behoved him to be made *like* unto his brethren —Hebrews 2:17(Authorized Version)⟩ **b** *chiefly British* **:** closely resembling the subject or original ⟨the portrait is very *like*⟩
2 : LIKELY ⟨the importance of statistics as the one discipline *like* to give accuracy of mind —H. J. Laski⟩

[4]like *preposition* (13th century)
1 a : having the characteristics of **:** similar to ⟨his house is *like* a barn⟩ ⟨it's *like* when we were kids⟩ **b :** typical of ⟨was *like* him to do that⟩ **c :** comparable to **:** APPROXIMATING ⟨costs something *like* fifty cents⟩
2 : in the manner of **:** similarly to ⟨acts *like* a fool⟩
3 : as though there would be ⟨looks *like* rain⟩
4 : such as ⟨a subject *like* physics⟩
5 — used to form intensive or ironic phrases ⟨fought *like* hell⟩ ⟨*like* fun he did⟩ ⟨laughed *like* anything⟩

[5]like *noun* (13th century)
1 a : one that is similar **:** COUNTERPART, EQUAL ⟨have . . . never seen the *like* before —Sir Winston Churchill⟩ **b :** KIND 4a ⟨put him and his *like* to some job —J. R. R. Tolkien⟩ — used with *the* and often followed by *of* and a substantive, or with a preceding possessive
2 : one of many that are similar to each other — used chiefly in proverbial expressions ⟨*like* breeds like⟩
— **and the like :** ET CETERA
— **the likes of 1 :** such people as **:** such things as ⟨reads *the likes of* Austen and Browning⟩ **2 :** such a one as and perhaps others similar to — often used with a singular object and usually with disparaging overtones ⟨have no use for *the likes of* you⟩

[6]like *adverb* (14th century)
1 *archaic* **:** EQUALLY
2 : LIKELY, PROBABLY ⟨you'll try it, some day, *like* enough —Mark Twain⟩
3 a : to some extent **:** RATHER, ALTOGETHER ⟨saunter over nonchalantly *like* —Walter Karig⟩ **b** — used interjectionally in informal speech often to emphasize a word or phrase (as in "He was, like, gorgeous") or for an apologetic, vague, or unassertive effect (as in "I need to, like, borrow some money")
4 : NEARLY **:** APPROXIMATELY ⟨the actual interest is more *like* 18 percent⟩ — used interjectionally in informal speech with expressions of measurement ⟨it was, *like,* five feet long⟩ ⟨goes there every day, *like*⟩
— **as like as not** *or* **like as not :** PROBABLY

[7]like *conjunction* (14th century)
1 a : AS IF ⟨middle-aged men who looked *like* they might be out for their one night of the year —Norman Mailer⟩ **b** — used in intensive phrases ⟨drove *like* mad⟩ ⟨hurts *like* crazy⟩
2 : in the same way that **:** AS ⟨they raven down scenery *like* children do sweetmeats —John Keats⟩
3 a : in the way or manner that ⟨the violin sounds *like* an old masterpiece should⟩ ⟨did it *like* you told me⟩ **b** — used interjectionally in informal speech often with the verb *be* to introduce a quotation, paraphrase, or thought expressed by or imputed to the subject of the verb, or with *it's* to report a generally held opinion ⟨so I'm *like,* "Give me a break"⟩ ⟨it's *like,* "Who cares what he thinks?"⟩
4 : such as ⟨a bag *like* a doctor carries⟩ ⟨when your car has trouble — *like* when it won't start⟩ — used interjectionally in informal speech ⟨often stays up late, until *like* three in the morning⟩ ■

[8]like *or* **liked** \ˈlīkt\ *verbal auxiliary* (15th century) *chiefly dialect* **:** came near **:** was near ⟨so loud I *like* to fell out of bed —Helen Eustis⟩

-like *adjective combining form*
: resembling or characteristic of ⟨bell-*like*⟩ ⟨lady*like*⟩

like·li·hood \ˈlī-klē-ˌhu̇d\ *noun* (14th century)
: PROBABILITY ⟨a strong *likelihood* that he is correct —T. D. Anderson⟩

[1]like·ly \ˈlī-klē\ *adjective* **like·li·er; -est** [Middle English, from Old Norse *glīkligr,* from *glīkr* like; akin to Old English *gelīc*] (14th century)
1 : having a high probability of occurring or being true **:** very probable ⟨rain is *likely* today⟩
2 : apparently qualified **:** SUITABLE ⟨a *likely* place⟩
3 : RELIABLE, CREDIBLE ⟨a *likely* enough story⟩
4 : PROMISING ⟨a *likely* candidate⟩
5 : ATTRACTIVE ⟨a *likely* child⟩

[2]likely *adverb* (14th century)
: in all probability **:** PROBABLY ⟨those who seek power will most *likely* wind up exercising it —Halton Arp⟩

like–mind·ed \ˈlīk-ˈmīn-dəd\ *adjective* (1526)
: having a like disposition or purpose **:** of the same mind or habit of thought
— **like–mind·ed·ly** *adverb*
— **like–mind·ed·ness** *noun*

lik·en \ˈlī-kən\ *transitive verb* **lik·ened; lik·en·ing** \ˈlī-kə-niŋ, ˈlīk-niŋ\ (14th century)
: COMPARE

like·ness \ˈlīk-nəs\ *noun* (before 12th century)
1 : COPY, PORTRAIT
2 : APPEARANCE, SEMBLANCE
3 : the quality or state of being like **:** RESEMBLANCE ☆

like·wise \ˈlīk-ˌwīz\ *adverb* (15th century)
1 : in like manner **:** SIMILARLY ⟨go and do *likewise*⟩

☆ SYNONYMS
Likeness, similarity, resemblance, similitude, analogy mean agreement or correspondence in details. LIKENESS implies a closer correspondence than SIMILARITY which often implies that things are merely somewhat alike ⟨a remarkable *likeness* to his late father⟩ ⟨some *similarity* between the two cases⟩. RESEMBLANCE implies similarity chiefly in appearance or external qualities ⟨statements that bear little *resemblance* to the truth⟩. SIMILITUDE applies chiefly to correspondence between abstractions ⟨two schools of social thought showing points of *similitude*⟩. ANALOGY implies likeness or parallelism in relations rather than in appearance or qualities ⟨pointed out *analogies* to past wars⟩.

□ USAGE
like *Like* has been used as a conjunction since the 14th century. In the 14th, 15th, and 16th centuries it was used in serious literature, but not often; in the 17th and 18th centuries it grew more frequent but less literary. It became markedly more frequent in literary use again in the 19th century. By mid-century it was coming under critical fire, but not from grammarians, oddly enough, who were wrangling over whether it could be called a preposition or not. There is no doubt that, after 600 years of use, conjunctive *like* is firmly established. It has been used by many prestigious literary figures of the past, though perhaps not in their most elevated works; in modern use it may be found in literature, journalism, and scholarly writing. While the present objection to it is perhaps more heated than rational, someone writing in a formal prose style may well prefer to use *as, as if, such as,* or an entirely different construction instead.

2 : in addition
3 : similarly so with me ⟨answered "*likewise*" to "Pleased to meet you"⟩

lik·ing \'lī-kiŋ\ *noun* (14th century)
: favorable regard **:** FONDNESS, TASTE ⟨had a greater *liking* for law —E. M. Coulter⟩ ⟨took a *liking* to the newcomer⟩

li·ku·ta \li-'kü-tə\ *noun, plural* **ma·ku·ta** \mä-\ [of Bantu origin] (1967)
— see *zaire* at MONEY table

li·lac \'lī-ˌläk, -ˌlak, -lək\ *noun* [obsolete French (now *lilas*), from Arabic *līlak*, from Persian *nīlak* bluish, from *nīl* blue, from Sanskrit *nīla* dark blue] (1625)
1 a : a European shrub (*Syringa vulgaris*) of the olive family that is often an escape in North America and has cordate ovate leaves and large panicles of fragrant pink-purple or white flowers **b :** a tree or shrub congeneric with the lilac
2 : a variable color averaging a moderate purple

lil·an·ge·ni \ˌli-lən-'ge-nē\ *noun, plural* **em·a·lan·ge·ni** \ˌe-mə-lən-'ge-nē\ [Swazi] (circa 1976)
— see MONEY table

lil·ied \'li-lēd\ *adjective* (1614)
1 *archaic* **:** resembling a lily in fairness
2 : full of or covered with lilies

Lil·ith \'li-ləth\ *noun* [Late Hebrew *līlīth*, from Hebrew, a female demon]
1 : a female figure who in rabbinic legend is Adam's first wife, is supplanted by Eve, and becomes an evil spirit
2 : a famous witch in medieval demonology

Lil·li·put \'li-li-(ˌ)pət\ *noun*
: an island in Swift's *Gulliver's Travels* where the inhabitants are six inches tall

lil·li·pu·tian \ˌli-lə-'pyü-shən\ *adjective, often capitalized* (1726)
1 : of, relating to, or characteristic of the Lilliputians or the island of Lilliput
2 a : SMALL, MINIATURE **b :** PETTY

Lilliputian *noun*
1 : an inhabitant of Lilliput
2 *often not capitalized* **:** one resembling a Lilliputian; *especially* **:** an undersized individual

[1]lilt \'lilt\ *noun* (circa 1680)
1 : a spirited and usually cheerful song or tune
2 : a rhythmical swing, flow, or cadence
3 : a springy buoyant movement

[2]lilt *verb* [Middle English *lulten* to sound an alarm] (1722)
transitive verb
: to sing or play in a lively cheerful manner
intransitive verb
1 : to sing or speak rhythmically and with fluctuating pitch
2 : to move in a lively springy manner

lilt·ing \'lil-tiŋ\ *adjective* (1800)
1 : characterized by a rhythmical swing or cadence ⟨a *lilting* stride⟩
2 : CHEERFUL, BUOYANT ⟨a *lilting* comedy⟩
— **lilt·ing·ly** \-lē\ *adverb*
— **lilt·ing·ness** *noun*

[1]lily \'li-lē\ *noun, plural* **lil·ies** [Middle English *lilie*, from Old English, from Latin *lilium*] (before 12th century)
1 : any of a genus (*Lilium* of the family Liliaceae, the lily family) of erect perennial leafy-stemmed bulbous herbs that are native to the northern hemisphere and are widely cultivated for their showy flowers; *broadly* **:** any of various plants of the lily family or of the related amaryllis or iris families
2 : any of various plants with showy flowers: as **a :** a scarlet anemone (*Anemone coronaria*) of the Mediterranean region **b :** WATER LILY **c :** CALLA LILY
3 : FLEUR-DE-LIS 2

[2]lily *adjective* (circa 1553)
: resembling a lily in fairness, purity, or fragility ⟨my lady's *lily* hand —John Keats⟩

lily–liv·ered \'li-lē-'li-vərd\ *adjective* (1605)
: lacking courage **:** COWARDLY ◇

lily of the valley (1563)
: a low perennial herb (*Convallaria majalis*) of the lily family that has usually two large oblong lanceolate leaves and a raceme of fragrant nodding bell-shaped white flowers

lily pad *noun* (1814)
: a floating leaf of a water lily

[1]lily–white \'li-lē-'hwīt, -'wīt\ *adjective* (14th century)
1 : white as a lily
2 : characterized by or favoring the exclusion of blacks especially from politics
3 : IRREPROACHABLE, PURE

[2]lily–white *noun* (circa 1903)
: a member of a lily-white political organization

Li·ma \'lē-mə\ (circa 1952)
— a communications code word for the letter *l*

li·ma bean \'lī-mə-\ *noun* [*Lima*, Peru] (1756)
1 a : a bushy or tall-growing tropical American bean (*Phaseolus limensis*) that is widely cultivated for its flat edible usually pale green or whitish seeds **b :** SIEVA BEAN
2 : the seed of a lima bean

li·ma·çon \ˌlē-mə-'sōⁿ\ *noun* [French, literally, snail, from Old French, diminutive of *limaz* slug, snail, from Latin *limax*; akin to Russian *slimak* snail and probably to Old English *līm* birdlime — more at LIME] (1874)
: a curve that is the locus of a variable point on a line as the line revolves about a fixed point of intersection with a circle while the variable point is always a fixed distance along the line from its second and variable point of intersection with the circle

[1]limb \'lim\ *noun* [Middle English *lim*, from Old English; akin to Old Norse *limr* limb and perhaps to Old English *lith* limb] (before 12th century)
1 a : one of the projecting paired appendages (as wings) of an animal body used especially for movement and grasping but sometimes modified into sensory or sexual organs **b :** a leg or arm of a human being
2 : a large primary branch of a tree
3 : an active member or agent
4 : EXTENSION, BRANCH
5 : a mischievous child
— **limb·less** \'lim-ləs\ *adjective*
— **limby** \'li-mē\ *adjective*
— **out on a limb :** in an exposed or dangerous position with little chance of retreat

[2]limb *transitive verb* (1674)
1 : DISMEMBER
2 : to cut off the limbs of (a felled tree)

[3]limb *noun* [Latin *limbus* border] (circa 1677)
1 : the outer edge of the apparent disk of a celestial body
2 : the expanded portion of an organ or structure; *especially* **:** the spreading upper portion of a gamopetalous corolla as distinguished from the lower tubular portion

lim·ba \'lim-bə\ *noun* [origin unknown] (1943)
1 : a tall whitish-trunked West African tree (*Terminalia superba*) with straight-grained wood
2 : the wood of a limba

lim·beck \'lim-ˌbek\ *noun* [Middle English *lembike*, from Medieval Latin *alembicum*] (14th century)
: ALEMBIC

limbed \'limd\ *adjective* (14th century)
: having limbs especially of a specified kind or number — usually used in combination ⟨strong-*limbed*⟩

[1]lim·ber \'lim-bər\ *noun* [Middle English *lymour*] (15th century)
: a two-wheeled vehicle to which a gun or caisson may be attached

[2]limber *adjective* [origin unknown] (1565)
1 : capable of being shaped **:** FLEXIBLE
2 : having a supple and resilient quality (as of mind or body) **:** AGILE, NIMBLE
— **lim·ber·ly** *adverb*
— **lim·ber·ness** *noun*

[3]limber *verb* **lim·bered; lim·ber·ing** \-b(ə-)riŋ\ (1748)
transitive verb
: to cause to become limber ⟨*limber* up his fingers⟩
intransitive verb
: to become limber ⟨*limber* up by running⟩

lim·bic \'lim-bik\ *adjective* [New Latin *limbicus* of a border or margin, from Latin *limbus*] (1882)
: of, relating to, or being the limbic system of the brain

limbic system *noun* (1952)
: a group of subcortical structures (as the hypothalamus, the hippocampus, and the amygdala) of the brain that are concerned especially with emotion and motivation

[1]lim·bo \'lim-(ˌ)bō\ *noun, plural* **limbos** [Middle English, from Medieval Latin, ablative of *limbus* limbo, from Latin, border] (14th century)
1 *often capitalized* **:** an abode of souls that are according to Roman Catholic theology barred from heaven because of not having received Christian baptism
2 a : a place or state of restraint or confinement **b :** a place or state of neglect or oblivion ⟨proposals kept in *limbo*⟩ **c :** an intermediate or transitional place or state **d :** a state of uncertainty

[2]limbo *noun, plural* **limbos** [English of Trinidad & Barbados; akin to Jamaican English *limba* to bend, from English [3]*limber*] (circa 1950)
: a West Indian acrobatic dance originally for men that involves bending over backwards and passing under a horizontal pole lowered slightly for each successive pass

Lim·burg·er \'lim-ˌbər-gər\ *noun* [Flemish, one from Limburg, from *Limburg*, Belgium] (1817)
: a pungent semisoft surface-ripened cheese

lim·bus \'lim bəs\ *noun* [Latin, border] (1877)
: the marginal region of the cornea of the eye by which it is continuous with the sclera

[1]lime \'līm\ *noun* [Middle English, from Old English *līm*; akin to Old High German *līm* birdlime, Latin *limus* mud, slime, and perhaps to Latin *linere* to smear] (before 12th century)
1 : BIRDLIME
2 a : a caustic highly infusible solid that consists of calcium oxide often together with magnesium oxide, that is obtained by calcining forms of calcium carbonate (as shells or limestone), and that is used in building (as in mortar and plaster) and in agriculture — called also *quicklime* **b :** a dry white powder consisting essentially of calcium hydroxide that is made by treating quicklime with water **c :** CALCIUM ⟨carbonate of *lime*⟩

◇ WORD HISTORY

lily–livered The basis of the word *lily-livered* lies in the former belief that a person's health and temperament are the product of a balance or imbalance of four bodily fluids, or humors: blood, phlegm, black bile and yellow bile. It was thought that a deficiency of yellow bile or choler, the humor that governed anger, spirit, and courage, would leave a person's liver colorless or white. Someone with this deficiency, and so white-livered, would be spiritless and a coward. *Lily-livered* and *white-livered* have been used synonymously since the 16th century; but *lily-livered*, probably because of its alliteration, is now the more common expression. See in addition CHOLERA, HUMOR.

²lime *transitive verb* **limed; lim·ing** (13th century)
1 : to smear with a sticky substance (as birdlime)
2 : to entangle with or as if with birdlime
3 : to treat or cover with lime ⟨*lime* the lawn in the spring⟩

³lime *adjective* (15th century)
: of, relating to, or containing lime or limestone

⁴lime *noun* [alteration of Middle English *lind,* from Old English; akin to Old High German *linta* linden] (1625)
: LINDEN 1a

⁵lime *noun* [French, from Spanish *lima,* from Arabic *līma, līm*] (1638)
1 : the small globose yellowish green fruit of a lime with an acid juicy pulp used as a flavoring agent and as a source of vitamin C
2 : a spiny tropical citrus tree (*Citrus aurantifolia*) with elliptical oblong narrowly winged leaves

lime·ade \ˌlīm-ˈād, ˈlī-ˌmād\ *noun* (1892)
: a beverage of sweetened lime juice mixed with plain or carbonated water

lime glass *noun* (circa 1909)
: glass containing a substantial proportion of lime

lime–juic·er \ˈlīm-ˌjü-sər\ *noun* [from the use of lime juice on British ships as a beverage to prevent scurvy] (1859)
1 *slang* **:** ENGLISHMAN
2 *slang* **a :** a British ship **b :** a British sailor

lime·kiln \-ˌkil, -ˌkiln\ *noun* (13th century)
: a kiln or furnace for reducing limestone or shells to lime by burning

¹lime·light \-ˌlīt\ *noun* (1826)
1 a : a stage lighting instrument producing illumination by means of an oxyhydrogen flame directed on a cylinder of lime and usually equipped with a lens to concentrate the light in a beam **b :** the white light produced by such an instrument **c** *British* **:** SPOTLIGHT
2 : the center of public attention

²limelight *transitive verb* (1909)
: to center attention on **:** SPOTLIGHT

li·men \ˈlī-mən\ *noun* [Latin *limin-, limen* transverse beam in a door frame, threshold; probably akin to Latin *limus* transverse] (1895)
: THRESHOLD 3a

lim·er·ick \ˈli-mə-rik, ˈlim-rik\ *noun* [*Limerick,* Ireland] (1896)
: a light or humorous verse form of 5 chiefly anapestic verses of which lines 1, 2, and 5 are of 3 feet and lines 3 and 4 are of 2 feet with a rhyme scheme of *aabba*

lime·stone \ˈlīm-ˌstōn\ *noun* (14th century)
: a rock that is formed chiefly by accumulation of organic remains (as shells or coral), consists mainly of calcium carbonate, is extensively used in building, and yields lime when burned

lime–twig \ˈlīm-ˌtwig\ *noun* (15th century)
1 : a twig covered with birdlime to catch birds
2 : SNARE

lime·wa·ter \-ˌwȯ-tər, -ˌwä-\ *noun* (circa 1500)
: an alkaline water solution of calcium hydroxide used as an antacid

¹lim·ey \ˈlī-mē\ *noun, plural* **limeys** *often capitalized* [*lime*-juicer + ⁴-y] (1918)
1 *slang* **:** a British sailor
2 *slang* **:** ENGLISHMAN

²limey *variant of* LIMY

lim·i·nal \ˈli-mə-nᵊl\ *adjective* [Latin *limin-, limen* threshold] (1884)
1 : of or relating to a sensory threshold
2 : barely perceptible

¹lim·it \ˈli-mət\ *noun* [Middle English, from Middle French *limite,* from Latin *limit-, limes* boundary] (14th century)
1 a : something that bounds, restrains, or confines **b :** the utmost extent
2 a : a geographical or political boundary **b** *plural* **:** the place enclosed within a boundary **:** BOUNDS
3 : LIMITATION
4 : a determining feature or differentia in logic
5 : a prescribed maximum or minimum amount, quantity, or number: as **a :** the maximum quantity of game or fish that may be taken legally in a specified period **b :** a maximum established for a gambling bet, raise, or payoff
6 a : a number whose numerical difference from a mathematical function is arbitrarily small for all values of the independent variables that are sufficiently close to but not equal to given prescribed numbers or that are sufficiently large positively or negatively **b :** a number that for an infinite sequence of numbers is such that ultimately each of the remaining terms of the sequence differs from this number by less than any given positive amount
7 : something that is exasperating or intolerable
— lim·it·less \-ləs\ *adjective*
— lim·it·less·ly *adverb*
— lim·it·less·ness *noun*

²limit *transitive verb* (14th century)
1 : to assign certain limits to **:** PRESCRIBE ⟨reserved the right to *limit* use of the land⟩
2 a : to restrict the bounds or limits of ⟨the specialist can no longer *limit* himself to his specialty⟩ **b :** to curtail or reduce in quantity or extent ⟨we must *limit* the power of aggressors⟩ ☆
— lim·it·able \-mə-tə-bəl\ *adjective*
— lim·it·er *noun*

lim·i·tary \ˈli-mə-ˌter-ē\ *adjective* (1620)
1 *archaic* **:** subject to limits
2 a *archaic* **:** of or relating to a boundary **b :** LIMITING, ENCLOSING

lim·i·ta·tion \ˌli-mə-ˈtā-shən\ *noun* (14th century)
1 : an act or instance of limiting
2 : the quality or state of being limited
3 : something that limits **:** RESTRAINT
4 : a certain period limited by statute after which actions, suits, or prosecutions cannot be brought in the courts
— lim·i·ta·tion·al \-shnəl, -shə-nᵊl\ *adjective*

lim·i·ta·tive \ˈli-mə-ˌtā-tiv\ *adjective* (1530)
: LIMITING, RESTRICTIVE

lim·it·ed *adjective* (1610)
1 a : confined within limits **:** RESTRICTED ⟨*limited* success⟩ **b** *of a train* **:** offering faster service especially by making a limited number of stops
2 : characterized by enforceable limitations prescribed (as by a constitution) upon the scope or exercise of powers ⟨a *limited* monarchy⟩
3 : lacking breadth and originality ⟨a bit *limited*; a bit thick in the head —Virginia Woolf⟩
— lim·it·ed·ly *adverb*
— lim·it·ed·ness *noun*

limited–access *adjective* (1944)
of a road **:** having access restricted to a relatively small number of points

limited edition *noun* (1903)
: an issue of something collectible (as books, prints, or medals) that is advertised to be limited to a relatively small number of copies

limited liability *noun* (1855)
: liability (as of a stockholder or shipowner) limited by statute or treaty

limited partner *noun* (1907)
: a partner in a venture who has no management authority and whose liability is restricted to the amount of his investment — compare GENERAL PARTNER
— limited partnership *noun*

limited war *noun* (1939)
: a war whose objective is less than the total defeat of the enemy

lim·it·ing *adjective* (1849)
1 a : functioning as a limit **:** RESTRICTIVE ⟨*limiting* value⟩ **b :** being an environmental factor (as a nutrient) that limits the population size of an organism
2 : serving to specify the application of the modified noun ⟨*this* in "this book" is a *limiting* word⟩
— lim·it·ing·ly *adverb*

limit point *noun* (1905)
: a point that is related to a set of points in such a way that every neighborhood of the point no matter how small contains another point belonging to the set — called also *point of accumulation*

lim·i·trophe \ˈli-mə-ˌtrōf, -ˌtrȯf\ *adjective* [French, from Late Latin *limitrophus* bordering upon, literally, providing subsistence for frontier troops, irregular from Latin *limit-, limes* boundary + Greek *trophos* feeder, from *trephein* to nourish] (1763)
: situated on a border or frontier **:** ADJACENT

lim·mer \ˈli-mər\ *noun* [Middle English (Scots)] (15th century)
1 *chiefly Scottish* **:** SCOUNDREL
2 *chiefly Scottish* **:** PROSTITUTE

limn \ˈlim\ *transitive verb* **limned; limn·ing** \ˈli-miŋ, ˈlim-niŋ\ [Middle English *luminen, limnen* to illuminate (a manuscript), from Middle French *enluminer,* from Latin *illuminare* to illuminate] (1592)
1 : to draw or paint on a surface
2 : to outline in clear sharp detail **:** DELINEATE
3 : DESCRIBE
— limn·er \ˈli-mər, ˈlim-nər\ *noun*

lim·net·ic \lim-ˈne-tik\ *adjective* [International Scientific Vocabulary, from Greek *limnē* pool, marshy lake; perhaps akin to Latin *limus* mud — more at LIME] (1899)
: of, relating to, or inhabiting the open water of a body of fresh water ⟨*limnetic* environment⟩

lim·nol·o·gy \lim-ˈnä-lə-jē\ *noun* [Greek *limnē* + International Scientific Vocabulary *-logy*] (circa 1888)
: the scientific study of bodies of fresh water (as lakes)
— lim·no·log·i·cal \ˌlim-nə-ˈlä-ji-kəl\ *also* **lim·no·log·ic** \-ˈlä-jik\ *adjective*
— lim·nol·o·gist \lim-ˈnä-lə-jist\ *noun*

limo \ˈli-(ˌ)mō\ *noun, plural* **lim·os** (1968)
: LIMOUSINE

Li·moges \li-ˈmōzh\ *noun* [*Limoges,* France] (1844)
: enamelware or porcelain made at Limoges

lim·o·nene \ˈli-mə-ˌnēn\ *noun* [International Scientific Vocabulary, from French *limon* lemon, from Middle French] (1845)
: a widely distributed terpene hydrocarbon $C_{10}H_{16}$ that occurs in essential oils (as of oranges or lemons) and has a lemon odor

li·mo·nite \ˈlī-mə-ˌnīt\ *noun* [German *Limonit,* from Greek *leimōn* wet meadow; akin to Greek *limnē* pool] (1823)
: a native hydrous ferric oxide of variable composition that is a major ore of iron
— li·mo·nit·ic \ˌlī-mə-ˈni-tik\ *adjective*

☆ SYNONYMS
Limit, restrict, circumscribe, confine mean to set bounds for. LIMIT implies setting a point or line (as in time, space, speed, or degree) beyond which something cannot or is not permitted to go ⟨visits are *limited* to 30 minutes⟩. RESTRICT suggests a narrowing or tightening or restraining within or as if within an encircling boundary ⟨laws intended to *restrict* the freedom of the press⟩. CIRCUMSCRIBE stresses a restriction on all sides and by clearly defined boundaries ⟨the work of the investigating committee was carefully *circumscribed*⟩. CONFINE suggests severe restraint and a resulting cramping, fettering, or hampering ⟨our freedom of choice was *confined* by finances⟩.

Lim·ou·sin \li-ˈmü-sin, ˈli-mə-ˌzēn, lē-mü-ˈzaⁿ\ *noun* [*Limousin,* France] (1920)
: any of a French breed of medium-sized yellow-red cattle bred especially for meat

lim·ou·sine \ˈli-mə-ˌzēn, ˌli-mə-ˈ\ *noun* [French, literally, cloak, from *Limousin,* France] (1902)
1 : a large luxurious often chauffeur-driven sedan that sometimes has a glass partition separating the driver's seat from the passenger compartment
2 : a large vehicle for transporting passengers to and from an airport

limousine liberal *noun* (1969)
: a wealthy political liberal

¹limp \ˈlimp\ *intransitive verb* [probably from Middle English *lympen* to fall short; akin to Old English *limpan* to happen, *lemp*healt lame] (circa 1570)
1 a : to walk lamely; *especially* **:** to walk favoring one leg **b :** to go unsteadily **:** FALTER
2 : to proceed slowly or with difficulty ⟨the ship *limped* back to port⟩
— limp·er *noun*

²limp *noun* (1818)
: a limping movement or gait

³limp *adjective* [akin to ¹*limp*] (circa 1706)
1 a : lacking firm texture, substance, or structure ⟨*limp* curtains⟩ ⟨her hair hung *limp* about her shoulders⟩ **b :** not stiff or rigid ⟨a book in a *limp* binding⟩
2 a : WEARY, EXHAUSTED ⟨*limp* with fatigue⟩ **b :** lacking in strength, vigor, or firmness **:** SPIRITLESS
— limp·ly *adverb*
— limp·ness *noun*

lim·pa \ˈlim-pə\ *noun* [Swedish] (1948)
: rye bread made with molasses or brown sugar

lim·pet \ˈlim-pət\ *noun* [Middle English *lempet,* from Old English *lempedu,* from Medieval Latin *lampreda* lamprey] (before 12th century)
1 : a marine gastropod mollusk (especially families Acmaeidae and Patellidae) that has a low conical shell broadly open beneath, browses over rocks or timbers in the littoral area, and clings very tightly when disturbed

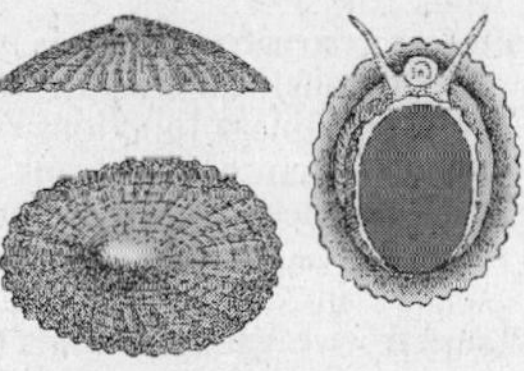
limpet 1

2 : one that clings tenaciously to someone or something
3 : an explosive device designed to cling magnetically to a metallic surface (as the hull of a ship)

lim·pid \ˈlim-pəd\ *adjective* [French or Latin; French *limpide,* from Latin *limpidus;* perhaps from *lympha* water — more at LYMPH] (1613)
1 a : marked by transparency **:** PELLUCID ⟨*limpid* streams⟩ **b :** clear and simple in style ⟨*limpid* prose⟩
2 : absolutely serene and untroubled
synonym see CLEAR
— lim·pid·i·ty \lim-ˈpi-də-tē\ *noun*
— lim·pid·ly \ˈlim-pəd-lē\ *adverb*
— lim·pid·ness *noun*

limp·kin \ˈlim(p)-kən\ *noun* [perhaps from ¹*limp*] (1871)
: a large brown wading bird (*Aramus guarauna*) of southern Georgia, Florida, and Central and South America that resembles a bittern but has a longer slightly curved bill, longer neck and legs, and white stripes on head and neck

limp–wrist·ed \ˈlimp-ˌris-təd\ *adjective* (circa 1960)
1 : EFFEMINATE
2 : WEAK

lim·u·lus \ˈlim-yə-ləs\ *noun, plural* **-li** \-ˌlī, -ˌlē\ [New Latin, genus name, from Latin *limus* oblique, transverse — more at LIMEN] (1837)
: HORSESHOE CRAB

limy \ˈlī-mē\ *adjective* **lim·i·er; -est** (circa 1552)
1 : smeared with or consisting of lime **:** VISCOUS
2 : containing lime or limestone
3 : resembling or having the qualities of lime

lin·ac \ˈli-ˌnak\ *noun* (1950)
: LINEAR ACCELERATOR

lin·age \ˈlī-nij\ *noun* (1884)
: the number of lines of printed or written matter

lin·al·o·ol \lə-ˈna-lə-ˌwȯl, lī-, -ˌwōl\ *noun* [International Scientific Vocabulary, from Mexican Spanish *lináloe,* tree yielding perfume, from Medieval Latin *lignum aloes,* literally, wood of the aloe] (1891)
: a fragrant liquid alcohol $C_{10}H_{18}O$ that occurs both free and in the form of esters in many essential oils and is used in perfumes, soaps, and flavoring materials

linch·pin \ˈlinch-ˌpin\ *noun* [Middle English *lynspin,* from *lyns* linchpin (from Old English *lynis*) + *pin;* akin to Middle High German *luns* linchpin] (13th century)
1 : a locking pin inserted crosswise (as through the end of an axle or shaft)
2 : one that serves to hold together the elements of a complex ⟨the *linchpin* in the defense's case⟩

Lin·coln \ˈliŋ-kən\ *noun* [*Lincoln*shire, England] (1837)
: any of an English breed of long-wooled mutton-type sheep

Lin·coln·i·a·na \(ˌ)liŋ-ˌkō-nē-ˈä-nə, -ˈa-nə, -ˈā-nə\ *noun plural* (1921)
: materials relating to Abraham Lincoln

Lincoln's Birthday \ˈliŋ-kənz-\ *noun* (1898)
1 : February 12 observed as a legal holiday in many states of the U.S.
2 : the first Monday in February observed as a legal holiday by some states of the U.S.

lin·co·my·cin \ˌliŋ-kə-ˈmī-sᵊn\ *noun* [New Latin *lincol*nensis (specific epithet of *Streptomyces lincolnensis*) + English *-mycin*] (1963)
: an antibiotic obtained from an actinomycete (*Streptomyces lincolnensis*) and effective especially against gram-positive bacteria

lin·dane \ˈlin-ˌdān\ *noun* [T. van der *Linden,* 20th century Dutch chemist] (circa 1949)
: an insecticide that consists chiefly of the gamma isomer of BHC and is biodegraded very slowly

lin·den \ˈlin-dən\ *noun* [Middle English, made of linden wood, from Old English, from *lind* linden tree; probably akin to Old English *līthe* gentle — more at LITHE] (1577)
1 : any of a genus (*Tilia* of the family Tiliaceae, the linden family) of trees of temperate regions that are planted as shade trees and are distinguished by having cordate leaves and a winglike bract attached to the peduncle of the flower and fruit: as **a :** a European tree (*T. europaea*) much used for ornamental planting **b :** a tall forest tree (*T. americana*) chiefly of the central and eastern U.S. — called also *basswood, whitewood*

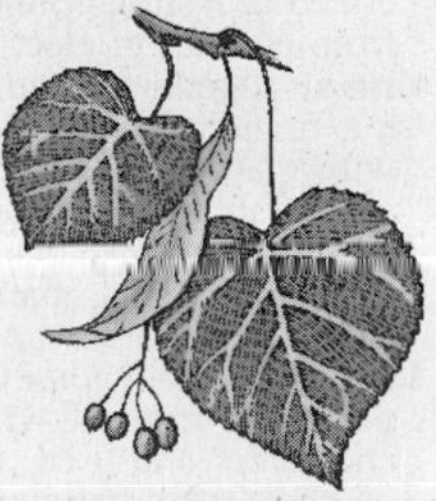
linden 1: branch with bract and fruit

2 : the light fine-grained white wood of a linden; *especially* **:** BASSWOOD 2

lin·dy \ˈlin-dē\ *noun* [probably from *Lindy,* nickname of Charles A. Lindbergh] (1931)
: a jitterbug dance originating in Harlem and later developing many local variants

¹line \ˈlīn\ *noun, often attributive* [Middle English; partly from Old French *ligne,* from Latin *linea,* from feminine of *lineus* made of flax, from *linum* flax; partly from Old English *līne;* akin to Old English *līn* flax — more at LINEN] (before 12th century)
1 a : THREAD, STRING, CORD, ROPE: as (1) **:** a comparatively strong slender cord (2) **:** CLOTHESLINE (3) **:** a rope used on shipboard **b** (1) **:** a device for catching fish consisting of a cord with hooks and other fishing gear (2) **:** scope for activity **:** ROPE **c :** a length of material used in measuring and leveling **d :** piping for conveying a fluid (as steam) **e** (1) **:** a wire or pair of wires connecting one telegraph or telephone station with another or a whole system of such wires; *also* **:** any circuit in an electronic communication system (2) **:** a telephone connection ⟨tried to get a *line*⟩; *also* **:** an individual telephone extension ⟨a call on *line* 2⟩ (3) **:** the principal circuits of an electric power system
2 a (1) **:** a horizontal row of written or printed characters; *also* **:** a blank row in lieu of such characters (2) **:** a unit in the rhythmic structure of verse formed by the grouping of a number of the smallest units of the rhythm (as metrical feet) (3) **:** an often numbered section of a computer program containing a single command or a small number of commands **b :** a short letter **:** NOTE **c** *plural* **:** a certificate of marriage **d :** the words making up a part in a drama — usually used in plural **e :** any of the successive horizontal rows of picture elements on the screen of a cathode-ray tube (as a television screen)
3 a : something (as a ridge or seam) that is distinct, elongated, and narrow **b :** a narrow crease (as on the face) **:** WRINKLE **c :** the course or direction of something in motion **:** ROUTE **d** (1) **:** a state of agreement or conformity **:** ACCORDANCE (2) **:** a state of order, control, or obedience ⟨you're getting out of *line*⟩ **e :** a boundary of an area ⟨the state *line*⟩ **f :** the track and roadbed of a railway **g :** an amount of cocaine that is arranged in a line to be inhaled through the nose
4 a : a course of conduct, action, or thought **b :** a field of activity or interest **c :** a glib often persuasive way of talking
5 a : LIMIT, RESTRAINT **b** *archaic* **:** position in life **:** LOT
6 a (1) **:** FAMILY, LINEAGE (2) **:** a strain produced and maintained especially by selective breeding or biological culture (3) **:** a chronological series **b :** dispositions made to cover extended military positions and presenting a front to the enemy — usually used in plural **c :** a military formation in which the different elements are abreast of each other **d :** naval ships arranged in a regular order **e** (1) **:** the combatant forces of an army distinguished from the staff corps and supply services (2) **:** the force of a regular navy **f** (1) **:** officers of the navy eligible for command at sea distinguished from officers of the staff (2) **:** officers of the army belonging to a combatant branch **g :** an arrangement or placement of persons or objects of one kind in an orderly series ⟨a *line* of trees⟩ ⟨stand on *line*⟩ ⟨waiting in *line*⟩; *also* **:** the persons or objects so positioned ⟨the *line* moved slowly at the bank⟩ **h** (1) **:** a group of public conveyances plying regularly under one management over a route (2) **:** a system of transportation together with its equipment, routes, and appurtenances; *also* **:** the company owning or operating it **i :** a succession of musical notes especially considered in melodic

phrases **j** (1) **:** an arrangement of operations in manufacturing permitting sequential occurrence on various stages of production (2) **:** the personnel of an organization that are responsible for its stated objective **k** (1) **:** the 7 players including center, 2 guards, 2 tackles, and 2 ends who in offensive football play line up on or within one foot of the line of scrimmage (2) **:** the players who in defensive play line up within one yard of the line of scrimmage
7 : a narrow elongated mark drawn or projected: as **a** (1) **:** a circle of latitude or longitude on a map (2) **:** EQUATOR **b :** a mark (as on a map) recording a boundary, division, or contour **c :** any of the horizontal parallel strokes on a music staff on or between which notes are placed — compare SPACE **d :** a mark (as by pencil) that forms part of the formal design of a picture distinguished from the shading or color **e :** a division on a bridge score dividing the score for bonuses from that for tricks **f** (1) **:** a demarcation of a limit with reference to which the playing of some game or sport is regulated — usually used in combination (2) **:** a marked or imaginary line across a playing area (as a football field) parallel to the end line (3) **:** LINE OF SCRIMMAGE
8 : a straight or curved geometric element that is generated by a moving point and that has extension only along the path of the point **:** CURVE
9 a : a defining outline **:** CONTOUR **b :** a general plan **:** MODEL — usually used in plural
10 a *chiefly British* **:** PICA — used to indicate the size of large type **b :** the unit of fineness of halftones expressed as the number of screen lines to the linear inch
11 : merchandise or services of the same general class for sale or regularly available
12 a : a source of information **:** INSIGHT **b :** betting odds offered by a bookmaker especially on a sporting event
13 : a complete game of 10 frames in bowling — called also *string*
14 : LINE DRIVE
— **liny** *also* **lin·ey** \'lī-nē\ *adjective*
— **between the lines 1 :** by implication **:** in an indirect way **2 :** by way of inference
— **down the line :** all the way **:** FULLY
— **in line for :** due or in a position to receive
— **on line :** in or into operation
— **on the line 1 :** in complete commitment and at great risk ⟨puts his future *on the line* by backing that policy⟩ **2 :** on the border between two categories **3 :** IMMEDIATELY ⟨paid cash *on the line*⟩

²line *verb* **lined; lin·ing** (1530)
transitive verb
1 : to mark or cover with a line or lines ⟨*lined* paper⟩
2 : to depict with lines **:** DRAW
3 : to place or form a line along ⟨pedestrians *line* the walks⟩
4 : to form into a line or lines **:** ALIGN ⟨*line* up troops⟩
5 : to hit (as a baseball) hard and in a usually straight line
intransitive verb
1 : to hit a line drive in baseball
2 : to come into the correct relative position **:** ALIGN

³line *transitive verb* **lined; lin·ing** [Middle English, from *line* flax, from Old English *līn*] (14th century)
1 : to cover the inner surface of ⟨*line* a cloak with silk⟩
2 : to put something in the inside of **:** FILL
3 : to serve as the lining of ⟨tapestries *lined* the walls⟩
4 *obsolete* **:** FORTIFY
— **line one's pockets :** to take money freely and especially dishonestly

¹lin·e·age \'li-nē-ij *also* 'li-nij\ *noun* (14th century)
1 a : descent in a line from a common progenitor **b :** DERIVATION
2 : a group of individuals tracing descent from a common ancestor; *especially* **:** such a group of persons whose common ancestor is regarded as its founder

²line·age \'lī-nij\ *variant of* LINAGE

lin·e·al \'li-nē-əl\ *adjective* (14th century)
1 : LINEAR
2 : composed of or arranged in lines
3 a : consisting of or being in a direct male or female line of ancestry — compare COLLATERAL 2 **b :** relating to or derived from ancestors **:** HEREDITARY **c :** descended in a direct line
4 a : belonging to one lineage ⟨*lineal* relatives⟩ **b :** of, relating to, or dealing with a lineage
— **lin·e·al·i·ty** \ˌli-nē-'a-lə-tē\ *noun*
— **lin·e·al·ly** \'li-nē-ə-lē\ *adverb*

lin·e·a·ment \'li-nē-ə-mənt\ *noun* [Middle English, from Latin *lineamentum,* from *lineare* to draw a line, from *linea*] (15th century)
1 a : an outline, feature, or contour of a body or figure and especially of a face — usually used in plural **b :** a linear topographic feature (as of the earth or a planet) that reveals a characteristic (as a fault or the subsurface structure)
2 : a distinguishing or characteristic feature — usually used in plural
— **lin·e·a·men·tal** \ˌli-nē-ə-'men-tᵊl\ *adjective*

lin·e·ar \'li-nē-ər\ *adjective* (circa 1656)
1 a (1) **:** of, relating to, resembling, or having a graph that is a line and especially a straight line **:** STRAIGHT (2) **:** involving a single dimension **b** (1) **:** of the first degree with respect to one or more variables (2) **:** of, relating to, based on, or being linear equations, linear differential equations, linear functions, linear transformations, or linear algebra **c** (1) **:** characterized by an emphasis on line ⟨*linear* art⟩ (2) **:** composed of simply drawn lines with little attempt at pictorial representation ⟨*linear* script⟩ **d :** consisting of a straight chain of atoms
2 : elongated with nearly parallel sides ⟨*linear* leaf⟩ — see LEAF illustration
3 : having or being a response or output that is directly proportional to the input
4 : of, relating to, or based or depending on sequential development ⟨*linear* thinking⟩ ⟨a *linear* narrative⟩
— **lin·e·ar·i·ty** \ˌli-nē-'ar-ə-tē\ *noun*
— **lin·e·ar·ly** \'li-nē-ər-lē\ *adverb*

Linear A \-'ā\ *noun* (1948)
: a linear form of writing used in Crete from the 18th to the 15th centuries B.C.

linear accelerator *noun* (1945)
: a device in which charged particles are accelerated in a straight line by successive impulses from a series of electric fields

linear algebra *noun* (circa 1884)
: a branch of mathematics that is concerned with mathematical structures closed under the operations of addition and scalar multiplication and that includes the theory of systems of linear equations, matrices, determinants, vector spaces, and linear transformations

Linear B \-'bē\ *noun* (1950)
: a linear form of writing employing syllabic characters and used at Knossos on Crete and on the Greek mainland from the 15th to the 12th centuries B.C. for documents in the Mycenaean language

linear combination *noun* (1960)
: a mathematical entity (as $4x + 5y + 6z$) which is composed of sums and differences of elements (as variables, matrices, or functions) especially when the coefficients are not all zero

linear dependence *noun* (1955)
: the property of one set (as of matrices or vectors) of having at least one linear combination of its elements equal to zero when the coefficients are taken from another given set and at least one of its coefficients is not equal to zero
— **linearly dependent** *adjective*

linear equation *noun* (1816)
: an equation of the first degree in any number of variables

linear function *noun* (circa 1889)
1 : a mathematical function in which the variables appear only in the first degree, are multiplied by constants, and are combined only by addition and subtraction
2 : LINEAR TRANSFORMATION

linear independence *noun* (1967)
: the property of a set (as of matrices or vectors) of having no linear combination of all its elements equal to zero when coefficients are taken from a given set unless the coefficient of each element is zero
— **linearly independent** *adjective*

lin·e·ar·ise *British variant of* LINEARIZE

lin·e·ar·ize \'li-nē-ə-ˌrīz\ *transitive verb* **-ized; -iz·ing** (1895)
: to give a linear form to; *also* **:** to project in linear form
— **lin·e·ar·i·za·tion** \ˌli-nē-ə-rə-'zā-shən\ *noun*

linear measure *noun* (circa 1890)
1 : a measure of length
2 : a system of measures of length

linear motor *noun* (1957)
: a motor that produces thrust in a straight line by direct induction rather than with the use of gears — called also *linear induction motor*

linear perspective *noun* (circa 1656)
: PERSPECTIVE 1a

linear programming *noun* (1949)
: a mathematical method of solving practical problems (as the allocation of resources) by means of linear functions where the variables involved are subject to constraints

linear space *noun* (circa 1889)
: VECTOR SPACE

linear transformation *noun* (circa 1889)
1 : a transformation in which the new variables are linear functions of the old variables
2 : a function that maps the vectors of one vector space onto the vectors of the same or another vector space with the same field of scalars in such a way that the image of the sum of two vectors equals the sum of their images and the image of the product of a scalar and a vector equals the product of the scalar and the image of the vector

lin·e·a·tion \ˌli-nē-'ā-shən\ *noun* [Middle English *lineacion* outline, from Latin *lineation-, lineatio,* from *lineare* to mark with lines, from *linea*] (14th century)
1 a : the action of marking with lines **:** DELINEATION **b :** OUTLINE
2 : an arrangement of lines

line·back·er \'līn-ˌba-kər\ *noun* (1949)
: a defensive football player who lines up immediately behind the line of scrimmage to make tackles on running plays through the line or defend against short passes

line·back·ing \-ˌba-kiŋ\ *noun* (1953)
: the action or art of playing linebacker

line·breed·ing \-ˌbrē-diŋ\ *noun* (circa 1879)
: the interbreeding of individuals within a particular line of descent usually to perpetuate desirable characters
— **line·bred** \-ˌbred\ *adjective*

line·cast·er \-ˌkas-tər\ *noun* (1964)
: a machine that casts metal type in lines
— **line·cast·ing** \-tiŋ\ *noun*

line·cut \'līn-ˌkət\ *noun* (circa 1909)
: a photoengraving of a line drawing

line drawing *noun* (1891)
: a drawing made in solid lines

line drive *noun* (1912)
: a batted baseball hit in a nearly straight line usually not far above the ground

line engraving *noun* (1802)
: an engraving cut by hand directly in the plate

line graph *noun* (circa 1924)
: a graph in which points representing values of a variable for suitable values of an independent variable are connected by a broken line

line–haul \ˈlīn-ˌhȯl\ *noun* (circa 1923)
: the transporting of items or persons between terminals

line item *noun* (1962)
: an appropriation that is itemized on a separate line in a budget
— **line–item** \ˈlīn-ˈī-təm\ *adjective*

line judge *noun* (1970)
: a football linesman whose duties include keeping track of the official time for the game

line•man \ˈlīn-mən\ *noun* (1876)
1 : one who sets up or repairs electric wire communication or power lines — called also *linesman*
2 : a player in the forward line of a team; *specifically* **:** a football player in the line

¹lin•en \ˈli-nən\ *adjective* [Middle English, from Old English *līnen,* from *līn* flax, from Latin *linum* flax; akin to Greek *linon* flax, thread] (before 12th century)
1 : made of flax
2 : made of or resembling linen

²linen *noun* (14th century)
1 a : cloth made of flax and noted for its strength, coolness, and luster **b :** thread or yarn spun from flax
2 : clothing or household articles made of linen cloth or similar fabric
3 : paper made from linen fibers or with a linen finish

line of credit (1917)
: the maximum credit allowed a buyer or borrower

line of duty (circa 1918)
: all that is authorized, required, or normally associated with some field of responsibility

line officer *noun* (1850)
: a commissioned officer assigned to the line of the army or navy — compare STAFF OFFICER

line of force (1873)
: a line in a field of force (as a magnetic or electric field) whose tangent at any point gives the direction of the field at that point

line of scrimmage (circa 1909)
: an imaginary line in football that is parallel to the goal lines and tangent to the nose of the ball laid on the ground and that marks the position of the ball at the start of each down

line of sight (1559)
1 : a line from an observer's eye to a distant point
2 : the straight path between a radio or television transmitting antenna and receiving antenna when unobstructed by the horizon

line out (1618)
transitive verb
1 : to indicate with or as if with lines **:** OUTLINE ⟨*line out* a route⟩
2 : to arrange in an extended line
3 : BELT ⟨*line out* a song⟩
intransitive verb
1 : to move rapidly ⟨*lined out* for home⟩
2 : to make an out by hitting a baseball in a line drive that is caught

line printer *noun* (1955)
: a high-speed printing device (as for a computer) that prints each line as a unit rather than character by character

¹lin•er \ˈlī-nər\ *noun* (15th century)
1 : one that makes, draws, or uses lines
2 a : a ship belonging to a regular line **b :** an airplane belonging to an airline
3 : LINE DRIVE
4 : something with which lines are made

²liner *noun* (1611)
1 : one that lines or is used to line or back something
2 : JACKET 3c(2)
— **lin•er•less** \-ləs\ *adjective*

li•ner•board \ˈlī-nər-ˌbōrd, -ˌbȯrd\ *noun* (1948)
: a thin paperboard used for the flat facings of corrugated containerboard

liner notes *noun plural* (1955)
: comments or explanatory notes about a recording printed on the jacket or an insert

line score *noun* (1946)
: a score of a baseball game giving the runs, hits, and errors made by each team — compare BOX SCORE

lines•man \ˈlīnz-mən\ *noun* (1883)
1 : LINEMAN 1
2 : an official who assists a referee in various games (as football or hockey) especially in determining if a ball, puck, or player is out-of-bounds or offside

line squall *noun* (1887)
: a squall or thunderstorm occurring along a cold front

line storm *noun* (1850)
: an equinoctial storm

line•up \ˈlī-ˌnəp\ *noun* (1889)
1 a : a list of players taking part in a game (as of baseball) **b :** the players on such a list
2 a : an alignment (as in entertainment or politics) of persons or things having a common purpose, distinction, or bond ⟨the show's star-studded *lineup*⟩ **b :** LINE 11 **c :** a television programming schedule
3 : a line of persons arranged especially for inspection or for identification by police

line up (1864)
intransitive verb
1 : to assume an orderly linear arrangement ⟨*line up* for inspection⟩
2 : to align oneself ⟨he *lined up* with the liberals against the bill⟩
transitive verb
1 : to put into alignment
2 : to arrange for ⟨*line up* support for a candidate⟩

¹ling \ˈliŋ\ *noun* [Middle English; akin to Dutch *leng* ling, Old English *lang* long] (13th century)
1 : any of various fishes (as a hake or burbot) of the cod family
2 : LINGCOD

²ling *noun* [Middle English, from Old Norse *lyng*] (13th century)
: a heath plant; *especially* **:** a common Old World heather (*Calluna vulgaris*)

¹-ling *noun suffix* [Middle English, from Old English; akin to Old English *-ing*]
1 : one connected with or having the quality of ⟨hire*ling*⟩
2 : young, small, or inferior one ⟨duck*ling*⟩

²-ling *or* **-lings** *adverb suffix* [Middle English *-ling* (from Old English), *-linges* (from *-ling* + *-es* -s); akin to Old High German *-lingūn* -ling, Old English *lang* long]
: in (such) a direction or manner ⟨side*ling*⟩ ⟨flat*ling*⟩

Lin•ga•la \liŋ-ˈgä-lə\ *noun* (1922)
: a Bantu language widely used in trade and public affairs in the Congo River area

lin•gam \ˈliŋ-gəm\ *or* **lin•ga** \-gə\ *noun* [Sanskrit *liṅga* (nominative *liṅgam*), literally, characteristic] (1719)
: a stylized phallic symbol of the masculine cosmic principle and of the Hindu god Siva — compare YONI

Lin•ga•yat \liŋ-ˈgä-yət\ *noun* [Kannada *lingāyata*] (1901)
: a member of a Saiva sect of southern India marked by wearing of the lingam and characterized by denial of caste distinctions

ling•cod \ˈliŋ-ˌkäd\ *noun* (1940)
: a large greenish-fleshed fish (*Ophiodon elongatus*) of the Pacific coast of North America that is an important food and sport fish and belongs to the same family as the greenlings

lin•ger \ˈliŋ-gər\ *verb* **lin•gered; lin•ger•ing** \-g(ə-)riŋ\ [Middle English (northern dialect) *lengeren* to dwell, frequentative of *lengen* to prolong, from Old English *lengan;* akin to Old English *lang* long] (14th century)
intransitive verb
1 : to be slow in parting or in quitting something **:** TARRY
2 a : to remain alive although gradually dying **b :** to remain existent although often waning in strength, importance, or influence ⟨*lingering* doubts⟩
3 : to be slow to act **:** PROCRASTINATE
4 : to move slowly **:** SAUNTER
transitive verb
1 *obsolete* **:** DELAY
2 : to pass (as a period of time) slowly
— **lin•ger•er** \-gər-ər\ *noun*
— **lin•ger•ing•ly** \-g(ə-)riŋ-lē\ *adverb*

lin•ge•rie \ˌlän-jə-ˈrā, ˌläⁿ-zhə-, -ˈrē; ˈlaⁿ-zhə-(ˌ)rē, ˈlän-jə-, ˈläⁿ-zhə-, -ˌrā\ *noun* [French, from Middle French, from *linge* linen, from Latin *lineus* made of linen — more at LINE] (1835)
1 *archaic* **:** linen articles or garments
2 : women's intimate apparel
— **lingerie** *adjective*

lin•go \ˈliŋ-(ˌ)gō\ *noun, plural* **lingoes** [probably from Lingua Franca, language, tongue, from Provençal, from Latin *lingua* — more at TONGUE] (1660)
: strange or incomprehensible language or speech: as **a :** a foreign language **b :** the special vocabulary of a particular field of interest **c :** language characteristic of an individual

ling•on•ber•ry \ˈliŋ-ən-ˌber-ē\ *noun* [Swedish *lingon* mountain cranberry; akin to Old Norse *lyng* ling] (1920)
: the fruit of the mountain cranberry; *also* **:** MOUNTAIN CRANBERRY

lin•gua \ˈliŋ-gwə\ *noun, plural* **lin•guae** \-ˌgwē, -ˌgwī\ [Latin — more at TONGUE] (circa 1826)
: a tongue or an organ resembling a tongue

lin•gua fran•ca \ˈliŋ-gwə-ˈfraŋ-kə\ *noun, plural* **lingua francas** *or* **lin•guae fran•cae** \-gwē-ˈfraŋ-(ˌ)kē\ [Italian, literally, Frankish language] (1619)
1 *often capitalized* **:** a common language consisting of Italian mixed with French, Spanish, Greek, and Arabic that was formerly spoken in Mediterranean ports
2 : any of various languages used as common or commercial tongues among peoples of diverse speech
3 : something resembling a common language

lin•gual \ˈliŋ-gwəl *also* ˈliŋ-gyə-wəl\ *adjective* [Latin *lingua*] (1650)
1 a : of, relating to, or resembling the tongue **b :** lying near or next to the tongue; *especially* **:** relating to or being the surface of tooth next to the tongue **c :** produced by the tongue
2 : LINGUISTIC
— **lin•gual•ly** *adverb*

lin•gui•ne *or* **lin•gui•ni** \liŋ-ˈgwē-nē\ *noun* [Italian, plural of *linguina,* diminutive of *lingua* tongue, from Latin] (circa 1948)
: narrow flat pasta

lin•guist \ˈliŋ-gwist\ *noun* [Latin *lingua* language, tongue] (1591)
1 : a person accomplished in languages; *especially* **:** one who speaks several languages
2 : a person who specializes in linguistics

lin•guis•tic \liŋ-ˈgwis-tik\ *also* **lin•guis•ti•cal** \-ti-kəl\ *adjective* (1846)
: of or relating to language or linguistics
— **lin•guis•ti•cal•ly** \-ti-k(ə-)lē\ *adverb*

linguistic atlas *noun* (1923)
: a publication containing a set of maps on which speech variations are recorded — called also *dialect atlas*

linguistic form *noun* (1921)
: a meaningful unit of speech (as a morpheme, word, or sentence) — called also *speech form*

\ə\ **abut** \ᵊ\ **kitten** \ər\ **further** \a\ **ash** \ā\ **ace** \ä\ **mop, mar** \au̇\ **out** \ch\ **chin** \e\ **bet** \ē\ **easy** \g\ **go** \i\ **hit** \ī\ **ice** \j\ **job** \ŋ\ **sing** \ō\ **go** \ȯ\ **law** \ȯi\ **boy** \th\ **thin** \t͟h\ **the** \ü\ **loot** \u̇\ **foot** \y\ **yet** \zh\ **vision** *see also* Guide to Pronunciation

linguistic geography *noun* (1926)
: local or regional variations of a language or dialect studied as a field of knowledge — called also *dialect geography*
— linguistic geographer *noun*

lin·guis·ti·cian \ˌliŋ-gwə-ˈsti-shən\ *noun* (1895)
: LINGUIST 2

lin·guis·tics \liŋ-ˈgwis-tiks\ *noun plural but singular in construction* (circa 1847)
: the study of human speech including the units, nature, structure, and modification of language

lin·i·ment \ˈli-nə-mənt\ *noun* [Middle English, from Late Latin *linimentum,* from Latin *linere* to smear — more at LIME] (15th century)
: a liquid or semiliquid preparation that is applied to the skin as an anodyne or a counterirritant

lin·ing \ˈlī-niŋ\ *noun* (14th century)
1 : material that lines or that is used to line especially the inner surface of something (as a garment)
2 : the act or process of providing something with a lining

¹link \ˈliŋk\ *noun* [Middle English, of Scandinavian origin; akin to Old Norse *hlekkr* chain; akin to Old English *hlanc* lank] (15th century)
1 : a connecting structure: as **a** (1) **:** a single ring or division of a chain (2) **:** one of the standardized divisions of a surveyor's chain that is 7.92 inches (20.1 centimeters) long and serves as a measure of length **b :** CUFF LINK **c :** BOND 3c **d :** an intermediate rod or piece for transmitting force or motion; *especially* **:** a short connecting rod with a hole or pin at each end **e :** the fusible member of an electrical fuse
2 : something analogous to a link of chain: as **a :** a segment of sausage in a chain **b :** a connecting element or factor ⟨found a *link* between smoking and cancer⟩ **c :** a unit in a communication system **d :** an identifier attached to an element (as an index term) in a system in order to indicate or permit connection with other similarly identified elements

²link (15th century)
transitive verb
: to couple or connect by or as if by a link
intransitive verb
: to become connected by or as if by a link
synonym see JOIN
— link·er *noun*

³link *noun* [perhaps modification of Medieval Latin *linchinus* candle, alteration of Latin *lychnus,* from Greek *lychnos;* akin to Greek *leukos* white — more at LIGHT] (1526)
: a torch formerly used to light a person's way through the streets

⁴link *intransitive verb* [origin unknown] (1715)
Scottish **:** to skip smartly along

link·age \ˈliŋ-kij\ *noun* (1874)
1 : the manner or style of being united: as **a :** the manner in which atoms or radicals are linked in a molecule **b :** BOND 3c
2 : the quality or state of being linked; *especially* **:** the relationship between genes on the same chromosome that causes them to be inherited together — compare MENDEL'S LAW 2
3 : a system of links; *especially* **:** a system of links or bars which are jointed together and more or less constrained by having a link or links fixed and by means of which straight or nearly straight lines or other point paths may be traced
4 : LINK 2b

linkage group *noun* (1921)
: a set of linked genes at different loci on the same chromosome

link·boy \ˈliŋk-ˌbȯi\ *noun* (1660)
: an attendant formerly employed to bear a light for a person on the streets at night

linked \ˈliŋ(k)t\ *adjective* (15th century)
1 : marked by linkage and especially genetic linkage ⟨*linked* genes⟩
2 : having or provided with links ⟨a *linked* list⟩

linking verb *noun* (1923)
: a word or expression (as a form of *be, become, feel,* or *seem*) that links a subject with its predicate

link·man \ˈliŋk-mən\ *noun* (1716)
1 : LINKBOY
2 *British* **:** a broadcasting moderator or anchorman

links \ˈliŋ(k)s\ *noun plural* [Middle English, from Old English *hlincas,* plural of *hlinc* ridge; akin to Old English *hlanc* lank] (15th century)
1 *Scottish* **:** sand hills especially along the seashore
2 : GOLF COURSE

links·man \ˈliŋ(k)s-mən\ *noun* (1937)
: one who plays golf

link·up \ˈliŋ-ˌkəp\ *noun* (1945)
1 : establishment of contact **:** MEETING ⟨the *linkup* of two spacecraft⟩
2 a : something that serves as a linking device or factor **b :** a functional whole resulting from the linking up of separate elements ⟨an instructional TV *linkup*⟩

linn \ˈlin\ *noun* [Scottish Gaelic *linne* pool] (1513)
1 *chiefly Scottish* **:** WATERFALL
2 *chiefly Scottish* **:** PRECIPICE

Lin·nae·an *or* **Lin·ne·an** \lə-ˈnē-ən, -ˈnā-; ˈli-nē-\ *adjective* [Carolus *Linnaeus*] (1753)
: of, relating to, or following the systematic methods of the Swedish botanist Linnaeus who established the system of binomial nomenclature

lin·net \ˈli-nət\ *noun* [Middle French *linette,* from *lin* flax, from Latin *linum;* from its feeding on flax seeds] (circa 1530)
: a common small brownish Old World finch (*Acanthis cannabina*) of which the male has red on the breast and crown during breeding season

li·no \ˈlī-(ˌ)nō\ *noun, plural* **linos** (1907)
chiefly British **:** LINOLEUM

li·no·cut \ˈlī-nō-ˌkət\ *noun* (1907)
: a print made from a design cut into a mounted piece of linoleum

li·no·le·ate \lə-ˈnō-lē-ˌāt\ *noun* (circa 1865)
: a salt or ester of linoleic acid

lin·o·le·ic acid \ˌli-nə-ˈlē-ik-, -ˈlā-\ *noun* [Greek *lin*on flax + International Scientific Vocabulary *oleic* (*acid*)] (1857)
: a liquid unsaturated fatty acid $C_{18}H_{32}O_2$ found especially in semidrying oils (as peanut oil) and essential for the nutrition of some animals

lin·o·le·nic acid \-ˈlē-nik-, -ˈlā-\ *noun* [International Scientific Vocabulary, irregular from *linoleic*] (1887)
: a liquid unsaturated fatty acid $C_{18}H_{30}O_2$ found especially in drying oils (as linseed oil) and essential for the nutrition of some animals

li·no·leum \lə-ˈnō-lē-əm, -ˈnōl-yəm\ *noun, often attributive* [Latin *lin*um flax + *oleum* oil — more at OIL] (1878)
1 : a floor covering made by laying on a burlap or canvas backing a mixture of solidified linseed oil with gums, cork dust or wood flour or both, and usually pigments
2 : a material similar to linoleum

Li·no·type \ˈlī-nə-ˌtīp\ *trademark*
— used for a keyboard-operated typesetting machine that uses circulating matrices and produces each line of type in the form of a solid metal slug

lin·sang \ˈlin-ˌsaŋ\ *noun* [Javanese *lingsang*] (1821)
: either of two nocturnal chiefly forest-dwelling Asian mammals (*Prionodon pardicolor* and *P. linsang*) that resemble and are related to the mongooses, civets, and genets; *also* **:** a related mammal (*Poiana richardsoni*) of Africa

lin·seed \ˈlin-ˌsēd\ *noun* [Middle English, from Old English *līnsǣd,* from *līn* flax + *sǣd* seed — more at LINEN] (before 12th century)
: FLAXSEED

linseed oil *noun* (15th century)
: a yellowish drying oil obtained from flaxseed and used especially in paint, varnish, printing ink, and linoleum

lin·sey–wool·sey \ˌlin-zē-ˈwu̇l-zē\ *noun* [Middle English *lynsy wolsye*] (15th century)
: a coarse sturdy fabric of wool and linen or cotton

lin·stock \ˈlin-ˌstäk\ *noun* [Dutch *lontstok,* from *lont* match + *stok* stick] (1575)
: a staff having a pointed foot (as for sticking into the ground) and a forked tip and formerly used to hold a lighted match for firing cannon

lint \ˈlint\ *noun* [Middle English] (14th century)
1 a : a soft fleecy material made from linen usually by scraping **b :** fuzz consisting especially of fine ravelings and short fibers of yarn and fabric
2 : a fibrous coat of thick convoluted hairs borne by cotton seeds that yields the cotton staple
— linty \ˈlin-tē\ *adjective*

lin·tel \ˈlin-tᵊl\ *noun* [Middle English, from Middle French, from Late Latin *limitaris* threshold, from Latin, constituting a boundary, from *limit-, limes* boundary] (14th century)
: a horizontal architectural member spanning and usually carrying the load above an opening

lint·er \ˈlin-tər\ *noun* (circa 1889)
1 : a machine for removing linters
2 *plural* **:** the fuzz of short fibers that adheres to cottonseed after ginning

1 lintel

lint·white \ˈlint-ˌhwīt, -ˌwīt\ *noun* [alteration of Middle English *lynkwhyt,* by folk etymology from Old English *līnetwige*] (1513)
: LINNET

li·on \ˈlī-ən\ *noun, plural* **lions** [Middle English, from Old French, from Latin *leon-, leo,* from Greek *leōn*] (12th century)
1 a *or plural* **lion :** a large heavily-built social cat (*Panthera leo* synonym *Leo leo*) of open or rocky areas chiefly of sub-Saharan Africa though once widely distributed throughout Africa and southern Asia that has a tawny body with a tufted tail and a shaggy blackish or dark brown mane in the male **b :** any of several large wildcats; *especially* **:** COUGAR **c** *capitalized* **:** LEO
2 a : a person felt to resemble a lion (as in courage or ferocity) **b :** a person of outstanding interest or importance
3 *capitalized* [*Lions* (*club*)] **:** a member of a major national and international service club
— li·on·like \-ˌlīk\ *adjective*

li·on·ess \ˈlī-ə-nəs\ *noun* (14th century)
: a female lion
usage see -ESS

li·on·fish \ˈlī-ən-ˌfish\ *noun* (circa 1907)
: any of several scorpion fishes (genus *Pterois*) of the Indian Ocean and the tropical Pacific that are brilliantly striped and barred with elongated fins and venomous dorsal spines

li·on·heart·ed \ˈlī-ən-ˌhär-təd\ *adjective* (1708)
: COURAGEOUS, BRAVE

li·on·ise *British variant of* LIONIZE

li·on·ize \ˈlī-ə-ˌnīz\ *transitive verb* **-ized; -iz·ing** (1809)
1 : to treat as an object of great interest or importance
2 *British* **:** to show the sights of a place to
— li·on·i·za·tion \ˌlī-ə-nə-ˈzā-shən\ *noun*
— li·on·iz·er \ˈlī-ə-ˌnī-zər\ *noun*

lion's share *noun* (1790)
: the largest portion ⟨received the *lion's share* of the research money⟩

¹lip \ˈlip\ *noun* [Middle English, from Old English *lippa;* akin to Old High German *leffur* lip and probably to Latin *labium, labrum* lip] (before 12th century)

1 : either of two fleshy folds that surround the mouth in humans and many other vertebrates and are the organs of human speech; *also* **:** the red or pinkish margin of the human lip
2 *slang* **:** BACK TALK
3 a : a fleshy edge or margin (as of a wound) **b :** LABIUM **c :** LABELLUM 1 **d :** a limb of a labiate corolla
4 a : the edge of a hollow vessel or cavity **b :** a projecting edge: as (1) **:** the beveled upper edge of the mouth of an organ flue pipe (2) **:** the sharp cutting edge on the end of a tool (as an auger) (3) **:** a short spout (as on a pitcher)
5 : EMBOUCHURE
— lip·less \-ləs\ *adjective*
— lip·like \-ˌlīk\ *adjective*

²lip *adjective* (1558)
1 : INSINCERE ⟨*lip* praise⟩
2 : produced with the participation of the lips **:** LABIAL ⟨*lip* consonants⟩

³lip *transitive verb* **lipped; lip·ping** (1604)
1 : to touch with the lips; *especially* **:** KISS
2 : UTTER
3 : to lap against **:** LICK
4 : to hit (a putt) so that the ball hits the edge of the cup but fails to drop in

lip- *or* **lipo-** *combining form* [New Latin, from Greek, from *lipos* — more at LEAVE]
: fat **:** fatty tissue **:** fatty ⟨*lip*oid⟩ ⟨*lipo*protein⟩

li·pase \ˈlī-ˌpās, -ˌpāz\ *noun* [International Scientific Vocabulary] (1897)
: an enzyme that hydrolyzes glycerides

lip·id \ˈli-pəd\ *also* **lip·ide** \-ˌpīd\ *noun* [International Scientific Vocabulary] (1912)
: any of various substances that are soluble in nonpolar organic solvents (as chloroform and ether), that with proteins and carbohydrates constitute the principal structural components of living cells, and that include fats, waxes, phosphatides, cerebrosides, and related and derived compounds
— li·pid·ic \li-ˈpi-dik\ *adjective*

Lip·iz·zan \ˌli-pət-ˈsän\ *or* **Lip·iz·zan·er** \-ˈsä-nər\ *also* **Lip·pi·zan** \-ˈsän\ *or* **Lip·pi·zan·er** \-ˈsä-nər\ *noun* [German *Lipizzaner, Lippizaner,* from *Lipizza, Lippiza,* former site of the Austrian Imperial Stud near Trieste, Italy] (1928)
: any of a breed of spirited horses developed from Spanish, Italian, Danish, and Arab stock that are usually born with a dark coat that lightens to white with age

li·po·gen·e·sis \ˌlī-pə-ˈje-nə-səs\ *noun* [New Latin] (1882)
: the formation of fatty acids from acetyl coenzyme A in the living body

li·po·ic acid \lī-ˈpō-ik-, li-\ *noun* [*lip-, lipo-*] (circa 1951)
: any of several microbial growth factors; *especially* **:** a crystalline compound $C_8H_{14}O_2S_2$ that is essential for the oxidation of alpha-keto acids (as pyruvic acid) in metabolism

¹li·poid \ˈlī-ˌpȯid, ˈli-\ *or* **li·poi·dal** \lī-ˈpȯi-dᵊl, li-\ *adjective* [International Scientific Vocabulary] (1876)
: resembling fat

²lipoid *noun* [International Scientific Vocabulary] (1906)
: LIPID

li·pol·y·sis \lī-ˈpä-lə-səs, li-\ *noun* [New Latin] (circa 1903)
: the hydrolysis of fat
— li·po·lyt·ic \ˌlī-pə-ˈli-tik, ˌli-\ *adjective*

li·po·ma \lī-ˈpō-mə, li-\ *noun, plural* **-mas** *or* **-ma·ta** \-mə-tə\ [New Latin] (1830)
: a tumor of fatty tissue
— li·po·ma·tous \-mə-təs\ *adjective*

li·po·phil·ic \ˌlī-pə-ˈfi-lik, ˌli-\ *adjective* (1939)
: having an affinity for lipids (as fats) ⟨a *lipophilic* metabolite⟩

li·po·poly·sac·cha·ride \ˌlī-pō-ˌpä-li-ˈsa-kə-ˌrīd, ˌli-\ *noun* (1950)
: a large molecule consisting of lipids and sugars joined by chemical bonds

li·po·pro·tein \-ˈprō-ˌtēn, -ˈprō-tē-ən\ *noun* (1909)
: a conjugated protein that is a complex of protein and lipid

li·po·some \ˈlī-pə-ˌsōm, ˈli-\ *noun* (1968)
: a vesicle composed of one or more concentric phospholipid bilayers and used medically especially to deliver a drug into the body
— li·po·so·mal \ˌlī-pə-ˈsō-məl, ˌli-\ *adjective*

li·po·suc·tion \ˈli-pə-ˌsək-shən, ˈlī-\ *noun* (1986)
: surgical removal of local fat deposits (as in the thighs) especially for cosmetic purposes

li·po·tro·pic \ˌlī-pō-ˈtrō-pik, ˌli-, -ˈträ-\ *adjective* [International Scientific Vocabulary] (1935)
: promoting the physiological utilization of fat ⟨*lipotropic* dietary factors⟩

li·po·tro·pin \-ˈtrō-pən\ *noun* (1964)
: either of two protein hormones of the pituitary gland that function in the mobilization of fat reserves

lipped \ˈlipt\ *adjective* (14th century)
: having a lip or lips especially of a specified kind or number — often used in combination ⟨tight-*lipped*⟩

lip·pen \ˈli-pən\ *verb* [Middle English *lipnien*] (12th century)
intransitive verb
chiefly Scottish **:** TRUST, RELY
transitive verb
chiefly Scottish **:** ENTRUST

Lippes loop \ˈli-pəs-, -pēz-\ *noun* [Jack *Lippes* (born 1924) American physician] (1964)
: an S-shaped plastic intrauterine device

lip·ping \ˈli-piŋ\ *noun* (1894)
1 : outgrowth of bone in liplike form at a joint margin
2 : a piece of wood set in an archer's bow where a flaw has been cut out
3 : EMBOUCHURE 1

lip·py \ˈli-pē\ *adjective* **lip·pi·er; -est** (circa 1875)
: given to back talk

lip–read \ˈlip-ˌrēd\ *verb* **-read** \-ˌred\; **-read·ing** \-ˌrē-diŋ\ (1892)
transitive verb
: to understand by lipreading
intransitive verb
: to use lipreading
— lip–read·er \-ˌrē-dər\ *noun*

lipreading *noun* (1874)
: the interpreting of speech by watching the speaker's lip and facial movements without hearing the voice

lip service *noun* (1644)
: an avowal of advocacy, adherence, or allegiance expressed in words but not backed by deeds — usually used with *pay*

lip·stick \ˈlip-ˌstik\ *noun* (1880)
: a waxy solid usually colored cosmetic in stick form for the lips; *also* **:** a stick of such cosmetic with its case
— lip·sticked *adjective*

lip–synch *or* **lip–sync** \ˈlip-ˌsiŋk\ (circa 1961)
transitive verb
: to pretend to sing or say in synchronization with recorded sound
intransitive verb
: to lip-synch something
— lip sync *noun*

li·quate \ˈlī-ˌkwāt\ *transitive verb* **li·quat·ed; li·quat·ing** [Latin *liquatus,* past participle of *liquare* to make liquid; akin to Latin *liquēre*] (circa 1859)
: to cause (a more fusible substance) to separate out of a combination or mixture by the application of heat ⟨*liquate* metallic lead from its ore⟩
— li·qua·tion \lī-ˈkwā-shən\ *noun*

liq·ue·fac·tion \ˌli-kwə-ˈfak-shən\ *noun* [Middle English, from Late Latin *liquefaction-, liquefactio,* from Latin *liquefacere,* from *liquēre* to be fluid + *facere* to make — more at DO] (15th century)
1 : the process of making or becoming liquid
2 : the state of being liquid

liquefied petroleum gas *noun* (1925)
: a compressed gas that consists of flammable hydrocarbons (as propane and butane) and is used especially as fuel or as raw material for chemical synthesis

liq·ue·fy *also* **liq·ui·fy** \ˈli-kwə-ˌfī\ *verb* **-fied; -fy·ing** [Middle English *liquefien,* from Middle French *liquefier,* from Latin *liquefacere*] (15th century)
transitive verb
: to reduce to a liquid state
intransitive verb
: to become liquid
— liq·ue·fi·er \-ˌfī(-ə)r\ *noun*

li·ques·cent \li-ˈkwe-sᵊnt\ *adjective* [Latin *liquescent-, liquescens,* present participle of *liquescere* to become fluid, inchoative of *liquēre*] (circa 1727)
: being or tending to become liquid **:** MELTING

li·queur \li-ˈkər, -ˈku̇r, -ˈkyu̇r\ *noun* [French, from Old French *licour* liquid — more at LIQUOR] (1729)
: a usually sweetened alcoholic liquor (as brandy) flavored with fruit, spices, nuts, herbs, or seeds

¹liq·uid \ˈli-kwəd\ *adjective* [Middle English, from Middle French *liquide,* from Latin *liquidus,* from *liquēre* to be fluid; akin to Latin *lixa* water, lye, and perhaps to Old Irish *fliuch* damp] (14th century)
1 : flowing freely like water
2 : having the properties of a liquid **:** being neither solid nor gaseous
3 a : shining and clear ⟨large *liquid* eyes⟩ **b :** being musical and free of harshness in sound **c :** smooth and unconstrained in movement **d :** articulated without friction and capable of being prolonged like a vowel ⟨a *liquid* consonant⟩
4 : consisting of or capable of ready conversion into cash ⟨*liquid* assets⟩
— li·quid·i·ty \li-ˈkwi-də-tē\ *noun*
— liq·uid·ly \ˈli-kwəd-lē\ *adverb*
— liq·uid·ness *noun*

²liquid *noun* (1530)
1 : a liquid consonant
2 : a fluid (as water) that has no independent shape but has a definite volume and does not expand indefinitely and that is only slightly compressible

liq·uid·am·bar \ˌli-kwə-ˈdam-bər\ *noun* [New Latin, from Latin *liquidus* + Medieval Latin *ambar, ambra* amber] (circa 1577)
1 : any of a genus (*Liquidambar*) of trees of the witch-hazel family with monoecious flowers and a globose fruit of many woody carpels
2 : STORAX 1b

liq·ui·date \ˈli-kwə-ˌdāt\ *verb* **-dat·ed; -dat·ing** [Late Latin *liquidatus,* past participle of *liquidare* to melt, from Latin *liquidus*] (circa 1575)
transitive verb
1 a (1) **:** to determine by agreement or by litigation the precise amount of (indebtedness, damages, or accounts) (2) **:** to determine the liabilities and apportion assets toward discharging the indebtedness of **b :** to settle (a debt) by payment or other settlement
2 *archaic* **:** to make clear
3 : to do away with
4 : to convert (assets) into cash
intransitive verb
1 : to liquidate debts, damages, or accounts
2 : to determine liabilities and apportion assets toward discharging indebtedness
— liq·ui·da·tion \ˌli-kwə-ˈdā-shən\ *noun*

liq·ui·da·tor \'li-kwə-ˌdā-tər\ *noun* (circa 1828)
: one that liquidates; *especially* **:** an individual appointed by law to liquidate assets
liquid crystal *noun* (1891)
: an organic liquid whose physical properties resemble those of a crystal in the formation of loosely ordered molecular arrays similar to a regular crystalline lattice and the anisotropic refraction of light
liquid crystal display *noun* (1968)
: LCD
liq·uid·ize \'li-kwə-ˌdīz\ *transitive verb* **-ized; -iz·ing** (1837)
: to cause to be liquid
liquid measure *noun* (circa 1855)
: a unit or series of units for measuring liquid capacity — see METRIC SYSTEM table, WEIGHT table
[1]li·quor \'li-kər\ *noun* [Middle English *licour,* from Old French, from Latin *liquor,* from *liquēre*] (13th century)
: a liquid substance: as **a :** a usually distilled rather than fermented alcoholic beverage **b :** a watery solution of a drug **c :** BATH 2b(1)
[2]liquor *verb* **li·quored; li·quor·ing** \'li-k(ə-)riŋ\ (1502)
transitive verb
1 : to dress (as leather) with oil or grease
2 : to make drunk with alcoholic liquor — usually used with *up*
intransitive verb
: to drink alcoholic liquor especially to excess — usually used with *up*
li·quo·rice *chiefly British variant of* LICORICE
[1]li·ra \'lir-ə, 'lē-rə\ *noun, plural* **li·re** \'lē-(ˌ)rā\ *also* **liras** [Italian, from Latin *libra,* a unit of weight] (1617)
— see MONEY table
[2]lira *noun, plural* **liras** [Turkish, from Italian] (1871)
— see MONEY table
[3]lira *noun, plural* **li·roth** *or* **li·rot** \'lē-ˌrōt, -ˌrōth\ [New Hebrew, from Italian] (circa 1946)
: the former Israeli pound
[4]lira *noun, plural* **li·ri** \'lē-(ˌ)rē\ [Maltese, from Italian] (circa 1985)
— see MONEY table
lir·i·pipe \'lir-ə-ˌpīp\ *noun* [Medieval Latin *liripipium*] (1614)
: a pendent part of a tippet; *also* **:** TIPPET, SCARF
lisente *plural of* SENTE
lisle \'lī(ə)l\ *noun, often attributive* [*Lisle* Lille, France] (1851)
: a smooth tightly twisted thread usually made of long-staple cotton
[1]lisp \'lisp\ *verb* [Middle English, from Old English *-wlyspian;* akin to Old High German *lispen* to lisp] (before 12th century)
intransitive verb
1 : to pronounce the sibilants \s\ and \z\ imperfectly especially by turning them into \th\ and \t͟h\
2 : to speak falteringly, childishly, or with a lisp
transitive verb
: to utter falteringly or with a lisp
— lisp·er *noun*
[2]lisp *noun* (circa 1625)
1 : a speech defect or affectation characterized by lisping
2 : a sound resembling a lisp
LISP \'lisp\ *noun* [*list processing*] (1959)
: a computer programming language that is designed for easy manipulation of data strings and is used extensively for work in artificial intelligence
lis·some *also* **lis·som** \'li-səm\ *adjective* [alteration of *lithesome*] (circa 1800)
1 a : easily flexed **b :** LITHE 2
2 : NIMBLE
— lis·some·ly *adverb*
— lis·some·ness *noun*

[1]list \'list\ *verb* [Middle English *lysten,* from Old English *lystan;* akin to Old English *lust* desire, lust] (before 12th century)
transitive verb
archaic **:** PLEASE, SUIT
intransitive verb
archaic **:** WISH, CHOOSE
[2]list *noun* [Middle English, probably from *lysten*] (13th century)
archaic **:** INCLINATION, CRAVING
[3]list *verb* [Middle English, from Old English *hlystan,* from *hlyst* hearing; akin to Old English *hlysnan* to listen] (before 12th century)
intransitive verb
archaic **:** LISTEN
transitive verb
archaic **:** to listen to **:** HEAR
[4]list *noun* [Middle English, from Old English *līste;* akin to Old High German *līsta* edge, Albanian *leth*] (before 12th century)
1 : a band or strip of material: as **a :** LISTEL **b :** SELVAGE **c :** a narrow strip of wood cut from the edge of a board
2 *plural but singular or plural in construction* **a :** an arena for combat (as jousting) **b :** a field of competition or controversy
3 *obsolete* **:** LIMIT, BOUNDARY
4 : STRIPE
[5]list *transitive verb* (1635)
1 : to cut away a narrow strip from the edge of
2 : to prepare or plant (land) in ridges and furrows with a lister
[6]list *noun* [French *liste,* from Italian *lista,* of Germanic origin; akin to Old High German *līsta* edge] (1602)
1 a : a simple series of words or numerals (as the names of persons or objects) ⟨a guest *list*⟩ **b :** an official roster **:** ROLL
2 : CATALOG, CHECKLIST
3 : the total number to be considered or included ⟨a situation that heads their *list* of troubles⟩
[7]list (1614)
transitive verb
1 a : to make a list of **:** ENUMERATE **b :** to include on a list **:** REGISTER
2 : to place (oneself) in a specified category ⟨*lists* himself as a political liberal⟩
3 *archaic* **:** RECRUIT
intransitive verb
1 *archaic* **:** ENLIST
2 : to become entered in a catalog with a selling price ⟨a car that *lists* for $12,000⟩
— list·ee \li-'stē\ *noun*
[8]list *verb* [origin unknown] (1626)
intransitive verb
: to tilt to one side; *especially, of a boat or ship* **:** to tilt to one side in a state of equilibrium (as from an unbalanced load) — compare HEEL
transitive verb
: to cause to list
[9]list *noun* (1633)
: a deviation from the vertical **:** TILT; *also* **:** the extent of such a deviation
lis·tel \'lis-tᵊl, lis-'tel\ *noun* [French, from Italian *listello,* diminutive of *lista* fillet, roster] (1598)
: a narrow band in architecture **:** FILLET
[1]lis·ten \'li-sᵊn\ *verb* **lis·tened; lis·ten·ing** \'lis-niŋ, 'li-sᵊn-iŋ\ [Middle English *listnen,* from Old English *hlysnan;* akin to Sanskrit *śroṣati* he hears, Old English *hlūd* loud] (before 12th century)
transitive verb
archaic **:** to give ear to **:** HEAR
intransitive verb
1 : to pay attention to sound ⟨*listen* to music⟩
2 : to hear something with thoughtful attention **:** give consideration ⟨*listen* to a plea⟩
3 : to be alert to catch an expected sound ⟨*listen* for his step⟩
— lis·ten·er \'lis-nər, 'li-sᵊn-ər\ *noun*
[2]listen *noun* (1788)
: an act of listening

lis·ten·able \'lis-nə-bəl, 'li-sᵊn-ə-\ *adjective* (1942)
: agreeable to listen to
lis·ten·er·ship \'lis-nər-ˌship, 'li-sᵊn-ər-\ *noun* (1943)
: the audience for a radio program or recording; *also* **:** the number or kind of that audience
listen in *intransitive verb* (1905)
1 : to tune in to or monitor a broadcast
2 : to listen to a conversation without participating in it; *especially* **:** EAVESDROP
— lis·ten·er–in \ˌlis-nər-'in, ˌli-sᵊn-ər-\ *noun*
listening post *noun* (1942)
: a center for monitoring electronic communications (as of an enemy)
[1]list·er \'lis-tər\ *noun* (1682)
: one that lists or catalogs
[2]lister *noun* [[5]*list*] (1887)
: a double-moldboard plow often equipped with a subsoiling attachment and used mainly where rainfall is limited
lis·te·ri·o·sis \lis-ˌtir-ē-'ō-səs\ *noun, plural* **-o·ses** \-ˌsēz\ [New Latin, from *Listeria,* from Joseph *Lister*] (1941)
: a serious commonly fatal encephalitic disease of a great variety of wild and domestic mammals and birds and occasionally humans that is caused by a bacterium (*Listeria monocytogenes*)
list·ing *noun* (1659)
1 : an act or instance of making or including in a list
2 : something that is listed
list·less \'list-ləs\ *adjective* [Middle English *listles,* from [2]*list*] (15th century)
: characterized by lack of interest, energy, or spirit **:** LANGUID ⟨a *listless* melancholy attitude⟩
— list·less·ly *adverb*
— list·less·ness *noun*
list price *noun* (1871)
: the basic price of an item as published in a catalog, price list, or advertisement before any discounts are taken
[1]lit \'lit\ *past and past participle of* LIGHT
[2]lit *noun* [by shortening] (1850)
: LITERATURE
[3]lit *adjective* [past participle of [3]*light*] (1904)
: affected by alcohol **:** DRUNK
lit·a·ny \'li-tᵊn-ē, 'lit-nē\ *noun, plural* **-nies** [Middle English *letanie,* from Old French & Late Latin; Old French, from Late Latin *litania,* from Late Greek *litaneia,* from Greek, entreaty, from *litanos* suppliant] (13th century)
1 : a prayer consisting of a series of invocations and supplications by the leader with alternate responses by the congregation
2 a : a resonant or repetitive chant ⟨a *litany* of cheering phrases —Herman Wouk⟩ **b :** a usually lengthy recitation or enumeration ⟨a familiar *litany* of complaints⟩
li·tchi \'lē-(ˌ)chē, 'lī-\ *noun* [Chinese (Beijing) *lìzhī*] (1588)
1 : the oval fruit of a tree (*Litchi chinensis*) of the soapberry family having a hard scaly reddish outer covering and sweet whitish edible flesh that surrounds a single large seed — called also *litchi nut*
2 : a tree bearing litchis
lit crit \'lit-ˌkrit\ *noun* (1963)
: literary criticism
lite \'līt\ *variant of* [4]LIGHT 9a
-lite *noun combining form* [French, alteration of *-lithe,* from Greek *lithos* stone]
: mineral ⟨rhodo*lite*⟩ **:** rock ⟨aero*lite*⟩ **:** fossil ⟨stromato*lite*⟩
li·ter \'lē-tər\ *noun* [French *litre,* from Medieval Latin *litra,* a measure, from Greek, a weight] (1797)
: a metric unit of capacity equal to one cubic decimeter — see METRIC SYSTEM table
lit·er·a·cy \'li-t(ə-)rə-sē\ *noun* (1883)
: the quality or state of being literate
[1]lit·er·al \'li-t(ə-)rəl\ *adjective* [Middle English, from Middle French, from Medieval

Latin *litteralis,* from Latin, of a letter, from *littera* letter] (14th century)
1 a : according with the letter of the scriptures **b :** adhering to fact or to the ordinary construction or primary meaning of a term or expression **:** ACTUAL ⟨liberty in the *literal* sense is impossible —B. N. Cardozo⟩ **c :** free from exaggeration or embellishment ⟨the *literal* truth⟩ **d :** characterized by a concern mainly with facts ⟨a very *literal* man⟩
2 : of, relating to, or expressed in letters
3 : reproduced word for word **:** EXACT, VERBATIM ⟨a *literal* translation⟩
— lit·er·al·i·ty \ˌli-tə-ˈra-lə-tē\ *noun*
— lit·er·al·ness \ˈli-t(ə-)rəl-nəs\ *noun*

[2]literal *noun* (1622)
: a small error usually of a single letter (as in writing)

lit·er·al·ism \ˈli-t(ə-)rə-ˌliz-əm\ *noun* (1644)
1 : adherence to the explicit substance of an idea or expression ⟨biblical *literalism*⟩
2 : fidelity to observable fact **:** REALISM
— lit·er·al·ist \-list\ *noun*
— lit·er·al·is·tic \ˌli-t(ə-)rə-ˈlis-tik\ *adjective*

lit·er·al·ize \ˈli-t(ə-)rə-ˌlīz\ *transitive verb* **-ized; -iz·ing** (1826)
: to make literal
— lit·er·al·i·za·tion \ˌli-t(ə-)rə-lə-ˈzā-shən\ *noun*

lit·er·al·ly \ˈli-tə-rə-lē, ˈli-trə-lē, ˈli-tər-lē\ *adverb* (1533)
1 : in a literal sense or manner **:** ACTUALLY ⟨took the remark *literally*⟩ ⟨was *literally* insane⟩
2 : in effect **:** VIRTUALLY ⟨will *literally* turn the world upside down to combat cruelty or injustice —Norman Cousins⟩ □

lit·er·ary \ˈli-tə-ˌrer-ē\ *adjective* (1749)
1 a : of, relating to, or having the characteristics of humane learning or literature **b :** BOOKISH 2 **c :** of or relating to books
2 a : WELL-READ **b :** of or relating to authors or scholars or to their professions
— lit·er·ar·i·ly \ˌli-tə-ˈrer-ə-lē\ *adverb*
— lit·er·ar·i·ness \ˈli-tə-ˌrer-ē-nəs\ *noun*

literary executor *noun* (1868)
: a person entrusted with the management of the papers and unpublished works of a deceased author

[1]lit·er·ate \ˈli-tə-rət *also* ˈli-trət\ *adjective* [Middle English *literat,* from Latin *litteratus* marked with letters, literate, from *litterae* letters, literature, from plural of *littera*] (15th century)
1 a : EDUCATED, CULTURED **b :** able to read and write
2 a : versed in literature or creative writing **:** LITERARY **b :** LUCID, POLISHED ⟨a *literate* essay⟩ **c :** having knowledge or competence ⟨computer-*literate*⟩ ⟨politically *literate*⟩
— lit·er·ate·ly *adverb*
— lit·er·ate·ness *noun*

[2]literate *noun* (circa 1550)
1 : an educated person
2 : a person who can read and write

li·te·ra·ti \ˌli-tə-ˈrä-(ˌ)tē\ *noun plural* [obsolete Italian *litterati,* from Latin, plural of *litteratus*] (1621)
1 : the educated class; *also* **:** INTELLIGENTSIA
2 : persons interested in literature or the arts

lit·er·a·tim \ˌli-tə-ˈrā-təm, -ˈrä-\ *adverb or adjective* [Medieval Latin, from Latin *littera*] (1643)
: letter for letter ⟨printed *literatim* from the manuscript —I. A. Gordon⟩

lit·er·a·tion \ˌli-tə-ˈrā-shən\ *noun* [Latin *littera* + English *-ation*] (circa 1889)
: the representation of sound or words by letters

lit·er·a·tor \ˈli-tə-ˌrā-tər, ˌli-tə-ˈrä-ˌtȯr\ *noun* (1791)
: LITTERATEUR

lit·er·a·ture \ˈli-tə-rə-ˌchu̇r, ˈli-trə-ˌchu̇r, ˈli-tə(r)-ˌchu̇r, -chər, -ˌtyu̇r, -ˌtu̇r\ *noun* [Middle English, from Latin *litteratura* writing, grammar, learning, from *litteratus*] (14th century)
1 *archaic* **:** literary culture
2 : the production of literary work especially as an occupation
3 a (1) **:** writings in prose or verse; *especially* **:** writings having excellence of form or expression and expressing ideas of permanent or universal interest (2) **:** an example of such writings ⟨what came out, though rarely *literature,* was always a roaring good story —*People*⟩ **b :** the body of written works produced in a particular language, country, or age **c :** the body of writings on a particular subject ⟨scientific *literature*⟩ **d :** printed matter (as leaflets or circulars) ⟨campaign *literature*⟩
4 : the aggregate of a usually specified type of musical compositions ⟨Brahms piano *literature*⟩

lit·er·a·tus \ˌli-tə-ˈrä-təs\ *noun* [New Latin, back-formation from English *literati* (taken as Latin)] (1704)
: a member of the literati

lith- *or* **litho-** *combining form* [Latin, from Greek, from *lithos*]
: stone ⟨*litho*logy⟩

-lith *noun combining form* [New Latin *-lithus* & French *-lithe,* from Greek *lithos*]
1 : structure or implement of stone ⟨mega*lith*⟩ ⟨eo*lith*⟩
2 : calculus ⟨uro*lith*⟩
3 : -LITE ⟨lacco*lith*⟩

li·tharge \ˈli-ˌthärj, li-ˈ\ *noun* [Middle English, from Middle French, from Latin *lithargyrus,* from Greek *lithargyros,* from *lithos* + *argyros* silver — more at ARGENT] (14th century)
: a fused lead monoxide; *broadly* **:** LEAD MONOXIDE

lithe \ˈlīth, ˈlīth\ *adjective* [Middle English, from Old English *līthe* gentle; akin to Old High German *lindi* gentle, Latin *lentus* slow] (14th century)
1 : easily bent or flexed ⟨*lithe* steel⟩ ⟨a *lithe* vine⟩
2 : characterized by easy flexibility and grace ⟨a *lithe* dancer⟩ ⟨treading with a *lithe* silent step⟩; *also* **:** athletically slim ⟨the most *lithe* and graspable of waists —R. P. Warren⟩
— lithe·ly *adverb*
— lithe·ness *noun*

lithe·some \ˈlīth-səm, ˈlīth-\ *adjective* (circa 1774)
: LISSOME

li·thi·a·sis \li-ˈthī-ə-səs\ *noun, plural* **-a·ses** \-ˌsēz\ [New Latin, from Greek, from *lithos*] (circa 1657)
: the formation of stony concretions in the body (as in the gallbladder)

lith·ic \ˈli-thik\ *adjective* [Greek *lithikos,* from *lithos*] (1797)
1 : STONY 1
2 : of, relating to, or being a stone tool

-lithic *adjective combining form* [*lithic*]
: relating to or characteristic of a (specified) stage in humankind's use of stone as a cultural tool ⟨Neo*lithic*⟩

lith·i·fy \ˈli-thə-ˌfī\ *verb* **-fied; -fy·ing** (1877)
transitive verb
: to change to stone **:** PETRIFY; *especially* **:** to convert (unconsolidated sediment) into solid rock
intransitive verb
: to become changed into stone
— lith·i·fi·ca·tion \ˌli-thə-fə-ˈkā-shən\ *noun*

lith·i·um \ˈli-thē-əm\ *noun* [New Latin, from *lithia* oxide of lithium, from Greek *lithos*] (1818)
1 : a soft silver-white element of the alkali metal group that is the lightest metal known and that is used in chemical synthesis and storage batteries — see ELEMENT table
2 : a salt of lithium (as lithium carbonate) used in psychiatric medicine

lithium carbonate *noun* (1873)
: a crystalline salt Li_2CO_3 used in the glass and ceramic industries and in medicine in the treatment of manic-depressive psychosis

lithium fluoride *noun* (1944)
: a crystalline salt LiF used especially in making prisms and ceramics and as a flux

lithium ni·o·bate \-ˈnī-ə-ˌbāt\ *noun* [*niobium* + [1]*-ate*] (1966)
: a crystalline material $LiNbO_3$ whose physical properties change in response to pressure or the presence of an electric field and which is used in fiber optics and as a synthetic gemstone

litho \ˈlī-(ˌ)thō\ *noun, plural* **lith·os** (circa 1889)
1 : LITHOGRAPH
2 : LITHOGRAPHY 1

[1]lith·o·graph \ˈli-thə-ˌgraf\ *transitive verb* (1825)
: to produce, copy, or portray by lithography
— li·tho·gra·pher \li-ˈthä-grə-fər, ˈli-thə-ˌgra-fər\ *noun*

[2]lithograph *noun* (1828)
: a print made by lithography
— lith·o·graph·ic \ˌli-thə-ˈgra-fik\ *adjective*
— lith·o·graph·i·cal·ly \-fi-k(ə-)lē\ *adverb*

li·thog·ra·phy \li-ˈthä-grə-fē\ *noun* [German *Lithographie,* from *lith-* + *-graphie* -graphy] (1813)
1 : the process of printing from a plane surface (as a smooth stone or metal plate) on which the image to be printed is ink-receptive and the blank area ink-repellent
2 : the process of producing patterns on semiconductor crystals for use as integrated circuits

li·thol·o·gy \li-ˈthä-lə-jē\ *noun, plural* **-gies** (1716)
1 : the study of rocks
2 : the character of a rock formation; *also* **:** a rock formation having a particular set of characteristics
— lith·o·log·ic \ˌli-thə-ˈlä-jik\ *or* **lith·o·log·i·cal** \-ji-kəl\ *adjective*
— lith·o·log·i·cal·ly \-ji-k(ə-)lē\ *adverb*

lith·o·phane \ˈli-thə-ˌfān\ *noun* [probably from German *Lithophan,* from Greek *lithos* + German *diaphan* diaphanous] (circa 1889)
: porcelain impressed with figures that are made distinct by transmitted light; *also* **:** an object of this material

lith·o·phyte \ˈli-thə-ˌfīt\ *noun* [French, from *lith-* + *-phyte*] (1774)
: a plant that grows on rock

lith·o·pone \ˈli-thə-ˌpōn\ *noun* [International Scientific Vocabulary *lith-* + Greek *ponos* work] (circa 1884)
: a white pigment consisting essentially of zinc sulfide and barium sulfate

lith·o·sol \ˈli-thə-ˌsäl, -ˌsȯl\ *noun* [*lith-* + Latin *solum* soil] (circa 1938)
: any of a group of shallow azonal soils consisting of imperfectly weathered rock fragments

lith·o·sphere \ˈli-thə-ˌsfir\ *noun* [International Scientific Vocabulary] (1887)

□ USAGE
literally Since some people take sense 2 to be the opposite of sense 1, it has been frequently criticized as a misuse. Instead, the use is pure hyperbole intended to gain emphasis, but it often appears in contexts where no additional emphasis is necessary.

: the solid part of a celestial body (as the earth); *specifically* **:** the outer part of the solid earth composed of rock essentially like that exposed at the surface and usually considered to be about 50 miles (80 kilometers) in thickness

— **lith·o·spher·ic** \ˌli-thə-ˈsfir-ik, -ˈsfer-\ *adjective*

li·thot·o·my \li-ˈthä-tə-mē\ *noun, plural* **-mies** [Late Latin *lithotomia,* from Greek, from *lithotomein* to perform a lithotomy, from *lith-* + *temnein* to cut — more at TOME] (1721)
: surgical incision of the urinary bladder for removal of a stone

lith·o·trip·sy \ˈli-thə-ˌtrip-sē\ *noun, plural* **-sies** [*lith-* + Greek *tripsis* a rubbing, from *tribein* to rub — more at THROW] (1834)
: the breaking of a stone (as by shock waves or crushing with a surgical instrument) in the urinary system into pieces small enough to be voided or washed out

lith·o·trip·ter *also* **lith·o·trip·tor** \ˈli-thə-ˌtrip-tər\ *noun* [alteration of *lithontriptor,* from *lithontriptic* breaking up bladder stones, modification of Greek (*pharmaka tōn*) *lithōn thryptika* (drugs) capable of pulverizing stones] (1825)
: a device for performing lithotripsy; *especially* **:** a noninvasive device that pulverizes stones by focusing shock waves on a patient immersed in a water bath

Lith·u·a·nian \ˌli-thə-ˈwā-nē-ən, -nyən, *chiefly British* ˌlith-yü-\ *noun* (1607)
1 : a native or inhabitant of Lithuania
2 : the Baltic language of the Lithuanian people
— **Lithuanian** *adjective*

lit·i·gant \ˈli-ti-gənt\ *noun* (1659)
: one engaged in a lawsuit
— **litigant** *adjective*

lit·i·gate \ˈli-tə-ˌgāt\ *verb* **-gat·ed; -gat·ing** [Latin *litigatus,* past participle of *litigare,* from *lit-, lis* lawsuit + *agere* to drive — more at AGENT] (1615)
intransitive verb
: to carry on a legal contest by judicial process
transitive verb
1 *archaic* **:** DISPUTE
2 : to contest in law
— **lit·i·ga·ble** \ˈli-ti-gə-bəl\ *adjective*
— **lit·i·ga·tion** \ˌli-tə-ˈgā-shən\ *noun*
— **lit·i·ga·tor** \ˈli-tə-ˌgā-tər\ *noun*

li·ti·gious \lə-ˈti-jəs, li-\ *adjective* [Middle English, from Middle French *litigieux,* from Latin *litigiosus,* from *litigium* dispute, from *litigare*] (14th century)
1 a : DISPUTATIOUS, CONTENTIOUS **b :** prone to engage in lawsuits
2 : subject to litigation
3 : of, relating to, or marked by litigation
— **li·ti·gious·ly** *adverb*
— **li·ti·gious·ness** *noun*

lit·mus \ˈlit-məs\ *noun* [Middle English *litmose,* of Scandinavian origin; akin to Old Norse *litmosi* herbs used in dyeing, from *litr* color (akin to Old English *wlite* brightness) + *mosi* moss; akin to Old English *mōs* moss] (14th century)
1 : a coloring matter from lichens that turns red in acid solutions and blue in alkaline solutions and is used as an acid-base indicator
2 : the critical factor in a litmus test; *also* **:** LITMUS TEST

litmus paper *noun* (1803)
: unsized paper colored with litmus and used as an indicator

litmus test *noun* (1952)
: a test in which a single factor (as an attitude, event, or fact) is decisive

li·to·tes \ˈlī-tə-ˌtēz, ˈli-, lī-ˈtō-ˌtēz\ *noun, plural* **litotes** [Greek *litotēs,* from *litos* simple, perhaps from *lit-, lis* linen cloth] (1589)
: understatement in which an affirmative is expressed by the negative of the contrary (as in "not a bad singer" or "not unhappy")

li·tre \ˈlē-tər\ *variant of* LITER

lit·ten \ˈli-tᵊn\ *adjective* [alteration of *lit,* past participle of *light*] (circa 1849)
archaic **:** being lighted

¹lit·ter \ˈli-tər\ *noun* [Middle English, from Middle French *litiere,* from *lit* bed, from Latin *lectus* — more at LIE] (14th century)
1 a : a covered and curtained couch provided with shafts and used for carrying a single passenger **b :** a device (as a stretcher) for carrying a sick or injured person

litter 1a

2 a (1) **:** material used as bedding for animals (2) **:** material used to absorb the urine and feces of animals **b :** the uppermost slightly decayed layer of organic matter on the forest floor
3 : the offspring at one birth of a multiparous animal
4 a : trash, wastepaper, or garbage lying scattered about ⟨trying to clean up the roadside *litter*⟩ **b :** an untidy accumulation of objects ⟨a shabby writing-desk covered with a *litter* of yellowish dusty documents —Joseph Conrad⟩ ◆
— **lit·tery** \-tə-rē\ *adjective*

²litter (14th century)
transitive verb
1 : BED 1a
2 : to give birth to a litter of (young)
3 a : to strew with scattered articles **b :** to scatter about in disorder **c :** to lie about in disorder ⟨their upside-down hats *littered* the top of the bar —Michael Chabon⟩ **d :** to mark with objects scattered at random ⟨a book *littered* with misprints⟩
intransitive verb
1 : to give birth to a litter
2 : to strew litter

lit·te·rae hu·ma·ni·o·res \ˈli-tə-ˌrī-hü-ˌmä-nē-ˈōr-ˌās, -ˈȯr-\ *noun plural* [Medieval Latin, literally, more humane letters] (1747)
: HUMANITIES

lit·ter·a·teur *or* **lit·té·ra·teur** \ˌli-tə-rə-ˈtər, ˌli-trə-, -ˈtu̇r\ *noun* [French *littérateur,* from Latin *litterator* critic, from *litterae* letters, literature] (1806)
: a literary person; *especially* **:** a professional writer

lit·ter·bag \ˈli-tər-ˌbag\ *noun* (1955)
: a bag used (as in an automobile) for temporary refuse disposal

lit·ter·bug \-ˌbəg\ *noun* (1947)
: one who litters a public area

lit·ter·er \ˈli-tər-ər\ *noun* (1928)
: LITTERBUG

lit·ter·mate \ˈli-tər-ˌmāt\ *noun* (1921)
: one of the offspring in a litter in relation to the others

¹lit·tle \ˈli-tᵊl\ *adjective* **lit·tler** \ˈli-tᵊl-ər, ˈlit-lər\ *or* **less** \ˈles\ *or* **less·er** \ˈle-sər\; **lit·tlest** \ˈli-tᵊl-əst, ˈlit-ləst\ *or* **least** \ˈlēst\ [Middle English *littel,* from Old English *lȳtel;* akin to Old High German *luzzil* little] (before 12th century)
1 : not big: as **a :** small in size or extent **:** TINY ⟨has *little* feet⟩ **b :** YOUNG ⟨was too *little* to remember⟩ **c** *of a plant or animal* **:** small in comparison with related forms — used in vernacular names **d :** small in number **e :** small in condition, distinction, or scope ⟨big business trampling on the *little* fellow⟩ **f :** NARROW, MEAN ⟨the pettiness of *little* minds⟩ **g :** pleasingly small ⟨a cute *little* thing⟩ **h** — used as an intensive ⟨why, you *little* devil!⟩
2 : not much: as **a :** existing only in a small amount or to a slight degree ⟨has *little* money⟩ **b :** short in duration **:** BRIEF **c :** existing to an appreciable though not extensive degree or amount — used with *a* ⟨had a *little* money in the bank⟩
3 : small in importance or interest **:** TRIVIAL
synonym see SMALL
— **lit·tle·ness** \ˈli-tᵊl-nəs\ *noun*

²little *adverb* **less** \ˈles\; **least** \ˈlēst\ (before 12th century)
1 a : in only a small quantity or degree **:** SLIGHTLY ⟨facts that were *little* known at the time⟩ **b :** not at all ⟨cared *little* for their neighbors⟩
2 : RARELY, INFREQUENTLY

³little *noun* (before 12th century)
1 : a small amount, quantity, or degree; *also* **:** practically nothing ⟨*little* has changed⟩
2 a : a short time **b :** a short distance
— **a little :** SOMEWHAT, RATHER ⟨found the play *a little* dull⟩
— **in little :** on a small scale; *especially* **:** in miniature

Little Bear *noun*
: URSA MINOR

little bitty *adjective* (1905)
: SMALL, TINY

little bluestem *noun* (circa 1898)
: a forage grass (*Schizachyrium scoparium* synonym *Andropogon scoparius*) of eastern and central North America

little brown bat *noun* (1842)
: a small widely distributed insectivorous North American bat (*Myotis lucifugus*) with brown fur

little by little *adverb* (15th century)
: by small degrees or amounts **:** GRADUALLY

Little Dipper *noun*
: the seven principal stars in Ursa Minor

little finger *noun* (before 12th century)
: the fourth and smallest finger of the hand counting the forefinger as the first

little guy *noun* (1863)
: LITTLE MAN

Little Hours *noun plural* (circa 1872)
: the offices of prime, terce, sext, and none forming part of the canonical hours

little leaf *noun* (1916)
: a plant disorder characterized by small and often chlorotic and distorted foliage: as **a :** a zinc-deficiency disease of deciduous woody plants (as grape, peach, and pecan) **b :** a destructive disease of unknown cause that affects southern pines and especially the shortleaf pine — called also *little-leaf disease*

Little League *noun* (1952)
: a commercially sponsored baseball league for boys and girls from 8 to 12 years old
— **Little Leaguer** *noun*

◇ WORD HISTORY

litter The diverse senses of the word *litter* all grew out of the basic notion "bed." In Old French *litiere,* a derivative of *lit* "bed," could refer to a sleeping place in a general way, but it was usually applied more narrowly either to a curtained portable couch or to straw spread on the ground as a sleeping place for animals. In borrowing the word, English retained both usages and added new ones. The "bedding for animals" sense was extended to the offspring of an animal such as a dog or cat that gives birth to a number of young at once. In a different direction, *litter* became not just straw for animal bedding but straw or similar material used for any purpose, and by the 18th century any odds and ends of rubbish lying scattered about, and hence a near-synonym for "trash" or "refuse." A much more recent development, presumably arising from euphemism, has extended *litter* to absorbent clay that a cat is trained to use not for sleeping or giving birth but for depositing its feces and urine.

little magazine *noun* (1900)
: a literary usually noncommercial magazine that features works especially of writers who are not well-known
little man *noun* (1933)
: the ordinary individual
lit·tle·neck \ˈli-tᵊl-ˌnek\ *noun* [*Littleneck* Bay, Long Island, N.Y.] (1883)
: a young quahog suitable to be eaten raw — called also *littleneck clam*
Little Office *noun* (circa 1872)
: an office in honor of the Virgin Mary like but shorter than the Divine Office
little people *noun plural* (circa 1731)
1 : tiny imaginary beings (as fairies, elves, and leprechauns) of folklore
2 : CHILDREN
3 : MIDGETS
4 : common people
little slam *noun* (circa 1897)
: the winning of all tricks except one in bridge
little theater *noun* (1812)
: a small theater for low-cost dramatic productions designed for a relatively limited audience
little toe *noun* (before 12th century)
: the outermost and smallest digit of the foot
little woman *noun* (1795)
: WIFE
¹lit·to·ral \ˈli-tə-rəl; ˌli-tə-ˈral, -ˈräl\ *adjective* [Latin *litoralis,* from *litor-, litus* seashore] (circa 1656)
: of, relating to, or situated or growing on or near a shore especially of the sea
²littoral *noun* (1828)
: a coastal region; *especially* **:** the shore zone between high and low watermarks
lit up *adjective* (circa 1914)
: DRUNK 1a
li·tur·gi·cal \lə-ˈtər-ji-kəl, li-\ *adjective* (1641)
1 : of, relating to, or having the characteristics of liturgy
2 : using or favoring the use of liturgy ⟨*liturgical* churches⟩
— li·tur·gi·cal·ly \-k(ə-)lē\ *adverb*
li·tur·gics \-jiks\ *noun plural but singular or plural in construction* (circa 1855)
: the practice or study of formal public worship
li·tur·gi·ol·o·gist \-ˌtər-jē-ˈä-lə-jist\ *noun* (1866)
: LITURGIST 2
li·tur·gi·ol·o·gy \-jē\ *noun* (1863)
: LITURGICS
lit·ur·gist \ˈli-tər-jist\ *noun* (1649)
1 : one who adheres to, compiles, or leads a liturgy
2 : a specialist in liturgics
lit·ur·gy \ˈli-tər-jē\ *noun, plural* **-gies** [Late Latin *liturgia,* from Greek *leitourgia* public service, from Greek (Attic) *leïton* public building (from Greek *laos* — Attic *lēōs* — people) + *-ourgia* -urgy] (1560)
1 *often capitalized* **:** a eucharistic rite
2 : a rite or body of rites prescribed for public worship
3 : a customary repertoire of ideas, phrases, or observances
liv·abil·i·ty *also* **live·abil·i·ty** \ˌli-və-ˈbi-lə-tē\ *noun* (1914)
1 : survival expectancy **:** VIABILITY — used especially of poultry and livestock
2 : suitability for human living
liv·able *also* **live·able** \ˈli-və-bəl\ *adjective* (1814)
1 : suitable for living in or with
2 : ENDURABLE
— liv·able·ness *noun*
¹live \ˈliv\ *verb* **lived; liv·ing** [Middle English, from Old English *libban;* akin to Old High German *lebēn* to live] (before 12th century)
intransitive verb
1 : to be alive **:** have the life of an animal or plant
2 : to continue alive
3 : to maintain oneself **:** SUBSIST
4 : to occupy a home **:** DWELL ⟨*living* in a shabby room⟩ ⟨they had always *lived* in the country⟩
5 : to attain eternal life ⟨though he die, yet shall he *live* —John 11:25 (Revised Standard Version)⟩
6 : to conduct or pass one's life ⟨*lived* only for his work⟩
7 : to remain in human memory or record ⟨the past *lives* in us all —W. R. Inge⟩
8 : to have a life rich in experience
9 : COHABIT
transitive verb
1 : to pass through or spend the duration of ⟨*lived* their lives alone⟩
2 : ACT OUT, PRACTICE — often used with *out* ⟨to *live* out their fantasies⟩
3 : to exhibit vigor, gusto, or enthusiasm in ⟨*lived* life to the fullest⟩
— live it up : to live with gusto and usually fast and loose ⟨*lived it up* with wine and song —*Newsweek*⟩
— live up to : to act or be in accordance with ⟨had no intention of *living up to* his promise⟩
²live \ˈlīv\ *adjective* [short for *alive*] (1542)
1 : having life **:** LIVING
2 : exerting force or containing energy: as **a :** AFIRE, GLOWING ⟨*live* coals⟩ **b :** connected to electric power **c :** charged with explosives and containing shot or a bullet ⟨*live* ammunition⟩; *also* **:** armed but not exploded ⟨a *live* bomb⟩ **d :** imparting or driven by power ⟨a *live* axle⟩ **e :** being in operation ⟨a *live* microphone⟩
3 : abounding with life **:** VIVID
4 : being in a pure native state
5 : of bright vivid color
6 : of continuing or current interest ⟨*live* issues⟩
7 a : not yet printed from or plated ⟨*live* type⟩ **b :** not yet typeset ⟨*live* copy⟩
8 a : of or involving a presentation (as a play or concert) in which both the performers and an audience are physically present ⟨a *live* record album⟩ ⟨a nightclub with *live* entertainment⟩ **b :** broadcast directly at the time of production ⟨a *live* radio program⟩
9 : being in play ⟨a *live* ball⟩
³live \ˈlīv\ *adverb* (1946)
: at the actual time of occurrence **:** during, from, or at a live production ⟨the programming originated *live* from New York City —*Current Biography*⟩
live–bear·er \ˈlīv-ˌbar-ər, -ˌber-\ *noun* (1934)
: a fish that brings forth living young rather than eggs; *especially* **:** any of a family (Poeciliidae) of numerous small surface-feeding fishes
live–box \-ˌbäks\ *noun* (1862)
: a box or pen suspended in water to keep aquatic animals alive
-lived \ˈlīvd, ˈlivd\ *adjective combining form* [Middle English, from *lif* life]
: having a life of a specified kind or length ⟨long-*lived*⟩
lived–in \ˈliv-ˌdin\ *adjective* (1873)
: of or suggesting long term human habitation or use **:** COMFORTABLE
live down *transitive verb* (1842)
: to live so as to wipe out the memory or effects of ⟨made a mistake and couldn't *live* it *down*⟩
live–for·ev·er \ˈliv-fə-ˌre-vər\ *noun* (1597)
: SEDUM
live–in \ˈliv-ˌin\ *adjective* (1953)
1 : living in one's place of employment ⟨a *live-in* maid⟩
2 : involving or involved with cohabitation ⟨a *live-in* relationship⟩ ⟨a *live-in* partner⟩
— live–in *noun*
live in *intransitive verb* (1890)
: to live in one's place of employment **:** live in another's home
live·li·hood \ˈlīv-lē-ˌhu̇d\ *noun* [alteration of Middle English *livelode* course of life, from Old English *līflād,* from *līf* + *lād* course — more at LODE] (15th century)
1 : means of support or subsistence
2 *obsolete* **:** the quality or state of being lively
live load *noun* (1866)
: the load to which a structure is subjected in addition to its own weight
live·long \ˈliv-ˌlȯŋ\ *adjective* [Middle English *lef long,* from *lef* dear + *long* — more at LIEF] (15th century)
: WHOLE, ENTIRE ⟨the *livelong* day⟩
live·ly \ˈlīv-lē\ *adjective* **live·li·er; -est** [Middle English, from Old English *līflīc,* from *līf* life] (before 12th century)
1 *obsolete* **:** LIVING
2 : briskly alert and energetic **:** VIGOROUS, ANIMATED ⟨a *lively* discussion⟩ ⟨*lively* children racing for home⟩
3 : ACTIVE, INTENSE ⟨takes a *lively* interest in politics⟩
4 : BRILLIANT, FRESH ⟨a *lively* wit⟩
5 : imparting spirit or vivacity **:** STIMULATING ⟨many a peer of England brews *livelier* liquor than the Muse —A. E. Housman⟩
6 : quick to rebound **:** RESILIENT
7 : responding readily to the helm ⟨a *lively* boat⟩
8 : full of life, movement, or incident ⟨*lively* streets at carnival time⟩ ☆
— live·li·ly \ˈlīv-lə-lē\ *adverb*
— live·li·ness \ˈlīv-lē-nəs\ *noun*
— lively *adverb*
liv·en \ˈlī-vən\ *verb* **liv·ened; liv·en·ing** \ˈlī-və-niŋ, ˈlīv-niŋ\ (1884)
transitive verb
: ENLIVEN — often used with *up* ⟨he . . . *livened* up the editorial page —*Current Biography*⟩
intransitive verb
: to become lively
live oak \ˈlīv-\ *noun* (1610)
: any of several American evergreen oaks: as **a :** a medium-sized oak (*Quercus virginiana*) of southeastern North America often cultivated as a shelter and shade tree and noted for its extremely hard tough durable wood **b :** any of several oaks of the western U.S. with evergreen foliage and hard durable wood
¹liv·er \ˈli-vər\ *noun* [Middle English, from Old English *lifer;* akin to Old High German *lebra* liver] (before 12th century)
1 a : a large very vascular glandular organ of vertebrates that secretes bile and causes important changes in many of the substances contained in the blood (as by converting sugars into glycogen which it stores up until required and by forming urea) **b :** any of various large compound glands associated with the digestive tract of invertebrate animals and probably concerned with the secretion of digestive enzymes
2 *archaic* **:** a determinant of the quality or temper of a man

☆ SYNONYMS
Lively, animated, vivacious, sprightly, gay mean keenly alive and spirited. LIVELY suggests briskness, alertness, or energy ⟨a *lively* debate on the issues⟩. ANIMATED applies to what is spirited and active ⟨an *animated* discussion of current events⟩. VIVACIOUS suggests an activeness of gesture and wit, often playful or alluring ⟨a *vivacious* party host⟩. SPRIGHTLY suggests lightness and spirited vigor of manner or wit ⟨a tuneful, *sprightly* musical⟩. GAY stresses complete freedom from care and overflowing spirits ⟨the *gay* spirit of Paris in the 1920s⟩.

\ə\ **abut** \ᵊ\ kitten \ər\ **further** \a\ ash \ā\ ace \ä\ mop, mar \au̇\ **out** \ch\ **chin** \e\ bet \ē\ **easy** \g\ go \i\ hit \ī\ ice \j\ job \ŋ\ sing \ō\ go \ȯ\ law \ȯi\ boy \th\ **thin** \t͟h\ **the** \ü\ **loot** \u̇\ **foot** \y\ yet \zh\ vision *see also* Guide to Pronunciation

3 : the liver of an animal (as a calf or chicken) eaten as food
4 : a grayish reddish brown — called also *liver brown, liver maroon*

[2]liv·er \ˈli-vər\ *noun* (14th century)
1 : one that lives especially in a specified way ⟨a fast *liver*⟩
2 : RESIDENT

-livered *adjective combining form*
: expressing vigor or courage considered suggestive of one with (such) a liver ⟨chicken-*livered*⟩ ⟨lily-*livered*⟩

liver fluke *noun* (circa 1798)
: any of various trematode worms (as *Fasciola hepatica*) that invade the mammalian liver

liv·er·ied \ˈli-və-rēd, ˈliv-rēd\ *adjective* (1634)
: wearing a livery ⟨a *liveried* chauffeur⟩

liv·er·ish \ˈli-və-rish, ˈliv-rish\ *adjective* (1740)
1 : resembling liver especially in color
2 a : suffering from liver disorder **:** BILIOUS **b :** PEEVISH, IRASCIBLE
— liv·er·ish·ness *noun*

liver sausage *noun* (1855)
: a sausage containing cooked ground liver and pork trimmings — called also *liver pudding*

liver spots *noun plural* (circa 1859)
: spots of dark pigmentation on the skin (as from exposure to sun) occurring especially among older people

liv·er·wort \ˈli-vər-ˌwərt, -ˌwȯrt\ *noun* (before 12th century)
1 : any of a class (Hepaticae) of bryophytic plants characterized by a thalloid gametophyte or sometimes an upright leafy gametophyte that resembles a moss
2 : HEPATICA

liv·er·wurst \ˈli-və(r)-ˌwərst *also* -ˌwu̇(r)st; *sometimes* -ˌvu̇sht\ *noun* [part translation of German *Leberwurst,* from *Leber* liver + *Wurst* sausage] (1869)
: LIVER SAUSAGE

[1]liv·ery \ˈli-və-rē, ˈliv-rē\ *noun, plural* **-er·ies** [Middle English, from Middle French *livree,* literally, delivery, from *livrer* to deliver, from Latin *liberare* to free — more at LIBERATE] (14th century)
1 *archaic* **:** the apportioning of provisions especially to servants **:** ALLOWANCE
2 a : the distinctive clothing or badge formerly worn by the retainers of a person of rank **b :** a servant's uniform **c :** distinctive dress **:** GARB **d** *chiefly British* **:** an identifying design (as on a vehicle) that designates ownership
3 *archaic* **a :** one's retainers or retinue **b :** the members of a British livery company
4 : the act of delivering legal possession of property
5 a : the feeding, stabling, and care of horses for pay **b :** LIVERY STABLE **c :** a concern offering vehicles (as boats) for rent

[2]livery *adjective* (1778)
1 : resembling liver
2 : suggesting liver disorder

livery company *noun* (1766)
: any of various London craft or trade associations that are descended from medieval guilds

liv·ery·man \ˈli-və-rē-mən, ˈliv-rē-\ *noun* (1682)
1 : a freeman of the City of London entitled to wear the livery of the company to which he belongs
2 *archaic* **:** a liveried retainer
3 : the keeper of a vehicle-rental service

livery stable *noun* (1705)
: a stable where horses and vehicles are kept for hire and where stabling is provided — called also *livery barn*

lives *plural of* LIFE

live steam *noun* (circa 1875)
: steam direct from a boiler and under full pressure

live·stock \ˈlīv-ˌstäk\ *noun* (1742)
: animals kept or raised for use or pleasure; *especially* **:** farm animals kept for use and profit

live·trap \-ˌtrap\ *transitive verb* (1944)
: to capture (an animal) in a live trap

live trap *noun* (circa 1875)
: a trap for catching an animal alive and uninjured

live wire *noun* (1903)
: an alert, active, or aggressive person

liv·id \ˈli-vəd\ *adjective* [French *livide,* from Latin *lividus,* from *livēre* to be blue; akin to Welsh *lliw* color and probably to Russian *sliva* plum] (1622)
1 : discolored by bruising **:** BLACK-AND-BLUE ⟨the *livid* traces of the sharp scourges —Abraham Cowley⟩
2 : ASHEN, PALLID ⟨this cross, thy *livid* face, thy pierced hands and feet —Walt Whitman⟩
3 : REDDISH ⟨a fan of gladiolas blushed *livid* under the electric letters —Truman Capote⟩
4 : very angry **:** ENRAGED ⟨was *livid* at his son's disobedience⟩
— li·vid·i·ty \li-ˈvi-də-tē\ *noun*
— liv·id·ness \ˈli-vəd-nəs\ *noun*

[1]liv·ing \ˈli-viŋ\ *adjective* (before 12th century)
1 a : having life **b :** ACTIVE, FUNCTIONING ⟨*living* languages⟩
2 a : exhibiting the life or motion of nature **:** NATURAL ⟨the wilderness is a *living* museum . . . of natural history —*NEA Journal*⟩ **b :** [2]LIVE 2a
3 a : full of life or vigor **b :** true to life **:** VIVID ⟨televised in *living* color⟩ **c :** suited for living ⟨the *living* area⟩
4 : involving living persons
5 : VERY — used as an intensive ⟨scared the *living* daylights out of me⟩
— liv·ing·ness *noun*

[2]living *noun* (14th century)
1 : the condition of being alive
2 a : means of subsistence **:** LIVELIHOOD ⟨earning a *living*⟩ **b** *archaic* **:** ESTATE, PROPERTY **c** *British* **:** BENEFICE 1
3 : conduct or manner of life ⟨the collegiate way of *living* —J. B. Conant⟩

living death *noun* (1671)
: life emptied of joys and satisfactions

living fossil *noun* (1922)
: an organism (as a horseshoe crab or a ginkgo tree) that has remained essentially unchanged from earlier geologic times and whose close relatives are usually extinct

liv·ing·ly \ˈli-viŋ-lē\ *adverb* (15th century)
: in a vital manner **:** REALISTICALLY

living room *noun* (1857)
1 : a room in a residence used for the common social activities of the occupants
2 : LEBENSRAUM — called also *living space*

living standard *noun* (1944)
: STANDARD OF LIVING

living unit *noun* (circa 1937)
: an apartment or house for use by one family

living wage *noun* (1888)
1 : a subsistence wage
2 : a wage sufficient to provide the necessities and comforts essential to an acceptable standard of living

living will *noun* (1972)
: a document in which the signer requests to be allowed to die rather than be kept alive by artificial means if disabled beyond a reasonable expectation of recovery

li·vre \lēvrᵊ\ *noun* [French, from Latin *libra,* a unit of weight] (1553)
1 : an old French monetary unit equal to 20 sols
2 : a coin representing one livre
3 : the pound of Lebanon

lix·iv·i·ate \lik-ˈsi-vē-ˌāt\ *transitive verb* **-at·ed; -at·ing** [Latin *lixivium* lye, from *lixivius* made of lye, from *lixa* lye — more at LIQUID] (1758)
: to extract a soluble constituent from (a solid mixture) by washing or percolation
— lix·iv·i·a·tion \(ˌ)lik-ˌsi-vē-ˈā-shən\ *noun*

liz·ard \ˈli-zərd\ *noun* [Middle English *liserd,* from Middle French *laisarde,* from Latin *lacerta*] (14th century)
1 : any of a suborder (Lacertilia) of reptiles distinguished from the snakes by a fused inseparable lower jaw, a single temporal opening, two pairs of well differentiated functional limbs which may be lacking in burrowing forms, external ears, and eyes with movable lids; *broadly* **:** any relatively long-bodied reptile (as a crocodile or dinosaur) with legs and tapering tail
2 : leather made from lizard skin

lizard's tail *noun* (circa 1753)
: a perennial herb (*Saururus cernuus* of the family Saururaceae) of North American wetlands having spikes of tiny white flowers

'll \l, əl, ᵊl\ *verb* (1578)
: WILL ⟨you*'ll* be late⟩

lla·ma \ˈlä-mə, ˈyä-mə\ *noun* [Spanish, from Quechua] (1600)
: any of a genus (*Lama*) of wild or domesticated South American ruminants related to the camels but smaller and without a hump; *especially* **:** the domesticated guanaco used especially in the Andes as a pack animal and a source of wool

llama

lla·no \ˈlä-(ˌ)nō, ˈla-\ *noun, plural* **llanos** [Spanish, plain, from Latin *planum* — more at PLAIN] (1604)
: an open grassy plain in Spanish America or the southwestern U.S.

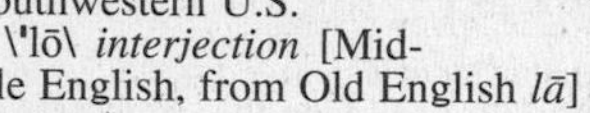

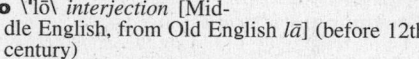

lo \ˈlō\ *interjection* [Middle English, from Old English *lā*] (before 12th century)
— used to call attention or to express wonder or surprise ⟨*lo* these many years⟩

loach \ˈlōch\ *noun* [Middle English *loche,* from Middle French] (14th century)
: any of a family (Cobitidae) of small Old World freshwater fishes related to the carps

[1]load \ˈlōd\ *noun* [Middle English *lod,* from Old English *lād* support, carrying — more at LODE] (12th century)
1 a : the quantity that can be carried at one time by a specified means; *especially* **:** a measured quantity of a commodity fixed for each type of carrier — often used in combination ⟨a boat*load* of tourists⟩ **b :** whatever is put on a person or pack animal to be carried **:** PACK **c :** whatever is put in a ship or vehicle or airplane for conveyance **:** CARGO; *especially* **:** a quantity of material assembled or packed as a shipping unit
2 a : a mass or weight supported by something ⟨branches bent low by their *load* of fruit⟩ **b :** the forces to which a structure is subjected due to superposed weight or to wind pressure on the vertical surfaces; *broadly* **:** the forces to which a given object is subjected
3 a : something that weighs down the mind or spirits ⟨took a *load* off her mind⟩ **b :** a burdensome or laborious responsibility ⟨always carried his share of the *load*⟩
4 *slang* **:** an intoxicating amount of liquor drunk
5 : a large quantity **:** LOT — usually used in plural
6 a : a charge for a firearm **b :** the quantity of material loaded into a device at one time
7 : external resistance overcome by a machine or prime mover
8 a : power output (as of a power plant) or power consumption (as by a device) **b :** a device to which power is delivered
9 a (1) **:** the amount of work that a person carries or is expected to carry (2) **:** the amount of authorized work to be performed by a machine, a group, a department, or a factory **b :** the demand on the operating resources of a

system (as a telephone exchange or a refrigerating apparatus)
10 *slang* **:** EYEFUL — used in the phrase *get a load of*
11 : an amount added (as to the price of a security or the net premium in insurance) to represent selling expense and profit to the distributor
12 : the decrease in capacity for survival of the average individual in a population due to the presence of deleterious genes in the gene pool ⟨the mutational *load* is the genetic *load* caused by mutation⟩

²load (15th century)
transitive verb
1 a : to put a load in or on ⟨*load* a truck⟩ **b :** to place in or on a means of conveyance ⟨*load* freight⟩
2 a : to encumber or oppress with something heavy, laborious, or disheartening **:** BURDEN ⟨a company *loaded* down with debts⟩ **b :** to place as a burden or obligation ⟨*load* more work on him⟩
3 a : to increase the weight of by adding something heavy **b :** to add a conditioning substance (as a mineral salt) to for body **c :** to weight or shape (dice) to fall unfairly **d :** to pack with one-sided or prejudicial influences **:** BIAS **e :** to charge with multiple meanings (as emotional associations or hidden implications) **f :** to weight (as a test) with factors influencing validity or outcome
4 a : to supply in abundance or excess **:** HEAP, PACK **b :** to put runners on (first, second, and third bases) in baseball
5 a : to put a load or charge in (a device or piece of equipment) ⟨*load* a gun⟩ **b :** to place or insert especially as a load in a device or piece of equipment ⟨*load* film in a camera⟩ ⟨*load* a program into a computer⟩
6 : to alter (as an alcoholic drink) by adding an adulterant or drug
7 a : to add a load to (an insurance premium) **b :** to add a sum to after profits and expenses are accounted for ⟨*loaded* prices⟩
intransitive verb
1 : to receive a load
2 : to put a load on or in a carrier, device, or container; *especially* **:** to insert the charge or cartridge in the chamber of a firearm
3 : to go or go in as a load ⟨sightseers *loading* onto a bus⟩
— load·er *noun*

load·ed *adjective* (1886)
1 *slang* **:** HIGH 12b
2 : having a large amount of money

load factor *noun* (1943)
: the percentage of available seats paid for and occupied in an aircraft

load·ing *noun* (15th century)
1 : a cargo, weight, or stress placed on something
2 : LOAD 11
3 : material used to load something **:** FILLER

load line *noun* (circa 1859)
: the line on a ship indicating the depth to which it sinks in the water when properly loaded — see PLIMSOLL MARK illustration

load·mas·ter \ˈlōd-ˌmas-tər\ *noun* (1961)
: a crew member of a transport aircraft who is in charge of the cargo

load·star *variant of* LODESTAR

load·stone *variant of* LODESTONE

¹loaf \ˈlōf\ *noun, plural* **loaves** \ˈlōvz\ [Middle English *lof,* from Old English *hlāf;* akin to Old High German *hleib* loaf] (before 12th century)
1 : a shaped or molded mass of bread
2 : a shaped or molded often symmetrical mass of food
3 *slang British* **:** HEAD, MIND

²loaf *intransitive verb* [probably back-formation from *loafer*] (1835)
: to spend time in idleness

loaf·er \ˈlō-fər\ *noun* [perhaps short for *landloafer,* from German *Landläufer* tramp, from *Land* + *Läufer* runner] (1830)
: one that loafs **:** IDLER

Loafer *trademark*
— used for a low step-in shoe

loam \ˈlōm, *chiefly Northern & Midland* ˈlüm, *New England also* ˈlu̇m\ *noun* [Middle English *lom,* from Old English *lām* clay, mud; akin to Old English *līm* lime] (12th century)
1 a : a mixture (as for plastering) composed chiefly of moistened clay **b :** a coarse molding sand used in founding
2 : SOIL; *specifically* **:** a soil consisting of a friable mixture of varying proportions of clay, silt, and sand
— loamy \ˈlō-mē, ˈlü-, ˈlu̇-\ *adjective*

¹loan \ˈlōn\ *noun* [Middle English *lon,* from Old Norse *lān;* akin to Old English *lǣn* loan, *lēon* to lend, Latin *linquere* to leave, Greek *leipein*] (12th century)
1 a : money lent at interest **b :** something lent usually for the borrower's temporary use
2 a : the grant of temporary use **b :** the temporary duty of a person transferred to another job for a limited time
3 : LOANWORD

²loan *transitive verb* (13th century)
: LEND ■
— loan·able \ˈlō-nə-bəl\ *adjective*

lo and behold *interjection* (1808)
— used to express wonder or surprise

loan·er \ˈlō-nər\ *noun* (1926)
: one (as a car or a watch) that is lent especially as a replacement for something being repaired

loan·ing \ˈlō-niŋ\ *noun* [Middle English *loning,* from *lone,* alteration of *lane*] (14th century)
1 *dialect British* **:** LANE
2 *dialect British* **:** a milking yard

loan shark *noun* (1905)
: one who lends money to individuals at exorbitant rates of interest

loan–shark·ing \-ˌshär-kiŋ\ *noun* (1914)
: the practice of lending money at exorbitant rates of interest

loan translation *noun* (circa 1933)
: a compound, derivative, or phrase that is introduced into a language through translation of the constituents of a term in another language (as *superman* from German *Übermensch*)

loan·word \ˈlōn-ˌwərd\ *noun* (1874)
: a word taken from another language and at least partly naturalized

loath \ˈlōth, ˈlō*th*\ *also* **loathe** \ˈlō*th*, ˈlōth\ *adjective* [Middle English *loth* loathsome, from Old English *lāth;* akin to Old High German *leid* loathsome, Old Irish *lius* loathing] (12th century)
: unwilling to do something contrary to one's ways of thinking **:** RELUCTANT
synonym see DISINCLINED
— loath·ness *noun*

loathe \ˈlō*th*\ *transitive verb* **loathed; loath·ing** [Middle English *lothen,* from Old English *lāthian* to dislike, be hateful, from *lāth*] (12th century)
: to dislike greatly and often with disgust or intolerance **:** DETEST
synonym see HATE
— loath·er *noun*

loathing *noun* (14th century)
: extreme disgust **:** DETESTATION

¹loath·ly \ˈlō*th*-lē, ˈlōth-\ *adjective* (before 12th century)
: LOATHSOME, REPULSIVE

²loath·ly \ˈlōth-lē, ˈlō*th*-\ *adverb* (15th century)
: not willingly **:** RELUCTANTLY

loath·some \ˈlōth-səm, ˈlō*th*-\ *adjective* [Middle English *lothsum,* from *loth* evil, from Old English *lāth,* from *lāth,* adjective] (14th century)
: giving rise to loathing **:** DISGUSTING
— loath·some·ly *adverb*
— loath·some·ness *noun*

¹lob \ˈläb\ *noun* [probably of Low German origin; akin to Low German *lubbe* coarse person] (1508)
dialect British **:** a dull heavy person **:** LOUT

²lob *verb* **lobbed; lob·bing** [*lob* a loosely hanging object] (1599)
transitive verb
1 : to let hang heavily **:** DROOP
2 : to throw, hit, or propel easily or in a high arc
intransitive verb
1 a : to move slowly and heavily **b :** to move in an arc
2 : to hit a tennis ball easily in a high arc

³lob *noun* (1851)
: a soft high-arching shot, throw, or kick

lob- *or* **lobo-** *combining form* [*lobe*]
: lobe ⟨*lobar*⟩ ⟨*lobotomy*⟩

lo·bar \ˈlō-bər, -ˌbär\ *adjective* (circa 1856)
: of or relating to a lobe

lo·bate \ˈlō-ˌbāt\ *also* **lo·bat·ed** \-ˌbā-təd\ *adjective* [New Latin *lobatus,* from Late Latin *lobus*] (circa 1760)
1 : LOBED
2 : resembling a lobe
— lo·ba·tion \lō-ˈbā-shən\ *noun*

¹lob·by \ˈlä-bē\ *noun, plural* **lobbies** [Medieval Latin *lobium* gallery, of Germanic origin; akin to Old High German *louba* porch] (1593)
1 : a corridor or hall connected with a larger room or series of rooms and used as a passageway or waiting room: as **a :** an anteroom of a legislative chamber; *especially* **:** one of two anterooms of a British parliamentary chamber to which members go to vote during a division **b :** a large hall serving as a foyer (as of a hotel or theater)
2 : a group of persons engaged in lobbying especially as representatives of a particular interest group

²lobby *verb* **lob·bied; lob·by·ing** (1837)
intransitive verb
: to conduct activities aimed at influencing public officials and especially members of a legislative body on legislation
transitive verb
1 : to promote (as a project) or secure the passage of (as legislation) by influencing public officials
2 : to attempt to influence or sway (as a public official) toward a desired action
— lob·by·er *noun*
— lob·by·ism \-ˌi-zəm\ *noun*
— lob·by·ist \-ist\ *noun*

lob·by·gow \ˈlä-bē-ˌgau̇\ *noun* [origin unknown] (1906)

☐ USAGE
loan The verb *loan* is one of the words English settlers brought to America and continued to use after it had died out in Britain. Its use was soon noticed by British visitors and somewhat later by the New England literati, who considered it a bit provincial. It was flatly declared wrong in 1870 by a popular commentator, who based his objection on etymology. A later scholar showed that the commentator was ignorant of Old English and thus unsound in his objection, but by then it was too late, as the condemnation had been picked up by many other commentators. Although a surprising number of critics still voice objections, *loan* is entirely standard as a verb. You should note that it is used only literally; *lend* is the verb used for figurative expressions, such as "lending a hand" or "lending enchantment."

: an errand boy

lobe \ˈlōb\ *noun* [Middle French, from Late Latin *lobus,* from Greek *lobos*] (1541)
: a curved or rounded projection or division; *specifically* **:** a usually somewhat rounded projection or division of a bodily organ or part

lo·bec·to·my \lō-ˈbek-tə-mē\ *noun, plural* **-mies** [International Scientific Vocabulary] (circa 1911)
: surgical removal of a lobe of an organ (as a lung) or gland (as the thyroid)

lobed \ˈlōbd\ *adjective* (1787)
: having lobes ⟨palmately *lobed* leaves⟩

lobe–fin \ˈlōb-ˌfin\ *noun* (1941)
: CROSSOPTERYGIAN
— lobe–finned \-ˈfind\ *adjective*

lo·be·lia \lō-ˈbēl-yə, -ˈbē-lē-ə\ *noun* [New Latin, from Matthias de *Lobel* (died 1616) Flemish botanist] (1739)
1 : any of a genus (*Lobelia* of the family Lobeliaceae, the lobelia family) of widely cultivated plants having terminal clusters of showy lipped flowers
2 : the leaves and tops of Indian tobacco

lo·be·line \ˈlō-bə-ˌlēn\ *noun* [New Latin *Lobelia* + English [2]*-ine*] (1852)
: a crystalline alkaloid $C_{22}H_{27}NO_2$ that is obtained from Indian tobacco and is used chiefly as a respiratory stimulant and as a smoking deterrent

lob·lol·ly \ˈläb-ˌlä-lē\ *noun, plural* **-lies** [probably from English dialect *lob* to boil + obsolete English dialect *lolly* broth] (1597)
1 *dialect* **a :** a thick gruel **b :** MIRE, MUDHOLE
2 *dialect* **:** LOUT
3 : LOBLOLLY PINE

loblolly pine *noun* (1760)
: a pine (*Pinus taeda*) of the southeastern U.S. with flaky bark, long needles in groups of three, and cones having spine-tipped scales; *also* **:** its coarse-grained wood

lo·bo \ˈlō-(ˌ)bō\ *noun, plural* **lobos** [Spanish, wolf, from Latin *lupus* — more at WOLF] (1839)
: GRAY WOLF

lo·bot·o·mise *British variant of* LOBOTOMIZE

lo·bot·o·mize \lō-ˈbä-tə-ˌmīz\ *transitive verb* **-mized; -miz·ing** (1943)
1 : to perform a lobotomy on
2 : to deprive of sensitivity, intelligence, or vitality ⟨fear of prosecution was causing the press to *lobotomize* itself —Tony Eprile⟩

lo·bot·o·my \lō-ˈbä-tə-mē\ *noun, plural* **-mies** [International Scientific Vocabulary] (1936)
: surgical severance of nerve fibers connecting the frontal lobes to the thalamus for the relief of some mental disorders

lob·scouse \ˈläb-ˌskau̇s\ *noun* [origin unknown] (1706)
: a sailor's dish of stewed or baked meat with vegetables and hardtack

lob·ster \ˈläb-stər\ *noun, often attributive* [Middle English, from Old English *loppestre,* from *loppe* spider] (before 12th century)
1 : any of a family (Nephropidae and especially *Homarus americanus*) of large edible marine decapod crustaceans that have stalked eyes, a pair of large claws, and a long abdomen and that include species from coasts on both sides of the North Atlantic and from the Cape of Good Hope
2 : SPINY LOBSTER
— lob·ster·like \-ˌlīk\ *adjective*

lobster 1

lob·ster·ing \ˈläb-st(ə-)riŋ\ *noun* (1881)
: the activity or business of catching lobsters

lob·ster·man \-mən\ *noun* (1881)
: one whose business is lobstering

lobster pot *noun* (1764)
: an oblong case with slat sides and a funnel-shaped net used to trap lobsters — called also *lobster trap*

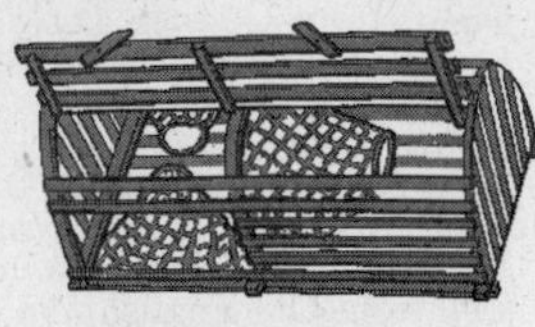
lobster pot

lobster shift *noun* (circa 1933)
: a work shift (as on a newspaper) that covers the late evening and early morning hours — called also *lobster trick*

lobster ther·mi·dor \-ˈthər-mə-ˌdȯr\ *noun* [*thermidor,* from French, from *Thermidor,* drama (1891) by Victorien Sardou] (1930)
: cooked lobster meat in a rich wine sauce stuffed into a lobster shell and browned

lob·u·lar \ˈlä-byə-lər\ *adjective* (1826)
: of, relating to, affecting, or resembling a lobule

lob·u·lat·ed \ˈlä-byə-ˌlā-təd\ *also* **lob·u·late** \-ˌlāt\ *adjective* (1783)
: made up of or having lobules ⟨the pancreas is a *lobulated* organ⟩
— lob·u·la·tion \ˌlä-byə-ˈlā-shən\ *noun*

lob·ule \ˈlä-(ˌ)byü(ə)l\ *noun* (1682)
: a small lobe; *also* **:** a subdivision of a lobe

[1]**lo·cal** \ˈlō-kəl\ *adjective* [Middle English *locale,* from Middle French *local,* from Late Latin *localis,* from Latin *locus* place — more at STALL] (15th century)
1 : characterized by or relating to position in space **:** having a definite spatial form or location
2 a : of, relating to, or characteristic of a particular place **:** not general or widespread **b :** of, relating to, or applicable to part of a whole
3 a : primarily serving the needs of a particular limited district **b** *of a public conveyance* **:** making all the stops on a route
4 : involving or affecting only a restricted part of the organism **:** TOPICAL
5 : of or relating to telephone communication within a specified area

[2]**local** *noun* (circa 1824)
: a local person or thing: as **a :** a local public conveyance (as a train or an elevator) **b :** a local or particular branch, lodge, or chapter of an organization (as a labor union) **c** *British* **:** a nearby or neighborhood pub

local area network *noun* (1981)
: a network of personal computers in a small area (as an office) that are linked by cable, can communicate directly with other devices in the network, and can share resources

local color *noun* (1884)
: the presentation of the features and peculiarities of a particular locality and its inhabitants in writing

lo·cale \lō-ˈkal\ *noun* [modification of French *local,* from *local,* adjective] (1772)
1 : a place or locality especially when viewed in relation to a particular event or characteristic
2 : SITE, SCENE ⟨the *locale* of a story⟩

local government *noun* (1844)
: the government of a specific local area constituting a subdivision of a major political unit (as a nation or state); *also* **:** the body of persons constituting such a government

lo·cal·ise *British variant of* LOCALIZE

lo·cal·ism \ˈlō-kə-ˌli-zəm\ *noun* (1823)
1 a : a local idiom **b :** a local peculiarity of speaking or acting
2 : affection or partiality for a particular place **:** SECTIONALISM

lo·cal·ite \ˈlō-kə-ˌlīt\ *noun* (1951)
: a native or resident of the locality under consideration **:** LOCAL

lo·cal·i·ty \lō-ˈka-lə-tē\ *noun, plural* **-ties** (1628)
1 : the fact or condition of having a location in space or time
2 : a particular place, situation, or location

lo·cal·ize \ˈlō-kə-ˌlīz\ *verb* **-ized; -iz·ing** (1792)
transitive verb
1 : to make local **:** orient locally
2 : to assign to or keep within a definite locality
intransitive verb
: to accumulate in or be restricted to a specific or limited area ⟨an infection that *localizes* in the ear⟩
— lo·cal·iz·abil·i·ty \ˌlō-kə-ˌlī-zə-ˈbi-lə-tē\ *noun*
— lo·cal·iz·able \ˈlō-kə-ˌlī-zə-bəl\ *adjective*
— lo·cal·i·za·tion \ˌlō-kə-lə-ˈzā-shən\ *noun*

lo·cal·ly \ˈlō-k(ə-)lē\ *adverb* (15th century)
1 : with respect to a particular place or situation
2 : NEARBY
3 : in the region of origin

local option *noun* (1878)
: the power granted by a legislature to a political subdivision to determine by popular vote the local applicability of a law on a controversial issue (as the sale of liquor)

local time *noun* (1833)
: time based on the meridian through a particular place as contrasted with that of a time zone

lo·cate \ˈlō-ˌkāt, lō-ˈ\ *verb* **lo·cat·ed; lo·cat·ing** [Latin *locatus,* past participle of *locare* to place, from *locus*] (1652)
intransitive verb
: to establish oneself or one's business **:** SETTLE
transitive verb
1 : to determine or indicate the place, site, or limits of
2 : to set or establish in a particular spot **:** STATION
3 : to seek out and determine the location of
4 : to find or fix the place of especially in a sequence **:** CLASSIFY
— lo·cat·able \-ˌkā-tə-bəl, -ˈkā-\ *adjective*

lo·ca·tion \lō-ˈkā-shən\ *noun* (1597)
1 a : a position or site occupied or available for occupancy or marked by some distinguishing feature **:** SITUATION **b** (1) **:** a tract of land designated for a purpose (2) *Australian* **:** FARM, STATION **c :** a place outside a motion-picture studio where a picture or part of it is filmed — usually used in the phrase *on location*
2 : the act or process of locating
— lo·ca·tion·al \-shnəl, -shə-nᵊl\ *adjective*
— lo·ca·tion·al·ly *adverb*

[1]**loc·a·tive** \ˈlä-kə-tiv\ *noun* [Latin *locus* + English *-ative* (as in *vocative*)] (1804)
: the locative case; *also* **:** a word in that case

[2]**locative** *adjective* (1841)
: of or being a grammatical case that denotes place or the place where or wherein

lo·ca·tor *also* **lo·cat·er** \ˈlō-ˌkā-tər, lō-ˈ\ *noun* (1784)
: one that locates something (as a mining claim or the course of a road)

loch \ˈläk, ˈläk\ *noun* [Middle English (Scots) *louch,* from Scottish Gaelic *loch;* akin to Latin *lacus* lake — more at LAKE] (14th century)
1 *Scottish* **:** LAKE
2 *Scottish* **:** a bay or arm of the sea especially when nearly landlocked

loch·an \ˈlä-kən\ *noun* [Scottish Gaelic, diminutive of *loch*] (1670)
Scottish **:** a small lake

loci *plural of* LOCUS

[1]**lock** \ˈläk\ *noun* [Middle English *lok,* from Old English *locc;* akin to Old High German *loc* lock, Greek *lygos* withe, Latin *luxus* dislocated] (before 12th century)
1 a : a tuft, tress, or ringlet of hair **b** *plural* **:** the hair of the head
2 : a cohering bunch (as of wool, cotton, or flax) **:** TUFT

²lock *noun* [Middle English *lok,* from Old English *loc;* akin to Old High German *loh* enclosure and perhaps to Old English *locc* lock of hair] (before 12th century)
1 a : a fastening (as for a door) operated by a key or a combination **b :** the mechanism for exploding the charge or cartridge of a firearm
2 a : an enclosure (as in a canal) with gates at each end used in raising or lowering boats as they pass from level to level **b :** AIR LOCK
3 a : a locking or fastening together **b :** an intricate mass of objects impeding each other (as in a traffic jam) **c :** a hold in wrestling secured on one part of the body; *broadly* **:** a controlling hold ⟨his paper . . . had a *lock* on a large part of the state —John Corry⟩
4 : one that is assured of success or favorable outcome

³lock (14th century)
transitive verb
1 a : to fasten the lock of **b :** to make fast with or as if with a lock ⟨*lock* up the house⟩
2 a : to fasten in or out or to make secure or inaccessible by or as if by means of locks ⟨*locked* himself away from the curious world⟩ ⟨*locked* her husband out⟩ **b :** to fix in a particular situation or method of operation ⟨a team firmly *locked* in last place⟩
3 a : to make fast, motionless, or inflexible especially by the interlacing or interlocking of parts ⟨*lock* wheels⟩ ⟨*lock* a knee⟩ **b :** to hold in a close embrace **c :** to grapple in combat; *also* **:** to bind closely ⟨administration and students were *locked* in conflict⟩
4 : to invest (capital) without assurance of easy convertibility into money
5 : to move or permit to pass (as a ship) by raising or lowering in a lock
intransitive verb
1 a : to become locked **b :** to be capable of being locked
2 : INTERLACE, INTERLOCK
3 : to go or pass by means of a lock (as in a canal)
— lock·able \ˈlä-kə-bəl\ *adjective*
— lock horns : to come into conflict
— lock on *also* **lock onto :** to sight and follow (a target) automatically using a sensor (as radar)

lock·box \ˈläk-ˌbäks\ *noun* (1872)
: a box (as a post-office box, strongbox, or safe-deposit box) that locks

lock·down \-ˌdau̇n\ *noun* (1977)
: the confinement of prisoners to their cells for all or most of the day as a temporary security measure

locked–in \ˈläkt-ˈin\ *adjective* (1952)
1 : not subject to adjustment **:** FIXED ⟨*locked-in* interest rates⟩
2 : unable or unwilling to shift invested funds because of the tax effect of realizing capital gains

lock·er \ˈlä-kər\ *noun* (14th century)
1 a : a drawer, cupboard, or compartment that may be closed with a lock; *especially* **:** one for individual storage use **b :** a chest or compartment on shipboard for compact stowage of articles **c :** a refrigerated compartment or room for the storage of fresh or frozen foods ⟨a meat *locker*⟩
2 : one that locks

lock·er–room \ˈlä-kə(r)-ˌrüm\ *adjective* (1946)
: of, relating to, or suitable for use in a locker room; *especially* **:** of an earthy or sexual nature ⟨*locker-room* talk⟩

locker room *noun* (circa 1896)
: a room for changing clothes and for storing clothing and equipment in lockers; *especially* **:** one for use by sports players

lock·et \ˈlä-kət\ *noun* [Middle French *loquet* latch, from Middle Dutch *loke;* akin to Old English *loc*] (1679)
: a small case usually of precious metal that has space for a memento and that is worn typically suspended from a chain or necklace

lock·jaw \ˈläk-ˌjȯ\ *noun* (1803)
: an early symptom of tetanus characterized by spasm of the jaw muscles and inability to open the jaws; *also* **:** TETANUS

lock·keep·er \ˈläk-ˌkē-pər\ *noun* (1794)
: a person in charge of a lock (as on a canal)

lock·nut \-ˌnət\ *noun* (circa 1864)
1 : a nut screwed down hard on another to prevent it from slacking back
2 : a nut so constructed that it locks itself when screwed tight against another part

lock·out \ˈläk-ˌau̇t\ *noun* (1854)
: the withholding of employment by an employer and the whole or partial closing of his business establishment in order to gain concessions from or resist demands of employees

lock out *transitive verb* (1868)
: to subject (a body of employees) to a lockout

lock·ram \ˈlä-krəm\ *noun* [Middle English *lokerham,* from *Locronan,* town in Brittany] (14th century)
: a coarse plain-woven linen formerly used in England

lock·smith \ˈläk-ˌsmith\ *noun* (13th century)
: a person who makes or repairs locks

lock·smith·ing \-ˌsmi-thiŋ\ *noun* (circa 1909)
: the work or business of a locksmith

lock·step \ˈläk-ˌstep\ *noun* (circa 1802)
1 : a mode of marching in step by a body of men going one after another as closely as possible
2 : a standard method or procedure that is mindlessly adhered to or that minimizes individuality
— lockstep *adjective*
— in lockstep : in perfect or rigid often mindless conformity or unison ⟨politicians marching *in lockstep* with the party line⟩

lock·stitch \ˈläk-ˌstich\ *noun* (circa 1859)
: a sewing machine stitch formed by the looping together of two threads one on each side of the material being sewn
— lockstitch *verb*

lock, stock, and barrel *adverb* [from the principal parts of a flintlock] (1842)
: WHOLLY, COMPLETELY ⟨the only thing which had not been sold *lock, stock, and barrel* with the . . . house was this piano —Marcia Davenport⟩

lock·up \ˈläk-ˌəp\ *noun* (1839)
1 : JAIL; *especially* **:** a local jail where persons are detained prior to court hearing
2 : an act of locking **:** the state of being locked

¹lo·co \ˈlō-(ˌ)kō\ *adverb or adjective* [Italian dialect, there, from Latin *in loco* in the place] (circa 1801)
: in the register as written — used as a direction in music

²loco *noun, plural* **locos** *or* **locoes** [Mexican Spanish, from Spanish, crazy] (1844)
1 : LOCOWEED
2 : LOCOISM

³loco *transitive verb* (1884)
1 : to poison with locoweed
2 : to make frenzied or crazy

⁴loco *adjective* [Spanish] (1887)
slang **:** mentally disordered **:** CRAZY, FRENZIED

Lo·co·fo·co \ˌlō-kə-ˈfō-(ˌ)kō\ *noun, plural* **-focos** [*locofoco,* a kind of friction match, probably from *locomotive* + Italian *fuoco, foco* fire, from Latin *focus* hearth] (1835)
1 : a member of a radical group of New York Democrats organized in 1835 in opposition to the regular party organization
2 : DEMOCRAT 2

lo·co·ism \ˈlō-(ˌ)kō-ˌi-zəm\ *noun* (1900)
: a disease of horses, cattle, and sheep caused by chronic poisoning with locoweeds

lo·co·mote \ˈlō-kə-ˌmōt\ *intransitive verb* **-mot·ed; -mot·ing** [back-formation from *locomotion*] (1834)
: to move about

lo·co·mo·tion \ˌlō-kə-ˈmō-shən\ *noun* [Latin *locus* + English *motion*] (1646)
1 : an act or the power of moving from place to place
2 : TRAVEL ⟨interest in free *locomotion* and choice of occupation —Zechariah Chafee Jr.⟩

¹lo·co·mo·tive \ˌlō-kə-ˈmō-tiv\ *adjective* (1612)
1 : LOCOMOTORY
2 : of or relating to travel
3 : of, relating to, or being a machine that moves about by operation of its own mechanism

²locomotive *noun* (1829)
1 : a self-propelled vehicle that runs on rails and is used for moving railroad cars
2 : a school or college cheer characterized by a slow beginning and a progressive increase in speed

lo·co·mo·tor \ˌlō-kə-ˈmō-tər\ *adjective* (1870)
1 : of, relating to, or functioning in locomotion
2 : affecting or involving the locomotor organs

locomotor ataxia *noun* (1878)
: TABES DORSALIS

lo·co·mo·to·ry \ˌlō-kə-ˈmō-tə-rē\ *adjective* (circa 1836)
1 : LOCOMOTOR ⟨*locomotory* appendages⟩
2 : capable of moving independently from place to place ⟨*locomotory* animals⟩

lo·co·weed \ˈlō-(ˌ)kō-ˌwēd\ *noun* (1879)
: any of several leguminous plants (genera *Astragalus* and *Oxytropis*) of western North America that cause locoism in livestock

loc·u·lar \ˈlä-kyə-lər\ *adjective* (1784)
: having or composed of loculi — often used in combination ⟨uni*locular*⟩

loc·ule \ˈlä-(ˌ)kyü(ə)l\ *noun* [French, from Latin *loculus*] (circa 1888)
: LOCULUS; *especially* **:** any of the cells of a compound ovary of a plant
— loc·uled \-(ˌ)kyü(ə)ld\ *adjective*

loc·u·li·ci·dal \ˌlä-kyə-lə-ˈsī-dᵊl\ *adjective* [New Latin *loculus* + Latin *-cidere* to cut, from *caedere*] (circa 1819)
: dehiscing longitudinally so as to bisect each loculus ⟨*loculicidal* fruit⟩

loc·u·lus \ˈlä-kyə-ləs\ *noun, plural* **-li** \-ˌlī, -ˌlē\ [New Latin, from Latin, diminutive of *locus*] (1858)
: a small chamber or cavity especially in a plant or animal body

lo·cum \ˈlō-kəm\ *noun* (1901)
chiefly British **:** LOCUM TENENS

locum te·nens \-ˈtē-ˌnenz, -ˈte-, -nənz\ *noun, plural* **locum te·nen·tes** \-ti-ˈnen-ˌtēz\ [Medieval Latin, literally, (one) holding a place] (1641)
: one filling an office for a time or temporarily taking the place of another — used especially of a doctor or clergyman

lo·cus \ˈlō-kəs\ *noun, plural* **lo·ci** \ˈlō-ˌsī, -ˌkī, -ˌkē\ [Latin — more at STALL] (1715)
1 a : the place where something is situated or occurs **:** SITE, LOCATION ⟨was the culture of medicine in the beginning dispersed from a single focus or did it arise in several *loci*? —S. C. Harvey⟩ **b :** a center of activity, attention, or concentration ⟨in democracy the *locus* of power is in the people —H. G. Rickover⟩
2 : the set of all points whose location is determined by stated conditions
3 : the position in a chromosome of a particular gene or allele

locus clas·si·cus \-ˈkla-si-kəs\ *noun, plural* **loci clas·si·ci** \-ˈkla-sə-ˌsī, -ˌkī, -ˌkē\ [New Latin] (1853)
1 : a passage that has become a standard for the elucidation of a word or subject
2 : a classic case or example

locus coe·ru·le·us *also* **locus ce·ru·le·us** \-si-ˈrü-lē-əs\ *noun* [New Latin, literally, dark blue place] (circa 1889)

: a blue area of the brain stem with many norepinephrine-containing neurons

lo·cust \'lō-kəst\ *noun* [Middle English, from Latin *locusta*] (14th century)
1 : SHORT-HORNED GRASSHOPPER; *especially* **:** a migratory grasshopper often traveling in vast swarms and stripping the areas passed of all vegetation
2 : CICADA
3 a : any of various leguminous trees: as (1) **:** CAROB 1 (2) **:** BLACK LOCUST (3) **:** HONEY LOCUST **b :** the wood of a locust tree

locust bean *noun* (1847)
: CAROB

lo·cu·tion \lō-'kyü-shən\ *noun* [Middle English *locucioun,* from Latin *locution-, locutio,* from *loqui* to speak] (15th century)
1 : a particular form of expression or a peculiarity of phrasing; *especially* **:** a word or expression characteristic of a region, group, or cultural level
2 : style of discourse **:** PHRASEOLOGY

lode \'lōd\ *noun* [Middle English, from Old English *lād* course, support; akin to Old English *līthan* to go — more at LEAD] (before 12th century)
1 *dialect English* **:** WATERWAY
2 : an ore deposit
3 : something that resembles a lode **:** an abundant store

lo·den \'lō-dᵊn\ *noun* [German, from Old High German *lodo* coarse cloth; akin to Old English *lotha* mantle] (1911)
1 : a thick woolen cloth used for outer clothing
2 : a variable color averaging a dull grayish green

lode·star \'lōd-,stär\ *noun* [Middle English *lode sterre,* from *lode* course, from Old English *lād*] (14th century)
1 *archaic* **:** a star that leads or guides; *especially* **:** NORTH STAR
2 : one that serves as an inspiration, model, or guide

lode·stone \-,stōn\ *noun* [obsolete *lode* course, from Middle English] (circa 1515)
1 : magnetite possessing polarity
2 : something that strongly attracts

¹lodge \'läj\ *verb* **lodged; lodg·ing** (13th century)
transitive verb
1 a (1) **:** to provide temporary quarters for (2) **:** to rent lodgings to **b :** to establish or settle in a place
2 : to serve as a receptacle for **:** CONTAIN
3 : to beat (as a crop) flat to the ground
4 : to bring to an intended or a fixed position (as by throwing or thrusting)
5 : to deposit for safeguard or preservation
6 : to place or vest especially in a source, means, or agent
7 : to lay (as a complaint) before a proper authority **:** FILE
intransitive verb
1 a : to occupy a place temporarily **:** SLEEP **b** (1) **:** to have a residence **:** DWELL (2) **:** to be a lodger
2 : to come to a rest
3 : to fall or lie down — used especially of hay or grain crops

²lodge *noun* [Middle English *loge,* from Old French, of Germanic origin; akin to Old High German *louba* porch] (13th century)
1 *chiefly dialect* **:** a rude shelter or abode
2 a : a house set apart for residence in a particular season (as the hunting season) **b :** a resort hotel **:** INN
3 a : a house on an estate originally for the use of a gamekeeper, caretaker, or porter **b :** a shelter for an employee (as a gatekeeper)
4 : a den or lair especially of gregarious animals
5 a : the meeting place of a branch of an organization and especially a fraternal organization **b :** the body of members of such a branch
6 a : WIGWAM **b :** a family of North American Indians

lodge·pole pine \'läj-,pōl-\ *noun* (1859)
: either of two pines of western North America with needles in pairs and short ovoid usually asymmetric cones: **a :** a scrubby chiefly coastal pine (*Pinus contorta*) with thick deeply furrowed bark and hard strong coarse-grained medium-light wood **b :** a tall straight pine (*P. contorta* variety *latifolia* synonym *P. murrayana*) with thin and little furrowed bark and soft weak fine-grained lightweight wood

lodg·er \'lä-jər\ *noun* (1596)
: ROOMER

lodging *noun* (14th century)
1 a : a place to live **:** DWELLING **b :** LODGMENT 3b
2 a (1) **:** sleeping accommodations ⟨found *lodging* in the barn⟩ (2) **:** a temporary place to stay ⟨a *lodging* for the night⟩ **b :** a room in the house of another used as a residence — usually used in plural
3 : the act of lodging

lodging house *noun* (1766)
: ROOMING HOUSE

lodg·ment *or* **lodge·ment** \'läj-mənt\ *noun* (1598)
1 a : a lodging place **:** SHELTER **b :** ACCOMMODATIONS, LODGINGS ⟨found *lodgment* in the city⟩
2 a : the act, fact, or manner of lodging ⟨a hut for temporary *lodgment* of cattlemen⟩ **b :** a placing, depositing, or coming to rest
3 a : an accumulation or collection deposited in a place or remaining at rest **b :** a place of rest or deposit

lod·i·cule \'lä-di-,kyü(ə)l\ *noun* [Latin *lodicula,* diminutive of *lodic-, lodix* cover] (1864)
: one of usually two delicate membranous hyaline scales at the base of the ovary of a grass that by their swelling assist in anthesis

loess \'les, 'ləs, 'lō-əs, 'lərs\ *noun* [German *Löss*] (1833)
: an unstratified usually buff to yellowish brown loamy deposit found in North America, Europe, and Asia and believed to be chiefly deposited by the wind
— loess·ial \'le-sē-əl, 'lə-, lō-'e-, 'lər-\ *adjective*

¹loft \'lȯft\ *noun* [Middle English, air, sky, from Old Norse *lopt;* akin to Old High German *luft* air] (13th century)
1 : an upper room or floor **:** ATTIC
2 a : a gallery in a church or hall **b :** one of the upper floors of a warehouse or business building especially when not partitioned ⟨living in a converted *loft*⟩ **c :** HAYLOFT
3 a : the backward slant of the face of a golf-club head **b :** the act of lofting
4 : the thickness of a fabric or insulating material (as goose down)
— loft·like \-,līk\ *adjective*

²loft (1518)
transitive verb
1 : to place, house, or store in a loft
2 : to propel through the air or into space ⟨*lofted* a long hit to center⟩ ⟨instruments *lofted* by a powerful rocket⟩
3 : to lay out a full-sized working drawing of the lines and contours of (as a ship's hull)
intransitive verb
1 : to propel a ball high into the air
2 : to rise high

lofty \'lȯf-tē\ *adjective* **loft·i·er; -est** (15th century)
1 a : elevated in character and spirit **:** NOBLE **b :** elevated in status **:** SUPERIOR
2 : having a haughty overbearing manner **:** SUPERCILIOUS
3 a : rising to a great height **:** impressively high ⟨*lofty* mountains⟩ **b :** REMOTE, ESOTERIC
synonym see HIGH
— loft·i·ly \-tə-lē\ *adverb*
— loft·i·ness \-tē-nəs\ *noun*

¹log \'lȯg, 'läg\ *noun, often attributive* [Middle English *logge,* probably of Scandinavian origin; akin to Old Norse *lāg* fallen tree; akin to Old English *licgan* to lie — more at LIE] (14th century)
1 : a usually bulky piece or length of unshaped timber; *especially* **:** a length of a tree trunk ready for sawing and over six feet (1.8 meters) long
2 : an apparatus for measuring the rate of a ship's motion through the water that consists of a block fastened to a line and run out from a reel
3 a : the record of the rate of a ship's speed or of her daily progress; *also* **:** the full nautical record of a ship's voyage **b :** the full record of a flight by an aircraft
4 : a record of performance, events, or day-to-day activities ⟨a computer *log*⟩

²log *verb* **logged; log·ging** (1699)
transitive verb
1 a : to cut (trees) for lumber **b :** to clear (land) of trees in lumbering — often used with *off*
2 : to make a note or record of **:** enter details of or about in a log
3 a : to move (an indicated distance) or attain (an indicated speed) as noted in a log **b** (1) **:** to sail a ship or fly an airplane for (an indicated distance or period of time) (2) **:** to have (an indicated record) to one's credit **:** ACHIEVE
intransitive verb
: ³LUMBER 1

³log *noun, often attributive* [by shortening] (1631)
: LOGARITHM

log- *or* **logo-** *combining form* [Greek, from *logos* — more at LEGEND]
: word **:** thought **:** speech **:** discourse ⟨*logo*gram⟩ ⟨*logo*rrhea⟩

-log — see -LOGUE

lo·gan·ber·ry \'lō-gən-,ber-ē\ *noun* [James H. *Logan* (died 1928) American lawyer + English *berry*] (1893)
: a red-fruited upright-growing dewberry regarded as a variety (*Rubus ursinus loganobaccus*) of a western dewberry or as a hybrid of a western dewberry and a red raspberry; *also* **:** its berry

log·a·oe·dic \,lä-gə-'ē-dik\ *adjective* [Late Latin *logaoedicus,* from Late Greek *logaoidikos,* from Greek *log-* + *aeidein* to sing; from the resemblance of such rhythm to prose — more at ODE] (1844)
: marked by the mixture of several meters; *specifically* **:** having a rhythm that uses both dactyls and trochees or anapests and iambs
— logaoedic *noun*

log·a·rithm \'lȯ-gə-,ri-thəm, 'lä-\ *noun* [New Latin *logarithmus,* from *log-* + Greek *arithmos* number — more at ARITHMETIC] (circa 1616)
: the exponent that indicates the power to which a number is raised to produce a given number ⟨the *logarithm* of 100 to the base 10 is 2⟩
— log·a·rith·mic \,lȯ-gə-'rith-mik, ,lä-\ *adjective*
— log·a·rith·mi·cal·ly \-mi-k(ə-)lē\ *adverb*

logarithmic function *noun* (1927)
: a function (as $y = \log_a x$ or $y = \ln x$) that is the inverse of an exponential function (as $y = a^x$ or $y = e^x$) so that the independent variable appears in a logarithm

log·book \'lȯg-,bu̇k, 'läg-\ *noun* (circa 1679)
: LOG 3, 4

loge \'lōzh\ *noun* [French, from Old French, a shelter — more at LODGE] (1749)
1 a : a small compartment **:** BOOTH **b :** a box in a theater
2 a : a small partitioned area **b :** a separate forward section of a theater mezzanine or balcony

logged \'lȯgd, 'lägd\ *adjective* (circa 1820)
1 : HEAVY, SLUGGISH
2 : sodden especially with water

log·ger \'lȯ-gər, 'lä-\ *noun* (1732)
: one engaged in logging

log·ger·head \'lȯ-gər-ˌhed, 'lä-\ *noun* [probably from English dialect *logger* block of wood + English *head*] (1588)
1 *chiefly dialect* **a :** BLOCKHEAD **b :** HEAD; *especially* **:** a disproportionately large head
2 a : a very large carnivorous sea turtle (*Caretta caretta*) of subtropical and temperate waters **b :** ALLIGATOR SNAPPER
3 : an iron tool consisting of a long handle terminating in a ball or bulb that is heated and used to melt tar or to heat liquids
— at loggerheads : in or into a state of quarrelsome disagreement

log·gets *or* **log·gats** \'lȯ-gəts, 'lä-\ *noun plural but singular or plural in construction* [probably from [1]*log* + *-et*] (1581)
: a game formerly played in England in which participants throw pieces of wood at a stake

log·gia \'lō-jē-ə, 'lȯ-(ˌ)jä\ *noun, plural* **log·gias** \'lō-jē-əz, 'lȯ-(ˌ)jäz\ *also* **log·gie** \'lȯ-(ˌ)jā\ [Italian, from Old French *loge* lodge] (1742)
: a roofed open gallery especially at an upper story overlooking an open court

loggia

log·ic \'lä-jik\ *noun* [Middle English *logik,* from Middle French *logique,* from Latin *logica,* from Greek *logikē,* from feminine of *logikos* of reason, from *logos* reason — more at LEGEND] (12th century)
1 a (1) **:** a science that deals with the principles and criteria of validity of inference and demonstration **:** the science of the formal principles of reasoning (2) **:** a branch or variety of logic ⟨modal *logic*⟩ ⟨Boolean *logic*⟩ (3) **:** a branch of semiotic; *especially* **:** SYNTACTICS (4) **:** the formal principles of a branch of knowledge **b** (1) **:** a particular mode of reasoning viewed as valid or faulty (2) **:** RELEVANCE, PROPRIETY **c :** interrelation or sequence of facts or events when seen as inevitable or predictable **d :** the arrangement of circuit elements (as in a computer) needed for computation; *also* **:** the circuits themselves
2 : something that forces a decision apart from or in opposition to reason ⟨the *logic* of war⟩
— lo·gi·cian \lō-'ji-shən\ *noun*

log·i·cal \'lä-ji-kəl\ *adjective* (15th century)
1 a (1) **:** of, relating to, involving, or being in accordance with logic (2) **:** skilled in logic **b :** formally true or valid **:** ANALYTIC, DEDUCTIVE
2 : capable of reasoning or of using reason in an orderly cogent fashion ⟨a *logical* thinker⟩
— log·i·cal·i·ty \ˌlä-jə-'ka-lə-tē\ *noun*
— log·i·cal·ly \'lä-ji-k(ə-)lē\ *adverb*
— log·i·cal·ness \-kəl-nəs\ *noun*

logical positivism *noun* (1931)
: a 20th century philosophical movement that holds characteristically that all meaningful statements are either analytic or conclusively verifiable or at least confirmable by observation and experiment and that metaphysical theories are therefore strictly meaningless — called also *logical empiricism*
— logical positivist *noun*

log in *intransitive verb* (1962)
: LOG ON
— log–in \'lȯg-ˌin, 'läg-\ *noun*

lo·gi·on \'lō-gē-ˌän\ *noun, plural* **lo·gia** \-gē-ˌä\ *or* **logions** [Greek, diminutive of *logos*] (1875)
: SAYING; *especially* **:** a saying attributed to Jesus

[1]**lo·gis·tic** \lō-'jis-tik, lə-\ *or* **lo·gis·ti·cal** \-ti-kəl\ *adjective* (1628)
1 a : of or relating to symbolic logic **b :** of or relating to the philosophical attempt to reduce mathematics to logic
2 : of or relating to logistics
3 *logistic* **:** of, represented by, or relating to a logistic curve ⟨a *logistic* process⟩
— lo·gis·ti·cal·ly \-ti-k(ə-)lē\ *adverb*

[2]**logistic** *noun* (1905)
: SYMBOLIC LOGIC

logistic curve *noun* (circa 1903)
: an S-shaped curve that represents an exponential function and is used in mathematical models of growth processes

lo·gis·ti·cian \ˌlō-jəs-'ti-shən\ *noun* (1932)
: a specialist in logistics

lo·gis·tics \lō-'jis-tiks, lə-\ *noun plural but singular or plural in construction* [French *logistique* art of calculating, logistics, from Greek *logistikē* art of calculating, from feminine of *logistikos* of calculation, from *logizein* to calculate, from *logos* reason] (circa 1861)
1 : the aspect of military science dealing with the procurement, maintenance, and transportation of military matériel, facilities, and personnel
2 : the handling of the details of an operation

log·jam \'lȯg-ˌjam, 'läg-\ *noun* (1885)
1 : a jumble of logs jammed together in a watercourse
2 a : DEADLOCK, IMPASSE ⟨trying to break the *logjam* in negotiations⟩ **b :** BLOCKAGE

log·nor·mal \ˌlȯg-'nȯr-məl, ˌläg-\ *adjective* (1945)
: relating to or being a normal distribution that is the distribution of the logarithm of a random variable; *also* **:** relating to or being such a random variable
— log·nor·mal·i·ty \-nȯr-'ma-lə-tē\ *noun*
— log·nor·mal·ly \-'nȯr-mə-lē\ *adverb*

logo \'lō-(ˌ)gō *also* 'lä-\ *noun, plural* **log·os** \-(ˌ)gōz\ (1937)
1 : LOGOTYPE
2 : an identifying statement **:** MOTTO

Lo·go \'lō-(ˌ)gō\ *noun* [modification of Greek *logos* word] (1972)
: a computer programming language that employs simple English commands and is used especially for introducing school children to computers

logo- — see LOG-

logo·gram \'lȯ-gə-ˌgram, 'lä-\ *noun* (1840)
: a letter, symbol, or sign used to represent an entire word
— logo·gram·mat·ic \ˌlȯ-gə-grə-'ma-tik, ˌlä-\ *adjective*

logo·graph \'lȯ-gə-ˌgraf, 'lä-\ *noun* (circa 1888)
: LOGOGRAM

logo·graph·ic \ˌlȯ-gə-'gra-fik, ˌlä-\ *adjective* (1801)
: of, relating to, or marked by the use of logographs **:** consisting of logographs
— logo·graph·i·cal·ly \-fi-k(ə-)lē\ *adverb*

logo·griph \'lȯ-gə-ˌgrif, 'lä-\ *noun* [*log-* + Greek *griphos* reed basket, riddle — more at CRIB] (circa 1598)
: a word puzzle (as an anagram)

lo·gom·a·chy \lō-'gä-mə-kē\ *noun, plural* **-chies** [Greek *logomachia,* from *log-* + *machesthai* to fight] (1569)
1 : a dispute over or about words
2 : a controversy marked by verbiage

log on *intransitive verb* (1977)
: to establish communication and initiate interaction with a time-shared computer or network — often used with *to*
— log–on \'lȯg-ˌȯn, 'läg-ˌän\ *noun*

log·or·rhea \ˌlȯ-gə-'rē-ə, ˌlä-\ *noun* [New Latin] (circa 1892)
: excessive and often incoherent talkativeness or wordiness
— log·or·rhe·ic \-'rē-ik\ *adjective*

Lo·gos \'lō-ˌgäs, -ˌgōs\ *noun, plural* **Lo·goi** \-ˌgȯi\ [Greek, speech, word, reason — more at LEGEND] (1587)
1 : the divine wisdom manifest in the creation, government, and redemption of the world and often identified with the second person of the Trinity
2 : reason that in ancient Greek philosophy is the controlling principle in the universe

logo·type \'lȯ-gə-ˌtīp, 'lä-\ *noun* (circa 1816)
1 : a single piece of type or a single plate faced with a term (as the name of a newspaper or a trademark)
2 : an identifying symbol (as for advertising)

log·roll \'lȯg-ˌrōl, 'läg-\ *verb* [back-formation from *logrolling*] (1835)
intransitive verb
: to take part in logrolling
transitive verb
: to promote passage of by logrolling
— log·roll·er *noun*

log·roll·ing \-ˌrō-liŋ\ *noun* (1812)
1 [from a former American custom of neighbors assisting one another in rolling logs into a pile for burning] **:** the exchanging of assistance or favors; *specifically* **:** the trading of votes by legislators to secure favorable action on projects of interest to each one
2 : the rolling of logs in water by treading; *also* **:** a sport in which contestants treading logs try to dislodge one another

-logue *or* **-log** *noun combining form* [Middle English *-logue,* from Old French, from Latin *-logus,* from Greek *-logos,* from *legein* to speak — more at LEGEND]
1 : discourse **:** talk ⟨duo*logue*⟩
2 : student **:** specialist ⟨sino*logue*⟩

log·wood \'lȯg-ˌwu̇d, 'läg-\ *noun* (1581)
1 a : a Mexican and West Indian leguminous tree (*Haematoxylon campechianum*) **b :** the very hard brown or brownish red heartwood of logwood
2 : a dye extracted from the heartwood of logwood — compare HEMATOXYLIN

lo·gy \'lō-gē\ *also* **log·gy** \'lȯ-gē, 'lä-\ *adjective* **lo·gi·er; -est** [perhaps from Dutch *log* heavy; akin to Middle Low German *luggich* lazy] (1847)
: marked by sluggishness and lack of vitality **:** GROGGY

-logy *noun combining form* [Middle English *-logie,* from Old French, from Latin *-logia,* from Greek, from *logos* word]
1 : oral or written expression ⟨phraseo*logy*⟩
2 : doctrine **:** theory **:** science ⟨ethno*logy*⟩

Lo·hen·grin \'lō-ən-ˌgrin\ *noun* [German]
: a son of Parsifal and knight of the Holy Grail in Germanic legend

loin \'lȯin\ *noun* [Middle English *loyne,* from Middle French *loigne,* from (assumed) Vulgar Latin *lumbea,* from Latin *lumbus;* akin to Old English *lendenu* loins, Old Church Slavonic *lędvije*] (14th century)
1 a : the part of a human being or quadruped on each side of the spinal column between the hipbone and the false ribs **b :** a cut of meat comprising this part of one or both sides of a carcass with the adjoining half of the vertebrae included but without the flank
2 *plural* **a :** the upper and lower abdominal regions and the region about the hips **b** (1) **:** the pubic region (2) **:** the generative organs

loin·cloth \-ˌklȯth\ *noun* (1859)
: a cloth worn about the loins often as the sole article of clothing in warm climates

loi·ter \'lȯi-tər\ *intransitive verb* [Middle English] (14th century)
1 : to delay an activity with aimless idle stops and pauses **:** DAWDLE
2 a : to remain in an area for no obvious reason **:** HANG AROUND **b :** to lag behind
synonym see DELAY
— loiter *noun*
— loi·ter·er \-tər-ər\ *noun*

Lo·ki \'lō-kē\ *noun* [Old Norse]
: a Norse god who contrives evil and mischief for his fellow gods

Lo·li·ta \lō-ˈlē-tə\ *noun* [from *Lolita,* character in the novel *Lolita* (1955) by Vladimir Nabokov] (1959)
: a precociously seductive girl

[1]loll \ˈläl\ *verb* [Middle English] (14th century)
intransitive verb
1 : to hang loosely or laxly **:** DROOP
2 : to act or move in a lax, lazy, or indolent manner **:** LOUNGE
transitive verb
: to let droop or dangle
— loll·er \ˈlä-lər\ *noun*

[2]loll *noun* (1709)
archaic **:** the act of lolling **:** a relaxed posture

Lol·lard \ˈlä-lərd\ *noun* [Middle English, from Middle Dutch *lollaert,* from *lollen* to mutter] (14th century)
: one of the followers of Wycliffe who traveled in the 14th and 15th centuries as lay preachers throughout England and Scotland
— Lol·lard·ism \-lər-ˌdi-zəm\ *noun*
— Lol·lardy \-lər-dē\ *noun*

lol·li·pop *or* **lol·ly·pop** \ˈlä-li-ˌpäp\ *noun* [perhaps from English dialect *lolly* tongue + [2]*pop*] (1784)
1 : a lump of hard candy on the end of a stick
2 *British* **:** a round stop sign on a pole used to stop traffic (as at a school crossing)

lol·lop \ˈlä-ləp\ *intransitive verb* [[1]*loll* + *-op* (as in *gallop*)] (1745)
1 *dialect English* **:** LOLL
2 : to proceed with a bounding or bobbing motion

lol·ly \ˈlä-lē\ *noun, plural* **lollies** [short for *lollipop*] (1854)
1 *British* **:** a piece of candy; *especially* **:** HARD CANDY
2 *British* **:** MONEY

lol·ly·gag \ˈlä-lē-ˌgag\ *intransitive verb* **-gagged; -gag·ging** [origin unknown] (1868)
: FOOL AROUND 1 **:** DAWDLE

Lom·bard \ˈläm-ˌbärd, -bərd\ *noun* [Middle English *Lumbarde,* from Middle French *lombard,* from Old Italian *lombardo,* from Latin *Langobardus*] (14th century)
1 a : a member of a Germanic people that invaded Italy in A.D. 568, settled in the Po valley, and established a kingdom **b :** a native or inhabitant of Lombardy
2 [from the prominence of Lombards as moneylenders] **:** BANKER, MONEYLENDER
— Lom·bar·di·an \läm-ˈbär-dē-ən\ *adjective*
— Lom·bar·dic \läm-ˈbär-dik\ *adjective*

Lom·bar·dy poplar \ˈläm-ˌbär-dē-, -bər-\ *noun* [*Lombardy,* Italy] (1766)
: a poplar of a staminate variety (*Populus nigra italica*) of a European poplar having a columnar shape and strongly ascending branches

lo mein \ˈlō-ˈmān, ˌlō-\ *noun* [Chinese (Guangdong) *lòu-mihn* stirred noodles] (1970)
: a Chinese dish consisting of sliced vegetables, soft noodles, and usually meat or shrimp in bite-size pieces stir-fried in a seasoned sauce

lo·ment \ˈlō-ˌment, -mənt\ *noun* [New Latin *lomentum,* from Latin, wash made from bean meal, from *lavere* to wash — more at LYE] (circa 1830)
: a dry indehiscent one-celled fruit that is produced from a single superior ovary and breaks transversely into numerous segments at maturity

Lon·don broil \ˈlən-dən-\ *noun* [*London,* England] (1946)
: a boneless cut of beef (as from the shoulder or flank) usually served sliced diagonally across the grain

lone \ˈlōn\ *adjective* [Middle English, short for *alone*] (14th century)
1 a : having no company **:** SOLITARY **b :** preferring solitude
2 : ONLY, SOLE
3 : situated by itself **:** ISOLATED
synonym see ALONE
— lone·ness \ˈlōn-nəs\ *noun*

lone·li·ness \ˈlōn-lē-nəs\ *noun* (circa 1586)
: the quality or state of being lonely

lone·ly \ˈlōn-lē\ *adjective* **lone·li·er; -est** (1607)
1 a : being without company **:** LONE **b :** cut off from others **:** SOLITARY
2 : not frequented by human beings **:** DESOLATE
3 : sad from being alone **:** LONESOME
4 : producing a feeling of bleakness or desolation
synonym see ALONE
— lone·li·ly \-lə-lē\ *adverb*

lonely hearts *adjective* (1931)
: of or relating to lonely persons who are seeking companions or spouses

lon·er \ˈlō-nər\ *noun* (1947)
: one that avoids others; *especially* **:** INDIVIDUALIST

lone ranger *noun, often L&R capitalized* [*Lone Ranger,* hero of an American radio and television western] (1969)
: one who acts alone and without consultation or the approval of others; *broadly* **:** LONER

[1]lone·some \ˈlōn(t)-səm\ *adjective* (1647)
1 a : sad or dejected as a result of lack of companionship or separation from others ⟨don't be *lonesome* while we are gone⟩ **b :** causing a feeling of loneliness ⟨the empty house seemed so *lonesome*⟩
2 a : REMOTE, UNFREQUENTED ⟨look down, look down that *lonesome* road —Gene Austin⟩ **b :** LONE
synonym see ALONE
— lone·some·ly *adverb*
— lone·some·ness *noun*

[2]lonesome *noun* (1899)
: SELF ⟨sat all by his *lonesome*⟩

lone wolf *noun* (1909)
: a person who prefers to work, act, or live alone

[1]long \ˈlȯŋ\ *adjective* **lon·ger** \ˈlȯŋ-gər *also* -ər\; **lon·gest** \ˈlȯŋ-gəst *also* -əst\ [Middle English *long, lang,* from Old English; akin to Old High German *lang* long, Latin *longus*] (before 12th century)
1 a : extending for a considerable distance **b :** having greater length than usual **c :** having greater height than usual **:** TALL **d :** having a greater length than breadth **:** ELONGATED **e :** having a greater length than desirable or necessary **f :** FULL-LENGTH ⟨*long* pants⟩
2 a : having a specified length ⟨six feet *long*⟩ **b :** forming the chief linear dimension ⟨the *long* side of the room⟩
3 a : extending over a considerable time ⟨a *long* friendship⟩ **b :** having a specified duration ⟨two hours *long*⟩ **c :** prolonged beyond the usual time ⟨a *long* look⟩
4 a : containing many items in a series ⟨a *long* list⟩ **b :** having a specified number of units ⟨300 pages *long*⟩ **c :** consisting of a greater number or amount than usual **:** LARGE
5 a *of a speech sound* **:** having a relatively long duration **b :** being the member of a pair of similarly spelled vowel or vowel-containing sounds that is descended from a vowel long in duration ⟨*long a* in *fate*⟩ ⟨*long i* in *sign*⟩ **c** *of a syllable in prosody* (1) **:** of relatively extended duration (2) **:** bearing a stress or accent
6 : having the capacity to reach, extend, or travel a considerable distance ⟨a *long* left jab⟩ ⟨tried to hit the *long* ball⟩
7 : larger or longer than the standard ⟨a *long* count by the referee⟩
8 a : extending far into the future ⟨the thoughts of youth are *long, long* thoughts —H. W. Longfellow⟩ **b :** extending beyond what is known ⟨a *long* guess⟩ **c :** payable after a considerable period ⟨a *long* note⟩
9 : possessing a high degree or a great deal of something specified **:** STRONG ⟨*long* on common sense⟩
10 a : of an unusual degree of difference between the amounts wagered on each side ⟨*long* odds⟩ **b :** of or relating to the larger amount wagered ⟨take the *long* end of the bet⟩
11 : subject to great odds
12 : owning or accumulating securities or goods especially in anticipation of an advance in prices ⟨they are now *long* on wheat⟩ ⟨take a *long* position in steel⟩
— long·ness \ˈlȯŋ-nəs\ *noun*
— before long : in a short time **:** SOON
— long in the tooth : past one's best days **:** OLD
— not long for : having little time left to do or enjoy something

[2]long *adverb* (before 12th century)
1 : for or during a long time ⟨*long* a popular hangout⟩
2 : at or to a long distance **:** FAR ⟨*long*-traveled⟩
3 : for the duration of a specified period
4 : at a point of time far before or after a specified moment or event ⟨was excited *long* before the big day⟩
5 : after or beyond a specified or implied time ⟨didn't stay *longer* than midnight⟩ ⟨said it was no *longer* possible⟩
6 : for a considerable distance ⟨faded back and threw the ball *long*⟩
7 : in or into a long position (as on a market)
— so long : GOOD-BYE

[3]long *noun* (before 12th century)
1 : a long period of time
2 : a long syllable
3 : one taking a long position especially in a security or commodity market
4 a *plural* **:** long trousers **b :** a size in clothing for tall men
— the long and short *or* **the long and the short :** GIST

[4]long *intransitive verb* **longed; long·ing** \ˈlȯŋ-iŋ\ [Middle English, from Old English *langian;* akin to Old High German *langēn* to long, Old English *lang* long] (before 12th century)
: to feel a strong desire or craving especially for something not likely to be attained ⟨they *long* for peace but are driven to war⟩ ☆
— long·er \ˈlȯŋ-ər\ *noun*

[5]long *intransitive verb* [Middle English, from *along* (*on*) because (of)] (13th century)
archaic **:** to be suitable or fitting

long–ago \ˈlȯŋ-ə-ˈgō\ *adjective* (circa 1834)
: of or relating to the past ⟨*long-ago* leaders⟩

long ago *noun* (1851)
: the distant past

lon·gan \ˈlȯŋ-gən, -ən\ *noun* [Chinese (Beijing) *lóngyǎn,* literally, dragon's eye] (1732)
1 : a pulpy fruit related to the litchi and produced by an Indian evergreen tree (*Euphoria longana*)
2 : a tree that bears the longan

lon·ga·nim·i·ty \ˌlȯŋ-gə-ˈni-mə-tē\ *noun* [Middle English *longanymyte,* from Late Latin *longanimitat-, longanimitas,* from *longanimis* patient, from Latin *longus* long + *animus* soul — more at ANIMATE] (15th century)

> ☆ SYNONYMS
> **Long, yearn, hanker, pine, hunger, thirst** mean to have a strong desire for something. LONG implies a wishing with one's whole heart and often a striving to attain ⟨*longed* for some rest⟩. YEARN suggests an eager, restless, or painful longing ⟨*yearned* for a stage career⟩. HANKER suggests the uneasy promptings of unsatisfied appetite or desire ⟨always *hankering* for money⟩. PINE implies a languishing or a fruitless longing for what is impossible ⟨*pined* for a lost love⟩. HUNGER and THIRST imply an insistent or impatient craving or a compelling need ⟨*hungered* for a business of his own⟩ ⟨*thirsted* for absolute power⟩.

: a disposition to bear injuries patiently **:** FORBEARANCE

long·boat \'lȯŋ-ˌbōt\ *noun* (15th century)
: a large oared boat usually carried by a merchant sailing ship

long bone *noun* (circa 1860)
: any of the elongated bones supporting a vertebrate limb and consisting of an essentially cylindrical shaft that contains marrow and ends in enlarged heads for articulation with other bones

long·bow \'lȯŋ-ˌbō\ *noun* (14th century)
: a hand-drawn wooden bow held vertically and used especially by medieval English archers

long·bow·man \-mən\ *noun* (1925)
: an archer who uses a longbow

long·case clock \'lȯŋ-ˌkās-\ *noun* (1892)
: GRANDFATHER CLOCK

long–chain *adjective* (1930)
: having a relatively long chain of atoms and especially carbon atoms in the molecule ⟨*long-chain* hydrocarbons⟩

long–day *adjective* (1920)
: responding to or relating to a long photoperiod — used especially of a plant; compare DAY-NEUTRAL, SHORT-DAY

[1]long–distance *adjective* (1884)
1 : of or relating to telephone communication with a distant point
2 a : situated a long distance away **b :** going or covering a long distance ⟨*long-distance* roads⟩ ⟨a *long-distance* runner⟩ **c :** conducted or effective over long distance ⟨a *long-distance* courtship⟩ ⟨*long-distance* listening devices⟩

[2]long–distance *adverb* (circa 1961)
1 : by long-distance telephone
2 : over or from a long distance

long distance *noun* (1904)
1 : communication by long-distance telephone
2 : a telephone operator or exchange that gives long-distance connections

long division *noun* (1827)
: arithmetical division in which the several steps involved in the division are indicated in detail

long–drawn–out *or* **long–drawn** *adjective* (1646)
: extended to a great length

lon·ge·ron \'län-jə-ˌrän\ *noun* [French] (1912)
: a fore-and-aft framing member of an airplane fuselage

lon·gev·i·ty \län-'je-və-tē, lȯn-\ *noun* [Late Latin *longaevitas,* from Latin *longaevus* long-lived, from *longus* long + *aevum* age — more at AYE] (1615)
1 a : a long duration of individual life **b :** length of life ⟨a study of *longevity*⟩
2 : long continuance **:** PERMANENCE, DURABILITY

lon·ge·vous \-'jē-vəs\ *adjective* (1680)
: LONG-LIVED

long face *noun* (1786)
: a facial expression of sadness or melancholy

long green *noun* (circa 1891)
slang **:** MONEY

long·hair \'lȯŋ-ˌhar, -ˌher\ *noun* (1920)
1 : an impractical intellectual
2 : a person of artistic gifts or interests; *especially* **:** a lover of classical music
3 : a person with long hair; *especially* **:** HIPPIE
4 : a domestic cat having long outer fur
— long·haired \-'hard, -'herd\ *or* **long·hair** *adjective*

long·hand \-ˌhand\ *noun* (1666)
: HANDWRITING: **a :** characters or words written out fully by hand **b :** cursive writing

long haul *noun* (1936)
1 : a long distance
2 : a considerable period of time; *especially* **:** LONG RUN
— long–haul *adjective*

long·head \'lȯŋ-ˌhed\ *noun* (1650)
: a dolichocephalic person

long·head·ed \-ˌhe-dəd\ *adjective* (circa 1700)
1 : having unusual foresight
2 : DOLICHOCEPHALIC
— long·head·ed·ness *noun*

long·horn \-ˌhȯrn\ *noun* (1834)
1 a : any of the long-horned cattle of Spanish derivation formerly common in southwestern U.S. **b :** TEXAS LONGHORN 2
2 : a firm-textured usually mild cheese (as cheddar or Colby)

long–horned beetle \-ˌhȯrn(d)-\ *noun* (1840)
: any of a family (Cerambycidae synonym Longicornia) of beetles usually distinguished by their very long antennae — called also *longhorn beetle*

long–horned grasshopper *noun* (1893)
: any of various grasshoppers (family Tettigoniidae) distinguished by their very long antennae

long horse *noun* (circa 1934)
: VAULTING HORSE

long·house \'lȯŋ-ˌhau̇s, -ˈhau̇s\ *noun* (1643)
: a long communal dwelling of some North American Indians (as the Iroquois)

long hundredweight *noun* (circa 1934)
British **:** HUNDREDWEIGHT 2

lon·gi·corn \'län-jə-ˌkȯrn\ *adjective* [ultimately from Latin *longus* long + *cornu* horn — more at HORN] (circa 1848)
1 : of, relating to, or being long-horned beetles
2 : having long antennae
— longicorn *noun*

long·ing \'lȯŋ-iŋ\ *noun* (before 12th century)
: a strong desire especially for something unattainable **:** CRAVING
— long·ing·ly \-iŋ-lē\ *adverb*

long·ish \'lȯŋ-ish\ *adjective* (1611)
: somewhat long **:** moderately long

lon·gi·tude \'län-jə-ˌtüd, -ˌtyüd, *British also* 'läŋ-gə-\ *noun* [Middle English, from Latin *longitudin-, longitudo,* from *longus*] (14th century)
1 a : angular distance measured on a great circle of reference from the intersection of the adopted zero meridian with this reference circle to the similar intersection of the meridian passing through the object **b :** the arc or portion of the earth's equator intersected between the meridian of a given place and the prime meridian and expressed either in degrees or in time
2 *archaic* **:** long duration

longitude 1a: hemisphere marked with meridians of longitude

lon·gi·tu·di·nal \ˌlän-jə-'tüd-nəl, -'tyüd-, -'t(y)ü-dᵊn-əl, *British also* ˌläŋ-gə-\ *adjective* (15th century)
1 : placed or running lengthwise
2 : of or relating to length or the lengthwise dimension
3 : involving the repeated observation or examination of a set of subjects over time with respect to one or more study variables
— lon·gi·tu·di·nal·ly *adverb*

longitudinal wave *noun* (circa 1931)
: a wave (as a sound wave) in which the particles of the medium vibrate in the direction of the line of advance of the wave

long johns \'lȯŋ-ˌjänz\ *noun plural* (1943)
: long underwear

long jump *noun* (1882)
: a track-and-field event in which a jump for distance is made usually from a running start
— long jumper *noun*

long·leaf pine \'lȯŋ-ˌlēf-\ *noun* (1796)
: a tall pine (*Pinus palustris*) of the southeastern U.S. with long needles in bundles of three and long cones that is a major timber tree; *also* **:** its tough coarse-grained reddish orange wood

long–leaved pine \-ˌlēv(d)-\ *noun* (1765)
: LONGLEAF PINE

long·line \'lȯŋ-ˌlīn, -'līn\ *noun* (1876)
: a heavy fishing line that may be several miles long and that has baited hooks in series

long–lin·er \-ˌlī-nər\ *noun* (1909)
: one that fishes with a longline; *also* **:** a fishing vessel used in long-lining

long–lin·ing \-ˌlī-niŋ\ *noun* (1877)
: fishing with a longline

long–lived \'lȯŋ-'livd *also* -'līvd\ *adjective* (14th century)
1 : having a long life **:** characterized by long life ⟨a *long-lived* family⟩
2 : lasting a long time **:** ENDURING ■

long meter *noun* (1718)
: a quatrain in iambic tetrameter in which the second and fourth lines and often the first and third lines rhyme — called also *long measure*

Lon·go·bard \'lȯŋ-gə-ˌbärd, 'läŋ-\ *noun, plural* **Longobards** *also* **Lon·go·bar·di** \ˌlȯŋ-gə-'bär-ˌdī, ˌläŋ-, -dē\ [Middle English *Longobardes,* plural, from Latin *Langobardi, Longobardi*] (14th century)
: LOMBARD 1a
— Lon·go·bar·dic \ˌlȯŋ-gə-'bär-dik, ˌläŋ-\ *adjective*

long play *noun* (1952)
: a long-playing record

long–play·ing \'lȯŋ-'plā-iŋ\ *adjective* (1929)
: designed to be played at 33⅓ revolutions per minute — used of a microgroove record

long–range \-'rānj\ *adjective* (1854)
1 : relating to or fit for long distances ⟨*long-range* rockets⟩
2 : involving or taking into account a long period of time ⟨*long-range* planning⟩

long run *noun* (1627)
: a relatively long period of time — usually used in the phrase *in the long run*
— long–run \'lȯŋ-'rən\ *adjective*

long·shore·man \'lȯŋ-ˌshōr-mən, -ˌshȯr-, ˌlȯŋ-'\ *noun* [*longshore,* short for *alongshore*] (1811)
: a person who loads and unloads ships at a seaport

long·shor·ing \'lȯŋ-ˌshōr-iŋ, -ˌshȯr-, ˌlȯŋ-'\ *noun* (1926)
: the act or occupation of working as a longshoreman

long shot \'lȯŋ-ˌshät\ *noun* (1867)
1 : a venture involving great risk but promising a great reward if successful; *also* **:** a venture unlikely to succeed
2 : an entry (as in a horse race) given little chance of winning

□ USAGE
long–lived Americans were once split almost evenly between the \-'livd\ and the \-'līvd\ variants of *long-lived* and *short-lived,* but today the preference in American and British speech is for \-'livd\. Objections to this variant have surfaced over the years, though. Some critics contend that the *-lived* element is formed from the noun *life* plus the suffix *-ed,* and so *-lived* should have the \ī\ vowel of *life.* In modern English, though, *-lived* has been reanalyzed as the past participle of the verb *live.* Different kinds of reanalysis have shaped English over the centuries, bringing about for example the second *r* in *bridegroom* and the *n* in *newt* and *notch,* as a check of their etymologies will reveal. The pronunciations of such reanalyzed forms, as in the case of *-lived,* have become fully acceptable over time.

3 : a bet in which the chances of winning are slight but the possible winnings great
— **by a long shot :** by a great deal
long·sight·ed \-ˌsī-təd\ *adjective* (circa 1790)
: FARSIGHTED
— **long·sight·ed·ness** *noun*
long since *adverb* (14th century)
1 : long ago ⟨promises *long since* forgotten⟩
2 : for a long time ⟨has *long since* been a devoted friend⟩
long·some \ˈlȯŋ-səm\ *adjective* (before 12th century)
: tediously long
— **long·some·ly** *adverb*
— **long·some·ness** *noun*
long·spur \ˈlȯŋ-ˌspər\ *noun* (1831)
: any of several long-clawed finches (especially genus *Calcarius*) of the arctic regions and the Great Plains of North America
long–stand·ing \-ˈstan-diŋ\ *adjective* (1814)
: of long duration
long–suf·fer·ing \-ˌsə-f(ə-)riŋ, -ˈsə-\ *noun* (1526)
: long and patient endurance of offense
— **long–suffering** *adjective*
— **long–suf·fer·ing·ly** \-lē\ *adverb*
long suit *noun* (circa 1876)
1 : a holding of more than the average number of cards in a suit
2 : the activity or quality in which one excels
long–term \ˈlȯŋ-ˈtərm\ *adjective* (1904)
1 : occurring over or involving a relatively long period of time
2 a : of, relating to, or constituting a financial operation or obligation based on a considerable term and especially one of more than 10 years ⟨*long-term* bonds⟩ **b :** generated by assets held for longer than six months ⟨a *long-term* capital gain⟩
long·time \ˈlȯŋ-ˈtīm\ *adjective* (1584)
: having been so for a long time **:** LONG-STANDING ⟨a *longtime* friend⟩ ⟨a *longtime* friendship⟩
Long Tom \ˈlȯŋ-ˈtäm\ *noun* [from the name *Tom*] (1832)
1 : a large land gun having a long range
2 : a trough for washing gold-bearing earth
long ton *noun* (1829)
— see WEIGHT table
lon·gueur \lōⁿ-ˈgœr\ *noun, plural* **longueurs** \-ˈgœr(z)\ [French, literally, length] (1791)
: a dull and tedious passage or section (as of a book)
long view *noun* (1912)
: an approach to a problem or situation that emphasizes long-range factors
long–wind·ed \ˌlȯŋ-ˈwin-dəd, ˈlȯŋ-ˌwin-\ *adjective* (1589)
1 : tediously long in speaking or writing
2 : not easily subject to loss of breath
— **long–wind·ed·ly** *adverb*
— **long–wind·ed·ness** *noun*
¹loo \ˈlü\ *noun* [short for obsolete English *lanterloo,* from French *lanturelu* twaddle] (1675)
1 : an old card game in which the winner of each trick or a majority of tricks takes a portion of the pool while losing players are obligated to contribute to the next pool
2 : money staked at loo
²loo *transitive verb* (1680)
: to obligate to contribute to a new pool at loo for failing to win a trick
³loo *noun* [origin unknown] (1940)
chiefly British **:** TOILET 3
loo·by \ˈlü-bē\ *noun, plural* **loobies** [Middle English *loby*] (14th century)
: an awkward clumsy fellow **:** LUBBER
loo·fah \ˈlü-fə\ *noun* [New Latin *luffa,* from Arabic *lūf*] (1887)
1 : any of a genus (*Luffa*) of Old World tropical plants of the gourd family with large fruits; *also* **:** its fruit
2 : a sponge consisting of the fibrous skeleton of the fruit of a loofah (especially *Luffa aegyptiaca*)
¹look \ˈlu̇k\ *verb* [Middle English, from Old English *lōcian;* akin to Old Saxon *lōcōn* to look] (before 12th century)
transitive verb
1 : to make sure or take care (that something is done)
2 : to ascertain by the use of one's eyes
3 a : to exercise the power of vision upon **:** EXAMINE **b** *archaic* **:** to search for
4 a : EXPECT, ANTICIPATE ⟨we *look* to have a good year⟩ **b :** to have in mind as an end ⟨*looking* to win back some lost profits⟩
5 *archaic* **:** to bring into a place or condition by the exercise of the power of vision
6 : to express by the eyes or facial expression
7 : to have an appearance that befits or accords with
intransitive verb
1 a : to exercise the power of vision **:** SEE **b :** to direct one's attention ⟨*look* upon the future with hope⟩ **c :** to direct the eyes
2 : to have the appearance or likelihood of being **:** SEEM ⟨it *looks* unlikely⟩ ⟨*looks* to be hard work⟩
3 : to have a specified outlook ⟨the house *looked* east⟩
4 : to gaze in wonder or surprise **:** STARE
5 : to show a tendency ⟨the evidence *looks* to acquittal⟩ ■
synonym see EXPECT
— **look after :** to take care of
— **look at :** CONSIDER 1 ⟨*looking at* the possibility of relocating⟩
— **look down one's nose :** to view something with arrogance, disdain, or disapproval
— **look for :** to await with hope or anticipation
— **look forward :** to anticipate with pleasure or satisfaction ⟨*looking forward* to your visit⟩
— **look into :** EXPLORE 1a
— **look the other way :** to direct one's attention away from something unpleasant or troublesome
— **look to 1 :** to direct one's attention to ⟨*looking to* the future⟩ **2 :** to rely upon ⟨*looks to* reading for relaxation⟩
²look *noun* (13th century)
1 a : the act of looking **b :** GLANCE
2 a : the expression of the countenance **b :** physical appearance; *especially* **:** attractive physical appearance — usually used in plural **c :** a combination of design features giving a unified appearance ⟨a new *look* in design⟩
3 : the state or form in which something appears
look–alike \ˈlu̇-kə-ˌlīk\ *noun* (1947)
: one that looks like another **:** DOUBLE
— **look–alike** *adjective*
look·down \ˈlu̇k-ˌdau̇n\ *noun* (circa 1882)
: a silvery carangid fish (*Selene vomer*) chiefly of the Atlantic having a laterally compressed deep body and steeply sloping facial profile
look down *intransitive verb* (14th century)
1 : to be in a position that affords a downward view
2 : to regard with contempt **:** DESPISE — used with *on* or *upon*
look·er \ˈlu̇-kər\ *noun* (14th century)
1 : one that looks
2 a : one having an appearance of a specified kind **b :** one that has an attractive appearance **:** BEAUTY
look·er–on \ˌlu̇-kər-ˈȯn, -ˈän\ *noun, plural* **lookers–on** (1539)
: ONLOOKER
look–in \ˈlu̇k-ˌin\ *noun* (1870)
1 : a chance of success
2 : a quick pass in football to a receiver running diagonally toward the center of the field
looking glass *noun* (1562)
: MIRROR
look·out \ˈlu̇k-ˌau̇t\ *noun* (1699)
1 : one engaged in keeping watch **:** WATCHMAN
2 : an elevated place or structure affording a wide view for observation
3 : a careful looking or watching ⟨on the *lookout*⟩
4 : VIEW, OUTLOOK
5 : a matter of care or concern
look out *intransitive verb* (1602)
: to take care or concern oneself — used with *for* ⟨*looking out* for number one⟩
look over *transitive verb* (14th century)
: to inspect or examine especially in a cursory way
look–see \ˈlu̇k-ˈsē, -ˌsē\ *noun* (1883)
: a general survey **:** EVALUATION, INSPECTION
look·up \ˈlu̇k-ˌəp\ *noun* (1936)
: the process or an instance of looking something up; *especially* **:** the process of matching by computer the words of a text with material stored in memory
look up (14th century)
intransitive verb
1 : to cheer up ⟨*look up*—things are not all bad⟩
2 : to improve in prospects or conditions ⟨business is *looking up*⟩
transitive verb
1 : to search for in or as if in a reference work ⟨*look up* a phone number⟩
2 : to seek out especially for a brief visit
¹loom \ˈlüm\ *noun* [Middle English *lome* tool, loom, from Old English *gelōma* tool; akin to Middle Dutch al*lame* tool] (15th century)
: a frame or machine for interlacing at right angles two or more sets of threads or yarns to form a cloth
²loom *intransitive verb* [origin unknown] (circa 1541)
1 : to come into sight in enlarged or distorted and indistinct form often as a result of atmospheric conditions
2 a : to appear in an impressively great or exaggerated form ⟨deficits *loomed* large⟩ **b :** to take shape as an impending occurrence
³loom *noun* (1836)
: the indistinct and exaggerated appearance of something seen on the horizon or through fog or darkness; *also* **:** a looming shadow or reflection
¹loon \ˈlün\ *noun* [Middle English *loun*] (15th century)
1 : LOUT, IDLER
2 *chiefly Scottish* **:** BOY
3 a : a crazy person **b :** SIMPLETON
²loon *noun* [of Scandinavian origin; akin to Old Norse *lōmr* loon] (1634)

□ USAGE
look Some people are offended or made uneasy by the construction in which *look* is followed by *to* and an infinitive. This construction, usually expressing expectation (*transitive verb* sense 4a), has been in use since the 17th century ⟨good morrow . . . I *look'd* to have found you with your headache and morning qualms —Thomas Shadwell⟩ ⟨is elderly now and . . . *looks* to be gathered to his fathers before the times get any worse —Rudyard Kipling⟩. In the 19th century the present participle began to turn up in this construction ⟨two lovers *looking* to be wed —A. E. Housman⟩. These uses carry the notion of purpose and are the beginning of sense 4b. This construction is very common at present ⟨had been *looking* to change jobs for several years —Curry Kirkpatrick⟩ ⟨*looking* to blaze her own trails as an actress —*Town & Country*⟩ ⟨a number of states are now *looking* to circumvent the national limit —Daniel Seligman⟩. The linking verb sense of *look* (*intransitive verb* sense 2) is commonly followed by *to be* ⟨sure *looked* to be the hero of this game —Rick Telander⟩ ⟨the baby *looks* to be about nine or ten months old —Léon Bing⟩. All of these constructions are standard.

: any of several large birds (genus *Gavia*) of Holarctic regions that feed on fish by diving and have their legs placed far back under the body for optimal locomotion underwater

loo·ny *also* **loo·ney** \ˈlü-nē\ *adjective* **loo·ni·er; -est** [by shortening & alteration from *lunatic*] (1872)
: CRAZY, FOOLISH
— loo·ni·ness *noun*
— loony *noun*

loony bin *noun* (1919)
: an insane asylum **:** MADHOUSE

[1]**loop** \ˈlüp\ *noun* [Middle English *loupe;* perhaps akin to Middle Dutch *lupen* to watch, peer] (14th century)
archaic **:** LOOPHOLE 1a

[2]**loop** *noun* [Middle English *loupe,* of unknown origin] (14th century)
1 a : a curving or doubling of a line so as to form a closed or partly open curve within itself through which another line can be passed or into which a hook may be hooked **b :** such a fold of cord or ribbon serving as an ornament
2 a : something shaped like a loop **b :** a circular airplane maneuver executed in the vertical plane
3 : a ring or curved piece used to form a fastening, handle, or catch
4 : a closed electric circuit
5 : a piece of film or magnetic tape whose ends are spliced together so as to project or play back the same material continuously
6 : a series of instructions (as for a computer) that is repeated until a terminating condition is reached
7 : a sports league
8 : a select well-informed inner circle that is influential in decision making ⟨out of the policy *loop*⟩
— for a loop : into a state of amazement, confusion, or distress

[3]**loop** (1832)
intransitive verb
1 : to make or form a loop
2 : to execute a loop in an airplane
3 : to move in loops or in an arc
transitive verb
1 a : to make a loop in, on, or about **b :** to fasten with a loop
2 : to join (two courses of loops) in knitting
3 : to connect (electric conductors) so as to complete a loop
4 : to cause to move in an arc

looped \ˈlüpd\ *adjective* (1513)
1 : having, formed in, or characterized by loops ⟨*looped* fabrics⟩
2 : DRUNK 1a

loop·er \ˈlü-pər\ *noun* (1731)
1 : any of the usually rather small hairless caterpillars that are mostly larvae of moths (families Geometridae and Noctuidae) and move with a looping movement in which the anterior legs and the posterior prolegs are alternately made fast and released — called also *inchworm*
2 : one that loops

[1]**loop·hole** \ˈlüp-ˌhōl\ *noun* [[1]*loop*] (1591)
1 a : a small opening through which small arms may be fired **b :** a similar opening to admit light and air or to permit observation
2 : a means of escape; *especially* **:** an ambiguity or omission in the text through which the intent of a statute, contract, or obligation may be evaded

[2]**loophole** *transitive verb* (1664)
: to make loopholes in

loop of Hen·le \-ˈhen-lē\ [F. G. J. *Henle* (died 1885) German pathologist] (1890)
: a U-shaped part of the nephron of birds and mammals that lies between and is continuous with the proximal and distal convoluted tubules and that functions in water resorption

loopy \ˈlü-pē\ *adjective* **loop·i·er; -est** (1856)
1 : having or characterized by loops
2 : CRAZY, BIZARRE

[1]**loose** \ˈlüs\ *adjective* **loos·er; loos·est** [Middle English *lous,* from Old Norse *lauss;* akin to Old High German *lōs* loose — more at -LESS] (13th century)
1 a : not rigidly fastened or securely attached **b** (1) **:** having worked partly free from attachments ⟨a *loose* tooth⟩ (2) **:** having relative freedom of movement **c :** produced freely and accompanied by raising of mucus ⟨a *loose* cough⟩ **d :** not tight-fitting
2 a : free from a state of confinement, restraint, or obligation ⟨a lion *loose* in the streets⟩ ⟨spend *loose* funds wisely⟩ **b :** not brought together in a bundle, container, or binding **c** *archaic* **:** DISCONNECTED, DETACHED
3 : not dense, close, or compact in structure or arrangement
4 a : lacking in restraint or power of restraint ⟨a *loose* tongue⟩ ⟨*loose* bowels⟩ **b :** lacking moral restraint **:** UNCHASTE
5 a : not tightly drawn or stretched **:** SLACK **b :** being flexible or relaxed ⟨stay *loose*⟩
6 a : lacking in precision, exactness, or care ⟨*loose* brushwork⟩ ⟨*loose* usage⟩ **b :** permitting freedom of interpretation
7 : not in the possession of either of two competing teams ⟨a *loose* ball⟩ ⟨a *loose* puck⟩
— loose·ly *adverb*
— loose·ness *noun*

[2]**loose** *verb* **loosed; loos·ing** (13th century)
transitive verb
1 a : to let loose **:** RELEASE **b :** to free from restraint
2 : to make loose **:** UNTIE ⟨*loose* a knot⟩
3 : to cast loose **:** DETACH
4 : to let fly **:** DISCHARGE
5 : to make less rigid, tight, or strict **:** RELAX
intransitive verb
: to let fly a missile (as an arrow) **:** FIRE

[3]**loose** *adverb* (15th century)
: in a loose manner **:** LOOSELY

loose box *noun* (1849)
British **:** BOX STALL

loose cannon *noun* (1981)
: a dangerously uncontrollable person or thing

loose end *noun* (1546)
1 : something left hanging loose
2 : a fragment of unfinished business — usually used in plural

loose–joint·ed \ˈlüs-ˈjȯin-təd\ *adjective* (1859)
1 : having joints apparently not closely articulated
2 : characterized by unusually free movements
— loose–joint·ed·ness *noun*

loose–leaf \ˈlüs-ˈlēf\ *adjective* (1902)
1 : having leaves secured in book form in a cover whose backbone may be opened for adding, arranging, or removing leaves ⟨*loose-leaf* notebook⟩
2 : of, relating to, or used with a loose-leaf binding ⟨*loose-leaf* paper⟩

loos·en \ˈlü-sᵊn\ *verb* **loos·ened; loos·en·ing** \ˈlüs-niŋ, ˈlü-sᵊn-iŋ\ (14th century)
transitive verb
1 : to release from restraint
2 : to make looser
3 : to relieve (the bowels) of constipation
4 : to cause or permit to become less strict — often used with *up*
intransitive verb
: to become loose or looser

loosen up *intransitive verb* (1938)
: to become less tense **:** RELAX

loose sentence *noun* (circa 1890)
: a sentence in which the principal clause comes first and subordinate modifiers or trailing elements follow

loose smut *noun* (1890)
: a smut disease of grains in which the entire head is transformed into a dusty mass of spores

loose·strife \ˈlü(s)-ˌstrīf\ *noun* [intended as translation of Greek *lysimacheios* loosestrife (as if from *lysis* act of loosing + *machesthai* to fight) — more at LYSIS] (1548)
1 : any of a genus (*Lysimachia*) of plants of the primrose family with leafy stems and usually yellow or white flowers
2 : any of a genus (*Lythrum,* family Lythraceae, the loosestrife family) of herbs having entire leaves and including some with showy spikes of purple flowers; *especially* **:** PURPLE LOOSESTRIFE

loos·ey–goos·ey \ˌlü-sē-ˈgü-sē\ *adjective* (1964)
: notably loose or relaxed **:** not tense

[1]**loot** \ˈlüt\ *noun* [Hindi *lūṭ;* akin to Sanskrit *luṇṭati* he plunders] (circa 1788)
1 : goods usually of considerable value taken in war **:** SPOILS
2 : something held to resemble goods of value seized in war: as **a :** something appropriated illegally often by force or violence **b :** illicit gains by public officials **c :** MONEY
3 : the action of looting
synonym see SPOIL

[2]**loot** (1845)
transitive verb
1 a : to plunder or sack in war **b :** to rob especially on a large scale and usually by violence or corruption
2 : to seize and carry away by force especially in war
intransitive verb
: to engage in robbing or plundering especially in war
— loot·er *noun*

[1]**lop** \ˈläp\ *noun* [Middle English *loppe*] (14th century)
: material cut away from a tree; *especially* **:** parts discarded in lumbering

[2]**lop** *transitive verb* **lopped; lop·ping** (1519)
1 a (1) **:** to cut off branches or twigs from (2) **:** to sever from a woody plant **b** (1) *archaic* **:** to cut off the head or limbs of (2) **:** to cut from a person
2 a : to remove superfluous parts from **b :** to eliminate as unnecessary or undesirable — usually used with *off*
— lop·per *noun*

[3]**lop** *intransitive verb* **lopped; lop·ping** [perhaps imitative] (1578)
: to hang downward **:** DROOP

[1]**lope** \ˈlōp\ *noun* [Middle English *loup, lope* leap, from Old Norse *hlaup;* akin to Old English *hlēapan* to leap — more at LEAP] (1809)
1 : an easy natural gait of a horse resembling a canter
2 : an easy usually bounding gait capable of being sustained for a long time

[2]**lope** *intransitive verb* **loped; lop·ing** (circa 1825)
: to move or ride at a lope
— lop·er *noun*

lop–eared \ˈläp-ˌird\ *adjective* (1687)
: having ears that droop

loph·o·phore \ˈlä-fə-ˌfōr, -ˌfȯr\ *noun* [Greek *lophos* crest + English *-phore*] (1850)
: a circular or horseshoe-shaped organ about the mouth especially of a brachiopod or bryozoan that bears tentacles and functions especially in food-getting

lop·sid·ed \ˈläp-ˌsī-dəd\ *adjective* (1711)
1 : leaning to one side
2 : lacking in balance, symmetry, or proportion **:** disproportionately heavy on one side ⟨a *lopsided* vote of 99–1⟩
— lop·sid·ed·ly *adverb*
— lop·sid·ed·ness *noun*

lo·qua·cious \lō-ˈkwā-shəs\ *adjective* [Latin *loquac-, loquax,* from *loqui* to speak] (1667)
1 : given to fluent or excessive talk **:** GARRULOUS
2 : full of excessive talk **:** WORDY
synonym see TALKATIVE

— **lo·qua·cious·ly** *adverb*
— **lo·qua·cious·ness** *noun*

lo·quac·i·ty \-ˈkwa-sə-tē\ *noun* (13th century)
: the quality or state of being very talkative

lo·quat \ˈlō-ˌkwät\ *noun* [Chinese (Guangdong) *làuh-gwāt*] (1820)
: an Asian evergreen tree (*Eriobotrya japonica*) of the rose family often cultivated for its fruit; *also* **:** its small yellow edible fruit used especially for preserves

lo·ran \ˈlōr-ˌan, ˈlȯr-\ *noun* [*lo*ng-*ra*nge *n*avigation] (1932)
: a system of long-range navigation in which pulsed signals sent out by two pairs of radio stations are used to determine the geographical position of a ship or an airplane

[1]**lord** \ˈlȯrd\ *noun* [Middle English *loverd, lord,* from Old English *hlāford,* from *hlāf* loaf + *weard* keeper — more at LOAF, WARD] (before 12th century)
1 : one having power and authority over others: **a :** a ruler by hereditary right or preeminence to whom service and obedience are due **b :** one of whom a fee or estate is held in feudal tenure **c :** an owner of land or other real property **d** *obsolete* **:** the male head of a household **e :** HUSBAND **f :** one that has achieved mastery or that exercises leadership or great power in some area ⟨a drug *lord*⟩
2 *capitalized* **a :** GOD 1 **b :** JESUS
3 : a man of rank or high position: as **a :** a feudal tenant whose right or title comes directly from the king **b :** a British nobleman: as (1) **:** BARON 2a (2) **:** an hereditary peer of the rank of marquess, earl, or viscount (3) **:** the son of a duke or a marquess or the eldest son of an earl (4) **:** a bishop of the Church of England **c** *plural, capitalized* **:** HOUSE OF LORDS
4 — used as a British title: as **a** — used as part of an official title ⟨*Lord* Advocate⟩ **b** — used informally in place of the full title for a marquess, earl, or viscount **c** — used for a baron **d** — used by courtesy before the name and surname of a younger son of a duke or a marquess
5 : a person chosen to preside over a festival
◆

[2]**lord** *intransitive verb* (14th century)
: to act like a lord; *especially* **:** to put on airs — usually used with *it* ⟨*lords* it over his friends⟩

lord chancellor *noun, plural* **lords chancellor** (15th century)
: a British officer of state who presides over the House of Lords in both its legislative and judicial capacities, serves as the head of the British judiciary, and is usually a leading member of the cabinet

lord·ing \ˈlȯr-diŋ\ *noun* (13th century)
1 *archaic* **:** LORD
2 *obsolete* **:** LORDLING

lord·ling \ˈlȯrd-liŋ\ *noun* (13th century)
: a little or insignificant lord

lord·ly \-lē\ *adjective* **lord·li·er; -est** (before 12th century)
1 a : of, relating to, or having the characteristics of a lord **:** DIGNIFIED **b :** GRAND, NOBLE
2 : exhibiting such pride and assurance as could only be felt as appropriate to one of the highest birth or rank
synonym see PROUD
— **lord·li·ness** *noun*
— **lordly** *adverb*

lord of misrule (15th century)
: a master of Christmas revels in England especially in the 15th and 16th centuries

lor·do·sis \lȯr-ˈdō-səs\ *noun* [New Latin, from Greek *lordōsis,* from *lordos* curving forward; akin to Old English *belyrtan* to deceive] (1704)
1 : abnormal curvature of the spine forward
2 : a mating posture of some sexually receptive female mammals (as rats) in which the head and rump are raised and the back is arched downward
— **lor·dot·ic** \-ˈdä-tik\ *adjective*

Lord Protector of the Commonwealth (circa 1653)
: PROTECTOR 2b

Lord's day *noun, often D capitalized* (12th century)
: SUNDAY

lord·ship \ˈlȯrd-ˌship\ *noun* (before 12th century)
1 a : the rank or dignity of a lord — used as a title **b :** the authority or power of a lord **:** DOMINION
2 : the territory under the jurisdiction of a lord **:** SEIGNIORY

Lord's Prayer *noun* (circa 1549)
: the prayer with variant versions in Matthew and Luke that according to the Lucan account Christ taught his disciples

Lord's Supper *noun* (14th century)
: COMMUNION 2a

Lord's table *noun, often T capitalized* (1535)
: ALTAR 2

Lordy \ˈlȯr-dē\ *interjection* [[1]*lord* (God) + [4]*-y*] (1853)
— used to express surprise or strength of feeling

[1]**lore** \ˈlōr, ˈlȯr\ *noun* [Middle English, from Old English *lār;* akin to Old High German *lēra* doctrine, Old English *leornian* to learn] (before 12th century)
1 *archaic* **:** something that is taught **:** LESSON
2 : something that is learned: **a :** knowledge gained through study or experience **b :** traditional knowledge or belief
3 : a particular body of knowledge or tradition
word history see LEARN

[2]**lore** *noun* [New Latin *lorum,* from Latin, thong, rein; akin to Greek *eulēra* reins] (1828)
: the space between the eye and bill in a bird or the corresponding region in a reptile or fish
— **lo·re·al** \ˈlōr-ē-əl, ˈlȯr-\ *adjective*

Lo·re·lei \ˈlōr-ə-ˌlī, ˈlȯr-\ *noun* [German]
: a siren of Germanic legend whose singing lures Rhine River boatmen to destruction on a reef

lor·gnette \lȯrn-ˈyet\ *noun* [French, from *lorgner* to take a sidelong look at, from Middle French, from *lorgne* squinting] (1803)
: a pair of eyeglasses or opera glasses with a handle

lor·gnon \lȯrn-ˈyō^{n}\ *noun* [French, from *lorgner*] (1846)
: LORGNETTE

lo·ri·ca \lə-ˈrī-kə\ *noun, plural* **-cae** \-kē, -ˌsē\ [Latin] (circa 1706)
1 : a Roman cuirass of leather or metal
2 [New Latin, from Latin] **:** a hard protective case or shell (as of a rotifer)

lor·i·keet \ˈlȯr-ə-ˌkēt, ˈlär-\ *noun* [*lory* + *-keet* (as in *parakeet*)] (1770)
: any of numerous small arboreal chiefly Australasian parrots (family Loriidae) that usually have long slender tongue papillae which form an organ resembling a brush

lo·ris \ˈlōr-əs, ˈlȯr-\ *noun* [French, probably from obsolete Dutch *loeris* simpleton] (1774)
: any of several nocturnal slow-moving tailless arboreal primates (family Lorisidae): as **a :** a slim-bodied primate (*Loris tardigradus*) of southern India and Sri Lanka **b :** either of two larger related primates (*Nycticebus pygmaeus* or *N. coucang*) of southeastern Asia that are heavier limbed and slower moving

loris a

lorn \ˈlȯrn\ *adjective* [Middle English, from *loren,* past participle of *lesen* to lose, from Old English *lēosan* — more at LOSE] (14th century)
: DESOLATE, FORSAKEN

Lor·raine cross \lə-ˈrān-, lȯ-\ *noun* (1898)
: CROSS OF LORRAINE

lor·ry \ˈlȯr-ē, ˈlär-\ *noun, plural* **lorries** [origin unknown] (1908)
chiefly British **:** MOTORTRUCK

lo·ry \ˈlōr-ē, ˈlȯr-\ *noun, plural* **lories** [Dutch, from Malay *nuri, luri*] (1682)
: any of numerous parrots (family Loriidae) of Australia, New Guinea, and adjacent islands related to the lorikeets and usually having the tongue papillose at the tip and the mandibles less toothed than in other parrots

lose \ˈlüz\ *verb* **lost** \ˈlȯst\; **los·ing** \ˈlü-ziŋ\ [Middle English, from Old English *losian* to perish, lose, from *los* destruction; akin to Old English *lēosan* to lose; akin to Old Norse *losa* to loosen, Latin *luere* to atone for, Greek *lyein* to loosen, dissolve, destroy] (before 12th century)
transitive verb
1 a : to bring to destruction — used chiefly in passive construction ⟨the ship was *lost* on the reef⟩ **b :** DAMN ⟨if he shall gain the whole world and *lose* his own soul —Matthew 16:26 (Authorized Version)⟩
2 : to miss from one's possession or from a customary or supposed place
3 : to suffer deprivation of **:** part with especially in an unforeseen or accidental manner
4 a : to suffer loss through the death or removal of or final separation from (a person) **b :** to fail to keep control of or allegiance of ⟨*lose* votes⟩ ⟨*lost* his temper⟩
5 a : to fail to use **:** let slip by **:** WASTE ⟨*lose* the tide⟩ **b** (1) **:** to fail to win, gain, or obtain ⟨*lose* a prize⟩ ⟨*lose* a contest⟩ (2) **:** to undergo defeat in ⟨*lost* every battle⟩ **c :** to fail to catch with the senses or the mind ⟨*lost* part of what she said⟩
6 : to cause the loss of ⟨one careless statement *lost* him the election⟩
7 : to fail to keep, sustain, or maintain ⟨*lost* my balance⟩
8 a : to cause to miss one's way or bearings ⟨*lost* himself in the maze of streets⟩ **b :** to make (oneself) withdrawn from immediate reality ⟨*lost* herself in daydreaming⟩
9 a : to wander or go astray from ⟨*lost* his way⟩ **b :** to draw away from **:** OUTSTRIP ⟨*lost* his pursuers⟩
10 : to fail to keep in sight or in mind
11 : to free oneself from **:** get rid of ⟨dieting to *lose* some weight⟩
intransitive verb
1 : to undergo deprivation of something of value
2 : to undergo defeat ⟨*lose* with good grace⟩
3 *of a timepiece* **:** to run slow
— **los·able** \ˈlü-zə-bəl\ *adjective*

◇ WORD HISTORY
lord *Hlāford,* the Old English predecessor of *lord,* contains the word *hlāf* "bread, loaf," as does *hlǣfdige,* the ancestor of *lady.* The second element of *hlāford* was *weard* "keeper, guard" (hence Modern English *ward*), and the fuller form *hlāfweard* is actually found in one Old English manuscript. At first this compound undoubtedly alluded to the head of a household in his role as "the keeper of the bread," the provider for his family, servants, and retainers, but at a relatively early period it was applied more broadly to anyone in a position of power or authority, from king on down. This generalization may have contributed to the phonetic attrition of *-weard* to *-ord,* as the original sense of the compound lost relevance. Early Middle English *loverd* became *lord* through loss of the second vowel and merger of the \v\ into the preceding vowel; words that underwent a similar alteration include Modern English *head, hawk, auger,* and *lark,* from Old English *hēafod, hafoc, nafogār,* and *lāwerce.* See in addition LADY.

— **los·able·ness** *noun*
— **lose ground :** to suffer loss or disadvantage **:** fail to advance or improve
— **lose one's heart :** to fall in love

lo·sel \ˈlō-zəl\ *noun* [Middle English, from *losen* (past participle of *lesen* to lose), alteration of *loren* — more at LORN] (14th century)
: a worthless person

lose out *intransitive verb* (circa 1858)
: to fail to win in competition **:** fail to receive an expected reward or gain

los·er \ˈlü-zər\ *noun* (1548)
1 : one that loses especially consistently
2 : one who is incompetent or unable to succeed; *also* **:** something doomed to fail or disappoint

loss \ˈlȯs\ *noun* [Middle English *los,* probably back-formation from *lost,* past participle of *losen* to lose] (13th century)
1 : DESTRUCTION, RUIN
2 a : the act of losing possession **b :** the harm or privation resulting from loss or separation **c :** an instance of losing
3 : a person or thing or an amount that is lost: as **a** *plural* **:** killed, wounded, or captured soldiers **b :** the power diminution of a circuit or circuit element corresponding to conversion of electrical energy into heat by resistance
4 a : failure to gain, win, obtain, or utilize **b :** an amount by which the cost of an article or service exceeds the selling price
5 : decrease in amount, magnitude, or degree
6 : the amount of an insured's financial detriment by death or damage that the insurer becomes liable for
— **at a loss :** UNCERTAIN, PUZZLED
— **for a loss :** into a state of distress

loss leader *noun* (1917)
: something (as merchandise) sold at a loss in order to draw customers
— **loss–leader** *adjective*

loss ratio *noun* (1926)
: the ratio between insurance losses incurred and premiums earned during a given period

lossy \ˈlȯ-sē\ *adjective* (1946)
: causing attenuation or dissipation of electrical energy ⟨a *lossy* transmission line⟩ ⟨a *lossy* dielectric⟩

lost \ˈlȯst\ *adjective* [past participle of *lose*] (15th century)
1 : not made use of, won, or claimed
2 a : no longer possessed **b :** no longer known
3 : ruined or destroyed physically or morally **:** DESPERATE
4 a : taken away or beyond reach or attainment **:** DENIED ⟨regions *lost* to the faith⟩ **b :** INSENSIBLE, HARDENED ⟨*lost* to shame⟩
5 a : unable to find the way **b :** no longer visible **c :** lacking assurance or self-confidence **:** HELPLESS
6 : RAPT, ABSORBED ⟨*lost* in reverie⟩
7 : not appreciated or understood **:** WASTED ⟨their jokes were *lost* on me⟩
— **lost·ness** \ˈlȯs(t)-nəs\ *noun*

lost wax *noun* (1909)
: a process used in metal casting that consists of making a wax model, coating it with a refractory to form a mold, heating until the wax melts and runs out of the mold, and then pouring metal into the vacant mold

¹lot \ˈlät\ *noun* [Middle English, from Old English *hlot;* akin to Old High German *hlōz*] (before 12th century)
1 : an object used as a counter in determining a question by chance
2 a : the use of lots as a means of deciding something **b :** the resulting choice
3 a : something that comes to one upon whom a lot has fallen **:** SHARE **b :** one's way of life or worldly fate **:** FORTUNE
4 a : a portion of land **b :** a measured parcel of land having fixed boundaries and designated on a plot or survey **c :** a motion-picture studio and its adjoining property
5 a : a number of units of an article, a single article, or a parcel of articles offered as one item (as in an auction sale) **b :** all the members of a present group, kind, or quantity — used with *the*
6 a : a number of associated persons **:** SET **b :** KIND, SORT
7 : a considerable quantity or extent ⟨a *lot* of money⟩ ⟨*lots* of friends⟩ ■
synonym see FATE
— **a lot 1 :** to a considerable degree or extent ⟨this is *a lot* nicer⟩ **2 :** OFTEN, FREQUENTLY ⟨runs *a lot* every day⟩ **3 :** LOTS

²lot *transitive verb* **lot·ted; lot·ting** (15th century)
1 : ALLOT, APPORTION
2 : to form or divide into lots

Lot \ˈlät\ *noun* [Hebrew *Lōṭ*]
: a nephew of Abraham who according to the account in Genesis escaped from the doomed city of Sodom with his wife who turned into a pillar of salt when she looked back

lo·ta *also* **lo·tah** \ˈlō-tə\ *noun* [Hindi *loṭā*] (1809)
: a small usually spherical water vessel of brass or copper used in India

loth \ˈlōth, ˈlōth\ *variant of* LOATH

lo·thar·io \lō-ˈthar-ē-ˌō, -ˈther-, -ˈthär-\ *noun, plural* **-i·os** *often capitalized* [*Lothario,* seducer in the play *The Fair Penitent* (1703) by Nicholas Rowe] (1756)
: a man whose chief interest is seducing women ◆

lo·ti \ˈlō-tē\ *noun, plural* **ma·lo·ti** \mə-ˈlō-tē\ [Sesotho] (1980)
— see MONEY table

lo·tic \ˈlō-tik\ *adjective* [Latin *lotus,* past participle of *lavere*] (1916)
: of, relating to, or living in actively moving water ⟨a *lotic* habitat⟩ — compare LENTIC

lo·tion \ˈlō shən\ *noun* [Middle English *loscion,* from Latin *lotion-, lotio* act of washing, from *lavere* to wash — more at LYE] (14th century)
: a liquid preparation for cosmetic or external medicinal use

lots \ˈläts\ *adverb* [plural of ¹*lot*] (1891)
: MUCH ⟨feeling *lots* better⟩

lotte \ˈlät, ˈlȯt\ *noun* [French, from Middle French] (1977)
: MONKFISH

lot·tery \ˈlä-tə-rē *also* ˈlä-trē\ *noun, plural* **-ter·ies** *often attributive* [Middle French *loterie,* from Middle Dutch, from *lot* lot; akin to Old English *hlot* lot] (1567)
1 a : a drawing of lots in which prizes are distributed to the winners among persons buying a chance **b :** a drawing of lots used to decide something
2 : an event or affair whose outcome is or seems to be determined by chance

lot·to \ˈlä-(ˌ)tō\ *noun* [Italian, lottery, lotto, from French *lot* lot, of Germanic origin; akin to Old English *hlot* lot] (1778)
: a game of chance resembling bingo

lo·tus \ˈlō-təs\ *noun* [Latin & Greek; Latin *lotus,* from Greek *lōtos*] (circa 1541)
1 *also* **lo·tos** \ˈlō-təs\ **:** a fruit eaten by the lotus-eaters and considered to cause indolence and dreamy contentment; *also* **:** a tree (as *Zizyphus lotus* of the buckthorn family) reputed to bear this fruit
2 : any of various water lilies including several represented in ancient Egyptian and Hindu art and religious symbolism
3 [New Latin, from Latin] **a :** any of a genus (*Lotus*) of widely distributed upright leguminous herbs or subshrubs **b :** SWEET CLOVER

lotus 2

lo·tus–eat·er \-ˌē-tər\ *noun* (1832)
1 : any of a people in Homer's Odyssey subsisting on the lotus and living in the dreamy indolence it induces
2 : an indolent person

lo·tus·land \-ˌland\ *noun* [from the Homeric land of lotus-eaters] (1842)
1 : a place inducing contentment especially through offering an idyllic living situation
2 : a state or an ideal marked by contentment often achieved through self-indulgence

lotus position *noun* [from the supposed resemblance of the position to a lotus blossom] (1953)
: a cross-legged sitting position used in yoga in which the right foot is on the left thigh and the left foot is on the right thigh

louche \ˈlüsh\ *adjective* [French, literally, cross-eyed, squint-eyed, from Latin *luscus* blind in one eye] (1819)
: not reputable or decent

loud \ˈlau̇d\ *adjective* [Middle English, from Old English *hlūd;* akin to Old High German *hlūt* loud, Latin *inclutus* famous, Greek *klytos,* Sanskrit *śṛṇoti* he hears] (before 12th century)
1 a : marked by intensity or volume of sound **b :** producing a loud sound
2 : CLAMOROUS, NOISY

□ USAGE
lot The phrases *a lot of* and *lots of* have been stigmatized as colloquial and unworthy of being used in writing since at least Ambrose Bierce did it in 1909. A surprising number of handbooks are still repeating his opinion even though the expressions are clearly well established and standard on all levels of usage except perhaps the most elevated. Literary use is common ⟨a *lot* of my old St. Louis chums will be out in the Spring —Mark Twain⟩ ⟨the Seurat paintings want a *lot* of seeing to appreciate —Arnold Bennett⟩ ⟨there were *lots* of people in Oxford like that —John Galsworthy⟩ ⟨I have sat around in bars with a *lot* of other ex-newspapermen —James Thurber⟩. Similarly the adverbial uses of *a lot* and *lots* are standard ⟨gave him something *a lot* more important —William Faulkner⟩ ⟨I hope you're *lots* better now —F. Scott Fitzgerald⟩. The spelling *alot* is rare in print and is generally considered a mistake, especially by editors and English teachers, who, along with computer spelling checkers, are evidently responsible for its rarity in print. It is by all accounts very common in spontaneously produced writing—notes, letters, drafts, memoranda, and the like. And it is well attested in the world of interactive computer networks, where no editors or spelling checkers are found.

◇ WORD HISTORY
lothario The word *lothario* derives from a character in a now seldom-performed tragedy, *The Fair Penitent* (1703) by Nicholas Rowe (1674–1718). In the play Lothario is a notorious seducer who is described as "haughty, gallant, and gay." He seduces Calista, an unfaithful wife and later the fair penitent of the title. Rowe's Lothario became a pattern for later characterizations in English literature, most notably Lovelace in Samuel Richardson's novel *Clarissa* (1747–48). Since the 18th century *lothario* has been used generically for a foppish, unscrupulous rake.

3 : obtrusive or offensive in appearance or smell **:** OBNOXIOUS ☆
— **loud** *adverb*
— **loud·ly** *adverb*

loud·en \'laù-dᵊn\ *verb* **loud·ened; loud·en·ing** \'laùd-niŋ, 'laù-dᵊn-iŋ\ (circa 1848)
intransitive verb
: to become loud
transitive verb
: to make loud

loud–hail·er \,laùd-'hā-lər\ *noun* (1941)
chiefly British **:** BULLHORN

loud·mouth \'laùd-,maùth\ *noun* (1914)
: a loudmouthed person

loud·mouthed \-,maùthd, -,maùtht\ *adjective* (1628)
: given to loud offensive talk

loud·ness *noun* (before 12th century)
: the attribute of a sound that determines the magnitude of the auditory sensation produced and that primarily depends on the amplitude of the sound wave involved

loud·speak·er \'laùd-,spē-kər\ *noun* (1920)
: a device that changes electrical signals into sounds loud enough to be heard at a distance

Lou Geh·rig's disease \,lü-'ger-igz-, -'gar-\ *noun* [*Lou Gehrig* (died 1941) American baseball player who suffered from the disease] (1958)
: AMYOTROPHIC LATERAL SCLEROSIS

lough \'läk, 'läk\ *noun* [Middle English, of Celtic origin; akin to Old Irish *loch* lake; akin to Latin *lacus* lake — more at LAKE] (14th century)
1 *chiefly Irish* **:** LAKE
2 *chiefly Irish* **:** a bay or inlet of the sea

lou·is d'or \,lü-ē-'dòr\ *noun, plural* **louis d'or** [French, from *Louis* XIII of France + *d'or* of gold] (1665)
1 : a French gold coin first struck in 1640 and issued up to the Revolution
2 : the French 20-franc gold piece issued after the Revolution

Lou·is Qua·torze \,lü-ē-kə-'tòrz\ *adjective* [French, Louis XIV] (1848)
: of, relating to, or characteristic of the architecture or furniture of the reign of Louis XIV of France

Louis Quinze \-'kaⁿz\ *adjective* [French, Louis XV] (1855)
: of, relating to, or characteristic of the architecture or furniture of the reign of Louis XV of France

Louis Seize \-'säz, -'sez\ *adjective* [French, Louis XVI] (1882)
: of, relating to, or characteristic of the architecture or furniture of the reign of Louis XVI of France

Louis Treize \-'träz, -'trez\ *adjective* [French, Louis XIII] (1883)
: of, relating to, or characteristic of the architecture or furniture of the reign of Louis XIII of France

¹lounge \'laùnj\ *verb* **lounged; loung·ing** [origin unknown] (1508)
intransitive verb
: to act or move idly or lazily **:** LOAF
transitive verb
: to pass (time) idly

²lounge *noun* (1775)
1 : a place for lounging: as **a :** a room in a private home or public building for leisure occupations **:** LIVING ROOM; *also* **:** LOBBY **b :** a room in a usually public building or vehicle often combining lounging, smoking, and toilet facilities
2 : a long couch
3 : an article of clothing or furniture designed for comfort and leisure use

lounge car *noun* (1947)
: CLUB CAR

lounge lizard *noun* (1918)
1 : LADIES' MAN
2 : FOP
3 : a social parasite

loung·er \'laùn-jər\ *noun* (1508)
1 : one that lounges; *especially* **:** IDLER
2 : an article of clothing or furniture designed for comfort and leisure use

lounge suit *noun* (1901)
chiefly British **:** BUSINESS SUIT

lounge·wear \'laùnj-,war, -,wer\ *noun* (circa 1957)
: informal clothing usually designed to be worn at home

loup \'laùp, 'lōp\ *verb* [Middle English, from Old Norse *hlaupa;* akin to Old English *hlēapan* to leap — more at LEAP] (14th century)
chiefly Scottish **:** LEAP
— **loup** *noun*

loupe \'lüp\ *noun* [French] (circa 1775)
: a small magnifier used especially by jewelers and watchmakers

loup–ga·rou \,lü-gə-'rü\ *noun, plural* **loups–garous** \,lü-gə-'rü(z)\ [Middle French, from Old French *leu garoul,* from *leu* (from Latin *lupus*) wolf + *garoul* werewolf, of Germanic origin; akin to Old High German *werwolf* werewolf] (circa 1580)
: WEREWOLF

lour \'laù(-ə)r\, **loury** \'laù(ə)r-ē\ *variant of* LOWER, LOWERY

¹louse \'laùs\ *noun* [Middle English *lous,* from Old English *lūs;* akin to Old High German *lūs* louse, Welsh *llau* lice] (before 12th century)
1 *plural* **lice** \'līs\ **a :** any of various small wingless usually flattened insects (orders Anoplura and Mallophaga) parasitic on warm-blooded animals **b :** a small usually sluggish arthropod that lives on other animals or on plants and sucks their blood or juices — usually used in combination ⟨plant *louse*⟩ **c :** any of several small arthropods that are not parasitic — usually used in combination ⟨book *louse*⟩ ⟨wood *louse*⟩
2 *plural* **lous·es** \'laù-səz\ **:** a contemptible person **:** HEEL

²louse \'laùs, 'laùz\ *transitive verb* **loused; lous·ing** (14th century)
: to pick lice from **:** DELOUSE

louse up (1934)
transitive verb
: FOUL UP, SNARL
intransitive verb
: to make a mess

louse·wort \'laùs-,wərt, -,wòrt\ *noun* (1597)
: any of a genus (*Pedicularis*) of herbs of the snapdragon family typically having pinnatifid leaves and bilabiate flowers in terminal spikes

lousy \'laù-zē\ *adjective* **lous·i·er; -est** (14th century)
1 : infested with lice
2 a : totally repulsive **:** CONTEMPTIBLE **b :** miserably poor or inferior ⟨got *lousy* grades⟩ ⟨felt *lousy* after dinner⟩ **c :** amply supplied **:** REPLETE ⟨*lousy* with money⟩
3 *of silk* **:** fuzzy and specked because of splitting of the fiber
— **lous·i·ly** \-zə-lē\ *adverb*
— **lous·i·ness** \-zē-nəs\ *noun*

¹lout \'laùt\ *intransitive verb* [Middle English, from Old English *lūtan;* akin to Old Norse *lūta* to bow down] (before 12th century)
1 : to bow in respect
2 : SUBMIT, YIELD

²lout *noun* [perhaps from Old Norse *lūtr* bent down, from *lūta*] (1542)
: an awkward brutish person

³lout *transitive verb* (circa 1530)
: to treat as a lout **:** SCORN

lout·ish \'laù-tish\ *adjective* (1542)
: resembling or befitting a lout
synonym see BOORISH
— **lout·ish·ly** *adverb*
— **lout·ish·ness** *noun*

lou·ver *or* **lou·vre** \'lü-vər\ *noun* [Middle English *lover,* from Middle French *lovier*] (14th century)
1 : a roof lantern or turret often with slatted apertures for escape of smoke or admission of light in a medieval building
2 a : an opening provided with one or more slanted fixed or movable fins to allow flow of air but to exclude rain or sun or to provide privacy **b :** a finned or vaned device for controlling a flow of air or the radiation of light **c :** a fin or shutter of a louver
— **lou·vered** *also* **lou·vred** \-vərd\ *adjective*

lov·able *also* **love·able** \'lə-və-bəl\ *adjective* (14th century)
: having qualities that attract affection
— **lov·abil·i·ty** \,lə-və-'bi-lə-tē\ *noun*
— **lov·able·ness** *noun*
— **lov·ably** \-blē\ *adverb*

lov·age \'lə-vij\ *noun* [Middle English *lovache,* from Anglo-French, from Late Latin *levisticum,* alteration of Latin *ligusticum,* from neuter of *ligusticus* Ligurian, from *Ligur-, Ligus,* noun, Ligurian] (14th century)
: any of several aromatic perennial herbs of the carrot family; *especially* **:** a European herb (*Levisticum officinale*) sometimes cultivated as a domestic remedy, flavoring agent, or potherb

lov·a·stat·in \'lō-və-,sta-tən, 'lə-; ,lō-və-', ,lə-\ *noun* [*lov-* (perhaps from *mevinolin,* an earlier name) + connective *-a-* + *-stat* + ¹*-in*] (1987)
: a drug $C_{24}H_{36}O_8$ used to lower cholesterol levels in the blood

lov·at \'lə-vət\ *noun* [probably from T. A. Fraser, Lord *Lovat* (died 1875) Scottish nobleman who popularized muted tweeds] (1907)
: a predominantly dusty color mixture (as of green) in fabrics

¹love \'ləv\ *noun* [Middle English, from Old English *lufu;* akin to Old High German *luba* love, Old English *lēof* dear, Latin *lubēre, libēre* to please] (before 12th century)
1 a (1) **:** strong affection for another arising out of kinship or personal ties ⟨maternal *love* for a child⟩ (2) **:** attraction based on sexual desire **:** affection and tenderness felt by lovers (3) **:** affection based on admiration, benevolence, or common interests ⟨*love* for his old schoolmates⟩ **b :** an assurance of love ⟨give her my *love*⟩
2 : warm attachment, enthusiasm, or devotion ⟨*love* of the sea⟩
3 a : the object of attachment, devotion, or admiration ⟨baseball was his first *love*⟩ **b** (1) **:** a beloved person **:** DARLING — often used as a term of endearment (2) *British* — used as an informal term of address
4 a : unselfish loyal and benevolent concern for the good of another: as (1) **:** the fatherly concern of God for humankind (2) **:** brotherly concern for others **b :** a person's adoration of God
5 : a god or personification of love
6 : an amorous episode **:** LOVE AFFAIR
7 : the sexual embrace **:** COPULATION
8 : a score of zero (as in tennis)
9 *capitalized, Christian Science* **:** GOD
— **at love :** holding one's opponent scoreless in tennis
— **in love :** inspired by affection

²love *verb* **loved; lov·ing** (before 12th century)
transitive verb
1 : to hold dear **:** CHERISH

☆ SYNONYMS
Loud, stentorian, earsplitting, raucous, strident mean marked by intensity or volume of sound. LOUD applies to any volume above normal and may suggest undue vehemence or obtrusiveness ⟨*loud* shouts of protest⟩. STENTORIAN implies great power and range ⟨an actor with a *stentorian* voice⟩. EARSPLITTING implies loudness that is physically discomforting ⟨the *earsplitting* sound of a siren⟩. RAUCOUS implies a loud harsh grating tone, especially of voice, and may suggest rowdiness ⟨the *raucous* shouts of drunken revelers⟩. STRIDENT implies a rasping discordant but insistent quality, especially of voice ⟨the *strident* voices of hecklers⟩.

2 a : to feel a lover's passion, devotion, or tenderness for **b** (1) **:** CARESS (2) **:** to fondle amorously (3) **:** to copulate with
3 : to like or desire actively **:** take pleasure in ⟨*loved* to play the violin⟩
4 : to thrive in ⟨the rose *loves* sunlight⟩
intransitive verb
: to feel affection or experience desire

love affair *noun* (1591)
1 : a romantic attachment or episode between lovers
2 : a lively enthusiasm ⟨America's *love affair* with baseball⟩

love apple *noun* [probably translation of French *pomme d'amour*] (1578)
: TOMATO

love beads *noun plural* (1968)
: beads worn as a symbol of love and peace

love•bird \ˈləv-ˌbərd\ *noun* (1595)
: any of various small usually gray or green parrots (especially genus *Agapornis* of Africa) that show great affection for their mates

love•bug \-ˌbəg\ *noun* [from the fact that it is usually seen copulating] (circa 1966)
: a small black fly (*Plecia nearctica*) with a red thorax that swarms along highways in the Gulf states of the U.S.

love child *noun* (1805)
: an illegitimate child

love feast *noun* (1580)
1 : a meal eaten in common by a Christian congregation in token of brotherly love
2 : a gathering held to promote reconciliation and good feeling or show someone affectionate honor

love grass *noun* (1702)
: any of a genus (*Eragrostis*) of grasses that resemble the bluegrasses but have flattened spikelets and deciduous lemmas

love handles *noun plural* (1975)
: fatty bulges along the sides at the waist

love–in \ˈləv-ˌin\ *noun* (1967)
: a gathering of people especially for the expression of their mutual love

love–in–a–mist \ˈləv-ə-nə-ˌmist\ *noun* (circa 1760)
: a European and African annual herb (*Nigella damascena*) of the buttercup family having whitish flowers enveloped in numerous finely dissected bracts

love knot *noun* (14th century)
: a stylized knot sometimes used as an emblem of love

love•less \ˈləv-ləs\ *adjective* (14th century)
1 : having no love
2 : not loved
— love•less•ly *adverb*
— love•less•ness *noun*

love•lock \-ˌläk\ *noun* (1592)
: a long lock of hair variously worn (as over the front of the shoulder) especially by men in the 17th and 18th centuries

love•lorn \-ˌlȯrn\ *adjective* (1634)
: bereft of love or of a lover
— love•lorn•ness \-ˌlȯrn-nəs\ *noun*

[1]love•ly \ˈləv-lē\ *adjective* **love•li•er; -est** (before 12th century)
1 *obsolete* **:** LOVABLE
2 : delightful for beauty, harmony, or grace **:** ATTRACTIVE
3 : GRAND, SWELL
4 : eliciting love by moral or ideal worth
synonym see BEAUTIFUL
— love•li•ly \ˈləv-lə-lē\ *adverb*
— love•li•ness \ˈləv-lē-nəs\ *noun*
— lovely *adverb*

[2]lovely *noun, plural* **lovelies** (1652)
1 : a beautiful woman
2 : a lovely object

love•mak•ing \ˈləv-ˌmā-kiŋ\ *noun* (15th century)
1 : COURTSHIP
2 : sexual activity; *especially* **:** COPULATION

lov•er \ˈlə-vər\ *noun* (13th century)
1 a : a person in love; *especially* **:** a man in love with a woman **b** *plural* **:** two persons in love with each other
2 : an affectionate or benevolent friend
3 : DEVOTEE
4 a : PARAMOUR **b :** a person with whom one has sexual relations

lov•er•ly \-lē\ *adjective* (1875)
: resembling or befitting a lover

love seat *noun* (1904)
: a double chair, sofa, or settee for two persons

love•sick \ˈləv-ˌsik\ *adjective* (15th century)
1 : languishing with love **:** YEARNING
2 : expressing a lover's longing
— love•sick•ness *noun*

love•some \-səm\ *adjective* (before 12th century)
1 : WINSOME, LOVELY
2 : AFFECTIONATE, AMOROUS

lovey–dovey \ˌlə-vē-ˈdə-vē\ *adjective* (1886)
: expressing much love or sentimentality; *also* **:** MUSHY
— lovey–dovey•ness \-nəs\ *noun*

lov•ing \ˈlə-viŋ\ *adjective* (before 12th century)
1 : AFFECTIONATE
2 : PAINSTAKING
— lov•ing•ly \-viŋ-lē\ *adverb*
— lov•ing•ness *noun*

loving cup *noun* [from its former use in ceremonial drinking] (1812)
1 : a large ornamental drinking vessel with two or more handles
2 : a loving cup given as a token or trophy

lov•ing–kind•ness \ˌlə-viŋ-ˈkīn(d)-nəs\ *noun* (1535)
: tender and benevolent affection

[1]low \ˈlō\ *intransitive verb* [Middle English *loowen,* from Old English *hlōwan;* akin to Old High German *hluoen* to moo, Latin *calare* to call, summon, Greek *kalein*] (before 12th century)
: MOO

[2]low *noun* (1549)
: the deep sustained sound characteristic especially of a cow

[3]low *adjective* **low•er** \ˈlō(-ə)r\; **low•est** \ˈlō-əst\ [Middle English *lah, low,* from Old Norse *lāgr;* akin to Middle High German *læge* low, flat; probably akin to Old English *licgan* to lie] (12th century)
1 a : having a small upward extension or elevation ⟨a *low* wall⟩ **b :** situated or passing little above a reference line, point, or plane ⟨*low* bridges⟩ **c** (1) **:** having a low-cut neckline (2) **:** not extending as high as the ankle ⟨*low* oxfords⟩
2 a : situated or passing below the normal level, surface, or base of measurement, or the mean elevation ⟨*low* ground⟩ **b :** marking a nadir or bottom ⟨the *low* point of his career⟩
3 : DEAD — used as a predicate adjective
4 a : not loud **:** SOFT **b :** FLAT 8a **c :** characterized by being toward the bottom of the range of pitch attainable (as by an instrument)
5 a : being near the equator ⟨*low* northern latitudes⟩ **b :** being near the horizon
6 : socially or economically humble in character or status ⟨*low* birth⟩
7 a : lacking strength, health, or vitality **:** WEAK, PROSTRATE ⟨very *low* with pneumonia⟩ **b :** lacking spirit or vivacity **:** DEPRESSED ⟨a *low* frame of mind⟩
8 a : of lesser degree, size, or amount than average or ordinary ⟨*low* energy⟩ **b** (1) **:** small in number or amount (2) **:** SUBSTANDARD, INADEQUATE ⟨a *low* level of employment⟩ ⟨a *low* income group⟩ (3) **:** CHEAP ⟨*low* prices⟩ (4) **:** SHORT, DEPLETED ⟨oil is in *low* supply⟩ **c :** of lesser position, rank, or order
9 : falling short of some standard: as **a :** lacking dignity or elevation ⟨a *low* style of writing⟩ **b :** morally reprehensible **:** BASE ⟨a *low* trick⟩ **c :** COARSE, VULGAR ⟨*low* language⟩
10 a : not advanced in complexity, development, or elaboration ⟨*low* organisms⟩ **b** *often capitalized* **:** LOW CHURCH
11 : UNFAVORABLE, DISPARAGING ⟨had a *low* opinion of him⟩
12 : designed for slow and usually the slowest speed ⟨*low* gear⟩
13 : articulated with a wide opening between the relatively flat tongue and the palate **:** OPEN ⟨\ä\ is a *low* vowel⟩
14 : intended to attract little attention ⟨kept a *low* profile⟩
synonym see BASE
— low *adverb*
— low•ness *noun*

[4]low *noun* (12th century)
1 : something that is low: as **a :** DEPTH **b :** a region of low barometric pressure
2 : the transmission gear of an automotive vehicle giving the lowest ratio of driveshaft to crankshaft speed

[5]low *or* **lowe** \ˈlō\ *noun* [Middle English, from Old Norse *logi, log;* akin to Old English *lēoht* light — more at LIGHT] (13th century)
chiefly Scottish **:** FLAME, BLAZE

[6]low *or* **lowe** *verb* **lowed; low•ing** (14th century)
Scottish **:** FLAME, BLAZE

low•ball \ˈlō-ˌbȯl\ *transitive verb* (circa 1961)
: to give (a customer) a deceptively low price or cost estimate
— lowball *noun*

low beam *noun* (circa 1952)
: the short-range focus of a vehicle headlight

low blood pressure *noun* (1924)
: HYPOTENSION

low blow *noun* (1952)
: an unprincipled attack ⟨gossip column that landed one *low blow* after another —James Fallows⟩

low•born \ˈlō-ˈbȯrn\ *adjective* (13th century)
: born in a low condition or rank

low•boy \-ˌbȯi\ *noun* (circa 1891)
: a chest or side table about three feet (one meter) high with drawers and usually with cabriole legs

lowboy

low•bred \-ˈbred\ *adjective* (1757)
: RUDE, VULGAR

low•brow \-ˌbrau̇\ *noun* (1906)
: a person with little taste or intellectual interest
— lowbrow *adjective*

Low Church *adjective* (1710)
: tending especially in Anglican worship to minimize emphasis on the priesthood, sacraments, and ceremonial in worship and often to emphasize evangelical principles

Low Churchman *noun* (1702)
: a person holding or advocating Low Church views

low comedy *noun* (1608)
: comedy employing burlesque, horseplay, or the representation of low life — compare HIGH COMEDY

low country *noun, often L&C capitalized* (15th century)
: a low-lying country or region; *especially* **:** the part of a southern state extending from the seacoast inland to the fall line
— low–country *adjective, often L&C capitalized*

low–density lipoprotein *noun* (1951)
: LDL

low•down \ˈlō-ˌdau̇n\ *noun* (1915)
: the inside facts **:** DOPE

low–down \ˈlō-(ˌ)dau̇n\ *adjective* (1850)
1 : CONTEMPTIBLE, BASE
2 : deeply emotional ⟨*low-down* blues⟩

low earth orbit *noun* (1963)
: a usually circular orbit from about 90 to 600 miles (144 to 960 kilometers) above the earth

low–end \'lō-ˌend\ *adjective* (1926)
: of, relating to, or being the lowest priced merchandise in a manufacturer's line

[1]low·er \'lau̇(-ə)r, 'lō(-ə)r\ *intransitive verb* [Middle English *louren;* akin to Middle High German *lūren* to lie in wait] (13th century)
1 : to look sullen **:** FROWN
2 : to be or become dark, gloomy, and threatening

[2]lower *noun* (14th century)
: FROWN

[3]low·er \'lō(-ə)r\ *adjective* [[3]*low*] (13th century)
1 : relatively low in position, rank, or order
2 : SOUTHERN ⟨*lower* New York State⟩
3 : less advanced in the scale of evolutionary development
4 a : situated or held to be situated beneath the earth's surface **b** *capitalized* **:** being an earlier epoch or series of the period or series named ⟨*Lower* Cretaceous⟩ ⟨*Lower* Paleolithic⟩
5 : constituting the popular and often the larger and more representative branch of a bicameral legislative body ⟨*lower* house⟩

[4]low·er \'lō(-ə)r\ (1606)
intransitive verb
: to move down **:** DROP; *also* **:** DIMINISH
transitive verb
1 a : to let descend **:** LET DOWN **b :** to depress as to direction ⟨*lower* your aim⟩ **c :** to reduce the height of
2 a : to reduce in value, number, or amount **b** (1) **:** to bring down in quality or character **:** DEGRADE (2) **:** ABASE, HUMBLE **c :** to reduce the objective of

[1]low·er·case \ˌlō(-ə)r-'kās\ *adjective* [from the compositor's practice of keeping such types in the lower of a pair of type cases] (1683)
of a letter **:** having as its typical form a f g or *b n i* rather than A F G or *B N I*
— lowercase *noun*

[2]lowercase *transitive verb* **-cased; -cas·ing** (1908)
: to print or set in lowercase letters

low·er–class \ˌlō(-ə)r-'klas\ *adjective* (1892)
1 : of, relating to, or characteristic of the lower class
2 : being an inferior or low-ranking specimen of its kind

lower class *noun* (1772)
: a social class occupying a position below the middle class and having the lowest status in a society

lower criticism *noun* (circa 1889)
: criticism concerned with the recovery of original texts especially of Scripture through collation of extant manuscripts — compare HIGHER CRITICISM

lower fungus *noun* (1900)
: a fungus with hyphae absent or rudimentary and nonseptate

low·er·ing \'lau̇-(ə-)riŋ, 'lō-\ *adjective* (15th century)
: dark and threatening **:** GLOOMY

low·er·most \'lō(-ə)r-ˌmōst\ *adjective* (1547)
: LOWEST

low·ery \'lau̇-(ə-)rē, 'lō-\ *adjective* (15th century)
: GLOOMY, LOWERING

lowest common denominator *noun* (1924)
1 : LEAST COMMON DENOMINATOR
2 : something of small intellectual content designed to appeal to a lowbrow audience; *also* **:** such an audience

lowest common multiple *noun* (circa 1924)
: LEAST COMMON MULTIPLE

lowest terms *noun plural* (1806)
: the form of a fraction in which the numerator and denominator have no factor in common except 1 ⟨reduce a fraction to *lowest terms*⟩

low frequency *noun* (circa 1868)
: a radio frequency between medium frequency and very low frequency — see RADIO FREQUENCY table

Low German *noun* (1838)
1 : the German dialects of northern Germany especially as used since the end of the medieval period **:** PLATTDEUTSCH
2 : the West Germanic languages other than High German

low–grade \'lō-'grād\ *adjective* (1878)
1 : of inferior grade or quality
2 : being near that extreme of a specified range which is lowest, least intense, or least competent ⟨a *low-grade* fever⟩ ⟨a *low-grade* infection⟩

low–key \-'kē\ *also* **low–keyed** \-'kēd\ *adjective* (1907)
1 : having or producing dark tones only with little contrast
2 : of low intensity

[1]low·land \'lō-lənd, -ˌland\ *noun* (15th century)
: low or level country

[2]lowland *adjective* (1508)
1 *capitalized* **:** of or relating to the Lowlands of Scotland
2 : of or relating to a lowland

low·land·er \-lən-dər, -ˌlan-\ *noun* (1692)
1 *capitalized* **:** an inhabitant of the Lowlands of Scotland
2 : a native or inhabitant of a lowland region

Low Latin *noun* (1872)
: postclassical Latin in its later stages

low–lev·el \'lō-'le-vəl\ *adjective* (1881)
1 : occurring, done, or placed at a low level
2 : being of low importance or rank

low·life \'lō-ˌlīf\ *noun, plural* **low·lifes** \-ˌlīfs\ *also* **low·lives** \-ˌlīvz\ (1911)
1 : a person of low social status
2 : a person of low moral character
— low–life *adjective*

low·light \'lō-ˌlīt\ *noun* (1941)
: a particularly bad or unpleasant event, detail, or part

low·li·head \'lō-lē-ˌhed\ *noun* [Middle English *lowliheed,* from *lowly* + *-hed* -hood; akin to Middle English *-hod* -hood] (15th century)
archaic **:** lowly state

[1]low·ly \'lō-lē\ *adverb* (14th century)
1 : in a humble or meek manner
2 : in a low position, manner, or degree
3 : not loudly

[2]lowly *adjective* **low·li·er; -est** (14th century)
1 : humble in manner or spirit **:** free from self-assertive pride
2 : not lofty or sublime **:** PROSAIC
3 : ranking low in some hierarchy
4 : of or relating to a low social or economic rank
5 : low in the scale of biological or cultural evolution
— low·li·ness *noun*

low–ly·ing \'lō-'lī-iŋ\ *adjective* (1856)
1 : rising relatively little above the base of measurement ⟨*low-lying* hills⟩
2 : lying below the normal level, surface, or the base of measurement or mean elevation ⟨*low-lying* clouds⟩

low mass *noun, often L&M capitalized* (1568)
: a mass that is recited without singing by the celebrant, without a deacon, subdeacon, or choir assisting the celebrant, and without the use of incense

low–mind·ed \'lō-'mīn-dəd\ *adjective* (circa 1746)
: inclined to low or unworthy things
— low–mind·ed·ly *adverb*
— low–mind·ed·ness *noun*

lown \'lau̇n, 'lün\ *adjective* [Middle English (Scots) *lowne*] (15th century)
dialect **:** CALM, QUIET

low–pres·sure \'lō-'pre-shər\ *adjective* (1827)
1 : having, exerting, or operating under a relatively small pressure
2 : EASYGOING

low relief *noun* (1711)
: BAS-RELIEF

low–rent \'lō-'rent\ *adjective* (1957)
: low in character, cost, or prestige ⟨*low-rent* thugs⟩ ⟨a *low-rent* movie⟩ ⟨a *low-rent* literary form⟩

low–rise \'lō-'rīz\ *adjective* (1957)
1 : having few stories and not equipped with elevators ⟨a *low-rise* classroom building⟩
2 : of, relating to, or characterized by low-rise buildings ⟨a *low-rise* housing development⟩

low–slung \'lō-ˌsləŋ\ *adjective* (1943)
: relatively low to the ground or floor ⟨a *low-slung* convertible⟩ ⟨a *low-slung* modern building⟩ ⟨a *low-slung* sofa⟩

low–spir·it·ed \'lō-'spir-ə-təd\ *adjective* (1693)
: DEJECTED, DEPRESSED
— low–spir·it·ed·ly *adverb*
— low–spir·it·ed·ness *noun*

Low Sunday *noun* (15th century)
: the Sunday following Easter

low–tech \'lō-'tek\ *adjective* (1981)
: technologically simple or unsophisticated ⟨*low-tech* industries⟩

low tide *noun* (1843)
: the farthest ebb of the tide

low water *noun* (15th century)
: a low stage of the water in a river or lake; *also* **:** LOW TIDE

[1]lox \'läks\ *noun* [*l*iquid *ox*ygen] (1923)
: liquid oxygen

[2]lox *noun, plural* **lox** *or* **lox·es** [Yiddish *laks,* from Middle High German *lahs* salmon, from Old High German; akin to Old English *leax* salmon] (1941)
: smoked salmon

lox·o·drome \'läk-sə-ˌdrōm\ *noun* [back-formation from *loxodromic* of a rhumb line, from French *loxodromique,* from Greek *loxos* oblique + *dromos* course — more at DROMEDARY] (1880)
: RHUMB LINE

loy·al \'lȯi(-ə)l\ *adjective* [Middle French, from Old French *leial, leel,* from Latin *legalis* legal] (1531)
1 : unswerving in allegiance: as **a :** faithful in allegiance to one's lawful sovereign or government **b :** faithful to a private person to whom fidelity is due **c :** faithful to a cause, ideal, custom, institution, or product
2 : showing loyalty
3 *obsolete* **:** LAWFUL, LEGITIMATE
synonym see FAITHFUL
— loy·al·ly \'lȯi-ə-lē\ *adverb*

loy·al·ist \'lȯi-ə-list\ *noun* (1647)
: one who is or remains loyal especially to a political cause, party, government, or sovereign

loy·al·ty \'lȯi(-ə)l-tē\ *noun, plural* **-ties** [Middle English *loyaltee,* from Middle French *loialté,* from Old French *leialté,* from *leial*] (15th century)
: the quality or state or an instance of being loyal
synonym see FIDELITY

loz·enge \'lä-zᵊnj *also* -sᵊnj\ *noun* [Middle English *losenge,* from Middle French *losange*] (14th century)
1 : a figure with four equal sides and two acute and two obtuse angles **:** DIAMOND
2 : something shaped like a lozenge
3 : a small often medicated candy

LP \ˌel-'pē\ *noun* [*l*ong-*p*laying] (1948)
: a microgroove phonograph record designed to be played at 33⅓ revolutions per minute

LPN \ˌel-(ˌ)pē-'en\ *noun* (1948)
: LICENSED PRACTICAL NURSE

LSD \ˌel-(ˌ)es-'dē\ *noun* [German *L*yserg*s*äure-*D*iäthylamid lysergic acid diethylamide] (1950)
: an organic compound $C_{20}H_{25}N_3O$ that induces psychotic symptoms similar to those of schizophrenia — called also *lysergic acid diethylamide*

lu·au \'lü-,au̇\ *noun* [Hawaiian *lū'au*] (1853)
: an Hawaiian feast

Lu·ba·vitch·er \'lü-bə-,vi-chər, lü-'bä-\ *noun* [Yiddish *lyubavitsher,* from *Lyubavitsh,* town in Belarus] (1954)
: a member of a Hasidic sect founded by Schneour Zalman of Lyady in the late 18th century
— **Lubavitcher** *adjective*

lub·ber \'lə-bər\ *noun* [Middle English *lobre, lobur*] (14th century)
1 : a big clumsy fellow
2 : a clumsy seaman
— **lub·ber·li·ness** \-lē-nəs\ *noun*
— **lub·ber·ly** \-lē\ *adjective or adverb*

lubber line *noun* (1889)
: a fixed line on the compass of a ship or airplane that is aligned with the longitudinal axis of the vehicle

lubber's hole *noun* (circa 1784)
: a hole in a square-rigger's top near the mast through which one may go farther aloft without going over the rim by the futtock shrouds

lube \'lüb\ *noun* [short for *lubricating oil*] (1926)
1 : LUBRICANT
2 : an application of a lubricant **:** LUBRICATION

lu·bric \'lü-brik\ *adjective* [Middle French *lubrique,* from Medieval Latin *lubricus*] (15th century)
archaic **:** LUBRICIOUS
— **lu·bri·cal** \-bri-kəl\ *adjective*

lu·bri·cant \'lü-bri-kənt\ *noun* (circa 1828)
1 : a substance (as grease) capable of reducing friction, heat, and wear when introduced as a film between solid surfaces
2 : something that lessens or prevents friction or difficulty
— **lubricant** *adjective*

lu·bri·cate \'lü-brə-,kāt\ *verb* **-cat·ed; -cat·ing** [Latin *lubricatus,* past participle of *lubricare,* from *lubricus* slippery — more at SLEEVE] (circa 1623)
transitive verb
1 : to make smooth or slippery
2 : to apply a lubricant to
intransitive verb
: to act as a lubricant
— **lu·bri·ca·tion** \,lü-brə-'kā-shən\ *noun*
— **lu·bri·ca·tive** \'lü-brə-,kā-tiv\ *adjective*
— **lu·bri·ca·tor** \-,kā-tər\ *noun*

lu·bri·cious \lü-'bri-shəs\ *or* **lu·bri·cous** \'lü-bri-kəs\ *adjective* [Medieval Latin *lubricus,* from Latin, slippery, easily led astray] (1535)
1 : marked by wantonness **:** LECHEROUS; *also* **:** SALACIOUS
2 [Latin *lubricus*] **:** having a smooth or slippery quality ⟨a *lubricious* skin⟩
— **lu·bri·cious·ly** *adverb*

lu·bric·i·ty \lü-'bri-sə-tē\ *noun, plural* **-ties** (15th century)
: the property or state of being lubricious; *also* **:** the capacity for reducing friction

Lu·can \'lü-kən\ *adjective* [Late Latin *lucanus,* from *Lucas* Luke, from Greek *Loukas*] (1890)
: of or relating to Luke or the Gospel ascribed to him

lu·carne \lü-'kärn\ *noun* [French] (circa 1825)
: DORMER

Lu·ca·yo \lü-'kī-(,)ō\ *also* **Lu·ca·yan** \-'kī-ən\ *noun* (1929)
: a member of an Arawakan people of the Bahamas

lu·cen·cy \'lü-sᵊn(t)-sē\ *noun* (1656)
: the quality or state of being lucent

lu·cent \'lü-sᵊnt\ *adjective* [Middle English, from Latin *lucent-, lucens,* present participle of *lucēre* to shine — more at LIGHT] (15th century)
1 : glowing with light **:** LUMINOUS
2 : marked by clarity or translucence **:** CLEAR
— **lu·cent·ly** *adverb*

lu·cern *noun* [probably modification of German *lüchsern* of a lynx, from *Luchs* lynx] (circa 1533)
obsolete **:** LYNX

lu·cerne *also* **lu·cern** \lü-'sərn\ *noun* [French *luzerne,* from Provençal *luserno*] (1626)
chiefly British **:** ALFALFA

lu·cid \'lü-səd\ *adjective* [Latin *lucidus,* from *lucēre*] (1591)
1 a : suffused with light **:** LUMINOUS **b :** TRANSLUCENT
2 : having full use of one's faculties **:** SANE
3 : clear to the understanding **:** INTELLIGIBLE
synonym see CLEAR
— **lu·cid·ly** *adverb*
— **lu·cid·ness** *noun*

lu·cid·i·ty \lü-'si-də-tē\ *noun* (1851)
1 : clearness of thought or style
2 : a presumed capacity to perceive the truth directly and instantaneously **:** CLAIRVOYANCE

Lu·ci·fer \'lü-sə-fər\ *noun* [Middle English, the morning star, a fallen rebel archangel, the Devil, from Old English, from Latin, the morning star, from *lucifer* light-bearing, from *luc-, lux* light + *-fer* -ferous — more at LIGHT] (before 12th century)
1 — used as a name of the devil
2 : the planet Venus when appearing as the morning star
3 *not capitalized* **:** a friction match having as active substances antimony sulfide and potassium chlorate
— **Lu·ci·fe·ri·an** \,lü-sə-'fir-ē-ən\ *adjective*

lu·cif·er·ase \lü-'si-fə-,rās, -,rāz\ *noun* [International Scientific Vocabulary, from *luciferin*] (1888)
: an enzyme that catalyzes the oxidation of luciferin

lu·cif·er·in \-f(ə-)rən\ *noun* [International Scientific Vocabulary, from Latin *lucifer* light-bearing] (1888)
: any of various organic substances in luminescent organisms (as fireflies) that upon oxidation produce a virtually heatless light

lu·cif·er·ous \lü-'si-f(ə-)rəs\ *adjective* [Latin *lucifer*] (1648)
: bringing light or insight **:** ILLUMINATING

Lu·ci·na \lü-'sī-nə\ *noun* [Latin, Roman goddess of childbirth] (1658)
archaic **:** MIDWIFE

Lu·cite \'lü-,sīt\ *trademark*
— used for an acrylic resin or plastic consisting essentially of polymerized methyl methacrylate

[1]luck \'lək\ *noun* [Middle English *lucke,* from Middle Dutch *luc;* akin to Middle High German *gelücke* luck] (15th century)
1 a : a force that brings good fortune or adversity **b :** the events or circumstances that operate for or against an individual
2 : favoring chance; *also* **:** SUCCESS
— **luck·less** \-ləs\ *adjective*

[2]luck *intransitive verb* (circa 1584)
1 : to prosper or succeed especially through chance or good fortune — usually used with *out*
2 : to come upon something desirable by chance — usually used with *out, on, onto,* or *into*

luck·i·ly \'lə-kə-lē\ *adverb* (1530)
1 : in a lucky manner
2 : FORTUNATELY 2 ⟨*luckily,* we were on time⟩
usage see HOPEFULLY

lucky \'lə-kē\ *adjective* **luck·i·er; -est** (15th century)
1 : having good luck
2 : happening by chance **:** FORTUITOUS
3 : producing or resulting in good by chance **:** FAVORABLE
4 : seeming to bring good luck ⟨a *lucky* rabbit's foot⟩ ☆
— **luck·i·ness** \'lə-kē-nəs\ *noun*

lucky dip *noun* (1925)
British **:** GRAB BAG

lu·cra·tive \'lü-krə-tiv\ *adjective* [Middle English *lucratif,* from Middle French, from Latin *lucrativus,* from *lucratus,* past participle of *lucrari* to gain, from *lucrum*] (15th century)
: producing wealth **:** PROFITABLE
— **lu·cra·tive·ly** *adverb*
— **lu·cra·tive·ness** *noun*

lu·cre \'lü-kər\ *noun* [Middle English, from Latin *lucrum;* probably akin to Old English *lēan* reward, Old High German *lōn,* Greek *apolauein* to enjoy] (14th century)
: monetary gain **:** PROFIT; *also* **:** MONEY

lu·cu·bra·tion \,lü-kyə-'brā-shən, ,lü-kə-\ *noun* [Latin *lucubration-, lucubratio* study by night, work produced at night, from *lucubrare* to work by lamplight; akin to Latin *luc-, lux*] (1595)
: laborious or intensive study; *also* **:** the product of such study — usually used in plural

lu·cu·lent \'lü-kyə-lənt\ *adjective* [Latin *luculentus,* from *luc-, lux* light] (circa 1548)
: clear in thought or expression **:** LUCID
— **lu·cu·lent·ly** *adverb*

Lu·cul·lan \lü-'kə-lən\ *also* **Lu·cul·li·an** \-'kə-lē-ən\ *adjective* [Latin *lucullanus* of Licinius *Lucullus;* from his reputation for luxurious banquets] (1861)
: LAVISH, LUXURIOUS ⟨a *Lucullan* feast⟩

Ludd·ite \'lə-,dīt\ *noun* [perhaps from Ned *Ludd,* 18th century Leicestershire workman who destroyed machinery] (1811)
: one of a group of early 19th century English workmen destroying laborsaving machinery as a protest; *broadly* **:** one who is opposed to especially technological change ◆

☆ SYNONYMS
Lucky, fortunate, happy, providential mean meeting with unforeseen success. LUCKY stresses the agency of chance in bringing about a favorable result ⟨won because of a *lucky* bounce⟩. FORTUNATE suggests being rewarded beyond one's deserts ⟨*fortunate* in my investments⟩. HAPPY combines the implications of LUCKY and FORTUNATE with stress on being blessed ⟨a series of *happy* accidents⟩. PROVIDENTIAL more definitely implies the help or intervention of a higher power ⟨a *providential* change in the weather⟩.

◇ WORD HISTORY
Luddite Since *Luddite* first appeared in the fall of 1811, it has never gone out of use as a label for those opposed to new technology and its consequences. But introduction of labor-saving machinery was only one of the origins of the rampages in the Midlands and North of England that gave rise to the word. The initial attacks by discontented stocking and lace makers in Nottinghamshire, when the possibly fictitious Ned Ludd's name first appears in anonymous letters as the movement's leader, were directed against stocking-making frames of a type dating back to Elizabethan times. The first Luddites' raids were a labor protest at a time when trade unions were proscribed by law and sabotage of machinery, whether old or new, was an available form of collective action. Moreover, the disturbances took place against a background of crop failures and economic depression caused by Britain's blockade of continental Europe during the Napoleonic Wars. Concerning the origin of the name Ned Ludd, nothing is known for certain. Tradition holds that Ned Ludd (or Ludlam) was a Leicestershire knitter who out of anger at his employer destroyed his knitting frame, and that subsequently his name was proverbially connected with frame-breaking.

— **Luddite** *adjective*

lu·dic \'lü-dik\ *adjective* [French *ludique,* from Latin *ludus*] (1940)
: of, relating to, or characterized by play : PLAYFUL ⟨*ludic* behavior⟩ ⟨a *ludic* novel⟩

lu·di·crous \'lü-də-krəs\ *adjective* [Latin *ludicrus,* from *ludus* play, sport; perhaps akin to Greek *loidoros* abusive] (1782)
1 : amusing or laughable through obvious absurdity, incongruity, exaggeration, or eccentricity
2 : meriting derisive laughter or scorn as absurdly inept, false, or foolish
synonym see LAUGHABLE
— **lu·di·crous·ly** *adverb*
— **lu·di·crous·ness** *noun*

lu·es \'lü-(,)ēz\ *noun, plural* **lues** [New Latin, from Latin, plague; akin to Greek *lyein* to loosen, destroy — more at LOSE] (1634)
: SYPHILIS
— **lu·et·ic** \lü-'e-tik\ *adjective*

[1]**luff** \'ləf\ *noun* [Middle English, weather side of a ship, luff, from Middle French *lof* weather side of ship] (14th century)
1 : the act of sailing a ship nearer the wind
2 : the forward edge of a fore-and-aft sail

[2]**luff** *intransitive verb* (14th century)
: to turn the head of a ship toward the wind

luf·fa *variant of* LOOFAH

luft·mensch \'lu̇ft-,men(t)sh\ *noun, plural* **luft·mensch·en** \-,men(t)-shən\ [Yiddish *luftmentsh,* from *luft* air + *mentsh* human being] (1907)
: an impractical contemplative person having no definite business or income

[1]**lug** \'ləg\ *verb* **lugged; lug·ging** [Middle English *luggen* to pull by the hair or ear, drag, probably of Scandinavian origin; akin to Norwegian *lugga* to pull by the hair] (14th century)
transitive verb
1 : DRAG, PULL
2 : to carry laboriously
3 : to introduce in a forced manner ⟨*lugs* my name into the argument⟩
intransitive verb
1 : to pull with effort : TUG
2 : to move heavily or by jerks ⟨the car *lugs* on hills⟩
3 *of a racehorse* **:** to swerve from the course toward or away from the inside rail

[2]**lug** *noun* (1616)
1 *archaic* **a :** an act of lugging **b :** something that is lugged **c :** a shipping container for produce
2 : LUGSAIL
3 *plural* **:** superior airs or affectations ⟨put on *lugs*⟩
4 *slang* **:** an exaction of money — used in the phrase *put the lug on*

[3]**lug** *noun* [Middle English (Scots) *lugge,* perhaps from Middle English *luggen*] (15th century)
1 : something (as a handle) that projects like an ear: as **a :** a leather loop on a harness saddle through which the shaft passes **b :** a metal fitting to which electrical wires are soldered or connected
2 *chiefly British* **:** EAR
3 : a ridge (as on the bottom of a shoe) to increase traction
4 : a nut used to secure a wheel on an automotive vehicle — called also *lug nut*
5 : BLOCKHEAD, LOUT

Lu·gan·da \lü-'gän-də, -'gan-\ *noun* (1889)
: the Bantu language of the Ganda people

luge \'lüzh\ *noun* [French] (1905)
: a small sled that is ridden in a supine position and used especially in competition; *also* : the competition itself
— **luge** *intransitive verb*
— **lug·er** \'lü-zhər\ *noun*

lug·gage \'lə-gij\ *noun* (1596)
: something that is lugged; *especially* **:** suitcases for a traveler's belongings : BAGGAGE

lug·ger \'lə-gər\ *noun* [*lugsail*] (1757)
: a small fishing or coasting boat that carries one or more lugsails

lugger

lug·sail \'ləg-,sāl, -səl\ *noun* [perhaps from [3]*lug*] (1677)
: a 4-sided sail bent to an obliquely hanging yard that is hoisted and lowered with the sail

lu·gu·bri·ous \lu̇-'gü-brē-əs *also* -'gyü-\ *adjective* [Latin *lugubris,* from *lugēre* to mourn; akin to Greek *lygros* mournful] (1601)
1 : MOURNFUL; *especially* **:** exaggeratedly or affectedly mournful ⟨dark, dramatic and *lugubrious* brooding —V. S. Pritchett⟩
2 : DISMAL ⟨a *lugubrious* landscape⟩
— **lu·gu·bri·ous·ly** *adverb*
— **lu·gu·bri·ous·ness** *noun*

lug·worm \'ləg-,wərm\ *noun* [origin unknown] (1802)
: any of a genus (*Arenicola*) of marine polychaete worms that have a row of tufted gills along each side of the back and are used for bait

Luk·an \'lü-kən\ *variant of* LUCAN

Luke \'lük\ *noun* [Latin *Lucas,* from Greek *Loukas*]
1 : a Gentile physician and companion of the apostle Paul traditionally identified as the author of the third Gospel in the New Testament and of the book of Acts
2 : the third Gospel in the New Testament — see BIBLE table

luke·warm \'lük-'wȯrm\ *adjective* [Middle English, from *luke* lukewarm + *warm;* probably akin to Old High German *lāo* lukewarm — more at LEE] (14th century)
1 : moderately warm : TEPID
2 : lacking conviction : HALFHEARTED
— **luke·warm·ly** *adverb*
— **luke·warm·ness** *noun*

[1]**lull** \'ləl\ *transitive verb* [Middle English; probably of imitative origin] (14th century)
1 : to cause to sleep or rest : SOOTHE
2 : to cause to relax vigilance

[2]**lull** *noun* (1719)
1 *archaic* **:** something that lulls; *especially* : LULLABY
2 : a temporary pause or decline in activity ⟨the early morning *lull* in urban noise⟩: as **a :** a temporary calm before or during a storm **b :** a temporary drop in business activity

[1]**lul·la·by** \'lə-lə-,bī\ *noun, plural* **-bies** [obsolete English *lulla,* interjection used to lull a child (from Middle English) + *bye,* interjection used to lull a child, from Middle English *by*] (1588)
: a soothing refrain; *specifically* **:** a song to quiet children or lull them to sleep

[2]**lullaby** *transitive verb* **-bied; -by·ing** (1592)
: to quiet with or as if with a lullaby

lu·lu \'lü-(,)lü\ *noun* [probably from *Lulu,* nickname from *Louise*] (1886)
slang **:** one that is remarkable or wonderful

lum \'ləm\ *noun* [origin unknown] (circa 1628)
chiefly Scottish **:** CHIMNEY

lum·ba·go \,ləm-'bā-(,)gō\ *noun* [Latin, from *lumbus*] (circa 1693)
: usually painful muscular rheumatism involving the lumbar region

lum·bar \'ləm-bər, -,bär\ *adjective* [New Latin *lumbaris,* from Latin *lumbus* loin — more at LOIN] (circa 1656)
: of, relating to, or constituting the loins or the vertebrae between the thoracic vertebrae and sacrum ⟨*lumbar* region⟩

[1]**lum·ber** \'ləm-bər\ *intransitive verb* **lumbered; lum·ber·ing** \-b(ə-)riŋ\ [Middle English *lomeren*] (14th century)
1 : to move ponderously
2 : RUMBLE

[2]**lumber** *noun* [perhaps from *Lombard;* from the use of pawnshops as storehouses of disused property] (1552)
1 : surplus or disused articles (as furniture) that are stored away
2 a : timber or logs especially when dressed for use **b :** any of various structural materials prepared in a form similar to lumber
— **lumber** *adjective*

[3]**lumber** *verb* **lum·bered; lum·ber·ing** \-b(ə-)riŋ\ (1642)
transitive verb
1 : to clutter with or as if with lumber : ENCUMBER
2 : to heap together in disorder
3 : to log and saw the timber of
intransitive verb
1 : to cut logs for lumber
2 : to saw logs into lumber for the market
— **lum·ber·er** \-bər-ər\ *noun*

lum·ber·jack \'ləm-bər-,jak\ *noun* (1831)
: LOGGER

lum·ber·man \-mən\ *noun* (circa 1817)
: a person who is engaged in or oversees the business of cutting, processing, and marketing lumber

lumber room *noun* (1741)
: STOREROOM 1

lum·ber·yard \-,yärd\ *noun* (1786)
: a yard where a stock of lumber is kept for sale

lum·bo·sa·cral \,ləm-bō-'sa-krəl, -'sā-\ *adjective* (1840)
: relating to the lumbar and sacral regions or parts

lu·men \'lü-mən\ *noun, plural* **lumens** *also* **lu·mi·na** \-mə-nə\ [New Latin *lumin-, lumen,* from Latin, light, air shaft, opening] (1873)
1 : the cavity of a tubular organ ⟨the *lumen* of a blood vessel⟩
2 : the bore of a tube (as of a hollow needle or catheter)
3 : a unit of luminous flux equal to the light emitted in a unit solid angle by a uniform point source of one candle intensity
— **lu·mi·nal** *also* **lu·men·al** \-mə-nᵊl\ *adjective*

lumin- *or* **lumini-** *combining form* [Middle English *lumin-,* from Latin *lumin-, lumen*]
: light ⟨*lumini*ferous⟩

lu·mi·naire \,lü-mə-'nar, -'ner\ *noun* [French, lamp, lighting] (1921)
: a complete lighting unit

lu·mi·nance \'lü-mə-nən(t)s\ *noun* (1880)
1 : the quality or state of being luminous
2 : the luminous intensity of a surface in a given direction per unit of projected area

lu·mi·nar·ia \,lü-mə-'ner-ē-ə\ *noun, plural* **-nar·i·as** [Spanish, decorative light, from Late Latin] (1949)
: a traditional Mexican Christmas lantern originally consisting of a candle set in sand inside a paper bag

lu·mi·nary \'lü-mə-,ner-ē\ *noun, plural* **-naries** [Middle English *luminarye,* from Middle French & Late Latin; Middle French *luminaire* lamp, from Late Latin *luminaria,* plural of *luminare* lamp, heavenly body, from Latin, window, from *lumin-, lumen* light; akin to Latin *lucēre* to shine — more at LIGHT] (15th century)
1 : a person of prominence or brilliant achievement
2 : a body that gives light; *especially* **:** one of the celestial bodies
— **luminary** *adjective*

lu·mi·nesce \,lü-mə-'nes\ *intransitive verb* **-nesced; -nesc·ing** [back-formation from *luminescent*] (1896)
: to exhibit luminescence

lu·mi·nes·cence \-ˈne-sᵊn(t)s\ *noun* (1889)
: the low-temperature emission of light (as by a chemical or physiological process); *also* : light produced by luminescence
— **lu·mi·nes·cent** \-sᵊnt\ *adjective*

lu·mi·nif·er·ous \ˌlü-mə-ˈni-f(ə-)rəs\ *adjective* (1801)
: transmitting, producing, or yielding light

lu·mi·nism \ˈlü-mə-ˌni-zəm\ *noun, often capitalized* (circa 1974)
: a theory or practice of realist landscape and seascape painting developed in the U.S. in the mid 19th century and concerned with the study and depiction of effects of light and atmosphere
— **lu·mi·nist** \-nist\ *noun, often capitalized*

lu·mi·nos·i·ty \ˌlü-mə-ˈnä-sə-tē\ *noun, plural* **-ties** (1634)
1 a : the quality or state of being luminous **b :** something luminous
2 a : the relative quantity of light **b :** relative brightness of something
3 : the relative quantity of radiation emitted by a celestial source (as a star)

lu·mi·nous \ˈlü-mə-nəs\ *adjective* [Middle English, from Latin *luminosus,* from *lumin-, lumen*] (15th century)
1 a : emitting or reflecting usually steady, suffused, or glowing light **b :** of or relating to light or to luminous flux
2 : bathed in or exposed to steady light ⟨*luminous* with sunlight⟩
3 : CLEAR, ENLIGHTENING
synonym see BRIGHT
— **lu·mi·nous·ly** *adverb*
— **lu·mi·nous·ness** *noun*

luminous energy *noun* (circa 1931)
: energy transferred in the form of visible radiation

luminous flux *noun* (1925)
: radiant flux in the visible-wavelength range usually expressed in lumens instead of watts

luminous paint *noun* (circa 1889)
: a paint containing a phosphor (as zinc sulfide activated with copper) and so able to glow in the dark

lum·mox \ˈlə-məks, -miks\ *noun* [origin unknown] (circa 1825)
: a clumsy person

¹lump \ˈləmp\ *noun* [Middle English] (14th century)
1 : a piece or mass of indefinite size and shape
2 a : AGGREGATE, TOTALITY ⟨taken in the *lump*⟩ **b :** MAJORITY
3 : PROTUBERANCE; *especially* **:** an abnormal swelling
4 : a person who is heavy and awkward; *also* **:** one who is stupid or dull
5 *plural* **a :** BEATINGS, BRUISES ⟨had taken a lot of *lumps* growing up in the city⟩ **b :** DEFEAT, LOSS ⟨can cheerfully take his *lumps* on losers, because the payout is big on the winners —Martin Mayer⟩
— **lump in one's throat :** a constriction of the throat caused by emotion

²lump (1624)
transitive verb
1 : to group indiscriminately
2 : to make into lumps; *also* **:** to make lumps on or in
3 : to move noisily and clumsily
intransitive verb
1 : to become formed into lumps
2 : to move oneself noisily and clumsily

³lump *adjective* (circa 1700)
: not divided into parts **:** ENTIRE ⟨*lump* sum⟩

⁴lump *transitive verb* [origin unknown] (1791)
: to put up with ⟨like it or *lump* it⟩

lump·ec·to·my \ˌləm-ˈpek-tə-mē\ *noun, plural* **-mies** (1972)
: excision of a breast tumor with a limited amount of associated tissue

¹lum·pen \ˈlüm-pən, ˈləm-\ *adjective* [German *Lumpenproletariat* degraded section of the proletariat, from *Lump* contemptible person (from *Lumpen* rags) + *Proletariat*] (1936)
: of or relating to dispossessed and uprooted individuals cut off from the economic and social class with which they might normally be identified ⟨*lumpen* proletariat⟩ ⟨*lumpen* intellectuals⟩

²lumpen *noun, plural* **lumpen** *also* **lumpens** (1941)
: a member of the crude and uneducated lowest class of society

lump·er \ˈləm-pər\ *noun* (circa 1785)
1 : a laborer employed to handle freight or cargo
2 : one who classifies organisms into large often variable taxonomic groups based on major characters — compare SPLITTER

lump·fish \ˈləmp-ˌfish\ *noun* [obsolete English *lump* lumpfish (probably from Dutch *lomp* blenny, loach) + English *fish*] (circa 1620)
: a northern Atlantic usually greenish fish (*Cyclopterus lumpus* of the family Cyclopteridae) having rows of nodules on the body and eggs used as a caviar

lump·ish \ˈləm-pish\ *adjective* (1528)
1 : DULL, SLUGGISH
2 *obsolete* **:** low in spirits
3 : HEAVY, AWKWARD
4 : LUMPY 1a
5 : tediously slow or dull **:** BORING; *also* **:** LUMPY 3
— **lump·ish·ly** *adverb*
— **lump·ish·ness** *noun*

lumpy \ˈləm-pē\ *adjective* **lump·i·er; -est** (1707)
1 a : filled or covered with lumps **b :** characterized by choppy waves
2 : having a heavy clumsy appearance
3 : uneven and often crude in style
— **lump·i·ly** \-pə-lē\ *adverb*
— **lump·i·ness** \-pē-nəs\ *noun*

lumpy jaw *noun* (1890)
: ACTINOMYCOSIS; *especially* **:** actinomycosis of the head in cattle

lu·na·cy \ˈlü-nə-sē\ *noun, plural* **-cies** [*lunatic*] (1541)
1 : any of various forms of insanity: as **a :** intermittent insanity once believed to be related to phases of the moon **b :** insanity amounting to lack of capacity or of responsibility in the eyes of the law
2 : wild foolishness **:** extravagant folly
3 : a foolish act

lu·na moth \ˈlü-nə-\ *noun* [New Latin *luna* (specific epithet of *Actias luna*), from Latin, moon] (1869)
: a large mostly pale green American saturniid moth (*Actias luna*) with long tails on the hind wings

luna moth

lu·nar \ˈlü-nər *also* -ˌnär\ *adjective* [Middle English, from Latin *lunaris,* from *luna* moon; akin to Latin *lucēre* to shine — more at LIGHT] (15th century)
1 : CRESCENT, LUNATE
2 a : of or relating to the moon **b :** designed for use on the moon ⟨*lunar* vehicles⟩
3 : measured by the moon's revolution ⟨*lunar* month⟩

lunar caustic *noun* [obsolete *luna* silver, from Medieval Latin, from Latin, moon] (1771)
: silver nitrate especially when fused and molded into sticks for use as a caustic

lunar eclipse *noun* (circa 1737)
: an eclipse in which the moon near the full phase passes partially or wholly through the umbra of the earth's shadow — see ECLIPSE illustration

lunar module *noun* (1967)
: a space vehicle module designed to carry astronauts from the command module to the surface of the moon and back — called also *lunar excursion module;* abbreviation *LEM*

lu·nate \ˈlü-ˌnāt\ *adjective* [Latin *lunatus,* past participle of *lunare* to bend in a crescent, from *luna*] (circa 1777)
: shaped like a crescent

lu·na·tic \ˈlü-nə-ˌtik\ *adjective* [Middle English *lunatik,* from Old French or Late Latin; Old French *lunatique,* from Late Latin *lunaticus,* from Latin *luna;* from the belief that lunacy fluctuated with the phases of the moon] (14th century)
1 a : affected with lunacy **:** INSANE **b :** designed for the care of insane persons ⟨*lunatic* asylum⟩
2 : wildly foolish ◆
— **lunatic** *noun*

lunatic fringe *noun* (1913)
: the members of a usually political or social movement espousing extreme, eccentric, or fanatical views

lu·na·tion \lü-ˈnā-shən\ *noun* [Middle English *lunacioun,* from Medieval Latin *lunation-, lunatio,* from Latin *luna*] (14th century)
: the period of time averaging 29 days, 12 hours, 44 minutes, and 2.8 seconds elapsing between two successive new moons

¹lunch \ˈlənch\ *noun* [probably short for *luncheon*] (1812)
1 : a usually light meal; *especially* **:** one taken in the middle of the day
2 : the food prepared for a lunch
— **out to lunch** *slang* **:** out of touch with reality

²lunch (1823)
intransitive verb
: to eat lunch
transitive verb
: to treat to lunch
— **lunch·er** *noun*

lunch counter *noun* (1869)
1 : a long counter at which lunches are sold
2 : LUNCHEONETTE

lun·cheon \ˈlən-chən\ *noun* [perhaps alteration of *nuncheon* light snack] (circa 1652)
: LUNCH; *especially* **:** a formal usually midday meal as part of a meeting or for entertaining a guest

lun·cheon·ette \ˌlən-chə-ˈnet\ *noun* (1924)
: a small restaurant serving light lunches

◇ WORD HISTORY

lunatic The word *lunatic* is based ultimately on an old folk-medical belief, that human behavior is affected by the phases of the moon. The full moon, in particular, was once thought to trigger phenomena as diverse as epilepsy and schizophrenia. Greek authors of late antiquity and the early Middle Ages used the adjectives *selēnoblēktos* or *selēnoplēktos,* literally "moonstruck," to describe people influenced by the moon. (The word *moonstruck* is itself probably a loan translation of these words, introduced into English in the 17th century.) The Late Latin adjective *lunaticus,* a derivative of Latin *luna* "moon" (with the suffix of *fanaticus* "fanatic"), was coined to express the same notion. *Lunaticus* was borrowed into Old French as *lunatique* and from there into Middle English as *lunatik.* One peculiarity of *lunatic* that is due to its early appearance in English is its stress, which falls on the first syllable. All other English words ending in *-atic* are stressed on the *a.* Some words that were once stressed on the first syllable such as *phlegmatic* and *schismatic* have, in the last two centuries, joined the common pattern, so that *lunatic* now remains alone.

lunch·room \ˈlənch-ˌrüm, -ˌru̇m\ *noun* (1830)
1 : LUNCHEONETTE
2 : a room (as in a school) where lunches supplied on the premises or brought from home may be eaten

lunch·time \-ˌtīm\ *noun* (1859)
: the time at which lunch is usually eaten **:** NOON

lune \ˈlün\ *noun* [Latin *luna* moon — more at LUNAR] (circa 1704)
: the part of a plane surface bounded by two intersecting arcs or of a spherical surface bounded by two great circles

lunes \ˈlünz\ *noun plural* [French, plural of *lune* crazy whim, from Middle French, moon, crazy whim, from Latin *luna*] (1602)
: fits of lunacy

lu·nette \lü-ˈnet\ *noun* [French, from Old French *lunete* small object shaped like the moon, from *lune* moon] (circa 1673)
1 : something that has the shape of a crescent or half-moon: as **a :** an opening in a vault especially for a window **b :** the surface at the upper part of a wall that is partly surrounded by a vault which the wall intersects and that is often filled by windows or by mural painting **c :** a low crescentic mound (as of sand) formed by the wind
2 : the figure or shape of a crescent moon

lung \ˈləŋ\ *noun* [Middle English *lunge,* from Old English *lungen;* akin to Old High German *lungun* lung, *līhti* light in weight — more at LIGHT] (before 12th century)
1 a : one of the usually paired compound saccular thoracic organs that constitute the basic respiratory organ of air-breathing vertebrates **b :** any of various respiratory organs of invertebrates
2 a : a device enabling individuals abandoning a submarine to rise to the surface **b :** a mechanical device for regularly introducing fresh air into and withdrawing stale air from the lung **:** RESPIRATOR
— lung·ful \-ˌfu̇l\ *noun*

[1]lunge \ˈlənj\ *noun* [modification of French *allonge* extension, reach, from Old French *alonge,* from *alongier* to lengthen, from (assumed) Vulgar Latin *allongare,* from Latin *ad-* ad- + Late Latin *longare,* from Latin *longus* long] (1748)
1 : a quick thrust or jab (as of a sword) usually made by leaning or striding forward
2 : a sudden forward rush or reach ⟨made a *lunge* to catch the ball⟩

[2]lunge *verb* **lunged; lung·ing** (1821)
intransitive verb
: to make a lunge **:** move with or as if with a lunge
transitive verb
: to thrust or propel (as a blow) in a lunge

lunged \ˈləŋd\ *adjective* (1693)
1 : having lungs **:** PULMONATE
2 : having a lung or lungs of a specified kind or number — used in combination ⟨one-lunged⟩

[1]lung·er \ˈlən-jər\ *noun* (1842)
: one that lunges

[2]lung·er \ˈləŋ-ər\ *noun* (1893)
: a person suffering from a chronic disease of the lungs; *especially* **:** one who is tubercular

lung·fish \ˈləŋ-ˌfish\ *noun* (1883)
: any of a subclass (Dipneusti) of bony fishes that breathe by a modified swim bladder as well as gills

lung·worm \-ˌwərm\ *noun* (1882)
: any of various nematodes that infest the lungs and air passages of mammals

lung·wort \-ˌwərt, -ˌwȯrt\ *noun* (before 12th century)
: any of several plants (as a mullein) formerly used in the treatment of respiratory disorders; *especially* **:** a European herb (*Pulmonaria officinalis*) of the borage family with hispid leaves and usually bluish flowers

lu·ni·so·lar \ˌlü-ni-ˈsō-lər *also* -ˌlär\ *adjective* [Latin *luna* moon + English *-i-* + *solar*] (1691)
: relating or attributed to the moon and the sun

lun·ker \ˈləŋ-kər\ *noun* [origin unknown] (1912)
: something large of its kind — used especially of a game fish

lunk·head \ˈləŋk-ˌhed\ *noun* [probably alteration of *lump* + *head*] (1852)
: a stupid person **:** DOLT
— lunk·head·ed \-ˌhe-dəd\ *adjective*

lu·nule \ˈlü-(ˌ)nyü(ə)l\ *noun* [New Latin *lunula,* from Latin, crescent-shaped ornament, from diminutive of *luna* moon] (1828)
: a crescent-shaped body part or marking (as the whitish mark at the base of a fingernail)

lu·ny \ˈlü-nē\ *variant of* LOONY

lu·pa·nar \lü-ˈpā-nər, -ˈpä-\ *noun* [Latin, from *lupa* prostitute, literally, she-wolf, feminine of *lupus*] (1864)
: BROTHEL

Lu·per·ca·lia \ˌlü-pər-ˈkā-lē-ə, -ˈkāl-yə\ *noun* [Latin, plural, from *Lupercus,* god of flocks] (1600)
: an ancient Roman festival celebrated February 15 to ensure fertility for the people, fields, and flocks
— Lu·per·ca·lian \-ˈkā-lē-ən, -ˈkāl-yən\ *adjective*

[1]lu·pine *also* **lu·pin** \ˈlü-pən\ *noun* [Middle English, from Latin *lupinus, lupinum,* from *lupinus,* adjective] (14th century)
: any of a genus (*Lupinus*) of leguminous herbs including some poisonous forms and others cultivated for their long showy racemes of usually blue, purple, white, or yellow flowers or for green manure, fodder, or their edible seeds; *also* **:** an edible lupine seed

[2]lu·pine \-ˌpīn\ *adjective* [Latin *lupinus,* from *lupus* wolf — more at WOLF] (1660)
: WOLFISH

lu·pus \ˈlü-pəs\ *noun* [Middle English, from Medieval Latin, from Latin, wolf] (14th century)
: any of several diseases characterized by skin lesions; *especially* **:** SYSTEMIC LUPUS ERYTHEMATOSUS

lupus er·y·the·ma·to·sus \-ˌer-ə-ˌthē-mə-ˈtō-səs\ *noun* [New Latin, literally, erythematous lupus] (1860)
: a disorder characterized by skin inflammation; *especially* **:** SYSTEMIC LUPUS ERYTHEMATOSUS

[1]lurch \ˈlərch\ *verb* [Middle English *lorchen,* probably alteration of *lurken* to lurk] (15th century)
intransitive verb
dialect chiefly English **:** to loiter about a place furtively **:** PROWL
transitive verb
1 *obsolete* **:** STEAL
2 *archaic* **:** CHEAT

[2]lurch *noun* [Middle French *lourche,* adjective, defeated by a lurch, deceived] (1598)
: a decisive defeat in which an opponent wins a game by more than double the defeated player's score especially in cribbage
— in the lurch : in a vulnerable and unsupported position

[3]lurch *transitive verb* (circa 1651)
1 *archaic* **:** to leave in the lurch
2 : to defeat by a lurch (as in cribbage)

[4]lurch *noun* [origin unknown] (1819)
1 : a sudden roll of a ship to one side
2 : a jerking or swaying movement; *also* **:** STAGGER 3

[5]lurch *intransitive verb* (circa 1828)
: to roll or tip abruptly **:** PITCH; *also* **:** STAGGER

lurch·er \ˈlər-chər\ *noun* [[1]*lurch*] (1528)
1 *archaic* **:** a petty thief **:** PILFERER
2 *British* **:** a crossbred dog; *especially* **:** one that resembles a greyhound
3 *archaic* **:** one who lurks; *also* **:** SPY

lur·dane \ˈlər-dᵊn\ *noun* [Middle English *lurdan,* from Middle French *lourdin* dullard, from *lourd* dull, stupid, from Latin *luridus* lurid] (14th century)
archaic **:** a lazy stupid person
— lurdane *adjective*

[1]lure \ˈlu̇r\ *noun* [Middle English, from Middle French *loire,* of Germanic origin; akin to Middle High German *luoder* bait; perhaps akin to Old English *lathian* to invite, Old High German *ladōn*] (14th century)
1 : an object usually of leather or feathers attached to a long cord and used by a falconer to recall or exercise a hawk
2 a : an inducement to pleasure or gain **:** ENTICEMENT **b :** APPEAL, ATTRACTION
3 : a decoy for attracting animals to capture: as **a :** artificial bait used for catching fish **b :** an often luminous structure on the head of pediculate fishes that is used to attract prey

[2]lure *transitive verb* **lured; lur·ing** (14th century)
1 : to recall or exercise (a hawk) by means of a lure
2 : to draw with a hint of pleasure or gain **:** attract actively and strongly ☆

Lur·ex \ˈlu̇r-eks\ *trademark*
— used for metallic yarn or thread

lu·rid \ˈlu̇r-əd\ *adjective* [Latin *luridus* pale yellow, sallow] (circa 1656)
1 a : wan and ghastly pale in appearance **b :** of any of several light or medium grayish colors ranging in hue from yellow to orange
2 : shining with the red glow of fire seen through smoke or cloud
3 a : causing horror or revulsion **:** GRUESOME **b :** MELODRAMATIC, SENSATIONAL; *also* **:** SHOCKING ⟨paperbacks in the usual *lurid* covers —T. R. Fyvel⟩
synonym see GHASTLY
— lu·rid·ly *adverb*
— lu·rid·ness *noun*

lurk \ˈlərk\ *intransitive verb* [Middle English; akin to Middle High German *lūren* to lie in wait — more at LOWER] (14th century)
1 a : to lie in wait in a place of concealment especially for an evil purpose **b :** to move furtively or inconspicuously **c :** to persist in staying
2 a : to be concealed but capable of being discovered; *specifically* **:** to constitute a latent threat **b :** to lie hidden ☆
— lurk·er *noun*

☆ SYNONYMS

Lure, entice, inveigle, decoy, tempt, seduce mean to lead astray from one's true course. LURE implies a drawing into danger, evil, or difficulty through attracting and deceiving ⟨*lured* naive investors with get-rich-quick schemes⟩. ENTICE suggests drawing by artful or adroit means ⟨advertising designed to *entice* new customers⟩. INVEIGLE implies enticing by cajoling or flattering ⟨fund-raisers *inveigling* wealthy alumni⟩. DECOY implies a luring into entrapment by artifice ⟨attempting to *decoy* the enemy into an ambush⟩. TEMPT implies the presenting of an attraction so strong that it overcomes the restraints of conscience or better judgment ⟨*tempted* by the offer of money⟩. SEDUCE implies a leading astray by persuasion or false promises ⟨*seduced* by assurances of assistance⟩.

Lurk, skulk, slink, sneak mean to behave so as to escape attention. LURK implies a lying in wait in a place of concealment and often suggests an evil intent ⟨suspicious men *lurking* in alleyways⟩. SKULK suggests more strongly cowardice or fear or sinister intent ⟨something *skulking* in the shadows⟩. SLINK implies moving stealthily often merely to escape attention ⟨*slunk* around the corner⟩. SNEAK may add an implication of entering or leaving a place or evading a difficulty by furtive, indirect, or underhanded methods ⟨*sneaked* out early⟩.

lus·cious \'lə-shəs\ *adjective* [Middle English *lucius,* perhaps alteration of *licius,* short for *delicious*] (15th century)
1 a : having a delicious taste or smell **:** SWEET **b** *archaic* **:** excessively sweet **:** CLOYING
2 : sexually attractive **:** SEDUCTIVE, SEXY
3 a : richly luxurious or appealing to the senses **b :** excessively ornate
— **lus·cious·ly** *adverb*
— **lus·cious·ness** *noun*

[1]**lush** \'ləsh\ *adjective* [Middle English *lusch* soft, tender] (1610)
1 a : growing vigorously especially with luxuriant foliage ⟨*lush* grass⟩ **b :** lavishly productive: as (1) **:** FERTILE (2) **:** THRIVING (3) **:** characterized by abundance **:** PLENTIFUL (4) **:** PROSPEROUS, PROFITABLE
2 a : SAVORY, DELICIOUS **b :** appealing to the senses ⟨the *lush* sounds of the orchestra⟩ **c :** OPULENT, SUMPTUOUS
synonym see PROFUSE
— **lush·ly** *adverb*
— **lush·ness** *noun*

[2]**lush** *noun* [origin unknown] (circa 1790)
1 *slang* **:** intoxicating liquor **:** DRINK
2 : an habitual heavy drinker **:** DRUNKARD

[3]**lush** *verb* (circa 1811)
slang **:** DRINK

Lu·so- \,lü-(,)sō\ *combining form* [Portuguese, from *lusitano* Portuguese, from Latin *Lusitanus* of Lusitania (ancient region corresponding approximately to modern Portugal)]
: Portuguese and ⟨*Luso*-Brazilian⟩

[1]**lust** \'ləst\ *noun* [Middle English, from Old English; akin to Old High German *lust* pleasure and perhaps to Latin *lascivus* wanton] (before 12th century)
1 *obsolete* **a :** PLEASURE, DELIGHT **b :** personal inclination **:** WISH
2 : usually intense or unbridled sexual desire **:** LASCIVIOUSNESS
3 a : an intense longing **:** CRAVING **b :** ENTHUSIASM, EAGERNESS

[2]**lust** *intransitive verb* (12th century)
: to have an intense desire or need **:** CRAVE; *specifically* **:** to have a sexual urge

[1]**lus·ter** *or* **lus·tre** \'ləs-tər\ *noun* [Middle English *lustre,* from Latin *lustrum*] (14th century)
: a period of five years **:** LUSTRUM 2

[2]**luster** *or* **lustre** *noun* [Middle French *lustre,* from Old Italian *lustro,* from *lustrare* to brighten, from Latin, to purify ceremonially, from *lustrum*] (circa 1522)
1 : a glow of reflected light **:** SHEEN; *specifically* **:** the appearance of the surface of a mineral dependent upon its reflecting qualities
2 a : a glow of light from within **:** LUMINOSITY **b :** an inner beauty **:** RADIANCE
3 : a superficial attractiveness or appearance of excellence
4 a : a glass pendant used especially to ornament a candlestick or chandelier **b :** a decorative object (as a chandelier) hung with glass pendants
5 *chiefly British* **:** a fabric with cotton warp and a filling of wool, mohair, or alpaca
6 : LUSTERWARE
— **lus·ter·less** \-tər-ləs\ *adjective*

[3]**luster** *or* **lustre** *verb* **lus·tered** *or* **lus·tred; lus·ter·ing** *or* **lus·tring** \-t(ə-)riŋ\ (1582)
intransitive verb
: to have luster **:** GLEAM
transitive verb
1 : to give luster or distinction to
2 : to coat or treat with a substance that imparts luster

lus·ter·ware \'ləs-tər-,war, -,wer\ *noun* (1825)
: pottery with an iridescent metallic sheen in the glaze

lust·ful \'ləst-fəl\ *adjective* (14th century)
: excited by lust **:** LECHEROUS
— **lust·ful·ly** \-fə-lē\ *adverb*
— **lust·ful·ness** *noun*

lust·i·hood \'ləs-tē-,hu̇d\ *noun* (1599)
1 : vigor of body or spirit **:** ROBUSTNESS
2 : sexual inclination or capacity

lus·tral \'ləs-trəl\ *adjective* [Latin *lustralis,* from *lustrum*] (1533)
: PURIFICATORY

lus·trate \'ləs-,trāt\ *transitive verb* **lus·trat·ed; lus·trat·ing** [Latin *lustratus,* past participle of *lustrare*] (1653)
: to purify ceremonially
— **lus·tra·tion** \,ləs-'trā-shən\ *noun*

[1]**lus·tring** \'ləs-triŋ\ *noun* [modification of Italian *lustrino*] (1697)
archaic **:** LUTESTRING

[2]**lus·tring** \'ləs-t(ə-)riŋ\ *noun* [*lustring,* gerund of [3]*luster*] (circa 1889)
: a finishing process (as calendering) for giving a gloss to yarns and cloth

lus·trous \'ləs-trəs\ *adjective* (1601)
1 : reflecting light evenly and efficiently without glitter or sparkle ⟨a *lustrous* satin⟩ ⟨the *lustrous* glow of an opal⟩
2 : radiant in character or reputation **:** ILLUSTRIOUS
synonym see BRIGHT
— **lus·trous·ly** *adverb*
— **lus·trous·ness** *noun*

lus·trum \'ləs-trəm\ *noun, plural* **lustrums** *or* **lus·tra** \-trə\ [Latin] (1590)
1 : a period of five years
2 a : a purification of the whole Roman people made in ancient times after the census every five years **b :** the Roman census

lusty \'ləs-tē\ *adjective* **lust·i·er; -est** (13th century)
1 *archaic* **:** MERRY, JOYOUS
2 : LUSTFUL ⟨*lusty* passion⟩
3 a : full of strength and vitality **:** HEALTHY, VIGOROUS ⟨a young, *lusty,* growing country —Helen Harris⟩ **b :** HEARTY, ROBUST ⟨a *lusty* beef stew⟩ **c :** ENTHUSIASTIC, ROUSING ⟨a *lusty* rendition of the song⟩
synonym see VIGOROUS
— **lust·i·ly** \-tə-lē\ *adverb*
— **lust·i·ness** \-tē-nəs\ *noun*

[1]**lute** \'lüt\ *noun* [Middle English, from Middle French *lut,* from Old Provençal *laut,* from Arabic *al-'ūd,* literally, the wood] (13th century)
: a stringed instrument having a large pear-shaped body, a vaulted back, a fretted fingerboard, and a head with tuning pegs which is often angled backward from the neck

[2]**lute** *transitive verb* **lut·ed; lut·ing** [Middle English, from Latin *lutare,* from *lutum* mud — more at POLLUTE] (14th century)
: to seal or cover (as a joint or surface) with lute

lute

[3]**lute** *noun* (15th century)
: a substance (as cement or clay) for packing a joint or coating a porous surface to make it impervious to gas or liquid

lute- *or* **luteo-** *combining form* [New Latin (*corpus*) *luteum*]
: corpus luteum ⟨*lute*al⟩

lu·te·al \'lü-tē-əl\ *adjective* (1920)
: of, relating to, characterized by, or involving the corpus luteum ⟨the *luteal* phase of the menstrual cycle⟩

lute·fisk \'lüt-,fisk, 'lü-tə-\ *noun* [Norwegian, from *lute* to wash in lye solution + *fisk* fish] (1924)
: dried codfish that has been soaked in a water and lye solution before cooking

lu·tein \'lü-tē-ən, 'lü-,tēn\ *noun* (1869)
: an orange xanthophyll $C_{40}H_{56}O_2$ occurring in plants, animal fat, egg yolk, and the corpus luteum

lu·tein·i·za·tion \,lü-tē-ə-nə-'zā-shən, ,lü-,tē-nə-\ *noun* (1929)
: the process of forming corpora lutea
— **lu·tein·ize** \'lü-tē-ə-,nīz, 'lü-,tē-,nīz\ *verb*

luteinizing hormone *noun* (1931)
: a hormone from the anterior lobe of the pituitary gland that in the female stimulates especially the development of corpora lutea and in the male the development of interstitial tissue in the testis

luteinizing hormone–releasing hormone *noun* (1970)
: a hormone secreted by the hypothalamus that stimulates the pituitary gland to release gonadotropins (as luteinizing hormone and follicle-stimulating hormone) — called also *luteinizing hormone-releasing factor*

lu·te·nist *or* **lu·ta·nist** \'lü-tᵊn-ist, 'lüt-nist\ *noun* [Medieval Latin *lutanista,* from *lutana* lute, probably from Middle French *lut* lute] (1600)
: a lute player

lu·teo·tro·pic \,lü-tē-ə-'trō-pik, -'trä-\ *or* **lu·teo·tro·phic** \-'trō-fik, -'trä-\ *adjective* (1941)
: acting on the corpora lutea

luteotropic hormone *or* **luteotrophic hormone** *noun* (1949)
: PROLACTIN

lu·teo·tro·pin \,lü-tē-ə-'trō-pən\ *or* **lu·teo·tro·phin** \-fən\ *noun* (1941)
: PROLACTIN

lu·te·ous \'lü-tē-əs\ *adjective* [Latin *luteus* yellow, from *lutum,* a plant (*Reseda luteola*) used for dyeing yellow] (1657)
: yellow tinged with green or brown

lute·string \'lüt-(,)striŋ\ *noun* [by folk etymology from Italian *lustrino* glossy fabric, from *lustro* luster] (1661)
: a plain glossy silk formerly much used for women's dresses and ribbons

lu·te·tium *also* **lu·te·cium** \lü-'tē-sh(ē-)əm\ *noun* [New Latin, from Latin *Lutetia,* ancient name of Paris] (1907)
: a metallic element of the rare-earth group — see ELEMENT table

[1]**Lu·ther·an** \'lü-th(ə-)rən\ *noun* (1521)
: a member of a Lutheran church

[2]**Lutheran** *adjective* (1530)
1 : of or relating to religious doctrines (as justification by faith alone) developed by Martin Luther or his followers
2 : of or relating to the Protestant churches adhering to Lutheran doctrines, liturgy, and polity
— **Lu·ther·an·ism** \-rə-,ni-zəm\ *noun*

lu·thi·er \'lü-tē-ər, -thē-ər\ *noun* [French, from *luth* lute (from Middle French *lut*)] (1879)
: one who makes stringed musical instruments (as violins or guitars)

lutz \'ləts\ *noun* [probably irregular from Gustave *Lussi* (born 1898) Swiss figure skater] (1938)
: a figure-skating jump from the outer edge of one skate with a full turn in the air and a return to the outer edge of the other skate

Lu·wi·an \'lü-ē-ən\ *noun* [*Luwi,* an ancient people of the southern coast of Asia Minor] (1924)
: an Anatolian language of the Indo-European language family — see INDO-EUROPEAN LANGUAGES table
— **Luwian** *adjective*

lux \'ləks\ *noun, plural* **lux** *or* **lux·es** [Latin, light — more at LIGHT] (1889)
: a unit of illumination equal to the direct illumination on a surface that is everywhere one meter from a uniform point source of one candle intensity or equal to one lumen per square meter

lux·a·tion \,lək-'sā-shən\ *noun* [Late Latin *luxation-, luxatio,* from Latin *luxare* to dislocate, from *luxus* dislocated — more at LOCK] (1552)

\ə\ abut \ᵊ\ kitten \ər\ further \a\ ash \ā\ ace
\ä\ mop, mar \au̇\ out \ch\ chin \e\ bet \ē\ easy
\g\ go \i\ hit \ī\ ice \j\ job \ŋ\ sing \ō\ go
\ȯ\ law \ȯi\ boy \th\ thin \t͟h\ the \ü\ loot \u̇\ foot
\y\ yet \zh\ vision *see also* Guide to Pronunciation

: dislocation of an anatomical part (as a bone at a joint or the lens of the eye)

luxe \'lu̇ks, 'ləks, 'lüks\ *noun* [French, from Latin *luxus* — more at LUXURY] (1558)
: LUXURY
— **luxe** *adjective*

lux·u·ri·ance \(,)ləg-'zhu̇r-ē-ən(t)s, (,)lək-'shu̇r-\ *noun* (circa 1746)
: the quality or state of being luxuriant

lux·u·ri·ant \-ē-ənt\ *adjective* (circa 1540)
1 a : yielding abundantly **:** FERTILE, FRUITFUL **b :** characterized by abundant growth **:** LUSH
2 : abundantly and often extravagantly rich and varied **:** PROLIFIC
3 : characterized by luxury **:** LUXURIOUS
synonym see PROFUSE
— **lux·u·ri·ant·ly** *adverb*

lux·u·ri·ate \-ē-,āt\ *intransitive verb* **-at·ed; -at·ing** [Latin *luxuriatus,* past participle of *luxuriare,* from *luxuria*] (1621)
1 a : to grow profusely **:** THRIVE **b :** to develop extensively
2 : to indulge oneself luxuriously **:** REVEL

lux·u·ri·ous \(,)ləg-'zhu̇r-ē-əs, (,)lək-'shu̇r-\ *adjective* (14th century)
1 : LECHEROUS
2 : marked by or given to self-indulgence ⟨*luxurious* tastes⟩ ⟨*luxurious* feeling⟩
3 : of, relating to, or marked by luxury ⟨*luxurious* accommodations⟩
4 : of the finest and richest kind ⟨*luxurious* cashmeres⟩ ⟨a *luxurious* chocolate sauce⟩
synonym see SENSUOUS
— **lux·u·ri·ous·ly** *adverb*
— **lux·u·ri·ous·ness** *noun*

lux·u·ry \'lək-sh(ə-)rē, -zh(ə-)rē\ *noun, plural* **-ries** [Middle English *luxurie,* from Middle French, from Latin *luxuria* rankness, luxury, excess; akin to Latin *luxus* luxury, excess] (14th century)
1 *archaic* **:** LECHERY, LUST
2 : a condition of abundance or great ease and comfort **:** sumptuous environment ⟨lived in *luxury*⟩
3 a : something adding to pleasure or comfort but not absolutely necessary **b :** an indulgence in something that provides pleasure, satisfaction, or ease ⟨had the *luxury* of rejecting a handful of job offers . . . before accepting a choice assignment —Terri Minsky⟩
— **luxury** *adjective*

lwei \lə-'wā\ *noun, plural* **lwei** *also* **lweis** [probably from *Lwei* (Lué), name of several rivers in Angola] (circa 1979)
— see *kwanza* at MONEY table

ly- *or* **lyo-** *combining form* [International Scientific Vocabulary, from Greek *lyein* to loosen, dissolve — more at LOSE]
1 : degrading **:** reduction ⟨*ly*ase⟩
2 : dispersed state **:** dispersion ⟨*lyo*philic⟩

¹-ly \lē; *in some dialects, especially British, Southern, New England, often* li *but not shown at individual entries*\ *adjective suffix* [Middle English, from Old English *-līc, -lic;* akin to Old High German *-līh, -lic,* Old English *līc* body — more at LIKE]
1 : like in appearance, manner, or nature **:** having the characteristics of ⟨queen*ly*⟩ ⟨father*ly*⟩
2 : characterized by regular recurrence in (specified) units of time **:** every ⟨hour*ly*⟩

²-ly *adverb suffix* [Middle English, from Old English *-līce, -lice,* from *-līc,* adjective suffix]
1 a : in a (specified) manner ⟨slow*ly*⟩ **b :** at a (specified) time interval ⟨annual*ly*⟩
2 : from a (specified) point of view ⟨eschatological*ly*⟩
3 : with respect to ⟨part*ly*⟩
4 : to a (specified) degree ⟨relative*ly*⟩
5 : in a (specified) place in a series ⟨second*ly*⟩

ly·am–hound \'lī-əm-,hau̇nd\ *or* **lyme–hound** \'līm-,hau̇nd\ *noun* [obsolete *lyam* leash] (1527)
archaic **:** BLOODHOUND

ly·art \'lī-ərt\ *adjective* [Middle English, from Middle French *liart*] (14th century)
chiefly Scottish **:** streaked with gray **:** GRAY

ly·ase \'lī-,ās, -,āz\ *noun* (1965)
: an enzyme (as a decarboxylase) that forms double bonds by removing groups from a substrate other than by hydrolysis or that adds groups to double bonds

ly·can·thro·py \lī-'kan(t)-thrə-pē\ *noun* [New Latin *lycanthropia,* from Greek *lykanthrōpia,* from *lykanthrōpos* werewolf, from *lykos* wolf + *anthrōpos* human being — more at WOLF] (1594)
1 : a delusion that one has become a wolf
2 : the assumption of the form and characteristics of a wolf held to be possible by witchcraft or magic

ly·cée \lē-'sā\ *noun* [French, from Middle French, lyceum, from Latin *Lyceum*] (1865)
: a French public secondary school that prepares students for the university

ly·ce·um \lī-'sē-əm, 'lī-sē-\ *noun* [Latin *Lyceum,* gymnasium near Athens where Aristotle taught, from Greek *Lykeion,* from neuter of *lykeios,* epithet of Apollo] (1786)
1 : a hall for public lectures or discussions
2 : an association providing public lectures, concerts, and entertainments
3 : LYCÉE

ly·chee *variant of* LITCHI

lych–gate \'lich-,gāt\ *noun* [Middle English *lycheyate,* from *lich* body, corpse (from Old English *līc*) + *gate, yate* gate] (15th century)
: a roofed gate in a churchyard under which a bier rests during the initial part of the burial service

lych·nis \'lik-nəs\ *noun* [New Latin, from Latin, a red flower, from Greek; akin to Greek *lychnos* lamp, Latin *lux* light — more at LIGHT] (1601)
: any of a genus (*Lychnis*) of herbs of the pink family with terminal cymes of showy mostly red or white flowers having 5 or rarely 4 styles

Ly·cian \'li-sh(ē-)ən\ *noun* (1598)
1 : a native or inhabitant of Lycia
2 : an Anatolian language of the Indo-European language family — see INDO-EUROPEAN LANGUAGES table
— **Lycian** *adjective*

ly·co·pene \'lī-kə-,pēn\ *noun* [International Scientific Vocabulary *lycop-* (from New Latin *Lycopersicon,* genus of herbs) + *-ene*] (circa 1929)
: a carotenoid pigment $C_{40}H_{56}$ that is the red coloring matter of the tomato

ly·co·pod \'lī-kə-,päd\ *noun* [New Latin *Lycopodium*] (1861)
: LYCOPODIUM 1; *broadly* **:** CLUB MOSS

ly·co·po·di·um \,lī-kə-'pō-dē-əm\ *noun* [New Latin, from Greek *lykos* wolf + *podion,* diminutive of *pod-, pous* foot — more at FOOT] (circa 1706)
1 : any of a large genus (*Lycopodium*) of erect or creeping club mosses with reduced or scale-like evergreen leaves
2 : a fine yellowish flammable powder composed of lycopodium spores and used especially in pharmacy

Ly·cra \'lī-krə\ *trademark*
— used for a spandex synthetic fiber

lydd·ite \'li-,dīt\ *noun* [*Lydd,* England] (1888)
: a high explosive composed chiefly of picric acid

Lyd·i·an \'li-dē-ən\ *noun* (15th century)
1 : a native or inhabitant of Lydia
2 : an Anatolian language of the Indo-European language family — see INDO-EUROPEAN LANGUAGES table
— **Lydian** *adjective*

lye \'lī\ *noun* [Middle English, from Old English *lēag;* akin to Old High German *louga* lye, Latin *lavare, lavere* to wash, Greek *louein*] (before 12th century)
1 : a strong alkaline liquor rich in potassium carbonate leached from wood ashes and used especially in making soap and washing; *broadly* **:** a strong alkaline solution (as of sodium hydroxide or potassium hydroxide)
2 : a solid caustic (as sodium hydroxide)

ly·gus bug \'lī-gəs-\ *noun* [New Latin *Lygus*] (1940)
: any of various small sucking bugs (genus *Lygus*) including some pests of cultivated plants

¹ly·ing \'lī-iŋ\ *present participle of* LIE

²lying *adjective* [Middle English *leghynge,* present participle of *lien* to lie] (14th century)
: marked by or containing falsehoods **:** FALSE ⟨a *lying* account of the accident⟩

ly·ing–in \,lī-iŋ-'in\ *noun, plural* **lyings–in** *or* **lying–ins** (15th century)
: the state attending and consequent to childbirth **:** CONFINEMENT

Lyme disease \'līm-\ *noun* [*Lyme,* Connecticut, where it was first reported] (1980)
: an acute inflammatory disease that is caused by a spirochete (*Borrelia burgdorferi*) transmitted by ticks (genus *Ixodes* and especially *I. dammini*), that is often characterized initially by a spreading red annular skin lesion at the site of the infection, fever, and chills, and that may result in joint pain, arthritis, and cardiac and neurological disorders

lymph \'lim(p)f\ *noun* [Latin *lympha,* water goddess, water, perhaps modification of Greek *nymphē* nymph — more at NUPTIAL] (circa 1673)
1 *archaic* **:** the sap of plants
2 [New Latin *lympha,* from Latin, water] **:** a pale coagulable fluid that bathes the tissues, passes into lymphatic channels and ducts, is discharged into the blood by way of the thoracic duct, and consists of a liquid portion resembling blood plasma and containing white blood cells but normally no red blood cells

lymph- *or* **lympho-** *combining form* [New Latin *lympha*]
: lymph **:** lymphatic tissue ⟨*lympho*granuloma⟩

lymph·ad·e·ni·tis \,lim-,fa-dᵊn-'ī-təs\ *noun* [New Latin, from *lymphaden* lymph gland, from *lymph-* + Greek *adēn* gland — more at ADEN-] (1860)
: inflammation of lymph nodes

lymph·ade·nop·a·thy \,lim-,fa-dᵊn-'ä-pə-thē\ *noun, plural* **-thies** (1920)
: abnormal enlargement of the lymph nodes

lymph·an·gi·og·ra·phy \,lim-,fan-jē-'ä-grə-fē\ *noun* (circa 1941)
: X-ray depiction of lymph vessels and nodes after use of a radiopaque material — called also *lymphography*
— **lymph·an·gio·gram** \(,)lim-'fan-jē-ə-,gram\ *noun*
— **lymph·an·gio·graph·ic** \,lim-,fan-jē-ə-'gra-fik\ *adjective*

¹lym·phat·ic \lim-'fa-tik\ *adjective* (1649)
1 a : of, relating to, or produced by lymph, lymphoid tissue, or lymphocytes **b :** conveying lymph
2 : lacking physical or mental energy **:** SLUGGISH
— **lym·phat·i·cal·ly** \-ti-k(ə-)lē\ *adverb*

²lymphatic *noun* (1667)
: a vessel that contains or conveys lymph — called also *lymph vessel*

lymph gland *noun* (circa 1858)
: LYMPH NODE

lymph node *noun* (1892)
: any of the rounded masses of lymphoid tissue that are surrounded by a capsule of connective tissue, are distributed along the lymphatic vessels, and contain numerous lymphocytes which filter the flow of lymph

lym·pho·blast \'lim(p)-fə-,blast\ *noun* [International Scientific Vocabulary] (circa 1909)
: a lymphocyte that has enlarged following stimulation by an antigen, has the capacity to recognize the stimulating antigen, and is undergoing proliferation and differentiation either to an effector state in which it functions to eliminate the antigen or to a memory state in which it functions to recognize the future reappearance of the antigen

— **lym·pho·blas·tic** \ˌlim(p)-fə-ˈblas-tik\ *adjective*

lym·pho·cyte \ˈlim(p)-fə-ˌsīt\ *noun* [International Scientific Vocabulary] (1890)
: any of the colorless weakly motile cells originating from stem cells and differentiating in lymphoid tissue (as of the thymus or bone marrow) that are the typical cellular elements of lymph, include the cellular mediators of immunity, and constitute 20 to 30 percent of the leukocytes of normal human blood — compare B CELL, T CELL
— **lym·pho·cyt·ic** \ˌlim(p)-fə-ˈsi-tik\ *adjective*

lymphocytic cho·rio·men·in·gi·tis \-ˌkȯr-ē-ō-ˌme-nən-ˈjī-təs\ *noun* [New Latin *choriomeningitis* cerebral meningitis, from *chorio-* of a membrane resembling the chorion] (1934)
: an acute virus disease that is characterized by fever, nausea and vomiting, headache, stiff neck, and slow pulse, is marked by the presence of numerous lymphocytes in the cerebrospinal fluid, and is transmitted especially by rodents and bloodsucking insects

lym·pho·cy·to·sis \ˌlim(p)-fə-ˌsī-ˈtō-səs, -fə-sə-\ *noun* [New Latin, from International Scientific Vocabulary *lymphocyte*] (1896)
: an increase in the number of lymphocytes in the blood usually associated with chronic infections or inflammations

lym·pho·gran·u·lo·ma \ˈlim(p)-fō-ˌgran-yə-ˈlō-mə\ *noun, plural* **-mas** *or* **-ma·ta** \-mə-tə\ [New Latin] (1924)
: LYMPHOGRANULOMA VENEREUM

lymphogranuloma in·gui·na·le \-ˌiŋ-gwə-ˈnä-lē, -ˈna-, -ˈnā-\ *noun* [New Latin, inguinal lymphogranuloma] (1932)
: LYMPHOGRANULOMA VENEREUM

lym·pho·gran·u·lo·ma·to·sis \-ˌlō-mə-ˈtō-səs\ *noun, plural* **-to·ses** \-ˌsēz\ [New Latin *lymphogranulomat-*, *lymphogranuloma* + *-osis*] (1911)
: the development of benign or malignant nodular swellings of lymph nodes in various parts of the body; *also* **:** a condition characterized by these

lymphogranuloma ve·ne·re·um \-və-ˈnir-ē-əm\ *noun* [New Latin, venereal lymphogranuloma] (1938)
: a contagious venereal disease caused by various strains of a chlamydia (*Chlamydia trachomatis*) and marked by swelling and ulceration of lymphatic tissue in the iliac and inguinal regions

lym·phog·ra·phy \lim-ˈfä-grə-fē\ *noun* (circa 1935)
: LYMPHANGIOGRAPHY
— **lym·pho·gram** \ˈlim(p)-fə-ˌgram\ *noun*
— **lym·pho·graph·ic** \ˌlim-fə-ˈgra-fik\ *adjective*

lym·phoid \ˈlim-ˌfȯid\ *adjective* (1867)
1 : of, relating to, or being tissue (as of the lymph nodes or thymus) containing lymphocytes
2 : of, relating to, or resembling lymph

lym·pho·kine \ˈlim(p)-fə-ˌkīn\ *noun* [*lymph-* + *-kine*, from Greek *kinein* to move about — more at -KINESIS] (1969)
: any of various substances (as interleukin-2) of low molecular weight that are not immunoglobulins, are secreted by T cells in response to stimulation by antigens, and have a role (as the activation of macrophages or the enhancement or inhibition of antibody production) in cell-mediated immunity

lym·pho·ma \lim-ˈfō-mə\ *noun, plural* **-mas** *or* **-ma·ta** \-mə-tə\ [New Latin] (1873)
: a tumor of lymphoid tissue
— **lym·pho·ma·tous** \-mə-təs\ *adjective*

lym·pho·ma·to·sis \(ˌ)lim-ˌfō-mə-ˈtō-səs\ *noun, plural* **-to·ses** \-ˌsēz\ [New Latin *lymphomat-*, *lymphoma* + *-osis*] (circa 1900)
: the presence of multiple lymphomas in the body

lym·pho·sar·co·ma \ˌlim(p)-fə-sär-ˈkō-mə\ *noun, plural* **-mas** *or* **-ma·ta** \-mə-tə\ [New Latin] (1874)
: a malignant lymphoma that tends to metastasize freely especially along the regional lymphatic drainage

lynch \ˈlinch\ *transitive verb* [*lynch law*] (1836)
: to put to death (as by hanging) by mob action without legal sanction ◆
— **lynch·er** *noun*

lynch law *noun* [William *Lynch* (died 1820) American vigilante] (1811)
: the punishment of presumed crimes or offenses usually by death without due process of law

lynch·pin *variant of* LINCHPIN

lynx \ˈliŋ(k)s\ *noun, plural* **lynx** *or* **lynx·es** [Middle English, from Latin, from Greek; akin to Old English *lox* lynx and probably to Greek *leukos* white — more at LIGHT] (14th century)
: any of several wildcats (genus *Lynx*) with relatively long legs, a short stubby tail, mottled coat, and often tufted ears: as **a :** a lynx (*L. lynx*) of northern Europe and Asia **b :** BOBCAT **c :** a North American lynx (*L. canadensis*) distinguished from the bobcat by its larger size, longer tufted ears, and wholly black tail tip — called also *Canadian lynx*

lynx–eyed \ˈliŋ(k)s-ˌīd\ *adjective* (1597)
: SHARP-SIGHTED

lyo- — see LY-

ly·on·naise \ˌlī-ə-ˈnāz\ *adjective* [French (*à la*) *lyonnaise* in the manner of Lyons, France] (1846)
: prepared with onions ⟨*lyonnaise* potatoes⟩

Ly·on·nesse \ˌlī-ə-ˈnes\ *noun*
: a country that according to Arthurian legend was contiguous to Cornwall before sinking beneath the sea

lyo·phile \ˈlī-ə-ˌfīl\ *adjective* [International Scientific Vocabulary] (1934)
1 : of or relating to freeze-drying
2 *or* **lyo·philed** \-ˌfīld\ **:** obtained by freeze-drying

lyo·phil·ic \ˌlī-ə-ˈfi-lik\ *adjective* (1911)
: marked by strong affinity between a dispersed phase and the liquid in which it is dispersed ⟨a *lyophilic* colloid⟩

ly·oph·i·lise *British variant of* LYOPHILIZE

ly·oph·i·lize \lī-ˈä-fə-ˌlīz\ *transitive verb* **-lized; -liz·ing** (1938)
: FREEZE-DRY
— **ly·oph·i·li·za·tion** \-ˌä-fə-lə-ˈzā-shən\ *noun*
— **ly·oph·i·liz·er** \-ˈä-fə-ˌlī-zər\ *noun*

lyo·pho·bic \ˌlī-ə-ˈfō-bik\ *adjective* (1911)
: marked by lack of strong affinity between a dispersed phase and the liquid in which it is dispersed ⟨a *lyophobic* colloid⟩

Ly·ra \ˈlī-rə\ *noun* [Latin (genitive *Lyrae*), literally, lyre]
: a northern constellation representing the lyre of Orpheus or Mercury and containing Vega

ly·rate \ˈlī-ˌrāt\ *adjective* (circa 1760)
: having or suggesting the shape of a lyre ⟨the *lyrate* horns of the impala⟩

lyre \ˈlīr\ *noun* [Middle English *lire*, from Old French, from Latin *lyra*, from Greek] (13th century)
1 : a stringed instrument of the harp class used by the ancient Greeks especially to accompany song and recitation
2 *capitalized* **:** LYRA

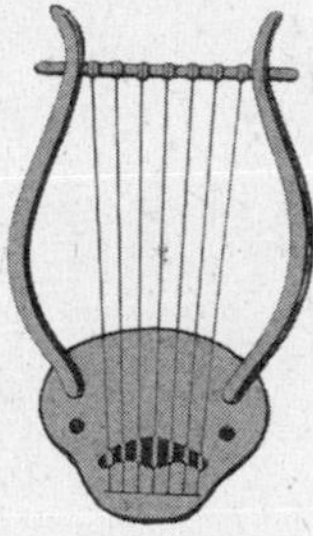
lyre 1

lyre·bird \-ˌbərd\ *noun* (1834)
: either of two Australian passerine birds (genus *Menura*) distinguished in the male by very long tail feathers displayed in the shape of a lyre during courtship

lyrebird

[1]lyr·ic \ˈlir-ik\ *noun* (1581)
1 : a lyric composition; *specifically* **:** a lyric poem
2 : the words of a song — often used in plural

[2]lyric *adjective* [Middle French or Latin; Middle French *lyrique*, from Latin *lyricus*, from Greek *lyrikos*, from *lyra*] (1589)
1 a : suitable for singing to the lyre or for being set to music and sung **b :** of, relating to, or being drama set to music; *especially* **:** OPERATIC ⟨*lyric* stage⟩
2 a : expressing direct usually intense personal emotion especially in a manner suggestive of song ⟨*lyric* poetry⟩ **b :** EXUBERANT, RHAPSODIC
3 *of an opera singer* **:** having a light voice and a melodic style — compare DRAMATIC

lyr·i·cal \ˈlir-i-kəl\ *adjective* (1581)
: LYRIC
— **lyr·i·cal·ly** \-i-k(ə-)lē\ *adverb*
— **lyr·i·cal·ness** \-kəl-nəs\ *noun*

lyr·i·cism \ˈlir-ə-ˌsi-zəm\ *noun* (1760)
1 : the quality or state of being lyric **:** SONGFULNESS
2 a : an intense personal quality expressive of feeling or emotion in an art (as poetry or music) **b :** EXUBERANCE ⟨the sort of author who inspires *lyricism* or invective, not judicious interpretation —*Time*⟩

lyr·i·cist \-sist\ *noun* (1881)
: a writer of lyrics

lyr·ism \ˈlir-ˌi-zəm\ *noun* (1859)
: LYRICISM

lyr·ist *noun* (circa 1656)
1 \ˈlīr-ist\ **:** a player on the lyre
2 \ˈlir-ist\ **:** LYRICIST

lys- *or* **lysi-** *or* **lyso-** *combining form* [New Latin, from Greek *lys-*, *lysi-* loosening, from *lysis*]
: lysis ⟨*lys*in⟩

ly·sate \ˈlī-ˌsāt\ *noun* (1922)
: a product of lysis

lyse \ˈlīs, ˈlīz\ *verb* **lysed; lys·ing** [back-formation from New Latin *lysis*] (1924)
transitive verb
: to cause to undergo lysis
intransitive verb
: to undergo lysis

-lyse *chiefly British variant of* -LYZE

Ly·sen·ko·ism \lə-ˈseŋ-kō-ˌi-zəm\ *noun* [Trofim *Lysenko*] (1948)
: a biological doctrine asserting the fundamental influence of somatic and environmental factors on heredity in contradiction of orthodox genetics

◇ WORD HISTORY
lynch Despite the claims made on behalf of numerous other Lynches, the weight of evidence for the eponymous word *lynch* falls on the side of Captain William Lynch (1742–1820). Lynch served with the Virginia militia and presided over a self-created tribunal that was organized to rid Pittsylvania County of a band of troublesome ruffians that had eluded the civil authorities. On 22 September 1780, Lynch and others entered into a compact stating their goals, reasons, and methods. Captain Lynch and his vigilantes became known as "lynch-men," and by 1782 their judicial code had become known as *lynch's law* and subsequently *lynch law*. By 1836 *lynch law* had given rise to the verb *lynch* in its modern meaning.

ly·ser·gic acid \lə-ˈsər-jik-, (ˌ)lī-\ *noun* [*lys-* + *ergot*] (1934)
: a crystalline acid $C_{16}H_{16}N_2O_2$ from ergotic alkaloids; *also* **:** LSD

lysergic acid di·eth·yl·am·ide \-ˌdī-ˌe-thə-ˈla-ˌmīd\ *noun* (1944)
: LSD

ly·sim·e·ter \lī-ˈsi-mə-tər\ *noun* (1879)
: a device for measuring the percolation of water through soils and for determining the soluble constituents removed in the drainage
— **ly·si·met·ric** \ˌlī-sə-ˈme-trik\ *adjective*

ly·sin \ˈlī-sᵊn\ *noun* (1900)
: a substance (as an antibody) capable of causing lysis

ly·sine \ˈlī-ˌsēn\ *noun* (1892)
: a crystalline essential amino acid $C_6H_{14}N_2O_2$ obtained from the hydrolysis of various proteins

ly·sis \ˈlī-səs\ *noun, plural* **ly·ses** \-ˌsēz\ [New Latin, from Greek, act of loosening, dissolution, remission of fever, from *lyein* to loosen — more at LOSE] (1543)
1 : the gradual decline of a disease process (as fever)
2 : a process of disintegration or dissolution (as of cells)

-lysis *noun combining form, plural* **-lyses** [New Latin, from Latin & Greek; Latin, loosening, from Greek, from *lysis*]
1 : decomposition ⟨electro*lysis*⟩
2 : disintegration **:** breaking down ⟨auto*lysis*⟩

ly·so·gen \ˈlī-sə-jən\ *noun* (circa 1934)
: a lysogenic bacterium or bacterial strain

ly·so·gen·ic \ˌlī-sə-ˈje-nik\ *adjective* [from the capacity of the prophage to lyse other bacteria] (1899)
1 : harboring a prophage as hereditary material ⟨*lysogenic* bacteria⟩
2 : TEMPERATE 3 ⟨*lysogenic* viruses⟩
— **ly·so·ge·nic·i·ty** \-jə-ˈni-sə-tē\ *noun*

ly·sog·e·nise *British variant of* LYSOGENIZE

ly·sog·e·nize \lī-ˈsä-jə-ˌnīz\ *transitive verb* **-nized; -niz·ing** (1953)
: to render lysogenic
— **ly·sog·e·ni·za·tion** \-ˌsä-jə-nə-ˈzā-shən\ *noun*

ly·sog·e·ny \lī-ˈsä-jə-nē\ *noun* (1956)
: the state of being lysogenic

ly·so·lec·i·thin \ˌlī-sə-ˈle-sə-thən\ *noun* (1923)
: a hydrolytic substance formed by the enzymatic hydrolysis (as by some snake venoms) of a lecithin

ly·so·some \ˈlī-sə-ˌsōm\ *noun* [International Scientific Vocabulary *lys-* + ³*-some*] (1955)
: a saclike cellular organelle that contains various hydrolytic enzymes — see CELL illustration
— **ly·so·som·al** \ˌlī-sə-ˈsō-məl\ *adjective*

ly·so·zyme \ˈlī-sə-ˌzīm\ *noun* (1922)
: a basic bacteriolytic protein that hydrolyzes peptidoglycan and is present in egg white and in human tears and saliva

-lyte *noun combining form* [Greek *lytos* that may be untied, soluble, from *lyein*]
: substance capable of undergoing (such) decomposition ⟨electro*lyte*⟩

lyt·ic \ˈli-tik\ *adjective* [Greek *lytikos* able to loose, from *lyein*] (1889)
: of or relating to lysis or a lysin; *also* **:** productive of or effecting lysis (as of cells)
— **ly·ti·cal·ly** \-ti-k(ə-)lē\ *adverb*

-lytic *adjective suffix* [Greek *lytikos*]
: of, relating to, or effecting (such) decomposition ⟨hydro*lytic*⟩

-lyze *verb combining form* **-lyzed; -lyzing** [International Scientific Vocabulary, probably irregular from New Latin *-lysis*]
: produce or undergo lytic disintegration or dissolution ⟨electro*lyze*⟩

M *is the thirteenth letter of the English alphabet. It comes through the Latin, via Etruscan, from the Greek* mu, *which took it from the Phoenician* mêm. *Its sound is that of a bilabial continuant, a sound formed by stopping the oral passage of air at the lips and allowing it to continue through the nose, giving the sound a nasal resonance. (For this reason* M *is also classified as a nasal consonant).* M *is silent only when initial before* n *in several words from Greek, as* mnemonic. *In related languages, English* m *corresponds to German* m, *Latin* m *or* n, *Greek* m *or* n, *and Sanskrit* m, *as in English* mouse, *German* Maus, *Latin* mus, *Greek* mys, *Sanskrit* mus; *and English* come, *German* kommen, *Latin* venire, *Greek* bainein, *Sanskrit* gamati *(goes). Small* m *developed by a rounding of the sharp angles and a reduction in size of the capital form of the letter.*

m \'em\ *noun, plural* **m's** *or* **ms** \'emz\ *often capitalized, often attributive* (before 12th century)
1 a : the 13th letter of the English alphabet **b :** a graphic representation of this letter **c :** a speech counterpart of orthographic *m*
2 : one thousand — see NUMBER table
3 : a graphic device for reproducing the letter *m*
4 : one designated *m* especially as the 13th in order or class
5 : something shaped like the letter M
6 a : EM 2 **b :** PICA 2

'm \m\ *verb* (circa 1594)
: AM ⟨I'*m* going⟩

ma \'mä, 'mȯ\ *noun* [short for *mama*] (1829)
: MOTHER

ma'am \'mam, *after "yes" often* əm\ *noun* (1668)
: MADAM

ma–and–pa \ˌmä-ən-'pä, ˌmȯ-ən-'pȯ\ *adjective* (1965)
: MOM-AND POP

Mab \'mab\ *noun*
: a queen of fairies in English literature

mabe pearl \'māb-\ *noun* [origin unknown] (1952)
: a cultured pearl essentially hemispherical in form — called also *mabe*

mac \'mak\ *noun* (1901)
British **:** MACKINTOSH

Mac \'mak\ *noun* [*Mac-, Mc-,* patronymic prefix in Scottish and Irish surnames] (circa 1937)
: FELLOW — used informally to address a man whose name is not known

ma·ca·bre \mə-'käb; -'kä-brə, -bər; -'käbrᵊ\ *adjective* [French, from (*danse*) *macabre* dance of death, from Middle French (*danse de*) *Macabré*] (1889)
1 : having death as a subject **:** comprising or including a personalized representation of death
2 : dwelling on the gruesome
3 : tending to produce horror in a beholder
synonym see GHASTLY

mac·ad·am \mə-'ka-dəm\ *noun* [John L. *McAdam* (died 1836) British engineer] (1824)
: macadamized roadway or pavement especially with a bituminous binder ◆

mac·a·da·mia nut \ˌma-kə-'dā-mē-ə-\ *noun* [New Latin *Macadamia,* from John *Macadam* (died 1865) Australian chemist] (1929)
: a hard-shelled nut somewhat resembling a filbert and produced by an Australian evergreen tree (*Macadamia integrifolia*) of the protea family that is cultivated extensively in Hawaii — called also *macadamia*

mac·ad·am·ize \mə-'ka-də-ˌmīz\ *transitive verb* **-ized; -iz·ing** (1826)
: to construct or finish (a road) by compacting into a solid mass a layer of small broken stone on a convex well-drained roadbed and using a binder (as cement or asphalt) for the mass

ma·caque \mə-'kak, -'käk\ *noun* [French, from Portuguese *macaco*] (1757)
: any of a genus (*Macaca*) of chiefly Asian monkeys typically having a sturdy build and including some short-tailed or tailless forms; *especially* **:** RHESUS MONKEY

mac·a·ro·ni \ˌma-kə-'rō-nē\ *noun* [Italian *maccheroni,* plural of *maccherone,* from Italian dialect *maccarone* dumpling, macaroni] (1599)
1 : pasta made from semolina and shaped in the form of slender tubes
2 *plural* **macaronis** *or* **macaronies a :** a member of a class of traveled young Englishmen of the late 18th and early 19th centuries who affected foreign ways **b :** an affected young man **:** FOP

mac·a·ron·ic \-'rä-nik\ *adjective* [New Latin *macaronicus,* from Italian dialect *maccarone* macaroni] (1638)
1 : characterized by a mixture of vernacular words with Latin words or with non-Latin words having Latin endings
2 : characterized by a mixture of two languages
— macaronic *noun*

mac·a·roon \ˌma-kə-'rün\ *noun* [French *macaron,* from Italian dialect *maccarone*] (circa 1611)
: a small cookie composed chiefly of egg whites, sugar, and ground almonds or coconut

ma·caw \mə-'kȯ\ *noun* [Portuguese *macau*] (1668)
: any of numerous parrots (especially genus *Ara*) of South and Central America including some of the largest and showiest of parrots

macaw

Mac·beth \mək-'beth, mak-\ *noun*
: a Scottish general who is the protagonist of Shakespeare's tragedy *Macbeth*

Mac·ca·bees \'ma-kə-(ˌ)bēz\ *noun plural* [Greek *Makkabaioi,* from plural of *Makkabaios,* surname of Judas Maccabaeus 2d century B.C. Jewish patriot] (1702)
1 : a priestly family leading a Jewish revolt begun in 168 B.C. against Hellenism and Syrian rule and reigning over Palestine from 142 B.C. to 63 B.C.
2 *singular in construction* **:** either of two narrative and historical books included in the Roman Catholic canon of the Old Testament and in the Protestant Apocrypha — see BIBLE table
— Mac·ca·be·an \ˌma-kə-'bē-ən\ *adjective*

[1]mace \'mās\ *noun* [Middle English, from Middle French *macis,* from Latin *macir,* an East Indian spice, from Greek *makir*] (13th century)
: an aromatic spice consisting of the dried external fibrous covering of a nutmeg

[2]mace *noun* [Middle English, from Old French, from (assumed) Vulgar Latin *mattia;* akin to Latin *mateola* mallet] (14th century)
1 a : a heavy often spiked staff or club used especially in the Middle Ages for breaking armor **b :** a club used as a weapon
2 a : an ornamental staff borne as a symbol of authority before a public official or a legislative body **b :** one who carries a mace

[3]mace *transitive verb* **maced; mac·ing** (1968)
: to attack with the liquid Mace

Mace \'mās\ *trademark*
— used for a temporarily disabling liquid usually used as a spray

ma·cé·doine \ˌma-sə-'dwän\ *noun* [French, from *Macédoine* Macedonia; perhaps from the mixture of ethnic groups in Macedonia] (1820)
1 : a confused mixture **:** MEDLEY
2 : a mixture of fruits or vegetables served as a salad or cocktail or in a jellied dessert or used in a sauce or as a garnish

Mac·e·do·nian \ˌma-sə-'dō-nyən, -nē-ən\ *noun* (1582)
1 : a native or inhabitant of Macedonia
2 : the Slavic language of modern Macedonia
3 : the language of ancient Macedonia of uncertain affinity but generally assumed to be Indo-European

mac·er·ate \'ma-sə-ˌrāt\ *verb* **-at·ed; -at·ing** [Latin *maceratus,* past participle of *macerare* to soften, steep] (1547)
transitive verb
1 : to cause to waste away by or as if by excessive fasting
2 : to cause to become soft or separated into constituent elements by or as if by steeping in fluid; *broadly* **:** STEEP, SOAK
intransitive verb
: to soften and wear away especially as a result of being wetted or steeped
— mac·er·a·tion \ˌma-sə-'rā-shən\ *noun*
— mac·er·a·tor \'ma-sə-ˌrā-tər\ *noun*

◇ WORD HISTORY
macadam *Macadam* is named after the Scottish engineer John Loudon McAdam (1756–1836). Concerned about the poor condition of Scotland's roads at the end of the 18th century, McAdam studied the problem for a decade and a half. He concluded that roads constructed of small, similarly sized fragments of stone could withstand years of heavy traffic. Upon his appointment as surveyor-general of roads in Bristol in 1815, he began to put his theories into practice. By 1823 the superiority of his paving process was generally recognized, and four years later he was named surveyor-general of roads for Great Britain and was even offered a knighthood. His name evolved into a lowercase, generic noun, the spelling eventually undergoing alteration to produce a more conventional-looking word. McAdam's name also survives, albeit rather covertly, in *tarmac,* shortened from *tarmacadam,* which designates a road surfacing that incorporates a tar binder.

Mach \'mäk\ *noun* [*Mach number*] (1946)
: a usually high speed expressed by a Mach number ⟨an airplane flying at *Mach* 2⟩

Mach·a·bees \'ma-kə-(,)bēz\ *noun plural but singular in construction* [Middle English, from Late Latin *Machabaei*, modification of Greek *Makkabaioi*] (14th century)
: MACCABEES

mâche \'mäsh\ *noun* [French, perhaps alteration of French dialect *pomache*, from (assumed) Vulgar Latin *pomasca*, from Latin *pomum* fruit] (1961)
: CORN SALAD

ma·chete \mə-'she-tē, -'che-; -'shet\ *noun* [Spanish] (1598)
: a large heavy knife used for cutting sugarcane and underbrush and as a weapon

Ma·chi·a·vel·lian \,ma-kē-ə-'ve-lē-ən, -'vel-yən\ *adjective* [Niccolo *Machiavelli*] (1579)
1 : of or relating to Machiavelli or Machiavellianism
2 : suggesting the principles of conduct laid down by Machiavelli; *specifically* **:** marked by cunning, duplicity, or bad faith
— Machiavellian *noun*

Ma·chi·a·vel·lian·ism \-'ve-lē-ə-,ni-zəm, -'vel-yə-,ni-zəm\ *noun* (1626)
: the political theory of Machiavelli; *especially* **:** the view that politics is amoral and that any means however unscrupulous can justifiably be used in achieving political power

ma·chic·o·la·tion \mə-,chi-kə-'lā-shən\ *noun* [Medieval Latin *machicolare* to furnish with machicolations, from Middle French *machicoller*, from *machicoleis* machicolation, from *macher* to crush + *col* neck, from Latin *collum* — more at COLLAR] (1788)
1 a : an opening between the corbels of a projecting parapet or in the floor of a gallery or roof of a portal for discharging missiles upon assailants below — see BATTLEMENT illustration **b :** a gallery or parapet containing such openings
2 : construction imitating medieval machicolation
— ma·chic·o·lat·ed \mə-'chi-kə-,lā-təd\ *adjective*

mach·i·nate \'ma-kə-,nāt, 'ma-shə-\ *verb* **-nat·ed; -nat·ing** [Latin *machinatus*, past participle of *machinari*, from *machina* machine, contrivance] (1600)
intransitive verb
: to plan or plot especially to do harm
transitive verb
: to scheme or contrive to bring about **:** PLOT
— mach·i·na·tor \-,nā-tər\ *noun*

mach·i·na·tion \,ma-kə-'nā-shən, ,ma-shə-\ *noun* (15th century)
1 : an act of machinating
2 : a scheming or crafty action or artful design intended to accomplish some usually evil end ⟨backstage *machinations* and power plays that have dominated the film industry —Peter Bogdanovich⟩
synonym see PLOT

¹ma·chine \mə-'shēn\ *noun, often attributive* [Middle French, from Latin *machina*, from Greek *mēchanē* (Doric dialect *machana*), from *mēchos* means, expedient — more at MAY] (circa 1545)
1 a *archaic* **:** a constructed thing whether material or immaterial **b :** CONVEYANCE, VEHICLE; *especially* **:** AUTOMOBILE **c** *archaic* **:** a military engine **d :** any of various apparatuses formerly used to produce stage effects **e** (1) **:** an assemblage of parts that transmit forces, motion, and energy one to another in a predetermined manner (2) **:** an instrument (as a lever) designed to transmit or modify the application of power, force, or motion **f :** a mechanically, electrically, or electronically operated device for performing a task ⟨a calculating *machine*⟩ ⟨a card-sorting *machine*⟩ **g :** a coin-operated device ⟨a cigarette *machine*⟩ **h :** MACHINERY — used with *the* or in plural
2 a : a living organism or one of its functional systems **b :** a person or organization that resembles a machine (as in being methodical, tireless, or unemotional) **c** (1) **:** a combination of persons acting together for a common end along with the agencies they use (2) **:** a highly organized political group under the leadership of a boss or small clique
3 : a literary device or contrivance introduced for dramatic effect

²machine *transitive verb* **ma·chined; ma·chin·ing** (circa 1864)
: to process by or as if by machine; *especially* **:** to reduce or finish by or as if by turning, shaping, planing, or milling by machine-operated tools
— ma·chin·abil·i·ty *also* **ma·chine·abil·i·ty** \-,shē-nə-'bi-lə-tē\ *noun*
— ma·chin·able *also* **ma·chine·able** \-'shē-nə-bəl\ *adjective*

ma·chine–gun \mə-'shēn-,gən\ *adjective* (1906)
: characterized by rapidity and sharpness **:** RAPID-FIRE ⟨a comic's *machine-gun* delivery⟩

machine gun *noun* (1870)
: a gun for sustained rapid fire that uses bullets; *broadly* **:** an automatic weapon
— machine–gun *verb*
— machine gun·ner *noun*

machine language *noun* (1949)
1 : the set of symbolic instruction codes usually in binary form that is used to represent operations and data in a machine (as a computer)
2 : ASSEMBLY LANGUAGE

ma·chine·like \mə-'shēn-,līk\ *adjective* (circa 1712)
: resembling a machine especially in regularity of action or stereotyped uniformity of product

machine pistol *noun* (1940)
: a usually small submachine gun

machine–readable *adjective* (1961)
: directly usable by a computer ⟨*machine-readable* text⟩

ma·chin·ery \mə-'shē-nə-rē, -'shēn-rē\ *noun, plural* **-er·ies** (1687)
1 a : machines in general or as a functioning unit **b :** the working parts of a machine
2 : the means or system by which something is kept in action or a desired result is obtained ⟨the *machinery* of government⟩ ⟨genetic *machinery* of cells⟩

machine shop *noun* (1827)
: a workshop in which work is machined to size and assembled

machine tool *noun* (1861)
: a machine designed for shaping solid work

ma·chin·ist \mə-'shē-nist\ *noun* (circa 1706)
1 a : a worker who fabricates, assembles, or repairs machinery **b :** a craftsman skilled in the use of machine tools **c :** one who operates a machine
2 *archaic* **:** a person in charge of the mechanical aspects of a theatrical production
3 : a warrant officer who supervises machinery and engine operation

ma·chis·mo \mä-'chēz-(,)mō, mə-, -'kēz-, -'kiz-, -'chiz-\ *noun* [Spanish, from *macho*] (1948)
1 : a strong sense of masculine pride **:** an exaggerated masculinity
2 : an exaggerated or exhilarating sense of power or strength

Mach number \'mäk-\ *noun* [Ernst *Mach* (died 1916) Austrian physicist] (1937)
: a number representing the ratio of the speed of a body (as an aircraft) to the speed of sound in a surrounding medium (as air)

¹ma·cho \'mä-(,)chō\ *adjective* [Spanish, literally, male, from Latin *masculus* — more at MASCULINE] (1928)
: characterized by machismo **:** aggressively virile

²macho *noun, plural* **machos** (1951)
1 : one who exhibits machismo
2 : MACHISMO

mack *variant of* MAC

mack·er·el \'ma-k(ə-)rəl\ *noun, plural* **mackerel** *or* **mackerels** [Middle English *makerel*, from Middle French] (14th century)
1 : a scombroid bony fish (*Scomber scombrus*) of the North Atlantic that is green above with dark blue bars and silvery below and is one of the most important food fishes
2 : any of a family (Scombridae) of fishes including the common mackerel; *especially* **:** a comparatively small member of this group as distinguished from a bonito or tuna

mackerel 1

mackerel shark *noun* (1819)
: any of a family (Lamnidae) of large pelagic sharks including the great white shark and mako sharks; *especially* **:** PORBEAGLE

mackerel sky *noun* (1669)
: a sky covered with rows of altocumulus or cirrocumulus clouds resembling the patterns on a mackerel's back

mack·i·naw \'ma-kə-,nȯ\ *noun* [*Mackinaw* (Mackinac), trading post at site of Mackinaw City, Michigan] (1836)
1 : a heavy woolen blanket formerly distributed by the U.S. government to the Indians
2 a : a heavy cloth of wool or wool and other fibers often with a plaid design and usually heavily napped and felted **b :** a short coat of mackinaw or similar heavy fabric

Mackinaw trout *noun* (1840)
: a large dark North American char (*Salvelinus namaycush*) that is an important commercial food fish in northern lakes

mack·in·tosh *also* **mac·in·tosh** \'ma-kən-,täsh\ *noun* [Charles *Macintosh* (died 1843) Scottish chemist & inventor] (1836)
1 *chiefly British* **:** RAINCOAT
2 : a lightweight waterproof fabric originally of rubberized cotton

Mac·lau·rin's series \mə-'klȯr-ən(z)-\ *noun* [Colin *Maclaurin* (died 1746) Scottish mathematician] (circa 1909)
: a Taylor's series of the form

$$f(x) = f(0) + \frac{f'(0)}{1!}x + \frac{f''(0)}{2!}x^2 + \ldots + \frac{f^{[n]}(0)}{n!}x^n + \ldots$$

in which the expansion is about the reference point zero — called also *Maclaurin series*

ma·cle \'ma-kəl\ *noun* [French, wide-meshed net, lozenge voided, macle, from Old French, mesh, lozenge voided, of Germanic origin; akin to Old High German *masca* mesh — more at MESH] (1801)
1 : a twin crystal
2 : a flat often triangular diamond that is usually a twin crystal
— ma·cled \'ma-kəld\ *adjective*

ma·con *also* **mâ·con** \ma-'kōⁿ\ *noun, often capitalized* [French *mâcon*, from *Mâcon*, France] (1863)
: a dry red or white wine produced in the area around Mâcon, France

Mac·Pher·son strut \mək-'fir-sᵊn-, -'fər-\ *noun* [Earle S. *MacPherson* (died 1960) American engineer] (1975)
: a component of an automobile suspension consisting of a shock absorber mounted within a coil spring

macr- *or* **macro-** *combining form* [French & Latin, from Greek *makr-*, *makro-* long, from *makros* — more at MEAGER]
1 : long ⟨*macro*biotic⟩
2 : large ⟨*macro*molecule⟩

mac·ra·mé *also* **mac·ra·me** \'ma-krə-,mā\ *noun* [French or Italian; French *macramé*,

from Italian *macramè,* from Turkish *makrama* napkin, towel, from Arabic *miqramah* coverlet] (1869)
: a coarse lace or fringe made by knotting threads or cords in a geometrical pattern; *also* **:** the art of tying knots in patterns

¹mac·ro \'ma-(ˌ)krō\ *adjective* [*macr-*] (1923)
1 : being large, thick, or exceptionally prominent
2 a : of, involving, or intended for use with relatively large quantities or on a large scale **b :** of or relating to macroeconomics
3 : GROSS 1c

²macro *noun, plural* **macros** [short for *macroinstruction*] (1959)
: a single computer instruction that stands for a sequence of operations

mac·ro·ag·gre·gate \ˌma-krō-'a-gri-gət\ *noun* (1926)
: a relatively large particle (as of soil or a protein)
— mac·ro·ag·gre·gat·ed \-ˌgā-təd\ *adjective*

mac·ro·bi·ot·ic \-bī-'ä-tik, -bē-\ *adjective* (1965)
: of, relating to, or being an extremely restricted diet (as one containing chiefly whole grains) that is held by its advocates to promote health and well-being although it may actually be deficient in essential nutrients (as fats)

mac·ro·cosm \'ma-krə-ˌkä-zəm\ *noun* [French *macrocosme,* from Medieval Latin *macrocosmos,* from Latin *macr-* + Greek *kosmos* order, universe] (1600)
1 : the great world **:** UNIVERSE
2 : a complex that is a large-scale reproduction of one of its constituents
— mac·ro·cos·mic \ˌma-krə-'käz-mik\ *adjective*
— mac·ro·cos·mi·cal·ly \-mi-k(ə-)lē\ *adverb*

mac·ro·cy·clic \ˌma-krō-'sī-klik, -'si-\ *adjective* (1936)
: containing or being a chemical ring that consists usually of 15 or more atoms

mac·ro·cyte \'ma-krə-ˌsīt\ *noun* [International Scientific Vocabulary] (circa 1889)
: an exceptionally large red blood cell occurring chiefly in anemias
— mac·ro·cyt·ic \ˌma-krə-'si-tik\ *adjective*

mac·ro·cy·to·sis \ˌma-krə-sī-'tō-səs, -krə-sə-\ *noun, plural* **-to·ses** \-ˌsēz\ [New Latin] (circa 1893)
: the occurrence of macrocytes in the blood

mac·ro·eco·nom·ics \'ma-krō-ˌe-kə-'nä-miks, -ˌē-kə-\ *noun plural but usually singular in construction* (1948)
: a study of economics in terms of whole systems especially with reference to general levels of output and income and to the interrelations among sectors of the economy — compare MICROECONOMICS
— mac·ro·eco·nom·ic \-mik\ *adjective*

mac·ro·evo·lu·tion \'ma-krō-ˌe-və-'lü-shən *also* -ˌē-və-\ *noun* (1939)
: evolution that results in relatively large and complex changes (as in species formation)
— mac·ro·evo·lu·tion·ary \-shə-ˌner-ē\ *adjective*

mac·ro·fos·sil \'ma-krō-ˌfä-səl\ *noun* (1937)
: a fossil large enough to be observed by direct inspection

mac·ro·ga·mete \ˌma-krō-'ga-ˌmēt *also* -gə-'mēt\ *noun* [International Scientific Vocabulary] (1899)
: the larger and usually female gamete of a heterogamous organism

mac·ro·glob·u·lin \-'glä-byə-lən\ *noun* [International Scientific Vocabulary] (1952)
: a highly polymerized globulin (as IgM) of high molecular weight

mac·ro·glob·u·li·ne·mia \-ˌglä-byə-lə-'nē-mē-ə\ *noun* [New Latin] (1949)
: a disorder characterized by increased blood serum viscosity and the presence of macroglobulins in the serum
— mac·ro·glob·u·li·ne·mic \-mik\ *adjective*

mac·ro·in·struc·tion \ˌma-krō-in-'strək-shən\ *noun* (1959)
: MACRO

macro lens *noun* [*macr-,* from the fact that the focal length is greater than normal] (1961)
: a camera lens designed to focus at very short distances with up to life-size magnification of the image

mac·ro·lep·i·dop·tera \'ma-krō-ˌle-pə-'däp-tə-rə\ *noun plural* [New Latin] (1882)
: lepidoptera (as butterflies, skippers, saturniids, noctuids, and geometrids) that include most of the large forms and none of the minute ones

mac·ro·mere \'ma-krə-ˌmir\ *noun* (1877)
: a large blastomere — see BLASTULA illustration

mac·ro·mol·e·cule \ˌma-krō-'mä-li-ˌkyü(ə)l\ *noun* [International Scientific Vocabulary] (circa 1929)
: a very large molecule (as of a protein or rubber)
— mac·ro·mo·lec·u·lar \-mə-'le-kyə-lər\ *adjective*

ma·cron \'mā-ˌkrän, 'ma-, -krən\ *noun* [Greek *makron,* neuter of *makros* long] (1851)
: a mark ¯ placed over a vowel to indicate that the vowel is long or placed over a syllable or used alone to indicate a stressed or long syllable in a metrical foot

mac·ro·nu·cle·us \ˌma-krō-'nü-klē-əs, -'nyü-\ *noun* [New Latin] (1892)
: a relatively large densely staining nucleus of most ciliate protozoans that is derived from micronuclei and controls various nonreproductive functions
— mac·ro·nu·cle·ar \ˌmak-rō-'nü-klē-ər, -'nyü-, ÷-kyə-lər\ *adjective*

mac·ro·nu·tri·ent \ 'nü-trē-ənt, -'nyü-\ *noun* (1942)
: a chemical element (as nitrogen, phosphorus, or potassium) of which relatively large quantities are essential to the growth and health of a plant

mac·ro·phage \'ma-krə-ˌfāj\ *noun* [International Scientific Vocabulary] (1890)
: a phagocytic tissue cell of the reticuloendothelial system that may be fixed or freely motile, is derived from a monocyte, and functions in the protection of the body against infection and noxious substances — called also *histiocyte*
— mac·ro·phag·ic \ˌma-krə-'fa-jik\ *adjective*

mac·ro·pho·tog·ra·phy \ˌma-krō-fə-'tä-grə-fē\ *noun* (1889)
: the making of photographs in which the object is either unmagnified or slightly magnified up to a limit often of about 10 diameters
— mac·ro·pho·to·graph \-'fō-tə-ˌgraf\ *noun*

mac·ro·phyte \'ma-krə-ˌfīt\ *noun* (circa 1909)
: a member of the macroscopic plant life especially of a body of water
— mac·ro·phyt·ic \ˌma-krə-'fi-tik\ *adjective*

mac·rop·ter·ous \ma-'kräp-tə-rəs\ *adjective* [Greek *makropteros,* from *makr-* + *pteron* wing — more at FEATHER] (circa 1836)
: having long or large wings

mac·ro·scale \'ma-krō-ˌskāl\ *noun* (1931)
: a large often macroscopic scale

mac·ro·scop·ic \ˌma-krə-'skä-pik\ *adjective* [International Scientific Vocabulary *macr-* + *-scopic* (as in *microscopic*)] (1872)
1 : large enough to be observed by the naked eye
2 : considered in terms of large units or elements
— mac·ro·scop·i·cal·ly \-pi-k(ə-)lē\ *adverb*

mac·ro·struc·ture \'ma-krō-ˌstrək-chər\ *noun* (circa 1899)
: the structure (as of metal, a body part, or the soil) revealed by visual examination with little or no magnification
— mac·ro·struc·tur·al \ˌma-krō-'strək-chə-rəl, -'strək-shə-rəl\ *adjective*

mac·u·la \'ma-kyə-lə\ *noun, plural* **-lae** \-ˌlē, -ˌlī\ *also* **-las** [Middle English, from Latin] (14th century)
1 : SPOT, BLOTCH; *especially* **:** MACULE
2 : an anatomical structure (as the macula lutea) having the form of a spot differentiated from surrounding tissues
— mac·u·lar \-lər\ *adjective*

macula lu·tea \-'lü-tē-ə\ *noun, plural* **maculae lu·te·ae** \-tē-ˌē, -tē-ˌī\ [New Latin, literally, yellow spot] (1848)
: a small yellowish area lying slightly lateral to the center of the retina that constitutes the region of maximum visual acuity — called also *yellow spot*

mac·u·late \'ma-kyə-lət\ *or* **mac·u·lat·ed** \-ˌlā-təd\ *adjective* [Latin *maculatus,* past participle of *maculare* to stain, from *macula*] (15th century)
1 : marked with spots **:** BLOTCHED
2 : IMPURE, BESMIRCHED

mac·u·la·tion \ˌma-kyə-'lā-shən\ *noun* (15th century)
1 *archaic* **:** the state of being spotted
2 a : a blemish in the form of a discrete spot ⟨acne scars and *maculations*⟩ **b :** the arrangement of spots and markings on an animal or plant

mac·ule \'ma-(ˌ)kyü(ə)l\ *noun* [French, from Latin *macula*] (1863)
: a patch of skin that is altered in color but usually not elevated and that is a characteristic feature of various diseases (as smallpox)

ma·cum·ba \mə-'küm-bə\ *noun* [Brazilian Portuguese] (1939)
: a polytheistic religion of African origin involving syncretistic elements and practiced mainly by Brazilian blacks in urban areas

¹mad \'mad\ *adjective* **mad·der; mad·dest** [Middle English *medd, madd,* from Old English *gemǣd,* past participle of (assumed) *gemǣdan* to madden, from *gemād* silly, mad; akin to Old High German *gimeit* foolish, crazy] (before 12th century)
1 : disordered in mind **:** INSANE
2 a : completely unrestrained by reason and judgment ⟨driven *mad* by the pain⟩ **b :** incapable of being explained or accounted for ⟨a *mad* decision⟩
3 : carried away by intense anger **:** FURIOUS ⟨*mad* at myself⟩ ⟨*mad* about the delay⟩
4 : carried away by enthusiasm or desire ⟨*mad* about horses⟩ ⟨*mad* for the boy next door⟩
5 : affected with rabies **:** RABID
6 : marked by wild gaiety and merriment **:** HILARIOUS
7 : intensely excited **:** FRANTIC ⟨*mad* with jealousy⟩
8 : marked by intense and often chaotic activity **:** WILD ⟨a *mad* scramble⟩
— mad·dish \'ma-dish\ *adjective*

²mad *verb* **mad·ded; mad·ding** (14th century)
: MADDEN

³mad *noun* (1834)
1 : a fit or mood of bad temper
2 : ANGER, FURY

Mad·a·gas·car periwinkle \ˌma-də-'gas-kər-\ *noun* [*Madagascar,* Africa] (1821)
: ¹PERIWINKLE 1b

mad·am \'ma-dəm\ *noun, plural* **madams** [Middle English, from Middle French *ma dame,* literally, my lady] (14th century)

1 *plural* **mes·dames** \mā-'däm, -'dam\ **:** LADY — used without a name as a form of respectful or polite address to a woman
2 : MISTRESS 1 — used as a title formerly with the given name but now with the surname or especially with a designation of rank or office ⟨*Madam* Chairman⟩ ⟨*Madam* President⟩
3 : the female head of a house of prostitution
4 : the female head of a household **:** WIFE

ma·dame \mə-'dam, ma-', *before a surname also* 'ma-dəm\ *noun* [French, from Old French *ma dame*] (1674)
1 *plural* **mes·dames** \mā-'däm, -'dam\ — used as a title equivalent to *Mrs.* for a married woman not of English-speaking nationality
2 *plural* **madames :** MADAM 3

mad–brained \'mad-'brānd\ *adjective* (1562) **:** RASH, HOTHEADED

mad·cap \'mad-ˌkap\ *adjective* (1588) **:** marked by capriciousness, recklessness, or foolishness
— **madcap** *noun*

mad·den \'ma-dᵊn\ *verb* **mad·dened; mad·den·ing** \'mad-niŋ, 'ma-dᵊn-iŋ\ (1735)
intransitive verb **:** to become or act as if mad
transitive verb
1 : to drive mad **:** CRAZE
2 : to make intensely angry **:** ENRAGE

maddening *adjective* (1822)
1 : tending to craze
2 a : tending to infuriate **b :** tending to vex **:** IRRITATING
— **mad·den·ing·ly** \-lē\ *adverb*

mad·der \'ma-dər\ *noun* [Middle English, from Old English *mædere;* akin to Old High German *matara* madder] (before 12th century)
1 : a Eurasian herb (*Rubia tinctorum* of the family Rubiaceae, the madder family) with whorled leaves and small yellowish panicled flowers succeeded by berries; *broadly* **:** any of several related herbs (genus *Rubia*)
2 a : the root of the Eurasian madder used formerly in dyeing; *also* **:** an alizarin dye prepared from it **b :** a moderate to strong red

mad·ding \'ma-diŋ\ *adjective* (1579) **:** acting in a frenzied manner ⟨the *madding* crowd⟩

made \'mād\ *adjective* [Middle English, from past participle of *maken* to make] (14th century)
1 a : FICTITIOUS, INVENTED ⟨a *made* excuse⟩ **b :** artificially produced **c :** put together of various ingredients ⟨a *made* dish⟩
2 : assured of success ⟨a *made* man⟩ — usually used in the phrase *have it made*

Ma·dei·ra \mə-'dir-ə, -'der-\ *noun* [Portuguese, from *Madeira* Islands] (1596) **:** an amber-colored fortified wine from Madeira; *also* **:** a similar wine made elsewhere

mad·e·leine \'ma-dᵊl-ən, ˌma-dᵊl-'ān\ *noun* [French, perhaps from *Madeleine* Paumier, 19th century French pastry cook] (1845)
1 : a small rich shell-shaped cake
2 : one that evokes a memory

ma·de·moi·selle \ˌmad-mwə-'zel, ˌma-də-, -mə-'zel, *sometimes* mam-'zel\ *noun, plural* **ma·de·moi·selles** \-'zelz\ *or* **mes·de·moi·selles** \ˌmād-mwə-'zel, ˌmā-də-, -mə-'zel\ [Middle English *madamoiselle,* from Middle French, from Old French *ma damoisele,* literally, my (young) lady] (15th century)
1 : an unmarried French girl or woman — used as a title equivalent to *Miss* for an unmarried woman not of English-speaking nationality
2 : a French governess
3 : SILVER PERCH a

made–to–order *adjective* (circa 1908)
1 : produced to supply a special or an individual demand **:** CUSTOM-MADE
2 : ideally suited (as to a particular purpose) ⟨a ground ball *made-to-order* for a double play⟩

made–up \'mā-'dəp\ *adjective* (1607)
1 : fancifully conceived or falsely devised
2 : fully manufactured
3 : marked by the use of makeup

mad·house \'mad-ˌhau̇s\ *noun* (1687)
1 : a place where insane persons are detained and treated
2 : a place of uproar or confusion

Mad·i·son Avenue \'ma-də-sən-\ *noun* [*Madison Avenue,* New York City, former center of the American advertising business] (1952) **:** the American advertising industry

mad·ly \'mad-lē\ *adverb* (13th century)
1 : in a mad manner
2 : to an extreme or excessive degree ⟨*madly* in love⟩

mad·man \'mad-ˌman, -mən\ *noun* (14th century) **:** a man who is or acts as if insane

mad money *noun* (1922) **:** money that a woman carries to pay her fare home in case a date ends in a quarrel; *also* **:** money set aside for an emergency or personal use

mad·ness \'mad-nəs\ *noun* (14th century)
1 : the quality or state of being mad: as **a :** RAGE **b :** INSANITY **c :** extreme folly **d :** ECSTASY, ENTHUSIASM
2 : any of several ailments of animals marked by frenzied behavior; *specifically* **:** RABIES

Ma·don·na \mə-'dä-nə\ *noun* [Italian, from Old Italian *ma donna,* literally, my lady] (1584)
1 *archaic* **:** LADY — used as a form of respectful address
2 *obsolete* **:** an Italian lady
3 a : VIRGIN MARY **b :** an artistic depiction (as a painting or statue) of the Virgin Mary

Madonna lily *noun* (1877) **:** a widely cultivated lily (*Lilium candidum*) with bell-shaped to broad funnel-shaped white flowers

ma·dras \'ma-drəs; mə-'dras, -'dräs\ *noun* [*Madras,* India] (circa 1830)
1 : a large silk or cotton kerchief usually of bright colors that is often worn as a turban
2 a : a fine plain-woven shirting and dress fabric usually of cotton with varied designs (as plaid) in bright colors or in white **b :** a light open usually cotton fabric with a heavy design used for curtains

mad·re·pore \'ma-drə-ˌpōr, -ˌpȯr\ *noun* [French *madrépore,* from Italian *madrepora,* from *madre* mother (from Latin *mater*) + *poro* pore (from Latin *porus*) — more at MOTHER] (1751) **:** any of various stony reef-building corals (order Madreporaria) of tropical seas that assume a variety of branching, encrusting, or massive forms
— **mad·re·po·ri·an** \ˌma-drə-'pōr-ē-ən, -'pȯr-\ *adjective or noun*
— **mad·re·por·ic** \-'pōr-ik, -'pȯr-\ *adjective*

mad·re·por·ite \'ma-drə-ˌpōr-ˌīt, -ˌpȯr-\ *noun* [International Scientific Vocabulary *madrepore* + [1]*-ite* (segment); from the resemblances of the perforations to those of a madrepore] (1877) **:** a perforated or porous body that is situated at the distal end of the stone canal in echinoderms

mad·ri·gal \'ma-dri-gəl\ *noun* [Italian *madrigale,* probably from Medieval Latin *matricale,* from neuter of (assumed) *matricalis* simple, from Late Latin, of the womb, from Latin *matric-, matrix* womb, from *mater* mother] (1588)
1 : a medieval short lyrical poem in a strict poetic form
2 a : a complex polyphonic unaccompanied vocal piece on a secular text developed especially in the 16th and 17th centuries **b :** PARTSONG; *especially* **:** GLEE
— **mad·ri·gal·ian** \ˌma-drə-'ga-lē-ən, -'gā-\ *adjective*
— **mad·ri·gal·ist** \'ma-dri-gə-list\ *noun*

ma·dri·lène \ˌma-drə-'len, -'lān\ *noun* [French (*consommé*) *madrilène,* literally, Madrid consommé] (1907) **:** a consommé flavored with tomato

ma·dro·na *or* **ma·dro·ne** *or* **ma·dro·no** \mə-'drō-nə\ *noun* [Spanish *madroño*] (1841) **:** any of several evergreen trees (genus *Arbutus*) of the heath family; *especially* **:** one (*A. menziesii*) of the Pacific coast of North America with smooth bark, thick shining leaves, and edible red berries

ma·du·ro \mə-'du̇r-(ˌ)ō\ *noun, plural* **-ros** [Spanish, from *maduro* ripe, from Latin *maturus* — more at MATURE] (1850) **:** a dark-colored relatively strong cigar

mad·wom·an \'mad-ˌwu̇-mən\ *noun* (15th century) **:** a woman who is or acts as if insane

mad·wort \-ˌwərt, -ˌwȯrt\ *noun* (1597)
1 : ALYSSUM 1
2 : a low hairy annual herb (*Asperugo procumbens*) of the borage family with blue flowers and a root used as a substitute for madder

Mae·ce·nas \mi-'sē-nəs\ *noun* [Latin, from Gaius *Maecenas* (died 8 B.C.) Roman statesman & patron of literature] (1542) **:** a generous patron especially of literature or art

mael·strom \'māl(ə)l-strəm, -ˌsträm\ *noun* [obsolete Dutch (now *maalstroom*), from *malen* to grind + *strom* stream] (1682)
1 : a powerful often violent whirlpool sucking in objects within a given radius
2 : something resembling a maelstrom in turbulence

mae·nad \'mē-ˌnad\ *noun* [Latin *maenad-, maenas,* from Greek *mainad-, mainas,* from *mainesthai* to be mad; akin to Greek *menos* spirit — more at MIND] (1579)
1 : a woman participant in orgiastic Dionysian rites **:** BACCHANTE
2 : an unnaturally excited or distraught woman
— **mae·nad·ic** \mē-'na-dik\ *adjective*

mae·sto·so \mī-'stō-(ˌ)sō, -(ˌ)zō\ *adjective or adverb* [Italian, from Latin *majestosus,* from *majestas* majesty] (circa 1724) **:** majestic and stately — used as a direction in music

mae·stro \'mī-(ˌ)strō\ *noun, plural* **mae·stros** *or* **mae·stri** \-ˌstrē\ [Italian, literally, master, from Latin *magister* — more at MASTER] (1724) **:** a master usually in an art; *especially* **:** an eminent composer, conductor, or teacher of music

Mae West \'mā-'west\ *noun* [*Mae West* (died 1980) American actress noted for her full figure] (1940) **:** an inflatable life jacket in the form of a collar extending down the chest that was worn by fliers in World War II

maf·fick \'ma-fik\ *intransitive verb* [back-formation from *Mafeking Night,* English celebration of the lifting of the siege of Mafeking, South Africa, May 17, 1900] (1900) **:** to celebrate with boisterous rejoicing and hilarious behavior

Ma·fia \'mä-fē-ə, 'ma-\ *noun* [*Mafia, Maffia,* a Sicilian secret criminal society, from Italian dialect (Sicily)] (1875)
1 a : a secret criminal society of Sicily or Italy **b :** a similarly conceived criminal organization in the U.S.; *also* **:** a similar organization elsewhere ⟨the Japanese *Mafia*⟩ **c :** a criminal organization associated with a particular traffic ⟨the cocaine *Mafia*⟩
2 *often not capitalized* **:** a group of people likened to the Mafia; *especially* **:** a group of people of similar interests or backgrounds prominent in a particular field or enterprise **:** CLIQUE

maf·ic \'ma-fik\ *adjective* [New Latin *magnesium* + Latin *ferrum* iron + English *-ic*] (1912) **:** of, relating to, or being a group of usually dark-colored minerals rich in magnesium and iron

ma·fi·o·so \ˌmä-fē-ˈō-(ˌ)sō, ˌma-, -(ˌ)zō\ *noun, plural* **-si** \-(ˌ)sē, -(ˌ)zē\ [Italian, from *Mafia*] (1875)
: a member of the Mafia or a mafia

mag \ˈmag\ *noun* (1796)
: MAGAZINE

mag·a·zine \ˈma-gə-ˌzēn, ˌma-gə-ˈ\ *noun* [Middle French, from Old Provençal, from Arabic *makhāzin,* plural of *makhzan* storehouse] (1583)
1 : a place where goods or supplies are stored **:** WAREHOUSE
2 : a room in which powder and other explosives are kept in a fort or a ship
3 : the contents of a magazine: as **a :** an accumulation of munitions of war **b :** a stock of provisions or goods
4 a : a periodical containing miscellaneous pieces (as articles, stories, poems) often illustrated **b :** a similar section of a newspaper usually appearing on Sunday **c :** a radio or television program presenting usually several short segments on a variety of topics
5 : a supply chamber: as **a :** a holder in or on a gun for cartridges to be fed into the gun chamber **b :** a lightproof chamber for films or plates on a camera or for film on a motion-picture projector ◆

mag·a·zin·ist \-ˌzē-nist, -ˈzē-\ *noun* (1821)
: a person who writes for or edits a magazine

mag·da·len \ˈmag-də-lən\ *or* **mag·da·lene** \-ˌlēn\ *noun, often capitalized* [Mary *Magdalen* or *Magdalene* woman healed by Jesus of evil spirits (Luke 8:2), considered identical with a reformed prostitute (Luke 7:36–50)] (1697)
1 : a reformed prostitute
2 : a house of refuge or reformatory for prostitutes

Mag·da·le·ni·an \ˌmag-də-ˈlē-nē-ən\ *adjective* [French *magdalénien,* from *La Madeleine,* rock shelter in southwest France] (1885)
: of or relating to an Upper Paleolithic culture characterized by flint, bone, and ivory implements, carving, and paintings

mage \ˈmāj\ *noun* [Middle English, from Latin *magus*] (14th century)
: MAGUS

Mag·el·lan·ic Cloud \ˌma-jə-ˈla-nik-, *chiefly British* ˌma-gə-\ *noun* [Ferdinand *Magellan*] (circa 1686)
: either of the two small galaxies that appear as conspicuous patches of light near the south celestial pole and are companions to the Milky Way galaxy

Ma·gen Da·vid \ˈmò-gən-ˈdò-vid, -ˈdə-; ˈmō-gən-ˈdā-vəd\ *noun* [Hebrew *māghēn Dāwīdh,* literally, shield of David] (circa 1904)
: a hexagram used as a symbol of Judaism

ma·gen·ta \mə-ˈjen-tə\ *noun* [*Magenta,* Italy] (1860)
1 : FUCHSIN
2 : a deep purplish red

mag·got \ˈma-gət\ *noun* [Middle English *mathek, magotte,* of Scandinavian origin; akin to Old Norse *mathkr* maggot; akin to Old English *matha* maggot] (14th century)
1 : a soft-bodied legless grub that is the larva of a dipterous insect (as the housefly)
2 : a fantastic or eccentric idea **:** WHIM
— mag·goty \-gə-tē\ *adjective*

magi *plural of* MAGUS

[1]Ma·gi·an \ˈmā-jē-ən\ *noun* (1578)
: MAGUS

[2]Ma·gi·an \-jē-ən, -ˌjī-\ *adjective* (1716)
: of or relating to the Magi
— Ma·gi·an·ism \-ə-ˌni-zəm\ *noun*

[1]mag·ic \ˈma-jik\ *noun* [Middle English *magique,* from Middle French, from Latin *magice,* from Greek *magikē,* feminine of *magikos* Magian, magical, from *magos* magus, sorcerer, of Iranian origin; akin to Old Persian *maguš* sorcerer] (14th century)
1 a : the use of means (as charms or spells) believed to have supernatural power over natural forces **b :** magic rites or incantations
2 a : an extraordinary power or influence seemingly from a supernatural source **b :** something that seems to cast a spell **:** ENCHANTMENT
3 : the art of producing illusions by sleight of hand

[2]magic *adjective* (14th century)
1 : of or relating to magic
2 a : having seemingly supernatural qualities or powers **b :** giving a feeling of enchantment
— mag·i·cal \ˈma-ji-kəl\ *adjective*
— mag·i·cal·ly \-ji-k(ə-)lē\ *adverb*

[3]magic *transitive verb* **mag·icked; mag·ick·ing** (1906)
: to produce, remove, or influence by magic

magic bullet *noun* (1940)
: a substance or therapy capable of destroying pathogens (as bacteria or cancer cells) without deleterious side effects

ma·gi·cian \mə-ˈji-shən\ *noun* (14th century)
1 : one skilled in magic; *especially* **:** SORCERER
2 : one who performs tricks of illusion and sleight of hand

magic lantern *noun* (1696)
: an early form of optical projector of still pictures using a transparent slide

Magic Marker *trademark*
— used for a felt-tipped pen

magic realism *noun* (1954)
: painting in a meticulously realistic style of imaginary or fantastic scenes or images
— magic realist *noun*

magic square *noun* (circa 1704)
: a square containing a number of integers arranged so that the sum of the numbers is the same in each row, column, and main diagonal and often in some or all of the other diagonals

4	9	2
3	5	7
8	1	6

6	3	10	15
9	16	5	4
7	2	11	14
12	13	8	1

magic squares

Ma·gi·not Line \ˈma-zhə-ˌnō-, ˈma-jə-\ *noun* [André *Maginot* (died 1932) French minister of war] (1936)
: a line of defensive fortifications built before World War II to protect the eastern border of France but easily outflanked by German invaders

mag·is·te·ri·al \ˌma-jə-ˈstir-ē-əl\ *adjective* [Late Latin *magisterialis* of authority, from *magisterium* office of a master, from *magister*] (1632)
1 a (1) **:** of, relating to, or having the characteristics of a master or teacher **:** AUTHORITATIVE (2) **:** marked by an overbearingly dignified or assured manner or aspect **b :** of, relating to, or required for a master's degree
2 : of or relating to a magistrate or a magistrate's office or duties
synonym see DICTATORIAL
— mag·is·te·ri·al·ly \-ē-ə-lē\ *adverb*

mag·is·te·ri·um \ˌma-jə-ˈstir-ē-əm\ *noun* [Latin] (1866)
: teaching authority especially of the Roman Catholic Church

mag·is·tra·cy \ˈma-jə-strə-sē\ *noun, plural* **-cies** (circa 1585)
1 : the state of being a magistrate
2 : the office, power, or dignity of a magistrate
3 : a body of magistrates
4 : the district under a magistrate

ma·gis·tral \ˈma-jə-strəl, mə-ˈjis-trəl\ *adjective* [Late Latin *magistralis,* from Latin *magistr-, magister*] (1605)
: MAGISTERIAL 1a
— ma·gis·tral·ly \-ē\ *adverb*

mag·is·trate \ˈma-jə-ˌstrāt, -strət\ *noun* [Middle English *magestrat,* from Latin *magistratus* magistracy, magistrate, from *magistr-, magister* master, political superior — more at MASTER] (14th century)
: an official entrusted with administration of the laws: as **a :** a principal official exercising governmental powers over a major political unit (as a nation) **b :** a local official exercising administrative and often judicial functions **c :** a local judiciary official having limited original jurisdiction especially in criminal cases
— mag·is·trat·i·cal \ˌma-jə-ˈstra-ti-kəl\ *adjective*
— mag·is·trat·i·cal·ly \-k(ə-)lē\ *adverb*

magistrate's court *noun* (1867)
1 : POLICE COURT
2 : a court that has minor civil and criminal jurisdiction

mag·is·tra·ture \ˈma-jə-ˌstrā-chər, -strə-ˌchu̇r\ *noun* (1672)
: MAGISTRACY

mag·lev \ˈmag-lev\ *noun, often attributive* [*mag*netic *lev*itation] (1969)
1 : the use of the physical properties of magnetic fields generated by superconducting magnets to cause an object (as a vehicle) to float above a solid surface
2 : a train utilizing maglev technology

mag·ma \ˈmag-mə\ *noun* [Middle English, from Latin *magmat-, magma,* from Greek, thick unguent, from *massein* to knead — more at MINGLE] (15th century)
1 *archaic* **:** DREGS, SEDIMENT
2 : a thin pasty suspension (as of a precipitate in water)
3 : molten rock material within the earth from which igneous rock results by cooling
— mag·mat·ic \mag-ˈma-tik\ *adjective*

Mag·na Car·ta *or* **Mag·na Char·ta** \ˈmag-nə-ˈkär-tə\ *noun* [Middle English, from Medieval Latin, literally, great charter] (15th century)
1 : a charter of liberties to which the English barons forced King John to give his assent in June 1215 at Runnymede
2 : a document constituting a fundamental guarantee of rights and privileges

mag·na cum lau·de \ˈmäg-nə-(ˌ)ku̇m-ˈlau̇-də, -ˈlau̇-dē; ˈmag-nə-ˌkəm-ˈlò-dē\ *adverb or adjective* [Latin] (1900)
: with great distinction ⟨graduated *magna cum laude*⟩ — compare CUM LAUDE, SUMMA CUM LAUDE

mag·na·nim·i·ty \ˌmag-nə-ˈni-mə-tē\ *noun, plural* **-ties** (14th century)
1 : the quality of being magnanimous **:** loftiness of spirit enabling one to bear trouble

◇ WORD HISTORY

magazine In Arabic *makhzan* denotes a storehouse. From the plural form *makhāzin* medieval French borrowed *magasin* in the 14th century, most likely via Old Provençal. English in turn borrowed the word from French in the 16th century, in the same sense. The term then underwent various specializations of meaning, depending on what was being stored. An early development was the application of the word to storage places for weapons or gunpowder; a further extension applied *magazine* not to a building but to a chamber or clip to hold cartridges in a firearm. *Magazine* was also used in the titles of books with the implication "storehouse of knowledge." The word was thus used in 1731 for the first issue of *The Gentleman's Magazine: or, Monthly Intelligencer,* the suggestion being that the publication would present the general reader each month with a collection or "storehouse" of stories and articles. Other periodical publications began using *magazine* in their titles, and the word was soon used generically for a certain kind of periodical.

calmly, to disdain meanness and pettiness, and to display a noble generosity
2 : a magnanimous act

mag•nan•i•mous \mag-'na-nə-məs\ *adjective* [Latin *magnanimus,* from *magnus* great + *animus* spirit — more at MUCH, ANIMATE] (1584)
1 : showing or suggesting a lofty and courageous spirit ⟨the irreproachable lives and *magnanimous* sufferings of their followers —Joseph Addison⟩
2 : showing or suggesting nobility of feeling and generosity of mind ⟨too sincere for dissimulation, too *magnanimous* for resentment —Ellen Glasgow⟩
— mag•nan•i•mous•ly *adverb*
— mag•nan•i•mous•ness *noun*

mag•nate \'mag-ˌnāt, -nət\ *noun* [Middle English *magnates,* plural, from Late Latin, from Latin *magnus*] (15th century)
: a person of rank, power, influence, or distinction often in a specified area

mag•ne•sia \mag-'nē-shə, -zhə\ *noun* [New Latin, from *magnes carneus,* a white earth, literally, flesh magnet] (1755)
: MAGNESIUM OXIDE — compare MILK OF MAGNESIA
— mag•ne•sian \-shən, -zhən\ *adjective*

mag•ne•site \'mag-nə-ˌsīt\ *noun* (1815)
: native magnesium carbonate used especially in making refractories and magnesium oxide

mag•ne•sium \mag-'nē-zē-əm, -zhəm\ *noun* [New Latin, from *magnesia*] (1812)
: a silver-white light malleable ductile metallic element that occurs abundantly in nature and is used in metallurgical and chemical processes, in photography, signaling, and pyrotechnics because of the intense white light it produces on burning, and in construction especially in the form of light alloys — see ELEMENT table

magnesium carbonate *noun* (1903)
: a carbonate of magnesium; *especially* **:** a white crystalline salt $MgCO_3$ that occurs naturally as dolomite and magnesite

magnesium chloride *noun* (circa 1910)
: a bitter deliquescent salt $MgCl_2$ used especially as a source of magnesium metal

magnesium hydroxide *noun* (circa 1909)
: a slightly alkaline crystalline compound $Mg(OH)_2$ used especially as a laxative and gastric antacid

magnesium oxide *noun* (circa 1909)
: a white highly infusible compound MgO used especially in refractories, cements, insulation, and fertilizers, in rubber manufacture, and in medicine as an antacid and mild laxative

magnesium sulfate *noun* (circa 1890)
: a sulfate of magnesium: as **a :** a white salt $MgSO_4$ used in medicine and in industry **b :** EPSOM SALTS

mag•net \'mag-nət\ *noun* [Middle English *magnete,* from Middle French, from Latin *magnet-, magnes,* from Greek *magnēs* (*lithos*), literally, stone of Magnesia, ancient city in Asia Minor] (15th century)
1 a : LODESTONE **b :** a body having the property of attracting iron and producing a magnetic field external to itself; *specifically* **:** a mass of iron, steel, or alloy that has this property artificially imparted
2 : something that attracts ⟨a box-office *magnet*⟩

magnet- *or* **magneto-** *combining form* [Latin *magnet-, magnes*]
1 : magnetic force ⟨*magneto*meter⟩
2 : magnetism **:** magnetic ⟨*magneto*electric⟩ ⟨*magnet*on⟩
3 : magnetoelectric ⟨*magneto*resistance⟩
4 : magnetosphere ⟨*magneto*pause⟩

¹mag•net•ic \mag-'ne-tik\ *adjective* (1611)
1 : possessing an extraordinary power or ability to attract ⟨a *magnetic* personality⟩
2 a : of or relating to a magnet or to magnetism **b :** of, relating to, or characterized by the earth's magnetism **c :** magnetized or capable of being magnetized **d :** actuated by magnetic attraction
— mag•net•i•cal•ly \-ti-k(ə-)lē\ *adverb*

²magnetic *noun* (1654)
: a magnetic substance

magnetic bubble *noun* (1969)
: a tiny movable magnetized cylindrical volume in a thin amorphous or crystalline magnetic material that along with other like volumes can be used to represent a bit of information (as in a computer)

magnetic disk *noun* (circa 1960)
: DISK 4c

magnetic equator *noun* (1832)
: an imaginary line roughly parallel to the geographical equator and passing through those points where a magnetic needle has no dip

magnetic field *noun* (1845)
: the portion of space near a magnetic body or a current-carrying body in which the magnetic forces due to the body or current can be detected

magnetic flux *noun* (1896)
: lines of force used to represent magnetic induction

magnetic mirror *noun* (1952)
: a magnetic field that confines a plasma by reflecting ions of the plasma back toward the main plasma concentration

magnetic moment *noun* (1865)
: a vector quantity that is a measure of the torque exerted on a magnetic system (as a bar magnet or dipole) when placed in a magnetic field and that for a magnet is the product of the distance between its poles and the strength of either pole

magnetic north *noun* (1812)
: the northerly direction in the earth's magnetic field indicated by the north-seeking pole of a compass needle

magnetic pole *noun* (1701)
1 : either of two small regions which are located respectively in the polar areas of the northern and southern hemispheres and toward which a compass needle points from any direction throughout adjacent regions; *also* **:** either of two comparable regions on a celestial body
2 : either of the poles of a magnet

magnetic quantum number *noun* (1923)
: an integer that expresses the component of the quantized angular momentum of an electron, atom, or molecule in the direction of an externally applied magnetic field

magnetic recording *noun* (1945)
: the process of recording sound, data (as for a computer), or a television program by producing varying local magnetization of a moving tape or disc
— magnetic recorder *noun*

magnetic resonance *noun* (1903)
: the response of electrons, atoms, molecules, or nuclei to various discrete radiation frequencies as a result of space quantization in a magnetic field

magnetic resonance imaging *noun* (1984)
: a noninvasive diagnostic technique that produces computerized images of internal body tissues and is based on nuclear magnetic resonance of atoms within the body induced by the application of radio waves — abbreviation *MRI*

magnetic storm *noun* (1860)
: a marked temporary disturbance of the earth's magnetic field held to be related to sunspots

magnetic tape *noun* (1937)
: a thin ribbon (as of plastic) coated with a magnetic material on which information (as sound or television images) may be stored

mag•ne•tise *British variant of* MAGNETIZE

mag•ne•tism \'mag-nə-ˌti-zəm\ *noun* (1616)
1 a : a class of physical phenomena that include the attraction for iron observed in lodestone and a magnet, are inseparably associated with moving electricity, are exhibited by both magnets and electric currents, and are characterized by fields of force **b :** a science that deals with magnetic phenomena
2 : an ability to attract or charm

mag•ne•tite \'mag-nə-ˌtīt\ *noun* (1851)
: a black isometric mineral of the spinel group that is an oxide of iron and an important iron ore

mag•ne•ti•za•tion \ˌmag-nə-tə-'zā-shən\ *noun* (1801)
: an instance of magnetizing or the state of being magnetized; *also* **:** the degree to which a body is magnetized

mag•ne•tize \'mag-nə-ˌtīz\ *transitive verb* **-tized; -tiz•ing** (1801)
1 : to induce magnetic properties in
2 : to attract like a magnet **:** CHARM
— mag•ne•tiz•able \-ˌtī-zə-bəl\ *adjective*
— mag•ne•tiz•er *noun*

mag•ne•to \mag-'nē-(ˌ)tō\ *noun, plural* **-tos** (1882)
: a magnetoelectric machine; *especially* **:** an alternator with permanent magnets used to generate current for the ignition in an internal-combustion engine

mag•ne•to•elec•tric \-ˌnē-tō-ə-'lek-trik, -ˌne-\ *adjective* (1831)
: relating to or characterized by electromotive forces developed by magnetic means

mag•ne•to•flu•id•dy•nam•ics \-ˌnē-tō-ˌflü-ə-dī-'na-miks, -ˌne-, -də-'\ *noun plural but singular or plural in construction* (1962)
: the study of magnetohydrodynamic phenomena

mag•ne•to•graph \-ˌgraf\ *noun* (1847)
: an automatic instrument for recording measurements of a magnetic field (as of the earth or the sun)

mag•ne•to•hy•dro•dy•nam•ic \mag-ˌnē-tō-ˌhī-drə-dī-'na-mik, -'ne-, -də-'\ *adjective* (1943)
: of, relating to, or being phenomena arising from the motion of electrically conducting fluids (as plasmas) in the presence of electric and magnetic fields
— mag•ne•to•hy•dro•dy•nam•ics \-miks\ *noun plural but singular or plural in construction*

mag•ne•tom•e•ter \ˌmag-nə-'tä-mə-tər\ *noun* (1827)
: an instrument used to detect the presence of a metallic object or to measure the intensity of a magnetic field
— mag•ne•to•met•ric \mag-ˌnē-tə-'me-trik, -ˌne-\ *adjective*
— mag•ne•tom•e•try \ˌmag-nə-'tä-mə-trē\ *noun*

mag•ne•to•mo•tive force \mag-ˌnē-tə-'mō-tiv-, -ˌne-\ *noun* (1883)
: a force that is the cause of a flux of magnetic induction

mag•ne•ton \'mag-nə-ˌtän\ *noun* [International Scientific Vocabulary *magnet-* + ²*-on*] (1911)
: a unit of the quantized magnetic moment of a particle (as an atom)

mag•ne•to–op•tic \mag-ˌnē-tō-'äp-tik, -ˌne-\ *also* **mag•ne•to–op•ti•cal** \-ti-kəl\ *adjective* (1848)
: of or relating to the influence of a magnetic field upon light
— mag•ne•to–op•tics \-tiks\ *noun plural but singular or plural in construction*

mag•ne•to•pause \mag-'nē-tə-ˌpȯz, -'ne-\ *noun* (1962)
: the outer boundary of a magnetosphere

mag•ne•to•re•sis•tance \-ˌnē-tō-ri-'zis-tən(t)s, -ˌne-\ *noun* (1927)
: a change in electrical resistance due to the presence of a magnetic field

mag•ne•to•sphere \mag-'nē-tə-ˌsfir, -'ne-\ *noun* (1959)

: a region of space surrounding a celestial object (as the earth or a star) that is dominated by the object's magnetic field so that charged particles are trapped in it
— **mag·ne·to·spher·ic** \-ˌnē-tə-ˈsfir-ik, -ˈsfer-\ *adjective*

mag·ne·to·stat·ic \mag-ˌnē-tō-ˈsta-tik, -ˌne-\ *adjective* (1893)
: of, relating to, or being a stationary magnetic field

mag·ne·to·stric·tion \-ˈstrik-shən\ *noun* [International Scientific Vocabulary *magnet-* + *-striction* (as in *constriction*)] (1896)
: the change in the dimensions of a ferromagnetic body caused by a change in its state of magnetization
— **mag·ne·to·stric·tive** \-ˈstrik-tiv\ *adjective*
— **mag·ne·to·stric·tive·ly** *adverb*

mag·ne·tron \ˈmag-nə-ˌträn\ *noun* [blend of *magnet* and *-tron*] (1924)
: a diode vacuum tube in which the flow of electrons is controlled by an externally applied magnetic field to generate power at microwave frequencies

magnet school *noun* (1968)
: a school with superior facilities and staff and specialized curricula designed to attract pupils from all segments of the community

mag·nif·ic \mag-ˈni-fik\ *adjective* [Middle French *magnifique,* from Latin *magnificus*] (15th century)
1 : MAGNIFICENT 2
2 : imposing in size or dignity
3 a : SUBLIME, EXALTED **b :** characterized by grandiloquence **:** POMPOUS ⟨commenced the conversation in the most *magnific* style —S. T. Coleridge⟩
— **mag·nif·i·cal** \-fi-kəl\ *adjective*
— **mag·nif·i·cal·ly** \-k(ə-)lē\ *adverb*

mag·nif·i·cat \mag-ˈni-fi-ˌkat, -ˌkät; män-ˈyi-fi-ˌkät\ *noun* [Middle English, from Latin, magnifies, from *magnificare* to magnify; from the first word of the canticle] (13th century)
1 *capitalized* **a :** the canticle of the Virgin Mary in Luke 1:46–55 **b :** a musical setting for the Magnificat
2 : an utterance of praise

mag·ni·fi·ca·tion \ˌmag-nə-fə-ˈkā-shən\ *noun* (15th century)
1 : the act of magnifying
2 a : the state of being magnified **b :** the apparent enlargement of an object by an optical instrument — called also *power*

mag·nif·i·cence \mag-ˈni-fə-sən(t)s, məg-\ *noun* [Middle English, from Middle French, from Latin *magnificentia,* from *magnificus* noble in character, magnificent, from *magnus* great — more at MUCH] (14th century)
1 : the quality or state of being magnificent
2 : splendor of surroundings

mag·nif·i·cent \-sənt\ *adjective* (15th century)
1 : great in deed or exalted in place — used only of former famous rulers ⟨Lorenzo the *Magnificent*⟩
2 : marked by stately grandeur and lavishness ⟨a *magnificent* way of life⟩
3 : sumptuous in structure and adornment ⟨a *magnificent* cathedral⟩; *broadly* **:** strikingly beautiful or impressive ⟨a *magnificent* physique⟩
4 : impressive to the mind or spirit **:** SUBLIME ⟨*magnificent* prose⟩
5 : exceptionally fine ⟨a *magnificent* day⟩
synonym see GRAND
— **mag·nif·i·cent·ly** *adverb*

mag·nif·i·co \mag-ˈni-fi-ˌkō\ *noun, plural* **-coes** *or* **-cos** [Italian, from *magnifico,* adjective, magnificent, from Latin *magnificus*] (1573)
1 : a nobleman of Venice
2 : a person of high position

mag·ni·fi·er \ˈmag-nə-ˌfī(-ə)r\ *noun* (1550)
: one that magnifies; *especially* **:** a lens or combination of lenses that makes something appear larger

mag·ni·fy \ˈmag-nə-ˌfī\ *verb* **-fied; -fy·ing** [Middle English *magnifien,* from Middle French *magnifier,* from Latin *magnificare,* from *magnificus*] (14th century)
transitive verb
1 a : EXTOL, LAUD **b :** to cause to be held in greater esteem or respect
2 a : to increase in significance **:** INTENSIFY **b :** EXAGGERATE
3 : to enlarge in fact or in appearance
intransitive verb
: to have the power of causing objects to appear larger than they are

mag·nil·o·quence \mag-ˈni-lə-kwən(t)s\ *noun* [Latin *magniloquentia,* from *magniloquus* magniloquent, from *magnus* + *loqui* to speak] (circa 1623)
: the quality or state of being magniloquent

mag·nil·o·quent \-kwənt\ *adjective* [back-formation from *magniloquence*] (circa 1656)
: speaking in or characterized by a high-flown often bombastic style or manner
— **mag·nil·o·quent·ly** *adverb*

mag·ni·tude \ˈmag-nə-ˌtüd, -ˌtyüd\ *noun* [Middle English, from Latin *magnitudo,* from *magnus*] (15th century)
1 a : great size or extent **b** (1) **:** spatial quality **:** SIZE (2) **:** QUANTITY, NUMBER
2 : the importance, quality, or caliber of something
3 : a number representing the intrinsic or apparent brightness of a celestial body on a logarithmic scale in which an increase of one unit corresponds to a reduction in the brightness of light by a factor of 2.512
4 : a numerical quantitative measure expressed usually as a multiple of a standard unit

mag·no·lia \mag-ˈnōl-yə\ *noun* [New Latin, from Pierre *Magnol* (died 1715) French botanist] (1748)
: any of a genus (*Magnolia* of the family Magnoliaceae, the magnolia family) of American and Asian shrubs and trees with entire evergreen or deciduous leaves and usually showy white, yellow, rose, or purple flowers usually appearing in early spring

magnolia

mag·num \ˈmag-nəm\ *noun* [Latin, neuter of *magnus* great] (1788)
: a large wine bottle holding about 1.5 liters

mag·num opus \ˈmag-nəm-ˈō-pəs\ *noun* [Latin] (1791)
: a great work; *especially* **:** the greatest achievement of an artist or writer

^{1}**mag·pie** \ˈmag-ˌpī\ *noun* [*Mag* (nickname for *Margaret*) + ^{1}*pie*] (1598)
1 : any of various birds (especially *Pica pica*) related to the jays but having a long graduated tail and black-and-white or brightly colored plumage
2 : a person who chatters noisily
3 : one who collects indiscriminately

^{2}**magpie** *adjective* (1808)
1 : collected indiscriminately **:** MISCELLANEOUS ⟨*magpie* compilations of unrelated tidbits —Helen R. Cross⟩
2 : given to indiscriminate collecting **:** ACQUISITIVE ⟨what possible *magpie* instinct had impelled me to retain them —S. J. Perelman⟩

mag tape \ˈmag-\ *noun* (1966)
: MAGNETIC TAPE

ma·guey \mə-ˈgā\ *noun* [Spanish, from Taino] (1589)
1 : any of various fleshy-leaved agaves (as the century plant)
2 : any of several hard fibers derived from magueys; *especially* **:** CANTALA

ma·gus \ˈmā-gəs\ *noun, plural* **ma·gi** \ˈmā-ˌjī\ [Latin, from Greek *magos* — more at MAGIC] (1621)
1 a : a member of a hereditary priestly class among the ancient Medes and Persians **b** *often capitalized* **:** one of the traditionally three wise men from the East paying homage to the infant Jesus
2 : MAGICIAN, SORCERER

Mag·yar \ˈmag-ˌyär, ˈmäg-; ˈmä-ˌjär\ *noun* [Hungarian] (1797)
1 : a member of the dominant people of Hungary
2 : HUNGARIAN 2
— **Magyar** *adjective*

ma·ha·ra·ja *or* **ma·ha·ra·jah** \ˌmä-hə-ˈrä-jə, -ˈrä-zhə\ *noun* [Sanskrit *mahārāja,* from *mahat* great + *rājan* raja; akin to Latin *rex* king — more at MUCH, ROYAL] (1698)
: a Hindu prince ranking above a raja

ma·ha·ra·ni *or* **ma·ha·ra·nee** \-ˈrä-nē\ *noun* [Hindi *mahārānī,* from *mahā* great (from Sanskrit *mahat*) + *rānī* rani] (circa 1855)
1 : the wife of a maharaja
2 : a Hindu princess ranking above a rani

ma·ha·ri·shi \ˌmä-hə-ˈrē-shē, -ˈhä-, -ˈri-\ *noun* [Sanskrit *mahārṣi,* from *mahat* + *ṛṣi* sage and poet] (1785)
: a Hindu teacher of mystical knowledge

ma·hat·ma \mə-ˈhät-mə, -ˈhat-\ *noun* [Sanskrit *mahātman,* from *mahātman* great-souled, from *mahat* + *ātman* soul — more at ATMAN] (1923)
1 : a person to be revered for high-mindedness, wisdom, and selflessness
2 : a person of great prestige in a field of endeavor

Ma·ha·ya·na \ˌmä-hə-ˈyä-nə\ *noun* [Sanskrit *mahāyāna,* literally, great vehicle] (1855)
: a liberal and theistic branch of Buddhism comprising sects chiefly in China and Japan, recognizing a large body of scripture in addition to the Pali canon, and teaching social concern and universal salvation — compare THERAVADA
— **Ma·ha·ya·nist** \-ˈyä-nist\ *noun or adjective*
— **Ma·ha·ya·nis·tic** \-yä-ˈnis-tik\ *adjective*

Mah·di \ˈmä-dē\ *noun* [Arabic *mahdīy,* literally, one rightly guided] (1800)
1 : the expected messiah of Muslim tradition
2 : a Muslim leader who assumes a messianic role
— **Mah·dism** \ˈmä-ˌdi-zəm\ *noun*
— **Mah·dist** \ˈmä-dist\ *noun*

Ma·hi·can \mə-ˈhē-kən\ *noun, plural* **Mahican** *or* **Mahicans** [Mahican] (circa 1614)
1 : a member of an American Indian people of the upper Hudson River valley
2 : the extinct Algonquian language of the Mahican people

ma·hi·ma·hi \ˌmä-hē-ˈmä-(ˌ)hē\ *noun* [Hawaiian, Tahitian, & Marquesan] (1943)
: the flesh of a dolphin (*Coryphaena hippurus*) used for food; *also* **:** the fish

mah–jongg *or* **mah·jong** \ˌmä-ˈzhäŋ, -ˈjäŋ, -ˈzhȯŋ, -ˈjȯŋ, ˈmä-ˌ\ *noun* [from *Mah-Jongg,* a trademark] (1920)
: a game of Chinese origin usually played by four persons with 144 tiles that are drawn and discarded until one player secures a winning hand

mahl·stick \ˈmȯl-\ *variant of* MAULSTICK

ma·hoe \mə-ˈhō, ˈmä-ˌ\ *noun* [French *maho,* from Taino] (1666)
: either of two tropical hibiscus trees (*Hibiscus elatus* and *H. tiliaceus*)

ma·hog·a·ny \mə-ˈhä-gə-nē\ *noun, plural* **-nies** [origin unknown] (circa 1660)

\ə\ **abut** \ᵊ\ **kitten** \ər\ **further** \a\ **ash** \ā\ **ace** \ä\ **mop, mar** \au̇\ **out** \ch\ **chin** \e\ **bet** \ē\ **easy** \g\ **go** \i\ **hit** \ī\ **ice** \j\ **job** \ŋ\ **sing** \ō\ **go** \ȯ\ **law** \ȯi\ **boy** \th\ **thin** \t͟h\ **the** \ü\ **loot** \u̇\ **foot** \y\ **yet** \zh\ **vision** *see also* Guide to Pronunciation

1 : the wood of any of various chiefly tropical trees (family Meliaceae, the mahogany family): **a** (1) **:** the durable yellowish brown to reddish brown usually moderately hard and heavy wood of a West Indian tree (*Swietenia mahagoni*) that is widely used for cabinetwork and fine finish work (2) **:** a wood similar to mahogany from a congeneric tree (especially *S. macrophylla*) **b** (1) **:** the rather hard heavy usually odorless wood of any of several African trees (genus *Khaya*) (2) **:** the rather lightweight cedar-scented wood of any of several African trees (genus *Entandrophragma*) that varies in color from pinkish to deep reddish brown
2 : any of various woods resembling or substituted for mahogany obtained from trees of the mahogany family
3 : a tree that yields mahogany
4 : a moderate reddish brown

ma·ho·nia \mə-ˈhō-nē-ə\ *noun* [New Latin, from Bernard Mc*Mahon* (died 1816) American botanist] (1829)
: any of a genus (*Mahonia*) of American and Asian shrubs (as the Oregon grape) of the barberry family

ma·hout \mə-ˈhau̇t\ *noun* [Hindi *mahāwat, mahāut*] (1662)
: a keeper and driver of an elephant

Mah·rat·ta *variant of* MARATHA

maid \ˈmād\ *noun* [Middle English *maide*, short for *maiden*] (13th century)
1 : an unmarried girl or woman especially when young **:** VIRGIN
2 a : MAIDSERVANT **b :** a woman or girl employed to do domestic work

[1]**maid·en** \ˈmā-d^{ə}n\ *noun* [Middle English, from Old English *mægden, mæden*, diminutive of *mægeth;* akin to Old High German *magad* maiden, Old Irish *mug* serf] (before 12th century)
1 : an unmarried girl or woman **:** MAID
2 : a former Scottish beheading device resembling the guillotine
3 : a horse that has never won a race

[2]**maiden** *adjective* (14th century)
1 a (1) **:** not married ⟨*maiden* aunt⟩ (2) **:** VIRGIN **b** *of a female animal* (1) **:** never yet mated (2) **:** never having borne young
2 : of, relating to, or befitting a maiden
3 : FIRST, EARLIEST ⟨*maiden* voyage⟩ ⟨*maiden* flight of a spacecraft⟩

maid·en·hair fern \-ˌhar-, -ˌher-\ *noun* (1833)
: any of a genus (*Adiantum*) of ferns with delicate palmately branched fronds — called also *maidenhair*

maidenhair fern

maidenhair tree *noun* (1773)
: GINKGO

maid·en·head \ˈmā-d^{ə}n-ˌhed\ *noun* [Middle English *maidenhed*, from *maiden* + *-hed* -hood; akin to Middle English *-hod* -hood] (13th century)
1 : the quality or state of being a maiden **:** VIRGINITY
2 : HYMEN

maid·en·hood \-ˌhu̇d\ *noun* (before 12th century)
: the quality, state, or time of being a maiden

maid·en·li·ness \-lē-nəs\ *noun* (1555)
: conduct or traits befitting a maiden

maid·en·ly \-lē\ *adjective* (15th century)
: of, resembling, or suitable to a maiden

maiden name *noun* (1689)
: the surname of a woman before she marries

maid·hood \ˈmād-ˌhu̇d\ *noun* (before 12th century)
: MAIDENHOOD

maid–in–wait·ing \ˈmā-d^{ə}n-ˈwā-tiŋ\ *noun, plural* **maids–in–wait·ing** \ˈmādz-\ (1953)
: a young woman of a queen's or princess's household appointed to attend her

Maid Mar·i·an \-ˈmer-ē-ən, -ˈmar-\ *noun*
: a companion of Robin Hood in some forms of his legend

maid of honor (circa 1586)
1 : an unmarried lady usually of noble birth whose duty it is to attend a queen or a princess
2 : a bride's principal unmarried wedding attendant

maid·ser·vant \ˈmād-ˌsər-vənt\ *noun* (14th century)
: a female servant

ma·ieu·tic \mā-ˈyü-tik, mī-\ *adjective* [Greek *maieutikos* of midwifery] (1655)
: relating to or resembling the Socratic method of eliciting new ideas from another

[1]**mail** \ˈmā(ə)l\ *noun* [Middle English *male, maille*, from Old English *māl* agreement, pay, from Old Norse *māl* speech, agreement; akin to Old English *mǣl* speech] (before 12th century)
chiefly Scottish **:** PAYMENT, RENT

[2]**mail** *noun, often attributive* [Middle English *male*, from Old French, of Germanic origin; akin to Old High German *malaha* bag] (13th century)
1 *chiefly Scottish* **:** BAG, WALLET
2 a : something sent or carried in the postal system **b :** a conveyance that transports mail **c :** messages sent electronically to an individual (as through a computer system)
3 : a nation's postal system — often used in plural

[3]**mail** *transitive verb* (1827)
: to send by mail **:** POST
— mail·abil·i·ty \ˌmā-lə-ˈbi-lə-tē\ *noun*
— mail·able \ˈmā-lə-bəl\ *adjective*

[4]**mail** *noun* [Middle English *maille*, from Middle French, from Latin *macula* spot, mesh] (14th century)
1 : armor made of metal links or sometimes plates
2 : a hard enclosing covering of an animal (as a tortoise)
— mailed \ˈmā(ə)ld\ *adjective*

mail·bag \ˈmā(ə)l-ˌbag\ *noun* (1812)
1 : a letter carrier's shoulder bag
2 : a pouch used in the shipment of mail

mail·box \-ˌbäks\ *noun* (1872)
1 : a public box for deposit of outgoing mail
2 : a box at or near a dwelling for the occupant's mail
3 : a computer file in which electronic mail is collected

mail carrier *noun* (1790)
: LETTER CARRIER

mail drop *noun* (1945)
1 : an address used in transmitting secret communications
2 : a receptacle or a slot for deposit of mail

mai·le \ˈmī-lē\ *noun* [Hawaiian] (1903)
: a Pacific island vine (*Alyxia olivaeformis*) of the dogbane family with fragrant leaves and bark that are used for decoration and in Hawaii for leis

mailed fist *noun* (1897)
: a threat of armed or overbearing force

mail·er \ˈmā-lər\ *noun* (1884)
1 : one that mails
2 : a container for mailing something
3 : something (as an advertisement) sent by mail

Mail·gram \ˈmāl-ˌgram\ *trademark*
— used for a message sent by wire to a post office that delivers it to the addressee

mail·ing \ˈmā-liŋ\ *noun* (1946)
: the mail dispatched at one time by a sender

mail·lot \mī-ˈō, mä-ˈyō\ *noun* [French] (1888)
1 : tights for dancers or gymnasts
2 : JERSEY 2
3 : a woman's one-piece bathing suit

mail·man \ˈmā(ə)l-ˌman\ *noun* (1881)
: a man who delivers mail — called also *postman*

mail order *noun* (1867)
: an order for goods that is received and filled by mail
— mail–or·der \-ˌȯr-dər\ *adjective*

[1]**maim** \ˈmām\ *transitive verb* [Middle English *maynhen, maymen*, from Old French *maynier* — more at MAYHEM] (14th century)
1 : to commit the felony of mayhem upon
2 : to mutilate, disfigure, or wound seriously ☆
— maim·er *noun*

[2]**maim** *noun* (14th century)
1 *obsolete* **:** serious physical injury; *especially* **:** loss of a member of the body
2 *obsolete* **:** a serious loss

[1]**main** \ˈmān\ *noun* [in sense 1, from Middle English, from Old English *mægen;* akin to Old High German *magan* strength, Old English *magan* to be able; in other senses, from [2]*main* or by shortening — more at MAY] (before 12th century)
1 : physical strength **:** FORCE — used in the phrase *with might and main*
2 a : MAINLAND **b :** HIGH SEA
3 : the chief part **:** essential point ⟨they are in the *main* well-trained⟩
4 : a pipe, duct, or circuit which carries the combined flow of tributary branches of a utility system
5 a : MAINMAST **b :** MAINSAIL

[2]**main** *adjective* [Middle English, from Old English *mægen-*, from *mægen* strength] (15th century)
1 : CHIEF, PRINCIPAL ⟨the *main* idea⟩
2 : fully exerted **:** SHEER ⟨*main* force⟩ ⟨by *main* strength⟩
3 *obsolete* **:** of or relating to a broad expanse (as of sea)
4 : connected with or located near the mainmast or mainsail
5 : expressing the chief predication in a complex sentence ⟨the *main* clause⟩

main chance *noun* (1584)
: the best chance for personal or financial gain ⟨kept an eye on the *main chance*⟩

Maine coon *noun* (1935)
: any of a breed of large long-haired domestic cats that have a very full tapered tail — called also *coon cat, Maine cat*

Maine coon

main·frame \ˈmān-ˌfrām\ *noun* (circa 1964)
: a computer with its cabinet and internal circuits; *also* **:** a large fast computer that can handle multiple tasks concurrently

main·land \ˈmān-ˌland, -lənd\ *noun* (14th century)
: a continent or the main part of a continent as distinguished from an offshore island or sometimes from a cape or peninsula

☆ SYNONYMS
Maim, cripple, mutilate, batter, mangle mean to injure so severely as to cause lasting damage. MAIM implies the loss or injury of a bodily member through violence ⟨*maimed* by a shark⟩. CRIPPLE implies the loss or serious impairment of an arm or leg ⟨*crippled* for life in an accident⟩. MUTILATE implies the cutting off or removal of an essential part of a person or thing thereby impairing its completeness, beauty, or function ⟨a tree *mutilated* by inept pruning⟩. BATTER implies a series of blows that bruise deeply, deform, or mutilate ⟨an old ship *battered* by fierce storms⟩. MANGLE implies a tearing or crushing that leaves deep extensive wounds ⟨a soldier's leg *mangled* by shrapnel⟩.

— **main·land·er** \-ˌlan-dər, -lən-\ *noun*
¹main·line \ˈmān-ˌlīn\ (1938)
transitive verb
slang **:** to take by or as if by injecting into a principal vein
intransitive verb
slang **:** to mainline a narcotic drug
²mainline *adjective* (1941)
: being part of an established group; *also* **:** being in the mainstream
main line *noun* (1841)
1 : a principal highway or railroad line
2 *slang* **:** a principal vein of the circulatory system
main·ly \ˈmān-lē\ *adverb* (13th century)
1 *obsolete* **:** in a forceful manner
2 : for the most part **:** CHIEFLY
main man *noun* (1967)
1 : best male friend
2 : a man whose character or work is most admired
3 : the most reliable or effective performer
main·mast \ˈmān-ˌmast, -məst\ *noun* (15th century)
: a sailing ship's principal mast usually second from the bow
mains \ˈmānz\ *adjective* (circa 1927)
British **:** of or relating to utility distribution mains ⟨*mains* voltage⟩ ⟨*mains* water⟩
main·sail \ˈmān-ˌsāl, ˈmān(t)-səl\ *noun* (15th century)
: the principal sail on the mainmast — see SAIL illustration
main sequence *noun* (1925)
: the group of stars that on a graph of spectrum versus luminosity forms a band comprising 90% of stellar types and that includes stars believed to be representative of the stages a normal star passes through for the majority of its lifetime
main·sheet \ˈmān-ˌshēt\ *noun* (15th century)
: a rope by which the mainsail is trimmed and secured
main·spring \ˈmān-ˌspriŋ\ *noun* (1591)
1 : the chief spring in a mechanism especially of a watch or clock
2 : the chief or most powerful motive, agent, or cause
main squeeze *noun* (circa 1968)
slang **:** one's principal romantic partner
main·stay \ˈmān-ˌstā\ *noun* (15th century)
1 : a ship's stay extending from the maintop forward usually to the foot of the foremast
2 : a chief support
main stem *noun* (1832)
: a main trunk or channel: as **a :** the main course of a river or stream **b :** the main street of a city or town
¹main·stream \ˈmān-ˌstrēm\ *noun* (1831)
: a prevailing current or direction of activity or influence
— **mainstream** *adjective*
²main·stream \ˈmān-ˈstrēm\ *transitive verb* (1974)
: to place (as a handicapped child) in regular school classes
Main Street *noun* (circa 1743)
1 : the principal street of a small town
2 a : the sections of a country centering about its small towns **b :** a place or environment characterized by materialistic self-complacent provincialism **c :** MIDDLE AMERICA 3
— **Main Street·er** \ˈmān-ˌstrē-tər\ *noun*
main·tain \mān-ˈtān, mən-\ *transitive verb* [Middle English *mainteinen,* from Old French *maintenir,* from Medieval Latin *manutenēre,* from Latin *manu tenēre* to hold in the hand] (14th century)
1 : to keep in an existing state (as of repair, efficiency, or validity) **:** preserve from failure or decline ⟨*maintain* machinery⟩
2 : to sustain against opposition or danger **:** uphold and defend ⟨*maintain* a position⟩
3 : to continue or persevere in **:** CARRY ON, KEEP UP ⟨couldn't *maintain* his composure⟩
4 a : to support or provide for ⟨has a family to *maintain*⟩ **b :** SUSTAIN ⟨enough food to *maintain* life⟩
5 : to affirm in or as if in argument **:** ASSERT ⟨*maintained* that the earth is flat⟩ ☆
— **main·tain·abil·i·ty** \-ˌtā-nə-ˈbi-lə-tē\ *noun*
— **main·tain·able** \-ˈtā-nə-bəl\ *adjective*
— **main·tain·er** *noun*
main·te·nance \ˈmānt-nən(t)s, ˈmān-tᵊn-ən(t)s\ *noun* [Middle English, from Middle French, from Old French, from *maintenir*] (14th century)
1 : the act of maintaining **:** the state of being maintained **:** SUPPORT
2 : something that maintains
3 : the upkeep of property or equipment
4 : an officious or unlawful intermeddling in a legal suit by assisting either party with means to carry it on
main·top \ˈmān-ˌtäp\ *noun* (15th century)
: a platform about the head of the mainmast of a square-rigged ship
main–top·mast \mān-ˈtäp-ˌmast, -məst\ *noun* (15th century)
: a mast next above the mainmast
main yard *noun* (15th century)
: the yard of a mainsail
mair \ˈmār\ *chiefly Scottish variant of* MORE
mai·son·ette \ˌmā-zᵊn-ˈet, -sᵊn-\ *noun* [French *maisonnette,* from Old French, diminutive of *maison* house, from Latin *mansion-, mansio* dwelling place — more at MANSION] (1793)
1 : a small house
2 : an apartment often on two floors
maî·tre d' *or* **mai·tre d'** \ˌmā-trə-ˈdē, ˌme-, -tər-\ *noun, plural* **maître d's** *or* **maitre d's** \-ˈdēz\ (1950)
: MAÎTRE D'HÔTEL
maî·tre d'hô·tel \ˌmā-trə-(ˌ)dō-ˈtel, ˌme-; ˌmāt-dō-, ˌmet-\ *noun, plural* **maîtres d'hôtel** *same*\ [French, literally, master of house] (1538)
1 a : MAJORDOMO **b :** HEADWAITER
2 : a sauce of butter, parsley, salt, pepper, and lemon juice — called also *maître d'hôtel butter*
maize \ˈmāz\ *noun* [Spanish *maíz,* from Taino *mahiz*] (1555)
: INDIAN CORN
ma·jes·tic \mə-ˈjes-tik\ *adjective* (1601)
: having or exhibiting majesty **:** STATELY
synonym see GRAND
— **ma·jes·ti·cal·ly** \-ti-k(ə-)lē\ *adverb*
maj·es·ty \ˈma-jə-stē\ *noun, plural* **-ties** [Middle English *maieste,* from Middle French *majesté,* from Latin *majestat-, majestas;* akin to Latin *major* greater] (14th century)
1 : sovereign power, authority, or dignity
2 — used in addressing or referring to reigning sovereigns and their consorts ⟨Your *Majesty*⟩ ⟨Her *Majesty's* Government⟩
3 a : royal bearing or aspect **:** GRANDEUR **b :** greatness or splendor of quality or character
ma·jol·i·ca \mə-ˈjä-li-kə\ *also* **ma·iol·i·ca** \-ˈyä-\ *noun* [Italian *maiolica,* from Old Italian *Maiolica, Maiorica* Majorca] (1555)
1 : earthenware covered with an opaque tin glaze and decorated on the glaze before firing; *especially* **:** an Italian ware of this kind
2 : a 19th century earthenware modeled in naturalistic shapes and glazed in lively colors
¹ma·jor \ˈmā-jər\ *adjective* [Middle English *maiour,* from Latin *major,* comparative of *magnus* great, large — more at MUCH] (15th century)
1 : greater in dignity, rank, importance, or interest ⟨one of the *major* poets⟩
2 : greater in number, quantity, or extent ⟨the *major* part of his work⟩
3 : having attained majority
4 : notable or conspicuous in effect or scope **:** CONSIDERABLE ⟨a *major* improvement⟩
5 : involving grave risk **:** SERIOUS ⟨a *major* illness⟩
6 : of or relating to a subject of academic study chosen as a field of specialization
7 a : having half steps between the third and fourth and the seventh and eighth degrees ⟨*major* scale⟩ **b :** based on a major scale ⟨*major* key⟩ **c :** equivalent to the distance between the keynote and another tone (except the fourth and fifth) of a major scale ⟨*major* third⟩ **d :** having a major third above the root ⟨*major* triad⟩
²major *noun* (1616)
1 : a person who has attained majority
2 a : one that is superior in rank, importance, size, or performance ⟨economic power of the oil *majors*⟩ **b :** a major musical interval, scale, key, or mode
3 : a commissioned officer in the army, air force, or marine corps ranking above a captain and below a lieutenant colonel
4 a : an academic subject chosen as a field of specialization **b :** a student specializing in such a field ⟨a history *major*⟩
5 *plural* **:** major league baseball — used with *the*
³major *intransitive verb* (1913)
: to pursue an academic major
major axis *noun* (1854)
: the axis passing through the foci of an ellipse
ma·jor·do·mo \ˌmā-jər-ˈdō-(ˌ)mō\ *noun, plural* **-mos** [Spanish *mayordomo* or obsolete Italian *maiordomo,* from Medieval Latin *major domus,* literally, chief of the house] (1589)
1 : a head steward of a large household (as a palace)
2 : BUTLER, STEWARD
3 : a person who speaks, makes arrangements, or takes charge for another
ma·jor·ette \ˌmā-jə-ˈret\ *noun* (1940)
: DRUM MAJORETTE 2
major general *noun* [French *major général,* from *major,* noun + *général,* adjective, general] (1642)
: a commissioned officer in the army, air force, or marine corps who ranks above a brigadier general and whose insignia is two stars
major histocompatibility complex *noun* (1975)
: a group of genes in mammals that function especially in determining the histocompatibility antigens found on cell surfaces
ma·jor·i·tar·i·an \mə-ˌjȯr-ə-ˈter-ē-ən, -ˌjär-\ *noun* (1942)
: a person who believes in or advocates majoritarianism
— **majoritarian** *adjective*
ma·jor·i·tar·i·an·ism \-ē-ə-ˌni-zəm\ *noun* (1942)
: the philosophy or practice according to which decisions of an organized group should be made by a numerical majority of its members

☆ SYNONYMS
Maintain, assert, defend, vindicate, justify mean to uphold as true, right, just, or reasonable. MAINTAIN stresses firmness of conviction ⟨steadfastly *maintained* his innocence⟩. ASSERT suggests determination to make others accept one's claim ⟨*asserted* her rights⟩. DEFEND implies maintaining in the face of attack or criticism ⟨*defended* his voting record⟩. VINDICATE implies successfully defending ⟨his success *vindicated* our faith in him⟩. JUSTIFY implies showing to be true, just, or valid by appeal to a standard or to precedent ⟨the action was used to *justify* military intervention⟩.

ma·jor·i·ty \mə-ˈjȯr-ə-tē, -ˈjär-\ *noun, plural* **-ties** (1552)
1 *obsolete* **:** the quality or state of being greater
2 a : the age at which full civil rights are accorded **b :** the status of one who has attained this age
3 a : a number greater than half of a total **b :** the excess of a majority over the remainder of the total **:** MARGIN **c :** the preponderant quantity or share
4 : the group or political party whose votes preponderate
5 : the military office, rank, or commission of a major
— majority *adjective*

majority leader *noun* (1952)
: a leader of the majority party in a legislative body (as the U.S. Senate)

majority rule *noun* (1893)
: a political principle providing that a majority usually constituted by fifty percent plus one of an organized group will have the power to make decisions binding upon the whole

major league *noun* (1906)
1 : a league of highest classification in U.S. professional baseball; *broadly* **:** a league of major importance in any of various sports
2 : BIG TIME 2
— major–league *adjective*

ma·jor·ly \ˈmā-jər-lē\ *adverb* (1956)
: in a major way **a :** PRIMARILY 1 ⟨was *majorly* a poet⟩ **b :** EXTREMELY 1 ⟨was *majorly* annoyed⟩

major–medical *adjective* (circa 1955)
: of, relating to, or being a form of insurance designed to pay all or part of the medical bills of major illnesses usually after deduction of a fixed initial sum

major order *noun* (circa 1741)
: one of the Roman Catholic or Eastern clerical orders that are sacramentally conferred and have a sacred character that implies major religious obligations (as clerical celibacy) — usually used in plural; compare MINOR ORDER

major party *noun* (1950)
: a political party having electoral strength sufficient to permit it to win control of a government usually with comparative regularity and when defeated to constitute the principal opposition to the party in power

major penalty *noun* (circa 1936)
: a 5-minute suspension of a player in ice hockey or lacrosse

major premise *noun* (1860)
: the premise of a syllogism containing the major term

major seminary *noun* (1945)
: a Roman Catholic seminary giving usually the entire six years of senior college and theological training required for major orders

major suit *noun* (1916)
: either of the suits hearts or spades having superior scoring value in bridge

major term *noun* (1847)
: the term of a syllogism constituting the predicate of the conclusion

ma·jus·cule \ˈma-jəs-ˌkyü(ə)l, mə-ˈjəs-\ *noun* [French, from Latin *majusculus* rather large, diminutive of *major*] (circa 1825)
: a large letter (as a capital)
— ma·jus·cu·lar \mə-ˈjəs-kyə-lər\ *adjective*
— majuscule *adjective*

mak·able *or* **make·able** \ˈmā-kə-bəl\ *adjective* (15th century)
: capable of being made

mak·ar \ˈmä-kər, ˈmā-\ *noun* [Middle English *maker*] (14th century)
chiefly Scottish **:** POET

[1]make \ˈmāk\ *verb* **made** \ˈmād\; **mak·ing** [Middle English, from Old English *macian;* akin to Old High German *mahhōn* to prepare, make, Greek *magenai* to be kneaded, Old Church Slavonic *mazati* to anoint, smear] (before 12th century)
transitive verb
1 a *obsolete* **:** BEHAVE, ACT **b :** to seem to begin (an action) ⟨*made* to go⟩
2 a : to cause to happen to or be experienced by someone ⟨*made* trouble for us⟩ **b :** to cause to exist, occur, or appear **:** CREATE ⟨*make* a disturbance⟩ **c :** to favor the growth or occurrence of ⟨haste *makes* waste⟩ **d :** to fit, intend, or destine by or as if by creating ⟨was *made* to be an actor⟩
3 a : to bring into being by forming, shaping, or altering material **:** FASHION ⟨*make* a dress⟩ **b :** COMPOSE, WRITE ⟨*make* verses⟩ **c :** to lay out and construct ⟨*make* a road⟩
4 : to frame or formulate in the mind ⟨*make* plans⟩
5 : to put together from components **:** CONSTITUTE ⟨houses *made* of stone⟩
6 a : to compute or estimate to be **b :** to form and hold in the mind ⟨*make* no doubt of it⟩
7 a : to assemble and set alight the materials for (a fire) **b :** to set in order ⟨*make* beds⟩ **c :** PREPARE, FIX ⟨*make* dinner⟩ **d :** to shuffle (a deck of cards) in preparation for dealing
8 : to prepare (hay) by cutting, drying, and storing
9 a : to cause to be or become ⟨*made* them happy⟩ **b :** APPOINT ⟨*made* him bishop⟩
10 a : ENACT, ESTABLISH ⟨*make* laws⟩ **b :** to execute in an appropriate manner ⟨*make* a will⟩ **c :** SET, NAME ⟨*make* a price⟩
11 a *chiefly dialect* **:** SHUT ⟨the doors are *made* against you —Shakespeare⟩ **b :** to cause (an electric circuit) to be completed
12 a : to conclude as to the nature or meaning of something ⟨what do you *make* of this development?⟩ **b :** to regard as being ⟨not the fool some *make* him⟩
13 a : to carry out (an action indicated or implied by the object) ⟨*make* war⟩ ⟨*make* a speech⟩ **b :** to perform with a bodily movement ⟨*make* a sweeping gesture⟩ **c :** to achieve by traversing ⟨*make* a detour⟩ ⟨*making* the rounds⟩
14 a : to produce as a result of action, effort, or behavior with respect to something ⟨*make* a mess of the job⟩ ⟨tried to *make* a thorough job of it⟩ **b** *archaic* **:** to turn into another language by translation
15 : to cause to act in a certain way **:** COMPEL ⟨*make* her give it back⟩
16 : to cause or assure the success or prosperity of ⟨can either *make* you or break you⟩
17 a : to amount to in significance ⟨*makes* a great difference⟩ **b :** to form the essential being of ⟨clothes *make* the man⟩ **c :** to form by an assembling of individuals ⟨*make* a quorum⟩ **d :** to count as ⟨that *makes* the third time you've said it⟩
18 a : to be or be capable of being changed or fashioned into ⟨rags *make* the best paper⟩ **b :** to develop into ⟨she will *make* a fine judge⟩ **c :** FORM 6b
19 a : REACH, ATTAIN — often used with *it* ⟨you'll never *make* it that far⟩; *also* **:** SURVIVE ⟨half the cubs won't *make* it through their first year⟩ **b :** to gain the rank of ⟨*make* major⟩ **c :** to gain a place on or in ⟨*make* the team⟩ ⟨the story *made* the papers⟩
20 : to gain (as money) by working, trading, or dealing ⟨*make* a living⟩
21 a : to act so as to win or acquire ⟨*makes* friends easily⟩ **b :** to score (points) in a game or sport **c :** to convert (a split) into a spare in bowling
22 a : to fulfill (a contract) in a card game **b :** to win a trick with (a card)
23 a : to include in a route or itinerary ⟨*make* New York on the return trip⟩ ⟨*make* it to the party⟩ **b :** CATCH 6b ⟨*made* the bus just in time⟩
24 : to persuade to consent to sexual intercourse **:** SEDUCE
25 : to provide the most enjoyable or satisfying experience of ⟨meeting the star of the show really *made* our day⟩
intransitive verb
1 *archaic* **:** to compose poetry
2 a : BEHAVE, ACT **b :** to begin or seem to begin a certain action ⟨*made* as though to hand it to me⟩ **c :** to act so as to be or to seem to be ⟨*make* merry⟩ **d** *slang* **:** to play a part — usually used with *like*
3 : SET OUT, HEAD ⟨*made* after the fox⟩ ⟨*made* straight for home⟩
4 : to increase in height or size ⟨the tide is *making* now⟩
5 : to reach or extend in a certain direction
6 : to have considerable effect ⟨courtesy *makes* for safer driving⟩
7 : to undergo manufacture or processing ⟨the silk *makes* up beautifully⟩
— make a face : to distort one's features **:** GRIMACE
— make a mountain out of a molehill : to treat a trifling matter as of great importance
— make away with 1 : to carry off **:** STEAL **2 :** KILL
— make believe : PRETEND, FEIGN
— make bold : VENTURE, DARE
— make book : to accept bets at calculated odds on all the entrants in a race or contest
— make common cause : to unite to achieve a shared goal
— make do : to get along or manage with the means at hand
— make ends meet : to make one's means adequate to one's needs
— make eyes : OGLE
— make fun of : to make an object of amusement or laughter **:** RIDICULE, MOCK
— make good 1 : to make valid or complete: as **a :** to make up for (a deficiency) **b :** INDEMNIFY ⟨*make good* the loss⟩ **c :** to carry out successfully ⟨*made good* his promise⟩ ⟨*made good* their escape⟩ **d :** PROVE ⟨*made good* a charge⟩ **2 :** to prove to be capable; *also* **:** SUCCEED
— make hay : to make use of offered opportunity especially in gaining an early advantage
— make head 1 : to make progress especially against resistance **2 :** to rise in armed revolt
— make it 1 : to be successful ⟨trying to *make it* in the big time as a fashion photographer —Joe Kane⟩ ⟨the crème brûlée didn't quite *make it*⟩ **2 :** to have sexual intercourse
— make light of : to treat as of little account
— make love 1 : WOO, COURT **2 a :** NECK, PET **b :** to engage in sexual intercourse
— make much of 1 : to treat as of importance **2 :** to treat with obvious affection or special consideration
— make no bones : to be straightforward, unhesitating, or sure ⟨*makes no bones* about the seriousness of the matter⟩
— make one's mark : to achieve success or fame
— make public : DISCLOSE
— make sail 1 : to raise or spread sail **2 :** to set out on a voyage
— make shift : to manage with difficulty
— make sport of : RIDICULE, MOCK
— make the grade : to measure up to some standard **:** be successful
— make the most of : to show or use to the best advantage
— make the scene : to be present at or participate in a usually specified activity or event
— make time 1 : to travel fast **2 :** to gain time **3 :** to make progress toward winning favor ⟨trying to *make time* with the waitress⟩
— make tracks 1 : to proceed at a walk or run **2 :** to go in a hurry **:** RUN AWAY, FLEE
— make use of : to put to use **:** EMPLOY
— make water 1 *of a boat* **:** LEAK **2 :** URINATE
— make waves : to create a stir or disturbance

— **make way 1 :** to give room for passing, entering, or occupying **2 :** to make progress
— **make with** *slang* **:** PRODUCE, PERFORM — usually used with *the*

[2]make *noun* (14th century)
1 a : the manner or style in which a thing is constructed **b :** BRAND 4a
2 : the physical, mental, or moral constitution of a person ⟨men of his *make* are rare⟩
3 a : the action of producing or manufacturing **b :** the actual yield or amount produced over a specified period **:** OUTPUT
4 : the act of shuffling cards; *also* **:** turn to shuffle
— **on the make 1 :** in the process of forming, growing, or improving **2 :** in quest of a higher social or financial status **3 :** in search of sexual adventure

make·bate \'māk-ˌbāt\ *noun* [[1]*make* + obsolete *bate* strife] (1529)
archaic **:** one that excites contention and quarrels

[1]make–be·lieve \'māk-bə-ˌlēv\ *also* **make–be·lief** \-ˌlēf\ *noun* (1811)
: a pretending to believe

[2]make–believe *adjective* (1824)
: IMAGINARY, PRETENDED

make–do \'māk-'dü\ *adjective* (1923)
: MAKESHIFT
— **make–do** *noun*

make·fast \-ˌfast\ *noun* (1898)
: something (as a post or buoy) to which a boat can be fastened

make off *intransitive verb* (1709)
: to leave in haste
— **make off with :** to take away; *especially* **:** GRAB, STEAL

make–or–break \'mā-kər-'brāk\ *adjective* (1919)
: allowing no middle ground between success and failure

make out (15th century)
transitive verb
1 : to complete (as a printed form) by supplying required information ⟨*make out* a check⟩
2 : to find or grasp the meaning of ⟨tried to *make out* what had really happened⟩
3 : to form an opinion or idea about **:** CONCLUDE ⟨how do you *make* that *out*⟩
4 a : to represent as being ⟨*made* them *out* to be losers⟩ **b :** to pretend to be true ⟨*made out* that he had never heard of me⟩
5 : to represent or delineate in detail
6 : to see and identify with difficulty or effort **:** DISCERN ⟨*make out* a ship through the fog⟩
intransitive verb
1 : GET ALONG, FARE ⟨how are you *making out* with the new job⟩
2 a : to engage in sexual intercourse **b :** NECK 1

make·over \'mā-ˌkō-vər\ *noun* (1927)
: an act or instance of making over; *especially* **:** a changing of a person's appearance (as by the use of cosmetics or a different hairdo)

make over *transitive verb* (1546)
1 : to transfer the title of (property)
2 : REMAKE, REMODEL ⟨*made* the whole house *over*⟩

mak·er \'mā-kər\ *noun* (14th century)
: one that makes; as **a** *capitalized* **:** GOD 1 **b** *archaic* **:** POET **c :** a person who borrows money on a promissory note **d :** MANUFACTURER

make·ready \'māk-ˌre-dē\ *noun, plural* **-readies** (1887)
: final preparation (as of a form on a printing press) for running

make·shift \'māk-ˌshift\ *noun* (circa 1812)
: a usually crude and temporary expedient **:** SUBSTITUTE
synonym see RESOURCE
— **makeshift** *adjective*

make·up \'mā-ˌkəp\ *noun* (1821)
1 a : the way in which the parts or ingredients of something are put together **:** COMPOSITION **b :** physical, mental, and moral constitution
2 a : the operation of making up especially pages for printing **b :** design or layout of printed matter
3 a (1) **:** cosmetics used to color and beautify the face (2) **:** a cosmetic applied to other parts of the body **b :** materials (as wigs and cosmetics) used in making up or in special costuming (as for a play)
4 : REPLACEMENT; *specifically* **:** material added (as in a manufacturing process) to replace material that has been used up ⟨*makeup* water⟩
5 : a special examination in which a student may make up for absence or previous failure

make up (14th century)
transitive verb
1 a : to form by fitting together or assembling ⟨*make up* a train of cars⟩ **b :** to arrange typeset matter in (as pages) for printing
2 a : to combine to produce (a sum or whole) **b :** CONSTITUTE, COMPOSE ⟨10 chapters *make up* this volume⟩
3 : to make good (a deficiency)
4 : SETTLE, DECIDE ⟨*made up* my mind to depart⟩
5 : to wrap or fasten up ⟨*make* the books *up* into a parcel⟩
6 a : to prepare in physical appearance for a role **b :** to apply cosmetics to
7 a : INVENT, IMPROVISE ⟨*make up* a story⟩ **b :** to set in order ⟨rooms are *made up* daily⟩
intransitive verb
1 : to become reconciled ⟨quarreled but later *made up*⟩
2 a : to act ingratiatingly and flatteringly ⟨*made up* to his aunt for a new bicycle⟩ **b :** to make advances **:** COURT
3 : COMPENSATE ⟨*make up* for lost time⟩
4 a : to put on costumes or makeup (as for a play) **b :** to apply cosmetics

make·weight \'māk-ˌwāt\ *noun* (1695)
1 a : something thrown into a scale to bring the weight to a desired value **b :** something of little independent value thrown in to fill a gap
2 : COUNTERWEIGHT, COUNTERPOISE

make–work \'māk-ˌwərk\ *noun* (1923)
: work assigned or done chiefly to keep one busy

ma·ki·mo·no \ˌmä-ki-'mō-(ˌ)nō\ *noun, plural* **-nos** [Japanese, scroll, from *maki* roll + *mono* thing] (1882)
: a horizontal Japanese ornamental pictorial or calligraphic scroll

mak·ing \'mā-kiŋ\ *noun* [Middle English, from Old English *macung,* from *macian* to make] (12th century)
1 : the act or process of forming, causing, doing, or coming into being ⟨spots problems in the *making*⟩
2 : a process or means of advancement or success
3 : something made; *especially* **:** a quantity produced at one time **:** BATCH
4 a : POTENTIALITY — often used in plural ⟨had the *makings* of a great artist⟩ **b** *plural* **:** the material from which something is to be made; *especially* *usually* 'mā-kənz\ **:** paper and tobacco for rolling cigarettes by hand

ma·ko shark \'mā-(ˌ)kō-, 'mä-\ *noun* [Maori *mako* mako shark] (1926)
: either of two mackerel sharks (*Isurus paucus* and *I. oxyrinchus*) that are notable sport fish and are considered dangerous to humans — called also *mako;* see SHARK illustration

makuta *plural of* LIKUTA

mal- *combining form* [Middle English, from Middle French, from Old French, from *mal* bad (from Latin *malus*) & *mal* badly, from Latin *male,* from *malus*]
1 a : bad ⟨*mal*practice⟩ **b :** badly ⟨*mal*odorous⟩
2 a : abnormal ⟨*mal*formation⟩ **b :** abnormally ⟨*mal*formed⟩
3 a : inadequate ⟨*mal*adjustment⟩ **b :** inadequately ⟨*mal*nourished⟩

mal·ab·sorp·tion \ˌma-ləb-'sȯrp-shən, -'zȯrp-\ *noun* (circa 1929)
: faulty absorption especially of nutrient materials from the alimentary canal

ma·lac·ca \mə-'la-kə\ *adjective* [*Malacca,* Malaya] (1844)
: made or comprised of the cane of an Asian rattan palm (*Calamus rotang*) ⟨an umbrella with a *malacca* handle⟩
— **malacca** *noun*

Mal·a·chi \'ma-lə-ˌkī\ *noun* [Hebrew *Mal'ākhī*]
1 — used as the conventional name for the unidentified 5th century B.C. writer of the book of Malachi
2 : a prophetic book of canonical Jewish and Christian Scripture — see BIBLE table

Mal·a·chi·as \ˌma-lə-'kī-əs\ *noun* [Late Latin, from Greek, from Hebrew *Mal'ākhī*]
: MALACHI

mal·a·chite \'ma-lə-ˌkīt\ *noun* [alteration of Middle English *melochites,* from Latin *molochites,* from Greek *molochitēs,* from *molochē, malachē* mallow] (1656)
: a mineral that is a green basic carbonate of copper used as an ore and for making ornamental objects

mal·a·col·o·gy \ˌma-lə-'kä-lə-jē\ *noun* [French *malacologie,* contraction of *malacozoologie,* from New Latin *Malacozoa,* zoological group including soft-bodied animals (from Greek *malakos* soft + New Latin *-zoa*) + French *-logie* -logy] (1836)
: a branch of zoology dealing with mollusks
— **mal·a·co·log·i·cal** \ˌma lə-kə-'lä-ji-kəl\ *adjective*
— **mal·a·col·o·gist** \ˌma-lə-'kä-lə-jist\ *noun*

mal·a·cos·tra·can \ˌma-lə-'käs-tri-kən\ *noun* [ultimately from Greek *malakostrakos* soft-shelled, from *malakos* soft + *ostrakon* shell — more at MOLLIFY, OYSTER] (1835)
: any of a large subclass (Malacostraca) of crustaceans having a thorax consisting of eight segments usually covered by a carapace and including the decapods and isopods
— **malacostracan** *adjective*

mal·ad·ap·ta·tion \ˌma-ˌla-ˌdap-'tā-shən\ *noun* (1877)
: poor or inadequate adaptation

mal·adapt·ed \ˌma-lə-'dap-təd\ *adjective* (1943)
: unsuited or poorly suited (as to a particular use, purpose, or situation)

mal·adap·tive \-tiv\ *adjective* (1931)
1 : marked by poor or inadequate adaptation
2 : not conducive to adaptation

mal·ad·just·ed \ˌma-lə-'jəs-təd\ *adjective* (1886)
: poorly or inadequately adjusted; *specifically* **:** lacking harmony with one's environment from failure to adjust one's desires to the conditions of one's life

mal·ad·jus·tive \-'jəs-tiv\ *adjective* (1928)
: not conducive to adjustment

mal·ad·just·ment \-'jəs(t)-mənt\ *noun* (1833)
: poor, faulty, or inadequate adjustment

mal·ad·min·is·tra·tion \ˌma-ləd-ˌmi-nə-'strā-shən\ *noun* (1644)
: administration that is corrupt or incompetent (as that of a public office) or incorrect (as that of a drug)
— **mal·ad·min·is·ter** \-'mi-nə-stər\ *transitive verb*

mal·adroit \ˌma-lə-'drȯit\ *adjective* [French, from Middle French, from *mal-* + *adroit*] (1685)
: lacking adroitness **:** INEPT
synonym see AWKWARD
— **mal·adroit·ly** *adverb*
— **mal·adroit·ness** *noun*

mal·a·dy \'ma-lə-dē\ *noun, plural* **-dies** [Middle English *maladie,* from Old French, from *malade* sick, from Latin *male habitus* in bad condition] (13th century)
1 : a disease or disorder of the animal body
2 : an unwholesome or disordered condition

ma·la fi·de \ˌma-lə-'fī-dē, -də\ *adverb or adjective* [Late Latin] (1561)
: with or in bad faith

Ma·la·ga \'ma-lə-gə\ *noun* (1608)
: a sweet brown fortified wine from Málaga, Spain; *also* **:** a similar wine made elsewhere

Mal·a·gasy \ˌma-lə-'ga-sē; ˌmä-lə-'gä-sē, -shē\ *noun, plural* **Malagasy** *also* **Mal·a·gas·ies** (1839)
1 : a member of a people of Indonesian and African origin who inhabit Madagascar
2 : the Austronesian language of the Malagasy people
— **Malagasy** *adjective*

ma·la·gue·na \ˌma-lə-'gān-yə, ˌmä-\ *noun* [Spanish *malagueña,* from feminine of *malagueño* of Málaga, from *Málaga*] (circa 1883)
1 : a folk tune native to Málaga that is similar to a fandango
2 : a Spanish dance for couples that is similar to a fandango

mal·aise \mə-'lāz, ma-, -'lez\ *noun* [French *malaise,* from Old French, from *mal-* + *aise* comfort — more at EASE] (circa 1768)
1 : an indefinite feeling of debility or lack of health often indicative of or accompanying the onset of an illness
2 : a vague sense of mental or moral ill-being ⟨a *malaise* of cynicism and despair —Malcolm Boyd⟩

mal·a·mute \'ma-lə-ˌmyüt, -ˌmüt\ *noun* [*Malemute,* an Alaskan Eskimo people] (1898)
: a sled dog of northern North America; *especially* **:** ALASKAN MALAMUTE

mal·a·pert \ˌma-lə-'pərt\ *adjective* [Middle English, from Middle French, unskillful, from *mal-* + *apert* skillful, literally, open — more at PERT] (14th century)
: impudently bold **:** SAUCY
— **mal·a·pert·ly** *adverb*
— **mal·a·pert·ness** *noun*

mal·ap·por·tioned \ˌma-lə-'pōr-shənd, -'pȯr-\ *adjective* (1965)
: characterized by an inequitable or unsuitable apportioning of representatives to a legislative body
— **mal·ap·por·tion·ment** \-shən-mənt\ *noun*

¹mal·a·prop \'ma-lə-ˌpräp\ *noun* [Mrs. *Malaprop*] (1823)
: an example of malapropism ⟨was famed for *malaprops:* he always said "polo bears" and "Remember Pearl Island" and "neon stockings" —*Time*⟩

²malaprop *or* **mal·a·prop·ian** \ˌma-lə-'prä-pē-ən\ *adjective* (1840)
: using or marked by the use of malapropisms

mal·a·prop·ism \'ma-lə-ˌprä-ˌpi-zəm\ *noun* [Mrs. *Malaprop,* character noted for her misuse of words in R. B. Sheridan's comedy *The Rivals* (1775)] (1849)
1 : the usually unintentionally humorous misuse or distortion of a word or phrase; *especially* **:** the use of a word sounding somewhat like the one intended but ludicrously wrong in the context
2 : MALAPROP
— **mal·a·prop·ist** \-ˌprä-pist\ *noun*

mal·ap·ro·pos \ˌma-ˌla-prə-'pō\ *adverb* [French *mal à propos*] (1668)
: in an inappropriate or inopportune way
— **malapropos** *adjective*

¹ma·lar \'mā-lər, -ˌlär\ *adjective* [New Latin *malaris,* from Latin *mala* jawbone, cheek] (1782)
: of or relating to the cheek or the side of the head

²malar *noun* (circa 1828)
: ZYGOMATIC BONE — called also *malar bone*

ma·lar·ia \mə-'ler-ē-ə\ *noun* [Italian, from *mala aria* bad air] (1740)
1 *archaic* **:** air infected with a noxious substance capable of causing disease; *especially* **:** MIASMA
2 a : a human disease that is caused by sporozoan parasites (genus *Plasmodium*) in the red blood cells, is transmitted by the bite of anopheline mosquitoes, and is characterized by periodic attacks of chills and fever **b :** any of various diseases of birds and mammals caused by blood protozoans
— **ma·lar·i·al** \-əl\ *adjective*
— **ma·lar·i·ous** \-əs\ *adjective*

ma·lar·i·ol·o·gy \-ˌler-ē-'ä-lə-jē\ *noun* (circa 1923)
: the scientific study of malaria
— **ma·lar·i·ol·o·gist** \-jist\ *noun*

ma·lar·key *also* **ma·lar·ky** \mə-'lär-kē\ *noun* [origin unknown] (1929)
: insincere or foolish talk **:** BUNKUM

ma·late \'ma-ˌlāt, 'mā-\ *noun* (1794)
: a salt or ester of malic acid

mal·a·thi·on \ˌma-lə-'thī-ən, -ˌän\ *noun* [from *Malathion,* a trademark] (1953)
: an insecticide $C_{10}H_{19}O_6PS_2$ with a lower mammalian toxicity than parathion

Ma·lay \mə-'lā, 'mā-(ˌ)lā\ *noun* [obsolete Dutch *Malayo* (now *Maleier*), from Malay *Mĕlayu*] (1598)
1 : a member of a people of the Malay Peninsula, eastern Sumatra, parts of Borneo, and some adjacent islands
2 : the Austronesian language of the Malays
— **Malay** *adjective*
— **Ma·lay·an** \mə-'lā-ən, mā-; 'mā-ˌlā-\ *noun or adjective*

Mal·a·ya·lam \ˌma-lə-'yä-ləm, ˌmä-\ *noun* (1837)
: the Dravidian language of Kerala, southwest India, closely related to Tamil

¹mal·con·tent \ˌmal-kən-'tent\ *noun* (1581)
: a discontented person: **a :** one who bears a grudge from a sense of grievance or thwarted ambition **b :** one who is in active opposition to an established order or government **:** REBEL

²malcontent *adjective* [Middle French, from Old French, from *mal-* + *content* content] (1586)
: dissatisfied with the existing state of affairs **:** DISCONTENTED

mal·con·tent·ed \-'ten-təd\ *adjective* (1586)
: MALCONTENT
— **mal·con·tent·ed·ly** *adverb*
— **mal·con·tent·ed·ness** *noun*

mal de mer \ˌmal-də-'mer\ *noun* [French] (1778)
: SEASICKNESS

mal·dis·tri·bu·tion \ˌmal-ˌdis-trə-'byü-shən\ *noun* (1895)
: bad or faulty distribution **:** undesirable inequality or unevenness of placement or apportionment (as of population, resources, or wealth) over an area or among members of a group

¹male \'mā(ə)l\ *noun* [Middle English, from Middle French *masle, male,* adjective & noun, from Latin *masculus* — more at MASCULINE] (14th century)
1 a : a male person **:** a man or a boy **b :** an individual that produces small usually motile gametes (as spermatozoa or spermatozoids) which fertilize the eggs of a female
2 : a staminate plant

²male *adjective* (14th century)
1 a (1) **:** of, relating to, or being the sex that produces gametes which fertilize the eggs of a female (2) **:** STAMINATE; *especially* **:** having only staminate flowers and not producing fruit or seeds ⟨a *male* holly⟩ **b** (1) **:** of, relating to, or characteristic of the male sex ⟨a deep *male* voice⟩ (2) **:** made up of usually adult male individuals ⟨a *male* choir⟩
2 : MASCULINE 3a
3 : designed for fitting into a corresponding female part ⟨*male* hose coupling⟩
— **male·ness** \-nəs\ *noun*

male alto *noun* (circa 1879)
: COUNTERTENOR

ma·le·ate \'mā-lē-ˌāt, -lē-ət\ *noun* (1853)
: a salt or ester of maleic acid

¹mal·e·dict \ˌma-lə-'dikt\ *adjective* [Late Latin *maledictus*] (circa 1550)
archaic **:** ACCURSED

²maledict *transitive verb* (1623)
: CURSE, EXECRATE

mal·e·dic·tion \ˌma-lə-'dik-shən\ *noun* [Middle English *malediccioun,* from Late Latin *malediction-, maledictio,* from *maledicere* to curse, from Latin, to speak evil of, from *male* badly + *dicere* to speak, say — more at MAL-, DICTION] (14th century)
: CURSE, EXECRATION
— **mal·e·dic·to·ry** \-'dik-t(ə-)rē\ *adjective*

mal·e·fac·tion \ˌma-lə-'fak-shən\ *noun* (15th century)
: an evil deed **:** CRIME

mal·e·fac·tor \'ma-lə-ˌfak-tər\ *noun* [Middle English *malefactour,* from Latin *malefactor,* from *malefacere* to do evil, from *male* + *facere* to do — more at DO] (15th century)
1 : one who commits an offense against the law; *especially* **:** FELON
2 : one who does ill toward another

male fern *noun* (1562)
: a fern (*Dryopteris filix-mas*) producing an oleoresin used in expelling tapeworms

ma·lef·ic \mə-'le-fik\ *adjective* [Latin *maleficus* wicked, mischievous, from *male* + *-ficus* -fic] (1652)
1 : having malignant influence **:** BALEFUL
2 : MALICIOUS

ma·lef·i·cence \mə-'le-fə-sən(t)s\ *noun* [Italian *maleficenza,* from Latin *maleficentia,* from *maleficus*] (1598)
1 a : the act of committing harm or evil **b :** a harmful or evil act
2 : the quality or state of being maleficent

ma·lef·i·cent \-sənt\ *adjective* [back-formation from *maleficence*] (1678)
: working or productive of harm or evil **:** BALEFUL

ma·le·ic acid \mə-'lē-ik-, -'lā-\ *noun* [French *acide maléique,* alteration of *acide malique* malic acid; from its formation by dehydration of malic acid] (1857)
: a crystalline dicarboxylic acid $C_4H_4O_4$ that is isomeric with fumaric acid and is used especially in organic synthesis

maleic anhydride *noun* (1857)
: a caustic crystalline cyclic anhydride $C_4H_2O_3$ used especially in making resins

maleic hydrazide *noun* (1949)
: a crystalline cyclic hydrazide $C_4H_4N_2O_2$ used to retard plant growth

mal·e·mute *variant of* MALAMUTE

male–ster·ile \'mā(ə)l-'ster-əl\ *adjective* (1921)
: having male gametes lacking or nonfunctional

ma·lev·o·lence \mə-'le-və-lən(t)s\ *noun* (15th century)
1 : the quality or state of being malevolent
2 : malevolent behavior
synonym see MALICE

ma·lev·o·lent \-lənt\ *adjective* [Latin *malevolent-, malevolens,* from *male* badly + *volent-, volens,* present participle of *velle* to wish — more at MAL-, WILL] (1509)
1 : having, showing, or arising from intense often vicious ill will, spite, or hatred
2 : productive of harm or evil
— **ma·lev·o·lent·ly** *adverb*

mal·fea·sance \ˌmal-'fē-zᵊn(t)s\ *noun* [*mal-* + obsolete *feasance* doing, execution] (1696)
: wrongdoing or misconduct especially by a public official

mal·for·ma·tion \ˌmal-fȯr-ˈmā-shən, -fər-\ *noun* (1800)
: irregular, anomalous, abnormal, or faulty formation or structure

mal·formed \ˌmal-ˈfȯrmd\ *adjective* (1817)
: characterized by malformation **:** badly or imperfectly formed **:** MISSHAPEN

mal·func·tion \ˌmal-ˈfəŋ(k)-shən\ *intransitive verb* (1928)
: to function imperfectly or badly **:** fail to operate normally
— malfunction *noun*

mal·gré \mal-ˈgrā, ˈmal-ˌ\ *preposition* [French, from Old French *maugré* — more at MAUGRE] (1608)
: DESPITE

ma·lic \ˈma-lik, ˈmā-\ *adjective* (circa 1909)
: involved in and especially catalyzing a reaction in which malic acid participates ⟨*malic* dehydrogenase⟩

malic acid *noun* [French *acide malique,* ultimately from Latin *malum* apple, from Greek *mēlon, malon*] (1797)
: a crystalline dicarboxylic acid $C_4H_6O_5$; *especially* **:** the optical isomer of malic acid that is found in various fruits (as apples) and is formed as an intermediate in the Krebs cycle

mal·ice \ˈma-ləs\ *noun* [Middle English, from Middle French, from Latin *malitia,* from *malus* bad] (14th century)
1 : desire to cause pain, injury, or distress to another
2 : intent to commit an unlawful act or cause harm without legal justification or excuse ☆

ma·li·cious \mə-ˈli-shəs\ *adjective* (13th century)
: given to, marked by, or arising from malice
— ma·li·cious·ly *adverb*
— ma·li·cious·ness *noun*

malicious mischief *noun* (1769)
: willful, wanton, or reckless damage to or destruction of another's property

¹ma·lign \mə-ˈlīn\ *adjective* [Middle English *maligne,* from Middle French, from Latin *malignus,* from *male* badly + *gignere* to beget — more at MAL-, KIN] (14th century)
1 a : evil in nature, influence, or effect **:** INJURIOUS **b :** MALIGNANT, VIRULENT
2 : having or showing intense often vicious ill will **:** MALEVOLENT
synonym see SINISTER
— ma·lign·ly *adverb*

²malign *transitive verb* [Middle English, from Middle French *maligner* to act maliciously, from Late Latin *malignari,* from Latin *malignus*] (15th century)
: to utter injuriously misleading or false reports about **:** speak evil of ☆

ma·lig·nance \mə-ˈlig-nən(t)s\ *noun* (circa 1604)
: MALIGNANCY

ma·lig·nan·cy \-nən(t)-sē\ *noun, plural* **-cies** (1601)
1 : the quality or state of being malignant
2 a : exhibition (as by a tumor) of malignant qualities **:** VIRULENCE **b :** a malignant tumor

ma·lig·nant \mə-ˈlig-nənt\ *adjective* [Late Latin *malignant-, malignans,* present participle of *malignari*] (circa 1545)
1 a *obsolete* **:** MALCONTENT, DISAFFECTED **b :** evil in nature, influence, or effect **:** INJURIOUS **c :** passionately and relentlessly malevolent **:** aggressively malicious
2 : tending to produce death or deterioration ⟨*malignant* malaria⟩; *especially* **:** tending to infiltrate, metastasize, and terminate fatally ⟨*malignant* tumor⟩
— ma·lig·nant·ly *adverb*

ma·lig·ni·ty \mə-ˈlig-nə-tē\ *noun* (14th century)
1 : MALIGNANCY, MALEVOLENCE
2 : an instance of malignant or malicious behavior or nature
synonym see MALICE

ma·li·hi·ni \ˌmä-li-ˈhē-nē\ *noun* [Hawaiian] (1914)
: a newcomer or stranger among the people of Hawaii

ma·lines \mə-ˈlēn\ *noun, plural* **ma·lines** \-ˈlēn(z)\ [French, from *Malines* (Mechelen), Belgium] (1833)
1 : MECHLIN
2 *also* **ma·line :** a fine stiff net with a hexagonal mesh that is usually made of silk or rayon and that is often used for veils

ma·lin·ger \mə-ˈliŋ-gər\ *intransitive verb* **ma·lin·gered; ma·lin·ger·ing** \-g(ə-)riŋ\ [French *malingre* sickly] (1820)
: to pretend incapacity (as illness) so as to avoid duty or work
— ma·lin·ger·er \-gər-ər\ *noun*

Ma·lin·ke \mə-ˈliŋ-kē\ *noun, plural* **Malinke** *or* **Malinkes** (1883)
1 : a member of a people of Mandingo affiliation widespread in the western part of Africa
2 : the language of the Malinke people

Ma·li·nois \ˌma-lən-ˈwä\ *noun* [French, one from Malines, from *Malines* (Mechelen), Belgium] (1929)
: BELGIAN MALINOIS

mal·i·son \ˈma-lə-sən, -zən\ *noun* [Middle English, from Old French *maleïçon,* from Late Latin *malediction-, maledictio*] (13th century)
: CURSE, MALEDICTION

mal·kin \ˈmȯ(l)-kən, ˈmal-\ *noun* [Middle English *malkyn* servant woman, from *Malkyn,* diminutive of the name *Maud*] (1586)
1 *dialect chiefly British* **:** an untidy woman **:** SLATTERN
2 *dialect chiefly British* **a :** CAT **b :** HARE

¹mall \ˈmȯl\ *variant of* MAUL

²mall \ˈmȯl, *especially British & for 1* ˈmal\ *noun* [short for obsolete *pall-mall* mallet used in pall-mall] (1644)
1 : an alley used for pall-mall
2 [The *Mall,* promenade in London, originally a pall-mall alley] **a :** a usually public area often set with shade trees and designed as a promenade or as a pedestrian walk **b :** a usually paved or grassy strip between two roadways
3 a : an urban shopping area featuring a variety of shops surrounding a usually open-air concourse reserved for pedestrian traffic **b :** a usually large suburban building or group of buildings containing various shops with associated passageways ◇

mal·lard \ˈma-lərd\ *noun, plural* **mallard** *or* **mallards** [Middle English, from Middle French *mallart*] (14th century)
: a common and widely distributed wild duck (*Anas platyrhynchos*) of the northern hemisphere the males of which have a green head and white-ringed neck and which is the source of the domestic ducks

mallard

mal·lea·ble \ˈma-lē-ə-bəl, ˈmal-yə-bəl, ˈma-lə-bəl\ *adjective* [Middle English *malliable,* from Middle French or Medieval Latin; Middle French *malleable,* from Medieval Latin *malleabilis,* from *malleare* to hammer, from Latin *malleus* hammer — more at MAUL] (14th century)
1 : capable of being extended or shaped by beating with a hammer or by the pressure of rollers
2 a : capable of being altered or controlled by outside forces or influences **b :** having a capacity for adaptive change
synonym see PLASTIC
— mal·lea·bil·i·ty \ˌma-lē-ə-ˈbi-lə-tē, ˌmal-yə-, ˌma-lə-\ *noun*

mal·lee \ˈma-lē\ *noun* [Wuywurung (Australian aboriginal language of the Melbourne area, Victoria) *mali*] (1845)
1 : any of various low-growing shrubby Australian eucalypts (as *Eucalyptus dumosa* and *E. oleosa*)
2 : a dense thicket or growth of mallees; *also* **:** land covered by such growth

☆ SYNONYMS

Malice, malevolence, ill will, spite, malignity, spleen, grudge mean the desire to see another experience pain, injury, or distress. MALICE implies a deep-seated often unexplainable desire to see another suffer ⟨felt no *malice* toward their former enemies⟩. MALEVOLENCE suggests a bitter persistent hatred that is likely to be expressed in malicious conduct ⟨a look of dark *malevolence*⟩. ILL WILL implies a feeling of antipathy of limited duration ⟨*ill will* provoked by a careless remark⟩. SPITE implies petty feelings of envy and resentment that are often expressed in small harassments ⟨petty insults inspired by *spite*⟩. MALIGNITY implies deep passion and relentlessness ⟨a life consumed by motiveless *malignity*⟩. SPLEEN suggests the wrathful release of latent spite or persistent malice ⟨venting his *spleen* against politicians⟩. GRUDGE implies a harbored feeling of resentment or ill will that seeks satisfaction ⟨never one to harbor a *grudge*⟩.

Malign, traduce, asperse, vilify, calumniate, defame, slander mean to injure by speaking ill of. MALIGN suggests specific and often subtle misrepresentation but may not always imply deliberate lying ⟨the most *maligned* monarch in British history⟩. TRADUCE stresses the resulting ignominy and distress to the victim ⟨so *traduced* the governor that he was driven from office⟩. ASPERSE implies continued attack on a reputation often by indirect or insinuated detraction ⟨both candidates *aspersed* the other's motives⟩. VILIFY implies attempting to destroy a reputation by open and direct abuse ⟨no criminal was more *vilified* in the press⟩. CALUMNIATE imputes malice to the speaker and falsity to the assertions ⟨falsely *calumniated* as a traitor⟩. DEFAME stresses the actual loss of or injury to one's good name ⟨sued them for *defaming* her reputation⟩. SLANDER stresses the suffering of the victim ⟨town gossips *slandered* their good name⟩.

◇ WORD HISTORY

mall In 16th century Italy a popular alley game resembling croquet was known as *pallamaglio,* from *palla* "ball" and *maglio* "mallet." The game was adopted by the French as *pallemaille* and in the 17th century by the English as *pall-mall.* The alley on which the game was played came to be known as a *mall.* One of the best known of these alleys was located in London's St. James's Park and known as "The Mall." When still used as a playing field, the alley in St. James's had a hard sandy base covered with powdered cockleshells. After the game lost favor, The Mall at St. James's, as it continued to be called, was landscaped into a fashionable promenade lined with trees and flowers. Other similar, open-air places came to be called *malls* also. In the mid 20th century the word was applied to a variety of public spaces, ranging from an outdoor concourse between buildings to an enclosed shopping complex in which the stores and restaurants are linked by passageways.

mal·let \'ma-lət\ *noun* [Middle English *maillet,* from Middle French, from Old French, diminutive of *mail* maul — more at MAUL] (15th century)
: a hammer with a typically barrel-shaped head: as **a :** a tool with a large head for driving another tool or for striking a surface without marring it **b :** a long-handled wooden implement used for striking a ball (as in polo or croquet) **c :** a light hammer with a small rounded or spherical usually padded head used in playing certain musical instruments (as a vibraphone)

mal·le·us \'ma-lē-əs\ *noun, plural* **mal·lei** \-lē-,ī, -lē-,ē\ [New Latin, from Latin, hammer] (1669)
: the outermost of a chain of three small bones of the mammalian middle ear — see EAR illustration

mal·low \'ma-(,)lō\ *noun* [Middle English *malwe,* from Old English *mealwe,* from Latin *malva*] (before 12th century)
: any of a genus (*Malva* of the family Malvaceae, the mallow family) of herbs with palmately lobed or dissected leaves, usually showy flowers, and a disk-shaped fruit

malm·sey \'mäm-zē, 'mälm-, *New England also* 'mȧm-\ *noun, often capitalized* [Middle English *malmesey,* from Medieval Latin *Malmasia* Monemvasia, village in Greece where a sweet wine was produced] (15th century)
: the sweetest variety of Madeira wine

mal·nour·ished \,mal-'nər-isht, -'nə-risht\ *adjective* (1927)
: UNDERNOURISHED

mal·nu·tri·tion \,mal-nü-'tri-shən, -nyü-\ *noun* (1862)
: faulty or inadequate nutrition

mal·oc·clu·sion \,ma-lə-'klü-zhən\ *noun* (1888)
: improper occlusion; *especially* **:** abnormality in the coming together of teeth

mal·odor \,mal-'ō-dər\ *noun* (1825)
: an offensive odor

mal·odor·ous \-'ō-də-rəs\ *adjective* (1850)
1 : having a bad odor
2 : highly improper ⟨*malodorous* practices and chicanery in high financial places —*New Republic*⟩ ☆
— mal·odor·ous·ly *adverb*
— mal·odor·ous·ness *noun*

ma·lo·lac·tic \,ma-lō-'lak-tik, ,mā-\ *adjective* [*malic* + *-o-* + *lactic*] (1965)
: relating to or involved in the bacterial conversion of malic acid to lactic acid in wine ⟨*malolactic* fermentation⟩

maloti *plural of* LOTI

Mal·pi·ghi·an corpuscle \mal-'pi-gē-ən-, -'pē-\ *noun* [Marcello *Malpighi*] (1848)
: the part of a nephron that consists of a glomerulus and Bowman's capsule — called also *Malpighian body*

Malpighian layer *noun* (1878)
: the deeper part of the epidermis consisting of cells whose protoplasm has not yet changed into horny material

Malpighian tubule *noun* (1877)
: any of a group of long blind vessels opening into the posterior part of the alimentary canal in most insects and some other arthropods and functioning primarily as excretory organs — called also *Malpighian tube*

mal·po·si·tion \,mal-pə-'zi-shən\ *noun* (circa 1839)
: wrong or faulty position

mal·prac·tice \,mal-'prak-təs\ *noun* (1671)
1 : a dereliction from professional duty or a failure to exercise an accepted degree of professional skill or learning by one (as a physician) rendering professional services which results in injury, loss, or damage
2 : an injurious, negligent, or improper practice **:** MALFEASANCE

mal·prac·ti·tio·ner \,mal-prak-'ti-sh(ə-)nər\ *noun* (1800)
: one who engages in or commits malpractice

¹malt \'mȯlt\ *noun* [Middle English, from Old English *mealt;* akin to Old High German *malz* malt, Old English *meltan* to melt] (before 12th century)
1 : grain (as barley) softened by steeping in water, allowed to germinate, and used especially in brewing and distilling
2 : MALT LIQUOR
3 : MALTED MILK
— malty \'mȯl-tē\ *adjective*

²malt (15th century)
transitive verb
1 : to convert into malt
2 : to make or treat with malt or malt extract
intransitive verb
1 : to become malt
2 : to make grain into malt

malt·ase \'mȯl-,tās, -,tāz\ *noun* (1890)
: an enzyme that catalyzes the hydrolysis of maltose to glucose

malted milk *noun* (1887)
1 : a soluble powder prepared from dried milk and malted cereals
2 : a beverage made by dissolving malted milk in milk and usually adding ice cream and flavoring — called also *malted*

Mal·tese \mȯl-'tēz, -'tēs\ *noun, plural* **Maltese** (1615)
1 : a native or inhabitant of Malta
2 : the Semitic language of the Maltese people
3 : any of a breed of toy dogs with a long silky white coat, a black nose, and very dark eyes
— Maltese *adjective*

Maltese 3

Maltese cross *noun* (1877)
1 a : a cross formée **b :** a cross that resembles the cross formée but has the outer face of each arm indented in a V — see CROSS illustration
2 : a Eurasian perennial (*Lychnis chalcedonica*) having scarlet or rarely white flowers in dense terminal heads

Mal·thu·sian \mal-'thü-zhən, mȯl-, -'thyü-\ *adjective* [Thomas R. *Malthus*] (1821)
: of or relating to Malthus or to his theory that population tends to increase at a faster rate than its means of subsistence and that unless it is checked by moral restraint or disaster (as disease, famine, or war) widespread poverty and degradation inevitably result
— Malthusian *noun*
— Mal·thu·sian·ism \-zhə-,ni-zəm\ *noun*

malt liquor *noun* (1693)
: a fermented liquor (as beer) made with malt

malt·ose \'mȯl-,tōs, -,tōz\ *noun* [French, from English ¹*malt*] (1862)
: a crystalline dextrorotatory fermentable sugar $C_{12}H_{22}O_{11}$ formed especially from starch by amylase

mal·treat \,mal-'trēt\ *transitive verb* [part translation of French *maltraiter,* from Middle French, from *mal-* + *traiter* to treat, from Old French *traitier* — more at TREAT] (1708)
: to treat cruelly or roughly **:** ABUSE
— mal·treat·er \-'trē-tər\ *noun*
— mal·treat·ment \-'trēt-mənt\ *noun*

malt·ster \'mȯlt-stər\ *noun* (14th century)
: a maker of malt

malt sugar *noun* (1862)
: MALTOSE

malt whiskey *noun* (circa 1839)
: SCOTCH 3

mal·ver·sa·tion \,mal-vər-'sā-shən\ *noun* [Middle French, from *malverser* to be corrupt, from *mal* + *verser* to turn, handle, from Latin *versare,* frequentative of *vertere* to turn — more at WORTH] (1549)
1 : misbehavior and especially corruption in an office, trust, or commission
2 : corrupt administration

ma·ma *or* **mam·ma** \'mä-mə, *chiefly British* mə-'mä\ *noun* [baby talk] (1579)
1 : MOTHER
2 *slang* **:** WIFE, WOMAN

mama's boy *noun* (1850)
: a usually polite or timid boy or man who is extremely or excessively close to and solicitous of his mother

mam·ba \'mäm-bə, 'mam-\ *noun* [Zulu *imamba*] (1862)
: any of several chiefly arboreal venomous green or black elapid snakes (genus *Dendroaspis* synonym *Dendraspis*) of sub-Saharan Africa

mam·bo \'mäm-(,)bō\ *noun, plural* **mambos** [American Spanish] (1948)
: a ballroom dance of Cuban origin that resembles the rumba and the cha-cha; *also* **:** the music for this dance
— mambo *intransitive verb*

ma·mey \ma-'mē\ *noun* [Spanish, from Taino] (1604)
: a tropical American tree (*Mammea americana* of the family Guttiferae) having an ovoid fruit with thick russet leathery rind and yellow or reddish juicy sweet flesh; *also* **:** the fruit

Mam·luk \'mam-,lük\ *or* **Mam·e·luke** \'ma-mə-,lük\ *noun* [Arabic *mamlūk,* literally, slave] (circa 1506)
1 : a member of a politically powerful Egyptian military class occupying the sultanate from 1250 to 1517
2 *often not capitalized* **:** a white or east Asian slave in Muslim countries

mam·ma \'ma-mə\ *noun, plural* **mam·mae** \'ma-,mē, -,mī\ [Latin, mother, breast, of baby-talk origin] (circa 1693)
: a mammary gland and its accessory parts

mam·mal \'ma-məl\ *noun* [New Latin *Mammalia,* from Late Latin, neuter plural of *mammalis* of the breast, from Latin *mamma* breast] (1826)
: any of a class (Mammalia) of warm-blooded higher vertebrates (as placentals, marsupials, or monotremes) that nourish their young with milk secreted by mammary glands, have the skin usually more or less covered with hair, and include humans
— mam·ma·li·an \mə-'mā-lē-ən, ma-\ *adjective or noun*

mam·mal·o·gy \mə-'ma-lə-jē, ma-'ma-, -'mä-\ *noun* [International Scientific Vocabulary, blend of *mammal* and *-logy*] (1835)
: a branch of zoology dealing with mammals
— mam·mal·o·gist \-jist\ *noun*

mam·ma·ry \'ma-mə-rē\ *adjective* (1682)
: of, relating to, lying near, or affecting the mammae

mammary gland *noun* (1831)
: any of the large compound modified sebaceous glands that in female mammals are modified to secrete milk, are situated ventrally in pairs, and usually terminate in a nipple

☆ SYNONYMS
Malodorous, stinking, fetid, noisome, putrid, rank, fusty, musty mean bad-smelling. MALODOROUS may range from the unpleasant to the strongly offensive ⟨*malodorous* fertilizers⟩. STINKING and FETID suggest the foul or disgusting ⟨prisoners were held in *stinking* cells⟩ ⟨the *fetid* odor of skunk cabbage⟩. NOISOME adds a suggestion of being harmful or unwholesome as well as offensive ⟨a stagnant, *noisome* sewer⟩. PUTRID implies particularly the sickening odor of decaying organic matter ⟨the *putrid* smell of rotting fish⟩. RANK suggests a strong unpleasant smell ⟨*rank* cigar smoke⟩. FUSTY and MUSTY suggest lack of fresh air and sunlight, FUSTY also implying prolonged uncleanliness, MUSTY stressing the effects of dampness, mildew, or age ⟨a *fusty* attic⟩ ⟨the *musty* odor of a damp cellar⟩.

mam·mer *intransitive verb* [Middle English *mameren* to stammer, of imitative origin] (circa 1555)
obsolete **:** WAVER, HESITATE

mam·mil·la·ry \ˈma-mə-ˌler-ē, ma-ˈmi-lə-rē\ *adjective* [Latin *mammilla* breast, nipple, diminutive of *mamma*] (1669)
1 : of, relating to, or resembling the breasts
2 : studded with breast-shaped protuberances

mam·mil·lat·ed \ˈma-mə-ˌlā-təd\ *adjective* [Late Latin *mammillatus,* from Latin *mammilla*] (1741)
1 : having nipples or small protuberances
2 : having the form of a bluntly rounded protuberance

[1]mam·mock \ˈma-mək\ *noun* [origin unknown] (circa 1529)
chiefly dialect **:** a broken piece **:** SCRAP

[2]mammock *transitive verb* (1607)
chiefly dialect **:** to tear into fragments **:** MANGLE

mam·mo·gram \ˈma-mə-ˌgram\ *noun* [Latin *mamma* + English *-o-* + *-gram*] (1937)
: a photograph of the breasts made by X rays

mam·mog·ra·phy \ma-ˈmä-grə-fē\ *noun* (1937)
: X-ray examination of the breasts (as for early detection of cancer)
— mam·mo·graph·ic \ˌma-mə-ˈgra-fik\ *adjective*

mam·mon \ˈma-mən\ *noun, often capitalized* [Late Latin *mammona,* from Greek *mamōna,* from Aramaic *māmōnā* riches] (15th century)
: material wealth or possessions especially as having a debasing influence ⟨you cannot serve God and *mammon* —Matthew 6:24 (Revised Standard Version)⟩
— mam·mon·ism \-mə-ˌni-zəm\ *noun*

mam·mon·ist \-mə-nist\ *noun* (1550)
archaic **:** one devoted to the ideal or pursuit of wealth

[1]mam·moth \ˈma-məth\ *noun* [Russian *mamont, mamot*] (1706)
1 : any of a genus (*Mammuthus*) of extinct Pleistocene elephants distinguished from recent elephants by highly ridged molars, usually large size, very long tusks that curve upward, and well-developed body hair
2 : something immense of its kind **:** GIANT ⟨a company that is a *mammoth* of the industry⟩

[2]mammoth *adjective* (1802)
: of very great size
synonym see ENORMOUS

mam·my \ˈma-mē\ *noun, plural* **mammies** [alteration of *mamma*] (1523)
1 : MAMA
2 : a black woman serving as a nurse to white children especially formerly in the southern U.S.

[1]man \ˈman, *in compounds* ˌman *or* mən\ *noun, plural* **men** \ˈmen, *in compounds* ˌmen *or* mən\ [Middle English, from Old English *man, mon* human being, male human; akin to Old High German *man* human being, Sanskrit *manu*] (before 12th century)
1 a (1) **:** an individual human; *especially* **:** an adult male human (2) **:** a man belonging to a particular category (as by birth, residence, membership, or occupation) — usually used in combination ⟨council*man*⟩ (3) **:** HUSBAND (4) **:** LOVER **b :** the human race **:** MANKIND **c :** a bipedal primate mammal (*Homo sapiens*) that is anatomically related to the great apes but distinguished especially by notable development of the brain with a resultant capacity for articulate speech and abstract reasoning, is usually considered to form a variable number of freely interbreeding races, and is the sole representative of a natural family (Hominidae); *broadly* **:** any living or extinct member of this family **d** (1) **:** one possessing in high degree the qualities considered distinctive of manhood (2) *obsolete* **:** the quality or state of being manly **:** MANLINESS **e :** FELLOW, CHAP **f** — used interjectionally to express intensity of feeling ⟨*man,* what a game⟩
2 a : INDIVIDUAL, PERSON ⟨a *man* could get killed there⟩ **b :** the individual who can fulfill or who has been chosen to fulfill one's requirements ⟨she's your *man*⟩
3 a : a feudal tenant **:** VASSAL **b :** an adult male servant **c** *plural* **:** the working force as distinguished from the employer and usually the management
4 a : one of the distinctive objects moved by each player in various board games **b :** one of the players on a team
5 : an alumnus of or student at a college or university ⟨a Bowdoin *man*⟩
6 *Christian Science* **:** the compound idea of infinite Spirit **:** the spiritual image and likeness of God **:** the full representation of Mind
7 *often capitalized* **:** POLICE ⟨when I heard the siren, I knew it was the *Man* —*American Speech*⟩
8 *often capitalized* **:** the white establishment **:** white society ⟨surprise that any black . . . should take on so about The *Man* —Peter Goldman⟩
9 : one extremely fond of or devoted to something specified ⟨a real vanilla ice cream *man*⟩
— man·less \ˈman-ləs\ *adjective*
— man·like \-ˌlīk\ *adjective*
— as one man : with the agreement and consent of all **:** UNANIMOUSLY
— one's own man : free from interference or control **:** INDEPENDENT
— to a man : without exception

[2]man *transitive verb* **manned; man·ning** (12th century)
1 a : to supply with people (as for service) ⟨*man* a fleet⟩ **b :** to station members of a ship's crew at ⟨*man* the capstan⟩ **c :** to serve in the force or complement of ⟨we'll *man* the concession stand while you sell tickets⟩
2 : to accustom (as a hawk) to humans and the human environment
3 : to furnish with strength or powers of resistance **:** BRACE

ma·na \ˈmä-nə\ *noun* [of Polynesian origin; akin to Hawaiian & Maori *mana* mana] (circa 1843)
1 : the power of the elemental forces of nature embodied in an object or person
2 : moral authority **:** PRESTIGE

man–about–town \ˌman-ə-ˌbau̇t-ˈtau̇n\ *noun, plural* **men–about–town** \ˌmen-\ (1734)
: a worldly and socially active man

[1]man·a·cle \ˈma-ni-kəl\ *noun* [Middle English *manicle,* from Middle French, from Latin *manicula,* diminutive of *manus* hand — more at MANUAL] (14th century)
1 : a shackle for the hand or wrist **:** HANDCUFF — usually used in plural
2 : something used as a restraint

[2]manacle *transitive verb* **man·a·cled; man·a·cling** \-k(ə-)liŋ\ (14th century)
1 : to confine (the hands) with manacles
2 : to make fast or secure **:** BIND; *broadly* **:** to restrain from movement, progress, or action
synonym see HAMPER

[1]man·age \ˈma-nij\ *verb* **man·aged; man·ag·ing** [Italian *maneggiare,* from *mano* hand, from Latin *manus*] (1561)
transitive verb
1 : to handle or direct with a degree of skill: as **a :** to make and keep compliant ⟨can't *manage* her child⟩ **b :** to treat with care **:** HUSBAND ⟨*managed* his resources carefully⟩ **c :** to exercise executive, administrative, and supervisory direction of ⟨*manage* a business⟩ ⟨*manage* a bond issue⟩ ⟨*manage* a baseball team⟩
2 : to work upon or try to alter for a purpose ⟨*manage* the press⟩ ⟨*manage* stress⟩
3 : to succeed in accomplishing **:** CONTRIVE ⟨*managed* to escape from prison⟩
4 : to direct the professional career of ⟨an agency that *manages* entertainers⟩
intransitive verb
1 a : to direct or carry on business or affairs; *also* **:** to direct a baseball team **b :** to admit of being carried on
2 : to achieve one's purpose
synonym see CONDUCT

[2]manage *noun* [Italian *maneggio* management, training of a horse, from *maneggiare*] (circa 1587)
1 a *archaic* **:** the action and paces of a trained riding horse **b :** the schooling or handling of a horse **c :** a riding school
2 *obsolete* **:** MANAGEMENT

man·age·able \ˈma-ni-jə-bəl\ *adjective* (1598)
: capable of being managed
— man·age·abil·i·ty \ˌma-ni-jə-ˈbi-lə-tē\ *noun*
— man·age·able·ness \ˈma-ni-jə-bəl-nəs\ *noun*
— man·age·ably \-blē\ *adverb*

man·age·ment \ˈma-nij-mənt\ *noun* (1598)
1 : the act or art of managing **:** the conducting or supervising of something (as a business)
2 : judicious use of means to accomplish an end
3 : the collective body of those who manage or direct an enterprise
— man·age·men·tal \ˌma-nij-ˈmen-tᵊl\ *adjective*

man·ag·er \ˈma-ni-jər\ *noun* (1588)
: one that manages: as **a :** a person who conducts business or household affairs **b :** a person whose work or profession is management **c** (1) **:** a person who directs a team or athlete (2) **:** a student who in scholastic or collegiate sports supervises equipment and records under the direction of a coach
— man·a·ge·ri·al \ˌma-nə-ˈjir-ē-əl\ *adjective*
— man·a·ge·ri·al·ly \-ē-ə-lē\ *adverb*
— man·ag·er·ship \ˈma-ni-jər-ˌship\ *noun*

man·ag·er·ess \ˈma-ni-jə-rəs\ *noun* (1797)
: a woman who is a manager
usage see -ESS

managing editor *noun* (1865)
: an editor in executive and supervisory charge of all editorial activities of a publication (as a newspaper)

[1]ma·ña·na \mən-ˈyä-nə\ *noun* [Spanish, literally, tomorrow, from (assumed) Vulgar Latin *maneana,* from feminine of *maneanus* early, from Latin *mane* early in the morning] (1845)
: an indefinite time in the future

[2]mañana *adverb* (1879)
: at an indefinite time in the future

man ape *noun* (circa 1864)
1 : GREAT APE
2 : any of various fossil primates intermediate in characters between recent humans and the great apes

Ma·nas·seh \mə-ˈna-sə\ *noun* [Hebrew *Mĕnashsheh*]
1 : a son of Joseph and the traditional eponymous ancestor of one of the tribes of Israel
2 : a king of Judah reigning in the 7th century B.C. and noted for his attempt to establish polytheism

man–at–arms \ˌman-ət-ˈärmz\ *noun, plural* **men–at–arms** \ˌmen-\ (1581)
: SOLDIER; *especially* **:** a heavily armed and usually mounted soldier

man·a·tee \ˈma-nə-ˌtē\ *noun* [Spanish *manatí*] (1555)
: any of a genus (*Trichechus*) of chiefly tropical aquatic herbivorous mammals that differ from the related dugong especially in having the tail rounded

manatee

\ə\ **abut** \ᵊ\ **kitten** \ər\ **further** \a\ **ash** \ā\ **ace** \ä\ **mop, mar** \au̇\ **out** \ch\ **chin** \e\ **bet** \ē\ **easy** \g\ **go** \i\ **hit** \ī\ **ice** \j\ **job** \ŋ\ **sing** \ō\ **go** \ȯ\ **law** \ȯi\ **boy** \th\ **thin** \t͟h\ **the** \ü\ **loot** \u̇\ **foot** \y\ **yet** \zh\ **vision** *see also* Guide to Pronunciation

Man·ches·ter terrier \ˈman-ˌches-tər-, -chə-stər-\ *noun* [*Manchester,* England] (1891)
: any of a breed of small short-haired black-and-tan terriers developed in England

man–child \ˈman-ˌchīld\ *noun, plural* **men·chil·dren** \ˈmen-ˌchil-drən, -dərn\ (14th century)
: a male child **:** SON

man·chi·neel \ˌman-chə-ˈnē(ə)l\ *noun* [French *mancenille,* from Spanish *manzanilla,* from diminutive of *manzana* apple] (1630)
: a poisonous tropical American tree (*Hippomane mancinella*) of the spurge family having a blistering milky juice and apple-shaped fruit

Man·chu \ˈman-(ˌ)chü, man-ˈ\ *noun, plural* **Manchu** *or* **Manchus** (1697)
1 : a member of an indigenous people of Manchuria who conquered China and established a dynasty there in 1644
2 : the Tungusic language of the Manchu people
— Manchu *adjective*

man·ci·ple \ˈman(t)-sə-pəl\ *noun* [Middle English, from Medieval Latin *mancipium* office of steward, from Latin, act of purchase, from *mancip-, manceps* purchaser — more at EMANCIPATE] (13th century)
: a steward or purveyor especially for a college or monastery

-mancy *noun combining form* [Middle English *-mancie,* from Old French, from Latin *-mantia,* from Greek *-manteia,* from *manteia,* from *mantis* diviner, prophet — more at MANTIS]
: divination ⟨oneiro*mancy*⟩

Man·dae·an \man-ˈdē-ən\ *noun* [Mandaean *mandayyā* having knowledge] (1875)
1 : a member of a Gnostic sect of the lower Tigris and Euphrates regions
2 : a form of Aramaic found in documents written by Mandaeans
— Mandaean *adjective*

man·da·la \ˈmən-də-lə\ *noun* [Sanskrit *maṇḍala* circle] (1859)
1 : a Hindu or Buddhist graphic symbol of the universe; *specifically* **:** a circle enclosing a square with a deity on each side
2 : a graphic and often symbolic pattern usually in the form of a circle divided into four separate sections or bearing a multiple projection of an image
— man·dal·ic \ˌmən-ˈda-lik\ *adjective*

man·da·mus \man-ˈdā-məs\ *noun* [Latin, we enjoin, from *mandare*] (1535)
: a writ issued by a superior court commanding the performance of a specified official act or duty

Man·dan \ˈman-ˌdan, -dən\ *noun, plural* **Mandan** *or* **Mandans** (1805)
1 : a member of an American Indian people of the Missouri River Valley in North Dakota
2 : the Siouan language of the Mandans

¹man·da·rin \ˈman-d(ə-)rən\ *noun* [Portuguese *mandarim,* from Malay *mĕntĕri,* from Sanskrit *mantrin* counselor, from *mantra* counsel — more at MANTRA] (1589)
1 a : a public official in the Chinese Empire of any of nine superior grades **b** (1) **:** a pedantic official (2) **:** BUREAUCRAT **c :** a person of position and influence often in intellectual or literary circles; *especially* **:** an elder and often traditionalist or reactionary member of such a circle
2 *capitalized* **a :** a form of spoken Chinese used by the court and the official classes of the Empire **b :** the group of closely related Chinese dialects that are spoken in about four fifths of the country and have a standard variety centering about Beijing
3 [French *mandarine,* from Spanish *mandarina,* probably from *mandarín* mandarin, from Portuguese *mandarim;* probably from the color of a mandarin's robes] **a :** a small spiny orange tree (*Citrus reticulata*) of southeastern Asia with yellow to reddish orange loose-rinded fruits; *also* **:** a derivative of this tree developed in cultivation by artificial selection or hybridization **b :** the fruit of a mandarin
— man·da·rin·ic \ˌman-də-ˈri-nik\ *adjective*
— man·da·rin·ism \ˈman-d(ə-)rə-ˌni-zəm\ *noun*

²mandarin *adjective* (1604)
1 : of, relating to, or typical of a mandarin ⟨*mandarin* graces⟩
2 : marked by polished ornate complexity of language ⟨*mandarin* prose⟩

man·da·rin·ate \ˈman-d(ə-)rə-ˌnāt\ *noun* [probably from French *mandarinat,* from *mandarin* mandarin, from Portuguese *mandarim*] (circa 1741)
1 : the office or status of a mandarin
2 : a body of mandarins
3 : rule by mandarins

mandarin collar *noun* (1947)
: a narrow stand-up collar usually open in front

mandarin orange *noun* (1771)
: MANDARIN 3

man·da·tary \ˈman-də-ˌter-ē\ *noun, plural* **-tar·ies** (15th century)
: MANDATORY

¹man·date \ˈman-ˌdāt\ *noun* [Middle French & Latin; Middle French *mandat,* from Latin *mandatum,* from neuter of *mandatus,* past participle of *mandare* to entrust, enjoin, probably irregular from *manus* hand + *-dere* to put — more at MANUAL, DO] (1501)
1 : an authoritative command; *especially* **:** a formal order from a superior court or official to an inferior one
2 : an authorization to act given to a representative ⟨accepted the *mandate* of the people⟩
3 a : an order or commission granted by the League of Nations to a member nation for the establishment of a responsible government over a former German colony or other conquered territory **b :** a mandated territory

²mandate *transitive verb* **man·dat·ed; man·dat·ing** (1919)
1 : to administer or assign (as a territory) under a mandate
2 : to make mandatory **:** ORDER; *also* **:** DIRECT, REQUIRE

man·da·tor \ˈman-ˌdā-tər\ *noun* (1681)
: one that gives a mandate

¹man·da·to·ry \ˈman-də-ˌtȯr-ē, -ˌtȯr-\ *adjective* (1576)
1 : containing or constituting a command **:** OBLIGATORY ⟨*mandatory* retirement age⟩
2 : of, by, relating to, or holding a League of Nations mandate
— man·da·tor·i·ly \-ˌtōr-ə-lē, -ˌtȯr-\ *adverb*

²mandatory *noun, plural* **-ries** (1661)
: one given a mandate; *especially* **:** a nation holding a mandate from the League of Nations

man–day \ˈman-ˈdā\ *noun* (1925)
: a unit of one day's work by one person

Man·de \ˈmän-ˌdā, män-ˈ\ *noun* (1883)
1 : MANDINGO
2 : a branch of the Niger-Congo language family spoken primarily in Sierra Leone, Liberia, Guinea, Ivory Coast, Mali, and Burkina Faso

man·di·ble \ˈman-də-bəl\ *noun* [Middle French, from Late Latin *mandibula,* from Latin *mandere* to chew; probably akin to Greek *masasthai* to chew] (15th century)
1 a : JAW 1a; *especially* **:** a lower jaw consisting of a single bone or of completely fused bones **b :** the lower jaw with its investing soft parts **c :** either the upper or lower segment of the bill of a bird
2 : any of various invertebrate mouthparts serving to hold or bite food materials; *especially* **:** either member of the anterior pair of mouth appendages of an arthropod often forming strong biting jaws
— man·dib·u·lar \man-ˈdi-byə-lər\ *adjective*
— man·dib·u·late \-lət\ *adjective*

Man·din·go \man-ˈdiŋ-(ˌ)gō\ *noun, plural* **Mandingo** *or* **Mandingoes** *or* **Mandingos** (1623)
1 : a member of a people of western Africa centering in the area of the upper Niger valley
2 : the language of the Mandingo people

Man·din·ka \man-ˈdiŋ-kə\ *noun, plural* **Mandinka** *or* **Mandinkas** (1939)
: MALINKE

man·di·o·ca \ˌman-dē-ˈō-kə\ *variant of* MANIOC

man·do·la \man-ˈdō-lə\ *noun* [Italian, from French *mandore,* modification of Late Latin *pandura* 3-stringed lute — more at BANDORE] (1758)
: a 16th and 17th century lute that is the ancestor of the smaller mandolin

man·do·lin \ˌman-də-ˈlin, ˈman-dᵊl-ən\ *also* **man·do·line** \ˌman-də-ˈlēn, ˈman-dᵊl-ən\ *noun* [Italian *mandolino,* diminutive of *mandola*] (1707)
1 : a musical instrument of the lute family that has a usually pear-shaped body and fretted neck and four to six pairs of strings
2 *usually* *mandoline* [French, from Italian *mandolino* mandolin] **:** a kitchen utensil with a blade for slicing and shredding
— man·do·lin·ist \ˌman-də-ˈli-nist\ *noun*

mandolin 1

man·drag·o·ra \man-ˈdra-gə-rə\ *noun* [Middle English, from Old English, from Latin *mandragoras,* from Greek] (before 12th century)
: MANDRAKE 1

man·drake \ˈman-ˌdrāk\ *noun* [Middle English, probably alteration of *mandragora*] (14th century)
1 a : a Mediterranean herb (*Mandragora officinarum*) of the nightshade family with ovate leaves, yellowish or purple flowers, and a large forked root traditionally credited with human attributes **b :** the root of a mandrake formerly used especially to promote conception, as a cathartic, or as a narcotic and soporific
2 : MAYAPPLE

mandrake 1

man·drel *also* **man·dril** \ˈman-drəl\ *noun* [probably modification of French *mandrin*] (1665)
1 a : a usually tapered or cylindrical axle, spindle, or arbor inserted into a hole in a piece of work to support it during machining **b :** a metal bar that serves as a core around which material (as metal) may be cast, molded, forged, bent, or otherwise shaped
2 : the shaft and bearings on which a tool (as a circular saw) is mounted

man·drill \ˈman-drəl\ *noun* [probably from ¹*man* + ³*drill*] (1744)
: a large baboon (*Papio sphinx* synonym *Mandrillus sphinx*) of central Africa west of the Congo River with the male having a bright red and blue muzzle

mandrill

mane \ˈmān\ *noun* [Middle English, from Old English *manu;* akin to Old High German *mana* mane, Latin *monile* necklace] (before 12th century)
1 : long and heavy hair growing about the neck and head of some mammals (as a horse or lion)
2 : long heavy hair on a person's head

— **maned** \'mānd\ *adjective*
man–eat·er \'man-,ē-tər\ *noun* (1600)
: one that has or is thought to have an appetite for human flesh: as **a :** CANNIBAL **b :** MACKEREL SHARK; *especially* **:** GREAT WHITE SHARK — called also *man-eater shark, man-eating shark* **c :** a large feline (as a lion or tiger) that has acquired the habit of feeding on human flesh
— **man–eat·ing** \-,ē-tiŋ\ *adjective*
maned wolf *noun* (1903)
: a yellowish red wild canid (*Chrysocyon brachyurus*) inhabiting South American grasslands and having black coloration on the nape and lower legs
ma·nège *also* **ma·nege** \ma-'nezh, mə-, -'nāzh\ *noun* [French *manège,* from Italian *maneggio* training of a horse — more at MANAGE] (1644)
1 : a school for teaching horsemanship and for training horses
2 : the art of horsemanship or of training horses
3 : the movements or paces of a trained horse
ma·nes \'mä-,nās, 'mā-,nēz\ *noun plural* [Latin]
1 *often capitalized* **:** the deified spirits of the ancient Roman dead honored with graveside sacrifices
2 : the venerated or appeased spirit of a dead person
¹ma·neu·ver \mə-'nü-vər, -'nyü-\ *noun* [French *manœuvre,* from Old French *maneuvre* work done by hand, from Medieval Latin *manuopera,* from Latin *manu operare* to work by hand] (1758)
1 a : a military or naval movement **b :** an armed forces training exercise; *especially* **:** an extended and large-scale training exercise involving military and naval units separately or in combination — often used in plural
2 : a procedure or method of working usually involving expert physical movement
3 a : evasive movement or shift of tactics **b :** an intended and controlled variation from a straight and level flight path in the operation of an airplane
4 a : an action taken to gain a tactical end **b :** an adroit and clever management of affairs often using trickery and deception
synonym see TRICK
²maneuver *verb* **ma·neu·vered; ma·neu·ver·ing** \-'nü-və-riŋ, -'nyü-; -'n(y)üv-riŋ\ (1777)
intransitive verb
1 a : to perform a movement in military or naval tactics in order to secure an advantage **b :** to make a series of changes in direction and position for a specific purpose
2 : to use stratagems **:** SCHEME
transitive verb
1 : to cause to execute tactical movements
2 : to manage into or out of a position or condition **:** MANIPULATE
3 a : to guide with adroitness and design **b :** to bring about or secure as a result of skillful management ◆
— **ma·neu·ver·abil·i·ty** \-,nü-və-rə-'bi-lə-tē, -,nyü-; -,n(y)üv-rə-\ *noun*
— **ma·neu·ver·able** \-'nü-və-rə-bəl, -'nyü-; -'n(y)üv-rə-\ *adjective*
— **ma·neu·ver·er** \-'nü-vər-ər, -'nyü-\ *noun*
man–for–man \,man-fər-'man\ *adjective* (circa 1949)
: MAN-TO-MAN 2
man Friday *noun* [*Friday,* servant in *Robinson Crusoe* (1719), novel by Daniel Defoe] (1887)
: an efficient and devoted aide or employee **:** a right-hand man
man·ful \'man-fəl\ *adjective* (14th century)
: having or showing courage and resolution
— **man·ful·ly** \-fə-lē\ *adverb*
— **man·ful·ness** *noun*

man·ga·bey \'maŋ-gə-(,)bē\ *noun, plural* **-beys** [French, from *Mangabey,* region of Madagascar] (1774)
: any of a genus (*Cercocebus*) of slender long-tailed African monkeys
mangan- *combining form* [German *Mangan,* from French *manganèse*]
: manganese ⟨*mangan*ate⟩
man·ga·nate \'maŋ-gə-,nāt\ *noun* (1839)
1 : a salt containing manganese in the anion MnO_4
2 : MANGANITE
man·ga·nese \'maŋ-gə-,nēz, -,nēs\ *noun* [French *manganèse,* from Italian *manganese* manganese dioxide] (1783)
: a grayish white usually hard and brittle metallic element that resembles iron but is not magnetic — see ELEMENT table
— **man·ga·ne·sian** \,maŋ-gə-'nē-zhən, -shən\ *adjective*
manganese dioxide *noun* (1882)
: a dark insoluble compound MnO_2 used especially as an oxidizing agent, as a depolarizer of dry cells, and in making glass and ceramics
man·gan·ic \man-'ga-nik, maŋ-\ *adjective* (circa 1828)
: of, relating to, or derived from manganese; *especially* **:** containing this element with a valence of three or six
man·ga·nite \'maŋ-gə-,nīt\ *noun* (1827)
1 : an ore of manganese MnO(OH) that is a hydroxide of manganese usually in brilliant gray crystals
2 : any of various unstable salts made by reaction of manganese dioxide with a base
man·ga·nous \-nəs\ *adjective* (1842)
: of, relating to, or derived from manganese; *especially* **:** containing this element with a valence of two
mange \'mānj\ *noun* [alteration of Middle English *manjewe,* from Middle French *mangene* itching, from *mangier* to eat] (1540)
: any of various persistent contagious skin diseases marked especially by eczematous inflammation and loss of hair, affecting domestic animals or sometimes humans, and caused by a minute parasitic mite — compare SARCOPTIC MANGE
man·gel \'maŋ-gəl\ *noun* [short for *mangelwurzel*] (1877)
: a large coarse yellow to reddish orange beet extensively grown as food for cattle
man·gel–wur·zel \-,wər-zəl\ *noun* [German, alteration of *Mangoldwurzel,* from *Mangold* beet + *Wurzel* root] (1767)
: MANGEL
man·ger \'mān-jər\ *noun* [Middle English *mangeour, manger,* from Middle French *maingeure,* from *mangier* to eat, from Latin *manducare* to chew, devour, from *manducus* glutton, from *mandere* to chew — more at MANDIBLE] (14th century)
: a trough or open box in a stable designed to hold feed or fodder for livestock
¹man·gle \'maŋ-gəl\ *transitive verb* **man·gled; man·gling** \-g(ə-)liŋ\ [Middle English, from Anglo-French *mangler*] (15th century)
1 : to injure with deep disfiguring wounds by cutting, tearing, or crushing ⟨people . . . *mangled* by sharks —V. G. Heiser⟩
2 : to spoil, injure, or make incoherent especially through ineptitude ⟨a story *mangled* beyond recognition⟩
synonym see MAIM
— **man·gler** \-g(ə-)lər\ *noun*
²mangle *noun* [Dutch *mangel,* from German, from Middle High German, diminutive of *mange* mangonel, mangle, from Latin *manganum*] (1774)
: a machine for ironing laundry by passing it between heated rollers
³mangle *transitive verb* **man·gled; man·gling** \-g(ə-)liŋ\ (1775)
: to press or smooth (as damp linen) with a mangle

— **man·gler** \-g(ə-)lər\ *noun*
man·go \'maŋ-(,)gō\ *noun, plural* **mangoes** *also* **mangos** [Portuguese *manga,* probably from Malayalam *māṅṅa*] (1582)
1 : a tropical fruit commonly with a firm yellowish red skin, hard central stone, and juicy aromatic pulp; *also* **:** an evergreen tree (*Mangifera indica*) of the cashew family that bears mangoes
2 : SWEET PEPPER
man·go·nel \'maŋ-gə-,nel\ *noun* [Middle English, from Old French, probably from Medieval Latin *manganellus,* diminutive of Late Latin *manganum* philter, mangonel, from Greek *manganon*] (13th century)
: a military engine formerly used to throw missiles
man·go·steen \'maŋ-gə-,stēn\ *noun* [modification of Malay *manggisutan*] (1598)
: a dark reddish brown fruit of southeastern Asia with a thick rind and juicy flesh having a flavor suggestive of both peach and pineapple; *also* **:** a tree (*Garcinia mangostana*) of the Saint-John's-wort family that bears mangosteens
man·grove \'man-,grōv, 'maŋ-\ *noun* [probably from Portuguese *mangue* mangrove (from Spanish *mangle,* probably from Taino) + English *grove*] (1613)
1 : any of a genus (*Rhizophora,* especially *R. mangle* of the family Rhizophoraceae) of tropical maritime trees or shrubs that send out many prop roots and form dense masses important in coastal land building
2 : any of numerous trees (as of the genera *Avicennia* of the vervain family or *Sonneratia* of the family Sonneratiaceae) with growth habits like those of the true mangroves
mangy \'mān-jē\ *adjective* **mang·i·er; -est** (1540)
1 : affected with or resulting from mange
2 : having many worn or bare spots **:** SEEDY, SHABBY
— **man·gi·ness** \-jē-nəs\ *noun*
man·han·dle \'man-,han-dᵊl\ *transitive verb* (circa 1865)
1 : to handle roughly
2 : to move or manage by human force ⟨*manhandled* the posts into place⟩
man·hat·tan \man-'ha-tᵊn, mən-\ *noun, often capitalized* [*Manhattan,* borough of New York City] (1890)

◇ WORD HISTORY
maneuver As peculiar as the connection seems, we owe both *maneuver* and *manure* to the same French source. The Old French verb *manovrer,* meaning "to work" or "to place with the hand," developed from *manu operare,* literally, "to work by hand," in the spoken Latin of Gaul. From an Anglo-French variant of this Old French verb such as *meinourer,* Middle English adopted *maynouren* or *manouren,* which had the senses "to take in hand, manage" and "to cultivate (land)." In the meantime, English derived on the basis of the latter sense the noun *manure,* meaning "material that fertilizes land." In the meantime, *manovrer* continued to go its own way in French. By the middle of the 18th century, its Modern French outcome *manœuvrer* (influenced in form by the noun *manœuvre*) had developed a new specific meaning "to perform a movement in military or naval tactics" and English reborrowed the word, quickly adapting it to several nonmilitary uses as well.

: a cocktail consisting of vermouth, whiskey, and sometimes a dash of bitters

man·hole \'man-ˌhōl\ *noun* (1793)
: a hole through which one may go especially to gain access to an underground or enclosed structure

man·hood \'man-ˌhu̇d\ *noun* (13th century)
1 : the condition of being a human being
2 : qualities associated with men
3 : the condition of being an adult male as distinguished from a child or female
4 : adult males **:** MEN

man–hour \'man-'au̇(-ə)r\ *noun* (1912)
: a unit of one hour's work by one person that is used especially as a basis for cost accounting and wages

man·hunt \'man-ˌhənt\ *noun* (1846)
: an organized and usually intensive hunt for a person and especially for one charged with a crime

ma·nia \'mā-nē-ə, -nyə\ *noun* [Middle English, from Late Latin, from Greek, from *mainesthai* to be mad; akin to Greek *menos* spirit — more at MIND] (14th century)
1 : excitement manifested by mental and physical hyperactivity, disorganization of behavior, and elevation of mood; *specifically* **:** the manic phase of manic-depressive psychosis
2 : excessive or unreasonable enthusiasm ⟨a *mania* for saving things⟩; *also* **:** the object of such enthusiasm

ma·ni·ac \'mā-nē-ˌak\ *noun* [Late Latin *maniacus* maniacal, from Greek *maniakos,* from *mania*] (circa 1763)
1 : MADMAN, LUNATIC
2 : a person characterized by an inordinate or ungovernable enthusiasm for something

ma·ni·a·cal \mə-'nī-ə-kəl\ *also* **ma·ni·ac** \'mā-nē-ˌak\ *adjective* (1604)
1 : affected with or suggestive of madness
2 : characterized by ungovernable excitement or frenzy **:** FRANTIC
— ma·ni·a·cal·ly \mə-'nī-ə-k(ə-)lē\ *adverb*

man·ic \'ma-nik\ *adjective* (1902)
: affected with, relating to, or resembling mania
— manic *noun*
— man·i·cal·ly \-ni-k(ə-)lē\ *adverb*

man·ic–de·pres·sive \ˌma-nik-di-'pre-siv\ *adjective* (1902)
: characterized either by mania or by psychotic depression or by alternating mania and depression
— manic–depressive *noun*

Man·i·chae·an *or* **Man·i·che·an** \ˌma-nə-'kē-ən\ *or* **Man·i·chee** \'man-ə-ˌkē\ *noun* [Late Latin *manichaeus,* from Late Greek *manichaios,* from *Manichaios* Manes (died about 276 A.D.) Persian founder of the sect] (1556)
1 : a believer in a syncretistic religious dualism originating in Persia in the 3d century A.D. and teaching the release of the spirit from matter through asceticism
2 : a believer in religious or philosophical dualism
— Manichaean *adjective*
— Man·i·chae·an·ism \ˌma-nə-'kē-ə-ˌni-zəm\ *noun*
— Man·i·chae·ism \'ma-nə-(ˌ)kē-ˌi-zəm\ *noun*

man·i·cot·ti \ˌma-nə-'kä-tē\ *noun, plural* **manicotti** [Italian, plural of *manicotto* muff, from *manica* sleeve, from Latin, from *manus* hand] (1948)
: tubular pasta shells that may be stuffed with ricotta or a meat mixture; *also* **:** a dish of stuffed manicotti usually with tomato sauce

¹man·i·cure \'ma-nə-ˌkyu̇r\ *noun* [French, from Latin *manus* hand + French *-icure* (as in *pédicure* pedicure) — more at MANUAL] (1880)
1 : MANICURIST
2 : a treatment for the care of the hands and fingernails

²manicure *transitive verb* **-cured; -cur·ing** (circa 1890)
1 : to do manicure work on; *especially* **:** to trim and polish the fingernails of
2 a : to trim closely and evenly ⟨*manicured* lawns⟩ **b :** GROOM 2 ⟨*manicured* flower beds⟩

man·i·cur·ist \-ˌkyu̇r-ist\ *noun* (1889)
: a person who gives manicures

¹man·i·fest \'ma-nə-ˌfest\ *adjective* [Middle English, from Middle French or Latin; Middle French *manifeste,* from Latin *manifestus* caught in the act, flagrant, obvious, perhaps from *manus* + *-festus* (akin to Latin in*festus* hostile)] (14th century)
1 : readily perceived by the senses and especially by the sight
2 : easily understood or recognized by the mind **:** OBVIOUS
synonym see EVIDENT
— man·i·fest·ly *adverb*

²manifest *transitive verb* (14th century)
: to make evident or certain by showing or displaying
synonym see SHOW
— man·i·fest·er *noun*

³manifest *noun* (1561)
1 : MANIFESTATION, INDICATION
2 : MANIFESTO
3 : a list of passengers or an invoice of cargo for a vehicle (as a ship or plane)

man·i·fes·tant \ˌma-nə-'fes-tənt\ *noun* (1880)
: one who makes or participates in a manifestation

man·i·fes·ta·tion \ˌma-nə-fə-'stā-shən, -ˌfe-'stā-\ *noun* (15th century)
1 a : the act, process, or an instance of manifesting **b :** something that manifests or is manifest **c :** one of the forms in which an individual is manifested **d :** an occult phenomenon; *specifically* **:** MATERIALIZATION
2 : a public demonstration of power and purpose

manifest destiny *noun, often M&D capitalized* (1845)
: a future event accepted as inevitable ⟨in the mid-19th century expansion to the Pacific was regarded as the *Manifest Destiny* of the United States⟩; *broadly* **:** an ostensibly benevolent or necessary policy of imperialistic expansion

¹man·i·fes·to \ˌma-nə-'fes-(ˌ)tō\ *noun, plural* **-tos** *or* **-toes** [Italian, denunciation, manifest, from *manifestare* to manifest, from Latin, from *manifestus*] (1647)
: a written statement declaring publically the intentions, motives, or views of its issuer

²manifesto *intransitive verb* (1748)
: to issue a manifesto

¹man·i·fold \'ma-nə-ˌfōld\ *adjective* [Middle English, from Old English *manigfeald,* from *manig* many + *-feald* -fold] (before 12th century)
1 a : marked by diversity or variety **b :** MANY
2 : comprehending or uniting various features **:** MULTIFARIOUS
3 : rightfully so-called for many reasons ⟨a *manifold* liar⟩
4 : consisting of or operating many of one kind combined ⟨a *manifold* bellpull⟩
— man·i·fold·ly \-ˌfōl(d)-lē\ *adverb*
— man·i·fold·ness \-ˌfōl(d)-nəs\ *noun*

²manifold *adverb* (before 12th century)
: many times **:** a great deal ⟨will increase your blessings *manifold*⟩

³manifold (before 12th century)
transitive verb
1 : to make manifold **:** MULTIPLY
2 : to make several or many copies of
intransitive verb
: to make several or many copies

⁴manifold *noun* (1855)
: something that is manifold: as **a :** a whole that unites or consists of many diverse elements ⟨the *manifold* of aspirations, passions, frustrations — Harry Slochower⟩ **b :** a pipe fitting with several lateral outlets for connecting one pipe with others; *also* **:** a fitting on an internal-combustion engine that directs a fuel and air mixture to or receives the exhaust gases from several cylinders **c :** SET 21 **d :** a topological space in which every point has a neighborhood that is homeomorphic to the interior of a sphere in Euclidean space of the same number of dimensions

man·i·kin *or* **man·ni·kin** \'ma-ni-kən\ *noun* [Dutch *mannekijn* little man, from Middle Dutch, diminutive of *man;* akin to Old English *man*] (circa 1536)
1 : MANNEQUIN
2 : a little man **:** DWARF, PYGMY

ma·nila *also* **ma·nil·la** \mə-'ni-lə\ *adjective* (1834)
1 *capitalized* **:** made from Manila hemp
2 : made of manila paper
— manila *noun*

Manila hemp *noun* [*Manila,* Philippine Islands] (circa 1847)
: ABACA

manila paper *noun, often M capitalized* (1873)
: a strong and durable paper of a brownish or buff color and smooth finish made originally from Manila hemp

ma·nille \mə-'nil\ *noun* [modification of Spanish *malilla*] (1674)
: the second highest trump in various card games (as ombre)

man in the street (1831)
: an average or ordinary person

man·i·oc \'ma-nē-ˌäk\ *noun* [French *manioc* & Spanish & Portuguese *mandioca,* from Tupi *manioca*] (1568)
: CASSAVA

man·i·ple \'ma-nə-pəl\ *noun* [Middle English, from Medieval Latin *manipulus,* from Latin, handful, from *manus* hand + *-pulus* (perhaps akin to Latin *plēre* to fill); from its having been originally held in the hand — more at MANUAL, FULL] (15th century)
1 : a long narrow strip of silk worn at mass over the left arm by clerics of or above the order of subdeacon
2 [Latin *manipulus,* from *manipulus* handful] **:** a subdivision of the Roman legion consisting of either 120 or 60 men

ma·nip·u·la·ble \mə-'ni-pyə-lə-bəl\ *adjective* (1881)
: capable of being manipulated
— ma·nip·u·la·bil·i·ty \-ˌni-pyə-lə-'bi-lə-tē\ *noun*

ma·nip·u·lar \mə-'ni-pyə-lər\ *adjective* (1623)
1 : of or relating to the ancient Roman maniple
2 : of, relating to, or performed by manipulation **:** MANIPULATIVE

ma·nip·u·late \mə-'ni-pyə-ˌlāt\ *transitive verb* **-lat·ed; -lat·ing** [back-formation from *manipulation,* from French, from *manipuler* to handle an apparatus in chemistry, ultimately from Latin *manipulus*] (1831)
1 : to treat or operate with the hands or by mechanical means especially in a skillful manner
2 a : to manage or utilize skillfully **b :** to control or play upon by artful, unfair, or insidious means especially to one's own advantage
3 : to change by artful or unfair means so as to serve one's purpose **:** DOCTOR
— ma·nip·u·lat·able \-ˌlā-tə-bəl\ *adjective*
— ma·nip·u·la·tion \-ˌni-pyə-'lā-shən\ *noun*
— ma·nip·u·la·tive \-'ni-pyə-ˌlā-tiv, -lə-\ *adjective*
— ma·nip·u·la·tive·ly *adverb*
— ma·nip·u·la·tive·ness *noun*
— ma·nip·u·la·tor \-ˌlā-tər\ *noun*
— ma·nip·u·la·to·ry \-lə-ˌtōr-ē, -ˌtȯr-\ *adjective*

man·i·tou *or* **man·i·tu** \'ma-nə-ˌtü\ *also* **man·i·to** \-ˌtō\ *noun* [Ojibwa *manito·*] (1671)
: a supernatural force that according to an Algonquian conception pervades the natural world

man jack \'man-'jak\ *noun* (1840)
: individual man ⟨every *man jack*⟩

man·kind *noun singular but singular or plural in construction* (13th century)
1 \'man-'kīnd, -ˌkīnd\ : the human race : the totality of human beings
2 \-ˌkīnd\ : men especially as distinguished from women

¹man·ly \'man-lē\ *adverb* (before 12th century)
: in a manly manner

²manly *adjective* **man·li·er; -est** (13th century)
1 : having qualities generally associated with a man : STRONG, VIRILE
2 : appropriate in character to a man ⟨*manly* sports⟩
— **man·li·ness** *noun*

man–made \'man-'mād, -ˌmād\ *adjective* (circa 1718)
: manufactured, created, or constructed by human beings; *specifically* : SYNTHETIC ⟨*man-made* fibers⟩

man·na \'ma-nə\ *noun* [Middle English, from Old English, from Late Latin, from Greek, from Hebrew *mān*] (before 12th century)
1 a : food miraculously supplied to the Israelites in their journey through the wilderness **b** : divinely supplied spiritual nourishment **c** : a usually sudden and unexpected source of gratification, pleasure, or gain
2 a : the sweetish dried exudate of a European ash (especially *Fraxinus ornus*) that contains mannitol and has been used as a laxative and demulcent **b** : a similar product excreted by a scale insect (*Trabutina mannipara*) feeding on the tamarisk

manna grass *noun* (1597)
: any of a genus (*Glyceria*) of chiefly North American perennial grasses of wetland or aquatic habitats

man·nan \'ma-ˌnan, -nən\ *noun* [International Scientific Vocabulary *mannose* + ³-*an*] (1895)
: any of several polysaccharides that are polymers of mannose and occur especially in plant cell walls

manned \'mand\ *adjective* (1617)
: carrying or performed by a human being ⟨*manned* space-flight⟩

man·ne·quin \'ma-ni-kən\ *noun* [French, from Dutch *mannekijn* little man — more at MANIKIN] (1730)
1 : an artist's, tailor's, or dressmaker's lay figure; *also* : a form representing the human figure used especially for displaying clothes
2 : one employed to model clothing

man·ner \'ma-nər\ *noun* [Middle English *manere,* from Old French *maniere* way of acting, from (assumed) Vulgar Latin *manuaria,* from Latin, feminine of *manuarius* of the hand, from *manus* hand — more at MANUAL] (12th century)
1 a : KIND, SORT ⟨what *manner* of man is he⟩ **b** : KINDS, SORTS ⟨all *manner* of problems⟩
2 a (1) : a characteristic or customary mode of acting : CUSTOM (2) : a mode of procedure or way of acting : FASHION (3) : method of artistic execution or mode of presentation : STYLE **b** *plural* : social conduct or rules of conduct as shown in the prevalent customs ⟨Victorian *manners*⟩ **c** : characteristic or distinctive bearing, air, or deportment ⟨his poised gracious *manner*⟩ **d** *plural* (1) : habitual conduct or deportment : BEHAVIOR ⟨mind your *manners*⟩ (2) : good manners **e** : a distinguished or stylish air
synonym see BEARING, METHOD
— **man·ner·less** \-ləs\ *adjective*

man·nered \'ma-nərd\ *adjective* (14th century)
1 : having manners of a specified kind ⟨well-*mannered*⟩
2 a : having or displaying a particular manner **b** : having an artificial or stilted character ⟨passages . . . so *mannered* as to be unintelligible —R. G. G. Price⟩

man·ner·ism \'ma-nə-ˌri-zəm\ *noun* (1803)
1 a : exaggerated or affected adherence to a particular style or manner : ARTIFICIALITY, PRECIOSITY ⟨refined almost to the point of *mannerism* —Winthrop Sargeant⟩ **b** *often capitalized* : an art style in late 16th century Europe characterized by spatial incongruity and excessive elongation of the human figures
2 : a characteristic and often unconscious mode or peculiarity of action, bearing, or treatment
synonym see POSE
— **man·ner·ist** \-rist\ *noun*
— **man·ner·is·tic** \ˌma-nə-'ris-tik\ *adjective*

man·ner·ly \'ma-nər-lē\ *adjective* (circa 1529)
: showing good manners
— **man·ner·li·ness** *noun*
— **mannerly** *adverb*

man·nish \'ma-nish\ *adjective* (14th century)
1 : resembling or suggesting a man rather than a woman
2 : generally associated with or characteristic of a man rather than a woman ⟨her *mannish* clothes⟩
— **man·nish·ly** *adverb*
— **man·nish·ness** *noun*

man·nite \'ma-ˌnīt\ *noun* [French, from *manna,* from Late Latin] (1830)
: MANNITOL

man·ni·tol \'ma-nə-ˌtȯl, -ˌtōl\ *noun* [International Scientific Vocabulary] (1879)
: a slightly sweet crystalline alcohol $C_6H_{14}O_6$ found in many plants and used especially as a diuretic and in testing kidney function

man·nose \'ma-ˌnōs, -ˌnōz\ *noun* [International Scientific Vocabulary *mannite* + ²-*ose*] (1888)
: an aldose $C_6H_{12}O_6$ whose dextrorotatory enantiomorph occurs especially as a structural unit of mannans from which it can be recovered by hydrolysis

ma·no \'mä-(ˌ)nō\ *noun, plural* **manos** [Spanish, literally, hand, from Latin *manus* — more at MANUAL] (circa 1892)
: a stone used as the upper millstone for grinding foods (as Indian corn) by hand in a metate

ma·noeu·vre *chiefly British variant of* MANEUVER

man of God (1748)
: CLERGYMAN

man of letters (1645)
1 : SCHOLAR
2 : AUTHOR

man of straw (1624)
: STRAW MAN

man of the house (circa 1904)
: the chief male in a household

man of the world (15th century)
: a practical or worldly-wise man of wide experience

man–of–war \ˌma-nə(v)-'wȯr\ *noun, plural* **men–of–war** \ˌmen-\ (15th century)
: a combatant warship of a recognized navy

ma·nom·e·ter \mə-'nä-mə-tər\ *noun* [French *manomètre,* from Greek *manos* sparse, loose, rare (akin to Armenian *manr* small) + French *-mètre*] (circa 1730)
1 : an instrument (as a pressure gauge) for measuring the pressure of gases and vapors
2 : SPHYGMOMANOMETER
— **man·o·met·ric** \ˌma-nə-'me-trik\ *adjective*
— **man·o·met·ri·cal·ly** \-tri-k(ə-)lē\ *adverb*
— **ma·nom·e·try** \mə-'nä-mə-trē\ *noun*

man on horseback (1860)
1 : a usually military figure whose ambitions and popularity mark him as a potential dictator
2 : DICTATOR

man·or \'ma-nər\ *noun* [Middle English *maner,* from Old French *manoir,* from *manoir* to sojourn, dwell, from Latin *manēre* — more at MANSION] (14th century)
1 a : the house or hall of an estate : MANSION **b** : a landed estate
2 a : a unit of English rural territorial organization; *especially* : such a unit in the Middle Ages consisting of an estate under a lord enjoying a variety of rights over land and tenants including the right to hold court **b** : a tract of land in North America occupied by tenants who pay a fixed rent in money or kind to the proprietor
— **ma·no·ri·al** \mə-'nōr-ē-əl, -'nȯr-\ *adjective*
— **ma·no·ri·al·ism** \-ə-ˌli-zəm\ *noun*

manor house *noun* (1575)
: the house of the lord of a manor

man–o'–war bird \ˌma-nə-'wȯr-\ *noun* (1707)
: FRIGATE BIRD

man·pack \'man-ˌpak\ *adjective* (1965)
: designed to be carried by one person

man power *noun* (1862)
1 : power available from or supplied by the physical effort of human beings
2 *usually* **man·pow·er** : the total supply of persons available and fitted for service

man·qué \mäⁿ-'kā\ *adjective* [French, from past participle of *manquer* to lack, fail, from Italian *mancare,* from *manco* lacking, left-handed, from Latin, having a crippled hand, probably from *manus*] (1778)
: short of or frustrated in the fulfillment of one's aspirations or talents — used postpositively ⟨a poet *manqué*⟩

man·rope \'man-ˌrōp\ *noun* (1769)
: a side rope (as to a ship's gangway or ladder) used as a handrail

man·sard \'man-ˌsärd, -sərd\ *noun* [French *mansarde,* from François *Mansart* (died 1666) French architect] (circa 1734)
: a roof having two slopes on all sides with the lower slope steeper than the upper one — see ROOF illustration
— **man·sard·ed** \-ˌsär-dəd, -sər-\ *adjective*

manse \'man(t)s\ *noun* [Middle English *manss,* from Medieval Latin *mansa, mansus, mansum,* from Latin *mansus* lodging, from *manēre*] (15th century)
1 *archaic* : the dwelling of a householder
2 : the residence of a clergyman; *especially* : the house of a Presbyterian clergyman
3 : a large imposing residence

man·ser·vant \'man-ˌsər-vənt\ *noun, plural* **men·ser·vants** \'men-ˌsər-vən(t)s\ (14th century)
: a male servant

-manship *noun suffix* [*-man* + *-ship* (as in *horsemanship*)]
1 : art or practice of a competitive nature ⟨brink*manship*⟩
2 : skilled engagement in a competitive activity ⟨grants*manship*⟩
word history see GAMESMANSHIP

man·sion \'man(t)-shən\ *noun* [Middle English, from Middle French, from Latin *mansion-, mansio,* from *manēre* to remain, dwell; akin to Greek *menein* to remain] (14th century)
1 a *obsolete* : the act of remaining or dwelling : STAY **b** *archaic* : DWELLING, ABODE
2 a (1) : the house of the lord of a manor (2) : a large imposing residence **b** : a separate apartment or lodging in a large structure
3 a : HOUSE 3b **b** : one of the 28 parts into which the moon's monthly course through the heavens is divided

man–size \'man-ˌsīz\ *or* **man–sized** \-ˌsīzd\ *adjective* (1913)
1 : suitable for or requiring a man ⟨a *man-size* job⟩
2 : larger than others of its kind ⟨constructed a *man-size* model⟩

man·slaugh·ter \'man-ˌslȯ-tər\ *noun* (15th century)

: the unlawful killing of a human being without express or implied malice

man·slay·er \-ˌslā-ər\ *noun* (14th century)
: one who commits homicide

man's man *noun* (1897)
: a man noted or admired for traditionally masculine interests and activities

man·sue·tude \ˈman(t)-swi-ˌtüd, man-ˈsü-ə-, -ˌtyüd\ *noun* [Middle English, from Latin *mansuetudo,* from *mansuescere* to tame, from *manus* hand + *suescere* to accustom; akin to Greek *ēthos* custom — more at MANUAL, SIB] (14th century)
: the quality or state of being gentle **:** MEEKNESS, TAMENESS

man·ta \ˈman-tə\ *noun* [Spanish, alteration of *manto* cloak, from Late Latin *mantus,* probably back-formation from Latin *mantellum* mantle] (1697)
1 : a square piece of cloth or blanket used in southwestern U.S. and Latin America usually as a cloak or shawl
2 [American Spanish, from Spanish; from its shape] **:** DEVILFISH 1

man–tai·lored \ˈman-ˌtā-lərd\ *adjective* (1922)
: made with the severe simplicity associated with men's coats and suits

manta ray *noun* (1936)
: DEVILFISH 1

man·teau \man-ˈtō, ˈman-ˌ\ *noun* [French, from Old French *mantel*] (1671)
: a loose cloak, coat, or robe

man·tel \ˈman-t^{ə}l\ *noun* [Middle English, from Middle French, from Old French, mantle] (15th century)
1 a : a beam, stone, or arch serving as a lintel to support the masonry above a fireplace **b :** the finish around a fireplace
2 : a shelf above a fireplace

man·tel·et \ˈmant-lət, ˈman-t^{ə}l-ət, ˌman-t^{ə}l-ˈet\ *noun* [Middle English, from Middle French *mantlet,* diminutive of *mantel*] (14th century)
1 : a very short cape or cloak
2 *or* **mant·let** \ˈmant-lət\ **:** a movable shelter formerly used by besiegers as a protection when attacking

man·tel·piece \ˈman-t^{ə}l-ˌpēs\ *noun* (1686)
1 : a mantel with its side elements
2 : MANTEL 2

man·tel·shelf \-ˌshelf\ *noun* (circa 1828)
: MANTEL 2

man·tic \ˈman-tik\ *adjective* [Greek *mantikos,* from *mantis*] (1850)
: of or relating to the faculty of divination **:** PROPHETIC

man·ti·core \ˈman-ti-ˌkōr, -ˌkȯr\ *noun* [Middle English, from Latin *mantichora,* from Greek *mantichōras*] (14th century)
: a legendary animal with the head of a man, the body of a lion, and the tail of a dragon or scorpion

man·tid \ˈman-təd\ *noun* [New Latin *Mantidae,* group name, from *Mantis,* genus name] (1895)
: MANTIS

man·til·la \man-ˈtē-yə, -ˈti-lə\ *noun* [Spanish, diminutive of *manta*] (1717)
1 : a light scarf worn over the head and shoulders especially by Spanish and Latin-American women
2 : a short light cape or cloak

man·tis \ˈman-təs\ *noun, plural* **man·tis·es** *or* **man·tes** \ˈman-ˌtēz\ [New Latin, from Greek, literally, diviner, prophet; akin to Greek *mainesthai* to be mad — more at MANIA] (1658)
: any of an order or suborder (Mantodea and especially family Mantidae) of large usually green insects that feed on other insects and clasp their prey in forelimbs held up as if in prayer

man·tis·sa \man-ˈti-sə\ *noun* [Latin *mantisa, mantissa* makeweight, from Etruscan] (circa 1847)
: the part of a logarithm to the right of the decimal point

mantis

¹man·tle \ˈman-t^{ə}l\ *noun* [Middle English *mantel,* from Old French, from Latin *mantellum*] (13th century)
1 a : a loose sleeveless garment worn over other clothes **:** CLOAK **b :** a mantle regarded as a symbol of preeminence or authority ⟨invested his people with the *mantle* of universal champions of justice —Denis Goulet⟩
2 a : something that covers, enfolds, or envelops **b** (1) **:** a fold or lobe or pair of lobes of the body wall of a mollusk or brachiopod that in shell-bearing forms lines the shell and bears shell-secreting glands — see CLAM illustration (2) **:** the soft external body wall that lines the test or shell of a tunicate or barnacle **c :** the outer wall and casing of a blast furnace above the hearth; *broadly* **:** an insulated support or casing in which something is heated
3 : the back, scapulars, and wings of a bird
4 : a lacy hood or sheath of some refractory material that gives light by incandescence when placed over a flame
5 a : REGOLITH **b :** the part of the interior of a terrestrial planet and especially the earth that lies beneath the lithosphere and above the central core
6 : MANTEL

²mantle *verb* **man·tled; man·tling** \ˈmant-liŋ, ˈman-t^{ə}l-iŋ\ (13th century)
transitive verb
: to cover with or as if with a mantle **:** CLOAK ⟨the encroaching jungle growth that *mantled* the building —Sanka Knox⟩
intransitive verb
1 : to become covered with a coating
2 : to spread over a surface
3 : BLUSH ⟨her rich face *mantling* with emotion —Benjamin Disraeli⟩

man–to–man \ˈman-tə-ˈman\ *adjective* (1902)
1 : characterized by frankness and honesty ⟨a *man-to-man* talk⟩
2 : of, relating to, or being a system of defense (as in football or basketball) in which each defensive player guards a specified opponent

Man·toux test \ˌman-ˈtü-, ˌmä^{n}-\ *noun* [Charles *Mantoux* (died 1947) French physician] (circa 1923)
: an intracutaneous test for hypersensitivity to tuberculin that indicates past or present infection with tubercle bacilli

man·tra \ˈmän-trə *also* ˈman- *or* ˈmən-\ *noun* [Sanskrit, sacred counsel, formula, from *manyate* he thinks; akin to Latin *mens* mind — more at MIND] (1808)
: a mystical formula of invocation or incantation (as in Hinduism); *also* **:** WATCHWORD 2
— man·tric \-trik\ *adjective*

man·trap \ˈman-ˌtrap\ *noun* (1788)
: a trap for catching humans **:** SNARE

man·tua \ˈman(t)-sh(ə-)wə, ˈman-tə-wə\ *noun* [modification of French *manteau* mantle] (1678)
: a usually loose-fitting gown worn especially in the 17th and 18th centuries

Manu \ˈma-(ˌ)nü\ *noun* [Sanskrit]
: the progenitor of the human race and giver of the religious laws of Manu according to Hindu mythology

¹man·u·al \ˈman-yə-wəl, -yəl\ *adjective* [Middle English *manuel,* from Middle French, from Latin *manualis,* from *manus* hand; akin to Old English *mund* hand and perhaps to Greek *marē* hand] (15th century)
1 a : of, relating to, or involving the hands ⟨*manual* dexterity⟩ **b :** worked or done by hand and not by machine ⟨a *manual* transmission⟩ ⟨*manual* computation⟩ ⟨*manual* indexing⟩
2 : requiring or using physical skill and energy ⟨*manual* labor⟩ ⟨*manual* workers⟩
— man·u·al·ly *adverb*

²manual *noun* (15th century)
1 : a book that is conveniently handled; *especially* **:** HANDBOOK
2 : the prescribed movements in the handling of a weapon or other military item during a drill or ceremony ⟨the *manual* of arms⟩
3 a : a keyboard for the hands; *specifically* **:** one of the several keyboards of an organ or harpsichord that controls a separate division of the instrument **b :** a device or apparatus intended for manual operation

manual alphabet *noun* (circa 1864)
: an alphabet especially for the deaf in which the letters are represented by finger positions

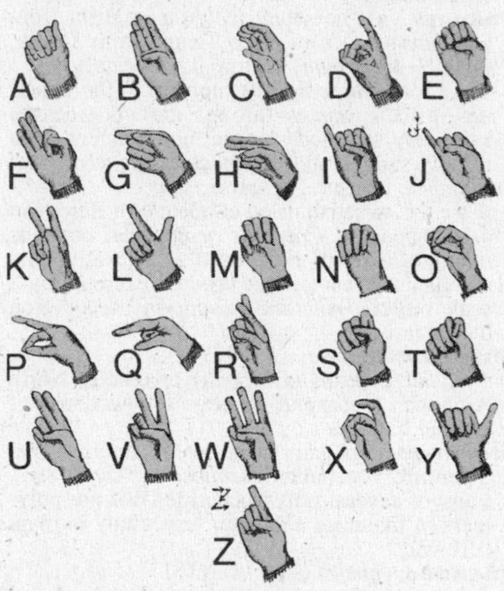

manual alphabet

manual training *noun* (1880)
: a course of training to develop skill in using the hands and to teach practical arts (as woodworking and metalworking)

ma·nu·bri·um \mə-ˈnü-brē-əm, -ˈnyü-\ *noun, plural* **-bria** \-brē-ə\ *also* **-bri·ums** [New Latin, from Latin, handle, from *manus*] (circa 1848)
: an anatomical process or part shaped like a handle: as **a :** the cephalic segment of the sternum of humans and many other mammals **b :** the process that bears the mouth of a hydrozoan **:** HYPOSTOME

man·u·fac·to·ry \ˌman-yə-ˈfak-t(ə-)rē, ˌma-nə-\ *noun* (1647)
: FACTORY 2a

¹man·u·fac·ture \ˌman-yə-ˈfak-chər, ˌma-nə-\ *noun* [Middle French, from Medieval Latin *manufactura,* from Latin *manu factus,* literally, made by hand] (1567)
1 : something made from raw materials by hand or by machinery
2 a : the process of making wares by hand or by machinery especially when carried on systematically with division of labor **b :** a productive industry using mechanical power and machinery
3 : the act or process of producing something

²manufacture *verb* **-tured; -tur·ing** \-ˈfak-chə-riŋ, -ˈfak-shriŋ\ (1683)
transitive verb
1 : to make into a product suitable for use
2 a : to make from raw materials by hand or by machinery **b :** to produce according to an organized plan and with division of labor
3 : INVENT, FABRICATE

4 : to produce as if by manufacturing **:** CREATE ⟨writers who *manufacture* stories for television⟩
intransitive verb
: to engage in manufacture
— manufacturing *noun*

man·u·fac·tur·er \-ˈfak-chər-ər, -ˈfak-shrər\ *noun* (1719)
: one that manufactures; *especially* **:** an employer of workers in manufacturing

man·u·mis·sion \ˌman-yə-ˈmi-shən\ *noun* [Middle English, from Middle French, from Latin *manumission-, manumissio,* from *manumittere*] (15th century)
: the act or process of manumitting; *especially* **:** formal emancipation from slavery

man·u·mit \ˌman-yə-ˈmit\ *transitive verb* **-mit·ted; -mit·ting** [Middle English *manumitten,* from Middle French *manumitter,* from Latin *manumittere,* from *manus* hand + *mittere* to let go, send] (15th century)
: to release from slavery
synonym see FREE

[1]ma·nure \mə-ˈnu̇r, -ˈnyu̇r\ *transitive verb* **ma·nured; ma·nur·ing** [Middle English *manouren,* from Middle French *manouvrer,* literally, to do work by hand, from Latin *manu operare*] (15th century)
1 *obsolete* **:** CULTIVATE
2 : to enrich (land) by the application of manure
word history see MANEUVER
— ma·nur·er *noun*

[2]manure *noun* (1549)
: material that fertilizes land; *especially* **:** refuse of stables and barnyards consisting of livestock excreta with or without litter
— ma·nu·ri·al \-ˈn(y)u̇r-ē-əl\ *adjective*

ma·nus \ˈmā-nəs, ˈmä-\ *noun, plural* **ma·nus** \-nəs, -ˌnüs\ [New Latin, from Latin, hand] (1826)
: the distal segment of the vertebrate forelimb from carpus to terminus

[1]man·u·script \ˈman-yə-ˌskript\ *adjective* [Latin *manu scriptus*] (1597)
: written by hand or typed ⟨*manuscript* letters⟩

[2]manuscript *noun* (1600)
1 : a written or typewritten composition or document as distinguished from a printed copy; *also* **:** a document submitted for publication
2 : writing as opposed to print

[1]Manx \ˈmaŋ(k)s\ *adjective* [alteration of *Maniske,* from (assumed) Old Norse *manskr,* from *Mana* Isle of Man] (1630)
: of, relating to, or characteristic of the Isle of Man, its people, or the Manx language

[2]Manx *noun* (1672)
1 : the Celtic language of the Manx people almost completely displaced by English
2 *plural in construction* **:** the people of the Isle of Man
3 : MANX CAT

Manx cat *noun* (1859)
: any of a breed of shorthaired tailless domestic cats

[1]many \ˈme-nē\ *adjective* **more** \ˈmōr, ˈmȯr\; **most** \ˈmōst\ [Middle English, from Old English *manig;* akin to Old High German *manag* many, Old Church Slavonic *mŭnogŭ* much] (before 12th century)
1 : consisting of or amounting to a large but indefinite number ⟨worked for *many* years⟩
2 : being one of a large but indefinite number ⟨*many* a man⟩ ⟨*many* another student⟩
— as many : the same in number ⟨saw three plays in *as many* days⟩

[2]many *pronoun, plural in construction* (before 12th century)
: a large number of persons or things ⟨*many* are called⟩

[3]many *noun, plural in construction* (12th century)
1 : a large but indefinite number ⟨a good *many* of them⟩
2 : the great majority of people ⟨the *many*⟩

man–year \ˈman-ˈyir\ *noun* (1916)
: a unit of the work done by one person in a year composed of a standard number of working days

many·fold \ˌme-nē-ˈfōld\ *adverb* (14th century)
: by many times ⟨aid to research has increased *manyfold*⟩

many–sid·ed \ˌme-nē-ˈsī-dəd\ *adjective* (1570)
1 : having many sides or aspects
2 : having many interests or aptitudes
— many–sid·ed·ness *noun*

many–val·ued \ˌme-nē-ˈval-(ˌ)yüd, -yəd\ *adjective* (1934)
1 : possessing more than the customary two truth-values of truth and falsehood
2 : MULTIPLE-VALUED

Man·za·nil·la \ˌman-zə-ˈnē-yə, -ˈni-lə\ *noun* [Spanish, diminutive of *manzana* apple] (1843)
: a pale very dry Spanish sherry

man·za·ni·ta \ˌman-zə-ˈnē-tə\ *noun* [American Spanish, diminutive of Spanish *manzana* apple] (1846)
: any of various western North American evergreen shrubs (genus *Arctostaphylos*) of the heath family with alternate leaves

Mao·ism \ˈmau̇-ˌi-zəm\ *noun* (1950)
: the theory and practice of Marxism-Leninism developed in China chiefly by Mao Tse-tung
— Mao·ist \ˈmau̇-ist\ *noun or adjective*

Mao·ri \ˈmau̇(ə)r-ē\ *noun, plural* **Maori** *or* **Maoris** (1843)
1 : a member of a Polynesian people native to New Zealand
2 : the Austronesian language of the Maori

mao–tai \ˈmau̇-ˈtī, -ˈdī\ *noun* [*Maotai,* town in China] (1943)
: a strong Chinese liquor made from sorghum

[1]map \ˈmap\ *noun* [Medieval Latin *mappa,* from Latin, napkin, towel] (1527)
1 a : a representation usually on a flat surface of the whole or a part of an area **b :** a representation of the celestial sphere or a part of it
2 : something that represents with a clarity suggestive of a map
3 : the arrangement of genes on a chromosome — called also *genetic map*
4 : FUNCTION 5a
— map·like \-ˌlīk\ *adjective*
— on the map : in a position of prominence or fame ⟨had put the fledgling university *on the map* —Lon Tinkle⟩

[2]map *verb* **mapped; map·ping** (1586)
transitive verb
1 a : to make a map of ⟨*map* the surface of the moon⟩ **b :** to delineate as if on a map ⟨sorrow was *mapped* on her face⟩ **c :** to make a survey of for or as if for the purpose of making a map **d :** to assign (as a set or element) in a mathematical correspondence ⟨*map* a set onto itself⟩ ⟨*map* picture elements to video memory⟩
2 : to plan in detail — often used with *out* ⟨*map* out a program⟩
3 : to locate (a gene) on a chromosome
intransitive verb
of a gene **:** to be located
— map·pa·ble \ˈma-pə-bəl\ *adjective*
— map·per *noun*

ma·ple \ˈmā-pəl\ *noun* [Middle English, from Old English *mapul-;* akin to Old Norse *mǫpurr* maple] (14th century)
: any of a genus (*Acer* of the family Aceraceae, the maple family) of chiefly deciduous trees or shrubs with opposite leaves and a fruit of two united samaras; *also* **:** the hard light-colored close-grained wood of a maple used especially for flooring and furniture

maple sugar *noun* (1720)
: sugar made by boiling maple syrup

maple syrup *noun* (1849)
: syrup made by concentrating the sap of maple trees and especially the sugar maple

map·mak·er \ˈmap-ˌmā-kər\ *noun* (1775)
: CARTOGRAPHER
— map·mak·ing \-kiŋ\ *noun*

map·ping \ˈma-piŋ\ *noun* (circa 1775)
1 : the act or process of making a map
2 : FUNCTION 5a ⟨a one-to-one continuous *mapping*⟩

ma·quette \ma-ˈket\ *noun* [French, from Italian *macchietta* sketch, diminutive of *macchia,* ultimately from Latin *macula* spot] (1903)
: a usually small preliminary model (as of a sculpture or a building)

ma·qui·la·do·ra \mə-ˌkē-lə-ˈdōr-ə, -ˈt͟hōr-\ *noun* [Mexican Spanish (*planta*) *maquiladora,* from *maquilar* to process ore for a fee, from *maquila* processing fee, multure, from Spanish, multure, from Arabic dialect *makīla* measure of grain] (1976)
: a foreign-owned factory in Mexico at which imported parts are assembled by lower-paid workers into products for export

ma·quil·lage \ˌma-kē-ˈyäzh\ *noun* [French] (1892)
: MAKEUP 3

ma·quis \ma-ˈkē, mä-\ *noun, plural* **ma·quis** \-ˈkē(z)\ [French, from Italian *macchie,* plural of *macchia* thicket, sketch, spot] (1858)
1 : thick scrubby underbrush of Mediterranean shores and especially of the island of Corsica; *also* **:** an area of such underbrush
2 *often capitalized* **a :** a guerrilla fighter in the French underground during World War II **b :** a band of maquis

[1]mar \ˈmär\ *transitive verb* **marred; mar·ring** [Middle English *marren,* from Old English *mierran* to obstruct, waste; akin to Old High German *merren* to obstruct] (before 12th century)
1 : to detract from the perfection or wholeness of **:** SPOIL
2 *archaic* **a :** to inflict serious bodily harm on **b :** DESTROY
synonym see INJURE

[2]mar *noun* (1551)
: something that mars **:** BLEMISH

mar·a·bou *also* **mar·a·bout** \ˈmar-ə-ˌbü\ *noun* [French *marabout,* literally, marabout] (1823)
1 a : a soft feathery fluffy material prepared from turkey feathers or the coverts of marabous and used especially for trimming women's hats or clothes **b** *marabou* **:** a large dark gray African stork (*Leptoptilos crumeniferus*) that has a distensible pouch of pink skin at the front of the neck and feeds especially on carrion — called also *marabou stork*

marabou 1b

2 a : a thrown silk usually dyed in the gum **b :** a fabric made of this silk

mar·a·bout \ˈmar-ə-bü\ *noun, often capitalized* [French, from Portuguese *marabuto,* from Arabic *murābiṭ*] (1621)
: a dervish in Muslim Africa believed to have supernatural power

ma·ra·ca \mə-ˈrä-kə, -ˈra-\ *noun* [Portuguese *maracá,* from Tupi] (1824)
: a rattle usually made from a gourd that is used as a percussion instrument

mar·ag·ing steel \ˈmär-ˌā-jiŋ-\ *noun* [*martensite* + *aging*] (1962)
: a strong tough low-carbon martensitic steel which contains up to 25 percent nickel and in which hardening precipitates are formed by aging

mar·a·schi·no \ˌmar-ə-ˈskē-(ˌ)nō, -ˈshē-\ *noun, plural* **-nos** *often capitalized* [Italian, from *marasca* bitter wild cherry, alteration of *amarasca,* from *amaro* bitter — more at AMARETTO] (circa 1793)
1 : a sweet liqueur distilled from the fermented juice of a bitter wild cherry
2 : a usually large cherry preserved in true or imitation maraschino

ma·ras·mus \mə-ˈraz-məs\ *noun* [Late Latin, from Greek *marasmos,* from *marainein* to waste away] (1656)
: a condition of chronic undernourishment occurring especially in children and usually caused by a diet deficient in calories and proteins
— ma·ras·mic \-ˈraz-mik\ *adjective*

Ma·ra·tha \mə-ˈrä-tə\ *noun* [Marathi *Marāṭhā* & Hindi *Marhaṭṭā,* from Sanskrit *Mahārāṣṭra* Maharashtra] (1748)
: a member of a people of the south central part of the subcontinent of India

Ma·ra·thi \mə-ˈrä-tē\ *noun* [Marathi *marāṭhī*] (1698)
: the chief Indo-Aryan language of the state of Maharashtra in India

mar·a·thon \ˈmar-ə-ˌthän\ *noun, often attributive* [*Marathon,* Greece, site of a victory of Greeks over Persians in 490 B.C., the news of which was carried to Athens by a long-distance runner] (1896)
1 : a long-distance race: **a :** a footrace run on an open course usually of 26 miles 385 yards (42.2 kilometers) **b :** a race other than a footrace marked especially by great length
2 a : an endurance contest **b :** something (as an event, activity, or session) characterized by great length or concentrated effort ◆

mar·a·thon·er \-ˌthä-nər\ *noun* (1923)
: one (as a runner) who takes part in a marathon
— mar·a·thon·ing \-niŋ\ *noun*

ma·raud \mə-ˈrȯd\ *verb* [French *marauder*] (1711)
intransitive verb
: to roam about and raid in search of plunder
transitive verb
: RAID, PILLAGE
— ma·raud·er *noun*

¹mar·ble \ˈmär-bəl\ *noun* [Middle English, from Old French *marbre,* from Latin *marmor,* from Greek *marmaros*] (12th century)
1 a : limestone that is more or less crystallized by metamorphism, that ranges from granular to compact in texture, that is capable of taking a high polish, and that is used especially in architecture and sculpture **b :** something (as a piece of sculpture) composed of or made from marble **c :** something suggesting marble (as in hardness, coldness, or smoothness) ⟨a heart of *marble*⟩
2 a : a little ball made of a hard substance (as glass) and used in various games **b** *plural but singular in construction* **:** any of several games played with these little balls
3 : MARBLING
4 *plural* **:** elements of common sense; *especially* **:** SANITY ⟨persons who are born without all their *marbles* —Arthur Miller⟩
— marble *adjective*

²marble *transitive verb* **mar·bled; mar·bling** \-b(ə-)liŋ\ (1683)
: to give a veined or mottled appearance to ⟨*marble* the edges of a book⟩

marble cake *noun* (1871)
: a cake made with light and dark batter so as to have a mottled appearance

mar·bled \ˈmär-bəld\ *adjective* (1599)
1 [¹*marble*] **a :** made of or covered with marble or marbling **b :** marked by an extensive use of marble as an architectural or decorative feature ⟨ancient *marbled* cities⟩
2 [²*marble*] **:** marked by an intermixture of fat and lean ⟨well-*marbled* beef⟩

mar·ble·ise *British variant of* MARBLEIZE

mar·ble·ize \ˈmär-bə-ˌlīz\ *transitive verb* **-ized; -iz·ing** (circa 1859)
: MARBLE

marbling *noun* (circa 1752)
1 : the action or process of making like marble especially in coloration
2 : coloration or markings resembling or suggestive of marble
3 : an intermixture of fat and lean especially in a cut of meat when evenly distributed

mar·bly \-b(ə-)lē\ *adjective* (15th century)
: resembling or suggestive of marble

marc \ˈmärk\ *noun* [French, from Middle French, from *marchier* to trample, march] (1601)
1 : the residue remaining after a fruit has been pressed; *broadly* **:** the organic residue from an extraction process ⟨the protein-rich cottonseed *marc*⟩
2 : brandy made from the residue of wine grapes after pressing

mar·ca·site \ˈmär-kə-ˌsīt, -ˌzīt; ˌmär-kə-ˈzēt\ *noun* [Middle English *marchasite,* from Medieval Latin *marcasita*] (15th century)
1 a : crystallized iron pyrites **b :** a mineral of the same composition and appearance as iron pyrites but of different crystalline organization and lower specific gravity
2 : a piece of marcasite used in jewelry

mar·ca·to \mär-ˈkä-(ˌ)tō\ *adverb or adjective* [Italian, past participle of *marcare* to mark, accent, of Germanic origin; akin to Old High German *marcōn* to mark] (circa 1840)
: with strong accentuation — used as a direction in music

¹mar·cel \mär-ˈsel\ *noun* [*Marcel* Grateau (died 1936) French hairdresser] (1895)
: a deep soft wave made in the hair by the use of a heated curling iron

²marcel *verb* **mar·celled; mar·cel·ling** (1906)
transitive verb
: to make a marcel in
intransitive verb
: to make a marcel

¹march \ˈmärch\ *noun* [Middle English *marche,* from Old French, of Germanic origin; akin to Old High German *marha* boundary — more at MARK] (14th century)
: a border region **:** FRONTIER; *especially* **:** a district originally set up to defend a boundary — usually used in plural

²march *intransitive verb* (14th century)
: to have common borders or frontiers ⟨a region that *marches* with Canada in the north and the Pacific in the west⟩

³march \ˈmärch, *imperatively often* ˈhärch *in the military*\ *verb* [Middle English, from Middle French *marchier* to trample, march, from Old French, to trample, probably of Germanic origin; akin to Old High German *marcōn* to mark] (15th century)
intransitive verb
1 : to move along steadily usually with a rhythmic stride and in step with others
2 a : to move in a direct purposeful manner **:** PROCEED **b :** to make steady progress **:** ADVANCE ⟨time *marches* on⟩
3 : to stand in orderly array suggestive of marching
transitive verb
1 : to cause to march ⟨*marched* the children off to bed⟩
2 : to cover by marching **:** TRAVERSE ⟨*marched* 10 miles⟩

⁴march \ˈmärch\ *noun* (circa 1572)
1 : a musical composition that is usually in duple or quadruple time with a strongly accentuated beat and that is designed or suitable to accompany marching
2 a (1) **:** the action of marching (2) **:** the distance covered within a specific period of time by marching (3) **:** a regular measured stride or rhythmic step used in marching **b :** forward movement **:** PROGRESS ⟨the *march* of a movie toward the climax⟩
3 : an organized procession of demonstrators who are supporting or protesting something
— march·like \-ˌlīk\ *adjective*
— on the march : moving steadily **:** ADVANCING

March \ˈmärch\ *noun* [Middle English, from Old French, from Latin *martius,* from *martius* of Mars, from *Mart-, Mars*] (13th century)
: the 3d month of the Gregorian calendar

mär·chen \ˈmer-k̲ən\ *noun, plural* **märchen** [German] (1871)
: TALE; *especially* **:** FOLKTALE

¹march·er \ˈmär-chər\ *noun* (14th century)
: one who inhabits a border region

²marcher *noun* (1589)
: one that marches; *especially* **:** one that marches for a specific cause ⟨a peace *marcher*⟩

mar·che·sa \mär-ˈkā-zə\ *noun, plural* **-se** \-(ˌ)zā\ [Italian, feminine of *marchese*] (1797)
: an Italian woman holding the rank of a marchese **:** MARCHIONESS

mar·che·se \-(ˌ)zā\ *noun, plural* **-si** \-(ˌ)zē\ [Italian, from Medieval Latin *marcensis,* from *marca* border region, of Germanic origin; akin to Old High German *marha*] (1517)
: an Italian nobleman next in rank above a count **:** MARQUIS

marching orders *noun plural* (1780)
: authoritative orders or instructions especially to set out on or as if on a march

mar·chio·ness \ˈmär-sh(ə-)nəs\ *noun* [Medieval Latin *marchionissa,* from *marchion-, marchio* marquess, from *marca*] (circa 1599)
1 : the wife or widow of a marquess
2 : a woman who holds the rank of marquess in her own right

march·pane \ˈmärch-ˌpān\ *noun* [Middle French *marcepain,* from Italian *marzapane*] (circa 1556)
: MARZIPAN

march–past \ˈmärch-ˌpast\ *noun* (circa 1876)
: a filing by **:** PROCESSION

Mar·cion·ism \ˈmär-shə-ˌni-zəm, -sē-ə-, -shē-ə-\ *noun* [*Marcion* 2d century A.D. Christian Gnostic] (1882)
: the doctrinal system of a sect of the 2d and 3d centuries A.D. accepting some parts of the New Testament but denying Christ's corporality and humanity and condemning the Creator God of the Old Testament
— Mar·cion·ite \-ˌnīt\ *noun*

Mar·co·ni \mär-ˈkō-nē\ *adjective* [probably from the resemblance of the complex arrangement of stays and struts to that used to support

◇ WORD HISTORY

marathon Marathon is the name of a plain in Greece located about 25 miles (40 kilometers) from the city of Athens. It is the site of an almost miraculous victory by the Athenians over a much stronger force of invading Persians that occurred in 490 B.C. Before the battle the Athenian Pheidippides had run from Athens to Sparta (a round-trip distance of 150 miles) in an unsuccessful attempt to enlist the aid of the Spartans. According to a story now regarded as apocryphal, after the battle Pheidippides also ran the 25 miles from Marathon to Athens to announce the stunning victory. When the Olympic Games were revived at Athens in 1896, the marathon footrace was intended as a re-creation of that legendary 25 mile run. The founding of the annual Boston Marathon the following year greatly helped to establish the marathon as a major sporting event. The marathon's eccentric distance of 26 miles, 385 yards, standard since 1924, stems from a decision made during the 1908 Olympics held in London. In order for the race to finish at the royal box in the London stadium, an extra 385 yards was added to the 26 mile distance then in effect.

the antennae used in wireless telegraphy, invented by Guglielmo *Marconi*] (1912)
: of, having, or being a Bermuda rig

Mar·di Gras \'mär-dē-ˌgrä, *in New Orleans commonly* -ˌgrȯ\ *noun* [French, literally, fat Tuesday] (1699)
1 a : Shrove Tuesday often observed (as in New Orleans) with parades and festivities **b :** a carnival period climaxing on Shrove Tuesday
2 : a festive occasion resembling a pre-Lenten Mardi Gras

[1]mare *noun* [Middle English, from Old English; akin to Old High German *mara* incubus, Serbo-Croatian *mora* nightmare] (before 12th century)
obsolete **:** an evil preternatural being causing nightmares

[2]mare \'mar, 'mer\ *noun* [Middle English, from Old English *mere;* akin to Old High German *merha* mare, Old English *mearh* horse, Welsh *march*] (before 12th century)
: a female horse or other equine animal especially when fully mature or of breeding age

[3]ma·re \'mär-(ˌ)ā\ *noun, plural* **ma·ria** \'mär-ē-ə\ *also* **ma·res** \'mär-(ˌ)āz\ [New Latin, from Latin, sea — more at MARINE] (1860)
: any of several mostly flat dark areas of considerable extent on the surface of the moon or Mars

ma·re clau·sum \'mär-(ˌ)ā-'klau̇-səm, -'klȯ-\ *noun* [New Latin, literally, closed sea] (1652)
: a navigable body of water (as a sea) that is under the jurisdiction of one nation and is closed to other nations

Mar·ek's disease \'mar-iks-, 'mer-\ *noun* [József *Marek* (died 1952) Hungarian veterinarian] (1947)
: a highly contagious viral disease of poultry that is characterized especially by proliferation of lymphoid cells and is caused by a herpesvirus

ma·re li·be·rum \'mär-(ˌ)ā-'lē-bə-ˌru̇m\ *noun* [New Latin, literally, free sea] (1652)
1 : a navigable body of water (as a sea) that is open to all nations
2 : FREEDOM OF THE SEAS

ma·ren·go \mə-'reŋ-(ˌ)gō\ *adjective, often capitalized* [French, from *Marengo,* village in northwest Italy] (circa 1924)
: of, consisting of, or served with a sauce of mushrooms, tomatoes, olives, oil, and wine ⟨veal *marengo*⟩

ma·re no·strum \'mär-(ˌ)ā-'nōs-trəm\ *noun* [New Latin, literally, our sea] (1941)
: a navigable body of water (as a sea) that belongs to a single nation or is mutually shared by two or more nations

mare's nest *noun, plural* **mare's nests** *or* **mares' nests** (1576)
1 : a false discovery, illusion, or deliberate hoax
2 : a place, condition, or situation of great disorder or confusion ⟨a *mare's nest* of spurious ambiguities to bewilder the simpleminded —J. H. Sledd⟩

mare's tail *noun, plural* **mare's tails** *or* **mares' tails** (circa 1762)
1 a : a common aquatic plant (*Hippuris vulgaris* of the family Hippuridaceae) with elongated shoots having dense whorls of narrow finely tapered leaves **b :** HORSEWEED 1
2 : a cirrus cloud that has a long slender flowing appearance

Mar·fan's syndrome \'mär-ˌfanz-\ *noun* [Antonin Bernard Jean *Marfan* (died 1942) French pediatrician] (circa 1923)
: a hereditary disorder of connective tissue that is characterized by abnormal elongation of the bones and often by ocular and circulatory defects — called also *Mar·fan syndrome* \'mär-ˌfan-\

mar·ga·rine \'mär-jə-rən, -ˌrēn; 'märj-rən\ *noun* [French, ultimately from Greek *margaron* pearl, probably back-formation from *margaritēs*] (1873)
: a food product made usually from vegetable oils churned with ripened skim milk to a smooth emulsion and used like butter

mar·ga·ri·ta \ˌmär-gə-'rē-tə\ *noun* [from the Spanish feminine name *Margarita*] (1963)
: a cocktail consisting of tequila, lime or lemon juice, and an orange-flavored liqueur

mar·ga·rite \'mär-gə-ˌrīt\ *noun* [Middle English, from Middle French, from Latin *margarita,* from Greek *margaritēs*] (13th century)
archaic **:** PEARL

mar·gay \'mär-ˌgā, mär-'\ *noun* [French, from Tupi *maracaja*] (1781)
: a small American spotted cat (*Felis wiedii*) resembling the ocelot and ranging from southernmost Texas to Argentina

margay

[1]marge \'märj\ *noun* [Middle French, from Latin *margo*] (1551)
archaic **:** MARGIN

[2]marge *noun* (1922)
British **:** MARGARINE

mar·gent \'mär-jənt\ *noun* (15th century)
archaic **:** MARGIN

[1]mar·gin \'mär-jən\ *noun* [Middle English, from Latin *margin-, margo* border — more at MARK] (14th century)
1 : the part of a page or sheet outside the main body of printed or written matter
2 : the outside limit and adjoining surface of something **:** EDGE ⟨at the *margin* of the woods⟩ ⟨continental *margin*⟩
3 a : a spare amount or measure or degree allowed or given for contingencies or special situations ⟨left no *margin* for error⟩ **b** (1) **:** a bare minimum below which or an extreme limit beyond which something becomes impossible or is no longer desirable ⟨on the *margin* of good taste⟩ (2) **:** the limit below which economic activity cannot be continued under normal conditions
4 a : the difference which exists between net sales and the cost of merchandise sold and from which expenses are usually met or profit derived **b :** the excess market value of collateral over the face of a loan **c** (1) **:** cash or collateral that is deposited by a client with a commodity or securities broker to protect the broker from loss on a contract (2) **:** the client's equity in securities bought with the aid of credit obtained specifically (as from a broker) for that purpose **d :** a range about a specified figure within which a purchase is to be made
5 : measure or degree of difference ⟨the bill passed by a one-vote *margin*⟩
— mar·gined \-jənd\ *adjective*

[2]margin *transitive verb* (1715)
1 a : to provide with an edging or border **b :** to form a margin to **:** BORDER
2 a : to add margin to ⟨*margin* up an account⟩ **b** (1) **:** to use as margin ⟨*margin* bonds to buy stock⟩ (2) **:** to provide margin for ⟨*margin* a transaction⟩ **c :** to buy (securities) on margin

mar·gin·al \'märj-nəl, 'mär-jə-nᵊl\ *adjective* [Medieval Latin *marginalis,* from Latin *margin-, margo*] (1573)
1 : written or printed in the margin of a page or sheet ⟨*marginal* notes⟩
2 a : of, relating to, or situated at a margin or border **b :** not of central importance ⟨regards violence as a *marginal* rather than a central problem⟩ **c** (1) **:** occupying the borderland of a relatively stable territorial or cultural area ⟨*marginal* tribes⟩ (2) **:** characterized by the incorporation of habits and values from two divergent cultures and by incomplete assimilation in either ⟨the *marginal* cultural habits of new immigrant groups⟩ (3) **:** excluded from or existing outside the mainstream of society, a group, or a school of thought ⟨*marginal* voters⟩
3 : located at the fringe of consciousness ⟨*marginal* sensations⟩
4 a : close to the lower limit of qualification, acceptability, or function **:** barely exceeding the minimum requirements ⟨a semiliterate person of *marginal* ability⟩ **b** (1) **:** having a character or capacity fitted to yield a supply of goods which when marketed at existing price levels will barely cover the cost of production ⟨*marginal* land⟩ (2) **:** of, relating to, or derived from goods produced and marketed with such result ⟨*marginal* profits⟩
5 : relating to or being a function of a random variable that is obtained from a function of several random variables by integrating or summing over all possible values of the other variables ⟨a *marginal* probability function⟩
— mar·gin·al·i·ty \ˌmär-jə-'na-lə-tē\ *noun*
— mar·gin·al·ly \'märj-nə-lē, 'mär-jə-nᵊl-ē\ *adverb*

mar·gi·na·lia \ˌmär-jə-'nā-lē-ə\ *noun plural* [New Latin, from Medieval Latin, neuter plural of *marginalis*] (1832)
1 : marginal notes (as in a book)
2 : nonessential items ⟨the meat and *marginalia* of American politics —*Saturday Review*⟩

mar·gin·al·ize \'märj-nə-ˌlīz, 'mär-jə-nᵊl-ˌīz\ *transitive verb* **-ized; -iz·ing** (1970)
: to relegate to a marginal position within a society or group
— mar·gin·al·i·za·tion \ˌmärj-nə-lə-'zā-shən, ˌmär-jə-nᵊl-ə-\ *noun*

marginal utility *noun* (1890)
: the amount of additional utility provided by an additional unit of an economic good or service

mar·gin·ate \'mär-jə-ˌnāt\ *transitive verb* **-at·ed; -at·ing** (1623)
1 : MARGIN 1, 2a
2 : MARGINALIZE
— mar·gin·ation \ˌmär-jə-'nā-shən\ *noun*

mar·gin·at·ed \-ˌnā-təd\ *adjective* (circa 1727)
: having a distinct margin

mar·gra·vate \'mär-grə-ˌvāt\ *or* **mar·gra·vi·ate** \mär-'grā-vē-ət, -ˌāt\ *noun* (1802)
: the territory of a margrave

mar·grave \'mär-ˌgrāv\ *noun* [Dutch *markgraaf,* from Middle Dutch *marcgrave;* akin to Old High German *marha* boundary and to Old High German *grāvo* count — more at MARK] (1551)
1 : the military governor especially of a German border province
2 : a member of the German nobility corresponding in rank to a British marquess
— mar·gra·vi·al \mär-'grā-vē-əl\ *adjective*

mar·gra·vine \'mär-grə-ˌvēn, ˌmär-grə-'\ *noun* (1692)
: the wife of a margrave

mar·gue·rite \ˌmär-gə-'rēt, -gyə-\ *noun* [French, from Middle French *margarite* pearl, daisy — more at MARGARITE] (circa 1866)
1 : DAISY 1b
2 : any of various single-flowered chrysanthemums; *especially* **:** a chrysanthemum (*Chrysanthemum frutescens*) of the Canary Islands

maria *plural of* MARE

ma·ri·a·chi \ˌmär-ē-'ä-chē, ˌmar-\ *noun* [Mexican Spanish, perhaps modification of French *mariage* marriage] (1927)
1 : a Mexican street band; *also* **:** a musician belonging to such a band
2 : the music performed by a mariachi

Mar·i·an \'mer-ē-ən, 'mar-ē-, 'mā-rē-\ *adjective* (1608)

1 : of or relating to Mary Tudor or her reign
2 : of or relating to the Virgin Mary

Mar•i•an•ist \-ə-nist\ *noun* (circa 1899)
: a member of the Roman Catholic Society of Mary of Paris founded by William Joseph Chaminade in France in 1817 and devoted especially to education

Ma•ria The•re•sa dollar \mə-'rē-ə-tə-'rā-sə-, -'rā-zə-\ *noun* (circa 1883)
: a 1780 silver trade coin used in the Middle East

mari•cul•ture \'mar-ə-,kəl-chər\ *noun* [Latin *mare* sea + English *-culture* (as in *agriculture*)] (circa 1909)
: the cultivation of marine organisms in their natural environment
— mari•cul•tur•ist \,mar-ə-'kəl-chə-rist, -'kəlch-rist\ *noun*

mari•gold \'mar-ə-,gōld, 'mer-\ *noun* [Middle English, from *Mary,* mother of Jesus + Middle English *gold*] (14th century)
1 : POT MARIGOLD
2 : any of a genus (*Tagetes*) of composite herbs with showy yellow, orange, or maroon flower heads

mar•i•jua•na *also* **mar•i•hua•na** \,mar-ə-'wä-nə *also* -'hwä-\ *noun* [Mexican Spanish *mariguana, marihuana*] (1894)
1 : HEMP 1a, c
2 : the dried leaves and flowering tops of the pistillate hemp plant that yield THC and are smoked in cigarettes for their intoxicating effect — compare BHANG, CANNABIS, HASHISH

ma•rim•ba \mə-'rim-bə\ *noun* [of Bantu origin; akin to Kimbundu *marimba* xylophone] (1704)
: a xylophone of southern Africa and Central America with resonators beneath each bar; *also* **:** a modern form of this instrument
— ma•rim•bist \-bist\ *noun*

ma•ri•na \mə-'rē-nə\ *noun* [Italian & Spanish, seashore, from feminine of *marino,* adjective, marine, from Latin *marinus*] (1924)
: a dock or basin providing secure moorings for pleasure boats and often offering supply, repair, and other facilities

¹mar•i•nade \,mar-ə-'nād\ *noun* [French, from *mariner* to pickle, marinate, probably from Italian *marinare*] (1725)
: a savory usually acidic sauce in which meat, fish, or a vegetable is soaked to enrich its flavor or to tenderize it

²marinade *transitive verb* **-nad•ed; -nad•ing** (1727)
: MARINATE

mar•i•na•ra \,mar-ə-'nar-ə, ,mer-ə-'ner-, -'när-\ *adjective* [Italian (*alla*) *marinara,* literally, in sailor style] (circa 1948)
: made with tomatoes, onions, garlic, and spices ⟨*marinara* sauce⟩; *also* **:** served with marinara sauce ⟨spaghetti *marinara*⟩

mar•i•nate \'mar-ə-,nāt\ *verb* **-nat•ed; -nat•ing** [probably from Italian *marinato,* past participle of *marinare* to marinate, from *marino*] (circa 1645)
transitive verb
: to steep (meat, fish, or vegetables) in a marinade
intransitive verb
: to become marinated
— mar•i•na•tion \,mar-ə-'nā-shən\ *noun*

¹ma•rine \mə-'rēn\ *adjective* [Middle English, from Latin *marinus,* from *mare* sea; akin to Old English *mere* sea, pool, Old High German *meri* sea, Old Church Slavonic *morje*] (15th century)
1 a : of or relating to the sea ⟨*marine* life⟩ **b :** of or relating to the navigation of the sea **:** NAUTICAL ⟨a *marine* chart⟩ **c :** of or relating to the commerce of the sea **:** MARITIME ⟨*marine* law⟩ **d :** depicting the sea, seashore, or ships ⟨a *marine* painter⟩
2 : of or relating to marines ⟨*marine* barracks⟩

²marine *noun* (1669)
1 a : the mercantile and naval shipping of a country **b :** seagoing ships especially in relation to nationality or class
2 : one of a class of soldiers serving on shipboard or in close association with a naval force; *specifically* **:** a member of the U.S. Marine Corps
3 : an executive department (as in France) having charge of naval affairs
4 : a marine picture **:** SEASCAPE

marine architect *noun* (1949)
: NAVAL ARCHITECT
— marine architecture *noun*

mar•i•ner \'mar-ə-nər\ *noun* [Middle English, from Anglo-French *marinier,* from Medieval Latin *marinarius,* from *marinus*] (14th century)
: a person who navigates or assists in navigating a ship **:** SEAMAN, SAILOR

mariner's compass *noun* (1627)
: a compass used in navigation that consists of parallel magnetic needles or bundles of needles permanently attached to a card marked to indicate direction and degrees of a circle

Mar•i•ol•a•try \,mer-ē-'ä-lə-trē, ,mar-ē-, ,mā-rē-\ *noun* (1612)
: excessive veneration of the Virgin Mary
— Mar•i•ol•a•ter \-'ä-lə-tər\ *noun*

Mar•i•ol•o•gy \-'ä-lə-jē\ *noun* (1857)
: study or doctrine relating to the Virgin Mary
— Mar•i•o•log•i•cal \-ə-'lä-ji-kəl\ *adjective*

mar•i•o•nette \,mar-ē-ə-'net, ,mer-\ *noun* [French *marionnette,* from Middle French *maryonete,* from *Marion,* diminutive of *Marie* Mary] (circa 1620)
: a small-scale usually wooden figure (as of a person) with jointed limbs that is moved from above by manipulation of the attached strings or wires ◆

mar•i•po•sa lily \,mar-ə-'pō-zə-, -sə-\ *noun* [probably from American Spanish *mariposa,* from Spanish, butterfly] (1882)
: any of a genus (*Calochortus*) of western North American plants of the lily family with showy flowers having three petals and three sepals — called also *mariposa tulip;* compare SEGO LILY

mar•ish \'mar-ish\ *noun* (15th century)
archaic **:** MARSH

Mar•ist \'mar-ist, 'mer-\ *noun* [French *mariste,* from *Marie* Mary] (circa 1872)
: a member of the Roman Catholic Society of Mary founded by Jean Claude Colin in France in 1816 and devoted to education

mar•i•tal \'mar-ə-tᵊl\ *adjective* [Latin *maritalis,* from *maritus* married] (1603)
1 : of or relating to marriage or the married state ⟨*marital* vows⟩
2 : of or relating to a husband and his role in marriage
— mar•i•tal•ly \-tᵊl-ē\ *adverb*

mar•i•time \'mar-ə-,tīm\ *adjective* [Latin *maritimus,* from *mare*] (1550)
1 : of, relating to, or bordering on the sea ⟨a *maritime* province⟩
2 : of or relating to navigation or commerce on the sea
3 : having the characteristics of a mariner

mar•jo•ram \'mär-jə-rəm, 'märj-rəm\ *noun* [alteration of Middle English *mageram, majorane,* from Middle French *majorane,* from Medieval Latin *majorana*] (circa 1550)
: any of various usually fragrant and aromatic Old World mints (genus *Origanum*) often used in cookery

¹mark \'märk\ *noun* [Middle English, from Old English *mearc* boundary, march, sign; akin to Old High German *marha* boundary, Latin *margo*] (before 12th century)
1 : a boundary land
2 a (1) **:** a conspicuous object serving as a guide for travelers (2) **:** something (as a line, notch, or fixed object) designed to record position **b :** one of the bits of leather or colored bunting placed on a sounding line at intervals **c :** TARGET **d :** the starting line or position in a track event **e** (1) **:** GOAL, OBJECT (2) **:** an object of attack, ridicule, or abuse; *specifically* **:** a victim or prospective victim of a swindle (3) **:** the point under discussion (4) **:** condition of being correct or accurate ⟨her observations are on the *mark*⟩ **f :** a standard of performance, quality, or condition **:** NORM ⟨not feeling up to the *mark* lately⟩
3 a (1) **:** SIGN, INDICATION ⟨gave her the necklace as a *mark* of his esteem⟩ (2) **:** an impression (as a scratch, scar, or stain) made on something (3) **:** a distinguishing trait or quality **:** CHARACTERISTIC ⟨the *marks* of an educated person⟩ **b :** a symbol used for identification or indication of ownership **c :** a cross made in place of a signature **d** (1) **:** TRADEMARK (2) *capitalized* — used with a numeral to designate a particular model of a weapon or machine ⟨*Mark* II⟩ **e :** a written or printed symbol (as a comma or colon) **f :** POSTMARK **g :** a symbol used to represent a teacher's estimate of a student's work or conduct; *especially* **:** GRADE **h :** a figure registering a point or level reached or achieved ⟨the halfway *mark* in the first period of play⟩; *especially* **:** RECORD
4 a : ATTENTION, NOTICE ⟨nothing worthy of *mark*⟩ **b :** IMPORTANCE, DISTINCTION ⟨stands out as a person of *mark*⟩ **c :** a lasting or strong impression ⟨worked at several jobs but didn't make much of a *mark*⟩ **d :** an assessment of merits **:** RATING ⟨got high *marks* for honesty⟩
synonym see SIGN

²mark *verb* [Middle English, from Old English *mearcian;* akin to Old High German *marcōn* to mark, determine the boundaries of, Old English *mearc* boundary] (before 12th century)
transitive verb
1 a (1) **:** to fix or trace out the bounds or limits of (2) **:** to plot the course of **:** CHART **b :** to set apart by or as if by a line or boundary — usually used with *off*
2 a (1) **:** to designate as if by a mark ⟨*marked* for greatness⟩ (2) **:** to make or leave a mark on (3) **:** to furnish with natural marks ⟨wings *marked* with white⟩ (4) **:** to label so as to indicate price or quality (5) **:** to make notations in or on **b** (1) **:** to make note of in writing **:** JOT ⟨*marking* the date in his journal⟩ (2) **:** to indicate by a mark or symbol ⟨*mark* an accent⟩ (3) **:** REGISTER, RECORD (4) **:** to determine the value of by means of marks or symbols **:** GRADE ⟨*mark* term papers⟩ **c** (1) **:** CHARACTERIZE, DISTINGUISH ⟨the flamboyance that *marks* her stage appearance⟩ (2) **:** SIGNALIZE ⟨this year *marks* our 50th anniversary⟩
3 : to take notice of **:** OBSERVE ⟨*mark* my words⟩

◇ WORD HISTORY
marionette Puppet shows have been widespread in the world since ancient times, but the elaborately jointed and intricately wired figure known as the marionette was not perfected until the 19th century. At the time the word entered English in the 17th century, French *marionette* denoted a much more rudimentary hand- or string-operated puppet. Nonetheless performances involving these figures were quite popular in France in the 16th century, when *marionette* is first attested in this sense in French. Puppet plays were performed in particular at the Feast of the Assumption of the Virgin Mary. This led to the puppets being called by a diminutive form of *Marie,* the French name of the Virgin Mary. (*Marionette* may be altered from a diminutive form of *mariole,* itself a diminutive of *Marie,* or formed directly from another diminutive, *Marion.*) Similar derivatives, such as *mariotte* and *mariole,* had been used to designate other objects associated with the Virgin, such as a 15th century coin bearing her image.

4 : to pick up (one's golf ball) from a putting green and substitute a marker
intransitive verb
: to take careful notice
— mark time 1 : to keep the time of a marching step by moving the feet alternately without advancing **2 :** to maintain a static state of readiness ⟨the House was *marking time* while the Senate talked —F. L. Paxson⟩

³mark *noun* [Middle English, from Old English *marc,* probably of Scandinavian origin; akin to Old Norse *mǫrk* mark; akin to Old English *mearc* sign] (before 12th century)
1 : any of various old European units of weight used especially for gold and silver; *especially* **:** a unit equal to about 8 ounces (248 grams)
2 : a unit of value: **a :** an old English unit equal to 13*s* 4*d* **b :** any one of various old Scandinavian or German units of value; *specifically* **:** a unit and corresponding silver coin of the 16th century worth ½ taler **c** (1) **:** DEUTSCHE MARK (2) **:** the basic monetary unit of East Germany replaced in 1990 by the West German deutsche mark **d :** MARKKA

Mark \'märk\ *noun* [Latin *Marcus*]
1 a : an early Jewish Christian traditionally identified as the writer of the Gospel of Mark — called also *John Mark* **b :** the second Gospel in the New Testament — see BIBLE table
2 : a king of Cornwall, uncle of Tristram, and husband of Isolde

mark·down \'märk-ˌdau̇n\ *noun* (1880)
1 : a lowering of price
2 : the amount by which an original selling price is reduced

mark down *transitive verb* (1859)
: to put a lower price on

marked \'märkt\ *adjective* (before 12th century)
1 : having an identifying mark ⟨a *marked* card⟩
2 *or* 'mär-kəd\ **:** having a distinctive or emphasized character **:** NOTICEABLE ⟨has a *marked* drawl⟩
3 a : enjoying fame or notoriety **b :** being an object of attack, suspicion, or vengeance ⟨a *marked* man⟩
4 : overtly signaled by a linguistic feature ⟨with most English nouns the plural is the *marked* number⟩
— mark·ed·ly \'mär-kəd-lē\ *adverb*
— mark·ed·ness \'mär-kəd-nəs\ *noun*

mark·er \'mär-kər\ *noun* (15th century)
1 : one that marks
2 : something used for marking
3 : SCORE 7; *specifically* **:** RUN
4 : PROMISSORY NOTE, IOU
5 : GENETIC MARKER

¹mar·ket \'mär-kət\ *noun, often attributive* [Middle English, from Old North French, from Latin *mercatus* trade, marketplace, from *mercari* to trade, from *merc-, merx* merchandise] (12th century)
1 a (1) **:** a meeting together of people for the purpose of trade by private purchase and sale and usually not by auction (2) **:** the people assembled at such a meeting **b** (1) **:** a public place where a market is held; *especially* **:** a place where provisions are sold at wholesale ⟨a farmers' *market*⟩ (2) **:** a retail establishment usually of a specified kind ⟨a fish *market*⟩
2 *archaic* **:** the act or an instance of buying and selling
3 : the rate or price offered for a commodity or security
4 a (1) **:** a geographical area of demand for commodities or services ⟨the foreign *market* for consulting firms⟩ (2) **:** a specified category of potential buyers ⟨the youth *market*⟩ **b :** the course of commercial activity by which the exchange of commodities is effected **:** extent of demand ⟨the *market* is dull⟩ **c** (1) **:** an opportunity for selling ⟨a good *market* for used cars⟩ (2) **:** the available supply of or potential demand for specified goods or services ⟨the labor *market*⟩ **d :** the area of economic activity in which buyers and sellers come together and the forces of supply and demand affect prices ⟨producing goods for *market* rather than for consumption⟩
— in the market : in the position of being a potential buyer ⟨*in the market* for a house⟩
— on the market : available for purchase; *also* **:** up for sale ⟨put their house *on the market*⟩

²market (1635)
intransitive verb
: to deal in a market
transitive verb
1 : to expose for sale in a market
2 : SELL

mar·ket·able \'mär-kə-tə-bəl\ *adjective* (1600)
1 a : fit to be offered for sale in a market ⟨food that is not *marketable*⟩ **b :** wanted by purchasers or employers **:** SALABLE ⟨*marketable* securities⟩ ⟨*marketable* skills⟩
2 : of or relating to buying or selling
— mar·ket·abil·i·ty \ˌmär-kə-tə-'bi-lə-tē\ *noun*

market basket *noun* (1943)
: a variety of consumer goods and services used to calculate a consumer price index

mar·ke·teer \ˌmär-kə-'tir\ *noun* (1832)
: a specialist in promoting or selling a product or service

mar·ket·er \'mär-kə-tər\ *noun* (1787)
: one that deals in a market; *specifically* **:** one that promotes or sells a product or service

market garden *noun* (1811)
British **:** TRUCK FARM
— market gardener *noun, British*
— market gardening *noun, British*

mar·ket·ing *noun* (1561)
1 a : the act or process of selling or purchasing in a market **b :** the process or technique of promoting, selling, and distributing a product or service
2 : an aggregate of functions involved in moving goods from producer to consumer

marketing research *noun* (circa 1937)
: research into the means of promoting, selling, and distributing a product or service

market order *noun* (circa 1920)
: an order to buy or sell securities or commodities immediately at the best price obtainable in the market

mar·ket·place \'mär-kət-ˌplās\ *noun* (14th century)
1 a : an open square or place in a town where markets or public sales are held **b :** MARKET ⟨the *marketplace* is the interpreter of supply and demand⟩
2 : the world of trade or economic activity **:** the everyday world ⟨a conviction that religion belongs in the *marketplace* —*Current Biography*⟩
3 : a sphere in which intangible values compete for acceptance ⟨the *marketplace* of ideas⟩

market price *noun* (15th century)
: a price actually given in current market dealings

market research *noun* (1926)
: research into the size, location, and makeup of a product market
— market researcher *noun*

market share *noun* (1970)
: the percentage of the market for a product or service that a company supplies

mark·ing *noun* (14th century)
1 : the act, process, or an instance of making or giving a mark
2 a : a mark made **b :** arrangement, pattern, or disposition of marks

mark·ka \'mär-ˌkä\ *noun, plural* **mark·kaa** \'mär-ˌkä\ *also* **mark·kas** \-ˌkäz\ [Finnish, from Swedish *mark,* a unit of value; akin to Old Norse *mǫrk* mark] (circa 1896)
— see MONEY table

Markov chain *noun* [A. A. *Markov* (died 1922) Russian mathematician] (1942)
: a usually discrete stochastic process (as a random walk) in which the probabilities of occurrence of various future states depend only on the present state of the system or on the immediately preceding state and not on the path by which the present state was achieved — called also *Markoff chain*

Mar·kov·ian \mär-'kō-vē-ən, -'kö-\ *or* **Mar·kov** \'mär-ˌkȯf, -ˌkȯv\ *also* **Mar·koff** \'mär-ˌkȯf\ *adjective* (1950)
: of, relating to, or resembling a Markov process or Markov chain especially by having probabilities defined in terms of transition from the possible existing states to other states

Markov process *noun* (1939)
: a stochastic process (as Brownian motion) that resembles a Markov chain except that the states are continuous; *also* **:** MARKOV CHAIN — called also *Markoff process*

marks·man \'märks-mən\ *noun* (1660)
: a person skilled in shooting at a mark or target
— marks·man·ship \-ˌship\ *noun*

marks·wom·an \'märks-ˌwu̇-mən\ *noun* (1802)
: a woman skilled in shooting at a mark or target

mark·up \'mär-ˌkəp\ *noun* (1916)
1 : an amount added to the cost price to determine the selling price; *broadly* **:** PROFIT
2 : a U.S. Congressional committee session at which a bill is put into final form before it is reported out

mark up *transitive verb* (1869)
: to put a markup on

marl \'mär(-ə)l\ *noun* [Middle English, from Middle French *marle,* from Medieval Latin *margila,* diminutive of Latin *marga* marl, from Gaulish] (14th century)
: a loose or crumbling earthy deposit (as of sand, silt, or clay) that contains a substantial amount of calcium carbonate and is used especially as a fertilizer for soils deficient in lime
— marly \'mär-lē\ *adjective*

mar·lin \'mär-lən\ *noun* [short for *marlinspike;* from the appearance of its beak] (1917)
: any of several large marine billfishes (genera *Makaira* and *Tetrapturus*) that are notable sport fishes

mar·line *also* **mar·lin** \'mär-lən\ *noun* [Middle English *merlyn,* probably from Middle Low German *marlinc, merlinc,* from *mēren* to tie, moor] (15th century)
: a small usually tarred line of two strands twisted loosely left-handed that is used especially for seizing and as a covering for wire rope

mar·line·spike *also* **mar·lin·spike** \'mär-lən-ˌspīk\ *noun* (1626)
: a tool (as of wood or iron) that tapers to a point and is used to separate strands of rope or wire (as in splicing)

marl·stone \'mär(-ə)l-ˌstōn\ *noun* (circa 1839)
: a rock that consists of a mixture of clay materials and calcium carbonate and often contains kerogen

mar·ma·lade \'mär-mə-ˌlād\ *noun* [Portuguese *marmelada* quince conserve, from *marmelo* quince, from Latin *melimelum,* a sweet apple, from Greek *melimēlon,* from *meli* honey + *mēlon* apple — more at MELLIFLUOUS] (1524)
: a clear sweetened jelly in which pieces of fruit and fruit rind are suspended

mar·mo·re·al \mär-'mōr-ē-əl, -'mȯr-\ *also* **mar·mo·re·an** \-ē-ən\ *adjective* [Latin *marmoreus,* from *marmor* marble] (1656)
: of, relating to, or suggestive of marble or a marble statue especially in coldness or aloofness

— **mar·mo·re·al·ly** \-ē-ə-lē\ *adverb*

mar·mo·set \'mär-mə-ˌset, -ˌzet\ *noun* [Middle English *marmusette* kind of monkey, from Middle French *marmoset* grotesque figure, from *marmouser* to mumble, of imitative origin] (1679)
: any of numerous small soft-furred South and Central American monkeys (family Callithricidae) with claws instead of nails on all the digits except the great toe

marmoset

mar·mot \'mär-mət\ *noun* [French *marmotte*] (1607)
: any of a genus (*Marmota*) of stout-bodied short-legged chiefly herbivorous burrowing rodents with coarse fur, a short bushy tail, and very small ears — compare WOODCHUCK

marmot

mar·o·cain \'mar-ə-ˌkān\ *noun* [French (*crêpe*) *marocain,* literally, Moroccan crepe] (1922)
: a ribbed crepe fabric used in women's clothing

Mar·o·nite \'mar-ə-ˌnīt\ *noun* [Medieval Latin *maronita,* from *Maron-, Maro* 5th century A.D. Syrian monk] (circa 1511)
: a member of a Uniate church chiefly in Lebanon having a Syriac liturgy and married clergy

[1]**ma·roon** \mə-'rün\ *noun* [French *maron, marron,* modification of American Spanish *cimarrón,* from *cimarrón* wild, savage] (1666)
1 *capitalized* **:** a fugitive black slave of the West Indies and Guiana in the 17th and 18th centuries; *also* **:** a descendant of such a slave
2 : a person who is marooned ◆

[2]**maroon** *transitive verb* (circa 1709)
1 : to put ashore on a desolate island or coast and leave to one's fate
2 : to place or leave in isolation or without hope of ready escape

[3]**maroon** *noun* [French *marron* Spanish chestnut] (1791)
: a dark red

mar·plot \'mär-ˌplät\ *noun* (1764)
: one who frustrates or ruins a plan or undertaking by meddling

[1]**marque** \'märk\ *noun* [Middle English, from Middle French, from Old Provençal *marca,* from *marcar* to mark, seize as pledge, of Germanic origin; akin to Old High German *marcōn* to mark] (15th century)
1 *obsolete* **:** REPRISAL, RETALIATION
2 : LETTERS OF MARQUE

[2]**marque** *noun* [French, mark, brand, from Middle French, from *marquer* to mark, of Germanic origin; akin to Old High German *marcōn* to mark] (1906)
: a brand or make of a product (as a sports car)

mar·quee \mär-'kē\ *noun* [modification of French *marquise,* literally, marchioness] (1690)
1 *chiefly British* **:** a large tent set up for an outdoor party, reception, or exhibition
2 : a permanent canopy often of metal and glass projecting over an entrance (as of a hotel or theater)
3 : BOX OFFICE 2 ⟨*marquee* value⟩

Mar·que·san \mär-'kā-z^{ə}n, -s^{ə}n\ *noun* (1799)
1 : a native or inhabitant of the Marquesas Islands
2 : the Austronesian language of the Marquesans
— **Marquesan** *adjective*

mar·quess \'mär-kwəs\ *or* **mar·quis** \'mär-kwəs, mär-'kē\ *noun, plural* **mar·quess·es** *or* **mar·quis·es** \-kwə-səz\ *or* **mar·quis** \-'kē(z)\ [Middle English *marquis, markis,* from Middle French *marquis,* alteration of *marchis,* from *marche* march] (14th century)
1 : a nobleman of hereditary rank in Europe and Japan
2 : a member of the British peerage ranking below a duke and above an earl
— **mar·quess·ate** \'mär-kwə-sət\ *or* **mar·quis·ate** \'mär-kwə-zət, -sət\ *noun*

mar·que·try *also* **mar·que·terie** \'mär-kə-trē\ *noun* [Middle French *marqueterie,* from *marqueter* to checker, inlay, from *marque* mark] (1563)
: decorative work in which elaborate patterns are formed by the insertion of pieces of material (as wood, shell, or ivory) into a wood veneer that is then applied to a surface (as of a piece of furniture)

mar·quise \mär-'kēz\ *noun, plural* **marquises** \-'kēz, -'kē-zəz\ [French, feminine of *marquis*] (1783)
1 : MARQUEE
2 : MARCHIONESS
3 : a gem or a ring setting or bezel usually elliptical in shape but with pointed ends

mar·qui·sette \ˌmär-kwə-'zet, -kə-\ *noun* [*marquise* + *-ette*] (1908)
: a sheer meshed fabric used for clothing, curtains, and mosquito nets

mar·ram grass \'mar-əm-\ *noun* [of Scandinavian origin; akin to Old Norse *maralmr,* a beach grass] (1834)
: any of several beach grasses (genus *Ammophila* and especially *A. arenaria*)

Mar·ra·no \mə-'rä-(ˌ)nō\ *noun, plural* **-nos** [Spanish, literally, pig] (1583)
: a Christianized Jew of medieval Spain

mar·riage \'mar-ij *also* 'mer-\ *noun* [Middle English *mariage,* from Old French, from *marier* to marry] (14th century)
1 a : the state of being married **b :** the mutual relation of husband and wife **:** WEDLOCK **c :** the institution whereby men and women are joined in a special kind of social and legal dependence for the purpose of founding and maintaining a family
2 : an act of marrying or the rite by which the married status is effected; *especially* **:** the wedding ceremony and attendant festivities or formalities
3 : an intimate or close union ⟨the *marriage* of painting and poetry —J. T. Shawcross⟩
— **mar·riage·abil·i·ty** \ˌmar-i-jə-'bi-lə-tē *also* ˌmer-\ *noun*
— **mar·riage·able** \'mar-i-jə-bəl *also* 'mer-\ *adjective*

marriage of convenience (1711)
: a marriage contracted for social, political, or economic advantage rather than for mutual affection; *broadly* **:** a union or cooperation formed solely for pragmatic reasons

[1]**mar·ried** \'mar-ēd *also* 'mer-\ *adjective* (14th century)
1 a : being in the state of matrimony **:** WEDDED **b :** of or relating to marriage **:** CONNUBIAL
2 : UNITED, JOINED

[2]**married** *noun, plural* **marrieds** *or* **married** (1890)
: a married person ⟨young *marrieds* are paid undue . . . attention —Paul Goodman⟩

mar·ron \ma-'rō^{n}\ *noun* [French] (1594)
1 : a large Mediterranean chestnut (*Castanea sativa*) or its sweet edible nut — called also *Spanish chestnut*
2 mar·rons \-'rō^{n}(z)\ *plural* **:** chestnuts preserved in vanilla-flavored syrup

mar·rons gla·cés \ma-'rō^{n}-gla-'sā\ *noun plural* [French, literally, glazed marrons] (1871)
: MARRON 2

[1]**mar·row** \'mar-(ˌ)ō\ *noun* [Middle English *marowe,* from Old English *mearg;* akin to Old High German *marag* marrow, Sanskrit *majjan*] (before 12th century)
1 a : a soft highly vascular modified connective tissue that occupies the cavities and cancellous part of most bones — compare RED MARROW **b :** the substance of the spinal cord
2 a : the choicest of food **b :** the seat of animal vigor **c :** the inmost, best, or essential part **:** CORE ⟨personal liberty is the *marrow* of the American tradition —Clinton Rossiter⟩
3 *chiefly British* **:** VEGETABLE MARROW
— **mar·rowy** \'mar-ə-wē\ *adjective*

[2]**marrow** *noun* [Middle English *marwe, marrow*] (1516)
chiefly Scottish **:** one of a pair

mar·row·bone \'mar-ə-ˌbōn, -ō-ˌbōn\ *noun* (14th century)
1 : a bone (as a shinbone) rich in marrow
2 *plural* **:** KNEES

mar·row·fat \-ə-ˌfat, -ō-ˌfat\ *noun* (1733)
: any of several wrinkled-seeded garden peas

[1]**mar·ry** \'mar-ē *also* 'mer-\ *verb* **mar·ried; mar·ry·ing** [Middle English *marien,* from Old French *marier,* from Latin *maritare,* from *maritus* married] (14th century)
transitive verb
1 a : to join as husband and wife according to law or custom **b :** to give in marriage ⟨*married* his daughter to his partner's son⟩ **c :** to take as spouse **:** WED ⟨*married* the girl next door⟩ **d :** to perform the ceremony of marriage for ⟨a priest will *marry* them⟩ **e :** to obtain by marriage ⟨*marry* wealth⟩
2 : to unite in close and usually permanent relation
intransitive verb
1 : to take a spouse **:** WED
2 : COMBINE, UNITE ⟨seafood *marries* with other flavors⟩
— **marry into :** to become a member of by marriage ⟨*married into* a prominent family⟩

[2]**marry** *interjection* [Middle English *marie,* from *Marie,* the Virgin Mary] (14th century)
archaic — used for emphasis and especially to express amused or surprised agreement

Mars \'märz\ *noun* [Latin *Mart-, Mars*]
1 : the Roman god of war — compare ARES
2 : the planet fourth in order from the sun and conspicuous for its red color — see PLANET table

mar·sa·la \mär-'sä-lə\ *noun, often capitalized* [*Marsala,* town in Sicily] (1806)
: a fortified Sicilian wine that varies from dry to sweet

marse \'märs\ *noun* [by shortening & alteration] (1869)
Southern **:** MASTER

Mar·seilles \mär-'sā(ə)lz\ *noun* [*Marseilles,* France] (1762)

◇ WORD HISTORY

maroon The modern Maroons of the Caribbean region are descendants of African slaves who fled from servitude into remote areas, such as the Cockpit Country of Jamaica and the interior of Suriname. In the 17th and 18th centuries, however, the word *maron* or *maroon* was used more or less generically to mean "fugitive slave." It is a loan from *maron* or *marron* in the local French of the Caribbean, originally used of a domestic animal and meaning "escaped into the wild." The French word in its turn was borrowed, directly or indirectly, from Spanish *cimarrón* "wild, savage," also applied in the 16th century to both animals and humans. In the late 17th century, the verbal form *marooned* was used by the English buccaneer William Dampier to mean "lost," and early in the next century *maroon* first clearly appears in the still-surviving sense "to put ashore on a desolate island or coast." In effect, the word now meant to reduce someone to the status of a fugitive slave, dependent on wits and knowledge of the wilderness for survival.

: a firm cotton fabric that is similar to piqué

marsh \'märsh\ *noun, often attributive* [Middle English *mersh,* from Old English *merisc, mersc;* akin to Middle Dutch *mersch* marsh, Old English *mere* sea, pool — more at MARINE] (before 12th century)
: a tract of soft wet land usually characterized by monocotyledons (as grasses or cattails)

[1]**mar·shal** *also* **mar·shall** \'mär-shəl\ *noun* [Middle English, from Old French *mareschal,* of Germanic origin; akin to Old High German *marahscalc* marshal, from *marah* horse + *scalc* servant] (13th century)
1 a : a high official in the household of a medieval king, prince, or noble originally having charge of the cavalry but later usually in command of the military forces **b :** a person who arranges and directs the ceremonial aspects of a gathering
2 a : FIELD MARSHAL **b :** a general officer of the highest military rank
3 a : an officer having charge of prisoners **b** (1) **:** a ministerial officer appointed for a judicial district (as of the U.S.) to execute the process of the courts and perform various duties similar to those of a sheriff (2) **:** a city law officer entrusted with particular duties **c :** the administrative head of a city police department or fire department ◆
— mar·shal·cy \-sē\ *noun*
— mar·shal·ship \-ˌship\ *noun*

[2]**marshal** *also* **marshall** *verb* **-shaled** *or* **-shalled; -shal·ing** *or* **-shal·ling** \'märsh-(ə-)liŋ\ (15th century)
transitive verb
1 : to place in proper rank or position ⟨*marshaling* the troops⟩
2 : to bring together and order in an appropriate or effective way ⟨*marshal* arguments⟩
3 : to lead ceremoniously or solicitously **:** USHER ⟨*marshaling* her little group of children down the street⟩
intransitive verb
: to take form or order ⟨ideas *marshaling* neatly⟩
synonym see ORDER

marshal of the Royal Air Force (1947)
: the highest ranking officer in the British air force

marsh elder *noun* (circa 1775)
: any of a genus (*Iva*) of coarse shrubby composite plants of moist areas in eastern and central North America

marsh gas *noun* (1848)
: METHANE

marsh hawk *noun* (1772)
: NORTHERN HARRIER

marsh hen *noun* (1709)
1 : any of various American rails
2 : BITTERN

marsh·land \'märsh-ˌland\ *noun* (12th century)
: a marshy tract or area **:** MARSH

marsh·mal·low \'märsh-ˌme-lō, -ˌma-\ *noun* (before 12th century)
1 : a pink-flowered European perennial herb (*Althaea officinalis*) of the mallow family that is naturalized in the eastern U.S. and has a mucilaginous root sometimes used in confectionery and in medicine
2 : a confection made from the root of the marshmallow or from corn syrup, sugar, albumen, and gelatin beaten to a light spongy consistency; *also* **:** a piece of partially dried marshmallow ⟨a bag of *marshmallows*⟩
— marsh·mal·lowy \-ˌme-lə-wē, -ˌma-\ *adjective*

marsh marigold *noun* (1578)
: a swamp herb (*Caltha palustris*) of the buttercup family that occurs in Europe and North America and has bright yellow flowers — called also *cowslip, kingcup*

marshy \'mär-shē\ *adjective* **marsh·i·er; -est** (14th century)
1 : resembling or constituting a marsh ⟨*marshy* ground⟩
2 : relating to or occurring in marshes ⟨*marshy* vegetation⟩
— marsh·i·ness *noun*

[1]**mar·su·pi·al** \mär-'sü-pē-əl\ *adjective* (1819)
1 : of, relating to, or being a marsupial
2 : of, relating to, or forming a marsupium

[2]**marsupial** *noun* [New Latin *Marsupialia,* from *marsupium*] (circa 1835)
: any of an order (Marsupialia) of mammals comprising kangaroos, wombats, bandicoots, opossums, and related animals that do not develop a true placenta and that usually have a pouch on the abdomen of the female which covers the teats and serves to carry the young

mar·su·pi·um \mär-'sü-pē-əm\ *noun, plural* **-pia** \-pē-ə\ [New Latin, from Latin, purse, pouch, from Greek *marsypion*] (1698)
1 : an abdominal pouch that is formed of a fold of the skin and encloses the mammary glands of most marsupials
2 : any of several structures in various invertebrates (as a bryozoan or mollusk) for enclosing or carrying eggs or young

[1]**mart** \'märt\ *noun* [Middle English, from Middle Dutch *marct, mart,* probably from Old North French *market*] (15th century)
1 *archaic* **:** a coming together of people to buy and sell **:** [5]FAIR 1
2 *obsolete* **:** the activity of buying and selling; *also* **:** BARGAIN
3 : MARKET

[2]**mart** *transitive verb* (1589)
: to deal in **:** SELL

mar·tel·lo tower \mär-'te-lō-\ *noun, often M capitalized* [Cape *Mortella,* Corsica] (1803)
: a circular masonry fort or blockhouse

mar·ten \'mär-tᵊn\ *noun, plural* **marten** *or* **martens** [Middle English *martryn,* from Middle French *martrine* marten fur, from Old French, from *martre* marten, of Germanic origin; akin to Old English *mearth* marten] (13th century)
1 : any of several semiarboreal slender-bodied carnivorous mammals (genus *Martes*) chiefly of the northern hemisphere that are larger than the related weasels
2 : the fur or pelt of a marten

mar·tens·ite \'mär-tᵊn-ˌzīt\ *noun* [Adolf *Martens* (died 1914) German metallurgist] (1898)
: the hard constituent that is the chief component of quenched steel
— mar·tens·it·ic \ˌmar-tᵊn-'zi-tik, -'si-\ *adjective*
— mar·tens·it·i·cal·ly \-ti-k(ə-)lē\ *adverb*

Mar·tha \'mär-thə\ *noun* [Late Latin, from Greek]
: a sister of Lazarus and Mary and friend of Jesus

mar·tial \'mär-shəl\ *adjective* [Middle English, from Latin *martialis* of Mars, from *Mart-, Mars*] (14th century)
1 : of, relating to, or suited for war or a warrior
2 : relating to an army or to military life
3 : experienced in or inclined to war **:** WARLIKE
— mar·tial·ly \-shə-lē\ *adverb*

martial art *noun* (1933)
: any of several arts of combat and self-defense (as karate and judo) that are widely practiced as sport
— martial artist *noun*

martial law *noun* (1568)
1 : the law applied in occupied territory by the military authority of the occupying power
2 : the law administered by military forces that is invoked by a government in an emergency when the civilian law enforcement agencies are unable to maintain public order and safety

mar·tian \'mär-shən\ *adjective, often capitalized* (1880)
: of or relating to the planet Mars or its hypothetical inhabitants
— martian *noun, often capitalized*

mar·tin \'mär-tᵊn\ *noun* [probably from Saint *Martin*] (1589)
1 : a small Eurasian swallow (*Delichon urbica*) with a forked tail, bluish black head and back, and white rump and underparts
2 : any of various swallows and flycatchers other than the martin

mar·ti·net \ˌmär-tᵊn-'et\ *noun* [Jean *Martinet,* 17th century French army officer] (1779)
1 : a strict disciplinarian
2 : a person who stresses a rigid adherence to the details of forms and methods

mar·tin·gale \'mär-tᵊn-ˌgāl, -tiŋ-\ *noun* [Middle French] (1589)
1 : a device for steadying a horse's head or checking its upward movement that typically consists of a strap fastened to the girth, passing between the forelegs, and bifurcating to end in two rings through which the reins pass

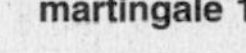

martingale 1

2 a : a lower stay of rope or chain for the jibboom used to sustain the strain of the forestays and fastened to or rove through the dolphin striker **b :** DOLPHIN STRIKER
3 : any of several systems of betting in which a player increases the stake usually by doubling each time a bet is lost

mar·ti·ni \mär-'tē-nē\ *noun* [probably alteration of *Martinez (cocktail),* from the name *Martinez*] (1894)
: a cocktail made of gin and dry vermouth; *also* **:** VODKA MARTINI

Mar·tin Lu·ther King Day \'mär-tᵊn-'lü-thər-'kiŋ-\ *noun* (1977)
: the third Monday in January observed as a legal holiday in some states of the U.S.

Mar·tin·mas \'mär-tᵊn-məs, -ˌmas\ *noun* [Middle English *martinmasse,* from Saint *Martin* + Middle English *masse* mass] (14th century)

◇ WORD HISTORY

marshal The French language is a direct descendant of Latin, but it has included since its earliest days a number of Germanic words. The Franks, a group of West Germanic peoples, entered Gaul in large numbers during the 5th century A.D. and left their name and some traces of their own speech to France. Old French *mareschal* is one of these words: etymologically it is a compound of two Frankish words, one meaning "horse" (and akin to Old English *mearh* "horse" and *mere* "mare"), and the other meaning "servant" (and akin to Old English *scealc* "servant"). The Old French word referred not only to a worker who shoed horses (a sense that survives in Modern French *maréchal-ferrant* "farrier"), but also to the officer of a noble household in charge of horses and by the 13th century to a royal officer in command of an army. This ascent in status of a man in charge of horses is paralleled by the history of *constable,* and Late Latin *comes stabuli,* the ultimate source of the word *constable,* may in fact be a loose translation of the Germanic compound underlying *marshal.* See in addition CONSTABLE.

: November 11 celebrated as the feast of Saint Martin

mart·let \'märt-lət\ *noun* [alteration of *martinet,* from Middle French, probably from Saint *Martin*] (1538)
: MARTIN 1

¹mar·tyr \'mär-tər\ *noun* [Middle English, from Old English, from Late Latin, from Greek *martyr-, martys,* literally, witness] (before 12th century)
1 : a person who voluntarily suffers death as the penalty of witnessing to and refusing to renounce a religion
2 : a person who sacrifices something of great value and especially life itself for the sake of principle
3 : VICTIM; *especially* **:** a great or constant sufferer ⟨a *martyr* to asthma all his life —A. J. Cronin⟩
— mar·tyr·i·za·tion \ˌmär-tə-rə-'zā-shən\ *noun*
— mar·tyr·ize \'mär-tə-ˌrīz\ *transitive verb*

²martyr *transitive verb* (before 12th century)
1 : to put to death for adhering to a belief, faith, or profession
2 : to inflict agonizing pain on **:** TORTURE

mar·tyr·dom \'mär-tər-dəm\ *noun* (before 12th century)
1 : the suffering of death on account of adherence to a cause and especially to one's religious faith
2 : AFFLICTION, TORTURE

mar·tyr·ol·o·gist \ˌmär-tə-'rä-lə-jist\ *noun* (1676)
: a writer of or a specialist in martyrology

mar·tyr·ol·o·gy \-jē\ *noun* (1599)
1 : a catalog of Roman Catholic martyrs and saints arranged by the dates of their feasts
2 : ecclesiastical history treating the lives and sufferings of martyrs

mar·tyry \'mär-tə-rē\ *noun, plural* **-tyr·ies** [Late Latin *martyrium,* from Late Greek *martyrion,* from Greek *martyr-, martys*] (circa 1722)
: a shrine erected in honor of a martyr

¹mar·vel \'mär-vəl\ *noun* [Middle English *mervel,* from Old French *merveille,* from Late Latin *mirabilia* marvels, from Latin, neuter plural of *mirabilis* wonderful, from *mirari* to wonder] (14th century)
1 : something that causes wonder or astonishment
2 : intense surprise or interest **:** ASTONISHMENT

²marvel *verb* **mar·veled** *or* **mar·velled; mar·vel·ing** *or* **mar·vel·ling** \'märv-liŋ, 'mär-və-\ (14th century)
intransitive verb
: to become filled with surprise, wonder, or amazed curiosity ⟨*marveled* at the magician's skill⟩
transitive verb
: to feel astonishment or perplexity at or about ⟨*marveled* that they had escaped⟩

mar·vel·ous *or* **mar·vel·lous** \'märv-(ə-)ləs\ *adjective* (14th century)
1 : causing wonder **:** ASTONISHING
2 : MIRACULOUS, SUPERNATURAL ⟨Gothic tales of the *marvelous* and the bizarre⟩
3 : of the highest kind or quality **:** notably superior ⟨has a *marvelous* way with children⟩
— mar·vel·ous·ly *adverb*
— mar·vel·ous·ness *noun*

Marx·ian \'märk-sē-ən\ *adjective* (1887)
: of, developed by, or influenced by the doctrines of Marx ⟨*Marxian* socialism⟩

Marx·ism \'märk-ˌsi-zəm\ *noun* (1897)
: the political, economic, and social principles and policies advocated by Marx; *especially* **:** a theory and practice of socialism including the labor theory of value, dialectical materialism, the class struggle, and dictatorship of the proletariat until the establishment of a classless society
— Marx·ist \-sist\ *noun or adjective*

Marx·ism–Le·nin·ism \'märk-ˌsi-zəm-'le-nə-ˌni-zəm\ *noun* (1932)
: a theory and practice of communism developed by Lenin from doctrines of Marx
— Marx·ist–Len·in·ist \'märk-sist-'le-nə-nist\ *noun or adjective*

Mary \'mer-ē, 'mar-ē, 'mā-rē\ *noun* [Late Latin *Maria,* from Greek *Mariam, Maria,* from Hebrew *Miryām* Miriam]
1 : the mother of Jesus
2 : a sister of Lazarus and Martha and a friend of Jesus

Mary Jane \-'jān\ *noun* [by folk etymology (influenced by Spanish *Juana* Jane)] (1928)
slang **:** MARIJUANA

Mary·knoll·er \-ˌnō-lər\ *noun* (1943)
: a member of the Catholic Foreign Mission Society of America founded by T. F. Price and J. A. Walsh at Maryknoll, N.Y. in 1911

Mary Mag·da·lene \-'mag-də-lən, -ˌlēn; -ˌmag-də-'lē-nē\ *noun* [Late Latin *Magdalene,* from Greek *Magdalēnē*]
: a woman who was healed of evil spirits by Jesus and who saw the risen Christ near his sepulchre

mar·zi·pan \'märt-sə-ˌpän, -ˌpan; 'mär-zə-ˌpan\ *noun* [German, from Italian *marzapane*] (1542)
: a confection of crushed almonds or almond paste, sugar, and egg whites that is often shaped into various forms

Ma·sai \mä-'sī, 'mä-ˌ\ *noun, plural* **Masai** *or* **Masais** (1857)
1 : a member of a pastoral and hunting people of Kenya and Tanzania
2 : the Nilotic language of the Masai people

mas·cara \ma-'skar-ə\ *noun* [probably from Italian *maschera* mask] (1886)
: a cosmetic especially for coloring the eyelashes

mas·car·po·ne \ˌmas-kär-'pō-(ˌ)nā\ *noun* [Italian, from Italian dialect (Lombardy) *mascarpón,* augmentative of *mascarpa* cream cheese] (1932)
: an Italian cream cheese

mas·con \'mas-ˌkän\ *noun* [²*mass* + *concentration*] (1968)
: any of the high-density regions below the surface of lunar maria that are held to perturb the motion of spacecraft in lunar orbit

mas·cot \'mas-ˌkät *also* -kət\ *noun* [French *mascotte,* from Provençal *mascoto,* from *masco* witch, from Medieval Latin *masca*] (1881)
: a person, animal, or object adopted by a group as a symbolic figure especially to bring them good luck ⟨the team had a mountain lion as their *mascot*⟩ ◆

¹mas·cu·line \'mas-kyə-lən\ *adjective* [Middle English *masculin,* from Middle French, from Latin *masculinus,* from *masculus,* noun, male, diminutive of *mas* male] (14th century)
1 a : MALE **b :** having qualities appropriate to or usually associated with a man
2 : of, relating to, or constituting the gender that ordinarily includes most words or grammatical forms referring to males
3 a : having or occurring in a stressed final syllable ⟨*masculine* rhyme⟩ **b :** having the final chord occurring on a strong beat ⟨*masculine* cadence⟩
4 : of or forming the formal, active, or generative principle of the cosmos
— mas·cu·line·ly *adverb*
— mas·cu·lin·i·ty \ˌmas-kyə-'li-nə-tē\ *noun*

²masculine *noun* (14th century)
1 : the masculine gender
2 : a noun, pronoun, adjective, or inflectional form or class of the masculine gender
3 : a male person

mas·cu·lin·ise *British variant of* MASCULINIZE

mas·cu·lin·ize \'mas-kyə-lə-ˌnīz\ *transitive verb* **-ized; -iz·ing** (1912)
: to give a chiefly masculine character to; *especially* **:** to cause (a female) to take on male characteristics
— mas·cu·lin·i·za·tion \ˌmas-kyə-lə-nə-'zā-shən\ *noun*

ma·ser \'mā-zər\ *noun* [*m*icrowave *a*mplification by *s*timulated *e*mission of *r*adiation] (1955)
: a device or object that emits coherent microwave radiation produced by the natural oscillations of atoms or molecules between energy levels

¹mash \'mash\ *transitive verb* (13th century)
1 a : to reduce to a soft pulpy state by beating or pressure **b :** CRUSH, SMASH ⟨*mash* a finger⟩
2 : to subject (as crushed malt) to the action of water with heating and stirring in preparing wort

²mash *noun* [Middle English *mash-,* from Old English *māx-;* akin to Middle High German *meisch* mash] (1577)
1 : a mixture of ground feeds for livestock
2 : crushed malt or grain meal steeped and stirred in hot water to ferment (as for the production of beer or whiskey)
3 : a soft pulpy mass
4 *British* **:** mashed potatoes

³mash *transitive verb* [origin unknown] (1879)
: to flirt with or seek the affection of

⁴mash *noun* (1880)
: CRUSH 4

¹mash·er \'ma-shər\ *noun* (circa 1500)
: one that mashes ⟨a potato *masher*⟩

²masher *noun* (1875)
: a man who makes passes at women

¹mask \'mask\ *noun* [Middle French *masque,* from Old Italian *maschera*] (1534)
1 a (1) **:** a cover or partial cover for the face used for disguise (2) **:** a person wearing a mask **:** MASKER **b** (1) **:** a figure of a head worn on the stage in antiquity to identify the character and project the voice (2) **:** a grotesque false face worn at carnivals or in rituals **c :** an often grotesque carved head or face used as an ornament (as on a keystone) **d :** a sculptured face or a copy of a face made by means of a mold
2 a : something that serves to conceal or disguise **:** PRETENSE, CLOAK ⟨aware of the *masks,* facades and defenses people erect to protect themselves —Kenneth Keniston⟩ **b :** something that conceals from view **c :** a translucent or opaque screen to cover part of the sensitive surface in taking or printing a photograph **d :** a pattern of opaque material used to shield selected areas of a surface (as of a semiconductor) in deposition or etching (as in producing an integrated circuit)

◇ WORD HISTORY
mascot The spoken Latin of Gaul, northern Italy, and the Iberian Peninsula had a root *mask-,* unattested in classical Latin. It overlapped two semantic fields, the color black and malevolent magic, that were strongly associated with each other in popular superstition. One outcome of this root in Occitan or Provençal, the Romance speech of southern France, was a feminine noun *masco* "sorceress, witch." *Mascoto* "enchantment, charm," a derivative of *masco,* was borrowed into French as *mascotte* and shortly thereafter provided English with the word *mascot.* Contemporary with the root *mask-* in spoken Latin was an extended form *maskar-,* which appears on the one hand in words such as Catalan *mascara* "lampblack, soot," and on the other in medieval Italian *maschera* "mask." The Italian word is the ultimate source of English *mask* and *masquerade,* both borrowed directly from French, and probably also of our *mascara,* which was originally theater argot.

3 a : a protective covering for the face **b :** GAS MASK **c :** a device covering the mouth and nose to facilitate inhalation **d :** a comparable device to prevent exhalation of infective material **e :** a cosmetic preparation for the skin of the face that produces a tightening effect as it dries
4 a : the head or face of an animal (as a fox or dog) **b :** an area (as the one around the eyes) of an animal's face that is distinguished by usually darker coloring

²mask (circa 1562)
intransitive verb
1 : to take part in a masquerade
2 a : to assume a mask **b :** to disguise one's true character or intentions
transitive verb
1 : to provide or conceal with a mask: as **a :** to conceal from view ⟨*mask* a gun battery⟩ **b :** to make indistinct or imperceptible ⟨*masks* undesirable flavors⟩ **c :** to cover up ⟨*masked* his real purpose⟩
2 : to cover for protection
3 : to modify the size or shape of (as a photograph) by means of an opaque border
synonym see DISGUISE
— mask·like \ˌlīk\ *adjective*

masked \ˈmaskt\ *adjective* (1585)
1 : marked by the use of masks ⟨a *masked* ball⟩
2 : failing to present or produce the usual symptoms **:** LATENT ⟨*masked* infection⟩ ⟨a *masked* virus⟩

mask·er \ˈmas-kər\ *noun* (circa 1548)
: a person who wears a mask; *especially* **:** a participant in a masquerade

mask·ing tape \ˈmas-kiŋ-\ *noun* (1936)
: a tape with adhesive on one side that has a variety of uses (as to cover a surface when painting an adjacent surface)

mas·och·ism \ˈma-sə-ˌki-zəm, ˈma-zə- *also* ˈmā-\ *noun* [International Scientific Vocabulary, from Leopold von Sacher-*Masoch* (died 1895) German novelist] (circa 1893)
1 : a sexual perversion characterized by pleasure in being subjected to pain or humiliation especially by a love object — compare SADISM
2 : pleasure in being abused or dominated **:** a taste for suffering
— mas·och·ist \-kist\ *noun*
— mas·och·is·tic \ˌma-sə-ˈkis-tik, ˌma-zə- *also* ˌmā-\ *adjective*
— mas·och·is·ti·cal·ly \-ˈkis-ti-k(ə-)lē\ *adverb*

ma·son \ˈmā-sᵊn\ *noun* [Middle English, from Old French *maçon,* of Germanic origin; akin to Old English *macian* to make] (13th century)
1 : a skilled worker who builds by laying units of substantial material (as stone or brick)
2 *capitalized* **:** FREEMASON

Ma·son·ic \mə-ˈsä-nik\ *adjective* (1797)
: of, relating to, or characteristic of Freemasons or Freemasonry

Ma·son·ite \ˈmā-sᵊn-ˌīt\ *trademark*
— used for fiberboard made from steam-exploded wood fiber

ma·son jar \ˈmā-sᵊn-\ *noun, often M capitalized* [John L. *Mason,* 19th century American inventor] (1888)
: a widemouthed jar used especially for home canning

ma·son·ry \ˈmā-sᵊn-rē\ *noun, plural* **-ries** (13th century)
1 a : something constructed of materials used by masons **b :** the art, trade, or occupation of a mason **c :** work done by a mason
2 *capitalized* **:** FREEMASONRY

mason wasp *noun* (1792)
: any of various solitary wasps that construct nests of hardened mud

Ma·so·ra *or* **Ma·so·rah** \mə-ˈsōr-ə, -ˈsȯr-\ *noun* [New Hebrew *mĕsōrāh,* from Late Hebrew *māsōreth* tradition, from Hebrew, bond] (1613)
: a body of notes on the textual traditions of the Hebrew Old Testament compiled by scribes during the 1st millennium of the Christian era

Mas·o·rete *or* **Mas·so·rete** \ˈma-sə-ˌrēt\ *noun* [Middle French *massoreth,* from Late Hebrew *māsōreth*] (1587)
: one of the scribes who compiled the Masora
— Mas·o·ret·ic \ˌma-sə-ˈre-tik\ *adjective*

masque *also* **mask** \ˈmask\ *noun* [Middle French *masque* — more at MASK] (1514)
1 : MASQUERADE
2 : a short allegorical dramatic entertainment of the 16th and 17th centuries performed by masked actors

masqu·er \ˈmas-kər\ *variant of* MASKER

¹mas·quer·ade \ˌmas-kə-ˈrād\ *noun* [Middle French, from Old Italian dialect *mascarada,* from Old Italian *maschera* mask] (1587)
1 a : a social gathering of persons wearing masks and often fantastic costumes **b :** a costume for wear at such a gathering
2 : an action or appearance that is mere disguise or outward show

²masquerade *intransitive verb* **-ad·ed; -ad·ing** (1692)
1 a : to disguise oneself, *also* **:** to go about disguised **b :** to take part in a masquerade
2 : to assume the appearance of something one is not
— mas·quer·ad·er *noun*

¹mass \ˈmas\ *noun* [Middle English, from Old English *mæsse,* modification of (assumed) Vulgar Latin *messa,* literally, dismissal at the end of a religious service, from Late Latin *missa,* from Latin, feminine of *missus,* past participle of *mittere* to send] (before 12th century)
1 *capitalized* **:** the liturgy of the Eucharist especially in accordance with the traditional Latin rite
2 *often capitalized* **:** a celebration of the Eucharist ⟨Sunday *masses* held at three different hours⟩
3 : a musical setting for the ordinary of the Mass

²mass *noun* [Middle English *masse,* from Middle French, from Latin *massa,* from Greek *maza;* akin to Greek *massein* to knead — more at MINGLE] (15th century)
1 a : a quantity or aggregate of matter usually of considerable size **b** (1) **:** EXPANSE, BULK (2) **:** massive quality or effect ⟨impressed me with such *mass* and such vividness —F. M. Ford⟩ (3) **:** the principal part or main body ⟨the great *mass* of the continent is buried under an ice cap —Walter Sullivan⟩ (4) **:** AGGREGATE, WHOLE ⟨men in the *mass*⟩ **c :** the property of a body that is a measure of its inertia and that is commonly taken as a measure of the amount of material it contains and causes it to have weight in a gravitational field
2 : a large quantity, amount, or number ⟨a great *mass* of material⟩
3 a : a large body of persons in a compact group **:** a body of persons regarded as an aggregate **b :** the great body of the people as contrasted with the elite — often used in plural ⟨the underprivileged and disadvantaged *masses* —C. A. Buss⟩
synonym see BULK

³mass (14th century)
transitive verb
: to form or collect into a mass
intransitive verb
: to assemble in a mass ⟨three thousand students had *massed* in the plaza —A. E. Neville⟩

⁴mass *adjective* (1733)
1 a : of or relating to the mass of the people ⟨*mass* market⟩ ⟨*mass* education⟩; *also* **:** being one of or at one with the mass **:** AVERAGE, COMMONPLACE ⟨*mass* man⟩ **b :** participated in by or affecting a large number of individuals ⟨*mass* destruction⟩ ⟨*mass* demonstrations⟩ **c :** having a large-scale character **:** WHOLESALE ⟨*mass* production⟩
2 : viewed as a whole **:** TOTAL ⟨the *mass* effect of a design⟩

mas·sa \ˈma-sə\ *noun* [by alteration] (1774)
Southern **:** MASTER

Mas·sa·chu·set *or* **Mas·sa·chu·sett** \ˌma-sə-ˈchü-sət, -zət *also* ˌmas-ˈchü-\ *noun, plural* **Massachusets** *or* **Massachuset** *or* **Massachusetts** *or* **Massachusett** [Massachuset, a locality, literally, at the big hill] (1616)
1 : a member of an American Indian people of Massachusetts
2 : the extinct Algonquian language of the Massachuset people

¹mas·sa·cre \ˈma-si-kər\ *noun* [Middle French] (circa 1578)
1 : the act or an instance of killing a number of usually helpless or unresisting human beings under circumstances of atrocity or cruelty
2 : a cruel or wanton murder
3 : a wholesale slaughter of animals
4 : an act of complete destruction ⟨the author's *massacre* of traditional federalist presuppositions —R. G. McCloskey⟩

²massacre *transitive verb* **mas·sa·cred; mas·sa·cring** \-k(ə-)riŋ\ (1581)
1 : to kill by massacre **:** SLAUGHTER
2 : MANGLE 2 ⟨words were misspelled and syntax *massacred* —Bice Clemow⟩
— mas·sa·crer \-kər-ər, -krər\ *noun*

¹mas·sage \mə-ˈsäzh, -ˈsäj\ *noun* [French, from *masser* to massage, from Arabic *massa* to stroke] (circa 1860)
: manipulation of tissues (as by rubbing, kneading, or tapping) with the hand or an instrument for therapeutic purposes

²massage *transitive verb* **mas·saged; mas·sag·ing** (1887)
1 : to subject to massage
2 a : to treat flatteringly **:** BLANDISH **b :** MANIPULATE 3, DOCTOR 2b ⟨researchers *massaged* the data to support their thesis⟩
— mas·sag·er *noun*

massage parlor *noun* (1913)
: an establishment that provides massage treatments; *also* **:** one offering sexual services in addition to or in lieu of massage

mas·sa·sau·ga \ˌma-sə-ˈsȯ-gə\ *noun* [*Missisauga* River, Ontario, Canada] (1835)
: a small North American rattlesnake (*Sistrurus catenatus*)

mass card *noun* (1948)
: a card notifying the recipient (as a bereaved family) that a mass is to be offered for the repose of the soul of a specified deceased person

mass defect *noun* (circa 1923)
: the difference between the mass of an isotope and its mass number

mass driver *noun* (1977)
: a large electromagnetic catapult designed to hurl material (as from an asteroid) into space

mas·sé \ma-ˈsā\ *noun* [French, from past participle of *masser* to make a massé shot, from *masse* sledgehammer, from Middle French *mace* mace] (1873)
: a shot in billiards or pool made by hitting the cue ball vertically or nearly vertically on the side to drive it around one ball in order to strike another

mas·se·ter \mə-ˈsē-tər, ma-\ *noun* [New Latin, from Greek *masētēr,* from *masasthai* to chew — more at MANDIBLE] (1666)
: a large muscle that raises the lower jaw and assists in mastication
— mas·se·ter·ic \ˌma-sə-ˈter-ik\ *adjective*

mas·seur \ma-ˈsər, mə-\ *noun* [French, from *masser*] (1876)
: a man who practices massage

mas·seuse \-ˈsə(r)z, -ˈsüz\ *noun* [French, feminine of *masseur*] (1876)

: a woman who practices massage

mas•sif \ma-'sēf\ *noun* [French, from *massif,* adjective, from Middle French] (1885)

1 : a principal mountain mass

2 : a block of the earth's crust bounded by faults or flexures and displaced as a unit without internal change

mas•sive \'ma-siv\ *adjective* [Middle English *massiffe,* from Middle French *massif,* from *masse* mass] (15th century)

1 : forming or consisting of a large mass: **a :** BULKY **b :** WEIGHTY, HEAVY ⟨*massive* walls⟩ ⟨a *massive* volume⟩ **c :** impressively large or ponderous **d :** having no regular form but not necessarily lacking crystalline structure ⟨*massive* sandstone⟩

2 a : large, solid, or heavy in structure ⟨*massive* jaw⟩ **b :** large in scope or degree ⟨the feeling of frustration, of being ineffectual, is *massive* —David Halberstam⟩ **c** (1) **:** large in comparison to what is typical ⟨*massive* dose of penicillin⟩ (2) **:** being extensive and severe ⟨*massive* hemorrhage⟩ ⟨*massive* collapse of a lung⟩ (3) **:** imposing in excellence or grandeur **:** MONUMENTAL ⟨*massive* simplicity⟩

3 : having mass ⟨a *massive* boson⟩

— mas•sive•ly *adverb*

— mas•sive•ness *noun*

mass•less \'mas-ləs\ *adjective* (1879)

: having no mass ⟨a *massless* particle⟩

mass–mar•ket \'mas-'mär-kət\ *adjective* (1952)

: sold through such retail outlets as supermarkets and drugstores as well as through bookstores ⟨a *mass-market* paperback⟩; *also* **:** of, relating to, or publishing mass-market materials

mass medium *noun, plural* **mass media** (1923)

: a medium of communication (as newspapers, radio, or television) that is designed to reach the mass of the people — usually used in plural

mass noun *noun* (1933)

: a noun (as *sand* or *water*) that characteristically denotes in many languages a homogeneous substance or a concept without subdivisions and that in English is preceded in indefinite singular constructions by *some* rather than *a* or *an* — compare COUNT NOUN

mass number *noun* (1923)

: an integer that approximates the mass of an isotope and designates the number of nucleons in the nucleus

mass–pro•duce \,mas-prə-'düs, -'dyüs\ *transitive verb* [back-formation from *mass production*] (1923)

: to produce in quantity usually by machinery

— mass production *noun*

mass spectrograph *noun* (1920)

: an instrument used to separate and often to determine the masses of isotopes

mass spectrometry *noun* (1943)

: an instrumental method for identifying the chemical constitution of a substance by means of the separation of gaseous ions according to their differing mass and charge — called also *mass spectroscopy*

— mass spectrometric *adjective*

— mass spectrometer *noun*

mass spectrum *noun* (1920)

: the spectrum of a stream of gaseous ions separated according to their mass and charge

massy \'ma-sē\ *adjective* (14th century)

: MASSIVE

¹mast \'mast\ *noun* [Middle English, from Old English *mæst;* akin to Old High German *mast* mast, Latin *malus*] (before 12th century)

1 : a long pole or spar rising from the keel or deck of a ship and supporting the yards, booms, and rigging

2 : a slender vertical or nearly vertical structure (as an upright post in various cranes)

3 : a disciplinary proceeding at which the commanding officer of a naval unit hears and disposes of cases against enlisted men — called also *captain's mast*

— mast•ed \'mas-təd\ *adjective*

— before the mast 1 : forward of the foremast **2 :** as a common sailor

²mast *transitive verb* (1627)

: to furnish with a mast

³mast *noun* [Middle English, from Old English *mæst;* akin to Old High German *mast* food, mast, and probably to Old English *mete* food — more at MEAT] (before 12th century)

: nuts (as acorns) accumulated on the forest floor and often serving as food for animals (as hogs)

mas•ta•ba \'mas-tə-bə\ *noun* [Arabic *maṣṭabah* stone bench] (1882)

: an Egyptian tomb of the time of the Memphite dynasties that is oblong in shape with sloping sides and a flat roof

mast cell \'mast-\ *noun* [part translation of German *Mastzelle,* from *Mast* food, mast (from Old High German) + *Zelle* cell] (circa 1890)

: a large cell that occurs especially in connective tissue and has basophilic granules containing substances (as histamine and heparin) which mediate allergic reactions

mas•tec•to•my \ma-'stek-tə-mē\ *noun, plural* **-mies** [Greek *mastos* breast + English *-ectomy*] (circa 1923)

: excision of the breast

¹mas•ter \'mas-tər\ *noun* [Middle English, from Old English *magister* & Old French *maistre,* both from Latin *magister;* akin to Latin *magnus* large — more at MUCH] (before 12th century)

1 a (1) **:** a male teacher (2) **:** a person holding an academic degree higher than a bachelor's but lower than a doctor's **b** *often capitalized* **:** a revered religious leader **c :** a worker or artisan qualified to teach apprentices **d** (1) **:** an artist, performer, or player of consummate skill (2) **:** a great figure of the past (as in science or art) whose work serves as a model or ideal

2 a : one having authority over another **:** RULER, GOVERNOR **b :** one that conquers or masters **:** VICTOR, SUPERIOR ⟨in this young, obscure challenger the champion found his *master*⟩ **c :** a person licensed to command a merchant ship **d** (1) **:** one having control (2) **:** an owner especially of a slave or animal **e :** the employer especially of a servant **f** (1) *dialect* **:** HUSBAND (2) **:** the male head of a household

3 a (1) *archaic* **:** MR. (2) **:** a youth or boy too young to be called *mister* — used as a title **b :** the eldest son of a Scottish viscount or baron

4 a : a presiding officer in an institution or society (as a college) **b :** any of several officers of court appointed to assist (as by hearing and reporting) a judge

5 a : a master mechanism or device **b :** an original from which copies can be made; *especially* **:** a master phonograph record or magnetic tape

— mas•ter•ship \-,ship\ *noun*

²master *adjective* (12th century)

: being or relating to a master: as **a :** having chief authority **:** DOMINANT **b :** SKILLED, PROFICIENT ⟨a prosperous *master* builder —*Current Biography*⟩ **c :** PRINCIPAL, PREDOMINANT **d :** SUPERLATIVE — often used in combination ⟨a *master*-liar⟩ **e :** being a device or mechanism that controls the operation of another mechanism or that establishes a standard (as a dimension or weight) **f :** being or relating to a master from which duplicates are made

³master *transitive verb* **mas•tered; mas•ter•ing** \-t(ə-)riŋ\ (13th century)

1 : to become master of **:** OVERCOME

2 a : to become skilled or proficient in the use of ⟨*master* a foreign language⟩ **b :** to gain a thorough understanding of ⟨had *mastered* every aspect of publishing —*Current Biography*⟩

3 : to produce a master phonograph record or magnetic tape of (as a musical rendition)

master–at–arms *noun, plural* **masters–at–arms** (1748)

: a petty officer charged with maintaining discipline aboard ship

master bedroom *noun* (1925)

: a large or principal bedroom

master chief petty officer *noun* (1958)

: an enlisted man in the navy or coast guard ranking above a senior chief petty officer

master chief petty officer of the coast guard (1966)

: the ranking petty officer in the coast guard serving as adviser to the commandant

master chief petty officer of the navy (1966)

: the ranking petty officer in the navy serving as adviser to the chief of naval operations

master class *noun* (1952)

: a seminar for advanced music students conducted by a master musician

mas•ter•ful \'mas-tər-fəl\ *adjective* (15th century)

1 a : inclined and usually competent to play the master **b :** suggestive of a domineering nature

2 : having or reflecting the power and skill of a master ☆ ■

— mas•ter•ful•ly \-fə-lē\ *adverb*

— mas•ter•ful•ness *noun*

master gunnery sergeant *noun* (1958)

: a noncommissioned officer in the marine corps ranking above a master sergeant

master key *noun* (1576)

: a key designed to open several different locks

mas•ter•ly \'mas-tər-lē\ *adjective* (15th century)

: suitable to or resembling that of a master; *especially* **:** indicating thorough knowledge or superior skill and power ⟨a *masterly* performance⟩

☆ SYNONYMS

Masterful, domineering, imperious, peremptory, imperative mean tending to impose one's will on others. MASTERFUL implies a strong personality and ability to act authoritatively ⟨her *masterful* personality soon dominated the movement⟩. DOMINEERING suggests an overbearing or arbitrary manner and an obstinate determination to enforce one's will ⟨children controlled by *domineering* parents⟩. IMPERIOUS implies a commanding nature or manner and often suggests arrogant assurance ⟨an *imperious* executive used to getting his own way⟩. PEREMPTORY implies an abrupt dictatorial manner coupled with an unwillingness to brook disobedience or dissent ⟨given a *peremptory* dismissal⟩. IMPERATIVE implies peremptoriness arising more from the urgency of the situation than from an inherent will to dominate ⟨an *imperative* appeal for assistance⟩.

□ USAGE

masterful Some commentators insist that use of *masterful* should be limited to sense 1 in order to preserve a distinction between it and *masterly.* The distinction is a modern one, excogitated by a 20th century pundit in disregard of the history of the word. Both words developed in a parallel manner but the earlier sense of *masterly,* equivalent to *masterful* 1, dropped out of use. Since *masterly* had but one sense, the pundit opined that it would be tidy if *masterful* were likewise limited to one sense and he forthwith condemned use of *masterful* 2 as an error. Sense 2 of *masterful,* which is slightly older than the sense of *masterly* intended to replace it, has continued in reputable use all along; it cannot rationally be called an error.

usage see MASTERFUL
— **mas·ter·li·ness** *noun*
— **masterly** *adverb*

[1]**mas·ter·mind** \'mas-tər-ˌmīnd, ˌmas-tər-'\ *noun* (1720)
: a person who supplies the directing or creative intelligence for a project

[2]**mastermind** *transitive verb* (1940)
: to be the mastermind of

master of arts *often M&A capitalized* (15th century)
1 : the recipient of a master's degree that usually signifies that the recipient has passed an integrated course of study in one or more of the humanities and sometimes has completed a thesis involving research or a creative project and that typically requires two years of work beyond a bachelor's degree
2 : the degree making one a master of arts — abbreviation *M.A., A.M.*

master of ceremonies (1662)
1 : a person who determines the forms to be observed on a public occasion
2 : a person who acts as host at a formal event
3 : a person who acts as host for a program of entertainment (as on television)

master of science *often M&S capitalized* (circa 1905)
1 : the recipient of a master's degree that usually signifies that the recipient has passed an integrated course of study in one or more of the sciences and sometimes has completed a thesis involving research and that typically requires two years of work beyond a bachelor's degree
2 : the degree making one a master of science — abbreviation *M.S., M.Sc.*

mas·ter·piece \'mas-tər-ˌpēs\ *noun* (1605)
1 : a work done with extraordinary skill; *especially* **:** a supreme intellectual or artistic achievement
2 : a piece of work presented to a medieval guild as evidence of qualification for the rank of master

master plan *noun* (circa 1930)
: a plan giving overall guidance

master race *noun* (1937)
: a people held to be racially preeminent and hence fitted to rule or enslave other peoples

master's *noun* (1939)
: a master's degree

master sergeant *noun* (circa 1934)
: a noncommissioned officer ranking in the army above a sergeant first class and below a staff sergeant major, in the air force above a technical sergeant and below a senior master sergeant, and in the marine corps above a gunnery sergeant and below a master gunnery sergeant

mas·ter·sing·er \'mas-tər-ˌsiŋ-ər\ *noun* (1810)
: MEISTERSINGER

mas·ter·stroke \-ˌstrōk\ *noun* (1679)
: a masterly performance or move

mas·ter·work \-ˌwərk\ *noun* (1617)
: MASTERPIECE

mas·tery \'mas-t(ə-)rē\ *noun* [Middle English *maistrie,* from Old French, from *maistre* master] (13th century)
1 a : the authority of a master **:** DOMINION **b :** the upper hand in a contest or competition **:** SUPERIORITY, ASCENDANCY
2 a : possession or display of great skill or technique **b :** skill or knowledge that makes one master of a subject **:** COMMAND

mast·head \'mast-ˌhed\ *noun* (1748)
1 : the top of a mast
2 a : the printed matter in a newspaper or periodical that gives the title and pertinent details of ownership, advertising rates, and subscription rates **b :** the name of a publication (as a newspaper) displayed on the top of the first page

mas·tic \'mas-tik\ *noun* [Middle English *mastik,* from Latin *mastiche,* from Greek *mastichē,* probably back-formation from *mastichan*] (14th century)
1 : an aromatic resinous exudate from mastic trees used chiefly in varnishes
2 : any of various pasty materials used as protective coatings or cements

mas·ti·cate \'mas-tə-ˌkāt\ *verb* **-cat·ed; -cat·ing** [Late Latin *masticatus,* past participle of *masticare,* from Greek *mastichan* to gnash the teeth; akin to Greek *masasthai* to chew — more at MANDIBLE] (1649)
transitive verb
1 : to grind or crush (food) with or as if with the teeth in preparation for swallowing **:** CHEW
2 : to soften or reduce to pulp by crushing or kneading
intransitive verb
: CHEW
— **mas·ti·ca·tion** \ˌmas-tə-'kā-shən\ *noun*
— **mas·ti·ca·tor** \'mas-tə-ˌkā-tər\ *noun*

[1]**mas·ti·ca·to·ry** \'mas-ti-kə-ˌtōr-ē, -ˌtȯr-\ *adjective* (1694)
1 : used for or adapted to chewing ⟨*masticatory* limbs of an arthropod⟩
2 : of, relating to, or involving the organs of mastication ⟨*masticatory* paralysis⟩

[2]**masticatory** *noun, plural* **-ries** (circa 1611)
: a substance chewed to increase saliva

mastic tree *noun* (15th century)
: a small southern European evergreen tree (*Pistacia lentiscus*) of the cashew family that yields mastic

mas·tiff \'mas-təf\ *noun* [Middle English *mastif,* modification of Middle French *mastin,* from (assumed) Vulgar Latin *mansuetinus,* from Latin *mansuetus* tame — more at MANSUETUDE] (14th century)
: any of a breed of very large massive powerful smooth-coated dogs that are apricot, fawn, or brindle and are often used as guard dogs

mastiff

mas·ti·goph·o·ran \ˌmas-tə-'gä-fə-rən\ *noun* [ultimately from Greek *mastig-, mastix* whip + *pherein* to carry — more at BEAR] (circa 1911)
: any of a subphylum (Mastigophora) of protozoans comprising forms with flagella and including many often treated as algae
— **mastigophoran** *adjective*

mas·ti·tis \ma-'stī-təs\ *noun, plural* **-tit·i·des** \-'sti-tə-ˌdēz\ [New Latin, from Greek *mastos* breast] (circa 1842)
: inflammation of the breast or udder usually caused by infection
— **mas·tit·ic** \-'sti-tik\ *adjective*

mas·to·don \'mas-tə-ˌdän, -dən\ *noun* [New Latin *mastodont-, mastodon,* from Greek *mastos* + *odont-, odōn, odous* tooth — more at TOOTH] (1813)
1 : any of numerous extinct mammals (genus *Mastodon* synonym *Mammut*) that differ from the related mammoths and existing elephants chiefly in the form of the molar teeth
2 : one that is unusually large
— **mas·to·don·ic** \ˌmas-tə-'dä-nik\ *adjective*
— **mas·to·dont** \'mas-tə-ˌdänt\ *adjective or noun*

[1]**mas·toid** \'mas-ˌtȯid\ *adjective* [New Latin *mastoides* resembling a nipple, mastoid, from Greek *mastoeidēs,* from *mastos* breast] (1732)
1 : being the process of the temporal bone behind the ear; *also* **:** being any of several bony elements that occupy a similar position in the skull of lower vertebrates
2 : of, relating to, or occurring in the region of the mastoid process

[2]**mastoid** *noun* (1842)
: a mastoid bone or process

mastoid cell *noun* (1800)
: one of the small cavities in the mastoid process that develop after birth and are filled with air

mas·toid·ec·to·my \ˌmas-ˌtȯi-'dek-tə-mē\ *noun, plural* **-mies** [International Scientific Vocabulary] (1898)
: surgical removal of part of the mastoid process of the temporal bone

mas·toid·itis \ˌmas-ˌtȯi-'dī-təs\ *noun* [New Latin] (circa 1890)
: inflammation of the mastoid and especially of the mastoid cells

mas·tur·bate \'mas-tər-ˌbāt\ *verb* **-bat·ed; -bat·ing** [Latin *masturbatus,* past participle of *masturbari*] (1857)
transitive verb
: to practice masturbation on
intransitive verb
: to practice masturbation
— **mas·tur·ba·tor** \-ˌbā-tər\ *noun*

mas·tur·ba·tion \ˌmas-tər-'bā-shən\ *noun* (1766)
: erotic stimulation especially of one's own genital organs commonly resulting in orgasm and achieved by manual or other bodily contact exclusive of sexual intercourse, by instrumental manipulation, occasionally by sexual fantasies, or by various combinations of these agencies

mas·tur·ba·to·ry \'mas-tər-bə-ˌtōr-ē, -ˌtȯr-\ *adjective* (1864)
1 : of, relating to, or involving masturbation ⟨*masturbatory* fantasies⟩
2 : excessively self-absorbed or self-indulgent ⟨write tedious, *masturbatory* books . . . about themselves for people to read . . . with envy —D. R. Katz⟩

[1]**mat** \'mat\ *noun* [Middle English, from Old English *meatte,* from Late Latin *matta,* of Semitic origin; akin to Hebrew *miṭṭāh* bed] (before 12th century)
1 a (1) **:** a piece of coarse, woven, plaited, or felted fabric used especially as a floor covering or a support (2) **:** a piece of material placed at a door for wiping soiled shoe soles **b :** a decorative piece of material used under a small item (as a dish) especially for support or protection **c :** a large thick pad or cushion used as a surface for wrestling, tumbling, and gymnastics
2 : something made up of densely tangled or adhering strands especially of organic matter ⟨algal *mat*⟩ ⟨a *mat* of unkempt hair⟩
3 : a large slab usually of reinforced concrete used as the supporting base of a building

[2]**mat** *verb* **mat·ted; mat·ting** (1549)
transitive verb
1 : to provide with a mat or matting
2 a : to form into a tangled mass **b :** to pack down so as to form a dense mass
intransitive verb
: to become matted

[3]**mat** \'mat\ *transitive verb* **mat·ted; mat·ting** (1602)
1 : to make (as a metal, glass, or color) matte
2 : to provide (a picture) with a mat

[4]**mat** *variant of* [2]MATTE

[5]**mat** *noun* [French *mat* dull color, unpolished surface, from *mat,* adjective — more at MATTE] (1845)
: a border going around a picture between picture and frame or serving as the frame

[6]**mat** *noun* (1904)
: MATRIX 2a

mat·a·dor \'ma-tə-ˌdȯr\ *noun* [Spanish, from *matar* to kill] (1681)
: a bullfighter who has the principal role and who kills the bull in a bullfight

[1]**match** \'mach\ *noun* [Middle English *macche,* from Old English *gemæcca* mate,

\ə\ **abut** \ᵊ\ **kitten** \ər\ **further** \a\ **ash** \ā\ **ace** \ä\ **mop, mar** \au̇\ **out** \ch\ **chin** \e\ **bet** \ē\ **easy** \g\ **go** \i\ **hit** \ī\ **ice** \j\ **job** \ŋ\ **sing** \ō\ **go** \ȯ\ **law** \ȯi\ **boy** \th\ **thin** \t͟h\ **the** \ü\ **loot** \u̇\ **foot** \y\ **yet** \zh\ **vision** *see also* Guide to Pronunciation

equal; akin to Old English *macian* to make — more at MAKE] (before 12th century)
1 a : a person or thing equal or similar to another **b :** one able to cope with another **c :** an exact counterpart
2 : a pair suitably associated ⟨carpet and curtains are a *match*⟩
3 a : a contest between two or more parties ⟨a golf *match*⟩ ⟨a soccer *match*⟩ ⟨a shouting *match*⟩ **b :** a contest (as in tennis or volleyball) completed when one player or side wins a specified number of sets or games
4 a : a marriage union **b :** a prospective partner in marriage

²match (14th century)
transitive verb
1 a : to encounter successfully as an antagonist **b** (1) **:** to set in competition or opposition (2) **:** to provide with a worthy competitor **c :** to set in comparison
2 : to join or give in marriage
3 a (1) **:** to put in a set possessing equal or harmonizing attributes (2) **:** to cause to correspond **:** SUIT **b** (1) **:** to be the counterpart of; *also* **:** to compare favorably with (2) **:** to harmonize with **c :** to provide with a counterpart **d :** to provide funds complementary to
4 : to fit together or make suitable for fitting together
5 a : to flip or toss (coins) and compare exposed faces **b :** to toss coins with
intransitive verb
: to be a counterpart
— match·able \ˈma-chə-bəl\ *adjective*
— match·er *noun*

³match *noun* [Middle English *macche*, from Middle French *meiche*] (1549)
1 : a chemically prepared wick or cord formerly used in firing firearms or powder
2 : a short slender piece of flammable material (as wood) tipped with a combustible mixture that bursts into flame when slightly heated through friction (as by being scratched against a rough surface)

match·board \ˈmach-ˌbōrd, -ˌbȯrd\ *noun* (circa 1858)
: a board with a groove cut along one edge and a tongue along the other so as to fit snugly with the edges of similarly cut boards

match·book \-ˌbu̇k\ *noun* (1944)
: a small folder containing rows of paper matches

match·box \-ˌbäks\ *noun* (1786)
: a box for matches

match·less \-ləs\ *adjective* (circa 1530)
: having no equal **:** PEERLESS
— match·less·ly *adverb*

match·lock \-ˌläk\ *noun* (1637)
1 : a slow-burning match lowered over a hole in the breech of a musket to ignite the charge
2 : a musket equipped with a matchlock

match·mak·er \-ˌmā-kər\ *noun* (circa 1639)
: one that arranges a match; *especially* **:** one who tries to bring two unmarried individuals together in an attempt to promote a marriage
— match·mak·ing \-kiŋ\ *noun*

match play *noun* (1893)
: golf competition in which the winner is the person or team winning the greater number of holes — compare STROKE PLAY

match point *noun* (1921)
: a situation (as in tennis) in which one player or side will win the match by winning the next point; *also* **:** the point itself

match·stick \ˈmach-ˌstik\ *noun* (1791)
1 : a slender piece especially of wood from which a match is made
2 : something resembling a matchstick especially in slenderness

match·up \-ˌəp\ *noun* (1964)
: ¹MATCH

match·wood \-ˌwu̇d\ *noun* (1838)
: small pieces of wood **:** SPLINTERS

¹mate \ˈmāt\ *transitive verb* **mat·ed; mat·ing** [Middle English, from Middle French *mater*, from Old French *mat*, noun, checkmate, from Arabic *māt* (in *shāh māt*)] (14th century)
: CHECKMATE 2

²mate *noun* (14th century)
: CHECKMATE 1

³mate *noun* [Middle English, probably from Middle Low German *māt*; akin to Old English *gemetta* guest at one's table, *mete* food — more at MEAT] (14th century)
1 a (1) **:** ASSOCIATE, COMPANION (2) *chiefly British* **:** an assistant to a more skilled worker **:** HELPER (3) *chiefly British* **:** FRIEND, BUDDY — often used as a familiar form of address **b** *archaic* **:** MATCH, PEER
2 : a deck officer on a merchant ship ranking below the captain
3 : one of a pair: as **a :** either member of a couple and especially a married couple **b :** either member of a breeding pair of animals **c :** either of two matched objects

⁴mate *verb* **mat·ed; mat·ing** (1509)
transitive verb
1 *archaic* **:** EQUAL, MATCH
2 : to join or fit together **:** COUPLE
3 a : to join together as mates **b :** to provide a mate for
intransitive verb
1 : to become mated ⟨gears that *mate* well⟩
2 : COPULATE

ma·té *or* **ma·te** \ˈmä-ˌtā\ *noun* [French & American Spanish; French *maté*, from American Spanish *mate* maté, vessel for drinking it, from Quechua *mati* vessel] (1758)
1 : a tealike beverage drunk especially in South America
2 : a South American shrub or tree (*Ilex paraguariensis*) of the holly family whose leaves and shoots are used in making maté; *also* **:** these leaves and shoots

mate·lot \ˈmat-ˌlō, ˈma-tə-\ *noun* [French, from Middle French, from Middle Dutch *mattenoot*, literally, bedmate] (1911)
British **:** SAILOR

ma·te·lote \ˌma-tᵊl-ˈōt, mat-ˈlōt\ *noun* [French, literally, sailor's wife, from *matelot*] (circa 1736)
: a stew made usually of fish in a seasoned wine sauce

ma·ter \ˈmā-tər\ *noun* [Latin — more at MOTHER] (circa 1859)
chiefly British **:** MOTHER

ma·ter·fa·mil·i·as \ˌmā-tər-fə-ˈmi-lē-əs, ˌmä-\ *noun* [Latin, from *mater* + *familias*, archaic genitive of *familia* household — more at FAMILY] (1756)
: a woman who is head of a household

¹ma·te·ri·al \mə-ˈtir-ē-əl\ *adjective* [Middle English *materiel*, from Middle French & Late Latin; Middle French, from Late Latin *materialis*, from Latin *materia* matter — more at MATTER] (14th century)
1 a (1) **:** relating to, derived from, or consisting of matter; *especially* **:** PHYSICAL ⟨the *material* world⟩ (2) **:** BODILY ⟨*material* needs⟩ **b** (1) **:** of or relating to matter rather than form ⟨*material* cause⟩ (2) **:** of or relating to the subject matter of reasoning; *especially* **:** EMPIRICAL ⟨*material* knowledge⟩
2 : having real importance or great consequences ⟨facts *material* to the investigation⟩
3 a : being of a physical or worldly nature **b :** relating to or concerned with physical rather than spiritual or intellectual things ⟨*material* progress⟩ ☆
— ma·te·ri·al·ly \-ē-ə-lē\ *adverb*
— ma·te·ri·al·ness *noun*

²material *noun* (1556)
1 a (1) **:** the elements, constituents, or substances of which something is composed or can be made (2) **:** matter that has qualities which give it individuality and by which it may be categorized ⟨sticky *material*⟩ ⟨explosive *materials*⟩ **b** (1) **:** something (as data) that may be worked into a more finished form ⟨*material* for a biography⟩ (2) **:** something used for or made the object of study ⟨*material* for the next semester⟩ (3) **:** a performer's repertoire ⟨a comedian's *material*⟩ **c :** MATTER 3b **d :** CLOTH **e :** a person potentially suited to some pursuit ⟨varsity *material*⟩ ⟨leadership *material*⟩
2 a : apparatus necessary for doing or making something ⟨writing *materials*⟩ **b :** MATÉRIEL

ma·te·ri·al·ise *British variant of* MATERIALIZE

ma·te·ri·al·ism \mə-ˈtir-ē-ə-ˌli-zəm\ *noun* (1748)
1 a : a theory that physical matter is the only or fundamental reality and that all being and processes and phenomena can be explained as manifestations or results of matter **b :** a doctrine that the only or the highest values or objectives lie in material well-being and in the furtherance of material progress **c :** a doctrine that economic or social change is materially caused — compare HISTORICAL MATERIALISM
2 : a preoccupation with or stress upon material rather than intellectual or spiritual things
— ma·te·ri·al·ist \-list\ *noun or adjective*
— ma·te·ri·al·is·tic \-ˌtir-ē-ə-ˈlis-tik\ *adjective*
— ma·te·ri·al·is·ti·cal·ly \-ti-k(ə-)lē\ *adverb*

ma·te·ri·al·i·ty \mə-ˌtir-ē-ˈa-lə-tē\ *noun, plural* **-ties** (1570)
1 : the quality or state of being material
2 : something that is material

ma·te·ri·al·i·za·tion \mə-ˌtir-ē-ə-lə-ˈzā-shən\ *noun* (1843)
1 : the action of materializing or becoming materialized
2 : something that has been materialized; *especially* **:** APPARITION

ma·te·ri·al·ize \mə-ˈtir-ē-ə-ˌlīz\ *verb* **-ized; -iz·ing** (1710)
transitive verb
1 a : to make material **:** OBJECTIFY **b :** to cause to appear in bodily form ⟨*materialize* the spirits of the dead⟩
2 : to cause to be materialistic
intransitive verb
1 : to assume bodily form
2 a : to appear especially suddenly **b :** to come into existence
— ma·te·ri·al·iz·er *noun*

materials science *noun* (1961)
: the scientific study of the properties and applications of materials of construction or manufacture (as ceramics, metals, polymers, and composites)
— materials scientist *noun*

☆ SYNONYMS
Material, physical, corporeal, phenomenal, sensible, objective mean of or belonging to actuality. MATERIAL implies formation out of tangible matter; used in contrast with *spiritual* or *ideal* it may connote the mundane, crass, or grasping ⟨*material* values⟩. PHYSICAL applies to what is perceived directly by the senses and may contrast with *mental, spiritual,* or *imaginary* ⟨the *physical* benefits of exercise⟩. CORPOREAL implies having the tangible qualities of a body such as shape, size, or resistance to force ⟨artists have portrayed angels as *corporeal* beings⟩. PHENOMENAL applies to what is known or perceived through the senses rather than by intuition or rational deduction ⟨scientists concerned with the *phenomenal* world⟩. SENSIBLE stresses the capability of readily or forcibly impressing the senses ⟨the earth's rotation is not *sensible* to us⟩. OBJECTIVE may stress material or independent existence apart from a subject perceiving it ⟨no *objective* evidence of damage⟩. See in addition RELEVANT.

ma·te·ria med·i·ca \mə-ˈtir-ē-ə-ˈme-di-kə\ *noun* [New Latin, literally, medical matter] (1699)
1 : substances used in the composition of medical remedies **:** DRUGS, MEDICINE
2 a : a branch of medical science that deals with the sources, nature, properties, and preparation of drugs **b :** a treatise on materia medica

ma·té·ri·el *or* **ma·te·ri·el** \mə-ˌtir-ē-ˈel\ *noun* [French *matériel,* from *matériel,* adjective] (1814)
: equipment, apparatus, and supplies used by an organization or institution

ma·ter·nal \mə-ˈtər-nᵊl\ *adjective* [Middle English, from Middle French *maternel,* from Latin *maternus,* from *mater* mother — more at MOTHER] (15th century)
1 : of, relating to, belonging to, or characteristic of a mother **:** MOTHERLY
2 a : related through a mother ⟨his *maternal* aunt⟩ **b :** inherited or derived from the female parent ⟨*maternal* genes⟩
— ma·ter·nal·ly \-nᵊl-ē\ *adverb*

¹ma·ter·ni·ty \mə-ˈtər-nə-tē\ *noun, plural* **-ties** (1611)
1 a : the quality or state of being a mother **:** MOTHERHOOD **b :** the qualities of a mother **:** MOTHERLINESS
2 : a hospital facility designed for the care of women before and during childbirth and for the care of newborn babies

²maternity *adjective* (1893)
1 : designed for wear during pregnancy ⟨a *maternity* dress⟩
2 : effective for the period close to and including childbirth ⟨*maternity* leave⟩

mat·ey \ˈmā-tē\ *adjective* (1915)
chiefly British **:** COMPANIONABLE
— mat·ey·ness \-nəs\ *noun, chiefly British*

math \ˈmath\ *noun* (circa 1878)
: MATHEMATICS

math·e·mat·i·cal \ˌmath-ˈma-ti-kəl, ˌma-thə-\ *also* **math·e·mat·ic** \-tik\ *adjective* [Middle English *mathematicalle,* from Latin *mathematicus,* from Greek *mathēmatikos,* from *mathēmat-, mathēma* learning, mathematics, from *manthanein* to learn; probably akin to Gothic *mundon* to pay attention] (15th century)
1 : of, relating to, or according with mathematics
2 a : rigorously exact **:** PRECISE **b :** CERTAIN
3 : possible but highly improbable ⟨only a *mathematical* chance⟩
— math·e·mat·i·cal·ly \-ti-k(ə-)lē\ *adverb*

mathematical expectation *noun* (1838)
: EXPECTED VALUE

mathematical induction *noun* (1838)
: INDUCTION 2b

mathematical logic *noun* (1858)
: SYMBOLIC LOGIC

math·e·ma·ti·cian \ˌmath-mə-ˈti-shən, ˌma-thə-\ *noun* (15th century)
: a specialist or expert in mathematics

math·e·mat·ics \ˌmath-ˈma-tiks, ˌma-thə-\ *noun plural but usually singular in construction* (1581)
1 : the science of numbers and their operations, interrelations, combinations, generalizations, and abstractions and of space configurations and their structure, measurement, transformations, and generalizations
2 : a branch of, operation in, or use of mathematics ⟨the *mathematics* of physical chemistry⟩

math·e·ma·ti·za·tion \ˌmath-mə-tə-ˈzā-shən, ˌma-thə-\ *noun* (1928)
: reduction to mathematical form
— math·e·ma·tize \ˈmath-mə-ˌtiz, ˈma-thə-\ *verb*

maths \ˈmaths\ *noun plural* (1911)
chiefly British **:** MATHEMATICS

mat·in \ˈma-tᵊn\ *adjective* [Middle English, from Old French] (14th century)
: of or relating to matins or to early morning

mat·in·al \ˈma-tᵊn-əl\ *adjective* (1803)
1 : of or relating to matins
2 : EARLY

mat·i·nee *or* **mat·i·née** \ˌma-tᵊn-ˈā\ *noun* [French *matinée,* literally, morning, from Old French, from *matin* morning, from Latin *matutinum,* from neuter of *matutinus* of the morning, from *Matuta,* goddess of morning; akin to Latin *maturus* ripe — more at MATURE] (1858)
: a musical or dramatic performance or social or public event held in the daytime and especially the afternoon

matinee idol *noun* (1902)
: a handsome male performer

mat·ins \ˈma-tᵊnz\ *noun plural but singular or plural in construction, often capitalized* [Middle English *matines,* from Old French, from Late Latin *matutinae,* from Latin, feminine plural of *matutinus*] (14th century)
1 : the night office forming with lauds the first of the canonical hours
2 : MORNING PRAYER

matr- *or* **matri-** *or* **matro-** *combining form* [Latin *matr-, matri-,* from *matr-, mater*]
: mother ⟨*matri*arch⟩ ⟨*matro*nymic⟩

ma·tri·arch \ˈmā-trē-ˌärk\ *noun* (1606)
: a female who rules or dominates a family, group, or state; *specifically* **:** a mother who is head and ruler of her family and descendants
— ma·tri·ar·chal \ˌmā-trē-ˈär-kəl\ *adjective*

ma·tri·ar·chate \ˈmā-trē-ˌär-kət, -ˌkāt\ *noun* (1885)
: MATRIARCHY 1

ma·tri·ar·chy \ˈmā-trē-ˌär-kē\ *noun, plural* **-chies** (1885)
1 : a family, group, or state governed by a matriarch
2 : a system of social organization in which descent and inheritance are traced through the female line

ma·tri·cide \ˈma-trə-ˌsīd, ˈmā-\ *noun* (1594)
1 [Latin *matricidium,* from *matr-* + *-cidium* -cide] **:** murder of a mother by her son or daughter
2 [Latin *matricida,* from *matr-* + *-cida* -cide] **:** one that murders his or her mother
— ma·tri·cid·al \ˌma-trə-ˈsī-dᵊl, ˌmā-\ *adjective*

ma·tric·u·late \mə-ˈtri-kyə-ˌlāt\ *verb* **-lat·ed; -lat·ing** [Medieval Latin *matriculatus,* past participle of *matriculare,* from Late Latin *matricula* public roll, diminutive of *matric-, matrix* list, from Latin, breeding female] (1577)
transitive verb
: to enroll as a member of a body and especially of a college or university
intransitive verb
: to become matriculated
— ma·tric·u·lant \-lənt\ *noun*
— ma·tric·u·la·tion \-ˌtri-kyə-ˈlā-shən\ *noun*

ma·tri·lin·eal \ˌma-trə-ˈli-nē-əl, ˌmā-\ *adjective* (1904)
: relating to, based on, or tracing descent through the maternal line ⟨*matrilineal* society⟩
— ma·tri·lin·eal·ly \-nē-ə-lē\ *adverb*

mat·ri·mo·nial \ˌma-trə-ˈmō-nē-əl, -nyəl\ *adjective* (15th century)
: of or relating to marriage, the married state, or married persons
— mat·ri·mo·nial·ly *adverb*

mat·ri·mo·ny \ˈma-trə-ˌmō-nē\ *noun* [Middle English, from Middle French *matremoine,* from Latin *matrimonium,* from *matr-, mater* mother, matron — more at MOTHER] (14th century)
: the union of man and woman as husband and wife **:** MARRIAGE

matrimony vine *noun* (circa 1818)
: a shrub or vine (genus *Lycium*) of the nightshade family with often showy flowers and usually red berries

ma·trix \ˈmā-triks\ *noun, plural* **ma·tri·ces** \ˈmā-trə-ˌsēz, ˈma-\ *or* **ma·trix·es** \ˈmā-trik-səz\ [Latin, female animal used for breeding, parent plant, from *matr-, mater*] (1555)
1 : something within or from which something else originates, develops, or takes form
2 a : a mold from which a relief surface (as a piece of type) is made **b :** DIE 3a(1) **c :** an engraved or inscribed die or stamp **d :** an electroformed impression of a phonograph record used for mass-producing duplicates of the original
3 a : the natural material (as soil or rock) in which something (as a fossil or crystal) is embedded **b :** material in which something is enclosed or embedded (as for protection or study)
4 a : the intercellular substance in which tissue cells (as of connective tissue) are embedded **b :** the thickened epithelium at the base of a fingernail or toenail from which new nail substance develops
5 a : a rectangular array of mathematical elements (as the coefficients of simultaneous linear equations) that can be combined to form sums and products with similar arrays having an appropriate number of rows and columns **b :** something resembling a mathematical matrix especially in rectangular arrangement of elements into rows and columns **c :** an array of circuit elements (as diodes and transistors) for performing a specific function
6 : a main clause that contains a subordinate clause

ma·tron \ˈmā-trən\ *noun* [Middle English *matrone,* from Middle French, from Latin *matrona,* from *matr-, mater*] (14th century)
1 a : a married woman usually marked by dignified maturity or social distinction **b :** a woman who supervises women or children (as in a school or police station) **c :** the chief officer in a women's organization
2 : a female animal kept for breeding

ma·tron·ly \ˈmā-trən-lē\ *adjective* (1656)
: having the character of or suitable to a matron

matron of honor (1903)
: a bride's principal married wedding attendant

mat·ro·nym·ic \ˌma-trə-ˈni-mik\ *noun* [*matr-* + *-onymic* (as in *patronymic*)] (1794)
: a name derived from that of the mother or a maternal ancestor

¹matte *or* **matt** \ˈmat\ *variant of* ³MAT

²matte *also* **matt** \ˈmat\ *adjective* [French *mat,* from Old French, faded, defeated] (circa 1648)
: lacking or deprived of luster or gloss: as **a :** having a usually smooth even surface free from shine or highlights ⟨*matte* metals⟩ ⟨a *matte* finish⟩ **b :** having a rough or granular surface

³matte \ˈmat\ *noun* [French, from Middle French, crude metal, curdled milk, from feminine of *mat* thick, dull, matte] (1839)
1 : a crude mixture of sulfides formed in smelting sulfide ores of metals (as copper, lead, or nickel)
2 : a motion-picture effect in which part of a scene is blocked out and later replaced by footage containing other material (as a background painting)

¹mat·ter \ˈma-tər\ *noun* [Middle English *matere,* from Old French, from Latin *materia* matter, physical substance, from *mater*] (13th century)

1 a : a subject under consideration **b :** a subject of disagreement or litigation **c** *plural* **:** the events or circumstances of a particular situation **d :** the subject or substance of a discourse or writing **e :** something of an indicated kind or having to do with an indicated field or situation ⟨this is a serious *matter*⟩ ⟨as a *matter* of policy⟩ ⟨*matters* of faith⟩ **f :** something to be proved in law **g** *obsolete* **:** sensible or serious material as distinguished from nonsense or drollery **h** (1) *obsolete* **:** REASON, CAUSE (2) **:** a source especially of feeling or emotion **i :** PROBLEM, DIFFICULTY
2 a : the substance of which a physical object is composed **b :** material substance that occupies space, has mass, and is composed predominantly of atoms consisting of protons, neutrons, and electrons, that constitutes the observable universe, and that is interconvertible with energy **c :** a material substance of a particular kind or for a particular purpose ⟨vegetable *matter*⟩ **d** (1) **:** material (as feces or urine) discharged from the living body (2) **:** material discharged by suppuration **:** PUS
3 a : the indeterminate subject of reality; *especially* **:** the element in the universe that undergoes formation and alteration **b :** the formless substratum of all things which exists only potentially and upon which form acts to produce realities
4 : a more or less definite amount or quantity ⟨cooks in a *matter* of minutes⟩
5 : something written or printed
6 : MAIL
7 *Christian Science* **:** the illusion that the objects perceived by the physical senses have the reality of substance
— for that matter : so far as that is concerned
— no matter : without regard to **:** irrespective of ⟨points in the same direction *no matter* how it is tilted⟩
— the matter : WRONG ⟨nothing's *the matter* with me⟩

²matter *intransitive verb* (1530)
1 : to form or discharge pus **:** SUPPURATE ⟨*mattering* wound⟩
2 : to be of importance **:** SIGNIFY

matter of course (1739)
: something that is to be expected as a natural or logical consequence

mat·ter–of–fact \ˌma-tə-rə(v)-ˈfakt\ *adjective* (1712)
: adhering to the unembellished facts; *also* **:** being plain, straightforward, or unemotional
— mat·ter–of–fact·ly \-ˈfak(t)-lē\ *adverb*
— mat·ter–of–fact·ness \-ˈfak(t)-nəs\ *noun*

mat·tery \ˈma-tə-rē\ *adjective* (14th century)
: producing or containing pus or material resembling pus ⟨eyes all *mattery*⟩

Mat·the·an *or* **Mat·thae·an** \ma-ˈthē-ən, mə-\ *adjective* [Late Latin *Matthaeus*] (1897)
: of, relating to, or characteristic of the evangelist Matthew or the gospel ascribed to him

Mat·thew \ˈma-(ˌ)thyü *also* -(ˌ)thü\ *noun* [French *Mathieu,* from Late Latin *Matthaeus,* from Greek *Matthaios,* from Hebrew *Mattithyāh*]
1 : an apostle traditionally identified as the author of the first Gospel in the New Testament
2 : the first Gospel in the New Testament — see BIBLE table

¹mat·ting \ˈma-tiŋ\ *noun* (circa 1847)
1 : material for mats
2 : MATS

²matting *noun* [from gerund of ⁴*mat*] (1854)
: a dull lusterless surface (as on gilding, metalwork, or satin)

mat·tins *often capitalized, chiefly British variant of* MATINS

mat·tock \ˈma-tək\ *noun* [Middle English *mattok,* from Old English *mattuc*] (before 12th century)
: a digging and grubbing tool with features of an adze and an ax or pick

mattock

mat·tress \ˈma-trəs\ *noun* [Middle English *materas,* from Old French, from Arabic *maṭraḥ* place where something is thrown] (14th century)
1 a : a fabric case filled with resilient material (as cotton, hair, feathers, foam rubber, or an arrangement of coiled springs) used either alone as a bed or on a bedstead **b :** an inflatable airtight sack for use as a mattress
2 : a device (as of interwoven brush and poles) used to protect a shoreline, bank, or streambed from erosion

mat·u·rate \ˈma-chə-ˌrāt\ *verb* **-rat·ed; -rat·ing** (1622)
: MATURE

mat·u·ra·tion \ˌma-chə-ˈrā-shən\ *noun* (1541)
1 a : the process of becoming mature **b :** the emergence of personal and behavioral characteristics through growth processes **c :** the final stages of differentiation of cells, tissues, or organs
2 a : the entire process by which diploid gonocytes are transformed into haploid gametes that includes both meiosis and physiological and structural changes **b :** SPERMIOGENESIS
— mat·u·ra·tion·al \-shnəl, -shə-nᵊl\ *adjective*

¹ma·ture \mə-ˈtu̇r, -ˈtyu̇r *also* -ˈchu̇r\ *adjective* **ma·tur·er; -est** [Middle English, from Latin *maturus* ripe; akin to Latin *mane* in the morning, *manus* good] (15th century)
1 : based on slow careful consideration ⟨a *mature* judgment⟩
2 a (1) **:** having completed natural growth and development **:** RIPE (2) **:** having undergone maturation **b :** having attained a final or desired state ⟨*mature* wine⟩ **c :** having achieved a low but stable growth rate ⟨paper is a *mature* industry⟩
3 a : of or relating to a condition of full development **b :** characteristic of or suitable to a mature individual ⟨*mature* outlook⟩
4 : due for payment ⟨a *mature* loan⟩
5 : belonging to the middle portion of a cycle of erosion
— ma·ture·ly *adverb*

²mature *verb* **ma·tured; ma·tur·ing** (15th century)
transitive verb
: to bring to maturity or completion
intransitive verb
1 : to become fully developed or ripe
2 : to become due

ma·tu·ri·ty \mə-ˈtu̇r-ə-tē, -ˈtyu̇r- *also* -ˈchu̇r-\ *noun* (15th century)
1 : the quality or state of being mature; *especially* **:** full development
2 : termination of the period that an obligation has to run

ma·tu·ti·nal \ˌma-chu̇-ˈtī-nᵊl; mə-ˈtüt-nəl, -ˈtyüt-, -ˈtü-tᵊn-əl, -ˈtyü-\ *adjective* [Late Latin *matutinalis,* from Latin *matutinus* — more at MATINEE] (circa 1656)
: of, relating to, or occurring in the morning **:** EARLY
— ma·tu·ti·nal·ly *adverb*

mat·zo *or* **mat·zoh** \ˈmät-sə *also* -(ˌ)sō\ *noun, plural* **mat·zoth** \-ˌsōt, -ˌsōth, -sōs\ *or* **mat·zos** *or* **mat·zohs** \-səz, -səs *also* -ˌsōz\ [Yiddish *matse,* from Hebrew *maṣṣāh*] (circa 1846)
1 : unleavened bread eaten especially at the Passover
2 : a wafer of matzo

matzo ball *noun* (1952)
: a small ball-shaped dumpling made from matzo meal

maud·lin \ˈmȯd-lən\ *adjective* [alteration of Mary *Magdalene;* from her depiction as a weeping penitent] (1509)
1 : drunk enough to be emotionally silly
2 : weakly and effusively sentimental ◆

mau·gre \ˈmȯ-gər\ *preposition* [Middle English, from Middle French *maugré,* from *maugré* displeasure, from *mau, mal* evil + *gré* pleasure] (13th century)
archaic **:** in spite of

¹maul \ˈmȯl\ *noun* [Middle English *malle* mace, maul, from Old French *mail,* from Latin *malleus;* akin to Old Church Slavonic *mlatŭ* hammer, Latin *molere* to grind — more at MEAL] (13th century)
: a heavy often wooden-headed hammer used especially for driving wedges; *also* **:** a tool like a sledgehammer with one wedge-shaped end that is used to split wood

²maul *transitive verb* (13th century)
1 : BEAT, BRUISE
2 : MANGLE 1
3 : to handle roughly
— maul·er *noun*

maul·stick \ˈmȯl-stik\ *noun* [part translation of Dutch *maalstok,* from obsolete Dutch *malen* to paint + Dutch *stok* stick, stock] (circa 1658)
: a stick used by painters as a rest for the hand while working

mau–mau \ˈmau̇-ˌmau̇\ *verb* **mau–maued; mau–mau·ing** *often both Ms capitalized* [*Mau Mau,* anti-European secret society in colonial Kenya] (1970)
transitive verb
: to intimidate (as an official) by hostile confrontation or threats
intransitive verb
: to engage in mau-mauing someone

maun \ˈmȯn, ˈmän, mən\ *verbal auxiliary* [Middle English *man,* from Old Norse, present of *munu* shall, will; akin to Old English *gemynd* mind — more at MIND] (13th century)
chiefly Scottish **:** MUST

maun·der \ˈmȯn-dər, ˈmän-\ *intransitive verb* **maun·dered; maun·der·ing** \-d(ə-)riŋ\ [probably imitative] (1621)
1 *dialect British* **:** GRUMBLE
2 : to wander slowly and idly
3 : to speak indistinctly or disconnectedly
— maun·der·er \-dər-ər\ *noun*

Maun·dy Thursday \ˈmȯn-dē-, ˈmän-\ *noun* [Middle English *maunde* ceremony of washing the feet of the poor on Maundy Thursday, from Old French *mandé,* from Latin *mandatum* command; from Jesus' words in John 13:34 — more at MANDATE] (15th century)
: the Thursday before Easter observed in commemoration of the institution of the Eucharist

mau·so·le·um \ˌmȯ-sə-ˈlē-əm, ˌmȯ-zə-\ *noun, plural* **-leums** *or* **-lea** \-ˈlē-ə\ [Middle English, from Latin, from Greek *mausōleion,* from *Mausōlos* Mausolus (died about 353 B.C.) ruler of Caria] (15th century)

◇ WORD HISTORY
maudlin In the New Testament of the Bible one of the most devoted followers of Jesus is Mary Magdalene, who was so called because she was thought to have come from the Palestinian city of Magdala. The popular pronunciation of *Magdalene* in early Modern English is indicated by spellings such as *Maudlen* and *Mawdlin,* which represent borrowings from medieval French *Madelaine,* the French development of Late Latin *Magdalena.* By the 16th century it had become an artistic tradition in scenes of the crucifixion and burial of Jesus to portray Mary Magdalene as weeping. So identified with weeping did she become that by the 17th century *maudlin* had come to mean "tearful, lachrymose." Later the word was used to describe a tearful show of emotion and especially the overwrought display of emotion brought on by drunkenness.

1 : a large tomb; *especially* **:** a usually stone building with places for entombment of the dead above ground
2 : a large gloomy building or room ◆
mauve \'mōv, 'mȯv\ *noun* [French, mallow, from Latin *malva*] (1859)
1 a : a moderate purple, violet, or lilac color **b :** a strong purple
2 : a dyestuff that produces a mauve color
— mauve *adjective*
ma·ven *or* **ma·vin** \'mā-vən\ *noun* [Yiddish *meyvn,* from Late Hebrew *mēbhīn*] (circa 1952)
: one who is experienced or knowledgeable **:** EXPERT; *also* **:** FREAK 4
[1]**mav·er·ick** \'mav-rik, 'ma-və-\ *noun* [Samuel A. *Maverick* (died 1870) American pioneer who did not brand his calves] (1867)
1 : an unbranded range animal; *especially* **:** a motherless calf
2 : an independent individual who does not go along with a group or party ◆
[2]**maverick** *adjective* (1886)
: characteristic of, suggestive of, or inclined to be a maverick
ma·vis \'mā-vəs\ *noun* [Middle English, from Middle French *mauvis*] (14th century)
: SONG THRUSH
ma·vour·neen \mə-'vu̇r-,nēn\ *noun* [Irish *mo mhuirnín*] (1800)
Irish **:** my darling
maw \'mȯ\ *noun* [Middle English, from Old English *maga;* akin to Old High German *mago* stomach, Lithuanian *makas* purse] (before 12th century)
1 : the receptacle into which food is taken by swallowing: **a :** STOMACH **b :** CROP
2 a : the throat, gullet, or jaws especially of a voracious animal **b :** something suggestive of a gaping maw
mawk·ish \'mȯ-kish\ *adjective* [Middle English *mawke* maggot, from Old Norse *mathkr* — more at MAGGOT] (circa 1697)
1 : having an insipid often unpleasant taste
2 : sickly or puerilely sentimental
— mawk·ish·ly *adverb*
— mawk·ish·ness *noun*
max \'maks\ *noun* (1968)
: MAXIMUM 1, 2
— to the max : to the greatest extent possible
maxi \'mak-sē\ *noun, plural* **max·is** [*maxi-*] (1967)
: a long skirt, dress, or coat
maxi- *combining form* [*maximum*]
1 : extra long ⟨*maxi*skirt⟩
2 : extra large ⟨*maxi*-problems⟩
max·il·la \mak-'si-lə\ *noun, plural* **max·il·lae** \-'si-(,)lē, -,lī\ *or* **maxillas** [Latin, diminutive of *mala* jaw] (1676)
1 a : JAW 1a **b** (1) **:** an upper jaw especially of humans and other mammals in which the bony elements are closely fused (2) **:** either of the two bones that lie with one on each side of the upper jaw lateral to the premaxilla and that in higher vertebrates bear most of the teeth
2 : one of the first or second pair of mouth parts posterior to the mandibles in insects, myriapods, crustaceans, and closely related arthropods
— max·il·lary \'mak-sə-,ler-ē, *chiefly British* mak-'si-lə-rē\ *adjective or noun*
max·il·li·ped \mak-'si-lə-,ped\ *noun* [International Scientific Vocabulary *maxilli-* (from Latin *maxilla*) + *-ped*] (1846)
: any of the crustacean appendages that comprise the first pair or first three pairs situated next behind the maxillae
max·il·lo·fa·cial \mak-,si-(,)lō-'fā-shəl, ,mak-sə-(,)lō-\ *adjective* (circa 1923)
: of, relating to, or treating the maxilla and the face ⟨*maxillofacial* surgeons⟩
max·im \'mak-səm\ *noun* [Middle English *maxime,* from Middle French, from Medieval Latin *maxima,* from Latin, feminine of *maximus,* superlative of *magnus* large — more at MUCH] (1567)
1 : a general truth, fundamental principle, or rule of conduct
2 : a saying of proverbial nature
max·i·mal \'mak-s(ə-)məl\ *adjective* (1882)
1 : being an upper limit **:** HIGHEST
2 : most comprehensive **:** COMPLETE
— max·i·mal·ly *adverb*
max·i·mal·ist \-s(ə-)mə-list\ *noun* (1907)
: one who advocates immediate and direct action to secure the whole of a program or set of goals
maxi·min \'mak-sə-,min\ *noun* [*maxi*mum + *min*imum] (1951)
: the maximum of a set of minima; *especially* **:** the largest of a set of minimum possible gains each of which occurs in the least advantageous outcome of a strategy followed by a participant in a situation governed by the theory of games — compare MINIMAX
max·i·mise *British variant of* MAXIMIZE
max·i·mize \'mak-sə-,mīz\ *transitive verb* **-mized; -miz·ing** (1802)
1 : to increase to a maximum
2 : to make the most of
3 : to find a maximum value of
— max·i·mi·za·tion \,mak-sə-mə-'zā-shən\ *noun*
— max·i·miz·er \'mak-sə-,mī-zər\ *noun*
max·i·mum \'mak-s(ə-)məm\ *noun, plural* **max·i·ma** \-sə-mə\ *or* **maximums** \-s(ə-)məmz\ [Latin, neuter of *maximus*] (1740)
1 a : the greatest quantity or value attainable or attained **b :** the period of highest, greatest, or utmost development
2 : an upper limit allowed (as by a legal authority) or allowable (as by the circumstances of a particular case)
3 : the largest of a set of numbers; *specifically* **:** the largest value assumed by a real-valued continuous function defined on a closed interval
— maximum *adjective*
maximum likelihood *noun* (1959)
: a statistical method for estimating population parameters (as the mean and variance) from sample data that selects as estimates those parameter values maximizing the probability of obtaining the observed data
ma·xixe \mə-'shēsh, -'shē-shə\ *noun, plural* **ma·xi·xes** \-'shē-shəz\ [Brazilian Portuguese] (1914)
: a ballroom dance of Brazilian origin that resembles the two-step
max·well \'maks-,wel, -wəl\ *noun* [James Clerk *Maxwell*] (1900)
: the centimeter-gram-second electromagnetic unit of magnetic flux equal to the flux per square centimeter of normal cross section in a region where the magnetic induction is one gauss **:** 10^{-8} weber
[1]**may** \'mā\ *verbal auxiliary, past* **might** \'mīt\; *present singular & plural* **may** [Middle English (1st & 3d singular present indicative), from Old English *mæg;* akin to Old High German *mag* (1st & 3d singular present indicative) have power, am able (infinitive *magan*), and perhaps to Greek *mēchos* means, expedient] (before 12th century)
1 a *archaic* **:** have the ability to **b :** have permission to ⟨you *may* go now⟩ **:** be free to ⟨a rug on which children *may* sprawl —C. E. Silberman⟩ — used nearly interchangeably with *can* **c** — used to indicate possibility or probability ⟨you *may* be right⟩ ⟨things you *may* need⟩; sometimes used interchangeably with *can* ⟨one of those slipups that *may* happen from time to time —Jessica Mitford⟩ ⟨copula *may* optionally be deleted —J. D. McCawley⟩; sometimes used where *might* would be expected ⟨you *may* think from a little distance that the country was solid woods —Robert Frost⟩
2 — used in auxiliary function to express a wish or desire especially in prayer, imprecation, or benediction ⟨long *may* he reign⟩
3 — used in auxiliary function expressing purpose or expectation ⟨I laugh that I *may* not weep⟩ or contingency ⟨she'll do her duty come what *may*⟩ or concession ⟨he *may* be slow but he is thorough⟩ or choice ⟨the angler *may* catch them with a dip net, or he *may* cast a large, bare treble hook —Nelson Bryant⟩
4 : SHALL, MUST — used in law where the sense, purpose, or policy requires this interpretation
usage see CAN
[2]**may** \'mā\ *noun* [Middle English, from Old English *mǣg* kinsman, kinswoman, maiden] (before 12th century)
archaic **:** MAIDEN
May \'mā\ *noun* [Middle English, from Old French & Latin; Old French *mai,* from Latin *Maius,* from *Maia,* Roman goddess] (12th century)
1 : the 5th month of the Gregorian calendar
2 *often not capitalized* **:** the early vigorous blooming part of human life **:** PRIME
3 : the festivities of May Day

◇ WORD HISTORY

mausoleum In the 4th century B.C. Mausolus was a satrap (governor) of the Persian empire who ruled over Caria, a region in what is now southwestern Turkey. Probably his most significant decision was moving his capital to Halicarnassus, where he conceived a building project that would assure the immortality of his name. This edifice, which he planned and his wife Artemisia completed, was his own tomb. It towered about 135 feet, had a rectangular base of white marble with a periphery of 411 feet, and was topped by an array of colossal statues, two of which portrayed Mausolus and Artemisia. Named the *Mausōleion* by the Greeks for its creator and occupant, this exercise in self-glorification was acclaimed as one of the Seven Wonders of the Ancient World. The Greek word later came to be used generically for any large and imposing burial structure.

maverick Samuel Augustus Maverick (1803–70) was a South Carolina-born lawyer who emigrated to Texas in 1835 and was a prominent figure in the early years of the state's history. Maverick's consuming passion was not law or politics, however, but the acquisition of land; at his death he owned over 300,000 acres spread over 32 Texas counties. He was not a rancher and resided much of his life in San Antonio, but a minor episode in his career made his name part of every cowboy's vocabulary and a fixture of American English. In 1847 he bought a farm near Matagorda Bay with about 450 cattle that were left in charge of a single slave with no experience as a cowboy. Most of the calves were never penned and branded, and loss of animals led Maverick to move the herd to a ranch southeast of San Antonio in 1853 and finally to sell them in 1856. Presumably in the 1850s cowboys finding unbranded yearlings on land adjoining Maverick's began to call the animals *mavericks.* The vast extent of his land holdings and later legends about Maverick as a great cattle baron fostered the spread of the word throughout Texas and the West.

4 *not capitalized* **a :** green or flowering branches used for May Day decorations **b :** a plant that yields may: as (1) **:** HAWTHORN (2) **:** a spring-flowering spirea

ma•ya \'mä-yə, 'mī-ə\ *noun* [Sanskrit *māyā*] (1788)
: the sense-world of manifold phenomena held in Vedanta to conceal the unity of absolute being; *broadly* **:** ILLUSION

Ma•ya \'mī-ə\ *noun, plural* **Maya** *or* **Mayas** [Spanish] (1825)
1 a : a Mayan language of the ancient Maya peoples recorded in inscriptions **b :** YUCATEC; *especially* **:** the older form of that language known from documents of the Spanish period
2 : a member of a group of Indian peoples chiefly of Yucatán, Belize, and Guatemala whose languages are Mayan

Ma•yan \'mī-ən\ *noun* (1900)
1 : a member of the peoples speaking Mayan languages
2 : an extensive language family of Central America and Mexico
— **Mayan** *adjective*

Mayanist \'mī-ə-nist\ *noun* (1950)
: a specialist in Mayan culture and history

may•ap•ple \'mā-,a-pəl\ *noun* [*May*] (circa 1733)
: a North American herb (*Podophyllum peltatum*) of the barberry family with a poisonous rootstock, one or two large-lobed peltate leaves, and a single large white flower followed by a yellow egg-shaped edible fruit; *also* **:** its fruit

[1]may•be \'mā-bē *also* 'me-\ *adverb* (15th century)
: PERHAPS

[2]maybe *noun* (circa 1586)
: UNCERTAINTY

May•day \'mā-,dā\ [French *m'aider* help me] (1927)
— an international radio-telephone signal word used as a distress call

May Day \'mā-,dā\ *noun* (13th century)
: May 1 celebrated as a springtime festival and in some countries as Labor Day

may•est *or* **mayst** \'mā-əst, 'māst\ *archaic present 2d singular of* MAY

may•flow•er \'mā-,flau̇(-ə)r\ *noun* (1594)
: any of various spring-blooming plants; *especially* **:** ARBUTUS 2

may•fly \'mā-flī\ *noun* (circa 1653)
: any of an order (Ephemeroptera) of insects with an aquatic nymph and a short-lived fragile adult having membranous wings and two or three long caudal styles — called also *ephemerid*

mayfly

may•hap \'mā-,hap, mā-'\ *adverb* [from the phrase *may hap*] (circa 1531)
: PERHAPS

may•hem \'mā-,hem, 'mā-əm\ *noun* [Middle English *mayme,* from Anglo-French *mahaim,* from Old French, loss of a limb, from *maynier* to maim, probably of Germanic origin; akin to Middle High German *meiden* gelding, Old Norse *meitha* to injure] (15th century)
1 a : willful and permanent deprivation of a bodily member resulting in the impairment of a person's fighting ability **b :** willful and permanent crippling, mutilation, or disfigurement of any part of the body
2 : needless or willful damage or violence

may•ing \'mā-iŋ\ *noun, often capitalized* (14th century)
: the celebrating of May Day

mayn't \'mā-ənt, 'mānt\ (circa 1631)
: may not

mayo \'mā-(,)ō\ *noun* (circa 1960)
: MAYONNAISE

may•on•naise \'mā-ə-,nāz, ,mā-ə-'\ *noun* [French] (1841)
: a dressing made of egg yolks, vegetable oils, and vinegar or lemon juice

may•or \'mā-ər, 'me(-ə)r, *especially before names* (,)mer\ *noun* [Middle English *maire,* from Old French, from Latin *major* greater — more at MAJOR] (14th century)
: an official elected or appointed to act as chief executive or nominal head of a city, town, or borough
— **may•or•al** \'mā-ə-rəl, 'me-ə-; ,mā-'ȯr-əl\ *adjective*

may•or•al•ty \'mā-ə-rəl-tē, 'me-; 'mer-əl-\ *noun* [Middle English *mairaltee,* from Middle French *mairalté,* from Old French, from *maire*] (14th century)
: the office or term of office of a mayor

may•or•ess \'mā-ə-rəs, 'me-\ *noun* (15th century)
chiefly British
1 : the wife or official hostess of a mayor
2 : a woman holding the office of mayor
usage see -ESS

may•pole \'mā-,pōl\ *noun, often capitalized* (1554)
: a tall flower-wreathed pole forming a center for May Day sports and dances

may•pop \'mā-,päp\ *noun* [alteration of *maracock,* perhaps from Virginia Algonquian] (1851)
: a climbing perennial passionflower (*Passiflora incarnata*) of the southern U.S. with a large ovoid yellow edible but insipid berry; *also* **:** its fruit

May•time \'mā-,tīm\ *noun* (14th century)
: the month of May

maz•ard \'ma-zərd\ *noun* [obsolete English *mazard* mazer, alteration of English *mazer*] (1602)
chiefly dialect **:** HEAD, FACE

[1]maze \'māz\ *transitive verb* **mazed; maz•ing** [Middle English] (13th century)
1 *chiefly dialect* **:** STUPEFY, DAZE
2 : BEWILDER, PERPLEX

[2]maze *noun* (14th century)
1 a : a confusing intricate network of passages **b :** something intricately or confusingly elaborate or complicated ⟨a *maze* of regulations⟩
2 *chiefly dialect* **:** a state of bewilderment
— **maze•like** \-,līk\ *adjective*

ma•zer \'mā-zər\ *noun* [Middle English, from Middle French *mazere,* of Germanic origin; akin to Old High German *masar* gnarled excrescence on a tree] (14th century)
: a large drinking bowl originally of a hard wood

ma•zur•ka \mə-'zər-kə, -'zu̇r-\ *also* **ma•zour•ka** \-'zu̇r-\ *noun* [Russian, from Polish *mazurek,* from *Mazury* Masuria, region of northeast Poland] (1818)
1 : a Polish folk dance in moderate triple measure
2 : music for the mazurka or in its rhythm usually in moderate 3/4 or 3/8 time

mazy \'mā-zē\ *adjective* (1579)
: resembling a maze

[1]maz•zard \'ma-zərd\ *noun* [origin unknown] (1578)
: SWEET CHERRY; *especially* **:** wild or seedling sweet cherry used as a rootstock for grafting

[2]mazzard *variant of* MAZARD

mbi•ra \em-'bir-ə\ *noun* [Shona] (circa 1911)
: an African musical instrument that consists of a wooden or gourd resonator and a varying number of tuned metal or wooden strips that vibrate when plucked

MC \,em-'sē\ *noun* (1790)
: MASTER OF CEREMONIES

Mc•Car•thy•ism \mə-'kär-thē-,i-zəm *also* -'kär-tē-\ *noun* [Joseph R. *McCarthy*] (1950)
: a mid-20th century political attitude characterized chiefly by opposition to elements held to be subversive and by the use of tactics involving personal attacks on individuals by means of widely publicized indiscriminate allegations especially on the basis of unsubstantiated charges
— **Mc•Car•thy•ite** \-,īt\ *noun or adjective*

Mc•Coy \mə-'kȯi\ *noun* [alteration of *Mackay* (in the phrase *the real Mackay*), of unknown origin] (1922)
: something that is neither imitation nor substitute — often used in the phrase *the real McCoy*

Mc•In•tosh \'ma-kən-,täsh\ *noun* [John *McIntosh* (flourished 1796) Canadian settler] (1878)
: a juicy bright red eating apple with a thin skin, white flesh, and aromatic slightly tart flavor

M–day \'em-,dā\ *noun* [*m*obilization *day*] (1924)
: a day on which a military mobilization is to begin

me \'mē\ *pronoun, objective case of* I [Middle English, from Old English *mē;* akin to Old High German *mīh* me, Latin *me,* Greek *me,* Sanskrit *mā*] ▢

mea cul•pa \,mā-ə-'ku̇l-pə, ,mā-ä-, -'ku̇l-(,)pä\ *noun* [Latin, through my fault] (1602)
: a formal acknowledgment of personal fault or error

[1]mead \'mēd\ *noun* [Middle English *mede,* from Old English *medu;* akin to Old High German *metu* mead, Greek *methy* wine] (before 12th century)
: a fermented beverage made of water and honey, malt, and yeast

[2]mead *noun* [Middle English *mede,* from Old English *mǣd*] (before 12th century)
archaic **:** MEADOW

mead•ow \'me-(,)dō\ *noun, often attributive* [Middle English *medwe,* from Old English *mǣdwe,* oblique case form of *mǣd;* akin to Old English *māwan* to mow — more at MOW] (before 12th century)
: land in or predominantly in grass; *especially* **:** a tract of moist low-lying usually level grassland
— **mead•owy** \'me-də-wē\ *adjective*

meadow beauty *noun* (1840)
: any of a genus (*Rhexia* of the family Melastomaceae, the meadow-beauty family) of low perennial American herbs with showy solitary or cymose flowers

meadow fescue *noun* (1794)
: a tall vigorous perennial European fescue grass (*Festuca pratensis*) with broad flat leaves widely cultivated for permanent pasture and hay

meadow grass *noun* (13th century)
: any of various grasses (as of the genus *Poa*) that thrive in the presence of abundant moisture; *especially* **:** KENTUCKY BLUEGRASS

mead•ow•land \'me-dō-,land, -də-\ *noun* (1530)
: land that is or is used for meadow

▢ USAGE
me *Me* is used in many constructions where strict grammarians prescribe *I.* This usage is not so much ungrammatical as indicative of the shrinking range of the nominative form: *me* began to replace *I* sometime around the 16th century largely because of the pressure of word order. *I* is now chiefly used as the subject of an immediately following verb. *Me* occurs in every other position: absolutely ⟨who, *me*?⟩, emphatically ⟨*me* too⟩, and after prepositions, conjunctions, and verbs, including *be* ⟨come with *me*⟩ ⟨you're as big as *me*⟩ ⟨it's *me*⟩. Almost all usage books recognize the legitimacy of *me* in these positions, especially in speech; some recommend *I* in formal and especially written contexts after *be* and after *as* and *than* when the first term of the comparison is the subject of a verb.

mead·ow·lark \ˈme-dō-ˌlärk, -də-\ *noun* (1611)
: any of several American songbirds (genus *Sturnella*) that are streaked brown above and in northernmost forms have a yellow breast marked with a black crescent

meadowlark

meadow mouse *noun* (1801)
: any of various voles (especially genus *Microtus*) that frequent open fields — called also *meadow vole*

meadow mushroom *noun* (1884)
: a common edible brown-spored agaric (*Agaricus campestris*) that occurs in moist open organically rich soil and is often cultivated

meadow nematode *noun* (1946)
: any of numerous plant-parasitic nematode worms (especially genus *Pratylenchus*) that destructively invade the roots of plants

meadow rue *noun* (1668)
: any of a genus (*Thalictrum*) of widely distributed plants of the buttercup family with decompound or compound leaves

meadow saffron *noun* (1578)
: AUTUMN CROCUS

meadow spittlebug *noun* (1942)
: a North American spittlebug (*Philaenus spumarius*) that does severe damage especially to grasses

mead·ow·sweet \ˈme-dō-ˌswēt, -də-\ *noun* (1530)
1 : SPIREA 1; *especially* **:** a North American native or naturalized spirea (as *Spiraea alba* and *S. latifolia*)
2 : any of a genus (*Filipendula*) of herbs closely related to the spireas

mea·ger *or* **mea·gre** \ˈmē-gər\ *adjective* [Middle English *megre,* from Middle French *maigre,* from Latin *macr-, macer* lean; akin to Old English *mæger* lean, Greek *makros* long] (14th century)
1 : having little flesh **:** THIN
2 a : lacking desirable qualities (as richness or strength) ⟨leading a *meager* life⟩ **b :** deficient in quality or quantity ⟨a *meager* diet⟩ ☆
— mea·ger·ly *adverb*
— mea·ger·ness *noun*

¹meal \ˈmē(ə)l\ *noun* [Middle English *meel* appointed time, meal, from Old English *mǣl;* akin to Old High German *māl* time, Latin *metiri* to measure — more at MEASURE] (before 12th century)
1 : an act or the time of eating a portion of food to satisfy appetite
2 : the portion of food eaten at a meal ◆

²meal *noun* [Middle English *mele,* from Old English *melu;* akin to Old High German *melo* meal, Latin *molere* to grind, Greek *mylē* mill] (before 12th century)
1 : the usually coarsely ground and unbolted seeds of a cereal grass or pulse; *especially* **:** CORNMEAL
2 : a product resembling seed meal especially in particle size or texture

-meal *adverb combining form* [Middle English *-mele,* from Old English *-mǣlum,* from *mǣlum,* dative plural of *mǣl*]
: by a (specified) portion or measure at a time ⟨piece*meal*⟩

mea·lie \ˈmē-lē\ *noun* [Afrikaans *mielie*] (1855)
South African **:** INDIAN CORN; *also* **:** an ear of Indian corn

meals–on–wheels *noun plural but singular in construction* (1961)
: a service that delivers daily hot meals to the homes of elderly or disabled people

meal ticket *noun* (circa 1899)
: one that serves as the ultimate source of one's income ⟨an advanced degree was his *meal ticket*⟩

meal·time \ˈmē(ə)l-ˌtīm\ *noun* (12th century)
: the usual time for serving a meal

meal·worm \-ˌwərm\ *noun* (1658)
: the larva of various beetles (genus *Tenebrio*) that infests and pollutes grain products but is often raised as food for insectivorous animals, for laboratory use, or as fishing bait

mealy \ˈmē-lē\ *adjective* **meal·i·er; -est** (1533)
1 : soft, dry, and friable
2 : containing meal **:** FARINACEOUS
3 a : covered with meal or with fine granules **b :** flecked with another color **c :** SPOTTY, UNEVEN **d :** PALLID, BLANCHED ⟨a *mealy* complexion⟩
4 : MEALYMOUTHED

mealy·bug \ˈmē-lē-ˌbəg\ *noun* (1824)
: any of a family (Pseudococcidae) of scale insects that have a white powdery covering and are destructive pests especially of fruit trees

mealy·mouthed \ˈmē-lē-ˌmau̇thd, -ˌmau̇tht\ *adjective* (circa 1572)
: not plain and straightforward **:** DEVIOUS ⟨a *mealymouthed* politician⟩

¹mean \ˈmēn\ *verb* **meant** \ˈment\; **mean·ing** \ˈmē-niŋ\ [Middle English *menen,* from Old English *mǣnan;* akin to Old High German *meinen* to have in mind, Old Church Slavonic *mĕniti* to mention] (before 12th century)
transitive verb
1 a : to have in the mind as a purpose **:** INTEND ⟨she *means* to win⟩ — sometimes used interjectionally with *I,* chiefly in informal speech for emphasis ⟨he throws, I *mean*, hard⟩ or to introduce a phrase restating the point of a preceding phrase ⟨we try to answer what we can, but I *mean* we're not God —Bobbie Ann Mason⟩ **b :** to design for or destine to a specified purpose or future ⟨I was *meant* to teach⟩
2 : to serve or intend to convey, show, or indicate **:** SIGNIFY ⟨a red sky *means* rain⟩
3 : to have importance to the degree of ⟨health *means* everything⟩
4 : to direct to a particular individual
intransitive verb
: to have an intended purpose ⟨he *means* well⟩
— mean·er \ˈmē-nər\ *noun*
— mean business : to be in earnest

²mean \ˈmēn\ *adjective* [Middle English *mene,* from *imene* common, shared, from Old English *gemǣne;* akin to Old High German *gimeini* common, Latin *communis* common, *munus* service, gift, Sanskrit *mayate* he exchanges] (14th century)
1 : lacking distinction or eminence **:** HUMBLE
2 : lacking in mental discrimination **:** DULL
3 a : of poor shabby inferior quality or status ⟨*mean* city streets⟩ **b :** worthy of little regard **:** CONTEMPTIBLE — often used in negative constructions as a term of praise ⟨no *mean* feat⟩
4 : lacking dignity or honor **:** BASE
5 a : PENURIOUS, STINGY **b :** characterized by petty selfishness or malice **c :** causing trouble or bother **:** VEXATIOUS **d :** EXCELLENT, EFFECTIVE ⟨plays a *mean* trumpet⟩ ⟨a lean, *mean* athlete⟩
6 : ASHAMED 1b ☆
— mean·ness \ˈmēn-nəs\ *noun*

³mean *adjective* [Middle English *mene,* from Middle French *meien,* from Latin *medianus* — more at MEDIAN] (14th century)
1 : occupying a middle position **:** intermediate in space, order, time, kind, or degree
2 : occupying a position about midway between extremes; *especially* **:** being the mean of a set of values **:** AVERAGE ⟨the *mean* temperature⟩
3 : serving as a means **:** INTERMEDIARY
synonym see AVERAGE

⁴mean *noun* (14th century)
1 a (1) **:** something intervening or intermediate (2) **:** a middle point between extremes **b :** a value that lies within a range of values and is computed according to a prescribed law: as (1) **:** ARITHMETIC MEAN (2) **:** EXPECTED VALUE **c :** either of the middle two terms of a proportion
2 *plural but singular or plural in construction* **:** something useful or helpful to a desired end
3 *plural* **:** resources available for disposal; *especially* **:** material resources affording a secure life
— by all means : most assuredly **:** CERTAINLY

☆ SYNONYMS

Meager, scanty, scant, skimpy, spare, sparse mean falling short of what is normal, necessary, or desirable. MEAGER implies the absence of elements, qualities, or numbers necessary to a thing's richness, substance, or potency ⟨a *meager* portion of meat⟩. SCANTY stresses insufficiency in amount, quantity, or extent ⟨supplies too *scanty* to last the winter⟩. SCANT suggests a falling short of what is desired or desirable rather than of what is essential ⟨in January the daylight hours are *scant*⟩. SKIMPY usually suggests niggardliness or penury as the cause of the deficiency ⟨tacky housing developments on *skimpy* lots⟩. SPARE may suggest a slight falling short of adequacy or merely an absence of superfluity ⟨a *spare,* concise style of writing⟩. SPARSE implies a thin scattering of units ⟨a *sparse* population⟩.

Mean, ignoble, abject, sordid mean being below the normal standards of human decency and dignity. MEAN suggests having such repellent characteristics as small-mindedness, ill temper, or cupidity ⟨*mean* and petty satire⟩. IGNOBLE suggests a loss or lack of some essential high quality of mind or spirit ⟨an *ignoble* scramble after material possessions⟩. ABJECT may imply degradation, debasement, or servility ⟨*abject* poverty⟩. SORDID is stronger than all of these in stressing physical or spiritual degradation and abjectness ⟨a *sordid* story of murder and revenge⟩.

◇ WORD HISTORY

meal It may be natural to suppose that the *meal* of "three meals a day" has some relation to the *meal* of "ground meal," since both words pertain to food, but in fact they are unrelated. The *meal* of "ground meal" is descended from Old English *melu*, a word akin to Old High German *malan* "to grind," as well as to Latin *molere* "to grind." Derived from *molere* is Latin *molina* "mill," the ultimate source of the English noun and verb *mill.* The *meal* of "three meals a day," on the other hand, is from Old English *mǣl* or *mēl* "measure, appointed time, mealtime." It is akin to Old High German *mal* "time," and much more distantly to Latin *metiri* "to measure." As *mēl* became *meel* in Middle English, the word became more closely associated with food than with time or measurement. The modern senses and spelling were established by the end of the 17th century. A relative of this *meal* can be seen in *piecemeal,* where *-meal* is a suffix having the sense "by a specified portion or measure at a time."

— **by means of :** through the use of
— **by no means :** in no way **:** not at all
[1]**me·an·der** \mē-ˈan-dər\ *noun* [Latin *maeander,* from Greek *maiandros,* from *Maiandros* (now *Menderes*), river in Asia Minor] (1576)
1 : a winding path or course; *especially* **:** LABYRINTH
2 : a turn or winding of a stream
— **me·an·drous** \-drəs\ *adjective*
[2]**meander** *intransitive verb* **-dered; -der·ing** \-d(ə-)riŋ\ (circa 1612)
1 : to follow a winding or intricate course
2 : to wander aimlessly or casually without urgent destination **:** RAMBLE
synonym see WANDER
mean deviation *noun* (1858)
: the mean of the absolute values of the numerical differences between the numbers of a set (as statistical data) and their mean or median
mean distance *noun* (circa 1889)
: the arithmetical mean of the maximum and minimum distances of an orbiting celestial object from its primary
mean free path *noun* (1879)
: the average distance traversed between collisions by particles (as molecules of a gas or free electrons in metal) in a system of agitated particles
mean·ie *also* **meany** \ˈmē-nē\ *noun, plural* **meanies** (1910)
: a mean or spiteful person
mean·ing \ˈmē-niŋ\ *noun* (14th century)
1 a : the thing one intends to convey especially by language **:** PURPORT **b :** the thing that is conveyed especially by language **:** IMPORT
2 : something meant or intended **:** AIM ⟨a mischievous *meaning* was apparent⟩
3 : significant quality; *especially* **:** implication of a hidden or special significance ⟨a glance full of *meaning*⟩
4 a : the logical connotation of a word or phrase **b :** the logical denotation or extension of a word or phrase
— **meaning** *adjective*
— **mean·ing·ly** \-niŋ-lē\ *adverb*
mean·ing·ful \-fəl\ *adjective* (1852)
1 a : having a meaning or purpose **b :** full of meaning **:** SIGNIFICANT ⟨a *meaningful* life⟩
2 : having an assigned function in a language system ⟨*meaningful* propositions⟩
— **mean·ing·ful·ly** \-fə-lē\ *adverb*
— **mean·ing·ful·ness** *noun*
mean·ing·less \ˈmē-niŋ-ləs\ *adjective* (1797)
1 : having no meaning; *especially* **:** lacking any significance
2 : having no assigned function in a language system
— **mean·ing·less·ly** *adverb*
— **mean·ing·less·ness** *noun*
[1]**mean·ly** \ˈmēn-lē\ *adverb* (14th century)
obsolete **:** fairly well **:** MODERATELY
[2]**meanly** *adverb* (15th century)
: in a mean manner: as **a :** in a lowly manner **:** HUMBLY **b :** in an inferior manner **:** BADLY **c :** in a base or ungenerous manner
mean proportional *noun* (1571)
: GEOMETRIC MEAN; *especially* **:** the square root (as x) of the product of two numbers (as a and b) when expressed as the means of a proportion (as $a/x = x/b$)
mean solar day *noun* (1816)
: the interval between successive transits of a given meridian by the mean sun
mean–spir·it·ed \ˈmēn-ˈspir-ə-təd, ˌmēn-\ *adjective* (1694)
: exhibiting or characterized by meanness of spirit
— **mean–spir·it·ed·ness** \-nəs\ *noun*
mean square *noun* (1845)
: the mean of the squares of a set of values
mean square deviation *noun* (1948)
1 : VARIANCE 5
2 : STANDARD DEVIATION
means test \ˈmēnz-\ *noun* (1930)
: an examination into the financial state of a person to determine his eligibility for public assistance
— **means–test·ed** \-ˌtes-təd\ *adjective*
mean sun *noun* (circa 1890)
: a fictitious sun used for timekeeping that moves uniformly along the celestial equator and maintains a constant rate of apparent motion equal to the average rate of apparent motion of the real sun
[1]**mean·time** \ˈmēn-ˌtīm\ *noun* (14th century)
: the intervening time
[2]**meantime** *adverb* (1588)
: MEANWHILE
mean time *noun* (circa 1864)
: time that is based on the motion of the mean sun — called also *mean solar time*
mean value theorem *noun* (1902)
1 : a theorem in differential calculus: if a function of one variable is continuous on a closed interval and differentiable on the interval minus its endpoints there is at least one point where the derivative of the function is equal to the slope of the line joining the endpoints of the curve representing the function on the interval
2 : a theorem in integral calculus: if a function of one variable is continuous on a closed interval and differentiable on the interval minus its endpoints, there is at least one point in the interval where the product of the value of the function and the length of the interval is equal to the integral of the function over the interval
[1]**mean·while** \ˈmēn-ˌhwīl, -ˌwīl\ *noun* (14th century)
: MEANTIME
[2]**meanwhile** *adverb* (14th century)
1 : during the intervening time
2 : at the same time
mea·sle \ˈmē-zəl\ *noun* [singular of *measles*] (1863)
: a cysticercus tapeworm larva; *specifically* **:** one found in the muscles of a domesticated mammal
mea·sles \ˈmē-zəlz\ *noun plural but singular or plural in construction* [Middle English *meseles,* plural of *mesel* measles, spot characteristic of measles; akin to Middle Dutch *masel* spot characteristic of measles] (14th century)
1 a : an acute contagious viral disease marked by an eruption of distinct red circular spots **b :** any of various eruptive diseases (as German measles)
2 [Middle English *mesel* infested with tapeworms, literally, leprous, from Old French, from Medieval Latin *misellus* leper, from Latin, wretch, from *misellus,* diminutive of *miser* miserable] **:** infestation with or disease caused by larval tapeworms in the muscles and tissues
mea·sly \ˈmēz-lē, ˈmē-zə-\ *adjective* **mea·sli·er; -est** (1687)
1 : infected with measles
2 a : containing larval tapeworms **b :** infested with trichinae
3 : contemptibly small
[1]**mea·sure** \ˈme-zhər, ˈmā-\ *noun* [Middle English *mesure,* from Old French, from Latin *mensura,* from *mensus,* past participle of *metiri* to measure; akin to Old English *mǣth* measure, Greek *metron*] (13th century)
1 a (1) **:** an adequate or due portion (2) **:** a moderate degree; *also* **:** MODERATION, TEMPERANCE (3) **:** a fixed or suitable limit **:** BOUNDS ⟨rich beyond *measure*⟩ **b :** the dimensions, capacity, or amount of something ascertained by measuring **c :** an estimate of what is to be expected (as of a person or situation) **d** (1) **:** a measured quantity (2) **:** AMOUNT, DEGREE
2 a : an instrument or utensil for measuring **b** (1) **:** a standard or unit of measurement — see WEIGHT table (2) **:** a system of standard units of measure ⟨metric *measure*⟩
3 : the act or process of measuring
4 a (1) **:** MELODY, TUNE (2) **:** DANCE; *especially* **:** a slow and stately dance **b :** rhythmic structure or movement **:** CADENCE: as (1) **:** poetic rhythm measured by temporal quantity or accent; *specifically* **:** METER (2) **:** musical time **c** (1) **:** a grouping of a specified number of musical beats located between two consecutive vertical lines on a staff (2) **:** a metrical unit **:** FOOT
5 : an exact divisor of a number
6 : a basis or standard of comparison ⟨wealth is not a *measure* of happiness⟩
7 : a step planned or taken as a means to an end; *specifically* **:** a proposed legislative act
— **for good measure :** in addition to the minimum required **:** as an extra
[2]**measure** *verb* **mea·sured; mea·sur·ing** \ˈme-zhə-riŋ, ˈmā-; ˈmezh-riŋ, ˈmāzh-\ (14th century)
transitive verb
1 a : to choose or control with cautious restraint **:** REGULATE ⟨*measure* his acts⟩ **b :** to regulate by a standard **:** GOVERN
2 : to allot or apportion in measured amounts ⟨*measure* out 3 cups⟩
3 : to lay off by making measurements
4 : to ascertain the measurements of
5 : to estimate or appraise by a criterion ⟨*measures* his skill against his rival⟩
6 *archaic* **:** to travel over **:** TRAVERSE
7 : to serve as a means of measuring ⟨a thermometer *measures* temperature⟩
intransitive verb
1 : to take or make a measurement
2 : to have a specified measurement
— **mea·sur·abil·i·ty** \ˌme-zhə-rə-ˈbi-lə-tē, ˌmā-; ˌmezh-rə-, ˌmāzh-\ *noun*
— **mea·sur·able** \ˈme-zhə-rə-bəl, ˈmā-; ˈmezh-rə-, ˈmāzh-\ *adjective*
— **mea·sur·ably** \-blē\ *adverb*
— **mea·sur·er** \-zhər-ər\ *noun*
mea·sured \ˈme-zhərd, ˈmā-\ *adjective* (14th century)
1 : marked by due proportion
2 a : marked by rhythm **:** regularly recurrent ⟨a *measured* gait⟩ **b :** METRICAL
3 : DELIBERATE, CALCULATED ⟨a *measured* response⟩
— **mea·sured·ly** *adverb*
mea·sure·less \-zhər-ləs\ *adjective* (14th century)
1 : having no observable limit **:** IMMEASURABLE ⟨the *measureless* universe⟩
2 : very great ⟨had *measureless* energy⟩
mea·sure·ment \ˈme-zhər-mənt, ˈmā-\ *noun* (1751)
1 : the act or process of measuring
2 : a figure, extent, or amount obtained by measuring **:** DIMENSION
3 : MEASURE 2b
measurement ton *noun* (circa 1934)
: TON 1c
measure up *intransitive verb* (1910)
1 : to have necessary or fitting qualifications — often used with *to*
2 : to be the equal (as in ability) — used with *to*
measuring worm *noun* (1843)
: LOOPER 1
meat \ˈmēt\ *noun* [Middle English *mete,* from Old English; akin to Old High German *maz* food] (before 12th century)
1 a : FOOD; *especially* **:** solid food as distinguished from drink **b :** the edible part of something as distinguished from its covering (as a husk or shell)
2 : animal tissue considered especially as food: **a :** FLESH 2b **b :** FLESH 1a; *specifically* **:** flesh of domesticated animals
3 *archaic* **:** [1]MEAL 1; *especially* **:** DINNER
4 a : the core of something **:** HEART **b :** PITH 2b ⟨a novel with *meat*⟩
5 : favorite pursuit or interest
— **meat·ed** \ˈmē-təd\ *adjective*
— **meat·less** *adjective*

meat–and–potatoes *adjective* (1949)
1 : of fundamental importance **:** BASIC; *also* **:** concerned with or emphasizing the basic aspects of something
2 : PRACTICAL; *also* **:** EVERYDAY
3 : providing or preferring simple food (as meat and potatoes)

meat and potatoes *noun plural but singular or plural in construction* (1951)
: the most interesting or fundamental part **:** MEAT 4

meat–ax \'mēt-ˌaks\ *noun* (1834)
1 : CLEAVER 1
2 : an extreme or heavy-handed method of cutting or altering something

meat•ball \-ˌbȯl\ *noun* (circa 1838)
: a small ball of chopped or ground meat often mixed with bread crumbs and spices

meat•head \-ˌhed\ *noun* (1945)
: a stupid or bungling person

meat loaf *noun* (1899)
: a dish of ground meat seasoned and baked in the form of a loaf

meat market *noun* (1896)
: a depersonalizing environment in which people are treated as sexual or economic resources

meat•pack•ing \'mēt-ˌpa-kiŋ\ *noun* (1873)
: the wholesale meat industry

me•a•tus \mē-'ā-təs\ *noun, plural* **me•a•tus•es** \-tə-səz\ *or* **me•a•tus** \-'ā-təs, -ˌtüs\ [Late Latin, from Latin, going, passage, from *meatus,* past participle of *meare* to go — more at PERMEATE] (1665)
: a natural body passage

meaty \'mē-tē\ *adjective* **meat•i•er; -est** (circa 1787)
1 a : full of meat **b :** having the character of meat
2 : rich especially in matter for thought **:** SUBSTANTIAL ⟨actors looking for *meaty* roles⟩
— meat•i•ness *noun*

mec•a•myl•amine \ˌme-kə-'mi-lə-ˌmēn\ *noun* [from *Mecamylamine,* a trademark] (1955)
: a drug that in the hydrochloride $C_{11}H_{21}N{\cdot}HCl$ is used orally as a ganglionic blocking agent to effect a rapid lowering of severely elevated blood pressure

mec•ca \'me-kə\ *noun, often capitalized* [*Mecca,* Saudi Arabia, a destination of pilgrims in the Islamic world] (1850)
: a center of activity sought as a goal by people sharing a common interest

mechan- *or* **mechano-** *combining form* [Greek, from *mēchanē* machine — more at MACHINE]
: mechanical ⟨*mechano*receptor⟩

¹me•chan•ic \mi-'ka-nik\ *adjective* [Middle English, probably from Middle French *mechanique,* adjective & noun, from Latin *mechanicus,* from Greek *mēchanikos,* from *mēchanē*] (14th century)
1 : of or relating to manual work or skill
2 : MECHANICAL 3a

²mechanic *noun* (1562)
1 : a manual worker **:** ARTISAN
2 : MACHINIST; *especially* **:** one who repairs machines

¹me•chan•i•cal \mi-'ka-ni-kəl\ *adjective* (15th century)
1 a (1) **:** of or relating to machinery or tools ⟨*mechanical* applications of science⟩ ⟨a *mechanical* genius⟩ ⟨*mechanical* aptitude⟩ (2) **:** produced or operated by a machine or tool ⟨*mechanical* power⟩ ⟨a *mechanical* refrigerator⟩ ⟨a *mechanical* saw⟩ **b :** of or relating to manual operations
2 : of or relating to artisans or machinists ⟨the *mechanical* trades⟩
3 a : done as if by machine **:** seemingly uninfluenced by the mind or emotions **:** AUTOMATIC ⟨her singing was cold and *mechanical*⟩ **b :** of or relating to technicalities or petty matters
4 a : relating to, governed by, or in accordance with the principles of mechanics ⟨*mechanical* work⟩ ⟨*mechanical* energy⟩ **b :** relating to the quantitative relations of force and matter ⟨*mechanical* pressure of wind on a tower⟩
5 : caused by, resulting from, or relating to a process that involves a purely physical as opposed to a chemical change ⟨*mechanical* erosion of rock⟩
synonym see SPONTANEOUS
— me•chan•i•cal•ly \-ni-k(ə-)lē\ *adverb*

²mechanical *noun* (1590)
1 *obsolete* **:** MECHANIC 1
2 : a piece of finished copy consisting typically of type proofs and artwork positioned and mounted for photomechanical reproduction

mechanical advantage *noun* (1894)
: the advantage gained by the use of a mechanism in transmitting force; *specifically* **:** the ratio of the force that performs the useful work of a machine to the force that is applied to the machine

mechanical drawing *noun* (circa 1890)
1 : drawing done with the aid of instruments
2 : a drawing made with instruments

mechanical engineering *noun* (circa 1890)
: a branch of engineering concerned primarily with the industrial application of mechanics and with the production of tools, machinery, and their products

mech•a•ni•cian \ˌme-kə-'ni-shən\ *noun* (1570)
: MECHANIC, MACHINIST

me•chan•ics \mi-'ka-niks\ *noun plural but singular or plural in construction* (1648)
1 : a branch of physical science that deals with energy and forces and their effect on bodies
2 : the practical application of mechanics to the design, construction, or operation of machines or tools
3 : mechanical or functional details or procedure

mech•a•nism \'me-kə-ˌni-zəm\ *noun* (1662)
1 a : a piece of machinery **b :** a process or technique for achieving a result
2 : mechanical operation or action **:** WORKING 2
3 : a doctrine that holds natural processes (as of life) to be mechanically determined and capable of complete explanation by the laws of physics and chemistry
4 : the fundamental physical or chemical processes involved in or responsible for an action, reaction, or other natural phenomenon (as organic evolution)

mech•a•nist \-nist\ *noun* (1606)
1 *archaic* **:** MECHANIC
2 : an adherent of the doctrine of mechanism

mech•a•nis•tic \ˌme-kə-'nis-tik\ *adjective* (1884)
1 : mechanically determined ⟨*mechanistic* universe⟩
2 : of or relating to a mechanism or the doctrine of mechanism
3 : MECHANICAL
— mech•a•nis•ti•cal•ly \-ti-k(ə-)lē\ *adverb*

mech•a•nize \'me-kə-ˌnīz\ *transitive verb* **-nized; -niz•ing** (1678)
1 : to make mechanical; *especially* **:** to make automatic or routine
2 a : to equip with machinery especially to replace human or animal labor **b :** to equip with armed and armored motor vehicles **c :** to provide with mechanical power
3 : to produce by or as if by machine
— mech•a•niz•able \-ˌnī-zə-bəl\ *adjective*
— mech•a•ni•za•tion \ˌme-kə-nə-'zā-shən\ *noun*
— mech•a•niz•er \'me-kə-ˌnī-zər\ *noun*

mech•a•no•chem•is•try \ˌme-kə-nō-'ke-mə-strē\ *noun* (1928)
: chemistry that deals with the conversion of chemical energy into mechanical work (as in the contraction of a muscle)
— mech•a•no•chem•i•cal \-'ke-mi-kəl\ *adjective*

mech•a•no•re•cep•tor \-ri-'sep-tər\ *noun* (1927)
: a neural end organ (as a tactile receptor) that responds to a mechanical stimulus (as a change in pressure)
— mech•a•no•re•cep•tion \-'sep-shən\ *noun*
— mech•a•no•re•cep•tive \-'sep-tiv\ *adjective*

Mech•lin \'me-klən\ *noun* [*Mechlin,* Belgium] (1699)
: a delicate bobbin lace used for dresses and millinery

mec•li•zine \'me-klə-ˌzēn\ *noun* [*me*thyl + *chl*or- + *-izine* (alteration of *azine*)] (1954)
: a drug $C_{25}H_{27}ClN_2$ used usually in the form of its hydrochloride to treat nausea and vertigo

me•co•ni•um \mi-'kō-nē-əm\ *noun* [Latin, literally, poppy juice, from Greek *mēkōnion,* from *mēkōn* poppy; akin to Old High German *mago* poppy] (1706)
: a dark greenish mass that accumulates in the bowel during fetal life and is discharged shortly after birth

med \'med\ *adjective* (1891)
: MEDICAL ⟨*med* school⟩

me•da•ka \mə-'dä-kə\ *noun* [Japanese] (1933)
: a small Japanese freshwater fish (*Oryzias latipes*) usually silvery brown in the wild but from pale yellow to deep red in aquarium strains

med•al \'me-dᵊl\ *noun* [Middle French *medaille,* from Old Italian *medaglia* coin worth half a denarius, medal, from (assumed) Vulgar Latin *medalis* half, alteration of Late Latin *medialis* middle, from Latin *medius* — more at MID] (circa 1578)
1 : a small usually metal object bearing a religious emblem or picture
2 : a piece of metal often resembling a coin and having a stamped design that is issued to commemorate a person or event or awarded for excellence or achievement

Medal for Merit (1942)
: a U.S. decoration awarded to civilians for exceptionally meritorious conduct in the performance of outstanding services

med•al•ist *or* **med•al•list** \'me-dᵊl-ist\ *noun* (circa 1757)
1 : a designer, engraver, or maker of medals
2 : a recipient of a medal as an award

me•dal•lic \mə-'da-lik\ *adjective* (1702)
: of, relating to, or shown on a medal

me•dal•lion \mə-'dal-yən\ *noun* [French *médaillon,* from Italian *medaglione,* augmentative of *medaglia*] (1658)
1 : a large medal
2 : something resembling a large medal; *especially* **:** a tablet or panel in a wall or window bearing a figure in relief, a portrait, or an ornament
3 *also* **me•dail•lon** \mā-dà-yōⁿ\ **:** a small round or oval serving (as of meat)

Medal of Freedom (1945)
: a U.S. decoration awarded to civilians for meritorious achievement in any of various fields

Medal of Honor (1898)
: a U.S. military decoration awarded in the name of the Congress for conspicuous intrepidity at the risk of life in action with an enemy

medal play *noun* (1899)
: STROKE PLAY

med•dle \'me-dᵊl\ *intransitive verb* **med•dled; med•dling** \'med-liŋ, 'me-dᵊl-iŋ\ [Middle English *medlen,* from Middle French *mesler, medler,* from (assumed) Vulgar Latin

misculare, from Latin *miscēre* to mix — more at MIX] (14th century)
: to interest oneself in what is not one's concern **:** interfere without right or propriety
— **med·dler** \'med-lər, 'me-dᵊl-ər\ *noun*

med·dle·some \'me-dᵊl-səm\ *adjective* (1615)
: given to meddling
synonym see IMPERTINENT
— **med·dle·some·ness** *noun*

Mede \'mēd\ *noun* [Middle English, from Latin *Medus,* from Greek *Mēdos*] (14th century)
: a native or inhabitant of ancient Media in Persia

Me·dea \mə-'dē-ə\ *noun* [Latin, from Greek *Mēdeia*]
: an enchantress noted in Greek mythology for helping Jason gain the Golden Fleece and for repeatedly resorting to murder to gain her ends

med·e·vac \'me-də-,vak, -dē-\ *noun* [*med*ical *evac*uation] (1966)
1 : emergency evacuation of the sick or wounded (as from a combat area)
2 : a helicopter used for medevac

med·fly \'med-,flī\ *noun, often capitalized* (1935)
: MEDITERRANEAN FRUIT FLY

¹me·dia \'mē-dē-ə\ *noun, plural* **me·di·ae** \-dē-,ē\ (1841)
1 [Late Latin, from Latin, feminine of *medius;* from the voiced stops' being regarded as intermediate between the tenues and the aspirates] **:** a voiced stop
2 [New Latin, from Latin] **:** the middle coat of the wall of a blood or lymph vessel consisting chiefly of circular muscle fibers

²media *noun, plural* **me·di·as** *often attributive* [plural of *medium*] (1923)
1 : a medium of cultivation, conveyance, or expression; *especially* **:** MEDIUM 2b
2 a *singular or plural in construction* **:** MASS MEDIA **b** *plural* **:** members of the mass media
■

me·di·ad \'mē-dē-,ad\ *adverb* (1878)
: toward the median line or plane of a body or part

media event *noun* (1972)
: a publicity event staged for coverage by the news media

me·dia·gen·ic \,mē-dē-ə-'je-nik\ *adjective* (1971)
: attractive or well-suited to the communications media

me·di·al \'mē-dē-əl\ *adjective* [Late Latin *medialis,* from Latin *medius*] (1570)
1 : MEAN, AVERAGE
2 a : being or occurring in the middle **b :** extending toward the middle
3 : situated between the extremes of initial and final in a word or morpheme
— **medial** *noun*
— **me·di·al·ly** \-ə-lē\ *adverb*

¹me·di·an \'mē-dē-ən\ *noun* (15th century)
1 : a medial part (as a vein or nerve)
2 a : a value in an ordered set of values below and above which there is an equal number of values or which is the arithmetic mean of the two middle values if there is no one middle number **b :** a vertical line that divides the histogram of a frequency distribution into two parts of equal area **c :** a value of a random variable for which all greater values make the distribution function greater than one half and all lesser values make it less than one half
3 a : a line from a vertex of a triangle to the midpoint of the opposite side **b :** a line joining the midpoints of the nonparallel sides of a trapezoid
4 : MEDIAN STRIP
synonym see AVERAGE

²median *adjective* [Middle French or Latin; Middle French, from Latin *medianus,* from *medius* middle — more at MID] (1645)
1 : being in the middle or in an intermediate position **:** MEDIAL
2 : lying in the plane dividing a bilateral animal into right and left halves
3 : relating to or constituting a statistical median
4 : produced without occlusion along the lengthwise middle line of the tongue
— **me·di·an·ly** *adverb*

median strip *noun* (1948)
: a paved or planted strip dividing a highway into lanes according to direction of travel

me·di·ant \'mē-dē-ənt\ *noun* [Italian *mediante,* from Late Latin *mediant-, medians,* present participle of *mediare* to be in the middle] (circa 1741)
: the third tone of a diatonic scale

me·di·as·ti·num \,mē-dē-ə-'stī-nəm\ *noun, plural* **-na** \-nə\ [New Latin, from Medieval Latin, neuter of *mediastinus* medial, from Latin *medius*] (1541)
: the space in the chest between the pleural sacs of the lungs that contains all the viscera of the chest except the lungs and pleurae; *also* **:** this space with its contents
— **me·di·as·ti·nal** \-'stī-nᵊl\ *adjective*

¹me·di·ate \'mē-dē-ət\ *adjective* [Middle English, from Late Latin *mediatus* intermediate, from past participle of *mediare*] (15th century)
1 : occupying a middle position
2 a : acting through an intervening agency **b :** exhibiting indirect causation, connection, or relation
— **me·di·a·cy** \-dē-ə-sē\ *noun*
— **me·di·ate·ly** *adverb*

²me·di·ate \'mē-dē-,āt\ *verb* **-at·ed; -at·ing** [Medieval Latin *mediatus,* past participle of *mediare,* from Late Latin, to be in the middle, from Latin *medius* middle] (1568)
transitive verb
1 a : to effect by action as an intermediary **b :** to bring accord out of by action as an intermediary
2 a : to act as intermediary agent in bringing, effecting, or communicating **:** CONVEY **b :** to transmit as intermediate mechanism or agency
intransitive verb
1 : to interpose between parties in order to reconcile them
2 : to reconcile differences
synonym see INTERPOSE
— **me·di·a·tive** \-,ā-tiv\ *adjective*
— **me·di·a·to·ry** \-ə-,tōr-ē, -,tȯr-\ *adjective*

me·di·a·tion \,mē-dē-'ā-shən\ *noun* (14th century)
: the act or process of mediating; *especially* **:** intervention between conflicting parties to promote reconciliation, settlement, or compromise
— **me·di·a·tion·al** \-shnəl, -shə-nᵊl\ *adjective*

me·di·a·tor \'mē-dē-,ā-tər\ *noun* [Middle English, from Late Latin, from *mediare*] (14th century)
1 : one that mediates; *especially* **:** one that mediates between parties at variance
2 : a mediating agent in a physical, chemical, or biological process

me·di·a·trix \-'ā-triks\ *noun* [Middle English, from Late Latin, feminine of *mediator*] (15th century)
: a woman who is a mediator

¹med·ic \'me-dik\ *noun* [Middle English *medike,* from Latin *medica,* from Greek *mēdikē,* from feminine of *mēdikos* of Media, from *Mēdia* Media] (15th century)
: any of a genus (*Medicago*) of leguminous herbs (as alfalfa)

²medic *noun* [Latin *medicus*] (1659)
: one engaged in medical work or study; *especially* **:** CORPSMAN

med·i·ca·ble \'me-di-kə-bəl\ *adjective* (circa 1616)
: CURABLE, REMEDIABLE

med·ic·aid \'me-di-,kād\ *noun, often capitalized* [*medical aid*] (1966)
: a program of medical aid designed for those unable to afford regular medical service and financed by the state and federal governments

med·i·cal \'me-di-kəl\ *adjective* [French or Late Latin; French *médical,* from Late Latin *medicalis,* from Latin *medicus* physician, from *mederi* to remedy, heal; akin to Avestan vī-*mad-* healer, Greek *medesthai* to be mindful of — more at METE] (1646)
1 : of, relating to, or concerned with physicians or the practice of medicine
2 : requiring or devoted to medical treatment
— **med·i·cal·ly** \-k(ə-)lē\ *adverb*

medical examiner *noun* (1877)
: a public officer who conducts autopsies of bodies to find the cause of death

me·di·ca·ment \mi-'di-kə-mənt, 'me-di-kə-\ *noun* [Middle French, from Latin *medicamentum,* from *medicare*] (1541)
: a substance used in therapy
— **me·di·ca·men·tous** \mi-,di-kə-'men-təs, ,me-di-kə-\ *adjective*

medi·care \'me-di-,ker, -,kar\ *noun, often capitalized* [blend of *medical* and *care*] (1955)
: a government program of medical care especially for the aged

med·i·cate \'me-də-,kāt\ *transitive verb* **-cat·ed; -cat·ing** [Latin *medicatus,* past participle of *medicare* to heal, from *medicus*] (circa 1623)
1 : to treat medicinally
2 : to impregnate with a medicinal substance ⟨*medicated* soap⟩

med·i·ca·tion \,me-də-'kā-shən\ *noun* (15th century)
1 : the act or process of medicating
2 : a medicinal substance **:** MEDICAMENT

me·dic·i·na·ble \mi-'dis-nə-bəl, -'di-sᵊn-ə-; *in Shakespeare* 'med-sə-nə-\ *adjective* (14th century)
: MEDICINAL

me·dic·i·nal \mə-'dis-nəl, -'di-sᵊn-əl; *in Shakespeare & Milton* ,me-di-'sī-nᵊl & 'med-sə-nəl\ *adjective* (14th century)
1 : tending or used to cure disease or relieve pain
2 : SALUTARY
— **medicinal** *noun*
— **me·dic·i·nal·ly** *adverb*

medicinal leech *noun* (circa 1890)
: a large European freshwater leech (*Hirudo medicinalis*) formerly used by physicians for bleeding patients

med·i·cine \'me-də-sən, *British usually* 'med-sən\ *noun* [Middle English, from Old French, from Latin *medicina,* from feminine of *medicinus* of a physician, from *medicus*] (13th century)
1 a : a substance or preparation used in treating disease **b :** something that affects well-being

☐ USAGE
media The singular *media* and its plural *medias* seem to have originated in the field of advertising over 50 years ago; they are apparently still so used without stigma in that specialized field. In most other applications *media* is used as a plural of *medium.* The great popularity of the word in references to the agencies of mass communication is leading to the formation of a mass noun, construed as a singular ⟨there's no basis for it. You know, the news *media* gets on to something —Edwin Meese 3d⟩ ⟨the *media* is less interested in the party's policies —James Lewis, *Guardian Weekly*⟩. This use is not as well established as the mass-noun use of *data* and is likely to incur criticism especially in writing.

2 a : the science and art dealing with the maintenance of health and the prevention, alleviation, or cure of disease **b :** the branch of medicine concerned with the nonsurgical treatment of disease
3 : a substance (as a drug or potion) used to treat something other than disease
4 : an object held in traditional American Indian belief to give control over natural or magical forces; *also* **:** magical power or a magical rite
— medicine *transitive verb*

medicine ball *noun* (1895)
: a heavy stuffed leather-covered ball used for conditioning exercises

medicine dropper *noun* (1898)
: DROPPER 2

medicine man *noun* (1801)
: a priestly healer or sorcerer especially among the American Indians **:** SHAMAN

medicine show *noun* (circa 1906)
: a traveling show using entertainers to attract a crowd among which remedies or nostrums are sold

med·i·co \'me-di-ˌkō\ *noun, plural* **-cos** [Italian *medico* or Spanish *médico,* both from Latin *medicus*] (1689)
: a medical practitioner **:** PHYSICIAN; *also* **:** a medical student

med·i·co·le·gal \ˌme-di-kō-'lē-gəl\ *adjective* [New Latin *medicolegalis,* from Latin *medicus* medical + *-o-* + *legalis* legal] (1835)
: of or relating to both medicine and law

[1]**me·di·e·val** *or* **me·di·ae·val** \mē-'dē-vəl, mi-, me-, -dē-'ē-vəl\ *adjective* [New Latin *medium aevum* Middle Ages] (1827)
1 : of, relating to, or characteristic of the Middle Ages
2 : extremely outmoded or antiquated
— me·di·e·val·ly *adverb*

[2]**medieval** *or* **mediaeval** *noun* (1856)
: a person of the Middle Ages

me·di·e·val·ism \-və-ˌli-zəm\ *noun* (1853)
1 : medieval quality, character, or state
2 : devotion to the institutions, arts, and practices of the Middle Ages

me·di·e·val·ist \-'dēv-list, -'dē-və-, -dē-'ēv-, -dē-'ē-və-\ *noun* (circa 1859)
1 : a specialist in medieval history and culture
2 : a connoisseur or devotee of medieval arts and culture

Medieval Latin *noun* (circa 1889)
: the Latin used especially for liturgical and literary purposes from the 7th to the 15th centuries inclusive

me·di·na \mə-'dē-nə\ *noun* [Arabic *madīna* city] (1906)
: the non-European part of a northern African city

me·di·o·cre \ˌmē-dē-'ō-kər\ *adjective* [Middle English, from Middle French, from Latin *mediocris,* from *medius* middle + Old Latin *ocris* stony mountain; akin to Latin *acer* sharp — more at EDGE] (circa 1586)
: of moderate or low quality, value, ability, or performance **:** ORDINARY, SO-SO

me·di·oc·ri·ty \ˌmē-dē-'ä-krə-tē\ *noun, plural* **-ties** (1531)
1 a : the quality or state of being mediocre **b :** moderate ability or value
2 : a mediocre person

med·i·tate \'me-də-ˌtāt\ *verb* **-tat·ed; -tat·ing** [Latin *meditatus,* past participle of *meditari,* frequentative of *medēri* to remedy — more at MEDICAL] (1560)
intransitive verb
: to engage in contemplation or reflection
transitive verb
1 : to focus one's thoughts on **:** reflect on or ponder over
2 : to plan or project in the mind **:** INTEND, PURPOSE
synonym see PONDER
— med·i·ta·tor \-ˌtā-tər\ *noun*

med·i·ta·tion \ˌme-də-'tā-shən\ *noun* (13th century)
1 : a discourse intended to express its author's reflections or to guide others in contemplation
2 : the act or process of meditating

med·i·ta·tive \'me-də-ˌtā-tiv\ *adjective* (circa 1656)
1 : disposed or given to meditation
2 : marked by or conducive to meditation
— med·i·ta·tive·ly *adverb*
— med·i·ta·tive·ness *noun*

Med·i·ter·ra·nean \ˌme-də-tə-'rā-nē-ən, -nyən\ *adjective* (15th century)
1 a : of, relating to, or characteristic of the Mediterranean Sea **b :** of, relating to, or characteristic of the peoples, lands, or cultures bordering the Mediterranean Sea
2 *not capitalized* [Latin *mediterraneus,* from *medius* middle + *terra* land — more at TERRACE] **:** enclosed or nearly enclosed with land
3 : of or relating to a group or physical type of the Caucasian race characterized by medium or short stature, slender build, dolichocephaly, and dark complexion

Mediterranean flour moth *noun* (1895)
: a small largely gray and black nearly cosmopolitan pyralid moth (*Anagasta kuehniella*) whose larva destroys processed grain products

Mediterranean fruit fly *noun* (1907)
: a small widely distributed dipteran fly (*Ceratitis capitata*) with black-and-white markings whose larva lives and feeds in ripening fruit

[1]**me·di·um** \'mē-dē-əm\ *noun, plural* **mediums** *or* **me·dia** \-dē-ə\ [Latin, from neuter of *medius* middle — more at MID] (1593)
1 a : something in a middle position **b :** a middle condition or degree **:** MEAN
2 : a means of effecting or conveying something: as **a** (1) **:** a substance regarded as the means of transmission of a force or effect (2) **:** a surrounding or enveloping substance (3) **:** the tenuous material (as gas and dust) in space that exists outside large agglomerations of matter (as stars) ⟨interstellar *medium*⟩ **b** *plural usually media* (1) **:** a channel or system of communication, information, or entertainment — compare MASS MEDIUM (2) **:** a publication or broadcast that carries advertising (3) **:** a mode of artistic expression or communication (4) **:** something (as a magnetic disk) on which information may be stored **c :** GO-BETWEEN, INTERMEDIARY **d** *plural mediums* **:** an individual held to be a channel of communication between the earthly world and a world of spirits **e :** material or technical means of artistic expression
3 a : a condition or environment in which something may function or flourish **b** *plural media* (1) **:** a nutrient system for the artificial cultivation of cells or organisms and especially bacteria (2) **:** a fluid or solid in which organic structures are placed (as for preservation or mounting) **c :** a liquid with which pigment is mixed by a painter
usage see MEDIA

[2]**medium** *adjective* (1711)
: intermediate in quantity, quality, position, size, or degree

medium frequency *noun* (1920)
: a radio frequency between high frequency and low frequency — see RADIO FREQUENCY table

me·di·um·is·tic \ˌmē-dē-ə-'mis-tik\ *adjective* (1868)
: of, relating to, or having the qualities of a spiritualistic medium

medium of exchange (1714)
: something commonly accepted in exchange for goods and services and recognized as representing a standard of value

me·di·um·ship \'mē-dē-əm-ˌship\ *noun* (1868)
: the capacity, function, or profession of a spiritualistic medium

med·lar \'med-lər\ *noun* [Middle English *medeler,* from Middle French *medlier,* from *medle* medlar fruit, from Latin *mespilum,* from Greek *mespilon*] (14th century)
: a small Eurasian tree (*Mespilus germanica*) of the rose family whose fruit resembles a crab apple and is used in preserves; *also* **:** its fruit

[1]**med·ley** \'med-lē\ *noun, plural* **medleys** [Middle English *medle,* from Middle French *medlee,* from feminine of *medlé,* past participle of *medler* to mix — more at MEDDLE] (14th century)
1 *archaic* **:** MELEE
2 : a diverse assortment or mixture; *especially* **:** HODGEPODGE
3 : a musical composition made up of a series of songs or short musical pieces

[2]**medley** *adjective* (14th century)
: MIXED, MOTLEY

medley relay *noun* (1949)
: a relay race in swimming in which each member of a team uses a different stroke

me·dul·la \mə-'də-lə\ *noun, plural* **-las** *or* **-lae** \-(ˌ)lē, -ˌlī\ [Middle English, from Latin] (15th century)
1 *plural medullae* **a :** MARROW 1 **b :** MEDULLA OBLONGATA
2 a : the inner or deep part of an animal or plant structure ⟨the adrenal *medulla*⟩ **b :** MYELIN SHEATH

medulla ob·lon·ga·ta \-ˌä-ˌbloŋ-'gä-tə\ *noun, plural* **medulla oblongatas** *or* **me·dullae ob·lon·ga·tae** \-'gä-tē, -ˌtī\ [New Latin, literally, oblong medulla] (1676)
: the part of the vertebrate brain that is continuous posteriorly with the spinal cord and that contains the centers controlling involuntary vital functions — see BRAIN illustration

med·ul·lary \'me-d^{ə}l-ˌer-ē, 'me-jə-ˌler-; mə-'də-lə-rē\ *adjective* (1620)
1 : of or relating to the pith of a plant
2 : of or relating to a medulla and especially the medulla oblongata

medullary ray *noun* (1830)
1 : a primary tissue composed of radiating bands of parenchyma cells extending between the vascular bundles of herbaceous dicotyledonous stems and connecting the pith with the cortex
2 : VASCULAR RAY

medullary sheath *noun* (circa 1846)
: MYELIN SHEATH

med·ul·lat·ed \'me-d^{ə}l-ˌā-təd, 'me-jə-ˌlā-\ *adjective* (1867)
1 : MYELINATED
2 : having a medulla — used of fibers other than nerve fibers

me·dul·lo·blas·to·ma \mə-ˌdə-lō-ˌblas-'tō-mə\ *noun, plural* **-to·mas** *also* **-to·ma·ta** \-'tō-mə-tə\ [New Latin, from *medulla* + *-o-* + *blast-* + *-oma*] (1925)
: a malignant tumor of the central nervous system arising in the cerebellum especially in children

me·du·sa \mi-'dü-sə, -'dyü-, -zə\ *noun*
1 *capitalized* [Latin, from Greek *Medousa*] **:** a mortal Gorgon who is slain when decapitated by Perseus
2 *plural* **me·du·sae** \-ˌsē, -ˌzē, -ˌsī, -ˌzī\ [New Latin, from Latin] **:** JELLYFISH 1a
— me·du·san \-s^{ə}n, -z^{ə}n\ *adjective or noun*
— me·du·soid \-ˌsȯid, -ˌzȯid\ *adjective or noun*

meed \'mēd\ *noun* [Middle English, from Old English *mēd;* akin to Old High German *miata* reward, Greek *misthos*] (before 12th century)
1 *archaic* **:** an earned reward or wage
2 : a fitting return or recompense

meek \'mēk\ *adjective* [Middle English, of Scandinavian origin; akin to Old Norse *mjūkr* gentle; akin to Welsh *mwyth* soft] (13th century)

1 : enduring injury with patience and without resentment **:** MILD
2 : deficient in spirit and courage **:** SUBMISSIVE
3 : not violent or strong **:** MODERATE
— **meek·ly** *adverb*
— **meek·ness** *noun*

meer·kat \ˈmir-ˌkat\ *noun* [Afrikaans, from Dutch, a kind of monkey, from Middle Dutch *meercatte* monkey, from *meer* sea + *catte* cat] (1801)
: any of several African mongooses; *especially* **:** a burrowing highly social mammal (*Suricata suricatta*) of southern Africa that is chiefly grayish with inconspicuous black markings

meer·schaum \ˈmir-shəm, -ˌshȯm\ *noun* [German, from *Meer* sea + *Schaum* foam] (1784)
1 : a fine light white clayey mineral that is a hydrous magnesium silicate found chiefly in Asia Minor and is used especially for tobacco pipes
2 : a tobacco pipe of meerschaum

¹meet \ˈmēt\ *verb* **met** \ˈmet\; **meet·ing** [Middle English *meten,* from Old English *mētan;* akin to Old English *gemōt* assembly — more at MOOT] (before 12th century)
transitive verb
1 a : to come into the presence of **:** FIND **b :** to come together with especially at a particular time or place ⟨I'll *meet* you at the station⟩ **c :** to come into contact or conjunction with **:** JOIN **d :** to appear to the perception of
2 : to encounter as antagonist or foe **:** OPPOSE
3 : to enter into conference, argument, or personal dealings with
4 : to conform to especially with exactitude and precision ⟨a concept to *meet* all requirements⟩
5 : to pay fully **:** SETTLE
6 : to cope with ⟨was able to *meet* every social situation⟩
7 : to provide for ⟨enough money to *meet* our needs⟩
8 : to become acquainted with
9 : ENCOUNTER, EXPERIENCE
10 : to receive or greet in an official capacity
intransitive verb
1 a : to come face-to-face **b :** to come together for a common purpose **:** ASSEMBLE **c :** to come together as contestants, opponents, or enemies
2 : to form a junction or confluence ⟨the lines *meet* in a point⟩
3 : to occur together
— **meet·er** *noun*
— **meet halfway :** to compromise with
— **meet with :** to be subjected to **:** ENCOUNTER ⟨the proposal *met with* opposition⟩

²meet *noun* (circa 1834)
1 : the act of assembling for a hunt or for competitive sports
2 : a competition in which individuals (as athletes) match skills

³meet *adjective* [Middle English *mete,* from Old English *gemǣte;* akin to Old English *metan* to mete] (14th century)
: precisely adapted to a particular situation, need, or circumstance **:** very proper
synonym see FIT
— **meet·ly** *adverb*

meet·ing \ˈmē-tiŋ\ *noun* (14th century)
1 : an act or process of coming together: as **a :** an assembly for a common purpose (as worship) **b :** a session of horse or dog racing
2 : a permanent organizational unit of the Society of Friends
3 : INTERSECTION, JUNCTION

meet·ing·house \-ˌhau̇s\ *noun* (1632)
: a building used for public assembly and especially for Protestant worship

meeting of minds (1939)
: AGREEMENT, CONCORD

mef·e·nam·ic acid \ˌme-fə-ˈna-mik-\ *noun* [di*me*thyl- + *fen-* (by shortening & alteration from *phenyl*) + *am*inobenzo*ic acid*] (circa 1964)
: a drug $C_{15}H_{15}NO_2$ used as an anti-inflammatory

mega- *or* **meg-** *combining form* [Greek, from *megas* large — more at MUCH]
1 a : great **:** large ⟨*mega*spore⟩ **b :** greatly surpassing others of its kind ⟨*mega*hit⟩
2 : million **:** multiplied by one million ⟨*meg*ohm⟩ ⟨*mega*cycle⟩

mega·bar \ˈme-gə-ˌbär\ *noun* [International Scientific Vocabulary] (1903)
: a unit of pressure equal to one million bars

mega·bit \-ˌbit\ *noun* (1956)
: one million bits

mega·buck \-ˌbək\ *noun* (1946)
: one million dollars; *also* **:** an indeterminately large sum of money — usually used in plural

mega·byte \-ˌbīt\ *noun* [from the fact that 1,048,576 (2^{20}) is the power of 2 closest to one million] (1970)
: a unit of computer information storage capacity equal to 1,048,576 bytes

mega·city \-ˌsi-tē\ *noun* (1968)
: MEGALOPOLIS 1

mega·cor·po·ra·tion \ˌme-gə-ˌkȯr-pə-ˈrā-shən\ *noun* (1973)
: a huge and powerful corporation

mega·cy·cle \ˈme-gə-ˌsī-kəl\ *noun* (1926)
: one million cycles; *especially* **:** MEGAHERTZ

mega·deal \-ˌdēl\ *noun* (1981)
: a business deal involving a lot of money

mega·death \-ˌdeth\ *noun* (1953)
: one million deaths — usually used as a unit in reference to nuclear warfare

mega·dose \-ˌdōs\ *noun* (1973)
: a large dose (as of a vitamin)

mega·fau·na \-ˌfȯ-nə, -ˌfä-\ *noun* (1935)
: fauna consisting of individuals large enough to be visible to the naked eye
— **mega·fau·nal** \-nᵊl\ *adjective*

mega·ga·mete \ˌme-gə-ˈga-ˌmēt *also* -gə-ˈmēt\ *noun* (1891)
: MACROGAMETE

mega·ga·me·to·phyte \-gə-ˈmē-tə-ˌfīt\ *noun* (1933)
: the female gametophyte produced by a megaspore

mega·hertz \ˈme-gə-ˌhərts, -ˌherts\ *noun* [International Scientific Vocabulary] (1941)
: a unit of frequency equal to one million hertz — abbreviation *MHz*

mega·hit \-ˌhit\ *noun* (1982)
: something (as a motion picture) that is extremely successful

mega·kar·yo·cyte \ˌme-gə-ˈkar-ē-ō-ˌsīt\ *noun* (1890)
: a large cell that has a lobulated nucleus, is found especially in the bone marrow, and is the source of blood platelets
— **mega·kar·yo·cyt·ic** \-ˌkar-ē-ō-ˈsi-tik\ *adjective*

megal- *or* **megalo-** *combining form* [New Latin, from Greek, from *megal-, megas* — more at MUCH]
: large **:** of giant size ⟨*megalo*polis⟩ **:** grandiose ⟨*megalo*mania⟩

mega·lith \ˈme-gə-ˌlith\ *noun* (1853)
: a very large usually rough stone used in prehistoric cultures as a monument or building block
— **mega·lith·ic** \ˌme-gə-ˈli-thik\ *adjective*

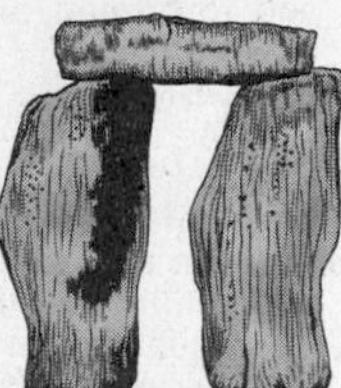
megalith

meg·a·lo·blast \ˈme-gə-lō-ˌblast\ *noun* (1899)
: a large erythroblast that appears in the blood especially in pernicious anemia
— **meg·a·lo·blas·tic** \ˌme-gə-lō-ˈblas-tik\ *adjective*

meg·a·lo·ma·nia \ˌme-gə-lō-ˈmā-nē-ə, -nyə\ *noun* [New Latin] (circa 1890)
1 : a mania for great or grandiose performance
2 : a delusional mental disorder that is marked by infantile feelings of personal omnipotence and grandeur
— **meg·a·lo·ma·ni·ac** \-ˈmā-nē-ˌak\ *adjective or noun*
— **meg·a·lo·ma·ni·a·cal** \-mə-ˈnī-ə-kəl\ *also* **meg·a·lo·man·ic** \-ˈma-nik\ *adjective*
— **meg·a·lo·ma·ni·a·cal·ly** \-mə-ˈnī-ə-k(ə-)lē\ *adverb*

meg·a·lop·o·lis \ˌme-gə-ˈlä-pə-ləs\ *noun* (circa 1828)
1 : a very large city
2 : a thickly populated region centering in a metropolis or embracing several metropolises
— **meg·a·lo·pol·i·tan** \-lō-ˈpä-lə-tᵊn\ *noun or adjective*

-megaly *noun combining form* [New Latin *-megalia,* from Greek *megal-, megas*]
: abnormal enlargement ⟨hepato*megaly*⟩

mega·par·sec \ˌme-gə-ˈpär-ˌsek\ *noun* [International Scientific Vocabulary] (1933)
: a unit of measure for distances in intergalactic space equal to one million parsecs

¹mega·phone \ˈme-gə-ˌfōn\ *noun* (1878)
: a cone-shaped device used to intensify or direct the voice
— **mega·phon·ic** \ˌme-gə-ˈfä-nik\ *adjective*

²megaphone (1901)
transitive verb
: to transmit or address through or as if through a megaphone
intransitive verb
: to speak through or as if through a megaphone

mega·proj·ect \-ˌprä-ˌjekt, -jikt *also* -ˌprō-\ *noun* (1982)
: a major project or undertaking (as in business or construction)

Me·gar·i·an \mə-ˈgar-ē-ən, me-\ *adjective* (1603)
: of or relating to a Socratic school of philosophy founded by Euclid of Megara and noted for its subtle attention to logic
— **Megarian** *noun*

Me·gar·ic \-ˈgar-ik\ *adjective* (1656)
: MEGARIAN
— **Megaric** *noun*

mega·scop·ic \ˌme-gə-ˈskä-pik\ *adjective* [*mega-* + *-scopic* (as in *microscopic*)] (1879)
1 : MACROSCOPIC 1 ⟨*megascopic* features of leaves⟩
2 : based on or relating to observations made with the unaided eye
— **mega·scop·i·cal·ly** \-pi-k(ə-)lē\ *adverb*

mega·spo·ran·gi·um \ˌme-gə-spə-ˈran-jē-əm\ *noun* [New Latin] (1886)
: a sporangium that develops only megaspores

mega·spore \ˈme-gə-ˌspōr, -ˌspȯr\ *noun* [International Scientific Vocabulary] (1858)
: a spore in heterosporous plants that gives rise to female gametophytes and is generally larger than a microspore
— **mega·spor·ic** \ˌme-gə-ˈspōr-ik, -ˈspȯr-\ *adjective*

mega·spo·ro·gen·e·sis \ˌme-gə-ˌspōr-ə-ˈje-nə-səs, -ˌspȯr-\ *noun* [New Latin] (circa 1928)
: the formation and maturation of a megaspore

mega·spo·ro·phyll \-ˈspōr-ə-ˌfil, -ˈspȯr-\ *noun* (circa 1899)
: a sporophyll that develops only megasporangia

mega·star \ˈme-gə-ˌstär\ *noun* (1975)
: SUPERSTAR

mega·ton \ˈme-gə-ˌtən\ *noun* (1952)
: an explosive force equivalent to that of one million tons of TNT

mega·ton·nage \-ˌtə-nij\ *noun* (1955)
: the destructive capability expressed in megatons especially of a collection of nuclear weapons

mega·vi·ta·min \ˌme-gə-ˈvī-tə-mən, *British usually* -ˈvi-\ *adjective* (1970)
: relating to or consisting of very large doses of vitamins ⟨*megavitamin* therapy⟩

mega·vi·ta·mins \-mənz\ *noun plural* (1974)
: a large quantity of vitamins

mega·watt \'me-gə-ˌwät\ *noun* [International Scientific Vocabulary] (circa 1900)
: one million watts

me·gil·lah \mə-'gi-lə\ *noun* [Yiddish *megile,* from Hebrew *mĕgillāh* scroll, volume (used especially of the Book of Esther, read aloud at the Purim celebration)] (circa 1952)
slang : a long involved story or account ⟨the whole *megillah*⟩

me·gilp \mə-'gilp\ *noun* [origin unknown] (1768)
: a gelatinous preparation commonly of linseed oil and mastic varnish that is used by artists as a vehicle for oil colors

meg·ohm \'meg-ˌōm\ *noun* [International Scientific Vocabulary] (circa 1868)
: one million ohms

me·grim \'mē-grəm\ *noun* [Middle English *migreime,* from Middle French *migraine* — more at MIGRAINE] (14th century)
1 a : MIGRAINE **b :** VERTIGO, DIZZINESS
2 a : FANCY, WHIM **b** *plural* **:** low spirits

Mei·ji \'mā-(ˌ)jē\ *noun* [Japanese, literally, enlightened rule] (1873)
: the period of the reign (1868–1912) of Emperor Mutsuhito of Japan

mei·kle \'mē-kəl\ *variant of* MICKLE

mei·ny *noun, plural* **meinies** [Middle English *meynie* — more at MENIAL] (13th century)
1 \'mā-nē\ *archaic* **:** RETINUE, COMPANY
2 \'men-yē\ *chiefly Scottish* **:** MULTITUDE

mei·o·sis \mī-'ō-səs\ *noun* [New Latin, from Greek *meiōsis* diminution, from *meioun* to diminish, from *meiōn* less; akin to Sanskrit *mīyate* he diminishes] (1550)
1 : the presentation of a thing with underemphasis especially in order to achieve a greater effect **:** UNDERSTATEMENT
2 : the cellular process that results in the number of chromosomes in gamete-producing cells being reduced to one half and that involves a reduction division in which one of each pair of homologous chromosomes passes to each daughter cell and a mitotic division — compare MITOSIS
— **mei·ot·ic** \mī-'ä-tik\ *adjective*
— **mei·ot·i·cal·ly** \-ti-k(ə-)lē\ *adverb*

Meis·sen \'mī-sᵊn\ *noun* [*Meissen,* Saxony, Germany] (1863)
: a ceramic ware made at Meissen near Dresden; *especially* **:** a European porcelain developed under the patronage of the king of Saxony about 1715 and used for both ornamental and table wares — called also *Meissen china, Meissen ware*

Mei·ster·sing·er \'mīs-tər-ˌsiŋ-ər, -ˌziŋ-\ *noun, plural* **Meistersinger** *or* **Meistersingers** [German, from Middle High German, from *meister* master + *singer* singer] (1845)
: a member of any of various German guilds formed chiefly in the 15th and 16th centuries by workingmen and craftsmen for the cultivation of poetry and music

mel·a·mine \'me-lə-ˌmēn\ *noun* [German *Melamin*] (circa 1835)
1 : a white crystalline organic base $C_3H_6N_6$ with a high melting point that is used especially in melamine resins
2 : a melamine resin or a plastic made from such a resin

melamine resin *noun* (1939)
: a thermosetting resin made from melamine and an aldehyde and used especially in molded or laminated products, adhesives, and coatings

melan- *or* **melano-** *combining form* [Middle English, from Middle French, from Late Latin, from Greek, from *melan-, melas;* perhaps akin to Lithuanian *mėlynas* blue, Sanskrit *malina* dirty]
1 : black **:** dark ⟨*melan*ic⟩ ⟨*melan*in⟩
2 : melanin ⟨*melano*phore⟩

mel·an·cho·lia \ˌme-lən-'kō-lē-ə\ *noun* [New Latin, from Late Latin, melancholy] (circa 1693)
: a mental condition characterized by extreme depression, bodily complaints, and often hallucinations and delusions; *especially* **:** a manic-depressive psychosis
— **mel·an·cho·li·ac** \-lē-ˌak\ *noun*

mel·an·chol·ic \ˌme-lən-'kä-lik\ *adjective* (14th century)
1 : of, relating to, or subject to melancholy **:** DEPRESSED
2 : of or relating to melancholia
3 : tending to depress the spirits **:** SADDENING
— **melancholic** *noun*

[1]**mel·an·choly** \'me-lən-ˌkä-lē\ *noun, plural* **-chol·ies** [Middle English *malencolie,* from Middle French *melancolie,* from Late Latin *melancholia,* from Greek, from *melan-* + *cholē* bile — more at GALL] (14th century)
1 a : an abnormal state attributed to an excess of black bile and characterized by irascibility or depression **b :** BLACK BILE **c :** MELANCHOLIA
2 a : depression of spirits **:** DEJECTION **b :** a pensive mood

[2]**melancholy** *adjective* (14th century)
1 a : suggestive or expressive of melancholy ⟨sang in a *melancholy* voice⟩ **b :** causing or tending to cause sadness or depression of mind or spirit **:** DISMAL ⟨a *melancholy* thought⟩
2 a : depressed in spirits **:** DEJECTED, SAD **b :** PENSIVE

Mel·a·ne·sian \ˌme-lə-'nē-zhən, -shən\ *noun* (1849)
1 : a language group consisting of the Austronesian languages of Melanesia
2 : a member of the dominant native group of Melanesia
— **Melanesian** *adjective*

mé·lange \mā-'läⁿzh, -'länj\ *noun* [French, from Middle French, from *mesler, meler* to mix — more at MEDDLE] (1653)
: a mixture often of incongruous elements

[1]**me·lan·ic** \mə-'la-nik\ *adjective* (1826)
1 : MELANOTIC
2 : affected with or characterized by melanism

[2]**melanic** *noun* (1952)
: a melanic individual

mel·a·nin \'me-lə-nən\ *noun* (1843)
: a dark brown or black animal or plant pigment

mel·a·nism \'me-lə-ˌni-zəm\ *noun* (1843)
1 : an increased amount of black or nearly black pigmentation (as of skin, feathers, or hair) of an individual or kind of organism
2 : intense human pigmentation of the skin, eyes, and hair
— **mel·a·nis·tic** \ˌme-lə-'nis-tik\ *adjective*

mel·a·nite \'me-lə-ˌnīt\ *noun* [German *Melanit,* from *melan-*] (circa 1807)
: a black andradite garnet
— **mel·a·nit·ic** \ˌme-lə-'ni-tik\ *adjective*

mel·a·nize \'me-lə-ˌnīz\ *transitive verb* **-nized; -niz·ing** (1885)
1 : to convert into or infiltrate with melanin
2 : to make dark or black
— **mel·a·ni·za·tion** \ˌme-lə-nə-'zā-shən\ *noun*

me·la·no·blast \mə-'la-nə-ˌblast, 'me-lə-nō-\ *noun* [International Scientific Vocabulary] (1901)
: a cell that is a precursor of a melanocyte or melanophore

me·la·no·cyte \mə-'la-nə-ˌsīt, 'me-lə-nō-\ *noun* [International Scientific Vocabulary] (circa 1890)
: an epidermal cell that produces melanin

melanocyte–stimulating hormone *noun* (1953)
: either of two vertebrate hormones of the pituitary gland especially of reptiles and amphibians that darken the skin by stimulating melanin dispersion in pigment-containing cells — called also *melanophore-stimulating hormone*

me·la·no·gen·e·sis \mə-ˌla-nə-'je-nə-səs, ˌme-lə-nō-\ *noun* [New Latin] (circa 1928)
: the formation of melanin

mel·a·no·ma \ˌme-lə-'nō-mə\ *noun, plural* **-mas** *also* **-ma·ta** \-mə-tə\ [New Latin] (1830)
: a usually malignant tumor containing dark pigment

me·la·no·phore \mə-'la-nə-ˌfōr, 'me-lə-nə-, -ˌfȯr\ *noun* (1903)
: a melanin-containing cell especially of fishes, amphibians, and reptiles

me·la·no·some \-ˌsōm\ *noun* (1940)
: a melanin-producing granule in a melanocyte

mel·a·not·ic \ˌme-lə-'nä-tik\ *adjective* (1829)
: having or characterized by black pigmentation

mel·a·to·nin \ˌme-lə-'tō-nən\ *noun* [Greek *melas* black + *-tonin* (as in *serotonin*)] (1958)
: a vertebrate hormone of the pineal gland that produces lightening of the skin by causing concentration of melanin in pigment-containing cells

mel·ba toast \'mel-bə-\ *noun* [Nellie *Melba*] (1925)
: very thin crisp toast

Mel·chite *or* **Mel·kite** \'mel-ˌkīt\ *noun* [Medieval Latin *Melchita,* from Middle Greek *Melchitēs,* literally, royalist, from Syriac *malkā* king] (1619)
1 : an Eastern Christian chiefly of Syria and Egypt adhering to Chalcedonian orthodoxy in preference to Monophysitism
2 : a member of a Uniate body derived from the Melchites

[1]**Mel·chiz·e·dek** \mel-'ki-zə-ˌdek\ *noun* [Greek *Melchisedek,* from Hebrew *Malkīṣedheq*]
: a priest-king of Jerusalem who prepared a ritual meal for Abraham and received tithes from him

[2]**Melchizedek** *adjective* (1842)
: of or relating to the higher order of the Mormon priesthood

[1]**meld** \'meld\ *verb* [German *melden* to announce, from Old High German *meldōn;* akin to Old English *meldian* to announce, Lithuanian *malda* prayer] (1897)
transitive verb
: to declare or announce (a card or combination of cards) for a score in a card game especially by placing face up on the table
intransitive verb
: to declare a card or combination of cards as a meld

[2]**meld** *noun* (1897)
: a card or combination of cards that is or can be melded in a card game

[3]**meld** *verb* [blend of *melt* and *weld*] (1936)
: MERGE, BLEND

[4]**meld** *noun* (1954)
: BLEND, MIXTURE

me·lee *also* **mê·lée** \'mā-ˌlā, mā-'\ *noun* [French *mêlée,* from Old French *meslee,* from *mesler* to mix — more at MEDDLE] (circa 1648)
: a confused struggle; *especially* **:** a hand-to-hand fight among several people

mel·ic \'me-lik\ *adjective* [Latin *melicus,* from Greek *melikos,* from *melos* song — more at MELODY] (1699)
: of or relating to song **:** LYRIC; *especially* **:** of or relating to Greek lyric poetry of the 7th and 6th centuries B.C.

mel·i·lot \'me-lə-ˌlät\ *noun* [Middle English *mellilot,* from Old French *melilot,* from Latin *melilotos,* from Greek *melilōtos,* from *meli* honey + *lōtos* clover, lotus — more at MELLIFLUOUS] (14th century)

: SWEET CLOVER; *especially* : a yellow-flowered sweet clover (*Melilotus officinalis*)

me·lio·rate \'mēl-yə-ˌrāt, 'mē-lē-ə-\ *verb* **-rat·ed; -rat·ing** [Late Latin *mélioratus,* past participle of *meliorare,* from Latin *melior* better; akin to Latin *multus* much, Greek *mala* very] (1542)
: AMELIORATE
— **me·lio·ra·tion** \ˌmēl-yə-'rā-shən, ˌmē-lē-ə-\ *noun*
— **me·lio·ra·tive** \'mēl-yə-ˌrā-tiv, 'mē-lē-ə-\ *adjective*
— **me·lio·ra·tor** \-ˌrā-tər\ *noun*

me·lio·rism \'mēl-yə-ˌri-zəm, 'mē-lē-ə-\ *noun* (1877)
: the belief that the world tends to become better and that humans can aid its betterment
— **me·lio·rist** \-rist\ *adjective or noun*
— **me·lio·ris·tic** \ˌmēl-yə-'ris-tik, ˌmē-lē-ə-\ *adjective*

me·lis·ma \mi-'liz-mə\ *noun, plural* **-ma·ta** \-mə-tə\ [New Latin, from Greek, song, melody, from *melizein* to sing, from *melos* song] (circa 1880)
1 : a group of notes or tones sung on one syllable in plainsong
2 : melodic embellishment or ornamentation
3 : CADENZA
— **mel·is·mat·ic** \ˌme-ləz-'ma-tik\ *adjective*

mell \'mel\ *verb* [Middle English, from Middle French *mesler*] (14th century)
: MIX

mel·lif·lu·ent \me-'li-flə-wənt\ *adjective* [Late Latin *mellifluent-, mellifluens,* from Latin *mell-, mel* + *fluent-, fluens,* present participle of *fluere*] (1601)
: MELLIFLUOUS
— **mel·lif·lu·ent·ly** *adverb*

mel·lif·lu·ous \me-'li-flə-wəs, mə-\ *adjective* [Middle English *mellyfluous,* from Late Latin *mellifluus,* from Latin *mell-, mel* honey + *fluere* to flow; akin to Gothic *milith* honey, Greek *melit-, meli*] (15th century)
1 : having a smooth rich flow ⟨a *mellifluous* voice⟩
2 : filled with something (as honey) that sweetens ◆
— **mel·lif·lu·ous·ly** *adverb*
— **mel·lif·lu·ous·ness** *noun*

mel·lo·phone \'me-lə-ˌfōn\ *noun* [¹*mellow* + *-phone*] (1926)
: a valved brass instrument similar in form and range to the French horn

mel·lo·tron \'me-lə-ˌträn\ *noun* [from *Mellotron,* a trademark] (1967)
: an electronic keyboard instrument programmed to produce the tape-recorded sounds usually of orchestral instruments

¹**mel·low** \'me-(ˌ)lō\ *adjective* [Middle English *melowe*] (15th century)
1 a *of a fruit* **:** tender and sweet because of ripeness **b** *of a wine* **:** well aged and pleasingly mild
2 a : made gentle by age or experience **b :** rich and full but free from garishness or stridency **c :** warmed and relaxed by or as if by liquor **d :** PLEASANT, AGREEABLE **e :** LAID-BACK
3 *of soil* **:** having a soft and loamy consistency
— **mel·low·ly** *adverb*
— **mel·low·ness** *noun*

²**mellow** (1572)
transitive verb
: to make mellow
intransitive verb
: to become mellow

me·lo·de·on \mə-'lō-dē-ən\ *noun* [German *Melodion,* from *Melodie* melody, from Old French] (1847)
: a small reed organ in which a suction bellows draws air inward through the reeds

me·lod·ic \mə-'lä-dik\ *adjective* (1823)
: of or relating to melody : MELODIOUS
— **me·lod·i·cal·ly** \-di-k(ə-)lē\ *adverb*

me·lo·di·ous \mə-'lō-dē-əs\ *adjective* (14th century)
1 : having a pleasing melody
2 : of, relating to, or producing melody
— **me·lo·di·ous·ly** *adverb*
— **me·lo·di·ous·ness** *noun*

mel·o·dist \'me-lə-dist\ *noun* (1789)
1 : SINGER
2 : a composer of melodies

mel·o·dize \'me-lə-ˌdīz\ *verb* **-dized; -diz·ing** (1662)
intransitive verb
: to compose a melody
transitive verb
: to make melodious : set to melody
— **mel·o·diz·er** *noun*

melo·dra·ma \'me-lə-ˌdrä-mə, -ˌdra-\ *noun* [modification of French *mélodrame,* from Greek *melos* song + French *drame* drama, from Late Latin *drama*] (1809)
1 a : a work (as a movie or play) characterized by extravagant theatricality and by the predominance of plot and physical action over characterization **b :** the genre of dramatic literature constituted by such works
2 : something resembling a melodrama or having a melodramatic quality
— **melo·dra·ma·tist** \ˌme-lə-'dra-mə-tist, -'drä-\ *noun*

melo·dra·mat·ic \ˌme-lə-drə-'ma-tik\ *adjective* (1816)
1 : of, relating to, or characteristic of melodrama
2 : appealing to the emotions : SENSATIONAL
synonym see DRAMATIC
— **melo·dra·mat·i·cal·ly** \-ti-k(ə-)lē\ *adverb*

melo·dra·mat·ics \-tiks\ *noun plural but singular or plural in construction* (1915)
: melodramatic conduct or writing

melo·dra·ma·tise *British variant of* MELODRAMATIZE

melo·dra·ma·tize \ˌme-lə-'dra-mə-ˌtīz, -'drä-\ *transitive verb* (1820)
1 : to make melodramatic ⟨*melodramatize* a situation⟩
2 : to make a melodrama of (as a novel)
— **melo·dra·ma·ti·za·tion** \-ˌdra-mə-tə-'zā-shən, -ˌdrä-\ *noun*

mel·o·dy \'me-lə-dē\ *noun, plural* **-dies** [Middle English *melodie,* from Old French, from Late Latin *melodia,* from Greek *melōidia* chanting, music, from *melos* limb, musical phrase, song (probably akin to Breton *mell* joint) + *aeidein* to sing — more at ODE] (13th century)
1 : a sweet or agreeable succession or arrangement of sounds : TUNEFULNESS
2 : a rhythmic succession of single tones organized as an aesthetic whole

mel·on \'me-lən\ *noun, often attributive* [Middle English, from Middle French, from Late Latin *melon-, melo,* short for Latin *melopepon-, melopepo,* from Greek *mēlopepōn,* from *mēlon* apple + *pepōn,* an edible gourd — more at PUMPKIN] (14th century)
1 : any of various gourds (as a muskmelon or watermelon) usually eaten raw as fruits
2 : something rounded like a melon: as **a :** the rounded organ in the front of the head of some cetaceans **b** *plural, slang* **:** large breasts
3 a : a surplus of profits available for distribution to stockholders **b :** a financial windfall

mel·pha·lan \'mel-fə-ˌlan\ *noun* [probably from *me*thanol + *ph*enyl*alan*ine] (circa 1964)
: an antineoplastic drug $C_{13}H_{18}Cl_2N_2O_2$

Mel·pom·e·ne \mel-'pä-mə-(ˌ)nē\ *noun* [Latin, from Greek *Melpomenē*]
: the Greek Muse of tragedy

¹**melt** \'melt\ *verb* [Middle English, from Old English *meltan;* akin to Old Norse *melta* to digest, Greek *meldein* to melt] (before 12th century)
intransitive verb
1 : to become altered from a solid to a liquid state usually by heat
2 a : DISSOLVE, DISINTEGRATE ⟨the sugar *melted* in the coffee⟩ **b :** to disappear as if by dissolving ⟨her anger *melted* at his kind words⟩
3 *obsolete* **:** to become subdued or crushed (as by sorrow)
4 : to become mild, tender, or gentle
5 : to lose outline or distinctness : BLEND
transitive verb
1 : to reduce from a solid to a liquid state usually by heat
2 : to cause to disappear or disperse
3 : to make tender or gentle : SOFTEN
— **melt·abil·i·ty** \ˌmel-tə-'bi-lə-tē\ *noun*
— **melt·able** \'mel-tə-bəl\ *adjective*
— **melt·er** *noun*

²**melt** *noun* (1854)
1 a : material in the molten state **b :** the mass melted at a single operation or the quantity melted during a specified period
2 a : the action or process of melting or the period during which it occurs ⟨the spring *melt*⟩ **b :** the condition of being melted
3 : a sandwich with melted cheese ⟨a tuna *melt*⟩

³**melt** *noun* [Middle English *milte,* from Old English; akin to Old High German *miltzi* spleen] (before 12th century)
: SPLEEN; *especially* : spleen of slaughtered animals for use as feed or food

melt·down \'melt-ˌdau̇n\ *noun* (1963)
1 : the accidental melting of the core of a nuclear reactor
2 : a rapid or disastrous decline or collapse

melt·ing \'mel-tiŋ\ *adjective* (1565)
: TENDER, DELICATE ⟨a love song's *melting* lyric⟩
— **melt·ing·ly** *adverb*

melting point *noun* (1823)
: the temperature at which a solid melts

melting pot *noun* (1912)
1 a : a place where racial amalgamation and social and cultural assimilation are going on ⟨long cherished the myth of the public school as the *melting pot* —M. R. Berube⟩ **b :** the population of such a place
2 : a process of blending that often results in invigoration or novelty

mel·ton \'mel-tᵊn\ *noun* [*Melton* Mowbray, town in England] (1823)
: a heavy smooth woolen fabric with short nap

melt·wa·ter \'melt-ˌwȯ-tər, -ˌwä-\ *noun* (1923)
: water derived from the melting of ice and snow

mem \'mem\ *noun* [Hebrew *mēm,* literally, water] (circa 1823)
: the 13th letter of the Hebrew alphabet — see ALPHABET table

mem·ber \'mem-bər\ *noun, often attributive* [Middle English *membre,* from Old French, from Latin *membrum;* akin to Gothic *mimz* flesh, Greek *mēros* thigh] (14th century)
1 : a body part or organ: as **a :** LIMB **b :** PENIS **c :** a unit of structure in a plant body
2 : one of the individuals composing a group
3 : a person baptized or enrolled in a church
4 : a constituent part of a whole: as **a :** a syntactic or rhythmic unit of a sentence : CLAUSE

◇ WORD HISTORY
mellifluous Having a much longer history as a sweetener than cane sugar, honey has since ancient times been a symbol of sweetness. In English we continue to speak of those halcyon days following a wedding as a *honeymoon* and to refer to an eloquent orator as *honey-tongued.* Less obviously, we are using the same metaphor when we speak of a voice as being *mellifluous.* In Latin *mel* meant "honey" and *fluere* meant to "to flow," hence the Late Latin word *mellifluus* literally meant "flowing like honey."

b : one of the propositions of a syllogism **c :** one of the elements of a set or class **d :** either of the equated elements in a mathematical equation
synonym see PART

mem·bered \'mem-bərd\ *adjective* (14th century)
: made up of or divided into members

mem·ber·ship \'mem-bər-ˌship\ *noun* (1647)
1 : the state or status of being a member
2 : the body of members ⟨an organization with a large *membership*⟩
3 : the relation between an element of a set or class and the set or class — compare INCLUSION 3

mem·brane \'mem-ˌbrān\ *noun* [Middle English, from Latin *membrana* skin, parchment, from *membrum*] (15th century)
1 : a thin soft pliable sheet or layer especially of animal or plant origin
2 : a piece of parchment forming part of a roll
— mem·braned \'mem-ˌbrānd\ *adjective*

mem·bra·nous \'mem-brə-nəs\ *adjective* (1597)
1 : of, relating to, or resembling membrane
2 : thin, pliable, and often somewhat transparent ⟨*membranous* leaves⟩
3 : characterized or accompanied by the formation of a usually abnormal membrane or membranous layer ⟨*membranous* croup⟩
— mem·bra·nous·ly *adverb*

membranous labyrinth *noun* (1840)
: the sensory structures of the inner ear

me·men·to \mə-'men-(ˌ)tō, ÷mō-\ *noun, plural* **-tos** *or* **-toes** [Middle English, from Latin, remember, imperative of *meminisse* to remember; akin to Latin *ment-, mens* mind — more at MIND] (1580)
: something that serves to warn or remind; *also* **:** SOUVENIR ■

me·men·to mo·ri \mə 'men-tō-'mōr-ē, -'mȯr-ē\ *noun, plural* **memento mori** [Latin, remember that you must die] (1596)
: a reminder of mortality; *especially* **:** DEATH'S-HEAD

Mem·non \'mem-ˌnän\ *noun* [Greek *Memnōn*]
: an Ethiopian king slain by Achilles at a late stage of the Trojan War

memo \'me-(ˌ)mō\ *noun, plural* **mem·os** (1889)
: MEMORANDUM

mem·oir \'mem-ˌwär, -ˌwȯr\ *noun* [Middle French *memoire,* from *memoire* memory, from Latin *memoria*] (1567)
1 : an official note or report **:** MEMORANDUM
2 a : a narrative composed from personal experience **b :** AUTOBIOGRAPHY — usually used in plural **c :** BIOGRAPHY
3 a : an account of something noteworthy **:** REPORT **b** *plural* **:** the record of the proceedings of a learned society
— mem·oir·ist \-ist\ *noun*

mem·o·ra·bil·ia \ˌme-mə-rə-'bi-lē-ə, -'bē-lē-ə, -'bil-yə\ *noun plural* [Latin, from neuter plural of *memorabilis*] (circa 1807)
: things that are remarkable and worthy of remembrance; *also* **:** things that stir recollection **:** MEMENTOS

mem·o·ra·bil·i·ty \-'bi-lə-tē\ *noun* (circa 1661)
: the quality or state of being easy to remember or worth remembering

mem·o·ra·ble \'mem-rə-bəl, 'mə-mə-rə-, 'me-mər-\ *adjective* [Middle English, from Latin *memorabilis,* from *memorare* to remind, mention, from *memor* mindful] (15th century)
: worth remembering **:** NOTABLE
— mem·o·ra·ble·ness *noun*
— mem·o·ra·bly \-blē\ *adverb*

mem·o·ran·dum \ˌme-mə-'ran-dəm\ *noun, plural* **-dums** *or* **-da** \-də\ [Middle English, to be remembered, from Latin, neuter of *memorandus,* gerundive of *memorare*] (circa 1543)
1 : an informal record; *also* **:** a written reminder
2 : an informal written note of a transaction or proposed instrument
3 a : an informal diplomatic communication **b :** a usually brief communication written for interoffice circulation **c :** a communication that contains directive, advisory, or informative matter ■

[1]me·mo·ri·al \mə-'mōr-ē-əl, -'mȯr-\ *adjective* [Middle English, from Latin *memorialis,* from *memoria* memory] (14th century)
1 : serving to preserve remembrance **:** COMMEMORATIVE
2 : of or relating to memory
— me·mo·ri·al·ly \-ə-lē\ *adverb*

[2]memorial *noun* (14th century)
1 : something that keeps remembrance alive: as **a :** MONUMENT **b :** something (as a speech or ceremony) that commemorates **c :** KEEPSAKE, MEMENTO
2 a : RECORD, MEMOIR ⟨language and literature . . . the *memorials* of another age —J. H. Fisher⟩ **b :** MEMORANDUM, NOTE; *specifically* **:** a legal abstract **c :** a statement of facts addressed to a government and often accompanied by a petition or remonstrance

Memorial Day *noun* (1869)
1 : May 30 formerly observed as a legal holiday in most states of the U.S. in remembrance of war dead
2 : the last Monday in May observed as a legal holiday in most states of the U.S.
3 : CONFEDERATE MEMORIAL DAY

me·mo·ri·al·ise *British variant of* MEMORIALIZE

me·mo·ri·al·ist \mə-'mōr-ē-ə-list, -'mȯr-\ *noun* (1706)
1 : a person who writes or signs a memorial
2 : a person who writes a memoir

me·mo·ri·al·ize \-ˌlīz\ *transitive verb* **-ized; -iz·ing** (1798)
1 : to address or petition by a memorial
2 : COMMEMORATE

memorial park *noun* (circa 1928)
: CEMETERY

me·mo·ri·ter \mə-'mȯr-ə-ˌter, -'mär-\ *adjective* [Latin, adverb, by memory, from *memor*] (circa 1812)
: marked by emphasis on memorization

mem·o·rise *British variant of* MEMORIZE

mem·o·rize \'me-mə-ˌrīz\ *transitive verb* **-rized; -riz·ing** (1838)
: to commit to memory **:** learn by heart
— mem·o·riz·able \ˌme-mə-'rī-zə-bəl\ *adjective*
— mem·o·ri·za·tion \ˌme-mə-rə-'zā-shən, ˌme-mə-(ˌ)rī-, ˌmem-rə-\ *noun*
— mem·o·riz·er *noun*

mem·o·ry \'mem-rē, 'me-mə-\ *noun, plural* **-ries** [Middle English *memorie,* from Middle French *memoire,* from Latin *memoria,* from *memor* mindful; akin to Old English *gemimor* well-known, Greek *mermēra* care, Sanskrit *smarati* he remembers] (14th century)
1 a : the power or process of reproducing or recalling what has been learned and retained especially through associative mechanisms **b :** the store of things learned and retained from an organism's activity or experience as evidenced by modification of structure or behavior or by recall and recognition
2 a : commemorative remembrance ⟨erected a statue in *memory* of the hero⟩ **b :** the fact or condition of being remembered ⟨days of recent *memory*⟩
3 a : a particular act of recall or recollection **b :** an image or impression of one that is remembered ⟨fond *memories* of her youth⟩ **c :** the time within which past events can be or are remembered ⟨within the *memory* of living men⟩
4 a : a device or a component of a device in which information especially for a computer can be inserted and stored and from which it may be extracted when wanted **b :** capacity for storing information ⟨four megabytes of *memory*⟩
5 : a capacity for showing effects as the result of past treatment or for returning to a former condition — used especially of a material (as metal or plastic) ☆

memory lane *noun* (1954)
: an imaginary path through the nostalgically remembered past — usually used in such phrases as *a walk down memory lane*

memory trace *noun* (1923)
: ENGRAM

mem·sa·hib \'mem-ˌsä-ˌ(h)ib, -ˌsäb\ *noun* [Hindi, from English *ma'am* + Hindi *sahib* sahib] (1857)
: a white foreign woman of high social status living in India; *especially* **:** the wife of a British official

men *plural of* MAN

men- *or* **meno-** *combining form* [New Latin, from Greek *mēn* month — more at MOON]
: menstruation ⟨*meno*rrhagia⟩

[1]men·ace \'me-nəs\ *noun* [Middle English, from Middle French, from Latin *minacia,* from *minac-, minax* threatening, from *minari* to threaten — more at MOUNT] (14th century)
1 : a show of intention to inflict harm **:** THREAT

☆ SYNONYMS
Memory, remembrance, recollection, reminiscence mean the capacity for or the act of remembering, or the thing remembered. MEMORY applies both to the power of remembering and to what is remembered ⟨gifted with a remarkable *memory*⟩ ⟨that incident was now just a distant *memory*⟩. REMEMBRANCE applies to the act of remembering or the fact of being remembered ⟨any *remembrance* of his deceased wife was painful⟩. RECOLLECTION adds an implication of consciously bringing back to mind often with some effort ⟨after a moment's *recollection* he produced the name⟩. REMINISCENCE suggests the recalling of usually pleasant incidents, experiences, or feelings from a remote past ⟨recorded my grandmother's *reminiscences* of her Iowa girlhood⟩.

□ USAGE
memento A fairly rare variant of *memento,* spelled with an *o* as its first vowel, is found just often enough in edited prose ⟨a nostalgic *momento* of an earlier century —Joseph Wechsberg (*New Yorker*)⟩ to have been entered in a few dictionaries. Our first evidence for it is in an 1853 letter written by Chauncey A. Goodrich, professor at Yale, Noah Webster's son-in-law, and first editor in chief of Merriam-Webster dictionaries. It has also been found in the writings of George Eliot and Dylan Thomas. Although our file of evidence keeps growing, *momento* is still infrequently used, and many people think it an error. It is best to use *memento.*

memorandum Although some commentators warn against the use of *memoranda* as a singular and condemn the plural *memorandas,* our evidence indicates that these forms are rarely encountered in print. We have a little evidence of the confusion of forms, including use of *memorandum* as a plural, in speech (as at congressional hearings). As plurals *memoranda* and *memorandums* are about equally frequent.

2 a : one that represents a threat **:** DANGER **b :** a person who causes annoyance

²menace *verb* **men·aced; men·ac·ing** (14th century)
transitive verb
1 : to make a show of intention to harm
2 : to represent or pose a threat to **:** ENDANGER
intransitive verb
: to act in a threatening manner
— men·ac·ing·ly \-nə-siŋ-lē\ *adverb*

me·nad *variant of* MAENAD

men·a·di·one \ˌme-nə-ˈdī-ˌōn, -dī-ˈ\ *noun* [*me*thyl + *na*pthalene + *di-* + ket*one*] (1941)
: a yellow crystalline compound $C_{11}H_8O_2$ with the biological activity of natural vitamin K

mé·nage \mā-ˈnäzh\ *noun* [French, from Old French *mesnage* dwelling, from (assumed) Vulgar Latin *mansionaticum,* from Latin *mansion-, mansio* mansion] (1698)
: a domestic establishment **:** HOUSEHOLD; *also* **:** HOUSEKEEPING

mé·nage à trois \-ä-ˈtrwä\ *noun* [French, literally, household for three] (1891)
: an arrangement in which three persons (as a married pair and the lover of one of the pair) share sexual relations especially while living together

me·nag·er·ie \mə-ˈnaj-rē, -ˈna-jə- *also* -ˈnazh-rē, -ˈna-zhə-\ *noun* [French *ménagerie,* from Middle French, management of a household or farm, from *menage*] (1676)
1 a : a place where animals are kept and trained especially for exhibition **b :** a collection of wild or foreign animals kept especially for exhibition
2 : a varied mixture ⟨a wonderful *menagerie* of royal hangers-on —V. S. Pritchett⟩

men·ar·che \ˈme-ˌnär-kē\ *noun* [New Latin, from *men-* + Greek *archē* beginning] (1900)
: the beginning of the menstrual function; *especially* **:** the first menstrual period of an individual
— men·ar·che·al \ˌme-när-ˈkē-əl\ *adjective*

¹mend \ˈmend\ *verb* [Middle English, short for *amenden* — more at AMEND] (13th century)
transitive verb
1 : to free from faults or defects: as **a :** to improve in manners or morals **:** REFORM **b :** to set right **:** CORRECT **c :** to put into good shape or working order again **:** patch up **:** REPAIR **d :** to improve or strengthen (as a relationship) by negotiation or conciliation — used chiefly in the phrase *mend fences* ⟨spends the weekend *mending* political fences —E. O. Hauser⟩ **e :** to restore to health **:** CURE
2 : to make amends or atonement for ⟨least said, soonest *mended*⟩
intransitive verb
1 : to improve morally **:** REFORM
2 : to become corrected or improved
3 : to improve in health; *also* **:** HEAL ☆
— mend·able \ˈmen-də-bəl\ *adjective*
— mend·er *noun*

²mend *noun* (14th century)
1 : an act of mending **:** REPAIR
2 : a mended place
— on the mend : getting better **:** IMPROVING

men·da·cious \men-ˈdā-shəs\ *adjective* [Latin *mendac-, mendax* — more at AMEND] (1616)
: given to or characterized by deception or falsehood or divergence from absolute truth ⟨*mendacious* tales of his adventures⟩
synonym see DISHONEST
— men·da·cious·ly *adverb*
— men·da·cious·ness *noun*

men·dac·i·ty \men-ˈda-sə-tē\ *noun, plural* **-ties** (1646)
1 : the quality or state of being mendacious
2 : LIE

Men·de \ˈmen-dē, -dā\ *noun, plural* **Mende** *or* **Mendes** (1732)
1 : a Mande language of southern Sierra Leone and eastern Liberia
2 : a member of a people speaking Mende

men·de·le·vi·um \ˌmen-də-ˈlē-vē-əm, -ˈlā-\ *noun* [New Latin, from Dmitry *Mendeleyev*] (1955)
: a radioactive element that is artificially produced — see ELEMENT table

Men·de·lian \men-ˈdē-lē-ən, -ˈdēl-yən\ *adjective* (1901)
: of, relating to, or according with Mendel's laws or Mendelism
— Mendelian *noun*

Mendelian factor *noun* (1927)
: GENE

Mendelian inheritance *noun* (circa 1923)
: PARTICULATE INHERITANCE

Men·del·ism \ˈmen-dᵊl-ˌi-zəm\ *noun* (1903)
: the principles or the operations of Mendel's laws; *also* **:** PARTICULATE INHERITANCE
— Men·del·ist \-dᵊl-ist\ *adjective or noun*

Men·del's law \ˈmen-dᵊlz-\ *noun* [Gregor *Mendel*] (1903)
1 : a principle in genetics: hereditary units occur in pairs that separate during gamete formation so that every gamete receives but one member of a pair — called also *law of segregation*
2 : a principle in genetics limited and modified by the subsequent discovery of the phenomenon of linkage: the different pairs of hereditary units are distributed to the gametes independently of each other, the gametes combine at random, and the various combinations of hereditary pairs occur in the zygotes according to the laws of chance — called also *law of independent assortment*
3 : a principle in genetics proved subsequently to be subject to many limitations: because one of each pair of hereditary units dominates the other in expression, characters are inherited alternatively on an all-or-nothing basis — called also *law of dominance*

men·di·can·cy \ˈmen-di-kən(t)-sē\ *noun* (1790)
1 : the condition of being a beggar
2 : the practice of begging

men·di·cant \ˈmen-di-kənt\ *noun* [Middle English, from Latin *mendicant-, mendicans,* present participle of *mendicare* to beg, from *mendicus* beggar — more at AMEND] (14th century)
1 : BEGGAR 1
2 *often capitalized* **:** a member of a religious order (as the Franciscans) combining monastic life and outside religious activity and originally owning neither personal nor community property **:** FRIAR
— mendicant *adjective*

men·dic·i·ty \men-ˈdi-sə-tē\ *noun* [Middle English *mendicite,* from Middle French *mendicité,* from Latin *mendicitat-, mendicitas,* from *mendicus*] (15th century)
: MENDICANCY

Men·e·la·us \ˌme-nᵊl-ˈā-əs\ *noun* [Latin, from Greek *Menelaos*]
: a king of Sparta, brother of Agamemnon, and husband of the abducted Helen of Troy

men·folk \ˈmen-ˌfōk\ *or* **men·folks** \-ˌfōks\ *noun plural* (1802)
1 : men in general
2 : the men of a family or community

men·ha·den \men-ˈhā-dᵊn, mən-\ *noun, plural* **-den** *also* **-dens** [of Algonquian origin; akin to Narraganset *munnawhatteaûg* menhaden] (1792)
: a marine fish (*Brevoortia tyrannus*) of the herring family abundant along the Atlantic coast of the U.S. where it is used for bait or converted into oil and fertilizer; *also* **:** any of several congeneric fishes

men·hir \ˈmen-ˌhir\ *noun* [French, from Breton, from *men* stone + *hir* long] (1840)
: a single upright rude monolith usually of prehistoric origin

¹me·nial \ˈmē-nē-əl, -nyəl\ *noun* (14th century)
: a person doing menial work; *specifically* **:** a domestic servant or retainer

²menial *adjective* [Middle English *meynial,* from *meynie* household, retinue, from Old French *mesnie,* from (assumed) Vulgar Latin *mansionata,* from Latin *mansion-, mansio* dwelling — more at MANSION] (15th century)
1 : of or relating to servants **:** LOWLY
2 a : appropriate to a servant **:** HUMBLE, SERVILE ⟨answered in *menial* tones⟩ **b :** lacking interest or dignity ⟨a *menial* task⟩
— me·nial·ly *adverb*

Mé·nière's disease \mən-ˈyerz-, ˈmen-yərz-\ *noun* [Prosper *Ménière* (died 1862) French physician] (1876)
: a disorder of the membranous labyrinth of the inner ear that is marked by recurrent attacks of dizziness, tinnitus, and deafness — called also *Ménière's syndrome*

mening- *or* **meningo-** *also* **meningi-** *combining form* [New Latin, from *mening-, meninx*]
1 : meninges ⟨*meningo*coccus⟩ ⟨*mening*itis⟩
2 : meninges and ⟨*meningo*encephalitis⟩

men·in·ge·al \ˌme-nən-ˈjē-əl\ *adjective* (1829)
: of, relating to, or affecting the meninges

meninges *plural of* MENINX

me·nin·gi·o·ma \mə-ˌnin-jē-ˈō-mə\ *noun, plural* **-mas** *or* **-ma·ta** \-ˈō-mə-tə\ [New Latin] (1922)
: a slow-growing encapsulated tumor arising from the meninges and often causing damage by pressing upon the brain and adjacent parts

men·in·gi·tis \ˌme-nən-ˈjī-təs\ *noun, plural* **-git·i·des** \-ˈji-tə-ˌdēz\ [New Latin] (1828)
1 : inflammation of the meninges and especially of the pia mater and the arachnoid
2 : a usually bacterial disease in which inflammation of the meninges occurs
— men·in·git·ic \-ˈji-tik\ *adjective*

me·nin·go·coc·cus \mə-ˌniŋ-gə-ˈkä-kəs, -ˌnin-jə-\ *noun, plural* **-coc·ci** \-ˈkäk-ˌsī, -ˌsē; -ˈkä-ˌkī, -ˌkē\ [New Latin] (circa 1893)
: the bacterium (*Neisseria meningitidis*) that causes cerebrospinal meningitis
— me·nin·go·coc·cal \-ˈkä-kəl\ *also* **me·nin·go·coc·cic** \-ˈkäk-sik, -ˈkä-kik\ *adjective*

me·nin·go·en·ceph·a·li·tis \-gō-ən-ˌse-fə-ˈlī-təs, -jō-\ *noun, plural* **-lit·i·des** \-ˈli-tə-ˌdēz\ [New Latin] (circa 1860)
: inflammation of the brain and meninges
— me·nin·go·en·ceph·a·lit·ic \-ˈli-tik\ *adjective*

me·ninx \ˈmē-niŋ(k)s, ˈme-\ *noun, plural* **me·nin·ges** \mə-ˈnin-(ˌ)jēz\ [New Latin, from Greek *mēning-, mēninx* membrane] (1543)
: any of the three membranes that envelop the brain and spinal cord

me·nis·cus \mə-ˈnis-kəs\ *noun, plural* **me·nis·ci** \-ˈni-ˌskī, -ˌskē, -ˌsī\ *also* **me·nis·cus·es** [New Latin, from Greek *mēniskos,* from diminutive of *mēnē* moon, crescent — more at MOON] (1693)
1 : a concavo-convex lens
2 : a crescent or crescent-shaped body
3 : the curved upper surface of a column of liquid
4 : a fibrous cartilage within a joint especially of the knee

☆ SYNONYMS
Mend, repair, patch, rebuild mean to put into good order something that has been injured, damaged, or defective. MEND implies making whole or sound something broken, torn, or injured ⟨*mended* the torn dress⟩. REPAIR applies to the fixing of more extensive damage or dilapidation ⟨*repaired* the back steps⟩. PATCH implies an often temporary fixing of a hole or break with new material ⟨*patch* worn jeans⟩. REBUILD suggests making like new without completely replacing ⟨a *rebuilt* automobile engine⟩.

Men·no·nite \'me-nə-,nīt\ *noun* [German *Mennonit,* from *Menno* Simons] (1565)
: a member of any of various Protestant groups derived from the Anabaptist movement in Holland and characterized by congregational autonomy and rejection of military service

meno- — see MEN-

me·no mos·so \,mā-nō-'mȯ(s)-(,)sō\ *adverb* [Italian] (circa 1854)
: less rapid — used as a direction in music

men·o·pause \'me-nə-,pȯz, 'mē-\ *noun* [French *ménopause,* from *méno-* men- + *pause* stop, pause] (1872)
: the period of natural cessation of menstruation occurring usually between the ages of 45 and 50
— **men·o·paus·al** \,me-nə-'pȯ-zəl, ,mē-\ *adjective*

me·no·rah \mə-'nōr-ə, -'nȯr-\ *noun* [Hebrew *mĕnōrāh* candlestick] (1888)
: a candelabrum with seven or nine candles that is used in Jewish worship

menorah

men·or·rha·gia \,me-nə-'rā-j(ē-)ə, -'rā-zhə; -'rä-jə, -zhə\ *noun* [New Latin] (circa 1784)
: abnormally profuse menstrual flow

men·sal \'men(t)-səl\ *adjective* [Middle English, from Late Latin *mensalis,* from Latin *mensa* table] (15th century)
: of, relating to, or done at the table

mensch \'men(t)sh\ *noun* [Yiddish *mentsh* human being, from Middle High German *mensch,* from Old High German *mennisco;* akin to Old English *man* human being, man] (1953)
: a person of integrity and honor

[1]mense \'men(t)s\ *noun* [Middle English *menske* honor, from Old Norse *mennska* humanity; akin to Old English *man*] (circa 1500)
chiefly Scottish **:** PROPRIETY
— **mense·ful** \-fəl\ *adjective*
— **mense·less** \-ləs\ *adjective*

[2]mense *transitive verb* **mensed; mens·ing** (1540)
chiefly Scottish **:** to do honor to **:** GRACE

men·ses \'men-,sēz\ *noun plural but singular or plural in construction* [Latin, literally, months, plural of *mensis* month — more at MOON] (1597)
: the menstrual flow

Men·she·vik \'men(t)-shə-,vik, -,vēk\ *noun, plural* **Mensheviks** *or* **Men·she·vi·ki** \,men(t)-shə-'vi-kē, -'vē-kē\ [Russian *men'shevik,* from *men'she* less; from their forming the minority group of the party] (1907)
: a member of a wing of the Russian Social Democratic party before and during the Russian Revolution believing in the gradual achievement of socialism by parliamentary methods in opposition to the Bolsheviks
— **Men·she·vism** \'men(t)-shə-,vi-zəm\ *noun*
— **Men·she·vist** \-vist\ *noun or adjective*

mens rea \'menz-'rē-ə\ *noun* [New Latin, literally, guilty mind] (1861)
: criminal intent

men's room *noun* (1929)
: a room equipped with one or more sinks, toilets, and usually urinals for the use of men and boys

men·stru·al \'men(t)-strü-əl, -strəl\ *adjective* (14th century)
: of or relating to menstruation

men·stru·ate \'men(t)-strü-,wāt, 'men-,strāt\ *intransitive verb* **-at·ed; -at·ing** [Late Latin *menstruatus,* past participle of *menstruari,* from Latin *menstrua* menses, from neuter plural of *menstruus* monthly, from *mensis*] (1800)
: to undergo menstruation

men·stru·a·tion \,men(t)-strü-'wā-shən, men-'strā-\ *noun* (circa 1784)
: a discharging of blood, secretions, and tissue debris from the uterus that recurs in nonpregnant breeding-age primate females at approximately monthly intervals and that is considered to represent a readjustment of the uterus to the nonpregnant state following proliferative changes accompanying the preceding ovulation; *also* **:** PERIOD 6c

men·stru·um \'men(t)-strü-əm, -strəm\ *noun, plural* **-stru·ums** *or* **-strua** \-strü-ə, -strə\ [Medieval Latin, literally, menses, alteration of Latin *menstrua*] (1612)
: a substance that dissolves a solid or holds it in suspension **:** SOLVENT

men·su·ra·ble \'men(t)s-rə-bəl, 'men(t)sh-; 'men(t)-sə-rə-, -shə-\ *adjective* [Late Latin *mensurabilis,* from *mensurare* to measure, from Latin *mensura* measure — more at MEASURE] (1604)
1 : capable of being measured **:** MEASURABLE
2 : MENSURAL 1
— **men·sur·abil·i·ty** \,men(t)s-rə-'bi-lə-tē, ,men(t)sh-; ,men(t)-sə-rə-, -shə-\ *noun*

men·su·ral \'men(t)s-rəl, 'men(t)sh-; 'men(t)-sə-rəl, -shə-\ *adjective* [Late Latin *mensuralis* measurable, from Latin *mensura*] (1609)
1 : of, relating to, or being polyphonic music originating in the 13th century with each note having a definite and exact time value
2 : of or relating to measure

men·su·ra·tion \,men(t)-sə-'rā-shən, -shə-\ *noun* (1571)
1 : the act of measuring **:** MEASUREMENT
2 : geometry applied to the computation of lengths, areas, or volumes from given dimensions or angles

mens·wear \'menz-,war, -,wer\ *noun* (1908)
: clothing for men

-ment \mənt; *in verbs derived by functional shift (as* [2]ORNAMENT) ,ment *also* mənt\ *noun suffix* [Middle English, from Old French, from Latin *-mentum;* akin to Latin *-men,* suffix denoting concrete result, Greek *-mat-, -ma*]
1 a : concrete result, object, or agent of a (specified) action ⟨embank*ment*⟩ ⟨entangle*ment*⟩ **b :** concrete means or instrument of a (specified) action ⟨entertain*ment*⟩
2 a : action **:** process ⟨encircle*ment*⟩ ⟨develop*ment*⟩ **b :** place of a (specified) action ⟨encamp*ment*⟩
3 : state or condition resulting from (a specified action) ⟨involve*ment*⟩

[1]men·tal \'men-tᵊl\ *adjective* [Middle English, from Middle French, from Late Latin *mentalis,* from Latin *ment-, mens* mind — more at MIND] (15th century)
1 a : of or relating to the mind; *specifically* **:** of or relating to the total emotional and intellectual response of an individual to external reality ⟨*mental* health⟩ **b :** of or relating to intellectual as contrasted with emotional activity **c :** of, relating to, or being intellectual as contrasted with overt physical activity **d :** occurring or experienced in the mind **:** INNER ⟨*mental* anguish⟩ **e :** relating to the mind, its activity, or its products as an object of study **:** IDEOLOGICAL **f :** relating to spirit or idea as opposed to matter
2 a (1) **:** of, relating to, or affected by a psychiatric disorder ⟨a *mental* patient⟩ ⟨*mental* illness⟩ (2) **:** mentally disordered **:** MAD, CRAZY **b :** intended for the care or treatment of persons affected by psychiatric disorders ⟨*mental* hospitals⟩
3 : of or relating to telepathic or mind-reading powers
— **men·tal·ly** \-tᵊl-ē\ *adverb*

[2]mental *adjective* [Latin *mentum* chin — more at MOUTH] (circa 1727)
: of or relating to the chin **:** GENIAL

mental age *noun* (1912)
: a measure used in psychological testing that expresses an individual's mental attainment in terms of the number of years it takes an average child to reach the same level

mental deficiency *noun* (1856)
: MENTAL RETARDATION

men·tal·ist \'men-tᵊl-ist\ *noun* (1930)
: MIND READER

men·tal·is·tic \,men-tᵊl-'is-tik\ *adjective* (1882)
1 : of or relating to any school of psychology or psychiatry that in contrast to behaviorism values subjective data (as those gained by introspection) in the study and explanation of behavior
2 : of or relating to mental phenomena
— **men·tal·ism** \'men-tᵊl-,i-zəm\ *noun*

men·tal·i·ty \men-'ta-lə-tē\ *noun, plural* **-ties** (1691)
1 : mental power or capacity **:** INTELLIGENCE
2 : mode or way of thought **:** OUTLOOK

mental retardation *noun* (1914)
: subaverage intellectual ability that is equivalent to or less than an IQ of 70, is present from birth or infancy, and is manifested especially by abnormal development, by learning difficulties, and by problems in social adjustment

men·ta·tion \men-'tā-shən\ *noun* [Latin *ment-, mens* + English *-ation*] (1850)
: mental activity

men·thol \'men-,thȯl, -,thōl\ *noun* [German, ultimately from Latin *mentha* mint] (1876)
: a crystalline alcohol $C_{10}H_{20}O$ that occurs especially in mint oils and has the odor and cooling properties of peppermint

men·tho·lat·ed \'men(t)-thə-,lā-təd\ *adjective* (1922)
: containing or impregnated with menthol ⟨a *mentholated* salve⟩

[1]men·tion \'men(t)-shən\ *noun* [Middle English *mencioun,* from Middle French *mention,* from Latin *mention-, mentio;* akin to Latin *meminisse* to remember, *ment-, mens* mind] (14th century)
1 : the act or an instance of citing or calling attention to someone or something especially in a casual or incidental manner
2 : formal citation for outstanding achievement

[2]mention *transitive verb* **men·tioned; men·tion·ing** \'men(t)-sh(ə-)niŋ\ (1530)
: to make mention of **:** refer to; *also* **:** to cite for outstanding achievement
— **men·tion·able** \'men(t)-sh(ə-)nə-bəl\ *adjective*
— **men·tion·er** \-sh(ə-)nər\ *noun*
— **not to mention :** to say nothing of

[1]men·tor \'men-,tȯr, -tər\ *noun* [Latin, from Greek *Mentōr*]
1 *capitalized* **:** a friend of Odysseus entrusted with the education of Odysseus' son Telemachus
2 a : a trusted counselor or guide **b :** TUTOR, COACH
— **men·tor·ship** \-,ship\ *noun*

[2]mentor *transitive verb* (1983)
: to serve as a mentor for

men·tum \'men-təm\ *noun, plural* **men·ta** \-tə\ [Latin, chin — more at MOUTH] (1826)
: a median plate of the labium of an insect

menu \'men-(,)yü, 'mān-\ *noun, plural* **menus** [French, from *menu* small, detailed, from Old French — more at MINUET] (1837)
1 a : a list of the dishes that may be ordered (as in a restaurant) or that are to be served (as at a banquet) **b** (1) **:** a comparable list or assortment of offerings ⟨a *menu* of television programs⟩ (2) **:** a list shown on the display of a computer from which a user can select the operation the computer is to perform

\ə\ **abut** \ᵊ\ **kitten** \ər\ **further** \a\ **ash** \ā\ **ace** \ä\ **mop, mar** \au̇\ **out** \ch\ **chin** \e\ **bet** \ē\ **easy** \g\ **go** \i\ **hit** \ī\ **ice** \j\ **job** \ŋ\ **sing** \ō\ **go** \ȯ\ **law** \ȯi\ **boy** \th\ **thin** \t͟h\ **the** \ü\ **loot** \u̇\ **foot** \y\ **yet** \zh\ **vision** *see also* Guide to Pronunciation

2 : the dishes available for or served at a meal; *also* **:** the meal itself ◆

menu–driv·en \-ˌdri-vən\ *adjective* (1977)
: relating to or being a computer program in which options are offered to the user via menus

me·ow \mē-ˈau̇\ *noun* [imitative] (1600)
1 : the cry of a cat
2 : a spiteful or malicious remark
— meow *intransitive verb*

me·per·i·dine \mə-ˈper-ə-ˌdēn\ *noun* [*me*thyl + *pi*peridine] (1947)
: a synthetic narcotic drug $C_{15}H_{21}NO_2$ used in the form of its hydrochloride as an analgesic, sedative, and antispasmodic

Meph·is·toph·e·les \ˌme-fə-ˈstä-fə-ˌlēz\ *noun* [German]
: a chief devil in the Faust legend
— Me·phis·to·phe·lian \ˌmef-ə-stə-ˈfēl-yən, mə-ˌfis-tə-\ *or* **Me·phis·to·phe·lean** *same, or* ˌme-fə-ˌstä-fə-ˈlē-ən\ *adjective*

me·phit·ic \mə-ˈfi-tik\ *adjective* (circa 1623)
: of, relating to, or resembling mephitis **:** foul-smelling

me·phi·tis \mə-ˈfī-təs\ *noun* [Latin] (circa 1706)
: a noxious, pestilential, or foul exhalation from the earth; *also* **:** STENCH

mep·ro·bam·ate \ˌme-prō-ˈba-ˌmāt\ *noun* [*me*thyl + *pro*pyl + dicar*bamate*] (1955)
: a bitter carbamate $C_9H_{18}N_2O_4$ used as a tranquilizer

-mer *noun combining form* [International Scientific Vocabulary, from Greek *meros* part — more at MERIT]
: member of a (specified) class ⟨mono*mer*⟩

mer·bro·min \ˌmər-ˈbrō-mən\ *noun* [*mer*curic + *brom-* + fluoresce*in*] (1945)
: a green crystalline mercurial compound $C_{20}H_8Br_2HgNa_2O_6$ used as a topical antiseptic and germicide in the form of its red solution

Mer·cal·li scale \mer-ˈkä-lē-, (ˌ)mər-\ *noun* [Giuseppe *Mercalli* (died 1914) Italian priest and geologist] (1921)
: a scale of earthquake intensity ranging from I for an earthquake detected only by seismographs to XII for one causing total destruction of all buildings

mer·can·tile \ˈmər-kən-ˌtēl, -ˌtīl\ *adjective* [French, from Italian, from *mercante* merchant, from Latin *mercant-, mercans,* from present participle of *mercari* to trade — more at MARKET] (1642)
1 : of or relating to merchants or trading
2 : of, relating to, or having the characteristics of mercantilism ⟨*mercantile* system⟩

mer·can·til·ism \-ˌtē-ˌli-zəm, -ˌtī-, -tə-\ *noun* (1873)
1 : the theory or practice of mercantile pursuits **:** COMMERCIALISM
2 : an economic system developing during the decay of feudalism to unify and increase the power and especially the monetary wealth of a nation by a strict governmental regulation of the entire national economy usually through policies designed to secure an accumulation of bullion, a favorable balance of trade, the development of agriculture and manufactures, and the establishment of foreign trading monopolies
— mer·can·til·ist \-list\ *noun or adjective*
— mer·can·til·is·tic \ˌmər-kən-ˌtē-ˈlis-tik, -ˌtī-, -tə-\ *adjective*

mer·cap·tan \(ˌ)mər-ˈkap-ˌtan\ *noun* [German, from Danish, from Medieval Latin *mercurium captans,* literally, seizing mercury] (1835)
: any of various compounds that contain a thiol functional group **:** THIOL 1

mer·cap·to·pu·rine \(ˌ)mər-ˌkap-tə-ˈpyu̇r-ˌēn\ *noun* [*mercapt*an + *-o-* + *purine*] (circa 1952)
: an antimetabolite $C_5H_4N_4S$ that interferes especially with the metabolism of purine bases and the biosynthesis of nucleic acids and that is sometimes useful in the treatment of acute leukemia

Mer·ca·tor \(ˌ)mər-ˈkā-tər\ *adjective* (circa 1876)
: of, relating to, or drawn on the Mercator projection

Mercator projection *noun* [Gerardus *Mercator*] (circa 1881)
: a conformal map projection in the usual case of which the meridians are drawn parallel to each other and the parallels of latitude are straight lines whose distance from each other increases with their distance from the equator

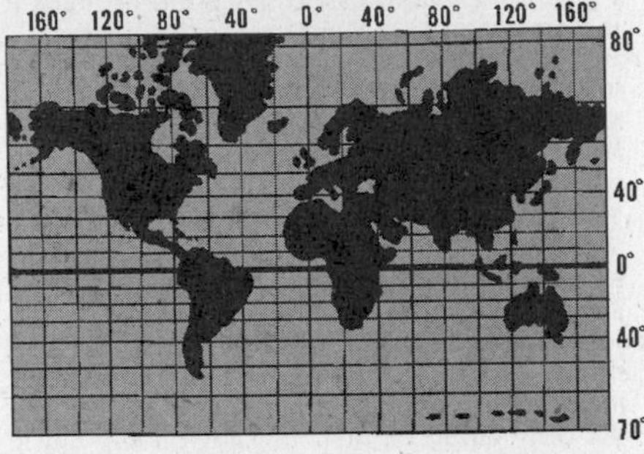

Mercator projection

[1]mer·ce·nary \ˈmər-sᵊn-ˌer-ē\ *noun, plural* **-nar·ies** [Middle English, from Latin *mercenarius,* irregular from *merced-, merces* wages — more at MERCY] (14th century)
: one that serves merely for wages; *especially* **:** a soldier hired into foreign service

[2]mercenary *adjective* (1532)
1 : serving merely for pay or sordid advantage **:** VENAL; *also* **:** GREEDY
2 : hired for service in the army of a foreign country
— mer·ce·nar·i·ly \ˌmər-sᵊn-ˈer-ə-lē\ *adverb*
— mer·ce·nar·i·ness \ˈmər-sᵊn-ˌer-ē-nəs\ *noun*

mer·cer \ˈmər-sər\ *noun* [Middle English, from Old French *mercier* merchant, from *mers* merchandise, from Latin *merc-, merx*] (13th century)
British **:** a dealer in usually expensive fabrics

mer·cer·ise *British variant of* MERCERIZE

mer·cer·ize \ˈmər-sə-ˌrīz\ *transitive verb* **-ized; -iz·ing** [John *Mercer* (died 1866) English calico printer] (1859)
: to give (as cotton yarn) luster, strength, and receptiveness to dyes by treatment under tension with caustic soda
— mer·cer·i·za·tion \ˌmərs-rə-ˈzā-shən, ˌmər-sə-\ *noun*

mer·cery \ˈmərs-rē, ˈmər-sə-\ *noun, plural* **-cer·ies** (14th century)
British **:** a mercer's wares, shop, or occupation

[1]mer·chan·dise \ˈmər-chən-ˌdīz, -ˌdīs\ *noun* [Middle English *marchaundise,* from Old French *marcheandise,* from *marcheant*] (13th century)
1 *archaic* **:** the occupation of a merchant **:** TRADE
2 : the commodities or goods that are bought and sold in business **:** WARES

[2]mer·chan·dise *also* **mer·chan·dize** \-ˌdīz\ *verb* **-dised** *also* **-dized; -dis·ing** *also* **diz·ing** (14th century)
intransitive verb
archaic **:** to carry on commerce **:** TRADE
transitive verb
1 : to buy and sell in business
2 : to promote for or as if for sale ⟨*merchandise* a movie star⟩
— mer·chan·dis·er *noun*

mer·chan·dis·ing *also* **mer·chan·diz·ing** \-ˌdī-ziŋ\ *noun* (1932)
: sales promotion as a comprehensive function including market research, development of new products, coordination of manufacture and marketing, and effective advertising and selling

[1]mer·chant \ˈmər-chənt\ *noun* [Middle English *marchant,* from Old French *marcheant,* from (assumed) Vulgar Latin *mercatant-, mercatans,* from present participle of *mercatare* to trade, frequentative of Latin *mercari* — more at MARKET] (13th century)
1 : a buyer and seller of commodities for profit **:** TRADER
2 : the operator of a retail business **:** STOREKEEPER
3 : one that is noted for a particular quality or activity **:** SPECIALIST ⟨a speed *merchant* on the base paths⟩
— merchant *adjective*

[2]merchant (14th century)
intransitive verb
archaic **:** to deal or trade as a merchant
transitive verb
: to deal or trade in

mer·chant·able \ˈmər-chən-tə-bəl\ *adjective* (15th century)
: of commercially acceptable quality **:** SALABLE
— mer·chant·abil·i·ty \ˌmər-chən-tə-ˈbi-lə-tē\ *noun*

merchant bank *noun* (1930)
chiefly British **:** a bank that specializes in bankers' acceptances and in underwriting or syndicating equity or bond issues
— merchant banker *noun*
— merchant banking *noun*

mer·chant·man \ˈmər-chənt-mən\ *noun* (15th century)
1 *archaic* **:** MERCHANT
2 : a ship used in commerce

merchant marine *noun* (1855)
1 : the privately or publicly owned commercial ships of a nation
2 : the personnel of a merchant marine

merchant ship *noun* (15th century)
: MERCHANTMAN 2

Mer·cian \ˈmər-sh(ē-)ən\ *noun* (1513)
1 : a native or inhabitant of Mercia
2 : the Old English dialect of Mercia
— Mercian *adjective*

mer·ci·ful \ˈmər-si-fəl\ *adjective* (14th century)
: full of mercy **:** COMPASSIONATE; *also* **:** providing relief
— mer·ci·ful·ness \-fəl-nəs\ *noun*

mer·ci·ful·ly \-f(ə-)lē\ *adverb* (14th century)
1 : in a merciful manner
2 : FORTUNATELY 2 ⟨*mercifully* we didn't have to attend the meeting⟩

mer·ci·less \ˈmər-si-ləs\ *adjective* (14th century)
: having or showing no mercy **:** PITILESS
— mer·ci·less·ly *adverb*
— mer·ci·less·ness *noun*

mer·cu·rate \ˈmər-kyə-ˌrāt\ *transitive verb* **-rat·ed; -rat·ing** (1923)
: to combine or treat with mercury or a mercury salt
— mer·cu·ra·tion \ˌmər-kyə-ˈrā-shən\ *noun*

[1]mer·cu·ri·al \(ˌ)mər-ˈkyu̇r-ē-əl\ *adjective* (14th century)

◇ WORD HISTORY

menu A restaurant that describes the dishes on the menu in lavish detail, right down to the source of the ingredients, is following a practice more in keeping with the origin of *menu* than one might suspect. *Menu* can be traced to the Latin objective *minutus,* meaning "small." From *minutus* came the French adjective *menu,* the various meanings of which include "small" and "detailed." The use of *menu* as a noun meaning "a list of dishes" came from the "detailed" sense of the adjective, since menus are to varying degrees detailed lists.

1 : of, relating to, or born under the planet Mercury
2 : having qualities of eloquence, ingenuity, or thievishness attributed to the god Mercury or to the influence of the planet Mercury
3 : characterized by rapid and unpredictable changeableness of mood
4 : of, relating to, containing, or caused by mercury
synonym see INCONSTANT
— mer•cu•ri•al•ly \-ē-ə-lē\ *adverb*
— mer•cu•ri•al•ness *noun*

²mercurial *noun* (1676)
: a pharmaceutical or chemical containing mercury

mer•cu•ric \(ˌ)mər-ˈkyu̇r-ik\ *adjective* (circa 1828)
: of, relating to, or containing mercury; *especially* **:** containing mercury with a valence of two

mercuric chloride *noun* (1874)
: a heavy crystalline poisonous compound $HgCl_2$ used as a disinfectant and fungicide and in photography

Mer•cu•ro•chrome \(ˌ)mər-ˈkyu̇r-ə-ˌkrōm\ *trademark*
— used for merbromin

mer•cu•rous \(ˌ)mər-ˈkyu̇r-əs, ˈmər-kyə-rəs\ *adjective* (circa 1865)
: of, relating to, or containing mercury; *especially* **:** containing mercury with a valence of one

mercurous chloride *noun* (circa 1885)
: CALOMEL

mer•cu•ry \ˈmər-kyə-rē, -k(ə-)rē\ *noun, plural* **-ries** [Latin *Mercurius,* Roman god and the planet]
1 a *capitalized* **:** a Roman god of commerce, eloquence, travel, cunning, and theft who serves as messenger to the other gods — compare HERMES **b** *often capitalized, archaic* **:** a bearer of messages or news or a conductor of travelers
2 [Middle English *mercurie,* from Medieval Latin *mercurius,* from Latin, the god] **a :** a heavy silver-white poisonous metallic element that is liquid at ordinary temperatures and is used especially in scientific instruments — called also *quicksilver;* see ELEMENT table **b :** the mercury in a thermometer or barometer
3 *capitalized* **:** the planet nearest the sun — see PLANET table

mercury chloride *noun* (circa 1885)
: a chloride of mercury: as **a :** CALOMEL **b :** MERCURIC CHLORIDE

mercury–vapor lamp *noun* (1904)
: an electric lamp in which the discharge takes place through mercury vapor — called also *mercury lamp*

mer•cy \ˈmər-sē\ *noun, plural* **mercies** [Middle English, from Old French *merci,* from Medieval Latin *merced-, merces,* from Latin, price paid, wages, from *merc-, merx* merchandise] (13th century)
1 a : compassion or forbearance shown especially to an offender or to one subject to one's power; *also* **:** lenient or compassionate treatment ⟨begged for *mercy*⟩ **b :** imprisonment rather than death imposed as penalty for first-degree murder
2 a : a blessing that is an act of divine favor or compassion **b :** a fortunate circumstance ⟨it was a *mercy* they found her before she froze⟩
3 : compassionate treatment of those in distress ⟨works of *mercy* among the poor⟩ ☆ ◆
— mercy *adjective*
— at the mercy of : wholly in the power of **:** with no way to protect oneself against

mercy killing *noun* (1935)
: EUTHANASIA

merde \ˈmerd, ˈmard\ *noun* [French, from Old French, from Latin *merda;* perhaps akin to Lithuanian *smirdėti* to stink] (1920)
: ²CRAP 1a, 2 — sometimes considered vulgar

¹mere \ˈmir\ *noun* [Middle English, from Old English — more at MARINE] (before 12th century)
chiefly British **:** an expanse of standing water **:** LAKE, POOL

²mere *noun* [Middle English, from Old English *mǣre;* akin to Old Norse landa*mæri* borderland] (before 12th century)
: BOUNDARY; *also* **:** LANDMARK

³mere \ˈmir\ *adjective, superlative* **mer•est** [Middle English, from Latin *merus;* akin to Old English *āmerian* to purify and perhaps to Greek *marmairein* to sparkle — more at MORN] (1536)
1 : having no admixture **:** PURE
2 *obsolete* **:** being nothing less than **:** ABSOLUTE
3 : being nothing more than ⟨a *mere* mortal⟩
— mere•ly *adverb*

-mere *noun combining form* [French *-mère,* from Greek *meros* part — more at MERIT]
: part **:** segment ⟨meta*mere*⟩

me•ren•gue \mə-ˈreŋ-(ˌ)gā\ *noun* [American Spanish *merengue* & Haitian Creole *méringue*] (1936)
: a ballroom dance of Haitian and Dominican origin in which one foot is dragged on every step; *also* **:** the music for a merengue

mer•e•tri•cious \ˌmer-ə-ˈtri-shəs\ *adjective* [Latin *meretricius,* from *meretric-, meretrix* prostitute, from *merēre* to earn — more at MERIT] (circa 1626)
1 : of or relating to a prostitute **:** having the nature of prostitution ⟨*meretricious* relationship⟩
2 a : tawdrily and falsely attractive ⟨the paradise they found was a piece of *meretricious* trash —Carolyn See⟩ **b :** superficially significant **:** PRETENTIOUS ⟨scholarly names to provide fig-leaves of respectability for *meretricious* but stylish books —*Times Literary Supplement*⟩
synonym see GAUDY
— mer•e•tri•cious•ly *adverb*
— mer•e•tri•cious•ness *noun*

mer•gan•ser \(ˌ)mər-ˈgan(t)-sər\ *noun* [New Latin, from Latin *mergus,* a waterfowl (from *mergere*) + *anser* goose — more at GOOSE] (1752)
: any of various fish-eating diving ducks (especially genus *Mergus*) with a slender bill hooked at the end and serrated along the margins and usually a crested head

merganser

merge \ˈmərj\ *verb* **merged; merg•ing** [Latin *mergere;* akin to Sanskrit *majjati* he dives] (1636)
transitive verb
1 *archaic* **:** to plunge or engulf in something **:** IMMERSE
2 : to cause to combine, unite, or coalesce
3 : to blend gradually by stages that blur distinctions
intransitive verb
1 : to become combined into one
2 : to blend or come together without abrupt change ⟨*merging* traffic⟩
synonym see MIX
— mer•gence \ˈmər-jən(t)s\ *noun*

merg•er \ˈmər-jər\ *noun* [*merge* + *-er* (as in *waiver*)] (1728)
1 *law* **:** the absorption of an estate, a contract, or an interest in another, of a minor offense in a greater, or of an obligation into a judgment
2 a : the act or process of merging **b :** absorption by a corporation of one or more others; *also* **:** any of various methods of combining two or more organizations (as business concerns)

me•rid•i•an \mə-ˈri-dē-ən\ *noun* [Middle English, from Middle French *meridien,* from *meridien* of noon, from Latin *meridianus,* from *meridies* noon, south, irregular from *medius* mid + *dies* day — more at MID, DEITY] (14th century)
1 *archaic* **:** the hour of noon **:** MIDDAY
2 : a great circle of the celestial sphere passing through its poles and the zenith of a given place
3 : a high point
4 a (1) **:** a great circle on the surface of the earth passing through the poles (2) **:** the half of such a circle included between the poles **b :** a representation of such a circle or half circle numbered for longitude on a map or globe — see LONGITUDE illustration
— meridian *adjective*

¹me•rid•i•o•nal \mə-ˈri-dē-ə-nᵊl\ *adjective* [Middle English, from Middle French *meridional,* from Late Latin *meridionalis,* from Latin *meridies*] (14th century)
1 : of, relating to, or situated in the south **:** SOUTHERN
2 : of, relating to, or characteristic of people living in the south especially of France
3 : of, relating to, or situated on or along a meridian
— me•rid•i•o•nal•ly \-nᵊl-ē\ *adverb*

²meridional *noun* (1591)
: a native or inhabitant of southern Europe and especially southern France

me•ringue \mə-ˈraŋ\ *noun* [French] (1706)

☆ SYNONYMS
Mercy, charity, clemency, grace, leniency mean a disposition to show kindness or compassion. MERCY implies compassion that forbears punishing even when justice demands it ⟨threw himself on the *mercy* of the court⟩. CHARITY stresses benevolence and goodwill shown in broad understanding and tolerance of others ⟨show a little *charity* for the less fortunate⟩. CLEMENCY implies a mild or merciful disposition in one having the power or duty of punishing ⟨the judge refused to show *clemency*⟩. GRACE implies a benign attitude and a willingness to grant favors or make concessions ⟨by the *grace* of God⟩. LENIENCY implies lack of severity in punishing ⟨criticized the courts for excessive *leniency*⟩.

◇ WORD HISTORY
mercy Mercy is not something that can be bought or sold; yet it does have a connection with the marketplace. The word was borrowed into Middle English from Old French *mercit* or *merci,* which had about the same range of senses as English *mercy.* The Old French word is descended from Latin *merces* "price," "payment," or "reward," a derivative of *merx* "commodity, piece of merchandise." The basis of what is now the primary sense of *mercy* is to be found in the Latin of Christian writers, who began to use *merces* for the spiritual reward that derives from kindness to those who do not necessarily have a direct claim to it and from whom no recompense is to be expected. Thus, in a sense, the dispensation of mercy is a figurative form of trade, transacted in a less mundane currency than our daily negotiations. While English has retained most of the sense of Old French *merci,* by and large these senses have not survived in French itself, where today *merci* is used primarily to express thanks.

1 : a dessert topping baked from a mixture of stiffly beaten egg whites and sugar
2 : a shell made of meringue and filled with fruit or ice cream

me·ri·no \mə-'rē-(,)nō\ *noun, plural* **-nos** [Spanish] (1810)
1 : any of a breed of fine-wooled white sheep originating in Spain and producing a heavy fleece of exceptional quality
2 : a soft wool or wool and cotton clothing fabric resembling cashmere
3 : a fine wool and cotton yarn used for hosiery and knitwear
— **merino** *adjective*

-merism *noun combining form* [International Scientific Vocabulary, from Greek *meros* part — more at MERIT]
: possession of (such) an arrangement of or relation among constituent chemical units ⟨tau*tomerism*⟩

mer·i·stem \'mer-ə-,stem\ *noun* [Greek *meristos* divided (from *merizein* to divide, from *meros*) + English *-em* (as in *system*)] (1874)
: a formative plant tissue usually made up of small cells capable of dividing indefinitely and giving rise to similar cells or to cells that differentiate to produce the definitive tissues and organs
— **mer·i·ste·mat·ic** \,mer-əs-tə-'ma-tik\ *adjective*
— **mer·i·ste·mat·i·cal·ly** \-ti-k(ə-)lē\ *adverb*

me·ris·tic \mə-'ris-tik\ *adjective* [Greek *meristos*] (1894)
1 : SEGMENTAL
2 : involving modification in number or in geometrical relation of body parts ⟨*meristic* variation in flower petals⟩
— **me·ris·ti·cal·ly** \-ti-k(ə-)lē\ *adverb*

[1]mer·it \'mer-ət\ *noun* [Middle English, from Middle French *merite,* from Latin *meritum,* from neuter of *meritus,* past participle of *merēre* to deserve, earn; akin to Greek *meiresthai* to receive as one's portion, *meros* part] (14th century)
1 a *obsolete* **:** reward or punishment due **b :** the qualities or actions that constitute the basis of one's deserts **c :** a praiseworthy quality **:** VIRTUE **d :** character or conduct deserving reward, honor, or esteem; *also* **:** ACHIEVEMENT
2 : spiritual credit held to be earned by performance of righteous acts and to ensure future benefits
3 a *plural* **:** the intrinsic nature of a legal case apart from considerations of circumstance, jurisdiction, or procedure **b :** individual significance or justification

[2]merit (1526)
transitive verb
: to be worthy of or entitled or liable to **:** EARN
intransitive verb
1 *obsolete* **:** to be entitled to reward or honor
2 : DESERVE

mer·i·toc·ra·cy \,mer-ə-'tä-krə-sē\ *noun, plural* **-cies** [[1]*merit* + *-o-* + *-cracy*] (1958)
1 : a system in which the talented are chosen and moved ahead on the basis of their achievement
2 : leadership selected on the basis of intellectual criteria
— **mer·it·o·crat·ic** \,mer-ə-tə-'kra-tik\ *adjective*

mer·it·o·crat \'mer-ə-tə-,krat\ *noun* (1960)
chiefly British **:** a person who advances through a meritocratic system

mer·i·to·ri·ous \,mer-ə-'tōr-ē-əs, -'tȯr-\ *adjective* (15th century)
: deserving of honor or esteem
— **mer·i·to·ri·ous·ly** *adverb*
— **mer·i·to·ri·ous·ness** *noun*

merit system *noun* (1879)
: a system by which appointments and promotions in the civil service are based on competence rather than political favoritism

[1]merle *also* **merl** \'mər(-ə)l\ *noun* [Middle English, from Middle French, from Latin *merulus;* akin to Old English *ōsle* blackbird, Old High German *amsla*] (15th century)
: BLACKBIRD 1a

[2]merle *noun* [origin unknown] (1905)
: a bluish or reddish gray mixed with splotches of black that is the color of the coats of some dogs

mer·lin \'mər-lən\ *noun* [Middle English *merlioun,* from Anglo-French *merilun,* from Old French *esmerillon,* augmentative of *esmeril,* of Germanic origin; akin to Old High German *smiril* merlin] (14th century)
: a small compact Holarctic falcon (*Falco columbarius*) which has a broad dark terminal band on the tail and of which the upperparts are slate blue in males and dark brown in females — compare PIGEON HAWK 1

Mer·lin \'mər-lən\ *noun* [Medieval Latin *Merlinus,* from Middle Welsh *Myrddin*]
: a prophet and magician in Arthurian legend

mer·lon \'mər-lən\ *noun* [French, from Italian *merlone,* augmentative of *merlo* battlement, from Medieval Latin *merulus,* from Latin, merle] (circa 1704)
: any of the solid intervals between crenellations of a battlement — see BATTLEMENT illustration

mer·lot \mer-'lō\ *noun, often capitalized* [French] (circa 1941)
: a dry red wine made from a widely grown grape originally used in the Bordeaux region of France for blending

mer·maid \'mər-,mād\ *noun* [Middle English *mermayde,* from *mere* sea (from Old English) + *mayde* maid — more at MARINE] (14th century)
: a fabled marine creature with the head and upper body of a woman and the tail of a fish

mer·man \-,man, -mən\ *noun* (1601)
: a fabled marine creature with the head and upper body of a man and the tail of a fish

mero- *combining form* [International Scientific Vocabulary, from Greek, from *meros* part — more at MERIT]
: part **:** partial ⟨*mero*blastic⟩

mer·o·blas·tic \,mer-ə-'blas-tik\ *adjective* [International Scientific Vocabulary] (1870)
: characterized by incomplete cleavage as a result of the presence of a mass of yolk material — compare HOLOBLASTIC
— **mer·o·blas·ti·cal·ly** \-ti-k(ə-)lē\ *adverb*

mer·o·crine \'mer-ə-krən, -,krīn, -,krēn\ *adjective* [International Scientific Vocabulary, from *mero-* + Greek *krinein* to separate — more at CERTAIN] (circa 1905)
: producing a secretion that is discharged without major damage to the secretory cells; *also* **:** produced by a merocrine gland

mer·o·mor·phic \,mer-ə-'mȯr-fik\ *adjective* (circa 1890)
: relating to or being a function of a complex variable that is analytic everywhere in a region except for singularities at each of which infinity is the limit and each of which is contained in a neighborhood where the function is analytic except for the singular point itself

mer·o·my·o·sin \,mer-ə-'mī-ə-sən\ *noun* (1952)
: either of two structural subunits of myosin that are obtained especially by tryptic digestion

-merous *adjective combining form* [New Latin *-merus,* from Greek *-merēs,* from *meros* — more at MERIT]
: having (such or so many) parts ⟨di*merous*⟩

Mer·o·vin·gian \,mer-ə-'vin-j(ē-)ən\ *adjective* [French *mérovingien,* from Medieval Latin *Merovingi* Merovingians, from *Merovaeus* Merowig (died 458) Frankish founder of the dynasty] (circa 1694)
: of or relating to the first Frankish dynasty reigning from about A.D. 500 to 751
— **Merovingian** *noun*

mer·o·zo·ite \,mer-ə-'zō-,īt\ *noun* [International Scientific Vocabulary, from *mero-* + *zo-* + *-ite*] (1900)
: a sporozoan trophozoite produced by schizogony that is capable of initiating a new sexual or asexual cycle of development

mer·ri·ment \'mer-i-mənt\ *noun* (1576)
1 : lighthearted gaiety or fun-making **:** HILARITY
2 : a lively celebration or party **:** FESTIVITY

mer·ry \'mer-ē\ *adjective* **mer·ri·er; -est** [Middle English *mery,* from Old English *myrge, merge;* akin to Old High German *murg* short — more at BRIEF] (before 12th century)
1 *archaic* **:** giving pleasure **:** DELIGHTFUL
2 : full of gaiety or high spirits **:** MIRTHFUL
3 : marked by festivity or gaiety
4 : QUICK, BRISK ⟨a *merry* pace⟩ ☆
— **mer·ri·ly** \'mer-ə-lē\ *adverb*
— **mer·ri·ness** \'mer-ē-nəs\ *noun*

mer·ry–an·drew \,mer-ē-'an-(,)drü\ *noun, often M&A capitalized* [*merry* + *Andrew,* proper name] (1673)
: a person who clowns publicly

mer·ry–go–round \'mer-ē-gō-,rau̇nd, -gə-\ *noun* (1729)
1 : an amusement park ride with seats often in the form of animals (as horses) revolving about a fixed center
2 : a cycle of activity that is complex, fast-paced, or difficult to break out of ⟨the corporate *merry-go-round*⟩

mer·ry·mak·er \'mer-ē-,mā-kər\ *noun* (1827)
: REVELER

mer·ry·mak·ing \-kiŋ\ *noun* (1714)
1 : gay or festive activity **:** CONVIVIALITY
2 : a convivial occasion **:** FESTIVITY

mer·ry·thought \'mer-ē-,thȯt\ *noun* (1607)
chiefly British **:** WISHBONE

merry widow *noun, often M&W capitalized* [*The Merry Widow,* operetta (1905) by Franz Lehár] (1964)
: a strapless corset or bustier usually having garters attached

Mer·thi·o·late \(,)mər-'thī-ə-,lāt, -lət\ *trademark*
— used for thimerosal

mes- *or* **meso-** *combining form* [Latin, from Greek, from *mesos* — more at MID]
1 : mid **:** in the middle ⟨*meso*carp⟩
2 : intermediate (as in size or type) ⟨*meso*morph⟩ ⟨*mes*on⟩

me·sa \'mā-sə\ *noun* [Spanish, literally, table, from Latin *mensa*] (1759)
: an isolated relatively flat-topped natural elevation usually more extensive than a butte and less extensive than a plateau; *also* **:** a broad terrace with an abrupt slope on one side **:** BENCH

més·al·liance \,mā-,zal-'yä^{n}s, ,mā-zə-'lī-ən(t)s\ *noun, plural* **mésalliances** \-'yä^{n}s(-əz), -'lī-ən(t)-səz\ [French, from *més-* mis- + *alliance*] (1782)
: a marriage with a person of inferior social position

mes·arch \'me-,zärk, 'mē-, -,särk\ *adjective* (1891)
: having metaxylem developed both internal and external to the protoxylem

☆ SYNONYMS
Merry, blithe, jocund, jovial, jolly mean showing high spirits or lightheartedness. MERRY suggests cheerful, joyous, uninhibited enjoyment of frolic or festivity ⟨a *merry* group of revelers⟩. BLITHE suggests carefree, innocent, or even heedless gaiety ⟨arrived late in his usual *blithe* way⟩. JOCUND stresses elation and exhilaration of spirits ⟨singing, dancing, and *jocund* feasting⟩. JOVIAL suggests the stimulation of conviviality and good fellowship ⟨dinner put them in a *jovial* mood⟩. JOLLY suggests high spirits expressed in laughing, bantering, and jesting ⟨our *jolly* host enlivened the party⟩.

mes•cal \me-'skal, mə-\ *noun* [American Spanish *mezcal, mescal,* from Nahuatl *mexcalli* mescal liquor] (1702)
1 : a small cactus (*Lophophora williamsii*) with rounded stems covered with jointed tubercles that are used as a stimulant and antispasmodic especially among the Mexican Indians
2 a : a usually colorless Mexican liquor distilled especially from the central leaves of maguey plants **b :** a plant from which mescal is produced; *especially* **:** MAGUEY

mescal 1

mescal button *noun* (1888)
: one of the dried discoid tops of the mescal

Mes•ca•le•ro \,mes-kə-'ler-(,)ō\ *noun, plural* **Mescalero** *or* **Mescaleros** [American Spanish, from *mezcal, mescal* maguey, mescal liquor] (1844)
: a member of an Apache people of Texas and New Mexico

mes•ca•line \'mes-kə-lən, -,lēn\ *noun* (1896)
: a hallucinatory crystalline alkaloid $C_{11}H_{17}NO_3$ that is the chief active principle in mescal buttons

mesdames *plural of* MADAM, *or of* MADAME, *or of* MRS.

mesdemoiselles *plural of* MADEMOISELLE

me•seems \mi-'sēmz\ *impersonal verb, past* **me•seemed** \-'sēmd\ (15th century)
archaic **:** it seems to me

me•sem•bry•an•the•mum \mə-,zem-brē-'an(t)-thə-məm\ *noun* [New Latin, irregular from Greek *mesēmbria* midday (from *mes-* + *hēmera* day) + *anthemon* flower, from *anthos* — more at ANTHOLOGY] (1753)
: any of a genus (*Mesembryanthemum*) of chiefly southern African fleshy-leaved herbs or subshrubs of the carpetweed family

mes•en•ceph•a•lon \,me-,zen-'se-fə-,län, ,mē-, -zᵊn-, -,sen-, -sᵊn-, -lən\ *noun* [New Latin] (1846)
: MIDBRAIN
— mes•en•ce•phal•ic \-,zen(t)-sə-'fa-lik, -zᵊn(t)-, -,sen(t)-, -sᵊn(t)-\ *adjective*

mes•en•chy•mal \mə-'zeŋ-kə-məl, -'seŋ-; ,me-zᵊn-'kī-məl, ,mē-, -sᵊn-\ *adjective* [International Scientific Vocabulary] (1886)
: of, resembling, or being mesenchyme

mes•en•chyme \'me-zᵊn-,kīm, 'mē-, -sᵊn-\ *noun* [German *Mesenchym,* from *mes-* + New Latin *-enchyma*] (1888)
: loosely organized undifferentiated mostly mesodermal cells that give rise to such structures as connective tissues, blood, lymphatics, bone, and cartilage

mes•en•ter•on \(,)me-'zen-tə-,rän, ,mē-, -'sen-, -rən\ *noun, plural* **-tera** \-tə-rə\ [New Latin] (1877)
: the part of the alimentary canal that is developed from the archenteron and is lined with hypoblast

mes•en•tery \'me-zᵊn-,ter-ē, -sᵊn-\ *noun, plural* **-ter•ies** [Middle English *mesenterie,* from Middle French & Medieval Latin; Middle French *mesenterie,* from Medieval Latin *mesenterion,* from Greek, from *mes-* + *enteron* intestine — more at INTER-] (15th century)
1 a : one or more vertebrate membranes that consist of a double fold of the peritoneum and invest the intestines and their appendages and connect them with the dorsal wall of the abdominal cavity **b :** a fold of membrane comparable to a mesentery and supporting a viscus (as the heart) that is not a part of the digestive tract
2 : a support or partition in an invertebrate like the vertebrate mesentery
— mes•en•ter•ic \,me-zᵊn-'ter-ik, -sᵊn-\ *adjective*

¹mesh \'mesh\ *noun* [Middle English, probably from Middle Dutch *maesche;* akin to Old High German *masca* mesh, Lithuanian *mazgos* knot] (14th century)
1 : one of the openings between the threads or cords of a net; *also* **:** one of the similar spaces in a network — often used to designate screen size as the number of openings per linear inch
2 a : the fabric of a net **b :** a woven, knit, or knotted material of open texture with evenly spaced holes **c :** an arrangement of interlocking metal links used especially for jewelry
3 a : an interlocking or intertwining arrangement or construction **:** NETWORK **b :** WEB, SNARE — usually used in plural
4 : working contact (as of the teeth of gears) ⟨in *mesh*⟩
— meshed \'mesht\ *adjective*

²mesh (circa 1547)
transitive verb
1 a : to catch in the openings of a net **b :** ENMESH, ENTANGLE
2 : to cause to resemble network
3 a : to cause (as gears) to engage **b :** to coordinate closely **:** INTERLOCK
intransitive verb
1 : to become entangled in or as if in meshes
2 : to be in or come into mesh — used especially of gears
3 : to fit or work together properly **:** COORDINATE

me•shuga *or* **me•shug•ge** *also* **me•shug•ah** *or* **me•shug•gah** \mə-'shu̇-gə\ *adjective* [Yiddish *meshuge,* from Hebrew *mĕshuggā'*] (1892)
: CRAZY, FOOLISH

me•shug•gen•er \-'shu̇-gə-nər\ *noun* [Yiddish *meshugener,* from *meshuge*] (1900)
: a foolish or crazy person

mesh•work \'mesh-,wərk\ *noun* (1830)
: NETWORK ⟨a vascular *meshwork*⟩

me•si•al \'mē-zē-əl, -sē-\ *adjective* (1803)
1 : MIDDLE, MEDIAN
2 : of, relating to, or being the surface of a tooth that is next to the tooth in front of it or that is closest to the middle of the front of the jaw — compare DISTAL 2
— me•si•al•ly \-ə-lē\ *adverb*

¹me•sic \'me-zik, 'mē-, -sik\ *adjective* [*mes-* + *-ic*] (1926)
: characterized by, relating to, or requiring a moderate amount of moisture ⟨a *mesic* habitat⟩ ⟨a *mesic* plant⟩ — compare HYDRIC, XERIC

²mesic *adjective* [*meson* + *-ic*] (1939)
: of or relating to a meson

mes•mer•ic \mez-'mer-ik *also* mes-\ *adjective* (1829)
1 : of, relating to, or induced by mesmerism
2 : FASCINATING, IRRESISTIBLE
— mes•mer•i•cal•ly \-i-k(ə-)lē\ *adverb*

mes•mer•ise *British variant of* MESMERIZE

mes•mer•ism \'mez-mə-,ri-zəm *also* 'mes-\ *noun* [F. A. *Mesmer*] (1784)
1 : hypnotic induction held to involve animal magnetism; *broadly* **:** HYPNOTISM
2 : hypnotic appeal
— mes•mer•ist \-rist\ *noun*

mes•mer•ize \-mə-,rīz\ *transitive verb* **-ized; -iz•ing** (1829)
1 : to subject to mesmerism; *also* **:** HYPNOTIZE
2 : SPELLBIND, FASCINATE ◆
— mes•mer•iz•er *noun*

mesne \'mēn\ *adjective* [Anglo-French, alteration of Middle French *meien* — more at MEAN] (1548)
: INTERMEDIATE, INTERVENING — used in law

mesne lord *noun* (1614)
: a feudal lord who holds land as tenant of a superior (as a king) but who is lord to his own tenant

meso- — see MES-

me•so•carp \'me-zə-,kärp, 'mē-, -sə-\ *noun* (1849)
: the middle layer of a pericarp — see ENDOCARP illustration

me•so•cy•clone \,me-zə-'sī-,klōn, ,mē-, -sə-'sī-\ *noun* (1975)
: a rapidly rotating air mass within a thunderstorm that often gives rise to a tornado

me•so•derm \'me-zə-,dərm, 'mē-, -sə-\ *noun* [International Scientific Vocabulary] (1873)
: the middle of the three primary germ layers of an embryo that is the source of many bodily tissues and structures (as bone, muscle, connective tissue, and dermis); *broadly* **:** tissue derived from this germ layer
— me•so•der•mal \,me-zə-'dər-məl, ,mē-, -sə-\ *adjective*

me•so•glea *or* **me•so•gloea** \,me-zə-'glē-ə, ,mē-, -sə-\ *noun* [New Latin, from *mes-* + Late Greek *gloia, glia* glue — more at CLAY] (1886)
: a gelatinous substance between the endoderm and ectoderm of sponges or coelenterates

Me•so•lith•ic \-'li-thik\ *adjective* [International Scientific Vocabulary] (1866)
: of, relating to, or being a transitional period of the Stone Age between the Paleolithic and the Neolithic

me•so•mere \'me-zə-,mir, 'mē-, -sə-\ *noun* (circa 1900)
: a blastomere of medium size; *also* **:** an intermediate part of the mesoderm

me•so•morph \'me-zə-,mȯrf, 'mē-, -sə-\ *noun* [*meso*derm + *-morph*] (1940)
: a mesomorphic body or person

me•so•mor•phic \,me-zə-'mȯr-fik, ,mē-, -sə-\ *adjective* [*meso*derm + *-morphic;* from the predominance in such types of structures developed from the mesoderm] (1940)
1 : of or relating to the component in W. H. Sheldon's classification of body types that measures especially the degree of muscularity and bone development
2 : having a husky muscular body build
— me•so•mor•phy \'me-zə-,mȯr-fē, -sə-\ *noun*

me•son \'me-,zän, 'mā-, 'mē-, -,sän\ *noun* [International Scientific Vocabulary *mes-* + ²*-on*] (1939)
: any of a group of fundamental particles (as the pion and kaon) made up of a quark and an antiquark that are subject to the strong force

◇ WORD HISTORY

mesmerize Franz Anton Mesmer (1734–1815), from whose surname *mesmerize* is derived, was an Austrian physician. A graduate of a traditional medical school in Vienna, Mesmer early on formulated unorthodox medical beliefs. He claimed that a mysterious fluid, which he eventually called "animal magnetism," permeates all matter. Forced to leave Vienna for practicing "magic," Mesmer settled in Paris and soon put his theories of animal magnetism into practice. His therapy basically consisted of having patients sit around a tub of dilute sulfuric acid from which iron bars protruded. As the patients touched the previously "magnetized" bars, the outflowing current of animal magnetism, as well as a tap from Mesmer's magic wand, caused them to become highly agitated. Once the necessary crises had been produced, Mesmer effected the individual cures by having the animal magnetism flow from his body to the patient's. A royal commission's finding that the "cures" were entirely due to the imagination failed to dampen the rage for Mesmer's treatment. What Mesmer had discovered—or rediscovered—was the technique once called "artificial somnambulism" and now known as hypnotism.

and have zero or an integer number of quantum units of spin
— **me·son·ic** \me-'zä-nik, mā-, mē-, -'sä-\ *adjective*

me·so·neph·ros \ˌme-zə-'ne-frəs, ˌmē-, -sə-, -ˌfräs\ *noun, plural* **-neph·roi** \-ˌfrȯi\ [New Latin, from *mes-* + Greek *nephros* kidney — more at NEPHRITIS] (1887)
: either member of the second and midmost of the three paired vertebrate renal organs that functions in adult fishes and amphibians but functions only in the embryo of reptiles, birds, and mammals in which it is replaced by a metanephros in the adult — compare METANEPHROS, PRONEPHROS
— **me·so·neph·ric** \-frik\ *adjective*

me·so·pause \'me-zə-ˌpȯz, 'mē-, -sə-\ *noun* [*meso*sphere + *pause*] (1950)
: the upper boundary of the mesosphere where the temperature of the atmosphere reaches its lowest point

me·so·pe·lag·ic \ˌme-zə-pə-'la-jik, ˌmē-, -sə-\ *adjective* (1947)
: of or relating to oceanic depths from about 600 feet to 3000 feet (200 to 1000 meters)

me·so·phyll \'me-zə-ˌfil, 'mē-, -sə-\ *noun* [New Latin *mesophyllum,* from *mes-* + Greek *phyllon* leaf — more at BLADE] (1839)
: the parenchyma between the epidermal layers of a foliage leaf
— **me·so·phyl·lic** \ˌme-zə-'fi-lik, ˌmē-, -sə-\ *or* **me·so·phyl·lous** \-ləs\ *adjective*

me·so·phyte \'me-zə-ˌfīt, 'mē-, -sə-\ *noun* [International Scientific Vocabulary] (1899)
: a plant that grows under medium conditions of moisture
— **me·so·phyt·ic** \ˌme-zə-'fi-tik, ˌmē-, -sə-\ *adjective*

me·so·scale \'me-zə-ˌskāl, 'mē-, -sə-\ *adjective* (1956)
: of intermediate size; *especially* **:** of or relating to a meteorological phenomenon approximately 10 to 1000 kilometers in horizontal extent ⟨*mesoscale* cloud pattern⟩

me·so·some \-ˌsōm\ *noun* (1960)
: an organelle of bacteria that appears as an invagination of the plasma membrane and functions either in DNA replication and cell division or excretion of exoenzymes

me·so·sphere \-ˌsfir\ *noun* (1950)
: the part of the earth's atmosphere between the stratosphere and the thermosphere in which temperature decreases with altitude to the atmosphere's absolute minimum of about −112°F (−80°C)
— **me·so·spher·ic** \ˌme-zə-'sfir-ik, ˌmē-, -sə-, -'sfer-\ *adjective*

me·so·the·li·o·ma \ˌme-zə-ˌthē-lē-'ō-mə, ˌmē-, -sə-\ *noun, plural* **-mas** *or* **-ma·ta** \-mə-tə\ [New Latin] (circa 1899)
: a tumor derived from mesothelial tissue (as that lining the peritoneum or pleura)

me·so·the·li·um \-'thē-lē-əm\ *noun, plural* **-lia** \-lē-ə\ [New Latin, from *mes-* + *epithelium*] (1886)
: epithelium derived from mesoderm that lines the body cavity of a vertebrate embryo and gives rise to epithelia (as of the peritoneum, pericardium, and pleurae), striated muscle, heart muscle, and several minor structures
— **me·so·the·li·al** \-lē-əl\ *adjective*

me·so·tho·rac·ic \-thə-'ra-sik\ *adjective* (1839)
: of or relating to the mesothorax

me·so·tho·rax \-'thōr-ˌaks, -'thȯr-\ *noun* [New Latin] (circa 1826)
: the middle of the three segments of the thorax of an insect — see INSECT illustration

me·so·tro·phic \ˌme-zə-'trō-fik, ˌmē-, -sə-, -'trä-fik\ *adjective* (1940)
of a body of water **:** having a moderate amount of dissolved nutrients — compare EUTROPHIC, OLIGOTROPHIC

Me·so·zo·ic \-'zō-ik\ *adjective* (1840)
: of, relating to, or being an era of geological history comprising the interval between the Permian and the Tertiary or the corresponding system of rocks that was marked by the presence of dinosaurs, marine and flying reptiles, ammonites, ferns, and gymnosperms and the appearance of angiosperms, mammals, and birds — see GEOLOGIC TIME table
— **Mesozoic** *noun*

mes·quite \mə-'skēt, me-\ *noun* [American Spanish, from Nahuatl *mizquitl*] (1759)
: any of several spiny leguminous trees or shrubs (genus *Prosopis* and especially *P. glandulosa*) chiefly of the southwestern U.S. that often form extensive thickets and have sweet pods eaten by livestock; *also* **:** the wood of the mesquite used especially in grilling food

¹mess \'mes\ *noun* [Middle English *mes,* from Middle French, from Late Latin *missus* course at a meal, from *missus,* past participle of *mittere* to put, from Latin, to send — more at SMITE] (14th century)
1 : a quantity of food: **a** *archaic* **:** food set on a table at one time **b :** a prepared dish of soft food; *also* **:** a mixture of ingredients cooked or eaten together **c :** enough food of a specified kind for a dish or a meal ⟨picked a *mess* of peas for dinner⟩
2 a : a group of persons who regularly take their meals together; *also* **:** a meal so taken **b :** a place where meals are regularly served to a group **:** MESS HALL
3 a : a disordered, untidy, offensive, or unpleasant state or condition ⟨your room is in a *mess*⟩ **b :** one that is disordered, untidy, offensive, or unpleasant usually because of blundering, laxity, or misconduct ⟨[the movie] is a *mess,* as sloppy in concept as it is in execution —Judith Crist⟩ ⟨made a *mess* of his life⟩

²mess (14th century)
transitive verb
1 : to provide with meals at a mess
2 a : to make dirty or untidy **:** DISARRANGE ⟨warned not to *mess* up your room⟩ **b :** to mix up **:** BUNGLE ⟨really *messed* up my life⟩
3 : to interfere with ⟨magnetic storms that *mess* up communications —*Time*⟩
4 : to rough up **:** MANHANDLE ⟨*mess* him up good so he won't double-cross us again⟩
intransitive verb
1 : to take meals with a mess
2 : to make a mess
3 a : PUTTER, TRIFLE ⟨small boys and girls who like to *mess* around with paints⟩ **b :** to handle or play with something especially carelessly ⟨don't *mess* with my camera⟩ — often used with *around* **c :** to take an active interest in something or someone ⟨*messing* around with new video techniques⟩; *also* **:** INTERFERE, MEDDLE ⟨*messing* in other people's affairs⟩ ⟨you'd better not *mess* with me⟩
4 : to become confused or make an error — usually used with *up* ⟨got another chance and didn't want to *mess* up again⟩

¹mes·sage \'me-sij\ *noun* [Middle English, from Old French, from Medieval Latin *missaticum,* from Latin *missus,* past participle of *mittere*] (14th century)
1 : a communication in writing, in speech, or by signals
2 : a messenger's errand or function
3 : an underlying theme or idea

²message *verb* **mes·saged; mes·sag·ing** (1583)
transitive verb
1 : to send as a message or by messenger
2 : to send a message to
intransitive verb
: to communicate by message

mes·sa·line \ˌme-sə-'lēn\ *noun* [French] (circa 1890)
: a soft lightweight silk dress fabric with a satin weave

mes·san \'me-sᵊn\ *noun* [Scottish Gaelic *measan*] (15th century)
chiefly Scottish **:** LAPDOG 1

mess around *intransitive verb* (circa 1932)
1 : to waste time **:** DAWDLE, IDLE
2 a : ASSOCIATE ⟨don't *mess around* with admirals much —K. M. Dodson⟩ **b :** FLIRT, PHILANDER ⟨caught him *messing around* with my wife⟩

messeigneurs *plural of* MONSEIGNEUR

mes·sen·ger \'me-sᵊn-jər\ *noun* [Middle English *messangere,* from Old French *messagier,* from *message*] (14th century)
1 : one who bears a message or does an errand: as **a** *archaic* **:** FORERUNNER, HERALD **b :** a dispatch bearer in government or military service **c :** an employee who carries messages
2 : a light line used in hauling a heavier line (as between ships)
3 : a substance (as a hormone) that mediates a biological effect
4 : MESSENGER RNA

messenger RNA *noun* (1961)
: an RNA produced by transcription that carries the code for a particular protein from the nuclear DNA to a ribosome in the cytoplasm and acts as a template for the formation of that protein — compare TRANSFER RNA

mess hall *noun* (1862)
: a hall or building (as on an army post) in which mess is served

mes·si·ah \mə-'sī-ə\ *noun* [Hebrew *māshīaḥ* & Aramaic *mĕshīḥā,* literally, anointed]
1 *capitalized* **a :** the expected king and deliverer of the Jews **b :** JESUS 1
2 : a professed or accepted leader of some hope or cause
— **mes·si·ah·ship** \-ˌship\ *noun*

mes·si·an·ic \ˌme-sē-'a-nik\ *adjective* [probably from French *messianique,* from *messianisme*] (circa 1834)
1 : of or relating to a messiah
2 : marked by idealism and an aggressive crusading spirit ⟨a *messianic* sense of historic mission —Edmond Taylor⟩

mes·si·a·nism \'me-sē-ə-ˌni-zəm; mə-'sī-ə-, me-\ *noun* [French *messianisme,* from *messie* messiah + *-anisme* (as in *christianisme* Christianity)] (1876)
1 : belief in a messiah as the savior of mankind
2 : religious devotion to an ideal or cause

Mes·si·as \mə-'sī-əs\ *noun* [Middle English, from Late Latin, from Greek, from Aramaic *mĕshīḥā*]
: MESSIAH 1

messieurs *plural of* MONSIEUR

mess jacket *noun* (1891)
: a fitted waist-length man's jacket worn especially as part of a dress uniform

mess kit *noun* (circa 1877)
: a compact kit of nested cooking and eating utensils for use by soldiers and campers

mess·mate \'mes-ˌmāt\ *noun* (1746)
: a person with whom one regularly takes mess (as on a ship)

mess over *transitive verb* (1965)
slang **:** to treat harshly or unfairly **:** ABUSE

Messrs. \'me-sərz\ *plural of* MR.
⟨*Messrs.* Jones, Brown, and Robinson⟩

mes·suage \'mes-wij\ *noun* [Middle English, from Anglo-French, probably alteration of Old French *mesnage* — more at MÉNAGE] (14th century)
: PREMISE 3b

messy \'me-sē\ *adjective* **mess·i·er; -est** (1843)
1 : marked by confusion, disorder, or dirt **:** UNTIDY ⟨a *messy* room⟩
2 : lacking neatness or precision **:** CARELESS, SLOVENLY ⟨*messy* thinking⟩
3 : extremely unpleasant or trying ⟨*messy* lawsuits⟩
— **mess·i·ly** \'me-sə-lē\ *adverb*
— **mess·i·ness** \'me-sē-nəs\ *noun*

mes·ti·za \me-'stē-zə\ *noun* [Spanish, feminine of *mestizo*] (circa 1582)
: a woman who is a mestizo

mes·ti·zo \-(ˌ)zō\ *noun, plural* **-zos** [Spanish, from *mestizo,* adjective, mixed, from Late Latin *mixticius,* from Latin *mixtus,* past participle of *miscēre* to mix — more at MIX] (1582)
: a person of mixed blood; *specifically* **:** a person of mixed European and American Indian ancestry

mes·tra·nol \ˈmes-trə-ˌnȯl, -ˌnōl\ *noun* [*meth-* + *estrogen* + pregn*ane* ($C_{21}H_{36}$) + [1]*-ol*] (1962)
: a synthetic estrogen $C_{21}H_{26}O_2$ used in oral contraceptives

met *past and past participle of* MEET

meta- *or* **met-** *prefix* [New Latin & Medieval Latin, from Latin or Greek; Latin, from Greek, among, with, after, from *meta* among, with, after; akin to Old English *mid, mith* with, Old High German *mit*]
1 a : occurring later than or in succession to **:** after ⟨*met*estrus⟩ **b :** situated behind or beyond ⟨*met*encephalon⟩ ⟨*meta*carpus⟩ **c :** later or more highly organized or specialized form of ⟨*meta*xylem⟩
2 : change **:** transformation
3 [*metaphysics*] **:** more comprehensive **:** transcending ⟨*meta*psychology⟩ — used with the name of a discipline to designate a new but related discipline designed to deal critically with the original one ⟨*meta*mathematics⟩
4 a : involving substitution at or characterized by two positions in the benzene ring that are separated by one carbon atom ⟨*meta*-xylene⟩ **b :** derived from by loss of water ⟨*meta*phosphoric acid⟩

met·a·bol·ic \ˌme-tə-ˈbä-lik\ *adjective* (1845)
: of, relating to, or based on metabolism
— met·a·bol·i·cal·ly \-li-k(ə-)lē\ *adverb*

me·tab·o·lism \mə-ˈta-bə-ˌli-zəm\ *noun* [International Scientific Vocabulary, from Greek *metabolē* change, from *metaballein* to change, from *meta-* + *ballein* to throw — more at DEVIL] (1872)
1 a : the sum of the processes in the buildup and destruction of protoplasm; *specifically* **:** the chemical changes in living cells by which energy is provided for vital processes and activities and new material is assimilated **b :** the sum of the processes by which a particular substance is handled in the living body **c :** the sum of the metabolic activities taking place in a particular environment ⟨the *metabolism* of a lake⟩
2 : METAMORPHOSIS 2 — usually used in combination ⟨holo*metabolism*⟩

me·tab·o·lite \-ˌlīt\ *noun* (1884)
1 : a product of metabolism
2 : a substance essential to the metabolism of a particular organism or to a particular metabolic process

me·tab·o·lize \-ˌlīz\ *verb* **-lized; -liz·ing** (1887)
transitive verb
: to subject to metabolism
intransitive verb
: to perform metabolism
— me·tab·o·liz·able \mə-ˌta-bə-ˈlī-zə-bəl\ *adjective*

[1]**meta·car·pal** \ˌme-tə-ˈkär-pəl\ *adjective* (1739)
: of, relating to, or being the metacarpus or a metacarpal

[2]**metacarpal** *noun* (1854)
: a bone of the part of the hand or forefoot between the carpus and the phalanges that typically contains five more or less elongated bones when all the digits are present

meta·car·pus \ˌme-tə-ˈkär-pəs\ *noun* [New Latin] (1676)
: the part of the hand or forefoot that contains the metacarpals

meta·cen·ter \ˈme-tə-ˌsen-tər\ *noun* [French *métacentre,* from *méta-* meta- + *centre* center] (1794)
: the point of intersection of the vertical through the center of buoyancy of a floating body with the vertical through the new center of buoyancy when the body is displaced

meta·cen·tric \ˌme-tə-ˈsen-trik\ *adjective* (1798)
1 : of or relating to a metacenter
2 : having the centromere medially situated so that the two chromosomal arms are of roughly equal length
— metacentric *noun*

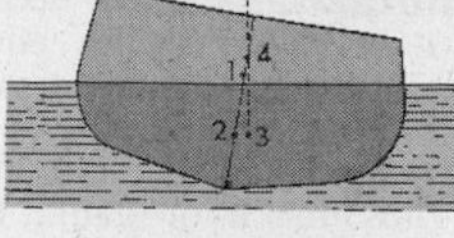

metacenter: *1* center of gravity, *2* center of buoyancy, *3* new center of buoyancy when floating body is displaced, *4* point of intersection

meta·cer·car·ia \ˌme-tə-(ˌ)sər-ˈkar-ē-ə, -ˈker-\ *noun* [New Latin] (1928)
: a tailless encysted late larva of a digenetic trematode that is usually the form which is infective for the definitive host
— meta·cer·car·i·al \-ē-əl\ *adjective*

meta·chro·mat·ic \-krō-ˈma-tik\ *adjective* (1876)
1 : staining or characterized by staining in a different color or shade from what is typical ⟨*metachromatic* granules in a bacterium⟩
2 : having the capacity to stain different elements of a cell or tissue in different colors or shades ⟨*metachromatic* stains⟩

meta·eth·ics \-ˈe-thiks\ *noun plural but usually singular in construction* (1949)
: the study of the meanings of ethical terms, the nature of ethical judgments, and the types of ethical arguments
— meta·eth·i·cal \-thi-kəl\ *adjective*

meta·fic·tion \-ˈfik-shən\ *noun* (1978)
: fiction which refers to or takes as its subject fictional writing and its conventions
— meta·fic·tion·al \-shnəl, -shə-nᵊl\ *adjective*
— meta·fic·tion·ist \-sh(ə-)nist\ *noun*

meta·gal·axy \-ˈga-lək sē\ *noun* [International Scientific Vocabulary] (1930)
: the entire system of galaxies **:** UNIVERSE
— meta·ga·lac·tic \-gə-ˈlak-tik\ *adjective*

meta·gen·e·sis \-ˈje-nə-səs\ *noun* [New Latin] (circa 1864)
: alternation of generations in animals; *especially* **:** regular alteration of a sexual and an asexual generation
— meta·ge·net·ic \-jə-ˈne-tik\ *adjective*

[1]**met·al** \ˈme-tᵊl\ *noun, often attributive* [Middle English, from Old French, from Latin *metallum* mine, metal, from Greek *metallon*] (14th century)
1 : any of various opaque, fusible, ductile, and typically lustrous substances that are good conductors of electricity and heat, form cations by loss of electrons, and yield basic oxides and hydroxides; *especially* **:** one that is a chemical element as distinguished from an alloy
2 a : METTLE 1a **b :** the material or substance out of which a person or thing is made
3 : glass in its molten state
4 a : printing type metal **b :** matter set in metal type
5 : ROAD METAL
6 : HEAVY METAL

[2]**metal** *transitive verb* **-aled** *or* **-alled; -al·ing** *or* **-al·ling** (1610)
: to cover or furnish with metal

meta·lan·guage \ˈme-tə-ˌlaŋ-gwij\ *noun* (1936)
: a language used to talk about language

meta·lin·guis·tic \ˈme-tə-liŋ-ˌgwis-tik\ *adjective* (1944)
: of or relating to a metalanguage or to metalinguistics

meta·lin·guis·tics \-tiks\ *noun plural but singular in construction* (1949)
: a branch of linguistics that deals with the relation between language and other cultural factors in a society

metall- *or* **metallo-** *combining form* [New Latin, from Latin *metallum*]
: metal ⟨*metallo*phone⟩

[1]**me·tal·lic** \mə-ˈta-lik\ *adjective* (15th century)
1 a : of, relating to, or being a metal **b :** made of or containing a metal **c :** having properties of a metal
2 : yielding metal
3 : resembling metal: as **a :** having iridescent and reflective properties ⟨*metallic* blond hair⟩ **b :** having an acrid quality like that of metal ⟨the tea has a *metallic* taste⟩
4 a : having a harsh resonance **:** GRATING ⟨a *metallic* voice⟩ **b :** having an impersonal or mechanical quality ⟨a *metallic* smile⟩
— me·tal·li·cal·ly \-li-k(ə-)lē\ *adverb*

[2]**metallic** *noun* (1952)
: a fiber or yarn made of or coated with metal; *also* **:** a fabric made with this

met·al·lif·er·ous \ˌme-tᵊl-ˈi-f(ə-)rəs\ *adjective* [Latin *metallifer,* from *metallum* + *-fer* -ferous] (circa 1656)
: yielding or containing metal

met·al·lize *also* **met·al·ize** \ˈme-tᵊl-ˌīz\ *transitive verb* **met·al·lized** *also* **met·al·ized; met·al·liz·ing** *also* **met·al·iz·ing** (1594)
: to coat, treat, or combine with a metal
— met·al·li·za·tion \ˌme-tᵊl-ə-ˈzā-shən\ *noun*

met·al·log·ra·phy \ˌme-tᵊl-ˈä-grə-fē\ *noun* [International Scientific Vocabulary] (circa 1864)
: a study of the structure of metals especially with the microscope
— met·al·log·ra·pher \ˌme-tᵊl-ˈä-grə-fər\ *noun*
— me·tal·lo·graph·ic \mə-ˌta-lə-ˈgra-fik\ *adjective*
— me·tal·lo·graph·i·cal·ly \-ˈgra-fi-k(ə-)lē\ *adverb*

[1]**met·al·loid** \ˈme-tᵊl-ˌȯid\ *noun* (1832)
1 : a nonmetal that can combine with a metal to form an alloy
2 : an element intermediate in properties between the typical metals and nonmetals

[2]**metalloid** *also* **met·al·loi·dal** \ˌme-tᵊl-ˈȯi-dᵊl\ *adjective* (circa 1850)
1 : resembling a metal
2 : of, relating to, or being a metalloid

me·tal·lo·phone \mə-ˈta-lə-ˌfōn\ *noun* (circa 1883)
: a percussion musical instrument consisting of a series of metal bars of varying pitch struck with hammers

met·al·lur·gy \ˈme-tᵊl-ˌər-jē, *especially British* mə-ˈta-lər-\ *noun* [New Latin *metallurgia,* from *metall-* + *-urgia* -urgy] (circa 1704)
: the science and technology of metals
— met·al·lur·gi·cal \ˌme-tᵊl-ˈər-ji-kəl\ *adjective*
— met·al·lur·gi·cal·ly \-k(ə-)lē\ *adverb*
— met·al·lur·gist \ˈme-tᵊl-ˌər-jist, *especially British* mə-ˈta-lər-\ *noun*

met·al·mark \ˈme-tᵊl-ˌmärk\ *noun* (circa 1909)
: any of a family (Riodinidae) of small or medium-sized usually brightly colored chiefly tropical butterflies that often have metallic coloration on the wings

met·al·smith \-ˌsmith\ *noun* (14th century)
: a person skilled in metalworking

met·al·ware \-ˌwar, -ˌwer\ *noun* (1896)
: ware made of metal; *especially* **:** metal utensils for household use

met·al·work \-ˌwərk\ *noun* (circa 1850)
: the product of metalworking; *especially* **:** a metal object of artistic merit
— met·al·work·er \-ˌwər-kər\ *noun*

met·al·work·ing \-ˌwər-kiŋ\ *noun* (1882)
: the act or process of shaping things out of metal

meta·math·e·mat·ics \ˈme-tə-ˌmath-ˈma-tiks, -ma-thə-\ *noun plural but usually singular in construction* (circa 1890)
: a field of study concerned with the formal structure and properties (as the consistency and completeness of axioms) of mathematical systems
— meta·math·e·mat·i·cal \-ti-kəl\ *adjective*

meta·mere \ˈme-tə-ˌmir\ *noun* [International Scientific Vocabulary] (1877)
: any of a linear series of primitively similar segments into which the body of a higher invertebrate or vertebrate is divisible
— meta·mer·ic \ˌme-tə-ˈmer-ik, -ˈmir-\ *adjective*
— meta·mer·i·cal·ly \-i-k(ə-)lē\ *adverb*

me·tam·er·ism \mə-ˈta-mə-ˌri-zəm\ *noun* (1877)
: the condition of having or the stage of evolutionary development characterized by a body made up of metameres

meta·mor·phic \ˌme-tə-ˈmȯr-fik\ *adjective* (1816)
1 : of or relating to metamorphosis
2 *of a rock* **:** of, relating to, or produced by metamorphism
— meta·mor·phi·cal·ly \-fi-k(ə-)lē\ *adverb*

meta·mor·phism \-ˈmȯr-ˌfi-zəm\ *noun* (1845)
: a change in the constitution of rock; *specifically* **:** a pronounced change effected by pressure, heat, and water that results in a more compact and more highly crystalline condition

meta·mor·phose \-ˌfōz, -ˌfōs\ *verb* **-phosed; -phos·ing** [probably from Middle French *metamorphoser,* from *metamorphose* metamorphosis, from Latin *metamorphosis*] (1576)
transitive verb
1 a : to change into a different physical form especially by supernatural means **b :** to change strikingly the appearance or character of **:** TRANSFORM ⟨you are so *metamorphosed* I can hardly think you my master —Shakespeare⟩
2 : to cause (rock) to undergo metamorphism
intransitive verb
1 : to undergo metamorphosis
2 : to become transformed
synonym see TRANSFORM

meta·mor·pho·sis \ˌme-tə-ˈmȯr-fə-səs\ *noun, plural* **-pho·ses** \-ˌsēz\ [Latin, from Greek *metamorphōsis,* from *metamorphoun* to transform, from *meta-* + *morphē* form] (1533)
1 a : change of physical form, structure, or substance especially by supernatural means **b :** a striking alteration in appearance, character, or circumstances
2 : a marked and more or less abrupt developmental change in the form or structure of an animal (as a butterfly or a frog) occurring subsequent to birth or hatching

met·anal·y·sis \ˌme-tə-ˈna-lə-səs\ *noun* (1914)
: a reanalysis of the division between sounds or words resulting in different constituents (as in the development of *an apron* from *a napron*)

meta·neph·ros \-ˈne-frəs, -ˌfräs\ *noun, plural* **-roi** \-ˌfrȯi\ [New Latin, from *meta-* + Greek *nephros* kidney — more at NEPHRITIS] (1884)
: either member of the final and most caudal pair of the three successive pairs of vertebrate renal organs that functions as a permanent adult kidney in reptiles, birds, and mammals but is not present at all in lower forms — compare MESONEPHROS, PRONEPHROS
— meta·neph·ric \-frik\ *adjective*

meta·phase \ˈme-tə-ˌfāz\ *noun* [International Scientific Vocabulary] (1887)
: the stage of mitosis and meiosis in which the chromosomes become arranged in the equatorial plane of the spindle

metaphase plate *noun* (1939)
: a section in the equatorial plane of the metaphase spindle having the chromosomes oriented upon it

met·a·phor \ˈme-tə-ˌfȯr *also* -fər\ *noun* [Middle French or Latin; Middle French *metaphore,* from Latin *metaphora,* from Greek, from *metapherein* to transfer, from *meta-* + *pherein* to bear — more at BEAR] (1533)
1 : a figure of speech in which a word or phrase literally denoting one kind of object or idea is used in place of another to suggest a likeness or analogy between them (as in *drowning in money*); *broadly* **:** figurative language — compare SIMILE
2 : an object, activity, or idea treated as a metaphor **:** SYMBOL 2
— met·a·phor·ic \ˌme-tə-ˈfȯr-ik, -ˈfär-\ *or* **met·a·phor·i·cal** \-i-kəl\ *adjective*
— met·a·phor·i·cal·ly \-i-k(ə-)lē\ *adverb*

meta·phos·phate \ˌme-tə-ˈfäs-ˌfāt\ *noun* [International Scientific Vocabulary] (1833)
: a salt or ester of a metaphosphoric acid

meta·phos·pho·ric acid \-ˌfäs-ˈfȯr-ik-, -ˈfär-; -ˈfäs-f(ə-)rik-\ *noun* (1833)
: a glassy solid acid HPO_3 or $(HPO_3)_n$ formed by heating orthophosphoric acid

meta·phrase \ˈme-tə-ˌfrāz\ *noun* (1640)
: a literal translation

meta·phys·ic \ˌme-tə-ˈfi-zik\ *noun* [Middle English *metaphesyk,* from Medieval Latin *metaphysica*] (14th century)
1 a : METAPHYSICS **b :** a particular system of metaphysics
2 : the system of principles underlying a particular study or subject **:** PHILOSOPHY 3b
— metaphysic *adjective*

meta·phys·i·cal \-ˈfi-zi-kəl\ *adjective* (15th century)
1 : of or relating to metaphysics
2 a : of or relating to the transcendent or to a reality beyond what is perceptible to the senses **b :** SUPERNATURAL
3 : highly abstract or abstruse; *also* **:** THEORETICAL
4 *often capitalized* **:** of or relating to poetry especially of the early 17th century that is highly intellectual and philosophical and marked by unconventional imagery
— meta·phys·i·cal·ly \-k(ə-)lē\ *adverb*

Metaphysical *noun* (1898)
: a metaphysical poet of the 17th century

meta·phy·si·cian \ˌme-tə-fə-ˈzi-shən\ *noun* (15th century)
: a student of or specialist in metaphysics

meta·phys·ics \-ˈfi-ziks\ *noun plural but singular in construction* [Medieval Latin *Metaphysica,* title of Aristotle's treatise on the subject, from Greek *(ta) meta (ta) physika,* literally, the (works) after the physical (works); from its position in his collected works] (1569)
1 a (1) **:** a division of philosophy that is concerned with the fundamental nature of reality and being and that includes ontology, cosmology, and often epistemology (2) **:** ONTOLOGY 2 **b :** abstract philosophical studies **:** a study of what is outside objective experience
2 : METAPHYSIC 2

meta·pla·sia \-ˈplā-zh(ē-)ə\ *noun* [New Latin] (1890)
1 : transformation of one tissue into another
2 : abnormal replacement of cells of one type by cells of another
— meta·plas·tic \-ˈplas-tik\ *adjective*

meta·psy·chol·o·gy \-sī-ˈkä-lə-jē\ *noun* [International Scientific Vocabulary] (circa 1909)
: speculative psychology concerned with postulating the structure (as the ego and id) and processes (as cathexis) of the mind which usually cannot be demonstrated objectively
— meta·psy·cho·log·i·cal \-ˌsī-kə-ˈlä-ji-kəl\ *adjective*

meta·se·quoia \-si-ˈkwȯi-ə\ *noun* [New Latin] (1948)
: any of a genus (*Metasequoia*) of fossil and living deciduous coniferous trees of the bald cypress family that have leaves, buds, and branches arranged oppositely and flat leaves resembling needles

meta·so·ma·tism \-ˈsō-mə-ˌti-zəm\ *noun* (1886)
: metamorphism that involves changes in the chemical composition as well as in the texture of rock
— meta·so·mat·ic \-sō-ˈma-tik\ *adjective*

meta·sta·ble \-ˈstā-bəl\ *adjective* [International Scientific Vocabulary] (1897)
: having or characterized by only a slight margin of stability ⟨a *metastable* compound⟩
— meta·sta·bil·i·ty \-stə-ˈbi-lə-tē\ *noun*
— meta·sta·bly \-ˈstā-b(ə-)lē\ *adverb*

me·tas·ta·sis \mə-ˈtas-tə-səs\ *noun, plural* **-ta·ses** \-ˌsēz\ [New Latin, from Late Latin, transition, from Greek, from *methistanai* to change, from *meta-* + *histanai* to set — more at STAND] (1663)
: change of position, state, or form: as **a :** transfer of a disease-producing agency from the site of disease to another part of the body **b :** a secondary metastatic growth of a malignant tumor
— met·a·stat·ic \ˌme-tə-ˈsta-tik\ *adjective*
— met·a·stat·i·cal·ly \-ti-k(ə-)lē\ *adverb*

me·tas·ta·size \mə-ˈtas-tə-ˌsīz\ *intransitive verb* **-sized; -siz·ing** (1907)
: to spread by or as if by metastasis

[1]meta·tar·sal \ˌme-tə-ˈtär-səl\ *adjective* (1739)
: of, relating to, or being the part of the human foot or of the hind foot in quadrupeds between the tarsus and the phalanges

[2]metatarsal *noun* (1854)
: a metatarsal bone

meta·tar·sus \ˌme-tə-ˈtär-səs\ *noun* [New Latin] (1676)
: the metatarsal part of a human foot or of a hind foot in quadrupeds

me·ta·te \mə-ˈtä-tē\ *noun* [Spanish, from Nahuatl *metatl*] (1834)
: a stone with a concave upper surface used as the lower millstone for grinding grains and especially maize

me·tath·e·sis \mə-ˈta-thə-səs\ *noun, plural* **-e·ses** \-ˌsēz\ [Late Latin, from Greek, from *metatithenai* to transpose, from *meta-* + *tithenai* to place — more at DO] (1577)
: a change of place or condition: as **a :** transposition of two phonemes in a word (as in the development of *crud* from *curd* or the pronunciation \ˈpu̇r-tē\ for *pretty*) **b :** a chemical reaction in which different kinds of molecules exchange parts to form other kinds of molecules
— met·a·thet·i·cal \ˌme-tə-ˈthe-ti-kəl\ *or* **met·a·thet·ic** \-tik\ *adjective*
— met·a·thet·i·cal·ly \-ti-k(ə-)lē\ *adverb*

meta·tho·rac·ic \ˌme-tə-thə-ˈra-sik\ *adjective* (circa 1839)
: of, relating to, or situated in or on the metathorax ⟨*metathoracic* legs⟩

meta·tho·rax \-ˈthōr-ˌaks, -ˈthȯr-\ *noun* [New Latin] (1816)
: the posterior segment of the thorax of an insect — see INSECT illustration

meta·xy·lem \-ˈzī-ləm, -ˌlem\ *noun* (1902)
: the part of the primary xylem that differentiates after the protoxylem and that is distinguished typically by broader tracheids and vessels with pitted or reticulate walls

meta·zo·al \-ˈzō-əl\ *adjective* [New Latin *Metazoa*] (1928)
: of or relating to the metazoans

meta·zo·an \-ˈzō-ən\ *noun* [New Latin *Metazoa,* from *meta-* + *-zoa*] (1884)
: any of a group (*Metazoa*) that comprises all animals having the body composed of cells differentiated into tissues and organs and usually a digestive cavity lined with specialized cells
— metazoan *adjective*

[1]mete \'mēt\ *transitive verb* **met·ed; met·ing** [Middle English, from Old English *metan;* akin to Old High German *mezzan* to measure, Latin *modus* measure, Greek *medesthai* to be mindful of] (before 12th century)
1 *archaic* **:** MEASURE
2 : to give out by measure **:** DOLE — usually used with *out* ⟨*mete* out punishment⟩

[2]mete *noun* [Middle English, from Anglo-French, from Latin *meta*] (15th century)
: BOUNDARY ⟨*metes* and bounds⟩

me·tem·psy·cho·sis \mə-ˌtem(p)-si-'kō-səs, ˌme-təm-ˌsī-\ *noun* [Late Latin, from Greek *metempsychōsis,* from *metempsychousthai* to undergo metempsychosis, from *meta-* + *empsychos* animate, from *en-* + *psychē* soul — more at PSYCH-] (1591)
: the passing of the soul at death into another body either human or animal

met·en·ceph·a·lon \ˌmet-ˌen-'se-fə-ˌlän, -lən\ *noun* [New Latin] (circa 1871)
: the anterior segment of the developing vertebrate hindbrain or the corresponding part of the adult brain composed of the cerebellum and pons
— met·en·ce·phal·ic \-ˌen(t)-sə-'fa-lik\ *adjective*

me·te·or \'mē-tē-ər, -ˌȯr\ *noun* [Middle English, from Middle French *meteore,* from Medieval Latin *meteorum,* from Greek *meteōron,* from neuter of *meteōros* high in air, from *meta-* + *-eōros,* from *aeirein* to lift] (15th century)
1 : an atmospheric phenomenon (as lightning or a snowfall)
2 a : any of the small particles of matter in the solar system that are directly observable only by their incandescence from frictional heating on entry into the atmosphere **b :** the streak of light produced by the passage of a meteor

me·te·or·ic \ˌmē-tē-'ȯr-ik, -'är-\ *adjective* (1789)
1 : of, relating to, or derived from the earth's atmosphere ⟨*meteoric* water⟩
2 a : of or relating to a meteor **b :** resembling a meteor in speed or in sudden and temporary brilliance ⟨a *meteoric* rise to fame⟩
— me·te·or·i·cal·ly \-i-k(ə-)lē\ *adverb*

me·te·or·ite \'mē-tē-ə-ˌrīt\ *noun* (1824)
: a meteor that reaches the surface of the earth without being completely vaporized
— me·te·or·it·ic \ˌmē-tē-ə-'ri-tik\ *also* **me·te·or·it·i·cal** \-ti-kəl\ *adjective*

me·te·or·it·ics \ˌmē-tē-ə-'ri-tiks\ *noun plural but singular in construction* (circa 1930)
: a science that deals with meteors
— me·te·or·it·i·cist \-'ri-tə-sist\ *noun*

me·te·or·oid \'mē-tē-ə-ˌrȯid\ *noun* (1865)
1 : a meteor particle itself without relation to the phenomena it produces when entering the earth's atmosphere
2 : a meteor in orbit around the sun
— me·te·or·oi·dal \ˌmē-tē-ə-'rȯi-dᵊl\ *adjective*

me·te·o·rol·o·gy \ˌmē-tē-ə-'rä-lə-jē\ *noun* [French or Greek; French *météorologie,* from Middle French, from Greek *meteōrologia,* from *meteōron* + *-logia* -logy] (1620)
1 : a science that deals with the atmosphere and its phenomena and especially with weather and weather forecasting
2 : the atmospheric phenomena and weather of a region
— me·te·o·ro·log·ic \-rə-'lä-jik\ *or* **me·te·o·ro·log·i·cal** \-ji-kəl\ *adjective*
— me·te·o·ro·log·i·cal·ly \-ji-k(ə-)lē\ *adverb*
— me·te·o·rol·o·gist \-'rä-lə-jist\ *noun*

[1]me·ter \'mē-tər\ *noun* [Middle English, from Old English & Middle French; Old English *mēter,* from Latin *metrum,* from Greek *metron* measure, meter; Middle French *metre,* from Old French, from Latin *metrum* — more at MEASURE] (before 12th century)
1 a : systematically arranged and measured rhythm in verse: (1) **:** rhythm that continuously repeats a single basic pattern ⟨iambic *meter*⟩ (2) **:** rhythm characterized by regular recurrence of a systematic arrangement of basic patterns in larger figures ⟨ballad *meter*⟩ **b :** a measure or unit of metrical verse — usually used in combination ⟨penta*meter*⟩; compare FOOT 4 **c :** a fixed metrical pattern **:** verse form
2 : the basic recurrent rhythmical pattern of note values, accents, and beats per measure in music

[2]met·er \'mē-tər\ *noun* [Middle English, from *meten* to mete] (14th century)
: one that measures; *especially* **:** an official measurer of commodities

[3]me·ter *noun* [French *mètre,* from Greek *metron* measure] (1797)
: the base unit of length in the International System of Units that is equal to the distance traveled by light in a vacuum in 1/299,792,458 second or to about 39.37 inches — see METRIC SYSTEM table

[4]me·ter *noun* [*-meter*] (1815)
1 : an instrument for measuring and sometimes recording the time or amount of something ⟨a parking *meter*⟩ ⟨a gas *meter*⟩
2 : POSTAGE METER; *also* **:** a postal marking printed by a postage meter

[5]me·ter *transitive verb* (1884)
1 : to measure by means of a meter
2 : to supply in a measured or regulated amount
3 : to print postal indicia on by means of a postage meter

-meter *noun combining form* [French *-mètre,* from Greek *metron* measure]
: instrument or means for measuring ⟨*barometer*⟩

meter–kilogram–second *adjective* (circa 1940)
: of, relating to, or being a system of units using the meter, kilogram, and second as its base units — abbreviation *mks*

meter maid *noun* (1957)
: a woman assigned by a police or traffic department to write tickets for parking violations

me·ter·stick \'mē-tər-ˌstik\ *noun* (1931)
: a measuring stick one meter long that is marked off in centimeters and usually millimeters

met·es·trus \ˌmet-'es-trəs\ *noun* [New Latin] (1900)
: the period of regression that follows estrus

meth- *or* **metho-** *combining form* [International Scientific Vocabulary, from *methyl*]
: methyl ⟨*meth*acrylic⟩

meth·ac·ry·late \ˌme-'tha-krə-ˌlāt\ *noun* [International Scientific Vocabulary] (1865)
1 : a salt or ester of methacrylic acid
2 : an acrylic resin or plastic made from a derivative of methacrylic acid

meth·acryl·ic acid \ˌme-thə-'kri-lik-\ *noun* [International Scientific Vocabulary] (1865)
: an acid $C_4H_6O_2$ used especially in making acrylic resins or plastics

meth·a·done \'me-thə-ˌdōn\ *also* **meth·a·don** \-ˌdän\ *noun* [*meth*yl + *a*mino + *d*iphenyl + *-one*] (1947)
: a synthetic addictive narcotic drug $C_{21}H_{27}NO$ used especially in the form of its hydrochloride for the relief of pain and as a substitute narcotic in the treatment of heroin addiction

meth·am·phet·amine \ˌme-tham-'fe-tə-ˌmēn, -thəm-, -mən\ *noun* (1949)
: an amine $C_{10}H_{15}N$ used medically in the form of its crystalline hydrochloride especially in the treatment of obesity and often used illicitly as a stimulant — called also *methedrine*

metha·na·tion \ˌme-thə-'nā-shən\ *noun* (1926)
: the production of methane especially from carbon monoxide and hydrogen

meth·ane \'me-ˌthān, *British usually* 'mē-\ *noun* [International Scientific Vocabulary] (circa 1868)
: a colorless odorless flammable gaseous hydrocarbon CH_4 that is a product of decomposition of organic matter and of the carbonization of coal, is used as a fuel and as a starting material in chemical synthesis, and is the simplest of the alkanes

meth·a·nol \'me-thə-ˌnȯl, -ˌnōl\ *noun* [International Scientific Vocabulary] (1894)
: a light volatile flammable poisonous liquid alcohol CH_3OH used especially as a solvent, antifreeze, or denaturant for ethyl alcohol and in the synthesis of other chemicals

meth·aqua·lone \me-'tha-kwə-ˌlōn\ *noun* [*meth-* + *-a-* (of unknown origin) + *qu*inoline + *azole* + *-one*] (1961)
: a sedative and hypnotic nonbarbiturate drug $C_{16}H_{14}N_2O$ that is habit-forming — compare QUAALUDE

meth·e·drine \'me-thə-ˌdrēn, -drən\ *noun* [from *Methedrine,* a trademark] (1939)
: METHAMPHETAMINE

me·theg·lin \mə-'the-glən\ *noun* [Welsh *meddyglyn*] (1533)
: [1]MEAD

met·he·mo·glo·bin \ˌmet-'hē-mə-ˌglō-bən\ *noun* [International Scientific Vocabulary] (1870)
: a soluble brown crystalline basic blood pigment that differs from hemoglobin in containing ferric iron and in being unable to combine reversibly with molecular oxygen

met·he·mo·glo·bi·ne·mia \ˌmet-ˌhē-mə-ˌglō-bə-'nē-mē-ə\ *noun* [New Latin] (1888)
: the presence of methemoglobin in the blood

me·the·na·mine \mə-'thē-nə-ˌmēn, -mən\ *noun* [*methene* (methylene) + *amine*] (1926)
: hexamethylenetetramine especially when used as a urinary antiseptic

meth·i·cil·lin \ˌme-thə-'si-lən\ *noun* [*meth-* + *penicillin*] (1961)
: a semisynthetic penicillin $C_{17}H_{19}N_2O_6NaS$ that is especially effective against penicillinase-producing staphylococci

me·thinks \mi-'thiŋ(k)s\ *impersonal verb, past* **me·thought** \-'thȯt\ [Middle English *me thinketh,* from Old English *mē thincth,* from *mē* (dative of *ic* I) + *thincth* seems, from *thyncan* to seem — more at I, THINK] (before 12th century)
archaic **:** it seems to me

me·thi·o·nine \mə-'thī-ə-ˌnēn\ *noun* [International Scientific Vocabulary, from *methyl* + *thion-* + [2]*-ine*] (1928)
: a crystalline sulfur-containing essential amino acid $C_5H_{11}NO_2S$

meth·od \'me-thəd\ *noun* [Middle French or Latin; Middle French *methode,* from Latin *methodus,* from Greek *methodos,* from *meta-* + *hodos* way] (1541)
1 : a procedure or process for attaining an object: as **a** (1) **:** a systematic procedure, technique, or mode of inquiry employed by or proper to a particular discipline or art (2) **:** a systematic plan followed in presenting material for instruction **b** (1) **:** a way, technique, or process of or for doing something (2) **:** a body of skills or techniques
2 : a discipline that deals with the principles and techniques of scientific inquiry
3 a : orderly arrangement, development, or classification **:** PLAN **b :** the habitual practice of orderliness and regularity
4 *capitalized* **:** a dramatic technique by which an actor seeks to gain complete identification

with the inner personality of the character being portrayed ☆

me·thod·i·cal \mə-ˈthä-di-kəl\ *also* **me·thod·ic** \-dik\ *adjective* (1570)
1 : arranged, characterized by, or performed with method or order ⟨a *methodical* treatment of the subject⟩
2 : habitually proceeding according to method **:** SYSTEMATIC ⟨*methodical* in his daily routine⟩
— **me·thod·i·cal·ly** \-di-k(ə-)lē\ *adverb*
— **me·thod·i·cal·ness** \-di-kəl-nəs\ *noun*

meth·od·ise *British variant of* METHODIZE

meth·od·ism \ˈme-thə-ˌdi-zəm\ *noun* (1739)
1 *capitalized* **a :** the doctrines and practice of Methodists **b :** the Methodist churches
2 : methodical procedure

meth·od·ist \-dist\ *noun* (1593)
1 : a person devoted to or laying great stress on method
2 *capitalized* **:** a member of one of the denominations deriving from the Wesleyan revival in the Church of England, having Arminian doctrine and in the U.S. modified episcopal polity, and stressing personal and social morality
— **methodist** *adjective, often capitalized*
— **meth·od·is·tic** \ˌme-thə-ˈdis-tik\ *adjective*

meth·od·ize \ˈme-thə-ˌdīz\ *transitive verb* **-ized; -iz·ing** (1589)
: to reduce to method **:** SYSTEMATIZE
synonym see ORDER

method of fluxions (circa 1719)
: DIFFERENTIAL CALCULUS

meth·od·o·log·i·cal \ˌme-thə-dᵊl-ˈä-ji-kəl\ *adjective* (1849)
: of or relating to method or methodology
— **meth·od·o·log·i·cal·ly** \-k(ə-)lē\ *adverb*

meth·od·ol·o·gist \-thə-ˈdä-lə-jist\ *noun* (1865)
: a student of methodology

meth·od·ol·o·gy \ˌme-thə-ˈdä-lə-jē\ *noun, plural* **-gies** [New Latin *methodologia,* from Latin *methodus* + *-logia* -logy] (1800)
1 : a body of methods, rules, and postulates employed by a discipline **:** a particular procedure or set of procedures
2 : the analysis of the principles or procedures of inquiry in a particular field

meth·o·trex·ate \ˌme-thə-ˈtrek-ˌsāt\ *noun* [*meth-* + *-trexate,* of unknown origin] (1955)
: a toxic anticancer drug $C_{20}H_{22}N_8O_5$ that is an analogue of folic acid and an antimetabolite

me·thoxy·chlor \me-ˈthäk-si-ˌklōr, -ˌklȯr\ *noun* [*meth-* + *oxy-* + *chlor-*] (1947)
: a chlorinated hydrocarbon insecticide $C_{16}H_{15}Cl_3O_2$

me·thoxy·flu·rane \me-ˌthäk-sē-ˈflu̇r-ˌān\ *noun* [*meth-* + *oxy-* + *fluor-* + eth*ane*] (1962)
: a potent nonexplosive inhalational general anesthetic $C_3H_4Cl_2F_2O$ administered as a vapor

Me·thu·se·lah \mə-ˈthü-zə-lə, -ˈthyü-; -ˈth(y)üz-lə\ *noun* [Hebrew *Mĕthūselaḥ*]
1 : an ancestor of Noah held to have lived 969 years
2 : an oversize wine bottle holding about 6 liters

meth·yl \ˈme-thəl\ *noun* [International Scientific Vocabulary, back-formation from *methylene*] (circa 1844)
: an alkyl group CH_3 derived from methane by removal of one hydrogen atom
— **me·thyl·ic** \mə-ˈthi-lik\ *adjective*

methyl acetate *noun* (1885)
: a flammable fragrant liquid $C_3H_6O_2$ used as a solvent and paint remover and in organic synthesis

methyl alcohol *noun* (circa 1847)
: METHANOL

me·thyl·amine \ˌme-thə-lə-ˈmēn, -ˈla-mən; mə-ˈthi-lə-ˌmēn\ *noun* [International Scientific Vocabulary] (circa 1850)
: a flammable explosive gas CH_3NH_2 with a strong ammoniacal odor used especially in organic synthesis (as of dyes and insecticides)

meth·yl·ase \ˈme-thə-ˌlās, -ˌlāz\ *noun* (circa 1952)
: an enzyme that catalyzes methylation (as of RNA or DNA)

meth·yl·ate \ˈme-thə-ˌlāt\ *transitive verb* **-at·ed; -at·ing** (1861)
: to introduce the methyl group into
— **meth·yl·a·tion** \ˌme-thə-ˈlā-shən\ *noun*
— **meth·yl·a·tor** \ˈme-thə-ˌlā-tər\ *noun*

methyl bromide *noun* (circa 1897)
: a poisonous gaseous compound CH_3Br used chiefly as a fumigant against rodents, worms, and insects

meth·yl·cel·lu·lose \ˌme-thəl-ˈsel-yə-ˌlōs, -ˌlōz\ *noun* (1921)
: any of various gummy products of cellulose methylation that swell in water and are used especially as emulsifiers, adhesives, thickeners, and bulk laxatives

meth·yl·cho·lan·threne \-kə-ˈlan-ˌthrēn\ *noun* [*methyl* + *chol*ic acid + *anthracene*] (1937)
: a potent carcinogenic hydrocarbon $C_{21}H_{16}$

meth·yl·do·pa \ˌme-thəl-ˈdō-pə\ *noun* (1954)
: a drug $C_{10}H_{13}NO_4$ used to lower blood pressure

meth·y·lene \ˈme-thə-ˌlēn, -lən\ *noun* [French *méthylène,* from Greek *methy* wine + *hylē* wood — more at MEAD] (1835)
: a bivalent hydrocarbon group CH_2 derived from methane by removal of two hydrogen atoms

methylene blue *noun* (circa 1890)
: a basic thiazine dye $C_{16}H_{18}ClN_3S{\cdot}3H_2O$ used especially as a biological stain, an antidote in cyanide poisoning, and an oxidation-reduction indicator

methylene chloride *noun* (1880)
: a nonflammable liquid CH_2Cl_2 used especially as a solvent, paint remover, and aerosol propellant

methyl ethyl ketone *noun* (1876)
: a flammable liquid compound C_4H_8O similar to acetone and used chiefly as a solvent — abbreviation *MEK*

methyl isocyanate *noun* (circa 1894)
: an extremely toxic chemical CH_3NCO used especially in the manufacture of pesticides — abbreviation *MIC*

meth·yl·mer·cury \ˌme-thəl-ˈmər-kyə-rē, -ˈmər-k(ə-)rē\ *noun* (1919)
: any of various toxic compounds of mercury containing the complex CH_3Hg- that often occur as pollutants which accumulate in living organisms (as fish) especially in higher levels of a food chain

methyl methacrylate *noun* (1933)
: a volatile flammable liquid $C_5H_8O_2$ that polymerizes readily and is used especially as a monomer for resins

meth·yl·naph·tha·lene \ˌme-thəl-ˈnaf-thə-ˌlēn, -ˈnap-\ *noun* (circa 1885)
: either of two isomeric hydrocarbons $C_{11}H_{10}$; *especially* **:** an oily liquid used in determining cetane numbers

methyl orange *noun* (1881)
: an alkaline dye used as a chemical indicator that in dilute solution is yellow when neutral and pink when acid

methyl parathion *noun* (1957)
: a potent synthetic organophosphate insecticide $C_8H_{10}NO_5PS$ that is more toxic than parathion

meth·yl·phe·ni·date \ˌme-thəl-ˈfe-nə-ˌdāt, -ˈfē-\ *noun* [*methyl* + *phen*yl + piper*idine* + acet*ate*] (1958)
: a mild stimulant $C_{14}H_{19}NO_2$ of the central nervous system used in the form of its hydrochloride to treat narcolepsy and hyperkinetic behavior disorders in children

meth·yl·pred·nis·o·lone \-pred-ˈni-sə-ˌlōn\ *noun* (1957)
: a glucocorticoid $C_{22}H_{30}O_5$ that is a derivative of prednisolone and is used as an anti-inflammatory agent; *also* **:** any of several of its salts (as an acetate) used similarly

meth·yl·xan·thine \-ˈzan-ˌthēn\ *noun* (1949)
: a methylated xanthine derivative (as caffeine, theobromine, or theophylline)

meth·y·ser·gide \ˌme-thə-ˈsər-ˌjīd\ *noun* [*methyl* + ly*serg*ic acid + am*ide*] (1962)
: a serotonin antagonist $C_{21}H_{27}N_3O_2$ used in the form of its maleate especially in the treatment and prevention of migraine headaches

met·i·cal \ˈme-ti-kəl\ *noun, plural* **met·i·cais** \-(ˌ)kī\ *also* **meticals** [Portuguese, from Arabic *mithqāl*] (1981)
— see MONEY table

me·tic·u·lous \mə-ˈti-kyə-ləs\ *adjective* [Latin *meticulosus* fearful, irregular from *metus* fear] (1827)
: marked by extreme or excessive care in the consideration or treatment of details
synonym see CAREFUL
— **me·tic·u·los·i·ty** \-ˌti-kyə-ˈlä-sə-tē\ *noun*
— **me·tic·u·lous·ly** \-ˈti-kyə-ləs-lē\ *adverb*
— **me·tic·u·lous·ness** \-nəs\ *noun*

mé·tier *also* **me·tier** \ˈme-ˌtyā, me-ˈ\ *noun* [French, from (assumed) Vulgar Latin *misterium,* alteration of Latin *ministerium* work, ministry] (1792)
1 : VOCATION, TRADE
2 : an area of activity in which one excels **:** FORTE
synonym see WORK

mé·tis \mā-ˈtē(s)\ *noun, plural* **mé·tis** \-ˈtē(s), -ˈtēz\ [French, from Late Latin *mixticius* mixed — more at MESTIZO] (1816)
: one of mixed blood; *especially, often capitalized* **:** the offspring of an American Indian and a person of European ancestry

met·o·nym \ˈme-tə-ˌnim\ *noun* [back-formation from *metonymy*] (1862)
: a word used in metonymy

me·ton·y·my \mə-ˈtä-nə-mē\ *noun, plural* **-mies** [Latin *metonymia,* from Greek *metōnymia,* from *meta-* + *-ōnymon* -onym] (1547)
: a figure of speech consisting of the use of the name of one thing for that of another of which it is an attribute or with which it is associated (as "crown" in "lands belonging to the crown")
— **met·o·nym·ic** \ˌme-tə-ˈni-mik\ *or* **met·o·nym·i·cal** \-mi-kəl\ *adjective*

me–too \ˈmē-ˈtü\ *adjective* (1926)
1 : marked by similarity to or by adoption of successful or persuasive policies or practices used or promoted by someone else
2 : similar or identical to an established product (as a drug) with no significant advantage over it
— **me–too·er** \-ər\ *noun*
— **me–too·ism** \-ˌi-zəm\ *noun*

met·o·pe \ˈme-tə-(ˌ)pē\ *noun* [Greek *metopē,* from *meta-* + *opē* opening; akin to Greek *ōps* eye, face — more at EYE] (1563)

☆ SYNONYMS
Method, mode, manner, way, fashion, system mean the means taken or procedure followed in achieving an end. METHOD implies an orderly logical effective arrangement usually in steps ⟨effective teaching *methods*⟩. MODE implies an order or course followed by custom, tradition, or personal preference ⟨the preferred *mode* of transportation⟩. MANNER is close to MODE but may imply a procedure or method that is individual or distinctive ⟨an odd *manner* of conducting⟩. WAY is very general and may be used for any of the preceding words ⟨has her own *way* of doing things⟩. FASHION may suggest a peculiar or characteristic way of doing something ⟨rushing about in his typical *fashion*⟩. SYSTEM suggests a fully developed or carefully formulated method often emphasizing the idea of rational orderliness ⟨a filing *system*⟩.

METRIC SYSTEM[1]

LENGTH

Unit	Abbreviation	Number of Meters	Approximate U.S. Equivalent
kilometer	km	1,000	0.62 mile
hectometer	hm	100	328.08 feet
dekameter	dam	10	32.81 feet
meter	m	1	39.37 inches
decimeter	dm	0.1	3.94 inches
centimeter	cm	0.01	0.39 inch
millimeter	mm	0.001	0.039 inch
micrometer	μm	0.000001	0.000039 inch

AREA

Unit	Abbreviation	Number of Square Meters	Approximate U.S. Equivalent
square kilometer	sq km *or* km²	1,000,000	0.3861 square miles
hectare	ha	10,000	2.47 acres
are	a	100	119.60 square yards
square centimeter	sq cm *or* cm²	0.0001	0.155 square inch

VOLUME

Unit	Abbreviation	Number of Cubic Meters	Approximate U.S. Equivalent
cubic meter	m³	1	1.307 cubic yards
cubic decimeter	dm³	0.001	61.023 cubic inches
cubic centimeter	cu cm *or* cm³ *also* cc	0.000001	0.061 cubic inch

CAPACITY

Unit	Abbreviation	Number of Liters	Approximate U.S. Equivalent		
			cubic	*dry*	*liquid*
kiloliter	kl	1,000	1.31 cubic yards		
hectoliter	hl	100	3.53 cubic feet	2.84 bushels	
dekaliter	dal	10	0.35 cubic foot	1.14 pecks	2.64 gallons
liter	l	1	61.02 cubic inches	0.908 quart	1.057 quarts
cubic decimeter	dm³	1	61.02 cubic inches	0.908 quart	1.057 quarts
deciliter	dl	0.10	6.1 cubic inches	0.18 pint	0.21 pint
centiliter	cl	0.01	0.61 cubic inch		0.338 fluid ounce
milliliter	ml	0.001	0.061 cubic inch		0.27 fluid dram
microliter	μl	0.000001	0.000061 cubic inch		0.00027 fluid dram

MASS AND WEIGHT

Unit	Abbreviation	Number of Grams	Approximate U.S. Equivalent
metric ton	t	1,000,000	1.102 short tons
kilogram	kg	1,000	2.2046 pounds
hectogram	hg	100	3.527 ounces
dekagram	dag	10	0.353 ounce
gram	g	1	0.035 ounce
decigram	dg	0.10	1.543 grains
centigram	cg	0.01	0.154 grain
milligram	mg	0.001	0.015 grain
microgram	μg	0.000001	0.000015 grain

[1]For metric equivalents of U.S. units, see Weights and Measures table

: the space between two triglyphs of a Doric frieze often adorned with carved work

met·o·pon \'me-tə-,pän\ *noun* [*meth-* + hydr*o-* + morphine + *-one*] (1941)
: a narcotic drug $C_{18}H_{21}NO_3$ that is derived from morphine and is used in the form of the hydrochloride to relieve pain

me·tre \'mē-tər\ *chiefly British variant of* METER

¹met·ric \'me-trik\ *noun* [Greek *metrikē,* from feminine of *metrikos* in meter, by measure, from *metron* measure — more at MEASURE] (1760)
1 *plural* **:** a part of prosody that deals with metrical structure
2 : a standard of measurement ⟨no *metric* exists that can be applied directly to happiness —*Scientific Monthly*⟩
3 : a mathematical function that associates with each pair of elements of a set a real nonnegative number with the general properties of distance such that the number is zero only if the two elements are identical, the number is the same regardless of the order in which the two elements are taken, and the number associated with one pair of elements plus that associated with one member of the pair and a third element is equal to or greater than the number associated with the other member of the pair and the third element

²metric *adjective* [French *métrique,* from *mètre* meter] (1864)
: of, relating to, or using the metric system ⟨a *metric* study⟩
— met·ri·cal·ly \-tri-k(ə-)lē\ *adverb*

-metric *or* **-metrical** *adjective combining form*
1 : of, employing, or obtained by (such) a meter ⟨galvano*metric*⟩
2 : of or relating to (such) an art, process, or science of measuring ⟨geo*metrical*⟩

met·ri·cal \'me-tri-kəl\ *or* **met·ric** \-trik\ *adjective* (15th century)
1 : of, relating to, or composed in meter
2 : of or relating to measurement
— met·ri·cal·ly \-tri-k(ə-)lē\ *adverb*

met·ri·ca·tion \,me-tri-'kā-shən\ *noun* (1965)
: the act or process of metricizing; *specifically*
: conversion of an existent system of units into the metric system

met·ri·cize \'me-trə-,sīz\ *transitive verb* **-cized; -ciz·ing** (1873)
: to change into or express in the metric system

metric space *noun* (1927)
: a mathematical set for which a metric is defined for any pair of elements

metric system *noun* (1864)
: a decimal system of weights and measures based on the meter and on the kilogram

metric ton *noun* (circa 1890)
— see METRIC SYSTEM table

me·trist \'me-trist, 'mē-\ *noun* (1535)
1 : a maker of verses
2 : one skillful in handling meter
3 : a student of meter or metrics

¹met·ro \'me-(,)trō, *in French context also* mā-'\ *noun, plural* **metros** [French *métro,* short for (*chemin de fer*) *métropolitain* metropolitan railroad] (1904)
: SUBWAY b

²met·ro \'me-(,)trō\ *adjective* (1950)
: METROPOLITAN 2

me·trol·o·gy \me-'trä-lə-jē\ *noun* [French *métrologie,* from Greek *metrologia* theory of ratios, from *metron* measure — more at MEASURE] (1816)
1 : the science of weights and measures or of measurement
2 : a system of weights and measures
— me·tro·log·i·cal \,me-trə-'lä-ji-kəl\ *adjective*
— me·trol·o·gist \me-'trä-lə-jist\ *noun*

\ə\ **abut** \ᵊ\ **kitten** \ər\ **further** \a\ **ash** \ā\ **ace** \ä\ **mop, mar** \au̇\ **out** \ch\ **chin** \e\ **bet** \ē\ **easy** \g\ **go** \i\ **hit** \ī\ **ice** \j\ **job** \ŋ\ **sing** \ō\ **go** \ȯ\ **law** \ȯi\ **boy** \th\ **thin** \t͟h\ **the** \ü\ **loot** \u̇\ **foot** \y\ **yet** \zh\ **vision** *see also* Guide to Pronunciation

met·ro·ni·da·zole \ˌme-trə-ˈnī-də-ˌzōl\ *noun* [*me*thyl + *-tron-* (probably alteration of *nitro-*) + im*ide* + *azole*] (1962)
: a drug $C_6H_9N_3O_3$ used especially in treating vaginal trichomoniasis

met·ro·nome \ˈme-trə-ˌnōm\ *noun* [Greek *metron* + *-nomos* controlling, from *nomos* law — more at NIMBLE] (1816)
: an instrument designed to mark exact time by a regularly repeated tick

metronome

met·ro·nom·ic \ˌme-trə-ˈnä-mik\ *also* **met·ro·nom·i·cal** \-mi-kəl\ *adjective* (1866)
: mechanically regular (as in action or tempo)
— **met·ro·nom·i·cal·ly** \-mi-k(ə-)lē\ *adverb*

me·trop·o·lis \mə-ˈträ-p(ə-)ləs\ *noun* [Middle English, from Late Latin, from Greek *mētropolis,* from *mētr-, mētēr* mother + *polis* city — more at MOTHER, POLICE] (14th century)
1 : the chief or capital city of a country, state, or region
2 : the city or state of origin of a colony (as of ancient Greece)
3 a : a city regarded as a center of a specified activity ⟨a great business *metropolis*⟩ **b :** a large important city ⟨the world's great *metropolises* —P. E. James⟩

¹met·ro·pol·i·tan \ˌme-trə-ˈpä-lə-tᵊn\ *noun* (14th century)
1 : the primate of an ecclesiastical province
2 : one who lives in a metropolis or displays metropolitan manners or customs

²metropolitan *adjective* [Middle English, from Late Latin *metropolitanus* of the see of a metropolitan, from *metropolita,* noun, metropolitan, from Late Greek *mētropolitēs,* from *mētropolis* see of a metropolitan, from Greek, capital] (15th century)
1 : of or constituting a metropolitan or his see
2 : of, relating to, or characteristic of a metropolis and sometimes including its suburbs
3 : of, relating to, or constituting a mother country as distinguished from a colony

me·tror·rha·gia \ˌmē-trə-ˈrā-j(ē-)ə, -ˈrā-zhə; -ˈrä-jə, -zhə\ *noun* [New Latin, from *metro-* womb (from Greek *mētra,* from *mētr-, mētēr* mother) + *-rrhagia* — more at MOTHER] (1879)
: profuse uterine bleeding especially between menstrual periods

-metry *noun combining form* [Middle English *-metrie,* from Middle French, from Latin *-metria,* from Greek, from *metrein* to measure, from *metron* — more at MEASURE]
: art, process, or science of measuring ⟨chrono*metry*⟩ ⟨photo*metry*⟩

met·tle \ˈme-tᵊl\ *noun* [alteration of *metal*] (1581)
1 a : vigor and strength of spirit or temperament ⟨suspected to have more tongue in his head than *mettle* in his bosom —Sir Walter Scott⟩ **b :** staying quality **:** STAMINA ⟨trucks had proved their *mettle* in army transport —*Pioneer & Pacemaker*⟩
2 : quality of temperament or disposition ⟨gentlemen of brave *mettle* —Shakespeare⟩
synonym see COURAGE
— **met·tled** \-tᵊld\ *adjective*
— **on one's mettle :** aroused to do one's best

met·tle·some \ˈme-tᵊl-səm\ *adjective* (1662)
: full of mettle **:** SPIRITED

meu·nière \(ˌ)mə(r)n-ˈyer, mœ̄n-\ *adjective* [French (*à la*) *meunière,* literally, in the manner of a miller's wife] (1903)
: rolled lightly in flour and sautéed in butter ⟨sole *meunière*⟩

Meur·sault \mə(r)-ˈsō, mœ̄-\ *noun* [French, from *Meursault,* commune in France] (1833)
: a dry white burgundy wine

¹mew \ˈmyü\ *noun* [Middle English, from Old English *mǣw;* akin to Old Norse *mār* gull] (before 12th century)
: GULL; *especially* **:** the common European gull (*Larus canus*)

²mew *verb* [Middle English *mewen,* of imitative origin] (14th century)
intransitive verb
: to utter a mew or similar sound ⟨gulls *mewed* over the bay⟩
transitive verb
: to utter by mewing **:** MEOW

³mew *noun* (1596)
: MEOW

⁴mew *noun* [Middle English *mewe,* from Middle French *mue,* from *muer* to molt, from Latin *mutare* to change — more at MUTABLE] (14th century)
1 : an enclosure for trained hawks — usually used in plural
2 : a place for hiding or retirement
3 *plural but singular or plural in construction, chiefly British* **a** (1) **:** stables usually with living quarters built around a court (2) **:** living quarters adapted from such stables **b :** back street **:** ALLEY

⁵mew *transitive verb* (15th century)
: to shut up **:** CONFINE — often used with *up*

mewl \ˈmyü(ə)l\ *intransitive verb* [imitative] (1600)
: to cry weakly **:** WHIMPER

Mex·i·can \ˈmek-si-kən\ *noun* (1604)
1 a : a native or inhabitant of Mexico **b :** a person of Mexican descent **c** *Southwest* **:** a person of mixed Spanish and Indian descent
2 : NAHUATL
— **Mexican** *adjective*

Mexican bean beetle *noun* (1921)
: a spotted ladybug (*Epilachna varivestis*) that feeds on the leaves of beans

Mexican jumping bean *noun* (circa 1929)
: JUMPING BEAN

Mexican Spanish *noun* (1945)
: the Spanish used in Mexico

Mexican standoff *noun* (1891)
: a situation in which no one emerges a clear winner; *also* **:** DEADLOCK

me·ze \me-ˈzā, ˈmā-(ˌ)zā\ *noun, plural* **mezes** *also* **meze** [New Greek & Turkish; New Greek *mezes,* from Turkish *meze,* perhaps from Arabic dialect *mazza, māzza*] (1926)
: an appetizer in Greek or Middle Eastern cuisine often served with an aperitif

me·ze·re·on \mə-ˈzir-ē-ən\ *noun* [Middle English *mizerion,* from Medieval Latin *mezereon,* from Arabic *māzariyūn,* from Persian] (15th century)
: a small European shrub (*Daphne mezereum* of the family Thymelaeaceae, the mezereon family) with fragrant lilac purple flowers and poisonous emetic leaves, fruit, and bark

me·zu·zah *or* **me·zu·za** \mə-ˈzu̇-zə\ *noun, plural* **-zahs** *or* **-zas** *or* **-zot** \-ˈzu̇-ˌzōt\ [Hebrew *mĕzūzāh* doorpost] (1650)
: a small parchment scroll inscribed with Deuteronomy 6:4–9 and 11:13–21 and the name Shaddai and placed in a case fixed to the doorpost by some Jewish families as a sign and reminder of their faith

mez·za·nine \ˈme-zᵊn-ˌēn, ˌme-zᵊn-ˈ\ *noun* [French, from Italian *mezzanino,* from *mezzano* middle, from Latin *medianus* middle, median] (1711)
1 : a low-ceilinged story between two main stories of a building; *especially* **:** an intermediate story that projects in the form of a balcony
2 a : the lowest balcony in a theater **b :** the first few rows of such a balcony

mez·za vo·ce \ˌmet-sä-ˈvō-(ˌ)chā, ˌmed-zä-\ *adverb or adjective* [Italian, half voice] (1775)
: with medium or half volume — used as a direction in music

mez·zo \ˈmet-(ˌ)sō, ˈmed-(ˌ)zō\ *noun, plural* **mezzos** [Italian, literally, middle, moderate, half, from Latin *medius* — more at MID] (1832)
: MEZZO-SOPRANO

mez·zo for·te \ˌmet-(ˌ)sō-ˈfȯr-ˌtā, ˌmed-(ˌ)zō-, -ˈfȯr-tē\ *adjective or adverb* [Italian] (1811)
: moderately loud — used as a direction in music

mez·zo pia·no \-pē-ˈä-(ˌ)nō\ *adjective or adverb* [Italian] (1811)
: moderately soft — used as a direction in music

mez·zo–re·lie·vo *or* **mez·zo–ri·lie·vo** \-ri-ˈlē-(ˌ)vō, -rēl-ˈyā-(ˌ)vō\ *noun, plural* **-vos** [Italian *mezzorilievo,* from *mezzo* + *rilievo* relief] (1598)
: sculptural relief intermediate between bas-relief and high relief

mez·zo–so·pra·no \-sə-ˈpra-(ˌ)nō, -ˈprä-\ *noun* [Italian *mezzosoprano,* from *mezzo* + *soprano* soprano] (circa 1753)
1 : a woman's voice with a range between that of the soprano and contralto
2 : a singer having a mezzo-soprano voice

mez·zo·tint \ˈmet-sō-ˌtint, ˈmed-zō-\ *noun* [modification of Italian *mezzatinta,* from *mezza* (feminine of *mezzo*) + *tinta* tint] (1738)
1 : a manner of engraving on copper or steel by scraping or burnishing a roughened surface to produce light and shade
2 : an engraving produced by mezzotint

mho \ˈmō\ *noun, plural* **mhos** [backward spelling of *ohm*] (1883)
: a unit of conductance equal to the reciprocal of the ohm **:** SIEMENS

mi \ˈmē\ *noun* [Medieval Latin, from the syllable sung to this note in a medieval hymn to Saint John the Baptist] (15th century)
: the 3d tone of the diatonic scale in solmization

MIA \ˌem-(ˌ)ī-ˈā\ *noun* [*m*issing *i*n *a*ction] (1944)
: a member of the armed forces whose whereabouts following a combat mission are unknown and whose death cannot be established beyond reasonable doubt

Mi·ami \mī-ˈa-mē, -mə\ *noun, plural* **Mi·ami** *or* **Mi·am·is** (1722)
: a member of an American Indian people originally of Wisconsin and Indiana

mi·aow \mē-ˈau̇\ *variant of* MEOW

mi·as·ma \mī-ˈaz-mə, mē-\ *noun, plural* **-mas** *also* **-ma·ta** \-mə-tə\ [New Latin, from Greek, defilement, from *miainein* to pollute] (1665)
1 : a vaporous exhalation formerly believed to cause disease; *also* **:** a heavy vaporous emanation or atmosphere ⟨a *miasma* of tobacco smoke⟩
2 : an influence or atmosphere that tends to deplete or corrupt ⟨freed from the *miasma* of poverty —Sir Arthur Bryant⟩; *also* **:** an atmosphere that obscures **:** FOG ⟨retreated into an asexual mental *miasma* —*Times Literary Supplement*⟩
— **mi·as·mal** \-məl\ *adjective*
— **mi·as·mat·ic** \ˌmī-əz-ˈma-tik\ *adjective*
— **mi·as·mic** \mī-ˈaz-mik, mē-\ *adjective*
— **mi·as·mi·cal·ly** \-mi-k(ə-)lē\ *adverb*

mi·ca \ˈmī-kə\ *noun* [New Latin, from Latin, grain, crumb; perhaps akin to Greek *mikros* small] (1777)
: any of various colored or transparent mineral silicates crystallizing in monoclinic forms that readily separate into very thin leaves
— **mi·ca·ceous** \mī-ˈkā-shəs\ *adjective*

Mi·cah \ˈmī-kə\ *noun* [Hebrew *Mīkhāh,* short for *Mīkhāyāh*]
1 : a Hebrew prophet of the 8th century B.C.
2 : a prophetic book of canonical Jewish and Christian Scripture — see BIBLE table

Mi·caw·ber \mi-ˈkȯ-bər, -ˈkä-\ *noun* [Wilkins *Micawber,* character in the novel *David Copperfield* (1849–50) by Charles Dickens] (1852)
: one who is poor but lives in optimistic expectation of better fortune
— **Mi·caw·ber·ish** \-bə-rish\ *adjective*

mice *plural of* MOUSE

mi·celle \mī-ˈsel\ *noun* [New Latin *micella,* from Latin *mica*] (1881)
: a unit of structure built up from polymeric molecules or ions: as **a :** an ordered region in a fiber (as of cellulose or rayon) **b :** a molecular aggregate that constitutes a colloidal particle
— **mi·cel·lar** \-ˈse-lər\ *adjective*

Mi·chael \ˈmī-kəl\ *noun* [Hebrew *Mīkhāʼēl*]
: one of the four archangels named in Hebrew tradition

Mi·chae·lis constant \mī-ˈkā-ləs-, mə-\ *noun* [Leonor *Michaelis* (died 1949) American biochemist] (1949)
: a constant that is a measure of the kinetics of an enzyme reaction and that is equivalent to the concentration of substrate at which the reaction takes place at one half its maximum rate

Mich·ael·mas \ˈmi-kəl-məs\ *noun* [Middle English *mychelmesse,* from Old English *Michaeles mæsse* Michael's mass] (before 12th century)
: September 29 celebrated as the feast of Saint Michael the Archangel

Michaelmas daisy *noun* (1785)
: a wild aster; *especially* **:** one blooming about Michaelmas

Mi·che·as \ˈmī-kē-əs, mī-ˈ\ *noun* [Late Latin *Michaeas,* from Greek *Michaias,* from Hebrew *Mīkhāyāh*]
: MICAH

mick \ˈmik\ *noun, often capitalized* [*Mick,* nickname for *Michael,* common Irish given name] (1856)
: IRISHMAN — often taken to be offensive

Mick·ey Finn \ˌmi-kē-ˈfin\ *noun* [probably from the name *Mickey Finn*] (1928)
: a drink of liquor doctored with a purgative or a drug

Mickey Mouse \ˈmi-kē-ˈmau̇s\ *adjective* [*Mickey Mouse,* cartoon character created by Walt Disney] (1936)
1 *often not capitalized* **:** being or performing insipid or corny popular music
2 : lacking importance **:** INSIGNIFICANT ⟨*Mickey Mouse* courses, where you don't work too hard —Willie Cager⟩
3 : annoyingly petty ⟨*Mickey Mouse* regulations⟩

mick·le \ˈmi-kəl\ *adjective* [Middle English *mikel,* from Old English *micel* — more at MUCH] (before 12th century)
chiefly Scottish **:** GREAT, MUCH
— **mickle** *adverb, chiefly Scottish*

Mic·mac \ˈmik-ˌmak\ *noun, plural* **Micmac** *or* **Micmacs** [Micmac *mìkmaw*] (1830)
1 : a member of an American Indian people of eastern Canada
2 : the Algonquian language of the Micmac people

micr- *or* **micro-** *combining form* [Middle English *micro-,* from Latin, from Greek *mikr-, mikro-,* from *mikros, smikros* small, short; perhaps akin to Old English *smēalīc* careful, exquisite]
1 a : small **:** minute ⟨*micro*capsule⟩ **b :** used for or involving minute quantities or variations ⟨*micro*barograph⟩
2 : one millionth part of a (specified) unit ⟨*micro*gram⟩
3 a : using microscopy ⟨*micro*dissection⟩ **:** used in microscopy **b :** revealed by or having the structure discernible only by microscopic examination ⟨*micro*organism⟩
4 : abnormally small ⟨*micro*cyte⟩
5 : of or relating to a small area ⟨*micro*climate⟩
6 : employed in or connected with microphotographing or microfilming ⟨*micro*copy⟩

[1]**mi·cro** \ˈmī-(ˌ)krō\ *adjective* [*micr-*] (1923)
1 : very small; *especially* **:** MICROSCOPIC
2 : involving minute quantities or variations

[2]**micro** *noun, plural* **micros** (1971)
1 : MICROCOMPUTER
2 : MICROPROCESSOR

mi·cro·am·pere \ˌmī-krō-ˈam-ˌpir\ *noun* (circa 1890)
: a unit of current equal to one millionth of an ampere

mi·cro·anal·y·sis \-ə-ˈna-lə-səs\ *noun* (1856)
: chemical analysis on a small or minute scale that usually requires special, very sensitive, or small-scale apparatus
— **mi·cro·an·a·lyst** \-ˈa-nᵊl-ist\ *noun*
— **mi·cro·an·a·lyt·i·cal** \-ˌa-nᵊl-ˈi-ti-kəl\ *also* **mi·cro·an·a·lyt·ic** \-ˌa-nᵊl-ˈi-tik\ *adjective*

mi·cro·anat·o·my \-ə-ˈna-tə-mē\ *noun* (circa 1899)
: HISTOLOGY
— **mi·cro·an·a·tom·i·cal** \-ˌa-nə-ˈtä-mi-kəl\ *adjective*

mi·cro·bal·ance \ˈmī-krō-ˌba-lən(t)s\ *noun* (1903)
: a balance designed to measure very small weights

mi·cro·baro·graph \ˌmī-krō-ˈbar-ə-ˌgraf\ *noun* [International Scientific Vocabulary] (1904)
: a barograph for recording small and rapid changes

mi·crobe \ˈmī-ˌkrōb\ *noun* [International Scientific Vocabulary *micr-* + Greek *bios* life — more at QUICK] (1881)
: MICROORGANISM, GERM
— **mi·cro·bi·al** \mī-ˈkrō-bē-əl\ *also* **mi·cro·bic** \-bik\ *adjective*

mi·cro·beam \ˈmī-krō-ˌbēm\ *noun* (1950)
: a beam of radiation of small cross section ⟨a focused laser *microbeam*⟩ ⟨a *microbeam* of electrons⟩

mi·cro·bi·ol·o·gy \ˌmī-krō-bī-ˈä-lə-jē\ *noun* [International Scientific Vocabulary] (1888)
: a branch of biology dealing especially with microscopic forms of life
— **mi·cro·bi·o·log·i·cal** \-ˌbī-ə-ˈlä-ji-kəl\ *also* **mi·cro·bi·o·log·ic** \-ˈlä-jik\ *adjective*
— **mi·cro·bi·o·log·i·cal·ly** \-ji-k(ə-)lē\ *adverb*
— **mi·cro·bi·ol·o·gist** \-bī-ˈä-lə-jist\ *noun*

mi·cro·brew \ˈmī-krō-ˌbrü\ *noun* (1987)
: a beer produced by a microbrewery

mi·cro·brew·ery \ˌmī-krō-ˈbrü-ə-rē, -ˈbru̇(-ə)r-ē\ *noun* (1984)
: a small brewery making specialty beer in limited quantities
— **mi·cro·brew·er** \ˈmī-krō-ˌbrü-ər, -ˌbru̇(-ə)r\ *noun*
— **mi·cro·brew·ing** \-ˌbrü-iŋ\ *noun*

mi·cro·burst \ˈmī-krō-ˌbərst\ *noun* (1982)
: a violent short-lived localized downdraft that creates extreme wind shears at low altitudes and is usually associated with thunderstorms

mi·cro·bus \-ˌbəs\ *noun* (1945)
: a station wagon shaped like a bus

mi·cro·cal·o·rim·e·ter \-ˌka-lə-ˈri-mə-tər\ *noun* (1911)
: an instrument for measuring very small quantities of heat
— **mi·cro·ca·lo·ri·met·ric** \-ˌka-lə-rə-ˈme-trik; -kə-ˌlȯr-ə-, -ˌlōr-, -ˌlär-\ *adjective*
— **mi·cro·cal·o·rim·e·try** \-ˌka-lə-ˈri-mə-trē\ *noun*

mi·cro·cap·sule \ˈmī-krō-ˌkap-səl, -(ˌ)sül *also* -ˌsyü(ə)l\ *noun* (1961)
: a tiny capsule containing material (as an adhesive or a medicine) that is released when the capsule is broken, melted, or dissolved

mi·cro·cas·sette \ˌmī-krō-kə-ˈset, -ka-\ *noun* (1979)
: a small cassette of magnetic tape that is used especially for dictation

[1]**mi·cro·ce·phal·ic** \-sə-ˈfa-lik\ *adjective* (circa 1856)
: having a small head; *specifically* **:** having an abnormally small head

[2]**microcephalic** *noun* (circa 1873)
: one that is microcephalic

mi·cro·ceph·a·ly \-ˈse-fə-lē\ *noun* [New Latin *microcephalia,* from *microcephalus* microcephalic, from *micr-* + Greek *kephalē* head — more at CEPHALIC] (1863)
: a condition of abnormal smallness of the head usually associated with mental defects

mi·cro·chip \ˈmī-krō-ˌchip\ *noun* (1969)
: INTEGRATED CIRCUIT

mi·cro·cir·cuit \-ˌsər-kət\ *noun* (1959)
: a compact electronic circuit **:** INTEGRATED CIRCUIT
— **mi·cro·cir·cuit·ry** \ˌmī-krō-ˈsər-kə-trē\ *noun*

mi·cro·cir·cu·la·tion \ˌmī-krō-ˌsər-kyə-ˈlā-shən\ *noun* (1959)
: blood circulation in the microvascular system; *also* **:** the microvascular system itself
— **mi·cro·cir·cu·la·to·ry** \-ˈsər-kyə-lə-ˌtȯr-ē, -ˌtȯr-\ *adjective*

mi·cro·cli·mate \ˈmī-krō-ˌklī-mət\ *noun* [International Scientific Vocabulary] (1925)
: the essentially uniform local climate of a usually small site or habitat
— **mi·cro·cli·mat·ic** \ˌmī-krō-klī-ˈma-tik\ *adjective*

mi·cro·cline \ˈmī-krō-ˌklīn\ *noun* [German *Mikroklin,* from *mikr-* micr- + Greek *klinein* to lean — more at LEAN] (1849)
: a triclinic mineral of the feldspar group that is like orthoclase in composition

mi·cro·coc·cus \ˌmī-krō-ˈkä-kəs\ *noun, plural* **-coc·ci** \-ˈkäk-(ˌ)sī, -(ˌ)sē; -ˈkä-(ˌ)kī, -(ˌ)kē\ [New Latin] (1870)
: a small spherical bacterium; *especially* **:** any of a genus (*Micrococcus*) of gram-positive chiefly harmless bacteria that typically occur in irregular clusters
— **mi·cro·coc·cal** \-ˈkä-kəl\ *adjective*

mi·cro·code \ˈmī-krə-ˌkōd\ *noun* (circa 1962)
: the microinstructions especially of a microprocessor

mi·cro·com·put·er \ˈmī-krō-kəm-ˌpyü-tər\ *noun* (1971)
1 : a very small computer that uses a microprocessor to handle information
2 : MICROPROCESSOR

mi·cro·copy \ˈmī-krō-ˌkä-pē\ *noun* [International Scientific Vocabulary] (1935)
: a photographic copy in which graphic matter is reduced in size (as on microfilm)

mi·cro·cosm \ˈmī-krə-ˌkä-zəm\ *noun* [Middle English, from Medieval Latin *microcosmus,* modification of Greek *mikros kosmos*] (15th century)
1 : a little world; *especially* **:** the human race or human nature seen as an epitome of the world or the universe
2 : a community or other unity that is an epitome of a larger unity
— **mi·cro·cos·mic** \ˌmī-krə-ˈkäz-mik\ *adjective*
— **mi·cro·cos·mi·cal·ly** \-mi-k(ə-)lē\ *adverb*
— **in microcosm :** in a greatly diminished size, form, or scale

microcosmic salt *noun* (1783)
: a white crystalline salt $NaNH_5PO_4 \cdot 4H_2O$ used as a flux in testing for metallic oxides and salts

mi·cro·cos·mos \ˈmī-krə-ˌkäz-məs, -ˌmōs, -ˌmäs\ *noun* [Middle English *mycrocossmos,* from Medieval Latin *microcosmus*] (13th century)
1 : MICROCOSM
2 : the microscopic or submicroscopic world

mi·cro·crys·tal \ˈmī-krō-ˌkris-tᵊl\ *noun* (1886)
: a crystal visible only under the microscope
— **mi·cro·crys·tal·line** \ˌmī-krō-ˈkris-tə-lən *also* -ˌlīn *or* -ˌlēn\ *adjective*
— **mi·cro·crys·tal·lin·i·ty** \-ˌkris-tə-ˈli-nə-tē\ *noun*

mi·cro·cul·ture \ˈmī-krō-ˌkəl-chər\ *noun* (1892)

1 : a microscopic culture of cells or organisms **2 :** the culture of a small group of human beings with limited perspective
— **mi·cro·cul·tur·al** \ˌmī-krō-ˈkəlch-rəl, -ˈkəl-chə-\ *adjective*

mi·cro·cu·rie \ˈmī-krō-ˌkyu̇r-ē, ˌmī-krō-kyu̇-ˈrē\ *noun* (1911)
: a unit of quantity or of radioactivity equal to one millionth of a curie

mi·cro·cyte \ˈmī-krə-ˌsīt\ *noun* [International Scientific Vocabulary] (1876)
: a small red blood cell present especially in some anemias
— **mi·cro·cyt·ic** \ˌmī-krə-ˈsi-tik\ *adjective*

mi·cro·den·si·tom·e·ter \ˈmī-krō-ˌden(t)-sə-ˈtä-mə-tər\ *noun* (1935)
: a densitometer for measuring the densities of very small areas of a photographic film or plate
— **mi·cro·den·si·to·met·ric** \-sə-tə-ˈme-trik\ *adjective*
— **mi·cro·den·si·tom·e·try** \-sə-ˈtä-mə-trē\ *noun*

mi·cro·dis·sec·tion \ˌmī-krō-di-ˈsek-shən, -dī-\ *noun* (1915)
: dissection under the microscope; *specifically* **:** dissection of cells and tissues by means of fine needles that are precisely manipulated by levers

mi·cro·dot \ˈmī-krə-ˌdät, -krō-\ *noun* (1946)
: a photographic reproduction of printed matter reduced to the size of a dot for ease or security of transmittal

mi·cro·earth·quake \ˌmī-krō-ˈərth-ˌkwāk\ *noun* (1965)
: an earthquake of low intensity

mi·cro·eco·nom·ics \-ˌe-kə-ˈnä-miks, -ˌē-kə-\ *noun plural but usually singular in construction* (1947)
: a study of economics in terms of individual areas of activity (as a firm, household, or prices) — compare MACROECONOMICS
— **mi·cro·eco·nom·ic** \-ˈnä-mik\ *adjective*

mi·cro·elec·trode \ˌmī-krō-i-ˈlek-ˌtrōd\ *noun* (1917)
: a minute electrode; *especially* **:** one that is inserted in a living biological cell or tissue in studying its electrical characteristics

mi·cro·elec·tron·ics \-i-ˌlek-ˈträ-niks\ *noun plural* (1958)
1 *singular in construction* **:** a branch of electronics that deals with the miniaturization of electronic circuits and components
2 : devices, equipment, or circuits produced using the methods of microelectronics
— **mi·cro·elec·tron·ic** \-nik\ *adjective*
— **mi·cro·elec·tron·i·cal·ly** \-ni-k(ə-)lē\ *adverb*

mi·cro·elec·tro·pho·re·sis \-i-ˌlek-trə-fə-ˈrē-səs\ *noun* [New Latin] (1936)
: electrophoresis in which the movement of single particles is observed in a microscope; *also* **:** electrophoresis in which micromethods are used
— **mi·cro·elec·tro·pho·ret·ic** \-ˈre-tik\ *adjective*
— **mi·cro·elec·tro·pho·ret·i·cal·ly** \-ti-k(ə-)lē\ *adverb*

mi·cro·el·e·ment \-ˈe-lə-mənt\ *noun* (1936)
: TRACE ELEMENT

mi·cro·en·cap·su·late \-in-ˈkap-sə-ˌlāt\ *transitive verb* (1963)
: to enclose in a microcapsule ⟨*microencapsulated* aspirin⟩
— **mi·cro·en·cap·su·la·tion** \-in-ˌkap-sə-ˈlā-shən\ *noun*

mi·cro·en·vi·ron·ment \-in-ˈvī-rən-mənt, -ˈvī(-ə)rn-\ *noun* (1938)
: a small or relatively small usually distinctly specialized and effectively isolated habitat (as a forest canopy) or environment (as of a nerve cell)
— **mi·cro·en·vi·ron·men·tal** \-ˌvī-rən-ˈmen-tᵊl\ *adjective*

mi·cro·evo·lu·tion \-ˌe-və-ˈlü-shən *also* -ˌē-və-\ *noun* (1940)
: comparatively minor evolutionary change involving the accumulation of variations in populations usually below the species level
— **mi·cro·evo·lu·tion·ary** \-shə-ˌner-ē\ *adjective*

mi·cro·far·ad \ˈmī-krō-ˌfar-ˌad, -ˌfar-əd\ *noun* (1873)
: a unit of capacitance equal to one millionth of a farad

mi·cro·fau·na \ˌmī-krō-ˈfȯ-nə, -ˈfä-\ *noun* [New Latin] (1902)
1 : minute animals; *especially* **:** those invisible to the naked eye ⟨the soil *microfauna*⟩
2 : a small or strictly localized fauna (as of a microenvironment)
— **mi·cro·fau·nal** \-ˈfȯ-nᵊl, -ˈfä-\ *adjective*

mi·cro·fi·bril \-ˈfī-brəl, -ˈfi-\ *noun* (1938)
: a fine fibril; *especially* **:** one of the submicroscopic elongated bundles of cellulose of a plant cell wall
— **mi·cro·fi·bril·lar** \-brə-lər\ *adjective*

mi·cro·fiche \ˈmī-krō-ˌfēsh, -ˌfish\ *noun, plural* **-fiche** *or* **-fiches** \-ˌfē-shəz, -ˌfēsh, -ˌfi-shəz, -ˌfish\ [French, from *micr-* micr- + *fiche* fiche] (1950)
: a sheet of microfilm containing rows of microimages of pages of printed matter

mi·cro·fil·a·ment \ˌmī-krō-ˈfi-lə-mənt\ *noun* (1963)
: any of the minute actin-containing protein filaments of eukaryotic cytoplasm that function in maintaining structure and in intracellular movement

mi·cro·fi·lar·ia \-fə-ˈlar-ē-ə, -ˈler-\ *noun* [New Latin] (1878)
: a minute larval filaria
— **mi·cro·fi·lar·i·al** \-ē-əl\ *adjective*

¹mi·cro·film \ˈmī-krə-ˌfilm\ *noun* [International Scientific Vocabulary] (1927)
: a film bearing a photographic record on a reduced scale of printed or other graphic matter

²microfilm (1937)
transitive verb
: to reproduce on microfilm
intransitive verb
: to make microfilms
— **mi·cro·film·able** \ˌmī-krə-ˈfil-mə-bəl\ *adjective*
— **mi·cro·film·er** \ˈmī-krə-ˌfil-mər\ *noun*

mi·cro·flo·ra \ˌmī-krə-ˈflōr-ə, -ˈflȯr-\ *noun* [New Latin] (1904)
1 : minute plants; *especially* **:** those invisible to the naked eye
2 : a small or strictly localized flora (as of a microenvironment)
— **mi·cro·flo·ral** \-əl\ *adjective*

mi·cro·form \ˈmī-krə-ˌfȯrm\ *noun* (1958)
1 : a process for reproducing printed matter in a much reduced size ⟨documents in *microform*⟩
2 a : matter reproduced by microform **b :** MICROCOPY

mi·cro·fos·sil \ˌmī-krō-ˈfä-səl\ *noun* (1924)
: a fossil that can be studied only microscopically and that may be either a fragment of a larger organism or an entire minute organism

mi·cro·fun·gus \-ˈfəŋ-gəs\ *noun* [New Latin] (1874)
: a fungus (as a mold) with a microscopic fruiting body

mi·cro·ga·mete \-ˈga-ˌmēt *also* -gə-ˈmēt\ *noun* [International Scientific Vocabulary] (circa 1891)
: the smaller and usually male gamete of a heterogamous organism

mi·cro·ga·me·to·cyte \-gə-ˈmē-tə-ˌsīt\ *noun* [International Scientific Vocabulary] (1902)
: a gametocyte producing microgametes

mi·cro·gram \ˈmī-krə-ˌgram\ *noun* [International Scientific Vocabulary] (circa 1890)
: one millionth of a gram — see METRIC SYSTEM table

mi·cro·graph \-ˌgraf\ *noun* [International Scientific Vocabulary] (1904)
: a graphic reproduction of the image of an object formed by a microscope
— **micrograph** *transitive verb*

mi·cro·graph·ics \ˌmī-krə-ˈgra-fiks\ *noun plural but singular in construction* (1969)
: the industry concerned with the manufacture and sale of graphic material in microform; *also* **:** the production of such material
— **mi·cro·graph·ic** \-fik\ *adjective*
— **mi·cro·graph·i·cal·ly** \-fi-k(ə-)lē\ *adverb*

mi·cro·grav·i·ty \-ˈgra-və-tē\ *noun* (1974)
: a condition in space in which only minuscule forces are experienced **:** virtual absence of gravity; *broadly* **:** a condition of weightlessness

mi·cro·groove \ˈmī-krō-ˌgrüv\ *noun* (1948)
: a narrow continuous V-shaped spiral track that has closely spaced turns and that is used on long-playing records

mi·cro·hab·i·tat \ˌmī-krō-ˈha-bə-ˌtat\ *noun* (1933)
: the microenvironment in which an organism lives ⟨decaying wood creates a *microhabitat* for insects⟩

mi·cro·im·age \-ˈi-mij\ *noun* (1950)
: an image (as on a microfilm) that is greatly reduced in size

mi·cro·inch \-ˈinch\ *noun* (1941)
: one millionth of an inch

mi·cro·in·jec·tion \ˌmī-krō-in-ˈjek-shən\ *noun* (1921)
: injection under the microscope; *specifically* **:** injection into tissues by means of a fine mechanically controlled capillary tube
— **mi·cro·in·ject** \-ˈjekt\ *transitive verb*

mi·cro·in·struc·tion \-ˈstrək-shən\ *noun* (1959)
: a computer instruction that activates the circuits necessary to perform a single machine operation usually as part of the execution of a machine-language instruction

mi·cro·lep·i·dop·tera \ˌmī-krō-ˌle-pə-ˈdäp-tə-rə\ *noun plural* [New Latin] (1852)
: lepidopterous insects (as tortricids) that belong to families of minute or medium-sized moths
— **mi·cro·lep·i·dop·ter·ous** \-tə-rəs\ *adjective*

mi·cro·li·ter \ˈmī-krō-ˌlē-tər\ *noun* [International Scientific Vocabulary] (circa 1890)
: a unit of capacity equal to one millionth of a liter — see METRIC SYSTEM table

mi·cro·lith \ˈmī-krə-ˌlith\ *noun* [International Scientific Vocabulary] (1908)
: a tiny blade tool especially of the Mesolithic usually in a geometric shape (as that of a triangle) and often set in a bone or wooden haft
— **mi·cro·lith·ic** \ˌmī-krə-ˈli-thik\ *adjective*

mi·cro·man·age \ˌmī-krō-ˈma-nij\ *transitive verb* (1979)
: to manage with great or excessive control or attention to details
— **mi·cro·man·age·ment** \-mənt\ *noun*
— **mi·cro·man·ag·er** \-ˈma-ni-jər\ *noun*

mi·cro·ma·nip·u·la·tion \ˌmī-krō-mə-ˌni-pyə-ˈlā-shən\ *noun* (1921)
: the technique or practice of microdissection and microinjection

mi·cro·ma·nip·u·la·tor \-ˈni-pyə-ˌlā-tər\ *noun* (1921)
: an instrument for micromanipulation

mi·cro·mere \ˈmī-krō-ˌmir\ *noun* [International Scientific Vocabulary] (1877)
: a small blastomere — see BLASTULA illustration

mi·cro·me·te·or·ite \ˌmī-krō-ˈmē-tē-ə-ˌrīt\ *noun* (1949)
1 : a meteorite so small that it can pass through the earth's atmosphere without becoming intensely heated
2 : a very small particle in interplanetary space
— **mi·cro·me·te·or·it·ic** \-ˌmē-tē-ə-ˈri-tik\ *adjective*

mi·cro·me·te·or·oid \-ˈmē-tē-ə-ˌrȯid\ *noun* (1954)
: MICROMETEORITE 2

mi·cro·me·te·o·rol·o·gy \-ˌmē-tē-ə-ˈrä-lə-jē\ *noun* (1930)
: meteorology that deals with small-scale weather systems ranging up to several kilometers in diameter and confined to the lower troposphere
— **mi·cro·me·te·o·ro·log·i·cal** \-ˌmē-tē-ˌȯr-ə-ˈlä-ji-kəl, -ˌär-ə-, -ə-rə-\ *adjective*
— **mi·cro·me·te·o·rol·o·gist** \-ˌmē-tē-ə-ˈrä-lə-jist\ *noun*

1mi·crom·e·ter \mī-ˈkrä-mə-tər\ *noun* [French *micromètre,* from *micr-* + *-mètre* -meter] (1670)
1 : an instrument used with a telescope or microscope for measuring minute distances
2 : a caliper for making precise measurements that has a spindle moved by a finely threaded screw

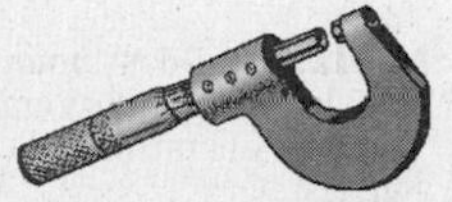
micrometer 2

2mi·cro·me·ter \ˈmī-krō-ˌmē-tər\ *noun* [International Scientific Vocabulary *micr-* + 3*meter*] (1880)
: a unit of length equal to one millionth of a meter — called also *micron*; see METRIC SYSTEM table

mi·cro·meth·od \ˈmī-krō-ˌme-thəd\ *noun* (1919)
: a method (as of microanalysis) that requires only very small quantities of material or that involves the use of the microscope

mi·cro·mini \ˌmī-krō-ˈmi-nē\ *noun* (1966)
: a very short miniskirt

mi·cro·min·i·a·ture \-ˈmi-nē-ə-ˌchu̇r, -ˈmi-ni-ˌchu̇r, -ˈmin-yə-, -chər, -ˌtyu̇r, -ˌtu̇r\ *adjective* (1958)
1 : MICROMINIATURIZED
2 : suitable for use with microminiaturized parts

mi·cro·min·i·a·tur·i·za·tion \-ˌmi-nē-ə-ˌchu̇r-ə-ˈzā-shən, -ˌmi-ni-ˌchu̇r-, -ˌmin-yə-ˌchu̇r-, -chər-, -ˌtyu̇r-, -ˌtu̇r-\ *noun* (1955)
: the process of producing microminiaturized things

mi·cro·min·i·a·tur·ized \-ˈmi-nē-ə-chə-ˌrīzd, -ˈmi-ni-chə-, -ˈmin-yə-chə-, -ˌtyu̇-, -ˌtu̇-\ *adjective* (1959)
: reduced to or produced in a very small size and especially in a size smaller than one considered miniature

mi·cro·mole \ˈmī-krə-ˌmōl\ *noun* [International Scientific Vocabulary] (1936)
: one millionth of a mole
— **mi·cro·mo·lar** \ˌmī-krə-ˈmō-lər\ *adjective*

mi·cro·mor·phol·o·gy \ˌmī-krō-mȯr-ˈfä-lə-jē\ *noun* (1945)
: MICROSTRUCTURE
— **mi·cro·mor·pho·log·i·cal** \-fə-ˈlä-ji-kəl\ *adjective*

mi·cron \ˈmī-ˌkrän\ *noun* [New Latin, from Greek *mikron,* neuter of *mikros* small — more at MICR-] (1885)
: 2MICROMETER

Mi·cro·ne·sian \ˌmī-krə-ˈnē-zhən, -shən\ *noun* (1847)
1 : a native or inhabitant of Micronesia
2 : a group of Austronesian languages spoken in the Micronesian islands
— **Micronesian** *adjective*

mi·cron·ize \ˈmī-krə-ˌnīz\ *transitive verb* **-ized; -iz·ing** [*micron*] (1940)
: to pulverize especially into particles a few micrometers in diameter

mi·cro·nu·cle·us \ˌmī-krō-ˈnü-klē-əs, -ˈnyü-\ *noun* [New Latin] (1892)
: a minute nucleus; *specifically* **:** one that is primarily concerned with reproductive and genetic functions in most ciliated protozoans

mi·cro·nu·tri·ent \-ˈnü-trē-ənt, -ˈnyü-\ *noun* (1939)
1 : TRACE ELEMENT
2 : an organic compound (as a vitamin) essential in minute amounts to the growth and health of an animal

mi·cro·or·gan·ism \-ˈȯr-gə-ˌni-zəm\ *noun* [International Scientific Vocabulary] (1880)
: an organism of microscopic or ultramicroscopic size

mi·cro·pa·le·on·tol·o·gy \-ˌpā-lē-ˌän-ˈtä-lə-jē, -lē-ən-, *especially British* -ˌpa-\ *noun* [International Scientific Vocabulary] (1883)
: the study of microscopic fossils
— **mi·cro·pa·le·on·to·log·i·cal** \-ˌän-tᵊl-ˈä-ji-kəl\ *also* **mi·cro·pa·le·on·to·log·ic** \-jik\ *adjective*
— **mi·cro·pa·le·on·tol·o·gist** \-ˌän-ˈtä-lə-jist, -ən-\ *noun*

mi·cro·par·ti·cle \-ˈpär-ti-kəl\ *noun* (1929)
: a very small particle; *especially* **:** one that is microscopic in size

mi·cro·phage \ˈmī-krə-ˌfāj *also* -ˌfäzh\ *noun* [International Scientific Vocabulary] (1890)
: a small phagocyte

mi·cro·phone \ˈmī-krə-ˌfōn\ *noun* [International Scientific Vocabulary] (1878)
: an instrument whereby sound waves are caused to generate or modulate an electric current usually for the purpose of transmitting or recording sound (as speech or music)
— **mi·cro·phon·ic** \ˌmī-krə-ˈfä-nik\ *adjective*

mi·cro·phon·ics \ˌmī-krə-ˈfä-niks\ *noun plural* (1929)
: noises in a loudspeaker caused by mechanical shock or vibration of the electronic components

mi·cro·pho·to·graph \-ˈfō-tə-ˌgraf\ *noun* [International Scientific Vocabulary] (1858)
1 : a small photograph that is normally magnified for viewing **:** MICROCOPY
2 : PHOTOMICROGRAPH
— **microphotograph** *transitive verb*
— **mi·cro·pho·tog·ra·pher** \-fə-ˈtä-grə-fər\ *noun*
— **mi·cro·pho·to·graph·ic** \-ˌfō-tə-ˈgra-fik\ *adjective*
— **mi·cro·pho·tog·ra·phy** \-fə-ˈtä-grə-fē\ *noun*

mi·cro·pho·tom·e·ter \-fō-ˈtä-mə-tər\ *noun* [International Scientific Vocabulary] (1899)
: an instrument for measuring the amount of light transmitted or reflected by small areas or for measuring the relative densities of spectral lines on a photographic film or plate
— **mi·cro·pho·to·met·ric** \-ˌfō-tə-ˈme-trik\ *adjective*
— **mi·cro·pho·to·met·ri·cal·ly** \-tri-k(ə-)lē\ *adverb*
— **mi·cro·pho·tom·e·try** \-fō-ˈtä-mə-trē\ *noun*

mi·cro·phyll \ˈmī-krə-ˌfil\ *noun* [International Scientific Vocabulary] (1935)
1 : a small leaf
2 : a leaf (as of a club moss) with single unbranched veins and no demonstrable gap around the leaf trace
— **mi·cro·phyl·lous** \ˌmī-krə-ˈfi-ləs\ *adjective*

mi·cro·phys·ics \ˌmī-krō-ˈfi-ziks\ *noun* (1885)
: the physics of molecules, atoms, and elementary particles
— **mi·cro·phys·i·cal** \-ˈfi-zi-kəl\ *adjective*
— **mi·cro·phys·i·cal·ly** \-k(ə-)lē\ *adverb*

mi·cro·pi·pette *or* **mi·cro·pi·pet** \-pī-ˈpet\ *noun* (1918)
1 : a pipette for the measurement of minute volumes
2 : a small and extremely fine-pointed pipette used in making microinjections

mi·cro·plank·ton \-ˈplaŋ(k)-tən, -ˌtän\ *noun* [International Scientific Vocabulary] (1903)
: microscopic plankton

mi·cro·pore \ˈmī-krə-ˌpōr, -ˌpȯr\ *noun* [International Scientific Vocabulary] (1884)
: a very fine pore
— **mi·cro·po·ros·i·ty** \-pə-ˈrä-sə-tē, -pȯ-\ *noun*
— **mi·cro·po·rous** \ˌmī-krə-ˈpōr-əs, -ˈpȯr-\ *adjective*

mi·cro·prism \ˈmī-krə-ˌpri-zəm\ *noun* (1966)
: a usually circular area on the focusing screen of a camera that is made up of tiny prisms and that causes the image in the viewfinder to blur if the subject is not in focus

mi·cro·probe \-ˌprōb\ *noun* (1944)
: a device for microanalysis that operates by exciting radiation in a minute area of material so that the composition may be determined from the emission spectrum

mi·cro·pro·ces·sor \ˌmī-krō-ˈprä-ˌse-sər, -ˈprō-\ *noun* (1970)
: a computer processor contained on an integrated-circuit chip; *also* **:** such a processor with memory and associated circuits

mi·cro·pro·gram \-ˈprō-ˌgram, -grəm\ *noun* (1953)
: a routine composed of microinstructions used in microprogramming

mi·cro·pro·gram·ming \-ˌgra-miŋ\ *noun* (1953)
: the use of routines stored in memory rather than specialized circuits to control a device (as a computer)

mi·cro·pro·jec·tor \-prə-ˈjek-tər\ *noun* (1927)
: a projector utilizing a compound microscope for projecting on a screen a greatly enlarged image of a microscopic object
— **mi·cro·pro·jec·tion** \-ˈjek-shən\ *noun*

mi·cro·pub·lish·ing \-ˈpə-bli-shiŋ\ *noun* (1966)
: publishing in microform
— **mi·cro·pub·lish·er** \-bli-shər\ *noun*

mi·cro·pul·sa·tion \-ˌpəl-ˈsā-shən\ *noun* (1949)
: a pulsation having a short period ⟨a *micropulsation* of the earth's magnetic field with a period in the range from a fraction of a second to several hundred seconds⟩

mi·cro·punc·ture \-ˈpəŋ(k)-chər\ *noun* (1948)
: an extremely small puncture (as of a nephron); *also* **:** an act of making a micropuncture

mi·cro·pyle \ˈmī-krə-ˌpīl\ *noun* [French, from *micr-* + Greek *pylē* gate] (1821)
1 : a minute opening in the integument of an ovule of a seed plant
2 : a differentiated area of surface in an egg through which a sperm enters
— **mi·cro·py·lar** \ˌmī-krə-ˈpī-lər\ *adjective*

mi·cro·quake \ˈmī-krō-ˌkwāk\ *noun* (1967)
: MICROEARTHQUAKE

mi·cro·ra·di·og·ra·phy \ˌmī-krō-ˌrā-dē-ˈä-grə-fē\ *noun* (1913)
: radiography in which an X-ray photograph is prepared showing minute internal structure
— **mi·cro·ra·dio·graph** \-ˈrā-dē-ə-ˌgraf\ *noun*
— **mi·cro·ra·dio·graph·ic** \-ˌrā-dē-ə-ˈgra-fik\ *adjective*

mi·cro·read·er \ˈmī-krō-ˌrē-dər\ *noun* (1949)
: an apparatus that gives an enlarged image of a microphotograph especially for reading

mi·cro·re·pro·duc·tion \ˌmī-krō-ˌrē-prə-ˈdək-shən\ *noun* (1938)
: the reproduction of written or printed matter in microform; *also* **:** an item so reproduced

mi·cro·scale \ˈmī-krō-ˌskāl\ *noun* (1931)
: a very small scale

mi·cro·scope \ˈmī-krə-ˌskōp\ *noun* [New Latin *microscopium,* from *micr-* + *-scopium* -scope] (1654)
1 : an optical instrument consisting of a lens or combination of lenses for making enlarged

images of minute objects; *especially* **:** COMPOUND MICROSCOPE
2 : an instrument using radiations other than light or using vibrations for making enlarged images of minute objects ⟨acoustic *microscope*⟩
mi·cro·scop·ic \ˌmī-krə-ˈskä-pik\ *or* **mi·cro·scop·i·cal** \-pi-kəl\ *adjective* (1732)
1 : resembling a microscope especially in perception
2 a : invisible or indistinguishable without the use of a microscope **b :** very small or fine or precise
3 : of, relating to, or conducted with the microscope or microscopy
— **mi·cro·scop·i·cal·ly** \-pi-k(ə-)lē\ *adverb*
mi·cros·co·py \mī-ˈkräs-kə-pē\ *noun* (circa 1665)
: the use of or investigation with the microscope
— **mi·cros·co·pist** \-pist\ *noun*
mi·cro·sec·ond \ˈmī-krō-ˌse-kənd, -kənt\ *noun* [International Scientific Vocabulary] (1906)
: one millionth of a second
mi·cro·seism \ˈmī-krə-ˌsī-zəm\ *noun* [International Scientific Vocabulary *micr-* + Greek *seismos* earthquake — more at SEISMIC] (1887)
: a feeble rhythmically and persistently recurring earth tremor
— **mi·cro·seis·mic** \ˌmī-krə-ˈsīz-mik, -ˈsīs-\ *adjective*
— **mi·cro·seis·mic·i·ty** \-sīz-ˈmi-sə-tē, -sīs-\ *noun*
mi·cro·some \ˈmī-krə-ˌsōm\ *noun* [German *Mikrosom,* from *mikr-* micr- + *-som* -some] (1885)
1 : any of various minute cellular structures (as a ribosome)
2 : a particle in a particulate fraction that is obtained by heavy centrifugation of broken cells and consists of various amounts of ribosomes, fragmented endoplasmic reticulum, and mitochondrial cristae
— **mi·cro·som·al** \ˌmī-krə-ˈsō-məl\ *adjective*
mi·cro·spec·tro·pho·tom·e·ter \ˌmī-krə-ˌspek-trə-fō-ˈtä-mə-tər\ *noun* (1949)
: a spectrophotometer adapted to the examination of light transmitted by a very small specimen (as a single organic cell)
— **mi·cro·spec·tro·pho·to·met·ric** \-ˌfō-tə-ˈme-trik\ *adjective*
— **mi·cro·spec·tro·pho·tom·e·try** \-fō-ˈtä-mə-trē\ *noun*
mi·cro·sphere \ˈmī-krə-ˌsfir\ *noun* (1894)
: a minute sphere
— **mi·cro·spher·i·cal** \ˌmī-krə-ˈsfir-i-kəl, -ˈsfer-\ *adjective*
mi·cro·spo·ran·gi·um \ˌmī-krō-spə-ˈran-jē-əm\ *noun* [New Latin] (1881)
: a sporangium that develops only microspores
— **mi·cro·spo·ran·gi·ate** \-jē-ət\ *adjective*
mi·cro·spore \ˈmī-krə-ˌspōr, -ˌspȯr\ *noun* [International Scientific Vocabulary] (1858)
: any of the spores in heterosporous plants that give rise to male gametophytes and are generally smaller than the megaspore
— **mi·cro·spo·rous** \ˌmī-krə-ˈspōr-əs, -ˈspȯr-; mī-ˈkräs-pə-rəs\ *adjective*
mi·cro·spo·ro·cyte \-ˈspōr-ə-ˌsīt, -ˈspȯr-\ *noun* (1940)
: a microspore mother cell
mi·cro·spo·ro·gen·e·sis \ˌmī-krə-ˌspōr-ə-ˈje-nə-səs, -ˌspȯr-\ *noun* [New Latin] (1921)
: the formation and maturation of microspores
mi·cro·spo·ro·phyll \-ˌfil\ *noun* (circa 1890)
: a sporophyll that develops only microsporangia
mi·cro·state \ˈmī-krō-ˌstāt\ *noun* (1962)
: a nation that is extremely small in area and population
mi·cro·struc·ture \ˈmī-krō-ˌstrək-chər\ *noun* [International Scientific Vocabulary] (1885)
: the microscopic structure of a material (as a mineral or a biological cell)
— **mi·cro·struc·tur·al** \ˌmī-krō-ˈstrək-chə-rəl, -ˈstrək-shrəl\ *adjective*
mi·cro·sur·gery \ˌmī-krō-ˈsərj-rē, -ˈsər-jə-\ *noun* (1926)
: minute dissection or manipulation (as by a micromanipulator or laser beam) of living structures or tissue
— **mi·cro·sur·gi·cal** \-ˈsər-ji-kəl\ *adjective*
mi·cro·switch \-ˌswich\ *noun* (1940)
: a very small switch that is sensitive to minute motions and is used especially in automatic devices
mi·cro·tech·nique \ˌmī-krō-tek-ˈnēk\ *also* **mi·cro·tech·nic** \-ˈtek-nik, -tek-ˈnēk\ *noun* [International Scientific Vocabulary] (1892)
: any of various methods of handling and preparing material for microscopic observation and study
mi·cro·tome \ˈmī-krə-ˌtōm\ *noun* [International Scientific Vocabulary] (1856)
: an instrument for cutting sections (as of organic tissues) for microscopic examination
mi·cro·tone \ˈmī-krə-ˌtōn\ *noun* (1920)
: a musical interval smaller than a halftone
— **mi·cro·ton·al** \ˌmī-krə-ˈtō-nᵊl\ *adjective*
— **mi·cro·to·nal·i·ty** \-tō-ˈna-lə-tē\ *noun*
— **mi·cro·ton·al·ly** \-ˈtō-nᵊl-ē\ *adverb*
mi·cro·tu·bule \ˌmī-krō-ˈtü-(ˌ)byü(ə)l, -ˈtyü-\ *noun* (1961)
: any of the minute tubules in eukaryotic cytoplasm that are composed of the protein tubulin and form an important component of the cytoskeleton, mitotic spindle, cilia, and flagella
— **mi·cro·tu·bu·lar** \-byə-lər\ *adjective*
mi·cro·vas·cu·lar \-ˈvas-kyə-lər\ *adjective* (1959)
: of, relating to, or constituting the part of the circulatory system made up of minute vessels (as venules or capillaries) that average less than 0.3 millimeters in diameter
— **mi·cro·vas·cu·la·ture** \-lə-ˌchu̇r, -ˌtyu̇r, -ˌtu̇r\ *noun*
mi·cro·vil·lus \-ˈvi-ləs\ *noun* [New Latin] (1953)
: a microscopic projection of a tissue, cell, or cell organelle; *especially* **:** any of the fingerlike outward projections of some cell surfaces
— **mi·cro·vil·lar** \-ˈvi-lər\ *adjective*
— **mi·cro·vil·lous** \-ˈvi-ləs\ *adjective*
mi·cro·volt \ˈmī-krə-ˌvōlt\ *noun* (1868)
: one millionth of a volt
mi·cro·watt \-ˌwät\ *noun* (circa 1909)
: one millionth of a watt
¹mi·cro·wave \-ˌwāv\ *noun, often attributive* (1931)
1 : a comparatively short electromagnetic wave; *especially* **:** one between about 1 millimeter and 1 meter in wavelength
2 : MICROWAVE OVEN
²microwave *transitive verb* (1973)
: to cook or heat in a microwave oven
— **mi·cro·wav·able** *or* **mi·cro·wave·able** \ˌmī-krə-ˈwā-və-bəl\ *adjective*
microwave oven *noun* (1963)
: an oven in which food is cooked by the heat produced by the absorption of microwave energy by water molecules in the food
mi·cro·world \-ˌwər(-ə)ld\ *noun* (1955)
: a small universe; *specifically* **:** the natural universe observed at the microscopic or submicroscopic level
mic·tu·rate \ˈmik-chə-ˌrāt, ˈmik-tə-\ *intransitive verb* **-rat·ed; -rat·ing** [Latin *micturire* to desire to urinate, from *meiere* to urinate; akin to Old English *mīgan* to urinate, Greek *omeichein*] (1842)
: URINATE
— **mic·tu·ri·tion** \ˌmik-chə-ˈri-shən, ˌmik-tə-\ *noun*
¹mid \ˈmid\ *adjective* [Middle English, from Old English *midde;* akin to Old High German *mitti* middle, Latin *medius,* Greek *mesos*] (before 12th century)
1 : being the part in the middle or midst ⟨in *mid* ocean⟩ — often used in combination ⟨*mid*-August⟩
2 : occupying a middle position ⟨the *mid* finger⟩
3 *of a vowel* **:** articulated with the arch of the tongue midway between its highest and its lowest elevation
— **mid** *adverb*
²mid *preposition* (1808)
: AMID
mid·air \ˈmid-ˈar, -ˈer\ *noun* (1667)
: a point or region in the air not immediately adjacent to the ground ⟨planes collided in *midair*⟩
Mi·das \ˈmī-dəs\ *noun* [Latin, from Greek]
: a legendary Phrygian king who is given the power of turning everything he touches to gold
Midas touch *noun* (1883)
: an uncanny ability for making money in every venture
mid·brain \ˈmid-ˌbrān\ *noun* (1875)
: the middle of the three primary divisions of the developing vertebrate brain or the corresponding part of the adult brain — called also *mesencephalon;* see BRAIN illustration
mid·course \-ˈkōrs, -ˈkȯrs\ *adjective* (circa 1956)
: being or occurring in the middle part of a course (as of a spacecraft) ⟨a *midcourse* correction⟩
mid·day \ˈmid-ˌdā, -ˈdā\ *noun, often attributive* (before 12th century)
: the middle of the day
mid·den \ˈmi-dᵊn\ *noun* [Middle English *midding,* of Scandinavian origin; akin to Old Norse *myki* dung & Old Norse *dyngja* manure pile — more at DUNG] (14th century)
1 : DUNGHILL
2 a : a refuse heap; *especially* **:** KITCHEN MIDDEN **b :** a small pile (as of seeds, bones, or leaves) gathered by a rodent (as a pack rat)
¹mid·dle \ˈmi-dᵊl\ *adjective* [Middle English *middel,* from Old English; akin to Old English *midde*] (before 12th century)
1 : equally distant from the extremes **:** MEDIAL, CENTRAL ⟨the *middle* house in the row⟩
2 : being at neither extreme **:** INTERMEDIATE
3 *capitalized* **a :** constituting a division intermediate between those prior and later or upper and lower ⟨*Middle* Paleozoic⟩ **b :** constituting a period of a language or literature intermediate between one called *Old* and one called *New* or *Modern* ⟨*Middle* Dutch⟩
4 *of a verb form or voice* **:** typically asserting that a person or thing both performs and is affected by the action represented
²middle *noun* (before 12th century)
1 : a middle part, point, or position
2 : the central portion of the human body **:** WAIST
3 : the position of being among or in the midst of something
4 : something intermediate between extremes **:** MEAN
5 : the center of an offensive or defensive formation; *especially* **:** the area between the second baseman and the shortstop
middle age *noun* (14th century)
: the period of life from about 40 to about 60
— **mid·dle–aged** \ˌmi-dᵊl-ˈājd\ *adjective*
— **mid·dle–ag·er** \-ˈā-jər\ *noun*
Middle Ages *noun plural* (1616)
: the period of European history from about A.D. 500 to about 1500
Middle America *noun* (1898)
1 : the region of the western hemisphere including Mexico, Central America, often the West Indies, and sometimes Colombia and Venezuela
2 : the midwestern section of the U.S.

3 : the middle-class segment of the U.S. population; *especially* **:** the traditional or conservative element of the middle class
— **middle–American** *adjective*
— **Middle American** *noun*

mid·dle·brow \ˈmi-dᵊl-ˌbrau̇\ *noun* (1925)
: a person who is moderately but not highly cultivated
— **middlebrow** *adjective*

middle C *noun* (1840)
: the note designated by the first ledger line below the treble staff and the first above the bass staff

mid·dle–class \ˌmi-dᵊl-ˈklas\ *adjective* (1836)
: of or relating to the middle class; *especially* **:** characterized by a high material standard of living, sexual morality, and respect for property
— **mid·dle–class·ness** \-nəs\ *noun*

middle class *noun* (1766)
: a class occupying a position between the upper class and the lower class; *especially* **:** a fluid heterogeneous socioeconomic grouping composed principally of business and professional people, bureaucrats, and some farmers and skilled workers sharing common social characteristics and values

middle distance *noun* (1813)
1 : a part of a pictorial representation or scene between the foreground and the background
2 : any footrace distance usually from 800 to 1500 meters or from 880 yards to one mile

Middle Dutch *noun* (circa 1959)
: the Dutch language in use from about 1100 to about 1500 — see INDO-EUROPEAN LANGUAGES table

middle ear *noun* (1852)
: a small membrane-lined cavity that is separated from the outer ear by the eardrum and that transmits sound waves from the eardrum to the partition between the middle and inner ears through a chain of tiny bones

Middle English *noun* (1830)
: the English in use from the 12th to 15th centuries — see INDO-EUROPEAN LANGUAGES table

middle finger *noun* (before 12th century)
: the midmost of the five digits of the hand

Middle French *noun* (1889)
: the French in use from the 14th to 16th centuries — see INDO-EUROPEAN LANGUAGES table

middle game *noun* (1894)
: the middle phase of a board game; *specifically* **:** the part of a chess game after the pieces have been developed when players attempt to gain and exploit positional and material superiority — compare ENDGAME, OPENING

Middle Greek *noun* (1889)
: the Greek language used in the 7th to 15th centuries

middle ground *noun* (1801)
1 : a standpoint midway between extremes
2 : MIDDLE DISTANCE 1

Middle High German *noun* (1889)
: the High German in use from about 1100 to 1500 — see INDO-EUROPEAN LANGUAGES table

Middle Irish *noun* (1952)
: the Irish in use between the 10th and 13th centuries — see INDO-EUROPEAN LANGUAGES table

middle lamella *noun* (circa 1886)
: a layer of pectinaceous intercellular material lying between the walls of adjacent plant cells — see CELL illustration

Middle Low German *noun* (circa 1889)
: the Low German in use from about 1100 to 1500 — see INDO-EUROPEAN LANGUAGES table

mid·dle·man \ˈmi-dᵊl-ˌman\ *noun* (1795)
: an intermediary or agent between two parties; *especially* **:** a dealer or agent intermediate between the producer of goods and the retailer or consumer

middle management *noun* (circa 1948)
: management personnel intermediate between operational supervisors and policy-making administrators
— **middle manager** *noun*

middle name *noun* (1835)
: a name between one's first name and surname

middle–of–the–road *adjective* (1894)
: standing for or following a course of action midway between extremes; *especially* **:** being neither liberal nor conservative in politics
— **mid·dle–of–the–road·er** \-ˈrō-dər\ *noun*
— **mid·dle–of–the–road·ism** \-ˈrō-ˌdi-zəm\ *noun*

middle of the road *noun* (1918)
: a course of action or a standpoint midway between extremes

Middle Persian *noun* (1930)
: any of the varieties of Persian in use from about 200 B.C. to about A.D. 1000

mid·dler \ˈmid-lər, ˈmi-dᵊl-ər\ *noun* (1882)
: one belonging to an intermediate group, division, or class: as **a :** a student in the second-year class of a three-year program (as at a seminary or law school) **b :** a student in the second- or third-year class in some private secondary schools having a four-year course **c :** a student in a division in some private schools that corresponds approximately to junior high school

middle school *noun* (1945)
: a school usually including grades 5 to 8 or 6 to 8
— **middle school·er** \-ˌskü-lər\ *noun*

Middle Scots *noun* (circa 1903)
: the Scots language in use between the latter half of the 15th and the early decades of the 17th centuries

middle term *noun* (1685)
: the term of a syllogism that occurs in both premises

mid·dle·weight \ˈmi-dᵊl-ˌwāt\ *noun* (1889)
: one of average weight; *specifically* **:** a boxer in a weight division having a maximum limit of 160 pounds for professionals and 165 pounds for amateurs — compare LIGHT HEAVYWEIGHT, WELTERWEIGHT

Middle Welsh *noun* (circa 1922)
: the Welsh in use from about 1150 to 1500 — see INDO-EUROPEAN LANGUAGES table

[1]mid·dling \ˈmid-liŋ, -lən\ *noun* (1543)
1 : any of various commodities of intermediate size, quality, or position
2 *plural but singular or plural in construction* **:** a granular product of grain milling; *especially* **:** a wheat milling by-product used in animal feeds

[2]middling *adjective* (1550)
1 : of middle, medium, or moderate size, degree, or quality
2 : MEDIOCRE, SECOND-RATE
— **middling** *adverb*
— **mid·dling·ly** \-lē\ *adverb*

mid·dor·sal \ˈmid-ˈdȯr-səl\ *adjective* (1879)
: of, relating to, or situated in the middle part or median line of the back

mid·dy \ˈmi-dē\ *noun, plural* **middies** [by shortening & alteration] (1818)
1 : MIDSHIPMAN
2 : a loosely fitting blouse with a sailor collar worn by women and children

mid·field \ˈmid-ˌfēld, ˌmid-ˈ\ *noun* (15th century)
1 : the middle portion of a field; *especially* **:** the portion of a playing field (as in football) that is midway between goals
2 : the players on a team (as in lacrosse or soccer) that normally play in the midfield

mid·field·er \-ˌfēl-dər, -ˈfēl-\ *noun* (1938)
: a member of a midfield

Mid·gard \ˈmid-ˌgärd\ *noun* [Old Norse *mithgarthr*]
: the abode of human beings in Norse mythology

midge \ˈmij\ *noun* [Middle English *migge,* from Old English *mycg;* akin to Old High German *mucka* midge, Greek *myia* fly, Latin *musca*] (before 12th century)
: a tiny dipteran fly (as a chironomid)

midg·et \ˈmi-jət\ *noun, often attributive* [*midge*] (1865)
1 : a very small person; *specifically* **:** a person of unusually small size who is physically well-proportioned
2 : something (as an animal) much smaller than usual
3 : a front-engine, single-seat, open-wheel racing car smaller and of less engine displacement than standard cars of the type

mid·gut \ˈmid-ˌgət\ *noun* (1875)
: the middle part of an alimentary canal

midi \ˈmi-dē\ *noun* [[1]*mid* + *-i* (as in *mini*)] (1967)
: a dress, skirt, or coat that usually extends to the mid-calf

Mid·i·an·ite \ˈmi-dē-ə-ˌnīt\ *noun* [*Midian,* son of Abraham] (1560)
: a member of an ancient northern Arabian people

mid·land \ˈmid-lənd, -ˌland\ *noun* (1555)
1 : the interior or central region of a country
2 *capitalized* **a :** the dialect of English spoken in the midland counties of England **b :** the dialect of English spoken in an area of the east central U.S. often divided into north Midland extending westward from an area including southern New Jersey; northern Delaware and Maryland; central and southern Pennsylvania; and central Ohio, Indiana, and Illinois and south Midland extending westward and southwestward from an area including the Appalachian regions of Virginia, North Carolina, South Carolina, and Georgia; Tennessee, Kentucky, West Virginia; and southern Ohio, Indiana, and Illinois
— **midland** *adjective, often capitalized*
— **Mid·land·er** \-lən-dər, -ˌlan-\ *noun*

mid·lat·i·tudes \ˈmid-ˈla-tə-ˌtüdz, ˌmid-, -ˌtyüdz\ *noun plural* (1925)
: latitudes of the temperate zones or from about 30 to 60 degrees north or south of the equator
— **mid·lat·i·tude** \-ˌtüd, -ˌtyüd\ *adjective*

mid·life \ˈmid-ˈlīf\ *noun* (1898)
: MIDDLE AGE

midlife crisis *noun* (1965)
: a period of emotional turmoil in middle age characterized especially by a strong desire for change

mid·line \ˈmid-ˌlīn, -ˈlīn\ *noun* (circa 1859)
: a median line; *especially* **:** the median line or median plane of the body or some part of the body

mid·most \-ˌmōst\ *adjective* (before 12th century)
1 : being in or near the exact middle
2 : most intimate **:** INNERMOST
— **midmost** *adverb or noun*

mid·night \ˈmid-ˌnīt\ *noun* (before 12th century)
1 : the middle of the night; *specifically* **:** 12 o'clock at night
2 : deep or extended darkness or gloom
— **midnight** *adjective*
— **mid·night·ly** *adverb or adjective*

midnight blue *noun* (1916)
: a deep blackish blue

midnight sun *noun* (1857)
: the sun above the horizon at midnight in the arctic or antarctic summer

mid·point \ˈmid-ˌpȯint, -ˈpȯint\ *noun* (14th century)
: a point at or near the center or middle

mid·rash \ˈmi-ˌdräsh\ *noun, plural* **mid·rash·im** \mi-ˈdrä-shəm\ [Hebrew *midhrāsh* exposition, explanation] (1613)
1 : a haggadic or halakic exposition of the underlying significance of a Bible text
2 : a collection of midrashim

\ə\ **abut** \ᵊ\ **kitten** \ər\ **further** \a\ **ash** \ā\ **ace** \ä\ **mop, mar** \au̇\ **out** \ch\ **chin** \e\ **bet** \ē\ **easy** \g\ **go** \i\ **hit** \ī\ **ice** \j\ **job** \ŋ\ **sing** \ō\ **go** \ȯ\ **law** \ȯi\ **boy** \th\ **thin** \t͟h\ **the** \ü\ **loot** \u̇\ **foot** \y\ **yet** \zh\ **vision** *see also* Guide to Pronunciation

3 *capitalized* **:** the midrashic literature written during the first Christian millennium
— **mid·rash·ic** \mi-ˈdrä-shik\ *adjective, often capitalized*

mid·rib \ˈmid-ˌrib\ *noun* (1794)
: the central vein of a leaf

mid·riff \ˈmi-ˌdrif\ *noun* [Middle English *midrif,* from Old English *midhrif,* from *midde* mid + *hrif* belly; akin to Old High German *href* body, and probably to Latin *corpus* body] (before 12th century)
1 : DIAPHRAGM 1
2 : the mid-region of the human torso
3 a : a section of a woman's garment that covers the midriff **b :** a woman's garment that exposes the midriff

mid–rise \ˈmid-ˌrīz, -ˈrīz\ *adjective* (1967)
: being approximately 5 to 10 stories high ⟨*mid-rise* condominiums⟩

mid·sag·it·tal \ˈmid-ˈsa-jə-tᵊl\ *adjective* (1947)
: median and sagittal

mid·sec·tion \-ˌsek-shən\ *noun* (1936)
: a section midway between the extremes; *especially* **:** MIDRIFF 2

mid·ship·man \ˈmid-ˌship-mən, ˌmid-ˈ\ *noun* (1685)
: a person in training for a naval commission; *especially* **:** a student in a naval academy

mid·ships \ˈmid-ˌships\ *adverb* (circa 1828)
: AMIDSHIPS

mid·size \-ˌsīz\ *also* **mid·sized** \-ˌsīzd\ *adjective* (1970)
: of intermediate size ⟨*midsize* car⟩

mid·sole \-ˌsōl\ *noun* (1926)
: a layer (as of leather or rubber) between the insole and the outsole of a shoe

midst \ˈmidst, ˈmitst\ *noun* [Middle English *middest,* alteration of *middes,* short for *amiddes* amid] (15th century)
1 : the interior or central part or point **:** MIDDLE ⟨in the *midst* of the forest⟩
2 : a position of proximity to the members of a group ⟨a traitor in our *midst*⟩
3 : the condition of being surrounded or beset ⟨in the *midst* of his troubles⟩
4 : a period of time about the middle of a continuing act or condition ⟨in the *midst* of a meal⟩
— **midst** *preposition*

mid·stream \ˈmid-ˈstrēm, -ˌstrēm\ *noun* (1669)
1 : the middle of a stream
2 : an intermediate stage in an act or process ⟨the tone changes in *midstream*⟩

mid·sum·mer \-ˈsə-mər, -ˌsəm-\ *noun* (before 12th century)
1 : the middle of summer
2 : the summer solstice
— **midsummer** *adjective*

Midsummer Day *noun* (before 12th century)
: June 24 celebrated as the feast of the nativity of John the Baptist

mid·term \ˈmid-ˌtərm (*usual for 1b*), -ˈtərm\ *noun* (1906)
1 a : the middle of an academic term **b :** an examination at midterm
2 : the approximate middle of a term of office

mid·town \ˈmid-ˌtau̇n, -ˈtau̇n\ *noun* (1926)
: a central section of a city; *especially* **:** one situated between sections conventionally called *downtown* and *uptown*
— **midtown** *adjective*

¹mid·way \ˈmid-ˌwā, -ˈwā\ *adverb* (13th century)
: in the middle of the way or distance **:** HALFWAY

²mid·way \-ˌwā\ *noun* [*Midway* (*Plaisance*), Chicago, site of the amusement section of the Columbian Exposition 1893] (1893)
: an avenue at a fair, carnival, or amusement park for concessions and amusements

mid·week \-ˌwēk\ *noun* (1707)
: the middle of the week
— **midweek** *adjective or adverb*
— **mid·week·ly** \-ˌwē-klē, -ˈwē-\ *adjective or adverb*

¹mid·wife \ˈmid-ˌwīf\ *noun* [Middle English *midwif,* from *mid* with (from Old English) + *wif* woman] (14th century)
1 : a person who assists women in childbirth
2 : one that helps to produce or bring forth something

²midwife *transitive verb* **mid·wifed** \-ˌwīft\ *or* **mid·wived** \-ˌwīvd\; **mid·wif·ing** \-ˌwī-fiŋ\ *or* **mid·wiv·ing** \-ˌwī-viŋ\ (1638)
: to assist in producing, bringing forth, or bringing about

mid·wife·ry \ˌmid-ˈwi-f(ə-)rē, -ˈwī-; ˈmid-ˌwī-\ *noun* (15th century)
1 : the art or act of assisting at childbirth; *also* **:** OBSTETRICS
2 : the art, act, or process of producing, bringing forth, or bringing about

mid·win·ter \ˈmid-ˈwin-tər, -ˌwin-\ *noun* (before 12th century)
1 : the winter solstice
2 : the middle of winter
— **midwinter** *adjective*

mid·year \-ˌyir\ *noun* (1896)
1 a : an examination at the middle of an academic year **b** *plural* **:** the set of examinations at the middle of an academic year; *also* **:** the period of such examinations
2 a : the middle or middle portion of a calendar year **b :** the middle of an academic year
— **midyear** *adjective*

mien \ˈmēn\ *noun* [by shortening & alteration from ¹*demean*] (1513)
1 : air or bearing especially as expressive of attitude or personality **:** DEMEANOR ⟨of aristocratic *mien*⟩
2 : APPEARANCE, ASPECT ⟨the inherent dangers of government encroachment . . . presented such a distasteful *mien* —H. W. Baldwin⟩
synonym see BEARING

¹miff \ˈmif\ *noun* [origin unknown] (1623)
1 : a fit of ill humor
2 : a trivial quarrel

²miff *transitive verb* (1811)
: to put into an ill humor **:** OFFEND

¹might \ˈmīt\ [Middle English, from Old English *meahte, mihte;* akin to Old High German *mahta, mohta* could] *past of* MAY (before 12th century)
— used in auxiliary function to express permission, liberty, probability, possibility in the past ⟨the president *might* do nothing without the board's consent⟩ or a present condition contrary to fact ⟨if you were older you *might* understand⟩ or less probability or possibility than *may* ⟨*might* get there before it rains⟩ or as a polite alternative to *may* ⟨*might* I ask who is calling⟩ or to *ought* or *should* ⟨you *might* at least apologize⟩

²might *noun* [Middle English, from Old English *miht;* akin to Old High German *maht* might, *magan* to be able — more at MAY] (before 12th century)
1 a : the power, authority, or resources wielded (as by an individual or group) ⟨the *might* of the armed forces⟩ **b** (1) **:** bodily strength (2) **:** the power, energy, or intensity of which one is capable ⟨striving with *might* and main⟩
2 *dialect* **:** a great deal
synonym see POWER

might·i·ly \ˈmī-tᵊl-ē\ *adverb* (before 12th century)
1 : in a mighty manner **:** VIGOROUSLY ⟨applauded *mightily*⟩
2 : very much ⟨depressed me *mightily*⟩

might·i·ness \ˈmī-tē-nəs\ *noun* (14th century)
: the quality or state of being mighty

mightn't \ˈmī-tᵊnt\ (1889)
: might not

¹mighty \ˈmī-tē\ *adjective* **might·i·er; -est** (before 12th century)
1 : possessing might **:** POWERFUL
2 : accomplished or characterized by might ⟨a *mighty* thrust⟩
3 : great or imposing in size or extent **:** EXTRAORDINARY

²mighty *adverb* (14th century)
: EXTREMELY, VERY ⟨a *mighty* handy gadget⟩ ■

mi·gnon \mēn-ˈyōⁿ, min-ˈyȯn\ *noun* (circa 1919)
: FILET MIGNON

mi·gnon·ette \ˌmin-yə-ˈnet\ *noun* [French *mignonnette,* from obsolete French, feminine of *mignonnet* dainty, from Middle French, from *mignon* darling] (1798)
: any of a genus (*Reseda* of the family Resedaceae, the mignonette family) of herbs; *especially* **:** a garden annual (*R. odorata*) bearing racemes of fragrant whitish flowers

mi·graine \ˈmī-ˌgrān, *British often* ˈmē-\ *noun* [French, modification of Late Latin *hemicrania* pain in one side of the head, from Greek *hēmikrania,* from *hēmi-* hemi- + *kranion* cranium] (15th century)
1 : a condition marked by recurrent severe headache often with nausea and vomiting
2 : an episode or attack of migraine
— **mi·grain·ous** \-ˌgrā-nəs\ *adjective*

mi·grant \ˈmī-grənt\ *noun* [Latin *migrant-, migrans,* present participle of *migrare*] (1760)
: one that migrates: as **a :** a person who moves regularly in order to find work especially in harvesting crops **b :** an animal that shifts from one habitat to another
— **migrant** *adjective*

mi·grate \ˈmī-ˌgrāt, mī-ˈ\ *intransitive verb* **mi·grat·ed; mi·grat·ing** [Latin *migratus,* past participle of *migrare;* perhaps akin to Greek *ameibein* to change] (1697)
1 : to move from one country, place, or locality to another
2 : to pass usually periodically from one region or climate to another for feeding or breeding
3 : to change position in an organism or substance ⟨filarial worms *migrate* within the human body⟩
— **mi·gra·tion** \mī-ˈgrā-shən\ *noun*
— **mi·gra·tion·al** \-shnəl, -shə-nᵊl\ *adjective*
— **mi·gra·tor** \ˈmī-ˌgrā-tər, mī-ˈ\ *noun*

mi·gra·to·ry \ˈmī-grə-ˌtōr-ē, -ˌtȯr-\ *adjective* (1753)
1 : of, relating to, or characterized by migration
2 : WANDERING, ROVING

mih·rab \ˈmē-rəb\ *noun* [Arabic *miḥrāb*] (1816)
: a niche or chamber in a mosque indicating the direction of Mecca

mi·ka·do \mə-ˈkä-(ˌ)dō\ *noun, plural* **-dos** [Japanese] (1727)
: an emperor of Japan

¹mike \ˈmīk\ *noun* [by shortening & alteration] (1924)
: MICROPHONE

²mike *transitive verb* **miked; mik·ing** (1939)
: to supply with a microphone

Mike \ˈmīk\ (1942)

□ USAGE

mighty The use of *mighty* as an intensive usually conveys a folksy down-home feeling ⟨plain and simple fare . . . but *mighty* filling and *mighty* satisfying —*Asheville (N.C.) Citizen-Times*⟩. It is used especially to create a chatty style ⟨turnip greens, corn bread and biscuits. That sounds *mighty* good to me —Julia Child⟩ or to stress a rural atmosphere ⟨a man must be *mighty* serious about his squirrel hunting —Stuart Williams, *Field & Stream*⟩. In a more formal context, *mighty* is used to create additional emphasis by drawing attention to itself ⟨the chairman made sure that there were *mighty* few of them —Mollie Panter-Downes⟩.

— a communications code word for the letter *m*

¹mil \'mil\ *noun* [Latin *mille* thousand] (1721)
1 : THOUSAND ⟨found a salinity of 38.4 per *mil*⟩
2 : a monetary unit formerly used in Cyprus equal to 1/1000 pound
3 : a unit of length equal to 1/1000 inch used especially in measuring thickness (as of plastic films)
4 : a unit of angular measurement equal to 1/6400 of 360 degrees and used especially in artillery

²mil *noun, plural* **military** [short for *million*] (circa 1955)
slang **:** a million dollars

mi·la·dy \mi-'lā-dē, *US also* mī-\ *noun* [French, from English *my lady*] (1839)
1 : an Englishwoman of noble or gentle birth
2 : a woman of fashion

milch \'milk, 'milch, 'milks\ *adjective* [Middle English *milche,* from Old English *-milce;* akin to Old English *melcan* to milk — more at EMULSION] (14th century)
: MILK

mil·chig \'mil-kik\ *adjective* [Yiddish *milkhik,* from *milkh* milk, from Middle High German, from Old High German *miluh* — more at MILK] (circa 1928)
: made of or derived from milk or dairy products — compare FLEISHIG, PAREVE

mild \'mī(ə)ld\ *adjective* [Middle English, from Old English *milde;* akin to Greek *malthakos* soft, Latin *mollis* — more at MELT] (before 12th century)
1 : gentle in nature or behavior ⟨has a *mild* disposition⟩
2 a (1) **:** moderate in action or effect ⟨a *mild* cigar⟩ (2) **:** not sharp or bitter ⟨*mild* cheese⟩ ⟨*mild* ale⟩ **b :** not being or involving what is extreme ⟨an analysis under *mild* conditions⟩ ⟨a *mild* slope⟩
3 : not severe **:** TEMPERATE ⟨a *mild* climate⟩ ⟨*mild* symptoms of disease⟩
— **mild·ly** \mī(ə)l(d)-lē\ *adverb*
— **mild·ness** \'mī(ə)l(d) nəs\ *noun*

¹mil·dew \'mil-,dü, -,dyü\ *noun* [Middle English, from Old English *meledēaw* honeydew; akin to Old High German *militou* honeydew] (14th century)
1 a : a superficial usually whitish growth produced especially on organic matter or living plants by fungi (as of the families Erysiphaceae and Peronosporaceae) **b :** a fungus producing mildew
2 : a discoloration caused by fungi
— **mil·dewy** \-ē\ *adjective*

²mildew (circa 1552)
transitive verb
: to affect with or as if with mildew
intransitive verb
: to become affected with mildew

mild steel *noun* (1868)
: a low-carbon structural steel that is easily worked

mile \'mī(ə)l\ *noun* [Middle English, from Old English *mīl,* from Latin *milia* miles, from *milia passuum,* literally, thousands of paces, from *milia,* plural of *mille* thousand] (before 12th century)
1 : any of various units of distance: as **a :** a unit equal to 5280 feet — see WEIGHT table **b :** NAUTICAL MILE
2 : a race of a mile
3 : a relatively great distance or interval — used chiefly adverbially in plural ⟨was *miles* too small⟩

mile·age \'mī-lij\ *noun* (1754)
1 : an allowance for traveling expenses at a certain rate per mile
2 : aggregate length or distance in miles: as **a :** the total miles traveled especially in a given period of time **b :** the amount of service that something will yield especially as expressed in terms of miles of travel **c :** the average number of miles a car will travel on a gallon of gasoline that is used as a measure of fuel economy ⟨gets good *mileage*⟩
3 a : USEFULNESS ⟨got a lot of *mileage* left in her⟩ **b :** benefit derived from something ⟨got good political *mileage* from the debates⟩

mile·post \'mī(ə)l-,pōst\ *noun* (1768)
1 : a post indicating the distance in miles from or to a given point; *also* **:** a post placed a mile from a similar post
2 : MILESTONE 2

mil·er \'mī-lər\ *noun* (1889)
1 : one that competes in mile races — often qualified in combination ⟨a quarter-*miler*⟩
2 : one that is a specified number of miles in length — used in combination ⟨a 15-*miler*⟩

mi·les glo·ri·o·sus \'mē-,lās-,glōr-ē-'ō-səs, -,glòr-\ *noun, plural* **mi·li·tes glo·ri·o·si** \'mē-lə-,tās-,glōr-ē-'ō-(,)sē, -,glòr-\ [Latin] (1917)
: a boastful soldier; *especially* **:** a stock character of this type in comedy

mile·stone \'mī(ə)l-,stōn\ *noun* (circa 1746)
1 : a stone serving as a milepost
2 : a significant point in development

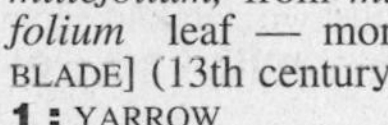

milestone 1

mil·foil \'mil-,fòil\ *noun* [Middle English, from Old French, from Latin *millefolium,* from *mille* + *folium* leaf — more at BLADE] (13th century)
1 : YARROW
2 : WATER MILFOIL

mil·i·ar·ia \,mi-lē-'ar-ē-ə, -'er-\ *noun* [New Latin, from Latin, feminine of *miliarius*] (1807)
: an inflammatory disorder of the skin characterized by redness, eruption, burning or itching, and the release of sweat in abnormal ways (as by the eruption of vesicles) due to blockage of the ducts of the sweat glands; *especially* **:** PRICKLY HEAT
— **mil·i·ar·i·al** \-əl\ *adjective*

mil·i·ary \'mi-lē-,er-ē\ *adjective* [Latin *miliarius* of millet, from *milium* millet — more at MILLET] (1685)
: having or made up of many small projections or lesions ⟨*miliary* tubercles⟩

mi·lieu \mēl-'yə(r), -'yü; 'mēl-,yü, mē-lyœ\ *noun, plural* **milieus** *or* **mi·lieux** \-'yə(r)(z), -'yüz; -,yü(z), -lyœ(z)\ [French, from Old French, midst, from *mi* middle (from Latin *medius*) + *lieu* place, from Latin *locus* — more at MID, STALL] (1854)
: the physical or social setting in which something occurs or develops **:** ENVIRONMENT
synonym see BACKGROUND

mil·i·tance \'mi-lə-tən(t)s\ *noun* (circa 1947)
: MILITANCY

mil·i·tan·cy \-tən(t)-sē\ *noun* (1648)
: the quality or state of being militant

mil·i·tant \-tənt\ *adjective* (15th century)
1 : engaged in warfare or combat **:** FIGHTING
2 : aggressively active (as in a cause) **:** COMBATIVE ⟨*militant* conservationists⟩ ⟨a *militant* attitude⟩
synonym see AGGRESSIVE
— **militant** *noun*
— **mil·i·tant·ly** *adverb*
— **mil·i·tant·ness** *noun*

mil·i·tar·ia \,mi-lə-'ter-ē-ə\ *noun plural* (1964)
: military objects (as firearms and uniforms) of historical value or interest

mil·i·tari·ly \,mi-lə-'ter-ə-lē\ *adverb* (1660)
1 : in a military manner
2 : from a military standpoint

mil·i·ta·rise *British variant of* MILITARIZE

mil·i·ta·rism \'mi-lə-tə-,ri-zəm\ *noun* (1864)
1 a : predominance of the military class or its ideals **b :** exaltation of military virtues and ideals
2 : a policy of aggressive military preparedness
— **mil·i·ta·rist** \-rist\ *noun or adjective*
— **mil·i·ta·ris·tic** \,mi-lə-tə-'ris-tik\ *adjective*
— **mil·i·ta·ris·ti·cal·ly** \-ti-k(ə-)lē\ *adverb*

mil·i·ta·rize \'mi-lə-tə-,rīz\ *transitive verb* **-rized; -riz·ing** (1880)
1 : to equip with military forces and defenses
2 : to give a military character to
3 : to adapt for military use
— **mil·i·ta·ri·za·tion** \,mi-lə-t(ə-)rə-'zā-shən\ *noun*

¹mil·i·tary \'mi-lə-,ter-ē\ *adjective* [Middle English, from Middle French *militaire,* from Latin *militaris,* from *milit-, miles* soldier] (15th century)
1 a : of or relating to soldiers, arms, or war **b :** of or relating to armed forces; *especially* **:** of or relating to ground or sometimes ground and air forces as opposed to naval forces
2 a : performed or made by armed forces **b :** supported by armed force
3 : of or relating to the army

²military *noun, plural* **military** *also* **mil·i·tar·ies** (1736)
1 : military persons; *especially* **:** army officers
2 : ARMED FORCES

military–industrial complex *noun* (1961)
: an informal alliance of the military and related government departments with defense industries that is held to influence government policy

military police *noun* (1827)
: a branch of an army that exercises guard and police functions

military press *noun* (1939)
: ¹PRESS 9

military science *noun* (circa 1830)
: the principles of military conflict

military time *noun* (1955)
: time measured in hours numbered to twenty-four (as 0100 or 2300) from one midnight to the next

mil·i·tate \'mi-lə-,tāt\ *intransitive verb* **-tat·ed; -tat·ing** [Latin *militatus,* past participle of *militare* to engage in warfare, from *milit-, miles*] (1642)
: to have weight or effect ⟨his boyish appearance *militated* against his getting an early promotion⟩
usage see MITIGATE

mi·li·tia \mə-'li-shə\ *noun* [Latin, military service, from *milit-, miles*] (circa 1660)
1 a : a part of the organized armed forces of a country liable to call only in emergency **b :** a body of citizens organized for military service
2 : the whole body of able-bodied male citizens declared by law as being subject to call to military service

mi·li·tia·man \-mən\ *noun* (1780)
: a member of a militia

mil·i·um \'mi-lē-əm\ *noun, plural* **mil·ia** \-lē-ə\ [New Latin, from Latin, millet — more at MILLET] (1856)
: a small whitish lump in the skin due to retention of keratin in an oil gland duct — called also *whitehead*

¹milk \'milk\ *noun* [Middle English, from Old English *meolc, milc;* akin to Old High German *miluh* milk, Old English *melcan* to milk — more at EMULSION] (before 12th century)
1 a : a fluid secreted by the mammary glands of females for the nourishment of their young; *especially* **:** cow's milk used as a food by humans **b :** LACTATION ⟨cows in *milk*⟩
2 : a liquid resembling milk in appearance: as **a :** the latex of a plant **b :** the juice of a coconut **c :** the contents of an unripe kernel of grain

²milk (before 12th century)
transitive verb

1 a (1) **:** to draw milk from the breasts or udder of (2) *obsolete* **:** SUCKLE 2 **b :** to draw (milk) from the breast or udder **c :** SUCKLE 1 — used of lower mammals
2 : to draw something from as if by milking: as **a :** to induce (a snake) to eject venom **b :** to draw or coerce profit or advantage from illicitly or to an extreme degree **:** EXPLOIT
intransitive verb
: to draw or yield milk

³milk *adjective* (14th century)
: giving milk; *specifically* **:** bred or suitable primarily for milk production ⟨*milk* cows⟩

milk–and–water *adjective* (1783)
: WEAK, INSIPID

milk chocolate *noun* (1904)
: chocolate made with milk solids

milk·er \'mil-kər\ *noun* (15th century)
1 : one that milks an animal
2 : one that yields milk

milk fever *noun* (1758)
1 : a febrile disorder following parturition
2 : a disease of fresh cows, sheep, or goats that is caused by excessive drain on the body mineral reserves during the establishment of the milk flow

milk·fish \'milk-ˌfish\ *noun* (circa 1890)
: a large fork-tailed silvery herbivorous food fish (*Chanos chanos*) of warm parts of the Pacific and Indian oceans that is the sole living representative of its family (Chanidae)

milk glass *noun* (1874)
: an opaque and typically milky white glass used especially for novelty and ornamental objects

milk house *noun* (1589)
: a building for the cooling, handling, or bottling of milk

milk leg *noun* (circa 1860)
: a painful swelling of the leg caused by inflammation and clotting in the veins and affecting some postpartum women

milk–liv·ered \'milk-ˌli-vərd\ *adjective* (1605)
archaic **:** COWARDLY, TIMOROUS

milk·maid \-ˌmād\ *noun* (1552)
: DAIRYMAID

milk·man \-ˌman, -mən\ *noun* (1589)
: one who sells or delivers milk and milk products

milk of magnesia (1880)
: a milky white suspension of magnesium hydroxide in water used as an antacid and laxative

milk punch *noun* (1704)
: a mixed drink of alcoholic liquor, milk, and sugar

milk run *noun* [from the resemblance in regularity and uneventfulness to the morning delivery of milk] (1925)
: a short, routine, or uneventful flight

milk shake *noun* (1889)
: a thoroughly shaken or blended drink made of milk, a flavoring syrup, and often ice cream

milk sickness *noun* (1823)
1 : an acute disease characterized by weakness, vomiting, and constipation and caused by eating dairy products or meat from cattle poisoned by various plants
2 : TREMBLE 2

milk snake *noun* (1800)
: a common harmless grayish or tan American colubrid snake (*Lampropeltis triangulum*) having an arrow-shaped occipital marking and brown blotches on the body bordered with black or rings usually of black, red, and yellow; *broadly* **:** KING SNAKE

milk·sop \'milk-ˌsäp\ *noun* [Middle English, literally, bread soaked in milk] (14th century)
: an unmanly man **:** MOLLYCODDLE

milk sugar *noun* (1846)
: LACTOSE

milk tooth *noun* (circa 1752)
: a deciduous tooth of a young mammal; *especially* **:** one of the human dentition including four incisors, two canines, and four molars in each JAW

milk vein *noun* (1844)
: a large subcutaneous vein that extends along the lower side of the abdomen of a cow and returns blood from the udder — see COW illustration

milk vetch *noun* [from the popular belief that it increases the milk yield of goats] (1597)
: any of a genus (*Astragalus*) of leguminous herbs; *especially* **:** an Old World perennial (*A. glycyphyllos*) that has sulfur yellow flowers in dense spikes

milk·weed \'milk-ˌwēd\ *noun* (circa 1598)
: any of various plants that secrete latex; *especially* **:** any of a genus (*Asclepias* of the family Asclepiadaceae, the milkweed family) of erect chiefly perennial herbs with milky juice and umbellate flowers

milkweed bug *noun* (1905)
: a large black red-marked bug (*Oncopeltus fasciatus*) cultured as a research organism

milkweed butterfly *noun* (1880)
: any of a family (Danaidae) or nymphalid subfamily (Danaine) of large butterflies feeding on plants of the milkweed family as larvae; *especially* **:** MONARCH BUTTERFLY

milk·wort \'milk-ˌwərt, -ˌwȯrt\ *noun* (1578)
: any of a genus (*Polygala* of the family Polygalaceae, the milkwort family) of plants typically having showy flowers with three sometimes crested petals united below into a tube and an irregular calyx with two petaloid sepals

milky \'mil-kē\ *adjective* **milk·i·er; -est** (14th century)
1 : resembling milk in color or consistency
2 : MILD, TIMOROUS
3 a : consisting of, containing, or abounding in milk **b :** yielding milk; *specifically* **:** having the characteristics of a good milk producer
— milk·i·ness *noun*

milky disease *noun* (circa 1940)
: a destructive bacterial disease of scarab beetle grubs and especially Japanese beetle larvae

Milky Way *noun*
1 : a broad luminous irregular band of light that stretches completely around the celestial sphere and is caused by the light of myriads of faint stars
2 : MILKY WAY GALAXY
word history see GALAXY

Milky Way galaxy *noun*
: the galaxy of which the sun and the solar system are a part and which contains the myriads of stars that comprise the Milky Way

¹mill \'mil\ *noun* [Middle English *mille,* from Old English *mylen,* from Late Latin *molina, molinum,* from feminine and neuter of *molinus* of a mill, of a millstone, from Latin *mola* mill, millstone; akin to Latin *molere* to grind — more at MEAL] (before 12th century)
1 : a building provided with machinery for grinding grain into flour
2 a : a machine or apparatus (as a quern) for grinding grain **b :** a machine for crushing or comminuting
3 : a machine that manufactures by the continuous repetition of some simple action
4 : a building or collection of buildings with machinery for manufacturing
5 a : a machine formerly used for stamping coins **b :** a machine for expelling juice from vegetable tissues by pressure or grinding
6 : MILLING MACHINE, MILLING CUTTER
7 a : a slow, laborious, or mechanical process or routine **b :** one that produces or processes people or things mechanically or in large numbers ⟨a diploma *mill*⟩ ⟨a rumor *mill*⟩
8 : a difficult and often educational experience — used in the phrase *through the mill*
9 *slang* **:** the engine of an automobile or boat

²mill (1552)
transitive verb
1 : to subject to an operation or process in a mill: as **a :** to grind into flour, meal, or powder **b :** to shape or dress by means of a rotary cutter **c :** to mix and condition (as rubber) by passing between rotating rolls
2 : to give a raised rim or a ridged or corrugated edge to (a coin)
3 : to cut grooves in the metal surface of (as a knob)
intransitive verb
1 : to hit out with the fists
2 : to move in a circle or in an eddying mass; *also* **:** WANDER
3 : to undergo milling

³mill *noun* [Latin *mille* thousand] (1786)
: a money of account equal to 1/10 cent

⁴mill *variant of* ²MIL

mill·age \'mi-lij\ *noun* (1891)
: a rate (as of taxation) expressed in mills per dollar

mill·dam \'mil-ˌdam\ *noun* (12th century)
: a dam to make a millpond; *also* **:** MILLPOND

mille \'mil\ *noun* [Latin] (1894)
: THOUSAND

mille–feuille \mēl-'fwē\ *noun* [French, from *mille feuilles* a thousand leaves] (1895)
: a dish composed of puff pastry layered with a filling (as salmon or cream)

mil·le·fi·o·ri \ˌmi-lə-fē-'ōr-ē, -'ȯr-\ *noun* [Italian, from *mille fiori* a thousand flowers] (1849)
: ornamental glass produced by cutting cross sections of fused bundles of glass rods of various colors and sizes

mille·fleur *or* **mille·fleurs** \(ˌ)mēl-'flər, -'flu̇r\ *adjective* [French *mille-fleurs,* from *mille fleurs* a thousand flowers] (1908)
: having an allover pattern of small flowers and plants ⟨*millefleur* tapestry⟩

¹mil·le·nar·i·an \ˌmi-lə-'ner-ē-ən\ *adjective* (1631)
1 a : of or relating to belief in a millennium **b :** APOCALYPTIC 2
2 : of or relating to 1000 years

²millenarian *noun* (circa 1674)
: one that believes in a millennium

mil·le·nar·i·an·ism \-ē-ə-ˌni-zəm\ *noun* (circa 1847)
1 : belief in the millennium of Christian prophecy
2 : belief in a coming ideal society and especially one created by revolutionary action

¹mil·le·na·ry \'mi-lə-ˌner-ē, mə-'le-nə-rē\ *noun, plural* **-ries** [Late Latin *millenarium,* from neuter of *millenarius* of a thousand, from Latin *milleni* one thousand each, from *mille*] (1550)
1 a : a group of 1000 units or things **b :** 1000 years **:** MILLENNIUM
2 : MILLENARIAN

²millenary *adjective* [Latin *millenarius*] (circa 1641)
1 : relating to or consisting of 1000
2 : suggesting a millennium

mil·len·ni·al \mə-'le-nē-əl\ *adjective* (1664)
: of or relating to a millennium

mil·len·ni·al·ism \-ē-ə-ˌli-zəm\ *noun* (1906)
: MILLENARIANISM

mil·len·ni·al·ist \-list\ *noun* (circa 1841)
: MILLENARIAN

mil·len·ni·um \mə-'le-nē-əm\ *noun, plural* **-nia** \-nē-ə\ *or* **-niums** [New Latin, from Latin *mille* thousand + New Latin *-ennium* (as in *biennium*)] (circa 1638)
1 a : the thousand years mentioned in Revelation 20 during which holiness is to prevail and Christ is to reign on earth **b :** a period of great happiness or human perfection
2 a : a period of 1000 years **b :** a 1000th anniversary or its celebration

mill·er \'mi-lər\ *noun* (14th century)
1 : one that operates a mill; *specifically* **:** one that grinds grain into flour
2 : any of various moths having powdery wings
3 a : MILLING MACHINE **b :** a tool for use in a milling machine

mil·ler·ite \'mi-lə-ˌrīt\ *noun* [German *Millerit,* from William H. *Miller* (died 1880) English mineralogist] (1854)

: sulfide of nickel NiS usually occurring as a mineral in capillary yellow crystals

mill·er's–thumb \ˌmi-lərz-ˈthəm\ *noun* (15th century)
: any of several small freshwater spiny-finned sculpins (genus *Cottus*) of Europe and North America

mil·les·i·mal \mə-ˈle-sə-məl\ *noun* [Latin *millesimus,* adjective, thousandth, from *mille*] (1719)
: the quotient of a unit divided by 1000 **:** one of 1000 equal parts
— millesimal *adjective*
— mil·les·i·mal·ly \-mə-lē\ *adverb*

mil·let \ˈmi-lət\ *noun* [Middle English *milet,* from Middle French, diminutive of *mil,* from Latin *milium;* akin to Greek *melinē* millet] (15th century)
1 : any of various small-seeded annual cereal and forage grasses: **a :** a grass (*Panicum miliaceum*) cultivated for its grain which is used for food **b :** any of several grasses related to common millet
2 : the seed of a millet

milli- *combining form* [French, from Latin *milli-* thousand, from *mille*]
: thousandth ⟨*milli*ampere⟩

mil·li·am·pere \ˌmi-lē-ˈam-ˌpir\ *noun* [International Scientific Vocabulary] (1885)
: one thousandth of an ampere

mil·liard \ˈmil-ˌyärd, ˈmi-lē-ˌärd\ *noun* [French, from Middle French *miliart,* from *mili-* (from *milion* million)] (1793)
British **:** a thousand millions — see NUMBER table

mil·li·ary \ˈmi-lē-ˌer-ē\ *adjective* [Latin *milliarius, miliarius* consisting of a thousand, one mile long, from *mille* thousand, mile] (1644)
: marking the distance of a Roman mile

mil·li·bar \ˈmi-lə-ˌbär\ *noun* [International Scientific Vocabulary] (1910)
: a unit of atmospheric pressure equal to 1/1000 bar or 1000 dynes per square centimeter

mil·li·cu·rie \ˌmi-lə-ˈkyu̇r-(ˌ)ē, -kyu̇-ˈrē\ *noun* [International Scientific Vocabulary] (1910)
: one thousandth of a curie

mil·li·de·gree \-di ˈgrē\ *noun* (1942)
: one thousandth of a degree

mil·lieme \mē(l)-ˈyem\ *noun, plural* **mil·liemes** \-ˈyem(z)\ [French *millième* thousandth, from Middle French, from *mille* thousand, from Latin] (1902)
: a unit of value of Egypt and Sudan equal to 1/1000 pound

mil·li·gal \ˈmi-lə-ˌgal\ *noun* [International Scientific Vocabulary] (1914)
: a unit of acceleration equivalent to 1/1000 gal

mil·li·gram \-ˌgram\ *noun* [French *milligramme,* from *milli-* + *gramme* gram] (circa 1810)
— see METRIC SYSTEM table

mil·li·hen·ry \-ˌhen-rē\ *noun* [International Scientific Vocabulary] (1897)
: one thousandth of a henry

mil·li·lam·bert \-ˌlam-bərt\ *noun* (1916)
: one thousandth of a lambert

mil·li·li·ter \ˈmi-lə-ˌlē-tər\ *noun* [French *millilitre,* from *milli-* + *litre* liter] (circa 1810)
— see METRIC SYSTEM table

mil·lime \mə-ˈlēm\ *noun* [modification of Arabic *mallim,* from French *millième*] (circa 1919)
— see *dinar* at MONEY table

mil·li·me·ter \ˈmi-lə-ˌmē-tər\ *noun* [French *millimètre,* from *milli-* + *mètre* meter] (1807)
— see METRIC SYSTEM table

mil·li·mi·cron \ˌmi-lə-ˈmī-ˌkrän\ *noun* [International Scientific Vocabulary] (1904)
: NANOMETER

mil·li·mole \ˈmi-lə-ˌmōl\ *noun* [International Scientific Vocabulary *milli-* + ⁵*mole*] (circa 1904)
: one thousandth of a mole (as of a substance)
— mil·li·mo·lar \-ˌmō-lər\ *adjective*

mil·li·ner \ˈmi-lə-nər\ *noun* [irregular from *Milan,* Italy; from the importation of women's finery from Italy in the 16th century] (1530)
: a person who designs, makes, trims, or sells women's hats ◆

mil·li·nery \ˈmi-lə-ˌner-ē\ *noun* (circa 1688)
1 : women's apparel for the head
2 : the business or work of a milliner

mill·ing \ˈmi-liŋ\ *noun* (1817)
: a corrugated edge on a coin

milling cutter *noun* (1884)
: a rotary tool-steel cutter used in a milling machine for shaping and dressing metal surfaces

milling machine *noun* (1876)
: a machine tool on which work usually of metal secured to a carriage is shaped by rotating milling cutters

mil·lion \ˈmi(l)-yən\ *noun, plural* **millions** *or* **million** [Middle English *milioun,* from Middle French *milion,* from Old Italian *milione,* augmentative of *mille* thousand, from Latin] (14th century)
1 — see NUMBER table
2 : a very large number ⟨*millions* of cars on the road⟩
3 : the mass of common people ⟨someone who writes for the *millions* —Bergen Evans⟩
— million *adjective*
— mil·lion·fold \-ˌfōld\ *adverb or adjective*
— mil·lionth \ˈmi(l)-yən(t)th\ *adjective or noun*

mil·lion·aire \ˌmi(l)-yə-ˈnar, -ˈner, ˈmi(l)-yə-ˌ\ *noun* [French *millionnaire,* from *million,* from Middle French *milion*] (1826)
: a person whose wealth is estimated at a million or more (as of dollars or pounds)

mil·lion·air·ess \-ˈar-əs, -ˈer-, -ˌar-, -ˌer-\ *noun* (1881)
1 : a woman who is a millionaire
2 : the wife of a millionaire

mil·li·os·mol \ˌmi-lē-ˈäz-ˌmōl, -ˈäs-\ *noun* (1939)
: one thousandth of an osmol

mil·li·pede \ˈmi-lə-ˌpēd\ *noun* [Latin *millepeda,* a small crawling animal, from *mille* thousand + *ped-, pes* foot — more at FOOT] (1601)
: any of a class (Diplopoda) of myriapod arthropods having usually a cylindrical segmented body covered with hard integument, two pairs of legs on most apparent segments, and unlike centipedes no poison fangs

mil·li·ra·di·an \ˌmi-lə-ˈrā-dē-ən\ *noun* [International Scientific Vocabulary] (1954)
: one thousandth of a radian

mil·li·rem \ˈmi-lə-ˌrem\ *noun* (1947)
: one thousandth of a rem

mil·li·roent·gen \ˌmi-lə-ˈrent-gən, -ˈrənt-, -jən, -shən\ *noun* [International Scientific Vocabulary] (1947)
: one thousandth of a roentgen

mil·li·sec·ond \ˈmi-lə-ˌse-kənd, -kənt\ *noun* [International Scientific Vocabulary] (1909)
: one thousandth of a second

mil·li·volt \-ˌvōlt\ *noun* [International Scientific Vocabulary] (1861)
: one thousandth of a volt

mil·li·watt \-ˌwät\ *noun* [International Scientific Vocabulary] (circa 1914)
: one thousandth of a watt

mill·pond \ˈmil-ˌpänd\ *noun* (14th century)
: a pond created by damming a stream to produce a head of water for operating a mill

mill·race \-ˌrās\ *noun* (15th century)
: a canal in which water flows to and from a mill wheel; *also* **:** the current that drives the wheel

mill·stone \ˈmil-ˌstōn\ *noun* (before 12th century)
1 : either of two circular stones used for grinding (as grain)
2 a : something that grinds or crushes **b :** a heavy burden

mill·stream \-ˌstrēm\ *noun* (before 12th century)
1 : a stream whose flow is utilized to run a mill
2 : MILLRACE

mill wheel *noun* (before 12th century)
: a waterwheel that drives a mill

mill·work \ˈmil-ˌwərk\ *noun* (1899)
: woodwork (as doors, sashes, or trim) manufactured at a mill

mill·wright \-ˌrīt\ *noun* (14th century)
1 : one whose occupation is planning and building mills or setting up their machinery
2 : one who maintains and cares for mechanical equipment (as of a mill or factory)

mi·lo \ˈmī-(ˌ)lō\ *noun, plural* **milos** [perhaps from Sesotho *maili*] (1882)
: a small usually early and drought-resistant grain sorghum with compact bearded heads of large yellow or whitish seeds

mi·lord \mi-ˈlȯr(d)\ *noun* [French, from English *my lord*] (1596)
: an Englishman of noble or gentle birth

mil·pa \ˈmil-pə\ *noun* [Mexican Spanish, from Nahuatl *mīlpan*] (1844)
1 a : a small field in Mexico or Central America that is cleared from the forest, cropped for a few seasons, and abandoned for a fresh clearing **b :** a maize field in Mexico or Central America
2 : the maize plant

Milque·toast \ˈmilk-ˌtōst\ *noun* [Caspar *Milquetoast,* comic strip character created by H. T. Webster (died 1952) American cartoonist] (1935)
: a timid, meek, or unassertive person

mil·reis \mil-ˈrāsh, -ˈrās\ *noun, plural* **mil·reis** *same or* -ˈrāz, -ˈrāzh\ [Portuguese *mil-réis*] (1589)
1 : a Portuguese unit of value equal before 1911 to 1000 reis
2 : the basic monetary unit of Brazil until 1942
3 : a coin representing one milreis

milt \ˈmilt\ *noun* [probably from Middle Dutch *milte* milt of fish, spleen; akin to Old English *milte* spleen — more at MELT] (15th century)
: the sperm-containing fluid of a male fish

mim \ˈmim\ *adjective* [imitative of the act of pursing the lips] (1641)
dialect **:** affectedly shy or modest

¹mime \ˈmīm *also* ˈmēm\ *noun* [Latin *mimus,* from Greek *mimos*] (1603)
1 : an ancient dramatic entertainment representing scenes from life usually in a ridiculous manner
2 a : an actor in a mime **b :** one that practices mime
3 : MIMIC 2
4 : PANTOMIME 3a,b

²mime *verb* **mimed; mim·ing** (1616)

◇ WORD HISTORY

milliner The Italian city of Milan was a significant source of luxury goods for 16th century England. Milan bonnets, Milan gloves, Milan ribbons, Milan point lace, Milan jewelry—all represented the best in Renaissance finery. The purveyors of these goods imported from Milan were called—regardless of their actual nationality—*milaners* or *millaners* or, in the variant that survives today, *milliners.* Eventually, *milliner* was extended to all merchants specializing in fancy accessories, the actual origin of the merchandise notwithstanding. For several centuries *milliner* could refer to a purveyor of luxury items of any kind. Apparently, not until the 19th century was *milliner* reserved exclusively for makers or retailers of women's hats.

intransitive verb
: to act a part with mimic gesture and action usually without words
transitive verb
1 : MIMIC
2 : to act out in the manner of a mime
— mim·er *noun*
mim·eo·graph \'mi-mē-ə-ˌgraf\ *noun* [from *Mimeograph,* a trademark] (1889)
: a duplicator for making many copies that utilizes a stencil through which ink is pressed
— mimeograph *transitive verb*
mi·me·sis \mə-'mē-səs, mī-\ *noun* [Late Latin, from Greek *mimēsis,* from *mimeisthai*] (1550)
: IMITATION, MIMICRY
mi·met·ic \-'me-tik\ *adjective* [Late Latin *mimeticus,* from Greek *mimētikos,* from *mimeisthai* to imitate, from *mimos* mime] (1637)
1 : IMITATIVE
2 : relating to, characterized by, or exhibiting mimicry ⟨*mimetic* coloring of a butterfly⟩
— mi·met·i·cal·ly \-ti-k(ə-)lē\ *adverb*
1mim·ic \'mi-mik\ *noun* (1590)
1 : MIME 2
2 : one that mimics
2mimic *adjective* [Latin *mimicus,* from Greek *mimikos,* from *mimos* mime] (1598)
1 a : IMITATIVE **b :** IMITATION, MOCK ⟨a *mimic* battle⟩
2 : of or relating to mime or mimicry
3mimic *transitive verb* **mim·icked** \-mikt\; **mim·ick·ing** (1687)
1 : to imitate closely **:** APE
2 : to ridicule by imitation
3 : SIMULATE
4 : to resemble by biological mimicry
synonym see COPY
mim·ic·ry \'mi-mi-krē\ *noun, plural* **-ries** (1687)
1 a : an instance of mimicking **b :** the action, practice, or art of mimicking
2 : a superficial resemblance of one organism to another or to natural objects among which it lives that secures it a selective advantage (as protection from predation)
mi·mo·sa \mə-'mō-sə, mī-, -zə\ *noun* [New Latin, from Latin *mimus* mime] (1751)
1 : any of a genus (*Mimosa*) of leguminous trees, shrubs, and herbs of tropical and warm regions with usually bipinnate often prickly leaves and globular heads of small white or pink flowers
2 : SILK TREE
3 : a mixed drink consisting of champagne and orange juice
mi·na \'mī-nə\ *noun* [Latin, from Greek *mna,* of Semitic origin; akin to Hebrew *māneh* mina] (circa 1580)
: an ancient unit of weight and value equal to 1/60 talent
min·able *or* **mine·able** \'mī-nə-bəl\ *adjective* (circa 1570)
: capable of being mined
min·a·ret \ˌmi-nə-'ret, 'mi-nə-ˌ\ *noun* [French, from Turkish *minare,* from Arabic *manārah* lighthouse] (1682)
: a tall slender tower of a mosque having one or more balconies from which the summons to prayer is cried by the muezzin

minaret

mi·na·to·ry \'mi-nə-ˌtōr-ē, 'mī-, -ˌtȯr-\ *adjective* [Late Latin *minatorius,* from Latin *minari* to threaten — more at MOUNT] (1532)
: having a menacing quality **:** THREATENING
min·au·dière \ˌmē-nōd-'yer\ *noun* [French, feminine of *minaudier* affected, coquettish, from *minauder* to mince] (1940)
: a small decorative case for carrying small articles (as cosmetics or jewelry)
1mince \'min(t)s\ *verb* **minced; minc·ing** [Middle English, from Middle French *mincer,* from (assumed) Vulgar Latin *minutiare,* from Latin *minutia* smallness — more at MINUTIA] (14th century)
transitive verb
1 a : to cut or chop into very small pieces **b :** to subdivide minutely; *especially* **:** to damage by cutting up
2 : to utter or pronounce with affectation
3 a *archaic* **:** MINIMIZE **b :** to restrain (words) within the bounds of decorum
intransitive verb
: to walk with short steps in a prim affected manner
— minc·er *noun*
2mince *noun* (1600)
1 : small chopped bits (as of food); *specifically* **:** MINCEMEAT
2 *British* **:** HAMBURGER 1a
mince·meat \'min(t)s-ˌmēt\ *noun* (1663)
1 : minced meat
2 : a finely chopped mixture (as of raisins, apples, and spices) sometimes with meat that is often used as pie filling
3 : a state of destruction or annihilation — used in the phrase *make mincemeat of*
minc·ing \'min(t)-siŋ\ *adjective* (1530)
: affectedly dainty or delicate
— minc·ing·ly \-siŋ-lē\ *adverb*
1mind \'mīnd\ *noun* [Middle English, from Old English *gemynd;* akin to Old High German *gimunt* memory, Latin *ment-, mens* mind, *monēre* to remind, warn, Greek *menos* spirit, *mnasthai, mimnēskesthai* to remember] (before 12th century)
1 : RECOLLECTION, MEMORY ⟨keep that in *mind*⟩ ⟨time out of *mind*⟩
2 a : the element or complex of elements in an individual that feels, perceives, thinks, wills, and especially reasons **b :** the conscious mental events and capabilities in an organism **c :** the organized conscious and unconscious adaptive mental activity of an organism
3 : INTENTION, DESIRE ⟨I changed my *mind*⟩
4 : the normal or healthy condition of the mental faculties
5 : OPINION, VIEW
6 : DISPOSITION, MOOD
7 a : a person or group embodying mental qualities ⟨the public *mind*⟩ **b :** intellectual ability
8 *capitalized, Christian Science* **:** GOD 1b
9 : a conscious substratum or factor in the universe
2mind (14th century)
transitive verb
1 *chiefly dialect* **:** REMIND
2 *chiefly dialect* **:** REMEMBER
3 : to attend to closely
4 a (1) **:** to become aware of **:** NOTICE (2) **:** to regard with attention **:** consider important — often used in the imperative with following *you* for emphasis ⟨I'm not against inspiration, *mind* you; I simply refuse to sit and stare at a blank page waiting for it —Dennis Whitcomb⟩ **b** *chiefly dialect* **:** INTEND, PURPOSE
5 a : to give heed to attentively in order to obey **b :** to follow the orders or instructions of
6 a : to be concerned about **b :** DISLIKE ⟨I don't *mind* going⟩
7 a : to be careful **:** SEE ⟨*mind* you finish it⟩ **b :** to be cautious about ⟨*mind* the broken rung⟩
8 : to give protective care to **:** TEND
intransitive verb
1 : to be attentive or wary
2 : to become concerned **:** CARE
3 : to pay obedient heed or attention
— mind·er *noun*
mind–bend·ing \'mīn(d)-ˌben-diŋ\ *adjective* (1965)
: MIND-BLOWING
mind–blow·ing \-ˌblō-iŋ\ *adjective* (1967)
1 : PSYCHEDELIC 1a
2 : MIND-BOGGLING
— mind·blow·er \-ˌblō(-ə)r\ *noun*
mind–bog·gling \-ˌbä-g(ə-)liŋ\ *adjective* (1964)
: mentally or emotionally exciting or overwhelming
— mind–bog·gling·ly \-lē\ *adverb*
mind·ed \'mīn-dəd\ *adjective* (15th century)
1 : INCLINED, DISPOSED
2 : having a mind especially of a specified kind or concerned with a specified thing — usually used in combination ⟨narrow-*minded*⟩ ⟨health-*minded*⟩
— mind·ed·ness \-dəd-nəs\ *noun*
mind–ex·pand·ing \'mīnd-ik-ˌspan-diŋ\ *adjective* (1963)
: PSYCHEDELIC 1a
mind·ful \'mīn(d)-fəl\ *adjective* (14th century)
1 : bearing in mind **:** AWARE
2 : inclined to be aware
— mind·ful·ly \-fə-lē\ *adverb*
— mind·ful·ness *noun*
mind·less \-ləs\ *adjective* (before 12th century)
1 a : marked by a lack of mind or consciousness ⟨a *mindless* sleep⟩ **b** (1) **:** marked by or displaying no use of the powers of the intellect ⟨*mindless* violence⟩ (2) **:** requiring little attention or thought; *especially* **:** not intellectually challenging or stimulating ⟨*mindless* work⟩ ⟨a *mindless* movie⟩
2 : not mindful **:** HEEDLESS ⟨*mindless* of the consequences⟩
— mind·less·ly *adverb*
— mind·less·ness *noun*
mind reader *noun* (1887)
: one that professes or is held to be able to perceive another's thought without normal means of communication
— mind reading *noun*
mind–set \'mīn(d)-ˌset\ *noun* (1926)
1 : a mental attitude or inclination
2 : a fixed state of mind
mind's eye *noun* (15th century)
: the mental faculty of conceiving imaginary or recollected scenes; *also* **:** the mental picture so conceived ⟨in the *mind's eye* one sees dinosaurs, mammoths, and sabertoothed tigers —F. P. Brooks Jr.⟩
1mine \'mīn\ *adjective* [Middle English *min,* from Old English *mīn* — more at MY] (before 12th century)
: MY — used before a word beginning with a vowel or *h* ⟨this treasure in *mine* arms —Shakespeare⟩ or sometimes as a modifier of a preceding noun; archaic except in an elevated style
2mine *pronoun singular or plural in construction* (before 12th century)
: that which belongs to me — used without a following noun as a pronoun equivalent in meaning to the adjective *my*
3mine *noun* [Middle English, from Middle French, from (assumed) Vulgar Latin *mina,* probably of Celtic origin; akin to Welsh *mwyn* ore] (14th century)
1 a : a pit or excavation in the earth from which mineral substances are taken **b :** an ore deposit
2 : a subterranean passage under an enemy position
3 : an encased explosive that is placed in the ground or in water and set to explode when disturbed
4 : a rich source of supply
4mine *verb* **mined; min·ing** (14th century)
transitive verb
1 a : to dig under to gain access or cause the collapse of (an enemy position) **b :** UNDERMINE
2 a : to get (as ore) from the earth **b :** to extract from a source ⟨information *mined* from the files⟩
3 : to burrow beneath the surface of ⟨larva that *mines* leaves⟩
4 : to place military mines in, on, or under ⟨*mine* a harbor⟩

5 a : to dig into for ore or metal **b :** to process for obtaining a natural constituent ⟨*mine* the air for nitrogen⟩ **c :** to seek valuable material in
intransitive verb
: to dig a mine
— min•er *noun*
mine•field \'mīn-ˌfēld\ *noun* (1886)
1 : an area (as of water or land) set with mines
2 : something resembling a minefield especially in having many dangers or requiring extreme caution
mine•lay•er \-ˌlā-ər, -ˌle(-ə)r\ *noun* (1909)
: a naval vessel for laying underwater mines
[1]min•er•al \'min-rəl, 'mi-nə-\ *noun* [Middle English, from Medieval Latin *minerale,* from neuter of *mineralis*] (15th century)
1 : ORE
2 : an inorganic substance (as in the ash of calcined tissue)
3 *obsolete* **:** MINE
4 : something neither animal nor vegetable
5 a : a solid homogeneous crystalline chemical element or compound that results from the inorganic processes of nature; *broadly* **:** any of various naturally occurring homogeneous substances (as stone, coal, salt, sulfur, sand, petroleum, water, or natural gas) obtained usually from the ground **b :** a synthetic substance having the chemical composition and crystalline form and properties of a naturally occurring mineral
6 *plural, British* **:** MINERAL WATER
[2]mineral *adjective* [Middle English, from Medieval Latin *mineralis,* from *minera* mine, ore, from Old French *miniere,* from *mine*] (15th century)
1 : of or relating to minerals; *also* **:** INORGANIC
2 : impregnated with mineral substances
min•er•al•ise *British variant of* MINERALIZE
min•er•al•ize \'min-rə-ˌlīz, 'mi-nə-\ *transitive verb* **-ized; -iz•ing** (1655)
1 : to transform (a metal) into an ore
2 a : to impregnate or supply with minerals or an inorganic compound **b :** to convert into mineral or inorganic form
3 : PETRIFY
— min•er•al•iz•able \ˌmin-rə-'lī-zə-bəl, ˌmi-nə-\ *adjective*
— min•er•al•i•za•tion \-rə-lə-'zā-shən\ *noun*
— min•er•al•iz•er \'min-rə-ˌlī-zər, 'mi-nə-\ *noun*
mineral kingdom *noun* (circa 1691)
: a basic group of natural objects that includes inorganic objects — compare ANIMAL KINGDOM, PLANT KINGDOM
min•er•al•o•cor•ti•coid \ˌmin-rə-lō-'kȯr-tə-ˌkȯid, ˌmi-nə-\ *noun* (1946)
: a corticosteroid (as aldosterone) that affects chiefly the electrolyte and fluid balance in the body — compare GLUCOCORTICOID
min•er•al•o•gy \ˌmi-nə-'rä-lə-jē, -'ra-\ *noun* [probably from (assumed) New Latin *mineralogia,* irregular from Medieval Latin *minerale* + Latin *-logia* -logy] (1690)
1 : a science dealing with minerals, their crystallography, properties, classification, and the ways of distinguishing them
2 : the minerological characteristics of an area, a rock, or a rock formation
— min•er•al•og•i•cal \ˌmi-nə-rə-'lä-ji-kəl, ˌmin-rə-\ *also* **min•er•al•og•ic** \-'lä-jik\ *adjective*
— min•er•al•og•i•cal•ly \-'lä-ji-k(ə-)lē\ *adverb*
— min•er•al•o•gist \ˌmi-nə-'rä-lə-jist, -'ra-\ *noun*
mineral oil *noun* (1805)
: an oil of mineral origin; *especially* **:** a refined petroleum oil used especially as a laxative
mineral spirits *noun plural but singular or plural in construction* (1927)
: a petroleum distillate that is used especially as a paint or varnish thinner
mineral water *noun* (1562)
: water naturally or artificially infused with mineral salts or gases (as carbon dioxide)
mineral wax *noun* (circa 1864)
: a wax of mineral origin; *especially* **:** OZOKERITE
mineral wool *noun* (circa 1881)
: any of various lightweight vitreous fibrous materials used especially in heat and sound insulation
Mi•ner•va \mə-'nər-və\ *noun* [Latin]
: the Roman goddess of wisdom — compare ATHENA
min•e•stro•ne \ˌmi-nə-'strō-nē, -'strōn\ *noun* [Italian, augmentative of *minestra,* from *minestrare* to serve, dish up, from Latin *ministrare,* from *minister* servant — more at MINISTER] (1891)
: a rich thick vegetable soup usually with dried beans and pasta (as macaroni or vermicelli)
mine•sweep•er \'mīn-ˌswē-pər\ *noun* (1905)
: a warship designed for removing or neutralizing mines by dragging
— mine•sweep•ing \-piŋ\ *noun*
Ming \'miŋ\ *noun* [Chinese (Beijing) *míng* luminous] (1795)
: a Chinese dynasty dated 1368–1644 and marked by restoration of earlier traditions and in the arts by perfection of established techniques
min•gle \'miŋ-gəl\ *verb* **min•gled; min•gling** \-g(ə-)liŋ\ [Middle English *menglen,* frequentative of *mengen* to mix, from Old English *mengan;* akin to Middle High German *mengen* to mix, Greek *massein* to knead] (15th century)
transitive verb
1 : to bring or mix together or with something else usually without fundamental loss of identity **:** INTERMIX
2 *archaic* **:** to prepare by mixing **:** CONCOCT
intransitive verb
1 : to become mingled
2 a : to come into contact **:** ASSOCIATE **b :** to move about (as in a group) ⟨*mingled* with the guests⟩
synonym see MIX
ming tree \'miŋ-\ *noun* [perhaps from *Ming*] (1948)
: a dwarfed evergreen conifer grown as bonsai; *also* **:** an artificial plant resembling this
min•gy \'min-jē\ *adjective* **min•gi•er; -est** [perhaps blend of [1]*mean* and *stingy*] (1912)
: MEAN, STINGY
[1]mini \'mi-nē\ *adjective* [*mini-*] (1954)
1 : small in relation to others of the same kind
2 : of short length or duration **:** BRIEF
[2]mini *noun, plural* **min•is** (1961)
: something small of its kind: as **a :** MINICAR **b :** MINISKIRT **c :** MINICOMPUTER
mini- *combining form* [*miniature*]
: smaller or briefer than usual, normal, or standard ⟨*mini*course⟩ ⟨*mini*bus⟩
[1]min•i•a•ture \'mi-nē-ə-ˌchu̇r, 'mi-ni-ˌchu̇r, 'min-yə-, -chər, -ˌtyu̇r, -ˌtu̇r\ *noun* [Italian *miniatura* art of illuminating a manuscript, from Medieval Latin, from Latin *miniatus,* past participle of *miniare* to color with minium, from *minium* red lead] (circa 1586)
1 a : a copy on a much reduced scale **b :** something small of its kind
2 : a painting in an illuminated book or manuscript
3 : the art of painting miniatures
4 : a very small portrait or other painting (as on ivory or metal) ◆
— min•i•a•tur•ist \-ˌchu̇r-ist, -chər-, -ˌtyu̇r, -ˌtu̇r\ *noun*
— min•i•a•tur•is•tic \ˌmi-nē-ə-chə-'ris-tik, ˌmi-ni-, ˌmin-yə-, -ˌtyu̇-, -ˌtu̇-\ *adjective*
— in miniature : in a greatly diminished size, form, or scale
[2]miniature *adjective* (1714)
: being or represented on a small scale
synonym see SMALL
miniature golf *noun* (1915)
: a novelty golf game played with a putter on a miniature course usually having tunnels, bridges, sharp corners, and obstacles
miniature pin•scher \-'pin(t)-shər\ *noun* (1929)
: any of a breed of toy dogs that suggest a small Doberman pinscher and are 10 to 12½ inches (25 to 32 centimeters) in height at the withers
miniature schnauzer *noun* (circa 1929)
: any of a breed of schnauzers that are 12 to 14 inches (30 to 36 centimeters) in height at the withers and are classified as terriers

miniature schnauzer

min•i•a•tur•ize \'mi-nē-ə-chə-ˌrīz, 'mi-ni-chə-, 'min-yə-chə-, -ˌtyu̇-, -ˌtu̇-\ *transitive verb* **-ized; -iz•ing** (1946)
: to design or construct in small size
— min•i•a•tur•i•za•tion \ˌmi-nē-ə-ˌchu̇r-ə-'zā-shən, ˌmi-ni-, ˌmin-yə-, -chər-, -ˌtyu̇r-, -ˌtu̇r-\ *noun*
mini•bike \'mi-nē-ˌbīk\ *noun* (1962)
: a small one-passenger motorcycle with a low frame and raised handlebars
— mini•bik•er *noun*
mini•bus \-ˌbəs\ *noun* (1958)
: a small bus or van
Mini•cam \-ˌkam\ *trademark*
— used for a portable television camera
mini•camp \-ˌkamp\ *noun* (1979)
: a special abbreviated training camp for football players held especially in the spring or early summer
mini•car \-ˌkär\ *noun* (1949)
: a very small automobile; *especially* **:** SUBCOMPACT
mini•com•put•er \ˌmi-nē-kəm-'pyü tər\ *noun* (1968)
: a small computer that is intermediate between a microcomputer and a mainframe in size, speed, and capacity, that can support time-sharing, and that is often dedicated to a single application
mini•course \'mi-nē-ˌkōrs, -ˌkȯrs\ *noun* (1970)
: a brief course of study usually lasting less than a semester

◇ WORD HISTORY

miniature The Latin name for the red coloring now commonly known as cinnabar or red lead was *minium.* Cinnabar, or *minium,* was used to embellish manuscripts in the era before the printing press. Titles, large initial letters, and decorative drawings were done in red as a contrast to the black ink of the regular text. The Latin verb meaning to color with minium was *miniare.* In early Italian, its meaning was broadened to the point where it simply meant "to decorate a manuscript," and the noun *miniature* was used to refer to any manuscript illustration, regardless of the colors used. Since the illustrations in manuscripts (called illuminations) are small by comparison with most other paintings, the word *miniature,* borrrowed into English from Italian *miniatura,* came to mean not only a manuscript illumination but any small portrait or painting, and eventually anything very small.

min·ié ball \'mi-nē-, ˌmi-nē-'ā-\ *noun* [Claude Étienne *Minié* (died 1879) French army officer] (1859)
: a rifle bullet with a conical head used in muzzle-loading firearms

min·i·fy \'mi-nə-ˌfī\ *transitive verb* **-fied; -fy·ing** [Latin *min*us less + English *-ify*] (1676)
: LESSEN

min·i·kin \'mi-ni-kən\ *noun* [obsolete Dutch *minneken* darling, ultimately from Middle Dutch *minne* love, beloved; akin to Old English *gemynd* mind, memory — more at MIND] (1761)
archaic **:** a small or dainty creature
— minikin *adjective*

mini·lab \'mi-nē-ˌlab\ *noun* (1982)
: a retail outlet offering rapid on-site film development and printing

min·im \'mi-nəm\ *noun* [Latin *minimus* least] (15th century)
1 : HALF NOTE
2 : something very minute
3 — see WEIGHT table
— minim *adjective*

min·i·mal \'mi-nə-məl\ *adjective* (1666)
1 : relating to or being a minimum: as **a :** the least possible ⟨a victory won with *minimal* loss of life⟩ **b :** barely adequate ⟨a *minimal* standard of living⟩ **c :** very small or slight ⟨a *minimal* interest in art⟩
2 *often capitalized* **:** of, relating to, or being minimal art or minimalism
— min·i·mal·ly \-mə-lē\ *adverb*

minimal art *noun* (1965)
: abstract art consisting primarily of simple geometric forms executed in an impersonal style

minimal brain dysfunction *noun* (1972)
: ATTENTION DEFICIT DISORDER

min·i·mal·ism \'mi-nə-mə-ˌli-zəm\ *noun* (1969)
1 : MINIMAL ART
2 : a style or technique (as in music, literature, or design) that is characterized by extreme spareness and simplicity

[1]**min·i·mal·ist** \-list\ *noun* (1907)
1 : one who favors restricting the functions and powers of a political organization or the achievement of a set of goals to a minimum
2 a : a minimal artist **b :** an adherent of minimalism

[2]**minimalist** *adjective* (1967)
: of, relating to, or done in the style of minimalism

minimal pair *noun* (1942)
: two linguistic units that differ in a single distinctive feature or constituent (as voice in the initial consonants of *bat* and *pat*)

mini·max \'mi-ni-ˌmaks\ *noun* [*mini*mum + *max*imum] (1918)
: the minimum of a set of maxima; *especially* **:** the smallest of a set of maximum possible losses each of which occurs in the most unfavorable outcome of a strategy followed by a participant in a situation governed by the theory of games — compare MAXIMIN

mini·mill \'mi-nē-ˌmil\ *noun* (1969)
: a relatively small-scale steel mill that uses scrap metal as starting material

min·i·mise *British variant of* MINIMIZE

min·i·mize \'mi-nə-ˌmīz\ *transitive verb* **-mized; -miz·ing** (1802)
1 : to reduce or keep to a minimum
2 : to underestimate intentionally **:** play down **:** SOFT-PEDAL ⟨*minimizing* losses in our own forces while maximizing those of the enemy⟩
— min·i·mi·za·tion \ˌmi-nə-mə-'zā-shən\ *noun*
— min·i·miz·er \'mi-nə-ˌmī-zər\ *noun*

min·i·mum \'mi-nə-məm\ *noun, plural* **-i·ma** \-nə-mə\ *or* **-i·mums** [Latin, neuter of *minimus* smallest; akin to Latin *minor* smaller — more at MINOR] (1674)
1 : the least quantity assignable, admissible, or possible
2 : the least of a set of numbers; *specifically* **:** the smallest value assumed by a continuous function defined on a closed interval
3 a : the lowest degree or amount of variation (as of temperature) reached or recorded **b :** the lowest speed allowed on a highway
— minimum *adjective*

minimum wage *noun* (1860)
1 : LIVING WAGE
2 : the lowest wage paid or permitted to be paid; *specifically* **:** a wage fixed by legal authority or by contract as the least that may be paid either to employed persons generally or to a particular category of employed persons

min·ing \'mī-niŋ\ *noun* (1523)
: the process or business of working mines

min·ion \'min-yən\ *noun* [Middle French *mignon* darling] (1501)
1 : a servile dependent, follower, or underling
2 : one highly favored **:** IDOL
3 : a subordinate or petty official

mini·park \'mi-nē-ˌpärk\ *noun* (1967)
: a small city park

mini·school \-ˌskül\ *noun* (1968)
: an experimental school offering specialized or individual instruction to its students

min·is·cule \'mi-nəs-ˌkyü(ə)l\ *variant of* MINUSCULE

mini·se·ries \'mi-nē-ˌsir-(ˌ)ēz\ *noun* (1973)
: a television production of a story presented in sequential episodes

mini·skirt \'mi-nē-ˌskərt\ *noun* (1965)
: a woman's short skirt with the hemline several inches above the knee
— mini·skirt·ed *adjective*

mini·state \-ˌstāt\ *noun* (1966)
: a small independent nation

[1]**min·is·ter** \'mi-nə-stər\ *noun* [Middle English *ministre,* from Old French, from Latin *minister* servant; akin to Latin *minor* smaller] (14th century)
1 : AGENT
2 a : one officiating or assisting the officiant in church worship **b :** a clergyman especially of a Protestant communion
3 a : the superior of one of several religious orders — called also *minister-general* **b :** the assistant to the rector or the bursar of a Jesuit house
4 : a high officer of state entrusted with the management of a division of governmental activities
5 a : a diplomatic representative (as an ambassador) accredited to the court or seat of government of a foreign state **b :** a diplomatic representative ranking below an ambassador

[2]**minister** *intransitive verb* **-tered; -ter·ing** \-st(ə-)riŋ\ (14th century)
1 : to perform the functions of a minister of religion
2 : to give aid or service ⟨*minister* to the sick⟩

min·is·te·ri·al \ˌmi-nə-'stir-ē-əl\ *adjective* (1561)
1 : of, relating to, or characteristic of a minister or the ministry
2 a : being or having the characteristics of an act or duty prescribed by law as part of the duties of an administrative office **b :** relating to or being an act done after ascertaining the existence of a specified state of facts in obedience to a legal order without exercise of personal judgment or discretion
3 : acting or active as an agent **:** INSTRUMENTAL
— min·is·te·ri·al·ly \-ē-ə-lē\ *adverb*

minister plenipotentiary *noun, plural* **ministers plenipotentiary** (1796)
: a diplomatic agent ranking below an ambassador but possessing full power and authority

minister resident *noun, plural* **ministers resident** (1848)
: a diplomatic agent resident at a foreign court or seat of government and ranking below a minister plenipotentiary

[1]**min·is·trant** \'mi-nə-strənt\ *adjective* (1667)
archaic **:** performing service in attendance on someone

[2]**ministrant** *noun* (1818)
: one that ministers

min·is·tra·tion \ˌmi-nə-'strā-shən\ *noun* (14th century)
: the act or process of ministering

min·is·try \'mi-nə-strē\ *noun, plural* **-tries** (14th century)
1 : MINISTRATION
2 : the office, duties, or functions of a minister
3 : the body of ministers of religion **:** CLERGY
4 : AGENCY 2, INSTRUMENTALITY
5 : the period of service or office of a minister or ministry
6 *often capitalized* **a :** the body of ministers governing a nation or state from which a smaller cabinet is sometimes selected **b :** the group of ministers constituting a cabinet
7 a : a government department presided over by a minister **b :** the building in which the business of a ministry is transacted

mini·van \'mi-nē-ˌvan\ *noun* (1960)
: a small van

min·i·ver \'mi-nə-vər\ *noun* [Middle English *meniver,* from Middle French *menu vair* small vair] (13th century)
: a white fur worn originally by medieval nobles and used chiefly for robes of state

mink \'miŋk\ *noun, plural* **mink** *or* **minks** [Middle English] (15th century)
1 : soft fur or pelt of the mink varying in color from white to dark brown
2 : either of two slender-bodied semiaquatic carnivorous mammals (*Mustela vison* of North America and *M. lutreola* of Eurasia) that resemble the related weasels and have partially webbed feet, a rather short bushy tail, and a soft thick coat

mink 2

min·ke whale \'miŋ-kə-\ *noun* [part translation of Norwegian *minkehval,* from *minke-* (perhaps from *Meincke,* a crewman of Svend Foyn (died 1894) Norwegian whaler) + *hval* whale] (1939)
: a small grayish baleen whale (*Balaenoptera acutorostrata*) with a whitish underside — called also *minke*

min·ne·sing·er \'mi-nə-ˌsiŋ-ər, -ˌziŋ-\ *noun* [German, from Middle High German, from *minne* love + *singer* singer] (1825)
: any of a class of German poets and musicians of the 12th to the 14th centuries

Min·ne·so·ta Multiphasic Personality Inventory \ˌmi-nə-'sō-tə-ˌməl-ti-'fā-zik-, -ˌməl-ˌtī-\ *noun* [University of *Minnesota*] (1943)
: a test of personal and social adjustment based on a complex scaling of the answers to an elaborate true or false test

min·now \'mi-(ˌ)nō\ *noun, plural* **minnows** *also* **minnow** [Middle English *menawe;* akin to Old English *myne* minnow, Old High German *munewa,* a kind of fish] (15th century)
1 a : a small cyprinid, killifish, or topminnow **b :** any of various small fish that are less than a designated size and are not game fish
2 : a live or artificial minnow used as bait

[1]**Mi·no·an** \mə-'nō-ən, mī-\ *adjective* [Latin *minous* of Minos, from Greek *minōios,* from *Minōs* Minos] (1894)
: of or relating to a Bronze Age culture of Crete (3000 B.C.–1100 B.C.)

[2]**Minoan** *noun* (1902)
: a native or inhabitant of ancient Crete

[1]**mi·nor** \'mī-nər\ *adjective* [Latin, smaller, inferior; akin to Old High German *minniro* smaller, Latin *minuere* to lessen] (1526)
1 : inferior in importance, size, or degree **:** comparatively unimportant
2 : not having reached majority
3 a : having half steps between the second and third, the fifth and sixth, and sometimes the

seventh and eighth degrees ⟨*minor* scale⟩ **b :** based on a minor scale ⟨*minor* key⟩ **c :** less by a semitone than the corresponding major interval ⟨*minor* third⟩ **d :** having a minor third above the root ⟨*minor* triad⟩
4 : not serious or involving risk to life ⟨*minor* illness⟩
5 : of or relating to an academic subject requiring fewer courses than a major

²minor *noun* (1612)
1 : a person who has not attained majority
2 : a minor musical interval, scale, key, or mode
3 a : a minor academic subject **b :** a student taking a specified minor
4 : a determinant or matrix obtained from a given determinant or matrix by eliminating the row and column in which a given element lies
5 *plural* **:** minor league baseball — used with *the*

³minor *intransitive verb* (1926)
: to take courses in a minor subject

minor axis *noun* (1862)
: the chord of an ellipse passing through the center and perpendicular to the major axis

minor element *noun* (circa 1945)
: TRACE ELEMENT

Mi·nor·ite \'mī-nə-ˌrīt\ *noun* [from *Friar Minor* Franciscan] (circa 1587)
: FRANCISCAN

mi·nor·i·ty \mə-'nȯr-ə-tē, mī-, -'när-\ *noun, plural* **-ties** *often attributive* (1547)
1 a : the period before attainment of majority **b :** the state of being a legal minor
2 : the smaller in number of two groups constituting a whole; *specifically* **:** a group having less than the number of votes necessary for control
3 a : a part of a population differing from others in some characteristics and often subjected to differential treatment **b :** a member of a minority group ⟨an effort to hire more *minorities*⟩

minority leader *noun* (1949)
: the leader of the minority party in a legislative body

minor league *noun* (1889)
: a league of professional clubs in a sport other than the recognized major leagues

minor order *noun* (1844)
: one of the Roman Catholic or Eastern clerical orders that are lower in rank and less sacred in character than major orders — usually used in plural

minor party *noun* (1949)
: a political party whose electoral strength is so small as to prevent its gaining control of a government except in rare and exceptional circumstances

minor penalty *noun* (1936)
: a two-minute suspension of a player in ice hockey with no substitute allowed

minor planet *noun* (1861)
: ASTEROID

minor premise *noun* (circa 1741)
: the premise of a syllogism that contains the minor term

minor seminary *noun* (circa 1948)
: a Roman Catholic seminary giving all or part of high school and junior college training with emphasis on preparing candidates for a major seminary

minor suit *noun* (1916)
: either of the suits diamonds or clubs having inferior scoring value in bridge

minor term *noun* (1843)
: the term of a syllogism that forms the subject of the conclusion

Mi·nos \'mī-nəs\ *noun* [Latin, from Greek *Minōs*]
: a son of Zeus and Europa and king of Crete who for his just rule is made supreme judge in the underworld after his death

Mi·no·taur \'mi-nə-ˌtȯr, 'mī-\ *noun* [Middle English, from Middle French, from Latin *Minotaurus,* from Greek *Minōtauros,* from *Minōs* + *tauros* bull]
: a monster shaped half like a man and half like a bull, confined in the labyrinth built by Daedalus for Minos, and given a periodic tribute of youths and maidens as food until slain by Theseus

mi·nox·i·dil \mə-'näk-sə-ˌdil\ *noun* [perhaps from *amino* + *oxi-* (alteration of *oxy-*) + *piperidine* + *-yl*] (1972)
: a peripheral vasodilator $C_9H_{15}N_5O$ used orally to treat hypertension and topically in a propylene glycol solution to promote hair regrowth in some forms of baldness

min·ster \'min(t)-stər\ *noun* [Middle English, monastery, church attached to a monastery, from Old English *mynster,* from Late Latin *monasterium* monastery] (before 12th century)
: a large or important church often having cathedral status

min·strel \'min(t)-strəl\ *noun* [Middle English *menestrel,* from Middle French, official, servant, minstrel, from Late Latin *ministerialis* imperial household officer, from Latin *ministerium* service, from *minister* servant — more at MINISTER] (14th century)
1 : one of a class of medieval musical entertainers; *especially* **:** a singer of verses to the accompaniment of a harp
2 a : MUSICIAN **b :** POET
3 a : any of a troupe of performers typically giving a program of black American melodies, jokes, and impersonations and usually wearing blackface **b :** a performance by a troupe of minstrels

min·strel·sy \-sē\ *noun* [Middle English *minstralcie,* from Middle French *menestralsie,* from *menestrel*] (14th century)
1 : the singing and playing of a minstrel
2 : a body of minstrels
3 : a group of songs or verse

¹mint \'mint\ *noun* [Middle English *minte,* from Old English, from Latin *mentha, menta;* akin to Greek *minthē* mint] (before 12th century)
1 : any of a family (Labiatae, the mint family) of aromatic plants with a square stem and a 4-lobed ovary which produces four one-seeded nutlets in fruit; *especially* **:** any of a genus (*Mentha*) of mints that have white, purple, or pink verticillate flowers with a nearly regular corolla and four equal stamens and that include some used in flavoring and cookery
2 : a confection flavored with mint
— minty \'min-tē\ *adjective*

²mint *noun* [Middle English *mynt* coin, money, from Old English *mynet,* from Latin *moneta* mint, coin, from *Moneta,* epithet of Juno; from the fact that the Romans coined money in the temple of Juno Moneta] (15th century)
1 : a place where coins, medals, or tokens are made
2 : a place where something is manufactured
3 : a vast sum or amount

³mint *transitive verb* (1546)
1 : to make (as coins) out of metal **:** COIN
2 : CREATE, PRODUCE
— mint·er *noun*

⁴mint *adjective* (1902)
: unmarred as if fresh from a mint ⟨in *mint* condition⟩

mint·age \'min-tij\ *noun* (circa 1570)
1 : the action or process of minting coins
2 : an impression placed upon a coin
3 : coins produced by minting or in a single period of minting

mint julep *noun* (1809)
: JULEP 2

min·u·end \'min-yə-ˌwend\ *noun* [Latin *minuendum,* neuter of *minuendus,* gerundive of *minuere* to lessen — more at MINOR] (1706)
: a number from which the subtrahend is to be subtracted

min·u·et \ˌmin-yə-'wet\ *noun* [French *menuet,* from obsolete French, tiny, from Old French, from *menu* small, from Latin *minutus*] (1673)
1 : a slow graceful dance in ¾ time characterized by forward balancing, bowing, and toe pointing
2 : music for or in the rhythm of a minuet

¹mi·nus \'mī-nəs\ *preposition* [Middle English, from Latin *minus,* adverb, less, from neuter of *minor* smaller — more at MINOR] (15th century)
1 : diminished by **:** LESS ⟨seven *minus* four is three⟩
2 : deprived of **:** WITHOUT ⟨*minus* his hat⟩

²minus *noun* (1654)
1 : a negative quantity
2 : DEFICIENCY, DEFECT

³minus *adjective* (1800)
1 : algebraically negative ⟨a *minus* quantity⟩
2 : having negative qualities
3 : relating to or being a particular one of the two mating types that are required for successful fertilization in sexual reproduction in some lower plants (as a fungus)
4 : falling low in a specified range ⟨B *minus*⟩

¹mi·nus·cule \'mi-nəs-ˌkyü(ə)l *also* mi-'nəs-\ *noun* [French, from Latin *minusculus* rather small, diminutive of *minor* smaller] (1705)
1 a : one of several ancient and medieval writing styles developed from cursive and having simplified and small forms **b :** a letter in this style
2 : a lowercase letter

²minuscule *adjective* (circa 1741)
1 : written in or in the size or style of minuscules
2 : very small

minus sign *noun* (1668)
: a sign − used in mathematics to indicate subtraction (as in $8-6=2$) or a negative quantity (as in $-10°$)

¹min·ute \'mi-nət\ *noun* [Middle English, from Middle French, from Late Latin *minuta,* from Latin *minutus* small, from past participle of *minuere* to lessen — more at MINOR] (14th century)
1 : a 60th part of an hour of time or of a degree **:** 60 seconds
2 : the distance one can traverse in a minute
3 : a short space of time **:** MOMENT
4 a : a brief note (as of summary or recommendation) **b :** MEMORANDUM, DRAFT **c** *plural* **:** the official record of the proceedings of a meeting

²minute *transitive verb* **min·ut·ed; min·ut·ing** (circa 1648)
: to make notes or a brief summary of

³mi·nute \mī-'nüt, mə-, -'nyüt\ *adjective* **mi·nut·er; -est** [Latin *minutus*] (circa 1626)
1 : very small **:** INFINITESIMAL
2 : of small importance **:** TRIFLING
3 : marked by close attention to details
synonym see SMALL, CIRCUMSTANTIAL
— mi·nute·ness *noun*

minute hand *noun* (1726)
: the long hand that marks the minutes on the face of a watch or clock

¹mi·nute·ly \mī-'nüt-lē, mə-, -'nyüt-\ *adverb* (1599)
1 : into very small pieces
2 : in a minute manner or degree

²min·ute·ly \'mi-nət-lē\ *adjective* (1605)
archaic **:** minute by minute

min·ute·man \'mi-nət-ˌman\ *noun* (1774)
: a member of a group of armed men pledged to take the field at a minute's notice during and immediately before the American Revolution

min·ute steak \'mi-nət-\ *noun* (1921)
: a small thin steak that can be quickly cooked

mi·nu·tia \mə-'nü-sh(ē-)ə, mī-, -'nyü-\ *noun, plural* **-ti·ae** \-shē-,ē, -,ī\ [Latin *minutiae* trifles, details, from plural of *minutia* smallness, from *minutus*] (1751)
: a minute or minor detail — usually used in plural

minx \'miŋ(k)s\ *noun* [origin unknown] (1592)
1 : a pert girl
2 *obsolete* **:** a wanton woman

min·yan \'min-yən\ *noun, plural* **-ya·nim** \,min-yə-'nēm\ *or* **-yans** [Hebrew *minyān,* literally, number, count] (1753)
: the quorum of 10 adult Jews required for communal worship

Mio·cene \'mī-ə-,sēn\ *adjective* [*mio-* (from Greek *meiōn* less) + *-cene* — more at MEIOSIS] (1831)
: of, relating to, or being an epoch of the Tertiary between the Pliocene and the Oligocene or the corresponding system of rocks — see GEOLOGIC TIME table
— Miocene *noun*

mi·o·sis \mī-'ō-səs, mē-\ *noun, plural* **mi·o·ses** \-,sēz\ [New Latin, from Greek *myein* to be closed (of the eyes) + New Latin *-osis*] (1819)
: excessive smallness or contraction of the pupil of the eye

¹mi·ot·ic \-'ä-tik\ *noun* (1864)
: an agent that causes miosis

²miotic *adjective* (1864)
: relating to or characterized by miosis

mi·que·let \,mi-kə-'let, ,mē-\ *noun* [Spanish *miquelete*] (1827)
: a Spanish or French irregular soldier during the Peninsular War

mir \'mir\ *noun* [Russian] (1877)
: a village community in czarist Russia in which land was owned jointly but cultivated by individual families

mi·ra·bi·le dic·tu \mə-,rä-bə-lē-'dik-(,)tü\ [Latin] (1831)
: wonderful to relate

mi·ra·cid·i·um \,mir-ə-'si-dē-əm, ,mī-rə-\ *noun, plural* **-cid·ia** \-dē-ə\ [New Latin, from Greek *meirak-, meirax* youth, stripling + New Latin *-idium*] (1898)
: the free-swimming ciliated first larva of a digenetic trematode that seeks out and penetrates a suitable snail intermediate host in which it develops into a sporocyst
— mi·ra·cid·i·al \-dē-əl\ *adjective*

mir·a·cle \'mir-i-kəl\ *noun* [Middle English, from Old French, from Late Latin *miraculum,* from Latin, a wonder, marvel, from *mirari* to wonder at] (12th century)
1 : an extraordinary event manifesting divine intervention in human affairs
2 : an extremely outstanding or unusual event, thing, or accomplishment
3 *Christian Science* **:** a divinely natural phenomenon experienced humanly as the fulfillment of spiritual law

miracle drug *noun* (1950)
: a drug usually newly discovered that elicits a dramatic response in a patient's condition — called also *wonder drug*

miracle fruit *noun* (1964)
: a tropical African shrub (*Synsepalum dulcificum*) of the sapodilla family whose fruit contains a glycoprotein that when applied to the tongue causes sour substances to taste sweet; *also* **:** its fruit

miracle play *noun* (circa 1852)
1 : a medieval drama based on episodes from the life of a saint or martyr
2 : MYSTERY PLAY

mi·rac·u·lous \mə-'ra-kyə-ləs\ *adjective* [Middle English, from Middle French *miraculeux,* from Medieval Latin *miraculosus,* from Latin *miraculum*] (15th century)
1 : of the nature of a miracle **:** SUPERNATURAL ⟨a *miraculous* event⟩
2 : suggesting a miracle **:** MARVELOUS ⟨gave proof of a *miraculous* memory —*Time*⟩
3 : working or able to work miracles ⟨*miraculous* power⟩
— mi·rac·u·lous·ly *adverb*
— mi·rac·u·lous·ness *noun*

mir·a·dor \'mir-ə-,dȯr, ,mir-ə-'\ *noun* [Spanish, from Catalan, from *mirar* to look at, from Latin *mirari*] (1797)
: a turret, window, or balcony designed to command an extensive outlook

mi·rage \mə-'räzh\ *noun* [French, from *mirer* to look at, from Latin *mirari*] (1803)
1 : an optical effect that is sometimes seen at sea, in the desert, or over a hot pavement, that may have the appearance of a pool of water or a mirror in which distant objects are seen inverted, and that is caused by the bending or reflection of rays of light by a layer of heated air of varying density
2 : something illusory and unattainable like a mirage

Mi·ran·da \mə-'ran-də\ *adjective* [from *Miranda v. Arizona,* the U.S. Supreme Court ruling establishing such rights] (1972)
: of, relating to, or being the legal rights of an arrested person to have an attorney and to remain silent so as to avoid self-incrimination ⟨*Miranda* warnings⟩

¹mire \'mīr\ *noun* [Middle English, from Old Norse *mȳrr;* akin to Old English *mōs* marsh — more at MOSS] (14th century)
1 : wet spongy earth (as of a bog or marsh)
2 : heavy often deep mud or slush
3 : a troublesome or intractable situation ⟨found themselves in a *mire* of debt⟩
— miry \'mīr-ē\ *adjective*

²mire *verb* **mired; mir·ing** (15th century)
transitive verb
1 a : to cause to stick fast in or as if in mire **b :** to hamper or hold back as if by mire **:** ENTANGLE
2 : to cover or soil with mire
intransitive verb
: to stick or sink in mire

mi·rex \'mī-,reks\ *noun* [origin unknown] (1962)
: an organochlorine insecticide $C_{10}Cl_{12}$ used especially against ants

mirk, mirky *variant of* MURK, MURKY

mir·li·ton \,mir-lə-'tōⁿ\ *noun* [French] (circa 1909)
: CHAYOTE

¹mir·ror \'mir-ər\ *noun* [Middle English *mirour,* from Old French, from *mirer* to look at, from Latin *mirari* to wonder at] (13th century)
1 : a polished or smooth surface (as of glass) that forms images by reflection
2 a : something that gives a true representation **b :** an exemplary model
— mir·rored \-ə(r)d\ *adjective*
— mir·ror·like \-,līk\ *adjective*

²mirror *transitive verb* (1593)
: to reflect in or as if in a mirror

mirror image *noun* (1885)
: something that has its parts reversely arranged in comparison with another similar thing or that is reversed with reference to an intervening axis or plane

mirth \'mərth\ *noun* [Middle English, from Old English *myrgth,* from *myrge* merry — more at MERRY] (before 12th century)
: gladness or gaiety as shown by or accompanied with laughter
— mirth·ful \-fəl\ *adjective*
— mirth·ful·ly \-fə-lē\ *adverb*
— mirth·ful·ness *noun*
— mirth·less \-ləs\ *adjective*
— mirth·less·ly \-lē\ *adverb*

¹MIRV \'mərv\ *noun* [*m*ultiple *i*ndependently targeted *r*eentry *v*ehicle] (1967)
: a missile with two or more warheads designed to strike separate enemy targets; *also* **:** any of the warheads of such a missile

²MIRV *verb* **MIRVed; MIRV·ing** (1969)
transitive verb
: to equip with MIRV warheads
intransitive verb
: to arm one's forces with MIRVs

mis- *prefix* [partly from Middle English, from Old English; partly from Middle English *mes-, mis-,* from Old French *mes-,* of Germanic origin; akin to Old English *mis-;* akin to Old English *missan* to miss]
1 a : badly **:** wrongly ⟨*mis*judge⟩ **b :** unfavorably ⟨*mis*esteem⟩ **c :** in a suspicious manner ⟨*mis*doubt⟩
2 : bad **:** wrong ⟨*mis*deed⟩
3 : opposite or lack of ⟨*mis*trust⟩
4 : not ⟨*mis*know⟩

mis·act
mis·ad·dress
mis·ad·just
mis·ad·min·is·tra·tion
mis·ad·vise
mis·aim
mis·align
mis·align·ment
mis·al·lo·cate
mis·al·lo·ca·tion
mis·anal·y·sis
mis·ap·pli·ca·tion
mis·ap·ply
mis·ap·prais·al
mis·ar·tic·u·late
mis·as·sem·ble
mis·as·sump·tion
mis·at·trib·ute
mis·at·tri·bu·tion
mis·bal·ance
mis·be·have
mis·be·hav·er
mis·be·hav·ior
mis·bound
mis·but·ton
mis·cal·cu·late
mis·cal·cu·la·tion
mis·cap·tion
mis·cat·a·log
mis·chan·nel
mis·char·ac·ter·i·za·tion
mis·char·ac·ter·ize
mis·charge
mis·choice
mis·ci·ta·tion
mis·clas·si·fi·ca·tion
mis·clas·si·fy
mis·code
mis·com·pre·hen·sion
mis·com·pu·ta·tion
mis·com·pute
mis·con·ceive
mis·con·ceiv·er
mis·con·cep·tion
mis·con·nect
mis·con·nec·tion
mis·con·struc·tion
mis·con·strue
mis·copy
mis·cor·re·la·tion
mis·cre·ate
mis·cre·a·tion
mis·cut
mis·date
mis·deem
mis·de·fine
mis·de·scribe
mis·de·scrip·tion
mis·de·vel·op
mis·di·ag·nose
mis·di·ag·no·sis
mis·dial
mis·dis·tri·bu·tion
mis·di·vi·sion
mis·draw
mis·ed·u·cate
mis·ed·u·ca·tion
mis·em·pha·sis
mis·em·pha·size
mis·em·ploy
mis·em·ploy·ment
mis·es·ti·mate
mis·es·ti·ma·tion
mis·eval·u·ate
mis·eval·u·a·tion
mis·field
mis·file
mis·fo·cus
mis·func·tion
mis·gauge
mis·gov·ern
mis·gov·ern·ment
mis·grade
mis·iden·ti·fi·ca·tion
mis·iden·ti·fy
mis·in·form
mis·in·for·ma·tion
mis·kick
mis·la·bel
mis·learn
mis·lo·cate
mis·lo·ca·tion
mis·man·age
mis·man·age·ment
mis·mark
mis·mar·riage
mis·match
mis·mate
mis·or·der
mis·ori·ent
mis·ori·en·ta·tion
mis·pack·age
mis·per·ceive
mis·per·cep·tion
mis·plan
mis·po·si·tion
mis·print
mis·pro·gram
mis·quo·ta·tion
mis·quote
mis·rec·ol·lec·tion
mis·re·cord
mis·ref·er·ence
mis·reg·is·ter
mis·reg·is·tra·tion
mis·re·late
mis·re·mem·ber
mis·ren·der
mis·re·port
mis·route
mis·set
mis·shape
mis·shap·en
mis·shap·en·ly
mis·sort
mis·strike
mis·throw
mis·time
mis·ti·tle
mis·train
mis·tran·scribe
mis·tran·scrip·tion
mis·trans·late
mis·trans·la·tion
mis·truth
mis·tune
mis·type
mis·uti·li·za·tion
mis·vo·cal·i·za·tion
mis·write

mis·ad·ven·ture \ˌmi-səd-ˈven-chər\ *noun* [Middle English *mesaventure,* from Old French, from *mesavenir* to chance badly, from *mes-* mis- + *avenir* to chance, happen, from Latin *advenire* — more at ADVENTURE] (14th century)
: MISFORTUNE, MISHAP

mis·al·li·ance \ˌmi-sə-ˈlī-ən(t)s\ *noun* [modification of French *mésalliance*] (1738)
1 : an improper alliance
2 a : MÉSALLIANCE **b :** a marriage between persons unsuited to each other

mis·an·thrope \ˈmi-sᵊn-ˌthrōp\ *noun* [Greek *misanthrōpos* hating mankind, from *misein* to hate + *anthrōpos* human being] (1683)
: a person who hates or distrusts mankind

mis·an·throp·ic \ˌmi-sᵊn-ˈthrä-pik\ *adjective* (1762)
1 : of, relating to, or characteristic of a misanthrope
2 : marked by a hatred or contempt for mankind
synonym see CYNICAL
— mis·an·throp·i·cal·ly \-pi-k(ə-)lē\ *adverb*

mis·an·thro·py \mi-ˈsan(t)-thrə-pē\ *noun* (circa 1656)
: a hatred or distrust of mankind

mis·ap·pre·hend \(ˌ)mis-ˌa-pri-ˈhend\ *transitive verb* (circa 1629)
: to apprehend wrongly **:** MISUNDERSTAND
— mis·ap·pre·hen·sion \-ˈhen(t)-shən\ *noun*

mis·ap·pro·pri·ate \ˌmi-sə-ˈprō-prē-ˌāt\ *transitive verb* (1857)
: to appropriate wrongly (as by theft or embezzlement)
— mis·ap·pro·pri·a·tion \-ˌprō-prē-ˈā-shən\ *noun*

mis·be·come \ˌmis-bi-ˈkəm\ *transitive verb* **-came** \-ˈkām\; **-come; -com·ing** (1530)
: to be inappropriate or unbecoming to

mis·be·got·ten \-bi-ˈgä-tᵊn\ *adjective* (1554)
1 : unlawfully conceived **:** ILLEGITIMATE ⟨a *misbegotten* child⟩
2 a : having a disreputable or improper origin **:** ill-conceived ⟨antiquated and *misbegotten* tax laws —R. M. Blough⟩ **b :** CONTEMPTIBLE, DEFORMED ⟨a *misbegotten* scoundrel⟩

mis·be·lief \ˌmis-bə-ˈlēf\ *noun* (13th century)
: erroneous or false belief

mis·be·lieve \-ˈlēv\ *intransitive verb* (14th century)
obsolete **:** to hold a false or unorthodox belief

mis·be·liev·er \-ˈlē-vər\ *noun* (15th century)
: HERETIC, INFIDEL

mis·brand \ˌmis-ˈbrand\ *transitive verb* (circa 1930)
: to brand falsely or in a misleading way; *specifically* **:** to label in violation of statutory requirements

mis·call \ˌmis-ˈkȯl\ *transitive verb* (14th century)
: to call by a wrong name **:** MISNAME

mis·car·riage \ˌmis-ˈkar-ij, ˈmis-ˌ\ *noun* (circa 1652)
1 : corrupt or incompetent management; *especially* **:** a failure in the administration of justice
2 : spontaneous expulsion of a human fetus before it is viable and especially between the 12th and 28th weeks of gestation

mis·car·ry \ˌmis-ˈkar-ē, ˈmis-ˌkar-ē\ *intransitive verb* (14th century)
1 *obsolete* **:** to come to harm
2 : to suffer miscarriage of a fetus
3 : to fail to achieve the intended purpose **:** go wrong or amiss ⟨the plan *miscarried*⟩

mis·cast \ˌmis-ˈkast\ *transitive verb* **-cast; -cast·ing** (1925)
: to cast in an unsuitable role ⟨life had *miscast* her in the role of wife and mother —Edna Ferber⟩

mis·ce·ge·na·tion \(ˌ)mi-ˌse-jə-ˈnā-shən, ˌmi-si-jə-ˈnā-\ *noun* [irregular from Latin *miscēre* to mix + *genus* race — more at MIX, KIN] (1864)
: a mixture of races; *especially* **:** marriage or cohabitation between a white person and a member of another race
— mis·ce·ge·na·tion·al \-shnəl, -shə-nᵊl\ *adjective*

mis·cel·la·nea \ˌmi-sə-ˈlā-nē-ə, -nyə\ *noun plural* [Latin, from neuter plural of *miscellaneus*] (1571)
: a collection of miscellaneous objects or writings

mis·cel·la·neous \ˌmi-sə-ˈlā-nē-əs, -nyəs\ *adjective* [Latin *miscellaneus,* from *miscellus* mixed] (1637)
1 : consisting of diverse things or members **:** HETEROGENEOUS
2 a : having various traits **b :** dealing with or interested in diverse subjects ⟨as a writer I was too *miscellaneous* —George Santayana⟩
— mis·cel·la·neous·ly *adverb*
— mis·cel·la·neous·ness *noun*

mis·cel·la·nist \ˈmi-sə-ˌlā-nist, *chiefly British* mi-ˈse-lə-nist\ *noun* (1810)
: a writer of miscellanies

mis·cel·la·ny \-nē\ *noun, plural* **-nies** [probably modification of French *miscellanées,* plural, from Latin *miscellanea*] (1615)
1 a *plural* **:** separate writings collected in one volume **b :** a collection of writings on various subjects
2 : a mixture of various things

mis·chance \ˌmis-ˈchan(t)s\ *noun* [Middle English *mischaunce,* from Old French *meschance,* from *mes-* mis- + *chance* chance] (14th century)
1 : bad luck
2 : a piece of bad luck **:** MISHAP
synonym see MISFORTUNE

mis·chief \ˈmis-chəf, ˈmish-\ *noun* [Middle English *meschief,* from Old French, calamity, from *mes-* + *chief* head, end — more at CHIEF] (14th century)
1 : a specific injury or damage attributed to a particular agent
2 : a cause or source of harm, evil, or irritation; *especially* **:** a person who causes mischief
3 a : action that annoys or irritates **b :** the quality or state of being mischievous **:** MISCHIEVOUSNESS ⟨had *mischief* in his eyes⟩

mis·chie·vous \ˈmis-chə-vəs, ˈmish-; ÷mis-ˈchē-vē-əs, mish-\ *adjective* (14th century)
1 : HARMFUL, INJURIOUS ⟨*mischievous* gossip⟩
2 a : able or tending to cause annoyance, trouble, or minor injury **b :** irresponsibly playful ⟨*mischievous* behavior⟩ ■
— mis·chie·vous·ly *adverb*
— mis·chie·vous·ness *noun*

misch metal \ˈmish-\ *noun* [German *Mischmetall,* from *mischen* to mix + *Metall* metal] (1916)
: a complex alloy of rare earth metals used especially in tracer bullets and as a flint in lighters

mis·ci·ble \ˈmi-sə-bəl\ *adjective* [Medieval Latin *miscibilis,* from Latin *miscēre* to mix — more at MIX] (1570)
: capable of being mixed; *specifically* **:** capable of mixing in any ratio without separation of two phases ⟨*miscible* liquids⟩
— mis·ci·bil·i·ty \ˌmi-sə-ˈbi-lə-tē\ *noun*

mis·com·mu·ni·ca·tion \ˌmis-kə-ˌmyü-nə-ˈkā-shən\ *noun* (1964)
: failure to communicate clearly

mis·con·duct \-ˈkän-(ˌ)dəkt\ *noun* (1710)
1 : mismanagement especially of governmental or military responsibilities
2 : intentional wrongdoing; *specifically* **:** deliberate violation of a law or standard especially by a government official **:** MALFEASANCE
3 a : improper behavior **b :** ADULTERY
— mis·con·duct \-kən-ˈdəkt\ *transitive verb*

mis·count \-ˈkau̇nt\ *verb* [Middle English *misconten,* from Middle French *mesconter,* from *mes-* + *conter* to count] (14th century)
intransitive verb
: to make a wrong count
transitive verb
: to count wrongly **:** MISCALCULATE
— miscount \-ˈkau̇nt, ˈmis-ˌ\ *noun*

¹mis·cre·ant \ˈmis-krē-ənt\ *adjective* [Middle English *miscreaunt,* from Middle French *mescreant,* present participle of *mescroire* to disbelieve, from *mes-* + *croire* to believe, from Latin *credere* — more at CREED] (14th century)
1 : UNBELIEVING, HERETICAL
2 : DEPRAVED, VILLAINOUS

²miscreant *noun* (14th century)
1 : INFIDEL, HERETIC
2 : one who behaves criminally or viciously

¹mis·cue \ˌmis-ˈkyü, ˈmis-ˌkyü\ *noun* (1873)
1 : a faulty stroke in billiards in which the cue slips
2 : MISTAKE, SLIP

²miscue *intransitive verb* (1894)
: to make a miscue

mis·deal \ˌmis-ˈdē(ə)l\ *verb* **-dealt** \-ˈdelt\; **-deal·ing** (1850)
intransitive verb
: to deal cards incorrectly
transitive verb
: to deal incorrectly
— misdeal *noun*

mis·deed \-ˈdēd\ *noun* (before 12th century)
: a wrong deed **:** OFFENSE

mis·de·mean·ant \-di-ˈmē-nənt\ *noun* (1819)
: a person convicted of a misdemeanor

mis·de·mean·or \-di-ˈmē-nər\ *noun* (15th century)
1 : a crime less serious than a felony
2 : MISDEED

mis·di·rect \-də-ˈrekt, -(ˌ)dī-\ *transitive verb* (1603)
1 : to give a wrong direction to
2 : to direct wrongly ⟨*misdirected* their energies⟩

mis·di·rec·tion \-ˈrek-shən\ *noun* (1768)
1 a : the act or an instance of misdirecting **b :** the state of being misdirected
2 : a wrong direction

mis·do \ˌmis-ˈdü\ *verb* **-did** \-ˈdid\; **-done** \-ˈdən\; **-do·ing** \-ˈdü-iŋ\; **-does** \-ˈdəz\ (before 12th century)
intransitive verb
obsolete **:** to act wrongly **:** transgress the laws of God ⟨the erring soul not wilfully *misdoing* —John Milton⟩
transitive verb
: to do (something) incorrectly or poorly
— mis·do·er \-ˈdü-ər\ *noun*

misdoing *noun* (15th century)
: the act or an instance of misbehaving **:** MISCONDUCT

mis·doubt \ˌmis-ˈdau̇t\ *transitive verb* (circa 1540)
1 : DOUBT
2 : SUSPECT, FEAR
— misdoubt *noun*

□ USAGE
mischievous A pronunciation \mis-ˈchē-vē-əs\ and a consequent spelling *mischievious* are of long standing: evidence for the spelling goes back to the 16th century. Our pronunciation files contain modern attestations ranging from dialect speakers to Herbert Hoover. But both the pronunciation and the spelling are still considered nonstandard.

mise–en–scène \ˌmē-ˌzäⁿ-ˈsen, -ˈsän\ *noun, plural* **mise–en–scènes** \-ˈsen(z), -ˈsän(z)\ [French *mise en scène*] (1833)
1 a : the arrangement of actors and scenery on a stage for a theatrical production **b :** stage setting
2 a : the physical setting of an action **:** CONTEXT **b :** ENVIRONMENT, MILIEU
synonym see BACKGROUND

mi·ser \ˈmī-zər\ *noun* [Latin *miser* miserable] (circa 1560)
: a mean grasping person; *especially* **:** one who is extremely stingy with money

mis·er·a·ble \ˈmi-zər-bəl, ˈmiz-rə-, ˈmi-zə-rə-\ *adjective* [Middle English, from Middle French, from Latin *miserabilis* wretched, pitiable, from *miserari* to pity, from *miser*] (15th century)
1 : being in a pitiable state of distress or unhappiness (as from want or shame) ⟨*miserable* refugees⟩
2 a : wretchedly inadequate or meager ⟨a *miserable* hovel⟩ **b :** causing extreme discomfort or unhappiness ⟨a *miserable* situation⟩
3 : being likely to discredit or shame ⟨his *miserable* neglect of his wife⟩ ⟨it was *miserable* of you to make fun of him⟩
— miserable *noun*
— mis·er·a·ble·ness *noun*
— mis·er·a·bly \-blē\ *adverb*

mi·se·re·re \ˌmi-zə-ˈrir-ē, -ˈrer-; ˌmē-zə-ˈrā-(ˌ)rā\ *noun* [Middle English, from Latin, be merciful, from *misereri* to be merciful, from *miser* wretched; from the first word of the Psalm] (13th century)
1 *capitalized* **:** the 50th Psalm in the Vulgate
2 : MISERICORD
3 : a vocal complaint or lament

mi·ser·i·cord *also* **mi·ser·i·corde** \mə-ˈzer-ə-ˌkȯrd, -ˈser-\ *noun* [Medieval Latin *misericordia* seat in church, from Latin, mercy, from *misericord-, misericors* merciful, from *misereri* + *cord-, cor* heart — more at HEART] (circa 1515)
: a small projection on the bottom of a hinged church seat that gives support to a standing worshiper when the seat is turned up

mi·ser·ly \ˈmī-zər-lē\ *adjective* (1593)
: of, relating to, or characteristic of a miser; *especially* **:** marked by grasping meanness and penuriousness
synonym see STINGY
— mi·ser·li·ness *noun*

mis·ery \ˈmi-zə-rē, ˈmiz-rē\ *noun, plural* **-er·ies** (14th century)
1 : a state of suffering and want that is the result of poverty or affliction
2 : a circumstance, thing, or place that causes suffering or discomfort
3 : a state of great unhappiness and emotional distress
synonym see DISTRESS

mis·es·teem \ˌmi-sə-ˈstēm\ *transitive verb* (1611)
: to esteem wrongly; *especially* **:** to hold in too little regard

mis·fea·sance \mis-ˈfē-zᵊn(t)s\ *noun* [Middle French *mesfaisance,* from *mesfaire* to do wrong, from *mes-* mis- + *faire* to make, do, from Latin *facere* — more at DO] (1596)
: TRESPASS; *specifically* **:** the performance of a lawful action in an illegal or improper manner
— mis·fea·sor \-ˈfē-zər, -ˌzȯr\ *noun*

¹mis·fire \ˌmis-ˈfīr\ *intransitive verb* (1752)
1 : to have the explosive or propulsive charge fail to ignite at the proper time ⟨the engine *misfired*⟩
2 : to fail to fire ⟨the gun *misfired*⟩
3 : to miss an intended effect or objective

²mis·fire \ˌmis-ˈfīr, ˈmis-ˌ\ *noun* (1839)
1 : a failure (as of a cartridge or firearm) to fire
2 : something that misfires

mis·fit \ˈmis-ˌfit *also* ˌmis-ˈfit\ *noun* (circa 1823)
1 : something that fits badly
2 : one who is poorly adapted to a situation or environment ⟨social *misfits*⟩

mis·for·tune \ˌmis-ˈfȯr-chən\ *noun* (15th century)
1 a : an event or conjunction of events that causes an unfortunate or distressing result **:** bad luck ⟨by *misfortune* he fell into bad company⟩ ⟨had the *misfortune* to break his leg⟩ **b :** an unhappy situation ⟨always ready to help people in *misfortune*⟩
2 : a distressing or unfortunate incident or event ⟨*misfortunes* never come singly⟩ ☆

mis·give \ˌmis-ˈgiv\ *verb* **-gave** \-ˈgāv\; **-giv·en** \-ˈgi-vən\; **-giv·ing** (1513)
transitive verb
: to suggest doubt or fear to
intransitive verb
: to be fearful or apprehensive

mis·giv·ing \-ˈgi-viŋ\ *noun* (1601)
: a feeling of doubt or suspicion especially concerning a future event

mis·guid·ance \ˌmis-ˈgī-dᵊn(t)s\ *noun* (1640)
: MISDIRECTION

mis·guide \-ˈgīd\ *transitive verb* (14th century)
: to lead astray **:** MISDIRECT ⟨prejudice *misguides* our minds⟩
— mis·guid·er *noun*

mis·guid·ed \(ˌ)mis-ˈgī-dəd\ *adjective* (1659)
: led or prompted by wrong or inappropriate motives or ideals ⟨well-meaning but *misguided* do-gooders⟩
— mis·guid·ed·ly \-lē\ *adverb*
— mis·guid·ed·ness \-nəs\ *noun*

mis·han·dle \-ˈhan-dᵊl\ *transitive verb* (1530)
1 : to treat roughly **:** MALTREAT
2 : to deal with or manage wrongly or ignorantly

mi·shan·ter \mi-ˈshän-tər\ *noun* [Middle English *misaunter,* alteration of *mesaventure*] (1742)
chiefly Scottish **:** MISADVENTURE

mis·hap \ˈmis-ˌhap, mis-ˈ\ *noun* (14th century)
1 : an unfortunate accident
2 : bad luck **:** MISFORTUNE
synonym see MISFORTUNE

mis·hear \ˌmis-ˈhir\ *verb* **-heard** \-ˈhərd\; **-hear·ing** \-ˈhir-iŋ\ (before 12th century)
transitive verb
: to hear wrongly
intransitive verb
: to misunderstand what is heard

mis·hit \ˌmis-ˈhit\ *transitive verb* **-hit; -hit·ting** (1904)
: to hit in a faulty manner
— mis·hit \ˌmis-ˈhit, ˈmis-ˌ\ *noun*

mish·mash \ˈmish-ˌmash, -ˌmäsh\ *noun* [Middle English & Yiddish; Middle English *mysse masche,* perhaps reduplication of *mash* mash; Yiddish *mish-mash,* perhaps reduplication of *mishn* to mix] (15th century)
: HODGEPODGE, JUMBLE

Mish·nah *or* **Mish·na** \ˈmish-nə\ *noun* [Hebrew *mishnāh* instruction, oral law] (1610)
: the collection of mostly halakic Jewish traditions compiled about A.D. 200 and made the basic part of the Talmud
— Mish·na·ic \mish-ˈnā-ik\ *adjective*

mis·im·pres·sion \ˌmi-sim-ˈpre-shən\ *noun* (1670)
: a mistaken impression

mis·in·ter·pret \ˌmi-sᵊn-ˈtər-prət, -pət\ *transitive verb* (1547)
1 : to explain wrongly
2 : to understand wrongly
— mis·in·ter·pre·ta·tion \-ˌtər-prə-ˈtā-shən, -pə-\ *noun*

mis·join·der \ˌmis-ˈjȯin-dər\ *noun* (circa 1847)
: an improper union of parties or of causes of action in a single legal proceeding

mis·judge \ˌmis-ˈjəj\ (15th century)
intransitive verb
: to be mistaken in judgment
transitive verb
1 : to estimate wrongly
2 : to have an unjust opinion of
— mis·judg·ment \-ˈjəj-mənt\ *noun*

Mi·ski·to \mis-ˈkē-(ˌ)tō\ *noun, plural* **Miskito** *or* **Miskitos** (1789)
1 : a member of an American Indian people of the Atlantic coast of Nicaragua and Honduras
2 : the language of the Miskito people

mis·know \ˌmis-ˈnō\ *transitive verb* **-knew** \-ˈnü, -ˈnyü\; **-known** \-ˈnōn\; **-know·ing** (1535)
: MISUNDERSTAND
— mis·knowl·edge \-ˈnä-lij\ *noun*

mis·lay \ˌmis-ˈlā\ *transitive verb* **-laid** \-ˈlād\; **-lay·ing** (1614)
: to put in an unremembered place **:** LOSE

mis·lead \ˌmis-ˈlēd\ *verb* **-led** \-ˈled\; **-lead·ing** (before 12th century)
transitive verb
: to lead in a wrong direction or into a mistaken action or belief often by deliberate deceit
intransitive verb
: to lead astray
synonym see DECEIVE
— mis·lead·er *noun*
— mis·lead·ing·ly \-ˈlē-diŋ-lē\ *adverb*

mis·leared \-ˈlird, -ˈlerd\ *adjective* [¹*mis-* + *lear* to learn] (1560)
chiefly Scottish **:** UNMANNERLY, ILL-BRED

mis·like \-ˈlīk\ *transitive verb* (before 12th century)
1 *archaic* **:** DISPLEASE
2 : DISLIKE
— mislike *noun*

mis·name \-ˈnām\ *transitive verb* (1537)
: to name incorrectly **:** MISCALL

mis·no·mer \ˌmis-ˈnō-mər\ *noun* [Middle English *misnoumer,* from Middle French *mesnommer* to misname, from *mes-* mis- + *nommer* to name, from Latin *nominare* — more at NOMINATE] (15th century)
1 : the misnaming of a person in a legal instrument
2 a : a use of a wrong name **b :** a wrong name or designation
— mis·no·mered \-mərd\ *adjective*

mi·so \ˈmē-(ˌ)sō\ *noun* [Japanese] (1727)
: a high-protein food paste consisting chiefly of soybeans, salt, and usually fermented grain (as barley or rice) and ranging in taste from very salty to very sweet

mi·sog·a·my \mi-ˈsä-gə-mē, mī-\ *noun* [Greek *misein* to hate + English *-gamy*] (circa 1656)
: a hatred of marriage
— mi·sog·a·mist \-mist\ *noun*

mi·sog·y·ny \mə-ˈsä-jə-nē\ *noun* [Greek *misogynia,* from *misein* to hate + *gynē* woman — more at QUEEN] (circa 1656)
: a hatred of women
— miso·gy·nic \ˌmi-sə-ˈji-nik, -ˈgī-\ *adjective*
— mi·sog·y·nist \mə-ˈsä-jə-nist\ *noun or adjective*
— mi·sog·y·nis·tic \mə-ˌsä-jə-ˈnis-tik\ *adjective*

mi·sol·o·gy \mə-ˈsä-lə-jē\ *noun* [Greek *misologia,* from *misein* + *-logia* -logy] (1833)

☆ SYNONYMS
Misfortune, mischance, adversity, mishap mean adverse fortune or an instance of this. MISFORTUNE may apply to either the incident or conjunction of events that is the cause of an unhappy change of fortune or to the ensuing state of distress ⟨never lost hope even in the depths of *misfortune*⟩. MISCHANCE applies especially to a situation involving no more than slight inconvenience or minor annoyance ⟨took the wrong road by *mischance*⟩. ADVERSITY applies to a state of grave or persistent misfortune ⟨had never experienced great *adversity*⟩. MISHAP applies to an often trivial instance of bad luck ⟨the usual *mishaps* of a family vacation⟩.

: a hatred of argument, reasoning, or enlightenment

mi·so·ne·ism \ˌmi-sə-ˈnē-ˌi-zəm\ *noun* [Italian *misoneismo,* from Greek *misein* + *neos* new + Italian *-ismo* -ism — more at NEW] (1886)
: a hatred, fear, or intolerance of innovation or change

mis·place \ˌmis-ˈplās\ *transitive verb* (1594)
1 a : to put in a wrong or inappropriate place ⟨*misplace* a comma⟩ **b :** MISLAY ⟨*misplaced* the keys⟩
2 : to set on a wrong object or eventuality ⟨his trust had been *misplaced*⟩
— mis·place·ment \-ˈplās-mənt\ *noun*

mis·play \ˌmis-ˈplā\ *noun* (1867)
: a wrong or unskillful play **:** ERROR
— mis·play \ˌmis-ˈplā, ˈmis-ˌ\ *transitive verb*

[1]**mis·pri·sion** \(ˌ)mis-ˈpri-zhən\ *noun* [Middle English, from Middle French *mesprison* error, wrongdoing, from Old French, from *mesprendre* to make a mistake, from *mes-* mis- + *prendre* to take, from Latin *prehendere* to seize — more at GET] (15th century)
1 a : neglect or wrong performance of official duty **b :** concealment of treason or felony by one who is not a participant in the treason or felony **c :** seditious conduct against the government or the courts
2 : MISUNDERSTANDING, MISINTERPRETATION

[2]**misprision** *noun* [*misprize*] (1586)
: CONTEMPT, SCORN

mis·prize \ˌmis-ˈprīz\ *transitive verb* [Middle French *mesprisier,* from *mes-* mis- + *prisier* to appraise — more at PRIZE] (14th century)
1 : to hold in contempt **:** DESPISE
2 : UNDERVALUE

mis·pro·nounce \ˌmis-prə-ˈnau̇n(t)s\ *transitive verb* (1593)
: to pronounce incorrectly or in a way regarded as incorrect

mis·pro·nun·ci·a·tion \-ˌnən(t)-sē-ˈā-shən *also* -ˌnau̇n(t)-\ *noun* (1530)
: the act or an instance of mispronouncing

mis·read \ˌmis-ˈrēd\ *transitive verb* **-read** \-ˈred\; **-read·ing** \-ˈrē-diŋ\ (1809)
1 : to read incorrectly
2 : to misinterpret in or as if in reading ⟨totally *misread* the lesson of history —Christopher Hollis⟩

mis·reck·on \-ˈre-kən\ *verb* (circa 1525)
: MISCOUNT, MISCALCULATE

mis·rep·re·sent \(ˌ)mis-ˌre-pri-ˈzent\ *transitive verb* (1647)
1 : to give a false or misleading representation of usually with an intent to deceive or be unfair ⟨*misrepresented* the facts⟩
2 : to serve badly or improperly as a representative of
— mis·rep·re·sen·ta·tion \(ˌ)mis-ˌre-pri-ˌzen-ˈtā-shən, -zən-\ *noun*
— mis·rep·re·sen·ta·tive \-ˈzen-tə-tiv\ *adjective*

[1]**mis·rule** \ˌmis-ˈrül\ *transitive verb* (14th century)
: to rule incompetently **:** MISGOVERN

[2]**misrule** *noun* (14th century)
1 : the action of misruling **:** the condition of being misruled
2 : DISORDER, ANARCHY

[1]**miss** \ˈmis\ *verb* [Middle English, from Old English *missan;* akin to Old High German *missan* to miss] (before 12th century)
transitive verb
1 : to fail to hit, reach, or contact
2 : to discover or feel the absence of
3 : to fail to obtain
4 : ESCAPE, AVOID ⟨just *missed* hitting the other car⟩
5 : to leave out **:** OMIT
6 : to fail to comprehend, sense, or experience ⟨*missed* the point of the speech⟩
7 : to fail to perform or attend ⟨had to *miss* school for a week⟩
intransitive verb
1 *archaic* **:** to fail to get, reach, or do something
2 : to fail to hit something
3 a : to be unsuccessful **b :** MISFIRE ⟨the engine *missed*⟩
— miss·able \ˈmi-sə-bəl\ *adjective*
— miss a beat : to deviate from regular smooth performance ⟨the company changed ownership without *missing a beat*⟩
— miss out on : to lose a good opportunity for ⟨*missed out on* a better job⟩
— miss the boat : to fail to take advantage of an opportunity

[2]**miss** *noun* (12th century)
1 *chiefly dialect* **:** disadvantage or regret resulting from loss ⟨we know the *miss* of you, and even hunger . . . to see you —Samuel Richardson⟩
2 a : a failure to hit **b :** a failure to attain a desired result
3 : MISFIRE

[3]**miss** *noun* [short for *mistress*] (circa 1667)
1 *capitalized* **a** — used as a title prefixed to the name of an unmarried woman or girl **b** — used before the name of a place or of a line of activity or before some epithet to form a title for a usually young unmarried female who is representative of the thing indicated ⟨*Miss* America⟩
2 : young lady — used without a name as a conventional term of address to a young woman
3 : a young unmarried woman or girl

mis·sa can·ta·ta \ˌmi-sə-kən-ˈtä-tə\ *noun* [New Latin, sung mass] (circa 1903)
: HIGH MASS

mis·sal \ˈmi-səl\ *noun* [Middle English *messel,* from Middle French & Medieval Latin; Middle French, from Medieval Latin *missale,* from neuter of *missalis* of the mass, from Late Latin *missa* mass — more at MASS] (14th century)
: a book containing all that is said or sung at mass during the entire year

mis·send \ˌmis-ˈsend\ *transitive verb* **-sent** \-ˈsent\; **-send·ing** (15th century)
: to send incorrectly ⟨*missent* mail⟩

mis·sense \ˈmis-ˌsen(t)s\ *noun* [[1]*mis-* + *-sense* (as in *nonsense*)] (1961)
: genetic mutation involving alteration of one or more codons so that different amino acids are determined — compare ANTISENSE, NONSENSE

[1]**mis·sile** \ˈmi-səl, *chiefly British* -ˌsīl\ *adjective* [Latin *missilis,* from *mittere* to throw, send] (1611)
1 : capable of being thrown or projected to strike a distant object
2 : adapted for throwing or hurling missiles

[2]**missile** *noun* (circa 1656)
: an object (as a weapon) thrown or projected usually so as to strike something at a distance ⟨stones, artillery shells, bullets, and rockets are *missiles*⟩: as **a :** GUIDED MISSILE **b :** BALLISTIC MISSILE

mis·sil·eer \ˌmi-sə-ˈlir\ *noun* (1958)
: MISSILEMAN

mis·sile·man \ˈmi-səl-mən\ *noun* (1951)
: one engaged in designing, building, or operating guided missiles

mis·sile·ry \ˈmi-səl-rē\ *noun* (1880)
: MISSILES; *especially* **:** GUIDED MISSILES

miss·ing \ˈmi-siŋ\ *adjective* (circa 1530)
: ABSENT; *also* **:** LOST ⟨*missing* in action⟩

missing link *noun* (1851)
1 : an absent member needed to complete a series or resolve a problem
2 : a hypothetical intermediate form between humans and their presumed simian progenitors

mis·si·ol·o·gy \ˌmi-sē-ˈä-lə-jē\ *noun* [*mission* + *-logy*] (1924)
: the study of the church's mission especially with respect to missionary activity

[1]**mis·sion** \ˈmi-shən\ *noun* [New Latin, Medieval Latin, & Latin; New Latin *mission-, missio* religious mission, from Medieval Latin, task assigned, from Latin, act of sending, from *mittere* to send] (1606)
1 *obsolete* **:** the act or an instance of sending
2 a : a ministry commissioned by a religious organization to propagate its faith or carry on humanitarian work **b :** assignment to or work in a field of missionary enterprise **c** (1) **:** a mission establishment (2) **:** a local church or parish dependent on a larger religious organization for direction or financial support **d** *plural* **:** organized missionary work **e :** a course of sermons and services given to convert the unchurched or quicken Christian faith
3 : a body of persons sent to perform a service or carry on an activity: as **a :** a group sent to a foreign country to conduct diplomatic or political negotiations **b :** a permanent embassy or legation **c :** a team of specialists or cultural leaders sent to a foreign country
4 a : a specific task with which a person or a group is charged **b** (1) **:** a definite military, naval, or aerospace task ⟨a bombing *mission*⟩ ⟨a space *mission*⟩ (2) **:** a flight operation of an aircraft or spacecraft in the performance of a mission ⟨a *mission* to Mars⟩
5 : CALLING, VOCATION

[2]**mission** *transitive verb* **mis·sioned; mis·sion·ing** \ˈmi-sh(ə-)niŋ\ (1692)
1 : to send on or entrust with a mission
2 : to carry on a religious mission among or in

[3]**mission** *adjective* (1904)
1 : of or relating to a style used in the early Spanish missions of the southwestern U.S. ⟨*mission* architecture⟩
2 : of, relating to, or having the characteristics of a style of plain heavy usually oak furniture originating in the U.S. in the early part of the 20th century

[1]**mis·sion·ary** \ˈmi-shə-ˌner-ē\ *adjective* (1644)
1 : relating to, engaged in, or devoted to missions
2 : characteristic of a missionary

[2]**missionary** *noun, plural* **-ar·ies** (circa 1656)
: a person undertaking a mission and especially a religious mission

missionary position *noun* [perhaps from the notion that missionaries insisted that this coital position is the only acceptable one] (1948)
: a coital position in which the female lies on her back with the male on top and with his face opposite hers

mis·sion·er \ˈmi-sh(ə-)nər\ *noun* (1654)
: MISSIONARY

mis·sion·ize \ˈmi-shə-ˌnīz\ *verb* **-ized; -iz·ing** (1826)
intransitive verb
: to carry on missionary work
transitive verb
: to do missionary work among
— mis·sion·i·za·tion \ˌmi-shə-nə-ˈzā-shən\ *noun*
— mis·sion·iz·er \ˈmi-shə-ˌnī-zər\ *noun*

Mis·sis·sip·pi·an \ˌmi-sə-ˈsi-pē-ən, (ˌ)mis-ˈsi-\ *adjective* [*Mississippi* River] (1835)
1 : of or relating to Mississippi, its people, or the Mississippi River
2 : of, relating to, or being the period of the Paleozoic era in North America following the Devonian and preceding the Pennsylvanian or the corresponding system of rocks — see GEOLOGIC TIME table
— Mississippian *noun*

mis·sive \ˈmi-siv\ *noun* [Middle French *lettre missive,* literally, letter intended to be sent] (1501)
: a written communication **:** LETTER

miss·out \ˈmis-ˌau̇t\ *noun* (1945)

: a throw of dice that loses the main bet
miss out *transitive verb* (1870)
British : to leave out : OMIT
mis·speak \ˌmis-ˈspēk\ *verb* **-spoke** \-ˈspōk\; **-spo·ken** \-ˈspō-kən\; **-speak·ing** (14th century)
transitive verb
1 : to speak (as a word) incorrectly
2 : to express (oneself) imperfectly or incorrectly ⟨claims now that he *misspoke* himself⟩
intransitive verb
: to speak incorrectly : misspeak oneself
mis·spell \ˌmis-ˈspel\ *transitive verb* (1655)
: to spell incorrectly
mis·spell·ing \-ˈspe-liŋ\ *noun* (circa 1696)
: an incorrect spelling
mis·spend \ˌmis-ˈspend\ *transitive verb* **-spent** \-ˈspent\; **-spend·ing** (14th century)
: to spend wrongly : SQUANDER ⟨a *misspent* life⟩
mis·state \ˌmis-ˈstāt\ *transitive verb* (1650)
: to state incorrectly : give a false account of
— **mis·state·ment** \-mənt\ *noun*
mis·step \-ˈstep\ *noun* (circa 1800)
1 : a mistake in judgment or action : BLUNDER
2 : a wrong step
mis·sus *or* **mis·sis** \ˈmi-səz, -səs, *especially Southern* -zəz\ *noun* [alteration of *mistress*] (1790)
1 *dialect* : MISTRESS 1a
2 : WIFE ⟨men spend money on themselves, but argue over every dime the *missus* wants —W. A. Lydgate⟩
missy \ˈmi-sē\ *noun* (1676)
: a young girl : MISS
¹mist \ˈmist\ *noun* [Middle English, from Old English; akin to Middle Dutch *mist* mist, Greek *omichlē*] (before 12th century)
1 : water in the form of particles floating or falling in the atmosphere at or near the surface of the earth and approaching the form of rain
2 : something that obscures understanding ⟨*mists* of antiquity⟩
3 : a film before the eyes
4 a : a cloud of small particles or objects suggestive of a mist **b :** a suspension of a finely divided liquid in a gas **c :** a fine spray
5 : a drink of liquor served over cracked ice
²mist (before 12th century)
intransitive verb
1 : to be or become misty
2 : to become moist or blurred
transitive verb
: to cover or spray with or convert to mist
mis·tak·able \mə-ˈstā-kə-bəl\ *adjective* (1646)
: capable of being misunderstood or mistaken
¹mis·take \mə-ˈstāk\ *verb* **mis·took** \-ˈstu̇k\; **mis·tak·en** \-ˈstā-kən\; **mis·tak·ing** [Middle English, from Old Norse *mistaka* to take by mistake, from *mis-* + *taka* to take — more at TAKE] (14th century)
transitive verb
1 : to blunder in the choice of ⟨*mistook* her way in the dark⟩
2 a : to misunderstand the meaning or intention of : MISINTERPRET ⟨don't *mistake* me, I mean exactly what I said⟩ **b :** to make a wrong judgment of the character or ability of
3 : to identify wrongly : confuse with another ⟨I *mistook* him for his brother⟩
intransitive verb
: to be wrong ⟨you *mistook* when you thought I laughed at you —Thomas Hardy⟩
— **mis·tak·en·ly** *adverb*
— **mis·tak·er** *noun*
²mistake *noun* (1638)
1 : a misunderstanding of the meaning or implication of something
2 : a wrong action or statement proceeding from faulty judgment, inadequate knowledge, or inattention
synonym see ERROR
¹mis·ter \ˈmis-tər, *for 1* ˌmis-tər *or in rapid speech* (ˌ)mis(t)\ *noun* [alteration of ¹*master*] (1551)
1 *capitalized* : MR. — used sometimes in writing instead of *Mr.*
2 : SIR — used without a name as a generalized term of direct address of a man who is a stranger ⟨hey, *mister*, do you want to buy a paper⟩
3 : a man not entitled to a title of rank or an honorific or professional title ⟨though he was only a *mister*, he was a greater scholar in his field than any PhD⟩
4 : HUSBAND
²mist·er \ˈmis-tər\ *noun* [²*mist* + ²*-er*] (1973)
: a device for spraying a mist
mis·think \ˌmis-ˈthiŋk\ *verb* **-thought** \-ˈtho̊t\; **-think·ing** (circa 1530)
intransitive verb
archaic : to think mistakenly or unfavorably
transitive verb
archaic : to think badly or unfavorably of
mis·tle·toe \ˈmi-səl-ˌtō, *chiefly British* ˈmi-zəl-\ *noun* [Middle English *mistilto,* from Old English *misteltān,* from *mistel* mistletoe + *tān* twig; akin to Old High German & Old Saxon *mistil* mistletoe and to Old High German *zein* twig] (before 12th century)
: a European semiparasitic green shrub (*Viscum album* of the family Loranthaceae, the mistletoe family) with thick leaves, small yellowish flowers, and waxy-white glutinous berries; *broadly* : any of various plants of the mistletoe family (as of an American genus *Phoradendron*) resembling the true mistletoe
mis·tral \ˈmis-trəl, mi-ˈsträl\ *noun* [French, from Provençal, from *mistral* masterful, from Late Latin *magistralis* of a teacher — more at MAGISTRAL] (1604)
: a strong cold dry northerly wind of southern France
mis·treat \ˌmis-ˈtrēt\ *transitive verb* [Middle English *mistreten,* probably from Middle French *mestraitier,* from Old French, from *mes-* mis- + *traitier* to treat — more at TREAT] (15th century)
: to treat badly : ABUSE
— **mis·treat·ment** \-mənt\ *noun*
mis·tress \ˈmis-trəs\ *noun* [Middle English *maistresse,* from Middle French, from Old French, feminine of *maistre* master — more at MASTER] (15th century)
1 : a woman who has power, authority, or ownership: as **a :** the female head of a household **b :** a woman who employs or supervises servants **c :** a woman who is in charge of a school or other establishment **d :** a woman of the Scottish nobility having a status comparable to that of a master
2 a *chiefly British* : a female teacher or tutor **b :** a woman who has achieved mastery in some field
3 : something personified as female that rules, directs, or dominates ⟨when Rome was *mistress* of the world⟩
4 a : a woman other than his wife with whom a married man has a continuing sexual relationship **b** *archaic* : SWEETHEART
5 a — used archaically as a title prefixed to the name of a married or unmarried woman **b** *chiefly Southern & Midland* : MRS. 1a
mistress of ceremonies (1952)
: a woman who presides at a public ceremony or who acts as hostess of a stage, radio, or television show
mis·tri·al \ˈmis-ˌtrī(-ə)l\ *noun* (1628)
: a trial that has no legal effect by reason of some error or serious prejudicial misconduct in the proceedings
¹mis·trust \ˌmis-ˈtrəst\ *noun* (14th century)
: a lack of confidence : DISTRUST
synonym see UNCERTAINTY
— **mis·trust·ful** \-fəl\ *adjective*
— **mis·trust·ful·ly** \-fə-lē\ *adverb*
— **mis·trust·ful·ness** *noun*
²mistrust (14th century)
transitive verb
1 : to have no trust or confidence in : SUSPECT ⟨*mistrusted* his neighbors⟩
2 : to doubt the truth, validity, or effectiveness of ⟨*mistrusted* his own judgment⟩
3 : SURMISE ⟨your mind *mistrusted* there was something wrong —Robert Frost⟩
intransitive verb
: to be suspicious
misty \ˈmis-tē\ *adjective* **mist·i·er; -est** (before 12th century)
1 a : obscured by mist **b :** consisting of or marked by mist
2 a : INDISTINCT ⟨a *misty* recollection of the event⟩ **b :** VAGUE, CONFUSED ⟨avoided the large, vague, *misty* issues —Reuben Abel⟩
3 : TEARFUL
— **mist·i·ly** \-tə-lē\ *adverb*
— **mist·i·ness** \-tē-nəs\ *noun*
misty–eyed \ˈmis-tē-ˌīd\ *adjective* (1928)
1 : having tearful eyes
2 : DREAMY, SENTIMENTAL ⟨*misty-eyed* recollections⟩
mis·un·der·stand \(ˌ)mi-ˌsən-dər-ˈstand\ *transitive verb* **-stood** \-ˈstu̇d\; **-stand·ing** (13th century)
1 : to fail to understand
2 : to interpret incorrectly
mis·un·der·stand·ing *noun* (15th century)
1 : a failure to understand : MISINTERPRETATION
2 : QUARREL, DISAGREEMENT
mis·us·age \ˌmis-ˈyü-sij, ˌmish-, -zij\ *noun* [Middle French *mesusage,* from *mes-* mis- + *usage* usage] (1532)
1 : bad treatment : ABUSE
2 : wrong or improper use (as of words)
¹mis·use \-ˈyüz\ *transitive verb* [Middle English, partly from *mis-* + *usen* to use; partly from Middle French *mesuser* to abuse, from Old French, from *mes-* + *user* to use] (14th century)
1 : to use incorrectly : MISAPPLY ⟨*misused* his talents⟩
2 : ABUSE, MISTREAT ⟨*misused* his servants⟩
— **mis·us·er** *noun*
²mis·use \-ˈyüs\ *noun* (14th century)
: incorrect or improper use : MISAPPLICATION
¹mite \ˈmīt\ *noun* [Middle English, from Old English *mīte;* akin to Middle Dutch *mite* mite, small copper coin and perhaps to Old High German *meizan* to cut] (before 12th century)
: any of numerous small acarid arachnids that often infest animals, plants, and stored foods and include important disease vectors
²mite *noun* [Middle English, from Middle French or Middle Dutch; Middle French, small Flemish copper coin, from Middle Dutch] (14th century)
1 : a small coin or sum of money
2 a : a very little : BIT **b :** a very small object or creature
— **a mite :** SOMEWHAT, RATHER ⟨could be that I am *a mite* prejudiced —John Fischer⟩
¹mi·ter *or* **mi·tre** \ˈmī-tər\ *noun* [Middle English *mitre,* from Middle French, from Latin *mitra* headband, turban, from Greek] (14th century)
1 : a liturgical headdress worn by bishops and abbots
2 a : a surface forming the beveled end or edge of a piece where a joint is made by cutting two pieces at an angle and fitting them together **b :** MITER JOINT

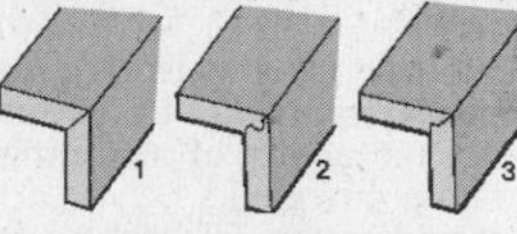

miter 2a: *1* **plain,** *2* **milled,** *3* **rabbeted**

²miter *or* **mitre** *transitive verb* **mi·tered** *or* **mi·tred; mi·ter·ing** *or* **mi·tring** \ˈmī-tə-riŋ\ (14th century)
1 : to confer a miter on
2 a : to match or fit together in a miter joint **b :** to bevel the ends of for making a miter joint
— **mi·ter·er** \ˈmī-tər-ər\ *noun*
miter box *noun* (1678)
: a device for guiding a handsaw at the proper angle in making a miter joint in wood

miter joint *noun* (1688)
: a joint made by fastening together usually perpendicularly parts with the ends cut at an angle

miter square *noun* (1678)
: a bevel with an immovable arm at an angle of 45 degrees for striking marking lines; *also* **:** a square with an arm adjustable to any angle

Mith•ra•ic \mi-ˈthrā-ik\ *adjective* [Late Greek *mithraikos* of Mithras, ancient Persian god of light, from Greek *Mithras,* from Old Persian *Mithra*] (1678)
: of or relating to an oriental mystery cult for men flourishing in the late Roman empire
— **Mith•ra•ism** \ˈmith-rə-ˌi-zəm, -(ˌ)rā-\ *noun*
— **Mith•ra•ist** \mi-ˈthrā-ist\ *noun or adjective*

mith•ri•date \ˈmith-rə-ˌdāt\ *noun* [Medieval Latin *mithridatum,* from Late Latin *mithridatium,* from Latin, dogtooth violet (used as an antidote), from Greek *mithridation,* from *Mithridatēs* Mithridates VI] (1528)
: an antidote against poison; *especially* **:** an electuary held to be effective against poison

mi•ti•cide \ˈmī-tə-ˌsīd\ *noun* [*mite*] (circa 1946)
: an agent used to kill mites
— **mi•ti•cid•al** \ˌmī-tə-ˈsī-dᵊl\ *adjective*

mit•i•gate \ˈmi-tə-ˌgāt\ *transitive verb* **-gat•ed; -gat•ing** [Middle English, from Latin *mitigatus,* past participle of *mitigare* to soften, from *mitis* soft + *-igare* (akin to Latin *agere* to drive); akin to Old Irish *moíth* soft — more at AGENT] (15th century)
1 : to cause to become less harsh or hostile **:** MOLLIFY ⟨aggressiveness may be *mitigated* or . . . channeled —Ashley Montagu⟩
2 a : to make less severe or painful **:** ALLEVIATE **b :** EXTENUATE ■
synonym see RELIEVE
— **mit•i•ga•tion** \ˌmi-tə-ˈgā-shən\ *noun*
— **mit•i•ga•tive** \ˈmi-tə-ˌgā-tiv\ *adjective*
— **mit•i•ga•tor** \-ˌgā-tər\ *noun*
— **mit•i•ga•to•ry** \ˈmi-ti-gə-ˌtōr-ē, -ˌtȯr-\ *adjective*

mi•to•chon•dri•on \ˌmī-tə-ˈkän-drē-ən\ *noun, plural* **-dria** \-drē-ə\ [New Latin, from Greek *mitos* thread + *chondrion,* diminutive of *chondros* grain] (1901)
: any of various round or long cellular organelles of most eukaryotes that are found outside the nucleus, produce energy for the cell through cellular respiration, and are rich in fats, proteins, and enzymes — see CELL illustration
— **mi•to•chon•dri•al** \-drē-əl\ *adjective*

mi•to•gen \ˈmī-tə-jən\ *noun* [*mitosis* + *-gen*] (circa 1951)
: a substance that induces mitosis
— **mi•to•gen•ic** \ˌmī-tə-ˈje-nik\ *adjective*
— **mi•to•ge•nic•i•ty** \-jə-ˈni-sə-tē\ *noun*

mi•to•my•cin \ˌmī-tə-ˈmī-sᵊn\ *noun* [International Scientific Vocabulary *mito-* (probably from New Latin *mitosis*) + *-mycin*] (1956)
: a complex of antibiotic substances which is produced by a Japanese streptomyces (*Streptomyces caespitosus*) and one form of which acts directly on DNA and shows promise as an anticancer agent

mi•to•sis \mī-ˈtō-səs\ *noun, plural* **-to•ses** \-ˌsēz\ [New Latin, from Greek *mitos* thread] (1887)
1 : a process that takes place in the nucleus of a dividing cell, involves typically a series of steps consisting of prophase, metaphase, anaphase, and telophase, and results in the formation of two new nuclei each having the same number of chromosomes as the parent nucleus — compare MEIOSIS
2 : cell division in which mitosis occurs
— **mi•tot•ic** \-ˈtä-tik\ *adjective*
— **mi•tot•i•cal•ly** \-ti-k(ə-)lē\ *adverb*

mi•tral \ˈmī-trəl\ *adjective* (1610)
1 : resembling a miter
2 : of, relating to, being, or adjoining a mitral valve or orifice

mitral valve *noun* (1705)
: BICUSPID VALVE

mi•tre•wort *also* **mi•ter•wort** \ˈmī-tər-ˌwərt, -ˌwȯrt\ *noun* (circa 1818)
: any of a genus (*Mitella*) of rhizomatous perennial herbs of the saxifrage family that bear a capsule resembling a bishop's miter

mitt \ˈmit\ *noun* [short for *mitten*] (1765)
1 a : a woman's glove that leaves the fingers uncovered **b :** MITTEN 1 **c :** a baseball catcher's or first baseman's glove made in the style of a mitten
2 *slang* **:** HAND

mit•ten \ˈmi-tᵊn\ *noun* [Middle English *mitain,* from Middle French *mitaine,* from Old French, from *mite* mitten] (14th century)
1 : a covering for the hand and wrist having a separate section for the thumb only
2 : MITT 1a

mit•ti•mus \ˈmi-tə-məs\ *noun* [Latin, we send, from *mittere* to send] (circa 1591)
: a warrant of commitment to prison

mitz•vah \ˈmits-və\ *noun, plural* **mitz•voth** \-ˌvōt, -ˌvōth, -ˌvōs\ *or* **mitz•vahs** [Hebrew *miṣwāh*] (1650)
1 : a commandment of the Jewish law
2 : a meritorious or charitable act

¹mix \ˈmiks\ *verb* [Middle English, back-formation from *mixte* mixed, from Middle French, from Latin *mixtus,* past participle of *miscēre* to mix; akin to Greek *mignynai* to mix] (15th century)
transitive verb
1 a (1) **:** to combine or blend into one mass (2) **:** to combine with another **b :** to bring into close association ⟨*mix* business with pleasure⟩
2 a : to form by mixing components ⟨*mix* a drink at the bar⟩ **b :** to produce (a sound recording) by electronically combining or adjusting sounds from more than one source
3 : CONFUSE — often used with *up* ⟨*mixes* things up in his eagerness to speak out —Irving Howe⟩
intransitive verb
1 a : to become mixed **b :** to be capable of mixing
2 : to enter into relations **:** ASSOCIATE
3 : CROSSBREED
4 : to become involved **:** PARTICIPATE ⟨decided not to *mix* in politics⟩ ☆
— **mix•able** \ˈmik-sə-bəl\ *adjective*
— **mix it up :** to engage in a fight, contest, or dispute

²mix *noun* (circa 1586)
1 : an act or process of mixing
2 : a product of mixing: as **a :** a commercially prepared mixture of food ingredients ⟨a cake *mix*⟩ **b :** a combination of different kinds ⟨the right *mix* of jobs, people and amenities —*London Times*⟩
3 : MIXER 2b

mixed \ˈmikst\ *adjective* [Middle English *mixte*] (15th century)
1 : combining characteristics of more than one kind; *specifically* **:** combining features of two or more systems of government ⟨a *mixed* constitution⟩
2 : made up of or involving individuals or items of more than one kind: as **a :** made up of or involving persons differing in race, national origin, religion, or class **b :** made up of or involving individuals of both sexes ⟨*mixed* company⟩ ⟨a *mixed* school⟩
3 : including or accompanied by inconsistent, incompatible, or contrary elements ⟨*mixed* emotions⟩ ⟨received *mixed* reviews⟩ ⟨a *mixed* blessing⟩
4 : deriving from two or more races or breeds ⟨a stallion of *mixed* blood⟩

mixed alphabet *noun* (1931)
: an alphabet (as in a cryptographic system) that has been rearranged or disordered systematically or randomly

mixed bag *noun* (1926)
: a miscellaneous collection **:** ASSORTMENT

mixed bud *noun* (1900)
: a bud that produces a branch and leaves as well as flowers

mixed drink *noun* (1943)
: an alcoholic beverage prepared from two or more ingredients

mixed farming *noun* (1872)
: the growing of food or cash crops, feed crops, and livestock on the same farm

mixed grill *noun* (1913)
: meats (as lamb chop, kidney, and bacon) and vegetables broiled together and served on one plate

mixed marriage *noun* (1829)
: a marriage between persons of different races or religions

mixed–media *adjective* (1962)
: MULTIMEDIA

mixed metaphor *noun* (1800)
: a figure of speech combining inconsistent or incongruous metaphors

mixed nerve *noun* (1878)
: a nerve containing both sensory and motor fibers

mixed number *noun* (1542)
: a number (as 5⅔) composed of an integer and a fraction

mixed–up \ˈmikst-ˈəp\ *adjective* (1862)
: marked by bewilderment, perplexity, or disorder **:** CONFUSED

mix•er \ˈmik-sər\ *noun* (circa 1611)
1 : one that mixes: as **a** (1) **:** one whose work is mixing the ingredients of a product (2) **:** one who balances and controls the dialogue, music, and sound effects to be recorded for or

☆ SYNONYMS

Mix, mingle, commingle, blend, merge, coalesce, amalgamate, fuse mean to combine into a more or less uniform whole. MIX may or may not imply loss of each element's identity ⟨*mix* the salad greens⟩ ⟨*mix* a drink⟩. MINGLE usually suggests that the elements are still somewhat distinguishable or separately active ⟨fear *mingled* with anticipation in my mind⟩. COMMINGLE implies a closer or more thorough mingling ⟨a sense of duty *commingled* with a fierce pride drove her⟩. BLEND implies that the elements as such disappear in the resulting mixture ⟨*blended* several teas to create a balanced flavor⟩. MERGE suggests a combining in which one or more elements are lost in the whole ⟨in his mind reality and fantasy *merged*⟩. COALESCE implies an affinity in the merging elements and usually a resulting organic unity ⟨telling details that *coalesce* into a striking portrait⟩. AMALGAMATE implies the forming of a close union without complete loss of individual identities ⟨refugees who were readily *amalgamated* into the community⟩. FUSE stresses oneness and indissolubility of the resulting product ⟨a building in which modernism and classicism are *fused*⟩.

□ USAGE

mitigate *Mitigate* is sometimes used as an intransitive (followed by *against*) where *militate* might be expected. Even though Faulkner used it ⟨some intangible and invisible social force that *mitigates* against him —William Faulkner⟩ and one critic thinks it should be called an American idiom, it is usually considered a mistake.

with a motion picture or television **b :** a container, device, or machine for mixing **c :** a game, stunt, or dance used at a get-together to give members of the group an opportunity to meet one another in a friendly and informal atmosphere

2 : one that mixes with others: as **a :** a person considered in regard to casual sociability ⟨was shy and a poor *mixer*⟩ **b :** a nonalcoholic beverage (as ginger ale) used in a mixed drink

mix·ol·o·gy \mik-'sä-lə-jē\ *noun* (1948)
: the art or skill of preparing mixed drinks
— **mix·ol·o·gist** \-jist\ *noun*

Mix·tec \mēs-'tek, mis-, mēsh-, mish-\ *noun, plural* **Mixtec** *or* **Mixtecs** [American Spanish *mixteco*] (1850)
1 : the language of the Mixtec people
2 : a member of an American Indian people of the state of Oaxaca, Mexico

mix·ture \'miks-chər\ *noun* [Middle English, from Middle French, from Old French *misture,* from Latin *mixtura,* from *mixtus*] (15th century)
1 a : the act, the process, or an instance of mixing **b** (1) **:** the state of being mixed (2) **:** the relative proportions of constituents; *especially* **:** the proportion of fuel to air produced in a carburetor
2 : a product of mixing **:** COMBINATION: as **a :** a portion of matter consisting of two or more components in varying proportions that retain their own properties **b :** a fabric woven of variously colored threads **c :** a combination of several different kinds

mix–up \'miks-,əp\ *noun* (1841)
1 : a state or instance of confusion
2 : MIXTURE
3 : CONFLICT, FIGHT

[1]**miz·zen** *also* **miz·en** \'mi-zᵊn\ *noun* [Middle English *meson,* from Middle French *misaine* foremast sail, probably ultimately from Latin *medianus* of the middle — more at MEDIAN] (15th century)
1 : a fore-and-aft sail set on the mizzenmast
2 : MIZZENMAST

[2]**mizzen** *also* **mizen** *adjective* (15th century)
: of or relating to the mizzenmast

miz·zen·mast \-,mast, -məst\ *noun* (15th century)
: the mast aft or next aft of the mainmast in a ship

[1]**miz·zle** \'mi-zəl\ *intransitive verb* **miz·zled; miz·zling** \'mi-zə-liŋ, 'miz-liŋ\ [Middle English *misellen;* akin to Flemish *mizzelen* to drizzle, Middle Dutch *mist* fog, mist] (15th century)
: to rain in very fine drops **:** DRIZZLE
— **mizzle** *noun*
— **miz·zly** \'mi-zə-lē, 'miz-lē\ *adjective*

[2]**mizzle** *intransitive verb* **miz·zled; miz·zling** \'mi-zə-liŋ, 'miz-liŋ\ [origin unknown] (1781)
chiefly British **:** to depart suddenly

[1]**mne·mon·ic** \ni-'mä-nik\ *adjective* [Greek *mnēmonikos,* from *mnēmōn* mindful, from *mimnēskesthai* to remember — more at MIND] (1753)
1 : assisting or intended to assist memory; *also* **:** of or relating to mnemonics
2 : of or relating to memory
— **mne·mon·i·cal·ly** \-ni-k(ə-)lē\ *adverb*

[2]**mnemonic** *noun* (1858)
: a mnemonic device or code

mne·mon·ics \ni-'mä-niks\ *noun plural but singular in construction* (circa 1721)
: a technique of improving the memory

Mne·mos·y·ne \ni-'mä-sᵊn-ē, -zᵊn-\ *noun* [Latin, from Greek *Mnēmosynē*]
: the Greek goddess of memory and mother of the Muses by Zeus

-mo *noun suffix* [duodecim*o*]
— after numerals or their names to indicate the number of leaves made by folding a sheet of paper ⟨sixteen*mo*⟩ ⟨16*mo*⟩

moa \'mō-ə\ *noun* [Maori] (1842)
: any of various usually very large extinct flightless birds of New Zealand of a ratite family (Dinornithidae) including one (*Dinornis giganteus*) about 12 feet (3.7 meters) in height

Mo·ab·ite \'mō-ə-,bīt\ *noun* [Middle English, from Late Latin *Moabita, Moabites,* from Greek *Mōabitēs,* from *Mōab* Moab, ancient kingdom in Syria] (14th century)
: a member of an ancient Semitic people related to the Hebrews
— **Moabite** *or* **Mo·ab·it·ish** \-,bī-tish\ *adjective*

[1]**moan** \'mōn\ *noun* [Middle English *mone,* from (assumed) Old English *mān*] (13th century)
1 : LAMENTATION, COMPLAINT
2 : a low prolonged sound of pain or of grief

[2]**moan** (14th century)
transitive verb
1 : to bewail audibly **:** LAMENT
2 : to utter with moans
intransitive verb
1 : LAMENT, COMPLAIN
2 a : to make a moan **:** GROAN **b :** to emit a sound resembling a moan ⟨the wind *moaned* in the trees⟩
— **moan·er** \'mō-nər\ *noun*

moat \'mōt\ *noun* [Middle English *mote,* probably from Middle French *motte* hill, mound] (14th century)
1 : a deep and wide trench around the rampart of a fortified place (as a castle) that is usually filled with water
2 : a channel resembling a moat (as about a seamount or for confinement of animals in a zoo)

moat 1

— **moat·ed** \'mō-təd\ *adjective*
— **moat·like** \-,līk\ *adjective*

[1]**mob** \'mäb\ *noun* [Latin *mobile vulgus* vacillating crowd] (1688)
1 : a large or disorderly crowd; *especially* **:** one bent on riotous or destructive action
2 : the lower classes of a community **:** MASSES, RABBLE
3 *chiefly Australian* **:** a flock, drove, or herd of animals
4 : a criminal set **:** GANG; *especially, often capitalized* **:** MAFIA 1
5 *chiefly British* **:** a group of people **:** CROWD
synonym see CROWD
— **mob·bish** \'mä-bish\ *adjective*

[2]**mob** *transitive verb* **mobbed; mob·bing** (1709)
1 : to crowd about and attack or annoy ⟨*mobbed* by autograph hunters⟩ ⟨a crow *mobbed* by songbirds⟩
2 : to crowd into or around ⟨customers *mob* the stores on sale days⟩

mob·cap \'mäb-,kap\ *noun* [*mob* woman's cap + *cap*] (1795)
: a woman's fancy indoor cap made with a high full crown and often tied under the chin

mo·be pearl *or* **mo·bé pearl** \'mō-,bā-, ,mō-'\ *noun, often M capitalized* [origin unknown] (1955)
: MABE

[1]**mo·bile** \'mō-bəl, -,bīl *also* -,bēl\ *adjective* [Middle English *mobyll,* from Middle French *mobile,* from Latin *mobilis,* from *movēre* to move] (15th century)
1 : capable of moving or being moved **:** MOVABLE ⟨a *mobile* missile launcher⟩
2 a : changeable in appearance, mood, or purpose ⟨*mobile* face⟩ **b :** ADAPTABLE, VERSATILE
3 : MIGRATORY
4 a : characterized by the mixing of social groups **b :** having the opportunity for or undergoing a shift in status within the hierarchical social levels of a society ⟨socially *mobile* workers⟩
5 : marked by the use of vehicles for transportation ⟨*mobile* warfare⟩
6 : of or relating to a mobile
— **mo·bil·i·ty** \mō-'bi-lə-tē\ *noun*

[2]**mo·bile** \'mō-,bēl\ *noun* (1936)
: a construction or sculpture frequently of wire and sheet metal shapes with parts that can be set in motion by air currents; *also* **:** a similar structure (as of paper or plastic) suspended so that it moves in a current of air

-mobile *combining form* [auto*mobile*]
1 : motorized vehicle ⟨snow*mobile*⟩
2 : automotive vehicle bringing services to the public ⟨blood*mobile*⟩ ⟨book*mobile*⟩

mobile home *noun* (1949)
: a dwelling structure built on a steel chassis and fitted with wheels that is intended to be hauled to a usually permanent site — compare MOTOR HOME

mo·bi·lise *chiefly British variant of* MOBILIZE

mo·bi·li·za·tion \,mō-bə-lə-'zā-shən\ *noun* (1799)
1 : the act of mobilizing
2 : the state of being mobilized

mo·bi·lize \'mō-bə-,līz\ *verb* **-lized; -liz·ing** (1838)
transitive verb
1 a : to put into movement or circulation ⟨*mobilize* financial assets⟩ **b :** to release (something stored in the organism) for bodily use
2 a : to assemble and make ready for war duty **b :** to marshal (as resources) for action ⟨*mobilize* support for a proposal⟩
intransitive verb
: to undergo mobilization

Mö·bi·us strip \'mœ-bē-əs-, 'mə(r)-, 'mō-\ *noun* [August F. *Möbius* (died 1868) German mathematician] (1904)
: a one-sided surface that is constructed from a rectangle by holding one end fixed, rotating the opposite end through 180 degrees, and joining it to the first end

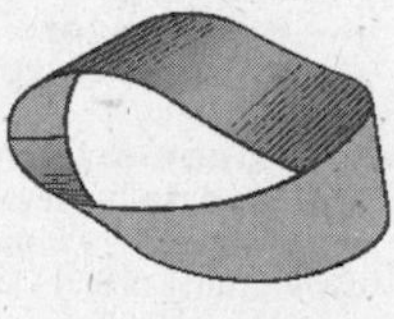
Möbius strip

mo·bled \'mä-bəld\ *adjective* [past participle of *moble* to muffle, probably frequentative of *mob* to muffle, of unknown origin] (1603)
: being wrapped or muffled in or as if in a hood

mob·oc·ra·cy \mä-'bä-krə-sē\ *noun* (1754)
1 : rule by the mob
2 : the mob as a ruling class
— **mob·o·crat** \'mä-bə-,krat\ *noun*
— **mob·o·crat·ic** \,mä-bə-'kra-tik\ *adjective*

mob·ster \'mäb-stər\ *noun* (1917)
: a member of a criminal gang

moc \'mäk\ *noun* (1948)
: MOCCASIN 1

moc·ca·sin \'mä-kə-sən\ *noun* [Virginia Algonquian *mockasin*] (circa 1612)
1 a : a soft leather heelless shoe or boot with the sole brought up the sides of the foot and over the toes where it is joined with a puckered seam to a U-shaped piece lying on top of the foot **b :** a regular shoe having a seam on the forepart of the vamp imitating the seam of a moccasin
2 a : WATER MOCCASIN **b :** a snake (as of the genus *Natrix*) resembling a water moccasin

moccasin flower *noun* (1680)
: any of several lady's slippers (genus *Cypripedium*); *especially* **:** a once common woodland orchid (*C. acaule*) of eastern North America with pink or white moccasin-shaped flowers

mo·cha \'mō-kə\ *noun* [*Mocha,* Yemen] (1773)

1 a (1) **:** a superior Arabian coffee consisting of small green or yellowish beans (2) **:** a coffee of superior quality **b :** a flavoring made of a strong coffee infusion or of a mixture of cocoa or chocolate with coffee
2 : a pliable suede-finished glove leather from African sheepskins
3 : a dark chocolate-brown color

¹mock \'mäk, 'mȯk\ *verb* [Middle English, from Middle French *mocquer*] (15th century)
transitive verb
1 : to treat with contempt or ridicule **:** DERIDE
2 : to disappoint the hopes of
3 : DEFY, CHALLENGE
4 a : to imitate (as a mannerism) closely **:** MIMIC **b :** to mimic in sport or derision
intransitive verb
: JEER, SCOFF
synonym see RIDICULE, COPY
— mock·er *noun*
— mock·ing·ly \'mä-kiŋ-lē, 'mȯ-\ *adverb*

²mock *noun* (15th century)
1 : an act of ridicule or derision **:** JEER
2 : one that is an object of derision or scorn
3 : MOCKERY
4 a : an act of imitation **b :** something made as an imitation

³mock *adjective* (1548)
: of, relating to, or having the character of an imitation **:** SIMULATED, FEIGNED ⟨the *mock* solemnity of the parody⟩

⁴mock *adverb* (circa 1619)
: in an insincere or counterfeit manner — usually used in combination ⟨*mock*-serious⟩

mock·ery \'mä-k(ə-)rē, 'mȯ-\ *noun, plural* **-er·ies** (15th century)
1 : insulting or contemptuous action or speech **:** DERISION
2 : a subject of laughter, derision, or sport
3 a : a counterfeit appearance **:** IMITATION **b :** an insincere, contemptible, or impertinent imitation ⟨makes a *mockery* of justice⟩
4 : something ridiculously or impudently unsuitable

¹mock-he·ro·ic \ˌmäk-hi-'rō-ik, ˌmȯk-\ *adjective* (circa 1712)
: ridiculing or burlesquing heroic style, character, or action ⟨a *mock-heroic* poem⟩
— mock-he·ro·ical·ly \-i-k(ə-)lē\ *adverb*

²mock-heroic *noun* (1728)
: a mock-heroic composition — called also *mock-epic*

mock·ing·bird \'mä-kiŋ-ˌbərd, 'mȯ-\ *noun* (1676)
: a common grayish bird (*Mimus polyglottos*) especially of the southern U.S. that is remarkable for its exact imitations of the notes of other birds

mock orange *noun* (1731)
1 : any of a genus (*Philadelphus*) of ornamental shrubs of the saxifrage family of which several are widely grown for their showy white flowers — called also *philadelphus, syringa*
2 : any of several usually shrubby plants considered to resemble the orange

mock turtle soup *noun* (1783)
: a soup made of meat (as calf's head or veal), wine, and spices in imitation of green turtle soup

mock-up \'mäk-ˌəp, 'mȯk-\ *noun* (1920)
: a full-sized structural model built accurately to scale chiefly for study, testing, or display

¹mod \'mäd\ *noun* [probably from ²*mod*] (1960)
: one who wears mod clothes

²mod *adjective* [short for *modern*] (1964)
1 : of, relating to, or being the characteristic style of 1960s British youth culture
2 : HIP, TRENDY

mod·acryl·ic fiber \ˌmä-də-'kri-lik-\ *noun* [*modified acrylic*] (1960)
: any of various synthetic textile fibers that are long-chain polymers composed of 35 to 85 percent by weight of acrylonitrile units

mod·al \'mō-dᵊl\ *adjective* [Medieval Latin *modalis,* from Latin *modus*] (1569)
1 : of or relating to modality in logic
2 : containing provisions as to the mode of procedure or the manner of taking effect — used of a contract or legacy
3 : of or relating to a musical mode
4 : of or relating to structure as opposed to substance
5 : of, relating to, or constituting a grammatical form or category characteristically indicating predication of an action or state in some manner other than as a simple fact
6 : of or relating to a statistical mode
— mod·al·ly \-dᵊl-ē\ *adverb*

modal auxiliary *noun* (circa 1904)
: an auxiliary verb (as *can, must, might, may*) that is characteristically used with a verb of predication and expresses a modal modification and that in English differs formally from other verbs in lacking *-s* and *-ing* forms

mo·dal·i·ty \mō-'da-lə-tē\ *noun, plural* **-ties** (circa 1617)
1 a : the quality or state of being modal **b :** a modal quality or attribute **:** FORM
2 : the classification of logical propositions according to their asserting or denying the possibility, impossibility, contingency, or necessity of their content
3 : one of the main avenues of sensation (as vision)
4 : a usually physical therapeutic agency

mod con \ˌmäd-'kän\ *noun* [from *mod. con.,* abbreviation for *modern convenience*] (1934)
chiefly British **:** a modern convenience — usually used in plural

¹mode \'mōd\ *noun* [Middle English *moede,* from Latin *modus* measure, manner, musical mode — more at METE] (14th century)
1 a : an arrangement of the eight diatonic notes or tones of an octave according to one of several fixed schemes of their intervals **b :** a rhythmical scheme (as in 13th and 14th century music)
2 : ²MOOD 2
3 [Late Latin *modus,* from Latin] **a :** ²MOOD 1 **b :** the modal form of the assertion or denial of a logical proposition
4 a : a particular form or variety of something **b :** a form or manner of expression **:** STYLE
5 : a possible, customary, or preferred way of doing something ⟨explained in the usual solemn *mode*⟩
6 a : a manifestation, form, or arrangement of being; *specifically* **:** a particular form or manifestation of an underlying substance **b :** a particular functioning arrangement or condition **:** STATUS ⟨a spacecraft in reentry *mode*⟩ ⟨a computer operating in parallel *mode*⟩
7 a : the most frequent value of a set of data **b :** a value of a random variable for which a function of probabilities defined on it achieves a relative maximum
8 : any of various stationary vibration patterns of which an elastic body or oscillatory system is capable ⟨the vibration *mode* of an airplane propeller blade⟩ ⟨the vibrational *modes* of a molecule⟩
synonym see METHOD

²mode *noun* [French, from Latin *modus*] (circa 1645)
: a prevailing fashion or style (as of dress or behavior)
synonym see FASHION

¹mod·el \'mä-dᵊl\ *noun* [Middle French *modelle,* from Old Italian *modello,* from (assumed) Vulgar Latin *modellus,* from Latin *modulus* small measure, from *modus*] (1575)
1 *obsolete* **:** a set of plans for a building
2 *dialect British* **:** COPY, IMAGE
3 : structural design ⟨a home on the *model* of an old farmhouse⟩
4 : a usually miniature representation of something; *also* **:** a pattern of something to be made
5 : an example for imitation or emulation
6 : a person or thing that serves as a pattern for an artist; *especially* **:** one who poses for an artist
7 : ARCHETYPE
8 : an organism whose appearance a mimic imitates
9 : one who is employed to display clothes or other merchandise **:** MANNEQUIN
10 a : a type or design of clothing **b :** a type or design of product (as a car)
11 : a description or analogy used to help visualize something (as an atom) that cannot be directly observed
12 : a system of postulates, data, and inferences presented as a mathematical description of an entity or state of affairs
13 : VERSION ☆

²model *verb* **mod·eled** *or* **mod·elled; mod·el·ing** *or* **mod·el·ling** \'mäd-liŋ, 'mä-dᵊl-iŋ\ (1625)
transitive verb
1 : to plan or form after a pattern **:** SHAPE
2 *archaic* **:** to make into an organization (as an army, government, or parish)
3 a : to shape or fashion in a plastic material **b :** to produce a representation or simulation of ⟨using a computer to *model* a problem⟩
4 : to construct or fashion in imitation of a particular model ⟨*modeled* its constitution on that of the U.S.⟩
5 : to display by wearing, using, or posing with ⟨*modeled* gowns⟩
intransitive verb
1 : to design or imitate forms **:** make a pattern ⟨enjoys *modeling* in clay⟩
2 : to work or act as a fashion model
— mod·el·er \'mäd-lər, 'mä-dᵊl-ər\ *noun*

³model *adjective* (1844)
1 : serving as or capable of serving as a pattern ⟨a *model* student⟩
2 : being a usually miniature representation of something ⟨a *model* airplane⟩

mo·dem \'mō-ˌdəm, -ˌdem\ *noun* [*mo*dulator + *dem*odulator] (circa 1952)
: a device that converts signals produced by one type of device (as a computer) to a form compatible with another (as a telephone)

¹mod·er·ate \'mä-d(ə-)rət\ *adjective* [Middle English, from Latin *moderatus,* from past participle of *moderare* to moderate; akin to Latin *modus* measure] (15th century)
1 a : avoiding extremes of behavior or expression **:** observing reasonable limits ⟨a *moderate* drinker⟩ **b :** CALM, TEMPERATE
2 a : tending toward the mean or average amount or dimension **b :** having average or less than average quality **:** MEDIOCRE

☆ SYNONYMS
Model, example, pattern, exemplar, ideal mean someone or something set before one for guidance or imitation. MODEL applies to something taken or proposed as worthy of imitation ⟨a decor that is a *model* of good taste⟩. EXAMPLE applies to a person to be imitated or in some contexts on no account to be imitated but to be regarded as a warning ⟨children tend to follow the *example* of their parents⟩. PATTERN suggests a clear and detailed archetype or prototype ⟨American industry set a *pattern* for others to follow⟩. EXEMPLAR suggests either a faultless example to be emulated or a perfect typification ⟨cited Joan of Arc as the *exemplar* of courage⟩. IDEAL implies the best possible exemplification either in reality or in conception ⟨never found a job that matched his *ideal*⟩.

3 : professing or characterized by political or social beliefs that are not extreme
4 : limited in scope or effect
5 : not expensive **:** reasonable or low in price
6 *of a color* **:** of medium lightness and medium chroma
— **mod•er•ate•ly** *adverb*
— **mod•er•ate•ness** *noun*

²mod•er•ate \'mä-də-ˌrāt\ *verb* **-at•ed; -at•ing** (15th century)
transitive verb
1 : to lessen the intensity or extremeness of ⟨the sun *moderated* the chill⟩
2 : to preside over or act as chairman of
intransitive verb
1 : to act as a moderator
2 : to become less violent, severe, or intense
— **mod•er•a•tion** \ˌmä-də-'rā-shən\ *noun*

³mod•er•ate \'mä-d(ə-)rət\ *noun* (1794)
: one who holds moderate views or who belongs to a group favoring a moderate course or program

moderate breeze *noun* (circa 1881)
: wind having a speed of 13 to 18 miles (21 to 29 kilometers) per hour — see BEAUFORT SCALE table

moderate gale *noun* (1704)
: wind having a speed of 32 to 38 miles (51 to 61 kilometers) per hour — see BEAUFORT SCALE table

mo•der•a•to \ˌmä-də-'rä-(ˌ)tō\ *adverb or adjective* [Italian, from Latin *moderatus*] (circa 1724)
: MODERATE — used as a direction in music to indicate tempo

mod•er•a•tor \'mä-də-ˌrā-tər\ *noun* (circa 1560)
1 : one who arbitrates **:** MEDIATOR
2 : one who presides over an assembly, meeting, or discussion: as **a :** the presiding officer of a Presbyterian governing body **b :** the nonpartisan presiding officer of a town meeting **c :** the chairman of a discussion group
3 : a substance (as graphite) used for slowing down neutrons in a nuclear reactor
— **mod•er•a•tor•ship** \-ˌship\ *noun*

¹mod•ern \'mä-dərn, ÷'mä-d(ə-)rən\ *adjective* [Late Latin *modernus,* from Latin *modo* just now, from *modus* measure — more at METE] (1585)
1 a : of, relating to, or characteristic of the present or the immediate past **:** CONTEMPORARY **b :** of, relating to, or characteristic of a period extending from a relevant remote past to the present time
2 : involving recent techniques, methods, or ideas **:** UP-TO-DATE
3 *capitalized* **:** of, relating to, or having the characteristics of the present or most recent period of development of a language
4 : of or relating to modernism **:** MODERNIST
— **mo•der•ni•ty** \mə-'dər-nə-tē, mä- *also* -'der-\ *noun*
— **mod•ern•ly** \'mä-dərn-lē\ *adverb*
— **mod•ern•ness** \-dərn-nəs\ *noun*

²modern *noun* (1585)
1 a : a person of modern times or views **b :** an adherent of modernism **:** MODERNIST
2 : a style of printing type distinguished by regularity of shape, precise curves, straight hairline serifs, and heavy downstrokes

mo•derne \mō-'dern, mə-\ *noun, often capitalized* [French, modern] (1955)
: ART DECO

Modern Hebrew *noun* (1949)
: NEW HEBREW

mod•ern•i•sa•tion, mod•ern•ise *British variant of* MODERNIZATION, MODERNIZE

mod•ern•ism \'mä-dər-ˌni-zəm\ *noun* (1737)
1 : a practice, usage, or expression peculiar to modern times
2 *often capitalized* **:** a tendency in theology to accommodate traditional religious teaching to contemporary thought and especially to devalue supernatural elements
3 : modern artistic or literary philosophy and practice; *especially* **:** a self-conscious break with the past and a search for new forms of expression
— **mod•ern•ist** \-nist\ *noun or adjective*
— **mod•ern•is•tic** \ˌmä-dər-'nis-tik\ *adjective*

mod•ern•i•za•tion \ˌmä-dər-nə-'zā-shən\ *noun* (1770)
1 : the act of modernizing **:** the state of being modernized
2 : something modernized **:** a modernized version

mod•ern•ize \'mä-dər-ˌnīz\ *verb* **-ized; -iz•ing** (1748)
transitive verb
: to make modern (as in taste, style, or usage)
intransitive verb
: to adopt modern ways
— **mod•ern•iz•er** *noun*

modern pentathlon *noun* (circa 1912)
: a composite contest in which all contestants compete in a 300-meter freestyle swim, a 4000-meter cross-country run, a 5000-meter 30-jump equestrian steeplechase, épée fencing, and target shooting at 25 meters

mod•est \'mä-dəst\ *adjective* [Latin *modestus* moderate; akin to Latin *modus* measure] (1565)
1 a : placing a moderate estimate on one's abilities or worth **b :** neither bold nor self-assertive **:** tending toward diffidence
2 : arising from or characteristic of a modest nature
3 : observing the proprieties of dress and behavior **:** DECENT
4 a : limited in size, amount, or scope **b :** UNPRETENTIOUS ⟨a *modest* cottage⟩
synonym see SHY, CHASTE
— **mod•est•ly** *adverb*

mod•es•ty \'mä-də-stē\ *noun* (1531)
1 : freedom from conceit or vanity
2 : propriety in dress, speech, or conduct

mo•di•cum \'mä-di-kəm *also* 'mō-\ *noun* [Middle English, from Latin, neuter of *modicus* moderate, from *modus* measure] (15th century)
: a small portion **:** a limited quantity

mod•i•fi•ca•tion \ˌmä-də-fə-'kā-shən\ *noun* (1603)
1 : the limiting of a statement **:** QUALIFICATION
2 : ¹MODE 6a
3 a : the making of a limited change in something; *also* **:** the result of such a change **b :** a change in an organism caused by environmental factors
4 : a limitation or qualification of the meaning of a word by another word, by an affix, or by internal change

mod•i•fi•er \'mä-də-ˌfī(-ə)r\ *noun* (1583)
1 : one that modifies
2 : a word or phrase that makes specific the meaning of another word or phrase
3 : a gene that modifies the effect of another

mod•i•fy \'mä-də-ˌfī\ *verb* **-fied; -fy•ing** [Middle English *modifien,* from Middle French *modifier,* from Latin *modificare* to measure, moderate, from *modus*] (14th century)
transitive verb
1 : to make less extreme **:** MODERATE
2 a : to limit or restrict the meaning of especially in a grammatical construction **b :** to change (a vowel) by umlaut
3 a : to make minor changes in **b :** to make basic or fundamental changes in often to give a new orientation to or to serve a new end ⟨the wing of a bird is an arm *modified* for flying⟩
intransitive verb
: to undergo change
synonym see CHANGE
— **mod•i•fi•abil•i•ty** \ˌmä-də-ˌfī-ə-'bi-lə-tē\ *noun*
— **mod•i•fi•able** \'mä-də-ˌfī-ə-bəl\ *adjective*

mo•dil•lion \mō-'dil-yən\ *noun* [Italian *modiglione*] (1563)
: an ornamental block or bracket under the corona of the cornice (as in the Corinthian order)

mod•ish \'mō-dish\ *adjective* (1660)
: FASHIONABLE, STYLISH ⟨a *modish* hat⟩ ⟨a *modish* writer⟩
— **mod•ish•ly** *adverb*
— **mod•ish•ness** *noun*

mo•diste \mō-'dēst\ *noun* [French, from *mode* style, mode] (circa 1840)
: one who makes and sells fashionable dresses and hats for women

Mo•dred \'mō-dred\ *noun*
: a knight of the Round Table and nephew of King Arthur

mod•u•la•bil•i•ty \ˌmä-jə-lə-'bi-lə-tē\ *noun* (1928)
: the capability of being modulated

mod•u•lar \'mä-jə-lər\ *adjective* (1798)
1 : of, relating to, or based on a module or a modulus
2 : constructed with standardized units or dimensions for flexibility and variety in use
— **mod•u•lar•i•ty** \ˌmä-jə-'lar-ə-tē\ *noun*
— **mod•u•lar•ly** \'mä-jə-lər-lē\ *adverb*

modular arithmetic *noun* (1959)
: arithmetic that deals with whole numbers where the numbers are replaced by their remainders after division by a fixed number ⟨in a *modular arithmetic* with modulus 5, 3 multiplied by 4 is 2⟩

mod•u•lar•ized \'mä-jə-lə-ˌrīzd\ *adjective* (1959)
1 : containing or consisting of modules
2 : produced in the form of modules

mod•u•late \'mä-jə-ˌlāt\ *verb* **-lat•ed; -lat•ing** [Latin *modulatus,* past participle of *modulari* to play, sing, from *modulus* small measure, rhythm, diminutive of *modus* measure — more at METE] (1615)
transitive verb
1 : to tune to a key or pitch
2 : to adjust to or keep in proper measure or proportion **:** TEMPER
3 : to vary the amplitude, frequency, or phase of (a carrier wave or a light wave) for the transmission of intelligence (as by radio); *also* **:** to vary the velocity of electrons in an electron beam
intransitive verb
1 : to play or sing with modulation
2 : to pass from one musical key into another by means of intermediary chords or notes that have some relation to both keys
3 : to pass gradually from one state to another
— **mod•u•la•tor** \-ˌlā-tər\ *noun*
— **mod•u•la•to•ry** \-lə-ˌtōr-ē, -ˌtȯr-\ *adjective*

mod•u•la•tion \ˌmä-jə-'lā-shən\ *noun* (1531)
1 : a regulating according to measure or proportion **:** TEMPERING
2 : an inflection of the tone or pitch of the voice; *specifically* **:** the use of stress or pitch to convey meaning
3 : a change from one musical key to another by modulating
4 : the process of modulating a carrier or signal (as in radio); *also* **:** the result of this process

mod•ule \'mä-(ˌ)jü(ə)l\ *noun* [Latin *modulus*] (circa 1628)
1 : a standard or unit of measurement
2 : the size of some one part taken as a unit of measure by which the proportions of an architectural composition are regulated
3 a : any in a series of standardized units for use together: as (1) **:** a unit of furniture or architecture (2) **:** an educational unit which covers a single subject or topic **b :** a usually packaged functional assembly of electronic components for use with other such assemblies
4 : an independently-operable unit that is a part of the total structure of a space vehicle
5 a : a subset of an additive group that is also a group under addition **b :** a mathematical set that is a commutative group under addition and that is closed under multiplication which

is distributive from the left or right or both by elements of a ring and for which $a(bx) = (ab)x$ or $(xb)a = x(ba)$ or both where a and b are elements of the ring and x belongs to the set

mod·u·lo \ˈmä-jə-ˌlō\ *preposition* [New Latin, ablative of *modulus*] (1897)
: with respect to a modulus of ⟨19 and 54 are congruent *modulo* 7⟩

mod·u·lus \ˈmä-jə-ləs\ *noun, plural* **-li** \-ˌlī, -ˌlē\ [New Latin, from Latin, small measure] (1753)
1 a : the factor by which a logarithm of a number to one base is multiplied to obtain the logarithm of the number to a new base **b :** ABSOLUTE VALUE 2 **c** (1) **:** the number (as a positive integer) or other mathematical entity (as a polynomial) in a congruence that divides the difference of the two congruent members without leaving a remainder — compare RESIDUE b (2) **:** the number of different numbers used in a system of modular arithmetic
2 : a constant or coefficient that expresses usually numerically the degree to which a body or substance possesses a particular property (as elasticity)

mo·dus op·e·ran·di \ˌmō-dəs-ˌä-pə-ˈran-dē, -ˌdī\ *noun, plural* **mo·di operandi** \ˈmō-ˌdē-, ˈmō-ˌdī-\ [New Latin] (1654)
: a method of procedure

mo·dus vi·ven·di \ˌmō-dəs-vi-ˈven-dē, -ˌdī\ *noun, plural* **mo·di vivendi** \ˈmō-ˌdē-, ˈmō-ˌdī-\ [New Latin, manner of living] (circa 1879)
1 : a feasible arrangement or practical compromise; *especially* **:** one that bypasses difficulties
2 : a manner of living **:** a way of life

Mogen David *variant of* MAGEN DAVID

mog·gy *also* **mog·gie** \ˈmä-gē\ *noun, plural* **mog·gies** [probably from *Moggy,* from *Mog,* nickname from the name *Margaret*] (circa 1911)
British **:** CAT

[1]mo·gul \ˈmō-(ˌ)gəl\ *noun* [Persian *Mughul,* from Mongolian *mongγol* Mongol] (1588)
1 *or* **mo·ghul** *capitalized* **:** an Indian Muslim of or descended from one of several conquering groups of Mongol, Turkish, and Persian origin; *especially* **:** GREAT MOGUL
2 : a great personage **:** MAGNATE
— **mogul** *adjective,* often capitalized

[2]mo·gul \ˈmō-gəl\ *noun* [German dialect; akin to German dialect (Viennese) *mugl* small hill] (1959)
: a bump in a ski run

mo·hair \ˈmō-ˌhar, -ˌher\ *noun* [modification of obsolete Italian *mocaiarro,* from Arabic *mukhayyar,* literally, choice] (1619)
: a fabric or yarn made wholly or in part of the long silky hair of the Angora goat; *also* **:** this hair

Mo·ham·med·an *variant of* MUHAMMADAN

Mo·hawk \ˈmō-ˌhȯk\ *noun, plural* **Mohawk** *or* **Mohawks** [of Algonquian origin; akin to Narraganset or Massachuset *Mohowawog* Mohawk, literally, cannibal] (1634)
1 : a member of an American Indian people of the Mohawk River valley, New York
2 : the Iroquoian language of the Mohawk people
3 : a hairstyle with a narrow center strip of upright hair and the sides shaved ◆

Mo·he·gan \mō-ˈhē-gən, mə-\ *or* **Mo·hi·can** \-ˈhē-kən\ *noun, plural* **Mohegan** *or* **Mohegans** *or* **Mohican** *or* **Mohicans** (1660)
: a member of an American Indian people of southeastern Connecticut

Mo·hi·can \mō-ˈhē-kən, mə-\ *variant of* MAHICAN

Mo·ho \ˈmō-ˌhō\ *noun* [short for *Mohorovicic discontinuity,* from Andrija *Mohorovičić* (died 1936) Yugoslavian geologist] (1952)
: the boundary layer between the earth's crust and mantle whose depth varies from about 3 miles (5 kilometers) beneath the ocean floor to about 25 miles (40 kilometers) beneath the continents

Mo·hock \ˈmō-ˌhäk\ *noun* [alteration of *Mohawk*] (circa 1712)
: one of a gang of aristocratic ruffians who assaulted and otherwise maltreated people in London streets in the early 18th century
— **Mo·hock·ism** \-ˌhä-ˌki-zəm\ *noun*

Mo·ho·ro·vi·cic discontinuity \ˌmō-hə-ˈrō-və-ˌchich-\ *noun* (1936)
: MOHO

Mohs' scale \ˈmōz-, ˈmōs-, ˈmō-səz-\ *noun* [Friedrich *Mohs* (died 1839) German mineralogist] (1879)
: a scale of hardness for minerals that ranges from a value of 1 for talc to 10 for diamond

mo·hur \ˈmō(-ə)r, mə-ˈhu̇r\ *noun* [Hindi *muhr* gold coin, seal, from Persian, from Middle Persian; akin to Sanskrit *mudrā* seal] (1690)
: a former gold coin of India and Persia equal to 15 rupees

moi·e·ty \ˈmȯi-ə-tē\ *noun, plural* **-ties** [Middle English *moite,* from Middle French *moité,* from Late Latin *medietat-, medietas,* from Latin *medius* middle — more at MID] (15th century)
1 a : one of two equal parts **:** HALF **b :** one of two approximately equal parts
2 : one of the portions into which something is divided **:** COMPONENT, PART
3 : one of two basic complementary tribal subdivisions

[1]moil \ˈmȯi(ə)l\ *verb* [Middle English *moillen,* from Middle French *moillier,* from (assumed) Vulgar Latin *molliare,* from Latin *mollis* soft — more at MOLLIFY] (15th century)
transitive verb
chiefly dialect **:** to make wet or dirty
intransitive verb
1 : to work hard **:** DRUDGE
2 : to be in continuous agitation **:** CHURN, SWIRL
— **moil·er** *noun*

[2]moil *noun* (1612)
1 : hard work **:** DRUDGERY
2 : CONFUSION, TURMOIL

moil·ing \ˈmȯi-liŋ\ *adjective* (1603)
1 a : requiring hard work **b :** INDUSTRIOUS ⟨*moiling* workers⟩
2 : violently agitated **:** TURBULENT
— **moil·ing·ly** \-lē\ *adverb*

Moi·rai \ˈmȯi-ˌrī\ *noun plural* [Greek, from plural of *moira* lot, fate; akin to Greek *meros* part — more at MERIT]
: FATE 4

moire \ˈmȯi(-ə)r, ˈmȯr, ˈmwär\ *noun* [French, from English *mohair*] (1660)
archaic **:** a watered mohair

moi·ré \mȯ-ˈrā, mwä-\ *or* **moire** *same or* ˈmȯi(-ə)r, ˈmȯr, ˈmwär\ *noun* [French *moiré,* from *moiré* like moire, from *moire*] (1818)
1 a : an irregular wavy finish on a fabric **b :** a ripple pattern on a stamp
2 : a fabric having a wavy watered appearance
3 : an independent usually shimmering pattern seen when two geometrically regular patterns (as two sets of parallel lines or two halftone screens) are superimposed especially at an acute angle
— **moiré** *or* **moire** *adjective*

moist \ˈmȯist\ *adjective* [Middle English *moiste,* from Middle French, perhaps from (assumed) Vulgar Latin *muscidus,* alteration of Latin *mucidus* slimy, from *mucus* nasal mucus] (14th century)
1 : slightly or moderately wet **:** DAMP
2 : TEARFUL
3 : characterized by high humidity
synonym see WET
— **moist·ly** *adverb*
— **moist·ness** \ˈmȯis(t)-nəs\ *noun*

moist·en \ˈmȯi-sᵊn\ *verb* **moist·ened; moist·en·ing** \ˈmȯis-niŋ, ˈmȯi-sᵊn-iŋ\ (1580)
transitive verb
: to make moist
intransitive verb
: to become moist
— **moist·en·er** \ˈmȯis-nər, ˈmȯi-sᵊn-ər\ *noun*

mois·ture \ˈmȯis-chər, ˈmȯish-\ *noun* [Middle English, modification of Middle French *moistour,* from *moiste*] (14th century)
: liquid diffused or condensed in relatively small quantity

mois·tur·ise *British variant of* MOISTURIZE

mois·tur·ize \-chə-ˌrīz\ *transitive verb* **-ized; -iz·ing** (1945)
: to add moisture to ⟨*moisturize* the air⟩
— **mois·tur·iz·er** *noun*

mo·jo \ˈmō-(ˌ)jō\ *noun, plural* **mojoes** *or* **mojos** [probably of African origin; akin to Fulani *moco'o* medicine man] (1926)
: a magic spell, hex, or charm; *broadly* **:** magical power

moke \ˈmōk\ *noun* [origin unknown] (1848)
1 *slang British* **:** DONKEY
2 *slang Australian* **:** [1]NAG

mol·al \ˈmō-ləl\ *adjective* [[5]*mole*] (1905)
: of, relating to, or containing a mole of solute per 1000 grams of solvent
— **mo·lal·i·ty** \mō-ˈla-lə-tē\ *noun*

[1]mo·lar \ˈmō-lər\ *noun* [Middle English *molares,* plural, from Latin *molaris,* from *molaris* of a mill, from *mola* millstone — more at MILL] (14th century)
: a tooth with a rounded or flattened surface adapted for grinding; *specifically* **:** one of the cheek teeth in mammals behind the incisors and canines — see TOOTH illustration

[2]molar *adjective* (1626)
1 : pulverizing by friction **:** GRINDING
2 : of, relating to, or located near the molar teeth

[3]molar *adjective* [[5]*mole*] (1902)
1 : of or relating to a mole of a substance ⟨the *molar* volume of a gas⟩
2 : containing one mole of solute in one liter of solution
— **mo·lar·i·ty** \mō-ˈlar-ə-tē\ *noun*

mo·las·ses \mə-ˈla-səz\ *noun* [modification of Portuguese *melaço,* from Late Latin *mellaceum* grape juice, from Latin *mell-, mel* honey — more at MELLIFLUOUS] (1582)
1 : the thick dark to light brown syrup that is separated from raw sugar in sugar manufacture

◇ WORD HISTORY
Mohawk The Mohawks are the easternmost of the five Iroquois tribes that inhabited upstate New York in colonial times. These five tribes, joined in the 18th century by the Tuscarora, formed a remarkable confederacy, the League of the Iroquois, that ensured peace among its members; some have alleged that the example of the League influenced the framers of the U.S. Constitution. The English name *Mohawk,* like the names of a number of other Indian peoples, is based not on the Mohawks' own name for themselves, but rather on a designation applied to them by another people, in this case the New England Algonquian tribes who lived to the east of the Mohawks. This Algonquian name, which the colonist Roger Williams recorded from the Narragansetts as *Mohowawog,* means literally "man-eater." In part it was probably a literal accusation of cannibalism, and in part a reflection of the terror the Mohawks and other Iroquois inspired in their enemies. The Mohawks' self-designation is a word usually interpreted as "place of the gunflint."

2 : a syrup made from boiling down sweet vegetable or fruit juice ⟨citrus *molasses*⟩

[1]**mold** \'mōld\ *noun* [Middle English, from Old English *molde;* akin to Old High German *molta* soil, Latin *molere* to grind — more at MEAL] (before 12th century)
1 : crumbling soft friable earth suited to plant growth **:** SOIL; *especially* **:** soil rich in humus — compare LEAF MOLD
2 *dialect British* **a :** the surface of the earth **:** GROUND **b :** the earth of the burying ground
3 *archaic* **:** earth that is the substance of the human body ⟨be merciful great Duke to men of *mold* —Shakespeare⟩

[2]**mold** *noun* [Middle English, modification of Old French *modle,* from Latin *modulus,* diminutive of *modus* measure — more at METE] (13th century)
1 : distinctive nature or character **:** TYPE
2 : the frame on or around which an object is constructed
3 a : a cavity in which a substance is shaped: as (1) **:** a matrix for casting metal (2) **:** a form in which food is given a decorative shape **b :** a molded object
4 : MOLDING
5 a *obsolete* **:** an example to be followed **b :** PROTOTYPE **c :** a fixed pattern **:** DESIGN

[3]**mold** *transitive verb* (14th century)
1 *archaic* **:** to knead (dough) into a desired consistency or shape
2 : to give shape to ⟨the wind *molds* the waves⟩
3 : to form in a mold ⟨*mold* candles⟩
4 : to determine or influence the quality or nature of ⟨*mold* public opinion⟩
5 : to fit the contours of
6 : to ornament with molding or carving ⟨*molded* picture frames⟩
— **mold·able** \'mōl-də-bəl\ *adjective*
— **mold·er** *noun*

[4]**mold** *noun* [Middle English *mowlde,* perhaps alteration of *mowle,* from *moulen* to grow moldy, of Scandinavian origin; akin to Old Norse *mygla* to grow moldy] (14th century)
1 : a superficial often woolly growth produced especially on damp or decaying organic matter or on living organisms
2 : a fungus (as of the order Mucorales) that produces mold

[5]**mold** *intransitive verb* (1530)
: to become moldy

mold·board \'mōl(d)-ˌbōrd, -ˌbȯrd\ *noun* (1508)
1 : a curved iron plate attached above a plowshare to lift and turn the soil
2 : the flat or curved blade (as of a bulldozer) that pushes material to one side as the machine advances

mold·er \'mōl-dər\ *intransitive verb* **mold·ered; mold·er·ing** \-d(ə-)riŋ\ [frequentative of [5]*mold*] (1531)
: to crumble into particles **:** DISINTEGRATE, DECAY

mold·ing \'mōl-diŋ\ *noun* (14th century)
1 a : the art or occupation of a molder **b :** an object produced by molding
2 a : a decorative recessed or relieved surface **b :** a decorative plane or curved strip used for ornamentation or finishing

molding 2a: *1* **fillet and fascia,** *2* **torus,** *3* **reeding,** *4* **cavetto,** *5* **scotia,** *6* **congé,** *7* **beak**

moldy \'mōl-dē\ *adjective* **mold·i·er; -est** (14th century)
1 : of, resembling, or covered with a mold-producing fungus
2 a : being old and moldering **:** CRUMBLING **b :** ANTIQUATED, FUSTY ⟨*moldy* tradition⟩
— **mold·i·ness** *noun*

[1]**mole** \'mōl\ *noun* [Middle English, from Old English *māl;* akin to Old High German *meil* spot] (14th century)
: a pigmented spot, mark, or small permanent protuberance on the human body; *especially* **:** NEVUS

[2]**mole** *noun* [Middle English; akin to Middle Low German *mol*] (14th century)
1 : any of numerous burrowing insectivores (especially family Talpidae) with tiny eyes, concealed ears, and soft fur
2 : one who works in the dark
3 : a machine for tunneling
4 : a spy (as a double agent) who establishes a cover long before beginning espionage

[3]**mole** *noun* [Middle English, from Latin *mola* mole, literally, mill, millstone — more at MILL] (15th century)
: an abnormal mass in the uterus especially when containing fetal tissues

[4]**mole** *noun* [Middle French, from Old Italian *molo,* from Late Greek *mōlos,* from Latin *moles,* literally, mass, exertion; akin to Greek *mōlos* exertion] (circa 1548)
1 : a massive work formed of masonry and large stones or earth laid in the sea as a pier or breakwater
2 : the harbor formed by a mole

[5]**mole** *also* **mol** \'mōl\ *noun* [German *Mol,* short for *Molekulargewicht* molecular weight, from *molekular* molecular + *Gewicht* weight] (1902)
: the base unit of amount of pure substance in the International System of Units that contains the same number of elementary entities as there are atoms in exactly 12 grams of the isotope carbon 12

[6]**mo·le** \'mō-lā\ *noun* [Mexican Spanish, from Nahuatl *mōlli* sauce] (1927)
: a spicy sauce made with chilies and usually chocolate and served with meat

mo·lec·u·lar \mə-'le-kyə-lər\ *adjective* (1823)
1 : of, relating to, or produced by molecules ⟨*molecular* oxygen⟩
2 : of or relating to simple or elementary organization
— **mo·lec·u·lar·ly** \-lē\ *adverb*

molecular biology *noun* (1938)
: a branch of biology dealing with the ultimate physicochemical organization of living matter and especially with the molecular basis of inheritance and protein synthesis
— **molecular biologist** *noun*

molecular formula *noun* (circa 1903)
: a chemical formula that is based on both analysis and molecular weight and gives the total number of atoms of each element in a molecule — compare STRUCTURAL FORMULA

molecular mass *noun* (1970)
: the mass of a molecule that is equal to the sum of the masses of all the atoms contained in the molecule

molecular orbital *noun* (1932)
: a solution of the Schrödinger equation that describes the probable location of an electron relative to the nuclei in a molecule and so indicates the nature of any bond in which the electron is involved

molecular sieve *noun* (1926)
: a crystalline substance (as a zeolite) characterized by uniformly sized pores of molecular dimension that can adsorb small molecules and is used especially in separations

molecular weight *noun* (1880)
: the average mass of a molecule of a compound compared to $\frac{1}{12}$ the mass of carbon 12 and calculated as the sum of the atomic weights of the constituent atoms

mol·e·cule \'mä-li-ˌkyü(ə)l\ *noun* [French *molécule,* from New Latin *molecula,* diminutive of Latin *moles* mass] (1794)
1 : the smallest particle of a substance that retains all the properties of the substance and is composed of one or more atoms
2 : a tiny bit **:** PARTICLE

mole·hill \'mōl-ˌhil\ *noun* (15th century)
: a little mound or ridge of earth pushed up by a mole

mole·skin \-ˌskin\ *noun* (1668)
1 : the skin of the mole used as fur
2 a : a heavy durable cotton fabric with a short thick velvety nap on one side **b :** a garment made of moleskin — usually used in plural

mo·lest \mə-'lest\ *transitive verb* [Middle English, from Middle French *molester,* from Latin *molestare,* from *molestus* burdensome, annoying; akin to Latin *moles* mass] (14th century)
1 : to annoy, disturb, or persecute especially with hostile intent or injurious effect
2 : to make annoying sexual advances to; *especially* **:** to force physical and usually sexual contact on
— **mo·les·ta·tion** \ˌmō-ˌles-'tā-shən, ˌmä-, -ləs-\ *noun*
— **mo·lest·er** \mə-'les-tər\ *noun*

mo·line \mə-'lēn, -'līn\ *adjective* [(assumed) Anglo-French *moliné,* from Old French *molin* mill, from Late Latin *molinum* — more at MILL] (1562)
of a heraldic cross **:** having the end of each arm forked and recurved — see CROSS illustration

moll \'mäl, 'mȯl\ *noun* [probably from *Moll,* nickname for *Mary*] (1604)
1 : PROSTITUTE
2 a : DOLL 2 **b :** a gangster's girlfriend

mol·li·fy \'mä-lə-ˌfī\ *verb* **-fied; -fy·ing** [Middle English *mollifien,* from Middle French *mollifier,* from Late Latin *mollificare,* from Latin *mollis* soft; akin to Greek *amaldynein* to soften, Sanskrit *mṛdu* soft, and probably to Greek *malakos* soft, *amblys* dull, Old English *meltan* to melt] (15th century)
transitive verb
1 : to soothe in temper or disposition **:** APPEASE ⟨*mollified* the staff with a raise⟩
2 : to reduce the rigidity of **:** SOFTEN
3 : to reduce in intensity **:** ASSUAGE, TEMPER
intransitive verb
archaic **:** SOFTEN, RELENT
synonym see PACIFY
— **mol·li·fi·ca·tion** \ˌmä-lə-fə-'kā-shən\ *noun*

mol·lus·ci·cide \mə-'ləs-kə-ˌsīd, -'lə-sə-\ *noun* [New Latin *Mollusca* + English *-i-* + *-cide*] (circa 1947)
: an agent for destroying mollusks (as snails)
— **mol·lus·ci·cid·al** \-ˌləs-kə-'sī-dᵊl, -ˌlə-sə-\ *adjective*

mol·lusk *or* **mol·lusc** \'mä-ləsk\ *noun* [French *mollusque,* from New Latin *Mollusca,* from Latin, neuter plural of *molluscus* thin-shelled (of a nut), from *mollis*] (1783)
: any of a large phylum (Mollusca) of invertebrate animals (as snails, clams, or squids) with a soft unsegmented body usually enclosed in a calcareous shell; *broadly* **:** SHELLFISH
— **mol·lus·can** *also* **mol·lus·kan** \mə-'ləs-kən, mä-\ *adjective*

mol·ly *also* **mol·lie** \'mä-lē\ *noun, plural* **mollies** [by shortening from New Latin *Mollienisia,* former genus name, from Comte François N. *Mollien* (died 1850) French statesman] (circa 1933)
: any of several brightly colored live-bearers (genus *Poecilia* of the family Poeciliidae) highly valued as aquarium fishes

[1]**mol·ly·cod·dle** \'mä-lē-ˌkä-dᵊl\ *noun* [*Molly,* nickname for *Mary*] (1833)
: a pampered or effeminate man or boy

[2]**mollycoddle** *transitive verb* **-cod·dled; -cod·dling** \-ˌkäd-liŋ, -ˌkä-dᵊl-iŋ\ (1864)
: to surround with an excessive or absurd degree of indulgence and attention
synonym see INDULGE
— **mol·ly·cod·dler** \-ˌkäd-lər, -ˌkä-dᵊl-ər\ *noun*

Mo·loch \'mä-lək, 'mō-ˌläk\ *noun* [Late Latin, from Greek, from Hebrew *Mōlekh*]

: a Semitic god to whom children were sacrificed

Mo·lo·tov cocktail \'mä-lə-ˌtȯf-, 'mȯ-, 'mō-, -ˌtȯv-\ *noun* [Vyacheslav M. *Molotov*] (1939)
: a crude bomb made of a bottle filled with a flammable liquid (as gasoline) and usually fitted with a wick (as a saturated rag) that is ignited just before the bottle is hurled

¹molt \'mōlt\ *verb* [alteration of Middle English *mouten,* from Old English *-mūtian* to change, from Latin *mutare* — more at MUTABLE] (1591)
intransitive verb
: to shed hair, feathers, shell, horns, or an outer layer periodically
transitive verb
: to cast off (an outer covering) periodically; *specifically* **:** to throw off (the old cuticle) — used of arthropods
— molt·er *noun*

²molt *noun* (1815)
: the act or process of molting; *specifically* **:** ECDYSIS

mol·ten \'mōl-tᵊn\ *adjective* [Middle English, from past participle of *melten* to melt] (14th century)
1 *obsolete* **:** made by melting and casting
2 : fused or liquefied by heat **:** MELTED ⟨*molten* lava⟩
3 : having warmth or brilliance **:** GLOWING ⟨the *molten* sunlight of warm skies —T. B. Costain⟩

mol·to \'mōl-(ˌ)tō, 'mȯl-\ *adverb* [Italian, from Latin *multum,* from neuter of *multus* much] (circa 1801)
: MUCH, VERY — used in music directions ⟨*molto* vivace⟩

mo·ly \'mō-lē\ *noun* [Latin, from Greek *mōly*]
: a mythical herb with a black root, white blossoms, and magical powers

mo·lyb·date \mə-'lib-ˌdāt\ *noun* (1794)
: a salt of molybdenum containing the group MoO_4 or Mo_2O_7

mo·lyb·de·nite \mə-'lib-də-ˌnīt\ *noun* [New Latin *molybdena*] (1796)
: a blue usually foliated mineral consisting of molybdenum disulfide that is a source of molybdenum

mo·lyb·de·num \-nəm\ *noun* [New Latin, from *molybdena,* a lead ore, molybdenite, molybdenum, from Latin *molybdaena* galena, from Greek *molybdaina,* from *molybdos* lead] (1816)
: a metallic element that resembles chromium and tungsten in many properties, is used especially in strengthening and hardening steel, and is a trace element in plant and animal metabolism — see ELEMENT table

molybdenum disulfide *noun* (circa 1931)
: a compound MoS_2 used especially as a lubricant in grease

mom \'mäm, 'məm\ *noun* [short for *momma*] (circa 1894)
: MOTHER

mom-and-pop \'mäm-ən(d)-'päp\ *adjective* (1951)
: being a small owner-operated business

mome \'mōm\ *noun* [origin unknown] (1553)
archaic **:** BLOCKHEAD, FOOL

mo·ment \'mō-mənt\ *noun* [Middle English, from Middle French, from Latin *momentum* movement, particle sufficient to turn the scales, moment, from *movēre* to move] (14th century)
1 a : a minute portion or point of time **:** INSTANT **b :** a comparatively brief period of time
2 a : present time ⟨at the *moment* she is working on a novel⟩ **b :** a time of excellence or conspicuousness ⟨he has his *moments*⟩
3 : importance in influence or effect ⟨a matter of great *moment*⟩
4 *obsolete* **:** a cause or motive of action
5 : a stage in historical or logical development
6 a : tendency or measure of tendency to produce motion especially about a point or axis **b :** the product of quantity (as a force) and the distance to a particular axis or point
7 a : the mean of the *n*th powers of the deviations of the observed values in a set of statistical data from a fixed value **b :** the expected value of a power of the deviation of a random variable from a fixed value
synonym see IMPORTANCE

mo·men·tar·i·ly \ˌmō-mən-'ter-ə-lē\ *adverb* (circa 1666)
1 : for a moment
2 *archaic* **:** INSTANTLY
3 : at any moment **:** in a moment ■

mo·men·tary \'mō-mən-ˌter-ē\ *adjective* (15th century)
1 a : continuing only a moment **:** FLEETING **b :** having a very brief life
2 : operative or recurring at every moment
synonym see TRANSIENT
— mo·men·tar·i·ness *noun*

mo·ment·ly \'mō-mənt-lē\ *adverb* (1676)
1 : from moment to moment
2 : at any moment
3 : for a moment
usage see MOMENTARILY

mo·men·to *variant of* MEMENTO
usage see MEMENTO

moment of inertia (1830)
: a measure of the resistance of a body to angular acceleration about a given axis that is equal to the sum of the products of each element of mass in the body and the square of the element's distance from the axis

moment of truth (1932)
1 : the final sword thrust in a bullfight
2 : a moment of crisis on whose outcome much or everything depends

mo·men·tous \mō-'men-təs, mə-\ *adjective* (1652)
: IMPORTANT, CONSEQUENTIAL
— mo·men·tous·ly *adverb*
— mo·men·tous·ness *noun*

mo·men·tum \mō-'men-təm, mə-\ *noun, plural* **mo·men·ta** \-'men-tə\ *or* **momentums** [New Latin, from Latin, movement] (1610)
1 : a property of a moving body that the body has by virtue of its mass and motion and that is equal to the product of the body's mass and velocity; *broadly* **:** a property of a moving body that determines the length of time required to bring it to rest when under the action of a constant force or moment
2 : strength or force gained by motion or through the development of events **:** IMPETUS ⟨the campaign gained *momentum*⟩

mom·ma \'mä-mə, 'mə-mə\ *variant of* MAMA

mom·my \'mä-mē, 'mə-mə\ *noun, plural* **mom·mies** [alteration of *mammy*] (1899)
: MOTHER

mommy track *noun* (1989)
: a career path that allows a mother flexible or reduced work hours but tends to slow or block advancement

Mo·mus \'mō-məs\ *noun* [Latin, from Greek *Mōmos*]
: the Greek god of censure and mockery

mon \'män\ *dialect chiefly British variant of* MAN

Mon \'mōn\ *noun, plural* **Mon** *or* **Mons** (1798)
1 : a member of the dominant native people of Pegu division, Myanmar (Burma)
2 : the Mon-Khmer language of the Mon people

mon- *or* **mono-** *combining form* [Middle English, from Middle French & Latin; Middle French, from Latin, from Greek, from *monos* alone, single — more at MONK]
1 : one **:** single **:** alone ⟨*mono*plane⟩ ⟨*mono*drama⟩
2 a : containing one (usually specified) atom, radical, or group ⟨*mono*hydroxy⟩ **b :** monomolecular ⟨*mono*layer⟩

mon·a·chal \'mä-ni-kəl\ *adjective* [Middle French or Late Latin; Middle French, from Late Latin *monachalis,* from *monachus* monk — more at MONK] (1587)
: MONASTIC
— mon·a·chism \'mä-nə-ˌki-zəm\ *noun*

mo·nad \'mō-ˌnad\ *noun* [Late Latin *monad-, monas,* from Greek, from *monos*] (1615)
1 a : UNIT, ONE **b :** ATOM 1 **c :** an elementary individual substance which reflects the order of the world and from which material properties are derived
2 : a flagellated protozoan (as of the genus *Monas*)
— mo·nad·ic \mō-'na-dik, mə-\ *adjective*
— mo·nad·ism \'mō-ˌna-ˌdi-zəm\ *noun*

mon·adel·phous \ˌmä-nə-'del-fəs\ *adjective* (1806)
of stamens **:** united by the filaments into one group usually forming a tube around the gynoecium

mo·nad·nock \mə-'nad-ˌnäk\ *noun* [Mount *Monadnock,* N.H.] (1893)
: INSELBERG

mon·an·dry \'mä-ˌnan-drē\ *noun, plural* **-dries** [*mon-* + *-andry* (as in *polyandry*)] (1855)
: a marriage form or custom in which a woman has only one husband at a time

mon·arch \'mä-nərk, -ˌnärk\ *noun* [Late Latin *monarcha,* from Greek *monarchos,* from *mon-* + *-archos* -arch] (15th century)
1 : a person who reigns over a kingdom or empire: as **a :** a sovereign ruler **b :** a constitutional king or queen
2 : one that holds preeminent position or power
3 : MONARCH BUTTERFLY
— mo·nar·chal \mə-'när-kəl, mä-\ *or* **mo·nar·chi·al** \-kē-əl\ *adjective*

□ USAGE

momentarily The question about what sense *momentarily* can properly be used in originated in 1926, when an English commentator opined that *momentarily* should be restricted in meaning to "for a moment" (its most common use) and *momently* to "from moment to moment." The first problem with this decree was that both words were already being used in both senses. A second problem, one the commentator could not have foreseen, was that *momentarily* would become the usual word and *momently* would recede sharply in use. Later commentators shifted their objection to the sense of *momentarily* meaning "at any moment"; this sense was assigned to *momently.* Now *momently* has indeed been so used ⟨an embittered man . . . *momently* expecting death from heart disease —*Time* (1950)⟩. But *momentarily* had already had the same meaning for some time ⟨during the early part of the morning, I *momentarily* expected his coming —Charlotte Brontë (1847)⟩ ⟨additional arrests were expected *momentarily* —*Sun* (Baltimore) (1928)⟩, and its use continues ⟨was expecting to be drafted *momentarily* —Robert A. Caro (1989)⟩. We have no very recent use of *momently* in this meaning, but it is still used occasionally in the sense "from moment to moment," indicating a continuing process over time ⟨the stink grew *momently* more revolting —Mervyn Peake⟩. The critics remain at odds, some disapproving *momentarily* in sense 3, others approving. Our evidence clearly shows that it is standard.

monarch butterfly *noun* (1890)
: a large migratory American butterfly (*Danaus plexippus*) that has orange-brown wings with black veins and borders and a larva that feeds on milkweed

monarch butterfly

Mo·nar·chi·an \mə-ˈnär-kē-ən, mä-\ *noun* (1765)
: an adherent of one of two anti-Trinitarian groups of the 2d and 3d centuries A.D. teaching that God is one person as well as one being
— **Mo·nar·chi·an·ism** \-kē-ə-ˌni-zəm\ *noun*

mo·nar·chi·cal \mə-ˈnär-ki-kəl, mä-\ *also* **mo·nar·chic** \-kik\ *adjective* (1576)
: of, relating to, suggestive of, or characteristic of a monarch or monarchy
— **mo·nar·chi·cal·ly** \-ki-k(ə-)lē\ *adverb*

mon·ar·chism \ˈmä-nər-ˌki-zəm, -ˌnär-\ *noun* (1838)
: monarchical government or principles
— **mon·ar·chist** \-kist\ *noun or adjective*

mon·ar·chy \ˈmä-nər-kē *also* -ˌnär-\ *noun, plural* **-chies** (14th century)
1 : undivided rule or absolute sovereignty by a single person
2 : a nation or state having a monarchical government
3 : a government having an hereditary chief of state with life tenure and powers varying from nominal to absolute

mo·nar·da \mə-ˈnär-də\ *noun* [New Latin, from Nicolás *Monardes* (died 1588) Spanish botanist] (1712)
: any of a genus (*Monarda*) of coarse North American mints with a tubular many-nerved calyx and whorls of showy flowers

mon·as·tery \ˈmä-nə-ˌster-ē\ *noun, plural* **-ter·ies** [Middle English *monasterie,* from Late Latin *monasterium,* from Late Greek *monastērion,* from Greek, hermit's cell, from *monazein* to live alone, from *monos* single — more at MONK] (15th century)
: a house for persons under religious vows; *especially* **:** an establishment for monks

mo·nas·tic \mə-ˈnas-tik\ *adjective* (circa 1563)
1 : of or relating to monasteries or to monks or nuns
2 : resembling (as in seclusion or ascetic simplicity) life in a monastery
— **monastic** *noun*
— **mo·nas·ti·cal·ly** \-ti-k(ə-)lē\ *adverb*
— **mo·nas·ti·cism** \-tə-ˌsi-zəm\ *noun*

mon·atom·ic \ˌmä-nə-ˈtä-mik\ *adjective* (1848)
: consisting of one atom; *especially* **:** having but one atom in the molecule

mon·au·ral \(ˌ)mä-ˈnȯr-əl\ *adjective* (1931)
: MONOPHONIC 2
— **mon·au·ral·ly** \-ə-lē\ *adverb*

mon·a·zite \ˈmä-nə-ˌzīt\ *noun* [German *Monazit,* from Greek *monazein*] (1836)
: a mineral that is a yellow, red, or brown phosphate of cerium and other rare earth elements and thorium and is found often in sand and gravel deposits

Mon·day \ˈmən-dē, -(ˌ)dā\ *noun* [Middle English, from Old English *mōnandæg;* akin to Old High German *mānatag* Monday; akin to Old English *mōna* moon and to Old English *dæg* day] (before 12th century)
: the second day of the week
— **Mon·days** \-dēz\ *adverb*

Monday–morning quarterback *noun* [from a fan's usually critical rehashing of the weekend football game strategy] (1941)
: one who second-guesses
— **Monday–morning quarterbacking** *noun*

mon·e·cious *variant of* MONOECIOUS

mo·ner·an \mō-ˈnir-ən, mə-\ *noun* [New Latin *Monera,* kingdom comprising prokaryotes, ultimately from Greek *monērēs* single] (1869)
: PROKARYOTE
— **moneran** *adjective*

M1 rifle \ˈem-ˈwən-\ *noun* (1938)
: a .30 caliber gas-operated clip-fed semiautomatic rifle used by U.S. troops in World War II

mon·es·trous \(ˌ)mä-ˈnes-trəs\ *adjective* (1900)
: experiencing estrus once each year or breeding season

mon·e·tar·ism \ˈmä-nə-tə-ˌri-zəm *also* ˈmən-\ *noun* (1969)
: a theory in economics that stable economic growth can be assured only by control of the rate of increase of the money supply to match the capacity for growth of real productivity
— **mon·e·tar·ist** \-rist\ *noun or adjective*

mon·e·tary \ˈmä-nə-ˌter-ē *also* ˈmə-\ *adjective* [Late Latin *monetarius* of a mint, of money, from Latin *moneta*] (circa 1812)
: of or relating to money or to the mechanisms by which it is supplied to and circulates in the economy
— **mon·e·tar·i·ly** \ˌmä-nə-ˈter-ə-lē *also* ˌmə-\ *adverb*

monetary aggregate *noun* (1979)
: one of the formal categories of money (as cash and demand deposits or bank credits) in a national economy that is used as a measure in predictions of economic growth

monetary unit *noun* (circa 1864)
: the standard unit of value of a currency

mon·e·tize \ˈmä-nə-ˌtīz *also* ˈmə-\ *transitive verb* **-tized; -tiz·ing** [Latin *moneta*] (circa 1879)
1 : to coin into money; *also* **:** to establish as legal tender
2 : to purchase (public or private debt) and thereby free for other uses moneys that would have been devoted to debt service
— **mon·e·ti·za·tion** \ˌmä-nə-tə-ˈzā-shən *also* ˌmə-\ *noun*

¹mon·ey \ˈmə-nē\ *noun, plural* **moneys** *or* **mon·ies** \ˈmə-nēz\ *often attributive* [Middle English *moneye,* from Middle French *moneie,* from Latin *moneta* mint, money — more at MINT] (14th century)
1 : something generally accepted as a medium of exchange, a measure of value, or a means of payment: as **a :** officially coined or stamped metal currency **b :** MONEY OF ACCOUNT **c :** PAPER MONEY
2 a : wealth reckoned in terms of money **b :** an amount of money **c** *plural* **:** sums of money **:** FUNDS
3 : a form or denomination of coin or paper money
4 a : the first, second, and third place winners (as in a horse or dog race) — usually used in the phrases *in the money* or *out of the money* **b :** prize money ⟨his horse took third *money*⟩
5 a : persons or interests possessing or controlling great wealth **b :** a position of wealth ⟨born into *money*⟩ ◆
— **for one's money :** according to one's preference or opinion
— **on the money :** exactly right or accurate

²money *adjective* (1935)
: involving or reliable in a crucial situation ⟨a *money* player⟩ ⟨a *money* pitch⟩

mon·ey–back \-ˈbak\ *adjective* (1922)
: providing that the consumer is entitled to a refund if the purchased product is unsatisfactory ⟨a *money-back* guarantee⟩

mon·ey·bags \ˈmə-nē-ˌbagz\ *noun plural but singular or plural in construction* (1596)
1 : WEALTH
2 : a wealthy person

money changer *noun* (15th century)
1 : one whose business is the exchanging of kinds or denominations of currency
2 : a device for holding and dispensing sorted change

mon·eyed *also* **mon·ied** \ˈmə-nēd\ *adjective* (15th century)
1 : having money **:** WEALTHY
2 : consisting in or derived from money

mon·ey·er \ˈmə-nē-ər\ *noun* [Middle English, from Old French *monier,* from Late Latin *monetarius* master of a mint, coiner, from *monetarius* of a mint — more at MONETARY] (15th century)
: an authorized coiner of money **:** MINTER

mon·ey–grub·ber \ˈmə-nē-ˌgrə-bər\ *noun* (1840)
: a person bent on accumulating money
— **mon·ey·grub·bing** \-biŋ\ *adjective or noun*

mon·ey·lend·er \-ˌlen-dər\ *noun* (circa 1780)
: one whose business is lending money; *specifically* **:** PAWNBROKER

mon·ey·mak·er \-ˌmā-kər\ *noun* (1834)
1 : one that accumulates wealth
2 : one that produces profit
— **mon·ey·mak·ing** \-kiŋ\ *adjective or noun*

mon·ey·man \-ˌman\ *noun* (circa 1585)
: FINANCIER

money market *noun* (1950)
: the trade in short-term negotiable instruments (as certificates of deposit or U.S. Treasury securities)

money of account (1691)
: a denominator of value or basis of exchange which is used in keeping accounts and for which there may or may not be an equivalent coin or denomination of paper money

money order *noun* (1802)
: an order issued by a post office, bank, or telegraph office for payment of a specified sum of money usually at any branch of the organization

mon·ey–spin·ner \ˈmə-nē-ˌspi-nər\ *noun* (1859)
chiefly British **:** MONEYMAKER
— **mon·ey–spin·ning** \-niŋ\ *adjective or noun, chiefly British*

mon·ey·wort \-ˌwərt, -ˌwȯrt\ *noun* (1578)
: a trailing perennial herb (*Lysimachia nummularia*) of the primrose family with rounded opposite leaves and solitary yellow flowers in their axils

¹mon·ger \ˈməŋ-gər, ˈmäŋ-\ *noun* [Middle English *mongere,* from Old English *mangere,* from Latin *mangon-, mango,* of Greek origin; akin to Greek *manganon* charm, philter] (before 12th century)
1 : BROKER, DEALER — usually used in combination ⟨ale*monger*⟩

◇ WORD HISTORY
money One of the epithets of the goddess Juno, the wife of Jupiter in ancient Roman myth, was *Moneta.* When the Romans established a mint at the temple of Juno Moneta, this epithet became a generic Latin term for a place where money was made as well as for money itself. The English words *mint* and *money* are both ultimately borrowings from Latin *moneta.* The considerable difference between the two words may be accounted for by the widely different routes by which each came into English. *Mint,* continuing Old English *mynet,* as well as the German word *Münze* "coin" (continuing Old High German *munizza*), goes back to a prehistoric West Germanic borrowing from Latin *moneta.* *Money,* on the other hand, was borrowed from medieval French in the Middle English period as *moneye;* Old French *moneie,* in turn, is a Romance descendant of Latin *moneta.*

MONEY

NAME	SYMBOL	SUBDIVISION	COUNTRY
afghani	Af	100 puls	Afghanistan
baht *or* tical	B	100 satang	Thailand
balboa[1]		100 centesimos	Panama
birr		100 cents	Ethiopia
bolivar	B	100 centimos	Venezuela
boliviano	$b	100 centavos	Bolivia
cedi	₵	100 pesewas	Ghana
colón	C *or* ₵	100 centimos	Costa Rica
colón	₵	100 centavos	El Salvador
córdoba	C$	100 centavos	Nicaragua
dalasi		100 bututs	Gambia
denar		100 deni[2]	Republic of Macedonia
deutsche mark	DM	100 pfennig	Germany
dinar	DA	100 centimes	Algeria
dinar	BD	1000 fils	Bahrain
dinar	ID	1000 fils	Iraq
dinar	JD	1000 fils	Jordan
dinar	KD	1000 fils	Kuwait
dinar	LD	1000 dirhams	Libya
dinar			Sudan
dinar	D	1000 millimes	Tunisia
dinar		100 paras	Yugoslavia
dirham	DH	100 centimes	Morocco
dirham		100 fils	United Arab Emirates
dobra	Db	100 centimos	São Tomé and Príncipe
dollar	EC$[3] *or* $	100 cents	Antigua and Barbuda, Dominica, Grenada, St. Kitts-Nevis, St. Lucia, St. Vincent and the Grenadines
dollar	$A *or* $	100 cents	Australia
dollar	$ *or* B$	100 cents	Bahamas
dollar	$ *or* Bds$	100 cents	Barbados
dollar	$	100 cents	Belize
dollar	$	100 cents	Bermuda
dollar	B$	100 sen *or* cents	Brunei
dollar	$	100 cents	Canada
dollar	$ *or* F$	100 cents	Fiji
dollar	$ *or* G$	100 cents	Guyana
dollar	$ *or* HK$	100 cents	Hong Kong
dollar	$ *or* J$	100 cents	Jamaica
dollar	L$	100 cents	Liberia
dollar	N$	100 cents	Namibia
dollar	NZ$	100 cents	New Zealand
dollar	$ *or* S$	100 cents	Singapore
dollar *or* yuan	NT$	100 cents	Taiwan
dollar	$ *or* TT$	100 cents	Trinidad and Tobago
dollar	$	100 cents	United States
dollar	$ *or* Z$	100 cents	Zimbabwe
dollar—see RINGGIT, below			
dong		100 xu	Vietnam
drachma	Dr	100 lepta	Greece
dram		100 luma	Armenia
escudo		100 centavos	Cape Verde
escudo	$ *or* Esc	100 centavos	Portugal
florin—see GULDEN, below			
forint	Ft	100 filler	Hungary
franc	BF *or* F *or* FR	100 centimes	Belgium
franc	Fr *or* FR *or* F *or* CFA[4]	100 centimes	Benin, Burkina Faso, Cameroon, Central African Republic, Chad, Republic of the Congo, Equatorial Guinea, Gabon, Ivory Coast, Mali, Niger, Senegal, Togo
franc	F *or* FBu	100 centimes	Burundi
franc	Fr *or* F	100 centimes	Djibouti
franc	Fr *or* F	100 centimes	France
franc	GF	100 centimes	Guinea
franc	F *or* Flux	100 centimes	Luxembourg
franc	FMG	100 centimes	Madagascar
franc	FR *or* RF	100 centimes[2]	Rwanda
franc	Fr *or* FR	100 centimes *or* rappen	Switzerland
gourde	G	100 centimes	Haiti
guarani	G *or*	100 centimos	Paraguay
gulden *or* guilder *or* florin	f *or* Fl *or* G	100 cents	Netherlands
gulden *or* guilder *or* florin	Sf	100 cents	Surinam
hryvnia		100 kopiykas	Ukraine
kina	K	100 toea	Papua New Guinea
kip	K	100 at	Laos
koruna	Kčs	100 haleru	Czech Republic
koruna	Kčs	100 haleru	Slovakia
krona	KR	100 aurar (*sing* eyrir)	Iceland
krona	SKr *or* KR	100 ore	Sweden
krone	DKr	100 ore	Denmark
krone	KR *or* Nkr	100 ore	Norway
kroon	EEK	100 senti	Estonia
kuna		100 lipa	Croatia
kwacha	K	100 tambala	Malawi
kwacha	K	100 ngwee	Zambia
kwanza		100 lwei	Angola
kyat	K	100 pyas	Myanmar
lari		100 tetri	Republic of Georgia
lats (*pl* lati *or* latu)		100 santimi *or* santimu (*sing* santims)	Latvia
lek		100 qindarka	Albania
lempira	L	100 centavos	Honduras
leone	Le	100 cents	Sierra Leone
leu		100 bani (*sing* ban)	Moldova
leu		100 bani (*sing* ban)	Romania
lev		100 stotinki	Bulgaria
lilangeni (*pl* emalangeni)	E	100 cents	Swaziland
lira	L *or* Lit	100 centesimi[2]	Italy
lira *or* pound	£ *or* £M *or* Lm	100 cents	Malta
lira	TL	100 kurus	Turkey
litas		100 centai *or* centu (*sing* centas)	Lithuania
livre—see POUND, below			
loti (*pl* maloti)		100 licente *or* lisente (*sing* sente)	Lesotho
manat		100 gopik	Azerbaijan
mark—see DEUTSCHE MARK, above			
markka	Fmk	100 pennia	Finland
metical		100 centavos	Mozambique
naira	₦	100 kobo	Nigeria
nakfa	nfa	100 cents	Eritrea
ngultrum	Nu	100 chetrums	Bhutan
ouguiya		5 khoums	Mauritania
pa'anga	T$	100 seniti	Tonga
pataca		100 avos	Macao
peseta	PTA	100 centimos[2]	Spain
peso		100 centavos	Argentina
peso		100 centavos	Chile
peso		100 centavos	Colombia
peso		100 centavos	Cuba
peso	RD$	100 centavos	Dominican Republic
peso		100 centavos	Guinea-Bissau
peso	$	100 centavos	Mexico
peso *or* piso	P	100 sentimos *or* centavos	Philippines
peso	$	100 centesimos	Uruguay
pound	£	100 cents	Cyprus
pound	£E	100 piastres	Egypt
pound	£	100 pence (*sing* penny)	Ireland
pound *or* livre	L	100 piastres	Lebanon
pound		100 piastres	Syria
pound	£	100 pence (*sing* penny)	United Kingdom
pound—see LIRA, above			
pula	P	100 thebe	Botswana
quetzal	Q	100 centavos	Guatemala
rand	R	100 cents	South Africa
real		100 centavos	Brazil
rial	R *or* Rl	100 dinars	Iran
rial	RO	1000 baiza	Oman
rial *also* riyal	YR *or* R	100 fils	Yemen
rial—see RIYAL, below			
riel		100 sen	Cambodia
ringgit *or* dollar	$ *or* M$	100 sen	Malaysia
riyal	QR	100 dirhams	Qatar
riyal *also* rial		100 halala	Saudi Arabia
riyal—see RIAL, above			
ruble		100 kopecks	Russia
ruble		100 tanga	Tajikistan
rufiyaa	Rf	100 laari	Maldives
rupee	Re (*pl* Rs)	100 paise	India
rupee	Re (*pl* Rs)	100 cents	Mauritius
rupee	Re (*pl* Rs)	100 paisa	Nepal
rupee	Re (*pl* Rs)	100 paisa	Pakistan
rupee	SR	100 cents	Seychelles
rupee	Re (*pl* Rs)	100 cents	Sri Lanka
rupiah	Rp	100 sen	Indonesia
schilling	S *or* Sch	100 groschen	Austria
shekel *or* sheqel	IS	100 agorot	Israel
shilling	S *or* KSh	100 cents	Kenya
shilling		100 cents	Somalia
shilling	TSh	100 cents	Tanzania
shilling		100 cents	Uganda
sol		100 centimos	Peru
som		100 tyiyn	Kyrgyzstan
sucre[1]		100 centavos	Ecuador
taka	Tk	100 paisa *or* poisha	Bangladesh
tala	$	100 sene	Samoa
tenge		100 tyin	Kazakstan
tical—see BAHT, above			
tolar		100 stotinov	Slovenia
tugrik		100 mongo	Mongolia
vatu			Vanuatu
won		100 chon	North Korea
won		100 chon	South Korea
yen	¥	100 sen[2]	Japan
yuan		100 fen	China (mainland)
yuan—see DOLLAR, above			
zaire	Z	100 makuta (*sing* likuta)	Democratic Republic of the Congo
zloty	Zl	100 groszy	Poland

[1] A monetary unit in name only, replaced by the U.S. dollar in 2000.
[2] Now a subdivision in name only.
[3] Dollars issued by the *E*astern *C*aribbean Central Bank, established to promote economic cooperation among the member nations.
[4] Francs issued by the African Financial Community (*C*ommunauté *F*inancière *A*fricaine), established to promote economic cooperation among member nations.

2 : a person who attempts to stir up or spread something that is usually petty or discreditable — usually used in combination ⟨war*monger*⟩

²monger *transitive verb* **mon·gered; mon·ger·ing** \-g(ə-)riŋ\ (circa 1864)
: PEDDLE

mon·go \'mäŋ-(ˌ)gō\ *noun, plural* **mongo** [Mongolian *möngö*] (1935)
— see *tugrik* at MONEY table

Mon·gol \'mäŋ-gəl; 'män-ˌgōl, 'mäŋ-\ *noun* [Mongolian *mongγol*] (1698)
1 : a member of any of a group of traditionally pastoral peoples of Mongolia
2 : MONGOLIAN 1
3 : a person of Mongoloid racial stock
4 *often not capitalized* **:** one affected with Down's syndrome
— Mongol *adjective*

¹Mon·go·lian \män-'gōl-yən, mäŋ-, -'gō-lē-ən\ *adjective* (1706)
1 : of, relating to, or constituting Mongolia, the Mongolian People's Republic, the Mongols, or Mongolian
2 : MONGOLOID 2

²Mongolian *noun* (1846)
1 a : the language of the Mongol people **b :** a family of Altaic languages that includes the languages of the Mongols and the Kalmucks
2 a : MONGOL 1 **b :** a person of Mongoloid racial stock **c :** a native or inhabitant of Mongolia

Mongolian gerbil *noun* (1948)
: a gerbil (*Meriones unguiculatus*) of Mongolia and northern China that has an external resemblance to a rat, has a high capacity for temperature regulation, and is used as an experimental laboratory animal

Mon·gol·ic \män-'gä-lik, mäŋ-\ *adjective* (1834)
: MONGOLOID 1

mon·gol·ism \'mäŋ-gə-ˌli-zəm\ *noun* (1900)
: DOWN'S SYNDROME

Mon·gol·oid \'mäŋ-gə-ˌlȯid\ *adjective* (1868)
1 : of, constituting, or characteristic of a major racial stock native to Asia as classified according to physical features (as the presence of an epicanthic fold) that includes peoples of northern and eastern Asia, Malaysians, Eskimos, and often American Indians
2 *often not capitalized* **:** of, relating to, or affected with Down's syndrome
— Mongoloid *noun*

mon·goose \'män-ˌgüs, 'mäŋ-\ *noun, plural* **mon·goos·es** *also* **mon·geese** \-ˌgēs\ [Hindi & Marathi *māgūs,* from Prakrit *maṁgūsa*] (1698)
: any of the viverrid mammals that comprise two subfamilies (Herpestinae and Galidiinae) often grouped in a separate family (Herpestidae), that include agile ferret-sized mammals sometimes with bands or stripes, and that feed on small animals and fruits

mongoose

mon·grel \'mäŋ-grəl, 'məŋ-\ *noun* [Middle English, probably from *mong* mixture, short for *ymong,* from Old English *gemong* crowd — more at AMONG] (15th century)
1 : an individual resulting from the interbreeding of diverse breeds or strains; *especially* **:** one of unknown ancestry
2 : a cross between types of persons or things
— mongrel *adjective*
— mon·grel·i·za·tion \ˌmäŋ-grə-lə-'zā-shən, ˌməŋ-\ *noun*
— mon·grel·ize \'mäŋ-grə-ˌlīz, 'məŋ-\ *transitive verb*

monies *plural of* MONEY

mon·i·ker *also* **mon·ick·er** \'mä-ni-kər\ *noun* [probably from Shelta (language of Irish itinerants) *mŭnnik,* modification of Irish *ainm*] (1851)
: NAME, NICKNAME

mo·ni·li·a·sis \ˌmō-nə-'lī-ə-səs, ˌmä-\ *noun, plural* **-a·ses** \-ˌsēz\ [New Latin, from *Monilia,* genus of fungi, from Latin *monile* necklace] (1920)
: CANDIDIASIS

mo·nil·i·form \mə-'ni-lə-ˌfȯrm\ *adjective* [Latin *monile* necklace — more at MANE] (circa 1803)
: jointed or constricted at regular intervals so as to resemble a string of beads ⟨a *moniliform* root⟩ ⟨*moniliform* insect antennae⟩

mon·ish \'mä-nish\ *transitive verb* [Middle English *monesen,* alteration of *monesten,* from Old French *monester,* from (assumed) Vulgar Latin *monestare,* from Latin *monēre* to warn] (14th century)
: WARN

mo·nism \'mō-ˌni-zəm, 'mä-\ *noun* [German *Monismus,* from *mon-* + *-ismus* -ism] (1862)
1 a : a view that there is only one kind of ultimate substance **b :** the view that reality is one unitary organic whole with no independent parts
2 : MONOGENESIS
3 : a viewpoint or theory that reduces all phenomena to one principle
— mo·nist \'mō-nist, 'mä-\ *noun*
— mo·nis·tic \mō-'nis-tik, mä-\ *adjective*

mo·ni·tion \mō-'ni-shən, mə-\ *noun* [Middle English *monicioun,* from Middle French *monition,* from Latin *monition-, monitio,* from *monēre*] (14th century)
1 : WARNING, CAUTION
2 : an intimation of danger

¹mon·i·tor \'mä-nə-tər\ *noun* [Latin, one that warns, overseer, from *monēre* to warn — more at MIND] (1546)
1 a : a student appointed to assist a teacher **b :** one that warns or instructs **c :** one that monitors or is used in monitoring: as (1) **:** a cathode-ray tube used for display (as of television pictures or computer information) (2) **:** a device for observing a biological condition or function ⟨a heart *monitor*⟩
2 : any of various large tropical Old World lizards (genus *Varanus* of the family Varanidae) closely related to the iguanas
3 [*Monitor,* first ship of the type] **a :** a heavily armored warship formerly used in coastal operations having a very low freeboard and one or more revolving gun turrets **b :** a small modern warship with shallow draft for coastal bombardment
4 : a raised central portion of a roof having low windows or louvers for providing light and air
— mon·i·to·ri·al \ˌmä-nə-'tōr-ē-əl, -'tȯr-\ *adjective*
— mon·i·tor·ship \'mä-nə-tər-ˌship\ *noun*

²monitor *transitive verb* **mon·i·tored; mon·i·tor·ing** \'mä-nə-t(ə-)riŋ\ (1924)
: to watch, keep track of, or check usually for a special purpose

¹mon·i·to·ry \'mä-nə-ˌtōr-ē, -ˌtȯr-\ *adjective* [Middle English, from Latin *monitorius,* from *monēre*] (15th century)
: giving admonition **:** WARNING

²monitory *noun, plural* **-ries** (15th century)
: a letter containing an admonition or warning

¹monk \'məŋk\ *noun* [Middle English, from Old English *munuc,* from Late Latin *monachus,* from Late Greek *monachos,* from Greek, adjective, single, from *monos* single, alone] (before 12th century)
: a man who is a member of a religious order and lives in a monastery; *also* **:** FRIAR

²monk *noun* (1843)
: MONKEY

monk·ery \'məŋ-kə-rē\ *noun, plural* **-er·ies** (1536)
1 : monastic life or practice **:** MONASTICISM
2 : a monastic house **:** MONASTERY

¹mon·key \'məŋ-kē\ *noun, plural* **monkeys** [probably of Low German origin; akin to *Moneke,* name of an ape, probably of Romance origin; akin to Old Spanish *mona* monkey] (circa 1530)
1 : a nonhuman primate mammal with the exception usually of the lemurs and tarsiers; *especially* **:** any of the smaller longer-tailed primates as contrasted with the apes
2 a : a person resembling a monkey **b :** a ludicrous figure **:** DUPE
3 : any of various machines, implements, or vessels; *especially* **:** the falling weight of a pile driver
4 : a desperate desire for or addiction to drugs — often used in the phrase *monkey on one's back; broadly* **:** a persistent or annoying encumbrance or problem

²monkey *verb* **mon·keyed; mon·key·ing** (1859)
transitive verb
: MIMIC, MOCK
intransitive verb
1 : to act in a grotesque or mischievous manner
2 a : FOOL, TRIFLE — often used with *around*
b : TAMPER — usually used with *with*

monkey bars *noun plural* (1955)
: a three-dimensional framework of horizontal and vertical bars from which children can hang and swing

monkey business *noun* (1883)
: SHENANIGAN 2

monkey jacket *noun* (1830)
: MESS JACKET

mon·key·pod \'məŋ-kē-ˌpäd\ *noun* (1888)
1 : an ornamental tropical leguminous tree (*Samanea saman* synonym *Pithecolobium saman*) that has bipinnate leaves, globose clusters of flowers with crimson stamens, sweet-pulp pods eaten by cattle, and wood used in carving — called also *rain tree*
2 : the wood of a monkeypod

monkey puzzle *noun* (1866)
: a tall araucaria (*Araucaria araucana*) that is native to Chile and western Argentina but widely grown elsewhere — called also *monkey puzzle tree*

mon·key·shine \'məŋ-kē-ˌshīn\ *noun* (circa 1832)
: PRANK — usually used in plural

monkey wrench *noun* (circa 1858)
1 : a wrench with one fixed and one adjustable jaw at right angles to a straight handle
2 : something that disrupts ⟨threw a *monkey wrench* into the peace negotiations⟩

monk·fish \'məŋk-ˌfish\ *noun* (1666)
: either of two goosefishes (*Lophius americanus* of America and *L. piscatorius* of Europe) used for food

Mon–Khmer \ˌmōn-kə-'mer\ *noun* (1887)
: a language family containing Mon, Khmer, and a number of other languages of southeast Asia

monk·hood \'məŋk-ˌhu̇d\ *noun* (before 12th century)
1 : the character, condition, or profession of a monk **:** MONASTICISM
2 : monks as a class

monk·ish \'məŋ-kish\ *adjective* (1546)
1 : of, relating to, or resembling a monk; *also* **:** resembling that of a monk
2 : inclined to disciplinary self-denial

monk's cloth *noun* (circa 1847)
: a coarse heavy fabric in basket weave made originally of worsted and used for monk's habits but now chiefly of cotton or linen and used for draperies

monk seal *noun* (1841)
: any of a genus (*Monachus*) of hair seals of Hawaii, the Mediterranean Sea, and formerly the Caribbean Sea

monks·hood \'məŋ(k)s-ˌhu̇d\ *noun* (1578)
: any of a genus (*Aconitum*) of usually bluish flowered poisonous herbs of the buttercup

family; *especially* **:** a poisonous Eurasian herb (*A. napellus*) often cultivated for its showy terminal racemes of white or purplish flowers — compare WOLFSBANE

[1]mono \ˈmä-(ˌ)nō\ *noun, plural* **mon·os** [[2]*mono*] (1959)
: monophonic reproduction

[2]mono *adjective* (1961)
: MONOPHONIC 2

[3]mono *noun* (1962)
: INFECTIOUS MONONUCLEOSIS

mono- — see MON-

mono·ac·id \ˌmä-nō-ˈa-səd\ *noun* (1863)
: an acid having only one acid hydrogen atom

mono·acid·ic \-ə-ˈsi-dik\ *adjective* (circa 1929)
: able to react with one molecule of a monobasic acid to form a salt or ester — used especially of bases

mono·amine \ˌmä-nō-ə-ˈmēn\ *noun* [International Scientific Vocabulary] (1951)
: an amine RNH_2 that has one organic substituent attached to the nitrogen atom; *especially* **:** one (as serotonin) that is functionally important in neural transmission

monoamine oxidase *noun* (1951)
: an enzyme that deaminates monoamines oxidatively and that functions in the nervous system by breaking down monoamine neurotransmitters oxidatively

mono·am·in·er·gic \ˌmä-nō-ˌa-mə-ˈnər-jik\ *adjective* (1966)
: liberating or involving monoamines (as serotonin or norepinephrine) in neural transmission ⟨*monoaminergic* neurons⟩ ⟨*monoaminergic* mechanisms⟩

mono·ba·sic \ˌmä-nə-ˈbā-sik\ *adjective* [International Scientific Vocabulary] (1842)
of an acid **:** having only one replaceable hydrogen atom

mono·car·box·yl·ic \ ˌkär-(ˌ)bäk-ˈsi-lik\ *adjective* (circa 1909)
: containing one carboxyl group ⟨acetic acid is a *monocarboxylic* acid⟩

mono·car·pic \-ˈkär-pik\ *adjective* [probably from (assumed) New Latin *monocarpicus,* from New Latin *mon-* + *-carpicus* -carpic] (1849)
: bearing fruit but once and then dying

mono·cha·si·um \-ˈkā-zē-əm, -zhē-\ *noun, plural* **-sia** \-zē-ə, -zhē-\ [New Latin, from *mon-* + *-chasium* (as in *dichasium*)] (circa 1890)
: a cymose inflorescence that produces only one main axis
— **mono·cha·sial** \-zē-əl, -zhē-\ *adjective*

mono·chord \ˈmä-nə-ˌkȯrd\ *noun* [Middle English *monocorde,* from Middle French, from Medieval Latin *monochordum,* from Greek *monochordon,* from *mon-* + *chordē* string — more at YARN] (15th century)
: an instrument of ancient origin for measuring and demonstrating the mathematical relations of musical tones and that consists of a single string stretched over a sound box and a movable bridge set on a graduated scale

mono·chro·mat \ˈmä-nə-krō-ˌmat, ˌmä-nə-ˈ\ *noun* [*mon-* + Greek *chrōmat-, chrōma*] (1902)
: a completely color-blind individual

mono·chro·mat·ic \ˌmä-nə-krō-ˈma-tik\ *adjective* [Latin *monochromatos,* from Greek *monochrōmatos,* from *mon-* + *chrōmat-, chrōma* color] (1822)
1 a : having or consisting of one color or hue **b :** MONOCHROME 2
2 : consisting of radiation of a single wavelength or of a very small range of wavelengths
3 : of, relating to, or exhibiting monochromatism
— **mono·chro·mat·i·cal·ly** \-ti-k(ə-)lē\ *adverb*
— **mono·chro·ma·tic·i·ty** \-ˌkrō-mə-ˈti-sə-tē\ *noun*

mono·chro·ma·tism \-ˈkrō-mə-ˌti-zəm\ *noun* (circa 1930)
: complete color blindness in which all colors appear as shades of gray

mono·chro·ma·tor \ˌmä-nə-ˈkrō-ˌmā-tər\ *noun* [*monochromatic* + *illuminator*] (1909)
: a device for isolating a narrow portion of a spectrum

[1]mono·chrome \ˈmä-nə-ˌkrōm\ *noun* [Medieval Latin *monochroma,* from Latin, feminine of *monochromos* of one color, from Greek *monochrōmos,* from *mon-* + *-chrōmos* -chrome] (1662)
: a painting, drawing, or photograph in a single hue
— **mono·chro·mic** \ˌmä-nə-ˈkrō-mik\ *adjective*
— **mono·chrom·ist** \ˈmä-nə-ˌkrō-mist\ *noun*

[2]monochrome *adjective* (1849)
1 : of, relating to, or made with a single color or hue
2 : involving or producing visual images in a single color or in varying tones of a single color (as gray) ⟨*monochrome* film⟩ ⟨*monochrome* television monitor⟩

mon·o·cle \ˈmä-ni-kəl\ *noun* [French, from Late Latin *monoculus* having one eye, from Latin *mon-* + *oculus* eye — more at EYE] (circa 1858)
: an eyeglass for one eye
— **mon·o·cled** \-kəld\ *adjective*

monocle

mono·cline \ˈmä-nə-ˌklīn\ *noun* (1879)
: an oblique geologic fold

mono·clin·ic \ˌmä-nə-ˈkli-nik\ *adjective* [International Scientific Vocabulary] (circa 1864)
: having one oblique intersection of the crystallographic axes

monoclinic system *noun* (1869)
: a crystal system characterized by three unequal axes with one oblique intersection

mono·clo·nal \ˌmä-nə-ˈklō-nᵊl\ *adjective* (1914)
: produced by, being, or composed of cells derived from a single cell ⟨*monoclonal* antibodies⟩ ⟨a *monoclonal* tumor⟩
— **monoclonal** *noun*

mono·coque \ˈmä-nə-ˌkōk, -ˌkäk\ *noun* [French, from *mon-* + *coque* shell, probably from Latin *coccum* kermes — more at COCOON] (1913)
1 : a type of construction (as of a fuselage) in which the outer skin carries all or a major part of the stresses
2 : a type of vehicle construction (as of an automobile) in which the body is integral with the chassis

mono·cot \-ˌkät\ *noun* (1890)
: MONOCOTYLEDON

mono·cot·y·le·don \ˌmä-nə-ˌkä-tᵊl-ˈē-dᵊn\ *noun* [ultimately from New Latin *mon-* + *cotyledon* cotyledon] (circa 1727)
: any of a class or subclass (Liliopsida or Monocotyledoneae) of chiefly herbaceous seedplants having an embryo with a single cotyledon, usually parallel-veined leaves, and floral organs arranged in cycles of three — compare DICOTYLEDON
— **mono·cot·y·le·don·ous** \-dᵊn-əs\ *adjective*

mo·noc·ra·cy \mä-ˈnä-krə-sē, mə-\ *noun* (1651)
: government by a single person
— **mono·crat** \ˈmä-nə-ˌkrat\ *noun*
— **mono·crat·ic** \ˌmä-nə-ˈkra-tik\ *adjective*

mono·crys·tal \ˈmä-nə-ˌkris-tᵊl\ *noun* (1926)
: a single crystal
— **monocrystal** *adjective*
— **mono·crys·tal·line** \ˌmä-nə-ˈkris-tə-lən *also* -ˌlīn, -ˌlēn\ *adjective*

[1]mon·oc·u·lar \mä-ˈnä-kyə-lər, mə-\ *adjective* [Late Latin *monoculus* having one eye] (1640)
1 : of, involving, or affecting a single eye
2 : suitable for use with only one eye
— **mon·oc·u·lar·ly** *adverb*

[2]monocular *noun* (1936)
: a monocular device

mono·cul·ture \ˈmä-nə-ˌkəl-chər\ *noun* (1915)
1 : the cultivation or growth of a single crop or organism especially on agricultural or forest land
2 : a crop or a population of a single kind of organism grown on land in monoculture
— **mono·cul·tur·al** \ˌmä-nə-ˈkəlch-rəl, -ˈkəl-chə-\ *adjective*

mono·cy·clic \ˌmä-nə-ˈsī-klik, -ˈsi-\ *adjective* [International Scientific Vocabulary] (1910)
: containing one ring in the molecular structure

mono·cyte \ˈmä-nə-ˌsīt\ *noun* [International Scientific Vocabulary] (circa 1913)
: a large white blood cell with finely granulated chromatin dispersed throughout the nucleus that is formed in the bone marrow, enters the blood, and migrates into the connective tissue where it differentiates into a macrophage
— **mono·cyt·ic** \ˌmä-nə-ˈsi-tik\ *adjective*

mono·dis·perse \ˌmä-nō-dis-ˈpərs\ *adjective* [*mon-* + *disperse,* adjective, from *disperse,* verb] (1925)
: characterized by particles of uniform size in a dispersed phase

mon·o·dist \ˈmä-nə-dist\ *noun* (1751)
: a writer, singer, or composer of monody

mono·dra·ma \ˈmä-nə-ˌdrä-mə, -ˌdra-\ *noun* (1793)
: a drama acted or designed to be acted by a single person
— **mono·dra·mat·ic** \ˌmä-nə-drə-ˈma-tik\ *adjective*

mon·o·dy \ˈmä-nə-dē\ *noun, plural* **-dies** [Medieval Latin *monodia,* from Greek *monōidia,* from *monōidos* singing alone, from *mon-* + *aeidein* to sing — more at ODE] (circa 1623)
1 : an ode sung by one voice (as in a Greek tragedy)
2 : an elegy or dirge performed by one person
3 a : a monophonic vocal piece **b :** the monophonic style of 17th century opera
— **mo·nod·ic** \mə-ˈnä-dik\ *or* **mo·nod·i·cal** \-di-kəl\ *adjective*
— **mo·nod·i·cal·ly** \-di-k(ə-)lē\ *adverb*

mon·oe·cious \mə-ˈnē-shəs, mä-\ *adjective* [ultimately from Greek *mon-* + *oikos* house — more at VICINITY] (1753)
1 : having pistillate and staminate flowers on the same plant
2 : having male and female sex organs in the same individual **:** HERMAPHRODITIC

mon·oe·cism \-ˈnē-ˌsi-zəm\ *noun* (1875)
: the condition of being monoecious

mono·es·ter \ˈmä-nō-ˌes-tər\ *noun* (1927)
: an ester (as of a dibasic acid) that contains only one ester group

mono·fil·a·ment \ˌmä-nə-ˈfi-lə-mənt\ *noun* (1940)
: a single untwisted synthetic filament (as of nylon)

mo·nog·a·mist \mə-ˈnä-gə-mist\ *noun* (1651)
: one who practices or upholds monogamy

mo·nog·a·my \-mē\ *noun* [French *monogamie,* from Late Latin *monogamia,* from Greek, from *monogamos* monogamous, from *mon-* + *gamos* marriage, from *gamein* to marry] (1612)
1 *archaic* **:** the practice of marrying only once during a lifetime
2 : the state or custom of being married to one person at a time

3 : the condition or practice of having a single mate during a period of time ⟨*monogamy* is common among birds⟩
— **mo·nog·a·mous** \mə-ˈnä-gə-məs\ *also* **mono·gam·ic** \ˌmä-nə-ˈga-mik\ *adjective*
— **mo·nog·a·mous·ly** *adverb*

mono·gas·tric \ˌmä-nə-ˈgas-trik\ *adjective* (1814)
: having a stomach with only a single compartment ⟨swine, chicks, and human beings are *monogastric*⟩

mono·ge·ne·an \-ˈjē-nē-ən\ *noun* [New Latin *Monogenea,* group name] (1899)
: a monogenetic trematode
— **monogenean** *adjective*

mono·gen·e·sis \-ˈje-nə-səs\ *noun* [New Latin] (circa 1859)
: origin of diverse individuals or kinds (as of language) by descent from a single ancestral individual or kind

mono·ge·net·ic \-jə-ˈne-tik\ *adjective* (1873)
1 : relating to or involving monogenesis
2 : of, relating to, or being any of a subclass (Monogenea) of trematode worms that ordinarily live as ectoparasites on a single host (as a fish or amphibian) throughout their entire life cycle

mono·gen·ic \-ˈje-nik\ *adjective* [International Scientific Vocabulary] (1939)
: of, relating to, or controlled by a single gene and especially by either of an allelic pair
— **mono·gen·i·cal·ly** \-ni-k(ə-)lē\ *adverb*

mono·germ \ˈmä-nə-ˌjərm\ *adjective* [*mon-* + *germ*inate] (1950)
: producing or being a fruit that gives rise to a single plant ⟨a *monogerm* sugar beet⟩

mono·glot \-ˌglät\ *adjective* [*mono-* + *-glot* (as in *polyglot*)] (1830)
: MONOLINGUAL
— **monoglot** *noun*

mono·glyc·er·ide \ˌmä-nə-ˈgli-sə-ˌrīd\ *noun* (1860)
: any of various esters of glycerol in which only one of the three hydroxyl groups is esterified and which are often used as emulsifiers

¹mono·gram \ˈmä-nə-ˌgram\ *noun* [Late Latin *monogramma,* from Greek *mon-* + *gramma* letter — more at GRAM] (circa 1696)
: a sign of identity usually formed of the combined initials of a name
— **mono·gram·mat·ic** \ˌmä-nə-grə-ˈma-tik\ *adjective*

²monogram *transitive verb* **-grammed; -gram·ming** (1868)
: to mark with a monogram
— **mono·gram·mer** \-ˌgra-mər\ *noun*

¹mono·graph \ˈmä-nə-ˌgraf\ *noun* (1821)
: a learned treatise on a small area of learning; *also* **:** a written account of a single thing
— **mono·graph·ic** \ˌmä-nə-ˈgra-fik\ *adjective*

²monograph *transitive verb* (1876)
: to write a monograph on

mo·nog·y·nous \mə-ˈnä-jə-nəs, mä-\ *adjective* (circa 1890)
: of, relating to, or living in monogyny

mo·nog·y·ny \-nē\ *noun* [International Scientific Vocabulary] (1876)
: the state or custom of having only one wife at a time

mono·hull \ˈmä-nə-ˌhəl\ *noun* (1967)
: a vessel (as a sailboat) with a single hull — compare MULTIHULL

mono·hy·brid \ˌmä-nō-ˈhī-brəd\ *noun* (1903)
: an individual or strain heterozygous for one specified gene
— **monohybrid** *adjective*

mono·hy·dric \-ˈhī-drik\ *adjective* (1880)
: MONOHYDROXY

mono·hy·droxy \-(ˌ)hī-ˈdräk-sē\ *adjective* [International Scientific Vocabulary *monohydroxy-,* from *mon-* + *hydroxy-*] (circa 1934)
: containing one hydroxyl group in the molecule

mono·lay·er \ˈmä-nō-ˌlā-ər, -ˌle(-ə)r\ *noun* (1926)
: a single continuous layer or film that is one cell or molecule in thickness

mono·lin·gual \ˌmä-nə-ˈliŋ-gwəl, ˌmō-, -ˈliŋ-gyə-wəl\ *adjective* (1926)
: knowing or using only one language
— **monolingual** *noun*

mono·lith \ˈmä-nᵊl-ˌith\ *noun* [French *monolithe,* from *monolithe* consisting of a single stone, from Latin *monolithus,* from Greek *monolithos,* from *mon-* + *lithos* stone] (1848)
1 : a single great stone often in the form of an obelisk or column
2 : a massive structure
3 : an organized whole that acts as a single unified powerful or influential force

mono·lith·ic \ˌmä-nᵊl-ˈi-thik\ *adjective* (1825)
1 a : of, relating to, or resembling a monolith **:** HUGE, MASSIVE **b** (1) **:** formed from a single crystal ⟨a *monolithic* silicon chip⟩ (2) **:** produced in or on a monolithic chip ⟨a *monolithic* circuit⟩
2 a : cast as a single piece ⟨a *monolithic* concrete wall⟩ **b :** formed or composed of material without joints or seams ⟨a *monolithic* floor covering⟩ ⟨a *monolithic* furnace lining⟩ **c :** consisting of or constituting a single unit
3 a : constituting a massive undifferentiated and often rigid whole ⟨a *monolithic* society⟩ **b :** exhibiting or characterized by often rigidly fixed uniformity ⟨*monolithic* party unity⟩
— **mono·lith·i·cal·ly** \-thi-k(ə-)lē\ *adverb*

mono·logue *also* **mono·log** \ˈmä-nᵊl-ˌȯg, -ˌäg\ *noun* [Middle French *monologue,* from *mon-* + *-logue*] (1549)
1 a : SOLILOQUY 2 **b :** a dramatic sketch performed by one actor **c :** the routine of a stand-up comic
2 : a literary composition written in the form of a soliloquy
3 : a long speech monopolizing conversation
— **mono·logu·ist** \-ˌȯ-gist, -ˌä-\ *or* **mo·no·lo·gist** *same or* mə-ˈnä-lə-jist, -gist\ *noun*

mono·ma·nia \ˌmä-nə-ˈmā-nē-ə, -nyə\ *noun* [New Latin] (1823)
1 : mental illness especially when limited in expression to one idea or area of thought
2 : excessive concentration on a single object or idea
— **mono·ma·ni·ac** \-nē-ˌak\ *noun or adjective*
— **mono·ma·ni·a·cal** \-mə-ˈnī-ə-kəl\ *adjective*
— **mono·ma·ni·a·cal·ly** \-k(ə-)lē\ *adverb*

mono·mer \ˈmä-nə-mər\ *noun* [International Scientific Vocabulary] (1914)
: a chemical compound that can undergo polymerization
— **mo·no·mer·ic** \ˌmä-nə-ˈmer-ik, ˌmō-\ *adjective*

mono·me·tal·lic \ˌmä-nō-mə-ˈta-lik\ *adjective* (1877)
1 : of or relating to monometallism
2 : consisting of or employing one metal

mono·met·al·lism \-ˈme-tᵊl-ˌi-zəm\ *noun* [International Scientific Vocabulary *mon-* + *-metallism* (as in *bimetallism*)] (1879)
: the adoption of one metal only in a currency
— **mono·met·al·list** \-tᵊl-ist\ *noun*

mo·nom·e·ter \mə-ˈnä-mə-tər, mä-\ *noun* [Late Latin, from Greek *monometros,* from *mon-* + *metron* measure — more at MEASURE] (circa 1846)
: a line of verse consisting of a single metrical foot or dipody

mo·no·mi·al \mä-ˈnō-mē-əl, mə-\ *noun* [blend of *mon-* and *-nomial* (as in *binomial*)] (circa 1706)
1 : a mathematical expression consisting of a single term
2 : a taxonomic name consisting of a single word or term
— **monomial** *adjective*

mono·mo·lec·u·lar \ˌmä-nə-mə-ˈle-kyə-lər\ *adjective* (1917)
: being only one molecule thick ⟨a *monomolecular* film⟩
— **mono·mo·lec·u·lar·ly** *adverb*

mono·mor·phe·mic \-mȯr-ˈfē-mik\ *adjective* (1936)
: consisting of only one morpheme ⟨the word *talk* is *monomorphemic* but *talked* is not⟩

mono·mor·phic \-ˈmȯr-fik\ *adjective* (circa 1879)
: having but a single form, structural pattern, or genotype ⟨a *monomorphic* species of insect⟩
— **mono·mor·phism** \-ˌfi-zəm\ *noun*

mono·nu·cle·ar \-ˈnü-klē-ər, -ˈnyü-, ÷-kyə-lər\ *adjective* [International Scientific Vocabulary] (1886)
: having only one nucleus ⟨a *mononuclear* cell⟩
— **mononuclear** *noun*

mono·nu·cle·at·ed \-ˈnü-klē-ˌā-təd, -ˈnyü-\ *also* **mono·nu·cle·ate** \-klē-ət, -ˌāt\ *adjective* (1890)
: MONONUCLEAR

mono·nu·cle·o·sis \-ˌnü-klē-ˈō-səs, -ˌnyü-\ *noun* [New Latin, from International Scientific Vocabulary *mononucle*ar + New Latin *-osis*] (1920)
: an abnormal increase of mononuclear white blood cells in the blood; *specifically* **:** INFECTIOUS MONONUCLEOSIS

mono·nu·cle·o·tide \-ˈnü-klē-ə-ˌtīd, -ˈnyü-\ *noun* (1908)
: a nucleotide that is derived from one molecule each of a nitrogenous base, a sugar, and a phosphoric acid

mo·noph·a·gous \mə-ˈnä-fə-gəs, mä-\ *adjective* (circa 1868)
: feeding on or utilizing a single kind of food; *especially* **:** feeding on a single kind of plant or animal
— **mo·noph·a·gy** \-fə-jē\ *noun*

mono·pho·nic \ˌmä-nə-ˈfä-nik, -ˈfō-\ *adjective* (circa 1864)
1 : having a single unaccompanied melodic line
2 : of or relating to sound transmission, recording, or reproduction involving a single transmission path
— **mono·pho·ni·cal·ly** \-ni-k(ə-)lē\ *adverb*

mo·noph·o·ny \mə-ˈnä-fə-nē, mä-\ *noun* (circa 1890)
: monophonic music

mon·oph·thong \ˈmä-nə(f)-ˌthȯŋ\ *noun* [Late Greek *monophthongos* single vowel, from Greek *mon-* + *phthongos* sound] (1616)
: a vowel sound that throughout its duration has a single constant articulatory position
— **mon·oph·thon·gal** \ˌmä-nə(f)-ˈthȯŋ-(g)əl\ *adjective*

mono·phy·let·ic \ˌmä-nō-fī-ˈle-tik\ *adjective* [International Scientific Vocabulary] (1874)
: of or relating to a single stock; *specifically* **:** developed from a single common ancestral form
— **mono·phy·ly** \ˈmä-nə-ˌfī-lē\ *noun*

Mo·noph·y·site \mə-ˈnä-fə-ˌsīt\ *noun* [Medieval Latin *Monophysita,* from Middle Greek *Monophysitēs,* from Greek *mon-* + *physis* nature — more at PHYSICS] (1698)
: one holding the doctrine that Christ's nature remains altogether divine and not human even though he has taken on an earthly and human body with its cycle of birth, life, and death
— **Monophysite** *or* **Mo·noph·y·sit·ic** \-ˌnä-fə-ˈsi-tik\ *adjective*
— **Mo·noph·y·sit·ism** \-ˈnä-fə-ˌsī-ˌti-zəm\ *noun*

mono·plane \ˈmä-nə-ˌplān\ *noun* (1907)
: an airplane with only one main supporting surface

¹mono·ploid \ˈmä-nə-ˌplȯid\ *adjective* (1928)
1 : HAPLOID
2 : having or being the basic haploid number of chromosomes in a polyploid series of organisms

²**monoploid** *noun* [International Scientific Vocabulary] (1944)
: a monoploid individual or organism

mono·po·di·al \ˌmä-nə-ˈpō-dē-əl\ *adjective* [New Latin *monopodium* main axis, from *mon-* + *-podium*] (1876)
: having or involving the formation of offshoots from a main axis
— **mono·po·di·al·ly** \-dē-ə-lē\ *adverb*

mono·pole \ˈmä-nə-ˌpōl\ *noun* (1937)
1 : a single positive or negative electrical charge; *also* **:** a hypothetical north or south magnetic pole existing alone
2 : a radio antenna consisting of a single often straight element

mo·nop·o·lise *British variant of* MONOPOLIZE

mo·nop·o·list \mə-ˈnä-pə-list\ *noun* (1601)
: one who monopolizes
— **mo·nop·o·lis·tic** \-ˌnä-pə-ˈlis-tik\ *adjective*
— **mo·nop·o·lis·ti·cal·ly** \-ti-k(ə-)lē\ *adverb*

mo·nop·o·lize \mə-ˈnä-pə-ˌlīz\ *transitive verb* **-lized; -liz·ing** (1611)
: to get a monopoly of **:** assume complete possession or control of ⟨*monopolize* a conversation⟩
— **mo·nop·o·li·za·tion** \-ˌnä-pə-lə-ˈzā-shən\ *noun*
— **mo·nop·o·liz·er** \-ˈnä-pə-ˌlī-zər\ *noun*

mo·nop·o·ly \mə-ˈnä-p(ə-)lē\ *noun, plural* **-lies** [Latin *monopolium,* from Greek *monopōlion,* from *mon-* + *pōlein* to sell] (1534)
1 : exclusive ownership through legal privilege, command of supply, or concerted action
2 : exclusive possession or control
3 : a commodity controlled by one party
4 : one that has a monopoly

mono·pro·pel·lant \ˌmä-nō-prə-ˈpe-lənt\ *noun* (circa 1945)
: a rocket propellant containing both the fuel and the oxidizer in a single substance

mo·nop·so·ny \mə-ˈnäp-sə-nē\ *noun, plural* **-nies** [*mon-* + *-opsony* (as in *oligopsony*)] (1933)
: an oligopsony limited to one buyer
— **mo·nop·so·nis·tic** \-ˌnäp-sə-ˈnis-tik\ *adjective*

mono·rail \ˈmä nə-ˌrāl\ *noun* (1897)
: a single rail serving as a track for a wheeled vehicle; *also* **:** a vehicle traveling on such a track

monorail

mon·or·chid \mä-ˈnȯr-kəd\ *noun* [irregular from Greek *monorchis,* from *mon-* + *orchis* testicle — more at ORCHIS] (1874)
: an individual who has only one testis or only one descended into the scrotum
— **monorchid** *adjective*
— **mon·or·chi·dism** \-kə-ˌdi-zəm\ *noun*

mono·rhyme \ˈmä-nə-ˌrīm\ *noun* (1731)
: a strophe or poem in which all the lines have the same end rhyme
— **mono·rhymed** \-ˌrīmd\ *adjective*

mono·sac·cha·ride \ˌmä-nə-ˈsa-kə-ˌrīd\ *noun* [International Scientific Vocabulary] (1896)
: a sugar not decomposable to simpler sugars by hydrolysis

mono·so·di·um glu·ta·mate \ˌmä-nə-ˌsō-dē-əm-ˈglü-tə-ˌmāt\ *noun* (1929)
: a crystalline salt $C_5H_8O_4NaN$ used to enhance the flavor of food — abbreviation *MSG*

mono·some \ˈmä-nə-ˌsōm\ *noun* (circa 1909)
1 : a chromosome lacking a synaptic mate; *especially* **:** an unpaired X chromosome
2 : a single ribosome

mono·so·mic \ˌmä-nə-ˈsō-mik\ *adjective* (1926)
: having one less than the diploid number of chromosomes
— **monosomic** *noun*
— **mono·so·my** \ˈmä-nə-ˌsō-mē\ *noun*

mono·spe·cif·ic \ˌmä-nō-spə-ˈsi-fik\ *adjective* (1947)
: specific for a single antigen or receptor site on an antigen
— **mono·spec·i·fic·i·ty** \-ˌspe-sə-ˈfi-sə-tē\ *noun*

mono·stele \ˈmä-nə-ˌstēl, ˌmä-nə-ˈstē-lē\ *noun* (circa 1900)
: PROTOSTELE
— **mono·ste·lic** \ˌmä-nə-ˈstē-lik\ *adjective*
— **mono·ste·ly** \ˈmä-nə-ˌstē-lē\ *noun*

mono·syl·lab·ic \ˌmä-nə-sə-ˈla-bik\ *adjective* [probably from French *monosyllabique,* from *monosyllabe*] (1768)
1 : consisting of one syllable or of monosyllables
2 : using or speaking only monosyllables
3 : conspicuously brief in answering or commenting **:** TERSE
— **mono·syl·lab·i·cal·ly** \-bi-k(ə-)lē\ *adverb*
— **mono·syl·la·bic·i·ty** \-ˌsi-lə-ˈbi-sə-tē\ *noun*

mono·syl·la·ble \ˈmä-nə-ˌsi-lə-bəl, ˌmä-nə-ˈ\ *noun* [modification of Middle French or Late Latin; Middle French *monosyllabe,* from Late Latin *monosyllabon,* from Greek, from neuter of *monosyllabos* having one syllable, from *mon-* + *syllabē* syllable] (1533)
: a word of one syllable

mono·syn·ap·tic \ˌmä-nō-sə-ˈnap-tik\ *adjective* (1942)
: having or involving a single neural synapse
— **mono·syn·ap·ti·cal·ly** \-ti-k(ə-)lē\ *adverb*

mono·ter·pene \ˌmä-nə-ˈtər-ˌpēn\ *noun* (circa 1959)
: any of a class of terpenes $C_{10}H_{16}$ containing two isoprene units per molecule; *also* **:** a derivative of a monoterpene

mono·the·ism \ˈmä-nə-(ˌ)thē-ˌi-zəm\ *noun* (1660)
: the doctrine or belief that there is but one God
— **mono·the·ist** \-ˌthē-ist\ *noun*
— **mono·the·is·tic** \ˌmä-nə-thē-ˈis-tik\ *also* **mono·the·is·ti·cal** \-ti-kəl\ *adjective*
— **mono·the·is·ti·cal·ly** \-ti-k(ə-)lē\ *adverb*

¹**mono·tone** \ˈmä-nə-ˌtōn\ *noun* [Greek *monotonos* monotonous] (1644)
1 : a succession of syllables, words, or sentences in one unvaried key or pitch
2 : a single unvaried musical tone
3 : a tedious sameness or reiteration
4 : a person unable to produce or to distinguish between musical intervals

²**monotone** *adjective* (1769)
1 : MONOTONIC 2
2 : having a uniform color

mono·ton·ic \ˌmä-nə-ˈtä-nik\ *adjective* (1797)
1 : characterized by the use of or uttered in a monotone
2 : having the property either of never increasing or of never decreasing as the values of the independent variable or the subscripts of the terms increase ⟨*monotonic* functions⟩ ⟨a *monotonic* sequence⟩
— **mono·ton·i·cal·ly** \-ni-k(ə-)lē\ *adverb*
— **mono·to·nic·i·ty** \ˌmä-nə-tō-ˈni-sə-tē\ *noun*

mo·not·o·nous \mə-ˈnä-tᵊn-əs, -ˈnät-nəs\ *adjective* [Greek *monotonos,* from *mon-* + *tonos* tone] (1778)
1 : uttered or sounded in one unvarying tone **:** marked by a sameness of pitch and intensity
2 : tediously uniform or unvarying
— **mo·not·o·nous·ly** *adverb*
— **mo·not·o·nous·ness** *noun*

mo·not·o·ny \mə-ˈnä-tᵊn-ē, -ˈnät-nē\ *noun* (1706)
1 : tedious sameness
2 : sameness of tone or sound

mono·treme \ˈmä-nə-ˌtrēm\ *noun* [New Latin *Monotremata,* from Greek *mon-* + *trēmat-, trēma* hole — more at TREMATODE] (1835)
: any of an order (Monotremata) of egg-laying mammals comprising the platypuses and echidnas

mono·type \ˈmä-nə-ˌtīp\ *noun* (1882)
: an impression on paper of a design painted usually with the finger or a brush on a surface (as glass)

Monotype *trademark*
— used for a keyboard typesetting machine that casts and sets type in separate characters

mono·typ·ic \ˌmä-nə-ˈti-pik\ *adjective* [*mon-* + *type* + *-ic*] (circa 1859)
: including a single representative — used especially of a genus with only one species

mono·un·sat·u·rat·ed \ˌmä-nō-ˌən-ˈsa-chə-ˌrā-təd\ *adjective* (circa 1939)
: containing one double or triple bond per molecule — used especially of an oil or fatty acid; compare POLYUNSATURATED
— **mono·un·sat·u·rate** \-rət\ *noun*

mono·va·lent \ˌmä-nə-ˈvā-lənt\ *adjective* [International Scientific Vocabulary] (1869)
1 : having a valence of one
2 : containing antibodies specific for or antigens of a single strain of a microorganism ⟨a *monovalent* vaccine⟩

mon·ovu·lar \(ˌ)mä-ˈnäv-yə-lər, -ˈnōv-\ *adjective* (1929)
: MONOZYGOTIC

mon·ox·ide \mə-ˈnäk-ˌsīd\ *noun* [International Scientific Vocabulary] (1869)
: an oxide containing one atom of oxygen in a molecule

mono·zy·got·ic \ˌmä-nə-zī-ˈgä-tik\ *adjective* (1916)
: derived from a single egg ⟨*monozygotic* twins⟩

Mon·roe Doctrine \mən-ˈrō- *also* ˈmən- *or* ˈmän-\ *noun* [James *Monroe*] (1853)
: a statement of U.S. foreign policy expressing opposition to extension of European control or influence in the western hemisphere

mon·sei·gneur \ˌmōⁿ-ˌsān-ˈyər\ *noun, plural* **mes·sei·gneurs** \ˌmā ˌsān-ˈyər(z)\ [French, literally, my lord] (1602)
: a French dignitary (as a prince or prelate) — used as a title preceding a title of office or rank

mon·sieur \məs-ˈyə(r), məsh-; mə-ˈsir\ *noun, plural* **mes·sieurs** \məs-ˈyə(r)(z), məsh-, mās-; mə-ˈsir(z)\ [Middle French, literally, my lord] (1512)
: a Frenchman of high rank or station — used as a title equivalent to *Mister* and prefixed to the name of a Frenchman

mon·si·gnor \män-ˈsē-nyər, mən-\ *noun, plural* **monsignors** *or* **mon·si·gno·ri** \ˌmän-ˌsēn-ˈyōr-ē, -ˈyȯr-\ [Italian *monsignore,* from French *monseigneur*] (1641)
: a Roman Catholic prelate having a dignity or titular distinction (as of domestic prelate or protonotary apostolic) usually conferred by the pope — used as a title prefixed to the surname or to the given name and surname
— **mon·si·gno·ri·al** \ˌmän-ˌsēn-ˈyōr-ē-əl, -ˈyȯr-\ *adjective*

mon·soon \män-ˈsün, ˈmän-ˌ\ *noun* [obsolete Dutch *monssoen,* from Portuguese *monção,* from Arabic *mawsim* time, season] (1584)
1 : a periodic wind especially in the Indian Ocean and southern Asia
2 : the season of the southwest monsoon in India and adjacent areas that is characterized by very heavy rainfall
3 : rainfall that is associated with the monsoon
— **mon·soon·al** \-ˈsü-nᵊl, -ˌsü-\ *adjective*

\ə\ abut \ᵊ\ kitten \ər\ further \a\ ash \ā\ ace \ä\ mop, mar \au̇\ out \ch\ chin \e\ bet \ē\ easy \g\ go \i\ hit \ī\ ice \j\ job \ŋ\ sing \ō\ go \ȯ\ law \ȯi\ boy \th\ thin \t͟h\ the \ü\ loot \u̇\ foot \y\ yet \zh\ vision *see also* Guide to Pronunciation

mons pu·bis \'mänz-'pyü-bəs\ *noun, plural* **mon·tes pubis** \'män-tēz-\ [New Latin, pubic eminence] (circa 1903)
: a rounded eminence of fatty tissue upon the pubic symphysis especially of the human female

[1]**mon·ster** \'män(t)-stər\ *noun* [Middle English *monstre,* from Middle French, from Latin *monstrum* omen, monster, from *monēre* to warn — more at MIND] (14th century)
1 a : an animal or plant of abnormal form or structure **b :** one who deviates from normal or acceptable behavior or character
2 : a threatening force
3 a : an animal of strange or terrifying shape **b :** one unusually large for its kind
4 : something monstrous; *especially* **:** a person of unnatural or extreme ugliness, deformity, wickedness, or cruelty
5 : one that is highly successful

[2]**monster** *adjective* (1837)
: enormous or impressive in size, extent, or numbers

mon·strance \'män(t)-strən(t)s\ *noun* [Middle English *monstrans,* from Middle French *monstrance,* from Medieval Latin *monstrantia,* from Latin *monstrant-, monstrans,* present participle of *monstrare* to show, from *monstrum*] (15th century)
: a vessel in which the consecrated Host is exposed for the adoration of the faithful

monstrance

mon·stros·i·ty \män-'strä-sə-tē\ *noun, plural* **-ties** (15th century)
1 a : a malformation of a plant or animal **b :** something deviating from the normal **:** FREAK
2 : the quality or state of being monstrous
3 a : an object of great and often frightening size, force, or complexity **b :** an excessively bad or shocking example

mon·strous \'män(t)-strəs\ *adjective* (15th century)
1 *obsolete* **:** STRANGE, UNNATURAL
2 : having extraordinary often overwhelming size **:** GIGANTIC
3 a : having the qualities or appearance of a monster **b** *obsolete* **:** teeming with monsters
4 a : extraordinarily ugly or vicious **:** HORRIBLE **b :** shockingly wrong or ridiculous
5 : deviating greatly from the natural form or character **:** ABNORMAL
6 : very great — used as an intensive ☆
— mon·strous·ly *adverb*
— mon·strous·ness *noun*

mons ve·ne·ris \'mänz-'ve-nə-rəs\ *noun, plural* **mon·tes veneris** \'män-ˌtēz-\ [New Latin, literally, eminence of Venus or of venery] (1621)
: the mons pubis of the human female

mon·ta·dale \'män-tə-ˌdāl\ *noun* [*Monta*na state + *dale*] (1949)
: any of an American breed of white-faced hornless sheep noted for heavy fleece and good meat conformation

[1]**mon·tage** \män-'täzh, mō^{n}(n)-, -'tȧzh\ *noun* [French, from *monter* to mount] (1929)
1 : the production of a rapid succession of images in a motion picture to illustrate an association of ideas
2 a : a literary, musical, or artistic composite of juxtaposed more or less heterogeneous elements **b :** a composite picture made by combining several separate pictures
3 : a heterogeneous mixture **:** JUMBLE ⟨a *montage* of emotions⟩

[2]**montage** *transitive verb* **mon·taged; mon·tag·ing** (1944)
: to combine into or depict in a montage

mon·ta·gnard \ˌmō^{n}-ˌtän-'yär(d)\ *noun, often capitalized* [French, mountaineer, from *montagne* mountain, from Old French *montaigne*] (1842)
: a member of any of various peoples inhabiting the highlands of central and southern Vietnam
— montagnard *adjective, often capitalized*

Mon·ta·gue \'män-tə-ˌgyü\ *noun*
: the family of Romeo in Shakespeare's *Romeo and Juliet*

mon·tane \ˌmän-'tān, 'män-ˌ\ *adjective* [Latin *montanus* of a mountain — more at MOUNTAIN] (1863)
1 : of, relating to, growing in, or being the biogeographic zone of relatively moist cool upland slopes below timberline dominated by large coniferous trees
2 : of, relating to, or made up of montane plants or animals

Mon·ta·nist \'män-t^{ə}n-ist\ *noun* [*Montanus,* 2d century A.D. Phrygian schismatic] (1577)
: an adherent of a Christian sect arising in the late second century and stressing apocalyptic expectations, the continuing prophetic gifts of the Spirit, and strict ascetic discipline
— Mon·ta·nism \-t^{ə}n-ˌi-zəm\ *noun*

mon·tan wax \'män-t^{ə}n-\ *noun* [Latin *montanus* of a mountain] (1908)
: a hard brittle mineral wax obtained usually from lignites by extraction and used especially in polishes, carbon paper, and insulating compositions

mon·te \'män-tē\ *noun* [Spanish, bank, mountain, heap, from Italian, from Latin *mont-, mons* mountain] (1824)
1 : a card game in which players select any two of four cards turned face up in a layout and bet that one of them will be matched before the other as cards are dealt one at a time from the pack — called also *monte bank*
2 : THREE-CARD MONTE

Mon·te Car·lo \ˌmän-ti-'kär-(ˌ)lō\ *adjective* [*Monte Carlo,* Monaco, famous for its gambling casino] (1949)
: of, relating to, or involving the use of random sampling techniques and often the use of computer simulation to obtain approximate solutions to mathematical or physical problems especially in terms of a range of values each of which has a calculated probability of being the solution ⟨*Monte Carlo* calculations⟩

mon·teith \män-'tēth\ *noun* [*Monteith,* 17th century Scottish eccentric who wore a cloak with a scalloped hem] (1683)
: a large silver punch bowl with scalloped rim

Mon·te·rey cypress \'män-tə-ˌrā-\ *noun* [*Monterey,* California] (1873)
: a California cypress (*Cupressus macrocarpa*) that is endemic to the Monterey and Carmel seacoast and is often planted for ornament or in a windbreak

Monterey Jack *noun* (circa 1947)
: a semisoft whole-milk cheese with high moisture content

Monterey pine *noun* (1834)
: a pine (*Pinus radiata*) native to coastal California that is widely grown especially in the southern hemisphere for its wood

mon·te·ro \män-'ter-(ˌ)ō\ *noun, plural* **-ros** [Spanish, hunter, from *monte* mountain] (1622)
: a round hunter's cap with ear flaps

Mon·te·zu·ma's revenge \ˌmän-tə-'zü-məz-\ *noun* [*Montezuma* II] (1962)
: diarrhea contracted in Mexico especially by tourists

month \'mən(t)th\ *noun, plural* **months** \'mən(t)s, 'mən(t)ths\ [Middle English, from Old English *mōnath;* akin to Old High German *mānōd* month, Old English *mōna* moon] (before 12th century)
1 : a measure of time corresponding nearly to the period of the moon's revolution and amounting to approximately 4 weeks or 30 days or $1/12$ of a year
2 *plural* **:** an indefinite usually extended period of time ⟨he has been gone for *months*⟩
3 : one ninth of the typical duration of human pregnancy ⟨she was in her eighth *month*⟩

month·long \'mən(t)th-'lȯŋ\ *adjective* (1843)
: lasting a month

[1]**month·ly** \-lē\ *adverb* (circa 1534)
: once a month **:** by the month

[2]**monthly** *adjective* (1572)
1 a : of or relating to a month **b :** payable or reckoned by the month
2 : lasting a month
3 : occurring or appearing every month

[3]**monthly** *noun, plural* **monthlies** (1833)
1 : a monthly periodical
2 *plural* **:** a menstrual period

Monthly Meeting *noun* (circa 1772)
: a district unit of an organization of Friends

month's mind *noun* (15th century)
: a Roman Catholic requiem mass held a month after a person's death

Mont·mo·ren·cy \ˌmänt-mə-'ren(t)-sē\ *noun* [French, from *Montmorency,* France] (1924)
: a cherry that is grown commercially for its bright red sour fruit; *also* **:** the fruit

mont·mo·ril·lon·ite \ˌmänt-mə-'ri-lə-ˌnīt, -'rē-ə-ˌnīt\ *noun* [French, from *Montmorillon,* commune in western France] (1854)
: a soft clayey mineral that is a hydrous aluminum silicate with considerable capacity for exchanging part of the aluminum for magnesium and bases
— mont·mo·ril·lon·it·ic \-ˌri-lə-'ni-tik, -ˌrē-ə-'ni-\ *adjective*

Mon·tra·chet \ˌmō^{n}-trä-'she\ *noun* [French, from *Montrachet,* vineyard in Burgundy, France] (1833)
: a dry white burgundy wine

mon·u·ment \'män-yə-mənt\ *noun* [Middle English, from Latin *monumentum,* literally, memorial, from *monēre* to remind — more at MIND] (13th century)
1 *obsolete* **:** a burial vault **:** SEPULCHRE
2 : a written legal document or record **:** TREATISE
3 a (1) **:** a lasting evidence, reminder, or example of someone or something notable or great (2) **:** a distinguished person **b :** a memorial stone or a building erected in remembrance of a person or event
4 *archaic* **:** an identifying mark **:** EVIDENCE; *also* **:** PORTENT, SIGN
5 *obsolete* **:** a carved statue **:** EFFIGY
6 : a boundary or position marker (as a stone)
7 : NATIONAL MONUMENT
8 : a written tribute

mon·u·men·tal \ˌmän-yə-'men-t^{ə}l\ *adjective* (1604)
1 : of or relating to a monument
2 : serving as or resembling a monument **:** MASSIVE; *also* **:** highly significant **:** OUTSTANDING
3 : very great
— mon·u·men·tal·i·ty \-mən-'ta-lə-tē, -ˌmen-\ *noun*
— mon·u·men·tal·ly \-t^{ə}l-ē\ *adverb*

☆ SYNONYMS
Monstrous, prodigious, tremendous, stupendous mean extremely impressive. MONSTROUS implies a departure from the normal (as in size, form, or character) and often carries suggestions of deformity, ugliness, or fabulousness ⟨the *monstrous* waste of the project⟩. PRODIGIOUS suggests a marvelousness exceeding belief, usually in something felt as going far beyond a previous maximum (as of goodness, greatness, intensity, or size) ⟨made a *prodigious* effort and rolled the stone aside⟩. TREMENDOUS may imply a power to terrify or inspire awe ⟨the *tremendous* roar of the cataract⟩. STUPENDOUS implies a power to stun or astound, usually because of size, numbers, complexity, or greatness beyond description ⟨a *stupendous* volcanic eruption⟩.

MONTHS OF THE PRINCIPAL CALENDARS

GREGORIAN[1]	
Name	*Days*
January	
begins 10 days after the winter solstice	31
February	28
in leap years	29
March	31
April	30
May	31
June	30
July	31
August	31
September	30
October	31
November	30
December	31

JEWISH	
Name	*Days*
Tishri	
in year 5753 began Sept. 28, 1992	30
Heshvan	29 *or* 30
Kislev	29 *or* 30
Tebet	29
Shebat	30
Adar[2]	29 *or* 30
Nisan[3]	30
Iyar	29
Sivan	30
Tammuz	29
Ab	30
Elul	29

ISLAMIC	
Name	*Days*
Muharram[4]	
in A.H. 1413 began July 2, 1992	30
Safar	29
Rabi I	30
Rabi II	29
Jumada I	30
Jumada II	29
Rajab	30
Sha'ban	29
Ramadan	30
Shawwal	29
Dhu'l-Qa'dah	30
Dhu'l-Hijja	29
in leap years	30

[1] The equinoxes occur about March 21 and September 23, the solstices about June 22 and December 22.

[2] In leap years an intercalary month of 30 days takes the place of Adar, and is sometimes called Adar I. Adar then becomes the 29-day Veadar or Adar Sheni (sometimes called Adar II), and retains the usual festivals and holidays of Adar.

[3] The first month of the ecclesiastical year; anciently called Abib.

[4] Retrogresses through the seasons; the Islamic year is lunar and each month begins at the approximate new moon; the year 1 A.H. began on Friday, July 16, A.D. 622.

mon·u·men·tal·ize \-ˈmen-tᵊl-ˌīz\ *transitive verb* **-ized; -iz·ing** (1857)
: to record or memorialize lastingly by a monument

mon·u·ron \ˈmän-yə-ˌrän\ *noun* [*mon-* + *urea* + [1]*-on*] (circa 1957)
: a persistent herbicide $C_9H_{11}ClN_2O$ used especially to control broad-leaved weeds

mon·zo·nite \män-ˈzō-ˌnīt, ˈmän-zə-\ *noun* [French, from Mount *Monzoni,* Italy] (1895)
: a granular igneous rock composed of plagioclase and orthoclase in about equal quantities together with augite and a little biotite

moo \ˈmü\ *intransitive verb* [imitative] (1549)
: to make the throat noise of a cow
— **moo** *noun*

mooch \ˈmüch\ *verb* [probably from French dialect *muchier* to hide, lurk] (1851)
intransitive verb
1 : to wander aimlessly **:** AMBLE; *also* **:** SNEAK
2 : SPONGE, CADGE
transitive verb
1 : to take surreptitiously **:** STEAL
2 : BEG, CADGE
— **mooch·er** *noun*

[1]**mood** \ˈmüd\ *noun* [Middle English, from Old English *mōd;* akin to Old High German *muot* mood] (before 12th century)
1 : a conscious state of mind or predominant emotion **:** FEELING; *also* **:** the expression of mood especially in art or literature
2 *archaic* **:** a fit of anger **:** RAGE
3 a : a prevailing attitude **:** DISPOSITION **b :** a receptive state of mind predisposing to action **c :** a distinctive atmosphere or context **:** AURA

[2]**mood** *noun* [alteration of [1]*mode*] (1569)
1 : the form of a syllogism as determined by the quantity and quality of its constituent propositions
2 : distinction of form or a particular set of inflectional forms of a verb to express whether the action or state it denotes is conceived as fact or in some other manner (as command, possibility, or wish)
3 : MODE 1b

moody \ˈmü-dē\ *adjective* **mood·i·er; -est** (1593)
1 : subject to depression **:** GLOOMY
2 : subject to moods **:** TEMPERAMENTAL
3 : expressive of a mood
— **mood·i·ly** \ˈmü-dᵊl-ē\ *adverb*
— **mood·i·ness** \ˈmü-dē-nəs\ *noun*

moo·la *or* **moo·lah** \ˈmü-(ˌ)lä, -lə\ *noun* [origin unknown] (1939)
slang **:** MONEY

[1]**moon** \ˈmün\ *noun* [Middle English *mone,* from Old English *mōna;* akin to Old High German *māno* moon, Latin *mensis* month, Greek *mēn* month, *mēnē* moon] (before 12th century)
1 a *often capitalized* **:** the earth's natural satellite that shines by the sun's reflected light, revolves about the earth from west to east in about 29½ days with reference to the sun or about 27⅓ days with reference to the stars, and has a diameter of 2160 miles (3475 kilometers), a mean distance from the earth of about 238,900 miles (384,400 kilometers), and a mass about one eightieth that of the earth — usually used with *the* **b :** one complete moon cycle consisting of four phases **c :** SATELLITE 2; *specifically* **:** a natural satellite of a planet
2 : an indefinite usually extended period of time ⟨a labor of many *moons*⟩
3 : MOONLIGHT
4 : something that resembles a moon: as **a :** a highly translucent spot on old porcelain **b :** LUNULE **c** *slang* **:** naked buttocks
5 : something impossible or inaccessible ⟨reach for the *moon*⟩
— **moon·like** \-ˌlīk\ *adjective*
— **over the moon :** very pleased **:** in high spirits

[2]**moon** (1836)
transitive verb
1 : to spend in idle reverie **:** DREAM — used with *away*
2 *slang* **:** to expose one's naked buttocks to
intransitive verb
: to spend time in idle reverie **:** behave abstractedly

moon·beam \ˈmün-ˌbēm\ *noun* (1590)
: a ray of light from the moon

moon blindness *noun* (circa 1720)
: a recurrent inflammation of the eye of the horse

moon·calf \ˈmün-ˌkaf, -ˌkåf\ *noun* (1614)
: a foolish or absentminded person **:** SIMPLETON

moon·dust \-ˌdəst\ *noun* (circa 1959)
: fine dry particles of the moon's soil

moon·eye \-ˌī\ *noun* (1842)
: any of a genus (*Hiodon*) of silvery North American freshwater fishes that resemble shad

moon–eyed \-ˈīd\ *adjective* (1699)
: having the eyes wide open

moon·faced \-ˌfāst\ *adjective* (1855)
: having a round face

moon·fish \-ˌfish\ *noun, plural* **moonfish** *or* **moon·fish·es** (1646)
: any of various compressed often short deep-bodied silvery or yellowish marine fishes: as **a :** OPAH **b :** PLATY

moon·flow·er \-ˌflau̇(-ə)r\ *noun* (circa 1909)
: a tropical American morning glory (*Calonyction aculeatum* synonym *Ipomoea alba*) with fragrant flowers; *also* **:** any of several related plants

Moon·ie \ˈmü-nē\ *noun* [Sun Myung *Moon* (born 1920) Korean evangelist] (1974)
: a member of the Unification Church founded by Sun Myung Moon

moon·ish \ˈmü-nish\ *adjective* (15th century)
: influenced by the moon; *also* **:** CAPRICIOUS
— **moon·ish·ly** *adverb*

moon·less \ˈmün-ləs\ *adjective* (1508)
: lacking the light of the moon

moon·let \-lət\ *noun* (1832)
: a small natural or artificial satellite

[1]**moon·light** \-ˌlīt\ *noun* (14th century)
: the light of the moon

[2]**moonlight** *intransitive verb* **moon·light·ed; moon·light·ing** [back-formation from *moonlighter*] (1957)
: to hold a second job in addition to a regular one
— **moon·light·er** *noun*

moon·lit \-ˌlit\ *adjective* (circa 1827)
: lighted by the moon

moon·quake \-ˌkwāk\ *noun* (1946)
: a seismic event on the moon

moon·rise \-ˌrīz\ *noun* (1728)
1 : the rising of the moon above the horizon
2 : the time of the moon's rising

moon·scape \-ˌskāp\ *noun* (1916)
: the surface of the moon as seen or as depicted; *also* **:** a landscape resembling this surface
word history see LANDSCAPE

moon·seed \-ˌsēd\ *noun* (1739)
: a twining plant (*Menispermum canadense* of the family Menispermaceae, the moonseed family) of eastern North America that has crescent-shaped seeds and black fruits

moon·set \-ˌset\ *noun* (1845)
1 : the descent of the moon below the horizon
2 : the time of the moon's setting

moon shell *noun* (1936)
: any of a cosmopolitan family (Naticidae) of carnivorous marine snails having smooth globose shells

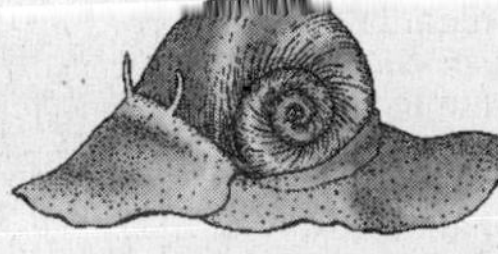

moon shell

moon·shine \ˈmün-ˌshīn\ *noun* (15th century)
1 : MOONLIGHT
2 : empty talk **:** NONSENSE

3 : intoxicating liquor; *especially* **:** illegally distilled corn whiskey

moon·shin·er \-ˌshī-nər\ *noun* (1860)
: a maker or seller of illicit whiskey

moon shot *noun* (1958)
: a spacecraft mission to the moon

moon·stone \-ˌstōn\ *noun* (1632)
: a transparent or translucent feldspar of pearly or opaline luster used as a gem

moon·struck \-ˌstrək\ *adjective* (1674)
: affected by or as if by the moon: as **a :** mentally unbalanced **b :** romantically sentimental **c :** lost in fantasy or reverie

moon·ward \-wərd\ *adverb* (1855)
: toward the moon

moony \ˈmü-nē\ *adjective* (circa 1586)
1 : of or relating to the moon
2 a : crescent-shaped **b :** resembling the full moon **:** ROUND
3 : MOONLIT
4 : DREAMY, MOONSTRUCK

¹moor \ˈmu̇r\ *noun* [Middle English *mor,* from Old English *mōr;* akin to Old High German *muor* moor] (before 12th century)
1 *chiefly British* **:** an expanse of open rolling infertile land
2 : a boggy area; *especially* **:** one that is peaty and dominated by grasses and sedges

²moor *verb* [Middle English *moren;* akin to Middle Dutch *meren, maren* to tie, moor] (15th century)
transitive verb
: to make fast with or as if with cables, lines, or anchors
intransitive verb
1 : to secure a boat by mooring **:** ANCHOR
2 : to be made fast

Moor \ˈmu̇r\ *noun* [Middle English *More,* from Middle French, from Latin *Maurus* inhabitant of Mauretania] (14th century)
1 : one of the Arab and Berber conquerors of Spain
2 : BERBER
— Moor·ish \-ish\ *adjective*

moor·age \ˈmu̇r-ij\ *noun* (1648)
1 : an act of mooring
2 : a place to moor

moor·hen \ˈmu̇r-ˌhen\ *noun* (14th century)
: the common gallinule (*Gallinula chloropus*) of the New World, Eurasia, and Africa

moor·ing \-iŋ\ *noun* (15th century)
1 : an act of making fast a boat or aircraft with lines or anchors
2 a : a place where or an object to which something (as a craft) can be moored **b :** a device (as a line or chain) by which an object is secured in place
3 : an established practice or stabilizing influence **:** ANCHORAGE 2 — usually used in plural

moor·land \-lənd, -ˌland\ *noun* (before 12th century)
: land consisting of moors **:** a stretch of moor

moose \ˈmüs\ *noun, plural* **moose** [of Algonquian origin; akin to Massachuset *moos* moose] (1603)
1 : a ruminant mammal (*Alces alces*) with humped shoulders, long legs, and broadly palmated antlers that is the largest existing member of the deer family and inhabits forested areas of Canada, the northern U.S., Europe, and Asia

moose 1

2 *capitalized* [Loyal Order of *Moose*] **:** a member of a major benevolent and fraternal order

¹moot \ˈmüt\ *noun* [Middle English, from Old English *mōt, gemōt;* akin to Middle High German *muoze* meeting] (before 12th century)
1 : a deliberative assembly primarily for the administration of justice; *especially* **:** one held by the freemen of an Anglo-Saxon community
2 *obsolete* **:** ARGUMENT, DISCUSSION

²moot *transitive verb* (before 12th century)
1 *archaic* **:** to discuss from a legal standpoint **:** ARGUE
2 a : to bring up for discussion **:** BROACH **b :** DEBATE

³moot *adjective* (circa 1587)
1 a : open to question **:** DEBATABLE **b :** subjected to discussion **:** DISPUTED
2 : deprived of practical significance **:** made abstract or purely academic

moot court *noun* (1788)
: a mock court in which law students argue hypothetical cases for practice

¹mop \ˈmäp\ *noun* [Middle English *mappe*] (15th century)
1 : an implement made of absorbent material fastened to a handle and used especially for cleaning floors
2 : something that resembles a mop; *especially* **:** a thick mass of hair

²mop *verb* **mopped; mop·ping** (1709)
transitive verb
: to use a mop on: as **a :** to clean or clear away by mopping ⟨*mop* the floors⟩ — often used with *up* ⟨*mop* up the spillage⟩ **b :** to wipe as if with a mop ⟨*mopped* his brow with a handkerchief⟩
intransitive verb
: to clean a surface (as a floor) with a mop
— mop·per *noun*

mop·board \ˈmäp-ˌbōrd, -ˌbȯrd\ *noun* (1853)
: BASEBOARD

¹mope \ˈmōp\ *intransitive verb* **moped; mop·ing** [probably from obsolete *mop, mope* fool] (1568)
1 *archaic* **:** to act in a dazed or stupid manner
2 : to give oneself up to brooding **:** become listless or dejected
3 : to move slowly or aimlessly **:** DAWDLE
— mop·er *noun*
— mop·ey \ˈmō-pē\ *adjective*

²mope *noun* (1693)
1 : one that mopes
2 *plural* **:** BLUES 1

mo·ped \ˈmō-ˌped\ *noun* [Swedish, from *mo*tor motor + *ped*al pedal] (1955)
: a lightweight low-powered motorbike that can be pedaled

mop·pet \ˈmä-pət\ *noun* [obsolete English *mop* fool, child] (1601)
1 *archaic* **:** BABY, DARLING
2 : CHILD

mop–up \ˈmäp-ˌəp\ *noun* (1900)
: a concluding action

mop up (circa 1811)
transitive verb
1 *British* **:** to consume eagerly
2 : to gather as if by absorbing **:** GARNER ⟨*mopped up* all the awards⟩
3 : to overcome decisively **:** TROUNCE
4 : to clear (an area) of remaining pockets of resistance in the wake of a military offensive
intransitive verb
: to complete a project or transaction

mo·quette \mō-ˈket\ *noun* [French] (1762)
: a carpet or upholstery fabric having a velvety pile

mor \ˈmȯr\ *noun* [Danish, literally, humus] (1931)
: forest humus that forms a layer of largely organic matter abruptly distinct from the mineral soil beneath

mo·ra \ˈmōr-ə, ˈmȯr-\ *noun, plural* **mo·rae** \ˈmōr-(ˌ)ē, ˈmȯr-, -ˌī\ *or* **mo·ras** [Latin, delay; akin to Old Irish *maraid* it lasts] (1832)
: the minimal unit of measure in quantitative verse equivalent to the time of an average short syllable

mo·raine \mə-ˈrān\ *noun* [French, from French dialect (Savoy) *morêna*] (1789)
: an accumulation of earth and stones carried and finally deposited by a glacier
— mo·rain·al \-ˈrā-nᵊl\ *adjective*
— mo·rain·ic \-ˈrā-nik\ *adjective*

¹mor·al \ˈmȯr-əl, ˈmär-\ *adjective* [Middle English, from Middle French, from Latin *moralis,* from *mor-, mos* custom] (14th century)
1 a : of or relating to principles of right and wrong in behavior **:** ETHICAL ⟨*moral* judgments⟩ **b :** expressing or teaching a conception of right behavior ⟨a *moral* poem⟩ **c :** conforming to a standard of right behavior **d :** sanctioned by or operative on one's conscience or ethical judgment ⟨a *moral* obligation⟩ **e :** capable of right and wrong action ⟨a *moral* agent⟩
2 : probable though not proved **:** VIRTUAL ⟨a *moral* certainty⟩
3 : having the effects of such on the mind, confidence, or will ⟨a *moral* victory⟩ ⟨*moral* support⟩ ☆
— mor·al·ly \-ə-lē\ *adverb*

²mor·al \ˈmȯr-əl, ˈmär-; *3 is* mə-ˈral\ *noun* (15th century)
1 a : the moral significance or practical lesson (as of a story) **b :** a passage pointing out usually in conclusion the lesson to be drawn from a story
2 *plural* **a :** moral practices or teachings **:** modes of conduct **b :** ETHICS
3 : MORALE

mo·rale \mə-ˈral\ *noun* [in sense 1, from French, from feminine of *moral,* adjective; in other senses, modification of French *moral* morale, from *moral,* adjective] (1752)
1 : moral principles, teachings, or conduct
2 a : the mental and emotional condition (as of enthusiasm, confidence, or loyalty) of an individual or group with regard to the function or tasks at hand **b :** a sense of common purpose with respect to a group **:** ESPRIT DE CORPS
3 : the level of individual psychological well-being based on such factors as a sense of purpose and confidence in the future

moral hazard *noun* (circa 1917)
: the possibility of loss to an insurance company arising from the character or circumstances of the insured

mor·al·ise *British variant of* MORALIZE

mor·al·ism \ˈmȯr-ə-ˌli-zəm, ˈmär-\ *noun* (1828)
1 a : the habit or practice of moralizing **b :** a conventional moral attitude or saying
2 : an often exaggerated emphasis on morality (as in politics)

mor·al·ist \-list\ *noun* (1621)
1 : one who leads a moral life
2 : a philosopher or writer concerned with moral principles and problems

☆ SYNONYMS
Moral, ethical, virtuous, righteous, noble mean conforming to a standard of what is right and good. MORAL implies conformity to established sanctioned codes or accepted notions of right and wrong ⟨the basic *moral* values of a community⟩. ETHICAL may suggest the involvement of more difficult or subtle questions of rightness, fairness, or equity ⟨committed to the highest *ethical* principles⟩. VIRTUOUS implies the possession or manifestation of moral excellence in character ⟨not a religious person, but *virtuous* nevertheless⟩. RIGHTEOUS stresses guiltlessness or blamelessness and often suggests the sanctimonious ⟨wished to be *righteous* before God and the world⟩. NOBLE implies moral eminence and freedom from anything petty, mean, or dubious in conduct and character ⟨had the *noblest* of reasons for seeking office⟩.

3 : one concerned with regulating the morals of others

mor·al·is·tic \ˌmȯr-ə-ˈlis-tik, ˌmär-\ *adjective* (1865)
1 : characterized by or expressive of a concern with morality
2 : characterized by or expressive of a narrow and conventional moral attitude
— mor·al·is·ti·cal·ly \-ti-k(ə-)lē\ *adverb*

mo·ral·i·ty \mə-ˈra-lə-tē, mȯ-\ *noun, plural* **-ties** (14th century)
1 a : a moral discourse, statement, or lesson **b :** a literary or other imaginative work teaching a moral lesson
2 a : a doctrine or system of moral conduct **b** *plural* **:** particular moral principles or rules of conduct
3 : conformity to ideals of right human conduct
4 : moral conduct **:** VIRTUE

morality play *noun* (1929)
1 : an allegorical play popular especially in the 15th and 16th centuries in which the characters personify abstract qualities or concepts (as virtues, vices, or death)
2 : something which involves a direct conflict between right and wrong or good and evil and from which a moral lesson may be drawn

mor·al·ize \ˈmȯr-ə-ˌlīz, ˈmär-\ *verb* **-ized; -iz·ing** (15th century)
transitive verb
1 : to explain or interpret morally
2 a : to give a moral quality or direction to **b :** to improve the morals of
intransitive verb
: to make moral reflections
— mor·al·i·za·tion \ˌmȯr-ə-lə-ˈzā-shən, ˌmär-\ *noun*
— mor·al·iz·er \ˈmȯr-ə-ˌlī-zər, ˈmär-\ *noun*

moral philosophy *noun* (1606)
: ETHICS; *also* **:** the study of human conduct and values

mo·rass \mə-ˈras, mȯ-\ *noun* [Dutch *moeras,* modification of Old French *maresc,* of Germanic origin; akin to Old English *mersc* marsh — more at MARSH] (1655)
1 : MARSH, SWAMP
2 : something that traps, confuses, or impedes ⟨a *morass* of troubles⟩
— mo·rassy \-ˈra-sē\ *adjective*

mor·a·to·ri·um \ˌmȯr-ə-ˈtȯr-ē-əm, ˌmär-, -ˈtȯr-\ *noun, plural* **-riums** *or* **-ria** \-ē-ə\ [New Latin, from Late Latin, neuter of *moratorius* dilatory, from Latin *morari* to delay, from *mora* delay] (1875)
1 a : a legally authorized period of delay in the performance of a legal obligation or the payment of a debt **b :** a waiting period set by an authority
2 : a suspension of activity

Mo·ra·vi·an \mə-ˈrā-vē-ən\ *noun* (1555)
1 a : a native or inhabitant of Moravia **b :** the group of Czech dialects spoken in Moravia
2 : a member of a Protestant denomination arising from a 15th century religious reform movement in Bohemia and Moravia
— Moravian *adjective*

mo·ray eel \mə-ˈrā-, ˈmȯr-(ˌ)ā-\ *noun* [Portuguese *moréia,* from Latin *muraena,* from Greek *myraina*] (1926)
: any of numerous often brightly colored eels (family Muraenidae) that have sharp teeth capable of inflicting a severe bite, that occur in warm seas, and that include a chiefly Mediterranean eel (*Muraena helena*) sometimes used for food — called also *moray*

mor·bid \ˈmȯr-bəd\ *adjective* [Latin *morbidus* diseased, from *morbus* disease] (1656)
1 a : of, relating to, or characteristic of disease ⟨*morbid* anatomy⟩ **b :** affected with or induced by disease ⟨a *morbid* condition⟩ **c :** productive of disease ⟨*morbid* substances⟩
2 : abnormally susceptible to or characterized by gloomy or unwholesome feelings
3 : GRISLY, GRUESOME ⟨*morbid* details⟩ ⟨*morbid* curiosity⟩
— mor·bid·ly *adverb*
— mor·bid·ness *noun*

mor·bid·i·ty \mȯr-ˈbi-də-tē\ *noun* (circa 1721)
1 : the quality or state of being morbid
2 : the relative incidence of disease

mor·ceau \mȯr-ˈsō\ *noun, plural* **mor·ceaux** \-ˈsō(z)\ [French, from Old French *morsel* morsel] (1751)
: a short literary or musical piece

mor·dan·cy \ˈmȯr-dᵊn(t)-sē\ *noun* (circa 1656)
1 : a biting and caustic quality of style **:** INCISIVENESS
2 : a sharply critical or bitter quality of thought or feeling **:** HARSHNESS

¹mor·dant \ˈmȯr-dᵊnt\ *adjective* [Middle English, from Middle French, present participle of *mordre* to bite, from Latin *mordēre;* perhaps akin to Sanskrit *mṛdnāti* he presses, rubs] (15th century)
1 : biting and caustic in thought, manner, or style **:** INCISIVE ⟨a *mordant* wit⟩
2 : acting as a mordant
3 : BURNING, PUNGENT
synonym see CAUSTIC
— mor·dant·ly *adverb*

²mordant *noun* (1791)
1 : a chemical that fixes a dye in or on a substance by combining with the dye to form an insoluble compound
2 : a corroding substance used in etching

³mordant *transitive verb* (1836)
: to treat with a mordant

Mor·de·cai \ˈmȯr-di-ˌkī\ *noun* [Hebrew *Mordĕkhai*]
: a relative of Esther who gives advice on saving the Jews from the destruction planned by Haman

mor·dent \ˈmȯr-dᵊnt, mȯr-ˈdent\ *noun* [Italian *mordente,* from Latin *mordent-, mordens,* present participle of *mordēre*] (1806)
: a musical ornament made by a quick alternation of a principal tone with the tone immediately below it

¹more \ˈmōr, ˈmȯr\ *adjective* [Middle English, from Old English *māra;* akin to Old English *mā,* adverb, more, Old High German *mēr,* Old Irish *mó* more] (before 12th century)
1 : GREATER ⟨something *more* than she expected⟩
2 : ADDITIONAL, FURTHER ⟨*more* guests arrived⟩

²more *adverb* (before 12th century)
1 a : in addition ⟨a couple of times *more*⟩ **b :** MOREOVER
2 : to a greater or higher degree — often used with an adjective or adverb to form the comparative ⟨*more* evenly matched⟩

³more *noun* (before 12th century)
1 : a greater quantity, number, or amount ⟨liked the idea better the *more* I thought about it⟩
2 : something additional **:** an additional amount
3 *obsolete* **:** persons of higher rank

⁴more *pronoun, singular or plural in construction* (13th century)
: additional persons or things or a greater amount ⟨*more* were found as the search continued⟩ ⟨*more* was spilled⟩

more and more *adverb* (13th century)
: to a progressively increasing extent

mo·reen \mə-ˈrēn, mȯ-\ *noun* [probably irregular from *moire*] (circa 1691)
: a strong fabric of wool, wool and cotton, or cotton with a plain glossy or moiré finish

mo·rel \mə-ˈrel, mȯ-\ *noun* [French *morille,* probably from (assumed) Vulgar Latin *mauricula,* from *maurus* brown, from Latin *Maurus* inhabitant of Mauretania] (1672)
: any of several edible fungi (genus *Morchella,* especially *M. esculenta*) having a cap with a highly pitted surface

mo·rel·lo \mə-ˈre-(ˌ)lō\ *noun, plural* **-los** [probably modification of Flemish *amarelle, marelle,* from Medieval Latin *amarellum,* a sour cherry, from Latin *amarus* bitter, sour] (1598)
: a cultivated sour cherry (as the Montmorency) having a dark-colored skin and juice

more or less *adverb* (13th century)
1 : to a varying or undetermined extent or degree **:** SOMEWHAT ⟨they were *more or less* willing to help⟩
2 : with small variations **:** APPROXIMATELY ⟨contains 16 acres *more or less*⟩

more·over \mōr-ˈō-vər, mȯr-, ˈmōr-ˌ, ˈmȯr-ˌ\ *adverb* (14th century)
: in addition to what has been said **:** BESIDES

mo·res \ˈmȯr-ˌāz, ˈmōr- *also* -(ˌ)ēz\ *noun plural* [Latin, plural of *mor-, mos* custom] (circa 1899)
1 : the fixed morally binding customs of a particular group
2 : moral attitudes
3 : HABITS, MANNERS ■

¹mo·resque \mȯ-ˈresk, mə-\ *adjective, often capitalized* [French, from Spanish *morisco,* from *moro* Moor, from Latin *Maurus*] (circa 1611)
: having the characteristics of Moorish art or architecture

²moresque *noun, often capitalized* (circa 1752)
: an ornament or decorative motif in Moorish style

mor·gan \ˈmȯr-gən\ *noun* [Thomas Hunt *Morgan*] (1919)
1 : a unit of inferred distance between genes on a chromosome that is used in constructing genetic maps and is equal to the distance for which the frequency of crossing over between specific pairs of genes is 100 percent
2 : CENTIMORGAN

Mor·gan \ˈmȯr-gən\ *noun* [Justin *Morgan* (died 1798) American teacher] (1841)
: any of an American breed of light strong horses originated in Vermont from the progeny of one prepotent stallion of uncertain ancestry

mor·ga·nat·ic \ˌmȯr-gə-ˈna-tik\ *adjective* [New Latin *matrimonium ad morganaticam,* literally, marriage with morning gift] (circa 1741)
: of, relating to, or being a marriage between a member of a royal or noble family and a person of inferior rank in which the rank of the

☐ USAGE
mores The pronunciation ending in \-(ˌ)ēz\ is now used by a minority of American and British speakers, but it was the original pronunciation of *mores* when the word first came into English about a century ago. Up until the middle of this century it was fashionable to anglicize words borrowed from another language, especially in British English. Thus, we do not pronounce *habeas corpus* according to the reconstructed Latin model as \ˈhä-bā-äs-ˈkōr-pūs\, but as a fully Englished form. One finds this tendency at work especially in the English of the Anglican Church, as in the church pronunciations of *Miserere, Venite, Kyrie,* and so forth. The word *mores* was thus also given a fully anglicized pronunciation at the start of its life in English, but when fashions changed and Latinate forms were given "Latin" pronunciations, the second syllable of *mores* became \-ˌāz\. Most of the evidence we have collected for the \-(ˌ)ēz\ pronunciation is from earlier decades, and it seems that among younger speakers the only recognized pronunciation ends in \-ˌāz\.

inferior partner remains unchanged and the children of the marriage do not succeed to the titles, fiefs, or entailed property of the parent of higher rank

— **mor·ga·nat·i·cal·ly** \-ti-k(ə-)lē\ *adverb*

mor·gan·ite \'mȯr-gə-ˌnīt\ *noun* [J. P. *Morgan* (died 1913)] (1911)
: a rose-colored gem variety of beryl

Morgan le Fay \-lə-'fā\ *noun* [Old French *Morgain la fee* Morgan the fairy]
: a sorceress and sister of King Arthur

mor·gen \'mȯr-gə(n)\ *noun, plural* **morgen** [Dutch, literally, morning] (1626)
: a Dutch and southern African unit of land area equal to 2.116 acres (0.856 hectare)

morgue \'mȯrg\ *noun* [French] (1821)
1 : a place where the bodies of persons found dead are kept until identified and claimed by relatives or are released for burial
2 : a collection of reference works and files of reference material in a newspaper or news periodical office

mor·i·bund \'mȯr-ə-(ˌ)bənd, 'mär-\ *adjective* [Latin *moribundus,* from *mori* to die — more at MURDER] (circa 1721)
1 : being in the state of dying **:** approaching death
2 : being in a state of inactivity or obsolescence

— **mor·i·bun·di·ty** \ˌmȯr-ə-'bən-də-tē, ˌmär-\ *noun*

¹mo·ri·on \'mōr-ē-ˌän, 'mȯr-\ *noun* [Middle French] (1563)
: a high-crested helmet with no visor

²morion *noun* [modification of Latin *mormorion*] (1748)
: a nearly black variety of smoky quartz

Mo·ris·co \mə-'ris-(ˌ)kō, mȯ-\ *noun, plural* **-cos** *or* **-coes** [Spanish, from *morisco,* adjective, from *moro* Moor] (1629)
: MOOR; *especially* **:** a Spanish Moor

— **Morisco** *adjective*

Mor·mon \'mȯr-mən\ *noun*
1 : the ancient redactor and compiler of the Book of Mormon presented as divine revelation by Joseph Smith
2 : LATTER-DAY SAINT; *especially* **:** a member of the Church of Jesus Christ of Latter-day Saints

— **Mor·mon·ism** \-mə-ˌni-zəm\ *noun*

Mormon cricket *noun* (1896)
: a large dark wingless katydid (*Anabrus simplex*) that resembles a cricket and is found in the arid parts of the western U.S. where it is occasionally an abundant pest of crops

morn \'mȯrn\ *noun* [Middle English, from Old English *morgen;* akin to Old High German *morgan* morning and perhaps to Greek *marmairein* to sparkle] (before 12th century)
1 : DAWN
2 : MORNING

Mor·nay sauce \mȯr-'nā-\ *noun* [Philippe de *Mornay*] (circa 1924)
: a cheese-flavored cream sauce

morn·ing \'mȯr-niŋ\ *noun* [Middle English, from *morn* + *-ing* (as in *evening*)] (13th century)
1 a : DAWN **b :** the time from sunrise to noon **c :** the time from midnight to noon
2 : a period of first development **:** BEGINNING

morn·ing–after pill \ˌmȯr-niŋ-'af-tər-\ *noun* [from its being taken after rather than before intercourse] (1966)
: an oral drug that interferes with pregnancy by blocking implantation of a fertilized egg in the human uterus

morning glory *noun* (1814)
: any of various usually twining plants (genus *Ipomoea* of the family Convolvulaceae, the morning-glory family) with showy trumpet-shaped flowers; *broadly* **:** a plant of the morning-glory family including herbs, shrubs, or trees with alternate leaves and regular pentamerous flowers

morning line *noun* (circa 1935)
: a bookmaker's list of entries for a race meet and the probable odds on each that is printed or posted before the betting begins

Morning Prayer *noun* (1552)
: a service of liturgical prayer used for regular morning worship in churches of the Anglican communion

morn·ings \'mȯr-niŋz\ *adverb* (1652)
: in the morning repeatedly **:** on any morning

morning sickness *noun* (1879)
: nausea and vomiting that occur on rising in the morning especially during the earlier months of pregnancy

morning star *noun* (1535)
: a bright planet (as Venus) seen in the eastern sky before or at sunrise

Mo·ro \'mōr-(ˌ)ō, 'mȯr-\ *noun, plural* **Moros** [Spanish, literally, Moor, from Latin *Maurus*] (1886)
1 : a member of any of several Muslim peoples of the southern Philippines
2 : any of the Austronesian languages of the Moro peoples

mo·roc·co \mə-'rä-(ˌ)kō\ *noun* [*Morocco,* Africa] (1634)
: a fine leather from goatskin tanned with sumac

mo·ron \'mōr-ˌän, 'mȯr-\ *noun* [irregular from Greek *mōros* foolish, stupid] (1910)
1 : a mentally retarded person who has a potential mental age of between 8 and 12 years and is capable of doing routine work under supervision
2 : a very stupid person

— **mo·ron·ic** \mə-'rä-nik, mȯ-\ *adjective*

— **mo·ron·i·cal·ly** \-ni-k(ə-)lē\ *adverb*

— **mo·ron·ism** \'mōr-ˌä-ˌni-zəm, 'mȯr-\ *noun*

— **mo·ron·i·ty** \mə-'rä-nə-tē, mȯ-\ *noun*

mo·rose \mə-'rōs, mȯ-\ *adjective* [Latin *morosus,* literally, capricious, from *mor-, mos* will] (1565)
1 : having a sullen and gloomy disposition
2 : marked by or expressive of gloom

synonym see SULLEN

— **mo·rose·ly** *adverb*

— **mo·rose·ness** *noun*

— **mo·ros·i·ty** \-'rä-sə-tē\ *noun*

morph \'mȯrf\ *noun* [back-formation from *morpheme*] (1947)
1 a : ALLOMORPH **b :** a distinctive collocation of phones (as a portmanteau form) that serves as the realization of more than one morpheme in a context (as the French *du* for the sequence of *de* and *le*)
2 a : a local population of a species that consists of interbreeding organisms and is distinguishable from other populations by morphology or behavior though capable of interbreeding with them **b :** a phenotypic variant of a species

morph- *or* **morpho-** *combining form* [German, from Greek, from *morphē*]
1 : form ⟨*morpho*genesis⟩
2 : morpheme ⟨*morpho*phonemics⟩

-morph *noun combining form* [International Scientific Vocabulary, from *-morphous*]
: one having (such) a form ⟨iso*morph*⟩

mor·phac·tin \mȯr-'fak-tən\ *noun* [probably from *morph-* + *act*ive + ¹*-in*] (1966)
: any of several synthetic fluorine-containing compounds that tend to produce morphological changes and suppress growth in plants

mor·phal·lax·is \ˌmȯr-fə-'lak-səs\ *noun, plural* **-lax·es** \-ˌsēz\ [New Latin, from *morph-* + Greek *allaxis* exchange, from *allassein* to change, exchange, from *allos* other — more at ELSE] (1901)
: regeneration of a part or organism from a fragment by reorganization without cell proliferation

mor·pheme \'mȯr-ˌfēm\ *noun* [French *morphème,* from Greek *morphē* form] (1926)
: a distinctive collocation of phonemes (as the free form *pin* or the bound form *-s* of *pins*) that contains no smaller meaningful parts

— **mor·phe·mic** \mȯr-'fē-mik\ *adjective*

— **mor·phe·mi·cal·ly** \-mi-k(ə-)lē\ *adverb*

mor·phe·mics \mȯr-'fē-miks\ *noun plural but singular in construction* (1947)
1 : a branch of linguistic analysis that consists of the study of morphemes
2 : the structure of a language in terms of morphemes

Mor·pheus \'mȯr-fē-əs, -ˌfyüs, -ˌfüs\ *noun* [Latin, from Greek]
: the Greek god of dreams

mor·phia \'mȯr-fē-ə\ *noun* [New Latin, from *Morpheus*] (1818)
: MORPHINE

-morphic *adjective combining form* [probably from French *-morphique,* from Greek *morphē*]
: having (such) a form ⟨endo*morphic*⟩

mor·phine \'mȯr-ˌfēn\ *noun* [French, from *Morpheus*] (1828)
: a bitter crystalline addictive narcotic base $C_{17}H_{19}NO_3$ that is the principal alkaloid of opium and is used in the form of a soluble salt (as a hydrochloride or a sulfate) as an analgesic and sedative ◆

mor·phin·ism \'mȯr-ˌfē-ˌni-zəm, -fə-\ *noun* (1882)
: a disordered condition of health produced by habitual use of morphine

-morphism *noun combining form* [Late Latin *-morphus* -morphous, from Greek *-morphos*]
: quality or state of having (such) a form ⟨hetero*morphism*⟩

mor·pho \'mȯr-(ˌ)fō\ *noun, plural* **morphos** [New Latin, from Greek *Morphō,* epithet of Aphrodite] (1853)
: any of a genus (*Morpho* of the family Morphoidae) of large showy tropical American butterflies that typically have a brilliant blue metallic luster on the upper surface of the wings

mor·pho·gen \'mȯr-fə-jən, -ˌjen\ *noun* (1950)
: a diffusible chemical substance that exerts control over morphogenesis especially by forming a gradient in concentration

mor·pho·gen·e·sis \ˌmȯr-fə-'je-nə-səs\ *noun* [New Latin] (circa 1890)
: the formation and differentiation of tissues and organs — compare ORGANOGENESIS

mor·pho·ge·net·ic \-jə-'ne-tik\ *adjective* (1884)
: relating to or concerned with the development of normal organic form ⟨*morphogenetic* movements of early embryonic cells⟩

— **mor·pho·ge·net·i·cal·ly** \-ti-k(ə-)lē\ *adverb*

mor·pho·gen·ic \-'je-nik\ *adjective* (circa 1890)
: MORPHOGENETIC

mor·phol·o·gy \mȯr-'fä-lə-jē\ *noun* [German *Morphologie,* from *morph-* + *-logie* -logy] (1830)

◇ WORD HISTORY

morphine The word *morphine* is based on classical mythology. Hypnos, the Greek god of sleep, was usually conceived as perpetually reclining on a couch, surrounded by his numerous sons, the Dreams. Prominent among the sons was Morpheus, who, when called upon by the chief gods to intervene in mortal affairs, had the power to make human forms appear to dreamers. (Greek *Morpheus* is appropriately a derivative of *morphē* "form, shape.") The drug morphine was first isolated in 1806 by the German pharmacologist Friedrich Wilhelm Adam Sertürner (1783–1841). Sertürner named this derivative of opium *morphium,* after the god of dreams, for the deep sleep it induces in addition to relieving pain. English borrowed the French version of this name, *morphine,* without change.

1 a : a branch of biology that deals with the form and structure of animals and plants **b :** the form and structure of an organism or any of its parts
2 a : a study and description of word formation (as inflection, derivation, and compounding) in language **b :** the system of word-forming elements and processes in a language
3 a : a study of structure or form **b :** STRUCTURE, FORM
4 : the external structure of rocks in relation to the development of erosional forms or topographic features
— mor·pho·log·i·cal \ˌmȯr-fə-ˈlä-ji-kəl\ *also* **mor·pho·log·ic** \-ˈlä-jik\ *adjective*
— mor·pho·log·i·cal·ly \-k(ə-)lē\ *adverb*
— mor·phol·o·gist \mȯr-ˈfä-lə-jist\ *noun*

mor·phom·e·try \mȯr-ˈfä-mə-trē\ *noun* (circa 1856)
: measurement of external form
— mor·pho·met·ric \ˌmȯr-fə-ˈme-trik\ *adjective*
— mor·pho·met·ri·cal·ly \-tri-k(ə-)lē\ *adverb*

mor·pho·pho·ne·mics \ˌmȯr-fō-fə-ˈnē-miks\ *noun plural but singular in construction* (1939)
1 : a study of the phonemic differences between allomorphs of the same morpheme
2 : the distribution of allomorphs in one morpheme
3 : the structure of a language in terms of morphophonemics

-morphous *adjective combining form* [Greek *-morphos,* from *morphē*]
: having (such) a form ⟨iso*morphous*⟩

-morphy *noun combining form* [International Scientific Vocabulary, from *-morphous*]
: quality or state of having (such) a form ⟨endo*morphy*⟩

mor·ris \ˈmȯr-əs, ˈmär-\ *noun* [Middle English *moreys daunce,* from *moreys* Moorish (from *More* Moor) + *daunce* dance] (1512)
: a vigorous English dance traditionally performed by men wearing costumes and bells

morris chair *noun* [William *Morris*] (1900)
: an easy chair with adjustable back and removable cushions

morris chair

mor·row \ˈmär-(ˌ)ō, ˈmȯr-\ *noun* [Middle English *morn, morwen* morn] (13th century)
1 *archaic* **:** MORNING
2 : the next day
3 : the time immediately after a specified event

Morse code \ˈmȯrs-\ *noun* [Samuel F. B. *Morse*] (1867)
: either of two codes consisting of variously spaced dots and dashes or long and short sounds used for transmitting messages by audible or visual signals

[1]mor·sel \ˈmȯr-səl\ *noun* [Middle English, from Old French, diminutive of *mors* bite, from Latin *morsus,* from *mordēre* to bite — more at MORDANT] (14th century)
1 : a small piece of food **:** BITE
2 : a small quantity **:** FRAGMENT
3 a : a tasty dish **b :** something delectable and pleasing
4 : a negligible person

[2]morsel *transitive verb* **-seled** *or* **-selled; -sel·ing** *or* **-sel·ling** (1598)
: to divide into or distribute in small pieces

[1]mort \ˈmȯrt\ *noun* [probably alteration of Middle English *mot* horn note, from Middle French, word, horn note — more at MOT] (circa 1500)
1 : a note sounded on a hunting horn when a deer is killed
2 : KILLING 1

INTERNATIONAL MORSE CODE

a	·–	á	·––·–
b	–···	ä	·–·–
c	–·–·	é	··–··
d	–··	ñ	––·––
e	·	ö	–––·
f	··–·	ü	··––
g	––·	1	·––––
h	····	2	··–––
i	··	3	···––
j	·–––	4	····–
k	–·–	5	·····
l	·–··	6	–····
m	––	7	––···
n	–·	8	–––··
o	–––	9	––––·
p	·––·	0	–––––
q	––·–	,	––··–– (comma)
r	·–·	.	·–·–·–
s	···	?	··––··
t	–	;	–·–·–·
u	··–	:	–––···
v	···–	’	·––––· (apostrophe)
w	·––	-	–····– (hyphen)
x	–··–	/	–··–·
y	–·––	(	–·––· (left parenthesis)
z	––··	)	–·––·– (right parenthesis)

[2]mort *noun* [perhaps back-formation from [1]*mortal*] (1694)
: a great quantity or number

mor·ta·del·la \ˌmȯr-tə-ˈde-lə\ *noun* [Italian, from Latin *murtatum* sausage seasoned with myrtle berries, from *murtus* myrtle] (1613)
: a large smoked sausage made of beef, pork, and pork fat and seasoned with pepper and garlic

[1]mor·tal \ˈmȯr-tᵊl\ *adjective* [Middle English, from Middle French, from Latin *mortalis,* from *mort-, mors* death — more at MURDER] (14th century)
1 : causing or having caused death **:** FATAL ⟨a *mortal* injury⟩
2 a : subject to death ⟨*mortal* man⟩ **b :** POSSIBLE, CONCEIVABLE ⟨have done every *mortal* thing⟩ **c :** DEADLY 3 ⟨waited three *mortal* hours⟩
3 : marked by unrelenting hostility **:** IMPLACABLE ⟨a *mortal* enemy⟩
4 : marked by great intensity or severity **:** EXTREME ⟨*mortal* fear⟩ ⟨a *mortal* shame⟩
5 : HUMAN ⟨*mortal* limitations⟩
6 : of, relating to, or connected with death ⟨*mortal* agony⟩
synonym see DEADLY

[2]mortal *adverb* (15th century)
chiefly dialect **:** MORTALLY

[3]mortal *noun* (1567)
: a human being

mor·tal·i·ty \mȯr-ˈta-lə-tē\ *noun* (14th century)
1 : the quality or state of being mortal
2 : the death of large numbers (as of people or animals)
3 *archaic* **:** DEATH
4 : the human race
5 a : the number of deaths in a given time or place **b :** the proportion of deaths to population **c :** the number lost or the rate of loss or failure

mortality table *noun* (1880)
: an actuarial table based on mortality statistics over a number of years

mor·tal·ly \ˈmȯr-tᵊl-ē\ *adverb* (14th century)
1 : in a deadly or fatal manner **:** to death ⟨*mortally* wounded⟩
2 : to an extreme degree **:** INTENSELY ⟨*mortally* afraid⟩

mortal mind *noun* (1875)
Christian Science **:** a belief that life, substance, and intelligence are in and of matter **:** ILLUSION

mortal sin *noun* (15th century)
: a sin (as murder) that is deliberately committed and is of such serious consequence according to Thomist theology that it deprives the soul of sanctifying grace — compare VENIAL SIN

[1]mor·tar \ˈmȯr-tər\ *noun* [Middle English *morter,* from Old English *mortere* & Middle French *mortier,* from Latin *mortarium*] (before 12th century)
1 : a strong vessel in which material is pounded or rubbed with a pestle
2 [Middle French *mortier*] **a :** a muzzle-loading cannon having a tube short in relation to its caliber that is used to throw projectiles with low muzzle velocities at high angles **b :** any of several similar firing devices

[2]mortar *noun* [Middle English *morter,* from Old French *mortier,* from Latin *mortarium*] (14th century)
: a plastic building material (as a mixture of cement, lime, or gypsum plaster with sand and water) that hardens and is used in masonry or plastering
— mor·tar·less *adjective*

[3]mortar *transitive verb* (14th century)
: to plaster or make fast with mortar

mor·tar·board \ˈmȯr-tər-ˌbōrd, -ˌbȯrd\ *noun* (1854)
1 a : HAWK 2 **b :** a board or platform about three feet (one meter) square for holding mortar
2 : an academic cap consisting of a closely fitting headpiece with a broad flat projecting square top

mortarboard 2

[1]mort·gage \ˈmȯr-gij\ *noun* [Middle English *morgage,* from Middle French, from Old French, from *mort* dead (from Latin *mortuus*) + *gage* gage — more at MURDER] (15th century)
1 : a conveyance of or lien against property that is defeated upon payment or performance according to stipulated terms
2 a : the instrument evidencing the mortgage **b :** the state of the property so mortgaged **c :** the interest of the mortgagee in such property

[2]mortgage *transitive verb* **mort·gaged; mort·gag·ing** (15th century)
1 : to grant or convey by a mortgage
2 : to subject to a claim or obligation **:** PLEDGE

mort·gag·ee \ˌmȯr-gi-ˈjē\ *noun* (1584)
: a person to whom property is mortgaged

mort·gag·or \ˌmȯr-gi-ˈjȯr\ *also* **mort·gag·er** \ˈmȯr-gi-jər\ *noun* (1584)
: a person who mortgages property

mor·ti·cian \mȯr-ˈti-shən\ *noun* [Latin *mort-, mors* death] (1895)
: UNDERTAKER 2

mor·ti·fi·ca·tion \ˌmȯr-tə-fə-ˈkā-shən\ *noun* (14th century)
1 : the subjection and denial of bodily passions and appetites by abstinence or self-inflicted pain or discomfort
2 : NECROSIS, GANGRENE
3 a : a sense of humiliation and shame caused by something that wounds one's pride or self-respect **b :** the cause of such humiliation

mor·ti·fy \ˈmȯr-tə-ˌfī\ *verb* **-fied; -fy·ing** [Middle English *mortifien,* from Middle French *mortifier,* from Late Latin *mortificare,* from Latin *mort-, mors*] (14th century)
transitive verb

\ə\ **abut** \ᵊ\ **kitten** \ər\ **further** \a\ **ash** \ā\ **ace** \ä\ **mop, mar** \au̇\ **out** \ch\ **chin** \e\ **bet** \ē\ **easy** \g\ **go** \i\ **hit** \ī\ **ice** \j\ **job** \ŋ\ **sing** \ō\ **go** \ȯ\ **law** \ȯi\ **boy** \th\ **thin** \t͟h\ **the** \ü\ **loot** \u̇\ **foot** \y\ **yet** \zh\ **vision** *see also* Guide to Pronunciation

1 *obsolete* **:** to destroy the strength, vitality, or functioning of
2 : to subdue or deaden (as the body or bodily appetites) especially by abstinence or self-inflicted pain or discomfort
3 : to subject to severe and vexing embarrassment **:** SHAME
intransitive verb
1 : to practice mortification
2 : to become necrotic or gangrenous

¹mor·tise *also* **mor·tice** \'mȯr-təs\ *noun* [Middle English *mortays,* from Middle French *mortaise*] (15th century)
: a hole, groove, or slot into or through which some other part of an arrangement of parts fits or passes; *especially* **:** a cavity cut into a piece of material (as timber) to receive a tenon — see DOVETAIL illustration

²mortise *also* **mortice** *transitive verb* **mor·tised** *also* **mor·ticed; mor·tis·ing** *also* **mor·tic·ing** (15th century)
1 : to join or fasten securely; *specifically* **:** to join or fasten by a tenon and mortise
2 : to cut or make a mortise in

mort·main \'mȯrt-ˌmān\ *noun* [Middle English *morte-mayne,* from Middle French *mortemain,* from Old French, from *morte* (feminine of *mort* dead) + *main* hand, from Latin *manus* — more at MANUAL] (15th century)
1 a : an inalienable possession of lands or buildings by an ecclesiastical or other corporation **b :** the condition of property or other gifts left to a corporation in perpetuity especially for religious, charitable, or public purposes
2 : the influence of the past regarded as controlling the present

¹mor·tu·ary \'mȯr-chə-ˌwer-ē\ *adjective* [Latin *mortuarius* of the dead, from *mortuus* dead] (1514)
1 : of or relating to the burial of the dead
2 : of, relating to, or characteristic of death

²mortuary *noun, plural* **-ar·ies** (1865)
: a place in which dead bodies are kept until burial; *especially* **:** FUNERAL HOME

mor·u·la \'mȯr-(y)ə-lə, 'mär-\ *noun, plural* **-lae** \-ˌlē, -ˌlī\ [New Latin, from Latin *morum* mulberry, from Greek *moron*] (1874)
: a globular solid mass of blastomeres formed by cleavage of a zygote that typically precedes the blastula
— mor·u·lar \-lər\ *adjective*
— mor·u·la·tion \ˌmȯr-(y)ə-'lā-shən, ˌmär-\ *noun*

¹mo·sa·ic \mō-'zā-ik\ *noun* [Middle English *musycke,* from Middle French *mosaique,* from Old Italian *mosaico,* from Medieval Latin *musaicum,* alteration of Late Latin *musivum,* from Latin *museum, musaeum*] (15th century)
1 : a surface decoration made by inlaying small pieces of variously colored material to form pictures or patterns; *also* **:** the process of making it
2 : a picture or design made in mosaic
3 : something resembling a mosaic ⟨a *mosaic* of visions and daydreams and memories —Lawrence Shainberg⟩
4 : an organism or one of its parts composed of cells of more than one genotype **:** CHIMERA 3
5 : a virus disease of plants characterized by diffuse light and dark green or yellow and green mottling of the foliage
6 : a composite map made of photographs taken by an aircraft or spacecraft
7 : the part of a television camera tube consisting of many minute photoelectric particles that convert light to an electric charge
— mo·sa·ic·like \-'zā-ik-ˌlīk\ *adjective*

²mosaic *adjective* (1585)
1 : of, relating to, produced by, or resembling a mosaic
2 : exhibiting mosaicism
3 : DETERMINATE 5
— mo·sa·i·cal·ly \-'zā-ə-k(ə-)lē\ *adverb*

³mosaic *transitive verb* **-icked; -ick·ing** (1839)
1 : to decorate with mosaics
2 : to form into a mosaic

Mo·sa·ic \mō-'zā-ik\ *adjective* [New Latin *Mosaicus,* from *Moses* Moses] (1662)
: of or relating to Moses or the institutions or writings attributed to him

mo·sa·i·cism \mō-'zā-ə-ˌsi-zəm\ *noun* (1926)
: the condition of possessing cells of two or more different genetic constitutions

mo·sa·i·cist \-sist\ *noun* (1847)
1 a : a designer of mosaics **b :** a worker who makes mosaics
2 : a dealer in mosaics

mo·sa·saur \'mō-zə-ˌsȯr\ *noun* [New Latin *Mosasaurus,* from Latin *Mosa* the river Meuse + Greek *sauros* lizard] (1841)
: any of a family (Mosasauridae) of very large extinct marine fish-eating lizards of the Upper Cretaceous with limbs modified into paddles that are related to the recent monitor lizards

Mo·selle \mō-'zel\ *noun* [German *Moselwein,* from *Mosel* Moselle, river in Germany + German *Wein* wine] (1693)
: a white wine from the Moselle valley

Mo·ses \'mō-zəz *also* -zəs\ *noun* [Latin, from Greek *Mōsēs,* from Hebrew *Mōsheh*]
: a Hebrew prophet who led the Israelites out of Egyptian slavery and at Mount Sinai delivered to them the Law establishing God's covenant with them

mo·sey \'mō-zē\ *intransitive verb* **mo·seyed; mo·sey·ing** [origin unknown] (1829)
1 : to hurry away
2 : to move in a leisurely or aimless manner **:** SAUNTER ⟨*moseyed* around the general store, testing the cheese straight off the round —Eric Sevareid⟩

mo·shav \mō-'shäv\ *noun, plural* **mo·sha·vim** \ˌmō-shə-'vēm\ [New Hebrew *mōshābh,* from Hebrew, dwelling] (1931)
: a cooperative settlement of small individual farms in Israel — compare KIBBUTZ

Mos·lem \'mäz-ləm *also* 'mäs-\ *variant of* MUSLIM

Mo·so·tho \mō-'sō-(ˌ)tō, -'sü-(ˌ)tü\ *noun, plural* **Ba·so·tho** \bä-\ (1952)
: a member of the Basotho people

mosque \'mäsk\ *noun* [Middle French *mosquee,* from Old Italian *moschea,* from Old Spanish *mezquita,* from Arabic *masjid* temple, from *sajada* to prostrate oneself, worship] (1711)
: a building used for public worship by Muslims

mos·qui·to \mə-'skē-(ˌ)tō\ *noun, plural* **-toes** *also* **-tos** [Spanish, diminutive of *mosca* fly, from Latin *musca* — more at MIDGE] (circa 1583)
: any of a family (Culicidae) of dipteran flies with females that have a set of slender organs in the proboscis adapted to puncture the skin of animals and to suck their blood and that are in some cases vectors of serious diseases
— mos·qui·to·ey \-'skē-tə-wē\ *adjective*

mosquito fish *noun* (1928)
: either of two North American live-bearers (*Gambusia affinis* and *Heterandria formosa* of the family Poeciliidae) used especially to exterminate mosquito larvae

mosquito hawk *noun* (1737)
: DRAGONFLY

mosquito net *noun* (1745)
: a net or screen for keeping out mosquitoes

¹moss \'mȯs\ *noun* [Middle English, from Old English *mos;* akin to Old High German *mos* moss, Latin *muscus*] (before 12th century)
1 *chiefly Scottish* **:** BOG, SWAMP; *especially* **:** a peat bog
2 a : any of a class (Musci) of bryophytic plants having a small leafy often tufted stem bearing sex organs at its tip; *also* **:** a clump or sward of these plants **b :** any of various plants resembling moss in appearance or habit of growth
3 : a mossy covering
— moss·like \-ˌlīk\ *adjective*

²moss *transitive verb* (15th century)
: to cover or overgrow with moss

moss agate *noun* (1798)
: an agate mineral containing brown, black, or green mosslike or dendritic markings

moss animal *noun* (1881)
: BRYOZOAN

moss·back \'mȯs-ˌbak\ *noun* (1872)
1 : a large sluggish fish (as a largemouth bass)
2 : an extremely old-fashioned or reactionary person **:** FOGY
— moss·backed \-ˌbakt\ *adjective*

moss green *noun* (1884)
: a variable color averaging a moderate yellow-green

moss–grown \'mȯs-ˌgrōn\ *adjective* (14th century)
1 : overgrown with moss
2 : ANTIQUATED

moss pink *noun* (circa 1856)
: a low-growing perennial phlox (*Phlox subulata*) widely cultivated for its abundant usually pink or white flowers

moss rose *noun* (1776)
: an old-fashioned garden rose (*Rosa centifolia mucosa*) that has a glandular mossy calyx and flower stalk

moss–troop·er \'mȯs-ˌtrü-pər\ *noun* (1645)
1 : one of a class of 17th century raiders in the marshy border country between England and Scotland
2 : PIRATE
— moss–troop·ing \-piŋ\ *adjective*

mossy \'mȯ-sē\ *adjective* **moss·i·er; -est** (15th century)
1 : resembling moss
2 : covered with moss or something like moss
3 : ANTIQUATED ⟨the *mossy* precepts of the . . . prescriptive grammarians —Thomas Pyles⟩

mossy zinc *noun* (1910)
: a granulated form of zinc made by pouring melted zinc into water

¹most \'mōst\ *adjective* [Middle English, from Old English *mǣst;* akin to Old High German *meist* most, Old English *māra* more — more at MORE] (before 12th century)
1 : greatest in quantity, extent, or degree ⟨the *most* ability⟩
2 : the majority of ⟨*most* people⟩

²most *adverb* (before 12th century)
1 : to the greatest or highest degree — often used with an adjective or adverb to form the superlative ⟨the *most* challenging job he ever had⟩
2 : to a very great degree ⟨was *most* persuasive⟩

³most *noun* (12th century)
: the greatest amount ⟨it's the *most* I can do⟩
— at most *or* **at the most :** as an extreme limit ⟨took him an hour *at most*⟩

⁴most *pronoun, singular or plural in construction* (13th century)
: the greatest number or part ⟨*most* become discouraged and quit⟩

⁵most *adverb* [by shortening] (circa 1584)
: ALMOST ⟨we'll be crossing the river *most* any time now —Hamilton Basso⟩ ■

-most *adjective suffix* [Middle English, alteration of *-mest* (as in *formest* foremost)]
: most ⟨inner*most*⟩ **:** most toward ⟨head*most*⟩

most·ly \'mōst-lē\ *adverb* (1594)
: for the greatest part **:** MAINLY

□ USAGE
most Although considered by some to be unacceptable in all cases, *most* is often used to mean "almost" in both spoken and, to a lesser extent, written English to modify the adjectives *all, every,* and *any;* the pronouns *all, everyone, everything, everybody, anyone, anything,* and *anybody;* and the adverbs *everywhere, anywhere,* and *always.* Other uses of this sense of *most* are dialectal.

Most Reverend (15th century)
— used as a title for an archbishop or a Roman Catholic bishop

mot \'mō\ *noun, plural* **mots** \'mō(z)\ [French, word, saying, from Late Latin *muttum* grunt — more at MOTTO] (1631)
: a pithy or witty saying

¹mote \'mōt\ *verbal auxiliary* [Middle English, from Old English *mōtan* to be allowed to — more at MUST] (before 12th century)
archaic : MAY, MIGHT

²mote \'mōt\ *noun* [Middle English *mot,* from Old English; akin to Middle Dutch & Frisian *mot* sand] (before 12th century)
: a small particle : SPECK

mo•tel \mō-'tel\ *noun* [blend of *motor* and *hotel*] (1925)
: an establishment which provides lodging and parking and in which the rooms are usually accessible from an outdoor parking area

mo•tet \mō-'tet\ *noun* [Middle English, from Middle French, diminutive of *mot*] (14th century)
: a polyphonic choral composition on a sacred text usually without instrumental accompaniment

moth \'mȯth\ *noun, plural* **moths** \'mȯthz, 'mȯths\ [Middle English *mothe,* from Old English *moththe;* akin to Middle High German *motte* moth] (before 12th century)
1 : CLOTHES MOTH
2 : any of various usually nocturnal lepidopteran insects with antennae that are often feathery, with a stouter body, duller coloring, and proportionately smaller wings than the butterflies, and with larvae that are plant-eating caterpillars
— **moth•like** \-,līk\ *adjective*
— **mothy** \'mȯ-thē\ *adjective*

¹moth•ball \'mȯth-,bȯl\ *noun* (1906)
1 : a ball made formerly of camphor but now often of naphthalene and used to keep moths from clothing
2 *plural* : a condition of protective storage ⟨put the ships in *mothballs* after the war⟩; *also* : a state of having been rejected for further use or dismissed from further consideration ⟨put that idea in *mothballs*⟩

²mothball *transitive verb* (1943)
1 : to deactivate (as a ship) and prevent deterioration chiefly by dehumidification
2 : to withdraw from use or service and keep in reserve

moth bean *noun* [probably by folk etymology from Marathi *math* moth bean] (1884)
: a bean (*Phaseolus aconitifolius*) that is cultivated especially in India for its cylindrical pods and small yellowish brown seeds

moth–eat•en \'mȯth-,ē-tᵊn\ *adjective* (14th century)
1 : eaten into by moth larvae ⟨*moth-eaten* clothes⟩
2 a : DILAPIDATED **b** : ANTIQUATED, OUTMODED

¹moth•er \'mə-thər\ *noun* [Middle English *moder,* from Old English *mōdor;* akin to Old High German *muoter* mother, Latin *mater,* Greek *mētēr,* Sanskrit *mātṛ*] (before 12th century)
1 a : a female parent **b** (1) : a woman in authority; *specifically* : the superior of a religious community of women (2) : an old or elderly woman
2 : SOURCE, ORIGIN ⟨necessity is the *mother* of invention⟩
3 : maternal tenderness or affection
4 [short for *motherfucker*] : one that is particularly impressive or contemptible — sometimes considered vulgar
— **moth•er•hood** \-,hu̇d\ *noun*
— **moth•er•less** \-ləs\ *adjective*
— **moth•er•less•ness** *noun*

²mother *adjective* (13th century)
1 a : of, relating to, or being a mother **b** : bearing the relation of a mother
2 : derived from or as if from one's mother
3 : acting as or providing parental stock — used without reference to sex

³mother *transitive verb* **moth•ered; moth•er•ing** \'mə-thə-riŋ, 'məth-riŋ\ (1548)
1 a : to give birth to **b** : to give rise to : PRODUCE
2 : to care for or protect like a mother

⁴mother *noun* [archaic *mother* dregs, lees; akin to Middle Dutch *moeder* dregs] (circa 1828)
: MOTHER OF VINEGAR

moth•er•board \-,bōrd, -,bȯrd\ *noun* (1972)
: the main circuit board especially of a microcomputer

Mother Car•ey's chicken \-'kar-ēz-, -'ker-\ *noun* [origin unknown] (1767)
: STORM PETREL

mother cell *noun* (1845)
: a cell that gives rise to other cells usually of a different sort

mother country *noun* (1587)
1 : the country from which the people of a colony or former colony derive their origin
2 : the country of one's parents or ancestors; *also* : FATHERLAND
3 : a country that is the origin of something

moth•er•fuck•er \'mə-thər-,fə-kər\ *noun* (1959)
: one that is formidable, contemptible, or offensive — usually considered obscene; usually used as a generalized term of abuse
— **moth•er•fuck•ing** \-kiŋ\ *adjective*

Mother Goose *noun*
: the legendary author of a collection of nursery rhymes first published in London about 1760

mother hen *noun* (1954)
: a person who assumes an overly protective maternal attitude

moth•er•house \-,hau̇s\ *noun* (1661)
1 : the convent in which the superior of a religious community resides
2 : the original convent of a religious community

Mother Hub•bard \-'hə-bərd\ *noun* [probably from *Mother Hubbard,* character in a nursery rhyme] (1884)
: a loose usually shapeless dress

moth•er–in–law \'mə-thər-ən-,lȯ, 'məth-rən-, 'mə-thərn-\ *noun, plural* **moth•ers–in–law** \'mə-thər-zən-\ (14th century)
1 : the mother of one's spouse
2 *archaic* : STEPMOTHER

moth•er•land \'mə-thər-,land\ *noun* (1711)
1 : a country regarded as a place of origin (as of an idea or a movement)
2 : MOTHER COUNTRY 2

mother lode *noun* (1874)
1 : the principal vein or lode of a region
2 : a principal source or supply

moth•er•ly \-lē\ *adjective* (before 12th century)
1 : of, relating to, or characteristic of a mother ⟨*motherly* advice⟩
2 : resembling a mother : MATERNAL
— **moth•er•li•ness** *noun*

moth•er–na•ked \,mə-thər-'nā-kəd, *especially Southern* -'ne-kəd\ *adjective* (14th century)
: stark naked

Mother Nature *noun* (1601)
: nature personified as a woman considered as the source and guiding force of creation

moth•er–of–pearl \,mə-thə-rə(v)-'pər(-ə)l\ *noun* (circa 1510)
: the hard pearly iridescent substance forming the inner layer of a mollusk shell

mother of vinegar (1601)
: a slimy membrane composed of yeast and bacterial cells that develops on the surface of alcoholic liquids undergoing acetous fermentation and is added to wine or cider to produce vinegar — called also *mother*

Mother's Day *noun* (1908)
: the 2d Sunday in May appointed for the honoring of mothers

mother tongue *noun* (14th century)
1 : one's native language
2 : a language from which another language derives

mother wit *noun* (15th century)
: natural wit or intelligence

¹moth•proof \'mȯth-,prüf\ *adjective* (1893)
: impervious to penetration by moths ⟨*mothproof* wool⟩

²mothproof *transitive verb* (1925)
: to make mothproof
— **moth•proof•er** *noun*

mo•tif \mō-'tēf\ *noun* [French, motive, motif, from Middle French — more at MOTIVE] (1848)
1 : a usually recurring salient thematic element (as in the arts); *especially* : a dominant idea or central theme
2 : a single or repeated design or color
— **mo•tif•ic** \-'tē-fik, -'ti-\ *adjective*

¹mo•tile \'mō-tᵊl, -,tīl\ *adjective* [Latin *motus,* past participle of *movēre*] (circa 1859)
: exhibiting or capable of movement
— **mo•til•i•ty** \mō-'ti-lə-tē\ *noun*

²motile *noun* (1886)
: a person whose prevailing mental imagery takes the form of inner feelings of action

¹mo•tion \'mō-shən\ *noun* [Middle English *mocioun,* from Middle French *motion,* from Latin *motion-, motio* movement, from *movēre* to move] (14th century)
1 a : an act, process, or instance of changing place : MOVEMENT **b** : an active or functioning state or condition ⟨set the divorce proceedings in *motion*⟩
2 : an impulse or inclination of the mind or will
3 a : a proposal for action; *especially* : a formal proposal made in a deliberative assembly **b** : an application made to a court or judge to obtain an order, ruling, or direction
4 *obsolete* **a** : a puppet show **b** : PUPPET
5 : MECHANISM
6 a : an act or instance of moving the body or its parts : GESTURE **b** *plural* : ACTIVITIES, MOVEMENTS
7 : melodic change of pitch
— **mo•tion•al** \'mo-shnəl, -shə-nᵊl\ *adjective*
— **mo•tion•less** \'mō-shən-ləs\ *adjective*
— **mo•tion•less•ly** *adverb*
— **mo•tion•less•ness** *noun*
— **in motion** *of an offensive football player* : running parallel to the line of scrimmage before the snap

²motion *verb* **mo•tioned; mo•tion•ing** \'mō-sh(ə-)niŋ\ (1747)
intransitive verb
: to signal by a movement or gesture ⟨the pitcher *motioned* to the catcher⟩
transitive verb
: to direct by a motion ⟨*motioned* me to the seat⟩

motion picture *noun* (1896)
1 : a series of pictures projected on a screen in rapid succession with objects shown in successive positions slightly changed so as to produce the optical effect of a continuous picture in which the objects move
2 : a representation (as of a story) by means of motion pictures : MOVIE

motion sickness *noun* (1942)
: sickness induced by motion (as in travel by air, car, or ship) and characterized by nausea

mo•ti•vate \'mō-tə-,vāt\ *transitive verb* **-vat•ed; -vat•ing** (1885)
: to provide with a motive : IMPEL ⟨questions that excite and *motivate* youth⟩
— **mo•ti•va•tive** \-,vā-tiv\ *adjective*
— **mo•ti•va•tor** \-,vā-tər\ *noun*

mo·ti·va·tion \ˌmō-tə-ˈvā-shən\ *noun* (1873)
1 a : the act or process of motivating **b :** the condition of being motivated
2 : a motivating force, stimulus, or influence **:** INCENTIVE, DRIVE
— **mo·ti·va·tion·al** \-shnəl, -shə-nᵊl\ *adjective*
— **mo·ti·va·tion·al·ly** *adverb*

[1]**mo·tive** \ˈmō-tiv, *2 is also* mō-ˈtēv\ *noun* [Middle English, from Middle French *motif,* from *motif,* adjective, moving, from Medieval Latin *motivus,* from Latin *motus,* past participle of *movēre* to move] (15th century)
1 : something (as a need or desire) that causes a person to act
2 : a recurrent phrase or figure that is developed through the course of a musical composition
3 : MOTIF ☆
— **mo·tive·less** \-ləs\ *adjective*
— **mo·tive·less·ly** *adverb*
— **mo·ti·vic** \mō-ˈtē-vik\ *adjective*

[2]**mo·tive** \ˈmō-tiv\ *adjective* [Middle French or Medieval Latin; Middle French *motif,* from Medieval Latin *motivus*] (1502)
1 : moving or tending to move to action
2 : of or relating to motion or the causing of motion ⟨*motive* energy⟩

[3]**mo·tive** \ˈmō-tiv\ *transitive verb* **mo·tived; mo·tiv·ing** (circa 1650)
: MOTIVATE

motive power *noun* (1889)
1 : an agency (as water or steam) used to impart motion especially to machinery
2 : something (as a locomotive or a motor) that provides motive power to a system

mo·tiv·i·ty \mō-ˈti-və-tē\ *noun* (circa 1687)
: the power of moving or producing motion

mot juste \mō-zhæst\ *noun, plural* **mots justes** *same*\ [French] (1912)
: the exactly right word or phrasing

[1]**mot·ley** \ˈmät-lē\ *adjective* [Middle English, perhaps from *mot* mote, speck] (14th century)
1 : variegated in color ⟨a *motley* coat⟩
2 : composed of diverse often incongruous elements ⟨a *motley* crowd⟩

[2]**motley** *noun* [Middle English, probably from [1]*motley*] (14th century)
1 : a woolen fabric of mixed colors made in England between the 14th and 17th centuries
2 : a garment made of motley; *especially* **:** the characteristic dress of the professional fool
3 : JESTER, FOOL
4 : a mixture especially of incongruous elements

mot·mot \ˈmät-ˌmät\ *noun* [New Latin *momot, motmot*] (1837)
: any of a family (Momotidae) of long-tailed mostly green nonpasserine birds of Central and South American tropical forests

mo·to·cross \ˈmō-tō-ˌkrȯs\ *noun* [French, from *moto* motorcycle (short for *motocyclette*) + *cross*-country, from English] (1951)
: a closed-course motorcycle race over natural or simulated rough terrain (as with steep inclines, hairpin turns, and mud); *also* **:** the sport of engaging in motocross races

mo·to·neu·ron \ˌmō-tə-ˈnü-ˌrän, -ˈnyü-; -ˈnu̇r-ˌän, -ˈnyu̇r-\ *noun* [*motor* + *neuron*] (1908)
: a neuron that passes from the central nervous system or a ganglion toward a muscle and conducts an impulse that causes movement — called also *motor neuron*
— **mo·to·neu·ro·nal** \-ˈnu̇r-ə-nᵊl, -ˈnyu̇r-; -nu̇-ˈrō-nᵊl, -nyu̇-\ *adjective*

[1]**mo·tor** \ˈmō-tər\ *noun* [Latin, from *movēre* to move] (1586)
1 : one that imparts motion; *specifically* **:** PRIME MOVER
2 : any of various power units that develop energy or impart motion: as **a :** a small compact engine **b :** INTERNAL COMBUSTION ENGINE; *especially* **:** a gasoline engine **c :** a rotating machine that transforms electrical energy into mechanical energy
3 : MOTOR VEHICLE; *especially* **:** AUTOMOBILE
— **mo·tor·dom** \-dəm\ *noun*
— **mo·tor·less** \-ləs\ *adjective*

[2]**motor** *adjective* (1840)
1 a : causing or imparting motion **b :** of, relating to, or being a motoneuron or a nerve containing motoneurons ⟨*motor* fiber⟩ **c :** of, relating to, concerned with, or involving muscular movement ⟨*motor* areas of the brain⟩
2 a : equipped with or driven by a motor **b :** of, relating to, or involving an automobile **c :** designed for motor vehicles or motorists

[3]**motor** (1896)
intransitive verb
1 : to travel by automobile **:** DRIVE
2 : to move or proceed at a vigorous steady pace ⟨*motored* down the field for a touchdown⟩
transitive verb
: to transport by automobile

mo·tor·bike \ˈmō-tər-ˌbīk\ *noun* (1903)
: a small usually lightweight motorcycle
— **motorbike** *intransitive verb*

mo·tor·boat \-ˌbōt\ *noun* (1902)
: a boat propelled usually by an internal combustion engine
— **mo·tor·boat·er** \-ˌbō-tər\ *noun*
— **mo·tor·boat·ing** \-tiŋ\ *noun*

motor bus *noun* (1901)
: BUS 1a — called also *motor coach*

mo·tor·cade \ˈmō-tər-ˌkād\ *noun* (1913)
: a procession of motor vehicles
— **motorcade** *intransitive verb*

mo·tor·car \-ˌkär\ *noun* (circa 1890)
1 : AUTOMOBILE
2 *usually* **motor car :** a railroad car containing motors for propulsion

motor court *noun* (1936)
: MOTEL

mo·tor·cy·cle \ˈmō-tər-ˌsī-kəl\ *noun* [*motor bicycle*] (1896)
: a 2-wheeled automotive vehicle having one or two saddles
— **motorcycle** *intransitive verb*
— **mo·tor·cy·clist** \-k(ə-)list\ *noun*

motor home *noun* (1965)
: a large motor vehicle equipped as living quarters — compare MOBILE HOME

mo·tor·ic \mō-ˈtȯr-ik, -ˈtär-\ *adjective* (1930)
: MOTOR 1c
— **mo·tor·i·cal·ly** \-i-k(ə-)lē\ *adverb*

motor inn *noun* (1951)
: MOTEL; *especially* **:** a large multistory motel — called also *motor hotel*

mo·tor·ise *British variant of* MOTORIZE

mo·tor·ist \ˈmō-tə-rist\ *noun* (1896)
: a person who travels by automobile

mo·tor·ize \ˈmō-tə-ˌrīz\ *transitive verb* **-ized; -iz·ing** (circa 1913)
1 : to equip with a motor
2 : to equip with automobiles
— **mo·tor·i·za·tion** \ˌmō-tə-rə-ˈzā-shən\ *noun*

motor lodge *noun* (1949)
: MOTEL

mo·tor·man \ˈmō-tər-mən\ *noun* (1890)
: an operator of a motor-driven vehicle (as a streetcar or subway train)

mo·tor·mouth \-ˌmau̇th\ *noun* (1971)
: a person who talks excessively

motor pool *noun* (1942)
: a group of motor vehicles centrally controlled (as by a governmental agency) and dispatched for use as needed

motor sailer *noun* (1923)
: a motorboat with sailing equipment

motor scooter *noun* (1919)
: a low 2- or 3-wheeled automotive vehicle resembling a child's scooter and having a seat so that the rider does not straddle the engine

motor ship *noun* (1915)
: a seagoing ship propelled by an internal combustion engine

motor torpedo boat *noun* (1940)
: PT BOAT

mo·tor·truck \ˈmō-tər-ˌtrək\ *noun* (1930)
: an automotive truck used especially for transporting freight

motor unit *noun* (1925)
: a motoneuron together with the muscle fibers on which it acts

motor vehicle *noun* (1890)
: an automotive vehicle not operated on rails; *especially* **:** one with rubber tires for use on highways

mo·tor·way \ˈmō-tər-ˌwā\ *noun* (1903)
chiefly British **:** SUPERHIGHWAY

motte \ˈmät\ *noun* [Middle English *mote,* from Old French *mote, motte*] (13th century)
: MOUND, HILL; *especially* **:** a hill serving as a site for a Norman castle in Britain

motte and bailey *noun* (1900)
: a medieval Norman castle consisting of two connecting ditched stockaded mounds with the higher mound surmounted by the keep and the lower one containing barracks and other buildings

[1]**mot·tle** \ˈmä-tᵊl\ *noun* [probably back-formation from *motley*] (1676)
1 : a colored spot
2 a : a surface having colored spots or blotches **b :** the arrangement of such spots or blotches on a surface
3 : MOSAIC 5
— **mot·tled** \-tᵊld\ *adjective*

[2]**mottle** *transitive verb* **mot·tled; mot·tling** \ˈmät-liŋ, ˈmä-tᵊl-iŋ\ (1676)
: to mark with spots or blotches of different color or shades of color as if stained
— **mot·tler** \ˈmät-lər, ˈmä-tᵊl-ər\ *noun*

mottled enamel *noun* (1928)
: spotted tooth enamel caused by drinking water containing excessive fluorides during the time the teeth are calcifying

mot·to \ˈmä-(ˌ)tō\ *noun, plural* **mottoes** *also* **mottos** [Italian, from Late Latin *muttum* grunt, from Latin *muttire* to mutter] (1588)
1 : a sentence, phrase, or word inscribed on something as appropriate to or indicative of its character or use
2 : a short expression of a guiding principle

moue \ˈmü\ *noun* [French, from Middle French — more at MOW] (1850)
: a little grimace **:** POUT

mou·flon *also* **mouf·flon** \mü-ˈflōⁿ\ *noun* [French *mouflon,* from Italian dialect *muvrone,* from Late Latin *mufron-, mufro*] (1774)
: either of two wild sheep (*Ovis orientalis* and *O. musimon*) of the mountains of Sardinia, Corsica and western Asia that have large curling horns in the males and are sometimes included in a single species

mouflon

☆ SYNONYMS
Motive, impulse, incentive, inducement, spur, goad mean a stimulus to action. MOTIVE implies an emotion or desire operating on the will and causing it to act ⟨a *motive* for the crime⟩. IMPULSE suggests a driving power arising from personal temperament or constitution ⟨buying on *impulse*⟩. INCENTIVE applies to an external influence (as an expected reward) inciting to action ⟨a bonus was offered as an *incentive*⟩. INDUCEMENT suggests a motive prompted by the deliberate enticements or allurements of another ⟨offered a watch as an *inducement* to subscribe⟩. SPUR applies to a motive that stimulates the faculties or increases energy or ardor ⟨fear was a *spur* to action⟩. GOAD suggests a motive that keeps one going against one's will or desire ⟨thought insecurity a *goad* to worker efficiency⟩.

mou·jik \mü-'zhēk, -'zhik\ *variant of* MUZHIK

mou·lage \mü-'läzh\ *noun* [French, molding, from Middle French, from *mouler* to mold, from Old French *modle* mold — more at MOLD] (1902)
1 : an impression or cast made for use especially as evidence in a criminal investigation
2 : the taking of an impression for use as evidence in a criminal investigation

mould \'mōld\ *variant of* MOLD

mould·ing \'mōl-diŋ\ *variant of* MOLDING

moult \'mōlt\ *variant of* MOLT

[1]**mound** \'maund\ *transitive verb* [origin unknown] (1515)
1 *archaic* **:** to enclose or fortify with a fence or a ridge of earth
2 : to form into a mound

[2]**mound** *noun, often attributive* [origin unknown] (1551)
1 *archaic* **:** HEDGE, FENCE
2 a (1) **:** an artificial bank or hill of earth or stones; *especially* **:** one constructed over a burial or ceremonial site (2) **:** the slightly elevated ground on which a baseball pitcher stands **b :** a rounded hill or natural formation
3 a : HEAP, PILE ⟨*mounds* of work⟩ **b :** a small rounded mass ⟨a *mound* of mashed potatoes⟩

Mound Builder *noun* (1838)
: a member of a prehistoric American Indian people whose extensive earthworks are found from the Great Lakes down the Mississippi valley to the Gulf of Mexico

[1]**mount** \'maunt\ *noun* [Middle English, from Old English *munt* & Old French *mont,* both from Latin *mont-, mons;* akin to Welsh *mynydd* mountain, Latin *minari* to project, threaten] (before 12th century)
1 : a high hill **:** MOUNTAIN — used especially before an identifying name ⟨*Mount* Everest⟩
2 *archaic* **:** EARTHWORK 1
3 : MOUND 2a(1)

[2]**mount** *verb* [Middle English, from Middle French *monter,* from (assumed) Vulgar Latin *montare,* from Latin *mont-, mons*] (14th century)
intransitive verb
1 : RISE, ASCEND
2 : to increase in amount or extent ⟨expenses began to *mount*⟩
3 : to get up on something above the level of the ground; *especially* **:** to seat oneself (as on a horse) for riding
transitive verb
1 a : to go up **:** CLIMB **b** (1) **:** to seat or place oneself on (2) **:** COVER 6a
2 a : to lift up **:** RAISE **b** (1) **:** to put or have (as artillery) in position (2) **:** to have as equipment **c** (1) **:** to organize and equip (an attacking force) ⟨*mount* an army⟩ (2) **:** to launch and carry out (as an assault or a campaign)
3 : to set on something that elevates
4 a : to cause to get on a means of conveyance **b :** to furnish with animals for riding
5 : to post or set up for defense or observation ⟨*mounted* some guards⟩
6 a : to attach to a support **b :** to arrange or assemble for use or display
7 a : to prepare (as a specimen) for examination or display **b :** to prepare and supply with materials needed for performance or execution ⟨*mount* an opera⟩
— mount·able \'maun-tə-bəl\ *adjective*
— mount·er *noun*

[3]**mount** *noun* (15th century)
1 : an act or instance of mounting; *specifically* **:** an opportunity to ride a horse in a race
2 : FRAME, SUPPORT: as **a :** the material (as cardboard) on which a picture is mounted **b :** a jewelry setting **c** (1) **:** an undercarriage or part on which a device (as a motor or an artillery piece) rests in service (2) **:** an attachment for an accessory **d :** a hinge, card, or acetate envelope for mounting a stamp **e :** a glass slide with its accessories on which objects are placed for examination with a microscope
3 : a means of conveyance; *especially* **:** SADDLE HORSE

moun·tain \'maun-tᵊn\ *noun, often attributive* [Middle English, from Old French *montaigne,* from (assumed) Vulgar Latin *montanea,* from feminine of *montaneus* of a mountain, alteration of Latin *montanus,* from *mont-, mons*] (13th century)
1 a : a landmass that projects conspicuously above its surroundings and is higher than a hill **b :** an elongated ridge
2 a : a great mass **b :** a vast number or quantity

mountain ash *noun* (1597)
: any of various trees or shrubs (genus *Sorbus*) of the rose family with pinnate leaves and red or orange-red fruits

mountain bike *noun* (1984)
: an all-terrain bicycle with wide knobby tires, straight handlebars, and typically 18 or 21 gears
— mountain biker *noun*
— mountain biking *noun*

mountain bluebird *noun* (1860)
: a bluebird (*Sialia currucoides*) of western North America having a blue-breasted rather than red-breasted male

mountain cranberry *noun* (1848)
: a low evergreen shrub (*Vaccinium vitis-idaea minus*) of north temperate uplands with red edible berries — called also *lingonberry*

mountain dew *noun* (1827)
: MOONSHINE 3

moun·tain·eer \,maun-tᵊn-'ir\ *noun* (1610)
1 : a native or inhabitant of a mountainous region
2 : a person who climbs mountains for sport

moun·tain·eer·ing *noun* (1803)
: the sport or technique of scaling mountains

mountain goat *noun* (1833)
: a ruminant mammal (*Oreamnos americanus*) of mountainous northwestern North America that has a thick white coat and slightly curved horns and resembles a goat

mountain goat

mountain gorilla *noun* (1939)
: a gorilla (*Gorilla gorilla beringei*) inhabiting the Virunga mountain range

mountain laurel *noun* (1759)
: a North American evergreen shrub or small tree (*Kalmia latifolia*) of the heath family with glossy leaves and umbels of rose-colored or white flowers

mountain lion *noun* (1859)
: COUGAR

mountain mahogany *noun* (1810)
: any of a genus (*Cercocarpus*) of western North American evergreen shrubs or small trees of the rose family

mountain man *noun* (1839)
: an American frontiersman (as a trapper) at home in the wilderness

moun·tain·ous \'maun-tᵊn-əs, 'maunt-nəs\ *adjective* (14th century)
1 : containing many mountains
2 : resembling a mountain **:** HUGE
— moun·tain·ous·ly *adverb*
— moun·tain·ous·ness *noun*

mountain sheep *noun* (1802)
: any of various wild sheep (as bighorn, argali, or Dall sheep) inhabiting high mountains

mountain sickness *noun* (1848)
: altitude sickness experienced especially above 10,000 feet (3000 meters) and caused by insufficient oxygen in the air

moun·tain·side \'maun-tᵊn-,sīd\ *noun* (14th century)
: the side of a mountain

mountain time *noun, often M capitalized* (1883)
: the time of the 7th time zone west of Greenwich that includes the Rocky mountain states of the U.S. — see TIME ZONE illustration

moun·tain·top \'maun-tᵊn-,täp\ *noun* (1593)
: the summit of a mountain

moun·tainy \'maun-tᵊn-ē, 'maunt-nē\ *adjective* (1613)
1 : MOUNTAINOUS
2 : of, relating to, or living in mountains

[1]**moun·te·bank** \'maun-ti-,baŋk\ *noun* [Italian *montimbanco,* from *montare* to mount (from—assumed—Vulgar Latin) + *in* in, on (from Latin) + *banco, banca* bench — more at BANK] (1577)
1 : a person who sells quack medicines from a platform
2 : a boastful unscrupulous pretender **:** CHARLATAN
— moun·te·bank·ery \-,baŋ-k(ə-)rē\ *noun*

[2]**mountebank** (1607)
transitive verb
obsolete **:** to beguile or transform by trickery ⟨I'll *mountebank* their loves —Shakespeare⟩
intransitive verb
: to play the mountebank

Mount·ie \'maun-tē\ *noun* [*mount*ed policeman] (1914)
: a member of the Royal Canadian Mounted Police

mount·ing \'maun-tiŋ\ *noun* (1563)
: [3]MOUNT 2

mourn \'mōrn, 'mȯrn\ *verb* [Middle English, from Old English *murnan;* akin to Old High German *mornēn* to mourn, Greek *mermēra* care — more at MEMORY] (before 12th century)
intransitive verb
1 : to feel or express grief or sorrow
2 : to show the customary signs of grief for a death; *especially* **:** to wear mourning
3 : to murmur mournfully — used especially of doves
transitive verb
1 : to feel or express grief or sorrow for
2 : to utter mournfully
— mourn·er *noun*
— mourn·ing·ly \'mōr-niŋ-lē, 'mȯr-\ *adverb*

mourn·ful \'mōrn-fəl, 'mȯrn-\ *adjective* (15th century)
1 : expressing sorrow **:** SORROWFUL
2 : full of sorrow **:** SAD
3 : causing sorrow or melancholy **:** GLOOMY
— mourn·ful·ly \-fə-lē\ *adverb*
— mourn·ful·ness *noun*

mourn·ing \'mōr-niŋ, 'mȯr-\ *noun* (13th century)
1 : the act of sorrowing
2 a : an outward sign (as black clothes or an armband) of grief for a person's death **b :** a period of time during which signs of grief are shown

mourning cloak *noun* (1898)
: a blackish brown nymphalid butterfly (*Nymphalis antiopa*) with a broad yellow border on the wings found in temperate parts of Europe, Asia, and North America

mourning dove *noun* (1833)
: a wild American dove (*Zenaida macroura*) with a pointed tail and a plaintive coo

[1]**mouse** \'maus\ *noun, plural* **mice** \'mīs\ [Middle English, from Old English *mūs;* akin to Old High German *mūs* mouse, Latin *mus,* Greek *mys* mouse, muscle] (before 12th century)
1 : any of numerous small rodents (as of the genus *Mus*) with pointed snout, rather small ears, elongated body, and slender tail
2 : a timid person

\ə\ **abut** \ᵊ\ **kitten** \ər\ **further** \a\ **ash** \ā\ **ace** \ä\ **mop, mar** \au\ **out** \ch\ **chin** \e\ **bet** \ē\ **easy** \g\ **go** \i\ **hit** \ī\ **ice** \j\ **job** \ŋ\ **sing** \ō\ **go** \ȯ\ **law** \ȯi\ **boy** \th\ **thin** \th\ **the** \ü\ **loot** \u\ **foot** \y\ **yet** \zh\ **vision** *see also* Guide to Pronunciation

3 : a dark-colored swelling caused by a blow; *specifically* **:** BLACK EYE
4 : a small mobile manual device that controls movement of the cursor and selection of functions on a computer display

²mouse \'mau̇z\ *verb* **moused; mous·ing** (13th century)
intransitive verb
1 : to hunt for mice
2 : to search or move stealthily or slowly
transitive verb
1 *obsolete* **a :** BITE, GNAW **b :** to toy with roughly
2 : to search for carefully — usually used with *out*

mouse–ear \'mau̇s-ˌir\ *noun* (13th century)
1 : a Eurasian hawkweed (*Hieracium pilosella*) that has soft hairy leaves and has been introduced into North America
2 : any of several plants other than mouse-ear that have soft hairy leaves

mouse–ear chickweed *noun* (1731)
: any of several hairy chickweeds (genus *Cerastium* and especially *C. vulgatum*)

mous·er \'mau̇-sər, *US also & chiefly British* -zər\ *noun* (15th century)
: a catcher of mice and rats; *especially* **:** a cat proficient at mousing

¹mouse·trap \'mau̇s-ˌtrap\ *noun* (15th century)
1 : a trap for mice
2 : a stratagem that lures one to defeat or destruction
3 : TRAP 2b

²mousetrap *transitive verb* (circa 1864)
: to snare in or as if in a mousetrap

Mous·que·taire \ˌmü-skə-'tar, -'ter\ *noun* [French — more at MUSKETEER] (1705)
: a French musketeer; *especially* **:** one of the royal musketeers of the 17th and 18th centuries conspicuous for their daring and their dandified dress

mous·sa·ka \ˌmü-sə-'kä\ *noun* [New Greek *mousakas,* from Turkish *musakka*] (1931)
: a Middle Eastern dish of ground meat (as lamb or beef) and sliced eggplant often topped with a seasoned sauce

¹mousse \'müs\ *noun* [French, literally, froth, moss, of Germanic origin; akin to Old High German *mos* moss — more at MOSS] (1892)
1 : a light spongy food usually containing cream or gelatin
2 : a molded chilled dessert made with sweetened and flavored whipped cream or egg whites and gelatin ⟨chocolate *mousse*⟩
3 : a foamy preparation used in styling hair

²mousse *transitive verb* **moussed; mouss·ing** (1984)
: to style (hair) with mousse

mous·se·line \ˌmü-sə-'lēn, ˌmüs-'lēn\ *noun* [French, literally, muslin — more at MUSLIN] (1696)
1 : a fine sheer fabric (as of rayon) that resembles muslin
2 a : a sauce (as hollandaise) to which whipped cream or beaten egg whites have been added **b :** MOUSSE 1 ⟨salmon *mousseline*⟩

mousseline de soie \-də-'swä\ *noun, plural* **mousselines de soie** *same*\ [French, literally, silk muslin] (1850)
: a silk muslin having a crisp finish

mous·tache *variant of* MUSTACHE

mous·ta·chio *variant of* MUSTACHIO

Mous·te·ri·an \mü-'stir-ē-ən\ *adjective* [French *moustérien,* from Le *Moustier,* cave in Dordogne, France] (1890)
: of or relating to a Middle Paleolithic culture that is characterized by well-made flake tools often considered the work of Neanderthal man

mousy *or* **mous·ey** \'mau̇-sē, -zē\ *adjective* **mous·i·er; -est** (1853)
: of, relating to, or resembling a mouse: as **a :** QUIET, STEALTHY **b :** TIMID, RETIRING **c :** grayish brown
— mous·i·ly \-sə-lē, -zə-\ *adverb*
— mous·i·ness \-sē-nəs, zē \ *noun*

¹mouth \'mau̇th\ *noun, plural* **mouths** \'mau̇t͟hz *also* 'mau̇z, 'mau̇ths; *in synecdochic compounds like "blabbermouths"* ths *more frequently*\ *often attributive* [Middle English, from Old English *mūth;* akin to Old High German *mund* mouth and perhaps to Latin *mentum* chin] (before 12th century)
1 a : the natural opening through which food passes into the body of an animal and which in vertebrates is typically bounded externally by the lips and internally by the pharynx and encloses the tongue, gums, and teeth **b :** GRIMACE ⟨made a *mouth*⟩ **c :** an individual requiring food ⟨had too many *mouths* to feed⟩
2 a : VOICE, SPEECH ⟨finally gave *mouth* to her feelings⟩ **b :** MOUTHPIECE 3a **c** (1) **:** a tendency to excessive talk (2) **:** saucy or disrespectful language **:** IMPUDENCE
3 : something that resembles a mouth especially in affording entrance or exit: as **a :** the place where a stream enters a larger body of water **b :** the surface opening of an underground cavity **c :** the opening of a container **d :** an opening in the side of an organ flue pipe
— mouth·like \'mau̇th-ˌlīk\ *adjective*
— down in the mouth : DEJECTED, SULKY

²mouth \'mau̇t͟h *also* 'mau̇th\ (14th century)
transitive verb
1 a : SPEAK, PRONOUNCE **b :** to utter bombastically **:** DECLAIM **c :** to repeat without comprehension or sincerity ⟨always *mouthing* platitudes⟩ **d :** to form soundlessly with the lips ⟨the librarian *mouthed* the word "quiet"⟩ **e :** to utter indistinctly **:** MUMBLE ⟨*mouthed* his words⟩
2 : to take into the mouth; *especially* **:** EAT
intransitive verb
1 a : to talk pompously **:** RANT — often used with *off* **b :** to talk insolently or impudently — usually used with *off*
2 : to move the mouth especially so as to make faces
— mouth·er *noun*

mouth·breed·er \'mau̇th-ˌbrē-dər\ *noun* (1927)
: any of several fishes that carry their eggs and young in the mouth; *especially* **:** a North African cichlid fish (*Haplochromis multicolor*) often kept in aquariums

mouthed \'mau̇t͟hd, 'mau̇tht\ *adjective* (14th century)
: having a mouth especially of a specified kind — often used in combination ⟨a soft-*mouthed* fish⟩

mouth·ful \'mau̇th-ˌfu̇l\ *noun* (15th century)
1 a : as much as a mouth will hold **b :** the quantity usually taken into the mouth at one time
2 : a small quantity
3 a : a very long word or phrase **b :** a comment or a statement rich in meaning or substance

mouth hook *noun* (1937)
: one of a pair of hooked larval mouthparts of some dipteran flies that function as jaws

mouth organ *noun* (1866)
: HARMONICA 2

mouth·part \'mau̇th-ˌpärt\ *noun* (1799)
: a structure or appendage near the mouth (as of an insect) especially when adapted for use in gathering or eating food

mouth·piece \-ˌpēs\ *noun* (1683)
1 : something placed at or forming a mouth
2 : a part (as of an instrument) that goes in the mouth or to which the mouth is applied
3 a : one that expresses or interprets another's views **:** SPOKESMAN **b** *slang* **:** a criminal lawyer

mouth–to–mouth *adjective* (1941)
: of, relating to, or being a method of artificial respiration in which the rescuer's mouth is placed tightly over the victim's mouth in order to force air into the victim's lungs by blowing forcefully enough every few seconds to inflate them

mouth·wash \'mau̇th-ˌwȯsh, -ˌwäsh\ *noun* (1840)
: a usually antiseptic liquid preparation for cleaning the mouth and teeth or freshening the breath

mouth·wa·ter·ing \-ˌwȯ-tə-riŋ, -ˌwä-\ *adjective* (1900)
: arousing the appetite **:** tantalizingly delicious or appealing ⟨a *mouthwatering* aroma⟩
— mouth·wa·ter·ing·ly \-lē\ *adverb*

mouthy \'mau̇-thē, -t͟hē\ *adjective* **mouth·i·er; -est** (1589)
1 : excessively talkative **:** GARRULOUS
2 : marked by bombast or back talk

mou·ton \'mü-ˌtän, mü-'\ *noun* [French, sheep, sheepskin, from Middle French, ram — more at MUTTON] (1944)
: processed sheepskin that has been sheared and dyed to resemble beaver or seal

¹mov·able *or* **move·able** \'mü-və-bəl\ *adjective* (14th century)
1 : capable of being moved
2 : changing date from year to year ⟨*movable* holidays⟩
— mov·abil·i·ty \ˌmü-və-'bi-lə-tē\ *noun*
— mov·able·ness \'mü-və-bəl-nəs\ *noun*
— mov·ably \-blē\ *adverb*

²movable *or* **moveable** *noun* (15th century)
: something (as an article of furniture) that can be removed or displaced

¹move \'müv\ *verb* **moved; mov·ing** [Middle English, from Middle French *movoir,* from Latin *movēre;* probably akin to Sanskrit *mīvati* he moves, pushes] (13th century)
intransitive verb
1 a (1) **:** to go or pass to another place or in a certain direction with a continuous motion ⟨*moved* into the shade⟩ (2) **:** to proceed toward a certain state or condition ⟨*moving* up the executive ladder⟩ ⟨*moved* into second place in the tournament⟩ (3) **:** to become transferred during play ⟨checkers *move* along diagonally adjacent squares⟩ (4) **:** to keep pace ⟨*moving* with the times⟩ **b :** to start away from some point or place **:** DEPART **c :** to change one's residence or location
2 : to carry on one's life or activities in a specified environment ⟨*moves* in the best circles⟩
3 : to change position or posture **:** STIR ⟨told him to be quiet and not to *move*⟩
4 : to take action **:** ACT
5 a : to begin operating or functioning or working in a usual way **b :** to show marked activity ⟨after a brief lull things really began to *move*⟩ **c :** to move a piece (as in chess or checkers) during one's turn
6 : to make a formal request, application, or appeal
7 : to change hands by being sold or rented ⟨goods that were *moving* slowly⟩
8 *of the bowels* **:** EVACUATE
transitive verb
1 a (1) **:** to change the place or position of (2) **:** to dislodge or displace from a fixed position **:** BUDGE **b :** to transfer (as a piece in chess) from one position to another
2 a (1) **:** to cause to go or pass from one place to another with a continuous motion ⟨*moved* the flag slowly up and down⟩ (2) **:** to cause to advance **b :** to cause to operate or function **:** ACTUATE ⟨this button *moves* the whole machine⟩ **c :** to put into activity or rouse up from inactivity
3 : to cause to change position or posture
4 : to prompt or rouse to the doing of something **:** PERSUADE ⟨the report *moved* the faculty to take action⟩
5 a : to stir the emotions, feelings, or passions of ⟨was deeply *moved* by such kindness⟩ **b :** to affect in such a way as to lead to an indicated show of emotion ⟨the story *moved* her to tears⟩
6 a *obsolete* **:** BEG **b :** to make a formal application to
7 : to propose formally in a deliberative assembly ⟨*moved* adjournment⟩
8 : to cause (the bowels) to void

9 : to cause to change hands through sale or rent ☆

²move *noun* (1656)
1 a : the act of moving a piece (as in chess) **b :** the turn of a player to move
2 a : a step taken so as to gain an objective **:** MANEUVER ⟨a *move* to end the dispute⟩ **b :** the action of moving from a motionless position **c :** a change of residence or location **d :** an agile or deceptive action especially in sports
— on the move 1 : in a state of moving about from place to place ⟨a salesman is constantly *on the move*⟩ **2 :** in a state of moving ahead or making progress ⟨said that civilization is always *on the move*⟩

move in *intransitive verb* (1898)
: to occupy a dwelling or place of work
— move in on : to make advances or aggressive movements toward

move·less \'müv-ləs\ *adjective* (1578)
: being without movement **:** FIXED, IMMOBILE
— move·less·ly *adverb*
— move·less·ness *noun*

move·ment \'müv-mənt\ *noun* (14th century)
1 a (1) **:** the act or process of moving; *especially* **:** change of place or position or posture (2) **:** a particular instance or manner of moving **b** (1) **:** a tactical or strategic shifting of a military unit **:** MANEUVER (2) **:** the advance of a military unit **c :** ACTION, ACTIVITY — usually used in plural
2 a : TENDENCY, TREND ⟨detected a *movement* toward fairer pricing⟩ **b :** a series of organized activities working toward an objective; *also* **:** an organized effort to promote or attain an end ⟨the civil rights *movement*⟩
3 : the moving parts of a mechanism that transmit a definite motion
4 a : MOTION 7 **b :** the rhythmic character or quality of a musical composition **c :** a distinct structural unit or division having its own key, rhythmic structure, and themes and forming part of an extended musical composition **d :** particular rhythmic flow of language **:** CADENCE
5 a : the quality (as in a painting or sculpture) of representing or suggesting motion **b :** the vibrant quality in literature that comes from elements that constantly hold a reader's interest (as a quickly moving action-filled plot)
6 a : an act of voiding the bowels **b :** matter expelled from the bowels at one passage

mov·er \'mü-vər\ *noun* (14th century)
: one that moves or sets something in motion; *especially* **:** one whose business or occupation is the moving of household goods from one residence to another

mover and shaker *noun, plural* **movers and shakers** (1951)
: a person who is active or influential in some field of endeavor

mov·ie \'mü-vē\ *noun* [*moving picture*] (1912)
1 : MOTION PICTURE
2 *plural* **:** a showing of a motion picture
3 *plural* **:** the motion-picture medium or industry

mov·ie·dom \'mü-vē-dəm\ *noun* (1916)
: FILMDOM

mov·ie·go·er \-,gō(-ə)r\ *noun* (1923)
: FILMGOER
— mov·ie·going \-,gō-iŋ, -,gȯ(-)iŋ\ *noun*

mov·ie·mak·er \-,mā-kər\ *noun* (1915)
: one who makes movies
— mov·ie·mak·ing \-,mā-kiŋ\ *noun*

mov·ing *adjective* (14th century)
1 a : marked by or capable of movement **b :** of or relating to a change of residence ⟨*moving* expenses⟩ **c :** used for transferring furnishings from one residence to another ⟨a *moving* van⟩ **d :** involving a motor vehicle that is in motion ⟨a *moving* violation⟩
2 a : producing or transferring motion or action **b :** stirring deeply in a way that evokes a strong emotional response ⟨a *moving* story of a faithful dog⟩ ☆
— mov·ing·ly \'mü-viŋ-lē\ *adverb*

moving picture *noun* (1896)
: MOTION PICTURE

Mov·i·ola \,mü-vē-'ō-lə\ *trademark*
— used for a device for editing motion-picture film and synchronizing the sound

¹mow \'maů\ *noun* [Middle English, heap, stack, from Old English *mūga;* akin to Old Norse *mūgi* heap] (before 12th century)
1 : a piled-up stack (as of hay or fodder); *also* **:** a pile of hay or grain in a barn
2 : the part of a barn where hay or straw is stored

²mow \'mō\ *verb* **mowed; mowed** *or* **mown** \'mōn\; **mow·ing** [Middle English, from Old English *māwan;* akin to Old High German *māen* to mow, Latin *metere* to reap, mow, Greek *aman*] (before 12th century)
transitive verb
1 a : to cut down with a scythe or sickle or machine **b :** to cut the standing herbage (as grass) of
2 a (1) **:** to kill or destroy in great numbers or mercilessly ⟨machine guns *mowed* down the enemy⟩ (2) **:** to cause to fall **:** KNOCK DOWN **b :** to overcome swiftly and decisively **:** ROUT ⟨*mowed* down the opposing team⟩
intransitive verb
: to cut down standing herbage (as grass)
— mow·er \'mō(-ə)r\ *noun*

³mow \'maů, 'mō\ *noun* [Middle English *mowe,* from Middle French *moue,* of Germanic origin; akin to Middle Dutch *mouwe* protruding lip] (14th century)
: GRIMACE

⁴mow \'maů, 'mō\ *intransitive verb* (15th century)
: to make grimaces

mox·ie \'mäk-sē\ *noun* [from *Moxie,* a trademark for a soft drink] (1930)
1 : ENERGY, PEP
2 : COURAGE, DETERMINATION
3 : KNOW-HOW, EXPERTISE

moyen–âge \mwȧ-ye-näzh\ *adjective* [French *moyen âge* Middle Ages] (1849)
: of or relating to medieval times

moz·za·rel·la \,mät-sə-'re-lə\ *noun* [Italian, diminutive of *mozza,* a kind of cheese, from *mozzare* to cut off, from *mozzo* cut off, docked, from (assumed) Vulgar Latin *mutius,* alteration of Latin *mutilus*] (1911)
: a moist white unsalted unripened cheese of mild flavor and a smooth rubbery texture

moz·zet·ta \mōt-'se-tə\ *noun* [Italian, probably from *mozzo* cut off] (1774)
: a short cape with a small ornamental hood worn over the rochet by Roman Catholic prelates

M phase *noun* [*m*itosis] (1945)
: the period in the cell cycle during which cell division takes place — compare G_1 PHASE, G_2 PHASE, S PHASE

Mr. \'mis-tər, *in rapid speech especially in sense 2* (,)mis(t)\ *noun, plural* **Messrs.** \'me-sərz\ [*Mr.* from Middle English, abbreviation of *maister* master; *Messrs.* abbreviation of *Messieurs,* from French, plural of *Monsieur*] (15th century)
1 — used as a conventional title of courtesy except when usage requires the substitution of a title of rank or an honorific or professional title before a man's surname ⟨spoke to *Mr.* Doe⟩
2 — used in direct address as a conventional title of respect before a man's title of office ⟨may I ask one more question, *Mr.* President⟩
3 — used before the name of a place (as a country or city) or of a profession or activity (as a sport) or before some epithet (as *clever*) to form a title applied to a male viewed or recognized as representative of the thing indicated ⟨*Mr.* Baseball⟩

Mr. Charlie \-'chär-lē\ *noun* [*Charlie,* from *Charles,* proper name] (circa 1941)
: a white man **:** white people — usually used disparagingly

mri·dan·ga \mri-'däŋ-gə, ,mər-i-\ *or* **mri·dan·gam** \-gəm\ *noun* [Sanskrit *mṛdaṅga*] (1921)
: a drum of India that is shaped like an elongated barrel and has tuned heads of different diameters

Mrs. \'mi-səz, -səs, *especially Southern* 'mi-zəz, -zəs, *or in rapid speech in sense 1* (,)miz, *or before given names* (,)mis\ *noun, plural* **Mes·dames** \mā-'däm, -'dam\ [*Mrs.* abbreviation of *mistress; Mesdames* from French, plural of *Madame*] (1615)
1 a — used as a conventional title of courtesy except when usage requires the substitution of a title of rank or an honorific or professional title before a married woman's surname ⟨spoke to *Mrs.* Doe⟩ **b** — used before the name of a place (as a country or city) or of a profession or activity (as a sport) or before some epithet (as *clever*) to form a title applied to a married woman viewed or recognized as representative of the thing indicated ⟨*Mrs.* Golf⟩
2 : WIFE ⟨took the *Mrs.* to dinner⟩

Mrs. Grun·dy \-'grən-dē\ *noun* [from a character alluded to in Thomas Morton's *Speed the Plough* (1798)] (1813)
: one marked by prudish conventionality in personal conduct

Ms. \'miz\ *noun, plural* **Mss.** *or* **Mses.** \'mi-zəz\ [probably blend of *Miss* and *Mrs.*] (1949)
— used instead of *Miss* or *Mrs.* (as when the marital status of a woman is unknown or irrelevant) ⟨*Ms.* Mary Smith⟩

M16 \,em-(,)sik-'stēn\ *noun* [*m*odel *16*] (1968)
: a .223 caliber (5.56 mm.) gas-operated magazine-fed rifle for semiautomatic or automatic operation used by U.S. troops since the mid 1960s — called also *M16 rifle*

mu \'myü, 'mü\ *noun* [Greek *my*] (1823)
: the 12th letter of the Greek alphabet — see ALPHABET table

☆ SYNONYMS

Move, actuate, drive, impel mean to set or keep in motion. MOVE is very general and implies no more than the fact of changing position ⟨*moved* the furniture⟩. ACTUATE stresses transmission of power so as to work or set in motion ⟨turbines *actuated* by waterpower⟩. DRIVE implies imparting forward and continuous motion and often stresses the effect rather than the impetus ⟨a ship *driven* aground by hurricane winds⟩. IMPEL is usually figurative and suggests a great motivating impetus ⟨a candidate *impelled* by ambition⟩.

Moving, impressive, poignant, affecting, touching, pathetic mean having the power to produce deep emotion. MOVING may apply to any strong emotional effect including thrilling, agitating, saddening, or calling forth pity or sympathy ⟨a *moving* appeal for contributions⟩. IMPRESSIVE implies compelling attention, admiration, wonder, or conviction ⟨an *impressive* list of achievements⟩. POIGNANT applies to what keenly or sharply affects one's sensitivities ⟨a *poignant* documentary on the homeless⟩. AFFECTING is close to MOVING but most often suggests pathos ⟨an *affecting* deathbed reunion⟩. TOUCHING implies arousing tenderness or compassion ⟨the *touching* innocence in a child's eyes⟩. PATHETIC implies moving to pity or sometimes contempt ⟨*pathetic* attempts to justify misconduct⟩.

muc- *or* **muci-** *or* **muco-** *combining form* [Latin *muc-,* from *mucus*]
1 : mucus ⟨*muco*protein⟩
2 : mucous and ⟨*muco*cutaneous⟩

¹much \'məch\ *adjective* **more** \'mōr, 'mȯr\; **most** \'mōst\ [Middle English *muche* large, much, from *michel, muchel,* from Old English *micel, mycel;* akin to Old High German *mihhil* great, large, Latin *magnus,* Greek *megas,* Sanskrit *mahat*] (13th century)
1 : great in quantity, amount, extent, or degree ⟨there is *much* truth in what you say⟩ ⟨taken too *much* time⟩
2 *obsolete* **:** many in number
3 : more than is expected or acceptable **:** more than enough ⟨eighty bedrooms, a bit *much* for a family of seven —Peter Dragadze⟩
— too much 1 : WONDERFUL, EXCITING **2 :** TERRIBLE, AWFUL

²much *adverb* **more; most** (13th century)
1 a (1) **:** to a great degree or extent **:** CONSIDERABLY ⟨*much* happier⟩ (2) **:** VERY **b** (1) **:** FREQUENTLY, OFTEN (2) **:** by or for a long time ⟨didn't get to work *much* before noon⟩ **c :** by far ⟨was *much* the brightest student⟩
2 : NEARLY, APPROXIMATELY ⟨looks *much* the way his father did⟩
— as much : the same in quantity

³much *noun* (13th century)
1 : a great quantity, amount, extent, or degree ⟨gave away *much*⟩
2 : something considerable or impressive ⟨was not *much* to look at⟩

mu·cha·cho \mü-'chä-(ˌ)chō\ *noun, plural* **-chos** [Spanish, probably from *mocho* cropped, shorn] (1591)
1 *chiefly Southwest* **:** a male servant
2 *chiefly Southwest* **:** a young man

much as *conjunction* (circa 1699)
: however much **:** even though

much less *conjunction* (1710)
: to say nothing of — used especially in negative contexts to add to one item another denoting something less likely ⟨can hardly grow, *much less* ripen, till the stock is in the earth —Jonathan Swift⟩ ⟨he is never going to get out of the eighth grade, *much less* ever make it to college —Thomas Meehan⟩

much·ness \'məch-nəs\ *noun* (14th century)
: the quality or state of being great **:** GREATNESS
— much of a muchness : very much the same

mu·ci·lage \'myü-s(ə-)lij\ *noun* [Middle English *muscilage,* from Late Latin *mucilago* mucus, musty juice, from Latin *mucus*] (15th century)
1 : a gelatinous substance of various plants (as legumes or seaweeds) that contains protein and polysaccharides and is similar to plant gums
2 : an aqueous usually viscid solution (as of a gum) used especially as an adhesive

mu·ci·lag·i·nous \ˌmyü-sə-'la-jə-nəs\ *adjective* [Middle English *muscilaginous,* from Late Latin *mucilaginosus,* from *mucilagin-, mucilago*] (15th century)
1 : STICKY, VISCID
2 : of, relating to, full of, or secreting mucilage
— mu·ci·lag·i·nous·ly *adverb*

mu·cin \'myü-sᵊn\ *noun* [International Scientific Vocabulary *muc-*] (1846)
: any of various mucoproteins that occur especially in secretions of mucous membranes
— mu·cin·ous \-sᵊn-əs, 'myüs-nəs\ *adjective*

¹muck \'mək\ *noun* [Middle English *muk,* perhaps from Old English *-moc;* akin to Old Norse *myki* dung] (13th century)
1 : soft moist farmyard manure
2 : slimy dirt or filth
3 : defamatory remarks or writings
4 a (1) **:** dark highly organic soil (2) **:** MIRE, MUD **b :** something resembling muck **:** GUNK
5 : material removed in the process of excavating or mining
— mucky \'mə-kē\ *adjective*

²muck (14th century)
transitive verb
1 a : to clean up; *especially* **:** to clear of manure or filth — usually used with *out* **b :** to clear of muck
2 : to dress (as soil) with muck
3 : to dirty with or as if with muck **:** SOIL
intransitive verb
1 : to move or load muck (as in a mine)
2 : to engage in aimless activity — usually used with *about* or *around*
— muck·er *noun*

muck–a–muck \'mə-kə-ˌmək\ *or* **muck·ety–muck** \'mə-kə-tē-\ *noun* (1912)
: HIGH-MUCK-A-MUCK

muck·luck *variant of* MUKLUK

muck·rake \'mək-ˌrāk\ *intransitive verb* [obsolete *muckrake,* noun, rake for dung] (1910)
: to search out and publicly expose real or apparent misconduct of a prominent individual or business ◆
— muck·rak·er *noun*

muck up *transitive verb* (1896)
: to make a mess of **:** BUNGLE, SPOIL

mu·co·cu·ta·ne·ous \ˌmyü-kō-kyu̇-'tā-nē-əs\ *adjective* (1898)
: made up of or involving both typical skin and mucous membrane

¹mu·coid \'myü-ˌkȯid\ *adjective* [International Scientific Vocabulary *muc-*] (1849)
: resembling mucus

²mucoid *noun* [International Scientific Vocabulary] (1900)
: MUCOPROTEIN

mu·co·lyt·ic \ˌmyü-kə-'li-tik\ *adjective* (circa 1923)
: hydrolyzing mucopolysaccharides **:** tending to break down or lower the viscosity of mucin-containing body secretions or components ⟨*mucolytic* enzymes⟩

mu·co·pep·tide \ˌmyü-kō-'pep-ˌtīd\ *noun* (1959)
: PEPTIDOGLYCAN

mu·co·poly·sac·cha·ride \'myü-kō-ˌpä-li-'sa-kə-ˌrīd\ *noun* [International Scientific Vocabulary] (1938)
: GLYCOSAMINOGLYCAN

mu·co·pro·tein \ˌmyü-kə-'prō-ˌtēn *also* -'prō-tē-ən\ *noun* (1925)
: any of various complex conjugated proteins (as mucins) that contain polysaccharides and occur in body fluids and tissues

mu·co·sa \myü-'kō-zə\ *noun, plural* **-sae** \-(ˌ)zē, -ˌzī\ *or* **-sas** [New Latin, from Latin, feminine of *mucosus* mucous] (1880)
: MUCOUS MEMBRANE
— mu·co·sal \-zəl\ *adjective*

mu·cous \'myü-kəs\ *adjective* [Latin *mucosus,* from *mucus*] (1646)
1 : of, relating to, or resembling mucus
2 : secreting or containing mucus
3 : covered with or as if with mucus **:** SLIMY

mucous membrane *noun* (1812)
: a membrane rich in mucous glands; *specifically* **:** one that lines body passages and cavities which communicate directly or indirectly with the exterior

mu·cro \'myü-ˌkrō\ *noun, plural* **mu·cro·nes** \myü-'krō-(ˌ)nēz\ [New Latin *mucron-, mucro,* from Latin, point, edge] (1646)
: an abrupt sharp terminal point or tip or process (as of a leaf)
— mu·cro·nate \'myü-krə-ˌnāt\ *adjective*

mu·cus \'myü-kəs\ *noun* [Latin, nasal mucus; akin to Greek *myxa* mucus] (1661)
: a viscid slippery secretion that is usually rich in mucins and is produced by mucous membranes which it moistens and protects

¹mud \'məd\ *noun* [Middle English *mudde,* probably from Middle Low German] (14th century)
1 : a slimy sticky mixture of solid material with a liquid and especially water; *especially* **:** soft wet earth
2 : abusive and malicious remarks or charges ⟨political campaigners slinging *mud* at each other⟩
3 : ANATHEMA 1b — usually used in the phrase *one's name is mud*

²mud *transitive verb* **mud·ded; mud·ding** (1593)
1 : to make muddy or turbid
2 : to treat or plaster with mud

mud dauber *noun* (1856)
: any of various wasps (especially family Sphecidae) that construct mud cells in which the female places an egg with spiders or insects paralyzed by a sting to serve as food for the larva

¹mud·dle \'mə-dᵊl\ *verb* **mud·dled; mud·dling** \'məd-liŋ, 'mə-dᵊl-iŋ\ [probably from obsolete Dutch *moddelen,* from Middle Dutch, from *modde* mud; akin to Middle Low German *mudde*] (1676)
transitive verb
1 : to make turbid or muddy
2 : to befog or stupefy especially with liquor
3 : to mix confusedly
4 : to make a mess of **:** BUNGLE
intransitive verb
: to think or act in a confused aimless way
— mud·dler \'məd-lər, 'mə-dᵊl-ər\ *noun*

²muddle *noun* (circa 1818)
1 : a state of especially mental confusion
2 : a confused mess
— mud·dly \'məd-lē, 'mə-dᵊl-ē\ *adjective*

mud·dle·head·ed \ˌmə-dᵊl-'he-dəd\ *adjective* (1759)
1 : mentally confused
2 : INEPT, BUNGLING
— mud·dle·head·ed·ly *adverb*
— mud·dle·head·ed·ness *noun*

muddle through *intransitive verb* (circa 1864)
: to achieve a degree of success without much planning or effort

¹mud·dy \'mə-dē\ *adjective* **mud·di·er; -est** (15th century)
1 : morally impure **:** BASE
2 a : full of or covered with mud **b :** characteristic or suggestive of mud ⟨a *muddy* flavor⟩ ⟨*muddy* colors⟩ **c :** turbid with sediment
3 a : lacking in clarity or brightness **:** CLOUDY, DULL ⟨a *muddy* recording⟩ ⟨eyes *muddy* with sleep⟩ **b :** obscure in meaning **:** MUDDLED, CONFUSED ⟨*muddy* thinking⟩ ⟨a *muddy* style⟩

◇ WORD HISTORY

muckrake John Bunyan's religious allegory *The Pilgrim's Progress* (1678) includes a character identified as "a man . . . with a Muckrake in his hand," who busies himself so much with raking the muck—that is, with attending to worldly things—that his gaze is always downward and he never sees a celestial crown held above him. The metaphor inspired some occasional uses of *muckrake* in the 19th century, most notably in political circles, but the word received its greatest boost when it was used by Theodore Roosevelt in an April 1906 speech criticizing the excesses of journalists who had achieved popularity by exposing the corruption of public figures and institutions. Roosevelt said, "The men with the muckrakes are often indispensable to the well-being of society; but only if they know when to stop raking the muck." The term *muckraker* quickly became established in popular use as a name for reform writers such as Lincoln Steffens, Ida M. Tarbell, Edwin Markham, and Upton Sinclair. Although originally meant to be pejorative, the word was adopted by the writers themselves, and it acquired connotations of courageous honesty and social conscience.

— **mud·di·ly** \'mə-dᵊl-ē\ *adverb*
— **mud·di·ness** \'mə-dē-nəs\ *noun*

[2]**muddy** *transitive verb* **mud·died; mud·dy·ing** (1601)
1 : to soil or stain with or as if with mud
2 : to make turbid
3 : to make cloudy or dull
4 : CONFUSE

Mu·de·jar \mü-'the-ˌhär, -ˌkär\ *noun, plural* **-ja·res** \-'the-hä-ˌrās, -kä-\ [Spanish *mudéjar,* from Arabic *mudajjan,* literally, allowed to remain] (1893)
: a Muslim living under a Christian king especially during the 8th to 11th centuries
— **Mudejar** *adjective*

mud·flat \'məd-ˌflat\ *noun* (1813)
: a level tract lying at little depth below the surface of water or alternately covered and left bare by the tide

mud·flow \-ˌflō\ *noun* (circa 1900)
: a moving mass of soil made fluid by rain or melting snow

mud·guard \-ˌgärd\ *noun* (1886)
1 a : FENDER d **b :** SPLASH GUARD
2 : a strip of material applied to a shoe upper just above the sole for protection against dampness or as an ornament

mud puppy *noun* (1882)
: a large American salamander (*Necturus maculosus*) that has external gills and is gray to rusty brown usually with bluish black spots

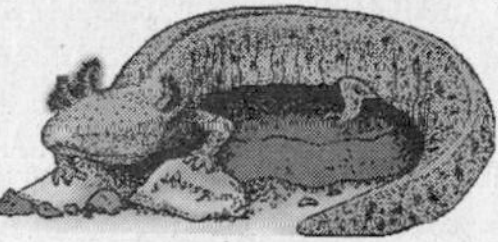
mud puppy

mu·dra \mə-'drä\ *noun* [Sanskrit *mudrā*] (1811)
: one of the symbolic hand gestures used in religious ceremonies and dances of India

mud·room \'məd-ˌrüm, -ˌrům\ *noun* (circa 1950)
: a room in a house designed especially for the shedding of dirty or wet footwear and clothing and located typically off the kitchen or in the basement

mud·sill \-ˌsil\ *noun* (1685)
1 : a supporting sill (as of a building or bridge) resting directly on a base and especially the earth
2 : a person of the lowest social level

mud·skip·per \-ˌski-pər\ *noun* (1860)
: any of several Asian and Polynesian gobies (genera *Periophthalmus* and *Boleophthalmus*) that are able to skip about actively over wet mud and sand

mud·sling·er \-ˌsliŋ-ər\ *noun* (circa 1890)
: one that uses offensive epithets and invective especially against a political opponent
— **mud·sling·ing** \-ˌsliŋ-iŋ\ *noun*

mud·stone \-ˌstōn\ *noun* (circa 1736)
: an indurated shale produced by the consolidation of mud

mud turtle *noun* (1785)
: any of a genus (*Kinosternon*) of bottom-dwelling freshwater turtles with two transverse hinges on the plastron

Muen·ster \'mən(t)-stər, 'mün(t)-, 'mǔn(t)-, 'min(t)-\ *noun* [*Münster, Munster,* France] (1902)
: a semisoft cheese that may be bland or sharp in flavor

mues·li \'myüs-lē, 'myüz-\ *noun* [German dialect (Swiss) *Müsli,* diminutive of German *Mus* soft food, mush, from Old High German *muos;* akin to Old English *mōs* food and probably to Old English *mete* food — more at MEAT] (1939)
: a breakfast cereal of Swiss origin consisting of rolled oats, nuts, and fruit

mu·ez·zin \mü-'e-zᵊn, myü-; 'mwe-zᵊn\ *noun* [Arabic *mu'adhdhin*] (1585)
: a Muslim crier who calls the hour of daily prayers

[1]**muff** \'məf\ *noun* [Dutch *mof,* from Middle French *moufle* mitten, from Medieval Latin *muffula*] (1599)
1 : a warm tubular covering for the hands
2 : a cluster of feathers on the side of the face of some domestic fowls

muff 1

[2]**muff** *verb* [probably from [1]*muff*] (1827)
transitive verb
1 : to handle awkwardly **:** BUNGLE
2 : to fail to hold (a ball) when attempting a catch
intransitive verb
1 : to act or do something stupidly or clumsily
2 : to muff a ball — compare FUMBLE

[3]**muff** *noun* (1871)
1 : a bungling performance
2 : a failure to hold a ball in attempting a catch

muf·fin \'mə-fən\ *noun* [probably from Low German *muffen,* plural of *muffe* cake] (1703)
: a quick bread made of batter containing egg and baked in a muffin pan

muf·fle \'mə-fəl\ *transitive verb* **muf·fled; muf·fling** \'mə-f(ə-)liŋ\ [Middle English *muflen*] (15th century)
1 : to wrap up so as to conceal or protect **:** ENVELOP
2 *obsolete* **:** BLINDFOLD
3 a : to wrap or pad with something to dull the sound ⟨*muffle* the oarlocks⟩ **b :** to deaden the sound of
4 : KEEP DOWN, SUPPRESS

muf·fler \'mə-flər\ *noun* (circa 1536)
1 a : a scarf worn around the neck **b :** something that hides or disguises
2 : a device to deaden noise; *especially* **:** one forming part of the exhaust system of an automotive vehicle
— **muf·flered** *adjective*

[1]**muf·ti** \'məf-tē, 'mǔf-\ *noun* [Arabic *muftī*] (1586)
: a professional jurist who interprets Muslim law

[2]**muf·ti** \'məf-tē\ *noun* [probably from [1]*mufti*] (1816)
: ordinary dress as distinguished from that denoting an occupation or station ⟨a priest in *mufti*⟩; *especially* **:** civilian clothes when worn by a person in the armed forces

[1]**mug** \'məg\ *noun* [origin unknown] (1664)
1 : a cylindrical drinking cup
2 a : the face or mouth of a person **b :** GRIMACE **c :** MUG SHOT
3 a *chiefly British* (1) **:** FOOL, BLOCKHEAD (2) **:** a person easily deceived **b :** PUNK, THUG
— **mug·ful** *noun*

[2]**mug** *verb* **mugged; mug·ging** (1855)
intransitive verb
: to pose or make faces especially to attract attention or for a camera
transitive verb
: PHOTOGRAPH

[3]**mug** *transitive verb* **mugged; mug·ging** [probably from earlier *mug* to strike in the face, perhaps from [1]*mug*] (circa 1864)
: to assault usually with intent to rob
— **mug·gee** \ˌmə-'gē\ *noun*

[1]**mug·ger** \'mə-gər\ *noun* [Hindi *magar,* from Sanskrit *makara* water monster] (1844)
: a common usually harmless freshwater crocodile (*Crocodylus palustris*) of southeastern Asia

[2]**mugger** *noun* [[3]*mug*] (1863)
: one who attacks with intent to rob

[3]**mugger** *noun* [[2]*mug*] (1892)
: one that grimaces especially before an audience

mug·gy \'mə-gē\ *adjective* **mug·gi·er; -est** [English dialect *mug* drizzle] (1746)
: being warm, damp, and close
— **mug·gi·ness** \'mə-gē-nəs\ *noun*

Mu·ghal *variant of* MOGUL

mu·gho pine \'mü-(ˌ)gō-, 'myü-\ *noun* [probably from French *mugho* mugho pine, from Italian *mugo*] (circa 1756)
: a shrubby spreading pine (*Pinus mugo mughus*) widely cultivated as an ornamental — called also *mugo pine*

mug's game \'məgz-\ *noun* (1910)
: a profitless or futile activity

mug shot *noun* (1950)
: a photograph of usually a person's head and especially face; *specifically* **:** a police photograph of a suspect's face or profile

mug up (circa 1860)
intransitive verb
British **:** to study intensively (as for an examination)
transitive verb
: to work up by study

mug·wump \'məg-ˌwəmp\ *noun* [obsolete slang *mugwump* kingpin, from Massachuset *mugquomp, muggumquomp* war leader] (1884)
1 : a bolter from the Republican party in 1884
2 : an independent in politics ◆

Mu·ham·mad·an \mō-'ha-mə-dən, -'hä- *also* mü-\ *adjective* (1681)
: of or relating to Muhammad or Islam
— **Muhammadan** *noun*
— **Mu·ham·mad·an·ism** \-də-ˌni-zəm\ *noun*

Muhammadan calendar *noun* (circa 1889)
: ISLAMIC CALENDAR

Muhammadan era *noun* (circa 1889)
: ISLAMIC ERA

Mu·har·ram \mü-'har əm\ *noun* [Arabic *Muḥarram*] (circa 1615)
1 : the 1st month of the Islamic year — see MONTH table
2 : a Muslim festival held during Muharram

mu·ja·hid·een *or* **mu·ja·hed·in** *also* **mu·ja·hed·een** \mü-ˌja-hi-'dēn, mǔ-, -jä-\ *noun plural* [Arabic *mujāhidīn,* plural of *mujāhid,* literally, person who wages jihad] (1922)
: Islamic guerrilla fighters especially in the Middle East

mu·jik \mü-'zhek, -'zhik\ *variant of* MUZHIK

muk·luk \'mək-ˌlək\ *noun* [Yupik *maklak* bearded seal] (1868)

◇ WORD HISTORY

mugwump When James G. Blaine received the Republican Party's nomination for the presidency in 1884, many Republicans refused to have anything to do with his candidacy, supporting instead the Democratic candidate, Grover Cleveland. Those Republicans who remained loyal to their party accused the bolters of a lofty and supercilious attitude and nicknamed them *mugwumps.* The word had previously been used as a jesting, slightly derogatory term for someone who considered himself an important person. The name given to Republican dissidents in 1884 came later to be used for anyone who took an independent, noncommittal attitude toward political issues; in the 20th century it was jocularly defined as someone who sits on the political fence "with his *mug* on one side and his *wump* on the other." Despite appearances, however, the word is not a compound with *mug* but rather a loan into colonial American English from Massachuset, the now extinct language of the Algonquian-speaking Indians of eastern Massachusetts. In a 17th century translation of the Bible into Massachuset, the missionary John Eliot used *mugquomp* to render "duke," "lord," or "chief."

1 : a sealskin or reindeer-skin boot worn by Eskimos
2 : a boot often of duck with a soft leather sole and worn over several pairs of socks

muk·tuk \ˈmək-ˌtək\ *noun* [Inuit *maktak*] (1835)
: whale skin used for food

mu·lat·to \mə-ˈla-(ˌ)tō, mu̇-, myu̇-, -ˈlä-\ *noun, plural* **-toes** *or* **-tos** [Spanish *mulato,* from *mulo* mule, from Latin *mulus*] (1593)
1 : the first-generation offspring of a black person and a white person
2 : a person of mixed white and black ancestry

mul·ber·ry \ˈməl-ˌber-ē, -b(ə-)rē\ *noun* [Middle English *murberie, mulberie,* from Middle French *moure* mulberry (from Latin *morum,* from Greek *moron*) + Middle English *berie* berry] (14th century)
1 : any of a genus (*Morus* of the family Moraceae, the mulberry family) of trees with an edible usually purple multiple fruit that is an aggregate of juicy one-seeded drupes; *also* **:** the fruit
2 : a dark purple or purplish black

mulch \ˈməlch\ *noun* [perhaps irregular from English dialect *melch* soft, mild] (1657)
: a protective covering (as of sawdust, compost, or paper) spread or left on the ground to reduce evaporation, maintain even soil temperature, prevent erosion, control weeds, enrich the soil, or keep fruit (as strawberries) clean
— mulch *transitive verb*

1mulct \ˈməlkt\ *noun* [Latin *multa, mulcta*] (1598)
: FINE, PENALTY

2mulct *transitive verb* (1611)
1 : to punish by a fine
2 a : to defraud especially of money **:** SWINDLE **b :** to obtain by fraud, duress, or theft

1mule \ˈmyü(ə)l\ *noun* [Middle English, from Old French *mul,* from Latin *mulus*] (13th century)
1 a : a hybrid between a horse and a donkey; *especially* **:** the offspring of a male donkey and a mare **b :** a self-sterile plant whether hybrid or not **c :** a usually sterile hybrid
2 : a very stubborn person
3 : a machine for simultaneously drawing and twisting fiber into yarn or thread and winding it into cops
4 *slang* **:** a person who smuggles or delivers illicit drugs

2mule *noun* [Middle French, a kind of slipper, from Latin *mulleus* shoe worn by magistrates] (1562)
: a shoe or slipper without quarter or heel strap — compare SCUFF

mule deer *noun* (1805)
: a long-eared deer (*Odocoileus hemionus*) of western North America that is larger and more heavily built than the common white-tail

mule deer

mule skinner *noun* (1870)
: MULETEER

mu·le·ta \mü-ˈlā-tə, myü-\ *noun* [Spanish, crutch, muleta, diminutive of *mula* she-mule, from Latin, feminine of *mulus* mule] (1838)
: a small cloth attached to a short tapered stick and used by a matador in place of the large cape during the final stage of a bullfight

mu·le·teer \ˌmyü-lə-ˈtir\ *noun* [French *muletier,* from *mulet,* from Old French, diminutive of *mul* mule] (1538)
: one who drives mules

mu·ley \ˈmyü-lē, ˈmu̇-, ˈmü-\ *adjective* [of Celtic origin; akin to Irish & Scottish Gaelic *maol* bald, hornless, Welsh *moel*] (1840)
: POLLED, HORNLESS; *especially* **:** naturally hornless

mu·li·eb·ri·ty \ˌmyü-lē-ˈe-brə-tē\ *noun* [Late Latin *muliebritat-, muliebritas,* from Latin *muliebris* of a woman, from *mulier* woman] (1592)
: FEMININITY

mul·ish \ˈmyü-lish\ *adjective* [1*mule*] (1751)
: unreasonably and inflexibly obstinate
synonym see OBSTINATE
— mul·ish·ly *adverb*
— mul·ish·ness *noun*

1mull \ˈməl\ *verb* [Middle English, from *mul, mol* dust, probably from Middle Dutch; akin to Old English *melu* meal — more at MEAL] (15th century)
transitive verb
1 : to grind or mix thoroughly **:** PULVERIZE
2 : to consider at length **:** PONDER — often used with *over*
intransitive verb
: MEDITATE, PONDER

2mull *transitive verb* [origin unknown] (1618)
: to heat, sweeten, and flavor (as wine or cider) with spices

3mull *noun* [by shortening & alteration from *mulmul* muslin, from Hindi *malmal*] (1798)
: a soft fine sheer fabric of cotton, silk, or rayon

4mull *noun* [German, from Danish *muld,* from Old Norse *mold* dust, soil; akin to Old High German *molta* dust, soil — more at MOLD] (1928)
1 : friable forest humus that forms a layer of mixed organic matter and mineral soil and merges gradually into the mineral soil beneath
2 : a finely powdered solid especially in a suspension

mul·lah \ˈmə-lə, ˈmu̇-\ *noun* [Turkish *molla* & Persian & Hindi *mulla,* from Arabic *mawlā*] (1613)
: an educated Muslim trained in traditional religious law and doctrine and usually holding an official post
— mul·lah·ism \-lə-ˌi-zəm\ *noun*

mul·lein *also* **mul·len** \ˈmə-lən\ *noun* [Middle English *moleyne,* from Anglo-French *moleine*] (14th century)
: any of a genus (*Verbascum*) of usually woolly-leaved Eurasian herbs of the snapdragon family including some that are naturalized in North America

mullein pink *noun* (circa 1850)
: an Old World herb (*Lychnis coronaria*) of the pink family cultivated chiefly for its woolly herbage and crimson flowers

mull·er \ˈmə-lər\ *noun* [alteration of Middle English *molour,* probably from *mullen* to grind] (1612)
: a stone or piece of wood, metal, or glass used as a pestle for pounding or grinding

Mül·le·ri·an \myü-ˈlir-ē-ən, mi-, ˌmə-\ *adjective* [Fritz Müller (died 1897) German zoologist] (1899)
: of, relating to, or being mimicry that exists between two or more inedible or dangerous species (as of butterflies) and that is considered in evolutionary theory to be a mechanism reducing loss to predation by simplification of the recognition process

mul·let \ˈmə-lət\ *noun, plural* **mullet** *or* **mullets** [Middle English *molet,* from Middle French *mulet,* from Latin *mullus* red mullet, from Greek *myllos*] (14th century)
: any of a family (Mugilidae) of valuable chiefly marine food fishes with an elongate rather stout body

mul·li·gan \ˈmə-li-gən\ *noun* [probably from the name *Mulligan*] (circa 1949)
: a free shot sometimes given a golfer in informal play when the previous shot was poorly played

mulligan stew *noun* [probably from the name *Mulligan*] (1904)
: a stew made from whatever ingredients are available

mul·li·ga·taw·ny \ˌmə-li-gə-ˈtȯ-nē, -ˈtä-\ *noun* [Tamil *miḷakutaṇṇi,* from *miḷaku* pepper + *taṇṇi* water] (1784)
: a rich soup usually of chicken stock seasoned with curry

mul·lion \ˈməl-yən\ *noun* [probably alteration of *monial* mullion] (1567)
: a slender vertical member that forms a division between units of a window, door, or screen or is used decoratively
— mullion *transitive verb*

mull·ite \ˈmə-ˌlīt\ *noun* [*Mull,* island of the Inner Hebrides] (1924)
: a mineral that is an orthorhombic silicate of aluminum which is resistant to corrosion and heat and is used as a refractory

multi- *combining form* [Middle English, from Middle French or Latin; Middle French, from Latin, from *multus* much, many — more at MELIORATE]
1 a : many **:** multiple **:** much ⟨*multi*valent⟩ **b :** more than two ⟨*multi*lateral⟩ **c :** more than one ⟨*multi*parous⟩
2 : many times over ⟨*multi*millionaire⟩

mul·ti·age
mul·ti·agen·cy
mul·ti·armed
mul·ti·at·om
mul·ti·au·thor
mul·ti·ax·i·al
mul·ti·band
mul·ti·bank
mul·ti·bar·rel
mul·ti·bar·reled
mul·ti·bil·lion
mul·ti·bil·lion·aire
mul·ti·blad·ed
mul·ti·branched
mul·ti·build·ing
mul·ti·cam·pus
mul·ti·car
mul·ti·car·bon
mul·ti·caus·al
mul·ti·cell
mul·ti·celled
mul·ti·cel·lu·lar
mul·ti·cel·lu·lar·i·ty
mul·ti·cen·ter
mul·ti·chain
mul·ti·cham·bered
mul·ti·chan·nel
mul·ti·char·ac·ter
mul·ti·city
mul·ti·cli·ent
mul·ti·coat·ed
mul·ti·col·or
mul·ti·col·ored
mul·ti·col·umn
mul·ti·com·po·nent
mul·ti·con·duc·tor
mul·ti·copy
mul·ti·coun·ty
mul·ti·course
mul·ti·cu·rie
mul·ti·cur·ren·cy
mul·ti·di·a·lec·tal
mul·ti·di·men·sion·al
mul·ti·di·men·sion·al·i·ty
mul·ti·di·rec·tion·al
mul·ti·dis·ci·plin·ary
mul·ti·dis·ci·pline
mul·ti·di·vi·sion·al
mul·ti·do·main
mul·ti·drug
mul·ti·elec·trode
mul·ti·el·e·ment
mul·ti·em·ploy·er
mul·ti·en·gine
mul·ti·eth·nic
mul·ti·fac·et·ed
mul·ti·fam·i·ly
mul·ti·fil·a·ment
mul·ti·flash
mul·ti·fo·cal
mul·ti·fre·quen·cy
mul·ti·func·tion
mul·ti·func·tion·al
mul·ti·gen·er·a·tion·al
mul·ti·gen·ic
mul·ti·grade
mul·ti·grain
mul·ti·grid
mul·ti·group
mul·ti·hand·i·capped
mul·ti·head·ed
mul·ti·hos·pi·tal
mul·ti·hued
mul·ti–in·dus·try
mul·ti–in·sti·tu·tion·al
mul·ti·lane
mul·ti·lev·el
mul·ti·lev·eled
mul·ti·line
mul·ti·lobed
mul·ti·manned
mul·ti·mega·ton
mul·ti·mega·watt
mul·ti·mem·ber
mul·ti·me·tal·lic
mul·ti·mil·len·ni·al
mul·ti·mil·lion
mul·ti·mil·lion·aire
mul·ti·mode
mul·ti·mo·lec·u·lar
mul·ti·na·tion
mul·ti·nu·cle·ar
mul·ti·nu·cle·ate
mul·ti·nu·cle·at·ed
mul·ti·or·gas·mic
mul·ti·page
mul·ti·paned
mul·ti·pa·ram·e·ter
mul·ti·part
mul·ti·par·ti·cle
mul·ti·par·ty
mul·ti·path
mul·ti·phase
mul·ti·phasic
mul·ti·pho·ton
mul·ti·pic·ture
mul·ti·piece
mul·ti·pi·on
mul·ti·pis·ton
mul·ti·plant
mul·ti·play·er
mul·ti·pole
mul·ti·pow·er
mul·ti·prob·lem
mul·ti·prod·uct
mul·ti·pur·pose
mul·ti·range
mul·ti·re·gion·al
mul·ti·re·li·gious

mul•ti•room
mul•ti•screen
mul•ti•ser•vice
mul•ti•sid•ed
mul•ti•site
mul•ti•size
mul•ti•skilled
mul•ti•source
mul•ti•spe•cies
mul•ti•speed
mul•ti•sport
mul•ti•stemmed
mul•ti•step
mul•ti•sto•ried
mul•ti•sto•ry
mul•ti•strand•ed
mul•ti•syl•lab•ic
mul•ti•sys•tem
mul•ti•tal•ent•ed
mul•ti•ter•mi•nal
mul•ti•tiered
mul•ti•ton
mul•ti•tone
mul•ti•tow•ered
mul•ti•track
mul•ti•tril•lion
mul•ti•union
mul•ti•unit
mul•ti•use
mul•ti•vi•ta•min
mul•ti•vol•ume
mul•ti•wall
mul•ti•war•head
mul•ti•wave•length
mul•ti•year

mul•ti•cul•tur•al \ˌməl-tē-ˈkəlch-rəl, -ˌtī-, -ˈkəl-chə-\ *adjective* (1941)
: of, relating to, reflecting, or adapted to diverse cultures ⟨a *multicultural* society⟩ ⟨*multicultural* education⟩ ⟨a *multicultural* menu⟩
— mul•ti•cul•tur•al•ism \-rə-ˌli-zəm\ *noun*

mul•ti•en•zyme \-ˈen-ˌzīm\ *adjective* (1961)
: composed of or involving two or more enzymes that function in a biosynthetic pathway ⟨*multienzyme* complex⟩

mul•ti•fac•to•ri•al \-fak-ˈtōr-ē-əl, -ˈtȯr-\ *adjective* (1920)
1 : having characters or a mode of inheritance dependent on a number of genes at different loci
2 *or* **mul•ti•fac•tor** \-ˈfak-tər\ **:** having, involving, or produced by a variety of elements or causes
— mul•ti•fac•to•ri•al•ly \-fak-ˈtōr-ē-ə-lē, -ˈtȯr-\ *adverb*

mul•ti•far•i•ous \ˌməl-tə-ˈfar-ē-əs, -ˈfer-\ *adjective* [Medieval Latin *multifarius,* from Latin *multifariam* in many places] (1593)
: having or occurring in great variety **:** DIVERSE
— mul•ti•far•i•ous•ness *noun*

mul•ti•flo•ra rose \ˌməl-tə-ˈflōr-ə-, -ˈflȯr-\ *noun* [New Latin *multiflora,* specific epithet, literally, having many flowers] (1829)
: a vigorous thorny rose (*Rosa multiflora*) with clusters of small flowers

mul•ti•fold \ˈməl-ti-ˌfōld\ *adjective* (1806)
: MANY, NUMEROUS

mul•ti•form \ˈməl-ti-ˌfȯrm\ *adjective* [French *multiforme,* from Latin *multiformis,* from *multi-* + *-formis* -form] (1603)
: having many forms or appearances
— mul•ti•for•mi•ty \ˌməl-ti-ˈfȯr-mə-tē\ *noun*

mul•ti•germ \ˌməl-ti-ˈjərm, -ˌtī-\ *adjective* [probably from *multi-* + *germinate*] (1950)
: producing or being a fruit cluster capable of giving rise to several plants ⟨a *multigerm* variety of sugar beet⟩

mul•ti•hull \ˈməl-ti-ˌhəl, -ˌtī-\ *noun* (1960)
: a vessel (as a catamaran or trimaran) with multiple side-by-side hulls — compare MONOHULL

mul•ti•lat•er•al \ˌməl-ti-ˈla-t(ə-)rəl, -ˌtī-\ *adjective* (1696)
1 : having many sides
2 : involving or participated in by more than two nations or parties ⟨*multilateral* agreements⟩
— mul•ti•lat•er•al•ism \-ˈla-t(ə-)rə-ˌli-zəm\ *noun*
— mul•ti•lat•er•al•ist \-list\ *noun*
— mul•ti•lat•er•al•ly *adverb*

mul•ti•lay•ered \-ˈlā-ərd, -ˈle(-ə)rd\ *or* **mul•ti•lay•er** \-ˈlā-ər, -ˈle(-ə)r\ *adjective* (1923)
: having or involving several distinct layers, strata, or levels

mul•ti•lin•gual \-ˈliŋ-gwəl, -ˈliŋ-gyə-wəl\ *adjective* (1838)
1 : of, containing, or expressed in several languages ⟨a *multilingual* sign⟩ ⟨*multilingual* dictionaries⟩
2 : using or able to use several languages ⟨*multilingual* translators⟩
— mul•ti•lin•gual•ism \-gwə-ˌli-zəm, -gyə-wə-\ *noun*
— mul•ti•lin•gual•ly \-gwə-lē, -gyə-wə-lē\ *adverb*

mul•ti•me•dia \-ˈmē-dē-ə\ *adjective* (1962)
: using, involving, or encompassing several media ⟨a *multimedia* approach to learning⟩
— multimedia *noun*

mul•ti•mod•al \-ˈmō-dᵊl\ *adjective* (1902)
: having or involving several modes, modalities, or maxima ⟨*multimodal* distributions⟩ ⟨*multimodal* responses⟩

mul•ti•na•tion•al \-ˈnash-nəl, -ˈna-shə-nᵊl\ *adjective* (1926)
1 : of or relating to more than two nationalities ⟨a *multinational* society⟩
2 a : of, relating to, or involving more than two nations ⟨a *multinational* alliance⟩ **b :** having divisions in more than two countries ⟨a *multinational* corporation⟩
— multinational *noun*

mul•ti•no•mi•al \-ˈnō-mē-əl\ *noun* [*multi-* + *-nomial* (as in *binomial*)] (1674)
: a mathematical expression that consists of the sum of several terms **:** POLYNOMIAL
— multinomial *adjective*

mul•tip•a•rous \ˌməl-ˈti-pə-rəs\ *adjective* [New Latin *multiparus,* from *multi-* + Latin *-parus* -parous] (1646)
1 : producing many or more than one at a birth
2 : having experienced one or more previous parturitions

mul•ti•par•tite \ˌməl-ti-ˈpär-ˌtīt\ *adjective* [Latin *multipartitus,* from *multi-* + *partitus,* past participle of *partire* to divide, from *part-, pars* part] (circa 1721)
1 : divided into several or many parts
2 : having numerous members or signatories ⟨a *multipartite* treaty⟩

¹mul•ti•ple \ˈməl-tə-pəl\ *adjective* [French, from Latin *multiplex,* from *multi-* + *-plex* -fold — more at -FOLD] (1647)
1 : consisting of, including, or involving more than one ⟨*multiple* births⟩
2 : MANY, MANIFOLD ⟨*multiple* achievements⟩
3 : shared by many ⟨*multiple* ownership⟩
4 : having numerous aspects or functions **:** VARIOUS
5 : being a group of terminals which make a circuit available at a number of points
6 : formed by coalescence of the ripening ovaries of several flowers ⟨a *multiple* fruit⟩

²multiple *noun* (1685)
1 a : the product of a quantity by an integer ⟨35 is a *multiple* of 7⟩ **b :** something in units of more than one or two
2 : PARALLEL 4b
3 *chiefly British* **:** CHAIN STORE

multiple allele *noun* (1938)
: an allele of a genetic locus having more than two allelic forms within a population

multiple–choice *adjective* (1926)
1 : having several answers from which one is to be chosen ⟨a *multiple-choice* question⟩
2 : composed of multiple-choice questions ⟨a *multiple-choice* test⟩

multiple factor *noun* (1915)
: one of a group of nonallelic genes that according to the multiple-factor hypothesis control various quantitative hereditary characters

multiple myeloma *noun* (1897)
: a disease of bone marrow that is characterized by the presence of numerous myelomas in various bones of the body

multiple personality *noun* (1901)
: an hysterical neurosis in which the personality becomes dissociated into two or more distinct but complex and socially and behaviorally integrated parts each of which becomes dominant and controls behavior from time to time to the exclusion of the others — compare SCHIZOPHRENIA, SPLIT PERSONALITY

multiple regression *noun* (1924)
: regression in which one variable is estimated by the use of more than one other variable

multiple sclerosis *noun* (1885)
: a demyelinating disease marked by patches of hardened tissue in the brain or the spinal cord and associated especially with partial or complete paralysis and jerking muscle tremor

multiple star *noun* (1850)
: several stars in close proximity that appear to form a single system

multiple store *noun* (1929)
chiefly British **:** CHAIN STORE

mul•ti•plet \ˈməl-tə-plət\ *noun* (1922)
1 : a spectrum line having several components
2 : a group of elementary particles that are different in charge but similar in other properties (as mass)

multiple–valued *adjective* (1882)
: having at least one and sometimes more of the values of the range associated with each value of the domain ⟨a *multiple-valued* function⟩ — compare SINGLE-VALUED

¹mul•ti•plex \ˈməl-tə-ˌpleks\ *adjective* [Latin] (1557)
1 : MANY, MULTIPLE
2 : being or relating to a system of transmitting several messages or signals simultaneously on the same circuit or channel

²multiplex (1907)
transitive verb
: to send (messages or signals) by a multiplex system
intransitive verb
: to multiplex messages or signals
— mul•ti•plex•er *also* **mul•ti•plex•or** \-ˌplek-sər\ *noun*

³multiplex *noun* (1985)
: a complex that houses several movie theaters

mul•ti•pli•cand \ˌməl-tə-plə-ˈkand\ *noun* [Latin *multiplicandus,* gerundive of *multiplicare*] (1594)
: the number that is to be multiplied by another

mul•ti•pli•ca•tion \ˌməl-tə-plə-ˈkā-shən\ *noun* [Middle English *multiplicacioun,* from Middle French *multiplication,* from Latin *multiplication-, multiplicatio,* from *multiplicare* to multiply] (14th century)
1 : the act or process of multiplying **:** the state of being multiplied
2 a : a mathematical operation that at its simplest is an abbreviated process of adding an integer to itself a specified number of times and that is extended to other numbers in accordance with laws that are valid for integers **b :** any of various mathematical operations that are analogous in some way to multiplication of the real numbers but are defined for other or larger sets of elements (as complex numbers, vectors, matrices, or functions)

multiplication sign *noun* (1907)
: a symbol used to indicate multiplication: **a :** TIMES SIGN **b :** DOT 2b

mul•ti•pli•ca•tive \ˌməl-tə-ˈpli-kə-tiv, ˈməl-tə-plə-ˌkā-tiv\ *adjective* (1653)
1 : tending or having the power to multiply
2 : of, relating to, or associated with a mathematical operation of multiplication ⟨the *multiplicative* property of 0 requires that $a \cdot 0 = 0$ and $0 \cdot a = 0$⟩
— mul•ti•pli•ca•tive•ly *adverb*

multiplicative identity *noun* (1958)
: an identity element (as 1 in the group of rational numbers without 0) that in a given mathematical system leaves unchanged any element by which it is multiplied

multiplicative inverse *noun* (1958)
: an element of a mathematical set that when multiplied by a given element yields the identity element — called also *reciprocal*

mul·ti·plic·i·ty \ˌməl-tə-ˈpli-sə-tē\ *noun, plural* **-ties** [Middle English, from Middle French *multiplicité,* from Late Latin *multiplicitat-, multiplicitas,* from Latin *multiplic-, multiplex*] (15th century)
1 a : the quality or state of being multiple or various **b :** the number of components in a system (as a multiplet or a group of energy levels)
2 : a great number
3 : the number of times a root of an equation or zero of a function occurs when there is more than one root or zero ⟨the *multiplicity* of $x = 2$ for the equation $(x - 2)^3 = 0$ is 3⟩

mul·ti·pli·er \ˈməl-tə-ˌplī(-ə)r\ *noun* (15th century)
: one that multiplies: as **a :** a number by which another number is multiplied **b :** an instrument or device for multiplying or intensifying some effect **c :** a machine, mechanism, or circuit that multiplies figures

[1]**mul·ti·ply** \ˈməl-tə-ˌplī\ *verb* **-plied; -ply·ing** [Middle English *multiplien,* from Old French *multiplier,* from Latin *multiplicare,* from *multiplic-, multiplex* multiple] (13th century)
transitive verb
1 : to increase in number especially greatly or in multiples **:** AUGMENT
2 a : to find the product of by multiplication ⟨*multiply* 7 and 8⟩ **b :** to use as a multiplicand in multiplication with another number ⟨*multiply* 7 by 8⟩
intransitive verb
1 a : to become greater in number **:** SPREAD **b :** BREED, PROPAGATE
2 : to perform multiplication
synonym see INCREASE

[2]**mul·ti·ply** \-plē\ *adverb* (1881)
: in a multiple manner **:** in several ways ⟨*multiply* handicapped children⟩

mul·ti–ply \ˌməl-ti-ˈplī, -ˌtī-\ *adjective* (1926)
: composed of several plies

mul·ti·po·lar \-ˈpō-lər\ *adjective* [International Scientific Vocabulary] (1859)
1 : having several poles ⟨a *multipolar* generator⟩ ⟨*multipolar* mitoses⟩
2 : having several dendrites ⟨*multipolar* nerve cells⟩
3 : characterized by more than two centers of power or interest ⟨a *multipolar* world⟩
— mul·ti·po·lar·i·ty \-pō-ˈlar-ə-tē\ *noun*

mul·ti·po·ten·tial \-pə-ˈten(t)-shəl\ *adjective* (1913)
: having the potential of becoming any of several mature cell types ⟨*multipotential* stem cell⟩

mul·ti·pro·cess·ing \-ˈprä-ˌse-siŋ, -ˈprō-, -sə-siŋ\ *noun* (1961)
: the processing of several computer programs at the same time especially by a computer system with two or more processors sharing a single memory
— mul·ti·pro·ces·sor \-sər\ *noun*

mul·ti·pro·gram·ming \-ˈprō-ˌgra-miŋ, -grə-\ *noun* (1959)
: the technique of utilizing several interleaved programs concurrently in a single computer system

mul·ti·pronged \-ˈprȯŋd\ *adjective* (1957)
1 : having several prongs ⟨*multipronged* fishing spears⟩
2 : having several distinct aspects or elements ⟨a *multipronged* attack on the problem⟩

mul·ti·ra·cial \-ˈrā-shəl\ *adjective* (1923)
: composed of, involving, or representing various races
— mul·ti·ra·cial·ism \-shə-ˌli-zəm\ *noun*

mul·ti·sense \ˈməl-ti-ˌsen(t)s\ *adjective* (1957)
: having several meanings ⟨*multisense* words⟩

mul·ti·sen·so·ry \ˌməl-ti-ˈsen(t)-sə-rē, -ˈsen(t)s-rē\ *adjective* (1949)
: relating to or involving several physiological senses ⟨*multisensory* teaching methods⟩ ⟨*multisensory* experience⟩

mul·ti·spec·tral \-ˈspek-trəl\ *adjective* (1965)
: of or relating to two or more ranges of frequencies or wavelengths in the electromagnetic spectrum

mul·ti·stage \ˈməl-ti-ˌstāj, -ˌtī-\ *adjective* (1904)
1 : having successive operating stages; *especially* **:** having propulsion units that operate in turn ⟨*multistage* rockets⟩
2 : conducted by or occurring in stages ⟨a *multistage* investigation⟩

mul·ti·state \-ˌstāt\ *adjective* (1944)
1 : having divisions in several states ⟨*multistate* enterprises⟩
2 : of, relating to, or involving several states ⟨a *multistate* attack on environmental pollution⟩

mul·ti·task·ing \-ˌtas-kiŋ\ *noun, often attributive* (1966)
: the concurrent performance of several jobs by a computer

mul·ti·tude \ˈməl-tə-ˌtüd, -ˌtyüd\ *noun* [Middle English, from Middle French or Latin; Middle French, from Latin *multitudin-, multitudo,* from *multus* much — more at MELIORATE] (14th century)
1 : the state of being many
2 : a great number **:** HOST
3 : a great number of people
4 : POPULACE, PUBLIC

mul·ti·tu·di·nous \ˌməl-tə-ˈtüd-nəs, -ˈtyüd-; -ˈtü-dᵊn-əs, -ˈtyü-\ *adjective* (1604)
1 : including a multitude of individuals **:** POPULOUS ⟨the *multitudinous* city⟩
2 : existing in a great multitude ⟨*multitudinous* opportunities⟩
3 : existing in or consisting of innumerable elements or aspects ⟨*multitudinous* applause⟩
— mul·ti·tu·di·nous·ly *adverb*
— mul·ti·tu·di·nous·ness *noun*

mul·ti·us·er \ˌməl-tī-ˈyü-zər, -tē-, -ti-\ *adjective* (1964)
: able to be used by more than one person simultaneously

mul·tiv·a·lence \ˌməl-ˈti-və-lən(t)s\ *noun* (circa 1881)
: the quality or state of having many values, meanings, or appeals

mul·ti·va·lent \ˌməl-ti-ˈvā-lənt, -ˌtī-, *especially in sense 3* ˌməl-ˈti-və-\ *adjective* [International Scientific Vocabulary] (1874)
1 : POLYVALENT
2 : represented more than twice in the somatic chromosome number ⟨*multivalent* chromosomes⟩
3 : having many values, meanings, or appeals
— multivalent *noun*

mul·ti·var·i·able \ˌməl-ti-ˈver-ē-ə-bəl, -ˌtī-, -ˈvar-\ *adjective* (1963)
: MULTIVARIATE

mul·ti·var·i·ate \-ˈver-ē-ət, -ˌāt\ *adjective* [*multi-* + *variable* + [3]*-ate*] (1951)
: having or involving a number of independent mathematical or statistical variables ⟨*multivariate* calculus⟩ ⟨*multivariate* data analysis⟩

mul·ti·ver·si·ty \-ˈvər-sə-tē, -stē\ *noun, plural* **-ties** [*multi-* + *-versity* (as in *university*)] (1963)
: a very large university with many component schools, colleges, or divisions and widely diverse functions

mul·ti·vol·tine \-ˈvōl-ˌtēn, -ˈvȯl-\ *adjective* [*multi-* + *-voltine* having a given number of broods (from French, from Italian *volta* time, turn) — more at VOLT] (1874)
: having several broods in a season ⟨*multivoltine* insects⟩

mul·ture \ˈməl-chər, *Scottish usually* ˈmü-tər\ *noun* [Middle English *multyr,* from Middle French *molture,* literally, grinding, from (assumed) Vulgar Latin *molitura,* from Latin *molitus,* past participle of *molere* to grind — more at MEAL] (14th century)
chiefly Scottish **:** a fee for grinding grain at a mill

[1]**mum** \ˈməm\ *adjective* [probably imitative of a sound made with closed lips] (14th century)
: SILENT ⟨keep *mum*⟩

[2]**mum** *intransitive verb* **mummed; mum·ming** [Middle English *mommen,* from Middle French *momer* to go masked] (1530)
1 : to perform in a pantomime
2 : to go about merrymaking in disguise during festivals

[3]**mum** *noun* [German *Mumme*] (1640)
: a strong ale or beer

[4]**mum** *chiefly British variant of* MOM

[5]**mum** *noun* (1917)
: CHRYSANTHEMUM

mum·ble \ˈməm-bəl\ *verb* **mum·bled; mum·bling** \-b(ə-)liŋ\ [Middle English *momelen,* of imitative origin] (14th century)
intransitive verb
: to utter words in a low confused indistinct manner **:** MUTTER
transitive verb
1 : to utter with a low inarticulate voice
2 : to chew or bite with or as if with toothless gums
— mumble *noun*
— mum·bler \-b(ə-)lər\ *noun*
— mum·bly \-b(ə-)lē\ *adjective*

mum·ble·ty–peg \ˈməm-bəl-(tē-)ˌpeg, -blē-ˌpeg\ *also* **mum·ble–the–peg** \-bəl-(thə-)ˌpeg, -blē-ˌpeg\ *or* **mumble peg** \-bəl-ˌpeg, -blē-\ *noun* [from the phrase *mumble the peg;* from the loser's originally having to pull out with his teeth a peg driven into the ground] (1627)
: a game in which the players try to flip a knife from various positions so that the blade will stick into the ground

mum·bo jum·bo \ˌməm-bō-ˈjəm-(ˌ)bō\ *noun* [*Mumbo Jumbo,* a masked figure among Mandingo peoples of western Africa] (1738)
1 : an object of superstitious homage and fear
2 a : a complicated often ritualistic observance with elaborate trappings **b :** complicated activity usually intended to obscure and confuse
3 : unnecessarily involved and incomprehensible language **:** GIBBERISH

mum·mer \ˈmə-mər\ *noun* [Middle French *momeur,* from *momer* to go masked] (1502)
1 : a performer in a pantomime; *broadly* **:** ACTOR
2 : one who goes merrymaking in disguise during festivals

mum·mery \ˈmə-mə-rē\ *noun, plural* **-mer·ies** (circa 1530)
1 : a performance by mummers
2 : a ridiculous, hypocritical, or pretentious ceremony or performance

mum·mi·chog \ˈmə-mi-ˌchȯg, -ˌchäg\ *noun* [Narraganset *moamitteaúg*] (1787)
: a common American killifish (*Fundulus heteroclitus* of the family Cyprinodontidae)

mum·mi·fy \ˈmə-mi-ˌfī\ *verb* **-fied; -fy·ing** (1628)
transitive verb
1 : to embalm and dry as or as if a mummy
2 a : to make into or like a mummy **b :** to cause to dry up and shrivel
intransitive verb
: to dry up and shrivel like a mummy
— mum·mi·fi·ca·tion \ˌmə-mi-fə-ˈkā-shən\ *noun*

[1]**mum·my** \ˈmə-mē\ *noun, plural* **mummies** [Middle English *mummie* powdered parts of a mummified body used as a drug, from Middle French *momie,* from Medieval Latin *mumia* mummy, powdered mummy, from Arabic *mūmiyah* bitumen, mummy, from Persian *mūm* wax] (1615)
1 a : a body embalmed or treated for burial with preservatives in the manner of the ancient

Egyptians **b :** a body unusually well preserved
2 : one resembling a mummy

²mummy *chiefly British variant of* MOMMY

mumps \'məmps\ *noun plural but singular or plural in construction* [from plural of obsolete *mump* grimace] (1598)
: an acute contagious virus disease marked by fever and by swelling especially of the parotid gland

munch \'mənch\ *verb* [Middle English *monchen,* probably of imitative origin] (14th century)
transitive verb
: to eat with a chewing action ⟨many a mouthful is *munched* in private —Washington Irving⟩; *also* **:** to snack on ⟨drank coffee and *munched* homemade cookies —Lady Bird Johnson⟩
intransitive verb
: to eat or chew something; *also* **:** SNACK — usually used with *on*
— munch·er *noun*

Mun·chau·sen syndrome \'mən-ˌchau̇-zən-, ˌmən-'chau̇-\ *noun* [Baron K. F. H. von *Münchausen* (died 1797) German soldier and proverbial teller of exaggerated tales] (1951)
: a condition characterized by the feigning of the symptoms of a disease or injury in order to undergo diagnostic tests, hospitalization, or medical or surgical treatment — called also *Munchausen's syndrome*

munch·ies \'mən-chēz\ *noun plural* (circa 1971)
1 : hunger pangs
2 : light snack foods

munch·kin \'mənch-ˌkin\ *noun* [the *Munchkins,* diminutive creatures in *The Wonderful Wizard of Oz* (1900) by L. Frank Baum] (1972)
: a person who is notably small and often endearing

Mun·da \'mu̇n-də\ *noun* (1901)
: a branch of the Austroasiatic language family spoken by tribal peoples of central and eastern India

mun·dane \ˌmən-'dān, 'mən-ˌ\ *adjective* [Middle English *mondeyne,* from Middle French *mondain,* from Late Latin *mundanus,* from Latin *mundus* world] (15th century)
1 : of, relating to, or characteristic of the world
2 : characterized by the practical, transitory, and ordinary **:** COMMONPLACE ⟨the *mundane* concerns of day-to-day life⟩
synonym see EARTHLY
— mun·dane·ly *adverb*
— mun·dane·ness \-'dān-nəs, -ˌdān-\ *noun*
— mun·dan·i·ty \ˌmən-'dā-nə-tē\ *noun*

mun·dun·gus \ˌmən-'dəŋ-gəs\ *noun* [modification of Spanish *mondongo* tripe] (1641)
archaic **:** foul-smelling tobacco

mung bean \'məŋ-\ *noun* [Hindi *mū̃g,* from Sanskrit *mudga*] (1910)
: an erect bushy annual bean (*Phaseolus aureus*) that is widely cultivated in warm regions for its edible usually green or yellow seeds, for forage, and as the chief source of bean sprouts; *also* **:** the seed of the mung bean

mun·go \'məŋ-(ˌ)gō\ *noun, plural* **mungos** [origin unknown] (1857)
: reclaimed wool of poor quality and very short staple

mu·ni \'myü-nē\ *noun* (1973)
: MUNICIPAL

¹mu·nic·i·pal \myu̇-'ni-s(ə-)pəl *also* myə-, mə-, -'ni-sə-bəl, ÷ˌmyü-nə-'si-pəl\ *adjective* [Latin *municipalis* of a municipality, from *municip-, municeps* inhabitant of a municipality, from *munus* duty, service + *capere* to take — more at MEAN, HEAVE] (circa 1540)
1 : of or relating to the internal affairs of a major political unit (as a nation)
2 a : of, relating to, or characteristic of a municipality **b :** having local self-government
3 : restricted to one locality

²municipal *noun* (1925)
: a security issued by a state or local government or by an authority set up by such a government — usually used in plural

municipal court *noun* (1828)
1 : a court that sits in some cities and larger towns and that usually has civil and criminal jurisdiction over cases arising within the municipality
2 : POLICE COURT

mu·nic·i·pal·i·ty \myu̇-ˌni-sə-'pa-lə-tē\ *noun, plural* **-ties** (1790)
1 : a primarily urban political unit having corporate status and usually powers of self-government
2 : the governing body of a municipality

mu·nic·i·pal·ize \myu̇-'ni-sə-pə-ˌlīz\ *transitive verb* **-ized; -iz·ing** (1880)
: to bring under municipal ownership or supervision
— mu·nic·i·pal·i·za·tion \-ˌni-s(ə-)pə-lə-'zā-shən\ *noun*

mu·nic·i·pal·ly \myu̇-'ni-sə-p(ə-)lē\ *adverb* (circa 1842)
: by or in terms of a municipality

mu·nif·i·cent \myu̇-'ni-fə-sənt\ *adjective* [back-formation from *munificence,* from Latin *munificentia,* from *munificus* generous, from *munus* service, gift — more at MEAN] (1583)
1 : very liberal in giving or bestowing **:** LAVISH
2 : characterized by great liberality or generosity
synonym see LIBERAL
— mu·nif·i·cence \-sən(t)s\ *noun*
— mu·nif·i·cent·ly *adverb*

mu·ni·ment \'myü-nə-mənt\ *noun* [Middle English, from Anglo-French, from Middle French, defense, from Latin *munimentum,* from *munire* to fortify] (15th century)
1 : the evidence (as documents) that enables one to defend the title to an estate or a claim to rights and privileges — usually used in plural
2 *archaic* **:** a means of defense

mu·ni·tion \myu̇-'ni-shən\ *noun* [Middle French, from Latin *munition-, munitio,* from *munire* to fortify, from *moenia* walls; akin to Latin *murus* wall and perhaps to Sanskrit *minoti* he builds, fastens] (1508)
1 *archaic* **:** RAMPART, DEFENSE
2 : ARMAMENT, AMMUNITION
— munition *transitive verb*

Mun·ster *variant of* MUENSTER

mun·tin \'mən-tᵊn\ *noun* [alteration of *montant* vertical dividing bar, from French, from present participle of *monter* to rise — more at MOUNT] (1774)
: a strip separating panes of glass in a sash

munt·jac \'mən(t)-ˌjak, 'mən-ˌchak\ *noun* [Sundanese (Austronesian language of western Java) *mənyčək*] (circa 1798)
: any of a genus (*Muntiacus*) of small deer of southeastern Asia and the East Indies — called also *barking deer*

muntjac

mu·on \'myü-ˌän\ *noun* [contraction of earlier *mu-meson,* from *mu*] (1952)
: an unstable lepton that is common in the cosmic radiation near the earth's surface, has a mass about 207 times the mass of the electron, and exists in negative and positive forms
— mu·on·ic \myü-'ä-nik\ *adjective*

mu·on·ium \myü-'ō-nē-əm, -'ä-\ *noun* [New Latin] (1957)
: a short-lived quasi-atom consisting of an electron and a positive muon

¹mu·ral \'myu̇r-əl\ *adjective* [Latin *muralis,* from *murus* wall — more at MUNITION] (1586)
1 : of, relating to, or resembling a wall
2 : applied to and made integral with a wall or ceiling surface

²mural *noun* (1916)
: a mural work of art (as a painting)
— mu·ral·ist \-ə-list\ *noun*

mu·ram·ic acid \myu̇-'ra-mik-\ *noun* [Latin *mur*us wall + English + glucos*amine* + ¹*-ic*] (1957)
: an amino sugar $C_9H_{17}NO_7$ that is a lactic acid derivative of glucosamine and is found especially in bacterial cell walls and in blue-green algae

¹mur·der \'mər-dər\ *noun* [partly from Middle English *murther,* from Old English *morthor;* partly from Middle English *murdre,* from Old French, of Germanic origin; akin to Old English *morthor;* akin to Old High German *mord* murder, Latin *mort-, mors* death, *mori* to die, *mortuus* dead, Greek *brotos* mortal] (before 12th century)
1 : the crime of unlawfully killing a person especially with malice aforethought
2 a : something very difficult or dangerous ⟨the traffic was *murder*⟩ **b :** something outrageous or blameworthy ⟨getting away with *murder*⟩

²murder *verb* **mur·dered; mur·der·ing** \'mər-d(ə-)riŋ\ (13th century)
transitive verb
1 : to kill (a human being) unlawfully and with premeditated malice
2 : to slaughter wantonly **:** SLAY
3 a : to put an end to **b :** TEASE, TORMENT **c :** MUTILATE, MANGLE ⟨*murders* French⟩ **d :** to defeat badly
intransitive verb
: to commit murder
synonym see KILL

mur·der·ee \ˌmər-də-'rē\ *noun* (1920)
: an actual or potential victim of a murder

mur·der·er \'mər-dər-ər\ *noun* (14th century)
: one who murders; *especially* **:** one who commits the crime of murder

mur·der·ess \'mər-də-rəs\ *noun* (14th century)
: a woman who murders

mur·der·ous \'mər-d(ə-)rəs\ *adjective* (1535)
1 a : having the purpose or capability of murder **b :** characterized by or causing murder or bloodshed
2 : having the ability or power to overwhelm **:** DEVASTATING ⟨*murderous* heat⟩
— mur·der·ous·ly *adverb*
— mur·der·ous·ness *noun*

mu·rein \'myu̇r-ē-ən, 'myu̇r-ˌēn\ *noun* [*muramic acid* + *-ein,* alteration of ²*-ine*] (1964)
: PEPTIDOGLYCAN

mu·rex \'myu̇r-ˌeks\ *noun, plural* **mu·ri·ces** \'myu̇r-ə-ˌsēz\ *or* **mu·rex·es** [New Latin, from Latin, mollusk yielding a purple dye; akin to Greek *myak-, myax* mussel] (1589)
: any of a genus (*Murex*) of marine gastropod mollusks having a rough and often spinose shell, abounding in tropical seas, and yielding a purple dye

mu·ri·ate \'myu̇r-ē-ˌāt\ *noun* [French, back-formation from (*acide*) *muriatique* muriatic acid] (1790)
: CHLORIDE

mu·ri·at·ic acid \ˌmyu̇r-ē-ˈa-tik-\ *noun* [French *muriatique,* from Latin *muriaticus* pickled in brine, from *muria* brine] (1790)
: HYDROCHLORIC ACID

mu·rid \ˈmyu̇r-əd\ *adjective* [ultimately from Latin *mur-, mus* mouse — more at MOUSE] (circa 1909)
: of or relating to a family (Muridae) comprising the typical mice and rats
— murid *noun*

mu·rine \ˈmyu̇r-ˌīn\ *adjective* [ultimately from Latin *mur-, mus*] (1607)
: of or relating to a murid genus (*Mus*) or its subfamily (Murinae) which includes the common household rats and mice; *also* **:** of, relating to, or involving these rodents and especially the house mouse
— murine *noun*

murine typhus *noun* (1933)
: a mild febrile disease that is marked by headache and rash, is caused by a rickettsia (*Rickettsia mooseri*), is widespread in nature in rodents, and is transmitted to humans by a flea

murk \ˈmərk\ *noun* [Middle English *mirke,* from Old English *mirce;* akin to Old Norse *myrkr* darkness] (before 12th century)
: GLOOM, DARKNESS; *also* **:** FOG
— murk *adjective, archaic*

murky \ˈmər-kē\ *adjective* **murk·i·er; -est** (14th century)
1 : characterized by a heavy dimness or obscurity caused by or like that caused by overhanging fog or smoke
2 : characterized by thickness and heaviness of air **:** FOGGY, MISTY
3 : darkly vague or obscure ⟨*murky* official rhetoric⟩
— murk·i·ly \-kə-lē\ *adverb*
— murk·i·ness \-kē-nəs\ *noun*

[1]**mur·mur** \ˈmər-mər\ *noun* [Middle English *murmure,* from Middle French, from Latin *murmur* murmur, roar, of imitative origin] (14th century)
1 : a half-suppressed or muttered complaint **:** GRUMBLING
2 a : a low indistinct but often continuous sound **b :** a soft or gentle utterance
3 : an atypical sound of the heart indicating a functional or structural abnormality

[2]**murmur** (14th century)
intransitive verb
1 : to make a murmur ⟨the breeze *murmured* in the pines⟩
2 : COMPLAIN, GRUMBLE
transitive verb
: to say in a murmur
— mur·mur·er *noun*

mur·mur·ous \ˈmər-mə-rəs, ˈmərm-rəs\ *adjective* (1582)
: filled with or characterized by murmurs **:** low and indistinct
— mur·mur·ous·ly *adverb*

Mur·phy \ˈmər-fē\ *noun* [Miss *Murphy,* nonexistent prostitute used to lure victims] (1959)
: any of various confidence games; *especially* **:** one in which the victim believes he is paying for sex

Murphy bed *noun* [William L. *Murphy* (died 1959) American inventor] (1925)
: a bed that may be folded or swung into a closet

Murphy's Law *noun* [Edward A. *Murphy* (born 1917) American engineer] (1958)
: an observation: anything that can go wrong will go wrong

mur·rain \ˈmər-ən, ˈmə-rən\ *noun* [Middle English *moreyne,* from Middle French *morine,* from *morir* to die, from Latin *mori* — more at MURDER] (14th century)
: a pestilence or plague especially affecting domestic animals

murre \ˈmər\ *noun* [origin unknown] (1602)
: any of a genus (*Uria*) of seabirds; *especially* **:** a common bird (*U. aalge*) of northern seas

mur·rey \ˈmər-ē, ˈmə-rē\ *noun* [Middle English, from Middle French *moré,* from Medieval Latin *moratum,* from neuter of *moratus* mulberry colored, from Latin *morum* mulberry — more at MULBERRY] (15th century)
: a purplish black **:** MULBERRY

mur·ther \ˈmər-t͟hər\ *chiefly dialect variant of* MURDER

mus·ca·det \ˌməs-kə-ˈdā, -ˈde\ *noun, often capitalized* [French, from Provençal, muscadet grape, from *musc* musk] (circa 1899)
: a dry white wine from the Loire valley of France

mus·ca·dine \ˈməs-kə-ˌdīn\ *noun* [probably alteration of *muscatel*] (circa 1785)
: a grape (*Vitis rotundifolia*) of the southern U.S. with musky fruits borne in small clusters; *also* **:** the fruit

mus·cae vo·li·tan·tes \ˈməs-ˌkē-ˌvä-lə-ˈtan-ˌtēz, ˈmə-ˌsē-\ *noun plural* [New Latin, literally, flying flies] (1797)
: spots before the eyes due to cells and cell fragments in the vitreous humor and lens

mus·ca·rine \ˈməs-kə-ˌrēn\ *noun* [German *Muskarin,* from New Latin (*Amanita*) *muscaria* fly agaric] (1872)
: a toxic ammonium base $[C_9H_{20}NO_2]^+$ that is biochemically related to acetylcholine, is found especially in fly agaric, and acts directly on smooth muscle

mus·ca·rin·ic \ˌməs-kə-ˈri-nik\ *adjective* (1936)
: relating to, resembling, producing, or mediating the parasympathetic effects (as a slowed heart rate and increased activity of smooth muscle) produced by muscarine ⟨*muscarinic* receptors⟩ — compare NICOTINIC

mus·cat \ˈməs-ˌkat, -kət\ *noun* [Middle French, from Old Provençal, from *muscat* musky, from *musc* musk, from Late Latin *muscus* — more at MUSK] (1548)
1 : MUSCATEL
2 : any of several cultivated grapes used in making wine and raisins

mus·ca·tel \ˌməs-kə-ˈtel\ *noun* [Middle English *muskadelle,* from Middle French *muscadel,* from Old Provençal, from *muscadel* resembling musk, from *muscat*] (15th century)
1 : a sweet fortified wine from muscat grapes
2 : a raisin from muscat grapes

[1]**mus·cle** \ˈmə-səl\ *noun, often attributive* [Middle English, from Middle French, from Latin *musculus,* from diminutive of *mus* mouse — more at MOUSE] (14th century)
1 a : a body tissue consisting of long cells that contract when stimulated and produce motion **b :** an organ that is essentially a mass of muscle tissue attached at either end to a fixed point and that by contracting moves or checks the movement of a body part
2 a : muscular strength **:** BRAWN **b :** effective strength **:** POWER ⟨political *muscle*⟩ ◆

[2]**muscle** *verb* **mus·cled; mus·cling** \ˈmə-s(ə-)liŋ\ (1913)
transitive verb
: to move or force by or as if by muscular effort ⟨*muscled* him out of office⟩
intransitive verb
: to make one's way by brute strength or by force

mus·cle–bound \ˈmə-səl-ˌbau̇nd\ *adjective* (1879)
1 : having some of the muscles tense and enlarged and of impaired elasticity sometimes as a result of excessive exercise
2 : lacking in flexibility **:** RIGID

muscle car \-ˌkär\ *noun* (1969)
: any of a group of American-made 2-door sports coupes with powerful engines designed for high-performance driving

mus·cled \ˈmə-səld\ *adjective* (1644)
: having muscles especially of a specified kind — often used in combination ⟨hard-*muscled* arms⟩

muscle spindle *noun* (1894)
: a sensory end organ in a muscle that is sensitive to stretch in the muscle, consists of small striated muscle fibers richly supplied with nerve fibers, and is enclosed in a connective tissue sheath — called also *stretch receptor*

mus·co·vite \ˈməs-kə-ˌvīt\ *noun* [Medieval Latin or New Latin *Muscovia, Moscovia* Moscow] (1535)
1 *capitalized* **a :** a native or resident of the ancient principality of Moscow or of the city of Moscow **b :** RUSSIAN
2 [*muscovy (glass)*] **:** a mineral that is a colorless to pale brown potassium mica
— Muscovite *adjective*

Mus·co·vy duck \ˈməs-ˌkō-vē-\ *noun* [*Muscovy,* principality of Moscow, Russia] (1657)
: a large dark crested duck (*Cairina moschata*) of Central and South America that is widely kept in domestication

muscul- *or* **musculo-** *combining form* [Late Latin *muscul-,* from Latin *musculus*]
1 : muscle ⟨*muscul*ar⟩
2 : muscular and ⟨*musculo*skeletal⟩

mus·cu·lar \ˈməs-kyə-lər\ *adjective* (1681)
1 a : of, relating to, or constituting muscle **b :** of, relating to, or performed by the muscles
2 : having well-developed musculature
3 a : of or relating to physical strength **:** BRAWNY **b :** having strength of expression or character **:** VIGOROUS ⟨*muscular* prose⟩
— mus·cu·lar·i·ty \ˌməs-kyə-ˈlar-ə-tē\ *noun*
— mus·cu·lar·ly \ˈməs-kyə-lər-lē\ *adverb*

muscular dystrophy *noun* (1886)
: any of a group of hereditary diseases characterized by progressive wasting of muscles

mus·cu·la·ture \ˈməs-kyə-lə-ˌchu̇r, -chər, -ˌtyu̇r, -ˌtu̇r\ *noun* [French, from Latin *musculus*] (1875)
: the muscles of all or a part of the animal body

mus·cu·lo·skel·e·tal \ˌməs-kyə-lō-ˈske-lə-tᵊl\ *adjective* (circa 1944)
: of, relating to, or involving both musculature and skeleton

[1]**muse** \ˈmyüz\ *verb* **mused; mus·ing** [Middle English, from Middle French *muser* to gape, idle, muse, from *muse* mouth of an animal, from Medieval Latin *musus*] (14th century)
intransitive verb
1 : to become absorbed in thought; *especially* **:** to turn something over in the mind meditatively and often inconclusively
2 *archaic* **:** WONDER, MARVEL

◇ WORD HISTORY
muscle The rippling muscles of a bodybuilder may not conjure in most people's imagination the image of a small rodent, but in a number of older Indo-European languages the word for "muscle" was identical with or a derivative of the word for "mouse." Hence Greek *mys* and Old English *mūs* mean both "mouse" and "muscle." Latin *musculus* (from which, via medieval French, comes English *muscle*) and Russian *myshtsa,* both words meaning "muscle," are derivatives of words meaning "mouse," respectively, *mus* and *mysh'.* (In some English dialects and in Scots, *mouse* continued to be applied to certain fleshy cuts of meat at least into the 19th century.) The usual explanation for this usage is that the lump of a contracting muscle in the upper arm or leg—perhaps more in the eyes of a child than an adult—has some resemblance to the furtive movement of a mouse. Alternatively, however, both the "mouse" and "muscle" senses of the word may reflect derivation from an Indo-European verbal base meaning "move," from which Latin *movēre* "to move" also descends.

transitive verb
: to think or say reflectively
synonym see PONDER
— **mus·er** *noun*

[2]**muse** *noun* (15th century)
: a state of deep thought or dreamy abstraction

[3]**muse** *noun* [Middle English, from Middle French, from Latin *Musa,* from Greek *Mousa*] (14th century)
1 *capitalized* **:** any of the nine sister goddesses in Greek mythology presiding over song and poetry and the arts and sciences
2 : a source of inspiration; *especially* **:** a guiding genius
3 : POET

mu·se·ol·o·gy \ˌmyü-zē-'ä-lə-jē\ *noun* [*museum* + *-logy*] (1885)
: the science or profession of museum organization and management
— **mu·seo·log·i·cal** \-ə-'lä-ji-kəl\ *adjective*
— **mu·se·ol·o·gist** \-'ä-lə-jist\ *noun*

mu·sette \myů-'zet\ *noun* [Middle English, from Middle French, diminutive of *muse* bagpipe, from *muser* to muse, play the bagpipe] (14th century)
1 : a bellows-blown bagpipe popular in France in the 17th and 18th centuries
2 : a small knapsack; *also* **:** a similar bag with one shoulder strap — called also *musette bag*

mu·se·um \myů-'zē-əm\ *noun* [Latin *Museum* place for learned occupation, from Greek *Mouseion,* from neuter of *Mouseios* of the Muses, from *Mousa*] (1672)
: an institution devoted to the procurement, care, study, and display of objects of lasting interest or value; *also* **:** a place where objects are exhibited

museum piece *noun* (1901)
1 : something preserved in or suitable for a museum
2 : one that is out-of-date **:** a thing of the past

[1]**mush** \'məsh, *especially 3 also* 'műsh\ *noun* [probably alteration of *mash*] (1671)
1 : a thick porridge made with cornmeal boiled in water or milk
2 : something soft and spongy or shapeless
3 a : weak sentimentality **:** DRIVEL **b :** mawkish amorousness

[2]**mush** (circa 1781)
transitive verb
chiefly dialect **:** to reduce to a crumbly mass
intransitive verb
of an airplane **:** to fly in a partly or nearly stalled condition
— **mush·er** *noun*

[3]**mush** *intransitive verb* [probably from French *marchons,* 1st plural imperative of *marcher* to move, march, from Middle French *marchier* — more at MARCH] (1897)
: to travel especially over snow with a sled drawn by dogs — often used as a command to a dog team
— **mush·er** *noun*

[4]**mush** *noun* (1902)
: a trip especially across snow with a dog team

[1]**mush·room** \'məsh-ˌrüm, -ˌrům; *chiefly Northern & Midland* -ˌrün; *dialect* 'mə-shə-ˌrüm, -ˌrům, -ˌrün\ *noun* [alteration of Middle English *musseroun,* from Middle French *mousseron,* from Late Latin *mussirion-, mussirio*] (1533)
1 a : an enlarged complex aboveground fleshy fruiting body of a fungus (as a basidiomycete) that consists typically of a stem bearing a pileus; *especially* **:** one that is edible **b :** FUNGUS
2 : UPSTART
3 : something resembling a mushroom

[2]**mushroom** *intransitive verb* (1894)
1 : to collect wild mushrooms
2 a : to well up and spread out laterally from a central source **b :** to become enlarged or extended **:** GROW
3 : to spring up suddenly or multiply rapidly

mushroom cloud *noun* (circa 1909)
: a mushroom-shaped cloud; *specifically* **:** one caused by the explosion of a nuclear weapon

mushy \'mə-shē, *especially 2 also* 'mů-\ *adjective* **mush·i·er; -est** (1839)
1 a : having the consistency of mush **:** SOFT **b :** lacking in definition or precision
2 : excessively tender or emotional; *especially* **:** mawkishly amorous
— **mush·i·ly** \'mə-shə-lē, 'mů-\ *adverb*
— **mush·i·ness** \'mə-shē-nəs, 'mů-\ *noun*

mu·sic \'myü-zik\ *noun, often attributive* [Middle English *musik,* from Old French *musique,* from Latin *musica,* from Greek *mousikē* any art presided over by the Muses, especially music, from feminine of *mousikos* of the Muses, from *Mousa* Muse] (13th century)
1 a : the science or art of ordering tones or sounds in succession, in combination, and in temporal relationships to produce a composition having unity and continuity **b :** vocal, instrumental, or mechanical sounds having rhythm, melody, or harmony
2 a : an agreeable sound **:** EUPHONY ⟨her voice was *music* to my ears⟩ **b :** musical quality ⟨the *music* of verse⟩
3 : a musical accompaniment ⟨a play set to *music*⟩
4 : the score of a musical composition set down on paper
5 : a distinctive type or category of music ⟨there is a *music* for everybody —Eric Salzman⟩

[1]**mu·si·cal** \'myü-zi-kəl\ *adjective* [Middle English, from Middle French, from Medieval Latin *musicalis,* from *musica*] (15th century)
1 a : of or relating to music **b :** having the pleasing harmonious qualities of music **:** MELODIOUS
2 : having an interest in or talent for music
3 : set to or accompanied by music
4 : of or relating to musicians or music lovers
— **mu·si·cal·ly** \-k(ə-)lē\ *adverb*

[2]**musical** *noun* (1823)
1 *archaic* **:** MUSICALE
2 : a film or theatrical production typically of a sentimental or humorous nature that consists of musical numbers and dialogue based on a unifying plot

musical box *noun* (1829)
chiefly British **:** MUSIC BOX

musical chairs *noun plural but singular in construction* (1877)
: a game in which players march to music around a row of chairs numbering one less than the players and scramble for seats when the music stops; *also* **:** a situation or series of events suggesting the game of musical chairs (as in rapid change or confusing activity)

musical comedy *noun* (1765)
: MUSICAL 2

mu·si·cale \ˌmyü-zi-'kal\ *noun* [French *soirée musicale,* literally, musical evening] (1872)
: a social entertainment with music as the leading feature

mu·si·cal·ise *British variant of* MUSICALIZE

mu·si·cal·i·ty \ˌmyü-zi-'ka-lə-tē\ *noun* (1853)
1 : sensitivity to, knowledge of, or talent for music
2 : the quality or state of being musical **:** MELODIOUSNESS

mu·si·cal·ize \'myü-zi-kə-ˌlīz\ *transitive verb* **-ized; -iz·ing** (1919)
: to set to music
— **mu·si·cal·i·za·tion** \ˌmyü-zi-kə-lə-'zā-shən\ *noun*

musical saw *noun* (1927)
: a handsaw made to produce melody by bending the blade while sounding it with a hammer or violin bow

music box *noun* (1844)
: a container enclosing an apparatus that reproduces music mechanically when activated by a clockwork

music drama *noun* (1877)
: an opera in which the action is not interrupted by formal song divisions (as recitatives or arias) and the music is determined solely by dramatic appropriateness

music hall *noun* (1842)
: a vaudeville theater; *also* **:** VAUDEVILLE

mu·si·cian \myů-'zi-shən\ *noun* (14th century)
: a composer, conductor, or performer of music; *especially* **:** INSTRUMENTALIST
— **mu·si·cian·ly** \-lē\ *adjective*
— **mu·si·cian·ship** \-ˌship\ *noun*

music of the spheres (1609)
: an ethereal harmony thought by the Pythagoreans to be produced by the vibration of the celestial spheres

mu·si·col·o·gy \ˌmyü-zi-'kä-lə-jē\ *noun* [Italian *musicologia,* from Latin *musica* music + *-logia* -logy] (1909)
: a study of music as a branch of knowledge or field of research as distinct from composition or performance
— **mu·si·co·log·i·cal** \-kə-'lä-ji-kəl\ *adjective*
— **mu·si·col·o·gist** \-'kä-lə-jist\ *noun*

[1]**mus·ing** \'myü-ziŋ\ *noun* (14th century)
: MEDITATION

[2]**musing** *adjective* (15th century)
: thoughtfully abstracted **:** MEDITATIVE
— **mus·ing·ly** \-lē\ *adverb*

mu·sique con·crète \myů-'zēk-kōn-'kret, mue-\ *noun* [French, literally, concrete music] (1952)
: a recorded montage of natural sounds often electronically modified and presented as a musical composition

musk \'məsk\ *noun* [Middle English *muske,* from Middle French *musc,* from Late Latin *muscus,* from Late Greek *moschos,* from (assumed) Middle Persian *mušk-,* from Sanskrit *muṣka* testicle, from diminutive of *mūṣ* mouse; akin to Old English *mūs* mouse] (14th century)
1 a : a substance with a penetrating persistent odor obtained from a sac beneath the abdominal skin of the male musk deer and used as a perfume fixative; *also* **:** a similar substance from another animal or a synthetic substitute **b :** the odor of musk; *also* **:** an odor resembling musk especially in heaviness or persistence
2 : any of various plants with musky odors; *especially* **:** MUSK PLANT

musk deer *noun* (1681)
: any of a genus (*Moschus*) of small heavy-limbed hornless deer of central Asian uplands with tusked musk-producing males

mus·keg \'məs-ˌkeg, -ˌkāg\ *noun* [Cree *maske·k*] (1806)
1 : BOG; *especially* **:** a sphagnum bog of northern North America often with tussocks
2 : a usually thick deposit of partially decayed vegetable matter of wet boreal regions

mus·kel·lunge \'məs-kə-ˌlənj\ *noun, plural* **muskellunge** [alteration of Canadian French *maskinongé,* from Ojibwa *ma·skino·še·*] (1789)
: a large North American pike (*Esox masquinongy*) that has dark markings, may weigh over 60 pounds (27 kilograms), and is a valuable sport fish

mus·ket \'məs-kət\ *noun* [Middle French *mousquet,* from Old Italian *moschetto* small artillery piece, sparrow hawk, from diminutive of *mosca* fly, from Latin *musca* — more at MIDGE] (circa 1587)

: a heavy large-caliber muzzle-loading usually smoothbore shoulder firearm; *broadly* **:** a shoulder gun carried by infantry ◆

mus•ke•teer \ˌməs-kə-ˈtir\ *noun* [modification of Middle French *mousquetaire,* from *mousquet*] (1590)
1 : a soldier armed with a musket
2 [from the musketeers' friendship in the novel *Les Trois Mousquetaires* (1844) by Alexandre Dumas] **:** a good friend **:** BUDDY

mus•ket•ry \ˈməs-kə-trē\ *noun* (1646)
1 : MUSKETS
2 : MUSKETEERS
3 a : musket fire **b :** the art or science of using small arms especially in battle

mus•kie *or* **mus•ky** \ˈməs-kē\ *noun, plural* **muskies** (1894)
: MUSKELLUNGE

musk•mel•on \ˈməsk-ˌme-lən\ *noun* (1573)
: a usually sweet musky-odored edible melon that is the fruit of a trailing or climbing Asian herbaceous vine (*Cucumis melo*): as **a :** any of various melons of small or moderate size with netted skin that include most of the muskmelons cultivated in North America **b :** CANTALOUPE 1 **c :** WINTER MELON

Mus•ko•ge•an *or* **Mus•kho•ge•an** \(ˌ)məs-ˈkō-gē-ən\ *noun* (1891)
: a language family of southeastern U.S. that includes Muskogee

Mus•ko•gee \mə-ˈskō-gē\ *noun, plural* **Muskogee** *or* **Muskogees** (1775)
1 : a member of an American Indian people of Georgia and eastern Alabama constituting the nucleus of the Creek confederacy
2 : the language of the Muskogees and of some of the Seminoles

musk ox *noun* (1744)
: a heavyset shaggy-coated wild ox (*Ovibos moschatus*) now confined to Greenland and the barren northern lands of North America

musk ox

musk plant *noun* (1852)
: a yellow-flowered North American herb (*Mimulus moschatus*) of the snapdragon family that has hairy foliage and sometimes a musky odor

musk•rat \ˈməs-ˌkrat\ *noun, plural* **muskrat** *or* **muskrats** [probably by folk etymology from a word of Algonquian origin; akin to Massachuset *musquash* muskrat] (1607)
: an aquatic rodent (*Ondatra zibethica*) of the U.S. and Canada with a long scaly laterally compressed tail, webbed hind feet, and dark glossy brown fur; *also* **:** its fur or pelt

musk rose *noun* (1577)
: a rose (*Rosa moschata*) of the Mediterranean region with white flowers having a musky odor

musk thistle *noun* (1731)
: a Eurasian thistle (*Carduus nutans*) that has nodding musky flower heads and is naturalized in eastern North America

musk turtle *noun* (1868)
: any of various small American freshwater turtles (genera *Sternotherus* and *Kinosternon*) that have musk glands; *especially* **:** a dark turtle (*S. odoratus*) having a strong musky odor

musky \ˈməs-kē\ *adjective* **musk•i•er; -est** (1613)
: having an odor of or resembling musk
— musk•i•ness *noun*

Mus•lim \ˈməz-ləm, ˈmu̇s-, ˈmu̇z-\ *noun* [Arabic *muslim,* literally, one who surrenders (to God)] (1615)
1 : an adherent of Islam
2 : BLACK MUSLIM
— Muslim *adjective*

Muslim era *noun* (1948)
: ISLAMIC ERA

mus•lin \ˈməz-lən\ *noun* [French *mousseline,* from Italian *mussolina,* from Arabic *mawṣilīy* of Mosul, from al-*Mawṣil* Mosul, Iraq] (1609)
: a plain-woven sheer to coarse cotton fabric

mus•quash \ˈməs-ˌkwäsh, -ˌkwȯsh\ *noun* [Massachuset] (1633)
: MUSKRAT

¹muss \ˈməs\ *noun* [origin unknown] (1591)
1 *obsolete* **a :** a game in which players scramble for small objects thrown to the ground **b :** SCRAMBLE
2 *slang* **:** a confused conflict **:** ROW
3 : a state of disorder **:** MESS

²muss *transitive verb* (1837)
: to make untidy **:** DISARRANGE

mus•sel \ˈmə-səl\ *noun* [Middle English *muscle,* from Old English *muscelle,* from (assumed) Vulgar Latin *muscula,* from Latin *musculus* muscle, mussel] (before 12th century)
1 : a marine bivalve mollusk (especially genus *Mytilus*) usually having a dark elongated shell
2 : a freshwater bivalve mollusk (as of *Unio, Anodonta,* or related genera) that is especially abundant in rivers of the central U.S. and has a shell with a lustrous nacreous lining

Mus•sul•man *also* **Mus•sal•man** \ˈmə-səl-mən\ *noun, plural* **Mus•sul•men** \-mən\ *or* **Mussulmans** [Turkish *müslüman* & Persian *musulmān,* modification of Arabic *muslim*] (circa 1583)
: MUSLIM

mussy \ˈmə-sē\ *adjective* **muss•i•er; -est** (circa 1859)
: characterized by clutter or muss **:** MESSY
— muss•i•ly \ˈmə-sə-lē\ *adverb*
— muss•i•ness \ˈmə-sē-nəs\ *noun*

¹must \məs(t), ˈməst\ *verb, present & past all persons* **must** [Middle English *moste,* from Old English *mōste,* past indicative & subjunctive of *mōtan* to be allowed to, have to; akin to Old High German *muozan* to be allowed to, have to]
verbal auxiliary
(before 12th century)
1 a : be commanded or requested to ⟨you *must* stop⟩ **b :** be urged to **:** ought by all means to ⟨you *must* read that book⟩
2 : be compelled by physical necessity to ⟨man *must* eat to live⟩ **:** be required by immediate or future need or purpose to ⟨we *must* hurry to catch the bus⟩
3 a : be obliged to **:** be compelled by social considerations to ⟨I *must* say you're looking well⟩ **b :** be required by law, custom, or moral conscience to ⟨we *must* obey the rules⟩ **c :** be determined to ⟨if you *must* go at least wait for me⟩ **d :** be unreasonably or perversely compelled to ⟨why *must* you be so stubborn⟩
4 : be logically inferred or supposed to ⟨it *must* be time⟩
5 : be compelled by fate or by natural law to ⟨what *must* be will be⟩
6 : was or were presumably certain to **:** was or were bound to ⟨if he really was there I *must* have seen him⟩
7 *dialect* **:** MAY, SHALL — used chiefly in questions
intransitive verb
archaic **:** to be obliged to go ⟨I *must* to Coventry —Shakespeare⟩

²must \ˈməst\ *noun* (1616)
1 : an imperative need or duty **:** REQUIREMENT
2 : an indispensable item **:** ESSENTIAL ⟨exercise is a *must*⟩

³must *noun* [Middle English, from Old English, from Latin *mustum*] (before 12th century)
: the expressed juice of fruit and especially grapes before and during fermentation; *also* **:** the pulp and skins of the crushed grapes

⁴must *noun* [Middle English (Scots) *moist,* from Middle French *must,* alteration of *musc* musk] (15th century)
1 : MUSK
2 : MOLD, MUSTINESS

mus•tache \ˈməs-ˌtash, (ˌ)mə-ˈstash\ *noun* [Middle French *moustache,* from Old Italian *mustaccio,* from Middle Greek *moustaki,* diminutive of Greek *mystak-, mystax* upper lip, mustache] (1585)
1 : the hair growing on the human upper lip; *especially* **:** such hair grown and often trimmed in a particular style
2 : hair or bristles about the mouth of a mammal
— mus•tached *adjective*

mus•ta•chio \(ˌ)mə-ˈsta-shē-ˌō, -ˈstä-, -shō\ *noun, plural* **-chios** [Spanish & Italian; Spanish *mostacho,* from Italian *mustaccio,* from Old Italian] (1588)
: MUSTACHE; *especially* **:** a large mustache
— mus•ta•chioed \-shē-ˌōd, -shōd\ *adjective*

mus•tang \ˈməs-ˌtaŋ\ *noun* [Mexican Spanish *mestengo,* from Spanish, stray, from *mesteño* strayed, from *mesta* annual roundup of cattle that disposed of strays, from Medieval Latin (*animalia*) *mixta* mixed animals] (1808)
1 : the small hardy naturalized horse of the western plains directly descended from horses brought in by the Spaniards; *also* **:** BRONC
2 *slang* **:** a commissioned officer (as in the U.S. Navy) who has risen from the ranks

mus•tard \ˈməs-tərd\ *noun* [Middle English, from Old French *mostarde,* from *moust* must, from Latin *mustum*] (13th century)
1 a : a pungent yellow powder of the seeds of any of several common mustards (*Brassica hirta, B. nigra,* or *B. juncea*) used as a condiment or in medicine as a stimulant and diuretic, an emetic, or a counterirritant **b** *slang* **:** ZEST
2 : any of several herbs (genus *Brassica* of the family Cruciferae, the mustard family) with lobed leaves, yellow flowers, and linear beaked pods
3 : a dark to moderate yellow
— mus•tardy \-tər-dē\ *adjective*

mustard gas *noun* (1917)
: an irritant vesicant oily liquid $C_4H_8Cl_2S$ used especially as a chemical weapon

mustard plaster *noun* (1810)
: a counterirritant and rubefacient plaster containing powdered mustard

¹mus•ter \ˈməs-tər\ *noun* [Middle English *mustre,* from Middle French *mostre, monstre,* from *monstrer*] (14th century)
1 : a representative specimen **:** SAMPLE
2 a : an act of assembling; *specifically* **:** formal military inspection **b :** critical examination **c :** an assembled group **:** COLLECTION **d :** INVENTORY

²muster *verb* **mus•tered; mus•ter•ing** \-t(ə-)riŋ\ [Middle English *mustren* to show,

◇ WORD HISTORY
musket In the early era of firearms, cannons of lesser size such as the falconet sometimes took their names from birds of prey. Following this pattern, Italians applied *moschetto* or *moschetta* "sparrow hawk" (a diminutive of *mosca* "fly") to a small-caliber piece of ordnance in the 16th century. Spaniards borrowed this word as *mosquete* and the French as *mosquet* (later *mousquet*), but applied it to a heavy shoulder firearm rather than a cannon; English *musket* was borrowed soon thereafter from French. The musket was not essentially different from the older harquebus, but unlike *harquebus,* which fell from use, the word *musket* was retained after the original matchlock firing mechanism was replaced by a wheel lock, and the wheel lock by the flintlock. As the practice of rifling firearms—incising the barrel with spiral grooves to improve the bullet's accuracy—became more common, *musket* gradually gave way to the newer word *rifle* in the 18th and 19th centuries.

muster, from Middle French *monstrer,* from Latin *monstrare* to show, from *monstrum* evil omen, monster — more at MONSTER] (15th century)
transitive verb
1 a : to cause to gather **:** CONVENE **b :** to enroll formally — usually used with *in* or *into* ⟨was *mustered* into the army⟩ **c :** to call the roll of
2 a : to bring together **:** COLLECT **b :** to call forth **:** ROUSE
3 : to amount to **:** COMPRISE
intransitive verb
: to come together **:** CONGREGATE
synonym see SUMMON
muster out *transitive verb* (1834)
: to discharge from service
muster roll *noun* (1605)
: INVENTORY, ROSTER; *specifically* **:** a register of the officers and men in a military unit or ship's company
musth *also* **must** \'məst\ *noun* [Hindi *mast* intoxicated, from Persian] (1878)
: a periodic state of the bull elephant characterized especially by aggressive behavior and usually connected with the rutting season
mustn't \'mə-sᵊnt\ (1741)
: must not
must–see \'məst-'sē\ *noun* (1946)
: something (as a film) that must or should be seen
— must–see *adjective*
musty \'məs-tē\ *adjective* **must·i·er; -est** (1530)
1 a : impaired by damp or mildew **:** MOLDY **b :** tasting of mold **c :** smelling of damp and decay **:** FUSTY
2 a : TRITE, STALE **b** (1) **:** ANTIQUATED (2) **:** SUPERANNUATED
synonym see MALODOROUS
— must·i·ly \'məs-tə-lē\ *adverb*
— must·i·ness \-tē-nəs\ *noun*
mu·ta·ble \'myü-tə-bəl\ *adjective* [Middle English, from Latin *mutabilis,* from *mutare* to change; akin to Old English *mīthan* to conceal, Sanskrit *mināti* he exchanges, deceives] (14th century)
1 : prone to change **:** INCONSTANT
2 a : capable of change or of being changed **b :** capable of or liable to mutation
— mu·ta·bil·i·ty \ˌmyü-tə-'bi-lə-tē\ *noun*
— mu·ta·bly \'myü-tə-blē\ *adverb*
mu·ta·gen \'myü-tə-jən\ *noun* [International Scientific Vocabulary *muta*tion + *-gen*] (1933)
: an agent (as mustard gas or various radiations) that tends to increase the frequency or extent of mutation
— mu·ta·gen·ic \ˌmyü-tə-'je-nik\ *adjective*
— mu·ta·gen·i·cal·ly \-ni-k(ə-)lē\ *adverb*
mu·ta·gen·e·sis \ˌmyü-tə-'je-nə-səs\ *noun* [New Latin] (1948)
: the occurrence or induction of mutation
mu·ta·ge·nic·i·ty \-jə-'ni-sə-tē\ *noun* (1956)
: the capacity to induce mutations
mu·tant \'myü-tᵊnt\ *adjective* [Latin *mutant-, mutans,* present participle of *mutare*] (1903)
: of, relating to, or produced by mutation
— mutant *noun*
mu·tase \'myü-ˌtās, -ˌtāz\ *noun* [International Scientific Vocabulary *mut-* (from Latin *mutare*) + *-ase*] (1938)
: any of various enzymes that catalyze molecular rearrangements and especially those involving the transfer of phosphate from one hydroxyl group to another in the same molecule
mu·tate \'myü-ˌtāt, myü-'\ *verb* **mu·tat·ed; mu·tat·ing** [Latin *mutatus,* past participle of *mutare*] (1818)
intransitive verb
: to undergo mutation
transitive verb
: to cause to undergo mutation
— mu·ta·tive \'myü-ˌtā-tiv, -tə-tiv\ *adjective*
mu·ta·tion \myü-'tā-shən\ *noun* (14th century)
1 : a significant and basic alteration **:** CHANGE
2 : UMLAUT
3 a : a relatively permanent change in hereditary material involving either a physical change in chromosome relations or a biochemical change in the codons that make up genes; *also* **:** the process of producing a mutation **b :** an individual strain or trait resulting from mutation
— mu·ta·tion·al \-shnəl, -shə-nᵊl\ *adjective*
— mu·ta·tion·al·ly *adverb*
mu·ta·tis mu·tan·dis \m(y)ü-'tä-təs-m(y)ü-'tän-dəs, -'tā-təs-, -'tan-\ *adverb* [Middle English, from Medieval Latin] (15th century)
1 : with the necessary changes having been made
2 : with the respective differences having been considered
mutch·kin \'məch-kən\ *noun* [Middle English (Scots) *muchekyn*] (15th century)
: a Scottish unit of liquid capacity equal to 0.90 pint (0.42 liter)
[1]**mute** \'myüt\ *adjective* **mut·er; mut·est** [alteration of Middle English *muet,* from Middle French, from Old French *mu,* from Latin *mutus,* probably from *mu,* representation of a muttered sound] (1513)
1 : unable to speak **:** DUMB
2 : characterized by absence of speech: as **a :** felt or experienced but not expressed ⟨touched her hand in *mute* sympathy⟩ **b :** refusing to plead directly or stand trial ⟨the prisoner stands *mute*⟩
3 : remaining silent, undiscovered, or unrecognized
4 a : contributing nothing to the pronunciation of a word ⟨the *b* in *plumb* is *mute*⟩ **b :** contributing to the pronunciation of a word but not representing the nucleus of a syllable ⟨the *e* in *mate* is *mute*⟩
— mute·ly *adverb*
— mute·ness *noun*
[2]**mute** *noun* (1530)
1 : STOP 9
2 : a person who cannot or does not speak
3 : a device attached to a musical instrument to reduce, soften, or muffle its tone
[3]**mute** *transitive verb* **mut·ed; mut·ing** (1883)
1 : to muffle, reduce, or eliminate the sound of
2 : to tone down **:** SOFTEN, SUBDUE ⟨*mute* a color⟩
[4]**mute** *intransitive verb* **mut·ed; mut·ing** [Middle English, from Middle French *meutir,* short for *esmeutir,* from Old French *esmeltir,* of Germanic origin; akin to Middle Dutch *smelten* to melt, defecate (used of birds)] (15th century)
of a bird **:** to evacuate the cloaca
mut·ed \'myü-təd\ *adjective* (1861)
1 : provided with or produced or modified by the use of a mute
2 a : being mute **:** SILENT **b :** toned down **:** LOW-KEY, SUBDUED
— mut·ed·ly *adverb*
mute swan *noun* (1785)
: a common white swan (*Cygnus olor*) that produces no loud notes, is native to Europe and western Asia, and has been introduced into parts of the U.S.
mu·ti·late \'myü-tᵊl-ˌāt\ *transitive verb* **-lat·ed; -lat·ing** [Latin *mutilatus,* past participle of *mutilare,* from *mutilus* truncated, maimed] (1534)
1 : to cut up or alter radically so as to make imperfect ⟨the child *mutilated* the book with his scissors⟩
2 : to cut off or permanently destroy a limb or essential part of **:** CRIPPLE
synonym see MAIM
— mu·ti·la·tion \ˌmyü-tᵊl-'ā-shən\ *noun*
— mu·ti·la·tor \'myü-tᵊl-ˌā-tər\ *noun*
mu·tine \'myü-tᵊn\ *intransitive verb* **mu·tined; mu·tin·ing** [Middle French (*se*) *mutiner*] (1555)
obsolete **:** REBEL, MUTINY
mu·ti·neer \ˌmyü-tᵊn-'ir\ *noun* (1610)
: one that mutinies
mu·ti·nous \'myü-tᵊn-əs, 'myüt-nəs\ *adjective* (1578)
1 a : disposed to or in a state of mutiny **:** REBELLIOUS **b :** TURBULENT, UNRULY
2 : of, relating to, or constituting mutiny
— mu·ti·nous·ly *adverb*
— mu·ti·nous·ness *noun*
mu·ti·ny \'myü-tᵊn-ē, 'myüt-nē\ *noun, plural* **-nies** [*mutine* to rebel, from Middle French (*se*) *mutiner,* from *mutin* mutinous, from *meute* revolt, from (assumed) Vulgar Latin *movita,* from feminine of *movitus,* alteration of Latin *motus,* past participle of *movēre* to move] (1567)
1 *obsolete* **:** TUMULT, STRIFE
2 : forcible or passive resistance to lawful authority; *especially* **:** concerted revolt (as of a naval crew) against discipline or a superior officer
synonym see REBELLION
— mutiny *intransitive verb*
mut·ism \'myü-ˌti-zəm\ *noun* [French *mutisme,* from Latin *mutus* mute] (1824)
: the condition of being mute
mutt \'mət\ *noun* [short for *muttonhead* dull-witted person] (1901)
1 : a stupid or insignificant person **:** FOOL
2 : a mongrel dog **:** CUR
mut·ter \'mə-tər\ *verb* [Middle English *muteren,* of imitative origin] (14th century)
intransitive verb
1 : to utter sounds or words indistinctly or with a low voice and with the lips partly closed
2 : to murmur complainingly or angrily **:** GRUMBLE
transitive verb
: to utter especially in a low or imperfectly articulated manner
— mutter *noun*
— mut·ter·er \-tər-ər\ *noun*
mut·ton \'mə-tᵊn\ *noun* [Middle English *motoun,* from Old French *moton* ram, wether, of Celtic origin; akin to Old Irish *molt* wether] (13th century)
: the flesh of a mature sheep used for food
— mut·tony \'mə-tᵊn-ē, 'mət-nē\ *adjective*
mut·ton·chops \'mə-tᵊn-ˌchäps\ *noun plural* (1865)
: side-whiskers that are narrow at the temple and broad and round by the lower jaws — called also *muttonchop whiskers*
mut·ton·fish \-ˌfish\ *noun* [from its flavor] (1735)
: a common snapper (*Lutjanus analis*) of the warmer parts of the western Atlantic that is usually olive green and sometimes nearly white or tinged with rosy red and that is a commercially important food and sport fish — called also *mutton snapper*
mu·tu·al \'myü-chə-wəl, -chəl; 'myüch-wəl\ *adjective* [Middle English, from Middle French *mutuel,* from Latin *mutuus* lent, borrowed, mutual, from *mutare* to change — more at MUTABLE] (15th century)
1 a : directed by each toward the other or the others ⟨*mutual* affection⟩ **b :** having the same feelings one for the other ⟨they had long been *mutual* enemies⟩ **c :** shared in common ⟨enjoying their *mutual* hobby⟩ **d :** JOINT
2 : characterized by intimacy
3 : of or relating to a plan whereby the members of an organization share in the profits and expenses; *specifically* **:** of, relating to, or taking the form of an insurance method in which

the policyholders constitute the members of the insuring company
— **mu·tu·al·ly** *adverb*

mutual fund *noun* (1934)
: an open-end investment company that invests money of its shareholders in a usually diversified group of securities of other corporations

mu·tu·al·ism \'myü-chə-wə-ˌli-zəm, 'myü-chə-ˌli-, 'myüch-wə-ˌli-\ *noun* (1849)
1 : the doctrine or practice of mutual dependence as the condition of individual and social welfare
2 : mutually beneficial association between different kinds of organisms
— **mu·tu·al·ist** \-list\ *noun*
— **mu·tu·al·is·tic** \ˌmyü-chə-wə-'lis-tik, ˌmyü-chə-'lis-, ˌmyüch-wə-'lis-\ *adjective*

mu·tu·al·i·ty \ˌmyü-chə-'wa-lə-tē\ *noun* (circa 1586)
1 : the quality or state of being mutual
2 : a sharing of sentiments **:** INTIMACY

mu·tu·al·ize \'myü-chə-wə-ˌlīz, 'myü-chə-ˌlīz, 'myüch-wə-ˌlīz\ *transitive verb* **-ized; -iz·ing** (1812)
: to make mutual
— **mu·tu·al·i·za·tion** \ˌmyü-chə-wə-lə-'zā-shən, ˌmyü-chə-lə-, ˌmyüch-wə-lə-\ *noun*

mutually exclusive *adjective* (1874)
: being related such that each excludes or precludes the other ⟨*mutually exclusive* events⟩; *also* **:** INCOMPATIBLE ⟨their outlooks were not *mutually exclusive*⟩

mutuel *noun* (1908)
: PARI-MUTUEL

muu·muu \'mü-ˌmü\ *noun* [Hawaiian *mu'umu'u,* from *mu'umu'u* cut off] (1923)
: a loose often long dress having bright colors and patterns and adapted from the dresses originally distributed by missionaries to the native women of Hawaii

Mu·zak \'myü-ˌzak\ *trademark*
— used for recorded background music transmitted by wire to the loudspeaker of a subscriber (as an office)

mu·zhik \mü-'zhēk, -'zhik\ *noun* [Russian] (1568)
: a Russian peasant

¹muz·zle \'mə-zəl\ *noun* [Middle English *musell,* from Middle French *musel,* from diminutive of *muse* mouth of an animal, from Medieval Latin *musus*] (15th century)
1 : the projecting jaws and nose of an animal **:** SNOUT
2 a : a fastening or covering for the mouth of an animal used to prevent eating or biting **b :** something (as censorship) that restrains normal expression
3 : the open end of an implement; *especially* **:** the discharging end of a weapon

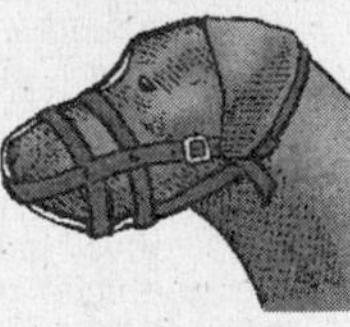
muzzle 2a

²muzzle *transitive verb* **muz·zled; muz·zling** \'məz-liŋ, 'mə-zə-\ (15th century)
1 : to fit with a muzzle
2 : to restrain from expression **:** GAG
— **muz·zler** \-lər\ *noun*

muz·zy \'mə-zē\ *adjective* **muz·zi·er; -est** [perhaps blend of *muddled* and *fuzzy*] (circa 1728)
1 a : deficient in brightness **:** DULL, GLOOMY ⟨a *muzzy* day⟩ **b :** lacking in clarity and precision ⟨his conclusions can be *muzzy* and naive —*Times Literary Supplement*⟩
2 : muddled or confused in mind
— **muz·zi·ly** \'mə-zə-lē\ *adverb*
— **muz·zi·ness** \'mə-zē-nəs\ *noun*

MX \ˌem-'eks\ *noun* [*m*issile, e*x*perimental] (1976)
: a mobile ICBM having up to 10 independently targeted nuclear warheads

my \'mī, mə\ *adjective* [Middle English, from Old English *mīn,* from *mīn,* suppletive genitive of *ic* I; akin to Old English *mē* me] (12th century)
1 : of or relating to me or myself especially as possessor, agent, or object of an action ⟨*my* car⟩ ⟨*my* promise⟩ ⟨*my* injuries⟩
2 — used interjectionally to express surprise and sometimes reduplicated ⟨*my* oh *my*⟩; used also interjectionally with names of various parts of the body to express doubt or disapproval ⟨*my* foot⟩

my- *or* **myo-** *combining form* [New Latin, from Greek, from *mys* mouse, muscle — more at MOUSE]
: muscle ⟨*myo*globin⟩ **:** muscle and ⟨*myo*neural⟩

my·al·gia \mī-'al-j(ē-)ə\ *noun* [New Latin] (1860)
: pain in one or more muscles
— **my·al·gic** \-jik\ *adjective*

my·as·the·nia \ˌmī-əs-'thē-nē-ə\ *noun* [New Latin] (circa 1856)
: muscular debility
— **my·as·then·ic** \-'the-nik\ *adjective or noun*

myasthenia gra·vis \-'gra-vəs, -'grä-\ *noun* [New Latin, literally, grave myasthenia] (1900)
: a disease characterized by progressive weakness and exhaustibility of voluntary muscles without atrophy or sensory disturbance and caused by an autoimmune attack on acetylcholine receptors at the neuromuscular junction

myc- *or* **myco-** *combining form* [New Latin, from Greek *mykēt-, mykēs* fungus; akin to Greek *myxa* nasal mucus]
: fungus ⟨*myco*logy⟩ ⟨*myco*sis⟩

my·ce·li·um \mī-'sē-lē-əm\ *noun, plural* **-lia** \-lē-ə\ [New Latin, from *myc-* + Greek *hēlos* nail, wart, callus] (1836)
: the mass of interwoven filamentous hyphae that forms especially the vegetative portion of the thallus of a fungus and is often submerged in another body (as of soil or organic matter or the tissues of a host); *also* **:** a similar mass of filaments formed by some bacteria (as streptomyces)
— **my·ce·li·al** \-əl\ *adjective*

My·ce·nae·an \ˌmī-sə-'nē-ən\ *also* **My·ce·ni·an** \mī-'sē-nē-ən\ *adjective* (1598)
1 : of, relating to, or characteristic of Mycenae, its people, or the period (1400 to 1100 B.C.) of Mycenae's political ascendancy
2 : characteristic of the Bronze Age Mycenaean culture of the eastern Mediterranean area
— **Mycenaean** *noun*

my·ce·to·ma \ˌmī-sə-'tō-mə\ *noun, plural* **-mas** *or* **-ma·ta** \-mə-tə\ [New Latin, from Greek *mykēt-, mykēs*] (1874)
: a condition marked by invasion of the deep subcutaneous tissues with fungi or actinomycetes; *also* **:** a tumorous mass occurring in such a condition
— **my·ce·to·ma·tous** \-mə-təs\ *adjective*

my·ce·toph·a·gous \ˌmī-sə-'tä-fə-gəs\ *adjective* [Greek *mykēt-, mykēs* + English *-phagous*] (circa 1890)
: feeding on fungi

my·ce·to·zo·an \mī-ˌsē-tə-'zō-ən\ *noun* [New Latin *Mycetozoa,* order of protozoans, from Greek *mykēt-, mykēs* + New Latin *-zoa*] (1881)
: SLIME MOLD
— **mycetozoan** *adjective*

-mycin *noun combining form* [*streptomycin*]
: substance obtained from a fungus-like bacterium ⟨erythro*mycin*⟩

my·co·bac·te·ri·um \ˌmī-kō-bak-'tir-ē-əm\ *noun* [New Latin] (1909)
: any of a genus (*Mycobacterium*) of nonmotile aerobic acid-fast bacteria that include numerous saprophytes and the pathogens causing tuberculosis and leprosy
— **my·co·bac·te·ri·al** \-ē-əl\ *adjective*

my·co·flo·ra \ˌmī-kə-'flōr-ə, -'flȯr-\ *noun* [New Latin] (1945)
: the fungi characteristic of a region or special environment

my·col·o·gy \mī-'kä-lə-jē\ *noun* [New Latin *mycologia,* from *myc-* + Latin *-logia* -logy] (1836)
1 : a branch of biology dealing with fungi
2 : fungal life
— **my·co·log·i·cal** \ˌmī-kə-'lä-ji-kəl\ *adjective*
— **my·co·log·i·cal·ly** \ˌmī-kə-'lä-ji-k(ə-)lē\ *adverb*
— **my·col·o·gist** \mī-'kä-lə-jist\ *noun*

my·coph·a·gist \mī-'kä-fə-jist\ *noun* [*mycophagy,* from *myc-* + *-phagy*] (1861)
: one that eats fungi (as mushrooms)
— **my·coph·a·gy** \-jē\ *noun*

my·coph·a·gous \-gəs\ *adjective* (circa 1909)
: feeding on fungi

my·co·phile \'mī-kō-ˌfīl\ *noun* (1953)
: a person whose hobby is hunting wild edible mushrooms

my·co·plas·ma \ˌmī-kō-'plaz-mə\ *noun, plural* **-mas** *or* **-ma·ta** \-mə-tə\ [New Latin] (1955)
: any of a genus (*Mycoplasma*) of minute pleomorphic gram-negative chiefly nonmotile microorganisms without cell walls that are intermediate in some respects between viruses and bacteria and are mostly parasitic usually in mammals — called also *pleuropneumonia-like organism*
— **my·co·plas·mal** \-məl\ *adjective*

mycoplasma–like organism *noun* (1968)
: any of a group of prokaryotic organisms that resemble mycoplasmas, cause disease in plants, are transmitted by insect vectors, and have proved extremely difficult to grow in artificial media

my·cor·rhi·za \ˌmī-kə-'rī-zə\ *noun, plural* **-zae** \-ˌzē\ *or* **-zas** [New Latin, from *myc-* + Greek *rhiza* root — more at ROOT] (1895)
: the symbiotic association of the mycelium of a fungus with the roots of a seed plant
— **my·cor·rhi·zal** \-zəl\ *adjective*

my·co·sis \mī-'kō-səs\ *noun, plural* **my·co·ses** \-ˌsēz\ [New Latin] (1876)
: infection with or disease caused by a fungus
— **my·cot·ic** \-'kä-tik\ *adjective*

my·co·tox·in \ˌmī-kə-'täk-sən\ *noun* (1962)
: a toxic substance produced by a fungus and especially a mold

my·dri·a·sis \mə-'drī-ə-səs\ *noun* [Latin, from Greek] (circa 1775)
: a long-continued or excessive dilatation of the pupil of the eye
— **myd·ri·at·ic** \ˌmi-drē-'a-tik\ *adjective or noun*

myel- *or* **myelo-** *combining form* [New Latin, from Greek, from *myelos,* probably from *myōn* cluster of muscles, from *mys* mouse, muscle — more at MOUSE]
: marrow **:** spinal cord ⟨*myel*encephalon⟩

my·e·len·ceph·a·lon \ˌmī-ə-len-'se-fə-ˌlän, -lən\ *noun* [New Latin] (1871)
: the posterior part of the developing vertebrate hindbrain or the corresponding part of the adult brain composed of the medulla oblongata
— **my·e·len·ce·phal·ic** \-ˌlen(t)-sə-'fa-lik\ *adjective*

my·e·lin \'mī-ə-lən\ *noun* [International Scientific Vocabulary] (1873)
: a soft white somewhat fatty material that forms a thick myelin sheath about the protoplasmic core of a myelinated nerve fiber
— **my·e·lin·ic** \ˌmī-ə-'li-nik\ *adjective*

my·e·lin·at·ed \'mī-ə-lə-ˌnā-təd\ *adjective* (1899)

: having a myelin sheath ⟨*myelinated* nerve fibers⟩

myelin sheath *noun* (1896)
: a layer of myelin surrounding some nerve fibers — called also *medullary sheath*

my·e·li·tis \ˌmī-ə-'lī-təs\ *noun, plural* **-lit·i·des** \-'li-tə-ˌdēz\ [New Latin] (1835)
: inflammation of the spinal cord or of the bone marrow

my·e·lo·blast \'mī-ə-lə-ˌblast\ *noun* [International Scientific Vocabulary] (circa 1904)
: a large mononuclear nongranular bone-marrow cell; *especially* : one that is a precursor of a myelocyte
— **my·e·lo·blas·tic** \ˌmī-ə-lə-'blas-tik\ *adjective*

my·e·lo·cyte \'mī-ə-lə-ˌsīt\ *noun* [International Scientific Vocabulary] (1891)
: a bone-marrow cell; *especially* : a motile cell with cytoplasmic granules that gives rise to the granulocytes of the blood and occurs abnormally in the circulating blood
— **my·e·lo·cyt·ic** \ˌmī-ə-lə-'si-tik\ *adjective*

my·e·lo·fi·bro·sis \ˌmī-ə-lō-fī-'brō-səs\ *noun* [New Latin] (1947)
: an anemic condition in which bone marrow becomes fibrotic and the liver and spleen usually exhibit a development of blood-cell precursors
— **my·e·lo·fi·brot·ic** \-'brä-tik\ *adjective*

my·e·log·e·nous \ˌmī-ə-'lä-jə-nəs\ *adjective* [International Scientific Vocabulary] (1876)
: of, relating to, originating in, or produced by the bone marrow ⟨*myelogenous* sarcoma⟩

myelogenous leukemia *noun* (1904)
: leukemia characterized by proliferation of myeloid tissue (as of the bone marrow and spleen) and an abnormal increase in the number of granulocytes, myelocytes, and myeloblasts in the circulating blood — called also *myelocytic leukemia, myeloid leukemia*

my·e·loid \'mī-ə-ˌlȯid\ *adjective* [International Scientific Vocabulary] (1857)
: of, relating to, or resembling bone marrow

my·e·lo·ma \ˌmī-ə-'lō-mə\ *noun* [New Latin] (circa 1857)
: a primary tumor of the bone marrow
— **my·e·lo·ma·tous** \-mə-təs\ *adjective*

my·e·lop·a·thy \-'lä-pə-thē\ *noun* [International Scientific Vocabulary] (circa 1891)
: a disease or disorder of the spinal cord or bone marrow
— **my·e·lo·path·ic** \ˌmī-ə-lō-'pa-thik\ *adjective*

my·e·lo·pro·lif·er·a·tive \'mī-ə-lō-prə-'li-f(ə-)rə-tiv, -fə-ˌrā-\ *adjective* (1951)
: of, relating to, or being a disorder (as leukemia) marked by excessive proliferation of bone marrow elements and especially blood cell precursors

my·ia·sis \mī-'ī-ə-səs, mē-\ *noun, plural* **my·ia·ses** \-ˌsēz\ [New Latin, from Greek *myia* fly — more at MIDGE] (1837)
: infestation with fly maggots

My·lar \'mī-ˌlär\ *trademark*
— used for a polyester film

my·nah *or* **my·na** \'mī-nə\ *noun* [Hindi *mainā* from Sanskrit *madana*, a kind of bird] (1769)
: any of various Asian starlings (especially genera *Acridotheres, Gracula,* and *Sturnus*); *especially* : a dark brown slightly crested bird (*A. tristis*) of southeastern Asia with a white tail tip and wing markings and bright yellow bill and feet — compare HILL MYNAH

myn·heer \mə-'ner\ *noun* [Dutch *mijnheer,* from *mijn* my + *heer* master, sir] (1652)
: a male Netherlander — used as a title equivalent to *Mr.*

myo- — see MY-

myo·blast \'mī-ə-ˌblast\ *noun* [International Scientific Vocabulary] (1884)
: an undifferentiated cell capable of giving rise to muscle cells

myo·car·di·tis \ˌmī-ə-(ˌ)kär-'dī-təs\ *noun* [New Latin] (1866)
: inflammation of the myocardium

myo·car·di·um \ˌmī-ə-'kär-dē-əm\ *noun* [New Latin, from *my-* + Greek *kardia* heart — more at HEART] (1879)
: the middle muscular layer of the heart wall
— **myo·car·di·al** \-dē-əl\ *adjective*

my·oc·lo·nus \ˌmī-'ä-klə-nəs\ *noun* [New Latin] (1883)
: irregular involuntary contraction of a muscle usually resulting from functional disorder of controlling motoneurons; *also* : a condition characterized by myoclonus
— **myo·clon·ic** \ˌmī-ə-'klä-nik\ *adjective*

myo·elec·tric \ˌmī-ō-i-'lek-trik\ *also* **myo·elec·tri·cal** \-tri-kəl\ *adjective* (circa 1919)
: of, relating to, or utilizing electricity generated by muscle

myo·fi·bril \ˌmī-ə-'fī-brəl, -'fi-\ *noun* [New Latin *myofibrilla,* from *my-* + *fibrilla* fibril] (1898)
: any of the longitudinal parallel contractile elements of a muscle cell that are composed of myosin and actin
— **myo·fi·bril·lar** \-brə-lər\ *adjective*

myo·fil·a·ment \-'fi-lə-mənt\ *noun* (1949)
: one of the individual filaments of actin or myosin that make up a myofibril

myo·gen·ic \-'je-nik\ *adjective* [International Scientific Vocabulary] (1904)
: taking place or functioning in ordered rhythmic fashion because of the inherent properties of cardiac muscle rather than specific neural stimuli ⟨a *myogenic* heartbeat⟩

myo·glo·bin \-'glō-bən, 'mī-ə-ˌ\ *noun* [International Scientific Vocabulary] (1925)
: a red iron containing protein pigment in muscles that is similar to hemoglobin

myo·ino·si·tol \ˌmī-ō-i-'nō-sə-ˌtȯl, -ˌtōl\ *noun* (1951)
: a biologically active inositol that is a component of the vitamin B complex and a lipotropic agent

my·ol·o·gy \mī-'ä-lə-jē\ *noun* [French or New Latin; French *myologie,* from New Latin *myologia,* from *my-* + Latin *-logia* -logy] (1649)
: a scientific study of muscles

my·o·ma \mī-'ō-mə\ *noun, plural* **-mas** *or* **-ma·ta** \-mə-tə\ [New Latin] (1875)
: a tumor consisting of muscle tissue
— **my·o·ma·tous** \-mə-təs\ *adjective*

myo·neu·ral \ˌmī-ə-'nu̇r-əl, -'nyu̇r-\ *adjective* (1905)
: of, relating to, or connecting muscles and nerves ⟨*myoneural* junctions⟩

my·op·a·thy \mī-'ä-pə-thē\ *noun* [International Scientific Vocabulary] (circa 1849)
: a disorder of muscle tissue or muscles
— **myo·path·ic** \ˌmī-ə-'pa-thik\ *adjective*

my·ope \'mī-ˌōp\ *noun* [French, from Late Latin *myops* myopic, from Greek *myōps,* from *myein* to be closed + *ōps* eye, face — more at EYE] (1728)
: a myopic person

my·o·pia \mī-'ō-pē-ə\ *noun* [New Latin, from Greek *myōpia,* from *myōp-, myōps*] (circa 1752)
1 : a condition in which the visual images come to a focus in front of the retina of the eye resulting especially in defective vision of distant objects
2 : a lack of foresight or discernment : a narrow view of something
— **my·o·pic** \-'ō-pik, -'ä-\ *adjective*
— **my·o·pi·cal·ly** \-pi-k(ə-)lē\ *adverb*

my·o·sin \'mī-ə-sən\ *noun* [International Scientific Vocabulary *my-* + ^{2}*-ose* + ^{1}*-in*] (1942)
: a fibrous globulin of muscle that can split ATP and that reacts with actin to form actomyosin

my·o·sis, my·ot·ic *variant of* MIOSIS, MIOTIC

myo·si·tis \ˌmī-ə-'sī-təs\ *noun* [New Latin, irregular from Greek *mys* muscle, mouse] (circa 1819)
: muscle soreness

myo·tome \'mī-ə-ˌtōm\ *noun* [International Scientific Vocabulary] (1894)
: the portion of an embryonic somite from which skeletal musculature is produced

myo·to·nia \ˌmī-ə-'tō-nē-ə\ *noun* [New Latin] (circa 1896)
: tonic spasm of one or more muscles; *also* : a condition characterized by such spasms
— **myo·ton·ic** \-'tä-nik\ *adjective*

^{1}**myr·i·ad** \'mir-ē-əd\ *noun* [Greek *myriad-, myrias,* from *myrioi* countless, ten thousand] (1555)
1 : ten thousand
2 : a great number ⟨a *myriad* of ideas⟩

^{2}**myriad** *adjective* (1791)
1 : INNUMERABLE ⟨those *myriad* problems⟩
2 : having innumerable aspects or elements ⟨the *myriad* activity of the new land —Meridel Le Sueur⟩

myr·ia·pod *also* **myr·io·pod** \'mir-ē-ə-ˌpäd\ *noun* [ultimately from Greek *myrioi* + *pod-, pous* foot — more at FOOT] (1826)
: any of a group (Myriapoda) of arthropods having the body made up of numerous similar segments nearly all of which bear true jointed legs and including the millipedes and centipedes
— **myriapod** *also* **myriopod** *adjective*

my·ris·tic acid \mə-'ris-tik-, mī-\ *noun* [International Scientific Vocabulary, from New Latin *Myristica,* genus of trees] (1848)
: a crystalline fatty acid $C_{14}H_{28}O_2$ occurring especially in the form of glycerides in most fats

myrmeco- *combining form* [Greek *myrmēko-,* from *myrmēk-, myrmēx* — more at PISMIRE]
: ant ⟨*myrmeco*philous⟩

myr·me·col·o·gy \ˌmər-mə-'kä-lə-jē\ *noun* [International Scientific Vocabulary] (circa 1902)
: the scientific study of ants
— **myr·me·co·log·i·cal** \-kə-'lä-ji-kəl\ *adjective*
— **myr·me·col·o·gist** \-'kä-lə-jist\ *noun*

myr·me·co·phile \'mər-mi-kə-ˌfīl\ *noun* [International Scientific Vocabulary] (1898)
: an organism that habitually shares an ant nest

myr·me·coph·i·lous \ˌmər-mə-'kä-fə-ləs\ *adjective* (1866)
: fond of, associated with, or benefited by ants

myr·mi·don \'mər-mə-ˌdän, -dən\ *noun* [Middle English *Mirmydon,* Latin *Myrmidon-, Myrmido,* from Greek *Myrmidōn*] (15th century)
1 *capitalized* : a member of a legendary Thessalian people who accompanied their king Achilles in the Trojan War
2 : a loyal follower; *especially* : a subordinate who executes orders unquestioningly or unscrupulously

my·rob·a·lan \mī-'rä-bə-lən, mə-\ *noun* [Middle French *mirobolan,* from Latin *myrobalanus,* from Greek *myrobalanos,* from *myron* unguent + *balanos* acorn — more at GLAND] (circa 1530)
: the dried astringent fruit of an East Indian tree (genus *Terminalia* of the family Combretaceae) used chiefly in tanning and in inks

myrrh \'mər\ *noun* [Middle English *myrre,* from Old English, from Latin *myrrha,* from Greek, of Semitic origin; akin to Arabic *murr* myrrh] (before 12th century)
: a yellowish brown to reddish brown aromatic gum resin with a bitter slightly pungent taste obtained from a tree (especially *Commiphora*

abyssinica of the family Burseraceae) of eastern Africa and Arabia; *also* **:** a mixture of myrrh and labdanum

myr·tle \'mər-tᵊl\ *noun, often attributive* [Middle English *mirtille,* from Middle French, from Medieval Latin *myrtillus,* from Latin *myrtus,* from Greek *myrtos*] (1562)
1 a : a common evergreen bushy shrub (*Myrtus communis*) of southern Europe with oval to lance-shaped shiny leaves, fragrant white or rosy flowers, and black berries **b :** any of a family (Myrtaceae, the myrtle family) of chiefly tropical shrubs or trees to which the common myrtle belongs
2 a : ¹PERIWINKLE 1a **b :** CALIFORNIA LAUREL

myrtle 1a

my·self \mī-'self, mə-, *Southern also* -'sef\ *pronoun* (before 12th century)
1 : that identical one that is I — used reflexively ⟨I'm going to get *myself* a new suit⟩, for emphasis ⟨I *myself* will go⟩, or in absolute constructions ⟨*myself* a tourist, I nevertheless avoided other tourists⟩
2 : my normal, healthy, or sane condition ⟨didn't feel *myself* yesterday⟩ ■

mys·ta·gogue \'mis-tə-ˌgäg\ *noun* [Latin *mystagogus,* from Greek *mystagōgos,* from *mystēs* initiate (perhaps akin to Greek *myein* to be closed) + *agein* to lead — more at AGENT] (circa 1550)
1 : one who initiates another into a mystery cult
2 : one who understands or teaches mystical doctrines
— mys·ta·go·gy \-ˌgä-jē, -ˌgō-\ *noun*

mys·te·ri·ous \mis-'tir-ē-əs\ *adjective* (1600)
1 a : of, relating to, or constituting mystery ⟨the *mysterious* ways of God⟩ **b :** exciting wonder, curiosity, or surprise while baffling efforts to comprehend or identify **:** MYSTIFYING ⟨heard a *mysterious* noise⟩ ⟨a *mysterious* stranger⟩
2 : stirred by or attracted to the inexplicable
— mys·te·ri·ous·ly *adverb*
— mys·te·ri·ous·ness *noun*

¹mys·tery \'mis-t(ə-)rē\ *noun, plural* **-ter·ies** [Middle English *mysterie,* from Latin *mysterium,* from Greek *mystērion,* from *mystēs* initiate] (14th century)
1 a : a religious truth that one can know only by revelation and cannot fully understand **b** (1) **:** any of the 15 events (as the Nativity, the Crucifixion, or the Assumption) serving as a subject for meditation during the saying of the rosary (2) *capitalized* **:** a Christian sacrament; *specifically* **:** EUCHARIST **c** (1) **:** a secret religious rite believed (as in Eleusinian and Mithraic cults) to impart enduring bliss to the initiate (2) **:** a cult devoted to such rites
2 a : something not understood or beyond understanding **:** ENIGMA **b** *obsolete* **:** a private secret **c :** the secret or specialized practices or ritual peculiar to an occupation or a body of people ⟨the *mysteries* of the tailor's craft⟩ **d :** a piece of fiction dealing usually with the solution of a mysterious crime
3 : profound, inexplicable, or secretive quality or character ⟨the *mystery* of her smile⟩ ☆

²mystery *noun, plural* **-ter·ies** [Middle English, from Late Latin *misterium, mysterium,* alteration of *ministerium* service, occupation, from *minister* servant — more at MINISTER] (14th century)
1 *archaic* **:** TRADE, CRAFT
2 *archaic* **:** a body of persons engaged in a particular trade, business, or profession **:** GUILD
3 : MYSTERY PLAY

mystery play *noun* [²*mystery*] (1852)
: a medieval drama based on scriptural incidents (as the creation of the world, the Flood, or the life, death, and resurrection of Christ) — compare MIRACLE PLAY

¹mys·tic \'mis-tik\ *adjective* [Middle English *mistik,* from Latin *mysticus* of mysteries, from Greek *mystikos,* from *mystēs* initiate] (14th century)
1 : MYSTICAL 1a
2 : of or relating to mysteries or esoteric rites **:** OCCULT
3 : of or relating to mysticism or mystics
4 a : MYSTERIOUS **b :** OBSCURE, ENIGMATIC **c :** inducing a feeling of awe or wonder **d :** having magical properties

²mystic *noun* (1679)
1 : a follower of a mystical way of life
2 : an advocate of a theory of mysticism

mys·ti·cal \'mis-ti-kəl\ *adjective* (15th century)
1 a : having a spiritual meaning or reality that is neither apparent to the senses nor obvious to the intelligence ⟨the *mystical* food of the sacrament⟩ **b :** involving or having the nature of an individual's direct subjective communion with God or ultimate reality ⟨the *mystical* experience of the Inner Light⟩
2 : MYSTERIOUS, UNINTELLIGIBLE
3 : MYSTIC 2, 3
— mys·ti·cal·ly \-k(ə-)lē\ *adverb*

mys·ti·cism \'mis-tə-ˌsi-zəm\ *noun* (1736)
1 : the experience of mystical union or direct communion with ultimate reality reported by mystics
2 : the belief that direct knowledge of God, spiritual truth, or ultimate reality can be attained through subjective experience (as intuition or insight)
3 a : vague speculation **:** a belief without sound basis **b :** a theory postulating the possibility of direct and intuitive acquisition of ineffable knowledge or power

mys·ti·fi·ca·tion \ˌmis-tə-fə-'kā-shən\ *noun* (1815)
1 : an act or instance of mystifying
2 : the quality or state of being mystified
3 : something designed to mystify

mys·ti·fy \'mis-tə-ˌfī\ *transitive verb* **-fied; -fy·ing** [French *mistifier,* from *mystère* mystery, from Latin *mysterium*] (circa 1734)
1 : to perplex the mind of **:** BEWILDER
2 : to make mysterious or obscure ⟨*mystify* an interpretation of a prophecy⟩
— mys·ti·fi·er \-ˌfī(-ə)r\ *noun*
— mys·ti·fy·ing·ly \-ˌfī-iŋ-lē\ *adverb*

mys·tique \mi-'stēk\ *noun* [French, from *mystique,* adjective, mystic, from Latin *mysticus*] (1891)
1 : an air or attitude of mystery and reverence developing around something or someone
2 : the special esoteric skill essential in a calling or activity

myth \'mith\ *noun* [Greek *mythos*] (1830)
1 a : a usually traditional story of ostensibly historical events that serves to unfold part of the world view of a people or explain a practice, belief, or natural phenomenon **b :** PARABLE, ALLEGORY
2 a : a popular belief or tradition that has grown up around something or someone; *especially* **:** one embodying the ideals and institutions of a society or segment of society ⟨seduced by the American *myth* of individualism —Orde Coombs⟩ **b :** an unfounded or false notion
3 : a person or thing having only an imaginary or unverifiable existence
4 : the whole body of myths

myth·i·cal \'mith-i-kəl\ *or* **myth·ic** \-ik\ *adjective* (1669)
1 : based on or described in a myth especially as contrasted with history
2 *usually mythical* **:** existing only in the imagination **:** FICTITIOUS, IMAGINARY ⟨sportswriters picked a *mythical* all-star team⟩
3 *usually mythic* **:** having qualities suitable to myth **:** LEGENDARY ⟨the twilight of a *mythic* professional career —Clayton Riley⟩
synonym see FICTITIOUS
— myth·i·cal·ly \-thi-k(ə-)lē\ *adverb*

myth·i·cize \'mi-thə-ˌsīz\ *transitive verb* **-cized; -ciz·ing** (1840)
1 : to turn into or envelop in myth
2 : to treat as myth
— myth·i·ciz·er *noun*

myth·mak·er \'mith-ˌmā-kər\ *noun* (1871)
: a creator of myths or of mythical situations or lore
— myth·mak·ing \-kiŋ\ *noun*

my·thog·ra·phy \mi-'thä-grə-fē\ *noun* [Greek *mythographia,* from *mythos* + *-graphia* -graphy] (1851)
1 : the representation of mythical subjects in art
2 : a critical compilation of myths
— my·thog·ra·pher \-fər\ *noun*

myth·o·log·i·cal \ˌmi-thə-'lä-ji-kəl\ *also* **myth·o·log·ic** \-jik\ *adjective* (1614)
1 : of or relating to mythology or myths **:** dealt with in mythology
2 : lacking factual basis or historical validity **:** MYTHICAL, FABULOUS
— myth·o·log·i·cal·ly \-ji-k(ə-)lē\ *adverb*

my·thol·o·gize \mi-'thä-lə-ˌjīz\ *verb* **-gized; -giz·ing** (1603)
transitive verb
1 *obsolete* **:** to explain the mythological significance of
2 : to build a myth around **:** MYTHICIZE
intransitive verb
1 : to relate, classify, and explain myths
2 : to create or perpetuate myths

☆ SYNONYMS
Mystery, problem, enigma, riddle, puzzle mean something which baffles or perplexes. MYSTERY applies to what cannot be fully understood by human reason or less strictly to whatever resists or defies explanation ⟨the *mystery* of the stone monoliths⟩. PROBLEM applies to any question or difficulty calling for a solution or causing concern ⟨the *problems* created by high technology⟩. ENIGMA applies to utterance or behavior that is very difficult to interpret ⟨his suicide was an *enigma* his family never understood⟩. RIDDLE suggests an enigma or problem involving paradox or apparent contradiction ⟨the *riddle* of the reclusive pop star⟩. PUZZLE applies to an enigma or problem that challenges ingenuity for its solution ⟨the mechanisms of heredity were long a *puzzle* for scientists⟩.

□ USAGE
myself *Myself* is often used where *I* or *me* might be expected: as subject ⟨to wonder what *myself* will say —Emily Dickinson⟩ ⟨others and *myself* continued to press for the legislation⟩, after *as, than,* or *like* ⟨an aversion to paying such people as *myself* to tutor⟩ ⟨was enough to make a better man than *myself* quail⟩ ⟨old-timers like *myself*⟩, and as object ⟨now here you see *myself* with the diver⟩ ⟨for my wife and *myself* it was a happy time⟩. Such uses almost always occur when the speaker or writer is referring to himself or herself as an object of discourse rather than as a participant in discourse. The other reflexive personal pronouns are similarly but less frequently used in the same circumstances. Critics have frowned on these uses since about the turn of the century, probably unaware that they serve a definite purpose. Users themselves are as unaware as the critics—they simply follow their instincts. These uses are standard.

— **my·thol·o·giz·er** *noun*
my·thol·o·gy \mi-ˈthä-lə-jē\ *noun, plural* **-gies** [French or Late Latin; French *mythologie,* from Late Latin *mythologia* interpretation of myths, from Greek, legend, myth, from *mythologein* to relate myths, from *mythos* + *logos* speech — more at LEGEND] (1603)
1 : an allegorical narrative
2 : a body of myths: as **a :** the myths dealing with the gods, demigods, and legendary heroes of a particular people **b :** MYTHOS 2 ⟨cold war *mythology*⟩
3 : a branch of knowledge that deals with myth
4 : a popular belief or assumption that has grown up around someone or something ⟨defective *mythologies* that ignore masculine depth of feeling —Robert Bly⟩
— **my·thol·o·ger** \-jər\ *noun*
— **my·thol·o·gist** \-jist\ *noun*
mytho·ma·nia \ˌmi-thə-ˈmā-nē-ə, -nyə\ *noun* [New Latin, from Greek *mythos* + Late Latin *mania* mania] (circa 1909)
: an excessive or abnormal propensity for lying and exaggerating
— **mytho·ma·ni·ac** \-nē-ˌak\ *noun or adjective*
mytho·poe·ia \ˌmi-thə-ˈpē-ə\ *noun* [Late Latin, from Greek *mythopoiia,* from *mythopoiein* to make a myth, from *mythos* + *poiein* to make — more at POET] (1846)
: a creating of myth **:** a giving rise to myths
— **mytho·poe·ic** \-ˈpē-ik\ *also* **mytho·po·et·ic** \-pō-ˈe-tik\ *or* **mytho·po·et·i·cal** \-ti-kəl\ *adjective*
my·thos \ˈmi-ˌthōs, -ˌthäs\ *noun, plural* **my·thoi** \-ˌthȯi\ [Greek] (1753)
1 a : MYTH 1a **b :** MYTHOLOGY 2a
2 : a pattern of beliefs expressing often symbolically the characteristic or prevalent attitudes in a group or culture
3 : THEME, PLOT
mythy \ˈmi-thē\ *adjective* (1931)
: resembling, concerned with, or of a subject for myth ⟨a *mythy* theme⟩
my word *interjection* (1841)
— used to express surprise or astonishment
myx·ede·ma \ˌmik-sə-ˈdē-mə\ *noun* [New Latin, from Greek *myxa* lamp wick, nasal mucus + New Latin *edema* edema — more at MUCUS] (1877)
: severe hypothyroidism characterized by firm inelastic edema, dry skin and hair, and loss of mental and physical vigor
— **myx·ede·ma·tous** \-ˈde-mə-təs, -ˈdē-\ *adjective*
myx·o·ma \mik-ˈsō-mə\ *noun, plural* **-mas** *or* **-ma·ta** \-mə-tə\ [New Latin, from Greek *myxa*] (1870)
: a soft tumor made up of gelatinous connective tissue like that of the umbilical cord
— **myx·o·ma·tous** \-mə-təs\ *adjective*
myx·o·ma·to·sis \mik-ˌsō-mə-ˈtō-səs\ *noun* [New Latin, from *myxomat-, myxoma*] (1927)
: a condition characterized by the presence of myxomas in the body; *specifically* **:** a severe virus disease of rabbits that is caused by a poxvirus transmitted by mosquitoes and that has been used in the biological control of rabbits in plague areas
myxo·my·cete \ˌmik-sō-ˈmī-ˌsēt, ˌmik-sō-(ˌ)mī-ˈ\ *noun* [ultimately from Greek *myxa* + *mykēt-, mykēs* fungus — more at MYC-] (1877)
: SLIME MOLD
myxo·vi·rus \ˈmik-sə-ˌvī-rəs\ *noun* [New Latin, from Greek *myxa* + New Latin *virus* virus; from its affinity for certain mucins] (1955)
: any of a group of rather large RNA-containing viruses that includes the influenza viruses
— **myxo·vi·ral** \ˌmik-sə-ˈvī-rəl\ *adjective*

N *is the fourteenth letter of the English alphabet. It came through the Latin by way of Etruscan from the Greek* nu, *which was borrowed from Phoenician* nun. *It represents either of two nasal consonant sounds: one, the* n *in* new, *is called an alveolar* n, *and is usually formed in English by bringing the tip of the tongue against the gums behind the upper front teeth; the other, called a velar* n, *occurs before* k *or* g *in such words as* sink *and* sing, *and is produced by pressing the back of the tongue against the velum (or rear roof of the mouth).* N *sometimes forms the nucleus of a syllable without a vowel, as in* hidden *and* listen. *It is silent in some words after* m, *as in* hymn. *Small* n *is believed to have evolved from an early Greek form of* nu *similar to the Roman capital* N *except that its cross-stroke slanted from the lower left to the upper right. Eventually this cross-stroke rose higher along the left side of the letter while the angle it formed on the right side rounded to a single curved line.*

n \'en\ *noun, plural* **n's** *or* **ns** \'enz\ *often capitalized, often attributive* (before 12th century)
1 a : the 14th letter of the English alphabet **b :** a graphic representation of this letter **c :** a speech counterpart of orthographic *n*
2 : a graphic device for reproducing the letter *n*
3 a : one designated *n* especially as the 14th in order or class **b :** an indefinite number; *especially* **:** a constant integer or a variable taking on integral values
4 : something shaped like the letter N
5 : the haploid or gametic number of chromosomes
6 : EN 2

-n — see -EN

'n \ən, ᵊn\ *conjunction* [by shortening]
: THAN

'n' *also* **'n** \ən, ᵊn\ *conjunction*
: AND ⟨fish *'n'* chips⟩

naan *variant of* NAN

nab \'nab\ *transitive verb* **nabbed; nab·bing** [perhaps alteration of English dialect *nap*] (1686)
1 : to catch or seize in arrest **:** APPREHEND
2 : to seize suddenly

nabe \'nāb\ *noun* [by shortening & alteration from *neighborhood*] (1935)
: a neighborhood theater — usually used in plural with *the*

na·bob \'nā-ˌbäb\ *noun* [Urdu *nawwāb,* from Arabic *nuwwāb,* plural of *nā'ib* governor] (1612)
1 : a provincial governor of the Mogul empire in India
2 : a person of great wealth or prominence

Na·both \'nā-ˌbäth\ *noun* [Hebrew *Nābhōth*]
: the owner of a vineyard coveted and seized by Ahab king of Israel

na·celle \nə-'sel\ *noun* [French, literally, small boat, from Late Latin *navicella,* diminutive of Latin *navis* ship — more at NAVE] (1904)
: a streamlined enclosure (as for an engine) on an aircraft

na·cho \'nä-(ˌ)chō\ *noun, plural* **nachos** [American Spanish, perhaps from Spanish *nacho* flat-nosed] (1969)
: a tortilla chip topped with melted cheese and often additional savory toppings (as hot peppers or refried beans)

na·cre \'nā-kər\ *noun* [Middle French, from Old Italian *naccara* drum, nacre, from Arabic *naqqārah* drum] (1718)
: MOTHER-OF-PEARL
— na·cre·ous \-krē-əs, -k(ə-)rəs\ *adjective*

NAD \ˌen-(ˌ)ā-'dē\ *noun* [*n*icotinamide *a*denine *d*inucleotide] (circa 1962)
: a coenzyme $C_{21}H_{27}N_7O_{14}P_2$ of numerous dehydrogenases that occurs in most cells and plays an important role in all phases of intermediary metabolism as an oxidizing agent or when in the reduced form as a reducing agent for various metabolites — called also *nicotinamide adenine dinucleotide, diphosphopyridine nucleotide, DPN*

na·da \'nä-də\ *noun* [Spanish, from Latin (*res*) *nata* situation, circumstance, literally, a thing come into being] (1914)
: NOTHING

Na–dene *also* **Na–dé·né** \nä-'dā-(ˌ)nā, -'de-, -nē\ *noun, often D capitalized* [*na-* (from Haida *na* to dwell & Tlingit *na* people) + *Déné*] (1915)
: a hypothetically related group of American Indian languages that includes the Athabascan family, Tlingit, and Haida

NADH \en-(ˌ)ā-(ˌ)dē-'āch\ *noun* [*NAD + H,* symbol for hydrogen] (1965)
: the reduced form of NAD

na·dir \'nā-ˌdir, 'nā-dər\ *noun* [Middle English, from Middle French, from Arabic *naẓīr* opposite] (15th century)
1 : the point of the celestial sphere that is directly opposite the zenith and vertically downward from the observer
2 : the lowest point

NADP \ˌen-(ˌ)ā-(ˌ)dē-'pē\ *noun* [*n*icotinamide *a*denine *d*inucleotide *p*hosphate] (circa 1962)
: a coenzyme $C_{21}H_{28}N_7O_{17}P_3$ of numerous dehydrogenases (as that acting on glucose-6-phosphate) that occurs especially in red blood cells and plays a role in intermediary metabolism similar to NAD but acting often on different metabolites — called also *nicotinamide adenine dinucleotide phosphate, TPN, triphosphopyridine nucleotide*

NADPH \ˌen-(ˌ)ā-(ˌ)dē-(ˌ)pē-'āch\ *noun* [*NADP + H,* symbol for hydrogen] (circa 1966)
: the reduced form of NADP

¹nag \'nag\ *noun* [Middle English *nagge;* akin to Dutch *negge* small horse] (15th century)
: HORSE; *especially* **:** one that is old or in poor condition

²nag *verb* **nagged; nag·ging** [probably of Scandinavian origin; akin to Old Norse *gnaga* to gnaw; akin to Old English *gnagan* to gnaw] (circa 1828)
intransitive verb
1 : to find fault incessantly **:** COMPLAIN
2 : to be a persistent source of annoyance or distraction
transitive verb
1 : to irritate by constant scolding or urging
2 : BADGER, WORRY
— nag·ger *noun*
— nagging *adjective*
— nag·ging·ly \'na-giŋ-lē\ *adverb*

³nag *noun* (1925)
: one who nags habitually

na·ga·na \nə-'gä-nə\ *noun* [Zulu *unakane, ulunakane*] (1895)
: trypanosomiasis (especially when caused by *Trypanosoma brucei*) of domestic animals

nah \'na, 'nä, 'nȧ\ *variant of* NO

Na·huatl \'nä-ˌwä-tᵊl\ *noun* [Spanish *náhuatl,* from Nahuatl *Nāhuatl*] (1822)
: a group of closely related Uto-Aztecan languages that includes the speech of several peoples (as the Aztecs) of central and southern Mexico and Central America
— Na·huat·lan \nä-'wät-lən\ *adjective or noun*

Na·hum \'nā-əm, -həm\ *noun* [Hebrew *Naḥūm*]
1 : a Hebrew prophet of the 7th century B.C.
2 : a prophetic book of canonical Jewish and Christian Scripture — see BIBLE table

na·iad \'nā-əd, 'nī-, -ˌad\ *noun, plural* **naiads** *or* **na·ia·des** \-ə-ˌdēz\ [Middle English, from Middle French or Latin; Middle French *naïade,* from Latin *naiad-, naias,* from Greek, from *nan* to flow — more at NOURISH] (14th century)
1 : any of the nymphs in classical mythology living in and giving life to lakes, rivers, springs, and fountains
2 : any of the aquatic young of a mayfly, dragonfly, damselfly, or stone fly — compare NYMPH 3
3 : any of a genus (*Najas* of the family Najadaceae) of submerged aquatic plants

¹na·ïf *or* **na·if** \nä-'ēf\ *adjective* [French] (1598)
: NAIVE

²naïf *or* **naif** *noun* (1893)
: a naive person

¹nail \'nā(ə)l\ *noun* [Middle English, from Old English *nægl;* akin to Old High German *nagal* nail, fingernail, Latin *unguis* fingernail, toenail, claw, Greek *onyx*] (before 12th century)
1 a : a horny sheath protecting the upper end of each finger and toe of humans and most other primates **b :** a structure (as a claw) that terminates a digit and corresponds to a nail
2 : a slender usually pointed and headed fastener designed to be pounded in

²nail *transitive verb* (before 12th century)
1 : to fasten with or as if with a nail
2 : to fix in steady attention ⟨*nailed* his eye on the crack⟩
3 a : CATCH, TRAP **b :** to detect and expose usually so as to discredit
4 a : STRIKE, HIT **b :** to put out (a runner) in baseball
5 : to perform or complete perfectly or impressively ⟨*nailed* a jump shot⟩
— nail·er *noun*

nail–bit·er \'nā(ə)l-ˌbī-tər\ *noun* (1971)
: something (as a close contest) that induces tension or anxiety ⟨lost a tough, 14-inning *nail-biter* —Steve Pate⟩

nail·brush \-ˌbrəsh\ *noun* (1802)
: a small firm-bristled brush for cleaning the hands and especially the fingernails

nail down *transitive verb* (1615)
1 : to gain or win decisively ⟨*nail down* his consent⟩
2 : to settle or establish clearly and unmistakably

nail file *noun* (1875)
: a small narrow instrument (as of metal or cardboard) with a rough or emery surface that is used for shaping fingernails

nain·sook \'nān-ˌsu̇k\ *noun* [Hindi *nainsukh,* from *nain* eye + *sukh* delight] (1790)
: a soft lightweight muslin

nai·ra \'nī-rə\ *noun* [alteration of *Nigeria*] (1972)
— see MONEY table

na·ive *or* **na·ïve** \nä-'ēv\ *adjective* **na·iv·er; -est** [French *naïve,* feminine of *naïf,* from Old French, inborn, natural, from Latin *nativus* native] (1654)
1 : marked by unaffected simplicity **:** ARTLESS, INGENUOUS

2 a : deficient in worldly wisdom or informed judgment; *especially* **:** CREDULOUS **b :** not previously subjected to experimentation or a particular experimental situation ⟨made the test with *naive* rats⟩; *also* **:** not having previously used a particular drug (as marijuana)
3 : SELF-TAUGHT, PRIMITIVE
synonym see NATURAL
— na•ive•ly *or* **na•ïve•ly** *adverb*
— na•ive•ness *noun*

na•ïve•té *also* **na•ive•te** *or* **na•ive•té** \(,)nä-,ē-və-'tā, -,ē-və-; nä-'ēv-,tā, -'ē-və-\ *noun* [French *naïveté,* from Old French, inborn character, from *naif*] (1673)
1 : a naive remark or action
2 : the quality or state of being naive

na•ive•ty *also* **na•ïve•ty** \nä-'ē-və-tē, -'ēv-tē\ *noun, plural* **-ties** (1708)
chiefly British **:** NAÏVETÉ

na•ked \'nā-kəd, *especially Southern* 'ne-kəd\ *adjective* [Middle English, from Old English *nacod;* akin to Old High German *nackot* naked, Latin *nudus,* Greek *gymnos*] (before 12th century)
1 : not covered by clothing **:** NUDE
2 : devoid of customary or natural covering **:** BARE: as **a :** not enclosed in a sheath or scabbard **b :** not provided with a shade **c** *of a plant or one of its parts* **:** lacking pubescence or enveloping or subtending parts **d :** lacking foliage or vegetation **e** *of an animal or one of its parts* **:** lacking an external covering (as of hair, feathers, or shell)
3 a : scantily supplied or furnished **b :** lacking embellishment **:** UNADORNED
4 : UNARMED, DEFENSELESS
5 : lacking confirmation or support
6 : devoid of concealment or disguise
7 : unaided by any optical device or instrument ⟨visible to the *naked* eye⟩
8 : not backed by the writer's ownership of the commodity contract or security
synonym see BARE
— na•ked•ly *adverb*
— na•ked•ness *noun*

na•led \'nā-,led\ *noun* [origin unknown] (circa 1962)
: a short-lived insecticide $C_4H_7Br_2Cl_2O_4P$ of relatively low toxicity to warm-blooded animals that is used especially to control crop pests and mosquitoes

na•li•dix•ic acid \,nā-lə-'dik-sik-\ *noun* [perhaps from *na*phthyri*di*ne ($C_8H_6N_2$ — from *naphth-* + *pyridine*) + carbox*ylic acid*] (1964)
: an antibacterial agent $C_{12}H_{12}N_2O_3$ that is used especially in the treatment of genitourinary infections

na•lor•phine \na-'lȯr-,fēn\ *noun* [*N-al*lyl + m*orphine*] (circa 1953)
: a white crystalline compound $C_{19}H_{21}NO_3$ that is derived from morphine and is used in the form of its hydrochloride as a respiratory stimulant to counteract poisoning by morphine and similar narcotic drugs

nal•ox•one \na-'läk-,sōn\ *noun* [*N-al*lyl + hydr*oxy-* + *-one*] (1964)
: a potent antagonist $C_{19}H_{21}NO_4$ of narcotic drugs and especially morphine that is administered especially as the hydrochloride

nal•trex•one \nal-'trek-,sōn\ *noun* [*N-al*lyl + *trex-* (as in *methotrexate*) + *-one*] (1973)
: a narcotic antagonist $C_{20}H_{23}NO_4$

nam•by–pam•by \,nam-bē-'pam-bē\ *adjective* [*Namby Pamby,* nickname given to Ambrose Philips] (1745)
1 : lacking in character or substance **:** INSIPID
2 : WEAK, INDECISIVE ◆
— namby–pamby *noun*

[1]name \'nām\ *noun* [Middle English, from Old English *nama;* akin to Old High German *namo* name, Latin *nomen,* Greek *onoma, onyma*] (before 12th century)
1 a : a word or phrase that constitutes the distinctive designation of a person or thing **b :** a word or symbol used in logic to designate an entity
2 : a descriptive often disparaging epithet ⟨called him *names*⟩
3 a : REPUTATION ⟨gave the town a bad *name*⟩ **b :** an illustrious record **:** FAME ⟨made a *name* for himself in golf⟩ **c :** a person or thing with a reputation
4 : FAMILY, CLAN
5 : appearance as opposed to reality ⟨a friend in *name* only⟩
6 : one referred to by a name ⟨praise his holy *name*⟩
— in the name of 1 : by authority of ⟨open *in the name of* the law⟩ **2 :** for the reason of **:** using the excuse of ⟨called for reforms *in the name of* progress⟩

[2]name *transitive verb* **named; nam•ing** (before 12th century)
1 : to give a name to **:** CALL
2 a : to mention or identify by name **b :** to accuse by name
3 : to nominate for office **:** APPOINT
4 : to decide on **:** CHOOSE ⟨*name* the day for the wedding⟩
5 : to mention explicitly **:** SPECIFY ⟨unwilling to *name* a price⟩
— nam•er *noun*

[3]name *adjective* (1598)
1 : of, relating to, or bearing a name ⟨*name* tags⟩
2 : appearing in the name of a literary or theatrical production
3 a : having an established reputation **b :** featuring celebrities

name•able *also* **nam•able** \'nā-mə-bəl\ *adjective* (1780)
1 : worthy of being named **:** MEMORABLE
2 : capable of being named **:** IDENTIFIABLE

name–call•ing \'nām-,kȯ-liŋ\ *noun* (1853)
: the use of offensive names especially to win an argument or to induce rejection or condemnation (as of a person or project) without objective consideration of the facts

name day *noun* (1721)
: the church feast day of the saint after whom one is named

name–drop•ping \-,drä-piŋ\ *noun* (1950)
: the practice of seeking to impress others by studied but apparently casual mention of prominent persons as associates
— name–drop•per \-pər\ *noun*

name•less \'nām-ləs\ *adjective* (14th century)
1 : OBSCURE, UNDISTINGUISHED
2 : not known by name **:** ANONYMOUS
3 : having no legal right to a name **:** ILLEGITIMATE
4 : not having been given a name **:** UNNAMED
5 : not marked with a name ⟨a *nameless* grave⟩
6 a : incapable of precise description **:** INDEFINABLE **b :** too repulsive or distressing to describe
— name•less•ly *adverb*
— name•less•ness *noun*

name•ly \'nām-lē\ *adverb* (14th century)
: that is to say **:** TO WIT

name of the game (1966)
1 : the essential quality or matter ⟨patience is the *name of the game* in coastal duck hunting —Dick Beals⟩
2 : the fundamental goal of an activity

name•plate \-,plāt\ *noun* (circa 1859)
: something (as a plate or plaque) bearing a name (as of a resident or manufacturer)

name•sake \-,sāk\ *noun* [probably from *name's sake*] (1646)
: one that has the same name as another; *especially* **:** one who is named after another or for whom another is named

nan \'nän, 'nan\ *noun* [Hindi & Persian; Hindi *nān,* from Persian] (1948)
: a round flat leavened bread especially of the Indian subcontinent

nana \'na-nə\ *noun* [probably of baby-talk origin] (circa 1844)
: GRANDMOTHER

nance \'nan(t)s\ *noun* [short for *nancy,* from the name *Nancy*] (1920)
1 : an effeminate male — often used disparagingly
2 : HOMOSEXUAL — often used disparagingly

NAND \'nand\ *noun* [*not AND*] (1958)
: a computer logic circuit that produces an output which is the inverse of that of an AND circuit

nan•di•na \nan-'dī-nə, -'dē-\ *noun* [New Latin, from Japanese *nanten*] (circa 1890)
: a widely cultivated Asian evergreen shrub (*Nandina domestica*) of the barberry family

nan•keen \,nan-'kēn\ *noun* [*Nanking* (Nanjing), China] (1755)
1 : a durable brownish yellow cotton fabric originally loomed by hand in China
2 *plural* **:** trousers made of nankeen

Nan•kin \'nan-'kin, 'nän-\ *or* **Nan•king** \-'kiŋ\ *noun* [*Nanking* (Nanjing), China] (1781)
: Chinese porcelain decorated in blue on a white ground

nan•no•plank•ton \,na-nō-'plaŋ(k)-tən, -,tän\ *noun* [New Latin, from Greek *nanos, nannos* dwarf + New Latin *plankton* plankton] (1912)
: the smallest plankton that consists of those organisms (as bacteria) passing through nets of very fine mesh silk cloth

nan•ny *also* **nan•nie** \'na-nē\ *noun, plural* **nannies** [probably of baby-talk origin] (1795)
: a child's nurse **:** NURSEMAID

nanny goat *noun* [*Nanny,* nickname for *Anne*] (1788)
: a female domestic goat

nano- *combining form* [International Scientific Vocabulary, from Greek *nanos* dwarf]
: one billionth (10^{-9}) part of ⟨*nano*second⟩

nano•gram \'na-nə-,gram\ *noun* [International Scientific Vocabulary] (1951)
: one billionth of a gram

nano•me•ter \'na-nə-,mē-tər\ *noun* [International Scientific Vocabulary] (1963)
: one billionth of a meter

nano•sec•ond \-,se-kənd, -kənt\ *noun* [International Scientific Vocabulary] (1959)
: one billionth of a second

nano•tech•nol•o•gy \,na-nō-tek-'nä-lə-jē\ *noun* (1987)
: the art of manipulating materials on an atomic or molecular scale especially to build microscopic devices (as robots)

nano•tes•la \'na-nō-,tes-lə\ *noun* (1968)
: a unit of magnetic flux density equal to 10^{-9} tesla

Nan•tua sauce \nä^{n}(n)-'twä-\ *noun* [*Nantua,* France] (circa 1961)

◇ WORD HISTORY

namby–pamby *Namby-pamby* mockingly perpetuates the memory of English poet Ambrose Philips (1674–1749). Once highly celebrated but now largely unread, he wrote poems praising public figures, the pastoral life, and precocious children. His poetry was little loved, however, by a number of his rivals, including Henry Carey, a close ally of Alexander Pope and author of a parody of Philips's poetry. Carey is credited with devising the rhyming pet name *Namby-Pamby* for Philips. Delighted by its aptness, Pope borrowed it for the 1733 edition of *The Dunciad,* his mock epic satirizing the popular authors of the day. The success of *The Dunciad* did much to ensure that the name *Namby-Pamby* would forever be associated with insipid sentimentality.

\ə\ **abut** \$^{ə}$\ **kitten** \ər\ **further** \a\ **ash** \ā\ **ace** \ä\ **mop, mar** \au̇\ **out** \ch\ **chin** \e\ **bet** \ē\ **easy** \g\ **go** \i\ **hit** \ī\ **ice** \j\ **job** \ŋ\ **sing** \ō\ **go** \ȯ\ **law** \ȯi\ **boy** \th\ **thin** \t͟h\ **the** \ü\ **loot** \u̇\ **foot** \y\ **yet** \zh\ **vision** *see also* Guide to Pronunciation

: a cream sauce flavored with shellfish (as lobster)

Na·o·mi \nā-ˈō-mē\ *noun* [Hebrew *Nā'ŏmī*]
: the mother-in-law of the Old Testament heroine Ruth

¹nap \ˈnap\ *intransitive verb* **napped; nap·ping** [Middle English *nappen,* from Old English *hnappian;* akin to Old High German *hnaffezen* to doze] (before 12th century)
1 : to sleep briefly especially during the day **:** DOZE
2 : to be off guard

²nap *noun* (14th century)
: a short sleep especially during the day **:** SNOOZE

³nap *noun* [Middle English *noppe,* from Middle Dutch, flock of wool, nap] (15th century)
: a hairy or downy surface (as on a fabric)
— **nap·less** \-ləs\ *adjective*
— **napped** \ˈnapt\ *adjective*

⁴nap *transitive verb* **napped; nap·ping** (1620)
: to raise a nap on (fabric or leather)

⁵nap *noun* [from to go *nap* (to make all the points in the card game Napoleon)] (1895)
British **:** a pick or recommendation as a good bet to win a contest (as a horse race); *also* **:** one named in a nap

⁶nap *transitive verb* **napped; nap·ping** (1927)
British **:** to pick or single out in a nap

napa cabbage \ˈna-pə-\ *noun* [perhaps from Japanese dialect *nappa* greens] (1980)
: CHINESE CABBAGE b

napa leather *noun* [*Napa,* California] (1897)
: a glove leather made by tawing sheepskins with a soap-and-oil mixture; *also* **:** a similarly soft leather

¹na·palm \ˈnā-ˌpäm, -ˌpälm *also* ˈna- *also* nə-ˈpä(l)m\ *noun* [*na*phthene + *palm*itate] (1942)
1 : a thickener consisting of a mixture of aluminum soaps used in jelling gasoline (as for incendiary bombs)
2 : fuel jelled with napalm

²napalm *transitive verb* (1950)
: to assault with napalm

nape \ˈnāp, ˈnap\ *noun* [Middle English] (14th century)
: the back of the neck

na·pery \ˈnā-p(ə-)rē\ *noun* [Middle English, from Middle French *naperie,* from *nappe, nape* tablecloth — more at NAPKIN] (14th century)
: household linen; *especially* **:** TABLE LINEN

Naph·ta·li \ˈnaf-tə-ˌlī\ *noun* [Hebrew *Naphtālī*]
: a son of Jacob and the traditional eponymous ancestor of one of the tribes of Israel

naph·tha \ˈnaf-thə, ÷ˈnap-\ *noun* [Latin, from Greek, of Iranian origin; akin to Persian *neft* naphtha] (1572)
1 : any of various volatile often flammable liquid hydrocarbon mixtures used chiefly as solvents and diluents
2 : PETROLEUM

naph·tha·lene \-ˌlēn\ *noun* [alteration of earlier *naphthaline,* irregular from *naphtha*] (1821)
: a crystalline aromatic hydrocarbon $C_{10}H_8$ usually obtained by distillation of coal tar and used especially in organic synthesis

naph·thene \ˈnaf-ˌthēn, ÷ˈnap-\ *noun* (1884)
: CYCLOPARAFFIN
— **naph·then·ic** \ˌnaf-ˈthē-nik, ÷ˌnap-, -ˈthe-\ *adjective*

naph·thol \ˈnaf-ˌthȯl, ÷ˈnap-, -ˌthōl\ *noun* [International Scientific Vocabulary] (1849)
1 : either of two isomeric derivatives $C_{10}H_8O$ of naphthalene used as antiseptics and in the manufacture of dyes
2 : any of various hydroxy derivatives of naphthalene that resemble the simpler phenols

naph·thyl·amine \naf-ˈthi-lə-ˌmēn, ÷nap-\ *noun* [International Scientific Vocabulary] (1857)
: either of two isomeric crystalline bases $C_{10}H_9N$ used especially as dye intermediates

na·pi·er grass \ˈnā-pē-ər-\ *noun* [*Napier,* town in South Africa] (1914)
: a tall stout African perennial grass (*Pennisetum purpureum*) that resembles sugarcane and is widely grown for forage — called also *elephant grass*

Na·pier·ian logarithm \nə-ˈpir-ē-ən-, nā-\ *noun* [John *Napier*] (1816)
: NATURAL LOGARITHM

Na·pi·er's bones \ˈnā-pē-ərz-\ *noun* (circa 1658)
: a set of graduated rods (as of wood or bone) invented by John Napier and used for multiplication and division based on the principles of logarithms

na·pi·form \ˈnā-pə-ˌfȯrm\ *adjective* [Latin *napus* turnip (perhaps from Greek *napy, sinapy* mustard) + International Scientific Vocabulary *-iform*] (circa 1846)
: globular at the top and tapering off abruptly — used especially of roots

nap·kin \ˈnap-kən\ *noun* [Middle English *nappekin,* from *nappe* tablecloth, from Middle French, from Latin *mappa* napkin] (14th century)
1 : a piece of material (as cloth or paper) used at table to wipe the lips or fingers and protect the clothes
2 : a small cloth or towel: as **a** *dialect British* **:** HANDKERCHIEF **b** *chiefly Scottish* **:** KERCHIEF **c** *chiefly British* **:** DIAPER 3
3 : SANITARY NAPKIN

na·po·leon \nə-ˈpōl-yən, -ˈpō-lē-ən\ *noun* [French *napoléon,* from *Napoléon* Napoleon I] (1814)
1 : a former French 20-franc gold coin
2 : an oblong pastry with a filling of cream, custard, or jelly
3 *capitalized* **:** one like Napoleon I (as in ambition)

nappe \ˈnap\ *noun* [French, tablecloth, sheet, nappe — more at NAPKIN] (1904)
1 : SHEET 6
2 : a large mass thrust over other rocks
3 : one of the two sheets that lie on opposite sides of the vertex and together make up a cone

¹nap·py \ˈna-pē\ *noun* [obsolete *nappy,* adjective, foaming] (1700)
chiefly Scottish **:** LIQUOR; *specifically* **:** ALE

²nappy *noun, plural* **nappies** [English dialect *nap* bowl, from Middle English, from Old English *hnæpp;* akin to Old High German *hnapf* bowl] (1864)
: a rimless shallow open serving dish

³nappy *noun, plural* **nappies** [*nap*kin + ⁴-y] (1927)
chiefly British **:** DIAPER 3

⁴nappy *adjective* **nap·pi·er; -est** [³*nap*] (1928)
: KINKY 1

na·prap·a·thy \nə-ˈpra-pə-thē\ *noun* [Czech *naprava* correction + English *-pathy*] (1909)
: a system of treatment by manipulation of connective tissue and adjoining structures (as ligaments, joints, and muscles) and by dietary measures that is held to facilitate the recuperative and regenerative processes of the body

narc \ˈnärk\ *noun* [short for *narcotics agent*] (1967)
slang **:** a person (as a government agent) who investigates narcotics violations

nar·cis·sism \ˈnär-sə-ˌsi-zəm\ *noun* [German *Narzissismus,* from *Narziss* Narcissus, from Latin *Narcissus*] (1822)
1 : EGOISM, EGOCENTRISM
2 : love of or sexual desire for one's own body ◆
— **nar·cis·sist** \ˈnär-sə-sist\ *noun or adjective*
— **nar·cis·sis·tic** \ˌnär-sə-ˈsis-tik\ *adjective*

nar·cis·sus \när-ˈsi-səs\ *noun* [Latin, from Greek *Narkissos*]
1 *capitalized* **:** a beautiful youth in Greek mythology who pines away for love of his own reflection and is then turned into the narcissus flower
2 *plural* **nar·cis·si** \-ˈsi-ˌsī, -(ˌ)sē\ *or* **nar·cis·sus·es** *or* **narcissus** [New Latin, genus name, from Latin, narcissus, from Greek *narkissos*] **:** DAFFODIL; *especially* **:** one whose flowers have a short corona and are usually borne separately

nar·co \ˈnär-(ˌ)kō\ *noun, plural* **narcos** (1955)
slang **:** NARC

nar·co·lep·sy \ˈnär-kə-ˌlep-sē\ *noun, plural* **-sies** [International Scientific Vocabulary, from Greek *narkē*] (1880)
: a condition characterized by brief attacks of deep sleep
— **nar·co·lep·tic** \ˌnär-kə-ˈlep-tik\ *adjective*

nar·co·lep·tic \ˌnär-kə-ˈlep-tik\ *noun* (1928)
: a person who is subject to attacks of narcolepsy

nar·co·sis \när-ˈkō-səs\ *noun, plural* **-co·ses** \-ˌsēz\ [New Latin, from Greek *narkōsis,* action of benumbing, from *narkoun*] (circa 1693)
: a state of stupor, unconsciousness, or arrested activity produced by the influence of narcotics or other chemicals

¹nar·cot·ic \när-ˈkä-tik\ *noun* [Middle English *narkotik,* from Middle French *narcotique,* from *narcotique,* adjective, from Medieval Latin *narcoticus,* from Greek *narkōtikos,* from *narkoun* to benumb, from *narkē* numbness — more at SNARE] (14th century)
1 a : a drug (as opium) that in moderate doses dulls the senses, relieves pain, and induces profound sleep but in excessive doses causes stupor, coma, or convulsions **b :** a drug (as marijuana or LSD) subject to restriction similar to that of addictive narcotics whether physiologically addictive and narcotic or not
2 : something that soothes, relieves, or lulls

²narcotic *adjective* (1526)
1 a : having the properties of or yielding a narcotic **b :** inducing mental lethargy **:** SOPORIFEROUS
2 : of, induced by, or concerned with narcotics
3 : of, involving, or intended for narcotic addicts
— **nar·cot·i·cal·ly** \-ti-k(ə-)lē\ *adverb*

nar·co·tize \ˈnär-kə-ˌtīz\ *verb* **-tized; -tiz·ing** [International Scientific Vocabulary] (1526)

◇ WORD HISTORY

narcissism The term *narcissism* represents the result of modern science finding a psychological truth in an ancient myth. In Greek mythology Narcissus was a youth so strikingly beautiful that he attracted many would-be lovers, both male and female. He spurned the attentions of all of them. Finally, Nemesis, the goddess of retribution, in response to the prayer of one rejected lover, contrived that Narcissus should catch sight of his own reflection in a pool on Mount Helicon. Narcissus became so entranced with his own beauty that he could not bear to tear himself away. He lay beside the pool day after day until he wasted away and finally expired. By an act of the gods his body was metamorphosed into the flower now known as the narcissus. Modern psychologists were reminded of the myth of Narcissus as they studied individuals morbidly interested in themselves or sexually attracted to their own bodies. In 1899 a German psychologist coined the term *Narzissismus* to describe this form of self-love. In 1905 Havelock Ellis, in his *Studies in the Psychology of Sex,* accepted this coinage and used *narcissism* as the English equivalent.

transitive verb
1 a : to treat with or subject to a narcotic **b :** to put into a state of narcosis
2 : to soothe to unconsciousness or unawareness
intransitive verb
: to act as a narcotizing agent

nard \'närd\ *noun* [Middle English *narde,* from Old English, from Latin *nardus,* from Greek *nardos,* of Semitic origin; akin to Hebrew *nērd* nard] (before 12th century)
: SPIKENARD 1b

na•res \'nar-(ˌ)ēz, 'ner-\ *noun plural* [Middle English, from Latin, plural of *naris;* akin to Latin *nasus* nose — more at NOSE] (14th century)
: the pair of openings of the nose or nasal cavity of a vertebrate

nar•ghi•le \'när-gə-lē\ *or* **nar•gi•leh** \-ˌle\ *noun* [Persian *nārgīla,* from *nārgīl* coconut, of Indo-Aryan origin; akin to Sanskrit *nārikela* coconut; from the original material used in making its bowl] (1758)
: a water pipe that originated in the Near East

[1]**nark** \'närk\ *noun* [perhaps from Romany *nak* nose] (circa 1860)
British **:** STOOL PIGEON

[2]**nark** *variant of* NARC

[3]**nark** *transitive verb* [origin unknown] (1888)
British **:** IRRITATE, ANNOY

Nar•ra•gan•set *or* **Nar•ra•gan•sett** \ˌnar-ə-'gan(t)-sət\ *noun, plural* **Narraganset** *or* **Narragansets** *or* **Narragansett** *or* **Narragansetts** (1622)
1 : a member of an American Indian people of Rhode Island
2 : the Algonquian language of the Narraganset people

nar•rate \'nar-ˌāt, na-'rāt\ *transitive verb* **nar•rat•ed; nar•rat•ing** [Latin *narratus,* past participle of *narrare,* from Latin *gnarus* knowing; akin to Latin *gnoscere, noscere* to know — more at KNOW] (1656)
: to tell (as a story) in detail; *also* **:** to provide spoken commentary for (as a movie or television show)
— **nar•ra•tor** \'nar-ˌā-tər; na-'rā-, nə-; 'nar-ə-\ *noun*

nar•ra•tion \na-'rā-shən, nə-\ *noun* (15th century)
1 : the act or process or an instance of narrating
2 : STORY, NARRATIVE
— **nar•ra•tion•al** \-shnəl, -shə-nᵊl\ *adjective*

nar•ra•tive \'nar-ə-tiv\ *noun* (1566)
1 : something that is narrated **:** STORY
2 : the art or practice of narration
3 : the representation in art of an event or story; *also* **:** an example of such a representation
— **narrative** *adjective*
— **nar•ra•tive•ly** *adverb*

nar•ra•tol•o•gy \ˌnar-ə-'tä-lə-jē\ *noun* (1976)
: the study of structure in narratives
— **nar•ra•to•log•i•cal** \-tə-'lä-ji-kəl\ *adjective*
— **nar•ra•tol•o•gist** \-'tä-lə-jist\ *noun*

[1]**nar•row** \'nar-(ˌ)ō\ *adjective* [Middle English *narowe,* from Old English *nearu;* akin to Old High German *narwa* scar] (before 12th century)
1 a : of slender width **b :** of less than standard width **c** *of a textile* **:** woven in widths less than 18 inches (46 centimeters)
2 : limited in size or scope
3 a : illiberal in views or disposition **:** PREJUDICED **b** *chiefly dialect* **:** STINGY, NIGGARDLY
4 a : barely sufficient **:** CLOSE **b :** barely successful
5 : minutely precise **:** METICULOUS
6 *of a ration* **:** relatively rich in protein as compared with carbohydrate and fat
7 : TENSE 3
— **nar•row•ly** *adverb*
— **nar•row•ness** *noun*

[2]**narrow** (before 12th century)
transitive verb
1 : to decrease the breadth or extent of **:** CONTRACT — often used with *down*
2 : to decrease the scope or sphere of **:** LIMIT — often used with *down*
intransitive verb
: to lessen in width or extent **:** CONTRACT — often used with *down*

[3]**narrow** *noun* (13th century)
: a narrow part or passage; *specifically* **:** a strait connecting two bodies of water — usually used in plural but singular or plural in construction

nar•row•band \'nar-ō-ˌband, 'nar-ə-\ *adjective* (1950)
: operating at, responsive to, or including a narrow range of frequencies

narrow boat *noun* (1949)
British **:** a barge with a beam of less than seven feet (2.1 meters)

nar•row•cast•ing \'nar-ō-ˌkas-tiŋ, 'nar-ə-\ *noun* (1928)
: radio or television transmission aimed at a narrowly defined area or audience (as paying subscribers or night workers)

nar•row–mind•ed \ˌnar-ō-'mīn-dəd, ˌnar-ə-\ *adjective* (1625)
: lacking in tolerance or breadth of vision **:** PETTY
— **nar•row–mind•ed•ly** *adverb*
— **nar•row–mind•ed•ness** *noun*

nar•thex \'när-ˌtheks\ *noun* [Late Greek *narthēx,* from Greek, giant fennel, cane, casket] (circa 1673)
1 : the portico of an ancient church
2 : a vestibule leading to the nave of a church

nar•whal \'när-ˌwäl, -ˌhwäl, -wəl\ *also* **nar•whale** \-ˌwāl, -ˌhwāl\ *noun* [Norwegian & Danish *narhval* & Swedish *narval,* probably modification of Icelandic *nárhvalur,* from Old Norse *nāhvalr,* from *nār* corpse + *hvalr* whale; from its color] (1646)
: an arctic cetacean (*Monodon monoceros*) about 20 feet (6 meters) long with the male having a long twisted ivory tusk

narwhal

nary \'nar-ē, 'ner-\ *adjective* [alteration of *ne'er a*] (1746)
: not any ⟨*nary* a person wanted to go⟩

[1]**na•sal** \'nā-zəl\ *noun* [Middle English, from Middle French, from Old French, from *nes* nose, from Latin *nasus* — more at NOSE] (14th century)
1 : the nosepiece of a helmet
2 : a nasal part
3 : a nasal consonant

[2]**nasal** *adjective* (1656)
1 : of or relating to the nose
2 a : uttered with the soft palate lowered and with passage of air through the nose (as with \m\, \n\, \ŋ\, \ō^{n}\, or \a^{n}\) **b :** characterized by resonance produced through the nose
3 *of a musical tone* **:** SHARP, PENETRATING
— **na•sal•i•ty** \nā-'za-lə-tē\ *noun*
— **na•sal•ly** \'nā-zə-lē, 'nāz-lē\ *adverb*

na•sal•ize \'nā-zə-ˌlīz\ *verb* **-ized; -iz•ing** (1817)
transitive verb
: to make nasal or pronounce as a nasal sound
— **na•sal•i•za•tion** \ˌnā-zə-lə-'zā-shən\ *noun*

na•scence \'na-sᵊn(t)s, 'nā-\ *noun* (1570)
: NASCENCY

na•scen•cy \-sᵊn(t)-sē\ *noun, plural* **-cies** (1682)
: BIRTH, ORIGIN

na•scent \'na-sᵊnt, 'nā-\ *adjective* [Latin *nascent-, nascens,* present participle of *nasci* to be born] (circa 1624)
: coming or having recently come into existence

naso- *combining form* [Latin *nasus* nose]
1 : nose and ⟨*naso*pharynx⟩
2 : nasal ⟨*naso*gastric⟩

na•so•gas•tric \ˌnā-zō-'gas-trik\ *adjective* (1942)
: being or performed by intubation of the stomach through the nasal passages ⟨*nasogastric* tube⟩ ⟨*nasogastric* suction⟩

na•so•pha•ryn•geal \-fə-'rin-j(ē-)əl, -ˌfar-ən-'jē-əl\ *adjective* (1872)
: of, relating to, or affecting the nose and pharynx or the nasopharynx

na•so•phar•ynx \-'far-iŋ(k)s\ *noun* [New Latin] (1877)
: the upper part of the pharynx continuous with the nasal passages

nas•tic \'nas-tik\ *adjective* [Greek *nastos* close-pressed, from *nassein* to press] (1908)
: of, relating to, or constituting a movement of a plant part caused by disproportionate growth or increase of turgor in one surface

nas•tur•tium \nə-'stər-shəm, na-\ *noun* [Latin, a cress] (1704)
: any of a genus (*Tropaeolum* of the family Tropaeolaceae, the nasturtium family) of herbs of Central and South America with showy spurred flowers and pungent seeds; *especially* **:** either of two widely cultivated ornamentals (*T. majus* and *T. minus*)

nasturtium

nas•ty \'nas-tē\ *adjective* **nas•ti•er; -est** [Middle English] (14th century)
1 a : disgustingly filthy **b :** physically repugnant
2 : INDECENT, OBSCENE
3 : MEAN, TAWDRY
4 a : extremely hazardous or harmful ⟨had a *nasty* climb to reach the summit⟩ **b :** causing severe pain or suffering ⟨a *nasty* wound⟩ **c :** sharply unpleasant **:** DISAGREEABLE ⟨*nasty* weather⟩
5 a : difficult to understand or deal with **:** VEXATIOUS ⟨a *nasty* problem⟩ **b :** psychologically unsettling **:** TRYING ⟨a *nasty* fear that she was lost⟩
6 : lacking in courtesy or sportsmanship ⟨a *nasty* trick⟩
synonym see DIRTY
— **nas•ti•ly** \-tə-lē\ *adverb*
— **nas•ti•ness** \-tē-nəs\ *noun*
— **nasty** *noun*

na•tal \'nā-tᵊl\ *adjective* [Middle English, from Latin *natalis,* from *natus,* past participle of *nasci* to be born — more at NATION] (15th century)
1 : NATIVE
2 : of, relating to, or present at birth; *especially* **:** associated with one's birth ⟨a *natal* star⟩

na•tal•i•ty \nā-'ta-lə-tē, nə-\ *noun, plural* **-ties** (1888)
: BIRTHRATE

na•tant \'nā-tᵊnt\ *adjective* [Middle English *natand,* from Latin *natant-, natans,* present participle of *natare* to swim; akin to Latin *nare* to swim; akin to Greek *nein, nēchein* to swim, Sanskrit *snāti* he bathes] (15th century)
: swimming or floating in water ⟨*natant* decapods⟩

na•ta•tion \nā-'tā-shən, na-\ *noun* (1542)
: the action or art of swimming

na·ta·to·ri·al \ˌnā-tə-ˈtōr-ē-əl, ˌna-, -ˈtȯr-\ *or* **na·ta·to·ry** \ˈnā-tə-ˌtōr-ē, ˈna-, -ˌtȯr-\ *adjective* (1816)
1 : of or relating to swimming
2 : adapted to or characterized by swimming ⟨a *natatorial* leg of an aquatic insect⟩

na·ta·to·ri·um \ˌnā-tə-ˈtōr-ē-əm, ˌna-, -ˈtȯr-\ *noun* [Late Latin, from Latin *natare*] (circa 1890)
: an indoor swimming pool

natch \ˈnach\ *adverb* [by shortening & alteration from *naturally*] (circa 1945)
slang **:** of course **:** NATURALLY

Natch·ez \ˈna-chiz\ *noun, plural* **Natchez** (1845)
1 : a member of an American Indian people of southwestern Mississippi
2 : the language of the Natchez people

na·tes \ˈnā-ˌtēz\ *noun plural* [Latin, plural of *natis* buttock] (circa 1706)
: BUTTOCKS

nathe·less \ˈnāth-ləs\ *or* **nath·less** \ˈnath-\ *adverb* [Middle English, from Old English *nā thē lǣs* not the less] (before 12th century)
archaic **:** NEVERTHELESS, NOTWITHSTANDING

na·tion \ˈnā-shən\ *noun* [Middle English *nacioun,* from Middle French *nation,* from Latin *nation-, natio* birth, race, nation, from *nasci* to be born; akin to Latin *gignere* to beget — more at KIN] (14th century)
1 a (1) **:** NATIONALITY 5a (2) **:** a politically organized nationality (3) **:** a non-Jewish nationality ⟨why do the *nations* conspire —Psalms 2:1 (Revised Standard Version)⟩ **b :** a community of people composed of one or more nationalities and possessing a more or less defined territory and government **c :** a territorial division containing a body of people of one or more nationalities and usually characterized by relatively large size and independent status
2 *archaic* **:** GROUP, AGGREGATION
3 : a tribe or federation of tribes (as of American Indians)

¹na·tion·al \ˈnash-nəl, ˈna-shə-nᵊl\ *adjective* (1597)
1 : of or relating to a nation
2 : NATIONALIST
3 : comprising or characteristic of a nationality
4 : belonging to or maintained by the federal government
5 : of, relating to, or being a coalition government formed by most or all major political parties usually in a crisis
— na·tion·al·ly *adverb*

²national *noun* (1887)
1 : one that owes allegiance to or is under the protection of a nation without regard to the more formal status of citizen or subject
2 : a competition that is national in scope — usually used in plural
synonym see CITIZEN

national bank *noun* (1790)
1 : CENTRAL BANK
2 : a bank operating under federal charter and supervision

national forest *noun* (circa 1905)
: a usually forested area of considerable extent that is preserved by government decree from private exploitation and is harvested only under supervision

National Guard *noun* (1847)
1 : a militia force recruited by each state of the U.S., equipped by the federal government, and jointly maintained subject to the call of either
2 *often not capitalized* **:** a military establishment serving as a national constabulary and defense force

national income *noun* (1878)
: the aggregate of earnings from a nation's current production including compensation of employees, interest, rental income, and profits of business after taxes

na·tion·al·ise *chiefly British variant of* NATIONALIZE

na·tion·al·ism \ˈnash-nə-ˌli-zəm, ˈna-shə-nᵊl-ˌi-zəm\ *noun* (1844)
: loyalty and devotion to a nation; *especially* **:** a sense of national consciousness exalting one nation above all others and placing primary emphasis on promotion of its culture and interests as opposed to those of other nations or supranational groups

¹na·tion·al·ist \-list, -ist\ *noun* (1715)
1 : an advocate of or believer in nationalism
2 *capitalized* **:** a member of a political party or group advocating national independence or strong national government

²nationalist *adjective* (1889)
1 : of, relating to, or advocating nationalism
2 *capitalized* **:** of, relating to, or being a political group advocating or associated with nationalism

na·tion·al·is·tic \ˌnash-nə-ˈlis-tik, ˌna-shə-nᵊl-ˈis-tik\ *adjective* (1866)
1 : of, favoring, or characterized by nationalism ⟨*nationalistic* election speeches⟩
2 : NATIONAL 1
— na·tion·al·is·ti·cal·ly \-ti-k(ə-)lē\ *adverb*

na·tion·al·i·ty \ˌna-shə-ˈna-lə-tē, ˌnash-ˈna-\ *noun, plural* **-ties** (1691)
1 : national character
2 : NATIONALISM
3 a : national status; *specifically* **:** a legal relationship involving allegiance on the part of an individual and usually protection on the part of the state **b :** membership in a particular nation
4 : political independence or existence as a separate nation
5 a : a people having a common origin, tradition, and language and capable of forming or actually constituting a nation-state **b :** an ethnic group constituting one element of a larger unit (as a nation)

na·tion·al·ize \ˈnash-nə-ˌlīz, ˈna-shə-nᵊl-ˌīz\ *transitive verb* **-ized; -iz·ing** (1800)
1 : to give a national character to
2 : to invest control or ownership of in the national government
— na·tion·al·i·za·tion \ˌnash-nə-lə-ˈzā-shən, ˌna-shə-nᵊl-ə-\ *noun*
— na·tion·al·iz·er \ˈnash-nə-ˌlī-zər, ˌna-shə-nᵊl-ˌī-zər\ *noun*

national monument *noun* (1916)
: a place of historic, scenic, or scientific interest set aside for preservation usually by presidential proclamation

national park *noun* (1868)
: an area of special scenic, historical, or scientific importance set aside and maintained by a national government and in the U.S. by an act of Congress

national seashore *noun* (1962)
: a recreational area adjacent to a seacoast and maintained by the federal government

national socialism *noun, often N&S capitalized* (1931)
: NAZISM
— national socialist *adjective, often N&S capitalized*

na·tion·hood \ˈnā-shən-ˌhu̇d\ *noun* (1850)
: NATIONALITY 1, 3a, 4

na·tion–state \ˈnā-shən-ˈstāt, -ˌstāt\ *noun* (1918)
: a form of political organization under which a relatively homogeneous people inhabits a sovereign state; *especially* **:** a state containing one as opposed to several nationalities

¹na·tion·wide \ˌnā-shən-ˈwīd\ *adjective* (1912)
: extending throughout a nation

²nationwide *adverb* (1926)
: throughout the nation

¹na·tive \ˈnā-tiv\ *adjective* [Middle English *natif,* from Middle French, from Latin *nativus,* from *natus,* past participle of *nasci* to be born — more at NATION] (14th century)
1 : INBORN, INNATE ⟨*native* talents⟩
2 : belonging to a particular place by birth ⟨*native* to Wisconsin⟩
3 *archaic* **:** closely related
4 : belonging to or associated with one by birth
5 : NATURAL, NORMAL
6 a : grown, produced, or originating in a particular place or in the vicinity **:** LOCAL **b :** living or growing naturally in a particular region **:** INDIGENOUS
7 : SIMPLE, UNAFFECTED
8 a : constituting the original substance or source **b :** found in nature especially in an unadulterated form ⟨mining *native* silver⟩
9 *chiefly Australian* **:** having a usually superficial resemblance to a specified English plant or animal ☆
— na·tive·ly *adverb*
— na·tive·ness *noun*

²native *noun* (1535)
1 : one born or reared in a particular place
2 a : an original or indigenous inhabitant **b :** something indigenous to a particular locality
3 : a local resident; *especially* **:** a person who has always lived in a place as distinguished from a visitor or a temporary resident

Native American *noun* (1925)
: AMERICAN INDIAN
— Native American *adjective*

na·tiv·ism \ˈnā-ti-ˌvi-zəm\ *noun* (1844)
1 : a policy of favoring native inhabitants as opposed to immigrants
2 : the revival or perpetuation of an indigenous culture especially in opposition to acculturation
— na·tiv·ist \-vist\ *noun or adjective*
— na·tiv·is·tic \ˌnā-ti-ˈvis-tik\ *adjective*

na·tiv·i·ty \nə-ˈti-və-tē, nā-\ *noun, plural* **-ties** [Middle English *nativite,* from Middle French *nativité,* from Medieval Latin *nativitat-, nativitas,* from Late Latin, birth, from Latin *nativus*] (14th century)
1 : the process or circumstances of being born **:** BIRTH; *especially, capitalized* **:** the birth of Jesus
2 : a horoscope at or of the time of one's birth
3 : the place of origin

na·tri·ure·sis \ˌnā-trē-yu̇-ˈrē-səs\ *noun* [New Latin, from *natrium* sodium (from International Scientific Vocabulary *natron*) + *uresis* urination, from Greek *ourēsis,* from *ourein* to urinate — more at URINE] (1957)
: excessive loss of sodium in the urine
— na·tri·uret·ic \-ˈre-tik\ *adjective or noun*

na·tro·lite \ˈnā-trə-ˌlīt\ *noun* [German *Natrolith,* from *Natron* (from French) + *-lith* -lite] (circa 1805)
: a hydrous silicate of sodium and aluminum that is related to zeolite

na·tron \ˈnā-ˌträn, -trən\ *noun* [French, from Spanish *natrón,* from Arabic *naṭrūn,* from Greek *nitron*] (1684)
: a hydrous native sodium carbonate used in ancient times in embalming, in ceramic pastes, and as a cleansing agent

¹nat·ter \ˈna-tər\ *intransitive verb* [probably imitative] (1942)
: CHATTER 2

²natter *noun* (1943)

☆ SYNONYMS
Native, indigenous, endemic, aboriginal mean belonging to a locality. NATIVE implies birth or origin in a place or region and may suggest compatibility with it ⟨*native* tribal customs⟩. INDIGENOUS applies to species or races and adds to NATIVE the implication of not having been introduced from elsewhere ⟨maize is *indigenous* to America⟩. ENDEMIC implies being peculiar to a region ⟨edelweiss is *endemic* in the Alps⟩. ABORIGINAL implies having no known race preceding in occupancy of the region ⟨the *aboriginal* peoples of Australia⟩.

chiefly British **:** idle talk or conversation **:** CHAT

nat·ty \'na-tē\ *adjective* **nat·ti·er; -est** [perhaps alteration of earlier *netty,* from obsolete *net* neat, clean] (1557)
: trimly neat and tidy **:** SMART
— **nat·ti·ly** \'na-tᵊl-ē\ *adverb*
— **nat·ti·ness** \'na-tē-nəs\ *noun*

¹nat·u·ral \'na-chə-rəl, 'nach-rəl\ *adjective* [Middle English, from Middle French, from Latin *naturalis* of nature, from *natura* nature] (14th century)
1 : based on an inherent sense of right and wrong ⟨*natural* justice⟩
2 a : being in accordance with or determined by nature **b :** having or constituting a classification based on features existing in nature
3 a (1) **:** begotten as distinguished from adopted; *also* **:** LEGITIMATE (2) **:** being a relation by actual consanguinity as distinguished from adoption ⟨*natural* parents⟩ **b :** ILLEGITIMATE ⟨a *natural* child⟩
4 : having an essential relation with someone or something **:** following from the nature of the one in question ⟨his guilt is a *natural* deduction from the evidence⟩
5 : implanted or being as if implanted by nature **:** seemingly inborn ⟨a *natural* talent for art⟩
6 : of or relating to nature as an object of study and research
7 : having a specified character by nature ⟨a *natural* athlete⟩
8 a : occurring in conformity with the ordinary course of nature **:** not marvelous or supernatural ⟨*natural* causes⟩ **b :** formulated by human reason alone rather than revelation ⟨*natural* religion⟩ ⟨*natural* rights⟩ **c :** having a normal or usual character ⟨events followed their *natural* course⟩
9 : possessing or exhibiting the higher qualities (as kindliness and affection) of human nature ⟨a noble . . . brother . . . ever most kind and *natural* —Shakespeare⟩
10 a : growing without human care; *also* **:** not cultivated ⟨*natural* prairie unbroken by the plow⟩ **b :** existing in or produced by nature **:** not artificial ⟨*natural* turf⟩ ⟨*natural* curiosities⟩ **c :** relating to or being natural food
11 a : being in a state of nature without spiritual enlightenment **:** UNREGENERATE ⟨*natural* man⟩ **b :** living in or as if in a state of nature untouched by the influences of civilization and society
12 a : having a physical or real existence as contrasted with one that is spiritual, intellectual, or fictitious ⟨a corporation is a legal but not a *natural* person⟩ **b :** of, relating to, or operating in the physical as opposed to the spiritual world ⟨*natural* laws describe phenomena of the physical universe⟩
13 a : closely resembling an original **:** true to nature **b :** marked by easy simplicity and freedom from artificiality, affectation, or constraint **c :** having a form or appearance found in nature
14 a : having neither flats nor sharps ⟨the *natural* scale of C major⟩ **b :** being neither sharp nor flat **c :** having the pitch modified by the natural sign
15 : of an off-white or beige color ☆
— **nat·u·ral·ness** \-nəs\ *noun*

²natural *noun* (1533)
1 : one born without the usual powers of reason and understanding
2 a : a sign ♮ placed on any degree of the musical staff to nullify the effect of a preceding sharp or flat **b :** a note or tone affected by the natural sign
3 : a result or combination that immediately wins the stake in a game; *specifically* **:** a throw of 7 or 11 on the first cast in craps
4 a : one having natural skills, talents, or abilities **b :** something that is likely to become an immediate success **c :** one that is obviously suitable for a specific purpose
5 : AFRO

natural childbirth *noun* (1933)
: a system of managing childbirth in which the mother receives preparatory education in order to remain conscious during and assist in delivery with minimal or no use of drugs or anesthetics

natural family planning *noun* (1975)
: a method of birth control that involves abstention from sexual intercourse during the period of ovulation which is determined through observation and measurement of bodily symptoms

natural food *noun* (1917)
: food that has undergone minimal processing and contains no preservatives or artificial additives

natural gas *noun* (1825)
1 : gas issuing from the earth's crust through natural openings or bored wells; *especially* **:** a combustible mixture of methane and higher hydrocarbons used chiefly as a fuel and raw material
2 : gas manufactured from organic matter (as coal)

natural history *noun* (1567)
1 : a treatise on some aspect of nature
2 : the natural development of something (as an organism or disease) over a period of time
3 : the study of natural objects especially in the field from an amateur or popular point of view

nat·u·ral·ise *British variant of* NATURALIZE

nat·u·ral·ism \'na-chə-rə-ˌli-zəm, 'nach-rə-\ *noun* (circa 1641)
1 : action, inclination, or thought based only on natural desires and instincts
2 : a theory denying that an event or object has a supernatural significance; *specifically* **:** the doctrine that scientific laws are adequate to account for all phenomena
3 : realism in art or literature; *specifically* **:** a theory in literature emphasizing scientific observation of life without idealization or the avoidance of the ugly

nat·u·ral·ist \-list\ *noun* (1587)
1 : one that advocates or practices naturalism
2 : a student of natural history; *especially* **:** a field biologist

nat·u·ral·is·tic \ˌna-chə-rə-'lis-tik, ˌnach-rə-\ *also* **nat·u·ral·ist** \'na-chə-rə-list, 'nach-rə-\ *adjective* (1838)
: of, characterized by, or according with naturalism
— **nat·u·ral·is·ti·cal·ly** \ˌna-chə-rə-'lis-ti-k(ə-)lē, ˌnach-rə-\ *adverb*

nat·u·ral·ize \'na-chə-rə-ˌlīz, 'nach-rə-\ *verb* **-ized; -iz·ing** (1593)
transitive verb
1 : to introduce into common use or into the vernacular
2 : to bring into conformity with nature
3 : to confer the rights of a national on; *especially* **:** to admit to citizenship
4 : to cause (as a plant) to become established as if native
intransitive verb
: to become established as if native
— **nat·u·ral·i·za·tion** \ˌna-chə-rə-lə-'zā-shən, ˌnach-rə-\ *noun*

natural killer cell *noun* (1975)
: a large granular lymphocyte capable of killing a tumor or microbial cell without prior exposure to the target cell and without having it presented with or marked by a histocompatibility antigen

natural law *noun* (15th century)
: a body of law or a specific principle held to be derived from nature and binding upon human society in the absence of or in addition to positive law

natural logarithm *noun* (1816)
: a logarithm with *e* as a base

nat·u·ral·ly \'na-chər-ə-lē, 'nach-rə-, 'na-chər-\ *adverb* (14th century)
1 : by nature **:** by natural character or ability ⟨*naturally* timid⟩
2 : according to the usual course of things **:** as might be expected ⟨we *naturally* dislike being hurt⟩
3 a : without artificial aid ⟨hair that curls *naturally*⟩ **b :** without affectation ⟨speak *naturally*⟩
4 : with truth to nature **:** REALISTICALLY

natural number *noun* (1763)
: the number 1 or any number (as 3, 12, 432) obtained by adding 1 to it one or more times **:** a positive integer

natural philosophy *noun* (14th century)
: NATURAL SCIENCE; *especially* **:** PHYSICAL SCIENCE
— **natural philosopher** *noun*

natural resource *noun* (1870)
1 *plural* **:** industrial materials and capacities (as mineral deposits and waterpower) supplied by nature
2 : RESOURCE 1b

natural science *noun* (14th century)
: any of the sciences (as physics, chemistry, or biology) that deal with matter, energy, and their interrelations and transformations or with objectively measurable phenomena
— **natural scientist** *noun*

natural selection *noun* (1857)
: a natural process that results in the survival and reproductive success of individuals or groups best adjusted to their environment and that leads to the perpetuation of genetic qualities best suited to that particular environment

natural theology *noun* (1677)
: theology deriving its knowledge of God from the study of nature independent of special revelation

na·ture \'nā-chər\ *noun* [Middle English, from Middle French, from Latin *natura,* from *natus,* past participle of *nasci* to be born — more at NATION] (14th century)
1 a : the inherent character or basic constitution of a person or thing **:** ESSENCE **b :** DISPOSITION, TEMPERAMENT
2 a : a creative and controlling force in the universe **b :** an inner force or the sum of such forces in an individual
3 : a kind or class usually distinguished by fundamental or essential characteristics ⟨documents of a confidential *nature*⟩ ⟨acts of a ceremonial *nature*⟩
4 : the physical constitution or drives of an organism; *especially* **:** an excretory organ or function — used in phrases like *the call of nature*
5 : a spontaneous attitude (as of generosity)

☆ SYNONYMS
Natural, ingenuous, naive, unsophisticated, artless mean free from pretension or calculation. NATURAL implies lacking artificiality and self-consciousness and having a spontaneousness suggesting the natural rather than the man-made world ⟨her unaffected, *natural* manner⟩. INGENUOUS implies inability to disguise or conceal one's feelings or intentions ⟨the *ingenuous* enthusiasm of children⟩. NAIVE suggests lack of worldly wisdom often connoting credulousness and unchecked innocence ⟨politically *naive*⟩. UNSOPHISTICATED implies a lack of experience and training necessary for social ease and adroitness ⟨*unsophisticated* adolescents⟩. ARTLESS suggests a naturalness resulting from unawareness of the effect one is producing on others ⟨*artless* charm⟩. See in addition REGULAR.

6 : the external world in its entirety
7 a : humankind's original or natural condition **b :** a simplified mode of life resembling this condition
8 : the genetically controlled qualities of an organism
9 : natural scenery
synonym see TYPE

na·tur·ism \'nā-chə-ˌri-zəm\ *noun* (1847)
1 : NATURALISM 1, 2
2 : the worship of the forces of nature
3 : NUDISM
— **na·tur·ist** \-rist\ *noun*

na·tu·rop·a·thy \ˌnā-chə-'rä-pə-thē\ *noun* (1901)
: a system of treatment of disease that avoids drugs and surgery and emphasizes the use of natural agents (as air, water, and sunshine) and physical means (as manipulation and electrical treatment)
— **na·tu·ro·path** \'nā-chər-ə-ˌpath, 'na-\ *noun*
— **na·tu·ro·path·ic** \ˌnā-chər-ə-'pa-thik, ˌna-; nə-ˌtyu̇r-ə-, -ˌtu̇r-\ *adjective*

Nau·ga·hyde \'nȯ-gə-ˌhīd, 'nä-\ *trademark*
— used for vinyl-coated fabrics

¹naught \'nȯt, 'nät\ *pronoun* [Middle English *nought,* from Old English *nāwiht,* from *nā* no + *wiht* creature, thing — more at NO, WIGHT] (before 12th century)
: NOTHING ⟨efforts came to *naught*⟩
word history see NAUGHTY

²naught *noun* (before 12th century)
1 : NOTHINGNESS, NONEXISTENCE
2 : the arithmetical symbol 0 **:** ZERO, CIPHER

naugh·ty \'nȯ-tē, 'nä-\ *adjective* **naugh·ti·er; -est** [Middle English *noughti,* from *nought*] (14th century)
1 a *archaic* **:** vicious in moral character **:** WICKED **b :** guilty of disobedience or misbehavior
2 : lacking in taste or propriety ◆
— **naugh·ti·ly** \'nȯ-tᵊl-ē, 'nä-\ *adverb*
— **naugh·ti·ness** \'nȯ-tē-nəs, 'nä-\ *noun*

nau·ma·chia \nȯ-'mā-kē-ə, -'ma-\ *noun, plural* **-chi·ae** \-kē-ˌē, -kē-ˌī\ *or* **-chi·as** [Latin, from Greek, naval battle, from *naus* ship + *machesthai* to fight — more at NAVE] (1596)
1 : an ancient Roman spectacle representing a naval battle
2 : a place for naumachiae

nau·pli·us \'nȯ-plē-əs\ *noun, plural* **-plii** \-plē-ˌī, -ˌē\ [New Latin, from Latin, a shellfish, from Greek *nauplios*] (1836)
: a crustacean larva in usually the first stage after leaving the egg and with three pairs of appendages, a median eye, and little or no segmentation

nau·sea \'nȯ-zē-ə, -sē-ə; 'nȯ-zhə, -shə\ *noun* [Latin, seasickness, nausea, from Greek *nautia, nausia,* from *nautēs* sailor] (1569)
1 : a stomach distress with distaste for food and an urge to vomit
2 : extreme disgust ◆
— **nau·se·ant** \-zhē-ənt, -shē-, -zē-, -sē-\ *noun or adjective*

nau·se·ate \'nȯ-zhē-ˌāt, -shē-, -zē-, -sē-\ *verb* **-at·ed; -at·ing** (1640)
intransitive verb
1 : to become affected with nausea
2 : to feel disgust
transitive verb
: to affect with nausea or disgust

nauseating *adjective* (1645)
: causing nausea or especially disgust
usage see NAUSEOUS
— **nau·se·at·ing·ly** *adverb*

nau·seous \'nȯ-shəs, 'nȯ-zē-əs\ *adjective* (1612)
1 : causing nausea or disgust **:** NAUSEATING
2 : affected with nausea or disgust ■
— **nau·seous·ly** *adverb*
— **nau·seous·ness** *noun*

nautch \'nȯch\ *noun* [Hindi *nāc,* from Sanskrit *nṛtya,* from *nṛtyati* he dances] (1809)
: an entertainment in India consisting chiefly of dancing by professional dancing girls

nau·ti·cal \'nȯ-ti-kəl, 'nä-\ *adjective* [Latin *nauticus,* from Greek *nautikos,* from *nautēs* sailor, from *naus* ship — more at NAVE] (1552)
: of, relating to, or associated with seamen, navigation, or ships
— **nau·ti·cal·ly** \-k(ə-)lē\ *adverb*

nautical mile *noun* (1834)
: any of various units of distance used for sea and air navigation based on the length of a minute of arc of a great circle of the earth and differing because the earth is not a perfect sphere: as **a :** a British unit equal to 6080 feet (1853.2 meters) **b :** an international unit equal to 6076.115 feet (1852 meters) used officially in the U.S. since July 1, 1959

nau·ti·loid \'nȯ-tᵊl-ˌȯid, 'nä-\ *noun* (circa 1847)
: any of a subclass (Nautiloidea) of shell-bearing cephalopods that were abundant in the Ordovician and especially the Silurian but are represented in the recent fauna only by the nautiluses
— **nautiloid** *adjective*

nau·ti·lus \'nȯ-tᵊl-əs, 'nä-\ *noun, plural* **-lus·es** *or* **-li** \-tᵊl-ˌī, -ˌē\ [New Latin, from Latin, paper nautilus, from Greek *nautilos,* literally, sailor, from *naus* ship] (1601)
1 : any of a genus (*Nautilus*) of cephalopod mollusks of the South Pacific and Indian Oceans with a spiral chambered shell that is pearly on the inside — called also *chambered nautilus*
2 : PAPER NAUTILUS

nautilus

nav·aid \'na-ˌvād\ *noun* [*na*vigation *aid*] (1956)
: a device or system (as a radar beacon) that provides a navigator with navigational data

Na·va·jo *also* **Na·va·ho** \'na-və-ˌhō, 'nä-\ *noun, plural* **-jo** *or* **-jos** *also* **-ho** *or* **-hos** [Spanish (*Apache de*) *Navajó,* literally, Apache of Navajó, from *Navajó,* area occupied by Navajos, probably from Tewa (Pueblo Indian language of northern New Mexico) *navahu·,* literally, arroyo with planted fields] (1780)
1 : a member of an American Indian people of northern New Mexico and Arizona
2 : the language of the Navajo people

na·val \'nā-vəl\ *adjective* [Middle English, from Latin *navalis,* from *navis* ship] (15th century)
1 *obsolete* **:** of or relating to ships or shipping
2 a : of or relating to a navy **b :** consisting of or involving warships

naval architect *noun* (circa 1885)
: one who designs ships

naval stores *noun plural* [from their former use in the construction and maintenance of wooden sailing vessels] (1678)
: products (as turpentine, pitch, and rosin) obtained from resinous conifers and especially pines

¹nave \'nāv\ *noun* [Middle English, from Old English *nafu;* akin to Old English *nafela* navel] (before 12th century)
: the hub of a wheel

²nave *noun* [Medieval Latin *navis,* from Latin, ship; akin to Old English *nōwend* sailor, Greek *naus* ship, Sanskrit *nau*] (1673)
: the main part of the interior of a church; *especially* **:** the long narrow central hall in a cruciform church that rises higher than the aisles flanking it to form a clerestory

na·vel \'nā-vəl\ *noun* [Middle English, from Old English *nafela;* akin to Old High German *nabalo* navel, Latin *umbilicus,* Greek *omphalos*] (before 12th century)
1 : a depression in the middle of the abdomen that marks the point of former attachment of the umbilical cord or yolk stalk
2 : the central point **:** MIDDLE

navel orange *noun* (1888)
: a seedless orange having a pit at the apex where the fruit encloses a small secondary fruit — called also *navel*

¹na·vic·u·lar \nə-'vi-kyə-lər\ *adjective* [Middle English *naviculare,* from Latin *navicula* boat, diminutive of *navis*] (15th century)
1 : shaped like a boat ⟨a *navicular* bone⟩
2 : of, relating to, or involving a navicular bone ⟨*navicular* disease⟩

²navicular *noun* [New Latin (*os*) *naviculare* navicular bone] (1816)
: a navicular bone; *especially* **:** one situated on the big-toe side of the tarsus

□ USAGE

nauseous Those who insist that *nauseous* can properly be used only in sense 1 and that in sense 2 it is an error for *nauseated* are mistaken. Current evidence shows these facts: *nauseous* is most frequently used to mean physically affected with nausea, usually after a linking verb such as *feel* or *become;* figurative use is quite a bit less frequent. Use of *nauseous* in sense 1 is much more often figurative than literal, and this use appears to be losing ground to *nauseating. Nauseated,* while not rare, is less common than *nauseous* in sense 2.

◇ WORD HISTORY

naughty The word *wight* "person, creature" is a more or less deliberate archaism in 20th century English, but a fragment of *wight* survives buried in a few much more familiar words. In Old English, *wiht,* the predecessor of *wight,* meant "thing" as well as "creature" and entered into indefinite pronouns such as *āwiht, ōwiht* "anything" and *nāwiht, nōwiht* "nothing." The latter shrank in Middle English to *naught* and *nought,* which by the end of the Middle English period had merged in pronunciation and become, in effect, spelling variants. A distinctive development of *nought* was its use as an adverb meaning "not at all, absolutely not." Reduced under weak stress to *not,* in Middle English it competed with and eventually ousted the older negative *ne* to become the ordinary negative adverb of Modern English. In a completely different direction, Old English *nāwiht, nōwiht* "nothing" came to be used adjectivally to mean "good for nothing, valueless." Then as *naught, nought* in Middle English they developed the meaning "bad, wicked." *Naught* was extended by the adjective suffix *-i* in late Middle English to give Modern English *naughty,* though the word now typically refers only to very mild wickedness, such as the pranks of misbehaving children or the improprieties of adults.

nausea The ancient Greeks were a seafaring people, so shipboard motion sickness was hardly unknown to them. In fact, the Greek word for it, *nausia* (or *neutia*) was a derivative of *naus,* meaning "ship." Since the dominant symptom of seasickness, as we call it, is an upset stomach and the concomitent urge to vomit, *nausia* became synonymous with similar stomach distress of any origin. The word was borrowed into Latin as *nausea* and from Latin into English. Ever since, many a landlubber has been in the throes of nausea without ever having gone near the ocean.

nav·i·ga·ble \ˈna-vi-gə-bəl\ *adjective* (15th century)
1 : deep enough and wide enough to afford passage to ships
2 : capable of being steered
— **nav·i·ga·bil·i·ty** \ˌna-vi-gə-ˈbi-lə-tē\ *noun*
— **nav·i·ga·bly** \ˈna-vi-gə-blē\ *adverb*

nav·i·gate \ˈna-və-ˌgāt\ *verb* **-gat·ed; -gat·ing** [Latin *navigatus,* past participle of *navigare,* from *navis* ship + *-igare* (from *agere* to drive) — more at AGENT] (1588)
intransitive verb
1 : to travel by water **:** SAIL
2 : to steer a course through a medium; *specifically* **:** to operate an airplane
3 : GET AROUND, MOVE
transitive verb
1 a : to sail over, on, or through **b :** to make one's way over or through **:** TRAVERSE
2 a : to steer or manage (a boat) in sailing **b :** to operate or control the course of (as an airplane)

nav·i·ga·tion \ˌna-və-ˈgā-shən\ *noun* (1547)
1 : the act or practice of navigating
2 : the science of getting ships, aircraft, or spacecraft from place to place; *especially* **:** the method of determining position, course, and distance traveled
3 : ship traffic or commerce
— **nav·i·ga·tion·al** \-shnəl, -shə-nᵊl\ *adjective*
— **nav·i·ga·tion·al·ly** *adverb*

nav·i·ga·tor \ˈna-və-ˌgā-tər\ *noun* (1590)
: one that navigates or is qualified to navigate

nav·vy \ˈna-vē\ *noun, plural* **navvies** [by shortening & alteration from *navigator* construction worker on a canal] (circa 1834)
chiefly British **:** an unskilled laborer

na·vy \ˈnā-vē\ *noun, plural* **navies** [Middle English *navie,* from Middle French, from Latin *navigia* ships, from *navigare*] (14th century)
1 : a group of ships **:** FLEET
2 : a nation's ships of war and of logistic support
3 *often capitalized* **:** the complete naval establishment of a nation including yards, stations, ships, and personnel
4 : a dark grayish purplish blue

navy bean *noun* (1856)
: a white-seeded kidney bean grown especially for its nutritious seeds

Navy Cross *noun* (1919)
: a U.S. Navy decoration awarded for extraordinary heroism in operations against an armed enemy

navy yard *noun* (1771)
: a yard where naval vessels are built or repaired

na·wab \nə-ˈwäb\ *noun* [Urdu *nawwāb*] (1758)
: NABOB

¹nay \ˈnā\ *adverb* [Middle English, from Old Norse *nei,* from *ne* not + *ei* ever — more at NO, AYE] (13th century)
: NO

²nay *noun* (14th century)
1 : DENIAL, REFUSAL
2 a : a negative reply or vote **b :** one who votes no

³nay *conjunction* (1560)
: not merely this but also **:** not only so but ⟨the letter made him happy, *nay,* ecstatic⟩

nay·say·er \ˈnā-ˌsā-ər, -ˌse(-ə)r\ *noun* (1721)
: one who denies, refuses, opposes, or is skeptical or cynical about something
— **nay–say·ing** \-ˌsā-iŋ\ *noun*

Naz·a·rene \ˌna-zə-ˈrēn\ *noun* [Middle English *Nazaren,* from Late Latin *Nazarenus,* from Greek *Nazarēnos,* from *Nazareth* Nazareth, Palestine] (13th century)
1 : a native or resident of Nazareth
2 a : CHRISTIAN 1a **b :** a member of the Church of the Nazarene that is a Protestant denomination deriving from the merging of three holiness groups, stressing sanctification, and following Methodist polity

Na·zi \ˈnät-sē, ˈnat-\ *noun* [German, by shortening & alteration from *Nationalsozialist,* from *national* national + *Sozialist* socialist] (1930)
1 : a member of a German fascist party controlling Germany from 1933 to 1945 under Adolf Hitler
2 *often not capitalized* **:** one who resembles a German Nazi
— **nazi** *adjective, often capitalized*
— **na·zi·fi·ca·tion** \ˌnät-si-fə-ˈkā-shən, ˌnat-\ *noun, often capitalized*
— **na·zi·fy** \ˈnät-si-ˌfī, ˈnat-\ *transitive verb, often capitalized*

Naz·i·rite *or* **Naz·a·rite** \ˈna-zə-ˌrīt\ *noun* [Late Latin *nazaraeus,* from Greek *naziraios, nazaraios,* from Hebrew *nāzīr,* literally, consecrated] (1560)
: a Jew of biblical times consecrated to God by a vow to avoid drinking wine, cutting the hair, and being defiled by the presence of a corpse
— **Naz·i·rit·ism** \-ˌrī-ˌti-zəm\ *noun*

Na·zism \ˈnät-ˌsi-zəm, ˈnat-\ *or* **Na·zi·ism** \-sē-ˌi-zəm\ *noun* [*Nazi* + *-ism*] (1934)
: the body of political and economic doctrines held and put into effect by the National Socialist German Workers' party in the Third German Reich including the totalitarian principle of government, state control of all industry, predominance of groups assumed to be racially superior, and supremacy of the führer

NCO \ˌen-(ˌ)sē-ˈō\ *noun* (circa 1810)
: NONCOMMISSIONED OFFICER

NC–17 \ˈen-ˈsē-ˌse-vən-ˈtēn\ *certification mark*
— used to certify that a motion picture is of such a nature that no one under the age of 17 can be admitted; compare G, PG, PG-13, R

-nd *symbol*
— used after the figure 2 to indicate the ordinal number *second* ⟨2*nd*⟩ ⟨72*nd*⟩

né \ˈnā\ *adjective* [French, literally, born — more at NÉE] (1905)
1 — used to indicate the original, former, or legal name of a man ⟨Robert Roe, *né* John Doe⟩
2 : originally or formerly called

ne- *or* **neo-** *combining form* [Greek, from *neos* new — more at NEW]
1 a : new **:** recent ⟨*Neo*gene⟩ **b :** new and different period or form of ⟨*Neo*platonism⟩ **:** in a new and different form or manner ⟨*Neo*platonic⟩ **c :** New World ⟨*Neo*tropical⟩ **d :** new and abnormal ⟨*neo*plasm⟩
2 : new chemical compound isomeric with or otherwise related to (such) a compound ⟨*neo*stigmine⟩

Ne·an·der·thal \nē-ˈan-dər-ˌtȯl, -ˌthȯl; nā-ˈän-dər-ˌtäl\ *adjective* (1861)
1 *also* **Ne·an·der·tal** \-ˌtȯl, -ˌtäl\ **:** being, relating to, or resembling Neanderthal man
2 : suggesting a caveman in appearance, mentality, or behavior
— **Neanderthal** *noun*

Neanderthal man *noun* [*Neanderthal,* valley in western Germany] (1863)
: a Middle Paleolithic hominid (*Homo sapiens neanderthalensis*) known from skeletal remains in Europe, northern Africa, and western Asia
— **Ne·an·der·thal·oid** \-ˌtȯ-ˌlȯid, -ˌthȯ-, -ˌtä-\ *adjective or noun*

¹neap \ˈnēp\ *adjective* [Middle English *neep,* from Old English *nēp* being at the stage of neap tide] (before 12th century)
: of, relating to, or constituting a neap tide

²neap *noun* (1584)
: NEAP TIDE

Ne·a·pol·i·tan \ˌnē-ə-ˈpä-lə-tᵊn\ *noun* [Middle English, from Latin *neapolitanus* of Naples, from Greek *neapolitēs* citizen of Naples, from *Neapolis* Naples] (15th century)
: a native or inhabitant of Naples, Italy
— **Neapolitan** *adjective*

Neapolitan ice cream *noun* (1895)
: a brick of from two to four layers of ice cream of different flavors

neap tide *noun* (circa 1548)
: a tide of minimum range occurring at the first and the third quarters of the moon

¹near \ˈnir\ *adverb* [Middle English *ner,* partly from *ner* nearer, from Old English *nēar,* comparative of *nēah* nigh; partly from Old Norse *nær* nearer, comparative of *nā-* nigh — more at NIGH] (13th century)
1 : at, within, or to a short distance or time
2 : ALMOST, NEARLY ⟨*near* dead⟩
3 : in a close or intimate manner **:** CLOSELY ⟨*near* related⟩
4 *archaic* **:** in a frugal manner

²near *preposition* (13th century)
: close to

³near *adjective* (14th century)
1 a : not far distant in time, place, or degree ⟨in the *near* future⟩ **b :** almost happening **:** narrowly missed or avoided ⟨a *near* win in the primary⟩ ⟨a *near* midair collision⟩ **c :** nearly not happening ⟨a *near* escape⟩
2 : closely related or intimately associated
3 a : being the closer of two ⟨the *near* side⟩ **b :** being the left-hand one of a pair ⟨the *near* wheel of a cart⟩
4 : DIRECT, SHORT ⟨the *nearest* road⟩
5 : STINGY, CLOSEFISTED
6 a : closely resembling the standard or typical ⟨a *near* desert⟩ **b :** approximating the genuine ⟨*near* silk⟩
— **near·ness** *noun*

⁴near *verb* (1513)
: APPROACH

near beer *noun* (1909)
: any of various malt liquors considered nonalcoholic because they contain less than a specified percentage of alcohol

near·by \nir-ˈbī, ˈnir-ˌ\ *adverb or adjective* (14th century)
: close at hand

Ne·arc·tic \ˌnē-ˈärk-tik, -ˈär-tik\ *adjective* (1858)
: of, relating to, or being the biogeographic subregion that includes Greenland and North America north of tropical Mexico

near gale *noun* (circa 1975)
: MODERATE GALE — see BEAUFORT SCALE table

near–in·fra·red \ˌnir-ˌin-frə-ˈred, -(ˌ)frä-\ *adjective* (1950)
: of or relating to the shorter wavelengths of radiation in the infrared spectrum and especially to those between 0.7 and 2.5 micrometers

near·ly \ˈnir-lē\ *adverb* (1561)
1 : in a close manner or relationship ⟨*nearly* related⟩
2 a : almost but not quite ⟨*nearly* identical⟩ ⟨*nearly* a year later⟩ **b :** to the least extent ⟨not *nearly* as good as we expected⟩

near miss *noun* (1940)
1 a : a miss (as with a bomb) close enough to cause damage **b :** something that falls just short of success
2 a : a near collision (as between aircraft) **b :** CLOSE CALL

near money *noun* (1942)
: assets (as savings accounts or government bonds) quickly and easily converted to cash

near point *noun* (1876)
: the point nearest the eye at which an object is accurately focused on the retina at full accommodation

near·shore \ˈnir-ˈshōr, -ˈshȯr, ˈnir-ˌ\ *adjective* (1896)
: extending outward an indefinite but usually short distance from shore ⟨*nearshore* sediments⟩

near·side \-ˌsīd\ *adjective* (circa 1840)
British **:** LEFT-HAND 1
— **nearside** *noun*

near·sight·ed \-ˌsī-təd\ *adjective* (1686)
: able to see near things more clearly than distant ones **:** MYOPIC
— **near·sight·ed·ly** *adverb*
— **near·sight·ed·ness** *noun*

near–ul·tra·vi·o·let \ˌnir-ˌəl-trə-'vī-(ə-)lət\ *adjective* (1951)
: of, relating to, or being the longest wavelengths of radiation especially between 300 and 400 nanometers

[1]**neat** \'nēt\ *noun, plural* **neat** *or* **neats** [Middle English *neet,* from Old English *nēat;* akin to Old High German *nōz* head of cattle, Old English *nēotan* to make use of, Lithuanian *nauda* use] (before 12th century)
: the common domestic bovine (*Bos taurus*)

[2]**neat** *adjective* [Middle French *net,* from Latin *nitidus* bright, neat, from *nitēre* to shine; probably akin to Middle Irish *níam* luster] (1542)
1 : free from dirt and disorder **:** habitually clean and orderly
2 a : free from admixture or dilution **:** STRAIGHT ⟨*neat* brandy⟩ ⟨*neat* cement⟩ **b :** free from irregularity **:** SMOOTH ⟨*neat* silk⟩
3 : marked by tasteful simplicity ⟨a *neat* outfit⟩
4 a : PRECISE, SYSTEMATIC **b :** marked by skill or ingenuity **:** ADROIT
5 : NET ⟨*neat* profit⟩
6 : FINE, ADMIRABLE
— **neat·ly** *adverb*
— **neat·ness** *noun*

[3]**neat** *adverb* (1649)
: without admixture or dilution **:** STRAIGHT

neat·en \'nē-t^{ə}n\ *transitive verb* **neat·ened; neat·en·ing** \'nēt-niŋ, 'nē-t^{ə}n-iŋ\ (1898)
1 : to set in order **:** make neat
2 : to finish (as a piece of sewing) carefully

neath \'nēth\ *preposition* (1787)
dialect **:** BENEATH

neat·herd \'nēt-ˌhərd\ *noun* (14th century)
: HERDSMAN

neat's–foot oil \'nēts-ˌfu̇t-\ *noun* (1787)
: a pale yellow fatty oil made especially from the bones of cattle and used chiefly as a leather dressing

neb \'neb\ *noun* [Middle English, from Old English; akin to Old Norse *nef* beak] (before 12th century)
1 a : the beak of a bird or tortoise **:** BILL **b** *chiefly dialect* **:** a person's mouth **c :** NOSE 1, SNOUT
2 : NIB, TIP

neb·bish \'ne-bish\ *noun* [Yiddish *nebekh* poor, unfortunate, from Czech *nebohý*] (1951)
: a timid, meek, or ineffectual person
— **neb·bishy** \-bi-shē\ *adjective*

ne·ben·kern \'nā-bən-ˌkərn, -ˌkern\ *noun* [German, literally, subsidiary nucleus] (1898)
: a 2-stranded helical structure of the proximal tail region of a spermatozoon that is derived from mitochondria

neb·u·la \'ne-byə-lə\ *noun, plural* **-lae** \-ˌlē, -ˌlī\ *also* **-las** [New Latin, from Latin, mist, cloud; akin to Old High German *nebul* fog, Greek *nephelē, nephos* cloud] (circa 1738)
1 : any of numerous clouds of gas or dust in interstellar space
2 : GALAXY; *especially* **:** a galaxy other than the Milky Way galaxy — not used technically
— **neb·u·lar** \-lər\ *adjective*

nebular hypothesis *noun* (1837)
: a hypothesis in astronomy: the solar system has evolved from a hot gaseous nebula

neb·u·lize \'ne-byə-ˌlīz\ *transitive verb* **-lized; -liz·ing** [Latin *nebula*] (1872)
: to reduce to a fine spray
— **neb·u·li·za·tion** \ˌne-byə-lə-'zā-shən\ *noun*
— **neb·u·liz·er** \'ne-byə-ˌlī-zər\ *noun*

neb·u·los·i·ty \ˌne-byə-'lä-sə-tē\ *noun, plural* **-ties** (1761)
1 : the quality or state of being nebulous
2 : nebulous matter; *also* **:** NEBULA 1

neb·u·lous \'ne-byə-ləs\ *adjective* [Latin *nebulosus* misty, from *nebula*] (1784)
1 : of, relating to, or resembling a nebula **:** NEBULAR
2 : INDISTINCT, VAGUE
— **neb·u·lous·ly** *adverb*
— **neb·u·lous·ness** *noun*

nec·es·sar·i·ly \ˌne-sə-'ser-ə-lē\ *adverb* (15th century)
1 : of necessity **:** UNAVOIDABLY
2 : as a logical result or consequence

[1]**nec·es·sary** \'ne-sə-ˌser-ē\ *adjective* [Middle English *necessarie,* from Latin *necessarius,* from *necesse* necessary, probably from *ne-* not + *cedere* to withdraw — more at NO] (14th century)
1 a : of an inevitable nature **:** INESCAPABLE **b** (1) **:** logically unavoidable (2) **:** that cannot be denied without contradiction **c :** determined or produced by the previous condition of things **d :** COMPULSORY
2 : absolutely needed **:** REQUIRED

[2]**necessary** *noun, plural* **-sar·ies** (14th century)
: an indispensable item **:** ESSENTIAL

necessary condition *noun* (1817)
1 : a proposition whose falsity assures the falsity of another
2 : a state of affairs that must prevail if another is to occur **:** PREREQUISITE

ne·ces·si·tar·i·an·ism \ni-ˌse-sə-'ter-ē-ə-ˌni-zəm\ *noun* (1854)
: the theory that results follow by invariable sequence from causes
— **ne·ces·si·tar·i·an** \-'ter-ē-ən\ *adjective or noun*

ne·ces·si·tate \ni-'se-sə-ˌtāt\ *transitive verb* **-tat·ed; -tat·ing** (1628)
1 : to make necessary **:** REQUIRE
2 : FORCE, COMPEL
— **ne·ces·si·ta·tion** \-ˌse-sə-'tā-shən\ *noun*

ne·ces·si·tous \ni-'se-sə-təs\ *adjective* (1611)
1 : NEEDY, IMPOVERISHED
2 : URGENT, PRESSING
3 : NECESSARY
— **ne·ces·si·tous·ly** *adverb*
— **ne·ces·si·tous·ness** *noun*

ne·ces·si·ty \ni-'se-sə-tē, -'ses-tē\ *noun, plural* **-ties** [Middle English *necessite,* from Middle French *necessité,* from Latin *necessitat-, necessitas,* from *necesse*] (14th century)
1 : the quality or state of being necessary
2 a : pressure of circumstance **b :** physical or moral compulsion **c :** impossibility of a contrary order or condition
3 : the quality or state of being in need; *especially* **:** POVERTY
4 a : something that is necessary **:** REQUIREMENT **b :** an urgent need or desire
— **of necessity :** in such a way that it cannot be otherwise; *also* **:** as a necessary consequence

[1]**neck** \'nek\ *noun* [Middle English *nekke,* from Old English *hnecca;* akin to Old High German *hnac* nape] (before 12th century)
1 a (1) **:** the part of an animal that connects the head with the body (2) **:** the siphon of a bivalve mollusk (as a clam) **b :** the part of a garment that covers or is next to the neck
2 : a relatively narrow part suggestive of a neck: as **a** (1) **:** the constricted end of a bottle (2) **:** the slender proximal end of a fruit **b :** CERVIX 2 **c :** the part of a stringed musical instrument extending from the body and supporting the fingerboard and strings **d :** a narrow stretch of land **e :** STRAIT 1b **f :** the part of a tooth between the crown and the root — see TOOTH illustration **g :** a column of solidified magma of a volcanic pipe or laccolith
3 : a narrow margin ⟨won by a *neck*⟩
4 : REGION, PART ⟨my *neck* of the woods⟩

[2]**neck** (1842)
transitive verb
1 : to kiss and caress amorously
2 : to reduce in diameter
intransitive verb
1 : to engage in amorous kissing and caressing
2 : to become constricted **:** NARROW

neck and neck *adverb or adjective* (1799)
: very close (as in a race)

necked \'nekt\ *adjective* (14th century)
: having a neck especially of a specified kind — often used in combination ⟨long-*necked*⟩

neck·er·chief \'ne-kər-chəf, -(ˌ)chif, -ˌchēf\ *noun, plural* **-chiefs** *also* **-chieves** *see* HANDKERCHIEF *plural*\ [Middle English *nekkerchef,* from *nekke* + *kerchef* kerchief] (14th century)
: a kerchief for the neck

neck·ing \'ne-kiŋ\ *noun* (1804)
1 : a narrow molding near the top of a column or pilaster
2 : the act or practice of kissing and caressing amorously

neck·lace \'ne-kləs\ *noun* (circa 1590)
: an ornament worn around the neck

neck·line \'nek-ˌlīn\ *noun* (1904)
: the line of the neck opening of a garment

neck–rein \-ˌrān\ (1940)
intransitive verb
of a saddle horse **:** to respond to the pressure of a rein on one side of the neck by turning in the opposite direction
transitive verb
: to direct (a horse) by pressures of the rein on the neck

neck·tie \-ˌtī\ *noun* (1838)
: a narrow length of material worn about the neck and tied in front; *especially* **:** FOUR-IN-HAND

neck·wear \-ˌwar, -ˌwer\ *noun* (circa 1879)
: articles of clothing (as ties and scarves) worn about the neck

necr- *or* **necro-** *combining form* [Late Latin, from Greek *nekr-, nekro-,* from *nekros* dead body — more at NOXIOUS]
1 : those that are dead ⟨*necro*philia⟩
2 : one that is dead ⟨*necr*opsy⟩

ne·crol·o·gy \nə-'krä-lə-jē, ne-\ *noun, plural* **-gies** [New Latin *necrologium,* from *necr-* + *-logium* (as in Medieval Latin *eulogium* eulogy)] (1799)
1 : OBITUARY
2 : a list of the recently dead
— **nec·ro·log·i·cal** \ˌne-krə-'lä-ji-kəl\ *adjective*
— **ne·crol·o·gist** \nə-'krä-lə-jist, ne-\ *noun*

nec·ro·man·cy \'ne-krə-ˌman(t)-sē\ *noun* [alteration of Middle English *nigromancie,* from Middle French, from Medieval Latin *nigromantia,* by folk etymology from Late Latin *necromantia,* from Late Greek *nekromanteia,* from Greek *nekr-* + *-manteia* -mancy] (1522)
1 : conjuration of the spirits of the dead for purposes of magically revealing the future or influencing the course of events
2 : MAGIC, SORCERY
— **nec·ro·man·cer** \-sər\ *noun*
— **nec·ro·man·tic** \ˌne-krə-'man-tik\ *adjective*
— **nec·ro·man·ti·cal·ly** \-ti-k(ə-)lē\ *adverb*

ne·croph·a·gous \nə-'krä-fə-gəs, ne-\ *adjective* (1835)
: feeding on corpses or carrion ⟨*necrophagous* insects⟩

nec·ro·phil·ia \ˌne-krə-'fi-lē-ə\ *noun* [New Latin] (1892)
: obsession with and usually erotic interest in or stimulation by corpses
— **nec·ro·phil·i·ac** \-'fi-lē-ˌak\ *adjective or noun*
— **nec·ro·phil·ic** \-'fi-lik\ *adjective*

ne·croph·i·lism \nə-'krä-fə-ˌli-zəm, ne-\ *noun* (1864)
: NECROPHILIA

ne·crop·o·lis \nə-ˈkrä-pə-ləs, ne-\ *noun, plural* **-lis·es** *or* **-les** \-ˌlēz\ *or* **-leis** \-ˌlās\ *or* **-li** \-ˌlī, -ˌlē\ [Late Latin, city of the dead, from Greek *nekropolis,* from *nekr-* + *-polis* -polis] (1819)
: CEMETERY; *especially* **:** a large elaborate cemetery of an ancient city

¹nec·rop·sy \ˈne-ˌkräp-sē\ *noun, plural* **-sies** (1856)
: AUTOPSY 1

²necropsy *transitive verb* **-sied; -sy·ing** (1927)
: to perform an autopsy on

ne·cro·sis \nə-ˈkrō-səs, ne-\ *noun, plural* **ne·cro·ses** \-ˌsēz\ [Late Latin, from Greek *nekrōsis,* from *nekroun* to make dead, from *nekros* dead body] (1665)
: usually localized death of living tissue
— ne·crot·ic \-ˈkrä-tik\ *adjective*

nec·ro·tiz·ing \ˈne-krə-ˌtī-ziŋ\ *adjective* [Greek *nekrōtikos* necrotic, from *nekroun*] (1899)
: causing, associated with, or undergoing necrosis ⟨*necrotizing* infections⟩ ⟨*necrotizing* tissue⟩

nec·tar \ˈnek-tər\ *noun* [Latin, from Greek *nektar*]
1 a : the drink of the Greek and Roman gods **b :** something delicious to drink **c :** a beverage of fruit juice and pulp ⟨apricot *nectar*⟩
2 : a sweet liquid that is secreted by the nectaries of a plant and is the chief raw material of honey
— nec·tar·ous \-t(ə-)rəs\ *adjective*

nec·tar·ine \ˌnek-tə-ˈrēn\ *noun* [obsolete *nectarine,* adjective, like nectar] (1611)
: a peach with a smooth-skinned fruit that is a frequent somatic mutation of the normal peach; *also* **:** its fruit

nec·tary \ˈnek-t(ə-)rē\ *noun, plural* **-tar·ies** [New Latin *nectarium,* irregular from Latin *nectar* + *-arium* -ary] (1759)
: a plant gland that secretes nectar

née *or* **nee** \ˈnā\ *adjective* [French *née,* feminine of *né,* literally, born, past participle of *naître* to be born, from Latin *nasci* — more at NATION] (1758)
1 — used to identify a woman by her maiden family name
2 : originally or formerly called ⟨the Brewers *née* Pilots who also are in their third year —Fred Ciampa⟩

¹need \ˈnēd\ *noun* [Middle English *ned,* from Old English *nīed, nēd;* akin to Old High German *nōt* distress, need, Old Prussian *nautin* need] (before 12th century)
1 : necessary duty **:** OBLIGATION
2 a : a lack of something requisite, desirable, or useful **b :** a physiological or psychological requirement for the well-being of an organism
3 : a condition requiring supply or relief
4 : lack of the means of subsistence **:** POVERTY

²need *verb* **need·ed; need·ing; needs** *or (auxiliary)* **need** (before 12th century)
intransitive verb
1 : to be needful or necessary
2 : to be in want
transitive verb
: to be in need of **:** REQUIRE
verbal auxiliary
: be under necessity or obligation to ⟨you *need* not answer⟩ ■

¹need·ful \ˈnēd-fəl\ *adjective* (12th century)
1 : being in need
2 : NECESSARY, REQUISITE
— need·ful·ly \-fə-lē\ *adverb*
— need·ful·ness *noun*

²needful *noun* (1709)
1 : something needed or requisite
2 : MONEY

¹nee·dle \ˈnē-dᵊl\ *noun* [Middle English *nedle,* from Old English *nǣdl;* akin to Old High German *nādala* needle, *nājan* to sew, Latin *nēre* to spin, Greek *nēn*] (before 12th century)
1 a : a small slender usually steel instrument that has an eye for thread or surgical sutures at one end and that is used for sewing **b :** any of various devices for carrying thread and making stitches (as in crocheting or knitting) **c** (1) **:** a slender hollow instrument for introducing material into or removing material from the body parenterally (2) **:** any of various slender hollow devices used to introduce matter (as air) into or remove it from an object (as a ball)
2 a : a slender bar of magnetized steel that when allowed to turn freely (as in a compass) indicates the direction of a magnetic field (as of the earth) **b :** a slender usually sharp-pointed indicator on a dial
3 a : a slender pointed object resembling a needle: as (1) **:** a pointed crystal (2) **:** a sharp rock (3) **:** OBELISK **b :** a needle-shaped leaf (as of a conifer) **c :** a slender rod (as of jewel or steel) with a rounded tip used in a phonograph to transmit vibrations from a record **:** STYLUS **d :** a slender pointed rod controlling a fine inlet or outlet (as in a valve)
4 : a teasing or gibing remark
— nee·dle·like \ˈnē-dᵊl-ˌ(l)īk\ *adjective*

²needle *verb* **nee·dled; nee·dling** \ˈnēd-liŋ, ˈnē-dᵊl-iŋ\ (circa 1715)
transitive verb
1 : to sew or pierce with or as if with a needle
2 a : TEASE, TORMENT **b :** to incite to action by repeated gibes ⟨*needled* the boy into a fight⟩
intransitive verb
: SEW, EMBROIDER
— nee·dler \ˈnēd-lər, ˈnē-dᵊl-ər\ *noun*

nee·dle·fish \ˈnē-dᵊl-ˌfish\ *noun* (1601)
1 : any of a family (Belonidae) of elongate green and silvery chiefly marine bony fishes with long slender jaws
2 : PIPEFISH

needlefish 1

nee·dle·point \-ˌpȯint\ *noun* (1865)
1 : lace worked with a needle over a paper pattern
2 : embroidery done on canvas usually in simple even stitches across counted threads
— needlepoint *adjective*

need·less \ˈnēd-ləs\ *adjective* (14th century)
: not needed **:** UNNECESSARY ⟨*needless* waste⟩ ⟨*needless* to say⟩
— need·less·ly *adverb*
— need·less·ness *noun*

nee·dle·wom·an \ˈnē-dᵊl-ˌwu̇-mən\ *noun* (1535)
: a woman who does needlework; *especially* **:** SEAMSTRESS

nee·dle·work \-ˌwərk\ *noun* (14th century)
1 : work done with a needle; *especially* **:** work (as embroidery) other than plain sewing
2 : the occupation of one who does needlework
— nee·dle·work·er \-ˌwər-kər\ *noun*

needn't \ˈnē-dᵊnt\ (1865)
: need not

needs \ˈnēdz\ *adverb* [Middle English *nedes,* from Old English *nēdes,* from genitive of *nēd* need] (before 12th century)
: of necessity **:** NECESSARILY ⟨must *needs* be recognized⟩

needy \ˈnē-dē\ *adjective* **need·i·er; -est** (12th century)
: being in want **:** POVERTY-STRICKEN ⟨*needy* families⟩
— need·i·ness *noun*

neem \ˈnēm\ *noun* [Hindi *nīm,* from Sanskrit *nimba*] (1813)
: a large East Indian tree (*Azadirachta indica* of the family Meliaceae) whose trunk exudes a tenacious gum and has a bitter bark used as a tonic and whose fruit and seeds yield a medicinal aromatic oil

neep \ˈnēp\ *noun* [Middle English *nepe,* from Old English *nǣp,* from Latin *napus* — more at NAPIFORM] (before 12th century)
chiefly Scottish **:** TURNIP

ne'er \ˈnar, ˈner\ *adverb* (13th century)
: NEVER

ne'er–do–well \-du̇-ˌwel\ *noun* (1736)
: an idle worthless person
— ne'er–do–well *adjective*

ne·far·i·ous \ni-ˈfar-ē-əs, -ˈfer-\ *adjective* [Latin *nefarius,* from *nefas* crime, from *ne-* not + *fas* right, divine law; perhaps akin to Greek *themis* law, *tithenai* to place — more at DO] (circa 1609)
: flagrantly wicked or impious **:** EVIL
synonym see VICIOUS
— ne·far·i·ous·ly *adverb*

ne·gate \ni-ˈgāt\ *transitive verb* **ne·gat·ed; ne·gat·ing** [Latin *negatus,* past participle of *negare* to say no, deny, from *neg-* no, not (akin to *ne-* not) — more at NO] (circa 1623)
1 : to deny the existence or truth of
2 : to cause to be ineffective or invalid
synonym see NULLIFY
— negate *noun*
— ne·ga·tor \-ˈgā-tər\ *noun*

ne·ga·tion \ni-ˈgā-shən\ *noun* (15th century)
1 a : the action or logical operation of negating or making negative **b :** a negative statement, judgment, or doctrine; *especially* **:** a logical proposition formed by asserting the falsity of a given proposition
2 a : something that is the absence of something actual **:** NONENTITY **b :** something considered the opposite of something regarded as positive
— ne·ga·tion·al \-shnəl, -shə-nᵊl\ *adjective*

¹neg·a·tive \ˈne-gə-tiv\ *adjective* (15th century)
1 a : marked by denial, prohibition, or refusal ⟨received a *negative* answer⟩; *also* **:** marked by absence, withholding, or removal of something positive ⟨the *negative* motivation of shame —Garrett Hardin⟩ **b** (1) **:** denying a predicate of a subject or a part of a subject ⟨"no A is B" is a *negative* proposition⟩ (2) **:** denoting the absence or the contradictory of something ⟨*nontoxic* is a *negative* term⟩ (3) **:** expressing negation ⟨*negative* particles such as *no* and *not*⟩ **c :** ADVERSE, UNFAVORABLE ⟨the reviews were mostly *negative*⟩
2 a : lacking positive qualities; *especially* **:** DISAGREEABLE **b :** marked by features of hostility, withdrawal, or pessimism that hinder or oppose constructive treatment or development ⟨a *negative* outlook⟩ ⟨*negative* criticism⟩

☐ USAGE

need Those who live in western Pennsylvania, central Ohio, Tennessee, or southeastern Colorado—somewhere in the area known as Midland to linguistic geographers—may have heard a curious idiom in which *need* is followed directly by a past participle, as in "the house needs painted," where the more widely used version would be "the house needs painting" or "the house needs to be painted." The idiom appears to be limited to the Midland area and, as far as we know now, is used almost exclusively in speech. A few people have told us that they have seen it in newspapers, but we have no printed evidence. One dictionary edited in England mentions the construction but says only that it is "widely disliked," which does not tell us much about possible dialectal restrictions. We do know that the construction exists in Scottish English, and it may be that the construction came from Scotland to this country. But 19th century use is only conjectural, so we cannot be sure.

3 a (1) **:** less than zero and opposite in sign to a positive number that when added to the given number yields zero ⟨−2 is a *negative* number⟩ (2) **:** having more outgo than income **:** constituting a loss ⟨*negative* cash flow⟩ ⟨*negative* worth⟩ **b :** extending or generated in a direction opposite to an arbitrarily chosen regular direction or position ⟨*negative* angle⟩
4 a : being, relating to, or charged with electricity of which the electron is the elementary unit **b :** having more electrons than protons ⟨a *negative* ion⟩ **c** (1) **:** having lower electric potential and constituting the part toward which the current flows from the external circuit ⟨the *negative* pole⟩ (2) **:** being the electron-emitting electrode of an electron tube
5 a : not affirming the presence of the organism or condition in question ⟨a *negative* TB test⟩ **b :** directed or moving away from a source of stimulation ⟨*negative* tropism⟩ **c :** less than the pressure of the atmosphere ⟨*negative* pressure⟩
6 : having the light and dark parts in approximately inverse order to those of the original photographic subject
7 *of a lens* **:** diverging light rays and forming a virtual inverted image
— neg·a·tive·ly *adverb*
— neg·a·tive·ness *noun*
— neg·a·tiv·i·ty \ˌne-gə-ˈti-və-tē\ *noun*

²negative *noun* (1571)
1 a (1) **:** a reply that indicates the withholding of assent **:** REFUSAL (2) *archaic* **:** a right of veto (3) *obsolete* **:** an adverse vote **:** VETO **b :** a proposition which denies or contradicts another; *especially* **:** the one of a pair of propositions in which negation is expressed
2 a : something that is the opposite or negation of something else **b :** DRAWBACK, LIABILITY
3 a : an expression (as the word *no*) of negation or denial **b :** a negative number
4 : the side that upholds the contradictory proposition in a debate
5 : a negative photographic image on transparent material used for printing positive pictures; *also* **:** the material that carries such an image
6 : a reverse impression taken from a piece of sculpture or ceramics

³negative *transitive verb* **-tived; -tiv·ing** (1706)
1 a : to refuse assent to **b :** to reject by or as if by a vote
2 : to demonstrate the falsity of
3 : to deny the truth, reality, or validity of
4 : NEUTRALIZE, COUNTERACT

negative feedback *noun* (1934)
: feedback that tends to stabilize a process by reducing its rate or output when its effects are too great

negative income tax *noun* (1966)
: a system of federal subsidy payments to families with incomes below a stipulated level

negative transfer *noun* (1921)
: the impeding of learning or performance in a situation by the carry-over of learned responses from another situation

neg·a·tiv·ism \ˈne-gə-ti-ˌvi-zəm\ *noun* (1824)
1 : an attitude of mind marked by skepticism especially about nearly everything affirmed by others
2 : a tendency to refuse to do, to do the opposite of, or to do something at variance with what is asked
— neg·a·tiv·ist \-vist\ *noun or adjective*
— neg·a·tiv·is·tic \ˌne-gə-ti-ˈvis-tik\ *adjective*

¹ne·glect \ni-ˈglekt\ *transitive verb* [Latin *neglectus,* past participle of *neglegere, neclegere,* from *nec-* not (akin to *ne-* not) + *legere* to gather — more at NO, LEGEND] (1529)
1 : to give little attention or respect to **:** DISREGARD
2 : to leave undone or unattended to especially through carelessness ☆
— ne·glect·er *noun*

²neglect *noun* (1588)
1 : an act or instance of neglecting something
2 : the condition of being neglected

ne·glect·ful \ni-ˈglek(t)-fəl\ *adjective* (1644)
: given to neglecting **:** CARELESS, HEEDLESS
synonym see NEGLIGENT
— ne·glect·ful·ly \-fə-lē\ *adverb*
— ne·glect·ful·ness *noun*

neg·li·gee *also* **neg·li·gé** \ˌne-glə-ˈzhā, ˈne-glə-ˌ\ *noun* [French *négligé,* from past participle of *négliger* to neglect, from Latin *neglegere*] (1756)
1 : a woman's long flowing usually sheer dressing gown
2 : carelessly informal or incomplete attire

neg·li·gence \ˈne-gli-jən(t)s\ *noun* (14th century)
1 a : the quality or state of being negligent **b :** failure to exercise the care that a prudent person usually exercises
2 : an act or instance of being negligent

neg·li·gent \-jənt\ *adjective* [Middle English, from Middle French & Latin; Middle French, from Latin *neglegent-, neglegens,* present participle of *neglegere*] (14th century)
1 : marked by or given to neglect especially habitually or culpably **:** not taking prudent care
2 : marked by a carelessly easy manner ☆
— neg·li·gent·ly *adverb*

neg·li·gi·ble \ˈne-gli-jə-bəl\ *adjective* [Latin *neglegere, negligere*] (1829)
: so small or unimportant or of so little consequence as to warrant little or no attention **:** TRIFLING
— neg·li·gi·bil·i·ty \ˌne-gli-jə-ˈbi-lə-tē\ *noun*
— neg·li·gi·bly \ˈne-gli-jə-blē\ *adverb*

ne·go·tia·ble \ni-ˈgō-sh(ē-)ə-bəl\ *adjective* (1758)
: capable of being negotiated: as **a :** transferable from one person to another by being delivered with or without endorsement so that the title passes to the transferee ⟨*negotiable* securities⟩ **b :** capable of being traversed, dealt with, or accomplished ⟨a difficult but *negotiable* road⟩ ⟨*negotiable* demands⟩ **c :** open to discussion or dispute
— ne·go·tia·bil·i·ty \-ˌgō-sh(ē-)ə-ˈbi-lə-tē\ *noun*

ne·go·ti·ant \-ˈgō-sh(ē-)ənt\ *noun* (circa 1611)
: one that negotiates

ne·go·ti·ate \ni-ˈgō-shē-ˌāt, ÷-sē-\ *verb* **-at·ed; -at·ing** [Latin *negotiatus,* past participle of *negotiari* to carry on business, from *negotium* business, from *neg-* not + *otium* leisure — more at NEGATE] (1599)
intransitive verb
: to confer with another so as to arrive at the settlement of some matter
transitive verb
1 a : to deal with (some matter or affair that requires ability for its successful handling) **:** MANAGE **b :** to arrange for or bring about through conference, discussion, and compromise ⟨*negotiate* a treaty⟩
2 a : to transfer (as a bill of exchange) to another by delivery or endorsement **b :** to convert into cash or the equivalent value ⟨*negotiate* a check⟩
3 a : to successfully travel along or over ⟨*negotiate* a turn⟩ **b :** COMPLETE, ACCOMPLISH ⟨*negotiate* the trip in two hours⟩
— ne·go·ti·a·tor \-ˌā-tər\ *noun*
— ne·go·ti·a·to·ry \-sh(ē-)ə-ˌtōr-ē, ÷-sē-, -ˌtȯr-\ *adjective*

ne·go·ti·a·tion \ni-ˌgō-shē-ˈā-shən, ÷-sē-\ *noun* (15th century)
: the action or process of negotiating or being negotiated — often used in plural

Ne·gress \ˈnē-grəs\ *noun* (1786)
: a black woman or girl — sometimes taken to be offensive

Ne·gril·lo \ni-ˈgri-(ˌ)lō, -ˈgrē-(ˌ)yō\ *noun, plural* **-los** *or* **-loes** [Spanish, diminutive of *negro*] (1853)
: a member of a people (as Pygmies) belonging to a group of dark-skinned peoples of small stature that live in Africa

Ne·gri·to \nə-ˈgrē-(ˌ)tō\ *noun, plural* **-tos** *or* **-toes** [Spanish, diminutive of *negro*] (1812)
: a member of a people (as the Andamanese) belonging to a group of dark-skinned peoples of small stature that live in Oceania and the southeastern part of Asia

ne·gri·tude \ˈne-grə-ˌtüd, ˈnē-, -ˌtyüd\ *noun* [French *négritude,* from *nègre* Negro + *-i-* + *-tude*] (1950)
1 : a consciousness of and pride in the cultural and physical aspects of the African heritage
2 : the state or condition of being black

Ne·gro \ˈnē-(ˌ)grō\ *noun, plural* **Negroes** [Spanish or Portuguese, from *negro* black, from Latin *nigr-, niger*] (1555)
1 : a member of the black race distinguished from members of other races by usually inherited physical and physiological characteristics without regard to language or culture; *especially* **:** a member of a people belonging to the African branch of the black race — sometimes taken to be offensive
2 : a person of Negro descent — sometimes taken to be offensive
— Negro *adjective*
— Ne·gro·ness \-grō-nəs\ *noun*

ne·groid \ˈnē-ˌgrȯid\ *adjective, often capitalized* (1859)
: of, resembling, related to, or characteristic of the Negro race
— negroid *noun, often capitalized*

ne·gro·phobe \ˈnē-grə-ˌfōb\ *noun, often capitalized* (1900)
: one who strongly dislikes or fears black people
— ne·gro·pho·bia \ˌnē-grə-ˈfō-bē-ə\ *noun, often capitalized*

☆ SYNONYMS
Neglect, disregard, ignore, overlook, slight, forget mean to pass over without giving due attention. NEGLECT implies giving insufficient attention to something that has a claim to one's attention ⟨habitually *neglected* his studies⟩. DISREGARD suggests voluntary inattention ⟨*disregarded* the wishes of other members⟩. IGNORE implies a failure to regard something obvious ⟨*ignored* the snide remark⟩. OVERLOOK suggests disregarding or ignoring through haste or lack of care ⟨in my rush I *overlooked* a key example⟩. SLIGHT implies contemptuous or disdainful disregarding or omitting ⟨*slighted* several major authors in her survey⟩. FORGET may suggest either a willful ignoring or a failure to impress something on one's mind ⟨*forget* what others say and listen to me⟩.

Negligent, neglectful, lax, slack, remiss mean culpably careless or indicative of such carelessness. NEGLIGENT implies inattention to one's duty or business ⟨*negligent* about writing a note of thanks⟩. NEGLECTFUL adds a more disapproving implication of laziness or deliberate inattention ⟨a society callously *neglectful* of the poor⟩. LAX implies a blameworthy lack of strictness, severity, or precision ⟨a reporter *lax* about accurate quotation⟩. SLACK implies want of due or necessary diligence or care ⟨*slack* workmanship and slipshod construction⟩. REMISS implies blameworthy carelessness shown in slackness, forgetfulness, or neglect ⟨had been *remiss* in her familial duties⟩.

¹ne·gus \'nē-gəs, ni-'güs\ *noun* [Amharic *nəgus,* from Ge'ez *něgūša nagašt* king of kings] (1594)
: KING — used as a title of the sovereign of Ethiopia
²ne·gus \'nē-gəs\ *noun* [Francis *Negus* (died 1732) English colonel] (1743)
: a beverage of wine, hot water, sugar, lemon juice, and spices
Ne·he·mi·ah \ˌnē-(h)ə-'mī-ə\ *noun* [Hebrew *Nĕḥemyāh*]
1 : a Jewish leader of the 5th century B.C. who supervised the rebuilding of the Jerusalem city walls and instituted religious reforms in the city
2 : a narrative and historical book of canonical Jewish and Christian Scripture — see BIBLE table
Ne·he·mi·as \-'mī-əs\ *noun* [Late Latin, from Hebrew *Nĕḥemyāh*]
: NEHEMIAH
neigh \'nā\ *intransitive verb* [Middle English *neyen,* from Old English *hnǣgan;* akin to Middle High German *nēgen* to neigh] (before 12th century)
: to make the prolonged cry of a horse
— neigh *noun*
¹neigh·bor \'nā-bər\ *noun* [Middle English, from Old English *nēahgebūr* (akin to Old High German *nāhgibūr*); akin to Old English *nēah* near and Old English *gebūr* dweller — more at NIGH, BOOR] (before 12th century)
1 : one living or located near another
2 : FELLOWMAN
²neighbor *adjective* (1530)
: being immediately adjoining or relatively near
³neighbor *verb* **neigh·bored; neigh·bor·ing** \-b(ə-)riŋ\ (circa 1586)
transitive verb
: to adjoin immediately or lie relatively near to
intransitive verb
1 : to live or be located as a neighbor
2 : to associate in a neighborly way
neigh·bor·hood \'nā-bər-ˌhu̇d\ *noun* (15th century)
1 : neighborly relationship
2 : the quality or state of being neighbors **:** PROXIMITY
3 a : a place or region near **:** VICINITY **b :** an approximate amount, extent, or degree ⟨cost in the *neighborhood* of $10⟩
4 a : the people living near one another **b :** a section lived in by neighbors and usually having distinguishing characteristics
5 : the set of all points belonging to a given set whose distances from a given point are less than a given positive number; *broadly* **:** a set that contains a neighborhood
neigh·bor·ly \-lē\ *adjective* (1558)
: of, relating to, or characteristic of congenial neighbors; *especially* **:** FRIENDLY
synonym see AMICABLE
— neigh·bor·li·ness *noun*
neigh·bour \-bər\ *chiefly British variant of* NEIGHBOR
¹nei·ther \'nē-t͟hər *also* 'nī-\ *conjunction* [Middle English, alteration (influenced by *either*) of *nauther, nother,* from Old English *nāhwæther, nōther,* from *nā, nō* not + *hwæther* which of two, whether] (12th century)
1 : not either ⟨*neither* black nor white⟩
2 : also not ⟨*neither* did I⟩ ▪
²neither *pronoun* (13th century)
: not the one or the other of two or more ▪
³neither *adjective* (14th century)
: not either ⟨*neither* hand⟩
⁴neither *adverb* (1551)
1 *chiefly dialect* **:** EITHER ⟨are not to be understood *neither* —Earl of Chesterfield⟩
2 : similarly not **:** also not ⟨just as the serf was not permitted to leave the land, so *neither* was his offspring —G. G. Coulton⟩
nek·ton \'nek-tən, -ˌtän\ *noun* [German *Nekton,* from Greek *nēkton,* neuter of *nēktos* swimming, from *nēchein* to swim — more at NATANT] (1893)
: free-swimming aquatic animals essentially independent of wave and current action
— nek·ton·ic \nek-'tä-nik\ *adjective*
nel·son \'nel-sən\ *noun* [probably from the name *Nelson*] (1889)
: a wrestling hold marked by the application of leverage against an opponent's arm, neck, and head — compare FULL NELSON, HALF NELSON
nemat- *or* **nemato-** *combining form* [New Latin, from Greek *nēmat-,* from *nēmat-, nēma,* from *nēn* to spin — more at NEEDLE]
1 : thread ⟨*nemato*cyst⟩
2 : nematode ⟨*nemato*logy⟩
ne·mat·ic \ni-'ma-tik\ *adjective* [International Scientific Vocabulary *nemat-* + ¹*-ic*] (1923)
: of, relating to, or being the phase of a liquid crystal characterized by arrangement of the long axes of the molecules in parallel lines but not layers — compare CHOLESTERIC, SMECTIC
nem·a·to·cide *or* **nem·a·ti·cide** \'ne-mə-tə-ˌsīd, ni-'ma-tə-\ *noun* (1898)
: a substance or preparation used to destroy nematodes
— nem·a·to·ci·dal *also* **nem·a·ti·ci·dal** \ˌne-mə-tə-'sī-dᵊl, ni-ˌma-tə-\ *adjective*
nem·a·to·cyst \-ˌsist\ *noun* [International Scientific Vocabulary] (1875)
: one of the stinging organelles of coelenterates used in catching prey
nem·a·tode \'ne-mə-ˌtōd\ *noun* [ultimately from Greek *nēmat-, nēma*] (1865)
: any of a phylum (Nematoda or Nemata) of elongated cylindrical worms parasitic in animals or plants or free-living in soil or water
nem·a·tol·o·gy \ˌne-mə-'tä-lə-jē\ *noun* (circa 1916)
: a branch of zoology that deals with nematodes
— nem·a·to·log·i·cal \ˌne-mə-tᵊl-'ä-ji-kəl\ *adjective*
— nem·a·tol·o·gist \ˌne-mə-'tä-lə-jist\ *noun*
Nem·bu·tal \'nem-byə-ˌtȯl\ *trademark*
— used for the sodium salt of pentobarbital
ne·mer·te·an \ni-'mər-tē-ən\ *noun* [ultimately from Greek *Nēmertēs* Nemertes, one of the Nereids] (1861)
: any of a phylum (Nemertea synonym Rhynchocoela) of often vividly colored marine worms most of which burrow in the mud or sand along seacoasts — called also *ribbon worm*
— nem·er·tine \'ne-mər-ˌtīn\ *adjective or noun*
nem·e·sis \'ne-mə-səs\ *noun* [Latin, from Greek]
1 *capitalized* **:** the Greek goddess of retributive justice
2 *plural* **-e·ses** \-ˌsēz\ **a :** one that inflicts retribution or vengeance **b :** a formidable and usually victorious rival or opponent
3 *plural* **-eses a :** an act or effect of retribution **b :** BANE 2
ne·moph·i·la \ni-'mä-fə-lə\ *noun* [New Latin, from Greek *nemos* wooded pasture + *philos* loving] (1830)
: any of a genus (*Nemophila*) of American annual herbs of the waterleaf family cultivated for their showy blue or white sometimes spotted flowers
ne·ne \'nā-(ˌ)nā\ *noun* [Hawaiian *nēnē*] (1902)
: an endangered goose (*Branta sandvicensis* synonym *Nesochen sandvicensis*) of the Hawaiian Islands that usually inhabits waterless uplands and feeds on berries and vegetation

nene

neo- — see NE-
neo·clas·sic \ˌnē-ō-'kla-sik\ *or* **neo·clas·si·cal** \-si-kəl\ *adjective* (1877)
: of, relating to, or constituting a revival or adaptation of the classical especially in literature, music, art, or architecture
— neo·clas·si·cism \-'kla-sə-ˌsi-zəm\ *noun*
— neo·clas·si·cist \-sist\ *noun or adjective*
neo·co·lo·nial·ism \ˌnē-ō-kə-'lōn-yə-ˌli-zəm, -'lō-nē-ə-ˌli-\ *noun* (1961)
: the economic and political policies by which a great power indirectly maintains or extends its influence over other areas or people
— neo·co·lo·nial \-'lōn-yəl, -'lō-nē-əl\ *adjective*
— neo·co·lo·nial·ist \-yə-list, -ə-list\ *noun or adjective*
neo·con·ser·va·tive \-kən-'sər-və-tiv\ *noun* (1955)
: a former liberal espousing political conservatism
— neo·con·ser·va·tism \-və-ˌti-zəm\ *noun*
— neoconservative *adjective*
neo·cor·tex \ˌnē-ō-'kȯr-ˌteks\ *noun* [New Latin; from its being the cortex of the phylogenetically most recently developed part of the brain] (1909)
: the dorsal region of the cerebral cortex that is unique to mammals
neo·cor·ti·cal \-'kȯr-ti-kəl\ *adjective* (1909)
: of or relating to the neocortex
neo–Dar·win·ian \-där-'wi-nē-ən\ *adjective, often N capitalized* (1895)
: of or relating to neo-Darwinism
neo–Dar·win·ism \-'där-wə-ˌni-zəm\ *noun, often N capitalized* (circa 1900)
: a theory of evolution that is a synthesis of Darwin's theory in terms of natural selection and modern population genetics
— neo–Dar·win·ist \-'där-wə-nist\ *noun, often N capitalized*
neo·dym·i·um \ˌnē-ō-'di-mē-əm\ *noun* [New Latin, from *ne-* + *-dymium* (from *didymium*)] (1885)
: a yellow metallic element of the rare-earth group that is used in magnets and lasers — see ELEMENT table
neo–Ex·pres·sion·ism \-ik-'spre-shə-ˌni-zəm\ *noun, often N capitalized* (1961)
: a revival of expressionism in art characterized by intense colors, dramatic usually figural forms, and emotive subject matter
— neo–Ex·pres·sion·ist \-nist\ *noun or adjective, often N capitalized*

☐ USAGE
¹neither Although use with *or* is neither archaic nor wrong, the conjunction *neither* is usually followed by *nor.* A few commentators think that *neither* must be limited in reference to two, but reference to more than two has been quite common since the 17th century ⟨rigid enforcement of antique decorum will help *neither* language, literature, nor literati —James Sledd⟩. See in addition EITHER.

²neither Some commentators insist that the pronoun *neither* must be used with a singular verb. It generally is, but especially when a prepositional phrase intervenes between it and the verb, a plural verb is quite common ⟨*neither* of those ideal solutions are in sight —C. P. Snow⟩.

neo–Freud·ian \-ˈfrói-dē-ən\ *adjective, often N capitalized* (1945)
: of or relating to a school of psychoanalysis that differs from Freudian orthodoxy in emphasizing the importance of social and cultural factors in the development of an individual's personality
— **neo–Freudian** *noun, often N capitalized*

Neo·gene \ˈnē-ə-ˌjēn\ *adjective* [International Scientific Vocabulary *ne-* + *-gene* (from Greek *-genēs* born) — more at -GEN] (1878)
: of, relating to, or being the later portion of the Tertiary including the Miocene and Pliocene or the corresponding system of rocks
— **Neogene** *noun*

neo–Goth·ic \ˌnē-ō-ˈgä-thik\ *adjective* (1892)
: of, relating to, or constituting a revival or adaptation of the Gothic especially in literature or architecture

neo–im·pres·sion·ism \ˌnē-ō-im-ˈpre-shə-ˌni-zəm\ *noun, often N&I capitalized* [French *néo-impressionisme,* from *né-* ne- + *impressionisme* impressionism] (1892)
: a late 19th century French art theory and practice characterized by an attempt to make impressionism more precise in form and the use of a pointillist painting technique
— **neo–im·pres·sion·ist** \-ˈpre-sh(ə-)nist\ *adjective or noun, often N&I capitalized*

Neo–Lat·in \-ˈla-tᵊn\ *noun* [International Scientific Vocabulary] (1850)
1 : NEW LATIN
2 : ROMANCE 5

neo·lib·er·al \-ˈli-b(ə-)rəl\ *noun* (1945)
: a liberal who de-emphasizes traditional liberal doctrines in order to seek progress by more pragmatic methods
— **neoliberal** *adjective*
— **neo·lib·er·al·ism** \-b(ə-)rə-ˌli-zəm\ *noun*

neo·lith \ˈnē-ə-ˌlith\ *noun* [back-formation from *neolithic*] (1882)
: a Neolithic stone implement

neo·lith·ic \ˌnē-ə-ˈli-thik\ *adjective* (1865)
1 *capitalized* **:** of or relating to the latest period of the Stone Age characterized by polished stone implements
2 : belonging to an earlier age and now outmoded

ne·ol·o·gism \nē-ˈä-lə-ˌji-zəm\ *noun* [French *néologisme,* from *ne-* + *log-* + *-isme* -ism] (1800)
1 : a new word, usage, or expression
2 : a meaningless word coined by a psychotic
— **ne·ol·o·gis·tic** \-ˌä-lə-ˈjis-tik\ *adjective*

neo–Mal·thu·sian \ˌnē-ō-mal-ˈthü-zhən, -ˈmól-, -ˈthyü-\ *adjective* (1896)
: advocating control of population growth (as by contraception)
— **neo–Malthusian** *noun*
— **neo–Mal·thu·sian·ism** \-zhə-ˌni-zəm\ *noun*

neo·my·cin \ˌnē-ə-ˈmī-sᵊn\ *noun* (1949)
: a broad-spectrum highly toxic antibiotic or mixture of antibiotics produced by a streptomyces (*Streptomyces fradiae*) and used medically especially to treat local infections

¹ne·on \ˈnē-ˌän\ *noun* [Greek, neuter of *neos* new — more at NEW] (1898)
1 : a colorless odorless mostly inert gaseous element that is found in minute amounts in air and is used in electric lamps — see ELEMENT table
2 a : a discharge lamp in which the gas contains a large proportion of neon **b :** a sign composed of such lamps **c :** the illumination provided by such lamps or signs
— **ne·oned** \-ˌänd\ *adjective*

²neon *adjective* (1904)
1 : of, relating to, or using neon
2 : extremely bright **:** FLUORESCENT ⟨*neon* yellow⟩

neo·na·tal \ˌnē-ō-ˈnā-tᵊl\ *adjective* (1902)
: of, relating to, or affecting the newborn and especially the human infant during the first month after birth ⟨*neonatal* mortality⟩
— **neo·na·tal·ly** \-tᵊl-ē\ *adverb*

ne·o·nate \ˈnē-ə-ˌnāt\ *noun* [New Latin *neonatus,* from *ne-* + Latin *natus,* past participle of *nasci* to be born — more at NATION] (1932)
: a newborn child; *especially* **:** a child less than a month old

neo·na·tol·o·gy \ˌnē-ə-nā-ˈtä-lə-jē\ *noun* (1960)
: a branch of medicine concerned with the care, development, and diseases of newborn infants
— **neo·na·tol·o·gist** \-jist\ *noun*

neo–Na·zi \ˌnē-ō-ˈnät-sē, -ˈnat-\ *noun* (1938)
: a member of a group espousing the programs and policies of Hitler's Nazis
— **neo–Nazi** *adjective*
— **neo–Na·zism** \-ˈnät-ˌsi-zəm, -ˈnat-\ *noun*

neo·or·tho·dox \ˌnē-ō-ˈór-thə-ˌdäks\ *adjective* (1946)
: of or relating to a 20th century movement in Protestant theology characterized by a reaction against liberalism and emphasis on various scripturally based Reformation doctrines
— **neo·or·tho·doxy** \-ˌdäk-sē\ *noun*

neo·phil·ia \ˌnē-ə-ˈfi-lē-ə\ *noun* (circa 1947)
: love of or enthusiasm for what is new or novel
— **neo·phil·i·ac** \-lē-ˌak\ *noun*

neo·phyte \ˈnē-ə-ˌfīt\ *noun* [Middle English, from Late Latin *neophytus,* from Greek *neophytos,* from *neophytos* newly planted, newly converted, from *ne-* + *phyein* to bring forth — more at BE] (14th century)
1 : a new convert **:** PROSELYTE
2 : NOVICE 1
3 : TYRO, BEGINNER

neo·pla·sia \ˌnē-ə-ˈplā-zh(ē-)ə\ *noun* [New Latin] (1890)
1 : the formation of tumors
2 : a tumorous condition

neo·plasm \ˈnē-ə-ˌpla-zəm\ *noun* [International Scientific Vocabulary] (1864)
: a new growth of tissue serving no physiological function **:** TUMOR

neo·plas·tic \ˌnē-ə-ˈplas-tik\ *adjective* [International Scientific Vocabulary] (circa 1890)
1 : of, relating to, or constituting a neoplasm or neoplasia
2 : of or relating to neoplasticism

neo·plas·ti·cism \-tə-ˌsi-zəm\ *noun* [French *néo-plasticisme,* from *ne-* + *plastique* plastic + *-isme* -ism] (1933)
: DE STIJL
— **neo·plas·ti·cist** \-sist\ *noun*

Neo·pla·to·nism \ˌnē-ō-ˈplā-tᵊn-ˌi-zəm\ *noun* (1845)
1 : Platonism modified in later antiquity to accord with Aristotelian, post-Aristotelian, and oriental conceptions that conceives of the world as an emanation from an ultimate indivisible being with whom the soul is capable of being reunited in trance or ecstasy
2 : a doctrine similar to ancient Neoplatonism
— **Neo·pla·ton·ic** \-plə-ˈtä-nik, -plā-\ *adjective*
— **Neo·pla·to·nist** \-ˈplā-tᵊn-ist\ *noun*

neo·prene \ˈnē-ə-ˌprēn\ *noun* [*ne-* + *chloroprene*] (1937)
: a synthetic rubber made by the polymerization of chloroprene, characterized by superior resistance (as to oils), and used especially for special-purpose clothing (as gloves and wet suits)

neo·re·al·ism \ˌnē-ō-ˈrē-ə-ˌli-zəm, -ˈri-ə-\ *noun* (1950)
: a movement especially in Italian filmmaking characterized by the simple direct depiction of lower-class life
— **neo·re·al·ist** \-list\ *adjective or noun*
— **neo·re·al·is·tic** \-ˌrē-ə-ˈlis-tik, -ˌri-ə-\ *adjective*

Neo·ri·can \ˌnē-ō-ˈrē-kən\ *noun* [American Spanish *neorriqueño* Puerto Rican living in New York City, blend of Spanish *neoyorquino* New Yorker and *puertorriqueño* Puerto Rican] (1970)
: a Puerto Rican who lives on the U.S. mainland or has lived there but has returned to Puerto Rico

neo–scho·las·ti·cism \ˌnē-ō-skə-ˈlas-tə-ˌsi-zəm\ *noun* (circa 1909)
: a movement among Catholic scholars aiming to restate medieval Scholasticism in a manner suited to present intellectual needs

neo·stig·mine \ˌnē-ə-ˈstig-ˌmēn\ *noun* [*ne-* + *-stigmine* (as in *physostigmine*)] (1941)
: a cholinergic drug used in the form of its bromide $C_{12}H_{19}BrN_2O_2$ or a methyl sulfate derivative $C_{13}H_{22}N_2O_6S$ especially in the treatment of some ophthalmic conditions and in the diagnosis and treatment of myasthenia gravis

ne·o·te·ny \nē-ˈä-tᵊn-ē\ *noun* [New Latin *neotenia,* from *ne-* + Greek *teinein* to stretch — more at THIN] (1901)
1 : retention of some larval or immature characters in adulthood
2 : attainment of sexual maturity during the larval stage
— **ne·o·ten·ic** \ˌnē-ə-ˈte-nik\ *adjective*

ne·o·ter·ic \ˌnē-ə-ˈter-ik\ *adjective* [Late Latin *neotericus,* from Late Greek *neōterikos,* from Greek, youthful, from *neōterios,* comparative of *neos* new, young — more at NEW] (1596)
: recent in origin **:** MODERN

Neo·trop·i·cal \ˌnē-ō-ˈträ-pi-kəl\ *also* **Neo·trop·ic** \-pik\ *adjective* [International Scientific Vocabulary] (1858)
: of, relating to, or constituting the biogeographic region that extends south, east, and west from the central plateau of Mexico

neo·trop·ics \-ˈträ-piks\ *noun plural* (1923)
: the Neotropical region

neo·type \ˈnē-ə-ˌtīp\ *noun* (1905)
: a type specimen that is selected subsequent to the description of a species to replace a preexisting type that has been lost or destroyed

Ne·pali \nə-ˈpó-lē, -ˈpä-, -ˈpa-\ *noun, plural* **Nepali** *also* **Ne·pal·is** [Hindi *naipālī* of Nepal, from Sanskrit *naipālīya,* from *Nepāla* Nepal] (circa 1885)
1 : a native or inhabitant of Nepal
2 : an Indo-Aryan language spoken in Nepal
— **Nepali** *adjective*

ne·pen·the \nə-ˈpen(t)-thē\ *noun* [Latin *nepenthes,* from Greek *nēpenthes,* neuter of *nēpenthēs* banishing pain and sorrow, from *nē-* not + *penthos* grief, sorrow; akin to Greek *pathos* suffering — more at NO, PATHOS] (1580)
1 : a potion used by the ancients to induce forgetfulness of pain or sorrow
2 : something capable of causing oblivion of grief or suffering
— **ne·pen·the·an** \-thē-ən\ *adjective*

neph·e·line \ˈne-fə-ˌlēn\ *also* **neph·e·lite** \-ˌlīt\ *noun* [French *néphéline,* from Greek *nephelē* cloud — more at NEBULA] (circa 1814)
: a hexagonal mineral that is a usually glassy crystalline silicate of sodium, potassium, and aluminum common in igneous rocks
— **neph·e·lin·ic** \ˌne-fə-ˈli-nik\ *adjective*

neph·e·lin·ite \ˈne-fə-lə-ˌnīt\ *noun* [International Scientific Vocabulary] (circa 1863)
: a silica-deficient igneous rock having nepheline as the predominant mineral
— **neph·e·lin·it·ic** \ˌne-fə-lə-ˈni-tik\ *adjective*

neph·e·lom·e·ter \ˌne-fə-ˈlä-mə-tər\ *noun* [Greek *nephelē* cloud + International Scientific Vocabulary *-meter*] (1884)
1 : an instrument for measuring the extent or degree of cloudiness
2 : an instrument for determining the concentration or particle size of suspensions by means of transmitted or reflected light
— **neph·e·lo·met·ric** \ˌne-fə-lō-ˈme-trik\ *adjective*

— **neph·e·lo·met·ri·cal·ly** \-tri-k(ə-)lē\ *adverb*

— **neph·e·lom·e·try** \-ˈlä-mə-trē\ *noun*

neph·ew \ˈne-(ˌ)fyü, *chiefly British* -(ˌ)vyü\ *noun* [Middle English *nevew,* from Old French *neveu,* from Latin *nepot-, nepos* grandson, nephew; akin to Old English *nefa* grandson, nephew, Sanskrit *napāt* grandson] (14th century)
1 a : a son of one's brother or sister or of one's brother-in-law or sister-in-law **b :** an illegitimate son of an ecclesiastic
2 *obsolete* **:** a lineal descendant; *especially* **:** GRANDSON

neph·o·scope \ˈne-fə-ˌskōp\ *noun* [Greek *nephos* cloud + International Scientific Vocabulary *-scope* — more at NEBULA] (1881)
: an instrument for observing the direction and velocity of clouds

nephr- *or* **nephro-** *combining form* [New Latin, from Greek, from *nephros* — more at NEPHRITIS]
: kidney ⟨*nephr*ic⟩ ⟨*nephro*logy⟩

ne·phrec·to·my \ni-ˈfrek-tə-mē\ *noun, plural* **-mies** [International Scientific Vocabulary] (1880)
: the surgical removal of a kidney

— **ne·phrec·to·mize** \-ˌmīz\ *transitive verb*

neph·ric \ˈne-frik\ *adjective* (1887)
: RENAL

ne·phrid·i·um \ni-ˈfri-dē-əm\ *noun, plural* **-ia** \-dē-ə\ [New Latin] (1877)
: a tubular glandular excretory organ characteristic of various invertebrates

— **ne·phrid·i·al** \-dē-əl\ *adjective*

neph·rite \ˈne-ˌfrīt\ *noun* [German *Nephrit,* from Greek *nephros;* from its formerly being worn as a remedy for kidney diseases] (1794)
: a compact tremolite or actinolite that is the commoner and less valuable kind of jade and that varies in color from white to dark green or black

ne·phrit·ic \ni-ˈfri-tik\ *adjective* (1580)
1 : RENAL
2 : of, relating to, or affected with nephritis

ne·phri·tis \ni-ˈfrī-təs\ *noun, plural* **ne·phrit·i·des** \-ˈfri-tə-ˌdēz\ [Late Latin, from Greek, from *nephros* kidney; probably akin to Middle English *nere* kidney] (1580)
: acute or chronic inflammation of the kidney caused by infection, degenerative process, or vascular disease

ne·phrol·o·gy \ni-ˈfrä-lə-jē\ *noun* (circa 1842)
: a branch of medicine concerned with the kidneys

— **ne·phrol·o·gist** \-jist\ *noun*

neph·ron \ˈne-ˌfrän\ *noun* [German, from Greek *nephros*] (1932)
: a single excretory unit of the vertebrate kidney

ne·phrop·a·thy \ni-ˈfrä-pə-thē\ *noun, plural* **-thies** [International Scientific Vocabulary] (circa 1900)
: an abnormal state of the kidney; *especially* **:** one associated with or secondary to some other pathological process

— **neph·ro·path·ic** \ˌne-frə-ˈpa-thik\ *adjective*

ne·phro·sis \ni-ˈfrō-səs\ *noun* [New Latin] (1916)
: a noninflammatory disease of the kidneys chiefly affecting function of the nephrons; *also* **:** NEPHROTIC SYNDROME

— **ne·phrot·ic** \-ˈfrä-tik\ *adjective or noun*

neph·ro·stome \ˈne-frə-ˌstōm\ *noun* [New Latin *nephrostoma,* from *nephr-* + *stoma* stoma] (1888)
: the ciliated funnel-shaped coelomic opening of a typical nephridium

nephrotic syndrome *noun* (1939)
: an abnormal condition that is marked by deficiency of albumin in the blood and its excretion in the urine due to altered permeability of the glomerular basement membranes

neph·ro·tox·ic \ˌne-frə-ˈtäk-sik\ *adjective* (1902)
: poisonous to the kidney ⟨*nephrotoxic* drugs⟩; *also* **:** resulting from or marked by poisoning of the kidney ⟨*nephrotoxic* effects⟩

— **neph·ro·tox·i·ci·ty** \-täk-ˈsi-sə-tē\ *noun*

ne plus ul·tra \ˌnā-ˌpləs-ˈəl-trə, ˌnē-\ *noun* [New Latin, (go) no more beyond] (1638)
1 : the highest point capable of being attained **:** ACME
2 : the most profound degree of a quality or state

nep·o·tism \ˈne-pə-ˌti-zəm\ *noun* [French *népotisme,* from Italian *nepotismo,* from *nepote* nephew, from Latin *nepot-, nepos* grandson, nephew — more at NEPHEW] (1670)
: favoritism (as in appointment to a job) based on kinship

— **nep·o·tis·tic** \ˌne-pə-ˈtis-tik\ *adjective*

Nep·tune \ˈnep-ˌtün, -ˌtyün\ *noun* [Latin *Neptunus*]
1 a : the Roman god of the sea — compare POSEIDON **b :** OCEAN
2 : the planet 8th in order from the sun — see PLANET table

— **Nep·tu·ni·an** \nep-ˈtü-nē-ən, -ˈtyü-\ *adjective*

nep·tu·ni·um \nep-ˈtü-nē-əm, -ˈtyü-\ *noun* [New Latin, from International Scientific Vocabulary *Neptune*] (1941)
: a radioactive metallic element that is chemically similar to uranium and is obtained in nuclear reactors as a by-product in the production of plutonium — see ELEMENT table

nerd \ˈnərd\ *noun* [perhaps from *nerd,* a creature in the children's book *If I Ran the Zoo* (1950) by Dr. Seuss (Theodor Geisel)] (1951)
: an unstylish, unattractive, or socially inept person; *especially* **:** one slavishly devoted to intellectual or academic pursuits ⟨computer *nerds*⟩ ◆

— **nerd·ish** \ˈnər-dish\ *adjective*

— **nerdy** \-dē\ *adjective*

ne·re·id \ˈnir-ē-əd\ *noun* [New Latin *Nereidae,* from *Nereis,* a genus, from Latin, Nereid] (1840)
: any of a family (Nereidae) of polychaete worms; *especially* **:** any of a genus (*Nereis*) of usually large often dimorphic and greenish mostly marine worms

— **nereid** *adjective*

Ne·re·id \ˈnir-ē-əd\ *noun* [Latin *Nereid-, Nereis,* from Greek *Nēreid-, Nēreis,* from *Nēreus* Nereus] (1680)
: any of the sea nymphs fathered by Nereus

Ne·re·us \ˈnir-ē-əs\ *noun* [Latin, from Greek *Nēreus*]
: a sea-god in Greek mythology

ne·rit·ic \nə-ˈri-tik\ *adjective* [International Scientific Vocabulary, perhaps from New Latin *Nerita,* genus of marine snails] (1891)
: of, relating to, inhabiting, or constituting the belt or region of shallow water adjoining the seacoast

ne·rol \ˈner-ˌȯl, ˈnir-\ *noun* [International Scientific Vocabulary *ner-* (from *neroli oil*) + [1]*-ol*] (1869)
: a liquid alcohol $C_{10}H_{18}O$ that has a rose scent and is used in perfumery

ner·o·li oil \ˈner-ə-lē-\ *noun* [French *néroli,* from Italian *neroli,* from Anna Maria de La Trémoille, princess of *Nerola* (flourished 1670)] (1849)
: a fragrant pale yellow essential oil obtained from orange flowers and used especially in cologne and as a flavoring

nerts \ˈnərts\ *noun plural* [alteration of *nuts*] (circa 1932)
slang **:** NONSENSE, NUTS — often used interjectionally

ner·va·tion \ˌnər-ˈvā-shən\ *noun* (1849)
: an arrangement or system of nerves; *also* **:** VENATION

[1]**nerve** \ˈnərv\ *noun* [Latin *nervus* sinew, nerve; akin to Greek *neuron* sinew, nerve, *nēn* to spin — more at NEEDLE] (14th century)
1 : SINEW, TENDON ⟨strain every *nerve*⟩
2 : any of the filamentous bands of nervous tissue that connect parts of the nervous system with the other organs, conduct nervous impulses, and are made up of axons and dendrites together with protective and supportive structures
3 a : power of endurance or control **:** FORTITUDE, STRENGTH **b :** ASSURANCE, BOLDNESS; *also* **:** presumptuous audacity **:** GALL
4 a : a sore or sensitive point **b** *plural* **:** nervous disorganization or collapse **:** NERVOUSNESS
5 : VEIN 3
6 : the sensitive pulp of a tooth
synonym see TEMERITY

[2]**nerve** *transitive verb* **nerved; nerv·ing** (circa 1749)
: to give strength or courage to **:** supply with physical or moral force

nerve cell *noun* (1858)
: NEURON; *also* **:** CELL BODY

nerve center *noun* (1868)
1 : CENTER 2c
2 : a source of leadership, organization, control, or energy ⟨the financial *nerve center* of the nation⟩

nerve cord *noun* (1877)
1 : the pair of closely united ventral longitudinal nerves with their segmental ganglia that is characteristic of many elongate invertebrates (as earthworms)
2 : the dorsal tubular cord of nervous tissue above the notochord of a chordate that comprises or develops into the central nervous system

nerved \ˈnərvd\ *adjective* (1800)
1 a : VEINED ⟨a *nerved* wing⟩ **b :** having veins or nerves especially of a specified kind or number — used in combination ⟨fan-*nerved* leaves⟩
2 : showing courage or strength

nerve ending *noun* (circa 1890)
: a structure forming the distal end of a nerve axon

nerve fiber *noun* (circa 1847)
: any of the processes (as axons or dendrites) of a neuron

nerve gas *noun* (1940)

◇ WORD HISTORY

nerd The word *nerd* was first noted in an article on teenage slang that appeared in the October 8, 1951, issue of *Newsweek:* "In Detroit, someone who would once be called a drip or a square is now, regrettably, a nerd, or, in a less severe case, a scurve." No one knows if Detroit was actually the birthplace of *nerd,* but the word went on to a remarkably successful career in American English. It provided a label for a new character in postwar America: the student wholly devoted to intellectual interests and unable or unwilling to mix with the masses of young people for whom higher education was a less intensely serious matter. While the collegiate *nerd* was supplanted by *dweeb, dork, wonk,* and newer words, *nerd* expanded into adult realms, and the word became an only slightly derogatory epithet for a narrowly focused specialist, especially in the common collocation *computer nerd.* The origin of *nerd* remains unclear; the children's author Theodore Geisel ("Dr. Seuss") used *nerd* for an imaginary creature in *If I Ran the Zoo* (1950), but his coinage may be unrelated to the appearance of the word in 1951 Detroit.

\ə\ **abut** \ᵊ\ **kitten** \ər\ **further** \a\ **ash** \ā\ **ace** \ä\ **mop, mar** \au̇\ **out** \ch\ **chin** \e\ **bet** \ē\ **easy** \g\ **go** \i\ **hit** \ī\ **ice** \j\ **job** \ŋ\ **sing** \ō\ **go** \ȯ\ **law** \ȯi\ **boy** \th\ **thin** \t͟h\ **the** \ü\ **loot** \u̇\ **foot** \y\ **yet** \zh\ **vision** *see also* Guide to Pronunciation

: an organophosphate chemical weapon that interferes with normal nerve transmission and induces intense bronchial spasm with resulting inhibition of respiration

nerve growth factor *noun* (1962)
: a protein that promotes development of the sensory and sympathetic nervous systems and is required for maintenance of sympathetic neurons

nerve impulse *noun* (1900)
: the progressive physicochemical change in the membrane of a nerve fiber that follows stimulation and serves to transmit a record of sensation from a receptor or an instruction to act to an effector — called also *nervous impulse*

nerve·less \'nərv-ləs\ *adjective* (1742)
1 : lacking strength or courage **:** FEEBLE
2 : exhibiting control or balance **:** POISED, COOL
— **nerve·less·ly** *adverb*
— **nerve·less·ness** *noun*

nerve net *noun* (1904)
: a network of nerve cells apparently continuous with one another and conducting impulses in all directions; *also* **:** a primitive nervous system (as in a jellyfish) consisting of such a network

nerve–rack·ing *or* **nerve–wrack·ing** \'nərv-,ra-kiŋ\ *adjective* (1812)
: extremely trying on the nerves ⟨a *nerve-racking* ordeal⟩

nerve trunk *noun* (1851)
: a bundle of nerve fibers enclosed in a connective tissue sheath

ner·vos·i·ty \,nər-'vä-sə-tē\ *noun* (1787)
: the quality or state of being nervous

ner·vous \'nər-vəs\ *adjective* (14th century)
1 *archaic* **:** SINEWY, STRONG
2 : marked by strength of thought, feeling, or style **:** SPIRITED ⟨a vibrant tight-packed *nervous* style of writing⟩
3 : of, relating to, or composed of neurons
4 a : of or relating to the nerves; *also* **:** originating in or affected by the nerves ⟨*nervous* energy⟩ **b :** easily excited or irritated **:** JUMPY **c :** TIMID, APPREHENSIVE ⟨a *nervous* smile⟩
5 a : tending to produce nervousness or agitation **:** UNEASY ⟨a *nervous* situation⟩ **b :** appearing or acting unsteady, erratic, or irregular — used of inanimate things
synonym see VIGOROUS
— **ner·vous·ly** *adverb*
— **ner·vous·ness** *noun*

nervous breakdown *noun* (1905)
: an attack of mental or emotional disorder especially when of sufficient severity to require hospitalization

nervous Nel·lie *or* **nervous Nel·ly** \-'ne-lē\ *noun, plural* **nervous Nellies** *often 1st N capitalized* [from the name *Nellie*] (1926)
: a timid or worrisome person

nervous system *noun* (1740)
: the bodily system that in vertebrates is made up of the brain and spinal cord, nerves, ganglia, and parts of the receptor organs and that receives and interprets stimuli and transmits impulses to the effector organs — compare CENTRAL NERVOUS SYSTEM; AUTONOMIC NERVOUS SYSTEM, PERIPHERAL NERVOUS SYSTEM

ner·vure \'nər-vyər\ *noun* [French, from *nerf* sinew, from Latin *nervus*] (1816)
: VEIN 3

nervy \'nər-vē\ *adjective* **nerv·i·er; -est** (1607)
1 *archaic* **:** SINEWY, STRONG
2 a : showing calm courage **:** BOLD **b :** marked by effrontery or presumption **:** BRASH
3 : EXCITABLE, NERVOUS
— **nerv·i·ly** \-və-lē\ *adverb*
— **nerv·i·ness** \-vē-nəs\ *noun*

ne·science \'ne-sh(ē-)ən(t)s, 'nē-, -sē-ən(t)s\ *noun* [Late Latin *nescientia,* from Latin *nescient-, nesciens,* present participle of *nescire* not to know, from *ne-* not + *scire* to know — more at NO, SCIENCE] (1612)
: lack of knowledge or awareness **:** IGNORANCE
— **ne·scient** \-sh(ē-)ənt, -sē-ənt\ *adjective*

ness \'nes\ *noun* [Middle English *nasse,* from Old English *næss;* akin to Old English *nasu* nose — more at NOSE] (before 12th century)
: CAPE, PROMONTORY

-ness \nəs\ *noun suffix* [Middle English *-nes,* from Old English; akin to Old High German *-nissa* -ness]
: state **:** condition **:** quality **:** degree ⟨good*ness*⟩

Nes·sel·rode \'ne-səl-,rōd\ *noun* [Count Karl R. *Nesselrode* (died 1862) Russian statesman] (1845)
: a mixture of candied fruits, nuts, and maraschino used in puddings, pies, and ice cream

Nes·sus \'ne-səs\ *noun* [Latin, from Greek *Nessos*]
: a centaur slain by Hercules for trying to carry away Hercules' wife but avenged by means of a poisoned garment that causes Hercules to die in torment

[1]nest \'nest\ *noun* [Middle English, from Old English; akin to Old High German *nest* nest, Latin *nidus*] (before 12th century)
1 a : a bed or receptacle prepared by an animal and especially a bird for its eggs and young **b :** a place or specially modified structure serving as an abode of animals and especially of their immature stages ⟨an ants' *nest*⟩ **c :** a receptacle resembling a bird's nest
2 a : a place of rest, retreat, or lodging **:** HOME ⟨grown children who have left the *nest*⟩ **b :** DEN, HANGOUT
3 : the occupants or frequenters of a nest
4 a : a group of similar things **:** AGGREGATION ⟨a *nest* of giant mountains —Helen MacInnes⟩ **b :** HOTBED 2 ⟨a *nest* of rebellion⟩
5 : a group of objects made to fit close together or one within another
6 : an emplaced group of weapons

[2]nest (13th century)
intransitive verb
1 : to build or occupy a nest **:** settle in or as if in a nest
2 : to fit compactly together or within one another **:** EMBED
transitive verb
1 : to form a nest for
2 : to pack compactly together
3 : to form a hierarchy, series, or sequence of with each member, element, or set contained in or containing the next ⟨*nested* subroutines in a computer program⟩

nest egg *noun* (14th century)
1 : a natural or artificial egg left in a nest to induce a fowl to continue to lay there
2 : a fund of money accumulated as a reserve

nest·er \'nes-tər\ *noun* (1880)
1 *West* **:** a homesteader or squatter who takes up land on open range for a farm
2 : one that nests

nes·tle \'ne-səl\ *verb* **nes·tled; nes·tling** \-s(ə-)liŋ\ [Middle English, from Old English *nestlian,* from *nest*] (before 12th century)
intransitive verb
1 *archaic* **:** NEST 1
2 : to settle snugly or comfortably
3 : to lie in an inconspicuous or sheltered manner
transitive verb
1 : to settle, shelter, or house in or as if in a nest ⟨the children were *nestled* all snug in their beds —Clement Moore⟩
2 : to press closely and affectionately ⟨*nestles* a kitten in her arms⟩
— **nes·tler** \-s(ə-)lər\ *noun*

nest·ling \'nest-liŋ\ *noun* (14th century)
: a young bird that has not abandoned the nest

Nes·tor \'nes-tər, -,tȯr\ *noun* [Latin, from Greek *Nestōr*]
1 : a king of Pylos who serves in his old age as a counselor to the Greeks at Troy
2 *often not capitalized* **:** one who is a patriarch or leader in a field

Nes·to·ri·an \ne-'stōr-ē-ən, -'stȯr-\ *adjective* (1565)
1 : of or relating to the doctrine ascribed to Nestorius and ecclesiastically condemned in 431 that divine and human persons remained separate in the incarnate Christ
2 : of or relating to a church separating from Byzantine Christianity after 431, centering in Persia, and surviving chiefly in Asia Minor
— **Nestorian** *noun*
— **Nes·to·ri·an·ism** \-ə-,ni-zəm\ *noun*

[1]net \'net\ *noun* [Middle English *nett,* from Old English; akin to Old High German *nezzi* net] (before 12th century)
1 a : an open-meshed fabric twisted, knotted, or woven together at regular intervals **b :** something made of net: as (1) **:** a device for catching fish, birds, or insects (2) **:** a fabric barricade which divides a court in half (as in tennis or volleyball) and over which a ball or shuttlecock must be hit to be in play (3) **:** the fabric that encloses the sides and back of the goal in various games (as soccer or hockey)
2 : an entrapping device or situation ⟨caught in the *net* of suspicious circumstances⟩
3 : something resembling a net in reticulation (as of lines, fibers, or figures)
4 a : a group of communications stations operating under unified control **b :** NETWORK 4
— **net·less** \-ləs\ *adjective*
— **net·like** \-,līk\ *adjective*
— **net·ty** \'ne-tē\ *adjective*

[2]net *transitive verb* **net·ted; net·ting** (1593)
1 : to cover or enclose with or as if with a net
2 : to catch in or as if in a net
3 : to cover with or as if with a network
4 a : to hit (a ball) into the net for the loss of a point in a racket game **b :** to hit (a ball or puck) into the goal for a score (as in hockey or soccer)
— **net·ter** *noun*

[3]net *adjective* [Middle English, clean, bright, from Middle French — more at NEAT] (15th century)
1 : free from all charges or deductions: as **a :** remaining after the deduction of all charges, outlay, or loss ⟨*net* earnings⟩ ⟨*net* worth⟩ — compare GROSS **b :** excluding all tare ⟨*net* weight⟩
2 : excluding all nonessential considerations **:** BASIC, FINAL ⟨the *net* result⟩ ⟨*net* effect⟩

[4]net *transitive verb* **net·ted; net·ting** (1758)
1 a : to receive by way of profit **:** CLEAR **b :** to produce by way of profit **:** YIELD
2 : to get possession of **:** GAIN

[5]net *noun* (circa 1904)
1 : a net amount, profit, weight, or price
2 : the score of a golfer in a handicap match after deducting his handicap from his gross
3 : ESSENCE, GIST

neth·er \'ne-thər\ *adjective* [Middle English, from Old English *nithera,* from *nither* down; akin to Old High German *nidar* down, Sanskrit *ni*] (before 12th century)
1 : situated down or below **:** LOWER ⟨the *nether* side⟩
2 : situated or believed to be situated beneath the earth's surface ⟨the *nether* regions⟩

neth·er·most \-,mōst\ *adjective* (14th century)
: farthest down **:** LOWEST

neth·er·world \-,wərld\ *noun* (1638)
1 : the world of the dead
2 : UNDERWORLD 4 ⟨the *netherworld* of deceit, subversion, and espionage —R. M. Nixon⟩

net·mind·er \'net-,mīn-dər\ *noun* (1937)
: GOALKEEPER

net·su·ke \'nets-(,)kā, -kē, -ke *also* 'net-sü-\ *noun, plural* **netsuke** *or* **netsukes** [Japanese] (1876)
: a small and often intricately carved toggle (as of wood, ivory, or metal) used to fasten a small container to a kimono sash

nett *British variant of* NET

net·ting \'ne-tiŋ\ *noun* (1567)
1 : NETWORK 1

2 : the act or process of making a net or network
3 : the act, process, or right of fishing with a net

[1]net·tle \ˈne-tᵊl\ *noun* [Middle English, from Old English *netel;* akin to Old High German *nazza* nettle, Greek *adikē*] (before 12th century)
1 : any of a genus (*Urtica* of the family Urticaceae, the nettle family) of chiefly coarse herbs armed with stinging hairs
2 : any of various prickly or stinging plants other than the true nettles (genus *Urtica*)

[2]nettle *transitive verb* **net·tled; net·tling** \ˈnet-liŋ, ˈne-tᵊl-iŋ\ (15th century)
1 : to strike or sting with or as if with nettles
2 : to arouse to sharp but transitory annoyance or anger
synonym see IRRITATE

nettle rash *noun* (1740)
: URTICARIA

net·tle·some \ˈne-tᵊl-səm\ *adjective* (1766)
: causing vexation **:** IRRITATING

net–veined \ˈnet-ˌvānd\ *adjective* (1861)
: having veins arranged in a fine network ⟨a *net-veined* leaf⟩ — compare PARALLEL-VEINED; see VENATION illustration
— net venation *noun*

net–winged \-ˌwiŋd\ *adjective* (circa 1890)
: having wings with a fine network of veins

[1]net·work \ˈnet-ˌwərk\ *noun* (1560)
1 : a fabric or structure of cords or wires that cross at regular intervals and are knotted or secured at the crossings
2 : a system of lines or channels resembling a network
3 a : an interconnected or interrelated chain, group, or system ⟨a *network* of hotels⟩ **b :** a system of computers, terminals, and databases connected by communications lines
4 a : a group of radio or television stations linked by wire or radio relay **b :** a radio or television company that produces programs for broadcast over such a network

[2]network (1887)
transitive verb
1 : to cover with or as if with a network ⟨a continent . . . so *networked* with navigable rivers and canals —*Lamp*⟩
2 *chiefly British* **:** to distribute for broadcast on a television network; *also* **:** BROADCAST 3
3 : to join (as computers) in a network
intransitive verb
: to engage in networking

net·work·ing *noun* (1966)
1 : the exchange of information or services among individuals, groups, or institutions
2 : the establishment or use of a computer network

Neuf·châ·tel \ˌnü-shä-ˈtel, ˌnyü-, ˌnə(r)-\ *noun* [French, from *Neufchâtel,* France] (circa 1865)
: a soft unripened cheese similar to cream cheese but containing less fat and more moisture

neume \ˈnüm, ˈnyüm\ *noun* [Middle English, from Medieval Latin *pneuma, neuma,* from Greek *pneuma* breath — more at PNEUMATIC] (14th century)
: any of various symbols used in the notation of Gregorian chant
— neu·mat·ic \nu̇-ˈma-tik, nyu̇-\ *adjective*

neur- *or* **neuro-** *combining form* [New Latin, from Greek, nerve, sinew, from *neuron* — more at NERVE]
1 : nerve ⟨*neural*⟩ ⟨*neurology*⟩
2 : neural **:** neural and ⟨*neuromuscular*⟩

neu·ral \ˈnu̇r-əl, ˈnyu̇r-\ *adjective* (circa 1847)
1 : of, relating to, or affecting a nerve or the nervous system
2 : situated in the region of or on the same side of the body as the brain and spinal cord **:** DORSAL
— neu·ral·ly \-ə-lē\ *adverb*

neural arch *noun* (circa 1860)
: the cartilaginous or bony arch enclosing the spinal cord on the dorsal side of a vertebra

neural crest *noun* (circa 1885)
: the ridge of one of the folds forming the neural tube that gives rise to the spinal ganglia and various structures of the autonomic nervous system

neu·ral·gia \nu̇-ˈral-jə, nyu̇-\ *noun* [New Latin] (circa 1834)
: acute paroxysmal pain radiating along the course of one or more nerves usually without demonstrable changes in the nerve structure
— neu·ral·gic \-jik\ *adjective*

neural net *noun* (1949)
: a computer architecture in which a number of processors are interconnected in a manner suggestive of the connections between neurons in a human brain and which is able to learn by a process of trial and error — called also *neural network*

neural plate *noun* (1888)
: a thickened plate of ectoderm along the dorsal midline of the early vertebrate embryo that gives rise to the neural tube and crests

neural tube *noun* (1888)
: the hollow longitudinal dorsal tube formed by infolding and subsequent fusion of the opposite ectodermal folds in the vertebrate embryo that gives rise to the brain and spinal cord

neur·amin·i·dase \ˌnu̇r-ə-ˈmi-nə-ˌdās, ˌnyu̇r-, -ˌdāz\ *noun* [*neuraminic* acid, an amino acid + *-ide* + *-ase*] (1956)
: a glycosidase that splits mucoproteins by breaking a glucoside link and occurs especially in influenza viruses as an antigen

neur·as·the·nia \ˌnu̇r əs-ˈthē-nē-ə, ˌnyu̇r-\ *noun* [New Latin] (1856)
: an emotional and psychic disorder that is characterized especially by easy fatigability and often by lack of motivation, feelings of inadequacy, and psychosomatic symptoms
— neur·as·then·ic \-ˈthe-nik, -ˈthē-\ *adjective or noun*
— neur·as·then·i·cal·ly \-ni-k(ə-)lē\ *adverb*

neu·ri·lem·ma \ˌnu̇r-ə-ˈle-mə, ˌnyu̇r-\ *noun* [New Latin, from *neur-* + Greek *eilēma* covering, coil, from *eilein* to wind; akin to Greek *eilyein* to wrap — more at VOLUBLE] (1852)
: the plasma membrane surrounding a Schwann cell of a myelinated nerve fiber and separating layers of myelin
— neu·ri·lem·mal \-ˈle-məl\ *adjective*

neu·ri·tis \nu̇-ˈrī-təs, nyu̇-\ *noun, plural* **-rit·i·des** \-ˈri-tə-ˌdēz\ *or* **-ri·tis·es** [New Latin] (1840)
: an inflammatory or degenerative lesion of a nerve marked especially by pain, sensory disturbances, and impaired or lost reflexes
— neu·rit·ic \-ˈri-tik\ *adjective or noun*

neu·ro·ac·tive \ˌnu̇r-ō-ˈak-tiv, ˌnyu̇r-\ *adjective* (1961)
: stimulating neural tissue

neu·ro·anat·o·my \-ə-ˈna-tə-mē\ *noun* (circa 1899)
: the anatomy of nervous tissue and the nervous system
— neu·ro·an·a·tom·i·cal \-ˌa-nə-ˈtä-mi-kəl\ *also* **neu·ro·an·a·tom·ic** \-mik\ *adjective*
— neu·ro·anat·o·mist \-ə-ˈna-tə-mist\ *noun*

neu·ro·bi·ol·o·gy \-bī-ˈä-lə-jē\ *noun* (1906)
: a branch of the life sciences that deals with the anatomy, physiology, and pathology of the nervous system
— neu·ro·bi·o·log·i·cal \-ˌbī-ə-ˈlä-ji-kəl\ *adjective*
— neu·ro·bi·ol·o·gist \-bī-ˈä-lə-jist\ *noun*

neu·ro·blas·to·ma \-blas-ˈtō-mə\ *noun, plural* **-mas** *or* **-ma·ta** \-mə-tə\ [New Latin, from International Scientific Vocabulary *neuroblast* embryonic ganglion cell, from *neur-* + *-blast* -blast] (1910)
: a malignant tumor formed of embryonic ganglion cells

neu·ro·chem·is·try \-ˈke-mə-strē\ *noun* (1924)
1 : the study of the chemical makeup and activities of nervous tissue
2 : chemical processes and phenomena related to the nervous system
— neu·ro·chem·i·cal \-ˈke-mi-kəl\ *adjective or noun*
— neu·ro·chem·ist \-ˈke-mist\ *noun*

neu·ro·de·gen·er·a·tive \-di-ˈje-nə-rə-tiv, -ˈjen-rə-; -ˈje-nə-ˌrā-; -dē-\ *adjective* (1907)
: relating to or characterized by degeneration of nervous tissue

neu·ro·en·do·crine \-ˈen-də-krən, -ˌkrīn, -ˌkrēn\ *adjective* (1922)
1 : of, relating to, or being a hormonal substance that influences the activity of nerves
2 : of, relating to, or functioning in neurosecretion

neu·ro·en·do·cri·nol·o·gy \-ˌen-də-kri-ˈnä-lə-jē, -(ˌ)krī-\ *noun* (1922)
: a branch of the life sciences dealing with neurosecretion and the physiological interaction between the central nervous system and the endocrine system
— neu·ro·en·do·cri·no·log·i·cal \-ˌkri-nᵊl-ˈä-ji-kəl, -ˌkrī-, -ˌkrē-\ *adjective*
— neu·ro·en·do·cri·nol·o·gist \-kri-ˈnä-lə-jist, -(ˌ)krī-\ *noun*

neu·ro·fi·bril \-ˈfī-brəl, -ˈfi-\ *noun* [New Latin *neurofibrilla,* from *neur-* + *fibrilla* fibril] (1898)
: a fine proteinaceous fibril that is found in cytoplasm (as of a neuron or a paramecium) and is capable of conducting excitation
— neu·ro·fi·bril·lary \-brə-ˌler-ē\ *adjective*

neu·ro·fi·bro·ma \-fī-ˈbrō-mə\ *noun* (1892)
: a fibroma composed of nervous and connective tissue and produced by proliferation of Schwann cells

neu·ro·fi·bro·ma·to·sis \-(ˌ)fī-ˌbrō-mə-ˈtō-səs\ *noun* [New Latin, from *neurofibromat-, neurofibroma*] (1896)
: a disorder inherited as an autosomal dominant and characterized especially by brown spots on the skin, neurofibromas of peripheral nerves, and deformities of subcutaneous tissue and bone

neu·ro·gen·ic \ˌnu̇r-ə-ˈje-nik, ˌnyu̇r-\ *adjective* (1901)
1 : originating in or controlled by nervous tissue ⟨*neurogenic* heartbeat⟩
2 : induced or modified by nervous factors; *especially* **:** disordered because of abnormally altered neural relations
— neu·ro·gen·i·cal·ly \-ni-k(ə-)lē\ *adverb*

neu·ro·glia \nu̇-ˈrō-glē-ə, nyu̇-, -ˈrä-; ˌn(y)u̇r-ə-ˈglē-ə, -ˈglī-\ *noun* [New Latin, from *neur-* + Middle Greek *glia* glue — more at CLAY] (1873)
: supporting tissue intermingled with the essential elements of nervous tissue especially in the brain, spinal cord, and ganglia
— neu·ro·gli·al \-əl\ *adjective*

neu·ro·hor·mon·al \ˌnu̇r-ō-hȯr-ˈmō-nᵊl, ˌnyu̇r-\ *adjective* (circa 1935)
1 : involving both neural and hormonal mechanisms
2 : of, relating to, or being a neurohormone

neu·ro·hor·mone \-ˈhȯr-ˌmōn\ *noun* [International Scientific Vocabulary] (1935)
: a hormone (as acetylcholine or norepinephrine) produced by or acting on nervous tissue

neu·ro·hu·mor \-ˈhyü-mər, -ˈyü-\ *noun* (1932)
: NEUROHORMONE; *especially* **:** NEUROTRANSMITTER
— neu·ro·hu·mor·al \-mə-rəl\ *adjective*

\ə\ **abut** \ᵊ\ **kitten** \ər\ **further** \a\ **ash** \ā\ **ace** \ä\ **mop, mar** \au̇\ **out** \ch\ **chin** \e\ **bet** \ē\ **easy** \g\ **go** \i\ **hit** \ī\ **ice** \j\ **job** \ŋ\ **sing** \ō\ **go** \ȯ\ **law** \ȯi\ **boy** \th\ **thin** \t͟h\ **the** \ü\ **loot** \u̇\ **foot** \y\ **yet** \zh\ **vision** *see also* Guide to Pronunciation

neu·ro·hy·poph·y·sis \-hī-ˈpä-fə-səs\ *noun* [New Latin] (1912)
: the portion of the pituitary gland that is composed of the infundibulum and posterior lobe and is concerned with the secretion of various hormones
— **neu·ro·hy·po·phy·se·al** *or* **neu·ro·hy·po·phy·si·al** \-(ˌ)hī-ˌpä-fə-ˈsē-əl, -ˌhī-pə-fə-, -ˈzē-, -ˌhī-pə-ˈfi-zē-\ *adjective*

neu·ro·lep·tic \ˌnu̇r-ə-ˈlep-tik, ˌnyu̇r-\ *noun* [French *neuroleptique,* from *neur-* + *-leptique* affecting, from Greek *lēptikos* seizing, from *lambanein* to take, seize — more at LATCH] (1958)
: any of the powerful tranquilizers (as the phenothiazines) used especially to treat psychosis and believed to act by blocking dopamine nervous receptors — called also *antipsychotic*
— **neuroleptic** *adjective*

neu·rol·o·gist \nu̇-ˈrä-lə-jist, nyu̇-\ *noun* (1832)
: one specializing in neurology; *especially* **:** a physician skilled in the diagnosis and treatment of disease of the nervous system

neu·rol·o·gy \-jē\ *noun* [New Latin *neurologia,* from *neur-* + *-logia* -logy] (circa 1681)
: the scientific study of the nervous system especially in respect to its structure, functions, and abnormalities
— **neu·ro·log·i·cal** \ˌnu̇r-ə-ˈlä-ji-kəl, ˌnyu̇r-\ *or* **neu·ro·log·ic** \-jik\ *adjective*
— **neu·ro·log·i·cal·ly** \-ji-k(ə-)lē\ *adverb*

neu·ro·ma \nu̇-ˈrō-mə, nyu̇-\ *noun, plural* **-mas** *or* **-ma·ta** \-mə-tə\ [New Latin] (circa 1847)
: a tumor or mass growing from a nerve and usually consisting of nerve fibers

neu·ro·mus·cu·lar \ˌnu̇r-ō-ˈməs-kyə-lər, ˌnyu̇r-\ *adjective* [International Scientific Vocabulary] (1864)
: of or relating to nerves and muscles; *especially* **:** jointly involving nervous and muscular elements ⟨a *neuromuscular* junction⟩

neu·ron \ˈnü-ˌrän, ˈnyü-; ˈnu̇r-ˌän, ˈnyu̇r-\ *also* **neu·rone** \-ˌrōn, -ˌōn\ *noun* [New Latin *neuron,* from Greek, nerve, sinew — more at NERVE] (1891)
: a grayish or reddish granular cell with specialized processes that is the fundamental functional unit of nervous tissue
— **neu·ro·nal** \ˈnu̇r-ə-nᵊl, ˈnyu̇r-; nu̇-ˈrō-nᵊl, nyu̇-\ *also* **neu·ron·ic** \nu̇-ˈrä-nik, nyu̇-\ *adjective*

neu·ro·pa·thol·o·gy \ˌnu̇r-ō-pə-ˈthä-lə-jē, ˌnyu̇r-, -pa-\ *noun* [International Scientific Vocabulary] (1853)
: pathology of the nervous system
— **neu·ro·path·o·log·ic** \-ˌpa-thə-ˈlä-jik\ *or* **neu·ro·path·o·log·i·cal** \-ji-kəl\ *adjective*
— **neu·ro·pa·thol·o·gist** \-pə-ˈthä-lə-jist, -pa-\ *noun*

neu·rop·a·thy \nu̇-ˈrä-pə-thē, nyu̇-\ *noun, plural* **-thies** [International Scientific Vocabulary] (1857)
: an abnormal and usually degenerative state of the nervous system or nerves; *also* **:** a systemic condition that stems from a neuropathy
— **neu·ro·path·ic** \ˌnu̇r-ə-ˈpa-thik, ˌnyu̇r-\ *adjective*
— **neu·ro·path·i·cal·ly** \-thi-k(ə-)lē\ *adverb*

neu·ro·pep·tide \ˌnu̇r-ə-ˈpep-ˌtīd, ˌnyu̇r-\ *noun* (1975)
: an endogenous peptide that influences neural activity or functioning

neu·ro·phar·ma·col·o·gy \ˌnu̇r-ō-ˌfär-mə-ˈkä-lə-jē, ˌnyu̇r-\ *noun* (1950)
1 : a branch of medical science dealing with the action of drugs on and in the nervous system
2 : the properties and reactions of a drug on and in the nervous system
— **neu·ro·phar·ma·co·log·i·cal** \-kə-ˈlä-ji-kəl\ *also* **neu·ro·phar·ma·co·log·ic** \-jik\ *adjective*
— **neu·ro·phar·ma·col·o·gist** \-ˈkä-lə-jist\ *noun*

neu·ro·phys·i·ol·o·gy \-ˌfi-zē-ˈä-lə-jē\ *noun* (1868)
: physiology of the nervous system
— **neu·ro·phys·i·o·log·i·cal** \-ə-ˈlä-ji-kəl\ *also* **neu·ro·phys·i·o·log·ic** \-jik\ *adjective*
— **neu·ro·phys·i·o·log·i·cal·ly** \-ji-k(ə-)lē\ *adverb*
— **neu·ro·phys·i·ol·o·gist** \-ˈä-lə-jist\ *noun*

neu·ro·psy·chi·a·try \-sə-ˈkī-ə-trē, -sī-\ *noun* (1918)
: a branch of medicine concerned with both neurology and psychiatry
— **neu·ro·psy·chi·at·ric** \-ˌsī-kē-ˈa-trik\ *adjective*
— **neu·ro·psy·chi·at·ri·cal·ly** \-tri-k(ə-)lē\ *adverb*
— **neu·ro·psy·chi·a·trist** \-sə-ˈkī-ə-trist, -sī-\ *noun*

neu·ro·psy·chol·o·gy \-sī-ˈkä-lə-jē\ *noun* (circa 1893)
: a science concerned with the integration of psychological observations on behavior and the mind with neurological observations on the brain and nervous system
— **neu·ro·psy·cho·log·i·cal** \-ˌsī-kə-ˈlä-ji-kəl\ *adjective*
— **neu·ro·psy·chol·o·gist** \-sī-ˈkä-lə-jist\ *noun*

neu·rop·ter·an \nu̇-ˈräp-tə-rən, nyu̇-\ *noun* [ultimately from Greek *neur-* + *pteron* wing — more at FEATHER] (circa 1842)
: any of an order (Neuroptera) of usually net-winged insects that include the lacewings and ant lions
— **neuropteran** *adjective*
— **neu·rop·ter·ous** \-rəs\ *adjective*

neu·ro·ra·di·ol·o·gy \ˌnu̇r-ō-ˌrā-dē-ˈä-lə-jē, ˌnyu̇r-\ *noun* (1938)
: radiology of the nervous system
— **neu·ro·ra·dio·log·i·cal** \-ə-ˈlä-ji-kəl\ *adjective*
— **neu·ro·ra·di·ol·o·gist** \-ˈä-lə-jist\ *noun*

neu·ro·sci·ence \-ˈsī-ən(t)s\ *noun* (1963)
: a branch (as neurophysiology) of the life sciences that deals with the anatomy, physiology, biochemistry, or molecular biology of nerves and nervous tissue and especially with their relation to behavior and learning
— **neu·ro·sci·en·tif·ic** \-ˌsī-ən-ˈti-fik\ *adjective*
— **neu·ro·sci·en·tist** \-ˈsī-ən-tist\ *noun*

neu·ro·se·cre·tion \-si-ˈkrē-shən\ *noun* (1941)
1 : the process of producing a secretion by nerve cells
2 : a secretion produced by neurosecretion
— **neu·ro·se·cre·to·ry** \-ˈkrē-tə-rē\ *adjective*

neu·ro·sen·so·ry \-ˈsen(t)-sə-rē, -ˈsen(t)s-rē\ *adjective* (1929)
: of or relating to afferent nerves

neu·ro·sis \nu̇-ˈrō-səs, nyu̇-\ *noun, plural* **-ro·ses** \-ˌsēz\ [New Latin] (circa 1784)
: a mental and emotional disorder that affects only part of the personality, is accompanied by a less distorted perception of reality than in a psychosis, does not result in disturbance of the use of language, and is accompanied by various physical, physiological, and mental disturbances (as visceral symptoms, anxieties, or phobias)

neu·ros·po·ra \nu̇-ˈräs-p(ə-)rə, nyu̇-\ *noun* [New Latin, from *neur-* + *spora* spore] (1928)
: any of a genus (*Neurospora* of the family Sordariaceae) of ascomycetous fungi which are used extensively in genetic research and have black perithecia and persistent asci and some of which have salmon-pink or orange spore masses and are severe pests in bakeries

neu·ro·sur·gery \-ˈsər-jə-rē, -ˈsərj-rē\ *noun* (1904)
: surgery of nervous structures (as nerves, the brain, or the spinal cord)
— **neu·ro·sur·geon** \-ˈsər-jən\ *noun*
— **neu·ro·sur·gi·cal** \-ˈsər-ji-kəl\ *adjective*

¹neu·rot·ic \nu̇-ˈrä-tik\ *adjective* (1873)
: of, relating to, constituting, or affected with neurosis
— **neu·rot·i·cal·ly** \-ti-k(ə-)lē\ *adverb*

²neurotic *noun* (1896)
1 : one affected with a neurosis
2 : an emotionally unstable individual

neu·rot·i·cism \nu̇-ˈrä-tə-ˌsi-zəm, nyu̇-\ *noun* (1900)
: a neurotic character, condition, or trait

neu·ro·tox·ic \ˌnu̇r-ō-ˈtäk-sik, ˌnyu̇r-\ *adjective* (circa 1903)
: toxic to the nerves or nervous tissue
— **neu·ro·tox·ic·i·ty** \-ˌtäk-ˈsi-sə-tē\ *noun*

neu·ro·tox·in \-ˈtäk-sən\ *noun* [International Scientific Vocabulary] (1902)
: a poisonous complex especially of protein that acts on the nervous system

neu·ro·trans·mis·sion \-tran(t)s-ˈmi-shən, -tranz-\ *noun* (1961)
: the transmission of nerve impulses across a synapse

neu·ro·trans·mit·ter \-tran(t)s-ˈmi-tər, -tranz-\ *noun* (1961)
: a substance (as norepinephrine or acetylcholine) that transmits nerve impulses across a synapse

neu·ro·trop·ic \-ˈträ-pik\ *adjective* [International Scientific Vocabulary] (1903)
: having an affinity for or localizing selectively in nerve tissue ⟨a *neurotropic* virus⟩

neu·ru·la \ˈnu̇r-ə-lə, ˈnyu̇r-, -yə-lə\ *noun, plural* **-lae** \-ˌlē\ *or* **-las** [New Latin, from *neur-* + Latin *-ula* -ule] (circa 1890)
: an early vertebrate embryo which follows the gastrula and in which nervous tissue begins to differentiate and the basic pattern of the vertebrate begins to emerge
— **neu·ru·la·tion** \ˌnu̇r-ə-ˈlā-shən, ˌnyu̇r-, -yə-\ *noun*

neus·ton \ˈnü-ˌstän, ˈnyü-\ *noun* [German, from Greek, neuter of *neustos* swimming, from *nein* to swim — more at NATANT] (1928)
: minute organisms that float in the surface film of water

¹neu·ter \ˈnü-tər, ˈnyü-\ *adjective* [Middle English *neutre,* from Middle French & Latin; Middle French *neutre,* from Latin *neuter,* literally, neither, from *ne-* not + *uter* which of two — more at NO, WHETHER] (14th century)
1 a : of, relating to, or constituting the gender that ordinarily includes most words or grammatical forms referring to things classed as neither masculine nor feminine **b :** neither active nor passive **:** INTRANSITIVE
2 : taking no side **:** NEUTRAL
3 : lacking or having imperfectly developed or nonfunctional generative organs ⟨the worker bee is *neuter*⟩

²neuter *noun* (15th century)
1 a : a noun, pronoun, adjective, or inflectional form or class of the neuter gender **b :** the neuter gender
2 : one that is neutral
3 a : WORKER 2 **b :** a spayed or castrated animal

³neuter *transitive verb* (1903)
: CASTRATE, ALTER

¹neu·tral \ˈnü-trəl, ˈnyü-\ *noun* (15th century)
1 : one that is neutral
2 : a neutral color
3 : a position of disengagement (as of gears)

²neutral *adjective* [Middle French, from (assumed) Medieval Latin *neutralis,* from Latin, of neuter gender, from *neutr-, neuter*] (1549)
1 : not engaged on either side; *specifically* **:** not aligned with a political or ideological grouping ⟨a *neutral* nation⟩
2 : of or relating to a neutral state or power ⟨*neutral* territory⟩

3 a : not decided or pronounced as to characteristics **:** INDIFFERENT **b** (1) **:** ACHROMATIC (2) **:** nearly achromatic **c** (1) **:** NEUTER 3 (2) **:** lacking stamens or pistils **d :** neither acid nor basic **e :** not electrically charged
4 : produced with the tongue in the position it has when at rest ⟨the *neutral* vowels of \ə-'bəv\ *above*⟩
— **neu·tral·ly** \-trə-lē\ *adverb*
— **neu·tral·ness** *noun*

neu·tral·ise *British variant of* NEUTRALIZE

neu·tral·ism \'nü-trə-ˌli-zəm, 'nyü-\ *noun* (1579)
1 : NEUTRALITY
2 : a policy or the advocacy of neutrality especially in international affairs
— **neu·tral·ist** \-list\ *noun*
— **neu·tral·is·tic** \ˌnü-trə-'lis-tik, ˌnyü-\ *adjective*

neu·tral·i·ty \nü-'tra-lə-tē, nyü-\ *noun* (15th century)
: the quality or state of being neutral; *especially* **:** refusal to take part in a war between other powers

neu·tral·i·za·tion \ˌnü-trə-lə-'zā-shən, ˌnyü-\ *noun* (1808)
1 : an act or process of neutralizing
2 : the quality or state of being neutralized

neu·tral·ize \'nü-trə-ˌlīz, 'nyü-\ *verb* **-ized; -iz·ing** (1759)
transitive verb
1 : to make chemically neutral
2 a : to counteract the activity or effect of **:** make ineffective ⟨propaganda that is difficult to *neutralize*⟩ **b :** KILL, DESTROY
3 : to make electrically inert by combining equal positive and negative quantities
4 : to invest (as a territory or a nation) with conventional or obligatory neutrality conferring inviolability during a war
5 : to make neutral by blending with the complementary color
6 : to give (as a pair of phonemes) a nondistinctive form or pronunciation ⟨\t\ and \d\ are *neutralized* when pronounced as flaps⟩
intransitive verb
: to undergo neutralization
— **neu·tral·iz·er** *noun*

neutral red *noun* (1890)
: a basic dye used chiefly as a biological stain and acid-base indicator

neutral spirits *noun plural but singular or plural in construction* (1919)
: ethyl alcohol of 190 or higher proof used especially for blending other alcoholic liquors

neu·tri·no \nü-'trē-(ˌ)nō, nyü-\ *noun, plural* **-nos** [Italian, from *neutro* neutral, neuter, from Latin *neutr-, neuter*] (1934)
: an uncharged elementary particle that is believed to be massless or to have a very small mass, that has any of three forms, and that interacts only rarely with other particles
— **neu·tri·no·less** \-ləs\ *adjective*

neu·tron \'nü-ˌträn, 'nyü-\ *noun* [probably from *neutral*] (1932)
: an uncharged elementary particle that has a mass nearly equal to that of the proton and is present in all known atomic nuclei except the hydrogen nucleus
— **neu·tron·ic** \nü-'trä-nik, nyü-\ *adjective*

neutron activation analysis *noun* (1951)
: an analytical method used to determine the chemical elements comprising a material by bombarding it with neutrons to produce radioactive atoms whose radiations are indicative of the elements present — called also *neutron activation*

neutron bomb *noun* (1959)
: a nuclear bomb designed to produce lethal neutrons but less blast and fire damage than other nuclear bombs

neutron star *noun* (1934)
: a hypothetical dense celestial object that consists primarily of closely packed neutrons and that results from the collapse of a much larger stellar body

¹neu·tro·phil \'nü-trə-ˌfil, 'nyü-\ *or* **neu·tro·phil·ic** \ˌnü-trə-'fi-lik, ˌnyü-\ *adjective* [International Scientific Vocabulary *neutro-* (from Latin *neutr-, neuter* neither) + *-phil*] (circa 1890)
: staining to the same degree with acid or basic dyes ⟨*neutrophil* granulocytes⟩

²neutrophil *noun* (1897)
: a granulocyte that is the chief phagocytic white blood cell of the blood

né·vé \nā-'vā\ *noun* [French dialect (Swiss), from (assumed) Vulgar Latin *nivatum,* from Latin *niv-, nix* snow — more at SNOW] (1843)
: the partially compacted granular snow that forms the surface part of the upper end of a glacier; *broadly* **:** a field of granular snow

nev·er \'ne-vər\ *adverb* [Middle English, from Old English *nǣfre,* from *ne* not + *ǣfre* ever — more at NO] (before 12th century)
1 : not ever **:** at no time ⟨*never* saw her before⟩
2 : not in any degree **:** not under any condition ⟨*never* the wiser for his experience⟩

never mind *conjunction* (1954)
: MUCH LESS, LET ALONE — used especially in negative contexts to add to one term another denoting something less likely ⟨with this knee I can hardly walk, *never mind* run⟩

nev·er·more \ˌne-vər-'mōr, -'mȯr\ *adverb* (12th century)
: never again

nev·er–nev·er land \ˌne-vər-'ne-vər-\ *noun* (1907)
: an ideal or imaginary place

nev·er·the·less \ˌne-vər-t͟hə-'les\ *adverb* (14th century)
: in spite of that **:** HOWEVER ⟨her childish but *nevertheless* real delight —Richard Corbin⟩

ne·vus \'nē-vəs\ *noun, plural* **ne·vi** \-ˌvī\ [New Latin, from Latin *naevus*] (circa 1836)
: a congenital pigmented area on the skin **:** BIRTHMARK

¹new \'nü, *chiefly British* 'nyü, *in place names usually* (ˌ)nü *or* nə *or* (ˌ)ni\ *adjective* [Middle English, from Old English *nīwe;* akin to Old High German *niuwi* new, Latin *novus,* Greek *neos*] (before 12th century)
1 : having existed or having been made but a short time **:** RECENT, MODERN
2 a (1) **:** having been seen, used, or known for a short time **:** NOVEL ⟨rice was a *new* crop for the area⟩ (2) **:** UNFAMILIAR ⟨visit *new* places⟩ **b :** being other than the former or old ⟨a steady flow of *new* money⟩
3 : having been in a relationship or condition but a short time ⟨*new* to the job⟩
4 a : beginning as the resumption or repetition of a previous act or thing ⟨a *new* day⟩ ⟨the *new* edition⟩ **b :** made or become fresh ⟨awoke a *new* person⟩
5 : different from one of the same category that has existed previously ⟨*new* realism⟩
6 : of dissimilar origin and usually of superior quality ⟨introducing *new* blood⟩
7 *capitalized* **:** MODERN 3; *especially* **:** having been in use after medieval times ☆
— **new·ish** \'nü-ish, 'nyü-\ *adjective*
— **new·ness** *noun*

²new *adverb* (before 12th century)
: NEWLY, RECENTLY — usually used in combination

New Age *adjective* (1956)
1 : of, relating to, or being a late 20th century social movement drawing on ancient concepts especially from Eastern and American Indian traditions and incorporating such themes as holism, concern for nature, spirituality, and metaphysics
2 : of, relating to, or being a soft soothing form of instrumental music often used to promote relaxation

¹new·born \-ˌbȯrn\ *adjective* (14th century)
1 : recently born
2 : born anew

²newborn *noun, plural* **newborn** *or* **newborns** (1879)
: a newborn individual

New·burg *or* **New·burgh** \'nü-ˌbərg, 'nyü-\ *adjective* [origin unknown] (1902)
: served with a sauce made of cream, butter, sherry, and egg yolks ⟨lobster *Newburg*⟩ ⟨shrimp *Newburg*⟩

New·cas·tle disease \'nü-ˌka-səl-, 'nyü-; n(y)ü-'\ *noun* [*Newcastle* upon Tyne, England] (1927)
: a destructive virus disease of birds and especially domestic fowl that involves respiratory and nervous symptoms

new·com·er \'nü-ˌkə-mər, 'nyü-\ *noun* (15th century)
1 : one recently arrived
2 : BEGINNER

New Criticism *noun* (1941)
: an analytic literary criticism that is marked by concentration on the language, imagery, and emotional or intellectual tensions in literary works
— **New Critic** *noun*
— **New Critical** *adjective*

New Deal *noun* [from the supposed resemblance to the situation of freshness and equality of opportunity afforded by a fresh deal in a card game] (1932)
: the legislative and administrative program of President F. D. Roosevelt designed to promote economic recovery and social reform during the 1930s; *also* **:** the period of this program
— **New Dealer** *noun*
— **New Deal·ish** \-'dē-lish\ *adjective*
— **New Deal·ism** \-'dē(ə)-ˌli-zəm\ *noun*

new drug *noun* (circa 1951)
: a drug that has not been declared safe and effective by qualified experts under the conditions prescribed, recommended, or suggested in the label and that may be a new chemical formula or an established drug prescribed for use in a new way

new economics *noun plural but usually singular in construction* (1928)
: an economic concept that is a logical extension of Keynesianism and that holds that appropriate fiscal and monetary maneuvering can maintain healthy economic growth and prosperity indefinitely

new·el \'nü-əl, 'nyü-\ *noun* [Middle English *nowell,* from Middle French *nouel* stone of a fruit, from Late Latin *nucalis* like a nut, from Latin *nuc-, nux* nut — more at NUT] (14th century)
1 : an upright post about which the steps of a circular staircase wind
2 : a post at the foot of a straight stairway or one at a landing

New English Bible *noun* (1957)
: a translation of the Bible by a British interdenominational committee first published in its entirety in 1970

new·fan·gled \'nü-'faŋ-gəld\ *adjective* [Middle English, from *newefangel,* from *new* + (assumed) Old English *-fangol,* from Old English

☆ SYNONYMS
New, novel, original, fresh mean having recently come into existence or use. NEW may apply to what is freshly made and unused ⟨*new* brick⟩ or has not been known before ⟨*new* designs⟩ or not experienced before ⟨starts the *new* job⟩. NOVEL applies to what is not only new but strange or unprecedented ⟨a *novel* approach to the problem⟩. ORIGINAL applies to what is the first of its kind to exist ⟨a man without one *original* idea⟩. FRESH applies to what has not lost its qualities of newness such as liveliness, energy, brightness ⟨*fresh* towels⟩ ⟨a *fresh* start⟩.

fōn (past participle *fangen*) to take, seize — more at PACT] (14th century)
1 : attracted to novelty
2 : of the newest style or kind ⟨had many *newfangled* gadgets in the kitchen⟩
— **new·fan·gled·ness** *noun*

new–fash·ioned \-ˈfa-shənd\ *adjective* (1592)
1 : made in a new fashion or form
2 : UP-TO-DATE

new·found \-ˈfau̇nd\ *adjective* (15th century)
: newly found ⟨a *newfound* friend⟩

New·found·land \ˈnü-fən(d)-lənd, ˈnyü-, -ˌland; n(y)ü-ˈfau̇nd-lənd\ *noun* [*Newfoundland,* Canada] (1773)
: any of a breed of very large heavy highly intelligent, black, black and white, or bronze dogs developed in Newfoundland

Newfoundland

New·gate \ˈnü-ˌgāt, ˈnyü-, -gət\ *noun* (14th century)
: a London prison razed in 1902

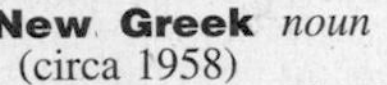

New Greek *noun* (circa 1958)
: Greek as used by the Greeks since the end of the medieval period

New Hebrew *noun* (circa 1959)
: the Hebrew language in use in present-day Israel

new·ie \ˈnü-ē, ˈnyü-\ *noun* (circa 1945)
: something new

New Jer·sey tea \nü-ˈjər-zē-, nyü-\ *noun* [*New Jersey,* state of U.S.; from the use of its leaves as a substitute for tea during the American Revolution] (1759)
: a low deciduous shrub (*Ceanothus americanus*) of the buckthorn family that is found in the eastern U.S. and has dull green leaves and small white flowers borne in large terminal panicles

New Je·ru·sa·lem \-jə-ˈrü-s(ə-)ləm, -zə-ləm; -ˈrüz-ləm\ *noun* [from the phrase "the holy city, *New Jerusalem*" (Revelation 21:2)] (1535)
1 : the final abode of souls redeemed by Christ
2 : an ideal earthly community

New Journalism *noun* (1967)
: journalism that features the author's subjective responses to people and events and that often includes fictional elements meant to illuminate and dramatize those responses

New Latin *noun* (circa 1889)
: Latin as used since the end of the medieval period especially in scientific description and classification

New Left *noun* (1960)
: a political movement originating especially among students in the 1960s, favoring confrontational tactics, often breaking with traditional leftist ideologies, and associated especially with antiwar, antinuclear, feminist, and ecological issues
— **new leftist** *noun, often N&L capitalized*

new·ly \ˈnü-lē, ˈnyü-\ *adverb* (before 12th century)
1 : LATELY, RECENTLY ⟨a *newly* married couple⟩
2 : ANEW, AFRESH

new·ly·wed \-ˌwed\ *noun* (1918)
: a person recently married

new·mar·ket \ˈnü-ˌmär-kət, ˈnyü-\ *noun* [*Newmarket,* England] (1837)
: a long close-fitting coat worn in the 19th century

new math *noun* (1964)
: basic mathematics taught with emphasis on abstraction and the principles of set theory — called also *new mathematics*

new moon *noun* (before 12th century)
1 : the moon's phase when it is in conjunction with the sun so that its dark side is toward the earth; *also* **:** the thin crescent moon seen shortly after sunset for a few days after the actual occurrence of the new moon phase
2 : the first day of each Jewish month marked by a special liturgy

New Right *noun* (1966)
: a political movement made up especially of Protestants, opposed especially to secular humanism, and concerned with issues especially of church and state, patriotism, laissez-faire economics, pornography, and abortion

news \ˈnüz, ˈnyüz\ *noun plural but singular in construction, often attributive* (15th century)
1 a : a report of recent events **b :** previously unknown information ⟨I've got *news* for you⟩
2 a : material reported in a newspaper or news periodical or on a newscast **b :** matter that is newsworthy
3 : NEWSCAST ■
— **news·less** \-ləs\ *adjective*

news agency *noun* (1873)
: an organization that supplies news to subscribing newspapers, periodicals, and newscasters

news·agent \ˈnüz-ˌā-jənt, ˈnyüz-\ *noun* (1851)
chiefly British **:** NEWSDEALER

news·boy \-ˌbȯi\ *noun* (1764)
: one who delivers or sells newspapers

news·break \-ˌbrāk\ *noun* (1944)
: a newsworthy event

news·cast \-ˌkast\ *noun* [*news* + broad*cast*] (circa 1939)
: a radio or television broadcast of news
— **news·cast·er** \-ˌkas-tər\ *noun*

news conference *noun* (1946)
: PRESS CONFERENCE

news·deal·er \-ˌdē-lər\ *noun* (1861)
: a dealer in newspapers, magazines, and often paperback books

news·hound \-ˌhau̇nd\ *noun* (1918)
: an aggressive journalist

news·let·ter \-ˌle-tər\ *noun* (1903)
: a small publication (as a leaflet or newspaper) containing news of interest chiefly to a special group

news·mag·a·zine \-ˌma-gə-ˌzēn, -ˈzēn\ *noun* (1923)
1 : a usually weekly magazine devoted chiefly to summarizing and analyzing news
2 : MAGAZINE 4c

news·man \-mən, -ˌman\ *noun* (1596)
: one who gathers, reports, or comments on the news **:** REPORTER, CORRESPONDENT

news·mon·ger \-ˌməŋ-gər, -ˌmäŋ-\ *noun* (1592)
: one who is active in gathering and repeating news; *especially* **:** GOSSIP

¹news·pa·per \ˈnüz-ˌpā-pər, ˈnyüz-, ˈn(y)üs-\ *noun* (1670)
1 : a paper that is printed and distributed usually daily or weekly and that contains news, articles of opinion, features, and advertising
2 : an organization that publishes a newspaper
3 : the paper of a newspaper **:** NEWSPRINT

²newspaper *intransitive verb* (1943)
: to do newspaper work

news·pa·per·man \-ˌman\ *noun* (1806)
: one who owns or is employed by a newspaper

news·pa·per·wom·an \-ˌwu̇-mən\ *noun* (1881)
: a woman who owns or is employed by a newspaper

new·speak \ˈnü-ˌspēk, ˈnyü-\ *noun, often capitalized* [*Newspeak,* a language "designed to diminish the range of thought," in the novel *1984* (1949) by George Orwell] (1950)
: propagandistic language marked by euphemism, circumlocution, and the inversion of customary meanings **:** DOUBLE-TALK 2

news·peo·ple \ˈnüz-ˌpē-pəl, ˈnyüz-\ *noun plural* (1972)
: REPORTERS

news·per·son \-ˌpər-sᵊn\ *noun* (1973)
: REPORTER

news·print \-ˌprint\ *noun* (1909)
: paper made chiefly from groundwood pulp and used mostly for newspapers

news·read·er \-ˌrē-dər\ *noun* (1925)
chiefly British **:** one who broadcasts the news

news·reel \-ˌrēl\ *noun* (1916)
: a short movie dealing with current events

news·room \-ˌrüm, -ˌru̇m\ *noun* (1929)
1 : a place (as an office) where news is prepared for publication or broadcast
2 : NEWSSTAND

news·stand \-ˌstand\ *noun* (1872)
: a place (as an outdoor stall) where newspapers and periodicals are sold

New Style *adjective* (1615)
: using or according to the Gregorian calendar

news·week·ly \-ˌwē-klē\ *noun* (1947)
: a weekly newspaper or newsmagazine

news·wom·an \-ˌwu̇-mən\ *noun* (1928)
: a woman who is a reporter

news·wor·thy \-ˌwər-t͟hē\ *adjective* (1932)
: sufficiently interesting to the general public to warrant reporting (as in a newspaper)
— **news·wor·thi·ness** \-t͟hē-nəs\ *noun*

news·writ·ing \-ˌrī-tiŋ\ *noun* (1916)
: JOURNALISM 1a

newsy \ˈnü-zē, ˈnyü-\ *adjective* **news·i·er; -est** (1832)
1 : containing or filled with news ⟨*newsy* letters⟩
2 : NEWSWORTHY
— **news·i·ness** *noun*

newt \ˈnüt, ˈnyüt\ *noun* [Middle English, alteration (resulting from misdivision of *an ewte*) of *ewte* — more at EFT] (15th century)
: any of various small semiaquatic salamanders (as of the genus *Triturus*)

newt

New Testament *noun* (14th century)
: the second part of the Christian Bible comprising the canonical Gospels and Epistles and also the book of Acts and book of Revelation — see BIBLE table

new·ton \ˈnü-tᵊn, ˈnyü-\ *noun* [Sir Isaac *Newton*] (1904)
: the unit of force in the meter-kilogram-second system equal to the force required to impart an acceleration of one meter per second per second to a mass of one kilogram

new town *noun* (1918)
: an urban development comprising a small to medium-sized city with a broad range of housing and planned industrial, commercial, and recreational facilities

new wave *noun, often N&W capitalized* (1960)
1 : a cinematic movement that is characterized by improvisation, abstraction, and subjective

□ USAGE

news The habit of sounding a \y\ in *news* is British, but it is also practiced by some Americans who insist that the more common American pronunciation \ˈnüz\ is incorrect. This opinion rests on the belief that the *ew* should be sounded \yü\, as in *few, hew,* and *pew.* At work in the usual American pronunciation of *news* is a type of dissimilation in which the sound \y\, formed with the tongue tip near the roof of the mouth, is lost when it follows a similarly formed sound such as \d,j,l,n,r,s,t,z,ch,sh,zh\. Even in modern British speech the pronunciation of words like *lewd, sewer,* and *suit* generally lacks a \y\ before the \ü\, showing that such dissimilation is not mere American carelessness. Similarly, the absence of \y\ in the American pronunciation of words like *dew, duty, astute,* and *presume* is standard and acceptable, not uncultivated.

symbolism and that often makes use of experimental photographic techniques
2 : a new movement in a particular field
3 : rock music characterized by cohesive ensemble playing and usually lyrics which express anger and social discontent
4 : DERNIER CRI; *especially* **:** fashion that is strikingly outrageous
— new–wave \ˈnü-ˈwāv, ˈnyü-, -ˌwāv\ *adjective*
— new wav•er \-ˈwā-vər\ *noun*

New World *noun* (1555)
: WESTERN HEMISPHERE; *especially* **:** the continental landmass of North and South America

New Year *noun* (13th century)
1 : the calendar year about to start or recently started
2 a : NEW YEAR'S DAY **b :** the first days of a calendar year
3 : ROSH HASHANAH

New Year's Day *noun* (13th century)
: the first day of the calendar year observed as a legal holiday in many countries

[1]**next** \ˈnekst\ *adjective* [Middle English, from Old English *nīehst,* superlative of *nēah* nigh — more at NIGH] (before 12th century)
1 : immediately adjacent (as in place, rank, or time)
2 : any other considered hypothetically ⟨knew it as well as the *next* man⟩

[2]**next** *preposition* (before 12th century)
: nearest or adjacent to

[3]**next** *adverb* (14th century)
1 : in the time, place, or order nearest or immediately succeeding ⟨*next* we drove home⟩ ⟨the *next* closest school⟩
2 : on the first occasion to come ⟨when *next* we meet⟩

next–door \ˈneks(t)-ˈdōr, -ˈdȯr\ *adjective* (1749)
: located or living in the next building, house, apartment, or room; *broadly* **:** NEARBY, ADJACENT

next door *adverb* (1579)
: in or to the next building, house, apartment, or room ⟨lives *next door*⟩; *broadly* **:** in or at an adjacent place
— next door to : NEXT TO

next friend *noun* (1579)
: a person admitted to or appointed by a court to act for the benefit of a person (as an infant) lacking full legal capacity

next of kin (1766)
: one or more persons in the nearest degree of relationship to another person

[1]**next to** *preposition* (1633)
1 : immediately following or adjacent to
2 : in comparison to ⟨*next to* you I'm wealthy⟩

[2]**next to** *adverb* (1667)
: very nearly **:** ALMOST ⟨it was *next to* impossible to see in the fog⟩

nex•us \ˈnek-səs\ *noun, plural* **nex•us•es** \-sə-səz\ *or* **nex•us** \-səs, -ˌsüs\ [Latin, from *nectere* to bind] (1663)
1 : CONNECTION, LINK; *also* **:** a causal link
2 : a connected group or series
3 : CENTER, FOCUS

Nez Percé *or* **Nez Perce** \ˈnez-ˈpərs, ˈnes-ˈpers, ÷ˈnā-per-ˈsā\ *noun, plural* **Nez Percé** *or* **Nez Percés** *same or* -ˈpər-səz, -ˈper-; ÷-per-ˈsā(z)\ *or* **Nez Perce** *or* **Nez Perces** *same or* -ˈpər-səz, -ˈper-\ [French, literally, pierced nose] (1812)
1 : a member of an American Indian people of Idaho, Washington, and Oregon
2 : the language of the Nez Percé people

ngul•trum \eŋ-ˈgu̇l-trəm, en-\ *noun* [Tibetan] (1973)
— see MONEY table

ngwee \eŋ-ˈgwē, en-\ *noun, plural* **ngwee** [Bemba, literally, bright] (1966)
— see *kwacha* at MONEY table

ni•a•cin \ˈnī-ə-sən\ *noun* [*ni*cotinic *ac*id + [1]*-in*] (1942)
: NICOTINIC ACID; *also* **:** NICOTINAMIDE

ni•a•cin•amide \ˌnī-ə-ˈsi-nə-ˌmīd\ *noun* (1942)
: NICOTINAMIDE

Ni•ag•a•ra \nī-ˈa-g(ə-)rə\ *noun* [*Niagara* Falls] (1841)
: an overwhelming flood **:** TORRENT ⟨a *Niagara* of protests⟩

ni•al•amide \nī-ˈa-lə-ˌmīd\ *noun* [*ni*cotinic acid + *amyl* + *amide*] (1959)
: a synthetic antidepressant drug $C_{16}H_{18}N_4O_2$ that is an inhibitor of monoamine oxidase

nib \ˈnib\ *noun* [probably alteration of *neb*] (1585)
1 : BILL, BEAK
2 a : the sharpened point of a quill pen **b :** PEN POINT; *also* **:** each of the two divisions of a pen point
3 : a small pointed or projecting part

[1]**nib•ble** \ˈni-bəl\ *verb* **nib•bled; nib•bling** \-b(ə-)liŋ\ [origin unknown] (circa 1512)
transitive verb
1 a : to bite gently **b :** to eat or chew in small bits
2 : to take away bit by bit ⟨waves *nibbling* the shore⟩
intransitive verb
1 : to take gentle, small, or cautious bites; *also* **:** SNACK
2 : to deal with something as if by nibbling
— nib•bler \-b(ə-)lər\ *noun*

[2]**nibble** *noun* (1658)
1 : an act of nibbling
2 : a very small quantity or portion (as of food); *also* **:** SNACK
3 : a tentative expression of interest

Ni•be•lung \ˈnē-bə-ˌlu̇ŋ\ *noun, plural* **-lungs** *also* **-lung•en** \-ˌlu̇ŋ-ən\ [German] (1861)
1 : a member of a race of dwarfs in Germanic legend
2 : any of the followers of Siegfried
3 : any of the Burgundian kings in the medieval German *Nibelungenlied*

nibs \ˈnibz\ *noun plural but singular in construction* [origin unknown] (circa 1821)
: an important or self-important person — usually used in the phrases *his nibs* or *her nibs* as if a title of honor

ni•cad \ˈnī-ˌkad\ *noun* [[1]*ni*ckel + *cad*mium] (1965)
: a rechargeable dry cell that has a nickel cathode and a cadmium anode

nic•co•lite \ˈni-kə-ˌlīt\ *noun* [New Latin *niccolum* nickel, probably from Swedish *nickel*] (1868)
: a pale copper-red usually massive mineral of metallic luster that is essentially a nickel arsenide

nice \ˈnīs\ *adjective* **nic•er; nic•est** [Middle English, foolish, wanton, from Old French, from Latin *nescius* ignorant, from *nescire* not to know — more at NESCIENCE] (14th century)
1 *obsolete* **a :** WANTON, DISSOLUTE **b :** COY, RETICENT
2 a : showing fastidious or finicky tastes **:** PARTICULAR **b :** exacting in requirements or standards **:** PUNCTILIOUS
3 : possessing, marked by, or demanding great or excessive precision and delicacy
4 *obsolete* **:** TRIVIAL
5 a : PLEASING, AGREEABLE ⟨a *nice* time⟩ ⟨a *nice* person⟩ **b :** well-executed ⟨*nice* shot⟩ **c :** APPROPRIATE, FITTING
6 a : socially acceptable **:** WELL-BRED **b :** VIRTUOUS, RESPECTABLE
7 : POLITE, KIND ⟨that's *nice* of you to say⟩ ◆
synonym see CORRECT
— nice *adverb*
— nice•ly *adverb*
— nice•ness *noun*

Ni•cene \ˈnī-ˌsēn, nī-ˈ\ *adjective* [Middle English, from Late Latin *nicaenus,* from Latin *Nicaea* Nicaea, from Greek *Nikaia*] (14th century)
1 : of or relating to Nicaea or the Nicaeans
2 : of or relating to the ecumenical church council held in Nicaea in A.D. 325 or to the Nicene Creed

Nicene Creed *noun* (circa 1569)
: a Christian creed expanded from a creed issued by the first Nicene Council, beginning "I believe in one God," and used in liturgical worship

nice–nel•ly \ˈnīs-ˈne-lē\ *adjective, often 2d N capitalized* [from the name *Nelly*] (1925)
1 : marked by euphemism
2 : PRUDISH
— nice nelly *noun, often 2d N capitalized*
— nice–nel•ly•ism \-ˌi-zəm\ *noun, often 2d N capitalized*

nice•ty \ˈnī-sə-tē, -stē\ *noun, plural* **-ties** [Middle English *nicete,* from Middle French *niceté* foolishness, from *nice,* adjective] (14th century)
1 : the quality or state of being nice
2 : an elegant, delicate, or civilized feature ⟨enjoy the *niceties* of life⟩
3 : a fine point or distinction **:** SUBTLETY ⟨the *niceties* of table manners⟩
4 : careful attention to details **:** delicate exactness **:** PRECISION
5 : delicacy of taste or feeling **:** FASTIDIOUSNESS

[1]**niche** \ˈnich, ÷ˈnēsh\ *noun* [French, from Middle French, from *nicher* to nest, from (assumed) Vulgar Latin *nidicare,* from Latin *nidus* nest — more at NEST] (1611)
1 a : a recess in a wall especially for a statue **b :** something that resembles a niche
2 a : a place, employment, status, or activity for which a person or thing is best fitted **b :** a habitat supplying the factors necessary for the existence of an organism or species **c :** the ecological role of an organism in a community especially in regard to food consumption **d :** a specialized market

niche 1a

[2]**niche** *transitive verb* **niched; nich•ing** (1752)
: to place in or as if in a niche

Ni•chrome \ˈnī-ˌkrōm\ *trademark*

◇ WORD HISTORY

nice This bland-looking, workaday word has undergone a remarkable semantic evolution over the course of its history. *Nice* was borrowed from Old French, in which it meant "foolish, simpleminded." (French had in turn borrowed the word at an early date from the Latin adjective *nescius* "ignorant, not knowing," a derivative of the verb *nescire* "to not know.") In Middle English, from the original sense "foolish, frivolous," the word developed an extraordinarily diverse range of meanings, from "fainthearted, cowardly" to "fastidious, scrupulous" to "lascivious." Of these only the "fastidious" sense has survived into the 19th and 20th centuries, in phrases like "a nice distinction." Not until the 18th century did *nice* come to be employed in the now common meaning "pleasing, agreeable"; in the 1864 revision of Noah Webster's *American Dictionary of the English Language,* this usage was still considered both improper and a peculiarity of the British, the American equivalent being *fine.* Succeeding Merriam-Webster dictionaries labeled the sense as colloquial until the 1934 publication of *Webster's New International Dictionary, Second Edition.*

— used for an alloy of nickel, chromium, and iron that is commonly used in the heating elements of electrical appliances

¹nick \'nik\ *noun* [Middle English *nyke,* probably alteration of *nocke* nock] (15th century)
1 a : a small notch or groove **b :** a break in one strand of two-stranded DNA caused by a missing phosphodiester bond **c :** a small cut incurred in shaving
2 : a final critical moment ⟨in the *nick* of time⟩
3 *slang British* **:** PRISON; *also* **:** POLICE STATION
4 *British* **:** CONDITION ⟨in good *nick*⟩

²nick (1523)
transitive verb
1 : to jot down **:** RECORD
2 a : to make a nick in **:** NOTCH, CHIP **b :** to cut into or wound slightly ⟨a bullet *nicked* his leg⟩
3 : to cut short ⟨cold weather, which *nicked* steel and automobile output —*Time*⟩
4 : to catch at the right point or time
5 : CHEAT, OVERCHARGE
6 a *slang British* **:** ARREST **b** *slang British* **:** STEAL
intransitive verb
1 : to make petty attacks **:** SNIPE
2 : to complement one another genetically and produce superior offspring

nick·el *also* **nick·le** \'ni-kəl\ *noun* [probably from Swedish *nickel,* from German *Kupfernickel* niccolite, probably from *Kupfer* copper + *Nickel* goblin; from the deceptive copper color of niccolite] (1755)
1 : a silver-white hard malleable ductile metallic element capable of a high polish and resistant to corrosion that is used chiefly in alloys and as a catalyst — see ELEMENT table
2 a (1) **:** the U.S. 5-cent piece regularly containing 25 percent nickel and 75 percent copper (2) **:** the Canadian 5-cent piece **b :** five cents
3 : a pass defense in football that employs five defensive backs

¹nick·el–and–dime \,ni-kəl-ən-'dīm\ *adjective* (1950)
1 : involving or offering only a small amount of money
2 : SMALL-TIME

²nickel–and–dime *transitive verb* **nick·eled–and–dimed** \,ni-kəld-ən-'dīmd\ *or* **nick·el–and–dimed** \-ən-'dīmd\; **nick·el·ing–and–dim·ing** \-'dī-miŋ\ *or* **nickel–and–diming** (1961)
: to impair, weaken, or defeat gradually (as through a series of small incursions or excessive attention to detail)

nick·el·if·er·ous \,ni-kə-'li-f(ə-)rəs\ *adjective* (1821)
: containing nickel

nick·el·ode·on \,ni-kə-'lō-dē-ən\ *noun* [probably from *nickel* + *-odeon* (as in *melodeon* music hall)] (1907)
1 : an early movie theater to which admission usually cost five cents
2 : JUKEBOX

nickel silver *noun* (1860)
: GERMAN SILVER

¹nick·er \'ni-kər\ *intransitive verb* **nick·ered; nick·er·ing** \-k(ə-)riŋ\ [perhaps alteration of *neigh*] (1641)
: NEIGH, WHINNY
— nicker *noun*

²nicker *noun* [perhaps from *nicker* one that nicks] (1910)
slang British **:** ¹POUND 2a

nick·nack *variant of* KNICKKNACK

¹nick·name \'nik-,nām\ *noun* [Middle English *nekename* additional name, alteration (resulting from misdivision of *an ekename*) of *ekename,* from *eke* eke, also + *name* name] (15th century)
1 : a usually descriptive name given instead of or in addition to the one belonging to a person, place, or thing
2 : a familiar form of a proper name (as of a person or a city) ◆

²nickname *transitive verb* (1536)
1 : MISNAME, MISCALL
2 : to give a nickname to
— nick·nam·er *noun*

ni·co·ti·a·na \ni-,kō-shē-'a-nə, -'ä-nə, -'ā-nə\ *noun* [New Latin, from *herba nicotiana,* literally, Nicot's herb, from Jean *Nicot* (died 1600) French diplomat and scholar] (1600)
: any of several tobaccos (as *Nicotiana alata*) grown for their showy flowers

nic·o·tin·amide \,ni-kə-'tē-nə-,mīd, -'ti-\ *noun* [International Scientific Vocabulary] (1895)
: a compound $C_6H_6N_2O$ of the vitamin B complex found especially as a constituent of coenzymes and used similarly to nicotinic acid

nicotinamide adenine dinucleotide *noun* (1961)
: NAD

nicotinamide adenine dinucleotide phosphate *noun* (1962)
: NADP

nic·o·tine \'ni-kə-,tēn\ *noun* [French, from New Latin *nicotiana*] (1819)
: a poisonous alkaloid $C_{10}H_{14}N_2$ that is the chief active principle of tobacco and is used as an insecticide

nic·o·tin·ic \,ni-kə-'tē-nik, -'ti-\ *adjective* [International Scientific Vocabulary] (1873)
: relating to, resembling, producing, or mediating the effects produced by nicotine on nerve fibers at autonomic ganglia and at the neuromuscular junctions of voluntary muscle which increases activity in small doses and inhibits it in larger doses ⟨*nicotinic* receptors⟩ — compare MUSCARINIC

nicotinic acid *noun* (1873)
: an acid $C_6H_5NO_2$ of the vitamin B complex found widely in animals and plants and used especially against pellagra — called also *niacin*

nic·ti·tate \'nik-tə-,tāt\ *intransitive verb* **-tat·ed; -tat·ing** [alteration of *nictate* to wink, from Latin *nictatus,* past participle of *nictare* — more at CONNIVE] (circa 1834)
: WINK

nictitating membrane *noun* (1713)
: a thin membrane found in many vertebrates at the inner angle or beneath the lower lid of the eye and capable of extending across the eyeball

ni·dic·o·lous \nī-'di-kə-ləs\ *adjective* [Latin *nidus* nest + English *-colous* — more at NEST] (circa 1902)
1 : reared for a time in a nest
2 : sharing the nest of another kind of animal

ni·di·fi·ca·tion \,ni-də-fə-'kā-shən, ,nī-\ *noun* [Medieval Latin *nidification-, nidificatio,* from Latin *nidificare* to build a nest, from *nidus* nest] (1658)
: the act, process, or technique of building a nest

ni·dif·u·gous \nī-'di-fyə-gəs\ *adjective* [Latin *nidus* nest + *fugere* to flee — more at FUGITIVE] (1896)
: leaving the nest soon after hatching

ni·dus \'nī-dəs\ *noun, plural* **ni·di** \'nī-,dī\ *or* **ni·dus·es** [New Latin, from Latin] (1742)
1 : a nest or breeding place; *especially* **:** a place or substance in an animal or plant where bacteria or other organisms lodge and multiply
2 : a place where something originates, develops, or is located

niece \'nēs\ *noun* [Middle English *nece* granddaughter, niece, from Old French *niece,* from Late Latin *neptia,* from Latin *neptis;* akin to Latin *nepot-, nepos* grandson, nephew — more at NEPHEW] (14th century)
1 : a daughter of one's brother, sister, brother-in-law, or sister-in-law
2 : an illegitimate daughter of an ecclesiastic

¹ni·el·lo \nē-'e-(,)lō\ *noun, plural* **ni·el·li** \-(,)lē\ *or* **niellos** [Italian, from Medieval Latin *nigellum,* from neuter of Latin *nigellus* blackish, diminutive of *niger* black] (1816)
1 : any of several black enamel-like alloys usually of sulfur with silver, copper, and lead
2 : the art or process of decorating metal with incised designs filled with niello
3 : a piece of metal or an object decorated with niello

²niello *transitive verb* (1866)
: to inlay or ornament with niello

ni·fed·i·pine \nə-'fe-də-,pēn, -pən\ *noun* [probably *nitr-* + *-fe-* (from *phenyl*) + *-dipine* (by alteration & shortening from *pyridine*)] (1974)
: a calcium channel blocker $C_{17}H_{18}N_2O_6$ that is a coronary vasodilator used especially in the treatment of angina pectoris

Nifl·heim \'ni-vəl-,hām\ *noun* [Old Norse *Niflheimr*]
: the abode of the dead in Norse mythology

¹nif·ty \'nif-tē\ *adjective* **nif·ti·er; -est** [origin unknown] (1865)
: very good **:** very attractive **:** FINE ⟨*nifty* clothes⟩
— nif·ti·ly \'nif-tə-lē\ *adverb*

²nifty *noun, plural* **nifties** (1923)
: something that is nifty; *especially* **:** a clever or neatly turned phrase or joke

Ni·ger–Con·go \,nī-jər-'käŋ-(,)gō\ *noun* [*Niger* (river) + *Congo* (river)] (circa 1950)
: a language family that includes the Mande and Kwa branches and that is spoken by most of the indigenous peoples of west, central, and south Africa

¹nig·gard \'ni-gərd\ *noun* [Middle English, of Scandinavian origin; akin to Old Norse *hnøggr* niggardly; akin to Old English *hnēaw* niggardly] (14th century)
: a meanly covetous and stingy person **:** MISER
— niggard *adjective*

²niggard (circa 1600)
intransitive verb
obsolete **:** to act niggardly
transitive verb
obsolete **:** to treat in a niggardly manner

nig·gard·ly \-lē\ *adjective* (1571)
1 : grudgingly mean about spending or granting **:** BEGRUDGING
2 : provided in meanly limited supply
synonym see STINGY
— nig·gard·li·ness *noun*
— niggardly *adverb*

nig·ger \'ni-gər\ *noun* [alteration of earlier *neger,* from Middle French *negre,* from Spanish or Portuguese *negro,* from *negro* black, from Latin *niger*] (1700)
1 : a black person — usually taken to be offensive
2 : a member of any dark-skinned race — usually taken to be offensive
3 : a member of a socially disadvantaged class of persons ⟨it's time for somebody to lead all of America's *niggers* . . . all the people who feel left out of the political process —Ron Dellums⟩ ☐

¹nig·gle \'ni-gəl\ *verb* **nig·gled; nig·gling**

◇ WORD HISTORY

nickname The earliest sense of *nickname* is a descriptive name given to a person, or even a place, in addition to the proper name, while today it can also refer to a shortened or familiar form of a proper name. In Middle English the word expressed the first of these senses very explicitly. The noun *eke,* meaning "addition or extension," was combined with *name,* and thus an *ekename* was an additional name. By the 15th century *an ekename,* by misperception of the boundary between article and noun, was taken to be *a nekename,* and both *ekename* and *nekename* begin to appear in writing. The latter, in the modern form *nickname,* eventually superseded the former. The noun *eke* involved in the history of *nickname* is related both to the verb *eke,* as in "to eke out a living," and to the once common but now archaic adverb *eke* "too, also."

\-g(ə-)liŋ\ [origin unknown] (circa 1616)
intransitive verb
1 a : TRIFLE **b :** to spend too much effort on minor details
2 : to find fault constantly in a petty way **:** CARP ⟨she haggles, she *niggles,* she wears out our patience —Virginia Woolf⟩
3 : GNAW
transitive verb
: to give stingily or in tiny portions
— **nig·gler** \-g(ə-)lər\ *noun*

²niggle *noun* (1886)
chiefly British **:** a trifling doubt, objection, or complaint

nig·gling \'ni-g(ə-)liŋ\ *adjective* (1599)
: PETTY; *also* **:** bothersome or persistent especially in a petty or tiresome way ⟨*niggling* injuries⟩
— **niggling** *noun*
— **nig·gling·ly** \-lē\ *adverb*

¹nigh \'nī\ *adverb* [Middle English, from Old English *nēah;* akin to Old High German *nāh,* adverb, nigh, preposition, nigh, after, Old Norse *nā-* nigh] (before 12th century)
1 : near in place, time, or relationship — often used with *on, onto,* or *unto*
2 : NEARLY, ALMOST

²nigh *adjective* (before 12th century)
1 : CLOSE, NEAR
2 *chiefly dialect* **:** DIRECT, SHORT
3 : being on the left side ⟨the *nigh* horse⟩

³nigh *preposition* (before 12th century)
: NEAR

⁴nigh (13th century)
transitive verb
: to draw or come near to **:** APPROACH
intransitive verb
: to draw near

¹night \'nīt\ *noun* [Middle English, from Old English *niht;* akin to Old High German *naht* night, Latin *noct-, nox,* Greek *nykt-, nyx*] (before 12th century)
1 : the time from dusk to dawn when no sunlight is visible
2 a : an evening or night taken as an occasion or point of time ⟨the opening *night*⟩ **b :** an evening set aside for a particular purpose
3 a : the quality or state of being dark **b :** a condition or period felt to resemble the darkness of night: as (1) **:** a period of dreary inactivity or affliction (2) **:** absence of moral values **c :** the beginning of darkness **:** NIGHTFALL
— **night·less** \-ləs\ *adjective*

²night *adjective* (before 12th century)
1 : of, relating to, or associated with the night ⟨*night* air⟩
2 : intended for use at night ⟨a *night* lamp⟩
3 a : existing, occurring, or functioning at night ⟨*night* baseball⟩ ⟨a *night* nurse⟩ **b :** active or functioning best at night ⟨*night* people⟩

night and day *adverb* (before 12th century)
: all the time **:** CONTINUALLY

night blindness *noun* (1754)
: reduced visual capacity in faint light (as at night)
— **night–blind** \'nīt-,blīnd\ *adjective*

night–blooming cereus *noun* (1832)
: any of several night-blooming cacti, *especially* **:** a slender sprawling or climbing cactus (*Selenicereus grandiflorus*) often cultivated for its large showy fragrant white flowers

night·cap \'nīt-,kap\ *noun* (14th century)
1 : a cloth cap worn with nightclothes
2 : a usually alcoholic drink taken at the end of the day
3 : the final race or contest of a day's sports; *especially* **:** the second game of a baseball doubleheader

night·clothes \-,klō(th)z\ *noun plural* (1602)
: garments for wear in bed

¹night·club \-,kləb\ *noun* (1894)
: a place of entertainment open at night usually serving food and liquor, having a floor show, and providing music and space for dancing

²nightclub *intransitive verb* (1936)
: to patronize nightclubs
— **night·club·ber** *noun*

night court *noun* (1934)
: a criminal court in a large city that sits at night (as for the summary disposition of criminal charges and the granting of bail)

night crawler *noun* (1924)
: EARTHWORM; *especially* **:** a large earthworm found on the soil surface at night

night·dress \'nīt-,dres\ *noun* (circa 1714)
1 : NIGHTGOWN
2 : NIGHTCLOTHES

night·fall \'nīt-,fól\ *noun* (1700)
: the close of the day **:** DUSK

night·glow \-,glō\ *noun* (1951)
: airglow seen during the night

night·gown \-,gau̇n\ *noun* (14th century)
1 *archaic* **:** DRESSING GOWN
2 : a loose garment for wear in bed

night·hawk \-,hók\ *noun* (1611)
1 a : any of a genus (*Chordeiles*) of American nightjars related to the whippoorwill **b :** a common European nightjar (*Caprimulgus europaeus*)
2 : a person who habitually is active late at night

nighthawk 1a

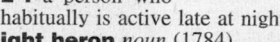

night heron *noun* (1784)
: any of various widely distributed nocturnal or crepuscular herons (especially genus *Nycticorax*)

night·ie \'nī-tē\ *or* **nighty** *noun, plural* **night·ies** [*night*gown + *-ie*] (1871)
: a nightgown for a woman or child

night·in·gale \'nī-tᵊn-,gāl, -tiŋ-\ *noun* [Middle English, alteration of Old English *nihtegale,* from *niht* + *galan* to sing — more at YELL] (13th century)
: any of several Old World thrushes (genus *Erithacus* and especially *E. megarhynchos*) noted for the sweet usually nocturnal song of the male; *also* **:** any of various other birds that sing at night

nightingale

night·jar \'nīt-,jär\ *noun* [from its harsh sound] (1630)
: any of a family (Caprimulgidae) of medium-sized long-winged crepuscular or nocturnal birds (as the whippoorwills and nighthawks) having a short bill, short legs, and soft mottled plumage and feeding on insects which they catch on the wing — called also *goatsucker*

night latch *noun* (1854)
: a door lock having a spring bolt operated from the outside by a key and from the inside by a knob

night letter *noun* (1910)
: a telegram sent at night at a reduced rate for delivery the following morning

night·life \'nīt-,līf\ *noun* (1852)
: the activity of or entertainment provided for pleasure-seekers at night (as in nightclubs); *also* **:** establishments providing nightlife

night–light \-,līt\ *noun* (1839)
: a light kept burning throughout the night

¹night·long \-,lóŋ\ *adjective* (1850)
: lasting the whole night ⟨*nightlong* festivities⟩

²night·long \-'lóŋ\ *adverb* (1870)
: through the whole night

¹night·ly \-lē\ *adjective* (before 12th century)
1 : happening, done, or used by night or every night
2 : of or relating to the night or every night

²nightly *adverb* (15th century)
: every night; *also* **:** at or by night

night·mare \'nīt-,mar, -,mer\ *noun* [Middle English, from ¹*night* + ¹*mare*] (14th century)
1 : an evil spirit formerly thought to oppress people during sleep
2 : a frightening dream that usually awakens the sleeper
3 : something (as an experience, situation, or object) having the monstrous character of a nightmare or producing a feeling of anxiety or terror ◆
— **nightmare** *adjective*
— **night·mar·ish** \-,mar-ish, -,mer-\ *adjective*
— **night·mar·ish·ly** *adverb*

night owl *noun* (circa 1846)
: a person who keeps late hours at night

night rail \-,rāl\ *noun* [*night* + *rail,* a woman's loose garment] (1554)
archaic **:** NIGHTGOWN

night raven *noun* (before 12th century)
: a bird that cries at night

night rider *noun* (1877)
: a member of a secret band who ride masked at night doing acts of violence for the purpose of punishing or terrorizing

nights \'nīts\ *adverb* (before 12th century)
: in the nighttime repeatedly **:** on any night ⟨works *nights*⟩

night·scope \'nīt-,skōp\ *noun* (1972)
: an optical device usually using infrared radiation that enables a person to see objects in the dark better

night·shade \-,shād\ *noun* (before 12th century)
1 : any of a genus (*Solanum* of the family Solanaceae, the nightshade family) of herbs, shrubs, and trees having alternate leaves, cymose flowers, and fruits that are berries and including some poisonous weeds, various ornamentals, and important crop plants (as the potato and eggplant)
2 : BELLADONNA 1

night·shirt \-,shərt\ *noun* (1857)
: a nightgown resembling a shirt

night·side \-,sīd\ *noun* (1848)

□ USAGE

nigger *Nigger* in senses 1 and 2 can be found in the works of such writers of the past as Joseph Conrad, Mark Twain, and Charles Dickens, but it now ranks as perhaps the most offensive and inflammatory racial slur in English. Its use by and among blacks is not always intended or taken as offensive, but, except in sense 3, it is otherwise a word expressive of racial hatred and bigotry.

◇ WORD HISTORY

nightmare During the Middle Ages there was a widespread belief in an evil spirit that visits and lies on people as they sleep. Known as a *mare* in Old English, this evil spirit, or incubus, became known in Middle English as a *nightmare* after the time of its supposed visitation. It was not until the 16th century that the word *nightmare* came to refer to a frightening dream. Again, evil spirits were involved, for they were credited with causing such dreams. *Mare* in the sense of "a female horse," is an entirely different, unrelated word, though it too is from Old English.

: the side of a celestial body (as the earth, the moon, or a planet) not in daylight

night soil *noun* (circa 1774)
: human excrement used especially for fertilizing the soil

night•spot \ˈnīt-ˌspät\ *noun* (1936)
: NIGHTCLUB

night•stand \-ˌstand\ *noun* (1892)
: NIGHT TABLE

night•stick \-ˌstik\ *noun* (1887)
: a police officer's club

night table *noun* (1788)
: a small bedside table or stand

night•time \-ˌtīm\ *noun, often attributive* (14th century)
: the time from dusk to dawn

night•walk•er \-ˌwȯ-kər\ *noun* (15th century)
1 : a person who roams about at night especially with criminal intent
2 : PROSTITUTE, STREETWALKER

ni•hil•ism \ˈnī-(h)ə-ˌli-zəm, ˈnē-\ *noun* [German *Nihilismus,* from Latin *nihil* nothing — more at NIL] (circa 1817)
1 a : a viewpoint that traditional values and beliefs are unfounded and that existence is senseless and useless **b** : a doctrine that denies any objective ground of truth and especially of moral truths
2 a (1) : a doctrine or belief that conditions in the social organization are so bad as to make destruction desirable for its own sake independent of any constructive program or possibility (2) *capitalized* : the program of a 19th century Russian party advocating revolutionary reform and using terrorism and assassination **b** : TERRORISM
— **ni•hil•ist** \-list\ *noun or adjective*
— **ni•hil•is•tic** \ˌnī-(h)ə-ˈlis-tik, ˌnē-\ *adjective*

-nik *noun suffix* [Yiddish, from Polish & Ukrainian]
: one connected with or characterized by being ⟨beat*nik*⟩

Ni•ke \ˈnī-kē\ *noun* [Greek *Nikē*]
: the Greek goddess of victory

nil \ˈnil\ *noun* [Latin, nothing, contraction of *nihil,* from Old Latin *nihilum,* from *ne-* not + *hilum* trifle — more at NO] (1833)
: NOTHING, ZERO
— **nil** *adjective*

nile green *noun, often N capitalized* [*Nile* River, Africa] (1871)
: a pale yellow green

Nile perch *noun* (1926)
: a large predaceous food fish (*Lates niloticus*) of the rivers and lakes of northern and central Africa that may exceed 200 pounds (91 kilograms) in weight

nill \ˈnil\ *verb* [Middle English *nilen,* from Old English *nyllan,* from *ne* not + *wyllan* to wish — more at NO, WILL] (before 12th century)
intransitive verb
archaic : to be unwilling : will not ⟨will you *nill* you, I will marry you —Shakespeare⟩
transitive verb
archaic : REFUSE

Ni•lot•ic \nī-ˈlä-tik\ *adjective* [Latin *Niloticus,* from Greek *Neilōtēs,* from *Neilos* Nile] (1653)
1 : of or relating to the Nile or the peoples of the Nile basin
2 : of, relating to, or being the languages of the Nilotic people

nil•po•tent \ˈnil-ˌpō-tᵊnt\ *adjective* [Latin *nil* nothing + *potent-, potens* having power — more at POTENT] (1870)
: equal to zero when raised to some power ⟨*nilpotent* matrices⟩

¹nim \ˈnim\ *verb* **nimmed; nim•ming** [earlier *nim* to take, from Middle English *nimen,* from Old English *niman*] (12th century)
transitive verb
archaic : STEAL, FILCH
intransitive verb
archaic : THIEVE

²nim *noun* [probably from ¹*nim*] (1901)
: any of various games in which counters are laid out in one or more piles and each player in turn draws one or more counters with the object of taking the last counter, forcing the opponent to take it, or taking the most or fewest counters

nim•ble \ˈnim-bəl\ *adjective* **nim•bler** \-b(ə-)lər\; **nim•blest** \-b(ə-)ləst\ [Middle English *nimel,* from Old English *numol* holding much, from *niman* to take; akin to Old High German *neman* to take, Greek *nemein* to distribute, manage, *nomos* pasture, *nomos* usage, custom, law] (14th century)
1 : quick and light in motion : AGILE ⟨*nimble* fingers⟩
2 a : marked by quick, alert, clever conception, comprehension, or resourcefulness ⟨a *nimble* mind⟩ **b** : RESPONSIVE, SENSITIVE ⟨a *nimble* listener⟩
— **nim•ble•ness** \-bəl-nəs\ *noun*
— **nim•bly** \-blē\ *adverb*

nim•bo•stra•tus \ˌnim-bō-ˈstrā-təs, -ˈstra-\ *noun* [New Latin, from Latin *nimbus* + New Latin *stratus* stratus] (circa 1909)
: a low dark gray rainy cloud layer — see CLOUD illustration

nim•bus \ˈnim-bəs\ *noun, plural* **nim•bi** \-ˌbī, -ˌbē\ *or* **nim•bus•es** [Latin, rainstorm, cloud; probably akin to Latin *nebula* cloud — more at NEBULA] (1616)
1 a : a luminous vapor, cloud, or atmosphere about a god or goddess when on earth **b** : a cloud or atmosphere (as of romance) about a person or thing
2 : an indication (as a circle) of radiant light or glory about the head of a drawn or sculptured divinity, saint, or sovereign
3 a : a rain cloud **b** : THUNDERHEAD; *also* : CUMULUS 2

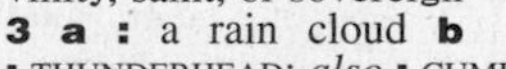
nimbus 2

ni•mi•e•ty \ni-ˈmī-ə-tē\ *noun, plural* **-ties** [Late Latin *nimietas,* from Latin *nimius* too much, adjective, from *nimis,* adverb] (circa 1564)
: EXCESS, REDUNDANCY

nim•i•ny–pim•i•ny \ˌni-mə-nē-ˈpi-mə-nē\ *adjective* [probably alteration of *namby-pamby*] (1786)
: affectedly refined : FINICKY

Nim•rod \ˈnim-ˌräd\ *noun* [Hebrew *Nimrōdh*]
1 : a descendant of Ham represented in Genesis as a mighty hunter and a king of Shinar
2 *not capitalized* : HUNTER

nin•com•poop \ˈnin-kəm-ˌpüp, ˈniŋ-\ *noun* [origin unknown] (1676)
: FOOL, SIMPLETON
— **nin•com•poop•ery** \-ˌpü-pə-rē\ *noun*

nine \ˈnīn\ *noun* [Middle English, from *nyne,* adjective, from Old English *nigon;* akin to Old High German *niun* nine, Latin *novem,* Greek *ennea*] (before 12th century)
1 — see NUMBER table
2 : the ninth in a set or series ⟨wears a *nine*⟩
3 : something having nine units or members: as **a** *capitalized* : the nine Muses **b** : a baseball team **c** : the first or last nine holes of an 18-hole golf course
— **nine** *adjective*
— **nine** *pronoun, plural in construction*
— **to the nines 1** : to perfection **2** : in a highly elaborate or showy manner ⟨dressed *to the nines*⟩

nine days' wonder *noun* (1594)
: something or someone that creates a short-lived sensation — called also *nine day wonder*

nine•fold \ˈnīn-ˌfōld, -ˈfōld\ *adjective* (before 12th century)
1 : being nine times as great or as many
2 : having nine units or members
— **nine•fold** \-ˈfōld\ *adverb*

nine•pin \-ˌpin\ *noun* (1580)
1 *plural but singular in construction* : a bowling game resembling tenpins played without the headpin
2 : a pin used in ninepins

nine•teen \nīn-ˈtēn, ˈnīn-ˌ\ *noun* [Middle English *nynetene,* adjective, from Old English *nigontēne,* from *nigon* + *-tīene* (akin to Old English *tīen* ten) — more at TEN] (before 12th century)
— see NUMBER table
— **nineteen** *adjective*
— **nineteen** *pronoun, plural in construction*
— **nine•teenth** \-ˈtēn(t)th, -ˌtēn(t)th\ *adjective or noun*

nine•ty \ˈnīn-tē\ *noun, plural* **nineties** [Middle English *ninety,* adjective, from Old English *nigontig,* short for *hundnigontig,* from *hundnigontig,* noun, group of 90, from *hund-,* literally, hundred + *nigon* nine + *-tig* group of 10; akin to Old English *tīen* ten] (before 12th century)
1 — see NUMBER table
2 *plural* : the numbers 90 to 99; *specifically* : the years 90 to 99 in a lifetime or century
— **nine•ti•eth** \-tē-əth\ *adjective or noun*
— **ninety** *adjective*
— **ninety** *pronoun, plural in construction*

nin•hy•drin \nin-ˈhī-drən\ *noun* [from *Ninhydrin,* a trademark] (1913)
: a poisonous crystalline oxidizing agent $C_9H_6O_4$ used especially as an analytical reagent

nin•ja \ˈnin-jə, -(ˌ)jä\ *noun, plural* **ninja** *also* **ninjas** [Japanese, from *nin-* persevere + *-ja* person] (1975)
: a person trained in ancient Japanese martial arts and employed especially for espionage and assassinations

nin•ny \ˈni-nē\ *noun, plural* **ninnies** [perhaps by shortening & alteration from *an innocent*] (1593)
: FOOL, SIMPLETON

nin•ny•ham•mer \ˈni-nē-ˌha-mər\ *noun* (1592)
: NINNY

ni•non \ˈnē-ˌnän\ *noun* [probably from French *Ninon,* nickname for *Anne*] (1911)
: a smooth sheer fabric

ninth \ˈnīn(t)th\ *noun, plural* **ninths** \ˈnīn(t)s, ˈnīn(t)ths\ (13th century)
1 — see NUMBER table
2 a : a musical interval embracing an octave and a second **b** : the tone at this interval **c** : a chord containing a ninth
— **ninth** *adjective or adverb*

ninth cranial nerve *noun* (circa 1961)
: GLOSSOPHARYNGEAL NERVE

ni•o•bate \ˈnī-ə-ˌbāt\ *noun* [New Latin *niobium* + English ¹*-ate*] (1845)
: a salt containing an anionic grouping of niobium and oxygen

Ni•o•be \ˈnī-ə-bē\ *noun* [Latin, from Greek *Niobē*]
: a daughter of Tantalus and wife of Amphion who while weeping for her slain children is turned into a stone from which her tears continue to flow

ni•o•bi•um \nī-ˈō-bē-əm\ *noun* [New Latin, from Latin *Niobe;* from its occurrence in tantalite] (1845)
: a lustrous light gray ductile metallic element that resembles tantalum chemically and is used in alloys — see ELEMENT table

¹nip \ˈnip\ *verb* **nipped; nip•ping** [Middle English *nippen;* akin to Old Norse *hnippa* to prod] (14th century)
transitive verb
1 a : to catch hold of and squeeze tightly between two surfaces, edges, or points : PINCH, BITE ⟨the dog *nipped* his ankle⟩ **b** : to pinch in (as a garment) ⟨a dress *nipped* at the waist⟩
2 a : to sever by or as if by pinching sharply **b** : to destroy the growth, progress, or fulfillment of ⟨*nipped* in the bud⟩
3 : to injure or make numb with cold : CHILL

4 : SNATCH, STEAL
5 : to defeat by a small margin
intransitive verb
1 : to move briskly, nimbly, or quickly
2 *chiefly British* **:** to make a quick trip

[2]nip *noun* (1549)
1 : something that nips: as **a** *archaic* **:** a sharp biting comment **b :** a sharp stinging cold ⟨a *nip* in the air⟩ **c :** a biting or pungent flavor **:** TANG ⟨cheese with a *nip*⟩
2 : the act of nipping **:** PINCH, BITE
3 : the region of a squeezing or crushing device (as a calender) where the rolls or jaws are closest together
4 : a small portion

[3]nip *noun* [probably from *nipperkin,* a liquor container] (circa 1796)
: a small quantity of liquor **:** SIP; *also* **:** a very small bottle of liquor

[4]nip *intransitive verb* **nipped; nip·ping** (1887)
: to take liquor in nips **:** TIPPLE

ni·pa \'nē-pə\ *noun* [probably from Italian, from Malay *nipah* nipa palm] (1779)
: thatch made of leaves of the nipa palm

nip and tuck \,nip-ən(d)-'tək\ *adjective or adverb* (1832)
: being so close that the lead or advantage shifts rapidly from one opponent to another

nipa palm *noun* (1882)
: a semiaquatic creeping palm (*Nipa fruticans*) found chiefly from India to Melanesia

nipa palm

nip·per \'ni-pər\ *noun* (1541)
1 : any of various devices (as pincers) for nipping — usually used in plural
2 a *chiefly British* **:** a boy employed as a helper (as of a carter or hawker) **b :** CHILD; *especially* **:** a small boy

nip·ping \'ni-piŋ\ *adjective* (1547)
: SHARP, CHILLING
— nip·ping·ly \-lē\ *adverb*

nip·ple \'ni pəl\ *noun* [earlier *neble, nible,* probably diminutive of *neb, nib*] (1530)
1 : the protuberance of a mammary gland upon which in the female the lactiferous ducts open and from which milk is drawn
2 a : an artificial teat through which a bottle-fed infant nurses **b :** a device with an orifice through which the discharge of a liquid can be regulated
3 a : a protuberance resembling or suggesting the nipple of a breast **b :** a small projection through which oil or grease is injected into machinery
4 : a pipe coupling consisting of a short piece of threaded tubing
— nip·pled \-pəld\ *adjective*

Nip·pon·ese \,ni-pə-'nēz, -'nēs\ *adjective* [*Nippon,* Japan] (1859)
: JAPANESE
— Nipponese *noun*

nip·py \'ni-pē\ *adjective* **nip·pi·er; -est** (1575)
1 : marked by a tendency to nip ⟨a *nippy* dog⟩
2 : brisk, quick, or nimble in movement **:** SNAPPY
3 : PUNGENT, SHARP
4 : CHILLY, CHILLING ⟨a *nippy* day⟩
— nip·pi·ly \'ni-pə-lē\ *adverb*
— nip·pi·ness \'ni-pē-nəs\ *noun*

nip–up \'nip-,əp\ *noun* (1938)
: a spring from a supine position to a standing position

nir·va·na \nir-'vä-nə, (,)nər-\ *noun, often capitalized* [Sanskrit *nirvāṇa,* literally, act of extinguishing, from *nis-* out + *vāti* it blows — more at WIND] (1801)
1 : the final beatitude that transcends suffering, karma, and samsara and is sought especially in Buddhism through the extinction of desire and individual consciousness
2 a : a place or state of oblivion to care, pain, or external reality; *also* **:** BLISS, HEAVEN **b :** a goal hoped for but apparently unattainable **:** DREAM
— nir·van·ic \-'vä-nik, -'va-\ *adjective*

Ni·san \'ni-sᵊn, nē-'sän\ *noun* [Hebrew *Nīsān*] (14th century)
: the 7th month of the civil year or the 1st month of the ecclesiastical year in the Jewish calendar — see MONTH table

ni·sei \nē-'sā, 'nē-,\ *noun, plural* **nisei** *often capitalized* [Japanese, literally, second generation, from *ni* second + *sei* generation] (1929)
: a son or daughter of Japanese immigrants who is born and educated in America and especially in the U.S.

ni·si \'nī-,sī\ *adjective* [Latin, unless, from *ne-* not + *si* if] (1836)
: taking effect at a specified time unless previously modified or avoided by cause shown, further proceedings, or a condition fulfilled ⟨a decree *nisi*⟩

Nis·sen hut \'ni-sᵊn-\ *noun* [Peter N. *Nissen* (died 1930) British mining engineer] (1932)
: a prefabricated shelter with a semicircular arching roof of corrugated iron sheeting and a concrete floor

ni·sus \'nī-səs\ *noun, plural* **ni·sus** \-səs, -,süs\ [Latin, from *niti* to lean, rely, strive; akin to Latin *nictare* to wink — more at CONNIVE] (1699)
: a mental or physical effort to attain an end **:** a perfective urge or endeavor

[1]nit \'nit\ *noun* [Middle English *nite,* from Old English *hnitu;* akin to Old High German *hniz* nit, Greek *konid-, konis*] (before 12th century)
: the egg of a louse or other parasitic insect; *also* **:** the insect itself when young

[2]nit *noun* (circa 1941)
chiefly British **:** NITWIT

nite *variant of* NIGHT

ni·ter \'nī-tər\ *noun* [Middle English *nitre* natron, from Middle French, from Latin *nitrum,* from Greek *nitron,* from Egyptian *ntry*] (1684)
1 : POTASSIUM NITRATE
2 *archaic* **:** CHILE SALTPETER

nit·ery *also* **nit·er·ie** \'nī-tə-rē\ *noun, plural* **ni·ter·ies** [*nite* + *-ery* (as in *eatery*); *niterie* from *nite* + French *-erie* -ery] (circa 1934)
: NIGHTCLUB

nit·id \'ni-təd\ *adjective* [Latin *nitidus* — more at NEAT] (circa 1656)
: BRIGHT, LUSTROUS

ni·ti·nol \'nī-tᵊn-,ȯl, -,ōl\ *noun* [*ni*ckel + *ti*tanium + *-nol* (from *N*aval *O*rdnance *L*aboratory, where it was created)] (1968)
: a nonmagnetic alloy of titanium and nickel that after being deformed returns to its original shape upon being reheated

nit·pick \'nit-,pik\ *verb* [back-formation from *nit-picking*] (1966)
intransitive verb
: to engage in nit-picking
transitive verb
: to criticize by nit-picking
— nitpick *noun*
— nit·pick·er *noun*
— nit·picky \-,pi-kē\ *adjective*

nit–pick·ing \'nit-,pi-kiŋ\ *noun* [[1]*nit*] (1956)
: minute and usually unjustified criticism

nitr- *or* **nitro-** *combining form* [*niter*]
1 : nitrogen ⟨*nitr*ide⟩
2 *usually nitro-* **:** containing the univalent group NO_2 composed of one nitrogen and two oxygen atoms ⟨*nitro*benzene⟩

ni·trate \'nī-,trāt, -trət\ *noun* [French, from *nitrique*] (1794)
1 : a salt or ester of nitric acid
2 : sodium nitrate or potassium nitrate used as a fertilizer

nitrate of soda (1843)
: sodium nitrate used as a fertilizer

ni·tra·tion \nī-'trā-shən\ *noun* (1887)
: the process of adding a nitro group to an organic compound
— ni·trate \'nī-,trāt\ *transitive verb*
— ni·tra·tor \-,trā-tər\ *noun*

ni·tre *chiefly British variant of* NITER

ni·tric acid \'nī-trik-\ *noun* [French *nitrique,* from *nitre* niter, from Middle French] (1794)
: a corrosive liquid inorganic acid HNO_3 used especially as an oxidizing agent, in nitrations, and in making organic compounds (as fertilizers, explosives, and dyes)

nitric oxide *noun* (1807)
: a colorless poisonous gas NO formed by oxidation of nitrogen or ammonia

[1]ni·tride \'nī-,trīd\ *noun* [International Scientific Vocabulary] (1850)
: a binary compound of nitrogen with a more electropositive element

[2]nitride *transitive verb* **ni·trid·ed; ni·trid·ing** (1928)
: to case-harden (as steel) by causing the surface to absorb nitrogen

ni·tri·fi·ca·tion \,nī-trə-fə-'kā-shən\ *noun* (1827)
: the oxidation (as by bacteria) of ammonium salts to nitrites and the further oxidation of nitrites to nitrates

ni·tri·fi·er \'nī-trə-,fī(-ə)r\ *noun* (1903)
: any of the nitrifying bacteria

ni·tri·fy·ing \'nī-trə-,fī-iŋ\ *adjective* [French *nitrifier* to convert into nitrite, from *nitr-*] (1827)
: active in or relating to nitrification ⟨*nitrifying* organisms⟩ ⟨*nitrifying* activities⟩

nitrifying bacteria *noun plural* (1925)
: bacteria of a family (Nitrobacteraceae) comprising gram-negative bacteria commonly found in the soil and obtaining energy through the process of nitrification

ni·trile \'nī-trəl, -,trīl\ *noun* [International Scientific Vocabulary *nitr-* + *-il, -ile* (from Latin *-ilis* [1]-ile)] (1848)
: an organic cyanide containing the group CN which on hydrolysis yields an acid with elimination of ammonia

ni·trite \'nī-,trīt\ *noun* (1800)
: a salt or ester of nitrous acid

[1]ni·tro \'nī-(,)trō\ *adjective* [*nitr-*] (1892)
: containing or being the univalent group NO_2 united through nitrogen

[2]nitro *noun, plural* **nitros** (1903)
: any of various nitrated products; *specifically* **:** NITROGLYCERIN

ni·tro·ben·zene \,nī-trō-'ben-,zēn, -ben-'\ *noun* [International Scientific Vocabulary] (1868)
: a poisonous yellow insoluble oil $C_6H_5NO_2$ with an almond odor that is used especially as a solvent, as a mild oxidizing agent, and in making aniline

ni·tro·cel·lu·lose \-'sel-yə-,lōs, -,lōz\ *noun* [International Scientific Vocabulary] (1882)
: any of several nitric-acid esters of cellulose used especially for making explosives, plastics, and varnishes

ni·tro·fu·ran \,nī-trō-'fyu̇r-,an, -fyu̇-'ran\ *noun* (1930)
: any of several nitro derivatives of furan used as bacteria-inhibiting agents

ni·tro·gen \'nī-trə-jən\ *noun, often attributive* [French *nitrogène,* from *nitre* niter + *-gène* -gen] (1794)
: a colorless tasteless odorless element that as a diatomic gas is relatively inert and constitutes 78 percent of the atmosphere by volume and that occurs as a constituent of all living tissues — see ELEMENT table
— ni·trog·e·nous \nī-'trä-jə-nəs\ *adjective*

ni·tro·ge·nase \nī-'trä-jə-,nās, 'nī-trə-jə-, -,nāz\ *noun* (1934)
: an enzyme of various nitrogen-fixing bacteria that catalyzes the reduction of molecular nitrogen to ammonia

\ə\ **abut** \ᵊ\ kitten \ər\ **further** \a\ **ash** \ā\ **ace** \ä\ mop, mar \au̇\ **out** \ch\ **chin** \e\ **bet** \ē\ **easy** \g\ **go** \i\ **hit** \ī\ **ice** \j\ **job** \ŋ\ **sing** \ō\ **go** \ȯ\ **law** \ȯi\ **boy** \th\ **thin** \t͟h\ **the** \ü\ **loot** \u̇\ **foot** \y\ **yet** \zh\ **vision** *see also* **Guide to Pronunciation**

nitrogen balance *noun* (1944)
: the difference between nitrogen intake and nitrogen loss in the body or the soil

nitrogen cycle *noun* (1908)
: a continuous series of natural processes by which nitrogen passes successively from air to soil to organisms and back to air or soil involving principally nitrogen fixation, nitrification, decay, and denitrification

nitrogen dioxide *noun* (1885)
: a toxic reddish brown gas NO_2 that is a strong oxidizing agent, is produced by combustion (as of fossil fuels), and is an atmospheric pollutant

nitrogen fixation *noun* (1895)
: the metabolic assimilation of atmospheric nitrogen into ammonia by soil microorganisms and especially rhizobia

nitrogen–fixer *noun* (1912)
: any of various soil microorganisms that are involved in nitrogen fixation

nitrogen–fixing *adjective* (1899)
: capable of nitrogen fixation ⟨*nitrogen-fixing* bacteria⟩

nitrogen mustard *noun* (1943)
: any of various toxic blistering compounds analogous to mustard gas but containing nitrogen instead of sulfur

nitrogen narcosis *noun* (1937)
: a state of euphoria and exhilaration that occurs when nitrogen in normal air enters the bloodstream at approximately seven times atmospheric pressure (as in deep-water diving) — called also *rapture of the deep*

nitrogen oxide *noun* (circa 1934)
: any of several oxides of nitrogen most of which are produced in combustion and are considered to be atmospheric pollutants: as **a** **:** NITRIC OXIDE **b :** NITROGEN DIOXIDE **c :** NITROUS OXIDE

nitrogen tetroxide *noun* (1885)
: a colorless toxic gas N_2O_4 that is a dimer of nitrogen dioxide and that in liquid form is used as an oxidizer in rocket engines

ni•tro•glyc•er•in *or* **ni•tro•glyc•er•ine** \ˌnī-trə-ˈglis-rən, -ˈgli-sə-\ *noun* [International Scientific Vocabulary] (1857)
: a heavy oily explosive poisonous liquid $C_3H_5N_3O_9$ used chiefly in making dynamites and in medicine as a vasodilator

ni•tro•meth•ane \ˌnī-trō-ˈme-ˌthān, *British usually* -ˈmē-\ *noun* (1872)
: a liquid nitroparaffin CH_3NO_2 that is used as an industrial solvent, in chemical synthesis, and as a fuel for rockets and high-performance engines

ni•tro•par•af•fin \-ˈpar-ə-fən\ *noun* [International Scientific Vocabulary] (1892)
: any of various nitro derivatives of alkanes

nitros- *or* **nitroso-** *combining form* [New Latin *nitrosus* nitrous]
: containing the group NO ⟨*nitros*amine⟩

ni•tro•sa•mine \nī-ˈtrō-sə-ˌmēn\ *noun* (1878)
: any of various organic compounds which are characterized by the grouping NNO and some of which are powerful carcinogens

ni•trous \ˈnī-trəs\ *adjective* [New Latin *nitrosus,* from Latin, full of natron, from *nitrum* natron — more at NITER] (1601)
archaic **:** of, relating to, or containing niter

nitrous acid *noun* (1676)
: an unstable acid HNO_2 known only in solution or in the form of its salts

nitrous oxide *noun* (1800)
: a colorless gas N_2O that when inhaled produces loss of sensibility to pain preceded by exhilaration and sometimes laughter and is used especially as an anesthetic in dentistry and that is an atmospheric pollutant produced by combustion and a suspected contributor to greenhouse warming — called also *laughing gas*

nit•ty–grit•ty \ˈni-tē-ˌgri-tē, ˌni-tē-ˈ\ *noun* [origin unknown] (1963)
: what is essential and basic **:** specific practical details ⟨get down to the *nitty-gritty* of the problem⟩
— nitty–gritty *adjective*

nit•wit \ˈnit-ˌwit\ *noun* [probably from German dialect *nit* not + English *wit*] (circa 1922)
: a scatterbrained or stupid person

[1]**nix** \ˈniks\ *noun* [German *nichts* nothing] (1789)
: NOTHING

[2]**nix** *transitive verb* (circa 1934)
: VETO, REJECT ⟨the court *nixed* the merger⟩

[3]**nix** *adverb* (circa 1909)
: NO — used to express disagreement or the withholding of permission; often used with *on* ⟨they said *nix* on our plan⟩

[4]**nix** *noun* [German, from Old High German *nihhus;* akin to Old English *nicor* water monster and perhaps to Greek *nizein* to wash] (1833)
: a water sprite of Germanic folklore

[1]**nix•ie** \ˈnik-sē\ *noun* [German *Nixe* female nix, from Old High German *nichessa,* feminine of *nihhus* nix] (1816)
: [4]NIX

[2]**nix•ie** *also* **nixy** \ˈnik-sē\ *noun, plural* **nix•ies** [[1]*nix* + *-ie*] (circa 1890)
: a piece of mail that is undeliverable because illegibly or incorrectly addressed

Nix•ie \ˈnik-sē\ *trademark*
— used for an electronic indicator tube

ni•zam \ni-ˈzäm, ˈnī-ˌzam, nī-ˈ\ *noun* [Hindi *nizām* order, governor, from Arabic *niẓām*] (1768)
: one of a line of sovereigns of Hyderabad, India, reigning from 1713 to 1950
— ni•zam•ate \ni-ˈzä-ˌmāt, nī-ˈza-\ *noun*

[1]**no** \ˈnō\ *adverb* [Middle English, from Old English *nā,* from *ne* not + *ā* always; akin to Old Norse & Old High German *ne* not, Latin *ne-,* Greek *nē-* — more at AYE] (before 12th century)
1 a *chiefly Scottish* **:** NOT **b** — used as a function word to express the negative of an alternative choice or possibility ⟨shall we go out to dinner or *no*⟩
2 : in no respect or degree — used in comparisons
3 : not so — used to express negation, dissent, denial, or refusal ⟨*no,* I'm not going⟩
4 — used with a following adjective to imply a meaning expressed by the opposite positive statement ⟨in *no* uncertain terms⟩
5 — used as a function word to emphasize a following negative or to introduce a more emphatic, explicit, or comprehensive statement
6 — used as an interjection to express surprise, doubt, or incredulity
7 — used in combination with a verb to form a compound adjective ⟨*no*-bake pie⟩

[2]**no** *adjective* (12th century)
1 a : not any ⟨*no* parking⟩ ⟨*no* disputing the decision⟩ **b :** hardly any **:** very little ⟨finished in *no* time⟩
2 : not a **:** quite other than a ⟨he's *no* expert⟩
3 — used in combination with a noun to form a compound adjective ⟨a *no*-nonsense realist⟩

[3]**no** \ˈnō\ *noun, plural* **noes** *or* **nos** \ˈnōz\ (1588)
1 : an act or instance of refusing or denying by the use of the word *no* **:** DENIAL
2 a : a negative vote or decision **b** *plural* **:** persons voting in the negative

No *or* **Noh** \ˈnō\ *noun, plural* **No** *or* **Noh** [Japanese *nō,* literally, talent] (1871)
: classic Japanese dance-drama having a heroic theme, a chorus, and highly stylized action, costuming, and scenery

no–account *adjective* (1845)
: of no account **:** TRIFLING ⟨her *no-account* relatives⟩

No•a•chi•an \nō-ˈā-kē-ən\ *adjective* [Hebrew *Nōaḥ* Noah] (1678)
1 : of or relating to the patriarch Noah or his time
2 : ANCIENT, ANTIQUATED

No•ah \ˈnō-ə\ *noun* [Hebrew *Nōaḥ*]
: an Old Testament patriarch who built the ark in which he, his family, and living creatures of every kind survived the Flood

[1]**nob** \ˈnäb\ *noun* [probably alteration of *knob*] (circa 1700)
1 : HEAD 1
2 : a jack of the same suit as the starter in cribbage that scores one point for the holder — usually used in the phrases *his nob* or *his nobs*

[2]**nob** *noun* [perhaps from [1]*nob*] (1703)
chiefly British **:** one in a superior position in life

nob•ble \ˈnä-bəl\ *transitive verb* **nob•bled; nob•bling** \-b(ə-)liŋ\ [perhaps irregular frequentative of *nab*] (1847)
1 *British* **:** to incapacitate (a racehorse) especially by drugging
2 *slang British* **a :** to win over to one's side **b :** STEAL **c :** SWINDLE, CHEAT **d :** to get hold of **:** CATCH
— nob•bler \-b(ə-)lər\ *noun*

nob•by \ˈnä-bē\ *adjective* **nob•bi•er; -est** (1788)
: CHIC, SMART

No•bel•ist \nō-ˈbe-list\ *noun* (1938)
: a winner of a Nobel prize

no•bel•i•um \nō-ˈbe-lē-əm\ *noun* [New Latin, from Alfred B. *Nobel*] (1957)
: a radioactive element produced artificially — see ELEMENT table

No•bel prize \nō-ˈbel-, ˈnō-ˌbel-\ *noun* (1900)
: any of various annual prizes (as in peace, literature, medicine) established by the will of Alfred Nobel for the encouragement of persons who work for the interests of humanity

no•bil•i•ty \nō-ˈbi-lə-tē\ *noun* [Middle English *nobilite,* from Middle French *nobilité,* from Latin *nobilitat-, nobilitas,* from *nobilis*] (14th century)
1 : the quality or state of being noble in character, quality, or rank
2 : the body of persons forming the noble class in a country or state **:** ARISTOCRACY

[1]**no•ble** \ˈnō-bəl\ *adjective* **no•bler** \-b(ə-)lər\; **no•blest** \-b(ə-)ləst\ [Middle English, from Old French, from Latin *nobilis* well-known, noble, from *noscere* to come to know — more at KNOW] (13th century)
1 a : possessing outstanding qualities **:** ILLUSTRIOUS **b :** FAMOUS, NOTABLE ⟨*noble* deeds⟩
2 : of high birth or exalted rank **:** ARISTOCRATIC
3 a : possessing very high or excellent qualities or properties ⟨*noble* wine⟩ **b :** very good or excellent
4 : grand or impressive especially in appearance ⟨*noble* edifice⟩
5 : possessing, characterized by, or arising from superiority of mind or character or of ideals or morals **:** LOFTY ⟨a *noble* ambition⟩
6 : chemically inert or inactive especially toward oxygen ⟨a *noble* metal such as platinum⟩ — compare BASE 6a
synonym see MORAL
— no•ble•ness \-bəl-nəs\ *noun*
— no•bly \-blē *also* -bə-lē\ *adverb*

[2]**noble** *noun* (14th century)
1 : a person of noble rank or birth
2 : an old English gold coin equivalent to 6*s* 8*d*

noble gas *noun* (1902)
: any of a group of rare gases that include helium, neon, argon, krypton, xenon, and sometimes radon and that exhibit great stability and extremely low reaction rates — called also *inert gas*

no•ble•man \ˈnō-bəl-mən\ *noun* (14th century)
: a man of noble rank **:** PEER

noble savage *noun* (1672)
: a mythic conception of people belonging to non-European cultures as having innate natural simplicity and virtue uncorrupted by European civilization; *also* **:** a person exemplifying this conception

no·blesse \nō-ˈbles\ *noun* [Middle English, from Old French *noblesce,* from *noble*] (13th century)
1 : noble birth or condition
2 : the members especially of the French nobility

no·blesse oblige \nō-ˈbles-ə-ˈblēzh\ *noun* [French, literally, nobility obligates] (1837)
: the obligation of honorable, generous, and responsible behavior associated with high rank or birth

no·ble·wom·an \ˈnō-bəl-ˌwu̇-mən\ *noun* (13th century)
: a woman of noble rank **:** PEERESS

¹no·body \ˈnō-bə-dē, -ˌbä-dē\ *pronoun* (14th century)
: no person **:** not anybody ■

²nobody *noun, plural* **no·bod·ies** (1581)
: a person of no influence or consequence

no–brain·er \ˈnō-ˈbrā-nər\ *noun* (1973)
: something that requires a minimum of thought

no·cent \ˈnō-sᵊnt\ *adjective* [Middle English, from Latin *nocent-, nocens,* from present participle of *nocēre* to harm, hurt — more at NOXIOUS] (15th century)
: HARMFUL

no·ci·cep·tive \ˌnō-si-ˈsep-tiv\ *adjective* [Latin *nocēre* + English *-i-* + *receptive*] (1904)
1 *of a stimulus* **:** PAINFUL, INJURIOUS
2 : of, induced by, or responding to a nociceptive stimulus — used especially of receptors or protective reflexes

¹nock \ˈnäk\ *noun* [Middle English *nocke* notched tip on the end of a bow; akin to Middle Dutch *nocke* summit] (14th century)
1 : one of the notches cut in either of two tips of horn fastened on the ends of a bow or in the bow itself for holding the string
2 a : the part of an arrow having a notch for the bowstring **b :** the notch itself

²nock *transitive verb* (14th century)
1 : to make a nock in (a bow or arrow)
2 : to fit (an arrow) against the bowstring

no contest *noun* (1952)
: NOLO CONTENDERE

noct·am·bu·list \näk-ˈtam-byə-list\ *noun* [Latin *noct-, nox* night + *-ambulist* (as in *somnambulist*) — more at NIGHT] (circa 1731)
: SLEEPWALKER

noc·ti·lu·cent cloud \ˌnäk-tə-ˈlü-sᵊnt-\ *noun* [*noctilucent* ultimately from Latin *noct-* + *lucent-, lucens* lucent] (1910)
: a luminous thin usually colored cloud seen especially at twilight at a height of about 50 miles (80 kilometers)

noc·tu·id \ˈnäk-chə-wəd, ˈnäk-tə-\ *noun* [New Latin *Noctuidae,* from *Noctua,* genus of moths, from Latin, the little owl (*Athene noctua*); akin to Latin *nox* night] (1894)
: any of a large family (Noctuidae) of medium-sized often dull-colored moths with larvae (as cutworms and armyworms) that are often destructive agricultural pests
— noctuid *adjective*

noc·turn \ˈnäk-ˌtərn\ *noun* [Middle English *nocturne,* from Middle French, from Medieval Latin *nocturna,* from Latin, feminine of *nocturnus*] (14th century)
: a principal division of the office of matins

noc·tur·nal \näk-ˈtər-nᵊl\ *adjective* [Middle English, from Middle French or Late Latin; Middle French, from Late Latin *nocturnalis,* from Latin *nocturnus* of night, nocturnal, from *noct-, nox* night] (15th century)
1 : of, relating to, or occurring in the night ⟨a *nocturnal* journey⟩
2 : active at night ⟨a *nocturnal* predator⟩
— noc·tur·nal·ly \-nᵊl-ē\ *adverb*

noc·turne \ˈnäk-ˌtərn\ *noun* [French, adjective, nocturnal, from Latin *nocturnus*] (1862)
: a work of art dealing with evening or night; *especially* **:** a dreamy pensive composition for the piano — compare AUBADE 3

noc·u·ous \ˈnä-kyə-wəs\ *adjective* [Latin *nocuus,* from *nocēre* to harm — more at NOXIOUS] (1635)
: HARMFUL
— noc·u·ous·ly *adverb*

¹nod \ˈnäd\ *verb* **nod·ded; nod·ding** [Middle English *nodden;* perhaps akin to Old High German *hnotōn* to shake] (14th century)
intransitive verb
1 : to make a quick downward motion of the head whether deliberately (as in expressing assent, salutation, or command) or involuntarily (as from drowsiness)
2 : to incline or sway from the vertical as though ready to fall
3 : to bend or sway the upper part gently downward or forward **:** bob gently
4 : to make a slip or error in a moment of abstraction
transitive verb
1 : to incline (as the head) downward or forward
2 : to bring, invite, or send by a nod ⟨*nodded* them into the room⟩
3 : to signify by a nod ⟨*nodded* their approval⟩
— nod·der *noun*

²nod *noun* (circa 1541)
1 : the act or an instance of nodding ⟨gave a *nod* of greeting⟩
2 : an indication especially of approval or recognition ⟨received the party's *nod* as candidate for governor⟩

nod·al \ˈnō-dᵊl\ *adjective* (1831)
: being, relating to, or located at or near a node
— no·dal·i·ty \nō-ˈda-lə-tē\ *noun*
— nod·al·ly \ˈnō-dᵊl-ē\ *adverb*

nodding *adjective* (1590)
1 : bending downward or forward **:** PENDULOUS, DROOPING ⟨a plant with *nodding* flowers⟩
2 : SLIGHT, SUPERFICIAL ⟨a *nodding* acquaintance⟩

nod·dle \ˈnä-dᵊl\ *noun* [Middle English *nodle* back of the head or neck] (1579)
: HEAD, PATE

nod·dy \ˈnä-dē\ *noun, plural* **noddies** [probably short for obsolete *noddypoll,* alteration of *hoddypoll* fumbling inept person] (circa 1530)
1 : a stupid person
2 : any of several stout-bodied terns (especially genus *Anous*) of warm seas

node \ˈnōd\ *noun* [Middle English, from Latin *nodus* knot, node; akin to Middle Irish *naidm* bond] (15th century)
1 a : a pathological swelling or enlargement (as of a rheumatic joint) **b :** a discrete mass of one kind of tissue enclosed in tissue of a different kind
2 : an entangling complication (as in a drama) **:** PREDICAMENT
3 : either of the two points where the orbit of a planet or comet intersects the ecliptic; *also* **:** either of the points at which the orbit of an earth satellite crosses the plane of the equator
4 a : a point, line, or surface of a vibrating body or system that is free or relatively free from vibratory motion **b :** a point at which a wave has an amplitude of zero
5 a : a point at which subsidiary parts originate or center **b :** a point on a stem at which a leaf or leaves are inserted **c :** a point at which a curve intersects itself in such a manner that the branches have different tangents **d :** VERTEX 1b

node of Ran·vier \-ˈräⁿ-vē-ˌā\ [Louis A. *Ranvier* (died 1922) French histologist] (circa 1885)
: a constriction in the myelin sheath of a myelinated nerve fiber

nod off *intransitive verb* (1914)
: to fall asleep

no·dose \ˈnō-ˌdōs\ *adjective* [Latin *nodosus,* from *nodus*] (circa 1721)
: having numerous or conspicuous protuberances
— no·dos·i·ty \nō-ˈdä-sə-tē\ *noun*

nod·u·lar \ˈnä-jə-lər\ *adjective* (1794)
: of, relating to, characterized by, or occurring in the form of nodules ⟨*nodular* lesions⟩

nod·u·la·tion \ˌnä-jə-ˈlā-shən\ *noun* (1872)
1 : the process of forming nodules and especially root nodules containing symbiotic bacteria
2 : NODULE

nod·ule \ˈnä-(ˌ)jü(ə)l\ *noun* [Middle English, from Latin *nodulus,* diminutive of *nodus*] (15th century)
: a small mass of rounded or irregular shape: as **a :** a small rounded lump of a mineral or mineral aggregate **b :** a swelling on a leguminous root that contains symbiotic bacteria **c :** a small abnormal knobby bodily protuberance (as a tumorous growth or a calcification near an arthritic joint)

no·dus \ˈnō-dəs\ *noun, plural* **no·di** \ˈnō-ˌdī, -ˌdē\ [Latin, knot, node] (circa 1738)
: COMPLICATION, DIFFICULTY

no·el \nō-ˈel\ *noun* [French *noël* Christmas, carol, from Latin *natalis* birthday, from *natalis* natal] (1811)
1 : a Christmas carol
2 *capitalized* **:** CHRISTMAS

noes *plural of* NO

no·et·ic \nō-ˈe-tik\ *adjective* [Greek *noētikos* intellectual, from *noein* to think, from *nous* mind] (1653)
: of, relating to, or based on the intellect

no–fault *adjective* (1967)
1 : of, relating to, or being a motor vehicle insurance plan under which someone involved in an accident is compensated usually up to a stipulated limit for actual losses (as for property damage, medical bills, and lost wages) by that person's own insurance company regardless of who is responsible for the accident
2 : of, relating to, or being a divorce law under which neither party is held responsible for the breakup of the marriage
3 : characterized by the absence of a prevailing sense of individual responsibility (as for behavior) ⟨a *no-fault* society⟩

no–frills \ˈnō-ˈfrilz\ *adjective* (1960)
: offering or providing only the essentials **:** not fancy, elaborate, or luxurious

nog \ˈnäg\ *noun* [origin unknown] (1693)
1 : a strong ale formerly brewed in Norfolk, England
2 [by shortening] **:** EGGNOG

nog·gin \ˈnä-gən\ *noun* [origin unknown] (1630)
1 : a small mug or cup
2 : a small quantity (as a gill) of drink
3 : a person's head

☐ USAGE
nobody The indefinite pronouns *nobody* and *no one* share with other indefinite pronouns the characteristic of taking a singular verb and, more often than not, a plural pronoun in reference. The use of the plural pronoun—*they, their, them*—has traditionally been disapproved by those who insist that agreement in number is more important than avoidance of gender-specific third-person singular pronouns. But the use of the plural pronouns has long been established ⟨*nobody* was in their right place —Jane Austen (1814)⟩ and increasing resistance to generic *he* may be easing their acceptance. Usage is still mixed ⟨*nobody* attains reality for my mother until he eats —Flannery O'Connor⟩ ⟨*nobody* naturally likes a mind quicker than their own —F. Scott Fitzgerald⟩.

nog·ging \'nä-gən, -giŋ\ *noun* [*nog* wooden block the size of a brick] (1825)
: rough brick masonry used to fill in the open spaces of a wooden frame

[1]**no–good** \'nō-'gu̇d\ *adjective* (1908)
: having no worth, virtue, use, or chance of success

[2]**no–good** \'nō-ˌgu̇d\ *noun* (1924)
: a no-good person or thing

Noh *variant of* NO

no–hit *adjective* (1916)
: of, relating to, or being a baseball game or a part of a game in which a pitcher allows the opposition no base hits

no–hit·ter \ˌnō-'hi-tər\ *noun* (1947)
: a no-hit game in baseball

no–holds–barred \ˌnō-ˌhōl(d)z-'bärd\ *adjective* (1942)
: free of restrictions or hampering conventions ⟨a *no-holds-barred* contest⟩

no–hop·er \'nō-'hō-pər\ *noun* (circa 1943) *chiefly British* **:** one that has no chance of success

no·how \'nō-ˌhau̇\ *adverb* (1775)
1 : in no manner or way **:** not at all ⟨was *no-how* equal to the task⟩
2 *dialect* **:** ANYHOW

noil \'nȯi(ə)l\ *noun* [origin unknown] (circa 1624)
: short fiber removed during the combing of a textile fiber and often separately spun into yarn

noir \'nwär\ *noun, often attributive* [short for *film noir*] (1980)
1 : crime fiction featuring hard-boiled cynical characters and bleak sleazy settings
2 : FILM NOIR
— noir·ish \-ish\ *adjective*

[1]**noise** \'nȯiz\ *noun* [Middle English, from Old French, strife, quarrel, noise, from Latin *nausea* nausea] (13th century)
1 : loud, confused, or senseless shouting or outcry
2 a : SOUND; *especially* **:** one that lacks agreeable musical quality or is noticeably unpleasant **b :** any sound that is undesired or interferes with one's hearing of something **c :** an unwanted signal or a disturbance (as static or a variation of voltage) in an electronic device or instrument (as radio or television); *broadly* **:** a disturbance interfering with the operation of a usually mechanical device or system **d :** electromagnetic radiation (as light or radio waves) that is composed of several frequencies and that involves random changes in frequency or amplitude **e :** irrelevant or meaningless data or output occurring along with desired information
3 : common talk **:** RUMOR; *especially* **:** SLANDER
4 : something that attracts attention ⟨the play . . . will make little *noise* in the world —Brendan Gill⟩
5 : something spoken or uttered
— noise·less \-ləs\ *adjective*
— noise·less·ly *adverb*

[2]**noise** *verb* **noised; nois·ing** (14th century)
intransitive verb
1 : to talk much or loudly
2 : to make a noise
transitive verb
: to spread by rumor or report — usually used with *about* or *abroad* ⟨the scandal was quickly *noised* about⟩

noise·mak·er \'nȯiz-ˌmā-kər\ *noun* (1574)
: one that makes noise; *especially* **:** a device (as a horn or rattle) used to make noise at parties
— noise·mak·ing \-kiŋ\ *noun or adjective*

noise pollution *noun* (1966)
: annoying or harmful noise (as of automobiles or jet airplanes) in an environment

noi·sette \nwə-'zet, nwä-\ *noun* [French, diminutive of *nois* choice cut of meat, literally, nut, from Old French, from Latin *nux* — more at NUT] (1891)
: a small piece of lean meat

noi·some \'nȯi-səm\ *adjective* [Middle English *noysome,* from *noy* annoyance, from Old French *enui, anoi* — more at ENNUI] (14th century)
1 : NOXIOUS, HARMFUL
2 a : offensive to the senses and especially to the sense of smell **b :** highly obnoxious or objectionable
synonym see MALODOROUS
— noi·some·ly *adverb*
— noi·some·ness *noun*

noisy \'nȯi-zē\ *adjective* **nois·i·er; -est** (1693)
1 : making noise
2 : full of or characterized by noise or clamor
3 : noticeably showy, gaudy, or bright **:** CONSPICUOUS
— nois·i·ly \'nȯi-zə-lē\ *adverb*
— nois·i·ness \-zē-nəs\ *noun*

no·li me tan·ge·re \ˌnō-lē-(ˌ)mē-'tan-jə-rē, -ˌlī-, -(ˌ)mā-\ *noun* [Latin, do not touch me; from Jesus' words to Mary Magdalene (John 20:17)] (1591)
: a warning against touching or interference

nol·le pro·se·qui \ˌnä-lē-'prä-sə-ˌkwī\ *noun* [Latin, to be unwilling to pursue] (1681)
: an entry on the record of a legal action denoting that the prosecutor or plaintiff will proceed no further in an action or suit either as a whole or as to some count or as to one or more of several defendants

no·lo \'nō-(ˌ)lō\ *noun* (1914)
: NOLO CONTENDERE

no–load \'nō-'lōd\ *adjective* (1963)
: charging no sales commission ⟨a *no-load* mutual fund⟩
— no–load \'nō-ˌlōd\ *noun*

no·lo con·ten·de·re \'nō-(ˌ)lō-kən-'ten-də-rē\ *noun* [Latin, I do not wish to contend] (1872)
: a plea in a criminal prosecution that without admitting guilt subjects the defendant to conviction but does not preclude denying the truth of the charges in a collateral proceeding

nol–pros \'näl-'präs\ *transitive verb* **nol–prossed; nol–pros·sing** [*nol*le *pros*equi] (circa 1878)
: to discontinue by entering a nolle prosequi

no·ma \'nō-mə\ *noun* [New Latin, from Greek *nomē,* from *nemein* to spread (of an ulcer), literally, to graze, pasture — more at NIMBLE] (1834)
: a spreading invasive gangrene chiefly of the lining of the cheek and lips that is usually fatal and occurs most often in persons severely debilitated by disease or profound nutritional deficiency

no·mad \'nō-ˌmad, *British also* 'nä-\ *noun* [Latin *nomad-, nomas* member of a wandering pastoral people, from Greek, from *nemein*] (1579)
1 : a member of a people who have no fixed residence but move from place to place usually seasonally and within a well-defined territory
2 : an individual who roams about aimlessly
— nomad *adjective*
— no·mad·ism \'nō-ˌma-ˌdi-zəm\ *noun*

no·mad·ic \nō-'ma-dik\ *adjective* (circa 1818)
1 : of, relating to, or characteristic of nomads ⟨a *nomadic* tribe⟩
2 : roaming about from place to place aimlessly, frequently, or without a fixed pattern of movement

no–man's–land \'nō-ˌmanz-ˌland\ *noun* (14th century)
1 a : an area of unowned, unclaimed, or uninhabited land **b :** an unoccupied area between opposing armies
2 : an anomalous, ambiguous, or indefinite area especially of operation, application, or jurisdiction ⟨the *no-man's-land* between art and science⟩

nom·bril \'näm-brəl\ *noun* [Middle French, literally, navel, ultimately from Latin *umbilicus*] (1562)
: the center point of the lower half of an armorial escutcheon

nom de guerre \ˌnäm-di-'ger\ *noun, plural* **noms de guerre** \ˌnäm(z)-di-\ [French, literally, war name] (1679)
: PSEUDONYM

nom de plume \-'plüm\ *noun, plural* **noms de plume** \ˌnäm(z)-di-\ [French, pen name; probably coined in English] (1823)
: PSEUDONYM, PEN NAME

nome \'nōm\ *noun* [Greek *nomos* pasture, district — more at NIMBLE] (circa 1727)
: a province of ancient Egypt

no·men \'nō-mən\ *noun, plural* **no·mi·na** \'nä-mə-nə, 'nō-\ [Latin *nomin-, nomen* name — more at NAME] (circa 1890)
: the second of the three usual names of an ancient Roman male

no·men·cla·tor \'nō-mən-ˌklā-tər\ *noun* [Latin, slave whose duty was to announce the names of persons met during a political campaign, from *nomen* + *calare* to call — more at LOW] (1585)
1 : a book containing collections or lists of words
2 *archaic* **:** one who announces the names of guests or of persons generally
3 : one who gives names to or invents names for things

no·men·cla·to·ri·al \ˌnō-mən-klə-'tōr-ē-əl, -'tȯr-\ *adjective* (1885)
: relating to or connected with nomenclature

no·men·cla·ture \'nō-mən-ˌklā-chər *also* nō-'men-klə-ˌchu̇r, -'meŋ-, -chər, -ˌtyu̇r, -ˌtu̇r\ *noun* [Latin *nomenclatura* assigning of names, from *nomen* + *calatus,* past participle of *calare*] (1610)
1 : NAME, DESIGNATION
2 : the act or process or an instance of naming
3 a : a system or set of terms or symbols especially in a particular science, discipline, or art **b :** an international system of standardized New Latin names used in biology for kinds and groups of kinds of animals and plants
— no·men·cla·tur·al \ˌnō-mən-'klāch-rəl, -'klā-chə-\ *adjective*

no·men con·ser·van·dum \'nō-mən-ˌkän(t)-sər-'van-dəm\ *noun, plural* **no·mi·na con·ser·van·da** \'nä-mə-nə-ˌkän(t)-sər-'van-də, 'nō-\ [New Latin, name to be kept] (circa 1925)
: a biological taxonomic name that is preserved by special sanction in exception to the usual rules

nomen du·bi·um \-'dü-bē-əm, -'dyü-\ *noun, plural* **nomina du·bia** \-bē-ə\ [New Latin, doubtful name] (1937)
: a taxonomic name that cannot be assigned with certainty to any taxonomic group because the description is insufficient for identification and the original specimens no longer exist

nomen nu·dum \-'nü-dəm, -'nyü-\ *noun, plural* **nomina nu·da** \-də\ [New Latin, bare name] (1900)
: a proposed taxonomic name that is invalid because the group designated is not described or illustrated sufficiently for recognition, that has no nomenclatural status, and that consequently can be used as though never previously proposed

[1]**nom·i·nal** \'nä-mə-nᵊl, 'näm-nəl\ *adjective* [Middle English *nominalle,* from Medieval Latin *nominalis,* from Latin, of a name, from *nomin-, nomen* name — more at NAME] (15th century)
1 : of, relating to, or being a noun or a word or expression taking a noun construction
2 a : of, relating to, or constituting a name **b :** bearing the name of a person
3 a : existing or being something in name or form only ⟨*nominal* head of his party⟩ **b :** of, being, or relating to a designated or theoretical

size that may vary from the actual **:** APPROXIMATE **c :** TRIFLING, INSIGNIFICANT
4 *of a rate of interest* **a :** equal to the annual rate of simple interest that would obtain if interest were not compounded when in fact it is compounded and paid for periods of less than a year **b :** equal to the percentage by which a repaid loan exceeds the principal borrowed with no adjustment made for inflation
5 : being according to plan **:** SATISFACTORY ⟨everything was *nominal* during the spacecraft launch⟩
— nom·i·nal·ly *adverb*

²nominal *noun* (1904)
: a word or word group functioning as a noun

nom·i·nal·ism \'nä-mə-nᵊl-ˌi-zəm, 'näm-nə-ˌli-zəm\ *noun* (1844)
1 : a theory that there are no universal essences in reality and that the mind can frame no single concept or image corresponding to any universal or general term
2 : the theory that only individuals and no abstract entities (as essences, classes, or propositions) exist — compare ESSENTIALISM, REALISM
— nom·i·nal·ist \-ist, -list\ *noun*
— nominalist *or* **nom·i·nal·is·tic** \ˌnä-mə-nᵊl-'is-tik, ˌnäm-nə-'lis-\ *adjective*

nominal value *noun* (circa 1901)
: PAR 1b

nominal wages *noun plural* (1898)
: wages measured in money as distinct from actual purchasing power

nom·i·nate \'nä-mə-ˌnāt\ *transitive verb* **-nat·ed; -nat·ing** [Latin *nominatus,* past participle of *nominare,* from *nomin-, nomen* name — more at NAME] (1545)
1 : DESIGNATE, NAME
2 a : to appoint or propose for appointment to an office or place **b :** to propose as a candidate for election to office **c :** to propose for an honor ⟨*nominate* her for player of the year⟩
3 : to enter (a horse) in a race
— nom·i·na·tor \-ˌnā-tər\ *noun*

nom·i·na·tion \ˌnä-mə-'nā-shən\ *noun* (15th century)
1 : the act, process, or an instance of nominating
2 : the state of being nominated

nom·i·na·tive \'näm-nə-tiv, 'nä-mə-; *2 & 3 are also* 'nä-mə-ˌnā-\ *adjective* [Middle English *nominatyf,* from Middle French or Latin; Middle French *nominatif,* from Latin *(casus) nominativus* nominative case, from *nominare;* from the traditional use of the nominative form in naming a noun] (14th century)
1 a : marking typically the subject of a verb especially in languages that have relatively full inflection ⟨*nominative* case⟩ **b :** of or relating to the nominative case ⟨a *nominative* ending⟩
2 : nominated or appointed by nomination
3 : bearing a person's name
— nominative *noun*

nom·i·nee \ˌnä-mə-'nē\ *noun* [*nominate*] (1688)
: a person who has been nominated

no·mo·gram \'nä-mə-ˌgram, 'nō-\ *noun* [Greek *nomos* law + International Scientific Vocabulary *-gram* — more at NIMBLE] (1908)
: a graphic representation that consists of several lines marked off to scale and arranged in such a way that by using a straightedge to connect known values on two lines an unknown value can be read at the point of intersection with another line

no·mo·graph \-ˌgraf\ *noun* (circa 1909)
: NOMOGRAM
— no·mo·graph·ic \ˌnä-mə-'gra-fik, ˌnō-\ *adjective*
— no·mog·ra·phy \nō-'mä-grə-fē\ *noun*

no·mo·log·i·cal \ˌnä-mə-'lä-ji-kəl, ˌnō-\ *adjective* [*nomology* science of physical and logical laws, from Greek *nomos* + English *-logy*] (1845)
: relating to or expressing basic physical laws or rules of reasoning ⟨*nomological* universals⟩

no·mo·thet·ic \-'the-tik\ *adjective* [Greek *nomothetikos* of legislation, from *nomothetēs* lawgiver, from *nomos* law + *-thetēs* one who establishes, from *tithenai* to put — more at DO] (1658)
: relating to, involving, or dealing with abstract, general, or universal statements or laws

-nomy *noun combining form* [Middle English *-nomie,* from Old French, from Latin *-nomia,* from Greek, from *nomos*]
: system of laws governing or sum of knowledge regarding a (specified) field ⟨agro*nomy*⟩

non- \(ˈ)nän *also* ˌnən *or* 'nən *before* '*-stressed syllable,* ˌnän *also* ˌnən *before* ˌ*-stressed or unstressed syllable; the variant with* ə *is also to be understood at pronounced entries, though not shown*\ *prefix* [Middle English, from Middle French, from Latin *non* not, from Old Latin *noenum,* from *ne-* not + *oinom,* neuter of *oinos* one — more at NO, ONE]
1 : not **:** other than **:** reverse of **:** absence of
2 : of little or no consequence **:** unimportant **:** worthless ⟨*non*issues⟩ ⟨*non*system⟩
3 : lacking the usual especially positive characteristics of the thing specified ⟨*non*celebration⟩ ⟨*non*art⟩

non·abra·sive
non·ab·sorb·able
non·ab·sor·bent
non·ab·sorp·tive
non·ab·stract
non·ac·a·dem·ic
non·ac·cep·tance
non·ac·count·able
non·ac·cred·it·ed
non·ac·cru·al
non·achieve·ment
non·ac·id
non·acid·ic
non·ac·quis·i·tive
non·act·ing
non·ac·tion
non·ac·ti·vat·ed
non·ac·tor
non·adap·tive
non·ad·dict
non·ad·dic·tive
non·ad·he·sive
non·adi·a·bat·ic
non·ad·ja·cent
non·ad·mir·er
non·ad·mis·sion
non·aes·thet·ic
non·af·fil·i·at·ed
non·af·flu·ent
non–Af·ri·can
non·ag·gres·sion
non·ag·gres·sive
non·ag·ri·cul·tur·al
non·al·co·hol·ic
non·al·ler·gen·ic
non·al·ler·gic
non·al·pha·bet·ic
non·alu·mi·num
non·am·big·u·ous
non–Amer·i·can
non·an·a·lyt·ic
non·an·a·tom·ic
non·an·i·mal
non·an·swer
non·an·tag·o·nis·tic
non·an·thro·po·log·i·cal
non·an·thro·pol·o·gist
non·an·ti·bi·ot·ic
non·an·ti·gen·ic
non·ap·pear·ance
non·aquat·ic
non·aque·ous
non·ar·a·ble
non·ar·bi·trar·i·ness
non·ar·bi·trary
non·ar·chi·tect
non·ar·chi·tec·ture
non·ar·gu·ment
non·aris·to·crat·ic
non·ar·o·mat·ic
non·art
non·art·ist
non·ar·tis·tic
non·as·cet·ic
non·as·pi·rin
non·as·ser·tive
non·as·so·ci·at·ed
non·as·tro·nom·i·cal
non·ath·lete
non·ath·let·ic
non·atom·ic
non·at·tached
non·at·tach·ment
non·at·ten·dance
non·at·tend·er
non·au·di·to·ry
non·au·thor
non·au·thor·i·tar·i·an
non·au·to·mat·ed
non·au·to·mat·ic
non·au·to·mo·tive
non·au·ton·o·mous
non·avail·abil·i·ty
non·bac·te·ri·al
non·bar·bi·tu·rate
non·ba·sic
non·bear·ing
non·be·hav·ior·al
non·be·ing
non·be·lief
non·be·liev·er
non·bel·lig·er·en·cy
non·bel·lig·er·ent
non·bet·ting
non·bib·lio·graph·ic
non·bi·na·ry
non·bind·ing
non·bio·de·grad·able
non·bio·graph·i·cal
non·bi·o·log·i·cal
non·bi·o·log·i·cal·ly
non·bi·ol·o·gist
non·bit·ing
non·black
non·body
non·bond·ed
non·bot·a·nist
non·brand
non·break·able
non·breath·ing
non·breed·er
non·breed·ing
non·broad·cast
non·build·ing
non·burn·able
non·buy·ing
non·cab·i·net
non·cak·ing
non·call·able
non·cam·pus
non·can·cel·able
non·can·cer·ous
non·can·ni·bal·is·tic
non·cap·i·tal
non·cap·i·tal·ist
non·car·cin·o·gen
non·car·ci·no·gen·ic
non·car·di·ac
non·ca·reer
non·car·ri·er
non·cash
non·ca·su·al
non–Cath·o·lic
non·caus·al
non·cel·e·bra·tion
non·ce·leb·ri·ty
non·cel·lu·lar
non·cel·lu·los·ic
non–Celt·ic
non·cen·tral
non·cer·tif·i·cat·ed
non·cer·ti·fied
non·char·ac·ter
non·char·is·mat·ic
non·chau·vin·ist
non·chem·i·cal
non–Chris·tian
non·chro·no·log·i·cal
non·church
non·church·go·er
non·cir·cu·lar
non·cir·cu·lat·ing
non·cit·i·zen
non·clan·des·tine
non·class
non·clas·si·cal
non·clas·si·fied
non·class·room
non·cler·i·cal
non·cling
non·clin·i·cal
non·clog·ging
non·co·er·cive
non·cog·ni·tive
non·co·her·ent
non·co·in·ci·dence
non·co·i·tal
non·cok·ing
non·co·la
non·col·lec·tor
non·col·lege
non·col·le·giate
non·col·lin·ear
non·col·or
non·col·ored
non·col·or·fast
non·com·bat
non·com·bat·ive
non·com·bus·ti·ble
non·com·mer·cial
non·com·mit·ment
non·com·mit·ted
non·com·mu·ni·cat·ing
non·com·mu·ni·ca·tion
non·com·mu·ni·ca·tive
non·com·mu·nist
non·com·mu·ni·ty
non·com·mu·ta·tive
non·com·mu·ta·tiv·i·ty
non·com·pa·ra·bil·i·ty
non·com·pa·ra·ble
non·com·pat·i·ble
non·com·pe·ti·tion
non·com·pet·i·tive
non·com·pet·i·tor
non·com·ple·men·ta·ry
non·com·plex
non·com·pli·ance
non·com·pli·cat·ed
non·com·ply·ing
non·com·pos·er
non·com·pound
non·com·pre·hen·sion
non·com·press·ible
non·com·put·er
non·com·put·er·ized
non·con·cep·tu·al
non·con·cern
non·con·clu·sion
non·con·cur·rent
non·con·dens·able
non·con·di·tioned
non·con·duct·ing
non·con·duc·tion
non·con·duc·tive
non·con·fer·ence
non·con·fi·dence
non·con·fi·den·tial
non·con·flict·ing
non·con·fron·ta·tion
non·con·fron·ta·tion·al
non·con·gru·ent
non·con·ju·gat·ed
non·con·nec·tion
non·con·scious
non·con·sec·u·tive
non·con·sen·su·al
non·con·ser·va·tion
non·con·ser·va·tive
non·con·sol·i·dat·ed
non·con·stant
non·con·sti·tu·tion·al
non·con·struc·tion
non·con·struc·tive
non·con·sum·er
non·con·sum·ing
non·con·sump·tion
non·con·sump·tive
non·con·tact
non·con·ta·gious
non·con·tem·po·rary
non·con·tig·u·ous
non·con·tin·gent
non·con·tin·u·ous
non·con·tract
non·con·trac·tu·al
non·con·tra·dic·tion
non·con·tra·dic·to·ry
non·con·trib·u·to·ry
non·con·trol·la·ble
non·con·trolled
non·con·trol·ling
non·con·tro·ver·sial
non·con·ven·tion·al
non·con·vert·ible
non·co·pla·nar
non·cor·po·rate
non·cor·re·la·tion
non·cor·rod·ible
non·cor·rod·ing
non·cor·ro·sive
non·coun·try

non•coun•ty
non•cov•er•age
non•cre•a•tive
non•cre•a•tiv•i•ty
non•cre•den•tialed
non•crime
non•crim•i•nal
non•cri•sis
non•crit•i•cal
non•crush•able
non•crys•tal•line
non•cu•li•nary
non•cul•ti•vat•ed
non•cul•ti•va•tion
non•cul•tur•al
non•cu•mu•la•tive
non•cur•rent
non•cus•tom•er
non•cy•clic
non•cy•cli•cal
non•dance
non•danc•er
non–Dar•win•i•an
non•de•cep•tive
non•de•ci•sion
non•de•creas•ing
non•de•duc•tive
non•de•fer•ra•ble
non•de•form•ing
non•de•gen•er•ate
non•de•grad•able
non•de•gree
non•del•e•gate
non•de•lib•er•ate
non•de•lin•quent
non•de•liv•ery
non•de•mand•ing
non•dem•o•crat•ic
non•de•nom•i•na-
tion•al
non•de•nom•i•na-
tion•al•ism
non•de•part•men•tal
non•de•pen•dent
non•de•plet•able
non•de•plet•ing
non•de•po•si•tion
non•de•pressed
non•de•riv•a•tive
non•de•scrip•tive
non•de•sert
non•de•tach•able
non•de•ter•min•is•tic
non•de•vel•op•ment
non•de•vi•ant
non•di•a•bet•ic
non•di•a•lyz•able
non•di•dac•tic
non•dif•fus•ible
non•di•men•sion•al
non•dip•lo•mat•ic
non•di•rect•ed
non•di•rec•tion•al
non•dis•abled
non•dis•clo•sure
non•dis•count
non•dis•cre•tion•ary
non•dis•crim•i•na-
tion
non•dis•crim•i•na-
to•ry
non•dis•cur•sive
non•dis•per•sive
non•dis•rup•tive
non•di•ver•si•fied
non•doc•tor
non•doc•tri•naire
non•doc•u•men•ta•ry
non•dog•mat•ic
non•dol•lar
non•do•mes•tic
non•dom•i•nant
non•dra•mat•ic
non•driv•er
non•drug
non•du•ra•ble
non•earn•ing
non•ec•cle•si•as•ti-
cal
non•econ•o•mist
non•ed•i•ble
non•ed•i•to•ri•al
non•ed•u•ca•tion
non•ed•u•ca•tion•al
non•ef•fec•tive
non•elas•tic
non•elect•ed
non•elec•tion
non•elec•tive
non•elec•tric
non•elec•tri•cal
non•elec•tron•ic
non•el•e•men•ta•ry
non•elite
non•emer•gen•cy
non•emo•tion•al
non•em•phat•ic
non•em•pir•i•cal
non•em•ploy•ee
non•em•ploy•ment
non•emp•ty
non•en•cap•su•lat•ed
non•end•ing
non•en•er•gy
non•en•force•abil•i•ty
non•en•force•ment
non•en•gage•ment
non•en•gi•neer•ing
non•en•ter•tain•ment
non•en•zy•mat•ic
non•en•zy•mic
non•equi•lib•ri•um
non•equiv•a•lence
non•equiv•a•lent
non•erot•ic
non•es•tab•lished
non•es•tab•lish•ment
non•es•ter•i•fied
non•eth•i•cal
non•eth•nic
non–Eu•ro•pe•an
non•eval•u•a•tive
non•ev•i•dence
non•ex•clu•sive
non•ex•ec•u•tive
non•ex•empt
non•ex•is•ten•tial
non•ex•ot•ic
non•ex•pend•able
non•ex•per•i•men•tal
non•ex•pert
non•ex•plan•a•to•ry
non•ex•ploi•ta•tion
non•ex•ploit•a•tive
non•ex•ploit•ive
non•ex•plo•sive
non•ex•posed
non•ex•tant
non•fact
non•fac•tor
non•fac•tu•al
non•fac•ul•ty
non•fad•ing
non•fa•mil•ial
non•fam•i•ly
non•fan
non•farm
non•farm•er
non•fa•tal
non•fat•ten•ing
non•fat•ty
non•fed•er•al
non•fed•er•at•ed
non•fem•i•nist
non•fil•a•men•tous
non•fil•ter•able
non•fi•nal
non•fi•nan•cial
non•fi•nite
non•fis•sion•able
non•flu•o•res•cent
non•fly•ing
non•food
non•for•feit•able
non•for•fei•ture
non•for•mal
non•fos•sil
non•frat•er•ni•za•tion
non•freez•ing
non•friv•o•lous
non•fro•zen
non•fuel
non•ful•fill•ment
non•func•tion•al
non•func•tion•ing
non•game
non•gas•eous
non•gay
non•ge•net•ic
non•gen•i•tal
non•geo•met•ri•cal
non•ghet•to
non•glam•or•ous
non•glare
non•golf•er
non•gov•ern•ment
non•gov•ern•men•tal
non•grad•ed
non•grad•u•ate
non•gram•mat•i•cal
non•gran•u•lar
non•grav•i•ta•tion•al
non•greasy
non•gre•gar•i•ous
non•grow•ing
non•growth
non•guest
non•ha•lo•ge•nat•ed
non•hand•i•capped
non•hap•pen•ing
non•har•dy
non•har•mon•ic
non•haz•ard•ous
non•heme
non•he•mo•lyt•ic
non•he•red•i•tary
non•hi•er•ar•chi•cal
non–His•pan•ic
non•his•tor•i•cal
non•home
non•ho•mo•ge•neous
non•ho•mol•o•gous
non•ho•mo•sex•u•al
non•hor•mon•al
non•hos•pi•tal
non•hos•pi•tal•ized
non•hos•tile
non•hous•ing
non•hu•man
non•hunt•er
non•hunt•ing
non•hy•gro•scop•ic
non•hys•ter•i•cal
non•ide•al
non•iden•ti•ty
non•ideo•log•i•cal
non•im•age
non•im•i•ta•tive
non•im•mi•grant
non•im•mune
non•im•pact
non•im•pli•ca•tion
non•im•por•ta•tion
non•in•clu•sion
non•in•creas•ing
non•in•cum•bent
non•in•de•pen•dence
non–In•di•an
non•in•dig•e•nous
non•in•di•vid•u•al
non–In•do–Eu•ro-
pe•an
non•in•dus•tri•al
non•in•dus•tri•al•ized
non•in•dus•try
non•in•fect•ed
non•in•fec•tious
non•in•fec•tive
non•in•fest•ed
non•in•flam•ma•ble
non•in•flam•ma•to•ry
non•in•fla•tion•ary
non•in•flec•tion•al
non•in•flu•ence
non•in•for•ma•tion
non•ini•tial
non•ini•ti•ate
non•in•ju•ry
non•in•sect
non•in•sec•ti•cid•al
non•in•stall•ment
non•in•sti•tu•tion•al
non•in•sti•tu•tion•al-
ized
non•in•struc•tion•al
non•in•stru•men•tal
non•in•sur•ance
non•in•sured
non•in•te•gral
non•in•te•grat•ed
non•in•tel•lec•tu•al
non•in•ter•act•ing
non•in•ter•ac•tive
non•in•ter•change-
able
non•in•ter•course
non•in•ter•est
non•in•ter•fer•ence
non•in•ter•sect•ing
non•in•tim•i•dat•ing
non•in•tox•i•cant
non•in•tox•i•cat•ing
non•in•tru•sive
non•in•tu•i•tive
non•ion•iz•ing
non•ir•ra•di•at•ed
non•ir•ri•gat•ed
non•ir•ri•tant
non•ir•ri•tat•ing
non•is•sue
non–Jap•a•nese
non–Jew
non–Jew•ish
non•join•er
non•ju•di•cial
non•ju•ry
non•jus•ti•cia•ble
non•ko•sher
non•la•bor
non•land•own•er
non•lan•guage
non•law•yer
non•lead•ed
non•league
non•le•gal
non•le•gume
non•le•gu•mi•nous
non•le•thal
non•lex•i•cal
non•li•brar•i•an
non•li•brary
non•life
non•lin•e•al
non•lin•e•ar
non•lin•e•ar•i•ty
non•lin•guis•tic
non•liq•uid
non•lit•er•al
non•lit•er•ary
non•liv•ing
non•lo•cal
non•log•i•cal
non•lu•mi•nous
non•mag•net•ic
non•main•stream
non•ma•jor
non•ma•lig•nant
non•mal•lea•ble
non•man•age•ment
non•man•a•ge•ri•al
non•man•u•al
non•man•u•fac•tur-
ing
non•mar•i•tal
non•mar•ket
non–Marx•ist
non•ma•te•ri•al
non•ma•te•ri•al•is•tic
non•math•e•mat•i-
cal
non•math•e•ma•ti-
cian
non•ma•tric•u•lat•ed
non•mean•ing•ful
non•mea•sur•able
non•meat
non•me•chan•i•cal
non•mech•a•nis•tic
non•med•i•cal
non•meet•ing
non•mem•ber
non•mem•ber•ship
non•men•tal
non•mer•cu•ri•al
non•meta•mer•ic
non•met•a•phor•i•cal
non•met•ric
non•met•ri•cal
non•met•ro
non•met•ro•pol•i•tan
non•mi•cro•bi•al
non•mi•grant
non•mi•gra•to•ry
non•mil•i•tant
non•mil•i•tary
non•mi•met•ic
non•mi•nor•i•ty
non•mo•bile
non•mo•lec•u•lar
non•mon•e•tar•ist
non•mon•e•tary
non•mon•ey
non•mo•nog•a•mous
non•mo•tile
non•mo•til•i•ty
non•mo•tor•ized
non•mov•ing
non•mu•nic•i•pal
non•mu•sic
non•mu•si•cal
non•mu•si•cian
non•mu•tant
non•my•e•lin•at•ed
non•mys•ti•cal
non•nar•ra•tive
non•na•tion•al
non•na•tive
non•nat•u•ral
non•ne•ces•si•ty
non•neg•li•gent
non•ne•go•tia•ble
non•net•work
non•news
non•ni•trog•e•nous
non•nor•ma•tive
non•nov•el
non•nu•cle•at•ed
non•nu•mer•i•cal
non•nu•tri•tious
non•nu•tri•tive
non•ob•scene
non•ob•ser•vance
non•ob•ser•vant
non•ob•vi•ous
non•oc•cu•pa•tion•al
non•oc•cur•rence
non•of•fi•cial
non•ohm•ic
non•oily
non•op•er•at•ic
non•op•er•at•ing
non•op•er•a•tion•al
non•op•er•a•tive
non•op•ti•mal
non•or•gan•ic
non•or•tho•dox
non•over•lap•ping
non•own•er
non•ox•i•diz•ing
non•paid
non•par•al•lel
non•par•a•sit•ic
non•par•tic•i•pant
non•par•tic•i•pat•ing
non•par•tic•i•pa•tion
non•par•tic•i•pa•to•ry
non•par•ty
non•pas•sive
non•past
non•pay•ing
non•pay•ment
non•per•for•mance
non•per•form•er
non•per•ish•able
non•per•mis•sive
non•per•son•al
non•pe•tro•leum
non•phi•los•o•pher
non•phil•o•soph•i-
cal
non•pho•ne•mic
non•pho•net•ic
non•phos•phate
non•pho•to•graph•ic
non•phys•i•cal
non•phy•si•cian
non•pla•nar
non•plas•tic
non•play
non•play•ing
non•po•et•ic
non•poi•son•ous
non•po•lar•iz•able
non•po•lice
non•po•lit•i•cal
non•po•lit•i•cal•ly
non•pol•i•ti•cian
non•pol•lut•ing
non•poor
non•po•rous
non•pos•ses•sion
non•prac•ti•cal
non•prac•tic•ing
non•preg•nant
non•print
non•prob•lem
non•pro•duc•ing
non•pro•fes•sion•al
non•pro•fes•sion-
al•ly
non•pro•fes•so•ri•al
non•pro•gram
non•pro•gram•mer
non•pro•gres•sive
non•pro•pri•etary
non•psy•chi•at•ric
non•psy•chi•a•trist
non•psy•cho•log•i-
cal
non•psy•chot•ic
non•pub•lic
non•pu•ni•tive
non•pur•po•sive
non•quan•ti•fi•able
non•quan•ti•ta•tive
non•ra•cial
non•ra•cial•ly
non•ra•dio•ac•tive
non•rail•road
non•ran•dom
non•ran•dom•ness
non•rat•ed
non•ra•tio•nal
non•re•ac•tive
non•re•ac•tor
non•re•al•is•tic
non•re•ap•point-
ment
non•re•ceipt
non•re•cip•ro•cal
non•rec•og•ni•tion
non•re•cy•cla•ble
non•re•duc•ing
non•re•dun•dant
non•re•fill•able
non•re•flect•ing

non•reg•u•lat•ed
non•reg•u•la•tion
non•rel•a•tive
non•rel•e•vant
non•re•li•gious
non•re•new•able
non•re•new•al
non•re•pay•able
non•rep•re•sen•ta•tive
non•re•pro•duc•tive
non•res•i•den•tial
non•res•o•nant
non•re•spon•dent
non•re•spond•er
non•re•sponse
non•re•spon•sive
non•re•strict•ed
non•re•trac•tile
non•ret•ro•ac•tive
non•re•us•able
non•re•vers•ible
non•rev•o•lu•tion•ary
non•ri•ot•er
non•ri•ot•ing
non•ro•tat•ing
non•rou•tine
non•roy•al
non•rub•ber
non•rul•ing
non•ru•mi•nant
non–Rus•sian
non•sal•able
non•sa•line
non•sa•pon•i•fi•able
non•schiz•o•phren•ic
non•school
non•sci•en•tif•ic
non•sci•en•tist
non•sea•son•al
non•sec•re•tory
non•se•cure
non•sed•i•ment•able
non•seg•re•gat•ed
non•seg•re•ga•tion
non•se•lect•ed
non•se•lec•tive
non–self–gov•ern•ing
non•sen•sa•tion•al
non•sen•si•tive
non•sen•su•ous
non•sen•tence
non•sep•tate
non•se•quen•tial
non•se•ri•ous
non•sex•ist
non•sex•u•al
non•shrink
non•shrink•able
non•sign•er
non•si•mul•ta•neous
non•sink•able
non•skat•er
non•skel•e•tal
non•ski•er
non•smok•er
non•smok•ing
non•so•cial•ist
non•so•lar
non•so•lu•tion
non•spa•tial
non•speak•er
non•speak•ing
non•spe•cial•ist
non•spe•cif•ic
non•spe•cif•i•cal•ly
non•spec•tac•u•lar
non•spec•u•la•tive
non•speech
non•spher•i•cal
non•sta•tion•ary
non•sta•tis•ti•cal
non•steady
non•sto•ry
non•stra•te•gic

non•struc•tur•al
non•struc•tured
non•stu•dent
non•style
non•sub•ject
non•sub•jec•tive
non•sub•si•dized
non•suc•cess
non•sug•ar
non•su•per•im•pos•able
non•su•per•vi•so•ry
non•sur•gi•cal
non•swim•mer
non•sym•bol•ic
non•sym•met•ric
non•sym•met•ri•cal
non•syn•chro•nous
non•sys•tem•at•ic
non•sys•tem•ic
non•tar•iff
non•tax•able
non•teach•ing
non•tech•ni•cal
non•tem•po•ral
non•ten•ured
non•ter•mi•nal
non•the•at•ri•cal
non•the•ist
non•the•is•tic
non•theo•log•i•cal
non•the•o•ret•i•cal
non•ther•a•peu•tic
non•think•ing
non•threat•en•ing
non•tid•al
non•to•bac•co
non•ton•al
non•to•tal•i•tar•i•an
non•tox•ic
non•tra•di•tion•al
non•trans•fer•able
non•treat•ment
non•triv•i•al
non•trop•i•cal
non•tur•bu•lent
non•typ•i•cal
non•unan•i•mous
non•uni•form
non•uni•for•mi•ty
non•union•ized
non•unique
non•unique•ness
non•uni•ver•sal
non•uni•ver•si•ty
non•ur•ban
non•ur•gent
non•util•i•tar•i•an
non•util•i•ty
non•uto•pi•an
non•val•id
non•va•lid•i•ty
non•vas•cu•lar
non•veg•e•tar•i•an
non•ven•om•ous
non•vet•er•an
non•vi•a•ble
non•view•er
non•vi•ral
non•vir•gin
non•vis•cous
non•vi•su•al
non•vo•cal
non•vo•ca•tion•al
non•vol•ca•nic
non•vol•un•tary
non•vot•er
non•vot•ing
non•war
non•win•ning
non•woody
non•work
non•work•er
non•work•ing
non•writ•er
non•yel•low•ing

non•ad•di•tive \ˈnän-ˈa-də-tiv\ *adjective* (1926)
1 : not having a numerical value equal to the sum of values for the component parts
2 : of, relating to, or being a genic effect that is not additive
— **non•ad•di•tiv•i•ty** \ˌnän-ˌa-də-ˈti-və-tē\ *noun*

non•age \ˈnä-nij, ˈnō-\ *noun* [Middle English, from Middle French, from *non-* + *age* age] (15th century)
1 : MINORITY 1
2 a : a period of youth **b :** lack of maturity

no•na•ge•nar•i•an \ˌnō-nə-jə-ˈner-ē-ən, ˌnä-\ *noun* [Latin *nonagenarius* containing ninety, from *nonageni* ninety each, from *nonaginta* ninety, from *nona-* (akin to *novem* nine) + *-ginta* (akin to *viginti* twenty) — more at NINE, VIGESIMAL] (1804)
: a person whose age is in the nineties
— **nonagenarian** *adjective*

no•na•gon \ˈnō-nə-ˌgän\ *noun* [Latin *nonus* ninth + English *-gon* — more at NOON] (circa 1639)
: a polygon of nine angles and nine sides

non•aligned \ˌnän-ə-ˈlīnd\ *adjective* (1960)
: not allied with other nations and especially with either the Communist or the non-Communist blocs
— **non•align•ment** \-mənt\ *noun*

non•al•le•lic \ˌnän-ə-ˈlē-lik, -ˈle-\ *adjective* (1945)
: not behaving as alleles toward one another ⟨*nonallelic* genes⟩

non–A, non–B hepatitis \ˈnän-ˈā-ˈnän-ˈbē-\ *noun* (1976)
: hepatitis clinically and immunologically similar to hepatitis A and hepatitis B but caused by different viruses

non•bank \ˈnän-ˈbaŋk\ *noun, often attributive* (1939)
: a business that is not an officially established bank but offers many similar services
— **non•bank•ing** \-ˈbaŋ-kiŋ\ *adjective*

non•bond•ing \-ˈbän-diŋ\ *adjective* (1952)
: being or occupied by electrons especially of an atom's valence shell that are not involved in a chemical bond ⟨a *nonbonding* atomic orbital⟩

¹non•book \-ˈbu̇k\ *adjective* (1949)
: being something other than a book; *especially* **:** being a library holding (as a microfilm) that is not a book

²non•book \-ˌbu̇k\ *noun* (1960)
: a book of little literary merit which is often a compilation (as of pictures, press clippings, or speeches)

non•busi•ness \-ˈbiz-nəs, -nəz\ *adjective* (1927)
: not related to business; *especially* **:** not related to one's primary business

non•ca•lo•ric \ˌnän-kə-ˈlȯr-ik, -ˈlōr-\ *adjective* (1950)
: free from or very low in calories

non•can•di•date \ˈnän-ˈkan-də-ˌdāt, -ˈka-nə-, -dət\ *noun* (1944)
: a person who is not a candidate; *especially* **:** one who has refused to be a candidate for a particular political office
— **non•can•di•da•cy** \-də-sē\ *noun*

¹nonce \ˈnän(t)s\ *noun* [Middle English *nanes,* alteration (from misdivision of *then anes* in such phrases as *to then anes* for the one purpose) of *anes* one purpose, irregular from *an, on* one — more at ONE] (13th century)
1 : the one, particular, or present occasion, purpose, or use ⟨for the *nonce*⟩
2 : the time being

²nonce *adjective* (1884)
: occurring, used, or made only once or for a special occasion ⟨a *nonce* word⟩

non•cha•lance \ˌnän-shə-ˈlän(t)s; ˈnän-shə-ˌlän(t)s, -lən(t)s\ *noun* (1678)
: the quality or state of being nonchalant

non•cha•lant \-ˈlänt, -ˌlänt, -lənt\ *adjective* [French, from Old French, from present participle of *nonchaloir* to disregard, from *non-* + *chaloir* to concern, from Latin *calēre* to be warm — more at LEE] (circa 1734)
: having an air of easy unconcern or indifference
synonym see COOL
— **non•cha•lant•ly** *adverb*

non•chro•mo•som•al \ˌnän-ˌkrō-mə-ˈsō-məl\ *adjective* (1960)
1 : not situated on a chromosome
2 : not involving chromosomes

non•com \ˈnän-ˌkäm\ *noun* (1883)
: NONCOMMISSIONED OFFICER

non•com•bat•ant \ˌnän-kəm-ˈba-tᵊnt *also* ˈnän-ˈkäm-bə-tənt\ *noun* (1811)
: one that does not engage in combat: as **a :** a member (as a chaplain) of the armed forces whose duties do not include fighting **b :** CIVILIAN
— **noncombatant** *adjective*

non•com•mis•sioned officer \ˌnän-kə-ˈmi-shənd-\ *noun* (1703)
: a subordinate officer (as a sergeant) in the army, air force, or marine corps appointed from among enlisted personnel

non•com•mit•tal \-kə-ˈmi-tᵊl\ *adjective* (1829)
1 : giving no clear indication of attitude or feeling
2 : having no clear or distinctive character
— **non•com•mit•tal•ly** \-tᵊl-ē\ *adverb*

non com•pos men•tis \ˌnän-ˌkäm-pəs-ˈmen-təs, ˌnōn-\ *adjective* [Latin, literally, not having mastery of one's mind] (1607)
: not of sound mind

non•con•cur \ˌnän-kən-ˈkər\ *intransitive verb* (circa 1847)
: to refuse or fail to concur
— **non•con•cur•rence** \-ˈkər-ən(t)s, -ˈkə-rən(t)s\ *noun*

non•con•duc•tor \-kən-ˈdək-tər\ *noun* (1751)
: a substance that conducts heat, electricity, or sound only in very small degree

non•con•form \-kən-ˈfȯrm\ *intransitive verb* [back-formation from *nonconformist*] (1681)
: to fail to conform
— **non•con•form•er** *noun*

non•con•for•mance \-ˈfȯr-mən(t)s\ *noun* (1843)
: failure to conform

non•con•form•ism \-ˈfȯr-ˌmi-zəm\ *noun* (1844)
: NONCONFORMITY

non•con•form•ist \-ˈfȯr-mist\ *noun* (1619)
1 *often capitalized* **:** a person who does not conform to an established church; *especially* **:** one who does not conform to the Church of England
2 : a person who does not conform to a generally accepted pattern of thought or action
— **nonconformist** *adjective, often capitalized*

non•con•for•mi•ty \-ˈfȯr-mə-tē\ *noun* (1618)
1 a : failure or refusal to conform to an established church **b** *often capitalized* **:** the movement or principles of English Protestant dissent **c** *often capitalized* **:** the body of English Nonconformists
2 : refusal to conform to an established or conventional creed, rule, or practice
3 : absence of agreement or correspondence

non•co•op•er•a•tion \ˌnän-kō-ˌä-pə-ˈrā-shən\ *noun* (1795)
: failure or refusal to cooperate; *specifically* **:** refusal through civil disobedience of a people to cooperate with the government of a country
— **non•co•op•er•a•tion•ist** \-sh(ə-)nist\ *noun*
— **non•co•op•er•a•tor** \-ˈä-pə-ˌrā-tər\ *noun*

non·co·op·er·a·tive \-'ä-p(ə-)rə-tiv, -pə-ˌrā-\ *adjective* (1922)
: of, relating to, or characterized by noncooperation

non·cred·it \'nän-'kre-dət\ *adjective* (1965)
: not offering credit toward a degree ⟨*noncredit* courses⟩

non·cross·over \-'krȯ-ˌsō-vər\ *adjective* (1919)
: having or being chromosomes that have not participated in genetic crossing-over ⟨*noncrossover* offspring⟩

non·cus·to·di·al \-kə-'stō-dē-əl\ *adjective* (1973)
: of or being a parent who does not have legal custody of a child

non·dairy \-'der-ē, -'dar-; 'nän-ˌ\ *adjective* (1968)
: containing no milk or milk products ⟨*nondairy* whipped topping⟩

non·de·duct·ible \ˌnän-di-'dək-tə-bəl\ *adjective* (1943)
: not deductible; *especially* : not deductible for income tax purposes
— **non·de·duct·ibil·i·ty** \-ˌdək-tə-'bi-lə-tē\ *noun*

non·de·fense \ˌnän-di-'fen(t)s\ *adjective* (1961)
: not used or intended for or associated with the military ⟨*nondefense* spending⟩

non·de·script \ˌnän-di-'skript\ *adjective* [*non-* + Latin *descriptus,* past participle of *describere* to describe] (circa 1807)
1 : belonging or appearing to belong to no particular class or kind : not easily described
2 : lacking distinctive or interesting qualities : DULL, DRAB
— **nondescript** *noun*

non·de·struc·tive \-di-'strək-tiv\ *adjective* (1926)
: not destructive; *specifically* : not causing destruction of material being investigated or treated ⟨*nondestructive* testing of metal⟩
— **non·de·struc·tive·ly** *adverb*
— **non·de·struc·tive·ness** *noun*

non·dia·paus·ing \ˌnän-ˌdī-ə-'pȯ-ziŋ\ *adjective* (1963)
1 : not having a diapause
2 : not being in a state of diapause

non·di·rec·tive \ˌnän-də-'rek-tiv, -(ˌ)dī-\ *adjective* (1931)
: of, relating to, or being psychotherapy, counseling, or interviewing in which the counselor refrains from interpretation or explanation but encourages the client (as by repeating phrases) to talk freely

non·dis·junc·tion \ˌnän-dis-'jəŋ(k)-shən\ *noun* [International Scientific Vocabulary] (1913)
: failure of homologous chromosomes or sister chromatids to separate subsequent to metaphase in meiosis or mitosis so that one daughter cell has both and the other neither of the chromosomes
— **non·dis·junc·tion·al** \-shnəl, -shə-nᵊl\ *adjective*

non·dis·tinc·tive \-di-'stiŋ(k)-tiv\ *adjective* (1916)
of a speech sound : having no signaling value

non·di·vid·ing \ˌnän-də-'vī-diŋ\ *adjective* (1945)
: not undergoing cell division

non·dor·mant \'nän-'dȯr-mənt\ *adjective* (1940)
1 : being in such a condition that germination is possible ⟨*nondormant* seeds⟩
2 : being in active vegetative growth ⟨*nondormant* plants⟩

non·drink·er \-'driŋ-kər\ *noun* (1899)
: a person who abstains from alcoholic beverages
— **non·drink·ing** \-kiŋ\ *adjective*

non·dry·ing oil \-'drī-iŋ-\ *noun* (1905)
: a highly saturated oil (as olive oil) that is unable to solidify when exposed in a thin film to air

¹none \'nən\ *pronoun, singular or plural in construction* [Middle English, from Old English *nān,* from *ne* not + *ān* one — more at NO, ONE] (before 12th century)
1 : not any
2 : not one : NOBODY
3 : not any such thing or person
4 : no part : NOTHING ■

²none *adjective* (before 12th century)
archaic : not any : NO

³none *adverb* (1651)
1 : by no means : not at all ⟨*none* too soon to begin⟩
2 : in no way : to no extent ⟨*none* the worse for wear⟩

⁴none \'nōn\ *noun, often capitalized* [Late Latin *nona,* from Latin, 9th hour of the day from sunrise — more at NOON] (1845)
: the fifth of the canonical hours
word history see NOON

non·eco·nom·ic \ˌnän-ˌe-kə-'nä-mik, -ˌē-kə-\ *adjective* (1920)
: not economic; *especially* : having no economic importance or implication

non·elec·tro·lyte \ˌnän-ə-'lek-trə-ˌlīt\ *noun* (1891)
: a substance that does not readily ionize when dissolved or melted and is a poor conductor of electricity

non·en·ti·ty \-'en-tə-tē, -'e-nə-\ *noun* (circa 1600)
1 : something that does not exist or exists only in the imagination
2 : NONEXISTENCE
3 : a person of little consequence or significance

nones \'nōnz\ *noun plural but singular or plural in construction* [Middle English *nonys,* from Latin *nonae,* from feminine plural of *nonus* ninth] (14th century)
1 : the ninth day before the ides according to ancient Roman reckoning
2 *often capitalized* : ⁴NONE
word history see NOON

non·es·sen·tial \ˌnän-i-'sen(t)-shəl\ *adjective* (1751)
1 : not essential
2 : synthesized by the body in sufficient quantity to satisfy dietary needs ⟨*nonessential* amino acids⟩
— **nonessential** *noun*

none·such \'nən-ˌsəch\ *noun* (1590)
: a person or thing without an equal
— **nonesuch** *adjective*

no·net \nō-'net\ *noun* [Italian *nonetto,* from *nono* ninth, from Latin *nonus* — more at NOON] (1865)
: a combination of nine instruments or voices; *also* : a musical composition for such a combination

none·the·less \ˌnən-thə-'les\ *adverb* (1847)
: NEVERTHELESS

non–eu·clid·e·an \ˌnän-yü-'kli-dē-ən\ *adjective, often E capitalized* (circa 1864)
: not assuming or in accordance with all the postulates of Euclid's *Elements* ⟨*non-euclidean* geometry⟩

non·event \'nän-i-ˌvent, ˌnän-i-'\ *noun* (1962)
1 a : an expected event that fails to take place or to satisfy expectations **b** : an often highly publicized event of little intrinsic interest or significance
2 : an occurrence that is officially ignored

non·ex·is·tence \ˌnän-ig-'zis-tən(t)s\ *noun* (1646)
: absence of existence : the negation of being
— **non·ex·is·tent** \-tənt\ *adjective*

non·fat \'nän-'fat\ *adjective* (1926)
: lacking fat solids : having fat solids removed ⟨*nonfat* milk⟩

non·fea·sance \'nän-'fē-zᵊn(t)s\ *noun* [*non-* + obsolete English *feasance* doing, execution] (1596)
: failure to act; *especially* : failure to do what ought to be done

non·fer·rous \-'fer-əs\ *adjective* (1887)
1 : not containing, including, or relating to iron
2 : of or relating to metals other than iron

non·fic·tion \-'fik-shən\ *noun* (1909)
: literature that is not fictional
— **non·fic·tion·al** \ˌnän-'fik-shnəl, -shə-nᵊl\ *adjective*

non·fig·u·ra·tive \'nän-'fi-gyə-rə-tiv, -'fi-gə-\ *adjective* (1927)
: NONOBJECTIVE 2

non·flam·ma·ble \-'fla-mə-bəl\ *adjective* (1915)
: not flammable; *specifically* : not easily ignited and not burning rapidly if ignited
— **non·flam·ma·bil·i·ty** \ˌnän-ˌfla-mə-'bi-lə-tē\ *noun*

non·flow·er·ing \'nän-'flau̇(-ə)r-iŋ\ *adjective* (circa 1934)
: producing no flowers; *specifically* : lacking a flowering stage in the life cycle

non·flu·en·cy \-'flü-ən(t)-sē\ *noun, plural* **-cies** (circa 1945)
1 : lack of fluency
2 : an instance of nonfluency

non·gon·o·coc·cal \ˌnän-ˌgä-nə-'kä-kəl\ *adjective* (1961)
: not caused by a gonococcus ⟨*nongonococcal* urethritis⟩

non gra·ta \ˌnän-'gra-tə, -'grä-\ *adjective* [*persona non grata*] (1925)
: not approved : UNWELCOME

non·green \'nän-'grēn\ *adjective* (1897)
: not green; *specifically* : containing no chlorophyll ⟨*nongreen* saprophytes⟩

non·he·ro \'nän-'hē-(ˌ)rō, -'hir-(ˌ)ō\ *noun* (1940)
: ANTI-HERO

non·his·tone \-'his-ˌtōn\ *adjective* (circa 1966)
: relating to or being any of the eukaryotic proteins (as DNA polymerase) that bind to nuclear DNA but are not histones

non–Hodg·kin's lymphoma \'nän-'häj-kənz-\ *noun* (1976)
: any of the numerous malignant lymphomas (as Burkitt's lymphoma) that are not classified as Hodgkin's disease

□ USAGE

none As the entry here indicates, this pronoun takes either a singular or a plural verb. Yet many people are convinced that *none* is singular only. We know about these people because they write to newspaper editors and to broadcasters, and a small number of them write usage books. The idea that *none* must be singular is rooted in etymology: *nān,* its Old English equivalent, was formed from a negative particle *ne-* "not" and *ān* "one." The first person we know of who combined knowledge of the etymology with disapproval of the plural verb was Charles Coote, an Englishman who published a grammar in 1788. He was apparently not so familiar with the grammar of Old English, where *nān* was both singular and plural. King Alfred the Great, in fact, used *nān* as a plural as far back as A.D. 888. Many of Coote's contemporaries knew that *none* was both singular and plural, and most writers on usage to this day know it, but the notion of its unalterable singularity has become part of the folklore of English usage. Here are a few examples of both constructions ⟨but *none* of these things move me —Acts 20:24 (Authorized Version)⟩ ⟨all agree in one judgment, and *none* ever varies his opinion —Samuel Johnson⟩ ⟨*none* are wretched but by their own fault —Samuel Johnson⟩ ⟨she has seed from four different people, but *none* of it comes up —Jane Austen⟩. A recent commentator thinks that where either construction can be chosen, the singular is a bit more emphatic, but either number can be an effective choice in context.

non·iden·ti·cal \ˌnän-(ˌ)ī-ˈden-ti-kəl, -ə-ˈden-\ *adjective* (1890)
1 : DIFFERENT
2 : FRATERNAL 2

no·nil·lion \nō-ˈnil-yən\ *noun, often attributive* [French, from Latin *nonus* ninth + French *-illion* (as in *million*) — more at NOON] (1690)
— see NUMBER table

non·in·duc·tive \ˌnän-in-ˈdək-tiv\ *adjective* (1896)
: not inductive; *especially* **:** having negligible inductance

non·in·ter·ven·tion \-ˌin-tər-ˈven(t)-shən\ *noun* (1831)
: the state or policy of not intervening ⟨*nonintervention* in the affairs of other countries⟩
— non·in·ter·ven·tion·ist \-ˈven(t)-sh(ə-)nist\ *noun or adjective*

non·in·va·sive \-in-ˈvā-siv, -ziv\ *adjective* (1971)
: not involving penetration (as by surgery or hypodermic needle) of the skin of the intact organism ⟨*noninvasive* diagnostic techniques⟩

non·in·volve·ment \-in-ˈvälv-mənt, -ˈvȯlv- *also* -ˈväv-, -ˈvȯv-\ *noun* (1936)
: absence of involvement or emotional attachment
— non·in·volved \-ˈvä(l)vd, -ˈvȯ(l)vd\ *adjective*

non·ion·ic \-(ˌ)ī-ˈä-nik\ *adjective* (1929)
: not ionic; *especially* **:** not dependent on a surface-active anion for effect ⟨*nonionic* surfactants⟩

non·join·der \ˈnän-ˈjȯin-dər\ *noun* (1833)
: failure to include a necessary party to a suit at law

non·judg·men·tal \ˌnän-ˌjəj-ˈmen-tᵊl\ *adjective* (1952)
: avoiding judgments based on one's personal and especially moral standards

non·jur·ing \ˈnän-ˈju̇r-iŋ\ *adjective* [*non-* + Latin *jurare* to swear — more at JURY] (1691)
: not swearing allegiance — used especially of a member of a party in Great Britain that would not swear allegiance to William and Mary or to their successors

non·ju·ror \-ˈju̇r-ər, -ˈju̇r-ˌȯr\ *noun* (1691)
: a person refusing to take an oath especially of allegiance, supremacy, or abjuration; *specifically* **:** one of the beneficed clergy in England and Scotland refusing to take an oath of allegiance to William and Mary or to their successors after the revolution of 1688

non·lit·er·ate \-ˈli-t(ə-)rət\ *adjective* (1947)
1 : not literate
2 : having no written language
— nonliterate *noun*

non·met·al \-ˈme-tᵊl\ *noun* (circa 1864)
: a chemical element (as boron, carbon, or nitrogen) that lacks the characteristics of a metal

non·me·tal·lic \ˌnän-mə-ˈta-lik\ *adjective* (1815)
1 : not metallic
2 : of, relating to, or being a nonmetal

non·mor·al \ˈnän-ˈmȯr-əl, -ˈmär-\ *adjective* (circa 1866)
: not falling into or existing in the sphere of morals or ethics

non·neg·a·tive \-ˈne-gə-tiv\ *adjective* (1885)
: not negative: as **a :** being either positive or zero **b :** taking on nonnegative values ⟨a *nonnegative* function⟩

non·nu·cle·ar \-ˈnü-klē-ər, -ˈnyü-, ÷-kyə-lər\ *adjective* (1953)
1 : not nuclear: as **a :** being a weapon whose destructive power is not derived from a nuclear reaction **b :** not operated by, using, or produced by nuclear energy **c :** not using or involving nuclear weapons
2 : not having nuclear weapons ⟨a *nonnuclear* country⟩

no–no \ˈnō-ˌnō\ *noun, plural* **no–no's** *or* **no–nos** (circa 1942)
: something unacceptable or forbidden

non·ob·jec·tive \ˌnän-əb-ˈjek-tiv\ *adjective* (1905)
1 : not objective
2 : representing or intended to represent no natural or actual object, figure, or scene ⟨*nonobjective* art⟩
— non·ob·jec·tiv·ism \-ti-ˌvi-zəm\ *noun*
— non·ob·jec·tiv·ist \-vist\ *noun*
— non·ob·jec·tiv·i·ty \-ˌäb-ˌjek-ˈti-və-tē, -əb-\ *noun*

non ob·stan·te \ˌnän-əb-ˈstan-tē, ˌnōn-\ *preposition* [Middle English, from Medieval Latin] (15th century)
: NOTWITHSTANDING

non–oil \ˈnän-ˈȯi(ə)l\ *adjective* (1979)
1 : not relating to, containing, or derived from oil
2 : being a net importer of petroleum or petroleum products ⟨*non-oil* nations⟩

no–nonsense *adjective* (1928)
: tolerating no nonsense **:** SERIOUS, BUSINESSLIKE

non·or·gas·mic \ˌnän-ȯr-ˈgaz-mik\ *adjective* (1973)
: not capable of experiencing orgasm

no·nox·y·nol–9 \nä-ˈnäk-si-ˌnȯl-ˈnīn, nə-, -ˌnōl-\ *noun* [*non*yl (the radical C_9H_{19}) + *oxy-* + phe*nol* + *9* (from the fact that the compounds it contains have an average of nine ethylene oxide groups per molecule)] (1980)
: a spermicide used in contraceptive products

non·para·met·ric \-ˌpar-ə-ˈme-trik\ *adjective* (1942)
: not involving the estimation of parameters of a statistical function ⟨*nonparametric* statistical tests⟩

¹non·pa·reil \ˌnän-pə-ˈrel\ *adjective* [Middle English *nonparaille*, from Middle French *nonpareil*, from *non-* + *pareil* equal, from (assumed) Vulgar Latin *pariculus*, from Latin *par* equal] (15th century)
: having no equal

²nonpareil *noun* (1593)
1 : an individual of unequaled excellence **:** PARAGON
2 a : a small flat disk of chocolate covered with white sugar pellets **b :** sugar in small pellets of various colors

non·par·ti·san \ˈnän-ˈpär-tə-zən, -sən\ *adjective* (1885)
: not partisan; *especially* **:** free from party affiliation, bias, or designation ⟨*nonpartisan* ballot⟩ ⟨a *nonpartisan* board⟩
— non·par·ti·san·ship \-ˌship\ *noun*

non·pas·ser·ine \-ˈpa-sə-ˌrīn\ *adjective* (circa 1909)
: not passerine; *especially* **:** of or relating to an order (Coraciiformes) of arboreal birds including the rollers, kingfishers, and hornbills

non·patho·gen·ic \ˌnän-ˌpa-thə-ˈje-nik\ *adjective* (1884)
: not capable of inducing disease — compare AVIRULENT

non·peak \ˈnän-ˈpēk\ *adjective* (circa 1914)
: OFF-PEAK

non·per·form·ing \ˌnän-pə(r)-ˈfȯr-miŋ\ *adjective* (1979)
: not bringing in the expected return ⟨*nonperforming* loans⟩ ⟨*nonperforming* assets⟩

non·per·sis·tent \-pər-ˈsis-tənt, -ˈzis-\ *adjective* (1900)
: not persistent: as **a :** decomposed rapidly by environmental action ⟨*nonpersistent* insecticides⟩ **b :** capable of being transmitted by a vector for only a relatively short time ⟨*nonpersistent* viruses⟩

non·per·son \ˈnän-ˈpər-sᵊn\ *noun* (circa 1909)
: a person who is regarded as nonexistent: as **a :** UNPERSON **b :** one having no social or legal status

non pla·cet \ˈnän-ˈplā-sət, ˈnōn-\ *noun* [Latin, it does not please] (1589)
: a negative vote

¹non·plus \ˌnän-ˈpləs\ *noun* [Latin *non plus* no more] (1582)
: a state of bafflement or perplexity **:** QUANDARY

²nonplus *transitive verb* **-plussed** *also* **-plused** \-ˈpləst\; **-plus·sing** *also* **-plus·ing** \-ˈplə-siŋ\ (1591)
: to cause to be at a loss as to what to say, think, or do **:** PERPLEX
synonym see PUZZLE

non·point \ˈnän-ˈpȯint\ *adjective* (1977)
: not occurring at a single well-defined site; *specifically* **:** being pollution that cannot be traced to a single location

non·po·lar \ˌnän-ˈpō-lər\ *adjective* (1892)
: not polar; *especially* **:** consisting of molecules not having a dipole ⟨a *nonpolar* solvent⟩

non pos·su·mus \ˈnän-ˈpä-sə-məs, ˈnōn-\ *noun* [Latin, we cannot] (1883)
: a statement expressing inability to do something

non·pre·scrip·tion \ˌnän-pri-ˈskrip-shən\ *adjective* (1958)
: capable of being bought without a doctor's prescription ⟨*nonprescription* drugs⟩

non·pro·duc·tive \-prə-ˈdək-tiv\ *adjective* (1901)
: not productive: as **a :** failing to produce or yield **:** UNPRODUCTIVE ⟨a *nonproductive* oil well⟩ **b :** not directly concerned with production ⟨the *nonproductive* labor of clerks and inspectors⟩ **c** *of a cough* **:** DRY
— non·pro·duc·tive·ness *noun*

non·prof·it \ˈnän-ˈprä-fət\ *adjective* (1903)
: not conducted or maintained for the purpose of making a profit ⟨a *nonprofit* organization⟩
— nonprofit *noun*

non·pro·lif·er·a·tion \ˌnän-prə-ˌli-fə-ˈrā-shən\ *adjective* (1964)
: providing for the stoppage of proliferation (as of nuclear arms) ⟨*nonproliferation* treaty⟩
— nonproliferation *noun*

non·pros \ˈnän-ˈpräs\ *transitive verb* **non·prossed; non·pros·sing** [*non pros*equitur] (1755)
: to enter a non prosequitur against

non pro·se·qui·tur \ˌnän-prə-ˈse-kwə-tər, ˌnōn-\ *noun* [Late Latin, he does not prosecute] (1768)
: a judgment entered against the plaintiff in a suit in which that party does not appear to prosecute

non·pro·tein \ˈnän-ˈprō-ˌtēn, -ˈprō-tē-ən\ *adjective* (1926)
: not being or derived from protein ⟨the *nonprotein* part of an enzyme⟩ ⟨*nonprotein* nitrogen⟩

non·read·er \-ˈrē-dər\ *noun* (1924)
1 : one who does not or cannot read
2 : a child who is very slow in learning to read
— non·read·ing \-diŋ\ *adjective*

non·re·com·bi·nant \ˌnän-(ˌ)rē-ˈkäm-bə-nənt\ *adjective* (1962)
: not exhibiting the results of genetic recombination
— nonrecombinant *noun*

non·re·course \ˈnän-ˈrē-ˌkōrs, -ˌkȯrs, -ri-ˈ\ *adjective* (1926)
: being or based on an agreement in which the lender has no right of recourse to the borrower's assets beyond stated limits ⟨a *nonrecourse* note⟩ ⟨a *nonrecourse* loan⟩

non·re·cur·rent \ˌnän-ri-ˈkər-ənt, -ˈkə-rənt\ *adjective* (circa 1864)
: not recurring

non·re·cur·ring \-ri-ˈkər-iŋ, -ˈkə-riŋ\ *adjective* (circa 1864)
: NONRECURRENT; *specifically* **:** unlikely to happen again — used of financial transactions that affect a profit and loss statement abnormally

non·re·fund·able \-ri-ˈfən-də-bəl\ *adjective* (1963)
: not subject to refunding or being refunded ⟨a *nonrefundable* bond⟩ ⟨a *nonrefundable* fee⟩
non·rel·a·tiv·is·tic \-ˌre-lə-ti-ˈvis-tik\ *adjective* (1930)
1 : not based on or involving the theory of relativity ⟨*nonrelativistic* equations⟩ ⟨*nonrelativistic* kinematics⟩
2 : of, relating to, or being a body moving at less than a relativistic velocity
— non·rel·a·tiv·is·ti·cal·ly \-ˈvis-ti-k(ə-)lē\ *adverb*
non·rep·re·sen·ta·tion·al \-ˌre-pri-ˌzen-ˈtā-shnəl, -zən-, -shə-nᵊl\ *adjective* (1923)
: NONOBJECTIVE 2
— non·rep·re·sen·ta·tion·al·ism \-shnə-ˌli-zəm, -shə-nᵊl-ˌi-zəm\ *noun*
non·res·i·dence \ˈnän-ˈre-zə-dən(t)s, -ˈrez-dən(t)s, -ˌden(t)s\ *noun* (1585)
: the state or fact of being nonresident
non·res·i·den·cy \-ˈre-zə-dən(t)-sē, -ˈrez-dən(t)-, -ˌden(t)-\ *noun* (1584)
: NONRESIDENCE
non·res·i·dent \-ˈre-zə-dənt, -ˈrez-dənt, -ˌdent\ *adjective* (1540)
: not residing in a particular place
— nonresident *noun*
non·re·sis·tance \ˌnän-ri-ˈzis-tən(t)s\ *noun* (1643)
: the principles or practice of passive submission to constituted authority even when unjust or oppressive; *also* **:** the principle or practice of not resisting violence by force
non·re·sis·tant \-tənt\ *adjective* (1702)
: not resistant; *specifically* **:** susceptible to the effects of a deleterious agent (as an insecticide, a pathogen, or an extreme environmental condition)
— nonresistant *noun*
non·re·stric·tive \-ri-ˈstrik-tiv\ *adjective* (1916)
: not restrictive; *specifically* **:** not limiting the reference of a modified word or phrase
nonrestrictive clause *noun* (1916)
: a descriptive clause that is not essential to the definiteness of the meaning of the word it modifies (as *who is retired* in "my father, who is retired, does volunteer work")
non·re·turn·able \-ri-ˈtər-nə-bəl\ *adjective* (1903)
: not returnable; *specifically* **:** not returnable to a dealer in exchange for a deposit ⟨*nonreturnable* bottles⟩
— nonreturnable *noun*
non·rig·id \ˈnän-ˈri-jəd\ *adjective* (1909)
: not rigid; *especially* **:** maintaining form by pressure of contained gas ⟨a *nonrigid* airship⟩
non·sched·uled \-ˈske-(ˌ)jü(ə)ld, -jəld\ *adjective* (1947)
: licensed to transport by air without a regular schedule ⟨*nonscheduled* airlines⟩
¹non·sci·ence \-ˈsī-ən(t)s\ *noun* (1855)
: something (as a discipline) that is not a science
²nonscience *adjective* (1944)
: of or relating to fields other than science
non·se·cre·tor \ˌnän-si-ˈkrē-tər\ *noun* (1944)
: an individual of blood group A, B, or AB who does not secrete the antigens characteristic of these blood groups in bodily fluids (as saliva)
non·sec·tar·i·an \-(ˌ)sek-ˈter-ē-ən\ *adjective* (1831)
: not having a sectarian character **:** not affiliated with or restricted to a particular religious group
non·self \ˈnän-ˈself, *Southern also* -ˈsef\ *noun* (1963)
: material that is foreign to the body of an organism
¹non·sense \ˈnän-ˌsen(t)s, ˈnän(t)-sən(t)s\ *noun* (1614)
1 a : words or language having no meaning or conveying no intelligible ideas **b** (1) **:** language, conduct, or an idea that is absurd or contrary to good sense (2) **:** an instance of absurd action
2 a : things of no importance or value **:** TRIFLES **b :** affected or impudent conduct ⟨took no *nonsense* from subordinates⟩
3 : genetic information consisting of one or more codons that do not code for any amino acid and usually cause termination of the molecular chain in protein synthesis — compare ANTISENSE, MISSENSE
— non·sen·si·cal \ˌnän-ˈsen(t)-si-kəl\ *adjective*
— non·sen·si·cal·ly \-k(ə-)lē\ *adverb*
— non·sen·si·cal·ness \-kəl-nəs\ *noun*
²nonsense *adjective* (1778)
1 : consisting of an arbitrary grouping of speech sounds or symbols ⟨\ˈshrȯg-ˌthī-əmpth\ is a *nonsense* word⟩ ⟨a *nonsense* syllable⟩
2 : consisting of one or more codons that are genetic nonsense
nonsense verse *noun* (1799)
: humorous or whimsical verse that features absurd characters and actions and often contains evocative but meaningless nonce words
non se·qui·tur \ˈnän-ˈse-kwə-tər *also* -ˌtu̇r\ *noun* [Latin, it does not follow] (1540)
1 : an inference that does not follow from the premises; *specifically* **:** a fallacy resulting from a simple conversion of a universal affirmative proposition or from the transposition of a condition and its consequent
2 : a statement (as a response) that does not follow logically from anything previously said
non·sig·nif·i·cant \ˌnän-sig-ˈni-fi-kənt\ *adjective* (1902)
: not significant: as **a :** INSIGNIFICANT **b :** MEANINGLESS **c :** having or yielding a value lying within limits between which variation is attributed to chance ⟨a *nonsignificant* statistical test⟩
— non·sig·nif·i·cant·ly *adverb*
non·sked \ˈnän-ˈsked\ *noun* [by shortening & alteration from *nonscheduled*] (1949)
: a nonscheduled transport plane or airline
non·skid \-ˈskid\ *adjective* (1904)
: designed or equipped to prevent skidding
non·slip \-ˈslip\ *adjective* (1903)
: designed to reduce or prevent slipping
non·so·cial \-ˈsō-shəl\ *adjective* (1902)
: not socially oriented **:** lacking a social component
non·sport·ing \-ˈspōr-tiŋ, -ˈspȯr-\ *adjective* (1852)
: lacking the qualities characteristic of a hunting dog
non·stan·dard \-ˈstan-dərd\ *adjective* (1923)
1 : not standard
2 : not conforming in pronunciation, grammatical construction, idiom, or word choice to the usage generally characteristic of educated native speakers of a language — compare SUBSTANDARD
non·start·er \-ˈstär-tər\ *noun* (1902)
1 : one that does not start
2 : someone or something that is not productive or effective ⟨his son has been, in politics a *nonstarter* —Anthony Lejeune⟩
non·ste·roi·dal \ˌnän-stə-ˈrȯi-dᵊl\ *also* **non·ste·roid** \ˈnän-ˈstir-ˌȯid, -ˈster-\ *adjective* (1964)
: of, relating to, or being a compound and especially a drug that is not a steroid
— nonsteroid *noun*
non·stick \ˈnän-ˈstik\ *adjective* [³*stick*] (1958)
: allowing of easy removal of cooked food particles ⟨a *nonstick* coating in a frying pan⟩
non·stop \ˈnän-ˈstäp\ *adjective* (1902)
: done, made, or held without a stop **:** not easing or letting up
— nonstop *adverb*
non·such \ˈnən-ˌsəch *also* ˈnän-\ *variant of* NONESUCH
non·suit \ˈnän-ˈsüt\ *noun* [Middle English, from Anglo-French *nounsuyte,* from *noun-* non- + Old French *siute* following, pursuit — more at SUIT] (14th century)
: a judgment against a plaintiff for failure to prosecute a case or inability to establish a prima facie case
— nonsuit *transitive verb*
non·sup·port \ˌnän-sə-ˈpōrt, -ˈpȯrt\ *noun* (1909)
: failure to support; *specifically* **:** failure (as of a parent) to honor a statutory or contractual obligation to provide maintenance
non·syl·lab·ic \-sə-ˈla-bik\ *adjective* (circa 1909)
: not constituting a syllable or the nucleus of a syllable ⟨the second vowel of a falling diphthong is *nonsyllabic* (as \i\ in \ȯi\)⟩
non·sys·tem \ˈnän-ˈsis-təm\ *noun* (1964)
: a system that lacks effective organization
non·tar·get \-ˈtär-gət\ *adjective* (1945)
: not being the intended object of action by a particular agent ⟨effect of insecticides on *nontarget* organisms⟩
non·ter·mi·nat·ing \-ˈtər-mə-ˌnā-tiŋ\ *adjective* (circa 1908)
: not terminating or ending; *especially* **:** being a decimal for which there is no place to the right of the decimal point such that all places farther to the right contain the entry 0 ⟨1/3 gives the *nonterminating* decimal .33333 . . .⟩
non·ther·mal \-ˈthər-məl\ *adjective* (circa 1964)
: not produced by heat; *specifically* **:** of, relating to, or being radiation having a spectrum that is not the spectrum of a blackbody
non·ti·tle \-ˈtī-tᵊl\ *adjective* (1968)
: of, relating to, or being an athletic contest in which a title is not at stake
non trop·po \ˈnän-ˈtrȯ-(ˌ)pō, ˈnōn-\ *adverb or adjective* [Italian, literally, not too much] (circa 1854)
: without excess — used to qualify a direction in music
non–U \ˈnän-ˈyü\ *adjective* (1954)
: not characteristic of the upper classes
non·union \-ˈyün-yən\ *adjective* (1863)
1 : not belonging to or connected with a trade union ⟨*nonunion* carpenters⟩
2 : not recognizing or favoring trade unions or their members
3 : not produced or worked on by members of a trade union ⟨*nonunion* lettuce⟩
non·use \ˈnän-ˈyüs\ *noun* (1542)
1 : failure to use ⟨*nonuse* of available material⟩
2 : the fact or condition of not being used
non·us·er \-ˈyü-zər\ *noun* (1926)
: one who does not make use of something (as an available public facility or a harmful drug)
non·van·ish·ing \-ˈva-ni-shiŋ\ *adjective* (1907)
: not zero or becoming zero
non·vec·tor \-ˈvek-tər\ *noun* (1956)
: an organism (as an insect) that does not transmit a particular pathogen (as a virus)
non·ver·bal \-ˈvər-bəl\ *adjective* (1924)
: not verbal: as **a :** being other than verbal ⟨*nonverbal* factors⟩ **b :** involving minimal use of language ⟨*nonverbal* tests⟩ **c :** ranking low in verbal skill
— non·ver·bal·ly \-bə-lē\ *adverb*
non·vin·tage \-ˈvin-tij\ *adjective* (1924)
: undated and usually blended to approximate a standard ⟨a *nonvintage* wine⟩
non·vi·o·lence \-ˈvī-ə-lən(t)s\ *noun* (1920)
1 : abstention from violence as a matter of principle; *also* **:** the principle of such abstention
2 a : the quality or state of being nonviolent **:** avoidance of violence **b :** nonviolent demonstrations for the purpose of securing political ends ⟨studied the history and techniques of *nonviolence*⟩
non·vi·o·lent \-lənt\ *adjective* (1920)
: abstaining or free from violence
— non·vi·o·lent·ly *adverb*
non·vol·a·tile \-ˈvä-lə-tᵊl\ *adjective* (1866)

: not volatile: as **a :** not vaporizing readily ⟨a *nonvolatile* solvent⟩ **b** *of a computer memory* **:** retaining data when power is shut off

non–West·ern \-ˈwes-tərn\ *adjective* (1902)
1 : not being part of the western tradition ⟨*non-Western* countries⟩
2 : of or relating to non-Western societies ⟨*non-Western* values⟩

non·white \-ˈhwīt, -ˈwīt\ *noun* (1927)
: a person whose features and especially whose skin color are distinctively different from those of peoples of northwestern Europe; *especially* **:** one who has African ancestors of the black race
— nonwhite *adjective*

non·word \-ˈwərd\ *noun* (1961)
: a word that has no meaning, is not known to exist, or is disapproved

non·wo·ven \-ˈwō-vən\ *adjective* (1945)
1 : made of fibers held together by interlocking or bonding (as by chemical or thermal means) **:** not woven, knitted, or felted ⟨*nonwoven* fabric⟩
2 : made of nonwoven fabric ⟨a *nonwoven* dress⟩
— nonwoven *noun*

non·ze·ro \-ˈzē-(ˌ)rō, -ˈzir-(ˌ)ō\ *adjective* (1905)
1 : being, having, or involving a value other than zero
2 : having phonetic content ⟨*nonzero* affixes⟩

¹noo·dle \ˈnü-dᵊl\ *noun* [perhaps alteration of *noddle*] (1753)
1 : a stupid person **:** SIMPLETON
2 : HEAD, NOGGIN

²noodle *noun* [German *Nudel*] (1779)
: a food paste made with egg and shaped typically in ribbon form

³noodle *intransitive verb* **noo·dled; noo·dling** \ˈnüd-liŋ, ˈnü-dᵊl-iŋ\ [imitative] (circa 1937)
: to improvise on an instrument in an informal or desultory manner

nook \ˈnu̇k\ *noun* [Middle English *noke, nok*] (14th century)
1 *chiefly Scottish* **:** a right-angled corner
2 a : an interior angle formed by two meeting walls **:** RECESS **b :** a secluded or sheltered place or part ⟨searched every *nook* and cranny⟩

nooky \ˈnu̇-kē\ *noun* [perhaps from *nook* + ⁴-*y*] (1928)
1 : SEXUAL INTERCOURSE — often considered vulgar
2 : the female partner in sexual intercourse — often considered vulgar

noon \ˈnün\ *noun* [Middle English, from Old English *nōn* ninth hour from sunrise, from Latin *nona,* from feminine of *nonus* ninth; akin to Latin *novem* nine — more at NINE] (13th century)
1 : MIDDAY; *specifically* **:** 12 o'clock at midday
2 *archaic* **:** MIDNIGHT — used chiefly in the phrase *noon of night*
3 : the highest point ◆

noon·day \-ˌdā\ *noun* (1535)
: MIDDAY

no one *pronoun* (before 12th century)
: no person **:** NOBODY
usage see NOBODY

noon·ing \ˈnü-niŋ, -nən\ *noun* (circa 1652)
1 *chiefly dialect* **:** a meal eaten at noon
2 *chiefly dialect* **:** a period at noon for eating or resting

noon·tide \ˈnün-ˌtīd\ *noun* (12th century)
1 : NOONTIME
2 : the culminating point

noon·time \-ˌtīm\ *noun* (14th century)
: the time of noon **:** MIDDAY

¹noose \ˈnüs, *British also* ˈnüz\ *noun* [Middle English *nose,* perhaps of Provençal origin; akin to Provençal *nous* knot, from Latin *nodus* — more at NODE] (15th century)
1 : a loop with a running knot that binds closer the more it is drawn
2 : something that snares like a noose

²noose *transitive verb* **noosed; noos·ing** (1600)
1 : to secure by a noose
2 : to make a noose in or of

noo·sphere \ˈnō-ə-ˌsfir\ *noun* [International Scientific Vocabulary *noo-* mind (from Greek *noos, nous*) + *sphere* sphere] (1945)
: the sphere of human consciousness and mental activity especially in regard to its influence on the biosphere and in relation to evolution

Noot·ka \ˈnu̇t-kə\ *noun, plural* **Nootka** *or* **Nootkas** (1841)
1 : a member of a Wakashan people inhabiting the West Coast of Vancouver Island
2 : the language of the Nootka people

no·pal \nō-ˈpäl, -ˈpal; ˈnō-pəl\ *noun* [Spanish, from Nahuatl *nohpalli*] (1730)
: any of a genus (*Nopalea*) of cacti of Mexico and Central America that differ from the prickly pears in having erect petals and scarlet flowers with the stamens much longer than the petals; *broadly* **:** PRICKLY PEAR

no–par *or* **no–par–value** *adjective* (1922)
: having no nominal value ⟨*no-par* stocks⟩

nope \ˈnōp, *or with glottal stop instead of* p\ *adverb* [by alteration] (1888)
: NO

¹nor \nər, ˈnȯr, *Southern also* ˈnär\ *conjunction* [Middle English, contraction of *nother* neither, nor, from *nother,* pronoun & adjective, neither — more at NEITHER] (14th century)
1 — used as a function word to introduce the second or last member or the second and each following member of a series of items each of which is negated ⟨neither here *nor* there⟩ ⟨not done by you *nor* me *nor* anyone⟩
2 — used as a function word to introduce and negate a following clause or phrase
3 *chiefly British* **:** NEITHER

²nor *conjunction* [Middle English, perhaps from ¹*nor*] (15th century)
dialect **:** THAN

nor- *combining form* [*nor*mal]
: homologue containing one less methyl group ⟨*nor*epinephrine⟩

NOR \ˈnȯr\ *noun* [not OR] (1957)
: a computer logic circuit that produces an output that is the inverse of that of an OR circuit

nor·adren·a·line *also* **nor·adren·a·lin** \ˌnȯr-ə-ˈdre-nᵊl-ən\ *noun* (1932)
: NOREPINEPHRINE

nor·ad·ren·er·gic \ˌnȯr-ˌa-drə-ˈnər-jik\ *adjective* [*noradren*aline + *-ergic*] (1963)
: liberating, activated by, or involving norepinephrine in the transmission of nerve impulses ⟨*noradrenergic* nerve endings⟩ ⟨*noradrenergic* nerve fibers⟩

¹Nor·dic \ˈnȯr-dik\ *adjective* [French *nordique,* from *nord* north, from Old English *north*] (1898)
1 : of or relating to the Germanic peoples of northern Europe and especially of Scandinavia
2 : of or relating to a group or physical type of the Caucasian race characterized by tall stature, long head, light skin and hair, and blue eyes
3 a : of or relating to competitive ski events involving cross-country racing, ski jumping, or biathlon — compare ALPINE **b :** of, relating to, or being cross-country skiing

²Nordic *noun* (1901)
1 : a native of northern Europe
2 : a person of Nordic physical type
3 : a member of the peoples of Scandinavia

nor·epi·neph·rine \ˈnȯr-ˌe-pə-ˈne-frən\ *noun* (1945)
: a catecholamine $C_8H_{11}NO_3$ that is the chemical means of transmission across synapses in postganglionic neurons of the sympathetic nervous system and in some parts of the central nervous system, is a vasopressor hormone of the adrenal medulla, and is a precursor of epinephrine in its major biosynthetic pathway

nor·eth·in·drone \ˌnȯ-ˈre-thən-ˌdrōn\ *noun* [*nor-* + *ethin*yl + hy*dr-* + *-one* (as in *progesterone*)] (1958)
: a synthetic progestational hormone $C_{20}H_{26}O_2$ used in oral contraceptives often in the form of its acetate

Nor·folk jacket \ˈnȯr-fək-, -ˌfȯk-\ *noun* [*Norfolk,* England] (1866)
: a loose-fitting belted single-breasted jacket with box pleats

Norfolk terrier *noun* (1964)
: any of a breed of dogs developed in England and resembling the Norwich terrier but having folded-over ears

no·ri \ˈnō-rē, ˈnȯr-ē\ *noun* [Japanese] (1892)
: dried laver seaweed pressed into thin sheets and used especially as a seasoning or as a wrapper for sushi

nor·land \ˈnȯr-lənd\ *noun* (circa 1578)
chiefly dialect **:** NORTHLAND

norm \ˈnȯrm\ *noun* [Latin *norma,* literally, carpenter's square] (1674)
1 : an authoritative standard **:** MODEL
2 : a principle of right action binding upon the members of a group and serving to guide, control, or regulate proper and acceptable behavior
3 : AVERAGE: as **a :** a set standard of development or achievement usually derived from the average or median achievement of a large group **b :** a pattern or trait taken to be typical in the behavior of a social group **c :** a widespread practice, procedure, or custom **:** RULE ⟨standing ovations became the *norm*⟩
4 a : a real-valued nonnegative function defined on a vector space and satisfying the conditions that the function is zero if and only if the vector is zero, the function of the product of a scalar and a vector is equal to the product of the absolute value of the scalar and the function of the vector, and the function of the sum of two vectors is less than or equal to the sum of the functions of the two vectors; *specifically* **:** the square root of the sum of the squares of the absolute values of the elements of a matrix or of the components of a vector **b :** the greatest distance between two successive points of a set of points that partition an interval into smaller intervals
synonym see AVERAGE

¹nor·mal \ˈnȯr-məl\ *adjective* [Latin *normalis,* from *norma*] (circa 1696)
1 : PERPENDICULAR; *especially* **:** perpendicular to a tangent at a point of tangency

◇ WORD HISTORY

noon *Noon* has not always indicated that time of day at which the sun is most nearly overhead. According to the Roman method of reckoning time, the hours of the day were counted from sunrise to sunset, so that *nona,* short for *nona hora* "ninth hour," denoted the ninth hour of daylight, about 3 P.M. in modern reckoning. *Nona* was borrowed into Old English as *nōn,* which denoted either the ninth hour of the day or the liturgical office of nones that was read at this time. In early Middle English, however, *non* or *none* (later *noon*) was applied to a meal eaten before nones was read, perhaps originally an afternoon snack. If this was the case, however, the meaning seems soon to have shifted to the major midday meal, and then to midday itself, a sense that Modern English *noon* has retained. The word *none* or *nones* was later borrowed from Latin to refer specifically to the liturgical office.

2 a : according with, constituting, or not deviating from a norm, rule, or principle **b :** conforming to a type, standard, or regular pattern
3 : occurring naturally ⟨*normal* immunity⟩
4 a : of, relating to, or characterized by average intelligence or development **b :** free from mental disorder **:** SANE
5 a *of a solution* **:** having a concentration of one gram equivalent of solute per liter **b :** containing neither basic hydroxyl nor acid hydrogen ⟨*normal* silver phosphate⟩ **c :** not associated ⟨*normal* molecules⟩ **d :** having a straight-chain structure ⟨*normal* pentane⟩ ⟨*normal* butyl alcohol⟩
6 *of a subgroup* **:** having the property that every coset produced by operating on the left by a given element is equal to the coset produced by operating on the right by the same element
7 : relating to, involving, or being a normal curve or normal distribution ⟨*normal* approximation to the binomial distribution⟩
8 *of a matrix* **:** having the property of commutativity under multiplication by the transpose of the matrix each of whose elements is a conjugate complex number with respect to the corresponding element of the given matrix ◆ ***synonym*** see REGULAR
— nor·mal·i·ty \nȯr-ˈma-lə-tē\ *noun*
— nor·mal·ly \ˈnȯr-mə-lē\ *adverb*

[2]**normal** *noun* (circa 1738)
1 a : a normal line **b :** the portion of a normal line to a plane curve between the curve and the x-axis
2 : one that is normal
3 : a form or state regarded as the norm **:** STANDARD

normal curve *noun* (1894)
: the symmetrical bell-shaped curve of a normal distribution

nor·mal·cy \ˈnȯr-məl-sē\ *noun* (1857)
: the state or fact of being normal

normal distribution *noun* (1897)
: a probability density function that approximates the distribution of many random variables (as the proportion of outcomes of a particular sort in a large number of independent repetitions of an experiment in which the probabilities remain constant from trial to trial) and that has the form

$$f(x) = \frac{1}{\sigma\sqrt{2\pi}}\, e^{-\frac{1}{2}\left(\frac{x-\mu}{\sigma}\right)^2}$$

where μ is the mean and σ is the standard deviation — compare NORMAL CURVE

nor·mal·ise *British variant of* NORMALIZE

nor·mal·ize \ˈnȯr-mə-ˌlīz\ *transitive verb* **-ized; -iz·ing** (1865)
1 : to make conform to or reduce to a norm or standard
2 : to make normal (as by a transformation of variables)
3 : to bring or restore (as relations between countries) to a normal condition
— nor·mal·iz·able \-ˌlī-zə-bəl\ *adjective*
— nor·mal·i·za·tion \ˌnȯr-mə-lə-ˈzā-shən\ *noun*

nor·mal·iz·er \ˈnȯr-mə-ˌlī-zər\ *noun* (1926)
1 : one that normalizes
2 a : a subgroup consisting of those elements of a group for which the group operation with regard to a given element is commutative **b :** the set of elements of a group for which the group operation with regard to every element of a given subgroup is commutative

normal school *noun* [translation of French *école normale;* from the fact that the first French school so named was intended to serve as a model] (1839)
: a usually 2-year school for training chiefly elementary teachers

Nor·man \ˈnȯr-mən\ *noun* [Middle English, from Old French *Normant,* from Old Norse *Northmann-, Northmathr* Norseman, from *northr* north + *mann-, mathr* man; akin to Old English *north* north and to Old English *man* man] (13th century)
1 : a native or inhabitant of Normandy: **a :** one of the Scandinavian conquerors of Normandy in the 10th century **b :** one of the Norman-French conquerors of England in 1066
2 : NORMAN-FRENCH
— Norman *adjective*

Norman architecture *noun* (1797)
: a Romanesque style first appearing in and near Normandy about A.D. 950; *also* **:** architecture resembling or imitating this style

nor·mande \nȯr-ˈmänd\ *adjective* [French, from feminine of *normand* Norman, from *Normandy,* France] (circa 1929)
: prepared with any of several foods traditionally associated with Normandy (as cream, apples, cider, or calvados) ⟨veal *normande*⟩

Norman–French *noun* (1605)
1 : the French language of the medieval Normans
2 : the modern dialect of Normandy

nor·ma·tive \ˈnȯr-mə-tiv\ *adjective* [French *normatif,* from *norme* norm, from Latin *norma*] (1878)
1 : of, relating to, or determining norms or standards ⟨*normative* tests⟩
2 : conforming to or based on norms ⟨*normative* behavior⟩ ⟨*normative* judgments⟩
3 : prescribing norms ⟨*normative* rules of ethics⟩ ⟨*normative* grammar⟩
— nor·ma·tive·ly *adverb*
— nor·ma·tive·ness *noun*

normed \ˈnȯrmd\ *adjective* (1935)
: being a mathematical entity upon which a norm is defined ⟨a *normed* vector space⟩

nor·mo·ten·sive \ˌnȯr-mō-ˈten(t)-siv\ *adjective* [*normal* + *-o-* + *-tensive* (as in *hypotensive*)] (circa 1941)
: having blood pressure typical of the group to which one belongs
— normotensive *noun*

nor·mo·ther·mia \-ˈthər-mē-ə\ *noun* [New Latin, from *normal*is normal + *-o-* + *-thermia* -thermy] (1960)
: normal body temperature
— nor·mo·ther·mic \-mik\ *adjective*

Norn \ˈnȯrn\ *noun* [Old Norse]
: any of the three Norse goddesses of fate

[1]**Norse** \ˈnȯrs\ *noun, plural* **Norse** [probably from obsolete Dutch *noorsch,* adjective, Norwegian, Scandinavian, alteration of obsolete Dutch *noordsch* northern, from Dutch *noord* north; akin to Old English *north* north] (circa 1688)
1 a : NORWEGIAN 2 **b :** any of the western Scandinavian dialects or languages **c :** the Scandinavian group of Germanic languages
2 *plural* **a :** SCANDINAVIANS **b :** NORWEGIANS

[2]**Norse** *adjective* (1768)
1 : of or relating to ancient Scandinavia or the language of its inhabitants
2 : NORWEGIAN

Norse·man \ˈnȯrs-mən\ *noun* (1817)
: any of the ancient Scandinavians

[1]**north** \ˈnȯrth\ *adverb* [Middle English, from Old English; akin to Old High German *nord* north and perhaps to Greek *nerteros* lower, infernal] (before 12th century)
: to, toward, or in the north

[2]**north** *adjective* (before 12th century)
1 : situated toward or at the north ⟨the *north* entrance⟩
2 : coming from the north ⟨a *north* wind⟩

[3]**north** *noun* (13th century)
1 a : the direction of the north terrestrial pole **:** the direction to the left of one facing east **b :** the compass point directly opposite to south
2 *capitalized* **a :** regions or countries lying to the north of a specified or implied point of orientation **b :** the industrially and economically developed nations of the world — compare SOUTH
3 *often capitalized* **a :** the one of four positions at 90-degree intervals that lies to the north or at the top of a diagram **b :** a person occupying this position in the course of a specified activity (as the game of bridge)

north·bound \ˈnȯrth-ˌbau̇nd\ *adjective* (1903)
: traveling or headed north

north by east (1725)
: a compass point that is one point east of due north **:** N11°15′E

north by west (circa 1771)
: a compass point that is one point west of due north **:** N11°15′W

[1]**north·east** \nȯr-ˈthēst, *nautical* nȯ-ˈrēst\ *adverb* (before 12th century)
: to, toward, or in the northeast

[2]**northeast** *adjective* (before 12th century)
1 : coming from the northeast ⟨a *northeast* wind⟩
2 : situated toward or at the northeast ⟨the *northeast* corner⟩

[3]**northeast** *noun* (12th century)
1 a : the general direction between north and east **b :** the point midway between the north and east compass points
2 *capitalized* **:** regions or countries lying to the northeast of a specified or implied point of orientation

northeast by east (circa 1771)
: a compass point that is one point east of due northeast **:** N56°15′E

northeast by north (circa 1771)
: a compass point that is one point north of due northeast **:** N33°45′E

north·east·er \nȯr-ˈthē-stər, nȯ-ˈrē-\ *noun* (1774)
1 : a strong northeast wind
2 : a storm with northeast winds

north·east·er·ly \nȯr-ˈthē-stər-lē\ *adverb or adjective* (1739)
1 : from the northeast
2 : toward the northeast

north·east·ern \-stərn\ *adjective* (14th century)
1 *often capitalized* **:** of, relating to, or characteristic of a region conventionally designated Northeast
2 : lying toward or coming from the northeast
— north·east·ern·most \-stərn-ˌmōst\ *adjective*

North·east·ern·er \-stə(r)-nər\ *noun* (1961)
: a native or inhabitant of a northeastern region (as of the U.S.)

[1]**north·east·ward** \nȯr-ˈthēs-twərd, nȯ-ˈrēs-\ *adverb or adjective* (1553)
: toward the northeast
— north·east·wards \-twərdz\ *adverb*

[2]**northeastward** *noun* (1892)
: NORTHEAST

north·er \ˈnȯr-thər\ *noun* (1820)
1 : a strong north wind
2 : a storm with north winds

[1]**north·er·ly** \-lē\ *adjective or adverb* [[3]*north* + *-erly* (as in *easterly*)] (1551)

◇ WORD HISTORY
normal In Latin *norma* was the word for a carpenter's square, the tool which ensures that a carpenter can regularly reproduce corners and edges that are straight and that form right angles. *Norma* also had the extended sense "standard" or "pattern." Similarly, *normalis,* the Latin adjective formed from *norma,* had the literal meaning "forming a right angle" and by the Late Latin period the extended sense "according to rule." It is from this extended sense that we derive most of the English senses of *normal.* Incidentally, *normal school,* as a name for a college devoted to teacher training, is a translation of the French term *école normale.* The term derives from the fact that the first school named *école normale* was intended to serve as a standard or model for other teaching schools.

1 : situated toward or belonging to the north ⟨the *northerly* border⟩
2 : coming from the north ⟨a *northerly* wind⟩
²northerly *noun, plural* **-lies** (1955)
: a wind from the north
¹north·ern \'nȯr-t͟hə(r)n\ *adjective* [Middle English *northerne,* from Old English; akin to Old High German *nordrōni* northern, Old English *north* north] (before 12th century)
1 *capitalized* **a :** of, relating to, or characteristic of a region conventionally designated North **b :** of, relating to, or constituting the northern dialect
2 a : lying toward the north **b :** coming from the north ⟨a *northern* storm⟩
— north·ern·most \-ˌmōst\ *adjective*
²northern *noun* (1950)
1 *capitalized* **:** the dialect of English spoken in the part of the U.S. north of a line running northwest through central New Jersey, below the northern tier of counties in Pennsylvania, through northern Ohio, Indiana, and Illinois, across central Iowa, and through the northwest corner of South Dakota
2 : ³PIKE 1a
northern corn rootworm *noun* (1952)
: a corn rootworm (*Diabrotica barberi* synonym *D. longicornis*) often destructive to Indian corn in the northern parts of the central and eastern U.S.
Northern Cross *noun*
: a cross formed by six stars in Cygnus
North·ern·er \'nȯr-t͟hə(r)-nər\ *noun* (1599)
: a native or inhabitant of the North; *especially* **:** a native or resident of the northern part of the U.S.
northern harrier *noun* (1980)
: a widely distributed brown or grayish hawk (*Circus cyaneus*) that inhabits open and marshy regions and has a conspicuous white patch on the rump. — called also *marsh hawk*

northern harrier

northern hemisphere *noun, often N&H capitalized* (circa 1771)
: the half of the earth that lies north of the equator
northern lights *noun plural* (14th century)
: AURORA BOREALIS
northern oriole *noun* (1974)
: an American oriole (*Icterus galbula*) that is characterized by the black and orange plumage of the male and that includes the Baltimore oriole of the eastern U.S. and a more western form (*I. g. bullockii*) with a black and orange head
northern pike *noun* (1856)
: ³PIKE 1a
northern white cedar *noun* (1926)
: an arborvitae (*Thuja occidentalis*) of eastern North America that has branchlets in horizontal planes; *also* **:** its wood — called also *white cedar*
North Germanic *noun* (circa 1930)
: a subdivision of the Germanic languages including Icelandic, Norwegian, Swedish, and Danish — see INDO-EUROPEAN LANGUAGES table
north·ing \'nȯr-thiŋ, -t͟hiŋ\ *noun* (1669)
1 : difference in latitude to the north from the last preceding point of reckoning
2 : northerly progress
north·land \'nȯrth-ˌland, -lənd\ *noun, often capitalized* (before 12th century)
: land in the north **:** the north of a country
North·man \'nȯrth-mən\ *noun* (before 12th century)
: NORSEMAN
north–north·east \'nȯrth-ˌnȯr-'thēst, -ˌnȯ-'rēst\ *noun* (14th century)
: a compass point that is two points east of due north **:** N22°30′E
north–north·west \'nȯrth-ˌnȯr(th)-'west\ *noun* (14th century)
: a compass point that is two points west of due north **:** N22°30′W
north pole *noun* (14th century)
1 a *often N&P capitalized* **:** the northernmost point of the earth; *broadly* **:** the corresponding point of a celestial body (as a planet) **b :** the zenith of the heavens as viewed from the north terrestrial pole
2 *of a magnet* **:** the pole that points toward the north when the magnet is freely suspended
north–seeking pole *noun* (circa 1920)
: NORTH POLE 2
North Star *noun*
: the star of the northern hemisphere toward which the axis of the earth points — called also *polestar*
¹North·um·bri·an \nȯr-'thəm-brē-ən\ *adjective* (1622)
1 : of, relating to, or characteristic of ancient Northumbria, its people, or its language
2 : of, relating to, or characteristic of Northumberland, its people, or its language
²Northumbrian *noun* (1752)
1 : a native or inhabitant of ancient Northumbria
2 : a native or inhabitant of Northumberland
3 a : the Old English dialect of Northumbria **b :** the Modern English dialect of Northumberland
¹north·ward \'nȯrth-wərd\ *adverb or adjective* (before 12th century)
: toward the north
— north·wards \-wərdz\ *adverb*
²northward *noun* (14th century)
: northward direction or part ⟨sail to the *northward*⟩
¹north·west \nȯrth-'west, *nautical* nȯr-'west\ *adverb* (before 12th century)
: to, toward, or in the northwest
²northwest *adjective* (before 12th century)
1 : coming from the northwest ⟨a *northwest* wind⟩
2 : situated toward or at the northwest ⟨the *northwest* corner⟩
³northwest *noun* (12th century)
1 a : the general direction between north and west **b :** the point midway between the north and west compass points
2 *capitalized* **:** regions or countries lying to the northwest of a specified or implied point of orientation
northwest by north (circa 1771)
: a compass point that is one point north of due northwest **:** N33°45′W
northwest by west (circa 1771)
: a compass point that is one point west of due northwest **:** N56°15′W
north·west·er \nȯr(th)-'wes-tər\ *noun* (1737)
: a strong northwest wind
north·west·er·ly \-lē\ *adverb or adjective* (circa 1611)
1 : from the northwest
2 : toward the northwest
north·west·ern \-'wes-tərn\ *adjective* (1612)
1 *often capitalized* **:** of, relating to, or characteristic of a region conventionally designated Northwest
2 : lying toward or coming from the northwest
— north·west·ern·most \-ˌmōst\ *adjective*
North·west·ern·er \-'wes-tə(r)-nər\ *noun* (1955)
: a native or inhabitant of the Northwest and especially of the northwestern part of the U.S.
¹north·west·ward \-'wes-twərd\ *adverb or adjective* (14th century)
: toward the northwest
— north·west·wards \-twərdz\ *adverb*
²northwestward *noun* (1796)
: NORTHWEST
nor·trip·ty·line \nȯr-'trip-tə-ˌlēn\ *noun* [*nor-* + *-tript-* (perhaps from *tri*cyclic + he*pt*a-) + *-yl* + ²*-ine*] (1964)
: a tricyclic antidepressant $C_{19}H_{21}N$
Nor·way maple \'nȯr-ˌwā-\ *noun* (1797)
: a European maple (*Acer platanoides*) with dark green or often reddish or red veined leaves that is much planted for shade in the U.S.
Norway rat *noun* (circa 1759)
: BROWN RAT
Norway spruce *noun* (1797)
: a widely cultivated spruce (*Picea abies*) that is native to northern Europe and has a pyramidal shape, spreading branches and pendulous branchlets, dark foliage, and long pendulous cones
Nor·we·gian \nȯr-'wē-jən\ *noun* [Medieval Latin *Norwegia* Norway] (1605)
1 a : a native or inhabitant of Norway **b :** a person of Norwegian descent
2 : the Germanic language of the Norwegian people
— Norwegian *adjective*
Norwegian elkhound *noun* (1930)
: any of a Norwegian breed of short-bodied medium-sized dogs with erect ears and a very heavy coat of gray hairs with black tips

Norwegian elkhound

Nor·wich terrier \'nȯr-(ˌ)wich-, *British* 'När-ich- *or* 'När-ij-\ *noun* [*Norwich,* England] (1931)
: any of an English breed of small active low-set terriers that have a long straight wiry coat and erect ears
¹nose \'nōz\ *noun* [Middle English, from Old English *nosu;* akin to Old High German *nasa* nose, Latin *nasus*] (before 12th century)
1 a : the part of the face that bears the nostrils and covers the anterior part of the nasal cavity; *broadly* **:** this part together with the nasal cavity **b :** the anterior part of the head at the top or end of the muzzle **:** SNOUT, PROBOSCIS
2 a : the sense of smell **:** OLFACTION **b :** AROMA, BOUQUET
3 : the vertebrate olfactory organ
4 a : the forward end or projection of something **b :** the projecting or working end of a tool
5 : the stem of a boat or its protective metal covering
6 a : the nose as a symbol of prying or meddling curiosity or interference **b :** a knack for discovery or understanding ⟨a keen *nose* for absurdity⟩
— on the nose 1 a : at or to a target point ⟨the bombs landed right *on the nose*⟩ **b** (1) **:** on target **:** ACCURATE (2) **:** ACCURATELY **2 :** to win — used of horse or dog racing bets
²nose *verb* **nosed; nos·ing** (circa 1507)
transitive verb
1 : to detect by or as if by smell **:** SCENT
2 a : to push or move with the nose **b :** to move (as a vehicle) ahead slowly or cautiously ⟨*nosed* my car into the parking space⟩
3 : to touch or rub with the nose **:** NUZZLE
intransitive verb
1 : to use the nose in examining, smelling, or showing affection
2 : to search impertinently **:** PRY
3 : to move ahead slowly or cautiously ⟨the boat *nosed* around the bend⟩

\ə\ **abut** \ᵊ\ **kitten** \ər\ **further** \a\ **ash** \ā\ **ace** \ä\ **mop, mar** \au̇\ **out** \ch\ **chin** \e\ **bet** \ē\ **easy** \g\ **go** \i\ **hit** \ī\ **ice** \j\ **job** \ŋ\ **sing** \ō\ **go** \ȯ\ **law** \ȯi\ **boy** \th\ **thin** \t͟h\ **the** \ü\ **loot** \u̇\ **foot** \y\ **yet** \zh\ **vision** *see also* Guide to Pronunciation

nose•band \ˈnōz-ˌband\ *noun* (1611)
: the part of a headstall that passes over a horse's nose

nose•bleed \-ˌblēd\ *noun* (1848)
: an attack of bleeding from the nose

nose cone *noun* (1949)
: a protective cone constituting the forward end of an aerospace vehicle

nosed \ˈnōzd\ *adjective* (1505)
: having a nose especially of a specified kind — usually used in combination ⟨snub-*nosed*⟩

nose•dive \ˈnōz-ˌdīv\ *noun* (1912)
1 : a downward nose-first plunge of a flying object (as an airplane)
2 : a sudden extreme drop ⟨stock prices took a *nosedive*⟩
— nose–dive *intransitive verb*

no–see–um \nō-ˈsē-əm\ *noun* [from the words (as supposedly spoken by American Indians) *no see um* you don't see them] (1848)
: BITING MIDGE

nose•gay \ˈnōz-ˌgā\ *noun* [Middle English, from *nose* nose + *gay* ornament, literally, something gay, from *gay*] (15th century)
: a small bunch of flowers **:** POSY

nose•guard \-ˌgärd\ *noun* (1950)
: a defensive lineman in football who plays opposite the offensive center

nose job *noun* (1963)
: RHINOPLASTY

nose out *transitive verb* (circa 1630)
1 : to discover often by prying
2 : to defeat by a narrow margin

nose•piece \-ˌpēs\ *noun* (1611)
1 : a piece of armor for protecting the nose
2 : the end piece of a microscope body to which an objective is attached
3 : the bridge of a pair of eyeglasses

nose tackle *noun* (1977)
: NOSEGUARD

nose•wheel \-ˌhwēl, -ˌwēl\ *noun* (1934)
: a landing-gear wheel under the nose of an airplane

nos•ey par•ker \ˈnō-zē-ˈpär-kər\ *noun, often N&P capitalized* [probably from the name *Parker*] (1907)
chiefly British **:** BUSYBODY

[1]nosh \ˈnäsh\ *noun* (1952)
: a light meal **:** SNACK

[2]nosh *verb* [Yiddish *nashn,* from Middle High German *naschen* to eat on the sly] (1956)
intransitive verb
: to eat a snack
transitive verb
: CHEW, MUNCH
— nosh•er *noun*

no–show \ˈnō-ˌshō, -ˈshō\ *noun* (1941)
1 : a person who reserves space (as on an airplane) but neither uses nor cancels the reservation
2 : a person who buys a ticket (as to a sporting event) but does not attend; *broadly* **:** a person who is expected but who does not show up
3 : failure to show up

nos•ing \ˈnō-ziŋ\ *noun* (circa 1775)
: the usually rounded edge of a stair tread that projects over the riser; *also* **:** a similar rounded projection

nos•o•co•mi•al \ˌnä-sə-ˈkō-mē-əl\ *adjective* [Late Latin *nosocomium* hospital, from Late Greek *nosokomeion,* from Greek *nosokomos* one who tends the sick, from *nosos* disease + *-komos;* akin to Greek *kamnein* to suffer, toil, Sanskrit *śāmyati* he tires] (circa 1843)
: originating or taking place in a hospital ⟨*nosocomial* infection⟩

no•sol•o•gy \nō-ˈsä-lə-jē, -ˈzä-\ *noun* [probably from New Latin *nosologia,* from Greek *nosos* disease + New Latin *-logia* -logy] (circa 1721)
1 : a classification or list of diseases
2 : a branch of medical science that deals with classification of diseases
— no•so•log•i•cal \ˌnō-sə-ˈlä-ji-kəl\ *or* **no•so•log•ic** \-jik\ *adjective*
— no•so•log•i•cal•ly \-ji-k(ə-)lē\ *adverb*

nos•tal•gia \nä-ˈstal-jə, nə- *also* nȯ-, nō-; nə-ˈstäl-\ *noun* [New Latin, from Greek *nostos* return home + New Latin *-algia;* akin to Greek *neisthai* to return, Old English *genesan* to survive, Sanskrit *nasate* he approaches] (1770)
1 : the state of being homesick **:** HOMESICKNESS
2 : a wistful or excessively sentimental yearning for return to or of some past period or irrecoverable condition; *also* **:** something that evokes nostalgia
— nos•tal•gic \-jik\ *adjective or noun*
— nos•tal•gi•cal•ly \-ji-k(ə-)lē\ *adverb*

nos•tal•gist \-jist\ *noun* (1953)
: a person given to nostalgia for the past

nos•toc \ˈnäs-ˌtäk\ *noun* [New Latin] (1650)
: any of a genus (*Nostoc*) of blue-green algae that fix nitrogen

nos•tril \ˈnäs-trəl\ *noun* [Middle English *nosethirl,* from Old English *nosthyrl,* from *nosu* nose + *thyrel* hole; akin to Old English *thurh* through] (before 12th century)
1 : either of the external nares; *broadly* **:** either of the nares with the adjoining passage on the same side of the septum
2 : either fleshy lateral wall of the nose
word history see THRILL

nos•trum \ˈnäs-trəm\ *noun* [Latin, neuter of *noster* our, ours, from *nos* we — more at US] (1602)
1 : a medicine of secret composition recommended by its preparer but usually without scientific proof of its effectiveness
2 : a usually questionable remedy or scheme **:** PANACEA

nosy *or* **nos•ey** \ˈnō-zē\ *adjective* **nos•i•er; -est** [[1]*nose*] (1882)
: of prying or inquisitive disposition or quality **:** INTRUSIVE
— nos•i•ly \ˈnō-zə-lē\ *adverb*
— nos•i•ness \-zē-nəs\ *noun*

not \ˈnät\ *adverb* [Middle English, alteration of *nought,* from *nought,* pronoun — more at NAUGHT] (13th century)
1 — used as a function word to make negative a group of words or a word
2 — used as a function word to stand for the negative of a preceding group of words ⟨is sometimes hard to see and sometimes *not*⟩

NOT \ˈnät\ *noun* [*not*] (1947)
: a logical operator that produces a statement that is the inverse of an input statement

nota *plural of* NOTUM

no•ta be•ne \ˌnō-tə-ˈbē-nē, -ˈbe-\ [Latin, mark well] (circa 1721)
— used to call attention to something important

no•ta•bil•i•ty \ˌnō-tə-ˈbi-lə-tē\ *noun, plural* **-ties** (1832)
: a notable or prominent person

[1]no•ta•ble \ˈnō-tə-bəl, *for 2 also* ˈnä-\ *adjective* (14th century)
1 a : worthy of note **:** REMARKABLE **b :** DISTINGUISHED, PROMINENT
2 *archaic* **:** efficient or capable in performance of housewifely duties
— no•ta•ble•ness *noun*

[2]no•ta•ble \ˈnō-tə-bəl\ *noun* (1815)
1 : a person of note **:** NOTABILITY
2 *plural, often capitalized* **:** a group of persons summoned especially in monarchical France to act as a deliberative body

no•ta•bly \ˈnō-tə-blē\ *adverb* (14th century)
1 : in a notable manner **:** to a high degree ⟨was *notably* impressed⟩
2 : ESPECIALLY, PARTICULARLY ⟨other powers, *notably* Britain and the United States —C. A. Fisher⟩

no•tar•i•al \nō-ˈter-ē-əl\ *adjective* (15th century)
1 : of, relating to, or characteristic of a notary public
2 : done or executed by a notary public
— no•tar•i•al•ly \-ə-lē\ *adverb*

no•ta•ri•za•tion \ˌnō-tə-rə-ˈzā-shən\ *noun* (1940)
1 : the act, process, or an instance of notarizing
2 : the notarial certificate appended to a document

no•ta•rize \ˈnō-tə-ˌrīz\ *transitive verb* **-rized; -riz•ing** (1926)
: to acknowledge or attest as a notary public

no•ta•ry public \ˈnō-tə-rē-\ *noun, plural* **notaries public** *or* **notary publics** [Middle English *notary* clerk, notary public, from Latin *notarius* clerk, secretary, from *notarius* of shorthand, from *nota* note, shorthand character] (15th century)
: a public officer who attests or certifies writings (as a deed) to make them authentic and takes affidavits, depositions, and protests of negotiable paper — called also *notary*

no•tate \ˈnō-ˌtāt\ *transitive verb* **no•tat•ed; no•tat•ing** [back-formation from *notation*] (1903)
: to put into notation

no•ta•tion \nō-ˈtā-shən\ *noun* [Latin *notation-, notatio,* from *notare* to note] (1584)
1 : ANNOTATION, NOTE
2 a : the act, process, method, or an instance of representing by a system or set of marks, signs, figures, or characters **b :** a system of characters, symbols, or abbreviated expressions used in an art or science or in mathematics or logic to express technical facts or quantities
— no•ta•tion•al \-shnəl, -shə-nᵊl\ *adjective*

[1]notch \ˈnäch\ *noun* [perhaps alteration (from misdivision of *an otch*) of (assumed) *otch,* from Middle French *oche*] (1577)
1 a : a V-shaped indentation **b :** a slit made to serve as a record **c :** a rounded indentation cut into the pages of a book on the edge opposite the spine
2 : a deep close pass **:** GAP
3 : DEGREE, STEP
— notched \ˈnächt\ *adjective*

[2]notch *transitive verb* (1600)
1 : to cut or make a notch in
2 a : to mark or record by a notch **b :** SCORE, ACHIEVE — sometimes used with *up*

notch•back \ˈnäch-ˌbak\ *noun* (1965)
: an automobile with a trunk whose lid forms a distinct deck; *also* **:** the back of such an automobile

[1]note \ˈnōt\ *transitive verb* **not•ed; not•ing** [Middle English, from Old French *noter,* from Latin *notare* to mark, note, from *nota*] (13th century)
1 a : to notice or observe with care **b :** to record or preserve in writing
2 a : to make special mention of **:** REMARK **b :** INDICATE, SHOW
— not•er *noun*

[2]note *noun* [Middle English, from Old French, from Latin *nota* mark, character, written note] (13th century)
1 a (1) *obsolete* **:** MELODY, SONG (2) **:** TONE 2a (3) **:** CALL, SOUND; *especially* **:** the musical call of a bird **b :** a written symbol used to indicate duration and pitch of a tone by its shape and position on the staff
2 a : a characteristic feature (as of odor or flavor) **b :** something (as an emotion or disposition) like a note in tone or resonance ⟨a *note* of sadness⟩ ⟨end on a high *note*⟩
3 a (1) **:** MEMORANDUM (2) **:** a condensed or informal record **b** (1) **:** a brief comment or explanation (2) **:** a printed comment or reference

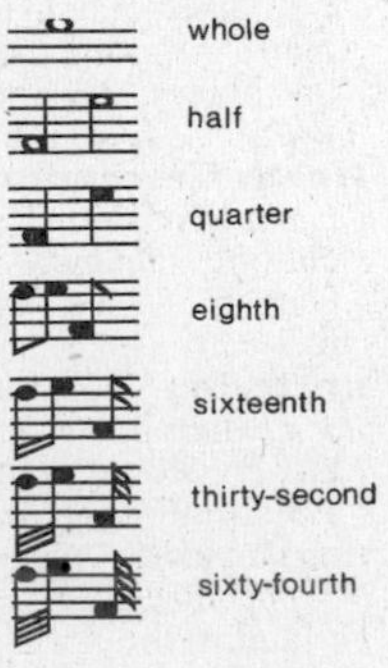

whole
half
quarter
eighth
sixteenth
thirty-second
sixty-fourth

note 1b

set apart from the text **c** (1) **:** a written promise to pay a debt (2) **:** a piece of paper money **d** (1) **:** a short informal letter (2) **:** a formal diplomatic communication **e :** a scholarly or technical essay shorter than an article and restricted in scope **f :** a sheet of notepaper
4 a : DISTINCTION, REPUTATION ⟨a figure of international *note*⟩ **b :** OBSERVATION, NOTICE ⟨took full *note* of the proceedings⟩ **c :** KNOWLEDGE, INFORMATION
synonym see SIGN

note·book \'nōt-ˌbu̇k\ *noun* (1579)
1 : a book for notes or memoranda
2 : a small portable microcomputer about the size of an ordinary loose-leaf binder

note·case \-ˌkās\ *noun* (1838)
British **:** WALLET 2a

not·ed \'nō-təd\ *adjective* (14th century)
: well-known by reputation **:** EMINENT, CELEBRATED
synonym see FAMOUS
— not·ed·ly *adverb*
— not·ed·ness *noun*

note·less \'nōt-ləs\ *adjective* (circa 1616)
: not noticed **:** UNDISTINGUISHED

note of hand (circa 1738)
: PROMISSORY NOTE

note·pad \'nōt-ˌpad\ *noun* (1922)
: PAD 4

note·pa·per \-ˌpā-pər\ *noun* (1849)
: writing paper suitable for notes

note·wor·thy \-ˌwər-t͟hē\ *adjective* (1552)
: worthy of or attracting attention especially because of some special excellence
— note·wor·thi·ly \-t͟hə-lē\ *adverb*
— note·wor·thi·ness \-t͟hē-nəs\ *noun*

not–for–profit *adjective* (circa 1973)
: NONPROFIT

noth·er *or* **'noth·er** \'nə-t͟hər\ *adjective* [alteration (from misdivision of *another*) of *other,* adjective] (1956)
: OTHER — used especially in the phrase *a whole nother;* used chiefly in speech or informal prose

¹noth·ing \'nə-thiŋ\ *pronoun* [Middle English, from Old English *nān thing, nāthing,* from *nān* no + *thing* thing — more at NONE] (before 12th century)
1 : not any thing **:** no thing ⟨leaves *nothing* to the imagination⟩
2 : no part
3 : one of no interest, value, or consequence ⟨they mean *nothing* to me⟩
— nothing doing : by no means **:** definitely no
— nothing for it : no alternative ⟨*nothing for it* but to start over⟩

²nothing *adverb* (12th century)
: not at all **:** in no degree
— nothing like : not nearly ⟨it's *nothing like* thorough enough⟩

³nothing *noun* (1535)
1 a : something that does not exist **b :** the absence of all magnitude or quantity; *also* **:** ZERO 1a **c :** NOTHINGNESS 3b
2 : someone or something of no or slight value or size

⁴nothing *adjective* (1611)
: of no account **:** WORTHLESS

noth·ing·ness \-nəs\ *noun* (circa 1631)
1 : the quality or state of being nothing: as **a :** NONEXISTENCE **b :** utter insignificance **c :** DEATH
2 : something insignificant or valueless
3 a : VOID, EMPTINESS **b :** a metaphysical entity opposed to and devoid of being and regarded by some existentialists as the ground of anxiety

¹no·tice \'nō-təs\ *noun* [Middle English, from Middle French, acquaintance, from Latin *notitia* knowledge, acquaintance, from *notus* known, from past participle of *noscere* to come to know — more at KNOW] (15th century)
1 a (1) **:** warning or intimation of something **:** ANNOUNCEMENT (2) **:** notification by one of the parties to an agreement or relation of intention of terminating it at a specified time (3) **:** the condition of being warned or notified — usually used in the phrase *on notice* **b :** INFORMATION, INTELLIGENCE
2 a : ATTENTION, HEED **b :** polite or favorable attention **:** CIVILITY
3 : a written or printed announcement
4 : a short critical account or examination

²notice *transitive verb* **no·ticed; no·tic·ing** (15th century)
1 : to give notice of
2 a : to comment upon **b :** REVIEW
3 a : to treat with attention or civility **b :** to take notice of **:** MARK
4 : to give a formal notice to
— no·tic·er *noun*

no·tice·able \'nō-tə-sə-bəl\ *adjective* (1796)
1 : worthy of notice
2 : capable of being noticed ☆
— no·tice·ably \-blē\ *adverb*

notice board *noun* (1854)
chiefly British **:** a board bearing a notice or on which notices may be posted; *especially* **:** BULLETIN BOARD

no·ti·fi·able \'nō-tə-ˌfī-ə-bəl, ˌnō-tə-'\ *adjective* (1889)
: required by law to be reported to official health authorities ⟨a *notifiable* disease⟩

no·ti·fi·ca·tion \ˌnō-tə-fə-'kā-shən\ *noun* (14th century)
1 : the act or an instance of notifying
2 : a written or printed matter that gives notice

no·ti·fy \'nō-tə-ˌfī\ *transitive verb* **-fied; -fy·ing** [Middle English *notifien,* from Middle French *notifier* to make known, from Late Latin *notificare,* from Latin *notus* known] (14th century)
1 *obsolete* **:** to point out
2 : to give notice of or report the occurrence of ⟨he *notified* his intention to sue⟩
3 : to give formal notice to ⟨*notify* a family of the death of a relation⟩
synonym see INFORM
— no·ti·fi·er \-ˌfī(-ə)r\ *noun*

no–till \'nō-'til\ *noun* (1968)
: NO-TILLAGE

no–till·age \-'ti-lij\ *noun* (1968)
: a system of farming that consists of planting a narrow slit trench without tillage and with the use of herbicides to suppress weeds

no·tion \'nō-shən\ *noun* [Latin *notion-, notio,* from *noscere*] (1537)
1 a (1) **:** an individual's conception or impression of something known, experienced, or imagined (2) **:** an inclusive general concept (3) **:** a theory or belief held by a person or group **b :** a personal inclination **:** WHIM
2 *obsolete* **:** MIND, INTELLECT
3 *plural* **:** small useful items **:** SUNDRIES
synonym see IDEA

no·tion·al \'nō-shnəl, -shə-nᵊl\ *adjective* (1597)
1 : THEORETICAL, SPECULATIVE
2 : existing in the mind only **:** IMAGINARY
3 : given to foolish or fanciful moods or ideas
4 a : of, relating to, or being a notion or idea **:** CONCEPTUAL **b** (1) **:** presenting an idea of a thing, action, or quality ⟨*has* is *notional* in *he has luck,* relational in *he has gone*⟩ (2) **:** of or representing what exists or occurs in the world of things as distinguished from syntactic categories
— no·tion·al·i·ty \ˌnō-shə-'na-lə-tē\ *noun*
— no·tion·al·ly \'nō-shnə-lē, -shə-nᵊl-ē\ *adverb*

no·to·chord \'nō-tə-ˌkȯrd\ *noun* [Greek *nōton, nōtos* back + Latin *chorda* cord — more at CORD] (1848)
: a longitudinal flexible rod of cells that in the lowest chordates (as a lancelet or a lamprey) and in the embryos of the higher vertebrates forms the supporting axis of the body
— no·to·chord·al \ˌnō-tə-'kȯr-dᵊl\ *adjective*

no·to·ri·e·ty \ˌnō-tə-'rī-ə-tē\ *noun, plural* **-ties** [Middle French or Medieval Latin; Middle French *notorieté,* from Medieval Latin *notorietat-, notorietas,* from *notorius*] (circa 1650)
1 : the quality or state of being notorious
2 : a notorious person

no·to·ri·ous \nō-'tōr-ē-əs, nə-, -'tȯr-\ *adjective* [Medieval Latin *notorius,* from Late Latin *notorium* information, indictment, from Latin *noscere* to come to know — more at KNOW] (1534)
: generally known and talked of; *especially* **:** widely and unfavorably known
synonym see FAMOUS

no·to·ri·ous·ly \-lē\ *adverb* (1512)
1 : in a notorious manner
2 : it is notorious **:** as is very well known

no–trump *adjective* (1899)
: being a bid, contract, or hand suitable to play without any suit being trumps
— no–trump *noun*

no·tum \'nō-təm\ *noun, plural* **no·ta** \'nō-tə\ [New Latin, from Greek *nōton* back] (1877)
: the dorsal surface of a thoracic segment of an insect

¹not·with·stand·ing \ˌnät-with-'stan-diŋ, -wit͟h-\ *preposition* [Middle English *notwithstanding,* from *not* + *withstonding,* present participle of *withstonden* to withstand] (14th century)
: DESPITE ⟨*notwithstanding* their lack of experience, they were an immediate success⟩ — often used after its object ⟨the motion passed, our objection *notwithstanding*⟩

²notwithstanding *adverb* (15th century)
: NEVERTHELESS, HOWEVER

³notwithstanding *conjunction* (15th century)
: ALTHOUGH

nou·gat \'nü gət, *especially British* ˌgä\ *noun* [French, from Provençal, from Old Provençal *nogat,* from *noga* nut, from (assumed) Vulgar Latin *nuca,* from Latin *nuc-, nux* — more at NUT] (1827)
: a confection of nuts or fruit pieces in a sugar paste

nought \'nȯt, 'nät\ *variant of* NAUGHT
word history see NAUGHTY

☆ SYNONYMS
Noticeable, remarkable, prominent, outstanding, conspicuous, salient, striking mean attracting notice or attention. NOTICEABLE applies to something unlikely to escape observation ⟨a piano recital with no *noticeable* errors⟩. REMARKABLE applies to something so extraordinary or exceptional as to invite comment ⟨a film of *remarkable* intelligence and wit⟩. PROMINENT applies to something commanding notice by standing out from its surroundings or background ⟨a doctor who occupies a *prominent* position in the town⟩. OUTSTANDING applies to something that rises above and excels others of the same kind ⟨honored for her *outstanding* contributions to science⟩. CONSPICUOUS applies to something that is obvious and unavoidable to the sight or mind ⟨*conspicuous* bureaucratic waste⟩. SALIENT applies to something of significance that merits the attention given it ⟨the *salient* points of the speech⟩. STRIKING applies to something that impresses itself powerfully and deeply upon the observer's mind or vision ⟨the region's *striking* poverty⟩.

nou·me·non \ˈnü-mə-ˌnän\ *noun, plural* **-na** \-nə, -ˌnä\ [German, from Greek *nooumenon* that which is apprehended by thought, from neuter of present passive participle of *noein* to think, conceive, from *nous* mind] (1796)
: a posited object or event as it appears in itself independent of perception by the senses
— nou·men·al \-mə-nᵊl\ *adjective*

noun \ˈnau̇n\ *noun* [Middle English *nowne,* from Anglo-French *noun* name, noun, from Old French *nom,* from Latin *nomen* — more at NAME] (14th century)
: any member of a class of words that typically can be combined with determiners to serve as the subject of a verb; can be interpreted as singular or plural, can be replaced with a pronoun, and refer to an entity, quality, state, action, or concept

noun phrase *noun, often N&P capitalized* (1923)
: a phrase formed by a noun and all its modifiers and determiners; *broadly* **:** any syntactic element (as a clause, clitic, pronoun, or zero element) with a noun's function (as the subject of a verb or the object of a verb or preposition) — abbreviation *NP*

nour·ish \ˈnər-ish, ˈnə-rish\ *transitive verb* [Middle English *nurishen,* from Old French *noriss-,* stem of *norrir,* from Latin *nutrire* to suckle, nourish; akin to Greek *nan* to flow, *noteros* damp, Sanskrit *snauti* it drips] (14th century)
1 : NURTURE, REAR
2 : to promote the growth of ⟨no occasions to exercise the feelings nor *nourish* passion —L. O. Coxe⟩
3 a : to furnish or sustain with nutriment **:** FEED **b :** MAINTAIN, SUPPORT ⟨their profits . . . *nourish* other criminal activities —Beverly Smith⟩
— nour·ish·er *noun*

nour·ish·ing *adjective* (14th century)
: giving nourishment **:** NUTRITIOUS

nour·ish·ment \ˈnər-ish-mənt, ˈnə-rish-\ *noun* (15th century)
1 a : FOOD, NUTRIMENT **b :** SUSTENANCE 3 ⟨books for intellectual *nourishment*⟩
2 : the act of nourishing **:** the state of being nourished

nous *noun* [Greek *noos, nous* mind] (1678)
1 \ˈnüs *also* ˈnau̇s\ **:** MIND, REASON: as **a :** an intelligent purposive principle of the world **b :** the divine reason regarded in Neoplatonism as the first emanation of God
2 \ˈnau̇s\ *chiefly British* **:** COMMON SENSE, ALERTNESS

nou·veau \nü-ˈvō\ *adjective* [French, from Middle French *novel*] (1828)
: newly arrived or developed

nou·veau riche \ˌnü-ˌvō-ˈrēsh\ *noun, plural* **nou·veaux riches** *same*\ [French, literally, new rich] (1813)
: a person newly rich **:** PARVENU

nou·velle \nü-ˈvel\ *adjective* [*nouvelle* cuisine] (1976)
: of or relating to nouvelle cuisine ⟨a *nouvelle* restaurant⟩

nouvelle cuisine *noun* [French, literally, new cuisine] (1975)
: a form of French cuisine that uses little flour or fat and stresses light sauces and the use of fresh seasonal produce; *also* **:** any national cuisine that stresses lightness and freshness in preparation ⟨American *nouvelle cuisine*⟩

nouvelle vague \-ˈväg, -ˈvȧg\ *noun* [French] (1959)
: NEW WAVE 1, 2

no·va \ˈnō-və\ *noun, plural* **novas** *or* **no·vae** \-(ˌ)vē, -ˌvī\ [New Latin, feminine of Latin *novus* new] (1927)
: a star that suddenly increases its light output tremendously and then fades away to its former obscurity in a few months or years
— no·va·like \-və-ˌlīk\ *adjective*

no·vac·u·lite \nō-ˈva-kyə-ˌlīt\ *noun* [Latin *novacula* razor] (1796)
: a very hard fine-grained siliceous rock used for whetstones and possibly of sedimentary origin

no·va·tion \nō-ˈvā-shən\ *noun* [Late Latin *novation-, novatio* renewal, legal novation, from Latin *novare* to make new, from *novus*] (1682)
: the substitution of a new legal obligation for an old one

¹nov·el \ˈnä-vəl\ *adjective* [Middle English, from Middle French, new, from Latin *novellus,* from diminutive of *novus* new — more at NEW] (15th century)
1 : new and not resembling something formerly known or used
2 : original or striking especially in conception or style ⟨a *novel* scheme to collect money⟩
synonym see NEW

²novel *noun* [Italian *novella*] (1639)
1 : an invented prose narrative that is usually long and complex and deals especially with human experience through a usually connected sequence of events
2 : the literary genre consisting of novels ◆
— nov·el·is·tic \ˌnä-və-ˈlis-tik\ *adjective*
— nov·el·is·ti·cal·ly \-ti-k(ə-)lē\ *adverb*

nov·el·ette \ˌnä-və-ˈlet\ *noun* (1814)
: NOVELLA 2

nov·el·ett·ish \-ˈle-tish\ *adjective* (1904)
: of, relating to, or characteristic of a novelette; *especially* **:** SENTIMENTAL

nov·el·ist \ˈnäv-list, ˈnä-və-\ *noun* (1728)
: a writer of novels

nov·el·ize \ˈnä-və-ˌlīz\ *transitive verb* **-ized; -iz·ing** (1828)
: to convert into the form of a novel ⟨*novelize* a play⟩
— nov·el·iza·tion \ˌnä-və-lə-ˈzā-shən\ *noun*

no·vel·la \nō-ˈve-lə\ *noun, plural* **novellas** *or* **no·vel·le** \-ˈve-lē\ [Italian, from feminine of *novello* new, from Latin *novellus*] (1898)
1 *plural novelle* **:** a story with a compact and pointed plot
2 *plural usually novellas* **:** a work of fiction intermediate in length and complexity between a short story and a novel

nov·el·ty \ˈnä-vəl-tē\ *noun, plural* **-ties** [Middle English *novelte,* from Middle French *noveleté,* from *novel*] (14th century)
1 : something new or unusual
2 : the quality or state of being novel **:** NEWNESS
3 : a small manufactured article intended mainly for personal or household adornment — usually used in plural

¹No·vem·ber \nō-ˈvem-bər, nə-\ *noun* [Middle English *Novembre,* from Old French, from Latin *November,* ninth month of the early Roman calendar, from *novem* nine — more at NINE] (13th century)
: the 11th month of the Gregorian calendar

²November (1956)
— a communications code word for the letter *n*

no·vem·de·cil·lion \ˌnō-ˌvem-di-ˈsil-yən\ *noun, often attributive* [Latin *novemdec*im nineteen (from *novem* + *decem* ten) + English *-illion* (as in *million*) — more at TEN] (circa 1934)
— see NUMBER table

no·ve·na \nō-ˈvē-nə\ *noun* [Medieval Latin, from Latin, feminine of *novenus* nine each, from *novem*] (1853)
: a Roman Catholic period of prayer lasting nine consecutive days

nov·ice \ˈnä-vəs\ *noun* [Middle English, from Middle French, from Medieval Latin *novicius,* from Latin, newly imported, from *novus* — more at NEW] (14th century)
1 : a person admitted to probationary membership in a religious community
2 : BEGINNER, TYRO

no·vi·tiate \nō-ˈvi-shət, nə-\ *noun* [French *noviciat,* from Medieval Latin *noviciatus,* from *novicius*] (1600)
1 : the period or state of being a novice
2 : a house where novices are trained
3 : NOVICE

no·vo·bi·o·cin \ˌnō-və-ˈbī-ə-sən\ *noun* [*novo-* (perhaps modification of Latin *niveus* snowy, specific epithet of the bacterium *Streptomyces niveus*) + *bi-* + *-mycin*] (1956)
: a weak dibasic acid $C_{31}H_{36}N_2O_{11}$ that is highly toxic to humans and is used as an antimicrobial drug in some serious cases of staphylococcic and urinary tract infection

No·vo·cain \ˈnō-və-ˌkān\ *trademark*
— used for a preparation containing the hydrochloride of procaine

no·vo·caine \-ˌkān\ *noun* [International Scientific Vocabulary *novo-* (from Latin *novus* new) + *cocaine*] (1910)
: PROCAINE; *also* **:** its hydrochloride

¹now \ˈnau̇\ *adverb* [Middle English, from Old English *nū;* akin to Old High German *nū* now, Latin *nunc,* Greek *nyn*] (before 12th century)
1 a : at the present time or moment **b :** in the time immediately before the present ⟨thought of them just *now*⟩ **c :** in the time immediately to follow **:** FORTHWITH ⟨come in *now*⟩
2 — used with the sense of present time weakened or lost to express command, request, or admonition ⟨*now* hear this⟩ ⟨*now* you be sure to write⟩
3 — used with the sense of present time weakened or lost to introduce an important point or indicate a transition (as of ideas)
4 : SOMETIMES ⟨*now* one and *now* another⟩
5 : under the present circumstances
6 : at the time referred to ⟨*now* the trouble began⟩

²now *conjunction* (before 12th century)
: in view of the fact that **:** SINCE — often followed by *that* ⟨*now* that we are here⟩

³now *noun* (12th century)
: the present time or moment ⟨been ill up to *now*⟩

⁴now *adjective* (14th century)
1 : of or relating to the present time **:** EXISTING ⟨the *now* president⟩
2 a : excitingly new ⟨*now* clothes⟩ **b :** constantly aware of what is new ⟨*now* people⟩ ⟨the *now* generation⟩

NOW account \ˈnau̇-\ *noun* [*n*egotiable *o*rder of *w*ithdrawal] (1974)
: a savings account on which checks may be drawn

now·a·days \ˈnau̇-(ə-)ˌdāz\ *adverb* [Middle English *now a dayes,* from ¹*now* + *a dayes* during the day] (14th century)
: at the present time

now and then *adverb* (15th century)

◇ WORD HISTORY

novel Italian *novella,* meaning literally "item of news," came to be applied in the 13th century to the short tales or fables that were becoming a popular literary form. Simple in plot and character development, these tales typically dealt with humorous, political, or amorous events, especially in a satiric way. Since similar tales were also becoming popular in England, English borrowed the word as *novel* in the 17th century. As the novel developed, however, the notion of brevity was abandoned and the narratives became considerably longer. During the 18th century the novel matured as a literary form, presenting psychologically developed characters in an extended, complex plot. At the end of the 19th century, English again borrowed the Italian word *novella,* this time intact to refer specifically to the medieval Italian stories of authors such as Boccaccio, and later to refer to a work of fiction intermediate in length between a short story and a novel.

: from time to time : OCCASIONALLY ⟨*now and then* we go off to the country⟩

no·way *adverb* (13th century)
1 \'nō-ˌwā\ *or* **no·ways** \-ˌwāz\ : NOWISE
2 *usually* **no way** \-'wā\ : NO — used emphatically

[1]no·where \'nō-ˌhwer, -ˌhwar, -hwər *or without* h\ *adverb* (before 12th century)
1 : not in or at any place
2 : to no place

[2]nowhere *noun* (1831)
1 : a nonexistent place
2 : an unknown, distant, or obscure place or state ⟨rose to fame out of *nowhere*⟩
— miles from nowhere : in an extremely remote place

nowhere near *adverb* (15th century)
: not nearly

no·wheres \'nō-ˌhwerz, -ˌhwarz, -hwərz *or without* h\ *adverb* (circa 1866)
chiefly dialect : NOWHERE

no·whith·er \'nō-ˌhwi-t͟hər, -ˌwi-t͟hər; ˌnō-'\ *adverb* (before 12th century)
: to or toward no place

no–win \'nō-'win, -ˌwin\ *adjective* (1962)
: not likely to give victory, success, or satisfaction : that cannot be won ⟨a *no-win* situation⟩ ⟨a *no-win* war⟩

no·wise \'nō-ˌwīz\ *adverb* (14th century)
: not at all

now·ness \'nau̇-nəs\ *noun* (1674)
: the quality or state of existing or occurring in or belonging to the present time

nowt \'nau̇t *also* 'nōt\ *dialect English variant of* NOUGHT

nox·ious \'näk-shəs\ *adjective* [Middle English *noxius,* from Latin, from *noxa* harm; akin to Latin *nocēre* to harm, *nec-, nex* violent death, Greek *nekros* dead body] (15th century)
1 a : physically harmful or destructive to living beings ⟨*noxious* wastes that poison our streams⟩ **b :** constituting a harmful influence on mind or behavior : morally corrupting ⟨*noxious* doctrines⟩
2 : DISTASTEFUL, OBNOXIOUS
synonym see PERNICIOUS
— nox·ious·ly *adverb*
— nox·ious·ness *noun*

noz·zle \'nä-zəl\ *noun* [diminutive of *nose*] (1683)
1 a : a projecting vent of something **b :** a short tube with a taper or constriction used (as on a hose) to speed up or direct a flow of fluid **c :** a part in a rocket engine that accelerates the exhaust gases from the combustion chamber to a high velocity
2 *slang* : NOSE

NSAID \'en-ˌsed *also* -ˌsād\ *noun* [*no*n*s*teroidal *a*nti-*i*nflammatory *d*rug] (1985)
: a nonsteroidal anti-inflammatory drug (as ibuprofen)

-n't *verb combining form*
: not ⟨is*n't*⟩

nth \'en(t)th\ *adjective* [*n* (indefinite number) + *-th*] (1852)
1 : numbered with an unspecified or indefinitely large ordinal number ⟨for the *nth* time⟩
2 : EXTREME, UTMOST ⟨to the *nth* degree⟩

n–type \'en-ˌtīp\ *adjective* [*negative* + *type*] (1946)
: relating to or being a semiconductor in which charge is carried by electrons — compare P-TYPE

nu \'nü, 'nyü\ *noun* [Greek *ny,* of Semitic origin; akin to Hebrew *nūn* nun] (circa 1823)
: the 13th letter of the Greek alphabet — see ALPHABET table

nu·ance \'nü-ˌän(t)s, 'nyü-, -ˌäⁿs; nu̇-', nyu̇-'\ *noun* [French, from Middle French, shade of color, from *nuer* to make shades of color, from *nue* cloud, from Latin *nubes;* perhaps akin to Welsh *nudd* mist] (1781)
1 : a subtle distinction or variation
2 : a subtle quality : NICETY
3 : sensibility to, awareness of, or ability to express delicate shadings (as of meaning, feeling, or value)
— nu·anced \-ˌän(t)st, -'än(t)st\ *adjective*

nub \'nəb\ *noun* [alteration of English dialect *knub,* probably from Low German *knubbe*] (1727)
1 : KNOB, LUMP
2 : NUBBIN
3 : GIST, POINT

nub·bin \'nə-bən\ *noun* [perhaps diminutive of *nub*] (1692)
1 : something (as an ear of Indian corn) that is small for its kind, stunted, undeveloped, or imperfect
2 : a small usually projecting part or bit
3 : NUB 3

nub·ble \'nə-bəl\ *noun* [diminutive of *nub*] (1818)
: a small knob or lump
— nub·bly \-b(ə-)lē\ *adjective*

nub·by \'nə-bē\ *adjective* **nub·bi·er; -est** [*nub* + [1]-y] (circa 1876)
1 : having or being like nubbles
2 : having nubs ⟨a *nubby* knit fabric⟩

Nu·bi·an \'nü-bē-ən, 'nyü-\ *noun* (15th century)
1 a : a native or inhabitant of Nubia **b :** a member of one of the group of dark-skinned peoples that formed a powerful empire between Egypt and Ethiopia from the 6th to the 14th centuries
2 : any of several languages spoken in central and northern Sudan
— Nubian *adjective*

nu·bile \'nü-ˌbīl, 'nyü-, -bəl\ *adjective* [French, from Latin *nubilis,* from *nubere* to marry — more at NUPTIAL] (circa 1642)
1 : of marriageable condition or age
2 : sexually attractive — used of a young woman
— nu·bil·i·ty \nü-'bi-lə-tē, nyü-\ *noun*

nu·cel·lus \nü-'se-ləs, nyü-\ *noun, plural* **nu·cel·li** \-'se-ˌlī\ [New Latin, from Latin *nucella* small nut, from *nuc-, nux* nut — more at NUT] (1882)
: the central and chief part of a plant ovule that encloses the female gametophyte
— nu·cel·lar \-'se-lər\ *adjective*

nu·chal \'nü-kəl, 'nyü-\ *adjective* [Medieval Latin *nucha* nape, from Arabic *nukhā'* spinal marrow] (1835)
: of, relating to, or lying in the region of the nape

nucle- *or* **nucleo-** *combining form* [French *nuclé-, nucléo-,* from New Latin *nucleus*]
1 : nucleus ⟨*nucleo*plasm⟩
2 : nucleic acid ⟨*nucleo*protein⟩

nu·cle·ar \'nü-klē-ər, 'nyü-, ÷-kyə-lər\ *adjective* (1846)
1 : of, relating to, or constituting a nucleus
2 a : of or relating to the atomic nucleus ⟨*nuclear* reaction⟩ ⟨*nuclear* physics⟩ **b :** used in or produced by a nuclear reaction (as fission) ⟨*nuclear* fuel⟩ ⟨*nuclear* waste⟩ ⟨*nuclear* energy⟩ **c** (1) **:** being a weapon whose destructive power derives from an uncontrolled nuclear reaction (2) **:** of, produced by, or involving nuclear weapons ⟨the *nuclear* age⟩ ⟨*nuclear* war⟩ (3) **:** armed with nuclear weapons ⟨*nuclear* powers⟩ **d :** of, relating to, or powered by nuclear energy ⟨a *nuclear* submarine⟩ ⟨the *nuclear* debate⟩ ⟨a *nuclear* plant⟩ ■

nuclear family *noun* (1947)
: a family group that consists only of father, mother, and children

nuclear magnetic resonance *noun* (1942)
: the magnetic resonance of an atomic nucleus

nuclear medicine *noun* (1952)
: a branch of medicine dealing with the use of radioactive materials in the diagnosis and treatment of disease

nuclear membrane *noun* (1888)
: a double membrane enclosing a cell nucleus and having its outer part continuous with the endoplasmic reticulum — see CELL illustration

nuclear–powered *adjective* (1948)
: powered by nuclear energy

nuclear resonance *noun* (1940)
: the resonance absorption of a gamma ray by a nucleus identical to the nucleus that emitted the gamma ray

nuclear sap *noun* (1877)
: the clear homogeneous ground substance of a cell nucleus — called also *karyolymph*

nuclear winter *noun* (1983)
: the chilling of climate that is hypothesized to be a consequence of nuclear war and to result from the prolonged blockage of sunlight by high-altitude dust clouds produced by nuclear explosions

nu·cle·ase \'nü-klē-ˌās, 'nyü-, -ˌāz\ *noun* (1902)
: any of various enzymes that promote hydrolysis of nucleic acids

nu·cle·ate \'nü-klē-ˌāt, 'nyü-\ *verb* **-at·ed; -at·ing** [Late Latin *nucleatus,* past participle of *nucleare* to become stony, from Latin *nucleus*] (circa 1864)
transitive verb
1 : to form into a nucleus : CLUSTER
2 : to act as a nucleus for
3 : to supply nuclei to
intransitive verb
1 : to form a nucleus
2 : to act as a nucleus
3 : to begin to form
— nu·cle·ation \ˌnü-klē-'ā-shən, ˌnyü-\ *noun*
— nu·cle·a·tor \'nü-klē-ˌā-tər, 'nyü-\ *noun*

nu·cle·at·ed \'nü-klē-ˌā-təd, 'nyü-\ *or* **nu·cle·ate** \-klē-ət\ *adjective* [Latin *nucleatus,* from *nucleus* kernel] (1845)
1 : having a nucleus or nuclei ⟨*nucleated* cells⟩
2 *usually nucleate* **:** originating or occurring at nuclei ⟨*nucleate* boiling⟩

nu·cle·ic acid \nu̇-'klē-ik-, -'klā-, nyu̇-\ *noun* [from their occurrence in cell nuclei] (1892)
: any of various acids (as DNA or RNA) that are composed of nucleotide chains

nu·cle·in \'nü-klē-ən, 'nyü-\ *noun* (1878)
1 : NUCLEOPROTEIN
2 : NUCLEIC ACID

nu·cle·o·cap·sid \ˌnü-klē-ō-'kap-səd, ˌnyü-\ *noun* (1963)
: the nucleic acid and surrounding protein coat in a virus

nu·cle·oid \'nü-klē-ˌȯid, 'nyü-\ *noun* (1938)
: the DNA-containing area of a prokaryotic cell (as a bacterium)

nu·cle·o·lus \nü-'klē-ə-ləs, nyü-\ *noun, plural* **-li** \-ˌlī\ [New Latin, from Latin, diminutive of *nucleus*] (1845)
: a spherical body of the nucleus of most eukaryotes that becomes enlarged during protein synthesis, is associated with a nucleolus organizer, and contains the DNA templates for ribosomal RNA — see CELL illustration

□ USAGE
nuclear Though disapproved of by many, pronunciations ending in \-kyə-lər\ have been found in widespread use among educated speakers including scientists, lawyers, professors, congressmen, U.S. cabinet members, and at least one U.S. president and one vice president. While most common in the U.S., these pronunciations have also been heard from British and Canadian speakers.

— **nu·cle·o·lar** \-lər\ *adjective*

nucleolus organizer *noun* (1939)
: the specific part of a chromosome with which a nucleolus is associated especially during its reorganization after nuclear division — called also *nucleolar organizer*

nu·cle·on \'nü-klē-ˌän, 'nyü-\ *noun* [International Scientific Vocabulary] (1923)
1 : a nuclear particle: **a :** PROTON **b :** NEUTRON
2 : a hypothetical single entity with one-half unit of isospin that can manifest itself as either a proton or a neutron
— **nu·cle·on·ic** \ˌnü-klē-'ä-nik, ˌnyü-\ *adjective*

nu·cle·on·ics \ˌnü-klē-'ä-niks, ˌnyü-\ *noun plural but singular or plural in construction* (1937)
: a branch of physical science that deals with nucleons or with all phenomena of the atomic nucleus

nu·cle·o·phile \'nü-klē-ə-ˌfīl, 'nyü-\ *noun* (1943)
: a nucleophilic substance (as an electron-donating reagent)

nu·cle·o·phil·ic \ˌnü-klē-ə-'fi-lik, ˌnyü-\ *adjective* (1933)
1 *of an atom, ion, or molecule* **:** having an affinity for atomic nuclei **:** being an electron donor
2 : involving a nucleophilic species ⟨a *nucleophilic* reaction⟩ — compare ELECTROPHILIC
— **nu·cleo·phil·i·cal·ly** \-li-k(ə-)lē\ *adverb*
— **nu·cle·o·phi·lic·i·ty** \-klē-ō-fi-'li-sə-tē\ *noun*

nu·cle·o·plasm \'nü-klē-ə-ˌpla-zəm, 'nyü-\ *noun* [International Scientific Vocabulary] (1888)
: the protoplasm of a nucleus; *especially* **:** NUCLEAR SAP
— **nu·cle·o·plas·mic** \-klē-ə-'plaz-mik\ *adjective*

nu·cle·o·pro·tein \ˌnü-klē-ō-'prō-ˌtēn, ˌnyü-, -'prō-tē-ən\ *noun* [International Scientific Vocabulary] (1907)
: a compound that consists of a protein (as a histone) conjugated with a nucleic acid (as a DNA) and that is the principal constituent of the hereditary material in chromosomes

nu·cle·o·side \'nü-klē-ə-ˌsīd, 'nyü-\ *noun* [International Scientific Vocabulary *nucle-* + [2]*-ose* + *-ide*] (1911)
: a compound (as guanosine or adenosine) that consists of a purine or pyrimidine base combined with deoxyribose or ribose and is found especially in DNA or RNA

nu·cle·o·some \-ˌsōm\ *noun* (1962)
: any of the repeating globular subunits of chromatin that consist of a complex of DNA and histone
— **nu·cle·o·so·mal** \ˌnü-klē-ə-'sō-məl, ˌnyü-\ *adjective*

nu·cle·o·syn·the·sis \ˌnü-klē-ō-'sin(t)-thə-səs, ˌnyü-\ *noun* [New Latin] (1960)
: the production of a chemical element from simpler nuclei (as of hydrogen) especially in a star
— **nu·cle·o·syn·thet·ic** \-sin-'the-tik\ *adjective*

nu·cle·o·tid·ase \ˌnü-klē-ə-'tī-ˌdās, ˌnyü-, -ˌdāz\ *noun* (1911)
: a phosphatase that promotes hydrolysis of a nucleotide (as into a nucleoside and phosphoric acid)

nu·cle·o·tide \'nü-klē-ə-ˌtīd, 'nyü-\ *noun* [International Scientific Vocabulary, irregular from *nucle-* + *-ide*] (1908)
: any of several compounds that consist of a ribose or deoxyribose sugar joined to a purine or pyrimidine base and to a phosphate group and that are the basic structural units of nucleic acids (as RNA and DNA) — compare NUCLEOSIDE

nu·cle·us \'nü-klē-əs, 'nyü-\ *noun, plural* **nu·clei** \-klē-ˌī\ *also* **nu·cle·us·es** [New Latin, from Latin, kernel, diminutive of *nuc-, nux* nut — more at NUT] (1704)
1 : the small brighter and denser portion of a galaxy or of the head of a comet
2 : a central point, group, or mass about which gathering, concentration, or accretion takes place: as **a :** a cellular organelle of eukaryotes that is essential to cell functions (as reproduction and protein synthesis), is composed of nuclear sap and a nucleoprotein-rich network from which chromosomes and nucleoli arise, and is enclosed in a definite membrane — see CELL illustration **b :** a mass of gray matter or group of nerve cells in the central nervous system **c :** a characteristic and stable complex of atoms or groups in a molecule; *especially* **:** RING ⟨the naphthalene *nucleus*⟩ **d :** the positively charged central portion of an atom that comprises nearly all of the atomic mass and that consists of protons and neutrons except in hydrogen which consists of one proton only
3 : the peak of sonority in the utterance of a syllable

nu·clide \'nü-ˌklīd, 'nyü-\ *noun* [*nucl*eus + Greek *eidos* form, species — more at IDOL] (1947)
: a species of atom characterized by the constitution of its nucleus and hence by the number of protons, the number of neutrons, and the energy content
— **nu·clid·ic** \nü-'kli-dik, nyü-\ *adjective*

[1]**nude** \'nüd, 'nyüd\ *adjective* **nud·er; nud·est** [Latin *nudus* naked — more at NAKED] (1531)
1 : lacking something essential especially to legal validity ⟨a *nude* contract⟩
2 a : devoid of a natural or conventional covering; *especially* **:** not covered by clothing or a drape **b** (1) **:** of the color of a white person's flesh (2) **:** giving the appearance of nudity ⟨a *nude* dress⟩ **c :** featuring nudes ⟨a *nude* movie⟩ **d :** frequented by naked people ⟨a *nude* beach⟩
synonym see BARE
— **nude** *adverb*
— **nude·ly** *adverb*
— **nude·ness** *noun*
— **nu·di·ty** \'nü-də-tē, 'nyü-\ *noun*

[2]**nude** *noun* (1708)
1 a : a representation of a nude human figure **b :** a nude person
2 : the condition of being nude ⟨in the *nude*⟩

nudge \'nəj\ *verb* **nudged; nudg·ing** [origin unknown] (1675)
transitive verb
1 : to touch or push gently; *especially* **:** to seek the attention of by a push of the elbow
2 : to prod lightly **:** urge into action
3 : APPROACH ⟨its circulation is *nudging* the four million mark —Bennett Cerf⟩
intransitive verb
: to give a nudge
— **nudge** *noun*
— **nudg·er** *noun*

nu·di·branch \'nü-də-ˌbraŋk, 'nyü-\ *noun, plural* **-branchs** [New Latin *Nudibranchia,* from Latin *nudus* + Greek *branchia* gills] (1817)
: any of an order (Nudibranchia) of marine gastropod mollusks without a shell in the adult state and without true gills
— **nudibranch** *adjective*

nud·ism \'nü-ˌdi-zəm, 'nyü-\ *noun* (1929)
: the practice of going nude especially in sexually mixed groups and during periods of time spent at specially secluded places
— **nud·ist** \'nü-dist, 'nyü-\ *adjective or noun*

nud·nick *or* **nud·nik** \'nu̇d-nik\ *noun* [Yiddish *nudnik,* from *nudyen* to bore, from Polish *nudzić,* from *nuda* boredom] (1947)
: a person who is a bore or nuisance

nu·ga·to·ry \'nü-gə-ˌtōr-ē, 'nyü-, -ˌtȯr-\ *adjective* [Latin *nugatorius,* from *nugari* to trifle, from *nugae* trifles] (1603)
1 : of little or no consequence **:** TRIFLING, INCONSEQUENTIAL
2 : having no force **:** INOPERATIVE
synonym see VAIN

nug·get \'nə-gət\ *noun* [origin unknown] (1852)
1 : a solid lump; *especially* **:** a native lump of precious metal
2 : TIDBIT 2 ⟨*nuggets* of wisdom⟩
3 : a small usually rounded piece of food ⟨chicken *nuggets*⟩

nui·sance \'nü-s^{ə}n(t)s, 'nyü-\ *noun* [Middle English *nusaunce,* from Anglo-French, from Old French *nuire* to harm, from Latin *nocēre* — more at NOXIOUS] (15th century)
1 : HARM, INJURY
2 : one that is annoying, unpleasant, or obnoxious **:** PEST

nuisance tax *noun* (1924)
: an excise tax collected in small amounts on a wide range of commodities directly from the consumer

[1]**nuke** \'nük, 'nyük\ *noun* [by shortening & alteration from *nuclear*] (1959)
1 : a nuclear weapon
2 : a nuclear-powered electric generating station

[2]**nuke** *transitive verb* **nuked; nuk·ing** (1967)
1 : to attack or destroy with or as if with nuclear bombs
2 : MICROWAVE

[1]**null** \'nəl\ *adjective* [Middle French *nul,* literally, not any, from Latin *nullus,* from *ne-* not + *ullus* any; akin to Latin *unus* one — more at NO, ONE] (circa 1567)
1 : having no legal or binding force **:** INVALID
2 : amounting to nothing **:** NIL
3 : having no value **:** INSIGNIFICANT
4 a : having no elements ⟨*null* set⟩ **b :** having zero as a limit ⟨*null* sequence⟩ **c** *of a matrix* **:** having all elements equal to zero
5 a : indicating usually by a zero reading on a scale when a given quantity (as current or voltage) is zero or when two quantities are equal — used of an instrument **b :** being or relating to a method of measurement in which an unknown quantity (as of electric current) is compared with a known quantity of the same kind and found equal by a null detector
6 : of, being, or relating to zero
7 : ZERO 1c

[2]**null** *noun* (1605)
1 : ZERO 3a(1)
2 a : a condition of a radio receiver when minimum or zero signal is received **b :** a minimum or zero value of an electric current or of a radio signal

[3]**null** *transitive verb* (1643)
: to make null

nul·lah \'nə-lə\ *noun* [Hindi *nālā*] (1776)
: GULLY, RAVINE

null and void *adjective* (1669)
: having no force, binding power, or validity

null hypothesis *noun* (1935)
: a statistical hypothesis to be tested and accepted or rejected in favor of an alternative; *specifically* **:** the hypothesis that an observed difference (as between the means of two samples) is due to chance alone and not due to a systematic cause

nul·li·fi·ca·tion \ˌnə-lə-fə-'kā-shən\ *noun* (1798)
1 : the act of nullifying **:** the state of being nullified
2 : the action of a state impeding or attempting to prevent the operation and enforcement within its territory of a law of the U.S.
— **nul·li·fi·ca·tion·ist** \-sh(ə-)nist\ *noun*

nul·li·fi·er \'nə-lə-ˌfī(-ə)r\ *noun* (1832)
: one that nullifies; *specifically* **:** one maintaining the right of nullification against the U.S. government

nul·li·fy \'nə-lə-ˌfī\ *transitive verb* **-fied; -fy·ing** [Late Latin *nullificare,* from Latin *nullus*] (1535)
1 : to make null; *especially* **:** to make legally null and void

2 : to make of no value or consequence ☆

nul·lip·a·rous \ˌnə-ˈli-pə-rəs\ *adjective* [New Latin *nullipara* one who has never borne an offspring, from Latin *nullus* not any + *-para* -para] (1859)
: of, relating to, or being a female that has not borne offspring

nul·li·ty \ˈnə-lə-tē\ *noun, plural* **-ties** (1570)
1 a : the quality or state of being null; *especially* **:** legal invalidity **b** (1) **:** NOTHINGNESS; *also* **:** INSIGNIFICANCE (2) **:** a mere nothing **:** NONENTITY
2 : one that is null; *specifically* **:** an act void of legal effect
3 : the number of elements in a basis of a null-space

null–space \ˈnəl-ˌspās\ *noun* (1884)
: a subspace of a vector space consisting of vectors that under a given linear transformation are mapped onto zero

numb \ˈnəm\ *adjective* [Middle English *nomen,* from past participle of *nimen* to take — more at NIM] (14th century)
1 : devoid of sensation especially as a result of cold or anesthesia
2 : devoid of emotion **:** INDIFFERENT
— numb *transitive verb*
— numb·ing·ly \ˈnə-miŋ-lē\ *adverb*
— numb·ly \ˈnəm-lē\ *adverb*
— numb·ness *noun*

[1]num·ber \ˈnəm-bər\ *noun* [Middle English *nombre,* from Old French, from Latin *numerus*] (14th century)
1 a (1) **:** a sum of units **:** TOTAL (2) **:** COMPLEMENT 1b (3) **:** an indefinite usually large total ⟨a *number* of members were absent⟩ ⟨the *number* of elderly is rising⟩ (4) *plural* **:** a numerous group **:** MANY (5) **:** a numerical preponderance **b** (1) **:** the characteristic of an individual by which it is treated as a unit or of a collection by which it is treated in terms of units (2) **:** an ascertainable total ⟨bugs beyond *number*⟩ **c** (1) **:** a unit belonging to an abstract mathematical system and subject to specified laws of succession, addition, and multiplication; *especially* **:** NATURAL NUMBER (2) **:** an element (as π) of any of many mathematical systems obtained by extension of or analogy with the natural number system (3) *plural* **:** ARITHMETIC
2 : a distinction of word form to denote reference to one or more than one; *also* **:** a form or group of forms so distinguished
3 *plural* **a** (1) **:** metrical structure **:** METER (2) **:** metrical lines **:** VERSES **b** *archaic* **:** musical sounds **:** NOTES
4 a : a word, symbol, letter, or combination of symbols representing a number **b :** a numeral or combination of numerals or other symbols used to identify or designate ⟨dialed the wrong *number*⟩ **c** (1) **:** a member of a sequence or collection designated by especially consecutive numbers (as an issue of a periodical) (2) **:** a position in a numbered sequence **d :** a group of one kind ⟨not of their *number*⟩
5 : one singled out from a group **:** INDIVIDUAL: as **a :** GIRL, WOMAN ⟨met an attractive *number* at the dance⟩ **b** (1) **:** a musical, theatrical, or literary selection or production (2) **:** ROUTINE, ACT **c :** an item of merchandise and especially clothing
6 : insight into a person's ability or character ⟨had my *number*⟩
7 *plural but singular or plural in construction* **a :** a form of lottery in which an individual wagers on the appearance of a certain combination of digits (as in regularly published numbers) — called also *numbers game* **b :** [2]POLICY 2a
8 *plural* **a :** figures representing amounts of money usually in dollars spent, earned, or involved **b** (1) **:** STATISTICS 2; *especially* **:** individual statistics (as of an athlete) (2) **:** RATING 3c
9 : a person represented by a number or considered without regard to individuality ⟨at the university I was just a *number*⟩
usage see AMOUNT
— by the numbers 1 : in unison to a specific count or cadence **2 :** in a systematic, routine, or mechanical manner
▶The Table of Numbers is on page 1254

[2]number *verb* **num·bered; num·ber·ing** \-b(ə-)riŋ\ (14th century)
transitive verb
1 : COUNT, ENUMERATE
2 : to claim as part of a total **:** INCLUDE
3 : to restrict to a definite number ⟨your days are *numbered*⟩
4 : to assign a number to
5 : to amount to in number **:** TOTAL ⟨the crew *numbers* 100⟩
intransitive verb
1 : to reach a total number
2 : to call off numbers in sequence ■
— num·ber·able \-b(ə-)rə-bəl\ *adjective*
— num·ber·er \-bər-ər\ *noun*

number cruncher *noun* (1966)
1 : a computer that performs fast numerical calculations especially on large amounts of data
2 : a person concerned with complex numerical data (as statistics)
— number crunching *noun*

num·ber·less \ˈnəm-bər-ləs\ *adjective* (1573)
: INNUMERABLE, COUNTLESS

number line *noun* (1960)
: a line of infinite extent whose points correspond to the real numbers according to their distance in a positive or negative direction from a point arbitrarily taken as zero

[1]number one *noun* (circa 1705)
1 : one's own interests or welfare **:** ONESELF ⟨looking out for *number one*⟩ — often written *No. 1*
2 : one that is first in rank, importance, or influence — often written *No. 1*

[2]number one *adjective* (1839)
1 : first in rank, importance, or influence **:** FOREMOST ⟨cancer is the country's *number one* killer⟩ — often written *No. 1*
2 : of highest or of high quality

Num·bers \ˈnəm-bərz\ *noun plural but singular in construction*
: the mainly narrative fourth book of canonical Jewish and Christian Scripture — see BIBLE table

number theory *noun* (1912)
: the study of the properties of integers
— number theoretic *adjective*
— number theorist *noun*

numb·skull \ˈnəm-ˌskəl\ *variant of* NUMSKULL

nu·men \ˈnü-mən, ˈnyü-\ *noun, plural* **nu·mi·na** \-mə-nə\ [Latin, nod, divine will, numen; akin to Latin *nutare* to nod, Greek *neuein*] (1628)
: a spiritual force or influence often identified with a natural object, phenomenon, or place

nu·mer·a·ble \ˈnüm-rə-bəl, ˈnü-mə-; ˈnyüm-, ˈnyü-mə-\ *adjective* [Latin *numerabilis,* from *numerare* to count] (1570)
: capable of being counted

[1]nu·mer·al \ˈnüm-rəl, ˈnü-mə-; ˈnyüm-, ˈnyü-mə-\ *adjective* [Middle English, from Middle French, from Late Latin *numeralis,* from Latin *numerus*] (14th century)
1 : of, relating to, or expressing numbers
2 : consisting of numbers or numerals
— nu·mer·al·ly *adverb*

[2]numeral *noun* (1686)
1 : a conventional symbol that represents a number
2 *plural* **:** numbers that designate by year a school or college class and that are awarded for distinction in an extracurricular activity

[1]nu·mer·ate \ˈnü-mə-ˌrāt, ˈnyü-\ *transitive verb* **-at·ed; -at·ing** [Latin *numeratus,* past participle of *numerare* to count, from *numerus*] (circa 1721)
: ENUMERATE

[2]nu·mer·ate \ˈnüm-rət, ˈnü-mə-; ˈnyüm-, ˈnyü-mə-\ *adjective* [Latin *numerus* number + English *-ate* (as in *literate*)] (1959)
: marked by the capacity for quantitative thought and expression
— nu·mer·acy \ˈn(y)üm-rə-sē, ˈn(y)ü-mə-\ *noun*

nu·mer·a·tion \ˌnü-mə-ˈrā-shən, ˌnyü-\ *noun* (15th century)
1 a : the act or process or an instance of counting or numbering; *also* **:** a system of counting or numbering **b :** an act or instance of designating by a number
2 : the art of reading in words numbers expressed by numerals

nu·mer·a·tor \ˈnü-mə-ˌrā-tər, ˈnyü-\ *noun* (1575)
1 : the part of a fraction that is above the line and signifies the number of parts of the denominator taken
2 : one that numbers

[1]nu·mer·ic \nu̇-ˈmer-ik, nyu̇-\ *adjective* (circa 1828)
: NUMERICAL; *especially* **:** denoting a number or a system of numbers ⟨*numeric* code⟩ ⟨a *numeric* sign⟩

[2]nu·mer·ic *noun* (1879)
: NUMBER, NUMERAL

nu·mer·i·cal \nu̇-ˈmer-i-kəl, nyu̇-\ *adjective* [Latin *numerus*] (1628)
1 : of or relating to numbers ⟨the *numerical* superiority of the enemy⟩
2 : expressed in or involving numbers or a number system ⟨*numerical* standing in a class⟩ ⟨a *numerical* code⟩
— nu·mer·i·cal·ly \-k(ə-)lē\ *adverb*

☆ SYNONYMS
Nullify, negate, annul, abrogate, invalidate mean to deprive of effective or continued existence. NULLIFY implies counteracting completely the force, effectiveness, or value of something ⟨a penalty *nullified* the touchdown⟩. NEGATE implies the destruction or canceling out of each of two things by the other ⟨the arguments *negate* each other⟩. ANNUL suggests making ineffective or nonexistent often by legal or official action ⟨the treaty *annuls* all previous agreements⟩. ABROGATE is like ANNUL but more definitely implies a legal or official purposeful act ⟨a law to *abrogate* trading privileges⟩. INVALIDATE implies making something powerless or unacceptable by declaration of its logical or moral or legal unsoundness ⟨the court *invalidated* the statute⟩.

□ USAGE
number There is a rule of thumb well-known in the newspaper business that *a number* takes a plural verb and *the number* a singular verb. Our files show that this rule is generally observed, even outside the newspaper business ⟨a *number* of celebrated modern manuscripts are available in published form —*New Yorker*⟩ ⟨the total *number* of emigrés is now estimated rather elastically —Barbara Tuchman⟩. When an adjective like *growing* or *increasing* is added, however, *a number* often takes a singular verb ⟨an increasing *number* of Ministers now believes the Government could be forced to go to the country —Ian Aitken⟩. We find some examples ⟨a large *number* of men was visible —Tom Clancy⟩ where the plural idea is belied by a singular verb; perhaps such instances are the result of uncritical copyediting. In this case, the rule of thumb is a good rule to follow.

TABLE OF NUMBERS

CARDINAL NUMBERS[1]

NAME[2]	SYMBOL: Hindu-Arabic	SYMBOL: Roman[3]
zero *or* naught *or* cipher	0	
one	1	I
two	2	II
three	3	III
four	4	IV
five	5	V
six	6	VI
seven	7	VII
eight	8	VIII
nine	9	IX
ten	10	X
eleven	11	XI
twelve	12	XII
thirteen	13	XIII
fourteen	14	XIV
fifteen	15	XV
sixteen	16	XVI
seventeen	17	XVII
eighteen	18	XVIII
nineteen	19	XIX
twenty	20	XX
twenty-one	21	XXI
twenty-two	22	XXII
twenty-three	23	XXIII
twenty-four	24	XXIV
twenty-five	25	XXV
twenty-six	26	XXVI
twenty-seven	27	XXVII
twenty-eight	28	XXVIII
twenty-nine	29	XXIX
thirty	30	XXX
thirty-one	31	XXXI
thirty-two *etc*	32	XXXII
forty	40	XL
forty-one *etc*	41	XLI
fifty	50	L
sixty	60	LX
seventy	70	LXX
eighty	80	LXXX
ninety	90	XC
one hundred	100	C
on hundred and one *or* one hundred one	101	CI
one hundred and two *etc*	102	CII
two hundred	200	CC
three hundred	300	CCC
four hundred	400	CD
five hundred	500	D
six hundred	600	DC
seven hundred	700	DCC
eight hundred	800	DCCC
nine hundred	900	CM
one thousand *or* ten hundred *etc*	1,000	M
two thousand *etc*	2,000	MM
five thousand	5,000	$\overline{V}$
ten thousand	10,000	$\overline{X}$
one hundred thousand	100,000	$\overline{C}$
one million	1,000,000	$\overline{M}$

ORDINAL NUMBERS[4]

NAME[5]	SYMBOL[6]
first	1st
second	2d *or* 2nd
third	3d *or* 3rd
fourth	4th
fifth	5th
sixth	6th
seventh	7th
eighth	8th
ninth	9th
tenth	10th
eleventh	11th
twelfth	12th
thirteenth	13th
fourteenth	14th
fifteenth	15th
sixteenth	16th
seventeenth	17th
eighteenth	18th
nineteenth	19th
twentieth	20th
twenty-first	21st
twenty-second	22d *or* 22nd
twenty-third	23d *or* 23rd
twenty-fourth	24th
twenty-fifth	25th
twenty-sixth	26th
twenty-seventh	27th
twenty-eighth	28th
twenty-ninth	29th
thirtieth	30th
thirty-first	31st
thirty-second *etc*	32d *or* 32nd
fortieth	40th
forty-first	41st
forty-second *etc*	42d *or* 42nd
fiftieth	50th
sixtieth	60th
seventieth	70th
eightieth	80th
ninetieth	90th
hundredth *or* one hundredth	100th
hundred and first *or* one hundred and first	101st
hundred and seconds *etc*	102d *or* 102nd
two hundredth	200th
three hundredth	300th
four hundredth	400th
five hundredth	500th
six hundredth	600th
seven hundredth	700th
eight hundredth	800th
nine hundredth	900th
thousandth *or* one thousandth	1,000th
two thousandth *etc*	2,000th
ten thousandth	10,000th
hundred thousandth *or* one hundred thousandth	100,000th
millionth *or* one millionth	1,000,000th

[1]The cardinal numbers are used in simple counting or in answer to "how many?" The words for these numbers may be used as nouns (he counted to *twelve*), as pronouns (*twelve* were found), or as adjectives (*twelve* boys).

[2]In formal contexts the numbers one to one hundred and in less formal contexts the numbers one to nine are commonly written out, while larger numbers are given in numerals. In nearly all contexts a number occurring at the beginning of a sentence is usually written out. Except in very formal contexts numerals are invariably used for dates. Hindu-Arabic numbers from 1,000 to 9,999 are often written without commas or spaces (1000, 9999). Year numbers are always written without commas (1783).

[3]The Roman numerals are written either in capitals or in lowercase letters.

[4]The ordinal numbers are used to show the order of succession in which such items as names, objects, and periods of time are considered (the *twelfth* month; the *fourth* row of seats; the *18th* century).

[5]Each of the terms for the ordinal numbers excepting *first* and *second* is used in designating one of a number of parts into which a whole may be divided (a *fourth*; a *sixth*; a *tenth*) and as the denominator in fractions designating the number of such parts constituting a certain portion of a whole (*one fourth*; *three fifths*). When used as nouns the fractions are usually written as two words, although they are regularly hyphenated as adjectives (a *two-thirds* majority). When fractions are written in numerals, the cardinal symbols are used (1/4, 3/5, 5/6).

[6]The Hindu-Arabic symbols for the cardinal numbers may be read as ordinals in certain contexts (January 1 = January first; 2 Samuel = Second Samuel). The Roman numerals are sometimes read as ordinals (Henry IV = Henry the Fourth); sometimes they are written with the ordinal suffixes (XIXth Dynasty).

DENOMINATIONS ABOVE ONE MILLION

American system[1]

NAME	VALUE IN POWERS OF TEN	NUMBER OF ZEROS[2]	NUMBER OF GROUPS OF THREE 0'S AFTER 1,000
billion	10^{9}	9	2
trillion	10^{12}	12	3
quadrillion	10^{15}	15	4
quintillion	10^{18}	18	5
sextillion	10^{21}	21	6
septillion	10^{24}	24	7
octillion	10^{27}	27	8
nonillion	10^{30}	30	9
decillion	10^{33}	33	10
undecillion	10^{36}	36	11
duodecillion	10^{39}	39	12
tredecillion	10^{42}	42	13
quattuordecillion	10^{45}	45	14
quindecillion	10^{48}	48	15
sexdecillion	10^{51}	51	16
septendecillion	10^{54}	54	17
octodecillion	10^{57}	57	18
novemdecillion	10^{60}	60	19
vigintillion	10^{63}	63	20
centillion	10^{303}	303	100

British system[1]

NAME	VALUE IN POWERS OF TEN	NUMBER OF ZEROS[2]	POWERS OF 1,000,000
milliard	10^{9}	9	—
billion	10^{12}	12	2
trillion	10^{18}	18	3
quadrillion	10^{24}	24	4
quintillion	10^{30}	30	5
sextillion	10^{36}	36	6
septillion	10^{42}	42	7
octillion	10^{48}	48	8
nonillion	10^{54}	54	9
decillion	10^{60}	60	10
undecillion	10^{66}	66	11
duodecillion	10^{72}	72	12
tredecillion	10^{78}	78	13
quattuordecillion	10^{84}	84	14
quindecillion	10^{90}	90	15
sexdecillion	10^{96}	96	16
septendecillion	10^{102}	102	17
octodecillion	10^{108}	108	18
novemdecillion	10^{114}	114	19
vigintillion	10^{120}	120	20
centillion	10^{600}	600	100

[1]The American system of numeration for denominations above one million was modeled on the French system but more recently the French system has been changed to correspond to the German and British systems. In the American system each of the denominations above 1,000 millions (the American *billion*) is 1,000 times the preceding one (one trillion = 1,000 billions; one quadrillion = 1,000 trillions). In the British system the first denomination above 1,000 millions (the British *milliard*) is 1,000 times the preceding one, but each of the denominations above 1,000 milliards (the British *billion*) is 1,000,000 times the preceding one (one trillion = 1,000,000 billions; one quadrillion = 1,000,000 trillions).

[2]For convenience in reading large numerals, the thousands, millions, etc., are usually separated by commas (21,530; 1,155,465) or especially in technical contexts by spaces (1 155 465). Serial numbers (as a social security number) are often written with hyphens (042-24-4705).

numerical analysis *noun* (1946)
: the study of quantitative approximations to the solutions of mathematical problems including consideration of the errors and bounds to the errors involved

numerical taxonomy *noun* (1963)
: taxonomy in which many quantitatively measured characters are given equal weight in the determination of taxa and the construction of diagrams indicating systematic relationships
— **numerical taxonomic** *adjective*
— **numerical taxonomist** *noun*

nu·mer·ol·o·gy \ˌnü-mə-ˈrä-lə-jē, ˌnyü-\ *noun* [Latin *numerus* + English *-o-* + *-logy*] (1911)
: the study of the occult significance of numbers
— **nu·mer·o·log·i·cal** \-mə-rə-ˈlä-ji-kəl\ *adjective*
— **nu·mer·ol·o·gist** \-mə-ˈrä-lə-jist\ *noun*

nu·me·ro uno \ˈnü-mə-rō-ˈü-(ˌ)nō, ˈnyü-\ *noun or adjective* [Italian *numero uno* or Spanish *número uno*] (1968)
: NUMBER ONE

nu·mer·ous \ˈnüm-rəs, ˈnü-mə-; ˈnyüm-, ˈnyü-mə-\ *adjective* [Middle English, from Middle French *numereux,* from Latin *numerosus,* from *numerus*] (15th century)
: consisting of great numbers of units or individuals ⟨born into a *numerous* family⟩; *also* : MANY ⟨received *numerous* complaints⟩
— **nu·mer·ous·ly** *adverb*
— **nu·mer·ous·ness** *noun*

nu·mi·nous \ˈnü-mə-nəs, ˈnyü-\ *adjective* [Latin *numin-, numen* numen] (1647)
1 : SUPERNATURAL, MYSTERIOUS
2 : filled with a sense of the presence of divinity : HOLY
3 : appealing to the higher emotions or to the aesthetic sense : SPIRITUAL
— **nu·mi·nous·ness** \-nəs\ *noun*

nu·mis·mat·ic \ˌnü-məz-ˈma-tik, -məs-, ˌnyü-\ *adjective* [French *numismatique,* from Latin *nomismat-, nomisma* coin, from Greek, current coin, from *nomizein* to use, from *nomos* custom, law — more at NIMBLE] (1792)
1 : of or relating to numismatics
2 : of or relating to currency : MONETARY
— **nu·mis·mat·i·cal·ly** \-ti-k(ə-)lē\ *adverb*

nu·mis·mat·ics \-tiks\ *noun plural but singular in construction* (circa 1828)
: the study or collection of coins, tokens, and paper money and sometimes related objects (as medals)
— **nu·mis·ma·tist** \nü-ˈmiz-mə-tist, nyü-\ *noun*

num·mu·lar \ˈnəm-yə-lər\ *adjective* [French *nummulaire,* from Latin *nummulus,* diminutive of *nummus* coin, probably from Greek *nomimos* customary, from *nomos*] (1846)
: characterized by circular or oval lesions or drops ⟨*nummular* dermatitis⟩

num·mu·lit·ic limestone \ˌnəm-yə-ˈli-tik-\ *noun* [New Latin *Nummulites,* genus of foraminifers, from Latin *nummulus*] (1833)
: the most widely distributed and distinctive formation of the Eocene in Europe, Asia, and northern Africa

num·skull \ˈnəm-ˌskəl\ *noun* [*numb* + *skull*] (1717)
1 : a thick or muddled head
2 : a dull or stupid person : DUNCE

[1]nun \ˈnən\ *noun* [Middle English, from Old English *nunne,* from Late Latin *nonna*] (before 12th century)
: a woman belonging to a religious order; *especially* : one under solemn vows of poverty, chastity, and obedience
— **nun·like** \-ˌlīk\ *adjective*

[2]nun \ˈnu̇n\ *noun* [Hebrew *nūn*] (circa 1823)
: the 14th letter of the Hebrew alphabet — see ALPHABET table

nun·a·tak \ˈnə-nə-ˌtak\ *noun* [Inuit (Greenland) *nunataq*] (1877)
: a hill or mountain completely surrounded by glacial ice

Nunc Di·mit·tis \ˌnəŋk-də-ˈmi-təs, ˌnu̇ŋk-\ *noun* [Latin, now lettest thou depart; from the first words of the canticle] (1552)
: the prayer of Simeon in Luke 2:29–32 used as a canticle

nun·cha·ku \ˈnən-ˌchək, ˌnən-ˈchä-kü\ *noun* [Japanese dialect (Okinawa)] (1970)
: a weapon that consists of two hardwood sticks joined at their ends by a short length of rawhide, cord, or chain

nun·ci·a·ture \ˈnən(t)-sē-ə-ˌchu̇r, ˈnu̇n(t)-, -chər, -ˌtyu̇r, -ˌtu̇r\ *noun* [Italian *nunciatura,* from *nuncio*] (1652)
1 : a papal diplomatic mission headed by a nuncio
2 : the office or period of office of a nuncio

nun·cio \ˈnən(t)-sē-ˌō, ˈnu̇n(t)-\ *noun, plural* **-ci·os** [Italian, from Latin *nuntius* messenger, message] (1528)
: a papal legate of the highest rank permanently accredited to a civil government

nun·cle \ˈnəŋ-kəl\ *noun* [by alteration (from misdivision of *an uncle*)] (circa 1589)
chiefly dialect : UNCLE

nun·cu·pa·tive \ˈnən-kyu̇-ˌpā-tiv, ˈnəŋ-; ˌnən-ˈkyü-pə-\ *adjective* [Medieval Latin *nuncupativus,* from Late Latin, so-called, from Latin *nuncupatus,* past participle of *nuncupare* to name, probably ultimately from *nomen* name + *capere* to take — more at NAME, HEAVE] (1546)
: not written : ORAL ⟨a *nuncupative* will⟩

nun·nery \ˈnən-rē, ˈnə-nə-\ *noun, plural* **-ner·ies** (14th century)
: a convent of nuns

Nu·pe \ˈnü-(ˌ)pā\ *noun, plural* **Nupe** *or* **Nupes** (1883)
1 : the language of the Nupe people
2 : a member of a people of west central Nigeria

[1]nup·tial \ˈnəp-shəl, -chəl, ÷-shə-wəl, ÷-chə-wəl\ *adjective* [Latin *nuptialis,* from *nuptiae,* plural, wedding, from *nubere* to marry; perhaps akin to Greek *nymphē* bride, nymph] (15th century)
1 : of or relating to marriage or the marriage ceremony
2 : characteristic of or occurring in the breeding season ⟨*nuptial* flight⟩

[2]nuptial *noun* (circa 1555)
: MARRIAGE, WEDDING — usually used in plural

nup·tial·i·ty \ˌnəp-shē-ˈa-lə-tē, -chē-\ *noun, plural* **-ties** (1899)
: the marriage rate

Nur·i·stani \ˌnu̇r-ə-ˈstä-nē, ˌnyu̇r-\ *noun* [*Nuristan,* Afghanistan] (1957)
: KAFIRI

[1]nurse \ˈnərs\ *noun* [Middle English, from Old French *nurice,* from Late Latin *nutricia,* from Latin, feminine of *nutricius* nourishing — more at NUTRITIOUS] (13th century)
1 a : a woman who suckles an infant not her own : WET NURSE **b :** a woman who takes care of a young child : DRY NURSE
2 : one that looks after, fosters, or advises
3 : a person who is skilled or trained in caring for the sick or infirm especially under the supervision of a physician
4 a : a member of an insect society that belongs to the worker caste and cares for the young **b :** a female mammal used to suckle the young of another

[2]nurse *verb* **nursed; nurs·ing** [Middle English *nurshen* to suckle, nourish, contraction of *nurishen*] (14th century)
transitive verb
1 a : to nourish at the breast : SUCKLE **b :** to take nourishment from the breast of
2 : REAR, EDUCATE
3 a : to promote the development or progress of **b :** to manage with care or economy ⟨*nursed* the business through hard times⟩ **c :** to take charge of and watch over
4 a : to care for and wait on (as a sick person) **b :** to attempt to cure by care and treatment
5 : to hold in one's memory or consideration ⟨*nurse* a grievance⟩
6 a : to use, handle, or operate carefully so as to conserve energy or avoid injury or pain ⟨*nurse* a sprained ankle⟩ **b :** to use sparingly **c :** to consume slowly or over a long period ⟨*nurse* a cup of coffee⟩
intransitive verb
1 a : to feed an offspring from the breast **b :** to feed at the breast : SUCK
2 : to act or serve as a nurse
— **nurs·er** *noun*

nurse·maid \ˈnərs-ˌmād\ *noun* (1657)
: a girl or woman who is regularly employed to look after children

nurse–mid·wife \-ˈmid-ˌwīf\ *noun* (1952)
: a registered nurse with additional training as a midwife who delivers infants and provides antepartum and postpartum care
— **nurse–mid·wife·ry** \-ˌmid-ˈwi-f(ə-)rē, -ˈwī-; -ˈmid-ˌwī-\ *noun*

nurse–prac·ti·tion·er \-prak-ˈti-sh(ə-)nər\ *noun* (1969)
: a registered nurse who is qualified through advanced training to assume some of the duties and responsibilities formerly assumed only by a physician

nurs·ery \ˈnərs-rē, ˈnər-sə-\ *noun, plural* **-er·ies** (14th century)
1 *obsolete* : attentive care : FOSTERAGE
2 a : a child's bedroom **b :** a place where children are temporarily cared for in their parents' absence **c :** DAY NURSERY
3 a : something that fosters, develops, or promotes **b :** a place in which persons are trained or educated
4 : an area where plants are grown for transplanting, for use as stocks for budding and grafting, or for sale
5 : a place where young animals grow or are cared for

nurs·ery·man \-mən\ *noun* (1672)
: one whose occupation is the cultivation of plants (as trees and shrubs) especially for sale

nursery rhyme *noun* (1832)
: a short rhyme for children that often tells a story

nursery school *noun* (1835)
: a school for children usually under five years

nurse's aide *noun* (1943)
: a worker who assists trained nurses in a hospital by performing unspecialized services (as giving baths)

nurse shark *noun* [alteration of Middle English *nusse*] (1851)
: any of various sharks of a widely distributed family (Orectolobidae); *especially* : a shark (*Ginglymostoma cirratum*) of the warmer parts of the Atlantic Ocean

nurs·ing *noun* (1860)
1 : the profession of a nurse ⟨schools of *nursing*⟩
2 : the duties of a nurse ⟨proper *nursing* is difficult work⟩

nursing home *noun* (1896)
: a privately operated establishment providing maintenance and personal or nursing care for persons (as the aged or the chronically ill) who are unable to care for themselves properly

nurs·ling \ˈnərs-liŋ\ *noun* (1557)
1 : one that is solicitously cared for
2 : a nursing child

nur·tur·ance \ˈnər-chə-rən(t)s\ *noun* (circa 1938)
: affectionate care and attention
— **nur·tur·ant** \-rənt\ *adjective*

¹nur·ture \'nər-chər\ *noun* [Middle English, from Middle French *norriture,* from Late Latin *nutritura* act of nursing, from Latin *nutritus,* past participle of *nutrire* to suckle, nourish — more at NOURISH] (14th century)
1 : TRAINING, UPBRINGING
2 : something that nourishes **:** FOOD
3 : the sum of the influences modifying the expression of the genetic potentialities of an organism

²nurture *transitive verb* **nur·tured; nur·tur·ing** \'nərch-riŋ, 'nər-chə-\ (15th century)
1 : to supply with nourishment
2 : EDUCATE
3 : to further the development of **:** FOSTER
— nur·tur·er \'nər-chər-ər\ *noun*

¹nut \'nət\ *noun* [Middle English *nute, note,* from Old English *hnutu;* akin to Old High German *nuz* nut and perhaps to Latin *nux* nut] (before 12th century)
1 a (1) **:** a hard-shelled dry fruit or seed with a separable rind or shell and interior kernel (2) **:** the kernel of a nut **b :** a dry indehiscent one-seeded fruit with a woody pericarp
2 a : a hard problem or undertaking **b :** CORE, HEART
3 : a perforated block usually of metal that has an internal screw thread and is used on a bolt or screw for tightening or holding something
4 : the ridge in a stringed instrument (as a violin) over which the strings pass on the upper end of the fingerboard
5 : a small lump (as of butter)
6 a : a foolish, eccentric, or crazy person **b :** ENTHUSIAST ⟨a movie *nut*⟩
7 *plural* **:** NONSENSE — often used interjectionally
8 *slang* **:** a person's head
9 : TESTIS — usually considered vulgar
10 : the amount of money that must be earned in order to break even
11 : EN 1
— nut·like \-,līk\ *adjective*

²nut *intransitive verb* **nut·ted; nut·ting** (1604)
: to gather or seek nuts

nu·tate \'nü-,tāt, 'nyü-\ *intransitive verb* **nu·tat·ed; nu·tat·ing** (1880)
: to exhibit or undergo nutation

nu·ta·tion \nü-'tā-shən, nyü-\ *noun* [Latin *nutation-, nutatio,* from *nutare* to nod, rock — more at NUMEN] (1612)
1 *archaic* **:** the act of nodding the head
2 : oscillatory movement of the axis of a rotating body (as the earth) **:** WOBBLE
3 : a spontaneous usually spiral movement of a growing plant part
— nu·ta·tion·al \-shnəl, -shə-nᵊl\ *adjective*

nut–brown \'nət-'brau̇n\ *adjective* (14th century)
: of the color of a brown nut

nut·case \-,kās\ *noun* (1959)
: NUT 6a

nut·crack·er \-,kra-kər\ *noun* (circa 1548)
: an implement for cracking nuts

nut·gall \-,gȯl\ *noun* (15th century)
: a gall that resembles a nut; *especially* **:** a gall produced on oak

nut grass *noun* (1775)
: a perennial sedge (*Cyperus rotundus*) of wide distribution that has slender rootstocks bearing small edible tubers resembling nuts; *also* **:** a related sedge (*C. esculentus*)

nut·hatch \'nət-,hach\ *noun* [Middle English *notehache,* from *note* nut + *-hache;* akin to Old English *tohaccian* to hack — more at HACK] (14th century)
: any of various small tree-climbing chiefly insectivorous birds (family Sittidae and especially genus *Sitta*) that have a compact body, a narrow bill, a short tail, and sometimes a black cap

nuthatch

nut·house \'nət-,hau̇s\ *noun* (1900)
slang **:** an insane asylum

nut·let \'nət-lət\ *noun* (1856)
1 a : a small nut **b :** a small fruit similar to a nut
2 : the stone of a drupelet

nut·meg \'nət-,meg, -,māg\ *noun* [Middle English *notemigge, notemuge,* ultimately from Old Provençal *noz muscada,* from *noz* nut (from Latin *nuc-, nux*) + *muscada,* feminine of *muscat* musky — more at MUSCAT] (15th century)
: an aromatic seed that is used as a spice and is produced by an evergreen tree (*Myristica fragrans* of the family Myristicaceae, the nutmeg family) native to the Moluccas; *also* **:** a tree yielding nutmeg

nut·pick \'nət-,pik\ *noun* (circa 1889)
: a small sharp-pointed implement for extracting the kernels from nuts

nu·tria \'nü-trē-ə, 'nyü-\ *noun* [American Spanish, from Spanish, otter, modification of Latin *lutra;* probably akin to Old English *oter* otter] (1820)
1 : the durable usually light brown fur of a nutria
2 : a large South American semiaquatic rodent (*Myocastor coypus*) with webbed hind feet that has been introduced into parts of Europe, Asia, and North America

¹nu·tri·ent \'nü-trē-ənt, 'nyü-\ *adjective* [Latin *nutrient-, nutriens,* present participle of *nutrire* to nourish — more at NOURISH] (1650)
: furnishing nourishment

²nutrient *noun* (circa 1828)
: a nutritive substance or ingredient

nu·tri·ment \'nü-trə-mənt, 'nyü-\ *noun* [Middle English, from Latin *nutrimentum,* from *nutrire*] (15th century)
: something that nourishes or promotes growth and repairs the natural wastage of organic life

nu·tri·tion \nu̇-'tri-shən, nyu̇-\ *noun* [Middle English *nutricioun,* from Middle French *nutrition,* from Late Latin *nutrition-, nutritio,* from Latin *nutrire*] (15th century)
: the act or process of nourishing or being nourished; *specifically* **:** the sum of the processes by which an animal or plant takes in and utilizes food substances
— nu·tri·tion·al \-'trish-nəl, -'tri-shə-nᵊl\ *adjective*
— nu·tri·tion·al·ly *adverb*

nu·tri·tion·ist \-'tri-sh(ə-)nist\ *noun* (1926)
: a specialist in the study of nutrition

nu·tri·tious \nu̇-'tri-shəs, nyu̇-\ *adjective* [Latin *nutricius,* from *nutric-, nutrix* nurse, from *nutrire* to nourish — more at NOURISH] (1665)
: NOURISHING
— nu·tri·tious·ly *adverb*
— nu·tri·tious·ness *noun*

nu·tri·tive \'nü-trə-tiv, 'nyü-\ *adjective* (14th century)
1 : of or relating to nutrition
2 : NOURISHING
— nu·tri·tive·ly *adverb*

nutritive ratio *noun* (1897)
: the ratio of digestible protein to other nutrients in a foodstuff or ration

nuts \'nəts\ *adjective* (1785)
1 : ENTHUSIASTIC, KEEN ⟨everyone seems *nuts* about it —Lois Long⟩
2 : INSANE, CRAZY ⟨said that it was a novel and all the people who said otherwise were *nuts* —Flannery O'Connor⟩ ◆

nuts and bolts *noun* (1967)
1 : the working parts or elements
2 : the practical workings of a machine or enterprise as opposed to theoretical considerations or speculative possibilities
— nuts–and–bolts *adjective*

nut·sedge \'nət-,sej\ *noun* (circa 1909)
: NUT GRASS

nut·shell \'nət-,shel\ *noun* (13th century)
1 : the hard external covering in which the kernel of a nut is enclosed
2 : something of small size, amount, or scope
— in a nutshell : in a very brief statement

nut·ter \'nə-tər\ *noun* (1958)
slang British **:** NUT 6a

nut·ty \'nə-tē\ *adjective* **nut·ti·er; -est** (15th century)
1 : having or producing nuts
2 : having a flavor like that of nuts
3 : ECCENTRIC, SILLY; *also* **:** mentally unbalanced
— nut·ti·ly \-tə-lē\ *adverb*
— nut·ti·ness *noun*

nux vom·i·ca \'nəks-'vä-mi-kə\ *noun, plural* **nux vomica** [New Latin, literally, emetic nut] (14th century)
1 : the poisonous seed of an Asian tree (*Strychnos nux-vomica* of the family Loganiaceae) that contains several alkaloids and especially strychnine and brucine; *also* **:** the tree yielding nux vomica
2 : a drug containing nux vomica

nuz·zle \'nə-zəl\ *verb* **nuz·zled; nuz·zling** \'nəz-liŋ, 'nə-zə-\ [Middle English *noselen* to bring the nose toward the ground, from *nose*] (1530)
intransitive verb
1 : to work with or as if with the nose; *especially* **:** to root, rub, or snuff something
2 : to lie close or snug **:** NESTLE
transitive verb
1 : to root, rub, or touch with or as if with the nose **:** NUDGE
2 : to rub or push gently (as one's face) against something ⟨*nuzzled* her face into the pillow⟩

ny·a·la \nē-'ä-lə\ *noun, plural* **nyalas** *or* **nyala** [of Bantu origin; akin to Venda *dzì-nyálà* nyala buck] (1894)
: an antelope (*Tragelaphus angasi*) of southeastern Africa with vertical white stripes on the sides of the body and with shaggy black hair along the male underside; *also* **:** a related antelope (*T. buxtoni*) of Ethiopia

nyala

nyc·ta·lo·pia \,nik-tə-'lō-pē-ə\ *noun* [New Latin, from

◇ WORD HISTORY

nuts *Nuts,* meaning "enthusiastic" or "crazy," takes us into an odd corner of English word derivation. It appears to be the plural of *nut,* though it is not apparent why plurality should make it an adjective. The sense "enthusiastic" is most likely based on the older idiom *nuts to* "a source of pleasure to," as in "This story would have been nuts to Mother Midnight" (Andrew Marvell). The sense "crazy," attested later, is probably dependent on *off one's nut,* where *nut* means figuratively "head." Synonyms of *nuts,* perhaps formed on its pattern, are *bonkers, crackers,* and *bananas,* though these words are much more difficult to account for semantically and their final *-s* seems to have even less reference to plurality. The element *-ers* in *bonkers* and *crackers* brings to mind the chiefly British colloquialisms *preggers* "pregnant" and *starkers* "naked." (There seems to be little to connect lunacy, pregnancy, and nudity, other than the fact that all are potentially embarrassing either to their subjects or to bystanders.) All of these adjectives are distinctive in that they cannot be used attributively: we would normally say "a nutty driver" rather than "a nuts driver."

Latin *nyctalops* suffering from night blindness, from Greek *nyktalops,* from *nykt-, nyx* night + *alaos* blind + *ōp-, ōps* eye — more at NIGHT, EYE] (1684)
: NIGHT BLINDNESS

ny·lon \ˈnī-ˌlän\ *noun* [coined word] (1938)
1 : any of numerous strong tough elastic synthetic polyamide materials that are fashioned into fibers, filaments, bristles, or sheets and used especially in textiles and plastics
2 *plural* **:** stockings made of nylon

nymph \ˈnim(p)f\ *noun* [Middle English *nimphe,* from Middle French, from Latin *nympha* bride, nymph, from Greek *nymphē* — more at NUPTIAL] (14th century)
1 : any of the minor divinities of nature in classical mythology represented as beautiful maidens dwelling in the mountains, forests, trees, and waters
2 : GIRL
3 : any of various immature insects; *especially* **:** a larva of an insect (as a grasshopper, true bug, or mayfly) with incomplete metamorphosis that differs from the imago especially in size and in its incompletely developed wings and genitalia — compare NAIAD 2
— nymph·al \ˈnim(p)-fəl\ *adjective*

nym·pha·lid \nim-ˈfa-ləd, ˈnim-fə-ləd\ *noun* [New Latin *Nymphalidae,* ultimately from Latin *nympha* nymph] (1897)
: any of a family (Nymphalidae) of butterflies (as a mourning cloak or fritillary) with the first pair of legs reduced in size in both sexes and useless for walking
— nymphalid *adjective*

nym·phet *also* **nym·phette** \nim-ˈfet, ˈnim(p)-fət\ *noun* (1955)
: a sexually precocious girl barely in her teens

nym·pho \ˈnim(p)-(ˌ)fō\ *noun, plural* **nym·phos** [short for *nymphomaniac*] (circa 1910)
: a person affected by nymphomania **:** NYMPHOMANIAC

nym·pho·lep·sy \ˈnim(p)-fə-ˌlep-sē\ *noun* [*nympholept,* from Greek *nympholēptos* frenzied, literally, caught by nymphs, from *nymphē* + *lambanein* to seize — more at LATCH] (1775)
1 : a demoniac enthusiasm held by the ancients to seize one bewitched by a nymph
2 : a frenzy of emotion
— nym·pho·lept \-ˌlept\ *noun*
— nym·pho·lep·tic \ˌnim(p)-fə-ˈlep-tik\ *adjective*

nym·pho·ma·nia \ˌnim(p)-fə-ˈmā-nē-ə, -nyə\ *noun* [New Latin, from *nymphae* inner lips of the vulva (from Latin, plural of *nympha*) + Late Latin *mania* mania] (1775)
: excessive sexual desire by a female
— nym·pho·ma·ni·ac \-nē-ˌak\ *noun or adjective*
— nym·pho·ma·ni·a·cal \-mə-ˈnī-ə-kəl\ *adjective*

Ny·norsk \nü-ˈnȯrsk, nyü-, nœ-\ *noun* [Norwegian, literally, new Norwegian] (1931)
: a literary form of Norwegian based on the spoken dialects of Norway — compare BOKMÅL

nys·tag·mus \nis-ˈtag-məs\ *noun* [New Latin, from Greek *nystagmos* drowsiness, from *nystazein* to doze; probably akin to Lithuanian *snusti* to doze] (1822)
: a rapid involuntary oscillation of the eyeballs (as from dizziness)
— nys·tag·mic \-mik\ *adjective*

nys·ta·tin \ˈnis-tə-tən\ *noun* [*N*ew *Y*ork *Stat*e (where it was developed) + [1]*-in*] (1952)
: an antibiotic that is derived from a soil actinomycete (*Streptomyces noursei*) and is used especially in the treatment of candidiasis

O is the fifteenth letter of the English alphabet. It comes through Latin, via Etruscan, from Greek. The Greeks in turn took the letter from Phoenician, where it represented a consonant sound called 'ayin, *sounding like an* h *pronounced harshly and with the vocal cords vibrating (rather like the sound one might make, for example, while being throttled). The Greeks, not having this consonant sound, assigned the letter-symbol to the short and long* o *vowel sounds. They later reassigned these two sounds to the separate letter-symbols* omicron *and* omega, *but the Romans continued to use one symbol for both.* O *stands for various sounds in English, the chief of which are its long sound, as in* bone, *its short sound, as in* nod, *and the sounds heard in the words* orb, son, do, *and* wolf. *With other vowels and with itself it forms several digraphs (*oa, œ, oi, oo, ou, *and* oy*), some representing simple vowels and others representing diphthongs. Small* o *developed by a simple reduction in the size of the capital form.*

o \'ō\ *noun, plural* **o's** *or* **os** \'ōz\ *often capitalized, often attributive* (before 12th century)
1 a : the 15th letter of the English alphabet **b :** a graphic representation of this letter **c :** a speech counterpart of orthographic *o*
2 : a graphic device for reproducing the letter *o*
3 : one designated *o* especially as the 15th in order or class
4 : something shaped like the letter O; *especially* **:** ZERO

O \'ō\ *variant of* OH

o- *or* **oo-** *combining form* [Greek *ōi-, ōio-,* from *ōion* — more at EGG]
: egg ⟨*o*ology⟩; *specifically* **:** ovum ⟨*oo*gonium⟩

-o- [Middle English, from Old French, from Latin, from Greek, thematic vowel of many nouns and adjectives in combination]
— used as a connective vowel originally to join word elements of Greek origin and now also to join word elements of Latin or other origin ⟨speed*o*meter⟩ ⟨elast*o*mer⟩

[1]-o *noun suffix* [perhaps from [1]*oh*]
: one that is, has the qualities of, or is associated with ⟨buck*o*⟩

[2]-o *interjection suffix* [probably from [1]*oh*]
— in interjections formed from other parts of speech ⟨cheeri*o*⟩ ⟨right*o*⟩

o' *also* **o** \ə, ō\ *preposition* [Middle English *o, o-*, contraction of *on & of*] (13th century)
1 *chiefly dialect* **:** ON
2 : OF ⟨one *o'*clock⟩

oaf \'ōf\ *noun, plural* **oafs** [of Scandinavian origin; akin to Old Norse *alfr* elf — more at ELF] (1625)
1 : a stupid person **:** BOOB
2 : a big clumsy slow-witted person
— oaf·ish \'ō-fish\ *adjective*
— oaf·ish·ly *adverb*
— oaf·ish·ness *noun*

oak \'ōk\ *noun, plural* **oaks** *or* **oak** *often attributive* [Middle English *ook,* from Old English *āc;* akin to Old High German *eih* oak and perhaps to Greek *aigilōps,* a kind of oak] (before 12th century)
1 a : any of a genus (*Quercus*) of trees or shrubs of the beech family that produce acorns; *also* **:** any of various plants related to or resembling the oaks **b :** the tough hard durable wood of an oak tree
2 : the leaves of an oak used as decoration
— oak·en \'ō-kən\ *adjective*

oak apple *noun* (15th century)
: a large round gall produced on oak leaves and twigs by a gall wasp (especially *Amphibolips confluenta* or *Andricus quercuscalifornicus*)

oak–leaf cluster *noun* (1918)
: a bronze or silver cluster of oak leaves and acorns added to various military decorations to signify a second or subsequent award of the basic decoration

oak·moss \'ōk-ˌmȯs\ *noun* (1921)
: any of several lichens (as *Evernia prunastri*) that grow on oak trees and yield a resin used in perfumery

oa·kum \'ō-kəm\ *noun* [Middle English *okum,* from Old English *ācumba* tow, from *ā-* (separative & perfective prefix) + *-cumba* (akin to Old English *camb* comb) — more at ABIDE] (15th century)
: loosely twisted hemp or jute fiber impregnated with tar or a tar derivative and used in caulking seams and packing joints

oak wilt *noun* (1942)
: a destructive disease of oak trees that is caused by a fungus (*Ceratocystis fagacearum*) and is characterized by wilting, discoloration, and defoliation

[1]oar \'ōr, 'ȯr\ *noun* [Middle English *oor,* from Old English *ār;* akin to Old Norse *ār* oar] (before 12th century)
1 : a long pole with a broad blade at one end used for propelling or steering a boat
2 : OARSMAN
— oared \'ōrd, 'ȯrd\ *adjective*

[2]oar (1610)
transitive verb
: to propel with or as if with oars **:** ROW
intransitive verb
: to progress by or as if by using oars

oar·fish \'ōr-ˌfish, 'ȯr-\ *noun* (1860)
: any of several sea fishes (genus *Regalecus* and especially *R. glesne*) with narrow soft bodies from 20 to 30 feet (6 to 9 meters) long, a red dorsal fin running the entire length of the body, and red-tipped anterior rays rising above the head

oar·lock \-ˌläk\ *noun* (before 12th century)
: a usually U-shaped device for holding an oar in place

oars·man \'ōrz-mən, 'ȯrz-\ *noun* (1701)
: one who rows especially in a racing crew
— oars·man·ship \-ˌship\ *noun*

oars·wom·an \-ˌwu̇-mən\ *noun* (1882)
: a woman who rows especially in a racing crew

oa·sis \ō-'ā-səs\ *noun, plural* **oa·ses** \-ˌsēz\ [Late Latin, from Greek] (1613)
1 : a fertile or green area in an arid region (as a desert)
2 : something that provides refuge, relief, or pleasant contrast

oast \'ōst\ *noun* [Middle English *ost,* from Old English *āst;* akin to Middle Dutch *eest* kiln, Latin *aestus* heat, *aestas* summer — more at EDIFY] (before 12th century)
: a usually conical kiln used for drying hops, malt, or tobacco — called also *oast·house* \-ˌhau̇s\

oat \'ōt\ *noun, often attributive* [Middle English *ote,* from Old English *āte*] (before 12th century)
1 a : any of several grasses (genus *Avena*); *especially* **:** a widely cultivated cereal grass (*A. sativa*) **b :** a crop or plot of the oat; *also* **:** oat seed — usually used in plural but singular or plural in construction
2 *archaic* **:** a reed instrument made of an oat straw

oat 1a

oat·cake \'ōt-ˌkāk\ *noun* (14th century)
: a thin flat oatmeal cake

oat–cell \-ˌsel\ *adjective* (1903)
: of, relating to, or being a highly malignant form of cancer especially of the lungs that is characterized by rapid proliferation of small anaplastic cells ⟨*oat-cell* carcinomas⟩

oat·en \'ō-tᵊn\ *adjective* (14th century)
: of or relating to oats, oat straw, or oatmeal

oat·er \'ō-tər\ *noun* (1946)
: WESTERN 2

oat grass *noun* (1578)
: WILD OAT 1a; *broadly* **:** one of several grasses resembling the oat

oath \'ōth\ *noun, plural* **oaths** \'ōt͟hz, 'ōths\ [Middle English *ooth,* from Old English *āth;* akin to Old High German *eid* oath, Middle Irish *oeth*] (before 12th century)
1 a (1) **:** a solemn usually formal calling upon God or a god to witness to the truth of what one says or to witness that one sincerely intends to do what one says (2) **:** a solemn attestation of the truth or inviolability of one's words **b :** something (as a promise) corroborated by an oath
2 : an irreverent or careless use of a sacred name; *broadly* **:** SWEARWORD

oat·meal \'ōt-ˌmēl, ōt-'mē(ə)l\ *noun* (14th century)
1 a : meal made from oats **b :** rolled oats
2 : porridge made from ground or rolled oats

ob- *prefix* [New Latin, from Latin, in the way, against, toward, from *ob* in the way of, on account of; akin to Old Church Slavonic *o, ob* on, around]
: inversely ⟨*ob*ovate⟩

Oba·di·ah \ˌō-bə-'dī-ə\ *noun* [Hebrew *'Ōbhadhyāh*]
1 : a Hebrew prophet
2 : a prophetic book of canonical Jewish and Christian Scripture — see BIBLE table

[1]ob·bli·ga·to \ˌä-blə-'gä-(ˌ)tō\ *adjective* [Italian, obligatory, from past participle of *obbligare* to oblige, from Latin *obligare* — more at OBLIGE] (1794)
: not to be omitted **:** OBLIGATORY — used as a direction in music; compare AD LIBITUM

[2]obbligato *noun, plural* **-tos** *also* **-ti** \-'gä-tē\ (1845)
1 : an elaborate especially melodic part accompanying a solo or principal melody and usually played by a single instrument ⟨a song with violin *obbligato*⟩
2 : ACCOMPANIMENT 2b; *especially* **:** an attendant background sound

ob·cor·date \ˌäb-'kȯr-ˌdāt\ *adjective* (1775)
: heart-shaped with the notch apical ⟨*obcordate* leaf⟩

ob·du·ra·cy \'äb-də-rə-sē, -dyə-; äb-'du̇r-ə-, əb-, -'dyu̇r-\ *noun, plural* **-cies** (1597)
: the quality or state of being obdurate

ob·du·rate \'äb-də-rət, -dyə-; äb-'du̇r-ət, əb-, -'dyu̇r-\ *adjective* [Middle English, from Latin *obduratus,* past participle of *obdurare* to harden, from *ob-* against + *durus* hard — more at DURING] (15th century)
1 a : stubbornly persistent in wrongdoing **b :** hardened in feelings

2 : resistant to persuasion or softening influences
synonym see INFLEXIBLE
— ob•du•rate•ly *adverb*
— ob•du•rate•ness *noun*

obe•ah \'ō-bē-ə\ *also* **obi** \'ō-bē\ *noun, often capitalized* [of African origin; akin to Twi *ɔ-bayifó* sorcerer, Ibo *díbìà* folk healer] (1760)
: a system of belief among blacks chiefly of the British West Indies, the Guianas, and the southeastern U.S. that is characterized by the use of sorcery and magic ritual

obe•di•ence \ō-'bē-dē-ən(t)s, ə-\ *noun* (13th century)
1 a : an act or instance of obeying **b :** the quality or state of being obedient
2 : a sphere of jurisdiction; *especially* **:** an ecclesiastical or sometimes secular dominion

obe•di•ent \-ənt\ *adjective* [Middle English, from Old French, from Latin *oboedient-, oboediens,* from present participle of *oboedire* to obey] (13th century)
: submissive to the restraint or command of authority **:** willing to obey ☆
— obe•di•ent•ly *adverb*

obei•sance \ō-'bē-sᵊn(t)s, ə-, -'bā-\ *noun* [Middle English *obeisaunce* obedience, obeisance, from Middle French *obeissance,* from *obeissant,* present participle of *obeir* to obey] (14th century)
1 : a movement of the body made in token of respect or submission **:** BOW
2 : DEFERENCE, HOMAGE
— obei•sant \-sᵊnt\ *adjective*
— obei•sant•ly *adverb*

obe•lia \ō-'bēl-yə\ *noun* [New Latin] (1868)
: any of a genus (*Obelia*) of small colonial marine hydroids with colonies branched like trees

obe•lisk \'ä-bə-ˌlisk *also* 'ō-\ *noun* [Middle French *obelisque,* from Latin *obeliscus,* from Greek *obeliskos,* from diminutive of *obelos*] (1569)
1 : an upright 4-sided usually monolithic pillar that gradually tapers as it rises and terminates in a pyramid
2 a : OBELUS **b :** DAGGER 2b

obelisk

obe•lize \-ˌlīz\ *transitive verb* **-lized; -liz•ing** (circa 1656)
: to designate or annotate with an obelus

obe•lus \-ləs\ *noun, plural* **ob•e•li** \-ˌlī, -ˌlē\ [Middle English, from Late Latin, from Greek *obelos* spit, pointed pillar, obelus] (14th century)
1 : a symbol — or ÷ used in ancient manuscripts to mark a questionable passage
2 : the symbol ÷

Ober•on \'ō-bə-ˌrän, -rən\ *noun* [French, from Old French *Auberon*]
: the king of the fairies in medieval folklore

obese \ō-'bēs\ *adjective* [Latin *obesus,* from *ob-* against + *esus,* past participle of *edere* to eat — more at OB-, EAT] (1651)
: excessively fat

obe•si•ty \ō-'bē-sə-tē\ *noun* (1611)
: a condition characterized by excessive bodily fat

obey \ō-'bā, ə-\ *verb* **obeyed; obey•ing** [Middle English *obeien,* from Old French *obeir,* from Latin *oboedire,* from *ob-* toward + *-oedire* (akin to *audire* to hear) — more at OB-, AUDIBLE] (14th century)
transitive verb
1 : to follow the commands or guidance of
2 : to conform to or comply with ⟨*obey* an order⟩ ⟨falling objects *obey* the laws of physics⟩
intransitive verb
: to behave obediently
— obey•er *noun*

ob•fus•cate \'äb-fə-ˌskāt; äb-'fəs-ˌkāt, əb-\ *transitive verb* **-cat•ed; -cat•ing** [Late Latin *obfuscatus,* past participle of *obfuscare,* from Latin *ob-* in the way + *fuscus* dark brown — more at OB-, DUSK] (1577)
1 a : DARKEN **b :** to make obscure
2 : CONFUSE
— ob•fus•ca•tion \ˌäb-(ˌ)fəs-'kā-shən\ *noun*
— ob•fus•ca•to•ry \äb-'fəs-kə-ˌtōr-ē, əb-, -ˌtȯr-\ *adjective*

obi \'ō-bē\ *noun* [Japanese] (1876)
: a broad sash worn especially with a Japanese kimono

Obie \'ō-bē\ *noun* [*O.B.,* abbreviation for *off Broadway*] (1965)
: an award presented annually by a professional organization for notable achievement in plays performed off-Broadway

obit \ō-'bit, 'ō-bət, *especially British* 'ä-bit\ *noun* [Middle English, from Middle French, from Latin *obitus* death, from *obire* to go to meet, die, from *ob-* in the way + *ire* to go — more at ISSUE] (15th century)
: OBITUARY

obi•ter dic•tum \ˌō-bə-tər-'dik-təm, ˌä-\ *noun, plural* **obiter dic•ta** \-tə\ [Late Latin, literally, something said in passing] (1812)
1 : an incidental and collateral opinion that is uttered by a judge but is not binding
2 : an incidental remark or observation

obit•u•ary \ə-'bi-chə-ˌwer-ē, ō-, -'bi-chə-rē\ *noun, plural* **-ar•ies** [Medieval Latin *obituarium,* from Latin *obitus* death] (1738)
: a notice of a person's death usually with a short biographical account
— obit•u•ar•ist \-'bi-chə-ˌwər-ist, -'bi-chə-rist\ *noun*
— obituary *adjective*

¹ob•ject \'äb-jikt, -(ˌ)jekt\ *noun* [Middle English, from Medieval Latin *objectum,* from Latin, neuter of *objectus,* past participle of *obicere* to throw in the way, present, hinder, from *ob-* in the way + *jacere* to throw — more at OB-, JET] (14th century)
1 a : something material that may be perceived by the senses ⟨I see an *object* in the distance⟩ **b :** something that when viewed stirs a particular emotion (as pity) ⟨look to the tragic loading of this bed . . . the *object* poisons sight; let it be hid —Shakespeare⟩
2 : something mental or physical toward which thought, feeling, or action is directed ⟨an *object* for study⟩ ⟨the *object* of my affection⟩ ⟨delicately carved art *objects*⟩
3 a : the goal or end of an effort or activity **:** PURPOSE, OBJECTIVE ⟨their *object* is to investigate the matter thoroughly⟩ **b :** a cause for attention or concern ⟨money is no *object*⟩
4 : a thing that forms an element of or constitutes the subject matter of an investigation or science
5 a : a noun or noun equivalent (as a pronoun, gerund, or clause) denoting the goal or result of the action of a verb **b :** a noun or noun equivalent in a prepositional phrase
synonym see INTENTION
— ob•ject•less \-ləs\ *adjective*
— ob•ject•less•ness *noun*

²ob•ject \əb-'jekt\ *verb* [Middle English, from Latin *objectus,* past participle of *obicere* to throw in the way, object] (15th century)
transitive verb
: to put forth in opposition or as an objection ⟨*objected* that the statement was misleading⟩
intransitive verb
1 : to oppose something firmly and usually with words or arguments
2 : to feel distaste for something
— ob•jec•tor \-'jek-tər\ *noun*

ob•ject ball \'äb-jik(t)-, -(ˌ)jek(t)-\ *noun* (1856)
: the ball first struck by the cue ball in pool or billiards; *also* **:** a ball hit by the cue ball

ob•jec•ti•fy \əb-'jek-tə-ˌfī\ *transitive verb* **-fied; -fy•ing** (circa 1837)
1 : to treat as an object or cause to have objective reality
2 : to give expression to (as an abstract notion, feeling, or ideal) in a form that can be experienced by others ⟨it is the essence of the fairy tale to *objectify* differing facets of the child's emotional experience —John Updike⟩
— ob•jec•ti•fi•ca•tion \-ˌjek-tə-fə-'kā-shən\ *noun*

ob•jec•tion \əb-'jek-shən\ *noun* (14th century)
1 : an act of objecting
2 a : a reason or argument presented in opposition **b :** a feeling or expression of disapproval

ob•jec•tion•able \-sh(ə-)nə-bəl\ *adjective* (1781)
: UNDESIRABLE, OFFENSIVE
— ob•jec•tion•able•ness *noun*
— ob•jec•tion•ably \-blē\ *adverb*

¹ob•jec•tive \əb-'jek-tiv, äb-\ *adjective* (1620)
1 a : relating to or existing as an object of thought without consideration of independent existence — used chiefly in medieval philosophy **b :** of, relating to, or being an object, phenomenon, or condition in the realm of sensible experience independent of individual thought and perceptible by all observers **:** having reality independent of the mind ⟨*objective* reality⟩ ⟨our reveries . . . are significantly and repeatedly shaped by our transactions with the *objective* world —Marvin Reznikoff⟩ — compare SUBJECTIVE 3a **c** *of a symptom of disease* **:** perceptible to persons other than the affected individual — compare SUBJECTIVE 4c **d :** involving or deriving from sense perception or experience with actual objects, conditions, or phenomena ⟨*objective* awareness⟩ ⟨*objective* data⟩
2 : relating to, characteristic of, or constituting the case of words that follow prepositions or transitive verbs
3 a : expressing or dealing with facts or conditions as perceived without distortion by personal feelings, prejudices, or interpretations ⟨*objective* art⟩ ⟨an *objective* history of the war⟩ ⟨an *objective* judgment⟩ **b** *of a test* **:** limited to choices of fixed alternatives and reducing subjective factors to a minimum
synonym see MATERIAL, FAIR
— ob•jec•tive•ly *adverb*
— ob•jec•tive•ness *noun*
— ob•jec•tiv•i•ty \ˌäb-ˌjek-'ti-və-tē, əb-\ *noun*

²objective *noun* (1835)
1 : a lens or system of lenses that forms an image of an object
2 a : something toward which effort is directed **:** an aim, goal, or end of action **b :** a strategic position to be attained or a purpose to be achieved by a military operation
synonym see INTENTION

☆ SYNONYMS
Obedient, docile, tractable, amenable mean submissive to the will of another. OBEDIENT implies compliance with the demands or requests of one in authority ⟨*obedient* to the government⟩. DOCILE implies a predisposition to submit readily to control or guidance ⟨a *docile* child⟩. TRACTABLE suggests having a character that permits easy handling or managing ⟨*tractable* beasts of burden⟩. AMENABLE suggests a willingness to yield or to cooperate either because of a desire to be agreeable or because of a natural open-mindedness ⟨*amenable* to new ideas⟩.

objective complement *noun* (1870)
: a noun, adjective, or pronoun used in the predicate as complement to a verb and as qualifier of its direct object (as *chairman* in "we elected him chairman")

objective correlative *noun* (1919)
: something (as a situation or chain of events) that symbolizes or objectifies a particular emotion and that may be used in creative writing to evoke a desired emotional response in the reader

ob·jec·tiv·ism \əb-ˈjek-ti-ˌvi-zəm, äb-\ *noun* (1854)
1 : any of various theories asserting the validity of objective phenomena over subjective experience; *especially* **:** REALISM 2a
2 : an ethical theory that moral good is objectively real or that moral precepts are objectively valid
3 : a 20th century movement in poetry growing out of imagism and putting stress on form
— ob·jec·tiv·ist \-vist\ *adjective or noun*
— ob·jec·tiv·is·tic \-ˌjek-ti-ˈvis-tik\ *adjective*

ob·ject language \ˈäb-jikt-, -(ˌ)jekt-\ *noun* (1935)
: TARGET LANGUAGE

ob·ject lesson \ˈäb-jikt-, -(ˌ)jekt-\ *noun* (1831)
: something that serves as a practical example of a principle or abstract idea

ob·jet d'art \ˌȯb-ˌzhā-ˈdär\ *noun, plural* **ob·jets d'art** *same*\ [French, literally, art object] (1865)
1 : an article of some artistic value
2 : CURIO — called also *objet*

ob·jet trou·vé \ˈȯb-ˌzhā-trü-ˈvā\ *noun, plural* **objets trouvés** *same*\ [French, literally, found object] (1937)
: a natural object (as a piece of driftwood) found by chance and held to have aesthetic value especially through the working of natural forces on it; *also* **:** an artifact not originally intended as art but held to have aesthetic value especially when displayed as a work of art

ob·jur·ga·tion \ˌäb-jər-ˈgā-shən\ *noun* [Middle English *objurgacyon,* from Middle French or Latin; Middle French *objurgation,* from Latin *objurgation-, objurgatio,* from *objurgare* to scold, blame, from *ob-* against + *jurgare* to quarrel, literally, to take to law, from *jur-, jus* law + *-igare* (from *agere* to lead) — more at OB-, JUST, AGENT] (15th century)
: a harsh rebuke
— ob·jur·gate \ˈäb-jər-ˌgāt\ *transitive verb*
— ob·jur·ga·to·ry \əb-ˈjər-gə-ˌtōr-ē, -ˌtȯr-\ *adjective*

ob·lan·ceo·late \(ˌ)äb-ˈlan(t)-sē-ə-ˌlāt\ *adjective* (1850)
: inversely lanceolate ⟨an *oblanceolate* leaf⟩ — see LEAF illustration

ob·last \ˈä-ˌblast, ˈō-bləst\ *noun, plural* **oblasts** *also* **ob·las·ti** \-ˌblas-tē, -bləs-\ [Russian *oblast'*] (circa 1886)
: a political subdivision of Imperial Russia or of a republic in the U.S.S.R.

¹ob·late \ä-ˈblāt, ˈä-ˌ\ *adjective* [probably from New Latin *oblatus,* from *ob-* + *-latus* (as in *prolatus* prolate)] (1705)
: flattened or depressed at the poles ⟨an *oblate* spheroid⟩
— ob·late·ness *noun*

²ob·late \ˈä-ˌblāt\ *noun* [Medieval Latin *oblatus,* literally, one offered up, from Latin, past participle of *offerre* — more at OFFER] (1864)
1 : a layman living in a monastery under a modified rule and without vows
2 : a member of one of several Roman Catholic communities of men or women

ob·la·tion \ə-ˈblā-shən, ō-\ *noun* [Middle English *oblacioun,* from Middle French *oblation,* from Late Latin *oblation-, oblatio,* from Latin *offerre*] (15th century)
1 : the act of making a religious offering; *specifically, capitalized* **:** the act of offering the eucharistic elements to God
2 : something offered in worship or devotion **:** a holy gift offered usually at an altar or shrine

¹ob·li·gate \ˈä-blə-ˌgāt\ *transitive verb* **-gat·ed; -gat·ing** [Latin *obligatus,* past participle of *obligare*] (1533)
1 : to bind legally or morally **:** CONSTRAIN
2 : to commit (as funds) to meet an obligation

²ob·li·gate \ˈä-bli-gət, -blə-ˌgāt\ *adjective* (1887)
1 : restricted to one particularly characteristic mode of life ⟨an *obligate* parasite⟩
2 : biologically essential for survival ⟨*obligate* mutualism⟩
— ob·li·gate·ly *adverb*

ob·li·ga·tion \ˌä-blə-ˈgā-shən\ *noun* (14th century)
1 : the action of obligating oneself to a course of action (as by a promise or vow)
2 a : something (as a formal contract, a promise, or the demands of conscience or custom) that obligates one to a course of action **b :** a debt security (as a mortgage or corporate bond) **c :** a commitment (as by a government) to pay a particular sum of money; *also* **:** an amount owed under such an obligation ⟨unable to meet its *obligations,* the company went into bankruptcy⟩
3 a : a condition or feeling of being obligated **b :** a debt of gratitude
4 : something one is bound to do **:** DUTY, RESPONSIBILITY

oblig·a·to·ry \ə-ˈbli-gə-ˌtōr-ē, ä-, -ˌtȯr- *also* ˈä-bli-gə-\ *adjective* (15th century)
1 : binding in law or conscience
2 : relating to or enforcing an obligation ⟨a writ *obligatory*⟩
3 : MANDATORY, REQUIRED ⟨*obligatory* military service⟩; *also* **:** so commonplace as to be a convention, fashion, or cliché ⟨the *obligatory* death scene in opera⟩
4 : OBLIGATE 1
— oblig·a·to·ri·ly \ə-ˌbli-gə-ˈtōr-ə-lē, ä-, -ˈtȯr- *also* ˌä-bli-gə-\ *adverb*

oblige \ə-ˈblīj\ *verb* **obliged; oblig·ing** [Middle English, from Old French *obliger,* from Latin *obligare,* literally, to bind to, from *ob-* toward + *ligare* to bind — more at LIGATURE] (14th century)
transitive verb
1 : to constrain by physical, moral, or legal force or by the exigencies of circumstance ⟨*obliged* to find a job⟩
2 a : to put in one's debt by a favor or service ⟨we are much *obliged* for your help⟩ **b :** to do a favor for ⟨always ready to *oblige* a friend⟩
intransitive verb
: to do something as or as if a favor
synonym see FORCE
— oblig·er *noun*

ob·li·gee \ˌä-blə-ˈjē\ *noun* (1574)
: one to whom another is obligated (as by a contract); *specifically* **:** one who is protected by a surety bond

oblig·ing \ə-ˈblī-jiŋ\ *adjective* (1632)
: willing to do favors **:** ACCOMMODATING
synonym see AMIABLE
— oblig·ing·ly \-jiŋ-lē\ *adverb*
— oblig·ing·ness *noun*

ob·li·gor \ˌä-blə-ˈgȯr, -ˈjȯr\ *noun* (1541)
: one who is bound by a legal obligation

¹oblique \ō-ˈblēk, ə-, -ˈblīk; *military usually* ī\ *adjective* [Middle English *oblike,* from Latin *obliquus*] (15th century)
1 a : neither perpendicular nor parallel **:** INCLINED **b :** having the axis not perpendicular to the base ⟨an *oblique* cone⟩ **c :** having no right angle ⟨an *oblique* triangle⟩
2 a : not straightforward **:** INDIRECT; *also* **:** OBSCURE **b :** DEVIOUS, UNDERHANDED
3 : situated obliquely and having one end not inserted on bone ⟨*oblique* muscles⟩
4 : taken from an airplane with the camera directed horizontally or diagonally downward ⟨an *oblique* photograph⟩
— oblique·ly *adverb*
— oblique·ness *noun*

²oblique *noun* (circa 1608)
1 : something (as a line) that is oblique
2 : any of several oblique muscles; *especially* **:** any of the thin flat muscles forming the middle and outer layers of the lateral walls of the abdomen

³oblique *adverb* (1667)
: at a 45 degree angle ⟨to the right *oblique,* march⟩

oblique angle *noun* (1695)
: an acute or obtuse angle

oblique case *noun* (1530)
: a grammatical case other than the nominative or vocative

obliq·ui·ty \ō-ˈbli-kwə-tē, ə-\ *noun, plural* **-ties** (15th century)
1 : deviation from moral rectitude or sound thinking
2 a : deviation from parallelism or perpendicularity; *also* **:** the amount of such deviation **b :** the angle between the planes of the earth's equator and orbit having a value of about 23°27′ ⟨*obliquity* of the ecliptic⟩
3 a : indirectness or deliberate obscurity of speech or conduct **b :** an obscure or confusing statement

oblit·er·ate \ə-ˈbli-tə-ˌrāt, ō-\ *transitive verb* **-at·ed; -at·ing** [Latin *oblitteratus,* past participle of *oblitterare,* from *ob-* ob- + *littera* letter] (1609)
1 : to make undecipherable or imperceptible by obscuring or wearing away
2 a : to remove utterly from recognition or memory **b :** to remove from existence **:** destroy utterly all trace, indication, or significance of **c :** to cause to disappear (as a bodily part or a scar) or collapse (as a duct conveying body fluid) **:** REMOVE 4 ⟨a blood vessel *obliterated* by inflammation⟩
3 : CANCEL 4
— oblit·er·a·tion \-ˌbli-tə-ˈrā-shən\ *noun*
— oblit·er·a·tor \-ˈbli-tə-ˌrā-tər\ *noun*

oblit·er·a·tive \ə-ˈbli-tə-ˌrā-tiv, ō-, -rə-tiv\ *adjective* (circa 1812)
: inducing or characterized by obliteration: as **a :** causing or accompanied by closure or collapse of a lumen ⟨*obliterative* arterial disease⟩ **b :** tending to make inconspicuous ⟨*obliterative* behavior⟩

obliv·i·on \ə-ˈbli-vē-ən, ō-, ä-\ *noun* [Middle English, from Middle French, from Latin *oblivion-, oblivio,* from *oblivisci* to forget, perhaps from *ob-* in the way + *levis* smooth — more at OB-, LEVIGATE] (14th century)
1 : the fact or condition of forgetting or having forgotten; *especially* **:** the condition of being oblivious
2 : the condition or state of being forgotten or unknown

obliv·i·ous \-vē-əs\ *adjective* (15th century)
1 : lacking remembrance, memory, or mindful attention
2 : lacking active conscious knowledge or awareness — usually used with *of* or *to*
— obliv·i·ous·ly *adverb*
— obliv·i·ous·ness *noun*

ob·long \ˈä-ˌblȯŋ\ *adjective* [Middle English, from Latin *oblongus,* from *ob-* toward + *longus* long — more at LONG] (15th century)
: deviating from a square, circular, or spherical form by elongation in one dimension ⟨an *oblong* piece of paper⟩ ⟨an *oblong* melon⟩
— oblong *noun*

ob·lo·quy \ˈä-blə-kwē\ *noun, plural* **-quies** [Middle English, from Late Latin *obloquium,* from *obloqui* to speak against, from *ob-* against + *loqui* to speak] (15th century)
1 : a strongly condemnatory utterance **:** abusive language
2 : the condition of one that is discredited **:** bad repute
synonym see ABUSE

ob•nox•ious \äb-ˈnäk-shəs, əb-\ *adjective* [Latin *obnoxius,* from *ob* in the way of, exposed to + *noxa* harm — more at NOXIOUS] (1597)
1 *archaic* **:** exposed to something unpleasant or harmful — used with *to*
2 *archaic* **:** deserving of censure
3 : odiously or disgustingly objectionable **:** highly offensive
— ob•nox•ious•ly *adverb*
— ob•nox•ious•ness *noun*

ob•nu•bi•late \äb-ˈnü-bə-ˌlāt, -ˈnyü-\ *transitive verb* **-lat•ed; -lat•ing** [Latin *obnubilatus,* past participle of *obnubilare,* from *ob-* in the way + *nubilare* to be cloudy, from *nubilus* cloudy, from *nubes* cloud — more at OB-, NUANCE] (1583)
: BECLOUD, OBSCURE
— ob•nu•bi•la•tion \-ˌnü-bə-ˈlā-shən, -ˌnyü-\ *noun*

oboe \ˈō-(ˌ)bō\ *noun* [Italian, from French *hautbois* — more at HAUTBOIS] (1794)
: a double-reed woodwind instrument having a conical tube, a brilliant penetrating tone, and a usual range from B flat below middle C upward for over 2½ octaves
— obo•ist \ˈō-(ˌ)bō-ist\ *noun*

oboe

obol \ˈä-bəl, ˈō-\ *noun* [Latin *obolus,* from Greek *obolos, obelos,* literally, spit] (1579)
: an ancient Greek coin or weight equal to ⅙ drachma

ob•ovate \(ˌ)äb-ˈō-ˌvāt\ *adjective* (1785)
: ovate with the narrower end basal ⟨*obovate* leaves⟩ — see LEAF illustration

ob•ovoid \-ˌvȯid\ *adjective* (1819)
: ovoid with the broad end toward the apex ⟨an *obovoid* fruit⟩

ob•scene \äb-ˈsēn, əb-\ *adjective* [Middle French, from Latin *obscenus, obscaenus*] (1593)
1 : disgusting to the senses **:** REPULSIVE
2 a : abhorrent to morality or virtue; *specifically* **:** designed to incite to lust or depravity **b :** containing or being language regarded as taboo in polite usage ⟨*obscene* lyrics⟩ **c :** repulsive by reason of crass disregard of moral or ethical principles ⟨an *obscene* misuse of power⟩ **d :** so excessive as to be offensive ⟨*obscene* wealth⟩ ⟨*obscene* waste⟩
synonym see COARSE
— ob•scene•ly *adverb*

ob•scen•i•ty \-ˈse-nə-tē *also* -ˈsē-\ *noun, plural* **-ties** (1589)
1 : the quality or state of being obscene
2 : something (as an utterance or act) that is obscene

ob•scur•ant \äb-ˈskyu̇r-ənt, əb-\ *or* **ob•scu•ran•tic** \ˌäb-skyə-ˈran-tik\ *adjective* (1878)
: tending to make obscure
— obscurant *noun*

ob•scu•ran•tism \äb-ˈskyu̇r-ən-ˌti-zəm, əb-; ˌäb-skyu̇-ˈran-\ *noun* (1834)
1 : opposition to the spread of knowledge **:** a policy of withholding knowledge from the general public
2 a : a style (as in literature or art) characterized by deliberate vagueness or abstruseness **b :** an act or instance of obscurantism
— ob•scu•ran•tist \-ən-tist, -ˈran-tist\ *noun or adjective*

¹ob•scure \äb-ˈskyu̇r, əb-\ *adjective* [Middle English, from Middle French *obscur,* from Latin *obscurus*] (15th century)
1 a : DARK, DIM **b :** shrouded in or hidden by darkness **c :** not clearly seen or easily distinguished **:** FAINT
2 : not readily understood or clearly expressed; *also* **:** MYSTERIOUS
3 : relatively unknown: as **a :** REMOTE, SECLUDED **b :** not prominent or famous ⟨an *obscure* poet⟩
4 : constituting the unstressed vowel \ə\ or having unstressed \ə\ as its value ☆
— ob•scure•ly *adverb*
— ob•scure•ness *noun*

²obscure *transitive verb* **ob•scured; ob•scur•ing** (15th century)
1 : to make dark, dim, or indistinct
2 : to conceal or hide by or as if by covering
3 : to reduce (a vowel) to the value \ə\
— ob•scu•ra•tion \ˌäb-skyu̇-ˈrā-shən\ *noun*

³obscure *noun* (1667)
: OBSCURITY

ob•scu•ri•ty \äb-ˈskyu̇r-ə-tē, əb-\ *noun, plural* **-ties** (14th century)
1 : one that is obscure
2 : the quality or state of being obscure

ob•se•qui•ous \əb-ˈsē-kwē-əs, äb-\ *adjective* [Middle English, compliant, from Latin *obsequiosus,* from *obsequium* compliance, from *obsequi* to comply, from *ob-* toward + *sequi* to follow — more at OB-, SUE] (15th century)
: marked by or exhibiting a fawning attentiveness
synonym see SUBSERVIENT
— ob•se•qui•ous•ly *adverb*
— ob•se•qui•ous•ness *noun*

ob•se•quy \ˈäb-sə-kwē\ *noun, plural* **-quies** [Middle English *obsequie,* from Middle French, from Medieval Latin *obsequiae* (plural), alteration of Latin *exsequiae,* from *exsequi* to follow out, execute — more at EXECUTION] (15th century)
: a funeral or burial rite — usually used in plural

ob•serv•able \əb-ˈzər-və-bəl\ *adjective* (1609)
1 : NOTEWORTHY
2 : capable of being observed **:** DISCERNIBLE
— ob•serv•abil•i•ty \-ˌzər-və-ˈbi-lə-tē\ *noun*
— observable *noun*
— ob•serv•ably \-ˈzər-və-blē\ *adverb*

ob•ser•vance \əb-ˈzər-vən(t)s\ *noun* (13th century)
1 a : a customary practice, rite, or ceremony ⟨Sabbath *observances*⟩ **b :** a rule governing members of a religious order
2 : an act or instance of following a custom, rule, or law ⟨*observance* of the speed limits⟩
3 : an act or instance of watching

¹ob•ser•vant \-vənt\ *adjective* (1602)
1 a : paying strict attention **:** WATCHFUL ⟨an *observant* spectator⟩ **b :** KEEN, PERCEPTIVE
2 : careful in observing (as rites, laws, or customs) **:** MINDFUL ⟨pious and religiously *observant* families —Sidney Hook⟩ ⟨always *observant* of the amenities⟩
— ob•ser•vant•ly *adverb*

²observant *noun* (1605)
obsolete **:** an assiduous or obsequious servant or attendant

ob•ser•va•tion \ˌäb-sər-ˈvā-shən, -zər-\ *noun* [Middle French, from Latin *observation-, observatio,* from *observare*] (1535)
1 : an act or instance of observing a custom, rule, or law **:** OBSERVANCE
2 a : an act of recognizing and noting a fact or occurrence often involving measurement with instruments ⟨weather *observations*⟩ **b :** a record or description so obtained
3 : a judgment on or inference from what one has observed; *broadly* **:** REMARK, STATEMENT
4 *obsolete* **:** attentive care **:** HEED
5 : the condition of one that is observed ⟨under *observation* at the hospital⟩
— ob•ser•va•tion•al \-shnəl, -shə-nᵊl\ *adjective*
— ob•ser•va•tion•al•ly *adverb*

ob•ser•va•to•ry \əb-ˈzər-və-ˌtōr-ē, -ˌtȯr-\ *noun, plural* **-ries** [probably from New Latin *observatorium,* from Latin *observare*] (1676)
1 : a building or place given over to or equipped for observation of natural phenomena (as in astronomy); *also* **:** an institution whose primary purpose is making such observations
2 : a situation or structure commanding a wide view

ob•serve \əb-ˈzərv\ *verb* **ob•served; ob•serv•ing** [Middle English, from Middle French *observer,* from Latin *observare* to guard, watch, observe, from *ob-* in the way, toward + *servare* to keep — more at CONSERVE] (14th century)
transitive verb
1 : to conform one's action or practice to (as a law, rite, or condition) **:** comply with
2 : to inspect or take note of as an augury, omen, or presage
3 : to celebrate or solemnize (as a ceremony or festival) in a customary or accepted way
4 a : to watch carefully especially with attention to details or behavior for the purpose of arriving at a judgment **b :** to make a scientific observation on or of
5 : to come to realize or know especially through consideration of noted facts
6 : to utter as a remark
intransitive verb
1 a : to take notice **b :** to make observations **:** WATCH
2 : REMARK, COMMENT
synonym see KEEP
— ob•serv•ing•ly \-ˈzər-viŋ-lē\ *adverb*

ob•serv•er \əb-ˈzər-vər\ *noun* (circa 1550)
: one that observes: as **a :** a representative sent to observe but not participate officially in an activity (as a meeting or war) **b :** an expert analyst and commentator in a particular field ⟨political *observers*⟩

ob•sess \əb-ˈses, äb-\ *verb* [Latin *obsessus,* past participle of *obsidēre* to frequent, besiege, from *ob-* against + *sedēre* to sit — more at OB-, SIT] (1531)
transitive verb
: to haunt or excessively preoccupy the mind of ⟨was *obsessed* with the idea⟩
intransitive verb
: to engage in obsessive thinking **:** become obsessed with an idea

ob•ses•sion \äb-ˈse-shən, əb-\ *noun* (1680)
1 : a persistent disturbing preoccupation with an often unreasonable idea or feeling; *broadly* **:** compelling motivation ⟨an *obsession* with profits⟩
2 : something that causes an obsession
— ob•ses•sion•al \-ˈsesh-nəl, -ˈse-shə-nᵊl\ *adjective*

☆ SYNONYMS
Obscure, dark, vague, enigmatic, cryptic, ambiguous, equivocal mean not clearly understandable. OBSCURE implies a hiding or veiling of meaning through some inadequacy of expression or withholding of full knowledge ⟨*obscure* poems⟩. DARK implies an imperfect or clouded revelation often with ominous or sinister suggestion ⟨muttered *dark* hints of revenge⟩. VAGUE implies a lack of clear formulation due to inadequate conception or consideration ⟨a *vague* sense of obligation⟩. ENIGMATIC stresses a puzzling, mystifying quality ⟨*enigmatic* occult writings⟩. CRYPTIC implies a purposely concealed meaning ⟨*cryptic* hints of hidden treasure⟩. AMBIGUOUS applies to language capable of more than one interpretation ⟨an *ambiguous* directive⟩. EQUIVOCAL applies to language left open to differing interpretations with the intention of deceiving or evading ⟨moral precepts with *equivocal* phrasing⟩.

— **ob·ses·sion·al·ly** *adverb*

ob·ses·sive \äb-ˈse-siv, əb-\ *adjective* (1901)
1 a : tending to cause obsession **b :** excessive often to an unreasonable degree
2 : of, relating to, or characterized by obsession **:** deriving from obsession
— **obsessive** *noun*
— **ob·ses·sive·ly** *adverb*
— **ob·ses·sive·ness** *noun*

obsessive–compulsive *adjective* (1927)
: relating to or characterized by recurring obsessions and compulsions especially as symptoms of a neurotic state
— **obsessive–compulsive** *noun*

ob·sid·i·an \əb-ˈsi-dē-ən\ *noun* [New Latin *obsidianus,* from Latin *obsidianus lapis,* false manuscript reading for *obsianus lapis,* literally, stone of Obsius, from *Obsius,* its supposed discoverer] (1796)
: a dark natural glass formed by the cooling of molten lava

ob·so·lesce \ˌäb-sə-ˈles\ *verb* **-lesced; -lesc·ing** [Latin *obsolescere*] (1873)
intransitive verb
: to be or become obsolescent
transitive verb
: to make obsolescent

ob·so·les·cence \-ˈle-sᵊn(t)s\ *noun* (circa 1841)
: the process of becoming obsolete or the condition of being nearly obsolete ⟨the gradual *obsolescence* of machinery⟩ ⟨reduced to *obsolescence*⟩

ob·so·les·cent \-sᵊnt\ *adjective* (1755)
: going out of use **:** becoming obsolete
— **ob·so·les·cent·ly** *adverb*

[1]**ob·so·lete** \ˌäb-sə-ˈlēt, ˈäb-sə-ˌ\ *adjective* [Latin *obsoletus,* from past participle of *obsolescere* to grow old, become disused, perhaps from *ob-* toward + *solēre* to be accustomed] (1579)
1 a : no longer in use or no longer useful **b :** of a kind or style no longer current **:** OLD-FASHIONED
2 *of a plant or animal part* **:** indistinct or imperfect as compared with a corresponding part in related organisms **:** VESTIGIAL
synonym see OLD
— **ob·so·lete·ly** *adverb*
— **ob·so·lete·ness** *noun*

[2]**obsolete** *transitive verb* **-let·ed; -let·ing** (1640)
: to make obsolete

ob·sta·cle \ˈäb-sti-kəl, -ˌsti-\ *noun* [Middle English, from Middle French, from Latin *obstaculum,* from *obstare* to stand in front of, from *ob-* in the way + *stare* to stand — more at OB-, STAND] (14th century)
: something that impedes progress or achievement

obstacle course *noun* (1943)
: a military training course filled with obstacles (as hurdles, fences, walls, and ditches) that must be negotiated; *broadly* **:** a series of obstacles that must be overcome

ob·stet·ric \əb-ˈste-trik, äb-\ *or* **ob·stet·ri·cal** \-tri-kəl\ *adjective* [modification of Latin *obstetricius,* from *obstetric-, obstetrix* midwife, from *obstare*] (1742)
: of, relating to, or associated with childbirth or obstetrics
— **ob·stet·ri·cal·ly** \-tri-k(ə-)lē\ *adverb*

ob·ste·tri·cian \ˌäb-stə-ˈtri-shən\ *noun* (circa 1828)
: a physician specializing in obstetrics

ob·stet·rics \əb-ˈste-triks, äb-\ *noun plural but singular or plural in construction* (circa 1819)
: a branch of medical science that deals with birth and with its antecedents and sequels

ob·sti·na·cy \ˈäb-stə-nə-sē\ *noun, plural* **-cies** (14th century)
1 a : the quality or state of being obstinate **:** STUBBORNNESS **b :** the quality or state of being difficult to remedy, relieve, or subdue ⟨the *obstinacy* of tuberculosis⟩
2 : an instance of being obstinate

ob·sti·nate \ˈäb-stə-nət\ *adjective* [Middle English, from Latin *obstinatus,* past participle of *obstinare* to be resolved, from *ob-* in the way + *-stinare* (akin to *stare* to stand)] (14th century)
1 : perversely adhering to an opinion, purpose, or course in spite of reason, arguments, or persuasion
2 : not easily subdued, remedied, or removed ⟨*obstinate* fever⟩ ☆
— **ob·sti·nate·ly** *adverb*
— **ob·sti·nate·ness** *noun*

ob·strep·er·ous \əb-ˈstre-p(ə-)rəs, äb-\ *adjective* [Latin *obstreperus,* from *obstrepere* to clamor against, from *ob-* against + *strepere* to make a noise] (circa 1600)
1 : marked by unruly or aggressive noisiness **:** CLAMOROUS ⟨*obstreperous* merriment⟩
2 : stubbornly resistant to control **:** UNRULY
synonym see VOCIFEROUS
— **ob·strep·er·ous·ly** *adverb*
— **ob·strep·er·ous·ness** *noun*

ob·struct \əb-ˈstrəkt, äb-\ *transitive verb* [Latin *obstructus,* past participle of *obstruere,* from *ob-* in the way + *struere* to build, heap up — more at OB-, STREW] (1590)
1 : to block or close up by an obstacle
2 : to hinder from passage, action, or operation **:** IMPEDE
3 : to cut off from sight ⟨a wall *obstructs* the view⟩
synonym see HINDER
— **ob·struc·tive** \-ˈstrək-tiv\ *adjective or noun*
— **ob·struc·tive·ness** *noun*
— **ob·struc·tor** \-tər\ *noun*

ob·struc·tion \əb-ˈstrək-shən, äb-\ *noun* (1533)
1 a : an act of obstructing **b :** the state of being obstructed; *especially* **:** a condition of being clogged or blocked
2 : something that obstructs

ob·struc·tion·ism \-shə-ˌni-zəm\ *noun* (1879)
: deliberate interference with the progress or business especially of a legislative body
— **ob·struc·tion·ist** \-sh(ə-)nist\ *noun or adjective*
— **ob·struc·tion·is·tic** \-ˌstrək-shə-ˈnis-tik\ *adjective*

ob·tain \əb-ˈtān, äb-\ *verb* [Middle English *obteinen,* from Middle French & Latin; Middle French *obtenir,* from Latin *obtinēre* to hold on to, possess, obtain, from *ob-* in the way + *tenēre* to hold — more at THIN] (15th century)
transitive verb
: to gain or attain usually by planned action or effort
intransitive verb
1 *archaic* **:** SUCCEED
2 : to be generally recognized or established **:** PREVAIL
— **ob·tain·abil·i·ty** \-ˌtā-nə-ˈbi-lə-tē\ *noun*
— **ob·tain·able** \-ˈtā-nə-bəl\ *adjective*
— **ob·tain·er** *noun*
— **ob·tain·ment** \-ˈtān-mənt\ *noun*

ob·tect \əb-ˈtekt, äb-\ *also* **ob·tect·ed** \-ˈtek-təd\ *adjective* [Latin *obtectus,* past participle of *obtegere* to cover over, from *ob-* in the way + *tegere* to cover — more at THATCH] (circa 1902)
: enclosed in or characterized by enclosure in a firm chitinous case or covering ⟨an *obtect* pupa⟩

ob·trude \əb-ˈtrüd, äb-\ *verb* **ob·trud·ed; ob·trud·ing** [Latin *obtrudere* to thrust at, from *ob-* in the way + *trudere* to thrust — more at OB-, THREAT] (circa 1609)
transitive verb
1 : to thrust out **:** EXTRUDE
2 : to force or impose (as oneself or one's ideas) without warrant or request
intransitive verb
: to become unduly prominent or interfering **:** INTRUDE
— **ob·trud·er** *noun*

ob·tru·sion \-ˈtrü-zhən\ *noun* [Late Latin *obtrusion-, obtrusio,* from Latin *obtrudere*] (1579)
1 : an act of obtruding
2 : something that is obtruded

ob·tru·sive \-ˈtrü-siv, -ziv\ *adjective* (1667)
1 a : forward in manner or conduct ⟨*obtrusive* behavior⟩ **b :** undesirably prominent
2 : thrust out **:** PROTRUDING
synonym see IMPERTINENT
— **ob·tru·sive·ly** *adverb*
— **ob·tru·sive·ness** *noun*

ob·tund \äb-ˈtənd\ *transitive verb* [Middle English, from Latin *obtundere*] (14th century)
: to reduce the edge or violence of **:** DULL ⟨*obtunded* reflexes⟩

ob·tu·ra·tion \ˌäb-tyə-ˈrā-shən, -tə-\ *noun* [Latin *obturation-, obturatio,* from *obturare* to obstruct] (1610)
: OBSTRUCTION, CLOSURE
— **ob·tu·rate** \ˈäb-tyə-ˌrāt, -tə-\ *transitive verb*

ob·tu·ra·tor \ˈäb-tyə-ˌrā-tər, -tə-\ *noun* [New Latin, from Latin *obturare*] (circa 1741)
: one that closes: as **a :** one (as a prosthetic device) that closes or blocks up an opening (as a fissure in the palate) **b :** a hooded swelling of the placenta that fits over the nucellus in some plants

ob·tuse \äb-ˈtüs, əb-, -ˈtyüs\ *adjective* **ob·tus·er; -est** [Middle English, from Latin *obtusus* blunt, dull, from past participle of *obtundere* to beat against, blunt, from *ob-* against + *tundere* to beat — more at OB-, CONTUSION] (15th century)
1 a : not pointed or acute **:** BLUNT **b** (1) *of an angle* **:** exceeding 90 degrees but less than 180 degrees (2) **:** having an obtuse angle ⟨an *obtuse* triangle⟩ — see TRIANGLE illustration **c** *of a leaf* **:** rounded at the free end
2 a : lacking sharpness or quickness of sensibility or intellect **:** INSENSITIVE, STUPID **b :** difficult to comprehend **:** not clear or precise in thought or expression
synonym see DULL
— **ob·tuse·ly** *adverb*
— **ob·tuse·ness** *noun*

[1]**ob·verse** \äb-ˈvərs, əb-, ˈäb-ˌ\ *adjective* [Latin *obversus,* from past participle of *obvertere* to turn toward, from *ob-* toward + *vertere* to turn — more at OB-, WORTH] (circa 1656)
1 : facing the observer or opponent
2 : having the base narrower than the top ⟨an *obverse* leaf⟩
3 : constituting the obverse of something **:** OPPOSITE
— **ob·verse·ly** *adverb*

[2]**ob·verse** \ˈäb-ˌvərs, äb-ˈ, əb-ˈ\ *noun* (1658)
1 : the side of a coin or currency note bearing the chief device and lettering; *broadly* **:** a front or principal surface
2 : a counterpart having the opposite orientation or force ⟨their rise was merely the *obverse* of the Empire's fall —A. J. Toynbee⟩; *also* **:** OPPOSITE 1 ⟨joy and its *obverse,* sorrow⟩

☆ SYNONYMS
Obstinate, dogged, stubborn, pertinacious, mulish mean fixed and unyielding in course or purpose. OBSTINATE implies usually an unreasonable persistence ⟨an *obstinate* proponent of conspiracy theories⟩. DOGGED suggests an admirable often tenacious and unwavering persistence ⟨pursued the story with *dogged* perseverance⟩. STUBBORN implies sturdiness in resisting change which may or may not be admirable ⟨a person too *stubborn* to admit error⟩. PERTINACIOUS suggests an annoying or irksome persistence ⟨a *pertinacious* salesclerk refusing to take no for an answer⟩. MULISH implies a thoroughly unreasonable obstinacy ⟨a *mulish* determination to have his own way⟩.

3 : a proposition inferred immediately from another by denying the opposite of what the given proposition affirms ⟨the *obverse* of "all *A* is *B*" is "no *A* is not *B*"⟩

ob·vi·ate \'äb-vē-,āt\ *transitive verb* **-at·ed; -at·ing** [Late Latin *obviatus,* past participle of *obviare* to meet, withstand, from Latin *obviam*] (1598)
: to anticipate and prevent (as a situation) or make unnecessary (as an action)
— ob·vi·a·tion \,äb-vē-'ā-shən\ *noun*

ob·vi·ous \'äb-vē-əs\ *adjective* [Latin *obvius,* from *obviam* in the way, from *ob* in the way of + *viam,* accusative of *via* way — more at OB-, VIA] (1603)
1 *archaic* **:** being in the way or in front
2 : easily discovered, seen, or understood
synonym see EVIDENT
— ob·vi·ous·ness *noun*

ob·vi·ous·ly \-lē\ *adverb* (1638)
1 : in an obvious manner ⟨showed his anger *obviously*⟩
2 : as is plainly evident ⟨*obviously,* something is wrong⟩

oca \'ō-kə\ *noun* [Spanish, from Quechua *oqa*] (1604)
: either of two South American wood sorrels (*Oxalis crenata* and *O. tuberosa*) cultivated for their edible tubers; *also* **:** the tuber of an oca

oc·a·ri·na \,ä-kə-'rē-nə\ *noun* [Italian, from Italian dialect, diminutive of *oca* goose, from Late Latin *auca,* from Latin *avis* bird — more at AVIARY] (1877)
: a simple wind instrument typically having an oval body with finger holes and a projecting mouthpiece

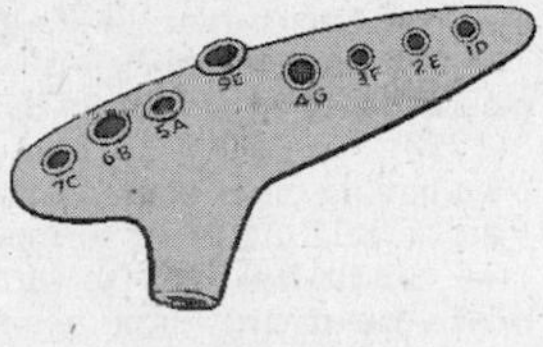
ocarina

Oc·cam's ra·zor \'ä-kəmz-\ *noun* [William of *Occam*] (circa 1837)
: a scientific and philosophic rule that entities should not be multiplied unnecessarily which is interpreted as requiring that the simplest of competing theories be preferred to the more complex or that explanations of unknown phenomena be sought first in terms of known quantities

[1]**oc·ca·sion** \ə-'kā-zhən\ *noun* [Middle English, from Middle French or Latin; Middle French, from Latin *occasion-, occasio,* from *occidere* to fall, fall down, from *ob-* toward + *cadere* to fall — more at OB-, CHANCE] (14th century)
1 : a favorable opportunity or circumstance ⟨did not have *occasion* to talk with them⟩
2 a : a state of affairs that provides a ground or reason ⟨the *occasion* of the discord was their mutual intolerance⟩ **b :** an occurrence or condition that brings something about; *especially* **:** the immediate inciting circumstance as distinguished from the fundamental cause ⟨his insulting remark was the *occasion* of a bitter quarrel⟩
3 a : HAPPENING, INCIDENT **b :** a time at which something happens **:** INSTANCE
4 a : a need arising from a particular circumstance **b** *archaic* **:** a personal want or need — usually used in plural
5 *plural* **:** AFFAIRS, BUSINESS
6 : a special event or ceremony **:** CELEBRATION
— on occasion : from time to time

[2]**occasion** *transitive verb* **-sioned; -sion·ing** \-'kāzh-niŋ, -'kā-zhə-\ (15th century)
: BRING ABOUT, CAUSE

oc·ca·sion·al \-'kāzh-nəl, -'kā-zhə-nᵊl\ *adjective* (circa 1631)
1 a : of or relating to a particular occasion ⟨a budget able to meet *occasional* demands as well as regular ones⟩ **b :** created for a particular occasion ⟨*occasional* verse⟩
2 : acting as the occasion or contributing cause of something
3 : encountered, occurring, appearing, or taken at irregular or infrequent intervals ⟨*occasional* visitors⟩ ⟨an *occasional* vacation⟩
4 : acting in a specified capacity from time to time ⟨an *occasional* lecturer⟩
5 : designed or constructed to be used as the occasion demands ⟨*occasional* furniture⟩

oc·ca·sion·al·ly \-'kāzh-nə-lē, -'kā-zhə-nᵊl-ē\ *adverb* (1630)
: on occasion **:** NOW AND THEN

Oc·ci·dent \'äk-sə-dənt, -,dent\ *noun* [Middle English, from Middle French, from Latin *occident-, occidens,* from present participle of *occidere* to fall, set (of the sun)] (14th century)
: WEST 2a

oc·ci·den·tal \,äk-sə-'den-tᵊl\ *adjective, often capitalized* (14th century)
1 : of, relating to, or situated in the Occident **:** WESTERN
2 : of or relating to Occidentals
— oc·ci·den·tal·ly *adverb*

Occidental *noun* (1857)
: a member of one of the occidental peoples; *especially* **:** a person of European ancestry

Oc·ci·den·tal·ism \,äk-sə-'den-tᵊl-,i-zəm\ *noun* (1839)
: the characteristic features of occidental peoples or culture

oc·ci·den·tal·ize \-tᵊl-,īz\ *transitive verb* **-ized; -iz·ing** *often capitalized* (1870)
: to make occidental (as in culture)

oc·cip·i·tal \äk-'si-pə-tᵊl\ *adjective* (1541)
: of, relating to, or located within or near the occiput or the occipital bone
— occipital *noun*
— oc·cip·i·tal·ly *adverb*

occipital bone *noun* (1679)
: a compound bone that forms the posterior part of the skull and bears a condyle by which the skull articulates with the atlas

occipital condyle *noun* (circa 1860)
: an articular surface on the occipital bone by which the skull articulates with the atlas

occipital lobe *noun* (circa 1890)
: the posterior lobe of each cerebral hemisphere that bears the visual areas and has the form of a 3-sided pyramid

oc·ci·put \'äk-sə-(,)pət\ *noun, plural* **occiputs** *or* **oc·cip·i·ta** \äk-'si-pə-tə\ [Middle English, from Latin *occipit-, occiput,* from *ob-* against + *capit-, caput* head — more at OB-, HEAD] (14th century)
: the back part of the head or skull

Oc·ci·tan \'äk-sə-,tan\ *noun* [French, from Medieval Latin *occitanus,* from Old Provençal *oc* yes (contrasted with Old French *oil* yes) + Medieval Latin *-itanus* (perhaps as in *aquitanus* of Aquitaine)] (1958)
: PROVENÇAL 2
— Occitan *adjective*

oc·clude \ə-'klüd, ä-\ *verb* **oc·clud·ed; oc·clud·ing** [Latin *occludere,* from *ob-* in the way + *claudere* to shut, close — more at CLOSE] (1597)
transitive verb
1 : to close up or block off **:** OBSTRUCT ⟨a thrombus *occluding* a coronary artery⟩; *also* **:** CONCEAL
2 : SORB
intransitive verb
1 : to come into contact with cusps of the opposing teeth fitting together ⟨his teeth do not *occlude* properly⟩
2 : to become occluded

occluded front *noun* (circa 1938)
: OCCLUSION 2

oc·clu·sal \ə-'klü-səl, ä-, -zəl\ *adjective* (1897)
: of or relating to the grinding or biting surface of a tooth or to occlusion of the teeth

oc·clu·sion \ə-'klü-zhən\ *noun* [Latin *occludere*] (circa 1645)
1 : the act of occluding **:** the state of being occluded: as **a :** the complete obstruction of the breath passage in the articulation of a speech sound **b :** the bringing of the opposing surfaces of the teeth of the two jaws into contact; *also* **:** the relation between the surfaces when in contact **c :** the inclusion or sorption of gas trapped during solidification of a material
2 : the front formed by a cold front overtaking a warm front and lifting the warm air above the earth's surface

oc·clu·sive \-siv, -ziv\ *adjective* [Latin *occlusus,* past participle of *occludere*] (1888)
1 : serving to occlude
2 : characterized by occlusion

[1]**oc·cult** \ə-'kəlt, ä-\ *transitive verb* [Latin *occultare,* frequentative of *occulere*] (1500)
: to shut off from view or exposure **:** COVER, ECLIPSE
— oc·cult·er *noun*

[2]**oc·cult** \ə-'kəlt, ä-; 'ä-,kəlt\ *adjective* [Latin *occultus,* from past participle of *occulere* to cover up, from *ob-* in the way + *-culere* (akin to *celare* to conceal) — more at OB-, HELL] (1567)
1 : not revealed **:** SECRET
2 : not easily apprehended or understood **:** ABSTRUSE, MYSTERIOUS
3 : hidden from view **:** CONCEALED
4 : of or relating to the occult
5 : not manifest or detectable by clinical methods alone ⟨*occult* carcinoma⟩; *also* **:** not present in macroscopic amounts ⟨*occult* blood in a stool⟩
— oc·cult·ly *adverb*

[3]**occult** *same as* [2]\ *noun* (1923)
: matters regarded as involving the action or influence of supernatural or supernormal powers or some secret knowledge of them — used with *the*

oc·cul·ta·tion \,ä-(,)kəl-'tā-shən\ *noun* (15th century)
1 : the state of being hidden from view or lost to notice
2 : the interruption of the light from a celestial body or of the signals from a spacecraft by the intervention of a celestial body; *especially* **:** an eclipse of a star or planet by the moon

oc·cult·ism \ə-'kəl-,ti-zəm, ä-; 'ä-,kəl-\ *noun* (1881)
: occult theory or practice **:** belief in or study of the action or influence of supernatural or supernormal powers
— oc·cult·ist \-tist\ *noun*

oc·cu·pan·cy \'ä-kyə-pən(t)-sē\ *noun, plural* **-cies** (1596)
1 : the fact or condition of holding, possessing, or residing in or on something ⟨*occupancy* of the estate⟩
2 : the act or fact of taking or having possession (as of unowned land) to acquire ownership
3 : the fact or condition of being occupied ⟨*occupancy* by more than 400 persons is unlawful⟩
4 : the use to which a property is put ⟨industrial *occupancy*⟩
5 : a building or part of a building intended to be occupied (as by a tenant) ⟨multiple *occupancy* buildings⟩

oc·cu·pant \-pənt\ *noun* (1596)
1 : one who occupies a particular place; *especially* **:** RESIDENT
2 : one who acquires title by occupancy

oc·cu·pa·tion \,ä-kyə-'pā-shən\ *noun* [Middle English *occupacioun,* from Middle French *occupation,* from Latin *occupation-, occupatio,* from *occupare*] (14th century)

1 a : an activity in which one engages ⟨in the first three grades learning to read is perhaps the major *occupation* of the pupil —J. B. Conant⟩ **b :** the principal business of one's life **:** VOCATION
2 a : the possession, use, or settlement of land **:** OCCUPANCY **b :** the holding of an office or position
3 a : the act or process of taking possession of a place or area **:** SEIZURE **b :** the holding and control of an area by a foreign military force **c :** the military force occupying a country or the policies carried out by it
synonym see WORK
— oc·cu·pa·tion·al \-shnəl, -shə-nᵊl\ *adjective*
— oc·cu·pa·tion·al·ly *adverb*

occupational therapy *noun* (1915)
: therapy by means of activity; *especially* **:** creative activity prescribed for its effect in promoting recovery or rehabilitation
— occupational therapist *noun*

oc·cu·py \ˈä-kyə-ˌpī\ *transitive verb* **-pied; -py·ing** [Middle English *occupien* to take possession of, occupy, modification of Middle French *occuper,* from Latin *occupare,* from *ob-* toward + *-cupare* (akin to *capere* to seize) — more at OB-, HEAVE] (14th century)
1 : to engage the attention or energies of
2 a : to take up (a place or extent in space) ⟨this chair is *occupied*⟩ ⟨the fireplace will *occupy* this corner of the room⟩ **b :** to take or fill (an extent in time) ⟨the hobby *occupies* all of my free time⟩
3 a : to take or hold possession or control of ⟨enemy troops *occupied* the ridge⟩ **b :** to fill or perform the functions of (an office or position)
4 : to reside in as an owner or tenant
— oc·cu·pi·er \-ˌpī(-ə)r\ *noun*

oc·cur \ə-ˈkər\ *intransitive verb* **oc·curred; oc·cur·ring** \-ˈkər-iŋ, -ˈkə-riŋ\ [Latin *occurrere,* from *ob-* in the way + *currere* to run — more at OB-, CAR] (1534)
1 : to be found or met with **:** APPEAR
2 : to come into existence **:** HAPPEN
3 : to come to mind

oc·cur·rence \ə-ˈkər-ən(t)s, -ˈkə-rən(t)s\ *noun* (1539)
1 : something that occurs ⟨a startling *occurrence*⟩
2 : the action or instance of occurring ⟨the repeated *occurrence* of petty theft in the locker room⟩ ☆

¹oc·cur·rent \ə-ˈkər-ənt, -ˈkə-rənt\ *adjective* [Middle English, from Middle French, from Latin *occurrent-, occurrens,* present participle of *occurrere*] (15th century)
1 : occurring at a particular time or place **:** CURRENT
2 : INCIDENTAL

²occurrent *noun* (1535)
: something that occurs as distinguished from something that continues to exist

ocean \ˈō-shən\ *noun, often attributive* [Middle English *occean,* from Latin *oceanus,* from Greek *Ōkeanos,* a river thought of as encircling the earth, ocean] (14th century)
1 a : the whole body of salt water that covers nearly three fourths of the surface of the globe **b :** any of the large bodies of water (as the Atlantic Ocean) into which the great ocean is divided
2 : a very large or unlimited space or quantity

ocean·ar·i·um \ˌō-shə-ˈnar-ē-əm, -ˈner-\ *noun, plural* **-iums** *or* **-ia** \-ē-ə\ (1938)
: a large marine aquarium

ocean·front \ˈō-shən-ˌfrənt\ *noun* (1919)
: a shore area on the ocean

ocean·go·ing \-ˌgō-iŋ\ *adjective* (1885)
: of, relating to, or designed for travel on the ocean

oce·an·ic \ˌō-shē-ˈa-nik\ *adjective* (1656)
1 a : of or relating to the ocean **b :** occurring in or frequenting the ocean and especially the open sea as distinguished from littoral or neritic waters
2 : VAST, GREAT

Oce·anid \ō-ˈsē-ə-nəd\ *noun* [Greek *ōkeanid-, ōkeanis,* from *Ōkeanos* Oceanus]
: any of the ocean nymphs that are daughters of Oceanus and Tethys in Greek mythology

ocean·og·ra·phy \ˌō-shə-ˈnä-grə-fē\ *noun* [International Scientific Vocabulary] (1859)
: a science that deals with the oceans and includes the delimitation of their extent and depth, the physics and chemistry of their waters, marine biology, and the exploitation of their resources
— ocean·og·ra·pher \-fər\ *noun*
— ocean·o·graph·ic \-nə-ˈgra-fik\ *also* **ocean·o·graph·i·cal** \-fi-kəl\ *adjective*
— ocean·o·graph·i·cal·ly \-fi-k(ə-)lē\ *adverb*

ocean·ol·o·gy \ˌō-shə-ˈnä-lə-jē\ *noun* (circa 1864)
: OCEANOGRAPHY; *specifically* **:** the science of marine resources and technology
— ocean·ol·o·gist \-ˈnä-lə-jist\ *noun*

ocean perch *noun* (1943)
: any of several marine scorpaenid food fishes (genus *Sebastes*): **a :** REDFISH a; *also* **:** a related food fish (*S. fasciatus*) **b :** one (*S. alutus*) abundant in the northeastern Pacific from Japan to the Bering Sea to southern California

ocean sunfish *noun* (1629)
: a large bony fish (*Mola mola* of the order Tetraodontiformes) having high dorsal and anal fins and a body nearly oval in outline and attaining a length of 10 feet (3 meters) and a weight in excess of 2 tons (1.8 metric tons)

Oce·anus \ō-ˈsē-ə-nəs\ *noun* [Latin, from Greek *Ōkeanos*]
: a Titan who rules over a great river encircling the earth in Greek mythology

ocel·lus \ō-ˈse-ləs\ *noun, plural* **ocel·li** \-ˈse-ˌlī, -(ˌ)lē\ [New Latin, from Latin, diminutive of *oculus* eye — more at EYE] (1819)
1 : a minute simple eye or eyespot of an invertebrate
2 : an eyelike colored spot (as on a peacock feather or the wings of some butterflies)
— ocel·lar \ō-ˈse-lər\ *adjective*

oce·lot \ˈä-sə-ˌlät, ˈō-\ *noun* [French, from Nahuatl *ōcēlōtl* jaguar] (1774)
: a medium-sized American wildcat (*Felis pardalis*) that ranges from Texas to northern Argentina and has a tawny yellow or grayish coat dotted and striped with black

ocelot

ocher *or* **ochre** \ˈō-kər\ *noun* [Middle English *oker,* from Middle French *ocre,* from Latin *ochra,* from Greek *ōchra,* from feminine of *ōchros* yellow] (14th century)
1 : an earthy usually red or yellow and often impure iron ore used as a pigment
2 : the color of ocher; *especially* **:** the color of yellow ocher
— ocher·ous \ˈō-k(ə-)rəs\ *or* **ochre·ous** \ˈō-k(ə-)rəs, -krē-əs\ *adjective*

och·loc·ra·cy \ä-ˈklä-krə-sē\ *noun* [Greek & Middle French; Middle French *ochlocratie,* from Greek *ochlokratia,* from *ochlos* mob + *-kratia* -cracy] (1584)
: government by the mob **:** mob rule
— och·lo·crat \ˈä-klə-ˌkrat\ *noun*
— och·lo·crat·ic \ˌä-klə-ˈkra-tik\ *or* **och·lo·crat·i·cal** \-ti-kəl\ *adjective*

-ock *noun suffix* [Middle English *-oc,* from Old English]
: small one ⟨hill*ock*⟩

Ock·ham's razor \ˈä-kəmz-\ *variant of* OCCAM'S RAZOR

o'·clock \ə-ˈkläk, ō-\ *adverb* [contraction of *of the clock*] (circa 1601)
1 : according to the clock ⟨the time is three *o'clock*⟩
2 — used for indicating position or direction as if on a clock dial that is oriented vertically or horizontally ⟨an airplane approaching at six *o'clock*⟩

oco·ti·llo \ˌō-kə-ˈtē-(ˌ)yō\ *noun, plural* **-llos** [Mexican Spanish] (1856)
: a thorny scarlet-flowered candlewood (*Fouquieria splendens* of the family Fouquieriaceae) of the southwestern U.S. and Mexico

ocotillo

octa- *or* **octo-** *also* **oct-** *combining form* [Greek *okta-, oktō-, okt-* (from *oktō*) & Latin *octo-, oct-,* from *octo* — more at EIGHT]
: eight ⟨*oct*ane⟩ ⟨*octo*roon⟩

oc·ta·gon \ˈäk-tə-ˌgän\ *noun* [Latin *octagonum,* from Greek *oktagōnon,* from *okta-* + *-gōnon* -gon] (1639)
: a polygon of eight angles and eight sides
— oc·tag·o·nal \äk-ˈta-gə-nᵊl\ *adjective*
— oc·tag·o·nal·ly \-nᵊl-ē\ *adverb*

oc·ta·he·dral \ˌäk-tə-ˈhē-drəl\ *adjective* (1758)
1 : having eight plane faces
2 : of, relating to, or formed in octahedrons
— oc·ta·he·dral·ly \-drə-lē\ *adverb*

oc·ta·he·dron \-drən\ *noun, plural* **-drons** *or* **-dra** \-drə\ [Greek *oktaedron,* from *okta-* + *-edron* -hedron] (1570)
: a solid bounded by eight plane faces

oc·tal \ˈäk-tᵊl\ *adjective* (1948)
: of, relating to, or being a number system with a base of eight

oc·tam·e·ter \äk-ˈta-mə-tər\ *noun* [Late Latin, having eight feet, from Late Greek *oktametros,* from *okta-* + *metron* measure — more at MEASURE] (1889)
: a line of verse consisting of eight metrical feet

oc·tane \ˈäk-ˌtān\ *noun* [International Scientific Vocabulary] (circa 1872)
1 : any of several isomeric liquid alkanes C_8H_{18}
2 : OCTANE NUMBER

octane number *noun* (1931)
: a number that is used to measure the antiknock properties of a liquid motor fuel (as gasoline) with a higher number indicating a

☆ SYNONYMS
Occurrence, event, incident, episode, circumstance mean something that happens or takes place. OCCURRENCE may apply to a happening without intent, volition, or plan ⟨an encounter that was a chance *occurrence*⟩. EVENT usually implies an occurrence of some importance and frequently one having antecedent cause ⟨the *events* following the assassination⟩. INCIDENT suggests an occurrence of brief duration or secondary importance ⟨a minor wartime *incident*⟩. EPISODE stresses the distinctiveness or apartness of an incident ⟨a brief romantic *episode* in a life devoted to work⟩. CIRCUMSTANCE implies a specific detail attending an action or event as part of its setting or background ⟨couldn't recall the exact *circumstances*⟩.

smaller likelihood of knocking — called also *octane rating*; compare CETANE NUMBER

oc·tant \'äk-tənt\ *noun* [Latin *octant-, octans* eighth of a circle, from *octo*] (1731)
1 : an instrument for observing altitudes of a celestial body from a moving ship or aircraft
2 : any of the eight parts into which a space is divided by three coordinate planes

oc·ta·pep·tide \ˌäk-tə-'pep-ˌtīd\ *noun* (1961)
: a protein fragment or molecule (as oxytocin or vasopressin) that consists of eight amino acids linked in a polypeptide chain

oc·tave \'äk-tiv, -təv, -ˌtāv\ *noun* [Middle English, from Medieval Latin *octava,* from Latin, feminine of *octavus* eighth, from *octo* eight — more at EIGHT] (14th century)
1 : an 8-day period of observances beginning with a festival day
2 a : a stanza of eight lines **:** OTTAVA RIMA **b :** the first eight lines of an Italian sonnet
3 a : a musical interval embracing eight diatonic degrees **b :** a tone or note at this interval **c :** the harmonic combination of two tones an octave apart **d :** the whole series of notes, tones, or digitals comprised within this interval and forming the unit of the modern scale **e :** an organ stop giving tones an octave above those corresponding to the digitals
4 : the interval between two frequencies (as in an electromagnetic spectrum) having a ratio of 2 to 1
5 : a group of eight

oc·ta·vo \äk-'tā-(ˌ)vō, -'tä-\ *noun, plural* **-vos** [Latin, ablative of *octavus* eighth] (1582)
: the size of a piece of paper cut eight from a sheet; *also* **:** a book, a page, or paper of this size

oc·tet \äk-'tet\ *noun* (1879)
1 : a group or set of eight: as **a :** OCTAVE 2b **b :** the performers of an octet
2 : a musical composition for eight instruments or voices

oc·til·lion \äk-'til-yən\ *noun* [French, from Middle French, from *oct-* octa- + *-illion* (as in *million*)] (1690)
— see NUMBER table

Oc·to·ber \äk-'tō-bər\ *noun* [Middle English *Octobre,* from Old English & Old French; Old English *October,* from Latin, 8th month of the early Roman calendar, from *octo;* Old French, from Latin *October*] (before 12th century)
: the 10th month of the Gregorian calendar

oc·to·de·cil·lion \ˌäk-tō-di-'sil-yən\ *noun* [Latin *octodecim* eighteen + English *-illion* (as in *million*)] (1939)
— see NUMBER table

oc·to·ge·nar·i·an \ˌäk-tə-jə-'ner-ē-ən\ *noun* [Latin *octogenarius* containing eighty, from *octogeni* eighty each, from *octoginta* eighty, from *octo* eight + *-ginta* (akin to *viginti* twenty) — more at VIGESIMAL] (1815)
: a person whose age is in the eighties
— octogenarian *adjective*

oc·to·ploid \'äk-tə-ˌplȯid\ *adjective* [International Scientific Vocabulary] (1925)
: having a chromosome number eight times the basic haploid chromosome number
— octoploid *noun*

oc·to·pod \'äk-tə-ˌpäd\ *noun* [ultimately from Greek *oktōpod-, oktōpous* scorpion, from *oktō-* octa- + *pod-, pous* foot — more at FOOT] (circa 1836)
: any of an order (Octopoda) of cephalopod mollusks (as an octopus or argonaut) that have eight arms bearing sessile suckers
— octopod *adjective*

oc·to·pus \'äk-tə-pəs, -ˌpu̇s\ *noun, plural* **-pus·es** *or* **-pi** \-ˌpī\ [New Latin *Octopod-, Octopus,* from Greek *oktōpous*] (1758)
1 : any of a genus (*Octopus*) of cephalopod mollusks that have eight muscular arms equipped with two rows of suckers; *broadly* **:** any octopod excepting the paper nautilus
2 : something that resembles an octopus especially in having many centrally directed branches

octopus 1

oc·to·roon \ˌäk-tə-'rün\ *noun* [*octa-* + *-roon* (as in *quadroon*)] (1861)
: a person of one-eighth black ancestry

oc·to·syl·lab·ic \ˌäk-tə-sə-'la-bik\ *adjective* [Late Latin *octosyllabus,* from Greek *oktasyllabos,* from *okta-* octa- + *syllabē* syllable] (circa 1771)
1 : consisting of eight syllables
2 : composed of verses of eight syllables
— octosyllabic *noun*

oc·to·syl·la·ble \'äk-tə-ˌsi-lə-bəl, ˌäk-tə-'\ *noun* (circa 1846)
: a word or line of eight syllables

oc·to·thorp \'äk-tə-ˌthȯrp, -tō-\ *noun* [*octo-* + *thorp,* of unknown origin; from the eight points on its circumference] (1971)
: the symbol #

¹oc·u·lar \'ä-kyə-lər\ *adjective* [Late Latin *ocularis* of eyes, from Latin *oculus* eye] (circa 1575)
1 a : done or perceived by the eye ⟨*ocular* inspection⟩ **b :** based on what has been seen ⟨*ocular* testimony⟩
2 a : of or relating to the eye ⟨*ocular* muscles⟩ **b :** resembling an eye in form or function

²ocular *noun* (1835)
: EYEPIECE

oc·u·lar·ist \'ä-kyə-lə-rist\ *noun* (1866)
: a person who makes and fits artificial eyes

oc·u·list \'ä-kyə-list\ *noun* [French *oculiste,* from Latin *oculus*] (1615)
1 : OPHTHALMOLOGIST
2 : OPTOMETRIST

oc·u·lo·mo·tor \ˌä-kyə-lə-'mō-tər\ *adjective* [Latin *oculus* eye + English *-o-* + *motor*] (circa 1890)
1 : moving or tending to move the eyeball
2 : of or relating to the oculomotor nerve

oculomotor nerve *noun* (1881)
: either of the pair of chiefly motor nerves that comprise the 3d pair of cranial nerves, arise from the midbrain, and supply four muscles of the eye

oc·u·lus \'ä-kyə-ləs\ *noun, plural* **oc·u·li** \-ˌlī, -ˌlē\ [Latin, literally, eye — more at EYE] (1848)
1 : a circular or oval window
2 : a circular opening at the top of a dome

od *or* **odd** \'äd\ *interjection, often capitalized* [euphemism for *God*] (1695)
archaic — used as a mild oath

¹OD \ˌō-'dē\ *noun* [overdose] (circa 1960)
1 : an overdose of a narcotic
2 : one who has taken an OD

²OD *intransitive verb* **OD'd** *or* **ODed; OD'·ing; OD's** (1966)
1 : to become ill or die of an OD
2 : to have or experience too much of something — used with *on* ⟨*OD* on television⟩

oda·lisque \'ō-dᵊl-ˌisk\ *noun* [French, from Turkish *odalık,* from *oda* room] (circa 1681)
1 : a female slave
2 : a concubine in a harem

odd \'äd\ *adjective* [Middle English *odde,* from Old Norse *oddi* point of land, triangle, odd number; akin to Old English *ord* point of a weapon] (14th century)
1 a : being without a corresponding mate ⟨an *odd* shoe⟩ **b** (1) **:** left over after others are paired or grouped (2) **:** separated from a set or series
2 a : somewhat more than the indicated approximate quantity, extent, or degree — usually used in combination ⟨300-*odd* pages⟩ **b** (1) **:** left over as a remainder ⟨had a few *odd* dollars for entertainment after paying his bills⟩ (2) **:** constituting a small amount ⟨had some *odd* change in her pocket⟩
3 a : being any of the integers (as −3, −1, +1, and +3) that are not exactly divisible by two **b :** marked by an odd number of units **c :** being a function such that $f(-x) = -f(x)$ where the sign is reversed but the absolute value remains the same if the sign of the independent variable is reversed
4 a : not regular, expected, or planned ⟨worked at *odd* jobs⟩ **b :** encountered or experienced from time to time **:** OCCASIONAL
5 : having an out-of-the-way location **:** REMOTE
6 : differing markedly from the usual or ordinary or accepted **:** PECULIAR ◆
synonym see STRANGE
— odd·ness *noun*

odd·ball \'äd-ˌbȯl\ *noun* (1948)
: one that is eccentric
— oddball *adjective*

Odd Fellow *noun* [Independent Order of *Odd Fellows*] (1795)
: a member of a major benevolent and fraternal order

odd·i·ty \'ä-də-tē\ *noun, plural* **-ties** (1713)
1 : an odd person, thing, event, or trait
2 : the quality or state of being odd

odd lot *noun* (1897)
: a number or quantity other than the usual unit in transactions; *especially* **:** a quantity of less than 100 shares of stock

odd·ly \'äd-lē\ *adverb* (1610)
1 : in an odd manner
2 : it is odd that ⟨was quite happy, *oddly* enough⟩

odd man out *noun* (1923)
: a person who differs from the other members of a group

odd·ment \'äd-mənt\ *noun* (1796)
1 a : something left over **:** REMNANT **b** *plural* **:** ODDS AND ENDS
2 : something odd **:** ODDITY

odd permutation *noun* (1929)
: a permutation that is produced by the successive application of an odd number of interchanges of pairs of elements

odd–pin·nate \'äd-'pi-ˌnāt\ *adjective* (circa 1890)
: having leaflets on each side of the petiole and having a single leaflet at the tip of the petiole
— odd–pin·nate·ly *adverb*

odds \'ädz\ *noun plural but singular or plural in construction* (circa 1520)
1 a *archaic* **:** INEQUALITIES **b** *obsolete* **:** degree of unlikeness

◇ WORD HISTORY

odd Old Norse *oddi* originally meant "point of land." From the roughly triangular shape of a point of land, *oddi* was generalized to mean "triangle." A point of land can be seen as the apex of a triangle, with the other two angles forming a pair and the apex the unpaired angle, so *oddi* was extended to mean "odd number," the item of a threesome left over after the other two items are paired off. During the time when Scandinavians dominated northern and central England in the Middle Ages, *oddi* was presumably borrowed into English, though it is first recorded only in 14th century Middle English as the adjective *odde* "without a corresponding mate." Already in Middle English, *odde* could mean "unconforming, irregular," though in reference to people it usually meant "outstanding, illustrious." The modern sense "peculiar, eccentric" did not become widely used before the 17th century.

2 a : an amount by which one thing exceeds or falls short of another ⟨won the election by considerable *odds*⟩ **b** (1) **:** a difference favoring one of two opposed things ⟨overwhelming *odds*⟩ (2) **:** a difference in terms of advantage or disadvantage ⟨what's the *odds*, if thinking so makes them happy —Flora Thompson⟩ **c** (1) **:** the probability that one thing is so or will happen rather than another **:** CHANCES ⟨the *odds* are against it⟩ (2) **:** the ratio of the probability of one event to that of an alternative event

3 : DISAGREEMENT, VARIANCE — usually used with *at* ⟨faculty and administration often are at *odds* on everything —W. E. Brock (born 1930)⟩

4 a : special favor **:** PARTIALITY **b :** an allowance granted by one making a bet to one accepting the bet and designed to equalize the chances favoring one of the bettors **c :** the ratio between the amount to be paid off for a winning bet and the amount of the bet

— by all odds : in every way **:** without question ⟨*by all odds* the best book of the year⟩

odds and ends *noun plural* (circa 1746)
1 a : miscellaneous articles **b :** miscellaneous small matters (as of business) to be attended to
2 : miscellaneous remnants or leftovers ⟨*odds and ends* of food⟩

odds•maker \ˈädz-ˌmā-kər\ *noun* (circa 1949)
: one who figures odds

odds–on \ˈädz-ˈȯn, -ˈän\ *adjective* (1890)
1 : having or viewed as having a better than even chance to win ⟨the *odds-on* favorite⟩
2 : not involving much risk **:** pretty sure ⟨an *odds-on* bet⟩

odd trick *noun* (1897)
: each trick in excess of six won by declarer's side at bridge — compare BOOK 9

ode \ˈōd\ *noun* [Middle French or Late Latin; Middle French, from Late Latin, from Greek *ōidē*, literally, song, from *aeidein, aidein* to sing; akin to Greek *audē* voice] (1588)
: a lyric poem usually marked by exaltation of feeling and style, varying length of line, and complexity of stanza forms
— od•ist \-ist\ *noun*

-ode *noun combining form* [Greek *-odos*, from *hodos*]
1 : way **:** path ⟨electr*ode*⟩
2 : electrode ⟨di*ode*⟩

ode•um \ō-ˈdē-əm, ˈō-dē-\ *noun, plural* **odea** \-ə\ [Latin & Greek; Latin, from Greek *ōideion*, from *ōidē* song] (1616)
1 : a small roofed theater of ancient Greece and Rome used chiefly for competitions in music and poetry
2 : a theater or concert hall

od•ic \ˈō-dik\ *adjective* (1863)
: of, relating to, or forming an ode

Odin \ˈō-dᵊn\ *noun* [Danish, from Old Norse *Ōthinn*]
: the supreme god and creator in Norse mythology

odi•ous \ˈō-dē-əs\ *adjective* [Middle English, from Middle French *odieus*, from Latin *odiosus*, from *odium*] (14th century)
: exciting or deserving hatred or repugnance ⟨*odious* associates⟩ ⟨an *odious* business⟩
— odi•ous•ly *adverb*
— odi•ous•ness *noun*

odi•um \ˈō-dē-əm\ *noun* [Latin, hatred, from *odisse* to hate; akin to Old English *atol* terrible, Greek *odyssasthai* to be angry] (1602)
1 : the state or fact of being subjected to hatred and contempt as a result of a despicable act or blameworthy circumstance
2 : hatred and condemnation accompanied by loathing or contempt **:** DETESTATION
3 : disrepute or infamy attached to something **:** OPPROBRIUM

odom•e•ter \ō-ˈdä-mə-tər\ *noun* [French *odomètre*, from Greek *hodometron*, from *hodos* way, road + *metron* measure — more at MEASURE] (1791)
: an instrument for measuring the distance traveled (as by a vehicle)

odo•nate \ˈō-dᵊn-ˌāt, ō-ˈdä-(ˌ)nāt\ *noun* [irregular from Greek *odous, odōn* tooth] (1947)
: any of an order (Odonata) of predaceous insects comprising the dragonflies and damselflies
— odonate *adjective*

odont- *or* **odonto-** *combining form* [French, from Greek, *odont-, odous* — more at TOOTH]
: tooth ⟨*odonto*blast⟩

-odont *adjective combining form* [Greek *odont-, odous* tooth]
: having teeth of a (specified) nature ⟨acr*odont*⟩

-odontia *noun combining form* [New Latin, from Greek *odont-, odous*]
: form, condition, or mode of treatment of the teeth ⟨orth*odontia*⟩

odon•to•blast \ō-ˈdän-tə-ˌblast\ *noun* [International Scientific Vocabulary] (1878)
: any of the elongated radially arranged cells on the surface of the dental pulp that secrete dentin
— odon•to•blas•tic \-ˌdän-tə-ˈblas-tik\ *adjective*

odon•to•glos•sum \ō-ˌdän-tə-ˈglä-səm\ *noun* [New Latin, from Greek *odont-* + *glōssa* tongue — more at GLOSS] (1880)
: any of a genus (*Odontoglossum*) of widely cultivated tropical American orchids

odon•toid process \ō-ˈdän-ˌtȯid-\ *noun* (circa 1819)
: a toothlike process projecting from the anterior end of the centrum of the axis vertebra on which the atlas vertebra rotates

odor \ˈō-dər\ *noun* [Middle English *odour*, from Middle French, from Latin *odor;* akin to Latin *olēre* to smell, Greek *ozein* to smell, *osmē* smell, odor] (13th century)
1 a : a quality of something that stimulates the olfactory organ **:** SCENT **b :** a sensation resulting from adequate stimulation of the olfactory organ **:** SMELL
2 a : a characteristic or predominant quality **:** FLAVOR ⟨the *odor* of sanctity⟩ **b :** REPUTE, ESTIMATION ⟨in bad *odor*⟩
3 *archaic* **:** something that emits a sweet or pleasing scent **:** PERFUME
synonym see SMELL
— odored \ˈō-dərd\ *adjective*
— odor•less \-dər-ləs\ *adjective*

odor•ant \ˈō-də-rənt\ *noun* (1935)
: an odorous substance; *especially* **:** one added to a dangerous odorless substance to warn of its presence

odor•if•er•ous \ˌō-də-ˈri-f(ə-)rəs\ *adjective* (15th century)
1 : yielding an odor **:** ODOROUS
2 : morally offensive
— odor•if•er•ous•ly *adverb*
— odor•if•er•ous•ness *noun*

odor•ize \ˈō-də-ˌrīz\ *transitive verb* **-ized; -iz•ing** (1884)
: to make odorous **:** SCENT

odor•ous \ˈō-də-rəs\ *adjective* (15th century)
: having an odor: as **a :** FRAGRANT **b :** MALODOROUS ☆
— odor•ous•ly *adverb*
— odor•ous•ness *noun*

odour *chiefly British variant of* ODOR

Odys•se•an \ō-ˈdi-sē-ən ("*Odysseus*"), ˌä-də-ˈsē-ən ("*journey*")\ *adjective* (circa 1711)
: of, relating to, or characteristic of Odysseus or his journey

Odys•seus \ō-ˈdi-sē-əs, -ˈdis-yəs, -ˈdi-shəs, -ˈdi-ˌshüs\ *noun* [Greek]
: a king of Ithaca and Greek leader in the Trojan War who after the war wanders 10 years before reaching home

od•ys•sey \ˈä-də-sē\ *noun, plural* **-seys** [the *Odyssey*, epic poem attributed to Homer recounting the long wanderings of Odysseus] (1889)
1 : a long wandering or voyage usually marked by many changes of fortune
2 : an intellectual or spiritual wandering or quest

oe•cu•men•i•cal *especially British* ˌē-\ *variant of* ECUMENICAL

oe•de•ma *chiefly British variant of* EDEMA

oe•di•pal \ˈe-də-pəl, ˈē-\ *adjective, often capitalized* (1939)
: of, relating to, or resulting from the Oedipus complex
— oe•di•pal•ly \-pə-lē\ *adverb, often capitalized*

¹Oe•di•pus \-pəs\ *noun* [Latin, from Greek *Oidipous*]
: the son of Laius and Jocasta who in fulfillment of an oracle unknowingly kills his father and marries his mother

²Oedipus *adjective* (1910)
: OEDIPAL

Oedipus complex *noun* (1910)
: the positive libidinal feelings of a child toward the parent of the opposite sex and hostile or jealous feelings toward the parent of the same sex that may be a source of adult personality disorder when unresolved

oeil–de–boeuf \ˌə(r)-də-ˈbəf, ˌəi-\ *noun, plural* **oeils–de–boeuf** *same*\ [French *œil-de-bœuf*, literally, ox's eye] (1849)
: OCULUS 1

oeil•lade \ˌə(r)-ˈyäd, œ-yȧd\ *noun* [Middle French *œillade*, from *oeil* eye, from Latin *oculus* — more at EYE] (1592)
: a glance of the eye; *especially* **:** OGLE

OEM \ˌō-(ˌ)ē-ˈem\ *noun* [*o*riginal *e*quipment *m*anufacturer] (1968)
: one that produces complex equipment (as a computer system) from components usually bought from other manufacturers

oe•nol•o•gy *variant of* ENOLOGY

Oe•no•ne \ē-ˈnō-nē\ *noun* [Latin, from Greek *Oinōnē*]
: a nymph who is abandoned by her husband Paris for Helen of Troy

oe•no•phile \ˈē-nə-ˌfīl\ *noun* [French *œnophile*, from *œno-* (from Greek *oinos* wine) + *-phile* -phile — more at WINE] (1930)
: a lover or connoisseur of wine

¹o'er \ˈōr, ˈȯr\ *adverb* (1592)
: OVER

²o'er *preposition* (1593)
: OVER

oer•sted \ˈər-stəd\ *noun* [Hans Christian *Oersted*] (1930)
: the unit of magnetic field strength in the centimeter-gram-second system

oe•soph•a•gus *chiefly British variant of* ESOPHAGUS

oestr- *or* **oestro-** *chiefly British variant of* ESTR-

oeu•vre \œvrᵊ\ *noun, plural* **oeuvres** *same*\ [French *œuvre*, literally, work, from Latin *opera* — more at OPERA] (1875)
: a substantial body of work constituting the lifework of a writer, an artist, or a composer

☆ SYNONYMS
Odorous, fragrant, redolent, aromatic mean emitting and diffusing scent. ODOROUS applies to whatever has a strong distinctive smell whether pleasant or unpleasant ⟨*odorous* cheeses should be tightly wrapped⟩. FRAGRANT applies to things (as flowers or spices) with sweet or agreeable odors ⟨a *fragrant* rose⟩. REDOLENT applies usually to a place or thing impregnated with odors ⟨the kitchen was *redolent* of garlic and tomatoes⟩. AROMATIC applies to things emitting pungent often fresh odors ⟨an *aromatic* blend of tobaccos⟩.

[1]of \əv, *before consonants also* ə; ˈəv, ˈäv\ *preposition* [Middle English, off, of, from Old English, adverb & preposition; akin to Old High German *aba* off, away, Latin *about* from, away, Greek *apo*] (before 12th century)
1 — used as a function word to indicate a point of reckoning ⟨north *of* the lake⟩
2 a — used as a function word to indicate origin or derivation ⟨a man *of* noble birth⟩ **b** — used as a function word to indicate the cause, motive, or reason ⟨died *of* flu⟩ **c :** BY ⟨plays *of* Shakespeare⟩ **d :** on the part of ⟨very kind *of* you⟩
3 — used as a function word to indicate the component material, parts, or elements or the contents ⟨throne *of* gold⟩ ⟨cup *of* water⟩
4 a — used as a function word to indicate the whole that includes the part denoted by the preceding word ⟨most *of* the army⟩ **b** — used as a function word to indicate a whole or quantity from which a part is removed or expended ⟨gave *of* his time⟩
5 a : relating to **:** ABOUT ⟨stories *of* her travels⟩ **b :** in respect to ⟨slow *of* speech⟩
6 a — used as a function word to indicate belonging or a possessive relationship ⟨king *of* England⟩ **b** — used as a function word to indicate relationship between a result determined by a function or operation and a basic entity (as an independent variable) ⟨a function *of x*⟩ ⟨the product *of* two numbers⟩
7 — used as a function word to indicate something from which a person or thing is delivered ⟨eased *of* her pain⟩ or with respect to which someone or something is made destitute ⟨robbed *of* all their belongings⟩
8 a — used as a function word to indicate a particular example belonging to the class denoted by the preceding noun ⟨the city *of* Rome⟩ **b** — used as a function word to indicate apposition ⟨that fool *of* a husband⟩
9 a — used as a function word to indicate the object of an action denoted or implied by the preceding noun ⟨love *of* nature⟩ **b** — used as a function word to indicate the application of a verb ⟨cheats him *of* a dollar⟩ or of an adjective ⟨fond *of* candy⟩
10 — used as a function word to indicate a characteristic or distinctive quality or possession ⟨a woman *of* courage⟩
11 a — used as a function word to indicate the position in time of an action or occurrence ⟨died *of* a Monday⟩ **b :** BEFORE ⟨quarter *of* ten⟩
12 *archaic* **:** ON ⟨a plague *of* all cowards —Shakespeare⟩

[2]of \əv, *before consonants also* ə\ *verbal auxiliary* [by alteration] (1837)
nonstandard **:** HAVE — used in place of the contraction *'ve* often in representations of uneducated speech ⟨I could *of* beat them easy —Ring Lardner⟩

ofay \ˈō-ˌfā, ō-ˈ\ *noun* [origin unknown] (1925)
: a white person — usually used disparagingly ◆

[1]off \ˈȯf\ *adverb* [Middle English *of*, from Old English — more at OF] (before 12th century)
1 a (1) **:** from a place or position ⟨march *off*⟩; *specifically* **:** away from land ⟨ship stood *off* to sea⟩ (2) **:** at a distance in space or time ⟨stood 10 paces *off*⟩ ⟨a long way *off*⟩ **b :** from a course **:** ASIDE ⟨turned *off* into a bypath⟩; *specifically* **:** away from the wind **c :** into an unconscious state ⟨dozed *off*⟩
2 a : so as to be separated from support ⟨rolled to the edge of the table and *off*⟩ or close contact ⟨blew the lid *off*⟩ ⟨the handle came *off*⟩ **b :** so as to be divided ⟨surface marked *off* into squares⟩
3 a : to a state of discontinuance or suspension ⟨shut *off* an engine⟩ **b** — used as an intensifier ⟨drink *off* a glass⟩ ⟨finish it *off*⟩
4 : in absence from or suspension of regular work or service ⟨take time *off* for lunch⟩
5 : OFFSTAGE

[2]off *preposition* (before 12th century)
1 a — used as a function word to indicate physical separation or distance from a position of rest, attachment, or union ⟨take it *off* the table⟩ ⟨a path *off* the main walk⟩ ⟨a shop just *off* the main street⟩ **b :** to seaward of ⟨two miles *off* shore⟩
2 — used as a function word to indicate the object of an action ⟨borrowed a dollar *off* him⟩ ⟨dined *off* oysters⟩
3 a — used as a function word to indicate the suspension of an occupation or activity ⟨*off* duty⟩ ⟨*off* liquor⟩ **b :** below the usual standard or level of ⟨*off* his game⟩

[3]off *adjective* (1666)
1 a : more removed or distant ⟨the *off* side of the building⟩ **b :** SEAWARD **c :** RIGHT
2 a : started on the way ⟨*off* on a spree⟩ **b :** not taking place or staying in effect **:** CANCELED ⟨the deal was *off*⟩ **c :** not operating **d :** not placed so as to permit operation
3 a : not corresponding to fact **:** INCORRECT ⟨*off* in his reckoning⟩ **b :** POOR, SUBNORMAL **c :** not entirely sane **:** ECCENTRIC **d :** REMOTE, SLIGHT ⟨an *off* chance⟩
4 a : spent off duty ⟨reading on his *off* days⟩ **b :** SLACK ⟨*off* season⟩
5 a : OFF-COLOR **b :** INFERIOR ⟨*off* grade of oil⟩; *also* **:** affected with putrefaction **c :** DOWN ⟨stocks were *off*⟩
6 : CIRCUMSTANCED ⟨worse *off*⟩

[4]off (1717)
intransitive verb
: to go away **:** DEPART — used chiefly as an imperative ⟨*off*, or I'll shoot⟩
transitive verb
slang **:** KILL, MURDER

of•fal \ˈȯ-fəl, ˈä-\ *noun* [Middle English, from *of* off + *fall*] (14th century)
1 : the waste or by-product of a process: as **a :** trimmings of a hide **b :** the by-products of milling used especially for stock feeds **c :** the viscera and trimmings of a butchered animal removed in dressing
2 : RUBBISH

off and on *adverb* (1535)
: with periodic cessation **:** INTERMITTENTLY ⟨rained *off and on* all day⟩

[1]off•beat \ˈȯf-ˌbēt\ *noun* (circa 1928)
: the unaccented beat of a musical measure

[2]off•beat \-ˈbēt\ *adjective* (1938)
: ECCENTRIC, UNCONVENTIONAL

off Broadway *noun, often O capitalized* [from its usually being produced in smaller theaters outside of the Broadway theatrical district] (1954)
: a part of the New York professional theater stressing fundamental and artistic values and formerly engaging in experimentation
— off–Broadway *adjective or adverb, often O capitalized*

off•cast \ˈȯf-ˌkast\ *adjective* (1571)
: cast off **:** DISCARDED
— offcast *noun*

off–col•or \ˈȯf-ˈkə-lər\ *or* **off–col•ored** \-lərd\ *adjective* (1860)
1 a : not having the right or standard color **b :** being out of sorts
2 a : of doubtful propriety **:** DUBIOUS **b :** verging on the indecent

off•cut \ˈȯf-ˌkət\ *noun* (circa 1664)
chiefly British **:** something that is cut off (as a waste piece of lumber)

of•fend \ə-ˈfend\ *verb* [Middle English, from Middle French *offendre*, from Latin *offendere* to strike against, offend, from *ob-* against + *-fendere* to strike — more at OB-, DEFEND] (14th century)
intransitive verb
1 a : to transgress the moral or divine law **:** SIN ⟨if it be a sin to covet honor, I am the most *offending* soul alive —Shakespeare⟩ **b :** to violate a law or rule **:** do wrong ⟨*offend* against the law⟩
2 a : to cause difficulty, discomfort, or injury ⟨took off his shoe and removed the *offending* pebble⟩ **b :** to cause dislike, anger, or vexation ⟨thoughtless words that *offend* needlessly⟩
transitive verb
1 a : VIOLATE, TRANSGRESS **b :** to cause pain to **:** HURT
2 *obsolete* **:** to cause to sin or fall
3 : to cause to feel vexation or resentment usually by violation of what is proper or fitting ⟨was *offended* by their language⟩ ☆
— of•fend•er *noun*

of•fense *or* **of•fence** \ə-ˈfen(t)s, *especially for 3* ˈä-ˌfen(t)s, ˈȯ-\ *noun* [Middle English, from Middle French, from Latin *offensa*, from feminine of *offensus*, past participle of *offendere*] (14th century)
1 a *obsolete* **:** an act of stumbling **b** *archaic* **:** a cause or occasion of sin **:** STUMBLING BLOCK
2 : something that outrages the moral or physical senses
3 a : the act of attacking **:** ASSAULT **b :** the means or method of attacking or of attempting to score **c :** the offensive team or members of a team playing offensive positions **d :** scoring ability
4 a : the act of displeasing or affronting **b :** the state of being insulted or morally outraged ⟨takes *offense* at the slightest criticism⟩

☆ SYNONYMS
Offend, outrage, affront, insult mean to cause hurt feelings or deep resentment. OFFEND need not imply an intentional hurting but it may indicate merely a violation of the victim's sense of what is proper or fitting ⟨hoped that my remarks had not *offended* her⟩. OUTRAGE implies offending beyond endurance and calling forth extreme feelings ⟨*outraged* by their accusations⟩. AFFRONT implies treating with deliberate rudeness or contemptuous indifference to courtesy ⟨deeply *affronted* by his callousness⟩. INSULT suggests deliberately causing humiliation, hurt pride, or shame ⟨managed to *insult* every guest at the party⟩.

◇ WORD HISTORY
ofay *Ofay* is a derogatory word for a white person, now perhaps obsolescent, recorded in urban African-American speech since the 1920s. Soon after it first saw print, theories about the word's origin proliferated. Probably the most common is that *ofay* is a Pig Latin permutation of *foe*. However, if a word meaning "enemy" were to be so disguised, the rather literary word *foe* is an unlikely choice, and this explanation is most likely an after-the-fact guess that suits popular notions of what the underlying meaning of the word *should* be. More serious have been attempts to find an African source for *ofay;* the word sounds as if it could come from any of a number of West African languages, but none of the suggestions thus far put forward is entirely compelling. There is also the absence of any documentation for *ofay* before it appeared in a New York City newspaper, the *Inter-State Tattler,* in March, 1925, and shortly thereafter in essays by the Harlem Renaissance authors Rudolph Fisher and George S. Schuyler published in the magazine *American Mercury.* Perhaps we should look no further than 1920s Harlem, culturally creative in many respects, for this piece of linguistic creativity.

5 a : a breach of a moral or social code **:** SIN, MISDEED **b :** an infraction of law; *especially* **:** MISDEMEANOR ☆☆

— of·fense·less \-ləs\ *adjective*

¹of·fen·sive \ə-ˈfen(t)-siv, *especially for 1* ˈä-ˌfen(t)-, ˈō-\ *adjective* (circa 1548)

1 a : making attack **:** AGGRESSIVE **b :** of, relating to, or designed for attack ⟨*offensive* weapons⟩ **c :** of or relating to an attempt to score in a game or contest; *also* **:** of or relating to a team in possession of the ball or puck

2 : giving painful or unpleasant sensations **:** NAUSEOUS, OBNOXIOUS ⟨*offensive* odor of garbage⟩

3 : causing displeasure or resentment

— of·fen·sive·ly *adverb*

— of·fen·sive·ness *noun*

²offensive *noun* (1720)

1 : the act of an attacking party

2 : ATTACK

¹of·fer \ˈō-fər, ˈä-\ *verb* **of·fered; of·fer·ing** \-f(ə-)riŋ\ [Middle English *offren,* in sense 1, from Old English *offrian,* from Late Latin *offerre,* from Latin, to present, tender, from *ob-* toward + *ferre* to carry; in other senses, from Old French *offrir,* from Latin *offerre* — more at BEAR] (before 12th century)

transitive verb

1 a : to present as an act of worship or devotion **:** SACRIFICE **b :** to utter (as a prayer) in devotion

2 a : to present for acceptance or rejection **:** TENDER ⟨was *offered* a job⟩ **b :** to present in order to satisfy a requirement ⟨candidates for degrees may *offer* French as one of their foreign languages⟩

3 a : PROPOSE, SUGGEST ⟨*offer* a solution to a problem⟩ **b :** to declare one's readiness or willingness ⟨*offered* to help me⟩

4 a : to try or begin to exert **:** PUT UP ⟨*offered* stubborn resistance⟩ **b :** THREATEN ⟨*offered* to strike him with his cane⟩

5 : to make available **:** AFFORD; *especially* **:** to place (merchandise) on sale

6 : to present in performance or exhibition

7 : to propose as payment **:** BID

intransitive verb

1 : to present something as an act of worship or devotion **:** SACRIFICE

2 *archaic* **:** to make an attempt

3 : to present itself

4 : to make a proposal (as of marriage)

²offer *noun* (15th century)

1 a : a presenting of something for acceptance ⟨considering job *offers* from several firms⟩ ⟨an *offer* of marriage⟩ **b :** an undertaking to do an act or give something on condition that the party to whom the proposal is made do some specified act or make a return promise

2 *obsolete* **:** OFFERING

3 : a price named by one proposing to buy **:** BID

4 a : ATTEMPT, TRY **b :** an action or movement indicating a purpose or intention

— on offer *chiefly British* **:** being offered especially for sale

of·fer·ing \ˈō-f(ə-)riŋ, ˈä-\ *noun* (before 12th century)

1 a : the act of one who offers **b :** something offered; *especially* **:** a sacrifice ceremonially offered as a part of worship **c :** a contribution to the support of a church

2 : something offered for sale or patronage ⟨latest *offerings* of the leading novelists⟩

3 : a course of instruction or study

of·fer·to·ry \ˈō-fə(r)-ˌtōr-ē, ˈä-, -ˌtȯr-\ *noun, plural* **-ries** [Medieval Latin *offertorium,* from Late Latin *offerre*] (1539)

1 *often capitalized* **a :** the eucharistic offering of bread and wine to God before they are consecrated at Communion **b :** a verse from a Psalm said or sung at the beginning of the offertory

2 a : the period of collection and presentation of the offerings of the congregation at public worship **b :** a musical composition played or sung during an offertory

¹off·hand \ˈōf-ˈhand, -ˌhand\ *adverb* (1694)

: without premeditation or preparation **:** EXTEMPORE ⟨couldn't give the figures *offhand*⟩

²offhand *adjective* (1708)

1 : CASUAL, INFORMAL ⟨a relaxed, *offhand* manner⟩

2 : done or made offhand ⟨*offhand* excuses⟩

off·hand·ed \-ˈhan-dəd, -ˌhan-\ *adjective* (1835)

: OFFHAND

— off·hand·ed·ly *adverb*

— off·hand·ed·ness *noun*

off–hour \ˈōf-ˌau̇(-ə)r\ *noun* (1932)

1 : a period of time other than a rush hour

2 : a period of time other than regular business hours

of·fice \ˈä-fəs, ˈō-\ *noun* [Middle English, from Middle French, from Latin *officium* service, duty, office, from *opus* work + *facere* to make, do — more at OPERATE, DO] (13th century)

1 a : a special duty, charge, or position conferred by an exercise of governmental authority and for a public purpose **:** a position of authority to exercise a public function and to receive whatever emoluments may belong to it **b :** a postion of responsibility or some degree of executive authority

2 [Middle English, from Old French, from Late Latin *officium,* from Latin] **:** a prescribed form or service of worship; *specifically, capitalized* **:** DIVINE OFFICE

3 : a religious or social ceremonial observance **:** RITE

4 a : something that one ought to do or must do **:** an assigned or assumed duty, task, or role **b :** the proper or customary action of something **:** FUNCTION **c :** something done for another **:** SERVICE

5 : a place where a particular kind of business is transacted or a service is supplied: as **a :** a place in which the functions of a public officer are performed **b :** the directing headquarters of an enterprise or organization **c :** the place in which a professional person conducts business

6 *plural, chiefly British* **:** the apartments, attached buildings, or outhouses in which the activities attached to the service of a house are carried on

7 a : a major administrative unit in some governments ⟨British Foreign *Office*⟩ **b :** a subdivision of some government departments ⟨Patent *Office*⟩

synonym see FUNCTION

office boy *noun* (1846)

: a boy or man employed for odd jobs in a business office

of·fice·hold·er \-ˌhōl-dər\ *noun* (1818)

: one holding a public office

¹of·fi·cer \ˈä-fə-sər, ˈō-\ *noun* [Middle English, from Middle French *officier,* from Medieval Latin *officiarius,* from Latin *officium*] (14th century)

1 a *obsolete* **:** AGENT **b :** one charged with police duties

2 : one who holds an office of trust, authority, or command ⟨the *officers* of the bank⟩

3 a : one who holds a position of authority or command in the armed forces; *specifically* **:** COMMISSIONED OFFICER **b :** the master or any of the mates of a merchant or passenger ship

²officer *transitive verb* (1670)

1 : to furnish with officers

2 : to command or direct as an officer

officer of arms (circa 1500)

: any of the officers (as king of arms, herald, or pursuivant) of a monarch or government responsible for devising and granting armorial bearings

¹of·fi·cial \ə-ˈfi-shəl *also* ō-\ *noun* (14th century)

1 : one who holds or is invested with an office **:** OFFICER ⟨government *officials*⟩

2 : one who administers the rules of a game or sport especially as a referee or umpire

²official *adjective* (circa 1604)

1 : of or relating to an office, position, or trust ⟨*official* duties⟩

2 : holding an office

3 a : AUTHORITATIVE, AUTHORIZED ⟨*official* statement⟩ **b :** prescribed or recognized as authorized ⟨an *official* language⟩ **c :** described by the U.S. Pharmacopeia or the National Formulary

4 : befitting or characteristic of a person in office ⟨extended an *official* greeting⟩

— of·fi·cial·ly \-ˈfi-sh(ə-)lē\ *adverb*

of·fi·cial·dom \-ˈfi-shəl-dəm\ *noun* (1863)

: officials as a class

of·fi·cial·ese \-ˌfi-shə-ˈlēz, -ˈlēs\ *noun* (1884)

: the characteristic language of official statements **:** wordy, pompous, or obscure language

of·fi·cial·ism \-ˈfi-shə-ˌli-zəm\ *noun* (1857)

: lack of flexibility and initiative combined with excessive adherence to regulations in the behavior of usually government officials

of·fi·ci·ant \ə-ˈfi-shē-ənt\ *noun* (1844)

: one (as a priest) that officiates at a religious rite

¹of·fi·ci·ary \ə-ˈfi-shē-ˌer-ē, ō-, ä-\ *noun, plural* **-ar·ies** [Medieval Latin *officiarius*] (1611)

1 : OFFICER, OFFICIAL

2 : a body of officers or officials

²officiary *adjective* (1612)

: connected with, derived from, or having a title or rank by virtue of holding an office ⟨*officiary* earl⟩

of·fi·ci·ate \ə-ˈfi-shē-ˌāt\ *verb* **-at·ed; -at·ing** (1631)

transitive verb

1 : to carry out (an official duty or function)

2 : to serve as a leader or celebrant of (a ceremony)

3 : to administer the rules of (a game or sport) especially as a referee or umpire

intransitive verb

1 : to perform a ceremony, function, or duty ⟨*officiate* at a wedding⟩

2 : to act in an official capacity **:** act as an official (as at a sports contest)

— of·fi·ci·a·tion \-ˌfi-shē-ˈā-shən\ *noun*

☆ SYNONYMS

Offense, resentment, umbrage, pique, dudgeon, huff mean an emotional response to or an emotional state resulting from a slight or indignity. OFFENSE implies hurt displeasure ⟨takes deep *offense* at racial slurs⟩. RESENTMENT suggests a longer lasting indignation or smoldering ill will ⟨harbored a lifelong *resentment* of his brother⟩. UMBRAGE may suggest hurt pride, resentment, or suspicion of another's motives ⟨took *umbrage* at the offer of advice⟩. PIQUE applies to a transient feeling of wounded vanity ⟨in a *pique* I foolishly declined the invitation⟩. DUDGEON suggests an angry fit of indignation ⟨stormed out of the meeting in high *dudgeon*⟩. HUFF implies a peevish short-lived spell of anger usually at a petty cause ⟨in a *huff* he slammed the door⟩.

Offense, sin, vice, crime, scandal mean a transgression of law. OFFENSE applies to the infraction of any law, rule, or code ⟨at that school no *offense* went unpunished⟩. SIN implies an offense against moral or religious law ⟨the *sin* of blasphemy⟩. VICE applies to a habit or practice that degrades or corrupts ⟨regarded gambling as a *vice*⟩. CRIME implies a serious offense punishable by the law of the state ⟨the *crime* of murder⟩. SCANDAL applies to an offense that outrages the public conscience ⟨a career ruined by a sex *scandal*⟩.

of·fi·ci·nal \ə-ˈfi-sᵊn-əl, ȯ-, ä-; ˌȯ-fə-ˈsī-nᵊl, ˌä-\ *adjective* [Medieval Latin *officinalis* of a storeroom, from *officina* storeroom, from Latin, workshop, from *opific-, opifex* workman, from *opus* work + *facere* to do] (circa 1720)
: MEDICINAL ⟨a monograph on *officinal* flora⟩

of·fi·cious \ə-ˈfi-shəs\ *adjective* [Latin *officiosus,* from *officium* service, office] (1565)
1 *archaic* **a :** KIND, OBLIGING **b :** DUTIFUL
2 : volunteering one's services where they are neither asked nor needed **:** MEDDLESOME
3 : INFORMAL, UNOFFICIAL
synonym see IMPERTINENT
— **of·fi·cious·ly** *adverb*
— **of·fi·cious·ness** *noun*

off·ing \ˈȯ-fiŋ, ˈä-\ *noun* [¹*off*] (1627)
1 : the part of the deep sea seen from the shore
2 : the near or foreseeable future ⟨in the *offing*⟩

off·ish \ˈȯ-fish\ *adjective* [¹*off*] (1831)
: STANDOFFISH
— **off·ish·ness** *noun*

off–key \ˈȯf-ˈkē\ *adjective or adverb* (1927)
1 : varying in pitch from the proper tone of a melody
2 : IRREGULAR, ANOMALOUS

off–kil·ter \-ˈkil-tər\ *adjective* (circa 1944)
: not in perfect balance **:** a bit askew

off–lim·its \ˈȯf-ˈli-məts\ *adjective* (1945)
: not to be entered or patronized by a designated class (as military personnel); *also* **:** not to be interfered with, considered, or spoken of ⟨the subject of sex was *off-limits* in her family⟩

off–line \-ˈlīn\ *adjective* (1950)
: not connected to or served by a system and especially a computer or telecommunications system; *also* **:** done independently of a system ⟨*off-line* computer storage⟩
— **off–line** *adverb*

off–load \(ˌ)ȯf-ˈlōd, ˈȯf-ˌ\ *verb* (1850)
: UNLOAD

off of *preposition* (1593)
: OFF ▢

off–off–Broadway *noun, often both Os capitalized* [from its relation to off Broadway being analogous to the relation of off Broadway to Broadway] (1965)
: an avant-garde theatrical movement in New York
— **off–off–Broadway** *adjective or adverb, often both Os capitalized*

off–peak \ˈȯf-ˈpēk\ *adjective* (1920)
: not being in the period of maximum use or business **:** not peak ⟨telephone rates during *off-peak* hours⟩

off–price \ˈȯf-ˈprīs\ *adjective* (1952)
: of, relating to, selling, or being discounted merchandise ⟨an *off-price* store⟩ ⟨*off-price* apparel⟩

off·print \ˈȯf-ˌprint\ *noun* (1885)
: a separately printed excerpt (as a magazine article)
— **offprint** *transitive verb*

off–put·ting \-ˌpu̇-tiŋ\ *adjective* (1828)
: that puts one off **:** REPELLENT, DISCONCERTING
— **off–put·ting·ly** \-lē\ *adverb*

off–ramp \-ˌramp\ *noun* (1954)
: a ramp by which one leaves a limited-access highway

off–road \ˈȯf-ˈrōd\ *adjective* (1968)
: of, relating to, done with, or being a vehicle designed especially to operate away from public roads

off·scour·ing \-ˌskau̇(-ə)r-iŋ\ *noun* (1526)
1 : someone rejected by society **:** OUTCAST
2 : something that is scoured off **:** REFUSE

off·screen \ˈȯf-ˈskrēn\ *adverb or adjective* (1935)
1 : out of sight of the motion picture or television viewer
2 : in private life

off–sea·son \ˈȯf-ˌsē-zᵊn\ *noun* (1848)
: a time of suspended or reduced activity; *especially* **:** the time during which an athlete is not training or competing

¹off·set \ˈȯf-ˌset\ *noun* (circa 1555)
1 a *archaic* **:** OUTSET, START **b :** CESSATION
2 a (1) **:** a short prostrate lateral shoot arising from the base of a plant (2) **:** a small bulb arising from the base of another bulb **b :** a lateral or collateral branch (as of a family or race) **:** OFFSHOOT **c :** a spur from a range of hills
3 a : a horizontal ledge on the face of a wall formed by a diminution of its thickness above **b :** DISPLACEMENT **c :** an abrupt change in the dimension or profile of an object or the part set off by such change
4 : something that sets off to advantage or embellishes something else **:** FOIL
5 : an abrupt bend in an object by which one part is turned aside out of line
6 : something that serves to counterbalance or to compensate for something else; *especially* **:** either of two balancing ledger items
7 a : unintentional transfer of ink (as from a freshly printed sheet) **b :** a printing process in which an inked impression from a plate is first made on a rubber-blanketed cylinder and then transferred to the paper being printed
— **offset** *adjective or adverb*

²off·set \ˈȯf-ˌset, *verb transitive senses are also* ȯf-ˈ\ *verb* **-set; -set·ting** (1792)
transitive verb
1 a : to place over against something **:** BALANCE ⟨credits *offset* debits⟩ **b :** to serve as a counterbalance for **:** COMPENSATE ⟨his speed *offset* his opponent's greater weight⟩
2 : to form an offset in ⟨*offset* a wall⟩
intransitive verb
: to become marked by offset

off·shoot \ˈȯf-ˌshüt\ *noun* (1710)
1 a : a collateral or derived branch, descendant, or member **:** OUTGROWTH **b :** a lateral branch (as of a mountain range)
2 : a branch of a main stem especially of a plant

¹off·shore \ˈȯf-ˈshōr, -ˈshȯr\ *adverb* (1720)
1 : from the shore **:** SEAWARD; *also* **:** at a distance from the shore
2 : outside the country **:** ABROAD

²off·shore \ˈȯf-ˌ\ *adjective* (1845)
1 : coming or moving away from the shore toward the water ⟨an *offshore* breeze⟩
2 a : situated off the shore but within waters under a country's control ⟨*offshore* fisheries⟩ **b :** distant from the shore — compare INSHORE
3 : situated or operating in a foreign country ⟨*offshore* mutual funds⟩ ⟨*offshore* banking⟩

³off·shore \ˈȯf-ˌ\ *preposition* (1965)
: off the shore of

off·side \ˈȯf-ˈsīd\ *adverb or adjective* (1867)
: illegally in advance of the ball or puck

off–site \-ˈsīt\ *adjective or adverb* (1946)
: not located or occurring at the site of a particular activity

off–speed \-ˈspēd\ *adjective* (1965)
: being slower than usual or expected ⟨throwing *off-speed* pitches⟩

off·spring \ˈȯf-ˌspriŋ\ *noun, plural* **offspring** *also* **offsprings** [Middle English *ofspring,* from Old English, from *of* off + *springan* to spring] (before 12th century)
1 a : the progeny of an animal or plant **:** YOUNG **b :** CHILD
2 a : PRODUCT, RESULT ⟨scholarly manuscripts —the labored *offsprings* of PhDs —Donna Martin⟩ **b :** OFFSHOOT 1a

off·stage \ˈȯf-ˈstāj, -ˌstāj\ *adverb or adjective* (1921)
1 : on a part of the stage not visible to the audience
2 : in private life ⟨known *offstage* as a kindly person⟩
3 : behind the scenes **:** out of the public view ⟨much of the important work of the conference was done *offstage*⟩

off–the–books *adjective* (1980)
: not reported or recorded ⟨*off-the-books* transactions⟩ ⟨*off-the-books* covert operations⟩
— **off the books** *adverb*

off–the–cuff *adjective* (1938)
: not prepared in advance **:** SPONTANEOUS, INFORMAL ⟨*off-the-cuff* remarks⟩
— **off–the–cuff** *adverb*

off–the–peg *adjective* (1959)
chiefly British **:** READY-MADE 1

off–the–rack *adjective* (1965)
: READY-MADE 1 ⟨*off-the-rack* suits⟩

off–the–record *adjective* (1933)
: given or made in confidence and not for publication ⟨*off-the-record* comments⟩

off–the–shelf *adjective* (1950)
: available as a stock item **:** not specially designed or custom-made ⟨made of *off-the-shelf* components⟩

off–the–wall *adjective* (circa 1966)
: highly unusual **:** BIZARRE ⟨an *off-the-wall* sense of humor⟩

off·track \ˈȯf-ˈtrak\ *adverb or adjective* (1944)
: away from a racetrack ⟨betting *offtrack*⟩ ⟨*offtrack* bookies⟩

off–white \ˈȯf-ˈhwīt, -ˈwīt\ *noun* (1927)
: a yellowish or grayish white

off year *noun* (1873)
1 : a year in which no major election is held
2 : a year of diminished activity or production ⟨an *off year* for auto sales⟩

oft \ˈȯft\ *adverb* [Middle English, from Old English; akin to Old High German *ofto* often] (before 12th century)
: OFTEN

of·ten \ˈȯ-fən, ÷ˈȯf-tən\ *adverb* [Middle English, alteration of *oft*] (14th century)
: many times **:** FREQUENTLY

of·ten·times \-ˌtīmz\ *or* **oft·times** \ˈȯf(t)-ˌtīmz\ *adverb* (14th century)
: OFTEN, REPEATEDLY

ogee *also* **OG** \ˈō-ˌjē\ *noun* [obsolete English *ogee* ogive; from the use of such moldings in ogives] (1677)
1 : a molding with an S-shaped profile
2 : a pointed arch having on each side a reversed curve near the apex — see ARCH illustration

ogham *or* **ogam** \ˈä-gəm, ˈō-; ˈō(-ə)m\ *noun* [Irish *ogham,* from Middle Irish *ogom, ogum*] (1729)
: the alphabetic system of 5th and 6th century Irish in which an alphabet of 20 letters is represented by notches for vowels and lines for consonants and which is known principally from inscriptions cut on the edges of rough standing tombstones
— **ogham·ic** \ä-ˈgä-mik, ō-; ˈō-(ə-)mik\ *adjective*
— **ogham·ist** \ˈä-gə-mist, ˈō-; ˈō-(ə-)mist\ *noun*

ogi·val \ō-ˈjī-vəl\ *adjective* (1841)
: of, relating to, or having the form of an ogive or an ogee

ogive \ˈō-ˌjīv\ *noun* [Middle English *oggif* stone comprising an arch, from Middle French *augive* diagonal arch] (1611)
1 a : a diagonal arch or rib across a Gothic vault **b :** a pointed arch
2 : a graph of a distribution function or a cumulative frequency distribution
3 : OGEE 1

¹ogle \ˈō-gəl *also* ˈä-\ *verb* **ogled; ogling** \-g(ə-)liŋ\ [probably from Low German *oegeln,* from *oog* eye; akin to Old High German *ouga* eye — more at EYE] (1682)

▢ USAGE
off of The *of* is often criticized as superfluous, a comment that is irrelevant because *off of* is an idiom. It is much more common in speech than in edited writing and is more common in American English than in British.

intransitive verb
: to glance with amorous invitation or challenge
transitive verb
1 : to eye amorously or provocatively
2 : to look at especially with greedy or interested attention
— ogler \-g(ə-)lər\ *noun*

[2]ogle *noun* (1711)
: an amorous or coquettish glance

ogre \ˈō-gər\ *noun* [French, probably ultimately from Latin *Orcus,* god of the underworld] (1713)
1 : a hideous giant of fairy tales and folklore that feeds on human beings **:** MONSTER
2 : a dreaded person or object
— ogre·ish \ˈō-g(ə-)rish\ *adjective*

ogress \ˈō-g(ə-)rəs\ *noun* (1713)
: a female ogre

[1]oh \(ˈ)ō\ *interjection* [Middle English *o*] (13th century)
1 — used to express an emotion (as surprise or desire) or in response to physical stimuli
2 — used in direct address ⟨*oh,* waiter! Will you come here, please?⟩
3 — used to express acknowledgment or understanding of a statement
4 — used to introduce an example or approximation

[2]oh \ˈō\ *noun* [*o;* from the similarity of the symbol for zero (0) to the letter *O*] (1936)
: ZERO

ohia \ō-ˈhē-ə\ *noun* [Hawaiian *ʻōhiʻa*] (1824)
: LEHUA

ohia lehua *noun* [Hawaiian *ʻōhiʻa-lehua*] (1888)
: LEHUA

ohm \ˈōm\ *noun* [Georg Simon *Ohm*] (1867)
: the practical meter-kilogram-second unit of electric resistance equal to the resistance of a circuit in which a potential difference of one volt produces a current of one ampere
— ohm·ic \ˈō-mik\ *adjective*
— ohm·i·cal·ly \-mi-k(ə-)lē\ *adverb*

ohm·me·ter \ˈō(m)-ˌmē-tər\ *noun* [International Scientific Vocabulary] (circa 1890)
: an instrument for indicating resistance in ohms directly

Ohm's law \ˈōmz-\ *noun* (1863)
: a law in electricity: the strength of a direct current is directly proportional to the potential difference and inversely proportional to the resistance of the circuit

-oic *adjective suffix* [French *-oïque* (as in *acide caproïque* caproic acid)]
: containing carboxyl or a derivative ⟨benz*oic* acid⟩

[1]-oid *noun suffix*
: something resembling a (specified) object or having a (specified) quality ⟨glob*oid*⟩

[2]-oid *adjective suffix* [Middle French & Latin; Middle French *-oïde,* from Latin *-oïdes,* from Greek *-oeidēs,* from *-o-* + *eidos* appearance, form — more at WISE]
: resembling **:** having the form or appearance of ⟨petal*oid*⟩

oid·i·um \ō-ˈi-dē-əm\ *noun, plural* **-ia** \-dē-ə\ [New Latin, from *o-* + *-idium*] (1857)
1 a : any of a genus (*Oidium* of the family Moniliaceae) of imperfect fungi many of which are now considered to be conidial stages of various powdery mildews **b :** one of the small conidia borne in chains by various fungi (as an oidium) — called also *arthrospore*
2 : a powdery mildew caused by an oidium especially in the grape

[1]oil \ˈȯi(ə)l\ *noun, often attributive* [Middle English *oile,* from Old French, from Latin *oleum* olive oil, from Greek *elaion,* from *elaia* olive] (13th century)
1 a : any of numerous unctuous combustible substances that are liquid or can be liquefied easily on warming, are soluble in ether but not in water, and leave a greasy stain on paper or cloth **b** (1) **:** PETROLEUM (2) **:** the petroleum industry
2 : a substance (as a cosmetic preparation) of oily consistency ⟨bath *oil*⟩
3 a : an oil color used by an artist **b :** a painting done in oil colors
4 : unctuous or flattering speech

[2]oil (15th century)
transitive verb
: to smear, rub over, furnish, or lubricate with oil
intransitive verb
: to take on fuel oil
— oil the hand *or* **oil the palm :** BRIBE, TIP

oil beetle *noun* (1658)
: any of various blister beetles (genus *Meloe* or a related genus) that emit a yellowish liquid from the leg joints when disturbed

oil·bird \ˈȯi(ə)l-ˌbərd\ *noun* (circa 1890)
: a nocturnal bird (*Steatornis caripensis*) of northern South America and Trinidad that is related to the nightjars, feeds chiefly on the fatty fruits of various palms, and has fatty young from which oil is extracted for use instead of butter — called also *guacharo*

oil cake *noun* (1743)
: the solid residue after extracting the oil from seeds (as of cotton)

oil·can \ˈȯi(ə)l-ˌkan\ *noun* (1839)
: a can for oil; *especially* **:** a spouted can designed to release oil drop by drop (as for lubricating machinery)

oil·cloth \-ˌklȯth\ *noun* (1796)
: cloth treated with oil or paint and used for table and shelf coverings

oil color *noun* (1539)
1 : a pigment used for oil paint
2 : OIL PAINT

oiled \ˈȯi(ə)ld\ *adjective* (1535)
1 : lubricated, treated, or covered with or as if with oil ⟨*oiled* paper⟩
2 *slang* **:** DRUNK 1a

oil·er \ˈȯi-lər\ *noun* (circa 1846)
1 : one (as a workman) that oils something
2 : a receptacle or device for applying oil
3 *plural* **:** OILSKIN 3
4 : an auxiliary naval vessel used for refueling at sea

oil field *noun* (1894)
: a region rich in petroleum deposits; *especially* **:** one that has been brought into production

oil gland *noun* (circa 1836)
: a gland (as of the skin) that produces an oily secretion; *specifically* **:** UROPYGIAL GLAND

oil·man \ˈȯi(ə)l-mən, -ˌman\ *noun* (1865)
1 : an oil company executive
2 : an oil field worker

oil of vitriol (1580)
: concentrated sulfuric acid

oil of wintergreen (1866)
: the methyl ester of salicylic acid that is used as a flavoring and as a counterirritant

oil paint *noun* (1790)
: paint in which a drying oil is the vehicle

oil painting *noun* (1782)
1 a : the act or art of painting in oil colors **b :** a picture painted in oils
2 : painting that uses pigments originally ground in oil

oil palm *noun* (circa 1864)
: an African pinnate-leaved palm (*Elaeis guineensis*) cultivated for its clustered fruit whose flesh and seeds yield oil

oil pan *noun* (1908)
: the lower section of the crankcase used as a lubricating-oil reservoir on an internal combustion engine

oil patch *noun* (circa 1952)
1 : OIL FIELD
2 : the petroleum industry

oil·seed \ˈȯi(ə)l-ˌsēd\ *noun* (1562)
: a seed or crop (as flaxseed) grown mainly for oil

oil shale *noun* (1873)
: a rock (as shale) from which oil can be recovered by distillation

oil·skin \ˈȯi(ə)l-ˌskin\ *noun* (1812)
1 : an oiled waterproof cloth used for coverings and garments
2 : an oilskin raincoat
3 *plural* **:** an oilskin suit of coat and trousers

oil slick *noun* (1889)
: a film of oil floating on water

oil·stone \ˈȯi(ə)l-ˌstōn\ *noun* (1585)
: a whetstone for use with oil

oil well *noun* (1847)
: a well from which petroleum is obtained

oily \ˈȯi-lē\ *adjective* **oil·i·er; -est** (14th century)
1 : of, relating to, or consisting of oil
2 a : covered or impregnated with oil ⟨*oily* rags⟩ **b :** relatively high in naturally secreted oils ⟨*oily* skin⟩ ⟨*oily* hair⟩
3 : excessively smooth or suave in manner
— oil·i·ly \ˈȯi-lə-lē\ *adverb*
— oil·i·ness \-lē-nəs\ *noun*

oink \ˈȯiŋk\ *noun* [imitative] (1941)
: the natural noise of a hog
— oink *intransitive verb*

oint·ment \ˈȯint-mənt\ *noun* [Middle English, alteration of *oignement,* from Old French, ultimately from Latin *unguentum,* from *unguere* to anoint; akin to Old High German *ancho* butter, Sanskrit *anakti* he salves] (14th century)
: a salve or unguent for application to the skin

oi·ti·ci·ca \ˌȯi-tə-ˈsē-kə\ *noun* [Portuguese, from Tupi] (1901)
: any of several South American trees; *especially* **:** a Brazilian tree (*Licania rigida*) with seeds that yield a drying oil similar to tung oil

Ojib·wa *or* **Ojib·way** \ō-ˈjib-(ˌ)wā\ *noun, plural* **Ojibwa** *or* **Ojibwas** *or* **Ojibway** *or* **Ojibways** [Ojibwa *očipweˑ,* an Ojibwa band] (1700)
1 : a member of an American Indian people of the region around Lake Superior and westward
2 : an Algonquian language of the Ojibwa people

[1]OK *or* **okay** \ō-ˈkā, *in assenting or agreeing also* ˈō-ˌkā\ *adverb or adjective* [abbreviation of *oll korrect,* facetious alteration of *all correct*] (1839)
: ALL RIGHT ◆

[2]OK *or* **okay** \ō-ˈkā\ *noun* (1841)
: APPROVAL, ENDORSEMENT

[3]OK *or* **okay** \ō-ˈkā\ *transitive verb* **OK'd** *or* **okayed; OK'·ing** *or* **okay·ing** (1888)
: APPROVE, AUTHORIZE

oka *variant of* OCA

◇ WORD HISTORY

OK *OK,* one of the most successful of all Americanisms, has long been the subject of etymological claims and counterclaims. According to the most plausible account, it arose in the late 1830s from a journalistic fad for reducing phrases to abbreviations. At the height of the fad it became fashionable to alter some of the abbreviations on the basis of pseudo-uneducated misspellings, so that *A.R.* for *all right* was transformed to *O.W.* on the basis of *oll wright,* and *N.G.* for *no go* to *K.G.* Hence *OK,* when it appeared in the Boston *Morning Post* on March 23, 1839—the earliest printed occurrence that anyone has yet found—is glossed as *oll korrect,* presumably a deliberate misspelling of *all correct. OK* might well have passed into oblivion with such other initialisms, but in the 1840 American presidential campaign it became associated with the Democratic O.K. Club, in which *O.K.* stood for *Old Kinderhook,* after Kinderhook, New York, the birthplace of the Democrats' candidate, Martin Van Buren. Humorous expansions of *OK,* including *oll korrect,* were bandied back and forth by both Democrats and Whigs. By 1841 *OK* had gone beyond the political arena to become an interjectional equivalent for "All right!"

oka·pi \ō-ˈkä-pē\ *noun* [Mvu'ba (language of northeast Zaire)] (1900)
: an ungulate mammal (*Okapia johnstoni*) of Zaire that is closely related to the giraffe but has a relatively short neck, a coat typically of solid reddish chestnut on the trunk, yellowish white on the cheeks, and purplish black and cream rings on the upper parts of the legs

okapi

okey·doke \ˌō-kē-ˈdōk\ *or* **okey·do·key** \-ˈdō-kē\ *adverb* [reduplication of *OK*] (circa 1932)
— used to express assent

Okie \ˈō-kē\ *noun* [*Ok*lahoma + *-ie*] (1938)
: a migrant agricultural worker; *especially* : one from Oklahoma in the 1930s — sometimes used disparagingly

okra \ˈō-krə, *Southern also* -krē\ *noun* [of African origin; akin to Ibo *ọ́kụ̀rụ̀* okra] (1679)
1 : a tall annual (*Abelmoschus esculentus*) of the mallow family that is cultivated for its mucilaginous green pods used especially in soups or stews; *also* : the pods of this plant
2 : ¹GUMBO 1

¹-ol *noun suffix* [International Scientific Vocabulary, from *alcohol*]
: chemical compound (as an alcohol or phenol) containing hydroxyl ⟨glycer*ol*⟩ ⟨cres*ol*⟩

²-ol — see -OLE

³-ol *noun combining form* [International Scientific Vocabulary, from Latin *oleum* oil — more at OIL]
: hydrocarbon chemically related to benzene ⟨xyl*ol*⟩

¹old \ˈōld; *for sense 9 usually* ˈōl\ *adjective* [Middle English, from Old English *eald;* akin to Old High German *alt* old, Latin *alere* to nourish, *alescere* to grow, *altus* high, deep] (before 12th century)
1 a : dating from the remote past : ANCIENT ⟨*old* traditions⟩ **b** : persisting from an earlier time ⟨an *old* ailment⟩ ⟨they brought up the same *old* argument⟩ **c** : of long standing ⟨an *old* friend⟩
2 a : distinguished from an object of the same kind by being of an earlier date ⟨many still used the *old* name⟩ **b** *capitalized* : belonging to an early period in the development of a language or literature ⟨*Old* Persian⟩
3 : having existed for a specified period of time ⟨a girl three years *old*⟩
4 : of, relating to, or originating in a past era ⟨*old* chronicles record the event⟩
5 a : advanced in years or age ⟨an *old* man⟩ **b** : showing the characteristics of age ⟨looked *old* at 20⟩
6 : EXPERIENCED ⟨an *old* trooper speaking of the last war⟩
7 : FORMER ⟨his *old* students⟩
8 a : showing the effects of time or use : WORN, AGED ⟨*old* shoes⟩ **b** : well advanced toward reduction by running water to the lowest level possible — used of topographic features **c** : no longer in use : DISCARDED ⟨*old* rags⟩ **d** : of a grayish or dusty color ⟨*old* mauve⟩ **e** : TIRESOME ⟨gets *old* fast⟩
9 a : long familiar ⟨same *old* story⟩ ⟨good *old* Joe⟩ **b** — used as an intensive ⟨a high *old* time⟩ **c** — used to express an attitude of affection or amusement ⟨a big *old* dog⟩ ⟨flex the *old* biceps⟩ ⟨any *old* time⟩ ☆

²old *noun* (13th century)
1 : one of a specified age — usually used in combination ⟨a 3-year-*old*⟩
2 : old or earlier time — used in the phrase *of old* ⟨mighty men of *old*⟩

old boy *noun* (1868)
1 *often O&B capitalized, British* : an alumnus especially of a boys' school
2 : a man who is a member of a long-standing and usually influential clique especially in a professional, business, or social sphere

Old Bulgarian *noun* (1861)
: OLD CHURCH SLAVONIC

Old Catholic *noun* (1871)
: a member of one of various hierarchical and liturgical churches separating from the Roman Catholic Church at various times since the 18th century

Old Christmas *noun* (1863)
chiefly Midland : EPIPHANY 1

Old Church Slavonic *noun* (circa 1929)
: the Slavic language used in the liturgical and Biblical translations of Cyril and Methodius as attested in manuscripts of the 10th and 11th centuries — called also *Old Church Slavic*; see INDO-EUROPEAN LANGUAGES table

old country *noun, often O&C capitalized* (1782)
: an emigrant's country of origin; *especially* : one in Europe — usually used with *the*

old·en \ˈōl-dən\ *adjective* (14th century)
: of or relating to a bygone era

Old English *noun* (13th century)
1 a : the language of the English people from the time of the earliest documents in the 7th century to about 1100 — see INDO-EUROPEAN LANGUAGES table **b** : English of any period before Modern English
2 : BLACK LETTER

Old English sheepdog *noun* (1891)
: any of a breed of tailless dogs developed in England and having a profuse blue-gray and white coat

Old English sheepdog

old-fan·gled \ˈōl(d)-ˈfaŋ-gəld\ *adjective* [*old* + *-fangled* (as in *newfangled*)] (1842)
: OLD-FASHIONED

¹old–fash·ioned \-ˈfa-shənd\ *adjective* (1596)
1 a : of, relating to, or characteristic of a past era ⟨wears an *old-fashioned* black bow tie —Green Peyton⟩ **b** : adhering to customs of a past era
2 : OUTMODED
— old–fash·ioned·ly \-shən(d)-lē\ *adverb*
— old–fash·ioned·ness \-nəs\ *noun*

²old–fashioned *noun* (1901)
: a cocktail usually made with whiskey, bitters, sugar, a twist of lemon peel, and a small amount of water or soda

Old French *noun* (1708)
: the French language from the 9th to the 16th century; *especially* : French from the 9th to the 13th century — see INDO-EUROPEAN LANGUAGES table

Old Glory *noun* (1862)
: the flag of the U.S.

old gold *noun* (1879)
: a dark yellow

old guard *noun, often O&G capitalized* (1850)
1 : a group of established prestige and influence
2 : the conservative members of an organization (as a political party)

old hand *noun* (circa 1785)
: HAND 10e

old hat *adjective* (1911)
1 : OLD-FASHIONED
2 : lacking in freshness : TRITE

Old High German *noun* (circa 1884)
: High German exemplified in documents prior to the 12th century — see INDO-EUROPEAN LANGUAGES table

old·ie \ˈōl-dē\ *noun* (1874)
: one that is old; *especially* : a popular song of an earlier day

Old Ionic *noun* (circa 1889)
: the Greek dialect of the Homeric epics

Old Iranian *noun* (1939)
: any Iranian language in use in the period B.C.

Old Irish *noun* (circa 1884)
: the Irish in use from the 7th century to about 950 — see INDO-EUROPEAN LANGUAGES table

old·ish \ˈōl-dish\ *adjective* (circa 1669)
: somewhat old or elderly

old lady *noun* (1836)
1 : WIFE
2 : MOTHER
3 : GIRLFRIEND; *especially* : one with whom a man cohabits

Old Latin *noun* (circa 1889)
: Latin used in the early inscriptions and in literature prior to the classical period

old–line \ˈōl(d)-ˈlīn\ *adjective* (1856)
1 a : having a reputation or authority based on length or proven quality of service ⟨an *old-line* firm⟩ **b** : of established prestige and influence ⟨*old-line* families⟩
2 : adhering to traditional policies or practices : CONSERVATIVE

old maid *noun* (circa 1530)
1 : SPINSTER 3
2 : a prim fussy person ⟨he was a real *old maid* about burning rubbish —R. C. Ruark⟩
3 : a simple card game in which cards are matched in pairs and the player holding the unmatched card at the end loses
— old–maid·ish \ˈōl(d)-ˈmā-dish\ *adjective*
— old–maid·ish·ness \-nəs\ *noun*

old man *noun* (1768)
1 a : HUSBAND **b** : FATHER
2 *capitalized* : one in authority; *especially* : COMMANDING OFFICER
3 : BOYFRIEND; *especially* : one with whom a woman cohabits

old–man's beard \ˈōl(d)-ˈmanz-\ *noun* (1742)
1 : any of several clematises (especially *Clematis vitalba*) having plumose styles
2 : a greenish gray pendulous lichen (*Usnea barbata*) growing on trees

old master *noun* (1824)
: a work of art by an established master and especially by any of the distinguished painters of the 16th, 17th, or early 18th century; *also* : such an artist

Old Nick \ˈōl(d)-ˈnik\ *noun* (1668)
— used as a name of the devil

☆ SYNONYMS

Old, ancient, venerable, antique, antiquated, archaic, obsolete mean having come into existence or use in the more or less distant past. OLD may apply to either actual or merely relative length of existence ⟨*old* houses⟩ ⟨an *old* sweater of mine⟩. ANCIENT applies to occurrence, existence, or use in or survival from the distant past ⟨*ancient* accounts of dragons⟩. VENERABLE stresses the impressiveness and dignity of great age ⟨the family's *venerable* patriarch⟩. ANTIQUE applies to what has come down from a former or ancient time ⟨collected *antique* Chippendale furniture⟩. ANTIQUATED implies being discredited or outmoded or otherwise inappropriate to the present time ⟨*antiquated* teaching methods⟩. ARCHAIC implies having the character or characteristics of a much earlier time ⟨the play used *archaic* language to convey a sense of period⟩. OBSOLETE may apply to something regarded as no longer acceptable or useful even though it is still in existence ⟨a computer that makes earlier models *obsolete*⟩.

Old Norse *noun* (1844)
: the North Germanic language of the Scandinavian peoples prior to about 1350 — see INDO-EUROPEAN LANGUAGES table
Old North French *noun* (circa 1930)
: the northern dialects of Old French including especially those of Normandy and Picardy
Old Persian *noun* (circa 1909)
: an ancient Iranian language known from cuneiform inscriptions from the 5th and 6th centuries B.C. — see INDO-EUROPEAN LANGUAGES table
Old Prussian *noun* (1872)
: a Baltic language used in East Prussia until the 17th century — see INDO-EUROPEAN LANGUAGES table
old rose *noun* (1893)
: a variable color averaging a grayish red
Old Saxon *noun* (1841)
: the language of the Saxons of northwest Germany until about the 12th century — see INDO-EUROPEAN LANGUAGES table
old school *noun* (1749)
: adherents of traditional policies and practices
old school tie *noun* (1932)
1 a : an attitude of conservatism, aplomb, and upper-class solidarity associated with English public school graduates **b :** a necktie displaying the colors of an English public school
2 : clannishness among members of an established clique
old–shoe \'ōl(d)-'shü\ *adjective* (1944)
: characterized by familiarity or freedom from restraint **:** COMFORTABLE, UNPRETENTIOUS
old sledge *noun* (1830)
: SEVEN-UP
old–squaw \'ōl(d)-'skwȯ\ *noun* (1838)
: a common sea duck (*Clangula hyemalis*) of the more northern parts of the northern hemisphere
old•ster \'ōl(d)-stər\ *noun* (1848)
: an old or elderly person
old style *noun* (1617)
1 *O&S capitalized* **:** a style of reckoning time used before the adoption of the Gregorian calendar
2 : a style of type distinguished by graceful irregularity among individual letters, bracketed serifs, and but slight contrast between light and heavy strokes
Old Style *adjective* (1678)
: using or according to the Julian calendar
Old Swedish *noun* (circa 1909)
: the Swedish language as exemplified in documents prior to about 1350
Old Testament *noun* (14th century)
: the first part of the Christian Bible containing the books of the Jewish canon of Scripture — see BIBLE table
old–time \'ōl(d)-'tīm\ *adjective* (1824)
1 : of, relating to, or characteristic of an earlier period ⟨*old-time* songs⟩
2 : of long standing ⟨*old-time* residents⟩
old–tim•er \-'tī-mər, -ˌtī-mər\ *noun* (1879)
1 a : VETERAN **b :** OLDSTER
2 : something that is old-fashioned **:** ANTIQUE
old–timey \'ōl(d)-'tī-mē\ *adjective* (1850)
: of a kind or style prevalent in or reminiscent of an earlier time ⟨*old-timey* music⟩
Old Welsh *noun* (1882)
: the Welsh language exemplified in documents prior to about 1150 — see INDO-EUROPEAN LANGUAGES table
old•wife \'ōl(d)-ˌwīf\ *noun* (1588)
1 : any of several marine fishes (as an alewife, menhaden, or triggerfish)
2 : OLD-SQUAW
old wives' tale *noun* (1656)
: an often traditional belief that is not based on fact **:** SUPERSTITION
old–world \'ōl(d)-'wər(-ə)ld\ *adjective* (1712)
: of, relating to, or characteristic of the Old World; *especially* **:** having the charm or picturesque qualities of the Old World ⟨narrow *old-world* streets⟩

Old World *noun* (circa 1596)
: the eastern hemisphere exclusive of Australia; *specifically* **:** Europe
ole \'ōl\ *adjective* [by alteration] (1844)
: OLD
ole- *or* **oleo-** *combining form* [French *olé-*, *oléo-*, from Latin *ole-*, from *oleum* — more at OIL]
: oil ⟨*oleo*graph⟩
-ole *also* **-ol** *noun combining form* [International Scientific Vocabulary, from Latin *oleum*]
1 : chemical compound containing a 5-membered usually heterocyclic ring ⟨pyrr*ole*⟩
2 : chemical compound not containing hydroxyl ⟨eucalypt*ol*⟩ — especially in names of ethers ⟨safr*ole*⟩
olé \ō-'lā\ *noun* [Spanish] (1922)
: [2]BRAVO
ole•ag•i•nous \ˌō-lē-'a-jə-nəs\ *adjective* [Middle English, from Middle French *oleagineux*, from Latin *oleagineus* of an olive tree, from *olea* olive tree, from Greek *elaia*] (15th century)
1 : resembling or having the properties of oil **:** OILY; *also* **:** containing or producing oil
2 : marked by an offensively ingratiating manner or quality
— **ole•ag•i•nous•ly** *adverb*
— **ole•ag•i•nous•ness** *noun*
ole•an•der \'ō-lē-ˌan-dər, ˌō-lē-'\ *noun* [Medieval Latin, alteration of *arodandrum, lorandrum,* perhaps alteration of Latin *rhododendron* — more at RHODODENDRON] (1548)
: a poisonous evergreen shrub (*Nerium oleander*) of the dogbane family with fragrant white to red flowers
ole•an•do•my•cin \ˌō-lē-ˌan-də-'mī-sᵊn\ *noun* [*oleand*rose, a sugar derived from oleandrin (a glycoside contained in oleander leaves) + *-o-* + *-mycin*] (1956)
: an antibiotic $C_{35}H_{61}NO_{12}$ produced by a streptomyces (*Streptomyces antibioticus*)
ole•as•ter \'ō-lē-ˌas-tər, ˌō-lē-'\ *noun* [Middle English, from Latin, from *olea*] (14th century)
: any of several plants (genus *Elaeagnus* of the family Elaeagnaceae, the oleaster family) having alternate leaves and perfect flowers with four stamens; *especially* **:** RUSSIAN OLIVE
ole•ate \'ō-lē-ˌāt\ *noun* (circa 1823)
: a salt or ester of oleic acid
olec•ra•non \ō-'le-krə-ˌnän\ *noun* [New Latin, from Greek *ōlekranon,* from *ōlenē* elbow + *kranion* skull — more at ELL, CRANIUM] (circa 1741)
: the process of the ulna projecting behind the elbow joint
ole•fin \'ō-lə-fən\ *noun* [International Scientific Vocabulary, from French (*gaz*) *oléfiant* ethylene, from Latin *oleum*] (1860)
1 : ALKENE
2 : a synthetic fiber (as polypropylene) derived from an alkene
— **ole•fin•ic** \ˌō-lə-'fi-nik\ *adjective*
oleic acid *noun* (1819)
: a monounsaturated fatty acid $C_{18}H_{34}O_2$ found in natural fats and oils
ole•in \'ō-lē-ən\ *noun* [French *oléine,* from Latin *oleum*] (1838)
1 : an ester of glycerol and oleic acid
2 : the liquid portion of a fat
oleo \'ō-lē-ˌō\ *noun* [short for *oleomargarine*] (1884)
: MARGARINE
oleo•graph \'ō-lē-ə-ˌgraf\ *noun* [International Scientific Vocabulary] (1873)
: a chromolithograph printed on cloth to imitate an oil painting
oleo•mar•ga•rine \ˌō-lē-ō-'mär-jə-rən, -ˌrēn; -'märj-rən\ *noun* [French *oléomargarine,* from *olé-* + *margarine* margarine] (1873)
: MARGARINE
oleo•res•in \ˌō-lē-ō-'re-zᵊn\ *noun* [International Scientific Vocabulary] (circa 1846)
1 : a natural plant product (as copaiba) containing chiefly essential oil and resin; *especially* **:** TURPENTINE 1b
2 : a preparation consisting essentially of oil holding resin in solution
— **oleo•res•in•ous** \-'re-zᵊn-əs, -'rez-nəs\ *adjective*
ole•um \'ō-lē-əm\ *noun* [Latin, olive oil — more at OIL] (circa 1823)
1 *plural* **olea** \-lē-ə\ **:** OIL
2 *plural* **oleums :** a heavy oily strongly corrosive solution of sulfur trioxide in anhydrous sulfuric acid
O level *noun* (1949)
1 : the lowest of three levels of standardized British examinations in a secondary school subject; *also* **:** successful completion of an O-level examination in a particular subject — called also *Ordinary level*; compare A LEVEL, S LEVEL
2 a : the level of education required to pass an O-level examination **b :** a course leading to an O-level examination
ol•fac•tion \äl-'fak-shən, ōl-\ *noun* (circa 1846)
1 : the sense of smell
2 : the act or process of smelling
ol•fac•tom•e•ter \ˌäl-ˌfak-'tä-mə-tər, ˌōl-\ *noun* (1889)
: an instrument for measuring the sensitivity of the sense of smell
ol•fac•to•ry \äl-'fak-t(ə-)rē, ōl-\ *adjective* [Latin *olfactorius,* from *olfacere* to smell, from *olēre* to smell + *facere* to do — more at ODOR, DO] (circa 1658)
: of, relating to, or connected with the sense of smell
olfactory bulb *noun* (circa 1860)
: a bulbous anterior projection of the olfactory lobe that is the place of termination of the olfactory nerves and is especially well developed in lower vertebrates (as fishes)
olfactory lobe *noun* (circa 1860)
: an anterior projection of each cerebral hemisphere that is continuous anteriorly with the olfactory nerve
olfactory nerve *noun* (1670)
: either of the pair of nerves that are the first cranial nerves and that arise in the olfactory neurosensory cells of the nasal mucous membrane and pass to the anterior part of the cerebrum
olig- *or* **oligo-** *combining form* [Medieval Latin, from Greek, from *oligos;* perhaps akin to Armenian *ałkat* scant]
: few ⟨*oligo*phagous⟩
ol•i•garch \'ä-lə-ˌgärk, 'ō-\ *noun* [Greek *oligarchēs,* from *olig-* + *-archēs* -arch] (circa 1610)
: a member or supporter of an oligarchy
ol•i•gar•chic \ˌä-lə-'gär-kik, ˌō-\ *or* **ol•i•gar•chi•cal** \-ki-kəl\ *adjective* (1586)
: of, relating to, or based on an oligarchy
ol•i•gar•chy \'ä-lə-ˌgär-kē, 'ō-\ *noun, plural* **-chies** (1542)
1 : government by the few
2 : a government in which a small group exercises control especially for corrupt and selfish purposes; *also* **:** a group exercising such control
3 : an organization under oligarchic control
Ol•i•go•cene \'ä-li-gō-ˌsēn, 'ō-; ə-'li-gə-\ *adjective* [International Scientific Vocabulary] (circa 1859)
: of, relating to, or being an epoch of the Tertiary between the Eocene and Miocene or the corresponding system of rocks — see GEOLOGIC TIME table
— **Oligocene** *noun*
ol•i•go•chaete \-ˌkēt\ *noun* [New Latin *Oligochaeta,* ultimately from Greek *olig-* + *chaitē* long hair] (1896)

: any of a class or order (Oligochaeta) of hermaphroditic terrestrial or aquatic annelids (as an earthworm) that lack a specialized head
— **oligochaete** *adjective*

ol•i•go•clase \'ä-li-gō-ˌklās, 'ō-, -ˌklāz; ə-'li-gə-\ *noun* [German *Oligoklas,* from *olig-* olig- + Greek *klasis* breaking, from *klan* to break — more at CLAST] (1832)
: a mineral of the plagioclase series

ol•i•go•den•dro•cyte \ˌä-li-gō-'den-drə-ˌsīt, ˌō-; ə-ˌli-gə-\ *noun* [International Scientific Vocabulary, from *olig-* + *dendr-* + *-cyte*] (1932)
: a neuroglial cell resembling an astrocyte but smaller with few and slender processes having few branches

ol•i•go•den•drog•lia \-den-'drä-glē-ə, -'drō-\ *noun* [New Latin, from *oligodendro*cyte + *-glia* (form of neuroglia)] (1924)
: neuroglia made up of oligodendrocytes that is held to function in myelin formation in the central nervous system
— **ol•i•go•den•drog•li•al** \-glē-əl\ *adjective*

olig•o•mer \ə-'li-gə-mər\ *noun* (1952)
: a polymer or polymer intermediate containing relatively few structural units
— **olig•o•mer•ic** \-ˌli-gə-'mer-ik\ *adjective*
— **olig•o•mer•i•za•tion** \-mə-rə-'zā-shən\ *noun*

ol•i•go•nu•cle•o•tide \-'nü-klē-ə-ˌtīd, -'nyü-\ *noun* (1942)
: a chain of usually up to 20 nucleotides

ol•i•goph•a•gous \ˌä-lə-'gä-fə-gəs, ˌō-\ *adjective* (1920)
: eating only a few specific kinds of food
— **ol•i•goph•a•gy** \-'gä-fə-jē\ *noun*

ol•i•gop•o•ly \-'gä-pə-lē\ *noun* [*olig-* + *-poly* (as in *monopoly*)] (1895)
: a market situation in which each of a few producers affects but does not control the market
— **ol•i•gop•o•lis•tic** \-ˌgä-pə-'lis-tik\ *adjective*

ol•i•gop•so•ny \-'gäp-sə-nē\ *noun* [*olig-* + Greek *opsōnia* purchase of victuals, from *opsōnein* to purchase victuals, from *opson* food + *ōneisthai* to buy — more at VENAL] (1942)
: a market situation in which each of a few buyers exerts a disproportionate influence on the market
— **ol•i•gop•so•nis•tic** \-ˌgäp-sə-'nis-tik\ *adjective*

ol•i•go•sac•cha•ride \ˌä-li-gō-'sa-kə-ˌrīd, ˌō-; ə-'li-gə-\ *noun* [International Scientific Vocabulary] (1930)
: a saccharide (as a disaccharide) that contains a known small number of monosaccharide units

ol•i•go•tro•phic \-'trō-fik\ *adjective* [International Scientific Vocabulary] (1928)
: deficient in plant nutrients ⟨*oligotrophic* boggy acid soils⟩; *especially* **:** having abundant dissolved oxygen ⟨an *oligotrophic* body of water⟩ — compare EUTROPHIC, MESOTROPHIC

olio \'ō-lē-ˌō\ *noun, plural* **oli•os** [modification of Spanish *olla*] (circa 1643)
1 : OLLA PODRIDA 1
2 a : a miscellaneous mixture **:** HODGEPODGE **b :** a miscellaneous collection (as of literary or musical selections)

ol•i•va•ceous \ˌä-lə-'vā-shəs\ *adjective* (1776)
: OLIVE 1

[1]**ol•ive** \'ä-liv, -ləv\ *noun* [Middle English, from Old French, from Latin *oliva,* from Greek *elaia*] (13th century)
1 a : a Mediterranean evergreen tree (*Olea europaea* of the family Oleaceae, the olive family) cultivated for its drupaceous fruit that is an important food and source of oil; *also* **:** the fruit **b :** any of various shrubs and trees resembling the olive
2 : any of several colors resembling that of the unripe fruit of the olive tree that are yellowish green
3 : an oval eminence on each ventrolateral aspect of the medulla oblongata

[2]**olive** *adjective* (1657)
1 : of the color olive or olive green
2 : approaching olive in color or complexion

olive branch *noun* (14th century)
1 : a branch of the olive tree especially when used as a symbol of peace
2 : an offer or gesture of conciliation or goodwill

olive drab *noun* (1897)
1 : a grayish olive
2 a : a wool or cotton fabric of an olive drab color **b :** a uniform of this fabric

olive green *noun* (circa 1757)
: a greenish olive

oliv•en•ite \ō-'li-və-ˌnīt\ *noun* [German *Olivenit,* from *oliven-, olive* olive] (1820)
: a mineral that is a basic arsenate of copper

Ol•i•ver \'ä-lə-vər\ *noun* [French *Olivier*]
: the close friend of Roland in the Charlemagne legends

ol•iv•ine \'ä-lə-ˌvēn\ *noun* [German *Olivin,* from Latin *oliva*] (1794)
: a usually greenish mineral that is a complex silicate of magnesium and iron used especially in refractories — compare PERIDOT
— **ol•iv•in•ic** \ˌä-lə-'vi-nik\ *or* **ol•iv•in•it•ic** \-və-'ni-tik\ *adjective*

ol•la \'ä-lə, 'ȯi-ə\ *noun* [Spanish, from Latin *olla, aulla* pot; akin to Sanskrit *ukhā* pot and probably to Gothic *auhns* oven] (1622)
: a large bulging widemouthed earthenware vessel sometimes with looped handles used (as by Pueblo Indians) for storage, cooking, or as a container for water

olla

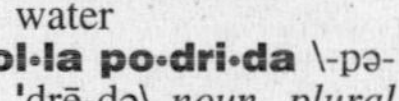

ol•la po•dri•da \-pə-'drē-də\ *noun, plural* **olla podridas** \-'drē-dəz\ *also* **ollas podridas** [Spanish, literally, rotten pot] (1599)
1 : a rich highly seasoned stew of meat and vegetables usually including sausage and chick-peas that is slowly simmered and is a traditional Spanish and Latin-American dish
2 : HODGEPODGE

olo•li•u•qui \ˌō-ˌlō-lē-'ü-kē\ *noun* [Spanish *ololiuque,* from Nahuatl *ololiuhqui,* literally, something rolled into a ball] (1915)
: a woody stemmed Mexican vine (*Rivea corymbosa*) of the morning glory family having small fleshy fruits with single seeds that are used especially by the Indians for medicinal, narcotic, and religious purposes

olo•ro•so \ˌō-lə-'rō-(ˌ)sō\ *noun, plural* **-sos** [Spanish, from *oloroso* fragrant, from *olor* odor, from Latin, from *olēre* to smell — more at ODOR] (1876)
: a dry full-bodied Spanish sherry

olym•pi•ad \ə-'lim-pē-ˌad, ō-\ *noun, often capitalized* [Middle English, from Middle French *Olympiade,* from Latin *Olympiad-, Olympias,* from Greek, from *Olympia,* site of ancient Olympic Games] (14th century)
1 : one of the 4-year intervals between Olympic Games by which time was reckoned in ancient Greece
2 : a quadrennial celebration of the modern Olympic Games

[1]**Olym•pi•an** \-pē-ən\ *adjective* (15th century)
1 : of or relating to Mount Olympus in Thessaly
2 : befitting or characteristic of an Olympian; *especially* **:** LOFTY ⟨his . . . formula of glib simplicity and *Olympian* arrogance —Richard Pollak⟩

[2]**Olympian** *adjective* (1593)
1 : of or relating to the ancient Greek region of Olympia
2 : of, relating to, or constituting the Olympic Games

[3]**Olympian** *noun* (1606)
: a participant in Olympic Games

[4]**Olympian** *noun* (1843)
1 : one of the ancient Greek deities dwelling on Olympus
2 : a being of lofty detachment or superior attainments

Olympian Games *noun plural* (1593)
: OLYMPIC GAMES 1

Olym•pia oyster \ə-'lim-pē-ə-, ō-\ *noun* [*Olympia,* Washington] (1908)
: a small flavorful native oyster (*Ostrea lurida*) of the Puget Sound area of the Pacific coast of North America — called also *Olympia*

Olym•pic \ə-'lim-pik, ō-\ *adjective* (1590)
1 : [1]OLYMPIAN
2 : of or relating to the Olympic Games

Olympic Games *noun plural* (circa 1610)
1 : an ancient Panhellenic festival held every fourth year and made up of contests of sports, music, and literature with the victor's prize a crown of wild olive
2 : a modified revival of the ancient Olympic Games consisting of international athletic contests that are held at separate winter and summer gatherings at four year intervals — called also *Olympics*

Olym•pus \ə-'lim-pəs, ō-\ *noun* [Latin, from Greek *Olympos*] (1580)
: a mountain in Thessaly that in Greek mythology is the abode of the gods

om \'ōm\ *noun* [Sanskrit] (1788)
: a mantra consisting of the sound \'ōm\ and used in contemplation of ultimate reality

-oma *noun suffix, plural* **-omas** *or* **-omata** [Latin *omat-, -oma,* from Greek *-ōmat-, -ōma,* from *-ō-* (stem vowel of causative verbs in *-oun*) + *-mat-, -ma,* suffix denoting result — more at -MENT]
: tumor ⟨aden*oma*⟩ ⟨fibr*oma*⟩

Oma•ha \'ō-mə-ˌhȯ, -ˌhä\ *noun, plural* **Omaha** *or* **Omahas** (1804)
: a member of an American Indian people of northeastern Nebraska

oma•sum \ō-'mā-səm\ *noun, plural* **oma•sa** \-sə\ [New Latin, from Latin, tripe of a bullock] (circa 1706)
: the third chamber of the ruminant stomach that is situated between the reticulum and the abomasum — compare RUMEN

om•bre \'äm-bər; 'äm-brē, 'əm-, -ˌbrā\ *noun* [French or Spanish; French *hombre,* from Spanish, literally, man — more at HOMBRE] (circa 1661)
: an old three-handed card game popular in Europe especially in the 17th and 18th centuries

om•bré \'äm-ˌbrā\ *adjective* [French, past participle of *ombrer* to shade, from Italian *ombrare,* from *ombra* shade, from Latin *umbra* — more at UMBRAGE] (circa 1896)
: having colors or tones that shade into each other — used especially of fabrics in which the color is graduated from light to dark
— **ombré** *noun*

om•buds•man \'äm-ˌbu̇dz-mən, 'ȯm-, -bədz-, -ˌman; äm-'bu̇dz-, ȯm-\ *noun, plural* **-men** \-mən\ [Swedish, literally, representative, from Old Norse *umbothsmathr,* from *umboth* commission + *mathr* man] (1959)
1 : a government official (as in Sweden or New Zealand) appointed to receive and investigate complaints made by individuals against abuses or capricious acts of public officials

2 : one that investigates reported complaints (as from students or consumers), reports findings, and helps to achieve equitable settlements ◆

— om·buds·man·ship \-ˌship\ *noun*

-ome *noun suffix* [New Latin *-oma,* from Latin, -oma]
: mass ⟨phyll*ome*⟩

ome·ga \ō-ˈmā-gə, -ˈmē-, -ˈme-\ *noun* [Middle English, from Middle French, from Greek *ō mega,* literally, large o] (15th century)
1 : the 24th and last letter of the Greek alphabet — see ALPHABET table
2 : LAST, ENDING
3 a : a negatively charged elementary particle that has a mass 3270 times the mass of an electron — called also *omega minus* **b :** a very short-lived unstable meson with mass 1532 times the mass of an electron — called also *omega meson*

ome·ga–3 \-ˈthrē\ *adjective* (1980)
: being or composed of polyunsaturated fatty acids that have the final double bond in the hydrocarbon chain between the third and fourth carbon atoms from one end of the molecule and that are found especially in fish, fish oils, vegetable oils, and green leafy vegetables

om·e·lette *or* **om·e·let** \ˈäm-lət, ˈä-mə-\ *noun* [French *omelette,* alteration of Middle French *alumelle,* literally, knife blade, modification of Latin *lamella,* diminutive of *lamina* thin plate] (circa 1611)
: beaten eggs cooked without stirring until set and served folded in half ◆

omen \ˈō-mən\ *noun* [Latin *omin-, omen*] (1582)
: an occurrence or phenomenon believed to portend a future event **:** AUGURY

omen·tum \ō-ˈmen-təm\ *noun, plural* **-ta** \-tə\ *or* **-tums** [Latin; perhaps akin to Latin *induere* to put on, *exuere* to take off — more at EXUVIAE] (1547)
: a fold of peritoneum connecting or supporting abdominal structures (as the viscera); *also* **:** a fold of peritoneum free at one end

— omen·tal \-ˈmen-t^{ə}l\ *adjective*

omer \ˈō-mər\ *noun* [Hebrew *ʿōmer*] (circa 1608)
1 : an ancient Hebrew unit of dry capacity equal to 1/10 ephah
2 a *often capitalized* **:** the sheaf of barley traditionally offered in Jewish Temple worship on the second day of the Passover **b** *capitalized* **:** a 7-week liturgical period of expectancy between the second day of the Passover and Shabuoth

om·i·cron \ˈä-mə-ˌkrän, ˈō-, *British* ō-ˈmī-(ˌ)krän\ *noun* [Middle English, from Middle French, from Greek *o mikron,* literally, small o] (15th century)
: the 15th letter of the Greek alphabet — see ALPHABET table

om·i·nous \ˈä-mə-nəs\ *adjective* (1587)
: being or exhibiting an omen **:** PORTENTOUS; *especially* **:** foreboding or foreshowing evil **:** INAUSPICIOUS ☆

— om·i·nous·ly *adverb*

— om·i·nous·ness *noun*

omis·si·ble \ō-ˈmi-sə-bəl\ *adjective* (1816)
: that may be omitted

omis·sion \ō-ˈmi-shən, ə-\ *noun* [Middle English *omissioun,* from Late Latin *omission-, omissio,* from Latin *omittere*] (14th century)
1 a : something neglected or left undone **b :** apathy toward or neglect of duty
2 : the act of omitting **:** the state of being omitted

omit \ō-ˈmit, ə-\ *transitive verb* **omit·ted; omit·ting** [Middle English *omitten,* from Latin *omittere,* from *ob-* toward + *mittere* to let go, send — more at OB-] (15th century)
1 : to leave out or leave unmentioned
2 : to fail to perform or make use of **:** FORBEAR
3 *obsolete* **:** DISREGARD
4 *obsolete* **:** GIVE UP

om·ma·tid·i·um \ˌä-mə-ˈti-dē-əm\ *noun, plural* **-tid·ia** \-dē-ə\ [New Latin, from Greek *ommat-, omma* eye; akin to Greek *ōps* eye — more at EYE] (1884)
: one of the elements corresponding to a small simple eye that make up the compound eye of an arthropod

— om·ma·tid·i·al \-dē-əl\ *adjective*

omni- *combining form* [Latin, from *omnis*]
: all **:** universally ⟨*omni*directional⟩

[1]om·ni·bus \ˈäm-ni-(ˌ)bəs\ *noun* [French, from Latin, for all, dative plural of *omnis*] (1829)
1 : a usually automotive public vehicle designed to carry a large number of passengers **:** BUS
2 : a book containing reprints of a number of works

[2]omnibus *adjective* (1842)
1 : of, relating to, or providing for many things at once
2 : containing or including many items

om·ni·com·pe·tent \ˌäm-ni-ˈkäm-pə-tənt\ *adjective* (1827)
: able to handle any situation; *especially* **:** having the authority or legal capacity to act in all matters

— om·ni·com·pe·tence \-tən(t)s\ *noun*

om·ni·di·rec·tion·al \ˌäm-ni-də-ˈrek-shnəl, -ˌnī-, -(ˌ)dī-, -shə-n^{ə}l\ *adjective* (1927)
: being in or involving all directions; *especially* **:** receiving or sending radio waves equally well in all directions ⟨*omnidirectional* antenna⟩

om·ni·far·i·ous \ˌäm-nə-ˈfar-ē-əs, -ˈfer-\ *adjective* [Late Latin *omnifarius,* from Latin *omni-* + *-farius* (as in *multifarius* diverse) — more at MULTIFARIOUS] (1653)
: of all varieties, forms, or kinds

om·nif·i·cent \äm-ˈni-fə-sənt\ *adjective* [Latin *omni-* + English *-ficent* (as in *magnificent*)] (1677)
: unlimited in creative power

om·nip·o·tence \äm-ˈni-pə-tən(t)s\ *noun* (15th century)
1 : the quality or state of being omnipotent
2 : an agency or force of unlimited power

[1]om·nip·o·tent \-tənt\ *adjective* [Middle English, from Middle French, from Latin *omnipotent-, omnipotens,* from *omni-* + *potent-, potens* potent] (14th century)
1 *often capitalized* **:** ALMIGHTY 1
2 : having virtually unlimited authority or influence
3 *obsolete* **:** ARRANT

— om·nip·o·tent·ly *adverb*

[2]omnipotent *noun* (1600)
1 : one who is omnipotent
2 *capitalized* **:** GOD 1

om·ni·pres·ence \ˌäm-ni-ˈpre-z^{ə}n(t)s\ *noun* (1601)
: the quality or state of being omnipresent **:** UBIQUITY

om·ni·pres·ent \-z^{ə}nt\ *adjective* (1609)
: present in all places at all times

om·ni·range \ˈäm-ni-ˌrānj\ *noun* (1946)
: a system of radio navigation in which any bearing relative to a special radio transmitter on the ground may be chosen and flown by an airplane pilot — called also *omnidirectional range*

om·ni·science \äm-ˈni-shən(t)s\ *noun* [Medieval Latin *omniscientia,* from Latin *omni-* + *scientia* knowledge — more at SCIENCE] (1612)
: the quality or state of being omniscient

om·ni·scient \-shənt\ *adjective* [New Latin *omniscient-, omnisciens,* back-formation from Medieval Latin *omniscientia*] (1604)
1 : having infinite awareness, understanding, and insight
2 : possessed of universal or complete knowledge

— om·ni·scient·ly *adverb*

om·ni·um–gath·er·um \ˌäm-nē-əm-ˈga-thə-rəm\ *noun, plural* **omnium–gatherums** [Latin *omnium* (genitive plural of *omnis*) + English *gather* + Latin *-um,* noun ending] (1530)
: a miscellaneous collection (as of things or persons)

om·ni·vore \ˈäm-ni-ˌvōr, -ˌvȯr\ *noun* [New Latin *omnivora,* neuter plural of *omnivorus,* from Latin] (1890)

☆ SYNONYMS

Ominous, portentous, fateful mean having a menacing or threatening aspect. OMINOUS implies having a menacing, alarming character foreshadowing evil or disaster ⟨*ominous* rumblings from a once-dormant volcano⟩. PORTENTOUS suggests being frighteningly big or impressive but now seldom definitely connotes forwarning of calamity ⟨an eerie and *portentous* stillness⟩. FATEFUL suggests being of momentous or decisive importance ⟨the *fateful* conference that led to war⟩.

◇ WORD HISTORY

ombudsman For a word that became current in late 20th century America, *ombudsman* has deep roots: it can be traced back to the Old Norse word *umbothsmathr.* Compounded from *umboth* "commission," and *mathr* "man," *umbothsmathr* had the meaning "commissary" or "manager." The word evolved into Modern Swedish *ombudsman,* acquiring the generic sense of "representative" or "commissioner." In 1809 the Swedish government established the office of *justitieombudsmannen,* or justice commissioner, who was empowered to investigate and redress the complaints of private citizens regarding abuse and maladministration on the part of public officials and bureaucrats. This functionary established a reputation as the one person to whom the victim of bureaucratic abuse could resort when other channels had failed. The concept of an ombudsman later spread to Finland, Denmark, and Norway. In 1962 New Zealand decided to appoint a comparably empowered official and to confer upon him the title of *ombudsman.* The establishment of the office of ombudsman in New Zealand inspired other parts of the English-speaking world to create their own officials with similar powers under this name.

omelette Although the word *omelette* bears little resemblance to the Latin word *lamina,* the shape of an omelette does resemble a thin plate, which is what *lamina,* the ultimate source of *omelette,* means. The Romans used the noun *lamella,* the diminutive form of *lamina,* to mean "thin metal plate." This diminutive became in Old French *lemelle* "blade of a knife." By a process somewhat similar to that which produced our *apron* from an earlier *napron, la lemelle* "the blade" was misinterpreted as *l'alemelle,* which accounts for the appearance of the initial vowel. In Middle French the word *alemelle* became *alumelle* and then was altered to *alumette,* by substitution of the suffix *-ette* for *-elle.* It also acquired the additional meaning "dish made with beaten eggs," since such a dish resembled a thin plate or blade. In turn, *alumette* was altered to *amelette* by transposition of the *l* and *m* sounds (a process called metathesis). It finally changed again (influenced by southern dialect words for "egg" descended from Old Provençal *ou*) to *omelette.* The word, now often written *omelet,* has been in English since the 17th century, when an omelette was described as "a pancake of egges."

: one that is omnivorous

om·niv·o·rous \äm-ˈniv-rəs, -ˈni-və-\ *adjective* [Latin *omnivorus,* from *omni-* + *-vorus* -vorous] (circa 1656)
1 : feeding on both animal and vegetable substances
2 : avidly taking in everything as if devouring or consuming
— **om·niv·o·rous·ly** *adverb*

om·pha·los \ˈäm(p)-fə-ˌläs, -ləs\ *noun* [Greek, navel — more at NAVEL] (1855)
: a central point **:** HUB 2 , FOCAL POINT

om·pha·lo·skep·sis \ˌäm(p)-fə-lō-ˈskep-səs\ *noun* [New Latin, from Greek *omphalos* + *skepsis* examination — more at SPY] (1925)
: contemplation of one's navel as an aid to meditation; *also* **:** INERTIA 2

[1]on \ˈȯn, ˈän\ *preposition* [Middle English *an, on,* preposition & adverb, from Old English; akin to Old High German *ana* on, Greek *ana* up, on] (before 12th century)
1 a — used as a function word to indicate position in contact with and supported by the top surface of ⟨the book is lying *on* the table⟩ **b** — used as a function word to indicate position in or in contact with an outer surface ⟨the fly landed *on* the ceiling⟩ ⟨I have a cut *on* my finger⟩ ⟨paint *on* the wall⟩ **c** — used as a function word to indicate position in close proximity with ⟨a village *on* the sea⟩ ⟨stay *on* your opponent⟩ **d** — used as a function word to indicate direction or location with respect to something ⟨*on* the south⟩ ⟨the garden is *on* the side of the house⟩
2 a — used as a function word to indicate a source of attachment or support ⟨*on* a string⟩ ⟨stand *on* one foot⟩ ⟨hang it *on* a nail⟩ **b** — used as a function word to indicate a source of dependence ⟨you can rely *on* me⟩ ⟨feeds *on* insects⟩ ⟨lives *on* a pension⟩ **c** — used as a function word to indicate means of conveyance ⟨*on* the bus⟩ or presence within the confines or in possession of ⟨had a knife *on* him⟩
3 — used as a function word to indicate a time frame during which something takes place ⟨a parade *on* Sunday⟩ or an instant, action, or occurrence when something begins or is done ⟨*on* cue⟩ ⟨*on* arriving home, I found your letter⟩ ⟨news *on* the hour⟩ ⟨cash *on* delivery⟩
4 *archaic* **:** OF
5 a — used as a function word to indicate manner of doing something; often used with *the* ⟨*on* the sly⟩ ⟨keep everything *on* the up-and-up⟩ **b** — used as a function word to indicate means or agency ⟨cut myself *on* a knife⟩ ⟨talk *on* the telephone⟩ **c** — used as a function word to indicate a medium of expression; used originally to refer to physical position ⟨best show *on* television⟩
6 a (1) — used as a function word to indicate active involvement in a condition or status ⟨*on* fire⟩ ⟨*on* the increase⟩ ⟨*on* the lookout⟩ (2) **:** regularly using or showing the effects of using ⟨*on* drugs⟩ **b** — used as a function word to indicate involvement with the activity, work, or function of ⟨*on* tour⟩ ⟨*on* the jury⟩ ⟨*on* duty⟩ **c** — used as a function word to indicate position or status in proper relationship with a standard or objective ⟨*on* schedule⟩
7 a — used as a function word to indicate reason, ground, or basis (as for an action, opinion, or computation) ⟨I have it *on* good authority⟩ ⟨*on* one condition⟩ ⟨the interest will be 10 cents *on* the dollar⟩ **b** — used as a function word to indicate the cause or source ⟨profited *on* the sale of stock⟩ **c** — used as a function word to indicate the focus of obligation or responsibility ⟨drinks are *on* the house⟩ ⟨put the blame *on* my actions⟩
8 a — used as a function word to indicate the object of collision, opposition, or hostile action ⟨bumped my head *on* a limb⟩ ⟨an attack *on* religion⟩ ⟨pulled a gun *on* me⟩ **b** — used as a function word to indicate the object with respect to some disadvantage, handicap, or detriment ⟨has three inches in height *on* me⟩ ⟨a 3-game lead *on* the second-place team⟩ ⟨the joke's *on* me⟩
9 a — used as a function word to indicate destination or the focus of some action, movement, or directed effort ⟨crept up *on* him⟩ ⟨feast your eyes *on* this⟩ ⟨working *on* my skiing⟩ ⟨made a payment *on* the loan⟩ **b** — used as a function word to indicate the focus of feelings, determination, or will ⟨have pity *on* me⟩ ⟨keen *on* sports⟩ ⟨a curse *on* you⟩ **c** — used as a function word to indicate the object with respect to some misfortune or disadvantageous event ⟨the crops died *on* them⟩ **d** — used as a function word to indicate the subject of study, discussion, or consideration ⟨a book *on* insects⟩ ⟨reflect *on* that a moment⟩ ⟨agree *on* price⟩
10 — used as a function word to indicate reduplication or succession in a series ⟨loss *on* loss⟩

[2]on *adverb* (before 12th century)
1 a : in or into a position of contact with an upper surface especially so as to be positioned for use or operation ⟨put the plates *on*⟩ **b :** in or into a position of being attached to or covering a surface; *especially* **:** in or into the condition of being worn ⟨put his new shoes *on*⟩
2 a : forward or at a more advanced point in space or time ⟨went *on* home⟩ ⟨later *on*⟩ **b :** in continuance or succession ⟨rambled *on*⟩ ⟨and so *on*⟩
3 : into operation or a position permitting operation ⟨switched the light *on*⟩

[3]on *adjective* (circa 1541)
1 : engaged in an activity or function (as a dramatic role)
2 a (1) **:** being in operation ⟨the radio is *on*⟩ (2) **:** placed so as to permit operation ⟨the switch is *on*⟩ **b :** taking place ⟨the game is *on*⟩
3 : aware of something — usually used with *to* ⟨my boss was *on* to me⟩
4 : INTENDED, PLANNED ⟨has nothing *on* for tonight⟩
5 *British* **:** talking or harping incessantly — used with *about*
6 *chiefly British* **:** regarded as possible or feasible — usually used in negative constructions
7 a : engaged in or as if in a performance ⟨the comedian was always *on*⟩ **b :** being at a high level of performance

[1]-on *noun suffix* [International Scientific Vocabulary, alteration of *-one*]
: chemical compound not a ketone or other oxo compound ⟨parathi*on*⟩

[2]-on *noun suffix* [from *-on* (in *ion*)]
1 : subatomic particle ⟨nucle*on*⟩
2 a : unit **:** quantum ⟨phot*on*⟩ ⟨magnet*on*⟩ **b :** basic hereditary component ⟨cistr*on*⟩ ⟨oper*on*⟩

[3]-on *noun suffix* [New Latin, from *-on* (in *argon*)]
: noble gas ⟨rad*on*⟩

on–again, off–again *adjective* (1948)
: existing briefly and then disappearing in an intermittent unpredictable way ⟨*on-again, off-again* fads⟩

on·a·ger \ˈä-ni-jər\ *noun* [Middle English, wild ass, from Latin, from Greek *onagros,* from *onos* ass + *agros* field — more at ACRE] (14th century)
1 : an Asian wild ass (*Equus hemionus onager* synonym *E. onager*) that usually has a broad dorsal stripe and is related to the kiang
2 [Late Latin, from Latin] **:** a heavy catapult used in ancient and medieval times

onager 2

on and off *adverb* (1855)
: OFF AND ON
— **on–and–off** *adjective*

onan·ism \ˈō-nə-ˌni-zəm\ *noun* [probably from New Latin *onanismus,* from *Onan,* son of Judah (Genesis 38:9)] (circa 1741)
1 : MASTURBATION
2 : COITUS INTERRUPTUS
3 : SELF-GRATIFICATION
— **onan·is·tic** \ˌō-nə-ˈnis-tik\ *adjective*

on·board \ˈȯn-ˈbōrd, ˈän-, -ˈbȯrd\ *adjective* (1960)
: carried within or occurring aboard a vehicle (as a satellite or spacecraft) ⟨an *onboard* computer⟩

[1]once \ˈwən(t)s\ *adverb* [Middle English *ones,* from genitive of *on* one] (12th century)
1 : one time and no more
2 : at any one time **:** under any circumstances **:** EVER
3 : at some indefinite time in the past **:** FORMERLY
4 : by one degree of relationship
— **once and for all 1 :** with finality **:** DEFINITIVELY **2 :** for the last time

[2]once *noun* (13th century)
: one single time **:** one time at least
— **at once 1 :** at the same time **:** SIMULTANEOUSLY **2 :** IMMEDIATELY **3 :** [2]BOTH

[3]once *adjective* (1691)
: that once was **:** FORMER

[4]once *conjunction* (1761)
: at the moment when **:** AS SOON AS

once–over \ˌwən(t)s-ˈō-vər, ˈwən(t)s-ˌ\ *noun* (1914)
: a swift examination or survey; *especially* **:** a swift comprehensive appraising glance

once that *conjunction* (1874)
: ONCE

on·cho·cer·ci·a·sis \ˌäŋ-kō-ˌsər-ˈkī-ə-səs\ *noun, plural* **-a·ses** \-ˌsēz\ [New Latin, from *Onchocerca,* genus of worms] (1911)
: infestation with or disease caused by filarial worms (genus *Onchocerca*); *especially* **:** a human disease caused by a worm (*O. volvulus*) that is native to Africa but now present in parts of tropical America and is transmitted by several blackflies

on·cid·i·um \än-ˈsi-dē-əm, äŋ-ˈki-\ *noun* [New Latin, from Greek *onkos* barbed hook — more at ANGLE] (circa 1868)
: any of a genus (*Oncidium*) of showy tropical American chiefly epiphytic orchids

onco- *combining form* [New Latin, from Greek *onkos* bulk, mass; akin to Greek *enenkein* to carry — more at ENOUGH]
: tumor ⟨*onco*logy⟩

on·co·gene \ˈäŋ-kō-ˌjēn\ *noun* (1969)
: a gene having the potential to cause a normal cell to become cancerous

on·co·gen·e·sis \ˌäŋ-kō-ˈje-nə-səs\ *noun* [New Latin] (circa 1932)
: the induction or formation of tumors

on·co·gen·ic \-ˈje-nik\ *adjective* (1936)
1 : relating to tumor formation
2 : tending to cause tumors

on·co·ge·nic·i·ty \-jə-ˈni-sə-tē\ *noun* (1944)
: the capacity to induce or form tumors

on·col·o·gy \än-ˈkä-lə-jē, äŋ-\ *noun* (circa 1857)
: the study of tumors
— **on·co·log·i·cal** \ˌäŋ-kə-ˈlä-ji-kəl\ *also* **on·co·log·ic** \-jik\ *adjective*
— **on·col·o·gist** \än-ˈkä-lə-jist, äŋ-\ *noun*

on·com·ing \ˈȯn-ˌkə-miŋ, ˈän-\ *adjective* (1844)
1 a : coming nearer in time or space ⟨the *oncoming* year⟩ ⟨an *oncoming* car⟩ **b :** FUTURE ⟨looked forward to his *oncoming* visit⟩
2 : EMERGENT, RISING ⟨the *oncoming* generation⟩

on·cor·na·vi·rus \ˌäŋ-ˌkȯr-nə-ˈvī-rəs\ *noun* [*onco-* + *RNA* + *virus*] (1970)
: any of a group of RNA-containing viruses that produce tumors

[1]**one** \ˈwən\ *adjective* [Middle English *on, an,* from Old English *ān;* akin to Old High German *ein* one, Latin *unus* (Old Latin *oinos*), Sanskrit *eka*] (before 12th century)
1 : being a single unit or thing ⟨*one* day at a time⟩
2 a : being one in particular ⟨early *one* morning⟩ **b :** being preeminently what is indicated ⟨*one* fine person⟩
3 a : being the same in kind or quality ⟨both of *one* species⟩ **b** (1) **:** constituting a unified entity of two or more components ⟨the combined elements form *one* substance⟩ (2) **:** being in agreement or union ⟨am *one* with you on this⟩
4 a : SOME 1 ⟨will see you again *one* day⟩ **b :** being a certain individual specified by name ⟨*one* John Doe made a speech⟩
5 : ONLY 2 ⟨the *one* person she wanted to marry⟩

[2]**one** *noun* (before 12th century)
1 — see NUMBER table
2 : the number denoting unity
3 a : the first in a set or series — often used with an attributive noun ⟨day *one*⟩ **b :** an article of clothing of a size designated *one* ⟨wears a *one*⟩
4 : a single person or thing ⟨has the *one* but needs the other⟩
5 : a one-dollar bill
— at one : at harmony **:** in a state of agreement
— for one : as one example ⟨I *for one* disagree⟩

[3]**one** *pronoun* (13th century)
1 : a certain indefinitely indicated person or thing ⟨saw *one* of his friends⟩
2 a : an individual of a vaguely indicated group **:** anyone at all ⟨*one* never knows⟩ **b** — used as a third person substitute for a first person pronoun ⟨I'd like to read more but *one* doesn't have the time⟩
3 : a single instance of a specified action ⟨felt like belting him *one* —John Casey⟩ ☐

-one *noun suffix* [International Scientific Vocabulary, alteration of *-ene*]
: ketone or related or analogous compound or class of compounds ⟨lact*one*⟩ ⟨quin*one*⟩

one another *pronoun* (1526)
: EACH OTHER
usage see EACH OTHER

one–armed bandit \ˈwən-ˈärm(d)-\ *also* **one–arm bandit** *noun* (1934)
: SLOT MACHINE 2

one–bag·ger \-ˈba-gər\ *noun* (1952)
: SINGLE 2

one–di·men·sion·al *adjective* (1883)
1 : having one dimension
2 : lacking depth **:** SUPERFICIAL ⟨*one-dimensional* characters⟩
— one–di·men·sion·al·i·ty *noun*

one·fold \ˈwən-ˌfōld, -ˈfōld\ *adjective* (before 12th century)
: constituting a single undivided whole

one–hand·ed \-ˈhan-dəd\ *adjective* (15th century)
1 : having or using only one hand ⟨could beat him up *one-handed*⟩
2 a : designed for or requiring the use of only one hand **b :** effected by the use of only one hand

one–horse \-ˈhȯrs\ *adjective* (1750)
1 : drawn or operated by one horse
2 : SMALL, SMALL-TIME ⟨a *one-horse* town⟩

Onei·da \ō-ˈnī-də\ *noun, plural* **Oneida** *or* **Oneidas** [Oneida *onęˀyóteˀ*, literally, standing rock] (1666)
1 : a member of an American Indian people originally of New York
2 : the Iroquoian language of the Oneida people

onei·ric \ō-ˈnī-rik\ *adjective* [Greek *oneiros* dream; akin to Armenian *anurǰ* dream] (1859)
: of or relating to dreams **:** DREAMY
— onei·ri·cal·ly \-ri-k(ə-)lē\ *adverb*

onei·ro·man·cy \ō-ˈnī-rə-ˌman(t)-sē\ *noun* [Greek *oneiros* + English *-mancy*] (1652)
: divination by means of dreams

one–line octave *noun* (1931)
: the musical octave that begins on middle C — see PITCH illustration

one–lin·er \ˌwən-ˈlī-nər\ *noun* (1967)
: a very succinct joke or witticism

one–man *adjective* (1842)
: of or relating to just one individual: as **a :** consisting of only one individual ⟨a *one-man* committee⟩ **b** (1) **:** done, presented, or produced by only one individual ⟨a *one-man* stage play⟩ (2) **:** featuring the work of a single artist (as a painter) ⟨a *one-man* show of oils⟩ **c :** designed for or limited to one individual

one·ness \ˈwən-nəs\ *noun* (1594)
: the quality or state or fact of being one: as **a :** SINGLENESS **b :** INTEGRITY, WHOLENESS **c :** HARMONY **d :** SAMENESS, IDENTITY **e :** UNITY, UNION

one–night·er \ˌwən-ˈnī-tər\ *noun* (circa 1937)
: ONE-NIGHT STAND

one–night stand *noun* (1880)
1 : a performance (as of a play or concert) given (as by a traveling group of actors or musicians) only once in each of a series of localities
2 a : a locality used for one-night stands **b :** a stopover for a one-night stand
3 : a sexual encounter limited to a single occasion; *also* **:** a partner in such an encounter

one–note \ˈwən-ˈnōt\ *adjective* (1973)
: unvarying in tone or emphasis **:** MONOTONOUS

one–off \ˌwən-ˈȯf\ *adjective* (1934)
British **:** limited to a single time, occasion, or instance **:** ONE-SHOT, UNIQUE
— one–off *noun*

one–on–one \ˌwən-ȯn-ˈwən, ˌwən-än-\ *adjective or adverb* (1967)
1 : playing directly against a single opposing player
2 : involving a direct encounter between one person and another

one–piece *adjective* (1880)
: consisting of or made in a single undivided piece ⟨a *one-piece* bathing suit⟩
— one–piec·er \ˈwən-ˈpē-sər\ *noun*

oner·ous \ˈä-nə-rəs, ˈō-\ *adjective* [Middle English, from Middle French *onereus,* from Latin *onerosus,* from *oner-, onus* burden; akin to Sanskrit *anas* cart] (14th century)
1 : involving, imposing, or constituting a burden **:** TROUBLESOME ⟨an *onerous* task⟩
2 : having legal obligations that outweigh the advantages ⟨*onerous* contract⟩ ☆
— oner·ous·ly *adverb*
— oner·ous·ness *noun*

one·self \(ˌ)wən-ˈself, *Southern also* -ˈsef\ *also* **one's self** \(ˌ)wən-, ˌwənz-\ *pronoun* (1548)
1 : a person's self **:** one's own self — used reflexively as object of a preposition or verb or for emphasis in various constructions
2 : one's normal, healthy, or sane condition or self
— be oneself : to conduct oneself in a usual or fitting manner

one–shot \ˈwən-ˌshät\ *adjective* (1927)
1 : that is complete or effective through being done or used or applied only once ⟨there is no easy *one-shot* answer to the problem⟩
2 : that is not followed by something else of the same kind ⟨a *one-shot* tax cut⟩
— one–shot *noun*

one–sid·ed \ˈwən-ˈsī-dəd\ *adjective* (1813)
1 a (1) **:** having one side prominent **:** LOPSIDED (2) **:** having or occurring on one side only **b :** limited to one side **:** PARTIAL ⟨a *one-sided* interpretation⟩
2 : UNILATERAL ⟨a *one-sided* decision⟩
— one–sid·ed·ly *adverb*
— one–sid·ed·ness *noun*

ones place *noun* (1976)
: UNITS PLACE

one–step \ˈwən-ˌstep\ *noun* (1911)
1 : a ballroom dance in 2/4 time marked by quick walking steps backward and forward
2 : music used for the one-step
— one–step *intransitive verb*

one–stop \-ˈstäp\ *adjective* (1934)
: being or relating to a business that provides a complete range of goods or services of a particular kind

one–tailed \ˈwən-ˈtāl(d)\ *also* **one–tail** \-ˈtāl\ *adjective* (1947)
: being a statistical test for which the critical region consists of all values of the test statistic greater than a given value or less than a given value but not both — compare TWO-TAILED

[1]**one·time** \ˈwən-ˈtīm\ *adjective* (1840)
1 : FORMER, SOMETIME ⟨a *onetime* actor⟩
2 : occurring only once **:** ONE-SHOT

[2]**onetime** *adverb* (1886)
: FORMERLY

one–to–one \ˌwən-tə-ˈwən, -də-\ *adjective* (1873)
1 : pairing each element of a set uniquely with an element of another set
2 : ONE-ON-ONE 2

one–track *adjective* (1926)
: marked by often narrowly restricted attention to or absorption in just one thing ⟨a *one-track* mind⟩

one–two \ˈwən-ˈtü, -ˌtü\ *noun* (1809)
1 : a combination of two quick blows in rapid succession in boxing; *especially* **:** a left jab followed at once by a hard blow with the right hand
2 *or* **one–two punch :** a combination of two forces acting against something

one–up \ˌwən-ˈəp, ˈwən-\ *transitive verb* [back-formation from *one-upmanship*] (1963)
: to practice one-upmanship on

one up *adjective* (1919)

☆ SYNONYMS
Onerous, burdensome, oppressive, exacting mean imposing hardship. ONEROUS stresses being laborious and heavy especially because distasteful ⟨the *onerous* task of cleaning up the mess⟩. BURDENSOME suggests causing mental as well as physical strain ⟨*burdensome* responsibilities⟩. OPPRESSIVE implies extreme harshness or severity in what is imposed ⟨the *oppressive* tyranny of a police state⟩. EXACTING implies rigor or sternness rather than tyranny or injustice in the demands made or in the one demanding ⟨an *exacting* employer⟩.

☐ USAGE
one Sense 2a is usually a sign of a formal style. A formal style excludes the participation of the reader or hearer; thus *one* is used where a less formal style might address the reader directly ⟨for the consequences of such choices, *one* has only oneself to thank —Walker Gibson⟩. This generic *one* has never been common in informal use in either British or American English, and people who start sentences with *one* often shift to another pronoun more natural to casual discourse ⟨when *one* is learning the river, he is not allowed to do or think about anything else —Mark Twain⟩. Use of *one* to replace a first-person pronoun—sense 2b—has occasionally been criticized. It is more common in British English than in American ⟨I'm watching this pretty carefully and I hope that the issue will come up in the Lords and *one* may be able to speak about it —Donald Coggan, Archbishop of Canterbury⟩. See in addition YOU

: being in a position of advantage — usually used with *on*

one–up·man·ship \ˌwən-ˈəp-mən-ˌship\ *also* **one–ups·man·ship** \-ˈəps-mən-\ *noun* (1952)
: the art or practice of outdoing or keeping one jump ahead of a friend or competitor ⟨engaged in a round of verbal *one-upmanship*⟩
word history see GAMESMANSHIP

one–way *adjective* (1824)
1 : that moves in or allows movement in only one direction ⟨*one-way* street⟩
2 : ONE-SIDED, UNILATERAL ⟨a *one-way* conversation⟩
3 : that functions in only one of two or more ways

on·go·ing \ˈȯn-ˌgō-iŋ, ˈän-, -ˌgȯ(-)iŋ\ *adjective* (1877)
1 a : being actually in process **b :** CONTINUING
2 : continuously moving forward **:** GROWING
— on·go·ing·ness \-nəs\ *noun*

on·ion \ˈən-yən\ *noun* [Middle English, from Middle French *oignon,* from Latin *union-, unio*] (14th century)
1 : a widely cultivated Asian herb (*Allium cepa*) of the lily family with pungent edible bulbs; *also* **:** its bulb
2 : any of various plants of the same genus as the onion
— on·iony \-yə-nē\ *adjective*

onion dome *noun* (1941)
: a dome (as of a church) having the general shape of an onion
— onion–domed *adjective*

onion ring *noun* (1946)
: a ring of sliced onion coated with batter or crumbs and fried

on·ion·skin \-ˌskin\ *noun* (1879)
: a thin strong translucent paper of very light weight

oni·um \ˈō-nē-əm\ *adjective* [*-onium*] (1905)
: being or characterized by a usually complex cation

-onium *noun suffix* [New Latin, from *ammonium*]
: an ion having a positive charge ⟨hydr*onium*⟩ — compare -IUM 1b

on–line *adjective or adverb* (1950)
: connected to, served by, or available through a system and especially a computer or telecommunications system ⟨an *on-line* database⟩; *also* **:** done while connected to a system ⟨*on-line* computer storage⟩

on·look·er \ˈȯn-ˌlu̇-kər, ˈän-\ *noun* (1606)
: one that looks on; *especially* **:** a passive spectator
— on·look·ing \-kiŋ\ *adjective*

¹on·ly \ˈōn-lē\ *adjective* [Middle English, from Old English *ānlīc,* from *ān* one — more at ONE] (before 12th century)
1 : unquestionably the best **:** PEERLESS
2 : alone in its class or kind **:** SOLE ⟨an *only* child⟩

²only *adverb* (14th century)
1 a : as a single fact or instance and nothing more or different **:** MERELY ⟨has *only* lost one election —George Orwell⟩ **b :** SOLELY, EXCLUSIVELY ⟨known *only* to him⟩
2 : at the very least ⟨it was *only* too true⟩
3 a : in the final outcome ⟨will *only* make you sick⟩ **b :** with nevertheless the final result ⟨won the battles, *only* to lose the war⟩
4 a : as recently as ⟨*only* last week⟩ **b :** in the immediate past ⟨*only* just talked to her⟩ ■

³only *conjunction* (14th century)
1 a : with the restriction that **:** BUT ⟨you may go, *only* come back early⟩ **b :** and yet **:** HOWEVER ⟨they look very nice, *only* we can't use them⟩
2 : were it not that **:** EXCEPT ⟨I'd introduce you to her, *only* you'd win her —Jack London⟩

on·o·mas·tic \ˌä-nə-ˈmas-tik\ *adjective* [Greek *onomastikos,* from *onomazein* to name, from *onoma* name — more at NAME] (1716)
: of, relating to, or consisting of a name or names
— on·o·mas·ti·cal·ly \-ti-k(ə-)lē\ *adverb*

on·o·mas·tics \-tiks\ *noun plural but singular or plural in construction* (1930)
1 a : the science or study of the origins and forms of words especially as used in a specialized field **b :** the science or study of the origin and forms of proper names of persons or places
2 : the system underlying the formation and use of words especially for proper names or of words used in a specialized field
— on·o·mas·ti·cian \ˌä-nə-mas-ˈti-shən\ *noun*

on·o·ma·tol·o·gy \ˌä-nə-mə-ˈtä-lə-jē\ *noun* [French *onomatologie,* from Greek *onomat-, onoma* name + French *-logie* -logy] (circa 1847)
: ONOMASTICS
— on·o·ma·tol·o·gist \-jist\ *noun*

on·o·mato·poe·ia \ˌä-nə-ˌmä-tə-ˈpē-ə, -ˌma-\ *noun* [Late Latin, from Greek *onomatopoiia,* from *onomat-, onoma* name + *poiein* to make — more at POET] (circa 1577)
1 : the naming of a thing or action by a vocal imitation of the sound associated with it (as *buzz, hiss*)
2 : the use of words whose sound suggests the sense
— on·o·mato·poe·ic \-ˈpē-ik\ *or* **on·o·mato·po·et·ic** \-pō-ˈe-tik\ *adjective*
— on·o·mato·poe·i·cal·ly \-ˈpē-ə-k(ə-)lē\ *or* **on·o·mato·po·et·i·cal·ly** \-pō-ˈe-ti-k(ə-)lē\ *adverb*

On·on·da·ga \ˌä-nə(n)-ˈdȯ-gə, -ˈdā-, -ˈdä-\ *noun, plural* **-ga** *or* **-gas** [Onondaga *onǫ́ˑtàʔke,* the chief Onondaga town] (1684)
1 : a member of an American Indian people of New York and Canada
2 : the Iroquoian language of the Onondaga people

on–ramp \ˈȯn-ˌramp, ˈän-\ *noun* (1958)
: a ramp by which one enters a limited-access highway

on·rush \-ˌrəsh\ *noun* (1844)
1 : a rushing forward or onward
2 : ONSET
— on·rush·ing \-ˌrə-shiŋ\ *adjective*

on–screen \ˈȯn-ˈskrēn, ˈän-\ *adverb or adjective* (1955)
: in a motion picture or a television program

on·set \-ˌset\ *noun* (1535)
1 : ATTACK, ASSAULT ⟨withstand the *onset* of the army⟩
2 : BEGINNING, COMMENCEMENT ⟨the *onset* of winter⟩

on·shore \ˈȯn-ˌshōr, ˈän-, -ˌshȯr\ *adjective* (1875)
1 : coming or moving from the water toward or onto the shore ⟨an *onshore* wind⟩
2 a : situated on or near the shore as distinguished from being in deep or open water **b :** situated on land
3 : DOMESTIC 2 ⟨*onshore* markets⟩
— on·shore \ˈȯn-ˈ, ˈän-ˈ\ *adverb*

on·side \-ˈsīd\ *adverb or adjective* (1871)
: not offside **:** in a position legally to play or receive the ball or puck

onside kick *noun* (1976)
: a kickoff in football in which the ball travels just far enough to be legally recoverable by the kicking team

on–site \-ˈsīt\ *adjective or adverb* (1946)
: carried out or located at the place connected with a particular activity ⟨*on-site* training in construction skills⟩

on·slaught \ˈän-ˌslȯt, ˈȯn-\ *noun* [modification of Dutch *aanslag* act of striking; akin to Old English *an* on and to Old English *slēan* to strike — more at SLAY] (circa 1625)
: an especially fierce attack; *also* **:** something resembling such an attack ⟨an *onslaught* of technological changes⟩

on·stage \ˈȯn-ˈstāj, ˈän-, -ˌstāj\ *adverb or adjective* (1925)
: on a part of the stage visible to the audience

on·stream \ˈȯn-ˈstrēm, ˈän-\ *adverb or adjective* (1930)
: in or into production ⟨plants scheduled to come *onstream*⟩

ont- *or* **onto-** *combining form* [New Latin, from Late Greek, from Greek *ont-, ōn,* present participle of *einai* to be — more at IS]
1 : being **:** existence ⟨*onto*logy⟩
2 : organism ⟨*onto*geny⟩

-ont *noun combining form* [Greek *ont-, ōn,* present participle]
: cell **:** organism ⟨dipl*ont*⟩

on–the–job *adjective* (1946)
: of, relating to, or being something (as training or experience) learned, gained, or done while working at a job and often under supervision

on·tic \ˈän-tik\ *adjective* (1942)
: of, relating to, or having real being
— on·ti·cal·ly \-ti-k(ə-)lē\ *adverb*

¹on·to \ˈȯn-(ˌ)tü, ˈän-\ *preposition* (1581)
1 : to a position on
2 : in or into a state of awareness about ⟨put me *onto* your methods⟩
3 — used as a function word to indicate a set each element of which is the image of at least one element of another set ⟨a function mapping the set *S onto* the set *T*⟩

²onto *adjective* (1942)
: mapping elements in such a way that every element in one set is the image of at least one element in another set ⟨a function that is one-to-one and *onto*⟩

on·to·gen·e·sis \ˌän-tə-ˈje-nə-səs\ *noun* [New Latin] (1875)
: ONTOGENY

on·to·ge·net·ic \-jə-ˈne-tik\ *adjective* [International Scientific Vocabulary] (1878)
1 : of, relating to, or appearing in the course of ontogeny
2 : based on visible morphological characters
— on·to·ge·net·i·cal·ly \-ti-k(ə-)lē\ *adverb*

on·tog·e·ny \än-ˈtä-jə-nē\ *noun* [International Scientific Vocabulary] (1872)
: the development or course of development especially of an individual organism

on·to·log·i·cal \ˌän-tᵊl-ˈä-ji-kəl\ *adjective* (1782)
1 : of or relating to ontology
2 : relating to or based upon being or existence
— on·to·log·i·cal·ly \-k(ə-)lē\ *adverb*

ontological argument *noun* (1877)
: an argument for the existence of God based upon the meaning of the term *God*

on·tol·o·gy \än-ˈtä-lə-jē\ *noun* [New Latin *ontologia,* from *ont-* + *-logia* -logy] (circa 1721)
1 : a branch of metaphysics concerned with the nature and relations of being
2 : a particular theory about the nature of being or the kinds of existents

□ USAGE
only The placement of *only* in a sentence has been a source of studious commentary since the 18th century, most of it intended to prove by force of argument that prevailing standard usage is wrong. After 200 years of preachment the following observations may be made: the position of *only* in standard spoken English is not fixed, since ambiguity is avoided through sentence stress; in casual prose that keeps close to the rhythms of speech *only* is often placed where it would be in speech; and in edited and more formal prose *only* tends to be placed immediately before the word or words it modifies.

— **on•tol•o•gist** \-jist\ *noun*
onus \'ō-nəs\ *noun* (circa 1640)
1 [Latin — more at ONEROUS] **a :** BURDEN **b :** a disagreeable necessity **:** OBLIGATION **c :** BLAME **d :** STIGMA
2 [New Latin *onus* (*probandi*), literally, burden of proving] **:** BURDEN OF PROOF
[1]**on•ward** \'ȯn-wərd, 'än-\ *also* **on•wards** \-wərdz\ *adverb* (1532)
: toward or at a point lying ahead in space or time **:** FORWARD
[2]**onward** *adjective* (1674)
: directed or moving onward **:** FORWARD
on•y•choph•o•ran \ˌä-ni-'kä-fə-rən\ *noun* [New Latin *Onychophora*, group name, from Greek *onych-*, *onyx* claw + *-phoros* -phore] (circa 1890)
: PERIPATUS
— **onychophoran** *adjective*
-onym *noun combining form* [Middle English, from Latin *-onymum*, from Greek *-ōnymon*, from *onyma* — more at NAME]
: name **:** word ⟨ant*onym*⟩
on•yx \'ä-niks\ *noun* [Middle English *onix*, from Middle French & Latin; Middle French, from Latin *onych-*, *onyx*, from Greek, literally, claw, nail — more at NAIL] (14th century)
: a translucent chalcedony in parallel layers of different colors
oo- — see O-
oo•cyst \'ō-ə-ˌsist\ *noun* [International Scientific Vocabulary] (1875)
: ZYGOTE; *specifically* **:** a sporozoan zygote undergoing sporogenous development
oo•cyte \'ō-ə-ˌsīt\ *noun* [International Scientific Vocabulary] (1895)
: an egg before maturation **:** a female gametocyte
oo•dles \'ü-dᵊlz\ *noun plural but singular or plural in construction* [origin unknown] (1869)
: a great quantity **:** LOT
oog•a•mous \ō-'ä-gə-məs\ *adjective* (1888)
: having or involving a small motile male gamete and a large immobile female gamete
— **oog•a•my** \-mē\ *noun*
oo•gen•e•sis \ˌō-ə-'je-nə-səs\ *noun* [New Latin] (circa 1879)
: formation and maturation of the egg
— **oo•ge•net•ic** \-jə-'ne-tik\ *adjective*
oo•go•ni•um \ˌō-ə-'gō-nē-əm\ *noun, plural* **-nia** \-nē-ə\ [New Latin] (1867)
1 : a female sexual organ in various algae and fungi that corresponds to the archegonium of ferns and mosses
2 : a descendant of a primordial germ cell that gives rise to oocytes
— **oo•go•ni•al** \-nē-əl\ *adjective*
[1]**ooh** \'ü\ *interjection* (1939)
— used to express amazement, joy, or surprise
[2]**ooh** *intransitive verb* (1951)
: to exclaim in amazement, joy, or surprise ⟨*oohing* and aahing over the new automobiles⟩
— **ooh** *noun*
oo•lite \'ō-ə-ˌlīt\ *noun* [probably from French *oolithe*, from *oo-* o- + *-lithe* -lite] (1785)
: a rock consisting of small round grains usually of calcium carbonate cemented together
— **oo•lit•ic** \ˌō-ə-'li-tik\ *adjective*
ool•o•gist \ō-'ä-lə-jist\ *noun* (1863)
1 : a person specializing in the study of birds' eggs
2 : a collector of birds' eggs
— **ool•o•gy** \-jē\ *noun*
oo•long \'ü-ˌlȯŋ\ *noun* [Chinese (Beijing) *wūlóng*, literally, black dragon] (1850)
: tea made from leaves that have been partially fermented before firing
oom•pah \ˌüm-(ˌ)pä, 'üm-\ *also* **oom•pah-pah** \ˌüm-(ˌ)pä-'pä, ˌu̇m-\ *noun* [imitative] (1877)
: a repeated rhythmic bass accompaniment especially in a band
— **oompah** *verb*

oomph \'u̇m(p)f\ *noun* [imitative of a sound made under exertion] (1936)
1 : personal charm or magnetism **:** GLAMOUR
2 : SEX APPEAL
3 : PUNCH, VITALITY
oo•pho•rec•to•my \ˌō-ə-fə-'rek-tə-mē\ *noun, plural* **-mies** [New Latin *oophoron* ovary (from *o-* + Greek *-phoron*, neuter of *-phoros* -phore) + English *-ectomy*] (1872)
: OVARIECTOMY
oops \'(w)u̇(ə)ps\ *interjection* (1933)
— used typically to express mild apology, surprise, or dismay
Oort cloud \'ōrt-, 'ȯrt-\ *noun* [Jan *Oort* (died 1992) Dutch astronomer] (1968)
: a spherical shell of small frozen bodies believed to surround the sun far beyond the orbit of Pluto and from which some are dislodged when perturbed (as by a passing star) to fall toward the sun as comets
oo•spore \'ō-ə-ˌspōr, -ˌspȯr\ *noun* [International Scientific Vocabulary] (1865)
: ZYGOTE; *especially* **:** a spore produced by heterogamous fertilization that yields a sporophyte
oo•the•ca \ˌō-ə-'thē-kə\ *noun, plural* **oo•the•cae** \-'thē-(ˌ)kē, -(ˌ)sē\ [New Latin] (circa 1856)
: a firm-walled and distinctive egg case (as of a cockroach)
— **oo•the•cal** \-'thē-kəl\ *adjective*
oo•tid \'ō-ə-ˌtid\ *noun* [irregular from *o-* + *-id*] (1904)
: an egg cell after meiosis
[1]**ooze** \'üz\ *noun* [Middle English *wose*, from Old English *wāse* mire; akin to Old Norse *veisa* stagnant water] (before 12th century)
1 : a soft deposit (as of mud, slime, or shells) on the bottom of a body of water
2 : a piece of soft wet plastic ground
[2]**ooze** *verb* **oozed; ooz•ing** [Middle English *wosen*, from *wose* sap] (14th century)
intransitive verb
1 : to pass or flow slowly through or as if through small openings or interstices
2 : to move slowly or imperceptibly ⟨the crowd began to *ooze* forward —Bruce Marshall⟩
3 a : to exude moisture **b :** to exude something often in a faintly repellent manner ⟨*ooze* with sympathy⟩
transitive verb
1 : to emit slowly
2 : EXUDE 2 ⟨*ooze* confidence⟩
[3]**ooze** *noun* [Middle English *wose* sap, juice, from Old English *wōs*; akin to Old High German *waso* damp] (circa 1581)
1 : a decoction of vegetable material used for tanning leather
2 : the act of oozing
3 : something that oozes
oozy \'ü-zē\ *adjective* **ooz•i•er; -est** (14th century)
1 : containing or composed of ooze **:** resembling ooze
2 : exuding moisture **:** SLIMY
op \'äp\ *noun* (1964)
: OPTICAL ART
opac•i•ty \ō-'pa-sə-tē\ *noun, plural* **-ties** [French *opacité* shadiness, from Latin *opacitat-*, *opacitas*, from *opacus* shaded, dark] (circa 1611)
1 : the quality or state of a body that makes it impervious to the rays of light; *broadly* **:** the relative capacity of matter to obstruct the transmission of radiant energy
2 a : obscurity of sense **:** UNINTELLIGIBLENESS **b :** the quality or state of being mentally obtuse **:** DULLNESS
3 : an opaque spot in a normally transparent structure (as the lens of the eye)
opah \'ō-pə, -ˌpä\ *noun* [perhaps from Ibo *úbà*] (1750)
: a large elliptical marine bony fish (*Lampris guttatus* of the family Lampridae) with brilliant colors

opal \'ō-pəl\ *noun* [Latin *opalus*, from Greek *opallios*, from Sanskrit *upala* stone, jewel] (1591)
: a mineral that is a hydrated amorphous silica softer and less dense than quartz and typically with definite and often marked iridescent play of colors
opal•es•cent \ˌō-pə-'le-sᵊnt\ *adjective* (circa 1813)
: reflecting an iridescent light
— **opal•es•cence** \-sᵊn(t)s\ *noun*
— **opal•es•cent•ly** \-sᵊnt-lē\ *adverb*
opal•ine \'ō-pə-ˌlīn, -ˌlēn\ *adjective* (1784)
: resembling opal
opaque \ō-'pāk\ *adjective* [Latin *opacus*] (1641)
1 : exhibiting opacity **:** blocking the passage of radiant energy and especially light
2 a : hard to understand or explain ⟨*opaque* prose⟩ **b :** OBTUSE, THICKHEADED
— **opaque** *noun*
— **opaque•ly** *adverb*
— **opaque•ness** *noun*
opaque projector *noun* (1951)
: a projector using reflected light for projecting an image of an opaque object or matter on an opaque support (as a photograph)
op art \'äp-\ *noun* (1964)
: OPTICAL ART
— **op artist** *noun*
ope \'ōp\ *verb* **oped; op•ing** (15th century)
archaic **:** OPEN
op–ed \'äp-'ed\ *noun, often O&E capitalized, often attributive* [short for *opposite editorial*] (1970)
: a page of special features usually opposite the editorial page of a newspaper
[1]**open** \'ō-pən, -pᵊm\ *adjective* **open•er** \'ōp-nər, 'ō-pə-\; **open•est** \'ōp-nəst, 'ō-pə-\ [Middle English, from Old English; akin to Old High German *offan* open, Old English *ūp* up] (before 12th century)
1 : having no enclosing or confining barrier **:** accessible on all or nearly all sides ⟨cattle grazing on an *open* range⟩
2 a (1) **:** being in a position or adjustment to permit passage **:** not shut or locked ⟨an *open* door⟩ (2) **:** having a barrier (as a door) so adjusted as to allow passage ⟨the house was *open*⟩ **b :** having the lips parted ⟨stood there with his mouth wide *open*⟩ **c :** not buttoned or zipped ⟨an *open* shirt⟩
3 a : completely free from concealment **:** exposed to general view or knowledge ⟨their hostilities eventually erupted with *open* war⟩ **b :** exposed or vulnerable to attack or question **:** SUBJECT ⟨*open* to doubt⟩
4 a : not covered with a top, roof, or lid ⟨an *open* car⟩ ⟨her eyes were *open*⟩ **b :** having no protective covering ⟨*open* wiring⟩ ⟨an *open* wound⟩
5 : not restricted to a particular group or category of participants ⟨*open* to the public⟩ ⟨*open* housing⟩: as **a :** enterable by both amateur and professional contestants ⟨an *open* tournament⟩ **b :** enterable by a registered voter regardless of political affiliation ⟨an *open* primary⟩
6 : fit to be traveled over **:** presenting no obstacle to passage or view ⟨the *open* road⟩ ⟨*open* country⟩
7 : having the parts or surfaces laid out in an expanded position **:** spread out **:** UNFOLDED ⟨an *open* book⟩
8 a (1) **:** LOW 13 (2) **:** formed with the tongue in a lower position ⟨Italian has an *open* and a close *e*⟩ **b** (1) **:** having clarity and resonance unimpaired by undue tension or constriction of the throat ⟨an *open* vocal tone⟩ (2) *of a tone* **:** produced by an open string or on a wind instrument by the lip without the use of slides, valves, or keys
9 a : available to follow or make use of ⟨the only course *open* to us⟩ **b :** not taken up with duties or engagements ⟨keep an hour *open* on Friday⟩ **c :** not finally decided **:** subject to further consideration ⟨the salary is *open*⟩ ⟨an

open question⟩ **d :** available for a qualified applicant **:** VACANT ⟨the job is still *open*⟩ **e :** remaining available for use or filling until canceled ⟨an *open* order for more items⟩ **f :** available for future purchase ⟨these items are in *open* stock⟩
10 a : characterized by ready accessibility and usually generous attitude: as (1) **:** generous in giving (2) **:** willing to hear and consider or to accept and deal with **:** RESPONSIVE (3) **:** free from reserve or pretense **:** FRANK **b :** accessible to the influx of new factors (as foreign goods) ⟨an *open* market⟩
11 a : having openings, interruptions, or spaces: as (1) **:** being porous and friable ⟨*open* soil⟩ (2) **:** sparsely distributed **:** SCATTERED ⟨*open* population⟩ (3) *of a compound* **:** having components separated by a space in writing or printing (as *opaque projector*) **b :** not made up of a continuous closed circuit of channels ⟨the insect circulatory system is *open*⟩
12 a *of an organ pipe* **:** not stopped at the top **b** *of a string on a musical instrument* **:** not stopped by the finger
13 : being in operation ⟨an *open* microphone⟩; *especially* **:** ready for business, patronage, or use ⟨the store is *open* from 9 to 5⟩ ⟨the new highway will be *open* next week⟩
14 a (1) **:** characterized by lack of effective regulation of various commercial enterprises ⟨an *open* town⟩ (2) **:** not repressed by legal controls ⟨*open* gambling⟩ **b :** free from checking or hampering restraints ⟨an *open* economy⟩ **c :** relatively unguarded by opponents ⟨passed to an *open* teammate⟩
15 : having been opened by a first ante, bet, or bid ⟨the bidding is *open*⟩
16 *of punctuation* **:** characterized by sparing use especially of the comma
17 a : containing none of its endpoints ⟨an *open* interval⟩ **b :** being a set or composed of sets each point of which has a neighborhood all of whose points are contained in the set ⟨the interior of a sphere is an *open* set⟩
18 a : being an incomplete electrical circuit **b :** not allowing the flow of electricity ⟨an *open* switch⟩
synonym see FRANK, LIABLE
— **open** *adverb*
— **open·ly** \'ō-pən-lē\ *adverb*
— **open·ness** \-pə(n)-nəs\ *noun*

²open *verb* **opened** \'ō-pənd, 'ō-pᵊmd\; **open·ing** \'ōp-niŋ, 'ō-pə-\ (before 12th century)
transitive verb
1 a : to move (as a door) from a closed position **b :** to make available for entry or passage by turning back (as a barrier) or removing (as a cover or an obstruction)
2 a : to make available for or active in a regular function ⟨*open* a new store⟩ **b :** to make accessible for a particular purpose ⟨*opened* new land for settlement⟩ ⟨*open* the way for changes⟩ **c :** to initiate access to (a computer file) prior to use
3 a : to disclose or expose to view **:** REVEAL **b :** to make more discerning or responsive **:** ENLIGHTEN ⟨must *open* our minds to the problems⟩ **c :** to bring into view or come in sight of by changing position
4 a : to make an opening in ⟨*opened* the boil⟩ **b :** to loosen and make less compact ⟨*open* the soil⟩
5 : to spread out **:** UNFOLD ⟨*opened* the book⟩
6 a : to enter upon **:** BEGIN ⟨*opened* the meeting⟩ **b :** to commence action in a card game by making (a first bid), putting a first bet in (the pot), or playing (a card or suit) as first lead
7 : to restore or recall (as an order) from a finally determined state to a state in which the parties are free to prosecute or oppose
intransitive verb
1 : to become open ⟨the office *opened* early⟩
2 a : to spread out **:** EXPAND ⟨the wound *opened* under the strain⟩ **b :** to become disclosed ⟨a beautiful vista *opened* before us⟩
3 : to become enlightened or responsive
4 : to give access ⟨the rooms *open* onto a hall⟩
5 : SPEAK OUT 2 ⟨finally he *opened* freely on the subject⟩
6 a : to begin a course or activity ⟨the play *opens* on Tuesday⟩ **b :** to make a bet, bid, or lead in commencing a round or hand of a card game
— **open·abil·i·ty** \ˌōp-nə-'bi-lə-tē, ˌō-pə-\ *noun*
— **open·able** \'ōp-nə-bəl, 'ō-pə-\ *adjective*

³open *noun* (13th century)
1 : OPENING
2 : open and unobstructed space: as **a :** OPEN AIR **b :** open water
3 : an open contest, competition, or tournament
4 : a public or unconcealed state or position

open admission *noun* (1969)
: OPEN ENROLLMENT 2

open–air *adjective* (1830)
: OUTDOOR

open air *noun* (15th century)
: the space where air is unconfined; *especially* **:** OUTDOORS

open–and–shut *adjective* (1841)
1 : perfectly simple **:** OBVIOUS
2 : easily settled ⟨an *open-and-shut* case⟩

open bar *noun* (1973)
: a bar (as at a wedding reception) at which drinks are served free — compare CASH BAR

open·cast \'ō-pən-ˌkast\ *adjective* (circa 1890)
chiefly British **:** worked from a surface open to the air ⟨an *opencast* mine⟩ ⟨*opencast* mining⟩

open chain *noun* (1884)
: an arrangement of atoms represented in a structural formula by a chain whose ends are not joined so as to form a ring

open city *noun* (1914)
: a city that is not occupied or defended by military forces and that is not allowed to be bombed under international law

open dating *noun* (1971)
: the marking of perishable food products with a clearly readable date indicating when the food was packaged or the last date on which it should be sold or used

open door *noun* (1526)
1 : a recognized right of admittance **:** freedom of access
2 : a policy giving opportunity for commercial relations with a country to all nations on equal terms
— **open–door** *adjective*

open–end *adjective* (1917)
: organized to allow for contingencies: as **a :** permitting additional debt to be incurred under the original indenture subject to specified conditions ⟨an *open-end* mortgage⟩ **b :** having a fluctuating capitalization of shares that are issued or redeemed at the current net asset value or at a figure in fixed ratio to this ⟨an *open-end* investment company⟩ — compare CLOSED-END

open–end·ed \ˌō-pən-'en-dəd\ *adjective* (1825)
: not rigorously fixed: as **a :** adaptable to the developing needs of a situation **b :** permitting or designed to permit spontaneous and unguided responses
— **open–end·ed·ness** *noun*

open enrollment *noun* (1964)
1 : the voluntary enrollment of a student in a public school other than the one assigned on the basis of residence
2 : enrollment on demand as a student in an institution of higher learning irrespective of formal qualifications

open·er \'ōp-nər, 'ō-pə-\ *noun* (15th century)
: one that opens ⟨a bottle *opener*⟩: as **a** *plural* **:** cards of sufficient value for a player to open the betting in a poker game **b :** the first item, contest, or event of a series
— **for openers :** to begin with

open–eyed \ˌō-pən-'īd\ *adjective* (1601)
1 : having the eyes open
2 : carefully observant **:** DISCERNING

open·hand·ed \-'han-dəd\ *adjective* (1593)
: GENEROUS, MUNIFICENT
— **open·hand·ed·ly** *adverb*
— **open·hand·ed·ness** *noun*

open–heart *adjective* (1960)
: of, relating to, or performed on a heart temporarily relieved of circulatory function and surgically opened for inspection and treatment ⟨*open-heart* surgery⟩

open·heart·ed \ˌō-pən-'här-təd\ *adjective* (1611)
1 : candidly straightforward **:** FRANK
2 : responsive to emotional appeal
— **open·heart·ed·ly** *adverb*
— **open·heart·ed·ness** *noun*

open–hearth *adjective* (1885)
: of, relating to, involving, or produced in the open-hearth process ⟨*open-hearth* steel⟩

open–hearth process *noun* (1882)
: a process of making steel from pig iron in a furnace of the regenerative reverberatory type

open house *noun* (15th century)
1 : ready and usually informal hospitality or entertainment for all comers
2 : a house or apartment open for inspection especially by prospective buyers or tenants

open·ing \'ōp-niŋ, 'ō-pə-\ *noun* (13th century)
1 a : an act or instance of making or becoming open **b :** an act or instance of beginning **:** COMMENCEMENT; *especially* **:** a formal and usually public event by which something new is put officially into operation
2 : something that is open: as **a** (1) **:** BREACH, APERTURE (2) **:** an open width **:** SPAN **b :** an area without trees or with scattered usually mature trees that occurs as a break in a forest **c :** two pages that face one another in a book
3 : something that constitutes a beginning: as **a :** a planned series of moves made at the beginning of a game of chess or checkers — compare ENDGAME, MIDDLE GAME **b :** a first performance
4 a : OCCASION, CHANCE **b :** an opportunity for employment

open letter *noun* (1878)
: a published letter of protest or appeal usually addressed to an individual but intended for the general public

open loop *noun* (1947)
: a control system for an operation or process in which there is no self-correcting action as there is in a closed loop

open marriage *noun* (1971)
: a marriage in which the partners agree to let each other have sexual partners outside the marriage

open–mind·ed \ˌō-pən-'mīn-dəd\ *adjective* (1828)
: receptive to arguments or ideas
— **open–mind·ed·ly** *adverb*
— **open–mind·ed·ness** *noun*

open·mouthed \ˌō-pən-'mau̇t͟hd, -'mau̇tht\ *adjective* (15th century)
1 : CLAMOROUS, VOCIFEROUS
2 : having the mouth wide open
3 : struck with amazement or wonder
— **open·mouth·ed·ly** \-'mau̇-t͟həd-lē, -thəd-\ *adverb*
— **open·mouth·ed·ness** \-'mau̇-t͟həd-nəs, -thəd-\ *noun*

open–pol·li·nat·ed \ˌō-pən-'pä-lə-ˌnā-təd\ *adjective* (1925)
: pollinated by natural agencies without human intervention

open season *noun* (circa 1890)
1 : a period when it is legal to kill or catch game or fish protected at other times by law

\ə\ **abut** \ᵊ\ **kitten** \ər\ **further** \a\ **ash** \ā\ **ace** \ä\ **mop, mar** \au̇\ **out** \ch\ **chin** \e\ **bet** \ē\ **easy** \g\ **go** \i\ **hit** \ī\ **ice** \j\ **job** \ŋ\ **sing** \ō\ **go** \ȯ\ **law** \ȯi\ **boy** \th\ **thin** \t͟h\ **the** \ü\ **loot** \u̇\ **foot** \y\ **yet** \zh\ **vision** *see also* Guide to Pronunciation

2 : a time during which someone or something is the object of sustained attack or criticism

open secret *noun* (1828)
: a supposedly secret but generally known matter

open sentence *noun* (1937)
: a statement (as in mathematics) that contains at least one blank or unknown and that becomes true or false when the blank is filled or a quantity is substituted for the unknown

open ses·a·me \-'se-sə-mē\ *noun* [from *open sesame,* the magical command used by Ali Baba to open the door of the robbers' den in *Ali Baba and the Forty Thieves*] (circa 1837)
: something that unfailingly brings about a desired end

open shop *noun* (1903)
: an establishment in which eligibility for employment and retention on the payroll are not determined by membership or nonmembership in a labor union though there may be an agreement by which a union is recognized as sole bargaining agent

open sight *noun* (1591)
: a firearm rear sight having an open notch

open stance *noun* (1948)
: a stance (as in golf) in which the forward foot is farther from the line of play than the back foot — compare CLOSED STANCE

open syllable *noun* (1891)
: a syllable ended by a vowel or diphthong

open up (1582)
transitive verb
1 : to make available
2 : to make plain or visible **:** DISCLOSE
3 : to open by cutting into
intransitive verb
1 : to spread out or come into view ⟨the road *opens up* ahead⟩
2 : to commence firing
3 : to become communicative ⟨tried to get the patient to *open up*⟩

open·work \'ō-pən-ˌwərk\ *noun, often attributive* (1598)
: work constructed so as to show openings through its substance **:** work that is perforated or pierced ⟨wrought-iron *openwork*⟩
— open–worked \-ˌwərkt\ *adjective*

¹opera *plural of* OPUS

²op·era \'ä-p(ə-)rə, *Southern also* 'ä-prē\ *noun* [Italian, work, opera, from Latin, work, pains; akin to Latin *oper-, opus* — more at OPERATE] (1644)
1 : a drama set to music and made up of vocal pieces with orchestral accompaniment and orchestral overtures and interludes; *specifically* **:** GRAND OPERA
2 : the score of a musical drama
3 : the performance of an opera; *also* **:** a house where operas are performed

op·er·a·ble \'ä-p(ə-)rə-bəl\ *adjective* (1646)
1 : fit, possible, or desirable to use **:** PRACTICABLE
2 : likely to result in a favorable outcome upon surgical treatment ⟨an *operable* cancer⟩
— op·er·a·bil·i·ty \ˌä-p(ə-)rə-'bi-lə-tē\ *noun*
— op·er·a·bly \'ä-pə-rə-blē\ *adverb*

opé·ra bouffe \'ō-pā-rä-'büf, 'ä-p(ə-)rə-\ *noun* [French, from Italian *opera buffa*] (1870)
: satirical comic opera

opera buf·fa \-'bü-fə\ *noun* [Italian, literally, comic opera] (1802)
: an 18th century farcical comic opera with dialogue in recitative

opéra co·mique \-kä-'mēk, -kō-\ *noun* [French, literally, comic opera] (1744)
: an opera characterized by spoken dialogue interspersed between the arias and ensemble numbers — compare GRAND OPERA

opera glass *noun* (1738)
: a small low-power binocular without prisms for use at the opera or theater — often used in plural

op·era·go·er \'ä-p(ə-)rə-ˌgō(-ə)r\ *noun* (1850)
: a person who frequently goes to operas
— op·era·go·ing \-ˌgō-iŋ, -ˌgȯ(-)iŋ\ *noun*

opera glass

opera hat *noun* (1810)
: a man's collapsible top hat

opera house *noun* (1720)
: a theater devoted principally to the performance of operas; *broadly* **:** THEATER

op·er·and \ˌä-pə-'rand\ *noun* [Latin *operandum,* neuter of gerundive of *operari*] (1886)
: something (as a quantity or data) that is operated on (as in a mathematical operation); *also* **:** the address in a computer instruction of data to be operated on

¹op·er·ant \'ä-pə-rənt\ *adjective* (15th century)
1 : functioning or tending to produce effects **:** EFFECTIVE ⟨an *operant* conscience⟩
2 : of or relating to the observable or measurable
3 : of, relating to, or being an operant or operant conditioning ⟨*operant* behavior⟩
— op·er·ant·ly *adverb*

²operant *noun* (1937)
: behavior (as bar pressing by a rat to obtain food) that operates on the environment to produce rewarding and reinforcing effects

operant conditioning *noun* (1941)
: conditioning in which the desired behavior or increasingly closer approximations to it are followed by a rewarding or reinforcing stimulus — compare CLASSICAL CONDITIONING

opera se·ria \-'ser-ē-ə, -'sir-\ *noun* [Italian, literally, serious opera] (circa 1854)
: an 18th century opera with a heroic or legendary subject

op·er·ate \'ä-pə-ˌrāt, 'ä-ˌprāt\ *verb* **-at·ed; -at·ing** [Latin *operatus,* past participle of *operari* to work, from *oper-, opus* work; akin to Old English *efnan* to perform, Sanskrit *apas* work] (1588)
intransitive verb
1 : to perform a function **:** exert power or influence ⟨factors *operating* against our success⟩
2 : to produce an appropriate effect ⟨the drug *operated* quickly⟩
3 a : to perform an operation or a series of operations **b :** to perform surgery **c :** to carry on a military or naval action or mission
4 : to follow a course of conduct that is often irregular ⟨crooked gamblers *operating* in the club⟩
transitive verb
1 : BRING ABOUT, EFFECT
2 a : to cause to function **:** WORK **b :** to put or keep in operation
3 : to perform an operation on; *especially* **:** to perform surgery on

op·er·at·ic \ˌä-pə-'ra-tik\ *adjective* (1749)
1 : of or relating to opera
2 : grand, dramatic, or romantic in style or effect
— op·er·at·i·cal·ly \-ti-k(ə-)lē\ *adverb*

operating *adjective* (1808)
: of, relating to, or used for or in operations ⟨*operating* expenses⟩ ⟨a hospital *operating* room⟩

operating system *noun* (1961)
: software that controls the operation of a computer and directs the processing of programs (as by assigning storage space in memory and controlling input and output functions)

op·er·a·tion \ˌä-pə-'rā-shən\ *noun* [Middle English *operacioun,* from Middle French *operation,* from Latin *operation-, operatio,* from *operari*] (14th century)
1 : performance of a practical work or of something involving the practical application of principles or processes
2 a : an exertion of power or influence ⟨the *operation* of a drug⟩ **b :** the quality or state of being functional or operative ⟨the plant is now in *operation*⟩ **c :** a method or manner of functioning ⟨a machine of very simple *operation*⟩
3 : EFFICACY, POTENCY — archaic except in legal usage
4 : a procedure carried out on a living body usually with instruments especially for the repair of damage or the restoration of health
5 : any of various mathematical or logical processes (as addition) of deriving one entity from others according to a rule
6 a : a usually military action, mission, or maneuver including its planning and execution **b** *plural* **:** the office on the flight line of an airfield where pilots file clearance for flights and where flying from the field is controlled **c** *plural* **:** the agency of an organization charged with carrying on the principal planning and operating functions of a headquarters and its subordinate units
7 : a business transaction especially when speculative
8 : a single step performed by a computer in the execution of a program

op·er·a·tion·al \-shnəl, -shə-nᵊl\ *adjective* (circa 1909)
1 : of or relating to operation or to an operation ⟨the *operational* gap between planning and production⟩
2 : of, relating to, or based on operations
3 a : of, engaged in, or connected with execution of military or naval operations in campaign or battle **b :** ready for or in condition to undertake a destined function
— op·er·a·tion·al·ly *adverb*

op·er·a·tion·al·ism \-shnə-ˌli-zəm, -shə-nᵊl-ˌi-zəm\ *noun* (1931)
: a view that the concepts or terms used in nonanalytic scientific statements must be definable in terms of identifiable and repeatable operations
— op·er·a·tion·al·ist \-list, -ist\ *noun*
— op·er·a·tion·al·is·tic \-ˌrā-shnə-'lis-tik, -shə-nᵊl-'is-tik\ *adjective*

op·er·a·tion·ism \ˌä-pə-'rā-shə-ˌni-zəm\ *noun* (1935)
: OPERATIONALISM
— op·er·a·tion·ist \-sh(ə-)nist\ *noun*

operations research *noun* (1945)
: the application of scientific and especially mathematical methods to the study and analysis of problems involving complex systems — called also *operational research*

¹op·er·a·tive \'ä-p(ə-)rə-tiv, 'ä-pə-ˌrā-\ *adjective* (15th century)
1 : producing an appropriate effect **:** EFFICACIOUS
2 : exerting force or influence **:** OPERATING
3 a : having to do with physical operations (as of machines) **b :** WORKING ⟨an *operative* craftsman⟩
4 : based on or consisting of an operation ⟨*operative* dentistry⟩
— op·er·a·tive·ly *adverb*
— op·er·a·tive·ness *noun*

²operative *noun* (circa 1810)
: OPERATOR: as **a :** ARTISAN, MECHANIC **b :** a secret agent **c :** PRIVATE DETECTIVE

op·er·a·tor \'ä-pə-ˌrā-tər, 'ä-ˌprā-\ *noun* (1611)
1 : one that operates: as **a :** one that operates a machine or device **b :** one that operates a business **c :** one that performs surgical operations **d :** one that deals in stocks or commodities
2 a : MOUNTEBANK, FRAUD **b :** a shrewd and skillful person who knows how to circumvent restrictions or difficulties
3 a : something and especially a symbol that denotes or performs a mathematical or logical operation **b :** a mathematical function
4 : a binding site in a DNA chain at which a genetic repressor binds to inhibit the initiation of transcription of messenger RNA by one or more nearby structural genes — called also *operator gene,* compare OPERON

— **op·er·a·tor·less** *adjective*

[1]**oper·cu·lar** \ō-ˈpər-kyə-lər\ *adjective* (1830)
: of, relating to, or constituting an operculum

[2]**opercular** *noun* (circa 1890)
: an opercular part (as a bone or scale)

oper·cu·late \ō-ˈpər-kyə-lət\ *also* **oper·cu·lat·ed** \-ˌlā-təd\ *adjective* (circa 1775)
: having an operculum

oper·cu·lum \ō-ˈpər-kyə-ləm\ *noun, plural* **-la** \-lə\ *also* **-lums** [New Latin, from Latin, cover, from *operire* to shut, cover] (1752)
1 : a body process or part that suggests a lid: as **a :** a horny or shelly plate on the posterior dorsal surface of the foot in many gastropod mollusks that closes the shell when the animal is retracted **b :** the covering of the gills of a fish — see FISH illustration
2 : a lid or covering flap (as of a moss capsule or a pyxidium in a seed plant)

op·er·et·ta \ˌä-pə-ˈre-tə\ *noun* [Italian, diminutive of *opera*] (1770)
: a usually romantic comic opera that includes songs and dancing
— **op·er·et·tist** \-ˈre-tist\ *noun*

op·er·on \ˈä-pə-ˌrän\ *noun* [French *opéron,* from *opérer* to bring about, effect (from Latin *operari*) + *-on* [2]-on] (1961)
: a group of closely linked genes that produces a single messenger RNA molecule in transcription and that consists of structural genes and regulating elements (as an operator and promoter)

op·er·ose \ˈä-pə-ˌrōs\ *adjective* [Latin *operosus,* from *oper-, opus* work — more at OPERATE] (1678)
: TEDIOUS, WEARISOME
— **op·er·ose·ly** *adverb*
— **op·er·ose·ness** *noun*

Ophe·lia \ō-ˈfēl-yə\ *noun*
: the daughter of Polonius in Shakespeare's *Hamlet*

ophid·i·an \ō-ˈfi-dē-ən\ *adjective* [ultimately from Greek *ophis*] (1883)
: of, relating to, or resembling snakes
— **ophidian** *noun*

Ophir \ˈō-fər\ *noun* [Hebrew *Ōphīr*]
: a biblical land of uncertain location but reputedly rich in gold

ophit·ic \ä-ˈfi-tik, ō-\ *adjective* [*ophite* serpentine (stone), from Latin *ophites,* from Greek *ophitēs* (*lithos*), from *ophitēs* snakelike, from *ophis* snake; akin to Sanskrit *ahi* snake and probably to Latin *anguis* snake, *anguilla* eel, Greek *enchelys* eel, *echidna* viper, *echinos* hedgehog, Old English *igil*] (1875)
: having or being a rock fabric in which lath-shaped plagioclase crystals are enclosed in later formed augite

ophi·u·roid \ˌō-fē-ˈyu̇r-ˌȯid, ˌä-\ *noun* [New Latin *Ophiuroidea,* group name, from *Ophiura,* genus name, from Greek *ophis* + *oura* tail — more at ASS] (circa 1879)
: BRITTLE STAR
— **ophiuroid** *adjective*

ophthalm- *or* **ophthalmo-** *combining form* [Greek, from *ophthalmos*]
: eye ⟨*ophthalmo*logy⟩

oph·thal·mia \äf-ˈthal-mē-ə, äp-\ *noun* [Middle English *obtalmia,* from Late Latin *ophthalmia,* from Greek, from *ophthalmos* eye; akin to Greek *ōps* eye — more at EYE] (14th century)
: inflammation of the conjunctiva or the eyeball

oph·thal·mic \-mik\ *adjective* (circa 1741)
1 : of, relating to, or situated near the eye
2 : supplying or draining the eye or structures in the region of the eye ⟨*ophthalmic* artery⟩

oph·thal·mol·o·gist \ˌäf-thə(l)-ˈmä-lə-jist, ˌäp-, -ˌthal-\ *noun* (1834)
: a physician that specializes in ophthalmology — compare OPTICIAN, OPTOMETRIST

oph·thal·mol·o·gy \-ˈmä-lə-jē\ *noun* (circa 1842)
: a branch of medical science dealing with the structure, functions, and diseases of the eye
— **oph·thal·mo·log·ic** \-mə-ˈlä-jik\ *or* **oph·thal·mo·log·i·cal** \-ji-kəl\ *adjective*
— **oph·thal·mo·log·i·cal·ly** \-ji-k(ə-)lē\ *adverb*

oph·thal·mo·scope \äf-ˈthal-mə-ˌskōp, äp-\ *noun* [International Scientific Vocabulary] (circa 1857)
: an instrument for use in viewing the interior of the eye and especially the retina
— **oph·thal·mo·scop·ic** \(ˌ)äf-ˌthal-mə-ˈskä-pik, (ˌ)äp-\ *adjective*
— **oph·thal·mos·co·py** \ˌäf-thəl-ˈmäs-kə-pē, ˌäp-, -ˌthal-\ *noun*

-opia *noun combining form* [New Latin, from Greek *-ōpia,* from *ōps*]
1 : condition of having (such) vision ⟨diplo*pia*⟩
2 : condition of having (such) a visual defect ⟨hyper*opia*⟩

[1]**opi·ate** \ˈō-pē-ət, -ˌāt\ *noun* (15th century)
1 : a preparation or derivative of opium; *broadly* **:** a narcotic or opioid peptide
2 : something that induces rest or inaction or quiets uneasiness

[2]**opiate** *adjective* (1543)
1 a : containing or mixed with opium **b :** of, relating to, binding, or being an opiate ⟨*opiate* receptors⟩
2 a : inducing sleep **:** NARCOTIC **b :** causing dullness or inaction

opine \ō-ˈpīn\ *verb* **opined; opin·ing** [Middle English, from Middle French *opiner,* from Latin *opinari* to have an opinion] (15th century)
intransitive verb
: to express opinions
transitive verb
: to state as an opinion

opin·ion \ə-ˈpin-yən\ *noun* [Middle English, from Middle French, from Latin *opinion-, opinio,* from *opinari*] (14th century)
1 a : a view, judgment, or appraisal formed in the mind about a particular matter **b :** APPROVAL, ESTEEM
2 a : belief stronger than impression and less strong than positive knowledge **b :** a generally held view
3 a : a formal expression of judgment or advice by an expert **b :** the formal expression (as by a judge, court, or referee) of the legal reasons and principles upon which a legal decision is based ☆
— **opin·ioned** \-yənd\ *adjective*

opin·ion·at·ed \-yə-ˌnā-təd\ *adjective* (1601)
: unduly adhering to one's own opinion or to preconceived notions
— **opin·ion·at·ed·ly** *adverb*
— **opin·ion·at·ed·ness** *noun*

opin·ion·a·tive \-ˌnā-tiv\ *adjective* (1536)
1 : of, relating to, or consisting of opinion **:** DOCTRINAL
2 : OPINIONATED
— **opin·ion·a·tive·ly** *adverb*
— **opin·ion·a·tive·ness** *noun*

opi·oid \ˈō-pē-ˌȯid\ *adjective* (1957)
1 : possessing some properties characteristic of opiate narcotics but not derived from opium
2 : of, involving, or induced by an opioid substance or an opioid peptide

opioid peptide *noun* (1976)
: any of a group of endogenous neural polypeptides (as an endorphin or enkephalin) that bind especially to opiate receptors and mimic some of the pharmacological properties of opiate drugs — called also *opioid*

opis·tho·branch \ə-ˈpis-thə-ˌbraŋk\ *noun, plural* **-branchs** [New Latin *Opisthobranchia,* from Greek *opisthen* behind (akin to Greek *epi* on) + *branchia* gills — more at EPI-] (circa 1856)
: any of a subclass (Opisthobranchia) of marine gastropod mollusks that have the gills when present posterior to the heart and often lack a shell
— **opisthobranch** *adjective*

opi·um \ˈō-pē-əm\ *noun* [Middle English, from Latin, from Greek *opion,* from diminutive of *opos* sap] (14th century)
1 : a bitter brownish addictive narcotic drug that consists of the dried juice of the opium poppy
2 : something having an effect like that of opium

opium poppy *noun* (1863)
: an annual Eurasian poppy (*Papaver somniferum*) cultivated since antiquity as the source of opium, for its edible oily seeds, or for its showy flowers

opos·sum \(ə-)ˈpä-səm\ *noun, plural* **opossums** *also* **opossum** [Virginia Algonquian, literally, white animal] (1610)
1 : any of a family (Didelphidae) of American marsupials that usually have a pointed snout and prehensile tail; *especially* **:** a common omnivorous largely nocturnal and arboreal mammal (*Didelphis virginiana*) of North America having grayish to blackish fur with white on the cheeks
2 : any of several Australian phalangers

opossum 1

[1]**op·po·nent** \ə-ˈpō-nənt\ *noun* [Latin *opponent-, opponens,* present participle of *opponere*] (1588)
1 : one that takes an opposite position (as in a debate, contest, or conflict)
2 : a muscle that opposes or counteracts and limits the action of another

[2]**opponent** *adjective* (1647)
1 : ANTAGONISTIC, OPPOSING
2 : situated in front

op·por·tune \ˌä-pər-ˈtün, -ˈtyün\ *adjective* [Middle English, from Middle French *opportun,* from Latin *opportunus,* from *ob-* toward + *portus* port, harbor — more at OB-, FORD] (15th century)
1 : suitable or convenient for a particular occurrence ⟨an *opportune* moment⟩
2 : occurring at an appropriate time ⟨an *opportune* offer of assistance⟩
— **op·por·tune·ly** *adverb*
— **op·por·tune·ness** \-ˈt(y)ün-nəs\ *noun*

op·por·tun·ism \-ˈtü-ˌni-zəm, -ˈtyü-\ *noun* (1870)
: the art, policy, or practice of taking advantage of opportunities or circumstances often with little regard for principles or consequences

☆ SYNONYMS
Opinion, view, belief, conviction, persuasion, sentiment mean a judgment one holds as true. OPINION implies a conclusion thought out yet open to dispute ⟨each expert seemed to have a different *opinion*⟩. VIEW suggests a subjective opinion ⟨very assertive in stating his *views*⟩. BELIEF implies often deliberate acceptance and intellectual assent ⟨a firm *belief* in her party's platform⟩. CONVICTION applies to a firmly and seriously held belief ⟨the *conviction* that animal life is as sacred as human⟩. PERSUASION suggests a belief grounded on assurance (as by evidence) of its truth ⟨was of the *persuasion* that everything changes⟩. SENTIMENT suggests a settled opinion reflective of one's feelings ⟨her feminist *sentiments* are well-known⟩.

\ə\ **abut** \ᵊ\ **kitten** \ər\ **further** \a\ **ash** \ā\ **ace** \ä\ **mop, mar** \au̇\ **out** \ch\ **chin** \e\ **bet** \ē\ **easy** \g\ **go** \i\ **hit** \ī\ **ice** \j\ **job** \ŋ\ **sing** \ō\ **go** \ȯ\ **law** \ȯi\ **boy** \th\ **thin** \t͟h\ **the** \ü\ **loot** \u̇\ **foot** \y\ **yet** \zh\ **vision** *see also* Guide to Pronunciation

— **op·por·tun·ist** \-'tü-nist, -'tyü-\ *noun or adjective*

op·por·tu·nis·tic \-tü-'nis-tik, -tyü-\ *adjective* (1892)
: taking advantage of opportunities as they arise: as **a :** exploiting opportunities with little regard to principle or consequences ⟨a politician considered *opportunistic*⟩ **b :** feeding on whatever food is available ⟨*opportunistic* feeders⟩ **c :** being or caused by a usually harmless microorganism that can become pathogenic when the host's resistance is impaired ⟨*opportunistic* infections⟩
— **op·por·tu·nis·ti·cal·ly** \-ti-k(ə-)lē\ *adverb*

op·por·tu·ni·ty \ˌä-pər-'tü-nə-tē, -'tyü-\ *noun, plural* **-ties** (14th century)
1 : a favorable juncture of circumstances ⟨the halt provided an *opportunity* for rest and refreshment⟩
2 : a good chance for advancement or progress

opportunity cost *noun* (1911)
: the cost of making an investment that is the difference between the return on one investment and the return on an alternative

op·pos·able \ə-'pō-zə-bəl\ *adjective* (1667)
1 : capable of being opposed or resisted
2 : capable of being placed against one or more of the remaining digits of a hand or foot ⟨the *opposable* human thumb⟩
— **op·pos·abil·i·ty** \-ˌpō-zə-'bi-lə-tē\ *noun*

op·pose \ə-'pōz\ *transitive verb* **op·posed; op·pos·ing** [French *opposer,* from Latin *opponere* (perfect indicative *opposui*), from *ob-* against + *ponere* to place — more at OB-, POSITION] (1579)
1 : to place opposite or against something
2 : to place over against something so as to provide resistance, counterbalance, or contrast
3 : to offer resistance to ☆
— **op·pos·er** *noun*

op·posed \-'pōzd\ *adjective* (1596)
: set or placed in opposition **:** CONTRARY ⟨with politicians, as *opposed* to soap, you cannot return what you have bought —Felix G. Rohatyn⟩

op·pose·less \ə-'pōz-ləs\ *adjective* (1605)
archaic **:** IRRESISTIBLE

[1]**op·po·site** \'ä-pə-zət, 'äp-sət\ *adjective* [Middle English, from Middle French, from Latin *oppositus,* past participle of *opponere*] (14th century)
1 a : set over against something that is at the other end or side of an intervening line or space ⟨*opposite* interior angles⟩ ⟨*opposite* ends of a diameter⟩ **b :** situated in pairs on an axis with each member being separated from the other by half the circumference of the axis ⟨*opposite* leaves⟩ — compare ALTERNATE
2 a : occupying an opposing and often antagonistic position ⟨*opposite* sides of the question⟩ **b :** diametrically different (as in nature or character) ⟨*opposite* meanings⟩
3 : contrary to one another or to a thing specified **:** REVERSE ⟨gave them *opposite* directions⟩
4 : being the other of a pair that are corresponding or complementary in position, function, or nature ⟨members of the *opposite* sex⟩
5 : of, relating to, or being the side of a baseball field that is near the first base line for a right-handed batter and near the third base line for a left-handed batter ☆
— **op·po·site·ly** *adverb*
— **op·po·site·ness** *noun*

[2]**opposite** *noun* (15th century)
1 : something that is opposed to some other often specified thing
2 : ANTONYM
3 : ADDITIVE INVERSE; *especially* **:** the additive inverse of a real number

[3]**opposite** *adverb* (1667)
: on or to an opposite side

[4]**opposite** *preposition* (1758)
1 : across from and usually facing or on the same level with ⟨sat *opposite* each other⟩
2 : in a role complementary to ⟨played *opposite* the leading man in the comedy⟩

opposite number *noun* (1906)
: a member of a system or class who holds relatively the same position as a particular member in a corresponding system or class **:** COUNTERPART

op·po·si·tion \ˌä-pə-'zi-shən\ *noun* (14th century)
1 : a configuration in which one celestial body is opposite another in the sky or in which the elongation is near or equal to 180 degrees
2 : the relation between two propositions having the same subject and predicate but differing in quantity or quality or both
3 : an act of setting opposite or over against **:** the condition of being so set
4 : hostile or contrary action or condition
5 a : something that opposes; *specifically* **:** a body of persons opposing something **b** *often capitalized* **:** a political party opposing and prepared to replace the party in power
— **op·po·si·tion·al** \-'zish-nəl, -'zi-shə-nᵊl\ *adjective*

op·po·si·tion·ist \-'zi-sh(ə-)nist\ *noun* (1773)
: a member of an opposition
— **oppositionist** *adjective*

op·press \ə-'pres\ *transitive verb* [Middle English, from Middle French *oppresser,* from Latin *oppressus,* past participle of *opprimere,* from *ob-* against + *premere* to press — more at OB-, PRESS] (14th century)
1 a *archaic* **:** SUPPRESS **b :** to crush or burden by abuse of power or authority
2 : to burden spiritually or mentally **:** weigh heavily upon
synonym see WRONG
— **op·pres·sor** \-'pre-sər\ *noun*

op·pres·sion \ə-'pre-shən\ *noun* (14th century)
1 a : unjust or cruel exercise of authority or power **b :** something that oppresses especially in being an unjust or excessive exercise of power
2 : a sense of being weighed down in body or mind **:** DEPRESSION

op·pres·sive \ə-'pre-siv\ *adjective* (circa 1677)
1 : unreasonably burdensome or severe ⟨*oppressive* legislation⟩
2 : TYRANNICAL
3 : overwhelming or depressing to the spirit or senses ⟨an *oppressive* climate⟩
synonym see ONEROUS
— **op·pres·sive·ly** *adverb*
— **op·pres·sive·ness** *noun*

op·pro·bri·ous \ə-'prō-brē-əs\ *adjective* (14th century)
1 : expressive of opprobrium **:** SCURRILOUS ⟨*opprobrious* language⟩
2 : deserving of opprobrium **:** INFAMOUS
— **op·pro·bri·ous·ly** *adverb*
— **op·pro·bri·ous·ness** *noun*

op·pro·bri·um \-brē-əm\ *noun* [Latin, from *opprobrare* to reproach, from *ob* in the way of + *probrum* reproach; akin to Latin *pro* forward and to Latin *ferre* to carry, bring — more at OB-, FOR, BEAR] (1656)
1 : something that brings disgrace
2 a : public disgrace or ill fame that follows from conduct considered grossly wrong or vicious **b :** CONTEMPT, REPROACH

op·pugn \ə-'pyün, ä-\ *transitive verb* [Middle English, from Latin *oppugnare,* from *ob-* against + *pugnare* to fight — more at OB-, PUNGENT] (15th century)
1 : to fight against
2 : to call in question
— **op·pugn·er** *noun*

Ops \'äps\ *noun* [Latin]
: the Roman goddess of abundance and the wife of Saturn

op·sin \'äp-sən\ *noun* [probably from *rhodopsin*] (1951)
: any of various colorless proteins that in combination with retinal or a related prosthetic group form a visual pigment (as rhodopsin) in a reaction which is reversed by light

-opsis *noun combining form, plural* **-opses** *or* **-opsides** [New Latin, from Greek, from *opsis* appearance, vision]
: structure resembling a (specified) thing ⟨cary*opsis*⟩

op·son·ic \äp-'sä-nik\ *adjective* (1903)
: of, relating to, or involving opsonin

op·so·nin \'äp-sə-nən\ *noun* [Latin *opsonare* to buy provisions, cater (from Greek *opsōnein*) + English [1]*-in* — more at OLIGOPSONY] (1903)
: an antibody of blood serum that makes foreign cells more susceptible to the action of the phagocytes

-opsy *noun combining form* [Greek *-opsia,* from *opsis*]
: examination ⟨necr*opsy*⟩

opt \'äpt\ *intransitive verb* [French *opter,* from Latin *optare*] (1877)
: to make a choice; *especially* **:** to decide in favor of something ⟨*opted* for a tax increase —Tom Wicker⟩

op·ta·tive \'äp-tə-tiv\ *adjective* (15th century)
1 a : of, relating to, or constituting a verbal mood that is expressive of wish or desire **b :** of, relating to, or constituting a sentence that is expressive of wish or hope
2 : expressing desire or wish
— **optative** *noun*
— **op·ta·tive·ly** *adverb*

[1]**op·tic** \'äp-tik\ *adjective* [Middle English, from Middle French *optique,* from Medieval Latin *opticus,* from Greek *optikos,* from *opsesthai* to be going to see; akin to Greek *opsis* appearance, *ōps* eye — more at EYE] (14th century)
: of or relating to vision or the eye

[2]**optic** *noun* (1600)
1 : EYE
2 a : any of the elements (as lenses, mirrors, or light guides) of an optical instrument or system — usually used in plural **b :** an optical instrument

op·ti·cal \'äp-ti-kəl\ *adjective* (1570)
1 : of or relating to the science of optics
2 a : of or relating to vision **:** VISUAL **b :** VISIBLE 1 ⟨*optical* wavelength⟩ **c :** of, relating to, or being objects that emit light in the visible

☆ SYNONYMS

Oppose, combat, resist, withstand mean to set oneself against someone or something. OPPOSE can apply to any conflict, from mere objection to bitter hostility or warfare ⟨*opposed* the plan⟩. COMBAT stresses the forceful or urgent countering of something ⟨*combat* disease⟩. RESIST implies an overt recognition of a hostile or threatening force and a positive effort to counteract or repel it ⟨*resisting* temptation⟩. WITHSTAND suggests a more passive resistance ⟨trying to *withstand* peer pressure⟩.

Opposite, contradictory, contrary, antithetical mean being so far apart as to be or seem irreconcilable. OPPOSITE applies to things in sharp contrast or in conflict ⟨*opposite* views on foreign aid⟩. CONTRADICTORY applies to two things that completely negate each other so that if one is true or valid the other must be untrue or invalid ⟨made *contradictory* predictions about whether the market would rise or fall⟩. CONTRARY implies extreme divergence or diametrical opposition ⟨*contrary* assessments of the war situation⟩. ANTITHETICAL stresses clear and unequivocal diametrical opposition ⟨a law that is *antithetical* to the very idea of democracy⟩.

range of frequencies ⟨an *optical* galaxy⟩ **d** **:** using the properties of light to aid vision ⟨an *optical* instrument⟩
3 a : of, relating to, or utilizing light especially instead of other forms of energy ⟨*optical* microscopy⟩ **b :** involving the use of light-sensitive devices to acquire information for a computer ⟨*optical* character recognition⟩
4 : of or relating to optical art
— op·ti·cal·ly \-k(ə-)lē\ *adverb*

optical activity *noun* (1877)
: ability of a chemical substance to rotate the plane of vibration of polarized light to the right or left

optical art *noun* (1964)
: nonobjective art characterized by the use of straight or curved lines or geometric patterns often for an illusory effect (as of motion)

optical bench *noun* (1883)
: an apparatus that is fitted for the convenient location and adjustment of light sources and optical devices and that is used for the observation and measurement of optical phenomena

optical disk *noun* (1980)
: a disk with a plastic coating on which information (as music or visual images) is recorded digitally (as in the form of tiny pits) and which is read by using a laser

optical fiber *noun* (1962)
: a single fiber-optic strand

optical glass *noun* (1840)
: flint or crown glass of well-defined characteristics used especially for making lenses

optical illusion *noun* (1794)
: ILLUSION 2a(1)

optically active *adjective* (1885)
: capable of rotating the plane of polarization of light to the right or left — used of compounds, molecules, or atoms

optical rotation *noun* (1895)
: the angle through which the plane of vibration of polarized light that traverses an optically active substance is rotated

optic axis *noun* (1664)
: a line in a doubly refracting medium that is parallel to the direction in which all components of plane-polarized light travel with the same speed

optic chiasma *noun* (1872)
: the X-shaped partial decussation on the undersurface of the hypothalamus through which the optic nerves are continuous with the brain — called also *optic chiasm*

optic cup *noun* (circa 1885)
: the optic vesicle after invaginating to form a 2-layered cup from which the retina and pigmented layer of the eye will develop — called also *eyecup*

optic disk *noun* (circa 1890)
: BLIND SPOT 1a

op·ti·cian \äp-'ti-shən\ *noun* (1687)
1 : a maker of or dealer in optical items and instruments
2 : a person who reads prescriptions for visual correction, orders lenses, and dispenses spectacles and contact lenses — compare OPHTHALMOLOGIST, OPTOMETRIST

optic lobe *noun* (1854)
: either of two prominences of the midbrain concerned with vision

optic nerve *noun* (1613)
: either of the pair of nerves that comprise the 2d pair of cranial nerves, arise from the ventral part of the diencephalon, supply the retina, and conduct visual stimuli to the brain — see EYE illustration

op·tics \'äp-tiks\ *noun plural but singular in construction* (1579)
: a science that deals with the genesis and propagation of light, the changes that it undergoes and produces, and other phenomena closely associated with it

optic vesicle *noun* (circa 1885)
: an evagination of each lateral wall of the embryonic vertebrate forebrain from which the nervous structures of the eye develop

op·ti·mal \'äp-tə-məl\ *adjective* (1890)
: most desirable or satisfactory **:** OPTIMUM
— op·ti·mal·i·ty \ˌäp-tə-'ma-lə-tē\ *noun*
— op·ti·mal·ly \-mə-lē\ *adverb*

op·ti·mi·sa·tion, op·ti·mise *British variant of* OPTIMIZATION, OPTIMIZE

op·ti·mism \'äp-tə-ˌmi-zəm\ *noun* [French *optimisme,* from Latin *optimum,* noun, best, from neuter of *optimus* best; akin to Latin *ops* power — more at OPULENT] (1759)
1 : a doctrine that this world is the best possible world
2 : an inclination to put the most favorable construction upon actions and events or to anticipate the best possible outcome
— op·ti·mist \-mist\ *noun*
— op·ti·mis·tic \ˌäp-tə-'mis-tik\ *adjective*
— op·ti·mis·ti·cal·ly \-ti-k(ə-)lē\ *adverb*

Op·ti·mist \'äp-tə-mist\ *noun* [*Optimist (Club)*] (1911)
: a member of a major international service club

op·ti·mi·za·tion \ˌäp-tə-mə-'zā-shən\ *noun* (1857)
: an act, process, or methodology of making something (as a design, system, or decision) as fully perfect, functional, or effective as possible; *specifically* **:** the mathematical procedures (as finding the maximum of a function) involved in this

op·ti·mize \'äp-tə-ˌmīz\ *transitive verb* **-mized; -miz·ing** (1857)
: to make as perfect, effective, or functional as possible
— op·ti·miz·er \-ˌmī-zər\ *noun*

op·ti·mum \'äp-tə-məm\ *noun, plural* **-ma** \-mə\ *also* **-mums** [Latin] (1879)
1 : the amount or degree of something that is most favorable to some end; *especially* **:** the most favorable condition for the growth and reproduction of an organism
2 : greatest degree attained or attainable under implied or specified conditions
— optimum *adjective*

[1]op·tion \'äp-shən\ *noun* [French, from Latin *option-, optio* free choice; akin to Latin *optare* to choose] (circa 1604)
1 : an act of choosing
2 a : the power or right to choose **:** freedom of choice **b :** a privilege of demanding fulfillment of a contract on any day within a specified time **c :** a contract conveying a right to buy or sell designated securities, commodities, or property interest at a specified price during a stipulated period; *also* **:** the right conveyed by an option **d :** a right of an insured person to choose the form in which payments due on a policy shall be made or applied
3 : something that may be chosen: as **a :** an alternative course of action ⟨didn't have many *options* open⟩ **b :** an item that is offered in addition to or in place of standard equipment
4 : an offensive football play in which a back may choose whether to pass or run with the ball — called also *option play*
synonym see CHOICE

[2]option *transitive verb* (1926)
: to grant or take an option on

op·tion·al \'äp-shnəl, -shə-nᵊl\ *adjective* (1792)
: involving an option **:** not compulsory
— op·tion·al·i·ty \ˌäp-shə-'na-lə-tē\ *noun*
— op·tion·al·ly *adverb*

opto- *combining form* [Greek *optos,* verbal of *opsesthai* — more at OPTIC]
1 : vision ⟨*opto*metry⟩
2 : optic and ⟨*opto*electronics⟩

op·to·elec·tron·ics \ˌäp-(ˌ)tō-i-lek-'trä-niks, -ē-lek-\ *noun plural but singular in construction* (1959)
: a branch of electronics that deals with electronic devices for emitting, modulating, transmitting, and sensing light
— op·to·elec·tron·ic \-nik\ *adjective*

op·to·ki·net·ic \ˌäp-tō-kə-'ne-tik, -kī-\ *adjective* (1925)
: of, relating to, or involving movements of the eyes

op·tom·e·trist \äp-'tä-mə-trist\ *noun* (1903)
: a specialist licensed to practice optometry — compare OPHTHALMOLOGIST, OPTICIAN

op·tom·e·try \-trē\ *noun* [International Scientific Vocabulary] (1886)
: the art or profession of examining the eye for defects and faults of refraction and prescribing correctional lenses or exercises
— op·to·met·ric \ˌäp-tə-'me-trik\ *adjective*

opt out *intransitive verb* (1951)
: to choose not to participate in something — often used with *of* ⟨*opted out* of the project⟩

op·u·lence \'ä-pyə-lən(t)s\ *noun* (circa 1510)
1 : WEALTH, AFFLUENCE
2 : ABUNDANCE, PROFUSION

op·u·lent \-lənt\ *adjective* [Latin *opulentus,* from *ops* power, help; akin to Latin *opus* work] (1601)
: exhibiting or characterized by opulence: as **a :** having a large estate or property **:** WEALTHY ⟨hoping to marry an *opulent* widow⟩ **b :** amply or plentifully provided or fashioned often to the point of ostentation ⟨living in *opulent* comfort⟩
synonym see RICH
— op·u·lent·ly *adverb*

opun·tia \ō-'pən(t)-sh(ē-)ə\ *noun* [Latin, a plant, from feminine of *opuntius* of Opus, from *Opunt-, Opus* Opus, ancient city in Greece] (1601)
: any of a large genus (*Opuntia*) of cacti with usually yellow flowers and flat or terete joints usually studded with tubercles bearing spines or prickly hairs — compare CHOLLA, PRICKLY PEAR

opus \'ō-pəs\ *noun, plural* **op·era** \'ō-pə-rə, 'ä-\ *also* **opus·es** \'ō-pə-səz\ [Latin *oper-, opus* — more at OPERATE] (1809)
: WORK; *especially* **:** a musical composition or set of compositions usually numbered in the order of its issue

opus·cule \ō-'pəs-(ˌ)kyü(ə)l\ *noun* [French, from Latin *opusculum,* diminutive of *opus*] (circa 1656)
: a small or petty work **:** OPUSCULUM

opus·cu·lum \ō-'pəs-kyə-ləm\ *noun, plural* **-la** \-lə\ [Latin] (1654)
: a minor work (as of literature) — usually used in plural

[1]or \ər, 'ȯr, *Southern also* 'är\ *conjunction* [Middle English, alteration of *other,* alteration of Old English *oththe;* akin to Old High German *eddo* or] (13th century)
1 — used as a function word to indicate an alternative ⟨coffee *or* tea⟩ ⟨sink *or* swim⟩, the equivalent or substitutive character of two words or phrases ⟨lessen *or* abate⟩, or approximation or uncertainty ⟨in five *or* six days⟩
2 *archaic* **:** EITHER
3 *archaic* **:** WHETHER
4 — used in logic as a sentential connective that forms a complex sentence which is true when at least one of its constituent sentences is true; compare DISJUNCTION

[2]or *preposition* [Middle English, from *or,* adverb, early, before, from Old Norse *ār,* akin to Old English *ǣr* early — more at ERE] (13th century)
archaic **:** BEFORE

[3]or *conjunction* (13th century)
archaic **:** BEFORE

[4]or \'ȯr\ *noun* [Middle English, from Middle French, gold, from Latin *aurum* — more at AUREUS] (15th century)
: the heraldic color gold or yellow

OR \'ȯr\ *noun* [[1]*or*] (1947)
: a logical operator that requires either of two inputs to be present or conditions to be met for

an output to be made or a statement to be executed ⟨*OR* gate in a computer⟩

¹-or *noun suffix* [Middle English, from Old French *-eur, -eor* & Latin *-or;* Old French *-eur,* from Latin *-or;* Old French *-eor,* from Latin *-ator,* from *-a-,* verb stem + *-tor,* agent suffix; akin to Greek *-tōr,* agent suffix, Sanskrit *-tā*]
: one that does a (specified) thing ⟨grant*or*⟩

²-or *noun suffix* [Middle English, from Old French *-eur,* from Latin *-or*]
: condition **:** activity ⟨demean*or*⟩

ora *plural of* os

or·ache *or* **or·ach** \'ȯr-ich, 'är-\ *noun* [Middle English *orage, arage,* from Middle French *arrache,* from (assumed) Vulgar Latin *atrapic-, atrapex,* from Greek *atraphaxys*] (13th century)
: any of various herbs (genus *Atriplex*) of the goosefoot family that include some (as *A. hortensis*) used as potherbs

or·a·cle \'ȯr-ə-kəl, 'är-\ *noun* [Middle English, from Middle French, from Latin *oraculum,* from *orare* to speak — more at ORATION] (15th century)
1 a : a person (as a priestess of ancient Greece) through whom a deity is believed to speak **b :** a shrine in which a deity reveals hidden knowledge or the divine purpose through such a person **c :** an answer or decision given by an oracle
2 a : a person giving wise or authoritative decisions or opinions **b :** an authoritative or wise expression or answer

orac·u·lar \ȯ-'ra-kyə-lər, ə-\ *adjective* [Latin *oraculum*] (1678)
1 : of, relating to, or being an oracle
2 : resembling an oracle (as in solemnity of delivery)
synonym see DICTATORIAL
— orac·u·lar·i·ty \-ˌra-kyə-'lar-ə-tē\ *noun*
— orac·u·lar·ly \-'ra-kyə-lər-lē\ *adverb*

¹oral \'ōr-əl, 'ȯr-, 'är-\ *adjective* [Latin *or-, os* mouth; akin to Old Norse *ōss* mouth of a river, Sanskrit *ās* mouth] (1628)
1 a : uttered by the mouth or in words **:** SPOKEN **b :** using speech or the lips especially in teaching the deaf
2 a : of, given through, or involving the mouth **b :** being on or relating to the same surface as the mouth
3 a : of, relating to, or characterized by the first stage of psychosexual development in psychoanalytic theory during which libidinal gratification is derived from intake (as of food), by sucking, and later by biting **b :** of, relating to, or characterized by personality traits of passive dependency and aggressiveness
usage see VERBAL
— oral·i·ty \ȯ-'ra-lə-tē, ō-\ *noun*
— oral·ly \'ōr-ə-lē, 'ȯr-, 'är-\ *adverb*

²oral *noun* (1876)
: an oral examination — usually used in plural

oral history *noun* (1955)
1 : tape-recorded historical information obtained in interviews concerning personal experiences and recollections; *also* **:** the study of such information
2 : a written work based on oral history
— oral historian *noun*

oral·ism \'ōr-ə-ˌli-zəm, 'ȯr-, 'är-\ *noun* (1883)
: advocacy or use of the oral method of teaching the deaf
— oral·ist \-list\ *noun*

orang \ə-'raŋ\ *noun* [by shortening] (1778)
: ORANGUTAN

¹or·ange \'är-inj, 'är(-ə)nj; *chiefly Northern & Midland* 'ȯr-inj, 'ȯr(-ə)nj\ *noun* [Middle English, from Middle French, from Old Provençal *auranja,* from Arabic *nāranj,* from Persian *nārang,* from Sanskrit *nāraṅga* orange tree] (14th century)
1 a : a globose berry with a yellowish to reddish orange rind and a sweet edible pulp **b :** any of various rather small evergreen trees (genus *Citrus*) with ovate unifoliolate leaves, hard yellow wood, fragrant white flowers, and fruits that are oranges
2 : any of several trees or fruits resembling the orange
3 : any of a group of colors that lie midway between red and yellow in hue

²orange *adjective* (1542)
1 : of or relating to the orange
2 : of the color orange

Orange *adjective* (1795)
: of, relating to, or sympathizing with Orangemen
— Or·ange·ism \'är-in-ˌji-zəm, 'är(ə-)n-, 'ȯr-in-, 'ȯr(ə-)n-\ *noun*

or·ange·ade \ˌär-in-'jād, ˌär(-ə)n-, ˌȯr-in-, ˌȯr(-ə)n-\ *noun* [French, from *orange* + *-ade*] (1706)
: a beverage of sweetened orange juice mixed with water

orange chromide *noun* [*chromide,* ultimately from Greek *chromis,* a sea fish] (1933)
: a brilliant orange or yellow red-spotted fish (*Etroplus maculatus*) often kept in tropical aquariums

orange hawkweed *noun* (circa 1900)
: a European hawkweed (*Hieracium aurantiacum*) that has flower heads with bright orange-red rays and is a troublesome weed especially in northeastern North America

Or·ange·man \'är-inj-mən, 'är(-ə)nj-, 'ȯr-inj-, 'ȯr(-ə)nj-\ *noun* [William III of England, prince of *Orange*] (1796)
1 : a member of a secret society organized in the north of Ireland in 1795 to defend the British sovereign and to support the Protestant religion
2 : a Protestant Irishman especially of Ulster

orange peel *noun* (circa 1909)
: a rough surface (as on porcelain) like that of an orange

orange pekoe *noun* (circa 1877)
: tea made from the smallest and youngest leaves of the shoot

or·ange·ry *also* **or·ange·rie** \'är-inj-rē, 'är(-ə)nj-, 'ȯr-inj-, 'ȯr(-ə)nj-\ *noun, plural* **-ries** (1664)
: a protected place and especially a greenhouse for raising oranges in cool climates

or·ange·wood \'är-inj-ˌwu̇d, 'är(-ə)nj-, 'ȯr-inj-, 'ȯr(-ə)nj-\ *noun* (1884)
: the wood of the orange tree used especially in turnery and carving

or·ang·ish \'är-in-jish, 'är(-ə)n-, 'ȯr-in-, 'ȯr(-ə)n-\ *adjective* (1967)
: somewhat orange

orang·u·tan \ə-'raŋ-ə-ˌtaŋ, -'raŋ-gə-, -ˌtan\ *noun* [Bazaar Malay (Malay-based pidgin), from Malay *orang* man + *hutan* forest] (1691)
: a largely herbivorous arboreal anthropoid ape (*Pongo pygmaeus*) of Borneo and Sumatra that is about two thirds as large as the gorilla and has brown skin, long sparse reddish brown hair, and very long arms ◆

orangutan

or·angy *or* **or·ang·ey** \'är-in-jē, 'är(-ə)n-, 'ȯr-in-, 'ȯr(-ə)n-\ *adjective* (1778)
: resembling or suggestive of an orange (as in flavor or color)

orate \ȯ-'rāt, ō-; 'ȯr-ˌāt, 'ōr-\ *intransitive verb* **orat·ed; orat·ing** [back-formation from *oration*] (circa 1600)
: to speak in an elevated and often pompous manner

ora·tion \ə-'rā-shən, ȯ-\ *noun* [Latin *oration-, oratio* speech, oration, from *orare* to plead, speak, pray; akin to Hittite *ariya-* to consult an oracle and perhaps to Greek *ara* prayer] (1502)
: an elaborate discourse delivered in a formal and dignified manner

or·a·tor \'ȯr-ə-tər, 'är-\ *noun* (15th century)
1 : one who delivers an oration
2 : one distinguished for skill and power as a public speaker

Or·a·to·ri·an \ˌȯr-ə-'tōr-ē-ən, ˌär-, -'tȯr-\ *noun* (circa 1656)
: a member of the Congregation of the Oratory of Saint Philip Neri founded in Rome in 1575 and comprising independent communities of secular priests under obedience but without vows
— Oratorian *adjective*

or·a·tor·i·cal \ˌȯr-ə-'tȯr-i-kəl, ˌär-ə-'tär-\ *adjective* (1589)
: of, relating to, or characteristic of an orator or oratory
— or·a·tor·i·cal·ly \-k(ə-)lē\ *adverb*

or·a·to·rio \ˌȯr-ə-'tōr-ē-ˌō, ˌär-, -'tȯr-\ *noun, plural* **-ri·os** [Italian, from the *Oratorio* di San Filippo Neri (Oratory of Saint Philip Neri) in Rome] (circa 1738)
: a lengthy choral work usually of a religious nature consisting chiefly of recitatives, arias, and choruses without action or scenery

¹or·a·to·ry \'ȯr-ə-ˌtōr-ē, 'är-, -ˌtȯr-\ *noun, plural* **-ries** [Middle English *oratorie,* from Late Latin *oratorium,* from Latin *orare*] (14th century)
1 : a place of prayer; *especially* **:** a private or institutional chapel
2 *capitalized* **:** an Oratorian congregation, house, or church

²oratory *noun* [Latin *oratoria,* from feminine of *oratorius* oratorical, from *orare*] (1593)
1 : the art of speaking in public eloquently or effectively
2 a : public speaking that employs oratory **b :** public speaking that is characterized by the use of stock phrases and that appeals chiefly to the emotions

¹orb \'ȯrb\ *noun* [Middle English, from Middle French *orbe,* from Latin *orbis* circle, disk, orb] (14th century)
1 : any of the concentric spheres in old astronomy surrounding the earth and carrying the celestial bodies in their revolutions
2 *archaic* **:** something circular **:** CIRCLE, ORBIT
3 : a spherical body; *especially* **:** a spherical celestial object
4 : EYE

◇ WORD HISTORY

orangutan The orangutan, native to Borneo and Sumatra, has been known in western Europe since the 17th century. The name for the animal, similar in most European languages to the English word, is borrowed from two Malay words, *orang* "man, person," and *hutan* "forest," but the Malay word for "orangutan" is *mawas,* and the precise origin of the compound *orangutan* is not known. Presumably Europeans borrowed it from Bazaar Malay, a group of restructured forms of Malay used as common languages by the many non-Malay speakers of the Indonesian archipelago. Queried by Europeans about the name of the animal and not knowing the actual Malay word, inhabitants of the islands perhaps responded by making up *orangutan* on the spot, the sense "forest man" fitting well enough as a descriptive epithet. There has been a tendency ever since *orangutan* was borrowed to rhyme the second and last syllables (as if \raŋ\ and \taŋ\) and in some European languages including English a *g* is sometimes added to the last syllable.

5 : a sphere surmounted by a cross symbolizing kingly power and justice

²orb (1600)
transitive verb
1 : to form into a disk or circle
2 *archaic* **:** ENCIRCLE, SURROUND, ENCLOSE
intransitive verb
archaic **:** to move in an orbit

or·bic·u·lar \ȯr-ˈbi-kyə-lər\ *adjective* [Middle English *orbiculer,* from Middle French or Late Latin; Middle French *orbiculaire,* from Late Latin *orbicularis,* from Latin *orbiculus,* diminutive of *orbis*] (15th century)
: SPHERICAL, CIRCULAR
— or·bic·u·lar·ly \-ˈbi-kyə-lər-lē\ *adverb*

or·bic·u·late \ȯr-ˈbi-kyə-lət\ *adjective* (circa 1760)
: circular or nearly circular in outline ⟨an *orbiculate* leaf⟩ — see LEAF illustration

¹or·bit \ˈȯr-bət\ *noun* [Middle English, from Medieval Latin *orbita,* from Latin, rut, track, probably from *orbis*] (15th century)
: the bony socket of the eye
— or·bit·al \ˈȯr-bə-ˈt^{ə}l\ *adjective*

²orbit *noun* [Latin *orbita* path, rut, orbit] (1696)
1 a : a path described by one body in its revolution about another (as by the earth about the sun or by an electron about an atomic nucleus); *also* **:** one complete revolution of a body describing such a path **b :** a circular path
2 : a range or sphere of activity or influence
synonym see RANGE
— or·bit·al \-t^{ə}l\ *adjective*

³orbit (1943)
transitive verb
1 : to revolve in an orbit around **:** CIRCLE
2 : to send up and make revolve in an orbit ⟨*orbit* a satellite⟩
intransitive verb
: to travel in circles

or·bit·al \ˈȯr-bə-t^{ə}l\ *noun* [*orbital,* adjective] (1932)
: a subdivision of a nuclear shell containing zero, one, or two electrons

or·bit·er \-bə-tər\ *noun* (1951)
: one that orbits: as **a :** a spacecraft designed to orbit a celestial body without landing on its surface **b :** SPACE SHUTTLE

orb weaver *noun* (1889)
: any of a family (Araneidae) of North American spiders that have eight similar eyes and typically spin a large elaborate web

or·ca \ˈȯr-kə\ *noun* [New Latin *Orca,* genus name, from Latin, a whale, probably modification of Greek *oryg-, oryx* — more at ORYX] (1866)
: KILLER WHALE

Or·ca·di·an \ȯr-ˈkā-dē-ən\ *noun* [Latin *Orcades* Orkney Islands] (1661)
: a native or inhabitant of the Orkney Islands
— Orcadian *adjective*

or·chard \ˈȯr-chərd\ *noun* [Middle English, from Old English *ortgeard,* from *ort-* (from Latin *hortus* garden) + *geard* yard — more at YARD] (before 12th century)
: a planting of fruit trees, nut trees, or sugar maples; *also* **:** the trees of such a planting

orchard grass *noun* (1765)
: a widely grown tall stout hay and pasture grass (*Dactylis glomerata*) that grows in tufts and has loose open panicles

or·chard·ist \ˈȯr-chər-dist\ *noun* (1794)
: an owner or supervisor of orchards

or·ches·tra \ˈȯr-kəs-trə, -ˌkes-\ *noun* [Latin, from Greek *orchēstra,* from *orcheisthai* to dance; perhaps akin to Sanskrit *ṛghāyati* he trembles, he rages] (1606)
1 a : the circular space used by the chorus in front of the proscenium in an ancient Greek theater **b :** a corresponding semicircular space in a Roman theater used for seating important persons
2 a : the space in front of the stage in a modern theater that is used by an orchestra **b :** the forward section of seats on the main floor of a theater **c :** the main floor of a theater
3 : a group of musicians including especially string players organized to perform ensemble music — compare BAND

or·ches·tral \ȯr-ˈkes-trəl\ *adjective* (circa 1811)
1 : of, relating to, or composed for an orchestra
2 : suggestive of an orchestra or its musical qualities
— or·ches·tral·ly \-trə-lē\ *adverb*

or·ches·trate \ˈȯr-kə-ˌstrāt\ *transitive verb* **-trat·ed; -trat·ing** (1880)
1 a : to compose or arrange (music) for an orchestra **b :** to provide with orchestration ⟨*orchestrate* a ballet⟩
2 : to arrange or combine so as to achieve a desired or maximum effect ⟨*orchestrated* preparations for the banquet⟩
— or·ches·tra·tor *also* **or·ches·trat·er** \-ˌstrā-tər\ *noun*

or·ches·tra·tion \ˌȯr-kə-ˈstrā-shən\ *noun* (circa 1859)
1 : the arrangement of a musical composition for performance by an orchestra; *also* **:** orchestral treatment of a musical composition
2 : harmonious organization ⟨develop a world community through *orchestration* of cultural diversities —L. K. Frank⟩
— or·ches·tra·tion·al \-shnəl, -shə-n^{ə}l\ *adjective*

or·chid \ˈȯr-kəd\ *noun* [irregular from New Latin *Orchis*] (1845)
1 : any of a large family (Orchidaceae, the orchid family) of perennial epiphytic or terrestrial monocotyledonous plants that usually have showy 3-petaled flowers with the middle petal enlarged into a lip and differing from the others in shape and color
2 : a light purple
— or·chid·like \-ˌlīk\ *adjective*

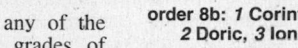

orchid 1

or·chi·da·ceous \ˌȯr-kə-ˈdā-shəs\ *adjective* [New Latin *Orchidaceae,* family name, from *Orchis*] (1838)
1 : of, relating to, or resembling the orchids
2 : SHOWY, OSTENTATIOUS

or·chis \ˈȯr-kəs\ *noun* [New Latin, from Latin, orchid, from Greek, testicle, orchid; akin to Middle Irish *uirgge* testicle] (1562)
: ORCHID; *especially* **:** any of a genus (*Orchis*) with fleshy roots and a spurred lip

or·dain \ȯr-ˈdān\ *verb* [Middle English *ordeinen,* from Old French *ordener,* from Late Latin *ordinare,* from Latin, to put in order, appoint, from *ordin-, ordo* order] (14th century)
transitive verb
1 : to invest officially (as by the laying on of hands) with ministerial or priestly authority
2 a : to establish or order by appointment, decree, or law **:** ENACT **b :** DESTINE, FOREORDAIN
intransitive verb
: to issue an order
— or·dain·er *noun*
— or·dain·ment \-ˈdān-mənt\ *noun*

or·deal \ȯr-ˈdē(-ə)l, ˈȯr-ˌ\ *noun* [Middle English *ordal,* from Old English *ordāl;* akin to Old High German *urteil* judgment, Old English *dāl* division — more at DEAL] (before 12th century)
1 : a primitive means used to determine guilt or innocence by submitting the accused to dangerous or painful tests believed to be under supernatural control ⟨*ordeal* by fire⟩
2 : a severe trial or experience

¹or·der \ˈȯr-dər\ *verb* **or·dered; or·der·ing** \ˈȯr-d(ə-)riŋ\ [Middle English, from *ordre,* noun] (13th century)
transitive verb
1 : to put in order **:** ARRANGE
2 a : to give an order to **:** COMMAND **b :** DESTINE, ORDAIN **c :** to command to go or come to a specified place **d :** to give an order for ⟨*order* a meal⟩
intransitive verb
1 : to bring about order **:** REGULATE
2 a : to issue orders **:** COMMAND **b :** to give or place an order ☆
— or·der·able \-ə-bəl\ *adjective*
— or·der·er \-dər-ər\ *noun*

²order *noun* [Middle English, from Middle French *ordre,* from Medieval Latin & Latin; Medieval Latin *ordin-, ordo* ecclesiastical order, from Latin, arrangement, group, class; akin to Latin *ordiri* to lay the warp, begin] (14th century)
1 a : a group of people united in a formal way: as (1) **:** a fraternal society ⟨the Masonic *Order*⟩ (2) **:** a community under a religious rule; *especially* **:** one requiring members to take solemn vows **b :** a badge or medal of such a society; *also* **:** a military decoration

1 2 3

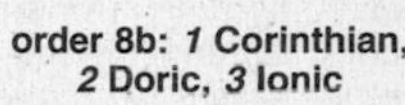
order 8b: *1* Corinthian, *2* Doric, *3* Ionic

2 a : any of the several grades of the Christian ministry **b** *plural* **:** the office of a person in the Christian ministry **c** *plural* **:** ORDINATION
3 a : a rank, class, or special group in a community or society **b :** a class of persons or things grouped according to quality, value, or natural characteristics: as (1) **:** a category of taxonomic classification ranking above the family and below the class (2) **:** the broadest category in soil classification
4 a (1) **:** RANK, LEVEL ⟨a statesman of the first *order*⟩ (2) **:** CATEGORY, CLASS ⟨in emergencies of this *order* —R. B. Westerfield⟩ **b** (1) **:** the arrangement or sequence of objects or of events in time ⟨listed the items in *order* of importance⟩ (2) **:** a sequential arrangement of mathematical elements **c :** DEGREE 12a, b **d** (1) **:** the number of times differentiation is applied

☆ SYNONYMS
Order, arrange, marshal, organize, systematize, methodize mean to put persons or things into their proper places in relation to each other. ORDER suggests a straightening out so as to eliminate confusion ⟨*ordered* her business affairs⟩. ARRANGE implies a setting in sequence, relationship, or adjustment ⟨*arranged* the files numerically⟩. MARSHAL suggests gathering and arranging in preparation for a particular operation or effective use ⟨*marshaling* the facts for argument⟩. ORGANIZE implies arranging so that the whole aggregate works as a unit with each element having a proper function ⟨*organized* the volunteers into teams⟩. SYSTEMATIZE implies arranging according to a predetermined scheme ⟨*systematized* billing procedures⟩. METHODIZE suggests imposing an orderly procedure rather than a fixed scheme ⟨*methodizes* every aspect of daily living⟩. See in addition COMMAND.

successively ⟨derivatives of higher *order*⟩ (2) *of a differential equation* **:** the order of the derivative of highest order **e :** the number of columns or rows or columns and rows in a magic square, determinant, or matrix ⟨the *order* of a matrix with 2 rows and 3 columns is 2 by 3⟩ **f :** the number of elements in a finite mathematical group
5 a (1) **:** a sociopolitical system ⟨was opposed to changes in the established *order*⟩ (2) **:** a particular sphere or aspect of a sociopolitical system ⟨the present economic *order*⟩ **b :** a regular or harmonious arrangement ⟨the *order* of nature⟩
6 a : a prescribed form of a religious service **:** RITE **b :** the customary mode of procedure especially in debate ⟨point of *order*⟩
7 a : the state of peace, freedom from confused or unruly behavior, and respect for law or proper authority ⟨promised to restore law and *order*⟩ **b :** a specific rule, regulation, or authoritative direction **:** COMMAND
8 a : a style of building **b :** a type of column and entablature forming the unit of a style
9 a : state or condition especially with regard to functioning or repair ⟨things were in terrible *order*⟩ **b :** a proper, orderly, or functioning condition ⟨their passports were in *order*⟩ ⟨the phone is out of *order*⟩
10 a : a written direction to pay money to someone **b :** a commission to purchase, sell, or supply goods or to perform work **c :** goods or items bought or sold **d :** an assigned or requested undertaking ⟨landing men on the moon was a large *order*⟩
11 : ORDER OF THE DAY ⟨flat roofs were the *order* in the small villages⟩
— or·der·less \-ləs\ *adjective*
— in order : APPROPRIATE, DESIRABLE ⟨an apology is *in order*⟩
— in order to : for the purpose of
— on order : in the process of being ordered
— on the order of 1 : after the fashion of **:** LIKE ⟨much *on the order of* Great Lakes bulk carriers —*Ships and the Sea*⟩ **2 :** ABOUT, APPROXIMATELY ⟨spent *on the order of* two million dollars⟩
— to order : according to the specifications of an order ⟨shoes made *to order*⟩

order arms *noun* [from the command *order arms!*] (1844)
1 : a position in the manual of arms in which the rifle is held vertically beside the right leg with the butt resting on the ground
2 : a command to return the rifle to order arms from present arms or to drop the hand from a hand salute

or·dered \'ȯr-dərd\ *adjective* (1579)
: characterized by order: as **a :** marked by regularity or discipline ⟨led an *ordered* life⟩ **b :** marked by regular or harmonious arrangement or disposition ⟨an *ordered* landscape⟩ ⟨the *ordered* crystal structure⟩ **c :** having elements arranged or identified according to a rule: as (1) **:** having the property that every pair of different elements is related by a transitive relationship that is not symmetric (2) **:** having elements labeled by ordinal numbers ⟨an *ordered* triple has a first, second, and third element⟩

or·der·li·ness \'ȯrd-ər-lē-nəs\ *noun* (1571)
: the quality or state of being orderly

¹or·der·ly \-lē\ *adjective* (circa 1577)
1 a (1) **:** arranged or disposed in some order or pattern **:** REGULAR ⟨*orderly* rows of houses⟩ (2) **:** not marked by disorder **:** TIDY ⟨keeps an *orderly* desk⟩ **b :** governed by law **:** REGULATED ⟨an *orderly* universe⟩ **c :** METHODICAL ⟨an *orderly* mind⟩
2 : well behaved **:** PEACEFUL ⟨an *orderly* crowd⟩
— orderly *adverb*

²orderly *noun, plural* **-lies** (1701)
1 : a soldier assigned to perform various services (as carrying messages) for a superior officer
2 : a hospital attendant who does routine or heavy work (as cleaning, carrying supplies, or moving patients)

order of business [*order of business* (predetermined sequence of matters to be dealt with by an assembly)] (circa 1890)
: a matter which must be dealt with **:** TASK ⟨the discipline problem was the first *order of business* at the meeting of the school board⟩

order of magnitude (1875)
: a range of magnitude extending from some value to ten times that value

order of the day (1698)
1 : the business or tasks appointed for an assembly for a given day
2 : the characteristic or dominant feature or activity ⟨growth and change are the *order of the day* in every field —Ruth G. Strickland⟩

¹or·di·nal \'ȯrd-nəl, 'ȯr-dᵊn-əl\ *noun* (14th century)
1 *capitalized* [Middle English, from Medieval Latin *ordinale,* from Late Latin, neuter of *ordinalis*] **:** a book of rites for the ordination of deacons, priests, and bishops
2 [Late Latin *ordinalis,* from *ordinalis,* adjective] **:** ORDINAL NUMBER

²ordinal *adjective* [Late Latin *ordinalis,* from Latin *ordin-, ordo*] (1599)
1 : of a specified order or rank in a series
2 : of or relating to a taxonomic order

ordinal number *noun* (1607)
1 : a number designating the place (as first, second, or third) occupied by an item in an ordered sequence — see NUMBER table
2 : a number assigned to an ordered set that designates both the order of its elements and its cardinal number

or·di·nance \'ȯrd-nən(t)s, 'ȯr-dᵊn-ən(t)s\ *noun* [Middle English, from Middle French & Medieval Latin; Middle French *ordenance,* literally, act of arranging, from Medieval Latin *ordinantia,* from Latin *ordinant-, ordinans,* present participle of *ordinare* to put in order — more at ORDAIN] (14th century)
1 a : an authoritative decree or direction **:** ORDER **b :** a law set forth by a governmental authority; *specifically* **:** a municipal regulation
2 : something ordained or decreed by fate or a deity
3 : a prescribed usage, practice, or ceremony
synonym see LAW

or·di·nand \ˌȯr-dᵊn-'and\ *noun* [Late Latin *ordinandus,* gerundive of *ordinare* to ordain] (circa 1842)
: a candidate for ordination

¹or·di·nary \'ȯr-dᵊn-ˌer-ē\ *noun, plural* **-naries** [Middle English *ordinarie,* from Anglo-French & Medieval Latin; Anglo-French, from Medieval Latin *ordinarius,* from Latin *ordinarius,* adjective] (14th century)
1 a (1) **:** a prelate exercising original jurisdiction over a specified territory or group (2) **:** a clergyman appointed formerly in England to attend condemned criminals **b :** a judge of probate in some states of the U.S.
2 *often capitalized* **:** the parts of the Mass that do not vary from day to day
3 : the regular or customary condition or course of things — usually used in the phrase *out of the ordinary*
4 a *British* **:** a meal served to all comers at a fixed price **b** *chiefly British* **:** a tavern or eating house serving regular meals
5 : a common heraldic charge (as the bend) of simple form

²ordinary *adjective* [Middle English *ordinarie,* from Latin *ordinarius,* from *ordin-, ordo* order] (15th century)
1 : of a kind to be expected in the normal order of events **:** ROUTINE, USUAL
2 : having or constituting immediate or original jurisdiction; *also* **:** belonging to such jurisdiction
3 a : of common quality, rank, or ability **b :** deficient in quality **:** POOR, INFERIOR
synonym see COMMON
— or·di·nar·i·ly \ˌȯr-dᵊn-'er-ə-lē\ *adverb*
— or·di·nar·i·ness \'ȯr-dᵊn-ˌer-ē-nəs\ *noun*

ordinary–language philosophy *noun* (1957)
: a trend in philosophical analysis that seeks to resolve philosophical perplexity by revealing sources of puzzlement in the misunderstanding of ordinary language

Ordinary level *noun* (1947)
: O LEVEL

ordinary seaman *noun* (1702)
: a seaman of some experience but not as skilled as an able seaman

ordinary share *noun* (1891)
British **:** a share of common stock

or·di·nate \'ȯrd-nət, 'ȯr-dᵊn-ət, -dᵊn-ˌāt\ *noun* [New Latin (*linea*) *ordinate* (*applicata*), literally, line applied in an orderly manner] (1676)
: the Cartesian coordinate obtained by measuring parallel to the y-axis — compare ABSCISSA

or·di·na·tion \ˌȯr-dᵊn-'ā-shən\ *noun* (14th century)
: the act or an instance of ordaining **:** the state of being ordained

ord·nance \'ȯrd-nən(t)s\ *noun* [Middle English *ordinaunce,* from Middle French *ordenance,* literally, act of arranging] (14th century)
1 a : military supplies including weapons, ammunition, combat vehicles, and maintenance tools and equipment **b :** a service of the army charged with the procuring, distributing, and safekeeping of ordnance
2 : CANNON, ARTILLERY

or·do \'ȯr-(ˌ)dō\ *noun, plural* **ordos** *or* **or·di·nes** \'ȯr-dᵊn-ˌēz\ [Medieval Latin, from Latin, order] (1849)
: a list of offices and feasts of the Roman Catholic Church for each day of the year

or·don·nance \ˌȯr-dᵊn-'äⁿs\ *noun* [French, alteration of Middle French *ordenance*] (1644)
: disposition of the parts (as of a literary composition) with regard to one another and the whole **:** ARRANGEMENT

Or·do·vi·cian \ˌȯr-də-'vi-shən\ *adjective* [Latin *Ordovices,* ancient people in northern Wales] (1879)
: of, relating to, or being the period between the Cambrian and the Silurian or the corresponding system of rocks — see GEOLOGIC TIME table
— Ordovician *noun*

or·dure \'ȯr-jər\ *noun* [Middle English, from Middle French, from *ord* filthy, from Latin *horridus* horrid] (14th century)
1 : EXCREMENT
2 : something that is morally degrading

¹ore \'ōr, 'ȯr\ *noun, often attributive* [Middle English *or, oor,* partly from Old English *ōra* ore; partly from Old English *ār* brass; akin to Old High German *ēr* bronze, Latin *aes* copper, bronze] (before 12th century)
1 : a mineral containing a valuable constituent (as metal) for which it is mined and worked
2 : a source from which valuable matter is extracted

²ore \'ər-ə\ *noun, plural* **ore** [Swedish *öre* & Danish & Norwegian *øre*] (1716)
— see *krona, krone* at MONEY table

ore·ad \'ōr-ē-ˌad, 'ȯr-, -ē-əd\ *noun* [Middle English *oreades,* plural, from Latin *oread-, oreas,* from Greek *oreiad-, oreias,* from *oreios* of a mountain, from *oros* mountain — more at ORIENT] (14th century)
: any of the nymphs of mountains and hills in Greek mythology

ore dressing *noun* (1862)
: mechanical preparation (as by crushing) and concentration (as by flotation) of ore

oreg·a·no \ə-ˈre-gə-ˌnō\ *noun* [American Spanish *orégano,* from Spanish, wild marjoram, from Latin *origanum* — more at ORIGANUM] (1771)
1 : a bushy perennial mint (*Origanum vulgare*) that is used as a seasoning and a source of aromatic oil — called also *origanum, wild marjoram*
2 : any of several plants (genera *Lippia* and *Coleus*) other than oregano of the vervain or mint families

Or·e·gon grape \ˈȯr-i-gən-, ˈär-, -ˌgän-\ *noun* [*Oregon,* state of the U.S.] (1869)
: an evergreen shrub (*Mahonia aquifolium*) of the barberry family that has yellow flowers, bears edible bluish black berries, and is native to western North America

Oregon grape

Or·eo \ˈȯr-ē-(ˌ)ō, ˈȯr-, ˈär-\ *noun* [from *Oreo,* trademark for a chocolate cookie with a white cream filling] (1969)
: a black person who adopts the characteristic mentality and behavior of white middle-class society — usually used disparagingly

Ores·tes \ə-ˈres-(ˌ)tēz, ȯ-\ *noun* [Latin, from Greek *Orestēs*]
: the son of Agamemnon and Clytemnestra who with his sister Electra avenges his father by killing his mother and her lover Aegisthus

or·gan \ˈȯr-gən\ *noun* [Middle English, partly from Old English *organa,* from Latin *organum,* from Greek *organon,* literally, tool, instrument; partly from Old French *organe,* from Latin *organum;* akin to Greek *ergon* work — more at WORK] (before 12th century)
1 a *archaic* **:** any of various musical instruments; *especially* **:** WIND INSTRUMENT **b** (1) **:** a wind instrument consisting of sets of pipes made to sound by compressed air and controlled by keyboards and producing a variety of musical effects — called also *pipe organ* (2) **:** REED ORGAN (3) **:** an instrument in which the sound and resources of the pipe organ are approximated by means of electronic devices (4) **:** any of various similar cruder instruments
2 a : a differentiated structure (as a heart, kidney, leaf, or stem) consisting of cells and tissues and performing some specific function in an organism **b :** bodily parts performing a function or cooperating in an activity ⟨the eyes and related structures that make up the visual *organs*⟩
3 : a subordinate group or organization that performs specialized functions ⟨the various *organs* of government⟩
4 : PERIODICAL

organ- *or* **organo-** *combining form* [Middle English, from Medieval Latin, from Latin *organum*]
1 : organ ⟨*organo*genesis⟩
2 : organic ⟨*organo*mercurial⟩

or·gan·dy *also* **or·gan·die** \ˈȯr-gən-dē\ *noun, plural* **-dies** [French *organdi*] (1835)
: a very fine transparent muslin with a stiff finish

or·gan·elle \ˌȯr-gə-ˈnel\ *noun* [New Latin *organella,* from Latin *organum*] (1920)
: a specialized cellular part (as a mitochondrion, lysosome, or ribosome) that is analogous to an organ

or·gan–grind·er \ˈȯr-gən-ˌgrīn-dər\ *noun* (circa 1807)
: one that cranks a hand organ; *especially* **:** a street musician who operates a barrel organ

[1]**or·gan·ic** \ȯr-ˈga-nik\ *adjective* (1517)
1 *archaic* **:** INSTRUMENTAL
2 a : of, relating to, or arising in a bodily organ **b :** affecting the structure of the organism
3 a (1) **:** of, relating to, or derived from living organisms (2) **:** of, relating to, yielding, or involving the use of food produced with the use of feed or fertilizer of plant or animal origin without employment of chemically formulated fertilizers, growth stimulants, antibiotics, or pesticides ⟨*organic* farming⟩ ⟨*organic* produce⟩ **b** (1) **:** of, relating to, or containing carbon compounds (2) **:** relating to, being, or dealt with by a branch of chemistry concerned with the carbon compounds of living beings and most other carbon compounds
4 a : forming an integral element of a whole **:** FUNDAMENTAL ⟨incidental music rather than *organic* parts of the action —Francis Fergusson⟩ **b :** having systematic coordination of parts **:** ORGANIZED ⟨an *organic* whole⟩ **c :** having the characteristics of an organism **:** developing in the manner of a living plant or animal ⟨society is *organic*⟩
5 : of, relating to, or constituting the law by which a government or organization exists
— **or·gan·i·cal·ly** \-ni-k(ə-)lē\ *adverb*
— **or·ga·nic·i·ty** \ˌȯr-gə-ˈni-sə-tē\ *noun*

[2]**organic** *noun* (1942)
: an organic substance: as **a :** a fertilizer of plant or animal origin **b :** a pesticide whose active component is an organic compound or a mixture of organic compounds

or·gan·i·cism \ȯr-ˈga-nə-ˌsi-zəm\ *noun* [International Scientific Vocabulary] (1883)
1 a : the explanation of life and living processes in terms of the levels of organization of living systems rather than in terms of the properties of their smallest components **b :** VITALISM
2 : any of various theories that attribute to society or the universe as a whole an existence or characteristics analogous to those of a biological organism
— **or·gan·i·cist** \-sist\ *noun or adjective*

or·ga·ni·sa·tion, or·ga·nise, or·ga·nis·er *British variant of* ORGANIZATION, ORGANIZE, ORGANIZER

or·gan·ism \ˈȯr-gə-ˌni-zəm\ *noun* (circa 1774)
1 : a complex structure of interdependent and subordinate elements whose relations and properties are largely determined by their function in the whole
2 : an individual constituted to carry on the activities of life by means of organs separate in function but mutually dependent **:** a living being
— **or·gan·is·mic** \ˌȯr-gə-ˈniz-mik\ *also* **or·gan·is·mal** \-məl\ *adjective*
— **or·gan·is·mi·cal·ly** \-mi-k(ə-)lē\ *adverb*

or·gan·ist \ˈȯr-gə-nist\ *noun* (1591)
: a person who plays the organ

[1]**or·ga·ni·za·tion** \ˌȯr-gə-nə-ˈzā-shən, ˌȯrg-nə-\ *noun* (15th century)
1 a : the act or process of organizing or of being organized **b :** the condition or manner of being organized
2 a : ASSOCIATION, SOCIETY ⟨charitable *organizations*⟩ **b :** an administrative and functional structure (as a business or a political party); *also* **:** the personnel of such a structure

[2]**organization** *adjective* (1949)
: characterized by complete conformity to the standards and requirements of an organization ⟨an *organization* man⟩

or·ga·ni·za·tion·al \-shnəl, -shə-nᵊl\ *adjective* (1881)
1 : of or relating to an organization **:** involving organization ⟨the *organizational* state of a crystal⟩
2 : ORGANIZATION
— **or·ga·ni·za·tion·al·ly** *adverb*

or·ga·nize \ˈȯr-gə-ˌnīz\ *verb* **-nized; -niz·ing** (15th century)
transitive verb
1 : to cause to develop an organic structure
2 : to form into a coherent unity or functioning whole **:** INTEGRATE ⟨trying to *organize* her thoughts⟩
3 a : to set up an administrative structure for **b :** to persuade to associate in an organization; *especially* **:** UNIONIZE
4 : to arrange by systematic planning and united effort
intransitive verb
1 : to undergo physical or organic organization
2 : to arrange elements into a whole of interdependent parts
3 : to form an organization; *specifically* **:** to form or persuade workers to join a union
synonym see ORDER
— **or·gan·iz·able** \ˌȯr-gə-ˈnī-zə-bəl\ *adjective*

organized *adjective* (1817)
1 : having a formal organization to coordinate and carry out activities ⟨*organized* baseball⟩ ⟨*organized* crime⟩
2 : affiliated by membership in an organization (as a union) ⟨*organized* steelworkers⟩

or·ga·niz·er \ˈȯr-gə-ˌnī-zər\ *noun* (1849)
1 : one that organizes
2 : a region of a developing embryo or a substance produced by such a region that is capable of inducing a specific type of development in undifferentiated tissue — called also *inductor*

or·gan·o·chlo·rine \ˌȯr-gə-nō-ˈklȯr-ˌēn, ȯr-ˌga-nə-, -ˈklȯr-, -ən\ *adjective* (1961)
: of, relating to, or belonging to the chlorinated hydrocarbon pesticides (as aldrin, DDT, or dieldrin)
— **organochlorine** *noun*

organ of Cor·ti \-ˈkȯr-tē\ [Alfonso *Corti* (died 1876) Italian anatomist] (1882)
: a complex epithelial structure in the cochlea that rests on the internal surface of the basilar membrane and in mammals is the chief part of the ear by which sound is directly perceived

or·gan·o·gen·e·sis \ˌȯr-gə-nō-ˈje-nə-səs, ȯr-ˌga-nə-\ *noun* [New Latin] (circa 1860)
: the origin and development of bodily organs — compare MORPHOGENESIS
— **or·gan·o·ge·net·ic** \-jə-ˈne-tik\ *adjective*

or·gan·o·lep·tic \ˌȯr-gə-nō-ˈlep-tik, ȯr-ˌga-nə-\ *adjective* [French *organoleptique,* from *organ-* + Greek *lēptikos* disposed to take, from *lambanein* to take — more at LATCH] (1852)
1 : being, affecting, or relating to qualities (as taste, color, odor, and feel) of a substance (as a food or drug) that stimulate the sense organs
2 : involving use of the sense organs ⟨*organoleptic* evaluation of foods⟩
— **or·gan·o·lep·ti·cal·ly** \-ti-k(ə-)lē\ *adverb*

or·ga·nol·o·gy \ˌȯr-gə-ˈnä-lə-jē\ *noun* [International Scientific Vocabulary] (circa 1842)
: the study of the organs of plants and animals

or·gan·o·mer·cu·ri·al \ˌȯr-gə-nō-(ˌ)mər-ˈkyu̇r-ē-əl, ȯr-ˌga-nə-\ *noun* (1938)
: an organic compound or a pharmaceutical preparation containing mercury

or·gan·o·me·tal·lic \-mə-ˈta-lik\ *adjective* [International Scientific Vocabulary] (1852)
: of, relating to, or being an organic compound that usually contains a metal or metalloid bonded directly to carbon
— **organometallic** *noun*

or·ga·non \ˈȯr-gə-ˌnän\ *noun* [Greek, literally, tool — more at ORGAN] (1610)
: an instrument for acquiring knowledge; *specifically* **:** a body of principles of scientific or philosophic investigation

or·gan·o·phos·phate \ˌȯr-gə-nō-ˈfäs-ˌfāt, ȯr-ˌga-nō-\ *noun* (1949)
: an organophosphorus compound (as a pesticide)
— **organophosphate** *adjective*

or·gan·o·phos·pho·rus \-ˈfäs-f(ə-)rəs\ *also* **or·gan·o·phos·pho·rous** \-fäs-ˈfōr-əs, -ˈfȯr-\ *adjective* (1950)
: of, relating to, or being a phosphorus-containing organic compound and especially a pesticide (as malathion) that acts by inhibiting cholinesterase
— **organophosphorus** *noun*

organ–pipe cactus *noun* (1908)
: any of several tall upright cacti (as *Lemaireocereus thurberi* or *L. marginatus*) of the southwestern U.S. and adjacent Mexico that usually branch at the base to form several upright stems

or·ga·num \ˈȯr-gə-nəm\ *noun* [Medieval Latin, from Latin, organ] (1614)
1 : ORGANON
2 : early polyphony of the late Middle Ages that consists of one or more voice parts accompanying the cantus firmus in parallel motion usually at a fourth, fifth, or octave above or below; *also* **:** a composition in this style

or·gan·za \ȯr-ˈgan-zə\ *noun* [probably alteration of *Lorganza,* a trademark] (1820)
: a sheer dress fabric (as of silk or nylon) resembling organdy

or·gan·zine \ˈȯr-gən-ˌzēn\ *noun* [French or Italian; French *organsin,* from Italian *organzino*] (1699)
: a raw silk yarn used for warp threads in fine fabrics

or·gasm \ˈȯr-ˌga-zəm\ *noun* [New Latin *orgasmus,* from Greek *orgasmos,* from *organ* to grow ripe, be lustful; probably akin to Sanskrit *ūrjā* sap, strength] (circa 1763)
: intense or paroxysmal excitement; *especially* **:** the climax of sexual excitement that is usually accompanied by the ejaculation of semen in the male and vaginal contractions in the female
— **or·gas·mic** \ȯr-ˈgaz-mik\ *also* **or·gas·tic** \-ˈgas-tik\ *adjective*

or·geat \ˈȯr-ˌzhä(t)\ *noun* [French, from Middle French, from *orge* barley, from Latin *hordeum;* akin to Old High German *gersta* barley] (1754)
: a sweet almond-flavored nonalcoholic syrup used as a cocktail ingredient or food flavoring

or·gi·as·tic \ˌȯr-jē-ˈas-tik\ *adjective* [Greek *orgiastikos,* from *orgiazein* to celebrate orgies, from *orgia*] (1698)
1 : of, relating to, or marked by orgies
2 : characterized by unrestrained emotion **:** FRENZIED
— **or·gi·as·ti·cal·ly** \-ti-k(ə-)lē\ *adverb*

or·gone \ˈȯr-ˌgōn\ *noun* [probably from *orgasm* + *-one* (as in *hormone*)] (1942)
: a vital energy held to pervade nature and be a factor in health in the theories of Wilhelm Reich

or·gu·lous \ˈȯr-gyə-ləs, -gə-\ *adjective* [Middle English, from Old French *orgueilleus,* from *orgueil* pride, of Germanic origin; akin to Old High German *urguol* distinguished] (13th century)
: PROUD

or·gy \ˈȯr-jē\ *noun, plural* **orgies** [Middle French *orgie,* from Latin *orgia,* plural, from Greek; akin to Greek *ergon* work — more at WORK] (circa 1561)
1 : secret ceremonial rites held in honor of an ancient Greek or Roman deity and usually characterized by ecstatic singing and dancing
2 a : drunken revelry **b :** a sexual encounter involving many people; *also* **:** an excessive sexual indulgence
3 : something that resembles an orgy in lack of control or moderation ⟨an *orgy* of destruction⟩

-oria *plural of* -ORIUM

-orial *adjective suffix* [Middle English, from Latin *-orius* -ory + Middle English *-al*]
: of, belonging to, or connected with ⟨combinatorial⟩

orib·a·tid \ō-ˈri-bə-təd, ˌȯr-ə-ˈba-təd\ *noun* [New Latin *Oribatidae* (coextensive with *Oribatoidea*), from *Oribata,* genus name, from Greek *oribatēs* walking the mountains, from *oros* mountain + *-batēs,* from *bainein* to go — more at ORIENT, COME] (1875)
: any of a superfamily (Oribatoidea) of small oval eyeless nonparasitic mites having a heavily sclerotized integument with a leathery appearance
— **oribatid** *adjective*

ori·el \ˈōr-ē-əl, ˈȯr-\ *noun* [Middle English, porch, oriel, from Middle French *oriol* porch] (14th century)
: a large bay window projecting from a wall and supported by a corbel or bracket

1**ori·ent** \ˈōr-ē-ənt, ˈȯr-, -ē-ˌent\ *noun* [Middle English, from Middle French, from Latin *orient-, oriens,* from present participle of *oriri* to rise; akin to Sanskrit *ṛnoti* he moves, arises, Greek *ornynai* to rouse, *oros* mountain] (14th century)
1 *archaic* **:** EAST 1b
2 *capitalized* **:** EAST 2
3 a : a pearl of great luster **b :** the luster of a pearl ◆

2**orient** *adjective* (15th century)
1 *archaic* **:** ORIENTAL 1
2 a : LUSTROUS, SPARKLING ⟨*orient* gems⟩ **b** *archaic* **:** RADIANT, GLOWING
3 *archaic* **:** rising in the sky

3**ori·ent** \ˈōr-ē-ˌent, ˈȯr-\ *transitive verb* [French *orienter,* from Middle French, from *orient*] (circa 1741)
1 a : to cause to face or point toward the east; *specifically* **:** to build (a church or temple) with the longitudinal axis pointing eastward and the chief altar at the eastern end **b :** to set or arrange in any determinate position especially in relation to the points of the compass **c :** to ascertain the bearings of
2 a : to set right by adjusting to facts or principles **b :** to acquaint with the existing situation or environment
3 : to direct (as a book or film) toward the interests of a particular group
4 : to cause the axes of the molecules of to assume the same direction
usage see ORIENTATE

ori·en·tal \ˌōr-ē-ˈen-tᵊl, ˌȯr-\ *adjective* (14th century)
1 *often capitalized* **:** of, relating to, or situated in the Orient
2 a : of superior grade, luster, or value **b :** being corundum or sapphire but simulating another gem in color
3 *often capitalized* **:** of, relating to, or having the characteristics of Orientals
4 *capitalized* **:** of, relating to, or constituting the biogeographic region that includes Asia south and southeast of the Himalayas and the Malay Archipelago west of Wallace's line
— **ori·en·tal·ly** \-tᵊl-ē\ *adverb*

Oriental *noun* (15th century)
: a member of one of the indigenous peoples of the Orient

oriental fruit moth *noun* (1921)
: a small nearly cosmopolitan moth (*Grapholita molesta*) probably of Japanese origin whose larva is injurious to the twigs and fruit of orchard trees and especially the peach — called also *oriental peach moth*

Ori·en·ta·lia \ˌōr-ē-ən-ˈtāl-yə, ˌȯr-, -ˌen-, -ˈtā-lē-ə\ *noun plural* [New Latin] (1903)
: materials concerning or characteristic of the Orient

ori·en·tal·ism \ˌōr-ē-ˈen-tᵊl-ˌi-zəm\ *noun, often capitalized* (1769)
1 : a trait, custom, or habit of expression characteristic of oriental peoples
2 : scholarship or learning in oriental subjects
— **ori·en·tal·ist** \-tᵊl-ist\ *noun, often capitalized*

ori·en·tal·ize \-tᵊl-ˌīz\ *verb* **-ized; -iz·ing** (1823)
transitive verb
often capitalized **:** to make oriental
intransitive verb
often capitalized **:** to become oriental

Oriental poppy *noun* (1731)
: an Asian perennial poppy (*Papaver orientale*) that is commonly cultivated for its large showy flowers

Oriental rug *noun* (1881)
: a handwoven or hand-knotted one-piece rug or carpet made in the Orient — called also *Oriental carpet*

ori·en·tate \ˈōr-ē-ən-ˌtāt, ˈȯr-, -ˌen-\ *verb* **-tat·ed; -tat·ing** (1849)
transitive verb
: ORIENT
intransitive verb
: to face or turn to the east ◻

ori·en·ta·tion \ˌōr-ē-ən-ˈtā-shən, ˌȯr-, -ˌen-\ *noun* (1839)
1 a : the act or process of orienting or of being oriented **b :** the state of being oriented; *broadly* **:** ARRANGEMENT, ALIGNMENT
2 : a usually general or lasting direction of thought, inclination, or interest
3 : change of position by organs, organelles, or organisms in response to external stimulus
— **ori·en·ta·tion·al** \-shnəl, -shə-nᵊl\ *adjective*
— **ori·en·ta·tion·al·ly** *adverb*

ori·ent·ed \ˈōr-ē-ˌen-təd, ˈȯr-\ *adjective* (1944)
: intellectually, emotionally, or functionally directed ⟨humanistically *oriented* scholars⟩

ori·en·teer \ˌōr-ē-ən-ˈtir, ˌȯr-, -ˌen-\ *noun* [back-formation from *orienteering*] (1965)
: one who engages in orienteering

ori·en·teer·ing \ˌōr-ē-ən-ˈtir-iŋ, ˌȯr-, -ˌen-\ *noun* [modification of Swedish *orientering,* from *orientera* to orient] (1948)
: a cross-country race in which each participant uses a map and compass to navigate between checkpoints along an unfamiliar course

◻ USAGE

orientate This back-formation from *orientation* had been in use for about a century when it was first criticized as a needlessly long variant of *orient.* A surprisingly large number of handbooks and other commentators have subsequently disparaged *orientate.* Since the whole controversy boils down to three letters and one syllable, one might think the matter too trivial to have drawn so much attention. Both *orient* and *orientate* are standard, but *orientate* is more common in British English ⟨where the conservatives tend to be clerically *orientated* —*Times Literary Supplement*⟩ ⟨more emotionally or culturally *orientated* toward Europe —Duke of Edinburgh⟩. *Orient* is the usual choice in American English.

◇ WORD HISTORY

orient As the ancient Romans did not have magnetic compasses, they depended upon the position of the rising sun to determine their daytime bearings. The direction from which the sun rose was called *oriens,* deriving from the present participle of the verb *oriri,* meaning "to rise." The word *oriens* also came to be used for the part of the world in the direction from which the sun rose, the East. After *oriens* was borrowed into English as *Orient,* it became the name for all the lands to the east of Europe, and especially eastern Asia. With the spread of Christianity across Europe it became customary to build churches with their longitudinal axes pointing eastward toward Jerusalem. This practice gave rise to the use of *orient* as a verb meaning "to cause to face or point toward the east." This sense served as the basis for development of the verb's other senses in English.

or·i·fice \'ȯr-ə-fəs, 'är-\ *noun* [Middle English, from Middle French, from Late Latin *orificium,* from Latin *or-, os* mouth + *facere* to make, do — more at ORAL, DO] (15th century)
: an opening (as a vent, mouth, or hole) through which something may pass
— or·i·fi·cial \,ȯr-ə-'fi-shəl, ,är-\ *adjective*

ori·flamme \'ȯr-ə-,flam, 'är-\ *noun* [Middle English *oriflamble,* the banner of Saint Denis, from Middle French, from Medieval Latin *aurea flamma,* literally, golden flame] (1600)
: a banner, symbol, or ideal inspiring devotion or courage

ori·ga·mi \,ȯr-ə-'gä-mē\ *noun* [Japanese, from *ori* fold + *kami* paper] (1956)
: the Japanese art or process of folding squares of paper into representational shapes

orig·a·num \ə-'ri-gə-nəm\ *noun* [Middle English, from Latin, wild marjoram, from Greek *origanon*] (14th century)
: any of several aromatic mints (especially genus *Origanum*) used as seasonings; *especially* **:** OREGANO 1

or·i·gin \'ȯr-ə-jən, 'är-\ *noun* [Middle English *origine,* probably from Middle French, from Latin *origin-, origo,* from *oriri* to rise — more at ORIENT] (15th century)
1 : ANCESTRY, PARENTAGE
2 a : rise, beginning, or derivation from a source **b :** the point at which something begins or rises or from which it derives ⟨the *origin* of the custom is forgotten⟩; *also* **:** something that creates, causes, or gives rise to another ⟨this spring is the *origin* of the brook⟩
3 : the more fixed, central, or larger attachment of a muscle
4 : the intersection of coordinate axes ☆

¹orig·i·nal \ə-'rij-ə-nᵊl, -'rij-nəl\ *noun* (14th century)
1 *archaic* **:** the source or cause from which something arises; *specifically* **:** ORIGINATOR
2 a : that from which a copy, reproduction, or translation is made **b :** a work composed firsthand
3 a : a person of fresh initiative or inventive capacity **b :** a unique or eccentric person

²original *adjective* (14th century)
1 : of, relating to, or constituting an origin or beginning **:** INITIAL ⟨the *original* part of the house⟩
2 a : not secondary, derivative, or imitative **b :** being the first instance or source from which a copy, reproduction, or translation is or can be made
3 : independent and creative in thought or action **:** INVENTIVE
synonym see NEW

orig·i·nal·i·ty \ə-,ri-jə-'na-lə-tē\ *noun* (1742)
1 : the quality or state of being original
2 : freshness of aspect, design, or style
3 : the power of independent thought or constructive imagination

orig·i·nal·ly \ə-'ri-jə-nᵊl-ē; -'rij-nə-lē, -'ri-jən-\ *adverb* (14th century)
1 *archaic* **:** by origin or derivation **:** INHERENTLY
2 : in the beginning **:** in the first place **:** INITIALLY
3 : in a fresh or original manner

original sin *noun* (14th century)
: the state of sin that according to Christian theology characterizes all human beings as a result of Adam's fall

orig·i·nate \ə-'ri-jə-,nāt\ *verb* **-nat·ed; -nat·ing** (1667)
transitive verb
: to give rise to **:** INITIATE
intransitive verb
: to take or have origin **:** BEGIN
synonym see SPRING
— orig·i·na·tion \-,ri-jə-'nā-shən\ *noun*
— orig·i·na·tor \-'ri-jə-,nā-tər\ *noun*

orig·i·na·tive \ə-'ri-jə-,nā-tiv, -nə-\ *adjective* (1827)
: having ability to originate **:** CREATIVE
— orig·i·na·tive·ly *adverb*

O-ring \'ō-,riŋ\ *noun* (1946)
: a ring (as of synthetic rubber) used as a gasket

ori·ole \'ōr-ē-,ōl, 'ȯr-, -ē-əl\ *noun* [French *oriol,* from Latin *aureolus,* diminutive of *aureus* golden — more at AUREUS] (1776)
1 : any of various usually brightly colored Old World passerine birds (family Oriolidae and especially genus *Oriolus*)
2 : any of various New World passerine birds (family Icteridae and especially genus *Icterus*) of which the males are usually black and yellow or black and orange

oriole 2

Ori·on \ə-'rī-ən, ō-\ *noun* [Latin, from Greek *Ōriōn*]
1 : a giant hunter slain by Artemis in Greek mythology
2 [Latin (genitive *Orionis*)] **:** a constellation on the equator east of Taurus represented on charts by the figure of a hunter with belt and sword

or·is·mol·o·gy \,ȯr-əz-'mä-lə-jē, ,är-\ *noun* [Greek *horismos* definition (from *horizein* to define) + English *-logy* — more at HORIZON] (1816)
: the science of defining technical terms
— or·is·mo·log·i·cal \,ȯr-əz-mə-'lä-ji-kəl, ,är-; ō-,riz-\ *adjective*

or·i·son \'ȯr-ə-sən, 'är-, -zən\ *noun* [Middle English, from Old French, from Late Latin *oration-, oratio,* from Latin, oration] (13th century)
: PRAYER

-orium *noun suffix, plural* **-oriums** *or* **-oria** [Latin, from neuter of *-orius* -ory]
: ¹-ORY ⟨haust*orium*⟩

Ori·ya \ȯ-'rē-ə\ *noun* (1801)
: the Indo-Aryan language of Orissa, India

Or·lean·ist \'ȯr-lē-ə-nist, ȯr-'lē-(ə-)nist\ *noun* (1834)
: a supporter of the Orleans family in its claim to the throne of France by descent from a younger brother of Louis XIV

Or·lon \'ȯr-,län\ *trademark*
— used for an acrylic fiber

or·lop deck \'ȯr-,läp-\ *noun* [Middle English *overlop* deck of a single decker, from Middle Low German *overlōp,* literally, something that overleaps] (1758)
: the lowest deck in a ship having four or more decks

Or·mazd \'ȯr-(,)məzd, -,mazd\ *noun* [Persian *Urmazd,* from Middle Persian, from Avestan *Ahuramazdāh-*] (1603)
: AHURA MAZDA

or·mo·lu \'ȯr-mə-,lü\ *noun, often attributive* [French *or moulu,* literally, ground gold] (1765)
: golden or gilded brass or bronze used for decorative purposes (as in mounts for furniture)

¹or·na·ment \'ȯr-nə-mənt\ *noun* [Middle English, from Old French *ornement,* from Latin *ornamentum,* from *ornare*] (13th century)
1 *archaic* **:** a useful accessory
2 a : something that lends grace or beauty **b :** a manner or quality that adorns
3 : one whose virtues or graces add luster to a place or society
4 : the act of adorning or being adorned
5 : an embellishing note not belonging to the essential harmony or melody — called also *embellishment, fioritura*

²or·na·ment \-,ment\ *transitive verb* (1720)
: to provide with ornament **:** EMBELLISH
synonym see ADORN

¹or·na·men·tal \,ȯr-nə-'men-tᵊl\ *adjective* (1646)
: of, relating to, or serving as ornament; *specifically* **:** grown as an ornamental
— or·na·men·tal·ly \-tᵊl-ē\ *adverb*

²ornamental *noun* (1650)
: a decorative object; *especially* **:** a plant cultivated for its beauty rather than for use

or·na·men·ta·tion \,ȯr-nə-mən-'tā-shən, -,men-\ *noun* (1851)
1 : something that ornaments **:** EMBELLISHMENT
2 : the act or process of ornamenting **:** the state of being ornamented

or·nate \ȯr-'nāt\ *adjective* [Middle English *ornat,* from Latin *ornatus,* past participle of *ornare* to furnish, embellish; akin to Latin *ordo* order — more at ORDER] (15th century)
1 : marked by elaborate rhetoric or florid style
2 : elaborately or excessively decorated
— or·nate·ly *adverb*
— or·nate·ness *noun*

or·nery \'ȯr-nə-rē, 'är-; 'ȯrn-rē, 'ärn-\ *adjective* **or·neri·er; -est** [alteration of *ordinary*] (1816)
: having an irritable disposition **:** CANTANKEROUS ◆
— or·neri·ness *noun*

ornith- *or* **ornitho-** *combining form* [Latin, from Greek, from *ornith-, ornis* — more at ERNE]
: bird ⟨*ornitho*logy⟩

☆ SYNONYMS
Origin, source, inception, root mean the point at which something begins its course or existence. ORIGIN applies to the things or persons from which something is ultimately derived and often to the causes operating before the thing itself comes into being ⟨an investigation into the *origin* of baseball⟩. SOURCE applies more often to the point where something springs into being ⟨the *source* of the Nile⟩ ⟨the *source* of recurrent trouble⟩. INCEPTION stresses the beginning of something without implying causes ⟨the business has been a success since its *inception*⟩. ROOT suggests a first, ultimate, or fundamental source often not easily discerned ⟨the real *root* of the violence⟩.

◇ WORD HISTORY
ornery The word *ornery* was originally a vernacular, nonstandard pronunciation of *ordinary,* produced by loss of the syllable coming just after the initial primary stress—a common process in polysyllabic English words. The first hint of such a pronunciation is a three-syllable phonetic spelling of *ordinary* in *The English-School Reformed,* a guide to English spelling and pronunciation by one R. Brown published in 1700. It is not until the 19th century that the spelling *ornery* actually appears, in both British and American writing, presumably at the point that its meaning had shifted far enough from the "commonplace" sense of *ordinary* that its link with this spelling was forgotten and *ornery* was perceived as a distinct word. At first *ornery* is attested in the senses "of poor quality" or "unpleasant," though it soon is found with the meaning "cantankerous" that it usually has today. Spellings such as *o'n'ry* recorded in the 20th century show further phonetic loss: dropping of the first \r\ and deletion of the syllable following the primary stress.

or·nith·ic \ȯr-ˈni-thik\ *adjective* [Greek *ornithikos,* from *ornith-, ornis*] (1854)
: of, relating to, or characteristic of birds

or·ni·thine \ˈȯr-nə-ˌthēn\ *noun* [International Scientific Vocabulary *ornith*uric acid (a compound of which ornithine is a component, found in the urine of birds) + ²*-ine*] (1881)
: a crystalline amino acid $C_5H_{12}N_2O_2$ that functions especially in urea production as a carrier by undergoing conversion to citrulline and then arginine in reaction with ammonia and carbon dioxide followed by recovery along with urea by enzymatic hydrolysis of arginine

or·nith·is·chi·an \ˌȯr-nə-ˈthis-kē-ən\ *noun* [New Latin *Ornithischia,* from *ornith-* + *ischium*] (1933)
: any of an order (Ornithischia) of herbivorous dinosaurs (as a stegosaurus) that have the pubis of the pelvis rotated backward to a position parallel and close to the ischium — compare SAURISCHIAN
— **ornithischian** *adjective*

or·ni·thol·o·gy \ˌȯr-nə-ˈthä-lə-jē\ *noun, plural* **-gies** [New Latin *ornithologia,* from *ornith-* + *-logia* -logy] (1678)
1 : a branch of zoology dealing with birds
2 : a treatise on ornithology
— **or·ni·tho·log·i·cal** \-thə-ˈlä-ji-kəl\ *also* **or·ni·tho·log·ic** \-jik\ *adjective*
— **or·ni·tho·log·i·cal·ly** \-ji-k(ə-)lē\ *adverb*
— **or·ni·thol·o·gist** \-ˈthä-lə-jist\ *noun*

or·nith·o·pod \ȯr-ˈni-thə-ˌpäd, ˈȯr-ni-thə-\ *noun* [ultimately from Greek *ornith-* + *pod-, pous* foot — more at FOOT] (circa 1890)
: any of a suborder (Ornithopoda) of bipedal ornithischian dinosaurs (as a hadrosaur) with digitigrade walking limbs usually having only three functional toes

or·ni·thop·ter \ˈȯr-nə-ˌthäp-tər\ *noun* [International Scientific Vocabulary *ornith-* + *-pter* (as in *helicopter*)] (1908)
: an aircraft designed to derive its chief support and propulsion from flapping wings

or·ni·tho·sis \ˌȯr-nə-ˈthō-səs\ *noun, plural* **-tho·ses** \-ˌsēz\ [New Latin] (1939)
: PSITTACOSIS

¹**oro-** *combining form* [Greek *oros* — more at ORIENT]
: mountain ⟨*oro*graphy⟩

²**oro-** *combining form* [Latin *or-, os* — more at ORAL]
: mouth ⟨*oro*pharynx⟩

oro·gen·e·sis \ˌōr-ə-ˈje-nə-səs, ˌȯr-\ *noun* [New Latin] (1886)
: OROGENY
— **oro·ge·net·ic** \-jə-ˈne-tik\ *adjective*

orog·e·ny \ȯ-ˈrä-jə-nē\ *noun* [International Scientific Vocabulary] (1890)
: the process of mountain formation especially by folding of the earth's crust
— **oro·gen·ic** \ˌōr-ə-ˈje-nik, ˌȯr-\ *adjective*

oro·graph·ic \ˌōr-ə-ˈgra-fik, ˌȯr-\ *also* **oro·graph·i·cal** \-fi-kəl\ *adjective* (circa 1803)
: of or relating to mountains; *especially* **:** associated with or induced by the presence of mountains ⟨*orographic* rainfall⟩

orog·ra·phy \ȯ-ˈrä-grə-fē\ *noun* [International Scientific Vocabulary] (circa 1846)
: a branch of physical geography that deals with mountains

Oro·mo \ȯ-ˈrō-(ˌ)mō, ō-\ *noun, plural* **Oromos** *or* **Oromo** (1893)
1 : a member of a Cushitic-speaking people of southern Ethiopia and adjacent parts of Kenya
2 : the Cushitic language of the Oromo

oro·pha·ryn·geal \ˌōr-ə-ˌfar-ən-ˈjē-əl, ˌȯr-, -fə-ˈrin-j(ē-)əl\ *adjective* (1885)
1 : of or relating to the oropharynx
2 : of or relating to the mouth and pharynx

oro·phar·ynx \-ˈfar-iŋ(k)s\ *noun* (1887)
: the part of the pharynx that is below the soft palate and above the epiglottis and is continuous with the mouth

oro·tund \ˈȯr-ə-ˌtənd, ˈär-, ˈōr-\ *adjective* [modification of Latin *ore rotundo,* literally, with round mouth] (circa 1799)
1 : marked by fullness, strength, and clarity of sound **:** SONOROUS
2 : POMPOUS, BOMBASTIC
— **oro·tun·di·ty** \ˌȯr-ə-ˈtən-də-tē, ˌär-, ˌōr-\ *noun*

¹**or·phan** \ˈȯr-fən\ *noun* [Middle English, from Late Latin *orphanus,* from Greek *orphanos;* akin to Old High German *erbi* inheritance, Latin *orbus* orphaned] (15th century)
1 : a child deprived by death of one or usually both parents
2 : a young animal that has lost its mother
3 : one deprived of some protection or advantage ⟨*orphans* of the storm⟩
— **orphan** *adjective*
— **or·phan·hood** \-ˌhu̇d\ *noun*

²**orphan** *transitive verb* **or·phaned; or·phan·ing** \ˈȯr-fə-niŋ, ˈȯrf-niŋ\ (1814)
: to cause to become an orphan

or·phan·age \ˈȯr-fə-nij, ˈȯrf-nij\ *noun* (circa 1580)
1 : the state of being an orphan
2 : an institution for the care of orphans

orphan drug *noun* (1981)
: a drug that is not developed or marketed because its extremely limited use makes it unprofitable

orphan's court *noun* (1713)
: a probate court which in some states has jurisdiction over the affairs of minors and the administration of estates

Or·pheus \ˈȯr-ˌfyüs, -fē-əs\ *noun* [Latin, from Greek]
: a poet and musician in Greek mythology who almost rescues his wife Eurydice from Hades by charming Pluto and Persephone with his lyre

or·phic \ˈȯr-fik\ *adjective* (1678)
1 *capitalized* **:** of or relating to Orpheus or the rites or doctrines ascribed to him
2 : MYSTIC, ORACULAR
3 : FASCINATING, ENTRANCING
— **or·phi·cal·ly** \-fi-k(ə-)lē\ *adverb*

Or·phism \ˈȯr-ˌfi-zəm\ *noun* [*Orpheus,* its reputed founder] (1880)
: a mystic Greek religion offering initiates purification of the soul from innate evil and release from the cycle of reincarnation

or·phrey \ˈȯr-frē\ *noun, plural* **orphreys** [Middle English *orfrey,* from Middle French *orfreis,* from Medieval Latin *aurifrigium,* from Latin *aurum* gold + *Phrygius* Phrygian — more at AUREUS] (13th century)
1 a : elaborate embroidery **b :** a piece of such embroidery
2 : an ornamental border or band especially on an ecclesiastical vestment

or·pi·ment \ˈȯr-pə-mənt\ *noun* [Middle English, from Middle French, from Latin *auripigmentum,* from *aurum* + *pigmentum* pigment] (14th century)
: native orange to lemon-yellow arsenic trisulfide

or·pine \ˈȯr-pən\ *noun* [Middle English *orpin,* from Middle French, from *orpiment*] (14th century)
: an herb (*Sedum telephium* of the family Crassulaceae, the orpine family) that has fleshy leaves and pink or purple flowers and was formerly used in folk medicine; *broadly* **:** SEDUM

Or·ping·ton \ˈȯr-piŋ-tən\ *noun* [*Orpington,* England] (1897)
: any of an English breed of large deep-chested domestic fowls

or·rery \ˈȯr-ər-ē, ˈär-\ *noun, plural* **or·rer·ies** [Charles Boyle (died 1731) 4th Earl of *Orrery*] (1713)
: an apparatus showing the relative positions and motions of bodies in the solar system by balls moved by a clockwork

orrery

or·ris \ˈȯr-əs, ˈär-\ *noun* [probably alteration of Middle English *ireos,* from Medieval Latin, alteration of Latin *iris* iris] (1545)
: ORRISROOT

or·ris·root \-ˌrüt, -ˌru̇t\ *noun* (1598)
: the fragrant rootstock of any of three European irises (*Iris florentina, I. germanica,* and *I. pallida*) used especially in perfumery

ort \ˈȯrt\ *noun* [Middle English, from Middle Low German *orte*] (15th century)
: a morsel left at a meal **:** SCRAP

orth- *or* **ortho-** *combining form* [Middle English, from Middle French, straight, right, true, from Latin, from Greek, from *orthos;* akin to Sanskrit *ūrdhva* high, upright]
1 : straight **:** upright **:** vertical ⟨*ortho*grade⟩
2 : perpendicular ⟨*ortho*rhombic⟩
3 : correct **:** corrective ⟨*ortho*dontia⟩
4 a : hydrated or hydroxylated to the highest degree ⟨*ortho*phosphoric acid⟩ **b :** involving substitution at or characterized by or having the relationship of two neighboring positions in the benzene ring ⟨*ortho*-xylene⟩

or·thi·con \ˈȯr-thi-ˌkän\ *noun* [International Scientific Vocabulary *orth-* + *icono*scope] (1939)
: a camera tube similar to but more sensitive than an iconoscope in which the charges are scanned by a low-velocity beam

or·tho \ˈȯr-(ˌ)thō\ *adjective* (1904)
: ORTHOCHROMATIC

or·tho·cen·ter \ˈȯr-thə-ˌsen-tər\ *noun* [International Scientific Vocabulary] (1869)
: the common intersection of the three altitudes of a triangle or their extensions or of the several altitudes of a polyhedron provided these latter exist and meet in a point

or·tho·chro·mat·ic \ˌȯr-thə-krō-ˈma-tik\ *adjective* [International Scientific Vocabulary] (1887)
1 : of, relating to, or producing tone values of light and shade in a photograph that correspond to the tones in nature
2 : sensitive to all colors except red

or·tho·clase \ˈȯr-thə-ˌklās, -ˌklāz\ *noun* [German *Orthoklas,* from *orth-* + Greek *klasis* breaking, from *klan* to break — more at CLAST] (1849)
: a mineral consisting of a monoclinic form of feldspar

or·tho·don·tia \ˌȯr-thə-ˈdän(t)-sh(ē-)ə\ *noun* [New Latin] (circa 1849)
: ORTHODONTICS

or·tho·don·tics \-ˈdän-tiks\ *noun plural but singular in construction* (1909)
: a branch of dentistry dealing with irregularities of the teeth and their correction (as by means of braces)
— **or·tho·don·tic** \-tik\ *adjective*
— **or·tho·don·ti·cal·ly** \-ti-k(ə-)lē\ *adverb*
— **or·tho·don·tist** \-ˈdän-tist\ *noun*

¹**or·tho·dox** \ˈȯr-thə-ˌdäks\ *adjective* [Middle English *orthodoxe,* from Middle French or Late Latin; Middle French *orthodoxe,* from Late Latin *orthodoxus,* from Late Greek *orthodoxos,* from Greek *orth-* + *doxa* opinion — more at DOXOLOGY] (15th century)
1 a : conforming to established doctrine especially in religion **b :** CONVENTIONAL
2 *capitalized* **:** of, relating to, or constituting any of various conservative religious or political groups: as **a :** EASTERN ORTHODOX **b :** of or relating to Orthodox Judaism
— **or·tho·dox·ly** *adverb*

²orthodox *noun, plural* **orthodox** *also* **or·tho·dox·es** (1587)
1 : one that is orthodox
2 *capitalized* **:** a member of an Eastern Orthodox church

Orthodox Judaism *noun* (1904)
: Judaism that adheres to the Torah and Talmud as interpreted in an authoritative rabbinic law code and applies their principles and regulations to modern living — compare CONSERVATIVE JUDAISM, REFORM JUDAISM

or·tho·doxy \ˈȯr-thə-ˌdäk-sē\ *noun, plural* **-dox·ies** (1630)
1 : the quality or state of being orthodox
2 : an orthodox belief or practice
3 *capitalized* **a :** Eastern Orthodox Christianity **b :** ORTHODOX JUDAISM

or·tho·epy \ˈȯr-thə-ˌwe-pē, ȯr-ˈthō-ə-pē\ *noun* [New Latin *orthoepia,* from Greek *orthoepeia,* from *orth-* + *epos* word — more at VOICE] (1668)
1 : the customary pronunciation of a language
2 : the study of the pronunciation of a language
— **or·tho·ep·ic** \ˌȯr-thə-ˈwe-pik\ *adjective*
— **or·tho·ep·i·cal·ly** \-pi-k(ə-)lē\ *adverb*
— **or·tho·epist** \ˈȯr-thə-ˌwe-pist, ȯr-ˈthō-ə-pist\ *noun*

or·tho·gen·e·sis \ˌȯr-thə-ˈje-nə-səs\ *noun* [New Latin] (1895)
: variation of organisms in successive generations that in some evolutionary theories takes place in some predestined direction and results in progressive evolutionary trends independent of external factors
— **or·tho·ge·net·ic** \-jə-ˈne-tik\ *adjective*
— **or·tho·ge·net·i·cal·ly** \-ti-k(ə-)lē\ *adverb*

or·thog·o·nal \ȯr-ˈthä-gə-nᵊl\ *adjective* [Middle French, from Latin *orthogonius,* from Greek *orthogōnios,* from *orth-* + *gōnia* angle — more at -GON] (1612)
1 a : intersecting or lying at right angles **b :** having perpendicular slopes or tangents at the point of intersection ⟨*orthogonal* curves⟩
2 : having a sum of products or an integral that is zero or sometimes one under specified conditions: as **a** *of real-valued functions* **:** having the integral of the product of each pair of functions over a specific interval equal to zero **b** *of vectors* **:** having the scalar product equal to zero **c** *of a square matrix* **:** having the sum of products of corresponding elements in any two rows or any two columns equal to one if the rows or columns are the same and equal to zero otherwise **:** having a transpose with which the product equals the identity matrix
3 *of a linear transformation* **:** having a matrix that is orthogonal **:** preserving length and distance
4 : composed of mutually orthogonal elements ⟨an *orthogonal* basis of a vector space⟩
5 : statistically independent
— **or·thog·o·nal·i·ty** \-ˌthä-gə-ˈna-lə-tē\ *noun*
— **or·thog·o·nal·ly** \-ˈthä-gə-nᵊl-ē\ *adverb*

or·thog·o·nal·ize \ȯr-ˈthä-gə-nᵊl-ˌīz\ *transitive verb* **-ized; -iz·ing** (1930)
: to make orthogonal
— **or·thog·o·nal·i·za·tion** \-ˌthä-gə-nᵊl-ə-ˈzā-shən\ *noun*

or·tho·grade \ˈȯr-thə-ˌgrād\ *adjective* (1902)
: walking with the body upright or vertical

or·tho·graph·ic \ˌȯr-thə-ˈgra-fik\ *also* **or·tho·graph·i·cal** \-fi-kəl\ *adjective* (1706)
1 : of, relating to, being, or prepared by orthographic projection ⟨an *orthographic* map⟩
2 a : of or relating to orthography **b :** correct in spelling
— **or·tho·graph·i·cal·ly** \-fi-k(ə-)lē\ *adverb*

orthographic projection *noun* (1668)
1 : projection of a single view of an object (as a view of the front) onto a drawing surface in which the lines of projection are perpendicular to the drawing surface
2 : the representation of related views of an object as if they were all in the same plane and projected by orthographic projection

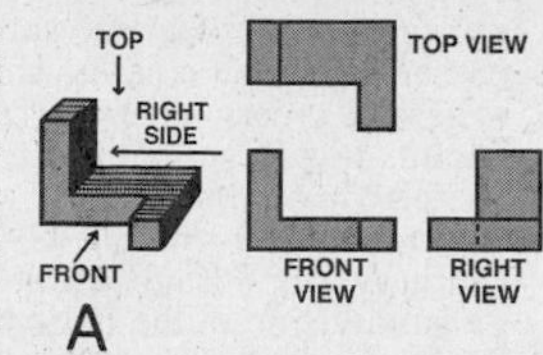

object *A* with top view, front view, and right view in orthographic projection

or·thog·ra·phy \ȯr-ˈthä-grə-fē\ *noun* [Middle English *ortografie,* from Middle French, from Latin *orthographia,* from Greek, from *orth-* + *graphein* to write — more at CARVE] (15th century)
1 a : the art of writing words with the proper letters according to standard usage **b :** the representation of the sounds of a language by written or printed symbols
2 : a part of language study that deals with letters and spelling

or·tho·mo·lec·u·lar \ˌȯr-thə-mə-ˈle-kyə-lər\ *adjective* (1968)
: relating to, based on, using, or being a theory according to which disease and especially mental illness may be cured by restoring the optimum amounts of substances normally present in the body ⟨*orthomolecular* therapy⟩ ⟨an *orthomolecular* psychiatrist⟩

or·tho·nor·mal \ˌȯr-thə-ˈnȯr-məl\ *adjective* (1932)
1 *of real-valued functions* **:** orthogonal with the integral of the square of each function over a specified interval equal to one
2 : being or composed of orthogonal elements of unit length ⟨*orthonormal* basis of a vector space⟩

or·tho·pe·dic *also* **or·tho·pae·dic** \ˌȯr-thə-ˈpē-dik\ *adjective* [French *orthopédique,* from *orthopédie* orthopedics, from *orth-* + Greek *paid-, pais* child — more at FEW] (1840)
1 : of, relating to, or employed in orthopedics
2 : marked by deformities or crippling
— **or·tho·pe·di·cal·ly** \-ˈpē-di-k(ə-)lē\ *adverb*

or·tho·pe·dics *also* **or·tho·pae·dics** \-ˈpē-diks\ *noun plural but singular or plural in construction* (circa 1853)
: a branch of medicine concerned with the correction or prevention of skeletal deformities
— **or·tho·pe·dist** \-ˈpē-dist\ *noun*

or·tho·phos·phate \ˌȯr-thə-ˈfäs-ˌfāt\ *noun* (1859)
: a salt or ester of orthophosphoric acid

or·tho·phos·pho·ric acid \ˌȯr-thə-ˌfäs-ˈfȯr-ik-, -ˈfär-; -ˈfäs-f(ə-)rik-\ *noun* [International Scientific Vocabulary] (1885)
: PHOSPHORIC ACID 1

or·tho·psy·chi·a·try \-sə-ˈkī-ə-trē, -(ˌ)sī-\ *noun* (circa 1927)
: prophylactic psychiatry concerned especially with incipient mental and behavioral disorders in youth
— **or·tho·psy·chi·at·ric** \-ˌsī-kē-ˈa-trik\ *adjective*
— **or·tho·psy·chi·a·trist** \-sə-ˈkī-ə-trist, -(ˌ)sī-\ *noun*

or·thop·tera \ȯr-ˈthäp-tə-rə\ *noun plural* [New Latin, order name, from *orth-* + Greek *pteron* wing — more at FEATHER] (1828)
: insects that are orthopterans

or·thop·ter·an \ȯr-ˈthäp-tə-rən\ *noun* [New Latin *Orthoptera*] (circa 1842)
: any of an order (Orthoptera) of insects (as crickets, grasshoppers, and sometimes mantises) that are characterized by biting mouthparts, two pairs of wings or none, and an incomplete metamorphosis
— **orthopteran** *adjective*
— **or·thop·ter·ist** \-rist\ *noun*
— **or·thop·ter·oid** \-ˌrȯid\ *noun or adjective*

or·tho·rhom·bic \ˌȯr-thə-ˈräm-bik\ *adjective* [International Scientific Vocabulary] (circa 1859)
: of, relating to, or constituting a system of crystallization characterized by three unequal axes at right angles to each other

or·tho·scop·ic \-ˈskä-pik\ *adjective* [International Scientific Vocabulary *orth-* + *-scopic* (as in *microscopic*)] (1853)
: giving an image in correct and normal proportions

or·tho·sis \ȯr-ˈthō-səs\ *noun, plural* **or·tho·ses** \-ˌsēz\ [New Latin, from Greek *orthōsis* straightening, from *orthoun* to straighten, from *orthos*] (1958)
: ORTHOTIC

or·tho·stat·ic \ˌȯr-thə-ˈsta-tik\ *adjective* (1902)
: of, relating to, or caused by erect posture ⟨*orthostatic* hypotension⟩

or·thot·ic \ȯr-ˈthä-tik\ *noun* [New Latin *orthosis*] (1955)
: a support or brace for weak or ineffective joints or muscles

or·thot·ics \-tiks\ *noun plural but singular in construction* (1957)
: a branch of mechanical and medical science that deals with the support and bracing of weak or ineffective joints or muscles
— **or·thot·ic** \-tik\ *adjective*
— **or·tho·tist** \ȯr-ˈthä-tist, ˈȯr-thə-tist\ *noun*

or·thot·ro·pous \ȯr-ˈthä-trə-pəs\ *adjective* [International Scientific Vocabulary] (1830)
: having the ovule straight so that the chalaza, hilum, and micropyle are in the same axial line

or·to·lan \ˈȯr-tᵊl-ən\ *noun* [French or Italian; French, from Italian *ortolano,* literally, gardener, from Latin *hortulanus,* from *hortulus,* diminutive of *hortus* garden — more at YARD] (1656)
: an Old World bunting (*Emberiza hortulana*) having a greenish gray head and breast, streaky brown back and wings, and a yellow throat

Or·vie·to \ȯr-ˈvyā-(ˌ)tō\ *noun* [*Orvieto,* city in central Italy] (1846)
: a usually dry Italian white wine

¹-ory *noun suffix* [Middle English *-orie,* from Latin *-orium,* from neuter of *-orius,* adjective suffix]
1 : place of or for ⟨observat*ory*⟩
2 : something that serves for ⟨cremat*ory*⟩

²-ory *adjective suffix* [Middle English *-orie,* from Middle French & Latin; Middle French, from Latin *-orius*]
1 : of, relating to, or characterized by ⟨gustat*ory*⟩
2 : serving for, producing, or maintaining ⟨justificat*ory*⟩

oryx \ˈōr-iks, ˈȯr-, ˈär-\ *noun, plural* **oryx** *or* **oryx·es** [New Latin, from Latin, a gazelle, from Greek, pickax, antelope, kind of whale, from *oryssein* to dig; akin to Latin *runcare* to grub up, weed, Sanskrit *luñcati* he plucks] (1535)
: any of a small genus (*Oryx*) of large heavily-built African and Arabian antelopes that have a light-colored coat with dark conspicuous markings especially on the face

oryx

or·zo \ˈȯrd-(ˌ)zō\ *noun* [Italian, literally, barley, from Latin *hordeum* — more at ORGEAT] (circa 1929)
: rice-shaped pasta

\ə\ abut \ᵊ\ kitten \ər\ further \a\ ash \ā\ ace \ä\ mop, mar \au̇\ out \ch\ chin \e\ bet \ē\ easy \g\ go \i\ hit \ī\ ice \j\ job \ŋ\ sing \ō\ go \ȯ\ law \ȯi\ boy \th\ thin \t͟h\ the \ü\ loot \u̇\ foot \y\ yet \zh\ vision *see also* Guide to Pronunciation

[1]os \'äs\ *noun, plural* **os·sa** \'ä-sə\ [Latin *oss-, os* — more at OSSEOUS] (15th century)
: BONE

[2]os \'ōs\ *noun, plural* **ora** \'ōr-ə, 'ȯr-ə\ [Latin *or-, os* — more at ORAL] (1737)
: MOUTH, ORIFICE

Osage \ō-'sāj, 'ō-ˌ\ *noun, plural* **Osag·es** *or* **Osage** (1698)
1 : a member of an American Indian people originally of Missouri
2 : the Siouan language of the Osage people

Osage orange *noun* (1817)
: an ornamental usually thorny American tree (*Maclura pomifera*) of the mulberry family with shiny ovate leaves and hard bright orange wood; *also* **:** its yellowish globose fruit

Os·can \'äs-kən\ *noun* [Latin *Oscus*] (1753)
1 : a member of a people of ancient Italy occupying Campania
2 : the language of the Oscan people — see INDO-EUROPEAN LANGUAGES table

[1]Os·car \'äs-kər\ *trademark*
— used especially for any of a number of golden statuettes awarded annually by a professional organization for notable achievement in motion pictures

[2]Oscar (1952)
— a communications code word for the letter *o*

os·cil·late \'ä-sə-ˌlāt\ *intransitive verb* **-lat·ed; -lat·ing** [Latin *oscillatus,* past participle of *oscillare* to swing, from *oscillum* swing] (1726)
1 a : to swing backward and forward like a pendulum **b :** to move or travel back and forth between two points
2 : to vary between opposing beliefs, feelings, or theories
3 : to vary above and below a mean value
synonym see SWING
— **os·cil·la·to·ry** \'ä-sə-lə-ˌtōr-ē, -ˌtȯr-\ *adjective*

os·cil·la·tion \ˌä-sə-'lā-shən\ *noun* (1658)
1 : the action or state of oscillating
2 : VARIATION, FLUCTUATION
3 : a flow of electricity changing periodically from a maximum to a minimum; *especially* **:** a flow periodically changing direction
4 : a single swing (as of an oscillating body) from one extreme limit to the other
— **os·cil·la·tion·al** \-shnəl, -shə-nᵊl\ *adjective*

os·cil·la·tor \'ä-sə-ˌlā-tər\ *noun* (1835)
1 : one that oscillates
2 : a device for producing alternating current; *especially* **:** a radio-frequency or audio-frequency generator

oscillo- *combining form* [International Scientific Vocabulary, from Latin *oscillare*]
: wave : oscillation ⟨*oscillo*scope⟩

os·cil·lo·gram \ä-'si-lə-ˌgram, ə-\ *noun* [International Scientific Vocabulary] (1903)
: a record made by an oscillograph or oscilloscope

os·cil·lo·graph \-ˌgraf\ *noun* [International Scientific Vocabulary] (1893)
: an instrument for recording alternating current wave forms or other electrical oscillations
— **os·cil·lo·graph·ic** \ä-ˌsi-lə-'gra-fik, ˌä-sə-lə-\ *adjective*
— **os·cil·lo·graph·i·cal·ly** \-fi-k(ə-)lē\ *adverb*
— **os·cil·log·ra·phy** \ˌä-sə-'lä-grə-fē\ *noun*

os·cil·lo·scope \ä-'si-lə-ˌskōp, ə-\ *noun* [International Scientific Vocabulary] (1906)
: an instrument in which the variations in a fluctuating electrical quantity appear temporarily as a visible wave form on the fluorescent screen of a cathode-ray tube
— **os·cil·lo·scop·ic** \ä-ˌsi-lə-'skä-pik, ˌä-sə-lə-\ *adjective*

os·cine \'ä-ˌsīn\ *adjective* [New Latin *Oscines,* suborder name, from Latin, plural of *oscin-, oscen* songbird, bird giving omens by its cry, from *obs-, ob-* in front of, in the way + *canere* to sing — more at OB-, CHANT] (1883)
: of or relating to a large suborder (Oscines) of passerine birds (as larks, shrikes, finches, orioles, and crows) characterized by a vocal apparatus highly specialized for singing
— **oscine** *noun*

Os·co–Um·bri·an \ˌäs-kō-'əm-brē-ən\ *noun* [Latin *Oscus* + English *Umbrian*] (1894)
: a subdivision of the Italic branch of the Indo-European language family containing Oscan and Umbrian

os·cu·late \'äs-kyə-ˌlāt\ *transitive verb* **-lat·ed; -lat·ing** [Latin *osculatus,* past participle of *osculari,* from *osculum* kiss, from diminutive of *os* mouth — more at ORAL] (circa 1656)
: KISS

os·cu·la·tion \ˌäs-kyə-'lā-shən\ *noun* (circa 1658)
: the act of kissing; *also* **:** KISS
— **os·cu·la·to·ry** \'äs-kyə-lə-ˌtōr-ē, -ˌtȯr-\ *adjective*

os·cu·lum \'äs-kyə-ləm\ *noun* [New Latin, from Latin, diminutive of *os* mouth] (1887)
: an excurrent opening of a sponge

[1]-ose *adjective suffix* [Middle English, from Latin *-osus*]
: full of : having : possessing the qualities of ⟨cym*ose*⟩

[2]-ose *noun suffix* [French, from *glucose*]
1 : carbohydrate ⟨amyl*ose*⟩; *especially* **:** sugar ⟨pent*ose*⟩
2 : primary hydrolysis product ⟨prote*ose*⟩

Osee \'ō-ˌzē, ō-'zā-ə\ *noun* [Late Latin, from Hebrew *Hōshēaʻ*]
: HOSEA

osier \'ō-zhər\ *noun* [Middle English, from Middle French, from Medieval Latin *auseria* osier bed] (14th century)
1 : any of various willows (especially *Salix viminalis*) whose pliable twigs are used for furniture and basketry
2 : a willow rod used in basketry
3 : any of several American dogwoods

Osi·ris \ō-'sī-rəs\ *noun* [Latin, from Greek, from Egyptian *Ws'r*]
: the Egyptian god of the underworld and husband and brother of Isis

-osis *noun suffix, plural* **-oses** *or* **-osises** [Middle English, from Latin, from Greek *-ōsis,* from *-ō-* (stem of causative verbs in *-oun*) + *-sis*]
1 a : action : process : condition ⟨hypn*osis*⟩ **b :** abnormal or diseased condition ⟨leuk*osis*⟩
2 : increase : formation ⟨leukocyt*osis*⟩

Os·man·li \äz-'man-lē\ *noun* [Turkish *osmanlı,* from *Osman,* founder of the Ottoman Empire] (1813)
1 : OTTOMAN 1
2 : TURKISH

os·me·te·ri·um \ˌäz-mə-'tir-ē-əm\ *noun, plural* **-ria** \-ē-ə\ [New Latin, irregular from Greek *osmē* odor] (1816)
: a protrusible forked glandular process that emits a disagreeable odor, is borne on the first thoracic segment of the larvae of many swallowtails and related butterflies, and is a defensive organ

osmic acid *noun* (1842)
: OSMIUM TETROXIDE

os·mi·rid·i·um \ˌäz-mə-'ri-dē-əm\ *noun* [Greek *osmē* + New Latin *iridium*] (1880)
: IRIDOSMINE

os·mi·um \'äz-mē-əm\ *noun* [New Latin, from Greek *osmē* odor] (1804)
: a hard brittle blue-gray or blue-black polyvalent metallic element of the platinum group with a high melting point that is the heaviest metal known and is used especially as a catalyst and in hard alloys — see ELEMENT table

osmium tetroxide *noun* (1876)
: a crystalline compound OsO_4 that is an oxide of osmium, has a poisonous irritating vapor, and is used as a catalyst, oxidizing agent, and biological fixative and stain

os·mol *or* **os·mole** \'äz-ˌmōl, 'äs-\ *noun* [blend of *osmosis* and *mol* (⁵mole)] (1942)
: a standard unit of osmotic pressure based on a one molal concentration of an ion in a solution

os·mo·lal·i·ty \ˌäz-mō-'la-lə-tē, ˌäs-\ *noun, plural* **-ties** [*osmol* + ¹*-al* + *-ity*] (circa 1944)
: the concentration of an osmotic solution especially when measured in osmols or milliosmols per 1000 grams of solvent
— **os·mo·lal** \äz-'mō-ləl, äs-\ *adjective*

os·mo·lar·i·ty \ˌäz-mō-'lar-ə-tē, ˌäs-\ *noun, plural* **-ties** [*osmol* + *-ar* + *-ity*] (1948)
: the concentration of an osmotic solution especially when measured in osmols or milliosmols per liter of solution
— **os·mo·lar** \äz-'mō-lər, äs-\ *adjective*

os·mom·e·ter \äz-'mä-mə-tər, äs-\ *noun* [*osmosis* + *-meter*] (1854)
: an apparatus for measuring osmotic pressure
— **os·mo·met·ric** \ˌäz-mə-'me-trik, ˌäs-\ *adjective*
— **os·mom·e·try** \äz-'mä-mə-trē\ *noun*

os·mo·reg·u·la·tion \'äz-mō-ˌre-gyə-'lā-shən, 'äs-\ *noun* [*osmosis* + *regulation*] (1927)
: regulation of osmotic pressure especially in the body of a living organism

os·mo·reg·u·la·to·ry \-'re-gyə-lə-ˌtōr-ē, -ˌtȯr-\ *adjective* (circa 1911)
: of, relating to, or concerned with the maintenance of constant osmotic pressure

os·mo·sis \äz-'mō-səs, äs-\ *noun* [New Latin, short for *endosmosis*] (1867)
1 : movement of a solvent through a semipermeable membrane (as of a living cell) into a solution of higher solute concentration that tends to equalize the concentrations of solute on the two sides of the membrane
2 : a process of absorption or diffusion suggestive of the flow of osmotic action; *especially* **:** a usually effortless often unconscious assimilation ⟨learned a number of languages by *osmosis* —Roger Kimball⟩

os·mot·ic \-'mä-tik\ *adjective* (1854)
: of, relating to, or having the properties of osmosis
— **os·mot·i·cal·ly** \-ti-k(ə-)lē\ *adverb*

osmotic pressure *noun* (1888)
: the pressure produced by or associated with osmosis and dependent on molar concentration and absolute temperature: as **a :** the maximum pressure that develops in a solution separated from a solvent by a membrane permeable only to the solvent **b :** the pressure that must be applied to a solution to just prevent osmosis

osmotic shock *noun* (1950)
: a rapid change in the osmotic pressure (as by transfer to a medium of different concentration) affecting a living system

os·mun·da \äz-'mən-də\ *noun* [New Latin, from Medieval Latin, from Old French *osmonde*] (1789)
: any of a genus (*Osmunda*) of rather large ferns (as the cinnamon fern) with pinnate or bipinnate fronds and fibrous creeping rhizomes

os·prey \'äs-prē, -ˌprā\ *noun, plural* **ospreys** [Middle English *ospray,* from (assumed) Middle French *osfraie,* from Latin *ossifraga,* a bird of prey] (15th century)
1 : a large fish-eating hawk (*Pandion haliaetus*) that is a dark brown color above and mostly pure white below
2 : a feather trimming used for millinery

osprey 1

ossa *plural of* OS

os·se·in \'ä-sē-ən\ *noun* [International Scientific Vocabulary, from Latin *oss-, os*] (1857)
: the collagen of bones

os·se·ous \'ä-sē-əs\ *adjective* [Latin *osseus,* from *oss-, os* bone; akin to Greek *osteon* bone, Sanskrit *asthi*] (1682)
: BONY 1

Os·sete \'ä-ˌsēt\ *also* **Os·set** \'ä-sət, -ˌset\ *noun* [Russian *osetin,* from *Osetiya* Ossetia, from Georgian *Oseti,* from *osi* Ossete] (1814)
: a member of a people of the central Caucasus
— **Os·se·tian** \ä-'sē-shən\ *adjective or noun*

Os·set·ic \ä-'se-tik\ *noun* (circa 1890)
: the Iranian language of the Ossetes

Os·si·an·ic \ˌä-sē-'a-nik, -shē-\ *adjective* (1808)
: of, relating to, or resembling the legendary Irish bard Ossian, the poems ascribed to him, or the rhythmic prose style used by James Macpherson in the poems he claimed to have translated from Ossian

os·si·cle \'ä-si-kəl\ *noun* [Latin *ossiculum,* diminutive of *oss-, os*] (1578)
: a small bone or bony structure (as the malleus, incus, or stapes)
— **os·sic·u·lar** \ä-'si-kyə-lər\ *adjective*

os·si·fi·ca·tion \ˌä-sə-fə-'kā-shən\ *noun* (1697)
1 a : the natural process of bone formation **b :** the hardening (as of muscular tissue) into a bony substance
2 : a mass or particle of ossified tissue
3 : a tendency toward or state of being molded into a rigid, conventional, sterile, or unimaginative condition

os·si·frage \'ä-sə-frij, -ˌfrāj\ *noun* [Latin *ossifraga,* a bird of prey, from feminine of *ossifragus* bone-breaking, from *oss-, os* + *frangere* to break — more at BREAK] (1601)
: LAMMERGEIER

os·si·fy \'ä-sə-ˌfī\ *verb* **-fied; -fy·ing** [Latin *oss-, os* + English *-ify*] (1713)
intransitive verb
1 : to change into bone
2 : to become hardened or conventional and opposed to change
transitive verb
1 : to change (as cartilage) into bone
2 : to make rigidly conventional and opposed to change

os·so bu·co *also* **os·so buc·co** \ˌō-sō-'bü-(ˌ)kō, ˌȯ-sō-\ *noun* [Italian *ossobuco* veal shank, literally, pierced bone] (1935)
: braised veal shanks

os·su·ary \'ä-shə-ˌwer-ē, -syə-, -sə-\ *noun, plural* **-ar·ies** [Late Latin *ossuarium,* from Latin, neuter of *ossuarius* of bones, from Old Latin *ossua,* plural of *oss-, os*] (1658)
: a depository for the bones of the dead

oste- *or* **osteo-** *combining form* [New Latin, from Greek, from *osteon* — more at OSSEOUS]
: bone ⟨*oste*al⟩ ⟨*osteo*myelitis⟩

os·te·al \'äs-tē-əl\ *adjective* [International Scientific Vocabulary] (1877)
: of, relating to, or resembling bone; *also* **:** affecting or involving bone or the skeleton

os·te·i·tis \ˌäs-tē-'ī-təs\ *noun* [New Latin] (circa 1847)
: inflammation of bone

os·ten·si·ble \ä-'sten(t)-sə-bəl, ə-\ *adjective* [French, from Latin *ostensus,* past participle of *ostendere* to show, from *obs-, ob-* in the way + *tendere* to stretch — more at OB-, THIN] (circa 1771)
1 : intended for display **:** open to view
2 : being such in appearance **:** plausible rather than demonstrably true or real ⟨the *ostensible* purpose for the trip⟩
synonym see APPARENT

os·ten·si·bly \-blē\ *adverb* (1765)
1 : in an ostensible manner
2 : to all outward appearances

os·ten·sive \ä-'sten(t)-siv\ *adjective* (1782)
1 : OSTENSIBLE 2
2 : of, relating to, or constituting definition by exemplifying the thing or quality being defined
— **os·ten·sive·ly** *adverb*

os·ten·so·ri·um \ˌäs-tən-'sōr-ē-əm, -ˌten-, -'sȯr-\ *noun, plural* **-ria** \-ē-ə\ [Medieval Latin, from Latin *ostendere*] (circa 1772)
: MONSTRANCE

os·ten·ta·tion \ˌäs-tən-'tā-shən\ *noun* [Middle English *ostentacion,* from Middle French *ostentation,* from Latin *ostentation-, ostentatio,* from *ostentare* to display, frequentative of *ostendere*] (15th century)
1 : excessive display **:** PRETENTIOUSNESS
2 *archaic* **:** an act of displaying

os·ten·ta·tious \-shəs\ *adjective* (1673)
: marked by or fond of conspicuous or vainglorious and sometimes pretentious display
synonym see SHOWY
— **os·ten·ta·tious·ly** *adverb*
— **os·ten·ta·tious·ness** *noun*

os·te·o·ar·thri·tis \ˌäs-tē-ō-är-'thrī-təs\ *noun* [New Latin] (1878)
: arthritis marked by degeneration of the cartilage and bone of joints
— **os·te·o·ar·thrit·ic** \-'thri-tik\ *adjective*

os·te·o·blast \'äs-tē-ə-ˌblast\ *noun* [International Scientific Vocabulary] (1875)
: a bone-forming cell
— **os·te·o·blas·tic** \ˌäs-tē-ə-'blas-tik\ *adjective*

os·te·o·clast \'äs-tē-ə-ˌklast\ *noun* [International Scientific Vocabulary *oste-* + Greek *klastos* broken — more at CLAST] (1872)
: any of the large multinucleate cells closely associated with areas of bone resorption
— **os·te·o·clas·tic** \ˌäs-tē-ə-'klas-tik\ *adjective*

os·te·o·cyte \'äs-tē-ə-ˌsīt\ *noun* (1942)
: a cell that is characteristic of adult bone and is isolated in a lacuna of the bone substance

os·te·o·gen·e·sis \ˌäs-tē-ə-'je-nə-səs\ *noun* [New Latin] (1830)
: development and formation of bone

osteogenesis im·per·fec·ta \-ˌim-(ˌ)pər-'fek-tə\ *noun* [New Latin, imperfect osteogenesis] (circa 1901)
: a hereditary disease marked especially by extreme brittleness of the long bones

os·te·o·gen·ic \ˌäs-tē-ə-'je-nik\ *adjective* (1867)
1 : producing bone
2 : originating in bone

osteogenic sarcoma *noun* (circa 1923)
: OSTEOSARCOMA

¹os·te·oid \'äs-tē-ˌȯid\ *adjective* [International Scientific Vocabulary] (1840)
: resembling bone

²osteoid *noun* (1934)
: uncalcified bone matrix

os·te·ol·o·gy \ˌäs-tē-'ä-lə-jē\ *noun* [New Latin *osteologia,* from Greek, description of bones, from *oste-* + *-logia* -logy] (1670)
1 : a branch of anatomy dealing with the bones
2 : the bony structure of an organism
— **os·te·o·log·i·cal** \-tē-ə-'lä-ji-kəl\ *adjective*
— **os·te·ol·o·gist** \-tē-'ä-lə-jist\ *noun*

os·te·o·ma \ˌäs-tē-'ō-mə\ *noun, plural* **-mas** *or* **-ma·ta** \-mə-tə\ [New Latin] (circa 1849)
: a benign tumor composed of bone tissue

os·te·o·ma·la·cia \ˌäs-tē-ō-mə-'lā-sh(ē-)ə\ *noun* [New Latin, from *oste-* + Greek *malakia* softness, from *malakos* soft — more at MOLLIFY] (circa 1834)
: a disease of adults that is characterized by softening of the bones and is analogous to rickets in the immature

os·te·o·my·e·li·tis \-ˌmī-ə-'lī-təs\ *noun* [New Latin] (1854)
: an infectious inflammatory disease of bone often of bacterial origin that is marked by local death and separation of tissue

os·te·o·path \'äs-tē-ə-ˌpath\ *noun* (1897)
: a practitioner of osteopathy

os·te·op·a·thy \ˌäs-tē-'ä-pə-thē\ *noun* [New Latin *osteopathia,* from *oste-* + Latin *-pathia* -pathy] (1899)
: a system of medical practice based on a theory that diseases are due chiefly to loss of structural integrity which can be restored by manipulation of the parts supplemented by therapeutic measures (as use of medicine or surgery)
— **os·te·o·path·ic** \ˌäs-tē-ə-'pa-thik\ *adjective*
— **os·te·o·path·i·cal·ly** \-thi-k(ə-)lē\ *adverb*

os·te·o·plas·ty \'äs-tē-ə-ˌplas-tē\ *noun* (circa 1860)
: plastic surgery on bone; *especially* **:** replacement of lost bone tissue or reconstruction of defective bony parts
— **os·te·o·plas·tic** \ˌäs-tē-ə-'plas-tik\ *adjective*

os·te·o·po·ro·sis \ˌäs-tē-ō-pə-'rō-səs\ *noun, plural* **-ro·ses** \-ˌsēz\ [New Latin, from *oste-* + *porosis* rarefaction, from *porus* pore + *-osis*] (1846)
: a condition that affects especially older women and is characterized by decrease in bone mass with decreased density and enlargement of bone spaces producing porosity and fragility
— **os·te·o·po·rot·ic** \-'rä-tik\ *adjective*

os·te·o·sar·co·ma \-sär-'kō-mə\ *noun, plural* **-mas** *or* **-ma·ta** \-mə-tə\ [New Latin] (circa 1826)
: a sarcoma derived from bone or containing bone tissue

osti·na·to \ˌäs-tə-'nä-(ˌ)tō, ˌȯs-\ *noun, plural* **-tos** *also* **-ti** \-tē\ [Italian, obstinate, from Latin *obstinatus*] (circa 1876)
: a musical figure repeated persistently at the same pitch throughout a composition — compare IMITATION, SEQUENCE

os·ti·ole \'äs-tē-ˌōl\ *noun* [New Latin *ostiolum,* from Latin, diminutive of *ostium*] (circa 1857)
: a small bodily aperture, orifice, or pore

os·ti·um \'äs-tē-əm\ *noun, plural* **os·tia** \-tē-ə\ [New Latin, from Latin, door, mouth of a river; akin to Latin *os* mouth — more at ORAL] (1828)
: a mouthlike opening in a bodily organ

os·tler *variant of* HOSTLER

ost·mark \'ōst-ˌmärk, 'ȯst-\ *noun* [German, literally, East mark] (1948)
: the former East German mark

os·to·my \'äs-tə-mē\ *noun, plural* **-mies** [*colostomy*] (1957)
: an operation (as a colostomy) to create an artificial passage for bodily elimination

-ostosis *noun combining form, plural* **-ostoses** *or* **-ostosises** [New Latin, from Greek *-ostōsis,* from *osteon* bone — more at OSSEOUS]
: ossification of a (specified) part or to a (specified) degree ⟨hyper*ostosis*⟩

os·tra·cise *British variant of* OSTRACIZE

os·tra·cism \'äs-trə-ˌsi-zəm\ *noun* (1588)
1 : a method of temporary banishment by popular vote without trial or special accusation practiced in ancient Greece
2 : exclusion by general consent from common privileges or social acceptance

os·tra·cize \-ˌsīz\ *transitive verb* **-cized; -ciz·ing** [Greek *ostrakizein* to banish by voting with potsherds, from *ostrakon* shell, potsherd — more at OYSTER] (1649)
1 : to exile by ostracism

2 : to exclude from a group by common consent ◆

os·tra·cod \ˈäs-trə-ˌkäd\ *also* **os·tra·code** \-ˌkōd\ *noun* [ultimately from Greek *ostrakon*] (1865)
: any of a subclass (Ostracoda) of very small active mostly freshwater crustaceans that have the body enclosed in a bivalve carapace, the body segmentation obscured, the abdomen rudimentary, and only seven pairs of appendages

os·tra·co·derm \ˈäs-trə-kō-ˌdərm, äs-ˈtra-kə-\ *noun* [ultimately from Greek *ostrakon* + *derma* skin — more at DERM-] (1891)
: any of the early fossil jawless fishes of the Lower Paleozoic usually having a bony covering of plates or scales

os·tra·con \ˈäs-trə-ˌkän\ *noun, plural* **-tra·ca** \-trə-kə\ [Greek *ostrakon* potsherd, shell — more at OYSTER] (1883)
: a fragment (as of pottery) containing an inscription — usually used in plural

os·trich \ˈäs-trich, ˈȯs- *also* -trij\ *noun* [Middle English, from Old French *ostrusce,* from (assumed) Vulgar Latin *avis struthio,* from Latin *avis* bird + Late Latin *struthio* ostrich — more at STRUTHIOUS] (13th century)
1 a : a swift-footed 2-toed flightless ratite bird (*Struthio camelus*) of Africa that is the largest of existing birds and often weighs 300 pounds (140 kilograms) **b :** RHEA **c :** leather made from ostrich skin
2 [from the belief that the ostrich when pursued hides its head in the sand and believes itself to be unseen] **:** one who attempts to avoid danger or difficulty by refusing to face it
— os·trich·like \-ˌlīk\ *adjective*

Os·tro·goth \ˈäs-trə-ˌgäth\ *noun* [Middle English, from Late Latin *Ostrogothi,* plural] (14th century)
: a member of the eastern division of the Goths
— Os·tro·goth·ic \ˌäs-trə-ˈgä-thik\ *adjective*

Os·we·go tea \ä-ˈswē-gō-\ *noun* [*Oswego* River, N. Y.] (1752)
: a North American mint (*Monarda didyma*) with showy scarlet irregular flowers

ot- *or* **oto-** *combining form* [Greek *ōt-, ōto-,* from *ōt-, ous* — more at EAR]
: ear ⟨*ot*itis⟩ **:** ear and ⟨*oto*laryngology⟩

Othel·lo \ə-ˈthe-(ˌ)lō, ō-\ *noun*
: a Moor in the military service of Venice, husband of Desdemona, and protagonist of Shakespeare's tragedy *Othello*

¹oth·er \ˈə-t͟hər\ *adjective* [Middle English, from Old English *ōther;* akin to Old High German *andar* other, Sanskrit *antara*] (before 12th century)
1 a : being the one (as of two or more) remaining or not included ⟨held on with one hand and waved with the *other* one⟩ **b :** being the one or ones distinct from that or those first mentioned or implied ⟨taller than the *other* boys⟩ **c :** SECOND ⟨every *other* day⟩
2 : not the same **:** DIFFERENT ⟨any *other* color would have been better⟩ ⟨something *other* than it seems to be⟩
3 : ADDITIONAL ⟨sold in the U.S. and 14 *other* countries⟩
4 a : recently past ⟨the *other* evening⟩ **b :** FORMER ⟨in *other* times⟩

²other *noun* (before 12th century)
1 a : one that remains of two or more **b :** a thing opposite to or excluded by something else ⟨went from one side to the *other*⟩
2 : a different or additional one ⟨the *others* came later⟩

³other *pronoun, sometimes plural in construction* (before 12th century)
1 *obsolete* **a :** one of two that remains **b :** each preceding one
2 : a different or additional one ⟨something or *other*⟩

⁴other *adverb* (13th century)
: OTHERWISE — used with *than* ⟨was unable to see them *other* than by going to their home⟩

oth·er–di·rect·ed \ˌə-t͟hər-də-ˈrek-təd, -dī-\ *adjective* (1950)
: directed in thought and action primarily by external norms rather than by one's own scale of values
— oth·er–di·rect·ed·ness *noun*

oth·er·guess \ˈə-t͟hər-ˌges\ *adjective* [alteration of English dialect *othergates*] (1632)
archaic **:** DIFFERENT

oth·er·ness \ˈə-t͟hər-nəs\ *noun* (1587)
1 : the quality or state of being other or different
2 : something that is other or different

other than *preposition* (1679)
: with the exception of **:** EXCEPT FOR, BESIDES ⟨*other than* that, nothing happened⟩

oth·er·where \-ˌhwer, -ˌhwar, -ˌwer, -ˌwar\ *adverb* (14th century)
: ELSEWHERE

oth·er·while \-ˌhwīl, -ˌwīl\ *also* **oth·er·whiles** \-ˌhwīlz, -ˌwīlz\ *adverb* (13th century)
chiefly dialect **:** at another time

¹oth·er·wise \-ˌwīz\ *pronoun* [Middle English, from Old English (*on*) *ōthre wīsan* in another manner] (before 12th century)
: something or anything else **:** something to the contrary ⟨do very little to enforce competition—and have never intended *otherwise* —Milton Viorst⟩

²otherwise *adverb* (13th century)
1 : in a different way or manner ⟨glossed over or *otherwise* handled —*Playboy*⟩
2 : in different circumstances ⟨might *otherwise* have left⟩
3 : in other respects ⟨an *otherwise* flimsy farce —*Current Biography*⟩
4 : if not ⟨do what I tell you, *otherwise* you'll be sorry⟩
5 : NOT — paired with an adjective, adverb, noun, or verb to indicate its contrary or to suggest an indefinite alternative ⟨people whose deeds, admirable or *otherwise* —John Fischer⟩ ⟨almost thirty thousand women, Irish and *otherwise* —J. M. Burns⟩ ⟨his opinion as to the success or *otherwise* of it —*Australian Dictionary of Biography*⟩

³otherwise *adjective* (14th century)
: DIFFERENT ⟨if conditions were *otherwise*⟩

other woman *noun* (1855)
: a woman with whom a married man has an affair — usually used with *the*

oth·er·world \ˈə-t͟hər-ˌwərld\ *noun* (13th century)
: a world beyond death or beyond present reality

oth·er·world·ly \ˌə-t͟hər-ˈwərl(d)-lē\ *adjective* (1879)
1 a : of, relating to, or resembling that of a world other than the actual world **b :** devoted to preparing for a world to come
2 : devoted to intellectual or imaginative pursuits
— oth·er·world·li·ness *noun*

¹-otic *adjective suffix* [Greek *-ōtikos,* from *-ōtos,* ending of verbals, from *-o-* (stem of causative verbs in *-oun*) + *-tos,* suffix forming verbals — more at -ED]
: of, relating to, or characterized by a (specified) action, process, or condition ⟨symbi*otic*⟩

²-otic *adjective combining form* [Greek *ōtikos* of the ear, from *ōt-, ous* ear]
: having (such) a relationship to the ear ⟨dich*otic*⟩

oti·ose \ˈō-shē-ˌōs, ˈō-tē-\ *adjective* [Latin *otiosus,* from *otium* leisure] (1794)
1 : producing no useful result **:** FUTILE
2 : being at leisure **:** IDLE
3 : lacking use or effect **:** FUNCTIONLESS
synonym see VAIN
— oti·ose·ly *adverb*
— oti·ose·ness *noun*
— oti·os·i·ty \ˌō-shē-ˈä-sə-tē, ˌō-tē-\ *noun*

oti·tis \ō-ˈtī-təs\ *noun* [New Latin] (circa 1799)
: inflammation of the ear

otitis me·dia \-ˈmē-dē-ə\ *noun* [New Latin] (1874)
: inflammation of the middle ear marked by pain, fever, dizziness, and abnormalities of hearing

oto·cyst \ˈō-tə-ˌsist\ *noun* [International Scientific Vocabulary; from its probable auditory function] (1877)
: a fluid-containing organ of many invertebrates that contains an otolith **:** STATOCYST
— oto·cys·tic \ˌō-tə-ˈsis-tik\ *adjective*

oto·lar·yn·gol·o·gy \ˌō-tō-ˌlar-ən-ˈgä-lə-jē\ *noun* (1897)
: a medical specialty concerned especially with the ear, nose, and throat
— oto·lar·yn·go·log·i·cal \-gə-ˈlä-ji-kəl\ *adjective*
— oto·lar·yn·gol·o·gist \-ˈgä-lə-jist\ *noun*

oto·lith \ˈō-tᵊl-ˌith\ *noun* [French *otolithe,* from *ot-* + *-lithe* -lith] (circa 1836)
: a calcareous concretion in the inner ear of a vertebrate or in the otocyst of an invertebrate
— oto·lith·ic \ˌō-tᵊl-ˈi-thik\ *adjective*

oto·rhi·no·lar·yn·gol·o·gy \ˌō-tō-ˌrī-nō-ˌlar-ən-ˈgä-lə-jē\ *noun* (circa 1900)
: OTOLARYNGOLOGY
— oto·rhi·no·lar·yn·go·log·i·cal \-gə-ˈlä-ji-kəl\ *adjective*
— oto·rhi·no·lar·yn·gol·o·gist \-ˈgä-lə-jist\ *noun*

oto·scle·ro·sis \ˌō-tō-sklə-ˈrō-səs\ *noun* [New Latin] (1901)
: growth of spongy bone in the inner ear that causes progressively increasing deafness

oto·tox·ic \ˌō-tə-ˈtäk-sik\ *adjective* (1951)
: producing, involving, or being adverse effects on organs or nerves involved in hearing or balance
— oto·tox·ic·i·ty \-täk-ˈsi-sə-tē\ *noun*

ot·ta·va \ō-ˈtä-və\ *adverb or adjective* [Italian, octave, from Medieval Latin *octava*] (1848)
: at an octave higher or lower than written — used as a direction in music

ottava ri·ma \-ˈrē-mə\ *noun, plural* **ottava rimas** [Italian, literally, eighth rhyme] (1820)
: a stanza of eight lines of heroic verse with a rhyme scheme of *abababcc*

Ot·ta·wa \ˈä-tə-wə, -ˌwä, -ˌwȯ\ *noun, plural* **-was** *or* **-wa** (1687)
: a member of an American Indian people of Michigan and southern Ontario

ot·ter \ˈä-tər\ *noun, plural* **otters** *also* **otter** [Middle English *oter,* from Old English *otor;* akin to Old High German *ottar* otter, Greek *hydōr* water — more at WATER] (before 12th century)
1 : any of various largely aquatic carnivorous mammals (as genus *Lutra* or *Enhydra*) that are

◇ WORD HISTORY

ostracize The Greek word *osteon* "bone" and its relatives *ostreon* "oyster" and *ostrakon* are names for hard, brittle objects. *Ostrakon* could refer to a seashell, an earthen vessel, or the broken fragment of such a vessel. Such potsherds served ancient Athenians as ballots in a particular kind of popular vote. Once a year the male citizens could gather in the agora or marketplace of Athens to decide who, if anyone, should be banished temporarily for the good of the city. Each voter wrote a name on his *ostrakon.* If at least six thousand votes were cast and if a majority of them named one man, then that man was banished. The verb to describe this banishment, and the source of English *ostracize,* was *ostrakizein,* a derivative of *ostrakos.* When the word *ostracism* first appeared in English in the 16th century, it was used only in historical reference to the ancient Athenian custom. Nonhistorical use began early in the 17th century, and nowadays *ostracism* refers more generally to exclusion from a social group by common consent.

related to the weasels and minks and usually have webbed and clawed feet and dark brown fur

otter 1

2 : the fur or pelt of an otter

otter hound *noun* [from its use in hunting otters] (1607)
: any of a British breed of large hounds that have a wiry water-resistant coat and a keen scent

ot·to \ä-(ˌ)tō\ *variant of* ATTAR

ot·to·man \ˈä-tə-mən\ *noun* (1605)
1 *capitalized* **a :** a member of a Turkish dynasty founded by Osman I that ruled the Ottoman Empire **b :** a citizen or functionary of the Ottoman Empire
2 [French *ottomane,* from feminine of *ottoman,* adjective] **a :** an upholstered often overstuffed seat or couch usually without a back **b :** an overstuffed footstool
3 : a heavy clothing fabric characterized by pronounced crosswise ribs

Ot·to·man \ˈä-tə-mən\ *adjective* [French, adjective & noun, probably from Italian *ottomano,* from Arabic *ʽothmānī,* from *ʽOthmān* Osman I, founder of the Ottoman Empire] (1603)
: of or relating to the Ottoman Empire, its rulers, or its citizens or functionaries

oua·bain \wä-ˈbā-ən, ˈwä-ˌbān\ *noun* [International Scientific Vocabulary, from French *ouabaïo,* an African tree, from Somali *waabayyo* arrow poison] (1893)
: a poisonous glycoside $C_{29}H_{44}O_{12}$ obtained from several African shrubs or trees (genera *Strophanthus* and *Acokanthera*) of the dogbane family and used medically like digitalis and in Africa as an arrow poison

ou·bli·ette \ˌü-blē-ˈet\ *noun* [French, from Middle French, from *oublier* to forget, from (assumed) Vulgar Latin *oblitare,* frequentative of Latin *oblivisci* to forget — more at OBLIVION] (1819)
: a dungeon with an opening only at the top

¹ouch \ˈau̇ch\ *noun* [Middle English, alteration (from misdivision of *a nouche*) of *nouche,* from Middle French, of Germanic origin; akin to Old High German *nusca* clasp] (14th century)
1 *obsolete* **:** CLASP, BROOCH
2 a : a setting for a precious stone **b :** JEWEL, ORNAMENT; *especially* **:** a buckle or brooch set with precious stones

²ouch *interjection* [origin unknown] (1838)
— used especially to express sudden pain

oud \ˈüd\ *noun* [Arabic *ʽūd,* literally, wood] (1738)
: a musical instrument of the lute family used in southwest Asia and northern Africa

¹ought \ˈȯt\ *verbal auxiliary* [Middle English *oughte* (1st & 3d singular present indicative), from *oughte,* 1st & 3d singular past indicative & subjunctive of *owen* to own, owe — more at OWE] (12th century)
— used to express obligation ⟨*ought* to pay our debts⟩, advisability ⟨*ought* to take care of yourself⟩, natural expectation ⟨*ought* to be here by now⟩, or logical consequence ⟨the result *ought* to be infinity⟩

²ought \ˈȯ(k)t\ *transitive verb* [Middle English *oughte,* 1st & 3d singular past indicative of *owen*] (13th century)
1 *chiefly Scottish* **:** POSSESS
2 *chiefly Scottish* **:** OWE

³ought \ˈȯt\ *noun* (1678)
: moral obligation **:** DUTY

⁴ought \ˈȯt, ˈät\ *variant of* AUGHT

oughtn't \ˈȯ-tᵊnt\ (1884)
: ought not

ou·gui·ya \ü-ˈgwē-ə, -ˈgē-ə\ *noun, plural* **ouguiya** [Arabic dialect *ūgīyah,* from Arabic *ūqīyah,* literally, ounce] (1973)
— see MONEY table

Oui·ja \ˈwē-jə, -jē\ *trademark*
— used for a board with the alphabet and other signs on it that is used with a planchette to seek spiritualistic or telepathic messages

¹ounce \ˈau̇n(t)s\ *noun* [Middle English, from Middle French *unce,* from Latin *uncia* 12th part, ounce, from *unus* one — more at ONE] (14th century)
1 a : a unit of weight equal to 1/12 troy pound — see WEIGHT table **b :** a unit of weight equal to 1/16 avoirdupois pound **c :** a small amount ⟨an *ounce* of common sense⟩
2 : FLUIDOUNCE ◆

²ounce *noun* [Middle English *once* lynx, from Middle French, alteration (by misdivision, as if *l'once* the ounce) of *lonce,* from (assumed) Vulgar Latin *lyncea,* from Latin *lync-, lynx*] (1774)
: SNOW LEOPARD

our \är, ˈau̇(-ə)r\ *adjective* [Middle English *oure,* from Old English *ūre;* akin to Old High German *unsēr* our, Old English *ūs* us] (before 12th century)
: of or relating to us or ourselves or ourself especially as possessors or possessor, agents or agent, or objects or object of an action ⟨*our* throne⟩ ⟨*our* actions⟩ ⟨*our* being chosen⟩

Our Father *noun* [from the opening words] (1882)
: LORD'S PRAYER

ours \ˈau̇(-ə)rz, ärz\ *pronoun, singular or plural in construction* (14th century)
: that which belongs to us — used without a following noun as a pronoun equivalent in meaning to the adjective *our*

our·self \är-ˈself, au̇(-ə)r-\ *pronoun* (14th century)
: MYSELF — used to refer to the single-person subject when *we* is used instead of *I* (as by a sovereign) ⟨will keep *ourself* till supper time alone —Shakespeare⟩

our·selves \-ˈselvz\ *pronoun plural* (15th century)
1 : those identical ones that are we — compare WE 1; used reflexively ⟨we're doing it solely for *ourselves*⟩, for emphasis ⟨we *ourselves* will never go⟩, or in absolute constructions ⟨*ourselves* no longer young, we can sympathize with those who are old⟩
2 : our normal, healthy, or sane condition ⟨just not *ourselves* today⟩

-ous *adjective suffix* [Middle English, from Old French *-ous, -eus, -eux,* from Latin *-osus*]
1 : full of **:** abounding in **:** having **:** possessing the qualities of ⟨clamor*ous*⟩ ⟨poison*ous*⟩
2 : having a valence lower than in compounds or ions named with an adjective ending in *-ic* ⟨mercur*ous*⟩

ou·sel \ˈü-zəl\ *variant of* OUZEL

Ou·shak \ü-ˈshäk\ *noun, often attributive* [from *Oushak, Ushak* (Uşak), town in Turkey] (1901)
: a heavy wool Oriental rug characterized especially by bright primary colors and an elaborate medallion pattern

oust \ˈau̇st\ *transitive verb* [Middle English, from Anglo-French *ouster,* from Old French *oster,* from Late Latin *obstare* to ward off, from Latin, to stand in the way, from *ob-* in the way + *stare* to stand — more at OB-, STAND] (15th century)
1 a : to remove from or dispossess of property or position by legal action, by force, or by the compulsion of necessity **b :** to take away (as a right or authority) **:** BAR, REMOVE
2 : to take the place of **:** SUPPLANT
synonym see EJECT

oust·er \ˈau̇s-tər\ *noun* [Anglo-French, to oust] (1531)
1 a : a wrongful dispossession **b :** a judgment removing an officer or depriving a corporation of a franchise
2 : EXPULSION

¹out \ˈau̇t\ *adverb* [Middle English, from Old English *ūt;* akin to Old High German *ūz* out, Greek *hysteros* later, Sanskrit *ud* up, out] (before 12th century)
1 a (1) **:** in a direction away from the inside or center ⟨went *out* into the garden⟩ (2) **:** OUTSIDE ⟨it's raining *out*⟩ **b :** from among others **c :** away from the shore **d :** away from home or work ⟨*out* to lunch⟩ **e :** away from a particular place
2 a : so as to be missing or displaced from the usual or proper place ⟨left a word *out*⟩ ⟨threw his shoulder *out*⟩ **b :** into the possession or control of another ⟨lend *out* money⟩ **c :** into a state of loss or defeat ⟨was voted *out*⟩ **d :** into a state of vexation ⟨they do not mark me, and that brings me *out* —Shakespeare⟩ **e :** into groups or shares ⟨sorted *out* her notes⟩ ⟨parceled *out* the farm⟩
3 a : to the point of depletion, extinction, or exhaustion ⟨the food ran *out*⟩ ⟨turn the light *out*⟩ ⟨all tuckered *out*⟩ **b :** to completion or satisfaction ⟨hear me *out*⟩ ⟨work the problem *out*⟩ **c :** to the full or a great extent or degree ⟨all decked *out*⟩ ⟨stretched *out* on the floor⟩
4 a : in or into the open ⟨the sun came *out*⟩ **b :** OUT LOUD ⟨cried *out*⟩ **c :** in or into public circulation ⟨the evening paper isn't *out* yet⟩ ⟨hand *out* pamphlets⟩ ⟨the library book is still *out*⟩
5 a : at an end ⟨before the day is *out*⟩ **b :** in or into an insensible or unconscious state ⟨she was *out* cold⟩ **c :** in or into a useless state ⟨landed the plane with one engine *out*⟩ **d :** so as to put a player or side out or to be put out in baseball
6 — used on a two-way radio circuit to indicate that a message is complete and no reply is expected

²out (before 12th century)
transitive verb
: EJECT, OUST
intransitive verb
: to become publicly known ⟨the truth will *out*⟩

³out *preposition* (13th century)
— used as a function word to indicate an outward movement ⟨ran *out* the door⟩ ⟨looked *out* the window⟩

⁴out *adjective* (13th century)
1 a : situated outside **:** EXTERNAL **b :** OUT-OF-BOUNDS

◇ WORD HISTORY
ounce The difficulty in determining and establishing standardized weights and measures is evidenced by the great variety of systems used in both ancient and modern times. The Romans used a system based on units divided into twelve parts. Thus Latin *uncia,* meaning "a twelfth part," was used to designate the twelfth part of a *pes* or "foot." A Germanic borrowing from *uncia* in this sense resulted in Old English *ince* or *ynce* and Modern English *inch.* The Roman pound, called *libra* in Latin (hence the English abbreviation *lb.*) was also divided into twelve parts similarly designated by the word *uncia.* In this sense *uncia* followed a different path, developing into *unce* in medieval French, which was borrowed into Middle English as *unce* or *ounce.* Why, then, do we have sixteen ounces in a pound? Unfortunately for the sake of consistency, two systems of weights were developed in the Middle Ages for weighing different sorts of material: one, the troy system, having twelve ounces, and the other, the avoirdupois system, which remains in use in the U.S. today, having sixteen ounces.

2 : situated at a distance **:** OUTLYING ⟨the *out* islands⟩
3 : not being in power
4 : ABSENT
5 : put out while at bat or baserunning in baseball
6 : directed outward or serving to direct something outward ⟨the *out* basket⟩
7 : not being in vogue or fashion
8 : not to be considered **:** out of the question
9 : DETERMINED 1 ⟨was *out* to get revenge⟩

[5]out *noun* (1717)
1 : OUTSIDE
2 : one who is out of office or power or on the outside ⟨a matter of *outs* versus ins⟩
3 a : an act or instance of putting a player out or of being put out in baseball **b :** a player that is put out
4 : a way of escaping from an embarrassing or difficult situation
— on the outs : on unfriendly terms **:** at variance

out- *prefix* [[1]*out*]
: in a manner that exceeds or surpasses and sometimes overpowers or defeats ⟨*out*maneuver⟩

out·achieve
out·act
out·bar·gain
out·bid
out·bitch
out·bluff
out·box
out·brag
out·brawl
out·bulk
out·buy
out·catch
out·charge
out·climb
out·coach
out·com·pete
out·dance
out·daz·zle
out·de·bate
out·de·liv·er
out·de·sign
out·drag
out·dress
out·drink
out·drive
out·du·el
out·earn
out·eat
out·fight
out·fig·ure
out·fish
out·fly
out·fum·ble
out·gain
out·glit·ter
out·gross
out·hit
out·ho·mer
out·hunt
out·hus·tle
out·in·trigue
out·jump
out·kick
out·kill
out·last
out·leap
out·learn
out·man
out·ma·neu·ver
out·ma·nip·u·late
out·march
out·mus·cle
out·or·ga·nize
out·pass
out·per·form
out·pitch
out·play
out·plot
out·pol·i·tick
out·poll
out·pop·u·late
out·pow·er
out·pray
out·preach
out·price
out·pro·duce
out·prom·ise
out·punch
out·rate
out·re·bound
out·re·pro·duce
out·ri·val
out·roar
out·row
out·rush
out·sail
out·scheme
out·scoop
out·score
out·shout
out·sing
out·sit
out·skate
out·soar
out·spar·kle
out·speed
out·sprint
out·stride
out·swear
out·swim
out·talk
out·think
out·throw
out·trade
out·vie
out·vote
out·wait
out·walk
out·watch
out·wres·tle
out·write
out·yell
out·yield

out·age \'au̇-tij\ *noun* (1899)
1 : a quantity or bulk of something lost in transportation or storage
2 a : a failure or interruption in use or functioning **b :** a period of interruption especially of electric current

out–and–out \ˌau̇-ᵊn(d)-'au̇t\ *adjective* (1813)
: being such completely at all times, in every way, or from every point of view ⟨an *out-and-out* fraud⟩

out–and–out·er \-'au̇-tər\ *noun* (circa 1812)
: one who goes to extremes

out·back \'au̇t-'bak, -ˌbak\ *noun* (1893)
: isolated rural country especially of Australia

out·bal·ance \ˌau̇t-'ba-lən(t)s\ *transitive verb* (1644)
: OUTWEIGH

[1]out·board \'au̇t-ˌbōrd, -ˌbȯrd\ *adjective* (circa 1823)
1 : situated outboard
2 : having, using, or limited to the use of an outboard motor

[2]outboard *adverb* (circa 1848)
1 : outside a ship's bulwarks **:** in a lateral direction from the hull
2 : in a position closer or closest to either of the wingtips of an airplane or to the sides of an automobile

[3]outboard *noun* (1935)
1 : OUTBOARD MOTOR
2 : a boat with an outboard motor

outboard motor *noun* (1909)
: a small internal combustion engine with propeller integrally attached for mounting at the stern of a small boat

out·bound \'au̇t-ˌbau̇nd\ *adjective* (1598)
: outward bound ⟨*outbound* traffic⟩

out·brave \ˌau̇t-'brāv\ *transitive verb* (1589)
1 : to face or resist defiantly
2 : to exceed in courage

out·break \'au̇t-ˌbrāk\ *noun* (1602)
1 a : a sudden or violent increase in activity or currency ⟨the *outbreak* of war⟩ **b :** a sudden rise in the incidence of a disease ⟨an *outbreak* of measles⟩ **c :** a sudden increase in numbers of a harmful organism and especially an insect within a particular area ⟨an *outbreak* of locusts⟩
2 : INSURRECTION, REVOLT

out·breed *transitive verb* **-bred** \-ˌbred, -'bred\; **-breed·ing** (circa 1909)
1 \'au̇t-ˌbrēd\ **:** to subject to outbreeding
2 \ˌau̇t-'\ **:** to breed faster than

out·breed·ing \'au̇t-ˌbrē-diŋ\ *noun* (1901)
: the interbreeding of individuals or stocks that are relatively unrelated

out·build·ing \'au̇t-ˌbil-diŋ\ *noun* (1626)
: a building (as a stable or a woodshed) separate from but accessory to a main house

out·burst \-ˌbərst\ *noun* (1657)
1 : a violent expression of feeling ⟨an *outburst* of anger⟩
2 : a surge of activity or growth ⟨new *outbursts* of creative power —C. E. Montague⟩
3 : ERUPTION ⟨volcanic *outbursts*⟩

out·bye *or* **out·by** \üt-'bī\ *adverb* [Middle English (Scots) *out-by,* from *out* + *by*] (15th century)
chiefly Scottish **:** a short distance away; *also* **:** OUTDOORS

out·cast \'au̇t-ˌkast\ *noun* (14th century)
1 : one that is cast out or refused acceptance (as by society) **:** PARIAH
2 [Scots *cast out* to quarrel] *Scottish* **:** QUARREL
— outcast *adjective*

out·caste \-ˌkast\ *noun* (1876)
1 : a Hindu who has been ejected from his caste for violation of its customs or rules
2 : one who has no caste

out·class \ˌau̇t-'klas\ *transitive verb* (1870)
: to excel or surpass so decisively as to be or appear to be of a higher class

out·come \'au̇t-ˌkəm\ *noun* (1788)
: something that follows as a result or consequence

[1]out·crop \'au̇t-ˌkräp\ *noun* (1805)
1 : a coming out of bedrock or of an unconsolidated deposit to the surface of the ground
2 : the part of a rock formation that appears at the surface of the ground

[2]out·crop \'au̇t-ˌkräp, ˌau̇t-'\ *intransitive verb* (circa 1847)
1 : to project from the surrounding soil ⟨ledges *outcropping* from the eroded slope⟩
2 : to come to the surface **:** APPEAR

out·crop·ping \'au̇t-ˌkrä-piŋ\ *noun* (1872)
: OUTCROP

[1]out·cross \'au̇t-ˌkrȯs\ *noun* (1890)
1 : a cross between relatively unrelated individuals
2 : the progeny of an outcross

[2]outcross *transitive verb* (1918)
: to cross with a relatively unrelated individual or strain

out·cry \'au̇t-ˌkrī\ *noun* (14th century)
1 a : a loud cry **:** CLAMOR **b :** a vehement protest
2 : AUCTION

out·dat·ed \ˌau̇t-'dā-təd\ *adjective* (1616)
: no longer current **:** OUTMODED
— out·dat·ed·ly *adverb*
— out·dat·ed·ness *noun*

out·dis·tance \ˌau̇t-'dis-tən(t)s\ *transitive verb* (1857)
: to go far ahead of (as in a race) **:** OUTSTRIP

out·do \-'dü\ *transitive verb* **-did** \-'did\; **-done** \-'dən\; **-do·ing** \-'dü-iŋ\; **-does** \-'dəz\ (1607)
1 : to go beyond in action or performance
2 : DEFEAT, OVERCOME
synonym see EXCEED

out·door \'au̇t-ˌdōr, -ˌdȯr; 'au̇t-'\ *also* **out·doors** \-ˌdōrz, -ˌdȯrz; 'au̇t-'\ *adjective* [*out (of) door, out (of) doors*] (1748)
1 : of or relating to the outdoors
2 a : performed outdoors ⟨*outdoor* sports⟩ **b :** OUTDOORSY ⟨an *outdoor* couple⟩
3 : not enclosed **:** having no roof ⟨an *outdoor* restaurant⟩

[1]out·doors \ˌau̇t-'dōrz, -'dȯrz; 'au̇t-'\ *adverb* (1817)
: outside a building **:** in or into the open air

[2]outdoors *noun plural but singular in construction* (1844)
1 : a place or location away from the confines of a building
2 : the world away from human habitations

out·doors·man \-mən\ *noun* (1918)
: one who spends much time in the outdoors or in outdoor activities
— out·doors·man·ship \-ˌship\ *noun*

out·doorsy \ˌau̇t-'dōr-zē, -'dȯr-\ *adjective* (1936)
1 : relating to, characteristic of, or appropriate for the outdoors ⟨*outdoorsy* clothing⟩
2 : fond of outdoor activities ⟨sounded rugged and *outdoorsy* —*N.Y. Times*⟩

out·draw \ˌau̇t-'drȯ\ *transitive verb* **-drew** \-'drü\; **-drawn** \-'drȯn\; **-draw·ing** (circa 1909)
1 : to attract a larger audience or following than
2 : to draw a handgun more quickly than

out·er \'au̇-tər\ *adjective* [Middle English, from [4]*out* + [1]*-er*] (13th century)
1 : existing independent of mind **:** OBJECTIVE
2 a : situated farther out ⟨the *outer* limits⟩ **b :** being away from a center **c :** situated or belonging on the outside ⟨the *outer* covering⟩

out·er·coat \'au̇-tər-ˌkōt\ *noun* (1948)
: COAT 1a

outer ear *noun* (1935)
: the outer visible portion of the ear that collects and directs sound waves toward the eardrum by way of a canal which extends inward through the temporal bone

out·er·most \'au̇-tər-ˌmōst\ *adjective* (14th century)
: farthest out

outer planet *noun* (1941)
: any of the planets Jupiter, Saturn, Uranus, Neptune, and Pluto whose orbits lie beyond the asteroid belt

outer space *noun* (1901)
: space immediately outside the earth's atmosphere; *broadly* **:** interplanetary or interstellar space

out·er·wear \'au̇-tər-ˌwar, -ˌwer\ *noun* (1921)

: clothing for outdoor wear

out·face \ˌau̇t-ˈfās\ *transitive verb* (circa 1529)
1 : to cause to waver or submit by or as if by staring
2 : to confront unflinchingly **:** DEFY

out·fall \ˈau̇t-ˌfȯl\ *noun* (1629)
: the outlet of a body of water (as a river or lake); *especially* **:** the mouth of a drain or sewer

out·field \-ˌfēld\ *noun* (1868)
1 : the part of a baseball field beyond the infield and between the foul lines
2 : the baseball defensive positions comprising right field, center field, and left field; *also* **:** the players who occupy these positions
— out·field·er \-ˌfēl-dər\ *noun*

[1]out·fit \ˈau̇t-ˌfit\ *noun* (circa 1769)
1 : the act of fitting out or equipping (as for a voyage or expedition)
2 a : the tools or equipment for the practice of a trade **b :** wearing apparel with accessories usually for a special occasion or activity **c :** physical, mental, or moral endowments or resources
3 : a group that works as a team **:** ORGANIZATION; *especially* **:** a military unit

[2]outfit *verb* **out·fit·ted; out·fit·ting** (1847)
transitive verb
1 : to furnish with an outfit
2 : SUPPLY ⟨*outfitting* every family with shoes —*American Guide Series: Vermont*⟩
intransitive verb
: to acquire an outfit
synonym see FURNISH

out·fit·ter \-ˌfi-tər\ *noun* (1846)
: one that outfits: as **a :** HABERDASHER **b :** a business providing equipment, supplies, and often trained guides (as for hunting trips); *also* **:** a guide working for such an outfitter

out·flank \ˌau̇t-ˈflaŋk\ *transitive verb* (1765)
1 : to get around the flank of (an opposing force)
2 : GET AROUND, CIRCUMVENT

[1]out·flow \ˈau̇t-ˌflō, ˌau̇t-ˈ\ *intransitive verb* (circa 1580)
: to flow out

[2]out·flow \ˈau̇t-ˌflō\ *noun* (circa 1864)
1 : a flowing out ⟨the *outflow* of dollars⟩
2 : something that flows out ⟨*outflow* of a sewage treatment plant⟩

out·foot \ˌau̇t-ˈfu̇t\ *transitive verb* (1737)
: to outdo in speed **:** OUTSTRIP

out·fox \-ˈfäks\ *transitive verb* (1924)
: OUTSMART

out–front \-ˈfrənt\ *adjective* (1968)
: FRANK, OPEN

out·gas \ˈau̇t-ˌgas, ˌau̇t-ˈ\ (1925)
transitive verb
1 : to remove occluded gases from usually by heating; *broadly* **:** to remove gases from
2 : to remove (gases) from a material or a space
intransitive verb
: to lose gases

out·gen·er·al \ˌau̇t-ˈjen-rəl, -ˈje-nə-\ *transitive verb* (1767)
: to surpass in generalship **:** OUTMANEUVER

[1]out·giv·ing \ˈau̇t-ˌgi-viŋ\ *noun* (1663)
: something that is given out; *especially* **:** a public statement or utterance

[2]outgiving *adjective* (1942)
: socially responsive and demonstrative

[1]out·go \ˌau̇t-ˈgō\ *transitive verb* (1530)
: to go beyond **:** OUTDO

[2]out·go \ˈau̇t-ˌgō\ *noun, plural* **outgoes** (circa 1640)
1 : something that goes out; *specifically* **:** EXPENDITURE
2 a : the act of going out **b :** DEPARTURE
3 : OUTLET 1a

out·go·ing \ˈau̇t-ˌgō-iŋ, -ˌgȯ(-)iŋ\ *adjective* (1633)
1 a : going away **:** DEPARTING ⟨an *outgoing* ship⟩ **b :** retiring or withdrawing from a place or position ⟨the *outgoing* president⟩ **c :** directed to an intended recipient ⟨*outgoing* mail⟩
2 : openly friendly and responsive **:** EXTROVERTED
— out·go·ing·ness *noun*

out·go·ings \-ˌgō-iŋz, -ˌgȯ(-)iŋz\ *noun plural* (1765)
British **:** costs incurred **:** EXPENSES

out–group \ˈau̇t-ˌgrüp\ *noun* (circa 1907)
: a group that is distinct from one's own and so usually an object of hostility or dislike — compare IN-GROUP 1

out·grow \ˌau̇t-ˈgrō\ *transitive verb* **-grew** \-ˈgrü\; **-grown** \-ˈgrōn\; **-grow·ing** (1594)
1 : to grow or increase faster than ⟨mankind is *outgrowing* food supplies —R. C. Murphy⟩
2 : to grow too large or too mature for ⟨*outgrew* his best suit⟩ ⟨the need to *outgrow* the habit of war —Norman Cousins⟩

out·growth \ˈau̇t-ˌgrōth\ *noun* (1837)
1 : a process or product of growing out ⟨an *outgrowth* of hair⟩
2 : CONSEQUENCE, BY-PRODUCT ⟨crime is often an *outgrowth* of poverty⟩

out·guess \ˌau̇t-ˈges\ *transitive verb* (1911)
: to anticipate the expectations, intentions, or actions of **:** OUTWIT

out·gun \-ˈgən\ *transitive verb* (1691)
: to surpass in firepower; *broadly* **:** OUTDO

out·haul \ˈau̇t-ˌhȯl\ *noun* (1840)
: a rope used to haul a sail taut along a spar

out–Her·od \ˌau̇t-ˈher-əd\ *transitive verb* [*out-* + *Herod* the Great, depicted in medieval mystery plays as a blustering tyrant] (1602)
: to exceed in violence or extravagance — usually used in the phrase *out-Herod Herod*

out·house \ˈau̇t-ˌhau̇s\ *noun* (14th century)
: OUTBUILDING; *especially* **:** PRIVY 1a

out·ing \ˈau̇-tiŋ\ *noun* (1821)
1 : a brief usually outdoor pleasure trip
2 : an athletic competition or race; *also* **:** an appearance therein
3 : a usually public presentation or appearance (as in a particular role) ⟨her first *outing* as a novelist⟩
4 : the public disclosure of the covert homosexuality of a prominent person by homosexual activists

outing flannel *noun* (1890)
: a flannelette sometimes having an admixture of wool

out·land \ˈau̇t-ˌland, -lənd\ *noun* (before 12th century)
1 : a foreign land
2 *plural* **:** the outlying regions of a country **:** PROVINCES
— outland *adjective*

out·land·er \-ˌlan-dər, -lən-\ *noun* (1598)
: a person who belongs to another region, culture, or group **:** FOREIGNER, STRANGER

out·land·ish \ˌau̇t-ˈlan-dish\ *adjective* (before 12th century)
1 : of or relating to another country **:** FOREIGN
2 a : strikingly out of the ordinary **:** BIZARRE ⟨an *outlandish* costume⟩ **b :** exceeding proper or reasonable limits or standards
3 : remote from civilization
synonym see STRANGE
— out·land·ish·ly *adverb*
— out·land·ish·ness *noun*

[1]out·law \ˈau̇t-ˌlȯ\ *noun* [Middle English *outlawe,* from Old English *ūtlaga,* from Old Norse *ūtlagi,* from *ūt* out (akin to Old English *ūt* out) + *lag-, lǫg* law — more at OUT, LAW] (before 12th century)
1 : a person excluded from the benefit or protection of the law
2 a : a lawless person or a fugitive from the law **b :** a person or organization under a ban or restriction **c :** one that is unconventional or rebellious
3 : an animal (as a horse) that is wild and unmanageable
— outlaw *adjective*

[2]outlaw *transitive verb* (before 12th century)
1 a : to deprive of the benefit and protection of law **:** declare to be an outlaw **b :** to make illegal ⟨*outlawed* dueling⟩
2 : to place under a ban or restriction
3 : to remove from legal jurisdiction or enforcement
— out·law·ry \ˈau̇t-ˌlȯr-ē\ *noun*

[1]out·lay \ˈau̇t-ˌlā, ˌau̇t-ˈ\ *transitive verb* **-laid** \-ˌlād, -ˈlād\; **-lay·ing** (1555)
: to lay out (money) **:** EXPEND

[2]out·lay \ˈau̇t-ˌlā\ *noun* (1798)
1 : the act of expending
2 : EXPENDITURE, PAYMENT ⟨*outlays* for national defense⟩

out·let \ˈau̇t-ˌlet, -lət\ *noun* [Middle English *utlete,* from *ut* out + *-lete* watercourse, from Old English *gelǣt,* from *lǣtan* to let] (13th century)
1 a : a place or opening through which something is let out **:** EXIT, VENT **b :** a means of release or satisfaction for an emotion or impulse ⟨sexual *outlets*⟩ **c :** a medium of expression or publication
2 : a stream flowing out of a lake or pond
3 a : a market for a commodity **b :** an agency (as a store or dealer) through which a product is marketed ⟨retail *outlets*⟩
4 : a receptacle for the plug of an electrical device

outlet pass *noun* (circa 1975)
: a pass made in basketball by the player taking a defensive rebound to a teammate to start a fast break

out·li·er \-ˌlī(-ə)r\ *noun* (1676)
1 : a person whose residence and place of business are at a distance
2 : something (as a geological feature) that is situated away from or classed differently from a main or related body

[1]out·line \ˈau̇t-ˌlīn\ *noun* (1662)
1 a : a line that marks the outer limits of an object or figure **:** BOUNDARY **b :** SHAPE
2 a : a style of drawing in which contours are marked without shading **b :** a sketch in outline
3 a : a condensed treatment of a particular subject ⟨an *outline* of world history⟩ **b :** a summary of a written work **:** SYNOPSIS
4 : a preliminary account of a project **:** PLAN
5 : a fishing line set out overnight **:** TROTLINE
☆

[2]outline *transitive verb* (circa 1790)
1 : to draw the outline of
2 : to indicate the principal features or different parts of ⟨*outlined* their responsibilities⟩

out·live \ˌau̇t-ˈliv\ *transitive verb* (15th century)
1 : to live beyond or longer than ⟨*outlived* most of his friends⟩ ⟨*outlive* its usefulness⟩
2 : to survive the effects of ⟨universities . . . *outlive* many political and social changes —J. B. Conant⟩

out·look \ˈau̇t-ˌlu̇k\ *noun* (1667)

☆ SYNONYMS
Outline, contour, profile, silhouette mean the line that bounds and gives form to something. OUTLINE applies to a line marking the outer limits or edges of a body or mass ⟨traced the *outline* of his hand⟩. CONTOUR stresses the quality of an outline or a bounding surface as being smooth, jagged, curving, or sharply angled ⟨a car with flowing *contours*⟩. PROFILE suggests a varied and sharply defined outline against a lighter background ⟨a portrait of her face in *profile*⟩. SILHOUETTE suggests a shape especially of a head or figure with all detail blacked out in shadow leaving only the outline clearly defined ⟨photograph in *silhouette* against a bright sky⟩.

1 a : a place offering a view **b :** a view from a particular place
2 : POINT OF VIEW ⟨a positive *outlook* on life⟩
3 : the act of looking out
4 : the prospect for the future ⟨the *outlook* for steel demand in the U.S. —*Wall Street Journal*⟩
synonym see PROSPECT

out loud *adverb* (1821)
: loudly enough to be heard **:** ALOUD

out·ly·ing \'au̇t-ˌlī-iŋ\ *adjective* (circa 1690)
: remote from a center or main body ⟨*outlying* areas⟩

out·match \ˌau̇t-'mach\ *transitive verb* (1603)
: to prove superior to **:** OUTDO

out–mi·grant \'au̇t-ˌmī-grənt\ *noun* (1945)
: one that out-migrates

out–mi·grate \-ˌgrāt\ *intransitive verb* (1953)
: to leave one region or community in order to settle in another especially as part of a large-scale and continuing movement of population — compare IN-MIGRATE
— out–mi·gra·tion \ˌau̇t-mī-'grā-shən\ *noun*

out·mode \ˌau̇t-'mōd\ *transitive verb* **out·mod·ed; out·mod·ing** [*out (of) mode*] (1906)
: to make unfashionable or obsolete

out·mod·ed \-'mō-dəd\ *adjective* (1903)
1 : not being in style
2 : no longer acceptable, current, or usable ⟨*outmoded* customs⟩

out·most \'au̇t-ˌmōst\ *adjective* (12th century)
: farthest out **:** OUTERMOST

out·num·ber \ˌau̇t-'nəm-bər\ *transitive verb* (1670)
: to exceed in number

out of *preposition* (before 12th century)
1 a (1) — used as a function word to indicate direction or movement from within to the outside of ⟨walked *out of* the room⟩ (2) — used as a function word to indicate a change in quality, state, or form ⟨woke up *out of* a deep sleep⟩ **b** (1) — used as a function word to indicate a position or situation beyond the range, limits, or sphere of ⟨*out of* control⟩ (2) — used as a function word to indicate a position or state away from the usual or expected ⟨*out of* practice⟩
2 — used as a function word to indicate origin, source, or cause ⟨a remarkable colt *out of* an ordinary mare⟩ ⟨built *out of* old lumber⟩ ⟨came *out of* fear⟩
3 — used as a function word to indicate exclusion from or deprivation of ⟨cheated him *out of* his savings⟩ ⟨*out of* breath⟩
4 — used as a function word to indicate choice or selection from a group ⟨one *out of* four survived⟩
— out of it 1 : not part of a group, activity, or fashion **2 :** in a dazed or confused state

out–of–body *adjective* (1970)
: relating to or involving a feeling of separation from one's body and of being able to view oneself and others from an external perspective ⟨an *out-of-body* experience⟩

out–of–bounds \ˌau̇t-ə(v)-'bau̇n(d)z\ *adverb or adjective* (1857)
: outside the prescribed boundaries or limits

out–of–date \-'dāt\ *adjective* (1628)
: OUTMODED, OBSOLETE

out–of–door \-'dōr, -'dȯr\ *or* **out–of–doors** \-'dōrz, -'dȯrz\ *adjective* (1800)
: OUTDOOR

out–of–doors *noun plural but singular in construction* (1819)
: OUTDOORS

out–of–pock·et \-'pä-kət\ *adjective* (1885)
: requiring an outlay of cash ⟨*out-of-pocket* expenses⟩

out–of–sight \ˌau̇-də-'sīt\ *adjective* (1893)
slang **:** WONDERFUL

out–of–the–way \ˌau̇t-ə(v)-t͟hə-'wā\ *adjective* (1704)
1 : UNUSUAL ⟨*out-of-the-way* information⟩
2 : being off the beaten track ⟨an *out-of-the-way* restaurant⟩

out·pace \ˌau̇t-'pās\ *transitive verb* (1611)
1 : to surpass in speed
2 : OUTDO

out·pa·tient \'au̇t-ˌpā-shənt\ *noun* (1715)
: a patient who is not hospitalized overnight but who visits a hospital, clinic, or associated facility for diagnosis or treatment — compare INPATIENT

out·place·ment \ˌau̇t-'plās-mənt, 'au̇t-ˌ\ *noun* (1970)
: the process of easing unwanted or unneeded executives out of a company by providing company-paid assistance in finding them new jobs

out·point \ˌau̇t-'pȯint\ *transitive verb* (1883)
1 : to sail closer to the wind than
2 : to win more points than (as in a boxing match)

out·port \'au̇t-ˌpōrt, -ˌpȯrt\ *noun* (1642)
1 *chiefly British* **:** a port other than the main port of a country
2 : a small fishing village especially in Newfoundland

out·post \'au̇t-ˌpōst\ *noun* (1757)
1 a : a security detachment dispatched by a main body of troops to protect it from enemy surprise **b :** a military base established by treaty or agreement in another country
2 a : an outlying or frontier settlement **b :** an outlying branch or position of a main organization or group

[1]out·pour \ˌau̇t-'pōr, -'pȯr, 'au̇t-ˌ\ *transitive verb* (1671)
: to pour out

[2]out·pour \'au̇t-ˌpōr, -ˌpȯr\ *noun* (1864)
: OUTPOURING

out·pour·ing \'au̇t-ˌpōr-iŋ, -ˌpȯr-\ *noun* (15th century)
1 : the act of pouring out
2 : something that pours out or is poured out **:** OUTFLOW

out·pull \ˌau̇t-'pu̇l\ *transitive verb* (1926)
: OUTDRAW 1

[1]out·put \'au̇t-ˌpu̇t\ *noun* (circa 1858)
1 : something produced: as **a :** mineral, agricultural, or industrial production ⟨steel *output*⟩ **b :** mental or artistic production ⟨literary *output*⟩ **c :** the amount produced by a person in a given time **d :** power or energy produced or delivered by a machine or system (as for storage or for conversion in kind or in characteristics) ⟨solar X-ray *output*⟩ **e :** the information produced by a computer
2 : the act, process, or an instance of producing
3 : the terminal for the output on an electrical device

[2]output *transitive verb* **out·put·ted** *or* **out·put; out·put·ting** (1858)
: to produce as output

out·race \ˌau̇t-'rās\ *transitive verb* (1657)
: OUTPACE

[1]out·rage \'au̇t-ˌrāj\ *noun* [Middle English, from Middle French, excess, outrage, from *outre* beyond, in excess, from Latin *ultra* — more at ULTRA-] (14th century)
1 : an act of violence or brutality
2 a : INJURY, INSULT ⟨do no *outrages* on silly women or poor passengers —Shakespeare⟩ **b :** an act that violates accepted standards of behavior or taste ⟨an *outrage* alike against decency and dignity —John Buchan⟩
3 : the anger and resentment aroused by injury or insult ◆

[2]outrage *transitive verb* **out·raged; out·rag·ing** (1590)
1 a : RAPE **b :** to violate the standards or principles of ⟨he has *outraged* respectability past endurance —John Braine⟩
2 : to arouse anger or resentment in usually by some grave offense
synonym see OFFEND

out·ra·geous \(ˌ)au̇t-'rā-jəs\ *adjective* (14th century)
1 a : exceeding the limits of what is usual **b :** not conventional or matter-of-fact **:** FANTASTIC
2 : VIOLENT, UNRESTRAINED
3 a : going beyond all standards of what is right or decent ⟨an *outrageous* disregard of human rights⟩ **b :** deficient in propriety or good taste ⟨*outrageous* language⟩ ⟨*outrageous* manners⟩
— out·ra·geous·ly *adverb*
— out·ra·geous·ness *noun*

ou·trance \ü-'träⁿs\ *noun* [Middle English, from Middle French, from *outrer* to pass beyond, carry to excess, from *outre*] (15th century)
: the last extremity

out·range \ˌau̇t-'rānj\ *transitive verb* (1858)
: to surpass in range

out·rank \-'raŋk\ *transitive verb* (1842)
1 : to rank higher than
2 : to exceed in importance

ou·tré \ü-'trā\ *adjective* [French, from past participle of *outrer* to carry to excess] (1722)
: violating convention or propriety **:** BIZARRE
word history see OUTRAGE

[1]out·reach \ˌau̇t-'rēch\ (circa 1568)
transitive verb
1 a : to surpass in reach **b :** EXCEED ⟨the demand *outreaches* the supply⟩
2 : to get the better of by trickery
intransitive verb
1 : to go too far
2 : to reach out

[2]out·reach \'au̇t-ˌrēch\ *noun* (1870)
1 : the act of reaching out
2 : the extent or limit of reach ⟨the *outreach* of the Ohio floods —Clifton Johnson⟩
3 : the extending of services or assistance beyond current or usual limits ⟨an *outreach* program⟩; *also* **:** the extent of such services or assistance

[1]out·ride \ˌau̇t-'rīd\ *transitive verb* **-rode** \-'rōd\; **-rid·den** \-'ri-dᵊn\; **-rid·ing** \-'rī-diŋ\ (1530)
1 : to ride better, faster, or farther than **:** OUTSTRIP
2 : to ride out (a storm)

[2]out·ride \'au̇t-ˌrīd\ *noun* (1880)
: an unstressed syllable or group of syllables added to a foot in sprung rhythm but not counted in the scansion

out·rid·er \-ˌrī-dər\ *noun* (1530)
1 : a mounted attendant
2 : one who escorts or clears the way for a vehicle or person
3 : FORERUNNER, HARBINGER

out·rig·ger \'au̇t-ˌri-gər\ *noun* (1748)
1 a : a projecting spar with a shaped log at the end attached to a canoe to prevent upsetting **b :** a spar or projecting beam run out from a ship's side to help secure the masts or from a mast to extend a rope or sail **c :** a projecting support for an oarlock; *also* **:** a boat equipped with these supports
2 : a projecting member run out from a main structure to provide additional stability or to

◇ WORD HISTORY

outrage Given the meaning of *outrage,* the assumption might be that the word is a compound of *out* and *rage.* Actually, such is not the case at all. The word's ultimate source is Latin *ultra* "beyond." *Ultra* developed into medieval French *outre.* To the adverb *outre* was appended the suffix *-age* to produce the noun *outrage.* Originally meaning "excess," *outrage* also developed the meaning "an act of violence or brutality." It was this sense that was borrowed into Middle English in the 14th century. In the 18th century English went back to a derivative of French *outre* once again, this time borrowing *outré* "bizarre," the past participle of the verb *outrer* "to carry to excess."

support something; *especially* **:** a projecting frame to support the elevator or tailplanes of an airplane or the rotor of a helicopter

[1]**out·right** \,au̇t-'rīt\ *adverb* (14th century)
1 *archaic* **:** straight ahead **:** DIRECTLY
2 : in entirety **:** COMPLETELY ⟨rejected the proposal *outright*⟩
3 : without restraint or reservation ⟨laughed *outright*⟩
4 : on the spot **:** INSTANTANEOUSLY ⟨was killed *outright*⟩
5 : without lien or encumbrance ⟨purchased the property *outright* for cash⟩

[2]**out·right** \'au̇t-,rīt\ *adjective* (1532)
1 a : being completely or exactly what is stated ⟨an *outright* lie⟩ **b :** given without reservation ⟨*outright* grants for research⟩ **c :** made without encumbrance or lien ⟨*outright* sale⟩
2 *archaic* **:** proceeding directly onward
— out·right·ly *adverb*

out·run \,au̇t-'rən\ *transitive verb* **-ran** \-'ran\; **-run; -run·ning** (1526)
1 : to run faster than
2 : EXCEED, SURPASS ⟨his ambitions *outrun* his abilities⟩

out·sell \-'sel\ *transitive verb* **-sold** \-'sōld\; **-sell·ing** (1609)
1 *archaic* **:** to exceed in value
2 : to exceed in number of items sold
3 : to surpass in selling or salesmanship

out·set \'au̇t-,set\ *noun* (1759)
: BEGINNING, START

out·shine \,au̇t-'shīn\ *verb* **-shone** \-'shōn, *especially British* -'shän\ *or* **-shined; -shin·ing** (1596)
transitive verb
1 a : to shine brighter than **b :** to excel in splendor or showiness
2 : OUTDO, SURPASS ⟨*outshone* most of the other films in quality —Kathleen Karr⟩
intransitive verb
: to shine out

out·shoot \,au̇t-'shüt\ *transitive verb* **-shot** \-'shät\; **-shoot·ing** (1530)
1 : to surpass in shooting or making shots
2 : to shoot or go beyond

[1]**out·side** \,au̇t-'sīd, 'au̇t-,\ *noun* (1505)
1 a : a place or region beyond an enclosure or boundary: as (1) **:** the world beyond the confines of an institution (as a prison) (2) *often capitalized, Alaska* **:** the world beyond the territory or state of Alaska; *especially* **:** the 48 contiguous states **b :** the area farthest from a specified point of reference: as (1) **:** the side of home plate farthest from the batter (2) **:** the part of a playing area toward the sidelines (3) **:** the part of a playing area away from the goal
2 : an outer side or surface
3 : an outer manifestation **:** APPEARANCE
4 : the extreme limit of a guess **:** MAXIMUM ⟨the crowd numbered 10,000 at the *outside*⟩

[2]**outside** *adjective* (1634)
1 a : of, relating to, or being on or toward the outer side or surface ⟨the *outside* edge⟩ **b :** of, relating to, or being on or toward the outer side of a curve or turn **c :** of, relating to, or being on or near the outside ⟨an *outside* pitch⟩
2 a : situated or performed outside a particular place **b :** connected with or giving access to the outside ⟨*outside* telephone line⟩
3 : MAXIMUM
4 a : not included or originating in a particular group or organization ⟨blamed the riot on *outside* agitators⟩ **b :** not belonging to one's regular occupation or duties ⟨*outside* interests⟩
5 : barely possible **:** REMOTE ⟨an *outside* chance⟩
6 : made or done from the outside ⟨borrowed a basketball and practiced his *outside* shot⟩

[3]**outside** *adverb* (1813)
1 : on or to the outside
2 : OUTDOORS

[4]**outside** *preposition* (1826)
1 — used as a function word to indicate movement to or position on the outer side of
2 : beyond the limits of ⟨*outside* the scope of this report⟩ ⟨*outside* the law⟩
3 : EXCEPT

outside of *preposition* (circa 1840)
1 : OUTSIDE
2 : ASIDE FROM

out·sid·er \,au̇t-'sī-dər, 'au̇t-,\ *noun* (1800)
1 : a person who does not belong to a particular group
2 *chiefly British* **:** a contender not expected to win
— out·sid·er·ness *noun*

out·sight \'au̇t-,sīt\ *noun* (1605)
: the power or act of perceiving external things ⟨the clear-eyed insight and *outsight* of the born writer —*New Yorker*⟩

[1]**out·size** \'au̇t-,sīz\ *noun* (1845)
: an unusual size; *especially* **:** a size larger than the standard

[2]**outsize** *also* **out·sized** \-,sīzd\ *adjective* (1880)
1 : unusually large or heavy
2 : exaggerated or extravagant in size or degree

out·skirt \'au̇t-,skərt\ *noun* (1596)
: a part remote from the center **:** BORDER — usually used in plural ⟨on the *outskirts* of town⟩

out·slick \,au̇t-'slik\ *transitive verb* (1926)
: to get the better of especially by trickery or cunning

out·smart \,au̇t-'smärt\ *transitive verb* (1924)
: to get the better of; *especially* **:** OUTWIT

out·sole \'au̇t-,sōl\ *noun* (1884)
: the outside sole of a boot or shoe

out·sourc·ing \-,sōr-siŋ, -,sȯr-\ *noun* (1982)
: the practice of subcontracting manufacturing work to outside and especially foreign or nonunion companies

out·speak \,au̇t-'spēk\ *transitive verb* **-spoke** \-'spōk\; **-spo·ken** \-'spō-kən\; **-speak·ing** (1603)
1 : to excel in speaking
2 : to declare openly or boldly

out·spend \-'spend\ *transitive verb* (1586)
1 : to exceed the limits of in spending ⟨*outspends* his income⟩
2 : to spend more than ⟨*outspent* the other candidates⟩

out·spent \-'spent\ *adjective* (1652)
: completely worn out **:** EXHAUSTED ⟨spurred him, like an *outspent* horse, to death —P. B. Shelley⟩

out·spo·ken \,au̇t-'spō-kən\ *adjective* (circa 1808)
1 : direct and open in speech or expression **:** FRANK ⟨*outspoken* in his criticism —*Current Biography*⟩
2 : spoken or expressed without reserve ⟨his *outspoken* advocacy of gun control⟩
— out·spo·ken·ly *adverb*
— out·spo·ken·ness \-kən-nəs\ *noun*

out·spread \,au̇t-'spred\ *transitive verb* **-spread; -spread·ing** (14th century)
: to spread out

out·stand \,au̇t-'stand\ *verb* **-stood; -stand·ing** (1571)
transitive verb
: to endure beyond ⟨I have *outstood* my time —Shakespeare⟩
intransitive verb
: STAND OUT

out·stand·ing \au̇t-'stan-diŋ, 'au̇t-,\ *adjective* (1611)
1 : standing out **:** PROJECTING
2 a : UNPAID ⟨left several bills *outstanding*⟩ **b :** continuing to exist **:** UNRESOLVED ⟨a long *outstanding* problem in astronomy⟩ **c** *of securities* **:** publicly issued and sold
3 a : standing out from a group **:** CONSPICUOUS **b :** marked by eminence and distinction
synonym see NOTICEABLE
— out·stand·ing·ly \-diŋ-lē\ *adverb*

out·stare \,au̇t-'star, -'ster\ *transitive verb* (1596)
: OUTFACE 1

out·sta·tion \'au̇t-,stā-shən\ *noun* (1844)
: a remote or outlying station

out·stay \,au̇t-'stā\ *transitive verb* (1600)
1 : OVERSTAY 1 ⟨*outstayed* their welcome⟩
2 : to surpass in staying power ⟨*outstayed* his competitors⟩

out·stretch \,au̇t-'strech\ *transitive verb* (15th century)
: to stretch out **:** EXTEND

out·strip \,au̇t-'strip\ *transitive verb* [*out-* + obsolete *strip* to move fast] (1580)
1 : to go faster or farther than
2 : to get ahead of **:** leave behind ⟨has civilization *outstripped* the ability of its users to use it? —Margaret Mead⟩
synonym see EXCEED

out·take \'au̇t-,tāk\ *noun* (1902)
1 : a passage outward **:** FLUE, VENT
2 : something that is taken out: as **a :** a take that is not used in an edited version of a film or video tape **b :** a recorded musical selection not included in a record album

out·turn \'au̇t-,tərn\ *noun* (1800)
: a quantity produced **:** OUTPUT

[1]**out·ward** \'au̇t-wərd\ *adjective* (before 12th century)
1 : moving, directed, or turned toward the outside or away from a center ⟨an *outward* flow⟩
2 : situated on the outside **:** EXTERIOR
3 : of or relating to the body or to appearances rather than to the mind or the inner life ⟨*outward* beauty⟩
4 : EXTERNAL

[2]**outward** *or* **out·wards** \-wərdz\ *adverb* (before 12th century)
1 : toward the outside
2 *obsolete* **:** on the outside **:** EXTERNALLY

[3]**outward** *noun* (1606)
: external form, appearance, or reality

out·ward–bound \,au̇t-wərd-'bau̇nd\ *adjective* (1602)
: bound in an outward direction or to foreign parts ⟨an *outward-bound* ship⟩

out·ward·ly \'au̇t-wərd-lē\ *adverb* (14th century)
1 a : on the outside **:** EXTERNALLY **b :** toward the outside
2 : in outward state, behavior, or appearance ⟨was *outwardly* friendly⟩

out·ward·ness \-nəs\ *noun* (1580)
1 : the quality or state of being external
2 : concern with or responsiveness to outward things

out·wash \'au̇t-,wȯsh, -,wäsh\ *noun* (1894)
: detritus chiefly consisting of gravel and sand carried by running water from the melting ice of a glacier and laid down in stratified deposits

out·wear \,au̇t-'war, -'wer\ *transitive verb* **-wore** \-'wōr, -'wȯr\; **-worn** \-'wōrn, -'wȯrn\; **-wear·ing** (circa 1541)
1 : WEAR OUT, EXHAUST
2 : to last longer than ⟨a fabric that *outwears* others⟩

out·weigh \-'wā\ *transitive verb* (1597)
: to exceed in weight, value, or importance ⟨the advantages *outweigh* the disadvantages⟩

out·wit \,au̇t-'wit\ *transitive verb* **-wit·ted; -wit·ting** (1652)
1 : to get the better of by superior cleverness **:** OUTSMART
2 *archaic* **:** to surpass in wisdom

[1]**out·work** \,au̇t-'wərk\ *transitive verb* (13th century)
1 : WORK OUT, COMPLETE
2 : to work harder, faster, or better than

[2]**out·work** \'au̇t-,wərk\ *noun* (circa 1615)
: a minor defensive position constructed outside a fortified area

out·work·er \-,wər-kər\ *noun* (1813)

chiefly British **:** a person who works at home for a business firm

out·worn \ˌau̇t-ˈwōrn, -ˈwȯrn\ *adjective* (1565)
: no longer useful or acceptable **:** OUTMODED ⟨an *outworn* social system⟩

out–year \ˈau̇t-ˌyir\ *noun* (1981)
: the year beyond a current fiscal year — usually used in plural except when attributive

ou·zel \ˈü-zəl\ *noun* [Middle English *ousel,* from Old English *ōsle* — more at MERLE] (before 12th century)
1 : BLACKBIRD 1a; *also* **:** a related bird (*Turdus torquatus*)
2 : DIPPER 1

ou·zo \ˈü-(ˌ)zō, -(ˌ)zȯ\ *noun* [New Greek] (1898)
: a colorless anise-flavored unsweetened Greek liqueur

ov- *or* **ovi-** *or* **ovo-** *combining form* [Latin *ov-, ovi-,* from *ovum* — more at EGG]
: egg ⟨*ovi*cide⟩ **:** ovum ⟨*ovi*duct⟩

ova *plural of* OVUM

¹oval \ˈō-vəl\ *noun* (1570)
1 : an oval figure or object
2 : a racetrack in the shape of an oval or a rectangle having rounded corners

²oval *adjective* [Medieval Latin *ovalis,* from Late Latin, of an egg, from Latin *ovum*] (1577)
: having the shape of an egg; *also* **:** broadly elliptical
— oval·i·ty \ō-ˈva-lə-tē\ *noun*
— oval·ly \-və-lē\ *adverb*
— oval·ness *noun*

ov·al·bu·min \ˌä-val-ˈbyü-mən, ˌō-\ *noun* (circa 1836)
1 : the principal albumin of white of egg; *especially* **:** the crystalline part of egg albumins
2 : dried whites of eggs

Oval Office *noun* [from the *Oval Office,* the U.S. president's office in the west wing of the White House] (1962)
: the seat of the executive department of the U.S. government

oval window *noun* (1683)
: FENESTRA 1a

Ovam·bo \ō-ˈvam-(ˌ)bō, -ˈväm-\ *noun, plural* **Ovambo** *or* **Ovambos** (1853)
1 : a member of a Bantu people of northern Namibia
2 : the Bantu language of the Ovambo people

ovar·i·an \ō-ˈvar-ē-ən, -ˈver-\ *also* **ovar·i·al** \-ē-əl\ *adjective* (circa 1834)
: of, relating to, or involving an ovary

ovari·ec·to·my \ō-ˌvar-ē-ˈek-tə-mē, -ˌver-\ *noun, plural* **-mies** (1889)
: the surgical removal of an ovary
— ovari·ec·to·mized \-ˌmīzd\ *adjective*

ovar·i·ole \ō-ˈvar-ē-ˌōl, -ˈver-\ *noun* [(assumed) New Latin *ovariolum,* diminutive of *ovarium*] (1877)
: one of the tubes of which the ovaries of most insects are composed

ovar·i·ot·o·my \ō-ˌvar-ē-ˈä-tə-mē, -ˌver-\ *noun, plural* **-mies** (1844)
1 : surgical incision of an ovary
2 : OVARIECTOMY

ova·ry \ˈō-və-rē, ˈōv-rē\ *noun, plural* **-ries** [New Latin *ovarium,* from Latin *ovum* egg] (1658)
1 : one of the typically paired essential female reproductive organs that produce eggs and in vertebrates female sex hormones
2 : the enlarged rounded usually basal portion of the pistil or gynoecium of an angiospermous plant that bears the ovules and consists of one or more carpels — see FLOWER illustration

ovate \ˈō-ˌvāt\ *adjective* (1760)
1 : shaped like an egg
2 : having an outline like a longitudinal section of an egg with the basal end broader ⟨*ovate* leaves⟩ — see LEAF illustration

ova·tion \ō-ˈvā-shən\ *noun* [Latin *ovation-, ovatio,* from *ovare* to exult; akin to Greek *euoi,* interjection used in bacchic revels] (1533)
1 : a ceremony attending the entering of Rome by a general who had won a victory of less importance than that for which a triumph was granted
2 : an expression or demonstration of popular acclaim especially by enthusiastic applause ⟨received a standing *ovation*⟩

ov·en \ˈə-vən\ *noun* [Middle English, from Old English *ofen;* akin to Old High German *ofan* oven and perhaps to Greek *ipnos* oven] (before 12th century)
: a chamber used for baking, heating, or drying

ov·en·bird \-ˌbərd\ *noun* [from the shape of its nest] (circa 1825)
1 : any of various South American small brown passerine birds (genus *Furnarius*)
2 : an American warbler (*Seiurus aurocapillus*) that builds a dome-shaped nest on the ground

ovenbird 2

ov·en·proof \-ˌprüf\ *adjective* (circa 1940)
: capable of withstanding the temperature range of a kitchen oven ⟨*ovenproof* dishes⟩

¹over \ˈō-vər\ *adverb* [Middle English, adverb & preposition, from Old English *ofer;* akin to Old High German *ubar* (preposition) above, beyond, over, Latin *super,* Greek *hyper*] (before 12th century)
1 a : across a barrier or intervening space; *especially* **:** across the goal line in football **b :** forward beyond an edge or brink and often down ⟨wandered too near the cliff and fell *over*⟩ **c :** across the brim ⟨soup boiled *over*⟩ **d :** so as to bring the underside up ⟨turned his cards *over*⟩ **e :** from a vertical to a prone or inclined position ⟨knocked the lamp *over*⟩ **f :** from one person or side to another ⟨hand it *over*⟩ **g :** ACROSS ⟨got his point *over*⟩ **h :** to one's home ⟨invite some friends *over*⟩ **i :** on the other side of an intervening space ⟨the next town *over*⟩ **j :** to agreement or concord ⟨won them *over*⟩
2 a (1) **:** beyond some quantity, limit, or norm often by a specified amount or to a specified degree ⟨show ran a minute *over*⟩ (2) **:** in an excessive manner **:** INORDINATELY **b :** till a later time (as the next day) **:** OVERNIGHT ⟨stay *over*⟩ ⟨sleep *over*⟩
3 a : ABOVE **b :** so as to cover the whole surface ⟨windows boarded *over*⟩
4 — used on a two-way radio circuit to indicate that a message is complete and a reply is expected
5 a : THROUGH ⟨read it *over*⟩; *also* **:** in an intensive or comprehensive manner **b :** once more **:** AGAIN ⟨do it *over*⟩

²over *preposition* (before 12th century)
1 — used as a function word to indicate motion or situation in a position higher than or above another ⟨towered *over* his mother⟩ ⟨flew *over* the lake⟩ ⟨rode *over* the old Roman road⟩
2 a — used as a function word to indicate the possession of authority, power, or jurisdiction in regard to some thing or person ⟨respected those *over* him⟩ **b** — used as a function word to indicate superiority, advantage, or preference ⟨a big lead *over* the others⟩ **c** — used as a function word to indicate one that is overcome, circumvented, or disregarded ⟨passed *over* the governor's veto⟩
3 a : more than ⟨cost *over* $5⟩ **b :** ABOVE 4
4 a — used as a function word to indicate position upon or movement down upon ⟨laid a blanket *over* the child⟩ ⟨hit him *over* the head⟩ **b** (1) **:** all through or throughout ⟨showed me *over* the house⟩ ⟨went *over* his notes⟩ (2) — used as a function word connecting one mathematical set and another whose elements are coefficients or values of parameters used to form elements of the first set ⟨polynomials *over* the field of real numbers⟩ **c** — used as a function word to indicate a particular medium or channel of communication ⟨*over* the radio⟩
5 — used as a function word to indicate position on or motion to the other side or beyond ⟨lives *over* the way⟩ ⟨fell *over* the edge⟩
6 a : THROUGHOUT, DURING ⟨*over* the past 25 years⟩ **b :** until the end of ⟨stay *over* Sunday⟩
7 a — used as a function word to indicate an object of solicitude, interest, consideration, or reference ⟨the Lord watches *over* his own⟩ **b** — used as a function word to indicate the object of an expressed or implied occupation, activity, or concern ⟨trouble *over* money⟩ ⟨met with advisers *over* lunch⟩ ■

³over *adjective* (before 12th century)
1 a : UPPER, HIGHER **b :** OUTER, COVERING **c :** EXCESSIVE ⟨*over* imagination⟩
2 a : not used up **:** REMAINING ⟨something *over* to provide for unusual requirements —J. A. Todd⟩ **b :** having or showing an excess or surplus
3 : being at an end ⟨the day is *over*⟩
4 : fried on both sides ⟨ordered two eggs *over*⟩
— over easy : fried on one side then turned and fried lightly on the other side ⟨eggs *over easy*⟩

⁴over *transitive verb* **overed; over·ing** \ˈō-və-riŋ, ˈōv-riŋ\ (1837)
: to leap over

over- *prefix*
1 : so as to exceed or surpass ⟨*over*achieve⟩
2 : EXCESSIVE ⟨*over*stimulation⟩
3 : to an excessive degree ⟨*over*thin⟩

over·ab·stract
over·abun·dance
over·abun·dant
over·ac·cen·tu·ate
over·ad·just·ment
over·ad·ver·tise
over·ag·gres·sive
over·alert
over·am·bi·tious
over·am·bi·tious·ness
over·am·pli·fied
over·anal·y·sis
over·an·a·lyt·i·cal
over·an·a·lyze
over·anx·i·ety
over·anx·ious
over·ap·pli·ca·tion
over·arous·al
over·ar·range
over·ar·tic·u·late
over·as·sert
over·as·ser·tion
over·as·ser·tive
over·as·sess·ment
over·at·ten·tion
over·bake
over·beat
over·be·jew·eled
over·bill
over·bleach
over·boil

□ USAGE

over There is a hallowed tradition among newspaper editors that the use of sense 3a of *over* is improper. The tradition may have begun with William Cullen Bryant when he was editor of the *New York Evening Post* in the 1870s. No particular reason for disliking this sense of *over* was given in the early objections, but later commentators seem to feel that *over* should be restricted to physical location. Now more than a dozen different uses of the preposition are given here, and many of them do not relate to physical location. Other writers, as distinct from journalists, have been using this sense of *over* for six and a half centuries ⟨I have now a library of nearly 900 volumes, *over* 700 of which I wrote myself —Henry David Thoreau⟩ ⟨I have been in show business man and boy for *over* forty years —Groucho Marx⟩ ⟨Mr. Kallman and I have been close friends for *over* thirty years —W. H. Auden⟩. There is no sound reason for you to avoid this sense of *over.*

over·bold
over·bor·row
over·breath·ing
over·brief
over·bright
over·broad
over·browse
over·bru·tal
over·burn
over·busy
over·care·ful
over·cau·tion
over·cau·tious
over·cen·tral·i·za·tion
over·cen·tral·ize
over·chill
over·civ·i·lized
over·claim
over·clas·si·fi·ca·tion
over·clas·si·fy
over·clean
over·clear
over·coach
over·com·mer·cial·i·za·tion
over·com·mer·cial·ize
over·com·mu·ni·cate
over·com·mu·ni·ca·tion
over·com·plex
over·com·pli·ance
over·com·pli·cate
over·com·pli·cat·ed
over·com·press
over·con·cen·tra·tion
over·con·cern
over·con·cerned
over·con·fi·dence
over·con·fi·dent
over·con·fi·dent·ly
over·con·sci·en·tious
over·con·scious
over·con·ser·va·tive
over·con·struct
over·con·sume
over·con·sump·tion
over·con·trol
over·cook
over·cool
over·cor·rect
over·count
over·cred·u·lous
over·crit·i·cal
over·cul·ti·va·tion
over·cure
over·dec·o·rate
over·dec·o·ra·tion
over·de·mand·ing
over·de·pen·dence
over·de·pen·dent
over·de·sign
over·dif·fer·en·ti·a·tion
over·di·rect
over·dis·count
over·di·ver·si·ty
over·doc·u·ment
over·dra·mat·ic
over·dra·ma·tize
over·drink
over·dry
over·ea·ger
over·ea·ger·ness
over·ear·nest
over·ed·it
over·ed·u·cate
over·ed·u·cat·ed
over·ed·u·ca·tion
over·elab·o·rate
over·elab·o·ra·tion
over·em·bel·lish
over·emote
over·emo·tion·al
over·em·pha·sis
over·em·pha·size
over·em·phat·ic
over·en·am·ored
over·en·cour·age
over·en·er·get·ic
over·en·gi·neer
over·en·rolled
over·en·ter·tained
over·en·thu·si·asm
over·en·thu·si·as·tic
over·equipped
over·es·ti·mate
over·es·ti·ma·tion
over·eval·u·a·tion
over·ex·ag·ger·ate
over·ex·ag·ger·a·tion
over·ex·cite
over·ex·cit·ed
over·ex·er·cise
over·ex·ert
over·ex·er·tion
over·ex·pand
over·ex·pan·sion
over·ex·pec·ta·tion
over·ex·plain
over·ex·plic·it
over·ex·ploit
over·ex·ploi·ta·tion
over·ex·trac·tion
over·ex·trap·o·la·tion
over·ex·trav·a·gant
over·ex·u·ber·ant
over·fac·ile
over·fa·mil·iar
over·fa·mil·iar·i·ty
over·fas·tid·i·ous
over·fat
over·fa·vor
over·fer·til·i·za·tion
over·fer·til·ize
over·fo·cus
over·fond
over·ful·fill
over·fund
over·fussy
over·gen·er·al·i·za·tion
over·gen·er·al·ize
over·gen·er·os·i·ty
over·gen·er·ous
over·gen·er·ous·ly
over·glam·or·ize
over·gov·ern
over·han·dle
over·har·vest
over·hasty
over·ho·mog·e·nize
over·hunt
over·hunt·ing
over·hype
over·ide·al·ize
over·iden·ti·fi·ca·tion
over·iden·ti·fy
over·imag·i·na·tive
over·im·press
over·in·debt·ed·ness
over·in·dulge
over·in·dul·gence
over·in·dul·gent
over·in·dus·tri·al·ize
over·in·flate
over·in·flat·ed
over·in·fla·tion
over·in·form
over·in·formed
over·in·ge·nious
over·in·ge·nu·ity
over·in·sis·tent
over·in·tel·lec·tu·al·i·za·tion
over·in·tel·lec·tu·al·ize
over·in·tense
over·in·ten·si·ty
over·in·ter·pre·ta·tion
over·in·vest·ment
over·la·bor
over·la·bored
over·lad·en
over·large
over·lav·ish
over·lend
over·length
over·length·en
over·light
over·lit·er·al
over·lit·er·ary
over·load
over·long
over·loud
over·lush
over·man·age
over·man·nered
over·ma·ture
over·ma·tu·ri·ty
over·med·i·cate
over·med·i·ca·tion
over·mighty
over·milk
over·mine
over·mix
over·mod·est
over·mod·est·ly
over·mus·cled
over·nice
over·nour·ish
over·nu·tri·tion
over·ob·vi·ous
over·op·er·ate
over·opin·ion·at·ed
over·op·ti·mism
over·op·ti·mist
over·op·ti·mis·tic
over·op·ti·mis·ti·cal·ly
over·or·ches·trate
over·or·ga·nize
over·or·ga·nized
over·or·na·ment
over·pack·age
over·par·tic·u·lar
over·pay
over·pay·ment
over·ped·al
over·peo·ple
over·plan
over·plant
over·plot
over·po·tent
over·praise
over·pre·cise
over·pre·scribe
over·pre·scrip·tion
over·priv·i·leged
over·prize
over·pro·cess
over·pro·duce
over·pro·duc·tion
over·pro·gram
over·prom·ise
over·pro·mote
over·pro·tect
over·pro·tec·tion
over·pro·tec·tive
over·pro·tec·tive·ness
over·pump
over·rate
over·re·act
over·re·ac·tion
over·re·fined
over·re·fine·ment
over·reg·u·late
over·reg·u·la·tion
over·re·li·ance
over·re·port
over·re·spond
over·rich
over·rig·id
over·salt
over·san·guine
over·sat·u·rate
over·sat·u·ra·tion
over·sauce
over·scru·pu·lous
over·se·cre·tion
over·sen·si·tive
over·sen·si·tive·ness
over·sen·si·tiv·i·ty
over·se·ri·ous
over·se·ri·ous·ly
over·ser·vice
over·sim·plis·tic
over·smoke
over·so·lic·i·tous
over·so·phis·ti·cat·ed
over·spe·cial·i·za·tion
over·spe·cial·ize
over·spec·u·late
over·spec·u·la·tion
over·sta·bil·i·ty
over·staff
over·stim·u·late
over·stim·u·la·tion
over·stock
over·strain
over·stress
over·stretch
over·struc·tured
over·sub·tle
over·suds
over·sup·ply
over·sus·pi·cious
over·sweet
over·sweet·en
over·sweet·ness
over·swing
over·talk
over·talk·a·tive
over·tax
over·tax·a·tion
over·thin
over·think
over·tight·en
over·tip
over·tired
over·train
over·treat
over·treat·ment
over·use
over·uti·li·za·tion
over·uti·lize
over·vi·o·lent
over·viv·id
over·wa·ter
over·wea·ry
over·wind
over·with·hold
over·zeal·ous
over·zeal·ous·ness

over·achiev·er \ˌō-və-rə-ˈchē-vər\ *noun* (1952)
: one who achieves success over and above the standard or expected level especially at an early age
— over·achieve \-ˈchēv\ *intransitive verb*
— over·achieve·ment \-mənt\ *noun*

over·act \ˌō-vər-ˈakt\ (1611)
intransitive verb
1 : to act more than is necessary
2 : to overact a part
transitive verb
: to exaggerate in acting
— over·ac·tion \-ˈak-shən\ *noun*

over·ac·tive \-ˈak-tiv\ *adjective* (1854)
: excessively or abnormally active
— over·ac·tiv·i·ty \-ak-ˈti-və-tē\ *noun*

over against *preposition* (1517)
: as opposed to **:** in contrast with

[1]**over·age** \ˌō-vər-ˈāj\ *also* **over·aged** \-ˈājd\ *adjective* [[2]*over* + *age*] (15th century)
1 : too old to be useful
2 : older than is normal for one's position, function, or grade

[2]**over·age** \ˈō-və-rij, ˈōv-rij\ *noun* [[3]*over* + *-age*] (1909)
: SURPLUS, EXCESS

[1]**over·all** \ˌō-vər-ˈȯl\ *adverb* (13th century)
1 : ALL OVER 1 ⟨the pattern used *overall*⟩
2 : from one end to the other ⟨600 feet long *overall*⟩
3 a : in view of all the circumstances or conditions ⟨*overall*, the sale was a success⟩ **b :** as a whole **:** GENERALLY ⟨doesn't do as well *overall*⟩ **c :** with everyone or everything taken into account ⟨was third *overall* in earnings⟩ ⟨got 31 miles to the gallon *overall*⟩

[2]**over·all** \ˈō-vər-ˌȯl\ *noun* (1815)
1 *plural* **a** *archaic* **:** loose protective trousers worn over regular clothes **b :** trousers of strong material usually with a bib and shoulder straps
2 *chiefly British* **:** a loose-fitting protective smock worn over regular clothing

[3]**over·all** \ˌō-vər-ˈȯl, ˈō-vər-ˌ\ *adjective* (1894)
1 : including everything
2 : viewed as a whole **:** GENERAL

over·alled \ˈō-vər-ˌȯld\ *adjective* (1908)
: wearing overalls

over and above *preposition* (15th century)
: in addition to **:** BESIDES

over and over *adverb* (15th century)
: REPEATEDLY

over·arch·ing \ˌō-vər-ˈär-chiŋ\ *adjective* (1720)
1 : forming an arch overhead
2 : dominating or embracing all else ⟨*overarching* goals⟩

over·arm \ˈō-vər-ˌärm\ *adjective* (1864)
1 : OVERHAND
2 *of a swimming stroke* **:** made with the arm lifted out of the water and stretched forward over the shoulder to begin the stroke

over·awe \ˌō-vər-ˈȯ\ *transitive verb* (1579)
: to restrain or subdue by awe

[1]**over·bal·ance** \ˌō-vər-ˈba-lən(t)s\ *transitive verb* (1608)
1 : OUTWEIGH
2 : to cause to lose balance

[2]**over·bal·ance** \ˈō-vər-ˌ\ *noun* (1659)
: something more than an equivalent

over·bear \ˌō-vər-ˈbar, -ˈber\ *transitive verb* **-bore** \-ˈbōr, -ˈbȯr\; **-borne** \-ˈbōrn, -ˈbȯrn\ *also* **-born** \-ˈbȯrn\; **-bear·ing** (1535)
1 : to bring down by superior weight or force **:** OVERWHELM
2 a : to domineer over **b :** to surpass in importance or cogency **:** OUTWEIGH

overbearing *adjective* (circa 1677)
1 a : tending to overwhelm **:** OVERPOWERING **b :** decisively important **:** DOMINANT
2 : harshly and haughtily arrogant
synonym see PROUD
— over·bear·ing·ly \-iŋ-lē\ *adverb*

over·bid \ˌō-vər-ˈbid\ *verb* **-bid; -bid·ding** (circa 1616)
intransitive verb
1 : to bid in excess of value

2 a : to bid more than the scoring capacity of a hand at cards **b** *British* **:** to make a higher bid than the preceding one
transitive verb
: to bid beyond or in excess of; *especially* **:** to bid more than the value of (one's hand at cards)
— over·bid \'ō-vər-ˌbid\ *noun*
over·bite \'ō-vər-ˌbīt\ *noun* (1887)
: the projection of the upper anterior teeth over the lower in the normal occlusal position of the jaws
over·blouse \-ˌblau̇s, -ˌblau̇z\ *noun* (1921)
: a usually fitted or belted blouse worn untucked
¹over·blown \ˌō-vər-'blōn\ *adjective* [³*blow*] (1616)
: past the prime of bloom ⟨*overblown* roses⟩
²overblown *adjective* [¹*blow*] (1864)
1 : excessively large in girth **:** PORTLY
2 : INFLATED ⟨*overblown* claims⟩ ⟨*overblown* rhetoric⟩; *also* **:** PRETENTIOUS
over·board \'ō-vər-ˌbōrd, -ˌbȯrd\ *adverb* (before 12th century)
1 : over the side of a ship or boat into the water
2 : to extremes of enthusiasm
3 : into discard **:** ASIDE
over·book \ˌō-vər-'bu̇k\ (1903)
transitive verb
: to issue reservations for (as an airplane flight) in excess of the space available
intransitive verb
: to issue reservations in excess of the space available
over·bought \-'bȯt\ *adjective* (1929)
: not likely to show an immediate rise in price because of prior heavy buying and accompanying price rises ⟨an *overbought* market⟩
over·build \-'bild\ *verb* **-built** \-'bilt\; **-building** (1601)
transitive verb
: to build beyond the actual demand of
intransitive verb
: to build houses or commercial developments in excess of demand
¹over·bur·den \-'bər-dᵊn\ *transitive verb* (1532)
: to place an excessive burden on
²over·bur·den \'ō-vər-ˌbər-dᵊn\ *noun* (1855)
: material overlying a deposit of useful geological materials or bedrock
over·buy \ˌō-vər-'bī\ *verb* **-bought** \-'bȯt\; **-buy·ing** (1745)
transitive verb
: to buy in excess of needs or demand
intransitive verb
: to make purchases beyond one's needs or in excess of one's ability to pay
over·call \-'kȯl\ (circa 1903)
transitive verb
: to make a higher bid than (the previous bid or bidder) in a card game
intransitive verb
: to bid over an opponent's bid in bridge when one's partner has not bid or doubled
— over·call \'ō-vər-ˌkȯl\ *noun*
over·ca·pac·i·ty \ˌō-vər-kə-'pa-sə-tē, -'pas-tē\ *noun* (1928)
: excessive capacity for production or services in relation to demand
over·cap·i·tal·ize \-'ka-pə-tᵊl-ˌīz, -'kap-tᵊl-\ *transitive verb* (1890)
1 : to put a nominal value on the capital of (a corporation) higher than actual cost or fair market value
2 : to capitalize beyond what the business or the profit-making prospects warrant
— over·cap·i·tal·i·za·tion \-ˌka-pə-tᵊl-ə-'zā-shən, -ˌkap-tᵊl-\ *noun*
¹over·cast *transitive verb* **-cast; -cast·ing** (14th century)
1 \ˌō-vər-'kast, 'ō-vər-ˌ\ **:** DARKEN, OVERSHADOW
2 \'ō-vər-ˌ\ **:** to sew (raw edges of a seam) with long slanting widely spaced stitches to prevent raveling
²over·cast \'ō-vər-ˌkast, ˌō-vər-'\ *adjective* (1536)
: clouded over ⟨an *overcast* day⟩
³over·cast \'ō-vər-ˌkast\ *noun* (1686)
: COVERING; *especially* **:** a covering of clouds over the sky
over·cast·ing \'ō-vər-ˌkas-tiŋ\ *noun* (1885)
: the act of stitching raw edges of fabric to prevent raveling; *also* **:** the stitching so done
overcast stitch *noun* (1891)
: a small close embroidery stitch sometimes done over a foundation thread and used to form outlines

overcast stitch

over·charge \ˌō-vər-'chärj\ (14th century)
transitive verb
1 : to charge too much or too fully
2 : to fill too full
3 : EXAGGERATE, OVERDRAW
intransitive verb
: to make an excessive charge
— over·charge \'ō-vər-ˌ\ *noun*
over·cloud \ˌō-vər-'klau̇d\ *transitive verb* (1592)
: to overspread with or as if with clouds
over·coat \'ō-vər-ˌkōt\ *noun* (1802)
1 : a warm coat worn over indoor clothing
2 : a protective coating (as of paint)
over·come \ˌō-vər-'kəm\ *verb* **-came** \-'kām\; **-come; -com·ing** [Middle English, from Old English *ofercuman*, from *ofer* over + *cuman* to come] (before 12th century)
transitive verb
1 : to get the better of **:** SURMOUNT ⟨*overcome* difficulties⟩
2 : OVERWHELM
intransitive verb
: to gain the superiority **:** WIN
synonym see CONQUER
— over·com·er *noun*
over·com·mit \-kə-'mit\ *transitive verb* (1951)
: to commit excessively: as **a :** to obligate (as oneself) beyond the ability for fulfillment **b :** to allocate (resources) in excess of the capacity for replenishment
— over·com·mit·ment \-mənt\ *noun*
over·com·pen·sa·tion \-ˌkäm-pən-'sā-shən, -ˌpen-\ *noun* (1912)
: excessive compensation; *specifically* **:** excessive reaction to a feeling of inferiority, guilt, or inadequacy leading to an exaggerated attempt to overcome the feeling
— over·com·pen·sate \-'käm-pən-ˌsāt\ *verb*
— over·com·pen·sa·to·ry \-kəm-'pen(t)-sə-ˌtōr-ē, -ˌtȯr-\ *adjective*
over·crowd \ˌō-vər-'krau̇d\ (1766)
transitive verb
: to cause to be too crowded
intransitive verb
: to crowd together too much
over·cut \-'kət\ *transitive verb* (1906)
: to cut excessively; *specifically* **:** to cut timber from (a forest) in excess of annual growth or an allotted annual amount
over·de·ter·mined \-di-'tər-mənd\ *adjective* (1915)
1 : excessively determined
2 : having more than one determining psychological factor
over·de·vel·op \-di-'ve-ləp\ *transitive verb* (1869)
: to develop excessively; *especially* **:** to subject (exposed photographic material) to a developing solution for excessive time or at excessive temperature, agitation, or concentration
— over·de·vel·op·ment \-mənt\ *noun*
over·do \ˌō-vər-'dü\ *verb* **-did** \-'did\; **-done** \-'dən\; **-do·ing** \-'dü-iŋ\; **-does** \-'dəz\ (before 12th century)
transitive verb
1 a : to do in excess **b :** to use to excess **c :** EXAGGERATE
2 : to cook too long
3 : EXHAUST
intransitive verb
: to go to extremes
over·dog \'ō-vər-ˌdȯg\ *noun* [³*over* + *underdog*] (1938)
: one that is dominant or victorious
over·dom·i·nance \ˌō-vər-'dä-mə-nən(t)s, -'däm-nən(t)s\ *noun* (1947)
: the condition wherein a heterozygote produces a phenotype more extreme or better adapted than that of the homozygote
— over·dom·i·nant \-nənt\ *adjective*
¹over·dose \'ō-vər-ˌdōs\ *noun* (1700)
1 : too great a dose (as of a therapeutic agent); *also* **:** a lethal or toxic amount (as of a drug)
2 : an excessive quantity or amount ⟨an *overdose* of sports⟩
— over·dos·age \ˌō-vər-'dō-sij\ *noun*
²over·dose \ˌō-vər-'dōs\ (1727)
transitive verb
: to give an overdose or too many doses to
intransitive verb
: to take or experience an overdose — usually used with *on*
over·draft \'ō-vər-ˌdraft\ *noun* (1878)
1 : an act of overdrawing at a bank **:** the state of being overdrawn; *also* **:** the sum overdrawn
2 : LINE OF CREDIT
over·draw \ˌō-vər-'drȯ\ *verb* **-drew** \-'drü\; **-drawn** \-'drȯn\; **-draw·ing** (1734)
transitive verb
1 : to draw checks on (a bank account) for more than the balance ⟨the account was *overdrawn*⟩
2 : EXAGGERATE, OVERSTATE
intransitive verb
: to make an overdraft
over·drawn *adjective* (1866)
: having an overdrawn account
¹over·dress \ˌō-vər-'dres\ (1706)
transitive verb
: to dress or adorn to excess
intransitive verb
: to dress oneself to excess
²over·dress \'ō-vər-ˌdres\ *noun* (1812)
: a dress worn over another
over·drive \'ō-vər-ˌdrīv\ *noun* (1926)
1 : an automotive transmission gear that transmits to the drive shaft a speed greater than engine speed
2 : a state of heightened activity ⟨going into rhetorical *overdrive*⟩
¹over·dub \'ō-vər-ˌdəb\ *noun* (circa 1965)
1 : the act or an instance of overdubbing
2 : recorded sound that is overdubbed ⟨vocal *overdubs*⟩
²over·dub \ˌō-vər-'dəb\ *transitive verb* (1967)
: to transfer (recorded sound) onto a recording that bears sound recorded earlier in order to produce a combined effect
over·due \-'dü, -'dyü\ *adjective* (1845)
1 a : unpaid when due **b :** delayed beyond an appointed time
2 : too great **:** EXCESSIVE
3 : more than ready
over·eat \ˌō-vər-'ēt\ *intransitive verb* **over·ate** \-'āt\; **over·eat·en** \-'ē-tᵊn\; **over·eat·ing** (1599)
: to eat to excess
— over·eat·er \ˌō-vər-'ē-tər, 'ō-vər-ˌ\ *noun*
over·ex·pose \ˌō-vər-ik-'spōz\ *transitive verb* (1869)
: to expose excessively; *especially* **:** to expose (as film) to excessive radiation (as light)
— over·ex·po·sure \-'spō-zhər\ *noun*
over·ex·tend \ˌō-vər-ik-'stend\ *transitive verb* (1937)

: to extend or expand beyond a safe or reasonable point; *especially* **:** to commit (oneself) financially beyond what can be paid
— **over·ex·ten·sion** \-ˈsten(t)-shən\ *noun*

over·fa·tigue \ˌō-vər-fə-ˈtēg\ *noun* (1727)
: excessive fatigue especially when carried beyond the recuperative capacity of the individual
— **over·fa·tigued** \-ˈtēgd\ *adjective*

over·feed \ˌō-vər-ˈfēd\ *verb* **-fed** \-ˈfed\; **-feed·ing** (1608)
transitive verb
: to feed to excess
intransitive verb
: to eat to excess

over·fill \-ˈfil\ (13th century)
transitive verb
: to fill to overflowing
intransitive verb
: to become full to overflowing

over·fish \-ˈfish\ *transitive verb* (1867)
: to fish to the detriment of (a fishing ground) or to the depletion of (a kind of organism)

over·flight \ˈō-vər-ˌflīt\ *noun* (1950)
: a passage over an area in an airplane

1over·flow \ˌō-vər-ˈflō\ (before 12th century)
transitive verb
1 : to cover with or as if with water **:** INUNDATE
2 : to flow over the brim of
3 : to cause to overflow
intransitive verb
1 : to flow over bounds
2 : to fill a space to capacity and spread beyond its limits ⟨the crowd *overflowed* into the street⟩

2over·flow \ˈō-vər-ˌflō\ *noun* (1589)
1 : a flowing over **:** INUNDATION
2 : something that flows over **:** SURPLUS
3 : an outlet or receptacle for surplus liquid

over·fly \ˌō-vər-ˈflī\ *transitive verb* **-flew** \-ˈflü\; **-flown** \-ˈflōn\; **-fly·ing** (14th century)
: to fly over; *especially* **:** to pass over in an airplane or spacecraft

over·gar·ment \ˈō-vər-ˌgär-mənt\ *noun* (15th century)
: an outer garment

over·glaze \-ˌglāz\ *adjective* (1879)
: applied or suitable for applying on top of a fired glaze ⟨*overglaze* enamels⟩
— **overglaze** *noun*

over·graze \ˌō-vər-ˈgrāz\ *transitive verb* (1919)
: to allow animals to graze (as a pasture) to the point of damaging vegetational cover

over·grow \ˌō-vər-ˈgrō\ *verb* **-grew** \-ˈgrü\; **-grown** \-ˈgrōn\; **-grow·ing** (14th century)
transitive verb
1 : to grow over so as to cover with herbage
2 : to grow beyond or rise above **:** OUTGROW
intransitive verb
1 : to grow excessively
2 : to become grown over
— **over·growth** \ˈō-vər-ˌgrōth\ *noun*

overgrown *adjective* (1604)
: grown abnormally or excessively large ⟨dismissed him as an *overgrown* adolescent⟩

1over·hand \ˈō-vər-ˌhand\ *adjective* (1656)
: made with the hand brought forward and down from above shoulder level
— **overhand** *adverb*
— **over·hand·ed** \ˌō-vər-ˈhan-dəd\ *adverb or adjective*

2overhand *transitive verb* (1871)
: to sew with short vertical stitches

3overhand *noun* (circa 1934)
: an overhand stroke (as in handball)

overhand knot *noun* (1840)
: a small knot often used to prevent the end of a cord from fraying — see KNOT illustration

1over·hang \ˈō-vər-ˌhaŋ, ˌō-vər-ˈ\ *verb* **-hung** \-ˌhəŋ, -ˈhəŋ\; **-hang·ing** (1592)
transitive verb
1 : to project over
2 : to impend over **:** THREATEN
intransitive verb
: to project so as to be over something

2over·hang \ˈō-vər-ˌhaŋ\ *noun* (1864)
1 : the part of the bow or stern of a ship that projects over the water above the waterline
2 : something that overhangs; *also* **:** the extent of the overhanging
3 : a projection of the roof or upper story of a building beyond the wall of the lower part
4 : an excess supply of a commodity that cannot be readily converted, sold, or disposed of ⟨dollar *overhang*⟩ ⟨inventory *overhang*⟩

over·haul \ˌō-vər-ˈhȯl\ *transitive verb* (1705)
1 a : to examine thoroughly **b** (1) **:** REPAIR (2) **:** to renovate, revise, or renew thoroughly
2 : to haul or drag over
3 : OVERTAKE
— **over·haul** \ˈō-vər-ˌhȯl\ *noun*

1over·head \ˌō-vər-ˈhed\ *adverb* (15th century)
: above one's head **:** ALOFT

2over·head \ˈō-vər-ˌhed\ *adjective* (1874)
1 a : operating, lying, or coming from above **b :** having the driving part above the part driven ⟨valves operated by an *overhead* camshaft⟩
2 : of or relating to overhead ⟨*overhead* costs⟩

3over·head \ˈō-vər-ˌhed\ *noun* (1914)
1 : business expenses (as rent, insurance, or heating) not chargeable to a particular part of the work or product
2 : CEILING; *especially* **:** the ceiling of a ship's compartment
3 : a stroke in a racket game made above head height **:** SMASH

overhead projector *noun* (1951)
: a projector for projecting onto a vertical screen magnified images of graphic material on a horizontal transparency illuminated from below — called also *overhead*

over·hear \ˌō-vər-ˈhir\ *verb* **-heard** \-ˈhərd\; **-hear·ing** \-ˈhir-iŋ\ (1549)
transitive verb
: to hear without the speaker's knowledge or intention
intransitive verb
: to overhear something

over·heat \-ˈhēt\ (14th century)
transitive verb
1 : to heat to excess
2 : to stimulate or agitate unduly
intransitive verb
: to become heated beyond a safe or desirable point

over·heat·ed \-ˈhē-təd\ *adjective* (1953)
: PERFERVID

over·is·sue \ˌō-vər-ˈi-(ˌ)shü\ *noun* (1803)
: an issue exceeding the limit of capital, credit, or authority
— **over·is·su·ance** \-ˈi-shə-wən(t)s\ *noun*
— **overissue** *transitive verb*

over·joyed \-ˈjȯid\ *adjective* (1594)
: feeling great joy

1over·kill \ˌō-vər-ˈkil\ *transitive verb* (1957)
: to obliterate (a target) with more nuclear force than required

2over·kill \ˈō-vər-ˌkil\ *noun* (1958)
1 : a destructive capacity greatly exceeding that required for a given target
2 : an excess of something (as a quantity or an action) beyond what is required or suitable for a particular purpose ⟨a propaganda *overkill*⟩ ⟨an *overkill* in weaponry⟩
3 : killing in excess of what is intended or required

1over·land \ˈō-vər-ˌland, -lənd\ *adverb* (12th century)
: by, on, or across land

2overland *adjective* (1800)
: going or accomplished over the land instead of by sea ⟨*overland* emigrants⟩ ⟨an *overland* route⟩

over·lap \ˌō-vər-ˈlap\ (1726)
transitive verb
1 : to extend over or past and cover a part of
2 : to have something in common with
intransitive verb
1 : to occupy the same area in part **:** lap over
2 : to have something in common
— **over·lap** \ˈō-vər-ˌlap\ *noun*

1over·lay \ˌō-vər-ˈlā\ *transitive verb* **-laid** \-ˈlād\; **-lay·ing** (14th century)
1 a : to lay or spread over or across **:** SUPERIMPOSE **b :** to prepare an overlay for
2 : OVERLIE 2

2over·lay \ˈō-vər-ˌlā\ *noun* (1794)
: a covering either permanent or temporary: as **a :** an ornamental veneer **b :** a decorative and contrasting design or article placed on top of a plain one **c :** a transparent sheet containing graphic matter to be superimposed on another sheet

over·leaf \ˈō-vər-ˌlēf, -ˈlēf\ *adverb* (1843)
: on the other side of a leaf (as of a book)

over·leap \ˌō-vər-ˈlēp\ *transitive verb* **-leaped** *or* **-leapt** \-ˈlēpt *also* -ˈlept\; **-leap·ing** \-ˈlē-piŋ\ (before 12th century)
1 : to leap over or across
2 : to defeat (oneself) by going too far

over·learn \-ˈlərn\ *transitive verb* (1874)
: to continue to study or practice after attaining proficiency

over·lie \-ˈlī\ *transitive verb* **-lay** \-ˈlā\; **-lain** \-ˈlān\; **-ly·ing** \-ˈlī-iŋ\ (13th century)
1 : to lie over or upon
2 : to cause the death of by lying upon

1over·look \-ˈlu̇k\ *transitive verb* (14th century)
1 : to look over **:** INSPECT
2 a : to look down upon from above **b :** to rise above or afford a view of
3 a : to look past **:** MISS **b :** IGNORE 1 **c :** EXCUSE 2
4 : SUPERINTEND, OVERSEE
5 : to look on with the evil eye **:** BEWITCH
synonym see NEGLECT

2over·look \ˈō-vər-ˌlu̇k\ *noun* (1861)
: a place from which one may look down on a scene below ⟨plenty of *overlooks* and trails —Thelma H. Bell⟩

over·lord \ˈō-vər-ˌlȯrd\ *noun* (13th century)
1 : a lord over other lords **:** a lord paramount
2 a : an absolute or supreme ruler **b :** one having great power or authority ⟨a corporate *overlord*⟩
— **over·lord·ship** \-ˌship, ˌō-vər-ˈ\ *noun*

over·ly \ˈō-vər-lē\ *adverb* (1821)
: to an excessive degree

1over·man \-mən, -ˌman\ *noun* (13th century)
1 : a man in authority over others; *specifically* **:** FOREMAN
2 \-ˌman\ [translation of German *Übermensch*] **:** SUPERMAN 1

2over·man \ˌō-vər-ˈman\ *transitive verb* (circa 1637)
: to have or get too many personnel for the needs of ⟨*overman* a ship⟩

over·man·tel \ˈō-vər-ˌman-tᵊl\ *noun* (1882)
: an ornamental structure (as a painting) above a mantelpiece
— **overmantel** *adjective*

over·mas·ter \ˌō-vər-ˈmas-tər\ *transitive verb* (14th century)
: OVERPOWER, SUBDUE

over·mas·ter·ing \-ˈmas-tə-riŋ\ *adjective* (1645)
: DOMINANT ⟨*overmastering* behavior⟩ ⟨the *overmastering* question⟩

over·match \-ˈmach\ *transitive verb* (14th century)
1 : to be more than a match for **:** DEFEAT
2 : to match with a superior opponent

1over·much \-ˈməch\ *adjective* (13th century)
: too much

2overmuch *adverb* (14th century)
: in too great a degree

\ə\ **abut** \ᵊ\ **kitten** \ər\ **further** \a\ **ash** \ā\ **ace** \ä\ **mop, mar** \au̇\ **out** \ch\ **chin** \e\ **bet** \ē\ **easy** \g\ **go** \i\ **hit** \ī\ **ice** \j\ **job** \ŋ\ **sing** \ō\ **go** \ȯ\ **law** \ȯi\ **boy** \th\ **thin** \t͟h\ **the** \ü\ **loot** \u̇\ **foot** \y\ **yet** \zh\ **vision** *see also* Guide to Pronunciation

[3]**over·much** \'ō-vər-ˌməch, ˌō-vər-'\ *noun* (14th century)
: too great an amount

[1]**over·night** \ˌō-vər-'nīt\ *adverb* (14th century)
1 a : on the evening before **b :** during the night ⟨stayed away *overnight*⟩
2 : very quickly or suddenly ⟨became famous *overnight*⟩

[2]**overnight** *adjective* (1824)
1 : of, lasting, or staying the night
2 : SUDDEN, RAPID ⟨an *overnight* sensation⟩

[3]**overnight** \'ō-vər-ˌnīt, ˌō-vər-'\ *intransitive verb* (1891)
: to stay overnight

[4]**over·night** \'ō-vər-ˌnīt\ *noun* (1959)
: an overnight stay

overnight bag *noun* (1925)
: a suitcase of a size to carry clothing and personal articles for an overnight trip — called also *overnight case*

over·night·er \ˌō-vər-'nī-tər\ *noun* (1949)
: OVERNIGHT BAG

[1]**over·pass** \ˌō-vər-'pas\ *transitive verb* (14th century)
1 : to pass across, over, or beyond **:** CROSS; *also* **:** SURPASS
2 : TRANSGRESS
3 : DISREGARD, IGNORE

[2]**over·pass** \'ō-vər-ˌpas\ *noun* (1929)
: a crossing of two highways or of a highway and pedestrian path or railroad at different levels where clearance to traffic on the lower level is obtained by elevating the higher level; *also* **:** the upper level of such a crossing

over·per·suade \ˌō-vər-pər-'swād\ *transitive verb* (1624)
: to persuade to act contrary to one's conviction or preference
— **over·per·sua·sion** \-'swā-zhən\ *noun*

over·plaid \'ō-vər-ˌplad\ *noun* (1926)
: a textile design consisting of a plaid pattern superimposed on another plaid or on a textured ground; *also* **:** a fabric with such a design
— **over·plaid·ed** *adjective*

over·play \ˌō-vər-'plā\ (1819)
transitive verb
1 a : to present (as a dramatic role) extravagantly **:** EXAGGERATE **b :** to place too much emphasis on
2 : to rely too much on the strength of — usually used in the phrase *overplay one's hand*
3 : to strike a golf ball beyond (a putting green)
intransitive verb
: to exaggerate a part or effect

over·plus \'ō-vər-ˌpləs\ *noun* [Middle English, part translation of Middle French *surplus*] (14th century)
: SURPLUS

over·pop·u·late \ˌō-vər-'päp-yə-ˌlāt\ (1868)
transitive verb
: to furnish or provide with more than the environment or market will bear
intransitive verb
: to become overly populous

over·pop·u·la·tion \ˌō-vər-ˌpä-pyə-'lā-shən\ *noun* (1823)
: the condition of having a population so dense as to cause environmental deterioration, an impaired quality of life, or a population crash

over·pow·er \ˌō-vər-'pau̇(-ə)r\ *transitive verb* (1593)
1 : to overcome by superior force **:** SUBDUE
2 : to affect with overwhelming intensity ⟨the stench *overpowered* us⟩
3 : to provide with more power than is needed or desirable ⟨a dangerously *overpowered* car⟩
— **over·pow·er·ing·ly** \-'pau̇(-ə)r-iŋ-lē\ *adverb*

over·pres·sure \'ō-vər-ˌpre-shər\ *noun* (1644)
: pressure significantly above what is usual or normal

over·price \ˌō-vər-'prīs\ *transitive verb* (1605)
: to price too high

[1]**over·print** \ˌō-vər-'print\ *transitive verb* (1863)
: to print over with something additional

[2]**over·print** \'ō-vər-ˌ\ *noun* (1876)
: something added by or as if by overprinting; *especially* **:** a printed marking added to a postage or revenue stamp especially to alter the original or to commemorate a special event

over·proof \ˌō-vər-'prüf\ *adjective* (1807)
: containing more alcohol than proof spirit

over·pro·por·tion \-prə-'pōr-shən, -'pȯr-\ *transitive verb* (1642)
: to make disproportionately large
— **overproportion** *noun*
— **over·pro·por·tion·ate** \-sh(ə-)nət\ *adjective*
— **over·pro·por·tion·ate·ly** *adverb*

over·qual·i·fied \-'kwä-lə-ˌfīd\ *adjective* (1954)
: having more education, training, or experience than a job calls for

over·reach \ˌō-və(r)-'rēch\ (14th century)
transitive verb
1 : to reach above or beyond **:** OVERTOP
2 : to defeat (oneself) by seeking to do or gain too much
3 : to get the better of especially in dealing and bargaining and typically by unscrupulous or crafty methods
intransitive verb
1 *of a horse* **:** to strike the forefoot with the front part of the hind foot
2 a : to go to excess **b :** EXAGGERATE
3 : to overreach oneself
— **over·reach** \'ō-və(r)-ˌrēch, ˌō-və(r)-'\ *noun*
— **over·reach·er** \-ˌrē-chər, -'rē-chər\ *noun*

over·rep·re·sent·ed \'ō-və(r)-ˌre-pri-'zen-təd\ *adjective* (1900)
: represented excessively; *especially* **:** having representatives in a proportion higher than the average
— **over·rep·re·sen·ta·tion** \-ˌre-pri-ˌzen-'tā-shən, -zən-\ *noun*

[1]**over·ride** \-'rīd\ *transitive verb* **-rode** \-'rōd\; **-rid·den** \-'ri-dᵊn\; **-rid·ing** \-'rī-diŋ\ (before 12th century)
1 : to ride over or across **:** TRAMPLE
2 : to ride (as a horse) too much or too hard
3 a : to prevail over **:** DOMINATE **b :** to set aside **:** ANNUL ⟨*override* a veto⟩ **c :** to neutralize the action of (as an automatic control)
4 : to extend or pass over; *especially* **:** OVERLAP

[2]**over·ride** \'o-və(r)-ˌrīd\ *noun* (1931)
1 : a commission paid to managerial personnel on sales made by subordinates
2 : ROYALTY 5a
3 : a device or system used to override a control
4 : an act or an instance of overriding

over·ripe \ˌō-və(r)-'rīp\ *adjective* (1671)
1 : passed beyond maturity or ripeness toward decay
2 a : DECADENT **b :** lacking originality or vigor
— **over–ripe·ness** \-nəs\ *noun*

over·rule \-'rül\ *transitive verb* (1576)
1 : to rule over **:** GOVERN
2 : to prevail over **:** OVERCOME
3 a : to rule against **b :** to set aside **:** REVERSE

[1]**over·run** \-'rən\ *transitive verb* **-ran** \-'ran\; **-run; -run·ning** (before 12th century)
1 a (1) **:** to defeat decisively and occupy the positions of (2) **:** to invade and occupy or ravage **b :** to spread or swarm over **:** INFEST
2 a : to run or go beyond or past ⟨the plane *overran* the runway⟩ **b :** EXCEED **c :** to readjust (set type) by shifting letters or words from one line into another
3 : to flow over

[2]**over·run** \'ō-və(r)-ˌrən\ *noun* (1898)
1 : an act or instance of overrunning; *especially* **:** an exceeding of the costs estimated in a contract for development and manufacture of new equipment
2 : the amount by which something overruns
3 : a run in excess of the quantity ordered by a customer

over·scale \'ō-vər-ˌskāl\ *or* **over·scaled** \-ˌskāld\ *adjective* (1953)
: OVERSIZE ⟨an *overscale* coat⟩ ⟨an *overscale* sofa⟩

over·sea \ˌo-vər-'sē, 'ō-vər-ˌ\ *adjective or adverb* (12th century)
chiefly British **:** OVERSEAS

[1]**over·seas** \ˌō-vər-'sēz\ *adverb* (1533)
: beyond or across the sea ⟨lived *overseas* for a time⟩

[2]**over·seas** \'ō-vər-ˌsēz\ *adjective* (1892)
1 : of or relating to movement, transport, or communication over the sea ⟨an *overseas* liner⟩
2 : situated, originating in, or relating to lands beyond the sea ⟨*overseas* installations⟩ ⟨*overseas* immigrants⟩

over·see \ˌō-vər-'sē\ *transitive verb* **-saw** \-'sȯ\; **-seen** \-'sēn\; **-see·ing** (before 12th century)
1 : SURVEY, WATCH
2 a : INSPECT, EXAMINE **b :** SUPERVISE

over·seer \'ō-və(r)-ˌsir, -ˌsē-ər, ˌō-və(r)-'\ *noun* (1523)
: SUPERVISOR, SUPERINTENDENT

over·sell \ˌō-vər-'sel\ *transitive verb* **-sold** \-'sōld\; **-sell·ing** (circa 1879)
1 a : to sell too much or too many to **b :** to sell too much or too many of
2 : to make excessive claims for
— **over·sell** \'ō-vər-ˌsel\ *noun*

over·set \-'set\ *transitive verb* **-set; -set·ting** (1583)
1 a : to disturb mentally or physically **:** UPSET **b :** to turn or tip over **:** OVERTURN
2 : to set too much type matter for
— **over·set** \'ō-vər-ˌset\ *noun*

over·sexed \ˌō-vər-'sekst\ *adjective* (1898)
: exhibiting an excessive sexual drive or interest

over·shad·ow \-'sha-(ˌ)dō\ *transitive verb* (before 12th century)
1 : to cast a shadow over
2 : to exceed in importance **:** OUTWEIGH

over·shirt \'ō-vər-ˌshərt\ *noun* (1805)
: a shirt usually worn over another shirt without being tucked in

over·shoe \-ˌshü\ *noun* (1823)
: an outer shoe; *especially* **:** GALOSH

over·shoot \ˌō-vər-'shüt\ *transitive verb* **-shot** \-'shät\; **-shoot·ing** (14th century)
1 : to pass swiftly beyond
2 : to shoot or pass over or beyond so as to miss
3 : to excel in shooting
— **over·shoot** \'ō-vər-ˌshüt\ *noun*

[1]**over·shot** \'ō-vər-ˌshät\ *adjective* (circa 1535)
1 : actuated by the weight of water passing over and flowing from above ⟨an *overshot* waterwheel⟩
2 a : having the upper jaw extending beyond the lower **b :** projecting beyond the lower jaw

[2]**overshot** *noun* (1945)
: a pattern or weave featuring filling threads which pass two or more warp yarns before reentering the fabric

over·sight \'ō-vər-ˌsīt\ *noun* (15th century)
1 a : watchful and responsible care **b :** regulatory supervision ⟨congressional *oversight*⟩
2 : an inadvertent omission or error

over·sim·ple \ˌō-vər-'sim-pəl\ *adjective* (15th century)
: too simple **:** not thoroughgoing or exhaustive ⟨*oversimple* theories⟩
— **over·sim·ply** \-plē\ *adverb*

over·sim·pli·fy \-'sim-plə-ˌfī\ (1923)
transitive verb
: to simplify to such an extent as to bring about distortion, misunderstanding, or error
intransitive verb
: to engage in undue or extreme simplification

— **over·sim·pli·fi·ca·tion** \-ˌsim-plə-fə-ˈkā-shən\ *noun*

over·size \ˌō-vər-ˈsīz\ *or* **over·sized** \-ˈsīzd\ *adjective* (1853)
: being of more than standard or ordinary size ⟨*oversize* pillows⟩ ⟨an *oversize* shirt⟩

over·skirt \ˈō-vər-ˌskərt\ *noun* (1870)
: a skirt worn over another skirt

over·slaugh \ˌō-vər-ˈslȯ\ *transitive verb* [Dutch *overslaan* to pass over, omit, from Middle Dutch *overslaen,* from *over-* over- + *slaen* to strike] (1846)
: to pass over for appointment or promotion in favor of another

over·sleep \ˌō-vər-ˈslēp\ *verb* **-slept** \-ˈslept\; **-sleep·ing** (14th century)
intransitive verb
: to sleep beyond the time for waking
transitive verb
: to allow (oneself) to oversleep

over·slip \ˌō-vər-ˈslip\ *transitive verb* (15th century)
obsolete **:** ESCAPE

over·sold \ˌō-vər-ˈsōld\ *adjective* (1926)
: likely to show a rise in price because of prior heavy selling and accompanying decline in price ⟨an *oversold* stock⟩

over·soul \ˈō-vər-ˌsōl\ *noun* (circa 1844)
: the absolute reality and basis of all existences conceived as a spiritual being in which the ideal nature imperfectly manifested in human beings is perfectly realized

over·spend \ˌō-vər-ˈspend\ *verb* **-spent** \-ˈspent\; **-spend·ing** (circa 1618)
transitive verb
1 : to spend or use to excess **:** EXHAUST
2 : to exceed in expenditure
intransitive verb
: to spend beyond one's means
— **over·spend·er** *noun*

over·spill \ˈō-vər-ˌspil\ *noun* (1884)
1 : the act or an instance of spilling over
2 *chiefly British* **:** the movement of excess urban population into less crowded areas

over·spread \ˌō-vər-ˈspred\ *transitive verb* **-spread; -spread·ing** (before 12th century)
: to spread over or above
— **over·spread** \ˈō-vər-ˌspred\ *noun*

over·state \-ˈstāt\ *transitive verb* (1803)
: to state in too strong terms **:** EXAGGERATE
— **over·state·ment** \-mənt\ *noun*

over·stay \-ˈstā\ *transitive verb* (1646)
: to stay beyond the time or the limits of

over·steer \ˈō-vər-ˌstir\ *noun* (1951)
: the tendency of an automobile to steer into a sharper turn than the driver intends sometimes with a thrusting of the rear to the outside; *also* **:** the action or an instance of oversteer

over·step \ˌō-vər-ˈstep\ *transitive verb* (before 12th century)
: EXCEED, TRANSGRESS

over·sto·ry \ˈō-vər-ˌstōr-ē, -ˌstȯr-\ *noun* (1925)
1 : the layer of foliage in a forest canopy
2 : the trees contributing to an overstory

over·strew \ˌō-vər-ˈstrü\ *transitive verb* **-strewed; -strewed** *or* **-strewn** \-ˈstrün\; **-strew·ing** (circa 1570)
1 : to strew or scatter about
2 : to cover here and there

over·stride \-ˈstrīd\ *transitive verb* **-strode** \-ˈstrōd\; **-strid·den** \-ˈstri-dᵊn\; **-strid·ing** \-ˈstrī-diŋ\ (13th century)
1 a : to stride over, across, or beyond **b :** BESTRIDE
2 : to stride faster than or beyond

over·strung \-ˈstrəŋ\ *adjective* (1810)
: too highly strung **:** too sensitive

over·stuff \-ˈstəf\ *transitive verb* (1904)
1 : to stuff too full
2 : to cover (as a chair or sofa) completely and deeply with upholstery

over·sub·scribe \-səb-ˈskrīb\ *transitive verb* (1891)
: to subscribe for more of than is available
— **over·sub·scrip·tion** \-ˈskrip-shən\ *noun*

overt \ō-ˈvərt, ˈō-(ˌ)vərt\ *adjective* [Middle English, from Middle French *ouvert, overt,* from past participle of *ouvrir* to open, from (assumed) Vulgar Latin *operire,* alteration of Latin *aperire*] (14th century)
: open to view **:** MANIFEST
— **overt·ly** *adverb*
— **overt·ness** *noun*

over·take \ˌō-vər-ˈtāk\ *transitive verb* **-took** \-ˈtu̇k\; **-tak·en** \-ˈtā-kən\; **-tak·ing** [Middle English, from [1]*over* + *taken* to take] (13th century)
1 a : to catch up with **b :** to catch up with and pass by
2 : to come upon suddenly

over–the–counter *adjective* (1921)
1 : not traded or effected on an organized securities exchange ⟨*over-the-counter* transactions⟩ ⟨*over-the-counter* securities⟩
2 : sold lawfully without prescription ⟨*over-the-counter* drugs⟩

over–the–hill *adjective* (1946)
1 : past one's prime
2 : advanced in age

over–the–top *adjective* (1985)
: extremely or excessively flamboyant or outrageous ⟨an *over-the-top* performance⟩

over–the–transom *adjective* (circa 1952)
: offered without prior arrangement especially for publication **:** UNSOLICITED ⟨an *over-the-transom* manuscript⟩

over·throw \ˌō-vər-ˈthrō\ *transitive verb* **-threw** \-ˈthrü\; **-thrown** \-ˈthrōn\; **-throw·ing** (14th century)
1 : OVERTURN, UPSET
2 : to cause the downfall of **:** BRING DOWN, DEFEAT
3 : to throw a ball over or past (as a base or a receiver)
— **over·throw** \ˈō-vər-ˌthrō\ *noun*

over·time \ˈō-vər-ˌtīm\ *noun* (1536)
1 : time in excess of a set limit: as **a :** working time in excess of a standard day or week **b :** an extra period of play in a contest
2 : the wage paid for overtime
— **overtime** *adverb*

over·tone \-ˌtōn\ *noun* (1867)
1 a : one of the higher tones produced simultaneously with the fundamental and that with the fundamental comprise a complex musical tone **:** HARMONIC 1a **b :** HARMONIC 2
2 : the color of the light reflected (as by a paint)
3 : a secondary effect, quality, or meaning **:** SUGGESTION 3, CONNOTATION

over·top \ˌō-vər-ˈtäp\ *transitive verb* (circa 1594)
1 : to rise above the top of
2 : to be superior to
3 : SURPASS

over·trade \-ˈtrād\ *intransitive verb* (1734)
: to trade beyond one's capital

over·trick \ˈō-vər-ˌtrik\ *noun* (1903)
: a card trick won in excess of the number bid

over·trump \ˌō-vər-ˈtrəmp\ (1746)
transitive verb
: to trump with a higher trump card than the highest previously played on the same trick
intransitive verb
: to play a higher trump card than the highest previously played on the same trick

[1]**over·ture** \ˈō-və(r)-ˌchu̇r, -chər, -ˌtyu̇r, -ˌtu̇r\ *noun* [Middle English, literally, opening, from Middle French, from (assumed) Vulgar Latin *opertura,* alteration of Latin *apertura* — more at APERTURE] (15th century)
1 a : an initiative toward agreement or action **:** PROPOSAL **b :** something introductory **:** PRELUDE
2 a : the orchestral introduction to a musical dramatic work **b :** an orchestral concert piece written especially as a single movement in sonata form

[2]**overture** *transitive verb* **-tured; -tur·ing** (circa 1650)
1 : to put forward as an overture
2 : to make or present an overture to

[1]**over·turn** \ˌō-vər-ˈtərn\ (13th century)
transitive verb
1 : to cause to turn over **:** UPSET
2 : INVALIDATE, DESTROY
intransitive verb
: UPSET, TURN OVER

[2]**over·turn** \ˈō-vər-ˌtərn\ *noun* (circa 1592)
1 : the act of overturning **:** the state of being overturned
2 : the sinking of surface water and rise of bottom water in a lake or sea that results from changes in temperature that commonly occur in spring and fall

over·val·ue \ˌō-vər-ˈval-(ˌ)yü\ *transitive verb* (1597)
: to assign an excessive or fictitious value to
— **over·valu·a·tion** \-ˌval-yə-ˈwā-shən\ *noun*

over·view \ˈō-vər-ˌvyü\ *noun* (1588)
: a general survey **:** SUMMARY

over·volt·age \ˌō-vər-ˈvōl-tij\ *noun* (1907)
1 : the excess potential required for the discharge of an ion at an electrode over and above the equilibrium potential of the electrode
2 : voltage in excess of the normal operating voltage of a device or circuit

over·wear \-ˈwar, -ˈwer\ *transitive verb* **-wore** \-ˈwōr, -ˈwȯr\; **-worn** \-ˈwōrn, -ˈwȯrn\; **-wear·ing** (1578)
: WEAR OUT, EXHAUST

over·ween·ing \-ˈwē-niŋ\ *adjective* [Middle English *overwening,* present participle of *overwenen* to be arrogant, from *over* + *wenen* to ween] (14th century)
1 : ARROGANT, PRESUMPTUOUS
2 : IMMODERATE, EXAGGERATED
— **over·ween·ing·ly** *adverb*

over·weigh \-ˈwā\ *transitive verb* (13th century)
1 : to exceed in weight
2 : OPPRESS 2

[1]**over·weight** \ˈō-vər-ˌwāt, *2 is usually* ˌō-vər-ˈ\ *noun* (1552)
1 : weight over and above what is required or allowed
2 : excessive or burdensome weight

[2]**over·weight** \ˌō-vər-ˈwāt\ *transitive verb* (1603)
1 : to give too much weight or consideration to
2 : to weight excessively

[3]**over·weight** \ˌō-vər-ˈwāt\ *adjective* (1638)
: exceeding expected, normal, or proper weight; *especially* **:** exceeding the bodily weight normal for one's age, height, and build

over·whelm \ˌō-vər-ˈhwelm, -ˈwelm\ *transitive verb* [Middle English, from [1]*over* + *whelmen* to turn over, cover up] (14th century)
1 : UPSET, OVERTHROW
2 a : to cover over completely **:** SUBMERGE **b :** to overcome by superior force or numbers **c :** to overpower in thought or feeling

over·whelm·ing *adjective* (1712)
: tending or serving to overwhelm ⟨an *overwhelming* majority⟩; *also* **:** EXTREME, GREAT ⟨*overwhelming* indifference⟩
— **over·whelm·ing·ly** *adverb*

[1]**over·win·ter** \ˌō-vər-ˈwin-tər\ *intransitive verb* (before 12th century)
: to survive the winter

[2]**overwinter** *adjective* (1900)
: occurring during the period spanning the winter

over with *adjective* (1915)
: being at an end **:** FINISHED, COMPLETED

over·work \ˌō-vər-ˈwərk\ (1818)
transitive verb
1 : to cause to work too hard, too long, or to exhaustion
2 : to decorate all over
3 a : to work too much on **:** OVERDO **b :** to make excessive use of
intransitive verb
: to work too much or too long **:** OVERDO
— overwork *noun*

over·write \ˌō-və(r)-ˈrīt\ *verb* **-wrote** \-ˈrōt\; **-writ·ten** \-ˈri-tᵊn\; **-writ·ing** \-ˈrī-tiŋ\ (1699)
transitive verb
1 : to write over the surface of
2 : to write in inflated or overly elaborate style
intransitive verb
: to write too much or in an overly elaborate style

over·wrought \-ˈrȯt\ *adjective* [past participle of *overwork*] (1670)
1 : extremely excited **:** AGITATED
2 : elaborated to excess **:** OVERDONE

ovi- *or* **ovo-**
— see OV-

ovi·cide \ˈō-və-ˌsīd\ *noun* [International Scientific Vocabulary] (1913)
: an agent that kills eggs; *especially* **:** an insecticide effective against the egg stage
— ovi·cid·al \ˌō-və-ˈsī-dᵊl\ *adjective*

ovi·duct \ˈō-və-ˌdəkt\ *noun* [New Latin *oviductus,* from *ov-* + *ductus* duct] (1672)
: a tube that serves exclusively or especially for the passage of eggs from an ovary
— ovi·duc·tal \ˌō-və-ˈdək-tᵊl\ *adjective*

ovine \ˈō-ˌvīn\ *adjective* [Late Latin *ovinus,* from Latin *ovis* sheep — more at EWE] (circa 1828)
: of, relating to, or resembling sheep
— ovine *noun*

ovip·a·rous \ō-ˈvi-p(ə-)rəs\ *adjective* [Latin *oviparus,* from *ov-* + *-parus* -parous] (1646)
: producing eggs that develop and hatch outside the maternal body; *also* **:** involving the production of such eggs

ovi·pos·it \ˈō-və-ˌpä-zət, ˌō-və-ˈ\ *intransitive verb* [probably back-formation from *ovipositor*] (1816)
: to lay eggs — used especially of insects
— ovi·po·si·tion \ˌō-və-pə-ˈzi-shən\ *noun*
— ovi·po·si·tion·al \-ˈzish-nəl, -ˈzi-shə-nᵊl\ *adjective*

ovi·pos·i·tor \ˈō-və-ˌpä-zə-tər, ˌō-və-ˈ\ *noun* [New Latin, from Latin *ov-* + *positor* one that places, from *ponere* to place — more at POSITION] (1816)
: a specialized organ (as of an insect) for depositing eggs — see INSECT illustration

ovoid \ˈō-ˌvȯid\ *also* **ovoi·dal** \ō-ˈvȯi-dᵊl\ *adjective* [French *ovoïde,* from Latin *ovum* egg — more at EGG] (circa 1828)
: resembling an egg in shape **:** OVATE
— ovoid *noun*

ovo·lo \ˈō-və-ˌlō\ *noun, plural* **-los** [Italian, diminutive of *uovo, ovo* egg, from Latin *ovum*] (1663)
: a rounded convex molding

Ovon·ics \ō-ˈvä-niks\ *noun plural but usually singular in construction* [Stanford R. *Ov*shinsky (born 1923) American inventor + electr*onics*] (1968)
: a branch of electronics that deals with applications of the change from an electrically nonconducting state to a semiconducting state shown by glasses of special composition upon application of a certain minimum voltage
— ovon·ic \-nik\ *adjective*

ovo·tes·tis \ˌō-vō-ˈtes-təs\ *noun* [New Latin] (1877)
: a hermaphrodite gonad (as in some scale insects)

ovo·vi·vip·a·rous \ˈō-vō-ˌvī-ˈvi-p(ə-)rəs\ *adjective* (1801)
: producing eggs that develop within the maternal body (as of various fishes or reptiles) and hatch within or immediately after extrusion from the parent
— ovo·vi·vip·a·rous·ly *adverb*
— ovo·vi·vip·a·rous·ness *noun*

ovu·late \ˈäv-yə-ˌlāt, ˈōv-, -lət\ *adjective* (1861)
: bearing an ovule

ovu·la·tion \ˌäv-yə-ˈlā-shən, ˌōv-\ *noun* (1848)
: the discharge of a mature ovum from the ovary
— ovu·late \ˈäv-yə-ˌlāt, ˈōv-\ *verb*

ovu·la·to·ry \ˈäv-yə-lə-ˌtōr-ē, ˈōv-, -ˌtȯr-\ *adjective* (1931)
: of, relating to, or involving ovulation

ovule \ˈäv-(ˌ)yü(ə)l, ˈōv-\ *noun* [New Latin *ovulum,* diminutive of Latin *ovum*] (1830)
1 : an outgrowth of the ovary of a seed plant that is a megasporangium and encloses an embryo sac within a nucellus
2 : a small egg; *especially* **:** one in an early stage of growth

ovum \ˈō-vəm\ *noun, plural* **ova** \-və\ [New Latin, from Latin, egg — more at EGG] (circa 1706)
: a female gamete **:** MACROGAMETE

ow \ˈau̇, ˈü\ *interjection* [from *ow,* interjection expressing surprise, from Middle English] (circa 1911)
— used especially to express sudden pain

owe \ˈō\ *verb* **owed; ow·ing** [Middle English, to possess, own, owe, from Old English *āgan;* akin to Old High German *eigun* (1st & 3d plural present indicative) possess, Sanskrit *īśe* he possesses] (before 12th century)
transitive verb
1 a *archaic* **:** POSSESS, OWN **b :** to have or bear (an emotion or attitude) to someone or something ⟨*owes* the boss a grudge⟩
2 a (1) **:** to be under obligation to pay or repay in return for something received **:** be indebted in the sum of ⟨*owes* me $5⟩ (2) **:** to be under obligation to render (as duty or service) ⟨I *owe* you a favor⟩ **b :** to be indebted to ⟨*owes* the grocer for supplies⟩
3 : to be indebted for ⟨*owed* his wealth to his father⟩ ⟨*owes* much to good luck⟩
intransitive verb
1 : to be in debt ⟨*owes* for his house⟩
2 : to be attributable ⟨an idea that *owes* to Greek philosophy⟩

owing *adjective* (15th century)
: due to be paid ⟨has bills *owing*⟩

owing to *preposition* (1695)
: BECAUSE OF ⟨delayed *owing to* a crash⟩

owl \ˈau̇(ə)l\ *noun* [Middle English *owle,* from Old English *ūle;* akin to Old High German *uwila* owl] (before 12th century)
: any of an order (Strigiformes) of chiefly nocturnal birds of prey with a large head and eyes, short hooked bill, strong talons, and soft fluffy often mottled brown plumage

owl·et \ˈau̇-lət\ *noun* (1542)
: a small or young owl

owl·ish \ˈau̇-lish\ *adjective* (1611)
: resembling or suggesting an owl
— owl·ish·ly *adverb*
— owl·ish·ness *noun*

[1]own \ˈōn\ *adjective* [Middle English *owen,* from Old English *āgen;* akin to Old High German *eigan* own, Old Norse *eiginn,* Old English *āgan* to possess — more at OWE] (before 12th century)
1 : belonging to oneself or itself — usually used following a possessive case or possessive adjective ⟨cooked my *own* dinner⟩
2 — used to express immediate or direct kinship ⟨an *own* son⟩ ⟨an *own* sister⟩

[2]own (before 12th century)
transitive verb
1 a : to have or hold as property **:** POSSESS **b :** to have power over **:** CONTROL ⟨wanted to *own* his own life⟩
2 : to acknowledge to be true, valid, or as claimed **:** ADMIT ⟨*own* a debt⟩
intransitive verb
: to acknowledge something to be true, valid, or as claimed — used with *to* or *up*
synonym see ACKNOWLEDGE
— own·er \ˈō-nər\ *noun*

[3]own *pronoun, singular or plural in construction* (before 12th century)
: one or ones belonging to oneself — used after a possessive and without a following noun ⟨gave out books so that each of us had our *own*⟩ ⟨a room of your *own*⟩
— on one's own : for or by oneself **:** independently of assistance or control

own·er·ship \ˈō-nər-ˌship\ *noun* (1583)
1 : the state, relation, or fact of being an owner
2 : a group or organization of owners

ox \ˈäks\ *noun, plural* **ox·en** \ˈäk-sən\ *also* **ox** [Middle English, from Old English *oxa;* akin to Old High German *ohso* ox, Sanskrit *ukṣā* bull, and perhaps to Sanskrit *ukṣati* he moistens, Greek *hygros* wet — more at HUMOR] (before 12th century)
1 : a domestic bovine mammal (*Bos taurus*); *broadly* **:** a bovine mammal
2 : an adult castrated male domestic ox

ox- *or* **oxo-** *combining form* [French, from *oxygène*]
: oxygen ⟨*ox*acillin⟩

ox·a·cil·lin \ˌäk-sə-ˈsi-lən\ *noun* [*ox-* + *az*ole + peni*cillin*] (1962)
: a semisynthetic penicillin that is especially effective in the control of infections caused by penicillin-resistant staphylococci

ox·a·late \ˈäk-sə-ˌlāt\ *noun* (1791)
: a salt or ester of oxalic acid

ox·al·ic acid \(ˌ)äk-ˈsa-lik-\ *noun* [French (*acide*) *oxalique,* from Latin *oxalis*] (1791)
: a poisonous strong acid $(COOH)_2$ or $H_2C_2O_4$ that occurs in various plants as oxalates and is used especially as a bleaching or cleaning agent and as a chemical intermediate

ox·al·is \äk-ˈsa-ləs\ *noun* [New Latin, genus name, from Latin, wood sorrel, from Greek, from *oxys* sharp — more at OXYGEN] (circa 1706)
: WOOD SORREL

ox·a·lo·ac·e·tate \ˌäk-sə-lō-ˈa-sə-ˌtāt\ *also* **ox·al·ac·e·tate** \ˌäk-sə-ˈla-\ *noun* (1891)
: a salt or ester of oxaloacetic acid

ox·a·lo·ace·tic acid \ˌäk-sə-lō-ə-ˈsē-tik-\ *also* **ox·al·ace·tic acid** \ˌäk-sə-lə-ˈsē-tik-\ *noun* [*oxal*ic + *acetic acid*] (1896)
: a crystalline acid $C_4H_4O_5$ that is formed by reversible oxidation of malic acid (as in carbohydrate metabolism via the Krebs cycle) and in reversible transamination reactions (as from aspartic acid)

ox·a·lo·suc·cin·ic acid \ˈäk-sə-lō-sək-ˌsi-nik-, äk-ˈsa-lō-\ *noun* [*oxal*ic + *succinic acid*] (1925)
: a tricarboxylic acid $C_6H_6O_7$ that is formed as an intermediate in the Krebs cycle

ox·az·e·pam \äk-ˈsa-zə-ˌpam\ *noun* [hydr*oxy*- + di*azepam*] (1964)
: a tranquilizing drug $C_{15}H_{11}ClN_2O_2$

ox·blood \ˈäks-ˌbləd\ *noun* (1705)
: a moderate reddish brown

ox·bow \ˈäks-ˌbō\ *noun* (14th century)
1 : a U-shaped frame forming a collar about an ox's neck and supporting the yoke
2 : something (as a bend in a river) resembling an oxbow
— oxbow *adjective*

oxbow 1

Ox·bridge \ˈäks-ˌbrij\ *adjective* [*Ox*ford + Cam*bridge*] (1960)
: of, relating to, or characteristic of Oxford and Cambridge Universities — compare PLATEGLASS, REDBRICK 2

ox·cart \-ˌkärt\ *noun* (1749)
: a cart drawn by oxen

ox·eye \-ˌī\ *noun* (15th century)

: any of several composite plants (as of the genera *Chrysanthemum* or *Heliopsis*) having heads with both disk and ray flowers; *especially* **:** DAISY 1b

oxeye daisy *noun* (circa 1763)
: DAISY 1b

ox·ford \'äks-fərd\ *noun* [*Oxford,* England] (circa 1890)
1 : a low shoe laced or tied over the instep
2 : a soft durable cotton or synthetic fabric made in plain or basket weaves — called also *oxford cloth*

Oxford down *noun, often D capitalized* [*Oxfordshire,* England] (1859)
: any of a Down breed of large hornless sheep developed by crossing Cotswolds and Hampshires — called also *Oxford*

Oxford movement *noun* (1841)
: a High Church movement within the Church of England begun at Oxford in 1833

ox·heart \'äks-ˌhärt\ *noun* (1870)
: any of various large sweet cherries

ox·i·dant \'äk-sə-dənt\ *noun* (1884)
: OXIDIZING AGENT
— **oxidant** *adjective*

ox·i·dase \'äk-sə-ˌdās, -ˌdāz\ *noun* [International Scientific Vocabulary] (1896)
: any of various enzymes that catalyze oxidations; *especially* **:** one able to react directly with molecular oxygen
— **ox·i·da·sic** \ˌäk-sə-'dā-sik, -zik\ *adjective*

ox·i·da·tion \ˌäk-sə-'dā-shən\ *noun* [French, from *oxider, oxyder* to oxidize, from *oxide*] (1791)
1 : the act or process of oxidizing
2 : the state or result of being oxidized
— **ox·i·da·tive** \'äk-sə-ˌdā-tiv\ *adjective*
— **ox·i·da·tive·ly** *adverb*

oxidation number *noun* (1926)
: a positive or negative number that represents the effective charge of an atom or element and that indicates the extent or possibility of its oxidation ⟨the usual *oxidation number* of sodium is +1 and of oxygen −2⟩ — called also *oxidation state*

oxidation–reduction *noun* (1909)
: a chemical reaction in which one or more electrons are transferred from one atom or molecule to another

oxidative phosphorylation *noun* (1954)
: the synthesis of ATP by phosphorylation of ADP for which energy is obtained by electron transport and which takes place in the mitochondria during aerobic respiration

ox·ide \'äk-ˌsīd\ *noun* [French *oxide, oxyde,* from *ox-* (from *oxygène* oxygen) + *-ide* (from *acide* acid)] (1790)
: a binary compound of oxygen with a more electropositive element or group
— **ox·id·ic** \äk-'si-dik\ *adjective*

ox·i·dize \'äk-sə-ˌdīz\ *verb* **-dized; -diz·ing** (1806)
transitive verb
1 : to combine with oxygen
2 : to dehydrogenate especially by the action of oxygen
3 : to change (a compound) by increasing the proportion of the electronegative part or change (an element or ion) from a lower to a higher positive valence **:** remove one or more electrons from (an atom, ion, or molecule)
intransitive verb
: to become oxidized
— **ox·i·diz·able** \ˌäk-sə-'dī-zə-bəl\ *adjective*

ox·i·diz·er \-ˌdī-zər\ *noun* (1875)
: OXIDIZING AGENT; *especially* **:** one used to support the combustion of a rocket propellant

oxidizing agent *noun* (1903)
: a substance that oxidizes something especially chemically (as by accepting electrons)

ox·i·do·re·duc·tase \'äk-sə-dō-ri-'dək-ˌtās, -ˌtāz\ *noun* [*oxid*ation + *-o-* + *reduction* + *-ase*] (1922)
: an enzyme that catalyzes an oxidation-reduction reaction

ox·ime \'äk-ˌsēm\ *noun* [International Scientific Vocabulary *ox-* + *-ime* (from *imide*)] (circa 1890)
: any of various compounds obtained chiefly by the action of hydroxylamine on aldehydes and ketones and characterized by the bivalent grouping C=NOH

ox·lip \'äk-ˌslip\ *noun* [(assumed) Middle English *oxeslippe,* from Old English *oxanslyppe,* literally, ox dung, from *oxa* ox + *slypa, slyppe* paste — more at SLIP] (before 12th century)
: a Eurasian primula (*Primula elatior*) having usually yellow flowers

oxo \'äk-(ˌ)sō\ *adjective* [*ox-*] (circa 1926)
: containing oxygen

oxo- — see OX-

Ox·o·ni·an \äk-'sō-nē-ən\ *noun* [Medieval Latin *Oxonia* Oxford] (circa 1540)
: a student or graduate of Oxford University
— **Oxonian** *adjective*

ox·tail \'äks-ˌtāl\ *noun* (15th century)
: the tail of a beef animal; *especially* **:** the skinned tail used for food (as in soup)

ox·ter \'äk-stər\ *noun* [Middle English (Scots), alteration of Old English *ōxta;* akin to Old English *eax* axis, axle — more at AXIS] (15th century)
1 *chiefly Scottish & Irish* **:** ARMPIT 1
2 *chiefly Scottish & Irish* **:** ARM

ox·tongue \'äks-ˌtəŋ\ *noun* (14th century)
: a European hawkweed (*Picris echioides*) that has yellow flowers and is now naturalized in the eastern U.S.

oxy \'äk-sē\ *adjective* [French, from *oxygène* oxygen] (1910)
: containing oxygen or additional oxygen — often used in combination ⟨*oxy*hemoglobin⟩

oxy·acet·y·lene \ˌäk-sē-ə-'se-tᵊl-ən, -tᵊl-ˌēn\ *adjective* [International Scientific Vocabulary] (1909)
: of, relating to, or utilizing a mixture of oxygen and acetylene ⟨an *oxyacetylene* torch⟩

oxy·ac·id \'äk-sē-ˌa-səd\ *noun* (circa 1841)
: an acid (as sulfuric acid) that contains oxygen — called also *oxygen acid*

ox·y·gen \'äk-si-jən\ *noun, often attributive* [French *oxygène,* from Greek *oxys,* adjective, acidic, literally, sharp + French *-gène* -gen; akin to Latin *acer* sharp — more at EDGE] (1790)
: a colorless tasteless odorless gaseous element that constitutes 21 percent of the atmosphere and is found in water, in most rocks and minerals, and in numerous organic compounds, that is capable of combining with all elements except the inert gases, that is active in physiological processes, and that is involved especially in combustion — see ELEMENT table
— **ox·y·gen·ic** \ˌäk-si-'je-nik\ *adjective*
— **ox·y·gen·less** \'äk-si-jən-ləs\ *adjective*

ox·y·gen·ate \'äk-si-jə-ˌnāt, äk-'si-jə-\ *transitive verb* **-at·ed; -at·ing** (1790)
: to impregnate, combine, or supply (as blood) with oxygen
— **ox·y·gen·a·tion** \ˌäk-si-jə-'nā-shən, äk-ˌsi-jə-\ *noun*

ox·y·gen·a·tor \'äk-si-jə-ˌnā-tər, äk-'si-jə-\ *noun* (circa 1864)
: one that oxygenates; *specifically* **:** an apparatus that oxygenates the blood extracorporeally (as during open-heart surgery)

oxygen cycle *noun* (1935)
: the cycle whereby atmospheric oxygen is converted to carbon dioxide in animal respiration and regenerated by green plants in photosynthesis

oxygen debt *noun* (1923)
: a cumulative deficit of oxygen available for oxidative metabolism that develops during periods of intense bodily activity and must be made good when the body returns to rest

oxygen demand *noun* (1950)
: BIOCHEMICAL OXYGEN DEMAND

oxygen mask *noun* (1920)
: a device worn over the nose and mouth (as by pilots at high altitudes) through which oxygen is supplied from a storage tank

oxygen tent *noun* (1925)
: a canopy which can be placed over a bedridden person and within which a flow of oxygen can be maintained

oxy·he·mo·glo·bin \ˌäk-si-'hē-mə-ˌglō-bən\ *noun* [International Scientific Vocabulary] (1873)
: hemoglobin loosely combined with oxygen that it releases to the tissues

oxy·hy·dro·gen \-'hī-drə-jən\ *adjective* (1827)
: of, relating to, or utilizing a mixture of oxygen and hydrogen ⟨*oxyhydrogen* torch⟩

ox·y·mo·ron \ˌäk-si-'mōr-ˌän, -'mȯr-\ *noun, plural* **-mo·ra** \-'mōr-ə, -'mȯr-\ [Late Greek *oxymōron,* from neuter of *oxymōros* pointedly foolish, from Greek *oxys* sharp, keen + *mōros* foolish] (1657)
: a combination of contradictory or incongruous words (as *cruel kindness*)
— **ox·y·mo·ron·ic** \-mə-'rä-nik, -mȯ-\ *adjective*
— **ox·y·mo·ron·i·cal·ly** \-ni-k(ə-)lē\ *adverb*

oxy·phen·bu·ta·zone \ˌäk-sē-ˌfen-'byü-tə-ˌzōn\ *noun* [*oxy* + *phenylbutazone*] (1961)
: a phenylbutazone derivative $C_{19}H_{20}N_2O_3$ used for its anti-inflammatory, analgesic, and antipyretic effects

oxy·phil·ic \ˌäk-si-'fi-lik\ *adjective* [Greek *oxys* acidic + English *-phil* — more at OXYGEN] (1901)
: ACIDOPHILIC

oxy·tet·ra·cy·cline \-ˌte-trə-'sī-ˌklēn\ *noun* (1953)
: a yellow crystalline broad-spectrum antibiotic $C_{22}H_{24}N_2O_9$ produced by a soil actinomycete (*Streptomyces rimosus*)

oxy·to·cic \ˌäk-si-'tō-sik\ *adjective* [International Scientific Vocabulary, from Greek *oxys* sharp, quick + *tokos* childbirth, from *tiktein* to bear — more at THANE] (1873)
: hastening parturition; *also* **:** inducing contraction of uterine smooth muscle
— **oxytocic** *noun*

oxy·to·cin \-'tō-sᵊn\ *noun* [International Scientific Vocabulary, from *oxytocic*] (1928)
: a pituitary octapeptide hormone $C_{43}H_{66}N_{12}O_{12}S_2$ that stimulates especially the contraction of uterine muscle and the secretion of milk

oxy·uri·a·sis \ˌäk-si-yu̇-'rī-ə-səs\ *noun* [New Latin, from *Oxyuris,* genus of worms] (circa 1909)
: infestation with or disease caused by pinworms (family Oxyuridae)

oy·er and ter·mi·ner \ˌȯi-ər-ən(d)-'tər-mə-nər\ *noun* [Middle English, part translation of Anglo-French *oyer et terminer,* literally, to hear and determine] (15th century)
1 : a commission authorizing a British judge to hear and determine a criminal case at the assizes
2 : a high criminal court in some U.S. states

¹oyez \ō-'yez, -'yā, -'yes, 'ō-ˌ\ *verb imperative* [Middle English, from Anglo-French, hear ye, imperative plural of *oir* to hear, from Latin *audire* — more at AUDIBLE] (15th century)
— used by a court or public crier to gain attention before a proclamation

²oyez *noun, plural* **oyes·ses** \-'ye-səz, -ˌye-səz\ (15th century)
: a cry of oyez

oys·ter \'ȯis-tər\ *noun, often attributive* [Middle English *oistre,* from Middle French, from Latin *ostrea,* from Greek *ostreon;* akin to Greek *ostrakon* shell, *osteon* bone — more at OSSEOUS] (13th century)
1 a : any of various marine bivalve mollusks (family Ostreidae) that have a rough irregular shell closed by a single adductor muscle and include commercially important shellfish **b :** any of various mollusks resembling or related to the oysters
2 : something that is or can be readily made to serve one's personal ends ⟨the world was her *oyster*⟩
3 : a small mass of muscle contained in a concavity of the pelvic bone on each side of the back of a fowl
4 : an extremely taciturn person
5 : a grayish-white color

oyster bed *noun* (1591)
: a place where oysters grow or are cultivated

oys·ter·catch·er \-ˌka-chər, -ˌke-\ *noun* (1731)
: any of a genus (*Haematopus*) of wading birds that have stout legs, a heavy wedge-shaped bill, and often black-and-white plumage

oyster crab *noun* (1756)
: a tiny crab (*Pinnotheres ostreum*) that lives as a commensal in the gill cavity of the oyster

oyster cracker *noun* (1873)
: a small salted usually round cracker

oyster drill *noun* (1925)
: DRILL 4a

oys·ter·ing \'ȯis-t(ə-)riŋ\ *noun* (1662)
: the act or business of taking oysters for the market or for food

oys·ter·man \'ȯis-tər-mən\ *noun* (1552)
: one who gathers, opens, breeds, or sells oysters

oyster mushroom *noun* (1875)
: an edible mushroom (*Pleurotus ostreatus*) that grows especially on deciduous trees and dead wood

oyster plant *noun* (1821)
: SALSIFY

oysters Rocke·fel·ler \-'rä-ki-ˌfe-lər\ *noun plural* [probably from John D. *Rockefeller* (died 1937)] (1939)
: a dish of oysters on the half shell cooked with various savory toppings typically including chopped spinach and a seasoned sauce

ozo·ke·rite \ˌō-zō-'kir-ˌīt\ *also* **ozo·ce·rite** \-'sir-\ *noun* [German *Ozokerit,* from Greek *ozein* to smell + *kēros* wax — more at CERUMEN] (circa 1837)
: a waxy mineral mixture of hydrocarbons that is colorless or white when pure and often of unpleasant odor and is used especially in making candles and in electrotyping

ozon·a·tion \ˌō-(ˌ)zō-'nā-shən\ *noun* (1854)
: the treatment or combination of a substance or compound with ozone
— ozon·ate \'ō-(ˌ)zō-ˌnāt, -zə-\ *transitive verb*

ozone \'ō-ˌzōn\ *noun* [German *Ozon,* from Greek *ozōn,* present participle of *ozein* to smell — more at ODOR] (circa 1840)
1 : a triatomic very reactive form of oxygen that is a bluish irritating gas of pungent odor, that is formed naturally in the atmosphere by a photochemical reaction and is a major air pollutant in the lower atmosphere but a beneficial component of the upper atmosphere, and that is used for oxidizing, bleaching, disinfecting, and deodorizing
2 : pure and refreshing air
— ozo·nic \ō-'zō-nik, -'zä-\ *adjective*

ozone hole *noun* (1986)
: an area of the ozone layer (as near the south pole) that is seasonally depleted of ozone

ozone layer *noun* (1929)
: an atmospheric layer at heights of about 20 to 30 miles (32 to 48 kilometers) that is normally characterized by high ozone content which blocks most solar ultraviolet radiation from entry into the lower atmosphere

ozon·ide \'ō-(ˌ)zō-ˌnīd\ *noun* (1867)
: a compound of ozone; *specifically* **:** a compound formed by the addition of ozone to the double or triple bond of an unsaturated organic compound

ozon·ize \-ˌnīz\ *transitive verb* **-ized; -iz·ing** (1850)
: to treat, impregnate, or combine with ozone
— ozon·i·za·tion \ˌō-(ˌ)zō-nə-'zā-shən\ *noun*
— ozon·iz·er \'ō-(ˌ)zō-ˌnī-zər\ *noun*

ozo·no·sphere \ō-'zō-nə-ˌsfir\ *noun* (1933)
: OZONE LAYER

P *is the sixteenth letter of the English alphabet. It comes through Latin, via Etruscan, from Greek* pi, *which in turn was borrowed from Phoenician* pê, *representing the sound which it has ever since retained, that of a voiceless bilabial stop (the* p *in English* pet*).* P *is sometimes silent in English, especially in the combinations* pn, ps, *and* pt *found at the beginning of certain words from Greek (for example,* pneumatic, psychic, *and* ptomaine*). With* h *it unites to form the independent digraph* ph, *which represents the Greek* phi *and has the value of* f *in words from that language (such as* phrase *and* graphic*). In related foreign words, English* p *corresponds to German* pf *or* f *and rarely to* b *in other Indo-European languages; compare, for example, English* plow *with German* Pflug *("plow") and English* deep *with German* tief *and Lithuanian* dubus *(both meaning "deep"). Small* p *was formed by reducing the size of the capital and placing the letter lower on the line.*

p \'pē\ *noun, plural* **p's** *or* **ps** \'pēz\ *often capitalized, often attributive* (before 12th century)
1 a : the 16th letter of the English alphabet **b :** a graphic representation of this letter **c :** a speech counterpart of orthographic *p*
2 : a graphic device for reproducing the letter *p*
3 : one designated *p* especially as the 16th in order or class
4 [abbreviation for *pass*] **a :** a grade rating a student's work as passing **b :** one graded or rated with a P
5 : something shaped like the letter P

pa \'pä, 'pȯ\ *noun* [short for *papa*] (1811)
: FATHER

PA \(,)pē-'ā\ *noun* (1970)
: PHYSICIAN'S ASSISTANT

pa·'an·ga \pä-'äŋ-gə, -'äŋ-ə\ *noun* [Tongan, literally, seed from a species of vine] (1966)
— see MONEY table

PABA \'pa-bə, 'pä-; ,pē-(,)ā-'bē-(,)ā\ *noun* [*para-aminobenzoic acid*] (1943)
: PARA-AMINOBENZOIC ACID

pab·lum \'pa-bləm\ *noun* [from *Pablum,* a trademark for an infant cereal] (1948)
: PABULUM 3

pab·u·lum \'pa-byə-ləm\ *noun* [Latin, food, fodder; akin to Latin *pascere* to feed — more at FOOD] (1733)
1 : FOOD; *especially* **:** a suspension or solution of nutrients in a state suitable for absorption
2 : intellectual sustenance
3 : something (as writing or speech) that is insipid, simplistic, or bland

pa·ca \'pä-kə, 'pa-\ *noun* [Portuguese & Spanish, from Tupi *páka*] (1657)
: either of two large chiefly Central and South American rodents that constitute a genus (*Agouti* synonym *Cuniculus*) and typically have a white-spotted brownish coat

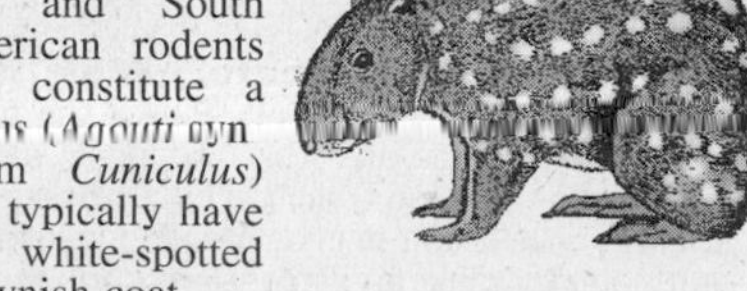
paca

¹pace \'pās\ *noun* [Middle English *pas,* from Old French, step, from Latin *passus,* from *pandere* to spread — more at FATHOM] (14th century)
1 a : rate of movement; *especially* **:** an established rate of locomotion **b :** rate of progress; *specifically* **:** parallel rate of growth or development ⟨supplies kept *pace* with demand⟩ **c :** an example to be emulated; *specifically* **:** first place in a competition ⟨three strokes off the *pace* —*Time*⟩ **d** (1) **:** rate of performance or delivery **:** TEMPO; *especially* **:** SPEED ⟨serves with great *pace*⟩ ⟨a *pace* bowler in cricket⟩ (2) **:** rhythmic animation **:** FLUENCY ⟨writes with color, with zest, and with *pace* —Amy Loveman⟩
2 : a manner of walking **:** TREAD
3 a : STEP 2a(1) **b :** any of various units of distance based on the length of a human step
4 a *plural* **:** an exhibition or test of skills or capacities ⟨the trainer put the tiger through its *paces*⟩ **b :** GAIT; *especially* **:** a fast 2-beat gait (as of the horse) in which the legs move in lateral pairs and support the animal alternately on the right and left legs

²pace *verb* **paced; pac·ing** (1513)
intransitive verb
1 a : to walk with often slow or measured tread **b :** to move along **:** PROCEED
2 : to go at a pace — used especially of a horse
transitive verb
1 a : to measure by pacing — often used with *off* ⟨*paced* off a 10-yard penalty⟩ **b :** to cover at a walk ⟨could hear him *pacing* the floor⟩
2 : to cover (a course) by pacing — used of a horse
3 a : to set or regulate the pace of ⟨taught them how to *pace* their solos for . . . impact —Richard Goldstein⟩; *also* **:** to establish a moderate or steady pace for (oneself) **b** (1) **:** to go before **:** PRECEDE (2) **:** to set an example for **:** LEAD **c :** to keep pace with

³pa·ce \'pā-(,)sē; 'pä-(,)chā, -(,)kā\ *preposition* [Latin, ablative of *pac-, pax* peace, permission — more at PACT] (1863)
: contrary to the opinion of — usually used as an expression of deference to someone's contrary opinion; usually italic ⟨easiness is a virtue in grammar, *pace* old-fashioned grammarians —Philip Howard⟩

pace car *noun* (1965)
: an automobile that leads the field of competitors through a pace lap but does not participate in the race

pace lap *noun* (1971)
: a lap of an auto racecourse by the entire field of competitors before the start of a race to allow the engines to warm up and to permit a flying start

pace·mak·er \'pās-,mā-kər\ *noun* (1884)
1 a : one that sets the pace for another **b :** one that takes the lead or sets an example
2 a : a body part (as the sinoatrial node of the heart) that serves to establish and maintain a rhythmic activity **b :** an electrical device for stimulating or steadying the heartbeat or reestablishing the rhythm of an arrested heart
— pace·mak·ing \-kiŋ\ *noun*

pac·er \'pā-sər\ *noun* (circa 1661)
1 : one that paces; *specifically* **:** a horse whose predominant gait is the pace
2 : PACEMAKER

pace·set·ter \'pās-,se-tər\ *noun* (1895)
: PACEMAKER 1
— pace·set·ting \-tiŋ\ *adjective*

pa·chi·si \pə-'chē-zē\ *noun* [Hindi *pacīsī,* from *pacīs* twenty-five] (1867)
: an ancient board game played with dice and counters on a cruciform board in which players attempt to be the first to reach the home square

pa·chu·co \pə-'chü-(,)kō\ *noun, plural* **-cos** [American Spanish] (1943)
: a young Mexican-American having a taste for flashy clothes and a special jargon and usually belonging to a neighborhood gang

pachy·derm \'pa-ki-,dərm\ *noun* [French *pachyderme,* from Greek *pachydermos* thick-skinned, from *pachys* thick + *derma* skin; akin to Sanskrit *bahu* dense, much — more at DERM-] (1838)
: any of various nonruminant hoofed mammals (as an elephant, a rhinoceros, or a pig) most of which have a thick skin

pachy·der·ma·tous \,pa-ki-'dər-mə-təs\ *adjective* [ultimately from Greek *pachys* + *dermat-, derma* skin] (1823)
1 : of or relating to the pachyderms
2 a : THICK, THICKENED ⟨*pachydermatous* skin⟩ **b :** CALLOUS, INSENSITIVE

pach·y·san·dra \,pa-ki-'san-drə\ *noun* [New Latin, irregular from Greek *pachys* + New Latin *-andrus* -androus] (1813)
: any of a genus (*Pachysandra*) of shrubby evergreen plants of the box family often used as a ground cover

pachy·tene \'pa-ki-,tēn\ *noun* [International Scientific Vocabulary *pachy-* (from Greek *pachys*) + *-tene*] (1912)
: the stage of meiotic prophase that immediately follows the zygotene and that is characterized by paired chromosomes thickened and visibly divided into chromatids and by the occurrence of crossing-over
— pachytene *adjective*

pa·cif·ic \pə-'si-fik\ *adjective* [Middle English *pacifique,* from Latin *pacificus,* from *pac-, pax* peace + *-i-* + *-ficus* -fic — more at PACT] (circa 1548)
1 a : tending to lessen conflict **:** CONCILIATORY **b :** rejecting the use of force as an instrument of policy
2 a : having a soothing appearance or effect ⟨mild *pacific* breezes⟩ **b :** mild of temper **:** PEACEABLE
3 *capitalized* **:** of, relating to, bordering on, or situated near the Pacific Ocean
— pa·cif·i·cal·ly \-fi-k(ə-)lē\ *adverb*

pac·i·fi·ca·tion \,pa-sə-fə-'kā-shən\ *noun* (15th century)
1 a : the act or process of pacifying **:** the state of being pacified **b :** the act of forcibly suppressing or eliminating a population considered to be hostile
2 : a treaty of peace

pa·cif·i·ca·tor \pə-'si-fə-,kā-tər\ *noun* (1539)
: PACIFIER 1

pa·cif·i·cism \pə-'si-fə-,si-zəm\ *noun* (1910)
: PACIFISM
— pa·cif·i·cist \-sist\ *noun*

Pacific salmon *noun* (1888)
: any of several anadromous salmonid fishes (genus *Oncorhynchus*) chiefly of the northern Pacific including the sockeye, coho, chum salmon, and chinook salmon

Pacific time \pə-'si-fik-\ *noun* [*Pacific* Ocean] (1883)
: the time of the 8th time zone west of Greenwich that includes the Pacific coastal region of the U.S. — see TIME ZONE illustration

pac·i·fi·er \'pa-sə-,fī(-ə)r\ *noun* (1533)

1 : one that pacifies
2 : a usually nipple-shaped device for babies to suck or bite on

pac·i·fism \'pa-sə-ˌfi-zəm\ *noun* [French *pacifisme,* from *pacifique* pacific] (1902)
1 : opposition to war or violence as a means of settling disputes; *specifically* **:** refusal to bear arms on moral or religious grounds
2 : an attitude or policy of nonresistance
— pac·i·fist \-fist\ *noun*

pac·i·fist \'pa-sə-fist\ *or* **pac·i·fis·tic** \ˌpa-sə-'fis-tik\ *adjective* (1908)
1 : of, relating to, or characteristic of pacifism or pacifists
2 : strongly and actively opposed to conflict and especially war
— pac·i·fis·ti·cal·ly \ˌpa-sə-'fis-ti-k(ə-)lē\ *adverb*

pac·i·fy \'pa-sə-ˌfī\ *transitive verb* **-fied; -fy·ing** [Middle English *pacifien,* from Latin *pacificare,* from *pac-, pax* peace] (15th century)
1 a : to allay the anger or agitation of **:** SOOTHE ⟨*pacify* a crying child⟩ **b :** APPEASE, PROPITIATE
2 a : to restore to a tranquil state **:** SETTLE ⟨made an attempt to *pacify* the commotion⟩ **b :** to reduce to a submissive state **:** SUBDUE ⟨forces moved in to *pacify* the country⟩ ☆
— pac·i·fi·able \ˌpa-sə-'fī-ə-bəl\ *adjective*

Pa·cin·i·an corpuscle \pə-'si-nē-ən-\ *noun* [Filippo *Pacini* (died 1883) Italian anatomist] (circa 1860)
: an oval capsule that terminates some sensory nerve fibers especially in the skin of the hands and feet

¹pack \'pak\ *noun, often attributive* [Middle English, of Low German or Dutch origin; akin to Middle Low German & Middle Dutch *pak* pack] (13th century)
1 a : a bundle arranged for convenience in carrying especially on the back **b :** a group or pile of related objects **c** (1) **:** a number of individual components packaged as a unit ⟨a *pack* of cigarettes⟩ (2) **:** CONTAINER (3) **:** a compact unitized assembly to perform a specific function (4) **:** a stack of magnetic disks in a container for use as a storage device
2 a : the contents of a bundle **b :** a large amount or number **:** HEAP ⟨a *pack* of lies⟩ **c :** a full set of playing cards
3 a : an act or instance of packing **b :** a method of packing
4 a : a set of persons with a common interest **:** CLIQUE **b :** an organized troop (as of Cub Scouts)
5 a (1) **:** a group of domesticated animals trained to hunt or run together (2) **:** a group of often predatory animals of the same kind ⟨a wolf *pack*⟩ (3) **:** a large group of individuals massed together (as in a race) **b :** WOLF PACK
6 : a concentrated or compacted mass (as of snow or ice)
7 : wet absorbent material for therapeutic application to the body
8 a : a cosmetic paste for the face **b :** an application or treatment of oils or creams for conditioning the scalp and hair
9 : material used in packing

²pack (14th century)
transitive verb
1 a : to make into a compact bundle **b :** to fill completely ⟨fans *packed* the stadium⟩ **c :** to fill with packing ⟨*pack* a joint in a pipe⟩ **d :** to load with a pack ⟨*pack* a mule⟩ **e :** to put in a protective container ⟨goods *packed* for shipment⟩
2 a : to crowd together **b :** to increase the density of **:** COMPRESS
3 a : to cause or command to go without ceremony ⟨*packed* him off to school⟩ **b :** to bring to an end **:** GIVE UP — used with *up* or *in* ⟨might *pack* up the assignment⟩; used especially in the phrase *pack it in*
4 : to gather into tight formation **:** make a pack of (as hounds)
5 : to cover or surround with a pack
6 a : to transport on foot or on the back of an animal ⟨*pack* a canoe overland⟩ **b :** to wear or carry as equipment ⟨*pack* a gun⟩ **c :** to be supplied or equipped with **:** POSSESS ⟨a storm *packing* hurricane winds⟩ **d :** to cause or be capable of making (an impact) ⟨a book that *packs* a man-sized punch —C. J. Rolo⟩
intransitive verb
1 a : to go away without ceremony **:** DEPART ⟨simply *packed* up and left⟩ **b :** QUIT, STOP — used with *up* or *in* ⟨why don't you *pack* in, before you kill yourself —Millard Lampell⟩
2 a : to stow goods and equipment for transportation **b :** to be suitable for packing ⟨a knit dress *packs* well⟩
3 a : to assemble in a group **:** CONGREGATE **b :** to crowd together
4 : to become built up or compacted in a layer or mass ⟨the ore *packed* into a stony mass⟩
5 a : to carry goods or equipment **b :** to travel with one's baggage (as by horse)
— pack·abil·i·ty \ˌpa-kə-'bi-lə-tē\ *noun*
— pack·able \'pa-kə-bəl\ *adjective*

³pack *transitive verb* [obsolete *pack* to make a secret agreement] (1587)
1 : to influence the composition of (as a political agency) so as to bring about a desired result ⟨*pack* a jury⟩
2 *archaic* **:** to arrange (the cards in a pack) so as to cheat

⁴pack *adjective* [perhaps from obsolete *pack* to make a secret agreement] (1701)
chiefly Scottish **:** INTIMATE

¹pack·age \'pa-kij\ *noun* (1611)
1 *archaic* **:** the act or process of packing
2 a : a small or moderate-sized pack **:** PARCEL **b :** a commodity or a unit of a product uniformly wrapped or sealed **c :** a preassembled unit
3 : a covering wrapper or container
4 : something that suggests a package: as **a :** PACKAGE DEAL **b :** a radio or television series offered for sale at a lump sum **c :** contract benefits gained through collective bargaining **d :** a ready-made computer program or collection of related software **e :** a travel arrangement contract that offers for a fixed price transportation, accommodations, and often sightseeing and entertainment **f :** a collection of related items; *especially* **:** one to be considered or acted on together ⟨presented his tax *package* to the nation⟩

²package *transitive verb* **pack·aged; pack·ag·ing** (1921)
1 a : to make into a package; *especially* **:** to produce as an entertainment package **b :** to present (as a product) in such a way as to heighten its appeal to the public
2 : to enclose in a package or covering
— pack·ag·er *noun*

package deal *noun* (circa 1948)
1 : an offer or agreement involving a number of related items or one making acceptance of one item dependent on the acceptance of another
2 : the items offered in a package deal

package store *noun* (circa 1918)
: a store that sells bottled or canned alcoholic beverages for consumption off the premises

pack animal *noun* (1847)
: an animal used for carrying packs

pack·board \'pak-ˌbōrd, -ˌbȯrd\ *noun* (1939)
: a usually canvas-covered light wood or metal frame with shoulder straps used for carrying goods and equipment

packed \'pakt\ *adjective* (1777)
1 a : COMPRESSED ⟨hard-*packed* snow⟩ **b :** that is crowded or stuffed — often used in combination ⟨an action-*packed* story⟩
2 : filled to capacity ⟨played to a *packed* house⟩

pack·er \'pa-kər\ *noun* (14th century)
1 : one that packs: as **a :** one engaged in processing food (as meat) and distributing it to retailers **b :** an automotive vehicle with a closed body and a compressing device (as for compacting rubbish) in the rear
2 : PORTER 1
3 : one that conveys goods by means of a pack

pack·et \'pa-kət\ *noun* [Middle English, from Middle French *pacquet,* of Germanic origin; akin to Middle Dutch *pak* pack] (15th century)
1 a : a small bundle or parcel **b :** a small thin package **c** *British* (1) **:** PAY ENVELOPE (2) **:** SALARY, PAYCHECK **d** *chiefly British* **:** a considerable amount ⟨that trip will cost you a *packet*⟩
2 a : a number of letters dispatched at one time **b :** a small group, cluster, or mass
3 : a passenger boat usually carrying mail and cargo
4 *British* **:** a pack of cigarettes
5 : a short fixed-length section of data that is transmitted as a unit in an electronic communications network

pack·horse \'pak-ˌhȯrs\ *noun* (circa 1500)
: a horse used as a pack animal

pack ice *noun* (1850)
: sea ice formed into a mass by the crushing together of pans, floes, and brash

pack·ing \'pa-kiŋ\ *noun* (14th century)
1 a : the action or process of packing something; *also* **:** a method of packing **b :** the processing of food and especially meat for future sale
2 : material (as a covering or stuffing) used to protect packed goods (as for shipping); *also* **:** material used for making airtight or watertight ⟨*packing* for a faucet⟩

pack·ing·house \-ˌhau̇s\ *noun* (1834)
: an establishment for slaughtering livestock and processing and packing meat, meat products, and by-products; *also* **:** one for processing and packing other foodstuffs — called also *packing plant*

pack·man \'pak-mən\ *noun* (circa 1625)
: PEDDLER

pack rat *noun* (1885)
1 : WOOD RAT; *especially* **:** a bushy-tailed rodent (*Neotoma cinerea*) of the Rocky Mountain area that has well-developed cheek pouches and hoards food and miscellaneous objects
2 : one who collects or hoards especially unneeded items

pack·sack \'pak-ˌsak\ *noun* (1851)
: a case (as of canvas) held on the back by shoulder straps and used to carry gear when traveling on foot

pack·sad·dle \-ˌsa-dᵊl\ *noun* (14th century)
: a saddle designed to support loads on the backs of pack animals

pack·thread \-ˌthred\ *noun* (14th century)
: strong thread or small twine used for sewing or tying packs or parcels

pact \'pakt\ *noun* [Middle English, from Middle French, from Latin *pactum,* from neuter of *pactus,* past participle of *pacisci* to agree, contract; akin to Old English *fōn* to seize, Latin

☆ SYNONYMS
Pacify, appease, placate, mollify, propitiate, conciliate mean to ease the anger or disturbance of. PACIFY suggests a soothing or calming ⟨*pacified* by a sincere apology⟩. APPEASE implies quieting insistent demands by making concessions ⟨*appease* their territorial ambitions⟩. PLACATE suggests changing resentment or bitterness to goodwill ⟨a move to *placate* local opposition⟩. MOLLIFY implies soothing hurt feelings or rising anger ⟨a speech that *mollified* the demonstrators⟩. PROPITIATE implies averting anger or malevolence especially of a superior being ⟨*propitiated* his parents by dressing up⟩. CONCILIATE suggests ending an estrangement by persuasion, concession, or settling of differences ⟨*conciliating* the belligerent nations⟩.

pax peace, *pangere* to fix, fasten, Greek *pēgnynai*] (15th century)
: [4]COMPACT; *especially* **:** an international treaty

[1]pad \ˈpad\ *verb* **pad•ded; pad•ding** [perhaps from Middle Dutch *paden* to follow a path, from *pad* path] (1553)
transitive verb
: to traverse on foot
intransitive verb
: to go on foot **:** WALK; *especially* **:** to walk with or as if with padded feet ⟨the dog *padded* along beside him⟩ ⟨*padding* around in bedroom slippers⟩

[2]pad *noun* [Middle Dutch *pad*] (1567)
1 *dialect British* **:** PATH
2 : a horse that moves along at an easy pace
3 *archaic* **:** FOOTPAD

[3]pad *noun* [origin unknown] (1570)
1 a : a thin flat mat or cushion: as (1) **:** a piece of soft stuffed material used as or under a saddle (2) **:** padding used to shape an article of clothing (3) **:** a guard worn to shield body parts against impact (4) **:** a piece of usually folded absorbent material (as gauze) used as a surgical dressing or protective covering (5) **:** frictional material that presses against the disks in a disk brake **b :** a piece of material saturated with ink for inking the surface of a rubber stamp
2 a : the foot of an animal **b :** the cushioned thickening of the underside of the toes of an animal
3 : a floating leaf of a water plant
4 : a collection of sheets of paper glued together at one end
5 a (1) **:** a section of an airstrip used for warm-ups or turnarounds (2) **:** an area used for helicopter takeoffs and landings **b :** LAUNCHPAD **c :** a horizontal concrete surface (as for parking a mobile home)
6 a : BED **b :** living quarters

[4]pad *transitive verb* **pad•ded; pad•ding** (1827)
1 a : to furnish with a pad or padding **b :** MUTE, MUFFLE
2 : to expand or increase especially with needless, misleading, or fraudulent matter ⟨*pad* an expense account⟩ — often used with *out* ⟨they *pad* out their bibliographies —J. P. Kenyon⟩

[5]pad *noun* [imitative] (1594)
: a soft muffled or slapping sound

pad•ding \ˈpa-diŋ\ *noun* (1828)
: material with which something is padded

[1]pad•dle \ˈpa-dᵊl\ *intransitive verb* **pad•dled; pad•dling** \ˈpad-liŋ, ˈpa-dᵊl-iŋ\ [origin unknown] (1530)
1 : to move the hands or feet about in shallow water
2 *archaic* **:** to use the hands or fingers in toying or caressing
3 : TODDLE
— pad•dler \ˈpad-lər, ˈpa-dᵊl-ər\ *noun*

[2]paddle *noun* [Middle English *padell* spade-shaped tool for cleaning a plow] (1624)
1 a : a usually wooden implement that has a long handle and a broad flattened blade and that is used to propel and steer a small craft (as a canoe) **b :** an implement often with a short handle and a broad flat blade that is used for stirring, mixing, or hitting; *especially* **:** one used to hit a ball in any of various games (as table tennis)
2 a : any of the broad boards at the circumference of a paddle wheel or waterwheel **b :** any of the broad blades attached to a shaft (as in an ice cream machine) and used for stirring
3 : a computer input device with a dial used to control linear movement of a cursor on a computer display

[3]paddle *verb* **pad•dled; pad•dling** \ˈpad-liŋ, ˈpa-dᵊl-iŋ\ (1677)
intransitive verb
: to go on or through water by or as if by means of a paddle or paddle wheel
transitive verb
1 a : to propel by a paddle **b :** to transport in a paddled craft ⟨*paddled* us to shore in his canoe⟩
2 a : to beat or stir with or as if with a paddle (as in washing or dyeing) **b :** to punish by or as if by beating with a paddle
— pad•dler \ˈpad-lər, ˈpa-dᵊl-ər\ *noun*

pad•dle•ball \ˈpa-dᵊl-ˌbȯl\ *noun* (1935)
: a game like handball played by hitting the ball with a paddle; *also* **:** the ball used in this game

pad•dle•board \-ˌbōrd, -ˌbȯrd\ *noun* (1938)
: a long narrow buoyant board used for riding the surf or in rescuing swimmers

pad•dle•boat \-ˌbōt\ *noun* (1874)
: a boat propelled by a paddle wheel

pad•dle•fish \-ˌfish\ *noun* (1807)
: any of a family (Polyodontidae) of ganoid fishes; *especially* **:** a large fish (*Polyodon spathula*) of the Mississippi valley that has a long paddle-shaped snout

paddlefish

paddle tennis *noun* (1925)
: a game like tennis that is played with a paddle and rubber ball on a small court

paddle wheel *noun* (1685)
: a wheel with paddles around its circumference used to propel a boat

paddle wheeler *noun* (1924)
: a steamer propelled by a paddle wheel

pad•dock \ˈpa-dək, -dik\ *noun* [alteration of Middle English *parrok,* from Old English *pearroc,* from Medieval Latin *parricus*] (1622)
1 a : a usually enclosed area used especially for pasturing or exercising animals; *especially* **:** an enclosure where racehorses are saddled and paraded before a race **b** *Australian & New Zealand* **:** an often enclosed field
2 : an area at an automobile racecourse where racing cars are parked

pad•dy *also* **padi** \ˈpa-dē\ *noun, plural* **pad•dies** *also* **pad•is** [Malay *padi*] (1623)
1 : RICE; *especially* **:** threshed unmilled rice
2 : wet land in which rice is grown

Pad•dy \ˈpa-dē\ *noun, plural* **Paddies** [from *Paddy,* Hiberno-English nickname for *Patrick*] (1780)
: IRISHMAN — often taken to be offensive

pad•dy wagon \ˈpa-dē-\ *noun* [probably from *Paddy*] (1930)
: an enclosed motortruck used by police to carry prisoners — called also *Black Maria, patrol wagon*

pad•lock \ˈpad-ˌläk\ *noun* [Middle English *padlok,* from *pad-* (of unknown origin) + *lok* lock] (15th century)
: a removable lock with a shackle that can be passed through a staple or link and then secured
— padlock *transitive verb*

pa•dre \ˈpä-(ˌ)drā, -drē\ *noun* [Spanish or Italian or Portuguese, literally, father, from Latin *pater* — more at FATHER] (1584)
1 : a Christian clergyman; *especially* **:** PRIEST
2 : a military chaplain

pa•dro•ne \pə-ˈdrō-nē\ *noun, plural* **-nes** *or* **-ni** \-nē\ [Italian, protector, owner, from Latin *patronus* patron] (1670)
1 a : MASTER **b :** an Italian innkeeper
2 : one that secures employment especially for Italian immigrants

pad•u•a•soy \ˈpa-jə-wə-ˌsȯi, ˈpaj-wə-\ *noun* [alteration of earlier *poudesoy,* from French *pou-de-soie*] (1663)
: a corded silk fabric; *also* **:** a garment made of it

pae•an \ˈpē-ən\ *noun* [Latin, hymn of thanksgiving especially addressed to Apollo, from Greek *paian, paiōn,* from *Paian, Paiōn,* epithet of Apollo in the hymn] (1589)
: a joyous song or hymn of praise, tribute, thanksgiving, or triumph; *broadly* **:** ENCOMIUM, TRIBUTE

paed- *or* **paedo-** — see PED-

pae•di•at•ric, pae•di•a•tri•cian, pae•di•at•rics *chiefly British variant of* PEDIATRIC, PEDIATRICIAN, PEDIATRICS

pae•do•gen•e•sis \ˌpē-dō-ˈje-nə-səs\ *noun* [New Latin] (circa 1871)
: reproduction by young or larval animals **:** NEOTENY
— pae•do•ge•net•ic \-jə-ˈne-tik\ *or* **pae•do•gen•ic** \-ˈje-nik\ *adjective*
— pae•do•ge•net•i•cal•ly \-jə-ˈne-ti-k(ə-)lē\ *adverb*

pae•do•mor•phic \ˌpē-də-ˈmȯr-fik\ *adjective* (1891)
: of, relating to, or involving paedomorphosis or paedomorphism

pae•do•mor•phism \-ˈmȯr-ˌfi-zəm\ *noun* (circa 1891)
: retention in the adult of infantile or juvenile characters

pae•do•mor•pho•sis \-ˈmȯr-fə-səs\ *noun* [New Latin, from *paed-* + Greek *morphōsis* formation, from *morphoun* to form, from *morphē* form] (1922)
: phylogenetic change that involves retention of juvenile characters by the adult

pa•el•la \pä-ˈe-lə, -ˈā-; -ˈāl-yə, -ˈā-yə\ *noun* [Catalan, literally, pot, pan, from Middle French *paelle,* from Latin *patella* small pan — more at PATELLA] (circa 1892)
: a saffron-flavored dish containing rice, meat, seafood, and vegetables

pae•on \ˈpē-ən, -ˌän\ *noun* [Latin, from Greek *paiōn,* from *paian, paiōn* paean] (1603)
: a metrical foot of four syllables with one long and three short syllables (as in classical prosody) or with one stressed and three unstressed syllables (as in English prosody)

pa•gan \ˈpā-gən\ *noun* [Middle English, from Late Latin *paganus,* from Latin, country dweller, from *pagus* country district; akin to Latin *pangere* to fix — more at PACT] (14th century)
1 : HEATHEN 1; *especially* **:** a follower of a polytheistic religion (as in ancient Rome)
2 : one who has little or no religion and who delights in sensual pleasures and material goods **:** an irreligious or hedonistic person
— pagan *adjective*
— pa•gan•ish \-gə-nish\ *adjective*

pa•gan•ism \ˈpā-gə-ˌni-zəm\ *noun* (15th century)
1 a : pagan beliefs or practices **b :** a pagan religion
2 : the quality or state of being a pagan

pa•gan•ize \-ˌnīz\ *verb* **-ized; -iz•ing** (1615)
transitive verb
: to make pagan
intransitive verb
: to become pagan
— pa•gan•iz•er *noun*

[1]page \ˈpāj\ *noun* [Middle English, from Middle French, from Old French] (14th century)
1 a (1) **:** a youth being trained for the medieval rank of knight and in the personal service of a knight (2) **:** a youth attendant on a person of rank especially in the medieval period **b :** a boy serving as an honorary attendant at a formal function (as a wedding)
2 : one employed to deliver messages, assist patrons, serve as a guide, or attend to similar duties
3 : an act or instance of paging ⟨a *page* came over the loudspeaker⟩

\ə\ **abut** \ᵊ\ **kitten** \ər\ **further** \a\ **ash** \ā\ **ace** \ä\ **mop, mar** \au̇\ **out** \ch\ **chin** \e\ **bet** \ē\ **easy** \g\ **go** \i\ **hit** \ī\ **ice** \j\ **job** \ŋ\ **sing** \ō\ **go** \ȯ\ **law** \ȯi\ **boy** \th\ **thin** \t͟h\ **the** \ü\ **loot** \u̇\ **foot** \y\ **yet** \zh\ **vision** *see also* Guide to Pronunciation

²page *transitive verb* **paged; pag·ing** (15th century)
1 : to wait on or serve in the capacity of a page
2 : to summon by repeatedly calling out the name of

³page *noun* [Middle French, from Latin *pagina;* akin to Latin *pangere* to fix, fasten — more at PACT] (1589)
1 a : one of the leaves of a publication or manuscript; *also* **:** a single side of one of these leaves **b :** the material printed or written on a page
2 a : a written record **b :** a noteworthy event or period ⟨one of the brightest *pages* of my life⟩
3 : a sizable subdivision of computer memory; *also* **:** a block of information that fills a page and can be transferred as a unit between the internal and external storage of a computer

⁴page *verb* **paged; pag·ing** (1628)
transitive verb
: to number or mark the pages of
intransitive verb
: to turn the pages (as of a book or magazine) especially in a steady or haphazard manner — usually used with *through*

pag·eant \'pa-jənt\ *noun* [Middle English *pagyn, padgeant,* literally, scene of a play, from Medieval Latin *pagina,* perhaps from Latin, page] (14th century)
1 a : a mere show **:** PRETENSE **b :** an ostentatious display
2 : SHOW, EXHIBITION; *especially* **:** an elaborate colorful exhibition or spectacle often with music that consists of a series of tableaux, of a loosely unified drama, or of a procession usually with floats
3 : PAGEANTRY 1

pag·eant·ry \'pa-jən-trē\ *noun* (1608)
1 : pageants and the presentation of pageants
2 : colorful, rich, or splendid display **:** SPECTACLE
3 : mere show **:** empty display

page boy *noun* (1874)
1 : a boy serving as a page
2 *usually* **pageboy :** an often shoulder-length hairdo with the ends of the hair turned under in a smooth roll

pag·er \'pā-jər\ *noun* (1901)
: one that pages; *especially* **:** BEEPER

Pag·et's disease \'pa-jəts-\ *noun* [Sir James *Paget* (died 1899) English surgeon] (1889)
1 : an eczematous inflammatory precancerous condition especially of the nipple and areola
2 : a chronic disease in which the bones become enlarged, weak, and deformed

page–turn·er \'pāj-ˌtər-nər\ *noun* (1972)
: an engrossing book or story

pag·i·nate \'pa-jə-ˌnāt\ *transitive verb* **-nat·ed; -nat·ing** [Latin *pagina* page] (1884)
: ⁴PAGE

pag·i·na·tion \ˌpa-jə-'nā-shən\ *noun* (1841)
1 : the action of paging **:** the condition of being paged
2 a : the numbers or marks used to indicate the sequence of pages (as of a book) **b :** the number and arrangement of pages or an indication of these

pa·go·da \pə-'gō-də\ *noun* [Portuguese *pagode* oriental idol, temple] (1588)
: a Far Eastern tower usually with roofs curving upward at the division of each of several stories and erected as a temple or memorial

pagoda

Pah·la·vi \'pa-lə-(ˌ)vē, 'pä-\ *noun* [Persian *pahlawī,* from *Pahlav* Parthia, from Old Persian *Parthava-*] (1773)
1 : the Iranian language of Sassanian Persia — see INDO-EUROPEAN LANGUAGES table
2 : a script used for writing Pahlavi

¹paid \'pād\ *past and past participle of* PAY

²paid *adjective* (1817)
1 : marked by the receipt of pay ⟨*paid* vacation time⟩
2 : being or having been paid or paid for ⟨a *paid* official⟩ ⟨a *paid* political announcement⟩

pail \'pā(ə)l\ *noun* [Middle English *payle, paille*] (14th century)
1 : a usually cylindrical container with a handle **:** BUCKET
2 : the quantity that a pail contains
— pail·ful \-ˌfu̇l\ *noun*

pail·lard \pī-'yär, pā-'yär\ *noun* [French *paillarde,* from *Paillard,* late 19th century French restaurateur] (1972)
: a piece of beef or veal usually pounded thin and grilled

pail·lette \pī-'yet, pā-'yet, pə-'let\ *noun* [French, from *paille* straw — more at PALLET] (circa 1890)
1 : a small shiny object (as a spangle) applied in clusters as a decorative trimming (as on women's clothing)
2 : a trimming made of paillettes

¹pain \'pān\ *noun* [Middle English, from Old French *peine,* from Latin *poena,* from Greek *poinē* payment, penalty; akin to Greek *tinein* to pay, *tinesthai* to punish, Avestan *kaēnā* revenge, Sanskrit *cayate* he revenges] (14th century)
1 : PUNISHMENT
2 a : usually localized physical suffering associated with bodily disorder (as a disease or an injury); *also* **:** a basic bodily sensation induced by a noxious stimulus, received by naked nerve endings, characterized by physical discomfort (as pricking, throbbing, or aching), and typically leading to evasive action **b :** acute mental or emotional distress or suffering **:** GRIEF
3 *plural* **:** the throes of childbirth
4 *plural* **:** trouble, care, or effort taken for the accomplishment of something ⟨was at *pains* to reassure us⟩
5 : one that irks or annoys or is otherwise troublesome — often used in such phrases as *pain in the neck*
— pain·less \-ləs\ *adjective*
— pain·less·ly *adverb*
— pain·less·ness *noun*
— on pain of *or* **under pain of :** subject to penalty or punishment of ⟨ordered not to leave the country *on pain of* death⟩

²pain (14th century)
transitive verb
1 : to make suffer or cause distress to **:** HURT
2 *archaic* **:** to put (oneself) to trouble or exertion
intransitive verb
1 *archaic* **:** SUFFER
2 : to give or have a sensation of pain

pained \'pānd\ *adjective* (14th century)
1 : feeling pain **:** HURT
2 : expressing or involving pain ⟨a *pained* expression⟩ ⟨with *pained* surprise⟩

pain·ful \'pān-fəl\ *adjective* **pain·ful·ler** \-fə-lər\; **pain·ful·lest** (14th century)
1 a : feeling or giving pain **b :** IRKSOME, ANNOYING
2 : requiring effort or exertion ⟨a long *painful* trip⟩
3 *archaic* **:** CAREFUL, DILIGENT
— pain·ful·ly \-f(ə-)lē\ *adverb*
— pain·ful·ness \-fəl-nəs\ *noun*

pain·kill·er \-ˌki-lər\ *noun* (1853)
: something (as a drug) that relieves pain
— pain·kill·ing \-liŋ\ *adjective*

¹pains·tak·ing \'pān-ˌstā-kiŋ\ *noun* (1538)
: the action of taking pains **:** diligent care and effort

²painstaking *adjective* (1696)
: taking pains **:** expending, showing, or involving diligent care and effort
— pains·tak·ing·ly \-kiŋ-lē\ *adverb*

¹paint \'pānt\ *verb* [Middle English, from Old French *peint,* past participle of *peindre,* from Latin *pingere* to tattoo, embroider, paint; akin to Old English *fāh* variegated, Greek *poikilos* variegated, *pikros* sharp, bitter] (13th century)
transitive verb
1 a (1) **:** to apply color, pigment, or paint to (2) **:** to color with a cosmetic **b** (1) **:** to apply with a movement resembling that used in painting (2) **:** to treat with a liquid by brushing or swabbing ⟨*paint* the wound with iodine⟩
2 a (1) **:** to produce in lines and colors on a surface by applying pigments (2) **:** to depict by such lines and colors **b :** to decorate, adorn, or variegate by applying lines and colors **c :** to produce or evoke as if by painting ⟨*paints* glowing pictures of the farm⟩
3 : to touch up or cover over by or as if by painting
4 : to depict as having specified or implied characteristics ⟨*paints* them whiter than the evidence justifies —Oliver La Farge⟩
intransitive verb
1 : to practice the art of painting
2 : to use cosmetics

²paint *noun* (1602)
1 : the action of painting **:** something produced by painting
2 : MAKEUP; *especially* **:** a cosmetic to add color
3 a (1) **:** a mixture of a pigment and a suitable liquid to form a closely adherent coating when spread on a surface in a thin coat (2) **:** the pigment used in this mixture especially when in the form of a cake ⟨a box of *paints*⟩ **b :** an applied coating of paint
4 : PINTO
5 : FREE THROW LANE

paint·brush \'pānt-ˌbrəsh\ *noun* (1827)
1 : a brush for applying paint
2 a : INDIAN PAINTBRUSH 1 **b :** ORANGE HAWKWEED

painted bunting *noun* (circa 1811)
: a brightly colored finch (*Passerina ciris*) that is found from the southern U.S. to Panama

painted cup *noun* (1787)
: INDIAN PAINTBRUSH 1

painted lady *noun* (1753)
1 : a migratory nymphalid butterfly (*Vanessa cardui*) with wings mottled in brown, orange, black, and white
2 : PROSTITUTE 1a

painted trillium *noun* (1855)
: a trillium (*Trillium undulatum*) of northeastern North America that has a solitary flower with white petals streaked with purple

painted turtle *noun* (1876)
: a North American turtle (*Chrysemys picta*) having a greenish to black carapace with yellow, red, or olive bordered scutes and a yellow plastron

¹paint·er \'pān-tər\ *noun* (14th century)
: one that paints: as **a :** an artist who paints **b :** one who applies paint especially as an occupation

²pain·ter \'pān-tər\ *noun* [Middle English *paynter,* probably from Middle French *pendoir, pentoir* clothesline, from *pendre* to hang — more at PENDANT] (14th century)
: a line used for securing or towing a boat

³pain·ter *noun* [alteration of *panther*] (circa 1764)
: COUGAR

paint·er·ly \'pān-tər-lē\ *adjective* (1586)
: of, relating to, or typical of a painter **:** ARTISTIC
— paint·er·li·ness *noun*

painter's colic *noun* (circa 1834)
: intestinal colic associated with obstinate constipation due to chronic lead poisoning

paint·ing \'pān-tiŋ\ *noun* (13th century)
1 : a product of painting; *especially* **:** a work produced through the art of painting
2 : the art or occupation of painting

paint·work \'pānt-ˌwərk\ *noun* (1888)
chiefly British
1 : work with paint
2 : PAINT 3b

[1]**pair** \ˈpar, ˈper\ *noun, plural* **pairs** *or* **pair** [Middle English *paire,* from Old French, from Latin *paria* equal things, from neuter plural of *par* equal] (14th century)
1 a (1) **:** two corresponding things designed for use together ⟨a *pair* of shoes⟩ (2) **:** two corresponding bodily parts or members ⟨a *pair* of hands⟩ **b :** something made up of two corresponding pieces ⟨a *pair* of trousers⟩
2 a : two similar or associated things: as (1) **:** two mated animals (2) **:** a couple in love, engaged, or married ⟨were a devoted *pair*⟩ (3) **:** two playing cards of the same value or denomination and especially of the same rank (4) **:** two horses harnessed side by side (5) **:** two members of a deliberative body that agree not to vote on a specific issue during a time agreed on; *also* **:** an agreement not to vote made by the two members **b :** a partnership especially of two players in a contest against another partnership
3 *chiefly dialect* **:** a set or series of small objects (as beads)

[2]**pair** (1606)
transitive verb
1 a : to make a pair of — often used with *off* or *up* ⟨*paired* off the animals⟩ **b :** to cause to be a member of a pair **c :** to arrange a voting pair between
2 : to arrange in pairs
intransitive verb
1 : to constitute a member of a pair ⟨a sock that didn't *pair*⟩
2 a : to become associated with another — often used with *off* or *up* ⟨*paired* up with an old friend⟩ **b :** to become grouped or separated into pairs — often used with *off* ⟨*paired* off for the next dance⟩

pair–bond \ˌbänd\ *noun* (1940)
: a monogamous relationship
— pair–bond·ing *noun*

paired–associate learning *noun* (1966)
: the learning of syllables, digits, or words in pairs (as in the study of a foreign language) so that one member of the pair evokes recall of the other

pair of compasses (1563)
: COMPASS 3c

pair of virginals (1542)
: VIRGINAL

pair production *noun* (1934)
: the transformation of a quantum of radiant energy simultaneously into an electron and a positron when the quantum interacts with the intense electric field near a nucleus

pai·sa \pī-ˈsä\ *noun* [ultimately from Hindi *paisā,* a quarter-anna coin] (1956)
1 *plural* **paisa** — see *rupee, taka* at MONEY table
2 *plural* **pai·se** \-ˈsā\ — see *rupee* at MONEY table

pais·ley \ˈpāz-lē\ *adjective, often capitalized* [*Paisley,* Scotland] (1824)
1 : made typically of soft wool and woven or printed with colorful curved abstract figures
2 : marked by designs, patterns, or figures typically used in paisley fabrics ⟨a *paisley* print⟩
— paisley *noun*

Pai·ute \ˈpī-ˌyüt\ *noun* (1827)
1 : a member of an American Indian people originally of Utah, Arizona, Nevada, and California
2 : either of the two Uto-Aztecan languages of the Paiute people

pa·ja·ma \pə-ˈjä-mə, -ˈja-\ *noun* [Hindi *pājāma,* from Persian *pā* leg + *jāma* garment] (1883)
: PAJAMAS
— pa·ja·maed \-məd\ *adjective*

pa·ja·mas \-məz\ *noun plural* [plural of *pajama*] (1800)
1 : loose lightweight trousers formerly much worn in the Near East
2 : a loose usually two-piece lightweight suit designed especially for sleeping or lounging

pak choi \ˈpäk-ˈchȯi, ˈpak-\ *variant of* BOK CHOY

pa·ke·ha \ˈpä-kə-ˌhä, ˈpa-kē-ə\ *noun, plural* **pakeha** *or* **pakehas** *often capitalized* [Maori] (1832)
chiefly New Zealand **:** a person who is not of Maori descent; *especially* **:** a white person

Paki \ˈpa-kē, ˈpä-\ *noun* [short for *Pakistani*] (1964)
chiefly British **:** a Pakistani immigrant — usually used disparagingly

[1]**pal** \ˈpal\ *noun* [Romany *phral, phal* brother, friend, from Sanskrit *bhrātṛ* brother; akin to Old English *brōthor* brother] (circa 1682)
: a close friend
— pal·ship \-ˌship\ *noun*

[2]**pal** *intransitive verb* **palled; pal·ling** (1879)
: to be or become pals **:** associate as pals ⟨they've *palled* around for years⟩

[1]**pal·ace** \ˈpa-ləs\ *noun* [Middle English *palais,* from Old French, from Latin *palatium,* from *Palatium,* the Palatine Hill in Rome where the emperors' residences were built] (13th century)
1 a : the official residence of a chief of state (as a monarch or a president) **b** *chiefly British* **:** the official residence of an archbishop or bishop
2 a : a large stately house **b :** a large public building **c :** a highly decorated place for public amusement or refreshment ⟨a movie *palace*⟩ ◆

[2]**palace** *adjective* (14th century)
1 : of or relating to a palace
2 : of, relating to, or involving the intimates of a chief executive ⟨a *palace* revolution⟩ ⟨*palace* politics⟩
3 : LUXURIOUS, DELUXE

pal·a·din \ˈpa-lə-dən\ *noun* [French, from Italian *paladino,* from Medieval Latin *palatinus* courtier, from Late Latin, imperial official — more at PALATINE] (1592)
1 : a trusted military leader (as for a medieval prince)
2 : a leading champion of a cause

palae- *or* **palaeo-** *chiefly British variant of* PALE-

pa·laes·tra \pə-ˈles-trə\ *noun, plural* **-trae** \-(ˌ)trē\ [Middle English *palestre* arena, from Latin *palaestra* place for wrestling, from Greek *palaistra,* from *palaiein* to wrestle] (1580)
1 : a school in ancient Greece or Rome for sports (as wrestling)
2 : GYMNASIUM

pa·lan·quin \ˌpa-lən-ˈkēn, -ˈkwin, -ˈkin, ˈpa-lən-ˌ; pə-ˈlaŋ-kwən\ *noun* [Portuguese *palanquim,* from Malay or Javanese *pelangki,* of Indo-Aryan origin; akin to Bengali *pālaṅka* bed] (1588)
: a conveyance formerly used especially in eastern Asia usually for one person that consists of an enclosed litter borne on the shoulders of men by means of poles

pal·at·able \ˈpa-lə-tə-bəl\ *adjective* (1669)
1 : agreeable to the palate or taste
2 : agreeable or acceptable to the mind ☆
pal·at·abil·i·ty \ˌpa-lə-tə-ˈbi-lə-tē\ *noun*
— pal·at·able·ness *noun*
— pal·at·ably \ˈpa-lə-tə-blē\ *adverb*

pal·a·tal \ˈpa-lə-t^{ə}l\ *adjective* (1728)
1 a : formed with some part of the tongue near or touching the hard palate posterior to the teethridge ⟨the \sh\ and \y\ in English and the \k̲\ of *ich* \ik̲\ in German are examples of *palatal* consonants⟩ **b** *of a vowel* **:** FRONT 2
2 : of, relating to, forming, or affecting the palate
— palatal *noun*
— pal·a·tal·ly \-t^{ə}l-ē\ *adverb*

pal·a·tal·i·za·tion \ˌpa-lə-t^{ə}l-ə-ˈzā-shən\ *noun* (1863)
1 : the quality or state of being palatalized
2 : an act or instance of palatalizing an utterance ⟨in rapid speech the \t\ in *got* undergoes *palatalization* before *you* to yield \ˈgächə\⟩

pal·a·tal·ize \ˈpa-lə-t^{ə}l-ˌīz\ *transitive verb* **-ized; -iz·ing** (1867)
: to pronounce as or change into a palatal sound

pal·ate \ˈpa-lət\ *noun* [Middle English, from Latin *palatum*] (14th century)
1 : the roof of the mouth separating the mouth from the nasal cavity
2 a : a usually intellectual taste or liking **b :** the sense of taste

pa·la·tial \pə-ˈlā-shəl\ *adjective* [Latin *palatium* palace] (1754)
1 : of, relating to, or being a palace
2 : suitable to a palace **:** MAGNIFICENT
— pa·la·tial·ly \-shə-lē\ *adverb*
— pa·la·tial·ness *noun*

pa·lat·i·nate \pə-ˈla-t^{ə}n-ət\ *noun* (circa 1580)
: the territory of a palatine

[1]**pal·a·tine** \ˈpa-lə-ˌtīn\ *adjective* [Middle English, from Latin *palatinus* imperial, from *palatium*] (15th century)
1 a : possessing royal privileges **b :** of or relating to a palatine or a palatinate
2 a : of or relating to a palace especially of a Roman or Holy Roman emperor **b :** PALATIAL

[2]**palatine** \-ˌtīn, *3 is also* -ˌtēn\ *noun* [Late Latin *palatinus* imperial official, from Latin *palatinus,* adjective] (1591)

☆ SYNONYMS
Palatable, appetizing, savory, tasty, toothsome mean agreeable or pleasant especially to the sense of taste. PALATABLE often applies to something that is found to be merely agreeable ⟨butterflies that birds find *palatable*⟩. APPETIZING suggests a whetting of the appetite and applies to aroma and appearance as well as taste ⟨*appetizing* hors d'oeuvres⟩. SAVORY applies to both taste and aroma and suggests piquancy and often spiciness ⟨dumplings with *savory* fillings⟩. TASTY implies a pronounced taste ⟨a tart and *tasty* pie⟩. TOOTHSOME stresses the notion of agreeableness and sometimes implies tenderness or daintiness ⟨an enticing array of *toothsome* desserts⟩.

◇ WORD HISTORY
palace The word *palace* conjures up images of kings and queens living in huge homes of splendor and luxury, often on high ground far above the common people. Although such palaces are the settings of fairy tales, they are not far removed from the first palaces of history. The *Palatium* or Palatine Hill, one of the seven hills of Rome, was the site of a number of large dwellings from about the 4th century B.C. Caesar Augustus took one of these homes as his residence, and the building of more imperial residences on the Palatine Hill by subsequent Roman emperors led to the use of *Palatium* to mean "the emperor's residence." In Late and Medieval Latin *palatium* became a generic noun denoting any royal or imperial residence, as well as the royal court and royal authority, and then any princely house or manor of a lord. When *paleis* or *palais,* the Old French outcome of *palatium,* first appeared, it referred simply to the luxurious residence of an important person, and this sense was carried into Middle English *palais* and Modern English *palace,* along with most of the other meanings of the Medieval Latin word.

1 a : a feudal lord having sovereign power within his domains **b :** a high officer of an imperial palace
2 *capitalized* **:** a native or inhabitant of the Palatinate
3 [French, from Elisabeth Charlotte of Bavaria (died 1722) Princess *Palatine*] **:** a fur cape or stole covering the neck and shoulders

³palatine \-ˌtīn\ *adjective* [French *palatin,* from Latin *palatum* palate] (circa 1656)
: of, relating to, or lying near the palate

⁴palatine \-ˌtīn\ *noun* (1854)
: either of a pair of bones that are situated behind and between the maxillae and in humans are of extremely irregular form

¹pa·lav·er \pə-ˈla-vər, -ˈlä-\ *noun* [Portuguese *palavra* word, speech, from Late Latin *parabola* parable, speech] (1735)
1 a : a long parley usually between persons of different cultures or levels of sophistication **b :** CONFERENCE, DISCUSSION
2 a : idle talk **b :** misleading or beguiling speech

²palaver *verb* **pa·lav·ered; pa·lav·er·ing** \pə-ˈla-və-riŋ, -ˈlä-; -ˈlav-riŋ, -ˈläv-\ (1773)
intransitive verb
1 : to talk profusely or idly
2 : PARLEY
transitive verb
: to use palaver to **:** CAJOLE

pa·laz·zo \pə-ˈlät-(ˌ)sō\ *noun, plural* **pa·laz·zi** \-(ˌ)sē\ [Italian, from Latin *palatium* palace] (circa 1666)
: a large imposing building (as a museum or a place of residence) especially in Italy

¹pale \ˈpā(ə)l\ *noun* [Middle English, from Middle French *pal* stake, from Latin *palus* — more at POLE] (12th century)
1 *archaic* **:** PALISADE, PALING
2 a : one of the stakes of a palisade **b :** PICKET
3 a : a space or field having bounds **:** ENCLOSURE **b :** a territory or district within certain bounds or under a particular jurisdiction
4 : an area or the limits within which one is privileged or protected (as from censure) ⟨conduct that was beyond the *pale*⟩
5 : a perpendicular stripe on a heraldic shield

²pale *transitive verb* **paled; pal·ing** (14th century)
: to enclose with pales **:** FENCE

³pale *adjective* **pal·er; pal·est** [Middle English, from Middle French, from Latin *pallidus,* from *pallēre* to be pale — more at FALLOW] (14th century)
1 : deficient in color or intensity of color **:** PALLID ⟨a *pale* complexion⟩
2 : not bright or brilliant **:** DIM ⟨a *pale* sun shining through the fog⟩
3 : FEEBLE, FAINT ⟨a *pale* imitation⟩
4 : deficient in chroma ⟨a *pale* pink⟩
— **pale·ly** \ˈpā(ə)l-lē\ *adverb*
— **pale·ness** \-nəs\ *noun*
— **pal·ish** \ˈpā-lish\ *adjective*

⁴pale *verb* **paled; pal·ing** (14th century)
intransitive verb
: to become pale
transitive verb
: to make pale

pale- *or* **paleo-** *combining form* [Greek *palai-, palaio-* ancient, from *palaios,* from *palai* long ago; probably akin to Greek *tēle* far off, Sanskrit *carama* last]
1 : involving or dealing with ancient forms or conditions ⟨*paleo*botany⟩
2 : early **:** primitive **:** archaic ⟨*Paleo*lithic⟩

pa·lea \ˈpā-lē-ə\ *noun, plural* **pa·le·ae** \-lē-ˌē\ [New Latin, from Latin, chaff — more at PALLET] (1753)
1 : one of the chaffy scales on the receptacle of many composite plants
2 : the upper bract that with the lemma encloses the flower in grasses
— **pa·le·al** \-lē-əl\ *adjective*

Pa·le·arc·tic \ˌpā-lē-ˈärk-tik, -ˈär-tik\ *adjective* (1858)
: of, relating to, or being a biogeographic region or subregion that includes Europe, Asia north of the Himalayas, northern Arabia, and Africa north of the Sahara

pale dry *adjective* (1953)
: dry and light colored ⟨*pale dry* ginger ale⟩

pale·face \ˈpā(ə)l-ˌfās\ *noun* (1822)
: a white person

pa·leo·an·thro·pol·o·gy \ˌpā-lē-ō-ˌan(t)-thrə-ˈpä-lə-jē, *especially British* ˌpa-\ *noun* (1916)
: a branch of anthropology dealing with fossil hominids
— **pa·leo·an·thro·po·log·i·cal** \-pə-ˈlä-ji-kəl\ *adjective*
— **pa·leo·an·thro·pol·o·gist** \-ˈpä-lə-jist\ *noun*

pa·leo·bi·ol·o·gy \-bī-ˈä-lə-jē\ *noun* (1893)
: a branch of paleontology concerned with the biology of fossil organisms
— **pa·leo·bi·o·log·i·cal** \-bī-ə-ˈlä-ji-kəl\ *also* **pa·leo·bi·o·log·ic** \-ˈlä-jik\ *adjective*
— **pa·leo·bi·ol·o·gist** \-bī-ˈä-lə-jist\ *noun*

pa·leo·bot·a·ny \-ˈbä-tᵊn-ē, -ˈbät-nē\ *noun* [International Scientific Vocabulary] (1872)
: a branch of botany dealing with fossil plants
— **pa·leo·bo·tan·i·cal** \-bə-ˈta-ni-kəl\ *or* **pa·leo·bo·tan·ic** \-ˈta-nik\ *adjective*
— **pa·leo·bo·tan·i·cal·ly** \-ni-k(ə-)lē\ *adverb*
— **pa·leo·bot·a·nist** \-ˈbä-tᵊn-ist, -ˈbät-nist\ *noun*

Pa·leo·cene \ˈpā-lē-ə-ˌsēn, *especially British* ˈpa-\ *adjective* [International Scientific Vocabulary] (1877)
: of, relating to, or being the earliest epoch of the Tertiary or the corresponding system of rocks — see GEOLOGIC TIME table
— **Paleocene** *noun*

pa·leo·cli·ma·tol·o·gy \ˌpā-lē-ō-ˌklī-mə-ˈtä-lə-jē, *especially British* ˌpa-\ *noun* [International Scientific Vocabulary] (circa 1909)
: a science dealing with the climate of past ages
— **pa·leo·cli·ma·tol·o·gist** \-jist\ *noun*

pa·leo·ecol·o·gy \-i-ˈkä-lə-jē, -e-ˈkä-\ *noun* (1898)
: a branch of ecology that is concerned with the characteristics of ancient environments and with their relationships to ancient plants and animals
— **pa·leo·eco·log·i·cal** \-ˌē-kə-ˈlä-ji-kəl, -ˌe-kə-\ *also* **pa·leo·eco·log·ic** \-jik\ *adjective*
— **pa·leo·ecol·o·gist** \-i-ˈkä-lə-jist, -e-ˈkä-\ *noun*

Pa·leo·gene \ˈpā-lē-ə-ˌjēn, *especially British* ˈpa-\ *adjective* [German *Paläogen,* from *paläpale-* + *-gen* (from Greek *-genēs* born) — more at -GEN] (1882)
: of, relating to, or being the earlier part of the Tertiary including the Paleocene, Eocene, and Oligocene or the corresponding system of rocks
— **Paleogene** *noun*

pa·leo·ge·og·ra·phy \ˌpā-lē-ō-jē-ˈä-grə-fē, *especially British* ˌpa-\ *noun* [International Scientific Vocabulary] (1881)
: the geography of ancient times or of a particular past geological epoch
— **pa·leo·geo·graph·ic** \-ˌjē-ə-ˈgra-fik\ *or* **pa·leo·geo·graph·i·cal** \-fi-kəl\ *adjective*
— **pa·leo·geo·graph·i·cal·ly** \-fi-k(ə-)lē\ *adverb*

pa·leo·graph·ic \-ə-ˈgra-fik\ *or* **pa·leo·graph·i·cal** \-fi-kəl\ *adjective* (circa 1842)
: relating to writings of former times
— **pa·leo·graph·i·cal·ly** \-fi-k(ə-)lē\ *adverb*

pa·le·og·ra·phy \ˌpā-lē-ˈä-grə-fē, *especially British* ˌpa-\ *noun* [New Latin *palaeographia,* from Greek *palai-* pale- + *-graphia* -graphy] (circa 1818)
1 : the study of ancient writings and inscriptions
2 a : an ancient manner of writing **b :** ancient writings
— **pa·le·og·ra·pher** \-fər\ *noun*

Pa·leo–In·di·an \ˌpā-lē-ō-ˈin-dē-ən, *especially British* ˌpa-\ *noun* (1940)
: one of the early American hunting people of Asian origin extant in the late Pleistocene
— **Paleo–Indian** *adjective*

Pa·leo·lith·ic \ˌpā-lē-ə-ˈli-thik, *especially British* ˌpa-\ *adjective* [International Scientific Vocabulary] (1865)
: of or relating to the earliest period of the Stone Age characterized by rough or chipped stone implements

pa·leo·mag·ne·tism \ˌpā-lē-ō-ˈmag-nə-ˌti-zəm, *especially British* ˌpa-\ *noun* (1854)
1 : the intensity and direction of residual magnetization in ancient rocks
2 : a science that deals with paleomagnetism
— **pa·leo·mag·net·ic** \-mag-ˈne-tik\ *adjective*
— **pa·leo·mag·net·i·cal·ly** \-ti-k(ə-)lē\ *adverb*
— **pa·leo·mag·ne·tist** \-ˈmag-nə-tist\ *noun*

pa·le·on·tol·o·gy \ˌpā-lē-ˌän-ˈtä-lə-jē, -ən-, *especially British* ˌpa-\ *noun* [French *paléontologie,* from *palé-* pale- + Greek *onta* existing things (from neuter plural of *ont-, ōn,* present participle of *einai* to be) + French *-logie* -logy — more at IS] (1838)
: a science dealing with the life of past geological periods as known from fossil remains
— **pa·le·on·to·log·i·cal** \-ˌän-tᵊl-ˈä-ji-kəl\ *or* **pa·le·on·to·log·ic** \-jik\ *adjective*
— **pa·le·on·tol·o·gist** \-ˌän-ˈtä-lə-jist, -ən-\ *noun*

pa·leo·pa·thol·o·gy \ˌpā-lē-ō-pə-ˈthä-lə-jē, -pa-ˈthä-, *especially British* ˌpa-lē-\ *noun* (1893)
: a branch of pathology concerned with diseases of former times as evidenced especially in fossil or other remains
— **pa·leo·patho·log·i·cal** \-ˌpa-thə-ˈlä-ji-kəl\ *adjective*
— **pa·leo·pa·thol·o·gist** \-pə-ˈthä-lə-jist, -pa-\ *noun*

Pa·leo·zo·ic \ˌpā-lē-ə-ˈzō-ik, *especially British* ˌpa-\ *adjective* (1838)
: of, relating to, originating in, or being an era of geological history that extends from the beginning of the Cambrian to the close of the Permian and is marked by the culmination of nearly all classes of invertebrates except the insects and in the later epochs by the appearance of terrestrial plants, amphibians, and reptiles; *also* **:** relating to the corresponding system of rocks — see GEOLOGIC TIME table
— **Paleozoic** *noun*

pa·leo·zo·ol·o·gy \-zō-ˈä-lə-jē, -zə-ˈwä-\ *noun* [International Scientific Vocabulary] (1857)
: a branch of paleontology dealing with ancient and fossil animals
— **pa·leo·zoo·log·i·cal** \-ˌzō-ə-ˈlä-ji-kəl\ *adjective*
— **pa·leo·zo·ol·o·gist** \-jist\ *noun*

pal·et \pā-ˈlet, ˈpā-lət\ *noun* [*pale* palea + ¹*-et*] (circa 1880)
: PALEA

pal·ette \ˈpa-lət\ *noun* [French, from Middle French, diminutive of *pale* spade, from Latin *pala;* probably akin to Latin *pangere* to fix — more at PACT] (1622)
1 : a thin oval or rectangular board or tablet that a painter holds and mixes pigments on
2 a : the set of colors put on the palette **b** (1) **:** a particular range, quality, or use of color (2) **:** a comparable range, quality, or use of available elements especially in another art (as music)

palette knife *noun* (1759)
: a knife with usually a flexible steel blade and no cutting edge used to mix colors or to apply colors (as to a painting)

pal·frey \'pòl-frē\ *noun, plural* **palfreys** [Middle English, from Old French *palefrei,* from Medieval Latin *palafredus,* from Late Latin *paraveredus* post-horse for secondary roads, from Greek *para-* beside, subsidiary + Late Latin *veredus* post-horse, from a Gaulish word akin to Welsh *gorwydd* horse; akin to Old Irish *réidid* he rides — more at PARA-, RIDE] (13th century)
archaic **:** a saddle horse other than a warhorse; *especially* **:** a lady's light easy-gaited horse

Pa·li \'pä-lē\ *noun* [Sanskrit *pāli* row, series of Buddhist sacred texts] (1800)
: an Indo-Aryan language used as the liturgical and scholarly language of Theravada Buddhism — see INDO-EUROPEAN LANGUAGES table

pal·i·mo·ny \'pa-lə-ˌmō-nē\ *noun* [blend of *pal* and *alimony*] (1979)
: a court-ordered allowance paid by one member of a couple formerly living together out of wedlock to the other

pa·limp·sest \'pa-ləm(p)-ˌsest, pə-'lim(p)-\ *noun* [Latin *palimpsestus,* from Greek *palimpsēstos* scraped again, from *palin* + *psēn* to rub, scrape; akin to Sanskrit *psāti, babhasti* he chews] (1825)
1 : writing material (as a parchment or tablet) used one or more times after earlier writing has been erased
2 : something having usually diverse layers or aspects apparent beneath the surface

pal·in·drome \'pa-lən-ˌdrōm\ *noun* [Greek *palindromos* running back again, from *palin* back, again + *dramein* to run; akin to Greek *polos* axis, pole — more at POLE, DROMEDARY] (circa 1629)
: a word, verse, or sentence (as "Able was I ere I saw Elba") or a number (as 1881) that reads the same backward or forward
— pal·in·drom·ic \ˌpa-lən-'drō-mik, -'drä-\ *adjective*
— pal·in·drom·ist \'pa-lən-ˌdrō-mist\ *noun*

pal·ing \'pā-liŋ\ *noun* (15th century)
1 : a fence of pales or pickets
2 : wood for making pales
3 : a pale or picket for a fence

pal·in·gen·e·sis \ˌpa-lən-'je-nə-səs\ *noun* [New Latin, from Greek *palin* again + Latin *genesis* genesis] (1818)
: METEMPSYCHOSIS

pal·in·ge·net·ic \-jə-'ne-tik\ *adjective* (1833)
: of or relating to palingenesis

pal·in·ode \'pa-lə-ˌnōd\ *noun* [Greek *palinōidia,* from *palin* + *aeidein* to sing — more at ODE] (1599)
1 : an ode or song recanting or retracting something in an earlier poem
2 : a formal retraction

¹pal·i·sade \ˌpa-lə-'sād\ *noun* [French *palissade,* ultimately from Latin *palus* stake — more at POLE] (1600)
1 a : a fence of stakes especially for defense **b :** a long strong stake pointed at the top and set close with others as a defense
2 : a line of bold cliffs

²palisade *transitive verb* **-sad·ed; -sad·ing** (1632)
: to fortify with palisades

palisade cell *noun* (1875)
: a cell of the palisade layer

palisade layer *noun* (1914)
: a layer of columnar cells rich in chloroplasts found beneath the upper epidermis of foliage leaves — called also *palisade mesophyll, palisade parenchyma, palisade tissue*; compare SPONGY PARENCHYMA

¹pall \'pòl\ *verb* [Middle English, short for *appallen* to become pale — more at APPALL] (14th century)
intransitive verb
1 : to lose strength or effectiveness
2 : to lose in interest or attraction ⟨his humor began to *pall* on us⟩
3 : to become tired of something
transitive verb
1 : to cause to become insipid
2 : to deprive of pleasure in something by satiating
synonym see SATIATE

²pall *noun* [Middle English, cloak, mantle, from Old English *pæll,* from Latin *pallium*] (14th century)
1 : PALLIUM 1a
2 a : a square of linen usually stiffened with cardboard that is used to cover the chalice **b** (1) **:** a heavy cloth draped over a coffin (2) **:** a coffin especially when holding a body
3 : something that covers or conceals; *especially* **:** an overspreading element that produces an effect of gloom ⟨a *pall* of thick black smoke⟩

³pall *transitive verb* (15th century)
: to cover with a pall **:** DRAPE

Pal·la·di·an \pə-'lā-dē-ən, -'lä-\ *adjective* (1731)
: of or relating to a revived classical style in architecture based on the works of Andrea Palladio
— Pal·la·di·an·ism \-ə-ˌni-zəm\ *noun*

¹pal·la·di·um \pə-'lā-dē-əm\ *noun* [Middle English, from Latin, from Greek *palladion,* from *Pallad-, Pallas*]
1 *capitalized* **:** a statue of Pallas whose preservation was believed to ensure the safety of Troy
2 *plural* **pal·la·dia** \-dē-ə\ **:** SAFEGUARD

²palladium *noun* [New Latin, from *Pallad-, Pallas,* an asteroid] (1803)
: a silver-white ductile malleable metallic element of the platinum group that is used especially in electrical contacts, as a catalyst, and in alloys — see ELEMENT table
— pal·la·dous \pə-'lā-dəs\ *adjective*

Pal·las \'pa-ləs\ *noun* [Latin *Pallad-, Pallas,* from Greek]
: ATHENA

pall·bear·er \'pòl-ˌbar-ər, -ˌber-\ *noun* [²*pall*] (1707)
: a person who helps to carry the coffin at a funeral; *also* **:** a member of the escort or honor guard of the coffin who does not actually help to carry it

¹pal·let \'pa-lət\ *noun* [Middle English *pailet,* from (assumed) Middle French *paillet,* from *paille* straw, from Latin *palea* chaff, straw; akin to Sanskrit *palāva* chaff] (14th century)
1 : a straw-filled tick or mattress
2 : a small, hard, or temporary bed

²pallet *noun* [Middle English *palette,* literally, small shovel, from Middle French — more at PALETTE] (1558)
1 : a wooden flat-bladed instrument
2 : a lever or surface in a timepiece that receives an impulse from the escapement wheel and imparts motion to a balance or pendulum
3 : a portable platform for handling, storing, or moving materials and packages (as in warehouses, factories, or vehicles)

pal·let·ise *British variant of* PALLETIZE

pal·let·ize \'pa-lə-ˌtīz\ *transitive verb* **-ized; -iz·ing** (1944)
: to place on, transport, or store by means of pallets
— pal·let·i·za·tion \ˌpa-lə-tə-'zā-shən\ *noun*
— pal·let·iz·er \'pa-lə-ˌtī-zər\ *noun*

pal·lette \pa-'let\ *noun* [alteration of *palette*] (1834)
: one of the plates at the armpits of a suit of armor — see ARMOR illustration

pal·li·al \'pa-lē-əl\ *adjective* [New Latin *pallium*] (1836)
1 : of, relating to, or produced by a mantle of a mollusk
2 : of or relating to the cerebral cortex

pal·liasse \pal-'yas\ *noun* [modification of French *paillasse,* from *paille* straw] (1763)
: a thin straw mattress used as a pallet

pal·li·ate \'pa-lē-ˌāt\ *transitive verb* **-at·ed; -at·ing** [Middle English, from Late Latin *palliatus,* past participle of *palliare* to cloak, conceal, from Latin *pallium* cloak] (15th century)
1 : to reduce the violence of (a disease) **:** ABATE
2 : to cover by excuses and apologies
3 : to moderate the intensity of ⟨trying to *palliate* the boredom⟩
— pal·li·a·tion \ˌpa-lē-'ā-shən\ *noun*
— pal·li·a·tor \'pa-lē-ˌā-tər\ *noun*

¹pal·li·a·tive \'pa-lē-ˌā-tiv, 'pal-yə-\ *adjective* (15th century)
: serving to palliate ⟨*palliative* surgery⟩
— pal·lia·tive·ly *adverb*

²palliative *noun* (1724)
: something that palliates

pal·lid \'pa-ləd\ *adjective* [Latin *pallidus* — more at PALE] (1590)
1 : deficient in color **:** WAN ⟨a *pallid* countenance⟩
2 : lacking sparkle or liveliness **:** DULL ⟨a *pallid* entertainment⟩
— pal·lid·ly *adverb*
— pal·lid·ness *noun*

pal·li·um \'pa-lē-əm\ *noun, plural* **-lia** \-lē-ə\ *or* **-li·ums** [Middle English, from Latin] (12th century)
1 a : a white woolen band with pendants in front and back worn over the chasuble by a pope or archbishop as a symbol of full episcopal authority **b :** a draped rectangular cloth worn as a cloak by men of ancient Greece and Rome
2 [New Latin, from Latin, cloak] **a :** CEREBRAL CORTEX **b :** the mantle of a mollusk, brachiopod, or bird

pall–mall \ˌpel-'mel, ˌpal-'mal, *US often* ˌpòl-'mòl\ *noun* [Middle French *pallemaille,* from Italian *pallamaglio,* from *palla* ball (of Germanic origin; akin to Old High German *balla* ball) + *maglio* mallet, from Latin *malleus* — more at BALL, MAUL] (1598)
: a 17th century game in which each player attempts to drive a wooden ball with a mallet down an alley and through a raised ring in as few strokes as possible; *also* **:** the alley in which it is played

pal·lor \'pa-lər\ *noun* [Middle English, from Latin, from *pallēre* to be pale — more at FALLOW] (15th century)
: deficiency of color especially of the face **:** PALENESS

pal·ly \'pa-lē\ *adjective* (1895)
: sharing the relationship of pals **:** INTIMATE

¹palm \'päm, 'pälm, 'pòm, 'pòlm, *New England also* 'pàm\ *noun* [Middle English, from Old English, from Latin *palma* palm of the hand, palm tree; from the resemblance of the tree's leaves to the outstretched hand; akin to Greek *palamē* palm of the hand, Old English *folm,* Old Irish *lám* hand] (before 12th century)
1 : any of a family (Palmae, the palm family) of mostly tropical or subtropical monocotyledonous trees, shrubs, or vines with usually a simple stem and a terminal crown of large pinnate or fan-shaped leaves
2 : a leaf of the palm as a symbol of victory or rejoicing; *also* **:** a branch (as of laurel) similarly used
3 : a symbol of triumph; *also* **:** VICTORY, TRIUMPH
4 : an addition to a military decoration in the form of a palm frond especially to indicate a second award of the basic decoration
— palm·like \-ˌlīk\ *adjective*

²palm *noun* [Middle English *paume,* from Middle French, from Latin *palma*] (14th century)
1 : the somewhat concave part of the human hand between the bases of the fingers and the

wrist or the corresponding part of the forefoot of a lower mammal
2 : a flat expanded part especially at the end of a base or stalk (as of an anchor)
3 [Latin *palmus,* from *palma*] **:** a unit of length based on the breadth or length of the hand
4 : something (as a part of a glove) that covers the palm of the hand
5 : an act of palming (as of cards)
³palm *transitive verb* (1673)
1 a : to conceal in or with the hand ⟨*palm* a card⟩ **b :** to pick up stealthily
2 : to impose by fraud ⟨a second imposter to be *palmed* upon you —Sir Walter Scott⟩
3 : to touch with the palm: as **a :** to stroke with the palm or hand **b :** to shake hands with **c :** to allow (a basketball) to come to rest momentarily in the hand while dribbling thus committing a violation
pal·mar \'pal-mər, 'pä-, 'päl-, 'pȯ-, 'pȯl-\ *adjective* (1656)
: of, relating to, or involving the palm of the hand
pal·ma·ry \'pal-mə-rē, 'pä-, 'päl-, 'pȯ-, 'pȯl-\ *adjective* [Latin *palmarius* deserving the palm, from *palma*] (1657)
: OUTSTANDING, BEST
pal·mate \'pal-ˌmāt, 'pä-, 'päl-, 'pȯ-, 'pȯl-\ *also* **pal·mat·ed** \-ˌmā-təd\ *adjective* (circa 1760)
: resembling a hand with the fingers spread: as **a :** having lobes radiating from a common point ⟨a *palmate* leaf⟩ — see LEAF illustration **b :** having the distal portion broad, flat, and lobed ⟨a *palmate* antler⟩
— pal·mate·ly *adverb*
— pal·ma·tion \pal-'mā-shən, pä-, 'päl-, 'pȯ-, 'pȯl-\ *noun*
palmed \'pämd, 'pälmd, 'pȯmd, 'pȯlmd, *New England also* 'pȧmd\ *adjective* (15th century)
: having a palm of a specified kind — used in combination ⟨leather-*palmed* gloves⟩
palm·er \'pä-mər, 'päl-, 'pȯ-, 'pȯl-\ *noun* (13th century)
: a person wearing two crossed palm leaves as a sign of a pilgrimage made to the Holy Land
palm·er·worm \-ˌwərm\ *noun* (1560)
: a caterpillar that suddenly appears in great numbers devouring herbage
pal·mette \ˌpal-'met\ *noun* [French, from *palme* palm, from Latin *palma*] (1850)
: a decorative motif suggestive of a palm
pal·met·to \pal-'me-(ˌ)tō *also* pä-, päl-, pȯ-, pȯl-\ *noun, plural* **-tos** *or* **-toes** [modification of Spanish *palmito,* from *palma* palm, from Latin] (circa 1746)
1 : any of several usually low-growing fan-leaved palms; *especially* **:** CABBAGE PALMETTO
2 : strips of the leaf blade of a palmetto used in weaving

palmetto 1

palm·ist \'pä-mist, 'päl-, 'pȯ-, 'pȯl-\ *noun* [probably back-formation from *palmistry*] (1886)
: one who practices palmistry
palm·ist·ry \'pä-mə-strē, 'päl-, 'pȯ-, 'pȯl-\ *noun* [Middle English *pawmestry,* probably from *paume* palm + *maistrie* mastery] (15th century)
: the art or practice of reading a person's character or future from the lines on the palms
pal·mi·tate \'pal-mə-ˌtāt, 'pä-, 'päl-, 'pȯ-, 'pȯl-\ *noun* (1873)
: a salt or ester of palmitic acid
pal·mit·ic acid \(ˌ)pal-'mi-tik-, (ˌ)pä-, (ˌ)päl-, (ˌ)pȯ-, (ˌ)pȯl-\ *noun* [International Scientific Vocabulary, from *palmitin*] (1857)
: a waxy crystalline saturated fatty acid $C_{16}H_{32}O_2$ occurring free or in the form of esters (as glycerides) in most fats and fatty oils and in several essential oils and waxes
pal·mi·tin \'pal-mə-tən, 'pä-, 'päl-, 'pȯ-, 'pȯl-\ *noun* [French *palmitine,* probably from *palmite* pith of the palm tree, from Spanish *palmito*] (1857)
: an ester of glycerol and palmitic acid; *especially* **:** a solid ester found in animal fats
palm off *transitive verb* (1822)
1 : to dispose of usually by trickery or guile
2 : PASS OFF 2
palm oil *noun* (1705)
: an edible fat obtained from the flesh of the fruit of several palms and used especially in soap and lubricating greases
Palm Sunday *noun* [from the palm branches strewn in Christ's way] (before 12th century)
: the Sunday before Easter celebrated in commemoration of Christ's triumphal entry into Jerusalem
palmy \'pä-mē, 'päl-, 'pȯ-, 'pȯl- *New England also* 'pȧ-mē\ *adjective* **palm·i·er; -est** (1602)
1 : marked by prosperity **:** FLOURISHING
2 : abounding in or bearing palms
pal·my·ra \pal-'mī-rə\ *noun* [Portuguese *palmeira,* from *palma* palm, from Latin] (1698)
: a tall fan-leaved palm (*Borassus flabellifer*) of India cultivated for its hard resistant wood, fiber, and sugar-rich sap — called also *palmyra palm*
pal·o·mi·no \ˌpa-lə-'mē-(ˌ)nō\ *noun, plural* **-nos** [American Spanish, from Spanish, like a dove, from Latin *palumbinus,* from *palumbes* ringdove; akin to Greek *peleia* dove, Latin *pallēre* to be pale — more at FALLOW] (1914)
: a horse that is pale cream to gold in color and has a flaxen or white mane and tail
pa·loo·ka \pə-'lü-kə\ *noun* [origin unknown] (1925)
1 : an inexperienced or incompetent boxer
2 : OAF, LOUT
pal·o·ver·de \ˌpa-lō-'ver-(ˌ)dā, ˌpä-, -'vər-\ *noun* [Mexican Spanish, literally, green tree] (1854)
1 : any of several small spiny leguminous trees or shrubs (genus *Cercidium*) that have greenish branches and are found chiefly in dry regions of the southwestern U.S. and Mexico
2 : JERUSALEM THORN
palp \'palp\ *noun* [New Latin *palpus*] (1842)
: PALPUS
pal·pa·ble \'pal-pə-bəl\ *adjective* [Middle English, from Late Latin *palpabilis,* from Latin *palpare* to stroke, caress — more at FEEL] (14th century)
1 : capable of being touched or felt **:** TANGIBLE
2 : easily perceptible **:** NOTICEABLE
3 : easily perceptible by the mind **:** MANIFEST
synonym see PERCEPTIBLE
— pal·pa·bil·i·ty \ˌpal-pə-'bi-lə-tē\ *noun*
— pal·pa·bly \'pal-pə-blē\ *adverb*
pal·pate \'pal-ˌpāt\ *transitive verb* **pal·pat·ed; pal·pat·ing** [probably back-formation from *palpation,* from Latin *palpation-, palpatio,* from *palpare*] (circa 1852)
: to examine by touch especially medically
— pal·pa·tion \pal-'pā-shən\ *noun*
pal·pe·bral \pal-'pē-brəl\ *adjective* [Late Latin *palpebralis,* from Latin *palpebra* eyelid; akin to Latin *palpare*] (1840)
: of, relating to, or located on or near the eyelids
pal·pi·tant \'pal-pə-tənt\ *adjective* (1837)
: marked by trembling or throbbing
pal·pi·tate \'pal-pə-ˌtāt\ *intransitive verb* **-tat·ed; -tat·ing** [Latin *palpitatus,* past participle of *palpitare,* frequentative of *palpare* to stroke] (circa 1623)
: to beat rapidly and strongly **:** THROB
— pal·pi·ta·tion \ˌpal-pə-'tā-shən\ *noun*
pal·pus \'pal-pəs\ *noun, plural* **pal·pi** \-ˌpī, -(ˌ)pē\ [New Latin, from Latin, caress, soft palm of the hand, from *palpare*] (1813)
: a segmented usually tactile or gustatory process on an arthropod mouthpart — see INSECT illustration
pals·grave \'pȯlz-ˌgrāv\ *noun* [Dutch *paltsgrave*] (1539)
: COUNT PALATINE 1a
pal·sied \'pȯl-zēd\ *adjective* (1550)
: affected with or as if with palsy
¹pal·sy \'pȯl-zē\ *noun, plural* **palsies** [Middle English *parlesie,* from Middle French *paralisie,* from Latin *paralysis*] (14th century)
1 : PARALYSIS
2 : a condition marked by uncontrollable tremor of the body or a part
²palsy *transitive verb* **pal·sied; pal·sy·ing** (1615)
: to affect with or as if with palsy
³palsy \'pal-zē\ *adjective* (1951)
: PALSY-WALSY
palsy–walsy \ˌpal-zē-'wal-zē\ *adjective* [reduplication of *palsy*] (1943)
slang **:** being or appearing to be very intimate
pal·ter \'pȯl-tər\ *intransitive verb* **pal·tered; pal·ter·ing** \-t(ə-)riŋ\ [origin unknown] (1601)
1 : to act insincerely or deceitfully **:** EQUIVOCATE
2 : HAGGLE, CHAFFER
synonym see LIE
— pal·ter·er \-tər-ər\ *noun*
pal·try \'pȯl-trē\ *adjective* **pal·tri·er; -est** [obsolete *paltry* trash, from dialect *palt, pelt* piece of coarse cloth, trash; akin to Middle Low German *palte* rag] (1570)
1 : INFERIOR, TRASHY
2 : MEAN, DESPICABLE
3 : TRIVIAL
4 : MEAGER, MEASLY ⟨made a *paltry* donation⟩
— pal·tri·ness *noun*
pa·lu·dal \pə-'lü-dᵊl, 'pal-yə-dᵊl\ *adjective* [Latin *palud-, palus* marsh; akin to Sanskrit *palvala* pond] (circa 1820)
: of or relating to marshes or fens **:** MARSHY
paly \'pā-lē\ *adjective* (1560)
archaic **:** somewhat pale **:** PALLID
pal·y·nol·o·gy \ˌpa-lə-'nä-lə-jē\ *noun* [Greek *palynein* to sprinkle, from *palē* fine meal] (1944)
: a branch of science dealing with pollen and spores
— pal·y·no·log·i·cal \-nə-'lä-ji-kəl\ *also* **pal·y·no·log·ic** \-jik\ *adjective*
— pal·y·no·log·i·cal·ly \-ji-k(ə-)lē\ *adverb*
— pal·y·nol·o·gist \-'nä-lə-jist\ *noun*
pam·pa \'pam-pə, 'päm-\ *noun, plural* **pampas** \-pəz, -pəs\ [American Spanish, from Quechua] (1704)
: an extensive generally grass-covered plain of temperate South America east of the Andes
: PRAIRIE
pam·pas grass \-pəz-, -pəs-\ *noun* (circa 1851)
: a South American grass (*Cortaderia selloana*) often grown for ornament that has showy white panicles borne on tall stems
pam·pe·an \'pam-pē-ən, 'päm-, pam-', päm-'\ *adjective* (1839)
: of or relating to the pampas of South America or their Indian inhabitants
pam·per \'pam-pər\ *transitive verb* **pam·pered; pam·per·ing** \-p(ə-)riŋ\ [Middle English, probably of Dutch origin; akin to Flemish *pamperen* to pamper] (14th century)
1 *archaic* **:** to cram with rich food **:** GLUT
2 a : to treat with extreme or excessive care and attention ⟨*pampered* their guests⟩ **b :** GRATIFY, HUMOR ⟨enabled him to *pamper* his wanderlust —*New Yorker*⟩
synonym see INDULGE
— pam·per·er \-pər-ər\ *noun*
pam·pe·ro \pam-'per-(ˌ)ō, päm-\ *noun, plural* **-ros** [American Spanish, from *pampa*] (1818)
: a strong cold wind from the west or southwest that sweeps over the pampas

pam·phlet \'pam(p)-flət\ *noun* [Middle English *pamflet* unbound booklet, from *Pamphilus seu De Amore* Pamphilus or On Love, popular Latin love poem of the 12th century] (14th century)
: an unbound printed publication with no cover or with a paper cover ◆

¹pam·phle·teer \,pam(p)-flə-'tir\ *noun* (1642)
: a writer of pamphlets attacking something or urging a cause

²pamphleteer *intransitive verb* (1698)
1 : to write and publish pamphlets
2 : to engage in partisan arguments indirectly in writings

¹pan \'pan\ *noun* [Middle English *panne,* from Old English (akin to Old High German *phanna* pan), from Latin *patina,* from Greek *patanē*] (before 12th century)
1 a : a usually broad, shallow, and open container for domestic use (as for warming, baking, or frying) **b :** any of various similar usually metal receptacles: as (1) **:** the hollow part of the lock in a firelock or flintlock gun that receives the priming (2) **:** either of the receptacles in a pair of scales (3) **:** a round shallow metal container for separating metal (as gold) from waste by washing
2 a (1) **:** a natural basin or depression in land (2) **:** a similar artificial basin (as for evaporating brine) **b :** a drifting fragment of the flat thin ice that forms in bays or along the shore
3 : HARDPAN 1
4 *slang* **:** FACE
5 : a harsh criticism

²pan *verb* **panned; pan·ning** (1839)
transitive verb
1 a : to wash in a pan for the purpose of separating heavy particles **b :** to separate (as gold) by panning **c :** to place in a pan
2 : to criticize severely ⟨the show was *panned*⟩
intransitive verb
1 : to wash material (as earth or gravel) in a pan in search of metal (as gold)
2 : to yield precious metal in the process of panning — usually used with *out*

³pan \'pän\ *noun* [Hindi *pān,* from Sanskrit *parṇa* wing, leaf — more at FERN] (1616)
1 : a betel leaf
2 : a masticatory of betel nut, mineral lime, and pan

⁴pan \'pan\ *noun* [short for *panorama*] (circa 1922)
1 : the process of panning a motion-picture or video camera
2 : a shot in which the camera is panned

⁵pan \'pan\ *verb* **panned; pan·ning** (1930)
transitive verb
: to rotate (as a motion-picture camera) so as to keep an object in the picture or secure a panoramic effect
intransitive verb
1 : to pan a motion-picture or video camera
2 *of a camera* **:** to undergo panning

Pan \'pan\ *noun* [Latin, from Greek]
: a Greek god of pastures, flocks, and shepherds usually represented as having the legs, horns, and ears of a goat

pan- *combining form* [Greek, from *pan,* neuter of *pant-, pas* all, every; akin to Tocharian B *pont-* all]
1 : all **:** completely ⟨*pan*chromatic⟩
2 a : involving all of a (specified) group ⟨*Pan*-American⟩ **b :** advocating or involving the union of a (specified) group ⟨*Pan*-Slavism⟩
3 : whole **:** general ⟨*pan*leukopenia⟩

pan·a·cea \,pa-nə-'sē-ə\ *noun* [Latin, from Greek *panakeia,* from *panakēs* all-healing, from *pan-* + *akos* remedy] (1548)
: a remedy for all ills or difficulties **:** CURE-ALL
— pan·a·ce·an \-'sē-ən\ *adjective*

pa·nache \pə-'nash, -'näsh\ *noun* [Middle French *pennache,* from Old Italian *pennacchio,* from Late Latin *pinnaculum* small wing — more at PINNACLE] (1553)
1 : an ornamental tuft (as of feathers) especially on a helmet
2 : dash or flamboyance in style and action **:** VERVE ◆

pa·na·da \pə-'nä-də\ *noun* [Spanish, from *pan* bread, from Latin *panis* — more at FOOD] (circa 1598)
: a paste of flour or bread crumbs and water or stock used as a base for sauce or a binder for forcemeat or stuffing

pan·a·ma \'pa-nə-,mä, -,mȯ\ *noun, often capitalized* [American Spanish *panamá,* from *Panama,* Central America] (1848)
: a lightweight hat of natural-colored straw hand-plaited of narrow strips from the young leaves of the jipijapa; *also* **:** a machine-made imitation of this

Panama Red *noun* (1967)
: marijuana of a reddish tint that is of Panamanian origin and is held to be very potent

Pan–Amer·i·can \,pa-nə-'mer-ə-kən\ *adjective* (1889)
: of, relating to, or involving the independent republics of North and South America

Pan–Amer·i·can·ism \-kə-,ni-zəm\ *noun* (1901)
: a movement for greater cooperation among the Pan-American nations

pan·a·tela \,pa-nə-'te-lə\ *noun* [Spanish, from American Spanish, a long thin biscuit, ultimately from Latin *panis* bread] (1847)
: a long slender straight-sided cigar

¹pan·cake \'pan-,kāk\ *noun* (14th century)
: a flat cake made of thin batter and cooked (as on a griddle) on both sides

²pancake *verb* **pan·caked; pan·cak·ing** (1911)
intransitive verb
: to make a pancake landing
transitive verb
: to cause to pancake

Pan–Cake \'pan-,kāk\ *trademark*
— used for a cosmetic in semimoist cake form

pancake landing *noun* (1928)
: a landing in which the airplane is stalled usually unintentionally above the landing surface causing it to drop abruptly in an approximately horizontal position with little forward motion

pan·cet·ta \(,)pan-'che-tə\ *noun* [Italian, from diminutive of *pancia* belly, paunch, from Latin *pantic-, pantex*] (1974)
: unsmoked bacon used especially in Italian cuisine

pan·chax \'pan-,kaks\ *noun* [New Latin] (1961)
: any of various small brilliantly colored Old World killifishes (genus *Aplocheilus*) often kept in the tropical aquarium

Pan·chen Lama \'pän-chən-\ *noun* [*Panchen* from Chinese (Beijing) *bānchán*] (1794)
: the lama who is the chief spiritual adviser of the Dalai Lama

pan·chro·mat·ic \,pan-krō-'ma-tik\ *adjective* [International Scientific Vocabulary] (1903)
: sensitive to light of all colors in the visible spectrum ⟨*panchromatic* film⟩

pan·cra·ti·um \pan-'krā-shē-əm\ *noun* [Latin, from Greek *pankration,* from *pan-* + *kratos* strength — more at HARD] (1603)
: an ancient Greek athletic contest involving both boxing and wrestling

pan·cre·as \'paŋ-krē-əs, 'pan-\ *noun* [New Latin, from Greek *pankreas* sweetbread, from *pan-* + *kreas* flesh, meat — more at RAW] (1578)
: a large lobulated gland of vertebrates that secretes digestive enzymes and the hormones insulin and glucagon
— pan·cre·at·ic \,paŋ-krē-'a-tik, ,pan-\ *adjective*

pancreat- *combining form* [New Latin, from Greek *pankreat-, pancreas*]
: pancreas ⟨*pancreat*ic⟩

pan·cre·a·tec·to·my \,paŋ-krē-ə-'tek-tə-mē, ,pan-\ *noun* (circa 1900)
: surgical removal of all or part of the pancreas
— pan·cre·a·tec·to·mized \-,mīzd\ *adjective*

pancreatic duct *noun* (circa 1860)
: a duct leading from the pancreas and opening into the duodenum

pancreatic juice *noun* (circa 1666)
: a clear alkaline secretion of pancreatic enzymes (as trypsin and lipase) that flows into the duodenum and acts on food already acted on by the gastric juice and saliva

pan·cre·a·tin \pan-'krē-ə-tən; 'paŋ-krē-, 'pan-\ *noun* (circa 1860)
: a mixture of enzymes from the pancreatic juice; *also* **:** a preparation containing such a mixture

pan·cre·a·ti·tis \,paŋ-krē-ə-'tī-təs, ,pan-\ *noun, plural* **-tit·i·des** \-'ti-tə-,dēz\ [New Latin] (circa 1842)
: inflammation of the pancreas

pan·cre·o·zy·min \-krē-ō-'zī-mən\ *noun* [*pancreas* + *-o-* + *zym-* + ¹*-in*] (1943)
: CHOLECYSTOKININ

pan·cy·to·pe·nia \,pan-,sī-tə-'pē-nē-ə\ *noun* [New Latin, from *pan-* + *cyt-* + *-penia*] (circa 1941)

◇ WORD HISTORY

pamphlet *Pamphilus, seu De Amore* ("Pamphilus, or About Love"), written in the late 12th century by an author now unknown, was a Latin poem describing a series of amusing amorous adventures. In its day *Pamphilus* enjoyed widespread popularity among those literate in Latin, especially university students. In the late Middle Ages, short compilations of popular classical authors were often given French diminutives based on the vernacular form of the author's name. Hence *Esopet* was a familiar title for the works of *Esope* (Aesop) and *Avionet* for Avianus (a Roman fabulist of around 400 A.D.). Similarly, *Pamphilus* became *Pamphilet* in Old French. Borrowed into Anglo-Latin as *panfletus* or *pamfletus* and into later Middle English as *pamfilet* or *pamflet,* the word denoted any short written work. In the 17th century *pamphlet* became associated particularly with polemics on controversial social, political, or religious issues, and it was borrowed back into French in the sense "diatribe."

panache *Pinnaculum,* a derivative of Latin *pinna* ("feather," "wing," or "battlement"), was a Late Latin word meaning "small wing" or "gable." The Italian outcome of *pinnaculum* was *pennachio,* which alluded to a plume of feathers decorating a warrior's helmet. This word was borrowed into 16th century French as *pennache* and later altered to *panache.* The figurative sense of French *panache* developed from the verve and swagger associated with anyone bold enough to wear such a distinctive mark as a colorful plume both in battle and among the ranks of polite society. This meaning of the French word apparently first appeared in English at the end of the 19th century in translations of Edmond Rostand's popular play *Cyrano de Bergerac*—Cyrano himself being the perfect example of panache. Another more straightforward development from Late Latin *pinnaculum* is English *pinnacle.*

: an abnormal reduction in the number of erythrocytes, white blood cells, and blood platelets in the blood; *also* **:** a disorder (as aplastic anemia) characterized by such a reduction

pan•da \'pan-də\ *noun* [French, perhaps from a language of the southeast Himalayas] (1835)
1 : a long-tailed largely herbivorous mammal (*Ailurus fulgens*) that is related to and resembles the American raccoon, has long rusty or chestnut fur, and is found from the Himalayas to China
2 : GIANT PANDA

***A* panda 1, *B* giant panda**

panda car *noun* [from its black and white coloration] (1967)
British **:** a police patrol car

pan•da•nus \pan-'dā-nəs, -'da-\ *noun, plural* **-ni** \-nī\ [New Latin, genus name, from Malay *pandan* screw pine] (1830)
: SCREW PINE; *also* **:** a fiber made from screw-pine leaves and used for woven products (as mats)

Pan•da•rus \'pan-də-rəs\ *noun* [Latin, from Greek *Pandaros*]
: a Lycian archer in the Trojan War who in medieval legend procures Cressida for Troilus

pan•dect \'pan-ˌdekt\ *noun* [Late Latin *Pandectae,* the Pandects, digest of Roman civil law (6th century A.D.), from Latin, plural of *pandectes* encyclopedic work, from Greek *pandektēs* all-receiving, from *pan-* + *dechesthai* to receive; akin to Greek *dokein* to seem, seem good — more at DECENT] (1533)
1 : a complete code of the laws of a country or system of law
2 : a treatise covering an entire subject

¹pan•dem•ic \pan-'de-mik\ *adjective* [Late Latin *pandemus,* from Greek *pandēmos* of all the people, from *pan-* + *dēmos* people — more at DEMAGOGUE] (1666)
: occurring over a wide geographic area and affecting an exceptionally high proportion of the population ⟨*pandemic* malaria⟩

²pandemic *noun* (circa 1853)
: a pandemic outbreak of a disease

Pan•de•mo•ni•um \ˌpan-də-'mō-nē-əm\ *noun* [New Latin, from Greek *pan-* + *daimōn* evil spirit — more at DEMON]
1 : the capital of Hell in Milton's *Paradise Lost*
2 : the infernal regions **:** HELL
3 *not capitalized* **:** a wild uproar **:** TUMULT

¹pan•der \'pan-dər\ *noun* [Middle English *Pandare* Pandarus, from Latin *Pandarus*] (1530)
1 a : a go-between in love intrigues **b :** PIMP
2 : someone who caters to or exploits the weaknesses of others ◇

²pander *intransitive verb* **pan•dered; pan•der•ing** \-d(ə-)riŋ\ (1602)
: to act as a pander; *especially* **:** to provide gratification for others' desires ⟨films that *pander* to the basest emotions⟩
— pan•der•er \-dər-ər\ *noun*

pan•dit \'pan-dət, 'pən-\ *noun* [Hindi *paṇḍit,* from Sanskrit *paṇḍita*] (circa 1828)
: a wise or learned man in India — often used as an honorary title

pan•do•ra \pan-'dōr-ə, -'dȯr-\ *noun* [Italian, from Late Latin *pandura* 3-stringed lute, from Greek *pandoura*] (1597)
: BANDORE

Pan•do•ra's box \pan-'dōr-əz-, -'dȯr-\ *noun* [from the box, sent by the gods to Pandora, which she was forbidden to open and which loosed a swarm of evils upon mankind when she opened it out of curiosity] (1579)
: a prolific source of troubles

pan•dow•dy \pan-'dau̇-dē\ *noun, plural* **-dies** [origin unknown] (1805)
: a deep-dish spiced apple dessert sweetened with sugar, molasses, or maple syrup and covered with a rich crust

pan•dy \'pan-dē\ *transitive verb* **pan•died; pan•dy•ing** [probably from Latin *pande,* imperative singular of *pandere* to spread out (the hand); command of the schoolmaster to the boy — more at FATHOM] (1863)
British **:** to punish (a schoolboy) with a blow on the palm of the hand especially with a ferule

pane \'pān\ *noun* [Middle English *pan, pane* strip of cloth, pane, from Middle French *pan,* from Latin *pannus* cloth, rag — more at VANE] (14th century)
: a piece, section, or side of something: as **a :** a framed sheet of glass in a window or door **b :** one of the sections into which a sheet of postage stamps is cut for distribution
— paned \'pānd\ *adjective*

pan•e•gy•ric \ˌpa-nə-'jir-ik, -'jī-rik\ *noun* [Latin *panegyricus,* from Greek *panēgyrikos,* from *panēgyrikos* of or for a festival assembly, from *panēgyris* festival assembly, from *pan-* + *agyris* assembly; akin to Greek *ageirein* to gather] (1603)
: a eulogistic oration or writing; *also* **:** formal or elaborate praise
synonym see ENCOMIUM
— pan•e•gy•ri•cal \-'jir-i-kəl, -'jī-ri-\ *adjective*
— pan•e•gy•ri•cal•ly \-k(ə-)lē\ *adverb*

pan•e•gy•rist \ˌpa-nə-'jir-ist, -'jī-rist\ *noun* (1605)
: EULOGIST

¹pan•el \'pa-nᵊl\ *noun* [Middle English, piece of cloth, slip of parchment, jury schedule, from Middle French, piece of cloth, piece, from (assumed) Vulgar Latin *pannellus,* diminutive of Latin *pannus*] (14th century)
1 a (1) **:** a schedule containing names of persons summoned as jurors (2) **:** the group of persons so summoned (3) **:** JURY 1 **b** (1) **:** a group of persons selected for some service (as investigation or arbitration) ⟨a *panel* of experts⟩ (2) **:** a group of persons who discuss before an audience a topic of public interest; *also* **:** PANEL DISCUSSION (3) **:** a group of entertainers or guests engaged as players in a quiz or guessing game on a radio or television program
2 : a separate or distinct part of a surface: as **a :** a fence section **:** HURDLE **b** (1) **:** a thin usually rectangular board set in a frame (as in a door) (2) **:** a usually sunken or raised section of a surface set off by a margin (3) **:** a flat usually rectangular piece of construction material (as plywood or precast masonry) made to form part of a surface **c :** a vertical section of fabric (as a gore) **d :** COMIC STRIP; *also* **:** a frame of a comic strip **e :** any of several units of construction of an airplane wing surface
3 : a thin flat piece of wood on which a picture is painted; *also* **:** a painting on such a surface
4 a : a section of a switchboard **b :** a flat often insulating support (as for computer hardware or parts of an electrical device) usually with control handles on one face **c :** a usually vertical mount for controls or dials (as of instruments of measurement)

²panel *transitive verb* **-eled** *or* **-elled; -el•ing** *or* **-el•ling** (15th century)
: to furnish or decorate with panels ⟨*paneled* the living room⟩

panel discussion *noun* (circa 1938)
: a formal discussion by a panel

paneling *also* **panelling** *noun* (1824)
: panels joined in a continuous surface; *especially* **:** decorative wood panels so joined

pan•el•ist \'pa-nᵊl-ist\ *noun* (1951)
: a member of a discussion or advisory panel or of a radio or television panel

panel truck *noun* (1937)
: a small motortruck with a fully enclosed body

pan•e•tela *variant of* PANATELA

pan•et•to•ne \ˌpä-nə-'tō-nē, ˌpa-\ *noun* [Italian, from *panetto* small loaf, diminutive of *pane* bread, from Latin *panis* — more at FOOD] (1922)
: a usually yeast-leavened bread containing raisins and candied fruit

pan•fish \'pan-ˌfish\ *noun* (1805)
: a small food fish (as a sunfish) usually taken with hook and line and not available on the market

pan•fry \'pan-ˌfrī, pan-'frī\ *transitive verb* (circa 1929)
: to cook in a frying pan with a small amount of fat

pan•ful \'pan-ˌfu̇l\ *noun* (circa 1864)
: as much or as many as a pan will hold

¹pang \'paŋ\ *noun* [origin unknown] (15th century)
1 : a brief piercing spasm of pain
2 : a sharp attack of mental anguish ⟨*pangs* of remorse⟩ ⟨a *pang* of jealousy struck me —Graham Greene⟩

²pang *transitive verb* (1502)
: to cause to have pangs **:** TORMENT

pan•ga \'päŋ-gə\ *noun* [Swahili] (1925)
: MACHETE

pan•gen•e•sis \ˌpan-'je-nə-səs\ *noun* [New Latin] (1868)
: a disproven hypothetical mechanism of heredity in which the cells throw off particles that collect in the reproductive products or in buds so that the egg or bud contains particles from all parts of the parent
— pan•ge•net•ic \-jə-'ne-tik\ *adjective*

Pan•gloss•ian \pan-'glä-sē-ən, paŋ-, -'glȯ-\ *adjective* [*Pangloss,* optimistic tutor in Voltaire's *Candide* (1759)] (1831)

◇ WORD HISTORY

pander In Greek legends of the Trojan War Pandarus was a Lycian archer who fought with the Trojans. He is remembered as the one who broke the truce between the Greeks and the Trojans by shooting an arrow at Menelaus, the king from Sparta. In medieval retellings of the story of Troy Pandarus becomes a player in the ill-fated love affair between Troilus, a Trojan prince, and Cressida, the daughter of a Greek priest. Pandarus acts as a benevolent go-between for the two lovers. Later, in Shakespeare's version of the story, Pandarus is portrayed in the less honorable role of a man procuring sexual favors for another. The meaning of *pander* in English mirrors the changing characterization of Pandarus. In the early 16th century *pander* referred to one who acts as a go-between for lovers; later the word took on the pejorative meaning "pimp" or "procurer." The use of *pander* as a verb meaning "to provide gratification for someone's baser desires" is attested by the early 17th century.

: marked by the view that all is for the best in this best of possible worlds **:** excessively optimistic

pan·go·la grass \pan-ˈgō-lə-, paŋ-\ *noun* [alteration of *Pongola grass,* from the *Pongola* River, South Africa] (1948)
: a rapid-growing perennial grass (*Digitaria decumbens*) of southern Africa that has been introduced into the southern U.S. as a pasture grass

pan·go·lin \ˈpaŋ-gə-lən, ˈpan-\ *noun* [Malay dialect *pĕngguling*] (1774)
: any of a family (Manidae) of Asian and African edentate mammals having the body covered with large imbricated horny scales

pangolin

[1]pan·han·dle \ˈpan-ˌhan-d^{ə}l\ *noun* (1856)
: a narrow projection of a larger territory (as a state) ⟨the Texas *panhandle*⟩

[2]panhandle *verb* **pan·han·dled; pan·han·dling** \-ˌhan(d)-liŋ, -ˌhan-d^{ə}l-iŋ\ [back-formation from *panhandler,* probably from *panhandle,* noun; from the extended forearm] (1903)
intransitive verb
: to stop people on the street and ask for food or money **:** BEG
transitive verb
1 : to accost on the street and beg from
2 : to get by panhandling
— pan·han·dler \-ˌhan(d)-lər, -ˌhan-d^{ə}l-ər\ *noun*

Pan·hel·len·ic \ˌpan-hə-ˈle-nik\ *adjective* (1847)
1 : of or relating to all Greece or all the Greeks
2 : of or relating to the Greek-letter sororities or fraternities in American colleges and universities or to an association representing them

pan·hu·man \ˌpan-ˈhyü-mən, -ˈyü-\ *adjective* (circa 1909)
: of or relating to all humanity ⟨the *panhuman* problem of evil — R. K. Merton⟩

[1]pan·ic \ˈpa-nik\ *adjective* [French *panique,* from Greek *panikos,* literally, of Pan, from *Pan*] (1603)
1 : of, relating to, or resembling the mental or emotional state believed induced by the god Pan ⟨*panic* fear⟩
2 : of, relating to, or arising from a panic ⟨a wave of *panic* buying⟩
3 : of or relating to the god Pan

[2]panic *noun* (1708)
1 : a sudden overpowering fright; *especially* **:** a sudden unreasoning terror often accompanied by mass flight
2 : a sudden widespread fright concerning financial affairs that results in a depression of values caused by extreme measures for protection of property (as securities)
3 *slang* **:** one that is very funny ◆
synonym see FEAR
— pan·icky \ˈpa-ni-kē\ *adjective*

[3]panic *verb* **pan·icked** \-nikt\; **pan·ick·ing** (1827)
transitive verb
1 : to affect with panic
2 : to produce demonstrative appreciation on the part of ⟨*panic* an audience with a gag⟩
intransitive verb
: to be affected with panic

panic button *noun* (circa 1950)
: something setting off a precipitous emergency response ⟨there was no pushing of *panic buttons* at the White House, no rushing of troops —J. C. Harsch⟩

pan·ic grass \ˈpa-nik-\ *noun* [Middle English *panik,* from Middle French or Latin; Middle French *panic* foxtail millet, from Latin *panicum,* from *panus* stalk of a panicle] (1597)
: any of various grasses (*Panicum* and related genera) including some important forage and cereal grasses

pan·i·cle \ˈpa-ni-kəl\ *noun* [Latin *panicula,* diminutive of *panus*] (1597)
1 : a compound racemose inflorescence — see INFLORESCENCE illustration
2 : a pyramidal loosely branched flower cluster
— pan·i·cled \-kəld\ *adjective*
— pa·nic·u·late \pa-ˈni-kyə-lət, pə-\ *adjective*

pan·ic–strick·en \ˈpa-nik-ˌstri-kən\ *adjective* (1804)
: overcome with panic

pan·i·cum \ˈpa-ni-kəm\ *noun* [New Latin, from Latin, foxtail millet] (circa 1864)
: any of a large and widely distributed genus (*Panicum*) of grasses that have a very diverse habit and 1- to 2-flowered spikelets arranged in a panicle

Pan·ja·bi \ˌpən-ˈjä-bē, -ˈja-\ *noun* [Hindi *pañjābī,* from *pañjābī* of Punjab] (1854)
1 : an Indo-Aryan language of the Punjab
2 : PUNJABI 2

pan·jan·drum \pan-ˈjan-drəm\ *noun, plural* **-drums** *also* **-dra** \-drə\ [Grand *Panjandrum,* burlesque title of an imaginary personage in some nonsense lines by Samuel Foote] (1755)
: a powerful personage or pretentious official

pan·leu·ko·pe·nia \ˌpan-ˌlü-kə-ˈpē-nē-ə\ *noun* [New Latin] (1940)
: an acute usually fatal viral epizootic disease especially of cats characterized by fever, diarrhea and dehydration, and extensive destruction of white blood cells

pan·mic·tic \ˌpan-ˈmik-tik\ *adjective* [*pan-* + Greek *miktos,* verbal of *mignynai* to mix] (1943)
: of, relating to, or exhibiting panmixia

pan·mix·ia \-ˈmik-sē-ə\ *noun* [New Latin, from *pan-* + Greek *mixis* act of mingling, mating, from *mignynai* to mix — more at MIX] (circa 1889)
: random mating within a breeding population

panne \ˈpan\ *noun* [French, from Old French *penne, panne* fur used for lining, from Latin *pinna* feather, wing — more at PEN] (circa 1794)
1 : a silk or rayon velvet with lustrous pile flattened in one direction — called also *panne velvet*
2 : a heavy silk or rayon satin with high luster and waxy smoothness

pan·nier *also* **pan·ier** \ˈpan-yər, ˈpa-nē-ər\ *noun* [Middle English *panier,* from Middle French, from Latin *panarium,* from *panis* bread — more at FOOD] (13th century)
1 : a large container: as **a :** a basket often carried on the back of an animal or the shoulders of a person **b :** one of a pair of packs or baskets hung over the rear wheel of a vehicle (as a bicycle)
2 a : one of a pair of hoops formerly used to expand women's skirts at the sides **b :** an overskirt draped at the sides of a skirt for an effect of fullness

pannier 2b

pan·ni·kin \ˈpa-ni-kən\ *noun* [[1]*pan* + *-nikin* (as in *cannikin*)] (1823)
British **:** a small pan or cup

pa·no·cha \pə-ˈnō-chə\ *or* **pa·no·che** \-chē\ *variant of* PENUCHE

pan·o·plied \ˈpa-nə-plēd\ *adjective* (1832)
: dressed in or having a panoply

pan·op·tic \ˌpa-ˈnäp-tik\ *adjective* [Greek *panoptēs* all-seeing, from *pan-* + *opsesthai* to be going to see — more at OPTIC] (1826)
: being or presenting a comprehensive or panoramic view

pan·o·ply \ˈpa-nə-plē\ *noun, plural* **-plies** [Greek *panoplia,* from *pan-* + *hopla* arms, armor, plural of *hoplon* tool, weapon — more at HOPLITE] (1632)
1 a : a full suit of armor **b :** ceremonial attire
2 : something forming a protective covering
3 a : a magnificent or impressive array ⟨the full *panoply* of a military funeral⟩ **b :** a display of all appropriate appurtenances ⟨has the *panoply* of science fiction . . . but it is not true science fiction —Isaac Asimov⟩

pan·o·rama \ˌpa-nə-ˈra-mə, -ˈrä-\ *noun* [*pan-* + Greek *horama* sight, from *horan* to see — more at WARY] (1796)
1 a : CYCLORAMA 1 **b :** a picture exhibited a part at a time by being unrolled before the spectator
2 a : an unobstructed or complete view of an area in every direction **b :** a comprehensive presentation of a subject ⟨a *panorama* of American history⟩ **c :** RANGE
3 : a mental picture of a series of images or events
— pan·o·ram·ic \-ˈra-mik\ *adjective*
— pan·o·ram·i·cal·ly \-mi-k(ə-)lē\ *adverb*

pan out *intransitive verb* [[2]*pan*] (circa 1868)
: TURN OUT; *especially* **:** SUCCEED ⟨the signs revealed that the experiment wasn't *panning out* —Ronald Reagan⟩

pan·pipe \ˈpan-ˌpīp\ *noun* [*Pan,* its traditional inventor] (1820)
: a primitive wind instrument consisting of a series of short vertical pipes of graduated length bound together with the mouthpieces in an even row — often used in plural

panpipe

pan·sex·u·al \ˌpan-ˈsek-sh(ə-)wəl, -shəl\ *adjective* (1926)
: exhibiting or implying many forms of sexual expression
— pan·sex·u·al·i·ty \ˌpan-ˌsek-shə-ˈwa-lə-tē\ *noun*

Pan–Slav·ism \ˌpan-ˈslä-ˌvi-zəm, -ˈsla-\ *noun* (1850)
: a political and cultural movement originally emphasizing the cultural ties between the Slavic peoples but later associated with Russian expansionism
— Pan–Slav·ic \-ˈsla-vik, -ˈslä-\ *adjective*
— Pan–Slav·ist \-ˈslä-vist, -ˈsla-\ *noun*

[1]pan·sy \ˈpan-zē\ *noun, plural* **pansies** [Middle English *pensee,* from Middle French *pensée,* from *pensée* thought, from feminine of *pensé,* past participle of *penser* to think, from Latin *pensare* to ponder — more at PENSIVE] (15th century)
1 : a garden plant (*Viola wittrockiana*) derived chiefly from the wild pansy (*Viola tricolor*) of Europe by hybridizing the latter with other wild violets; *also* **:** its flower

◇ WORD HISTORY

panic In ancient Greek religion Pan was the god of woodlands and pastures, flocks and shepherds. Although generally conceived as a pipe-playing, nymph-chasing pastoral deity, Pan was also recognized for his ability to induce terror. When accidentally awakened from his noontime slumber, he could give a great shout that would cause flocks to stampede. Athens became a seat of worship for Pan because its citizens believed that he had aided the Athenians against the Persians at the battle of Marathon. From this more formidable aspect of his nature later Greek authors derived the word *panikon* "groundless fear," the ultimate source of English *panic.*

2 a : an effeminate man or boy — usually used disparagingly **b :** a male homosexual — usually used disparagingly

²pansy *adjective* (1929)
: EFFEMINATE 1; *also* **:** HOMOSEXUAL — usually used disparagingly

¹pant \'pant\ *verb* [Middle English, modification of Middle French *pantaisier,* from (assumed) Vulgar Latin *phantasiare* to have hallucinations, from Greek *phantasioun,* from *phantasia* appearance, imagination — more at FANCY] (14th century)
intransitive verb
1 a : to breathe quickly, spasmodically, or in a labored manner **b :** to run panting ⟨*panting* along beside the bicycle⟩ **c :** to move with or make a throbbing or puffing sound
2 : to long eagerly **:** YEARN
3 : THROB, PULSATE
transitive verb
: to utter with panting **:** GASP

²pant *noun* (1513)
1 a : a panting breath **b :** the visible movement of the chest accompanying such a breath
2 : a throbbing or puffing sound

³pant *noun* [short for *pantaloons*] (1840)
1 : an outer garment covering each leg separately and usually extending from the waist to the ankle — usually used in plural
2 *plural, chiefly British* **:** men's underpants
3 *plural* **:** PANTIE
word history see PANTALOON
— with one's pants down : in an embarrassing position (as of being unprepared to act)

⁴pant *adjective* (1899)
: of or relating to pants ⟨a *pant* leg⟩

pant- *or* **panto-** *combining form* [Middle French, from Latin, from Greek, from *pant-, pas* — more at PAN-]
: all ⟨*pant*isocracy⟩

Pan·ta·gru·el \ˌpan-tə-'grü-əl; pan-'ta-grə-wəl, -ˌwel\ *noun* [French]
: the huge son of Gargantua in Rabelais's *Pantagruel*
— Pan·ta·gru·el·ian \ˌpan-tə-grü-'e-lē-ən, (ˌ)pan-ˌta-grə-'we-\ *adjective*
— Pan·ta·gru·el·ism \ˌpan-tə-'grü-ə-ˌli-zəm; pan-'ta-grə-wə-ˌli-zəm, -ˌwe-\ *noun*
— Pan·ta·gru·el·ist \-list\ *noun*

pan·ta·lets *or* **pan·ta·lettes** \ˌpan-tə-'lets\ *noun plural* [*pantaloons*] (1834)
: long drawers with a ruffle at the bottom of each leg worn especially by women and children in the first half of the 19th century

pan·ta·loon \ˌpan-tə-'lün\ *noun* [Middle French & Old Italian; Middle French *Pantalon,* from Old Italian *Pantaleone, Pantalone*] (circa 1590)
1 a *or* **pan·ta·lo·ne** \-tə-'lō-nē\ *capitalized* **:** a character in the commedia dell'arte that is usually a skinny old dotard who wears spectacles, slippers, and a tight-fitting combination of trousers and stockings **b :** a buffoon in pantomimes
2 *plural* **a :** wide breeches worn especially in England during the reign of Charles II **b :** close-fitting trousers usually having straps passing under the instep and worn especially in the 19th century
3 : loose-fitting usually shorter than ankle-length trousers ◆

pant·dress \'pant-ˌdres\ *noun* (1964)
: a dress having a divided skirt

pan·tech·ni·con \pan-'tek-ni-kən\ *noun* [short for *pantechnicon van,* from *pantechnicon* storage warehouse] (1891)
British **:** ³VAN 1

pan·the·ism \'pan(t)-thē-ˌi-zəm\ *noun* [French *panthéisme,* from *panthéiste* pantheist, from English *pantheist,* from *pan-* + Greek *theos* god] (1732)
1 : a doctrine that equates God with the forces and laws of the universe
2 : the worship of all gods of different creeds, cults, or peoples indifferently; *also* **:** toleration of worship of all gods (as at certain periods of the Roman empire)
— pan·the·ist \-thē-ist\ *noun*
— pan·the·is·tic \ˌpan(t)-thē-'is-tik\ *also* **pan·the·is·ti·cal** \-ti-kəl\ *adjective*
— pan·the·is·ti·cal·ly \-ti-k(ə-)lē\ *adverb*

pan·the·on \'pan(t)-thē-ˌän, -ən\ *noun* [Middle English *Panteon,* a temple at Rome, from Latin *Pantheon,* from Greek *pantheion* temple of all the gods, from neuter of *pantheios* of all gods, from *pan-* + *theos* god] (14th century)
1 : a temple dedicated to all the gods
2 : a building serving as the burial place of or containing memorials to the famous dead of a nation
3 : the gods of a people; *especially* **:** the officially recognized gods
4 : a group of illustrious persons

pan·ther \'pan(t)-thər\ *noun, plural* **panthers** *also* **panther** [Middle English *pantere,* from Old French, from Latin *panthera,* from Greek *panthēr*] (13th century)
1 : LEOPARD: as **a :** a leopard of a hypothetical exceptionally large fierce variety **b :** a leopard of the black color phase
2 : COUGAR
3 : JAGUAR

pant·ie *or* **panty** \'pan-tē\ *noun, plural* **panties** [³*pant*] (1908)
: a woman's or child's undergarment covering the lower trunk and made with closed crotch — usually used in plural
word history see PANTALOON

pantie girdle *noun* (1941)
: a woman's girdle having a sewed-in or detachable crotch and made with or without garters and bones

pan·tile \'pan-ˌtīl\ *noun* [¹*pan*] (1640)
1 : a roofing tile whose cross section is an ogee curve
2 : a roofing tile of which the cross section is an arc of a circle and which is laid with alternate convex and concave surfaces uppermost
— pan·tiled \-ˌtīld\ *adjective*

pant·i·soc·ra·cy \ˌpan-tə-'sä-krə-sē, ˌpan-ˌtī-\ *noun, plural* **-cies** [*pant-* + *isocracy* equal rule, from Greek *isokratia,* from *is-* + *-kratia* -cracy] (1794)
: a utopian community in which all rule equally
— pant·i·so·crat·ic \ˌpan-ˌtī-sə-'kra-tik\ *or* **pant·i·so·crat·i·cal** \-'kra-ti-kəl\ *adjective*
— pant·i·soc·ra·tist \ˌpan-tə-'sä-krə-tist, ˌpan-ˌtī-\ *noun*

pan·to \'pan-(ˌ)tō\ *noun, plural* **pantos** (1852)
British **:** PANTOMIME 2c

pan·to·fle \pan-'tō-fəl, -'tä-, -'tü-; 'pan-tə-fəl\ *noun* [Middle English *pantufle,* from Middle French *pantoufle*] (15th century)
: SLIPPER

pan·to·graph \'pan-tə-ˌgraf\ *noun* [French *pantographe,* from *pant-* + *-graphe* -graph] (1723)
1 : an instrument for copying (as a map) on a predetermined scale consisting of four light rigid bars jointed in parallelogram form; *also* **:** any of various extensible devices of similar construction (as for use as brackets or gates)

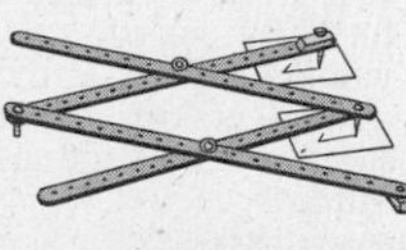

pantograph 1

2 : an electrical trolley carried by a collapsible and adjustable frame
— pan·to·graph·ic \ˌpan-tə-'gra-fik\ *adjective*

¹pan·to·mime \'pan-tə-ˌmīm\ *noun* [Latin *pantomimus,* from *pant-* + *mimus* mime] (1589)
1 : PANTOMIMIST
2 a : an ancient Roman dramatic performance featuring a solo dancer and a narrative chorus **b :** any of various dramatic or dancing performances in which a story is told by expressive bodily or facial movements of the performers **c :** a British theatrical entertainment of the Christmas season based on a nursery tale and featuring topical songs, tableaux, and dances
3 a : conveyance of a story by bodily or facial movements especially in drama or dance **b :** the art or genre of conveying a story by bodily movements only
— pan·to·mim·ic \ˌpan-tə-'mi-mik\ *adjective*

²pantomime *verb* **-mimed; -mim·ing** (1768)
intransitive verb
: to engage in pantomime
transitive verb
: to represent by pantomime

pan·to·mim·ist \'pan-tə-ˌmī-mist, -ˌmi-\ *noun* (1838)
1 : an actor or dancer in pantomimes
2 : a composer of pantomimes

pan·to·the·nate \ˌpan-tə-'the-ˌnāt, pan-'tä-thə-ˌnāt\ *noun* (circa 1934)
: a salt or ester of pantothenic acid

pan·to·then·ic acid \ˌpan-tə-'the-nik-, -'thē-\ *noun* [Greek *pantothen* from all sides, from *pant-, pas* all — more at PAN-] (1933)
: a viscous oily acid $C_9H_{17}NO_5$ of the vitamin B complex found in all living tissues

pan·trop·i·cal \ˌpan-'trä-pi-kəl\ *also* **pan·trop·ic** \-pik\ *adjective* (1936)
: occurring or distributed throughout the tropical regions of the earth

pan·try \'pan-trē\ *noun, plural* **pantries** [Middle English *panetrie,* from Middle French *paneterie,* from Old French, from *panetier* servant in charge of the pantry, from *pan* bread, from Latin *panis* — more at FOOD] (14th century)
1 : a room or closet used for storing (as provisions) or from which food is brought to the table
2 : a room (as in a hotel or hospital) for preparation of foods on order

pan·try·man \-mən\ *noun* (circa 1567)
: a man in charge of or working in a pantry (as in a hotel or hospital)

pants suit *noun* (1964)
: PANTSUIT

pant·suit \'pant-ˌsüt\ *noun* (1964)
: a woman's ensemble consisting usually of a long jacket and tailored pants of the same material
— pant·suit·ed *adjective*

panty hose *noun* (1963)

◇ WORD HISTORY

pantaloon A stock figure in the commedia dell'arte, a form of Italian comic theater popular throughout Europe from about the 16th to the mid-18th century, was Pantalone, an avaricious, lecherous, and scheming old man who often ended up being duped and humiliated. His costume consisted of a soft brimless hat; a pleated black cassock; slippers; and a vest, breeches, and stockings that were conspicuously red and tight-fitting. In later representations of the character, the breeches and stockings were replaced by long trousers. When trousers of a similar style became popular during the Restoration in England, they became known as *pantaloons, Pantaloon* being an anglicization of *Pantalone.* Even as fashions changed over the years, *pantaloons* remained the name for various types of trousers. The clipped form *pants,* an Americanism first current around 1840, became a standard word for the garment, serving as the basis for new formations (denoting new garments), such as *underpants* and *panties.*

: a one-piece undergarment for women that consists of hosiery made with a panty-style top — usually plural in construction

panty raid *noun* (circa 1952)
: a raid on a women's dormitory by college men usually to obtain panties as trophies

panty·waist \'pan-tē-ˌwāst\ *noun* (circa 1936)
: SISSY
— pantywaist *adjective*

Pan·urge \'pa-ˌnərj, pa-'nu̇rzh\ *noun* [French]
: a witty rascal and companion of Pantagruel in Rabelais's *Pantagruel*

pan·zer \'pan-zər, 'pän(t)-sər\ *noun* [German *Panzer* tank, armor, coat of mail, from Middle High German *panzier,* from Old French *pancière,* from *pance, panche* belly — more at PAUNCH] (circa 1939)
: TANK 3; *specifically* **:** a German tank of World War II

panzer division *noun* (circa 1939)
: a German armored division

[1]pap \'pap\ *noun* [Middle English *pappe;* of imitative origin] (13th century)
1 *chiefly dialect* **:** NIPPLE, TEAT
2 : something shaped like a nipple

[2]pap *noun* [Middle English] (15th century)
1 : a soft food for infants or invalids
2 : political patronage
3 : something lacking solid value or substance

pa·pa \'pä-pə, *chiefly British* pə-'pä, -'pä\ *noun* [French (baby talk)] (1681)
: FATHER

Pa·pa \pə-'pä, 'pä-pə\ (1952)
— a communications code word for the letter *p*

pa·pa·cy \'pā-pə-sē\ *noun, plural* **-cies** [Middle English *papacie,* from Medieval Latin *papatia,* from Late Latin *papa* pope — more at POPE] (14th century)
1 : the office of pope
2 : a succession or line of popes
3 : the term of a pope's reign
4 *capitalized* **:** the system of government of the Roman Catholic Church of which the pope is the supreme head

Pa·pa·go \'pä-pə-ˌgō, 'pa-\ *noun, plural* **Papago** *or* **Papagos** (1839)
1 : a member of an American Indian people of southwestern Arizona and northwestern Mexico
2 : the Uto-Aztecan language of the Papago people

pa·pa·in \pə-'pā-ən, -'pī-ən\ *noun* [International Scientific Vocabulary, from *papaya*] (circa 1890)
: a protease in the juice of unripe papaya that is used especially as a tenderizer for meat and in medicine

pa·pal \'pā-pəl\ *adjective* [Middle English, from Middle French, from Medieval Latin *papalis,* from Late Latin *papa*] (14th century)
: of or relating to a pope or to the Roman Catholic Church; *also* **:** resembling a pope or that of a pope
— pa·pal·ly \-pə-lē\ *adverb*

papal cross *noun* (circa 1889)
: a figure of a cross having a long upright shaft and three crossbars with the longest at or somewhat above its middle and the two other successively shorter crossbars above the longest one — see CROSS illustration

papal infallibility *noun* (1870)
: the Roman Catholic doctrine that the pope cannot err when speaking ex cathedra in defining a doctrine of Christian faith or morals

Pa·pa·ni·co·laou smear \ˌpä-pə-'nē-kə-ˌlau̇-, ˌpa-pə-'ni-\ *noun* [George N. *Papanicolaou* (died 1962) American medical scientist] (1950)
: PAP SMEAR

Papanicolaou test *noun* (1946)
: PAP SMEAR

pa·pa·raz·zo \ˌpä-pə-'rät-(ˌ)sō\ *noun, plural* **-raz·zi** \-(ˌ)sē\ [Italian] (1966)
: a free-lance photographer who aggressively pursues celebrities for the purpose of taking candid photographs

pa·pav·er·ine \pə-'pa-və-ˌrēn; -'pav-rən, -'pa-və-\ *noun* [International Scientific Vocabulary, from Latin *papaver* poppy] (1857)
: a crystalline alkaloid $C_{20}H_{21}NO_4$ that is found in opium and is used chiefly as an antispasmodic for its ability to relax smooth muscle

pa·paw *noun* [probably modification of Spanish *papaya*] (1624)
1 \pə-'pȯ\ **:** PAPAYA
2 \'pä-(ˌ)pȯ, 'pȯ-\ **:** a North American tree (*Asimina triloba*) of the custard-apple family with purple flowers and an edible fruit; *also* **:** its fruit

pa·pa·ya \pə-'pī-ə\ *noun* [Spanish, of American Indian origin; akin to Arawak *papáia*] (1598)
: a tropical American tree (*Carica papaya* of the family Caricaceae, the papaya family) with an oblong to globose yellow edible fruit; *also* **:** its fruit

[1]pa·per \'pā-pər\ *noun* [Middle English *papir,* from Middle French *papier,* from Latin *papyrus* papyrus, paper, from Greek *papyros* papyrus] (14th century)
1 a (1) **:** a felted sheet of usually vegetable fibers laid down on a fine screen from a water suspension (2) **:** a similar sheet of other material (as plastic) **b :** a piece of paper
2 a : a piece of paper containing a written or printed statement **:** DOCUMENT ⟨pedigree *papers*⟩ **b :** a piece of paper containing writing or print **c :** a formal written composition often designed for publication and often intended to be read aloud ⟨presented a scholarly *paper* at the meeting⟩ **d :** a piece of written schoolwork
3 : a paper container or wrapper
4 : NEWSPAPER
5 : the negotiable notes or instruments of commerce
6 : WALLPAPER
7 : TICKETS; *especially* **:** free passes
8 : PAPERBACK
— on paper 1 : in writing ⟨wants these promises *on paper*⟩ **2 :** in theory ⟨the plan looks good *on paper*⟩ **3 :** figured at face value ⟨*on paper* the stock was worth nearly a million dollars⟩

[2]paper *verb* **pa·pered; pa·per·ing** \'pā-p(ə-)riŋ\ (1594)
transitive verb
1 *archaic* **:** to put down or describe in writing
2 : to fold or enclose in paper
3 : to cover or line with paper; *especially* **:** to apply wallpaper to
4 : to fill by giving out free passes ⟨*paper* the theater for opening night⟩
5 : to cover (an area) with advertising bills, circulars, or posters
intransitive verb
: to hang wallpaper
— pa·per·er \-pər-ər\ *noun*

[3]paper *adjective* (1594)
1 a : made of paper, paperboard, or papier-mâché ⟨a *paper* bag⟩ **b :** PAPERY
2 : of or relating to clerical work or written communication
3 : existing only in theory **:** NOMINAL ⟨a *paper* blockade⟩
4 : admitted by free passes ⟨a *paper* audience⟩
5 : finished with a crisp smooth surface similar to that of paper ⟨*paper* taffeta⟩

pa·per·back \'pā-pər-ˌbak\ *noun* (1899)
: a book with a flexible paper binding
— paperback *also* **pa·per·backed** \-ˌbakt\ *adjective*

paper birch *noun* (1810)
: a North American birch (*Betula papyrifera*) with peeling white bark and toothed ovate leaves

pa·per·board \'pā-pər-ˌbōrd, -ˌbȯrd\ *noun* (1549)
: a material made from cellulose fiber (as wood pulp) like paper but usually thicker

pa·per·bound \-ˌbau̇nd\ *noun* (1950)
: PAPERBACK
— paperbound *adjective*

pa·per·boy \-ˌbȯi\ *noun* (1876)
: a boy who delivers newspapers **:** NEWSBOY

paper chromatography *noun* (1948)
: chromatography that uses paper strips or sheets as the adsorbent stationary phase through which a solution flows and that is used especially to separate amino acids

paper clip *noun* (1919)
: a length of wire bent into flat loops that is used to hold papers together

paper cutter *noun* (circa 1828)
1 : PAPER KNIFE
2 : a machine or device for cutting or trimming sheets of paper to required dimensions

pa·per·hang·er \'pā-pər-ˌhaŋ-ər\ *noun* (1796)
1 : one that applies wallpaper
2 *slang* **:** one who passes worthless checks

pa·per·hang·ing \-ˌhaŋ-iŋ\ *noun* (1873)
: the act of applying wallpaper

paper knife *noun* (circa 1807)
1 : a knife for slitting envelopes or uncut pages
2 : the knife of a paper cutter

pa·per·less \'pā-pər-ləs\ *adjective* (1969)
: recording or relaying information by electronic media rather than on paper ⟨*paperless* offices⟩

pa·per·mak·er \'pā-pər-ˌmā-kər\ *noun* (circa 1580)
: one that makes paper
— pa·per·mak·ing \-kiŋ\ *noun*

paper money *noun* (1691)
1 : money consisting of government notes and banknotes
2 : BANK MONEY

paper mulberry *noun* (1777)
: an Asian tree (*Broussonetia papyrifera*) of the mulberry family that is widely grown as a shade tree

paper nautilus *noun* (1835)
: a pelagic cephalopod (genus *Argonauta*) of which the female has a delicate papery shell

paper nautilus

paper over *transitive verb* (1955)
1 : to gloss over, explain away, or patch up (as major differences or disparities) especially in order to maintain a semblance of unity or agreement
2 : HIDE, CONCEAL

paper profit *noun* (1893)
: a profit that can be realized only by selling something (as a security) that has appreciated in market value

paper–thin *adjective* (1928)
: extremely thin ⟨*paper-thin* partitions⟩

paper tiger *noun* (1850)
: one that is outwardly powerful or dangerous but inwardly weak or ineffectual

paper trail *noun* (1965)
: documents (as financial records or published materials) from which a person's actions may be traced or opinions learned

pa·per–train \'pā-pər-ˌtrān\ *transitive verb* (1971)
: to train (as a dog) to defecate and urinate on paper indoors

pa·per·weight \-ˌwāt\ *noun* (circa 1858)

: a usually small heavy object used to hold down loose papers (as on a desk)

pa·per·work \-ˌwərk\ *noun* (1889)
: routine clerical or record-keeping work often incidental to a more important task

pa·pery \'pā-p(ə-)rē\ *adjective* (1627)
: resembling paper in thinness or consistency ⟨*papery* leaves⟩ ⟨*papery* silk⟩
— **pa·per·i·ness** *noun*

pa·pe·terie \'pa-pə-trē, ˌpa-pə-'\ *noun* [French] (circa 1847)
: packaged fancy stationery

[1]**Pa·phi·an** \'pā-fē-ən\ *noun* [Latin *paphius,* from Greek *paphios,* from *Paphos,* ancient city of Cyprus that was the center of worship of Aphrodite] (1598)
1 *often not capitalized* **:** PROSTITUTE
2 : a native or inhabitant of Paphos

[2]**Paphian** *adjective* (1605)
1 : of or relating to illicit love **:** WANTON
2 : of or relating to Paphos or its people

Pa·pia·men·to \ˌpä-pyə-'men-(ˌ)tō\ *also* **Pa·pia·men·tu** \-(ˌ)tü\ *noun* [Spanish, from Papiamento *papya* talk + *-mentu* -ment] (1923)
: a Spanish-based creole language of Netherlands Antilles

pa·pier col·lé \ˌpä-ˌpyā-(ˌ)kō-'lā, ˌpa-\ *noun, plural* **papiers col·lés** \-ˌpyā-(ˌ)kō-'lā(z)\ [French, glued paper] (1935)
: COLLAGE

[1]**pa·pier–mâ·ché** \ˌpā-pər-mə-'shā, ˌpa-ˌpyā-mə-, -(ˌ)ma-\ *noun* [French, literally, chewed paper] (1753)
: a light strong molding material of wastepaper pulped with glue and other additives

[2]**papier–mâché** *adjective* (1753)
1 : formed of papier-mâché
2 : UNREAL, ARTIFICIAL

pa·pil·i·o·na·ceous \pə-ˌpi-lē-ə-'nā-shəs\ *adjective* [Latin *papilion-, papilio* butterfly — more at PAVILION] (1668)
: having a corolla (as in the bean or pea) with usually five petals that include a large standard enclosing two lateral wings and a lower carina

pa·pil·la \pə-'pi-lə\ *noun, plural* **pa·pil·lae** \-'pi-(ˌ)lē, -ˌlī\ [Latin, nipple, from diminutive of *papula* pimple; akin to Lithuanian *papas* nipple] (1713)
: a small projecting body part similar to a nipple in form: **a :** a vascular process of connective tissue extending into and nourishing the root of a hair, feather, or developing tooth **b :** any of the vascular protuberances of the dermal layer of the skin extending into the epidermal layer and often containing tactile corpuscles **c :** any of the small protuberances on the upper surface of the tongue
— **pap·il·lary** \'pa-pə-ˌler-ē, *especially British* pə-'pi-lə-rē\ *adjective*
— **pa·pil·late** \'pa-pə-ˌlāt, pə-'pi-lət\ *adjective*
— **pa·pil·lose** \'pa-pə-ˌlōs, pə-'pi-ˌlōs\ *adjective*

pap·il·lo·ma \ˌpa-pə-'lō-mə\ *noun, plural* **-mas** *or* **-ma·ta** \-mə-tə\ [New Latin] (1866)
1 : a benign tumor (as a wart) due to overgrowth of epithelial tissue on papillae of vascular connective tissue (as of the skin)
2 : an epithelial tumor caused by a virus
— **pap·il·lo·ma·tous** \-'lō-mə-təs\ *adjective*

pap·il·lo·ma·vi·rus \-ˌvī-rəs\ *noun* [New Latin] (1960)
: any of a group of papovaviruses that cause papillomas

pa·pil·lon \ˌpä-pē-'yō^{n}, ˌpa-\ *noun* [French, literally, butterfly, from Latin *papilion-, papilio*] (1907)
: any of a breed of small slender toy spaniels having large erect heavily fringed ears

pa·pil·lote \ˌpä-pē-'yōt, ˌpa-\ *noun* [French, from *papillon* butterfly] (1818)
: a greased usually paper wrapper in which food (as meat or fish) is cooked

pa·pist \'pā-pist\ *noun, often capitalized* [Middle French or New Latin; Middle French *papiste,* from *pape* pope; New Latin *papista,* from Late Latin *papa* pope] (1534)
: ROMAN CATHOLIC — usually used disparagingly
— **papist** *adjective*

pa·pist·ry \'pā-pə-strē\ *noun* (1535)
: the Roman Catholic religion — usually used disparagingly

pa·poose \pa-'püs, pə-\ *noun* [Narraganset *papoòs*] (1634)
: a young child of American Indian parents

pa·po·va·vi·rus \pə-'pō-və-ˌvī-rəs\ *noun* [*pa*pilloma + *po*lyoma + *va*cuolating + *virus*] (1962)
: any of a group of viruses that have a capsid composed of 72 subunits and that are associated with or responsible for various neoplasms (as some warts) of mammals

pap·pose \'pa-ˌpōs\ *adjective* (1691)
: having or being a pappus

pap·pus \'pa-pəs\ *noun, plural* **pap·pi** \'pa-ˌpī, -ˌpē\ [Latin, from Greek *pappos*] (circa 1704)
: an appendage or tuft of appendages that crowns the ovary or fruit in various seed plants and functions in dispersal of the fruit

pap·py \'pa-pē\ *noun* (1763)
chiefly Southern & Midland **:** PAPA

pa·pri·ka \pə-'prē-kə, pa-\ *noun* [Hungarian, from Serbo-Croation, from *papar* ground pepper, ultimately from Latin *piper* — more at PEPPER] (circa 1896)
: a usually mild red condiment consisting of the dried finely ground pods of various cultivated sweet peppers; *also* **:** a sweet pepper used for making paprika

Pap smear \'pap-\ *noun* [George N. *Papanicolaou* (died 1962) American medical scientist] (1952)
: a method for the early detection of uterine cancer that involves the staining of exfoliated cells using a special technique which differentiates diseased tissue — called also *Papanicolaou smear, Papanicolaou test, Pap test*

Pap·u·an \'pa-pyə-wən, -pə-\ *noun* (1814)
1 : a native or inhabitant of Papua
2 : a member of any of the native peoples of New Guinea and adjacent areas of Melanesia
3 : any of a heterogeneous group of languages spoken in New Guinea, New Britain, and the Solomon islands
— **Papuan** *adjective*

pap·u·lar \'pa-pyə-lər\ *adjective* (circa 1820)
: consisting of or characterized by papules

pap·ule \'pa-(ˌ)pyü(ə)l\ *noun* [Latin *papula*] (1864)
: a small solid usually conical elevation of the skin

pap·y·rol·o·gy \ˌpa-pə-'rä-lə-jē\ *noun* [International Scientific Vocabulary] (1898)
: the study of papyrus manuscripts
— **pap·y·rol·o·gist** \-jist\ *noun*

pa·py·rus \pə-'pī-rəs\ *noun, plural* **pa·py·ri** \-'pī-(ˌ)rē, -rī\ *or* **pa·py·rus·es** [Middle English, from Latin — more at PAPER] (14th century)
1 : a tall sedge (*Cyperus papyrus*) of the Nile valley
2 : the pith of the papyrus plant especially when cut in strips and pressed into a material to write on
3 a : a writing on papyrus **b :** a written scroll made of papyrus

papyrus 1

[1]**par** \'pär\ *noun* [Latin, one that is equal, from *par* equal] (1622)
1 a : the established value of the monetary unit of one country expressed in terms of the monetary unit of another country using the same metal as the standard of value **b :** the face amount of an instrument of value (as a check or note): as (1) **:** the monetary value assigned to each share of stock in the charter of a corporation (2) **:** the principal of a bond
2 : common level **:** EQUALITY — usually used with *on* ⟨judged the recording to be on a *par* with previous ones⟩
3 a : an amount taken as an average or norm **b :** an accepted standard; *specifically* **:** a usual standard of physical condition or health
4 : the score standard for each hole of a golf course; *also* **:** a score equal to par
— **par** *adjective*
— **par for the course :** not unusual **:** NORMAL

[2]**par** *transitive verb* **parred; par·ring** (1950)
: to score par on (a hole)

pa·ra \'pär-ə\ *noun, plural* **paras** *or* **para** [Turkish, from Persian *pārah,* literally, piece] (1687)
1 a : any of several monetary units of the Turkish Empire **b :** a coin representing one para
2 [Serbo-Croatian, from Turkish] — see *dinar* at MONEY table

[1]**para-** *or* **par-** *prefix* [Middle English, from Middle French, from Latin, from Greek, from *para;* akin to Greek *pro* before — more at FOR]
1 : beside **:** alongside of **:** beyond **:** aside from ⟨*para*thyroid⟩ ⟨*par*enteral⟩
2 a : closely related to ⟨*par*aldehyde⟩ **b :** involving substitution at or characterized by two opposite positions in the benzene ring that are separated by two carbon atoms ⟨*para*dichlorobenzene⟩
3 a : faulty **:** abnormal ⟨*par*esthesia⟩ **b :** associated in a subsidiary or accessory capacity ⟨*para*medical⟩ **c :** closely resembling **:** almost ⟨*para*typhoid⟩

[2]**para-** *combining form* [*parachute*]
: parachute ⟨*para*trooper⟩

-para *noun combining form, plural* **-paras** *or* **-parae** [Latin, from *parere* to give birth to — more at PARE]
: woman delivered of (so many) children ⟨primi*para*⟩

para–ami·no·ben·zo·ic acid \ˌpar-ə-ə-ˌmē-ˌnō-ˌben-'zō-ik-, ˌpar-ə-ˌa-mə-(ˌ)nō-\ *noun* [International Scientific Vocabulary] (1906)
: a colorless para-substituted aminobenzoic acid that is a growth factor of the vitamin B complex — called also *PABA*

para–ami·no·sal·i·cyl·ic acid \-ˌsa-lə-'si-lik-\ *noun* (1946)
: the white crystalline para-substituted isomer of aminosalicylic acid that is made synthetically and is used in the treatment of tuberculosis

para·bi·o·sis \ˌpar-ə-(ˌ)bī-'ō-səs, -bē-\ *noun* [New Latin] (1903)
1 : reversible suspension of obvious vital activities
2 : anatomical and physiological union of two organisms
— **para·bi·ot·ic** \-'ä-tik\ *adjective*
— **para·bi·ot·i·cal·ly** \-ti-k(ə-)lē\ *adverb*

par·a·ble \'par-ə-bəl\ *noun* [Middle English, from Middle French, from Late Latin *parabola,* from Greek *parabolē* comparison, from *paraballein* to compare, from *para-* + *ballein* to throw — more at DEVIL] (14th century)
: EXAMPLE; *specifically* **:** a usually short fictitious story that illustrates a moral attitude or a religious principle

pa·rab·o·la \pə-'ra-bə-lə\ *noun* [New Latin, from Greek *parabolē,* literally, comparison] (1579)
1 : a plane curve generated by a point moving so that its distance from a fixed point is equal to its distance from a fixed line **:** the intersection of a right circular cone with a plane parallel to an element of the cone
2 : something bowl-shaped (as an antenna or microphone reflector)

par•a•bol•ic \ˌpar-ə-ˈbä-lik\ *adjective* [in sense 1, from Late Latin *parabola* parable; in sense 2, from New Latin *parabola*] (1669)
1 : expressed by or being a parable **:** ALLEGORICAL
2 : of, having the form of, or relating to a parabola ⟨motion in a *parabolic* curve⟩
— par•a•bol•i•cal•ly \-li-k(ə-)lē\ *adverb*

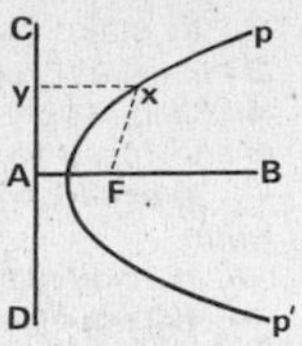

C, p, y, x, A, F, B, D, p′

parabola 1: *F* fixed point; *CD* fixed line; *x* moving point; *AB* axis; *xy* distance from *x* to *CD*; pp′ parabola

pa•rab•o•loid \pə-ˈra-bə-ˌlȯid\ *noun* (circa 1702)
: a surface all of whose intersections by planes are either parabolas and ellipses or parabolas and hyperbolas
— pa•rab•o•loi•dal \-ˌra-bə-ˈlȯi-d^{ə}l\ *adjective*

¹para•chute \ˈpar-ə-ˌshüt\ *noun* [French, from *para-* (as in *parasol*) + *chute* fall — more at CHUTE] (1785)
1 : a device for slowing the descent of a person or object through the air that consists of a usually hemispherical fabric canopy beneath which the person or object is suspended
2 : PATAGIUM
3 : a device or structure suggestive of a parachute in form, use, or operation
— para•chut•ic \ˌpar-ə-ˈshü-tik\ *adjective*

²parachute *verb* **-chut•ed; -chut•ing** (1809)
transitive verb
: to convey by means of a parachute
intransitive verb
: to descend by means of a parachute

para•chut•ist \ˈpar ə-ˌshü-tist\ *noun* (1888)
: one that parachutes: as **a :** PARATROOPER **b :** a person who parachutes as a sport

Par•a•clete \ˈpar-ə-ˌklēt\ *noun* [Middle English *Paraclyte,* from Late Latin *Paracletus, Paraclitus,* from Greek *Paraklētos,* literally, advocate, intercessor, from *parakalein* to invoke, from *para-* + *kalein* to call — more at LOW] (15th century)
: HOLY SPIRIT

¹pa•rade \pə-ˈrād\ *noun* [French, from Middle French, from *parer* to prepare — more at PARE] (circa 1656)
1 : a pompous show **:** EXHIBITION
2 a : the ceremonial formation of a body of troops before a superior officer **b :** a place where troops assemble regularly for parade
3 a : an informal procession **b :** a public procession **c :** a showy array or succession ⟨a *parade* of tycoons' castles —Gail Sheehy⟩
4 a : a place for strolling **b :** those who promenade

²parade *verb* **pa•rad•ed; pa•rad•ing** (1686)
transitive verb
1 : to cause to maneuver or march **:** MARSHAL
2 : PROMENADE
3 : to exhibit ostentatiously
intransitive verb
1 : to march in or as if in a procession
2 : PROMENADE
3 a : SHOW OFF **b :** MASQUERADE ⟨myths which *parade* as modern science —M. R. Cohen⟩
synonym see SHOW
— pa•rad•er *noun*

para•di•chlo•ro•ben•zene \ˌpar-ə-ˌdī-ˌklōr-ə-ˈben-ˌzēn, -ˌklȯr-, -ˌben-ˈ\ *noun* [International Scientific Vocabulary] (1876)
: a white crystalline compound $C_6H_4Cl_2$ made by chlorinating benzene and used chiefly as a fumigant against clothes moths

par•a•did•dle \ˈpar-ə-ˌdi-d^{ə}l\ *noun* [origin unknown] (1927)
: a quick succession of drumbeats slower than a roll and alternating left- and right-hand strokes in a typical L-R-L-L, R-L-R-R pattern

par•a•digm \ˈpar-ə-ˌdīm *also* -ˌdim\ *noun* [Late Latin *paradigma,* from Greek *paradeigma,* from *paradeiknynai* to show side by side, from *para-* + *deiknynai* to show — more at DICTION] (15th century)
1 : EXAMPLE, PATTERN; *especially* **:** an outstandingly clear or typical example or archetype
2 : an example of a conjugation or declension showing a word in all its inflectional forms
3 : a philosophical and theoretical framework of a scientific school or discipline within which theories, laws, and generalizations and the experiments performed in support of them are formulated
— par•a•dig•mat•ic \ˌpar-ə-dig-ˈma-tik\ *adjective*
— par•a•dig•mat•i•cal•ly \-ti-k(ə-)lē\ *adverb*

par•a•di•sa•ic \ˌpar-ə-ˌdī-ˈsā-ik, -ˈzā-\ *adjective* [*paradise* + *-aic* (as in *Hebraic*)] (1754)
: PARADISIACAL
— par•a•di•sa•i•cal \-ˈsā-ə-kəl, -ˈzā-\ *adjective*
— par•a•di•sa•i•cal•ly \-ə-k(ə-)lē\ *adverb*

par•a•dis•al \ˌpar-ə-ˈdī-səl, -zəl\ *adjective* (1560)
: PARADISIACAL

par•a•dise \ˈpar-ə-ˌdīs, -ˌdīz\ *noun* [Middle English *paradis,* from Old French, from Late Latin *paradisus,* from Greek *paradeisos,* literally, enclosed park, of Iranian origin; akin to Avestan *pairi-daēza-* enclosure; akin to Greek *peri* around and to Greek *teichos* wall — more at PERI-, DOUGH] (12th century)
1 a : EDEN 2 **b :** an intermediate place or state where the righteous departed await resurrection and judgment **c :** HEAVEN
2 : a place or state of bliss, felicity, or delight
— par•a•dis•i•al \ˌpar-ə-ˈdi-sē-əl, -zē-\ *also* **par•a•dis•i•cal** \-si-kəl, -zi-\ *adjective*

par•a•di•si•a•cal \ˌpar-ə-də-ˈsī-ə-kəl, -ˌdī-, -ˈzī-\ *or* **par•a•dis•i•ac** \-ˈdi-zē-ˌak, -sē-\ *adjective* [Late Latin *paradisiacus,* from *paradisus*] (1649)
: of, relating to, or resembling paradise
— par•a•di•si•a•cal•ly \-də-ˈsī-ə-k(ə-)lē, -ˌdī-, -ˈzī-\ *adverb*

pa•ra•dor \ˌpä-rä-ˈthȯr\ *noun* [Spanish, inn, from *parar* to stop, lodge for the night, from Latin *parare* to prepare — more at PARE] (1845)
: a usually government-operated hostelry found especially in Spain

par•a•dox \ˈpar-ə-ˌdäks\ *noun* [Latin *paradoxum,* from Greek *paradoxon,* from neuter of *paradoxos* contrary to expectation, from *para-* + *dokein* to think, seem — more at DECENT] (1540)
1 : a tenet contrary to received opinion
2 a : a statement that is seemingly contradictory or opposed to common sense and yet is perhaps true **b :** a self-contradictory statement that at first seems true **c :** an argument that apparently derives self-contradictory conclusions by valid deduction from acceptable premises
3 : something or someone with seemingly contradictory qualities or phases

par•a•dox•i•cal \ˌpar-ə-ˈdäk-si-kəl\ *adjective* (1581)
1 a : of the nature of a paradox **b :** inclined to paradoxes
2 : not being the normal or usual kind ⟨*paradoxical* pulse⟩
— par•a•dox•i•cal•i•ty \-ˌdäk-si-ˈka-lə-tē\ *noun*
— par•a•dox•i•cal•ly \-ˈdäk-si-k(ə-)lē\ *adverb*
— par•a•dox•i•cal•ness \-kəl-nəs\ *noun*

paradoxical sleep *noun* (1964)
: REM SLEEP

par•aes•the•sia *variant of* PARESTHESIA

par•af•fin \ˈpar-ə-fən\ *noun* [German, from Latin *parum* too little (akin to Greek *pauros* little, *paid-, pais* child) + *affinis* bordering on — more at FEW, AFFINITY] (1838)
1 a : a waxy crystalline flammable substance obtained especially from distillates of wood, coal, petroleum, or shale oil that is a complex mixture of hydrocarbons and is used chiefly in coating and sealing, in candles, in rubber compounding, and in pharmaceuticals and cosmetics **b :** any of various mixtures of similar hydrocarbons including mixtures that are semisolid or oily
2 : ALKANE
3 *chiefly British* **:** KEROSENE
— par•af•fin•ic \ˌpar-ə-ˈfi-nik\ *adjective*

para•for•mal•de•hyde \ˌpar-ə-fȯr-ˈmal-də-ˌhīd, -fər-\ *noun* (1894)
: a white powder $(CH_2O)_x$ that consists of a polymer of formaldehyde and is used especially as a fungicide

para•gen•e•sis \ˌpar-ə-ˈje-nə-səs\ *noun* [New Latin] (1853)
: the formation of minerals in contact in such a manner as to affect one another's development
— para•ge•net•ic \-jə-ˈne-tik\ *adjective*
— para•ge•net•i•cal•ly \-ti-k(ə-)lē\ *adverb*

¹par•a•gon \ˈpar-ə-ˌgän, -gən\ *noun* [Middle French, from Old Italian *paragone,* literally, touchstone, from *paragonare* to test on a touchstone, from Greek *parakonan* to sharpen, from *para-* + *akonē* whetstone, from *akē* point; akin to Greek *akmē* point — more at EDGE] (circa 1548)
: a model of excellence or perfection

²paragon *transitive verb* (circa 1586)
1 : to compare with **:** PARALLEL
2 : to put in rivalry **:** MATCH
3 *obsolete* **:** SURPASS

¹par•a•graph \ˈpar-ə-ˌgraf\ *noun* [Middle French & Medieval Latin; Middle French *paragraphe,* from Medieval Latin *paragraphus* sign marking a paragraph, from Greek *paragraphos* line used to mark change of persons in a dialogue, from *paragraphein* to write alongside, from *para-* + *graphein* to write — more at CARVE] (1525)
1 a : a subdivision of a written composition that consists of one or more sentences, deals with one point or gives the words of one speaker, and begins on a new usually indented line **b :** a short composition or note that is complete in one paragraph
2 : a character (as ¶) used to indicate the beginning of a paragraph and as a reference mark
— par•a•graph•ic \ˌpar-ə-ˈgra-fik\ *adjective*

²paragraph (1764)
transitive verb
1 : to write paragraphs about
2 : to divide into paragraphs
intransitive verb
: to write paragraphs

par•a•graph•er \ˈpar-ə-ˌgra-fər\ *noun* (1822)
: a writer of paragraphs especially for the editorial page of a newspaper

para•in•flu•en•za virus \ˌpar-ə-ˌin-flü-ˈen-zə-\ *noun* (1959)
: any of several paramyxoviruses that are associated with or responsible for some respiratory infections especially in children — called also *parainfluenza*

para•jour•nal•ism \ˌpar-ə-ˈjər-n^{ə}l-ˌi-zəm\ *noun* (1965)
: journalism that is heavily colored by the opinions of the reporter

par•a•keet \ˈpar-ə-ˌkēt\ *noun* [Spanish & Middle French; Spanish *periquito,* from Middle French *perroquet* parrot] (1581)
: any of numerous usually small slender parrots with a long graduated tail

para•lan•guage \ˈpar-ə-ˌlaŋ-gwij\ *noun* (circa 1958)

\ə\ abut \$^{ə}$\ kitten \ər\ further \a\ ash \ā\ ace \ä\ mop, mar \au̇\ out \ch\ chin \e\ bet \ē\ easy \g\ go \i\ hit \ī\ ice \j\ job \ŋ\ sing \ō\ go \ȯ\ law \ȯi\ boy \th\ thin \th\ the \ü\ loot \u̇\ foot \y\ yet \zh\ vision *see also* Guide to Pronunciation

: optional vocal effects (as tone of voice) that accompany or modify the phonemes of an utterance and that may communicate meaning

par·al·de·hyde \pə-ˈral-də-ˌhīd, pə-\ *noun* (1857)
: a colorless liquid polymeric modification $C_6H_{12}O_3$ of acetaldehyde used as a hypnotic

para·le·gal \ˌpar-ə-ˈlē-gəl\ *adjective* (1971)
: of, relating to, or being a paraprofessional who assists a lawyer
— **para·le·gal** \ˈpar-ə-ˌlē-gəl\ *noun*

para·lin·guis·tics \ˌpar-ə-liŋ-ˈgwis-tiks\ *noun* (circa 1958)
: the study of paralanguage
— **para·lin·guis·tic** \-tik\ *adjective*

Par·a·li·pom·e·non \ˌpar-ə-lə-ˈpä-mə-ˌnän, -lī-\ *noun* [Late Latin, from Greek *Paraleipomenōn,* genitive of *Paraleipomena,* literally, things left out, from neuter plural of passive present participle of *paraleipein* to leave out, from *para-* + *leipein* to leave; from its forming a supplement to Samuel and Kings — more at LOAN]
: CHRONICLES

par·al·lac·tic \ˌpar-ə-ˈlak-tik\ *adjective* [New Latin *parallacticus,* from Greek *parallaktikos,* from *parallaxis*] (1630)
: of, relating to, or due to parallax

par·al·lax \ˈpar-ə-ˌlaks\ *noun* [Middle French *parallaxe,* from Greek *parallaxis,* from *parallassein* to change, from *para-* + *allassein* to change, from *allos* other] (1580)
: the apparent displacement or the difference in apparent direction of an object as seen from two different points not on a straight line with the object; *especially* : the angular difference in direction of a celestial body as measured from two points on the earth's orbit

¹par·al·lel \ˈpar-ə-ˌlel, -ləl\ *adjective* [Latin *parallelus,* from Greek *parallēlos,* from *para* beside + *allēlōn* of one another, from *allos* . . . *allos* one . . . another, from *allos* other — more at ELSE] (1549)
1 a : extending in the same direction, everywhere equidistant, and not meeting ⟨*parallel* rows of trees⟩ **b :** everywhere equally distant ⟨concentric spheres are *parallel*⟩
2 a : having parallel sides ⟨a *parallel* reamer⟩ **b :** being or relating to an electrical circuit having a number of conductors in parallel **c :** arranged in parallel ⟨a *parallel* processor⟩ **d :** relating to or being a connection in a computer system in which the bits of a byte are transmitted over separate channels at the same time — compare SERIAL
3 a : similar, analogous, or interdependent in tendency or development **b :** readily compared : COMPANION **c :** having identical syntactical elements in corresponding positions; *also* : being such an element **d** (1) **:** having the same tonic — used of major and minor keys and scales (2) **:** keeping the same distance apart in musical pitch
4 : performed while keeping one's skis parallel ⟨*parallel* turns⟩
synonym see SIMILAR

²parallel *noun* (1551)
1 a : a parallel line, curve, or surface **b :** one of the imaginary circles on the surface of the earth paralleling the equator and marking the latitude; *also* : the corresponding line on a globe or map — see LATITUDE illustration **c :** a character ‖ used in printing especially as a reference mark
2 a : something equal or similar in all essential particulars : COUNTERPART **b :** SIMILARITY, ANALOGUE
3 : a comparison to show resemblance : a tracing of similarity
4 a : the state of being physically parallel : PARALLELISM **b :** an arrangement of electrical devices in a circuit in which the same potential difference is applied to two or more resistances with each resistance being on a different branch of the circuit — compare SERIES **c :** an arrangement of state that permits several operations or tasks to be performed simultaneously rather than consecutively

³parallel *transitive verb* (1598)
1 : to indicate analogy of : COMPARE
2 a : to show something equal to : MATCH **b :** to correspond to
3 : to place so as to be parallel in direction with something
4 : to extend, run, or move in a direction parallel to

⁴parallel *adverb* (1787)
: in a parallel manner

parallel bars *noun plural* (1868)
1 : a pair of wooden bars supported horizontally above the floor at the same height or at different heights usually by a common base and used in gymnastics
2 : an event in gymnastics competition in which even or uneven parallel bars are used

par·al·lel·e·pi·ped \ˌpar-ə-ˌle-lə-ˈpī-pəd *also* -ˌle-ˈle-pə-ˌped\ *noun* [Greek *parallēlepipedon,* from *parallēlos* + *epipedon* plane surface, from neuter of *epipedos* flat, from *epi-* epi- + *pedon* ground; akin to Latin *ped-, pes* foot — more at FOOT] (1570)
: a 6-faced polyhedron all of whose faces are parallelograms lying in pairs of parallel planes

par·al·lel·ism \ˈpar-ə-ˌle-ˌli-zəm, -lə-ˌli-\ *noun* (1610)
1 : the quality or state of being parallel
2 : RESEMBLANCE, CORRESPONDENCE
3 : repeated syntactical similarities introduced for rhetorical effect
4 : a theory that mind and matter accompany one another but are not causally related
5 : the development of similar new characters by two or more related organisms in response to similarity of environment — called also *parallel evolution*

par·al·lel·o·gram \ˌpar-ə-ˈle-lə-ˌgram\ *noun* [Late Latin or Greek; Late Latin *parallelogrammum,* from Greek *parallēlogrammon,* from neuter of *parallēlogrammos* bounded by parallel lines, from *parallēlos* + *grammē* line, from *graphein* to write — more at CARVE] (1570)
: a quadrilateral with opposite sides parallel and equal

par·al·lel–veined \ˌpar-ə-ˌlel-ˈvānd, -ləl-\ *adjective* (1861)
of a leaf : having veins arranged nearly parallel to one another — compare NET-VEINED

pa·ral·o·gism \pə-ˈra-lə-ˌji-zəm\ *noun* [Middle French *paralogisme,* from Late Latin *paralogismus,* from Greek *paralogismos,* from *paralogos* unreasonable, from *para-* + *logos* speech, reason — more at LEGEND] (1565)
: a fallacious argument

par·a·lyse *British variant of* PARALYZE

pa·ral·y·sis \pə-ˈra-lə-səs\ *noun, plural* **-y·ses** \-ˌsēz\ [Latin, from Greek, from *paralyein* to loosen, disable, from *para-* + *lyein* to loosen — more at LOSE] (1525)
1 : complete or partial loss of function especially when involving the motion or sensation in a part of the body
2 : loss of the ability to move
3 : a state of powerlessness or incapacity to act

paralysis agi·tans \-ˈa-jə-ˌtanz\ *noun* [New Latin, literally, shaking palsy] (1817)
: PARKINSON'S DISEASE

¹par·a·lyt·ic \ˌpar-ə-ˈli-tik\ *adjective* [Middle English *paralytyk,* from Middle French *paralitike,* from Latin *paralyticus,* from Greek *paralytikos,* from *paralyein*] (14th century)
1 : affected with or characterized by paralysis
2 : of, relating to, or resembling paralysis
— **par·a·lyt·i·cal·ly** \-ti-k(ə-)lē\ *adverb*

²paralytic *noun* (14th century)
: one affected with paralysis

par·a·lyze \ˈpar-ə-ˌlīz\ *transitive verb* **-lyzed; -lyz·ing** [French *paralyser,* back-formation from *paralysie* paralysis, from Latin *paralysis*] (1804)
1 : to affect with paralysis
2 : to make powerless or ineffective
3 : UNNERVE
4 : STUN, STUPEFY
5 : to bring to an end : PREVENT, DESTROY
— **par·a·ly·za·tion** \ˌpar-ə-lə-ˈzā-shən\ *noun*
— **par·a·lyz·er** \ˈpar-ə-ˌlī-zər\ *noun*
— **par·a·lyz·ing·ly** \-ˌlī-ziŋ-lē\ *adverb*

para·mag·net \ˈpar-ə-ˌmag-nət\ *noun* [back-formation from *paramagnetic*] (circa 1900)
: a paramagnetic substance

para·mag·net·ic \ˌpar-ə-mag-ˈne-tik\ *adjective* [International Scientific Vocabulary] (circa 1850)
: being or relating to a magnetizable substance (as aluminum) that has small but positive susceptibility which varies little with magnetizing force
— **para·mag·net·i·cal·ly** \-ti-k(ə-)lē\ *adverb*
— **para·mag·ne·tism** \-ˈmag-nə-ˌti-zəm\ *noun*

par·a·me·cium \ˌpar-ə-ˈmē-sh(ē-)əm, -sē-əm\ *noun, plural* **-cia** \-sh(ē-)ə, -sē-ə\ *also* **-ciums** [New Latin, from Greek *paramēkēs* oblong, from *para-* + *mēkos* length; akin to Greek *makros* long — more at MEAGER] (1752)
: any of a genus (*Paramecium*) of ciliate protozoans that have an elongate body rounded at the anterior end and an oblique funnel-shaped buccal groove bearing the mouth at the extremity

para·med·ic \ˌpar-ə-ˈme-dik\ *also* **para·med·i·cal** \-di-kəl\ *noun* (1967)
1 : a person who works in a health field in an auxiliary capacity to a physician (as by giving injections and taking X rays)
2 : a specially trained medical technician licensed to provide a wide range of emergency services (as defibrillation and the intravenous administration of drugs) before or during transportation to a hospital — compare EMT

para·med·i·cal \ˌpar-ə-ˈme-di-kəl\ *also* **para·med·ic** \-dik\ *adjective* (1921)
: concerned with supplementing the work of highly trained medical professionals ⟨*paramedical* aides and technicians⟩

par·a·ment \ˈpar-ə-mənt\ *noun* [Middle English, from Medieval Latin *paramentum,* from *parare* to adorn, from Latin, to prepare — more at PARE] (15th century)
: an ornamental ecclesiastical hanging or vestment

pa·ram·e·ter \pə-ˈram-ə-tər\ *noun* [New Latin, from *para-* + Greek *metron* measure — more at MEASURE] (1656)
1 a : an arbitrary constant whose value characterizes a member of a system (as a family of curves); *also* : a quantity (as a mean or variance) that describes a statistical population **b :** an independent variable used to express the coordinates of a variable point and functions of them — compare PARAMETRIC EQUATION
2 : any of a set of physical properties whose values determine the characteristics or behavior of something ⟨*parameters* of the atmosphere such as temperature, pressure, and density⟩
3 : something represented by a parameter : a characteristic element; *broadly* : CHARACTERISTIC, ELEMENT, FACTOR ⟨political dissent as a *parameter* of modern life⟩
4 : LIMIT, BOUNDARY — usually used in plural ⟨the *parameters* of science fiction⟩
— **para·met·ric** \ˌpar-ə-ˈme-trik\ *adjective*
— **para·met·ri·cal·ly** \-tri-k(ə-)lē\ *adverb*

pa·ram·e·ter·ize \pə-ˈra-mə-tə-ˌrīz, -mə-ˌtrīz\ *or* **pa·ram·e·trize** \-ˈra-mə-ˌtrīz\ *transitive verb* **-ter·ized** *or* **-trized; -ter·iz·ing** *or* **-triz·ing** (1940)
: to express in terms of parameters
— **pa·ram·e·ter·i·za·tion** \-ˌra-mə-tə-rə-ˈzā-shən, -mə-trə-\ *or* **pa·ram·e·tri·za·tion** \-mə-trə-\ *noun*

parametric equation *noun* (1909)

: any of a set of equations that express the coordinates of the points of a curve as functions of one parameter or that express the coordinates of the points of a surface as functions of two parameters

para·mil·i·tary \ˌpar-ə-ˈmi-lə-ˌter-ē\ *adjective* (1935)
: of, relating to, being, or characteristic of a force formed on a military pattern especially as a potential auxiliary military force ⟨a *paramilitary* border patrol⟩ ⟨*paramilitary* training⟩

par·am·ne·sia \ˌpar-ˌam-ˈnē-zhə, -əm-\ *noun* [New Latin, from *para-* + *-mnesia* (as in *amnesia*)] (1888)
: a disorder of memory; *especially* **:** DÉJÀ VU 1

[1]par·a·mount \ˈpar-ə-ˌmau̇nt\ *adjective* [Anglo-French *paramont,* from Old French *par* by (from Latin *per*) + *amont* above, from *a* to (from Latin *ad*) + *mont* mountain — more at FOR, AT, MOUNT] (1579)
: superior to all others **:** SUPREME
synonym see DOMINANT
— par·a·mount·cy \-ˌmau̇n(t)-sē\ *noun*
— par·a·mount·ly \-ˌmau̇nt-lē\ *adverb*

[2]paramount *noun* (1616)
: a supreme ruler

par·amour \ˈpar-ə-ˌmu̇r\ *noun* [Middle English, from *par amour* by way of love, from Middle French] (14th century)
: an illicit lover

par·am·y·lum \ˌpa-ˈra-mə-ləm\ *noun* [New Latin, from *para-* + Latin *amylum* starch — more at AMYL-] (1897)
: a reserve carbohydrate that is found in various protozoans and algae and resembles starch

para·myxo·vi·rus \ˌpar-ə-ˈmik-sə-ˌvī-rəs\ *noun* [New Latin] (1962)
: any of a group of RNA-containing viruses (as the mumps, measles, and parainfluenza viruses)

pa·rang \ˈpär-ˌaŋ\ *noun* [Malay] (1839)
: a short sword, cleaver, or machete common in Malaysia and Indonesia

para·noia \ˌpar-ə-ˈnȯi-ə\ *noun* [New Latin, from Greek, madness, from *paranous* demented, from *para-* + *nous* mind] (circa 1811)
1 : a psychosis characterized by systematized delusions of persecution or grandeur usually without hallucinations
2 : a tendency on the part of an individual or group toward excessive or irrational suspiciousness and distrustfulness of others
— para·noi·ac \-ˈnȯi-ˌak, -ˈnȯi-ik\ *also* **para·noic** \-ˈnȯi(-i)k, -ˈnō-ik\ *adjective or noun*
— para·noi·cal·ly \-ˈnȯi(-i)-k(ə-)lē, -ˈnō-i-k(ə-)lē\ *adverb*

para·noid \ˈpar-ə-ˌnȯid\ *also* **para·noi·dal** \ˌpar-ə-ˈnȯi-dᵊl\ *adjective* (1904)
1 : characterized by or resembling paranoia
2 : characterized by suspiciousness, persecutory trends, or megalomania
3 : extremely fearful
— paranoid *noun*

paranoid schizophrenia *noun* (1940)
: schizophrenia characterized especially by persecutory or grandiose delusions or hallucinations or by delusional jealousy

para·nor·mal \ˌpar-ə-ˈnȯr-məl, ˈpar-ə-ˌ\ *adjective* (circa 1920)
: not scientifically explainable **:** SUPERNATURAL
— paranormal *noun*
— para·nor·mal·i·ty \ˌpar-ə-ˌnȯr-ˈma-lə-tē\ *noun*
— para·nor·mal·ly \-ˈnȯr-mə-lē\ *adverb*

para·nymph \ˈpar-ə-ˌnim(p)f\ *noun* [Late Latin *paranymphus,* from Greek *paranymphos,* from *para-* + *nymphē* bride — more at NUPTIAL] (1600)
1 : a friend going with a bridegroom to fetch home the bride in ancient Greece; *also* **:** the bridesmaid conducting the bride to the bridegroom
2 a : BEST MAN **b :** BRIDESMAID

par·a·pet \ˈpar-ə-pət, -ˌpet\ *noun* [Italian *parapetto,* from *parare* to shield (from Latin, to prepare) + *petto* chest, from Latin *pectus* — more at PARE, PECTORAL] (1590)
1 : a wall, rampart, or elevation of earth or stone to protect soldiers **:** BREASTWORK
2 : a low wall or railing to protect the edge of a platform, roof, or bridge — called also *parapet wall*
— par·a·pet·ed \-ˌpe-təd\ *adjective*

pa·raph \ˈpar-əf, pə-ˈraf\ *noun* [Middle French, from Latin *paragraphus* paragraph] (1584)
: a flourish at the end of a signature

par·a·pher·na·lia \ˌpar-ə-fə(r)-ˈnāl-yə\ *noun plural but singular or plural in construction* [Medieval Latin, ultimately from Greek *parapherna* bride's property beyond her dowry, from *para-* + *phernē* dowry, from *pherein* to bear — more at BEAR] (1651)
1 : the separate real or personal property of a married woman that she can dispose of by will and sometimes according to common law during her life
2 : personal belongings
3 a : articles of equipment **:** FURNISHINGS **b :** accessory items **:** APPURTENANCES

[1]para·phrase \ˈpar-ə-ˌfrāz\ *noun* [Middle French, from Latin *paraphrasis,* from Greek, from *paraphrazein* to paraphrase, from *para-* + *phrazein* to point out] (1548)
1 : a restatement of a text, passage, or work giving the meaning in another form
2 : the use or process of paraphrasing in studying or teaching composition

[2]paraphrase *verb* **-phrased; -phras·ing** (1596)
intransitive verb
: to make a paraphrase
transitive verb
: to make a paraphrase of
— para·phras·able \ˌpar-ə-ˈfrā-zə-bəl\ *adjective*
— para·phras·er *noun*

para·phras·tic \ˌpar-ə-ˈfras-tik\ *adjective* [French *paraphrastique,* from Greek *paraphrastikos,* from *paraphrazein*] (circa 1623)
: having the nature of or being a paraphrase
— para·phras·ti·cal·ly \-ti-k(ə-)lē\ *adverb*

pa·raph·y·sis \pə-ˈra-fə-səs\ *noun, plural* **-y·ses** \-ˌsēz\ [New Latin, from Greek, sucker, offshoot, from *paraphyein* to produce at the side, from *para-* + *phyein* to bring forth — more at BE] (1857)
: one of the slender sterile filaments borne among the sporogenous or gametogenous organs in cryptogamic plants

para·ple·gia \ˌpar-ə-ˈplē-j(ē-)ə\ *noun* [New Latin, from Greek *paraplēgiē* hemiplegia, from *para-* + *-plēgia* -plegia] (circa 1657)
: paralysis of the lower half of the body with involvement of both legs
— para·ple·gic \-jik\ *adjective or noun*

para·po·di·um \-ˈpō-dē-əm\ *noun, plural* **-dia** \-dē-ə\ [New Latin] (1877)
1 : either of a pair of fleshy lateral processes borne by most segments of a polychaete worm
2 : a lateral expansion on each side of the foot usually forming a broad swimming organ in some gastropods
— para·po·di·al \-dē-əl\ *adjective*

para·pro·fes·sion·al \-prə-ˈfesh-nəl, -ˈfe-shə-nᵊl\ *noun* (1965)
: a trained aide who assists a professional person (as a teacher or doctor)
— paraprofessional *adjective*

para·psy·chol·o·gy \ˌpar-ə-(ˌ)sī-ˈkä-lə-jē\ *noun* [International Scientific Vocabulary] (1925)
: a field of study concerned with the investigation of evidence for paranormal psychological phenomena (as telepathy, clairvoyance, and psychokinesis)
— para·psy·cho·log·i·cal \-ˌsī-kə-ˈlä-ji-kəl\ *adjective*
— para·psy·chol·o·gist \-(ˌ)sī-ˈkä-lə-jist\ *noun*

para·quat \ˈpar-ə-ˌkwät\ *noun* [*para-* + *quaternary*] (circa 1961)
: a highly toxic contact herbicide containing a salt of a cation $[C_{12}H_{14}N_2]^{2+}$

para·ros·an·i·line \ˌpar-ə-ˌrō-ˈza-nᵊl-ən\ *noun* [International Scientific Vocabulary] (circa 1879)
: a white crystalline base $C_{19}H_{19}N_3O$ that is the parent compound of many dyes; *also* **:** its red chloride used especially as a biological stain

Pa·ra rubber \ˈpar-ə-, pə-ˈrä-\ *noun* [*Pará,* Brazil] (1857)
: native rubber from South American rubber trees (genus *Hevea* and especially *H. brasiliensis*)

Para rubber tree *noun* (1930)
: a South American rubber tree (*Hevea brasiliensis*)

para·sail·ing \ˈpar-ə-ˌsā-liŋ\ *noun* (1967)
: the recreational sport of soaring in a parachute while being towed usually by a motorboat

par·a·sang \ˈpar-ə-ˌsaŋ\ *noun* [Latin *parasanga,* from Greek *parasangēs,* of Iranian origin; akin to Persian *farsung* parasang] (1594)
: any of various Persian units of distance; *especially* **:** an ancient unit of about four miles (six kilometers)

para·sex·u·al \-ˈsek-sh(ə-)wəl, -shəl\ *adjective* (1954)
: relating to or being reproduction that results in recombination of genes from different individuals but does not involve meiosis and formation of a zygote by fertilization as in sexual reproduction ⟨the *parasexual* cycle in some fungi⟩
— para·sex·u·al·i·ty \-ˌsek-shə-ˈwa-lə-tē\ *noun*

pa·ra·shah \ˈpär-ə-ˌshä\ *noun* [Hebrew *pārāshāh,* literally, explanation] (1624)
: a passage in Jewish Scripture dealing with a single topic; *specifically* **:** a section of the Torah assigned for weekly reading in synagogue worship

par·a·site \ˈpar-ə-ˌsīt\ *noun* [Middle French, from Latin *parasitus,* from Greek *parasitos,* from *para-* + *sitos* grain, food] (1539)
1 : a person who exploits the hospitality of the rich and earns welcome by flattery
2 : an organism living in, with, or on another organism in parasitism
3 : something that resembles a biological parasite in dependence on something else for existence or support without making a useful or adequate return ☆

☆ SYNONYMS
Parasite, sycophant, toady, leech, sponge mean a usually obsequious flatterer or self-seeker. PARASITE applies to one who clings to a person of wealth, power, or influence or is useless to society ⟨a jet-setter with an entourage of *parasites*⟩. SYCOPHANT adds to this a strong suggestion of fawning, flattery, or adulation ⟨a powerful prince surrounded by *sycophants*⟩. TOADY emphasizes the servility and snobbery of the self-seeker ⟨cultivated leaders of society and became their *toady*⟩. LEECH stresses persistence in clinging to or bleeding another for one's own advantage ⟨a *leech* living off his family and friends⟩. SPONGE stresses the parasitic laziness, dependence, and opportunism of the cadger ⟨a shiftless *sponge,* always looking for a handout⟩.

— **par·a·sit·ic** \ˌpar-ə-ˈsi-tik\ *also* **par·a·sit·i·cal** \-ti-kəl\ *adjective*
— **par·a·sit·i·cal·ly** \-ti-k(ə-)lē\ *adverb*

par·a·sit·i·cid·al \ˌpar-ə-ˌsi-tə-ˈsī-dᵊl\ *adjective* (1892)
: destructive to parasites
— **par·a·sit·i·cide** \-ˈsi-tə-ˌsīd\ *noun*

par·a·sit·ise *British variant of* PARASITIZE

par·a·sit·ism \ˈpar-ə-sə-ˌti-zəm, -ˌsī-\ *noun* (1611)
1 : the behavior of a parasite
2 : an intimate association between organisms of two or more kinds; *especially* **:** one in which a parasite obtains benefits from a host which it usually injures
3 : PARASITOSIS

par·a·sit·ize \-sə-ˌtīz, -ˌsī-\ *transitive verb* **-ized; -iz·ing** (1890)
: to infest or live on or with as a parasite
— **par·a·sit·i·za·tion** \ˌpar-ə-sə-tə-ˈzā-shən, -ˌsī-\ *noun*

par·a·sit·oid \ˈpar-ə-sə-ˌtȯid, -ˌsī-\ *noun* (1922)
: an insect and especially a wasp that completes its larval development within the body of another insect eventually killing it and is free-living as an adult
— **parasitoid** *adjective*

par·a·si·tol·o·gy \ˌpar-ə-sə-ˈtä-lə-jē, -ˌsī-\ *noun* [Latin *parasitus* + International Scientific Vocabulary *-logy*] (1882)
: a branch of biology dealing with parasites and parasitism especially among animals
— **par·a·si·to·log·i·cal** \-ˌsi-tᵊl-ˈä-ji-kəl, -ˌsī-\ *also* **par·a·si·to·log·ic** \-jik\ *adjective*
— **par·a·si·to·log·i·cal·ly** \-ji-k(ə-)lē\ *adverb*
— **par·a·si·tol·o·gist** \-sə-ˈtä-lə-jist, -ˌsī-\ *noun*

par·a·sit·o·sis \-sə-ˈtō-səs, -ˌsī-\ *noun, plural* **-o·ses** \-ˌsēz\ [New Latin] (circa 1899)
: infestation with or disease caused by parasites

par·a·sol \ˈpar-ə-ˌsȯl, -ˌsäl\ *noun* [French, from Old Italian *parasole,* from *parare* to shield + *sole* sun, from Latin *sol* — more at PARAPET, SOLAR] (1660)
: a lightweight umbrella used as a sunshade especially by women

¹para·sym·pa·thet·ic \ˌpar-ə-ˌsim-pə-ˈthe-tik\ *adjective* [International Scientific Vocabulary] (1905)
: of, relating to, being, or acting on the parasympathetic nervous system

²parasympathetic *noun* (1925)
1 : a parasympathetic nerve
2 : PARASYMPATHETIC NERVOUS SYSTEM

parasympathetic nervous system *noun* (circa 1934)
: the part of the autonomic nervous system that contains chiefly cholinergic fibers, that tends to induce secretion, to increase the tone and contractility of smooth muscle, and to slow heart rate, and that consists of a cranial and a sacral part — compare SYMPATHETIC NERVOUS SYSTEM

para·sym·pa·tho·mi·met·ic \ˌpar-ə-ˈsim-pə-(ˌ)thō-mī-ˈme-tik, -mə-\ *adjective* [International Scientific Vocabulary] (1942)
: simulating parasympathetic nervous action in physiological effect

para·syn·the·sis \ˌpar-ə-ˈsin(t)-thə-səs\ *noun* [New Latin] (1862)
: the formation of words by adding a derivative ending and prefixing a particle (as in *denationalize*)
— **para·syn·thet·ic** \-sin-ˈthe-tik\ *adjective*

para·tac·tic \ˌpar-ə-ˈtak-tik\ *also* **para·tac·ti·cal** \-ti-kəl\ *adjective* (1871)
: of or relating to parataxis
— **para·tac·ti·cal·ly** \-ti-k(ə-)lē\ *adverb*

para·tax·is \ˌpar-ə-ˈtak-səs\ *noun* [New Latin, from Greek, act of placing side by side, from *paratassein* to place side by side, from *para-* + *tassein* to arrange] (circa 1842)
: the placing of clauses or phrases one after another without coordinating or subordinating connectives

para·thi·on \ˌpar-ə-ˈthī-ən, -ˌän\ *noun* [*para-* + *thio*phosphate + ¹*-on*] (1947)
: an extremely toxic insecticide $C_{10}H_{14}NO_5PS$

par·a·thor·mone \ˌpar-ə-ˈthȯr-ˌmōn\ *noun* [from *Parathormone,* a trademark] (1925)
: PARATHYROID HORMONE

¹para·thy·roid \-ˈthī-ˌrȯid\ *noun* (1897)
: PARATHYROID GLAND

²parathyroid *adjective* [International Scientific Vocabulary] (1902)
: of, relating to, or produced by the parathyroid glands

para·thy·roid·ec·to·my \-ˌrȯi-ˈdek-tə-mē\ *noun, plural* **-mies** (1903)
: partial or complete excision of the parathyroid glands
— **para·thy·roid·ec·to·mized** \-ˌmīzd\ *adjective*

parathyroid gland *noun* [International Scientific Vocabulary] (circa 1903)
: any of usually four small endocrine glands that are adjacent to or embedded in the thyroid gland and produce parathyroid hormone

parathyroid hormone *noun* (1953)
: a hormone of the parathyroid gland that regulates the metabolism of calcium and phosphorus in the body

para·troop·er \ˈpar-ə-ˌtrü-pər\ *noun* (1927)
: a member of the paratroops

para·troops \-ˌtrüps\ *noun plural* [²*para-*] (1940)
: troops trained and equipped to parachute from an airplane
— **para·troop** \-ˌtrüp\ *adjective*

¹para·ty·phoid \ˌpar-ə-ˈtī-ˌfȯid, -(ˌ)tī-ˈ\ *adjective* [International Scientific Vocabulary] (1902)
1 : resembling typhoid fever
2 : of or relating to paratyphoid or its causative organisms ⟨*paratyphoid* infection⟩

²paratyphoid *noun* (1903)
: a salmonellosis that resembles typhoid fever and is commonly contracted by eating contaminated food — called also *paratyphoid fever*

para·vane \ˈpar-ə-ˌvān\ *noun* [probably from French *para-* warding off (as in *parachute*) + English *vane*] (1919)
: a torpedo-shaped protective device with serrate teeth in its forward end used underwater by a ship in mined areas to sever the moorings of mines

par·boil \ˈpär-ˌbȯil\ *transitive verb* [Middle English, from *parboilen* to boil thoroughly, from Middle French *parboillir,* from Late Latin *perbullire,* from Latin *per-* thoroughly (from *per* through) + *bullire* to boil, from *bulla* bubble — more at FOR] (14th century)
: to boil briefly as a preliminary or incomplete cooking procedure

par·buck·le \ˈpär-ˌbə-kəl\ *noun* [origin unknown] (1626)
: a purchase for hoisting or lowering a cylindrical object by making fast the middle of a long rope aloft and looping both ends around the object which rests in the loops and rolls in them as the ends are hauled up or paid out; *also* **:** a single line made fast at one end and passed around an object that is used similarly
— **parbuckle** *transitive verb*

parbuckle

Par·cae \ˈpär-ˌkī, -ˌsē\ *noun plural* [Latin]
: FATE 4

¹par·cel \ˈpär-səl\ *noun* [Middle English, from Middle French, from (assumed) Vulgar Latin *particella,* from Latin *particula* small part — more at PARTICLE] (14th century)
1 a : FRAGMENT, PORTION **b :** a volume of a fluid (as air) considered as a single entity within a greater volume of the same fluid
2 : a tract or plot of land
3 : a company, collection, or group of persons, animals, or things **:** LOT ⟨the whole story was a *parcel* of lies⟩
4 a : a wrapped bundle **:** PACKAGE **b :** a unit of salable merchandise

²parcel *adverb* (15th century)
archaic **:** PARTLY

³parcel *adjective* (15th century)
: PART-TIME, PARTIAL

⁴parcel *transitive verb* **-celed** *or* **-celled; -cel·ing** *or* **-cel·ling** \ˈpär-s(ə-)liŋ\ (15th century)
1 : to divide into parts **:** DISTRIBUTE — often used with *out*
2 : to make up into a parcel **:** WRAP
3 : to cover (as a rope) with strips of canvas or tape

parcel post *noun* (1837)
1 : a mail service handling parcels
2 : packages handled by parcel post

par·ce·nary \ˈpär-sᵊn-ˌer-ē\ *noun* [Anglo-French *parcenarie,* from Old French *parçonerie,* from *parçon* portion, from Latin *partition-, partitio* partition] (1544)
: COPARCENARY 1

par·ce·ner \ˈpärs-nər, ˈpär-sᵊn-ər\ *noun* [Anglo-French, from Old French *parçonier,* from *parçon*] (1574)
: COPARCENER

parch \ˈpärch\ *verb* [Middle English] (14th century)
transitive verb
1 : to toast under dry heat
2 : to shrivel with heat
3 : to dry or shrivel with cold
intransitive verb
: to become dry or scorched

parched \ˈpärcht\ *adjective* (circa 1552)
: deprived of natural moisture; *also* **:** THIRSTY

Par·chee·si \pär-ˈchē-zē, pər-, *especially British* -sē\ *trademark*
— used for a board game adapted from pachisi

parch·ment \ˈpärch-mənt\ *noun* [Middle English *parchemin,* from Middle French, modification of Latin *pergamena,* from Greek *pergamēnē,* from feminine of *Pergamēnos* of Pergamum, from *Pergamon* Pergamum] (14th century)
1 : the skin of a sheep or goat prepared for writing on
2 : strong, tough, and often somewhat translucent paper made to resemble parchment
3 : a parchment manuscript; *also* **:** an academic diploma

¹pard \ˈpärd\ *noun* [Middle English *parde,* from Middle French, from Latin *pardus,* from Greek *pardos*] (14th century)
: LEOPARD

²pard *noun* [short for *pardner*] (1850)
chiefly dialect **:** PARTNER, CHUM

par·die *or* **par·di** *or* **par·dy** \pər-ˈdē, pär-\ *interjection* [Middle English *pardee,* from Old French *par Dé* by God] (14th century)
archaic — used as a mild oath

pard·ner \ˈpärd-nər\ *noun* [alteration of *partner*] (1795)
chiefly dialect **:** PARTNER, CHUM

¹par·don \ˈpär-dᵊn\ *noun* [Middle English, from Middle French, from *pardoner*] (14th century)
1 : INDULGENCE 1
2 : the excusing of an offense without exacting a penalty
3 a : a release from the legal penalties of an offense **b :** an official warrant of remission of penalty
4 : excuse or forgiveness for a fault, offense, or discourtesy ⟨I beg your *pardon*⟩

²pardon *transitive verb* **par·doned; par·don·ing** \'pärd-niŋ, 'pär-dᵊn-iŋ\ [Middle English, from Middle French *pardoner,* from Late Latin *perdonare* to grant freely, from Latin *per-* thoroughly + *donare* to give — more at PARBOIL, DONATION] (15th century)
1 a : to absolve from the consequences of a fault or crime **b :** to allow (an offense) to pass without punishment **:** FORGIVE **c :** to relieve of a penalty improperly assessed
2 : TOLERATE
synonym see EXCUSE

par·don·able \'pärd-nə-bəl, 'pär-dᵊn-ə-bəl\ *adjective* (15th century)
: admitting of being pardoned **:** EXCUSABLE ⟨*pardonable* offenses⟩
— par·don·able·ness *noun*
— par·don·ably \-blē\ *adverb*

par·don·er \'pärd-nər, 'pär-dᵊn-ər\ *noun* (14th century)
1 : a medieval preacher delegated to raise money for religious works by soliciting offerings and granting indulgences
2 : one that pardons

pare \'par, 'per\ *transitive verb* **pared; par·ing** [Middle English, from Middle French *parer* to prepare, trim, from Latin *parare* to prepare, acquire; akin to Latin *parere* to give birth to, produce, Lithuanian *perėti* to hatch] (14th century)
1 : to trim off an outside, excess, or irregular part of ⟨*pare* apples⟩ ⟨*paring* his nails⟩
2 : to diminish or reduce by or as if by paring ⟨*pare* expenses⟩ ⟨the novel was *pared* down to 200 pages⟩
— par·er *noun*

par·e·go·ric \ˌpar-ə-'gȯr-ik, -'gōr-, -'gär-\ *noun* [French *parégorique* mitigating pain, from Late Latin *paregoricus,* from Greek *parēgorikos,* from *parēgorein* to talk over, soothe, from *para-* + *agora* assembly, from *ageirein* to gather] (circa 1847)
: camphorated tincture of opium used especially to relieve pain

pa·ren·chy·ma \pə-'reŋ-kə-mə\ *noun* [New Latin, from Greek, visceral flesh, from *parenchein* to pour in beside, from *para-* + *en-* en- + *chein* to pour — more at FOUND] (1651)
1 : a tissue of higher plants that consists of thin-walled living photosynthetic or storage cells capable of division even when mature and that makes up much of the substance of leaves and roots, the pulp of fruits, and parts of stems and supporting structures
2 : the essential and distinctive tissue of an organ or an abnormal growth as distinguished from its supportive framework
— par·en·chy·ma·tous \ˌpar-ən-'kī-mə-təs, -'ki-\ *or* **pa·ren·chy·mal** \pə-'reŋ-kə-məl, ˌpar-ən-'kī-\ *adjective*

¹par·ent \'par-ənt, 'per-\ *noun* [Middle English, from Middle French, from Latin *parent-, parens;* akin to Latin *parere* to give birth to] (15th century)
1 a : one that begets or brings forth offspring **b :** a person who brings up and cares for another
2 a : an animal or plant that is regarded in relation to its offspring **b :** the material or source from which something is derived **c :** a group from which another arises and to which it usually remains subsidiary ⟨a *parent* company⟩
— parent *adjective*
— pa·ren·tal \pə-'ren-tᵊl\ *adjective*
— pa·ren·tal·ly \-tᵊl-ē\ *adverb*
— par·ent·less \'par-ənt-ləs, 'per-\ *adjective*

²parent *transitive verb* (1663)
: to be or act as the parent of **:** ORIGINATE, PRODUCE

par·ent·age \'par-ən-tij, 'per-\ *noun* [Middle English, from Middle French, from *parent*] (15th century)
1 a : descent from parents or ancestors **:** LINEAGE ⟨a person of noble *parentage*⟩ **b :** DERIVATION, ORIGIN ⟨a tradition of uncertain *parentage*⟩
2 : PARENTHOOD

par·en·ter·al \pə-'ren-tə-rəl\ *adjective* [International Scientific Vocabulary *para-* + *enteral*] (circa 1910)
: situated or occurring outside the intestine; *especially* **:** introduced otherwise than by way of the intestines
— par·en·ter·al·ly \-rə-lē\ *adverb*

pa·ren·the·sis \pə-'ren(t)-thə-səs\ *noun, plural* **-the·ses** \-ˌsēz\ [Late Latin, from Greek, literally, act of inserting, from *parentithenai* to insert, from *para-* + *en-* en- + *tithenai* to place — more at DO] (circa 1550)
1 a : an amplifying or explanatory word, phrase, or sentence inserted in a passage from which it is usually set off by punctuation **b :** a remark or passage that departs from the theme of a discourse **:** DIGRESSION
2 : INTERLUDE, INTERVAL
3 : one or both of the curved marks ( ) used in writing and printing to enclose a parenthetical expression or to group a symbolic unit in a logical or mathematical expression
— par·en·thet·i·cal \ˌpar-ən-'the-ti-kəl\ *also* **par·en·thet·ic** \-tik\ *adjective*
— par·en·thet·i·cal·ly \-ti-k(ə-)lē\ *adverb*

pa·ren·the·size \pə-'ren(t)-thə-ˌsīz\ *transitive verb* **-sized; -siz·ing** (1837)
: to make a parenthesis of **:** enclose within parentheses

par·ent·hood \'par-ənt-ˌhu̇d, 'per-\ *noun* (1856)
: the state of being a parent; *specifically* **:** the position, function, or standing of a parent

par·ent·ing \'par-ən-tiŋ, 'per-\ *noun* (1958)
1 : the raising of a child by its parents
2 : the act or process of becoming a parent
3 : the act of caring for someone in the manner of a parent

parent–teacher association *noun, often P&T&A capitalized* (1915)
: an organization of local groups of teachers and the parents of their pupils that works for the improvement of the schools and the benefit of the pupils

pa·re·sis \pə-'rē-səs, 'par-ə-\ *noun, plural* **pa·re·ses** \-ˌsēz\ [New Latin, from Greek, from *parienai* to let fall, from *para-* + *hienai* to let go, send — more at JET] (1693)
1 : slight or partial paralysis
2 : GENERAL PARESIS
— pa·ret·ic \pə-'re-tik\ *adjective or noun*

par·es·the·sia \ˌpar-əs-'thē-zhə\ *noun* [New Latin, from *para-* + *-esthesia* (as in *anesthesia*)] (circa 1860)
: a sensation of pricking, tingling, or creeping on the skin that has no objective cause
— par·es·thet·ic \-'the-tik\ *adjective*

pa·reu *or* **pa·reo** \'pär-ā-(ˌ)ü\ *noun* [Tahitian] (1860)
: a wraparound skirt usually made from a rectangular piece of printed cloth and worn by men and women throughout Polynesia

pa·re·ve \'pär-(ə-)və\ *adjective* [Yiddish *parev*] (1941)
: made without milk, meat, or their derivatives ⟨*pareve* margarine⟩ — compare FLEISHIG, MILCHIG

par ex·cel·lence \'pär-ˌek-sə-'läⁿs\ *adjective* [French, literally, by excellence] (1695)
: being the best of a kind **:** PREEMINENT ⟨a salesman *par excellence*⟩

par·fait \pär-'fā\ *noun* [French, literally, something perfect, from *parfait* perfect, from Latin *perfectus*] (1894)
1 : a flavored custard containing whipped cream and syrup frozen without stirring
2 : a cold dessert made of layers of fruit, syrup, ice cream, and whipped cream

parfait glass *noun* (circa 1951)
: a tall narrow glass with a short stem

par·fleche \'pär-ˌflesh\ *noun* [Canadian French *parflèche,* from French *parer* to ward off + *flèche* arrow] (1827)
1 : a raw hide soaked in lye to remove the hair and dried
2 : an article (as a bag or case) made of parfleche

par·fo·cal \ˌpär-'fō-kəl\ *adjective* [Latin *par* equal + English *focal*] (1886)
: being or having lenses or lens sets (as eyepieces) with the corresponding focal points all in the same plane
— par·fo·cal·i·ty \ˌpär-fō-'ka-lə-tē\ *noun*
— par·fo·cal·ize \ˌpär-'fō-kə-ˌlīz\ *transitive verb*

parge \'pärj\ *transitive verb* **parged; parg·ing** (1701)
: PARGET

¹par·get \'pär-jət\ *transitive verb* **-get·ed** *or* **-get·ted; -get·ing** *or* **-get·ting** [Middle English *pargetten,* from Middle French *parjeter* to throw on top of, from *par-* thoroughly (from Latin *per-*) + *jeter* to throw — more at JET] (14th century)
: to coat with plaster; *especially* **:** to apply ornamental or waterproofing plaster to

²parget *noun* (14th century)
1 : plaster, whitewash, or roughcast for coating a wall
2 : plasterwork especially in raised ornamental figures on walls

par·gy·line \'pär-jə-ˌlēn\ *noun* [pro*pargyl* (an alcohol) + ²*-ine*] (1961)
: a monoamine oxidase inhibitor $C_{11}H_{13}N$ that is used in the hydrochloride especially as an antihypertensive agent

parhelic circle *noun* (1890)
: a luminous circle or halo parallel to the horizon at the altitude of the sun — called also *parhelic ring*

par·he·lion \pär-'hēl-yən\ *noun, plural* **-lia** \-yə\ [Latin *parelion,* from Greek *parēlion,* from *para-* + *hēlios* sun — more at SOLAR] (1647)
: any of several bright spots often tinged with color that often appear on the parhelic circle
— par·he·lic \-'hē-lik\ *adjective*

pa·ri·ah \pə-'rī-ə\ *noun* [Tamil *paraiyan,* literally, drummer] (1613)
1 : a member of a low caste of southern India
2 : OUTCAST 1

par·i·an \'par-ē-ən, 'per-\ *noun* [*Parian,* adjective; from its suitability for making statuettes] (1850)
: a porcelain composed essentially of kaolin and feldspar and usually used unglazed in ornamental articles

Par·i·an \'par-ē-ən, 'per-\ *adjective* (1611)
: of or relating to the island of Paros noted for its marble used extensively for sculpture in ancient times

Parian ware *noun* (1894)
1 : PARIAN
2 : articles made of parian

¹pa·ri·e·tal \pə-'rī-ə-tᵊl\ *adjective* [Middle French, from New Latin *pariet-, paries* wall of a cavity or hollow organ, from Latin, wall] (15th century)
1 a : of or relating to the walls of a part or cavity **b :** of, relating to, or forming the upper posterior wall of the head
2 : attached to the main wall rather than the axis or a cross wall of a plant ovary — used of an ovule or a placenta
3 : of or relating to college living or its regulation; *especially* **:** of or relating to parietals

²parietal *noun* (15th century)
1 : a parietal part (as a bone, scale, or plate)

2 *plural* **:** the regulations governing the visiting privileges of members of the opposite sex in campus dormitories
parietal bone *noun* (15th century)
: either of a pair of bones of the roof of the skull between the frontal bones and the occipital bones
parietal cell *noun* (1875)
: any of the large oval cells of the gastric mucous membrane that secrete hydrochloric acid
parietal lobe *noun* (circa 1889)
: the middle division of each cerebral hemisphere that contains an area concerned with bodily sensations
pari–mu·tu·el \ˌpar-i-ˈmyü-chə-wəl, -chəl; -ˈmyüch-wəl\ *noun* [French *pari mutuel,* literally, mutual stake] (1888)
1 : a betting pool in which those who bet on competitors finishing in the first three places share the total amount bet minus a percentage for the management
2 : a machine for registering the bets and computing the payoffs in pari-mutuel betting
par·ing \ˈpar-iŋ, ˈper-\ *noun* (14th century)
1 : the act of cutting away an edge or surface
2 : something pared off ⟨apple *parings*⟩
paring knife *noun* (1591)
: a small short-bladed knife (as for paring fruit)
pa·ri pas·su \ˈpar-i-ˈpa-(ˌ)sü\ *adverb or adjective* [Latin, with equal step] (1567)
: at an equal rate or pace
Par·is \ˈpar-əs\ *noun* [Latin, from Greek]
: a son of Priam whose abduction of Helen leads to the Trojan War
Par·is green \ˈpar-əs-\ *noun* [*Paris,* France] (1868)
1 : a very poisonous copper-based bright green powder that is used as an insecticide and pigment
2 : a brilliant yellowish green
par·ish \ˈpar-ish\ *noun* [Middle English *parisshe,* from Middle French *parroche,* from Late Latin *parochia,* from Late Greek *paroikia,* from *paroikos* Christian, from Greek, stranger, from *para-* + *oikos* house — more at VICINITY] (14th century)
1 a (1) **:** the ecclesiastical unit of area committed to one pastor (2) **:** the residents of such an area **b** *British* **:** a subdivision of a county often coinciding with an original ecclesiastical parish and constituting the unit of local government
2 : a local church community composed of the members or constituents of a Protestant church
3 : a civil division of the state of Louisiana corresponding to a county in other states
pa·rish·ion·er \pə-ˈri-sh(ə-)nər\ *noun* [Middle English *parisshoner,* probably modification of Middle French *parrochien,* from *parroche*] (15th century)
: a member or inhabitant of a parish
¹par·i·ty \ˈpar-ə-tē\ *noun, plural* **-ties** [Latin *paritas,* from *par* equal] (circa 1609)
1 : the quality or state of being equal or equivalent
2 a : equivalence of a commodity price expressed in one currency to its price expressed in another **b :** equality of purchasing power established by law between different kinds of money at a given ratio
3 : an equivalence between farmers' current purchasing power and their purchasing power at a selected base period maintained by government support of agricultural commodity prices
4 a : the property of an integer with respect to being odd or even ⟨3 and 7 have the same *parity*⟩ **b** (1) **:** the state of being odd or even used as the basis of a method of detecting errors in binary-coded data (2) **:** PARITY BIT
5 : the property of oddness or evenness of a quantum mechanical function
6 : the symmetry of behavior in an interaction of a physical entity (as a subatomic particle) with that of its mirror image

²parity *noun* [*-parous*] (1878)
: the state or fact of having borne offspring; *also* **:** the number of children previously borne
parity bit *noun* (1957)
: a bit added to an array of bits (as on magnetic tape) to provide parity
¹park \ˈpärk\ *noun* [Middle English, from Old French *parc* enclosure, from Medieval Latin *parricus*] (13th century)
1 a : an enclosed piece of ground stocked with game and held by royal prescription or grant **b :** a tract of land that often includes lawns, woodland, and pasture attached to a country house and is used as a game preserve and for recreation
2 a : a piece of ground in or near a city or town kept for ornament and recreation **b :** an area maintained in its natural state as a public property
3 a *West* **:** a level valley between mountain ranges **b :** an open space and especially a grassland that is often all or partly surrounded by woodland and is suitable for cultivation or grazing
4 a : a space occupied by military animals, vehicles, or materials **b :** PARKING LOT
5 : an enclosed arena or stadium used especially for ball games
6 : an area designed for a specified industrial, commercial, or residential use ⟨amusement *park*⟩ ⟨industrial *park*⟩ ⟨mobile home *park*⟩
— park·like \ˈpärk-ˌlīk\ *adjective*
²park (1526)
transitive verb
1 : to enclose in a park
2 a (1) **:** to bring (a vehicle) to a stop and keep standing at the edge of a public way (2) **:** to leave temporarily on a public way or in a parking lot or garage **b :** to land and leave (as an airplane) **c :** to establish (as a satellite) in orbit
3 a : to set and leave temporarily ⟨*parked* his hat on the chair⟩ **b :** to place, settle, or establish especially for a considerable time ⟨kids *parked* in front of the TV⟩ ⟨*parked* her money in a savings account⟩
4 : to assemble (as equipment or stores) in a military dump or park
intransitive verb
: to park a vehicle
— park·er *noun*
par·ka \ˈpär-kə\ *noun* [Aleut, from Russian dialect, ultimately from Nenets (Samoyedic language of northern Russia)] (1780)
1 : a hooded fur pullover garment for arctic wear
2 : a usually lined fabric outerwear pullover or jacket
parking lot *noun* (1924)
: an area used for the parking of motor vehicles
parking meter *noun* (1936)
: a coin-operated device which registers the purchase of parking time for a motor vehicle
par·kin·so·nian \ˌpär-kən-ˈsō-nē-ən, -nyən\ *adjective* (1906)
1 : of or similar to that of parkinsonism
2 : affected with parkinsonism and especially Parkinson's disease
par·kin·son·ism \ˈpär-kən-sə-ˌni-zəm\ *noun* (circa 1923)
1 : PARKINSON'S DISEASE
2 : a nervous disorder that resembles Parkinson's disease
Par·kin·son's disease \ˈpär-kən-sənz-\ *noun* [James *Parkinson* (died 1824) English physician] (1877)
: a chronic progressive nervous disease chiefly of later life that is linked to decreased dopamine production in the substantia nigra and is marked by tremor and weakness of resting muscles and by a shuffling gait — called also *paralysis agitans, Parkinson's, Parkinson's syndrome*
Parkinson's Law *noun* [C. Northcote *Parkinson* (born 1909) English historian] (1955)
1 : an observation in office organization: the number of subordinates increases at a fixed rate regardless of the amount of work produced
2 : an observation in office organization: work expands so as to fill the time available for its completion
park·land \ˈpärk-ˌland\ *noun* (1907)
: land with clumps of trees and shrubs in cultivated condition used as or suitable for use as a park
park·way \ˈpärk-ˌwā\ *noun* (1887)
: a broad landscaped thoroughfare
par·lance \ˈpär-lən(t)s\ *noun* [Middle French, from Old French, from *parler*] (circa 1580)
1 : SPEECH; *especially* **:** formal debate or parley
2 : manner or mode of speech **:** IDIOM
par·lan·do \pär-ˈlän-(ˌ)dō\ *or* **par·lan·te** \-(ˌ)tā\ *adjective* [*parlando* from Italian, verbal of *parlare* to speak, from Medieval Latin *parabolare; parlante* from Italian, present participle of *parlare*] (circa 1854)
: delivered or performed in a style suggestive of speech — used as a direction in music
¹par·lay \ˈpär-ˌlā, -lē\ *transitive verb* [French *paroli,* noun, parlay, from Italian dialect, plural of *parolo,* perhaps from *paro* equal, from Latin *par*] (1828)
1 : to bet in a parlay
2 a : to exploit successfully **b :** to increase or otherwise transform into something of much greater value
²parlay *noun* (1904)
: a series of two or more bets so set up in advance that the original stake plus its winnings are risked on the successive wagers; *broadly* **:** the fresh risking of an original stake together with its winnings
parle \ˈpär(-ə)l\ *intransitive verb* **parled; parl·ing** [Middle English, to parley, from Middle French *parler*] (14th century)
archaic **:** PARLEY
— parle *noun, archaic*
¹par·ley \ˈpär-lē\ *noun, plural* **parleys** [Middle English *parlai* speech, probably from Middle French *parlee,* from feminine of *parlé,* past participle of *parler* to speak, from Medieval Latin *parabolare,* from Late Latin *parabola* speech, parable — more at PARABLE] (1581)
1 a : a conference for discussion of points in dispute **b :** a conference with an enemy
2 : DISCUSSION
²parley *intransitive verb* **par·leyed; par·ley·ing** (1591)
: to speak with another **:** CONFER; *specifically* **:** to discuss terms with an enemy
par·lia·ment \ˈpär-lə-mənt *also* ˈpärl-yə-\ *noun* [Middle English, from Old French *parlement,* from *parler*] (13th century)
1 : a formal conference for the discussion of public affairs; *specifically* **:** a council of state in early medieval England
2 a : an assemblage of the nobility, clergy, and commons called together by the British sovereign as the supreme legislative body in the United Kingdom **b :** a similar assemblage in another nation or state
3 a : the supreme legislative body of a usually major political unit that is a continuing institution comprising a series of individual assemblages **b :** the British House of Commons
4 : one of several principal courts of justice existing in France before the revolution of 1789
par·lia·men·tar·i·an \ˌpär-lə-ˌmen-ˈter-ē-ən, -mən- *also* ˌpärl-yə-\ *noun* (1644)
1 *often capitalized* **:** an adherent of the parliament in opposition to the king during the English Civil War
2 : an expert in the rules and usages of a deliberative assembly (as a parliament)
3 : a member of a parliament
par·lia·men·ta·ry \-ˈmen-t(ə-)rē\ *adjective* (1616)

1 a : of or relating to a parliament **b :** enacted, done, or ratified by a parliament
2 : of or adhering to the parliament as opposed to the king during the English Civil War
3 : of, based on, or having the characteristics of parliamentary government
4 : of or relating to members of a parliament
5 : of or according to parliamentary law ⟨*parliamentary* procedure⟩

parliamentary government *noun* (1858)
: a system of government having the real executive power vested in a cabinet composed of members of the legislature who are individually and collectively responsible to the legislature

parliamentary law *noun* (1893)
: the rules and precedents governing the proceedings of deliberative assemblies and other organizations

[1]par·lor \ˈpär-lər\ *noun* [Middle English *parlour,* from Old French, from *parler*] (13th century)
1 : a room used primarily for conversation or the reception of guests: as **a :** a room in a private dwelling for the entertainment of guests **b :** a conference chamber or private reception room **c :** a room in an inn, hotel, or club for conversation or semiprivate uses
2 : any of various business places ⟨a funeral *parlor*⟩ ⟨a beauty *parlor*⟩ ◆

[2]parlor *adjective* (1552)
1 : used in or suitable for a parlor ⟨*parlor* furniture⟩
2 a : fostered or advocated in comfortable seclusion without consequent action or application to affairs ⟨*parlor* bolshevism⟩ **b :** given to or characterized by fostering or advocating something (as a doctrine) in such a manner ⟨*parlor* socialist⟩

parlor car *noun* (1868)
: an extra-fare railroad passenger car for day travel equipped with individual chairs

parlor game *noun* (1872)
: a game suitable for playing indoors

parlor grand *noun* (1856)
: a grand piano intermediate in length between a concert grand and a baby grand

par·lour \ˈpär-lər\ *chiefly British variant of* PARLOR

par·lous \ˈpär-ləs\ *adjective* [Middle English, alteration of *perilous*] (14th century)
1 *obsolete* **:** dangerously shrewd or cunning
2 : full of danger or risk **:** HAZARDOUS
— par·lous·ly *adverb*

Par·me·san \ˈpär-mə-ˌzän, -ˌzhän, -zən, -ˌzan\ *noun* [*Parmesan* of Parma, from Middle French *parmesan,* from Old Italian *parmigiano*] (1538)
: a very hard dry sharply flavored cheese that is sold grated or in wedges

par·mi·gia·na \ˌpär-mi-ˈjä-nə, -ˈzhän; ˈpär-miˌzhän, -ˌzän\ *or* **par·mi·gia·no** \-ˈjä-(ˌ)nō\ *adjective* [Italian *parmigiana,* feminine of *parmigiano* of Parma, from *Parma*] (1943)
: made or covered with Parmesan cheese ⟨veal *parmigiana*⟩

Par·nas·si·an \pär-ˈna-sē-ən\ *adjective* (circa 1644)
1 [Latin *parnassius* of Parnassus, from Greek *parnasios,* from *Parnasos* Parnassus, mountain in Greece sacred to Apollo and the Muses] **:** of or relating to poetry
2 [French *parnassien,* from *Parnasse* Parnassus; from *Le Parnasse contemporain* (1866), an anthology of poetry] **:** of or relating to a school of French poets of the second half of the 19th century emphasizing metrical form rather than emotion
— Parnassian *noun*

pa·ro·chi·al \pə-ˈrō-kē-əl\ *adjective* [Middle English *parochiall,* from Middle French *parochial,* from Late Latin *parochialis,* from *parochia* parish — more at PARISH] (14th century)
1 : of or relating to a church parish
2 : of or relating to a parish as a unit of local government
3 : confined or restricted as if within the borders of a parish **:** limited in range or scope (as to a narrow area or region) **:** PROVINCIAL, NARROW
— pa·ro·chi·al·ly \-kē-ə-lē\ *adverb*

pa·ro·chi·al·ism \-kē-ə-ˌli-zəm\ *noun* (1847)
: the quality or state of being parochial; *especially* **:** selfish pettiness or narrowness (as of interests, opinions, or views)

parochial school *noun* (1755)
: a private school maintained by a religious body usually for elementary and secondary instruction

par·o·dist \ˈpar-ə-dist\ *noun* (1742)
: a writer of parodies

[1]par·o·dy \ˈpar-ə-dē\ *noun, plural* **-dies** [Latin *parodia,* from Greek *parōidia,* from *para-* + *aidein* to sing — more at ODE] (1598)
1 : a literary or musical work in which the style of an author or work is closely imitated for comic effect or in ridicule
2 : a feeble or ridiculous imitation
synonym see CARICATURE
— pa·rod·ic \pə-ˈrä-dik, pa-\ *adjective*
— par·o·dis·tic \ˌpar-ə-ˈdis-tik\ *adjective*

[2]parody *transitive verb* **-died; -dy·ing** (circa 1745)
1 : to compose a parody on ⟨*parody* a poem⟩
2 : to imitate in the manner of a parody

par·ol \ˈpar-əl\ *noun* [Middle French *parole*] (1590)
: WORD OF MOUTH
— parol *adjective*

[1]pa·role \pə-ˈrōl\ *noun* [French, speech, parole, from Middle French, from Late Latin *parabola* speech — more at PARABLE] (circa 1616)
1 : a promise made with or confirmed by a pledge of one's honor; *especially* **:** the promise of a prisoner of war to fulfill stated conditions in consideration of his release
2 : a watchword given only to officers of the guard and of the day
3 : a conditional release of a prisoner serving an indeterminate or unexpired sentence
4 a : language viewed as a specific individual usage **:** PERFORMANCE **b :** a linguistic act — compare LANGUE
— parole *adjective*

[2]parole *transitive verb* **pa·roled; pa·rol·ing** (1790)
: to release (a prisoner) on parole

pa·rol·ee \pə-ˌrō-ˈlē, -ˈrō-(ˌ); ˌpar-ə-ˈlē\ *noun* (1903)
: one released on parole

par·o·no·ma·sia \ˌpar-ə-nō-ˈmā-zh(ē-)ə, ˌparˌä-nə-ˈmā-\ *noun* [Latin, from Greek, from *paronomazein* to call with a slight change of name, from *para-* + *onoma* name — more at NAME] (1577)
: a play on words **:** PUN
— par·o·no·mas·tic \-ˈmas-tik\ *adjective*

par·o·nym \ˈpar-ə-ˌnim\ *noun* [Late Latin *paronymon,* from Greek *parōnymon,* neuter of *parōnymos*] (circa 1846)
: a paronymous word

par·on·y·mous \pə-ˈrä-nə-məs, pa-\ *adjective* [Greek *parōnymos,* from *para-* + *-ōnymos* (as in *homōnymos* homonymous)] (circa 1661)
1 : CONJUGATE 4
2 a : formed from a word in another language **b :** having a form similar to that of a cognate foreign word

pa·rot·id \pə-ˈrä-təd\ *adjective* [New Latin *parotid-, parotis* parotid gland, from Latin, tumor near the ear, from Greek *parōtid-, parōtis,* from *para-* + *ōt-, ous* ear — more at EAR] (1687)
: of or relating to the parotid gland

parotid gland *noun* (circa 1771)
: either of a pair of large serous salivary glands situated below and in front of the ear

par·o·ti·tis \ˌpar-ō-ˈtī-təs\ *noun* (1822)
: inflammation of the parotid glands; *also* **:** MUMPS

par·ous \ˈpar-əs, ˈper-\ *adjective* [*-parous*] (circa 1889)
: having produced offspring

-parous *adjective combining form* [Latin *-parus,* from *parere* to give birth to, produce — more at PARE]
: giving birth to **:** producing ⟨multi*parous*⟩

Par·ou·sia \ˌpär-ü-ˈsē-ə, pə-ˈrü-zē-ə\ *noun* [Greek, literally, presence, from *paront-, parōn,* present participle of *pareinai* to be present, from *para-* + *einai* to be — more at IS] (1875)
: SECOND COMING

par·ox·ysm \ˈpar-ək-ˌsi-zəm *also* pə-ˈräk-\ *noun* [French & Medieval Latin; French *paroxysme,* from Medieval Latin *paroxysmus,* from Greek *paroxysmos,* from *paroxynein* to stimulate, from *para-* + *oxynein* to provoke, from *oxys* sharp — more at OXYGEN] (15th century)
1 : a fit, attack, or sudden increase or recurrence of symptoms (as of a disease) **:** CONVULSION ⟨a *paroxysm* of coughing⟩
2 : a sudden violent emotion or action **:** OUTBURST ⟨a *paroxysm* of rage⟩
— par·ox·ys·mal \ˌpar-ək-ˈsiz-məl *also* pəˌräk-\ *adjective*

[1]par·quet \pär-ˈkā\ *transitive verb* **par·queted** \-ˈkād\; **par·quet·ing** \-ˈkā-iŋ\ (1678)
1 : to furnish with a floor of parquet
2 : to make of parquetry

[2]par·quet \ˈpär-ˌkā, pär-ˈ\ *noun* [French, from Middle French, small enclosure, from *parc* park] (1816)
1 a : a patterned wood surface (as flooring or paneling); *especially* **:** one made of parquetry **b :** PARQUETRY
2 : the main floor of a theater; *specifically* **:** the part from the front of the stage to the parquet circle

parquet circle *noun* (1854)
: the part of the main floor of a theater that is beneath the galleries

par·que·try \ˈpär-kə-trē\ *noun, plural* **-tries** (circa 1842)

◇ WORD HISTORY
parlor Within some medieval religious orders a rule of silence was enforced, and the members of a community belonging to the order could speak with each other or outsiders only in a special room. In 12th century Old French this room was called the *parleure,* a derivative of the verb *parler* "to speak." The Old French word was borrowed at an early date into Middle English as *parlure* or *parlour.* In both French and English the sense of the word was extended to rooms such as a chamber for private conversation off the main hall of a manor house, or a room in a municipal building used for conferences. In England the *parlour* in an ordinary middle-class home developed into the best-furnished room in the house, often facing the street, where a family received guests; the word eventually became virtually synonymous with *drawing room* or *living room.* A late 19th century development originally unique to the U.S. was the use of *parlor* with a preceding attributive noun to denote a well-appointed commercial establishment. Though the *Oxford English Dictionary* labeled such usages as "commercial cant," no stigma is attached to *ice cream parlor, beauty parlor,* or *funeral parlor* in 20th century American English.

: work in the form of usually geometrically patterned wood laid or inlaid especially for floors

parr \'pär\ *noun, plural* **parr** *also* **parrs** [origin unknown] (circa 1722)
: a young salmon actively feeding in freshwater

par·ra·keet *variant of* PARAKEET

par·rel *or* **par·ral** \'par-əl\ *noun* [Middle English *perell,* from alteration of *parail* apparel, short for *apparail,* from Middle French *apareil,* from *apareillier* to prepare — more at APPAREL] (15th century)
: a rope loop or sliding collar by which a yard or spar is held to a mast in such a way that it may be hoisted or lowered

par·ri·cid·al \ˌpar-ə-'sī-dᵊl\ *adjective* (1627)
: of, relating to, or guilty of parricide

par·ri·cide \'par-ə-ˌsīd\ *noun* (1554)
1 [Latin *parricida* killer of a close relative, from *parri-* (perhaps akin to Greek *pēos* kinsman by marriage) + *-cida* -cide] **:** one that murders his or her father, mother, or a close relative
2 [Latin *parricidium* murder of a close relative, from *parri-* + *-cidium* -cide] **:** the act of a parricide

¹par·rot \'par-ət\ *noun* [probably irregular from Middle French *perroquet*] (circa 1525)
1 : any of numerous widely distributed tropical zygodactyl birds (order Psittaciformes and especially family Psittacidae) that have a distinctive stout curved hooked bill, are often crested and brightly variegated, and include some excellent mimics
2 : a person who sedulously echoes another's words
— parrot *adjective*

²parrot *transitive verb* (1596)
: to repeat by rote

parrot fever *noun* (1930)
: PSITTACOSIS

parrot fish *noun* (1712)
: any of a family (Scaridae) of usually brightly colored chiefly tropical marine fishes that have the teeth in each jaw fused into a cutting plate like a beak

par·ry \'par-ē\ *verb* **par·ried; par·ry·ing** [probably from French *parez,* imperative of *parer* to parry, from Old Provençal *parar,* from Latin *parare* to prepare — more at PARE] (1672)
intransitive verb
1 : to ward off a weapon or blow
2 : to evade or turn aside something
transitive verb
1 : to ward off (as a blow)
2 : to evade especially by an adroit answer ⟨*parry* an embarrassing question⟩
— parry *noun*

¹parse \'pärs, *chiefly British* 'pärz\ *verb* **parsed; pars·ing** [Latin *pars orationis* part of speech] (1553)
transitive verb
1 a : to resolve (as a sentence) into component parts of speech and describe them grammatically **b :** to describe grammatically by stating the part of speech and explaining the inflection and syntactical relationships
2 : to examine in a minute way **:** analyze critically ⟨*parses* appellate court opinions⟩
intransitive verb
1 : to give a grammatical description of a word or a group of words
2 : to admit of being parsed

²parse *noun* (1963)
: a product or an instance of parsing

par·sec \'pär-ˌsek\ *noun* [*par*allax + *sec*ond] (1913)
: a unit of measure for interstellar space equal to the distance to an object having a parallax of one second or to 3.26 light-years

pars·er \'pär-sər\ *noun* (circa 1864)
: one that parses; *specifically* **:** a computer program that breaks down text into recognized strings of characters for further analysis

Par·si *also* **Par·see** \'pär-(ˌ)sē\ *noun* [Persian *pārsī,* from *Pārs* Persia] (1615)
1 : a Zoroastrian descended from Persian refugees settled principally at Bombay
2 : the Iranian dialect of the Parsi religious literature
— Par·si·ism \-ˌi-zəm\ *noun*

par·si·mo·ni·ous \ˌpär-sə-'mō-nē-əs\ *adjective* (1598)
: exhibiting or marked by parsimony; *especially* **:** frugal to the point of stinginess
synonym see STINGY
— par·si·mo·ni·ous·ly *adverb*

par·si·mo·ny \'pär-sə-ˌmō-nē\ *noun* [Middle English *parcimony,* from Latin *parsimonia,* from *parsus,* past participle of *parcere* to spare] (15th century)
1 a : the quality of being careful with money or resources **:** THRIFT **b :** the quality or state of being stingy
2 : economy in the use of means to an end; *especially* **:** economy of explanation in conformity with Occam's razor

pars·ley \'pär-slē\ *noun* [Middle English *persely,* from Old English *petersilie,* from (assumed) Vulgar Latin *petrosilium,* alteration of Latin *petroselinum,* from Greek *petroselinon,* from *petros* stone + *selinon* celery] (before 12th century)
: a European biennial herb (*Petroselinum crispum*) of the carrot family that is widely cultivated for its leaves which are used as a culinary herb or garnish

pars·leyed *also* **pars·lied** \-slēd\ *adjective* (1952)
: garnished or flavored with parsley ⟨*parsleyed* potatoes⟩

pars·nip \'pär-snəp\ *noun* [Middle English *pasnepe,* modification of Middle French *pasnaie,* from Latin *pastinaca,* from *pastinum* 2-pronged dibble] (14th century)
: a Eurasian biennial herb (*Pastinaca sativa*) of the carrot family with large pinnate leaves and yellow flowers that is cultivated for its long tapered edible root; *also* **:** the root

parsnip

par·son \'pär-sᵊn\ *noun* [Middle English *persone,* from Old French, from Medieval Latin *persona,* literally, person, from Latin] (13th century)
1 : RECTOR
2 : CLERGYMAN; *especially* **:** a Protestant pastor
word history see PERSON

par·son·age \'pär-s(ə-)nij, 'pär-sᵊn-ij\ *noun* (15th century)
: the house provided by a church for its pastor

Par·sons ta·ble \'pär-sᵊnz-\ *noun* [*Parsons* School of Design, New York City] (1967)
: a usually rectangular table having straight legs that are flush with the edge of the top

¹part \'pärt\ *noun* [Middle English, from Old French & Old English, both from Latin *part-, pars;* perhaps akin to Latin *parare* to prepare — more at PARE] (before 12th century)
1 a (1) **:** one of the often indefinite or unequal subdivisions into which something is or is regarded as divided and which together constitute the whole (2) **:** an essential portion or integral element **b :** one of several or many equal units of which something is composed or into which it is divisible **:** an amount equal to another amount ⟨mix one *part* of the powder with three *parts* of water⟩ **c** (1) **:** an exact divisor of a quantity **:** ALIQUOT (2) **:** PARTIAL FRACTION **d :** one of the constituent elements of a plant or animal body: as (1) **:** ORGAN, MEMBER (2) *plural* **:** PRIVATE PARTS **e :** a division of a literary work **f** (1) **:** a vocal or instrumental line or melody in concerted music or in harmony (2) **:** a particular voice or instrument in concerted music; *also* **:** the score for it **g :** a constituent member of a machine or other apparatus; *also* **:** a spare part
2 : something falling to one in a division or apportionment **:** SHARE
3 : one's share or allotted task (as in an action) ⟨one must do one's *part*⟩
4 : one of the opposing sides in a conflict or dispute
5 : a general area of indefinite boundaries — usually used in plural ⟨you're not from around these *parts*⟩ ⟨took off for *parts* unknown⟩
6 : a function or course of action performed
7 a : an actor's lines in a play **b :** the role of a character in a play
8 : a constituent of character or capacity **:** TALENT ⟨a man of many *parts*⟩
9 : the line where the hair is parted ☆
— for the most part : in general **:** on the whole ⟨*for the most part* the crowd was orderly⟩
— in part : in some degree **:** PARTIALLY
— on the part of : with regard to the one specified

²part *verb* [Middle English, from Old French *partir,* from Latin *partire* to divide, from *part-, pars*] (13th century)
intransitive verb
1 a : to separate from or take leave of someone **b :** to take leave of one another
2 : to become separated into parts
3 a : to go away **:** DEPART **b :** DIE
4 : to become separated, detached, or broken
5 : to relinquish possession or control ⟨hated to *part* with that money⟩
transitive verb
1 a : to divide into parts **b :** to separate by combing on each side of a line **c :** to break or suffer the breaking of (as a rope or anchor chain)
2 : to divide into shares and distribute **:** APPORTION
3 a : to remove from contact or association ⟨if aught but death *part* thee and me —Ruth 1:17 (Authorized Version)⟩ **b :** to keep separate ⟨the narrow channel that *parts* England from France⟩ **c :** to hold (as brawlers) apart **d :** to separate by a process of extraction, elimination, or secretion
4 a *archaic* **:** LEAVE, QUIT **b** *dialect British* **:** RELINQUISH, GIVE UP
synonym see SEPARATE

³part *adverb* (1513)
: PARTLY

⁴part *adjective* (1818)
: PARTIAL 1

par·take \pär-'tāk, pər-\ *verb* **-took** \-'tu̇k\; **-tak·en** \-'tā-kən\; **-tak·ing** [back-formation from *partaker,* alteration of *part taker*] (circa 1585)
intransitive verb
1 : to take part in or experience something along with others ⟨*partake* in the revelry⟩ ⟨*partake* of the good life⟩

☆ SYNONYMS
Part, portion, piece, member, division, section, segment, fragment mean something less than the whole. PART is a general term appropriate when indefiniteness is required ⟨they ran only *part* of the way⟩. PORTION implies an assigned or allotted part ⟨cut the pie into six *portions*⟩. PIECE applies to a separate or detached part of a whole ⟨a puzzle with 500 *pieces*⟩. MEMBER suggests one of the functional units composing a body ⟨a structural *member*⟩. DIVISION applies to a large or diversified part ⟨the manufacturing *division* of the company⟩. SECTION applies to a relatively small or uniform part ⟨the entertainment *section* of the newspaper⟩. SEGMENT applies to a part separated or marked out by or as if by natural lines of cleavage ⟨the retired *segment* of the population⟩. FRAGMENT applies to a part produced by or as if by breaking off or shattering ⟨only a *fragment* of the play still exists⟩.

2 : to have a portion (as of food or drink)
3 : to possess or share a certain nature or attribute ⟨the experience *partakes* of a mystical quality⟩
transitive verb
: to take part in
synonym see SHARE
— **par·tak·er** *noun*

part and parcel *noun* (15th century)
: an essential or integral component ⟨stress was *part and parcel* of the job⟩

part·ed \'pär-təd\ *adjective* (1590)
1 a : divided into parts **b :** cleft so that the divisions reach nearly but not quite to the base — usually used in combination ⟨a 3-*parted* corolla⟩
2 *archaic* **:** DEAD

par·terre \pär-'ter\ *noun* [French, from Middle French, from *par terre* on the ground] (circa 1639)
1 : an ornamental garden with paths between the beds
2 : the part of the main floor of a theater that is behind the orchestra; *especially* **:** PARQUET CIRCLE

par·the·no·car·py \'pär-thə-nō-,kär-pē\ *noun* [International Scientific Vocabulary, from Greek *parthenos* virgin + *karpos* fruit — more at HARVEST] (1911)
: the production of fruits without fertilization ⟨bananas set fruit by *parthenocarpy* and without pollination⟩
— **par·the·no·car·pic** \,pär-thə-nō-'kär-pik\ *adjective*

par·the·no·gen·e·sis \,pär-thə-nō-'je-nə-səs\ *noun* [New Latin, from Greek *parthenos* + Latin *genesis* genesis] (1849)
: reproduction by development of an unfertilized usually female gamete that occurs especially among lower plants and invertebrate animals

par·the·no·ge·net·ic \-jə-'ne-tik\ *adjective* (1872)
: of, characterized by, or produced by parthenogenesis
— **par·the·no·ge·net·i·cal·ly** \-ti-k(ə-)lē\ *adverb*

Par·the·non \'pär-thə-,nän\ *noun* [Latin, from Greek *Parthenōn*] (circa 1841)
: a Doric temple of Athena built on the acropolis at Athens in the 5th century B.C.

Par·thi·an \'pär-thē-ən\ *adjective* (1579)
1 : of, relating to, or characteristic of ancient Parthia or its people
2 : relating to, being, or having the effect of a shot fired while in real or feigned retreat
— **Parthian** *noun*

¹par·tial \'pär-shəl\ *adjective* [Middle English *parcial,* from Middle French *partial,* from Medieval Latin *partialis,* from Late Latin, of a part, from Latin *part-, pars* part] (14th century)
1 : of or relating to a part rather than the whole **:** not general or total ⟨a *partial* solution⟩
2 : inclined to favor one party more than the other **:** BIASED
3 : markedly fond of someone or something — used with *to* ⟨*partial* to pizza⟩

²partial *noun* (1880)
: OVERTONE 1a

partial denture *noun* (1860)
: a usually removable artificial replacement of one or more teeth

partial derivative *noun* (1889)
: the derivative of a function of several variables with respect to one of them and with the remaining variables treated as constants

partial differential equation *noun* (1889)
: a differential equation containing at least one partial derivative

partial differentiation *noun* (circa 1890)
: the process of finding a partial derivative

partial fraction *noun* (1816)
: one of the simpler fractions into the sum of which the quotient of two polynomials may be decomposed

par·tial·i·ty \,pär-shē-'a-lə-tē, pär-'sha-lə-tē\ *noun, plural* **-ties** (15th century)
1 : the quality or state of being partial **:** BIAS
2 : a special taste or liking

par·tial·ly \'pär-sh(ə-)lē\ *adverb* (15th century)
1 *archaic* **:** in a biased manner **:** with partiality
2 : to some extent **:** in some degree

partially ordered *adjective* (1941)
: having some or all elements connected by a relation that is reflexive, transitive, and antisymmetric

partial pressure *noun* (1857)
: the pressure exerted by a (specified) component in a mixture of gases

partial product *noun* (circa 1924)
: a product obtained by multiplying a multiplicand by one digit of a multiplier having more than one digit

par·ti·ble \'pär-tə-bəl\ *adjective* (14th century)
: capable of being parted **:** DIVISIBLE ⟨a *partible* inheritance⟩
— **par·ti·bil·i·ty** \,pär-tə-'bi-lə-tē\ *noun*

par·tic·i·pant \pär-'ti-sə-pənt, pər-\ *noun* (1562)
: one that participates
— **participant** *adjective*

par·tic·i·pate \pär-'ti-sə-,pāt, pər-\ *verb* **-pat·ed; -pat·ing** [Latin *participatus,* past participle of *participare,* from *particip-, particeps* participant, from *part-, pars* part + *capere* to take — more at HEAVE] (1531)
transitive verb
archaic **:** PARTAKE
intransitive verb
1 : to possess some of the attributes of a person, thing, or quality
2 a : to take part ⟨always tried to *participate* in class discussions⟩ **b :** to have a part or share in something
synonym see SHARE
— **par·tic·i·pa·tor** \-,pā-tər\ *noun*

par·tic·i·pa·tion \(,)pär-,ti-sə-'pā-shən, pər-\ *noun* (14th century)
1 : the act of participating
2 : the state of being related to a larger whole

par·tic·i·pa·tion·al \-'pā-shnəl, -'pā-shə-nᵊl\ *adjective* (1959)
: PARTICIPATORY

par·tic·i·pa·tive \pär-'ti-sə-pə-tiv, pər-, -,pā-tiv\ *adjective* (1951)
: relating to or involving participation; *especially* **:** of, relating to, or being a style of management in which subordinates participate in decision making

par·tic·i·pa·to·ry \pär-'ti-sə-pə-,tōr-ē, pər-, -,tȯr-\ *adjective* (1881)
: characterized by or involving participation; *especially* **:** providing the opportunity for individual participation ⟨*participatory* democracy⟩

par·ti·cip·i·al \,pär-tə-'si-pē-əl\ *adjective* [Latin *participialis,* from *participium*] (1591)
: of, relating to, or formed with or from a participle
— **par·ti·cip·i·al·ly** \-pē-ə-lē\ *adverb*

par·ti·ci·ple \'pär-tə-,si-pəl\ *noun* [Middle English, from Middle French, modification of Latin *participium,* from *particip-, particeps*] (14th century)
: a word having the characteristics of both verb and adjective; *especially* **:** an English verbal form that has the function of an adjective and at the same time shows such verbal features as tense and voice and capacity to take an object

par·ti·cle \'pär-ti-kəl\ *noun* [Middle English, from Latin *particula,* from diminutive of *part-, pars*] (14th century)
1 a : a minute quantity or fragment **b :** a relatively small or the smallest discrete portion or amount of something
2 *archaic* **:** a clause or article of a composition or document
3 : any of the basic units of matter and energy (as a molecule, atom, proton, electron, or photon)
4 : a unit of speech expressing some general aspect of meaning or some connective or limiting relation and including the articles, most prepositions and conjunctions, and some interjections and adverbs ⟨the *particle up* has a perfective meaning in phrases such as *beat up* and *cut up*⟩
5 : a small eucharistic wafer distributed to a Roman Catholic layman at Communion

particle accelerator *noun* (1946)
: ACCELERATOR d

par·ti·cle·board \-,bōrd, -,bȯrd\ *noun* (circa 1957)
: a composition board made of very small pieces of wood bonded together

particle physics *noun* (1946)
: a branch of physics dealing with the constitution, properties, and interactions of elementary particles especially as revealed in experiments using particle accelerators — called also *high-energy physics*
— **particle physicist** *noun*

par·ti–col·or \'pär-tē-,kə-lər\ *or* **par·ti–col·ored** \-lərd\ *adjective* [obsolete English *party* parti-color, from Middle English *parti,* from Middle French, striped, from past participle of *partir* to divide] (1535)
: showing different colors or tints; *especially* **:** having a predominant color broken by patches of one or more other colors ⟨*parti-color* setters⟩
— **parti–color** *noun*

¹par·tic·u·lar \pə(r)-'ti-kyə-lər, -k(ə-)lər\ *adjective* [Middle English *particuler,* from Middle French, from Late Latin *particularis,* from Latin *particula* small part] (14th century)
1 : of, relating to, or being a single person or thing ⟨the *particular* person I had in mind⟩
2 *obsolete* **:** PARTIAL
3 : of, relating to, or concerned with details ⟨gave us a very *particular* account of the trip⟩
4 a : distinctive among other examples or cases of the same general category **:** notably unusual ⟨suffered from measles of *particular* severity⟩ **b :** being one unit or element among others ⟨*particular* incidents in a story⟩
5 a : denoting an individual member or subclass in logic **b :** affirming or denying a predicate to a part of the subject — used of a proposition in logic ⟨"some men are wise" is a *particular* affirmative⟩
6 a : concerned over or attentive to details **:** METICULOUS ⟨a very *particular* gardener⟩ **b :** nice in taste **:** FASTIDIOUS **c :** hard to please **:** EXACTING
synonym see CIRCUMSTANTIAL, SPECIAL

²particular *noun* (15th century)
1 *archaic* **:** a separate part of a whole
2 a : an individual fact, point, circumstance, or detail **b :** a specific item or detail of information or news ⟨bill of *particulars*⟩
3 a : an individual or a specific subclass (as in logic) falling under some general concept or term **b :** a particular proposition in logic
synonym see ITEM
— **in particular :** in distinction from others **:** SPECIFICALLY

par·tic·u·lar·ise *British variant of* PARTICULARIZE

par·tic·u·lar·ism \pə(r)-'ti-k(yə-)lə-,ri-zəm *also* pär-\ *noun* (1824)
1 : exclusive or special devotion to a particular interest
2 : a political theory that each political group has a right to promote its own interests and especially independence without regard to the interests of larger groups
3 : a tendency to explain complex social phenomena in terms of a single causative factor

\ə\ **a**but \ᵊ\ kitt**e**n \ər\ **fur**ther \a\ **a**sh \ā\ **a**ce \ä\ m**o**p, m**a**r \au̇\ **ou**t \ch\ **ch**in \e\ b**e**t \ē\ **ea**sy \g\ **g**o \i\ h**i**t \ī\ **i**ce \j\ **j**ob \ŋ\ si**ng** \ō\ g**o** \ȯ\ l**aw** \ȯi\ b**oy** \th\ **th**in \t͟h\ **th**e \ü\ l**oo**t \u̇\ f**oo**t \y\ **y**et \zh\ vi**si**on *see also* Guide to Pronunciation

— **par·tic·u·lar·ist** \-rist\ *noun or adjective*
— **par·tic·u·lar·is·tic** \-ˌti-k(yə-)lə-ˈris-tik\ *adjective*

par·tic·u·lar·i·ty \pə(r)-ˌti-kyə-ˈlar-ə-tē *also* (ˌ)pär-\ *noun, plural* **-ties** (1528)
1 a : a minute detail **:** PARTICULAR **b :** an individual characteristic **:** PECULIARITY; *also* **:** SINGULARITY
2 : the quality or state of being particular as distinguished from universal
3 a : attentiveness to detail **:** EXACTNESS **b :** the quality or state of being fastidious in behavior or expression

par·tic·u·lar·i·za·tion \-ˌti-k(yə-)lə-rə-ˈzā-shən\ *noun* (1657)
: the act of particularizing **:** the condition of being particularized

par·tic·u·lar·ize \pə(r)-ˈti-k(yə-)lə-ˌrīz *also* pär-\ *verb* **-ized; -iz·ing** (1593)
transitive verb
: to state in detail **:** SPECIFY
intransitive verb
: to go into details

par·tic·u·lar·ly \pə(r)-ˈti-kyə-(lər-)lē, -kyə-lə-lē; pə(r)-ˈti-k(ə-)lē; *also* pär-\ *adverb* (14th century)
1 : in a particular manner **:** in detail
2 : to an unusual degree
3 : in particular **:** SPECIFICALLY

[1]**par·tic·u·late** \pär-ˈti-kyə-lət *also* -ˌlāt\ *adjective* [Latin *particula*] (1871)
: of or relating to minute separate particles

[2]**particulate** *noun* (1942)
: a particulate substance

particulate inheritance *noun* (1889)
: inheritance of characters specifically transmitted by genes in accord with Mendel's laws

[1]**part·ing** \ˈpär-tiŋ\ *noun* (15th century)
: a place or point where a division or separation occurs
— **parting of the ways 1 :** a point of separation or divergence **2 :** a place or time at which a choice must be made

[2]**parting** *adjective* (circa 1577)
: given, taken, or performed at parting ⟨a *parting* kiss⟩

par·ti pris \ˈpär-ˌtē-ˈprē\ *noun, plural* **partis pris** \-ˌtē-ˈprē(z)\ [French, literally, side taken] (1860)
: a preconceived opinion **:** PREJUDICE
— **parti pris** *adjective*

[1]**par·ti·san** *also* **par·ti·zan** \ˈpär-tə-zən, -sən, -ˌzan, *chiefly British* ˌpär-tə-ˈzan\ *noun* [Middle French *partisan,* from Old Italian *partigiano,* from *parte* part, party, from Latin *part-, pars* part] (1555)
1 : a firm adherent to a party, faction, cause, or person; *especially* **:** one exhibiting blind, prejudiced, and unreasoning allegiance
2 a : a member of a body of detached light troops making forays and harassing an enemy **b :** a member of a guerrilla band operating within enemy lines
synonym see FOLLOWER
— **partisan** *adjective*
— **par·ti·san·ly** \-lē\ *adverb*
— **par·ti·san·ship** \-ˌship\ *noun*

[2]**par·ti·san** *or* **par·ti·zan** \ˈpär-tə-zən, -sən\ *noun* [Middle French *partisane,* from Old Italian *partigiana,* feminine of *partigiano*] (1556)
: a weapon of the 16th and 17th centuries with long shaft and broad blade

par·ti·ta \pär-ˈtē-tə\ *noun* [Italian, from *partire* to divide, from Latin — more at PART] (1880)
1 : VARIATION 4
2 : SUITE 2b(1)

par·tite \ˈpär-ˌtīt\ *adjective* [Latin *partitus,* from past participle of *partire*] (circa 1570)
1 : divided into a usually specified number of parts
2 : PARTED 1b ⟨a *partite* leaf⟩

[1]**par·ti·tion** \pär-ˈti-shən, pər-\ *noun* (15th century)
1 : the action of parting **:** the state of being parted **:** DIVISION
2 : something that divides; *especially* **:** an interior dividing wall
3 : one of the parts or sections of a whole

[2]**partition** *transitive verb* (1741)
1 a : to divide into parts or shares **b :** to divide (as a country) into two or more territorial units having separate political status
2 : to separate or divide by a partition (as a wall) — often used with *off*
— **par·ti·tion·er** \-ˈti-sh(ə-)nər\ *noun*

par·ti·tion·ist \-ˈti-sh(ə-)nist\ *noun* (circa 1900)
: an advocate of political partition

par·ti·tive \ˈpär-tə-tiv\ *adjective* (14th century)
1 : serving to part or divide into parts
2 a : of, relating to, or denoting a part ⟨a *partitive* construction⟩ **b :** serving to indicate the whole of which a part is specified ⟨*partitive* genitive⟩
— **par·ti·tive·ly** *adverb*

part·let \ˈpärt-lət\ *noun* [Middle English (Scots) *patelet,* from Middle French *patelette,* from diminutive of *patte* paw] (1519)
: a 16th century chemisette with a band or collar

part·ly \ˈpärt-lē\ *adverb* (1523)
: in some measure or degree **:** PARTIALLY

[1]**part·ner** \ˈpärt-nər *also* ˈpärd-\ *noun* [Middle English *partener,* alteration of *parcener,* from Anglo-French, coparcener — more at PARCENER] (14th century)
1 *archaic* **:** one that shares **:** PARTAKER
2 a : one associated with another especially in an action **:** ASSOCIATE, COLLEAGUE **b :** either of two persons who dance together **c :** one of two or more persons who play together in a game against an opposing side **d :** either of two people living together; *especially* **:** SPOUSE
3 : a member of a partnership; *also* **:** such membership
4 : one of the heavy timbers that strengthen a ship's deck to support a mast — usually used in plural
— **part·ner·less** \-ləs\ *adjective*

[2]**partner** (1611)
transitive verb
1 : to join or associate with another as partner
2 : to provide with a partner
intransitive verb
: to act as a partner

partners desk *noun* (1946)
: a large desk with an open kneehole which allows use of the desk by two people seated opposite each other

part·ner·ship \-ˌship\ *noun* (1576)
1 : the state of being a partner **:** PARTICIPATION
2 a : a legal relation existing between two or more persons contractually associated as joint principals in a business **b :** the persons joined together in a partnership
3 : a relationship resembling a legal partnership and usually involving close cooperation between parties having specified and joint rights and responsibilities

part of speech (1509)
: a traditional class of words distinguished according to the kind of idea denoted and the function performed in a sentence

par·ton \ˈpär-ˌtän\ *noun* [*part*icle + [2]*-on*] (1969)
: a hypothetical particle (as a quark or gluon) that is held to be a constituent of hadrons

par·tridge \ˈpär-trij, *dialect* ˈpa-trij\ *noun, plural* **partridge** *or* **par·tridg·es** [Middle English *partrich,* modification of Old French *perdris,* modification of Latin *perdic-, perdix,* from Greek *perdik-, perdix*] (14th century)
1 : any of various typically medium-sized stout-bodied Old World gallinaceous game birds (*Perdix, Alectoris,* and related genera) with variegated plumage
2 : any of numerous gallinaceous birds (as the American ruffed grouse or bobwhite) somewhat like the Old World partridges in size, habits, or value as game

par·tridge·ber·ry \-ˌber-ē\ *noun* (1714)
: an American trailing evergreen plant (*Mitchella repens*) of the madder family with insipid scarlet berries; *also* **:** its fruit

partridge 1

part–song \ˈpärt-ˌsȯŋ\ *noun* (1850)
: a usually unaccompanied song consisting of two or more voice parts with one part carrying the melody

part–time \ˈpärt-ˈtīm\ *adjective* (1891)
: involving or working less than customary or standard hours ⟨a *part-time* job⟩ ⟨*part-time* students⟩
— **part–time** *adverb*
— **part–tim·er** \-ˌtī-mər, -ˈtī-\ *noun*

[1]**par·tu·ri·ent** \pär-ˈtu̇r-ē-ənt, -ˈtyu̇r-\ *adjective* [Latin *parturient-, parturiens,* present participle of *parturire* to be in labor, from *parere* to give birth to — more at PARE] (1592)
1 : bringing forth or about to bring forth young
2 : of or relating to parturition

[2]**parturient** *noun* (1947)
: a parturient individual

par·tu·ri·tion \ˌpär-chə-ˈri-shən, ˌpär-tyu̇-, ˌpär-tə-\ *noun* [Late Latin *parturition-, parturitio,* from Latin *parturire*] (1646)
: the action or process of giving birth to offspring

part·way \ˈpärt-ˈwā\ *adverb* (1859)
1 : to some extent **:** PARTIALLY, PARTLY
2 : at a point in the way or distance ⟨*partway* through the trip they met some friends⟩

[1]**par·ty** \ˈpär-tē\ *noun, plural* **parties** [Middle English *partie* part, party, from Old French, from *partir* to divide — more at PART] (14th century)
1 : a person or group taking one side of a question, dispute, or contest
2 : a group of persons organized for the purpose of directing the policies of a government
3 : a person or group participating in an action or affair ⟨mountain-climbing *party*⟩ ⟨a *party* to the transaction⟩
4 : a particular individual **:** PERSON ⟨an old *party* approaching 80⟩
5 : a detail of soldiers
6 : a social gathering; *also* **:** the entertainment provided for it
— **party** *adjective*

[2]**party** *intransitive verb* **par·tied; par·ty·ing** (1919)
: to attend or give parties; *broadly* **:** REVEL 1
— **par·ty·er** \-tē-ər\ *noun*

party line *noun* (1834)
1 : the policy or practice of a political party
2 : a single telephone circuit connecting two or more subscribers with the exchange — called also *party wire*
3 : the principles or policies of an individual or organization; *also* **:** the explanation or interpretation usually put forth ⟨the *party line* that her mother was a saint —Leslie Bennetts⟩
— **par·ty–lin·er** \ˌpär-tē-ˈlī-nər\ *noun*

party poop·er \-ˈpü-pər\ *noun* [[3]*poop* + *-er*] (1954)
: one who refuses to join in the fun of a party; *broadly* **:** one who refuses to go along with everyone else

party wall *noun* (circa 1798)
: a wall which divides two adjoining properties and in which each of the owners shares the rights

pa·rure \pə-ˈru̇r\ *noun* [French, literally, adornment, from Old French *pareure,* from *parer* to prepare, adorn — more at PARE] (1818)
: a matched set of ornaments (as jewelry)

par value *noun* (1807)
: PAR 1b(1)

par·ve \'pär-və\ *variant of* PAREVE

par·ve·nu \'pär-və-ˌnü, -ˌnyü\ *noun, plural* **-nus** \-ˌn(y)üz\ [French, from past participle of *parvenir* to arrive, from Latin *pervenire,* from *per* through + *venire* to come — more at FOR, COME] (1802)
: one that has recently or suddenly risen to an unaccustomed position of wealth or power and has not yet gained the prestige, dignity, or manner associated with it
— **parvenu** *adjective*

par·ve·nue \-ˌnü, -ˌnyü\ *noun* [French, from *parvenue,* feminine of *parvenum,* past participle] (1826)
: a woman who is a parvenu
— **parvenue** *adjective*

par·vis *also* **par·vise** \'pär-vəs\ *noun* [Middle English *parvis,* from Middle French, modification of Late Latin *paradisus* enclosed park — more at PARADISE] (14th century)
1 : a court or enclosed space before a building (as a church)
2 : a single portico or colonnade before a church

par·vo \'pär-(ˌ)vō\ *noun* (1980)
: PARVOVIRUS 2

par·vo·vi·rus \'pär-vō-ˌvī-rəs\ *noun* [New Latin, from Latin *parvus* small (akin to Greek *pauros* small, *paid-, pais* child) + New Latin *-o-* + *virus* — more at FEW] (1965)
1 : any of a group of small single-stranded DNA viruses pathogenic for vertebrates
2 : a highly contagious febrile disease of dogs that is caused by a parvovirus and marked by loss of appetite, lethargy, often bloody diarrhea and vomiting, and sometimes death

pas \'pä\ *noun, plural* **pas** \'pä(z)\ [French, from Latin *passus* step — more at PACE] (1707)
1 : the right of precedence
2 : a dance step or combination of steps

pas·cal \pas-'kal, päs-'käl\ *noun* [Blaise *Pascal*] (1956)
1 : a unit of pressure in the meter-kilogram-second system equivalent to one newton per square meter
2 *usually P capitalized or all capitals* **:** a structured computer programming language developed from Algol and designed to process both numerical and textual data

Pas·cal's triangle \pas-'kalz-, päs-'kälz-\ *noun* (1886)
: a system of numbers triangularly arranged in rows that consist of the coefficients in the expansion of $(a + b)^n$ for $n = 0, 1, 2, 3, \ldots$

Pasch \'pask\ *also* **Pas·cha** \'päs-kə\ *noun* [Middle English *pasche* Passover, Easter, from Old French, from Late Latin *pascha,* from Late Greek, from Greek, Passover, from Hebrew *pesaḥ*] (12th century)
1 : EASTER
2 : PASSOVER
— **pas·chal** \'pas-kəl\ *adjective*

paschal full moon *noun* (1892)
: the 14th day of a lunar month occurring on or next after March 21 according to a fixed set of ecclesiastical calendar rules and without regard to the real moon

Paschal Lamb *noun* (15th century)
: AGNUS DEI 2

pas de bour·rée \ˌpä-də-bü-'rā\ *noun, plural* **pas de bourrée** *same*\ *or* **pas de bourrées** \-'rā(z)\ [French, literally, bourrée step] (1897)
: a walking or running ballet step usually executed on the points of the toes

pas de chat \-'sha\ *noun, plural* **pas de chat** *same*\ [French, literally, cat's step] (1914)
: a forward leap in ballet

pas de deux \-'də(r), -'dü\ *noun, plural* **pas de deux** \-'dər(z), -'də(z), -'dü(z)\ [French, literally, step for two] (circa 1762)
1 : a dance or figure for two performers
2 : an intricate relationship or activity involving two parties or things

pas de qua·tre \-'kat, -'ka-trə, -'kätrᵊ\ *noun, plural* **pas de quatre** *same*\ [French, literally, step for four] (1884)
: a dance or figure for four performers

pas de trois \-'trwä, -'twä\ *noun, plural* **pas de trois** \-'trwä(z), -'twä(z)\ [French, literally, step for three] (circa 1762)
: a dance or figure for three performers

pa·se \'pä-(ˌ)sā\ *noun* [Spanish, literally, feint, from *pase* let him pass, from *pasar* to pass, from (assumed) Vulgar Latin *passare*] (1937)
: a movement of a cape by a matador in drawing a bull and taking his charge

pa·seo \pə-'sā-(ˌ)ō, pä-\ *noun, plural* **pa·se·os** [Spanish, from *pasear* to take a stroll, from *paso* passage, step, from Latin *passus*] (1832)
1 a : a leisurely stroll **:** PROMENADE **b :** a public walk or boulevard
2 : a formal entrance march of bullfighters into an arena

[1]**pash** \'pash\ *transitive verb* [Middle English *passhen*] (14th century)
dialect English **:** SMASH

[2]**pash** *noun* [origin unknown] (1611)
dialect English **:** HEAD

pa·sha \'pä-shə, 'pa-; pə-'shä, -'shȯ\ *noun* [Turkish *paşa*] (1646)
: a man of high rank or office (as in Turkey or northern Africa)

Pash·to \'pəsh-(ˌ)tō\ *noun* [Pashto *pašto*] (1784)
: the Iranian language of the Pashtuns

Pash·tun \ˌpəsh-'tün\ *noun, plural* **Pashtuns** *or* **Pashtun** (1815)
: a member of a people of eastern and southern Afghanistan and adjacent parts of Pakistan

Pa·siph·aë \pə-'si-fə-ˌē\ *noun* [Latin, from Greek *Pasiphaē*]
: the wife of Minos and mother of the Minotaur by a white bull

pasque·flow·er \'pask-ˌflau̇(-ə)r\ *noun* [modification of Middle French *passefleur,* from *passer* to pass (from Old French) + *fleur* flower, from Latin *flor-, flos* — more at BLOW] (1597)
: any of several low perennial herbs (genus *Anemone*) of the buttercup family with palmately compound leaves and large usually white or purple early spring flowers

pas·qui·nade \ˌpas-kwə-'nād\ *noun* [Middle French, from Italian *pasquinata,* from *Pasquino,* name given to a statue in Rome on which lampoons were posted] (1658)
1 : a lampoon posted in a public place
2 : satirical writing **:** SATIRE
— **pasquinade** *transitive verb*

[1]**pass** \'pas\ *verb* [Middle English, from Old French *passer,* from (assumed) Vulgar Latin *passare,* from Latin *passus* step — more at PACE] (13th century)
intransitive verb
1 : MOVE, PROCEED, GO
2 a : to go away **:** DEPART **b :** DIE — often used with *on*
3 a : to move in a path so as to approach and continue beyond something **:** move past; *especially* **:** to move past another vehicle going in the same direction **b :** to run the normal course — used of time or a period of time ⟨the hours *pass* quickly⟩
4 a : to go or make one's way through ⟨allow no one to *pass*⟩ **b :** to go uncensured, unchallenged, or seemingly unnoticed ⟨let the remark *pass*⟩
5 : to go from one quality, state, or form to another ⟨*passes* from a liquid to a gaseous state⟩
6 a : to sit in inquest or judgment **b** (1) **:** to render a decision, verdict, or opinion ⟨the court *passed* on the legality of wiretapping⟩ (2) **:** to become legally rendered ⟨judgment *passed* for the plaintiff⟩
7 : to go from the control, ownership, or possession of one person or group to that of another ⟨the throne *passed* to the king's son⟩ ⟨title *passes* to the buyer upon payment in full⟩
8 a : HAPPEN, OCCUR **b :** to take place or be exchanged as or in a social, personal, or business interaction ⟨words *passed*⟩
9 a : to become approved by a legislature or body empowered to sanction or reject ⟨the proposal *passed*⟩ **b :** to undergo an inspection, test, or course of study successfully
10 a : to serve as a medium of exchange **b :** to be accepted or regarded ⟨drivel that *passes* for literature⟩ **c :** to identify oneself or be identified as something one isn't ⟨tried to *pass* as an adult⟩ ⟨Mom could *pass* as my sister⟩
11 a *obsolete* **:** to make a pass in fencing **b :** to throw or hit a ball or puck to a teammate — often used with *off*
12 a : to decline to bid, double, or redouble in a card game **b :** to withdraw from the current poker pot
transitive verb
1 : to go beyond: as **a :** SURPASS, EXCEED ⟨*passes* all expectations⟩ **b :** to advance or develop beyond **c :** to go past (one moving in the same direction)
2 : to omit a regularly scheduled declaration and payment of (a dividend)
3 a : to go across, over, or through **:** CROSS **b :** to live through (as an experience or peril) **:** UNDERGO **c :** to go through (as a test) successfully
4 a : to secure the approval of ⟨the bill *passed* the Senate⟩ **b :** to cause or permit to win approval or legal or official sanction ⟨*pass* a law⟩ **c :** to give approval or a passing grade to ⟨*pass* the students⟩
5 a : to let (as time or a period of time) go by especially while involved in a leisure activity ⟨I'll read to *pass* the time⟩ **b :** to let go unnoticed **:** OVERLOOK, DISREGARD
6 a : PLEDGE **b :** to transfer the right to or property in ⟨*pass* title to a house⟩
7 a : to put in circulation ⟨*pass* bad checks⟩ **b** (1) **:** to transfer or transmit from one to another (2) **:** to relay or communicate (as information) to another **c :** to cause or enable to go **:** TRANSPORT **d :** to throw or hit (a ball or puck) especially to a teammate
8 a : to pronounce (as a sentence or opinion) especially judicially **b :** UTTER ⟨*passed* a cutting remark⟩
9 a : to cause or permit to go past or through a barrier **b :** to move or cause to move in a particular manner or direction ⟨*passed* my hand over my face⟩ ⟨*pass* the rope through the loop⟩ **c :** to cause to march or go by in order ⟨*pass* the troops in review⟩
10 : to emit or discharge from a bodily part and especially the bowels
11 a : to give a base on balls to **b :** to hit a ball past (an opponent) in a game (as tennis)
— **pass·er** *noun*
— **pass muster :** to gain approval or acceptance
— **pass the buck :** to shift a responsibility to someone else
— **pass the hat :** to take up a collection for money
— **pass the time of day :** to exchange greetings or engage in pleasant conversation

[2]**pass** *noun* [Middle English, from Middle French *pas,* from Latin *passus*] (14th century)
1 : a means (as an opening, road, or channel) by which a barrier may be passed or access to a place may be gained; *especially* **:** a low place in a mountain range
2 : a position to be held usually against odds

[3]**pass** *noun* [[1]*pass*] (1523)
1 : REALIZATION ⟨brought his dream to *pass*⟩
2 : the act or an instance of passing **:** PASSAGE

\ə\ **abut** \ᵊ\ **kitten** \ər\ **further** \a\ **ash** \ā\ **ace** \ä\ **mop, mar** \au̇\ **out** \ch\ **chin** \e\ **bet** \ē\ **easy** \g\ **go** \i\ **hit** \ī\ **ice** \j\ **job** \ŋ\ **sing** \ō\ **go** \ȯ\ **law** \ȯi\ **boy** \th\ **thin** \t͟h\ **the** \ü\ **loot** \u̇\ **foot** \y\ **yet** \zh\ **vision** *see also* Guide to Pronunciation

3 : a usually distressing or bad state of affairs ⟨what has brought you to such a *pass*?⟩
4 a : a written permission to move about freely in a place or to leave or enter it **b :** a written leave of absence from a military post or station for a brief period **c :** a permit or ticket allowing free transportation or free admission
5 *archaic* **:** a thrust or lunge in fencing
6 a : a transference of objects by sleight of hand or other deceptive means **b :** a moving of the hands over or along something
7 *archaic* **:** an ingenious sally (as of wit)
8 : the passing of an examination or course of study; *also* **:** the mark or certification of such passing
9 : a single complete mechanical operation; *also* **:** a single complete cycle of operations (as for processing, manufacturing, or printing)
10 a (1) **:** a transfer of a ball or a puck from one player to another on the same team (2) **:** a ball or puck so transferred **b :** PASSING SHOT
11 : BASE ON BALLS
12 : an election not to bid, bet, or draw an additional card in a card game
13 : a throw of dice in the game of craps that wins the bet for the shooter — compare [3]CRAP 2, MISSOUT
14 : a single passage or movement (as of an airplane) over a place or toward a target
15 a : EFFORT, TRY **b :** a sexually inviting gesture or approach
16 : PASE

pass·able \'pa-sə-bəl\ *adjective* (15th century)
1 a : capable of being passed, crossed, or traveled on ⟨*passable* roads⟩ **b :** capable of being freely circulated
2 : good enough **:** ADEQUATE
— pass·ably \-blē\ *adverb*

pas·sa·ca·glia \ˌpä-sə-'käl-yə, ˌpa-sə-'kal-yə\ *noun* [modification of Spanish *pasacalle,* from *pasar* to pass + *calle* street, from Latin *callis* path — more at PASE] (1659)
1 a : an old Italian or Spanish dance tune **b :** an instrumental musical composition consisting of variations usually on a ground bass in moderately slow triple time
2 : an old dance performed to a passacaglia

pas·sa·do \pə-'sä-(ˌ)dō\ *noun, plural* **-dos** *or* **-does** [modification of French *passade* (from Italian *passata*) or Italian *passata,* from *passare* to pass, from (assumed) Vulgar Latin] (1588)
archaic **:** a thrust in fencing with one foot advanced

[1]pas·sage \'pa-sij\ *noun* (13th century)
1 a : a way of exit or entrance **:** a road, path, channel, or course by which something passes **b :** a corridor or lobby giving access to the different rooms or parts of a building or apartment
2 a : the action or process of passing from one place, condition, or stage to another **b :** DEATH 1 **c :** a continuous movement or flow ⟨the *passage* of time⟩
3 a (1) **:** a specific act of traveling or passing especially by sea or air (2) **:** a privilege of conveyance as a passenger **:** ACCOMMODATIONS **b :** the passing of a legislative measure or law **:** ENACTMENT
4 : a right, liberty, or permission to pass
5 a : something that happens or is done **:** INCIDENT **b :** something that takes place between two persons mutually
6 a : a usually brief portion of a written work or speech that is relevant to a point under discussion or noteworthy for content or style **b :** a phrase or short section of a musical composition **c :** a detail of a work of art (as a painting)
7 : the act or action of passing something or undergoing a passing
8 : incubation of a pathogen (as a virus) in culture, a living organism, or a developing egg

[2]passage *verb* **pas·saged; pas·sag·ing** (1821)
intransitive verb
: to go past or across **:** CROSS
transitive verb
: to subject to passage ⟨*passaged* a virus⟩

pas·sage·way \-ˌwā\ *noun* (1649)
: a way that allows passage

pas·sage·work \-ˌwərk\ *noun* (1865)
: a section of a musical composition characteristically unimportant thematically and consisting especially of ornamental figures

pas·sant \'pa-sᵊnt\ *adjective* [Middle English *passaunt,* from Middle French *passant,* from present participle of *passer* to pass] (15th century)
: walking with the farther forepaw raised — used of a heraldic animal

pass away *intransitive verb* (13th century)
1 : to go out of existence
2 : DIE 1

pass·band \'pas-ˌband\ *noun* (1922)
: a band of frequencies (as in a radio circuit or a light filter) that is transmitted with maximum efficiency

pass·book \-ˌbu̇k\ *noun* (1828)
: BANKBOOK

pass degree *noun* (1868)
: a bachelor's degree without honors that is taken at a British university

pas·sé \pa-'sā\ *adjective* [French, from past participle of *passer*] (1775)
1 : past one's prime
2 a : OUTMODED **b :** behind the times

passed ball *noun* (1861)
: a baseball pitch not hit by the batter that passes the catcher when it should have been caught and allows a base runner to advance — compare WILD PITCH

passed pawn *noun* (1797)
: a chess pawn that has no enemy pawn in front of it on its own or an adjacent file

pas·sel \'pa-səl\ *noun* [alteration of *parcel*] (1835)
: a large number or amount ◆

passe·men·terie \pas-'men-t(ə-)rē\ *noun* [French, from *passement* ornamental braid, from *passer*] (1794)
: a fancy edging or trimming made of braid, cord, gimp, beading, or metallic thread

pas·sen·ger \'pa-sᵊn-jər\ *noun, often attributive* [Middle English *passager,* from Middle French, from *passager,* adjective, passing, from *passage* act of passing, from Old French, from *passer*] (14th century)
1 : WAYFARER
2 : a traveler in a public or private conveyance

passenger pigeon *noun* (1802)
: an extinct but formerly abundant North American migratory pigeon (*Ectopistes migratorius*)

passe–par·tout \ˌpas-pər-'tü, -ˌpär-\ *noun* [French, from *passe partout* pass everywhere] (1675)
1 : MASTER KEY
2 a : [5]MAT **b :** a method of framing in which a picture, a mat, a glass, and a back (as of cardboard) are held together by strips of paper or cloth pasted over the edges
3 : a strong paper gummed on one side and used especially for mounting pictures

pass·er·by \ˌpa-sər-'bī, 'pa-sər-ˌ\ *noun, plural* **pass·ers·by** \-sərz-\ (1568)
: one who passes by

pas·ser·ine \'pa-sə-ˌrīn\ *adjective* [Latin *passerinus* of sparrows, from *passer* sparrow] (1776)
: of or relating to the largest order (Passeriformes) of birds which includes more than half of all living birds and consists chiefly of altricial songbirds of perching habits — compare OSCINE
— passerine *noun*

pas seul \ˌpä-'səl, -'sər(-ə)l\ *noun* [French, literally, solo step] (1813)
: a solo dance or dance figure

pass–fail \'pas-'fā(ə)l\ *adjective* (1959)
: being a system of grading whereby the grades "pass" and "fail" replace the traditional letter grades
— pass–fail *noun*

pas·si·ble \'pa-sə-bəl\ *adjective* [Middle English, from Middle French, from Late Latin *passibilis,* from Latin *passus,* past participle of *pati* to suffer — more at PATIENT] (14th century)
: capable of feeling or suffering

pas·sim \'pa-səm; 'pa-ˌsim, 'pä-\ *adverb* [Latin, from *passus* scattered, from past participle of *pandere* to spread — more at FATHOM] (1803)
: HERE AND THERE

[1]pass·ing \'pa-siŋ\ *noun* (14th century)
: the act of one that passes or causes to pass; *especially* **:** DEATH 1
— in passing : by the way **:** INCIDENTALLY

[2]passing *adjective* (14th century)
1 : going by or past ⟨a *passing* pedestrian⟩
2 : having a brief duration ⟨a *passing* whim⟩
3 *obsolete* **:** SURPASSING
4 : SUPERFICIAL ⟨a *passing* acquaintance⟩ ⟨a *passing* resemblance⟩
5 a : of, relating to, or used in or for the act or process of passing ⟨*passing* lanes⟩ **b :** given on satisfactory completion of an examination or course of study ⟨a *passing* grade⟩

[3]passing *adverb* (14th century)
: to a surpassing degree **:** EXCEEDINGLY ⟨*passing* strange⟩

passing note *noun* (1730)
: a nonharmonic tone interposed between essential harmonic tones of adjacent chords — called also *passing tone*

passing shot *noun* (circa 1949)
: a stroke (as in tennis) that drives the ball to one side and beyond the reach of an opponent

pas·sion \'pa-shən\ *noun* [Middle English, from Old French, from Late Latin *passion-, passio* suffering, being acted upon, from Latin *pati* to suffer — more at PATIENT] (13th century)
1 *often capitalized* **a :** the sufferings of Christ between the night of the Last Supper and his death **b :** an oratorio based on a gospel narrative of the Passion
2 *obsolete* **:** SUFFERING
3 : the state or capacity of being acted on by external agents or forces
4 a (1) **:** EMOTION ⟨his ruling *passion* is greed⟩ (2) *plural* **:** the emotions as distinguished from reason **b :** intense, driving, or overmastering feeling or conviction **c :** an outbreak of anger

◇ WORD HISTORY

passel The loss of the sound \r\ before \s\ inside a word probably took place in most areas of England during the late Middle Ages. (This change is distinct from the more general loss of \r\ before all consonants and at the end of words that now characterizes most people's speech in England and local speech in parts of the U.S.) In some words, such as *bass,* the fish, the pronunciation without \r\ has become standard and is reflected in spelling. In most words containing this sequence, however, the influence of orthography and other factors have made the \r\ forms standard and the forms without \r\, such as *fust* for *first* and *hoss* for *horse,* a mark of nonstandard or folk speech. In yet another set of words the form without \r\ is associated with distinctive senses of the source word or has developed new senses. The word *passel,* for example, meaning "large number," is akin to the "group" sense of *parcel* (as in *a parcel of lies*), but *passel* is used for none of the other senses of *parcel,* and, though it occurs in all varieties of writing, still has a slightly folksy tinge.

5 a : ardent affection **:** LOVE **b :** a strong liking or desire for or devotion to some activity, object, or concept **c :** sexual desire **d :** an object of desire or deep interest ☆
— **pas·sion·less** \-ləs\ *adjective*
pas·sion·al \ˈpa-shə-nᵊl, ˈpash-nəl\ *adjective* (15th century)
: of, relating to, or marked by passion
pas·sion·ate \ˈpa-sh(ə-)nət\ *adjective* (15th century)
1 a : easily aroused to anger **b :** filled with anger **:** ANGRY
2 a : capable of, affected by, or expressing intense feeling **b :** ENTHUSIASTIC, ARDENT
3 : swayed by or affected with sexual desire
synonym see IMPASSIONED
— **pas·sion·ate·ly** *adverb*
— **pas·sion·ate·ness** *noun*
pas·sion·flow·er \ˈpa-shən-ˌflau̇(-ə)r\ *noun* [from the fancied resemblance of parts of the flower to the instruments of Christ's crucifixion] (1633)
: any of a genus (*Passiflora* of the family Passifloraceae, the passionflower family) of chiefly tropical woody tendriled climbing vines or erect herbs with usually showy flowers and pulpy often edible berries

passionflower

passion fruit *noun* (1752)
: the edible fruit of a passionflower
Pas·sion·ist \ˈpa-sh(ə-)nist\ *noun* [Italian *passionista,* from *passione* passion, from Late Latin *passion-, passio*] (1832)
: a member of a Roman Catholic mendicant order founded by Saint Paul of the Cross in Italy in 1720 and devoted chiefly to missionary work and retreats
passion play *noun, often 1st P capitalized* (1870)
: a dramatic representation of the scenes connected with the passion and crucifixion of Christ
Passion Sunday *noun* (14th century)
: the fifth Sunday in Lent
Pas·sion·tide \ˈpa-shən-ˌtīd\ *noun* (1847)
: the last two weeks of Lent
Passion Week *noun* (15th century)
1 : HOLY WEEK
2 : the week between Passion Sunday and Palm Sunday
pas·siv·ate \ˈpa-si-ˌvāt\ *transitive verb* **-at·ed; -at·ing** (1913)
1 : to make inactive or less reactive ⟨*passivate* the surface of steel by chemical treatment⟩
2 : to protect (as a solid-state device) against contamination by coating or surface treatment
— **pas·siv·ation** \ˌpa-si-ˈvā-shən\ *noun*
¹pas·sive \ˈpa-siv\ *adjective* [Middle English, from Latin *passivus,* from *passus,* past participle] (14th century)
1 a (1) **:** acted upon by an external agency (2) **:** receptive to outside impressions or influences **b** (1) **:** asserting that the grammatical subject of a verb is subjected to or affected by the action represented by that verb (2) **:** containing or yielding a passive verb form **c** (1) **:** lacking in energy or will **:** LETHARGIC (2) **:** tending not to take an active or dominant part **d :** induced by an outside agency ⟨*passive* exercise of a paralyzed leg⟩
2 a : not active or operating **:** INERT **b :** of, relating to, or making direct use of the sun's heat usually without the intervention of mechanical devices ⟨*passive* technique⟩ ⟨*passive* solar house⟩ **c :** LATENT **d** (1) **:** of, relating to, or characterized by a state of chemical inactivity; *especially* **:** resistant to corrosion (2) **:** not involving expenditure of chemical energy ⟨*passive* transport across a cell membrane⟩ **e** *of an electronic element* **:** exhibiting no gain or control **f :** operating solely by means of the power of an input signal ⟨a *passive* communication satellite that reflects television signals⟩ **g :** relating to the detection of an object through its emission of energy
3 a : receiving or enduring without resistance **:** SUBMISSIVE **b :** existing or occurring without being active, open, or direct ⟨*passive* support⟩
4 : of, relating to, or being business activity in which the investor does not have immediate control over income
synonym see INACTIVE
— **pas·sive·ly** *adverb*
— **pas·sive·ness** *noun*
— **pas·siv·i·ty** \pa-ˈsi-və-tē\ *noun*
²passive *noun* (1530)
1 : a passive verb form
2 : the passive voice of a language ■
passive immunity *noun* (1895)
: immunity acquired by transfer of antibodies (as by injection of serum from an individual with active immunity)
— **passive immunization** *noun*
passive resistance *noun* (1819)
: resistance especially to a government or an occupying power characterized mainly by noncooperation
passive restraint *noun* (1970)
: a restraint (as a self-locking seat belt) that acts automatically to protect an automobile rider during a crash
passive smoking *noun* (1971)
: the involuntary inhalation of tobacco smoke (as from another's cigarette) especially by a nonsmoker
passive transfer *noun* (1935)
: a local transfer of skin sensitivity from an allergic to a normal person by injection of serum from the former that is used especially for identifying specific allergens when a high degree of allergic sensitivity is suspected
pas·siv·ism \ˈpa-si-ˌvi-zəm\ *noun* (1872)
: a passive attitude, behavior, or way of life
— **pas·siv·ist** \-vist\ *noun*
pass·key \ˈpas-ˌkē\ *noun* (circa 1817)
1 : MASTER KEY
2 : SKELETON KEY
pass off *transitive verb* (1799)
1 : to make public or offer for sale with intent to deceive
2 : to give a false identity or character to
pass out (1899)
intransitive verb
1 : DIE 1
2 : to lose consciousness
transitive verb
: to reject (a deal in bridge) as unplayable because everyone has passed on the first round of bidding
Pass·over \ˈpas-ˌō-vər\ *noun* [from the exemption of the Israelites from the slaughter of the firstborn in Egypt (Exodus 12:23–27)] (1530)
: a Jewish holiday beginning on the 14th of Nisan and commemorating the Hebrews' liberation from slavery in Egypt
pass over *transitive verb* (14th century)
1 : to ignore in passing
2 : to pay no attention to the claims of **:** DISREGARD
pass·port \ˈpas-ˌpōrt, -ˌpȯrt\ *noun* [Middle English (Scots) *pasport,* from Middle French *passeport,* from *passer* to pass + *port* port, from Latin *portus* — more at FORD] (15th century)
1 a : a formal document issued by an authorized official of a country to one of its citizens that is usually necessary for exit from and reentry into the country, that allows the citizen to travel in a foreign country in accordance with visa requirements, and that requests protection for the citizen while abroad **b :** a license issued by a country permitting a foreign citizen to pass or take goods through its territory **:** SAFE-CONDUCT **c :** a document of identification required by law to be carried by persons residing or traveling within a country

☆ SYNONYMS
Passion, fervor, ardor, enthusiasm, zeal mean intense emotion compelling action. PASSION applies to an emotion that is deeply stirring or ungovernable ⟨was a slave to his *passions*⟩. FERVOR implies a warm and steady emotion ⟨read the poem aloud with great *fervor*⟩. ARDOR suggests warm and excited feeling likely to be fitful or short-lived ⟨the *ardor* of their honeymoon soon faded⟩. ENTHUSIASM applies to lively or eager interest in or admiration for a proposal, cause, or activity ⟨never showed much *enthusiasm* for sports⟩. ZEAL implies energetic and unflagging pursuit of an aim or devotion to a cause ⟨preaches with the *zeal* of the converted⟩. See in addition FEELING.

□ USAGE
passive A passive sentence is made from an active sentence with a transitive verb by a transformation which makes the direct object, or sometimes the indirect object, into the subject, changes the active verb to a past participle with a form of *be,* and relegates the original subject to the status of agent. This transformation throws emphasis on the original object and de-emphasizes the original subject. For instance we can change the active "I broke the vase" to the passive "The vase was broken by me." The agent is optional in a passive construction; we can shorten it to "The vase was broken." These characteristics are turned into liabilities by many critics, who disparage the passive as wordy, weak, evasive, or vague. But these are not faults of the construction; they are faults of the writer. When the agent is omitted, the passive sentence is shorter than the active. And there are many times when the agent is better omitted, as when the doer is unknown ⟨the bank *was robbed* last night⟩ or is unimportant ⟨lawn mowers should not *be left* outside in the rain⟩ or is too obvious to need mentioning ⟨Clinton *was elected* president⟩. Sometimes the agent wants to be omitted ⟨I am sorry to inform you that your request *has been denied*⟩; this is commonly true in scientific and business reports where the authors are not supposed to use first person pronouns. Perhaps the most important thing to keep in mind is focus. One can write either "The car hit a pedestrian" or "A pedestrian was hit by the car"; the choice should be governed by whether the reader is meant to focus on the car or the pedestrian. The passive is useful for generalizing ⟨people *are imprisoned* for years without trial, or *shot* in the back of the neck or *sent* to die of scurvy in Arctic lumber camps —George Orwell⟩. Had Orwell, who wrote "Never use the passive where you can use the active," meant to be specific, he would probably have begun with "The Soviet secret police imprison . . . " or something similar. The point to remember is that the passive need not be avoided—it is much less commonly used than the active anyway—but it should be used where it best serves the writer's purpose.

2 a : a permission or authorization to go somewhere **b :** something that secures admission, acceptance, or attainment ⟨education as a *passport* to success⟩

pass–through *noun* (1951)
1 : the act, action, or process of offsetting increased costs by raising prices
2 : an opening in a wall between two rooms through which something (as dishes) may be passed

pass up *transitive verb* (1896)
: to let go by without accepting or taking advantage of ⟨*pass up* a chance for promotion⟩; *also* **:** DECLINE, REJECT

pass•word \ˈpas-ˌwərd\ *noun* (circa 1817)
1 : something that enables one to pass or gain admission: as **a :** a spoken word or phrase required to pass by a guard **b :** a sequence of characters required for access to a computer system
2 : WATCHWORD

[1]past \ˈpast\ *adjective* [Middle English, from past participle of *passen* to pass] (14th century)
1 a : AGO ⟨12 years *past*⟩ **b :** just gone or elapsed ⟨for the *past* few months⟩
2 : having existed or taken place in a period before the present **:** BYGONE
3 : of, relating to, or constituting a verb tense that is expressive of elapsed time and that in English is usually formed by internal vowel change (as in *sang*) or by the addition of a suffix (as in *laughed*)
4 : having served as a specified officer in an organization ⟨*past* president⟩

[2]past *preposition* (14th century)
1 a : beyond the age for or of ⟨*past* playing with dolls⟩ **b :** AFTER ⟨half *past* two⟩
2 a : at the farther side of **:** BEYOND **b :** in a course or direction going close to and then beyond ⟨drove *past* the house⟩
3 *obsolete* **:** more than
4 : beyond the capacity, range, or sphere of ⟨*past* belief⟩

[3]past *noun* (1590)
1 a : time gone by **b :** something that happened or was done in the past ⟨regret the *past*⟩
2 a : the past tense of a language **b :** a verb form in the past tense
3 : a past life, history, or course of action; *especially* **:** one that is kept secret
— past•less \ˈpas(t)-ləs\ *adjective*

[4]past *adverb* (1784)
: so as to reach and go beyond a point near at hand

pas•ta \ˈpäs-tə *also* ˈpas-\ *noun* [Italian, from Late Latin] (1874)
1 : paste in processed form (as spaghetti) or in the form of fresh dough (as ravioli)
2 : a dish of cooked pasta

[1]paste \ˈpāst\ *noun* [Middle English, from Middle French, from Late Latin *pasta* dough, paste] (14th century)
1 a : a dough that contains a considerable proportion of fat and is used for pastry crust or fancy rolls **b :** a confection made by evaporating fruit with sugar or by flavoring a gelatin, starch, or gum arabic preparation **c :** a smooth food product made by evaporation or grinding ⟨tomato *paste*⟩ ⟨almond *paste*⟩ **d :** a shaped dough (as spaghetti or ravioli) prepared from semolina, farina, or wheat flour
2 : a soft plastic mixture or composition: as **a :** a preparation usually of flour or starch and water used as an adhesive or a vehicle for mordant or color **b :** clay or a clay mixture used in making pottery or porcelain
3 : a brilliant glass of high lead content used for the manufacture of artificial gems

[2]paste *transitive verb* **past•ed; past•ing** (1561)
1 : to cause to adhere by or as if by paste **:** STICK
2 : to cover with something pasted on

[3]paste *transitive verb* **past•ed; past•ing** [alteration of *baste*] (1846)
1 : to strike hard at
2 : to beat or defeat soundly

[1]paste•board \ˈpās(t)-ˌbōrd, -ˌbȯrd\ *noun* (1562)
1 : a solid paperboard with a paper facing; *broadly* **:** PAPERBOARD
2 : TICKET

[2]pasteboard *adjective* (1599)
1 : made of pasteboard
2 : SHAM, UNSUBSTANTIAL ⟨prefabricated plots and *pasteboard* heroes —Peter Andrews⟩

paste•down \-ˌdau̇n\ *noun* (circa 1888)
: the outer leaf of an endpaper that is pasted down to the inside of the front or back cover of a book

[1]pas•tel \pa-ˈstel\ *noun* [French, from Italian *pastello*, from Late Latin *pastellus* woad, from diminutive of *pasta*] (1662)
1 : a paste made of powdered pigment ranging from pale to deep colors and used for making crayons; *also* **:** a crayon made of such paste
2 a : a drawing in pastel **b :** the process or art of drawing with pastels
3 : a light literary sketch
4 : any of various pale or light colors

[2]pastel *adjective* (1884)
1 a : of or relating to a pastel **b :** made with pastels
2 : pale and light in color
3 : lacking in body or vigor

pas•tel•ist *or* **pas•tel•list** \-ˈste-list\ *noun* (1881)
: an artist who works with pastels

pas•tern \ˈpas-tərn\ *noun* [Middle French *pasturon*, from *pasture* tether attached to a horse's foot, alteration of (assumed) Old French *pastoire*, from Medieval Latin *pastoria*, from Latin, feminine of *pastorius* of a herdsman, from *pastor* herdsman — more at PASTOR] (circa 1530)
1 : a part of the foot of an equine extending from the fetlock to the top of the hoof — see HORSE illustration
2 : a part of the leg of an animal other than an equine that corresponds to the pastern

paste•up \ˈpāst-ˌəp\ *noun* (circa 1930)
: MECHANICAL; *also* **:** the process of making mechanicals

pas•teur•ise *British variant of* PASTEURIZE

pas•teur•i•za•tion \ˌpas-chə-rə-ˈzā-shən, ˌpas-tyə-, -tə-\ *noun* (1886)
1 : partial sterilization of a substance and especially a liquid (as milk) at a temperature and for a period of exposure that destroys objectionable organisms without major chemical alteration of the substance
2 : partial sterilization of perishable food products (as fruit or fish) with radiation (as gamma rays)

pas•teur•ize \ˈpas-chə-ˌrīz, ˈpas-tyə-, -tə-\ *transitive verb* **-ized; -iz•ing** [Louis *Pasteur*] (1881)
: to subject to pasteurization
— pas•teur•iz•er *noun*

Pasteur treatment *noun* (1926)
: a method of aborting rabies by stimulating production of antibodies through successive inoculations with attenuated virus of gradually increasing strength

pas•tic•cio \pa-ˈstē-ch(ē-ˌ)ō, pä-\ *noun, plural* **-ci** \-(ˌ)chē\ *or* **-cios** [Italian, literally, pasty, from (assumed) Vulgar Latin *pasticium*, from Late Latin *pasta*] (1752)
: PASTICHE

pas•tiche \pas-ˈtēsh, päs-\ *noun* [French, from Italian *pasticcio*] (1878)
1 : a literary, artistic, musical, or architectural work that imitates the style of previous work; *also* **:** such stylistic imitation
2 a : a musical, literary, or artistic composition made up of selections from different works **:** POTPOURRI **b :** HODGEPODGE
— pas•ti•cheur \ˌpas-tē-ˈshər, ˌpäs-\ *noun*

past•ies \ˈpās-tēz\ *noun plural* [[2]*paste*] (circa 1954)
: small round coverings for a woman's nipples worn especially by a stripteaser

pas•tille \pas-ˈtē(ə)l\ *also* **pas•til** \ˈpas-tᵊl\ *noun* [French *pastille*, from Latin *pastillus* small loaf, lozenge; akin to Latin *panis* bread — more at FOOD] (1658)
1 : a small mass of aromatic paste for fumigating or scenting the air of a room
2 : an aromatic or medicated lozenge **:** TROCHE

pas•time \ˈpas-ˌtīm\ *noun* (15th century)
: something that amuses and serves to make time pass agreeably **:** DIVERSION

pas•ti•na \ˈpäs-tē-nə\ *noun* [Italian, diminutive of *pasta* pasta] (circa 1948)
: very small bits of pasta used in soup or broth

pas•tis \pȧs-ˈtēs\ *noun* [French, from French dialect (Marseilles), literally, jumble, kind of pastry, from Provençal, from Old Provençal *pastitz* cake, from (assumed) Vulgar Latin *pasticium*] (1926)
: a French liqueur flavored with aniseed

past master *noun* (1762)
1 : one who has held the office of worshipful master in a lodge of Freemasons or of master in a guild, club, or society
2 [alteration of *passed master*] **:** one who is expert **:** ADEPT

past•ness \ˈpas(t)-nəs\ *noun* (1829)
1 : the quality or state of being past
2 : the subjective quality of something being remembered rather than immediately experienced

[1]pas•tor \ˈpas-tər\ *noun* [Middle English *pastour*, from Middle French, from Latin *pastor* herdsman, from *pascere* to feed — more at FOOD] (14th century)
: a spiritual overseer; *especially* **:** a clergyman serving a local church or parish
— pas•tor•ship \-ˌship\ *noun*

[2]pastor *transitive verb* **pas•tored; pas•tor•ing** \-t(ə-)riŋ\ (1623)
: to serve as pastor of (as a church)

[3]pas•tor \päs-ˈtȯr\ *noun* [Spanish, from Latin] (1849)
chiefly Southwest **:** HERDSMAN

[1]pas•to•ral \ˈpas-t(ə-)rəl\ *adjective* [Middle English, from Latin *pastoralis*, from *pastor* herdsman] (15th century)
1 a (1) **:** of, relating to, or composed of shepherds or herdsmen (2) **:** devoted to or based on livestock raising **b :** of or relating to the countryside **:** not urban **c :** portraying or expressive of the life of shepherds or country people especially in an idealized and conventionalized manner ⟨*pastoral* poetry⟩ **d :** pleasingly peaceful and innocent **:** IDYLLIC
2 a : of or relating to spiritual care or guidance especially of a congregation **b :** of or relating to the pastor of a church
— pas•to•ral•ly \-t(ə-)rə-lē\ *adverb*
— pas•to•ral•ness *noun*

[2]pastoral \ˈpas-t(ə-)rəl; *1d is often* ˌpas-tə-ˈräl, -ˈral\ *noun* (1584)
1 a : a literary work (as a poem or play) dealing with shepherds or rural life in a usually artificial manner and typically drawing a contrast between the innocence and serenity of the simple life and the misery and corruption of city and especially court life **b :** pastoral poetry or drama **c :** a rural picture or scene **d :** PASTORALE 1b
2 : CROSIER 1
3 : a letter of a pastor to his charge: as **a :** a letter addressed by a bishop to his diocese **b :** a letter of the house of bishops of the Protestant Episcopal Church to be read in each parish

pas•to•rale \ˌpas-tə-ˈräl, -ˈral *also* -ˈrä-lē\ *noun* [Italian, from *pastorale* of herdsmen, from Latin *pastoralis*] (1724)
1 a : an opera of the 16th or 17th centuries having a pastoral plot **b :** an instrumental or vocal composition having a pastoral theme
2 : PASTORAL 1a

Pastoral Epistle *noun* (1836)
: one of three New Testament letters including two addressed to Timothy and one to Titus that give advice on matters of church government and discipline

pas·to·ral·ism \ˈpas-t(ə-)rə-ˌli-zəm\ *noun* (1854)
1 : the quality or style characteristic of pastoral writing
2 a : livestock raising **b :** social organization based on livestock raising as the primary economic activity
— pas·to·ral·ist \-list\ *noun or adjective*

pas·tor·ate \ˈpas-t(ə-)rət\ *noun* (1795)
1 : the office, state, jurisdiction, or tenure of office of a pastor
2 : a body of pastors

past participle *noun* (1798)
: a participle that typically expresses completed action, that is traditionally one of the principal parts of the verb, and that is traditionally used in English in the formation of perfect tenses in the active voice and of all tenses in the passive voice

past perfect *adjective* (1889)
: of, relating to, or constituting a verb tense that is traditionally formed in English with *had* and denotes an action or state as completed at or before a past time spoken of
— past perfect *noun*

pas·tra·mi *also* **pas·tromi** \pə-ˈsträ-mē\ *noun* [Yiddish *pastrame,* from Romanian *pastramă* pressed and cured meat] (1936)
: a highly seasoned smoked beef prepared especially from shoulder cuts

pas·try \ˈpās-trē\ *noun, plural* **pastries** [[1]*paste*] (circa 1538)
1 a : PASTE 1a **b :** sweet baked goods made of dough having a high fat content
2 : a piece of pastry

past tense *noun* (1813)
: a verb tense expressing action or state in or as if in the past: **a :** a verb tense expressive of elapsed time (as *wrote* in "on arriving I wrote a letter") **b :** a verb tense expressing action or state in progress or continuance or habitually done or customarily occurring at a past time (as *was writing* in "I was writing while he dictated" or *loved* in "their sons loved fishing")

pas·tur·age \ˈpas-chə-rij\ *noun* (circa 1533)
: PASTURE

[1]pas·ture \ˈpas-chər\ *noun* [Middle English, from Middle French, from Late Latin *pastura,* from Latin *pastus,* past participle of *pascere* to feed — more at FOOD] (14th century)
1 : plants (as grass) grown for the feeding especially of grazing animals
2 : land or a plot of land used for grazing
3 : the feeding of livestock **:** GRAZING

[2]pasture *verb* **pas·tured; pas·tur·ing** (14th century)
intransitive verb
: GRAZE, BROWSE
transitive verb
1 : to feed (as cattle) on pasture
2 : to use as pasture

pas·ture·land \ˈpas-chər-ˌland\ *noun* (1591)
: PASTURE 2

[1]pas·ty \ˈpas-tē\ *noun, plural* **pasties** [Middle English *pastee,* from Middle French *pasté,* from *paste* dough, paste] (13th century)
1 : [2]PIE 1, 2; *especially* **:** a meat pie
2 : TURNOVER 5

[2]pasty \ˈpās-tē\ *adjective* **past·i·er; -est** (1659)
: resembling paste; *especially* **:** pallid and unhealthy in appearance
— past·i·ness \-nəs\ *noun*

PA system \ˌpē-ˈā-\ *noun* (circa 1936)
: PUBLIC-ADDRESS SYSTEM

[1]pat \ˈpat\ *noun* [Middle English *patte,* probably of imitative origin] (15th century)
1 : a light blow especially with the hand or a flat instrument
2 : a light tapping often rhythmical sound
3 : something (as butter) shaped into a small flat usually square individual portion
— pat on the back : an expression of approval

[2]pat *adverb* (1578)
: in a pat manner **:** APTLY, PERFECTLY

[3]pat *verb* **pat·ted; pat·ting** (1591)
transitive verb
1 : to strike lightly with a flat instrument
2 : to flatten, smooth, or put into place or shape with light blows
3 : to tap or stroke gently with the hand to soothe, caress, or show approval
intransitive verb
: to strike or beat gently

[4]pat *adjective* (1646)
1 a : exactly suited to the purpose or occasion **:** APT **b :** suspiciously appropriate **:** CONTRIVED
2 : learned, mastered, or memorized exactly
3 : FIRM, UNYIELDING
4 : reduced to a simple or mechanical form **:** STANDARD, TRITE

pa·ta·ca \pə-ˈtä-kə\ *noun* [Portuguese] (1830)
— see MONEY table

pat–a–cake \ˈpa-də-ˌkāk, ˈpat-ə-\ *variant of* PATTY-CAKE

pa·ta·gi·um \pə-ˈtā-jē-əm\ *noun, plural* **-gia** \-jē-ə\ [New Latin, from Latin, gold edging on a tunic] (1826)
1 : the fold of skin connecting the forelimbs and hind limbs of some tetrapods (as flying squirrels)
2 : the fold of skin in front of the main segments of a bird's wing

[1]patch \ˈpach\ *noun* [Middle English *pacche*] (14th century)
1 : a piece of material used to mend or cover a hole or a weak spot
2 : a tiny piece of black silk or court plaster worn on the face or neck especially by women to hide a blemish or to heighten beauty
3 a : a piece of material (as adhesive plaster) used medically usually to cover a wound or repair a defect **b :** a shield worn over the socket of an injured or missing eye
4 a : a small piece **:** SCRAP **b :** a part or area distinct from that about it ⟨cabbage *patch*⟩ **c :** a period of time **:** SPELL ⟨was going through a rough *patch*⟩
5 : a piece of cloth sewed on a garment as an ornament or insignia; *especially* **:** SHOULDER PATCH
6 : a temporary connection in a communication system (as a telephone hookup)
7 : a minor usually temporary correction or modification in a computer program

[2]patch *transitive verb* (15th century)
1 : to mend, cover, or fill up a hole or weak spot in
2 : to provide with a patch
3 a : to make of patches or fragments **b :** to mend or put together especially in hasty or shabby fashion — usually used with *up* **c :** to make a patch in (a computer program)
4 : to connect (as circuits) by a patch cord
synonym see MEND

[3]patch *noun* [perhaps by folk etymology from Italian dialect *paccio*] (1549)
: FOOL, DOLT

patch·board \ˈpach-ˌbōrd, -ˌbȯrd\ *noun* (1934)
: a switchboard in which circuits are interconnected by patch cords

patch cord *noun* (1926)
: a wire with a plug at each end that is used to connect electrical devices

patch·ou·li *also* **patch·ou·ly** \ˈpa-chə-lē, pə-ˈchü-lē\ *noun* [Tamil *paccuḷi*] (1845)
: a heavy perfume made from a fragrant essential oil from an East Indian shrubby mint (*Pogostemon cablin*); *also* **:** the plant itself

patch pocket *noun* (1895)
: a flat pocket applied to the outside of a garment

patch test *noun* (1933)
: a test for determining allergic sensitivity that is made by applying to the unbroken skin small pads soaked with the allergen to be tested

patch·work \ˈpach-ˌwərk\ *noun, often attributive* (1692)
1 : something composed of miscellaneous or incongruous parts **:** HODGEPODGE
2 : pieces of cloth of various colors and shapes sewn together to form a covering

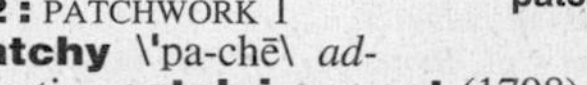
patchwork 2

patchwork quilt *noun* (1840)
1 : a quilt made of patchwork
2 : PATCHWORK 1

patchy \ˈpa-chē\ *adjective* **patch·i·er; -est** (1798)
1 : marked by, consisting of, or diversified with patches
2 : irregular in appearance, makeup, or quality
— patch·i·ly \ˈpa-chə-lē\ *adverb*
— patch·i·ness \ˈpa-chē-nəs\ *noun*

pate \ˈpāt\ *noun* [Middle English] (14th century)
1 : HEAD
2 : the crown of the head
3 : BRAIN — used chiefly disparagingly
— pat·ed \ˈpā-təd\ *adjective*

pâte \ˈpät\ *noun* [French, literally, paste, from Old French *paste*] (1863)
: PASTE 2b

pâ·té *also* **pate** \pä-ˈtā, pa-\ *noun* [French, from Middle French *pasté* — more at PASTY] (1706)
1 : a meat or fish pie or patty
2 : a spread of finely chopped or pureed seasoned meat ⟨chicken liver *pâté*⟩

pâ·té de foie gras \(ˌ)pä-ˌtā-də-ˌfwä-ˈgrä, (ˌ)pa-ˌtā-\ *noun, plural* **pâ·tés de foie gras** \-ˌtā(z)-\ [French, literally, pâté of fat liver] (1827)
: a pâté of fat goose liver and usually truffles sometimes with added fat pork

pa·tel·la \pə-ˈte-lə\ *noun, plural* **-lae** \-ˈte-(ˌ)lē, -ˌlī\ *or* **-las** [Latin, from diminutive of *patina* shallow dish] (1693)
: a thick flat triangular movable bone that forms the anterior point of the knee and protects the front of the joint — called also *kneecap*
— pa·tel·lar \-ˈte-lər\ *adjective*

pa·tel·li·form \pə-ˈte-lə-ˌfȯrm\ *adjective* [New Latin *Patella,* genus including the limpet, from Latin, small shallow dish] (1819)
1 : resembling a limpet or limpet shell
2 : disk-shaped with a narrow rim

pat·en \ˈpa-tᵊn\ *noun* [Middle English, from Middle French *patene,* from Medieval Latin & Latin; Medieval Latin *patina,* from Latin, shallow dish, from Greek *patanē*] (14th century)
1 : a plate usually made of precious metal and used to carry the bread at the Eucharist
2 a : PLATE **b :** something (as a metal disk) resembling a plate

pa·ten·cy \ˈpa-tᵊn(t)-sē, ˈpā-\ *noun* (1656)
: the quality or state of being patent

[1]pat·ent *1–3 are* ˈpa-tᵊnt, *chiefly British* ˈpā-; *4* ˈpā-; *5* ˈpā-, ˈpa-; *6–7* ˈpa-, ˈpā-, *British usually* ˈpā-\ *adjective* [Middle English, from Middle French, from Latin *patent-, patens,* from present participle of *patēre* to be open — more at FATHOM] (14th century)
1 a : open to public inspection — used chiefly in the phrase *letters patent* **b** (1) **:** secured by letters patent or by a patent to the exclusive

\ə\ **abut** \ᵊ\ kitten \ər\ **further** \a\ **ash** \ā\ **ace** \ä\ **mop, mar** \au̇\ **out** \ch\ **chin** \e\ **bet** \ē\ **easy** \g\ **go** \i\ **hit** \ī\ **ice** \j\ **job** \ŋ\ **sing** \ō\ **go** \ȯ\ **law** \ȯi\ **boy** \th\ **thin** \t͟h\ **the** \ü\ **loot** \u̇\ **foot** \y\ **yet** \zh\ **vision** *see also* Guide to Pronunciation

control and possession of a particular individual or party (2) **:** protected by a patent **:** made under a patent ⟨*patent* locks⟩ **c :** protected by a trademark or a trade name so as to establish proprietary rights analogous to those conveyed by letters patent or a patent **:** PROPRIETARY ⟨*patent* drugs⟩
2 : of, relating to, or concerned with the granting of patents especially for inventions ⟨a *patent* lawyer⟩
3 : making exclusive or proprietary claims or pretensions
4 : affording free passage **:** UNOBSTRUCTED ⟨a *patent* opening⟩
5 : PATULOUS, SPREADING ⟨a *patent* calyx⟩
6 *archaic* **:** ACCESSIBLE, EXPOSED
7 : readily visible or intelligible **:** OBVIOUS
synonym see EVIDENT
— pat·ent·ly *adverb*

[2]**pat·ent** \ˈpa-tᵊnt, *British also* ˈpā-\ *noun* (14th century)
1 : an official document conferring a right or privilege **:** LETTERS PATENT
2 a : a writing securing to an inventor for a term of years the exclusive right to make, use, or sell an invention **b :** the monopoly or right so granted **c :** a patented invention
3 : PRIVILEGE, LICENSE
4 : an instrument making a conveyance of public lands; *also* **:** the land so conveyed
5 : PATENT LEATHER

[3]**pat·ent** *transitive verb* (1675)
1 : to obtain or grant a patent right to
2 : to grant a privilege, right, or license to by patent
3 : to obtain or secure by patent; *especially* **:** to secure by letters patent exclusive right to make, use, or sell
— pat·ent·abil·i·ty \ˌpa-tᵊn-tə-ˈbi-lə-tē, *British also* ˌpā-\ *noun*
— pat·ent·able \ˈpa-tᵊn-tə-bəl, *British also* ˈpā-\ *adjective*

pat·ent·ed \ˈpa-tᵊn-təd, *British also* ˈpā-\ *adjective* (1951)
: originated by or peculiar to one person or group **:** INDIVIDUALIZED

pat·en·tee \ˌpa-tᵊn-ˈtē, *British also* ˌpā-\ *noun* (15th century)
: one to whom a grant is made or a privilege secured by patent

pat·ent flour \ˈpa-tᵊn(t)-, *British also* ˈpā-\ *noun* (1886)
: a high-grade wheat flour that consists solely of endosperm

pat·ent leather \ˈpa-tᵊn(t)-, *British usually* ˈpā-\ *noun* (1829)
: a leather with a hard smooth glossy surface

patent medicine *noun* (1770)
: a packaged nonprescription drug which is protected by a trademark and whose contents are incompletely disclosed; *also* **:** any drug that is a proprietary

patent office *noun* (1696)
: a government office for examining claims to patents and granting patents

pat·en·tor \ˈpa-tᵊn-tər, ˌpa-tᵊn-ˈtȯr, *British also* ˈpā-, ˌpā-\ *noun* (circa 1890)
: one that grants a patent

patent right *noun* (1805)
: a right granted by letters patent; *especially* **:** the exclusive right to an invention

pa·ter *noun* (14th century)
1 \ˈpä-ˌter\ *often capitalized* **:** PATERNOSTER
2 \ˈpā-tər\ [Latin] *chiefly British* **:** FATHER

pa·ter·fa·mil·i·as \ˌpa-tər-fə-ˈmi-lē-əs, ˌpä-, ˌpā-\ *noun, plural* **pa·tres·fa·mil·i·as** \ˌpā-ˌtrēz-, ˌpä-ˌtrās-\ [Middle English, from Latin, from *pater* father + *familias,* archaic genitive of *familia* household — more at FATHER, FAMILY] (15th century)
1 : the male head of a household
2 : the father of a family

pa·ter·nal \pə-ˈtər-nᵊl\ *adjective* [Middle English, from Late Latin *paternalis,* from Latin *paternus* paternal, from *pater*] (15th century)
1 a : of or relating to a father **b :** like that of a father ⟨*paternal* benevolence⟩
2 : received or inherited from one's male parent
3 : related through one's father ⟨*paternal* grandfather⟩
— pa·ter·nal·ly \-nᵊl-ē\ *adverb*

pa·ter·nal·ism \pə-ˈtər-nᵊl-ˌi-zəm\ *noun* (1881)
1 : a system under which an authority undertakes to supply needs or regulate conduct of those under its control in matters affecting them as individuals as well as in their relations to authority and to each other
2 : a policy or practice based on or characteristic of paternalism
— pa·ter·nal·ist \-nᵊl-ist\ *noun or adjective*
— pa·ter·nal·is·tic \-ˌtər-nᵊl-ˈis-tik\ *adjective*
— pa·ter·nal·is·ti·cal·ly \-ti-k(ə-)lē\ *adverb*

pa·ter·ni·ty \pə-ˈtər-nə-tē\ *noun* (1582)
1 : the quality or state of being a father
2 : origin or descent from a father

paternity test *noun* (1926)
: a test especially of DNA or genetic traits to determine whether a given man could be the biological father of a given child

pa·ter·nos·ter \ˌpä-tər-ˈnäs-tər, ˈpa-tər-ˌ, ˈpä-ˌter-ˈ, -ˈnäs-ˌter\ *noun* [Middle English, from Old English, from Medieval Latin, from Latin *pater noster* our father, from the opening words] (before 12th century)
1 *often capitalized* **:** LORD'S PRAYER
2 : a word formula repeated as a prayer or magical charm

path \ˈpath, ˈpȧth\ *noun, plural* **paths** \ˈpat͟hz, ˈpaths, ˈpȧt͟hz, ˈpȧths\ [Middle English, from Old English *pæth;* akin to Old High German *pfad* path] (before 12th century)
1 : a trodden way
2 : a track specially constructed for a particular use
3 a : COURSE, ROUTE **b :** a way of life, conduct, or thought
4 a : the continuous series of positions or configurations that can be assumed in any motion or process of change by a moving or varying system **b :** a sequence of arcs in a network that can be traced continuously without retracing any arc
5 : PATHWAY 2

path- *or* **patho-** *combining form* [New Latin, from Greek, from *pathos,* literally, suffering — more at PATHOS]
: pathological state **:** disease ⟨*path*ogen⟩

-path *noun combining form* [German, back-formation from *-pathie* -pathy]
1 : practitioner of a (specified) system of medicine that emphasizes one aspect of disease or its treatment ⟨naturo*path*⟩
2 [International Scientific Vocabulary, from Greek *-pathēs,* adjective, suffering, from *pathos*] **a :** one suffering from a disorder (of such a part or system) ⟨psycho*path*⟩ **b :** one perceiving ⟨tele*path*⟩

Pa·than \pə-ˈtän, ˌpä-\ *noun* [Hindi *Paṭhān,* from Pashto (eastern dialect) *Pakhtana,* plural of *Pakhtun*] (1638)
: PASHTUN

path·break·ing \ˈpath-ˌbrā-kiŋ\ *adjective* (1914)
: TRAILBLAZING

pa·thet·ic \pə-ˈthe-tik\ *adjective* [Middle French or Late Latin; Middle French *pathetique,* from Late Latin *patheticus,* from Greek *pathētikos* capable of feeling, pathetic, from *paschein* (aorist *pathein*) to experience, suffer — more at PATHOS] (1598)
1 : having a capacity to move one to either compassionate or contemptuous pity
2 : marked by sorrow or melancholy **:** SAD
synonym see MOVING
— pa·thet·i·cal \-ti-kəl\ *adjective*
— pa·thet·i·cal·ly \-ti-k(ə-)lē\ *adverb*

pathetic fallacy *noun* (1856)
: the ascription of human traits or feelings to inanimate nature (as in *cruel sea*)

path·find·er \ˈpath-ˌfīn-dər, ˈpȧth-\ *noun* (1840)
: one that discovers a way; *especially* **:** one that explores untraversed regions to mark out a new route
— path·find·ing \-diŋ\ *noun or adjective*

path·less \-ləs\ *adjective* (1591)
: UNTRODDEN, TRACKLESS
— path·less·ness *noun*

patho·bi·ol·o·gy \ˌpa-thō-bī-ˈä-lə-jē\ *noun* (circa 1909)
: PATHOLOGY

path·o·gen \ˈpa-thə-jən\ *noun* [International Scientific Vocabulary] (1880)
: a specific causative agent (as a bacterium or virus) of disease

path·o·gen·e·sis \ˌpa-thə-ˈje-nə-səs\ *noun* [New Latin] (1876)
: the origination and development of a disease

path·o·ge·net·ic \-jə-ˈne-tik\ *adjective* [International Scientific Vocabulary] (1838)
1 : of or relating to pathogenesis
2 : PATHOGENIC 2

path·o·gen·ic \-ˈje-nik\ *adjective* [International Scientific Vocabulary] (1852)
1 : PATHOGENETIC 1
2 : causing or capable of causing disease
— path·o·ge·nic·i·ty \-jə-ˈni-sə-tē\ *noun*

pa·tho·gno·mon·ic \ˌpa-thə(g)-nō-ˈmä-nik\ *adjective* [Greek *pathognōmonikos,* from *path-* + *gnōmonikos* fit to judge, from *gnōmōn* interpreter; akin to Greek *gignōskein* to know — more at KNOW] (1625)
: distinctively characteristic of a particular disease

path·o·log·i·cal \ˌpa-thə-ˈlä-ji-kəl\ *also* **path·o·log·ic** \-jik\ *adjective* (1688)
1 : of or relating to pathology
2 : altered or caused by disease
— path·o·log·i·cal·ly \-ji-k(ə-)lē\ *adverb*

pa·thol·o·gist \pə-ˈthä-lə-jist, pa-\ *noun* (1650)
: a specialist in pathology; *specifically* **:** one who interprets and diagnoses the changes caused by disease in tissues and body fluids

pa·thol·o·gy \-jē\ *noun, plural* **-gies** [New Latin *pathologia* & Middle French *pathologie,* from Greek *pathologia* study of the emotions, from *path-* + *-logia* -logy] (1611)
1 : the study of the essential nature of diseases and especially of the structural and functional changes produced by them
2 : something abnormal: **a :** the structural and functional deviations from the normal that constitute disease or characterize a particular disease **b :** deviation from propriety or from an assumed normal state of something nonliving or nonmaterial

path·o·phys·i·ol·o·gy \ˈpa-thō-ˌfi-zē-ˈä-lə-jē\ *noun* (1947)
: the physiology of abnormal states; *specifically* **:** the functional changes that accompany a particular syndrome or disease
— path·o·phys·i·o·log·i·cal \-zē-ə-ˈlä-ji-kəl\ *also* **path·o·phys·i·o·log·ic** \-jik\ *adjective*

pa·thos \ˈpā-ˌthäs, -ˌthȯs, -ˌthōs *also* ˈpa-\ *noun* [Greek, suffering, experience, emotion, from *paschein* (aorist *pathein*) to experience, suffer; perhaps akin to Lithuanian *kęsti* to suffer] (1591)
1 : an element in experience or in artistic representation evoking pity or compassion
2 : an emotion of sympathetic pity

path·way \ˈpath-ˌwā, ˈpȧth-\ *noun* (15th century)
1 : PATH, COURSE
2 : a line of communication over interconnecting neurons extending from one organ or center to another
3 : the sequence of enzyme catalyzed reactions by which an energy-yielding substance is utilized by protoplasm ⟨metabolic *pathways*⟩

-pathy *noun combining form* [Latin *-pathia,* from Greek *-patheia,* from *-pathēs* suffering, from *pathos*]
1 : feeling **:** suffering ⟨em*pathy*⟩ **:** perception ⟨tele*pathy*⟩
2 : disorder of (such) a part or kind ⟨neuro*pathy*⟩
3 : system of medicine based on (such) a factor ⟨osteo*pathy*⟩

pa·tience \'pā-shən(t)s\ *noun* (13th century)
1 : the capacity, habit, or fact of being patient
2 *chiefly British* **:** SOLITAIRE 2

[1]pa·tient \'pā-shənt\ *adjective* [Middle English *pacient,* from Middle French, from Latin *patient-, patiens,* from present participle of *pati* to suffer; perhaps akin to Greek *pēma* suffering] (14th century)
1 : bearing pains or trials calmly or without complaint
2 : manifesting forbearance under provocation or strain
3 : not hasty or impetuous
4 : steadfast despite opposition, difficulty, or adversity
5 a : able or willing to bear — used with *of* **b :** SUSCEPTIBLE, ADMITTING ⟨*patient* of one interpretation⟩
— pa·tient·ly *adverb*

[2]patient *noun* (14th century)
1 a : an individual awaiting or under medical care and treatment **b :** the recipient of any of various personal services
2 : one that is acted upon

pa·ti·na \pə-'tē-nə, 'pa-tə-nə\ *noun, plural* **pa·ti·nas** \-nəz\ *or* **pa·ti·nae** \-ˌnē, -ˌnī\ [Italian, from Latin, shallow dish — more at PATEN] (1748)
1 a : a usually green film formed naturally on copper and bronze by long exposure or artificially (as by acids) and often valued aesthetically for its color **b :** a surface appearance of something grown beautiful especially with age or use
2 : an appearance or aura that is derived from association, habit, or established character
3 : a superficial covering or exterior

pat·i·nate \'pa-tə-ˌnāt\ *verb* **-nat·ed; -nat·ing** (1880)
transitive verb
: to give a patina to
intransitive verb
: to take on a patina — usually used in the past participle ⟨*patinated* bronze⟩
— pat·i·na·tion \ˌpa-tə-'nā-shən\ *noun*

[1]pa·tine \pa-'tēn\ *noun* [French, from Italian *patina*] (1883)
: PATINA

[2]patine *transitive verb* **pa·tined; pa·tin·ing** (1896)
: to cover with a patina

pa·tio \'pa-tē-ˌō *also* 'pä-\ *noun, plural* **pa·ti·os** [Spanish] (1828)
1 : COURTYARD; *especially* **:** an inner court open to the sky
2 : a recreation area that adjoins a dwelling, is often paved, and is adapted especially to outdoor dining

pa·tis·se·rie *or* **pâ·tis·se·rie** \pə-'ti-sə-rē, pä-\ *noun* [French *pâtisserie,* from Middle French *pastiserie,* from *pasticier* to make pastry, from (assumed) Old French *pastitz* cake, from (assumed) Vulgar Latin *pasticium,* from Late Latin *pasta* dough] (1784)
1 : PASTRY
2 : a pastry shop

pâ·tis·sier *or* **pa·tis·sier** \pä-tis-'yā\ *noun* [French *pâtissier,* from Old French *pasticier,* from (assumed) Old French *pastitz* cake] (circa 1905)
: a pastry chef

pa·tois \'pa-ˌtwä, 'pä-\ *noun, plural* **pa·tois** \-ˌtwäz\ [French] (1643)
1 a : a dialect other than the standard or literary dialect **b :** uneducated or provincial speech
2 : the characteristic special language of an occupational or social group **:** JARGON

patr- *or* **patri-** *or* **patro-** *combining form* [*patr-, patri-* from Latin, from *patr-, pater; patr-, patro-* from Greek, from *patr-, patēr* — more at FATHER]
: father ⟨*patristic*⟩

pa·tri·arch \'pā-trē-ˌärk\ *noun* [Middle English *patriarche,* from Old French, from Late Latin *patriarcha,* from Greek *patriarchēs,* from *patria* lineage (from *patr-, patēr* father) + *-archēs* -arch — more at FATHER] (13th century)
1 a : one of the scriptural fathers of the human race or of the Hebrew people **b :** a man who is father or founder **c** (1) **:** the oldest member or representative of a group (2) **:** a venerable old man **d :** a man who is head of a patriarchy
2 a : any of the bishops of the ancient or Eastern Orthodox sees of Constantinople, Alexandria, Antioch, and Jerusalem or the ancient and Western see of Rome with authority over other bishops **b :** the head of any of various Eastern churches **c :** a Roman Catholic bishop next in rank to the pope with purely titular or with metropolitan jurisdiction
3 : a Mormon of the Melchizedek priesthood empowered to perform the ordinances of the church and pronounce blessings within a stake or prescribed jurisdiction

pa·tri·ar·chal \ˌpā-trē-'är-kəl\ (15th century)
: of, relating to, or being a patriarch or patriarchy

patriarchal cross *noun* (circa 1727)
: a chiefly heraldic cross denoting a cardinal's or archbishop's rank and having two crossbars of which the lower is the longer and intersects the upright above or at its center — see CROSS illustration

pa·tri·arch·ate \'pā-trē-ˌär-kət, -ˌkāt\ *noun* (1617)
1 a : the office, jurisdiction, or time in office of a patriarch **b :** the residence or headquarters of a patriarch
2 : PATRIARCHY

pa·tri·ar·chy \-ˌär-kē\ *noun, plural* **-chies** (1632)
1 : social organization marked by the supremacy of the father in the clan or family, the legal dependence of wives and children, and the reckoning of descent and inheritance in the male line; *broadly* **:** control by men of a disproportionately large share of power
2 : a society or institution organized according to the principles or practices of patriarchy

pa·tri·cian \pə-'tri-shən\ *noun* [Middle English *patricion,* from Middle French *patricien,* from Latin *patricius,* from *patres* senators, from plural of *pater* father — more at FATHER] (15th century)
1 : a member of one of the original citizen families of ancient Rome
2 a : a person of high birth **:** ARISTOCRAT **b :** a person of breeding and cultivation
— patrician *adjective*

pa·tri·ci·ate \-'tri-shē-ət, -ˌāt\ *noun* (circa 1656)
1 : the position or dignity of a patrician
2 : a patrician class

pat·ri·cide \'pa-trə-ˌsīd\ *noun* (1593)
1 [Latin *patricida,* from *patr-* + *-cida* -cide] **:** one who murders his or her own father
2 [Late Latin *patricidium,* from Latin *patr-* + *-cidium* -cide] **:** the murder of one's own father
— pat·ri·cid·al \ˌpa-trə-'sī-dᵊl\ *adjective*

pat·ri·lin·eal \ˌpa-trə-'li-nē-əl\ *adjective* (1904)
: relating to, based on, or tracing descent through the paternal line ⟨*patrilineal* society⟩

pat·ri·mo·ny \'pa-trə-ˌmō-nē\ *noun* [Middle English *patrimonie,* from Middle French, from Latin *patrimonium,* from *patr-, pater* father] (14th century)
1 a : an estate inherited from one's father or ancestor **b :** anything derived from one's father or ancestors **:** HERITAGE
2 : an estate or endowment belonging by ancient right to a church
— pat·ri·mo·ni·al \ˌpa-trə-'mō-nē-əl\ *adjective*

pa·tri·ot \'pā-trē-ət, -ˌät, *chiefly British* 'pa-trē-ət\ *noun* [Middle French *patriote* compatriot, from Late Latin *patriota,* from Greek *patriōtēs,* from *patria* lineage, from *patr-, patēr* father] (1605)
: one who loves his or her country and supports its authority and interests

pa·tri·ot·ic \ˌpā-trē-'ä-tik, *chiefly British* ˌpa-\ *adjective* (1757)
1 : inspired by patriotism
2 : befitting or characteristic of a patriot
— pa·tri·ot·i·cal·ly \-ti-k(ə-)lē\ *adverb*

pa·tri·ot·ism \'pā-trē-ə-ˌti-zəm, *chiefly British* 'pa-\ *noun* (circa 1726)
: love for or devotion to one's country

Patriots' Day *noun* (1897)
: the third Monday in April observed as a legal holiday in Maine and Massachusetts in commemoration of the battles of Lexington and Concord in 1775

pa·tris·tic \pə-'tris-tik\ *adjective* (circa 1828)
: of or relating to the church fathers or their writings
— pa·tris·ti·cal \-ti-kəl\ *adjective*

pa·tris·tics \-tiks\ *noun plural but singular in construction* (1847)
: the study of the writings and background of the church fathers

Pa·tro·clus \pə-'trō-kləs, -'trä-\ *noun* [Latin, from Greek *Patroklos*]
: a Greek hero and friend of Achilles slain by Hector at Troy

[1]pa·trol \pə-'trōl\ *noun* (1664)
1 a : the action of traversing a district or beat or of going the rounds along a chain of guards for observation or the maintenance of security **b :** the person performing such an action **c :** a unit of persons or vehicles employed for reconnaissance, security, or combat
2 : a subdivision of a Boy Scout troop or Girl Scout troop

[2]patrol *verb* **pa·trolled; pa·trol·ling** [French *patrouiller,* from Middle French, to tramp around in the mud, from *patte* paw — more at PATTEN] (1691)
intransitive verb
: to carry out a patrol
transitive verb
: to carry out a patrol of
— pa·trol·ler *noun*

pa·trol·man \pə-'trōl-mən\ *noun* (1867)
: one who patrols; *especially* **:** a police officer assigned to a beat

patrol wagon *noun* (1887)
: PADDY WAGON

pa·tron \'pā-trən, *for 6 also* pa-'trōⁿ\ *noun* [Middle English, from Middle French, from Medieval Latin & Latin; Medieval Latin *patronus* patron saint, patron of a benefice, pattern, from Latin, defender, from *patr-, pater*] (14th century)
1 a : a person chosen, named, or honored as a special guardian, protector, or supporter **b :** a wealthy or influential supporter of an artist or writer **c :** a social or financial sponsor of a social function (as a ball or concert)
2 : one that uses wealth or influence to help an individual, an institution, or a cause
3 : one who buys the goods or uses the services offered especially by an establishment
4 : the holder of the right of presentation to an English ecclesiastical benefice
5 : a master in ancient times who freed his slave but retained some rights over him

6 [French, from Middle French] **:** the proprietor of an establishment (as an inn) especially in France
7 : the chief male officer in some fraternal lodges having both men and women members
— pa·tron·al \'pā-trə-nᵊl; *British* pə-'trō-nᵊl, pa-\ *adjective*

pa·tron·age \'pa-trə-nij, 'pā-\ *noun* (14th century)
1 : ADVOWSON
2 : the support or influence of a patron
3 : kindness done with an air of superiority
4 : business or activity provided by patrons ⟨the new branch library is expected to have a heavy *patronage*⟩
5 a : the power to make appointments to government jobs especially for political advantage **b :** the distribution of jobs on the basis of patronage **c :** jobs distributed by patronage

pa·tron·ess \'pā-trə-nəs\ *noun* (15th century)
: a woman who is a patron

pa·tron·ise *British variant of* PATRONIZE

pa·tron·ize \'pā-trə-ˌnīz, 'pa-\ *transitive verb* **-ized; -iz·ing** (1589)
1 : to act as patron of **:** provide aid or support for
2 : to adopt an air of condescension toward **:** treat haughtily or coolly
3 : to be a frequent or regular customer or client of
— pa·tron·i·za·tion \ˌpā-trə-nə-'zā-shən, pa-\ *noun*
— pa·tron·iz·ing·ly \'pā-trə-ˌnī-ziŋ-lē, 'pa-\ *adverb*

patron saint *noun* (1717)
1 : a saint to whose protection and intercession a person, a society, a church, or a place is dedicated
2 : an original leader or prime exemplar

pat·ro·nym·ic \ˌpa-trə-'ni-mik\ *noun* [Late Latin *patronymicum,* from neuter of *patronymicus* of a patronymic, from Greek *patronymikos,* from *patronymia* patronymic, from *patr-* + *onyma* name — more at NAME] (1612)
: a name derived from that of the father or a paternal ancestor usually by the addition of an affix
— patronymic *adjective*

pa·troon \pə-'trün\ *noun* [French *patron* & Spanish *patrón,* from Medieval Latin *patronus,* from Latin, patron] (1743)
1 *archaic* **:** the captain or officer commanding a ship
2 [Dutch, from French *patron*] **:** the proprietor of a manorial estate especially in New York originally granted under Dutch rule but in some cases existing until the mid-19th century

pat·sy \'pat-sē\ *noun, plural* **pat·sies** [perhaps from Italian *pazzo* fool] (1903)
: a person who is easily manipulated or victimized **:** SUCKER

pat·ten \'pa-tᵊn\ *noun* [Middle English *patin,* from Middle French, from *patte* paw, hoof, from (assumed) Vulgar Latin *patta,* of imitative origin] (14th century)
: a clog, sandal, or overshoe often with a wooden sole or metal device to elevate the foot and increase the wearer's height or aid in walking in mud

¹pat·ter \'pa-tər\ *verb* [Middle English *patren,* from *paternoster*] (14th century)
transitive verb
: to say or speak in a rapid or mechanical manner
intransitive verb
1 : to recite prayers (as paternosters) rapidly or mechanically
2 : to talk glibly and volubly
3 : to speak or sing rapid-fire words in a theatrical performance
— pat·ter·er \-tər-ər\ *noun*

²patter *noun* (1758)
1 : a specialized lingo **:** CANT; *especially* **:** the jargon of criminals (as thieves)
2 : the spiel of a street hawker or of a circus barker
3 : empty chattering talk
4 a (1) **:** the rapid-fire talk of a comedian (2) **:** the talk with which an entertainer accompanies a routine **b :** the words of a comic song or of a rapidly spoken usually humorous monologue introduced into such a song

³patter *verb* [frequentative of ³*pat*] (1611)
intransitive verb
1 : to strike or pat rapidly and repeatedly
2 : to run with quick light-sounding steps
transitive verb
: to cause to patter

⁴patter *noun* (1844)
: a quick succession of light sounds or pats

¹pat·tern \'pa-tərn\ *noun* [Middle English *patron,* from Middle French, from Medieval Latin *patronus*] (14th century)
1 : a form or model proposed for imitation **:** EXEMPLAR
2 : something designed or used as a model for making things ⟨a dressmaker's *pattern*⟩
3 : a model for making a mold into which molten metal is poured to form a casting
4 : an artistic, musical, literary, or mechanical design or form
5 : a natural or chance configuration ⟨frost *pattern*⟩ ⟨the *pattern* of events⟩
6 : a length of fabric sufficient for an article (as of clothing)
7 a : the distribution of shrapnel, bombs on a target, or shot from a shotgun **b :** the grouping made on a target by bullets
8 : a reliable sample of traits, acts, tendencies, or other observable characteristics of a person, group, or institution ⟨behavior *pattern*⟩ ⟨spending *pattern*⟩
9 a : the flight path prescribed for an airplane that is coming in for a landing **b :** a prescribed route to be followed by a pass receiver in football
10 : TEST PATTERN
11 : a discernible coherent system based on the intended interrelationship of component parts ⟨foreign policy *patterns*⟩
12 : frequent or widespread incidence ⟨a *pattern* of dissent⟩
synonym see MODEL
— pat·terned \-tərnd\ *adjective*
— pat·tern·less *adjective*

²pattern (circa 1586)
transitive verb
1 *dialect chiefly English* **a :** MATCH **b :** IMITATE
2 : to make, adapt, or fashion according to a pattern
3 : to furnish, adorn, or mark with a design
intransitive verb
: to form a pattern

pat·tern·ing \'pa-tər-niŋ\ *noun* (1862)
1 : decoration, composition, or configuration according to a pattern
2 : physiotherapy that is designed to improve malfunctioning nervous control by means of feedback from muscular activity imposed by an outside source or induced by other muscles

pat·ty *also* **pat·tie** \'pa-tē\ *noun, plural* **patties** [French *pâté* pâté] (1710)
1 : a little pie
2 a : a small flat cake of chopped food ⟨a hamburger *patty*⟩ **b :** a small flat candy ⟨a peppermint *patty*⟩
3 : PATTY SHELL

pat·ty–cake \-ˌkāk\ *noun* [alteration of *pat-a-cake,* from the opening words of the rhyme] (1889)
: a game in which two participants (as mother and child) clap their hands together to the rhythm of an accompanying nursery rhyme

pat·ty·pan \-ˌpan\ *noun* [*pattypan* pan for baking patties] (1900)
: a summer squash having a scalloped edge — called also *cymling*

patty shell *noun* (1909)
: a shell of puff pastry made to hold a creamed meat, fish, or vegetable filling

pat·u·lous \'pa-chə-ləs\ *adjective* [Latin *patulus,* from *patēre* to be open — more at FATHOM] (1616)
: spreading widely from a center ⟨a tree with *patulous* branches⟩

pat·zer \'pät-sər, 'pat-\ *noun* [probably from German *Patzer* bungler, from *patzen* to blunder] (1959)
: an inept chess player

pau·ci·ty \'pȯ-sə-tē\ *noun* [Middle English *paucite,* from Middle French or Latin; Middle French *paucité,* from Latin *paucitat-, paucitas,* from *paucus* little — more at FEW] (15th century)
1 : smallness of number **:** FEWNESS
2 : smallness of quantity **:** DEARTH

Paul \'pȯl\ *noun* [Latin *Paulus,* from Greek *Paulos*]
: an early Christian apostle and missionary and author of several New Testament epistles

Paul Bun·yan \-'bən-yən\ *noun*
: a giant lumberjack of American folklore

Pau·li exclusion principle \'pau̇-lē-\ *noun* [Wolfgang *Pauli*] (1926)
: EXCLUSION PRINCIPLE — called also *Pauli principle*

Pau·line \'pȯ-ˌlīn\ *adjective* (1817)
: of or relating to the apostle Paul, his epistles, or the doctrine or theology implicit in his epistles

Paul·ist \'pȯ-list\ *noun* (circa 1883)
: a member of the Roman Catholic Congregation of the Missionary Priests of Saint Paul the Apostle founded by I. T. Hecker in the U.S. in 1858

pau·low·nia \pȯ-'lō-nē-ə\ *noun* [New Latin, from Anna *Pavlovna* (died 1865) Russian princess] (1843)
: any of a genus (*Paulownia*) of Chinese trees of the snapdragon family; *especially* **:** one (*P. tomentosa*) widely cultivated for its panicles of fragrant violet flowers

paunch \'pȯnch, 'pänch\ *noun* [Middle English, from Middle French *panche,* from Latin *pantic-, pantex*] (14th century)
1 a : the belly and its contents **b :** POTBELLY
2 : RUMEN

paunchy \'pȯn-chē, 'pän-\ *adjective* **paunch·i·er; -est** (1598)
: having a potbelly
— paunch·i·ness *noun*

pau·per \'pȯ-pər\ *noun* [Latin, poor — more at POOR] (1516)
1 : a person destitute of means except such as are derived from charity; *specifically* **:** one who receives aid from funds designated for the poor
2 : a very poor person
— pau·per·ism \-pə-ˌri-zəm\ *noun*

pau·per·ize \'pȯ-pə-ˌrīz\ *transitive verb* **-ized; -iz·ing** (1834)
: to reduce to poverty

pau·piette \pō-'pyet\ *noun* [French *paupiette,* from Italian *polpetta* meat croquette, diminutive of *polpa* pulp, flesh, from Latin *pulpa*] (1889)
: a thin slice of meat or fish wrapped around a forcemeat filling

¹pause \'pȯz\ *noun* [Middle English, from Latin *pausa,* from Greek *pausis,* from *pauein* to stop] (15th century)
1 : a temporary stop
2 a : a break in a verse **b :** a brief suspension of the voice to indicate the limits and relations of sentences and their parts
3 : temporary inaction especially as caused by uncertainty **:** HESITATION
4 a : the sign denoting a fermata **b :** a mark (as a period or comma) used in writing or printing to indicate or correspond to a pause of voice
5 : a reason or cause for pausing (as to reconsider) ⟨a thought that should give one *pause*⟩

²pause *verb* **paused; paus·ing** (15th century)
intransitive verb

1 : to stop temporarily
2 : to linger for a time
transitive verb
: to cause to pause **:** STOP

pa·vane \pə-ˈvän, -ˈvan\ *also* **pa·van** *same or* ˈpa-vən\ *noun* [Middle French *pavane,* from Italian dialect *pavana,* from feminine of *pavano* of Padua, from *Pava* (Tuscan *Padova*) Padua] (1535)
1 : a stately court dance by couples that was introduced from southern Europe into England in the 16th century
2 : music for the pavane; *also* **:** music having the slow duple rhythm of a pavane

pave \ˈpāv\ *transitive verb* **paved; pav·ing** [Middle English, from Middle French *paver,* from Latin *pavire* to strike, pound; perhaps akin to Greek *paiein* to strike] (14th century)
1 : to lay or cover with material (as asphalt or concrete) that forms a firm level surface for travel
2 : to cover firmly and solidly as if with paving material
3 : to serve as a covering or pavement of
— pav·er *noun*
— pave the way : to prepare a smooth easy way **:** facilitate development

pa·vé \pa-ˈvā\ *also* **pa·véed** *or* **pa·véd** \pa-ˈvād\ *or* **pa·ve** \pa-ˈvā\ *adjective* [*pavé* from French, from past participle of *paver* to pave] (1903)
of jewels **:** set as close together as possible to conceal a metal base

pave·ment \ˈpāv-mənt\ *noun* [Middle English, from Old French, from Latin *pavimentum,* from *pavire*] (13th century)
1 : a paved surface: as **a :** the artificially covered surface of a public thoroughfare **b** *chiefly British* **:** SIDEWALK
2 : the material with which something is paved
3 : something that suggests a pavement (as in flatness, hardness, and extent of surface)

pav·id \ˈpa-vəd\ *adjective* [Latin *pavidus,* from *pavēre* to be frightened; akin to Latin *pavire*] (circa 1656)
: TIMID

[1]pa·vil·ion \pə-ˈvil-yən\ *noun* [Middle English *pavilon,* from Old French *paveillon,* from Latin *papilion-, papilio* butterfly; perhaps akin to Old High German *fīfaltra* butterfly] (13th century)
1 a : a large often sumptuous tent **b :** something resembling a canopy or tent ⟨tree ferns spread their delicate *pavilions* —Blanche E. Baughan⟩
2 a : a part of a building projecting from the rest **b :** one of several detached or semidetached units into which a building is sometimes divided
3 a : a light sometimes ornamental structure in a garden, park, or place of recreation that is used for entertainment or shelter **b :** a temporary structure erected at an exposition by an individual exhibitor
4 : the lower faceted part of a brilliant below the girdle

[2]pavilion *transitive verb* (14th century)
: to furnish or cover with or put in a pavilion

pav·ing \ˈpā-viŋ\ *noun* (15th century)
: PAVEMENT

pav·ior *or* **pav·iour** \ˈpāv-yər\ *noun* [Middle English *pavier,* from *paven* to pave] (15th century)
British **:** one that paves

pav·lo·va \ˈpav-lə-və, pav-ˈlō-\ *noun, often capitalized* [Anna *Pavlova*] (1926)
: a dessert of Australian and New Zealand origin consisting of a meringue shell topped with whipped cream and usually fruit

Pav·lov·i·an \pav-ˈlō-vē-ən, -ˈlō-; -ˈlō-fē-\ *adjective* (1926)
1 : of or relating to Ivan Pavlov or to his work and theories ⟨*Pavlovian* conditioning⟩
2 : being or expressing a conditioned or predictable reaction **:** AUTOMATIC ⟨the candidates gave *Pavlovian* answers⟩

[1]paw \ˈpȯ\ *noun* [Middle English, from Middle French *poue*] (14th century)
1 : the foot of a quadruped (as a lion or dog) that has claws; *broadly* **:** the foot of an animal
2 : a human hand especially when large or clumsy

[2]paw (15th century)
transitive verb
1 : to touch or strike at with a paw
2 : to feel or touch clumsily, rudely, or sexually
3 : to scrape or beat with or as if with a hoof
4 : to flail at or grab for wildly
intransitive verb
1 : to beat or scrape something with or as if with a hoof
2 : to touch or strike with a paw
3 : to feel or touch someone or something clumsily, rudely, or sexually
4 : to flail or grab wildly

paw·ky \ˈpȯ-kē\ *adjective* [obsolete English dialect *pawk* trick] (1676)
chiefly British **:** artfully shrewd **:** CANNY

pawl \ˈpȯl\ *noun* [perhaps modification of Dutch *pal* pawl] (1626)
: a pivoted tongue or sliding bolt on one part of a machine that is adapted to fall into notches or interdental spaces on another part (as a ratchet wheel) so as to permit motion in only one direction — see RATCHET WHEEL illustration

[1]pawn \ˈpȯn, ˈpän\ *noun* [Middle English *pown,* from Middle French *poon,* from Medieval Latin *pedon-, pedo* foot soldier, from Late Latin, one with broad feet, from Latin *ped-, pes* foot — more at FOOT] (14th century)
1 : one of the chessmen of least value having the power to move only forward ordinarily one square at a time, to capture only diagonally forward, and to be promoted to any piece except a king upon reaching the eighth rank
2 : one that can be used to further the purposes of another

[2]pawn *noun* [Middle English *paun,* modification of Middle French *pan*] (15th century)
1 a : something delivered to or deposited with another as security for a loan **b :** HOSTAGE
2 : the state of being pledged
3 : something used as a pledge **:** GUARANTY
4 : the act of pawning

[3]pawn *transitive verb* (1570)
: to deposit in pledge or as security
— pawn·er \ˈpȯ-nər, ˈpä-\ *or* **paw·nor** *same or* pȯ-ˈnȯr, pä-\ *noun*

pawn·bro·ker \ˈpȯn-ˌbrō-kər, ˈpän-\ *noun* (1687)
: one who lends money on the security of personal property pledged in his keeping
— pawn·bro·king \-kiŋ\ *noun*

Paw·nee \pȯ-ˈnē, pä-\ *noun, plural* **Pawnee** *or* **Pawnees** (1770)
: a member of an American Indian people originally of Kansas and Nebraska

pawn off *transitive verb* (1832)
: to get rid of or pass off usually by deception **:** PALM OFF

pawn·shop \ˈpȯn-ˌshäp, ˈpän-\ *noun* (1849)
: a pawnbroker's shop

paw·paw *variant of* PAPAW

pax \ˈpaks, ˈpäks\ *noun* [Middle English, from Medieval Latin, from Latin, peace — more at PEACE] (14th century)
1 : a tablet decorated with a sacred figure (as of Christ) and sometimes ceremonially kissed by participants at mass
2 : the kiss of peace in the Mass
3 : PEACE; *especially, capitalized* **:** a period of general stability in international affairs under the influence of a dominant military power — usually used in combination with a latinized name ⟨*Pax* Americana⟩

[1]pay \ˈpā\ *verb* **paid** \ˈpād\ *also in sense 7* **payed; pay·ing** [Middle English, from Old French *paier,* from Latin *pacare* to pacify, from *pac-, pax* peace] (13th century)
transitive verb
1 a : to make due return to for services rendered or property delivered **b :** to engage for money **:** HIRE ⟨you couldn't *pay* me to do that⟩
2 a : to give in return for goods or service ⟨*pay* wages⟩ **b :** to discharge indebtedness for **:** SETTLE ⟨*pay* a bill⟩ **c :** to make a disposal or transfer of (money)
3 : to give or forfeit in expiation or retribution ⟨*pay* the penalty⟩
4 a : to make compensation for **b :** to requite according to what is deserved ⟨*pay* them back⟩
5 : to give, offer, or make freely or as fitting ⟨*pay* attention⟩ ⟨*pay* your respects⟩
6 a : to return value or profit to ⟨it *pays* you to stay open⟩ **b :** to bring in as a return ⟨an investment *paying* five percent⟩
7 : to slacken (as a rope) and allow to run out — used with *out*
intransitive verb
1 : to discharge a debt or obligation
2 : to be worth the expense or effort ⟨crime doesn't *pay*⟩
3 : to suffer the consequences of an act ☆ ◆
— pay one's dues 1 : to earn a right or position through experience, suffering, or hard work **2** *also* **pay dues :** PAY *intransitive verb* 3

☆ SYNONYMS
Pay, compensate, remunerate, satisfy, reimburse, indemnify, repay, recompense mean to give money or its equivalent in return for something. PAY implies the discharge of an obligation incurred ⟨*paid* their bills on time⟩. COMPENSATE implies a making up for services rendered or help given ⟨an attorney well *compensated* for her services⟩. REMUNERATE more clearly suggests paying for services rendered and may extend to payment that is generous or not contracted for ⟨promised to *remunerate* the searchers handsomely⟩. SATISFY implies paying a person what is demanded or required by law ⟨all creditors will be *satisfied* in full⟩. REIMBURSE implies a return of money that has been expended for another's benefit ⟨*reimbursed* employees for expenses⟩. INDEMNIFY implies making good a loss suffered through accident, disaster, warfare ⟨*indemnified* the families of the killed miners⟩. REPAY stresses paying back an equivalent in kind or amount ⟨*repay* a favor with a favor⟩. RECOMPENSE suggests due return in amends, friendly repayment, or reward ⟨hotel guests were *recompensed* for the inconvenience⟩.

◇ WORD HISTORY
pay Etymologically speaking, to *pay* is "to pacify." The Latin verb *pacare* "to pacify, impose a settlement on" is derived from *pax* "pact, settlement, peace." In the Middle Ages *pacare* was used specifically to mean "to appease (a creditor) by paying a debt" and eventually in the more general sense "to pay." Old French *paier,* descended from *pacare,* had both the original sense "to appease" and the later meaning "to pay." Middle English *payen,* which was borrowed from French in the late 12th or early 13th century, was also used in both senses. The original sense of *pay,* however, has long disappeared from English usage.

\ə\ **a**but \ᵊ\ kitt**e**n \ər\ f**ur**th**er** \a\ **a**sh \ā\ **a**ce \ä\ m**o**p, m**a**r \au̇\ **ou**t \ch\ **ch**in \e\ b**e**t \ē\ **ea**sy \g\ **g**o \i\ h**i**t \ī\ **i**ce \j\ **j**ob \ŋ\ si**ng** \ō\ g**o** \ȯ\ l**aw** \ȯi\ b**oy** \th\ **th**in \t͟h\ **th**e \ü\ l**oo**t \u̇\ f**oo**t \y\ **y**et \zh\ vi**si**on *see also* Guide to Pronunciation

— **pay one's way** *or* **pay one's own way :** to pay one's share of expenses
— **pay the piper :** to bear the cost of something
— **pay through the nose :** to pay exorbitantly or dearly

²pay *noun* (14th century)
1 : something paid for a purpose and especially as a salary or wage **:** REMUNERATION
2 a : the act or fact of paying or being paid **b :** the status of being paid by an employer **:** EMPLOY
3 : a person viewed with respect to reliability or promptness in paying debts or bills
4 a : ore or a natural situation that yields metal and especially gold in profitable amounts **b :** an oil-yielding stratum or zone
synonym see WAGE

³pay *adjective* (1856)
1 : containing or leading to something precious or valuable
2 : equipped with a coin slot for receiving a fee for use ⟨*pay* telephone⟩
3 : requiring payment

⁴pay *transitive verb* **payed** *also* **paid; paying** [obsolete French *peier,* from Latin *picare,* from *pic-, pix* pitch — more at PITCH] (1627)
: to coat with a waterproof composition

pay·able \ˈpā-ə-bəl\ *adjective* (14th century)
1 : that may, can, or must be paid
2 : PROFITABLE

pay–as–you–go *adjective* (1840)
: of or relating to a system or policy of paying bills when due or of paying for goods and services when purchased

pay·back \ˈpā-ˌbak\ *noun* (1955)
1 : a return on an investment equal to the original capital outlay; *also* **:** the period of time elapsed before an investment is recouped
2 : REQUITAL

pay–cable *noun* (1975)
: pay-TV utilizing a cable television system

pay·check \ˈpā-ˌchek\ *noun* (1899)
1 : a check in payment of wages or salary
2 : WAGES, SALARY

pay·day \-ˌdā\ *noun* (1529)
: a regular day on which wages are paid

pay dirt *noun* (1856)
1 : earth or ore that yields a profit to a miner
2 : a useful or remunerative discovery or object

pay·ee \(ˌ)pā-ˈē\ *noun* (1758)
: one to whom money is or is to be paid

pay envelope *noun* (1901)
: an envelope containing one's wages; *also* **:** WAGES

pay·er \ˈpā-ər\ *also* **pay·or** \ˈpā-ər, (ˌ)pā-ˈȯr\ *noun* (14th century)
: one that pays; *especially* **:** the person by whom a bill or note has been or should be paid

pay·load \ˈpā-ˌlōd\ *noun* (circa 1922)
1 : the load carried by a vehicle exclusive of what is necessary for its operation; *especially* **:** the load carried by an aircraft or spacecraft consisting of things (as passengers or instruments) necessary to the purpose of the flight
2 : the weight of a payload

pay·mas·ter \-ˌmas-tər\ *noun* (1550)
: an officer or agent whose duty it is to pay salaries or wages

pay·ment \ˈpā-mənt\ *noun* (14th century)
1 : the act of paying
2 : something that is paid **:** PAY
3 : REQUITAL

pay·nim \ˈpā-nəm\ *noun* [Middle English *painim,* from Old French *paienime* heathendom, from Late Latin *paganismus,* from *paganus* pagan] (13th century)
archaic **:** PAGAN; *especially* **:** MUSLIM

¹pay·off \ˈpā-ˌȯf\ *noun* (1905)
1 a : PROFIT, REWARD **b :** RETRIBUTION
2 : the act or occasion of receiving money or material gain especially as compensation or as a bribe
3 : the climax of an incident or enterprise; *specifically* **:** the denouement of a narrative
4 : a decisive fact or factor resolving a situation or bringing about a definitive conclusion

²payoff *adjective* (1934)
: yielding results in the final test **:** DECISIVE

pay off (1710)
transitive verb
1 a : to give all due wages to; *especially* **:** to pay in full and discharge (an employee) **b :** to pay (a debt or a creditor) in full **c :** BRIBE
2 : to inflict retribution on
3 : to allow (a thread or rope) to run off a spool or drum
intransitive verb
: to yield returns

pay·o·la \pā-ˈō-lə\ *noun* [¹*pay* + *-ola* (as in *Pianola,* trademark for a player piano)] (1938)
: undercover or indirect payment (as to a disc jockey) for a commercial favor (as for promoting a particular record)

pay·out \ˈpā-ˌau̇t\ *noun* (1943)
: the act of paying out **:** PAYOFF

pay–per–view *noun* (1980)
: a cable television service by which customers can order access to a single program for a fee

pay phone *noun* (1936)
: a usually coin-operated public telephone

pay·roll \ˈpā-ˌrōl\ *noun* (1740)
1 : a paymaster's or employer's list of those entitled to pay and of the amounts due to each
2 : the sum necessary for distribution to those on a payroll; *also* **:** the money to be distributed

pay station *noun* (1919)
: PAY PHONE

pay–TV *noun* (circa 1956)
: a service providing noncommercial television programming (as recent movies and entertainment specials) by means of a scrambled signal to subscribers who are provided with a decoder — called also *pay television;* compare PAY-CABLE, SUBSCRIPTION TV

pay up (15th century)
intransitive verb
: to pay what is due
transitive verb
: to pay in full

PBB \ˌpē-(ˌ)bē-ˈbē\ *noun* (circa 1975)
: POLYBROMINATED BIPHENYL

PBX \ˌpē-(ˌ)bē-ˈeks\ *noun* [*p*rivate *b*ranch e*x*change] (circa 1961)
: a private telephone switchboard

PC \ˌpē-ˈsē\ *noun, plural* **PCs** *or* **PC's** (1978)
: PERSONAL COMPUTER

PCB \ˌpē-(ˌ)sē-ˈbē\ *noun* (1966)
: POLYCHLORINATED BIPHENYL

PCP \ˌpē-(ˌ)sē-ˈpē\ *noun* [*p*henyl + *c*ycl- + *p*iperidine] (circa 1970)
: PHENCYCLIDINE

PDQ \ˌpē-(ˌ)dē-ˈkyü\ *adverb, often not capitalized* [abbreviation of *pretty damned quick*] (1875)
: IMMEDIATELY

pe \ˈpā\ *noun* [Hebrew *pē*] (1823)
: the 17th letter of the Hebrew alphabet — see ALPHABET table

pea \ˈpē\ *noun, plural* **peas** *also* **pease** \ˈpēz\ *often attributive* [back-formation from Middle English *pease* (taken as a plural), from Old English *pise,* from Latin *pisa,* plural of *pisum,* from Greek *pison*] (1611)
1 a : a variable annual leguminous Eurasian vine (*Pisum sativum*) that is cultivated especially for its rounded smooth or wrinkled edible protein-rich seeds **b :** the seed of the pea **c** *plural* **:** the immature pods of the pea with their included seeds
2 : any of various leguminous plants related to or resembling the pea — usually used in combination or with a qualifying term ⟨chick*pea*⟩ ⟨black-eyed *pea*⟩; *also* **:** the seed of such a plant
3 : something resembling a pea (as in size or shape)
word history see SHERRY

pea aphid *noun* (1925)
: a widely distributed aphid (*Acyrthosiphon pisum*) that is a serious pest on legumes (as alfalfa, pea, and clover)

pea bean *noun* (circa 1887)
: any of various kidney beans cultivated for their small white seeds which are used dried (as for baking)

¹peace \ˈpēs\ *noun* [Middle English *pees,* from Old French *pais,* from Latin *pac-, pax;* akin to Latin *pacisci* to agree — more at PACT] (12th century)
1 : a state of tranquillity or quiet: as **a :** freedom from civil disturbance **b :** a state of security or order within a community provided for by law or custom ⟨a breach of the *peace*⟩
2 : freedom from disquieting or oppressive thoughts or emotions
3 : harmony in personal relations
4 a : a state or period of mutual concord between governments **b :** a pact or agreement to end hostilities between those who have been at war or in a state of enmity
5 — used interjectionally to ask for silence or calm or as a greeting or farewell
— **at peace :** in a state of concord or tranquillity

²peace *intransitive verb* (14th century)
obsolete **:** to be, become, or keep silent or quiet

peace·able \ˈpē-sə-bəl\ *adjective* (14th century)
1 a : disposed to peace **:** not contentious or quarrelsome **b :** quietly behaved
2 : marked by freedom from strife or disorder
— **peace·able·ness** *noun*
— **peace·ably** \-blē\ *adverb*

peace corps *noun* (1960)
: a body of trained personnel sent as volunteers especially to assist underdeveloped nations

peace·ful \ˈpēs-fəl\ *adjective* (14th century)
1 : PEACEABLE 1
2 : untroubled by conflict, agitation, or commotion **:** QUIET, TRANQUIL
3 : of or relating to a state or time of peace
4 : devoid of violence or force
synonym see CALM
— **peace·ful·ly** \-fə-lē\ *adverb*
— **peace·ful·ness** *noun*

peace·keep·ing \ˈpēs-ˌkē-piŋ\ *noun* (1945)
: the preserving of peace; *especially* **:** international enforcement and supervision of a truce between hostile states or communities
— **peace·keep·er** \-pər\ *noun*

peace·mak·er \-ˌmā-kər\ *noun* (15th century)
: one who makes peace especially by reconciling parties at variance
— **peace·mak·ing** \-kiŋ\ *noun or adjective*

peace·nik \-ˌnik\ *noun* (1965)
: an opponent of the Vietnam War; *broadly* **:** PACIFIST

peace offering *noun* (circa 1530)
: a gift or service for the purpose of procuring peace or reconciliation

peace officer *noun* (1714)
: a civil officer (as a police officer) whose duty it is to preserve the public peace

peace pipe *noun* (1760)
: CALUMET

peace sign *noun* (1969)
1 : a sign made by holding the palm outward and forming a V with the index and middle fingers and used to indicate the desire for peace
2 : PEACE SYMBOL

peace symbol *noun* (1970)
: the symbol ☮ used to signify peace

peace·time \ˈpēs-ˌtīm\ *noun* (1551)
: a time when a nation is not at war

¹peach \ˈpēch\ *noun* [Middle English *peche,* from Middle French (the fruit), from Late Latin *persica,* from Latin (*malum*) *persicum,* literally, Persian fruit] (14th century)
1 a : a low spreading freely branching Chi-

nese tree (*Prunus persica*) of the rose family that is widely cultivated in temperate areas and has lanceolate leaves, sessile usually pink flowers borne on the naked twigs in early spring, and a fruit which is a single-seeded drupe with a hard endocarp, a pulpy white or yellow mesocarp, and a thin downy epicarp **b :** the edible fruit of the peach
2 : a moderate yellowish pink
3 : one resembling a peach (as in sweetness, beauty, or excellence) ◆

[2]peach *verb* [Middle English *pechen,* short for *apechen* to accuse, from (assumed) Anglo-French *apecher,* from Late Latin *impedicare* to entangle — more at IMPEACH] (1570)
transitive verb
: to inform against **:** BETRAY
intransitive verb
: to turn informer **:** BLAB

peach leaf curl *noun* (1888)
: leaf curl of the peach that is caused by a fungus (*Taphrina deformans*)

peach tree borer *noun* (1850)
: a blue-black orange-marked clearwing moth (*Synanthedon exitiosa* synonym *Sanninoidea exitiosa*) whose white brown-headed larva bores in the wood of stone fruit trees (as the peach)

peachy \ˈpē-chē\ *adjective* **peach·i·er; -est** (1599)
1 : resembling a peach
2 : unusually fine **:** DANDY

[1]pea·cock \ˈpē-ˌkäk\ *noun* [Middle English *pecok,* from *pe-* (from Old English *pēa* peafowl, from Latin *pavon-, pavo* peacock) + *cok* cock] (14th century)
1 : a male peafowl distinguished by a crest of upright plumules and by greatly elongated loosely webbed upper tail coverts which are mostly tipped with ocellated spots and can be erected and spread at will in a fan shimmering with iridescent color; *broadly* **:** PEAFOWL
2 : one making a proud display of himself **:** SHOW-OFF
— pea·cock·ish \-ˌkä-kish\ *adjective*
— pea·cocky \-kē\ *adjective*

[2]peacock (1586)
intransitive verb
: SHOW OFF

peacock blue *noun* (1881)
: a moderate greenish blue

peacock flower *noun* (1884)
: ROYAL POINCIANA

pea·fowl \ˈpē-ˌfau̇l\ *noun* [*pea-* (as in *peacock*) + *fowl*] (1804)
: either of two very large terrestrial pheasants (*Pavo cristatus* and *P. muticus*) of southeastern Asia and the East Indies often reared as ornamental fowls

peafowl: *A* female; *B* male

pea green *noun* (1752)
: a moderate yellow-green

pea·hen \ˈpē-ˌhen\ *noun* [Middle English *pehenne,* from *pe-* + *henne* hen] (15th century)
: a female peafowl

pea jacket \ˈpē-\ *noun* [by folk etymology from Dutch *pijjekker,* from *pij,* a kind of cloth + *jekker* jacket] (1721)
: a heavy woolen double-breasted jacket originally worn by sailors — called also *pea·coat* \ˈpē-ˌkōt\

[1]peak \ˈpēk\ *noun* [perhaps alteration of *pike*] (1530)
1 : a pointed or projecting part of a garment; *especially* **:** the visor of a cap or hat
2 : PROMONTORY
3 : a sharp or pointed end
4 a (1) **:** the top of a hill or mountain ending in a point (2) **:** a prominent mountain usually having a well-defined summit **b :** something resembling a mountain peak
5 a : the upper aftermost corner of a fore-and-aft sail **b :** the narrow part of a ship's bow or stern or the part of the hold in it
6 a : the highest level or greatest degree **b :** a high point in a course of development especially as represented on a graph
7 : a point formed by the hair on the forehead
synonym see SUMMIT

[2]peak (1577)
intransitive verb
: to reach a maximum (as of capacity, value, or activity) — often used with *out*
transitive verb
: to cause to come to a peak, point, or maximum

[3]peak *adjective* (1903)
: being at or reaching the maximum

[4]peak *intransitive verb* [origin unknown] (1605)
1 : to grow thin or sickly
2 : to dwindle away

[5]peak *transitive verb* [from *apeak* held vertically] (1626)
1 : to set (as a gaff) nearer the perpendicular
2 : to hold (oars) with blades well raised

[1]peaked \ˈpēkt *also* ˈpē-kəd\ *adjective* (15th century)
: having a peak **:** POINTED
— peaked·ness \ˈpēk(t)-nəs, ˈpē-kəd-nəs\ *noun*

[2]peak·ed \ˈpē-kəd *also* ˈpi-kəd\ *adjective* (1835)
: being pale and wan or emaciated **:** SICKLY

peaky \ˈpē-kē\ *adjective* (1821)
: [2]PEAKED

[1]peal \ˈpē(ə)l\ *noun* [Middle English, appeal, summons to church, short for *appel* appeal, from *appelen* to appeal] (14th century)
1 a : the loud ringing of bells **b :** a complete set of changes on a given number of bells **c :** a set of bells tuned to the tones of the major scale for change ringing
2 : a loud sound or succession of sounds ⟨*peals* of laughter⟩

[2]peal (1632)
intransitive verb
: to give out peals
transitive verb
: to utter or give forth loudly

pea·like \ˈpē-ˌlīk\ *adjective* (1774)
1 : resembling a garden pea especially in size, firmness, and shape ⟨a *pealike* lump under the skin⟩
2 *of a flower* **:** being showy and papilionaceous

[1]pea·nut \ˈpē-(ˌ)nət\ *noun* (1802)
1 : a low-branching widely cultivated leguminous annual herb (*Arachis hypogaea*) with showy yellow flowers having a peduncle which elongates and bends into the soil where the ovary ripens into a pod containing one to three oily edible seeds; *also* **:** its seed or seed-containing pod
2 : an insignificant or tiny person
3 *plural* **:** a trifling amount

[2]peanut *adjective* (1836)
: INSIGNIFICANT, PETTY ⟨*peanut* politics⟩

peanut oil *noun* (1882)
: a colorless to yellow fatty nondrying oil that is obtained from peanuts and is used chiefly as a salad oil, in margarine, in soap, and as a vehicle in pharmaceutical preparations and cosmetics

pear \ˈpar, ˈper\ *noun* [Middle English *pere,* from Old English *peru,* from (assumed) Vulgar Latin *pira,* from Latin, plural of *pirum*] (before 12th century)
1 : a fleshy pome fruit that is borne by a tree (genus *Pyrus,* especially *P. communis*) of the rose family and is usually larger at the apical end
2 : a tree bearing pears

[1]pearl \ˈpər(-ə)l\ *noun* [Middle English *perle,* from Middle French, probably from (assumed) Vulgar Latin *pernula,* diminutive of Latin *perna* upper leg, kind of sea mussel; akin to Old English *fiersn* heel, Greek *pternē*] (14th century)
1 a : a dense variously colored and usually lustrous concretion formed of concentric layers of nacre as an abnormal growth within the shell of some mollusks and used as a gem **b :** MOTHER-OF-PEARL
2 : one that is very choice or precious
3 : something resembling a pearl intrinsically or physically
4 : a nearly neutral slightly bluish medium gray

[2]pearl (14th century)
transitive verb
1 : to set or adorn with pearls
2 : to sprinkle or bead with pearly drops
3 : to form into small round grains
4 : to give a pearly color or luster to
intransitive verb
1 : to form drops or beads like pearls
2 : to fish or search for pearls
— pearl·er \ˈpər-lər\ *noun*

[3]pearl *adjective* (1610)
1 a : of, relating to, or resembling pearl **b :** made of or adorned with pearls
2 : having medium-sized grains

[4]pearl *noun or transitive verb* [alteration of *purl*] (1824)
British **:** PICOT

pearl·es·cent \ˌpər-ˈle-sᵊnt\ *adjective* (1936)
: having a pearly luster ⟨a *pearlescent* lacquer⟩
— pearl·es·cence \-sᵊn(t)s\ *noun*

pearl essence *noun* (circa 1909)
: a translucent substance that occurs in the silvery scales of various fish (as herring) and is used in making artificial pearls, lacquers, and plastics

pearl gray *noun* (1796)
1 : a yellowish to light gray
2 : a pale blue

Pearl Harbor *noun* [*Pearl Harbor,* Oahu, Hawaii, American naval station attacked without warning by the Japanese on December 7, 1941] (1942)
: a sneak attack often with devastating effect

pearl·ite \ˈpər(-ə)-ˌlīt\ *noun* [French *perlite,* from *perle* pearl] (1888)
: the lamellar mixture of ferrite and cementite in slowly cooled iron-carbon alloys occurring normally as a principal constituent of both steel and cast iron
— pearl·it·ic \ˌpər-ˈli-tik\ *adjective*

pearl·ized \ˈpər(-ə)-ˌlīzd\ *adjective* (1937)
: given a pearlescent surface or finish

pearl millet *noun* (circa 1890)

◇ WORD HISTORY

peach When the peach, which is native to China, was introduced to the classical world in the first century A.D., it was known in Greek as *mēlon Persikon* and in Latin as *malum Persicum.* Both names literally meant "Persian apple." The fruit may not have reached the Mediterranean directly from Persia, and its attribution to that country in the name simply reflects a general sense that it was eastern in origin. The Latin *malum Persicum* was shortened to *persicum,* which was superseded in Late Latin by the feminine form *persica. Persica* developed into Old French *pesche.* In a Middle French form this word provided later Middle English with *peche,* the antecedent of Modern English *peach.*

: a tall cereal grass (*Pennisetum glaucum* synonym *P. americanum*) that has large leaves and dense round spikes and is widely grown for its seeds and for forage

pearl onion *noun* (circa 1890)
: a very small usually pickled onion used especially in appetizers and as a garnish

pearly \ˈpər-lē\ *adjective* **pearl·i·er; -est** (15th century)
1 : resembling, containing, or adorned with pearls or mother-of-pearl
2 : highly precious

pearly everlasting *noun* (1857)
: an everlasting (*Anaphalis margaritacea*) that has herbage covered with white woolly hairs and corymbose heads with white scarious involucres

pearly nautilus *noun* (1822)
: NAUTILUS 1

pear psylla *noun* (1904)
: a yellowish or greenish jumping plant louse (*Psylla pyricola*) that is often destructive to the pear

pear–shaped \ˈpar-ˌshāpt, ˈper-\ *adjective* (1758)
1 : having an oval shape markedly tapering at one end
2 *of a vocal tone* **:** free from harshness, thinness, or nasality

peart \ˈpirt\ *adjective* [alteration of *pert*] (1578)
chiefly Southern & Midland **:** being in good spirits **:** LIVELY
— peart·ly *adverb*

peas·ant \ˈpe-zᵊnt\ *noun* [Middle English *paissaunt,* from Middle French *païsant,* from Old French, from *païs* country, from Late Latin *pagensis* inhabitant of a district, from Latin *pagus* district] (15th century)
1 : a member of a European class of persons tilling the soil as small landowners or as laborers; *also* **:** a member of a similar class elsewhere
2 : a usually uneducated person of low social status

peas·ant·ry \-zᵊn-trē\ *noun* (circa 1553)
1 : PEASANTS
2 : the position, rank, or behavior of a peasant

pease·cod *or* **peas·cod** \ˈpēz-ˌkäd\ *noun* [Middle English *pesecod,* from *pese* + *cod* bag, husk — more at CODPIECE] (14th century)
: a pea pod

pea·shoot·er \ˈpē-ˌshü-tər\ *noun* (1803)
: a toy blowgun that uses peas for projectiles

pea soup *noun* (1711)
1 : a thick purée made of dried peas
2 : a thick fog

1peat \ˈpēt\ *noun, often attributive* [Middle English *pete,* from Medieval Latin *peta*] (14th century)
1 : TURF 2b
2 : partially carbonized vegetable tissue formed by partial decomposition in water of various plants (as mosses of the genus *Sphagnum*)
— peaty \ˈpē-tē\ *adjective*

2peat *noun* [origin unknown] (1599)
: a bold lively woman

peat moss *noun* (1880)
: SPHAGNUM

pea·vey *or* **pea·vy** \ˈpē-vē\ *noun, plural* **peaveys** *or* **peavies** [probably from the name *Peavey*] (1870)
: a lumberman's lever that has a pivoting hooked arm and metal spike at one end — called also *cant dog;* compare CANT HOOK

peavey

1peb·ble \ˈpe-bəl\ *noun* [Middle English *pobble,* from Old English *papolstān,* from *papol-* (of unknown origin) + *stān* stone] (14th century)
1 : a small usually rounded stone especially when worn by the action of water
2 : transparent and colorless quartz **:** ROCK CRYSTAL
3 : an irregular, crinkled, or grainy surface
— peb·bly \-b(ə-)lē\ *adjective*

2pebble *transitive verb* **peb·bled; peb·bling** \-b(ə-)liŋ\ (1605)
1 : to pelt with pebbles
2 : to pave or cover with pebbles or something resembling pebbles
3 : to grain (as leather) so as to produce a rough and irregularly indented surface

pec \ˈpek\ *noun* (1966)
: PECTORAL MUSCLE

pe·can \pi-ˈkän, -ˈkan; ˈpē-ˌkan\ *noun* [American French *pacane,* of Algonquian origin; akin to Ojibwa *paka·n,* a hard-shelled nut] (1712)
1 : a large hickory (*Carya illinoensis*) that has roughish bark and hard but brittle wood and is widely grown in the warmer parts of the U.S. and in Mexico for its edible nut
2 : the wood of the pecan tree
3 : the smooth oblong thin-shelled nut of the pecan tree

pec·ca·dil·lo \ˌpe-kə-ˈdi-(ˌ)lō\ *noun, plural* **-loes** *or* **-los** [Spanish *pecadillo,* diminutive of *pecado* sin, from Latin *peccatum,* from neuter of *peccatus,* past participle of *peccare*] (1591)
: a slight offense

pec·cant \ˈpe-kənt\ *adjective* [Latin *peccant-, peccans,* present participle of *peccare* to stumble, sin] (circa 1604)
1 : guilty of a moral offense **:** SINNING
2 : violating a principle or rule **:** FAULTY
— pec·cant·ly *adverb*

pec·ca·ry \ˈpe-kə-rē\ *noun, plural* **-ries** [perhaps modification of Carib *baquira*] (1613)
: any of several largely nocturnal gregarious American mammals resembling the related pigs: as **a :** a grizzled animal (*Tayassu tajacu*) with an indistinct white collar **b :** a blackish animal (*Tayassu pecari*) with a whitish mouth region

pec·ca·vi \pe-ˈkä-(ˌ)wē, -(ˌ)vē; -ˈkā-ˌvī\ *noun* [Latin, I have sinned, from *peccare*] (1553)
: an acknowledgment of sin

1peck \ˈpek\ *noun* [Middle English *pek,* from Middle French] (13th century)
1 — see WEIGHT table
2 : a large quantity or number

2peck *verb* [Middle English, perhaps from Middle Low German *pekken*] (14th century)
transitive verb
1 a : to strike or pierce especially repeatedly with the bill or a pointed tool **b :** to make by pecking ⟨*peck* a hole⟩
2 : to pick up with the bill
intransitive verb
1 a : to strike, pierce, or pick up something with or as if with the bill **b :** CARP, NAG
2 : to eat reluctantly and in small bites ⟨*peck* at food⟩

3peck *noun* (circa 1591)
1 : an impression or hole made by pecking
2 : a quick sharp stroke

peck·er \ˈpe-kər\ *noun* (1587)
1 : one that pecks
2 *chiefly British* **:** COURAGE
3 : PENIS — often considered vulgar

peck·er·wood \ˈpe-kər-ˌwu̇d\ *noun* [probably inversion of *woodpecker*] (1904)
: a rural white Southerner — often used disparagingly

pecking order *also* **peck order** *noun* (1928)
1 : the basic pattern of social organization within a flock of poultry in which each bird pecks another lower in the scale without fear of retaliation and submits to pecking by one of higher rank; *broadly* **:** a dominance hierarchy in a group of social animals
2 : a social hierarchy

peck·ish \ˈpe-kish\ *adjective* [2*peck*] (1785)
1 *chiefly British* **:** HUNGRY
2 : CROTCHETY

Peck·sniff·ian \pek-ˈsni-fē-ən\ *adjective* [Seth *Pecksniff,* character in *Martin Chuzzlewit* (1843–44) by Charles Dickens] (1851)
: unctuously hypocritical **:** PHARISAICAL

pecky \ˈpe-kē\ *adjective* [3*peck*] (1848)
1 : marked by lenticular or finger-shaped pockets of decay caused by fungi ⟨*pecky* cypress⟩
2 : containing discolored or shriveled grains ⟨*pecky* rice⟩

pec·o·ri·no \ˌpe-kə-ˈrē-(ˌ)nō\ *noun, often capitalized* [Italian, from *pecorino* of sheep, from *pecora* sheep, ewe, from Latin, domestic animals, from plural of *pecus* cattle — more at FEE] (1912)
: any of various cheeses of Italian origin made from sheep's milk

pec·ten \ˈpek-tən\ *noun, plural* **pectens** [New Latin *pectin-, pecten,* from Latin, comb, scallop] (1682)
1 : SCALLOP 1a
2 *plural usually* **pec·ti·nes** \-tə-ˌnēz\ **:** a body part that resembles a comb; *especially* **:** a folded vascular pigmented membrane projecting into the vitreous humor in the eye of a bird or reptile

pec·tic \ˈpek-tik\ *adjective* [French *pectique,* from Greek *pēktikos* coagulating, from *pēgnynai* to fix, coagulate — more at PACT] (1831)
: of, relating to, or derived from pectin

pectic acid *noun* (1831)
: any of various water-insoluble substances formed by hydrolyzing the methyl ester groups of pectins

pec·tin \ˈpek-tən\ *noun* [French *pectine,* from *pectique*] (1838)
: any of various water-soluble substances that bind adjacent cell walls in plant tissues and yield a gel which is the basis of fruit jellies; *also* **:** a commercial product rich in pectins

pec·ti·na·ceous \ˌpek-tə-ˈnā-shəs\ *adjective* (circa 1844)
: of, relating to, or containing pectin

pec·ti·nate \ˈpek-tə-ˌnāt\ *adjective* [Latin *pectinatus,* from *pectin-, pecten* comb; akin to Greek *kten-, kteis* comb] (1793)
: having narrow parallel projections or divisions suggestive of the teeth of a comb ⟨*pectinate* antennae⟩
— pec·ti·na·tion \ˌpek-tə-ˈnā-shən\ *noun*

pec·tin·es·ter·ase \ˌpek-tə-ˈnes-tə-ˌrās, -ˌrāz\ *noun* (1945)
: an enzyme that catalyzes the hydrolysis of pectins into pectic acids and methanol

1pec·to·ral \ˈpek-t(ə-)rəl\ *noun* (15th century)
: something worn on the breast

2pectoral *adjective* [Middle French or Latin; Middle French, from Latin *pectoralis,* from *pector-, pectus* breast; akin to Tocharian A *päśśäm* breasts, Old Irish *ucht* breast] (1578)
1 : of, situated in or on, or worn on the chest
2 : coming from the breast or heart as the seat of emotion

pectoral cross *noun* (circa 1735)
: a cross worn on the breast especially by a prelate

pectoral fin *noun* (1769)
: either of the fins of a fish that correspond to the forelimbs of a quadruped

pectoral girdle *noun* (circa 1890)
: the bony or cartilaginous arch that supports the forelimbs of a vertebrate

pectoral muscle *noun* (1615)
: any of the muscles which connect the ventral walls of the chest with the bones of the upper arm and shoulder and of which there are two on each side of the human body

pec·u·late \ˈpe-kyə-ˌlāt\ *transitive verb* **-lat·ed; -lat·ing** [Latin *peculatus,* past participle of *peculari,* from *peculium*] (1802)
: EMBEZZLE
— pec·u·la·tion \ˌpe-kyə-ˈlā-shən\ *noun*
— pec·u·la·tor \ˈpe-kyə-ˌlā-tər\ *noun*

[1]**pe·cu·liar** \pi-ˈkyül-yər\ *adjective* [Middle English *peculier,* from Latin *peculiaris* of private property, special, from *peculium* private property, from *pecu* cattle; akin to Latin *pecus* cattle — more at FEE] (15th century)
1 : characteristic of only one person, group, or thing **:** DISTINCTIVE
2 : different from the usual or normal: **a :** SPECIAL, PARTICULAR **b :** ODD, CURIOUS **c :** ECCENTRIC, QUEER ◈
synonym see CHARACTERISTIC, STRANGE
— pe·cu·liar·ly *adverb*

[2]**peculiar** *noun* (1562)
: something exempt from ordinary jurisdiction; *especially* **:** a church or parish exempt from the jurisdiction of the ordinary in whose territory it lies

pe·cu·liar·i·ty \pi-ˌkyül-ˈyar-ə-tē, -ˌkyü-lē-ˈar-\ *noun, plural* **-ties** (1646)
1 : the quality or state of being peculiar
2 : a distinguishing characteristic
3 : ODDITY, QUIRK

pe·cu·ni·ary \pi-ˈkyü-nē-ˌer-ē\ *adjective* [Latin *pecuniarius,* from *pecunia* money — more at FEE] (1502)
1 : consisting of or measured in money
2 : of or relating to money **:** MONETARY
word history see PECULIAR
— pe·cu·ni·ar·i·ly \-ˌkyü-nē-ˈer-ə-lē\ *adverb*

ped \ˈped\ *noun* [Greek *pedon* ground; akin to Latin *ped-, pes* foot — more at FOOT] (1951)
: a natural soil aggregate

ped- *or* **pedo-** *or* **paed-** *or* **paedo-** *combining form* [Greek *paid-, paido-,* from *paid-, pais* child, boy — more at FEW]
: child ⟨*pedi*atric⟩ **:** childhood ⟨*paedo*genesis⟩

-ped *or* **-pede** *noun combining form* [Latin *ped-, pes*]
: foot ⟨maxilli*ped*⟩

ped·a·gog·i·cal \ˌpe-də-ˈgä-ji-kəl, -ˈgō-\ *also* **ped·a·gog·ic** \-jik\ *adjective* (1619)
: of, relating to, or befitting a teacher or education
— ped·a·gog·i·cal·ly \-ji-k(ə-)lē\ *adverb*

ped·a·gog·ics \-jiks\ *noun plural but singular in construction* (circa 1859)
: PEDAGOGY

ped·a·gogue *also* **ped·a·gog** \ˈpe-də-ˌgäg\ *noun* [Middle English *pedagoge,* from Middle French, from Latin *paedagogus,* from Greek *paidagōgos,* slave who escorted children to school, from *paid-* ped- + *agōgos* leader, from *agein* to lead — more at AGENT] (14th century)
: TEACHER, SCHOOLMASTER; *especially* **:** a dull, formal, or pedantic teacher

ped·a·go·gy \ˈpe-də-ˌgō-jē *also* -ˌgä-, *especially British* -ˌgä-gē\ *noun* (1583)
: the art, science, or profession of teaching; *especially* **:** EDUCATION 2

[1]**ped·al** \ˈpe-dᵊl\ *noun* [Middle French *pedale,* from Italian, from Latin *pedalis,* adjective] (1611)
1 : a lever pressed by the foot in the playing of a musical instrument (as an organ or piano)
2 : a foot lever or treadle by which a part is activated in a mechanism

[2]**ped·al** *adjective* [Latin *pedalis,* from *ped-, pes*] (1625)
1 \ˈpe-dᵊl *also* ˈpē-\ **:** of or relating to the foot
2 \ˈpe-\ **:** of, relating to, or involving a pedal

[3]**ped·al** \ˈpe-dᵊl\ *verb* **ped·aled** *also* **ped·alled; ped·al·ing** *also* **ped·al·ling** \ˈpe-dᵊl-iŋ, ˈped-liŋ\ (1888)
intransitive verb
1 : to ride a bicycle
2 : to use or work a pedal
transitive verb
: to work the pedals of

pedal bone *noun* (1881)
: COFFIN BONE

pe·dal·fer \pə-ˈdal-fər, -ˌfer\ *noun* [Greek *pedon* ground + English *al*umen + Latin *fer*rum iron] (1928)
: a soil that lacks a hardened layer of accumulated carbonates

ped·al–note \ˈpe-dᵊl-ˌnōt\ *noun* [from the playing of the lowest notes on the organ by means of pedals] (circa 1828)
1 : PEDAL POINT
2 : one of the lowest tones that can be sounded on a brass instrument being an octave below the normal usable range and representing the fundamental of the harmonic series

ped·alo \ˈpe-də-(ˌ)lō\ *noun, plural* **-los** [French *pédalo,* from *pédale* pedal + *-o* (perhaps as in *meccano* children's construction set)] (1945)
chiefly British **:** a small recreational paddleboat powered by pedals

pedal point *noun* (1852)
: a single tone usually the tonic or dominant that is normally sustained in the bass and sounds against changing harmonies in the other parts

pedal pushers *noun plural* (1944)
: women's and girls' calf-length trousers

pedal steel *noun* (1969)
: a box-shaped musical instrument with legs that has usually 10 strings which can be altered in pitch by the use of pedals and which are plucked while being pressed with a movable steel bar — called also *pedal steel guitar*

ped·ant \ˈpe-dᵊnt\ *noun* [Middle French, from Italian *pedante*] (1588)
1 *obsolete* **:** a male schoolteacher
2 a : one who makes a show of knowledge **b :** one who is unimaginative or who unduly emphasizes minutiae in the presentation or use of knowledge **c :** a formalist or precisionist in teaching

pe·dan·tic \pi-ˈdan-tik\ *adjective* (circa 1600)
1 : of, relating to, or being a pedant
2 : narrowly, stodgily, and often ostentatiously learned
3 : UNIMAGINATIVE, PEDESTRIAN
— pe·dan·ti·cal·ly \-ˈdan-ti-k(ə-)lē\ *adverb*

ped·ant·ry \ˈpe-dᵊn-trē\ *noun, plural* **-ries** (1612)
1 : pedantic presentation or application of knowledge or learning
2 : an instance of pedantry

ped·dle \ˈpe-dᵊl\ *verb* **ped·dled; ped·dling** \ˈped-liŋ, ˈpe-dᵊl-iŋ\ [back-formation from *peddler,* from Middle English *pedlere*] (1532)
intransitive verb
1 : to travel about with wares for sale; *broadly* **:** SELL
2 : to be busy with trifles **:** PIDDLE
transitive verb
1 : to sell or offer for sale from place to place **:** HAWK; *broadly* **:** SELL
2 : to deal out or seek to disseminate

ped·dler *also* **ped·lar** \ˈped-lər\ *noun* (14th century)
: one who peddles: as **a :** one who offers merchandise (as fresh produce) for sale along the street or from door to door **b :** one who deals in or promotes something intangible (as a personal asset or an idea) ⟨influence *peddlers*⟩

ped·er·ast \ˈpe-də-ˌrast\ *noun* [Greek *paiderastēs,* literally, lover of boys, from *paid-* ped- + *erastēs* lover, from *erasthai* to love — more at EROS] (circa 1736)
: one that practices anal intercourse especially with a boy
— ped·er·as·tic \ˌpe-də-ˈras-tik\ *adjective*
— ped·er·as·ty \ˈpe-də-ˌras-tē\ *noun*

[1]**ped·es·tal** \ˈpe-dəs-tᵊl\ *noun* [Middle French *piedestal,* from Old Italian *piedestallo,* from *pie di stallo* foot of a stall] (1563)
1 a : the support or foot of a late classic or neoclassic column — see COLUMN illustration **b :** the base of an upright structure
2 : BASE, FOUNDATION
3 : a position of esteem

[2]**pedestal** *transitive verb* **-taled** *or* **-talled; -tal·ing** *or* **-tal·ling** (1648)
: to place on or furnish with a pedestal

[1]**pe·des·tri·an** \pə-ˈdes-trē-ən\ *adjective* [Latin *pedestr-, pedester,* literally, going on foot, from *ped-, pes* foot — more at FOOT] (1716)
1 : COMMONPLACE, UNIMAGINATIVE
2 a : going or performed on foot **b :** of, relating to, or designed for walking ⟨a *pedestrian* mall⟩

[2]**pedestrian** *noun* (1793)
: a person going on foot **:** WALKER

pe·des·tri·an·ism \-ə-ˌni-zəm\ *noun* (1809)
1 a : the practice of walking **b :** fondness for walking for exercise or recreation
2 : the quality or state of being unimaginative or commonplace

pe·di·at·ric \ˌpē-dē-ˈa-trik\ *adjective* (1880)
: of or relating to pediatrics

pe·di·a·tri·cian \ˌpē-dē-ə-ˈtri-shən\ *or* **pe·di·a·trist** \ˌpē-dē-ˈa-trist, pē-ˈdī-ə-\ *noun* (circa 1903)
: a specialist in pediatrics

pe·di·at·rics \ˌpē-dē-ˈa-triks\ *noun plural but singular or plural in construction* (1884)
: a branch of medicine dealing with the development, care, and diseases of children

pedi·cab \ˈpe-di-ˌkab\ *noun* [Latin *ped-, pes* + English *cab*] (1945)
: a tricycle with a 2-seat passenger compartment covered by a usually folding top and a separate seat for a driver who pedals

ped·i·cel \ˈpe-də-ˌsel\ *noun* [New Latin *pedicellus,* diminutive of Latin *pediculus*] (1676)
: a slender basal part of an organism or one of its parts: as **a :** a plant stalk that supports a fruiting or spore-bearing organ **b :** a narrow basal attachment (as of the abdomen of an ant) of an animal organ or part
— ped·i·cel·late \ˌpe-də-ˈse-lət\ *adjective*

ped·i·cle \ˈpe-di-kəl\ *noun* [Latin *pediculus,* from diminutive of *ped-, pes*] (1626)
1 : PEDICEL b
2 : the part of a skin or tissue graft left attached to the original site during the preliminary stages of union
— ped·i·cled \-kəld\ *adjective*

pe·dic·u·late \pi-ˈdi-kyə-lət\ *adjective* [ultimately from Latin *pediculus* little foot, pedicel] (circa 1890)

◇ WORD HISTORY

peculiar The Latin words *pecunia* "property, wealth, money" and *peculium* "property managed by a person incapable of legal ownership, such as a slave or minor" have traditionally been explained as derivatives of *pecu* "flock" or "herd" and its plural *pecua* "farm animals, livestock." The assumption has been that wealth was measured by the Italic and Indo-European ancestors of the Romans in terms of cattle. This assumption is probably incorrect: although *pecunia* and *peculium* are certainly derivatives of *pecu,* the ancient sense of this word was more likely "movable property (that is, property other than land)" rather than "livestock." The latter is a specific application of the word that arose only subsequently, livestock being one major form of movable wealth. The relation between English *pecuniary* and Latin *pecunia* can be seen easily enough, but the relation of English *peculiar* to Latin *peculium* is less straightforward. The Latin derivative adjective *peculiaris* meant originally "pertaining to the property of a slave or minor" and by extension "one's own, personal, private," and hence "special, particular." The meanings "strange" and "eccentric" of English *peculiar* have developed since the word was borrowed from Latin.

\ə\ **abut** \ᵊ\ **kitten** \ər\ **further** \a\ **ash** \ā\ **ace** \ä\ **mop, mar** \aů\ **out** \ch\ **chin** \e\ **bet** \ē\ **easy** \g\ **go** \i\ **hit** \ī\ **ice** \j\ **job** \ŋ\ **sing** \ō\ **go** \ȯ\ **law** \ȯi\ **boy** \th\ **thin** \t͟h\ **the** \ü\ **loot** \ů\ **foot** \y\ **yet** \zh\ **vision** *see also* Guide to Pronunciation

: of or relating to an order (Lophiiformes synonym Pediculati) of marine bony fishes with pectoral fins at the end of an armlike process and part of the dorsal fin modified into a lure
— **pediculate** *noun*

pe·dic·u·lo·sis \pi-ˌdi-kyə-ˈlō-səs\ *noun* [New Latin, from Latin *pediculus* louse, diminutive of *pedis* louse] (circa 1890)
: infestation with lice

pe·dic·u·lous \pi-ˈdi-kyə-ləs\ *adjective* [Latin *pediculosus,* from *pediculus*] (circa 1550)
: infested with lice **:** LOUSY

ped·i·cure \ˈpe-di-ˌkyu̇r\ *noun* [French *pédicure,* from Latin *ped-, pes* foot + *curare* to take care, from *cura* care] (circa 1842)
1 : one who practices chiropody
2 a : care of the feet, toes, and nails **b :** a single treatment of these parts
— **ped·i·cur·ist** \-ˌkyu̇r-ist\ *noun*

ped·i·gree \ˈpe-də-ˌgrē\ *noun* [Middle English *pedegru,* from Middle French *pie de grue* crane's foot; from the shape made by the lines of a genealogical chart] (15th century)
1 : a register recording a line of ancestors
2 a : an ancestral line **:** LINEAGE **b :** the origin and the history of something
3 a : a distinguished ancestry **b :** the recorded purity of breed of an individual or strain
— **ped·i·greed** \-ˌgrēd\ *or* **pedigree** *adjective*

ped·i·ment \ˈpe-də-mənt\ *noun* [alteration of obsolete English *periment,* probably alteration of English *pyramid*] (1592)
1 : a triangular space that forms the gable of a low-pitched roof and that is usually filled with relief sculpture in classical architecture; *also* **:** a similar form used as a decoration
2 : a broad gently sloping bedrock surface with low relief that is situated at the base of a steeper slope and is usually thinly covered with alluvial gravel and sand
— **ped·i·men·tal** \ˌpe-də-ˈmen-tᵊl\ *adjective*
— **ped·i·ment·ed** \ˈpe-də-ˌmen-təd\ *adjective*

ped·i·palp \ˈpe-də-ˌpalp\ *noun* [New Latin *pedipalpus,* from *ped-, pes* foot + *palpus* palpus] (1826)
: either of the second pair of appendages of an arachnid (as a spider) that are borne near the mouth and are often modified for a special (as sensory) function

pedo- — see PED-

ped·o·cal \ˈpe-də-ˌkal\ *noun* [Greek *pedon* earth + Latin *calc-, calx* lime — more at PED, CHALK] (1928)
: a soil that includes a definite hardened layer of accumulated carbonates
— **ped·o·cal·ic** \ˌpe-də-ˈka-lik\ *adjective*

[1]**pe·do·gen·e·sis** \ˌpē-də-ˈje-nə-səs\ *variant of* PAEDOGENESIS

[2]**ped·o·gen·e·sis** \ˌpe-də-ˈje-nə-səs\ *noun* [New Latin, from Greek *pedon* + Latin *genesis*] (1936)
: the formation and development of soil
— **ped·o·gen·ic** \-ˈje-nik\ *also* **ped·o·ge·net·ic** \-jə-ˈne-tik\ *adjective*

pe·dol·o·gy \pi-ˈdä-lə-jē, pe-\ *noun* [Greek *pedon* + International Scientific Vocabulary *-logy*] (1912)
: SOIL SCIENCE
— **pe·do·log·i·cal** \ˌpe-dᵊl-ˈä-ji-kəl\ *also* **pe·do·log·ic** \-jik\ *adjective*
— **pe·dol·o·gist** \pi-ˈdä-lə-jist, pe-\ *noun*

pe·dom·e·ter \pi-ˈdä-mə-tər\ *noun* [French *pédomètre,* from Latin *ped-, pes* foot + French *-mètre* -meter — more at FOOT] (1723)
: an instrument usually in watch form that records the distance a person covers on foot by responding to the body motion at each step

pe·do·phile \ˈpē-də-ˌfīl\ *noun* (1951)
: one affected with pedophilia

pe·do·phil·ia \ˌpē-də-ˈfi-lē-ə\ *noun* [New Latin] (1906)
: sexual perversion in which children are the preferred sexual object
— **pe·do·phil·i·ac** \-ˈfi-lē-ˌak\ *or* **pe·do·phil·ic** \-ˈfi-lik\ *adjective*

pe·dun·cle \ˈpē-ˌdəŋ-kəl, pi-ˈ\ *noun* [New Latin *pedunculus,* diminutive of Latin *ped-, pes*] (circa 1753)
1 : a stalk bearing a flower or flower cluster or a fructification
2 : a narrow part by which some larger part or the whole body of an organism is attached **:** STALK, PEDICEL
3 : a narrow stalk by which a tumor or polyp is attached
— **pe·dun·cled** \-kəld\ *adjective*
— **pe·dun·cu·lar** \pi-ˈdəŋ-kyə-lər\ *adjective*

pe·dun·cu·lat·ed \pi-ˈdəŋ-kyə-ˌlā-təd\ *also* **pe·dun·cu·late** \-lət\ *adjective* [New Latin *pedunculus*] (1752)
: having, growing on, or being attached by a peduncle ⟨a *pedunculated* tumor⟩

[1]**pee** \ˈpē\ *noun* (1612)
1 : the letter *p*
2 *plural* **pee** *British* **:** PENNY

[2]**pee** *intransitive verb* **peed; pee·ing** [euphemism from the initial letter of *piss*] (circa 1880)
: URINATE — sometimes considered vulgar

[3]**pee** *noun* (1946)
1 : URINE — sometimes considered vulgar
2 : an act of urination — sometimes considered vulgar

[1]**peek** \ˈpēk\ *intransitive verb* [Middle English *piken*] (14th century)
1 a : to look furtively **b :** to peer through a crack or hole or from a place of concealment — often used with *in* or *out*
2 : to take a brief look **:** GLANCE

[2]**peek** *noun* (1843)
1 : a surreptitious look
2 : a brief look **:** GLANCE

[1]**peek·a·boo** \ˈpē-kə-ˌbü\ *noun* [[1]*peek* + [1]*boo*] (1599)
: a game for amusing a baby by repeatedly hiding one's face or body and popping back into view exclaiming "Peekaboo!"

[2]**peekaboo** *adjective* (1895)
1 : trimmed with eyelet embroidery ⟨a *peekaboo* blouse⟩
2 : made of a sheer or transparent fabric
3 : offering only limited display or disclosure especially of a teasing sort ⟨*peekaboo* publicity⟩

[1]**peel** \ˈpē(ə)l\ *verb* [Middle English *pelen,* from Middle French *peler,* from Latin *pilare* to remove the hair from, from *pilus* hair] (13th century)
transitive verb
1 : to strip off an outer layer of ⟨*peel* an orange⟩
2 : to remove by stripping ⟨*peel* the label off the can⟩
intransitive verb
1 a : to come off in sheets or scales **b :** to lose an outer layer (as of skin) ⟨his face is *peeling*⟩
2 : to take off one's clothes
3 : to break away from a group or formation — often used with *off*
— **peel·able** \ˈpē-lə-bəl\ *adjective*

[2]**peel** *noun* (14th century)
1 : the skin or rind of a fruit
2 : a thin layer of organic material that is embedded in a film of collodion and stripped from the surface of an object (as a plant fossil) for microscopic study

[3]**peel** *noun* [Middle English *pele,* from Middle French, from Latin *pala*] (14th century)
: a usually long-handled spade-shaped instrument that is used chiefly by bakers for getting something (as bread or pies) into or out of the oven

[4]**peel** *noun* [Middle English *pel* stockade, stake, from Anglo-French, stockade & Middle French, stake, from Latin *palus* stake — more at POLE] (1726)
: a medieval small massive fortified tower along the Scottish-English border — called also *peel tower*

[1]**peel·er** \ˈpē-lər\ *noun* (1597)
1 : one that peels
2 : a crab that is about to shed its shell
3 : a log of wood (as Douglas fir) suitable for cutting into rotary veneer — called also *peeler log*

[2]**peeler** *noun* [Sir Robert *Peel*] (1817)
British **:** POLICE OFFICER

peel·ing \ˈpē-liŋ\ *noun* (1597)
: a peeled-off piece or strip

peel off *intransitive verb* (1941)
1 : to veer away from an airplane formation especially for diving or landing
2 : DEPART, LEAVE

[1]**peen** \ˈpēn\ *transitive verb* (1513)
: to draw, bend, or flatten by or as if by hammering with a peen

[2]**peen** *noun* [probably of Scandinavian origin; akin to Norwegian *penn* peen] (1683)
: a usually hemispherical or wedge-shaped end of the head of a hammer that is opposite the face and is used especially for bending, shaping, or cutting the material struck

[1]**peep** \ˈpēp\ *intransitive verb* [Middle English *pepen,* of imitative origin] (15th century)
1 : to utter a feeble shrill sound as of a bird newly hatched **:** CHEEP
2 : to utter the slightest sound

[2]**peep** *noun* (15th century)
1 : a feeble shrill sound **:** CHEEP
2 : a slight utterance especially of complaint or protest ⟨don't let me hear another *peep* out of you⟩
3 : any of several small sandpipers

[3]**peep** *verb* [Middle English *pepen,* perhaps alteration of *piken* to peek] (15th century)
intransitive verb
1 a : to peer through or as if through a crevice **b :** to look cautiously or slyly
2 : to begin to emerge from or as if from concealment **:** show slightly
transitive verb
: to put forth or cause to protrude slightly

[4]**peep** *noun* (1530)
1 : a first glimpse or faint appearance ⟨at the *peep* of dawn⟩
2 a : a brief look **:** GLANCE **b :** a furtive look

[1]**peep·er** \ˈpē-pər\ *noun* (1591)
1 : one that makes a peeping sound
2 : any of various tree frogs that peep shrilly; *especially* **:** SPRING PEEPER

[2]**peeper** *noun* (1652)
1 : one that peeps; *specifically* **:** VOYEUR
2 : EYE

peep·hole \ˈpēp-ˌhōl\ *noun* (1681)
: a hole or crevice to peep through

Peeping Tom \-ˈtäm\ *noun* [*Peeping Tom,* legendary citizen of Coventry who watched Lady Godiva riding naked] (circa 1796)
: a pruriently prying person **:** VOYEUR
— **Peeping Tom·ism** \-ˈtä-ˌmi-zəm\ *noun*

peep show *noun* (1851)
: an entertainment (as a film) or object (as a small picture) that is viewed through a small opening or a magnifying glass and is usually sexually explicit

peep sight *noun* (1881)
: a rear sight for a gun having an adjustable metal piece pierced with a small hole to peep through in aiming

[1]**peer** \ˈpir\ *noun* [Middle English, from Middle French *per,* from *per,* adjective, equal, from Latin *par*] (13th century)
1 : one that is of equal standing with another **:** EQUAL; *especially* **:** one belonging to the same societal group especially based on age, grade, or status
2 *archaic* **:** COMPANION
3 a : a member of one of the five ranks (as duke, marquess, earl, viscount, or baron) of the British peerage **b :** NOBLE 1
— **peer** *adjective*

[2]**peer** *transitive verb* (14th century)

archaic **:** RIVAL, MATCH

[3]**peer** *intransitive verb* [perhaps by shortening & alteration from *appear*] (1591)
1 : to look narrowly or curiously; *especially* **:** to look searchingly at something difficult to discern
2 : to come slightly into view **:** emerge partly

peer·age \ˈpir-ij\ *noun* (15th century)
1 : the body of peers
2 : the rank or dignity of a peer
3 : a book containing a list of peers with their genealogy, history, and titles

peer·ess \ˈpir-əs\ *noun* (1689)
1 : the wife or widow of a peer
2 : a woman who holds in her own right the rank of a peer

peer·less \ˈpir-ləs\ *adjective* (14th century)
: MATCHLESS, INCOMPARABLE

[1]**peeve** \ˈpēv\ *transitive verb* **peeved; peev·ing** [back-formation from *peevish*] (1908)
: to make peevish or resentful **:** ANNOY
synonym see IRRITATE

[2]**peeve** *noun* (1911)
1 : a feeling or mood of resentment
2 : a particular grievance or source of aggravation

pee·vish \ˈpē-vish\ *adjective* [Middle English *pevish* spiteful] (circa 1530)
1 : querulous in temperament or mood **:** FRETFUL
2 : perversely obstinate
3 : marked by ill temper
— pee·vish·ly *adverb*
— pee·vish·ness *noun*

pee·wee \ˈpē-(ˌ)wē\ *noun* [imitative] (1806)
1 : PEWEE
2 : one that is diminutive or small; *especially* **:** a small child
— peewee *adjective*

pee·wit \ˈpē-ˌwit, ˈpyü-ət\ *noun* [imitative] (circa 1529)
: any of several birds; *especially* **:** LAPWING

[1]**peg** \ˈpeg\ *noun* [Middle English *pegge,* probably from Middle Dutch] (15th century)
1 a : a small usually cylindrical pointed or tapered piece (as of wood) used to pin down or fasten things or to fit into or close holes **:** PIN, PLUG **b** *British* **:** CLOTHESPIN **c :** a predetermined level at which something (as a price) is fixed
2 a : a projecting piece used as a support or boundary marker **b :** something (as a fact or opinion) used as a support, pretext, or reason
3 a : one of the movable wooden pegs set in the head of a stringed instrument (as a violin) that are turned to regulate the pitch of the strings — see VIOLIN illustration **b :** a step or degree especially in estimation
4 : a pointed prong or claw for catching or tearing
5 *British* **:** DRINK ⟨poured himself out a stiff *peg* —Dorothy Sayers⟩
6 : something (as a leg) resembling a peg
7 : THROW; *especially* **:** a hard throw in baseball made in an attempt to put out a base runner

[2]**peg** *verb* **pegged; peg·ging** (1543)
transitive verb
1 a : to put a peg into **b** *British* **:** to pin (laundry) on a clothesline
2 : to attach or fix as if with a peg: as **a :** to pin down **:** RESTRICT **b :** to fix or hold (as prices or wage increases) at a predetermined level or rate **c :** to place in a definite category **:** IDENTIFY
3 : to mark by pegs
4 : THROW
intransitive verb
1 : to work steadily and diligently — often used with *away*
2 : to move along vigorously or hastily **:** HUSTLE

[3]**peg** \ˈpeg\ *or* **pegged** \ˈpegd\ *adjective* (1681)
: wide at the top and narrow at the bottom ⟨*peg* pants⟩

Peg·a·sus \ˈpe-gə-səs\ *noun* [Latin (genitive *Pegasi*), from Greek *Pēgasos*]
1 : a winged horse that causes the stream Hippocrene to spring from Mount Helicon with a blow of his hoof
2 : poetic inspiration
3 : a northern constellation near the vernal equinoctial point

Peg–Board \ˈpeg-ˌbōrd, -ˌbȯrd\ *trademark*
— used for material (as fiberboard) with regularly spaced perforations into which hooks may be inserted for the storage or display of articles

peg leg *noun* [[3]*peg*] (1769)
: an artificial leg; *especially* **:** one fitted at the knee

peg·ma·tite \ˈpeg-mə-ˌtīt\ *noun* [French, from Greek *pēgmat-, pēgma* something fastened together, from *pēgnynai* to fasten together — more at PACT] (circa 1828)
: a coarse variety of granite occurring in dikes or veins
— peg·ma·tit·ic \ˌpeg-mə-ˈti-tik\ *adjective*

peg–top \ˈpeg-ˈtäp\ *or* **peg–topped** \-ˈtäpt\ *adjective* (1858)
: PEG

peg top *noun* (1788)
1 \ˈpeg-ˈtäp\ **:** a pear-shaped top that is made to spin on the sharp metal peg in its base by the unwinding of a string wound round its center
2 *plural* \-ˌtäps\ **:** peg trousers

pei·gnoir \pān-ˈwär, pen-\ *noun* [French, literally, garment worn while combing the hair, from Middle French, from *peigner* to comb the hair, from Latin *pectinare,* from *pectin-, pecten* comb — more at PECTINATE] (1835)
: a woman's loose negligee or dressing gown

pein *variant of* PEEN

[1]**pe·jo·ra·tive** \pi-ˈjȯr-ə-tiv, -ˈjär- *also* ˈpe-jə-rə-tiv *or* ˈpē- *or* -ˌrā- *or* ˈpej-rə- *or* ˈpēj-\ *noun* (1882)
: a pejorative word or phrase

[2]**pejorative** *adjective* [Late Latin *pejoratus,* past participle of *pejorare* to make or become worse, from Latin *pejor* worse; akin to Sanskrit *padyate* he falls, Latin *ped-, pes* foot — more at FOOT] (circa 1888)
: having negative connotations; *especially* **:** tending to disparage or belittle **:** DEPRECIATORY
— pe·jo·ra·tive·ly *adverb*

peke \ˈpēk\ *noun, often capitalized* (1915)
: PEKINGESE 2

Pe·kin \pi-ˈkin, ˈpē-ˌ\ *noun* [*Pekin, Peking* (Beijing), China] (1885)
: any of a breed of large white ducks of Chinese origin used for meat production

Pe·king duck \ˈpē-ˌkiŋ-\ *noun* (1955)
: a Chinese dish consisting of roasted duck meat and strips of crispy duck skin topped with scallions and sauce and wrapped in thin pancakes

Pe·king·ese *or* **Pe·kin·ese** \ˌpē-kə-ˈnēz, -ˈnēs; -kiŋ-ˈēz, -ˈēs\ *noun, plural* **Pekingese** *or* **Pekinese** (1849)
1 a : the Chinese dialect of Beijing **b :** a native or resident of Beijing
2 : any of a Chinese breed of small short-legged dogs with a broad flat face and a profuse long soft coat

Pekingese 2

Peking man *noun* (1926)
: an extinct Pleistocene hominid that is known from skeletal and cultural remains in cave deposits at Cho-k'ou-tien, China and is now classified with the pithecanthropines

pe·koe \ˈpē-(ˌ)kō *also* ˈpe-\ *noun* [Chinese (Xiamen) *pek-ho*] (1712)
: a tea made from young leaves that are slightly larger than those of orange pekoe

pel·age \ˈpe-lij\ *noun* [French, from Middle French, from *poil* hair, from Latin *pilus*] (circa 1828)
: the hairy covering of a mammal

[1]**Pe·la·gian** \pə-ˈlā-j(ē-)ən\ *noun* (15th century)
: one agreeing with Pelagius in denying original sin and consequently in holding that mankind has perfect freedom to do either right or wrong

[2]**Pelagian** *adjective* (15th century)
: of or relating to Pelagians or Pelagianism

Pe·la·gian·ism \-j(ē-)ə-ˌni-zəm\ *noun* (1583)
: the teaching of Pelagius or Pelagians

pe·lag·ic \pə-ˈla-jik\ *adjective* [Latin *pelagicus,* from Greek *pelagikos,* from *pelagos* sea — more at PLAGAL] (circa 1656)
: of, relating to, or living or occurring in the open sea **:** OCEANIC ⟨*pelagic* sediment⟩ ⟨*pelagic* birds⟩

pel·ar·go·ni·um \ˌpe-lär-ˈgō-nē-əm, -lər-\ *noun* [New Latin, from Greek *pelargos* stork (akin to Greek *pelios* livid, *polios* gray) + New Latin *-nium* (as in *Geranium*) — more at FALLOW] (circa 1819)
: any of a genus (*Pelargonium*) of southern African herbs (as a garden geranium) of the geranium family with showy flowers of various shades of red, pink, or white distinguished by a spurred calyx and irregular corolla

Pe·las·gian \pə-ˈlaz-j(ē-)ən, -ˈlaz-gē-ən\ *noun* [Greek *pelasgios,* adjective, Pelasgian, from *Pelasgoi* Pelasgians] (15th century)
: a member of an ancient people mentioned by classical writers as early inhabitants of Greece and the eastern islands of the Mediterranean
— Pelasgian *adjective*
— Pe·las·gic \-jik, -gik\ *adjective*

pe·lec·y·pod \pə-ˈle-sə-ˌpäd\ *noun* [New Latin *Pelecypoda,* group name, from Greek *pelekys* ax (akin to Sanskrit *paraśu* ax) + *pod-, pous* foot — more at FOOT] (circa 1890)
: LAMELLIBRANCH

pel·er·ine \ˌpe-lə-ˈrēn, ˈpe-lə-rən\ *noun* [obsolete French, neckerchief, from French *pèlerine,* feminine of *pèlerin* pilgrim, from Late Latin *pelegrinus* — more at PILGRIM] (1744)
: a woman's narrow cape made of fabric or fur and usually with long ends hanging down in front

Pe·leus \ˈpēl-ˌyüs, ˈpē-lē-əs\ *noun* [Latin, from Greek *Pēleus*]
: a son of Aeacus who becomes by the goddess Thetis the father of Achilles

pelf \ˈpelf\ *noun* [Middle English, from Middle French *pelfre* booty] (14th century)
: MONEY, RICHES

pel·i·can \ˈpe-li-kən\ *noun* [Middle English, from Old English *pellican,* from Late Latin *pelecanus,* from Greek *pelekan*] (before 12th century)
: any of a genus (*Pelecanus*) of large web-footed fish-eating birds with a very large bill and distensible gular pouch

pe·lisse \pə-ˈlēs, pe-\ *noun* [French, from Late Latin *pellicia,* from feminine of *pellicius* made of skin, from Latin *pellis* skin — more at FELL] (1717)
1 : a long cloak or coat made of fur or lined or trimmed with fur
2 : a woman's loose lightweight cloak with wide collar and fur trimming

pel·la·gra \pə-ˈla-grə, -ˈlā-, -ˈlä-\ *noun* [Italian, from *pelle* skin (from Latin *pellis*) + *-agra* (as in *podagra,* from Latin)] (circa 1811)
: a disease marked by dermatitis, gastrointestinal disorders, and central nervous symptoms and associated with a diet deficient in niacin

— **pel·la·grous** \-grəs\ *adjective*
pel·la·grin \-grən\ *noun* [irregular from *pellagra*] (1865)
: a person who is affected with pellagra
[1]**pel·let** \'pe-lət\ *noun* [Middle English *pelote*, from Middle French, from (assumed) Vulgar Latin *pilota*, diminutive of Latin *pila* ball] (14th century)
1 : a usually small rounded, spherical, or cylindrical body (as of food or medicine)
2 : any of various projectiles fired from a weapon (as an air rifle)
— **pel·let·al** \-lə-tᵊl\ *adjective*
[2]**pellet** *transitive verb* (1597)
1 : PELLETIZE
2 : to strike with pellets
pellet gun *noun* (1952)
: AIR GUN 1
pel·let·ise *British variant of* PELLETIZE
pel·let·ize \'pe-lə-ˌtīz\ *transitive verb* **-ized; -iz·ing** (1942)
: to form or compact into pellets ⟨*pelletize* ore⟩
— **pel·let·i·za·tion** \ˌpe-lə-tə-'zā-shən\ *noun*
— **pel·let·iz·er** \'pe-lə-ˌtī-zər\ *noun*
pel·li·cle \'pe-li-kəl\ *noun* [Middle French *pellicule*, from Medieval Latin *pellicula*, from Latin, diminutive of *pellis*] (1541)
: a thin skin or film: as **a :** an outer membrane of some protozoans (as euglenoids or paramecia) **b :** a film that reflects a part of the light falling upon it and transmits the rest and that is used for dividing a beam of light (as in a photographic device)
[1]**pel·li·to·ry** \'pe-lə-ˌtōr-ē, -ˌtȯr-\ *noun, plural* **-ries** [Middle English *peletre*, from Middle French *piretre*, from Latin *pyrethrum* — more at PYRETHRUM] (1533)
: a southern European composite plant (*Anacyclus pyrethrum*) resembling yarrow — called also *pellitory-of-Spain*
[2]**pellitory** *noun* [Middle English *paritorie*, from Middle French *paritaire*, from Late Latin *parietaria*, from feminine of *parietarius* of a wall, from Latin *pariet-, paries* wall] (1548)
: any of a genus (*Parietaria*) of herbs of the nettle family with alternate leaves and inconspicuous flowers — called also *pellitory-of-the-wall*
pell–mell \ˌpel-'mel\ *adverb* [Middle French *pelemele*] (1596)
1 : in mingled confusion or disorder
2 : in confused haste
— **pell–mell** *adjective or noun*
pel·lu·cid \pə-'lü-səd\ *adjective* [Latin *pellucidus*, from *per* through + *lucidus* lucid — more at FOR] (1619)
1 : admitting maximum passage of light without diffusion or distortion
2 : reflecting light evenly from all surfaces
3 : easy to understand
— **pel·lu·cid·ly** \pə-'lü-səd-lē\ *adverb*
pel·met \'pel-mət\ *noun* [probably modification of French *palmette* palmette] (1821)
: a short valance or small cornice for concealing curtain fixtures
Pe·lops \'pē-ˌläps, 'pe-\ *noun* [Latin, from Greek]
: a son of Tantalus served by his father to the gods for food but later restored to life by them
pe·lo·rus \pə-'lōr-əs, -'lȯr-\ *noun* [origin unknown] (1854)
: a navigational instrument resembling a mariner's compass without magnetic needles and having two sight vanes by which bearings are taken
pe·lo·ta \pə-'lō-tə\ *noun* [Spanish, from Old French *pelote* little ball — more at PELLET] (1844)
1 : a court game related to jai alai
2 : the ball used in jai alai
[1]**pelt** \'pelt\ *noun* [Middle English, perhaps back-formation from *peltry*] (15th century)
1 : a usually undressed skin with its hair, wool, or fur
2 : a skin stripped of hair or wool for tanning
[2]**pelt** *transitive verb* (1596)
: to strip off the skin or pelt of (an animal)
[3]**pelt** *verb* [Middle English] (15th century)
transitive verb
1 a : to strike with a succession of blows or missiles ⟨*pelted* him with stones⟩ **b :** to assail vigorously or persistently ⟨*pelted* her with accusations⟩
2 : HURL, THROW ⟨*pelted* snowballs at them⟩
3 : to beat or dash repeatedly against ⟨hailstones *pelting* the roof⟩
intransitive verb
1 : to deliver a succession of blows or missiles
2 : to beat incessantly
3 : to move rapidly and vigorously **:** HURRY
— **pelt·er** *noun*
[4]**pelt** *noun* (1513)
: BLOW, WHACK
pel·tate \'pel-ˌtāt\ *adjective* [Latin *pelta* small shield, from Greek *peltē*] (circa 1760)
: shaped like a shield; *specifically* **:** having the stem or support attached to the lower surface instead of at the base or margin — see LEAF illustration
pelt·ing \'pel-tiŋ\ *adjective* [probably from English dialect *pelt* piece of trash] (1540)
archaic **:** PALTRY, INSIGNIFICANT
pelt·ry \'pel-trē\ *noun, plural* **peltries** [Middle English, from Anglo-French *pelterie*] (15th century)
: PELTS, FURS; *especially* **:** raw undressed skins — often used in plural
pel·vic \'pel-vik\ *adjective* (1830)
: of, relating to, or located in or near the pelvis
— **pelvic** *noun*
pelvic fin *noun* (circa 1909)
: one of the paired fins of a fish that are homologous with the hind limbs of a quadruped
pelvic girdle *noun* (1883)
: a bony or cartilaginous arch that supports the hind limbs of a vertebrate
pelvic inflammatory disease *noun* (1974)
: inflammation of the female reproductive tract and especially the fallopian tubes that is caused especially by sexually transmitted disease, occurs more often in women using IUDs, and is a leading cause of female sterility
pel·vis \'pel-vəs\ *noun, plural* **pel·vis·es** \-və-səz\ *or* **pel·ves** \-ˌvēz\ [New Latin, from Latin, basin; perhaps akin to Old English & Old Norse *full* cup] (1615)
1 : a basin-shaped structure in the skeleton of many vertebrates that is formed by the pelvic girdle and adjoining bones of the spine
2 : the cavity of the pelvis
3 : the funnel-shaped cavity of the kidney into which urine is discharged
pel·y·co·saur \'pe-li-kə-ˌsȯr\ *noun* [ultimately from Greek *pelyc-, pelyx* bowl + *sauros* lizard] (1904)
: any of an order (Pelycosauria) of primitive Permian quadruped reptiles that resemble mammals and often have extreme development of the dorsal vertebral processes
Pem·broke table \'pem-ˌbrōk-, -ˌbrůk-\ *noun* [*Pembroke*, Wales] (1778)
: a small 4-legged table originating in the Georgian period and having two drop leaves and a drawer
Pembroke Welsh corgi *noun* (1938)
: any of a breed of Welsh corgis with pointed ears, straight forelegs, and a short tail — called also *Pembroke;* see WELSH CORGI illustration
pem·mi·can *also* **pem·i·can** \'pe-mi-kən\ *noun* [Cree *pimihkā·n*] (1791)
: a concentrated food used by North American Indians and consisting of lean meat dried, pounded fine, and mixed with melted fat; *also* **:** a similar preparation (as of dried beef, flour, molasses, suet) used for emergency rations
pem·o·line \'pe-mə-ˌlēn\ *noun* [origin unknown] (1961)
: a synthetic organic drug $C_9H_8N_2O_2$ that is a mild stimulant of the central nervous system
pem·phi·gus \'pem(p)-fi-gəs, pem-'fī-\ *noun* [New Latin, from Greek *pemphig-, pemphix* pustule] (circa 1779)
: a disease characterized by large blisters on skin and mucous membranes and often by itching or burning
[1]**pen** \'pen\ *transitive verb* **penned; pen·ning** [Middle English *pennen*, from Old English *-pennian*] (13th century)
: to shut in or as if in a pen
[2]**pen** *noun* [Middle English, perhaps from *pennen*] (14th century)
1 a : a small enclosure for animals **b :** the animals in a pen ⟨a *pen* of sheep⟩
2 : a small place of confinement or storage
3 : a protected dock or slip for a submarine
4 : BULL PEN 2
[3]**pen** *noun* [Middle English *penne*, from Middle French, feather, pen, from Latin *penna*, *pinna* feather; akin to Greek *pteron* wing — more at FEATHER] (14th century)
1 : an implement for writing or drawing with ink or a similar fluid: as **a :** QUILL **b :** PEN POINT **c :** a penholder containing a pen point **d :** FOUNTAIN PEN **e :** BALLPOINT
2 a : a writing instrument regarded as a means of expression ⟨enlisted the *pens* of the best writers —F. H. Chase⟩ **b :** WRITER
3 : the internal horny feather-shaped shell of a squid
[4]**pen** *transitive verb* **penned; pen·ning** (15th century)
: WRITE, INDITE ⟨*pen* a letter⟩
[5]**pen** *noun* [origin unknown] (1550)
: a female swan
[6]**pen** *noun* (1884)
: PENITENTIARY
pe·nal \'pē-nᵊl\ *adjective* [Middle English, from Middle French, from Latin *poenalis*, from *poena* punishment — more at PAIN] (15th century)
1 : of, relating to, or involving punishment, penalties, or punitive institutions
2 : liable to punishment ⟨a *penal* offense⟩
3 : used as a place of confinement and punishment ⟨a *penal* colony⟩
— **pe·nal·ly** \-nᵊl-ē\ *adverb*
penal code *noun* (1845)
: a code of laws concerning crimes and offenses and their punishment
pe·nal·ise *British variant of* PENALIZE
pe·nal·ize \'pē-nᵊl-ˌīz, 'pe-\ *transitive verb* **-ized; -iz·ing** (1868)
1 : to inflict a penalty on
2 : to put at a serious disadvantage
— **pe·nal·i·za·tion** \-nᵊl-ə-'zā-shən\ *noun*
pen·al·ty \'pe-nᵊl-tē\ *noun, plural* **-ties** [Medieval Latin *poenalitas*, from Latin *poenalis*] (15th century)
1 : the suffering in person, rights, or property that is annexed by law or judicial decision to the commission of a crime or public offense
2 : the suffering or the sum to be forfeited to which a person agrees to be subjected in case of nonfulfillment of stipulations
3 a : disadvantage, loss, or hardship due to some action **b :** a disadvantage (as loss of yardage, time, or possession of the ball or an addition to or subtraction from the score) imposed on a team or competitor for violation of the rules of a sport
4 : points scored in bridge by the side that defeats the opposing contract — usually used in plural
— **penalty** *adjective*
penalty box *noun* (1931)
: an area alongside an ice hockey rink to which penalized players are confined for the duration of their penalty
penalty kick (1889)
1 : a free kick in rugby
2 : a free kick at the goal in soccer made from a point 12 yards in front of the goal and allowed for certain violations within a designated area around the goal

penalty shot *noun* (circa 1948)
: an unhindered shot at the goal in ice hockey awarded to an individual for certain violations by an opponent

[1]pen·ance \ˈpe-nən(t)s\ *noun* [Middle English, from Old French, from Medieval Latin *poenitentia* penitence] (14th century)
1 : an act of self-abasement, mortification, or devotion performed to show sorrow or repentance for sin
2 : a sacramental rite that is practiced in Roman, Eastern, and some Anglican churches and that consists of private confession, absolution, and a penance directed by the confessor
3 : something (as a hardship or penalty) resembling an act of penance (as in compensating for an offense)

[2]penance *transitive verb* **pen·anced; pen·anc·ing** (1600)
: to impose penance on

Pe·na·tes \pə-ˈnā-tēz, -ˈnä-\ *noun plural* [Latin, from *penus* food, provisions]
: the Roman gods of the household worshiped in close connection with Vesta and with the Lares

pence \ˈpen(t)s\ *plural of* PENNY

pen·cel *or* **pen·cil** \ˈpen(t)-səl\ *noun* [Middle English *pencel,* modification of Old French *penoncel*] (13th century)
: PENNONCEL

pen·chant \ˈpen-chənt, *especially British* ˈpäⁿ-ˌshäⁿ\ *noun* [French, from present participle of *pencher* to incline, from (assumed) Vulgar Latin *pendicare,* from Latin *pendere* to weigh] (1672)
: a strong and continued inclination; *broadly* **:** LIKING
synonym see LEANING

[1]pen·cil \ˈpen(t)-səl\ *noun* [Middle English *pensel,* from Middle French *pincel,* from (assumed) Vulgar Latin *penicellus,* alteration of Latin *penicillus,* diminutive of *peniculus* brush, from diminutive of *penis* tail, penis] (14th century)
1 : an artist's brush
2 : an artist's individual skill or style
3 a : an implement for writing, drawing, or marking consisting of or containing a slender cylinder or strip of a solid marking substance **b :** a small medicated or cosmetic roll or stick for local applications
4 : a set of geometric objects each pair of which has a common property ⟨the lines in a plane through a point comprise a *pencil* of lines⟩
5 : something (as a beam of radiation) long and thin like a pencil

[2]pencil *transitive verb* **-ciled** *or* **-cilled; -cil·ing** *or* **-cil·ling** \-s(ə-)liŋ\ (circa 1532)
: to paint, draw, write, or mark with a pencil

pen·cil·ing *or* **pen·cil·ling** *noun* (1706)
: the work of the pencil or brush; *also* **:** a product of this

pencil pusher *noun* (1881)
: a person who does predominantly paperwork

pen·dant *also* **pen·dent** \ˈpen-dənt; *3 & 4 are also* ˈpe-nənt, *5 is also* päⁿ-däⁿ\ *noun* [Middle English *pendaunt,* from Middle French *pendant,* from present participle of *pendre* to hang, from (assumed) Vulgar Latin *pendere,* from Latin *pendēre;* akin to Latin *pendere* to weigh, estimate, pay, *pondus* weight] (14th century)
1 : something suspended: as **a :** an ornament (as on a necklace) allowed to hang free **b :** an electrical fixture suspended from the ceiling
2 : a hanging ornament of roofs or ceilings much used in the later styles of Gothic architecture
3 : a length of line usually used as a connector on a boat or ship; *especially* **:** a short rope hanging from a spar and having at its free end a block or spliced thimble
4 *chiefly British* **:** PENNANT 1a
5 a : COMPANION PIECE **b :** something secondary or supplementary

pen·den·cy \ˈpen-dən(t)-sē\ *noun* (1637)
: the state of being pending

pen·dent *or* **pen·dant** \ˈpen-dənt\ *adjective* [Middle English *pendaunt*] (14th century)
1 : jutting or leaning over **:** OVERHANGING ⟨a *pendent* cliff⟩
2 : supported from above **:** SUSPENDED ⟨icicles *pendent* from the eaves⟩
3 : remaining undetermined **:** PENDING

pen·den·tive \pen-ˈden-tiv\ *noun* [French *pendentif,* from Latin *pendent-, pendens,* present participle of *pendēre*] (circa 1741)
: one of the concave triangular members that supports a dome over a square space

1 pendentive

[1]pend·ing \ˈpen-diŋ\ *preposition* [French *pendant,* from present participle of *pendre*] (1642)
1 : DURING
2 : while awaiting

[2]pending *adjective* (1797)
1 : not yet decided **:** being in continuance
2 : IMMINENT, IMPENDING

pen·du·lar \ˈpen-jə-lər, ˈpen-dyə-, -də-\ *adjective* (1878)
: being or resembling the movement of a pendulum

pen·du·lous \-ləs\ *adjective* [Latin *pendulus,* from *pendēre* to hang] (circa 1605)
1 *archaic* **:** poised without visible support
2 a : suspended so as to swing freely ⟨branches hung with *pendulous* vines⟩ **b :** inclined or hanging downward ⟨*pendulous* jowls⟩
3 : marked by vacillation, indecision, or uncertainty
— pen·du·lous·ness *noun*

pen·du·lum \-ləm\ *noun* [New Latin, from Latin, neuter of *pendulus*] (1660)
1 : a body suspended from a fixed point so as to swing freely to and fro under the action of gravity and commonly used to regulate movements (as of clockwork)
2 : something (as a state of affairs) that alternates between opposites

Pe·nel·o·pe \pə-ˈne-lə-pē\ *noun* [Latin, from Greek *Pēnelopē*]
: the wife of Odysseus who waits faithfully for him during his 20 years' absence

pe·ne·plain *also* **pe·ne·plane** \ˈpē-ni-ˌplān, ˈpe-\ *noun* [Latin *paene, pene* almost + English *plain* or *plane*] (1889)
: a land surface of considerable area and slight relief shaped by erosion

pen·e·tra·ble \ˈpe-nə-trə-bəl\ *adjective* (1538)
: capable of being penetrated
— pen·e·tra·bil·i·ty \ˌpe-nə-trə-ˈbi-lə-tē\ *noun*

pen·e·tra·lia \ˌpe-nə-ˈtrā-lē-ə\ *noun plural* [Latin, neuter plural of *penetralis* inner, from *penetrare* to penetrate] (1668)
: the innermost or most private parts

pen·e·trance \ˈpe-nə-trən(t)s\ *noun* [International Scientific Vocabulary, from Latin *penetrare*] (1934)
: the proportion of individuals of a particular genotype that express its phenotypic effect in a given environment

[1]pen·e·trant \-trənt\ *adjective* (1543)
: PENETRATING

[2]penetrant *noun* (circa 1734)
: one that penetrates or is capable of penetrating

pen·e·trate \ˈpe-nə-ˌtrāt\ *verb* **-trat·ed; -trat·ing** [Latin *penetratus,* past participle of *penetrare,* from *penitus* deep within, far; akin to Latin *penus* provisions] (circa 1530)
transitive verb
1 a : to pass into or through **b :** to enter by overcoming resistance **:** PIERCE **c :** to gain entrance to
2 a : to see into or through **b :** to discover the inner contents or meaning of
3 : to affect profoundly with feeling
4 : to diffuse through or into
intransitive verb
1 a : to pass, extend, pierce, or diffuse into or through something **b :** to pierce something with the eye or mind
2 : to affect deeply the senses or feelings
synonym see ENTER

penetrating *adjective* (1593)
1 : having the power of entering, piercing, or pervading ⟨a *penetrating* shriek⟩
2 : ACUTE, DISCERNING ⟨*penetrating* insights into life⟩
— pen·e·trat·ing·ly \-ˌtrā-tiŋ-lē\ *adverb*

pen·e·tra·tion \ˌpe-nə-ˈtrā-shən\ *noun* (1605)
1 a : the power to penetrate; *especially* **:** the ability to discern deeply and acutely **b :** the depth to which something penetrates **c :** the extent to which a commercial product or agency is familiar or sells in a market
2 : the act or process of penetrating: as **a :** the act of entering a country so that actual establishment of influence is accomplished **b :** an attack that penetrates the enemy's front or territory
synonym see DISCERNMENT

pen·e·tra·tive \ˈpe-nə-ˌtrā-tiv\ *adjective* (15th century)
1 : tending to penetrate **:** PIERCING
2 : ACUTE ⟨*penetrative* observations⟩
3 : IMPRESSIVE ⟨a *penetrative* speaker⟩

pen·e·trom·e·ter \ˌpe-nə-ˈträ-mə-tər\ *noun* [Latin *penetrare* + International Scientific Vocabulary *-meter*] (1905)
: an instrument for measuring firmness or consistency (as of soil)

pen·gö \ˈpeŋ-ˌgə(r), -ˌgœ̄\ *noun, plural* **pengö** *or* **pengös** [Hungarian *pengő,* literally, jingling] (1925)
: the basic monetary unit of Hungary from 1925 to 1946

pen·guin \ˈpen-gwən, ˈpeŋ-\ *noun* [obsolete English *penguin* great auk, perhaps from Welsh *pen gwyn* white head (applied to the bird in winter plumage)] (1588)
: any of various erect short-legged flightless aquatic birds (family Spheniscidae) of the southern hemisphere

pen·hold·er \ˈpen-ˌhōl-dər\ *noun* (1815)
: a holder or handle for a pen point

-penia *combining form* [New Latin, from Greek *penia*]
: deficiency ⟨leuko*penia*⟩

pen·i·cil·la·mine \ˌpe-nə-ˈsi-lə-ˌmēn\ *noun* [*penicill*in + *amine*] (1943)
: an amino acid $C_5H_{11}NO_2S$ that is obtained from penicillins and is used especially in the treatment of poisoning by metals (as copper or lead) and of cystinuria

pen·i·cil·late \ˌpe-nə-ˈsi-lət, -ˌlāt\ *adjective* [Latin *penicillus* brush — more at PENCIL] (1819)
: furnished with a tuft of fine filaments ⟨a *penicillate* stigma⟩

pen·i·cil·lin \ˌpe-nə-ˈsi-lən\ *noun* [New Latin *Penicillium*] (1929)
1 : any of several relatively nontoxic antibiotic acids of the general constitution $C_9H_{11}N_2O_4SR$ that are produced by molds (genus *Penicillium* and especially *P. notatum* or *P. chrysogenum*) or synthetically and are used especially against cocci; *also* **:** a mixture of such acids
2 : a salt or ester of a penicillin or a mixture of such salts or esters

pen·i·cil·lin·ase \-ˈsi-lə-ˌnās, -ˌnāz\ *noun* (1940)

\ə\ **abut** \ᵊ\ kitten \ər\ **further** \a\ **ash** \ā\ **ace** \ä\ mop, **mar** \au̇\ **out** \ch\ **chin** \e\ **bet** \ē\ **easy** \g\ **go** \i\ **hit** \ī\ **ice** \j\ **job** \ŋ\ **sing** \ō\ **go** \ȯ\ **law** \ȯi\ **boy** \th\ **thin** \t͟h\ **the** \ü\ **loot** \u̇\ **foot** \y\ **yet** \zh\ **vision** *see also* Guide to Pronunciation

: an enzyme found especially in staphylococcal bacteria that inactivates the penicillins by hydrolyzing them

pen•i•cil•li•um \-ˈsi-lē-əm\ *noun, plural* **-lia** \-lē-ə\ [New Latin, from Latin *penicillus*] (1867)
: any of a genus (*Penicillium* of the family Moniliaceae) of fungi (as a blue mold) that are found chiefly on moist nonliving organic matter

pe•nile \ˈpē-ˌnīl\ *adjective* (circa 1861)
: of, relating to, or affecting the penis

pen•in•su•la \pə-ˈnin(t)-s(ə-)lə, -shə-lə\ *noun* [Latin *paeninsula,* from *paene* almost + *insula* island] (1538)
: a portion of land nearly surrounded by water and connected with a larger body by an isthmus; *also* **:** a piece of land jutting out into the water whether with or without a well-defined isthmus
— **pen•in•su•lar** \-s(ə-)lər, -shə-lər\ *adjective*

pe•nis \ˈpē-nəs\ *noun, plural* **pe•nes** \ˈpē-(ˌ)nēz\ *or* **pe•nis•es** [Latin, penis, tail; akin to Old High German *faselt* penis, Greek *peos*] (1676)
: a male organ of copulation that in male mammals including humans usually functions as the channel by which urine leaves the body

penis envy *noun* (1924)
: the supposed coveting of the penis by a young human female which is held in Freudian psychoanalytic theory to lead to feelings of inferiority and defensive or compensatory behavior

pen•i•tence \ˈpe-nə-tən(t)s\ *noun* [Middle English, from Old French, from Medieval Latin *poenitentia,* alteration of Latin *paenitentia* regret, from *paenitent-, paenitens,* present participle] (13th century)
: the quality or state of being penitent **:** sorrow for sins or faults ☆

[1]pen•i•tent \-tənt\ *adjective* [Middle English, from Middle French, from Latin *paenitent-, paenitens,* from present participle of *paenitēre* to cause regret, feel regret, perhaps from *paene* almost] (14th century)
: feeling or expressing humble or regretful pain or sorrow for sins or offenses **:** REPENTANT
— **pen•i•tent•ly** *adverb*

[2]penitent *noun* (14th century)
1 : a person who repents of sin
2 : a person under church censure but admitted to penance or reconciliation especially under the direction of a confessor

pen•i•ten•tial \ˌpe-nə-ˈten(t)-shəl\ *adjective* (1508)
: of or relating to penitence or penance
— **pen•i•ten•tial•ly** \-ˈten(t)-sh(ə-)lē\ *adverb*

[1]pen•i•ten•tia•ry \ˌpe-nə-ˈten(t)-sh(ə-)rē\ *noun, plural* **-ries** [Middle English *penitenciary,* from Medieval Latin *poenitentiarius,* from *poenitentia*] (15th century)
1 a : an officer in some Roman Catholic dioceses vested with power from the bishop to deal with cases of a nature normally handled only by the bishop **b** *capitalized* **:** a cardinal presiding over a tribunal of the Roman curia concerned with dispensations and indulgences
2 : a public institution in which offenders against the law are confined for detention or punishment; *specifically* **:** a state or federal prison in the U.S.

[2]pen•i•ten•tia•ry \ˌpe-nə-ˈten(t)-sh(ə-)rē, *1 also* -ˈten(t)-shē-ˌer-ē\ *adjective* (1577)
1 : PENITENTIAL
2 : of, relating to, or incurring confinement in a penitentiary

pen•knife \ˈpen-ˌnīf\ *noun* [Middle English; from its original use for mending quill pens] (15th century)
: a small pocketknife usually with only one blade

pen•light *also* **pen•lite** \-ˌlīt\ *noun* (1943)
: a small flashlight resembling a fountain pen in size or shape

pen•man \-mən\ *noun* (1539)
1 a : CALLIGRAPHER **b :** COPYIST, SCRIBE **c :** a person with a specified quality or kind of handwriting ⟨a poor *penman*⟩
2 : AUTHOR

pen•man•ship \-ˌship\ *noun* (1695)
1 : the art or practice of writing with the pen
2 : quality or style of handwriting

pen name *noun* (circa 1864)
: an author's pseudonym

pen•nant \ˈpe-nənt\ *noun* [alteration of *pendant*] (1698)
1 a : any of various nautical flags tapering usually to a point or swallowtail and used for identification or signaling **b :** a flag or banner longer in the fly than in the hoist; *especially* **:** one that tapers to a point
2 : a flag emblematic of championship (as in a professional baseball league); *also* **:** the championship itself

pen•nate \ˈpe-ˌnāt\ *adjective* [irregular from New Latin *Pennales*] (1938)
: of, relating to, or being diatoms of an order (Pennales) usually characterized by a raphe or a structure resembling a raphe and by ornamentation of the valves that is bilaterally symmetrical

pen•ne \ˈpe(n)-(ˌ)nā\ *noun, plural* **penne** [Italian, plural of *penna,* literally, quill, feather, pen, from Latin *pinna* feather & *penna* wing — more at PEN] (1974)
: short thick diagonally cut tubular pasta

pen•ni \ˈpe-nē\ *noun, plural* **pen•nia** \-nē-ə\ *also* **penni** *or* **pen•nis** \-nēz\ [Finnish] (circa 1893)
— see *markka* at MONEY table

pen•ni•less \ˈpe-ni-ləs, ˈpe-nᵊl-əs\ *adjective* (14th century)
: destitute of money

pen•non \ˈpe-nən\ *noun* [Middle English, from Middle French *penon,* diminutive of *penne* quill, wing feather — more at PEN] (14th century)
1 a : a long usually triangular or swallow-tailed streamer typically attached to the head of a lance as an ensign **b :** PENNANT 1a
2 : WING, PINION

pen•non•cel *or* **pen•on•cel** \ˈpe-nən-ˌsel\ *noun* [Middle English *penoncell,* from Middle French *penoncel,* diminutive of *penon*] (14th century)
: a small pennon used in late medieval or Renaissance times

Penn•syl•va•nia Dutch \ˌpen(t)-səl-ˈvā-nyə-, -nē-ə-\ *noun* (circa 1824)
1 : a people living mostly in eastern Pennsylvania whose characteristic cultural traditions go back to the German migrations of the 18th century
2 : a dialect of High German spoken in parts of Pennsylvania and Maryland
— **Pennsylvania Dutchman** *noun*

Pennsylvania German *noun* (1869)
: PENNSYLVANIA DUTCH

Penn•syl•va•nian \-ˈvā-nyən, -nē-ən\ *adjective* (1698)
1 : of or relating to Pennsylvania or its people
2 : of, relating to, or being the period of the Paleozoic era in North America between the Mississippian and Permian or the corresponding system of rocks — see GEOLOGIC TIME table
— **Pennsylvanian** *noun*

pen•ny \ˈpe-nē\ *noun, plural* **pennies** \-nēz\ *or* **pence** \ˈpen(t)s\ *often attributive* [Middle English, from Old English *penning, penig;* akin to Old High German *pfenning,* a coin] (before 12th century)
1 a : a monetary unit of the United Kingdom formerly equal to 1/240 pound but now equal to 1/100 pound **b :** a similar monetary unit of any of various other countries in or formerly in the Commonwealth — see *pound* at MONEY table **c :** a coin representing one penny
2 : DENARIUS
3 *plural pennies* **:** CENT
4 : a trivial amount
5 : a piece or sum of money ⟨that will cost a pretty *penny*⟩

-penny *adjective combining form* [*penny;* perhaps from the original price per hundred]
: being a (designated) nail size ⟨an eight*penny* nail⟩

pen•ny–an•te \ˈpe-nē-ˈan-tē\ *adjective* (1868)
: SMALL-TIME, TWO-BIT

penny ante *noun* (1855)
: poker played for very low stakes

penny arcade *noun* (1908)
: an amusement center having coin-operated devices for entertainment

pen•ny•cress \ˈpe-nē-ˌkres\ *noun* (1713)
: a Eurasian herb (*Thlaspi arvense*) with round flat pods that is widely naturalized in the New World

penny dreadful *noun* (circa 1873)
: a novel of violent adventure or crime

pen•ny–pinch•ing \-ˌpin-chiŋ\ *noun* (1935)
: FRUGALITY, PARSIMONY
— **pen•ny–pinch•er** \-chər\ *noun*
— **penny–pinching** *adjective*

pen•ny•roy•al \ˌpen-ē-ˈrȯi(-ə)l, ˈpe-ni-ˌrīl\ *noun* [probably by folk etymology from Middle French *poullieul,* modification of Latin *pulegium* fleabane, pennyroyal] (1530)
1 : a European perennial mint (*Mentha pulegium*) with small aromatic leaves
2 : an aromatic North American mint (*Hedeoma pulegioides*) that has blue or violet flowers borne in axillary tufts and yields an oil used in folk medicine or to drive away mosquitoes

penny stock *noun* (circa 1920)
: a usually unlisted highly speculative stock selling for a dollar or less

pen•ny•weight \ˈpe-nē-ˌwāt\ *noun* (14th century)
— see WEIGHT table

pen•ny•whistle \-ˌhwi-səl, -ˌwi-\ *noun* (1818)
1 : a small fipple flute
2 : a toy whistle

pen•ny–wise \-ˌwīz\ *adjective* [from the phrase *penny-wise and pound-foolish*] (1607)
: wise or prudent only in dealing with small sums or matters

pen•ny•wort \-ˌwərt, -ˌwȯrt\ *noun* (14th century)
: any of several usually round-leaved plants (as of the genus *Hydrocotyle* of the carrot family)

pen•ny•worth \ˈpe-nē-ˌwərth, *British often* ˈpe-nərth\ *noun, plural* **pennyworth** *or* **pennyworths** (before 12th century)
1 : a penny's worth
2 : value for the money spent **:** BARGAIN
3 : a small quantity **:** MODICUM

Pe•nob•scot \pə-ˈnäb-ˌskät, -skət\ *noun, plural* **-scot** *or* **-scots** (1624)

☆ SYNONYMS

Penitence, repentance, contrition, compunction, remorse mean regret for sin or wrongdoing. PENITENCE implies sad and humble realization of and regret for one's misdeeds ⟨absolution is dependent upon sincere *penitence*⟩. REPENTANCE adds the implication of a resolve to change ⟨*repentance* accompanied by a complete change of character⟩. CONTRITION stresses the sorrowful regret that constitutes true penitence ⟨tearful expressions of *contrition*⟩. COMPUNCTION implies a painful sting of conscience especially for contemplated wrongdoing ⟨had no *compunctions* about taking back what is mine⟩. REMORSE suggests prolonged and insistent self-reproach and mental anguish for past wrongs and especially for those whose consequences cannot be remedied ⟨thieves untroubled by feelings of *remorse*⟩.

: a member of an American Indian people of the Penobscot river valley and Penobscot Bay region of Maine

pe·no·che \pə-ˈnō-chē\ *variant of* PENUCHE

pe·nol·o·gy \pi-ˈnä-lə-jē\ *noun* [Greek *poinē* penalty + English *-logy* — more at PAIN] (1838)
: a branch of criminology dealing with prison management and the treatment of offenders
— **pe·no·log·i·cal** \ˌpē-nə-ˈlä-ji-kəl\ *adjective*
— **pe·nol·o·gist** \pi-ˈnä-lə-jist\ *noun*

pen pal *noun* (1938)
: a friend made and kept through correspondence

pen point *noun* (circa 1864)
: a small thin convex metal device that tapers to a split point, fits into a holder, and is used for writing or drawing

pen pusher *noun* (circa 1905)
: PENCIL PUSHER

¹pen·sion \ˈpen(t)-shən\ *noun* [Middle English, from Middle French, from Latin *pension-, pensio,* from *pendere* to pay — more at PENDANT] (14th century)
1 \ˈpen(t)-shən\ : a fixed sum paid regularly to a person: **a** *archaic* : WAGE **b** : a gratuity granted (as by a government) as a favor or reward **c** : one paid under given conditions to a person following retirement from service or to surviving dependents
2 \pänⁿs-yōⁿ\ [French, from Middle French] **a** : accommodations especially at a continental European hotel or boardinghouse : ROOM AND BOARD **b** *also* **pen·sio·ne** \pen(t)-ˈsyō-(ˌ)nā\ [*pensione,* from Italian] : a hotel or boardinghouse especially in continental Europe
— **pen·sion·less** \ˈpen(t)-shən-ləs\ *adjective*

²pension *transitive verb* **pen·sioned; pen·sion·ing** \ˈpen(t)-sh(ə-)niŋ\ (1702)
1 : to grant or pay a pension to
2 : to dismiss or retire from service with a pension ⟨*pensioned* off his faithful old servant⟩

pen·sion·able \ˈpen(t)-sh(ə-)nə-bəl\ *adjective* (1882)
chiefly British : of, relating to, qualified for, or qualifying for a pension ⟨*pensionable* employees⟩ ⟨a *pensionable* post⟩

pen·sion·ary \ˈpen(t)-shə-ˌner-ē\ *noun, plural* **-ar·ies** (1536)
: PENSIONER; *especially* : HIRELING
— **pensionary** *adjective*

pen·sion·er \ˈpen(t)-sh(ə-)nər\ *noun* (15th century)
1 : a person who receives or lives on a pension
2 *obsolete* **a** : GENTLEMAN-AT-ARMS **b** : RETAINER **c** : MERCENARY, HIRELING

pen·sive \ˈpen(t)-siv\ *adjective* [Middle English *pensif,* from Middle French, from *penser* to think, from Latin *pensare* to ponder, frequentative of *pendere* to weigh — more at PENDANT] (14th century)
1 : musingly or dreamily thoughtful
2 : suggestive of sad thoughtfulness
— **pen·sive·ly** *adverb*
— **pen·sive·ness** *noun*

pen·ste·mon \pen-ˈste-mən, ˈpen(t)-stə-\ *noun* [New Latin, from *penta-* + Greek *stēmōn* warp, thread — more at STAMEN] (1760)
: any of a genus (*Penstemon*) of chiefly American herbs of the snapdragon family with showy tubular flowers

pen·stock \ˈpen-ˌstäk\ *noun* (circa 1607)
1 : a sluice or gate for regulating a flow (as of water)
2 : a conduit or pipe for conducting water

pent \ˈpent\ *adjective* [probably from past participle of obsolete English *pend* to confine] (1550)
: shut up : CONFINED, REPRESSED ⟨a *pent* crowd⟩ ⟨*pent*-up feelings⟩

penta- *or* **pent-** *combining form* [Middle English, from Greek, from *pente* — more at FIVE]
1 : five ⟨*penta*hedron⟩
2 : containing five atoms or groups ⟨*pent*ane⟩

pen·ta·chlo·ro·phe·nol \ˌpen-tə-ˌklōr-ə-ˈfē-ˌnōl, -ˌklȯr-, -fi-ˈ\ *noun* (1879)
: a crystalline compound C_6Cl_5OH used especially as a wood preservative and fungicide and a disinfectant

pen·ta·cle \ˈpen-ti-kəl\ *noun* [(assumed) Medieval Latin *pentaculum,* probably from Greek *pente*] (1594)
: PENTAGRAM

pen·tad \ˈpen-ˌtad\ *noun* [Greek *pentad-, pentas,* from *pente*] (1653)
: a group of five

pen·ta·gon \ˈpen-tə-ˌgän\ *noun* [Greek *pentagōnon,* from neuter of *pentagōnos* pentagonal, from *penta-* + *gōnia* angle — more at -GON] (1571)
: a polygon of five angles and five sides

Pentagon *noun* [the *Pentagon* building, headquarters of the Department of Defense] (1941)
: the U.S. military leadership

pen·tag·o·nal \pen-ˈta-gə-nᵊl\ *noun* (1571)
1 : having five sides and five angles
2 : having a pentagon as a cross section or as a base ⟨a *pentagonal* pyramid⟩
— **pen·tag·o·nal·ly** \-nᵊl-ē\ *adverb*

Pen·ta·gon·ese \ˌpen-tə-ˌgä-ˈnēz, -ˈnēs\ *noun* (1951)
: military jargon

pen·ta·gram \ˈpen-tə-ˌgram\ *noun* [Greek *pentagrammon,* from *penta-* + *-grammon* (akin to *gramma* letter) — more at GRAM] (1833)
: a figure of a 5-pointed star usually made with alternate points connected by a continuous line and used as a magic or occult symbol; *also* : a similar 6-pointed star (as a Solomon's seal)

pen·ta·he·dron \ˌpen-tə-ˈhē-drən\ *noun* [New Latin] (circa 1775)
: a solid bounded by five faces
— **pen·ta·he·dral** \-drəl\ *adjective*

pen·tam·er·ous \pen-ˈta-mə-rəs\ *adjective* [New Latin *pentamerus,* from *penta-* (from Greek) + *-merus* -merous] (1826)
: divided into or consisting of five parts; *specifically* : having each floral whorl consisting of five or a multiple of five members

pen·tam·e·ter \pen-ˈta-mə-tər\ *noun* [Latin, from Greek *pentametros* having five metrical feet, from *penta-* + *metron* measure — more at MEASURE] (1589)
: a line of verse consisting of five metrical feet

pent·am·i·dine \pen-ˈta-mə-ˌdēn, -dən\ *noun* (1941)
: a drug $C_{19}H_{24}N_4O_2$ used especially to treat protozoal infections (as leishmaniasis) and to prevent AIDS-related pneumonia

pen·tane \ˈpen-ˌtān\ *noun* [International Scientific Vocabulary] (1877)
: any of three isomeric alkanes C_5H_{12} that occur especially in petroleum

pen·tan·gle \ˈpent-ˌaŋ-gəl, ˈpen-ˌtaŋ-\ *noun* (14th century)
: PENTAGRAM

pen·ta·pep·tide \ˌpen-tə-ˈpep-ˌtīd\ *noun* (1907)
: a polypeptide that contains five amino acid residues

pen·ta·ploid \ˈpen-tə-ˌplȯid\ *adjective* (1921)
: having or being a chromosome number that is five times the basic number
— **pentaploid** *noun*
— **pen·ta·ploi·dy** \-ˌplȯi-dē\ *noun*

pen·tar·chy \ˈpen-ˌtär-kē\ *noun* [Greek *pentarchia,* from *penta-* + *-archia* -archy] (1587)
: a group of five countries or districts each under its own ruler or government

Pen·ta·teuch \ˈpen-tə-ˌtük, -ˌtyük\ *noun* [Middle English *Penteteuke,* from Late Latin *Pentateuchus,* from Greek *Pentateuchos,* from *penta-* + *teuchos* tool, vessel, book, from *teuchein* to make — more at DOUGHTY] (15th century)
: the first five books of Jewish and Christian Scriptures

pen·tath·lete \pen-ˈtath-ˌlēt\ *noun* (1828)
: an athlete participating in a pentathlon

pen·tath·lon \pen-ˈtath-lən, -ˌlän\ *noun* [Greek, from *penta-* + *athlon* contest] (1603)
: an athletic contest involving participation by each contestant in five different events; *especially* : MODERN PENTATHLON

pen·ta·ton·ic \ˌpen-tə-ˈtä-nik\ *adjective* (1864)
: consisting of five tones; *specifically* : being or relating to a scale in which the tones are arranged like a major scale with the fourth and seventh tones omitted

pen·ta·va·lent \ˌpen-tə-ˈvā-lənt\ *adjective* (1871)
: having a valence of five

pen·taz·o·cine \pen-ˈta-zə-ˌsēn\ *noun* [probably from *penta-* + *az-* + *octa-* + ²*-ine*] (1963)
: an analgesic drug $C_{19}H_{27}NO$ that is less addictive than morphine

Pen·te·cost \ˈpen-ti-ˌkȯst, -ˌkäst\ *noun* [Middle English, from Old English *pentecosten,* from Late Latin *pentecoste,* from Greek *pentēkostē,* literally, fiftieth day, from *pentēkostos* fiftieth, from *pentēkonta* fifty, from *penta-* + *-konta* (akin to Latin *viginti* twenty) — more at VIGESIMAL] (before 12th century)
1 : SHABUOTH
2 : a Christian feast on the seventh Sunday after Easter commemorating the descent of the Holy Spirit on the apostles — called also *Whitsunday*

¹Pen·te·cos·tal \ˌpen-ti-ˈkäs-tᵊl, -ˈkȯs-\ *adjective* (1663)
1 : of, relating to, or suggesting Pentecost
2 : of, relating to, or constituting any of various Christian religious bodies that emphasize individual experiences of grace, spiritual gifts (as glossolalia and faith healing), expressive worship, and evangelism
— **Pen·te·cos·tal·ism** \-tə-ˌli-zəm\ *noun*
— **Pen·te·cos·tal·ist** \-tə-list\ *noun or adjective*

²Pentecostal *noun* (1904)
: a member of a Pentecostal religious body

pent·house \ˈpent-ˌhau̇s\ *noun* [alteration of Middle English *pentis,* from Middle French *apentiz,* from *apent,* past participle of *apendre, appendre* to attach, hang against — more at APPEND] (1530)
1 a : a shed or roof attached to and sloping from a wall or building **b** : a smaller structure joined to a building : ANNEX
2 : a structure or dwelling built on the roof of a building

pent·land·ite \ˈpent-lən-ˌdīt\ *noun* [French, from Joseph *Pentland* (died 1873) Irish scientist] (circa 1858)
: a bronzy yellow mineral that is an isometric nickel iron sulfide and the principal ore of nickel

pen·to·bar·bi·tal \ˌpen-tə-ˈbär-bə-ˌtȯl\ *noun* [*penta-* + *-o-* + *barbital*] (1931)
: a granular barbiturate $C_{11}H_{18}N_2O_3$ used especially in the form of its sodium or calcium salt as a sedative, hypnotic, and antispasmodic

pen·to·bar·bi·tone \-ˌtōn\ *noun* [*penta-* + *-o-* + *barbitone*] (1938)
British : PENTOBARBITAL

pen·to·san \ˈpen-tə-ˌsan\ *noun* [International Scientific Vocabulary] (1892)
: any of various polysaccharides that yield only pentoses on hydrolysis and occur widely in plants

pen·tose \ˈpen-ˌtōs, -ˌtōz\ *noun* [International Scientific Vocabulary] (1890)
: a monosaccharide $C_5H_{10}O_5$ (as ribose) that contains five carbon atoms in the molecule

Pen·to·thal \'pen-tə-ˌthȯl\ *trademark*
— used for thiopental

pent·ox·ide \pent-'äk-ˌsīd\ *noun* [International Scientific Vocabulary] (circa 1872)
: an oxide containing five atoms of oxygen in the molecule

pent·ste·mon *variant of* PENSTEMON

pen·tyl·ene·tet·ra·zol \ˌpen-tᵊl-ˌēn-'te-trə-ˌzȯl, -ˌzōl\ *noun* [*penta-* + methy*lene* + *tetrazole* (CH_2N_4)] (1949)
: an analeptic drug $C_6H_{10}N_4$

pe·nu·che \pə-'nü-chē\ *noun* [Mexican Spanish *panocha* raw sugar, from Spanish, ear of maize, ultimately from Latin *panicula* panicle — more at PANICLE] (1871)
: fudge made usually of brown sugar, butter, cream or milk, and nuts

pe·nult \'pē-ˌnəlt, pi-'\ *noun* [Latin *paenultima* penult, from feminine of *paenultimus* almost last, from *paene* almost + *ultimus* last — more at ULTIMATE] (1537)
: the next to the last member of a series; *especially* **:** the next to the last syllable of a word

pen·ul·ti·ma \pi-'nəl-tə-mə\ *noun* [Latin] (1589)
: PENULT

pen·ul·ti·mate \pi-'nəl-tə-mət\ *adjective* (1677)
1 : next to the last ⟨the *penultimate* chapter of a book⟩
2 : of or relating to a penult ⟨a *penultimate* accent⟩
— pen·ul·ti·mate·ly *adverb*

pen·um·bra \pə-'nəm-brə\ *noun, plural* **-brae** \-(ˌ)brē, -ˌbrī\ *or* **-bras** [New Latin, from Latin *paene* almost + *umbra* shadow — more at UMBRAGE] (1666)
1 a : a space of partial illumination (as in an eclipse) between the perfect shadow on all sides and the full light **b :** a shaded region surrounding the dark central portion of a sunspot
2 : a surrounding or adjoining region in which something exists in a lesser degree **:** FRINGE
3 : a body of rights held to be guaranteed by implication in a civil constitution
— pen·um·bral \-brəl\ *adjective*

pe·nu·ri·ous \pə-'nu̇r-ē-əs, -'nyu̇r-\ *adjective* (1590)
1 : marked by or suffering from penury
2 : given to or marked by extreme stinting frugality
synonym see STINGY
— pe·nu·ri·ous·ly *adverb*
— pe·nu·ri·ous·ness *noun*

pen·u·ry \'pen-yə-rē\ *noun* [Middle English, from Latin *penuria, paenuria* want; perhaps akin to Latin *paene* almost] (14th century)
1 : a cramping and oppressive lack of resources (as money); *especially* **:** severe poverty
2 : extreme and often niggardly frugality
synonym see POVERTY

pe·on \'pē-ˌän, -ən *also* pā-'ōn *for 2, British also* 'pyün *for 1*\ *noun, plural* **peons** *or* **pe·o·nes** \pā-'ō-nēz\ [Portuguese *peão* & French *pion,* from Medieval Latin *pedon-, pedo* foot soldier — more at PAWN] (1609)
1 : any of various workers in India, Sri Lanka, or Malaysia: as **a :** INFANTRYMAN **b :** ORDERLY
2 [Spanish *peón,* from Latin *pedon-, pedo*] **:** a member of the landless laboring class in Spanish America
3 *plural peons* **a :** a person held in compulsory servitude to a master for the working out of an indebtedness **b :** DRUDGE, MENIAL

pe·on·age \'pē-ə-nij\ *noun* (1844)
1 a : the use of laborers bound in servitude because of debt **b :** a system of convict labor by which convicts are leased to contractors
2 : the condition of a peon

pe·o·ny \'pē-ə-nē\ *noun, plural* **-nies** [Middle English *piony,* from Middle French *pioine,* from Latin *paeonia,* from Greek *paiōnia,* from *Paiōn* Paeon, physician of the gods] (14th century)
: any of a genus (*Paeonia* of the family Paeoniaceae) of chiefly Eurasian plants with large often double flowers

¹peo·ple \'pē-pəl\ *noun, plural* **people** [Middle English *peple,* from Old French *peuple,* from Latin *populus*] (13th century)
1 *plural* **:** human beings making up a group or assembly or linked by a common interest
2 *plural* **:** HUMAN BEINGS, PERSONS — often used in compounds instead of *persons* ⟨sales*people*⟩
3 *plural* **:** the members of a family or kinship
4 *plural* **:** the mass of a community as distinguished from a special class ⟨disputes between the *people* and the nobles⟩ — often used by Communists to distinguish Communists from other people
5 *plural* **peoples :** a body of persons that are united by a common culture, tradition, or sense of kinship, that typically have common language, institutions, and beliefs, and that often constitute a politically organized group
6 : lower animals usually of a specified kind or situation
7 : the body of enfranchised citizens of a state
— peo·ple·less \-pə(l)-ləs\ *adjective*

²people *transitive verb* **peo·pled; peo·pling** \-p(ə-)liŋ\ [Middle French *peupler,* from Old French, from *peuple*] (15th century)
1 : to supply or fill with people
2 : to dwell in **:** INHABIT

peo·ple·hood \'pē-pəl-ˌhu̇d\ *noun* (circa 1899)
1 : the quality or state of constituting a people
2 : the awareness of the underlying unity that makes the individual a part of a people

people mover *noun* (1968)
: any of various rapid-transit systems (as of moving sidewalks or automated driverless cars) for shuttling people

¹pep \'pep\ *noun* [short for *pepper*] (1912)
: brisk energy or initiative and high spirits

²pep *transitive verb* **pepped; pep·ping** (1925)
: to inject pep into ⟨*pep* him up⟩

pep·er·o·mia \ˌpe-pə-'rō-mē-ə\ *noun* [New Latin, from Greek *peperi* pepper + *homoios* like, similar — more at HOMEO-] (1882)
: any of a genus (*Peperomia*) of fleshy tropical herbs of the pepper family often cultivated for their showy foliage

pep·los \'pe-pləs, -ˌpläs\ *also* **pep·lus** \-pləs\ *noun* [Latin *peplus,* from Greek *peplos*] (1776)
: a garment worn like a shawl by women of ancient Greece

pep·lum \-pləm\ *noun* [Latin, from Greek *peplon* peplos] (1866)
: a short section attached to the waistline of a blouse, jacket, or dress
— pep·lumed \-pləmd\ *adjective*

peplum

pe·po \'pē-(ˌ)pō\ *noun, plural* **pepos** [Latin, a melon — more at PUMPKIN] (circa 1859)
: an indehiscent fleshy 1-celled or falsely 3-celled many-seeded berry (as a pumpkin, squash, melon, or cucumber) that has a hard rind and is the characteristic fruit of the gourd family

¹pep·per \'pe-pər\ *noun* [Middle English *peper,* from Old English *pipor,* from Latin *piper,* from Greek *peperi*] (before 12th century)
1 a : either of two pungent products from the fruit of an East Indian plant that are used as a condiment, carminative, or stimulant: (1) **:** BLACK PEPPER (2) **:** WHITE PEPPER **b :** any of a genus (*Piper* of the family Piperaceae, the pepper family) of tropical mostly jointed climbing shrubs with aromatic leaves; *especially* **:** a woody vine (*P. nigrum*) with spicate flowers that is widely cultivated in the tropics for its red berries from which black pepper and white pepper are prepared

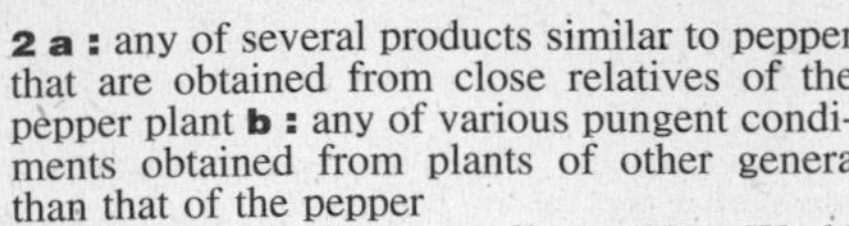

2 a : any of several products similar to pepper that are obtained from close relatives of the pepper plant **b :** any of various pungent condiments obtained from plants of other genera than that of the pepper
3 a : CAPSICUM 1; *especially* **:** a New World capsicum (*Capsicum annuum*) whose fruits are hot peppers or sweet peppers **b :** the fruit of a pepper that is usually red or yellow when ripe
— pepper *adjective*

²pepper *transitive verb* **pep·pered; pep·per·ing** \'pe-p(ə-)riŋ\ (1538)
1 a : to sprinkle, season, or cover with or as if with pepper **b :** to shower with or as if with shot or other missiles
2 : to hit with rapid repeated blows
3 : to sprinkle as if with pepper ⟨*peppered* the report with statistics⟩
— pep·per·er \-pər-ər\ *noun*

pep·per–and–salt \ˌpe-pər(-ə)n(d)-'sȯlt\ *adjective* (1774)
: SALT-AND-PEPPER

pep·per·box \'pe-pər-ˌbäks\ *noun* (1546)
1 : a small usually cylindrical box or bottle with a perforated top used for sprinkling ground pepper on food
2 : a pistol of the late 18th century with five or six revolving barrels

pep·per·corn \-ˌkȯrn\ *noun* (before 12th century)
: a dried berry of the black pepper

peppered moth *noun* (circa 1832)
: a European geometrid moth (*Biston betularia*) that typically has white wings with small black specks but also occurs as a solid black form especially in areas where the air is heavily polluted by industry

pep·per·grass \'pe-pər-ˌgras\ *noun* (15th century)
: any of a genus (*Lepidium*) of cresses; *especially* **:** GARDEN CRESS

pepper mill *noun* (circa 1858)
: a hand mill for grinding peppercorns

pep·per·mint \-ˌmint, -mənt, *in rapid speech* 'pep-mənt *or* -ᵊm-ənt\ *noun* (1696)
1 a : a pungent and aromatic mint (*Mentha piperita*) with dark green lanceolate leaves and whorls of small pink flowers in spikes **b :** any of several mints (as *M. arvensis*) that are related to the peppermint
2 : candy flavored with peppermint
— pep·per·minty \'pe-pər-ˌmin-tē\ *adjective*

pep·per·o·ni \ˌpe-pə-'rō-nē\ *noun* [Italian *peperoni* cayenne peppers, plural of *peperone,* augmentative of *pepe* pepper, from Latin *piper* — more at PEPPER] (1921)
: a highly seasoned beef and pork sausage

pepper pot *noun* (1679)
1 *British* **:** PEPPERBOX 1
2 a : a highly seasoned West Indian stew of vegetables and meat or fish **b :** a thick soup of tripe, meat, dumplings, and vegetables highly seasoned especially with crushed peppercorns — called also *Philadelphia pepper pot*

pepper shaker *noun* (1895)
: a container with a perforated top for sprinkling pepper

pep·per·tree \'pe-pər-ˌtrē\ *noun* (circa 1692)
: a Peruvian evergreen tree (*Schinus molle*) of the cashew family grown as a shade tree in mild regions

pep·pery \'pe-p(ə-)rē\ *adjective* (1699)
1 : of, relating to, or having the qualities of pepper **:** HOT, PUNGENT ⟨a *peppery* taste⟩
2 : having a hot temper **:** TOUCHY ⟨a *peppery* boss⟩
3 : FIERY, STINGING ⟨a *peppery* satire⟩
— pep·per·i·ness \-nəs\ *noun*

pep pill *noun* (1937)
: any of various stimulant drugs in pill or tablet form

pep·py \'pe-pē\ *adjective* **pep·pi·er; -est** (circa 1918)
: full of pep

— **pep·pi·ness** *noun*

pep·sin \ˈpep-sən\ *noun* [German, from Greek *pepsis* digestion, from *pessein*] (circa 1844)
1 : a protease of the stomach that breaks down most proteins to polypeptides
2 : a preparation containing pepsin that is obtained from the stomach especially of the hog and is used especially as a digestive

pep·sin·o·gen \pep-ˈsi-nə-jən\ *noun* [International Scientific Vocabulary] (1878)
: a granular zymogen of the gastric glands that is readily converted into pepsin in a slightly acid medium

pep talk *noun* (1925)
: a usually brief, intense, and emotional talk designed to influence or encourage an audience

pep·tic \ˈpep-tik\ *adjective* [Latin *pepticus,* from Greek *peptikos,* from *peptos* cooked, from *peptein, pessein* to cook, digest — more at COOK] (1651)
1 : relating to or promoting digestion **:** DIGESTIVE
2 : of, relating to, producing, or caused by pepsin ⟨*peptic* digestion⟩
3 : connected with or resulting from the action of digestive juices ⟨a *peptic* ulcer⟩

pep·ti·dase \ˈpep-tə-ˌdās, -ˌdāz\ *noun* (1918)
: an enzyme that hydrolyzes simple peptides or their derivatives

pep·tide \ˈpep-ˌtīd\ *noun* [International Scientific Vocabulary, from *peptone*] (1906)
: any of various amides that are derived from two or more amino acids by combination of the amino group of one acid with the carboxyl group of another and are usually obtained by partial hydrolysis of proteins
— **pep·tid·ic** \pep-ˈti-dik\ *adjective*

peptide bond *noun* (1935)
: the chemical bond between carbon and nitrogen in a peptide linkage

peptide linkage *noun* (1925)
: the bivalent group CONH that unites the amino acid residues in a peptide

pep·ti·do·gly·can \ˌpep-tə-dō-ˈglī-ˌkan\ *noun* (1966)
: a polymer that is composed of polysaccharide and peptide chains and is found especially in bacterial cell walls — called also *mucopeptide, murein*

pep·tone \ˈpep-ˌtōn\ *noun* [German *Pepton,* from Greek, neuter of *peptos* cooked] (1860)
: any of various water-soluble products of partial hydrolysis of proteins

Pe·quot \ˈpē-ˌkwät\ *noun* [Narraganset *Pequttôog*] (1631)
: a member of an American Indian people of what is now eastern Connecticut

[1]**per** \ˈpər\ *preposition* [Latin, through, by means of, by — more at FOR] (14th century)
1 : by the means or agency of **:** THROUGH ⟨*per* bearer⟩
2 : with respect to every member of a specified group **:** for each
3 : according to — often used with *as* ⟨*per* instructions⟩ ⟨as *per* usual⟩ □

[2]**per** *adverb* (1899)
: for each **:** APIECE ⟨a bargain at $3.50 *per*⟩

per- *prefix* [Latin, through, throughout, thoroughly, detrimental to, from *per*]
1 : throughout **:** thoroughly ⟨*per*use⟩
2 a : containing the largest possible or a relatively large proportion of a (specified) chemical element ⟨*per*chloroethylene⟩ **b :** containing an element in its highest or a high oxidation state ⟨*per*chloric acid⟩

[1]**per·ad·ven·ture** \ˈpər-əd-ˌven-chər, ˈper-; ˌpər-əd-ˈ, ˌper-\ *adverb* [Middle English *per aventure,* from Old French, by chance] (14th century)
archaic **:** PERHAPS, POSSIBLY

[2]**peradventure** *noun* (1627)
1 : DOUBT ⟨a fact established beyond *peradventure*⟩
2 : CHANCE 4a ⟨beyond *peradventure* of doubt⟩

per·am·bu·late \pə-ˈram-byə-ˌlāt\ *verb* **-lat·ed; -lat·ing** [Latin *perambulatus,* past participle of *perambulare,* from *per-* through + *ambulare* to walk] (1568)
transitive verb
1 : to travel over or through especially on foot **:** TRAVERSE
2 : to make an official inspection of (a boundary) on foot
intransitive verb
: STROLL
— **per·am·bu·la·tion** \-ˌram-byə-ˈlā-shən\ *noun*
— **per·am·bu·la·to·ry** \-ˈram-byə-lə-ˌtōr-ē, -ˌtȯr-\ *adjective*

per·am·bu·la·tor \pə-ˈram-byə-ˌlā-tər, *for 2 also* ˈpram-\ *noun* (1611)
1 : one that perambulates
2 *chiefly British* **:** a baby carriage

per an·num \(ˌ)pər-ˈa-nəm\ *adverb* [Medieval Latin] (1536)
: in or for each year

per·bo·rate \(ˌ)pər-ˈbōr-ˌāt, -ˈbȯr-\ *noun* [International Scientific Vocabulary] (1881)
: a salt that is a compound of a borate with hydrogen peroxide

per·cale \(ˌ)pər-ˈkā(ə)l, ˈpər-ˌ; (ˌ)pər-ˈkal\ *noun* [Persian *pargālah*] (1840)
: a fine closely woven cotton cloth variously finished for clothing, sheeting, and industrial uses

per·ca·line \ˌpər-kə-ˈlēn\ *noun* [French, from *percale*] (circa 1858)
: a lightweight cotton fabric; *especially* **:** a glossy fabric used for bookbindings

per cap·i·ta \(ˌ)pər-ˈka-pə-tə\ *adverb or adjective* [Medieval Latin, by heads] (1682)
1 : equally to each individual
2 : per unit of population **:** by or for each person ⟨the highest income *per capita* of any state in the union⟩

per·ceive \pər-ˈsēv\ *transitive verb* **per·ceived; per·ceiv·ing** [Middle English, from Middle French *perceivre,* from Latin *percipere,* from *per-* thoroughly + *capere* to take — more at HEAVE] (14th century)
1 a : to attain awareness or understanding of **b :** to regard as being such ⟨*perceived* threats⟩ ⟨was *perceived* as a loser⟩
2 : to become aware of through the senses; *especially* **:** SEE, OBSERVE
— **per·ceiv·able** \-ˈsē-və-bəl\ *adjective*
— **per·ceiv·ably** \-blē\ *adverb*
— **per·ceiv·er** *noun*

[1]**per·cent** \pər-ˈsent\ *adverb* [earlier *per cent,* from *per* + Latin *centum* hundred — more at HUNDRED] (1568)
: in the hundred **:** of each hundred

[2]**percent** *noun, plural* **percent** *or* **percents** (1667)
1 *plural percent* **a :** one part in a hundred **b :** PERCENTAGE ⟨a large *percent* of his income⟩
2 *percents plural, British* **:** securities bearing a specified rate of interest

[3]**percent** *adjective* (1888)
1 : reckoned on the basis of a whole divided into one hundred parts
2 : paying interest at a specified percent

per·cent·age \pər-ˈsen-tij\ *noun* (circa 1789)
1 a : a part of a whole expressed in hundredths **b :** the result obtained by multiplying a number by a percent
2 a : a share of winnings or profits **b :** ADVANTAGE, PROFIT ⟨no *percentage* in going around looking like an old sack of laundry —Wallace Stegner⟩
3 : an indeterminate part **:** PROPORTION
4 a : PROBABILITY **b :** favorable odds

per·cen·tile \pər-ˈsen-ˌtīl\ *noun* (1885)
: a value on a scale of one hundred that indicates the percent of a distribution that is equal to or below it ⟨a score in the 95th *percentile*⟩

per cen·tum \pər-ˈsen-təm\ *noun* [*per* + Latin *centum*] (circa 1565)
: PERCENT

per·cept \ˈpər-ˌsept\ *noun* [back-formation from *perception*] (1837)
: an impression of an object obtained by use of the senses **:** SENSE-DATUM

per·cep·ti·ble \pər-ˈsep-tə-bəl\ *adjective* (1603)
: capable of being perceived especially by the senses ⟨a *perceptible* change in her tone⟩ ⟨the light became increasingly *perceptible*⟩ ☆
— **per·cep·ti·bil·i·ty** \-ˌsep-tə-ˈbi-lə-tē\ *noun*
— **per·cep·ti·bly** \-blē\ *adverb*

per·cep·tion \pər-ˈsep-shən\ *noun* [Latin *perception-, perceptio* act of perceiving, from *percipere*] (14th century)
1 a : a result of perceiving **:** OBSERVATION **b :** a mental image **:** CONCEPT
2 *obsolete* **:** CONSCIOUSNESS
3 a : awareness of the elements of environment through physical sensation ⟨color *perception*⟩ **b :** physical sensation interpreted in the light of experience
4 a : quick, acute, and intuitive cognition **:** APPRECIATION **b :** a capacity for comprehension
synonym see DISCERNMENT
— **per·cep·tion·al** \-shnəl, -shə-nᵊl\ *adjective*

per·cep·tive \pər-ˈsep-tiv\ *adjective* (1656)
1 : responsive to sensory stimuli **:** DISCERNING ⟨a *perceptive* eye⟩
2 a : capable of or exhibiting keen perception **:** OBSERVANT ⟨a *perceptive* scholar⟩ **b :** characterized by sympathetic understanding or insight
— **per·cep·tive·ly** *adverb*
— **per·cep·tive·ness** *noun*
— **per·cep·tiv·i·ty** \(ˌ)pər ˌsep-ˈti-və-tē\ *noun*

☆ SYNONYMS
Perceptible, sensible, palpable, tangible, appreciable, ponderable mean apprehensible as real or existent. PERCEPTIBLE applies to what can be discerned by the senses often to a minimal extent ⟨a *perceptible* difference in sound to a careful listener⟩. SENSIBLE applies to whatever is clearly apprehended through the senses or impresses itself strongly on the mind ⟨an abrupt, *sensible* drop in temperature⟩. PALPABLE applies either to what has physical substance or to what is obvious and unmistakable ⟨the tension in the air was almost *palpable*⟩. TANGIBLE suggests what is capable of being handled or grasped both physically and mentally ⟨no *tangible* evidence of UFOs⟩. APPRECIABLE applies to what is distinctly discernible by the senses or definitely measurable ⟨an *appreciable* increase in income⟩. PONDERABLE suggests having definitely measurable weight or importance ⟨exerted a *ponderable* influence on world events⟩.

□ USAGE
per *Per* occurs most frequently in business contexts; its use outside such contexts is often criticized but is quite widespread, especially in sense 2. Its most common and natural nonbusiness uses always involve figures, usually in relation to price ⟨$150 *per* performance⟩, automobiles ⟨32 miles *per* gallon⟩ ⟨55 miles *per* hour⟩, or sports ⟨averages 15 points and 9 rebounds *per* game⟩.

per·cep·tu·al \(ˌ)pər-ˈsep-chə-wəl, -chəl, -shwəl\ *adjective* [*percept* + *-ual* (as in *conceptual*)] (1878)
: of, relating to, or involving perception especially in relation to immediate sensory experience
— per·cep·tu·al·ly *adverb*

Per·ce·val \ˈpər-sə-vəl\ *noun* [Old French]
: a knight of King Arthur who wins a sight of the Holy Grail

¹perch \ˈpərch\ *noun* [Middle English *perche,* from Middle French, from Latin *pertica* pole] (14th century)
1 : a bar or peg on which something is hung
2 a : a roost for a bird **b :** a resting place or vantage point **:** SEAT **c :** a prominent position ⟨his new *perch* as president⟩
3 *chiefly British* **:** ROD 2

²perch (14th century)
intransitive verb
: to alight, settle, or rest on a perch, a height, or a precarious spot
transitive verb
: to place on a perch, a height, or a precarious spot

³perch *noun, plural* **perch** *or* **perch·es** [Middle English *perche,* from Middle French, from Latin *perca,* from Greek *perkē;* akin to Old High German *faro* colored, Latin *porcus,* a spiny fish] (14th century)
1 a : a small European freshwater bony fish (*Perca fluviatilis* of the family Percidae, the perch family) **b :** YELLOW PERCH
2 : any of numerous bony fishes (as of the families Percidae, Centrarchidae, and Serranidae)

per·chance \pər-ˈchan(t)s\ *adverb* [Middle English *per chance,* from Middle French, by chance] (14th century)
: PERHAPS, POSSIBLY

Per·che·ron \ˈpər-chə-ˌrän, -shə-\ *noun* [French] (1875)
: any of a breed of powerful rugged draft horses that originated in the Perche region of France

Percheron

per·chlo·rate \(ˌ)pər-ˈklōr-ˌāt, -ˈklȯr-, -ət\ *noun* [International Scientific Vocabulary] (1826)
: a salt or ester of perchloric acid

per·chlo·ric acid \(ˌ)pər-ˈklōr-ik-, -ˈklȯr-\ *noun* (1818)
: a fuming corrosive strong acid $HClO_4$ that is the most highly oxidized acid of chlorine and a powerful oxidizing agent when heated

per·chlo·ro·eth·y·lene \(ˌ)pər-ˌklōr-ō-ˈe-thə-ˌlēn, -ˌklȯr-\ *noun* (1873)
: a colorless nonflammable liquid C_2Cl_4 used often as a solvent in dry cleaning and for removal of grease from metals

per·cip·i·ence \pər-ˈsi-pē-ən(t)s\ *noun* (circa 1774)
: PERCEPTION 4

¹per·cip·i·ent \-ənt\ *noun* [Latin *percipient-, percipiens,* present participle of *percipere* to perceive] (1662)
1 : one that perceives
2 : a person on whose mind a telepathic impulse or message is held to fall

²percipient *adjective* (1692)
: capable of or characterized by perception **:** DISCERNING
— per·cip·i·ent·ly *adverb*

per·co·late \ˈpər-kə-ˌlāt, ÷-kyə-\ *verb* **-lat·ed; -lat·ing** [Latin *percolatus,* past participle of *percolare,* from *per-* through + *colare* to sieve — more at PER-, COLANDER] (1626)
transitive verb
1 a : to cause (a solvent) to pass through a permeable substance (as a powdered drug) especially for extracting a soluble constituent **b :** to prepare (coffee) in a percolator
2 : to be diffused through **:** PENETRATE
intransitive verb
1 : to ooze or trickle through a permeable substance **:** SEEP
2 a : to become percolated **b :** to become lively or effervescent
3 : to spread gradually ⟨allow the sunlight to *percolate* into our rooms —Norman Douglas⟩
4 : SIMMER 2a
— per·co·la·tion \ˌpər-kə-ˈlā-shən\ *noun*

per·co·la·tor \ˈpər-kə-ˌlā-tər, ÷-kyə-\ *noun* (circa 1842)
: one that percolates; *specifically* **:** a coffeepot in which boiling water rising through a tube is repeatedly deflected downward through a perforated basket containing ground coffee beans to extract their essence

per con·tra \(ˌ)pər-ˈkän-trə\ *adverb* [Italian, by the opposite side (of the ledger)] (1554)
1 a : on the contrary **b :** by way of contrast
2 : as an offset

per cu·ri·am decision \(ˌ)pər-ˈkyu̇r-ē-ˌäm-, -ˈku̇r-\ *noun* [Medieval Latin *per curiam,* literally, by the court] (1927)
: a very brief usually unanimous decision of a court rendered without elaborate discussion

per·cuss \pər-ˈkəs\ *transitive verb* [Latin *percussus,* past participle of *percutere*] (1560)
: to tap sharply; *especially* **:** to practice percussion on

per·cus·sion \pər-ˈkə-shən\ *noun* [Middle English, from Latin *percussion-, percussio,* from *percutere* to beat, from *per-* thoroughly + *quatere* to shake] (15th century)
1 : the act of percussing: as **a :** the striking of a percussion cap so as to set off the charge in a firearm **b :** the beating or striking of a musical instrument **c :** the act or technique of tapping the surface of a body part to learn the condition of the parts beneath by the resultant sound
2 : the striking of sound on the ear
3 : percussion instruments that form a section of a band or orchestra
— percussion *adjective*

percussion cap *noun* (1823)
: CAP 5

percussion instrument *noun* (1872)
: a musical instrument (as a drum, xylophone, or maraca) sounded by striking, shaking, or scraping

per·cus·sion·ist \pər-ˈkə-sh(ə-)nist\ *noun* (1939)
: one skilled in the playing of percussion instruments

per·cus·sive \pər-ˈkə-siv\ *adjective* (1793)
1 : of or relating to percussion; *especially* **:** operative or operated by striking
2 : having powerful impact
— per·cus·sive·ly *adverb*
— per·cus·sive·ness *noun*

per·cu·ta·ne·ous \ˌpər-kyu̇-ˈtā-nē-əs\ *adjective* (1887)
: effected or performed through the skin
— per·cu·ta·ne·ous·ly *adverb*

per·die \(ˌ)pər-ˈdē, per-\ *variant of* PARDIE

¹per di·em \(ˌ)pər-ˈdē-əm, -ˈdī-\ *adverb* [Medieval Latin] (1520)
: by the day **:** for each day

²per diem *adjective* (1809)
1 : based on use or service by the day **:** DAILY
2 : paid by the day

³per diem *noun, plural* **per diems** (1812)
1 : a daily allowance
2 : a daily fee

per·di·tion \pər-ˈdi-shən\ *noun* [Middle English *perdicion,* from Late Latin *perdition-, perditio,* from Latin *perdere* to destroy, from *per-* through + *dare* to give — more at PER-, DATE] (14th century)
1 a *archaic* **:** utter destruction **b** *obsolete* **:** LOSS
2 a : eternal damnation **b :** HELL

¹per·du *or* **per·due** \ˈpər-(ˌ)dü, -(ˌ)dyü; (ˌ)pər-ˈ\ *noun* [French *sentinelle perdue,* literally, lost sentinel] (1605)
obsolete **:** a soldier assigned to extremely hazardous duty

²per·du *or* **per·due** \per-dœ\ *adjective* [French *perdu,* masculine, & *perdue,* feminine, from past participle of *perdre* to lose, from Latin *perdere*] (1624)
: remaining out of sight

per·du·ra·ble \(ˌ)pər-ˈdu̇r-ə-bəl, -ˈdyu̇r-; ˈpər-jə-rə-\ *adjective* [Middle English, from Middle French, from Late Latin *perdurabilis,* from Latin *perdurare* to endure, from *per-* throughout + *durare* to last — more at DURING] (14th century)
: very durable
— per·du·ra·bil·i·ty \(ˌ)pər-ˌdu̇r-ə-ˈbil-ət-ē, -ˌdyu̇r-; ˌpər-jə-rə-\ *noun*
— per·du·ra·bly \(ˌ)pər-ˈdu̇r-ə-blē, -ˈdyu̇r-; ˈpər-jə-rə-\ *adverb*

per·dure \(ˌ)pər-ˈdu̇r, -ˈdyu̇r\ *intransitive verb* **per·dured; per·dur·ing** [Middle English, from Latin *perdurare*] (15th century)
: to continue to exist **:** LAST

per·e·gri·nate \ˈper-ə-grə-ˌnāt\ *verb* **-nat·ed; -nat·ing** (1593)
intransitive verb
: to travel especially on foot **:** WALK
transitive verb
: to walk or travel over **:** TRAVERSE
— per·e·gri·na·tion \ˌper-ə-grə-ˈnā-shən\ *noun*

¹per·e·grine \ˈper-ə-grən, -ˌgrēn\ *adjective* [Middle English, from Medieval Latin *peregrinus,* from Latin, foreign — more at PILGRIM] (14th century)
: having a tendency to wander

²peregrine *noun* (1555)
: a swift nearly cosmopolitan falcon (*Falco peregrinus*) that is much used in falconry — called also *peregrine falcon*
word history see PILGRIM

peregrine

pe·rei·on \pə-ˈrī-ˌän\ *or* **pe·re·on** \-ˈrē-\ *noun* [New Latin, from Greek *peraiōn,* present participle of *peraioun* to transport, from *peraios* situated beyond, from *pera* beyond; akin to Greek *poros* passage — more at FARE] (1855)
: the thorax or the seven metameres comprising the thorax of some crustaceans (as a decapod)

pe·reio·pod \pə-ˈrī-ə-ˌpäd\ *or* **pe·reo·pod** \-ˈrē-\ *noun* [New Latin *perion* + English *-pod*] (1893)
: an appendage of the pereion

pe·remp·to·ry \pə-ˈrem(p)-t(ə-)rē\ *adjective* [Late Latin & Latin; Late Latin *peremptorius,* from Latin, destructive, from *perimere* to take entirely, destroy, from *per-* thoroughly + *emere* to take — more at REDEEM] (15th century)
1 a : putting an end to or precluding a right of action, debate, or delay ⟨a *peremptory* mandamus⟩ **b :** admitting of no contradiction ⟨a *peremptory* conclusion based on absolute evidence⟩
2 : expressive of urgency or command ⟨a *peremptory* call⟩
3 a : characterized by often imperious or arrogant self-assurance ⟨how insolent of late he is become, how proud, how *peremptory* —Shakespeare⟩ **b :** indicative of a peremptory attitude

or nature **:** HAUGHTY ⟨a *peremptory* tone⟩ ⟨*peremptory* disregard of an objection⟩
synonym see MASTERFUL
— **pe·remp·to·ri·ly** \-ˈrem(p)-t(ə-)rə-lē; -ˌrem(p)-ˈtōr-ə-lē, -ˈtȯr-\ *adverb*
— **pe·remp·to·ri·ness** \-ˈrem(p)-t(ə-)rē-nəs\ *noun*

peremptory challenge *noun* (circa 1531)
: a challenge (as of a juror) made as of right without assigning any cause

pe·ren·nate \ˈper-ə-ˌnāt, pə-ˈre-ˌnāt\ *intransitive verb* **-nat·ed; -nat·ing** [Latin *perennatus,* past participle of *perennare,* from *perennis*] (circa 1623)
: to live over from season to season ⟨a *perennating* rhizome⟩
— **per·en·na·tion** \ˌper-ə-ˈnā-shən\ *noun*

pe·ren·ni·al \pə-ˈre-nē-əl\ *adjective* [Latin *perennis,* from *per-* throughout + *annus* year — more at PER-, ANNUAL] (1644)
1 : present at all seasons of the year
2 : persisting for several years usually with new herbaceous growth from a perennating part ⟨*perennial* asters⟩
3 a : PERSISTENT, ENDURING ⟨*perennial* traditions⟩ **b :** continuing without interruption **:** CONSTANT, PERPETUAL ⟨the *perennial* quest for certainty⟩ ⟨a *perennial* student⟩ **c :** regularly repeated or renewed **:** RECURRENT ⟨death is a *perennial* literary theme⟩
synonym see CONTINUAL
— **perennial** *noun*
— **pe·ren·ni·al·ly** \-nē-ə-lē\ *adverb*

per·e·stroi·ka \ˌper-ə-ˈstrȯi-kə\ *noun* [Russian *perestroĭka,* literally, restructuring] (1986)
: the policy of economic and governmental reform instituted by Mikhail Gorbachev in the Soviet Union during the mid-1980s

[1]**per·fect** \ˈpər-fikt\ *adjective* [Middle English *parfit,* from Old French, from Latin *perfectus,* from past participle of *perficere* to carry out, perfect, from *per-* thoroughly + *facere* to make, do — more at DO] (14th century)
1 a : being entirely without fault or defect **:** FLAWLESS ⟨a *perfect* diamond⟩ **b :** satisfying all requirements **:** ACCURATE **c :** corresponding to an ideal standard or abstract concept ⟨a *perfect* gentleman⟩ **d :** faithfully reproducing the original; *specifically* **:** LETTER-PERFECT **e :** legally valid
2 : EXPERT, PROFICIENT ⟨practice makes *perfect*⟩
3 a : PURE, TOTAL **b :** lacking in no essential detail **:** COMPLETE **c** *obsolete* **:** SANE **d :** ABSOLUTE, UNEQUIVOCAL ⟨enjoys *perfect* happiness⟩ **e :** of an extreme kind **:** UNMITIGATED ⟨a *perfect* brat⟩ ⟨an act of *perfect* foolishness⟩
4 *obsolete* **:** MATURE
5 : of, relating to, or constituting a verb form or verbal that expresses an action or state completed at the time of speaking or at a time spoken of
6 *obsolete* **a :** CERTAIN, SURE **b :** CONTENTED, SATISFIED
7 *of a musical interval* **:** belonging to the consonances unison, fourth, fifth, and octave which retain their character when inverted and when raised or lowered by a half step become augmented or diminished
8 a : sexually mature and fully differentiated ⟨a *perfect* insect⟩ **b :** having both stamens and pistils in the same flower ⟨a *perfect* flower⟩ ☆ ▪
— **per·fect·ness** \-fik(t)-nəs\ *noun*

[2]**per·fect** \pər-ˈfekt *also* ˈpər-fikt\ *transitive verb* (14th century)
1 : to bring to final form
2 : to make perfect **:** IMPROVE, REFINE
— **per·fect·er** *noun*

[3]**per·fect** \ˈpər-fikt\ *noun* (1841)
: the perfect tense of a language; *also* **:** a verb form in the perfect tense

per·fec·ta \pər-ˈfek-tə\ *noun* [American Spanish *quiniela perfecta* perfect quiniela] (1967)
: a bet in which the bettor picks the first and second place finishers in order — compare QUINIELA, TRIFECTA

perfect binding *noun* (1926)
: a book binding in which a layer of adhesive holds the pages and cover together
— **per·fect–bound** \ˈpər-fik(t)-ˈbau̇nd\ *adjective*

perfect game *noun* (circa 1949)
: a baseball game in which a pitcher allows no hits, no runs, and no opposing batter to reach first base

per·fect·ible \pər-ˈfek-tə-bəl *also* ˈpər-fik-\ *adjective* (1635)
: capable of improvement or perfection
— **per·fect·ibil·i·ty** \pər-ˌfek-tə-ˈbi-lə-tē *also* ˌpər-fik-\ *noun*

per·fec·tion \pər-ˈfek-shən\ *noun* [Middle English *perfeccioun,* from Old French *perfection,* from Latin *perfection-, perfectio,* from *perficere*] (13th century)
1 : the quality or state of being perfect: as **a :** freedom from fault or defect **:** FLAWLESSNESS **b :** MATURITY **c :** the quality or state of being saintly
2 a : an exemplification of supreme excellence **b :** an unsurpassable degree of accuracy or excellence
3 : the act or process of perfecting

per·fec·tion·ism \-shə-ˌni-zəm\ *noun* (circa 1846)
1 a : the doctrine that the perfection of moral character constitutes a person's highest good **b :** the theological doctrine that a state of freedom from sin is attainable on earth
2 : a disposition to regard anything short of perfection as unacceptable
— **per·fec·tion·ist** \-sh(ə-)nist\ *noun or adjective*
— **per·fec·tion·is·tic** \-ˌfek-shə-ˈnis-tik\ *adjective*

per·fec·tive \pər-ˈfek-tiv *also* ˈpər-fik-\ *adjective* (1596)
1 *archaic* **a :** tending to make perfect **b :** becoming better
2 : expressing action as complete or as implying the notion of completion, conclusion, or result ⟨*perfective* verb⟩
— **perfective** *noun*
— **per·fec·tive·ly** *adverb*
— **per·fec·tive·ness** *noun*
— **per·fec·tiv·i·ty** \pər-ˌfek-ˈti-və-tē *also* ˌpər-fik-\ *noun*

per·fect·ly \ˈpər-fik(t)-lē\ *adverb* (14th century)
1 : in a perfect manner
2 : to a complete or adequate extent **:** QUITE ⟨was *perfectly* happy until now⟩

perfect number *noun* (14th century)
: an integer (as 6 or 28) the sum of whose integral factors including 1 but excluding itself is equal to itself

per·fec·to \pər-ˈfek-(ˌ)tō\ *noun, plural* **-tos** [Spanish, perfect, from Latin *perfectus*] (1894)
: a cigar that is thick in the middle and tapers at each end

perfect participle *noun* (1862)
: PAST PARTICIPLE

perfect pitch *noun* (1949)
: ABSOLUTE PITCH 2

perfect square *noun* (1856)
: an integer (as 9 or 36) whose square root is an integer

per·fer·vid \(ˌ)pər-ˈfər-vəd, ˈpər-\ *adjective* [New Latin *perfervidus,* from Latin *per-* thoroughly + *fervidus* fervid] (1856)
: marked by overwrought or exaggerated emotion **:** excessively fervent
synonym see IMPASSIONED

per·fid·i·ous \(ˌ)pər-ˈfi-dē-əs\ *adjective* (1598)
: of, relating to, or characterized by perfidy
synonym see FAITHLESS
— **per·fid·i·ous·ly** *adverb*
— **per·fid·i·ous·ness** *noun*

per·fi·dy \ˈpər-fə-dē\ *noun, plural* **-dies** [Latin *perfidia,* from *perfidus* faithless, from *per-* detrimental to + *fides* faith — more at PER-, FAITH] (1592)
1 : the quality or state of being faithless or disloyal **:** TREACHERY
2 : an act or an instance of disloyalty

per·fo·li·ate \ˌpər-ˈfō-lē-ət, ˈpər-\ *adjective* [New Latin *perfoliata,* an herb having leaves pierced by the stem, from Latin *per* through + *foliata,* feminine of *foliatus* foliate] (1687)
: having the basal part naturally united around the stem ⟨a *perfoliate* leaf of a honeysuckle⟩

☆ SYNONYMS
Perfect, whole, entire, intact mean not lacking or faulty in any particular. PERFECT implies the soundness and the excellence of every part, element, or quality of a thing frequently as an unattainable or theoretical state ⟨a *perfect* set of teeth⟩. WHOLE suggests a completeness or perfection that can be sought, gained, or regained ⟨felt like a *whole* person again after vacation⟩. ENTIRE implies perfection deriving from integrity, soundness, or completeness of a thing ⟨the *entire* Beethoven corpus⟩. INTACT implies retention of perfection of a thing in its natural or original state ⟨the boat survived the storm *intact*⟩.

□ USAGE
perfect Can something be more perfect than something else? Can something be the most perfect of its kind? Writers and speakers of English have long thought so ⟨our sight is the most *perfect* and most delightful of all our senses —Joseph Addison (1712)⟩ ⟨some of their years were much more *perfect* than others —Benjamin Franklin (1752)⟩ ⟨the most *perfect* model of Roman discipline —Edward Gibbon (1776)⟩ ⟨the most *perfect* and precious . . . of antiquity —Thomas Jefferson (1786)⟩ ⟨in order to form a more *perfect* union —*U.S. Constitution* (1787)⟩. *Perfect* had been used in the comparative and superlative since the 14th century and, as the examples show, was common in the 18th century. Yet toward the end of the 18th century some grammarians decided that such use must be wrong in spite of its considerable literary authority. What is perfect is perfect, they said; the word does not admit of comparison. The argument claims to be based on logic. But did logicians agree ⟨man and the more *perfect* animals —John Stuart Mill (*A System of Logic*)⟩? The upshot of all this is that we have now inherited two traditions: a 200-year-old tradition of insisting that *perfect* cannot be compared and a 600-year-old tradition of using *perfect* in the comparative and superlative ⟨I was feeling more *perfect* with every Rainbow [trout] —Norman MacLean⟩ ⟨he felt . . . that circular motion was the most *perfect* —Stephen W. Hawking⟩. Resolution of this conflict lies in the fact that few things in reality are perfect, and when most people use *more perfect* or *most perfect* they are suggesting an approach to perfection. In other words, *more perfect* is equivalent to *more nearly perfect.* Sometimes writers find an excess of perfection ⟨it was perhaps too *perfect* in its unity —W. B. Yeats⟩ ⟨he is too *perfect* to believe —Lady Bird Johnson⟩. All of these uses are perfectly standard, but those who are diffident about *more perfect* can write, as some authors do, *more nearly perfect.*

per·fo·rate \'pər-fə-ˌrāt\ *verb* **-rat·ed; -rat·ing** [Latin *perforatus,* past participle of *perforare* to bore through, from *per-* through + *forare* to bore — more at BORE] (1538)
transitive verb
1 : to make a hole through ⟨an ulcer *perforates* the duodenal wall⟩; *especially* **:** to make a line of holes in to facilitate separation
2 : to pass through or into by or as if by making a hole
intransitive verb
: to penetrate a surface
— **per·fo·rate** \'pər-f(ə-)rət, -fə-ˌrāt\ *adjective*
— **per·fo·ra·tor** \-fə-ˌrā-tər\ *noun*

per·fo·rat·ed \-fə-ˌrā-təd\ *adjective* (1597)
1 : having a hole or perforations; *especially* **:** having a specified number of perforations in 20 millimeters ⟨the stamps are *perforated* 10⟩
2 : characterized by perforation ⟨a *perforated* ulcer⟩

per·fo·ra·tion \ˌpər-fə-'rā-shən\ *noun* (15th century)
1 : the act or process of perforating
2 a : a hole or pattern made by or as if by piercing or boring **b :** one of the series of holes between rows of postage stamps in a sheet that serve as an aid in separation

per·force \pər-'fōrs, -'fȯrs\ *adverb* [Middle English *par force,* from Middle French, by force] (14th century)
1 *obsolete* **:** by physical coercion
2 : by force of circumstances

per·form \pə(r)-'fȯrm\ *verb* [Middle English, from Anglo-French *performer,* alteration of Old French *perfournir,* from *per-* thoroughly (from Latin) + *fournir* to complete — more at FURNISH] (14th century)
transitive verb
1 : to adhere to the terms of **:** FULFILL ⟨*perform* a contract⟩
2 : CARRY OUT, DO
3 a : to do in a formal manner or according to prescribed ritual **b :** to give a rendition of **:** PRESENT
intransitive verb
1 : to carry out an action or pattern of behavior **:** ACT, FUNCTION
2 : to give a performance **:** PLAY ☆
— **per·form·abil·i·ty** \-ˌfȯr-mə-'bi-lə-tē\ *noun*
— **per·form·able** \-'fȯr-mə-bəl\ *adjective*
— **per·form·er** \-'fȯr-mər\ *noun*

per·for·mance \pə(r)-'fȯr-mən(t)s\ *noun* (15th century)
1 a : the execution of an action **b :** something accomplished **:** DEED, FEAT
2 : the fulfillment of a claim, promise, or request **:** IMPLEMENTATION
3 a : the action of representing a character in a play **b :** a public presentation or exhibition ⟨a benefit *performance*⟩
4 a : the ability to perform **:** EFFICIENCY **b :** the manner in which a mechanism performs ⟨engine *performance*⟩
5 : the manner of reacting to stimuli **:** BEHAVIOR
6 : the linguistic behavior of an individual **:** PAROLE; *also* **:** the ability to speak a certain language — compare COMPETENCE 3

performance art *noun* (1971)
: a nontraditional art form that consists of or features a performance by the artist
— **performance artist** *noun*

per·for·ma·tive \-'fȯr-mə-tiv\ *adjective* (1955)
: being or relating to an expression that serves to effect a transaction or that constitutes the performance of the specified act by virtue of its utterance ⟨*performative* verbs such as *promise* and *congratulate*⟩ — compare CONSTATIVE
— **performative** *noun*

per·for·ma·to·ry \-mə-ˌtōr-ē, -ˌtȯr-\ *adjective* (1949)
: PERFORMATIVE; *also* **:** of or relating to performance

performing *adjective* (1889)
: of, relating to, or constituting an art (as drama) that involves public performance ⟨the *performing* arts⟩

¹per·fume \'pər-ˌfyüm, (ˌ)pər-'\ *noun* [Middle French *perfum,* probably from Old Provençal, from *perfumar* to perfume, from *per-* thoroughly (from Latin) + *fumar* to smoke, from Latin *fumare,* from *fumus* smoke — more at FUME] (1533)
1 : the scent of something sweet-smelling
2 : a substance that emits a pleasant odor; *especially* **:** a fluid preparation of natural essences (as from plants or animals) or synthetics and a fixative used for scenting
synonym see FRAGRANCE

²per·fume \(ˌ)pər-'fyüm, 'pər-ˌ\ *transitive verb* **per·fumed; per·fum·ing** (1538)
: to fill or imbue with an odor

per·fum·er \pə(r)-'fyü-mər, 'pər-ˌ\ *noun* (circa 1580)
: one that makes or sells perfumes

per·fum·ery \pə(r)-'fyü-mə-rē, -'fyüm-rē\ *noun, plural* **-er·ies** (1800)
1 a : the art or process of making perfume **b :** the products made by a perfumer
2 : an establishment where perfumes are made

per·func·to·ry \pər-'fəŋ(k)-t(ə-)rē\ *adjective* [Late Latin *perfunctorius,* from Latin *perfungi* to accomplish, get through with, from *per-* through + *fungi* to perform — more at PER-, FUNCTION] (1593)
1 : characterized by routine or superficiality **:** MECHANICAL ⟨a *perfunctory* smile⟩
2 : lacking in interest or enthusiasm
— **per·func·to·ri·ly** \-t(ə-)rə-lē\ *adverb*
— **per·func·to·ri·ness** \-t(ə-)rē-nəs\ *noun*

per·fus·ate \(ˌ)pər-'fyü-ˌzāt, -zət\ *noun* (1915)
: a fluid (as a solution pumped through the heart) that is perfused

per·fuse \(ˌ)pər-'fyüz\ *transitive verb* **per·fused; per·fus·ing** [Middle English, from Latin *perfusus,* past participle of *perfundere* to pour over, from *per-* through + *fundere* to pour — more at FOUND] (15th century)
1 : SUFFUSE
2 a : to cause to flow or spread **:** DIFFUSE **b :** to force a fluid through (an organ or tissue) especially by way of the blood vessels
— **per·fu·sion** \-'fyü-zhən\ *noun*

per·fu·sion·ist \(ˌ)pər-'fyü-zhə-nist\ *noun* (1964)
: a certified medical technician responsible for extracorporeal oxygenation of the blood during open-heart surgery and for the operation and maintenance of equipment (as a heart-lung machine) controlling it

per·go·la \'pər-gə-lə, pər-'gō-\ *noun* [Italian, from Latin *pergula*] (1675)
1 : ARBOR, TRELLIS
2 : a structure usually consisting of parallel colonnades supporting an open roof of girders and cross rafters

¹per·haps \pər-'haps, 'praps\ *adverb* [*per* + *hap*] (1528)
: possibly but not certainly **:** MAYBE

²perhaps *noun* (1534)
: something open to doubt or conjecture

pe·ri \'pir-ē\ *noun* [Persian *perī* fairy, genius, from Middle Persian *parīk;* akin to Avestan *pairikā* sorceress] (circa 1780)
1 : a supernatural being in Persian folklore descended from fallen angels and excluded from paradise until penance is accomplished
2 : a beautiful and graceful girl

peri- *prefix* [Latin, from Greek, around, in excess, from *peri;* akin to Greek *peran* to pass through — more at FARE]
1 : all around **:** about ⟨*peri*scope⟩
2 : near ⟨*peri*helion⟩
3 : enclosing **:** surrounding ⟨*peri*neurium⟩

peri·anth \'per-ē-ˌan(t)th\ *noun* [New Latin *perianthium,* from *peri-* + Greek *anthos* flower — more at ANTHOLOGY] (1828)
: the floral structure comprised of the calyx and corolla especially when the two whorls are fused — see FLOWER illustration

peri·apt \'per-ē-ˌapt\ *noun* [Middle French or Greek; Middle French *periapte,* from Greek *periapton,* from *periaptein* to fasten around (oneself), from *peri-* + *haptein* to fasten] (1584)
: AMULET

peri·car·di·al \ˌper-ə-'kär-dē-əl\ *adjective* (1654)
: of, relating to, or affecting the pericardium; *also* **:** situated around the heart

peri·car·di·tis \-ˌkär-'dī-təs\ *noun* [New Latin] (circa 1799)
: inflammation of the pericardium

peri·car·di·um \-'kär-dē-əm\ *noun, plural* **-dia** \-dē-ə\ [Middle English, from Medieval Latin, from Greek *perikardion,* neuter of *perikardios* around the heart, from *peri-* + *kardia* heart — more at HEART] (15th century)
1 : the conical sac of serous membrane that encloses the heart and the roots of the great blood vessels of vertebrates
2 : a cavity or space that contains the heart of an invertebrate and in arthropods is a part of the hemocoel

peri·carp \'per-ə-ˌkärp\ *noun* [New Latin *pericarpium,* from Greek *perikarpion* pod, from *peri-* + *-karpion* -carp] (1759)
: the ripened and variously modified walls of a plant ovary — see ENDOCARP illustration

peri·chon·dri·um \ˌper-ə-'kän-drē-əm\ *noun, plural* **-dria** \-drē-ə\ [New Latin, from *peri-* + Greek *chondros* grain, cartilage] (1741)
: the membrane of fibrous connective tissue that invests cartilage except at joints
— **peri·chon·dral** \-drəl\ *adjective*

pe·ric·o·pe \pə-'ri-kə-pē\ *noun* [Late Latin, from Greek *perikopē* section, from *peri-* + *kopē* act of cutting, from *koptein* to cut — more at CAPON] (1658)
: a selection from a book; *specifically* **:** LECTION 1

peri·cra·ni·um \ˌper-ə-'krā-nē-əm\ *noun, plural* **-nia** \-nē-ə\ [Middle English *pericraneum,* from Medieval Latin, from Greek *perikranion,* neuter of *perikranios* around the skull, from *peri-* + *kranion* skull — more at CRANIUM] (15th century)

☆ SYNONYMS

Perform, execute, discharge, accomplish, achieve, effect, fulfill mean to carry out or into effect. PERFORM implies action that follows established patterns or procedures or fulfills agreed-upon requirements and often connotes special skill ⟨*performed* gymnastics⟩. EXECUTE stresses the carrying out of what exists in plan or in intent ⟨*executed* the hit-and-run⟩. DISCHARGE implies execution and completion of appointed duties or tasks ⟨*discharged* his duties⟩. ACCOMPLISH stresses the successful completion of a process rather than the means of carrying it out ⟨*accomplished* everything they set out to do⟩. ACHIEVE adds to ACCOMPLISH the implication of conquered difficulties ⟨*achieve* greatness⟩. EFFECT adds to ACHIEVE an emphasis on the inherent force in the agent capable of surmounting obstacles ⟨*effected* sweeping reforms⟩. FULFILL implies a complete realization of ends or possibilities ⟨*fulfilled* their ambitions⟩.

: the external periosteum of the skull
— **peri·cra·ni·al** \-nē-əl\ *adjective*

peri·cy·cle \ˈper-ə-ˌsī-kəl\ *noun* [French *péricycle,* from Greek *perikyklos* spherical, from *peri-* + *kyklos* circle — more at WHEEL] (circa 1892)
: a thin layer of parenchymatous or sclerenchymatous cells that surrounds the stele in most vascular plants
— **peri·cy·clic** \ˌper-ə-ˈsī-klik, -ˈsi-\ *adjective*

peri·derm \ˈper-ə-ˌdərm\ *noun* [New Latin *peridermis,* from *peri-* + *-dermis*] (1849)
: an outer layer of tissue; *especially* **:** a cortical protective layer of many roots and stems that typically consists of phellem, phellogen, and phelloderm

pe·rid·i·um \pə-ˈri-dē-əm\ *noun, plural* **pe·rid·ia** \-dē-ə\ [New Latin, from Greek *pēridion,* diminutive of *pēra* leather bag] (circa 1823)
: the outer envelope of the sporophore of many fungi

per·i·dot \ˈper-ə-ˌdät, -ˌdō(t)\ *noun* [French *péridot,* from Old French *peritot*] (circa 1706)
: a deep yellowish green transparent olivine used as a gem
— **per·i·do·tic** \ˌper-ə-ˈdä-tik, -ˈdō-\ *adjective*

pe·ri·do·tite \pə-ˈri-də-ˌtīt\ *noun* [French *péridotite,* from *péridot*] (1878)
: any of a group of granitoid igneous rocks composed of ferromagnesian minerals and especially olivine
— **pe·ri·do·tit·ic** \pə-ˌri-də-ˈti-tik\ *adjective*

peri·gee \ˈper-ə-(ˌ)jē\ *noun* [Middle French, from New Latin *perigeum,* from Greek *perigeion,* from neuter of *perigeios* near the earth, from *peri-* + *gē* earth] (1594)
: the point in the orbit of an object (as a satellite) orbiting the earth that is nearest to the center of the earth; *also* **:** the point nearest a planet or a satellite (as the moon) reached by an object orbiting it — compare APOGEE
— **peri·ge·an** \ˌper-ə-ˈjē-ən\ *adjective*

pe·rig·y·nous \pə-ˈri-jə-nəs\ *adjective* [New Latin *perigynus,* from *peri-* + *-gynus* -gynous] (1807)
: borne on a ring or cup of the receptacle surrounding a pistil ⟨*perigynous* petals⟩; *also* **:** having perigynous stamens and petals ⟨*perigynous* flowers⟩
— **pe·rig·y·ny** \-nē\ *noun*

peri·he·lion \ˌper-ə-ˈhēl-yən\ *noun, plural* **-he·lia** \-ˈhēl-yə\ [New Latin, from *peri-* + Greek *hēlios* sun — more at SOLAR] (1666)
: the point in the path of a celestial body (as a planet) that is nearest to the sun — compare APHELION
— **peri·he·lial** \-ˈhēl-yəl\ *adjective*

peri·kar·y·on \-ˈkar-ē-ˌän, -ən\ *noun, plural* **-kar·ya** \-ē-ə\ [New Latin, from *peri-* + Greek *karyon* nut, kernel — more at CAREEN] (1897)
: CELL BODY
— **peri·kar·y·al** \-ē-əl\ *adjective*

[1]**per·il** \ˈper-əl\ *noun* [Middle English, from Old French, from Latin *periculum* — more at FEAR] (13th century)
1 : exposure to the risk of being injured, destroyed, or lost **:** DANGER ⟨fire put the city in *peril*⟩
2 : something that imperils **:** RISK ⟨lessen the *perils* of the streets⟩

[2]**peril** *transitive verb* **-iled** *also* **-illed; -il·ing** *also* **-il·ling** (1567)
: to expose to danger

pe·ril·la \pə-ˈri-lə\ *noun* [New Latin] (1900)
: any of a genus (*Perilla*) of Asian mints that have a bilabiate fruiting calyx and rugose nutlets

perilla oil *noun* (1917)
: a light yellow drying oil that is obtained from seeds of perillas

per·il·ous \ˈper-ə-ləs\ *adjective* (14th century)
: full of or involving peril **:** HAZARDOUS
synonym see DANGEROUS
— **per·il·ous·ly** *adverb*
— **per·il·ous·ness** *noun*

peri·lune \ˈper-ə-ˌlün\ *noun* [*peri-* + Latin *luna* moon — more at LUNAR] (1960)
: the point in the path of a body orbiting the moon that is nearest to the center of the moon — compare APOLUNE

peri·lymph \-ˌlim(p)f\ *noun* [International Scientific Vocabulary] (circa 1839)
: the fluid between the membranous and bony labyrinths of the ear

pe·rim·e·ter \pə-ˈri-mə-tər\ *noun* [Middle English *perimetre,* from Latin *perimetros,* from Greek, from *peri-* + *metron* measure — more at MEASURE] (15th century)
1 a : the boundary of a closed plane figure **b :** the length of a perimeter
2 : a line or strip bounding or protecting an area
3 : outer limits — often used in plural

peri·my·si·um \ˌper-ə-ˈmi-zhē-əm, -zē-\ *noun, plural* **-sia** \-zhē-ə, -zē-\ [New Latin, irregular from *peri-* + Greek *mys* mouse, muscle — more at MOUSE] (circa 1842)
: the connective-tissue sheath that surrounds a muscle and forms sheaths for the bundles of muscle fibers

peri·na·tal \-ˈnā-tᵊl\ *adjective* (1952)
: occurring in, concerned with, or being in the period around the time of birth ⟨*perinatal* mortality⟩ ⟨*perinatal* care⟩
— **peri·na·tal·ly** \-tᵊl-(l)ē\ *adverb*

per·i·ne·um \ˌper-ə-ˈnē-əm\ *noun, plural* **-nea** \-ˈnē-ə\ [Middle English, from Late Latin *perinaion,* from Greek, from *peri-* + *inan* to empty out; perhaps akin to Sanskrit *iṣṇāti* he sets in motion] (15th century)
: an area of tissue that marks externally the approximate boundary of the outlet of the pelvis and gives passage to the urinogenital ducts and rectum; *also* **:** the area between the anus and the posterior part of the external genitalia
— **per·i·ne·al** \-ˈne-əl\ *adjective*

peri·neu·ri·um \ˌper-ə-ˈnu̇r-ē-əm, -ˈnyu̇r-\ *noun, plural* **-ria** \-ē-ə\ [New Latin, from *peri-* + Greek *neuron* nerve — more at NERVE] (circa 1842)
: the connective-tissue sheath that surrounds a bundle of nerve fibers

[1]**pe·ri·od** \ˈpir-ē-əd\ *noun* [Middle English *pariode,* from Middle French *periode,* from Medieval Latin, Latin, & Greek; Medieval Latin *periodus* period of time, punctuation mark, from Latin & Greek; Latin, rhetorical period, from Greek *periodos* circuit, period of time, rhetorical period, from *peri-* + *hodos* way] (circa 1530)
1 : the completion of a cycle, a series of events, or a single action **:** CONCLUSION
2 a (1) **:** an utterance from one full stop to another **:** SENTENCE (2) **:** a well-proportioned sentence of several clauses (3) **:** PERIODIC SENTENCE **b :** a musical structure or melodic section usually composed of two or more contrasting or complementary phrases and ending with a cadence
3 a : the full pause with which the utterance of a sentence closes **b :** END, STOP
4 *obsolete* **:** GOAL, PURPOSE
5 a : a point . used to mark the end (as of a declarative sentence or an abbreviation) — often used interjectionally to emphasize that no more need be said ⟨I don't remember — *period*⟩ **b :** a rhythmical unit in Greek verse composed of a series of two or more cola
6 a : a portion of time determined by some recurring phenomenon **b** (1) **:** the interval of time required for a cyclic motion or phenomenon to complete a cycle and begin to repeat itself (2) **:** a number *k* that does not change the value of a periodic function *f* when added to the independent variable; *especially* **:** the smallest such number **c :** a single cyclic occurrence of menstruation
7 a : a chronological division **:** STAGE **b :** a division of geologic time longer than an epoch and included in an era **c :** a stage of culture having a definable place in time and space
8 a : one of the divisions of the academic day **b :** one of the divisions of the playing time of a game ☆

[2]**period** *adjective* (1905)
: of, relating to, or representing a particular historical period ⟨*period* furniture⟩ ⟨*period* costumes⟩

pe·ri·od·ic \ˌpir-ē-ˈä-dik\ *adjective* (1642)
1 : occurring or recurring at regular intervals
2 a : consisting of or containing a series of repeated stages, processes, or digits **:** CYCLIC ⟨*periodic* decimals⟩ ⟨a *periodic* vibration⟩ **b :** being a function any value of which recurs at regular intervals
3 : expressed in or characterized by periodic sentences

per·iod·ic acid \ˌpər-(ˌ)ī-ˈä-dik-\ *noun* [International Scientific Vocabulary *per-* + *iodic*] (1836)
: any of the strongly oxidizing acids (as H_5IO_6 or HIO_4) that are the most highly oxidized acids of iodine

[1]**pe·ri·od·i·cal** \ˌpir-ē-ˈä-di-kəl\ *adjective* (1601)
1 : PERIODIC 1
2 a : published with a fixed interval between the issues or numbers **b :** published in, characteristic of, or connected with a periodical

[2]**periodical** *noun* (1798)
: a periodical publication

periodical cicada *noun* (1890)
: SEVENTEEN-YEAR LOCUST

pe·ri·od·i·cal·ly \ˌpir-ē-ˈä-di-k(ə-)lē\ *adverb* (1646)
1 : at regular intervals of time
2 : from time to time **:** FREQUENTLY

pe·ri·od·ic·i·ty \ˌpir-ē-ə-ˈdi-sə-tē\ *noun* (1833)
: the quality, state, or fact of being regularly recurrent or having periods

periodic law *noun* (1872)
: a law in chemistry: the elements when arranged in the order of their atomic numbers show a periodic variation of atomic structure and of most of their properties

periodic sentence *noun* (circa 1928)
: a usually complex sentence that has no subordinate or trailing elements following its principal clause (as in "yesterday while I was walking down the street, I saw him")

> ☆ SYNONYMS
> **Period, epoch, era, age** mean a division of time. PERIOD may designate an extent of time of any length ⟨*periods* of economic prosperity⟩. EPOCH applies to a period begun or set off by some significant or striking quality, change, or series of events ⟨the steam engine marked a new *epoch* in industry⟩. ERA suggests a period of history marked by a new or distinct order of things ⟨the *era* of global communications⟩. AGE is used frequently of a fairly definite period dominated by a prominent figure or feature ⟨the *age* of Samuel Johnson⟩.

PERIODIC TABLE

This is a common long form of the table. Roman numerals and letters heading the vertical columns indicate the groups (there are differences of opinion regarding the letter designations, those given here being probably the most generally used). The horizontal rows represent the periods, with two series removed from the two very long periods and represented below the main table. Atomic numbers are given above the symbols for the elements. Compare ELEMENT table.

IA[1]	IIA[2]	IIIB	IVB	VB	VIB	VIIB	VIII	VIII	VIII	IB	IIB	IIIA	IVA	VA	VIA	VIIA[3]	Zero[4]
1 H																1 H	2 He
3 Li	4 Be											5 B	6 C	7 N	8 O	9 F	10 Ne
11 Na	12 Mg											13 Al	14 Si	15 P	16 S	17 Cl	18 Ar
19 K	20 Ca	21 Sc	22 Ti	23 V	24 Cr	25 Mn	26 Fe	27 Co	28 Ni	29 Cu	30 Zn	31 Ga	32 Ge	33 As	34 Se	35 Br	36 Kr
37 Rb	38 Sr	39 Y	40 Zr	41 Nb	42 Mo	43 Tc	44 Ru	45 Rh	46 Pd	47 Ag	48 Cd	49 In	50 Sn	51 Sb	52 Te	53 I	54 Xe
55 Cs	56 Ba	57 *La	72 Hf	73 Ta	74 W	75 Re	76 Os	77 Ir	78 Pt	79 Au	80 Hg	81 Tl	82 Pb	83 Bi	84 Po	85 At	86 Rn
87 Fr	88 Ra	89 #Ac	104 Rf	105 Db	106 Sg	107 Bh	108 Hs	109 Mt									

*LANTHANIDE SERIES	58 Ce	59 Pr	60 Nd	61 Pm	62 Sm	63 Eu	64 Gd	65 Tb	66 Dy	67 Ho	68 Er	69 Tm	70 Yb	71 Lu
#ACTINIDE SERIES	90 Th	91 Pa	92 U	93 Np	94 Pu	95 Am	96 Cm	97 Bk	98 Cf	99 Es	100 Fm	101 Md	102 No	103 Lr

[1] Group IA (excluding hydrogen) comprises the alkali metals.
[2] Group IIA comprises the alkaline-earth metals.
[3] Group VIIA (excluding hydrogen) comprises the halogens.
[4] Group Zero comprises the noble gases.

periodic table *noun* (1895)
: an arrangement of chemical elements based on the periodic law

pe·ri·od·i·za·tion \ˌpir-ē-ə-də-ˈzā-shən\ *noun* (1938)
: division (as of history) into periods

peri·odon·tal \ˌper-ē-ō-ˈdän-tᵊl\ *adjective* (1854)
1 : investing or surrounding a tooth
2 : of or affecting periodontal tissues or regions ⟨*periodontal* diseases⟩
— peri·odon·tal·ly \-tᵊl-ē\ *adverb*

periodontal membrane *noun* (circa 1903)
: the fibrous connective-tissue layer covering the cementum of a tooth and holding it in place in the jawbone

peri·odon·tics \-ˈdän-tiks\ *noun plural but singular or plural in construction* [New Latin *periodontium* periodontal tissue, from *peri-* + Greek *odont-, odous, odōn* tooth — more at TOOTH] (circa 1944)
: a branch of dentistry that deals with diseases of the supporting and investing structures of the teeth including the gums, cementum, periodontal membranes, and alveolar bone
— peri·odon·tist \-ˈdän-tist\ *noun*

peri·odon·tol·o·gy \-ˌdän-ˈtä-lə-jē\ *noun* (1914)
: PERIODONTICS

period piece *noun* (1940)
: a piece (as of fiction, art, furniture, or music) whose special value lies in its evocation of an historical period

peri·onych·i·um \ˌper-ē-ō-ˈni-kē-əm\ *noun, plural* **-ia** \-kē-ə\ [New Latin, from *peri-* + Greek *onych-, onyx* nail — more at NAIL] (circa 1879)
: the tissue bordering the root and sides of a fingernail or toenail

peri·os·te·al \ˌper-ē-ˈäs-tē-əl\ *adjective* (1830)
1 : situated around or produced external to bone
2 : of, relating to, or involving the periosteum

peri·os·te·um \-tē-əm\ *noun, plural* **-tea** \-tē-ə\ [New Latin, from Late Latin *periosteon,* from Greek, neuter of *periosteos* around the bone, from *peri-* + *osteon* bone — more at OSSEOUS] (1597)
: the membrane of connective tissue that closely invests all bones except at the articular surfaces

peri·os·ti·tis \-ˌäs-ˈtī-təs\ *noun* [New Latin] (1843)
: inflammation of the periosteum

¹peri·pa·tet·ic \ˌper-ə-pə-ˈte-tik\ *noun* (15th century)
1 *capitalized* **:** a follower of Aristotle or adherent of Aristotelianism
2 : PEDESTRIAN, ITINERANT
3 *plural* **:** movement or journeys hither and thither

²peripatetic *adjective* [Middle French & Latin; Middle French *peripatetique,* from Latin *peripateticus,* from Greek *peripatētikos,* from *peripatein* to walk up and down, discourse while pacing (as did Aristotle), from *peri-* + *patein* to tread; akin to Sanskrit *patha* path — more at FIND] (1566)
1 *capitalized* **:** ARISTOTELIAN
2 a : of, relating to, or given to walking **b :** moving or traveling from place to place **:** ITINERANT
— peri·pa·tet·i·cal·ly \-ti-k(ə-)lē\ *adverb*
— Peri·pa·tet·i·cism \-ˈte-tə-ˌsi-zəm\ *noun*

pe·rip·a·tus \pə-ˈri-pə-təs\ *noun* [New Latin, genus name, from Greek *peripatos* act of walking about, from *peri-* + *patein* to tread] (circa 1931)
: any of a class or phylum (Onychophora) of primitive tropical wormlike invertebrates that appear intermediate between annelid worms and arthropods

peri·pe·teia \ˌper-ə-pə-ˈtē-ə, -ˈtī-\ *noun* [Greek, from *peripiptein* to fall around, change suddenly, from *peri-* + *piptein* to fall — more at FEATHER] (1591)
: a sudden or unexpected reversal of circumstances or situation especially in a literary work

pe·rip·e·ty \pə-ˈri-pə-tē\ *noun, plural* **-ties** (1753)
: PERIPETEIA

¹pe·riph·er·al \pə-ˈri-f(ə-)rəl\ *adjective* (1808)
1 : of, relating to, involving, or forming a periphery or surface part
2 : of, relating to, or being part of the peripheral nervous system ⟨*peripheral* nerves⟩
3 : of, relating to, or being the outer part of the field of vision ⟨good *peripheral* vision⟩
4 : AUXILIARY, SUPPLEMENTARY ⟨*peripheral* equipment⟩; *also* **:** of or relating to computer peripherals
— pe·riph·er·al·ly *adverb*

²peripheral *noun* (1966)
: a device connected to a computer to provide communication (as input and output) or auxiliary functions (as additional storage)

peripheral nervous system *noun* (1935)
: the part of the nervous system that is outside the central nervous system and comprises the cranial nerves excepting the optic nerve, the spinal nerves, and the autonomic nervous system

pe·riph·ery \pə-ˈri-f(ə-)rē\ *noun, plural* **-er·ies** [Middle French *peripherie,* from Late

Latin *peripheria,* from Greek *periphereia,* from *peripherein* to carry around, from *peri-* + *pherein* to carry — more at BEAR] (1571)
1 : the perimeter of a circle or other closed curve; *also* **:** the perimeter of a polygon
2 : the external boundary or surface of a body
3 a : the outward bounds of something as distinguished from its internal regions or center **:** CONFINES **b :** an area lying beyond the strict limits of a thing

pe·riph·ra·sis \pə-ˈri-frə-səs\ *noun, plural* **-ra·ses** \-ˌsēz\ [Latin, from Greek, from *periphrazein* to express periphrastically, from *peri-* + *phrazein* to point out] (1533)
1 : use of a longer phrasing in place of a possible shorter form of expression
2 : an instance of periphrasis

peri·phras·tic \ˌper-ə-ˈfras-tik\ *adjective* (1805)
1 : of, relating to, or characterized by periphrasis
2 : formed by the use of function words or auxiliaries instead of by inflection ⟨*more fair* is a *periphrastic* comparative⟩
— peri·phras·ti·cal·ly \-ti-k(ə-)lē\ *adverb*

pe·riph·y·ton \pə-ˈri-fə-ˌtän\ *noun* [New Latin, from Greek *periphytos* (verbal of *periphyein* to grow around, from *peri-* + *phyein* to bring forth, grow) + *-on* (as in *plankton*) — more at BE] (1945)
: organisms (as some algae) that live attached to underwater surfaces
— peri·phyt·ic \ˌper-ə-ˈfi-tik\ *adjective*

peri·plast \ˈper-ə-ˌplast\ *noun* (1853)
: PLASMA MEMBRANE; *also* **:** a proteinaceous subcellular layer below the plasma membrane especially of a euglena

pe·rique \pə-ˈrēk\ *noun* [Louisiana French *périque*] (1882)
: an aromatic fermented Louisiana tobacco used in smoking mixtures

peri·sarc \ˈper-ə-ˌsärk\ *noun* [International Scientific Vocabulary *peri-* + Greek *sark-, sarx* flesh — more at SARCASM] (circa 1871)
: the outer usually chitinous integument of a hydroid

peri·scope \-ˌskōp\ *noun* [International Scientific Vocabulary] (1879)
: a tubular optical instrument containing lenses and mirrors by which an observer obtains an otherwise obstructed field of view

peri·scop·ic \ˌper-ə-ˈskä-pik\ *adjective* (1804)
1 : providing a view all around or on all sides ⟨*periscopic* lens⟩
2 : of or relating to a periscope

per·ish \ˈper-ish\ *verb* [Middle English *perisshen,* from Old French *periss-,* stem of *perir,* from Latin *perire,* from *per-* detrimentally + *ire* to go — more at PER-, ISSUE] (13th century)

periscope

intransitive verb
1 : to become destroyed or ruined **:** DIE ⟨recollection of a past already long since *perished* —Philip Sherrard⟩ ⟨guard against your mistakes or your attempts (*perish* the thought) to cheat —C. B. Davis⟩
2 *chiefly British* **:** DETERIORATE, SPOIL
transitive verb
1 *chiefly British* **:** to cause to die **:** DESTROY
2 : WEAKEN, BENUMB

per·ish·able \ˈper-i-shə-bəl\ *adjective* (circa 1611)
: liable to perish **:** liable to spoil or decay ⟨such *perishable* products as fruit, vegetables, butter, and eggs⟩
— per·ish·abil·i·ty \ˌper-i-shə-ˈbi-lə-tē\ *noun*
— perishable *noun*

pe·ris·so·dac·tyl \pə-ˈri-sə-ˌdak-tᵊl\ *noun* [New Latin *Perissodactyla,* from Greek *perissos* excessive, odd in number + *daktylos* finger, toe] (1849)
: any of an order (Perissodactyla) of nonruminant ungulate mammals (as a horse, a tapir, or a rhinoceros) that usually have an odd number of toes, molar teeth with transverse ridges on the grinding surface, and the posterior premolars resembling true molars
— perissodactyl *adjective*

peri·stal·sis \ˌper-ə-ˈstȯl-səs, -ˈstäl-, -ˈstal-\ *noun, plural* **-stal·ses** \-ˌsēz\ [New Latin, from Greek *peristaltikos* peristaltic] (1859)
: successive waves of involuntary contraction passing along the walls of a hollow muscular structure (as the esophagus or intestine) and forcing the contents onward

peri·stal·tic \-tik\ *adjective* [Greek *peristaltikos,* from *peristellein* to wrap around, from *peri-* + *stellein* to place] (1655)
1 : of, relating to, resulting from, or being peristalsis
2 : having an action suggestive of peristalsis

peristaltic pump *noun* (1962)
: a pump in which fluid is forced along by waves of contraction produced mechanically on flexible tubing

peri·stome \ˈper-ə-ˌstōm\ *noun* [New Latin *peristoma,* from *peri-* + Greek *stoma* mouth — more at STOMACH] (circa 1796)
1 : the fringe of teeth surrounding the orifice of a moss capsule
2 : the region around the mouth in various invertebrates
— peri·sto·mi·al \ˌper-ə-ˈstō-mē-əl\ *adjective*

peri·style \ˈper-ə-ˌstīl\ *noun* [French *péristyle,* from Latin *peristylum,* from Greek *peristylon,* from neuter of *peristylos* surrounded by a colonnade, from *peri-* + *stylos* pillar — more at STEER] (1612)
1 : a colonnade surrounding a building or court
2 : an open space enclosed by a colonnade

peri·the·ci·um \ˌper-ə-ˈthē-shē-əm, -sē-\ *noun, plural* **-cia** \-shē-ə, -sē-\ [New Latin, from *peri-* + Greek *thēkion,* diminutive of *thēkē* case — more at TICK] (circa 1832)
: a spherical, cylindrical, or flask-shaped hollow fruiting body in various ascomycetous fungi that contains the asci and usually opens by a terminal pore
— peri·the·cial \-ˈthē-sh(ē-)əl, -sē-əl\ *adjective*

peri·to·ne·um \ˌper-ə-tᵊn-ˈē-əm\ *noun, plural* **-ne·ums** \-ˈē-əmz\ *or* **-nea** \-ˈē-ə\ [Middle English, from Late Latin, from Greek *peritonaion,* neuter of *peritonaios* stretched around, from *peri-* + *teinein* to stretch — more at THIN] (15th century)
: the smooth transparent serous membrane that lines the cavity of the abdomen of a mammal and is folded inward over the abdominal and pelvic viscera
— peri·to·ne·al \-ˈē-əl\ *adjective*
— peri·to·ne·al·ly \-ə-lē\ *adverb*

peri·to·ni·tis \ˌper-ə-tᵊn-ˈī-təs\ *noun* [New Latin] (1776)
: inflammation of the peritoneum

pe·rit·ri·chous \pə-ˈri-tri-kəs\ *adjective* [*peri-* + Greek *trich-, thrix* hair] (1877)
1 : having flagella uniformly distributed over the body ⟨*peritrichous* bacteria⟩
2 : having a spiral line of modified cilia around the oral disk ⟨*peritrichous* protozoa⟩
— pe·rit·ri·chous·ly *adverb*

peri·wig \ˈper-i-ˌwig\ *noun* [modification of Middle French *perruque* — more at PERUKE] (1529)
: PERUKE
— peri·wigged \-ˌwigd\ *adjective*

[1]**per·i·win·kle** \ˈper-i-ˌwiŋ-kəl\ *noun* [Middle English *perwinke,* from Old English *perwince,* from (assumed) Vulgar Latin *pervinca,* short for Latin *vincapervinca*] (before 12th century)
1 : any of several trailing or woody evergreen herbs of the dogbane family: as **a :** a European creeper (*Vinca minor*) widely cultivated as a ground cover and for its blue or white flowers — called also *myrtle* **b :** a commonly cultivated subshrub (*Catharanthus roseus* synonym *Vinca rosea*) of the Old World tropics that is the source of several antineoplastic drugs — called also *Madagascar periwinkle*
2 : a light purplish blue — called also *periwinkle blue*

[2]**periwinkle** *noun* [(assumed) Middle English, alteration of Old English *pīnewincle,* from Latin *pina,* a kind of mussel (from Greek) + Old English *-wincle* (akin to Danish *vincle* snail shell)] (circa 1530)
: any of various gastropod mollusks: as **a :** any of a genus (*Littorina*) of edible littoral marine snails; *also* **:** any of various similar or related marine snails **b :** any of several North American freshwater snails

[2]periwinkle a

per·jure \ˈpər-jər\ *transitive verb* **per·jured; per·jur·ing** \ˈpər-jə-riŋ, ˈpərj-riŋ\ [Middle French *perjurer,* from Latin *perjurare,* from *per-* detrimentally, for the worse + *jurare* to swear — more at PER-, JURY] (1535)
1 : to make a perjurer of (oneself)
2 *obsolete* **:** to cause to commit perjury

per·jur·er \ˈpər-jər-ər\ *noun* (15th century)
: a person guilty of perjury

per·ju·ri·ous \(ˌ)pər-ˈju̇r-ē-əs\ *adjective* (1602)
: marked by perjury ⟨*perjurious* testimony⟩
— per·ju·ri·ous·ly *adverb*

per·ju·ry \ˈpər-jə-rē, ˈpərj-rē\ *noun* (14th century)
: the voluntary violation of an oath or vow either by swearing to what is untrue or by omission to do what has been promised under oath **:** false swearing

[1]**perk** \ˈpərk\ *verb* [Middle English] (14th century)
intransitive verb
1 a : to thrust up the head, stretch out the neck, or carry the body in a bold or insolent manner **b :** to stick up or out jauntily
2 : to gain in vigor or cheerfulness especially after a period of weakness or depression — usually used with *up* ⟨he's *perked* up noticeably⟩
transitive verb
1 : to make smart or spruce in appearance **:** FRESHEN, IMPROVE
2 : to thrust up quickly or impudently

[2]**perk** *intransitive verb* (1656)
: PERCOLATE

[3]**perk** *noun* (1824)
: PERQUISITE — usually used in plural

perky \ˈpər-kē\ *adjective* **perk·i·er; -est** (1855)
1 : briskly self-assured **:** COCKY
2 : JAUNTY ⟨a *perky* . . . waltz —*New Yorker*⟩
— perk·i·ly \-kə-lē\ *adverb*
— perk·i·ness \-kē-nəs\ *noun*

per·lite \ˈpər-ˌlīt\ *noun* [French, from *perle* pearl] (1833)
: volcanic glass that has a concentric shelly structure, appears as if composed of concretions, is usually grayish and sometimes spherulitic, and when expanded by heat forms a lightweight aggregate used especially in concrete and plaster and as a medium for potting plants
— per·lit·ic \ˌpər-ˈli-tik\ *adjective*

[1]**perm** \ˈpərm\ *noun* (1927)
: PERMANENT

[2]**perm** *transitive verb* (1928)
: to give (hair) a permanent wave

per·ma·frost \'pər-mə-ˌfrȯst\ *noun* [*perma*nent + *frost*] (1943)
: a permanently frozen layer at variable depth below the surface in frigid regions of a planet (as earth)

per·ma·nence \'pər-mə-nən(t)s, 'pərm-nən(t)s\ *noun* (15th century)
: the quality or state of being permanent **:** DURABILITY

per·ma·nen·cy \-nən(t)-sē\ *noun, plural* **-cies** (1555)
1 : PERMANENCE
2 : something permanent

[1]**per·ma·nent** \-nənt\ *adjective* [Middle English, from Middle French, from Latin *permanent-, permanens,* present participle of *permanēre* to endure, from *per-* throughout + *manēre* to remain — more at PER-, MANSION] (15th century)
: continuing or enduring without fundamental or marked change **:** STABLE
synonym see LASTING
— per·ma·nent·ly *adverb*
— per·ma·nent·ness *noun*

[2]**permanent** *noun* (1925)
: a long-lasting hair wave or straightening produced by mechanical and chemical means — called also *permanent wave*

permanent magnet *noun* (1828)
: a magnet that retains its magnetism after removal of the magnetizing force

permanent press *noun* (1964)
1 : the process of treating a fabric with a chemical (as a resin) and heat for setting the shape and for aiding wrinkle resistance
2 : material treated by permanent press
3 : the condition of material treated by permanent press
— permanent–press *adjective*

permanent tissue *noun* (circa 1928)
: plant tissue that has completed its growth and differentiation and is usually incapable of meristematic activity

permanent tooth *noun* (1836)
: any of the second set of teeth of a mammal that follow the milk teeth, typically persist into old age, and in humans are 32 in number

per·man·ga·nate \(ˌ)pər-'maŋ-gə-ˌnāt\ *noun* (1841)
1 : a salt containing the anion MnO_4^-; *especially* **:** POTASSIUM PERMANGANATE
2 : the anion MnO_4^- of a permanganate

per·me·abil·i·ty \ˌpər-mē-ə-'bi-lə-tē\ *noun* (1759)
1 : the quality or state of being permeable
2 : the property of a magnetizable substance that determines the degree in which it modifies the magnetic flux in the region occupied by it in a magnetic field

per·me·able \'pər-mē-ə-bəl\ *adjective* (15th century)
: capable of being permeated **:** PENETRABLE; *especially* **:** having pores or openings that permit liquids or gases to pass through ⟨a *permeable* membrane⟩ ⟨*permeable* limestone⟩

per·me·ase \'pər-mē-ˌās, -ˌāz\ *noun* [International Scientific Vocabulary *perme-* (from *permeate*) + *-ase*] (1957)
: a substance that catalyzes the transport of another substance across a cell membrane

per·me·ate \'pər-mē-ˌāt\ *verb* **-at·ed; -at·ing** [Latin *permeatus,* past participle of *permeare,* from *per-* through + *meare* to go, pass; akin to Middle Welsh *mynet* to go, Czech *míjet* to pass] (1656)
intransitive verb
: to diffuse through or penetrate something
transitive verb
1 : to spread or diffuse through ⟨a room *permeated* with tobacco smoke⟩
2 : to pass through the pores or interstices of
— per·me·ative \-ˌā-tiv\ *adjective*

per·me·ation \ˌpər-mē-'ā-shən\ *noun* (circa 1623)
1 : the quality or state of being permeated
2 : the action or process of permeating

per men·sem \(ˌ)pər-'men(t)-səm\ *adverb* [Medieval Latin] (1647)
: by the month

per·meth·rin \(ˌ)pər-'me-thrən\ *noun* [*per-* + *meth*yl + pyreth*rin*] (1976)
: a synthetic pyrethrin $C_{21}H_{20}Cl_2O_3$ used especially as an insecticide

Perm·ian \'pər-mē-ən, 'per-\ *adjective* [*Perm,* former province in eastern Russia] (1841)
: of, relating to, or being the last period of the Paleozoic era or the corresponding system of rocks — see GEOLOGIC TIME table
— Permian *noun*

per mill \(ˌ)pər-'mil\ *adverb* [*per* + Latin *mille* thousand] (1902)
: per thousand
— per·mil·lage \(ˌ)pər-'mi-lij\ *noun*

per·mis·si·ble \pər-'mi-sə-bəl\ *adjective* [Middle English, from Medieval Latin *permissibilis,* from Latin *permissus,* past participle of *permittere*] (15th century)
: that may be permitted **:** ALLOWABLE
— per·mis·si·bil·i·ty \-ˌmi-sə-'bi-lə-tē\ *noun*
— per·mis·si·ble·ness \-'mi-sə-bəl-nəs\ *noun*
— per·mis·si·bly \-blē\ *adverb*

per·mis·sion \pər-'mi-shən\ *noun* [Middle English, from Middle French, from Latin *permission-, permissio,* from *permittere*] (15th century)
1 : the act of permitting
2 : formal consent **:** AUTHORIZATION

per·mis·sive \pər-'mi-siv\ *adjective* [Middle English *permyssyf,* from Middle French *permissif,* from Latin *permissus*] (15th century)
1 *archaic* **:** granted on sufferance **:** TOLERATED
2 a : granting or tending to grant permission **:** TOLERANT **b :** deficient in firmness or control **:** INDULGENT, LAX
3 : allowing discretion **:** OPTIONAL ⟨reduced the *permissive* retirement age from 65 to 62⟩
— per·mis·sive·ly *adverb*
— per·mis·sive·ness *noun*

[1]**per·mit** \pər-'mit\ *verb* **per·mit·ted; per·mit·ting** [Middle English *permitten,* from Latin *permittere* to let through, permit, from *per-* through + *mittere* to let go, send] (15th century)
transitive verb
1 : to consent to expressly or formally ⟨*permit* access to records⟩
2 : to give leave **:** AUTHORIZE
3 : to make possible
intransitive verb
: to give an opportunity ⟨if time *permits*⟩
— per·mit·tee \pər-ˌmi(t)-'tē, ˌpər-mi(t)-\ *noun*
— per·mit·ter *noun*

[2]**per·mit** \'pər-ˌmit, pər-'\ *noun* (1682)
1 : a written warrant or license granted by one having authority ⟨a gun *permit*⟩
2 : PERMISSION

[3]**permit** *noun* [perhaps by folk etymology from Spanish *palometa,* a kind of pompano, from diminutive of *paloma* dove, from Latin *palumba, palumbes* — more at PALOMINO] (circa 1945)
: either of two pompanos (*Trachinotus falcatus* and *T. goodei*) that are important game fishes of temperate to tropical waters of the western Atlantic

per·mit·tiv·i·ty \ˌpər-ˌmi-'ti-və-tē, -mə-\ *noun* [[1]*permit* + *-ivity* (as in *selectivity*)] (1887)
: the ability of a dielectric to store electrical potential energy under the influence of an electric field measured by the ratio of the capacitance of a condenser with the material as dielectric to its capacitance with vacuum as dielectric — called also *dielectric constant*

per·mu·ta·tion \ˌpər-myu̇-'tā-shən\ *noun* [Middle English *permutacioun* exchange, transformation, from Middle French *permutation,* from Latin *permutation-, permutatio,* from *permutare*] (14th century)
1 : often major or fundamental change (as in character or condition) based primarily on rearrangement of existent elements ⟨land-owners and peasants . . . in the *permutations* of their tortured interdependence —P. E. Mosley⟩
2 a : the act or process of changing the lineal order of an ordered set of objects **b :** an ordered arrangement of a set of objects
— per·mu·ta·tion·al \-shnəl, -shə-nᵊl\ *adjective*

permutation group *noun* (1904)
: a group whose elements are permutations and in which the product of two permutations is a permutation whose effect is the same as the successive application of the first two

per·mute \pər-'myüt\ *transitive verb* **per·mut·ed; per·mut·ing** [Middle English, to exchange, from Middle French or Latin; Middle French *permuter,* from Latin *permutare,* from *per-* + *mutare* to change — more at MUTABLE] (1878)
: to change the order or arrangement of; *especially* **:** to arrange in all possible ways
— per·mut·able \-'myü-tə-bəl\ *adjective*

per·ni·cious \pər-'ni-shəs\ *adjective* [Middle English, from Middle French *pernicieus,* from Latin *perniciosus,* from *pernicies* destruction, from *per-* + *nec-, nex* violent death — more at NOXIOUS] (15th century)
1 : highly injurious or destructive **:** DEADLY
2 *archaic* **:** WICKED ☆
— per·ni·cious·ly *adverb*
— per·ni·cious·ness *noun*

pernicious anemia *noun* (1874)
: a severe megaloblastic anemia marked by a progressive decrease in the number of red blood cells and by pallor, weakness, and gastrointestinal and nervous disturbances and associated with malabsorption of vitamin B_{12} due to the absence of intrinsic factor

per·nick·e·ty \pər-'ni-kə-tē\ *adjective* [origin unknown] (circa 1818)
: PERSNICKETY

Per·nod \per-'nō, ˌpər-\ *trademark*
— used for a French liqueur

pe·ro·ne·al \ˌper-ō-'nē-əl, pə-'rō-nē-\ *adjective* [New Latin *peroneus,* from *perone* fibula, from Greek *peronē,* literally, pin, from *peirein* to pierce — more at DIAPIR] (1831)
: of, relating to, or located near the fibula

per·oral \(ˌ)pər-'ōr-əl, per-, -'ȯr-, -'är-\ *adjective* [International Scientific Vocabulary] (1908)
: occurring through or by way of the mouth
— per·oral·ly \-ə-lē\ *adverb*

per·o·rate \'per-ə-ˌrāt *also* 'pər-\ *intransitive verb* **-rat·ed; -rat·ing** [Latin *peroratus,* past participle of *perorare* to declaim at length, wind up an oration, from *per-* through + *orare* to speak — more at PER-, ORATION] (1603)
1 : to deliver a long or grandiloquent oration
2 : to make a peroration

per·o·ra·tion \ˌper-ə-'rā-shən *also* ˌpər-\ *noun* [Middle English *peroracyon,* from Latin *peroration-, peroratio,* from *perorare*] (15th century)

> ☆ SYNONYMS
> **Pernicious, baneful, noxious, deleterious, detrimental** mean exceedingly harmful. PERNICIOUS implies irreparable harm done through evil or insidious corrupting or undermining ⟨the claim that pornography has a *pernicious* effect on society⟩. BANEFUL implies injury through poisoning or destroying ⟨the *baneful* notion that discipline destroys creativity⟩. NOXIOUS applies to what is both offensive and injurious to the health of a body or mind ⟨*noxious* chemical fumes⟩. DELETERIOUS applies to what has an often unsuspected harmful effect ⟨a diet found to have *deleterious* effects⟩. DETRIMENTAL implies obvious harmfulness to something specified ⟨the *detrimental* effects of excessive drinking⟩.

1 : the concluding part of a discourse and especially an oration
2 : a highly rhetorical speech
— per·o·ra·tion·al \ˌper-ə-ˈrā-shnəl, ˌpər-, -shə-nᵊl\ *adjective*

pe·rov·skite \pə-ˈräv-ˌskīt, -ˈräf-\ *noun* [German *Perowskit,* from Count L. A. *Perovskiĭ* (died 1856) Russian statesman] (1840)
: a yellow, brown, or grayish black mineral sometimes containing rare earth elements

per·ox·i·dase \pə-ˈräk-sə-ˌdās, -ˌdāz\ *noun* (1900)
: an enzyme that catalyzes the oxidation of various substances by peroxides

¹per·ox·ide \pə-ˈräk-ˌsīd\ *noun* [International Scientific Vocabulary] (1804)
: an oxide containing a high proportion of oxygen; *especially* **:** a compound (as hydrogen peroxide) in which oxygen is visualized as joined to oxygen
— per·ox·id·ic \-ˌräk-ˈsi-dik\ *adjective*

²peroxide *transitive verb* **-id·ed; -id·ing** (1906)
: to treat with a peroxide; *especially* **:** to bleach (hair) with hydrogen peroxide

³peroxide *adjective* (1920)
: having or being bleached hair ⟨a *peroxide* blonde⟩

per·ox·i·some \pə-ˈräk-sə-ˌsōm\ *noun* [*peroxide* + ³*-some*] (1965)
: a cytoplasmic cell organelle containing enzymes (as catalase) which act in oxidative reactions and especially in the production and decomposition of hydrogen peroxide
— per·ox·i·som·al \-ˌräk-sə-ˈsō-məl\ *adjective*

peroxy- *combining form* [International Scientific Vocabulary *per-* + *oxy*]
: containing the bivalent group O—O ⟨*peroxy*acetyl nitrate⟩

per·oxy·ace·tyl nitrate \pə-ˌräk-sē-ə-ˈsē-tᵊl-; -ˈa-sə-tᵊl-, -ˌtēl-\ *noun* (1963)
: a toxic compound $C_2H_3O_5N$ found especially in smog

per·pend \(ˌ)pər-ˈpend\ *verb* [Middle English, from Latin *perpendere,* from *per-* thoroughly + *pendere* to weigh — more at PER-, PENDANT] (15th century)
transitive verb
: to reflect on carefully **:** PONDER
intransitive verb
: to be attentive **:** REFLECT

¹per·pen·dic·u·lar \ˌpər-pən-ˈdi-kyə-lər\ *adjective* [Middle English *perpendiculer,* from Middle French, from Latin *perpendicularis,* from *perpendiculum* plumb line, from *per-* + *pendēre* to hang — more at PENDANT] (14th century)
1 a : standing at right angles to the plane of the horizon **:** exactly upright **b :** being at right angles to a given line or plane
2 : extremely steep **:** PRECIPITOUS
3 *often capitalized* **:** of or relating to a medieval English Gothic style of architecture in which vertical lines predominate
4 : relating to, uniting, or consisting of individuals of dissimilar type or on different levels
synonym see VERTICAL
— per·pen·dic·u·lar·i·ty \-ˌdi-kyə-ˈlar-ə-tē\ *noun*
— per·pen·dic·u·lar·ly \-ˈdi-kyə-lər-lē\ *adverb*

²perpendicular *noun* (1571)
: a line at right angles to a line or plane (as of the horizon)

per·pe·trate \ˈpər-pə-ˌtrāt\ *transitive verb* **-trat·ed; -trat·ing** [Latin *perpetratus,* past participle of *perpetrare,* from *per-* through + *patrare* to accomplish, from *pater* father — more at FATHER] (1537)
: to bring about or carry out (as a crime or deception) **:** COMMIT
— per·pe·tra·tion \ˌpər-pə-ˈtrā-shən\ *noun*
— per·pe·tra·tor \ˈpər-pə-ˌtrā-tər\ *noun*

per·pet·u·al \pər-ˈpe-chə-wəl, -chəl; -ˈpech-wəl\ *adjective* [Middle English *perpetuel,* from Middle French, from Latin *perpetuus* uninterrupted, from *per-* through + *petere* to go to — more at FEATHER] (14th century)
1 a : continuing forever **:** EVERLASTING **b** (1) **:** valid for all time (2) **:** holding (as an office) for life or for an unlimited time
2 : occurring continually **:** indefinitely long-continued
3 : blooming continuously through the season
synonym see CONTINUAL
— per·pet·u·al·ly *adverb*

perpetual calendar *noun* (1895)
: a table for finding the day of the week for any one of a wide range of dates

perpetual check *noun* (circa 1909)
: an endless succession of checks to which an opponent's king may be subjected to force a draw in chess

per·pet·u·ate \pər-ˈpe-chə-ˌwāt\ *transitive verb* **-at·ed; -at·ing** [Latin *perpetuatus,* past participle of *perpetuare,* from *perpetuus*] (1530)
: to make perpetual or cause to last indefinitely ⟨*perpetuate* the species⟩
— per·pet·u·a·tion \-ˌpe-chə-ˈwā-shən\ *noun*
— per·pet·u·a·tor \-ˈpe-chə-ˌwā-tər\ *noun*

per·pe·tu·i·ty \ˌpər-pə-ˈtü-ə-tē, -ˈtyü-\ *noun, plural* **-ties** [Middle English *perpetuite,* from Middle French *perpetuité,* from Latin *perpetuitat-, perpetuitas,* from *perpetuus*] (15th century)
1 : ETERNITY 2
2 : the quality or state of being perpetual ⟨bequeathed to them in *perpetuity*⟩
3 a : the condition of an estate limited so that it will not take effect or vest within the period fixed by law **b :** an estate so limited
4 : an annuity payable forever

per·phe·na·zine \(ˌ)pər-ˈfe-nə-ˌzēn, -ˈfe-\ *noun* [blend of *piperazine* and *phen-*] (1957)
: a tranquilizing drug $C_{21}H_{26}ClN_3OS$ that is used to control tension, anxiety, and agitation especially in psychotic conditions

per·plex \pər-ˈpleks\ *transitive verb* [obsolete *perplex,* adjective, involved, perplexed, from Latin *perplexus,* from *per-* thoroughly + *plexus* involved, from past participle of *plectere* to braid, twine — more at PER-, PLY] (1593)
1 : to make unable to grasp something clearly or to think logically and decisively about something ⟨her attitude *perplexes* me⟩ ⟨a *perplexing* problem⟩
2 : to make intricate or involved **:** COMPLICATE
synonym see PUZZLE

per·plexed \-ˈplekst\ *adjective* (15th century)
1 : filled with uncertainty **:** PUZZLED
2 : full of difficulty
— per·plexed·ly \-ˈplek-səd-lē, -ˈplekst-lē\ *adverb*

per·plex·i·ty \pər-ˈplek-sə-tē\ *noun, plural* **-ties** [Middle English *perplexite,* from Middle French *perplexité,* from Late Latin *perplexitat-, perplexitas,* from Latin *perplexus*] (14th century)
1 : the state of being perplexed
2 : something that perplexes
3 : ENTANGLEMENT

per·qui·site \ˈpər-kwə-zət\ *noun* [Middle English, property acquired by means other than inheritance, from Medieval Latin *perquisitum,* from neuter of *perquisitus,* past participle of *perquirere* to purchase, acquire, from Latin, to search for thoroughly, from *per-* thoroughly + *quaerere* to seek] (15th century)
1 : a privilege, gain, or profit incidental to regular salary or wages; *especially* **:** one expected or promised
2 : GRATUITY, TIP
3 : something held or claimed as an exclusive right or possession ⟨concepts . . . not the *perquisites* of any particular groups —Gilbert Ryle⟩

per·ron \ˈper-ən, pe-rōⁿ\ *noun* [French, from Old French, from *perre, pierre* rock, stone, from Latin *petra,* from Greek] (1723)
: an outdoor stairway leading up to a building entrance; *also* **:** a platform at its top

per·ry \ˈper-ē\ *noun* [Middle English *peirrie,* from Middle French *peré,* from (assumed) Vulgar Latin *piratum,* from Latin *pirum* pear] (14th century)
: fermented pear juice often made sparkling

perse \ˈpərs\ *adjective* [Middle English *pers,* from Middle French, from Medieval Latin *persus*] (15th century)
: of a dark grayish blue resembling indigo

per se \(ˌ)pər-ˈsā *also* per-ˈsā *or* (ˌ)pər-ˈsē\ *adverb* [Latin] (1572)
: by, of, or in itself or oneself or themselves **:** as such **:** INTRINSICALLY

per second per second *adverb* (1922)
: per second every second — used of acceleration

per·se·cute \ˈpər-si-ˌkyüt\ *transitive verb* **-cut·ed; -cut·ing** [Middle English, from Middle French *persecuter,* back-formation from *persecuteur* persecutor, from Late Latin *persecutor,* from *persequi* to persecute, from Latin, to pursue, from *per-* through + *sequi* to follow — more at SUE] (15th century)
1 : to harass in a manner designed to injure, grieve, or afflict; *specifically* **:** to cause to suffer because of belief
2 : to annoy with persistent or urgent approaches (as attacks, pleas, or importunities) **:** PESTER
synonym see WRONG
— per·se·cu·tee \ˌpər-si-ˌkyü-ˈtē\ *noun*
— per·se·cu·tive \ˈpər-si-ˌkyü-tiv\ *adjective*
— per·se·cu·tor \-ˌkyü-tər\ *noun*
— per·se·cu·to·ry \-kyü-ˌtōr-ē, -ˌtȯr-; -ˌkyü-tə-rē\ *adjective*

per·se·cu·tion \ˌpər-si-ˈkyü-shən\ *noun* (14th century)
1 : the act or practice of persecuting especially those who differ in origin, religion, or social outlook
2 : the condition of being persecuted, harassed, or annoyed

Per·se·id \ˈpər-sē-əd\ *noun* [Latin *Perseus;* from their appearing to radiate from a point in Perseus]
: any of a group of meteors that appear annually about August 11

Per·seph·o·ne \pər-ˈse-fə-nē\ *noun* [Latin, from Greek *Persephonē*]
: a daughter of Zeus and Demeter abducted by Pluto to reign with him over the underworld

Per·seus \ˈpər-ˌsüs, -sē-əs\ *noun* [Latin, from Greek]
1 : a son of Zeus and Danaë and slayer of Medusa
2 [Latin (genitive *Persei*), from Greek] **:** a northern constellation between Taurus and Cassiopeia

per·se·ver·ance \ˌpər-sə-ˈvir-ən(t)s\ *noun* (14th century)
: the action or condition or an instance of persevering **:** STEADFASTNESS

per·sev·er·a·tion \pər-ˌse-və-ˈrā-shən\ *noun* [Latin *perseveration-, perseveratio,* from *perseverare*] (1910)
: continuation of something (as repetition of a word) usually to an exceptional degree or beyond a desired point
— per·sev·er·ate \-ˈse-və-ˌrāt\ *intransitive verb*

per·se·vere \ˌpər-sə-ˈvir\ *intransitive verb* **-vered; -ver·ing** [Middle English, from Middle French *perseverer,* from Latin *perseverare,* from *per-* through + *severus* severe] (14th century)
: to persist in a state, enterprise, or undertak-

ing in spite of counterinfluences, opposition, or discouragement
— **per·se·ver·ing·ly** *adverb*
Per·sian \'pər-zhən, *especially British* -shən\ *noun* (14th century)
1 : one of the people of Persia: as **a :** one of the ancient Iranians who under Cyrus and his successors founded an empire in southwest Asia **b :** a member of one of the peoples forming the modern Iranian nationality
2 a : any of several Iranian languages dominant in Persia at different periods **b :** the modern language of Iran and western Afghanistan — see INDO-EUROPEAN LANGUAGES table
3 : a thin soft silk formerly used especially for linings
4 : PERSIAN CAT
— **Persian** *adjective*
Persian cat *noun* (1824)
: any of a breed of stocky round-headed domestic cats that have a long silky coat and thick ruff
Persian lamb *noun* (1889)
1 : a pelt that is obtained from karakul lambs older than those yielding broadtail and that is characterized by very silky tightly curled fur
2 : the young of the karakul sheep that furnishes skins used in furriery

Persian cat

Persian lamb 2

per·si·flage \'pər-si-ˌfläzh, 'per-\ *noun* [French, from *persifler* to banter, from *per-* thoroughly + *siffler* to whistle, hiss, boo, ultimately from Latin *sibilare*] (1757)
: frivolous bantering talk **:** light raillery
per·sim·mon \pər-'si-mən\ *noun* [Virginia Algonquian *pessemmin*] (1612)
1 : any of a genus (*Diospyros*) of trees of the ebony family with hard fine wood, oblong leaves, and small bell-shaped flowers; *especially* **:** an American tree (*D. virginiana*) or a Japanese tree (*D. kaki*)
2 : the usually orange several-seeded globular berry of a persimmon that is edible when fully ripe but usually extremely astringent when unripe
per·sist \pər-'sist, -'zist\ *intransitive verb* [Middle French *persister,* from Latin *persistere,* from *per-* + *sistere* to take a stand, stand firm; akin to Latin *stare* to stand — more at STAND] (1538)
1 : to go on resolutely or stubbornly in spite of opposition, importunity, or warning
2 *obsolete* **:** to remain unchanged or fixed in a specified character, condition, or position
3 : to be insistent in the repetition or pressing of an utterance (as a question or an opinion)
4 : to continue to exist especially past a usual, expected, or normal time
synonym see CONTINUE
— **per·sist·er** *noun*
per·sis·tence \pər-'sis-tən(t)s, -'zis-\ *noun* (1546)
1 : the action or fact of persisting
2 : the quality or state of being persistent; *especially* **:** PERSEVERANCE
per·sis·ten·cy \-tən(t)-sē\ *noun* (1597)
: PERSISTENCE 2
per·sis·tent \-tənt\ *adjective* [Latin *persistent-, persistens,* present participle of *persistere*] (1826)
1 : existing for a long or longer than usual time or continuously: as **a :** retained beyond the usual period ⟨a *persistent* leaf⟩ **b :** continuing without change in function or structure ⟨*persistent* gills⟩ **c :** effective in the open for an appreciable time usually through slow volatilizing ⟨mustard gas is *persistent*⟩ **d :** degraded only slowly by the environment ⟨*persistent* pesticides⟩ **e :** remaining infective for a relatively long time in a vector after an initial period of incubation ⟨*persistent* viruses⟩
2 a : continuing or inclined to persist in a course **b :** continuing to exist in spite of interference or treatment ⟨a *persistent* cough⟩
— **per·sis·tent·ly** *adverb*
per·snick·e·ty \pər-'sni-kə-tē\ *adjective* [alteration of *pernickety*] (circa 1905)
1 a : fussy about small details **:** FASTIDIOUS ⟨a *persnickety* teacher⟩ **b :** having the characteristics of a snob
2 : requiring great precision ⟨a *persnickety* job⟩
— **per·snick·e·ti·ness** \-nəs\ *noun*
per·son \'pər-sᵊn\ *noun* [Middle English, from Old French *persone,* from Latin *persona* actor's mask, character in a play, person, probably from Etruscan *phersu* mask, from Greek *prosōpa,* plural of *prosōpon* face, mask — more at PROSOPOPOEIA] (13th century)
1 : HUMAN, INDIVIDUAL — sometimes used in combination especially by those who prefer to avoid *man* in compounds applicable to both sexes ⟨chair*person*⟩ ⟨spokes*person*⟩
2 : a character or part in or as if in a play
3 a : one of the three modes of being in the Trinitarian Godhead as understood by Christians **b :** the unitary personality of Christ that unites the divine and human natures
4 a *archaic* **:** bodily appearance **b :** the body of a human being; *also* **:** the body and clothing ⟨unlawful search of the *person*⟩
5 : the personality of a human being **:** SELF
6 : one (as a human being, a partnership, or a corporation) that is recognized by law as the subject of rights and duties
7 : reference of a segment of discourse to the speaker, to one spoken to, or to one spoken of as indicated by means of certain pronouns or in many languages by verb inflection ◆
— **per·son·hood** \-ˌhu̇d\ *noun*
— **in person :** in one's bodily presence
per·so·na \pər-'sō-nə, -ˌnä\ *noun, plural* **per·so·nae** \-(ˌ)nē, -ˌnī\ *or* **personas** [Latin] (1909)
1 : a character assumed by an author in a written work
2 a *plural personas* [New Latin, from Latin] **:** an individual's social facade or front that especially in the analytic psychology of C. G. Jung reflects the role in life the individual is playing — compare ANIMA **b :** the personality that a person (as an actor or politician) projects in public **:** IMAGE
3 *plural personae* **:** a character in a fictional presentation (as a novel or play) — usually used in plural ⟨comic *personae*⟩
per·son·able \'pərs-nə-bəl, 'pər-sᵊn-ə-bəl\ *adjective* (15th century)
: pleasant or amiable in person **:** ATTRACTIVE
— **per·son·able·ness** *noun*
per·son·age \'pərs-nij, 'pər-sᵊn-ij\ *noun* (15th century)
1 : a person of rank, note, or distinction; *especially* **:** one distinguished for presence and personal power
2 : a human individual **:** PERSON
3 : a dramatic, fictional, or historical character; *also* **:** IMPERSONATION
per·so·na gra·ta \pər-'sō-nə-'gra-tə, -'grä-\ *adjective* [New Latin, acceptable person] (1882)
: personally acceptable or welcome
¹per·son·al \'pərs-nəl, 'pər-sᵊn-əl\ *adjective* [Middle English, from Middle French, from Late Latin *personalis,* from Latin *persona*] (14th century)
1 : of, relating to, or affecting a person **:** PRIVATE, INDIVIDUAL ⟨*personal* ambition⟩
2 a : done in person without the intervention of another; *also* **:** proceeding from a single person **b :** carried on between individuals directly ⟨a *personal* interview⟩
3 : relating to the person or body
4 : relating to an individual or an individual's character, conduct, motives, or private affairs often in an offensive manner ⟨a *personal* insult⟩
5 a : being rational and self-conscious ⟨*personal,* responsive government is still possible —John Fischer⟩ **b :** having the qualities of a person rather than a thing or abstraction ⟨a *personal* devil⟩
6 : of, relating to, or constituting personal property ⟨a *personal* estate⟩
7 : denoting grammatical person
²personal *noun* (1861)
1 : a short newspaper paragraph relating to the activities of a person or a group or to personal matters
2 : a short personal communication in a special column of the classified ads section of a newspaper or periodical
3 : PERSONAL FOUL
personal computer *noun* (1977)
: MICROCOMPUTER 1
personal effects *noun plural* (1843)
: privately owned items (as clothing and jewelry) normally worn or carried on the person
personal equation *noun* (1845)
: variation (as in observation) occasioned by the personal peculiarities of an individual; *also* **:** a correction or allowance made for such variation
personal foul *noun* (circa 1829)
: a foul in a game (as basketball) involving usually physical contact with or deliberate roughing of an opponent — compare TECHNICAL FOUL
per·son·al·ise *British variant of* PERSONALIZE
per·son·al·ism \'pərs-nə-ˌli-zəm, 'pər-sᵊn-ə-\ *noun* (circa 1846)
: a doctrine emphasizing the significance, uniqueness, and inviolability of personality
— **per·son·al·ist** \-list\ *noun or adjective*
— **per·son·al·is·tic** \ˌpərs-nə-'lis-tik, ˌpər-sᵊn-ə-\ *adjective*
per·son·al·i·ty \ˌpər-sᵊn-'a-lə-tē, ˌpər-'sna-\ *noun, plural* **-ties** [Middle English *personalite,* from Late Latin *personalitat-, personalitas,* from *personalis*] (15th century)
1 a : the quality or state of being a person **b :** personal existence
2 a : the condition or fact of relating to a particular person; *specifically* **:** the condition of referring directly to or being aimed disparagingly or hostilely at an individual **b :** an offensively personal remark ⟨angrily resorted to *personalities*⟩

◇ WORD HISTORY
person The Latin noun *persona* originally referred to an actor's mask, but its sense was extended to include the character played by the actor as well as his role in the play. By Cicero's time *persona* was also being used outside dramatic contexts to refer to a role in life or to an individual human personality. In Medieval Latin the sense "individual, human being" was common, but *persona* also developed much more specific senses. One of these was "priest holding the lifetime rights to the tithes of a parish, rector" and this sense, along with older meanings of *persona,* was retained when the word was borrowed into Middle English as *person,* partly through the mediation of medieval French. By a regular sound change that shifted late Middle English \er\ to \ar\ before a consonant or at the end of a word, *person* changed to *parson,* and this spelling began to appear for the word. The influence of Latin spelling and the norms of school Latin pronunciation militated against the change, however, and the spelling was restored to the form with *e*—except in the sense "parish priest," which remained *parson.*

3 : the complex of characteristics that distinguishes an individual or a nation or group; *especially* **:** the totality of an individual's behavioral and emotional characteristics
4 a : distinction or excellence of personal and social traits; *also* **:** a person having such quality **b :** a person of importance, prominence, renown, or notoriety ⟨a TV *personality*⟩
synonym see DISPOSITION

personality inventory *noun* (1932)
: any of several tests that attempt to characterize the personality of an individual by objective scoring of replies to a large number of questions concerning his or her own behavior — compare MINNESOTA MULTIPHASIC PERSONALITY INVENTORY

personality test *noun* (1914)
: any of several tests that consist of standardized tasks designed to determine various aspects of the personality or the emotional status of the individual examined

per·son·al·ize \'pərs-nə-ˌlīz, 'pər-s^{ə}n-ə-\ *transitive verb* **-ized; -iz·ing** (circa 1741)
1 : PERSONIFY
2 : to make personal or individual; *specifically* **:** to mark as the property of a particular person ⟨*personalized* stationery⟩
— **per·son·al·i·za·tion** \ˌpərs-nə-lə-'zā-shən, ˌpər-s^{ə}n-ə-\ *noun*

per·son·al·ly \'pərs-nə-lē, 'pər-s^{ə}n-ə-\ *adverb* (14th century)
1 : in person ⟨attend to the matter *personally*⟩
2 : as a person **:** in personality ⟨*personally* attractive but not very trustworthy⟩
3 : for oneself **:** as far as oneself is concerned

personal pronoun *noun* (1668)
: a pronoun (as *I, you,* or *they*) that expresses a distinction of person

personal property *noun* (1838)
: property other than real property consisting of things temporary or movable **:** CHATTELS

personal tax *noun* (circa 1935)
: DIRECT TAX

per·son·al·ty \'pərs-nəl-tē, 'pər-s^{ə}n-əl-\ *noun, plural* **-ties** [Middle English, from Anglo-French *personalté,* from Late Latin *personalitat-, personalitas* personality] (15th century)
: PERSONAL PROPERTY

per·so·na non gra·ta \pər-'sō-nə-ˌnän-'gra-tə, -'grä-\ *adjective* [New Latin, unacceptable person] (1904)
: personally unacceptable or unwelcome

per·son·ate \'pər-s^{ə}n-ˌāt\ *transitive verb* **-at·ed; -at·ing** (1591)
1 a : IMPERSONATE, REPRESENT **b :** to assume without authority and with fraudulent intent (some character or capacity)
2 : to invest with personality or personal characteristics ⟨*personating* their gods ridiculous, and themselves past shame —John Milton⟩
— **per·son·a·tion** \ˌpər-s^{ə}n-'ā-shən\ *noun*
— **per·son·a·tive** \'pər-s^{ə}n-ˌā-tiv\ *adjective*
— **per·son·a·tor** \-ˌā-tər\ *noun*

per·son–hour \'pər-s^{ə}n-ˌau̇(-ə)r\ *noun* (1975)
: a unit of one hour's work by one person

per·son·i·fi·ca·tion \pər-ˌsä-nə-fə-'kā-shən\ *noun* (circa 1755)
1 : attribution of personal qualities; *especially* **:** representation of a thing or abstraction as a person or by the human form
2 : a divinity or imaginary being representing a thing or abstraction
3 : EMBODIMENT, INCARNATION

per·son·i·fy \pər-'sä-nə-ˌfī\ *transitive verb* **-fied; -fy·ing** (circa 1741)
1 : to conceive of or represent as a person or as having human qualities or powers
2 : to be the embodiment or personification of **:** INCARNATE ⟨a teacher who *personified* patience⟩
— **per·son·i·fi·er** \-ˌfī(-ə)r\ *noun*

per·son·nel \ˌpər-s^{ə}n-'el\ *noun* [French, from German *Personale, Personal,* from Medieval Latin *personale,* from Late Latin, neuter of *personalis* personal] (1837)
1 a : a body of persons usually employed (as in a factory, office, or organization) **b personnel** *plural* **:** PERSONS
2 : a division of an organization concerned with personnel

[1]**per·spec·tive** \pər-'spek-tiv\ *noun* [Middle English *perspectyf,* from Medieval Latin *perspectivum,* from neuter of *perspectivus* of sight, optical, from Latin *perspectus,* past participle of *perspicere* to look through, see clearly, from *per-* through + *specere* to look — more at PER-, SPY] (14th century)
archaic **:** an optical glass (as a telescope)

[2]**perspective** *noun* [Middle French, probably modification of Old Italian *prospettiva,* from *prospetto* view, prospect, from Latin *prospectus* — more at PROSPECT] (1563)
1 a : the technique or process of representing on a plane or curved surface the spatial relation of objects as they might appear to the eye; *specifically* **:** representation in a drawing or painting of parallel lines as converging in order to give the illusion of depth and distance **b :** a picture in perspective
2 a : the interrelation in which a subject or its parts are mentally viewed ⟨places the issues in proper *perspective*⟩; *also* **:** POINT OF VIEW **b :** the capacity to view things in their true relations or relative importance ⟨urge you to maintain your *perspective* and to view your own task in a larger framework —W. J. Cohen⟩
3 a : a visible scene; *especially* **:** one giving a distinctive impression of distance **:** VISTA **b :** a mental view or prospect ⟨to gain a broader *perspective* on the international scene —*Current Biography*⟩
4 : the appearance to the eye of objects in respect to their relative distance and positions
— **per·spec·tiv·al** \pər-'spek-ti-vəl, ˌpər-(ˌ)spek-'tī-vəl\ *adjective*

[3]**perspective** *adjective* [Middle English, optical, from Medieval Latin *perspectivus*] (1570)
1 *obsolete* **:** aiding the vision ⟨his eyes should be like unto the wrong end of a *perspective* glass —Alexander Pope⟩
2 : of, relating to, employing, or seen in perspective ⟨*perspective* drawing⟩
— **per·spec·tive·ly** *adverb*

Per·spex \'pər-ˌspeks\ *trademark*
— used for an acrylic plastic consisting essentially of polymethyl methacrylate

per·spi·ca·cious \ˌpər-spə-'kā-shəs\ *adjective* [Latin *perspicac-, perspicax,* from *perspicere*] (1640)
: of acute mental vision or discernment **:** KEEN
synonym see SHREWD
— **per·spi·ca·cious·ly** *adverb*
— **per·spi·ca·cious·ness** *noun*
— **per·spi·cac·i·ty** \-'ka-sə-tē\ *noun*

per·spic·u·ous \pər-'spi-kyə-wəs\ *adjective* [Latin *perspicuus* transparent, perspicuous, from *perspicere*] (1586)
: plain to the understanding especially because of clarity and precision of presentation ⟨a *perspicuous* argument⟩
synonym see CLEAR
— **per·spi·cu·ity** \ˌpər-spə-'kyü-ə-tē\ *noun*
— **per·spic·u·ous·ly** \pər-'spi-kyə-wəs-lē\ *adverb*
— **per·spic·u·ous·ness** *noun*

per·spi·ra·tion \ˌpər-spə-'rā-shən\ *noun* (1626)
1 : the action or process of perspiring
2 : a saline fluid secreted by the sweat glands **:** SWEAT

per·spi·ra·to·ry \pər-'spī-rə-ˌtōr-ē, 'pər-sp(ə-)rə-, -ˌtȯr-\ *adjective* (1725)
: of, relating to, secreting, or inducing perspiration

per·spire \pər-'spīr\ *intransitive verb* **per·spired; per·spir·ing** [French *perspirer,* from Middle French, from Latin *per-* through + *spirare* to blow, breathe — more at PER-] (circa 1682)
: to emit matter through the skin; *specifically* **:** to secrete and emit perspiration

per·suad·able \pər-'swā-də-bəl\ *adjective* (circa 1598)
: capable of being persuaded

per·suade \pər-'swād\ *transitive verb* **per·suad·ed; per·suad·ing** [Latin *persuadēre,* from *per-* thoroughly + *suadēre* to advise, urge — more at SWEET] (15th century)
1 : to move by argument, entreaty, or expostulation to a belief, position, or course of action
2 : to plead with **:** URGE
— **per·suad·er** *noun*

per·sua·si·ble \-'swā-zə-bəl, -'swā-sə-\ *adjective* [Middle French, from Latin *persuasibilis* persuasive, from *persuasus,* past participle of *persuadēre*] (1502)
: PERSUADABLE

per·sua·sion \pər-'swā-zhən\ *noun* [Middle English *persuasioun,* from Middle French or Latin; Middle French *persuasion,* from Latin *persuasion-, persuasio,* from *persuadēre*] (14th century)
1 a : the act or process or an instance of persuading **b :** a persuading argument **c :** the ability to persuade **:** PERSUASIVENESS
2 : the condition of being persuaded
3 a : an opinion held with complete assurance **b :** a system of religious beliefs; *also* **:** a group adhering to a particular system of beliefs
4 : KIND, SORT
synonym see OPINION

per·sua·sive \-'swā-siv, -ziv\ *adjective* (15th century)
: tending to persuade
— **per·sua·sive·ly** *adverb*
— **per·sua·sive·ness** *noun*

pert \'pərt\ *adjective* [Middle English, open, bold, pert, modification of Old French *apert,* from Latin *apertus* open, from past participle of *aperire* to open] (14th century)
1 a : saucily free and forward **:** flippantly cocky and assured **b :** being trim and chic **:** JAUNTY ⟨a *pert* little hat⟩ **c :** piquantly stimulating ⟨is a *pert* notion⟩
2 : LIVELY, VIVACIOUS
— **pert·ly** *adverb*
— **pert·ness** *noun*

per·tain \pər-'tān\ *intransitive verb* [Middle English *perteinen,* from Middle French *partenir,* from Latin *pertinēre* to reach to, belong, from *per-* through + *tenēre* to hold — more at THIN] (14th century)
1 a (1) **:** to belong as a part, member, accessory, or product (2) **:** to belong as an attribute, feature, or function ⟨the destruction and havoc *pertaining* to war⟩ (3) **:** to belong as a duty or right ⟨responsibilities that *pertain* to fatherhood⟩ **b :** to be appropriate to something ⟨the criteria . . . will be different from those that *pertain* elsewhere —J. B. Conant⟩
2 : to have reference ⟨books *pertaining* to birds⟩

per·ti·na·cious \ˌpər-t^{ə}n-'ā-shəs\ *adjective* [Latin *pertinac-, pertinax,* from *per-* thoroughly + *tenac-, tenax* tenacious, from *tenēre*] (1626)
1 a : adhering resolutely to an opinion, purpose, or design **b :** perversely persistent
2 : stubbornly unyielding or tenacious
synonym see OBSTINATE
— **per·ti·na·cious·ly** *adverb*
— **per·ti·na·cious·ness** *noun*
— **per·ti·nac·i·ty** \-'a-sə-tē\ *noun*

per·ti·nence \'pər-t^{ə}n-ən(t)s, 'pərt-nən(t)s\ *noun* (1659)
: the quality or state of being pertinent **:** RELEVANCE

per·ti·nen·cy \-t^{ə}n-ən(t)-sē, -nən(t)-sē\ *noun* (1598)
: PERTINENCE

per·ti·nent \ˈpər-tᵊn-ənt, ˈpərt-nənt\ *adjective* [Middle English, from Middle French, from Latin *pertinent-, pertinens,* present participle of *pertinēre*] (14th century)
: having a clear decisive relevance to the matter in hand
synonym see RELEVANT
— per·ti·nent·ly *adverb*

per·turb \pər-ˈtərb\ *transitive verb* [Middle English, from Middle French *perturber,* from Latin *perturbare* to throw into confusion, from *per-* + *turbare* to disturb — more at TURBID] (14th century)
1 : to disturb greatly in mind **:** DISQUIET
2 : to throw into confusion **:** DISORDER
3 : to cause to experience a perturbation
synonym see DISCOMPOSE
— per·turb·able \-ˈtər-bə-bəl\ *adjective*

per·tur·ba·tion \ˌpər-tər-ˈbā-shən, ˌpər-ˌtər-\ *noun* (14th century)
1 : the action of perturbing **:** the state of being perturbed
2 : a disturbance of motion, course, arrangement, or state of equilibrium; *especially* **:** a disturbance of the regular and usually elliptical course of motion of a celestial body that is produced by some force additional to that which causes its regular motion
— per·tur·ba·tion·al \-shnəl, -shə-nᵊl\ *adjective*

per·tus·sis \pər-ˈtə-səs\ *noun* [New Latin, from Latin *per-* thoroughly + *tussis* cough] (circa 1799)
: WHOOPING COUGH

pe·ruke \pə-ˈrük\ *noun* [Middle French *perruque,* from Old Italian *parrucca, perrucca* hair, wig] (circa 1573)
: WIG; *specifically* **:** one of a type popular from the 17th to the early 19th century
— pe·ruked \-ˈrükt\ *adjective*

peruke

pe·ruse \pə-ˈrüz\ *transitive verb* **pe·rused; pe·rus·ing** [Middle English, to use up, deal with in sequence, from Latin *per-* thoroughly + Middle English *usen* to use] (1532)
1 a : to examine or consider with attention and in detail **:** STUDY **b :** to look over or through in a casual or cursory manner
2 : READ; *especially* **:** to read over in an attentive or leisurely manner ■
— pe·rus·al \-ˈrü-zəl\ *noun*
— pe·rus·er *noun*

per·vade \pər-ˈvād\ *transitive verb* **per·vad·ed; per·vad·ing** [Latin *pervadere* to go through, pervade, from *per-* through + *vadere* to go — more at PER-, WADE] (1659)
: to become diffused throughout every part of

per·va·sion \pər-ˈvā-zhən\ *noun* (1661)
: the action of pervading or condition of being pervaded

per·va·sive \pər-ˈvā-siv, -ziv\ *adjective* (circa 1750)
: that pervades or tends to pervade ⟨a *pervasive* odor⟩
— per·va·sive·ly *adverb*
— per·va·sive·ness *noun*

per·verse \(ˌ)pər-ˈvərs, ˈpər-ˌ\ *adjective* [Middle English, from Latin *perversus,* from past participle of *pervertere*] (14th century)
1 a : turned away from what is right or good **:** CORRUPT **b :** IMPROPER, INCORRECT **c :** contrary to the evidence or the direction of the judge on a point of law ⟨*perverse* verdict⟩
2 a : obstinate in opposing what is right, reasonable, or accepted **:** WRONGHEADED **b :** arising from or indicative of stubbornness or obstinacy
3 : marked by peevishness or petulance **:** CRANKY
synonym see CONTRARY
— per·verse·ly *adverb*
— per·verse·ness *noun*
— per·ver·si·ty \pər-ˈvər-sə-tē, -stē\ *noun*

per·ver·sion \pər-ˈvər-zhən, -shən\ *noun* (14th century)
1 : the action of perverting **:** the condition of being perverted
2 : a perverted form; *especially* **:** an aberrant sexual practice especially when habitual and preferred to normal coitus

per·ver·sive \-ˈvər-siv, -ziv\ *adjective* (1817)
1 : that perverts or tends to pervert
2 : arising from or indicative of perversion

¹per·vert \pər-ˈvərt\ *transitive verb* [Middle English, from Middle French *pervertir,* from Latin *pervertere* to overturn, corrupt, pervert, from *per-* thoroughly + *vertere* to turn — more at PER-, WORTH] (14th century)
1 a : to cause to turn aside or away from what is good or true or morally right **:** CORRUPT **b :** to cause to turn aside or away from what is generally done or accepted **:** MISDIRECT
2 a : to divert to a wrong end or purpose **:** MISUSE **b :** to twist the meaning or sense of **:** MISINTERPRET
synonym see DEBASE
— per·vert·er *noun*

²per·vert \ˈpər-ˌvərt\ *noun* (circa 1661)
: one that has been perverted; *specifically* **:** one given to some form of sexual perversion

per·vert·ed \pər-ˈvər-təd\ *adjective* (14th century)
1 : CORRUPT
2 : marked by perversion
— per·vert·ed·ly *adverb*
— per·vert·ed·ness *noun*

per·vi·ous \ˈpər-vē-əs\ *adjective* [Latin *pervius,* from *per-* through + *via* way — more at PER-, WAY] (circa 1614)
1 : ACCESSIBLE ⟨*pervious* to reason⟩
2 : PERMEABLE ⟨*pervious* soil⟩
— per·vi·ous·ness *noun*

Pe·sach \ˈpā-ˌsäk\ *noun* [Hebrew *pesaḥ*] (1613)
: PASSOVER

pe·se·ta \pə-ˈsā-tə\ *noun* [Spanish, from diminutive of *peso*] (1811)
— see MONEY table

pe·se·wa \pə-ˈsā-wə\ *noun* [Twi *pésɛwa,* literally, penny, penny's worth of gold dust] (1965)
— see *cedi* at MONEY table

pes·ky \ˈpes-kē\ *adjective* **pes·ki·er; -est** [probably irregular from *pest* + *-y*] (1775)
: TROUBLESOME, VEXATIOUS ⟨*pesky* issues⟩

pe·so \ˈpā-(ˌ)sō, ˈpe-\ *noun, plural* **pesos** [Spanish, literally, weight, from Latin *pensum* — more at POISE] (1555)
1 : an old silver coin of Spain and Spanish America equal to eight reals
2 — see MONEY table

pes·sa·ry \ˈpe-sə-rē\ *noun, plural* **-ries** [Middle English *pessarie,* from Late Latin *pessarium,* from *pessus, pessum* pessary, from Greek *pessos* oval stone for playing checkers, pessary] (14th century)
1 : a vaginal suppository
2 : a device worn in the vagina to support the uterus, remedy a malposition, or prevent conception

pes·si·mism \ˈpe-sə-ˌmi-zəm *also* ˈpe-zə-\ *noun* [French *pessimisme,* from Latin *pessimus* worst — more at PEJORATIVE] (1815)
1 : an inclination to emphasize adverse aspects, conditions, and possibilities or to expect the worst possible outcome
2 a : the doctrine that reality is essentially evil **b :** the doctrine that evil overbalances happiness in life
— pes·si·mist \-mist\ *noun*

pes·si·mis·tic \ˌpe-sə-ˈmis-tik *also* ˌpe-zə-\ *adjective* (1868)
: of, relating to, or characterized by pessimism **:** GLOOMY
synonym see CYNICAL
— pes·si·mis·ti·cal·ly \-ti-k(ə-)lē\ *adverb*

pest \ˈpest\ *noun* [Middle French *peste,* from Latin *pestis*] (1553)
1 : an epidemic disease associated with high mortality; *specifically* **:** PLAGUE
2 : something resembling a pest in destructiveness; *especially* **:** a plant or animal detrimental to humans or human concerns (as agriculture or livestock production)
3 : one that pesters or annoys **:** NUISANCE ◆
— pesty \ˈpes-tē\ *adjective*

pes·ter \ˈpes-tər\ *transitive verb* **pes·tered; pes·ter·ing** \-t(ə-)riŋ\ [modification of Middle French *empestrer* to hobble, embarrass, from (assumed) Vulgar Latin *impastoriare,* from Latin *in-* + Medieval Latin *pastoria* hobble — more at PASTERN] (1533)
1 *obsolete* **:** OVERCROWD
2 : to harass with petty irritations **:** ANNOY
synonym see WORRY

pest·hole \ˈpest-ˌhōl\ *noun* (1903)
: a place liable to epidemic disease

pest·house \-ˌhau̇s\ *noun* (1611)
: a shelter or hospital for those infected with a pestilential or contagious disease

pes·ti·cide \ˈpes-tə-ˌsīd\ *noun* (circa 1925)
: an agent used to destroy pests

□ USAGE

peruse *Peruse* is a word with a literary flavor. It has been used in sense 2 since the time of Shakespeare and Marlowe, and it has been a useful alternative to the monosyllabic *read* for poets. It was probably a more ordinary word in the past than it is now, although it still has considerable use. About 1906 a writer on usage decided that *peruse* could mean only "to read with care and attention," for what reason we do not know. In time this opinion was echoed by a number of commentators, right down to the present. *Peruse* actually has been used in the "careful and attentive" sense, but writers almost always signal that meaning by a modifier ⟨have you with heed *perused* what I have written to you? —Shakespeare⟩ ⟨the lady in blue velvet, who so attentively *peruses* her book —Washington Irving⟩ ⟨each paragraph was carefully *perused* —Fred Allen⟩. And it has also been used, usually with a qualifier, to suggest casual or leisurely reading ⟨advertisements are now so numerous that they are very negligently *perused* —Samuel Johnson⟩ ⟨mirrors are gazed into, letters dreamily *perused* —Simon Schama⟩. But most of the time it simply means "to read" or "to read over or through" ⟨a magazine that the civilized reader may *peruse* without damage to his stomach —H. L. Mencken⟩ ⟨a dozen men, most of them huddled in dark suits at separate tables, *perusing* newspapers —A. N. Wilson⟩ ⟨these are what the bulk of the population *peruses* —Anthony Burgess⟩.

◇ WORD HISTORY

pest In contemporary usage a *pest* may be no more than a nuisance. Originally, however, the word meant something far more serious. In the 16th century when English borrowed *peste,* it was the French word for "plague." Although it could be used of any epidemic disease resulting in many deaths, the word customarily referred specifically to bubonic plague. The word *pest* had not been in English for very long before it was being used indiscriminately of anything destructive or troublesome. While French *peste* has retained all of its original force, the English word can be applied to persons or things that are only mildly irritating. *Pest* is now naturally associated with the verb *pester,* which is etymologically unrelated to it.

pes·tif·er·ous \pes-'ti-f(ə-)rəs\ *adjective* [Middle English, from Latin *pestifer* pestilential, noxious, from *pestis* + *-fer* -ferous] (15th century)
1 : dangerous to society **:** PERNICIOUS
2 a : carrying or propagating infection **:** PESTILENTIAL **b :** infected with a pestilential disease
3 : TROUBLESOME, ANNOYING
— pes·tif·er·ous·ly *adverb*
— pes·tif·er·ous·ness *noun*

pes·ti·lence \'pes-tə-lən(t)s\ *noun* (14th century)
1 : a contagious or infectious epidemic disease that is virulent and devastating; *especially* **:** BUBONIC PLAGUE
2 : something that is destructive or pernicious ⟨I'll pour this *pestilence* into his ear —Shakespeare⟩

pes·ti·lent \-lənt\ *adjective* [Middle English, from Latin *pestilent-, pestilens* pestilential, from *pestis*] (14th century)
1 : destructive of life **:** DEADLY
2 : injuring or endangering society **:** PERNICIOUS
3 : causing displeasure or annoyance
4 : INFECTIOUS, CONTAGIOUS ⟨*pestilent* disease⟩
— pes·ti·lent·ly *adverb*

pes·ti·len·tial \,pes-tə-'len(t)-shəl\ *adjective* (14th century)
1 a : causing or tending to cause pestilence **:** DEADLY **b :** of or relating to pestilence
2 : morally harmful **:** PERNICIOUS
3 : giving rise to vexation or annoyance **:** IRRITATING
— pes·ti·len·tial·ly \-'len(t)-sh(ə-)lē\ *adverb*

1pes·tle \'pe-səl, 'pes-tᵊl\ *noun* [Middle English *pestel,* from Middle French, from Latin *pistillum,* from *pinsere* to pound, crush; akin to Greek *ptissein* to crush, Sanskrit *pinaṣṭi* he pounds] (14th century)
1 : a usually club-shaped implement for pounding or grinding substances in a mortar
2 : any of various devices for pounding, stamping, or pressing

pestle 1 (with mortar)

2pestle *verb* **pes·tled; pes·tling** \'pe-s(ə-)liŋ, 'pes-tᵊl-iŋ\ (15th century)
transitive verb
archaic **:** to beat, pound, or pulverize with or as if with a pestle
intransitive verb
: to work with a pestle **:** use a pestle

pes·to \'pes-(,)tō\ *noun* [Italian, from *pesto,* adjective, pounded, from *pestare* to pound, from Late Latin *pistare,* frequentative of Latin *pinsere*] (1937)
: a sauce made especially of fresh basil, garlic, oil, pine nuts, and grated cheese

1pet \'pet\ *noun* [perhaps back-formation from Middle English *pety* small — more at PETTY] (1508)
1 a : a pampered and usually spoiled child **b :** a person who is treated with unusual kindness or consideration **:** DARLING
2 : a domesticated animal kept for pleasure rather than utility

2pet *adjective* (1584)
1 : kept or treated as a pet
2 : expressing fondness or endearment ⟨a *pet* name⟩
3 : FAVORITE ⟨a *pet* project⟩

3pet *verb* **pet·ted; pet·ting** (1629)
transitive verb
1 a : to treat as a pet **b :** to stroke in a gentle or loving manner
2 : to treat with unusual kindness and consideration **:** PAMPER
intransitive verb
: to engage in amorous embracing, caressing, and kissing **:** NECK
— pet·ter *noun*

4pet *noun* [origin unknown] (1590)
: a fit of peevishness, sulkiness, or anger

5pet *intransitive verb* **pet·ted; pet·ting** (1629)
: to take offense **:** SULK

peta- *combining form* [International Scientific Vocabulary, modification of Greek *penta-* penta-]
: quadrillion ⟨*peta*-electron volts⟩

pet·al \'pe-tᵊl\ *noun* [New Latin *petalum,* from Greek *petalon;* akin to Greek *petannynai* to spread out — more at FATHOM] (circa 1726)
: one of the modified often brightly colored leaves of the corolla of a flower — see FLOWER illustration
— pet·aled *or* **pet·alled** \-tᵊld\ *adjective*
— pet·al·like \-tᵊl-,(l)īk\ *adjective*

pet·al·oid \'pe-tᵊl-,ȯid\ *adjective* (1730)
1 : resembling a flower petal
2 : consisting of petaloid elements

pet·al·ous \'pe-tᵊl-əs\ *adjective* (circa 1736)
1 : having petals
2 : having (such or so many) petals — used in combination ⟨poly*petalous*⟩

pe·tard \pə-'tär(d)\ *noun* [Middle French, from *peter* to break wind, from *pet* expulsion of intestinal gas, from Latin *peditum,* from neuter of *peditus,* past participle of *pedere* to break wind; akin to Greek *bdein* to break wind] (1598)
1 : a case containing an explosive to break down a door or gate or breach a wall
2 : a firework that explodes with a loud report

pet·a·sos *or* **pet·a·sus** \'pe-tə-səs\ *noun* [Latin & Greek; Latin *petasus,* from Greek *petasos;* akin to Greek *petannynai* to spread out] (1599)
: a broad-brimmed low-crowned hat worn by ancient Greeks and Romans; *especially* **:** the winged hat of Hermes

pet·cock \'pet-,käk\ *noun* [*pet-* (perhaps from *petty*) + *cock*] (circa 1864)
: a small cock, faucet, or valve for releasing a gas (as air) or draining

pe·te·chia \pə-'tē-kē-ə\ *noun, plural* **-chi·ae** \-kē-,ī\ [New Latin, from Italian *petecchia,* ultimately from Latin *impetigo*] (circa 1796)
: a minute reddish or purplish spot containing blood that appears in skin or mucous membrane especially in some infectious diseases (as typhoid fever)
— pe·te·chi·al \-kē-əl\ *adjective*

1pe·ter \'pē-tər\ *intransitive verb* [origin unknown] (1846)
1 : to diminish gradually and come to an end **:** GIVE OUT — usually used with *out* ⟨novelists whose creative impetus seems largely to have *petered* out —*Times Literary Supplement*⟩
2 : to become exhausted — usually used with *out*

2peter *noun* [from the name *Peter*] (circa 1902)
: PENIS — often considered vulgar

Pe·ter \'pē-tər\ *noun* [Late Latin *Petrus,* from Greek *Petros,* from *petra* rock]
1 : a fisherman of Galilee and one of the twelve apostles
2 : either of two hortatory letters written to early Christians and included as books of the New Testament — see BIBLE table

Peter Pan \-'pan\ *noun*
1 : a boy in Sir James Barrie's play *Peter Pan* who lives without growing older in a never-never land
2 : an adult who doesn't want to grow up **:** one who hangs on to adolescent interests and attitudes

Peter Pan collar *noun* (1908)
: a usually small flat close-fitting collar with rounded ends that meet at the top in front

Peter Principle *noun* [Laurence J. *Peter* (born 1919) American (Canadian-born) educator] (1968)
: an observation: in a hierarchy employees tend to rise to the level of their incompetence

Peter's pence *noun plural but singular in construction* [Middle English; from the tradition that Saint Peter founded the papal see] (13th century)
1 : an annual tribute of a penny formerly paid by each householder in England to the papal see
2 : a voluntary annual contribution made by Roman Catholics to the pope

Peter Pan collar

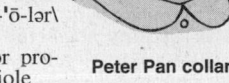

pet·i·o·lar \,pe-tē-'ō-lər\ *adjective* (1760)
: of, relating to, or proceeding from a petiole

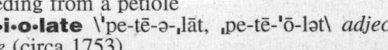

pet·i·o·late \'pe-tē-ə-,lāt, ,pe-tē-'ō-lət\ *adjective* (circa 1753)
: having a stalk or petiole

pet·i·ole \'pe-tē-,ōl\ *noun* [New Latin *petiolus,* from Latin *petiolus, peciolus* small foot, fruit stalk, probably alteration of (assumed) Latin *pediciolus,* diminutive of *pediculus,* diminutive of *ped-, pes* foot — more at FOOT] (1753)
1 : a slender stem that supports the blade of a foliage leaf
2 : PEDUNCLE; *specifically* **:** a slender abdominal segment joining the rest of the abdomen to the thorax in some insects
— pet·i·oled \-,ōld\ *adjective*

pet·i·o·lule \'pe-tē-ō-,lül, ,pe-tē-'ōl-(,)yü(ə)l\ *noun* [New Latin *petiolulus,* diminutive of *petiolus*] (1832)
: a stalk of a leaflet of a compound leaf

pet·it \'pe-tē\ *adjective* [Middle English, small, minor, from Middle French, small] (14th century)
: PETTY 1 — used chiefly in legal compounds

pe·tit bourgeois \pə-'tē-, ,pe-tē-\ *noun* [French, literally, small bourgeois] (1853)
1 : a member of the petite bourgeoisie
2 : PETITE BOURGEOISIE
— petit bourgeois *adjective*

1pe·tite \pə-'tēt\ *adjective* [French, feminine of *petit*] (1784)
: having a small trim figure — usually used of a woman
— pe·tite·ness *noun*

2petite *noun* (circa 1929)
: a clothing size for short women

petite bourgeoisie *noun* [French, literally, small bourgeoisie] (1916)
: the lower middle class including especially small shopkeepers and artisans

pe·tit four \,pe-tē-'fōr, pə-,tē-, -'fȯr; -'fu̇r\ *noun, plural* **petits fours** *or* **petit fours** \-'fōrz, -'fȯrz, -'fu̇r(z)\ [French, literally, small oven] (1884)
: a small cake cut from pound or sponge cake and frosted

1pe·ti·tion \pə-'ti-shən\ *noun* [Middle English, from Middle French, from Latin *petition-, petitio,* from *petere* to seek, request — more at FEATHER] (14th century)
1 : an earnest request **:** ENTREATY
2 a : a formal written request made to an official person or organized body (as a court) **b :** a document embodying such a formal written request
3 : something asked or requested
— pe·ti·tion·ary \-'ti-shə-,ner-ē\ *adjective*

2petition *verb* **pe·ti·tioned; pe·ti·tion·ing** \-'ti-sh(ə-)niŋ\ (1607)
transitive verb
: to make a request to **:** SOLICIT
intransitive verb
: to make a request; *especially* **:** to make a formal written request
— pe·ti·tion·er \-sh(ə-)nər\ *noun*

\ə\ **abut** \ᵊ\ **kitten** \ər\ **further** \a\ **ash** \ā\ **ace** \ä\ **mop, mar** \au̇\ **out** \ch\ **chin** \e\ **bet** \ē\ **easy** \g\ **go** \i\ **hit** \ī\ **ice** \j\ **job** \ŋ\ **sing** \ō\ **go** \ȯ\ **law** \ȯi\ **boy** \th\ **thin** \t͟h\ **the** \ü\ **loot** \u̇\ **foot** \y\ **yet** \zh\ **vision** *see also* Guide to Pronunciation

pe·ti·tio prin·ci·pii \pə-ˈtē-tē-ˌō-(ˌ)priŋ-ˈki-pē-ˌē\ *noun* [Medieval Latin, literally, postulation of the beginning, begging the question] (1531)
: a logical fallacy in which a premise is assumed to be true without warrant or in which what is to be proved is implicitly taken for granted

pet·it jury \ˈpe-tē-\ *noun* (15th century)
: a jury of 12 persons impaneled to try and to decide finally upon the facts at issue in causes for trial in a court

petit larceny *noun* (1587)
: larceny involving property of a value below a legally established minimum

pe·tit–maî·tre \pə-ˌtē-ˈmātrᵊ\ *noun, plural* **petits–maîtres** *same*\ [French, literally, small master] (1711)
: DANDY, FOP

pe·tit mal \ˈpe-tē-ˌmal, -ˌmäl\ *noun* [French, literally, small illness] (1874)
: epilepsy characterized by mild convulsive seizure with transient clouding of consciousness — compare GRAND MAL

pet·it point \ˈpe-tē-ˌpȯint\ *noun* [French, literally, small point] (1882)
: TENT STITCH; *also* **:** embroidery made with this stitch

pet peeve *noun* (circa 1919)
: a frequent subject of complaint

petr- *or* **petri-** *or* **petro-** *combining form* [New Latin, from Greek *petr-, petro-,* from *petros* stone & *petra* rock]
1 : stone **:** rock ⟨*petro*logy⟩
2 : petroleum ⟨*petro*dollar⟩

pe·tra·le sole \pə-ˈträ-lē-\ *noun* [*petrale* probably from Italian dialect, a flatfish] (1953)
: a flounder (*Eopsetta jordani*) chiefly of the Pacific waters of North America that is an important food fish — called also *petrale*

Pe·trar·chan sonnet \pi-ˈträr-kən-, ˌpē-, (ˌ)pe-\ *noun* [*Petrarch* (Francesco Petrarca)] (circa 1909)
: ITALIAN SONNET

pe·trel \ˈpe-trəl, ˈpē-\ *noun* [alteration of earlier *pitteral*] (1676)
: any of numerous seabirds (especially families Procellariidae and Hydrobatidae); *especially* **:** one of the smaller long-winged birds that fly far from land — compare STORM PETREL

pe·tri dish \ˈpē-trē-\ *noun* [Julius R. *Petri* (died 1921) German bacteriologist] (1892)
: a small shallow dish of thin glass or plastic with a loose cover used especially for cultures in bacteriology

pet·ri·fac·tion \ˌpe-trə-ˈfak-shən\ *noun* (15th century)
1 : the process of petrifying
2 : something petrified
3 : the quality or state of being petrified

pet·ri·fi·ca·tion \ˌpe-trə-fə-ˈkā-shən\ *noun* (circa 1611)
: PETRIFACTION

pet·ri·fy \ˈpe-trə-ˌfī\ *verb* **-fied; -fy·ing** [Middle French *petrifier,* from *petr-* + *-ifier* -ify] (1594)
transitive verb
1 : to convert (organic matter) into stone or a substance of stony hardness by the infiltration of water and the deposition of dissolved mineral matter
2 : to make rigid or inert like stone: **a :** to make lifeless or inactive **:** DEADEN ⟨slogans are apt to *petrify* a man's thinking —*Saturday Review*⟩ **b :** to confound with fear, amazement, or awe ⟨a novel about an airline pilot that will *petrify* you —Martin Levin⟩
intransitive verb
: to become stone or of stony hardness or rigidity

Pe·trine \ˈpē-ˌtrīn\ *adjective* [Late Latin *Petrus* Peter] (1846)
1 : of, relating to, or characteristic of the apostle Peter or the doctrines associated with his name
2 : of, relating to, or characteristic of Peter the Great or his reign

pet·ro·chem·i·cal \ˌpe-trō-ˈke-mi-kəl\ *noun* (1942)
: a chemical isolated or derived from petroleum or natural gas
— pet·ro·chem·is·try \-ˈke-mə-strē\ *noun*

pet·ro·dol·lar \ˈpe-trō-ˌdä-lər\ *noun* (1974)
: a dollar's worth of foreign exchange obtained by a petroleum-exporting country through sales abroad — usually used in plural

pet·ro·gen·e·sis \ˌpe-trō-ˈje-nə-səs\ *noun* [New Latin] (1901)
: the origin or formation of rocks
— pet·ro·ge·net·ic \-jə-ˈne-tik\ *adjective*

pet·ro·glyph \ˈpe-trə-ˌglif\ *noun* [French *pétroglyphe,* from *pétr-* petr- + *-glyphe* (as in *hiéroglyphe* hieroglyph)] (1870)
: a carving or inscription on a rock

pe·trog·ra·phy \pə-ˈträ-grə-fē, pe-\ *noun* [New Latin *petrographia,* from *petr-* + Latin *-graphia* -graphy] (1651)
: the description and systematic classification of rocks
— pe·trog·ra·pher \-fər\ *noun*
— pet·ro·graph·ic \ˌpe-trə-ˈgra-fik\ *or* **pet·ro·graph·i·cal** \-fi-kəl\ *adjective*
— pet·ro·graph·i·cal·ly \-fi-k(ə-)lē\ *adverb*

pet·rol \ˈpe-trəl, -ˌträl\ *noun* [French *essence de pétrole,* literally, essence of petroleum] (1895)
chiefly British **:** GASOLINE

pet·ro·la·tum \ˌpe-trə-ˈlā-təm, -ˈlä-\ *noun* [New Latin, from Medieval Latin *petroleum*] (1887)
: PETROLEUM JELLY

pe·tro·leum \pə-ˈtrō-lē-əm, -ˈtrōl-yəm\ *noun* [Middle English, from Medieval Latin, from Latin *petr-* + *oleum* oil — more at OIL] (15th century)
: an oily flammable bituminous liquid that may vary from almost colorless to black, occurs in many places in the upper strata of the earth, is a complex mixture of hydrocarbons with small amounts of other substances, and is prepared for use as gasoline, naphtha, or other products by various refining processes

petroleum jelly *noun* (1897)
: a neutral unctuous odorless tasteless substance obtained from petroleum and used especially in ointments and dressings

pe·trol·o·gy \pə-ˈträ-lə-jē, pe-\ *noun* [International Scientific Vocabulary] (1811)
: a science that deals with the origin, history, occurrence, structure, chemical composition, and classification of rocks
— pet·ro·log·ic \ˌpe-trə-ˈlä-jik\ *or* **pet·ro·log·i·cal** \-ji-kəl\ *adjective*
— pet·ro·log·i·cal·ly \-ji-k(ə-)lē\ *adverb*
— pe·trol·o·gist \pə-ˈträ-lə-jist, pe-\ *noun*

pet·ro·nel \ˌpe-trə-ˈnel\ *noun* [perhaps modification of Middle French *poitrinal, petrinal,* from *poitrinal* of the chest, from *poitrine* chest, derivative of Latin *pector-, pectus* — more at PECTORAL] (circa 1577)
: a portable firearm resembling a carbine of large caliber

pe·tro·sal \pə-ˈtrō-səl\ *adjective* [New Latin *petrosa* petrous portion of the temporal bone, from Latin, feminine of *petrosus*] (1741)
: of, relating to, or situated in the region of the petrous portion of the temporal bone or capsule of the inner ear

pe·trous \ˈpe-trəs, ˈpē-\ *adjective* [Middle English, from Middle French *petreux,* from Latin *petrosus,* from *petra* rock, from Greek] (14th century)
: of, relating to, or constituting the exceptionally hard and dense portion of the human temporal bone that contains the internal auditory organs

¹pet·ti·coat \ˈpe-tē-ˌkōt\ *noun* [Middle English *petycote* short tunic, petticoat, from *pety* small + *cote* coat] (15th century)
1 : a skirt worn by women, girls, or young children: as **a :** an outer skirt formerly worn by women and small children **b :** a fancy skirt made to show below a draped-up overskirt **c :** an underskirt usually a little shorter than outer clothing and often made with a ruffled, pleated, or lace edge **d** *archaic* **:** the skirt of a woman's riding habit
2 a : a garment characteristic or typical of women **b :** WOMAN
3 : something (as a valance) resembling a petticoat
— pet·ti·coat·ed \-ˌkō-təd\ *adjective*

²petticoat *adjective* (1660)
: of, relating to, or exercised by women **:** FEMALE

pet·ti·fog·ger \ˈpe-tē-ˌfȯ-gər, -ˌfä-\ *noun* [probably from *petty* + obsolete English *fogger* pettifogger] (1576)
1 : a lawyer whose methods are petty, underhanded, or disreputable **:** SHYSTER
2 : one given to quibbling over trifles
— pet·ti·fog·ging \-giŋ\ *adjective or noun*
— pet·ti·fog·gery \-g(ə-)rē\ *noun*

petting zoo *noun* (1968)
: a collection of farm animals or gentle exotic animals for children to pet and feed

pet·tish \ˈpe-tish\ *adjective* [probably from ⁴*pet*] (circa 1591)
: FRETFUL, PEEVISH
— pet·tish·ly *adverb*
— pet·tish·ness *noun*

pet·ti·toes \ˈpe-tē-ˌtōz\ *noun plural* [plural of obsolete *pettytoe* offal] (1555)
1 : the feet of a pig used as food
2 : TOES, FEET

pet·ty \ˈpe-tē\ *adjective* **pet·ti·er; -est** [Middle English *pety* small, minor, alteration of *petit*] (14th century)
1 : having secondary rank or importance **:** MINOR, SUBORDINATE
2 : having little or no importance or significance
3 : marked by or reflective of narrow interests and sympathies **:** SMALL-MINDED
— pet·ti·ly \ˈpe-tᵊl-ē\ *adverb*
— pet·ti·ness \ˈpe-tē-nəs\ *noun*

petty cash *noun* (1834)
: cash kept on hand for payment of minor items

petty larceny *noun* (1818)
: PETIT LARCENY

petty officer *noun* (1760)
: a subordinate officer in the navy or coast guard appointed from among the enlisted men — compare NONCOMMISSIONED OFFICER

petty officer first class *noun* (1942)
: an enlisted man in the navy or coast guard ranking above a petty officer second class and below a chief petty officer

petty officer second class *noun* (1942)
: an enlisted man in the navy or coast guard ranking above a petty officer third class and below a petty officer first class

petty officer third class *noun* (1942)
: an enlisted man in the navy or coast guard ranking above a seaman and below a petty officer second class

pet·u·lance \ˈpe-chə-lən(t)s\ *noun* (1610)
: the quality or state of being petulant **:** PEEVISHNESS

pet·u·lan·cy \-lən(t)-sē\ *noun* (1559)
archaic **:** PETULANCE

pet·u·lant \-lənt\ *adjective* [Latin or Middle French; Middle French, from Latin *petulant-, petulans;* akin to Latin *petere* to go to, attack, seek — more at FEATHER] (1605)
1 : insolent or rude in speech or behavior
2 : characterized by temporary or capricious ill humor **:** PEEVISH
— pet·u·lant·ly *adverb*

pe·tu·nia \pi-ˈtün-yə, -ˈtyün-\ *noun* [New Latin, from obsolete French *petun* tobacco, from Tupi *petima*] (1825)
: any of a genus (*Petunia*) of tropical American herbs of the nightshade family with flowers having funnel-shaped corollas

petunia

pew \ˈpyü\ *noun* [Middle English *pewe,* from Middle French *puie* balustrade, from Latin *podia,* plural of *podium* parapet, podium, from Greek *podion* base, diminutive of *pod-, pous* foot — more at FOOT] (14th century)
1 : a compartment in the auditorium of a church providing seats for several persons
2 : one of the benches with backs and sometimes doors fixed in rows in a church

pe·wee \ˈpē-(ˌ)wē\ *noun* [imitative] (1796)
: any of various small olivaceous flycatchers (genus *Contopus*)

pew·hold·er \ˈpyü-ˌhōl-dər\ *noun* (1845)
: a renter or owner of a pew

pe·wit *variant of* PEEWIT

pew·ter \ˈpyü-tər\ *noun* [Middle English, from Middle French *peutre,* from (assumed) Vulgar Latin *peltrum*] (14th century)
1 : any of various alloys having tin as chief component; *especially* **:** a dull alloy with lead formerly used for domestic utensils
2 : utensils of pewter
3 : a bluish gray
— pewter *adjective*

pew·ter·er \ˈpyü-tər-ər\ *noun* (14th century)
: one that makes pewter utensils

pey·o·te \pā-ˈō-tē\ *also* **pey·otl** \-ˈō-tᵊl\ *noun* [Mexican Spanish *peyote,* from Nahuatl *peyotl*] (1849)
1 : a stimulant drug derived from mescal buttons
2 : any of several American cacti (genus *Lophophora*); *especially* **:** MESCAL

pfen·nig \ˈfe-nig, -nik, *German* ˈ(p)fe-nik\ *noun, plural* **pfennig** *also* **pfen·nigs** \ˈfe-nigz, -niks\ *or* **pfen·ni·ge** \ˈ(p)fe-ni-gə, -ni-yə\ [German, from Old High German *pfenning* — more at PENNY] (1547)
— see *deutsche mark* at MONEY table

PG \ˈpē-ˈjē\ *certification mark*
— used to certify that a motion picture is of such a nature that all ages may be allowed admission but parental guidance is suggested; compare G, NC-17, PG-13, R

PG–13 \-ˌthər(t)-ˈtēn\ *certification mark*
— used to certify that a motion picture is of such a nature that persons of all ages may be admitted but parental guidance is suggested especially for children under 13; compare G, NC-17, PG, R

pH \ˈpē-ˈāch\ *noun* [German, from *P*otenz power + *H* (symbol for hydrogen)] (1909)
: a measure of acidity and alkalinity of a solution that is a number on a scale on which a value of 7 represents neutrality and lower numbers indicate increasing acidity and higher numbers increasing alkalinity and on which each unit of change represents a tenfold change in acidity or alkalinity and that is the negative logarithm of the effective hydrogen-ion concentration or hydrogen-ion activity in gram equivalents per liter of the solution; *also* **:** the condition represented by a pH number

Phae·dra \ˈfē-drə\ *noun* [Latin, from Greek *Phaidra*]
: a daughter of Minos who marries Theseus and falls in love with her stepson Hippolytus

Pha·ë·thon \ˈfā-ə-tᵊn, -tən, -ˌthän\ *noun* [Latin, from Greek *Phaethōn*]
: a son of Helios who drives his father's sun-chariot through the sky but loses control and is struck down by a thunderbolt of Zeus

pha·eton \ˈfā-ə-tᵊn\ *noun* [*Phaëthon*] (1742)
1 : any of various light four-wheeled horse-drawn vehicles
2 : TOURING CAR

phage \ˈfāj *also* ˈfäzh\ *noun* [by shortening] (1926)
: BACTERIOPHAGE

-phage *noun combining form* [Greek *-phagos* one that eats, from *-phagos* -phagous]
: virus or cell that destroys cells ⟨bacterio*phage*⟩ ⟨micro*phage*⟩

-phagia *noun combining form* [New Latin, from Greek]
: -PHAGY ⟨dys*phagia*⟩

phago·cyte \ˈfa-gə-ˌsīt\ *noun* [International Scientific Vocabulary, from Greek *phagein* + New Latin *-cyta* -cyte] (circa 1884)
: a cell (as a white blood cell) that engulfs and consumes foreign material (as microorganisms) and debris
— phago·cyt·ic \ˌfa-gə-ˈsi-tik\ *adjective*

phago·cy·tize \ˈfa-gə-sə-ˌtīz, -ˌsī-\ *transitive verb* **-tized; -tiz·ing** (1913)
: PHAGOCYTOSE

phago·cy·tose \-sə-ˌtōs, -sī-, -ˌtōz\ *transitive verb* **-tosed; -tos·ing** [back-formation from *phagocytosis*] (1912)
: to consume by phagocytosis

phago·cy·to·sis \ˌfa-gə-sə-ˈtō-səs, -sī-\ *noun, plural* **-to·ses** \-ˌsēz\ [New Latin] (1889)
: the engulfing and usually the destruction of particulate matter by phagocytes
— phago·cy·tot·ic \-ˈtä-tik\ *adjective*

-phagous *adjective combining form* [Greek *-phagos,* from *phagein* to eat — more at BAKSHEESH]
: eating ⟨sapro*phagous*⟩

-phagy *noun combining form, plural* **-phagies** [Greek *-phagia,* from *phagein*]
: eating of a (specified) type or substance ⟨geo*phagy*⟩

pha·lange \ˈfā-ˌlanj, fə-ˈ, fā-ˈ\ *noun* [French, from Greek *phalang-, phalanx*] (circa 1860)
: PHALANX 2

pha·lan·ge·al \ˌfā-lən-ˈjē-əl, ˌfa-; fə-ˈlan-jē-, fā-\ *adjective* (1831)
: of or relating to a phalanx or the phalanges

pha·lan·ger \fə-ˈlan-jər, ˈfā-ˌ\ *noun* [New Latin, from Greek *phalang-, phalanx*] (circa 1774)
: any of various small to medium-sized marsupial mammals (family Phalangeridae) of the Australian region that are chiefly arboreal and nocturnal and usually densely furred

phal·an·stery \ˈfa-lən-ˌster-ē\ *noun, plural* **-ster·ies** [French *phalanstère* dwelling of a Fourierist community, from Latin *phalang-, phalanx* + French *-stère* (as in *monastère* monastery)] (1846)
1 a : a Fourierist cooperative community **b :** a self-contained structure housing such a community
2 : something resembling a Fourierist phalanstery

pha·lanx \ˈfā-ˌlaŋ(k)s, *British usually* ˈfa-\ *noun, plural* **pha·lanx·es** *or* **pha·lan·ges** \fə-ˈlan-(ˌ)jēz, fā-, ˈfā-ˌ, *British usually* fa-\ [Latin *phalang-, phalanx,* from Greek, battle line, digital bone, literally, log — more at BALK] (1553)
1 : a body of heavily armed infantry in ancient Greece formed in close deep ranks and files; *broadly* **:** a body of troops in close array
2 *plural phalanges* **:** one of the digital bones of the hand or foot of a vertebrate
3 *plural usually phalanxes* **a :** a massed arrangement of persons, animals, or things **b :** an organized body of persons

phal·a·rope \ˈfa-lə-ˌrōp\ *noun, plural* **-ropes** *also* **-rope** [French, from New Latin *phalaropod-, phalaropus,* from Greek *phalaris* coot + *pod-, pous* foot; akin to Greek *phalios* having a white spot — more at BALD, FOOT] (circa 1776)
: any of a genus (*Phalaropus*) of small shorebirds that are related to sandpipers but have lobate toes and are good swimmers

phal·lic \ˈfa-lik\ *adjective* (1789)
1 : of or relating to phallicism ⟨a *phallic* cult⟩
2 : of, relating to, or resembling a phallus
3 : relating to or being the stage of psychosexual development in psychoanalytic theory during which a child becomes interested in his or her own sexual organs
— phal·li·cal·ly \-li-k(ə-)lē\ *adverb*

phal·li·cism \ˈfa-lə-ˌsi-zəm\ *noun* (1884)
: the worship of the generative principle as symbolized by the phallus

phal·lo·cen·tric \ˌfa-lə-ˈsen-trik\ *adjective* (1927)
: centered on or emphasizing the masculine point of view

phal·lus \ˈfa-ləs\ *noun, plural* **phal·li** \ˈfa-ˌlī, -ˌlē\ *or* **phal·lus·es** [Latin, from Greek *phallos* penis, representation of the penis; probably akin to Latin *flare* to blow — more at BLOW] (circa 1613)
1 : a symbol or representation of the penis
2 : PENIS

-phane *noun combining form* [Greek *phanēs* appearing, from *phainein* to show — more at FANCY]
: substance having a (specified) form, quality, or appearance ⟨hydro*phane*⟩

phan·er·o·gam \ˈfa-nə-rə-ˌgam, fə-ˈner-ə-\ *noun* [French *phanérogame,* ultimately from Greek *phaneros* visible (from *phainein*) + *gamos* marriage] (1861)
: a seed plant or flowering plant **:** SPERMATOPHYTE

phan·er·o·phyte \ˈfa-nə-rə-ˌfīt, fə-ˈner-ə-\ *noun* [Greek *phaneros* + International Scientific Vocabulary *-phyte*] (1913)
: a perennial plant that bears its perennating buds well above the surface of the ground

Phan·er·o·zo·ic \ˌfa-nə-rə-ˈzō-ik\ *adjective* [Greek *phaneros* + English ²*-zoic*] (1930)
: of, relating to, or being an eon of geologic history that comprises the Paleozoic, Mesozoic, and Cenozoic or the corresponding system of rocks — see GEOLOGIC TIME table
— Phanerozoic *noun*

phan·tasm \ˈfan-ˌta-zəm\ *noun* [Middle English *fantasme,* from Old French, from Latin *phantasma,* from Greek, from *phantazein* to present to the mind — more at FANCY] (13th century)
1 : a product of fantasy: as **a :** delusive appearance **:** ILLUSION **b :** GHOST, SPECTER **c :** a figment of the imagination
2 : a mental representation of a real object
— phan·tas·mal \fan-ˈtaz-məl\ *adjective*
— phan·tas·mic \-mik\ *adjective*

phan·tas·ma \fan-ˈtaz-mə\ *noun, plural* **-ma·ta** \-mə-tə\ [Latin] (1598)
: PHANTASM 1

phan·tas·ma·go·ria \(ˌ)fan-ˌtaz-mə-ˈgōr-ē-ə, -ˈgȯr-\ *noun* [French *phantasmagorie,* from *phantasme* phantasm (from Old French *fantasme*) + *-agorie* (perhaps from Greek *agora* assembly) — more at AGORA] (circa 1802)
1 : an exhibition or display of optical effects and illusions
2 a : a constantly shifting complex succession of things seen or imagined **b :** a scene that constantly changes
3 : a bizarre or fantastic combination, collection, or assemblage
— phan·tas·ma·gor·ic \-ˈgōr-ik, -ˈgȯr-, -ˈgär-\ *or* **phan·tas·ma·gor·i·cal** \-i-kəl\ *adjective*

phantasy *variant of* FANTASY

¹phan·tom \ˈfan-təm\ *noun* [Middle English *fantosme, fantome,* from Middle French *fantosme,* modification of Latin *phantasma*] (14th century)

1 a : something (as a specter) apparent to sense but with no substantial existence **:** APPARITION **b :** something elusive or visionary **:** WILL-O'-THE-WISP **c :** an object of continual dread or abhorrence **:** BUGBEAR ⟨the *phantom* of disease and want⟩
2 : something existing in appearance only
3 : a representation of something abstract, ideal, or incorporeal ⟨she was a *phantom* of delight —William Wordsworth⟩
— phan·tom·like \-ˌlīk\ *adverb or adjective*
²phantom *adjective* (15th century)
1 : of the nature of, suggesting, or being a phantom **:** ILLUSORY
2 : FICTITIOUS, DUMMY ⟨*phantom* voters⟩
pha·raoh \ˈfer-(ˌ)ō, ˈfar-(ˌ)ō, ˈfā-(ˌ)rō\ *noun, often capitalized* [Middle English *pharao,* from Old English, from Late Latin *pharaon-, pharao,* from Greek *pharaō,* from Hebrew *par'ōh,* from Egyptian *pr-ʿ'*,] (before 12th century)
1 : a ruler of ancient Egypt
2 : TYRANT
pharaoh ant *noun* (circa 1947)
: a little red ant (*Monomorium pharaonis*) that is a common household pest
phar·a·on·ic \ˌfer-ā-ˈä-nik, ˌfar-\ *adjective, often capitalized* [French *pharaonique,* from *pharaon* pharaoh, from Late Latin *Pharaon-, Pharao*] (circa 1828)
1 : of, relating to, or characteristic of a pharaoh or the pharaohs
2 : enormous in size or magnitude ⟨*pharaonic* construction projects⟩
phar·i·sa·ic \ˌfar-ə-ˈsā-ik\ *adjective* [Late Latin *pharisaicus,* from Late Greek *pharisaikos,* from Greek *pharisaios* Pharisee] (1618)
1 : PHARISAICAL
2 *capitalized* **:** of or relating to the Pharisees
phar·i·sa·ical \-ˈsā-ə-kəl\ *adjective* (1531)
: marked by hypocritical censorious self-righteousness
— phar·i·sa·ical·ly \-k(ə-)lē\ *adverb*
— phar·i·sa·ical·ness \-kəl-nəs\ *noun*
phar·i·sa·ism \ˈfar-ə-(ˌ)sā-ˌi-zəm\ *noun* [New Latin *pharisaismus,* from Greek *pharisaios*] (1610)
1 *capitalized* **:** the doctrines or practices of the Pharisees
2 *often capitalized* **:** pharisaical character, spirit, or attitude **:** HYPOCRISY
phar·i·see \ˈfar-ə-(ˌ)sē\ *noun* [Middle English *pharise,* from Old English *farise,* from Late Latin *pharisaeus,* from Greek *pharisaios,* from Aramaic *pĕrīshayyā,* plural of *pĕrīshā,* literally, separated] (before 12th century)
1 *capitalized* **:** a member of a Jewish sect of the intertestamental period noted for strict observance of rites and ceremonies of the written law and for insistence on the validity of their own oral traditions concerning the law
2 : a pharisaical person
¹phar·ma·ceu·ti·cal \ˌfär-mə-ˈsü-ti-kəl\ *adjective* [Late Latin *pharmaceuticus,* from Greek *pharmakeutikos,* from *pharmakeuein* to administer drugs — more at PHARMACY] (1648)
: of, relating to, or engaged in pharmacy or the manufacture and sale of pharmaceuticals ⟨a *pharmaceutical* company⟩
— phar·ma·ceu·ti·cal·ly \-ti-k(ə-)lē\ *adverb*
²pharmaceutical *noun* (1881)
: a medicinal drug
phar·ma·cist \ˈfär-mə-sist\ *noun* (1834)
: a person licensed to engage in pharmacy
pharmaco- *combining form* [Greek *pharmako-,* from *pharmakon*]
: medicine **:** drug ⟨*pharmacology*⟩
phar·ma·co·dy·nam·ics \ˌfär-mə-kō-dī-ˈna-miks, -də-\ *noun plural but singular in construction* (circa 1842)
: a branch of pharmacology dealing with the reactions between drugs and living systems
— phar·ma·co·dy·nam·ic \-mik\ *adjective*
— phar·ma·co·dy·nam·i·cal·ly \-mi-k(ə-)lē\ *adverb*
phar·ma·cog·no·sy \ˌfär-mə-ˈkäg-nə-sē\ *noun* [International Scientific Vocabulary, from Greek *pharmakon* + *-gnōsia* knowledge, from *gnōsis* — more at GNOSIS] (circa 1885)
: descriptive pharmacology dealing with crude drugs and simples
— phar·ma·cog·nos·tic \-ˌkäg-ˈnäs-tik\ *or* **phar·ma·cog·nos·ti·cal** \-ti-kəl\ *adjective*
phar·ma·co·ki·net·ics \-kō-kə-ˈne-tiks, -kō-kī-\ *noun plural but singular in construction* (1960)
1 : the study of the bodily absorption, distribution, metabolism, and excretion of drugs
2 : the characteristic interactions of a drug and the body in terms of its absorption, distribution, metabolism, and excretion
— phar·ma·co·ki·net·ic \-tik\ *adjective*
phar·ma·col·o·gy \ˌfär-mə-ˈkä-lə-jē\ *noun* (circa 1721)
1 : the science of drugs including materia medica, toxicology, and therapeutics
2 : the properties and reactions of drugs especially with relation to their therapeutic value
— phar·ma·co·log·i·cal \-kə-ˈlä-ji-kəl\ *also* **phar·ma·co·log·ic** \-jik\ *adjective*
— phar·ma·co·log·i·cal·ly \-ji-k(ə-)lē\ *adverb*
— phar·ma·col·o·gist \-ˈkä-lə-jist\ *noun*
phar·ma·co·poe·ia *also* **phar·ma·co·pe·ia** \-kə-ˈpē-ə\ *noun* [New Latin, from Late Greek *pharmakopoiia* preparation of drugs, from Greek *pharmako-* + *poiein* to make — more at POET] (1621)
1 : a book describing drugs, chemicals, and medicinal preparations; *especially* **:** one issued by an officially recognized authority and serving as a standard
2 : a collection or stock of drugs
— phar·ma·co·poe·ial *also* **phar·ma·co·pe·ial** \-əl\ *adjective*
phar·ma·co·ther·a·py \ˌfär-mə-kō-ˈther-ə-pē\ *noun* (circa 1909)
: the treatment of disease and especially mental illness with drugs
phar·ma·cy \ˈfär-mə-sē\ *noun, plural* **-cies** [Late Latin *pharmacia* administration of drugs, from Greek *pharmakeia,* from *pharmakeuein* to administer drugs, from *pharmakon* magic charm, poison, drug] (1651)
1 : the art, practice, or profession of preparing, preserving, compounding, and dispensing medical drugs
2 a : a place where medicines are compounded or dispensed **b :** DRUGSTORE
3 : PHARMACOPOEIA 2
pha·ryn·geal \ˌfar-ən-ˈjē-əl, fə-ˈrin-j(ē-)əl\ *adjective* [New Latin *pharyngeus,* from *pharyng-, pharynx*] (1828)
: relating to or located or produced in the region of the pharynx
phar·yn·gi·tis \ˌfar-ən-ˈjī-təs\ *noun, plural* **-git·i·des** \-ˈji-tə-ˌdēz\ (circa 1844)
: inflammation of the pharynx
phar·ynx \ˈfar-iŋ(k)s\ *noun, plural* **pha·ryn·ges** \fə-ˈrin-(ˌ)jēz\ *also* **phar·ynx·es** [New Latin *pharyng-, pharynx,* from Greek, throat, pharynx; akin to Old Norse *barki* throat and probably to Latin *ferire* to strike — more at BORE] (circa 1693)
1 : the part of the vertebrate alimentary canal between the cavity of the mouth and the esophagus
2 : a differentiated part of the alimentary canal in some invertebrates that may be thickened and muscular, eversible and toothed, or adapted as a suctorial organ
¹phase \ˈfāz\ *noun* [New Latin *phasis,* from Greek, appearance of a star, phase of the moon, from *phainein* to show (middle voice, to appear) — more at FANCY] (1812)
1 : a particular appearance or state in a regularly recurring cycle of changes ⟨*phases* of the moon⟩
2 a : a distinguishable part in a course, development, or cycle ⟨the early *phases* of her career⟩ **b :** an aspect or part (as of a problem) under consideration
3 : the point or stage in a period of uniform circular motion, harmonic motion, or the periodic changes of any magnitude varying according to a simple harmonic law to which the rotation, oscillation, or variation has advanced considered in its relation to a standard position or assumed instant of starting
4 : a homogeneous, physically distinct, and mechanically separable portion of matter present in a nonhomogeneous physicochemical system
5 : an individual or subgroup distinguishably different in appearance or behavior from the norm of the group to which it belongs; *also* **:** the distinguishing peculiarity
— pha·sic \ˈfā-zik\ *adjective*
— in phase : in a synchronized or correlated manner
— out of phase : in an unsynchronized manner **:** not in correlation
²phase *transitive verb* **phased; phas·ing** (1904)
1 : to adjust so as to be in a synchronized condition
2 : to conduct or carry out by planned phases
3 : to introduce in stages — often used with *in* ⟨*phase* in new models⟩
phase–contrast *adjective* (1934)
: of or employing the phase-contrast microscope
phase–contrast microscope *noun* (1947)
: a microscope that translates differences in phase of the light transmitted through or reflected by the object into differences of intensity in the image — called also *phase microscope*
phase·down \ˈfāz-ˌdau̇n\ *noun* (1964)
: a gradual reduction (as in operations)
phase modulation *noun* (1930)
: modulation of the phase of a radio carrier wave by voice or other signal
phase·out \ˈfā-ˌzau̇t\ *noun* (1958)
: a gradual stopping of operations or production **:** a closing down by phases
phase out (1940)
transitive verb
: to discontinue the practice, production, or use of by phases ⟨*phase out* the old machinery⟩
intransitive verb
: to stop production or operation by phases
-phasia *noun combining form* [New Latin, from Greek, speech, from *phasis* utterance, from *phanai* to speak, say — more at BAN]
: speech disorder of a (specified) type ⟨dys*phasia*⟩
phas·mid \ˈfaz-məd\ *noun* [New Latin *Phasmida,* group name, from *Phasma,* type genus, from Greek, apparition, from *phainein* to show — more at FANCY] (1872)
: any of an order (Phasmatodea synonym Phasmida) of large cylindrical or sometimes flattened chiefly tropical insects (as a walking stick) with long strong legs, strictly phytophagous habits, and slight metamorphosis
— phasmid *adjective*
phat·ic \ˈfa-tik\ *adjective* [Greek *phatos,* verbal of *phanai* to speak] (1923)
: of, relating to, or being speech used for social or emotive purposes rather than for communicating information
— phat·i·cal·ly \-ti-k(ə-)lē\ *adverb*
pheas·ant \ˈfe-zᵊnt\ *noun, plural* **pheasant** *or* **pheasants** [Middle English *fesaunt,* from Anglo-French, from Old French *fesan,* from Latin *phasianus,* from Greek *phasianos,* from *phasianos* of the Phasis River, from *Phasis,* river in Colchis] (13th century)

1 : any of numerous large often long-tailed and brightly colored Old World gallinaceous birds (*Phasianus* and related genera of the family Phasianidae) including many raised as ornamental or game birds
2 : any of various birds resembling a pheasant

phel·lem \'fe-ˌlem\ *noun* [Greek *phellos* cork + English *-em* (as in *phloem*)] (1887)
: a layer of usually suberized cells produced outwardly by a phellogen

phel·lo·derm \'fe-lə-ˌdərm\ *noun* [Greek *phellos* + International Scientific Vocabulary *-derm*] (1875)
: a layer of parenchyma produced inwardly by a phellogen

phel·lo·gen \'fe-lə-jən\ *noun* [Greek *phellos* + International Scientific Vocabulary *-gen*] (1875)
: a secondary meristem that initiates phellem and phelloderm in the periderm of a stem — called also *cork cambium*

phen- *or* **pheno-** *combining form* [obsolete *phene* benzene, from French *phène,* from Greek *phainein* to show; from its occurrence in illuminating gas — more at FANCY]
1 : related to or derived from benzene ⟨*phenol*⟩
2 : containing phenyl ⟨*pheno*barbital⟩

phe·na·caine \'fē-nə-ˌkān, 'fe-\ *noun* [probably from *phen*etidine + *acet-* + *-caine*] (1907)
: a crystalline base $C_{18}H_{22}N_2O_2$ or its hydrochloride used as a local anesthetic

phen·ac·e·tin \fi-'na-sə-tən\ *noun* [International Scientific Vocabulary] (1887)
: a white crystalline compound $C_{10}H_{13}NO_2$ formerly used to ease pain or fever but withdrawn because of its serious side effects — called also *acetophenetidin*

phen·a·kite \'fe-nə-ˌkīt, 'fē-\ *or* **phen·a·cite** \-ˌsīt\ *noun* [German *Phenakit,* from Greek *phenak-, phenax* deceiver; from its being easily mistaken for quartz] (circa 1834)
: a glassy mineral that consists of a beryllium silicate and occurs in rhombohedral crystals

phen·an·threne \fə-'nan-ˌthrēn\ *noun* [International Scientific Vocabulary *phen-* + *anthracene*] (1882)
: a crystalline aromatic hydrocarbon $C_{14}H_{10}$ of coal tar isomeric with anthracene

phen·a·zine \'fe-nə-ˌzēn\ *noun* [International Scientific Vocabulary] (circa 1900)
: a yellowish crystalline base $C_{12}H_8N_2$ used especially in organic synthesis

phen·cy·cli·dine \(ˌ)fen-'si-klə-ˌdēn, -'sī-, -dən\ *noun* [*phen-* + *cycl-* + *-idine*] (1959)
: a piperidine derivative $C_{17}H_{25}N$ used especially as a veterinary anesthetic and sometimes illicitly as a psychedelic drug — called also *angel dust, PCP*

phe·net·ic \fi-'ne-tik\ *adjective* [*phen*otype + *-etic* (as in *genetic*)] (1960)
: of or relating to taxonomic analysis that emphasizes the overall similarities of characteristics among biological taxa without regard to phylogenetic relationships

phe·net·ics \-tiks\ *noun plural but singular in construction* (circa 1960)
: a system of biological classification based on phenetic methods
— **phe·net·i·cist** \-'ne-tə-sist\ *noun*

phen·met·ra·zine \(ˌ)fen-'me-trə-ˌzēn\ *noun* [*phen*yl + *me*thyl + *tetra-* + *azine*] (1956)
: a sympathomimetic stimulant $C_{11}H_{15}NO$ used in the form of its hydrochloride as an appetite suppressant

phe·no·bar·bi·tal \ˌfē-nō-'bär-bə-ˌtȯl\ *noun* (1918)
: a crystalline barbiturate $C_{12}H_{12}N_2O_3$ used as a hypnotic and sedative

phe·no·bar·bi·tone \-bə-ˌtōn\ *noun* (circa 1932)
chiefly British **:** PHENOBARBITAL

phe·no·copy \'fē-nə-ˌkä-pē\ *noun* [International Scientific Vocabulary *phenotype* + *copy*] (1937)
: a phenotypic variation that is caused by unusual environmental conditions and resembles the normal expression of a genotype other than its own

phe·no·cryst \-ˌkrist\ *noun* [French *phénocryste,* from Greek *phainein* to show + *krystallos* crystal — more at FANCY] (circa 1893)
: one of the prominent embedded crystals of a porphyry
— **phe·no·crys·tic** \ˌfē-nə-'kris-tik\ *adjective*

phe·nol \'fē-ˌnōl, -ˌnȯl, fi-'\ *noun* [International Scientific Vocabulary *phen-* + [3]*-ol*] (circa 1852)
1 : a corrosive poisonous crystalline acidic compound C_6H_5OH present in coal tar and wood tar that in dilute solution is used as a disinfectant
2 : any of various acidic compounds analogous to phenol and regarded as hydroxyl derivatives of aromatic hydrocarbons

phe·no·late \'fē-nᵊl-ˌāt\ *noun* (1885)
: PHENOXIDE

phe·no·lat·ed \'fē-nᵊl-ˌā-təd\ *adjective* (1923)
: treated, mixed, or impregnated with phenol

[1]phe·no·lic \fi-'nō-lik, -'nä-\ *adjective* (1872)
1 a : of, relating to, or having the characteristics of a phenol **b :** containing or derived from a phenol
2 : of, relating to, or being a phenolic

[2]phenolic *noun* (1924)
: a usually thermosetting resin or plastic made by condensation of a phenol with an aldehyde and used especially for molding and insulating and in coatings and adhesives — called also *phenolic resin*

phe·nol·o·gy \fi-'nä-lə-jē\ *noun* [*pheno*mena + *-logy*] (circa 1884)
1 : a branch of science dealing with the relations between climate and periodic biological phenomena (as bird migration or plant flowering)
2 : periodic biological phenomena (as of a kind of organism) that are correlated with climatic conditions
— **phe·no·log·i·cal** \ˌfē-nᵊl-'ä-ji-kəl\ *adjective*
— **phe·no·log·i·cal·ly** \-k(ə-)lē\ *adverb*

phe·nol·phtha·lein \ˌfē-nᵊl-'tha-lē-ən, -'tha-ˌlēn, -'thā-\ *noun* [International Scientific Vocabulary] (circa 1875)
: a white or yellowish white crystalline compound $C_{20}H_{14}O_4$ used in analysis as an indicator because its solution is brilliant red in alkalies and is decolorized by acids and in medicine as a laxative

phenol red *noun* (1916)
: a red crystalline compound $C_{19}H_{14}O_5S$ used especially as an acid-base indicator

phe·nom \'fē-ˌnäm, fi-'näm\ *noun* (circa 1890)
: PHENOMENON; *especially* **:** a person of phenomenal ability or promise

phe·nom·e·na \fi-'nä-mə-nə, -ˌnä\ *noun, plural* **-nas** (1576)
nonstandard **:** PHENOMENON ■

phe·nom·e·nal \fi-'nä-mə-nᵊl\ *adjective* (1825)
: relating to or being a phenomenon: as **a :** known through the senses rather than through thought or intuition **b :** concerned with phenomena rather than with hypotheses **c :** EXTRAORDINARY, REMARKABLE
synonym see MATERIAL
— **phe·nom·e·nal·ly** \-nᵊl-ē\ *adverb*

phe·nom·e·nal·ism \-nᵊl-ˌi-zəm\ *noun* (circa 1865)
1 : a theory that limits knowledge to phenomena only
2 : a theory that all knowledge is of phenomena and that what is construed to be perception of material objects is simply perception of sense-data
— **phe·nom·e·nal·ist** \-ist\ *noun or adjective*
— **phe·nom·e·nal·is·tic** \-ˌnä-mə-nᵊl-'is-tik\ *adjective*
— **phe·nom·e·nal·is·ti·cal·ly** \-ti-k(ə-)lē\ *adverb*

phe·nom·e·no·log·i·cal \fi-ˌnä-mə-nᵊl-'ä-ji-kəl\ *adjective* (circa 1858)
1 : of or relating to phenomenology
2 : PHENOMENAL
3 : of or relating to phenomenalism
— **phe·nom·e·no·log·i·cal·ly** \-k(ə-)lē\ *adverb*

phe·nom·e·nol·o·gy \fi-ˌnä-mə-'nä-lə-jē\ *noun, plural* **-gies** [German *Phänomenologie,* from *Phänomenon* phenomenon + *-logie* -logy] (circa 1797)
1 : the study of the development of human consciousness and self-awareness as a preface to philosophy or a part of philosophy
2 a (1) **:** a philosophical movement that describes the formal structure of the objects of awareness and of awareness itself in abstraction from any claims concerning existence (2) **:** the typological classification of a class of phenomena ⟨the *phenomenology* of religion⟩ **b :** an analysis produced by phenomenological investigation
— **phe·nom·e·nol·o·gist** \-jist\ *noun*

phe·nom·e·non \fi-'nä-mə-ˌnän, -nən\ *noun, plural* **-na** \-nə, -ˌnä\ *or* **-nons** [Late Latin *phaenomenon,* from Greek *phainomenon,* from neuter of *phainomenos,* present participle of *phainesthai* to appear, middle voice of *phainein* to show — more at FANCY] (1605)
1 *plural phenomena* **:** an observable fact or event
2 *plural phenomena* **a :** an object or aspect known through the senses rather than by thought or intuition **b :** a temporal or spatiotemporal object of sensory experience as distinguished from a noumenon **c :** a fact or event of scientific interest susceptible of scientific description and explanation
3 a : a rare or significant fact or event **b** *plural phenomenons* **:** an exceptional, unusual, or abnormal person, thing, or occurrence
usage see PHENOMENA

phe·no·thi·a·zine \ˌfē-nō-'thī-ə-ˌzēn\ *noun* [International Scientific Vocabulary] (1894)
1 : a greenish yellow crystalline compound $C_{12}H_9NS$ used as an anthelmintic and insecticide especially in veterinary practice
2 : any of various phenothiazine derivatives (as chlorpromazine) that are used as tranquilizing agents especially in the treatment of schizophrenia

phe·no·type \'fē-nə-ˌtīp\ *noun* [German *Phänotypus,* from Greek *phainein* to show + *typos* type] (circa 1911)
: the visible properties of an organism that are produced by the interaction of the genotype and the environment
— **phe·no·typ·ic** \ˌfē-nə-'ti-pik\ *also* **phe·no·typ·i·cal** \-pi-kəl\ *adjective*
— **phe·no·typ·i·cal·ly** \-pi-k(ə-)lē\ *adverb*

□ USAGE
phenomena *Phenomena* has been in occasional use as a singular for more than 400 years and its plural *phenomenas* for more than 350. Our evidence shows that it is primarily a speech form used by poets, critics, and professors, among others, but one that sometimes turns up in edited prose ⟨the Borgia were, in modern terms, a media *phenomena —Economist*⟩. It is etymologically no more irregular than *stamina, agenda,* and *candelabra,* but it has nowhere near the frequency of use that they have, and while they are standard, *phenomena* is still rather borderline.

phen•ox•ide \fi-ˈnäk-ˌsīd\ *noun* (1888)
: a salt of a phenol especially in its capacity as a weak acid

phen•tol•amine \fen-ˈtä-lə-ˌmēn, -mən\ *noun* [*phen-* + *tol*uidine + *amine*] (1952)
: an adrenergic blocking agent $C_{17}H_{19}N_3O$ that is used especially in the diagnosis and treatment of hypertension due to pheochromocytoma

phe•nyl \ˈfe-nᵊl, ˈfē-\ *noun* [International Scientific Vocabulary] (circa 1850)
: a univalent group C_6H_5 that is an aryl group derived from benzene by removal of one hydrogen atom — often used in combination
— phe•nyl•ic \fi-ˈni-lik\ *adjective*

phe•nyl•al•a•nine \ˌfe-nᵊl-ˈa-lə-ˌnēn, ˌfē-\ *noun* [International Scientific Vocabulary] (1883)
: an essential amino acid $C_9H_{11}NO_2$ that is converted in the normal body to tyrosine

phen•yl•bu•ta•zone \ˌfe-nᵊl-ˈbyü-tə-ˌzōn\ *noun* [*phenyl* + *but*yric acid + pyr*azalone* ($C_3H_4N_2O$)] (1952)
: a drug $C_{19}H_{20}N_2O_2$ that is used for its analgesic and anti-inflammatory properties especially in the treatment of arthritis, gout, and bursitis

phen•yl•eph•rine \-ˈe-ˌfrēn, -frən\ *noun* [*phenyl* + epin*ephrine*] (1947)
: a sympathomimetic agent $C_9H_{13}NO_2$ that is used in the form of the hydrochloride as a vasoconstrictor, a mydriatic, and by injection to raise the blood pressure

phen•yl•eth•yl•amine \ˌfe-nᵊl-ˌe-thəl-ˈa-ˌmēn, ˌfē-\ *noun* (1939)
: a neurotransmitter $C_8H_{11}N$ that is an amine resembling amphetamine in structure and pharmacological properties; *also* **:** any of its derivatives

phe•nyl•ke•ton•uria \ˌfe-nᵊl-ˌkē-tᵊn-ˈu̇r-ē-ə, ˌfē-, -ˈyu̇r-\ *noun* [New Latin, from International Scientific Vocabulary *phenyl* + *ketone* + New Latin *-uria*] (1935)
: an inherited human metabolic disease that is characterized by inability to oxidize a metabolic product of phenylalanine and by severe mental deficiency — abbreviation *PKU*
— phe•nyl•ke•ton•uric \-ˈu̇r-ik, -ˈyu̇r-\ *noun or adjective*

phen•yl•pro•pa•nol•amine \ˌfe-nᵊl-ˌprō-pə-ˈnȯ-lə-ˌmēn, ˌfē-, -ˈnō-; -nȯ-ˈla-ˌmēn\ *noun* [*phenyl* + *propane* + [1]*-ol* + *amine*] (1947)
: a sympathomimetic drug $C_9H_{13}NO$ used in the form of its hydrochloride especially as a nasal and bronchial decongestant and as an appetite suppressant

phen•yl•thio•car•ba•mide \ˌfe-nᵊl-ˌthī-ō-ˈkär-bə-ˌmīd, ˌfē-\ *noun* (1879)
: a crystalline compound $C_7H_8N_2S$ that is extremely bitter or tasteless depending on the presence or absence of a single dominant gene in the taster — called also *phenylthiourea, PTC*

phen•yl•thio•urea \-ˌthī-ō-yu̇-ˈrē-ə\ *noun* (1896)
: PHENYLTHIOCARBAMIDE

phe•nyt•o•in \fə-ˈni-tə-wən\ *noun* [di*phenyl*hydan*toin*] (1941)
: a crystalline anticonvulsant compound $C_{15}H_{12}N_2O_2$ used in the form of its sodium salt in the treatment of epilepsy — called also *diphenylhydantoin*

pheo•chro•mo•cy•to•ma \ˌfē-ə-ˌkrō-mə-sə-ˈtō-mə, -sī-\ *noun, plural* **-mas** *or* **-ma•ta** \-mə-tə\ [New Latin, from International Scientific Vocabulary *pheochromocyte* chromaffin cell + New Latin *-oma*] (circa 1929)
: a tumor that is derived from chromaffin cells and is usually associated with paroxysmal or sustained hypertension

pher•o•mone \ˈfer-ə-ˌmōn\ *noun* [International Scientific Vocabulary *phero-* (from Greek *pherein* to carry) + *-mone* (as in *hormone*) — more at BEAR] (1959)
: a chemical substance that is produced by an animal and serves especially as a stimulus to other individuals of the same species for one or more behavioral responses
— pher•o•mon•al \ˌfer-ə-ˈmō-nᵊl\ *adjective*

phew *\a voiceless bilabial fricative usually followed by a voiceless* (y)ü *or* œ *sound; often read as* ˈf(y)ü\ *interjection* (1604)
1 — used to express relief or fatigue
2 — used to express disgust at or as if at an unpleasant odor

phi \ˈfī\ *noun* [Middle Greek, from Greek *phei*] (circa 1899)
: the 21st letter of the Greek alphabet — see ALPHABET table

phi•al \ˈfī(-ə)l\ *noun* [Middle English, from Latin *phiala,* from Greek *phialē*] (14th century)
: VIAL

Phi Be•ta Kap•pa \ˌfī-ˌbā-tə-ˈka-pə\ *noun* [*Phi Beta Kappa (Society)*, from *phi* + *beta* + *kappa,* initials of the society's Greek motto *philosophia biou kybernētēs* philosophy the guide of life] (1912)
: a person winning high scholastic distinction in an American college or university and being elected to membership in a national honor society founded in 1776

phil- *or* **philo-** *combining form* [Middle English, from Old French, from Latin, from Greek, from *philos* dear, friendly]
: loving **:** having an affinity for ⟨*philo*progenitive⟩

¹-phil *or* **-phile** *noun combining form* [French *-phile,* from Greek *-philos* -philous]
: lover **:** one having an affinity for or a strong attraction to ⟨acido*phil*⟩ ⟨Slavo*phile*⟩

²-phil *or* **-phile** *adjective combining form* [New Latin *-philus,* from Latin, from Greek *-philos*]
: loving **:** having a fondness or affinity for ⟨Franco*phile*⟩

Phil•a•del•phia lawyer \ˌfi-lə-ˈdel-fyə-, -fē-ə-\ *noun* [*Philadelphia,* Pa.] (1788)
: a lawyer knowledgeable in even the most minute aspects of the law

Philadelphia pepper pot *noun* (1929)
: PEPPER POT 2b

phil•a•del•phus \ˌfi-lə-ˈdel-fəs\ *noun* [New Latin, from Greek *philadelphos* brotherly, from *phil-* + *adelphos* brother — more at -ADELPHOUS] (1950)
: MOCK ORANGE 1

phi•lan•der \fə-ˈlan-dər\ *intransitive verb* **-dered; -der•ing** \-d(ə-)riŋ\ [from obsolete *philander* lover, philanderer, probably from the name *Philander*] (1737)
1 : to make love to someone with whom marriage is impossible (as because of an existing marriage) or with no intention of proposing marriage
2 : to have many love affairs
— phi•lan•der•er \-dər-ər\ *noun*

phil•an•throp•ic \ˌfi-lən-ˈthrä-pik\ *also* **phil•an•throp•i•cal** \-pi-kəl\ *adjective* (1789)
1 : of, relating to, or characterized by philanthropy **:** HUMANITARIAN
2 : dispensing or receiving aid from funds set aside for humanitarian purposes
— phil•an•throp•i•cal•ly \-pi-k(ə-)lē\ *adverb*

phi•lan•thro•pist \fə-ˈlan(t)-thrə-pist\ *noun* (circa 1736)
: one who practices philanthropy

phi•lan•thro•poid \-ˌpȯid\ *noun* [blend of *philanthropist* and *anthropoid*] (1945)
: a person who works for a philanthropic organization

phi•lan•thro•py \-pē\ *noun, plural* **-pies** [Late Latin *philanthropia,* from Greek *philanthrōpia,* from *philanthrōpos* loving people, from *phil-* + *anthrōpos* human being] (circa 1623)
1 : goodwill to fellowmen; *especially* **:** active effort to promote human welfare
2 a : a philanthropic act or gift **b :** an organization distributing or supported by philanthropic funds

phi•lat•e•list \fə-ˈla-tᵊl-ist\ *noun* (circa 1865)
: a specialist in philately **:** one who collects or studies stamps

phi•lat•e•ly \fə-ˈla-tᵊl-ē\ *noun* [French *philatélie,* from *phil-* + Greek *ateleia* tax exemption, from *atelēs* free from tax, from *a-* + *telos* tax; perhaps akin to Greek *tlēnai* to bear; from the fact that a stamped letter frees the recipient from paying the mailing charges — more at TOLERATE] (circa 1865)
: the collection and study of postage and imprinted stamps **:** stamp collecting
— phil•a•tel•ic \ˌfi-lə-ˈte-lik\ *adjective*
— phil•a•tel•i•cal•ly \-li-k(ə-)lē\ *adverb*

Phi•le•mon \fə-ˈlē-mən, fī-\ *noun* [Greek *Philēmōn*]
1 : a friend and probable convert of the apostle Paul
2 : a letter written by Saint Paul to a Christian living in the area of Colossae and included as a book in the New Testament — see BIBLE table
3 : a poor aged Phrygian in Greek mythology who with his wife Baucis treats a disguised Zeus hospitably and is rewarded by him with a splendid temple

phil•har•mon•ic \ˌfi-lər-ˈmä-nik, -(ˌ)lär-; ˌfil-(ˌ)här-\ *noun* [French *philharmonique,* literally, loving harmony, from Italian *filarmonico,* from *fil-* phil- + *armonia* harmony, from Latin *harmonia*] (1843)
: SYMPHONY ORCHESTRA

phil•hel•lene \(ˌ)fil-ˈhe-ˌlēn\ *or* **phil•hel•len•ic** \ˌfil-hə-ˈle-nik\ *adjective* [Greek *philellēn,* from *phil-* + *Hellēn* Hellene] (circa 1825)
: admiring Greece or the Greeks
— philhellene *noun*
— phil•hel•le•nism \(ˌ)fil-ˈhe-lə-ˌni-zəm\ *noun*
— phil•hel•le•nist \-nist\ *noun*

-philia *noun combining form* [New Latin, from Greek *philia* friendship, from *philos* dear]
1 : friendly feeling toward ⟨Franco*philia*⟩
2 : tendency toward ⟨hemo*philia*⟩
3 : abnormal appetite or liking for ⟨necro*philia*⟩

-philiac *noun combining form* [New Latin *-philia* + Greek *-akos,* adjective suffix]
1 : one having a tendency toward ⟨hemo*philiac*⟩
2 : one having an abnormal appetite or liking for ⟨copro*philiac*⟩

-philic *adjective combining form* [Greek *-philos* -philous]
: having an affinity for **:** loving ⟨acido*philic*⟩

Phi•lip•pi•ans \fə-ˈli-pē-ənz\ *noun plural but singular in construction* [short for *Epistle to the Philippians*]
: a hortatory letter written by Saint Paul to the Christians of Philippi and included as a book in the New Testament — see BIBLE table

phi•lip•pic \fə-ˈli-pik\ *noun* [Middle French *philippique,* from Latin & Greek; Latin *philippica, orationes philippicae,* speeches of Cicero against Mark Antony, translation of Greek *philippikoi logoi,* speeches of Demosthenes against Philip II of Macedon, literally, speeches relating to Philip] (1592)
: a discourse or declamation full of bitter condemnation **:** TIRADE

Phil•ip•pine mahogany \ˈfi-lə-ˌpēn-\ *noun* [*Philippine* Islands] (circa 1924)
: any of several Philippine dipterocarp timber trees (genera *Shorea, Parashorea,* and *Pentacme*) with wood resembling that of the true mahoganies; *also* **:** its wood

phi•lis•tia \fə-ˈlis-tē-ə\ *noun plural, often capitalized* [*Philistia,* ancient country of southwest Palestine] (1857)
: the class or world of cultural philistines

Phi•lis•tine \ˈfi-lə-ˌstēn; fə-ˈlis-tən, -ˌtēn; ˈfi-lə-stən\ *noun* (14th century)
1 : a native or inhabitant of ancient Philistia

2 *often not capitalized* **a :** a person who is guided by materialism and is usually disdainful of intellectual or artistic values **b :** one uninformed in a special area of knowledge ◆
— philistine *adjective, often capitalized*
— phi·lis·tin·ism \-lə-ˌstē-ˌni-zəm; -ˈlis-tə-, -ˌtē-, -lə-stə-\ *noun, often capitalized*

Phil·lips \ˈfi-ləps\ *adjective* [from *Phillips,* a trademark, from Henry M. *Phillips* 20th century American engineer] (1935)
: of, relating to, or being a screw having a head with a cross slot or the corresponding screwdriver

phil·lu·men·ist \fi-ˈlü-mə-nist\ *noun* [*phil-* + Latin *lumen* light — more at LUMINARY] (1943)
: one who collects matchbooks or matchbox labels

Phi·loc·te·tes \fi-ˈläk-tə-ˌtēz, ˌfi-läk-ˈtē-tēz\ *noun* [Greek *Philoktētēs*]
: a Greek archer who uses the bow of Hercules to slay Paris at Troy

phil·o·den·dron \ˌfi-lə-ˈden-drən\ *noun, plural* **-drons** *or* **-dra** \-drə\ [New Latin, from Greek, neuter of *philodendros* loving trees, from *phil-* + *dendron* tree — more at DENDR-] (1877)
: any of various aroid plants (as of the genus *Philodendron*) that are cultivated for their showy foliage

phi·lol·o·gy \fə-ˈlä-lə-jē *also* fī-\ *noun* [French *philologie,* from Latin *philologia* love of learning and literature, from Greek, from *philologos* fond of learning and literature, from *phil-* + *logos* word, speech — more at LEGEND] (1612)
1 : the study of literature and of disciplines relevant to literature or to language as used in literature
2 a : LINGUISTICS; *especially* **:** historical and comparative linguistics **b :** the study of human speech especially as the vehicle of literature and as a field of study that sheds light on cultural history
— phil·o·log·i·cal \ˌfi-lə-ˈlä-ji-kəl\ *adjective*
— phil·o·log·i·cal·ly \-k(ə-)lē\ *adverb*
— phi·lol·o·gist \fə-ˈlä-lə-jist *also* fī-\ *noun*

Phil·o·mel \ˈfi-lə-ˌmel\ *noun* [Latin *Philomela* Philomela, nightingale] (1579)
: NIGHTINGALE

Phil·o·me·la \ˌfi-lə-ˈmē-lə\ *noun* [Latin, from Greek *Philomēlē*]
: an Athenian princess in Greek mythology raped and deprived of her tongue by her brother-in-law Tereus, avenged by the killing of his son, and changed into a nightingale while fleeing from him

phil·o·pro·gen·i·tive \ˌfi-lə-prō-ˈje-nə-tiv\ *adjective* [*phil-* + Latin *progenitus,* past participle of *progignere* to beget — more at PROGENITOR] (1865)
1 : tending to produce offspring **:** PROLIFIC
2 : of, relating to, or characterized by love of offspring
— phil·o·pro·gen·i·tive·ness *noun*

phi·lo·sophe \ˌfē-lə-ˈzȯf\ *noun* [French, literally, philosopher] (1779)
: one of the deistic or materialistic writers and thinkers of the 18th century French Enlightenment

phi·los·o·pher \fə-ˈlä-s(ə-)fər\ *noun* [Middle English, modification of Middle French *philosophe,* from Latin *philosophus,* from Greek *philosophos,* from *phil-* + *sophia* wisdom, from *sophos* wise] (14th century)
1 a : a person who seeks wisdom or enlightenment **:** SCHOLAR, THINKER **b :** a student of philosophy
2 a : a person whose philosophical perspective makes meeting trouble with equanimity easier **b :** an expounder of a theory in a particular area of experience **c :** one who philosophizes

philosophers' stone *noun* (14th century)
: an imaginary stone, substance, or chemical preparation believed to have the power of transmuting baser metals into gold and sought by alchemists — called also *philosopher's stone*

phil·o·soph·i·cal \ˌfi-lə-ˈsä-fi-kəl *also* -ˈzä-\ *or* **phil·o·soph·ic** \-fik\ *adjective* (14th century)
1 a : of or relating to philosophers or philosophy **b :** based on philosophy
2 : characterized by the attitude of a philosopher; *specifically* **:** calm or unflinching in face of trouble, defeat, or loss
— phil·o·soph·i·cal·ly \-fi-k(ə-)lē\ *adverb*

philosophical analysis *noun* (1936)
: ANALYTIC PHILOSOPHY

phi·los·o·phise *British variant of* PHILOSOPHIZE

phi·los·o·phize \fə-ˈlä-sə-ˌfīz\ *verb* **-phized; -phiz·ing** (1594)
intransitive verb
1 : to reason in the manner of a philosopher
2 : to expound a moralizing and often superficial philosophy
transitive verb
: to consider from or bring into conformity with a philosophical point of view
— phi·los·o·phiz·er *noun*

phi·los·o·phy \fə-ˈlä-s(ə-)fē\ *noun, plural* **-phies** [Middle English *philosophie,* from Old French, from Latin *philosophia,* from Greek, from *philosophos* philosopher] (14th century)
1 a (1) **:** all learning exclusive of technical precepts and practical arts (2) **:** the sciences and liberal arts exclusive of medicine, law, and theology ⟨a doctor of *philosophy*⟩ (3) **:** the 4-year college course of a major seminary **b** (1) *archaic* **:** PHYSICAL SCIENCE (2) **:** ETHICS **c :** a discipline comprising as its core logic, aesthetics, ethics, metaphysics, and epistemology
2 a : pursuit of wisdom **b :** a search for a general understanding of values and reality by chiefly speculative rather than observational means **c :** an analysis of the grounds of and concepts expressing fundamental beliefs
3 a : a system of philosophical concepts **b :** a theory underlying or regarding a sphere of activity or thought ⟨the *philosophy* of war⟩ ⟨*philosophy* of science⟩
4 a : the most general beliefs, concepts, and attitudes of an individual or group **b :** calmness of temper and judgment befitting a philosopher

philosophy of life (1853)
1 : an overall vision of or attitude toward life and the purpose of life
2 [translation of German *Lebensphilosophie*] **:** any of various philosophies that emphasize human life or life in general

-philous *adjective combining form* [Greek *-philos,* from *philos* dear, friendly]
: loving **:** having an affinity for ⟨hygro*philous*⟩

phil·ter *or* **phil·tre** \ˈfil-tər\ *noun* [Middle French *philtre,* from Latin *philtrum,* from Greek *philtron;* akin to Greek *philos* dear] (circa 1587)
1 : a potion, drug, or charm held to have the power to arouse sexual passion
2 : a potion credited with magical power

phi phenomenon \ˈfī-\ *noun* (circa 1928)
: apparent motion resulting from an orderly sequence of stimuli (as lights flashed in rapid succession a short distance apart on a sign) without any actual motion being presented to the eye

phiz \ˈfiz\ *noun* [by shortening & alteration from *physiognomy*] (1688)
: FACE

phleb- *or* **phlebo-** *combining form* [Middle English *fleb-,* from Middle French, from Late Latin *phlebo-,* from Greek *phleb-, phlebo-,* from *phleb-, phleps;* perhaps akin to Greek *phlyein, phlyzein* to boil over — more at FLUID]
: vein ⟨*phleb*itis⟩

phle·bi·tis \fli-ˈbī-təs\ *noun* [New Latin] (circa 1834)
: inflammation of a vein

phle·bo·gram \ˈflē-bə-ˌgram\ *noun* [International Scientific Vocabulary] (1885)
: a figure of a vein or a record of its movements

phle·bog·ra·phy \fli-ˈbä-grə-fē\ *noun* [International Scientific Vocabulary] (1937)
: the process of making phlebograms
— phle·bo·graph·ic \ˌflē-bə-ˈgra-fik\ *adjective*

phle·bol·o·gy \fli-ˈbä-lə-jē\ *noun* [International Scientific Vocabulary] (1893)
: a branch of medicine concerned with the veins

phle·bo·to·mus fever \fli-ˈbä-tə-məs-\ *noun* [New Latin *Phlebotomus,* genus of sand flies] (circa 1923)
: SANDFLY FEVER

phle·bot·o·my \fli-ˈbä-tə-mē\ *noun, plural* **-mies** [Middle English *fleobotomie,* from Middle French *flebotomie,* from Late Latin *phlebotomia,* from Greek, from *phleb-* + *-tomia* -tomy] (14th century)
: the letting of blood for transfusion, diagnosis, or experiment, and especially formerly in the treatment of disease — called also *venesection*
— phle·bot·o·mist \-mist\ *noun*

Phleg·e·thon \ˈfle-gə-ˌthän\ *noun* [Latin, from Greek *Phlegethōn*]
: a river of fire in Hades

phlegm \ˈflem\ *noun* [Middle English *fleume,* from Middle French, from Late Latin *phlegmat-, phlegma,* from Greek, flame, inflammation, phlegm, from *phlegein* to burn — more at BLACK] (13th century)
1 : the one of the four humors in early physiology that was considered to be cold and moist and to cause sluggishness
2 : viscid mucus secreted in abnormal quantity in the respiratory passages
3 a : dull or apathetic coldness or indifference **b :** intrepid coolness or calm fortitude

◇ WORD HISTORY

Philistine In the Bible the *Philistines* are the perennial rivals and enemies of the neighboring Israelites. Biblical accounts portray the Philistines as a crude and warlike race. A figurative sense of *philistine* entered the English language as a result of a dispute between the townspeople and the university students in the German town of Jena in 1687. A bloody confrontation prompted a local clergyman to address the issue in a sermon to the townspeople on the value of education. For this sermon he chose his text from Judges 16:9 (Authorized Version): "The Philistines be upon these . . . ." *Philister,* the German word for *Philistine,* soon caught on with the students as an epithet first for the town militia and then for townsmen in general. Spreading more widely in German university slang, *Philister* came to be used for an outsider—someone who was not a member of the university community. Familiar with this usage, the English Victorian social and cultural critics Thomas Carlyle and Matthew Arnold, among others, used *Philistine* as an epithet for a materialist in various writings including Carlyle's *Sartor Resartus* (1831) and Arnold's *Culture and Anarchy* (1869). Since then, *philistine* has become widely used of any person who is indifferent or hostile to the arts and smugly accepts conventional values.

\ə\ **abut** \ᵊ\ **kitten** \ər\ **further** \a\ **ash** \ā\ **ace** \ä\ **mop, mar** \au̇\ **out** \ch\ **chin** \e\ **bet** \ē\ **easy** \g\ **go** \i\ **hit** \ī\ **ice** \j\ **job** \ŋ\ **sing** \ō\ **go** \ȯ\ **law** \ȯi\ **boy** \th\ **thin** \t͟h\ **the** \ü\ **loot** \u̇\ **foot** \y\ **yet** \zh\ **vision** *see also* Guide to Pronunciation

— **phlegmy** \'fle-mē\ *adjective*

phleg·mat·ic \fleg-'ma-tik\ *adjective* (14th century)
1 : resembling, consisting of, or producing the humor phlegm
2 : having or showing a slow and stolid temperament
synonym see IMPASSIVE
— **phleg·mat·i·cal·ly** \-ti-k(ə-)lē\ *adverb*

phlo·em \'flō-,em\ *noun* [German, from Greek *phloios, phloos* bark; perhaps akin to Greek *phlein* to teem, abound, *phlyein, phlyzein* to boil over — more at FLUID] (1875)
: a complex tissue in the vascular system of higher plants that consists mainly of sieve tubes and elongated parenchyma cells usually with fibers and that functions in translocation and in support and storage — compare XYLEM

phloem necrosis *noun* (1923)
: a pathological state in a plant characterized by brown discoloration and disintegration of the phloem; *especially* **:** a fatal disease of the American elm caused by a mycoplasma-like organism transmitted by a leafhopper (*Scaphoideus luteolus*)

phloem ray *noun* (1875)
: a vascular ray or part of a vascular ray that is located in phloem — compare XYLEM RAY

phlo·gis·tic \flō-'jis-tik\ *adjective* (1733)
1 [New Latin *phlogiston*] **:** of or relating to phlogiston
2 [Greek *phlogistos*] **:** of or relating to inflammations and fevers

phlo·gis·ton \-tən\ *noun* [New Latin, from Greek, neuter of *phlogistos* inflammable, from *phlogizein* to set on fire, from *phlog-, phlox* flame, from *phlegein*] (1733)
: the hypothetical principle of fire regarded formerly as a material substance

phlog·o·pite \'flä-gə-,pīt\ *noun* [German *Phlogopit,* from Greek *phlogōpos* fiery-looking, from *phlog-, phlox* + *ōps* face — more at EYE] (1850)
: a usually brown to red form of mica

phlox \'fläks\ *noun, plural* **phlox** *or* **phlox·es** [New Latin, from Latin, a flower, from Greek, flame, wallflower] (circa 1706)
: any of a genus (*Phlox* of the family Polemoniaceae, the phlox family) of American annual or perennial herbs that have red, purple, white, or variegated flowers, a salverform corolla with the stamens on its tube, and a 3-valved capsular fruit

phlox

-phobe *noun combining form* [Greek *-phobos* fearing]
: one fearing or averse to (something specified) ⟨Franco*phobe*⟩

pho·bia \'fō-bē-ə\ *noun* [*-phobia*] (1786)
: an exaggerated usually inexplicable and illogical fear of a particular object, class of objects, or situation

-phobia *noun combining form* [New Latin, from Late Latin, from Greek, from *-phobos* fearing, from *phobos* fear, flight, from *phebesthai* to flee; akin to Lithuanian *bėgti* to flee, Old Church Slavonic *běžati*]
1 : exaggerated fear of ⟨acro*phobia*⟩
2 : intolerance or aversion for ⟨photo*phobia*⟩

pho·bic \'fō-bik\ *adjective* (1897)
: of, relating to, affected with, or constituting phobia
— **phobic** *noun*

-phobic *adjective combining form* [French *-phobique,* from Late Latin *-phobicus,* from Greek *-phobikos,* from *-phobia*]
1 a : having an intolerance or aversion for ⟨photo*phobic*⟩ ⟨Anglo*phobic*⟩ **b :** exhibiting a phobia for ⟨claustro*phobic*⟩
2 : lacking affinity for ⟨hydro*phobic*⟩

phoe·be \'fē-(,)bē\ *noun* [imitative] (1700)
: any of a genus (*Sayornis*) of American flycatchers; *especially* **:** one (*S. phoebe*) of the eastern U.S. that has a slight crest and is plain grayish brown above and yellowish white below

Phoe·be \'fē-bē\ *noun* [Latin, from Greek *Phoibē,* from *phoibē,* feminine of *phoibos*]
: ARTEMIS

Phoe·bus \'fē-bəs\ *noun* [Latin, from Greek *Phoibos,* from *phoibos* radiant]
1 : APOLLO
2 *not capitalized* **:** SUN

Phoe·ni·cian \fi-'nē-shən, -'ni-\ *noun* (14th century)
1 : a native or inhabitant of ancient Phoenicia
2 : the Semitic language of ancient Phoenicia
— **Phoenician** *adjective*

phoe·nix \'fē-niks\ *noun* [Middle English *fenix,* from Old English, from Latin *phoenix,* from Greek *phoinix*] (before 12th century)
: a legendary bird which according to one account lived 500 years, burned itself to ashes on a pyre, and rose alive from the ashes to live another period; *also* **:** a person or thing likened to the phoenix
— **phoe·nix·like** \-,līk\ *adjective*

phon \'fän\ *noun* [International Scientific Vocabulary, from Greek *phōnē* voice, sound] (1932)
: the unit of loudness on a scale beginning at zero for the faintest audible sound and corresponding to the decibel scale of sound intensity with the number of phons of a given sound being equal to the decibels of a pure 1000-cycle tone judged by the average listener to be equal in loudness to the given sound

phon- *or* **phono-** *combining form* [Latin, from Greek *phōn-, phōno-,* from *phōnē* — more at BAN]
: sound **:** voice **:** speech ⟨*phon*ate⟩ ⟨*phono*graph⟩

pho·nate \'fō-,nāt\ *intransitive verb* **pho·nat·ed; pho·nat·ing** (1876)
: to produce vocal sounds and especially speech
— **pho·na·tion** \fō-'nā-shən\ *noun*

[1]phone \'fōn\ *noun* [Greek *phōnē*] (circa 1866)
: a speech sound considered as a physical event without regard to its place in the sound system of a language

[2]phone *noun* [by shortening] (1884)
1 : TELEPHONE
2 : EARPHONE

[3]phone *verb* **phoned; phon·ing** (1889)
: TELEPHONE

[1]-phone *noun combining form* [Greek *-phōnos* sounding, from *phōnē*]
1 : sound ⟨homo*phone*⟩ — often in names of musical instruments and sound-transmitting devices ⟨radio*phone*⟩ ⟨xylo*phone*⟩
2 : speaker of (a specified language) ⟨Franco*phone*⟩

[2]-phone *adjective combining form* [French, from Greek *-phōnos*]
: of or relating to a population that speaks (a specified language) ⟨Franco*phone*⟩

phone-in \'fōn-,in\ *noun* (1968)
: a call-in show (as on radio)

pho·ne·mat·ic \,fō-nē-'ma-tik\ *adjective* (1936)
: PHONEMIC

pho·neme \'fō-,nēm\ *noun* [French *phonème,* from Greek *phōnēmat-, phōnēma* speech sound, utterance, from *phōnein* to sound] (circa 1916)
: any of the abstract units of the phonetic system of a language that correspond to a set of similar speech sounds (as the velar \k\ of *cool* and the palatal \k\ of *keel*) which are perceived to be a single distinctive sound in the language

pho·ne·mic \fə-'nē-mik, fō-\ *adjective* (circa 1931)
1 : of, relating to, or having the characteristics of a phoneme
2 a : constituting members of different phonemes (as \n\ and \m\ in English) **b :** DISTINCTIVE 2
— **pho·ne·mi·cal·ly** \-mi-k(ə-)lē\ *adverb*

pho·ne·mics \-miks\ *noun plural but singular in construction* (1936)
1 : a branch of linguistic analysis that consists of the study of phonemes
2 : the structure of a language in terms of phonemes
— **pho·ne·mi·cist** \-mə-sist\ *noun*

pho·net·ic \fə-'ne-tik\ *adjective* [New Latin *phoneticus,* from Greek *phōnētikos,* from *phōnein* to sound with the voice, from *phōnē* voice] (1826)
1 a : of or relating to spoken language or speech sounds **b :** of or relating to the science of phonetics
2 : representing the sounds and other phenomena of speech: as **a :** constituting an alteration of ordinary spelling that better represents the spoken language, that employs only characters of the regular alphabet, and that is used in a context of conventional spelling **b :** representing speech sounds by means of symbols that have one value only **c :** employing for speech sounds more than the minimum number of symbols necessary to represent the significant differences in a speaker's speech
— **pho·net·i·cal·ly** \-ti-k(ə-)lē\ *adverb*

phonetic alphabet *noun* (1848)
1 : a set of symbols (as IPA) used for phonetic transcription
2 : any of various systems of identifying letters of the alphabet by means of code words in voice communication

pho·ne·ti·cian \,fō-nə-'ti-shən *also* ,fä-\ *noun* (1848)
: a specialist in phonetics

pho·net·ics \fə-'ne-tiks\ *noun plural but singular in construction* (1836)
1 : the system of speech sounds of a language or group of languages
2 a : the study and systematic classification of the sounds made in spoken utterance **b :** the practical application of this science to language study

pho·nic \'fä-nik *also* 'fō-\ *adjective* (1823)
1 : of, relating to, or producing sound **:** ACOUSTIC
2 a : of or relating to the sounds of speech **b :** of or relating to phonics
— **pho·ni·cal·ly** \-ni-k(ə-)lē\ *adverb*

pho·nics \'fä-niks, *1 is also* 'fō-\ *noun plural but singular in construction* (circa 1684)
1 : the science of sound **:** ACOUSTICS
2 : a method of teaching beginners to read and pronounce words by learning the phonetic value of letters, letter groups, and especially syllables

pho·no \'fō-(,)nō\ *noun, plural* **phonos** (1909)
: PHONOGRAPH

pho·no·car·dio·gram \,fō-nə-'kär-dē-ə-,gram\ *noun* [International Scientific Vocabulary] (1912)
: a graphic record of heart sounds made by means of a microphone, amplifier, and galvanometer

pho·no·car·di·og·ra·phy \-,kär-dē-'ä-grə-fē\ *noun* (1916)
: the process of producing a phonocardiogram
— **pho·no·car·dio·graph** \-'kär-dē-ə-,graf\ *noun*
— **pho·no·car·dio·graph·ic** \-,kär-dē-ə-'gra-fik\ *adjective*

pho·no·gram \'fō-nə-,gram\ *noun* [International Scientific Vocabulary] (1860)
1 : a character or symbol used to represent a word, syllable, or phoneme
2 : a succession of orthographic letters that occurs with the same phonetic value in several words (as the *ight* of *bright, fight,* and *flight*)
— **pho·no·gram·mic** *or* **pho·no·gram·ic** \,fō-nə-'gra-mik\ *adjective*

— **pho·no·gram·mi·cal·ly** *or* **pho·no·gram·i·cal·ly** \-mi-k(ə-)lē\ *adverb*

pho·no·graph \'fō-nə-ˌgraf\ *noun* (1877)
: an instrument for reproducing sounds by means of the vibration of a stylus or needle following a spiral groove on a revolving disc or cylinder

pho·nog·ra·pher \fə-'nä-grə-fər, fō-\ *noun* (1845)
: a specialist in phonography

pho·no·graph·ic \ˌfō-nə-'gra-fik, *1 is also* ˌfä-\ *adjective* (1828)
1 : of or relating to phonography
2 : of or relating to a phonograph
— **pho·no·graph·i·cal·ly** \-fi-k(ə-)lē\ *adverb*

pho·nog·ra·phy \fə-'nä-grə-fē, fō-\ *noun* (1701)
1 : spelling based on pronunciation
2 : a system of shorthand writing based on sound

pho·no·lite \'fō-nᵊl-ˌīt\ *noun* [French, from German *Phonolith,* from *phon-* + *-lith;* from its ringing sound when struck] (circa 1828)
: a gray or green volcanic rock consisting essentially of orthoclase and nepheline

pho·nol·o·gy \fə-'nä-lə-jē, fō-\ *noun* (1799)
1 : the science of speech sounds including especially the history and theory of sound changes in a language or in two or more related languages
2 : the phonetics and phonemics of a language at a particular time
— **pho·no·log·i·cal** \ˌfō-nᵊl-'ä-ji-kəl *also* ˌfä-\ *also* **pho·no·log·ic** \-jik\ *adjective*
— **pho·no·log·i·cal·ly** \-ji-k(ə-)lē\ *adverb*
— **pho·nol·o·gist** \fə-'nä-lə-jist, fō-\ *noun*

pho·non \'fō-ˌnän\ *noun* [*phon-* + ²*-on*] (1932)
: a quantum of vibrational energy (as in a crystal)

pho·no·tac·tics \ˌfo-nə-'tak-tiks\ *noun plural but singular in construction* (1956)
: the area of phonology concerned with the analysis and description of the permitted sound sequences of a language
— **pho·no·tac·tic** \-tik\ *adjective*

¹**pho·ny** *or* **pho·ney** \'fō-nē\ *adjective* **pho·ni·er; -est** [perhaps alteration of *fawney* gilded brass ring used in the fawney rig, a confidence game, from Irish *fáinne* ring, from Old Irish *ánne* — more at ANUS] (1900)
: not genuine or real: as **a** (1) **:** intended to deceive or mislead (2) **:** intended to defraud **:** COUNTERFEIT ⟨a *phony* $10 bill⟩ ⟨a *phony* check⟩ **b :** arousing suspicion **:** probably dishonest ⟨something *phony* about the story⟩ **c :** having no genuine existence **:** FICTITIOUS ⟨*phony* publicity stories⟩ **d :** FALSE, SHAM ⟨a *phony* name⟩ ⟨*phony* pearls⟩ **e :** making a false show: as (1) **:** HYPOCRITICAL (2) **:** SPECIOUS ⟨has a *phony* poetic elegance —*New Republic*⟩
— **pho·ni·ly** \'fō-ni-lē\ *adverb*
— **pho·ni·ness** \'fō-nē-nəs\ *noun*

²**phony** *or* **phoney** *noun, plural* **pho·nies** (1902)
: one that is phony

³**phony** *or* **phoney** *transitive verb* **pho·nied** *or* **pho·neyed; pho·ny·ing** *or* **pho·ney·ing** (circa 1942)
: COUNTERFEIT, FAKE — often used with *up* ⟨a paper *phonied* up on the spur of the moment —William Faulkner⟩

-phony *also* **-phonia** *noun combining form* [Middle English *-phonie,* from Old French, from Latin *-phonia,* from Greek *-phōnia,* from *-phōnos* sounding — more at -PHONE]
1 : sound ⟨tele*phony*⟩
2 *usually* *-phonia* **:** speech disorder of a (specified) type ⟨dys*phonia*⟩

phoo·ey \'fü-ē\ *interjection* (1929)
— used to express repudiation or disgust

pho·rate \'fōr-ˌāt, 'fȯr-\ *noun* [phos*phor*us + thio*ate* (salt of a thio acid)] (1959)
: a very toxic organophosphate systemic insecticide $C_7H_{17}O_2PS_3$ that is used especially in seed treatments

-phore *noun combining form* [New Latin *-phorus,* from Greek *-phoros,* from *-phoros* (adjective combining form) carrying, from *pherein* to carry — more at BEAR]
: carrier ⟨gameto*phore*⟩

-phoresis *noun combining form, plural* **-phoreses** [New Latin, from Greek *phorēsis* act of carrying, from *phorein* to carry, wear, frequentative of *pherein*]
: transmission ⟨electro*phoresis*⟩

phos·gene \'fäz-ˌjēn\ *noun* [Greek *phōs* light + *-genēs* born, produced — more at FANCY, -GEN; from its originally having been obtained by the action of sunlight] (1812)
: a colorless gas $COCl_2$ of unpleasant odor that is a severe respiratory irritant

phosph- *or* **phospho-** *combining form* [*phosphorus*]
1 : phosphorus ⟨*phosph*ide⟩
2 : phosphate ⟨*phospho*fructokinase⟩

phos·pha·tase \'fäs-fə-ˌtās, -ˌtāz\ *noun* (1912)
: an enzyme that accelerates the hydrolysis and synthesis of organic esters of phosphoric acid and the transfer of phosphate groups to other compounds: **a :** ALKALINE PHOSPHATASE **b :** ACID PHOSPHATASE

phos·phate \'fäs-ˌfāt\ *noun* [French, from *acide phosphorique* phosphoric acid] (1795)
1 a (1) **:** a salt or ester of a phosphoric acid (2) **:** the trivalent anion PO_4^{3-} derived from phosphoric acid H_3PO_4 **b :** an organic compound of phosphoric acid in which the acid group is bound to nitrogen or a carboxyl group in a way that permits useful energy to be released (as in metabolism)
2 : an effervescent drink of carbonated water with a small amount of phosphoric acid flavored with fruit syrup
3 : a phosphatic material used for fertilizers

phosphate rock *noun* (1870)
: a rock that consists largely of calcium phosphate usually together with other minerals (as calcium carbonate), is used in making fertilizers, and is a source of phosphorus compounds

phos·phat·ic \fäs-'fa-tik, -'fā-\ *adjective* (1843)
: of, relating to, or containing phosphoric acid or phosphates ⟨*phosphatic* fertilizers⟩

phos·pha·tide \'fäs-fə-ˌtīd\ *noun* [International Scientific Vocabulary] (1884)
: PHOSPHOLIPID
— **phos·pha·tid·ic** \ˌfäs-fə-'ti-dik\ *adjective*

phos·pha·ti·dyl \ˌfäs-fə-'tī-dᵊl, fäs-'fa-tə-dᵊl\ *noun* (1941)
: any of several univalent groups $(RCOO)_2C_3H_5OPO(OH)$ that are derived from phosphatidic acids

phos·pha·ti·dyl·cho·line \ˌfäs-fə-ˌtī-dᵊl-'kō-ˌlēn, (ˌ)fäs-ˌfa-tə-dᵊl-\ *noun* (1954)
: LECITHIN

phos·pha·ti·dyl·eth·a·nol·amine \-ˌe-thə-'nä-lə-ˌmēn, -'nō-\ *noun* (1942)
: any of a group of phospholipids that occur especially in blood plasma and the white matter of the central nervous system — called also *cephalin*

phos·pha·tize \'fäs-fə-ˌtīz\ *transitive verb* **-tized; -tiz·ing** (1883)
1 : to change to a phosphate or phosphates
2 : to treat with phosphoric acid or a phosphate
— **phos·pha·ti·za·tion** \ˌfäs-fə-tə-'zā-shən, -ˌfā-\ *noun*

phos·pha·tu·ria \ˌfäs-fə-'tu̇r-ē-ə, -'tyu̇r-\ *noun* [New Latin, from International Scientific Vocabulary *phosphate* + New Latin *-uria*] (1876)
: the excessive discharge of phosphates in the urine

phos·phene \'fäs-ˌfēn\ *noun* [International Scientific Vocabulary *phos-* + Greek *phainein* to show — more at FANCY] (circa 1860)
: a luminous impression due to excitation of the retina

phos·phide \-ˌfīd\ *noun* [International Scientific Vocabulary] (1849)
: a binary compound of phosphorus with a more electropositive element or group

phos·phine \-ˌfēn\ *noun* [International Scientific Vocabulary] (1873)
1 : a colorless poisonous flammable gas PH_3 that is a weaker base than ammonia and that is used especially to fumigate stored grain
2 : any of various derivatives of phosphine analogous to amines but weaker as bases

phos·phite \-ˌfīt\ *noun* (1799)
: a salt or ester of phosphorous acid

phos·pho·cre·a·tine \ˌfäs-(ˌ)fō-'krē-ə-ˌtēn\ *noun* [International Scientific Vocabulary] (1927)
: a compound $C_4H_{10}N_3O_5P$ of creatine and phosphoric acid that is found especially in vertebrate muscle where it is an energy source for muscle contraction

phos·pho·di·es·ter·ase \-dī-'es-tə-ˌrās, -ˌrāz\ *noun* (1936)
: a phosphatase (as from snake venom) that acts on diesters (as some nucleotides) to hydrolyze only one of the two ester groups

phos·pho·enol·pyr·uvate \ˌfäs-ˌfō-ə-ˌnȯl-pī-'rü-ˌvāt, -ˌnōl-, -ˌpīr-'yü-\ *noun* (1956)
: a salt or ester of phosphoenolpyruvic acid

phos·pho·enol·pyr·uvic acid \-pī-'rü-vik-, -ˌpīr-'yü-vik-\ *noun* (1959)
: the phosphate $H_2C{=}C(OPO_3H_2)COOH$ of the enol form of pyruvic acid that is formed as an intermediate in carbohydrate metabolism

phos·pho·fruc·to·ki·nase \ˌfäs-(ˌ)fō-ˌfrək-tō-'kī-ˌnās, -ˌfrük-, -ˌfru̇k-, -ˌnāz\ *noun* [*phosph-* + *fructo*se + *kinase*] (1947)
: an enzyme that functions in carbohydrate metabolism and especially in glycolysis by catalyzing the transfer of a second phosphate (as from ATP) to fructose

phos·pho·glu·co·mu·tase \-ˌglü-kō-'myü-ˌtās, -ˌtāz\ *noun* (1938)
: an enzyme that is found in all plant and animal cells and that catalyzes the reversible isomerization of glucose-1-phosphate to glucose-6-phosphate

phos·pho·glyc·er·al·de·hyde \-ˌgli-sə-'ral-də-ˌhīd\ *noun* (1941)
: a phosphate of glyceraldehyde $C_3H_5O_3(H_2PO_3)$ that is formed especially in anaerobic metabolism of carbohydrates by the splitting of a diphosphate of fructose

phos·pho·glyc·er·ate \ˌfäs-fō-'gli-sə-ˌrāt\ *noun* (1901)
: a salt or ester of phosphoglyceric acid

phos·pho·gly·cer·ic acid \-gli-'ser-ik-\ *noun* (1857)
: either of two isomeric phosphates $HOOCC_2H_3(OH)OPO_3H_2$ of glyceric acid that are formed as intermediates in photosynthesis and in carbohydrate metabolism

phos·pho·ki·nase \ˌfäs-fō-'kī-ˌnās, -ˌnāz\ *noun* (1946)
: KINASE

phos·pho·li·pase \-'lī-ˌpās, -ˌpāz\ *noun* (1945)
: any of several enzymes that hydrolyze lecithins or phosphatidylethanolamines — called also *lecithinase*

phos·pho·lip·id \-'li-pəd\ *noun* (1928)
: any of numerous lipids (as lecithins and phosphatidylethanolamines) in which phosphoric acid as well as a fatty acid is esterified to glycerol and which are found in all living cells and in the bilayers of plasma membranes

phos·pho·mono·es·ter·ase \-ˌmä-nō-'es-tə-ˌrās, -ˌrāz\ *noun* (1933)
: a phosphatase that acts on monoesters

phos·pho·ni·um \fäs-'fō-nē-əm\ *noun* [New Latin] (1871)

: a univalent cation PH_4^+ analogous to ammonium and derived from phosphine; *also* **:** an organic derivative of phosphonium (as $(C_2H_5)_4P^+$)

phos·pho·pro·tein \ˌfäs-fō-ˈprō-ˌtēn, -ˈprō-tē-ən\ *noun* (circa 1908)
: any of various proteins (as casein) that contain combined phosphoric acid

phos·phor \ˈfäs-fər, -ˌfȯr\ *also* **phos·phore** \-ˌfȯr, -ˌfȯr, -fər\ *noun* [Latin *phosphorus,* from Greek *phōsphoros,* literally, light bringer, from *phōsphoros* light-bearing, from *phōs* light + *pherein* to carry, bring — more at FANCY, BEAR] (1705)
: a phosphorescent substance; *specifically* **:** a substance that emits light when excited by radiation

phosphor bronze *noun* (1875)
: a bronze of great hardness, elasticity, and toughness that contains a small amount of phosphorus

phos·pho·resce \ˌfäs-fə-ˈres\ *intransitive verb* **-resced; -resc·ing** [probably back-formation from *phosphorescent*] (1794)
: to exhibit phosphorescence

phos·pho·res·cence \-ˈre-sᵊn(t)s\ *noun* (1796)
1 : luminescence that is caused by the absorption of radiation at one wavelength followed by delayed reradiation at a different wavelength and that continues for a noticeable time after the incident radiation stops — compare FLUORESCENCE
2 : an enduring luminescence without sensible heat

phos·pho·res·cent \-sᵊnt\ *adjective* (1766)
: exhibiting phosphorescence
— phos·pho·res·cent·ly *adverb*

phos·pho·ric \fäs-ˈfȯr-ik, -ˈfär-; ˈfäs-f(ə-)rik\ *adjective* (1800)
: of, relating to, or containing phosphorus especially with a valence higher than in phosphorous compounds

phosphoric acid *noun* (1791)
1 : a syrupy or deliquescent tribasic acid H_3PO_4 used especially in preparing phosphates (as for fertilizers), in rust-proofing metals, and especially formerly as a flavoring in soft drinks — called also *orthophosphoric acid*
2 : a compound (as pyrophosphoric acid or metaphosphoric acid) consisting of phosphate groups linked directly to each other by oxygen

phos·pho·rite \ˈfäs-fə-ˌrīt\ *noun* (1796)
1 : a fibrous concretionary apatite
2 : PHOSPHATE ROCK
— phos·pho·rit·ic \ˌfäs-fə-ˈri-tik\ *adjective*

phos·pho·rol·y·sis \ˌfäs-fə-ˈrä-lə-səs\ *noun* [New Latin] (1937)
: a reversible reaction analogous to hydrolysis in which phosphoric acid functions in a manner similar to that of water with the formation of a phosphate (as glucose-1-phosphate in the breakdown of liver glycogen)
— phos·pho·ro·lyt·ic \-rō-ˈli-tik\ *adjective*

phos·pho·rous \ˈfäs-f(ə-)rəs; fäs-ˈfōr-əs, -ˈfȯr-\ *adjective* (1815)
: of, relating to, or containing phosphorus especially with a valence lower than in phosphoric compounds

phosphorous acid *noun* (1794)
: a deliquescent crystalline acid H_3PO_3 used especially as a reducing agent and in making phosphites

phos·pho·rus \ˈfäs-f(ə-)rəs\ *noun, often attributive* [New Latin, from Greek *phōsphoros* light-bearing — more at PHOSPHOR] (1645)
1 : a phosphorescent substance or body; *especially* **:** one that shines or glows in the dark
2 : a nonmetallic element of the nitrogen family that occurs widely especially as phosphates — see ELEMENT table

phos·pho·ryl \ˈfäs-fə-ˌril\ *noun* [International Scientific Vocabulary] (1871)
: a usually trivalent group PO

phos·phor·y·lase \fäs-ˈfȯr-ə-ˌlās, -ˌlāz\ *noun* (1939)
: any of the enzymes that catalyze phosphorolysis with the formation of organic phosphates

phos·phor·y·late \-ˌlāt\ *transitive verb* **-lat·ed; -lat·ing** (1937)
: to cause (an organic compound) to take up or combine with phosphoric acid or a phosphorus-containing group
— phos·phor·y·la·tive \-ˌlā-tiv\ *adjective*

phos·phor·y·la·tion \ˌfäs-ˌfȯr-ə-ˈlā-shən\ *noun* (1925)
: the process of phosphorylating a chemical compound either by reaction with inorganic phosphate or by transfer of phosphate from another organic phosphate; *especially* **:** the enzymatic conversion of carbohydrates into their phosphoric esters in metabolic processes

phot- *or* **photo-** *combining form* [Greek *phōt-, phōto-,* from *phōt-, phōs* — more at FANCY]
1 : light **:** radiant energy ⟨*phot*on⟩ ⟨*photo*graphy⟩
2 : photograph **:** photographic ⟨*photo*engraving⟩
3 : photoelectric ⟨*photo*cell⟩

pho·tic \ˈfō-tik\ *adjective* (1843)
1 : of, relating to, or involving light especially in relation to organisms
2 : penetrated by light especially of the sun ⟨the *photic* zone of the ocean⟩
— pho·ti·cal·ly \ˈfō-ti-k(ə-)lē\ *adverb*

[1]pho·to \ˈfō-(ˌ)tō\ *noun, plural* **photos** (1860)
: PHOTOGRAPH

[2]photo *verb* (1868)
: PHOTOGRAPH

[3]photo *adjective* (1889)
: PHOTOGRAPHIC 1

pho·to·au·to·troph \ˌfō-tō-ˈȯ-tə-ˌtrōf\ *noun* (1949)
: a photoautotrophic organism

pho·to·au·to·tro·phic \-ˌȯ-tə-ˈtrō-fik\ *adjective* (1943)
: autotrophic and utilizing energy from light ⟨green plants are *photoautotrophic*⟩
— pho·to·au·to·tro·phi·cal·ly \-fi-k(ə-)lē\ *adverb*

pho·to·bi·ol·o·gy \ˌfō-tō-(ˌ)bī-ˈä-lə-jē\ *noun* [International Scientific Vocabulary] (1935)
: a branch of biology that deals with the effects on living beings of radiant energy (as light)
— pho·to·bi·o·log·i·cal \-ˌbī-ə-ˈlä-ji-kəl\ *also* **pho·to·bi·o·log·ic** \-ˈlä-jik\ *adjective*
— pho·to·bi·ol·o·gist \-(ˌ)bī-ˈä-lə-jist\ *noun*

pho·to·cath·ode \-ˈka-ˌthōd\ *noun* [International Scientific Vocabulary] (1930)
: a cathode that emits electrons when exposed to radiant energy and especially light

pho·to·cell \ˈfō-tə-ˌsel\ *noun* [International Scientific Vocabulary] (1891)
: PHOTOELECTRIC CELL

pho·to·chem·i·cal \ˌfō-tō-ˈke-mi-kəl\ *adjective* (1859)
1 : of, relating to, or resulting from the chemical action of radiant energy and especially light ⟨*photochemical* smog⟩
2 : of or relating to photochemistry ⟨*photochemical* studies⟩
— pho·to·chem·i·cal·ly \-k(ə-)lē\ *adverb*

pho·to·chem·is·try \-ˈke-mə-strē\ *noun* (1867)
1 : a branch of chemistry that deals with the effect of radiant energy in producing chemical changes
2 a : photochemical properties ⟨the *photochemistry* of gases⟩ **b :** photochemical processes ⟨*photochemistry* of vision⟩
— pho·to·chem·ist \-ˈke-mist\ *noun*

pho·to·chro·mic \ˌfō-tə-ˈkrō-mik\ *adjective* [*phot-* + *chrom-* + [1]*-ic*] (1953)
1 : capable of changing color on exposure to radiant energy (as light) ⟨*photochromic* glass⟩
2 : of, relating to, or utilizing the change of color shown by a photochromic substance ⟨a *photochromic* process⟩
— pho·to·chro·mism \-ˌmi-zəm\ *noun*

pho·to·co·ag·u·la·tion \-kō-ˌa-gyə-ˈlā-shən\ *noun* (1961)
: a surgical process of coagulating tissue by means of a precisely oriented high-energy light source (as a laser)

pho·to·com·pose \ˌfō-tō-kəm-ˈpōz\ *transitive verb* (1929)
: to set (as reading matter) by photocomposition
— pho·to·com·pos·er *noun*

pho·to·com·po·si·tion \-ˌkäm-pə-ˈzi-shən\ *noun* (1929)
: composition of text directly on film or photosensitive paper for reproduction

pho·to·con·duc·tive \-kən-ˈdək-tiv\ *adjective* (1929)
: having, involving, or operating by photoconductivity

pho·to·con·duc·tiv·i·ty \-ˌkän-ˌdək-ˈti-və-tē, -kən-\ *noun* (1929)
: electrical conductivity that is affected by exposure to electromagnetic radiation (as light)

[1]pho·to·copy \ˈfō-tə-ˌkä-pē\ *noun* [International Scientific Vocabulary] (circa 1909)
: a photographic reproduction of graphic matter

[2]photocopy (1944)
transitive verb
: to make a photocopy of
intransitive verb
: to make a photocopy
— pho·to·copi·er *noun*

pho·to·cur·rent \ˈfō-tō-ˌkər-ənt, -ˌkə-rənt\ *noun* (1913)
: a stream of electrons produced by photoelectric or photovoltaic effects

pho·to·de·com·po·si·tion \-ˌdē-ˌkäm-pə-ˈzi-shən\ *noun* (1888)
: chemical breaking down (as of a pesticide) by means of radiant energy

pho·to·de·grad·able \-di-ˈgrā-də-bəl\ *adjective* (1971)
: chemically degradable by the action of light ⟨*photodegradable* plastics⟩

pho·to·de·tec·tor \-di-ˈtek-tər\ *noun* (1947)
: any of various devices for detecting and measuring the intensity of radiant energy through photoelectric action

pho·to·di·ode \-ˈdī-ˌōd\ *noun* (1945)
: a photoelectric semiconductor device for detecting and often measuring radiant energy (as light)

pho·to·dis·in·te·gra·tion \-di-ˌsin-tə-ˈgrā-shən\ *noun* (1935)
: disintegration of the nucleus of an atom caused by absorption of radiant energy (as light)
— pho·to·dis·in·te·grate \-ˈsin-tə-ˌgrāt\ *transitive verb*

pho·to·dis·so·ci·a·tion \-di-ˌsō-sē-ˈā-shən, -shē-\ *noun* (1925)
: dissociation of the molecules of a substance (as water) caused by absorption of radiant energy
— pho·to·dis·so·ci·ate \-ˈsō-sē-ˌāt, -shē-\ *verb*

pho·to·du·pli·ca·tion \-ˌdü-plə-ˈkā-shən, -ˌdyü-\ *noun* (1941)
: the process of making photocopies
— pho·to·du·pli·cate \-ˈdü-plə-ˌkāt, -ˈdyü-\ *verb*
— pho·to·du·pli·cate \-pli-kət\ *noun*

pho·to·dy·nam·ic \-(ˌ)dī-ˈna-mik\ *adjective* [International Scientific Vocabulary] (1909)
: of, relating to, or having the property of intensifying or inducing a toxic reaction to light (as in the destruction of cancer cells stained with a light-sensitive dye) in a living system
— pho·to·dy·nam·i·cal·ly \-mi-k(ə-)lē\ *adverb*

pho·to·elec·tric \ˌfō-tō-i-ˈlek-trik\ *adjective* [International Scientific Vocabulary] (circa 1879)
: involving, relating to, or utilizing any of various electrical effects due to the interaction of radiation (as light) with matter
— **pho·to·elec·tri·cal·ly** \-tri-k(ə-)lē\ *adverb*

photoelectric cell *noun* (1891)
: an electronic device whose electrical properties are modified by the action of light

photoelectric effect *noun* (1892)
: the emission of free electrons from a metal surface when light strikes it

pho·to·elec·tron \ˌfō-tō-i-ˈlek-ˌträn\ *noun* [International Scientific Vocabulary] (1912)
: an electron released in photoemission
— **pho·to·elec·tron·ic** \-ˌlek-ˈträ-nik\ *adjective*

pho·to·emis·sion \-i-ˈmi-shən\ *noun* (1916)
: the release of electrons from a usually solid material (as a metal) by means of energy supplied by incidence of radiation and especially light
— **pho·to·emis·sive** \-ˈmi-siv\ *adjective*

pho·to·en·grave \-in-ˈgrāv\ *transitive verb* [back-formation from *photoengraving*] (1892)
: to make a photoengraving of
— **pho·to·en·grav·er** *noun*

pho·to·en·grav·ing *noun* (1872)
1 : a photomechanical process for making linecuts and halftone cuts by photographing an image on a metal plate and then etching
2 a : a plate made by photoengraving **b :** a print made from such a plate

pho·to–es·say \ˈfō-tō-ˌe-ˌsā\ *noun* (1944)
: a group of photographs (as in a book or magazine) arranged to explore a theme or tell a story

pho·to·ex·ci·ta·tion \ˌfō-tō-ˌek-ˌsī-ˈtā-shən, -ˌek-sə-\ *noun* (1918)
: the process of exciting the atoms or molecules of a substance by the absorption of radiant energy
— **pho·to·ex·cit·ed** \-ik-ˈsī-təd, -ek\ *adjective*

photo finish *noun* (1936)
1 : a race in which contestants are so close that a photograph of them as they cross the finish line has to be examined to determine the winner
2 : a close contest

pho·to·fin·ish·er \ˌfō-tō-ˈfi-ni-shər\ *noun* (circa 1934)
: one who develops and prints photographic film commercially
— **pho·to·fin·ish·ing** \-shiŋ\ *noun*

pho·to·flash \ˈfō-tō-ˌflash\ *noun* (1930)
: FLASH 6f

pho·to·flood \-ˌfləd\ *noun* (1933)
: an electric lamp that provides intense sustained illumination for taking photographs

pho·to·flu·o·rog·ra·phy \-flu̇-ˈrä-grə-fē, -flȯ-, -flō-\ *noun* (1941)
: the photography of the image produced on a fluorescent screen by X rays

pho·tog \fə-ˈtäg\ *noun* [short for *photographer*] (circa 1906)
: one who takes photographs **:** PHOTOGRAPHER

pho·to·ge·nic \ˌfō-tə-ˈje-nik, -ˈjē-\ *adjective* (1839)
1 : produced or precipitated by light ⟨*photogenic* dermatitis⟩
2 : producing or generating light **:** PHOSPHORESCENT ⟨*photogenic* bacteria⟩
3 : suitable for being photographed ⟨a *photogenic* smile⟩
— **pho·to·ge·ni·cal·ly** \-ni-k(ə-)lē\ *adverb*

pho·to·ge·ol·o·gy \ˌfō-tō-jē-ˈä-lə-jē\ *noun* (1941)
: a branch of geology concerned with the identification and study of geological features through the study of aerial or orbital photographs
— **pho·to·geo·log·ic** \-ˌjē-ə-ˈlä-jik\ *also* **pho·to·geo·log·i·cal** \-ji-kəl\ *adjective*
— **pho·to·ge·ol·o·gist** \-ˈä-lə-jist\ *noun*

pho·to·gram \ˈfō-tə-ˌgram\ *noun* [International Scientific Vocabulary] (1859)
: a shadowlike photograph made by placing objects between light-sensitive paper and a light source

pho·to·gram·me·try \ˌfō-tə-ˈgra-mə-trē\ *noun* [International Scientific Vocabulary *photogram* photograph (from *phot-* + *-gram*) + *-metry*] (1875)
: the science of making reliable measurements by the use of photographs and especially aerial photographs (as in surveying)
— **pho·to·gram·met·ric** \-grə-ˈme-trik\ *adjective*
— **pho·to·gram·me·trist** \-ˈgra-mə-trist\ *noun*

[1]pho·to·graph \ˈfō-tə-ˌgraf\ *noun* (1839)
: a picture or likeness obtained by photography

[2]photograph (1839)
transitive verb
: to take a photograph of
intransitive verb
1 : to take a photograph
2 : to appear as an image in a photograph ⟨an actress who *photographs* well⟩

pho·tog·ra·pher \fə-ˈtä-grə-fər\ *noun* (1847)
: one who practices photography; *especially* **:** one who makes a business of taking photographs

pho·to·graph·ic \ˌfō-tə-ˈgra-fik\ *adjective* (1839)
1 : relating to, obtained by, or used in photography
2 : representing nature and human beings with the exactness of a photograph
3 : capable of retaining vivid impressions ⟨a *photographic* mind⟩
— **pho·to·graph·i·cal·ly** \-fi-k(ə-)lē\ *adverb*

pho·tog·ra·phy \fə-ˈtä-grə-fē\ *noun* (1839)
: the art or process of producing images on a sensitized surface (as a film) by the action of radiant energy and especially light

pho·to·gra·vure \ˌfō-tə-grə-ˈvyu̇r\ *noun* [French, from *phot-* + *gravure*] (1879)
: a process for printing from an intaglio plate prepared by photographic methods

pho·to·in·duced \ˌfō-tō-in-ˈdüst, -ˈdyüst\ *adjective* (1947)
: induced by the action of light
— **pho·to·in·duc·tion** \-ˈdək-shən\ *noun*
— **pho·to·in·duc·tive** \-ˈdək-tiv\ *adjective*

pho·to·in·ter·pre·ta·tion \-in-ˌtər-prə-ˈtā-shən, -pə-\ *noun* (1923)
: the science of identifying and describing objects in photographs
— **pho·to·in·ter·pret·er** \-in-ˈtər-prə-tər, -pə-\ *noun*

pho·to·ion·i·za·tion \-ˌī-ə-nə-ˈzā-shən\ *noun* (1914)
: ionization (as in the ionosphere) of a molecule or atom caused by absorption of radiant energy
— **pho·to·ion·ize** \-ˈī-ə-ˌnīz\ *transitive verb*

pho·to·jour·nal·ism \ˌfō-tō-ˈjər-n^{ə}l-ˌi-zəm\ *noun* (1938)
: journalism in which written copy is subordinate to pictorial usually photographic presentation of news stories or in which a high proportion of pictorial presentation is used
— **pho·to·jour·nal·ist** \-n^{ə}l-ist\ *noun*
— **pho·to·jour·nal·is·tic** \-ˌjər-n^{ə}l-ˈis-tik\ *adjective*

pho·to·ki·ne·sis \-kə-ˈnē-səs, -kī-\ *noun* [New Latin] (1905)
: motion or activity induced by light
— **pho·to·ki·net·ic** \-ˈne-tik\ *adjective*

pho·to·li·thog·ra·phy \-li-ˈthä-grə-fē\ *noun* [International Scientific Vocabulary] (1856)
1 : lithography in which photographically prepared plates are used
2 : a process involving the photographic transfer of a pattern to a surface for etching (as in producing an integrated circuit)
— **pho·to·lith·o·graph** \-ˈli-thə-ˌgraf\ *noun or transitive verb*
— **pho·to·lith·o·graph·ic** \-ˌli-thə-ˈgra-fik\ *adjective*
— **pho·to·lith·o·graph·i·cal·ly** \-fi-k(ə-)lē\ *adverb*

pho·tol·y·sis \fō-ˈtä-lə-səs\ *noun* [New Latin] (1911)
: chemical decomposition by the action of radiant energy
— **pho·to·lyt·ic** \ˌfō-t^{ə}l-ˈi-tik\ *adjective*
— **pho·to·lyt·i·cal·ly** \-ti-k(ə-)lē\ *adverb*

pho·to·lyze \ˈfō-t^{ə}l-ˌīz\ *transitive verb* **-lyzed; -lyz·ing** (1936)
: to cause to undergo photolysis
— **pho·to·lyz·able** \-ˌī-zə-bəl\ *adjective*

pho·to·map \ˈfō-tō-ˌmap\ *noun* (1939)
: a photograph which is taken vertically from above (as from an airplane) and upon which a grid and data pertinent to maps have been added
— **photomap** *verb*

pho·to·mask \-ˌmask\ *noun* (1965)
: MASK 2d

pho·to·me·chan·i·cal \ˌfō-tō-mi-ˈka-ni-kəl\ *adjective* [International Scientific Vocabulary] (circa 1889)
: relating to or involving any of various processes for producing printed matter from a photographically prepared surface
— **pho·to·me·chan·i·cal·ly** \-ni-k(ə-)lē\ *adverb*

pho·tom·e·ter \fō-ˈtä-mə-tər\ *noun* [New Latin *photometrum,* from *phot-* + *-metrum* -meter] (1778)
: an instrument for measuring luminous intensity, luminous flux, illumination, or brightness

pho·to·met·ric \ˌfō-tə-ˈme-trik\ *adjective* (circa 1828)
: of or relating to photometry or the photometer
— **pho·to·met·ri·cal·ly** \-tri-k(ə-)lē\ *adverb*

pho·tom·e·try \fō-ˈtä-mə-trē\ *noun* [New Latin *photometria,* from *phot-* + *-metria* -metry] (1824)
: a branch of science that deals with measurement of the intensity of light; *also* **:** the practice of using a photometer

pho·to·mi·cro·graph \ˌfō-tə-ˈmī-krə-ˌgraf\ *noun* (1858)
: a photograph of a microscope image
— **pho·to·mi·cro·graph·ic** \-ˌmī-krə-ˈgra-fik\ *adjective*
— **pho·to·mi·crog·ra·phy** \-mī-ˈkrä-grə-fē\ *noun*

pho·to·mon·tage \-män-ˈtäzh, -mō^{n}(n)-, -ˈtȧzh\ *noun* [International Scientific Vocabulary] (1931)
: montage using photographic images; *also* **:** a picture made by photomontage

pho·to·mor·pho·gen·e·sis \ˌfō-tə-ˌmȯr-fə-ˈje-nə-səs\ *noun* [New Latin] (1959)
: plant morphogenesis controlled by radiant energy (as light)
— **pho·to·mor·pho·gen·ic** \-ˈje-nik\ *adjective*

pho·to·mo·sa·ic \-mō-ˈzā-ik\ *noun* (1942)
: a photographic mosaic; *especially* **:** one composed of aerial or orbital photographs

pho·to·mul·ti·pli·er tube \ˌfō-tō-ˈməl-tə-ˌplī(-ə)r-\ *noun* (1941)
: a vacuum tube that detects light especially from dim sources through the use of photoemission and successive instances of secondary emission to produce enough electrons to generate a useful current — called also *photomultiplier*

pho·to·mu·ral \-ˈmyu̇r-əl\ *noun* (1927)

\ə\ **abut** \ᵊ\ **kitten** \ər\ **further** \a\ **ash** \ā\ **ace** \ä\ **mop, mar** \au̇\ **out** \ch\ **chin** \e\ **bet** \ē\ **easy** \g\ **go** \i\ **hit** \ī\ **ice** \j\ **job** \ŋ\ **sing** \ō\ **go** \ȯ\ **law** \ȯi\ **boy** \th\ **thin** \t͟h\ **the** \ü\ **loot** \u̇\ **foot** \y\ **yet** \zh\ **vision** *see also* Guide to Pronunciation

: an enlarged photograph usually several yards long used on walls especially as decoration

pho·ton \'fō-ˌtän\ *noun [phot- +* ^{2}*-on]* (1916)
1 : a unit of intensity of light at the retina equal to the illumination received per square millimeter of a pupillary area from a surface having a brightness of one candle per square meter
2 : a quantum of electromagnetic radiation
— **pho·ton·ic** \fō-'tä-nik\ *adjective*

pho·to·neg·a·tive \ˌfō-tō-'ne-gə-tiv\ *adjective* (1914)
: exhibiting negative phototropism or phototaxis

pho·ton·ics \fō-'tä-niks\ *noun plural but singular in construction* (1952)
: a branch of physics that deals with the properties and applications of photons especially as a medium for transmitting information

pho·to·nu·cle·ar \ˌfō-tō-'nü-klē-ər, -'nyü-, ÷-kyə-lər\ *adjective* (1941)
: relating to or caused by the incidence of radiant energy (as gamma rays) on atomic nuclei

pho·to–off·set \-'ȯf-ˌset\ *noun* (1926)
: offset printing from photolithographic plates

photo opportunity *noun* (1972)
: a situation or event that lends itself to and is often arranged expressly for the taking of pictures that favorably publicize the individuals photographed

pho·to·ox·i·da·tion \-ˌäk-sə-'dā-shən\ *noun* (1888)
: oxidation under the influence of radiant energy (as light)
— **pho·to·ox·i·da·tive** \-'äk-sə-ˌdā-tiv\ *adjective*
— **pho·to·ox·i·dize** \-'äk-sə-ˌdīz\ *verb*

pho·to·pe·ri·od \-'pir-ē-əd\ *noun* (1920)
: a recurring cycle of light and dark periods of constant length; *also* **:** PHOTOPHASE 2
— **pho·to·pe·ri·od·ic** \-ˌpir-ē-'ä-dik\ *adjective*
— **pho·to·pe·ri·od·i·cal·ly** \-di-k(ə-)lē\ *adverb*

pho·to·pe·ri·od·ism \-'pir-ē-ə-ˌdi-zəm\ *noun* (circa 1911)
: a plant or animal's response or capacity to respond to photoperiod

pho·to·phase \'fō-tə-ˌfāz\ *noun* (1944)
1 : LIGHT REACTION
2 : the light period of a photoperiodic cycle of light and dark

pho·to·pho·bia \ˌfō-tə-'fō-bē-ə\ *noun* [New Latin] (circa 1799)
: intolerance to light; *especially* **:** painful sensitiveness to strong light

pho·to·pho·bic \-'fō-bik\ *adjective* (1858)
1 a : shunning or avoiding light **b :** growing best under reduced illumination
2 : of or relating to photophobia

pho·to·phore \'fō-tə-ˌfōr, -ˌfȯr\ *noun* [International Scientific Vocabulary] (1898)
: a light-emitting organ; *especially* **:** one of the luminous spots on various marine mostly deep-sea fishes

pho·to·phos·phor·y·la·tion \'fō-tō-ˌfäs-ˌfȯr-ə-'lā-shən\ *noun* (1956)
: the synthesis of ATP from ADP and phosphate that occurs in a plant using radiant energy absorbed during photosynthesis

phot·opic \fōt-'ō-pik, -'ä-pik\ *adjective* [New Latin *photopia,* from *phot-* + *-opia*] (1915)
: relating to or being vision in bright light with light-adapted eyes that is mediated by the cones of the retina

pho·to·play \'fō-tō-ˌplā\ *noun* (1910)
: MOTION PICTURE 2

pho·to·po·la·rim·e·ter \ˌfō-tō-ˌpō-lə-'ri-mə-tər\ *noun* (circa 1890)
: an instrument used to measure the intensity and polarization of reflected light (as from clouds enveloping a planet)

pho·to·poly·mer \ˌfō-tō-'pä-lə-mər\ *noun* (1932)
: a photosensitive plastic used especially in the manufacture of printing plates

pho·to·pos·i·tive \-'pä-zə-tiv, -'päz-tiv\ *adjective* (1914)
: exhibiting positive phototropism or phototaxis

pho·to·prod·uct \-'prä-(ˌ)dəkt\ *noun* (1926)
: a product of a photochemical reaction

pho·to·pro·duc·tion \-prə-'dək-shən\ *noun* (1949)
: the production of elementary particles (as mesons) as a result of the action of photons on atomic nuclei

pho·to·re·ac·tion \-rē-'ak-shən\ *noun* (1909)
: a photochemical reaction

pho·to·re·ac·ti·va·tion \-rē-ˌak-tə-'vā-shən\ *noun* (1949)
: repair of DNA (as of a bacterium) especially by a light-dependent enzymatic reaction after damage by ultraviolet irradiation
— **pho·to·re·ac·ti·vat·ing** \-'ak-tə-ˌvā-tiŋ\ *adjective*

photo–realism *noun* (1961)
: realism in painting characterized by extremely meticulous depiction of detail
— **photo–realist** *noun or adjective*

pho·to·re·cep·tion \-ri-'sep-shən\ *noun* (1902)
: perception of waves in the range of visible light; *specifically* **:** VISION
— **pho·to·re·cep·tive** \-'sep-tiv\ *adjective*

pho·to·re·cep·tor \-'sep-tər\ *noun* (1906)
: a receptor for light stimuli

pho·to·re·con·nais·sance \-ri-'kä-nə-zən(t)s *also* -sən(t)s\ *noun* (1940)
: reconnaissance in which aerial photographs are taken

pho·to·re·duc·tion \-ri-'dək-shən\ *noun* (1888)
: chemical reduction under the influence of radiant energy (as light) **:** photochemical reduction
— **pho·to·re·duce** \-ri-'düs, -'dyüs\ *transitive verb*

pho·to·re·pro·duc·tion \-ˌrē-prə-'dək-shən\ *noun* (1939)
: reproduction by photographic means; *also* **:** PHOTOCOPY

pho·to·re·sist \'fō-tō-ri-ˌzist, ˌfō-tō-ri-'\ *noun* (1953)
: a photosensitive resin that loses its resistance to chemical etching when exposed to radiation and is used especially in the transference of a circuit pattern to a semiconductor chip during the production of an integrated circuit

pho·to·res·pi·ra·tion \ˌfō-tō-ˌres-pə-'rā-shən\ *noun* (1945)
: oxidation involving production of carbon dioxide during photosynthesis

pho·to·sen·si·tive \-'sen(t)-s(ə-)tiv\ *adjective* (1886)
: sensitive or sensitized to the action of radiant energy
— **pho·to·sen·si·tiv·i·ty** \-ˌsen(t)-sə-'ti-və-tē\ *noun*

pho·to·sen·si·ti·za·tion \-ˌsen(t)-s(ə-)tə-'zā-shən\ *noun* (circa 1923)
1 : the process of photosensitizing
2 : the condition of being photosensitized; *especially* **:** the development of an abnormal capacity to react to sunlight typically by edematous swelling and dermatitis

pho·to·sen·si·tize \-'sen(t)-sə-ˌtīz\ *transitive verb* (circa 1923)
: to make sensitive to the influence of radiant energy and especially light
— **pho·to·sen·si·tiz·er** *noun*

pho·to·set \'fō-tō-ˌset\ *transitive verb* **-set; -set·ting** (1957)
: PHOTOCOMPOSE
— **pho·to·set·ter** *noun*

pho·to·sphere \'fō-tə-ˌsfir\ *noun* (1664)
1 : a sphere of light or radiance
2 : the luminous surface layer of the sun or a star
— **pho·to·spher·ic** \ˌfō-tə-'sfir-ik, -'sfer-\ *adjective*

pho·to·stat \'fō-tə-ˌstat\ *transitive verb* (1914)
: to copy by a Photostat device

Photostat *trademark*
— used for a device for making a photographic copy of graphic matter

pho·to·stat·ic \ˌfō-tə-'sta-tik\ *adjective* (1919)
: of, made by, or using a Photostat device ⟨a *photostatic* copy⟩ ⟨a *photostatic* process⟩

pho·to·syn·thate \ˌfō-tō-'sin-ˌthāt\ *noun* [*photosynth*esis + ^{1}*-ate*] (1913)
: a product of photosynthesis

pho·to·syn·the·sis \-'sin(t)-thə-səs\ *noun* [New Latin] (1898)
: synthesis of chemical compounds with the aid of radiant energy and especially light; *especially* **:** formation of carbohydrates from carbon dioxide and a source of hydrogen (as water) in the chlorophyll-containing tissues of plants exposed to light
— **pho·to·syn·the·size** \-ˌsīz\ *intransitive verb*
— **pho·to·syn·thet·ic** \-sin-'the-tik\ *adjective*
— **pho·to·syn·thet·i·cal·ly** \-ti-k(ə-)lē\ *adverb*

pho·to·sys·tem \'fō-tō-ˌsis-təm\ *noun* (1964)
: either of two photochemical reactions occurring in chloroplasts: **a :** one that proceeds best in long wavelength light — called also *photosystem I* **b :** one that proceeds best in short wavelength light — called also *photosystem II*

pho·to·tac·tic \ˌfō-tō-'tak-tik\ *adjective* [International Scientific Vocabulary] (1882)
: of, relating to, or exhibiting phototaxis
— **pho·to·tac·ti·cal·ly** \-ti-kə-lē\ *adverb*

pho·to·tax·is \-'tak-səs\ *noun* [New Latin] (circa 1890)
: a taxis in which light is the directive factor

pho·to·te·leg·ra·phy \-tə-'le-grə-fē\ *noun* [International Scientific Vocabulary] (1886)
: FACSIMILE 2

pho·to·tox·ic \-'täk-sik\ *adjective* (1942)
1 : rendering the skin susceptible to damage (as sunburn or blisters) upon exposure to light and especially ultraviolet light
2 : induced by a phototoxic substance
— **pho·to·tox·ic·i·ty** \-täk-'si-sə-tē\ *noun*

pho·to·tro·pic \ˌfō-tə-'trō-pik, -'trä-\ *adjective* (circa 1890)
: of, relating to, or capable of phototropism
— **pho·to·tro·pi·cal·ly** \-pi-k(ə-)lē\ *adverb*

pho·tot·ro·pism \fō-'tä-trə-ˌpi-zəm\ *noun* [International Scientific Vocabulary] (1899)
: a tropism in which light is the orienting stimulus

pho·to·tube \'fō-tō-ˌtüb, -ˌtyüb\ *noun* (1930)
: an electron tube having a photoemissive cathode whose released electrons are drawn to the anode by reason of its positive potential

pho·to·type·set·ting \ˌfō-tō-'tīp-ˌse-tiŋ\ *noun* (1931)
: PHOTOCOMPOSITION; *especially* **:** photocomposition done on a keyboard or tape-operated composing machine
— **pho·to·type·set·ter** *noun*

pho·to·vol·ta·ic \-väl-'tā-ik, -vōl-\ *adjective* [International Scientific Vocabulary] (circa 1890)
: of, relating to, or utilizing the generation of a voltage when radiant energy falls on the boundary between dissimilar substances (as two different semiconductors)
— **photovoltaic** *noun*

phrag·mo·plast \'frag-mō-ˌplast\ *noun* [International Scientific Vocabulary *phragmo-* (from Greek *phragmos* fence, from *phrassein* to enclose) + *-plast*] (1912)
: the enlarged barrel-shaped spindle that is characteristic of the later stages of plant mitosis and within which the cell plate forms

phras·al \'frā-zəl\ *adjective* (1871)

: of, relating to, or consisting of a phrase ⟨*phrasal* prepositions⟩
— **phras·al·ly** \-zə-lē\ *adverb*

[1]**phrase** \ˈfrāz\ *noun* [Latin *phrasis,* from Greek, from *phrazein* to point out, explain, tell] (1530)
1 : a characteristic manner or style of expression **:** DICTION
2 a : a brief expression; *especially* **:** CATCHWORD **b :** WORD
3 : a short musical thought typically two to four measures long closing with a cadence
4 : a word or group of words forming a syntactic constituent with a single grammatical function ⟨an adverbial *phrase*⟩
5 : a series of dance movements comprising a section of a pattern

[2]**phrase** *transitive verb* **phrased; phrasing** (1570)
1 a : to express in words or in appropriate or telling terms **b :** to designate by a descriptive word or phrase
2 : to divide into melodic phrases

phrase book *noun* (1594)
: a book containing idiomatic expressions of a foreign language and their translation

phrase·mak·er \ˈfrāz-ˌmā-kər\ *noun* (1822)
1 : one who coins impressive phrases
2 : one given to making fine-sounding but often hollow and meaningless phrases
— **phrase·mak·ing** \-kiŋ\ *noun*

phrase·mon·ger \-ˌməŋ-gər, -ˌmäŋ-\ *noun* (1815)
: PHRASEMAKER 2
— **phrase·mon·ger·ing** \-g(ə-)riŋ\ *noun*

phra·se·o·log·i·cal \ˌfrā-zē-ə-ˈlä-ji-kəl\ *adjective* (1664)
1 a : expressed in formal often sententious phrases **b :** marked by frequently insincere use of such phrases
2 : of or relating to phraseology

phra·se·ol·o·gist \ˌfrā-zē-ˈä-lə-jist, frā-ˈzä-\ *noun* (1713)
: one who uses sententious or insincere phrases

phra·se·ol·o·gy \-jē\ *noun, plural* **-gies** [New Latin *phraseologia,* irregular from Greek *phrasis* + *-logia* -logy] (1664)
1 : a manner of organizing words and phrases into longer elements **:** STYLE
2 : choice of words

phras·ing \ˈfrā-ziŋ\ *noun* (1611)
1 : style of expression **:** PHRASEOLOGY
2 : the act, method, or result of grouping notes into musical phrases

phra·try \ˈfrā-trē\ *noun, plural* **phratries** [Greek *phratria,* from *phratēr* member of the same clan, member of a phratry — more at BROTHER] (1833)
1 : a kinship group forming a subdivision of a Greek phyle
2 : a tribal subdivision; *specifically* **:** an exogamous group typically comprising several totemic clans

phre·at·ic \frē-ˈa-tik\ *adjective* [Greek *phreat-, phrear* well; akin to Armenian *albiwr* spring, Old High German *brunno* — more at BURN] (circa 1890)
1 : of, relating to, or being groundwater
2 : of, relating to, or being an explosion caused by steam derived from groundwater

phre·at·o·phyte \frē-ˈa-tə-ˌfīt\ *noun* [Greek *phreat-, phrear* well + English *-o-* + *-phyte*] (1920)
: a deep-rooted plant that obtains its water from the water table or the layer of soil just above it
— **phre·at·o·phyt·ic** \-ˌa-tə-ˈfi-tik\ *adjective*

phre·net·ic *variant of* FRENETIC

-phrenia *noun combining form* [New Latin, from Greek *phren-, phrēn* diaphragm, mind]
: disordered condition of mental functions ⟨hebe*phrenia*⟩

phren·ic \ˈfre-nik\ *adjective* [New Latin *phrenicus,* from Greek *phren-, phrēn*] (1704)
1 : of or relating to the diaphragm
2 : of or relating to the mind

phre·nol·o·gy \fri-ˈnä-lə-jē\ *noun* [Greek *phren-, phrēn*] (1805)
: the study of the conformation of the skull based on the belief that it is indicative of mental faculties and character
— **phre·no·log·i·cal** \ˌfre-nᵊl-ˈä-ji-kəl, ˌfrē-\ *adjective*
— **phre·nol·o·gist** \fri-ˈnä-lə-jist\ *noun*

phren·sy *variant of* FRENZY

Phry·gian \ˈfri-j(ē-)ən\ *noun* (15th century)
1 : a native or inhabitant of ancient Phrygia
2 : the extinct Indo-European language of the Phrygians — see INDO-EUROPEAN LANGUAGES table
— **Phrygian** *adjective*

phthal·ic acid \ˈtha-lik-\ *noun* [International Scientific Vocabulary, short for obsolete *naphthalic acid,* from *naphthalene*] (1857)
: any of three isomeric acids $C_8H_6O_4$ obtained by oxidation of various benzene derivatives

phthalic anhydride *noun* (1855)
: a crystalline cyclic acid anhydride $C_8H_4O_3$ used especially in making alkyd resins

phtha·lo·cy·a·nine \ˌtha-lō-ˈsī-ə-ˌnēn, ˌthā-\ *noun* [International Scientific Vocabulary *phthalic* acid + *-o-* + *cyanine*] (1933)
: a bright greenish blue crystalline compound $C_{32}H_{18}N_8$; *also* **:** any of several metal derivatives that are brilliant fast blue to green dyes or pigments

phthis·ic \ˈti-zik\ *noun* [Middle English *tisike,* from Middle French *tisique,* from *tisique* tubercular, from Latin *phthisicus,* from Greek *phthisikos,* from *phthisis*] (14th century)
: PHTHISIS
— **phthisic** *or* **phthis·i·cal** \-zi-kəl\ *adjective*

phthi·sis \ˈthī-səs, ˈtī-, ˈfthī- *or with* i *for* ī\ *noun, plural* **phthi·ses** \-ˌsēz\ [Latin, from Greek, from *phthinein* to waste away; akin to Sanskrit *kṣiṇoti* he destroys] (1526)
: a progressively wasting or consumptive condition; *especially* **:** pulmonary tuberculosis

phyco- *combining form* [Greek *phykos* seaweed]
: algae ⟨*phyco*logy⟩

phy·co·cy·a·nin \ˌfī-kō-ˈsī-ə-nən\ *noun* [International Scientific Vocabulary *phyco-* + *cyan-* + [1]*-in*] (1875)
: any of various bluish green protein pigments in the cells of blue-green algae

phy·co·er·y·thrin \-ˈer-i-thrən\ *noun* [International Scientific Vocabulary *phyco-* + *erythr-* + [1]*-in*] (circa 1868)
: any of the red protein pigments in the cells of red algae

phy·col·o·gy \fī-ˈkä-lə-jē\ *noun* (circa 1847)
: ALGOLOGY
— **phy·co·log·i·cal** \ˌfī-kə-ˈlä-ji-kəl\ *adjective*
— **phy·col·o·gist** \fī-ˈkä-lə-jist\ *noun*

phy·co·my·cete \ˌfī-kō-ˈmī-ˌsēt, -ˌmī-ˈsēt\ *noun* [ultimately from Greek *phykos* + *mykēt-, mykēs* fungus — more at MYC-] (circa 1900)
: any of a large class (Phycomycetes) of lower fungi that are in many respects similar to algae and are now often assigned to subdivisions (Mastigomycotina and Zygomycotina)
— **phy·co·my·ce·tous** \-ˌmī-ˈsē-təs\ *adjective*

phyl- *or* **phylo-** *combining form* [Latin, from Greek, from *phylē, phylon;* akin to Greek *phyein* to bring forth — more at BE]
: tribe **:** race **:** phylum ⟨*phylo*geny⟩

phy·lac·tery \fə-ˈlak-t(ə-)rē\ *noun, plural* **-ter·ies** [Middle English *philaterie,* from Medieval Latin *philaterium,* alteration of Late Latin *phylacterium,* from Greek *phylaktērion* amulet, phylactery, from *phylassein* to guard, from *phylak-, phylax* guard] (14th century)
1 : either of two small square leather boxes containing slips inscribed with scriptural passages and traditionally worn on the left arm and on the head by Jewish men during morning weekday prayers
2 : AMULET

phylactery 1

phy·le \ˈfī-(ˌ)lē\ *noun, plural* **phy·lae** \-ˌlē\ [Greek *phylē* tribe, phyle] (1863)
: the largest political subdivision among the ancient Athenians

phy·let·ic \fī-ˈle-tik\ *adjective* [International Scientific Vocabulary *phyl-* + *-etic* (as in *genetic*)] (1881)
: of or relating to evolutionary change in a single line of descent without branching
— **phy·let·i·cal·ly** \-ti-k(ə-)lē\ *adverb*

phyll- *or* **phyllo-** *combining form* [New Latin, from Greek, from *phyllon* — more at BLADE]
: leaf ⟨*phyll*ome⟩

-phyll *noun combining form* [New Latin *-phyllum,* from Greek *phyllon* leaf]
: leaf ⟨sporo*phyll*⟩

phyl·la·ry \ˈfi-lə-rē\ *noun, plural* **-ries** [New Latin *phyllarium,* from Greek *phyllarion,* diminutive of *phyllon* leaf] (1857)
: one of the involucral bracts subtending the flower head of a composite plant

phyl·lo \ˈfē-(ˌ)lō, ˈfī-\ *noun* [New Greek, sheet of pastry dough, literally, leaf, from Greek *phyllon*] (circa 1950)
: extremely thin pastry dough that is layered to produce a flaky pastry

phyl·lo·clade \ˈfi-lə-ˌklād\ *noun* [New Latin *phyllocladium,* from *phyll-* + Greek *klados* branch — more at HOLT] (1858)
: a flattened stem or branch (as a joint of a cactus) that functions as a leaf

phyl·lode \ˈfi-ˌlōd\ *noun* [New Latin *phyllodium,* from Greek *phyllōdēs* like a leaf, from *phyllon* leaf] (1848)
: a flat expanded petiole that replaces the blade of a foliage leaf, fulfills the same functions, and is analogous to a cladophyll

phyl·lo·di·um \fi-ˈlō-dē-əm\ *noun, plural* **-dia** \-dē-ə\ [New Latin] (circa 1847)
: PHYLLODE

phyl·lome \ˈfi-ˌlōm\ *noun* [International Scientific Vocabulary] (1875)
: a plant part that is a leaf or is phylogenetically derived from a leaf

phyl·lo·tac·tic \ˌfi-lə-ˈtak-tik\ *adjective* (1857)
: of or relating to phyllotaxis

phyl·lo·tax·is \ˌfi-lə-ˈtak-səs\ *also* **phyl·lo·taxy** \ˈfi-lə-ˌtak-sē\ *noun* [New Latin *phyllotaxis,* from *phyll-* + *-taxis*] (1857)
1 : the arrangement of leaves on a stem and in relation to one another
2 : the study of phyllotaxis and of the laws that govern it

-phyllous *adjective combining form* [New Latin *-phyllus,* from Greek *-phyllos,* from *phyllon* leaf — more at BLADE]
: having (such or so many) leaves, leaflets, or leaflike parts ⟨hetero*phyllous*⟩

phyl·lox·e·ra \ˌfi-ˌläk-ˈsir-ə, fə-ˈläk-sə-rə\ *noun* [New Latin, from *phyll-* + Greek *xēros* dry] (1868)
: any of various plant lice (especially genus *Phylloxera*) that differ from aphids especially in wing structure and in being continuously oviparous

phy·lo·ge·net·ic \ˌfī-lō-jə-ˈne-tik\ *adjective* [International Scientific Vocabulary, from New

Latin *phylogenesis* phylogeny, from *phyl-* + *genesis*] (1877)
1 : of or relating to phylogeny
2 : based on natural evolutionary relationships
3 : acquired in the course of phylogenetic development **:** RACIAL
— phy·lo·ge·net·i·cal·ly \-ti-k(ə-)lē\ *adverb*

phy·log·e·ny \fī-'lä-jə-nē\ *noun, plural* **-nies** [International Scientific Vocabulary] (1872)
1 : the evolutionary history of a kind of organism
2 : the evolution of a genetically related group of organisms as distinguished from the development of the individual organism
3 : the history or course of the development of something (as a word or custom)

phy·lum \'fī-ləm\ *noun, plural* **phy·la** \-lə\ [New Latin, from Greek *phylon* tribe, race — more at PHYL-] (1876)
1 a : a direct line of descent within a group **b :** a group that constitutes or has the unity of a phylum; *especially* **:** one of the usually primary divisions of the animal kingdom
2 : a group of languages related more remotely than those of a family or stock

phys ed \'fiz-'ed\ *noun* (1955)
: PHYSICAL EDUCATION

physi- *or* **physio-** *combining form* [Latin, from Greek, from *physis* — more at PHYSICS]
1 : nature ⟨*physio*graphy⟩
2 : physical ⟨*physio*therapy⟩

phys·iat·rist \,fi-zē-'a-trist\ *noun* [*physiatrics* physical medicine, from Greek *physis* + International Scientific Vocabulary *-iatrics*] (circa 1947)
: a physician who specializes in physical medicine

¹phys·ic \'fi-zik\ *noun* [Middle English *physik* natural science, art of medicine, from Old French *fisique,* from Latin *physica,* singular, natural science, from Greek *physikē,* from feminine of *physikos* — more at PHYSICS] (14th century)
1 a : the art or practice of healing disease **b :** the practice or profession of medicine
2 : a medicinal agent or preparation; *especially* **:** PURGATIVE
3 *archaic* **:** NATURAL SCIENCE

²physic *transitive verb* **phys·icked; phys·ick·ing** (14th century)
1 : to treat with or administer medicine to; *especially* **:** PURGE
2 : HEAL, CURE

¹phys·i·cal \'fi-zi-kəl\ *adjective* [Middle English *phisicale* medical, from Medieval Latin *physicalis,* from Latin *physica*] (1597)
1 a : having material existence **:** perceptible especially through the senses and subject to the laws of nature ⟨everything *physical* is measurable by weight, motion, and resistance —Thomas De Quincey⟩ **b :** of or relating to material things
2 a : of or relating to natural science **b** (1) **:** of or relating to physics (2) **:** characterized or produced by the forces and operations of physics
3 a : of or relating to the body **b :** concerned or preoccupied with the body and its needs **:** CARNAL **c :** characterized by especially rugged and forceful physical activity **:** ROUGH ⟨a *physical* hockey game⟩ ⟨a *physical* player⟩
synonym see MATERIAL
— phys·i·cal·ly \-k(ə-)lē\ *adverb*
— phys·i·cal·ness \-kəl-nəs\ *noun*

²physical *noun* (1934)
: PHYSICAL EXAMINATION

physical anthropology *noun* (1873)
: anthropology concerned with the comparative study of human evolution, variation, and classification especially through measurement and observation — compare CULTURAL ANTHROPOLOGY
— physical anthropologist *noun*

physical education *noun* (1830)
: instruction in the development and care of the body ranging from simple calisthenic exercises to a course of study providing training in hygiene, gymnastics, and the performance and management of athletic games

physical examination *noun* (1884)
: an examination of the bodily functions and condition of an individual

physical geography *noun* (1808)
: geography that deals with the exterior physical features and changes of the earth

phys·i·cal·ism \'fi-zi-kə-,li-zəm\ *noun* (circa 1931)
: a thesis that the descriptive terms of scientific language are reducible to terms which refer to spatiotemporal things or events or to their properties
— phys·i·cal·ist \-list\ *noun*
— phys·i·cal·is·tic \,fi-zi-kə-'lis-tik\ *adjective*

phys·i·cal·i·ty \,fi-zə-'ka-lə-tē\ *noun, plural* **-ties** (1660)
1 : intensely physical orientation **:** predominance of the physical usually at the expense of the mental, spiritual, or social
2 : a physical aspect or quality

physical medicine *noun* (1939)
: a branch of medicine concerned with the diagnosis and treatment of disease and disability by physical means (as radiation, heat, and electricity) — compare PHYSICAL THERAPY

physical science *noun* (1802)
: any of the natural sciences (as physics, chemistry, and astronomy) that deal primarily with nonliving materials
— physical scientist *noun*

physical therapy *noun* (1922)
: the treatment of disease by physical and mechanical means (as massage, regulated exercise, water, light, heat, and electricity)
— physical therapist *noun*

phy·si·cian \fə-'zi-shən\ *noun* [Middle English *fisicien,* from Old French, from *fisique* medicine] (13th century)
1 : a person skilled in the art of healing; *specifically* **:** a doctor of medicine
2 : one exerting a remedial or salutary influence

physician's assistant *noun* (1970)
: a person certified to provide basic medical services usually under the supervision of a licensed physician — called also *PA, physician assistant*

phys·i·cist \'fi-zə-sist, 'fiz-sist\ *noun* (1840)
1 : a specialist in physics
2 *archaic* **:** a person skilled in natural science

phys·i·co·chem·i·cal \,fi-zi-kō-'ke-mi-kəl\ *adjective* (1664)
1 : being physical and chemical
2 : of or relating to chemistry that deals with the physicochemical properties of substances
— phys·i·co·chem·i·cal·ly \-k(ə-)lē\ *adverb*

phys·ics \'fi-ziks\ *noun plural but singular or plural in construction* [Latin *physica,* plural, natural science, from Greek *physika,* from neuter plural of *physikos* of nature, from *physis* growth, nature, from *phyein* to bring forth — more at BE] (1715)
1 : a science that deals with matter and energy and their interactions
2 a : the physical processes and phenomena of a particular system **b :** the physical properties and composition of something

Phys·io·crat \'fi-zē-ə-,krat\ *noun* [French *physiocrate,* from *physi-* physi- + *-crate* -crat] (1798)
: a member of a school of political economists founded in 18th century France and characterized chiefly by a belief that government policy should not interfere with the operation of natural economic laws and that land is the source of all wealth
— phys·io·crat·ic \,fi-zē-ə-'kra-tik\ *adjective, often capitalized*

phys·i·og·nom·ic \,fi-zē-ə(g)-'nä-mik\ *also* **phys·i·og·nom·i·cal** \-mi-kəl\ *adjective* (1588)
: of, relating to, or characteristic of physiognomy or the physiognomy
— phys·i·og·nom·i·cal·ly \-mi-k(ə-)lē\ *adverb*

phys·i·og·no·my \,fi-zē-'ä(g)-nə-mē\ *noun, plural* **-mies** [Middle English *phisonomie,* from Middle French, from Late Latin *physiognomonia, physiognomia,* from Greek *physiognōmonia,* from *physiognōmōn* judging character by the features, from *physis* nature, physique, appearance + *gnōmōn* interpreter — more at GNOMON] (14th century)
1 : the art of discovering temperament and character from outward appearance
2 : the facial features held to show qualities of mind or character by their configuration or expression
3 : external aspect; *also* **:** inner character or quality revealed outwardly

phys·i·og·ra·phy \,fi-zē-'ä-grə-fē\ *noun* [probably from French *physiographie,* from *physi-* + *-graphie* -graphy] (circa 1828)
: PHYSICAL GEOGRAPHY
— phys·i·og·ra·pher \-fər\ *noun*
— phys·io·graph·ic \,fi-zē-ō-'gra-fik\ *also* **phys·io·graph·i·cal** \-fi-kəl\ *adjective*

phys·i·o·log·i·cal \,fi-zē-ə-'lä-ji-kəl\ *or* **phys·i·o·log·ic** \-jik\ *adjective* (1814)
1 : of or relating to physiology
2 : characteristic of or appropriate to an organism's healthy or normal functioning
3 : differing in, involving, or affecting physiological factors ⟨a *physiological* strain of bacteria⟩
— phys·i·o·log·i·cal·ly \-ji-k(ə-)lē\ *adverb*

physiological psychology *noun* (1888)
: a branch of psychology that deals with the effects of normal and pathological physiological processes on mental life — called also *psychophysiology*

physiological saline *noun* (1896)
: a solution of a salt or salts that is essentially isotonic with tissue fluids or blood

phys·i·ol·o·gy \,fi-zē-'ä-lə-jē\ *noun* [Latin *physiologia* natural science, from Greek, from *physi-* + *-logia* -logy] (1597)
1 : a branch of biology that deals with the functions and activities of life or of living matter (as organs, tissues, or cells) and of the physical and chemical phenomena involved — compare ANATOMY
2 : the organic processes and phenomena of an organism or any of its parts or of a particular bodily process
— phys·i·ol·o·gist \-jist\ *noun*

phys·io·pa·thol·o·gy \'fi-zē-ō-pə-'thä-lə-jē, -pa-\ *noun* (circa 1889)
: a branch of biology or medicine that combines physiology and pathology especially in the study of altered bodily function in disease
— phys·io·path·o·log·ic \-,pa-thə-'lä-jik\ *or* **phys·io·path·o·log·i·cal** \-ji-kəl\ *adjective*

phys·io·ther·a·py \,fi-zē-ō-'ther-ə-pē\ *noun* [New Latin *physiotherapia,* from *physi-* + *therapia* therapy] (circa 1903)
: PHYSICAL THERAPY
— phys·io·ther·a·pist \-pist\ *noun*

phy·sique \fə-'zēk\ *noun* [French, from *physique* physical, bodily, from Latin *physicus* of nature, from Greek *physikos*] (1813)
: the form or structure of a person's body **:** bodily makeup

phy·so·stig·mine \,fī-sə-'stig-,mēn\ *noun* [International Scientific Vocabulary, from New Latin *Physostigma,* genus of vines that bear the Calabar bean] (1864)
: a crystalline tasteless alkaloid $C_{15}H_{21}N_3O_2$ from the Calabar bean that is used in medicine especially in the form of its salicylate for its anticholinesterase activity

phyt- *or* **phyto-** *combining form* [New Latin, from Greek, from *phyton,* from *phyein* to bring forth — more at BE]
: plant ⟨*phyto*phagous⟩

phy·tane \'fī-ˌtān\ *noun* (1907)
: an isoprenoid hydrocarbon $C_{20}H_{42}$ that is found especially associated with fossilized plant remains from the Precambrian and later eras

-phyte *noun combining form* [International Scientific Vocabulary, from Greek *phyton* plant]
: plant having a (specified) characteristic or habitat ⟨xero*phyte*⟩

-phytic *adjective combining form* [International Scientific Vocabulary, from Greek *phyton* plant]
: like a plant ⟨holo*phytic*⟩

phy·to·alex·in \ˌfī-tō-ə-'lek-sən\ *noun* [International Scientific Vocabulary *phyt-* + *alexin* substance combating infection, from Greek *alexein* to ward off, protect; akin to Sanskrit *rakṣati* he protects] (1949)
: a chemical substance produced by a plant to combat infection by a pathogen (as a fungus)

phy·to·chem·i·cal \-'ke-mi-kəl\ *adjective* (circa 1858)
: of, relating to, or being phytochemistry
— **phy·to·chem·i·cal·ly** \-mi-k(ə-)lē\ *adverb*

phy·to·chem·is·try \-'ke-mə-strē\ *noun* (1837)
: the chemistry of plants, plant processes, and plant products
— **phy·to·chem·ist** \-'ke-mist\ *noun*

phy·to·chrome \'fī-tə-ˌkrōm\ *noun* (circa 1893)
: a chromoprotein that is present in traces in many plants and that plays a significant role in initiating floral and developmental processes when activated by red or near-infrared radiation

phy·to·fla·gel·late \ˌfī-tō-'fla-jə-lət, -ˌlāt; -flə-'je-\ *noun* (1935)
: any of various organisms (as dinoflagellates) that are considered a subclass (Phytomastigophora synonym Phytomastigina) usually of algae by botanists and of protozoans by zoologists and that have many characteristics in common with typical algae

phy·to·ge·og·ra·phy \ˌfī-tō-jē-'ä-grə-fē\ *noun* [International Scientific Vocabulary] (1847)
: the biogeography of plants
— **phy·to·ge·og·ra·pher** \-fər\ *noun*
— **phy·to·geo·graph·i·cal** \-ˌjē-ə-'gra-fi-kəl\ *or* **phy·to·geo·graph·ic** \-fik\ *adjective*
— **phy·to·geo·graph·i·cal·ly** \-fi-k(ə-)lē\ *adverb*

phy·to·he·mag·glu·ti·nin \-ˌhē-mə-'glü-tᵊn-ən\ *noun* (1949)
: a proteinaceous hemagglutinin of plant origin used especially to induce mitosis (as in lymphocytes)

phy·to·hor·mone \-'hȯr-ˌmōn\ *noun* [International Scientific Vocabulary] (1933)
: PLANT HORMONE

phy·ton \'fī-ˌtän\ *noun* [New Latin, from Greek, plant] (1848)
1 : a structural unit of a plant consisting of a leaf and its associated portion of stem
2 : the smallest part of a stem, root, or leaf that when severed may grow into a new plant
— **phy·ton·ic** \fī-'tä-nik\ *adjective*

phy·to·path·o·gen \ˌfī-tō-'pa-thə-jən\ *noun* (circa 1930)
: an organism parasitic on a plant host
— **phy·to·path·o·gen·ic** \-ˌpa-thə-'je-nik\ *adjective*

phy·to·pa·thol·o·gy \-pə-'thä-lə-jē, -pa-\ *noun* [International Scientific Vocabulary] (circa 1859)
: plant pathology
— **phy·to·path·o·log·i·cal** \-ˌpa-thə-'lä-ji-kəl\ *adjective*

phy·toph·a·gous \fī-'tä-fə-gəs\ *adjective* (1826)
: feeding on plants ⟨a *phytophagous* insect⟩

phy·to·plank·ter \ˌfī-tō-'plaŋ(k)-tər\ *noun* (1944)
: a planktonic plant

phy·to·plank·ton \-'plaŋ(k)-tən, -ˌtän\ *noun* [International Scientific Vocabulary] (1897)
: planktonic plant life
— **phy·to·plank·ton·ic** \-plaŋ(k)-'tä-nik\ *adjective*

phy·to·so·ci·ol·o·gy \-ˌsō-sē-'ä-lə-jē, -shē-\ *noun* (circa 1928)
: a branch of ecology concerned especially with the structure, composition, and interrelationships of plant communities
— **phy·to·so·cio·log·i·cal** \-sē-ə-'lä-ji-kəl\ *adjective*

phy·tos·ter·ol \fī-'täs-tə-ˌrȯl, -ˌrōl\ *noun* [International Scientific Vocabulary] (1898)
: any of various sterols derived from plants — compare ZOOSTEROL

phy·to·tox·ic \ˌfī-tə-'täk-sik\ *adjective* (1926)
: poisonous to plants
— **phy·to·tox·ic·i·ty** \-ˌtäk-'si-sə-tē\ *noun*

[1]pi *also* **pie** \'pī\ *noun, plural* **pies** [origin unknown] (circa 1659)
1 : type that is spilled or mixed
2 : a pi character or matrix

[2]pi *also* **pie** *verb* **pied; pi·ing** *or* **pie·ing** (1870)
transitive verb
: to spill or throw (type or type matter) into disorder
intransitive verb
: to become pied

[3]pi *adjective* (circa 1940)
1 : not intended to appear in final printing ⟨*pi* lines⟩
2 : capable of being inserted only by hand ⟨*pi* characters⟩

[4]pi *noun, plural* **pis** \'pīz\ [Middle Greek, from Greek *pei,* of Semitic origin; akin to Hebrew *pē* pe] (1823)
1 : the 16th letter of the Greek alphabet — see ALPHABET table
2 a : the symbol π denoting the ratio of the circumference of a circle to its diameter **b :** the ratio itself **:** a transcendental number having a value to eight decimal places of 3.14159265

pi·al \'pī-əl, 'pē-\ *adjective* (1889)
: of or relating to the pia mater

pia ma·ter \'pī-ə-ˌmā-tər, 'pē-ə-ˌmä-\ *noun* [Middle English, from Medieval Latin, from Latin, tender mother] (14th century)
: the thin vascular membrane that invests the brain and spinal cord internal to the arachnoid and dura mater

pi·a·nism \'pē-ə-ˌni-zəm\ *noun* (1844)
1 : the art or technique of piano playing
2 : the composition or adaptation of music for the piano

[1]pi·a·nis·si·mo \ˌpē-ə-'ni-sə-ˌmō\ *adverb or adjective* [Italian, from *piano* softly] (1724)
: very softly — used as a direction in music

[2]pianissimo *noun, plural* **-mi** \-(ˌ)mē\ *or* **-mos** (1883)
: a passage played, sung, or spoken very softly

pi·a·nist \pē-'a-nist, 'pē-ə-\ *noun* (circa 1828)
: one who plays the piano; *especially* **:** a skilled or professional performer on the piano

pi·a·nis·tic \ˌpē-ə-'nis-tik\ *adjective* (1881)
1 : of, relating to, or characteristic of the piano
2 : skilled in or well adapted to piano playing
— **pi·a·nis·ti·cal·ly** \-ti-k(ə-)lē\ *adverb*

[1]pi·a·no \pē-'ä-(ˌ)nō\ *adverb or adjective* [Italian, from Late Latin *planus* smooth, from Latin, level — more at FLOOR] (1683)
: at a soft volume **:** SOFT — used as a direction in music

[2]pi·ano \pē-'a-(ˌ)nō *also* -'ä-\ *noun, plural* **pi·anos** [Italian, short for *pianoforte,* from *gravicembalo col piano e forte,* literally, harpsichord with soft and loud; from the fact that its tones could be varied in loudness] (1803)
: a musical instrument having steel wire strings that sound when struck by felt-covered hammers operated from a keyboard ◆

piano accordion *noun* (1860)
: an accordion with a keyboard for the right hand resembling and corresponding to the middle register of a piano keyboard

pi·ano·forte \pē-'a-nə-ˌfōrt, -'ä-, -ˌfȯrt, -ˌfȯr-tē; -ˌa-nə-'fȯr-tē, -ˌä-\ *noun* [Italian] (1767)
1 : FORTEPIANO
2 : PIANO
word history see PIANO

piano hinge *noun* (1926)
: a hinge that has a thin pin joint and extends along the full length of the part to be moved

pi·as·sa·va \ˌpē-ə-'sä-və\ *noun* [Portuguese *piassaba,* from Tupi *piasawa*] (1835)
1 : any of several stiff coarse fibers obtained from palms and used especially in cordage or brushes
2 : a palm yielding piassava; *especially* **:** either of two Brazilian palms (*Attalia funifera* and *Leopoldinia piassaba*)

pi·as·tre *also* **pi·as·ter** \pē-'as-tər, -'äs-\ *noun* [French *piastre,* from Italian *piastra* thin metal plate, coin, from Latin *emplastra, emplastrum* plaster] (1592)
1 : PIECE OF EIGHT
2 — see *pound* at MONEY table

pi·az·za \pē-'a-zə, -'ä-, *1 is usually* -'at-sə, -'ät-\ *noun, plural* **piazzas** *or* **pi·az·ze** \-'at-(ˌ)sā, -'ät-\ [Italian, from Latin *platea* broad street — more at PLACE] (1563)
1 *plural piazze* **:** an open square especially in an Italian town
2 a : an arcaded and roofed gallery **b** *dialect* **:** VERANDA, PORCH

pi·broch \'pē-ˌbräk, -ˌbräk\ *noun* [Scottish Gaelic *piobaireachd* pipe music] (1719)
: a set of martial or mournful variations for the Scottish Highland bagpipe

[1]pic \'pik\ *noun, plural* **pics** *or* **pix** \'piks\ [short for *picture*] (1884)
1 : PHOTOGRAPH
2 : MOTION PICTURE

[2]pic \'pik, 'pēk\ *noun* [Spanish *pica,* from *picar* to prick] (1926)
: the picador's lance

[1]pi·ca \'pī-kə\ *noun* [New Latin, from Latin, magpie — more at PIE] (1563)
: an abnormal desire to eat substances (as chalk or ashes) not normally eaten

◇ WORD HISTORY

piano A harpsichord is played by means of a mechanism that plucks the strings, so it is not possible to achieve fine gradations of loudness. Feeling the need to overcome this drawback, Bartolomeo Cristofori (1655–1731), a Florentine instrument maker, invented a mechanism or action by means of which the strings of the instrument are struck by felt-covered hammers. This device allows the player to produce notes with varying degrees of loudness. In the first published notice of Cristofori's invention in 1711, it was called in Italian *gravicembalo col piano e forte,* "harpsichord with soft and loud." In English the terms *pianoforte* and *fortepiano* were used interchangeably in the 18th century. Eventually, both were largely replaced by *piano,* first attested in 1803, though scholars and performers continue to use *fortepiano* for 18th and early 19th century forms of the instrument.

²pica *noun* [probably from Medieval Latin, collection of church rules] (1588)
1 : 12-point type
2 : a unit of about ⅙ inch used in measuring typographic material
3 : a typewriter type providing 10 characters to the linear inch and six lines to the vertical inch

pic·a·dor \ˈpi-kə-ˌdȯr, ˌpi-kə-ˈ\ *noun, plural* **picadors** \-ˌdȯrz, -ˈdȯrz\ *or* **pic·a·do·res** \ˌpi-kə-ˈdōr-ēz, -ˈdȯr-\ [Spanish, from *picar* to prick, from (assumed) Vulgar Latin *piccare* — more at PIKE] (1797)
: a horseman in a bullfight who jabs the bull with a lance to weaken its neck and shoulder muscles

pi·ca·ra \ˈpē-kä-ˌrä\ *noun* [Spanish *pícara,* feminine of *pícaro*] (circa 1930)
: a woman who is a rogue

¹pi·ca·resque \ˌpi-kə-ˈresk, ˌpē-\ *adjective* [Spanish *picaresco,* from *pícaro*] (1810)
: of or relating to rogues or rascals; *also* **:** of, relating to, suggesting, or being a type of fiction dealing with the episodic adventures of a usually roguish protagonist ⟨a *picaresque* novel⟩

²picaresque *noun* (1895)
: one that is picaresque

pi·ca·ro \ˈpē-kä-ˌrō\ *noun, plural* **-ros** [Spanish *pícaro*] (1623)
: ROGUE, BOHEMIAN

¹pic·a·roon *or* **pick·a·roon** \ˌpi-kə-ˈrün\ *noun* [Spanish *picarón,* augmentative of *pícaro*] (1624)
1 : PIRATE
2 : PICARO

²picaroon *intransitive verb* (1675)
: to act as a pirate

¹pic·a·yune \ˌpi-kē-ˈyün\ *noun* [Provençal *picaioun,* a small coin, from *picaio* money, from *pica* to jingle, of imitative origin] (1804)
1 a : a Spanish half real piece formerly current in the South **b :** HALF DIME
2 : something trivial

²picayune *adjective* (1836)
: of little value **:** PALTRY; *also* **:** PETTY, SMALL-MINDED
— pic·a·yun·ish \-ˈyü-nish\ *adjective*

pic·ca·lil·li \ˌpi-kə-ˈli-lē\ *noun* [probably alteration of *pickle*] (1845)
: a relish of chopped vegetables and spices

¹pic·co·lo \ˈpi-kə-ˌlō\ *adjective* [Italian, small] (circa 1854)
: smaller than ordinary size ⟨a *piccolo* banjo⟩

²piccolo *noun, plural* **-los** [Italian, short for *piccolo flauto* small flute] (1856)
: a small shrill flute whose range is an octave higher than that of an ordinary flute
— pic·co·lo·ist \-ˌlō-ist\ *noun*

piccolo

pice \ˈpīs\ *noun, plural* **pice** [Hindi *paisā*] (1615)
: PAISA

pi·ce·ous \ˈpī-sē-əs\ *adjective* [Latin *piceus,* from *pic-, pix* pitch — more at PITCH] (1826)
: of, relating to, or resembling pitch; *especially* **:** glossy brownish black in color ⟨an insect with a *piceous* abdomen⟩

¹pick \ˈpik\ *verb* [Middle English *piken,* partly from (assumed) Old English *pīcian* (akin to Middle Dutch *picken* to prick); partly from Middle French *piquer* to prick — more at PIKE] (14th century)
transitive verb
1 : to pierce, penetrate, or break up with a pointed instrument ⟨*picked* the hard clay⟩
2 a : to remove bit by bit ⟨*pick* meat from bones⟩ **b :** to remove covering or adhering matter from ⟨*pick* the bones⟩
3 a : to gather by plucking ⟨*pick* apples⟩ **b :** CHOOSE, SELECT ⟨tried to *pick* the shortest route⟩ ⟨she *picked* out the most expensive dress⟩ **c :** to make (one's way) slowly and carefully ⟨*picked* his way through the rubble⟩
4 a : PILFER, ROB ⟨*pick* pockets⟩ **b :** to obtain useful information from by questioning — used in such phrases as *pick the brains of*
5 : PROVOKE ⟨*pick* a quarrel⟩
6 a : to dig into **:** PROBE ⟨*pick* his teeth⟩ **b :** to pluck (as a guitar) with a pick or with the fingers **c :** to loosen or pull apart with a sharp point ⟨*pick* wool⟩
7 : to unlock with a device (as a wire) other than the key ⟨*pick* a lock⟩
intransitive verb
1 : to use or work with a pick
2 : to gather or harvest something by plucking
3 : PILFER — used in the phrase *picking and stealing*
4 : to eat sparingly or mincingly ⟨*picking* listlessly at his dinner⟩
— pick and choose : to select with care and deliberation
— pick at : to criticize repeatedly especially for minor faults **:** NAG
— pick on : to single out for criticism, teasing, or bullying ⟨*picked on* smaller boys⟩; *also* **:** to single out for a particular purpose or for special attention

²pick *noun* (15th century)
1 : a blow or stroke with a pointed instrument
2 a : the act or privilege of choosing or selecting **:** CHOICE ⟨take your *pick*⟩ **b :** the best or choicest one ⟨the *pick* of the herd⟩
3 : the portion of a crop gathered at one time ⟨the first *pick* of peaches⟩
4 : a screen in basketball

³pick *noun* [Middle English *pik*] (14th century)
1 : a heavy wooden-handled iron or steel tool pointed at one or both ends — compare MATTOCK
2 a : TOOTHPICK **b :** PICKLOCK **c :** a small thin piece (as of plastic or metal) used to pluck the strings of a stringed instrument
3 : one of the points on the forepart of the blade of a skate used in figure skating
4 : a comb with long widely spaced teeth used to give height to a hair style

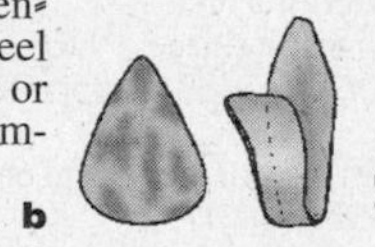

pick 2c

⁴pick *transitive verb* [Middle English *pykken* to pitch (a tent); akin to Middle English *picchen* to pitch] (1523)
1 *chiefly dialect* **:** to throw or thrust with effort **:** HURL
2 : to throw (a shuttle) across the loom

⁵pick *noun* (1627)
1 *dialect English* **a :** the act of pitching or throwing **b :** something thrown
2 a : a throw of the shuttle **b :** a filling thread

pick·a·back \ˈpi-gē-ˌbak, ˈpi-kə-\ *variant of* PIGGYBACK

pick–and–roll \ˈpik-ən(d)-ˈrōl\ *noun* (circa 1961)
: a basketball play in which a player sets a screen and then cuts toward the basket for a pass

pick–and–shovel *adjective* (1895)
: done with or as if with a pick and shovel **:** LABORIOUS

pick·a·nin·ny *or* **pic·a·nin·ny** \ˈpi-kə-ˌni-nē, ˌpi-kə-ˈ\ *noun, plural* **-nies** [probably ultimately from Portuguese *pequenino,* diminutive of *pequeno* small] (1653)
: a black child — often taken to be offensive

pick·ax \ˈpik-ˌaks\ *noun* [Middle English *pecaxe,* alteration of *pikois,* from Old French *picois,* from *pic* pick, from Latin *picus* woodpecker — more at PIE] (15th century)
: ³PICK 1

¹pick·ed \ˈpi-kəd\ *adjective* [Middle English, from ³*pick*] (14th century)
chiefly dialect **:** POINTED, PEAKED

²picked \ˈpikt\ *adjective* [¹*pick*] (circa 1548)
: CHOICE, PRIME

pick·eer \pi-ˈkir\ *intransitive verb* [probably modification of French *picorer* to maraud, perhaps from Middle French *pecore* sheep, from Old Italian *pecora* — more at PECORINO] (circa 1645)
archaic **:** to skirmish in advance of an army; *also* **:** SCOUT, RECONNOITER

pick·er \ˈpi-kər\ *noun* (14th century)
: one that picks: as **a :** a worker who picks something (as crops) **b :** a tool, implement, or machine used in picking something **c :** a musician who picks a stringed instrument (as a banjo)

pick·er·el \ˈpi-k(ə-)rəl\ *noun, plural* **-el** *or* **-els** [Middle English *pikerel,* diminutive of *pike*] (13th century)
1 a *dialect chiefly British* **:** a young or small pike **b :** either of two fishes resembling but smaller than the related northern pike: (1) **:** CHAIN PICKEREL (2) **:** one (*Esox americanus*) having green or red fins and a black bar below and slanting away from the eye
2 : WALLEYE 3

pick·er·el·weed \-ˌwēd\ *noun* (1836)
: an American shallow-water monocotyledonous plant (*Pontederia cordata* of the family Pontederiaceae) with large leaves and a spike of blue flowers

¹pick·et \ˈpi-kət\ *noun* [French *piquet,* from Middle French, from *piquer* to prick — more at PIKE] (circa 1702)
1 : a pointed or sharpened stake, post, or pale
2 a : a detached body of soldiers serving to guard an army from surprise **b :** a detachment kept ready in camp for such duty **c :** SENTRY
3 : a person posted by a labor organization at a place of work affected by a strike; *also* **:** a person posted for a demonstration or protest

²picket (1745)
transitive verb
1 : to enclose, fence, or fortify with pickets
2 a : to guard with a picket **b :** to post as a picket
3 : TETHER
4 a : to post pickets at **b :** to walk or stand in front of as a picket
intransitive verb
: to serve as a picket
— pick·et·er *noun*

pick·et·boat \ˈpi-kət-ˌbōt\ *noun* (1866)
: a craft used (as by the coast guard) for harbor patrol

picket line *noun* (1856)
1 : a position held by a line of military pickets
2 : a line of people picketing a business, organization, or institution

pick·ings \ˈpi-kiŋz, -kənz\ *noun plural* (1642)
: something that is picked or picked up: as **a :** gleanable or eatable fragments **:** SCRAPS **b :** yield or return for effort expended

¹pick·le \ˈpi-kəl\ *noun* [Middle English *pekille*] (15th century)
1 : a solution or bath for preserving or cleaning: as **a :** a brine or vinegar solution in which foods are preserved **b :** any of various baths used in industrial cleaning or processing
2 : a difficult situation **:** PLIGHT ⟨could see no way out of the *pickle* I was in —R. L. Stevenson⟩
3 : an article of food that has been preserved in brine or in vinegar; *specifically* **:** a cucumber that has been so preserved
4 *British* **:** a mischievous or troublesome person

²pickle *transitive verb* **pick·led; pick·ling** \-k(ə-)liŋ\ (1570)
1 : to treat, preserve, or clean in or with a pickle
2 : to give a light finish to (as furniture) by bleaching or painting and wiping

³pickle *noun* [perhaps from Scots *pickle* to trifle, pilfer] (1552)
1 *Scottish* **:** GRAIN, KERNEL
2 *Scottish* **:** a small quantity

pickled *adjective* (circa 1552)

1 : preserved in or cured with pickle ⟨*pickled* herring⟩
2 : DRUNK 1a ⟨gets thoroughly *pickled* before dinner —*New Yorker*⟩

pick·lock \'pik-,läk\ *noun* (1553)
1 : BURGLAR
2 : a tool for picking locks

pick–me–up \'pik-mē-,əp\ *noun* (1867)
: something that stimulates or restores **:** TONIC, BRACER

pick·off \'pik-,ȯf\ *noun* (1939)
: a baseball play in which a base runner is picked off

pick off *transitive verb* (1810)
1 : to shoot or bring down especially one by one
2 : to put out (a base runner who is off base) with a quick throw (as from the pitcher or catcher)
3 : INTERCEPT ⟨*picked off* a pass⟩

pick out *transitive verb* (1540)
1 : DISCERN, MAKE OUT
2 : to play the notes of by ear or one by one ⟨*picking out* tunes on the piano⟩

pick over *transitive verb* (1839)
: to examine in order to select the best or remove the unwanted

pick·pock·et \'pik-,pä-kət\ *noun* (1591)
: a thief who picks pockets

pick·proof \-,prüf\ *adjective* (1933)
: designed to prevent picking ⟨a *pickproof* lock⟩

pick·thank \-,thaŋk\ *noun* [from *pick a thank* to seek someone's favor] (15th century)
archaic **:** SYCOPHANT

[1]pick·up \'pik-,əp\ *noun* (1848)
1 : one that is picked up: as **a :** a hitchhiker who is given a ride **b :** a temporary chance acquaintance
2 : the act or process of picking up: as **a :** a revival of business activity **b :** ACCELERATION
3 : the conversion of mechanical movements into electrical impulses in the reproduction of sound; *also* **:** a device (as on a phonograph) for making such conversion
4 a (1) **:** the reception of sound or an image into a radio or television transmitting apparatus for conversion into electrical signals (2) **:** interference (as with such reception) from an adjacent electrical circuit or system **b :** a device (as a microphone or a television camera) for converting sound or the image of a scene into electrical signals **c :** the place where a broadcast originates **d :** the electrical system for connecting to a broadcasting station a program produced outside the studio
5 : a light truck having an enclosed cab and an open body with low sides and tailgate — called also *pickup truck*

[2]pickup *adjective* (1909)
: utilizing or comprising local or available personnel especially without formal organization ⟨a *pickup* basketball game⟩

pick up (14th century)
transitive verb
1 a : to take hold of and lift up **b :** to gather together **:** COLLECT ⟨*picked up* all the pieces⟩ **c :** to clean up **:** TIDY
2 : to take (passengers or freight) into a vehicle
3 a : to acquire casually or by chance ⟨*picked up* a valuable antique at an auction⟩ **b :** to acquire by study or experience **:** LEARN ⟨*picking up* a great deal of knowledge in the process —Robert Schleicher⟩ **c :** to obtain especially by payment **:** BUY ⟨*picked up* some groceries⟩ **d :** to acquire (a player) especially from another team through a trade or by financial recompense **e :** to accept for the purpose of paying ⟨offered to *pick up* the tab⟩ **f :** to come down with **:** CATCH ⟨*picked up* a cold⟩ **g :** GAIN ⟨*picked up* a few yards on the last play⟩
4 a : to enter informally into conversation or companionship with (a previously unknown person) ⟨had a brief affair with a girl he *picked up* in a bar⟩ **b :** to take into custody ⟨the police *picked up* the fugitive⟩
5 a : to catch sight of **:** PERCEIVE ⟨*picked up* the harbor lights⟩ **b :** to come to and follow ⟨*picked up* the outlaw's trail⟩ **c :** to bring within range of sight or hearing ⟨*pick up* distant radio signals⟩ **d :** UNDERSTAND, CATCH ⟨didn't *pick up* the hint⟩
6 a : REVIVE **b :** INCREASE
7 : to resume after a break **:** CONTINUE ⟨*pick up* the discussion tomorrow⟩
8 : to assume responsibility for guarding (an opponent) in an athletic contest
intransitive verb
1 : to recover speed, vigor, or activity **:** IMPROVE ⟨after the strike, business *picked up*⟩
2 : to put things in order ⟨was always *picking up* after her⟩
3 : to pack up one's belongings ⟨couldn't just *pick up* and leave⟩
— pick up on 1 a : UNDERSTAND, APPRECIATE **b :** to become aware of **:** NOTICE **2 :** to adopt as one's own **:** TAKE UP

Pick·wick·ian \(,)pik-'wi-kē-ən\ *adjective* [Samuel *Pickwick,* character in the novel *Pickwick Papers* (1836–37) by Charles Dickens] (1836)
1 : marked by simplicity and generosity
2 : intended or taken in a sense other than the obvious or literal one

picky \'pi-kē\ *adjective* **pick·i·er; -est** (1917)
: FUSSY, CHOOSY ⟨a *picky* eater⟩

pi·clo·ram \'pi-klə-,ram, 'pī-\ *noun* [*pic*oline + ch*lor-* + *amine*] (1965)
: a systemic herbicide $C_6H_3Cl_3N_2O_2$ that breaks down only very slowly in the soil

[1]pic·nic \'pik-(,)nik\ *noun, often attributive* [German or French; German *Picknick,* from French *pique-nique*] (1748)
1 : an excursion or outing with food usually provided by members of the group and eaten in the open; *also* **:** the food provided for a picnic
2 a : a pleasant or amusingly carefree experience ⟨I don't expect being married to be a *picnic* —Josephine Pinckney⟩ **b :** an easy task or feat
3 : a shoulder of pork with much of the butt removed
— pic·nic·ky \-(,)ni-kē\ *adjective*

[2]picnic *intransitive verb* **pic·nicked; pic·nick·ing** (1842)
: to go on a picnic **:** eat in picnic fashion
— pic·nick·er *noun*

pico- *combining form* [International Scientific Vocabulary, probably from Spanish *pico* small amount, literally, peak, beak]
1 : one trillionth (10^{-12}) part of ⟨*pico*gram⟩
2 : very small ⟨*pico*rnavirus⟩

pi·co·far·ad \,pē-kō-'far-,ad, -əd\ *noun* [International Scientific Vocabulary] (circa 1926)
: one trillionth of a farad

pi·co·gram \'pē-kō-,gram, -kə-\ *noun* [International Scientific Vocabulary] (1951)
: one trillionth of a gram

pic·o·line \'pi-kə-,lēn, 'pī-\ *noun* [Latin *pic-, pix* pitch + International Scientific Vocabulary [1]*-ol* + [2]*-ine* — more at PITCH] (1853)
: any of the three liquid pyridine bases C_6H_7N used chiefly as solvents and in organic synthesis

pi·co·mole \'pē-kō-,mōl, -kə-\ *noun* (1968)
: one trillionth of a mole

pi·cor·na·vi·rus \(,)pē-,kȯr-nə-'vī-rəs\ *noun* [*pico-* + *RNA* + *virus*] (1962)
: any of a group of small RNA-containing viruses that includes the enteroviruses and rhinoviruses

pi·co·sec·ond \,pē-kō-'se-kənd, -kənt\ *noun* [International Scientific Vocabulary] (circa 1962)
: one trillionth of a second

[1]pi·cot \'pē-(,)kō, pē-'\ *noun* [French, literally, small point, from Middle French, from *pic* prick, from *piquer* to prick — more at PIKE] (circa 1882)
: one of a series of small ornamental loops forming an edging on ribbon or lace

[2]picot *transitive verb* (1926)
: to finish with picots

pic·o·tee \,pi-kə-'tē\ *noun* [French *picoté* pointed, from *picoter* to mark with points, from *picot*] (1727)
: a flower (as some carnations or tulips) having one basic color with a margin of another color

picr- *or* **picro-** *combining form* [French, from Greek *pikr-, pikro-,* from *pikros* — more at PAINT]
: bitter ⟨*pic*ric acid⟩

pic·ric acid \'pi-krik-\ *noun* [International Scientific Vocabulary] (1852)
: a toxic explosive yellow crystalline strong acid $C_6H_3N_3O_7$ used especially in high explosives and as a dye

pic·ro·tox·in \,pi-krō-'täk-sən\ *noun* [International Scientific Vocabulary] (1815)
: a poisonous bitter crystalline stimulant and convulsive drug $C_{30}H_{34}O_{13}$ used intravenously as an antidote for barbiturate poisoning

Pict \'pikt\ *noun* [Middle English *Pictes,* plural, Picts, from Old English *Pihtas,* from Late Latin *Picti*] (before 12th century)
: a member of a people of the north of Scotland who are first noted in historical records in the late 3d century and who became amalgamated with the Scots in the mid-8th century
— Pict·ish \'pik-tish\ *adjective or noun*

pic·to·gram \'pik-tə-,gram\ *noun* [International Scientific Vocabulary *picto-* (from Latin *pictus*) + *-gram*] (1910)
: PICTOGRAPH

pic·to·graph \-,graf\ *noun* [Latin *pictus* + English *-o-* + *-graph*] (1851)
1 : an ancient or prehistoric drawing or painting on a rock wall
2 : one of the symbols belonging to a pictorial graphic system
3 : a diagram representing statistical data by pictorial forms
— pic·to·graph·ic \,pik-tə-'gra-fik\ *adjective*

pic·tog·ra·phy \pik-'tä-grə-fē\ *noun* (1851)
: use of pictographs **:** PICTURE WRITING 1

[1]pic·to·ri·al \pik-'tōr-ē-əl, -'tȯr-\ *adjective* [Late Latin *pictorius,* from Latin *pictor* painter] (1646)
1 : of or relating to a painter, a painting, or the painting or drawing of pictures ⟨*pictorial* perspective⟩
2 a : of, relating to, or consisting of pictures ⟨*pictorial* records⟩ **b :** illustrated by pictures ⟨*pictorial* weekly⟩ **c :** consisting of or displaying the characteristics of pictographs **d :** suggesting or conveying visual images ⟨*pictorial* poetry⟩
— pic·to·ri·al·ly \-ē-ə-lē\ *adverb*
— pic·to·ri·al·ness *noun*

[2]pictorial *noun* (1844)
: a periodical having much pictorial matter

pic·to·ri·al·ism \-ē-ə-,li-zəm\ *noun* (1869)
1 : the use or creation of pictures or visual images
2 : a movement or technique in photography emphasizing artificial often romanticized pictorial qualities
— pic·to·ri·al·ist \-list\ *adjective or noun*

pic·to·ri·al·ize \pik-'tōr-ē-ə-,līz, -'tȯr-\ *transitive verb* **-ized; -iz·ing** (1870)
: to represent by a picture or illustrate with pictures

\ə\ abut \ᵊ\ kitten \ər\ further \a\ ash \ā\ ace \ä\ mop, mar \au̇\ out \ch\ chin \e\ bet \ē\ easy \g\ go \i\ hit \ī\ ice \j\ job \ŋ\ sing \ō\ go \ȯ\ law \ȯi\ boy \th\ thin \t͟h\ the \ü\ loot \u̇\ foot \y\ yet \zh\ vision *see also* Guide to Pronunciation

— **pic·to·ri·al·i·za·tion** \-ˌtōr-ē-ə-lə-ˈzā-shən, -ˌtȯr-\ *noun*

¹pic·ture \ˈpik-chər\ *noun* [Middle English, from Latin *pictura,* from *pictus,* past participle of *pingere* to paint — more at PAINT] (15th century)
1 : a design or representation made by various means (as painting, drawing, or photography)
2 a : a description so vivid or graphic as to suggest a mental image or give an accurate idea of something ⟨the book gives a detailed *picture* of what is happening⟩ **b :** a mental image
3 : IMAGE, COPY ⟨the *picture* of his father⟩ ⟨the very *picture* of health⟩
4 a : a transitory visible image or reproduction **b :** MOTION PICTURE **c** *plural* **:** MOVIES
5 : TABLEAU 1, 2 ⟨stage *pictures*⟩
6 : SITUATION ⟨took a hard look at his financial *picture*⟩

²picture *transitive verb* **pic·tured; pic·tur·ing** \ˈpik-chə-riŋ, ˈpik-shriŋ\ (15th century)
1 : to paint or draw a representation, image, or visual conception of **:** DEPICT; *also* **:** ILLUSTRATE
2 : to describe graphically in words
3 : to form a mental image of **:** IMAGINE

picture–book *adjective* (1922)
: suitable for or suggestive of a picture book: as **a :** PICTURESQUE **b :** PICTURE-PERFECT

picture book *noun* (1847)
: a book that consists wholly or chiefly of pictures

picture hat *noun* (1887)
: a woman's dressy hat with a broad brim

picture–perfect *adjective* (1981)
: completely flawless **:** PERFECT ⟨made a *picture-perfect* landing⟩

pic·ture·phone \ˈpik-chər-ˌfōn\ *noun* (1956)
: VIDEOPHONE

picture–postcard *adjective* (1907)
: PICTURESQUE, PICTURE-BOOK ⟨a *picture-postcard* village⟩

picture puzzle *noun* (1898)
: JIGSAW PUZZLE

pic·tur·esque \ˌpik-chə-ˈresk\ *adjective* [French & Italian; French *pittoresque,* from Italian *pittoresco,* from *pittore* painter, from Latin *pictor,* from *pingere*] (1703)
1 a : resembling a picture **:** suggesting a painted scene **b :** charming or quaint in appearance
2 : evoking mental images **:** VIVID
synonym see GRAPHIC
— **pic·tur·esque·ly** *adverb*
— **pic·tur·esque·ness** *noun*

picture tube *noun* (1937)
: a cathode-ray tube on which the picture appears in a television receiver

picture window *noun* (1938)
: an outsize usually single-paned window designed to frame an exterior view

picture writing *noun* (1741)
1 : the recording of events or expression of messages by pictures representing actions or facts
2 : the record or message represented by picture writing

pic·tur·ize \ˈpik-chə-ˌrīz\ *transitive verb* **-ized; -iz·ing** (circa 1846)
: to make a picture of **:** present in pictures; *especially* **:** to make into a motion picture
— **pic·tur·i·za·tion** \ˌpik-chə-rə-ˈzā-shən\ *noun*

pid·dle \ˈpi-dᵊl\ *intransitive verb* **pid·dled; pid·dling** \ˈpid-liŋ, ˈpi-dᵊl-iŋ\ [origin unknown] (1545)
1 : DAWDLE, PUTTER
2 : URINATE

pid·dling \ˈpid-lən, -liŋ; ˈpi-dᵊl-ən, -iŋ\ *adjective* (1559)
: TRIVIAL, PALTRY

pid·dock \ˈpi-dək, -dik\ *noun* [origin unknown] (1851)
: a bivalve mollusk (genus *Pholas* or family Pholadidae) that bores holes in wood, clay, and rocks

pid·gin \ˈpi-jən\ *noun* [*pidgin English*] (1876)
: a simplified speech used for communication between people with different languages ◆
— **pid·gin·i·za·tion** \ˌpi-jə-nə-ˈzā-shən\ *noun*
— **pid·gin·ize** \ˈpi-jə-ˌnīz\ *transitive verb*

pidgin English *noun, often P capitalized* [Chinese Pidgin English *pidgin* business] (1859)
: an English-based pidgin; *especially* **:** one originally used in parts of the Orient

¹pie \ˈpī\ *noun* [Middle English, from Old French, from Latin *pica;* akin to Latin *picus* woodpecker, Old High German *speh*] (13th century)
: MAGPIE

²pie *noun* [Middle English] (14th century)
1 : a meat dish baked with biscuit or pastry crust — compare POTPIE
2 : a dessert consisting of a filling (as of fruit or custard) in a pastry shell or topped with pastry or both
3 a : AFFAIR, BUSINESS ⟨she wanted her finger . . . in every possible social *pie* —Mary Deasy⟩ **b :** a whole regarded as divisible into shares ⟨giving the less fortunate . . . a larger share of the economic *pie* —R. M. Hutchins⟩

³pie *variant of* PI

¹pie·bald \ˈpī-ˌbȯld\ *adjective* (1594)
1 : of different colors; *especially* **:** spotted or blotched with black and white
2 : composed of incongruous parts

²piebald *noun* (1765)
: a piebald animal (as a horse)

¹piece \ˈpēs\ *noun* [Middle English, from Old French, from (assumed) Vulgar Latin *pettia,* of Gaulish origin; akin to Welsh *peth* thing] (13th century)
1 : a part of a whole: as **a :** FRAGMENT ⟨*pieces* of broken glass⟩ **b :** any of the individual members comprising a unit — often used in combination ⟨a five-*piece* band⟩ ⟨a three-*piece* suit⟩
2 : an object or individual regarded as a unit of a kind or class ⟨a *piece* of fruit⟩
3 : a short distance ⟨down the road a *piece*⟩
4 : a standard quantity (as of length, weight, or size) in which something is made or sold
5 : a literary, journalistic, artistic, dramatic, or musical composition
6 : FIREARM
7 : COIN; *also* **:** TOKEN
8 : a man used in playing a board game; *specifically* **:** a chessman of superior rank
9 : OPINION, VIEW ⟨spoke his *piece*⟩
10 a : an act of copulation — usually considered vulgar **b :** the female partner in sexual intercourse — usually considered vulgar
synonym see PART
— **of a piece :** ALIKE, CONSISTENT
— **piece of one's mind :** a severe scolding **:** TONGUE-LASHING
— **piece of the action :** a share in activity or profit
— **to pieces 1 :** without reserve or restraint **:** COMPLETELY **2 :** into fragments; *also* **:** into component parts **3 :** out of control ⟨went *to pieces* from shock⟩

²piece *transitive verb* **pieced; piec·ing** (15th century)
1 : to repair, renew, or complete by adding pieces **:** PATCH
2 : to join into a whole — often used with *together* ⟨his new book . . . has been *pieced* together from talks —Merle Miller⟩
— **piec·er** *noun*

piece by piece *adverb* (1560)
: by degrees **:** PIECEMEAL

pièce de ré·sis·tance \pē-ˌes-də-rə-ˌzē-ˈstän(t)s, -rā-, -ˈstäⁿs\ *noun, plural* **pièces de ré·sis·tance** *same*\ [French, literally, piece of resistance] (1839)
1 : the chief dish of a meal
2 : an outstanding item or event **:** SHOWPIECE

piece–dye \ˈpēs-ˌdī\ *transitive verb* (1920)
: to dye after weaving or knitting

piece goods *noun plural* (1665)
: cloth fabrics sold from the bolt at retail in lengths specified by the customer — called also *yard goods*

¹piece·meal \ˈpēs-ˌmēl, -ˈmē(ə)l\ *adverb* (14th century)
1 : one piece at a time **:** GRADUALLY
2 : in pieces or fragments **:** APART

²piecemeal *adjective* (1600)
: done, made, or accomplished piece by piece or in a fragmentary way ⟨*piecemeal* reforms in the system⟩

piece of cake (1936)
: something easily done **:** CINCH, BREEZE

piece of eight (1610)
: an old Spanish peso of eight reals

piece of work (1928)
: a complicated, difficult, or eccentric person

piece·wise \ˈpēs-ˌwīz\ *adverb* (1674)
: with respect to a number of discrete intervals, sets, or pieces ⟨*piecewise* continuous functions⟩

piece·work \-ˌwərk\ *noun* (1549)
: work done by the piece and paid for at a set rate per unit
— **piece·work·er** \-ˌwər-kər\ *noun*

pie chart *noun* (1922)
: a circular chart cut by radii into segments illustrating relative magnitudes or frequencies — called also *circle graph*

pie·crust \ˈpī-ˌkrəst\ *noun* (1582)
: the pastry shell of a pie

¹pied \ˈpīd\ *adjective* (14th century)
: of two or more colors in blotches; *also* **:** wearing or having a parti-colored coat ⟨a *pied* horse⟩

²pied *past and past participle of* PI *or of* PIE

pied–à–terre \pē-ˌād-ə-ˈter, -ˌād-ä-; ˌpyā-dä-\ *noun, plural* **pieds–à–terre** *same*\ [French, literally, foot to the ground] (1829)
: a temporary or second lodging

pied·mont \ˈpēd-ˌmänt\ *adjective* [*Piedmont,* region of Italy] (1855)
: lying or formed at the base of mountains
— **piedmont** *noun*

pied piper *noun, often both Ps capitalized* [the *Pied Piper,* hero of a German folktale who charmed the rats of Hameln, Germany, into a river] (1925)
1 : one that offers strong but delusive enticement

◇ WORD HISTORY

pidgin *Pidgin,* alluding to a makeshift language used between people with no language in common, comes from *Pidgin English,* the name for a particular pidgin based on English that was formerly used along the China coast in commercial transactions between Chinese and foreigners. (Linguists usually call it *Chinese Pidgin English* to distinguish it from other pidgins based on English.) The origin of *pidgin* in *Pidgin English* is mysterious. In Chinese Pidgin English the word *pidgin* means "business, affair, occupation," so that *pidgin English* presumably meant simply "business English," the language used to carry on trade. Consequently, *pidgin* has been explained with little phonetic plausibility as a Chinese mispronunciation of *business* or, alternatively, of Portuguese *ocupação* "occupation." Another theory is that *pidgin* meant originally "pidgin language" and ultimately draws its origin from the *Pidians,* supposedly an American Indian group of Guiana with whom inhabitants of a short-lived English colony were in contact in 1605–06. But while the phonetic shift from *Pidian* to *pidgin* is unobjectionable, the shift of sense and the transfer from Guiana to China two centuries later are entirely hypothetical; moreover, *Pidian* in the sole text where the word is found may actually be a misprint for *Indian.*

2 : a leader who makes irresponsible promises
3 : a charismatic person who attracts followers
pie–eyed \ˈpī-ˌīd\ *adjective* (1904)
: INTOXICATED
pie–faced \-ˌfāst\ *adjective* (circa 1912)
: having a round, smooth, or blank face
pieing *present participle of* PI *or of* PIE
pie in the sky (1911)
: an unrealistic enterprise or prospect of prosperity
— pie–in–the–sky *adjective*
pie·plant \ˈpī-ˌplant\ *noun* (circa 1847)
: garden rhubarb
pier \ˈpir\ *noun* [Middle English *per,* from Old English, from Medieval Latin *pera*] (12th century)
1 : an intermediate support for the adjacent ends of two bridge spans
2 : a structure (as a breakwater) extending into navigable water for use as a landing place or promenade or to protect or form a harbor
3 : a vertical structural support: as **a :** the wall between two openings **b :** PILLAR, PILASTER **c :** a vertical member that supports the end of an arch or lintel **d :** an auxiliary mass of masonry used to stiffen a wall
4 : a structural mount (as for a telescope) usually of stonework, concrete, or steel
pierce \ˈpirs\ *verb* **pierced; pierc·ing** [Middle English *percen,* from Old French *percer,* from (assumed) Vulgar Latin *pertusiare,* from Latin *pertusus,* past participle of *pertundere* to perforate, from *per-* through + *tundere* to beat — more at PER-, CONTUSION] (14th century)
transitive verb
1 a : to run into or through as a pointed weapon does **:** STAB **b :** to enter or thrust into sharply or painfully
2 : to make a hole through **:** PERFORATE
3 : to force or make a way into or through
4 : to penetrate with the eye or mind **:** DISCERN
5 : to penetrate so as to move or touch the emotions of
intransitive verb
: to force a way into or through something
synonym see ENTER
pierced *adjective* (14th century)
1 : having holes; *especially* **:** decorated with perforations
2 : having the earlobe punctured for an earring ⟨*pierced* ears⟩
3 : designed for pierced ears ⟨*pierced* earrings⟩
piercing *adjective* (14th century)
: PENETRATING: as **a :** LOUD, SHRILL ⟨*piercing* cries⟩ **b :** PERCEPTIVE ⟨*piercing* eyes⟩ **c :** penetratingly cold **:** BITING ⟨a *piercing* wind⟩ **d :** CUTTING, INCISIVE ⟨*piercing* sarcasm⟩
— pierc·ing·ly \ˈpir-siŋ-lē\ *adverb*
pier glass *noun* (1703)
: a large high mirror; *especially* **:** one designed to occupy the wall space between windows — called also *pier mirror*
Pi·eri·an \pī-ˈir-ē-ən, -ˈer-\ *adjective* (1591)
1 : of or relating to the region of Pieria in ancient Macedonia or to the Muses who were once worshiped there
2 : of or relating to learning or poetry
pie·ro·gi *also* **pi·ro·gi** \pə-ˈrō-gē, pi-\ *noun, plural* **-gi** *also* **-gies** [Polish, plural of *pieróg* dumpling, pierogi] (1927)
: a case of dough filled with a savory filling (as of meat, cheese, or vegetables) and cooked by boiling and then panfrying
Pier·rot \ˈpē-ə-ˌrō\ *noun* [French, diminutive of *Pierre* Peter] (circa 1770)
: a stock comic character of old French pantomime usually having a whitened face and wearing loose white clothes
pier table *noun* (1803)
: a table to be placed under a pier glass
pies *plural of* PI *or of* PIE
pie safe *noun* (1951)
: a cupboard whose doors have decoratively pierced tin panels for ventilation
pie·tà \ˌpē-(ˌ)ā-ˈtä, pyā-\ *noun, often capitalized* [Italian, literally, pity, from Latin *pietat-, pietas*] (1644)
: a representation of the Virgin Mary mourning over the dead body of Christ
pi·e·tism \ˈpī-ə-ˌti-zəm\ *noun* (1697)
1 *capitalized* **:** a 17th century religious movement originating in Germany in reaction to formalism and intellectualism and stressing Bible study and personal religious experience
2 a : emphasis on devotional experience and practices **b :** affectation of devotion
— pi·e·tist \ˈpī-ə-tist\ *adjective or noun, often capitalized*
pi·e·tis·tic \ˌpī-ə-ˈtis-tik\ *adjective* (1830)
1 : of or relating to Pietism
2 a : of or relating to religious devotion or devout persons **b :** marked by overly sentimental or emotional devotion to religion **:** RELIGIOSE
— pi·e·tis·ti·cal·ly \-ti-k(ə-)lē\ *adverb*
pi·e·ty \ˈpī-ə-tē\ *noun, plural* **pi·e·ties** [French *piété* piety, pity, from Latin *pietat-, pietas,* from *pius* dutiful, pious] (1565)
1 : the quality or state of being pious: as **a :** fidelity to natural obligations (as to parents) **b :** dutifulness in religion **:** DEVOUTNESS
2 : an act inspired by piety
3 : a conventional belief or standard **:** ORTHODOXY
synonym see FIDELITY
piezo- *combining form* [Greek *piezein* to press; perhaps akin to Sanskrit *pīḍayati* he squeezes] **:** pressure ⟨*piezo*meter⟩
pi·e·zo·elec·tric \pē-ˌā-(ˌ)zō-ə-ˈlek-trik, pē-ˌāt-(ˌ)sō-\ *adjective* [International Scientific Vocabulary] (1883)
: of, relating to, marked by, or functioning by means of piezoelectricity
— pi·e·zo·elec·tri·cal·ly \-tri-k(ə-)lē\ *adverb*
pi·e·zo·elec·tric·i·ty \-ˌlek-ˈtri-s(ə-)tē\ *noun* [International Scientific Vocabulary] (1883)
: electricity or electric polarity due to pressure especially in a crystalline substance (as quartz)
pi·e·zom·e·ter \ˌpē-ə-ˈzä-mə-tər, pē-ˌāt-ˈsä-\ *noun* (1820)
: an instrument for measuring pressure or compressibility; *especially* **:** one for measuring the change of pressure of a material subjected to hydrostatic pressure
— pi·e·zo·met·ric \pē-ˌā-zə-ˈme-trik, pē-ˌāt-sə-\ *adjective*
[1]**pif·fle** \ˈpi-fəl\ *intransitive verb* **pif·fled; pif·fling** \-f(ə-)liŋ\ [perhaps blend of *piddle* and *trifle*] (circa 1878)
: to talk or act in a trivial, inept, or ineffective way
[2]**piffle** *noun* (1890)
: trivial nonsense ⟨pseudo-scientific *piffle*⟩
pif·fling \ˈpi-flən, -f(ə-)liŋ\ *adjective* (1894)
: of little worth or importance **:** TRIVIAL
[1]**pig** \ˈpig\ *noun, often attributive* [Middle English *pigge*] (13th century)
1 : a young swine not yet sexually mature; *broadly* **:** a wild or domestic swine
2 a : PORK **b :** the dressed carcass of a young swine weighing less than 130 pounds (60 kilograms) **c :** PIGSKIN
3 a : one that resembles a pig ⟨an unkempt . . . person is a *pig* —S. S. Hall⟩ **b :** an animal related to or resembling the pig
4 : a crude casting of metal (as iron)
5 *slang* **:** an immoral woman
6 *slang* **:** POLICE OFFICER — usually used disparagingly
— pig·like \-ˌlīk\ *adjective*
[2]**pig** *verb* **pigged; pig·ging** (15th century)
intransitive verb
1 : FARROW
2 : to live like a pig ⟨*pig* it⟩
transitive verb
: FARROW
pig·boat \ˈpig-ˌbōt\ *noun* (1921)
: SUBMARINE
[1]**pi·geon** \ˈpi-jən\ *noun* [Middle English, from Middle French *pijon,* from Late Latin *pipion-, pipio* young bird, from Latin *pipire* to chirp] (14th century)
1 : any of a widely distributed family (Columbidae, order Columbiformes) of birds with a stout body, rather short legs, and smooth and compact plumage; *especially* **:** a member of any of numerous varieties of the rock dove that exist in domestication and in the feral state in cities and towns throughout most of the world
2 : a young woman
3 : an easy mark **:** DUPE
4 : CLAY PIGEON
[2]**pigeon** *noun* [alteration of *pidgin*] (1938)
: an object of special concern **:** BUSINESS
pigeon breast *noun* (1842)
: a deformity of the chest marked by sharp projection of the sternum
pigeon hawk *noun* (circa 1728)
1 : a falcon of the North American population of the merlin
2 : SHARP-SHINNED HAWK
[1]**pi·geon·hole** \ˈpi-jən-ˌhōl\ *noun* (1577)
1 : a hole or small recess for pigeons to nest
2 : a small open compartment (as in a desk or cabinet) for keeping letters or documents
3 : a neat category which usually fails to reflect actual complexities
[2]**pigeonhole** *transitive verb* (1840)
1 a : to place in or as if in the pigeonhole of a desk **b :** to lay aside **:** SHELVE ⟨his reports continued to be *pigeonholed* and his advice not taken —Walter Mills⟩
2 : to assign to a category **:** CLASSIFY
— pi·geon·hol·er \-ˌhō-lər\ *noun*
pi·geon·ite \ˈpi-jə-ˌnīt\ *noun* [*Pigeon* Point, northeast Minn. + [1]*-ite*] (1900)
: a monoclinic mineral of the pyroxene group
pi·geon–liv·ered \ˈpi-jən-ˌli-vərd\ *adjective* (1602)
: GENTLE, MILD
pigeon pea *noun* (1725)
: a leguminous woody herb (*Cajanus cajan*) that has trifoliate leaves, yellow flowers, and somewhat flat pods and is much cultivated especially in the tropics; *also* **:** its small highly nutritious seed
pi·geon–toed \ˈpi-jən-ˌtōd\ *adjective* (1801)
: having the toes turned in
pi·geon·wing \-ˌwiŋ\ *noun* (circa 1808)
: a fancy dance step executed by jumping and striking the legs together
pig·fish \ˈpig-ˌfish\ *noun* (1860)
: a saltwater grunt (*Orthopristis chrysopterus*) that is a food fish found from Long Island southward
pig·gery \ˈpi-gə-rē\ *noun, plural* **-ger·ies** (1781)
1 : a place where swine are kept
2 : swinish behavior
pig·gin \ˈpi-gən\ *noun* [Middle English *pygyn*] (15th century)
: a small wooden pail with one stave extended upward as a handle
pig·gish \ˈpi-gish\ *adjective* (1792)
1 : of, relating to, or suggestive of a pig ⟨a *piggish* snort⟩
2 : having qualities (as greediness or stubbornness) associated with a pig
— pig·gish·ly *adverb*
— pig·gish·ness *noun*
pig·gy \ˈpi-gē\ *adjective* **pig·gi·er; -est** (circa 1845)
: PIGGISH
[1]**pig·gy·back** \ˈpi-gē-ˌbak\ *adverb* [alteration of earlier *a pick pack,* of unknown origin] (1565)
1 : up on the back and shoulders
2 : on or as if on the back of another; *especially* **:** on a railroad flatcar

[2]piggyback *noun* (circa 1590)
1 : the act of carrying piggyback
2 : the movement of loaded truck trailers on railroad flatcars

[3]piggyback *adjective* (1823)
1 : marked by being up on the shoulders and back ⟨a child needs hugging, tussling, and *piggyback* rides —Benjamin Spock⟩
2 : carried or transported piggyback: as **a :** of or relating to the hauling of truck trailers on railroad flatcars **b :** of, relating to, or being a radio or television commercial that is presented in addition to other commercials during one commercial break **c :** being or relating to something carried into space as an extra load by a vehicle (as a spacecraft)

[4]piggyback (1952)
transitive verb
1 : to carry up on the shoulders and back
2 : to haul (as a truck trailer) by railroad car
3 : to set up or cause to function in conjunction with something larger or more important
intransitive verb
1 : to haul truck trailers on railroad cars
2 : to function or be carried on or as if on the back of another

piggy bank *noun* (1941)
: a coin bank often in the shape of a pig

pig·head·ed \ˈpig-ˌhe-dəd\ *adjective* (1620)
: willfully or perversely unyielding **:** OBSTINATE
— pig·head·ed·ly *adverb*
— pig·head·ed·ness *noun*

pig in a poke (1562)
: something offered in such a way as to obscure its real nature or worth ⟨unwilling to buy a *pig in a poke*⟩

pig iron *noun* (1665)
: crude iron that is the direct product of the blast furnace and is refined to produce steel, wrought iron, or ingot iron

pig latin *noun, often L capitalized* (1931)
: a jargon that is made by systematic alteration of English (as *ipskay the ointjay* for *skip the joint*)

pig·let \ˈpi-glət\ *noun* (1883)
: a small usually young swine

[1]pig·ment \ˈpig-mənt\ *noun* [Latin *pigmentum,* from *pingere* to paint — more at PAINT] (14th century)
1 : a substance that imparts black or white or a color to other materials; *especially* **:** a powdered substance that is mixed with a liquid in which it is relatively insoluble and used especially to impart color to coating materials (as paints) or to inks, plastics, and rubber
2 : a coloring matter in animals and plants especially in a cell or tissue; *also* **:** any of various related colorless substances
— pig·men·tary \-mən-ˌter-ē\ *adjective*

[2]pig·ment \-ˌment, -mənt\ *transitive verb* (1900)
: to color with or as if with pigment

pig·men·ta·tion \ˌpig-mən-ˈtā-shən, -ˌmen-\ *noun* (1866)
: coloration with or deposition of pigment; *especially* **:** an excessive deposition of bodily pigment

pig·my *variant of* PYGMY

pi·gno·li \pēn-ˈyō-lē\ *or* **pi·gno·lia** \-lē-ə\ *noun* [*pignoli* from Italian, plural of *pignolo, pinolo,* from *pigna, pina* pine cone, from Latin *pinea; pignolia* perhaps modification of *pignoli* — more at PINEAL] (1898)
: PINE NUT

pig·nut \ˈpig-ˌnət\ *noun* (1666)
1 : any of several bitter-flavored hickory nuts
2 : a hickory (as *Carya glabra* and *C. cordiformis*) bearing pignuts

pig out *intransitive verb* (1978)
slang **:** to eat greedily **:** GORGE
— pig–out \ˈpig-ˌau̇t\ *noun*

pig·pen \-ˌpen\ *noun* (1803)
1 : a pen for pigs
2 : a dirty slovenly place

pig·skin \-ˌskin\ *noun* (1855)
1 : the skin of a swine or leather made of it
2 a : a jockey's saddle **b :** FOOTBALL 2a

pig·stick \-ˌstik\ *intransitive verb* (1891)
: to hunt the wild boar on horseback with a spear
— pig·stick·er *noun*

pig·sty \ˈpig-ˌstī\ *noun* (circa 1591)
: PIGPEN

pig·tail \-ˌtāl\ *noun* (1688)
1 : tobacco in small twisted strands or rolls
2 : a tight braid of hair

pig·tailed \-ˌtāld\ *adjective* (1754)
: wearing a pigtail ⟨*pigtailed* little girls⟩

pig·weed \-ˌwēd\ *noun* (circa 1801)
: any of various strongly growing weedy plants especially of the goosefoot or amaranth families

pi·ing *present participle of* PI *or of* PIE

pi·ka \ˈpē-kə, ˈpī-\ *noun* [perhaps from Evenki (Tungusic language of Siberia)] (1827)
: any of various short-eared small lagomorph mammals (family Ochotonidae) of rocky uplands of Asia and western North America with relatively short hind legs

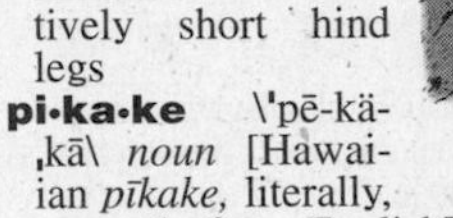
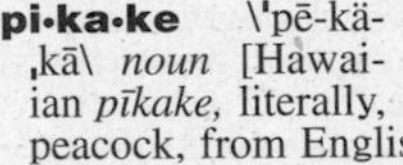
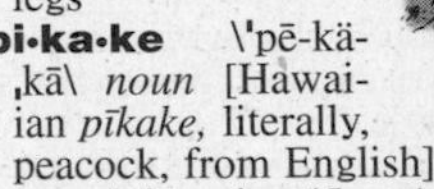
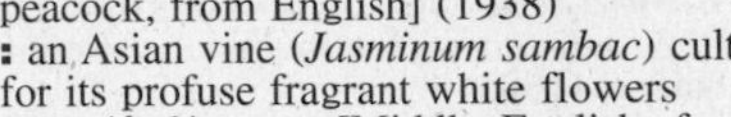

pika

pi·ka·ke \ˈpē-kä-ˌkā\ *noun* [Hawaiian *pīkake,* literally, peacock, from English] (1938)
: an Asian vine (*Jasminum sambac*) cultivated for its profuse fragrant white flowers

[1]pike \ˈpīk\ *noun* [Middle English, from Old English *pīc* pickax] (13th century)
1 : PIKESTAFF 1
2 : a sharp point or spike; *also* **:** the tip of a spear
— piked \ˈpīkt\ *adjective*

[2]pike *noun* [Middle English, perhaps of Scandinavian origin; akin to Norwegian dialect *pīk* pointed mountain] (13th century)
dialect English **:** a mountain or hill having a peaked summit — used especially in place names

[3]pike *noun, plural* **pike** *or* **pikes** [Middle English, from [1]*pike*] (14th century)
1 a : a large elongate long-snouted freshwater bony fish (*Esox lucius*) valued for food and sport and widely distributed in cooler parts of the northern hemisphere — called also *northern, northern pike* **b :** any of various fishes (family Esocidae) related to the pike: as (1) **:** MUSKELLUNGE (2) **:** PICKEREL
2 : any of various fishes resembling the pike in appearance or habits

[4]pike *noun* [Middle French *pique,* from *piquer* to prick, from (assumed) Vulgar Latin *piccare,* perhaps from Latin *picus* woodpecker — more at PIE] (circa 1511)
: a heavy spear with a very long shaft used by infantry especially in Europe from the Middle Ages to the 18th century

[5]pike *transitive verb* **piked; pik·ing** (1798)
: to pierce, kill, or wound with a pike

[6]pike *intransitive verb* **piked; pik·ing** [Middle English *pyken* (reflexive)] (1526)
1 : to leave abruptly ⟨get lonely and sore, and *pike* out —Sinclair Lewis⟩
2 : to make one's way ⟨*pike* along⟩

[7]pike *noun* (1812)
1 : TURNPIKE
2 : a railroad or railroad line or system
— down the pike 1 : in the course of events ⟨the greatest boxer to come *down the pike* in years⟩ **2 :** in the future ⟨today's advances only hint at what's *down the pike*⟩

[8]pike *noun* [perhaps from [3]*pike*] (1928)
: a body position (as in diving) in which the hips are bent, the knees are straight, the head is pressed forward, and the hands touch the toes or clasp the legs behind or just above the knees

pike·man \ˈpīk-mən\ *noun* (1566)
: a soldier armed with a pike

pike perch *noun* (1842)
: a fish (as the walleye) of the perch group that resembles the pike

pik·er \ˈpī-kər\ *noun* [*pike* to play cautiously, of unknown origin] (1872)
1 : one who gambles or speculates with small amounts of money
2 : one who does things in a small way; *also* **:** TIGHTWAD, CHEAPSKATE

pike·staff \ˈpīk-ˌstaf\ *noun* (14th century)
1 : a spiked staff for use on slippery ground
2 : the staff of a foot soldier's pike

pi·ki \ˈpē-kē\ *noun* [Hopi *píki*] (circa 1889)
: bread made especially from blue cornmeal and baked in thin sheets by the Indians of the southwestern U.S.

pi·laf *or* **pi·laff** \pi-ˈläf, -ˈlȯf; ˈpē-ˌ\ *or* **pi·lau** *also* **pi·law** \pi-ˈlō, -ˈlȯ, ˈpē-(ˌ)\; *Southern often* ˈpər-(ˌ)lü, -(ˌ)lō\ *noun* [Turkish & Persian; Turkish *pilav,* from Persian *pilāv*] (1612)
: a dish made of seasoned rice and often meat

pi·las·ter \pi-ˈlas-tər, ˈpī-ˌlas-\ *noun* [Middle French *pilastre,* from Italian *pilastro*] (1575)
: an upright architectural member that is rectangular in plan and is structurally a pier but architecturally treated as a column and that usually projects a third of its width or less from the wall

1 **pilaster**

pil·chard \ˈpil-chərd\ *noun* [origin unknown] (circa 1530)
1 : a fish (*Sardina pilchardus*) of the herring family resembling the herring and occurring in great schools along the coasts of Europe
2 : any of several sardines related to the European pilchard

[1]pile \ˈpī(ə)l\ *noun* [Middle English, dart, stake, from Old English *pīl,* from Latin *pilum* javelin] (12th century)
1 : a long slender column usually of timber, steel, or reinforced concrete driven into the ground to carry a vertical load
2 : a wedge-shaped heraldic charge usually placed vertically with the broad end up
3 a : a target-shooting arrowhead without cutting edges **b** [Latin *pilum*] **:** an ancient Roman foot soldier's heavy javelin

[2]pile *transitive verb* **piled; pil·ing** (15th century)
: to drive piles into

[3]pile *verb* **piled; pil·ing** [Middle English, from [4]*pile*] (14th century)
transitive verb
1 : to lay or place in a pile **:** STACK
2 a : to heap in abundance **:** LOAD ⟨*piled* potatoes on his plate⟩ **b :** to collect little by little into a mass — usually used with *up*
intransitive verb
1 : to form a pile or accumulation — usually used with *up*
2 : to move or press forward in or as if in a mass **:** CROWD ⟨*piled* into a car⟩

[4]pile *noun* [Middle English, from Middle French, from Latin *pila* pillar] (15th century)
1 a (1) **:** a quantity of things heaped together (2) **:** a heap of wood for burning a corpse or a sacrifice **b :** any great number or quantity **:** LOT
2 : a large building or group of buildings
3 : a great amount of money **:** FORTUNE
4 : REACTOR 3b

[5]pile *noun* [Middle English, from Latin *pilus* hair] (15th century)
1 : a coat or surface of usually short close fine furry hairs

2 : a velvety surface produced by an extra set of filling yarns that form raised loops which are cut and sheared
— **pile·less** \'pī(ə)l-ləs\ *adjective*

[6]**pile** *noun* [Middle English, from Latin *pila* ball] (15th century)
1 : a single hemorrhoid
2 *plural* **:** HEMORRHOIDS

pi·le·at·ed \'pī-lē-ˌā-təd, 'pi-\ *adjective* (circa 1728)
: having a crest covering the pileum

pileated woodpecker *noun* (1782)
: a large red-crested North American woodpecker (*Dryocopus pileatus*) that is black with white on the face, neck, and undersides of the wings

piled \'pī(ə)ld\ *adjective* (15th century)
: having a pile ⟨a deep-*piled* rug⟩

pile driver *noun* (1772)
: a machine for driving down piles with a drop hammer or a steam or air hammer

pi·le·um \'pī-lē-əm\ *noun, plural* **pi·lea** \-lē-ə\ [New Latin, from Latin *pileus, pileum* felt cap; akin to Greek *pilos* felt] (1874)
: the top of the head of a bird from the bill to the nape

pile·up \'pī(ə)l-ˌəp\ *noun* (circa 1929)
1 : a collision involving usually several motor vehicles
2 a : a jammed tangled mass or pile (as of motor vehicles or people) resulting from collision or accumulation **b :** ACCUMULATION

pi·le·us \'pī-lē-əs\ *noun, plural* **pi·lei** \-lē-ˌī\ [New Latin, from Latin] (1760)
1 : the convex, concave, or flattened spore-bearing structure of some basidiomycetes that is attached superiorly to the stem and typically is expanded with gills or pores on the underside — called also *cap*
2 [Latin] **:** a pointed or close-fitting cap worn by ancient Romans

pile·wort \'pī(ə)l-ˌwərt, -ˌwȯrt\ *noun* [Middle English *pyle wort;* from its use in treating piles] (15th century)
: LESSER CELANDINE

pil·fer \'pil-fər\ *verb* **pil·fered; pil·fer·ing** \-f(ə-)riŋ\ [Middle French *pelfrer,* from *pelfre* booty] (circa 1548)
intransitive verb
: STEAL; *especially* **:** to steal stealthily in small amounts and often again and again
transitive verb
: STEAL; *especially* **:** to steal in small quantities
synonym see STEAL
— **pil·fer·able** \-f(ə-)rə-bəl\ *adjective*
— **pil·fer·age** \-f(ə-)rij\ *noun*
— **pil·fer·er** \-fər-ər\ *noun*
— **pil·fer·proof** \-ˌprüf\ *adjective*

pil·gar·lic \pil-'gär-lik\ *noun* [*pilled garlic*] (circa 1529)
1 a : a bald head **b :** a bald-headed man
2 : a man looked upon with humorous contempt or mock pity

pil·grim \'pil-grəm\ *noun* [Middle English, from Old French *peligrin,* from Late Latin *pelegrinus,* alteration of Latin *peregrinus* foreigner, from *peregrinus,* adjective, foreign, from *peregri* abroad, from *per* through + *agr-, ager* land — more at FOR, ACRE] (13th century)
1 : one who journeys in foreign lands **:** WAYFARER
2 : one who travels to a shrine or holy place as a devotee
3 *capitalized* **:** one of the English colonists settling at Plymouth in 1620 ◆

[1]**pil·grim·age** \'pil-grə-mij\ *noun* (14th century)
1 : a journey of a pilgrim; *especially* **:** one to a shrine or a sacred place
2 : the course of life on earth

[2]**pilgrimage** *intransitive verb* **-aged; -ag·ing** (14th century)
: to go on a pilgrimage

pilgrim bottle *noun* (1874)
: COSTREL

pil·ing \'pī-liŋ\ *noun* (15th century)
: a structure of piles; *also* **:** PILES

Pi·li·pi·no \ˌpi-lə-'pē-(ˌ)nō, ˌpē-\ *noun* [Tagalog, literally, Philippine, from Spanish *Filipino*] (1936)
: the Tagalog-based official language of the Republic of the Philippines

[1]**pill** \'pil\ *verb* [Middle English *pilen, pillen,* partly from Old English *pilian* to peel, partly from Middle French *piller* to plunder] (12th century)
intransitive verb
dialect chiefly English **:** to come off in flakes or scales **:** PEEL
transitive verb
1 *archaic* **:** to subject to depredation or extortion
2 *dialect* **:** to peel or strip off

[2]**pill** *noun* [Middle English *pylle,* from Middle Dutch *pille,* ultimately from Latin *pilula,* from diminutive of *pila* ball] (14th century)
1 a : medicine in a small rounded mass to be swallowed whole **b** *often capitalized* **:** an oral contraceptive — usually used with *the*
2 : something repugnant or unpleasant that must be accepted or endured
3 : something resembling a pill in size or shape
4 : a disagreeable or tiresome person

[3]**pill** (1736)
transitive verb
1 : to dose with pills
2 : BLACKBALL
intransitive verb
: to become rough with or mat into little balls ⟨brushed woolens often *pill*⟩

[1]**pil·lage** \'pi-lij\ *noun* [Middle English, from Middle French, from *piller* to plunder, abuse] (14th century)
1 : the act of looting or plundering especially in war
2 : something taken as booty

[2]**pillage** *verb* **pil·laged; pil·lag·ing** (circa 1592)
transitive verb
: to plunder ruthlessly **:** LOOT
intransitive verb
: to take booty
synonym see RAVAGE
— **pil·lag·er** *noun*

[1]**pil·lar** \'pi-lər\ *noun* [Middle English *piler,* from Old French, from Medieval Latin *pilare,* from Latin *pila*] (13th century)
1 a : a firm upright support for a superstructure **:** POST **b :** a usually ornamental column or shaft; *especially* **:** one standing alone for a monument
2 : a chief supporter **:** PROP
3 : a solid mass of coal, rock, or ore left standing to support a mine roof
4 : a body part that resembles a column
— **pil·lar·less** *adjective*
— **from pillar to post :** from one place or one predicament to another

[2]**pillar** *transitive verb* (1607)
: to provide or strengthen with or as if with pillars

pil·lar–box \'pi-lər-ˌbäks\ *noun* (1858)
British **:** a pillar-shaped mailbox

pill·box \'pil-ˌbäks\ *noun* (1730)
1 : a box for pills; *especially* **:** a shallow round box of pasteboard
2 : a small low concrete emplacement for machine guns and antitank weapons
3 : a small round hat without a brim; *specifically* **:** a woman's shallow hat with a flat crown and straight sides

pill bug *noun* [[2]*pill;* from its rolling into a ball when disturbed] (1843)
: WOOD LOUSE

[1]**pil·lion** \'pil-yən\ *noun* [Scottish Gaelic or Irish; Scottish Gaelic *pillean,* diminutive of *peall* covering, couch; Irish *pillín,* diminutive of *peall* covering, couch] (1503)
1 a : a light saddle for women consisting chiefly of a cushion **b :** a pad or cushion put on behind a man's saddle chiefly for a woman to ride on
2 *chiefly British* **:** a motorcycle or bicycle saddle for a passenger

[2]**pillion** *adverb* (1926)
chiefly British **:** on or as if on a pillion ⟨ride *pillion*⟩

[1]**pil·lo·ry** \'pi-lə-rē, 'pil-rē\ *noun, plural* **-ries** [Middle English, from Old French *pilori*] (13th century)
1 : a device formerly used for publicly punishing offenders consisting of a wooden frame with holes in which the head and hands can be locked
2 : a means for exposing one to public scorn or ridicule

pillory 1

[2]**pillory** *transitive verb* **-ried; -ry·ing** (circa 1600)
1 : to set in a pillory as punishment
2 : to expose to public contempt, ridicule, or scorn

[1]**pil·low** \'pi-(ˌ)lō\ *noun* [Middle English *pilwe,* from Old English *pyle* (akin to Old High German *pfuliwi*), from Latin *pulvinus*] (before 12th century)
1 a : a support for the head of a reclining person; *especially* **:** one consisting of a cloth bag filled with feathers, down, sponge rubber, or plastic fiber **b :** something resembling a pillow especially in form
2 : a cushion or pad tightly stuffed and used as a support for the design and tools in making lace with a bobbin
— **pil·lowy** \'pi-lə-wē\ *adjective*

[2]**pillow** (1629)
transitive verb
1 : to rest or lay on or as if on a pillow
2 : to serve as a pillow for

◇ WORD HISTORY
pilgrim Latin *peregrinus* "foreigner, alien" is a derivative of *peregri* "abroad, away from home" itself formed from *per-* "through" and *ager* "field." In Late Latin and early Medieval Latin *peregrinus* was applied to wanderers directed by God, such as the Israelites after their departure from Egypt, prophets living in the wilderness, and those Church Fathers who like Saint Jerome left their homes for the desert or the Holy Land. From this use the word was extended to all wayfarers traveling to the Holy Land or the shrine of a saint as an act of faith or penance. *Peregrinus* became *pelegrinus* in Late Latin, the first of the two *r*s being altered by dissimilation. *Peregrinus* then became *pelerin* or *peligrin* in Old French. The latter form was borrowed into Middle English as *pilegrim* or *pilgrim.* Latin *peregrinus* reaches us more directly in *peregrine falcon* (in Medieval Latin *falco peregrinus*), so named because young birds were captured by falconers as "wanderers" from their nesting sites. Other birds of prey were taken directly from their nests, but the aerie of a peregrine falcon is typically on an inaccessible crag.

\ə\ abut \ᵊ\ kitten \ər\ further \a\ ash \ā\ ace \ä\ mop, mar \au̇\ out \ch\ chin \e\ bet \ē\ easy \g\ go \i\ hit \ī\ ice \j\ job \ŋ\ sing \ō\ go \ȯ\ law \ȯi\ boy \th\ thin \t͟h\ the \ü\ loot \u̇\ foot \y\ yet \zh\ vision *see also* Guide to Pronunciation

intransitive verb
: to lay or rest one's head on or as if on a pillow
pillow block *noun* (1844)
: a block or standard to support a journal (as of a shaft) **:** BEARING
pil·low·case \'pi-lə-ˌkās, -lō-\ *noun* (1724)
: a removable covering for a pillow
pillow lace *noun* [from its being worked over a pillow on which the pattern is marked] (circa 1858)
: lace made with a bobbin
pillow slip *noun* (circa 1828)
: PILLOWCASE
pillow talk *noun* (1939)
: intimate conversation between lovers in bed
pi·lo·car·pine \ˌpī-lə-'kär-ˌpēn\ *noun* [International Scientific Vocabulary, from New Latin *Pilocarpus jaborandi,* species of tropical shrubs] (1875)
: a miotic alkaloid $C_{11}H_{16}N_2O_2$ that is obtained from jaborandi and used especially in the treatment of glaucoma
pi·lose \'pī-ˌlōs\ *adjective* [Latin *pilosus,* from *pilus* hair] (1753)
: covered with usually soft hair
— pi·los·i·ty \pī-'lä-sə-tē\ *noun*
[1]**pi·lot** \'pī-lət\ *noun* [Middle French *pilote,* from Italian *pilota,* alteration of *pedota,* from (assumed) Middle Greek *pēdōtēs,* from Greek *pēda* steering oars, plural of *pēdon* oar; probably akin to Greek *pod-, pous* foot — more at FOOT] (1530)
1 a : one employed to steer a ship **:** HELMSMAN **b :** a person who is qualified and usually licensed to conduct a ship into and out of a port or in specified waters **c :** a person who flies or is qualified to fly an aircraft or spacecraft
2 : GUIDE, LEADER
3 : COWCATCHER
4 : a piece that guides a tool or machine part
5 : a television show produced and filmed or taped as a sample of a proposed series
6 : PILOT LIGHT 2
— pi·lot·less \-ləs\ *adjective*
[2]**pilot** *transitive verb* (1649)
1 : to act as a guide to **:** lead or conduct over a usually difficult course
2 a : to set and conn the course of ⟨*pilot* a ship⟩ **b :** to act as pilot of ⟨*pilot* a plane⟩
synonym see GUIDE
[3]**pilot** *adjective* (1802)
: serving as a guiding or tracing device, an activating or auxiliary unit, or a trial apparatus or operation ⟨a *pilot* study⟩
pi·lot·age \'pī-lə-tij\ *noun* (circa 1618)
1 : the action or business of piloting
2 : the compensation paid to a licensed ship's pilot
pilot biscuit *noun* (1836)
: HARDTACK — called also *pilot bread*
pilot engine *noun* (1838)
: a locomotive going in advance of a train to make sure that the way is clear
pilot fish *noun* (1634)
: a pelagic carangid fish (*Naucrates ductor*) that has dark stripes and often swims in company with a shark
pi·lot·house \'pī-lət-ˌhau̇s\ *noun* (1846)
: a deckhouse for a ship's helmsman containing the steering wheel, compass, and navigating equipment
pilot light *noun* (circa 1890)
1 : an indicator light showing where a switch or circuit breaker is located or whether a motor is in operation or power is on — called also *pilot lamp*
2 : a small permanent flame used to ignite gas at a burner
pilot officer *noun* (1919)
: a commissioned officer in the British air force who ranks with a second lieutenant in the army
pilot whale *noun* (1867)
: BLACKFISH 2
pil·sner *or* **pil·sen·er** \'pils-nər, 'pilz-, 'pil-zə-\ *noun* [German, literally, of Pilsen (Plzeň), city in the Czech Republic] (1877)
1 : a light beer with a strong flavor of hops
2 : a tall slender footed glass for beer
Pilt·down man \'pilt-ˌdau̇n-\ *noun* [*Piltdown,* East Sussex, England] (circa 1918)
: a supposedly very early hominid erroneously reconstructed from a combination of human and animal skeletal remains the latter of which were later found to have been planted by a hoaxer
pil·u·lar \'pil-yə-lər\ *adjective* (1802)
: of, relating to, or resembling a pill
pil·ule \'pil-(ˌ)yü(ə)l\ *noun* [Middle French, from Latin *pilula* pill — more at PILL] (1543)
: a little pill
pi·lus \'pī-ləs\ *noun, plural* **pi·li** \-ˌlī\ [Latin] (circa 1890)
: a hair or a structure (as of a bacterium) resembling a hair
Pi·ma \'pē-mə\ *noun, plural* **Pimas** *or* **Pima** (1850)
1 : a member of an American Indian people of southern Arizona and northern Mexico
2 : the Uto-Aztecan language of the Pima people
— Pi·man \-mən\ *adjective*
pi·ma cotton \'pē-mə-, 'pi-\ *noun* [*Pima* County, Arizona] (1925)
: a cotton that produces fiber of exceptional strength and firmness and that was developed in the southwestern U.S. by selection and breeding of Egyptian cottons
pi·men·to \pə-'men-(ˌ)tō\ *noun, plural* **-tos** *or* **-to** [Spanish *pimienta* allspice, pepper, from Late Latin *pigmenta,* plural of *pigmentum* plant juice, from Latin, pigment] (1660)
1 : ALLSPICE
2 : PIMIENTO 1
pimento cheese *noun* (1916)
: a Neufchâtel, process, cream, or occasionally cheddar cheese to which ground pimientos have been added
pi–me·son \'pī-'me-ˌzän, -'mā-, -'mē-, -ˌsän\ *noun* [[4]*pi*] (1948)
: PION
pi·mien·to \pə-'men-(ˌ)tō, pəm-'yen-\ *noun, plural* **-tos** [Spanish from *pimienta*] (1845)
1 : any of various bluntly conical thick-fleshed sweet peppers of European origin that have a distinctive mild sweet flavor and are used especially as a garnish, as a stuffing for olives, and as a source of paprika
2 : a plant that bears pimientos
[1]**pimp** \'pimp\ *noun* [origin unknown] (1600)
: a man who solicits clients for a prostitute
[2]**pimp** *intransitive verb* (1636)
: to work as a pimp
pim·per·nel \'pim-pər-ˌnel, -pər-nəl\ *noun* [Middle English *pimpernele,* from Middle French *pimprenelle,* from Late Latin *pimpinella,* a medicinal herb] (14th century)
: any of a genus (*Anagallis*) of herbs of the primrose family; *especially* **:** SCARLET PIMPERNEL
pimp·ing \'pim-pən, -piŋ\ *adjective* [origin unknown] (1687)
1 : PETTY, INSIGNIFICANT
2 *chiefly dialect* **:** PUNY, SICKLY
pim·ple \'pim-pəl\ *noun* [Middle English *pinple*] (14th century)
1 : a small inflamed elevation of the skin **:** PAPULE; *especially* **:** PUSTULE
2 : a swelling or protuberance like a pimple
— pim·pled \-pəld\ *adjective*
— pim·ply \-p(ə-)lē\ *adjective*
pimp·mo·bile \'pimp-mō-ˌbēl, -mə-\ *noun* (1971)
: an ostentatious luxury car of a kind characteristically used by a pimp
[1]**pin** \'pin\ *noun* [Middle English, from Old English *pinn* (akin to Old High German *pfinn* peg), perhaps from Latin *pinna* quill, feather — more at PEN] (before 12th century)
1 a : a piece of solid material (as wood or metal) used especially for fastening separate articles together or as a support by which one article may be suspended from another **b** *obsolete* **:** the center peg of a target; *also* **:** the center itself **c :** something that resembles a pin especially in slender elongated form ⟨an electrical connector *pin*⟩ **d** (1) **:** one of the wooden pieces constituting the target in various games (as bowling) (2) **:** the peg at which a quoit is pitched (3) **:** the staff of the flag marking a hole on a golf course **e :** a peg for regulating the tension of the strings of a musical instrument **f :** the part of a key stem that enters a lock **g** (1) **:** THOLE 1 (2) **:** a belaying pin
2 a (1) **:** a small pointed piece of wire with a head used especially for fastening cloth (2) **:** something of small value **:** TRIFLE **b :** an ornament or emblem fastened to clothing with a pin **c** (1) **:** BOBBY PIN (2) **:** HAIRPIN (3) **:** SAFETY PIN
3 : LEG — usually used in plural ⟨wobbly on his *pins*⟩
4 : a fall in wrestling
[2]**pin** *transitive verb* **pinned; pin·ning** (14th century)
1 a : to fasten, join, or secure with a pin **b :** to hold fast or immobile **c :** to present (a young woman) with a fraternity pin as a pledge of affection
2 a : ATTACH, HANG ⟨*pinned* his hopes on a miracle⟩ **b :** to assign the blame or responsibility for ⟨*pin* the robbery on a night watchman⟩ **c :** to define or determine clearly or precisely — usually used with *down* ⟨it is hard to *pin* down exactly when things changed —Katharine Whittemore⟩
3 a : to make (a chess opponent's piece) unable to move without exposing the king to check or a valuable piece to capture **b** *of a wrestler* **:** to secure a fall over (an opponent)
[3]**pin** *adjective* (1523)
1 : of or relating to a pin
2 *of leather* **:** having a grain suggesting the heads of pins
pi·ña cloth \'pēn-yə-\ *noun* [Spanish *piña* pineapple, pinecone, from Latin *pinea* pinecone — more at PINEAL] (circa 1858)
: a lustrous transparent cloth of Philippine origin that is woven of silky pineapple fibers
pi·ña co·la·da \'pēn-yə-kō-'lä-də, 'pē-nə-\ *noun* [Spanish, literally, strained pineapple] (1923)
: a tall drink made of rum, cream of coconut, and pineapple juice mixed with ice
pin·a·fore \'pi-nə-ˌfōr, -ˌfȯr\ *noun* [[2]*pin* + *afore*] (1782)
: a sleeveless usually low-necked garment fastened in the back and worn as an apron or dress
— pin·a·fored \-ˌfōrd, -ˌfȯrd\ *adjective*
pi·ña·ta *or* **pi·na·ta** \pēn-'yä-tə\ *noun* [Spanish *piñata,* literally, pot, from Italian *pignatta,* probably from *pigna* pinecone — more at PIGNOLI] (1887)
: a decorated vessel (as a pottery jar) filled with candies, fruits, and gifts and hung from the ceiling to be broken with sticks by blindfolded persons as part of especially Latin-American festivities (as at Christmas or for a birthday party)

piñata

pin·ball machine \'pin-ˌbȯl-\ *noun* (1936)
: an amusement device in which a ball propelled by a plunger scores points as it rolls down a slanting surface among pins and targets — called also *pinball game*
pin·bone \'pin-ˌbōn\ *noun* (1640)
: the hipbone especially of a quadruped — see COW illustration

pince–nez \paⁿs-ˈnā, pan(t)s-\ *noun, plural* **pince–nez** \-ˈnā(z)\ [French, from *pincer* to pinch + *nez* nose, from Latin *nasus* — more at NOSE] (1876)
: eyeglasses clipped to the nose by a spring

pince-nez

pin·cer \ˈpin(t)-sər, *especially for 1 US often* ˈpin-chər\ *noun* [Middle English *pinceour,* from (assumed) Anglo-French *pinceour,* from Middle French *pincier* to pinch, from (assumed) Vulgar Latin *pinctiare, punctiare,* from Latin *punctum* puncture — more at POINT] (14th century)
1 *plural but singular or plural in construction* **a :** an instrument having two short handles and two grasping jaws working on a pivot and used for gripping things **b :** a claw (as of a lobster) resembling a pair of pincers **:** CHELA
2 : one part of a double envelopment in which two military forces converge on opposite sides of an enemy position
— pin·cer·like \-ˌlīk\ *adjective*

¹pinch \ˈpinch\ *verb* [Middle English, from (assumed) Old North French *pinchier,* from (assumed) Vulgar Latin *pinctiare*] (14th century)
transitive verb
1 a : to squeeze between the finger and thumb or between the jaws of an instrument **b :** to prune the tip of (a plant or shoot) usually to induce branching **c :** to squeeze or compress painfully **d :** to cause physical or mental pain to **e** (1) **:** to cause to appear thin or shrunken (2) **:** to cause to shrivel or wither
2 a : to subject to strict economy or want **:** STRAITEN **b :** to confine or limit narrowly **:** CONSTRICT
3 a : STEAL **b :** ARREST
4 : to sail too close to the wind
intransitive verb
1 : COMPRESS, SQUEEZE
2 : to be miserly or closefisted
3 : to press painfully
4 : NARROW, TAPER ⟨the road *pinched* down to a trail —Cecelia Holland⟩
— pinch pennies : to practice strict economy

²pinch *noun* (15th century)
1 a : a critical juncture **:** EMERGENCY **b** (1) **:** PRESSURE, STRESS (2) **:** HARDSHIP, PRIVATION **c :** DEFICIT
2 a : an act of pinching **:** SQUEEZE **b :** as much as may be taken between the finger and thumb ⟨a *pinch* of snuff⟩ **c :** a very small amount
3 : a marked thinning of a vein or bed
4 a : THEFT **b :** a police raid; *also* **:** ARREST
synonym see JUNCTURE

³pinch *adjective* (1916)
1 : SUBSTITUTE ⟨*pinch* runner⟩
2 : hit by a pinch hitter ⟨*pinch* homer⟩

pinch bar *noun* (1837)
: a bar similar in form and use to a crowbar and sometimes having an end adapted for pulling spikes or inserting under a heavy wheel that is to be rolled

pinch·beck \ˈpinch-ˌbek\ *noun* [Christopher *Pinchbeck* (died 1732) English watchmaker] (1734)
1 : an alloy of copper and zinc used especially to imitate gold in jewelry
2 : something counterfeit or spurious
— pinchbeck *adjective*

pinch·er \ˈpin-chər\ *noun* (15th century)
1 : one that pinches
2 *plural* **:** PINCERS

pinch–hit \ˈpinch-ˈhit, ˌpinch-\ *intransitive verb* [back-formation from *pinch hitter*] (1915)
1 : to act or serve in place of another
2 : to bat in the place of another player especially when a hit is particularly needed

pinch hit *noun* (1927)
: a hit made by a pinch hitter

pinch hitter *noun* (1912)
: one that pinch-hits

pinch·pen·ny \ˈpinch-ˌpe-nē\ *adjective* (1582)
: STINGY, NIGGARDLY

pin curl *noun* (1896)
: a curl made usually by dampening a strand of hair with water or lotion, coiling it, and securing it by a hairpin or clip

pin·cush·ion \ˈpin-ˌku̇-shən\ *noun* (1632)
: a small cushion in which pins may be stuck ready for use

¹Pin·dar·ic \pin-ˈdar-ik\ *adjective* (1640)
1 : of or relating to the poet Pindar
2 : written in the manner or style characteristic of Pindar

²Pindaric *noun* (1685)
1 : a Pindaric ode
2 *plural* **:** loose irregular verses similar to those used in Pindaric odes

pin·dling \ˈpin(d)-lən, -liŋ; ˈpin-dᵊl-ən, -iŋ\ *adjective* [perhaps alteration of *spindling*] (1861)
dialect **:** PUNY, FRAIL

¹pine \ˈpīn\ *noun, often attributive* [Middle English, from Old English *pīn,* from Latin *pinus;* probably akin to Greek *pitys* pine] (before 12th century)
1 : any of a genus (*Pinus* of the family Pinaceae, the pine family) of coniferous evergreen trees that have slender elongated needles and include some valuable timber trees and ornamentals
2 : the straight-grained white or yellow usually durable and resinous wood of a pine varying from extreme softness in the white pine to hardness in the longleaf pine
3 : any of various Australian coniferous trees (as of the genera *Callitris* or *Araucaria*)
4 : PINEAPPLE
5 : BENCH 1c
— pin·ey *also* **piny** \ˈpī-nē\ *adjective*

²pine *intransitive verb* **pined; pin·ing** [Middle English, from Old English *pīnian* to suffer, from (assumed) Old English *pīn* punishment, from Latin *poena* — more at PAIN] (14th century)
1 : to lose vigor, health, or flesh (as through grief) **:** LANGUISH
2 : to yearn intensely and persistently especially for something unattainable ⟨they still *pined* for their lost wealth⟩
synonym see LONG

pi·ne·al \ˈpī-nē-əl, pī-ˈ\ *adjective* [French *pinéal,* from Middle French, from Latin *pinea* pinecone, from feminine of *pineus* of pine, from *pinus*] (1681)
: of, relating to, or being the pineal gland

pi·ne·al·ec·to·my \ˌpī-nē-ə-ˈlek-tə-mē, pī-ˌnē-\ *noun* (1915)
: surgical removal of the pineal gland
— pi·ne·al·ec·to·mize \-tə-ˌmīz\ *transitive verb*

pineal gland *noun* (1712)
: a small usually conical appendage of the brain of all craniate vertebrates that functions primarily as an endocrine organ and that in a few reptiles has the essential structure of an eye — called also *pineal, pineal body, pineal organ*

pine·ap·ple \ˈpī-ˌna-pəl\ *noun* (1664)
1 a : a tropical monocotyledonous plant (*Ananas comosus* of the family Bromeliaceae, the pineapple family) that has rigid spiny-margined recurved leaves and a short stalk with a dense oblong head of small abortive flowers **b :** the large edible multiple fruit of the pineapple that consists of the sweet succulent fleshy inflorescence
2 : a hand grenade

pineapple guava *noun* (circa 1924)
: FEIJOA

pine·cone \ˈpīn-ˌkōn\ *noun* (1695)
: a cone of a pine tree

pine·drops \ˈpīn-ˌdräps\ *noun plural but singular or plural in construction* (1857)
1 : a purplish brown leafless saprophytic plant (*Pterospora andromedea*) of the wintergreen family with racemose drooping white flowers
2 : BEECHDROPS

pine·land \ˈpīn-ˌland, -lənd\ *noun* (circa 1658)
: land naturally dominated by pine forests

pi·nene \ˈpī-ˌnēn\ *noun* [International Scientific Vocabulary, from Latin *pinus*] (1885)
: either of two liquid isomeric unsaturated bicyclic terpene hydrocarbons $C_{10}H_{16}$ of which one is a major constituent of wood turpentine

pine nut *noun* (before 12th century)
: the edible seed of any of several pines — compare PIÑON

pin·ery \ˈpī-nə-rē, ˈpīn-rē\ *noun, plural* **-er·ies** (1758)
1 : a hothouse or area where pineapples are grown
2 : a grove or forest of pine

pine·sap \ˈpīn-ˌsap\ *noun* (1840)
: any of several yellowish or reddish parasitic or saprophytic herbs (genus *Monotropa*) of the wintergreen family resembling the Indian pipe

pine siskin *noun* (1887)
: a North American finch (*Carduelis pinus*) with streaked plumage

pine snake *noun* (1791)
1 : a large constricting snake (*Pituophis melanoleucus*) of the eastern U.S. that is typically white and black and is found especially in coastal regions from New Jersey southward
2 : any of various snakes related to the pine snake

pine tar *noun* (1880)
: tar obtained by destructive distillation of the wood of the pine tree and used especially in roofing and soaps and in the treatment of skin diseases

pi·ne·tum \pī-ˈnē-təm\ *noun, plural* **pi·ne·ta** \-ˈnē-tə\ [Latin, from *pinus*] (1842)
: a plantation of pine trees; *especially* **:** a scientific collection of living coniferous trees

pine·wood \ˈpīn-ˌwu̇d\ *noun* (1673)
1 : a wood of pines — often used in plural but singular or plural in construction
2 : the wood of the pine tree

piney woods *noun plural* (1800)
: woodland of the southern U.S. in which pines are the dominant tree

pin·feath·er \ˈpin-ˌfe-t͟hər\ *noun* (circa 1775)
: a feather not fully developed; *especially* **:** a feather just emerging through the skin

pin·fish \-ˌfish\ *noun* (1878)
: a small compressed dark green grunt (*Lagodon rhomboides*) that has sharp dorsal spines and is found along the Atlantic coast

pin·fold \-ˌfōld\ *noun* [Middle English, from Old English *pundfald,* from *pund-* enclosure + *fald* fold] (13th century)
1 : ²POUND 1a
2 : a place of restraint

ping \ˈpiŋ\ *noun* [imitative] (1835)
1 : a sharp sound like that of a striking bullet
2 : KNOCK 2b
— ping *intransitive verb*

ping·er \ˈpiŋ-ər\ *noun* (1957)
: a device for producing pulses of sound (as for marking an underwater site or detecting an underwater object)

pin·go \ˈpiŋ-(ˌ)gō\ *noun, plural* **pingos** [Inuit *pinguq*] (1938)
: a low hill or mound forced up by hydrostatic pressure in an area underlain by permafrost

ping–pong \ˈpiŋ-ˌpäŋ, -ˌpȯŋ\ *verb* (1952)
: SHIFT, BOUNCE

Ping–Pong \ˈpiŋ-ˌpäŋ, -ˌpȯŋ\ *trademark*
— used for table tennis

pin·head \ˈpin-ˌhed\ *noun* (1593)

1 : the head of a pin
2 : something very small or insignificant
3 : a very dull or stupid person **:** FOOL

pin•head•ed \-ˌhe-dəd\ *adjective* (1901)
: lacking intelligence or understanding **:** DULL, STUPID
— pin•head•ed•ness *noun*

pin•hole \-ˌhōl\ *noun* (1676)
: a small hole made by, for, or as if by a pin

[1]**pin•ion** \'pin-yən\ *noun* [Middle English, from Middle French *pignon,* from (assumed) Vulgar Latin *pinnion-, pinnio,* from Latin *pinna* feather — more at PEN] (15th century)
1 : the terminal section of a bird's wing including the carpus, metacarpus, and phalanges; *broadly* **:** WING
2 : FEATHER, QUILL; *also* **:** FLIGHT FEATHERS
— pin•ioned \-yənd\ *adjective*

[2]**pinion** *transitive verb* (1558)
1 a : to disable or restrain by binding the arms **b :** to bind fast **:** SHACKLE
2 : to restrain (a bird) from flight especially by cutting off the pinion of one wing

[3]**pinion** *noun* [French *pignon,* from Middle French *peignon,* from *peigne* comb, from Latin *pecten* — more at PECTINATE] (1659)
1 : a gear with a small number of teeth designed to mesh with a larger wheel or rack
2 : the smaller of a pair or the smallest of a train of gear wheels

[1]**pink** \'piŋk\ *noun* [Middle English, from Middle Dutch *pinke*] (15th century)
: a ship with a narrow overhanging stern — called also *pinkie*

[2]**pink** *noun* [origin unknown] (1573)
1 : any of a genus (*Dianthus* of the family Caryophyllaceae, the pink family) of herbs having a cylindrical many-veined calyx with bracts at its base
2 a : the very embodiment **:** PARAGON **b** (1) **:** one dressed in the height of fashion (2) **:** ELITE **c :** highest degree possible **:** HEIGHT ⟨keep their house in the *pink* of repair —Rebecca West⟩
— in the pink : in the best of health or condition

[3]**pink** *noun* (1678)
1 : any of a group of colors bluish red to red in hue, of medium to high lightness, and of low to moderate saturation
2 a : the scarlet color of a fox hunter's coat; *also* **:** a fox hunter's coat of this color **b :** pink-colored clothing **c** *plural* **:** light-colored trousers formerly worn by army officers
3 : PINKO

[4]**pink** *adjective* (1720)
1 : of the color pink
2 : holding moderately radical and usually socialistic political or economic views
3 : emotionally moved **:** EXCITED — often used as an intensive ⟨tickled *pink*⟩
— pink•ness *noun*

[5]**pink** *transitive verb* [Middle English, to thrust] (1503)
1 a : to perforate in an ornamental pattern **b :** to cut a saw-toothed edge on
2 a : PIERCE, STAB **b :** to wound by irony, criticism, or ridicule

pink bollworm *noun* (1906)
: a small dark brown moth (*Pectinophora gossypiella*) whose pinkish larva bores into the flowers and bolls of cotton and is a destructive pest in most cotton-growing regions

pink–collar *adjective* (1977)
: of, relating to, or constituting a class of employees in occupations (as nursing and clerical jobs) traditionally held by women

pink elephants *noun plural* (1940)
: hallucinations arising especially from heavy drinking or use of narcotics ⟨began to see *pink elephants*⟩

pink•eye \'piŋk-ˌī\ *noun* (1855)
: an acute highly contagious conjunctivitis of humans and various domestic animals

[1]**pin•kie** \'piŋ-kē\ *noun* [probably from Dutch *pinkje* small pink, diminutive of *pink,* from Middle Dutch *pinke*] (1840)
: [1]PINK

[2]**pinkie** *or* **pin•ky** \'piŋ-kē\ *noun, plural* **pinkies** [probably from Dutch *pinkje,* diminutive of *pink* little finger] (circa 1808)
: LITTLE FINGER

pinking shears *noun plural* (circa 1939)
: shears with a saw-toothed inner edge on the blades for making a zigzag cut

pink•ish \'piŋ-kish\ *adjective* (1784)
: somewhat pink; *especially* **:** tending to be pink in politics
— pink•ish•ness *noun*

pink lady *noun* (1936)
: a cocktail consisting of gin, brandy, lemon juice, grenadine, and white of egg shaken with ice and strained

pink•ly \'piŋ-klē\ *adverb* (1836)
: in a pink manner **:** with a pink hue

pinko \'piŋ-(ˌ)kō\ *noun, plural* **pink•os** *also* **pink•oes** (1936)
: a person who holds advanced liberal or moderately radical political or economic views

pink•root \'piŋk-ˌrüt, -ˌru̇t\ *noun* (1763)
: any of several plants (genus *Spigelia*) related to the nux vomica and used as anthelmintics; *especially* **:** an American woodland herb (*S. marilandica*) sometimes cultivated for its showy red and yellow flowers

pink salmon *noun* (1905)
: a small spotted Pacific salmon (*Oncorhynchus gorbuscha*) native to the northern Pacific Ocean and adjacent rivers

pink slip *noun* (1915)
: a notice from an employer that a recipient's employment is being terminated
— pink–slip *transitive verb*

pin money *noun* (1697)
1 a : money given by a man to his wife for her own use **b :** money set aside for the purchase of incidentals
2 : a trivial amount of money ⟨worked for *pin money*⟩

pin•na \'pi-nə\ *noun, plural* **pin•nae** \'pi-ˌnē, -ˌnī\ *or* **pinnas** [New Latin, from Latin, feather, wing — more at PEN] (1785)
1 : a leaflet or primary division of a pinnate leaf or frond
2 a : a projecting body part (as a feather, wing, or fin) **b :** the largely cartilaginous projecting portion of the external ear — see EAR illustration

pin•nace \'pi-nəs\ *noun* [Middle French *pinace,* probably from Old Spanish *pinaza,* from *pino* pine, from Latin *pinus*] (1538)
1 : a light sailing ship; *especially* **:** one used as a tender
2 : any of various ship's boats

[1]**pin•na•cle** \'pi-ni-kəl\ *noun* [Middle English *pinacle,* from Middle French, from Late Latin *pinnaculum* small wing, gable, from Latin *pinna* wing, battlement] (14th century)
1 : an upright architectural member generally ending in a small spire and used especially in Gothic construction to give weight especially to a buttress
2 : a structure or formation suggesting a pinnacle; *specifically* **:** a lofty peak
3 : the highest point of development or achievement **:** ACME
synonym see SUMMIT

1 **pinnacle 1**

[2]**pinnacle** *transitive verb* **-cled; -cling** \-k(ə-)liŋ\ (14th century)
1 : to surmount with a pinnacle
2 : to raise or rear on a pinnacle

pin•nate \'pi-ˌnāt\ *adjective* [New Latin *pinnatus,* from Latin, feathered, from *pinna* feather, wing, fin] (circa 1727)
: resembling a feather especially in having similar parts arranged on opposite sides of an axis like the barbs on the rachis of a feather ⟨*pinnate* leaf⟩ — see LEAF illustration
— pin•nate•ly *adverb*

pin•nat•i•fid \pə-'na-tə-fəd, -ˌfid\ *adjective* [New Latin *pinnatifidus,* from *pinnatus* + Latin *-fidus* -fid] (circa 1753)
: cleft in a pinnate manner ⟨a *pinnatifid* leaf⟩

pin•ner \'pi-nər\ *noun* (1652)
1 : a woman's cap with long lappets worn in the 17th and 18th centuries
2 : one that pins

pin•ni•ped \'pi-nə-ˌped\ *noun* [ultimately from Latin *pinna* + *ped-, pes* foot — more at FOOT] (circa 1842)
: any of a suborder (Pinnipedia) of aquatic carnivorous mammals (as a seal or walrus) with all four limbs modified into flippers
— pinniped *adjective*

pin•nule \'pin-(ˌ)yü(ə)l\ *noun* [New Latin *pinnula,* from Latin, diminutive of *pinna*] (1748)
1 : any of the secondary branches of a plumose organ especially of a crinoid
2 : one of the ultimate divisions of a twice pinnate leaf

pin•ny \'pi-nē\ *noun, plural* **pinnies** [by shortening & alteration] (1851)
chiefly British **:** PINAFORE

pin oak *noun* (1813)
: a pyramidally-crowned oak (*Quercus palustris*) of wet regions of the eastern U.S. that has deeply cleft toothed leaves and rather small nearly hemispherical acorns

pi•noch•le \'pē-ˌnə-kəl\ *noun* [modification of German dialect *Binokel,* a game resembling bezique, from French dialect *binocle*] (1864)
: a card game played with a 48-card pack containing two each of A, K, Q, J, 10, 9 in each suit with the object to score points by melding certain combinations of cards or by winning tricks that contain scoring cards; *also* **:** the meld of queen of spades and jack of diamonds scoring 40 points in this game

pi•no•cy•to•sis \ˌpi-nə-sə-'tō-səs, ˌpī-, -ˌsī-\ *noun, plural* **-to•ses** \-ˌsēz\ [New Latin, from Greek *pinein* to drink + New Latin *cyt-* + *-osis* — more at POTABLE] (1895)
: the uptake of fluid by a cell by invagination and pinching off of the plasma membrane

pi•no•cy•tot•ic \-'tä-tik\ *or* **pi•no•cyt•ic** \ˌpi-nə-'sī-tik, ˌpī-\ *adjective* (1959)
: of, relating to, or being pinocytosis
— pi•no•cy•tot•i•cal•ly \-ti-k(ə-)lē\ *adverb*

pi•no•le \pi-'nō-lē\ *noun* [American Spanish, from Nahuatl *pinolli*] (1842)
1 : a finely ground flour made from parched corn
2 : any of various flours resembling pinole and ground from the seeds of other plants

pi•ñon *or* **pin•yon** \'pin-ˌyōn, -ˌyän, -yən; pin-'yōn\ *noun, plural* **pi•ñons** *or* **pin•yons** *or* **pi•ño•nes** \pin-'yō-nēz\ [American Spanish *piñón,* from Spanish, pine nut, from *piña* pinecone, from Latin *pinea* — more at PINEAL] (1831)
: any of various low-growing pines (as *Pinus quadrifolia, P. cembroides, P. edulis,* and *P. monophylla*) of western North America with edible seeds; *also* **:** the edible seed of a piñon

pi•not noir \'pē-(ˌ)nō-'nwär, pē-'nō-\ *noun, often P&N capitalized* [French, literally, black Pinot (a grape variety)] (1941)
: a dry red wine produced from the same grape as French burgundy

[1]**pin•point** \'pin-ˌpȯint\ *noun* (1849)
1 : something that is extremely small or insignificant
2 : the point of a pin
3 : an extremely small or sharp point

[2]**pinpoint** *adjective* (1899)
1 : extremely fine or precise
2 : located, fixed, or directed with extreme precision ⟨*pinpoint* targets⟩
3 : small as a pinpoint

[3]**pinpoint** *transitive verb* (1917)

1 : to locate or aim with great precision or accuracy
2 a : to fix, determine, or identify with precision **b :** to cause to stand out conspicuously **:** HIGHLIGHT

[1]pin·prick \ˈpin-ˌprik\ *noun* (1862)
1 : a small puncture made by or as if by a pin
2 : a petty irritation or annoyance

[2]pinprick (1899)
transitive verb
: to administer pinpricks to
intransitive verb
: to administer pinpricks

pins and needles *noun plural* (1813)
: a pricking tingling sensation in a limb growing numb or recovering from numbness
— on pins and needles : in a nervous or jumpy state of anticipation

pin·set·ter \ˈpin-ˌse-tər\ *noun* (1916)
: an employee or a mechanical device that spots pins in a bowling alley

pin·spot·ter \-ˌspä-tər\ *noun* (1946)
: PINSETTER

pin·stripe \-ˌstrīp\ *noun, often attributive* (1897)
: a very thin stripe especially on a fabric; *also* **:** a suit with such stripes — often used in plural
— pin–striped \-ˌstrīpt\ *adjective*

pint \ˈpīnt\ *noun* [Middle English *pinte,* from Middle French, from Medieval Latin *pincta,* probably from (assumed) Vulgar Latin, feminine of *pinctus,* past participle of Latin *pingere* to paint — more at PAINT] (14th century)
1 — see WEIGHT table
2 : a pint pot or vessel

[1]pin·ta \ˈpin-tə, -ˌtä\ *noun* [American Spanish, from Spanish, spot, mark, from (assumed) Vulgar Latin *pincta*] (1825)
: a chronic skin disease that is endemic in tropical America, that occurs successively as an initial papule, a generalized eruption, and a patchy loss of pigment, and that is caused by a spirochete (*Treponema careteum*) morphologically indistinguishable from the causative agent of syphilis

[2]pinta \ˈpīn-tə\ *noun* [*pint* + *-a* (as in *cuppa*)] (1959)
British **:** a pint of milk

pin–ta·ble \ˈpin-ˌtā-bəl\ *noun* (1936)
British **:** PINBALL MACHINE

pin·tail \ˈpin-ˌtāl\ *noun, plural* **pintail** *or* **pintails** (1768)
: a bird having elongated central tail feathers; *especially* **:** a slender duck (*Anas acuta*) of the northern hemisphere with the male having a brown head, a white breast with a white line continuing up the side of the neck, and chiefly gray upperparts

pin·tle \ˈpin-tᵊl\ *noun* [Middle English *pintel,* literally, penis, from Old English; akin to Middle Low German *pint* penis, Old English *pinn* pin] (15th century)
: a usually upright pivot pin on which another part turns

[1]pin·to \ˈpin-(ˌ)tō\ *noun, plural* **pintos** *also* **pintoes** [American Spanish, from *pinto* spotted, from obsolete Spanish, from (assumed) Vulgar Latin *pinctus*] (1860)
: a horse or pony marked with patches of white and another color — compare PIEBALD, SKEWBALD

pinto

[2]pinto *adjective* (1865)
: PIED, MOTTLED

pinto bean *noun* (1916)
: a mottled kidney bean that is grown chiefly in the southwestern U.S. for food and for stock feed

pint–size \ˈpīnt-ˌsīz\ *or* **pint–sized** \-ˌsīzd\ *adjective* (1936)
: SMALL

[1]pin–up \ˈpin-ˌəp\ *noun* (1943)
: something fastened to a wall: as **a :** a photograph or poster of a person considered to have glamorous qualities **b :** something (as a lamp) designed for wall attachment

[2]pinup *adjective* (1941)
1 : of or relating to pinups ⟨male *pinup* calendars⟩
2 : designed for hanging on a wall ⟨a *pinup* lamp⟩

pinup girl *noun* (1941)
: a girl or woman whose glamorous qualities make her a suitable subject for a pinup

pin·wale \ˈpin-ˌwāl\ *adjective* (1949)
of a fabric **:** made with narrow wales

pin·weed \-ˌwēd\ *noun* (1814)
: any of a genus (*Lechea*) of herbs of the rockrose family with slender stems, many small leaves, and tiny flowers

[1]pin·wheel \-ˌhwēl, -ˌwēl\ *noun* (1869)
1 : a fireworks device in the form of a revolving wheel of colored fire
2 : a toy consisting of lightweight vanes that revolve at the end of a stick
3 : something (as a galaxy) shaped like a pinwheel

[2]pinwheel *intransitive verb* (circa 1934)
: to move like a pinwheel

pin·worm \-ˌwərm\ *noun* (circa 1864)
1 : any of numerous small nematode worms (family Oxyuridae) that infest the intestines and especially the cecum of various vertebrates; *especially* **:** a worm (*Enterobius vermicularis*) parasitic in humans
2 : any of several rather slender insect larvae that burrow in plant tissue

pinx·ter flower \ˈpiŋ(k)-stər-\ *noun* [Dutch *pinkster* Whitsuntide] (1857)
: a deciduous pink-flowered azalea (*Rhododendron periclymenoides* synonym *R. nudiflorum*) that is native to rich moist woodlands of eastern North America

pin·yin \ˈpin-ˈyin\ *noun, often capitalized* [Chinese (Beijing) *pīnyīn* to spell phonetically, from *pīn* to arrange + *yīn* sound, pronunciation] (1963)
: a system for romanizing Chinese ideograms in which tones are indicated by diacritics and unaspirated consonants are transcribed as voiced — compare WADE-GILES

pi·on \ˈpī-ˌän\ *noun* [contraction of *pi-meson*] (1950)
: any of four mesons that consist of combinations of two up or down quarks, that may be positive, negative, or neutral, that have a mass about 270 times that of the electron, and that are subject to the strong force
— pi·on·ic \pī-ˈä-nik\ *adjective*

[1]pi·o·neer \ˌpī-ə-ˈnir\ *noun* [Middle French *pionier,* from Old French *peonier* foot soldier, from *peon* foot soldier, from Medieval Latin *pedon-, pedo* — more at PAWN] (1523)
1 : a member of a military unit usually of construction engineers
2 a : a person or group that originates or helps open up a new line of thought or activity or a new method or technical development **b :** one of the first to settle in a territory
3 : a plant or animal capable of establishing itself in a bare, barren, or open area and initiating an ecological cycle ◆

[2]pioneer (1780)
intransitive verb
: to act as a pioneer ⟨*pioneered* in the development of airplanes⟩
transitive verb
1 : to open or prepare for others to follow; *also* **:** SETTLE
2 : to originate or take part in the development of

[3]pioneer *adjective* (1840)
1 : ORIGINAL, EARLIEST
2 : relating to or being a pioneer; *especially* **:** of, relating to, or characteristic of early settlers or their time

pi·ous \ˈpī-əs\ *adjective* [Middle English, from Latin *pius*] (15th century)
1 a : marked by or showing reverence for deity and devotion to divine worship **b :** marked by conspicuous religiosity ⟨a hypocrite—a thing all *pious* words and uncharitable deeds —Charles Reade⟩
2 : sacred or devotional as distinct from the profane or secular **:** RELIGIOUS ⟨a *pious* opinion⟩
3 : showing loyal reverence for a person or thing **:** DUTIFUL
4 a : marked by sham or hypocrisy **b :** marked by self-conscious virtue **:** VIRTUOUS
5 : deserving commendation **:** WORTHY ⟨a *pious* effort⟩
— pi·ous·ly *adverb*
— pi·ous·ness *noun*

[1]pip \ˈpip\ *noun* [Middle English *pippe,* from Middle Dutch (akin to Old High German *pfiffīz*), from (assumed) Vulgar Latin *pipita,* alteration of Latin *pituita* phlegm, pip; perhaps akin to Greek *pitys* pine — more at PINE] (15th century)
1 a : a disorder of a bird marked by formation of a scale or crust on the tongue **b :** the scale or crust of this disorder
2 a : any of various human ailments; *especially* **:** a slight nonspecific disorder **b** *chiefly British* **:** a feeling of irritation or annoyance

[2]pip *verb* **pipped; pip·ping** [imitative] (1598)
intransitive verb
1 : [1]PEEP 1
2 : to break through the shell of the egg ⟨the chick *pipped*⟩
transitive verb
: to break open (the shell of an egg) in hatching

[3]pip *noun* [origin unknown] (1604)
1 a : one of the dots used on dice and dominoes to indicate numerical value **b :** SPOT 2c
2 a : SPOT, SPECK **b :** [1]SPIKE 6a; *also* **:** DLIP
3 : an individual rootstock of the lily of the valley
4 : a diamond-shaped insignia of rank worn by a second lieutenant, lieutenant, or captain in the British army

[4]pip *noun* [short for *pippin*] (1797)
1 : a small fruit seed; *especially* **:** one of a several-seeded fleshy fruit
2 : one extraordinary of its kind

◇ WORD HISTORY

pioneer Daniel Boone and the other pioneers who opened up the American West are heroic figures to many, but the origins of the word *pioneer* are somewhat less than grand. Medieval French *pēonier* or *pionier,* a derivative of *peon* "foot soldier," also originally meant "foot soldier," but later appeared in the sense "digger, excavator" and by the 14th century designated a solider who dug earthworks and prepared fortifications in advance of the troops who would occupy them. The word was borrowed by English in the latter sense as *pioner* in the 16th century, and with the modern spelling *pioneer* continues to refer to troops who build fortifications, roads, and bridges, though in the 20th century American army *engineer* has replaced it. In the figurative sense "forerunner" or "initiator," however, the English word has gone beyond its French source. Its specific application to a frontiersman first appeared in American English in the early 19th century, notably in the title of *The Pioneers,* a novel by James Fenimore Cooper published in 1823.

5pip *transitive verb* **pipped; pip·ping** [probably from *pip* to blackball, from *3pip* or *4pip*] (1880)
British **:** to beat by a narrow margin

6pip *noun* [imitative] (1907)
chiefly British **:** a short high-pitched tone

pip·age *or* **pipe·age** \ˈpī-pij\ *noun* (1612)
1 a : transportation by means of pipes **b :** the charge for such transportation
2 : material for pipelines **:** PIPING

pi·pal \ˈpē-(ˌ)pəl\ *noun* [Hindi *pīpal,* from Sanskrit *pippala*] (1788)
: a large long-lived fig (*Ficus religiosa*) of India

1pipe \ˈpīp\ *noun* [Middle English, from Old English *pīpa* (akin to Old High German *pfīfa* pipe), from (assumed) Vulgar Latin *pipa* pipe, from Latin *pipare* to peep, of imitative origin] (before 12th century)
1 a : a tubular wind instrument; *specifically* **:** a small fipple flute held in and played by the left hand **b :** one of the tubes of a pipe organ: (1) **:** FLUE PIPE (2) **:** REED PIPE **c :** BAGPIPE — usually used in plural **d** (1) **:** VOICE, VOCAL CORD — usually used in plural (2) **:** PIPING 1
2 : a long tube or hollow body for conducting a liquid, gas, or finely divided solid or for structural purposes
3 a : a tubular or cylindrical object, part, or passage **b :** a roughly cylindrical and vertical geological formation **c :** the eruptive channel opening into the crater of a volcano
4 a : a large cask of varying capacity used especially for wine and oil **b :** any of various units of liquid capacity based on the size of a pipe; *especially* **:** a unit equal to 2 hogsheads
5 : a device for smoking usually consisting of a tube having a bowl at one end and a mouthpiece at the other
6 : SNAP 2c, CINCH
— pipe·ful \-ˌfu̇l\ *noun*
— pipe·less \ˈpī-pləs\ *adjective*
— pipe·like \ˈpīp-ˌlīk\ *adjective*

2pipe *verb* **piped; pip·ing** (before 12th century)
intransitive verb
1 a : to play on a pipe **b :** to convey orders by signals on a boatswain's pipe
2 a : to speak in a high or shrill voice **b :** to emit a shrill sound
transitive verb
1 a : to play (a tune) on a pipe **b :** to utter in the shrill tone of a pipe
2 a : to lead or cause to go with pipe music **b** (1) **:** to call or direct by the boatswain's pipe (2) **:** to receive aboard or attend the departure of by sounding a boatswain's pipe
3 : to trim with piping
4 : to furnish or equip with pipes
5 : to convey by or as if by pipes; *especially* **:** to transmit by wire or coaxial cable
6 : NOTICE

pipe clay *noun* (1777)
: highly plastic grayish white clay used especially in making tobacco pipes and for whitening leather

pipe cleaner *noun* (1870)
: something used to clean the inside of a pipe; *specifically* **:** a piece of flexible wire in which tufted fabric is twisted and which is used to clean the stem of a tobacco pipe

pipe cutter

pipe cutter *noun* (circa 1890)
: a tool or machine for cutting pipe; *especially* **:** a hand tool comprising a grasping device and one or more sharp-edged wheels that cut into the pipe as the tool is rotated

pipe down *intransitive verb* [*2pipe*] (1850)
: to stop talking or making noise

pipe dream *noun* [from the fantasies brought about by the smoking of opium] (1890)
: an illusory or fantastic plan, hope, or story

pipe·fish \ˈpīp-ˌfish\ *noun* (1769)
: any of various fishes (family Syngnathidae) that are related to the sea horses and have a tube-shaped snout and a long slender body covered with bony plates

pipe fitter *noun* (circa 1883)
: a worker who installs and repairs piping

pipe fitting *noun* (circa 1890)
1 : a piece (as a coupling or elbow) used to connect pipes or as accessory to a pipe
2 : the work of a pipe fitter

pipe·line \ˈpīp-ˌlīn\ *noun* (1860)
1 : a line of pipe with pumps, valves, and control devices for conveying liquids, gases, or finely divided solids
2 : a direct channel for information
3 : the processes through which supplies pass from or as if from source to user
— in the pipeline : undergoing preparation, production, or completion **:** in the works

pipe of peace (1705)
: CALUMET

pipe organ *noun* (1885)
: ORGAN 1b(1)

pip·er \ˈpī-pər\ *noun* (before 12th century)
: one that plays on a pipe

pi·per·a·zine \pi-ˈper-ə-ˌzēn, pī-\ *noun* [International Scientific Vocabulary *piper*idine + *azine*] (1889)
: a crystalline heterocyclic base $C_4H_{10}N_2$ used especially as an anthelmintic

pi·per·i·dine \pi-ˈper-ə-ˌdēn, pī-\ *noun* [International Scientific Vocabulary *piper*ine + *-idine*] (1854)
: a liquid heterocyclic base $C_5H_{11}N$ that has a peppery ammoniacal odor and is obtained usually by hydrolysis of piperine

pip·er·ine \ˈpi-pə-ˌrēn\ *noun* [probably from French *pipérine,* from Latin *piper* pepper] (1820)
: a white crystalline alkaloid $C_{17}H_{19}NO_3$ that is the chief active constituent of pepper

pi·per·o·nal \pī-ˈper-ə-ˌnal\ *noun* [International Scientific Vocabulary *piperine* + *-one* + *3-al*] (1869)
: a crystalline aldehyde $C_8H_6O_3$ with an odor of heliotrope

pi·per·o·nyl bu·tox·ide \pī-ˈper-ə-ˌnil-byü-ˈtäk-ˌsīd, -nᵊl-\ *noun* [*piperon*al + *-yl* + *but*yl + *oxide*] (1945)
: an insecticide $C_{19}H_{30}O_5$; *especially* **:** an oily liquid containing this compound that is used chiefly as a synergist (as for pyrethrum insecticides)

pipe·stone \ˈpīp-ˌstōn\ *noun* (circa 1805)
: a pink or mottled pink-and-white argillaceous stone used especially by American Indians to make carved objects (as tobacco pipes)

pipe stop *noun* (circa 1909)
: an organ stop composed of flue pipes

pi·pette *also* **pi·pet** \pī-ˈpet\ *noun* [French *pipette,* diminutive of *pipe* pipe, cask, from (assumed) Vulgar Latin *pipa, pippa* pipe] (1839)
: a small piece of apparatus which typically consists of a narrow tube into which fluid is drawn by suction (as for dispensing or measurement) and retained by closing the upper end
— pipette *also* **pipet** *verb*

pipe up *intransitive verb* (1889)
: SPEAK UP

pipe wrench *noun* (circa 1875)
: a wrench for gripping and turning a cylindrical object (as a pipe) usually by use of two serrated jaws so designed as to grip the pipe when turning in one direction only

1pip·ing \ˈpī-piŋ\ *noun* (13th century)
1 a : a sound, note, or call like that of a pipe **b :** the music of a pipe
2 : a quantity or system of pipes
3 : trimming stitched in seams or along edges (as of clothing, slipcovers, or curtains)

2piping *adjective* (15th century)
1 : SHRILL ⟨a *piping* voice⟩
2 : TRANQUIL ⟨*piping* times of peace —Shakespeare⟩

piping hot *adjective* (14th century)
: very hot

pip·it \ˈpi-pət\ *noun* [imitative] (1768)
: any of various small singing birds (family Motacillidae and especially genus *Anthus*) resembling the lark

pip·kin \ˈpip-kən\ *noun* [perhaps from *pipe*] (1565)
: a small earthenware or metal pot usually with a horizontal handle

pip·pin \ˈpi-pən\ *noun* [Middle English *pepin,* from Middle French] (15th century)
1 : any of numerous apples that have usually yellow or greenish yellow skins strongly flushed with red and are used especially for cooking
2 : a highly admired or very admirable person or thing

pip–pip \ˌpi(p)-ˈpip\ *interjection* [probably from *pip-pip,* imitating the sound of a horn] (1907)
British — used to express farewell

pip·sis·se·wa \pip-ˈsi-sə-ˌwȯ\ *noun* [perhaps from Eastern Abenaki *kpi-pskʷáhsawe,* literally, flower of the woods] (1789)
: any of a genus (*Chimaphila* and especially *C. umbellata*) of evergreen herbs of the wintergreen family with astringent leaves used as a tonic and diuretic

pip–squeak \ˈpip-ˌskwēk\ *noun* (1910)
: one that is small or insignificant

pi·quance \ˈpē-kən(t)s, -kwən(t)s\ *noun* (1883)
: PIQUANCY

pi·quan·cy \ˈpē-kən(t)-sē, ˈpi-kwən(t)-\ *noun* (1664)
: the quality or state of being piquant

pi·quant \ˈpē-kənt, -ˌkänt; ˈpi-kwənt\ *adjective* [Middle French, from present participle of *piquer*] (1630)
1 : agreeably stimulating to the palate; *especially* **:** SPICY
2 : engagingly provocative; *also* **:** having a lively arch charm
synonym see PUNGENT
— pi·quant·ly *adverb*
— pi·quant·ness *noun*

1pique \ˈpēk\ *noun* (1592)
: a transient feeling of wounded vanity **:** RESENTMENT ⟨a fit of *pique*⟩
synonym see OFFENSE

2pique *transitive verb* **piqued; piqu·ing** [French *piquer,* literally, to prick — more at PIKE] (1669)
1 : to arouse anger or resentment in **:** IRRITATE ⟨what *piques* linguistic conservatives —T. H. Middleton⟩
2 a : to excite or arouse by a provocation, challenge, or rebuff ⟨sly remarks to *pique* their curiosity⟩ **b :** PRIDE ⟨he *piques* himself on his skill as a cook⟩
synonym see PROVOKE

pi·qué *or* **pi·que** \pi-ˈkā, ˈpē-ˌ\ *noun* [French *piqué,* from past participle of *piquer* to prick, quilt] (1852)
1 : a durable ribbed clothing fabric of cotton, rayon, or silk
2 : decoration of a tortoiseshell or ivory object with inlaid fragments of gold or silver

pi·quet \pi-ˈkā, -ˈket\ *noun* [French] (1646)
: a two-handed card game played with 32 cards

pi·ra·cy \ˈpī-rə-sē\ *noun, plural* **-cies** [Medieval Latin *piratia,* from Late Greek *peirateia,* from Greek *peiratēs* pirate] (1537)
1 : an act of robbery on the high seas; *also* **:** an act resembling such robbery
2 : robbery on the high seas
3 : the unauthorized use of another's production, invention, or conception especially in infringement of a copyright

pi·ra·gua \pə-ˈrä-gwə, -ˈra-\ *noun* [Spanish] (1609)
1 : DUGOUT 1

2 : a 2-masted flat-bottomed boat

pi·ra·nha \pə-'rä-nə; -'rän-yə, -'ran-\ *noun* [Portuguese, from Tupi *pirãya,* from *pira* fish + *ãya* tooth] (1869)
: any of various usually small South American characin fishes (genus *Serrasalmo*) having very sharp teeth and including some that may attack and inflict dangerous wounds upon humans and large animals — called also *caribe*

pi·ra·ru·cu \pi-,rär-ə-'kü\ *noun* [Portuguese, from Tupi *pirauruku,* from *pira* fish + *uruku* annatto] (1840)
: a very large edible bony fish (*Arapaima gigas* of the family Osteoglossidae) of the rivers of northern South America

¹pi·rate \'pī-rət\ *noun* [Middle English, from Middle French or Latin; Middle French, from Latin *pirata,* from Greek *peiratēs,* from *peiran* to attempt — more at FEAR] (14th century)
: one who commits or practices piracy
— pi·rat·i·cal \pə-'ra-ti-kəl, pī-\ *adjective*
— pi·rat·i·cal·ly \-k(ə-)lē\ *adverb*

²pirate *verb* **pi·rat·ed; pi·rat·ing** (1574)
transitive verb
1 : to commit piracy on
2 : to take or appropriate by piracy: as **a :** to reproduce without authorization especially in infringement of copyright **b :** to lure away from another employer by offers of betterment
intransitive verb
: to commit or practice piracy

pirn \'pərn, *2 is also* 'pirn\ *noun* [Middle English] (15th century)
1 : QUILL 1a(1)
2 *chiefly Scottish* **:** a device resembling a reel

pi·ro·gi *variant of* PIEROGI

pi·rogue \'pē-,rōg\ *noun* [French, from Spanish *piragua,* from Carib *piraua*] (1666)
1 : DUGOUT 1
2 : a boat like a canoe

piro·plasm \'pir-ə-,pla-zəm\ *or* **piro·plas·ma** \,pir-ə-'plaz-mə\ *noun, plural* **piro·plasms** *or* **piro·plas·ma·ta** \,pir-ə-'plaz-mə-tə\ [New Latin *Piroplasma,* genus of piroplasms] (1901)
: BABESIA

pi·rosh·ki *or* **pi·rozh·ki** \pi-'rōsh-kē, ,pir-əsh 'kē\ *noun plural* [Russian *pirozhki,* plural of *pirozhok,* diminutive of *pirog* pastry] (1912)
: small pastries with meat, cheese, or vegetable filling

pir·ou·ette \,pir-ə-'wet\ *noun* [French, literally, teetotum] (1706)
: a rapid whirling about of the body; *especially* **:** a full turn on the toe or ball of one foot in ballet
— pirouette *intransitive verb*

pis *plural of* PI

pis al·ler \,pē-za-'lā\ *noun, plural* **pis al·lers** \-'lā(z)\ [French, literally, to go worst] (1676)
: a last resource or device **:** EXPEDIENT

pis·ca·to·ri·al \,pis-kə-'tōr-ē-əl, -'tȯr-\ *adjective* (1828)
: PISCATORY

pis·ca·to·ry \'pis-kə-,tōr-ē, -,tȯr-\ *adjective* [Latin *piscatorius,* from *piscari* to fish, from *piscis*] (1633)
: of, relating to, or dependent on fish or fishing

Pi·sce·an \'pī-sē-ən, 'pi-sē-, 'pis-kē-\ *noun* (1925)
: PISCES 2b

Pi·sces \'pī-(,)sēz *also* 'pi-,sēz *or* 'pis-,kās\ *noun plural but singular in construction* [Middle English, from Latin (genitive *Piscium*), from plural of *piscis* fish — more at FISH]
1 : a zodiacal constellation directly south of Andromeda
2 a : the 12th sign of the zodiac in astrology — see ZODIAC table **b :** one born under this sign

pi·sci·cul·ture \'pī-sə-,kəl-chər, 'pi-sə-, 'pis-kə-\ *noun* [probably from French, from Latin *piscis* + French *culture* culture] (1859)
: fish culture

pi·sci·na \pə-'sē-nə, -'sī-\ *noun* [Medieval Latin, from Latin, fishpond, from *piscis*] (1793)
: a basin with a drain near the altar of a church for disposing of water from liturgical ablutions

pi·scine \'pī-,sēn, 'pi-,sīn, 'pis-,kīn\ *adjective* [Latin *piscinus,* from *piscis*] (1799)
: of, relating to, or characteristic of fish

pi·sciv·o·rous \pə-'si-və-rəs, pī-\ *adjective* [Latin *piscis* + English *-vorous*] (1668)
: feeding on fishes

pish \'pish\ *interjection* (1592)
— used to express disdain or contempt

¹pi·si·form \'pī-sə-,fȯrm\ *adjective* [Latin *pisum* pea + English *-iform* — more at PEA] (1767)
: resembling a pea in size or shape

²pisiform *noun* (circa 1771)
: a bone on the ulnar side of the carpus in most mammals

pis·mire \'pis-,mīr, 'piz-\ *noun* [Middle English *pissemire,* from *pisse* urine + *mire* ant, of Scandinavian origin; akin to Old Norse *maurr* ant; akin to Latin *formica* ant, Greek *myrmēx*] (14th century)
: ANT

pis·mo clam \'piz-(,)mō-\ *noun, often P capitalized* [*Pismo* Beach, Calif.] (1913)
: a thick-shelled clam (*Tivela stultorum*) of the southwest coast of North America used extensively for food

pi·so \'pē-(,)sō\ *noun* [probably from Tagalog, from Spanish *peso*] (circa 1975)
: the peso of the Philippines

pi·so·lite \'pī-sə-,līt\ *noun* [New Latin *pisolithus,* from Greek *pisos* pea + *-lithos* -lith] (1708)
: a limestone composed of pisiform concretions
— pi·so·lit·ic \,pī-sə-'li-tik\ *adjective*

¹piss \'pis\ *verb* [Middle English, from Old French *pissier,* from (assumed) Vulgar Latin *pissiare,* of imitative origin] (14th century)
intransitive verb
: URINATE — sometimes considered vulgar
transitive verb
: to urinate in or on — sometimes considered vulgar

²piss *noun* (14th century)
1 : URINE — sometimes considered vulgar
2 : an act of urinating — often used with *take;* sometimes considered vulgar

piss·ant \'pis-,ant\ *noun, often attributive* [*pissant* ant, from ¹*piss* + *ant*] (circa 1946)
: one that is insignificant — used as a generalized term of abuse; sometimes considered vulgar

piss away *transitive verb* (1949)
: to fritter away **:** SQUANDER — sometimes considered vulgar

pissed \'pist\ *adjective* (1929)
1 *chiefly British* **:** DRUNK 1a — sometimes considered vulgar
2 : ANGRY, IRRITATED — often used with *off;* sometimes considered vulgar

piss off (1953)
intransitive verb
British **:** to leave immediately **:** SCRAM — usually used as a command; sometimes considered vulgar
transitive verb
: ANGER, IRRITATE — sometimes considered vulgar

pis·soir \pi-'swär\ *noun* [French, from Middle French, from *pisser* to urinate, from Old French *pissier*] (1919)
: a public urinal usually located on the street in some European countries

pis·ta·chio \pə-'sta-sh(ē-,)ō, -'stä-\ *noun, plural* **-chios** [Italian *pistacchio,* from Latin *pistacium* pistachio nut, from Greek *pistakion,* from *pistakē* pistachio tree, of Iranian origin; akin to Persian *pistah* pistachio tree] (1598)
: a small Asian tree (*Pistacia vera*) of the cashew family whose drupaceous fruit contains a greenish edible seed; *also* **:** its seed

pis·ta·reen \,pis-tə-'rēn\ *noun* [probably modification of Spanish *peseta* peseta] (1744)
: an old Spanish silver piece circulating at a debased rate

piste \'pēst\ *noun* [French, from Middle French, from Old Italian *pista,* from *pistare* to trample down, pound — more at PISTON] (circa 1741)
: TRAIL; *especially* **:** a downhill ski trail

pis·til \'pis-tᵊl\ *noun* [New Latin *pistillum,* from Latin, pestle — more at PESTLE] (1749)
: a single carpel or group of fused carpels usually differentiated into an ovary, style, and stigma — see FLOWER illustration

pis·til·late \'pis-tə-,lāt\ *adjective* (circa 1828)
: having pistils; *specifically* **:** having pistils but no stamens

pis·tol \'pis-tᵊl\ *noun* [Middle French *pistole,* from German, from Middle High German *pischulle,* from Czech *píšt'ala,* literally, pipe, fife; akin to Czech *pištět* to squeak] (circa 1570)
: a handgun whose chamber is integral with the barrel; *broadly* **:** HANDGUN
— pistol *transitive verb*

pis·tole \pis-'tōl\ *noun* [Middle French] (1592)
: an old gold 2-escudo piece of Spain; *also* **:** any of several old gold coins of Europe of approximately the same value

pis·tol·eer \,pis-tə-'lir\ *noun* (1832)
: one who is armed with a pistol

pistol grip *noun* (1874)
1 : a grip of a shotgun or rifle shaped like a pistol stock
2 : a handle shaped like a pistol stock

pistol–whip *transitive verb* (circa 1942)
: to beat with a pistol

pis·ton \'pis-tən\ *noun* [French, from Italian *pistone,* from *pistare* to pound, from Old Italian, from Medieval Latin, from Latin *pistus,* past participle of *pinsere* to crush — more at PESTLE] (1704)
1 : a sliding piece moved by or moving against fluid pressure which usually consists of a short cylinder fitting within a cylindrical vessel along which it moves back and forth
2 a : a sliding valve moving in a cylinder in a brass instrument which when depressed by a finger knob serves to lower the instrument's pitch **b :** a button on an organ console to bring in a previously selected registration

piston pin *noun* (1897)
: WRIST PIN

piston ring *noun* (1867)
: a springy split metal ring for sealing the gap between a piston and the cylinder wall

piston rod *noun* (1786)
: CONNECTING ROD

¹pit \'pit\ *noun* [Middle English, from Old English *pytt* (akin to Old High German *pfuzzi* well), from Latin *puteus* well, pit] (before 12th century)
1 a (1) **:** a hole, shaft, or cavity in the ground (2) **:** MINE (3) **:** a scooped-out place used for burning something (as charcoal) **b :** an area often sunken or depressed below the adjacent floor area: as (1) **:** an enclosure in which animals are made to fight each other (2) **:** a space at the front of a theater for the orchestra (3) **:** an area in a securities or commodities exchange in which members trade (as stocks)
2 a : HELL — used with *the* **b :** a place or situation of futility, misery, or degradation **c** *plural* **:** WORST ⟨it's the *pits*⟩
3 : a hollow or indentation especially in the surface of an organism: as **a :** a natural hollow in the surface of the body **b :** one of the indented scars left in the skin by a pustular disease **:** POCKMARK **c :** a minute depression in the secondary wall of a plant cell functioning

in the intercellular movement of water and dissolved material
4 : any of the areas alongside an auto racecourse used for refueling and repairing the cars during a race — often used in plural with *the*

[2]pit *verb* **pit·ted; pit·ting** (15th century)
transitive verb
1 a : to place, cast, bury, or store in a pit **b :** to make pits in; *especially* **:** to scar or mark with pits
2 a : to set (as gamecocks) into or as if into a pit to fight **b :** to set into opposition or rivalry — usually used with *against*
intransitive verb
1 : to become marked with pits; *especially* **:** to preserve for a time an indentation made by pressure
2 : to make a pit stop

[3]pit *noun* [Dutch, from Middle Dutch — more at PITH] (1841)
: the stone of a drupaceous fruit

[4]pit *transitive verb* **pit·ted; pit·ting** (1913)
: to remove the pit from (a fruit)

[1]pi·ta \ˈpē-tə\ *noun* [Spanish & Portuguese] (1698)
1 : any of several fiber-yielding plants (as an agave)
2 : the fiber of a pita; *also* **:** any of several fibers from other sources

[2]pita *noun* [New Greek] (1951)
: a thin flat bread that can be separated easily into two layers to form a pocket — called also *pita bread*

pit–a–pat \ˌpit-i-ˈpat\ *noun* [imitative] (1582)
: PITTER-PATTER
— pit–a–pat *adverb or adjective*
— pit–a–pat *intransitive verb*

pit boss *noun* (1949)
: a person who supervises the gaming tables in a casino

pit bull *noun* (1930)
: a dog (as an American Staffordshire terrier) of any of several breeds or a real or apparent hybrid with one or more of these breeds that was developed and is now often trained for fighting and is noted for strength and stamina

pit bull terrier *noun* (1945)
1 : PIT BULL
2 : AMERICAN PIT BULL TERRIER

[1]pitch \ˈpich\ *noun* [Middle English *pich,* from Old English *pic,* from Latin *pic-, pix;* akin to Greek *pissa* pitch, Old Church Slavonic *pĭcĭlŭ*] (before 12th century)
1 : a black or dark viscous substance obtained as a residue in the distillation of organic materials and especially tars
2 : any of various bituminous substances
3 : resin obtained from various conifers and often used medicinally
4 : any of various artificial mixtures resembling resinous or bituminous pitches

[2]pitch *transitive verb* (before 12th century)
: to cover, smear, or treat with or as if with pitch

[3]pitch *verb* [Middle English *pichen*] (13th century)
transitive verb
1 : to erect and fix firmly in place ⟨*pitch* a tent⟩
2 : to throw usually with a particular objective or toward a particular point ⟨*pitch* hay onto a wagon⟩: as **a :** to throw (a baseball) to a batter **b :** to toss (as coins) so as to fall at or near a mark ⟨*pitch* pennies⟩ **c :** to put aside or discard by or as if by throwing ⟨*pitched* the trash into the bin⟩ ⟨decided to *pitch* the whole idea⟩
3 : to present or advertise for sale especially in a high-pressure way
4 a (1) **:** to cause to be at a particular level or of a particular quality (2) **:** to set in a particular musical key **b :** to cause to be set at a particular angle **:** SLOPE
5 : to utter glibly and insincerely
6 a : to use as a starting pitcher **b :** to play as pitcher
7 : to hit (a golf ball) in a high arc with backspin so that it rolls very little after striking the green
intransitive verb
1 a : to fall precipitately or headlong **b** (1) *of a ship* **:** to have the bow alternately plunge precipitately and rise abruptly (2) *of an aircraft, missile, or spacecraft* **:** to turn about a lateral axis so that the forward end rises or falls in relation to the after end **c :** BUCK 1
2 : ENCAMP
3 : to hit upon or happen upon something ⟨*pitch* upon the perfect gift⟩
4 : to incline downward **:** SLOPE
5 a : to throw a ball to a batter **b :** to play ball as a pitcher **c :** to pitch a golf ball
6 : to make a sales pitch
synonym see THROW
— pitch into 1 : ATTACK, ASSAIL **2 :** to set to work on energetically

[4]pitch *noun* (1542)
1 a : SLOPE; *also* **:** degree of slope **:** RAKE **b :** the distance between any of various things: as (1) **:** distance between one point on a gear tooth and the corresponding point on the next tooth (2) **:** distance from any point on the thread of a screw to the corresponding point on an adjacent thread measured parallel to the axis **c :** the theoretical distance a propeller would advance longitudinally in one revolution **d :** the number of teeth or of threads per inch **e :** a unit of width of type based on the number of times a letter can be set in a linear inch

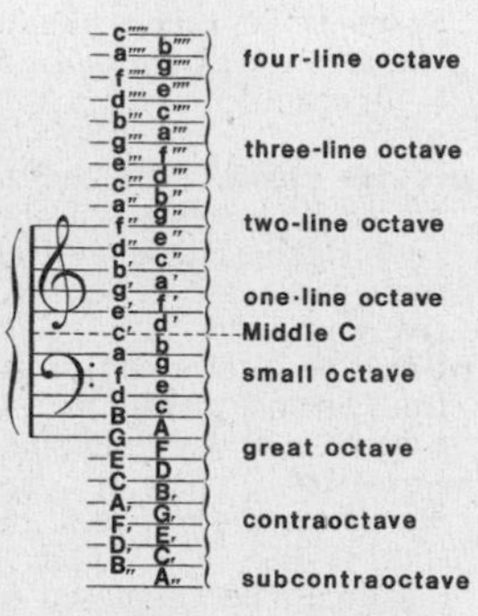

staff notation of pitch 4b(1)

2 : the action or a manner of pitching; *especially* **:** an up-and-down movement — compare YAW
3 *archaic* **:** TOP, ZENITH
4 a : the relative level, intensity, or extent of some quality or state ⟨tensions rose to a fever-ish *pitch*⟩ **b** (1) **:** the property of a sound and especially a musical tone that is determined by the frequency of the waves producing it **:** highness or lowness of sound (2) **:** a standard frequency for tuning instruments **c** (1) **:** the difference in the relative vibration frequency of the human voice that contributes to the total meaning of speech (2) **:** a definite relative pitch that is a significant phenomenon in speech
5 : a steep place **:** DECLIVITY
6 *chiefly British* **a :** an outdoor site (as for camping or doing business) **b :** PLAYING FIELD 1
7 : an all-fours game in which the first card led is a trump
8 a : an often high-pressure sales presentation **b :** ADVERTISEMENT
9 a : the delivery of a baseball by a pitcher to a batter **b :** a baseball so thrown **c :** PITCHOUT 2
— pitched \ˈpicht\ *adjective*

pitch–black \ˈpich-ˈblak\ *adjective* (1599)
: extremely dark or black

pitch·blende \ˈpich-ˌblend\ *noun* [part translation of German *Pechblende,* from *Pech* pitch + *Blende* blende] (1770)
: a brown to black mineral that consists of massive uraninite, has a distinctive luster, contains radium, and is the chief ore-mineral source of uranium

pitch–dark \ˈpich-ˈdärk\ *adjective* (1827)
: extremely dark **:** PITCH-BLACK

pitched battle *noun* (1607)
: an intensely fought battle in which the opposing forces are locked in close combat

[1]pitch·er \ˈpi-chər\ *noun* [Middle English *picher,* from Old French *pichier,* from Medieval Latin *bicarius* goblet] (13th century)
1 : a container for holding and pouring liquids that usually has a lip or spout and a handle
2 : a modified leaf of a pitcher plant in which the hollowed petiole and base of the blade form an elongated receptacle
— pitch·er·ful \-ˌfu̇l\ *noun*

[2]pitcher *noun* (circa 1722)
: one that pitches; *specifically* **:** the player that pitches in a game of baseball

pitcher plant *noun* (1819)
: a plant (especially family Sarraceniaceae, the pitcher-plant family) with leaves modified into pitchers in which insects are trapped and digested by means of liquids secreted by the leaves

pitch·fork \ˈpich-ˌfȯrk\ *noun* (13th century)
: a long-handled fork that has two or three long somewhat curved prongs and is used especially in pitching hay
— pitchfork *transitive verb*

pitch in *intransitive verb* (1843)
1 : to begin to work
2 : to contribute to a common endeavor

pitch·man \ˈpich-mən\ *noun* (circa 1926)
: one who makes a sales pitch: as **a :** one who sells merchandise on the streets or from a concession **b :** one who does radio or TV commercials

pitch·out \ˈpich-ˌau̇t\ *noun* (1912)
1 : a pitch in baseball deliberately out of reach of the batter to enable the catcher to put out a base runner especially with a throw
2 : a lateral pass in football between two backs behind the line of scrimmage
— pitch out *intransitive verb*

pitch pine *noun* (1676)
1 : any of several pines that yield pitch; *especially* **:** a 3-leaved pine (*Pinus rigida*) of eastern North America
2 : the wood of a pitch pine

pitch pipe *noun* (1711)
: a small reed pipe or flue pipe producing one or more tones to establish the pitch in singing or in tuning an instrument

pitch·pole \ˈpich-ˌpōl\ *verb* [*pitchpole* somersault, from [3]*pitch* + *pole, poll* head] (1851)
intransitive verb
: to turn end over end ⟨the catamaran *pitchpoled*⟩
transitive verb
: to cause to turn end over end

pitch·wom·an \-ˌwu̇-mən\ *noun* (1957)
: a woman who makes a sales pitch

pitchy \ˈpi-chē\ *adjective* (15th century)
1 a : full of pitch **:** TARRY **b :** of, relating to, or having the qualities of pitch
2 : PITCH-BLACK

pit·e·ous \ˈpi-tē-əs\ *adjective* (14th century)
: of a kind to move to pity or compassion
— pit·e·ous·ly *adverb*
— pit·e·ous·ness *noun*

pit·fall \ˈpit-ˌfȯl\ *noun* (14th century)
1 : TRAP, SNARE; *specifically* **:** a pit flimsily covered or camouflaged and used to capture and hold animals or men
2 : a hidden or not easily recognized danger or difficulty

[1]pith \ˈpith\ *noun* [Middle English, from Old English *pitha;* akin to Middle Dutch & Middle Low German *pit* pith, pit] (before 12th century)
1 a : a usually continuous central strand of spongy tissue in the stems of most vascular plants that probably functions chiefly in storage **b :** any of various loose spongy plant tissues that resemble true pith **c :** the soft or spongy interior of a part of the body
2 a : the essential part **:** CORE **b :** substantial quality (as of meaning)
3 : IMPORTANCE

[2]pith *transitive verb* (1805)

1 a : to kill (as cattle) by piercing or severing the spinal cord **b :** to destroy the spinal cord or central nervous system of (as a frog) usually by passing a wire or needle up and down the vertebral canal
2 : to remove the pith from (a plant stem)

pit·head \'pit-ˌhed\ *noun* (1839)
chiefly British **:** the top of a mining pit or coal shaft; *also* **:** the immediately adjacent ground and buildings

pith·ec·an·thro·pine \ˌpi-thi-'kan(t)-thrə-ˌpīn\ *noun* (1925)
: any of a group of Pleistocene hominids (as Java man, Peking man, and Heidelberg man) that have a smaller cranial capacity than modern humans (*Homo sapiens*) but a greater cranial capacity than the australopithecines, prominent eyebrow ridges, a receding forehead, constriction of the skull behind the eye sockets, and relatively large canine and incisor teeth and that were formerly considered to comprise a genus (*Pithecanthropus*) but are now grouped in a single species (*Homo erectus*)
— pithecanthropine *adjective*

pith·ec·an·thro·pus \-'kan(t)-thrə-pəs, -ˌkan-'thrō-\ *noun, plural* **-pi** \-ˌpī, -ˌpē\ [New Latin, from Greek *pithēkos* ape + *anthrōpos* human being] (1895)
: PITHECANTHROPINE

pith helmet *noun* (1889)
: TOPEE

pith ray *noun* (1902)
: MEDULLARY RAY

pithy \'pi-thē\ *adjective* **pith·i·er; -est** (1562)
1 : consisting of or abounding in pith
2 : having substance and point **:** tersely cogent
synonym see CONCISE
— pith·i·ly \'pi-thə-lē\ *adverb*
— pith·i·ness \'pi-thē-nəs\ *noun*

piti·able \'pi-tē-ə-bəl\ *adjective* (15th century)
1 : deserving or exciting pity **:** LAMENTABLE
2 : of a kind to evoke mingled pity and contempt especially because of inadequacy ⟨a *pitiable* excuse⟩
synonym see CONTEMPTIBLE
— piti·able·ness *noun*
— piti·ably \-blē\ *adverb*

piti·er \'pi-tē-ər\ *noun* (1601)
: one that pities

piti·ful \'pi-ti-fəl\ *adjective* (14th century)
1 *archaic* **:** full of pity **:** COMPASSIONATE
2 a : deserving or arousing pity or commiseration **b :** exciting pitying contempt (as by meanness or inadequacy)
— piti·ful·ly \-f(ə-)lē\ *adverb*
— piti·ful·ness \-fəl-nəs\ *noun*

piti·less \'pi-ti-ləs, 'pi-tᵊl-əs\ *adjective* (15th century)
: devoid of pity **:** MERCILESS
— piti·less·ly *adverb*
— piti·less·ness *noun*

pit·man \'pit-mən\ *noun* (1703)
1 *plural* **pit·men** \-mən\ **:** one who works in or near a pit (as in a coal mine)
2 *plural* **pitmans :** CONNECTING ROD

pi·ton \'pē-ˌtän\ *noun* [French] (1893)
: a spike, wedge, or peg that is driven into a rock or ice surface as a support (as for a mountain climber)

pi·tot–stat·ic tube \ˌpē-ˌtō-'sta-tik-\ *noun, often P capitalized* (1926)
: a device that consists of a pitot tube and a static tube and that measures pressures in such a way that the relative speed of a fluid can be determined

pi·tot tube \'pē-ˌtō-\ *noun, often P capitalized* [French (*tube de*) *Pitot,* from Henri *Pitot* (died 1771) French physicist] (circa 1859)
1 : a device that consists of a tube having a short right-angled bend which is placed vertically in a moving body of fluid with the mouth of the bent part directed upstream and that is used with a manometer to measure the velocity of fluid flow
2 : PITOT-STATIC TUBE

pit saw *noun* (1679)
: a handsaw worked by two persons one of whom stands on or above the log being sawed into planks and the other below it usually in a pit

pit stop *noun* (1932)
1 : a stop at the pits during an automobile race
2 a : a stop (as during a trip) for fuel, food, or rest or for use of a rest room **b** (1) **:** a place where a pit stop can be made (2) **:** an establishment providing food or drink

pit·tance \'pi-tᵊn(t)s\ *noun* [Middle English *pitance,* from Old French, piety, pity, from Medieval Latin *pietantia,* from *pietant-, pietans,* present participle of *pietari* to be charitable, from Latin *pietas* piety — more at PITY] (14th century)
: a small portion, amount, or allowance; *also* **:** a meager wage or remuneration

pit·ted \'pi-təd\ *adjective* (before 12th century)
: marked with pits

pit·ter–pat·ter \'pi-tər-ˌpa-tər, 'pi-tē-ˌ\ *noun* [reduplication of ⁴*patter*] (15th century)
: a rapid succession of light sounds or beats **:** PATTER
— pitter–patter \ˌpi-tər-', ˌpi-tē-'\ *adverb or adjective*
— pitter–patter *same as adverb*\ *intransitive verb*

pit·ting *noun* (1665)
1 : an arrangement of pits
2 : the action or process of forming pits
3 : the bringing of gamecocks together to fight

pit·tos·po·rum \pə-'täs-pə-rəm\ *noun* [New Latin, from Greek *pitta, pissa* pitch + *spora* seed — more at PITCH, SPORE] (1789)
: any of various Old World shrubs or trees (genus *Pittosporum* of the family Pittosporaceae) planted especially as ornamentals in warm regions

¹pi·tu·i·tary \pə-'tü-ə-ˌter-ē, -'tyü-\ *adjective* [Latin *pituita* phlegm; from the former belief that the pituitary gland secreted phlegm — more at PIP] (1615)
1 : of or relating to the pituitary gland
2 : caused or characterized by secretory disturbances of the pituitary gland ⟨a *pituitary* dwarf⟩

²pituitary *noun, plural* **-tar·ies** (1845)
: PITUITARY GLAND

pituitary gland *noun* (1825)
: a small oval endocrine organ that is attached to the infundibulum of the brain, consists of an epithelial anterior lobe joined by an intermediate part to a posterior lobe of nervous origin, and produces various internal secretions directly or indirectly impinging on most basic body functions — called also *hypophysis, pituitary body;* see BRAIN illustration

pit viper *noun* (circa 1885)
: any of various mostly New World venomous snakes (subfamily Crotalinae of the family Viperidae) with a sensory pit on each side of the head and hollow perforated fangs

¹pity \'pi-tē\ *noun, plural* **pit·ies** [Middle English *pite,* from Old French *pité,* from Latin *pietat-, pietas* piety, pity, from *pius* pious] (13th century)
1 a : sympathetic sorrow for one suffering, distressed, or unhappy **b :** capacity to feel pity
2 : something to be regretted ⟨it's a *pity* you can't go⟩ ☆

²pity *verb* **pit·ied; pity·ing** (15th century)
transitive verb
: to feel pity for
intransitive verb
: to feel pity

pitying *adjective* (1589)
: expressing or feeling pity
— pity·ing·ly \-iŋ-lē\ *adverb*

pit·y·ri·a·sis \ˌpi-ti-'rī-ə-səs\ *noun* [New Latin, from Greek, from *pityron* scurf] (circa 1693)
: a condition of humans or domestic animals marked by dry scaling or scurfy patches of skin

più \'pyü, pē-'ü\ *adverb* [Italian, from Latin *plus*] (1724)
: MORE — used to qualify an adverb or adjective used as a direction in music

Pi·ute *variant of* PAIUTE

¹piv·ot \'pi-vət\ *noun* [French] (1611)
1 : a shaft or pin on which something turns
2 a : a person, thing, or factor having a major or central role, function, or effect **b :** a key player or position; *specifically* **:** an offensive position of a basketball player standing usually with back to the basket to relay passes, shoot, or provide a screen for teammates
3 : the action of pivoting

²pivot *adjective* (1796)
1 : turning on or as if on a pivot
2 : PIVOTAL

³pivot (1841)
intransitive verb
: to turn on or as if on a pivot
transitive verb
1 : to provide with, mount on, or attach by a pivot
2 : to cause to pivot
— pivot·able \-və-tə-bəl\ *adjective*

piv·ot·al \'pi-və-tᵊl\ *adjective* (1844)
1 : of, relating to, or constituting a pivot
2 : vitally important **:** CRUCIAL
— piv·ot·al·ly \-tᵊl-ē\ *adverb*

piv·ot·man \'pi-vət-ˌman\ *noun* (circa 1814)
: one who plays the pivot; *specifically* **:** a center on a basketball team

pivot tooth *noun* (1842)
: an artificial crown attached to the root of a tooth by a usually metallic pin — called also *pivot crown*

pix *plural of* PIC

pix·el \'pik-səl, -ˌsel\ *noun* [*pix* + *el*ement] (1969)
1 : any of the small discrete elements that together constitute an image (as on a television screen)
2 : any of the detecting elements of a charge-coupled device used as an optical sensor

¹pix·ie *or* **pixy** \'pik-sē\ *noun, plural* **pix·ies** [origin unknown] (1746)
: FAIRY; *specifically* **:** a cheerful mischievous sprite
— pix·ie·ish \-sē-ish\ *adjective*

²pixie *or* **pixy** *adjective* (1943)
: playfully mischievous
— pixi·ness *noun*

pix·i·lat·ed *or* **pix·il·lat·ed** \'pik-sə-ˌlā-təd\ *adjective* [irregular from *pixie*] (1848)

☆ SYNONYMS

Pity, compassion, commiseration, condolence, sympathy mean the act or capacity for sharing the painful feelings of another. PITY implies tender or sometimes slightly contemptuous sorrow for one in misery or distress ⟨felt *pity* for the captives⟩. COMPASSION implies pity coupled with an urgent desire to aid or to spare ⟨treats the homeless with great *compassion*⟩. COMMISERATION suggests pity expressed outwardly in exclamations, tears, or words of comfort ⟨murmurs of *commiseration* filled the loser's headquarters⟩. CONDOLENCE applies chiefly to formal expression of grief to one who has suffered loss ⟨expressed their *condolences* to the widow⟩. SYMPATHY often suggests a tender concern but can also imply a power to enter into another's emotional experience of any sort ⟨went to my best friend for *sympathy*⟩ ⟨in *sympathy* with her desire to locate her natural parents⟩.

1 : somewhat unbalanced mentally; *also* **:** BEMUSED
2 : WHIMSICAL
— **pix·i·la·tion** \ˌpik-sə-ˈlā-shən\ *noun*

piz·za \ˈpēt-sə\ *noun* [Italian, perhaps of Germanic origin; akin to Old High German *bizzo, pizzo* bite, bit, *bīzan* to bite — more at BITE] (1935)
: an open pie made typically of flattened bread dough spread with a savory mixture usually including tomatoes and cheese and often other toppings and baked — called also *pizza pie*
— **pizza·like** \-ˌlīk\ *adjective*

piz·zazz *or* **pi·zazz** \pə-ˈzaz\ *noun* [origin unknown] (1937)
: the quality of being exciting or attractive: as **a :** GLAMOUR **b :** VITALITY
— **piz·zazzy** *or* **pi·zazzy** \-ˈza-zē\ *adjective*

piz·ze·ria \ˌpēt-sə-ˈrē-ə\ *noun* [Italian, from *pizza*] (1943)
: an establishment where pizzas are made or sold

¹piz·zi·ca·to \ˌpit-si-ˈkä-(ˌ)tō\ *noun, plural* **-ca·ti** \-ˈkä-(ˌ)tē\ (1845)
: a note or passage played by plucking strings

²pizzicato *adverb or adjective* [Italian, past participle of *pizzicare* to pluck] (circa 1854)
: by means of plucking instead of bowing — used as a direction in music; compare ARCO

piz·zle \ˈpi-zəl\ *noun* [probably from Flemish *pezel;* akin to Low German *pesel* pizzle] (1523)
1 : the penis of an animal
2 : a whip made of a bull's pizzle

pj's \ˈpē-ˌjāz\ *noun plural* [*pa*jamas] (1951)
: PAJAMAS

PK \ˌpē-ˈkā\ *noun* (1943)
: PSYCHOKINESIS

pla·ca·ble \ˈpla-kə-bəl, ˈplā-\ *adjective* (1586)
: easily placated **:** TOLERANT, TRACTABLE
— **pla·ca·bil·i·ty** \ˌpla-kə-ˈbi-lə-tē, ˌplā-\ *noun*
— **pla·ca·bly** \ˈpla-kə-blē, ˈplā-\ *adverb*

¹plac·ard \ˈpla-kərd, -ˌkärd\ *noun* [Middle English *placquart* formal document, from Middle French, from *plaquier* to plate — more at PLAQUE] (1560)
1 : a notice posted in a public place **:** POSTER
2 : a small card or metal plaque

²plac·ard \-ˌkärd, -kərd\ *transitive verb* (1813)
1 a : to cover with or as if with posters **b :** to post in a public place
2 : to announce by or as if by posting

pla·cate \ˈplā-ˌkāt, ˈpla-\ *transitive verb* **pla·cat·ed; pla·cat·ing** [Latin *placatus,* past participle of *placare* — more at PLEASE] (1678)
: to soothe or mollify especially by concessions **:** APPEASE
synonym see PACIFY
— **pla·cat·er** *noun*
— **pla·cat·ing·ly** \-ˌkā-tiŋ-lē\ *adverb*
— **pla·ca·tion** \plā-ˈkā-shən, pla-\ *noun*
— **pla·ca·tive** \ˈplā-ˌkā-tiv, ˈpla-\ *adjective*
— **pla·ca·to·ry** \ˈplā-kə-ˌtōr-ē, ˈpla-, -ˌtȯr-\ *adjective*

¹place \ˈplās\ *noun* [Middle English, from Old French, open space, from Latin *platea* broad street, from Greek *plateia* (*hodos*), from feminine of *platys* broad, flat; akin to Sanskrit *pṛthu* broad, Latin *planta* sole of the foot] (13th century)
1 a : physical environment **:** SPACE **b :** a way for admission or transit **c :** physical surroundings **:** ATMOSPHERE
2 a : an indefinite region or expanse ⟨all over the *place*⟩ **b :** a building or locality used for a special purpose ⟨a *place* of learning⟩ ⟨a fine eating *place*⟩ **c** *archaic* **:** the three-dimensional compass of a material object
3 a : a particular region, center of population, or location ⟨a nice *place* to visit⟩ **b :** a building, part of a building, or area occupied as a home ⟨our summer *place*⟩
4 : a particular part of a surface or body **:** SPOT
5 : relative position in a scale or series: as **a :** position in a social scale ⟨kept them in their *place*⟩ **b :** a step in a sequence ⟨in the first *place,* it's none of your business⟩ **c :** a position at the conclusion of a competition ⟨finished in last *place*⟩
6 a : a proper or designated niche ⟨the *place* of education in society⟩ **b :** an appropriate moment or point ⟨this is not the *place* to discuss compensation —Robert Moses⟩
7 a : an available seat or accommodation ⟨needs a *place* to stay⟩ **b :** an empty or vacated position ⟨new ones will take their *place*⟩
8 : the position of a figure in relation to others of a row or series; *especially* **:** the position of a digit within a numeral
9 a : remunerative employment **:** JOB **b :** prestige accorded to one of high rank **:** STATUS ⟨an endless quest for preferment and *place* —*Time*⟩
10 : a public square **:** PLAZA
11 : a small street or court
12 : second place at the finish (as of a horse race)
— **in place 1 :** in an original or proper position **2 :** in the same spot without forward or backward movement ⟨run *in place*⟩

²place *verb* **placed; plac·ing** (15th century)
transitive verb
1 a : to put in or as if in a particular place or position **:** SET **b :** to present for consideration ⟨a question *placed* before the group⟩ **c :** to put in a particular state ⟨*place* a performer under contract⟩ **d :** to direct to a desired spot **e :** to cause (the voice) to produce free and well resonated singing or speaking tones
2 a : to assign to a position in a series or category **:** RANK **b :** ESTIMATE ⟨*placed* the value of the estate too high⟩ **c :** to identify by connecting with an associated context ⟨couldn't quite *place* her face⟩ ⟨police *placed* them at the crime scene⟩
3 : to distribute in an orderly manner **:** ARRANGE
4 : to appoint to a position
5 : to find a place (as a home or employment) for
6 a : to give (an order) to a supplier **b :** to give an order for ⟨*place* a bet⟩ **c :** to try to establish a connection for ⟨*place* a telephone call⟩
intransitive verb
: to earn a given spot in a competition; *specifically* **:** to come in second (as in a horse race)
— **place·able** \ˈplā-sə-bəl\ *adjective*

pla·ce·bo *noun, plural* **-bos** (13th century)
1 \plä-ˈchā-(ˌ)bō\ [Middle English, from Latin, I shall please, from *placēre* to please — more at PLEASE] **:** the Roman Catholic vespers for the dead
2 \plə-ˈsē-\ [Latin, I shall please] **a** (1) **:** a medication prescribed more for the mental relief of the patient than for its actual effect on a disorder (2) **:** an inert or innocuous substance used especially in controlled experiments testing the efficacy of another substance (as a drug) **b :** something tending to soothe

placebo effect *noun* (1950)
: improvement in the condition of a sick person that occurs in response to treatment but cannot be considered due to the specific treatment used

place·hold·er \ˈplās-ˌhōl-dər\ *noun* (1958)
: a symbol in a mathematical or logical expression that may be replaced by the name of any element of a set

¹place·kick \ˈplās-ˌkik\ *noun* (1856)
: the kicking of a ball (as a football) placed or held in a stationary position on the ground

²placekick *transitive verb* (1856)
1 : to kick (a ball) from a stationary position
2 : to score by means of a placekick
— **place·kick·er** *noun*

place·less \-ləs\ *adjective* (14th century)
: lacking a fixed location
— **place·less·ly** *adverb*

place·man \-mən\ *noun* (1741)
: a political appointee to a public office especially in 18th century Britain

place mat *noun* (1928)
: a small often rectangular table mat on which a place setting is laid

place·ment \ˈplās-mənt\ *noun* (1844)
1 : an act or instance of placing: as **a :** an accurately hit ball (as in tennis) that an opponent cannot return **b :** the assignment of a person to a suitable place (as a job or a class in school)
2 : PLACEKICK

placement test *noun* (1928)
: a test usually given to a student entering an educational institution to determine specific knowledge or proficiency in various subjects for the purpose of assignment to appropriate courses or classes

place–name \ˈplās-ˌnām\ *noun* (1868)
: the name of a geographical locality

pla·cen·ta \plə-ˈsen-tə\ *noun, plural* **-centas** *or* **-cen·tae** \-ˈsen-(ˌ)tē\ [New Latin, from Latin, flat cake, from Greek *plakoenta,* accusative of *plakoeis,* from *plak-, plax* flat surface — more at FLUKE] (1691)
1 : the vascular organ in mammals except monotremes and marsupials that unites the fetus to the maternal uterus and mediates its metabolic exchanges through a more or less intimate association of uterine mucosal with chorionic and usually allantoic tissues; *also* **:** an analogous organ in another animal
2 : a sporangium-bearing surface; *especially* **:** the part of the carpel bearing ovules
— **pla·cen·tal** \-ˈsen-tᵊl\ *adjective or noun*

pla·cen·ta·tion \ˌpla-sᵊn-ˈtā-shən, plə-ˌsen-\ *noun* (1760)
1 : the arrangement of placentas and ovules in a plant ovary
2 a : the development of the placenta and attachment of the fetus to the uterus during pregnancy **b :** the morphological type of a placenta

¹plac·er \ˈplā-sər\ *noun* (1579)
: one that places: as **a :** one that deposits or arranges **b :** one of the winners in a competition

²plac·er \ˈpla-sər\ *noun* [Spanish, from Catalan, submarine plain, from *plaza* place, from Latin *platea* broad street — more at PLACE] (1848)
: an alluvial, marine, or glacial deposit containing particles of valuable mineral and especially of gold

place setting *noun* (1944)
: a table service for one person

place value *noun* (1911)
: the value of the place of a digit in a numeral

plac·id \ˈpla-səd\ *adjective* [Latin *placidus,* from *placēre* to please — more at PLEASE] (1626)
1 : serenely free of interruption or disturbance ⟨*placid* skies⟩ ⟨a *placid* disposition⟩
2 : COMPLACENT
synonym see CALM
— **pla·cid·i·ty** \pla-ˈsi-də-tē, plə-\ *noun*
— **plac·id·ly** \ˈpla-səd-lē\ *adverb*
— **plac·id·ness** *noun*

plack·et \ˈpla-kət\ *noun* [origin unknown] (1605)
1 a : a slit in a garment (as a skirt) often forming the closure **b** *archaic* **:** a pocket especially in a woman's skirt
2 *archaic* **a :** PETTICOAT **b :** WOMAN

plac·oid \ˈpla-ˌkȯid\ *adjective* [Greek *plak-, plax* flat surface] (1842)
: of, relating to, or being a scale of dermal origin with an enamel-tipped spine characteristic of the elasmobranchs

pla·gal \ˈplā-gəl\ *adjective* [Medieval Latin *plagalis,* ultimately from Greek *plagios* oblique, sideways, from *plagos* side; akin to Latin *plaga* net, region, Greek *pelagos* sea] (1597)
1 *of a church mode* **:** having the keynote on the 4th scale step — compare AUTHENTIC 4a

2 *of a cadence* **:** progressing from the subdominant chord to the tonic — compare AUTHENTIC 4b

plage \'pläzh\ *noun* [French, beach, luminous surface, from Italian *piaggia* beach, from Late Latin *plagia,* from Greek *plagios* oblique] (1888)
1 : the beach of a seaside resort
2 : a bright region on the sun caused by the light emitted by clouds of calcium or hydrogen and often associated with a sunspot

pla·gia·rise *British variant of* PLAGIARIZE

pla·gia·rism \'plā-jə-ˌri-zəm *also* -jē-ə-\ *noun* (1621)
1 : an act or instance of plagiarizing
2 : something plagiarized ◆
— **pla·gia·rist** \-rist\ *noun*
— **pla·gia·ris·tic** \ˌplā-jə-'ris-tik *also* -jē-ə-\ *adjective*

pla·gia·rize \'plā-jə-ˌrīz *also* -jē-ə-\ *verb* **-rized; -riz·ing** [*plagiary*] (1716)
transitive verb
: to steal and pass off (the ideas or words of another) as one's own **:** use (a created production) without crediting the source
intransitive verb
: to commit literary theft **:** present as new and original an idea or product derived from an existing source
— **pla·gia·riz·er** *noun*

pla·gia·ry \'plā-jē-ˌer-ē, -jə-rē\ *noun, plural* **-ries** [Latin *plagiarus,* literally, kidnapper, from *plagium* netting of game, kidnapping, from *plaga* net, trap] (1601)
1 *archaic* **:** one that plagiarizes
2 : PLAGIARISM

pla·gio·clase \'plā-j(ē-)ə-ˌklās, 'pla-, -ˌklāz\ *noun* [Greek *plagios* oblique + *klasis* breaking, from *klan* to break — more at CLAST] (circa 1868)
: a triclinic feldspar; *especially* **:** one having calcium or sodium in its composition

pla·gio·tro·pic \ˌplā-j(ē-)ə-'trō-pik, ˌpla-, -'trä-\ *adjective* [Greek *plagios* + International Scientific Vocabulary *-tropic*] (1882)
: having the longer axis inclined away from the vertical

[1]plague \'plāg\ *noun* [Middle English *plage,* from Middle French, from Late Latin *plaga,* from Latin, blow; akin to Latin *plangere* to strike — more at PLAINT] (14th century)
1 a : a disastrous evil or affliction **:** CALAMITY **b :** a destructively numerous influx ⟨a *plague* of locusts⟩
2 a : an epidemic disease causing a high rate of mortality **:** PESTILENCE **b :** a virulent contagious febrile disease that is caused by a bacterium (*Yersinia pestis*) and that occurs in bubonic, pneumonic, and septicemic forms — called also *black death*
3 a : a cause of irritation **:** NUISANCE **b :** a sudden unwelcome outbreak ⟨a *plague* of burglaries⟩

[2]plague *transitive verb* **plagued; plagu·ing** (15th century)
1 : to smite, infest, or afflict with or as if with disease, calamity, or natural evil
2 a : to cause worry or distress to **:** HAMPER, BURDEN **b :** to disturb or annoy persistently
synonym see WORRY
— **plagu·er** *noun*

plagu·ey *or* **plaguy** \'plā-gē, 'ple-\ *adjective* (1615)
: causing irritation or annoyance **:** TROUBLESOME
— **plaguey** *adverb*
— **plagu·i·ly** \-gə-lē\ *adverb*

plaice \'plās\ *noun, plural* **plaice** [Middle English *plaice,* from Middle French *plaïs,* from Late Latin *platessa*] (13th century)
: any of various flatfishes; *especially* **:** a large European flounder (*Pleuronectes platessa*) with red spots

plaid \'plad\ *noun* [Scottish Gaelic *plaide*] (1512)
1 : a rectangular length of tartan worn over the left shoulder as part of the Scottish national costume
2 a : a twilled woolen fabric with a tartan pattern **b :** a fabric with a pattern of tartan or an imitation of tartan
3 a : TARTAN 1 **b :** a pattern of unevenly spaced repeated stripes crossing at right angles
— **plaid** *adjective*
— **plaid·ed** \'pla-dəd\ *adjective*

[1]plain \'plān\ *intransitive verb* [Middle English, from Old French *plaindre,* from Latin *plangere* to lament — more at PLAINT] (14th century)
archaic **:** COMPLAIN

[2]plain *noun* [Middle English, from Old French, from Latin *planum,* from neuter of *planus* flat, plain — more at FLOOR] (14th century)
1 a : an extensive area of level or rolling treeless country **b :** a broad unbroken expanse
2 : something free from artifice, ornament, or extraneous matter

[3]plain *adjective* (14th century)
1 *archaic* **:** EVEN, LEVEL
2 : lacking ornament **:** UNDECORATED
3 : free of extraneous matter **:** PURE
4 : free of impediments to view **:** UNOBSTRUCTED
5 a (1) **:** evident to the mind or senses **:** OBVIOUS ⟨it's perfectly *plain* that they will resist⟩ (2) **:** CLEAR ⟨let me make my meaning *plain*⟩ **b :** marked by outspoken candor **:** free from duplicity or subtlety **:** BLUNT ⟨*plain* talk⟩
6 a : belonging to the great mass of humans **b :** lacking special distinction or affectation **:** ORDINARY
7 : characterized by simplicity **:** not complicated ⟨*plain* home-cooked meals⟩
8 : lacking beauty or ugliness
synonym see COMMON, EVIDENT, FRANK
— **plain·ly** *adverb*
— **plain·ness** \'plān-nəs\ *noun*

[4]plain *adverb* (14th century)
: in a plain manner ⟨saw them clearly and told you *plain* —*American Documentation*⟩

[5]plain *adverb* [partly from Middle English *plein* entire, complete, from Middle French, full, from Latin *plenus;* partly from [4]*plain* — more at FULL] (1535)
: ABSOLUTELY ⟨it *plain* galled me to pay fancy prices —F. R. Buckley⟩

plain·chant \'plān-ˌchant\ *noun* [French *plain-chant,* literally, plain song] (circa 1741)
: PLAINSONG

plain·clothes \'plān-'klō(t͟h)z\ *adjective* (1866)
: dressed in civilian clothes while on duty — used especially of a police officer

plain·clothes·man \'plān-'klō(t͟h)z-mən, -ˌman\ *noun* (1899)
: a plainclothes police officer

plain–Jane \'plān-'jān\ *adjective* [from the name *Jane*] (1912)
: not fancy or glamorous **:** ORDINARY

plain–laid \'plān-'lād\ *adjective* (1881)
of a rope **:** consisting of three strands laid right-handed

Plain People *noun* (1904)
: members of any of various Protestant groups (as Mennonites) especially in the U.S. who wear distinctively plain clothes and adhere to a simple and traditional style of life excluding many conveniences of modern technology

Plains \'plānz\ *adjective* (1844)
: of or relating to North American Indians of the Great Plains or to their culture

plain sailing *noun* (1756)
: easy progress over an unobstructed course

plains·man \'plānz-mən\ *noun* [Great *Plains* + *man*] (1870)
: an inhabitant of the plains

plain·song \'plān-ˌsȯŋ\ *noun* (1513)
: a monophonic rhythmically free liturgical chant of any of various Christian rites; *especially* **:** GREGORIAN CHANT

plain·spo·ken \-'spō-kən\ *adjective* (1678)
: CANDID, FRANK
— **plain·spo·ken·ness** \-kən-nəs\ *noun*

plaint \'plānt\ *noun* [Middle English, from Middle French, from Latin *planctus,* from *plangere* to strike, beat one's breast, lament; akin to Old High German *fluokhōn* to curse, Greek *plēssein* to strike] (13th century)
1 : LAMENTATION, WAIL
2 : PROTEST, COMPLAINT

plain·text \'plān-ˌtekst\ *noun* (1918)
: the intelligible form of an encrypted text or of its elements — compare CIPHERTEXT

plaint·ful \'plānt-fəl\ *adjective* (14th century)
: MOURNFUL

plain·tiff \'plān-təf\ *noun* [Middle English *plaintif,* from Middle French, from *plaintif,* adjective] (14th century)
: a person who brings a legal action — compare DEFENDANT

plain·tive \'plān-tiv\ *adjective* [Middle English *plaintif* grieving, from Middle French, from *plaint*] (1579)
: expressive of suffering or woe **:** MELANCHOLY
— **plain·tive·ly** *adverb*
— **plain·tive·ness** *noun*

plain–vanilla *adjective* (1975)
: lacking special features or qualities **:** BASIC

plain weave *noun* (1888)
: a weave in which the threads interlace alternately

plain–woven *adjective* (1925)
: made in plain weave

plais·ter \'plas-tər, 'plās-\ *variant of* PLASTER

[1]plait \'plāt, 'plat\ *noun* [Middle English *pleit,* from Middle French, from (assumed) Vulgar Latin *plicitum,* from neuter of Latin *plicitus,* past participle of *plicare* to fold — more at PLY] (14th century)
1 : PLEAT
2 : a braid of material (as hair or straw); *specifically* **:** PIGTAIL

[2]plait *transitive verb* (14th century)
1 : PLEAT 1
2 a : to interweave the strands or locks of **:** BRAID **b :** to make by plaiting
— **plait·er** *noun*

plait·ing *noun* (15th century)
: the interlacing of strands **:** BRAIDING

[1]plan \'plan\ *noun* [French, plane, foundation, ground plan; partly from Latin *planum* level ground, from neuter of *planus* level; partly from French *planter* to plant, fix in place, from Late Latin *plantare* — more at FLOOR, PLANT] (1706)
1 : a drawing or diagram drawn on a plane: as **a :** a top or horizontal view of an object **b :** a large-scale map of a small area
2 a : a method for achieving an end **b :** an often customary method of doing something

◇ WORD HISTORY

plagiarism The Latin noun *plaga* denoted a hunting net or snare used for capturing game. The netting of such animals was termed *plagium.* By extension, this word was also used for the crime of kidnapping children or freemen and selling them as slaves. The kidnapper was called by the derivative noun *plagiarius.* We know that by the time of the poet Martial, who died about A.D. 103, *plagiarius* was also being used to refer to a literary thief or plagiarist. When Latin *plagiarius* was borrowed into English as *plagiary* in the 17th century, both the original and literary senses were carried over. Only the literary sense of *plagiary* appears in the English derivatives *plagiarism* and *plagiarize.*

PLANETS

NAME	SYMBOL	MEAN DISTANCE FROM THE SUN *astronomical units*	*million miles*	PERIOD OF REVOLUTION IN DAYS OR YEARS	EQUATORIAL DIAMETER IN MILES
Mercury	☿	0.387	36.0	87.97 days	3,032
Venus	♀	0.723	67.2	224.70 days	7,523
Earth	⊕	1.000	92.9	365.26 days	7,928
Mars	♂	1.524	141.5	686.98 days	4,218
Jupiter	♃	5.203	483.4	11.86 years	88,900
Saturn	♄	9.522	884.6	29.46 years	74,900
Uranus	⛢	19.201	1783.8	84.01 years	31,800
Neptune	♆	30.074	2793.9	164.79 years	30,800
Pluto	♇	39.725	3690.5	247.69 years	1,400

: PROCEDURE **c :** a detailed formulation of a program of action **d :** GOAL, AIM
3 : an orderly arrangement of parts of an overall design or objective
4 : a detailed program (as for payment or the provision of some service) ⟨pension *plan*⟩ ☆
— plan·less \-ləs\ *adjective*
— plan·less·ly *adverb*
— plan·less·ness *noun*

²plan *verb* **planned; plan·ning** (1728)
transitive verb
1 : to arrange the parts of **:** DESIGN
2 : to devise or project the realization or achievement of ⟨*plan* a program⟩
3 : to have in mind **:** INTEND
intransitive verb
: to make plans
— plan·ner *noun*

plan- *or* **plano-** *combining form* [Latin *planus*]
1 : flat ⟨*plano*sol⟩
2 : flat and ⟨*plano*-concave⟩

pla·nar \ˈplā-nər, -ˌnär\ *adjective* (1850)
1 : of, relating to, or lying in a plane
2 : two-dimensional in quality
— pla·nar·i·ty \plā-ˈnar-ə-tē\ *noun*

pla·nar·ia \plə-ˈnar-ē-ə, -ˈner-\ *noun* [New Latin, from feminine of Late Latin *planarius* lying on a plane, from Latin *planum* plane] (circa 1909)
: PLANARIAN; *especially* **:** any of a genus (*Planaria*) of 2-eyed planarian worms

pla·nar·i·an \-ē-ən\ *noun* [New Latin *Planaria*] (circa 1858)
: any of various dark-colored freshwater triclad flatworms (family Planariidae) with two eyespots and a triangular head; *broadly* **:** TRICLAD

pla·na·tion \plā-ˈnā-shən\ *noun* (1877)
: the condition or process of becoming flattened

plan·chet \ˈplan-chət\ *noun* [diminutive of *planch* flat plate, from Middle French *planche*] (1611)
1 : a metal disk to be stamped as a coin
2 : a small metal or plastic disk

plan·chette \plan-ˈshet\ *noun* [French, from diminutive of *planche* plank, from Latin *planca*] (1860)
: a small triangular or heart-shaped board supported on casters at two points and a vertical pencil at a third and believed to produce automatic writing when lightly touched by the fingers; *also* **:** a similar board without a pencil

Planck's constant \ˈpläŋ(k)s-, ˈplaŋ(k)s-\ *noun* [Max K.E.L. *Planck*] (1910)
: a constant that gives the unvarying ratio of the energy of a quantum of radiation to its frequency and that has an approximate value of 6.626×10^{-34} J·S — symbol *h*

¹plane \ˈplān\ *verb* **planed; plan·ing** [Middle English, from Middle French *planer,* from Late Latin *planare,* from Latin *planus* level — more at FLOOR] (14th century)
transitive verb
1 a : to make smooth or even **:** LEVEL **b :** to make smooth or even by use of a plane
2 : to remove by planing — often used with *away* or *off*
intransitive verb
1 : to work with a plane
2 : to do the work of a plane
— plan·er *noun*

²plane *noun* [Middle English, from Middle French, from Latin *platanus,* from Greek *platanos;* probably akin to Greek *platys* broad — more at PLACE] (14th century)
: any of a genus (*Platanus* of the family Platanaceae, the plane-tree family) of trees with large palmately lobed leaves and flowers in globose heads — called also *buttonwood, plane tree, sycamore*

³plane *noun* [Middle English, from Middle French, from Late Latin *plana,* from *planare*] (14th century)
: a tool for smoothing or shaping a wood surface

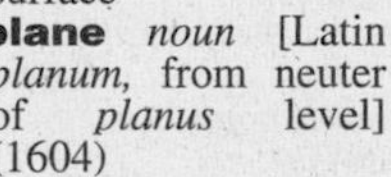
³plane

⁴plane *noun* [Latin *planum,* from neuter of *planus* level] (1604)
1 a : a surface of such nature that a straight line joining two of its points lies wholly in the surface **b :** a flat or level surface
2 : a level of existence, consciousness, or development ⟨on the intellectual *plane*⟩
3 a : one of the main supporting surfaces of an airplane **b** [by shortening] **:** AIRPLANE

⁵plane *adjective* [Latin *planus*] (1704)
1 : having no elevations or depressions **:** FLAT
2 a : of, relating to, or dealing with geometric planes **b :** lying in a plane ⟨a *plane* curve⟩
synonym see LEVEL

⁶plane *intransitive verb* **planed; plan·ing** [Middle English, from Middle French *planer,* from *plain* level, plain] (15th century)
1 a : to fly while keeping the wings motionless **b :** to skim across the surface of the water
2 : to travel by airplane

plane angle *noun* (1570)
: an angle that for a given dihedral angle is formed by two intersecting lines each of which lies on a face of the dihedral angle and is perpendicular to the edge of the face

plane geometry *noun* (1747)
: a branch of elementary geometry that deals with plane figures

plane·load \ˈplān-ˌlōd\ *noun* (1941)
: a load that fills an airplane

plane of polarization (1831)
: the plane in which the magnetic-vibration component of plane-polarized electromagnetic radiation lies

plane–polarized *adjective* (circa 1853)
: vibrating in a single plane ⟨*plane-polarized* light waves⟩

pla·ner tree \ˈplā-nər-\ *noun* [J. J. *Planer* (died 1789) German botanist] (circa 1810)
: a small-leaved North American tree (*Planera aquatica*) of the elm family with an oval ribbed fruit

plan·et \ˈpla-nət\ *noun* [Middle English *planete,* from Old French, from Late Latin *planeta,* modification of Greek *planēt-, planēs,* literally, wanderer, from *planasthai* to wander — more at FLOOR] (13th century)
1 a : any of the seven celestial bodies sun, moon, Venus, Jupiter, Mars, Mercury, and Saturn that in ancient belief have motions of their own among the fixed stars **b** (1) **:** any of the large bodies that revolve around the sun in the solar system (2) **:** a similar body associated with another star **c :** EARTH — usually used with *the*
2 : a celestial body held to influence the fate of human beings
3 : a person or thing of great importance **:** LUMINARY ◆
— plan·et·like \-ˌlīk\ *adjective*

plane table *noun* (1607)
: an instrument consisting essentially of a drawing board on a tripod with a ruler pointed at the object observed and used for plotting the lines of a survey directly from observation

plan·e·tar·i·um \ˌpla-nə-ˈter-ē-əm\ *noun, plural* **-i·ums** *or* **-ia** \-ē-ə\ (1860)
1 : a model or representation of the solar system
2 a : an optical device for projecting various celestial images and effects **b :** a building or room housing such a projector

plan·e·tary \ˈpla-nə-ˌter-ē\ *adjective* (1607)
1 a : of, relating to, being, or resembling a planet **b :** ERRATIC, WANDERING **c :** having a motion like that of a planet ⟨*planetary* electrons⟩ **d :** IMMENSE ⟨the scope of this project has reached *planetary* proportions⟩

☆ SYNONYMS
Plan, design, plot, scheme, project mean a method devised for making or doing something or achieving an end. PLAN always implies mental formulation and sometimes graphic representation ⟨*plans* for a house⟩. DESIGN often suggests a particular pattern and some degree of achieved order or harmony ⟨a *design* for a new dress⟩. PLOT implies a laying out in clearly distinguished sections with attention to their relations and proportions ⟨the *plot* of the play⟩. SCHEME stresses calculation of the end in view and may apply to a plan motivated by craftiness and self-interest ⟨a *scheme* to defraud the government⟩. PROJECT often stresses imaginative scope and vision ⟨a *project* to develop the waterfront⟩.

◇ WORD HISTORY
planet In scanning the nighttime sky, ancient astronomers observed that while the vast majority of stars maintained fixed relative positions, there were five heavenly bodies that looked like bright stars but quite obviously changed their position in relation to each other and to the fixed stars. These five—Mercury, Venus, Mars, Jupiter, and Saturn—were called by the Greeks *planētes asteres* "wandering stars," or simply *planētes* "wanderers." The corresponding Latin expression was *stellae errantes* or *erraticae,* which also means "wandering stars." By the second century A.D. the borrowed Greek words *planētes* and *planētai* (the latter from a different noun derivative of the verb *planasthai* "to wander") were also appearing in Latin as *planetes* and *planetae* respectively; *planetae,* the singular of which was *planeta,* became the usual word in Late and Medieval Latin. Through medieval French, Middle English borrowed this word in the 14th century, which ultimately became *planet* in Modern English. Of course, we now know that the planets "wander" because they revolve around the sun, as does the earth.

2 a : of, relating to, or belonging to the earth **:** TERRESTRIAL **b :** GLOBAL, WORLDWIDE
3 : having or consisting of an epicyclic train of gear wheels

planetary nebula *noun* (1785)
: a usually compact luminous ring-shaped nebula that is composed of matter which has been ejected from a hot star at its center

planetary science *noun* (1969)
: PLANETOLOGY
— **planetary scientist** *noun*

plan·e·tes·i·mal \ˌpla-nə-ˈte-sə-məl, -zə-məl\ *noun* [*planet* + *-esimal* (as in *infinitesimal*)] (1903)
: any of numerous small solid celestial bodies that may have existed at an early stage of the development of the solar system

planetesimal hypothesis *noun* (1904)
: a hypothesis in astronomy: the planets have evolved by aggregation from planetesimals

plan·et·oid \ˈpla-nə-ˌtȯid\ *noun* (1803)
: a small body resembling a planet; *especially* **:** ASTEROID
— **plan·et·oi·dal** \ˌpla-nə-ˈtȯi-dᵊl\ *adjective*

plan·et·ol·o·gy \ˌpla-nə-ˈtä-lə-jē\ *noun, plural* **-gies** (1907)
: a branch of astronomy that deals with the condensed matter of the solar system and especially with the planets and their moons
— **plan·et·o·log·i·cal** \ˌpla-nə-tᵊl-ˈä-ji-kəl\ *adjective*
— **plan·et·ol·o·gist** \ˌpla-nə-ˈtä-lə-jist\ *noun*

plan·et–strick·en \ˈpla-nət-ˌstri-kən\ *or* **plan·et–struck** \-ˌstrək\ *adjective* (1599)
1 *archaic* **:** affected by the influence of a planet
2 *archaic* **:** PANIC-STRICKEN

planet wheel *noun* (1827)
: a gear wheel that revolves around the wheel with which it meshes in an epicyclic train

plan·et·wide \ˈpla-nət-ˈwīd\ *adjective* (1969)
: extending throughout or involving an entire planet

plan·form \ˈplan-ˌfȯrm\ *noun* (1908)
: the contour of an object (as an airplane) or mass as viewed from above

plan·gen·cy \ˈplan-jən(t)-sē\ *noun* (1858)
: the quality or state of being plangent

plan·gent \-jənt\ *adjective* [Latin *plangent-, plangens,* present participle of *plangere* to strike, lament — more at PLAINT] (1858)
1 : having a loud reverberating sound
2 : having an expressive and especially plaintive quality
— **plan·gent·ly** *adverb*

pla·nim·e·ter \plā-ˈni-mə-tər, plə-\ *noun* [French *planimètre,* from Latin *planum* plane + French *-mètre* -meter] (circa 1858)
: an instrument for measuring the area of a plane figure by tracing its boundary line

pla·ni·met·ric \ˌplā-nə-ˈme-trik\ *adjective* (circa 1828)
1 : of, relating to, or made by means of a planimeter ⟨*planimetric* measurements⟩
2 *of a map* **:** having no indications of relief
— **pla·ni·met·ri·cal·ly** \-tri-k(ə-)lē\ *adverb*

plan·ish \ˈpla-nish\ *transitive verb* [Middle English *planysshen,* from Middle French *planiss-,* stem of *planir* to make smooth, from *plan* level, from Latin *planus* — more at FLOOR] (14th century)
: to smooth, toughen, and finish (metal) by hammering lightly
— **plan·ish·er** *noun*

pla·ni·sphere \ˈplā-nə-ˌsfir\ *noun* [alteration of Middle English *planisperie,* from Medieval Latin *planisphaerium,* from Latin *planum* plane + *sphaera* sphere] (1571)
: a representation of the circles of the sphere on a plane; *especially* **:** a polar projection of the celestial sphere and the stars on a plane with adjustable circles or other appendages for showing celestial phenomena for any given time
— **pla·ni·spher·ic** \ˌplā-nə-ˈsfir-ik, -ˈsfer-\ *adjective*

1plank \ˈplaŋk\ *noun* [Middle English, from Old North French *planke,* from Latin *planca*] (13th century)
1 a : a heavy thick board; *especially* **:** one 2 to 4 inches (5 to 10 centimeters) thick and at least 8 inches (20 centimeters) wide **b :** an object made of a plank or planking **c :** PLANKING
2 a : an article in the platform of a political party **b :** a principal item of a policy or program

2plank *transitive verb* (15th century)
1 : to cover, build, or floor with planks
2 : SET DOWN 1, 2 — usually used with *down*
3 : to cook and serve on a board ⟨*planked* salmon⟩ ⟨*planked* steak⟩

plank·ing *noun* (15th century)
1 : the act or process of covering or fitting with planks
2 : a quantity of planks

plank·ter \ˈplaŋ(k)-tər\ *noun* [Greek *planktēr* wanderer, from *plazesthai*] (1935)
: a planktonic organism

plank·ton \ˈplaŋ(k)-tən, -ˌtän\ *noun* [German, from Greek, neuter of *planktos* drifting, from *plazesthai* to wander, drift, middle voice of *plazein* to drive astray; akin to Latin *plangere* to strike — more at PLAINT] (1891)
: the passively floating or weakly swimming usually minute animal and plant life of a body of water
— **plank·ton·ic** \plaŋ(k)-ˈtä-nik\ *adjective*

Planned Parenthood *service mark*
— used for research and dissemination of information on contraception

plan·ning *noun* (1748)
: the act or process of making or carrying out plans; *specifically* **:** the establishment of goals, policies, and procedures for a social or economic unit ⟨city *planning*⟩ ⟨business *planning*⟩

plano- — see PLAN-

pla·no–con·cave \ˌplā-nō-(ˌ)kän-ˈkāv, -ˈkän-ˌ\ *adjective* (1693)
: flat on one side and concave on the other

pla·no–con·vex \-(ˌ)kän-ˈveks, -ˈkän-ˌ, -kən-ˈ\ *adjective* (1665)
: flat on one side and convex on the other

pla·nog·ra·phy \plā-ˈnä-grə-fē, plə-\ *noun* (circa 1909)
: a process (as lithography) for printing from a plane surface
— **pla·no·graph·ic** \ˌplā-nə-ˈgra-fik\ *adjective*

pla·no·sol \ˈplā-nə-ˌsäl, -ˌsȯl\ *noun* [*plan-* + Latin *solum* ground, soil] (1938)
: any of an intrazonal group of soils that have a strongly leached upper layer over a compacted clay or silt and occur on smooth flat uplands

plan position indicator *noun* (1932)
: a radarscope having a sweep synchronized with a usually rotating antenna so that the radar return can be used to find range and bearing

1plant \ˈplant\ *verb* [Middle English, from Old English *plantian,* from Late Latin *plantare* to plant, fix in place, from Latin, to plant, from *planta* plant] (before 12th century)
transitive verb
1 a : to put or set in the ground for growth ⟨*plant* seeds⟩ **b :** to set or sow with seeds or plants **c :** IMPLANT
2 a : ESTABLISH, INSTITUTE **b :** COLONIZE, SETTLE **c :** to place (animals) in a new locality **d :** to stock with animals
3 a : to place in or on the ground **b :** to place firmly or forcibly ⟨*planted* a hard blow on his chin⟩
4 a : CONCEAL **b :** to covertly place for discovery, publication, or dissemination
intransitive verb
: to plant something
— **plant·able** \ˈplan-tə-bəl\ *adjective*

2plant *noun* [Middle English *plante,* from Old English, from Latin *planta*] (before 12th century)
1 a : a young tree, vine, shrub, or herb planted or suitable for planting **b :** any of a kingdom (Plantae) of living things typically lacking locomotive movement or obvious nervous or sensory organs and possessing cellulose cell walls
2 a : the land, buildings, machinery, apparatus, and fixtures employed in carrying on a trade or an industrial business **b :** a factory or workshop for the manufacture of a particular product **c :** the total facilities available for production or service **d :** the buildings and other physical equipment of an institution
3 : an act of planting
4 : something or someone planted
— **plant·like** \-ˌlīk\ *adjective*

Plan·tag·e·net \plan-ˈta-jə-nət, -ˈtaj-nət\ *adjective* [*Plantagenet,* nickname of the family adopted as surname] (1868)
: of or relating to a royal house ruling England from 1154 to 1485
— **Plantagenet** *noun*

1plan·tain \ˈplan-tᵊn\ *noun* [Middle English, from Old French, from Latin *plantagin-, plantago,* from *planta* sole of the foot; from its broad leaves — more at PLACE] (13th century)
: any of a genus (*Plantago* of the family Plantaginaceae, the plantain family) of short-stemmed elliptic-leaved herbs with spikes of minute greenish flowers

2plantain *noun* [Spanish *plántano, plátano* plane tree, banana tree, from Medieval Latin *plantanus* plane tree, alteration of Latin *platanus* — more at PLANE] (1555)
1 : a banana plant (*Musa paradisiaca*)
2 : the angular greenish starchy fruit of the plantain that is a staple food in the tropics when cooked

plantain lily *noun* (1882)
: any of a genus (*Hosta*) of perennial plants of the lily family with densely growing basal leaves and racemose white or violet flowers — called also *hosta, funkia*

plan·tar \ˈplan-tər, -ˌtär\ *adjective* [Latin *plantaris,* from *planta* sole — more at PLACE] (circa 1706)
: of or relating to the sole of the foot ⟨*plantar* wart⟩

plan·ta·tion \plan-ˈtā-shən\ *noun* (1569)
1 : a usually large group of plants and especially trees under cultivation
2 : a settlement in a new country or region ⟨Plymouth *Plantation*⟩
3 a : a place that is planted or under cultivation **b :** an agricultural estate usually worked by resident labor

plant·er \ˈplan-tər\ *noun* (14th century)
1 : one that cultivates plants: as **a** (1) **:** FARMER (2) **:** one who owns or operates a plantation **b :** a planting machine or implement
2 : one who settles or founds a place and especially a new colony
3 : a container in which ornamental plants are grown

planter's punch *noun* (1924)
: a punch of rum, lime or lemon juice, sugar, water, and sometimes bitters

plant food *noun* (1869)
1 : FOOD 1b
2 : FERTILIZER

plant hormone *noun* (1935)
: an organic substance other than a nutrient that in minute amounts modifies a plant physiological process; *especially* **:** one produced by a plant and active elsewhere than at the site of production

\ə\ **abut** \ᵊ\ **kitten** \ər\ **further** \a\ **ash** \ā\ **ace** \ä\ **mop, mar** \au̇\ **out** \ch\ **chin** \e\ **bet** \ē\ **easy** \g\ **go** \i\ **hit** \ī\ **ice** \j\ **job** \ŋ\ **sing** \ō\ **go** \ȯ\ **law** \ȯi\ **boy** \th\ **thin** \t͟h\ **the** \ü\ **loot** \u̇\ **foot** \y\ **yet** \zh\ **vision** *see also* Guide to Pronunciation

plan·ti·grade \'plan-tə-,grād\ *adjective* [French, from Latin *planta* sole + French *-grade*] (1831)
: walking on the sole with the heel touching the ground ⟨humans are *plantigrade*⟩
— **plantigrade** *noun*

plant·ing *noun* (1632)
: an area where plants are grown for commercial or decorative purposes; *also* : the plants cultivated in such an area

plant kingdom *noun* (1884)
: a basic group of natural objects that includes all living and extinct plants — compare ANIMAL KINGDOM, MINERAL KINGDOM

plant·let \'plant-lət\ *noun* (1816)
: a small or young plant

plant louse *noun* (1805)
: APHID; *also* : any of various small insects (as a jumping plant louse) of similar habits

plan·to·cra·cy \plan-'tä-krə-sē\ *noun* [*planter* + *-o-* + *-cracy*] (circa 1846)
1 : a ruling class made up of planters
2 : government by planters

plants·man \'plants-mən, -,man\ *noun* (1881)
: a person skilled with plants : an expert gardener or horticulturist

plan·u·la \'plan-yə-lə\ *noun, plural* **-lae** \-,lē, -,lī\ [New Latin, from Latin *planus* level, flat — more at FLOOR] (1870)
: the very young usually flattened oval or oblong free-swimming ciliated larva of a coelenterate

plaque \'plak\ *noun* [French, from Middle French, metal sheet, from *plaquier* to plate, from Middle Dutch *placken* to piece, patch; akin to Middle Dutch *placke* piece, Middle High German *placke* patch] (1848)
1 a : an ornamental brooch; *especially* : the badge of an honorary order **b :** a flat thin piece (as of metal) used for decoration **c :** a commemorative or identifying inscribed tablet
2 a : a localized abnormal patch on a body part or surface **b :** a film of mucus that harbors bacteria on a tooth **c :** an atherosclerotic lesion **d :** a histopathologic lesion of brain tissue that is characteristic of Alzheimer's disease and consists of a cluster of degenerating nerve endings and dendrites around a core of amyloid
3 : a clear area in a bacterial culture produced by destruction of cells by a virus

¹plash \'plash\ *noun* [probably imitative] (1513)
: SPLASH

²plash (1542)
intransitive verb
: to cause a splashing or spattering effect
transitive verb
: to break the surface of (water) : SPLASH

-plasia *or* **-plasy** *noun combining form* [New Latin *-plasia,* from Greek *plasis* molding, from *plassein*]
: development : formation ⟨hyper*plasia*⟩ ⟨homo*plasy*⟩

plasm \'pla-zəm\ *noun* [Late Latin *plasma* something molded] (1747)
: PLASMA

plasm- *or* **plasmo-** *combining form* [French, from New Latin *plasma*]
: plasma ⟨*plasmo*dium⟩ ⟨*plasmo*lysis⟩

-plasm *noun combining form* [German *-plasma,* from New Latin *plasma*]
: formative or formed material (as of a cell or tissue) ⟨endo*plasm*⟩

plas·ma \'plaz-mə\ *noun* [German, from Late Latin, something molded, from Greek, from *plassein* to mold — more at PLASTER] (1772)
1 : a green faintly translucent quartz
2 [New Latin, from Late Latin] **a :** the fluid part of blood, lymph, or milk as distinguished from suspended material **b :** the juice that can be expressed from muscle
3 : PROTOPLASM
4 : a collection of charged particles (as in the atmospheres of stars or in a metal) containing about equal numbers of positive ions and electrons and exhibiting some properties of a gas but differing from a gas in being a good conductor of electricity and in being affected by a magnetic field
— **plas·mat·ic** \plaz-'ma-tik\ *adjective*

plasma cell *noun* (1888)
: a lymphocyte that is a mature antibody-secreting B cell

plas·ma·gel \'plaz-mə-,jel\ *noun* (1923)
: gelated cytoplasm; *especially* : the outer firm zone of a pseudopodium

plas·ma·gene \-,jēn\ *noun* [International Scientific Vocabulary] (1939)
: an extranuclear determiner of hereditary characteristics with a capacity for replication similar to that of a nuclear gene

plasma jet *noun* (1957)
: a stream of very hot ionized plasma; *also* : a device for producing such a stream

plas·ma·lem·ma \,plaz-mə-'le-mə\ *noun* [New Latin, from *plasma* + Greek *lemma* husk — more at LEMMA] (1923)
: PLASMA MEMBRANE

plasma membrane *noun* (1900)
: a semipermeable limiting layer of cell protoplasm — called also *cell membrane;* see CELL illustration

plas·ma·phe·re·sis \,plaz-mə-fə-'rē-səs, -'fer-ə-səs\ *noun* [New Latin, from *plasm-* + Greek *aphairesis* taking off — more at APHAERESIS] (1914)
: a process for obtaining blood plasma without depleting the donor or patient of other blood constituents (as red blood cells) by separating out the plasma from the whole blood and returning the rest to the donor's or patient's circulatory system

plas·ma·sol \'plaz-mə-,säl, -,sȯl, -,sōl\ *noun* (1923)
: cytoplasm in the form of a sol especially in a pseudopodium or amoeboid cell

plasma torch *noun* (1959)
: a device that heats a gas by electrical means to form a plasma for high-temperature operations (as melting metal)

plas·mid \'plaz-məd\ *noun* [*plasma* + ²*-id*] (1952)
: an extrachromosomal ring of DNA especially of bacteria that replicates autonomously

plas·min \-mən\ *noun* (circa 1866)
: a proteolytic enzyme that dissolves the fibrin of blood clots

plas·min·o·gen \plaz-'mi-nə-jən\ *noun* (1945)
: the precursor of plasmin that is found in blood plasma and serum

plas·mo·des·ma \,plaz-mə-'dez-mə\ *also* **plas·mo·desm** \'plaz-mə-,de-zəm\ *noun, plural* **-des·ma·ta** \-'dez-mə-tə\ *or* **-des·mas** \-'dez-məz\ [New Latin *plasmodesma,* from *plasma* + Greek *desmat-, desma* bond, from *dein* to bind — more at DIADEM] (1905)
: one of the cytoplasmic strands that pass through openings in some plant cell walls and provide living bridges between cells

plas·mo·di·um \plaz-'mō-dē-əm\ *noun, plural* **-dia** \-dē-ə\ [New Latin, from *plasm-* + *-odium* thing resembling, from Greek *-ōdēs* like] (1875)
1 a : a motile multinucleate mass of protoplasm resulting from fusion of uninucleate amoeboid cells; *also* : an organism (as a stage of a slime mold) that consists of such a structure **b :** SYNCYTIUM 1
2 : an individual malaria parasite

plas·mog·a·my \plaz-'mä-gə-mē\ *noun* [International Scientific Vocabulary] (1912)
: fusion of the cytoplasm of two or more cells as distinguished from fusion of nuclei

plas·mol·y·sis \plaz-'mä-lə-səs\ *noun* [New Latin] (1883)
: shrinking of the cytoplasm away from the wall of a living cell due to outward osmotic flow of water
— **plas·mo·lyt·ic** \,plaz-mə-'li-tik\ *adjective*

plas·mo·lyze \'plaz-mə-,līz\ *verb* **-lyzed; -lyz·ing** (1888)
transitive verb
: to subject to plasmolysis
intransitive verb
: to undergo plasmolysis

-plast *noun combining form* [Middle French *-plaste* thing molded, from Late Latin *-plastus,* from Greek *-plastos,* from *plastos* molded, from *plassein*]
: organized particle or granule : cell ⟨chromo*plast*⟩

¹plas·ter \'plas-tər\ *noun* [Middle English, from Old English, from Latin *emplastrum,* from Greek *emplastron,* from *emplassein* to plaster on, from *en-* + *plassein* to mold, plaster; perhaps akin to Latin *planus* level, flat — more at FLOOR] (before 12th century)
1 : a medicated or protective dressing that consists of a film (as of cloth or plastic) spread with a usually medicated substance ⟨adhesive *plaster*⟩; *broadly* : something applied to heal and soothe
2 : a pasty composition (as of lime, water, and sand) that hardens on drying and is used for coating walls, ceilings, and partitions
— **plas·tery** \-t(ə-)rē\ *adjective*

²plaster *verb* **plas·tered; plas·ter·ing** \-t(ə-)riŋ\ (14th century)
transitive verb
1 : to overlay or cover with plaster : COAT
2 : to apply a plaster to
3 a : to cover over or conceal as if with a coat of plaster **b :** to apply as a coating or incrustation **c :** to smooth down with a sticky or shiny substance ⟨*plastered* his hair down⟩
4 : to fasten or apply tightly to another surface
5 : to treat with plaster of paris
6 : to affix to or place on especially conspicuously or in quantity
7 : to inflict heavy damage or loss on especially by a concentrated or unremitting attack
intransitive verb
: to apply plaster
— **plas·ter·er** \-tər-ər\ *noun*

plas·ter·board \'plas-tər-,bōrd, -,bȯrd\ *noun* (1897)
: a board made of several plies of fiberboard, paper, or felt usually bonded to a hardened gypsum plaster core and used especially as wallboard

plaster cast *noun* (1825)
1 : a sculptor's model in plaster of paris
2 : a rigid dressing of gauze impregnated with plaster of paris

plas·tered \'plas-tərd\ *adjective* (1902)
: DRUNK, INTOXICATED

plastering *noun* (15th century)
1 : a coating of or as if of plaster
2 : a decisive defeat

plaster of par·is \-'par-əs\ *often 2d P capitalized* [*Paris,* France] (15th century)
: a white powdery slightly hydrated calcium sulfate $CaSO_4 \cdot \frac{1}{2} H_2O$ or $2CaSO_4 \cdot H_2O$ made by calcining gypsum and used chiefly for casts and molds in the form of a quick-setting paste with water

plaster saint *noun* (1890)
: a person without human failings

plas·ter·work \'plas-tər-,wərk\ *noun* (1600)
: plastering often ornate in design used to finish architectural constructions

¹plas·tic \'plas-tik\ *adjective* [Latin *plasticus* of molding, from Greek *plastikos,* from *plassein* to mold, form] (1632)
1 : FORMATIVE, CREATIVE ⟨*plastic* forces in nature⟩
2 a : capable of being molded or modeled ⟨*plastic* clay⟩ **b :** capable of adapting to varying conditions : PLIABLE ⟨ecologically *plastic* animals⟩
3 : SCULPTURAL
4 : made or consisting of a plastic
5 : capable of being deformed continuously and permanently in any direction without rupture

6 : of, relating to, or involving plastic surgery
7 : having a quality suggestive of mass-produced plastic goods; *especially* **:** ARTIFICIAL 4 ⟨*plastic* smiles⟩ ☆

^{2}**plastic** *noun* (1905)
1 : a plastic substance; *specifically* **:** any of numerous organic synthetic or processed materials that are mostly thermoplastic or thermosetting polymers of high molecular weight and that can be made into objects, films, or filaments
2 : credit cards used for payment
— **plas·ticky** \'plas-ti-kē\ *adjective*

-plastic *adjective combining form* [Greek *-plastikos,* from *plassein*]
1 : developing **:** forming ⟨thrombo*plastic*⟩
2 : of or relating to (something designated by a term ending in *-plasm, -plast, -plasty,* or *-plasy*) ⟨homo*plastic*⟩ ⟨neo*plastic*⟩

plas·ti·cal·ly \'plas-ti-k(ə-)lē\ *adverb* (1835)
1 : in a plastic manner
2 : with respect to plastic qualities

plastic art *noun* (circa 1907)
1 : art (as sculpture or bas-relief) characterized by modeling **:** three-dimensional art
2 : visual art (as painting, sculpture, or film) especially as distinguished from art that is written (as poetry or music) — often used in plural

plastic foam *noun* (1943)
: EXPANDED PLASTIC

plas·ti·cine *also* **plas·ti·cene** \'plas-tə-ˌsēn\ *noun* [from *Plasticine,* a trademark] (1897)
: a plastic paste used for models and sculptures

plas·tic·i·ty \pla-'sti-sə-tē\ *noun* (circa 1783)
1 : the quality or state of being plastic; *especially* **:** capacity for being molded or altered
2 : the ability to retain a shape attained by pressure deformation
3 : the capacity of organisms with the same genotype to vary in developmental pattern, in phenotype, or in behavior according to varying environmental conditions

plas·ti·cize \'plas-tə-ˌsīz\ *transitive verb* **-cized; -ciz·ing** (1919)
1 : to make plastic
2 : to treat with a plastic ⟨a *plasticized* mattress cover⟩
— **plas·ti·ci·za·tion** \ˌplas-tə-sə-'zā-shən\ *noun*

plas·ti·ciz·er \'plas-tə-ˌsī-zər\ *noun* (1925)
: one that plasticizes; *specifically* **:** a chemical added especially to rubbers and resins to impart flexibility, workability, or stretchability

plastic surgeon *noun* (1946)
: a specialist in plastic surgery

plastic surgery *noun* (1842)
: surgery concerned with the repair, restoration, or improvement of lost, injured, defective, or misshapen body parts

plas·tid \'plas-təd\ *noun* [German, from Greek *plastos* molded] (1885)
: any of various cytoplasmic organelles (as an amyloplast or chloroplast) of photosynthetic cells that serve in many cases as centers of special metabolic activities
— **plas·tid·i·al** \pla-'sti-dē-əl\ *adjective*

plas·ti·sol \'plas-tə-ˌsäl, -ˌsōl\ *noun* [*plastic* + ^{4}*sol*] (1946)
: a substance consisting of a mixture of a resin and a plasticizer that can be molded, cast, or made into a continuous film by application of heat

plas·to·cy·a·nin \ˌplas-tō-'sī-ə-nən\ *noun* [Greek *plastos* + English *cyan-* + ^{1}*-in*] (1961)
: a copper-containing protein that acts as an intermediary in photosynthetic electron transport

plas·to·qui·none \ˌplas-(ˌ)tō-kwi-'nōn, -'kwi-ˌ\ *noun* [Greek *plastos* + English *quinone*] (1958)
: a plant substance that is related to vitamin K and plays a role in photosynthetic phosphorylation

plas·tral \'plas-trəl\ *adjective* (1889)
: of or relating to a plastron

plas·tron \'plas-trən\ *noun* [Middle French, from Old Italian *piastrone,* augmentative of *piastra* thin metal plate — more at PIASTRE] (circa 1507)
1 a : a metal breastplate formerly worn under the hauberk **b :** a quilted pad worn in fencing practice to protect the chest, waist, and the side on which the weapon is held
2 : the ventral part of the shell of a tortoise or turtle consisting typically of nine symmetrically placed bones overlaid by horny plates
3 a : a trimming like a bib for a woman's dress **b :** DICKEY 1a
4 : a thin film of air held by water-repellent hairs of some aquatic insects

-plasty *noun combining form* [French *-plastie,* from Late Greek *-plastia* molding, from Greek *-plastēs* molder, from *plassein*]
: plastic surgery ⟨osteo*plasty*⟩

-plasy — see -PLASIA

^{1}**plat** \'plat\ *transitive verb* **plat·ted; plat·ting** [Middle English, alteration of *plaiten*] (14th century)
: PLAIT

^{2}**plat** *noun* (1535)
: PLAIT

^{3}**plat** *noun* [probably alteration of *plot*] (15th century)
1 : a small piece of ground (as a lot or quadrat) **:** PLOT
2 : a plan, map, or chart of a piece of land with actual or proposed features (as lots); *also* **:** the land represented

^{4}**plat** *transitive verb* **plat·ted; plat·ting** (1751)
: to make a plat of

plat·an \'pla-t^{ə}n\ *noun* [Middle English, from Latin *platanus* — more at PLANE] (14th century)
: ^{2}PLANE

plat du jour \ˌplä-də-'zhu̇r, ˌpla-\ *noun, plural* **plats du jour** *same*\ [French, literally, plate of the day] (1906)
: a dish that is featured by a restaurant on a particular day

^{1}**plate** \'plāt\ *noun* [Middle English, from Old French, from *plate,* feminine of *plat* flat, from (assumed) Vulgar Latin *plattus,* probably from Greek *platys* broad, flat — more at PLACE] (14th century)
1 a : a smooth flat thin piece of material **b** (1) **:** forged, rolled, or cast metal in sheets usually thicker than 1/4 inch (6 millimeters) (2) **:** a very thin layer of metal deposited on a surface of base metal by plating **c :** one of the broad metal pieces used in armor; *also* **:** armor of such plates **d** (1) **:** a lamina or plaque (as of bone or horn) that forms part of an animal body; *especially* **:** SCUTE (2) **:** the thin under portion of the forequarter of beef; *especially* **:** the fatty back part — see BEEF illustration **e :** HOME PLATE **f :** any of the large movable segments into which the earth's crust is divided according to the theory of plate tectonics
2 [Middle English; partly from Old French *plate* plate, piece of silver; partly from Old Spanish *plata* silver, from (assumed) Vulgar Latin *platta* metal plate, from feminine of *plattus* flat] **a** *obsolete* **:** a silver coin **b :** precious metal; *especially* **:** silver bullion
3 [Middle English, from Middle French *plat* dish, plate, from *plat* flat] **a :** domestic hollowware made of or plated with gold, silver, or base metals **b :** a shallow usually circular vessel from which food is eaten or served **c** (1) **:** a quantity to fill a plate **:** PLATEFUL (2) **:** a main course served on a plate (3) **:** food and service supplied to one person ⟨a dinner at $10 a *plate*⟩ **d** (1) **:** a prize given to the winner in a contest (2) *British* **:** a horse race in which the contestants compete for a prize of fixed value rather than stakes **e :** a dish or pouch passed in taking collections **f :** a flat glass dish used chiefly for culturing microorganisms
4 a : a prepared surface from which printing is done **b :** a sheet of material (as glass) coated with a light-sensitive photographic emulsion **c** (1) **:** the usually flat or grid-formed anode of an electron tube at which electrons collect (2) **:** a metallic grid with its interstices filled with active material that forms one of the structural units of a battery **d :** LICENSE PLATE
5 : a horizontal structural member that provides bearing and anchorage especially for the trusses of a roof or the rafters
6 : the part of a denture that fits to the mouth; *broadly* **:** DENTURE
7 : a full-page illustration often on different paper from the text pages
8 : a schedule of matters to deal with ⟨have a lot on my *plate* now⟩
— **plate·ful** \-ˌfu̇l\ *noun*
— **plate·like** \-ˌlīk\ *adjective*

^{2}**plate** *transitive verb* **plat·ed; plat·ing** (14th century)
1 : to cover or equip with plate: as **a :** to provide with armor plate **b :** to cover with an adherent layer mechanically, chemically, or electrically; *also* **:** to deposit (as a layer) on a surface
2 : to make a printing surface from or for
3 : to fix or secure with a plate
4 : to cause (as a runner) to score in baseball

^{1}**pla·teau** \pla-'tō, 'pla-ˌ\ *noun, plural* **pla·teaus** *or* **pla·teaux** \-'tōz, -ˌtōz\ [French, from Middle French, platter, from *plat* flat] (1796)
1 a : a usually extensive land area having a relatively level surface raised sharply above adjacent land on at least one side **:** TABLELAND **b :** a similar undersea feature
2 a : a region of little or no change in a graphic representation **b :** a relatively stable level, period, or condition
3 : a level of attainment or achievement ⟨the 500-point *plateau*⟩

^{2}**plateau** *intransitive verb* (1939)
: to reach a level, period, or condition of stability

plate·glass \'plāt-'glas, -ˌglas\ *adjective, often capitalized* (1968)
: of, relating to, or being the British universities founded in the latter half of the twentieth century — compare OXBRIDGE, REDBRICK

plate glass *noun* (circa 1741)
: rolled, ground, and polished sheet glass

plate·let \'plāt-lət\ *noun* (1895)
: a minute flattened body (as of ice or a mineral); *especially* **:** BLOOD PLATELET

plate·mak·er \'plāt-ˌmā-kər\ *noun* (1904)

☆ SYNONYMS
Plastic, pliable, pliant, ductile, malleable, adaptable mean susceptible of being modified in form or nature. PLASTIC applies to substances soft enough to be molded yet capable of hardening into the desired fixed form ⟨*plastic* materials allow the sculptor greater freedom⟩. PLIABLE suggests something easily bent, folded, twisted, or manipulated ⟨*pliable* rubber tubing⟩. PLIANT may stress flexibility and sometimes connote springiness ⟨an athletic shoe with a *pliant* sole⟩. DUCTILE applies to what can be drawn out or extended with ease ⟨*ductile* metals such as copper⟩. MALLEABLE applies to what may be pressed or beaten into shape ⟨the *malleable* properties of gold enhance its value⟩. ADAPTABLE implies the capability of being easily modified to suit other conditions, needs, or uses ⟨computer hardware that is *adaptable*⟩.

: a machine for making printing plates and especially offset printing plates
— **plate·mak·ing** \-kiŋ\ *noun*

plat·en \'pla-tᵊn\ *noun* [Middle French *plateine,* from *plate*] (1541)
1 : a flat plate; *especially* **:** one that exerts or receives pressure (as in a printing press)
2 : the roller of a typewriter or printer

plat·er \'plā-tər\ *noun* (1777)
1 : one that plates
2 a : a horse that runs chiefly in plate races **b :** a racehorse that competes in the lowest grade of races

plate rail *noun* (1902)
: a rail or narrow shelf along the upper part of a wall for holding plates or ornaments

plat·er·esque \ˌpla-tə-'resk\ *adjective, often capitalized* [Spanish *plateresco,* from *platero* silversmith, from *plata* silver] (circa 1864)
: of, relating to, or being a 16th century Spanish architectural style characterized by elaborate ornamentation suggestive of silver plate

plate tectonics *noun plural but singular in construction* (1969)
: a theory in geology: the lithosphere of the earth is divided into a small number of plates which float on and travel independently over the mantle and much of the earth's seismic activity occurs at the boundaries of these plates — compare CONTINENTAL DRIFT

plat·form \'plat-ˌfȯrm\ *noun, often attributive* [Middle French *plate-forme* diagram, map, literally, flat form] (1535)
1 : PLAN, DESIGN
2 : a declaration of the principles on which a group of persons stands; *especially* **:** a declaration of principles and policies adopted by a political party or a candidate
3 a (1) **:** a usually raised horizontal flat surface; *especially* **:** a raised flooring (as for speakers or performers) (2) **:** a device or structure incorporating or providing a platform; *especially* **:** such a structure on legs used for offshore drilling (as for oil) **b :** a place or opportunity for public discussion
4 a : a usually thick layer (as of cork) between the inner sole and outer sole of a shoe **b :** a shoe having such a sole
5 : a vehicle (as a satellite or aircraft) used for a particular activity or purpose or to carry a usually specified kind of equipment

platform balance *noun* (1811)
: a balance having a platform on which objects are weighed — called also *platform scale*

platform rocker *noun* (1944)
: a chair that rocks on a stable platform

platform tennis *noun* (1955)
: a variation of paddle tennis that is played on a platform enclosed by a wire fence

pla·ti·na \plə-'tē-nə\ *noun* [Spanish] (1750)
: PLATINUM

plat·ing \'plā-tiŋ\ *noun* (1831)
1 : the act or process of plating
2 a : a coating of metal plates **b :** a thin coating of metal

plat·i·nize \'pla-tᵊn-ˌīz\ *transitive verb* **-nized; -niz·ing** (1825)
: to cover, treat, or combine with platinum or a compound of platinum

plat·i·no·cy·a·nide \ˌpla-tᵊn-ō-'sī-ə-ˌnīd\ *noun* (1845)
: a fluorescent complex salt formed by the union of a compound of platinum and cyanide with another cyanide

plat·i·num \'plat-nəm, 'pla-tᵊn-əm\ *noun, often attributive* [New Latin, from Spanish *platina,* from diminutive of *plata* silver — more at PLATE] (1812)
1 : a heavy precious grayish white noncorroding ductile malleable metallic element that fuses with difficulty and is used especially in chemical ware and apparatus, as a catalyst, and in dental and jewelry alloys — see ELEMENT table
2 : a moderate gray

platinum blonde *noun* (1931)
1 : a person whose hair is of a pale silvery blonde color
2 : the color of the hair of a platinum blonde

plat·i·tude \'pla-tə-ˌtüd, -ˌtyüd\ *noun* [French, from *plat* flat, dull] (1812)
1 : the quality or state of being dull or insipid
2 : a banal, trite, or stale remark

plat·i·tu·di·nal \ˌpla-tə-'tüd-nəl, -'tyüd-; -'tü-dᵊn-əl, -'tyü-\ *adjective* (1870)
: PLATITUDINOUS

plat·i·tu·di·nar·i·an \-ˌtü-dᵊn-'er-ē-ən, -ˌtyü-\ *noun* (1855)
: one given to the use of platitudes

plat·i·tu·di·nize \-'tü-dᵊn-ˌīz, -'tyü-\ *intransitive verb* **-nized; -niz·ing** [*platitudinous*] (1885)
: to utter platitudes

plat·i·tu·di·nous \-'tüd-nəs, -'tyüd-; -'tü-dᵊn-əs, -'tyü-\ *adjective* [*platitude* + *-inous* (as in *multitudinous*)] (1862)
: having the characteristics of a platitude **:** full of platitudes ⟨*platitudinous* remarks⟩
— **plat·i·tu·di·nous·ly** *adverb*

pla·ton·ic \plə-'tä-nik, plā-\ *adjective* [Latin *platonicus,* from Greek *platōnikos,* from *Platōn* Plato] (1533)
1 *capitalized* **:** of, relating to, or characteristic of Plato or Platonism
2 a : relating to or based on platonic love; *also* **:** experiencing or professing platonic love **b :** of, relating to, or being a relationship marked by the absence of romance or sex
3 : NOMINAL, THEORETICAL
— **pla·ton·i·cal·ly** \-ni-k(ə-)lē\ *adverb*

platonic love *noun, often P capitalized* (1631)
1 : love conceived by Plato as ascending from passion for the individual to contemplation of the universal and ideal
2 : a close relationship between two persons in which sexual desire is nonexistent or has been suppressed or sublimated

Pla·to·nism \'plā-tᵊn-ˌi-zəm\ *noun* (circa 1570)
1 a : the philosophy of Plato stressing especially that actual things are copies of transcendent ideas and that these ideas are the objects of true knowledge apprehended by reminiscence **b :** NEOPLATONISM
2 : PLATONIC LOVE
— **Pla·to·nist** \-tᵊn-ist\ *noun*
— **Pla·to·nis·tic** \ˌplā-tᵊn-'is-tik\ *adjective*

Pla·to·nize \'plā-tᵊn-ˌīz\ *verb* **-nized; -niz·ing** (1608)
intransitive verb
: to adopt, imitate, or conform to Platonic opinions
transitive verb
: to explain in accordance with or adapt to Platonic doctrines; *especially* **:** IDEALIZE

[1]pla·toon \plə-'tün, pla-\ *noun* [French *peloton* small detachment, literally, ball, from Middle French *pelote* little ball — more at PELLET] (1637)
1 : a subdivision of a company-size military unit normally consisting of two or more squads or sections
2 : a group of persons sharing a common characteristic or activity ⟨a *platoon* of waiters⟩; *especially* **:** a group of football players who are trained for either offense or defense and are sent into or withdrawn from the game as a body

[2]platoon (1963)
transitive verb
: to play (one player) alternately with another player in the same position (as on a baseball team)
intransitive verb
1 : to alternate with another player at the same position
2 : to use alternate players at the same position

platoon sergeant *noun* (1915)
: a noncommissioned officer in the army ranking above a staff sergeant and below a first sergeant

Platt·deutsch \'plat-ˌdȯich, 'plät-\ *noun* [German, from Dutch *Platduitsch,* literally, Low German, from *plat* flat, low + *duitsch* German] (1677)
: the Low German speech of northern Germany comprising several dialects

plat·ter \'pla-tər\ *noun* [Middle English *plater,* from Anglo-French, from Middle French *plat* plate] (13th century)
1 a : a large plate used especially for serving meat **b :** PLATE 3c(2)
2 : a phonograph record
— **plat·ter·ful** \-ˌfu̇l\ *noun*
— **on a platter :** without effort **:** very easily ⟨can have the presidency *on a platter* —Jonathan Daniels⟩

[1]platy \'plā-tē\ *adjective* (1533)
: resembling a plate; *also* **:** consisting of plates or flaky layers — used chiefly of soil or mineral formations

[2]platy \'pla-tē\ *noun, plural* **platy** *or* **plat·ys** *or* **plat·ies** [New Latin *Platypoecilus,* former genus name of the fish] (1931)
: either of two live-bearers (*Xiphophorus maculatus* and *X. variatus* of the family Poeciliidae) that are popular aquarium fishes and are noted for variability and brilliant color — called also *platy·fish* \-ˌfish\

platy·hel·minth \ˌpla-ti-'hel-ˌmin(t)th\ *noun* [ultimately from Greek *platys* broad, flat + *helminth-, helmis* helminth] (circa 1890)
: any of a phylum (Platyhelminthes) of soft-bodied usually much flattened acoelomate worms (as the planarians, flukes, and tapeworms) — called also *flatworm*
— **platy·hel·min·thic** \-hel-'min(t)-thik, -'min-tik\ *adjective*

platy·pus \'pla-ti-pəs, -ˌpu̇s\ *noun, plural* **platy·pus·es** *also* **platy·pi** \-ˌpī, -ˌpē\ [New Latin, from Greek *platypous* flat-footed, from *platys* broad, flat + *pous* foot — more at PLACE, FOOT] (1799)
: a small carnivorous aquatic oviparous mammal (*Ornithorhynchus anatinus*) of eastern Australia and Tasmania that has a fleshy bill resembling that of a duck, dense fur, webbed feet, and a broad flattened tail

platypus

plat·yr·rhine \'pla-ti-ˌrīn\ *adjective* (1857)
1 [New Latin *Platyrrhina,* from Greek *platyrrhin-, platyrrhis* broad-nosed, from *platys* + *rhin-, rhis* nose] **:** of, relating to, or being any of a division (Platyrrhina) of monkeys all of which are New World monkeys and are characterized by a broad nasal septum, usually 36 teeth, and often a prehensile tail
2 [Greek *platyrrhin-, platyrrhis*] **:** having a short broad nose
— **platyrrhine** *noun*

plau·dit \'plȯ-dət\ *noun* [Latin *plaudite* applaud, plural imperative of *plaudere* to applaud] (1624)
1 : an act or round of applause
2 : enthusiastic approval — usually used in plural ⟨received the *plaudits* of the critics⟩

plau·si·bil·i·ty \ˌplȯ-zə-'bi-lə-tē\ *noun, plural* **-ties** (1649)
1 : the quality or state of being plausible
2 : something plausible

plau·si·ble \'plȯ-zə-bəl\ *adjective* [Latin *plausibilis* worthy of applause, from *plausus,* past participle of *plaudere*] (1565)
1 : superficially fair, reasonable, or valuable but often specious ⟨a *plausible* pretext⟩

2 : superficially pleasing or persuasive ⟨a swindler . . . , then a quack, then a smooth, *plausible* gentleman —R. W. Emerson⟩
3 : appearing worthy of belief ⟨the argument was both powerful and *plausible*⟩
— plau·si·ble·ness *noun*
— plau·si·bly \-blē\ *adverb*

plau·sive \ˈplȯ-ziv, -siv\ *adjective* [Latin *plausus,* past participle] (1600)
1 : manifesting praise or approval
2 *obsolete* **:** PLEASING
3 *archaic* **:** SPECIOUS

[1]play \ˈplā\ *noun* [Middle English, from Old English *plega;* akin to Old English *plegan* to play, Middle Dutch *pleyen*] (before 12th century)
1 a : SWORDPLAY **b** *archaic* **:** GAME, SPORT **c :** the conduct, course, or action of a game **d :** a particular act or maneuver in a game: as (1) **:** the action during an attempt to advance the ball in football (2) **:** the action in which a player is put out in baseball **e :** the action in which cards are played after bidding in a card game **f :** the moving of a piece in a board game (as chess) **g :** one's turn in a game ⟨it's your *play*⟩
2 a *obsolete* **:** SEXUAL INTERCOURSE **b :** amorous flirtation **:** DALLIANCE
3 a : recreational activity; *especially* **:** the spontaneous activity of children **b :** absence of serious or harmful intent **:** JEST ⟨said it in *play*⟩ **c :** the act or an instance of playing on words or speech sounds **d :** GAMING, GAMBLING
4 a (1) **:** an act, way, or manner of proceeding **:** MANEUVER ⟨that was a *play* to get your fingerprints —Erle Stanley Gardner⟩ (2) **:** DEAL, VENTURE **b** (1) **:** OPERATION, ACTIVITY ⟨other motives surely come into *play* —M. R. Cohen⟩ (2) **:** brisk, fitful, or light movement ⟨the gem presented a dazzling *play* of colors⟩ (3) **:** free or unimpeded motion (as of a part of a machine); *also* **:** the length or measure of such motion (4) **:** scope or opportunity for action
5 : emphasis or publicity especially in the news media ⟨wished the country received a better *play* in the American press —Hugh MacLennan⟩
6 : a move or series of moves calculated to arouse friendly feelings — usually used with *make* ⟨made a big *play* for the girl —Will Herman⟩
7 a : the stage representation of an action or story **b :** a dramatic composition **:** DRAMA
synonym see FUN
— in play : in condition or position to be legitimately played
— out of play : not in play

[2]play (before 12th century)
intransitive verb
1 a : to engage in sport or recreation **:** FROLIC **b :** to have sexual relations; *especially* **:** to have promiscuous or illicit sexual relations — usually used in the phrase *play around* **c** (1) **:** to move aimlessly about **:** TRIFLE (2) **:** to toy or fiddle around with something ⟨*played* with her food⟩ (3) **:** to deal or behave frivolously or mockingly **:** JEST (4) **:** to deal in a light, speculative, or sportive manner (5) **:** to make use of double meaning or of the similarity of sound of two words for stylistic or humorous effect
2 a : to take advantage ⟨*playing* on fears⟩ **b** (1) **:** FLUTTER, FRISK (2) **:** to move or operate in a lively, irregular, or intermittent manner **c :** to move or function freely within prescribed limits **d :** to produce a stream ⟨hoses *playing* on a fire⟩
3 a (1) **:** to perform music ⟨*play* on a violin⟩ (2) **:** to sound in performance ⟨the organ is *playing*⟩ (3) **:** to emit sounds ⟨the radio is *playing*⟩ (4) **:** to reproduce recorded sounds ⟨a record is *playing*⟩ **b** (1) **:** to act in a dramatic production (2) **:** SHOW, RUN ⟨what's *playing* at the theater⟩ **c :** to be suitable for dramatic performance **d :** to act with special consideration so as to gain favor, approval, or sympathy ⟨might *play* to popular prejudices to serve his political ends —V. L. Parrington⟩ — often used in the phrase *play up to*
4 a : to engage or take part in a game **b :** to perform in a position in a specified manner ⟨the outfielders were *playing* deep⟩ **c :** to perform an action during one's turn in a game **d :** GAMBLE **e** (1) **:** to behave or conduct oneself in a specified way ⟨*play* safe⟩ (2) **:** to feign a specified state or quality ⟨*play* dead⟩ (3) **:** to take part in or assent to some activity **:** COOPERATE ⟨*play* along with his scheme⟩ (4) **:** to act so as to prove advantageous to another — usually used in the phrase *play into the hands of*
5 : to gain approval **:** GO OVER ⟨her idea did not *play* well⟩
transitive verb
1 a (1) **:** to engage in or occupy oneself with ⟨*play* baseball⟩ (2) **:** to engage in (an activity) as a game (3) **:** to deal with, handle, or manage (4) **:** EXPLOIT, MANIPULATE **b :** to pretend to engage in the activities of ⟨*play* war⟩ ⟨children *playing* house⟩ **c** (1) **:** to amount to by one's efforts ⟨*played* an important role in their success⟩ (2) **:** to perform or execute for amusement or to deceive or mock ⟨*play* a trick⟩ (3) **:** WREAK ⟨*play* havoc⟩
2 a (1) **:** to put on a performance of (a play) (2) **:** to act in the character or part of (3) **:** to act or perform in ⟨*played* leading theaters⟩ **b :** to perform or act the part of ⟨*play* the fool⟩
3 a (1) **:** to contend against in or as if in a game (2) **:** to use as a contestant in a game ⟨the coach did not *play* him⟩ (3) **:** to perform the duties associated with (a certain position) ⟨*played* quarterback⟩ (4) **:** to guard or move into position to defend against (an opponent) in a specified manner **b** (1) **:** to wager in a game **:** STAKE (2) **:** to make wagers on ⟨*play* the races⟩ (3) **:** to speculate on or in ⟨*play* the stock market⟩ (4) **:** to operate on the basis of ⟨*play* a hunch⟩ **c :** to put into action in a game; *especially* **:** to remove (a playing card) from one's hand and place usually faceup on a table in one's turn either as part of a scoring combination or as one's contribution to a trick **d :** to catch or pick up (a batted ball) **:** FIELD ⟨*played* the ball bare-handed⟩ **e :** to direct the course of (as a ball) **:** HIT ⟨*played* a wedge shot to the green⟩; *also* **:** to cause (a ball or puck) to rebound ⟨*played* the ball off the backboard⟩
4 a : to perform (music) on an instrument ⟨*play* a waltz⟩ **b :** to perform music on ⟨*play* the violin⟩ **c :** to perform music of (a certain composer) **d** (1) **:** to cause (as a radio or phonograph) to emit sounds (2) **:** to cause the recorded sound or image of (as a record or a magnetic tape) to be reproduced
5 a : WIELD, PLY **b :** to discharge, fire, or set off with continuous effect ⟨*played* the hose on the burning building⟩ **c :** to cause to move or operate lightly and irregularly or intermittently **d :** to keep (a hooked fish) in action
— play·abil·i·ty \ˌplā-ə-ˈbi-lə-tē\ *noun*
— play·able \ˈplā-ə-bəl\ *adjective*
— play ball : COOPERATE
— play both ends against the middle : to set opposing interests against each other to one's own ultimate profit
— play by ear : to deal with something without previous planning or instructions
— play games : to try to hide the truth from someone by deceptive means
— play one's cards : to act with the means available to one
— play possum : to pretend to be asleep or dead
— play second fiddle : to take a subordinate position
— play the field : to date or have romantic connections with more than one person
— play the game : to act according to a code or set of standards
— play with fire : to do something risky or dangerous
— play with oneself : MASTURBATE

pla·ya \ˈplī-ə\ *noun* [Spanish, literally, beach, from Late Latin *plagia* — more at PLAGE] (1854)
: the flat-floored bottom of an undrained desert basin that becomes at times a shallow lake

play·act \ˈplā-ˌakt\ *verb* [back-formation from *playacting*] (1901)
transitive verb
: ACT OUT 1a
intransitive verb
1 a : to take part in theatrical performances especially as a professional **b :** to make believe
2 : to engage in theatrical or insincere behavior
— play·act·ing *noun*

play–action pass *noun* (1964)
: a pass play in football in which the quarterback fakes a handoff before passing the ball — called also *play-action*

play·back \ˈplā-ˌbak\ *noun* (1929)
: an act or instance of reproducing recorded sound or pictures often immediately after recording

play back *transitive verb* (1949)
: to perform a playback of (a usually recently recorded disc or tape)

play·bill \ˈplā-ˌbil\ *noun* (1616)
: a bill advertising a play and usually announcing the cast

Playbill *trademark*
— used for a theater program

play·book \-ˌbu̇k\ *noun* (1535)
1 : one or more plays in book form
2 : a notebook containing diagrammed football plays

play·boy \-ˌbȯi\ *noun* (1907)
: a man who lives a life devoted chiefly to the pursuit of pleasure

play–by–play \ˈpla-ˌbī-ˌplā, ˌplā-ˌbī-ˈ\ *adjective* (1931)
1 : being or giving a running commentary on a sports event
2 : relating each event as it occurs
— play–by–play *noun*

play down *transitive verb* (1930)
: to attach little importance to **:** MINIMIZE

played out *adjective* (circa 1859)
1 : worn out or used up
2 : tired out **:** SPENT

play·er \ˈplā-ər\ *noun* (14th century)
: one that plays: as **a :** a person who plays a game **b :** MUSICIAN **c :** ACTOR **d :** a device that reproduces recorded material (as video images or music) from a usually specified medium **e :** one actively involved especially in a competitive field or process **:** PARTICIPANT

player piano *noun* (1907)
: a piano containing a mechanism by which it plays automatically

play·fel·low \ˈplā-ˌfe-(ˌ)lō\ *noun* (1513)
: PLAYMATE

play·field \-ˌfēld\ *noun* (1883)
: a playground for outdoor athletics

play·ful \ˈplā-fəl\ *adjective* (13th century)
1 : full of play **:** FROLICSOME, SPORTIVE ⟨a *playful* kitten⟩
2 : HUMOROUS, JOCULAR
— play·ful·ly \-fə-lē\ *adverb*
— play·ful·ness *noun*

play·girl \-ˌgər(-ə)l\ *noun* (1938)
: a woman who lives a life devoted chiefly to the pursuit of pleasure

play·go·er \-ˌgō(-ə)r\ *noun* (1822)
: a person who frequently attends plays

play·ground \-ˌgrau̇nd\ *noun* (1794)
1 : a piece of land used for and usually equipped with facilities for recreation especially by children

2 : an area known or suited for activity of a specified sort ⟨a vacation *playground*⟩

play·house \-ˌhau̇s\ *noun* (before 12th century)
1 : THEATER
2 : a small house for children to play in

playing card *noun* (1543)
: one of a set of 24 to 78 thin rectangular pieces of paperboard or plastic marked on one side to show its rank and suit and used in playing any of numerous games

playing field *noun* (circa 1584)
1 : a field for various games; *especially* **:** the part of a field officially marked off for play
2 : a set of conditions for competition (as in business) — usually used in such phrases as *a level playing field*

play·land \ˈplā-ˌland\ *noun* (1918)
: PLAYGROUND

play·let \-lət\ *noun* (1884)
: a short play

play·list \-ˌlist\ *noun* (1972)
: a list of recordings to be played on the air by a radio station

play·mak·er \-ˌmā-kər\ *noun* (circa 1942)
: a player who leads the offense for a team (as in basketball or hockey)
— play·mak·ing \-kiŋ\ *noun or adjective*

play·mate \-ˌmāt\ *noun* (1642)
: a companion in play

play–off \ˈplā-ˌȯf\ *noun* (1895)
1 : a final contest or series of contests to determine the winner between contestants or teams that have tied
2 : a series of contests played after the end of the regular season to determine a championship

play off *transitive verb* (1807)
1 a : to set in opposition for one's own gain **b :** to set in contrast
2 : to complete the playing of (an interrupted contest)
3 : to break (a tie) by a play-off

play out (1596)
transitive verb
1 a : to perform to the end ⟨*play out* a role⟩ **b :** USE UP, FINISH
2 : UNREEL, UNFOLD ⟨*played out* a length of line —Gordon Webber⟩
intransitive verb
: to become spent or exhausted

play·pen \ˈplā-ˌpen\ *noun* (1931)
: a portable usually collapsible enclosure in which a baby or young child may play

play·room \-ˌrüm, -ˌru̇m\ *noun* (1819)
: RUMPUS ROOM

play·suit \-ˌsüt\ *noun* (1908)
: a sports and play outfit for women or children that consists usually of a blouse and shorts

play therapy *noun* (1939)
: psychotherapy in which a child is encouraged to reveal feelings and conflicts in play rather than by verbalization

play·thing \ˈplā-ˌthiŋ\ *noun* (1675)
: TOY

play·time \-ˌtīm\ *noun* (1661)
: a time for play or diversion

play up *transitive verb* (1909)
: EMPHASIZE; *also* **:** EXAGGERATE, OVEREMPHASIZE

play·wear \ˈplā-ˌwar, -ˌwer\ *noun* (1964)
: informal clothing worn for leisure activities

play·wright \ˈplā-ˌrīt\ *noun* [[1]*play* + obsolete *wright* maker — more at WRIGHT] (1687)
: a person who writes plays

play·writ·ing *also* **play·wright·ing** \-ˌrī-tiŋ\ *noun* (1809)
: the writing of plays

pla·za \ˈpla-zə, ˈplä-\ *noun* [Spanish, from Latin *platea* broad street — more at PLACE] (1683)
1 a : a public square in a city or town **b :** an open area usually located near urban buildings and often featuring walkways, trees and shrubs, places to sit, and sometimes shops
2 : a place on a thoroughfare (as a turnpike) at which all traffic must temporarily stop (as to pay tolls)
3 : an area adjacent to an expressway which has service facilities (as a restaurant, service station, and rest rooms)
5 : SHOPPING CENTER

plea \ˈplē\ *noun* [Middle English *plaid, plai,* from Old French *plait, plaid,* from Medieval Latin *placitum,* from Latin, decision, decree, from neuter of *placitus,* past participle of *placēre* to please, be decided — more at PLEASE] (13th century)
1 : a legal suit or action
2 : an allegation made by a party in support of a cause: as **a :** an allegation of fact — compare DEMURRER **b** (1) **:** a defendant's answer to a plaintiff's declaration in common-law practice (2) **:** an accused person's answer to a charge or indictment in criminal practice **c :** a plea of guilty to an indictment
3 : something offered by way of excuse or justification ⟨left early with the *plea* of a headache⟩
4 : an earnest entreaty **:** APPEAL ⟨their *plea* for understanding must be answered⟩
synonym see APOLOGY

plea bargaining *noun* (1964)
: the negotiation of an agreement between a prosecutor and a defendant whereby the defendant is permitted to plead guilty to a reduced charge
— plea–bargain *intransitive verb*
— plea bargain *noun*

pleach \ˈplēch, ˈplāch\ *transitive verb* [Middle English *plechen,* from Old North French *plechier,* from (assumed) Vulgar Latin *plactiare,* alteration of *plectere* to braid — more at PLY] (14th century)
: INTERLACE, PLAIT

plead \ˈplēd\ *verb* **plead·ed** \ˈplē-dəd\ *or* **pled** *also* **plead** \ˈpled\; **plead·ing** [Middle English *plaiden* to institute a lawsuit, from Middle French *plaidier,* from *plaid* plea] (13th century)
intransitive verb
1 : to argue a case or cause in a court of law
2 a : to make an allegation in an action or other legal proceeding; *especially* **:** to answer the previous pleading of the other party by denying facts therein stated or by alleging new facts **b :** to conduct pleadings
3 : to make a plea of a specified nature ⟨*plead* not guilty⟩
4 a : to argue for or against a claim **b :** to entreat or appeal earnestly
transitive verb
1 : to maintain (as a case or cause) in a court of law or other tribunal
2 : to allege in or by way of a legal plea
3 : to offer as a plea usually in defense, apology, or excuse
— plead·able \ˈplē-də-bəl\ *adjective*
— plead·er *noun*
— plead·ing·ly \ˈplē-diŋ-lē\ *adverb*

pleading *noun* (14th century)
1 : advocacy of a cause in a court of law
2 a : one of the formal usually written allegations and counter allegations made alternately by the parties in a legal action or proceeding **b :** the action or process performed by the parties in presenting such formal allegations until a single point at issue is produced **c :** the introduction of one of these allegations and especially the first one **d :** the body of rules according to which these allegations are framed
3 : the act or an instance of making a plea
4 : a sincere entreaty

pleas·ance \ˈple-zᵊn(t)s\ *noun* (14th century)
1 : a feeling of pleasure **:** DELIGHT
2 : a pleasant rest or recreation place usually attached to a mansion

pleas·ant \ˈple-zᵊnt\ *adjective* [Middle English *plesaunt,* from Middle French *plaisant,* from present participle of *plaisir*] (14th century)
1 : having qualities that tend to give pleasure **:** AGREEABLE ⟨a *pleasant* day⟩
2 : having or characterized by pleasing manners, behavior, or appearance
— pleas·ant·ly *adverb*
— pleas·ant·ness *noun*

pleas·ant·ry \-zᵊn-trē\ *noun, plural* **-ries** (1655)
1 : an agreeable playfulness in conversation **:** BANTER
2 : a humorous act or remark **:** JEST
3 : a polite social remark ⟨exchanged *pleasantries*⟩

[1]**please** \ˈplēz\ *verb* **pleased; pleas·ing** [Middle English *plesen,* from Middle French *plaisir,* from Latin *placēre;* akin to Latin *placare* to placate and perhaps to Greek *plak-, plax* flat surface — more at FLUKE] (14th century)
intransitive verb
1 : to afford or give pleasure or satisfaction
2 : LIKE, WISH ⟨do as you *please*⟩
3 *archaic* **:** to have the kindness ⟨will you *please* to enter the carriage —Charles Dickens⟩
transitive verb
1 : to give pleasure to **:** GRATIFY
2 : to be the will or pleasure of ⟨may it *please* your Majesty⟩
— pleas·er \ˈplē-zər\ *noun*

[2]**please** *adverb* (1667)
1 — used as a function word to express politeness or emphasis in a request ⟨*please* come in⟩
2 — used as a function word to express polite affirmation ⟨have some tea? *Please*⟩

pleas·ing \ˈplē-ziŋ\ *adjective* (14th century)
: giving pleasure **:** AGREEABLE ⟨the sun's *pleasing* warmth⟩
— pleas·ing·ly \-lē\ *adverb*
— pleas·ing·ness *noun*

plea·sur·able \ˈplezh-rə-bəl, ˈplāzh-; ˈple-zhə-, ˈplā-\ *adjective* (1579)
: PLEASANT, GRATIFYING
— plea·sur·abil·i·ty \ˌplezh-rə-ˈbi-lə-tē, ˌplāzh-; ˌple-zhə-, ˌplā-\ *noun*
— plea·sur·able·ness \ˈplezh-rə-bəl-nəs, ˈplāzh-; ˈple-zhə-, ˈplā-\ *noun*
— plea·sur·ably \-blē\ *adverb*

[1]**plea·sure** \ˈple-zhər, ˈplā-\ *noun* [Middle English *plesure,* alteration of *plesir,* from Middle French *plaisir,* from *plaisir* to please] (14th century)
1 : DESIRE, INCLINATION ⟨wait upon his *pleasure* —Shakespeare⟩
2 : a state of gratification
3 a : sensual gratification **b :** frivolous amusement
4 : a source of delight or joy

[2]**pleasure** *verb* **plea·sured; plea·sur·ing** \ˈplezh-riŋ, ˈplāzh-; ˈple-zhə-, ˈplā-\ (1537)
transitive verb
1 : to give pleasure to **:** GRATIFY
2 : to give sexual pleasure to
intransitive verb
1 : to take pleasure **:** DELIGHT
2 : to seek pleasure

pleasure dome *noun* (1797)
: a place of pleasurable entertainment or recreation **:** RESORT

plea·sure·less \ˈple-zhər-ləs, ˈplā-\ *adjective* (1814)
: giving no pleasure

pleasure principle *noun* (1912)
: a tendency for individual behavior to be directed toward immediate satisfaction of instinctual drives and immediate relief from pain or discomfort

[1]**pleat** \ˈplēt\ *transitive verb* [Middle English *pleten,* from *pleit, plete* plait] (14th century)
1 : FOLD; *especially* **:** to arrange in pleats ⟨*pleat* a skirt⟩
2 : PLAIT 2
— pleat·er *noun*

[2]**pleat** *noun* [Middle English *plete*] (15th century)

: a fold in cloth made by doubling material over on itself; *also* : something resembling such a fold
— **pleat•less** \-ləs\ *adjective*
pleb \'pleb\ *noun* (1865)
: PLEBEIAN
plebe \'plēb\ *noun* [obsolete *plebe* common people, from French *plèbe,* from Latin *plebs*] (1833)
: a freshman at a military or naval academy
[1]ple•be•ian \pli-'bē-ən\ *noun* [Latin *plebeius* of the common people, from *plebs* common people] (1533)
1 : a member of the Roman plebs
2 : one of the common people
— **ple•be•ian•ism** \-ə-,ni-zəm\ *noun*
[2]plebeian *adjective* (1566)
1 : of or relating to plebeians
2 : crude or coarse in manner or style **:** COMMON
— **ple•be•ian•ly** *adverb*
pleb•i•scite \'ple-bə-,sīt, -sət *also* -,sēt\ *noun* [Latin *plebis scitum* law voted by the comitia, literally, decree of the common people] (1860)
: a vote by which the people of an entire country or district express an opinion for or against a proposal especially on a choice of government or ruler
— **ple•bi•sci•ta•ry** \ple-'bi-sə-,ter-ē, pli-; ,ple-bə-'sī-tə-rē\ *adjective*
plebs \'plebz, 'pleps\ *noun, plural* **ple•bes** \'plē-(,)bēz, 'plā-,bās\ [Latin] (1647)
1 : the general populace
2 : the common people of ancient Rome
ple•cop•ter•an \pli-'käp-tə-rən\ *noun* [New Latin *Plecoptera,* group name, from Greek *plekein* to braid + *pteron* wing — more at PLY, FEATHER] (circa 1890)
: STONE FLY
— **plecopteran** *adjective*
plec•trum \'plek-trəm\ *noun, plural* **plec•tra** \-trə\ *or* **plectrums** [Latin, from Greek *plēktron,* from *plēssein* to strike — more at PLAINT] (1626)
: [3]PICK 2c
[1]pledge \'plej\ *noun* [Middle English, security, from Middle French *plege,* from Late Latin *plebium,* from (assumed) Late Latin *plebere* to pledge, probably of Germanic origin; akin to Old High German *pflegan* to take care of — more at PLIGHT] (14th century)
1 a : a bailment of a chattel as security for a debt or other obligation without involving transfer of title **b :** the chattel so delivered **c :** the contract incidental to such a bailment
2 a : the state of being held as a security or guaranty **b :** something given as security for the performance of an act
3 : a token, sign, or earnest of something else
4 : a gage of battle
5 : TOAST 3
6 a : a binding promise or agreement to do or forbear **b** (1) **:** a promise to join a fraternity, sorority, or secret society (2) **:** a person who has so promised
[2]pledge *transitive verb* **pledged; pledg•ing** (15th century)
1 : to make a pledge of; *especially* **:** PAWN
2 : to drink to the health of
3 : to bind by a pledge
4 : to promise the performance of by a pledge
— **pledg•er** \'ple-jər\ *or* **pled•gor** \'ple-jər, ple-'jȯr\ *noun*
pledg•ee \ple-'jē\ *noun* (1766)
: one to whom a pledge is given
pled•get \'ple-jət\ *noun* [origin unknown] (circa 1540)
: a compress or pad used to apply medication to or absorb discharges (as from a wound)
-plegia *noun combining form* [New Latin, from Greek *-plēgia,* from *plēssein* to strike — more at PLAINT]
: paralysis ⟨di*plegia*⟩
ple•iad \'plē-əd, 'plā-, -,ad, *chiefly British* 'plī-\ *noun* [French *Pléiade,* group of seven 16th century French poets, from Middle French, group of seven tragic poets of ancient Alexandria, from Greek *Pleiad-, Pleias,* from singular of *Pleiades*] (circa 1839)
: a group of usually seven illustrious or brilliant persons or things
Pleiad *noun*
: any of the Pleiades
Ple•ia•des \'plē-ə-,dēz, 'plā-, *chiefly British* 'plī-\ *noun plural* [Latin, from Greek]
1 : the seven daughters of Atlas turned into a group of stars in Greek mythology
2 : a conspicuous cluster of stars in the constellation Taurus that includes six stars in the form of a very small dipper
plein air \,plān-'ar, -'er; ple-'nar, -'ner\ *adjective* [French, open air] (1894)
1 : of or relating to painting in outdoor daylight
2 : of or relating to a branch of impressionism that attempts to represent outdoor light and air
— **plein•air•ism** \plā-'nar-,i-zəm, ple-, -'ner-\ *noun*
— **plein•air•ist** \-ist\ *noun*
pleio- *or* **pleo-** *or* **plio-** *combining form* [Greek *pleiōn, pleōn* — more at PLUS]
: more ⟨*pleio*tropic⟩ ⟨*pleo*morphic⟩ ⟨*Plio*cene⟩
pleio•tro•pic \,plī-ə-'trō-pik, -'trä-\ *adjective* (1938)
: producing more than one genic effect; *specifically* **:** having multiple phenotypic expressions ⟨a *pleiotropic* gene⟩
— **plei•ot•ro•py** \plī-'ä-trə-pē\ *noun*
Pleis•to•cene \'plīs-tə-,sēn\ *adjective* [Greek *pleistos* most + International Scientific Vocabulary *-cene;* akin to Greek *pleiōn* more] (1839)
: of, relating to, or being the earlier epoch of the Quaternary or the corresponding system of rocks — see GEOLOGIC TIME table
— **Pleistocene** *noun*
ple•na•ry \'plē-nə-rē, 'ple-\ *adjective* [Middle English, from Late Latin *plenarius,* from Latin *plenus* full — more at FULL] (15th century)
1 : complete in every respect **:** ABSOLUTE, UNQUALIFIED ⟨*plenary* power⟩
2 : fully attended or constituted by all entitled to be present ⟨a *plenary* session⟩
synonym see FULL
plenary indulgence *noun* (1675)
: a remission of the entire temporal punishment for sin
ple•nip•o•tent \pli-'ni-pə-tənt\ *adjective* [Late Latin *plenipotent-, plenipotens,* from Latin *plenus* + *potent-, potens* powerful — more at POTENT] (1658)
: PLENIPOTENTIARY
[1]plen•i•po•ten•tia•ry \,ple-nə-pə-'ten(t)-sh(ə-)rē, -shē-,er-ē\ *adjective* [Medieval Latin *plenipotentiarius,* adjective & noun, from Late Latin *plenipotent-, plenipotens*] (circa 1645)
1 : invested with full power
2 : of or relating to a plenipotentiary
[2]plenipotentiary *noun, plural* **-ries** (circa 1656)
: a person and especially a diplomatic agent invested with full power to transact business
plen•ish \'ple-nish\ *transitive verb* [Middle English (Scots) *plenyssen* to fill up, from Middle French *pleniss-,* stem of *plenir,* from *plen* full, from Latin *plenus*] (1513)
chiefly British **:** EQUIP
plen•i•tude \'ple-nə-,tüd, -,tyüd\ *noun* [Middle English *plenitude,* from Middle French or Latin; Middle French, from Latin *plenitudo,* from *plenus*] (15th century)
1 : the quality or state of being full **:** COMPLETENESS
2 : a great sufficiency **:** ABUNDANCE
plen•i•tu•di•nous \,ple-nə-'tüd-nəs, -'tyüd-; -'tü-dᵊn-əs, -'tyü-\ *adjective* [Latin *plenitudin-, plenitudo*] (1895)
: characterized by plenitude
plen•te•ous \'plen-tē-əs\ *adjective* [Middle English *plentevous, plenteous,* from Middle French *plentiveus,* from *plentif* abundant, from *plenté* plenty] (14th century)
1 : FRUITFUL, PRODUCTIVE ⟨a *plenteous* harvest —J. G. Frazer⟩ — usually used with *in* or *of* ⟨the seasons had been *plenteous* in corn —George Eliot⟩
2 : constituting or existing in plenty ⟨*plenteous* grace with thee is found —Charles Wesley⟩
— **plen•te•ous•ly** *adverb*
— **plen•te•ous•ness** *noun*
plen•ti•ful \'plen-ti-fəl\ *adjective* (15th century)
1 : containing or yielding plenty ⟨a *plentiful* land⟩
2 : characterized by, constituting, or existing in plenty ☆
— **plen•ti•ful•ly** \-fə-lē\ *adverb*
— **plen•ti•ful•ness** *noun*
plen•ti•tude \'plen-tə-,tüd, -,tyüd\ *noun* [by alteration (influenced by *plenty*)] (1615)
: PLENITUDE
[1]plen•ty \'plen-tē\ *noun* [Middle English *plente,* from Old French *plenté,* from Late Latin *plenitat-, plenitas,* from Latin, fullness, from *plenus* full — more at FULL] (13th century)
1 a : a full or more than adequate amount or supply ⟨had *plenty* of time to finish the job⟩ **b :** a large number or amount ⟨in *plenty* of trouble⟩
2 : the quality or state of being copious **:** PLENTIFULNESS
[2]plenty *adjective* (14th century)
1 : plentiful in amount, number, or supply ⟨if reasons were as *plenty* as blackberries —Shakespeare⟩
2 : AMPLE ⟨*plenty* work to be done —*Time*⟩ ■
[3]plenty *adverb* (1842)
: more than sufficiently **:** to a considerable degree ⟨the nights were *plenty* cold —F. B. Gipson⟩ ■

☆ SYNONYMS
Plentiful, ample, abundant, copious mean more than sufficient without being excessive. PLENTIFUL implies a great or rich supply ⟨peaches are *plentiful* this summer⟩. AMPLE implies a generous sufficiency to satisfy a particular requirement ⟨*ample* food to last the winter⟩. ABUNDANT suggests an even greater or richer supply than does PLENTIFUL ⟨streams *abundant* with fish⟩. COPIOUS stresses largeness of supply rather than fullness or richness ⟨*copious* examples of bureaucratic waste⟩.

□ USAGE
[2]plenty Many commentators object to the use of adjective sense 2 in writing; it appears to be limited chiefly to spoken English. Sense 1 is literary but is no longer in common use.

[3]plenty Many handbooks advise avoiding the adverb *plenty* in writing; "use *very, quite,* or a more precise word," they advise. Actually *plenty* is often a more precise word than its recommended replacements; *very, fully,* or *quite* will not work as well in these typical quotations ⟨it's already *plenty* hot for us in the kitchen without some dolt opening the oven —C. H. Bridges⟩ ⟨may not be rising quite as rapidly as other health costs, but it is going up *plenty* fast —*Changing Times*⟩. It is not used in more formal writing.

ple·num \'ple-nəm, 'plē-\ *noun, plural* **-nums** *or* **-na** \-nə\ [New Latin, from Latin, neuter of *plenus*] (1678)
1 a : a space or all space every part of which is full of matter **b** (1) **:** a condition in which the pressure of the air in an enclosed space is greater than that of the outside atmosphere (2) **:** an enclosed space in which such a condition exists
2 : a general assembly of all members especially of a legislative body
3 : the quality or state of being full

ple·och·ro·ism \plē-'ä-krə-,wi-zəm\ *noun* [International Scientific Vocabulary *pleochroic,* from *pleio-* + Greek *chrōs* skin, color] (1857)
: the property of a crystal of showing different colors when viewed by light that vibrates parallel to different axes
— pleo·chro·ic \,plē-ə-'krō-ik\ *adjective*

pleo·mor·phic \,plē-ə-'mȯr-fik\ *adjective* (1886)
: able to assume different forms **:** POLYMORPHIC ⟨*pleomorphic* bacteria⟩ ⟨a *pleomorphic* sarcoma⟩
— pleo·mor·phism \-,fi-zəm\ *noun*

ple·o·nasm \'plē-ə-,na-zəm\ *noun* [Late Latin *pleonasmus,* from Greek *pleonasmos,* from *pleonazein* to be excessive, from *pleiōn, pleōn* more — more at PLUS] (1610)
1 : the use of more words than those necessary to denote mere sense (as in *the man he said*) **:** REDUNDANCY
2 : an instance or example of pleonasm
— ple·o·nas·tic \,plē-ə-'nas-tik\ *adjective*
— ple·o·nas·ti·cal·ly \-ti-k(ə-)lē\ *adverb*

ple·o·pod \'plē-ə-,päd\ *noun* [Greek *plein* to sail + English *-o-* + *-pod;* from its use in swimming — more at FLOW] (circa 1890)
: an abdominal limb of a crustacean

ple·ro·cer·coid \,plir-ō-'sər-,kȯid\ *noun* [Greek *plērēs* full + *kerkos* tail — more at FULL] (1906)
: the solid elongate infective larva of some tapeworms usually occurring in the muscles of fishes

ple·si·o·saur \'plē-sē-ə-,sȯr, -zē-\ *noun* [ultimately from Greek *plēsios* close (akin to Greek *pelas* near) + *sauros* lizard] (1839)
: any of a suborder (Plesiosauria) of Mesozoic marine reptiles with dorsoventrally flattened bodies and limbs modified into paddles

pleth·o·ra \'ple-thə-rə\ *noun* [Medieval Latin, from Greek *plēthōra,* literally, fullness, from *plēthein* to be full — more at FULL] (1541)
1 : a bodily condition characterized by an excess of blood and marked by turgescence and a florid complexion
2 : EXCESS, SUPERFLUITY; *also* **:** PROFUSION, ABUNDANCE
— ple·tho·ric \plə-'thȯr-ik, ple-, -'thär-; 'ple-thə-rik\ *adjective*

ple·thys·mo·gram \ple-'thiz-mə-,gram, plə-\ *noun* (1894)
: a tracing made by a plethysmograph

ple·thys·mo·graph \-,graf\ *noun* [International Scientific Vocabulary, from Greek *plēthysmos* increase, from *plēthynein* to increase, from *plēthys* mass, quantity, from *plēthein*] (1872)
: an instrument for determining and registering variations in the size of an organ or limb resulting from changes in the amount of blood present or passing through it
— ple·thys·mo·graph·ic \-,thiz-mə-'gra-fik\ *adjective*
— ple·thys·mo·graph·i·cal·ly \-fi-k(ə-)lē\ *adverb*
— pleth·ys·mog·ra·phy \,ple-thiz-'mä-grə-fē\ *noun*

pleu·ra \'plu̇r-ə\ *noun, plural* **pleu·rae** \'plu̇r-,ē, -,ī\ *or* **pleuras** [Middle English, from Medieval Latin, from Greek, rib, side] (15th century)
: the delicate serous membrane that lines each half of the thorax of mammals and is folded back over the surface of the lung of the same side
— pleu·ral \'plu̇r-əl\ *adjective*

pleu·ri·sy \'plu̇r-ə-sē\ *noun* [Middle English *pluresie,* from Middle French *pleuresie,* from Late Latin *pleurisis,* alteration of Latin *pleuritis,* from Greek, from *pleura* side] (14th century)
: inflammation of the pleura usually with fever, painful and difficult respiration, cough, and exudation of fluid or fibrinous material into the pleural cavity
— pleu·rit·ic \plu̇-'ri-tik\ *adjective*

pleu·ro·pneu·mo·nia \,plu̇r-ō-nu̇-'mō-nyə, -ō-nyu̇-\ *noun* [New Latin] (1725)
1 : combined inflammation of the pleura and lungs
2 : an acute febrile and often fatal respiratory disorder of cattle, goats, sheep, and related animals caused by a mycoplasma (*Mycoplasma mycoides*)

pleuropneumonia–like organism *noun* (1935)
: MYCOPLASMA

pleus·ton \'plü-stən, -,stän\ *noun* [International Scientific Vocabulary *pleus-* (irregular from Greek *plein* to sail, float) + *-on* (as in *plankton*)] (1943)
: organisms living in the thin surface layer existing at the air-water interface of a body of fresh water
— pleus·ton·ic \plü-'stä-nik\ *adjective*

-plex *noun combining form* [partly from Latin *-plex* (as in *duplex*); partly from *complex*]
1 : a figure of a given power ⟨googol*plex*⟩
2 : a building divided into an often specified number of spaces (as apartments or movie theaters) ⟨four*plex*⟩ ⟨multi*plex*⟩

plex·i·form \'plek-sə-,fȯrm\ *adjective* [New Latin *plexus* + English *-iform*] (circa 1828)
: of, relating to, or having the form or characteristics of a plexus

Plex·i·glas \'plek-si-,glas\ *trademark*
— used for acrylic plastic sheets and molding powders

plex·us \'plek-səs\ *noun, plural* **plex·us·es** \-sə-səz\ [New Latin, from Latin *plectere* to braid — more at PLY] (1682)
1 : a network of anastomosing or interlacing blood vessels or nerves
2 : an interwoven combination of parts or elements in a structure or system

pli·able \'plī-ə-bəl\ *adjective* [Middle English, from Middle French, from *plier* to bend, fold — more at PLY] (14th century)
1 a : supple enough to bend freely or repeatedly without breaking **b :** yielding readily to others **:** COMPLAISANT
2 : adjustable to varying conditions
synonym see PLASTIC
— pli·abil·i·ty \,plī-ə-'bi-lə-tē\ *noun*
— pli·able·ness \'plī-ə-bəl-nəs\ *noun*
— pli·ably \-blē\ *adverb*

pli·an·cy \'plī-ən(t)-sē\ *noun* (1699)
: the quality or state of being pliant

pli·ant \'plī-ənt\ *adjective* (14th century)
1 : PLIABLE 1a
2 : easily influenced **:** YIELDING
3 : suitable for varied uses
synonym see PLASTIC
— pli·ant·ly *adverb*
— pli·ant·ness *noun*

pli·ca \'plī-kə\ *noun, plural* **pli·cae** \-,kē, -,sē\ [Medieval Latin, from Latin *plicare* to fold — more at PLY] (circa 1706)
: a fold or folded part; *especially* **:** a groove or fold of skin

pli·cate \'plī-,kāt\ *adjective* [Latin *plicatus,* past participle of *plicare*] (1760)
1 : folded lengthwise like a fan ⟨a *plicate* leaf⟩
2 : having the surface thrown up into or marked with parallel ridges ⟨*plicate* wing cases⟩

pli·ca·tion \plī-'kā-shən\ *noun* (14th century)
1 : the act or process of folding **:** the state of being folded
2 : FOLD

plié \plē-'ā\ *noun* [French, from past participle of *plier* to bend] (1892)
: a bending of the knees outward by a ballet dancer with the back held straight

pli·ers \'plī(-ə)rz\ *noun plural but singular or plural in construction* [[2]*ply*] (circa 1569)
: a small pincers for holding small objects or for bending and cutting wire

[1]plight \'plīt\ *transitive verb* [Middle English, from Old English *plihtan* to endanger, from *pliht* danger; akin to Old English *plēon* to expose to danger, Old High German *pflegan* to take care of] (13th century)
: to put or give in pledge **:** ENGAGE ⟨*plight* one's troth⟩
— plight·er *noun*

[2]plight *noun* (13th century)
: a solemnly given pledge **:** ENGAGEMENT

[3]plight *noun* [Middle English *plit* condition, from Anglo-French, from (assumed) Vulgar Latin *plicitum* fold — more at PLAIT] (13th century)
: an unfortunate, difficult, or precarious situation

plim·soll \'plim(p)-səl, 'plim-,sȯl\ *noun* [probably from the supposed resemblance of the upper edge of the shoe's mudguard to the Plimsoll mark on a ship] (1907)
British **:** SNEAKER 2

Plimsoll mark *noun* [Samuel *Plimsoll* (died 1898) English shipping reformer] (1884)
: a load line or a set of load-line markings on an oceangoing cargo ship — called also *Plimsoll line*

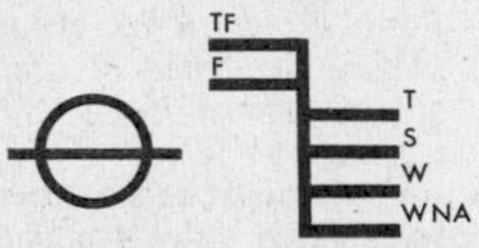

Plimsoll mark: *TF* tropical freshwater mark; *F* freshwater mark; *T* tropical load line; *S* summer load line; *W* winter load line; *WNA* winter load line, North Atlantic

[1]plink \'pliŋk\ *verb* [imitative] (1941)
intransitive verb
1 : to make a tinkling sound
2 : to shoot at random targets in an informal and noncompetitive manner
transitive verb
1 : to cause to make a tinkling sound
2 : to shoot at especially in a casual manner
— plink·er *noun*

[2]plink *noun* (1954)
: a tinkling metallic sound

plinth \'plin(t)th\ *noun* [Latin *plinthus,* from Greek *plinthos*] (1601)
1 a : the lowest member of a base **:** SUBBASE **b :** a block upon which the moldings of an architrave or trim are stopped at the bottom
2 : a usually square block serving as a base; *broadly* **:** any of various bases or lower parts — see BASE illustration
3 : a course of stones forming a continuous foundation or base course

plio- — see PLEIO-

Pli·o·cene \'plī-ə-,sēn\ *adjective* (1831)
: of, relating to, or being the latest epoch of the Tertiary or the corresponding system of rocks — see GEOLOGIC TIME table
— Pliocene *noun*

Plio·film \'plī-ə-,film\ *trademark*
— used for a glossy membrane made of rubber hydrochloride and used especially for packaging

plique–à–jour \,plēk-(,)ä-'zhu̇r\ *noun* [French, literally, braid letting in daylight] (1878)
: a style of enameling in which usually transparent enamels are fused into the openings of a metal filigree to produce an effect suggestive of stained glass

plis·kie *or* **plis·ky** \'plis-kē\ *noun, plural* **pliskies** [origin unknown] (1706)
chiefly Scottish **:** PRACTICAL JOKE, TRICK

plis·sé *or* **plis·se** \pli-ˈsā\ *noun* [French *plissé,* from past participle of *plisser* to pleat, from Middle French, from *pli* fold, from *plier* to fold — more at PLY] (1873)
1 : a textile finish of permanently puckered designs formed by treating with a sodium hydroxide solution
2 : a fabric with a plissé finish

plod \ˈpläd\ *verb* **plod·ded; plod·ding** [origin unknown] (1562)
intransitive verb
1 : to work laboriously and monotonously **:** DRUDGE
2 a : to walk heavily or slowly **:** TRUDGE **b :** to proceed slowly or tediously ⟨the movie just *plods* along⟩
transitive verb
: to tread slowly or heavily along or over
— plod *noun*
— plod·der *noun*
— plod·ding·ly \ˈplä-diŋ-lē\ *adverb*

-ploid *adjective combining form* [International Scientific Vocabulary, from *diploid & haploid*]
: having or being a chromosome number that bears (such) a relationship to or is (so many) times the basic chromosome number of a given group ⟨poly*ploid*⟩

ploi·dy \ˈplȯi-dē\ *noun* [from such words as *diploidy, hexaploidy*] (1939)
: degree of repetition of the basic number of chromosomes

PL/1 \ˌpē-(ˌ)el-ˈwən\ *noun* [*p*rogramming *l*anguage (version) *1*] (1965)
: a general-purpose language for programming a computer

[1]**plonk** \ˈpläŋk, ˈplȯŋk\ *variant of* PLUNK

[2]**plonk** *noun* [short for earlier *plink-plonk,* perhaps modification of French *vin blanc* white wine] (1930)
chiefly British **:** cheap or inferior wine

plop \ˈpläp\ *verb* **plopped; plop·ping** [imitative] (1821)
intransitive verb
1 : to fall, drop, or move suddenly with a sound like that of something dropping into water
2 : to allow the body to drop heavily ⟨*plopped* into a chair⟩
transitive verb
: to set, drop, or throw heavily
— plop *noun*

plo·sion \ˈplō-zhən\ *noun* (1899)
: EXPLOSION 3

plo·sive \ˈplō-siv\ *noun* [short for *explosive*] (1899)
: STOP 9
— plosive *adjective*

[1]**plot** \ˈplät\ *noun* [Middle English, from Old English] (before 12th century)
1 a : a small area of planted ground ⟨a vegetable *plot*⟩ **b :** a small piece of land in a cemetery **c :** a measured piece of land **:** LOT
2 : GROUND PLAN, PLAT
3 : the plan or main story of a literary work
4 : a secret plan for accomplishing a usually evil or unlawful end **:** INTRIGUE
5 : a graphic representation (as a chart) ☆
— plot·less \-ləs\ *adjective*
— plot·less·ness *noun*

[2]**plot** *verb* **plot·ted; plot·ting** (1588)
transitive verb
1 a : to make a plot, map, or plan of **b :** to mark or note on or as if on a map or chart
2 : to lay out in plots
3 a : to locate (a point) by means of coordinates **b :** to locate (a curve) by plotted points **c :** to represent (an equation) by means of a curve so constructed
4 : to plan or contrive especially secretly
5 : to invent or devise the plot of (a literary work)
intransitive verb
1 : to form a plot **:** SCHEME
2 : to be located by means of coordinates ⟨the data *plot* at a single point⟩

Plo·ti·nism \plō-ˈtī-ˌni-zəm, ˈplō-tᵊn-ˌi-zəm\ *noun* (circa 1890)
: the Neoplatonic ideas of the philosopher Plotinus
— Plo·ti·nist \-ˈtī-nist, -tᵊn-ist\ *noun*

plot·line \ˈplät-ˌlīn\ *noun* (1952)
: PLOT 3

plot·tage \ˈplä-tij\ *noun* (1936)
: the area included in a plot of land

plot·ter \ˈplä-tər\ *noun* (1588)
: one that plots: as **a :** a person who schemes or conspires **b :** a contriver of a literary plot **c :** a device for plotting; *specifically* **:** an instrument that graphs computer output

plot·ty \ˈplä-tē\ *adjective* **plot·ti·er; -est** (1897)
: marked by intricacy of plot or intrigue

plough·man's lunch \ˈplau̇-mənz-\ *noun* (1970)
: a cold lunch served especially in an English pub typically including bread, cheese, and pickled onions

plo·ver \ˈplə-vər, ˈplō-\ *noun, plural* **plover** *or* **plovers** [Middle English, from Middle French, from (assumed) Vulgar Latin *pluviarius,* from Latin *pluvia* rain — more at PLUVIAL] (14th century)
1 : any of a family (Charadriidae) of shore-inhabiting birds that differ from the sandpipers in having a short hard-tipped bill and usually a stouter more compact build
2 : any of various birds (as a turnstone or sandpiper) related to the plovers

[1]**plow** *or* **plough** \ˈplau̇\ *noun* [Middle English, from Old English *plōh* hide of land; akin to Old High German *pfluog* plow] (12th century)
1 : an implement used to cut, lift, and turn over soil especially in preparing a seedbed
2 : any of various devices (as a snowplow) operating like a plow

[2]**plow** *or* **plough** (15th century)
transitive verb
1 a : to turn, break up, or work with a plow **b :** to make (as a furrow) with a plow
2 : to cut into, open, or make furrows or ridges in with or as if with a plow
3 : to cleave the surface of or move through (water) ⟨whales *plowing* the ocean⟩
4 : to clear away snow from with a snowplow ⟨*plow* the street⟩
intransitive verb
1 a : to use a plow **b :** to bear or admit of plowing
2 a : to move in a way resembling that of a plow ⟨the car *plowed* into a fence⟩ **b :** to proceed steadily and laboriously ⟨had to *plow* through a stack of letters⟩
— plow·able \-ə-bəl\ *adjective*
— plow·er \ˈplau̇(-ə)r\ *noun*

plow back *transitive verb* (1930)
: to reinvest (profits) in a business
— plow·back \ˈplau̇-ˌbak\ *noun*

plow·boy \ˈplau̇-ˌbȯi\ *noun* (1596)
1 : a boy who leads the team drawing a plow
2 : a country youth

plow·man \-mən, -ˌman\ *noun* (13th century)
1 : a man who guides a plow
2 : a farm laborer

plow·share \ˈplau̇-ˌsher, -ˌshar\ *noun* [Middle English *ploughshare,* from *plough* plow + *schare* plowshare — more at SHARE] (14th century)
: a part of a plow that cuts the furrow

plow under *transitive verb* (1900)
: to cause to disappear **:** BURY, OVERWHELM

ploy \ˈplȯi\ *noun* [probably from *employ*] (1722)
1 : ESCAPADE, FROLIC
2 a : a tactic intended to embarrass or frustrate an opponent **b :** a devised or contrived move **:** STRATAGEM ⟨a *ploy* to get her to open the door —Robert B. Parker⟩

[1]**pluck** \ˈplək\ *verb* [Middle English, from Old English *pluccian;* akin to Middle High German *pflücken* to pluck] (before 12th century)
transitive verb
1 : to pull or pick off or out
2 a : to remove something (as hairs) from by or as if by plucking ⟨*pluck* one's eyebrows⟩ **b :** ROB, FLEECE
3 : to move, remove, or separate forcibly or abruptly ⟨*plucked* the child from the middle of the street⟩
4 a : to pick, pull, or grasp at **b :** to play by sounding the strings with the fingers or a pick
intransitive verb
: to make a sharp pull or twitch
— pluck·er *noun*

[2]**pluck** *noun* (15th century)
1 : an act or instance of plucking or pulling
2 : the heart, liver, lungs, and windpipe of a slaughtered animal especially as an item of food
3 : courageous readiness to fight or continue against odds **:** dogged resolution

plucky \ˈplə-kē\ *adjective* **pluck·i·er; -est** (1842)
: SPIRITED, BRAVE
— pluck·i·ly \ˈplə-kə-lē\ *adverb*
— pluck·i·ness \ˈplə-kē-nəs\ *noun*

[1]**plug** \ˈpləg\ *noun* [Dutch, from Middle Dutch *plugge;* akin to Middle High German *pfloc* plug] (1627)
1 a : a piece used to fill a hole **:** STOPPER **b :** an obtruding or obstructing mass of material resembling a stopper
2 : a flat compressed cake of tobacco
3 : a small core or segment removed from a larger object
4 : something inferior; *especially* **:** an inferior often aged or unsound horse; *also* **:** a quiet steady cold-blooded horse usually of light or moderate weight
5 a : HYDRANT, FIREPLUG **b :** SPARK PLUG
6 : an artificial angling lure used primarily for casting and made with one or more sets of gang hooks
7 : any of various devices resembling or functioning like a plug: as **a :** a male fitting for making an electrical connection to a live circuit by insertion in a receptacle (as an outlet) **b :** a device for connecting electric wires to a jack
8 : a piece of favorable publicity or a favorable mention usually incorporated in general matter

[2]**plug** *verb* **plugged; plug·ging** (1630)
transitive verb
1 : to stop, make tight, or secure by inserting a plug
2 : to hit with a bullet **:** SHOOT
3 : to advertise or publicize insistently
intransitive verb

☆ SYNONYMS
Plot, intrigue, machination, conspiracy, cabal mean a plan secretly devised to accomplish an evil or treacherous end. PLOT implies careful foresight in planning a complex scheme ⟨an assassination *plot*⟩. INTRIGUE suggests secret underhanded maneuvering in an atmosphere of duplicity ⟨backstairs *intrigue*⟩. MACHINATION implies a contriving of annoyances, injuries, or evils by indirect means ⟨the *machinations* of a party boss⟩. CONSPIRACY implies a secret agreement among several people usually involving treason or great treachery ⟨a *conspiracy* to fix prices⟩. CABAL typically applies to political intrigue involving persons of some eminence ⟨a *cabal* among powerful senators⟩. See in addition PLAN.

1 : to become plugged — usually used with *up*
2 : to work doggedly and persistently ⟨*plugged* away at her homework⟩
3 : to fire shots
— plug·ger *noun*
— plug into : to connect or become connected to by or as if by means of a plug ⟨the city was *plugged into* the new highway system⟩

plugged \ˈpləgd\ *adjective* (1694)
1 *of a coin* **:** altered by the insertion of a plug of base metal
2 : closed by or as if by a plug **:** OBSTRUCTED

plug hat *noun* (1863)
: a man's stiff hat (as a bowler or top hat)

plug–in \ˈpləg-ˌin\ *adjective* (1922)
: designed to be connected to an electric circuit by plugging in ⟨a *plug-in* toy⟩
— plug–in *noun*

plug in (1903)
intransitive verb
: to establish an electric circuit by inserting a plug
transitive verb
: to attach or connect to a service outlet

plug·o·la \(ˌ)plə-ˈgō-lə\ *noun* [*plug* + *-ola* (as in *payola*)] (1959)
: incidental advertising on radio or television that is not purchased like regular advertising

plug–ug·ly \ˈpləg-ˌəg-lē\ *noun* (1856)
: THUG, TOUGH; *especially* **:** one hired to intimidate

plum \ˈpləm\ *noun, often attributive* [Middle English, from Old English *plūme,* modification of Latin *prunum* plum, from Greek *proumnon*] (before 12th century)
1 a : any of numerous trees and shrubs (genus *Prunus*) with globular to oval smooth-skinned fruits that are drupes with oblong seeds **b :** the edible fruit of a plum
2 : any of various trees with edible fruits resembling plums; *also* **:** the fruit
3 a : a raisin when used in desserts **b :** SUGARPLUM
4 : something superior or very desirable; *especially* **:** something desirable given in return for a favor
5 : a dark reddish purple
— plum·like \-ˌlīk\ *adjective*

plum·age \ˈplü-mij\ *noun* [Middle English, from Middle French, from Old French, from *plume* feather — more at PLUME] (14th century)
: the feathers of a bird
— plum·aged \-mijd\ *adjective*

¹plumb \ˈpləm\ *noun* [Middle English, from (assumed) Old French *plomb* lead, from Latin *plumbum*] (14th century)
: a lead weight attached to a line and used to indicate a vertical direction
— out of plumb *or* **off plumb :** out of vertical or true

²plumb *adverb* (15th century)
1 : straight down or up **:** VERTICALLY
2 *chiefly dialect* **:** to a complete degree **:** ABSOLUTELY ⟨'you're *plumb* crazy', she remarked, with easy candor —*Harper's Weekly*⟩
3 : in a direct manner **:** EXACTLY; *also* **:** without interval of time **:** IMMEDIATELY

³plumb (15th century)
transitive verb
1 : to weight with lead
2 a : to measure the depth of with a plumb **b :** to examine minutely and critically ⟨*plumbing* the book's complexities⟩
3 : to adjust or test by a plumb line
4 : to seal with lead
5 [back-formation from *plumber*] **:** to supply with or install as plumbing
intransitive verb
: to work as a plumber

⁴plumb *adjective* (15th century)
1 : exactly vertical or true
2 : THOROUGH, COMPLETE
synonym see VERTICAL

plum·ba·go \ˌpləm-ˈbā-(ˌ)gō\ *noun, plural* **-gos** [Latin *plumbagin-, plumbago* galena, from *plumbum*] (1747)
1 [New Latin, from Latin] **:** any of a genus (*Plumbago* of the family Plumbaginaceae, the plumbago family) of woody chiefly tropical plants with alternate leaves and spikes of showy flowers
2 : GRAPHITE 1

plumb bob *noun* (circa 1840)
: the metal bob of a plumb line

plumb·er \ˈplə-mər\ *noun* [Middle English, from Middle French *plommier, plombier,* from Latin *plumbarius,* from *plumbum*] (15th century)
1 : a dealer or worker in lead
2 : one who installs, repairs, and maintains piping, fittings, and fixtures involved in the distribution and use of water in a building

plumber's helper *noun* (1952)
: PLUNGER d — called also *plumber's friend*

plumber's snake *noun* (1938)
: a long flexible rod or cable usually of steel that is used to free clogged pipes

plumb·ing \ˈplə-miŋ\ *noun* (1666)
1 : the act of using a plumb
2 : a plumber's occupation or trade
3 a : the apparatus (as pipes and fixtures) concerned in the distribution and use of water in a building **b :** an internal system that resembles plumbing; *especially* **:** one consisting of conduits or channels for conveying fluids

plum·bism \ˈpləm-ˌbi-zəm\ *noun* [Latin *plumbum* lead] (1876)
: lead poisoning especially when chronic

plumb line *noun* (15th century)
1 : a line (as of cord) that has at one end a weight (as a plumb bob) and is used especially to determine verticality
2 : a line directed to the center of gravity of the earth **:** a vertical line

¹plume \ˈplüm\ *noun* [Middle English, from Middle French, from Latin *pluma* small soft feather — more at FLEECE] (14th century)
1 : a feather of a bird: as **a :** a large conspicuous or showy feather **b :** CONTOUR FEATHER **c :** PLUMAGE **d :** a cluster of distinctive feathers
2 a : material (as a feather, cluster of feathers, or a tuft of hair) worn as an ornament **b :** a token of honor or prowess **:** PRIZE
3 : something resembling a feather (as in shape, appearance, or lightness): as **a :** a plumose appendage of a plant **b :** an elongated and usually open and mobile column or band (as of smoke, exhaust gases, or blowing snow) **c :** an animal structure having a main shaft bearing many hairs or filamentous parts; *especially* **:** a full bushy tail

²plume *transitive verb* **plumed; plum·ing** (15th century)
1 a : to provide or deck with feathers **b :** to array showily
2 : to indulge (oneself) in pride with an obvious or vain display of self-satisfaction
3 *of a bird* **a :** to preen and arrange the feathers of (itself) **b :** to preen and arrange (feathers)

plumed \ˈplümd\ *adjective* (15th century)
: provided with or adorned with or as if with a plume — often used in combination ⟨a white-*plumed* egret⟩

plume·let \ˈplüm-lət\ *noun* (circa 1847)
: a small tuft or plume

plu·me·ria \plü-ˈmir-ē-ə\ *noun* [New Latin, genus name, from Charles *Plumier* (died 1704) French botanist] (1753)
: FRANGIPANI

¹plum·met \ˈplə-mət\ *noun* [Middle English *plomet,* from Middle French *plombet* ball of lead, from *plomb* lead, from (assumed) Old French — more at PLUMB] (14th century)
: PLUMB; *also* **:** PLUMB LINE

²plummet *intransitive verb* (1937)
1 : to fall perpendicularly ⟨birds *plummeted* down⟩
2 : to drop sharply and abruptly ⟨prices *plummeted*⟩

plum·my \ˈplə-mē\ *adjective* **plum·mi·er; -est** (1759)
1 a : full of plums ⟨a rich *plummy* cake⟩ **b :** CHOICE, DESIRABLE ⟨got a *plummy* role in the movie⟩
2 a : having a plum color **b :** rich and mellow often to the point of affectation ⟨a *plummy* singing voice⟩

plu·mose \ˈplü-ˌmōs\ *adjective* (circa 1727)
1 : having feathers or plumes **:** FEATHERED
2 : FEATHERY

¹plump \ˈpləmp\ *verb* [Middle English] (14th century)
intransitive verb
1 : to drop, sink, or come in contact suddenly or heavily ⟨*plumped* down in the chair⟩
2 : to favor or decide in favor of someone or something strongly or emphatically — used with *for*
transitive verb
1 : to drop, cast, or place suddenly or heavily
2 : to give support and favorable publicity to

²plump *noun* (15th century)
: a sudden plunge, fall, or blow; *also* **:** the sound made by a plump

³plump *adverb* (1594)
1 : with a sudden or heavy drop
2 a : straight down **b :** straight ahead
3 : without qualification **:** DIRECTLY

⁴plump *noun* [Middle English *plumpe*] (15th century)
chiefly dialect **:** GROUP, FLOCK ⟨a *plump* of ducks rose at the same time —H. D. Thoreau⟩

⁵plump *verb* [⁶*plump*] (1533)
transitive verb
: to make plump
intransitive verb
: to become plump

⁶plump *adjective* [Middle English, dull, blunt, from Middle Dutch *plomp*] (1569)
1 : having a full rounded usually pleasing form ⟨a *plump* woman⟩
2 : AMPLE, ABUNDANT
— plump·ish \ˈpləm-pish\ *adjective*

plump·en \ˈpləm-pən\ *verb* (1687)
: ⁵PLUMP

¹plump·er \ˈpləm-pər\ *noun* [⁵*plump*] (1690)
: an object carried in the mouth to fill out the cheeks

²plumper *noun* [¹*plump*] (circa 1785)
chiefly British **:** a vote for only one candidate when two or more are to be elected to the same office

¹plump·ly \ˈpləm-plē\ *adverb* (1611)
: in a plump way ⟨a *plumply* pretty girl⟩

²plumply *adverb* [³*plump*] (1786)
: in a wholehearted manner and without hesitation or circumlocution **:** FORTHRIGHTLY

¹plump·ness \ˈpləmp-nəs\ *noun* (1545)
: the quality or state of being plump

²plumpness *noun* (1780)
: freedom from hesitation or circumlocution

plum pudding *noun* (1711)
: a rich boiled or steamed pudding containing fruits and spices

plum tomato *noun* (circa 1900)
: a small oblong tomato

plu·mule \ˈplü-(ˌ)myü(ə)l\ *noun* [New Latin *plumula,* from Latin, diminutive of *pluma* small soft feather — more at FLEECE] (circa 1741)
1 : the primary bud of a plant embryo usually situated at the apex of the hypocotyl and consisting of leaves and an epicotyl
2 : a down feather

plumule of morning glory seedling: *1* **hypocotyl,** *2* **plumule,** *3* **cotyledons**

plumy \ˈplü-mē\ *adjective* **plum·i·er; -est** (1582)
1 : DOWNY
2 : having or resembling plumes

[1]plun·der \ˈplən-dər\ *verb* **plun·dered; plun·der·ing** \-d(ə-)riŋ\ [German *plündern*] (1632)
transitive verb
1 a : to take the goods of by force (as in war) **:** PILLAGE, SACK ⟨invaders *plundered* the town⟩ **b :** to take by force or wrongfully **:** STEAL, LOOT
2 : to make extensive use of as if by plundering **:** use or use up wrongfully ⟨*plunder* the land⟩
intransitive verb
: to commit robbery or looting
— plun·der·er \-dər-ər\ *noun*

[2]plunder *noun* (1643)
1 : an act of plundering **:** PILLAGING
2 : something taken by force, theft, or fraud **:** LOOT
3 *chiefly dialect* **:** personal or household effects
synonym see SPOIL

plun·der·ous \-d(ə-)rəs\ *adjective* (1845)
: given to plundering

[1]plunge \ˈplənj\ *verb* **plunged; plung·ing** [Middle English, from Middle French *plonger,* from (assumed) Vulgar Latin *plumbicare,* from Latin *plumbum* lead] (14th century)
transitive verb
1 : to cause to penetrate or enter quickly and forcibly into something
2 : to cause to enter a state or course of action usually suddenly, unexpectedly, or violently
intransitive verb
1 : to thrust or cast oneself into or as if into water
2 a : to become pitched or thrown headlong or violently forward and downward; *also* **:** to move oneself in such a manner **b :** to act with reckless haste **:** enter suddenly or unexpectedly **c :** to bet or gamble heavily and recklessly
3 : to descend or dip suddenly

[2]plunge *noun* (15th century)
: an act or instance of plunging **:** DIVE; *also* **:** SWIM

plung·er \ˈplən-jər\ *noun* (1611)
: one that plunges: as **a :** DIVER **b :** a reckless gambler or speculator **c** (1) **:** a sliding reciprocating piece driven by or against fluid pressure; *especially* **:** PISTON (2) **:** a piece with a motion more or less like that of a ram or piston **d :** a rubber suction cup on a handle used to free plumbing traps and waste outlets of obstructions

plunk \ˈpləŋk\ *verb* [imitative] (1805)
transitive verb
1 : to pluck or hit so as to produce a quick, hollow, metallic, or harsh sound
2 : to set down suddenly **:** PLUMP
intransitive verb
1 : to make a plunking sound
2 : to drop abruptly **:** DIVE
3 : to come out in favor of someone or something — used with *for*
— plunk *noun*
— plunk·er *noun*

plunk down (1891)
intransitive verb
: to drop abruptly **:** settle into position
transitive verb
1 a : to put down usually firmly or abruptly ⟨*plunked* the money *down* on the counter⟩ **b :** to settle (oneself) into position ⟨*plunked* himself *down* on the bench⟩
2 : to pay out

plu·per·fect \ˌplü-ˈpər-fikt\ *adjective* [Middle English *pluperfyth,* modification of Late Latin *plusquamperfectus,* literally, more than perfect] (15th century)
1 : PAST PERFECT
2 : utterly perfect or complete
— pluperfect *noun*

plu·ral \ˈplu̇r-əl\ *adjective* [Middle English, from Middle French & Latin; Middle French *plurel,* from Latin *pluralis,* from *plur-, plus* more — more at PLUS] (14th century)
1 : of, relating to, or constituting a class of grammatical forms usually used to denote more than one or in some languages more than two ⟨*genetics* is *plural* in form but takes a singular verb⟩
2 : relating to, consisting of, or containing more than one or more than one kind or class ⟨a *plural* society⟩
— plural *noun*
— plu·ral·ly \-ə-lē\ *adverb*

plu·ral·ism \ˈplu̇r-ə-ˌli-zəm\ *noun* (1818)
1 : the holding of two or more offices or positions (as benefices) at the same time
2 : the quality or state of being plural
3 a : a theory that there are more than one or more than two kinds of ultimate reality **b :** a theory that reality is composed of a plurality of entities
4 a : a state of society in which members of diverse ethnic, racial, religious, or social groups maintain an autonomous participation in and development of their traditional culture or special interest within the confines of a common civilization **b :** a concept, doctrine, or policy advocating this state
— plu·ral·ist \-list\ *adjective or noun*
— plu·ral·is·tic \ˌplu̇r-ə-ˈlis-tik\ *adjective*
— plu·ral·is·ti·cal·ly \-ti-k(ə-)lē\ *adverb*

plu·ral·i·ty \plu̇-ˈra-lə-tē\ *noun, plural* **-ties** (14th century)
1 a : the state of being plural **b :** the state of being numerous **c :** a large number or quantity
2 : PLURALISM 1; *also* **:** a benefice held by pluralism
3 a : a number greater than another **b :** an excess of votes over those cast for an opposing candidate **c :** a number of votes cast for a candidate in a contest of more than two candidates that is greater than the number cast for any other candidate but not more than half the total votes cast

plu·ral·ize \ˈplu̇r-ə-ˌlīz\ *transitive verb* **-ized; -iz·ing** (1803)
: to make plural or express in the plural form
— plu·ral·i·za·tion \ˌplu̇r-ə-lə-ˈzā-shən\ *noun*

plu·rip·o·tent \plu̇-ˈri-pə-tənt\ *adjective* [Latin *plur-, plus* more + English *potent*] (1916)
: not fixed as to developmental potentialities **:** having developmental plasticity ⟨*pluripotent* stem cell⟩

[1]plus \ˈpləs\ *adjective* [Latin, adverb, more, from neuter of *plur-, plus,* adjective; akin to Greek *pleion* more, Latin *plenus* full — more at FULL] (1579)
1 : algebraically positive
2 : having, receiving, or being in addition to what is anticipated
3 a : falling high in a specified range ⟨a grade of C *plus*⟩ **b :** greater than that specified **c :** possessing a specified quality to a high degree
4 : electrically positive
5 : relating to or being a particular one of the two mating types that are required for successful fertilization in sexual reproduction in some lower plants (as a fungus)

[2]plus *noun, plural* **plus·es** \ˈplə-səz\ *also* **plus·ses** (1654)
1 : PLUS SIGN
2 : an added quantity
3 : a positive factor or quality
4 : SURPLUS

[3]plus *preposition* (1668)
1 : increased by **:** with the addition of ⟨four *plus* five⟩ ⟨principal *plus* interest⟩ ⟨made twelve dollars a week *plus* room and board —E. L. Doctorow⟩
2 : BESIDES — used chiefly in speech and casual writing ⟨*plus* all this, as a sedative it has no equal —Groucho Marx⟩

[4]plus *conjunction* (circa 1950)
1 : AND ⟨the Smyth Report, *plus* an idea and some knowledge of bureaucracy, were all I needed —Pat Frank⟩ ⟨a box-office *plus* critical and artistic success —G. J. Nathan⟩ ⟨eats alone, a hot beef sandwich *plus* a BLT *plus* apple pie —Garrison Keillor⟩
2 : in addition to which ⟨it was an achievement. *Plus,* I wrote the story and the musical score —Jackie Gleason⟩ ⟨it's also pretty on my open shelves, *plus* it smells good —Nikki Giovanni⟩ ▫

plus fours *noun plural* (1920)
: loose sports knickers made four inches longer than ordinary knickers

[1]plush \ˈpləsh\ *noun* [Middle French *peluche*] (1594)
: a fabric with an even pile longer and less dense than velvet pile

[2]plush *adjective* (circa 1645)
1 : relating to, resembling, or made of plush
2 : notably luxurious
— plush·ly *adverb*
— plush·ness *noun*

plushy \ˈplə-shē\ *adjective* **plush·i·er; -est** (1611)
1 : having the texture of or covered with plush
2 : LUXURIOUS, SHOWY
— plush·i·ness *noun*

plus·sage \ˈplə-sij\ *noun* (1924)
: an amount over and above another amount

plus sign *noun* (circa 1907)
: a sign + denoting addition or a positive quantity

Plu·to \ˈplü-(ˌ)tō\ *noun* [Latin *Pluton-, Pluto,* from Greek *Ploutōn*]
1 : the Greek god of the underworld — compare DIS
2 [New Latin] **:** the planet with the farthest mean distance from the sun — see PLANET table

plu·toc·ra·cy \plü-ˈtä-krə-sē\ *noun, plural* **-cies** [Greek *ploutokratia,* from *ploutos* wealth; akin to Greek *plein* to sail, float — more at FLOW] (1652)
1 : government by the wealthy
2 : a controlling class of the wealthy
— plu·to·crat \ˈplü-tə-ˌkrat\ *noun*
— plu·to·crat·ic \ˌplü-tə-ˈkra-tik\ *adjective*
— plu·to·crat·i·cal·ly \-ti-k(ə-)lē\ *adverb*

plu·ton \ˈplü-ˌtän\ *noun* [probably back-formation from *plutonic*] (1936)
: a typically large body of intrusive igneous rock

plu·to·ni·an \plü-ˈtō-nē-ən\ *adjective, often capitalized* (1667)
: of, relating to, or characteristic of Pluto or the lower world **:** INFERNAL

plu·ton·ic \plü-ˈtä-nik\ *adjective* [Latin *Pluton-, Pluto*] (1833)
1 : formed by solidification of magma deep within the earth and crystalline throughout ⟨*plutonic* rock⟩
2 *often capitalized* **:** PLUTONIAN

plu·to·ni·um \plü-ˈtō-nē-əm\ *noun* [New Latin, from *Pluton-, Pluto,* the planet Pluto] (1942)
: a radioactive metallic element similar chemically to uranium that is formed as the isotope 239 by decay of neptunium and found in minute quantities in pitchblende, that undergoes slow disintegration with the emission of an alpha particle to form uranium 235, and

☐ USAGE
plus The preposition *plus* has long been used with a meaning equivalent to *and* (as in "two *plus* two"); it is not, therefore, very surprising that in time people have begun to use it as a conjunction much like *and.* Sense 2 is considered to be an adverb by some commentators. It is used chiefly in speech and in informal writing.

that is fissionable with slow neutrons to yield atomic energy — see ELEMENT table

¹plu·vi·al \'plü-vē-əl\ *adjective* [Latin *pluvialis,* from *pluvia* rain, from feminine of *pluvius* rainy, from *pluere* to rain — more at FLOW] (circa 1656)
1 a : of or relating to rain **b :** characterized by abundant rain
2 *of a geologic change* **:** resulting from the action of rain

²pluvial *noun* (1929)
: a prolonged period of wet climate

¹ply \'plī\ *verb* **plied; ply·ing** [Middle English *plien,* short for *applien* to apply] (14th century)
transitive verb
1 a : to use or wield diligently ⟨busily *plying* his pen⟩ **b :** to practice or perform diligently ⟨*ply* a trade⟩
2 : to keep furnishing or supplying something to ⟨*plied* us with liquor⟩
3 a : to make a practice of rowing or sailing over or on ⟨the boat *plies* the river⟩ **b :** to go or travel regularly over, on, or through ⟨jets *plying* the skies⟩
intransitive verb
1 : to apply oneself steadily
2 : to go or travel regularly ⟨a steamer *plying* between the towns⟩

²ply *noun, plural* **plies** [³*ply*] (1532)
1 a : one of several layers (as of cloth) usually sewn or laminated together **b :** one of the strands in a yarn **c :** one of the veneer sheets forming plywood **d :** a layer of a paper or paperboard
2 : INCLINATION, BIAS

³ply *transitive verb* **plied; ply·ing** [Middle English *plien* to fold, from Middle French *plier,* from Latin *plicare;* akin to Old High German *flehtan* to braid, Latin *plectere,* Greek *plekein*] (circa 1909)
: to twist together ⟨*ply* two single yarns⟩

Plym·outh Rock \'pli-məth-\ *noun* [*Plymouth Rock,* on which the Pilgrims are supposed to have landed in 1620] (1849)
: any of an American breed of medium-sized single-combed dual-purpose domestic fowls

ply·wood \'plī-ˌwu̇d\ *noun* (1907)
: a structural material consisting of sheets of wood glued or cemented together with the grains of adjacent layers arranged at right angles or at a wide angle

-pnea *or* **-pnoea** *noun combining form* [New Latin, from Greek *-pnoia,* from *pnoia,* from *pnein* to breathe]
: breath **:** breathing ⟨hyper*pnea*⟩ ⟨a*pnoea*⟩

pneum- *or* **pneumo-** *combining form* [New Latin, partly from Greek *pneum-* (from *pneuma*); partly from Greek *pneumōn* lung]
1 : air **:** gas ⟨*pneumo*thorax⟩
2 : lung ⟨*pneumo*coniosis⟩
3 : respiration ⟨*pneumo*graph⟩
4 : pneumonia ⟨*pneumo*coccus⟩

pneu·ma \'nü-mə, 'nyü-\ *noun* [Greek] (1884)
: SOUL, SPIRIT

pneumat- *or* **pneumato-** *combining form* [Greek, from *pneumat-, pneuma*]
1 : air **:** vapor **:** gas ⟨*pneumato*lytic⟩
2 : respiration ⟨*pneumato*phore⟩

pneu·mat·ic \nu̇-'ma-tik, nyu̇-\ *adjective* [Latin *pneumaticus,* from Greek *pneumatikos,* from *pneumat-, pneuma* air, breath, spirit, from *pnein* to breathe — more at SNEEZE] (1659)
1 : of, relating to, or using gas (as air or wind): **a :** moved or worked by air pressure **b** (1) **:** adapted for holding or inflated with compressed air (2) **:** having air-filled cavities
2 : of or relating to the pneuma **:** SPIRITUAL
3 : having a well-proportioned feminine figure; *especially* **:** having a full bust
— pneu·mat·i·cal·ly \-ti-k(ə-)lē\ *adverb*
— pneu·ma·tic·i·ty \ˌnü-mə-'ti-sə-tē, ˌnyü-\ *noun*

pneu·ma·tol·o·gy \ˌnü-mə-'tä-lə-jē, ˌnyü-\ *noun* [New Latin *pneumatologia,* from Greek *pneumat-, pneuma* + New Latin *-logia* -logy] (1678)
: the study of spiritual beings or phenomena

pneu·ma·to·lyt·ic \ˌnü-mə-tᵊl-'i-tik, ˌnyü-; (ˌ)n(y)ü-ˌma-\ *adjective* [International Scientific Vocabulary] (1896)
: formed or forming by hot vapors or superheated liquids under pressure — used especially of minerals and ores

pneu·mat·o·phore \nu̇-'ma-tə-ˌfōr, nyu̇-, -ˌfȯr\ *noun* [International Scientific Vocabulary] (1859)
1 : a muscular gas-containing sac that serves as a float on a siphonophore colony
2 : a root often functioning as a respiratory organ in a wetland plant

pneu·mo·coc·cus \ˌnü-mə-'kä-kəs, ˌnyü-\ *noun, plural* **-coc·ci** \-'käk-ˌsī, -ˌsē; -'kä-ˌkī, -ˌkē\ [New Latin] (1890)
: a bacterium (*Streptococcus pneumoniae*) that causes an acute pneumonia involving one or more lobes of the lung
— pneu·mo·coc·cal \-'kä-kəl\ *adjective*

pneu·mo·co·ni·o·sis \ˌnü-mō-ˌkō-nē-'ō-səs, ˌnyü-\ *noun, plural* **-o·ses** \-ˌsēz\ [New Latin, from *pneum-* + Greek *konis* dust — more at INCINERATE] (1881)
: a disease of the lungs caused by the habitual inhalation of irritants (as mineral or metallic particles) — compare BLACK LUNG, SILICOSIS

Pneu·mo·cys·tis ca·ri·nii pneumonia \ˌnü-mə-'sis-təs-kə-'rī-nē-ˌē-, ˌnyü-\ *noun* [New Latin *Pneumocystis carinii,* species name] (1964)
: a pneumonia that affects individuals whose immunological defenses have been compromised, that is caused by a microorganism (*Pneumocystis carinii*), and that attacks especially the interstitial tissue of the lungs with marked thickening of the alveolar septa and alveoli

pneu·mo·graph \'nü-mə-ˌgraf, 'nyü-\ *noun* [International Scientific Vocabulary] (1878)
: an instrument for recording thoracic movements or volume change during respiration

pneu·mo·nec·to·my \ˌnü-mə-'nek-tə-mē, ˌnyü-\ *noun, plural* **-mies** [Greek *pneumōn* + International Scientific Vocabulary *-ectomy*] (circa 1895)
: excision of an entire lung or of one or more lobes of a lung

pneu·mo·nia \nu̇-'mō-nyə, nyu̇-\ *noun* [New Latin, from Greek, from *pneumōn* lung, alteration of *pleumōn* — more at PULMONARY] (1603)
: a disease of the lungs characterized by inflammation and consolidation followed by resolution and caused by infection or irritants

pneu·mon·ic \nu̇-'mä-nik, nyu̇-\ *adjective* [New Latin *pneumonicus,* from Greek *pneumonikos,* from *pneumōn*] (1675)
1 : of, relating to, or affecting the lungs **:** PULMONIC, PULMONARY
2 : of, relating to, or affected with pneumonia

pneu·mo·ni·tis \ˌnü-mə-'nī-təs, ˌnyü-\ *noun* [New Latin, from Greek *pneumōn*] (circa 1834)
: inflammation of the lungs

pneu·mo·tho·rax \ˌnü-mə-'thōr-ˌaks, ˌnyü-, -'thȯr-\ *noun* [New Latin] (1821)
: a condition in which air or other gas is present in the pleural cavity and which occurs spontaneously as a result of disease or injury of lung tissue or puncture of the chest wall or is induced as a therapeutic measure to collapse the lung

¹poach \'pōch\ *transitive verb* [Middle English *pochen,* from Middle French *pocher,* from Old French *pochier,* literally, to put into a bag, from *poche* bag, pocket, of Germanic origin; akin to Old English *pocca* bag] (15th century)
: to cook in simmering liquid

²poach *verb* [Middle French *pocher,* of Germanic origin; akin to Middle English *poken* to poke] (1611)
intransitive verb
1 : to encroach upon especially for the purpose of taking something
2 : to trespass for the purpose of stealing game; *also* **:** to take game or fish illegally
transitive verb
1 : to trespass on ⟨a field *poached* too frequently by the amateur —*Times Literary Supplement*⟩
2 a : to take (game or fish) by illegal methods **b :** to appropriate (something) as one's own

¹poach·er \'pō-chər\ *noun* [²*poach*] (1614)
1 : one that trespasses or steals
2 : one who kills or takes wild animals (as game or fish) illegally

²poacher *noun* [¹*poach*] (1861)
1 : a covered pan containing a plate with depressions or shallow cups in each of which an egg can be cooked over steam rising from boiling water in the bottom of the pan
2 : a baking dish in which food (as fish) can be poached

po'·boy \'pō-ˌbȯi\ *variant of* POOR BOY

po·chard \'pō-chərd\ *noun* [origin unknown] (1552)
: any of numerous rather heavy-bodied diving ducks (especially genus *Aythya*) with a large head and with feet and legs placed far back under the body

¹pock \'päk\ *noun* [Middle English *pokke,* from Old English *pocc;* akin to Middle Low German & Middle Dutch *pocke* pock] (before 12th century)
: a pustule in an eruptive disease (as smallpox); *also* **:** a spot suggesting such a pustule

²pock *transitive verb* (1841)
: to mark with or as if with pocks **:** PIT

¹pock·et \'pä-kət\ *noun* [Middle English *poket,* from Old North French *pokete,* diminutive of *poke* bag, of Germanic origin; akin to Old English *pocca* bag] (15th century)
1 a : a small bag carried by a person **:** PURSE **b :** a small bag that is sewed or inserted in a garment so that it is open at the top or side ⟨coat *pocket*⟩
2 : supply of money **:** MEANS
3 : RECEPTACLE, CONTAINER: as **a :** an opening at the corner or side of a billiard table **b :** a superficial pouch in some animals
4 : a small often isolated area or group ⟨*pockets* of unemployment⟩ : **a** (1) **:** a cavity containing a deposit (as of gold, water, or gas) (2) **:** a small body of ore **b :** AIR POCKET
5 : a place for a batten made by sewing a strip on a sail
6 a : BLIND ALLEY **b :** the position of a contestant in a race hemmed in by others **c :** an area formed by blockers from which a football quarterback attempts to pass
7 : the concave area at the base of the finger sections of a baseball glove or mitt in which the ball is normally caught
— pock·et·ful \-ˌfu̇l\ *noun*
— in one's pocket : in one's control or possession
— in pocket 1 : provided with funds **2 :** in the position of having made a profit
— out of pocket 1 : low on money or funds **2 :** having suffered a loss

²pocket *transitive verb* (1589)
1 a : to put or enclose in or as if in one's pocket ⟨*pocketed* the change⟩ **b :** to appropriate to one's own use **:** STEAL **c :** to refuse assent to (a bill) by a pocket veto
2 : to put up with **:** ACCEPT
3 : to set aside **:** SUPPRESS ⟨*pocketed* his pride⟩
4 a : to hem in **b :** to drive (a ball) into a pocket of a pool table
5 : to cover or supply with pockets
— pock·et·able \'pä-kə-tə-bəl\ *adjective*

³pocket *adjective* (1612)

1 a : small enough to be carried in the pocket **b :** SMALL, MINIATURE ⟨a *pocket* park⟩
2 a : of or relating to money **b :** carried in or paid from one's own pocket

pocket battleship *noun* (1930)
: a small German battleship built so as to come within treaty limitations of tonnage and armament

pocket billiards *noun plural but usually singular in construction* (1913)
: POOL 2b

[1]**pock·et·book** \ˈpä-kət-ˌbu̇k\ *noun* (1617)
1 *often* **pocket book :** a small especially paperback book that can be carried in the pocket
2 : a flat typically leather folding case for money or personal papers that can be carried in a pocket or handbag
3 a : PURSE **b :** HANDBAG 2
4 : financial resources **:** INCOME

[2]**pocketbook** *adjective* (1894)
: relating to or involving economic interests ⟨*pocketbook* issues⟩

pocket borough *noun* (1856)
: an English constituency controlled before parliamentary reform by a single person or family

pocket edition *noun* (1715)
1 : POCKETBOOK 1
2 : a miniature form of something

pocket gopher *noun* (1873)
: GOPHER 2a

pocket–handkerchief *noun* (1645)
: a handkerchief carried in the pocket

pock·et·knife \ˈpä-kət-ˌnīf\ *noun* (1727)
: a knife that has one or more blades that fold into the handle and that can be carried in the pocket

pocket money *noun* (1632)
: money for small personal expenses

pocket mouse *noun* (1884)
: any of various nocturnal burrowing rodents (family Heteromyidae) that resemble mice, live in arid parts of western North America, and have long hind legs and tail and fur-lined cheek pouches

pocket mouse

pock·et–size \ˈpä-kət-ˌsīz\ *also* **pock·et–sized** \-ˌsīzd\ *adjective* (1907)
1 : of a size convenient for carrying in the pocket
2 : SMALL ⟨a *pocket-size* country⟩

pocket veto *noun* (1842)
: an indirect veto of a legislative bill by an executive through retention of the bill unsigned until after adjournment of the legislature
— pocket veto *transitive verb*

[1]**pock·mark** \ˈpäk-ˌmärk\ *noun* (circa 1673)
: a mark, pit, or depressed scar caused by smallpox or acne; *also* **:** an imperfection or depression like a pockmark

[2]**pockmark** *transitive verb* (1756)
: to cover with or as if with pockmarks **:** PIT

pocky \ˈpä-kē\ *adjective* (14th century)
: covered with pocks

po·co \ˈpō-(ˌ)kō, ˈpȯ-\ *adverb* [Italian, little, from Latin *paucus* — more at FEW] (1724)
: to a slight degree **:** SOMEWHAT — used to qualify a direction in music ⟨*poco* allegro⟩

po·co a po·co \ˈpō-kō-(ˌ)ä-ˈpō-(ˌ)kō, ˈpȯ-kō-(ˌ)ä-ˈpȯ-\ *adverb* [Italian] (circa 1854)
: little by little **:** GRADUALLY — used as a direction in music

po·co·cu·ran·te \ˈpō-kō-kyu̇-ˈran-tē, -ku̇-\ *adjective* [Italian *poco curante* caring little] (1815)
: INDIFFERENT, NONCHALANT
— po·co·cu·ran·tism \-ˈran-ˌti-zəm\ *noun*

po·co·sin \pə-ˈkō-sᵊn\ *noun* [probably from Virginia or North Carolina Algonquian] (1634)
: an upland swamp of the coastal plain of the southeastern U.S.

[1]**pod** \ˈpäd\ *noun* [origin unknown] (1573)
1 : a bit socket in a brace
2 : a straight groove or channel in the barrel of an auger

[2]**pod** *noun* [probably alteration of *cod* bag — more at CODPIECE] (1688)
1 : a dry dehiscent pericarp or fruit that is composed of one or more carpels; *especially* **:** LEGUME
2 a : an anatomical pouch **b :** a grasshopper egg case
3 : a tapered and roughly cylindrical body of ore or mineral
4 : a usually protective container or housing: as **a :** a streamlined compartment (as for fuel) under the wings or fuselage of an aircraft **b :** a compartment (as for personnel, a power unit, or an instrument) on a ship or craft

[3]**pod** *intransitive verb* **pod·ded; pod·ding** (1734)
: to produce pods

[4]**pod** *noun* [origin unknown] (1832)
: a number of animals (as whales) clustered together

-pod *noun combining form* [Greek *-podos,* from *pod-, pous* foot — more at FOOT]
: foot **:** part resembling a foot ⟨pleo*pod*⟩

po·dag·ra \pə-ˈda-grə\ *noun* [Middle English, from Latin, from Greek, literally, foot trap, from *pod-, pous* + *agra* hunt, catch; probably akin to Greek *agein* to drive, lead — more at AGENT] (14th century)
: a painful condition of the big toe caused by gout

pod corn *noun* (1893)
: an Indian corn that has each kernel enclosed in a chaffy shell similar to that of other cereals

po·de·sta \ˌpō-də-ˈstä\ *noun* [Italian *podestà,* literally, power, from Latin *potestat-, potestas,* irregular from *potis* able — more at POTENT] (1548)
: a chief magistrate in a medieval Italian municipality

podgy \ˈpä-jē\ *adjective* **podg·i·er; -est** [*podge* something pudgy] (1846)
chiefly British **:** PUDGY

po·di·a·try \pə-ˈdī-ə-trē, pō-\ *noun* [Greek *pod-, pous* + English *-iatry*] (1914)
: the medical care and treatment of the human foot — called also *chiropody*
— po·di·at·ric \ˌpō-dē-ˈa-trik\ *adjective*
— po·di·a·trist \pə-ˈdī-ə-trist, pō-\ *noun*

po·di·um \ˈpō-dē-əm\ *noun, plural* **podiums** *or* **po·dia** \-dē-ə\ [Latin — more at PEW] (1743)
1 : a low wall serving as a foundation or terrace wall: as **a :** one around the arena of an ancient amphitheater serving as a base for the tiers of seats **b :** the masonry under the stylobate of a temple
2 a : a dais especially for an orchestral conductor **b :** LECTERN ■

-podium *noun combining form, plural* **-podia** [New Latin, from Greek *podion,* diminutive of *pod-, pous* foot — more at FOOT]
: foot **:** part resembling a foot ⟨pseudo*podium*⟩

podo·phyl·lin \ˌpä-də-ˈfi-lən\ *noun* [International Scientific Vocabulary, from New Latin *Podophyllum*] (1851)
: a resin obtained from podophyllum and used in medicine as a caustic

podo·phyl·lum \-ˈfi-ləm\ *noun, plural* **-phyl·li** \-ˈfi-ˌlī\ *or* **-phyllums** [New Latin, from *Podophyllum,* genus of herbs including the mayapple] (1842)
: the dried rhizome and rootlet of the mayapple that is used as a caustic or as a source of the more effective podophyllin

Po·dunk \ˈpō-ˌdəŋk\ *noun* [*Podunk,* village in Mass. or locality in Conn.] (1846)
: a small, unimportant, and isolated town

pod·zol *also* **pod·sol** \ˈpäd-ˌzȯl\ *noun* [Russian] (1908)
: any of a group of zonal soils that develop in a moist climate especially under coniferous or mixed forest and have an organic mat and a thin organic-mineral layer above a light gray leached layer resting on a dark illuvial horizon enriched with amorphous clay
— pod·zol·ic \päd-ˈzä-lik, -ˈzȯ-\ *adjective*

pod·zol·i·za·tion *also* **pod·sol·i·za·tion** \ˌpäd-ˌzȯ-lə-ˈzā-shən\ *noun* (1912)
: a process of soil formation especially in humid regions involving principally leaching of the upper layers with accumulation of material in lower layers and development of characteristic horizons; *specifically* **:** the development of a podzol
— pod·zol·ize \ˈpäd-ˌzȯ-ˌlīz\ *verb*

po·em \ˈpō-əm, -im, ˈpōm *also* ˈpȯ(-)im, ˈpō-ˌem\ *noun* [Middle French *poeme,* from Latin *poema,* from Greek *poiēma,* from *poiein*] (15th century)
1 : a composition in verse
2 : something suggesting a poem (as in expressiveness, lyricism, or formal grace) ⟨the house we stayed in . . . was itself a *poem* —H. J. Laski⟩

po·e·sy \ˈpō-ə-zē, -sē\ *noun, plural* **po·e·sies** [Middle English *poesie,* from Middle French, from Latin *poesis,* from Greek *poiēsis,* literally, creation, from *poiein*] (14th century)
1 a : a poem or body of poems **b :** POETRY **c :** artificial or sentimentalized poetic writing
2 : poetic inspiration

po·et \ˈpō-ət, -it *also* ˈpȯ(-)it\ *noun* [Middle English, from Middle French *poete,* from Latin *poeta,* from Greek *poiētēs* maker, poet, from *poiein* to make; akin to Sanskrit *cinoti* he gathers, heaps up] (14th century)
1 : one who writes poetry **:** a maker of verses
2 : one (as a creative artist) of great imaginative and expressive capabilities and special sensitivity to the medium

po·et·as·ter \ˈpō-ə-ˌtas-tər\ *noun* [New Latin, from Latin *poeta* + *-aster* -aster] (1599)
: an inferior poet

po·et·ess \ˈpō-ə-təs, ˈpō-i- *also* ˈpȯ(-)i-\ *noun* (1530)
: a girl or woman who is a poet
usage see -ESS

po·et·ic \pō-ˈe-tik\ *adjective* (1530)
1 a : of, relating to, or characteristic of poets or poetry **b :** given to writing poetry
2 : written in verse

po·et·i·cal \-ti-kəl\ *adjective* (14th century)
1 : POETIC
2 : being beyond or above the truth of history or nature **:** IDEALIZED ⟨had *poetical* ideas about love⟩
— po·et·i·cal·ly \-k(ə-)lē\ *adverb*
— po·et·i·cal·ness \-ti-kəl-nəs\ *noun*

po·et·i·cism \pō-ˈe-tə-ˌsi-zəm\ *noun* (1847)
: an archaic, trite, or strained expression in poetry

□ USAGE

podium Although critics sometimes complain that this or that writer does not know the difference between a podium and a lectern, sense 2b is standard ⟨the portable *podium* placed on a table at the end of the room —Stephen Coonts⟩ ⟨briskly seizes the *podium* in the still-humming law-school auditorium —*New England Monthly*⟩. It is a sort of in-joke among the critics that the man who seized the podium in the last quotation must be on his hands and knees. But the reading public never seems to be confused by the use. In general *on* is more likely to be used of musical conductors than of speakers; a speaker is often said to be standing *at* the podium ⟨Robert Frost, who was standing at this *podium* reading, suddenly slammed his book and threw it at me —Truman Capote⟩.

po·et·i·cize \-ˌsīz\ *transitive verb* **-cized; -ciz·ing** (1804)
: to give a poetic quality to
poetic justice *noun* (circa 1890)
: an outcome in which vice is punished and virtue rewarded usually in a manner peculiarly or ironically appropriate
poetic license *noun* (1819)
: LICENSE 4
po·et·ics \pō-ˈe-tiks\ *noun plural but singular or plural in construction* (circa 1741)
1 a : a treatise on poetry or aesthetics **b** *also* **po·et·ic** \-tik\ **:** poetic theory or practice
2 : poetic feelings or utterances
po·et·ize \ˈpō-ə-ˌtīz\ *verb* **-ized; -iz·ing** (1581)
intransitive verb
: to compose poetry
transitive verb
: POETICIZE
— po·et·iz·er *noun*
poet laureate *noun, plural* **poets laureate** *or* **poet laureates** (15th century)
1 : a poet honored for achievement in his art
2 : a poet appointed for life by an English sovereign as a member of the royal household and formerly expected to compose poems for court and national occasions
3 : one regarded by a country or region as its most eminent or representative poet
po·et·ry \ˈpō-ə-trē, -i-trē *also* ˈpȯ(-)i-trē\ *noun* (14th century)
1 a : metrical writing **:** VERSE **b :** the productions of a poet **:** POEMS
2 : writing that formulates a concentrated imaginative awareness of experience in language chosen and arranged to create a specific emotional response through meaning, sound, and rhythm
3 a : something likened to poetry especially in beauty of expression **b :** poetic quality or aspect ⟨the *poetry* of dance⟩
po–faced \ˈpō-ˌfāst\ *adjective* [perhaps from *po* chamber pot, toilet, from French *pot* pot] (1934)
British **:** having an assumed solemn, serious, or earnest expression or manner **:** piously or hypocritically solemn
pog·o·nip \ˈpäg-ə-ˌnip\ *noun* [Shoshone *payinappih* cloud] (1865)
: a dense winter fog containing frozen particles that is formed in deep mountain valleys of the western U.S.
po·go·noph·o·ran \ˌpō-gə-ˈnä-fə-rən\ *noun* [New Latin *Pogonophora,* from Greek *pōgōnophora,* neuter plural of *pōgōnophoros* wearing a beard, from *pōgōn* beard + *-phoros* -phore] (1963)
: any of a phylum (Pogonophora) of marine wormlike animals of uncertain systematic relationships that live in chitinous tubes, have obscure segmentation, and lack a mouth and digestive tract
— pogonophoran *adjective*
po·go stick \ˈpō-(ˌ)gō-\ *noun* [from *Pogo,* a trademark] (1921)
: a pole with a strong spring at the bottom and two footrests on which a person stands and moves along with a series of jumps
¹po·grom \ˈpō-grəm, ˈpä-; pō-ˈgräm, pə-\ *noun* [Yiddish, from Russian, literally, devastation] (1903)
: an organized massacre of helpless people; *specifically* **:** such a massacre of Jews
²pogrom *transitive verb* (1915)
: to massacre or destroy in a pogrom
po·grom·ist \ˈpō-grə-mist, ˈpä-; pō-ˈgrä-, pə-\ *noun* (1907)
: one who organizes or takes part in a pogrom
po·gy \ˈpō-gē\ *noun, plural* **pogies** [by shortening & alteration from *poghaden,* perhaps from Eastern Abenaki] (circa 1847)
: MENHADEN
poi \ˈpȯi\ *noun, plural* **poi** *or* **pois** [Hawaiian & Samoan] (1823)
: a Hawaiian food of taro root cooked, pounded, and kneaded to a paste and often allowed to ferment
-poiesis *noun combining form, plural* **-poieses** [New Latin, from Greek *poiēsis* creation — more at POESY]
: production **:** formation ⟨hemato*poiesis*⟩
-poietic *adjective combining form* [Greek *poiētikos* creative, from *poiētēs* maker — more at POET]
: productive **:** formative ⟨hemato*poietic*⟩
poi·gnance \ˈpȯi-nyən(t)s *sometimes* ˈpȯi(g)-nən(t)s\ *noun* (1769)
: POIGNANCY
poi·gnan·cy \ˈpȯi-nyən(t)-sē *sometimes* ˈpȯi(g)-nən(t)-sē\ *noun, plural* **-cies** (1730)
1 : the quality or state of being poignant
2 : an instance of poignancy
poi·gnant \ˈpȯi-nyənt *sometimes* ˈpȯi(g)-nənt\ *adjective* [Middle English *poinaunt,* from Middle French *poignant,* present participle of *poindre* to prick, sting, from Latin *pungere* — more at PUNGENT] (14th century)
1 : pungently pervasive ⟨a *poignant* perfume⟩
2 a (1) **:** painfully affecting the feelings **:** PIERCING (2) **:** deeply affecting **:** TOUCHING **b :** designed to make an impression **:** CUTTING ⟨*poignant* satire⟩
3 a : pleasurably stimulating **b :** being to the point **:** APT
synonym see PUNGENT, MOVING
— poi·gnant·ly *adverb*
poi·ki·lo·therm \pȯi-ˈkē-lə-ˌthərm, -ˈki-\ *noun* [Greek *poikilos* variegated + International Scientific Vocabulary *-therm* — more at PAINT] (1920)
: an organism (as a frog) with a variable body temperature that is usually slightly higher than the temperature of its environment **:** a cold-blooded organism
— poi·ki·lo·ther·mic \ˌpȯi-kə-lō-ˈthər-mik\ *adjective*
poi·lu \pwäl-ˈyü, pwä-ˈlü; ˈpwäl-ˌyü, ˈpwä-ˌlü; pwȧ-lᵫ\ *noun* [French, from *poilu* hairy, from Middle French, from *poil* hair, from Latin *pilus*] (1914)
: a French soldier; *especially* **:** a front-line soldier in World War I
poin·ci·ana \ˌpȯin(t)-sē-ˈa-nə, ˌp(w)än(t)-\ *noun* [New Latin, from De *Poinci,* 17th century governor of part of the French West Indies] (1731)
: any of several ornamental tropical leguminous trees or shrubs (genera *Caesalpinia* and *Delonix*) formerly placed in their own genus (*Poinciana*) — compare ROYAL POINCIANA
poin·set·tia \pȯin-ˈse-tē-ə, ÷pȯint-, ÷-ˈse-tə\ *noun* [New Latin, from Joel R. *Poinsett* (died 1851) American diplomat] (1836)
: any of several spurges (genus *Euphorbia*) with flower clusters subtended by showy involucral bracts; *especially* **:** a showy Mexican and Central American plant (*E. pulcherrima*) with tapering usually scarlet bracts that suggest petals and surround small yellow flowers

poinsettia

¹point \ˈpȯint\ *noun* [Middle English, partly from Old French, puncture, small spot, point in time or space, from Latin *punctum,* from neuter of *punctus,* past participle of *pungere* to prick; partly from Old French *pointe* sharp end, from (assumed) Vulgar Latin *puncta,* from Latin, feminine of *punctus,* past participle — more at PUNGENT] (13th century)
1 a (1) **:** an individual detail **:** ITEM (2) **:** a distinguishing detail ⟨tact is one of her strong *points*⟩ **b :** the most important essential in a discussion or matter ⟨missed the whole *point* of the joke⟩ **c :** COGENCY
2 *obsolete* **:** physical condition
3 : an end or object to be achieved **:** PURPOSE ⟨did not see what *point* there was in continuing the discussion⟩
4 a (1) **:** a geometric element of which it is postulated that at least two exist and that two suffice to determine a line (2) **:** a geometric element determined by an ordered set of coordinates **b** (1) **:** a narrowly localized place having a precisely indicated position ⟨walked to a *point* 50 yards north of the building⟩ (2) **:** a particular place **:** LOCALITY ⟨have come from distant *points*⟩ **c** (1) **:** an exact moment ⟨at this *point* I was interrupted⟩ (2) **:** a time interval immediately before something indicated **:** VERGE ⟨at the *point* of death⟩ **d** (1) **:** a particular step, stage, or degree in development ⟨had reached the *point* where nothing seemed to matter anymore⟩ (2) **:** a definite position in a scale
5 a : the terminal usually sharp or narrowly rounded part of something **:** TIP **b :** a weapon or tool having such a part and used for stabbing or piercing: as (1) **:** ARROWHEAD (2) **:** SPEARHEAD **c** (1) **:** the contact or discharge extremity of an electric device (as a spark plug or distributor) (2) *chiefly British* **:** an electric outlet
6 a : a projecting usually tapering piece of land or a sharp prominence **b** (1) **:** the tip of a projecting body part (2) **:** TINE 2 (3) *plural* **:** the extremities or markings of the extremities of an animal especially when of a color differing from the rest of the body **c** (1) **:** a railroad switch (2) **:** the tip of the angle between two rails in a railroad frog **d :** the head of the bow of a stringed instrument
7 : a short musical phrase; *especially* **:** a phrase in contrapuntal music
8 a : a very small mark **b** (1) **:** PUNCTUATION MARK; *especially* **:** PERIOD (2) **:** DECIMAL POINT
9 : a lace for tying parts of a garment together used especially in the 16th and 17th centuries
10 : one of usually 11 divisions of a heraldic shield that determines the position of a charge
11 a : one of the 32 equidistant spots of a compass card **b :** the difference of 11¼ degrees between two such successive points
12 : a small detachment ahead of an advance guard or behind a rear guard
13 a : NEEDLEPOINT 1 **b :** lace made with a bobbin
14 : one of 12 spaces marked off on each side of a backgammon board
15 : a unit of measurement: as **a** (1) **:** a unit of counting in the scoring of a game or contest (2) **:** a unit used in evaluating the strength of a bridge hand **b :** a unit of academic credit **c** (1) **:** a unit used in quoting prices (as of stocks, bonds, and commodities) (2) *plural* **:** a percentage of the face value of a loan often added as a placement fee or service charge (3) **:** a percentage of the profits of a business venture (as a motion-picture production) **d :** a unit of about 1/72 inch used especially to measure the size of type
16 : the action of pointing: as **a :** the rigidly intent attitude of a hunting dog marking game for a gunner **b :** the action in dancing of extending one leg and arching the foot so that only the tips of the toes touch the floor
17 : a position of a player in various games (as lacrosse); *also* **:** the player of such a position
18 : a number thrown on the first roll of the dice in craps which the player attempts to repeat before throwing a seven — compare MISSOUT, PASS 13
19 : credit accruing from creating a good impression ⟨scored *points* for hard work⟩
— beside the point : IRRELEVANT
— in point of : with regard to **:** in the matter of ⟨*in point of* law⟩ ⟨*in point of* fact⟩

— to the point : RELEVANT, PERTINENT ⟨a suggestion that was *to the point*⟩

[2]point (14th century)
transitive verb
1 a : to furnish with a point **:** SHARPEN ⟨*pointing* a pencil with a knife⟩ **b :** to give added force, emphasis, or piquancy to ⟨*point* up a remark⟩
2 : to scratch out the old mortar from the joints of (as a brick wall) and fill in with new material
3 a (1) **:** to mark the pauses or grammatical divisions in **:** PUNCTUATE (2) **:** to separate (a decimal fraction) from an integer by a decimal point — usually used with *off* **b :** to mark (as Hebrew words) with diacritics (as vowel points)
4 a (1) **:** to indicate the position or direction of especially by extending a finger ⟨*point* the way home⟩ (2) **:** to direct someone's attention to ⟨*point* the way to new knowledge —Elizabeth Hall⟩ — usually used with *out* or *up* ⟨*point* out a mistake⟩ ⟨*points* up the difference⟩ **b** *of a hunting dog* **:** to indicate the presence and place of (game) by a point
5 a : to cause to be turned in a particular direction ⟨*point* a gun⟩ ⟨*pointed* the boat upstream⟩ **b :** to extend (a leg) and arch (the foot) in executing a point in dancing
intransitive verb
1 a : to indicate the fact or probability of something specified ⟨everything *points* to a bright future⟩ **b :** to indicate the position or direction of something especially by extending a finger ⟨*point* at the map⟩ **c :** to direct attention ⟨can *point* with pride to their own traditions⟩ **d :** to point game ⟨a dog that *points* well⟩
2 a : to lie extended, aimed, or turned in a particular direction ⟨a directional arrow that *pointed* to the north⟩ **b :** to execute a point in dancing
3 *of a ship* **:** to sail close to the wind
4 : to train for a particular contest

point–blank \ˈpȯint-ˈblaŋk\ *adjective* (1591)
1 a : marked by no appreciable drop below initial horizontal line of flight **b :** so close to a target that a missile fired will travel in a straight line to the mark
2 : DIRECT, BLUNT ⟨a *point-blank* refusal⟩
— point–blank *adverb*

point count *noun* (1950)
: a method of evaluating the strength of a hand in bridge by counting points for each high card and usually for long or short suits; *also* **:** the value of a hand so evaluated

point d'ap·pui \ˌpwa^{n}(n)-da-ˈpwē\ *noun, plural* **points d'appui** *same*\ [French, literally, point of support] (1819)
: FOUNDATION, BASE

point–de·vice \ˌpȯint-di-ˈvīs\ *adjective* [Middle English *at point devis* at a fixed point] (1526)
archaic **:** marked by punctilious attention to detail **:** METICULOUS
— point–device *adverb, archaic*

pointe \ˈpwa^{n}(n)t\ *noun* [French *pointe (du pied)*, literally, tiptoe] (1846)
: a ballet position in which the body is balanced on the extreme tip of the toe

[1]point·ed \ˈpȯin-təd\ *adjective* (14th century)
1 a : having a point **b :** being an arch with a pointed crown; *also* **:** marked by the use of a pointed arch ⟨*pointed* architecture⟩
2 a : being to the point **:** PERTINENT **b :** aimed at a particular person or group
3 : CONSPICUOUS, MARKED ⟨*pointed* indifference⟩
— point·ed·ly *adverb*
— point·ed·ness *noun*

[2]pointed *adjective* [short for *appointed*] (1523)
obsolete **:** SET, FIXED

poin·telle \ˌpȯin-ˈtel\ *noun* [perhaps from [1]*point* + *-elle* (as in *dentelle* lace)] (1953)
: an openwork design (as in knitted fabric) typically in the shape of chevrons; *also* **:** a fabric with this design

point·er \ˈpȯin-tər\ *noun* (1574)
1 a *plural, capitalized* **:** the two stars in the Big Dipper a line through which points to the North Star **b :** one that points out; *especially* **:** a rod used to direct attention **c :** a computer memory address that contains another address (as of desired data)
2 : a large strong slender smooth-haired gundog that hunts by scent and indicates the presence of game by pointing
3 : one that furnishes with points
4 : a useful suggestion or hint **:** TIP

point estimate *noun* (1966)
: the single value assigned to a parameter in point estimation

point estimation *noun* (1962)
: estimation in which a single value is assigned to a parameter

point guard *noun* (1970)
: a guard in basketball who is chiefly responsible for running the offense

poin·til·lism \ˈpwa^{n}(n)-tē-ˌyi-zəm, ˈpȯin-t^{ə}l-ˌi-zəm\ *noun, often capitalized* [French *pointillisme*, from *pointiller* to stipple, from *point* spot — more at POINT] (1901)
: the theory or practice in art of applying small strokes or dots of color to a surface so that from a distance they blend together
— poin·til·list \ˌpwa^{n}(n)-tē-ˈyēst, ˈpȯin-t^{ə}l-ist\ *noun*

poin·til·lis·tic \ˌpwa^{n}(n)-tē-ˈyis-tik, ˌpȯin-t^{ə}l-ˈis-\ *also* **point·til·list** \ˌpwa^{n}(n)-tē-ˈyēst, ˈpȯin-t^{ə}l-ist\ *adjective* (1922)
1 : composed of many discrete details or parts
2 : of, relating to, or characteristic of pointillism or pointillists

point lace *noun* (1672)
: NEEDLEPOINT 1

point·less \ˈpȯint-ləs\ *adjective* (1726)
1 : devoid of meaning **:** SENSELESS ⟨a *pointless* remark⟩
2 : devoid of effectiveness **:** FLAT ⟨*pointless* attempts to be funny⟩
— point·less·ly *adverb*
— point·less·ness *noun*

point man *noun* (1903)
: a soldier who goes ahead of a patrol; *broadly* **:** one who is in the forefront (as on a political issue)

point mutation *noun* (1925)
: mutation due to intramolecular gene reorganization (as by substitution, addition, or deletion of a nucleotide)

point of accumulation (1929)
: LIMIT POINT

point of departure (1857)
: a starting point especially in a discussion

point of honor (1612)
: a matter seriously affecting one's honor

point of inflection (1743)
: INFLECTION POINT

point of no return (1941)
1 : the point in the flight of an aircraft beyond which the remaining fuel will be insufficient for a return to the starting point with the result that the craft must proceed
2 : a critical point at which turning back or reversal is not possible

point of view (1720)
: a position from which something is considered or evaluated **:** STANDPOINT

point set topology *noun* (1957)
: a branch of topology concerned with the properties and theory of topological spaces and metric spaces developed with emphasis on set theory

point source *noun* (1903)
: a source of radiation (as light) that is concentrated at a point and considered as having no spatial extension

point–to–point *noun* (1898)
: a cross-country steeplechase

pointy \ˈpȯin-tē\ *adjective* **point·i·er; -est** (1644)
1 : coming to a rather sharp point
2 : having parts that stick out sharply here and there

pointy–head \ˈpȯin-tē-ˌhed\ *noun* (1968)
: INTELLECTUAL — usually used disparagingly
— pointy–head·ed \-ˌhe-dəd\ *adjective*

[1]poise \ˈpȯiz\ *noun* [Middle English *poyse* weight, heaviness, from Middle French *pois*, from Latin *pensum*, from neuter of *pensus*, past participle of *pendere* to weigh — more at PENDANT] (1555)
1 : a stably balanced state **:** EQUILIBRIUM ⟨a *poise* between widely divergent impulses —F. R. Leavis⟩
2 a : easy self-possessed assurance of manner **:** gracious tact in coping or handling; *also* **:** the pleasantly tranquil interaction between persons of poise ⟨no angry outbursts marred the *poise* of the meeting⟩ **b :** a particular way of carrying oneself **:** BEARING, CARRIAGE
synonym see TACT

[2]poise *verb* **poised; pois·ing** [Middle English, to weigh, ponder, from Middle French *pois-*, stem of *peser*, from Latin *pensare* — more at PENSIVE] (1598)
transitive verb
1 a : BALANCE; *especially* **:** to hold or carry in equilibrium ⟨carried a water jar *poised* on her head⟩ **b :** to hold supported or suspended without motion in a steady position ⟨*poised* her fork and gave her guest a knowing look —Louis Bromfield⟩
2 : to hold or carry (the head) in a particular way
3 : to put into readiness **:** BRACE
intransitive verb
1 : to become drawn up into readiness
2 : HOVER

[3]poise \ˈpwäz\ *noun* [French, from Jean Louis Marie *Poiseuille* (died 1869) French physician and anatomist] (1913)
: a centimeter-gram-second unit of viscosity equal to the viscosity of a fluid that would require a shearing force of one dyne to move a square-centimeter area of either of two parallel layers of fluid one centimeter apart with a velocity of one centimeter per second relative to the other layer with the space between the layers being filled with the fluid

poised \ˈpȯizd\ *adjective* (circa 1643)
: having poise: **a :** marked by balance or equilibrium **b :** marked by easy composure of manner or bearing

poi·sha \ˈpȯi-shə\ *noun, plural* **poisha** [Bengali, from Hindi *paisā*] (circa 1976)
: the paisa of Bangladesh

[1]poi·son \ˈpȯi-z^{ə}n\ *noun* [Middle English, from Middle French, drink, poisonous drink, poison, from Latin *potion-*, *potio* drink — more at POTION] (13th century)
1 a : a substance that through its chemical action usually kills, injures, or impairs an organism **b** (1) **:** something destructive or harmful (2) **:** an object of aversion or abhorrence
2 : a substance that inhibits the activity of another substance or the course of a reaction or process ⟨a catalyst *poison*⟩
word history see VENOM

[2]poison *transitive verb* **poi·soned; poi·son·ing** \ˈpȯiz-niŋ, ˈpȯi-z^{ə}n-iŋ\ (14th century)
1 a : to injure or kill with poison **b :** to treat, taint, or impregnate with or as if with poison
2 : to exert a baneful influence on **:** CORRUPT ⟨*poisoned* their minds⟩
3 : to inhibit the activity, course, or occurrence of
— poi·son·er \ˈpȯiz-nər, ˈpȯi-z^{ə}n-ər\ *noun*

[3]poison *adjective* (1530)

1 **:** POISONOUS, VENOMOUS ⟨a *poison* plant⟩ ⟨a *poison* tongue⟩
2 **:** impregnated with poison **:** POISONED ⟨a *poison* arrow⟩

poison gas *noun* (1915)
: a poisonous gas or a liquid or a solid giving off poisonous vapors designed (as in chemical warfare) to kill, injure, or disable by inhalation or contact

poison hemlock *noun* (circa 1818)
1 **:** a large branching biennial poisonous herb (*Conium maculatum*) of the carrot family with finely divided leaves and white flowers
2 **:** WATER HEMLOCK

poison ivy *noun* (1784)
1 **a :** a climbing plant (*Rhus radicans*) of the cashew family that is especially common in the eastern and central U.S., that has ternate leaves, greenish flowers, and white berries, and that produces an acutely irritating oil causing a usually intensely itching skin rash **b :** any of several plants (as poison oak) closely related to poison ivy
2 **:** a skin rash caused by poison ivy

poison ivy 1a

poison oak *noun* (1743)
1 **:** any of several plants of the same genus (*Rhus*) as poison ivy that produce an oil with similar irritating properties: **a :** a bushy plant (*Rhus diversiloba*) of the Pacific coast **b :** a bushy plant (*Rhus toxicodendron* synonym *R. quercifolia*) of the southeastern U.S.
2 **:** POISON IVY 1

poi·son·ous \ˈpȯiz-nəs, ˈpȯi-zᵊn-əs\ *adjective* (circa 1580)
1 **:** having the properties or effects of poison **:** VENOMOUS
2 **:** DESTRUCTIVE, HARMFUL
3 **:** SPITEFUL, MALICIOUS
— **poi·son·ous·ly** *adverb*

poison–pen *adjective* (1925)
: written with malice and spite and usually anonymously ⟨*poison-pen* letter⟩

poison pill *noun* (1983)
: a financial tactic (as increasing indebtedness) used by a company to deter an unwanted takeover

poison sumac *noun* (1817)
: a smooth American swamp shrub (*Rhus vernix*) that has pinnate leaves, greenish flowers, and greenish white berries and produces an irritating oil — called also *poison dogwood*

poi·son·wood \ˈpȯi-zᵊn-ˌwu̇d\ *noun* (1721)
: a poisonous tree (*Metopium toxiferum*) of the cashew family that is native to Florida and the West Indies and has compound leaves, greenish paniculate flowers, and orange-yellow fruits

Pois·son distribution \pwä-ˈsōⁿ-\ *noun* [Siméon D. *Poisson* (died 1840) French mathematician] (1922)
: a probability density function that is often used as a mathematical model of the number of outcomes obtained in a suitable interval of time and space, that has its mean equal to its variance, that is used as an approximation to the binomial distribution, and that has the form

$$f(x) = \frac{e^{-\mu}\mu^x}{x!} \text{ where } \mu$$

is the mean and x takes on nonnegative integral values

Poisson's ratio *noun* [S. *Poisson*] (1886)
: the ratio of transverse to longitudinal strain in a material under tension

¹poke \ˈpōk\ *noun* [Middle English, from Old North French — more at POCKET] (13th century)
1 *chiefly Southern & Midland* **:** BAG, SACK
2 **a :** WALLET **b :** PURSE

²poke *verb* **poked; pok·ing** [Middle English; akin to Middle Dutch *poken* to poke] (14th century)
transitive verb
1 **a** (1) **:** PROD, JAB ⟨*poked* him in the ribs⟩ (2) **:** to urge or stir by prodding or jabbing (3) **:** to cause to prod **:** THRUST ⟨*poked* a stick at the snake⟩ **b** (1) **:** PIERCE, STAB (2) **:** to produce by piercing, stabbing, or jabbing ⟨*poke* a hole⟩ **c** (1) **:** HIT, PUNCH ⟨*poked* him in the nose⟩ (2) **:** to deliver (a blow) with the fist
2 **a :** to cause to project ⟨*poked* her head out of the window⟩ **b :** to make (one's way) by poking ⟨*poked* his way through the ruins⟩ **c :** to interpose or interject in a meddlesome manner ⟨asked him not to *poke* his nose into other people's business⟩
intransitive verb
1 **a :** to make a prodding, jabbing, or thrusting movement especially repeatedly **b :** to strike out at something
2 **a :** to look about or through something without system **:** RUMMAGE ⟨*poking* around in the attic⟩ **b :** MEDDLE
3 **:** to move or act slowly or aimlessly ⟨just *poked* around and didn't accomplish much⟩
4 **:** to become stuck out or forward **:** PROTRUDE
— **poke fun at :** RIDICULE, MOCK

³poke *noun* (circa 1796)
1 **a :** a quick thrust **:** JAB **b :** a blow with the fist **:** PUNCH
2 **:** a projecting brim on the front of a woman's bonnet

⁴poke *noun* [perhaps modification of Virginia Algonquian *pocone, poughkone* puccoon] (1708)
: POKEWEED

poke·ber·ry \ˈpōk-ˌber-ē\ *noun* (1774)
: the berry of the pokeweed; *also* **:** POKEWEED

poke bonnet *noun* (1820)
: a woman's bonnet with a projecting brim at the front

¹pok·er \ˈpō-kər\ *noun* (1534)
: one that pokes; *especially* **:** a metal rod for stirring a fire

²po·ker \ˈpō-kər\ *noun* [probably modification of French *poque,* a card game similar to poker] (1836)
: any of several card games in which a player bets that the value of his or her hand is greater than that of the hands held by others, in which each subsequent player must either equal or raise the bet or drop out, and in which the player holding the highest hand at the end of the betting wins the pot

poker hands in descending value: *1* five of a kind, *2* royal flush, *3* straight flush, *4* four of a kind, *5* full house, *6* flush, *7* straight, *8* three of a kind, *9* two pairs, *10* one pair

poker face *noun* [²*poker;* from the poker player's need to conceal emotions during play] (1885)
: an inscrutable face that reveals no hint of a person's thoughts or feelings
— **po·ker–faced** \ˈpō-kər-ˌfāst\ *adjective*

poke·weed \ˈpōk-ˌwēd\ *noun* (1751)
: a coarse American perennial herb (*Phytolacca americana* of the family Phytolaccaceae, the pokeweed family) with racemose white flowers, dark purple juicy berries, a poisonous root, and young shoots sometimes used as potherbs

po·key \ˈpō-kē\ *noun, plural* **pokeys** [origin unknown] (circa 1919)
slang **:** JAIL

poky *or* **pok·ey** \ˈpō-kē\ *adjective* **pok·i·er; -est** [²*poke*] (1849)
1 **:** small and cramped
2 **:** SHABBY, DULL
3 **:** annoyingly slow
— **pok·i·ly** \-kə-lē\ *adverb*
— **pok·i·ness** \-kē-nəs\ *noun*

pol \ˈpäl\ *noun* (circa 1942)
: POLITICIAN

Po·la·bi·an \pō-ˈlä-bē-ən, -ˈlā-\ *noun* [*Polab,* ultimately from Polabian *po* on + *Låbí,* the Elbe River] (1866)
1 *or* **Po·lab** \pō-ˈläb\ **:** a member of a Slavic people formerly dwelling in the basin of the Elbe and on the Baltic coast of Germany
2 **:** the extinct West Slavic language of the Polabians

Po·lack \ˈpō-ˌläk, -ˌlak\ *noun* [Polish *polak*] (1574)
1 *obsolete* **:** a native or inhabitant of Poland
2 **:** a person of Polish birth or descent — usually used disparagingly

Po·land Chi·na \ˈpō-lən(d)-ˈchī-nə\ *noun* [*Poland,* Europe + *China,* Asia] (1879)
: any of an American breed of large white-marked black swine

¹po·lar \ˈpō-lər\ *adjective* [New Latin *polaris,* from Latin *polus* pole] (1551)
1 **a :** of or relating to a geographical pole or the region around it **b :** coming from or having the characteristics of such a region **c** (1) **:** passing over a celestial body's north and south poles ⟨a satellite in a *polar* orbit⟩ (2) **:** traveling in a polar orbit ⟨a *polar* satellite⟩
2 **:** of or relating to one or more poles (as of a magnet)
3 **:** serving as a guide
4 **:** diametrically opposite
5 **:** exhibiting polarity; *especially* **:** having a dipole or characterized by molecules having dipoles ⟨a *polar* solvent⟩
6 **:** resembling a pole or axis around which all else revolves **:** PIVOTAL
7 **:** of, relating to, or expressed in polar coordinates ⟨*polar* equations⟩; *also* **:** of or relating to a polar coordinate system

²polar *noun* (1848)
: a straight line related to a point; *specifically* **:** the straight line joining the points of contact of the tangents from a point exterior to a conic section

polar bear *noun* (1781)
: a large creamy-white bear (*Thalarctos maritimus* synonym *Ursus maritimus*) that inhabits arctic regions

polar bear

polar body *noun* (1888)
: a cell that separates from an oocyte during meiosis and that contains a nucleus produced in the first or second meiotic division but very little cytoplasm

polar circle *noun* (1551)
: either of the two parallels of latitude each at a distance from a pole of the earth equal to about 23 degrees 27 minutes

polar coordinate *noun* (1816)
: either of two numbers that locate a point in a plane by its distance from a fixed point on a line and the angle this line makes with a fixed line

polar front *noun* (1920)
: the boundary between the cold air of a polar region and the warmer air of lower latitudes

po•lar•im•e•ter \ˌpō-lə-ˈri-mə-tər\ *noun* [International Scientific Vocabulary, from *polarization*] (circa 1859)
1 : an instrument for determining the amount of polarization of light or the proportion of polarized light in a partially polarized ray
2 : a polariscope for measuring the amount of rotation of the plane of polarization especially by liquids
— **po•lar•i•met•ric** \pō-ˌlar-ə-ˈme-trik\ *adjective*
— **po•lar•im•e•try** \ˌpō-lə-ˈri-mə-trē\ *noun*

Po•lar•is \pə-ˈlar-əs, -ˈlär-\ *noun* [New Latin, from *polaris* polar]
: NORTH STAR

po•lar•ise *British variant of* POLARIZE

po•lar•i•scope \pō-ˈlar-ə-ˌskōp\ *noun* [International Scientific Vocabulary, from *polarization*] (1829)
1 : an instrument for studying the properties of or examining substances in polarized light
2 : POLARIMETER 2
— **po•lar•i•scop•ic** \-ˌlar-ə-ˈskä-pik\ *adjective*

po•lar•i•ty \pō-ˈlar-ə-tē, pə-\ *noun, plural* **-ties** (1646)
1 : the quality or condition inherent in a body that exhibits opposite properties or powers in opposite parts or directions or that exhibits contrasted properties or powers in contrasted parts or directions **:** the condition of having poles
2 : attraction toward a particular object or in a specific direction
3 : the particular state either positive or negative with reference to the two poles or to electrification
4 a : diametrical opposition **b :** an instance of such opposition

po•lar•i•za•tion \ˌpō-lə-rə-ˈzā-shən\ *noun* (1812)
1 : the action of polarizing or state of being or becoming polarized: as **a** (1) **:** the action or process of affecting radiation and especially light so that the vibrations of the wave assume a definite form (2) **:** the state of radiation affected by this process **b :** an increase in the resistance of an electrolytic cell often caused by the deposition of gas on one or both electrodes **c :** MAGNETIZATION
2 a : division into two opposites **b :** concentration about opposing extremes of groups or interests formerly ranged on a continuum

po•lar•ize \ˈpō-lə-ˌrīz\ *verb* **-ized; -iz•ing** [French *polariser,* from New Latin *polaris* polar] (1811)
transitive verb
1 : to cause (as light waves) to vibrate in a definite pattern
2 : to give physical polarity to
3 : to break up into opposing factions or groupings ⟨a campaign that *polarized* the electorate⟩
4 : CONCENTRATE 1 ⟨recreate a cohesive rock community by *polarizing* . . . an amorphous, fragmented audience —Ellen Willis⟩
intransitive verb
: to become polarized
— **po•lar•iz•abil•i•ty** \ˌpō-lə-ˌrī-zə-ˈbi-lə-tē\ *noun*
— **po•lar•iz•able** \ˌpō-lə-ˈrī-zə-bəl\ *adjective*

polar nucleus *noun* (1882)
: either of the two nuclei of a seed plant embryo sac that are destined to form endosperm

po•lar•og•ra•phy \ˌpō-lə-ˈrä-grə-fē\ *noun* [International Scientific Vocabulary, from *polarization*] (1936)
: a method of qualitative or quantitative analysis based on current-voltage curves obtained during electrolysis of a solution with a steadily increasing electromotive force
— **po•lar•o•graph•ic** \pō-ˌlar-ə-ˈgra-fik\ *adjective*
— **po•lar•o•graph•i•cal•ly** \-fi-k(ə-)lē\ *adverb*

Po•lar•oid \ˈpō-lə-ˌrȯid\ *trademark*
— used especially for a light-polarizing material used especially in eyeglasses and lamps to prevent glare or for a camera that develops pictures instantly

po•lar•on \ˈpō-lə-ˌrän\ *noun* [International Scientific Vocabulary *polar* + ²*-on*] (1946)
: a conducting electron in an ionic crystal together with the induced polarization of the surrounding lattice

pol•der \ˈpōl-dər\ *noun* [Dutch] (1604)
: a tract of low land (as in the Netherlands) reclaimed from a body of water (as the sea)

¹pole \ˈpōl\ *noun* [Middle English, from Old English *pāl* stake, pole, from Latin *palus* stake; akin to Latin *pangere* to fix — more at PACT] (before 12th century)
1 a : a long slender usually cylindrical object (as a length of wood) **b :** a shaft which extends from the front axle of a wagon between wheelhorses and by which the wagon is drawn **:** TONGUE **c :** a long staff of wood, metal, or fiberglass used in the pole vault
2 a : a varying unit of length; *especially* **:** one measuring 16½ feet (5.03 meters) **b :** a unit of area equal to a square rod (25.293 square meters)
3 : a tree with a breast-high diameter of from 4 to 12 inches (10 to 30 centimeters)
4 : the inside front row position on the starting line for a race

²pole *verb* **poled; pol•ing** (1573)
transitive verb
1 : to act upon with a pole
2 : to impel or push with a pole
intransitive verb
1 : to propel a boat with a pole
2 : to use ski poles to gain speed

³pole *noun* [Middle English, from Latin *polus,* from Greek *polos* pivot, pole; akin to Greek *pelesthai* to become, Sanskrit *carati* he moves, wanders — more at WHEEL] (14th century)
1 : either extremity of an axis of a sphere and especially of the earth's axis
2 a : either of two related opposites **b :** a point of guidance or attraction
3 a : either of the two terminals of an electric cell, battery, generator, or motor **b :** one of two or more regions in a magnetized body at which the magnetic flux density is concentrated
4 : either of two morphologically or physiologically differentiated areas at opposite ends of an axis in an organism or cell — see BLASTULA illustration
5 a : the fixed point in a system of polar coordinates that serves as the origin **b :** the point of origin of two tangents to a conic section that determine a polar
— **poles apart :** as diametrically opposed as possible

Pole \ˈpōl\ *noun* [German, of Slavic origin; akin to Polish *polak* Pole] (1535)
1 : a native or inhabitant of Poland
2 : a person of Polish descent

¹pole•ax \ˈpō-ˌlaks\ *noun* [Middle English *polax, pollax,* from *pol, polle* poll + *ax*] (14th century)
1 : a battle-ax with a short handle and often a hook or spike opposite the blade; *also* **:** one with a long handle used as an ornamental weapon
2 : an ax used in slaughtering cattle

²poleax *transitive verb* (1882)
: to attack, strike, or fell with or as if with a poleax

pole bean *noun* (circa 1770)
: a cultivated bean that is usually trained to grow upright on supports

pole•cat \ˈpōl-ˌkat\ *noun, plural* **polecats** *or* **polecat** [Middle English *polcat,* probably from Middle French *poul, pol* cock + Middle English *cat;* probably from its preying on poultry — more at PULLET] (14th century)
1 : any of several carnivorous mammals (as of the genera *Mustela* or *Vormela*) of the weasel family; *especially* **:** a brown to black European mammal (*M. putorius*) from which the domesticated ferret is derived
2 : SKUNK
word history see FERRET

poleis *plural of* POLIS

pole•less \ˈpōl-ləs\ *adjective* (1647)
: having no pole

po•lem•ic \pə-ˈle-mik\ *noun* [French *polémique,* from Middle French, from *polemique* controversial, from Greek *polemikos* warlike, hostile, from *polemos* war; perhaps akin to Greek *pelemizein* to shake, Old English ea*lfelo* baleful] (1638)
1 a : an aggressive attack on or refutation of the opinions or principles of another **b :** the art or practice of disputation or controversy — usually used in plural but singular or plural in construction
2 : an aggressive controversialist **:** DISPUTANT
— **po•lem•i•cist** \-ˈle-mə-sist\ *noun*

po•lem•i•cal \-mi-kəl\ *also* **po•lem•ic** \-mik\ *adjective* (1640)
1 : of, relating to, or being a polemic **:** CONTROVERSIAL
2 : engaged in or addicted to polemics **:** DISPUTATIOUS
— **po•lem•i•cal•ly** \-mi-k(ə-)lē\ *adverb*

po•lem•i•cize \-ˈle-mə-ˌsīz\ *intransitive verb* **-cized; -ciz•ing** (1950)
: to engage in controversy **:** deliver a polemic

po•le•mist \pə-ˈle-mist, ˈpä-lə-mist\ *noun* (1825)
: one skilled in or given to polemics

pol•e•mize \ˈpä-lə-ˌmīz\ *intransitive verb* **-mized; -miz•ing** (1828)
: POLEMICIZE

pol•e•mo•ni•um \ˌpä-lə-ˈmō-nē-əm\ *noun* [New Latin, from Greek *polemōnion,* a plant] (1900)
: JACOB'S LADDER 1

po•len•ta \pō-ˈlen-tə, pə-, -ˌtä\ *noun* [Italian, from Latin, crushed and hulled barley; akin to Latin *pollen* fine flour] (1764)
: mush made of chestnut meal, cornmeal, semolina, or farina

pol•er \ˈpō-lər\ *noun* (1848)
: one that poles; *especially* **:** one that poles a boat

pole•star \ˈpōl-ˌstär\ *noun*
1 : NORTH STAR
2 a : a directing principle **:** GUIDE **b :** a center of attraction

pole vault *noun* (circa 1890)
: a vault with the aid of a pole; *specifically* **:** a field event consisting of a vault for height over a crossbar
— **pole–vault** *intransitive verb*
— **pole–vaulter** *noun*

pole•ward \ˈpōl-wərd\ *adverb or adjective* (1875)
: toward or in the direction of a pole of the earth ⟨as the sun moves *poleward*⟩ ⟨*poleward* variation in temperature⟩

¹po•lice \pə-ˈlēs\ *transitive verb* **po•liced; po•lic•ing** [in sense 1, from Middle French *policier,* from *police* conduct of public affairs; in other senses, from ²*police*] (1589)
1 *archaic* **:** GOVERN
2 : to control, regulate, or keep in order by use of police
3 : to make clean and put in order
4 a : to supervise the operation, execution, or administration of to prevent or detect and prosecute violations of rules and regulations **b :** to exercise such supervision over the policies and activities of
5 : to perform the functions of a police force in or over

²police *noun, plural* **police** *often attributive* [French, from Late Latin *politia* government, administration, from Greek *politeia,* from *politēs* citizen, from *polis* city, state; akin to Sanskrit *pur* rampart, Lithuanian *pilis* castle] (1716)
1 a : the internal organization or regulation of a political unit through exercise of governmental powers especially with respect to general comfort, health, morals, safety, or prosperity **b :** control and regulation of affairs affecting the general order and welfare of any unit or area **c :** the system of laws for effecting such control
2 a : the department of government concerned primarily with maintenance of public order, safety, and health and enforcement of laws and possessing executive, judicial, and legislative powers **b :** the department of government charged with prevention, detection, and prosecution of public nuisances and crimes
3 a : POLICE FORCE **b** *plural* **:** POLICE OFFICERS
4 a : a private organization resembling a police force ⟨campus *police*⟩ **b** *plural* **:** the members of a private police organization
5 a : the action or process of cleaning and putting in order **b :** military personnel detailed to perform this function

police action *noun* (1933)
: a localized military action undertaken without formal declaration of war by regular armed forces against persons (as guerrillas or aggressors) held to be violators of international peace and order

police court *noun* (1823)
: a court of record that has jurisdiction over various minor offenses (as breach of the peace) and the power to bind over for trial in a superior court or for a grand jury persons accused of more serious offenses

police dog *noun* (1908)
1 : a dog trained to assist police (as in drug detection)
2 : GERMAN SHEPHERD

police force *noun* (1838)
: a body of trained officers entrusted by a government with maintenance of public peace and order, enforcement of laws, and prevention and detection of crime

po·lice·man \pə-ˈlēs-mən\ *noun* (1801)
1 : a member of a police force
2 : one held to resemble a policeman ⟨making the United States the *policeman* for the whole wide world —R. B. Long⟩

police officer *noun* (1800)
: a member of a police force

police power *noun* (1827)
: the inherent power of a government to exercise reasonable control over persons and property within its jurisdiction in the interest of the general security, health, safety, morals, and welfare except where legally prohibited

police procedural *noun, plural* **police procedurals** (1967)
: a mystery story written from the point of view of the police investigating the crime

police reporter *noun* (1834)
: a reporter regularly assigned to cover police news (as crimes and arrests)

police state *noun* (1865)
: a political unit characterized by repressive governmental control of political, economic, and social life usually by an arbitrary exercise of power by police and especially secret police in place of regular operation of administrative and judicial organs of the government according to publicly known legal procedures

police station *noun* (1846)
: the headquarters of the police for a locality

po·lice·wom·an \pə-ˈlēs-ˌwu̇-mən\ *noun* (1853)
: a woman who is a member of a police force

¹pol·i·cy \ˈpä-lə-sē\ *noun, plural* **-cies** *often attributive* [Middle English *policie* government, policy, from Middle French, government, regulation, from Late Latin *politia*] (15th century)
1 a : prudence or wisdom in the management of affairs **b :** management or procedure based primarily on material interest
2 a : a definite course or method of action selected from among alternatives and in light of given conditions to guide and determine present and future decisions **b :** a high-level overall plan embracing the general goals and acceptable procedures especially of a governmental body

²policy *noun, plural* **-cies** [alteration of earlier *police,* from Middle French, certificate, from Old Italian *polizza,* modification of Medieval Latin *apodixa* receipt, from Middle Greek *apodeixis,* from Greek, proof, from *apodeiknynai* to demonstrate — more at APODICTIC] (1565)
1 : a writing whereby a contract of insurance is made
2 a : a daily lottery in which participants bet that certain numbers will be drawn from a lottery wheel **b :** NUMBER 7a

pol·i·cy·hold·er \ˈpä-lə-sē-ˌhōl-dər\ *noun* (1851)
: the owner of an insurance policy

policy science *noun* (1950)
: a social science dealing with the making of high-level policy (as in a government or business)

po·lio \ˈpō-lē-ˌō\ *noun* (1931)
: POLIOMYELITIS

po·lio·my·eli·tis \ˌpō-lē-(ˌ)ō-ˌmī-ə-ˈlī-təs\ *noun* [New Latin, from Greek *polios* gray + *myelos* marrow — more at FALLOW, MYEL-] (1878)
: an acute infectious virus disease characterized by fever, motor paralysis, and atrophy of skeletal muscles often with permanent disability and deformity and marked by inflammation of nerve cells in the anterior gray matter in each lateral half of the spinal cord — called also *infantile paralysis*

po·lio·vi·rus \ˈpō-lē-(ˌ)ō-ˌvī-rəs\ *noun* [New Latin, from *polio*myelitis + *virus*] (1953)
: an enterovirus that occurs in several antigenically distinct forms and is the causative agent of human poliomyelitis

po·lis \ˈpä-ləs\ *noun, plural* **po·leis** \ˈpä-ˌlās\ [Greek — more at POLICE] (1894)
: a Greek city-state; *broadly* **:** a state or society especially when characterized by a sense of community

-polis *noun combining form* [Late Latin, from Greek, from *polis*]
: city ⟨megalo*polis*⟩

¹pol·ish \ˈpä-lish\ *verb* [Middle English *polisshen,* from Middle French *poliss-,* stem of *polir,* from Latin *polire*] (14th century)
transitive verb
1 : to make smooth and glossy usually by friction **:** BURNISH
2 : to smooth, soften, or refine in manners or condition
3 : to bring to a highly developed, finished, or refined state **:** PERFECT
intransitive verb
: to become smooth or glossy by or as if by friction
— pol·ish·er *noun*

²polish *noun* (1704)
1 a : a smooth glossy surface **:** LUSTER **b :** freedom from rudeness or coarseness **:** CULTURE **c :** a state of high development or refinement
2 : the action or process of polishing
3 : a preparation that is used to produce a gloss and often a color for the protection and decoration of a surface ⟨furniture *polish*⟩ ⟨nail *polish*⟩

¹Pol·ish \ˈpō-lish\ *adjective* [*Pole*] (1674)
: of, relating to, or characteristic of Poland, the Poles, or Polish

²Polish *noun* (1784)
: the Slavic language of the Poles

polish off *transitive verb* (1829)
: to finish off or dispose of rapidly or completely

po·lit·bu·ro \ˈpä-lət-ˌbyu̇r-(ˌ)ō, ˈpō-lət-, pə-ˈlit-\ *noun* [Russian *politbyuro,* from *politicheskoe byuro* political bureau] (1925)
: the principal policy-making and executive committee of a Communist party

po·lite \pə-ˈlīt\ *adjective* **po·lit·er; -est** [Latin *politus,* from past participle of *polire*] (1501)
1 a : of, relating to, or having the characteristics of advanced culture **b :** marked by refined cultural interests and pursuits especially in arts and belles lettres
2 a : showing or characterized by correct social usage **b :** marked by an appearance of consideration, tact, deference, or courtesy **c :** marked by a lack of roughness or crudities ⟨*polite* literature⟩
synonym see CIVIL
— po·lite·ly *adverb*
— po·lite·ness *noun*

po·li·tesse \ˌpä-li-ˈtes, ˌpȯ-\ *noun* [French, from Middle French, cleanness, from Old Italian *pulitezza,* from *pulito,* past participle of *pulire* to polish, clean, from Latin *polire*] (1717)
: formal politeness **:** DECOROUSNESS

pol·i·tic \ˈpä-lə-ˌtik\ *adjective* [Middle English *politik,* from Middle French *politique,* from Latin *politicus,* from Greek *politikos,* from *politēs* citizen — more at POLICE] (15th century)
1 : POLITICAL
2 : characterized by shrewdness in managing, contriving, or dealing
3 : sagacious in promoting a policy
4 : shrewdly tactful
synonym see EXPEDIENT, SUAVE

po·lit·i·cal \pə-ˈli-ti-kəl\ *adjective* [Latin *politicus*] (1551)
1 a : of or relating to government, a government, or the conduct of government **b :** of, relating to, or concerned with the making as distinguished from the administration of governmental policy
2 : of, relating to, involving, or involved in politics and especially party politics
3 : organized in governmental terms ⟨*political* units⟩
4 : involving or charged or concerned with acts against a government or a political system ⟨*political* prisoners⟩
— po·lit·i·cal·ly \-k(ə-)lē\ *adverb*

political action committee *noun* (1944)
: a group formed (as by an industry or an issue-oriented organization) to raise and contribute money to the campaigns of candidates likely to advance the group's interests

political correctness *noun* (1990)
: conformity to a belief that language and practices which could offend political sensibilities should be eliminated

political economy *noun* (1740)
1 : ECONOMICS
2 : the theory or study of the role of public policy in influencing the economic and social welfare of a political unit
— political economist *noun*

po·lit·i·cal·ize \pə-ˈli-ti-kə-ˌlīz\ *transitive verb* **-ized; -izing** (1869)
: POLITICIZE
— po·lit·i·cal·i·za·tion \-ˌli-ti-kə-lə-ˈzā-shən\ *noun*

politically correct *adjective* (1936)
: conforming to a belief that language and practices which could offend political sensibilities (as in matters of sex or race) should be eliminated

political science *noun* (1779)
: a social science concerned chiefly with the description and analysis of political and especially governmental institutions and processes
— political scientist *noun*

pol·i·ti·cian \ˌpä-lə-ˈti-shən\ *noun* (1589)

1 : a person experienced in the art or science of government; *especially* **:** one actively engaged in conducting the business of a government
2 a : a person engaged in party politics as a profession **b :** a person primarily interested in political office for selfish or other narrow usually short-sighted reasons
po·lit·i·cise *British variant of* POLITICIZE
po·lit·i·cize \pə-'li-tə-ˌsīz\ *transitive verb* **-cized; -ciz·ing** (1846)
: to give a political tone or character to ⟨an attempt to *politicize* the civil service⟩
— po·lit·i·ci·za·tion \-ˌli-tə-sə-'zā-shən\ *noun*
pol·i·tick \'pä-lə-ˌtik\ *intransitive verb* [back-formation from *politicking,* noun, from *politics* + [3]*-ing*] (circa 1934)
: to engage in often partisan political discussion or activity
— pol·i·tick·er *noun*
po·lit·i·co \pə-'li-ti-ˌkō\ *noun, plural* **-cos** *also* **-coes** [Italian *politico* or Spanish *político,* ultimately from Latin *politicus* political] (1630)
: POLITICIAN 2
pol·i·tics \'pä-lə-ˌtiks\ *noun plural but singular or plural in construction* [Greek *politika,* from neuter plural of *politikos* political] (circa 1529)
1 a : the art or science of government **b :** the art or science concerned with guiding or influencing governmental policy **c :** the art or science concerned with winning and holding control over a government
2 : political actions, practices, or policies
3 a : political affairs or business; *especially* **:** competition between competing interest groups or individuals for power and leadership (as in a government) **b :** political life especially as a principal activity or profession **c :** political activities characterized by artful and often dishonest practices
4 : the political opinions or sympathies of a person
5 a : the total complex of relations between people living in society **b :** relations or conduct in a particular area of experience especially as seen or dealt with from a political point of view ⟨office *politics*⟩
pol·i·ty \'pä-lə-tē\ *noun, plural* **-ties** [Late Latin *politia* — more at POLICE] (1538)
1 : political organization
2 : a specific form of political organization
3 : a politically organized unit
4 a : the form or constitution of a politically organized unit **b :** the form of government of a religious denomination
pol·ka \'pōl-kə, 'pō-kə\ *noun* [Czech, from *Polka* Polish woman, feminine of *Polák* Pole] (1844)
1 : a vivacious couple dance of Bohemian origin in duple time with a basic pattern of hop-step-close-step
2 : a lively Bohemian dance tune in 2/4 time
— polka *intransitive verb*
pol·ka dot \'pō-kə-ˌdät\ *noun* (1884)
: a dot in a pattern of regularly distributed dots in textile design
— polka-dot *or* **pol·ka-dot·ted** \-ˌdä-təd\ *adjective*
[1]**poll** \'pōl\ *noun* [Middle English *pol, polle,* from Middle Low German] (14th century)
1 : HEAD
2 a : the top or back of the head **b :** NAPE
3 : the broad or flat end of a striking tool (as a hammer)
4 a (1) **:** the casting or recording of the votes of a body of persons (2) **:** a counting of votes cast **b :** the place where votes are cast or recorded — usually used in plural ⟨at the *polls*⟩ **c :** the period of time during which votes may be cast at an election **d :** the total number of votes recorded ⟨a heavy *poll*⟩
5 a : a questioning or canvassing of persons selected at random or by quota to obtain information or opinions to be analyzed **b :** a record of the information so obtained
[2]**poll** (14th century)
transitive verb
1 a : to cut off or cut short the hair or wool of **:** CROP, SHEAR **b :** to cut off or cut short (as wool)
2 a : to cut off or back the top of (as a tree); *specifically* **:** POLLARD **b :** to cut off or cut short the horns of (cattle)
3 a : to receive and record the votes of **b :** to request each member of to declare a vote individually ⟨*poll* the assembly⟩
4 : to receive (as votes) in an election
5 : to question or canvass in a poll
6 : to test (as several computer terminals sharing a single line) in sequence for messages to be transmitted
intransitive verb
: to cast one's vote at a poll
— poll·ee \pō-'lē\ *noun*
— poll·er \'pō-lər\ *noun*
pol·lack *or* **pol·lock** \'pä-lək\ *noun, plural* **pollack** *or* **pollock** [Middle English *poullok,* perhaps from Scottish Gaelic *pollag* or Irish *pollóg*] (15th century)
1 : a commercially important north Atlantic food fish (*Pollachius virens*) related to and resembling the cods but darker
2 : a commercially important northern Pacific food fish (*Theragra chalcogramma*) of the cod family that closely resembles the pollack — called also *walleye pollack*
[1]**pol·lard** \'pä-lərd\ *noun* [[2]*poll*] (1611)
: a tree cut back to the trunk to promote the growth of a dense head of foliage
[2]**pollard** *transitive verb* (1670)
: to make a pollard of (a tree)
polled \'pōld\ *adjective* (1584)
: having no horns
pol·len \'pä-lən\ *noun* [New Latin *pollin-, pollen,* from Latin, fine flour] (1760)
1 : a mass of microspores in a seed plant appearing usually as a fine dust
2 : a dusty bloom on the body of an insect
pollen basket *noun* (1860)
: a smooth area on each hind tibia of a bee that is edged by a fringe of stiff hairs and serves to collect and transport pollen — called also *corbicula*
pollen grain *noun* (1835)
: one of the granular microspores that occur in pollen and give rise to the male gametophyte of a seed plant
pol·len·iz·er *also* **pol·lin·iz·er** \'pä-lə-ˌnī-zər\ *noun* [*pollenize* to pollinate] (1897)
1 : a plant that is a source of pollen
2 : POLLINATOR a
pollen mother cell *noun* (1884)
: a cell that is derived from the hypodermis of the pollen sac and that gives rise by meiosis to four cells each of which develops into a pollen grain
pollen sac *noun* (1875)
: one of the pouches of a seed plant anther in which pollen is formed
pollen tube *noun* (1835)
: a tube that is formed by a pollen grain and conveys the sperm nuclei to the embryo sac of an angiosperm or the archegonium of a gymnosperm
pol·lex \'pä-ˌleks\ *noun, plural* **pol·li·ces** \'pä-lə-ˌsēz\ [New Latin *pollic-, pollex,* from Latin, thumb, big toe] (circa 1836)
: the first digit of the forelimb **:** THUMB
pollin- *combining form* [New Latin *pollin-, pollen*]
: pollen ⟨*pollinate*⟩
pol·li·nate \'pä-lə-ˌnāt\ *transitive verb* **-nat·ed; -nat·ing** (1875)
1 : to carry out the pollination of
2 : to mark or smudge with pollen
pol·li·na·tion \ˌpä-lə-'nā-shən\ *noun* (1875)
: the transfer of pollen from an anther to the stigma in angiosperms or from the microsporangium to the micropyle in gymnosperms
pol·li·na·tor \'pä-lə-ˌnā-tər\ *noun* (1903)
: one that pollinates: as **a :** an agent (as an insect) that pollinates flowers **b :** POLLENIZER 1
pol·lin·i·um \pä-'li-nē-əm\ *noun, plural* **-ia** \-nē-ə\ [New Latin, from *pollin-*] (1862)
: a coherent mass of pollen grains often with a stalk bearing an adhesive disk that clings to insects
pol·li·no·sis *or* **pol·len·osis** \ˌpä-lə-'nō-səs\ *noun* [New Latin *pollinosis,* from *pollin-*] (1925)
: an acute recurrent catarrhal disorder caused by allergic sensitivity to specific pollens
poll·ster \'pōl-stər\ *noun* (1939)
: one that conducts a poll or compiles data obtained by a poll
poll tax *noun* (1692)
: a tax of a fixed amount per person levied on adults
pol·lut·ant \pə-'lü-tᵊnt\ *noun* (1892)
: something that pollutes
pol·lute \pə-'lüt\ *transitive verb* **pol·lut·ed; pol·lut·ing** [Middle English, from Latin *pollutus,* past participle of *polluere,* from *por-* (akin to Latin *per* through) + *-luere* (akin to Latin *lutum* mud, Greek *lyma* dirt, defilement) — more at FOR] (14th century)
1 a : to make ceremonially or morally impure **:** DEFILE **b :** DEBASE 1 ⟨using language to deceive or mislead *pollutes* language —Linda C. Lederman⟩
2 a : to make physically impure or unclean **:** BEFOUL, DIRTY **b :** to contaminate (an environment) especially with man-made waste
synonym see CONTAMINATE
— pol·lut·er *noun*
— pol·lut·ive \-'lü-tiv\ *adjective*
pol·lu·tion \pə-'lü-shən\ *noun* (14th century)
1 : the action of polluting especially by environmental contamination with man-made waste; *also* **:** the condition of being polluted
2 : POLLUTANT
Pol·lux \'pä-ləks\ *noun* [Latin, modification of Greek *Polydeukēs*]
1 : one of the Dioscuri
2 : a first-magnitude star in the constellation Gemini
Pol·ly·an·na \ˌpä-lē-'a-nə\ *noun* [*Pollyanna,* heroine of the novel *Pollyanna* (1913) by Eleanor Porter (died 1920) American fiction writer] (1921)
: a person characterized by irrepressible optimism and a tendency to find good in everything
— Pollyanna *adjective*
— Pol·ly·an·na·ish \-'a-nə-ish\ *also* **Pol·ly·an·nish** \-'a-nish\ *adjective*
pol·ly·wog *or* **pol·li·wog** \'pä-lē-ˌwäg, -ˌwȯg\ *noun* [alteration of Middle English *polwygle,* probably from *pol* poll + *wiglen* to wiggle] (1832)
: TADPOLE
po·lo \'pō-(ˌ)lō\ *noun* [Balti, ball] (1872)
1 : a game played by teams of players on horseback using mallets with long flexible handles to drive a wooden ball through goalposts
2 : WATER POLO
— po·lo·ist \'pō-(ˌ)lō-ist\ *noun*
polo coat *noun* (1910)
: a tailored overcoat that is made especially of tan camel's hair and often has stitched edges and a half-belt on the back
po·lo·naise \ˌpä-lə-'nāz, ˌpō-\ *noun* [French, from feminine of *polonais* Polish, from *Pologne* Poland, from Medieval Latin *Polonia*] (1773)

1 : an elaborate overdress with a short-sleeved fitted waist and a draped cutaway overskirt
2 a : a stately Polish processional dance popular in 19th century Europe **b :** music for this dance in moderate 3/4 time

polonaise 1

Po·lo·nia \pə-ˈlō-nē-ə, -nyə\ *noun* [New Latin, Poland, from Medieval Latin] (1944)
: people of Polish descent living outside Poland

po·lo·ni·um \pə-ˈlō-nē-əm\ *noun* [New Latin, from Medieval Latin *Polonia*] (1898)
: a radioactive metallic element that is similar chemically to tellurium and bismuth, occurs especially in pitchblende and radium-lead residues, and emits an alpha particle to form an isotope of lead — see ELEMENT table

Po·lo·ni·us \pə-ˈlō-nē-əs\ *noun*
: a garrulous courtier and father of Ophelia and Laertes in Shakespeare's *Hamlet*

polo shirt *noun* (1920)
: a close-fitting pullover often knit shirt with short or long sleeves and turnover collar or banded neck

pol·ter·geist \ˈpōl-tər-ˌgīst\ *noun* [German, from *poltern* to knock + *Geist* spirit] (1848)
: a noisy usually mischievous ghost held to be responsible for unexplained noises (as rappings)

[1]pol·troon \päl-ˈtrün\ *noun* [Middle French *poultron,* from Old Italian *poltrone,* probably akin to *poltro* colt, ultimately from Latin *pullus* young of an animal — more at FOAL] (circa 1529)
: a spiritless coward **:** CRAVEN

[2]poltroon *adjective* (1645)
: characterized by complete cowardice

pol·troon·ery \-ˈtrü-nə-rē, -ˈtrün-rē\ *noun* (1590)
: mean pusillanimity **:** COWARDICE

poly \ˈpä-lē\ *noun, plural* **pol·ys** \-lēz\ *often attributive* [short for *polymer*] (1942)
: a polymerized plastic or something made of this; *especially* **:** a polyester fiber, fabric, or garment

poly- *combining form* [Middle English, from Latin, from Greek, from *polys;* akin to Old High German *filu* many, Sanskrit *puru,* Latin *plenus* full — more at FULL]
1 : many **:** several **:** much **:** MULTI- ⟨*poly*chotomous⟩ ⟨*poly*gyny⟩
2 a : containing an indefinite number more than one of a (specified) substance ⟨*poly*sulfide⟩ **b :** polymeric **:** polymer of a (specified) monomer ⟨*poly*ethylene⟩ ⟨*poly*adenylic acid⟩

poly·A \ˌpä-lē-ˈā\ *noun* [*poly-* + *a*denylic acid] (1957)
: RNA or a segment of RNA that is composed of a polynucleotide chain consisting entirely of adenylic acid residues and that codes for polylysine when functioning as messenger RNA in protein synthesis

poly·acryl·amide \ˌpä-lē-ə-ˈkri-lə-ˌmīd\ *noun* (1944)
: a polyamide of acrylic acid

poly·ac·ry·lo·ni·trile \ˌpä-lē-ˌa-krə-lō-ˈnī-trəl, -ˌtrēl\ *noun* (1935)
: a polymer of acrylonitrile used often as fibers

poly·ad·e·nyl·ic acid \ˌpä-lē-ˌa-dᵊn-ˈi-lik-\ *noun* (1956)
: POLYA

poly·al·co·hol \ˌpä-lē-ˈal-kə-ˌhȯl\ *noun* (1900)
: an alcohol (as a diol) that contains more than one hydroxyl group

poly·am·ide \ˌpä-lē-ˈa-ˌmīd, -məd\ *noun* [International Scientific Vocabulary] (1929)
: a compound characterized by more than one amide group; *especially* **:** a polymeric amide (as nylon)

poly·amine \ˈpä-lē-ə-ˌmēn, ˌpä-lē-ˈa-ˌmēn\ *noun* (1861)
: a compound characterized by more than one amino group

poly·an·dry \ˈpä-lē-ˌan-drē\ *noun* [Greek *polyandros,* adjective, having many husbands, from *poly-* + *andr-, anēr* man, husband — more at ANDR-] (1780)
: the state or practice of having more than one husband or male mate at one time — compare POLYGAMY, POLYGYNY
— poly·an·drous \ˌpä-lē-ˈan-drəs\ *adjective*

poly·an·tha \ˌpä-lē-ˈan(t)-thə\ *noun* [New Latin, from Greek *polyanthos* blooming] (1889)
: any of numerous dwarf hybrid roses characterized by many large clusters of small flowers

poly·an·thus \-ˈan(t)-thəs\ *noun, plural* **-an·thus·es** *or* **-an·thi** \-ˈan-ˌthī, -ˌthē\ [New Latin, from Greek *polyanthos* blooming, from *poly-* + *anthos* flower — more at ANTHOLOGY] (circa 1727)
1 : any of various hybrid primroses
2 : a narcissus (*Narcissus tazetta*) having small white or yellow flowers arranged in umbels and having a spreading perianth

poly·atom·ic \-ə-ˈtä-mik\ *adjective* [International Scientific Vocabulary] (1857)
: containing more than one and especially more than two atoms ⟨*polyatomic* molecules⟩

poly·bro·mi·nat·ed biphenyl \ˌpä-lē-ˈbrō-mə-ˌnā-təd-\ *noun* (1975)
: any of several compounds that are similar to polychlorinated biphenyls in environmental toxicity and in structure but that have various hydrogen atoms replaced by bromine rather than chlorine — called also *PBB*

poly·bu·ta·di·ene \-ˌbyü-tə-ˈdī-ˌēn, -ˌdī-ˈ\ *noun* (1939)
: a synthetic rubber that has a high resistance to wear and is used especially in the manufacture of tires

poly·car·bon·ate \-ˈkär-bə-ˌnāt, -nət\ *noun* (1930)
: any of various tough transparent thermoplastics characterized by high impact strength and high softening temperature

poly·cen·tric \-ˈsen-trik\ *adjective* (1887)
: having more than one center (as of development or control): as **a :** having several centromeres ⟨*polycentric* chromosomes⟩ ⟨*polycentric* cells⟩ **b :** characterized by polycentrism

poly·cen·trism \-ˈsen-ˌtri-zəm\ *noun* (1956)
: the existence of many centers of communist ideological thought; *especially* **:** the existence of a number of autonomous national communist movements

poly·chaete \ˈpä-lē-ˌkēt\ *noun* [ultimately from Greek *polychaitēs* having much hair, from *poly-* + *chaitē* long hair] (1896)
: any of a class (Polychaeta) of chiefly marine annelid worms (as clam worms) usually with paired segmental appendages, separate sexes, and a free-swimming trochophore larva
— polychaete *adjective*

poly·chlo·ri·nat·ed biphenyl \ˌpä-lē-ˈklōr-ə-ˌnā-təd-, -ˈklȯr-\ *noun* (1962)
: any of several compounds that are produced by replacing hydrogen atoms in biphenyl with chlorine, have various industrial applications, and are poisonous environmental pollutants which tend to accumulate in animal tissues — called also *PCB*

poly·chot·o·mous \-ˈkä-tə-məs\ *adjective* [*poly-* + *-chotomous* (as in *dichotomous*)] (circa 1858)
: dividing or marked by division into many parts, branches, or classes
— poly·chot·o·my \-mē\ *noun*

poly·chro·mat·ic \-krō-ˈma-tik\ *adjective* [Greek *polychrōmatos,* from *poly-* + *chrōmat-, chrōma* color — more at CHROMATIC] (circa 1841)
1 : showing a variety or a change of colors **:** MULTICOLORED
2 : being or relating to radiation that is composed of more than one wavelength

poly·chro·mato·phil·ic \-krō-ˌma-tə-ˈfi-lik\ *adjective* (1897)
: stainable with more than one type of stain and especially with both acid and basic dyes ⟨*polychromatophilic* erythroblasts⟩
— poly·chro·mato·phil·ia \-ˈfi-lē-ə\ *noun*

poly·chrome \ˈpä-lē-ˌkrōm\ *adjective* [Greek *polychrōmos,* from *poly-* + *chrōma*] (1837)
: relating to, made with, or decorated in several colors ⟨*polychrome* pottery⟩
— polychrome *transitive verb*
— poly·chro·my \-ˌkrō-mē\ *noun*

poly·cis·tron·ic \ˌpä-lē-sis-ˈträ-nik\ *adjective* (1963)
: containing the genetic information of a number of cistrons ⟨*polycistronic* messenger RNA⟩

poly·clin·ic \-ˈkli-nik\ *noun* [International Scientific Vocabulary] (circa 1890)
: a clinic or hospital treating diseases of many sorts

poly·clon·al \-ˈklō-nəl\ *adjective* (1914)
: produced by or being cells derived from two or more cells of different ancestry or genetic constitution ⟨*polyclonal* antibodies⟩

poly·con·den·sa·tion \-ˌkän-ˌden-ˈsā-shən, -dən-\ *noun* [International Scientific Vocabulary] (1936)
: a chemical condensation leading to the formation of a compound of high molecular weight

poly·con·ic projection \ˌpä-lē-ˈkä-nik-\ *noun* (circa 1864)
: a map projection consisting of a composite series of concentric cones each of which before being unrolled has been placed over a sphere so as to be tangent to a different parallel of latitude

poly·crys·tal·line \-ˈkris-tə-lən\ *adjective* (1918)
1 : consisting of crystals variously oriented
2 : composed of more than one crystal
— poly·crys·tal \ˈpä-lē-ˌkris-tᵊl\ *noun*

poly·cy·clic \ˌpä-lē-ˈsī-klik, -ˈsi-\ *adjective* [International Scientific Vocabulary] (1869)
: having more than one cyclic component; *especially* **:** having two or more rings in the molecule

poly·cys·tic \-ˈsis-tik\ *adjective* (1872)
: having or involving more than one cyst ⟨*polycystic* kidneys⟩ ⟨*polycystic* disease⟩

poly·cy·the·mia \-(ˌ)sī-ˈthē-mē-ə\ *noun* [New Latin, from *poly-* + *cyt-* + *-hemia*] (circa 1857)
: a condition marked by an abnormal increase in the number of circulating red blood cells; *specifically* **:** POLYCYTHEMIA VERA
— poly·cy·the·mic \-mik\ *adjective*

polycythemia ve·ra \-ˈvir-ə\ *noun* [New Latin, true polycythemia] (circa 1925)
: polycythemia of unknown cause that is characterized by increase in total blood volume and accompanied by nosebleed, distension of the circulatory vessels, and enlargement of the spleen — called also *erythremia*

poly·cyt·i·dyl·ic acid \ˌpä-lē-ˌsi-tə-ˈdi-lik-\ *noun* (1965)
: RNA or a segment of RNA that is composed of a polynucleotide chain consisting entirely of cytosine-containing nucleotides and that codes for a polypeptide chain consisting of proline residues when functioning as messenger RNA in protein synthesis

poly·dac·tyl \-ˈdak-tᵊl\ *adjective* [Greek *polydaktylos,* from *poly-* + *daktylos* digit] (circa 1890)
: having or causing polydactyly

poly·dac·ty·ly \-ˈdak-tᵊl-ē\ *noun* (1886)
: the condition of having more than the normal number of fingers or toes

poly·dip·sia \-ˈdip-sē-ə\ *noun* [New Latin, from *poly-* + Greek *dipsa* thirst] (1660)
: excessive or abnormal thirst

— **poly·dip·sic** \-sik\ *adjective*

poly·dis·perse \-dis-ˈpərs\ *adjective* [*poly-* + Latin *dispersus* dispersed, from past participle of *dispergere* to disperse] (1915)
: of, relating to, or characterized by or as particles of varied sizes in the dispersed phase of a disperse system
— **poly·dis·per·si·ty** \-ˈpər-sə-tē\ *noun*

poly·elec·tro·lyte \ˌpä-lē-ə-ˈlek-trə-ˌlīt\ *noun* (circa 1947)
: a substance of high molecular weight (as a protein) that is an electrolyte

poly·em·bry·o·ny \-ˈem-brē-ə-nē, -(ˌ)em-ˈbrī-\ *noun* [International Scientific Vocabulary *poly-* + *embryon-* + [2]*-y*] (1849)
1 : the condition of having several embryos
2 : the production of two or more embryos from one ovule or egg
— **poly·em·bry·on·ic** \-ˌem-brē-ˈä-nik\ *adjective*

poly·ene \ˈpä-lē-ˌēn\ *noun* [International Scientific Vocabulary] (1928)
: an organic compound containing many double bonds; *especially* **:** one having the double bonds in a long aliphatic hydrocarbon chain
— **poly·enic** \ˌpä-lē-ˈē-nik\ *adjective*

[1]**poly·es·ter** \ˈpä-lē-ˌes-tər, ˌpä-lē-ˈ\ *noun* [International Scientific Vocabulary] (1929)
: any of a group of polymers that consist basically of repeated units of an ester and are used especially in making fibers or plastics
— **poly·es·ter·i·fi·ca·tion** \-e-ˌster-ə-fə-ˈkā-shən\ *noun*

[2]**polyester** *adjective* (1975)
: characterized by inelegant or unsophisticated middle-class taste ⟨*polyester* suburbs⟩ ⟨*polyester* folks⟩

poly·es·trous \ˌpä-lē-ˈes-trəs\ *adjective* (1900)
: having more than one period of estrus in a year

poly·eth·yl·ene \-ˈe-thə-ˌlēn\ *noun* (circa 1862)
: a polymer of ethylene; *especially* **:** any of various partially crystalline lightweight thermoplastics $(CH_2CH_2)_x$ that are resistant to chemicals and moisture, have good insulating properties, and are used especially in packaging and insulation

po·lyg·a·la \pə-ˈli-gə-lə\ *noun* [New Latin, genus name, from Latin, milkwort, from Greek *polygalon,* from *poly-* + *gala* milk — more at GALAXY] (1578)
: MILKWORT

poly·gam·ic \ˌpä-lē-ˈga-mik\ *adjective* (1819)
: POLYGAMOUS

po·lyg·a·mous \pə-ˈli-gə-məs\ *adjective* [Greek *polygamos,* from *poly-* + *-gamos* -gamous] (1613)
1 a : relating to or practicing polygamy **b :** having more than one mate at one time ⟨baboons are *polygamous*⟩
2 : bearing both hermaphrodite and unisexual flowers on the same plant

po·lyg·a·my \-mē\ *noun* (circa 1591)
1 : marriage in which a spouse of either sex may have more than one mate at the same time — compare POLYANDRY, POLYGYNY
2 : the state of being polygamous
— **po·lyg·a·mist** \-mist\ *noun*
— **po·lyg·a·mize** \-ˌmīz\ *intransitive verb*

poly·gene \ˈpä-lē-ˌjēn\ *noun* [International Scientific Vocabulary] (1941)
: any of a group of nonallelic genes that collectively control the inheritance of a quantitative character or modify the expression of a qualitative character
— **poly·gen·ic** \ˌpä-lē-ˈjē-nik\ *adjective*

poly·gen·e·sis \ˌpä-lē-ˈje-nə-səs\ *noun* [New Latin] (circa 1882)
: development from more than one source

poly·ge·net·ic \-jə-ˈne-tik\ *adjective* (1861)
1 : POLYPHYLETIC
2 : having many distinct sources

[1]**poly·glot** \ˈpä-lē-ˌglät\ *noun* [[2]*polyglot*] (circa 1645)
1 : one who is polyglot
2 *capitalized* **:** a book containing versions of the same text in several languages; *especially* **:** the Scriptures in several languages
3 : a mixture or confusion of languages or nomenclatures

[2]**polyglot** *adjective* [Greek *polyglōttos,* from *poly-* + *glōtta* language — more at GLOSS] (1656)
1 a : speaking or writing several languages **:** MULTILINGUAL **b :** composed of numerous linguistic groups ⟨a *polyglot* population⟩
2 : containing matter in several languages ⟨a *polyglot* sign⟩
3 : composed of elements from different languages
4 : being widely diverse (as in ethnic or cultural origins) ⟨*polyglot* cuisine⟩

poly·glot·ism *or* **poly·glot·tism** \-ˌglä-ˌti-zəm\ *noun* (1882)
: the use of many languages **:** the ability to speak many languages

poly·gon \ˈpä-lē-ˌgän\ *noun* [Late Latin *polygonum,* from Greek *polygōnon,* from neuter of *polygōnos* polygonal, from *poly-* + *gōnia* angle — more at -GON] (1571)
1 : a closed plane figure bounded by straight lines
2 : a closed figure on a sphere bounded by arcs of great circles
— **po·lyg·o·nal** \pə-ˈli-gə-nᵊl\ *adjective*
— **po·lyg·o·nal·ly** \-nᵊl-ē\ *adverb*

po·lyg·o·num \pə-ˈli-gə-nəm\ *noun* [New Latin, from Greek *polygonon* knotgrass, from *poly-* + *gony* knee — more at KNEE] (circa 1706)
: any of a genus (*Polygonum*) of herbs of the buckwheat family with a prominent tubular sheath around the base of each petiole, thickened nodes, and flowers that are solitary and axillary or in spiked racemes — called also *knotweed*

poly·graph \ˈpä-lē-ˌgraf\ *noun* (1871)
: an instrument for recording variations of several different pulsations (as of physiological variables) simultaneously; *broadly* **:** LIE DETECTOR
— **poly·graph·ic** \ˌpä-lē-ˈgra-fik\ *adjective*

po·lyg·ra·pher \ˈpä-lē-ˌgra-fər, pə-ˈli-grə-fər\ *noun* (circa 1934)
: one who operates a polygraph

po·lyg·ra·phist \ˈpä-lē-ˌgra-fist, pə-ˈli-grə-fist\ *noun* (1954)
: POLYGRAPHER

po·lyg·y·nous \pə-ˈli-jə-nəs\ *adjective* (1874)
: relating to or practicing polygyny

po·lyg·y·ny \-nē\ *noun* (1780)
: the state or practice of having more than one wife or female mate at one time — compare POLYANDRY, POLYGAMY

polyhedral angle *noun* (circa 1864)
: a portion of space partly enclosed by three or more planes whose intersections meet in a vertex

poly·he·dron \ˌpä-lē-ˈhē-drən\ *noun, plural* **-drons** *or* **-dra** \-drə\ [New Latin] (1570)
: a solid formed by plane faces
— **poly·he·dral** \-drəl\ *adjective*

poly·he·dro·sis \ˌpä-lē-hē-ˈdrō-səs\ *noun, plural* **-dro·ses** \-ˌsēz\ [New Latin, from *polyhedron*] (1947)
: any of several virus diseases of insect larvae characterized by dissolution of tissues and accumulation of polyhedral granules in the resultant fluid

poly·his·tor \ˌpä-lē-ˈhis-tər\ *noun* [Greek *polyistōr* very learned, from *poly-* + *istōr, histōr* learned — more at HISTORY] (1588)
: POLYMATH
— **poly·his·tor·ic** \-his-ˈtȯr-ik, -ˈtär-\ *adjective*

poly·hy·droxy \-hī-ˈdräk-sē\ *adjective* (circa 1929)
: containing more than one hydroxyl group in the molecule

Poly·hym·nia \ˌpä-lē-ˈhim-nē-ə\ *noun* [Latin, from Greek *Polyymnia*]
: the Greek Muse of sacred song

poly I:C \ˌpä-lē-ˈī-ˈsē\ *or* **poly I·poly C** \ˌpä-lē-ˈī-ˌpä-lē-ˈsē\ *noun* [*poly-* + *i*nosinic acid + *c*ytidylic acid] (1969)
: a synthetic 2-stranded RNA composed of one strand of polyinosinic acid and one strand of polycytidylic acid that induces interferon formation and has been used experimentally as an anticancer and antiviral agent

poly·ino·sin·ic acid \ˌpä-lē-ˌi-nə-ˈsi-nik-, -ˌī-nə-\ *noun* [*poly-* + *inosinic acid* $(C_{10}H_{13}N_4O_8P)$, from Greek *in-, is* sinew + International Scientific Vocabulary [2]*-ose* + [1]*-in* + [1]*-ic* + *acid*] (1965)
: RNA or a segment of RNA that is composed of a polynucleotide chain consisting entirely of inosinic acid residues

poly·ly·sine \ˌpä-lē-ˈlī-ˌsēn\ *noun* (1947)
: a protein whose polypeptide chain consists entirely of lysine residues

poly·math \ˈpä-lē-ˌmath\ *noun* [Greek *polymathēs* very learned, from *poly-* + *manthanein* to learn — more at MATHEMATICAL] (1621)
: a person of encyclopedic learning
— **polymath** *or* **poly·math·ic** \ˌpä-lē-ˈma-thik\ *adjective*
— **po·ly·ma·thy** \pə-ˈli-mə-thē, ˈpä-lə-ˌma-thē\ *noun*

poly·mer \ˈpä-lə-mər\ *noun* [International Scientific Vocabulary, back-formation from *polymeric,* from Greek *polymerēs* having many parts, from *poly-* + *meros* part — more at MERIT] (1866)
: a chemical compound or mixture of compounds formed by polymerization and consisting essentially of repeating structural units
— **poly·mer·ic** \ˌpä-lə-ˈmer-ik\ *adjective*
— **poly·mer·ism** \pə-ˈli-mə-ˌri-zəm, ˈpä-lə-mə-\ *noun*

poly·mer·ase \ˈpä-lə-mə-ˌrās, -ˌrāz\ *noun* (1958)
: any of several enzymes that catalyze the formation of DNA or RNA from precursor substances in the presence of preexisting DNA or RNA acting as a template

polymerase chain reaction *noun* (1987)
: an in vitro technique for rapidly synthesizing large quantities of a given DNA segment that involves separating the DNA into its two complementary strands, using DNA polymerase to synthesize two-stranded DNA from each single strand, and repeating the process

po·ly·mer·i·sa·tion, po·ly·mer·ise *chiefly British variant of* POLYMERIZATION, POLYMERIZE

po·ly·mer·i·za·tion \pə-ˌli-mə-rə-ˈzā-shən, ˌpä-lə-mə-rə-\ *noun* [International Scientific Vocabulary] (1872)
1 : a chemical reaction in which two or more molecules combine to form larger molecules that contain repeating structural units — compare ASSOCIATION 5
2 : reduplication of parts in an organism

po·ly·mer·ize \pə-ˈli-mə-ˌrīz, ˈpä-lə-mə-\ *verb* **-ized; -iz·ing** (1865)
transitive verb
: to subject to polymerization
intransitive verb
: to undergo polymerization

poly·meth·yl methacrylate \ˈpä-lē-ˌme-thəl-\ *noun* (1936)
: a thermoplastic resin of polymerized methyl methacrylate which is characterized by its optical clarity (as in a contact lens)

poly·morph \ˈpä-lē-ˌmȯrf\ *noun* [International Scientific Vocabulary] (circa 1828)
1 : a polymorphic organism; *also* **:** one of the several forms of such an organism

2 : any of the crystalline forms of a polymorphic substance
poly·mor·phism \ˌpä-lē-ˈmȯr-ˌfi-zəm\ *noun* (1839)
: the quality or state of being able to assume different forms: as **a :** existence of a species in several forms independent of the variations of sex **b :** the property of crystallizing in two or more forms with distinct structure
— **poly·mor·phic** \-fik\ *adjective*
— **poly·mor·phi·cal·ly** \-fi-k(ə-)lē\ *adverb*
poly·mor·pho·nu·cle·ar \-ˌmȯr-fə-ˈnü-klē-ər, -ˈnyü-\ *adjective* (1897)
of a leukocyte **:** having the nucleus complexly lobed
— **polymorphonuclear** *noun*
poly·mor·phous \-ˈmȯr-fəs\ *adjective* [Greek *polymorphous,* from *poly-* + *-morphos* -morphous] (1785)
: having, assuming, or occurring in various forms, characters, or styles **:** POLYMORPHIC ⟨a *polymorphous* rash⟩ ⟨*polymorphous* sexuality⟩
— **poly·mor·phous·ly** *adverb*
polymorphous perverse *adjective* (1909)
: relating to or exhibiting infantile sexual tendencies in which the genitals are not yet identified as the sole or principal sexual organs nor coitus as the goal of erotic activity
poly·myx·in \ˌpä-lē-ˈmik-sən\ *noun* [International Scientific Vocabulary, from New Latin *polymyxa,* from *poly-* + Greek *myxa* mucus — more at MUCUS] (1947)
: any of several toxic antibiotics obtained from a soil bacterium (*Bacillus polymyxa*) and active against gram-negative bacteria
Poly·ne·sian \ˌpä-lə-ˈnē-zhən, -shən\ *noun* (1807)
1 : a member of any of the indigenous peoples of Polynesia
2 : a group of Austronesian languages spoken in Polynesia
— **Polynesian** *adjective*
poly·neu·ri·tis \ˌpä-lē-nu̇-ˈrī-təs, -nyu̇-\ *noun* [New Latin] (1886)
: neuritis of several peripheral nerves at the same time
Poly·ni·ces \ˌpä-lə-ˈnī-sēz\ *noun* [Latin, from Greek *Polyneikēs*]
: a son of Oedipus for whom the Seven against Thebes mount their expedition
¹poly·no·mi·al \ˌpä-lə-ˈnō-mē-əl\ *noun* [*poly-* + *-nomial* (as in *binomial*)] (1674)
: a mathematical expression of one or more algebraic terms each of which consists of a constant multiplied by one or more variables raised to a nonnegative integral power (as $a + bx + cx^2$)
²polynomial *adjective* (circa 1704)
: relating to, composed of, or expressed as one or more polynomials ⟨*polynomial* functions⟩ ⟨*polynomial* equations⟩
poly·nu·cle·ar \ˌpä-lē-ˈnü-klē-ər, -ˈnyü-, ÷-kyə-lər\ *adjective* [International Scientific Vocabulary] (1908)
: chemically polycyclic especially with respect to the benzene ring — used chiefly of aromatic hydrocarbons that are important as pollutants and possibly as carcinogens
poly·nu·cle·o·tide \-ˈnü-klē-ə-ˌtīd, -ˈnyü-\ *noun* [International Scientific Vocabulary] (1911)
: a polymeric chain of nucleotides
po·lyn·ya \ˌpä-lən-ˈyä\ *noun, plural* **polynyas** *also* **po·lyn·yi** \-ˈyē\ [Russian *polyn'ya*] (1853)
: an area of open water in sea ice
poly·ole·fin \ˌpä-lē-ˈō-lə-fən\ *noun* (1930)
: a polymer of an alkene (as polyethylene)
poly·oma virus \ˌpä-lē-ˈō-mə-\ *noun* [New Latin *polyoma,* from *poly-* + *-oma*] (1958)
: a papovavirus of rodents that is associated with various kinds of tumors — called also *polyoma*
poly·on·y·mous \ˌpä-lē-ˈä-nə-məs\ *adjective* [Greek *polyōnymos,* from *poly-* + *onoma, onyma* name — more at NAME] (1678)
: having or known by various names
pol·yp \ˈpä-ləp\ *noun* [Middle French *polype* octopus, nasal tumor, from Latin *polypus,* from Greek *polypous,* from *poly-* + *pous* foot — more at FOOT] (1742)
1 : a coelenterate that has typically a hollow cylindrical body closed and attached at one end and opening at the other by a central mouth surrounded by tentacles armed with nematocysts
2 : a projecting mass of swollen and hypertrophied or tumorous membrane
— **pol·yp·oid** \-lə-ˌpȯid\ *adjective*
poly·pep·tide \ˌpä-lē-ˈpep-ˌtīd\ *noun* [International Scientific Vocabulary] (1903)
: a molecular chain of amino acids
— **poly·pep·tid·ic** \-(ˌ)pep-ˈti-dik\ *adjective*
poly·pet·al·ous \-ˈpe-tᵊl-əs\ *adjective* [New Latin *polypetalus,* from *poly-* + *petalum* petal] (circa 1704)
: having or consisting of separate petals
poly·pha·gia \-ˈfā-j(ē-)ə\ *noun* [Greek *polyphagia,* from *polyphagos*] (circa 1693)
: excessive appetite or eating
po·lyph·a·gous \pə-ˈli-fə-gəs\ *adjective* [Greek *polyphagos* eating too much, from *poly-* + *-phagos* -phagous] (1815)
: feeding on or utilizing many kinds of food
— **po·lyph·a·gy** \-jē\ *noun*
poly·phase \ˈpä-lē-ˌfāz\ *adjective* [International Scientific Vocabulary] (1891)
: having or producing two or more phases ⟨a *polyphase* machine⟩ ⟨a *polyphase* current⟩
poly·pha·sic \ˌpä-lē-ˈfā-zik\ *adjective* (1922)
: consisting of two or more phases
Poly·phe·mus \ˌpä-lə-ˈfē-məs\ *noun* [Latin, from Greek *Polyphēmos*]
: a Cyclops whom Odysseus blinds in order to escape from his cave
poly·phe·nol \ˌpä-lē-ˈfē-ˌnōl, -fi-ˈ\ *noun* [International Scientific Vocabulary] (1894)
: a polyhydroxy phenol
— **poly·phe·no·lic** \-fi-ˈnō-lik, -ˈnä-\ *adjective*
poly·phil·o·pro·gen·i·tive \-ˌfi-lə-prō-ˈje-nə-tiv\ *adjective* (1919)
: extremely prolific **:** PHILOPROGENITIVE
poly·phone \ˈpä-lē-ˌfōn\ *noun* (1872)
: a symbol or sequence of symbols having more than one phonemic value (as *a* in English)
poly·phon·ic \ˌpä-lē-ˈfä-nik\ *or* **po·lyph·o·nous** \pə-ˈli-fə-nəs\ *adjective* (1782)
1 : of, relating to, or marked by polyphony
2 : being a polyphone
— **poly·phon·i·cal·ly** \ˌpä-lē-ˈfä-ni-k(ə-)lē\ *or* **po·lyph·o·nous·ly** *adverb*
polyphonic prose *noun* (1916)
: a freely rhythmical prose employing characteristic devices of verse (as alliteration and assonance)
po·lyph·o·ny \pə-ˈli-fə-nē\ *noun* [Greek *polyphōnia* variety of tones, from *polyphōnos* having many tones or voices, from *poly-* + *phōnē* voice — more at BAN] (circa 1864)
: a style of musical composition in which two or more independent melodies are juxtaposed in harmony
poly·phy·let·ic \ˌpä-lē-(ˌ)fī-ˈle-tik\ *adjective* [International Scientific Vocabulary, from Greek *polyphylos* of many tribes, from *poly-* + *phylē* tribe — more at PHYL-] (1875)
: of, relating to, or derived from different ancestral stocks; *specifically* **:** relating to or being a taxonomic group that includes members (as genera or species) from different ancestral lineages
— **poly·phy·let·i·cal·ly** \-ti-k(ə-)lē\ *adverb*
pol·yp·ide \ˈpä-lə-ˌpīd\ *noun* [*polyp* + Greek *-idēs,* patronymic suffix] (1850)
: one of the individual zooids of a bryozoan colony
poly·ploid \ˈpä lē ˌplȯid\ *adjective* [International Scientific Vocabulary] (1920)
: having or being a chromosome number that is a multiple greater than two of the monoploid number
— **polyploid** *noun*
— **poly·ploi·dy** \-ˌplȯi-dē\ *noun*
po·lyp·nea \pä-ˈlip-nē-ə, pə-\ *noun* [New Latin] (circa 1890)
: rapid or panting respiration
poly·po·dy \ˈpä-lə-ˌpō-dē\ *noun, plural* **-dies** [Middle English *polypodie,* from Latin *polypodium,* from Greek *polypodion,* from *poly-* + *pod-, pous* foot — more at FOOT] (14th century)
: any of a genus (*Polypodium*) of chiefly epiphytic ferns with creeping rhizomes; *especially* **:** either of two ferns (*P. vulgare* of Eurasia and *P. virginianum* of North America) with relatively narrow entire segments

polypody

poly·pro·pyl·ene \ˌpä-lē-ˈprō-pə-ˌlēn\ *noun* (1935)
: any of various thermoplastic plastics or fibers that are polymers of propylene
po·lyp·tych \ˈpä-ləp-ˌtik, pə-ˈlip-tik\ *noun* [Greek *polyptychos* having many folds, from *poly-* + *ptychē* fold, from *ptyssein* to fold] (1859)
: an arrangement of four or more panels (as of a painting) usually hinged and folding together
poly·rhythm \ˈpä-lē-ˌri-t͟həm\ *noun* (1929)
: the simultaneous combination of contrasting rhythms in music
— **poly·rhyth·mic** \ˌpä-lē-ˈrit͟h-mik\ *adjective*
— **poly·rhyth·mi·cal·ly** \-mi-k(ə-)lē\ *adverb*
poly·ri·bo·nu·cle·o·tide \ˌpä-lē-ˌrī-bō-ˈnü-klē-ə-ˌtīd, -ˈnyü-\ *noun* (1956)
: a polynucleotide in which the mononucleotides are ribonucleotides
poly·ri·bo·some \-ˈrī-bə-ˌsōm\ *noun* (1962)
: a cluster of ribosomes linked together by a molecule of messenger RNA and forming the site of protein synthesis
— **poly·ri·bo·som·al** \-ˌrī-bə-ˈsō-məl\ *adjective*
poly·sac·cha·ride \-ˈsa-kə-ˌrīd\ *noun* [International Scientific Vocabulary] (1892)
: a carbohydrate that can be decomposed by hydrolysis into two or more molecules of monosaccharides; *especially* **:** any of the more complex carbohydrates (as cellulose, starch, or glycogen)
po·ly·se·mous \ˌpä-lē-ˈsē-məs, pə-ˈli-sə-məs\ *adjective* [Late Latin *polysemus,* from Greek *polysēmos,* from *poly-* + *sēma* sign] (1884)
: having multiple meanings
— **po·ly·se·my** \-mē\ *noun*
poly·some \ˈpä-lē-ˌsōm\ *noun* (1962)
: POLYRIBOSOME
poly·sor·bate \ˌpä-lē-ˈsȯr-ˌbāt\ *noun* (1950)
: any of several emulsifiers used in the preparation of some pharmaceuticals or foods
po·lys·ti·chous \pə-ˈlis-ti-kəs\ *adjective* [Greek *polystichos,* from *poly-* + *stichos* row — more at DISTICH] (circa 1890)
: arranged in several rows
poly·sty·rene \ˌpä-lē-ˈstī-ˌrēn\ *noun* (1927)
: a polymer of styrene; *especially* **:** a rigid transparent thermoplastic that has good physical and electrical insulating properties and is used especially in molded products, foams, and sheet materials
poly·sul·fide \-ˈsəl-ˌfīd\ *noun* [International Scientific Vocabulary] (1849)
: a sulfide containing two or more atoms of sulfur in the molecule
poly·syl·lab·ic \-sə-ˈla-bik\ *adjective* [Medieval Latin *polysyllabus,* from Greek *polysyllabos,* from *poly-* + *syllabē* syllable] (1782)
1 : having more than one and usually more than three syllables

2 : characterized by words of many syllables
— **poly·syl·lab·i·cal·ly** \-bi-k(ə-)lē\ *adverb*

poly·syl·la·ble \'pä-lē-ˌsi-lə-bəl, ˌpä-lē-'\ *noun* [modification of Medieval Latin *polysyllaba,* from feminine of *polysyllabus*] (1570)
: a polysyllabic word

poly·syn·ap·tic \ˌpä-lē-sə-'nap-tik\ *adjective* (1964)
: involving two or more synapses in the central nervous system ⟨*polysynaptic* reflexes⟩
— **poly·syn·ap·ti·cal·ly** \-ti-k(ə-)lē\ *adverb*

poly·syn·de·ton \-'sin-də-ˌtän\ *noun* [New Latin, from Late Greek, neuter of *polysyndetos* using many conjunctions, from Greek *poly-* + *syndetos* bound together, conjunctive — more at ASYNDETON] (circa 1577)
: repetition of conjunctions in close succession (as in *we have ships and men and money and stores*)

[1]poly·tech·nic \-'tek-nik\ *adjective* [French *polytechnique,* from Greek *polytechnos* skilled in many arts, from *poly-* + *technē* art — more at TECHNICAL] (1805)
: relating to or devoted to instruction in many technical arts or applied sciences

[2]polytechnic *noun* (1836)
: a polytechnic school

poly·tene \'pä-lē-ˌtēn\ *adjective* [International Scientific Vocabulary] (1935)
: relating to, being, or having chromosomes each of which consists of many strands with the corresponding chromomeres in contact
— **poly·te·ny** \-ˌtē-nē\ *noun*

poly·the·ism \'pä-lē-(ˌ)thē-ˌi-zəm\ *noun* [French *polythéisme,* from Late Greek *polytheos* polytheistic, from Greek, of many gods, from *poly-* + *theos* god] (1613)
: belief in or worship of more than one god
— **poly·the·ist** \-ˌthē-ist\ *adjective or noun*
— **poly·the·is·tic** \ˌpä-lē-thē-'is-tik\ *also* **poly·the·is·ti·cal** \-'is-ti-kəl\ *adjective*

poly·thene \'pä-lə-ˌthēn\ *noun* [by contraction] (1939)
chiefly British **:** POLYETHYLENE

poly·to·nal·i·ty \ˌpä-lē-tō-'na-lə-tē\ *noun* (1923)
: the simultaneous use of two or more musical keys
— **poly·ton·al** \-'tō-nᵊl\ *adjective*
— **poly·ton·al·ly** \-nᵊl-ē\ *adverb*

poly·typ·ic \ˌpä-lē-'ti-pik\ *adjective* (1888)
: represented by several or many types or subdivisions ⟨a *polytypic* species of organism⟩

poly·un·sat·u·rat·ed \ˌpä-lē-ˌən-'sa-chə-ˌrā-təd\ *adjective* (1932)
of an oil or fatty acid **:** having many double or triple bonds in a molecule — compare MONOUNSATURATED

poly·ure·thane \ˌpä-lē-'yu̇r-ə-ˌthān\ *noun* [International Scientific Vocabulary] (1944)
: any of various polymers that contain NHCOO linkages and are used especially in flexible and rigid foams, elastomers, and resins (as for coatings)

poly·uria \-'yu̇r-ē-ə\ *noun* [New Latin] (circa 1842)
: excessive secretion of urine

poly·va·lent \-'vā-lənt *also* pə-'li-və-lənt\ *adjective* [International Scientific Vocabulary] (1881)
1 : having a chemical valence greater usually than two
2 : effective against, sensitive toward, or counteracting more than one exciting agent (as a toxin or antigen)
— **poly·va·lence** \-lən(t)s\ *noun*

poly·vi·nyl \ˌpä-lē-'vī-nᵊl\ *adjective* [International Scientific Vocabulary] (1927)
: of, relating to, or being a polymerized vinyl compound, resin, or plastic — often used in combination

polyvinyl chloride *noun* (1933)
: a polymer of vinyl chloride used especially for electrical insulation, films, and pipes — abbreviation *PVC*

poly·wa·ter \'pä-lē-ˌwȯ-tər, -ˌwä-\ *noun* [*polymeric water*] (1969)
: water condensed into a glass capillary tube and formerly held to be a stable form with special properties

Pom \'päm\ *noun* (1912)
Australian & New Zealand **:** POMMY — usually used disparagingly

pom·ace \'pə-məs, 'pä-\ *noun* [Middle English *pomys,* probably from Medieval Latin *pomacium* cider, from Late Latin *pomum* apple, from Latin, fruit] (15th century)
1 : the dry or pulpy residue of material (as fruit, seeds, or fish) from which a liquid (as juice or oil) has been pressed or extracted
2 : something crushed to a pulpy mass

po·ma·ceous \pō-'mā-shəs\ *adjective* [New Latin *pomaceus,* from Late Latin *pomum*] (1708)
1 : of or relating to apples
2 [*pome*] **:** resembling a pome

po·made \pō-'mād, -'mäd\ *noun* [Middle French *pommade* ointment formerly made from apples, from Italian *pomata,* from *pomo* apple, from Late Latin *pomum*] (1562)
: a perfumed ointment; *especially* **:** a fragrant hair dressing
— **pomade** *transitive verb*

po·man·der \'pō-ˌman-dər, pō-'\ *noun* [Middle English, modification of Middle French *pome d'ambre,* literally, apple or ball of amber] (15th century)
1 : a mixture of aromatic substances enclosed in a perforated bag or box and used to scent clothes and linens or formerly carried as a guard against infection; *also* **:** a clove-studded orange or apple used for the same purposes
2 : a box or hollow fruit-shaped ball for holding pomander

po·ma·tum \pō-'mā-təm, -'mä-\ *noun* [New Latin, from Late Latin *pomum* apple] (1562)
: POMADE

pome \'pōm\ *noun* [Middle English, from Middle French *pome, pomme* apple, pome, ball, ultimately from Late Latin *pomum*] (15th century)
: a fleshy fruit (as an apple or pear) consisting of an outer thickened fleshy layer and a central core with usually five seeds enclosed in a capsule

pome·gran·ate \'pä-mə-ˌgra-nət; 'päm-ˌgra-nət, 'pəm-\ *noun* [Middle English *poumgrenet,* from Middle French *pomme grenate,* literally, seedy apple] (14th century)
1 : a thick-skinned several-celled reddish berry that is about the size of an orange and has many seeds with pulpy crimson arils of tart flavor
2 : a widely cultivated tropical Old World tree (*Punica granatum* of the family Punicaceae) bearing pomegranates

pomegranate

pom·e·lo \'pä-mə-ˌlō\ *noun, plural* **-los** [alteration of earlier *pompelmous,* from Dutch *pompelmoes*] (1858)
1 : GRAPEFRUIT
2 : SHADDOCK

Pom·er·a·nian \ˌpä-mə-'rā-nē-ən, -nyən\ *noun* (1760)
1 : any of a breed of very small compact long-haired dogs
2 : a native or inhabitant of Pomerania
— **Pomeranian** *adjective*

pom·mée \pä-'mā, ˌpə-\ *adjective* [French, from Middle French *pomme* apple, ball] (1725)
of a heraldic cross **:** having the end of each arm terminating in a ball or disk — see CROSS illustration

[1]pom·mel \'pə-məl, 'pä-\ *noun* [Middle English *pomel,* from Middle French, from (assumed) Vulgar Latin *pomellum* ball, knob, from diminutive of Late Latin *pomum* apple] (14th century)
1 : the knob on the hilt of a sword or saber
2 : the protuberance at the front and top of a saddle
3 : either of a pair of removable rounded or U-shaped handles used on the top of a pommel horse

[2]pom·mel \'pə-məl\ *transitive verb* **-meled** *or* **-melled; -mel·ing** *or* **-mel·ling** \'pə-mə-liŋ, 'pəm-liŋ\ [[1]*pommel*] (1530)
: PUMMEL

pommel horse *noun* (1908)
1 : a gymnastics apparatus for swinging and balancing feats that consists of a padded rectangular or cylindrical form with two pommels on the top and that is supported in a horizontal position above the floor
2 : an event in which the pommel horse is used

pommel horse 1

Pom·my *or* **Pom·mie** \'pä-mē\ *noun, plural* **Pommies** [by shortening & alteration from *pomegranate,* alteration of *Jimmy Grant,* rhyming slang for *immigrant*] (1912)
Australian & New Zealand **:** BRITON; *especially* **:** an English immigrant — usually used disparagingly

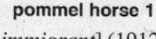

Po·mo \'pō-(ˌ)mō\ *noun, plural* **Pomo** *or* **Pomos** (1852)
: a member of an American Indian people of northern California

po·mol·o·gy \pō-'mä-lə-jē\ *noun* [New Latin *pomologia,* from Latin *pomum* fruit + *-logia* -logy] (1818)
: the science and practice of fruit growing
— **po·mo·log·i·cal** \ˌpō-mə-'lä-ji-kəl\ *adjective*
— **po·mol·o·gist** \pō-'mä-lə-jist\ *noun*

pomp \'pämp\ *noun* [Middle English, from Middle French *pompe,* from Latin *pompa* procession, pomp, from Greek *pompē* act of sending, escort, procession, pomp, from *pempein* to send] (14th century)
1 : a show of magnificence **:** SPLENDOR ⟨every day begins . . . in a *pomp* of flaming colours —F. D. Ommanney⟩
2 : a ceremonial or festival display (as a train of followers or a pageant)
3 a : ostentatious display **:** VAINGLORY **b :** an ostentatious gesture or act

pom·pa·dour \'päm-pə-ˌdōr, -ˌdȯr\ *noun* [Marquise de *Pompadour*] (1756)
1 a : a man's style of hairdressing in which the hair is combed into a high mound in front **b :** a woman's style of hairdressing in which the hair is brushed into a loose full roll around the face
2 : hair dressed in a pompadour
— **pom·pa·doured** \-ˌdōrd, -ˌdȯrd\ *adjective*

pom·pa·no \'päm-pə-ˌnō, 'pəm-\ *noun, plural* **-no** *or* **-nos** [American Spanish *pámpano,* from Spanish, a percoid fish (*Sparus auratus*), literally, vine leaf, from Latin *pampinus*] (1778)
1 : a carangid food fish (*Trachinotus carolinus*) of the southern Atlantic and Gulf coasts of North America; *broadly* **:** any of several related fishes
2 : a small bluish or greenish butterfish (*Peprilus simillimus*) of the Pacific coast

¹pom–pom \'päm-ˌpäm\ *noun* [alteration of *pompon*] (1897)
1 : an ornamental ball or tuft used especially on clothing, caps, or costumes
2 : a handheld usually brightly colored fluffy ball flourished by cheerleaders
²pom–pom *noun* [imitative; from the sound of its discharge] (1899)
: a multibarreled automatic antiaircraft gun of 20 to 40 millimeters mounted especially on ships
pom•pon \'päm-ˌpän\ *noun* [French, from Middle French *pompe* tuft of ribbons] (1753)
1 : ¹POM-POM 1
2 : a chrysanthemum or dahlia with small rounded flower heads
pom•pos•i•ty \päm-'pä-sə-tē\ *noun, plural* **-ties** (1620)
1 : pompous demeanor, speech, or behavior
2 : a pompous gesture, habit, or act
pomp•ous \'päm-pəs\ *adjective* (15th century)
1 : excessively elevated or ornate ⟨*pompous* rhetoric⟩
2 : having or exhibiting self-importance **:** ARROGANT ⟨a *pompous* politician⟩
3 : relating to or suggestive of pomp **:** MAGNIFICENT
— pomp•ous•ly *adverb*
— pomp•ous•ness *noun*
ponce \'pän(t)s\ *noun* [origin unknown] (1872)
British **:** PIMP; *also* **:** a male homosexual — usually used disparagingly
pon•cho \'pän-(ˌ)chō\ *noun, plural* **ponchos** [American Spanish, from Araucanian] (1717)
1 : a blanket with a slit in the middle so that it can be slipped over the head and worn as a sleeveless garment
2 : a waterproof garment resembling a poncho and having an integral hood
¹pond \'pänd\ *noun* [Middle English *ponde* artificially confined body of water, probably alteration of *pounde* enclosure — more at POUND] (14th century)
: a body of water usually smaller than a lake
²pond (1694)
transitive verb
: to block (as a stream) to create a pond
intransitive verb
: to collect in or form a pond
pon•der \'pän-dər\ *verb* **pon•dered; pon•der•ing** \-d(ə-)riŋ\ [Middle English, from Middle French *ponderer,* from Latin *ponderare* to weigh, ponder, from *ponder-, pondus* weight — more at PENDANT] (14th century)
transitive verb
1 : to weigh in the mind **:** APPRAISE ⟨*pondered* their chances of success⟩
2 : to think about **:** reflect on ⟨*pondered* the events of the day⟩
intransitive verb
: to think or consider especially quietly, soberly, and deeply ☆
— pon•der•er \-dər-ər\ *noun*
pon•der•a•ble \'pän-d(ə-)rə-bəl\ *adjective* [Late Latin *ponderabilis,* from *ponderare*] (1813)
: significant enough to be worth considering **:** APPRECIABLE
synonym see PERCEPTIBLE
pon•der•o•sa pine \ˌpän-də-'rō-sə-, -zə-\ *noun* [New Latin *ponderosa,* specific epithet of *Pinus ponderosa,* from Latin, feminine of *ponderosus* ponderous] (1878)
: a tall pine (*Pinus ponderosa*) of western North America with long needles usually in groups of 2 or 3; *also* **:** its strong reddish straight-grained wood — called also *ponderosa*
pon•der•ous \'pän-d(ə-)rəs\ *adjective* [Middle English, from Middle French *pondereux,* from Latin *ponderosus,* from *ponder-, pondus* weight] (15th century)
1 : of very great weight
2 : unwieldy or clumsy because of weight and size
3 : oppressively or unpleasantly dull **:** LIFELESS ⟨*ponderous* prose⟩
synonym see HEAVY
— pon•der•ous•ly *adverb*
— pon•der•ous•ness *noun*
pond lily *noun* (1748)
: WATER LILY
pond scum *noun* (circa 1890)
1 : SPIROGYRA; *also* **:** any of various related algae
2 : a mass of tangled algal filaments in stagnant waters
pond skater *noun* (1895)
: WATER STRIDER
pond•weed \'pänd-ˌwēd\ *noun* (1578)
: any of a genus (*Potamogeton* of the family Potamogetonaceae, the pondweed family) of aquatic plants with jointed usually rooting stems, 2-ranked floating or submerged leaves, and spikes of greenish flowers
pone \'pōn\ *noun* [modification of Virginia Algonquian *appone*] (1634)
Southern & Midland **:** CORN PONE
pon•gee \ˌpän-'jē, 'pän-ˌ\ *noun, often attributive* [Chinese (Beijing) *běnjī,* from *běn* own + *jī* loom] (1711)
: a thin soft fabric of Chinese origin woven from raw silk; *also* **:** an imitation of this fabric in cotton or a synthetic fiber (as of polyester or rayon)
pon•gid \'pän-jəd, 'päŋ-gəd\ *noun* [ultimately from Kongo *mpongi* ape] (1950)
: any of a family (Pongidae) of apes that includes the chimpanzee, gorilla, and orangutan
— pongid *adjective*
¹pon•iard \'pän-yərd\ *noun* [Middle French *poignard,* from *poing* fist, from Latin *pugnus* fist — more at PUNGENT] (1588)
: a dagger with a usually slender blade of triangular or square cross section
²poniard *transitive verb* (1601)
: to pierce or kill with a poniard
pons \'pänz\ *noun, plural* **pon•tes** \'pän-ˌtēz\ [New Latin, short for *pons Varolii*] (1831)
: a broad mass of chiefly transverse nerve fibers conspicuous on the ventral surface of the brain of man and lower mammals at the anterior end of the medulla oblongata — see BRAIN illustration
pons asi•no•rum \'pänz-ˌa-sə-'nōr-əm, -'nȯr-\ *noun* [New Latin, literally, asses' bridge, name applied to the proposition that the base angles of an isosceles triangle are equal] (1751)
: a critical test of ability or understanding; *also* **:** STUMBLING BLOCK
pons Va•ro•lii \-və-'rō-lē-ˌī, -lē-ˌē\ *noun* [New Latin, literally, bridge of Varoli, from Costanzo *Varoli* (died 1575) Italian surgeon and anatomist] (circa 1693)
: PONS
pon•ti•fex \'pän-tə-ˌfeks\ *noun, plural* **pon•tif•i•ces** \pän-'ti-fə-ˌsēz\ [Latin *pontific-, pontifex,* from *pont-, pons* bridge + *facere* to make — more at FIND, DO] (circa 1580)
: a member of the council of priests in ancient Rome
pon•tiff \'pän-təf\ *noun* [French *pontif,* from Latin *pontific-, pontifex*] (1626)
1 : PONTIFEX
2 : BISHOP; *specifically* **:** POPE
¹pon•tif•i•cal \pän-'ti-fi-kəl\ *noun* (14th century)
1 : episcopal attire; *specifically* **:** the insignia of the episcopal order worn by a prelate when celebrating a pontifical mass — usually used in plural
2 : a book containing the forms for sacraments and rites performed by a bishop
²pontifical *adjective* [Middle English, from Latin *pontificalis,* from *pontific-, pontifex*] (15th century)
1 a : of or relating to a pontiff or pontifex **b :** celebrated by a prelate of episcopal rank with distinctive ceremonies ⟨*pontifical* mass⟩
2 : POMPOUS
3 : pretentiously dogmatic
— pon•tif•i•cal•ly \-k(ə-)lē\ *adverb*
¹pon•tif•i•cate \pän-'ti-fi-kət, -ˌkāt\ *noun* [Middle English, from Latin *pontificatus,* from *pontific-, pontifex*] (15th century)
: the state, office, or term of office of a pontiff
²pon•tif•i•cate \pän-'ti-fə-ˌkāt\ *intransitive verb* **-cat•ed; -cat•ing** [Medieval Latin *pontificatus,* past participle of *pontificare,* from Latin *pontific-, pontifex*] (1818)
1 a : to officiate as a pontiff **b :** to celebrate pontifical mass
2 : to speak or express opinions in a pompous or dogmatic way
— pon•tif•i•ca•tion \(ˌ)pän-ˌti-fə-'kā-shən\ *noun*
— pon•tif•i•ca•tor \-ˌkā-tər\ *noun*
pon•til \'pän-tᵊl\ *noun* [French, perhaps from Italian *puntello,* diminutive of *punto* point, from Latin *punctus* — more at POINT] (1832)
: PUNTY — called also *pontil rod*
pon•tine \'pän-ˌtīn\ *adjective* [New Latin *pont-, pons* pons] (1889)
: of or relating to the pons
Pont l'É•vêque \ˌpōⁿ-lā-'vek\ *noun* [*Pont l'Évêque,* town in France] (circa 1889)
: a soft surface-ripened cheese firmer, yellower, and having less surface mold than Camembert
¹pon•toon \pän-'tün\ *noun* [French *ponton,* from Latin *ponton-, ponto*] (1690)
1 : a flat-bottomed boat (as a lighter); *especially* **:** a flat-bottomed boat or portable float used in building a floating temporary bridge
2 : a float especially of a seaplane
²pontoon *noun* [perhaps alteration of *vingt-et-un*] (circa 1917)
British **:** BLACKJACK 5
po•ny \'pō-nē\ *noun, plural* **ponies** [probably from obsolete French *poulenet,* diminutive of French *poulain* colt, from Medieval Latin *pullanus,* from Latin *pullus* young of an animal, foal — more at FOAL] (1659)
1 a : a small horse; *especially* **:** one of any of several breeds of very small stocky animals noted for their gentleness and endurance **b :** a bronco, mustang, or similar horse of the western U.S. **c :** RACEHORSE — usually used in plural
2 : something smaller than standard: as **a :** a small beer glass **b :** a small liqueur glass typically holding one ounce
3 : a literal translation of a foreign language text; *especially* **:** one used surreptitiously by students in preparing or reciting lessons
pony express *noun, often P&E capitalized* (1860)
: a rapid postal and express system that operated across the western U.S. in 1860–1861 by relays of horses and riders
po•ny•tail \'pō-nē-ˌtāl\ *noun* (1951)
: a hairstyle in which the hair is pulled together and banded at the back of the head so as to resemble a pony's tail
— po•ny•tailed \-ˌtāld\ *adjective*
po•ny up \'pō-nē-\ *verb* **po•nied up; po•ny•ing up** [origin unknown] (1824)
transitive verb

☆ SYNONYMS
Ponder, meditate, muse, ruminate mean to consider or examine attentively or deliberately. PONDER implies a careful weighing of a problem or, often, prolonged inconclusive thinking about a matter ⟨*pondered* the course of action⟩. MEDITATE implies a definite focusing of one's thoughts on something so as to understand it deeply ⟨*meditated* on the meaning of life⟩. MUSE suggests a more or less focused daydreaming as in remembrance ⟨*mused* upon childhood joys⟩. RUMINATE implies going over the same matter in one's thoughts again and again but suggests little of either purposive thinking or rapt absorption ⟨*ruminated* on past disappointments⟩.

: to pay (money) especially in settlement of an account ⟨*ponied up* $12.50 for the fine —*Newsweek*⟩
intransitive verb
: PAY

Pon·zi scheme \'pän-zē-\ *noun* [Charles A. *Ponzi* (died 1949) American (Italian-born) swindler] (1973)
: an investment swindle in which some early investors are paid off with money put up by later ones in order to encourage more and bigger risks

-poo \ˌpü, 'pü\ *suffix* [origin unknown]
— used as a disparaging diminutive ⟨cutesy-*poo*⟩

[1]pooch \'püch\ *verb* [alteration of [1]*pouch*] (circa 1923)
chiefly dialect **:** BULGE

[2]pooch *noun* [origin unknown] (1924)
: DOG

pood \'püd, 'püt\ *noun* [Russian *pud*, from Old Russian, from Old Norse *pund* pound — more at POUND] (1554)
: a Russian unit of weight equal to about 36.11 pounds (16.38 kilograms)

poo·dle \'pü-dᵊl\ *noun* [German *Pudel*, short for *Pudelhund*, from *pudeln* to splash + *Hund* dog] (1820)
1 : a dog of any of three breeds of active intelligent heavy-coated solid-colored dogs — compare TOY POODLE
2 : a fabric with a nubby or coarsely looped surface that resembles a poodle's coat — called also *poodle cloth*

[1]poof \'püf, 'pu̇f\ *interjection* (1824)
— used to express disdain or dismissal or to suggest instantaneous occurrence

[2]poof *also* **poove** \'püv, 'pu̇v\ *noun, plural* **poofs** *also* **pooves** [perhaps alteration of [2]*puff*] (circa 1860)
British **:** a male homosexual — usually used disparagingly

poof·ter \'püf-tər, 'pu̇f-\ *also* **poof·tah** \-tə\ *noun* [by alteration] (1903)
British **:** POOF — usually used disparagingly

pooh \'pü, 'pu̇\ *interjection* (1602)
— used to express contempt or disapproval

pooh–bah *also* **poo–bah** \'pü-ˌbä, -ˌbȯ\ *noun, often P&B capitalized* [*Pooh-Bah*, character in Gilbert and Sullivan's opera *The Mikado* (1885) bearing the title Lord-High-Everything-Else] (1888)
1 : a person holding many public or private offices
2 : a person in high position or of great influence

pooh–pooh \'pü-(ˌ)pü, pü-'\ *also* **pooh** \'pü\ *verb* [*pooh*] (1827)
intransitive verb
: to express contempt or impatience
transitive verb
: to express contempt for or make light of **:** PLAY DOWN, DISMISS

[1]pool \'pü(ə)l\ *noun* [Middle English, from Old English *pōl*; akin to Old High German *pfuol* pool] (before 12th century)
1 a (1) **:** a small and rather deep body of usually fresh water (2) **:** a quiet place in a stream (3) **:** a body of water forming above a dam **b :** something resembling a pool ⟨a *pool* of light⟩
2 : a small body of standing liquid
3 : a continuous area of porous sedimentary rock that yields petroleum or gas
4 : SWIMMING POOL

[2]pool *intransitive verb* (1626)
1 : to form a pool
2 *of blood* **:** to accumulate or become static (as in the veins of a bodily part)

[3]pool *noun* [French *poule*, literally, hen, from Old French, feminine of *poul* cock — more at PULLET] (circa 1712)
1 a : an aggregate stake to which each player of a game has contributed **b :** all the money bet by a number of persons on a particular event
2 a : a game played on an English billiard table in which each of the players stakes a sum and the winner takes all **b :** any of various games of billiards played on an oblong table having 6 pockets with usually 15 object balls
3 : an aggregation of the interests or property of different persons made to further a joint undertaking by subjecting them to the same control and a common liability
4 : a readily available supply: as **a :** the whole quantity of a particular material present in the body and available for function or the satisfying of metabolic demands **b :** a body product (as blood) collected from many donors and stored for later use **c :** a group of people available for some purpose ⟨a shrinking *pool* of applicants⟩ ⟨typing *pool*⟩
5 : GENE POOL

[4]pool *transitive verb* (1879)
: to combine (as resources) in a common fund or effort

pool·room \'pü(ə)l-ˌrüm, -ˌru̇m\ *noun* (1861)
1 : a room in which bookmaking is carried on
2 : a room for the playing of pool

pool·side \-ˌsīd\ *noun* (1921)
: the area surrounding a swimming pool

[1]poop \'püp\ *noun* [Middle English, from Middle French *poupe*, from Latin *puppis*] (15th century)
1 *obsolete* **:** STERN
2 : an enclosed superstructure at the stern of a ship above the main deck

[2]poop *transitive verb* (1748)
1 : to break over the stern of
2 : to ship (a sea or wave) over the stern

[3]poop *verb* [origin unknown] (circa 1932)
transitive verb
slang **:** to put out of breath; *also* **:** to tire out
intransitive verb
slang **:** to become exhausted ⟨*poop* out⟩

[4]poop *noun* [perhaps from [5]*poop*] (circa 1941)
slang **:** INFORMATION, SCOOP

[5]poop *noun* [*poop*, verb, to defecate, break wind, from Middle English *poupen* to make a gulping sound; of imitative origin] (1965)
: EXCREMENT

poop deck *noun* (1840)
: a partial deck above a ship's main afterdeck

poop·er–scoop·er \'pü-pər-ˌskü-pər\ *noun* [[5]*poop*] (1976)
: a device used for picking up the droppings of a pet (as a dog) for disposal

poor \'pu̇r, 'pōr\ *adjective* [Middle English *poure*, from Old French *povre*, from Latin *pauper*; akin to Latin *paucus* little and to Latin *parere* to give birth to, produce — more at FEW, PARE] (13th century)
1 a : lacking material possessions **b :** of, relating to, or characterized by poverty
2 a : less than adequate **:** MEAGER **b :** small in worth
3 : exciting pity ⟨you *poor* thing⟩
4 a : inferior in quality or value **b :** HUMBLE, UNPRETENTIOUS **c :** MEAN, PETTY
5 : LEAN, EMACIATED
6 : BARREN, UNPRODUCTIVE — used of land
7 : INDIFFERENT, UNFAVORABLE
8 : lacking a normal or adequate supply of something specified — often used in combination ⟨oil-*poor* countries⟩
— poor·ish \-ish\ *adjective*
— poor·ness *noun*

poor box *noun* (1621)
: a box (as in a church) for alms for the poor

poor boy \'pō(r)-ˌbȯi\ *noun* (circa 1941)
: SUBMARINE 2

Poor Clare \-'klar, -'kler\ *noun* (1608)
: a member of an austere order of nuns founded by Saint Clare under the direction of Saint Francis in Assisi, Italy, in 1212

poor farm \'pu̇r-ˌfärm, 'pōr-\ *noun* (1852)
: a farm maintained at public expense for the support and employment of needy persons

poor·house \-ˌhau̇s\ *noun* (1792)
: a place maintained at public expense to house needy or dependent persons

poor law *noun* (1752)
: a law providing for or regulating the public relief or support of the poor

[1]poor·ly \'pu̇r-lē, 'pōr-\ *adverb* (13th century)
: in a poor condition or manner; *especially* **:** in an inferior or imperfect way **:** BADLY ⟨sang *poorly*⟩

[2]poor·ly *adjective* (1750)
: somewhat ill **:** INDISPOSED

poor–mouth \'pu̇r-ˌmau̇t͟h, 'pōr-, -ˌmau̇th\ (1964)
intransitive verb
: to plead poverty as a defense or excuse
transitive verb
: to speak disparagingly of

poor mouth \-ˌmau̇th\ *noun* (1822)
: an exaggerated claim of poverty

poor–spir·it·ed \-'spir-ə-təd\ *adjective* (1670)
: lacking zest, confidence, or courage
— poor–spir·it·ed·ly *adverb*
— poor–spir·it·ed·ness *noun*

poor white *noun* (1819)
: a member of an inferior or underprivileged white social group — often taken to be offensive

[1]pop \'päp\ *verb* **popped; pop·ping** [Middle English *poppen*, of imitative origin] (15th century)
transitive verb
1 : to strike or knock sharply **:** HIT
2 : to push, put, or thrust suddenly and often deftly
3 : to cause to explode or burst open
4 : to fire at **:** SHOOT
5 : to take (pills) especially frequently or habitually
6 : to open with a pop ⟨*pop* a cold beer⟩
intransitive verb
1 a : to go, come, or appear suddenly — often used with *up* **b :** to escape or break away from something (as a point of attachment) usually suddenly or unexpectedly
2 : to make or burst with a sharp sound **:** EXPLODE
3 : to protrude from the sockets
4 : to shoot with a firearm
5 : to hit a pop fly — often used with *up* or *out*
— pop the question : to propose marriage

[2]pop *noun* (1591)
1 : a sharp explosive sound
2 : a shot from a gun
3 : SODA POP
4 : POP FLY
— a pop : for each one **:** APIECE ⟨tickets at $10 *a pop*⟩

[3]pop *adverb* (1621)
: like or with a pop **:** SUDDENLY

[4]pop *noun* [short for *poppa*] (1838)
: FATHER

[5]pop *adjective* [by shortening] (1880)
1 : POPULAR ⟨*pop* music⟩: as **a :** of or relating to popular music ⟨*pop* singer⟩ **b :** of or relating to the popular culture disseminated through the mass media ⟨*pop* psychology⟩ ⟨*pop* grammarians⟩ ⟨*pop* society⟩
2 a : of or relating to pop art ⟨*pop* painter⟩ **b :** having, using, or imitating themes or techniques characteristic of pop art ⟨*pop* movie⟩

[6]pop *noun* (1954)
1 : pop music
2 : POP ART
3 : pop culture

pop art *noun* (1957)
: art in which commonplace objects (as road signs, hamburgers, comic strips, or soup cans)

are used as subject matter and are often physically incorporated in the work
— **pop artist** *noun*

pop·corn \'päp-ˌkȯrn\ *noun* (1823)
: an Indian corn (*Zea mays praecox*) whose kernels on exposure to heat burst open to form a white starchy mass; *also* **:** the kernels especially after popping

pope \'pōp\ *noun* [Middle English, from Old English *pāpa,* from Late Latin *papa,* from Greek *pappas, papas,* title of bishops, literally, papa] (before 12th century)
1 *often capitalized* **:** a prelate who as bishop of Rome is the head of the Roman Catholic Church
2 : one that resembles a pope (as in authority)
3 a : the Eastern Orthodox or Coptic patriarch of Alexandria **b :** a priest of an Eastern church

pop·ery \'pō-p(ə-)rē\ *noun* (circa 1534)
: ROMAN CATHOLICISM — usually used disparagingly

pop eye \'päp-ˌī\ *noun* [back-formation from *pop-eyed*] (1828)
: an eye staring and bulging (as from excitement)
— **pop–eyed** \-ˌīd\ *adjective*

pop fly *noun* (1887)
: a high fly ball in baseball

pop·gun \'päp-ˌgən\ *noun* (1622)
: a toy gun that usually shoots a cork and produces a popping sound

pop·in·jay \'pä-pən-ˌjā\ *noun* [Middle English *papejay* parrot, from Middle French *papegai, papejai,* from Arabic *babghā'*] (1528)
: a strutting supercilious person

pop·ish \'pō-pish\ *adjective* [*pope*] (1528)
: ROMAN CATHOLIC — often used disparagingly

pop·lar \'pä-plər\ *noun* [Middle English *poplere,* from Middle French *pouplier,* from *pouple* poplar, from Latin *populus*] (14th century)
1 a : any of a genus (*Populus*) of slender catkin-bearing quick-growing trees (as an aspen or cottonwood) of the willow family **b :** the wood of a poplar
2 : TULIP TREE 1

pop·lin \'pä-plən\ *noun* [French *papeline*] (1710)
: a strong fabric in plain weave with crosswise ribs

pop·li·te·al \ˌpä-plə-'tē-əl *also* pä-'pli-tē-əl\ *adjective* [New Latin *popliteus,* from Latin *poplit-, poples* ham of the knee] (1786)
: of or relating to the back part of the leg behind the knee joint

pop off *intransitive verb* (1764)
1 a : to die unexpectedly **b :** to leave suddenly
2 : to talk thoughtlessly and often loudly or angrily

pop·over \'päp-ˌō-vər\ *noun* (1876)
: a hollow quick bread shaped like a muffin and made from a thin batter of eggs, milk, and flour

pop·pa \'pä-pə\ *variant of* PAPA

pop·per \'pä-pər\ *noun* (1750)
1 : one that pops; *especially* **:** a utensil for popping corn
2 *slang* **:** a vial of amyl nitrite or butyl nitrite especially when used illicitly as an aphrodisiac

pop·pet \'pä-pət\ *noun* [Middle English *popet* — more at PUPPET] (15th century)
1 a *Midland* **:** DOLL **b** *obsolete* **:** MARIONETTE
2 *chiefly British* **:** DEAR
3 a : an upright support or guide of a machine that is fastened at the bottom only **b :** a valve that rises perpendicularly to or from its seat

pop·pied \'pä-pēd\ *adjective* (1818)
1 *archaic* **:** growing or overgrown with poppies
2 : DROWSY

[1]pop·ple \'pä-pəl\ *noun* [Middle English *popul,* from Old English, from Latin *populus*] (14th century)
chiefly dialect **:** POPLAR 1

[2]popple *noun* [*popple,* verb, from Middle English *poplen* to bubble, ripple, probably of imitative origin] (1875)
: a choppy sea

pop·py \'pä-pē\ *noun, plural* **poppies** [Middle English *popi,* from Old English *popæg, popig,* modification of Latin *papaver*] (before 12th century)
1 a : any of a genus (*Papaver* of the family Papaveraceae, the poppy family) of chiefly annual or perennial herbs with milky juice, showy regular flowers, and capsular fruits including the opium poppy and several forms cultivated as ornamentals **b :** an extract or decoction of poppy used medicinally
2 : a strong reddish orange

pop·py·cock \'pä-pē-ˌkäk\ *noun* [Dutch dialect *pappekak,* literally, soft dung, from Dutch *pap* pap + *kak* dung] (1865)
: empty talk or writing **:** NONSENSE

pop·py·head \-ˌhed\ *noun* (1839)
: a raised ornament often in the form of a finial generally used on the tops of the upright ends of seats in Gothic churches

poppy seed *noun* (14th century)
: the seed of a poppy used chiefly as a topping or flavoring for baked goods

pop quiz *noun* (circa 1960)
: an unscheduled or unannounced quiz

Pop·si·cle \'päp-ˌsi-kəl, -sə-\ *trademark*
— used for flavored and colored water frozen on a stick

pop–top \'päp-ˌtäp\ *noun* (1965)
: a closure that can be pulled by hand to open a can

pop·u·lace \'pä-pyə-ləs\ *noun* [Middle French, from Italian *popolaccio* rabble, augmentative of *popolo* the people, from Latin *populus*] (1572)
1 : the common people **:** MASSES
2 : POPULATION

pop·u·lar \'pä-pyə-lər\ *adjective* [Latin *popularis,* from *populus* the people, a people] (1548)
1 : of or relating to the general public
2 : suitable to the majority: as **a :** adapted to or indicative of the understanding and taste of the majority ⟨a *popular* history of the war⟩ **b :** suited to the means of the majority **:** INEXPENSIVE ⟨sold at *popular* prices⟩
3 : frequently encountered or widely accepted
4 : commonly liked or approved ⟨a very *popular* girl⟩
synonym see COMMON
— **pop·u·lar·ly** *adverb*

popular front *noun, often P&F capitalized* (1936)
: a coalition especially of leftist political parties against a common opponent; *specifically* **:** one sponsored and dominated by Communists as a device for gaining power

pop·u·lar·ise *British variant of* POPULARIZE

pop·u·lar·i·ty \ˌpä-pyə-'lar-ə-tē\ *noun* (1601)
: the quality or state of being popular

pop·u·lar·ize \'pä-pyə-lə-ˌrīz\ *verb* **-ized; -iz·ing** (1593)
intransitive verb
: to cater to popular taste
transitive verb
: to make popular: as **a :** to cause to be liked or esteemed **b :** to present in generally understandable or interesting form
— **pop·u·lar·i·za·tion** \ˌpä-pyə-lə-rə-'zā-shən\ *noun*
— **pop·u·lar·iz·er** \'pä-pyə-lə-ˌrī-zər\ *noun*

popular sovereignty *noun* (1848)
1 : a doctrine in political theory that government is created by and subject to the will of the people
2 : a pre-Civil War doctrine asserting the right of the people living in a newly organized territory to decide by vote of their territorial legislature whether or not slavery would be permitted there

pop·u·late \'pä-pyə-ˌlāt\ *transitive verb* **-lat·ed; -lat·ing** [Medieval Latin *populatus,* past participle of *populare* to people, from Latin *populus* people] (1578)
1 : to have a place in **:** OCCUPY, INHABIT
2 a : to furnish or provide with inhabitants **:** PEOPLE **b :** to provide with members

pop·u·la·tion \ˌpä-pyə-'lā-shən\ *noun* [Late Latin *population-, populatio,* from Latin *populus*] (1612)
1 a : the whole number of people or inhabitants in a country or region **b :** the total of individuals occupying an area or making up a whole **c :** the total of particles at a particular energy level — used especially of atoms in a laser
2 : the act or process of populating
3 a : a body of persons or individuals having a quality or characteristic in common **b** (1) **:** the organisms inhabiting a particular locality (2) **:** a group of interbreeding organisms that represents the level of organization at which speciation begins
4 : a group of individual persons, objects, or items from which samples are taken for statistical measurement
— **pop·u·la·tion·al** \-shnəl, -shə-nᵊl\ *adjective*

population explosion *noun* (1953)
: a pyramiding of numbers of a biological population; *especially* **:** the recent great increase in human numbers resulting from increased survival and exponential population growth

[1]pop·u·list \'pä-pyə-list\ *noun* [Latin *populus* the people] (1892)
1 : a member of a political party claiming to represent the common people; *especially, often capitalized* **:** a member of a U.S. political party formed in 1891 primarily to represent agrarian interests and to advocate the free coinage of silver and government control of monopolies
2 : a believer in the rights, wisdom, or virtues of the common people
— **pop·u·lism** \-ˌli-zəm\ *noun*
— **pop·u·lis·tic** \ˌpä-pyə-'lis-tik\ *adjective*

[2]populist *adjective* (1893)
1 *often capitalized* **:** of, relating to, or characterized by populism
2 : POPULAR 1, 2

pop·u·lous \'pä-pyə-ləs\ *adjective* [Middle English, from Latin *populosus,* from *populus* people] (15th century)
1 a : densely populated **b :** having a large population
2 a : NUMEROUS **b :** filled to capacity
— **pop·u·lous·ly** *adverb*
— **pop·u·lous·ness** *noun*

pop–up \'päp-ˌəp\ *noun* (1906)
: POP FLY

por·bea·gle \'pȯr-ˌbē-gəl\ *noun* [Cornish *porgh-bugel*] (1758)
: a small viviparous shark (*Lamna nasus*) chiefly of the North Atlantic Ocean with a pointed nose and crescent-shaped tail

por·ce·lain \'pōr-s(ə-)lən, 'pȯr-\ *noun* [Middle French *porcelaine* cowrie shell, porcelain, from Italian *porcellana,* from *porcello* vulva, literally, little pig, from Latin *porcellus,* diminutive of *porcus* pig, vulva; from the shape of the shell — more at FARROW] (circa 1530)
: a hard, fine-grained, sonorous, nonporous, and usually translucent and white ceramic ware that consists essentially of kaolin, quartz, and feldspar and is fired at high temperatures
— **por·ce·lain·like** \-ˌlīk\ *adjective*
— **por·ce·la·ne·ous** *or* **por·cel·la·ne·ous** \ˌpōr-sə-'lā-nē-əs, ˌpȯr-\ *adjective*

porcelain enamel *noun* (1883)
: a fired-on opaque glassy coating on metal (as steel)

por·ce·lain·ize \'pōr-s(ə-)lə-ˌnīz, 'pȯr-\ *transitive verb* **-ized; -iz·ing** (1951)
: to fire a glassy coating on (as steel)

porch \'pōrch, 'pȯrch\ *noun* [Middle English *porche,* from Old French, from Latin *porticus*

portico, from *porta* gate; akin to Latin *portus* port — more at FORD] (14th century)
1 : a covered area adjoining an entrance to a building and usually having a separate roof
2 *obsolete* **:** PORTICO

por·cine \ˈpȯr-ˌsīn\ *adjective* [Latin *porcinus,* from *porcus* pig — more at FARROW] (circa 1656)
: of, relating to, or suggesting swine

por·ci·no \pȯr-ˈchē-(ˌ)nō\ *noun, plural* **-ni** \-(ˌ)nē\ [Italian, short for *fungo porcino,* literally, porcine mushroom] (1976)
: a wild edible boletus mushroom (especially *Boletus edulis*)

por·cu·pine \ˈpȯr-kyə-ˌpīn\ *noun, plural* **porcupines** *also* **porcupine** [Middle English *porkepin,* from Middle French *porc espin,* from Old Italian *porcospino,* from Latin *porcus* pig + *spina* spine, prickle] (15th century)
: any of various relatively large rodents having stiff sharp erectile bristles mingled with the hair and constituting an Old World terrestrial family (Hystricidae) and a New World arboreal family (Erethizontidae)

porcupine fish *noun* (1681)
: any of several bony fishes (family Diodontidae) having sharp spines covering the body; *especially* **:** a spotted chiefly tropical fish (*Diodon hystrix*) that is olive to brown above with white below

[1]**pore** \ˈpōr, ˈpȯr\ *intransitive verb* **pored; por·ing** [Middle English *pouren*] (13th century)
1 : to gaze intently
2 : to read studiously or attentively — usually used with *over*
3 : to reflect or meditate steadily

[2]**pore** *noun* [Middle English, from Middle French, from Latin *porus,* from Greek *poros* passage, pore — more at FARE] (14th century)
1 : a minute opening especially in an animal or plant; *especially* **:** one by which matter passes through a membrane
2 : a small interstice (as in soil) admitting absorption or passage of liquid
— **pored** \ˈpōrd, ˈpȯrd\ *adjective*

pore fungus *noun* (1922)
: a fungus (family Boletaceae or Polyporaceae) having the spore-bearing surface within tubes or pores

por·gy \ˈpȯr-gē\ *noun, plural* **porgies** *also* **porgy** [alteration of *pargo,* from Spanish & Portuguese, from Latin *phager,* from Greek *phagros*] (1671)
1 : a blue-spotted crimson food fish (*Pagrus pagrus* of the family Sparidae) of the eastern and western Atlantic; *also* **:** any of various fishes of the same family
2 [alteration of *pogy*] **:** any of various bony fishes (as a menhaden) of families other than that of the porgy

pork \ˈpōrk, ˈpȯrk\ *noun* [Middle English, from Old French *porc* pig, from Latin *porcus* — more at FARROW] (14th century)
1 : the fresh or salted flesh of swine when dressed for food
2 : government money, jobs, or favors used by politicians as patronage

pork barrel *noun* (1909)
: government projects or appropriations yielding rich patronage benefits; *also* **:** PORK 2

pork belly *noun* (circa 1950)
: an uncured side of pork

pork·er \ˈpōr-kər, ˈpȯr-\ *noun* (1657)
: HOG; *especially* **:** a young pig fattened for table use as fresh pork

pork·pie hat \ˈpōrk-ˌpī-, ˈpȯrk-\ *noun* [from its shape] (1860)
: a hat with a low telescoped crown, flat top, and brim turned up all around or up in back and down in front

[1]**porky** \ˈpōr-kē, ˈpȯr-\ *adjective* **pork·i·er; -est** (1852)
: resembling a pig **:** FAT

[2]**por·ky** \ˈpȯr-kē\ *noun, plural* **porkies** (1900)
: PORCUPINE

porn \ˈpȯrn\ *also* **por·no** \ˈpȯr-(ˌ)nō\ *noun, often attributive* (1962)
: PORNOGRAPHY

por·nog·ra·pher \pȯr-ˈnä-grə-fər\ *noun* (1850)
: one who produces pornography

por·nog·ra·phy \-fē\ *noun* [Greek *pornographos,* adjective, writing about prostitutes, from *pornē* prostitute + *graphein* to write; akin to Greek *pernanai* to sell, *poros* journey — more at FARE, CARVE] (circa 1864)
1 : the depiction of erotic behavior (as in pictures or writing) intended to cause sexual excitement
2 : material (as books or a photograph) that depicts erotic behavior and is intended to cause sexual excitement
3 : the depiction of acts in a sensational manner so as to arouse a quick intense emotional reaction ⟨the *pornography* of violence⟩
— **por·no·graph·ic** \ˌpȯr-nə-ˈgra-fik\ *adjective*
— **por·no·graph·i·cal·ly** \-fi-k(ə-)lē\ *adverb*

porny \ˈpȯr-ne\ *adjective* **porn·i·er; -est** (1961)
: of, relating to, involved in, or being pornography

po·ros·i·ty \pə-ˈrä-sə-tē, pȯ-\ *noun, plural* **-ties** (14th century)
1 a : the quality or state of being porous **b :** the ratio of the volume of interstices of a material to the volume of its mass
2 : PORE

po·rous \ˈpōr-əs, ˈpȯr-\ *adjective* (14th century)
1 a : possessing or full of pores **b :** containing vessels ⟨hardwood is *porous*⟩
2 a : permeable to fluids **b :** permeable to outside influences ⟨his imagination is astonishingly *porous* —Elizabeth Hardwick⟩
3 : capable of being penetrated ⟨*porous* national boundaries⟩ ⟨a *porous* defense⟩
— **po·rous·ly** *adverb*
— **po·rous·ness** *noun*

por·phyr·ia \pȯr-ˈfir-ē-ə\ *noun* [New Latin, from International Scientific Vocabulary *porphyrin*] (1923)
: any of several usually hereditary abnormalities of porphyrin metabolism characterized by excretion of excess porphyrins in the urine

por·phy·rin \ˈpȯr-fə-rən\ *noun* [International Scientific Vocabulary, from Greek *porphyra* purple] (1910)
: any of various compounds with a structure that consists essentially of four pyrrole rings joined by four =CH− groups; *especially* **:** one (as chlorophyll or hemoglobin) containing a central metal atom and usually exhibiting biological activity

por·phy·rit·ic \ˌpȯr-fə-ˈri-tik\ *adjective* [Medieval Latin *porphyriticus,* from Greek *porphyritikos,* from *porphyritēs* (*lithos*) porphyry] (15th century)
1 : of or relating to porphyry
2 : having distinct crystals (as of feldspar) in a relatively fine-grained base

por·phy·rop·sin \ˌpȯr-fə-ˈräp-sən\ *noun* [Greek *porphyra* purple + English *-opsin* (as in *rhodopsin*)] (1930)
: a purple pigment in the retinal rods of freshwater fishes that resembles rhodopsin

por·phy·ry \ˈpȯr-f(ə-)rē\ *noun, plural* **-ries** [Middle English *porphiri,* from Medieval Latin *porphyrium,* alteration of Latin *porphyrites,* from Greek *porphyritēs* (*lithos*), literally, stone like Tyrian purple, from *porphyra* purple] (15th century)
1 : a rock consisting of feldspar crystals embedded in a compact dark red or purple groundmass
2 : an igneous rock of porphyritic texture

por·poise \ˈpȯr-pəs\ *noun* [Middle English *porpoys,* from Middle French *porpois,* from Medieval Latin *porcopiscis,* from Latin *porcus* pig + *piscis* fish — more at FARROW, FISH] (14th century)
1 : any of a family (Phocoenidae) of small gregarious toothed whales; *especially* **:** a blunt-snouted usually largely black whale (*Phocoena phocoena*) of the North Atlantic and Pacific 5 to 8 feet (1.5 to 2.4 meters) long
2 : DOLPHIN 1a

por·rect \pə-ˈrekt, pä-\ *adjective* [Latin *porrectus,* past participle of *porrigere* to stretch out, from *por-* forward + *regere* to direct — more at PORTEND, RIGHT] (15th century)
: extended forward ⟨*porrect* antennae⟩

por·ridge \ˈpȯr-ij, ˈpär-\ *noun* [alteration of *pottage*] (circa 1643)
: a soft food made by boiling meal of grains or legumes in milk or water until thick
— **por·ridgy** \-i-jē\ *adjective*

por·rin·ger \-ən-jər\ *noun* [alteration of Middle English *poteger, potinger,* from Anglo-French *potageer,* from Middle French *potager* of pottage, from *potage* pottage] (1522)
: a low usually metal bowl with a single and usually flat and pierced handle

[1]**port** \ˈpōrt, ˈpȯrt\ *noun* [Middle English, from Old English & Old French, from Latin *portus* — more at FORD] (before 12th century)
1 : a place where ships may ride secure from storms **:** HAVEN
2 a : a harbor town or city where ships may take on or discharge cargo **b :** AIRPORT
3 : PORT OF ENTRY

[2]**port** *noun* [Middle English *porte,* from Middle French, gate, door, from Latin *porta* passage, gate; akin to Latin *portus* port] (before 12th century)
1 *chiefly Scottish* **:** GATE
2 a : an opening (as in a valve seat or valve face) for intake or exhaust of a fluid **b :** the area of opening in a cylinder face of a passageway for the working fluid in an engine; *also* **:** such a passageway
3 a : an opening in a vessel's side (as for admitting light or loading cargo) **b** *archaic* **:** the cover for a porthole
4 : a hole in an armored vehicle or fortification through which guns may be fired
5 : a hardware interface by which a computer communicates with another device or system

[3]**port** *noun* [Middle English, from Middle French, from *porter* to carry, from Latin *portare*] (14th century)
1 : the manner of bearing oneself
2 *archaic* **:** STATE 3
3 : the position in which a military weapon is carried at the command *port arms*

[4]**port** *transitive verb* [[5]*port*] (1580)
: to turn or put (a helm) to the left — used chiefly as a command

[5]**port** *noun* [probably from [1]*port* or [2]*port*] (circa 1625)
: the left side of a ship or aircraft looking forward — called also *larboard;* compare STARBOARD
word history see STARBOARD
— **port** *adjective*

[6]**port** *noun* [*Oporto,* Portugal] (1691)
: a sweet fortified wine of rich taste and aroma made in Portugal; *also* **:** a similar wine made elsewhere

[1]**por·ta·ble** \ˈpōr-tə-bəl, ˈpȯr-\ *adjective* [Middle English, from Middle French, from Late Latin *portabilis,* from Latin *portare* to carry — more at FARE] (15th century)
1 a : capable of being carried or moved about ⟨a *portable* TV⟩ ⟨a *portable* sawmill⟩ **b :** usable on many computers without modification ⟨*portable* software⟩

2 *archaic* **:** BEARABLE
— **por·ta·bil·i·ty** \ˌpōr-tə-ˈbi-lə-tē, ˌpȯr-\ *noun*
— **por·ta·bly** \ˈpōr-tə-blē, ˈpȯr-\ *adverb*

2portable *noun* (1883)
: something that is portable

1por·tage \ˈpōr-tij, ˈpȯr-, *3 is also* pȯr-ˈtäzh\ *noun* [Middle English, from Middle French, from *porter* to carry] (15th century)
1 : the labor of carrying or transporting
2 *archaic* **:** the cost of carrying **:** PORTERAGE
3 a : the carrying of boats or goods overland from one body of water to another or around an obstacle (as a rapids) **b :** the route followed in making such a transfer

2por·tage \ˈpōr-tij, ˈpȯr-; pȯr-ˈtäzh\ *verb* **por·taged; por·tag·ing** (1864)
transitive verb
: to carry over a portage
intransitive verb
: to move gear over a portage

1por·tal \ˈpōr-tᵊl, ˈpȯr-\ *noun* [Middle English, from Middle French, from Medieval Latin *portale* city gate, porch, from neuter of *portalis* of a gate, from Latin *porta* gate — more at PORT] (14th century)
1 : DOOR, ENTRANCE; *especially* **:** a grand or imposing one
2 : the whole architectural composition surrounding and including the doorways and porches of a church
3 : the approach or entrance to a bridge or tunnel
4 : a communicating part or area of an organism; *specifically* **:** the point at which something (as a pathogen) enters the body

2portal *adjective* [New Latin *porta* transverse fissure of the liver, from Latin, gate] (1845)
1 : of or relating to the transverse fissure on the underside of the liver where most of the vessels enter
2 : of, relating to, or being a portal vein

portal system *noun* [*portal vein*] (1851)
: a system of veins that begins and ends in capillaries

portal–to–portal *adjective* (1943)
: of or relating to the time spent by a worker in traveling between the entrance to an employer's property and the worker's actual job site (as in a mine) ⟨*portal-to-portal* pay⟩

portal vein *noun* [2*portal*] (1845)
: a vein that collects blood from one part of the body and distributes it in another through capillaries; *especially* **:** a vein carrying blood from the digestive organs and spleen to the liver

por·ta·men·to \ˌpōr-tə-ˈmen-(ˌ)tō, ˌpȯr-\ *noun, plural* **-men·ti** \-(ˌ)tē\ [Italian, literally, act of carrying, from *portare* to carry, from Latin] (1771)
: a continuous gliding movement from one tone to another (as by the voice)

por·ta·pak *or* **por·ta·pack** \ˈpōr-tə-ˌpak, ˈpȯr-\ *noun* [*porta*ble + 1*pack*] (1970)
: a small portable combined videotape recorder and camera

port arms *noun* [from the command *port arms!*] (circa 1890)
: a position in the manual of arms in which the rifle is held diagonally in front of the body with the muzzle pointing upward to the left; *also* **:** a command to assume this position

por·ta·tive \ˈpōr-tə-tiv, ˈpȯr-\ *adjective* [Middle English *portatif,* from Middle French, from Latin *portatus,* past participle of *portare*] (14th century)
: PORTABLE

port·cul·lis \pōrt-ˈkə-ləs, pȯrt-\ *noun* [Middle English *port colice,* from Middle French *porte coleïce,* literally, sliding door] (14th century)
: a grating of iron hung over the gateway of a fortified place and lowered between grooves to prevent passage

port de bras \ˌpȯr-də-ˈbrä\ *noun* [French, literally, carriage of the arm] (1912)
: the technique and practice of arm movement in ballet

portcullis

Port du Sa·lut \ˌpȯr-də-sə-ˈlü, -sa-; -səl-ˈyü, -sal-\ *noun* [French *port-du-salut, port-salut,* from *Port du Salut,* Trappist abbey in northwest France] (1881)
: a semisoft pressed ripened cheese of usually mild flavor originated by Trappist monks in France

Porte \ˈpōrt, ˈpȯrt\ *noun* [French, short for *Sublime Porte,* literally, sublime gate; from the gate of the sultan's palace where justice was administered] (15th century)
: the government of the Ottoman empire

porte co·chere \ˌpōrt-kō-ˈsher, ˌpȯrt-\ *noun* [French *porte cochère,* literally, coach door] (1698)
1 : a passageway through a building or screen wall designed to let vehicles pass from the street to an interior courtyard
2 : a roofed structure extending from the entrance of a building over an adjacent driveway and sheltering those getting in or out of vehicles

por·tend \pȯr-ˈtend, pōr-\ *transitive verb* [Middle English, from Latin *portendere,* from *por-* forward (akin to *per* through) + *tendere* to stretch — more at FOR, THIN] (15th century)
1 : to give an omen or anticipatory sign of **:** BODE
2 : INDICATE, SIGNIFY

por·tent \ˈpȯr-ˌtent, ˈpōr-\ *noun* [Latin *portentum,* from neuter of *portentus,* past participle of *portendere*] (circa 1587)
1 : something that foreshadows a coming event **:** OMEN
2 : prophetic indication or significance
3 : MARVEL, PRODIGY

por·ten·tous \pȯr-ˈten-təs, pōr-\ *adjective* (15th century)
1 : of, relating to, or constituting a portent
2 : eliciting amazement or wonder **:** PRODIGIOUS
3 a : being a grave or serious matter ⟨*portentous* decisions⟩ **b :** self-consciously solemn or important **:** POMPOUS **c :** ponderously excessive ⟨that discipline's overwrought, *portentous* phrases —R. M. Coles⟩
synonym see OMINOUS
— **por·ten·tous·ly** *adverb*
— **por·ten·tous·ness** *noun*

1por·ter \ˈpōr-tər, ˈpȯr-\ *noun* [Middle English, from Old French *portier,* from Late Latin *portarius,* from Latin *porta* gate — more at PORT] (13th century)
chiefly British **:** a person stationed at a door or gate to admit or assist those entering

2porter *noun* [Middle English *portour,* from Middle French *porteour,* from Late Latin *portator,* from Latin *portare* to carry — more at FARE] (14th century)
1 : a person who carries burdens; *especially* **:** one employed to carry baggage for patrons at a hotel or transportation terminal
2 : a parlor-car or sleeping-car attendant who waits on passengers and makes up berths
3 [short for *porter's beer*] **:** a heavy dark brown beer brewed from browned or charred malt
4 : a person who does routine cleaning (as in a hospital or office)

3porter (1609)
transitive verb
: to transport or carry as or as if by a porter
intransitive verb
: to act as a porter

por·ter·age \-tə-rij\ *noun* (15th century)
: a porter's work; *also* **:** the charge for it

por·ter·house \ˈpōr-tər-ˌhau̇s, ˈpȯr-\ *noun* (circa 1758)
1 *archaic* **:** a house where malt liquor (as porter) is sold
2 : a large steak cut from the thick end of the short loin to contain a T-shaped bone and a large piece of tenderloin — see BEEF illustration

port·fo·lio \pōrt-ˈfō-lē-ˌō, pȯrt-\ *noun, plural* **-li·os** [Italian *portafoglio,* from *portare* to carry (from Latin) + *foglio* leaf, sheet, from Latin *folium* — more at BLADE] (1722)
1 : a hinged cover or flexible case for carrying loose papers, pictures, or pamphlets
2 [from the use of such a case to carry documents of state] **:** the office and functions of a minister of state or member of a cabinet
3 : the securities held by an investor **:** the commercial paper held by a financial house (as a bank)
4 : a set of pictures (as drawings or photographs) either bound in book form or loose in a folder

port·hole \ˈpōrt-ˌhōl, ˈpȯrt-\ *noun* [2*port*] (circa 1591)
1 : an opening (as a window) with a cover or closure especially in the side of a ship or aircraft
2 : a port through which to shoot
3 : 2PORT 2

Por·tia \ˈpōr-shə, ˈpȯr-\ *noun*
: the heroine in Shakespeare's *The Merchant of Venice*

por·ti·co \ˈpōr-ti-ˌkō, ˈpȯr-\ *noun, plural* **-coes** *or* **-cos** [Italian, from Latin *porticus* — more at PORCH] (1605)
: a colonnade or covered ambulatory especially in classical architecture and often at the entrance of a building

por·tiere \pōr-ˈtyer, pȯr-, -ˈtir; ˈpōr-tē-ər, ˈpȯr-\ *noun* [French *portière,* from Old French, feminine of *portier* porter, doorkeeper] (1843)
: a curtain hanging across a doorway

1por·tion \ˈpōr-shən, ˈpȯr-\ *noun* [Middle English, from Old French, from Latin *portion-, portio;* akin to Latin *part-, pars* part] (14th century)
1 : an individual's part or share of something: as **a :** a share received by gift or inheritance **b :** DOWRY **c :** enough food especially of one kind to serve one person at one meal
2 : an individual's lot, fate, or fortune **:** one's share of good and evil
3 : an often limited part set off or abstracted from a whole ⟨give but that *portion* which yourself proposed —Shakespeare⟩
synonym see PART, FATE

2portion *transitive verb* **por·tioned; por·tion·ing** \-sh(ə-)niŋ\ (14th century)
1 : to divide into portions **:** DISTRIBUTE
2 : to allot a dowry to **:** DOWER

por·tion·less \-shən-ləs\ *adjective* (1782)
: having no portion; *especially* **:** having no dowry or inheritance

port·land cement \ˈpōrt-lən(d)-, ˈpȯrt-\ *noun* [Isle of *Portland,* England; from its resemblance to a limestone found there] (1824)
: a hydraulic cement made by finely pulverizing the clinker produced by calcining to incipient fusion a mixture of clay and limestone or similar materials

port·ly \ˈpōrt-lē, ˈpȯrt-\ *adjective* **port·li·er; -est** [3*port*] (15th century)
1 : DIGNIFIED, STATELY
2 : heavy or rotund of body **:** STOUT
— **port·li·ness** *noun*

1port·man·teau \pōrt-ˈman-(ˌ)tō, pȯrt-\ *noun, plural* **-teaus** *or* **-teaux** \-(ˌ)tōz\ [Middle French *portemanteau,* from *porter* to carry + *manteau* mantle, from Latin *mantellum* — more at PORT] (1579)
1 : a large suitcase
2 : a word or morpheme whose form and meaning are derived from a blending of two or more distinct forms (as *smog* from *smoke* and *fog*)

2portmanteau *adjective* (1909)
1 : combining more than one use or quality

2 : being a portmanteau ⟨a *portmanteau* word⟩

port of call (1884)
1 : an intermediate port where ships customarily stop for supplies, repairs, or transshipment of cargo
2 : a stop included on an itinerary

port of entry (1840)
1 : a place where foreign goods may be cleared through a customhouse
2 : a place where an alien may be permitted to enter a country

por·trait \'pōr-trət, 'pȯr-, -ˌtrāt\ *noun* [Middle French, from past participle of *portraire*] (1570)
1 : PICTURE; *especially* **:** a pictorial representation (as a painting) of a person usually showing the face
2 : a sculptured figure **:** BUST, STATUE
3 : a graphic portrayal in words

por·trait·ist \-trə-tist, -ˌtrā-\ *noun* (1866)
: a maker of portraits

por·trai·ture \'pōr-trə-ˌchu̇r, 'pȯr-, -chər, -ˌtyu̇r, -ˌtu̇r\ *noun* (14th century)
1 : the making of portraits **:** PORTRAYAL
2 : PORTRAIT

por·tray \pōr-'trā, pȯr-, pər-\ *transitive verb* [Middle English *portraien,* from Middle French *portraire,* from Latin *protrahere* to draw forth, reveal, expose — more at PROTRACT] (14th century)
1 : to make a picture of **:** DEPICT
2 a : to describe in words **b :** to play the role of **:** ENACT
— por·tray·er *noun*

por·tray·al \-'trā(-ə)l\ *noun* (circa 1847)
1 : the act or process or an instance of portraying **:** REPRESENTATION
2 : PORTRAIT

por·tress \'pōr-trəs, 'pȯr-\ *noun* (15th century)
: a female porter: as **a :** a doorkeeper in a convent or apartment house **b :** CHARWOMAN

Port Roy·al·ist \pōrt-'rȯi-ə-list, pȯrt-\ *noun* [French *port-royaliste,* from *Port-Royal,* a convent near Versailles, France] (circa 1741)
: a member or adherent of a 17th century French Jansenist lay community noted for its logicians and educators

Port Sa·lut \ˌpȯr-sə-'lü, -sa-; -səl-'yü, -sal-\ *noun* (1902)
: PORT DU SALUT

Por·tu·guese \'pōr-chə-ˌgēz, 'pȯr-, -ˌgēs; ˌpōr-chə-', ˌpȯr-\ *noun, plural* **Portuguese** [Portuguese *português,* adjective & noun, from *Portugal*] (1534)
1 a : a native or inhabitant of Portugal **b :** one who is of Portuguese descent
2 : the Romance language of Portugal and Brazil
— Portuguese *adjective*

Portuguese man-of-war *noun, plural* **Portuguese man-of-wars** *also* **Portuguese men-of-war** (1707)
: any of a genus (*Physalia*) of large tropical and subtropical pelagic siphonophores having a crested bladderlike float which bears the colony comprised of three types of zooids on the lower surface with one of the three having nematocyst-equipped tentacles

Portuguese man-of-war

por·tu·laca \ˌpōr-chə-'la-kə, ˌpȯr-\ *noun* [New Latin, from Latin, purslane, from *portula,* diminutive of *porta* gate; from the lid of its capsule — more at PORT] (1548)
: any of a genus (*Portulaca*) of mainly tropical succulent herbs of the purslane family; *especially* **:** a widely cultivated plant (*P. grandiflora*) with showy flowers and small conical leaves

port-wine stain \'pōrt-'wīn-, 'pȯrt-\ *noun* (circa 1909)
: a reddish purple superficial hemangioma of the skin commonly occurring as a birthmark

po·sa·da \pə-'sä-də\ *noun* [Spanish, from *posar* to lodge, from Late Latin *pausare*] (1763)
: an inn in Spanish-speaking countries

¹pose \'pōz\ *verb* **posed; pos·ing** [Middle English, from Middle French *poser,* from (assumed) Vulgar Latin *pausare,* from Late Latin, to stop, rest, pause, from Latin *pausa* pause] (14th century)
transitive verb
1 a : to present for attention or consideration ⟨let me *pose* a question⟩ **b :** to put or set forth **:** OFFER ⟨this attitude *poses* a threat to our hopes for peace⟩
2 a : to put or set in place **b :** to place (as a model) in a studied attitude
intransitive verb
1 : to assume a posture or attitude usually for artistic purposes
2 : to affect an attitude or character usually to deceive or impress

²pose *noun* (1818)
1 : a sustained posture; *especially* **:** one assumed for artistic effect
2 : an attitude, role, or characteristic assumed for effect ☆

³pose *transitive verb* **posed; pos·ing** [short for earlier *appose,* from Middle English *apposen,* alteration of *opposen* to oppose] (1593)
: PUZZLE, BAFFLE

Po·sei·don \pə-'sī-dᵊn\ *noun* [Latin, from Greek *Poseidōn*]
: the Greek god of the sea — compare NEPTUNE

¹pos·er \'pō-zər\ *noun* [³*pose*] (1793)
: a puzzling or baffling question

²poser *noun* [¹*pose*] (1888)
: a person who poses

po·seur \pō-'zər\ *noun* [French, literally, poser, from *poser*] (1872)
: a person who pretends to be what he or she is not **:** an affected or insincere person

posh \'päsh\ *adjective* [origin unknown] (1918)
: ELEGANT, FASHIONABLE ◆
— posh·ly *adverb*
— posh·ness *noun*

pos·it \'pä-zət\ *transitive verb* **pos·it·ed** \'pä-zə-təd, 'päz-təd\; **pos·it·ing** \'pä-zə-tiŋ, 'päz-tiŋ\ [Latin *positus,* past participle of *ponere*] (1647)
1 : to dispose or set firmly **:** FIX
2 : to assume or affirm the existence of **:** POSTULATE
3 : to propose as an explanation **:** SUGGEST

¹po·si·tion \pə-'zi-shən\ *noun* [Middle English *posycion,* from Middle French *position,* from Latin *position-, positio,* from *ponere* to lay down, put, place, from (assumed) Old Latin *posinere,* from *po-* away (akin to Old Church Slavonic *po-,* perfective prefix, Greek *apo* away) + Latin *sinere* to leave — more at OF] (14th century)
1 : an act of placing or arranging: as **a :** the laying down of a proposition or thesis **b :** an arranging in order
2 : a point of view adopted and held to ⟨made my *position* on the issue clear⟩
3 a : the point or area occupied by a physical object ⟨took her *position* at the head of the line⟩ **b :** a certain arrangement of bodily parts ⟨rose to a standing *position*⟩
4 : a market commitment in securities or commodities; *also* **:** the inventory of a market trader
5 a : relative place, situation, or standing ⟨is now in a *position* to make decisions on his own⟩ **b :** social or official rank or status **c :** an employment for which one has been hired **:** JOB ⟨a *position* with a brokerage firm⟩ **d :** a situation that confers advantage or preference

²position *transitive verb* **po·si·tioned; po·si·tion·ing** \pə-'zi-sh(ə-)niŋ\ (1817)
: to put in proper position; *also* **:** LOCATE

po·si·tion·al \pə-'zi-sh(ə-)nəl\ *adjective* (1571)
1 : of, relating to, or fixed by position ⟨*positional* astronomy⟩
2 : involving little movement ⟨*positional* warfare⟩
3 : dependent on position or environment or context ⟨the front-articulated \k\ in \kē\ *key* and the back-articulated \k\ in \kül\ *cool* are *positional* variants⟩
— po·si·tion·al·ly *adverb*

positional notation *noun* (1941)
: a system of expressing numbers in which the digits are arranged in succession, the position of each digit has a place value, and the number is equal to the sum of the products of each digit by its place value

position effect *noun* (1930)
: genetic effect that is due to interaction of adjacent genes and that is modified when the spatial relationships of the genes change (as by chromosomal inversion)

position paper *noun* (1949)
: a detailed report that recommends a course of action on a particular issue

¹pos·i·tive \'pä-zə-tiv, 'päz-tiv\ *adjective* [Middle English, from Middle French *positif,*

☆ SYNONYMS
Pose, air, airs, affectation, mannerism mean an adopted way of speaking or behaving. POSE implies an attitude deliberately assumed in order to impress others ⟨her shyness was just a *pose*⟩. AIR may suggest natural acquirement through environment or way of life ⟨a traveler's sophisticated *air*⟩. AIRS always implies artificiality and pretentiousness ⟨snobbish *airs*⟩. AFFECTATION applies to a trick of speech or behavior that strikes the observer as insincere ⟨the posh accent is an *affectation*⟩. MANNERISM applies to an acquired eccentricity that has become a habit ⟨gesturing with a cigarette was her most noticeable *mannerism*⟩.

◇ WORD HISTORY
posh The origin of the adjective *posh* "elegant, fashionable" is obscure, though dictionary editors have received hundreds of letters over the years claiming that *posh* comes from the first letters of the phrase "port out, starboard home." On the route from Suez to India taken by a British steamship line (the Peninsular and Oriental), the coolest cabins during the Red Sea passage were on the port side and the hottest cabins on the starboard side. On the return trip, the opposite was true. Hence the wealthiest and most important passengers supposedly took the cooler cabins and had the letters P.O.S.H. stamped on their tickets. Unfortunately, there is not a shred of evidence for this story from the P & O's archives, travelers' diaries, or any other source. Bookings for the homeward voyage were not issued when the outbound voyage was booked, so P.O.S.H. would not have appeared on tickets or documents, and there are no indications that first-class accommodations on one side of the ship were more desirable or expensive than accommodations on the other. Along with explanations of *cop* as "constable on patrol" and *news* as "north, east, west, south," the "port out, starboard home" theory for *posh* is most likely folklore.

from Latin *positivus,* from *positus,* past participle of *ponere*] (14th century)
1 a : formally laid down or imposed **:** PRESCRIBED ⟨*positive* laws⟩ **b :** expressed clearly or peremptorily ⟨her answer was a *positive no*⟩ **c :** fully assured **:** CONFIDENT
2 a : of, relating to, or constituting the degree of comparison that is expressed in English by the unmodified and uninflected form of an adjective or adverb and denotes no increase or diminution **b** (1) **:** independent of changing circumstances **:** UNCONDITIONED (2) **:** relating to or constituting a motion or device that is definite, unyielding, constant, or certain in its action ⟨a *positive* system of levers⟩ **c** (1) **:** INCONTESTABLE ⟨*positive* proof⟩ (2) **:** UNQUALIFIED ⟨a *positive* disgrace⟩
3 a : not fictitious **:** REAL ⟨a *positive* influence for good in the community⟩ **b :** active and effective in social or economic function rather than merely maintaining peace and order ⟨a *positive* government⟩
4 a : having or expressing actual existence or quality as distinguished from deprivation or deficiency ⟨*positive* change in temperature⟩: as (1) **:** capable of being constructively applied (2) **:** not speculative **:** EMPIRICAL **b :** having rendition of light and shade similar in tone to the tones of the original subject ⟨a *positive* photographic image⟩ **c :** that is or is generated in a direction arbitrarily or customarily taken as that of increase or progression ⟨*positive* rotation of the earth⟩ ⟨we are making some *positive* progress⟩ **d :** directed or moving toward a source of stimulation ⟨a *positive* taxis⟩ **e :** real and numerically greater than zero ⟨+2 is a *positive* integer⟩
5 a (1) **:** being, relating to, or charged with electricity of which the proton is the elementary unit and which predominates in a glass body after being rubbed with silk (2) **:** having more protons than electrons ⟨a *positive* ion⟩ **b** (1) **:** having higher electric potential and constituting the part from which the current flows to the external circuit ⟨the *positive* terminal of a discharging storage battery⟩ (2) **:** being an electron-collecting electrode of an electron tube
6 a : marked by or indicating acceptance, approval, or affirmation **b :** affirming the presence of that sought or suspected to be present ⟨a *positive* test for blood⟩
7 *of a lens* **:** converging light rays and forming a real inverted image
synonym see SURE
— pos·i·tive·ly \-lē, *for emphasis often* ˌpä-zə-ˈtiv-\ *adverb*
— pos·i·tive·ness \ˈpä-zə-tiv-nəs, ˈpäz-tiv-\ *noun*

²positive *noun* (1530)
: something positive: as **a** (1) **:** the positive degree of comparison in a language (2) **:** a positive form of an adjective or adverb **b :** something of which an affirmation can be made **:** REALITY **c :** a positive photograph or a print from a negative

positive definite *adjective* (1907)
1 : having a positive value for all values of the constituent variables ⟨*positive definite* quadratic forms⟩
2 *of a matrix* **:** having the characteristic roots real and positive

positive law *noun* (14th century)
: law established or recognized by governmental authority — compare NATURAL LAW

pos·i·tiv·ism \ˈpä-zə-ti-ˌvi-zəm, ˈpäz-ti-\ *noun* [French *positivisme,* from *positif* positive + *-isme* -ism] (1847)
1 a : a theory that theology and metaphysics are earlier imperfect modes of knowledge and that positive knowledge is based on natural phenomena and their properties and relations as verified by the empirical sciences **b :** LOGICAL POSITIVISM
2 : the quality or state of being positive
— pos·i·tiv·ist \-vist\ *adjective or noun*
— pos·i·tiv·is·tic \ˌpä-zə-ti-ˈvis-tik, ˌpäz-ti-\ *adjective*
— pos·i·tiv·is·ti·cal·ly \-ti-k(ə-)lē\ *adverb*

pos·i·tiv·i·ty \ˌpä-zə-ˈti-və-tē\ *noun, plural* **-ties** (1659)
1 : the quality or state of being positive
2 : something that is positive

pos·i·tron \ˈpä-zə-ˌträn\ *noun* [*posi*tive + *-tron* (as in *electron*)] (1933)
: a positively charged particle having the same mass and magnitude of charge as the electron and constituting the antiparticle of the electron

positron–emission tomography *noun* (1979)
: tomography in which a cross-sectional image of regional metabolism is obtained by a usually color-coded cathode-ray tube representation of the distribution of gamma radiation given off in the collision of electrons in cells with positrons emitted by radionuclides incorporated into metabolic substances

pos·i·tro·ni·um \ˌpä-zə-ˈtrō-nē-əm\ *noun* (1945)
: a short-lived system that consists of a positron and an electron bound together and is suggestive of a hydrogen atom

pos·se \ˈpä-sē\ *noun* [Medieval Latin *posse comitatus,* literally, power or authority of the county] (1645)
1 : a large group often with a common interest
2 : a body of persons summoned by a sheriff to assist in preserving the public peace usually in an emergency
3 : a group of people temporarily organized to make a search (as for a lost child)

pos·sess \pə-ˈzes *also* -ˈses\ *transitive verb* [Middle English, from Middle French *possesser* to have possession of, take possession of, from Latin *possessus,* past participle of *possidēre,* from *potis* able, having the power + *sedēre* to sit — more at POTENT, SIT] (14th century)
1 a : to have and hold as property **:** OWN **b :** to have as an attribute, knowledge, or skill
2 a : to take into one's possession **b :** to enter into and control firmly **:** DOMINATE ⟨was *possessed* by demons⟩ **c :** to bring or cause to fall under the influence, possession, or control of some emotional or intellectual reaction ⟨melancholy *possesses* her⟩
3 a *obsolete* **:** to instate as owner **b :** to make the owner or holder — used in passive construction to indicate simple possession ⟨*possessed* of riches⟩ ⟨*possessed* of knowledge and experience⟩
— pos·ses·sor *noun*

pos·sessed *adjective* (1534)
1 a (1) **:** influenced or controlled by something (as an evil spirit or a passion) (2) **:** MAD, CRAZED **b :** urgently desirous to do or have something
2 *obsolete* **:** held as a possession
3 : SELF-POSSESSED, CALM
— pos·sessed·ly \-ˈze-səd-lē, -ˈzest-lē *also* -ˈse-səd- *or* -ˈsest-\ *adverb*
— pos·sessed·ness \-ˈze-səd-nəs, -ˈzest-nəs *also* -ˈse-səd- *or* -ˈsest-\ *noun*

pos·ses·sion \-ˈze-shən *also* -ˈse-\ *noun* (14th century)
1 a : the act of having or taking into control **b :** control or occupancy of property without regard to ownership **c :** OWNERSHIP **d :** control of the ball or puck
2 : something owned, occupied, or controlled **:** PROPERTY
3 a : domination by something (as an evil spirit, a passion, or an idea) **b :** a psychological state in which an individual's normal personality is replaced by another **c :** the fact or condition of being self-controlled
— pos·ses·sion·al \-ˈzesh-nəl, -ˈze-shə-nᵊl *also* -ˈsesh-nəl *or* -ˈse-shə-nᵊl\ *adjective*
— pos·ses·sion·less \-ˈze-shən-ləs, -ˈse-\ *adjective*

¹pos·ses·sive \pə-ˈze-siv *also* -ˈse-\ *adjective* (15th century)
1 : of, relating to, or constituting a word, a word group, or a grammatical case that denotes ownership or a relation analogous to ownership
2 : manifesting possession or the desire to own or dominate
— pos·ses·sive·ly *adverb*
— pos·ses·sive·ness *noun*

²possessive *noun* (15th century)
1 : a possessive word or word group
2 a : the possessive case **b :** a word in the possessive case

possessive adjective *noun* (1870)
: a pronominal adjective expressing possession

possessive pronoun *noun* (15th century)
: a pronoun that derives from a personal pronoun and denotes possession and analogous relationships

pos·ses·so·ry \pə-ˈze-sə-rē, -ˈzes-rē *also* -ˈse-sə-rē *or* -ˈses-rē\ *adjective* (15th century)
1 : of, arising from, or having the nature of possession ⟨*possessory* rights⟩
2 : having possession
3 : characteristic of a possessor **:** POSSESSIVE

pos·set \ˈpä-sət\ *noun* [Middle English *poshet, possot*] (15th century)
: a hot drink of sweetened and spiced milk curdled with ale or wine

pos·si·bil·i·ty \ˌpä-sə-ˈbi-lə-tē\ *noun, plural* **-ties** (14th century)
1 : the condition or fact of being possible
2 *archaic* **:** one's utmost power, capacity, or ability
3 : something that is possible
4 : potential or prospective value — usually used in plural ⟨the house had great *possibilities*⟩

pos·si·ble \ˈpä-sə-bəl\ *adjective* [Middle English, from Middle French, from Latin *possibilis,* from *posse* to be able, from *potis, pote* able + *esse* to be — more at POTENT, IS] (14th century)
1 a : being within the limits of ability, capacity, or realization **b :** being what may be done or may occur according to nature, custom, or manners
2 a : being something that may or may not occur **b :** being something that may or may not be true or actual ⟨*possible* explanation⟩
3 : having an indicated potential ⟨a *possible* housing site⟩ ☆

pos·si·bly \-blē\ *adverb* (14th century)
1 : in a possible manner **:** by any possibility
2 : by merest chance **:** PERHAPS

pos·sum \ˈpä-səm\ *noun* (1613)
: OPOSSUM

¹post \ˈpōst\ *noun* [Middle English, from Old English, from Latin *postis;* probably akin to Latin *por-* forward and to Latin *stare* to stand — more at PORTEND, STAND] (before 12th century)
1 : a piece (as of timber or metal) fixed firmly in an upright position especially as a stay or support **:** PILLAR, COLUMN
2 : a pole or stake set up to mark or indicate something; *especially* **:** a pole that marks the starting or finishing point of a horse race
3 : a metallic fitting attached to an electrical device (as a storage battery) for convenience in making connections

☆ SYNONYMS
Possible, practicable, feasible mean capable of being realized. POSSIBLE implies that a thing may certainly exist or occur given the proper conditions ⟨a *possible* route up the west face of the mountain⟩. PRACTICABLE implies that something may be effected by available means or under current conditions ⟨a *practicable* route up the west face of the mountain⟩. FEASIBLE applies to what is likely to work or be useful in attaining the end desired ⟨commercially *feasible* for mass production⟩.

4 : GOALPOST
5 : the metal stem of a pierced earring

[2]post *transitive verb* (1633)
1 a : to publish, announce, or advertise by or as if by use of a placard **b :** to denounce by public notice **c :** to enter on a public listing **d :** to forbid (property) to trespassers under penalty of legal prosecution by notices placed along the boundaries **e :** SCORE ⟨*posted* a 70 in the final round⟩
2 : to affix to a usual place (as a wall) for public notices **:** PLACARD

[3]post *noun* [Middle French *poste* relay station, courier, from Old Italian *posta* relay station, from feminine of *posto,* past participle of *porre* to place, from Latin *ponere* — more at POSITION] (1507)
1 *obsolete* **:** COURIER
2 *archaic* **a :** one of a series of stations for keeping horses for relays **b :** the distance between any two such consecutive stations **:** STAGE
3 *chiefly British* **a :** a nation's organization for handling mail; *also* **:** the mail handled **b :** a single dispatch of mail **c :** POST OFFICE **d :** POSTBOX

[4]post (1533)
intransitive verb
1 : to travel with post-horses
2 : to ride or travel with haste **:** HURRY
3 : to rise from the saddle and return to it in rhythm with a horse's trot
transitive verb
1 *archaic* **:** to dispatch in haste
2 : MAIL ⟨*post* a letter⟩
3 a : to transfer or carry from a book of original entry to a ledger **b :** to make transfer entries in
4 : to make familiar with a subject **:** INFORM ⟨kept her *posted* on the latest gossip⟩

[5]post *adverb* (1549)
: with post-horses **:** EXPRESS

[6]post *noun* [Middle French *poste,* from Old Italian *posto,* from past participle of *porre* to place] (1598)
1 a : the place at which a soldier is stationed; *especially* **:** a sentry's beat or station **b :** a station or task to which one is assigned **c :** the place at which a body of troops is stationed **:** CAMP **d :** a local subdivision of a veterans' organization **e :** one of two bugle calls sounded (as in the British army) at tattoo
2 a : an office or position to which a person is appointed **b :** a player position in basketball that is the focal point of the offense; *specifically* **:** PIVOT 2b
3 a : TRADING POST, SETTLEMENT **b :** a trading station on the floor of a stock exchange

[7]post *transitive verb* (1683)
1 a : to station in a given place ⟨guards were *posted* at the doors⟩ **b :** to carry ceremoniously to a position ⟨*posting* the colors⟩
2 *chiefly British* **:** to assign to a unit, position, or location (as in the military or civil service)
3 : to put up (as bond)

post- *prefix* [Middle English, from Latin, from *post;* akin to Lithuanian *pas* at, Greek *apo* away from — more at OF]
1 a : after **:** subsequent **:** later ⟨*post*date⟩ **b :** behind **:** posterior **:** following after ⟨*post*lude⟩ ⟨*post*consonantal⟩
2 a : subsequent to **:** later than ⟨*post*operative⟩ **b :** posterior to ⟨*post*orbital⟩

post·abor·tion
post·ac·ci·dent
post·ad·o·les·cent
post·am·pu·ta·tion
post·apoc·a·lyp·tic
post·ar·rest
post·atom·ic
post·at·tack
post·bac·ca·lau·re·ate
post·base
post·bib·li·cal
post·bour·geois
post·burn
post·cap·i·tal·ist
post–Chris·tian
post·co·i·tal
post·col·lege
post·col·le·giate
post·co·lo·nial
post·con·cep·tion
post·con·cert
post·con·quest
post·con·so·nan·tal
post·con·ven·tion
post·cop·u·la·to·ry
post·cor·o·nary
post·coup
post·crash
post·cri·sis
post–Dar·win·i·an
post·dead·line
post·de·bate
post·deb·u·tante
post·de·liv·ery
post·de·po·si·tion·al
post·de·pres·sion
post·de·val·u·a·tion
post·dive
post·di·ves·ti·ture
post·di·vorce
post·drug
post·ed·it·ing
post–Ein·stein·ian
post·elec·tion
post·em·bry·o·nal
post·em·bry·on·ic
post·emer·gen·cy
post·en·ceph·a·lit·ic
post·ep·i·lep·tic
post·erup·tive
post·ex·er·cise
post·ex·pe·ri·ence
post·ex·per·i·men·tal
post·ex·po·sure
post·fault
post·fire
post·flight
post·frac·ture
post·freeze
post–Freud·ian
post·game
post·gla·cial
post·grad·u·a·tion
post·har·vest
post·hem·or·rhag·ic
post·hol·i·day
post·ho·lo·caust
post·hos·pi·tal
post·im·pact
post·im·pe·ri·al
post·in·au·gu·ral
post·in·de·pen·dence
post·in·dus·tri·al
post·in·fec·tion
post·in·jec·tion
post·in·oc·u·la·tion
post·ir·ra·di·a·tion
post·is·che·mic
post·iso·la·tion
post·land·ing
post·launch
post·lib·er·a·tion
post·mar·i·tal
post·mas·tec·to·my
post·mat·ing
post·me·di·e·val
post·mid·night
post·neo·na·tal
post·or·gas·mic
post·pol·li·na·tion
post·pres·i·den·tial
post·pri·ma·ry
post·pris·on
post·psy·cho·an·a·lyt·ic
post·pu·ber·ty
post·pu·bes·cent
post·race
post·re·ces·sion
post–Ref·or·ma·tion
post·res·ur·rec·tion
post·re·tire·ment
post·rev·o·lu·tion·ary
post·ri·ot
post·ro·man·tic
post·sea·son
post·sec·ond·ary
post·show
post·stim·u·la·tion
post·stim·u·la·to·ry
post·stim·u·lus
post·strike
post·sur·gi·cal
post·tax
post·teen
post·trau·mat·ic
post·treat·ment
post·trial
post·vac·ci·nal
post·vac·ci·na·tion
post·va·got·o·my
post·va·sec·to·my
post–Vic·to·ri·an
post·war
post·wean·ing
post·work·shop

post·age \ˈpōs-tij\ *noun* (1654)
1 : the fee for postal service
2 : adhesive stamps or printed indicia representing postal fees

postage–due stamp *noun* (1893)
: a special adhesive stamp that is applied by a post office to mail bearing insufficient postage

postage meter *noun* (1927)
: a machine that prints postal indicia on pieces of mail, records the amount of postage, and subtracts it from a total paid amount for which the machine has been set

postage–stamp *adjective* (1938)
: suggesting a postage stamp in size **:** very small ⟨*postage stamp* yards⟩

postage stamp *noun* (1840)
: a government adhesive stamp or imprinted stamp for use on mail as evidence of prepayment of postage

post·al \ˈpōs-tᵊl\ *adjective* (1843)
1 : of or relating to the mails or the post office
2 : conducted by mail ⟨*postal* chess⟩

postal card *noun* (1872)
1 : a card officially stamped and issued by the government for use in the mail
2 : POSTCARD

postal order *noun* (1883)
British **:** MONEY ORDER

postal service *noun* (circa 1920)
: POST OFFICE 1

postal union *noun* (1875)
: an association of governments setting up uniform regulations and practices for international mail

post·ax·i·al \ˌpōst-ˈak-sē-əl\ *adjective* (1872)
: of or relating to the ulnar side of the vertebrate forelimb or the fibular side of the hind limb; *also* **:** of or relating to the side of an animal or side of one of its limbs that is posterior to the axis of its body or limb

post·bag \ˈpōs(t)-ˌbag\ *noun* (1813)
1 *British* **:** MAILBAG
2 *British* **:** a single batch of mail **:** LETTERS

post·bel·lum \ˌpōs(t)-ˈbe-ləm\ *adjective* [Latin *post bellum* after the war] (1874)
: of, relating to, or characteristic of the period following a war and especially following the American Civil War

post·box \ˈpōs(t)-ˌbäks\ *noun* (1754)
: MAILBOX; *especially* **:** a public mailbox

post·boy \-ˌbȯi\ *noun* (1707)
: POSTILION

post·card \ˈpōs(t)-ˌkärd\ *noun* (1870)
1 : POSTAL CARD 1
2 : a card on which a message may be written for mailing without an envelope and to which the sender must affix a stamp
— post·card·like \-ˌlīk\ *adjective*

post·ca·va \ˌpōs(t)-ˈkā-və\ *noun* [New Latin] (1882)
: the inferior vena cava of vertebrates higher than fishes
— post·ca·val \-vəl\ *adjective*

post chaise *noun* (1712)
: a carriage usually having a closed body on four wheels and seating two to four persons

post·clas·si·cal \ˌpōs(t)-ˈkla-si-kəl\ *or* **post·clas·sic** \-sik\ *adjective* (1867)
: of or relating to a period (as in art, literature, or civilization) following a classical one

post·code \ˈpōs(t)-ˌkōd\ *noun* (1967)
: a code (as of numbers and letters) used similarly to the zip code especially in the United Kingdom and Australia

post–com·mu·nion \ˌpōs(t)-kə-ˈmyü-nyən\ *noun, often P&C capitalized* [Medieval Latin *postcommunion-, postcommunio,* from Latin *post-* + Late Latin *communio* communion] (15th century)
: a liturgically variable prayer following the communion at Mass

post·cra·ni·al \-ˈkrā-nē-əl\ *adjective* (1913)
: of or relating to the part of the body caudal to the head ⟨*postcranial* skeleton⟩ ⟨*postcranial* fossil remains⟩
— post·cra·ni·al·ly *adverb*

post·date \ˌpōs(t)-ˈdāt, ˈpōs(t)-ˌ\ *transitive verb* (1624)
1 a : to date with a date later than that of execution ⟨*postdate* a check⟩ **b :** to assign (an event) to a date subsequent to that of actual occurrence
2 : to follow in time

post·di·lu·vi·an \ˌpōs(t)-də-ˈlü-vē-ən, -dī-\ *adjective* [*post-* + Latin *diluvium* flood — more at DELUGE] (1680)
: of or relating to the period after the flood described in the Bible
— postdiluvian *noun*

[1]post·doc \ˈpōs(t)-ˌdäk\ *noun* (1968)
: one engaged in postdoctoral study or research

[2]postdoc *adjective* (1970)
: POSTDOCTORAL

post·doc·tor·al \ˌpōs(t)-ˈdäk-t(ə-)rəl, ˈpōst-ˌ\ *also* **post·doc·tor·ate** \-t(ə-)rət\ *adjective* (1936)
: being beyond the doctoral level: **a :** of or relating to advanced academic or professional work beyond a doctor's degree ⟨a *postdoctoral*

fellowship⟩ **b :** engaged in such work ⟨*post-doctoral* scholars⟩

post·emer·gence \ˌpōst-i-ˈmər-jən(t)s\ *adjective* (1940)
: used or occurring in the stage between the emergence of a seedling and the maturity of a crop plant ⟨*postemergence* herbicides⟩ ⟨*post-emergence* development⟩

[1]post·er \ˈpōs-tər\ *noun* [[4]*post*] (1605)
archaic **:** a swift traveler

[2]poster *noun* [[2]*post*] (1838)
: a bill or placard for posting often in a public place; *especially* **:** one that is decorative or pictorial

poster color *noun* (1925)
: an opaque watercolor paint with a gum or glue-size binder sold usually in jars — called also *poster paint*

poste res·tante \ˌpōst-ˌres-ˈtänt\ *noun* [French, literally, waiting mail] (1768)
chiefly British **:** GENERAL DELIVERY

[1]pos·te·ri·or \pō-ˈstir-ē-ər, pä-\ *adjective* [Latin, comparative of *posterus* coming after, from *post* after — more at POST-] (1534)
1 : later in time **:** SUBSEQUENT
2 : situated behind: as **a :** CAUDAL **b** *of the human body or its parts* **:** DORSAL
3 *of a plant part* **:** ADAXIAL, SUPERIOR
— pos·te·ri·or·ly *adverb*

[2]pos·te·ri·or \pä-ˈstir-ē-ər, pō-\ *noun* (circa 1616)
: the hinder parts of the body; *specifically* **:** BUTTOCKS

pos·te·ri·or·i·ty \(ˌ)pō-ˌstir-ē-ˈȯr-ə-tē, (ˌ)pä-, -ˈär-\ *noun* (14th century)
: the quality or state of being later or subsequent

pos·ter·i·ty \pä-ˈster-ə-tē\ *noun* [Middle English *posterite,* from Middle French *posterité,* from Latin *posteritat-, posteritas,* from *posterus* coming after] (14th century)
1 : the offspring of one progenitor to the furthest generation
2 : all future generations

pos·tern \ˈpōs-tərn, ˈpäs-\ *noun* [Middle English *posterne,* from Old French, alteration of *posterle,* from Late Latin *posterula,* diminutive of *postera* back door, from Latin, feminine of *posterus*] (14th century)
1 : a back door or gate
2 : a private or side entrance or way
— postern *adjective*

pos·tero·lat·er·al \ˌpäs-tə-rō-ˈla-t(ə-)rəl\ *adjective* [*poster*ior + *-o-* + *lateral*] (1852)
: posterior and lateral in position or direction

post exchange *noun* (1892)
: a store at a military installation that sells merchandise and services to military personnel and authorized civilians

post·ex·il·ic \ˌpōst-(ˌ)eg-ˈzi-lik\ *adjective* (1871)
: of or relating to the period of Jewish history between the end of the exile in Babylon in 538 B.C. and A.D. 1

post·face \ˈpōs(t)-fəs, -ˌfās; pȯs-fȧs\ *noun* [French, from *post-* + *-face* (as in *préface* preface)] (1782)
: a brief article or note (as of explanation) placed at the end of a publication

post·fem·i·nist \ˌpōst-ˈfe-mə-nist\ *adjective* (1983)
: of, relating to, occurring in, or being the period following widespread advocacy and acceptance of feminism

post·fix \ˈpōs(t)-ˌfiks\ *adjective* [*post-* + *-fix* (as in *prefix*)] (1973)
: characterized by placement of an operator after its operand or after its two operands if it is a binary operator — compare INFIX, PREFIX

post–free \ˈpōs(t)-ˈfrē\ *adjective* (1723)
chiefly British **:** POSTPAID

post·gan·gli·on·ic \ˌpōs(t)-ˌgaŋ-glē-ˈä-nik\ *adjective* (1897)
: distal to a ganglion; *specifically* **:** of, relating to, or being an axon arising from a cell body within an autonomic ganglion — compare PREGANGLIONIC

[1]post·grad·u·ate \-ˈgra-jə-wət, -ˌwāt, -ˈgraj-wət\ *adjective* (1858)
: GRADUATE 2

[2]postgraduate *noun* (circa 1890)
: a student continuing formal education after graduation from high school or college

[1]post·haste \ˈpōst-ˈhāst\ *noun* [[3]*post*] (1545)
archaic **:** great haste

[2]posthaste *adverb* (1593)
: with all possible speed

[3]posthaste *adjective* (1604)
obsolete **:** SPEEDY, IMMEDIATE ⟨requires your . . . *posthaste* appearance —Shakespeare⟩

post hoc \ˈpōst-ˈhäk\ *adjective* [New Latin *post hoc, ergo propter hoc* after this, therefore because of this] (1704)
1 : relating to or being the fallacy of arguing from temporal sequence to a causal relation
2 : formulated after the fact ⟨a *post hoc* rationalization⟩

post·hole \ˈpōst-ˌhōl\ *noun* (1703)
: a hole dug for a post (as to support a fence or wall)

post horn *noun* (1675)
: a simple straight or coiled brass or copper wind instrument with cupped mouthpiece used especially by guards of mail coaches of the 18th and 19th centuries

post horn

post–horse \ˈpōst-ˌhȯrs\ *noun* [[3]*post*] (1527)
: a horse for use especially by couriers or mail carriers

post·hu·mous \ˈpäs-chə-məs *also* -tə-, -tyə-, -thə-; päst-ˈhyü-məs, ˈpōst-, -ˈyü-\ *adjective* [Latin *posthumus,* alteration of *postumus* late-born, posthumous, from superlative of *posterus* coming after — more at POSTERIOR] (1619)
1 : born after the death of the father
2 : published after the death of the author
3 : following or occurring after death ⟨*posthumous* fame⟩
— post·hu·mous·ly *adverb*
— post·hu·mous·ness *noun*

post·hyp·not·ic \ˌpōst-hip-ˈnä-tik, -ip-\ *adjective* [International Scientific Vocabulary] (1890)
: of, relating to, or characteristic of the period following a hypnotic trance

pos·tiche \pȯs-tēsh\ *noun* [French, from Spanish *postizo*] (1886)
: WIG; *especially* **:** TOUPEE 2

pos·til·ion *or* **pos·til·lion** \pō-ˈstil-yən, pə-\ *noun* [Middle French *postillon* mail carrier using post-horses, from Italian *postiglione,* from *posta* post — more at POST] (circa 1611)
: one who rides as a guide on the near horse of one of the pairs attached to a coach or post chaise especially without a coachman

Post·im·pres·sion·ism \ˌpōst-im-ˈpre-shə-ˌni-zəm\ *noun* [French *postimpressionisme,* from *post-* + *impressionisme* impressionism] (1910)
: a theory or practice of art originating in France in the last quarter of the 19th century that in revolt against impressionism stresses variously volume, picture structure, or expressionism
— Post·im·pres·sion·ist \-ˈpre-sh(ə-)nist\ *adjective or noun*
— Post·im·pres·sion·is·tic \-ˌpre-shə-ˈnis-tik\ *adjective*

[1]post·ing \ˈpōs-tiŋ\ *noun* [[4]*post*] (1682)
1 : the act of transferring an entry or item from a book of original entry to the proper account in a ledger
2 : the record in a ledger account resulting from the transfer of an entry or item from a book of original entry

[2]posting *noun* [[7]*post*] (1880)
: appointment to a post or a command

post–Kant·ian \ˌpōs(t)-ˈkan-tē-ən, -ˈkän-\ *adjective* (1843)
: of or relating to the idealist philosophers (as Fichte, Schelling, and Hegel) following Kant and developing some of his ideas

post·lap·sar·i·an \-ˌlap-ˈser-ē-ən\ *adjective* [*post-* + Latin *lapsus* slip, fall — more at LAPSE] (1733)
: of, relating to, or characteristic of the time or state after the fall of mankind

post·lit·er·ate \-ˈli-tə-rət *also* -ˈli-trət\ *adjective* (1965)
: relating to or occurring after the introduction of the electronic media

post·lude \ˈpōst-ˌlüd\ *noun* [*post-* + *-lude* (as in *prelude*)] (1851)
1 : a closing piece of music; *especially* **:** an organ voluntary at the end of a church service
2 : a closing phase (as of an epoch or a literary work)

post·man \ˈpōs(t)-mən, -ˌman\ *noun* (1529)
: MAILMAN

[1]post·mark \-ˌmärk\ *noun* (1678)
: an official postal marking on a piece of mail; *specifically* **:** a mark showing the post office and date of mailing

[2]postmark *transitive verb* (1716)
: to put a postmark on

post·mas·ter \-ˌmas-tər\ *noun* (1513)
1 : one who has charge of a post office
2 : one who has charge of a station for the accommodation of travelers or who supplies post-horses
— post·mas·ter·ship \-ˌship\ *noun*

postmaster general *noun, plural* **postmasters general** (1626)
: an official in charge of a national post office department or agency

post·men·o·paus·al \ˌpōs(t)-ˌmę-nə-ˈpȯ-zəl\ *adjective* (1928)
1 : having undergone menopause
2 : occurring after menopause

post me·ri·di·em \-mə-ˈri-dē-əm, -ˌem\ *adjective* [Latin] (1647)
: being after noon — abbreviation *p.m.*

post·mil·le·nar·i·an·ism \-ˌmi-lə-ˈner-ē-ə-ˌni-zəm\ *noun* (circa 1890)
: POSTMILLENNIALISM
— postmillenarian *adjective or noun*

post·mil·len·ni·al \-mə-ˈle-nē-əl\ *adjective* (1851)
1 : coming after or relating to the period after the millennium
2 : holding or relating to postmillennialism

post·mil·len·ni·al·ism \-ə-ˌli-zəm\ *noun* (1879)
: the theological doctrine that the second coming of Christ will occur after the millennium
— post·mil·len·ni·al·ist \-ə-list\ *noun*

post·mis·tress \ˈpōs(t)-ˌmis-trəs\ *noun* (1697)
: a woman who is a postmaster

post·mod·ern \ˌpōs(t)-ˈmä-dərn, ÷-ˈmä-d(ə-)rən\ *adjective* (1949)
: of, relating to, or being any of several movements (as in art, architecture, or literature) that are reactions against the philosophy and practices of modern movements and are typically marked by revival of traditional elements and techniques
— post·mod·ern·ism \-dər-ˌni-zəm\ *noun*
— post·mod·ern·ist \-nist\ *adjective or noun*

[1]post·mor·tem \ˌpōs(t)-ˈmȯr-təm\ *adjective* [Latin *post mortem* after death] (circa 1836)
1 : done, occurring, or collected after death ⟨*postmortem* tissue specimens⟩
2 : following the event

[2]postmortem *noun* (1850)
1 : an analysis or discussion of an event after it is over
2 : AUTOPSY 1

postmortem examination *noun* (1837)
: AUTOPSY 1

post·na·sal drip \ˈpōst-ˌnā-zəl-\ *noun* (1949)

: flow of mucous secretion from the posterior part of the nasal cavity onto the wall of the pharynx occurring usually as a chronic accompaniment of an allergic state

post·na·tal \ˌpōs(t)-ˈnā-tᵊl\ *adjective* [International Scientific Vocabulary] (circa 1859)
: occurring or being after birth; *specifically* **:** of or relating to an infant immediately after birth ⟨*postnatal* care⟩
— **post·na·tal·ly** *adverb*

post·nup·tial \-ˈnəp-shəl, -chəl, ÷-chə-wəl\ *adjective* (1807)
: made or occurring after marriage or mating

post oak *noun* [[1]*post*] (1775)
: an oak (*Quercus stellata*) of the eastern and central U.S. having hard durable wood

[1]post–obit \pōst-ˈō-bət, *especially British* -ˈä-bit\ *noun* (1751)
: POST-OBIT BOND

[2]post–obit *adjective* [Latin *post obitum* after death] (circa 1834)
: occurring or taking effect after death

post–obit bond *noun* (1788)
: a bond made by a reversioner to secure a loan and payable out of his reversion

post office *noun* (1652)
1 : a government department or agency handling the transmission of mail
2 : a local branch of a national post office handling the mail for a particular place or area
3 : a game in which a player acting as postmaster or postmistress may exact a kiss from one of the opposite sex as payment for the pretended delivery of a letter

post·op·er·a·tive \ˌpōst-ˈä-p(ə-)rə-tiv, -pə-ˌrā-\ *adjective* [International Scientific Vocabulary] (circa 1890)
1 : following a surgical operation ⟨*postoperative* care⟩
2 : having undergone a surgical operation ⟨a *postoperative* patient⟩
— **post·op·er·a·tive·ly** *adverb*

post·or·bit·al \-ˈȯr-bə-tᵊl\ *adjective* (circa 1836)
: situated behind the eye socket

post·paid \ˈpōs(t)-ˈpād\ *adjective* (1653)
: having the postage paid by the sender and not chargeable to the receiver

post·par·tum \ˌpōs(t)-ˈpär-təm\ *adjective* [New Latin *post partum* after birth] (1846)
1 : following parturition ⟨*postpartum* period⟩
2 : being in the postpartum period ⟨*postpartum* mothers⟩
— **postpartum** *adverb*

post·pone \(ˌ)pōs(t)-ˈpōn\ *transitive verb* **post·poned; post·pon·ing** [Latin *postponere* to place after, postpone, from *post-* + *ponere* to place — more at POSITION] (circa 1520)
1 : to put off to a later time **:** DEFER
2 a : to place later (as in a sentence) than the normal position in English ⟨*postpone* an adjective⟩ **b :** to place later in order of precedence, preference, or importance
synonym see DEFER
— **post·pon·able** \-ˈpō-nə-bəl\ *adjective*
— **post·pone·ment** \-ˈpōn-mənt\ *noun*
— **post·pon·er** *noun*

post·po·si·tion \ˌpōs(t)-pə-ˈzi-shən, ˈpōs(t)-pə-ˌ\ *noun* [French, from *postposer* to place after, from Latin *postponere* (perfect indicative *postposui*)] (circa 1638)
: the placing of a grammatical element after a word to which it is primarily related in a sentence; *also* **:** such a word or particle especially when functioning as a preposition
— **post·po·si·tion·al** \ˌpōs(t)-pə-ˈzish-nəl, -ˈzi-shə-nᵊl\ *adjective*
— **post·po·si·tion·al·ly** *adverb*

post·pos·i·tive \-ˈpä-zə-tiv, -ˈpäz-tiv\ *adjective* (1786)
: placed after or at the end of another word
— **post·pos·i·tive·ly** *adverb*

post·pran·di·al \pōs(t)-ˈpran-dē-əl\ *adjective* (1820)
: occurring after a meal

post·pro·duc·tion \ˌpōs(t)-prə-ˈdək-shən, ˈpōs(t)-prə-ˌ, -prō-\ *noun* (1953)
: the period following filming or taping in which a motion picture or television show is readied for public presentation

post road *noun* (1657)
: a route over which mail is carried

post·script \ˈpōs(t)-ˌskript\ *noun* [New Latin *postscriptum*, from Latin, neuter of *postscriptus*, past participle of *postscribere* to write after, from *post-* + *scribere* to write — more at SCRIBE] (1551)
: a note or series of notes appended to a completed letter, article, or book

post·syn·ap·tic \ˌpōs(t)-sə-ˈnap-tik\ *adjective* (1909)
1 : occurring after synapsis ⟨a *postsynaptic* chromosome⟩
2 : relating to, occurring in, or being part of a nerve cell by which a wave of excitation is conveyed away from a synapse
— **post·syn·ap·ti·cal·ly** \-ti-k(ə-)lē\ *adverb*

post·ten·sion \-ˈten(t)-shən\ *transitive verb* (1950)
: to apply tension to (reinforcing steel) after concrete has set

post·test \ˈpōs(t)-ˌtest\ *noun* (circa 1951)
: a test given to students after completion of an instructional program or segment and often used in conjunction with a pretest to measure their achievement and the effectiveness of the program

post time *noun* [[1]*post*] (1941)
: the designated time for the start of a horse race

post·tran·scrip·tion·al \ˌpōs(t)-tran(t)-ˈskrip-shnəl, -shə-nᵊl\ *adjective* (1969)
: occurring, acting, or existing after genetic transcription

post·trans·fu·sion \-tran(t)s-ˈfyü-zhən\ *adjective* (1944)
1 : caused by transfused blood
2 : occurring after blood transfusion ⟨*posttransfusion* shock⟩

post·trans·la·tion·al \-tran(t)s-ˈlā-shnəl, -shə-nᵊl\ *adjective* (1975)
: occurring or existing after genetic translation

post–traumatic stress disorder *noun* (1980)
: a psychological reaction occurring after a highly stressing event that is usually characterized by depression, anxiety, flashbacks, recurrent nightmares, and avoidance of reminders of the event — called also *post-traumatic stress syndrome*

pos·tu·lan·cy \ˈpäs-chə-lən(t)-sē\ *noun, plural* **-cies** (circa 1883)
1 : the quality or state of being a postulant
2 : the period during which a person remains a postulant

pos·tu·lant \ˈpäs-chə-lənt\ *noun* [French, petitioner, candidate, postulant, from Middle French, from present participle of *postuler* to demand, solicit, from Latin *postulare*] (1759)
1 : a person admitted to a religious order as a probationary candidate for membership
2 : a person on probation before being admitted as a candidate for holy orders in the Episcopal Church

[1]pos·tu·late \ˈpäs-chə-ˌlāt\ *transitive verb* **-lat·ed; -lat·ing** [Latin *postulatus*, past participle of *postulare;* akin to Latin *poscere* to ask, Old High German *forscōn* to search, Sanskrit *pṛcchati* he asks — more at PRAY] (1593)
1 : DEMAND, CLAIM
2 a : to assume or claim as true, existent, or necessary **:** depend upon or start from the postulate of **b :** to assume as a postulate or axiom (as in logic or mathematics)
— **pos·tu·la·tion** \ˌpäs-chə-ˈlā-shən\ *noun*
— **pos·tu·la·tion·al** \-shnəl, -shə-nᵊl\ *adjective*

[2]pos·tu·late \ˈpäs-chə-lət, -ˌlāt\ *noun* [Medieval Latin *postulatum*, from neuter of *postulatus*, past participle of *postulare* to assume, from Latin, to demand] (1646)
1 : a hypothesis advanced as an essential presupposition, condition, or premise of a train of reasoning
2 : AXIOM 3

pos·tu·la·tor \-ˌlā-tər\ *noun* (1863)
: an official who presents a plea for beatification or canonization in the Roman Catholic Church — compare DEVIL'S ADVOCATE

pos·tur·al \ˈpäs-chə-rəl\ *adjective* (1857)
: of, relating to, or involving posture

[1]pos·ture \ˈpäs-chər\ *noun* [French, from Italian *postura*, from Latin *positura*, from *positus*, past participle of *ponere* to place — more at POSITION] (circa 1586)
1 a : the position or bearing of the body whether characteristic or assumed for a special purpose ⟨erect *posture*⟩ **b :** the pose of a model or artistic figure
2 : state or condition at a given time especially with respect to capability in particular circumstances ⟨maintain a competitive *posture* in the market⟩ ⟨put the country in a *posture* of defense⟩
3 : a conscious mental or outward behavioral attitude ⟨takes a neutral *posture* toward the discussions⟩

[2]posture *verb* **pos·tured; pos·tur·ing** (circa 1645)
transitive verb
: to cause to assume a given posture **:** POSE
intransitive verb
1 : to assume a posture; *especially* **:** to strike a pose for effect
2 : to assume an artificial or pretended attitude **:** ATTITUDINIZE
— **pos·tur·er** \-chər-ər\ *noun*

post·vo·cal·ic \ˌpōs(t)-vō-ˈka-lik, -və-\ *adjective* [International Scientific Vocabulary] (1892)
: immediately following a vowel

po·sy \ˈpō-zē\ *noun, plural* **posies** [alteration of *poesy*] (1533)
1 : a brief sentiment, motto, or legend
2 a : BOUQUET, NOSEGAY **b :** FLOWER

[1]pot \ˈpät\ *noun* [Middle English, from Old English *pott;* akin to Middle Low German *pot* pot] (before 12th century)
1 a : a usually rounded metal or earthen container used chiefly for domestic purposes (as in cooking or for holding liquids or growing plants); *also* **:** any of various technical or industrial vessels or enclosures resembling or likened to a household pot ⟨the *pot* of a still⟩ **b :** POTFUL ⟨a *pot* of coffee⟩
2 : an enclosed framework of wire, wood, or wicker for catching fish or lobsters
3 a : a large amount (as of money) **b** (1) **:** the total of the bets at stake at one time (2) **:** one round in a poker game **c :** the common fund of a group
4 : POTSHOT
5 : POTBELLY
6 : RUIN ⟨gone to *pot*⟩
7 *British* **:** a shot in snooker in which a ball is pocketed

[2]pot *verb* **pot·ted; pot·ting** (1616)
transitive verb
1 a : to place in a pot **b :** to pack or preserve (as cooked and chopped meat) in a sealed pot, jar, or can often with aspic
2 : to shoot with a potshot
3 : to make or shape (earthenware) as a potter
4 : to embed (as electronic components) in a container with an insulating or protective material (as plastic)
intransitive verb
: to take a potshot

\ə\ **abut** \ᵊ\ **kitten** \ər\ **further** \a\ **ash** \ā\ **ace** \ä\ **mop, mar** \au̇\ **out** \ch\ **chin** \e\ **bet** \ē\ **easy** \g\ **go** \i\ **hit** \ī\ **ice** \j\ **job** \ŋ\ **sing** \ō\ **go** \ȯ\ **law** \ȯi\ **boy** \th\ **thin** \t͟h\ **the** \ü\ **loot** \u̇\ **foot** \y\ **yet** \zh\ **vision** *see also* Guide to Pronunciation

³pot *noun* [perhaps modification of Mexican Spanish *potiguaya*] (1938)
: MARIJUANA

¹po·ta·ble \'pō-tə-bəl\ *adjective* [Middle English, from Late Latin *potabilis,* from Latin *potare* to drink; akin to Latin *bibere* to drink, Greek *pinein*] (15th century)
: suitable for drinking
— **po·ta·bil·i·ty** \ˌpō-tə-'bi-lə-tē\ *noun*
— **po·ta·ble·ness** \'pō-tə-bəl-nəs\ *noun*

²potable *noun* (1623)
: a liquid that is suitable for drinking; *especially* : an alcoholic beverage

po·tage \pȯ-'täzh\ *noun* [Middle French, from Old French, pottage] (1567)
: a thick soup

pot ale *noun* (1812)
: the residue of fermented wort left in a still after the distillation of whiskey or alcohol and used for feeding swine

pot·ash \'pät-ˌash\ *noun* [singular of *pot ashes*] (circa 1648)
1 : potassium carbonate especially from wood ashes
2 : potassium or a potassium compound especially as used in agriculture or industry

po·tas·sic \pə-'ta-sik\ *adjective* (1858)
: of, relating to, or containing potassium

po·tas·si·um \pə-'ta-sē-əm\ *noun, often attributive* [New Latin, from *potassa* potash, from English *potash*] (circa 1807)
: a silver-white soft light low-melting univalent metallic element of the alkali metal group that occurs abundantly in nature especially combined in minerals — see ELEMENT table

potassium–argon *adjective* (1953)
: being or relating to a method of dating paleontological or geological materials based on the radioactive decay of potassium to argon that has taken place in a specimen

potassium bromide *noun* (1873)
: a crystalline salt KBr with a saline taste that is used as a sedative and in photography

potassium carbonate *noun* (1885)
: a white salt K_2CO_3 that forms a strongly alkaline solution and is used in making glass and soap

potassium chlorate *noun* (1885)
: a crystalline salt $KClO_3$ that is used as an oxidizing agent in matches, fireworks, and explosives

potassium chloride *noun* (1885)
: a crystalline salt KCl occurring as a mineral and in natural waters and used especially as a fertilizer

potassium cyanide *noun* (1885)
: a very poisonous crystalline salt KCN used especially in gold and silver extraction from ore

potassium dichromate *noun* (1885)
: a soluble salt $K_2Cr_2O_7$ forming large orange-red crystals used especially in dyeing, in photography, and as an oxidizing agent

potassium hydroxide *noun* (1885)
: a white deliquescent solid KOH that dissolves in water with much heat to form a strongly alkaline and caustic liquid and is used chiefly in making soap and as a reagent

potassium nitrate *noun* (1885)
: a crystalline salt KNO_3 that occurs as a product of nitrification in arable soils, is a strong oxidizer, and is used especially in making gunpowder, as a fertilizer, and in medicine

potassium permanganate *noun* (1869)
: a dark purple salt $KMnO_4$ used as an oxidizer and disinfectant

potassium sorbate *noun* (1960)
: a potassium salt $C_6H_7KO_2$ of sorbic acid used especially as a food preservative

potassium sulfate *noun* (1885)
: a white crystalline compound K_2SO_4 used especially as a fertilizer

po·ta·tion \pō-'tā-shən\ *noun* [Middle English *potacioun,* from Middle French *potation,* from Latin *potation-, potatio* act of drinking, from *potare* to drink — more at POTABLE] (15th century)
1 : a usually alcoholic drink or brew
2 : the act or an instance of drinking or inhaling; *also* **:** the portion taken in one such act

po·ta·to \pə-'tā-(ˌ)tō, -tə, *dialect* pə-'dā-, bə-\ *noun, plural* **-toes** *often attributive* [Spanish *batata,* from Taino] (1565)
1 : SWEET POTATO
2 a : an erect South American herb (*Solanum tuberosum*) of the nightshade family widely cultivated as a vegetable crop **b :** the edible starchy tuber of a potato — called also *Irish potato, white potato*
usage see TOMATO

potato 2a

potato beetle *noun* (1821)
: COLORADO POTATO BEETLE

potato blight *noun* (1879)
: any of several destructive fungus diseases of the potato

potato bug *noun* (1799)
: COLORADO POTATO BEETLE

potato chip *noun* (1878)
: a thin slice of white potato that has been fried until crisp and then usually salted

potato leafhopper *noun* (1921)
: a small green white-spotted leafhopper (*Empoasca fabae*) of the eastern and southern U.S. that is a serious pest on many cultivated plants and especially on the potato

potato pancake *noun* (1935)
: a fried flat cake of grated potato mixed with raw egg and usually grated onion and spices

potato tu·ber·worm \-'tü-bər-ˌwərm, -'tyü-\ *noun* (1920)
: a grayish brown moth (*Phthorimaea operculella* of the family Gelechiidae) whose larva mines the leaves and bores in the stems especially of potato and tobacco plants and commonly overwinters in potato tubers

pot–au–feu \ˌpät-ō-'fə(r), pȯ-tō-fœ\ *noun, plural* **pot–au–feu** [French, literally, pot on the fire] (1791)
: a French boiled dinner of meat and vegetables

pot·bel·lied \'pät-ˌbe-lēd\ *adjective* (1657)
: having a potbelly

potbellied stove *noun* (1933)
: a stove with a rounded or bulging body — called also *potbelly stove*

pot·bel·ly \'pät-ˌbe-lē\ *noun* (circa 1714)
1 : an enlarged, swollen, or protruding abdomen
2 : POTBELLIED STOVE

pot·boil \-ˌbȯil\ *intransitive verb* (1867)
: to produce potboilers

pot·boil·er \-ˌbȯi-lər\ *noun* (1864)
: a usually inferior work (as of art or literature) produced chiefly for profit

pot–bound \'pät-ˌbau̇nd\ *adjective* (1850)
of a potted plant **:** having roots so densely matted as to allow little or no space for further growth

pot·boy \-ˌbȯi\ *noun* (1795)
: a boy who serves drinks in a tavern

pot cheese *noun* (1812)
: COTTAGE CHEESE

po·teen *also* **po·theen** \pə-'tēn, -'chēn, -'tyēn, -'thēn\ *noun* [Irish *poitín,* literally, small pot, diminutive of *pota* pot] (1812)
: whiskey illicitly distilled in Ireland

Po·tem·kin village \pə-'tem(p)-kən-\ *noun* [Grigori *Potëmkin,* who supposedly built impressive fake villages along a route Catherine the Great was to travel] (1937)
: an impressive facade or show designed to hide an undesirable fact or condition ◆

po·tence \'pō-tᵊn(t)s\ *noun* (15th century)
: POTENCY

po·ten·cy \'pō-tᵊn(t)-sē\ *noun, plural* **-cies** (15th century)
1 a : FORCE, POWER **b :** the quality or state of being potent **c :** the ability or capacity to achieve or bring about a particular result
2 : POTENTIALITY 1

¹po·tent \'pō-tᵊnt\ *adjective* [Middle English, from Latin *potent-, potens* (present participle of *posse* to be able), from Latin *potis, pote* able; akin to Gothic brūth*faths* bridegroom, Greek *posis* husband, Sanskrit *pati* master] (15th century)
1 : having or wielding force, authority, or influence **:** POWERFUL
2 : achieving or bringing about a particular result **:** EFFECTIVE
3 a : chemically or medicinally effective ⟨a *potent* vaccine⟩ **b :** rich in a characteristic constituent
4 : able to copulate — usually used of the male
— **po·tent·ly** *adverb*

²potent *adjective* [obsolete English *potent* crutch] (1610)
of a heraldic cross **:** having flat bars across the ends of the arms — see CROSS illustration

po·ten·tate \'pō-tᵊn-ˌtāt\ *noun* (15th century)
: RULER, SOVEREIGN; *broadly* **:** one who wields great power or sway

¹po·ten·tial \pə-'ten(t)-shəl\ *adjective* [Middle English *potencial,* from Late Latin *potentialis,* from *potentia* potentiality, from Latin, power, from *potent-, potens*] (14th century)
1 : existing in possibility **:** capable of development into actuality ⟨*potential* benefits⟩
2 : expressing possibility; *specifically* **:** of, relating to, or constituting a verb phrase expressing possibility, liberty, or power by the use of an auxiliary with the infinitive of the verb (as in "it may rain")
synonym see LATENT
— **po·ten·tial·ly** \-'ten(t)-sh(ə-)lē\ *adverb*

²potential *noun* (1817)
1 a : something that can develop or become actual ⟨a *potential* for violence⟩ **b :** PROMISE 2
2 a : any of various functions from which the intensity or the velocity at any point in a field may be readily calculated **b :** the work required to move a unit positive charge from a reference point (as at infinity) to a point in question **c :** POTENTIAL DIFFERENCE

potential difference *noun* (1896)
: the difference in potential between two points that represents the work involved or the energy released in the transfer of a unit quantity of electricity from one point to the other

potential energy *noun* (1853)
: the energy that a piece of matter has because of its position or because of the arrangement of parts

po·ten·ti·al·i·ty \pə-ˌten(t)-shē-'a-lə-tē\ *noun, plural* **-ties** (1625)
1 : the ability to develop or come into existence
2 : POTENTIAL 1

◇ WORD HISTORY

Potemkin village In 1787, according to an account by a Saxon diplomat named Helbig, the Russian Prince Grigory Aleksandrovich Potemkin arranged for the Empress Catherine the Great to take a grand tour of Ukraine and the newly conquered territory of Crimea. According to Helbig's account, Potemkin, in order to impress Catherine with the prosperity of her new acquisitions, erected whole villages along her traveling route. Fortunately, she was able to view these villages only from a distance, for they were mere facades. Though the story of the sham villages must be regarded as apocryphal, countless retellings have established *Potemkin village* as a general term for any imposing facade or display that masks an undesirable condition or fact.

po·ten·ti·ate \pə-ˈten(t)-shē-ˌāt\ *transitive verb* **-at·ed; -at·ing** (1817)
: to make effective or active or more effective or more active; *also* **:** to augment the activity of (as a drug) synergistically
— **po·ten·ti·a·tion** \-ˌten(t)-shē-ˈā-shən\ *noun*
— **po·ten·ti·a·tor** \-ˈten(t)-shē-ˌā-tər\ *noun*

po·ten·til·la \ˌpō-tᵊn-ˈti-lə\ *noun* [New Latin, from Medieval Latin, garden heliotrope, from Latin *potent-, potens*] (1548)
: CINQUEFOIL 1

po·ten·ti·om·e·ter \pə-ˌten(t)-shē-ˈä-mə-tər\ *noun* [International Scientific Vocabulary *potential* + *-o-* + *-meter*] (1881)
1 : an instrument for measuring electromotive forces
2 : VOLTAGE DIVIDER
— **po·ten·ti·o·met·ric** \-sh(ē-)ə-ˈme-trik\ *adjective*

pot·ful \ˈpät-ˌfu̇l\ *noun* (14th century)
1 : as much or as many as a pot will hold
2 : a large amount ⟨makes a *potful* of money —John Corry⟩

pot hat *noun* (1798)
: a hat with a stiff crown; *especially* **:** DERBY

pot·head \ˈpät-ˌhed\ *noun* (1959)
: a person who smokes marijuana

¹poth·er \ˈpä-t͟hər\ *noun* [origin unknown] (1591)
1 a : confused or fidgety flurry or activity **:** COMMOTION **b :** agitated talk or controversy usually over a trivial matter
2 : a choking cloud of dust or smoke
3 : mental turmoil

²pother *verb* **poth·ered; poth·er·ing** \ˈpä-t͟hə-riŋ, ˈpät͟h-riŋ\ (1692)
transitive verb
: to put into a pother
intransitive verb
: to be in a pother

pot·herb \ˈpät-ˌərb, -ˌhərb\ *noun* (1538)
: a usually leafy herb that is cooked for use as greens; *also* **:** one (as mint) used to season food

pot holder *noun* (1944)
: a small cloth pad used for handling hot cooking utensils or containers

pot·hole \ˈpät-ˌhōl\ *noun* (1826)
1 a : a circular hole formed in the rocky bed of a river by the grinding action of stones or gravel whirled round by the water **b :** a sizable rounded often water-filled depression in land
2 : a pot-shaped hole in a road surface
— **pot·holed** \-ˌhōld\ *adjective*

pot·hook \-ˌhu̇k\ *noun* (15th century)
1 : an S-shaped hook for hanging pots and kettles over an open fire
2 : a written character resembling a pothook

pot·house \-ˌhau̇s\ *noun* (1724)
: TAVERN 1

pot·hunt·er \-ˌhən-tər\ *noun* (1781)
1 : one who hunts game for food
2 : an amateur archaeologist
— **pot·hunt·ing** \-tiŋ\ *noun*

po·tion \ˈpō-shən\ *noun* [Middle English *pocioun,* from Middle French *potion,* from Latin *potion-, potio* drink, potion, from *potare* to drink — more at POTABLE] (14th century)
: a mixture of liquids (as liquor or medicine)

¹pot·latch \ˈpät-ˌlach\ *noun* [Chinook Jargon *patlač,* from Nootka *p̓aƛ̓p̓ač*] (circa 1861)
1 : a ceremonial feast of the American Indians of the northwest coast marked by the host's lavish distribution of gifts or sometimes destruction of property to demonstrate wealth and generosity with the expectation of eventual reciprocation
2 *Northwest* **:** a social event or celebration

²potlatch (1898)
transitive verb
1 : to give (as a gift) especially with the expectation of a gift in return
2 : to hold or give a potlatch for (as a tribe or group)
intransitive verb
: to hold or give a potlatch

pot lik·ker \-ˈli-kər\ *Southern & Midland variant of* POT LIQUOR

pot·line \ˈpät-ˌlīn\ *noun* (1944)
: a row of electrolytic cells used in the production of aluminum

pot liquor *noun* (1744)
: the liquid left in a pot after cooking something

pot·luck \ˈpät-ˈlək, *1b also* -ˌlək\ *noun* (1592)
1 a : the regular meal available to a guest for whom no special preparations have been made **b :** a communal meal to which people bring food to share — usually used attributively ⟨a *potluck* supper⟩
2 : whatever is offered or available in given circumstances or at a given time

pot marigold *noun* (1814)
: a calendula (*Calendula officinalis*) grown especially for ornament

po·tom·e·ter \pō-ˈtä-mə-tər\ *noun* [Greek *poton* drink (akin to Greek *pinein* to drink) + English *-meter* — more at POTABLE] (1884)
: an apparatus for measuring the rate of transpiration in a plant by determining the amount of water absorbed

pot·pie \ˈpät-ˈpī, -ˌpī\ *noun* (circa 1792)
: pastry-covered meat and vegetables cooked in a deep dish

pot·pour·ri \ˌpō-pu̇-ˈrē\ *noun* [French *pot pourri,* literally, rotten pot] (1749)
1 : a mixture of flowers, herbs, and spices that is usually kept in a jar and used for scent
2 : a miscellaneous collection **:** MEDLEY ⟨a *potpourri* of the best songs and sketches —*Current Biography*⟩ ◆

pot roast *noun* (1881)
: a piece of beef cooked by braising usually on top of the stove

pot·sherd \ˈpät-ˌshərd\ *noun* [Middle English *pot-sherd,* from *pot* + *sherd* shard] (14th century)
: a pottery fragment

¹pot·shot \-ˌshät\ *noun* [from the notion that such a shot is unsportsmanlike and worthy only of one whose object is to fill the cooking pot] (1858)
1 : a shot taken from ambush or at a random or easy target
2 : a critical remark made in a random or sporadic manner

²potshot *verb* **potshot; pot·shot·ting** (1918)
intransitive verb
: to take a potshot
transitive verb
: to attack or shoot with a potshot

pot still *noun* (1799)
: a still used especially in the distillation of Irish grain whiskey and Scotch malt whiskey in which the heat of the fire is applied directly to the pot containing the mash

pot·stone \ˈpät-ˌstōn\ *noun* (1771)
: a more or less impure steatite used especially in prehistoric times to make cooking vessels

pot·tage \ˈpä-tij\ *noun* [Middle English *potage,* from Old French, from *pot* pot, of Germanic origin; akin to Old English *pott* pot] (13th century)
: a thick soup of vegetables and often meat

pot·ted \ˈpä-təd\ *adjective* (1646)
1 *chiefly British* **:** preserved in a pot, jar, or can
2 : planted or grown in a pot
3 *chiefly British* **:** briefly and superficially summarized ⟨a dull, pedestrian *potted* history —*Times Literary Supplement*⟩
4 *slang* **:** DRUNK 1a

¹pot·ter \ˈpä-tər\ *noun* (before 12th century)
: one that makes pottery

²potter *intransitive verb* [probably frequentative of English dialect *pote* to poke] (1829)
: PUTTER
— **pot·ter·er** \ˈpä-tər-ər\ *noun*
— **pot·ter·ing·ly** \ˈpä-tə-riŋ-lē\ *adverb*

potter's clay *noun* (15th century)
: a plastic clay suitable for making pottery — called also *potter's earth*

potter's field *noun* [from the mention in Matthew 27:7 of the purchase of a potter's field for use as a graveyard] (1777)
: a public burial place for paupers, unknown persons, and criminals

potter's wheel *noun* (circa 1741)
: a usually horizontal disk revolving on a vertical spindle and carrying the clay being shaped by a potter

pot·tery \ˈpä-tə-rē\ *noun, plural* **-ter·ies** (15th century)
1 : a place where clayware is made and fired
2 a : the art or craft of the potter **b :** the manufacture of clayware
3 : CLAYWARE; *especially* **:** earthenware as distinguished on the one hand from porcelain and stoneware and on the other from brick and tile

pot·tle \ˈpä-tᵊl\ *noun* [Middle English *potel,* from Middle French, from *pot*] (14th century)
: a container holding a half gallon (1.9 liters)

pot·to \ˈpä-(ˌ)tō\ *noun, plural* **pottos** [perhaps from Wolof *pata,* a tailless monkey] (1705)
: any of several African primates (genera *Arctocebus* and *Perodicticus*); *especially* **:** a West African primate (*P. potto*) that has a vestigial index finger and tail

Pott's disease \ˈpäts-\ *noun* [Percivall *Pott* (died 1788) English surgeon] (1835)
: tuberculosis of the spine with destruction of bone resulting in curvature of the spine

¹pot·ty \ˈpä-tē\ *adjective* **pot·ti·er; -est** [probably from ¹*pot*] (circa 1860)
1 *British* **:** TRIVIAL, INSIGNIFICANT
2 *chiefly British* **:** slightly crazy
3 : SNOBBISH

²potty *noun, plural* **potties** (circa 1942)
: a small child's pot for urination or defecation

pot·ty–chair \-ˌcher, -ˌchar\ *noun* (1943)
: a child's chair having an open seat under which a receptacle in placed for toilet training

pot·zer \ˈpät-sər\ *variant of* PATZER

¹pouch \ˈpau̇ch\ *noun* [Middle English *pouche,* from Middle French, of Germanic origin; akin to Old English *pocca* bag] (14th century)
1 : a small drawstring bag carried on the person

◇ WORD HISTORY

potpourri A potpourri, which in ordinary usage alludes to a dried mixture of pleasant-smelling herbs or more broadly to any miscellaneous collection, is etymologically a "putrid pot." The original French *pot pourri* was a stew made of a variety of meats and vegetables cooked together. This was a traditional Spanish dish borrowed by the French, whose name for it is a loan translation of Spanish *olla podrida.* Just why the stew designated *olla podrida* was called "rotten" is not very clear. One explanation is that the Spanish adjective *podrido* has occasionally been used colloquially to mean "very full, teeming," a sense that would fit the nature of the dish. Though not common, similar uses of English *rotten* can be found (as, for example, in the passage "The country's rotten with money, if one could touch it" from a novel by Elizabeth Bowen). And *rotten* in the common idiom "to spoil a child rotten" means little more than "thoroughly."

2 a : a bag of small or moderate size for storing or transporting goods; *specifically* **:** a lockable bag for first-class mail or diplomatic dispatches **b** *chiefly Scottish* **:** POCKET **c :** PACKET
3 : an anatomical structure resembling a pouch
— pouched \'pau̇cht\ *adjective*

²pouch (circa 1566)
transitive verb
1 : to put or form into or as if into a pouch
2 : to transmit by pouch
intransitive verb
1 : to bulge or stick out or down in a manner suggesting a pouch ⟨*pouching* cheeks⟩
2 : to transmit mail or dispatches by pouch

pouchy \'pau̇-chē\ *adjective* **pouch·i·er; -est** (1828)
: having, tending to have, or resembling a pouch ⟨*pouchy* insomniac eyes —Graham Greene⟩

pouf *also* **pouffe** \'püf\ *noun* [French *pouf,* something inflated, of imitative origin] (1817)
1 : PUFF 3b(3)
2 : a bouffant or fluffy part of a garment or accessory
3 : OTTOMAN
— poufed *or* **pouffed** \'püft\ *adjective*

Pouil·ly–Fuis·sé \pü-'yē-fwē-'sā\ *noun* [Solutré-*Pouilly* and *Fuissé,* villages in France] (1927)
: a dry white burgundy from an area west of Mâcon, France

Pouil·ly–Fu·mé \pü-'yē-fü-'mā, -fᵫ-\ *noun* [French, from *Pouilly*-sur-Loire, village in France + *fumé,* past participle of *fumer* to smoke, from Latin *fumare,* from *fumus* smoke — more at FUME] (1935)
: a dry white wine from the Loire valley of France

pou·larde *also* **pou·lard** \pü-'lärd\ *noun* [French *poularde*] (1732)
: a pullet sterilized to produce fattening

poult \'pōlt\ *noun* [Middle English *polet, pulte* young fowl — more at PULLET] (15th century)
: a young fowl; *especially* **:** a young turkey

poul·ter·er \'pōl-tər-ər\ *noun* [alteration of Middle English *pulter,* from Middle French *pouletier*] (1638)
: one that deals in poultry

poul·ter's measure \'pōl-tərz-\ *noun* [obsolete *poulter* poulterer, from Middle English *pulter;* from the former practice of occasionally giving one or two extra when counting eggs by dozens] (1576)
: a meter in which lines of 12 and 14 syllables alternate

¹poul·tice \'pōl-təs\ *noun* [Middle English *pultes,* from Medieval Latin, literally, pap, from Latin, plural of *pult-, puls* porridge] (15th century)
: a soft usually heated and sometimes medicated mass spread on cloth and applied to sores or other lesions

²poultice *transitive verb* **-ticed; -tic·ing** (1730)
: to apply a poultice to

poul·try \'pōl-trē\ *noun* [Middle English *pultrie,* from Middle French *pouleterie,* from Old French, from *pouletier* poulterer, from *poulet* — more at PULLET] (14th century)
: domesticated birds kept for eggs or meat

poul·try·man \-mən\ *noun* (circa 1574)
1 : one who raises domestic fowls especially on a commercial scale for the production of eggs and meat
2 : one who deals in poultry or poultry products

¹pounce \'pau̇n(t)s\ *noun* [Middle English, talon] (15th century)
: the claw of a bird of prey

²pounce *intransitive verb* **pounced; pounc·ing** (1744)
1 a : to swoop upon and seize something with or as if with talons **b :** to seize upon and make capital of something (as another's blunder or an opportunity)
2 : to make a sudden assault or approach

³pounce *noun* (1841)
: the act of pouncing

⁴pounce *transitive verb* **pounced; pounc·ing** [Middle French *poncer,* from *ponce*] (1535)
: to dust, rub, finish, or stencil with pounce

⁵pounce *noun* [French *ponce* pumice, from Middle French, from Late Latin *pomic-, pomex,* alteration of Latin *pumic-, pumex* — more at FOAM] (1706)
1 : a fine powder formerly used to prevent ink from spreading
2 : a fine powder for making stenciled patterns

poun·cet–box \'pau̇n(t)-sət-\ *noun* [probably from (assumed) Middle French *poncette* small pounce bag] (1596)
archaic **:** a box for carrying pomander

¹pound \'pau̇nd\ *noun, plural* **pounds** *also* **pound** [Middle English, from Old English *pund,* from Latin *pondo* pound, from ablative of *pondus* weight — more at PENDANT] (before 12th century)
1 : any of various units of mass and weight; *specifically* **:** a unit now in general use among English-speaking peoples equal to 16 avoirdupois ounces or 7000 grains or 0.45359237 kilogram — see WEIGHT table
2 a : the basic monetary unit of the United Kingdom — called also *pound sterling* **b :** any of numerous basic monetary units of other countries — see MONEY table **c :** ²LIRA

²pound *noun* [Middle English, enclosure, from Old English *pund-*] (14th century)
1 a : an enclosure for animals; *especially* **:** a public enclosure for stray or unlicensed animals ⟨a dog *pound*⟩ **b :** a depot for holding impounded personal property until redeemed by the owner ⟨a car *pound*⟩
2 : a place or condition of confinement
3 : an enclosure within which fish are kept or caught; *especially* **:** the inner compartment of a fish trap or pound net

³pound *noun* (1562)
: an act or sound of pounding

⁴pound *verb* [alteration of Middle English *pounen,* from Old English *pūnian*] (1594)
transitive verb
1 : to reduce to powder or pulp by beating
2 a : to strike heavily or repeatedly **b :** to produce with or as if with repeated vigorous strokes — usually used with *out* ⟨*pound* out a story on the typewriter⟩ **c :** to inculcate by insistent repetition **:** DRIVE ⟨day after day the facts were *pounded* home to them —Ivy B. Priest⟩
3 : to move along heavily or persistently ⟨*pounded* the pavements looking for work⟩
intransitive verb
1 : to strike heavy repeated blows
2 : PULSATE, THROB ⟨my heart was *pounding*⟩
3 a : to move with or make a heavy repetitive sound **b :** to work hard and continuously — usually used with *away*

¹pound·age \'pau̇n-dij\ *noun* (circa 1500)
1 : a charge per pound of weight
2 : weight in pounds

²poundage *noun* (1554)
: IMPOUNDMENT 1

pound·al \'pau̇n-dᵊl\ *noun* [¹*pound* + *-al* (as in *quintal*)] (1879)
: a unit of force equal to the force that would give a free mass of one pound an acceleration of one foot per second per second

pound cake *noun* [from the original recipe prescribing a pound of each of the principal ingredients] (1747)
: a rich butter cake made with a large proportion of eggs and shortening

¹pound·er \'pau̇n-dər\ *noun* (before 12th century)
1 : one that pounds
2 : a tool used for pounding

²pounder *noun* (1684)
1 : a gun throwing a projectile of a specified weight — usually used in combination ⟨the ship was armed with six-*pounders*⟩
2 : one having a usually specified weight or value in pounds — usually used in combination ⟨caught a ten-*pounder*⟩

pound–fool·ish \'pau̇n(d)-'fü-lish\ *adjective* [from the phrase *penny-wise and pound-foolish*] (1607)
: imprudent in dealing with large sums or large matters

pound mile *noun* (1939)
: the transport of one pound of mail or express for one mile

pound net *noun* (1865)
: a fish trap consisting of a netting arranged into a directing wing and an enclosure with a narrow entrance

pound sign *noun* (1980)
1 : the symbol £
2 : the symbol #

¹pour \'pōr, 'pȯr\ *verb* [Middle English] (14th century)
transitive verb
1 a : to cause to flow in a stream **b :** to dispense from a container ⟨*poured* drinks for everyone⟩
2 : to supply or produce freely or copiously
3 : to give full expression to **:** VENT ⟨*poured* out his feelings⟩
intransitive verb
1 : to move with a continuous flow
2 : to rain hard
3 : to move or come continuously **:** STREAM ⟨complaints *poured* in⟩
— pour·able \'pōr-ə-bəl, 'pȯr-\ *adjective*
— pour·er \-ər\ *noun*
— pour·ing·ly \-iŋ-lē\ *adverb*

²pour *noun* (1790)
1 : the action of pouring **:** STREAM
2 a : an instance of pouring or an amount poured **b :** a heavy fall of rain **:** DOWNPOUR

pour·boire \pu̇r-'bwär\ *noun* [French, from *pour boire* for drinking] (1817)
: TIP, GRATUITY

pour·par·ler \ˌpu̇r-pär-'lā\ *noun* [French, from Middle French, from *pourparler* to discuss, from Old French, from *pour* for, before + *parler* to speak — more at PURCHASE, PARLEY] (1795)
: a discussion preliminary to negotiations

pour·point \'pu̇r-ˌpȯint, -ˌpwant\ *noun* [Middle English *purpoint,* from Middle French *pourpoint,* from Old French *porpoint,* from *porpoint,* adjective, quilted, alteration of (assumed) Vulgar Latin *perpunctus,* past participle of *perpungere* to perforate, from Latin *per* through + *pungere* to prick, pierce — more at PUNGENT] (14th century)
: a padded and quilted doublet

pour point \'pōr-ˌpȯint, 'pȯr-\ *noun* (1922)
: the lowest temperature at which a substance flows under specified conditions

pousse–ca·fé \ˌpüs-(ˌ)ka-'fā\ *noun* [French, literally, coffee chaser] (1880)
: an after-dinner drink consisting of several liqueurs of different colors and specific gravities poured so as to remain in separate layers

pous·sette \pü-'set\ *intransitive verb* **pous·sett·ed; pous·sett·ing** [French, game in which contestants cross pins with each attempting to get his pin on top, from *pousser* to push] (1812)
: to swing in a semicircle with hands joined with one's partner in a country-dance

¹pout \'pau̇t\ *verb* [Middle English] (14th century)
intransitive verb
1 a : to show displeasure by thrusting out the lips or wearing a sullen expression **b :** SULK
2 : PROTRUDE
transitive verb
: to cause to protrude ⟨*pouted* her lips⟩

²pout *noun* (1591)

1 : a protrusion of the lips expressive of displeasure
2 *plural* **:** a fit of pique

[3]pout *noun, plural* **pout** *or* **pouts** [probably from (assumed) Middle English *poute,* a fish with a large head, from Old English *-pūte;* akin to Middle English *pouten* to pout] (1591)
: any of several large-headed fishes (as a bullhead or eelpout)

pout·er \'pau̇-tər\ *noun* (1725)
1 : a domestic pigeon of a breed characterized by erect carriage and a dilatable crop
2 : one that pouts

pouty \'pau̇-tē\ *adjective* (1863)
1 : SULKY 1
2 : expressive of displeasure

pov·er·ty \'pä-vər-tē\ *noun, often attributive* [Middle English *poverte,* from Old French *poverté,* from Latin *paupertat-, paupertas,* from *pauper* poor — more at POOR] (12th century)
1 a : the state of one who lacks a usual or socially acceptable amount of money or material possessions **b :** renunciation as a member of a religious order of the right as an individual to own property
2 : SCARCITY, DEARTH
3 a : debility due to malnutrition **b :** lack of fertility ⟨*poverty* of the soil⟩ ☆

poverty line *noun* (1901)
: a level of personal or family income below which one is classified as poor according to governmental standards — called also *poverty level*

pov·er·ty–strick·en \-ˌstri-kən\ *adjective* (1803)
: very poor **:** DESTITUTE

[1]pow \'pō, 'pau̇\ *noun* [alteration of *poll*] (1724)
: HEAD, POLL

[2]pow \'pau̇\ *noun* [imitative] (1881)
: a sound of a blow or explosion

POW \ˌpē-(ˌ)ō-'də-bəl-(ˌ)yü, -bə-(ˌ)yü; -'dəb-(ˌ)yü\ *noun* (circa 1919)
: PRISONER OF WAR

[1]pow·der \'pau̇-dər\ *verb* **pow·dered; pow·der·ing** \'pau̇-d(ə-)riŋ\ (13th century)
transitive verb
1 : to sprinkle or cover with or as if with powder
2 : to reduce or convert to powder
3 : to hit (as a ball) very hard
intransitive verb
1 : to become powder
2 : to apply cosmetic powder
— pow·der·er \-dər-ər\ *noun*

[2]powder *noun, often attributive* [Middle English *poudre,* from Old French, from Latin *pulver-, pulvis* dust; probably akin to Sanskrit *palāva* chaff] (14th century)
1 : matter in a finely divided state **:** particulate matter
2 a : a preparation in the form of fine particles especially for medicinal or cosmetic use **b :** fine dry light snow
3 : any of various solid explosives used chiefly in gunnery and blasting
— pow·der·less \-ləs\ *adjective*
— pow·der·like \-ˌlīk\ *adjective*

powder blue *noun* (1896)
: a pale blue

powder horn *noun* (1533)
: a flask for carrying gunpowder; *especially* **:** one made of the horn of an ox or cow

powder keg *noun* (1855)
1 : a small usually metal cask for holding gunpowder or blasting powder
2 : something liable to explode

powder metallurgy *noun* (1933)
: a branch of science or an art concerned with the production of powdered metals or of metallic objects by compressing a powdered metal or alloy with or without other materials and heating without thoroughly melting to solidify and strengthen

powder monkey *noun* (1682)
: one who carries or has charge of explosives (as in blasting operations)

powder–puff *adjective* (1939)
: of, relating to, or being a competitive activity or event for women ⟨a *powder-puff* football game⟩

powder puff *noun* (circa 1704)
: a small fluffy device (as a pad) for applying cosmetic powder

powder room *noun* (circa 1937)
1 : a rest room for women
2 : a lavatory in the main living area of a house

pow·dery \'pau̇-də-rē\ *adjective* (15th century)
1 a : resembling or consisting of powder ⟨*powdery* snow⟩ **b :** easily reduced to powder **:** CRUMBLING
2 : covered with or as if with powder

powdery mildew *noun* (1889)
1 : an ascomycetous fungus (family Erysiphaceae) producing abundant powdery conidia on the host
2 : a plant disease caused by a powdery mildew

[1]pow·er \'pau̇(-ə)r\ *noun, often attributive* [Middle English, from Old French *poeir,* from *poeir* to be able, from (assumed) Vulgar Latin *potēre,* alteration of Latin *posse* — more at POTENT] (13th century)
1 a (1) **:** ability to act or produce an effect (2) **:** ability to get extra-base hits (3) **:** capacity for being acted upon or undergoing an effect **b :** legal or official authority, capacity, or right
2 a : possession of control, authority, or influence over others **b :** one having such power; *specifically* **:** a sovereign state **c :** a controlling group **:** ESTABLISHMENT — often used in the phrase *the powers that be* **d** *archaic* **:** a force of armed men **e** *chiefly dialect* **:** a large number or quantity
3 a : physical might **b :** mental or moral efficacy **c :** political control or influence
4 *plural* **:** an order of angels — see CELESTIAL HIERARCHY
5 a : the number of times as indicated by an exponent that a number occurs as a factor in a product; *also* **:** the product itself **b :** CARDINAL NUMBER 2
6 a : a source or means of supplying energy; *especially* **:** ELECTRICITY **b :** MOTIVE POWER **c :** the time rate at which work is done or energy emitted or transferred
7 : MAGNIFICATION 2b
8 : SCOPE, COMPREHENSIVENESS
9 : the probability of rejecting the null hypothesis in a statistical test when a particular alternative hypothesis happens to be true ☆ ☆

[2]power (1540)
transitive verb
1 : to supply with power and especially motive power
2 : to give impetus to
intransitive verb
1 : to move about by means of motive power
2 : to move with great speed or force

[3]power *adjective* (1949)
1 : relating to or utilizing power ⟨plays a *power* game⟩; *also* **:** POWERFUL 1 ⟨a *power* critic⟩
2 : of, relating to, or being a meal at which influential people discuss business or politics

power base *noun* (1959)
: a base of political support

pow·er·boat \'pau̇(-ə)r-ˌbōt\ *noun* (1908)
: MOTORBOAT

power broker *noun* (1961)
: a person (as in politics) able to exert strong influence through control of votes or individuals

pow·er–dive \-ˌdīv\ (1937)
intransitive verb
: to make a power dive
transitive verb
: to cause to power-dive

power dive *noun* (1930)
: a dive of an airplane accelerated by the power of the engine

pow·er·ful \'pau̇(-ə)r-fəl\ *adjective* (15th century)
1 : having great power, prestige, or influence
2 : leading to many or important deductions ⟨a *powerful* set of postulates⟩
— pow·er·ful·ly \-f(ə-)lē\ *adverb*

power function *noun* (1957)
1 : a function of a parameter under statistical test whose value for a particular value of the parameter is the probability of rejecting the null hypothesis if that value of the parameter happens to be true
2 : a function (as $f(x) = ax^k$) that equals the product of a constant and a power of the independent variable

pow·er·house \'pau̇(-ə)r-ˌhau̇s\ *noun* (circa 1890)

☆ SYNONYMS

Poverty, indigence, penury, want, destitution mean the state of one with insufficient resources. POVERTY may cover a range from extreme want of necessities to an absence of material comforts ⟨the extreme *poverty* of the slum dwellers⟩. INDIGENCE implies seriously straitened circumstances ⟨the *indigence* of her years as a graduate student⟩. PENURY suggests a cramping or oppressive lack of money ⟨a catastrophic illness that condemned them to years of *penury*⟩. WANT and DESTITUTION imply extreme poverty that threatens life itself through starvation or exposure ⟨lived in a perpetual state of *want*⟩ ⟨the widespread *destitution* in countries beset by famine⟩.

Power, authority, jurisdiction, control, command, sway, dominion mean the right to govern or rule or determine. POWER implies possession of ability to wield force, permissive authority, or substantial influence ⟨the *power* to mold public opinion⟩. AUTHORITY implies the granting of power for a specific purpose within specified limits ⟨gave her attorney the *authority* to manage her estate⟩. JURISDICTION applies to official power exercised within prescribed limits ⟨the bureau having *jurisdiction* over alcohol and firearms⟩. CONTROL stresses the power to direct and restrain ⟨you are responsible for the students under your *control*⟩. COMMAND implies the power to make arbitrary decisions and compel obedience ⟨the army officer in *command*⟩. SWAY suggests the extent or scope of exercised power or influence ⟨an empire that extended its *sway* over the known world⟩. DOMINION stresses sovereign power or supreme authority ⟨given *dominion* over all the animals⟩.

Power, force, energy, strength, might mean the ability to exert effort. POWER may imply latent or exerted physical, mental, or spiritual ability to act or be acted upon ⟨the awesome *power* of flowing water⟩. FORCE implies the actual effective exercise of power ⟨used enough *force* to push the door open⟩. ENERGY applies to power expended or capable of being transformed into work ⟨a worker with boundless *energy*⟩. STRENGTH applies to the quality or property of a person or thing that makes possible the exertion of force or the withstanding of strain, pressure, or attack ⟨use weight training to build your *strength*⟩. MIGHT implies great or overwhelming power or strength ⟨the belief that *might* makes right⟩.

1 a : POWER PLANT 1 **b :** a source of influence or inspiration
2 : one having great power: as **a :** one having great drive, energy, or ability **b :** an athletic team characterized by strong aggressive play
pow·er·less \-ləs\ *adjective* (15th century)
1 : devoid of strength or resources
2 : lacking the authority or capacity to act
— pow·er·less·ly *adverb*
— pow·er·less·ness *noun*
power mower *noun* (1940)
: a motor-driven lawn mower
power of attorney (1747)
: a legal instrument authorizing one to act as the attorney or agent of the grantor
power pack *noun* (1936)
: a unit for converting a power supply (as from a battery or household electrical circuit) to a voltage suitable for an electronic device
power plant *noun* (1890)
1 : an electric utility generating station
2 : an engine and related parts supplying the motive power of a self-propelled object (as a rocket or automobile)
power play *noun* (1947)
1 : a military, diplomatic, political, or administrative maneuver in which power is brought to bear
2 a : a concentrated attack in football in which the ballcarrier is preceded by a mass of blockers **b :** a situation in ice hockey in which one team temporarily has more players on the ice than the other team because of a penalty
power politics *noun plural but singular or plural in construction* (1926)
: politics based primarily on the use of power (as military and economic strength) as a coercive force rather than on ethical precepts
power series *noun* (1893)
: an infinite series whose terms are successive integral powers of a variable multiplied by constants
power shovel *noun* (1909)
: a power-operated excavating machine consisting of a boom or crane that supports a lever arm with a large bucket at the end of it
power station *noun* (1901)
: POWER PLANT 1
power steering *noun* (1932)
: automotive steering with engine power used to amplify the torque applied at the steering wheel by the driver
power structure *noun* (1950)
1 : a group of persons having control of an organization **:** ESTABLISHMENT
2 : the hierarchical interrelationships existing within a controlling group
power sweep *noun* (1964)
: SWEEP 3e
power take–off *noun* (1929)
: a supplementary mechanism (as on a tractor) enabling the engine power to be used to operate nonautomotive apparatus (as a pump or saw)
power train *noun* (1943)
: the intervening mechanism by which power is transmitted from an engine to a propeller or axle that it drives
power up *transitive verb* (1970)
: to cause to operate ⟨*power up* the computer⟩
— pow·er–up \ˈpau̇(-ə)r-ˌəp\ *noun*
¹pow·wow \ˈpau̇-ˌwau̇\ *noun* [Narraganset *powwaw* or Massachuset *pauwau*] (1625)
1 : an American Indian medicine man
2 a : an American Indian ceremony (as for victory in war) **b :** an American Indian social gathering or fair usually including competitive dancing
3 a : a social get-together **b :** a meeting for discussion
²powwow *intransitive verb* (1642)
: to hold a powwow
¹pox \ˈpäks\ *noun, plural* **pox** *or* **pox·es** [alteration of *pocks,* plural of *pock*] (1550)
1 a : a virus disease (as chicken pox) characterized by pustules or eruptions **b** *archaic* **:** SMALLPOX **c :** SYPHILIS
2 : a disastrous evil **:** PLAGUE, CURSE ⟨a *pox* on him⟩
²pox *transitive verb* (1601)
archaic **:** to infect with a pox and especially with syphilis
pox·vi·rus \ˈpäks-ˌvī-rəs\ *noun* (1941)
: any of a group of relatively large round, brick-shaped, or ovoid DNA-containing animal viruses (as the causative agent of smallpox) that have a fluffy appearance caused by a covering of tubules and threads
poz·zo·la·na \ˌpät-sə-ˈlä-nə\ *or* **poz·zo·lan** \ˈpät-sə-lən\ *noun* [Italian *pozzolana*] (1706)
: finely divided siliceous or siliceous and aluminous material that reacts chemically with slaked lime at ordinary temperature and in the presence of moisture to form a strong slow-hardening cement
— poz·zo·la·nic \ˌpät-sə-ˈla-nik, -ˈlä-\ *adjective*
PPO \ˌpē-(ˌ)pē-ˈō\ *noun, plural* **PPOs** [*p*referred *p*rovider *o*rganization] (1983)
: an organization providing health care that gives economic incentives to the individual purchaser of a health-care contract to patronize certain physicians, laboratories, and hospitals that agree to supervision and reduced fees — compare HMO
prac·ti·ca·ble \ˈprak-ti-kə-bəl\ *adjective* (1670)
1 : capable of being put into practice or of being done or accomplished **:** FEASIBLE
2 : capable of being used **:** USABLE
synonym see POSSIBLE
— prac·ti·ca·bil·i·ty \ˌprak-ti-kə-ˈbi-lə-tē\ *noun*
— prac·ti·ca·ble·ness \ˈprak-ti-kə-bəl-nəs\ *noun*
— prac·ti·ca·bly \-blē\ *adverb*
¹prac·ti·cal \ˈprak-ti-kəl\ *adjective* [Middle English, from Late Latin *practicus,* from Greek *praktikos,* from *prassein* to pass over, fare, do; akin to Greek *peran* to pass through — more at FARE] (15th century)
1 a : of, relating to, or manifested in practice or action **:** not theoretical or ideal ⟨a *practical* question⟩ ⟨for all *practical* purposes⟩ **b :** being such in practice or effect **:** VIRTUAL ⟨a *practical* failure⟩
2 : actively engaged in some course of action or occupation ⟨a *practical* farmer⟩
3 : capable of being put to use or account **:** USEFUL ⟨he had a *practical* knowledge of French⟩
4 a : disposed to action as opposed to speculation or abstraction **b** (1) **:** qualified by practice or practical training ⟨a good *practical* mechanic⟩ (2) **:** designed to supplement theoretical training by experience
5 : concerned with voluntary action and ethical decisions ⟨*practical* reason⟩
— prac·ti·cal·i·ty \ˌprak-ti-ˈka-lə-tē\ *noun*
— prac·ti·cal·ness \ˈprak-ti-kəl-nəs\ *noun*
²practical *noun* (1925)
: an examination requiring demonstration of some practical skill ⟨a zoology *practical*⟩
practical art *noun* (circa 1925)
: an art (as woodworking) that serves ordinary or material needs — usually used in plural
practical joke *noun* (circa 1847)
: a prank intended to trick or embarrass someone or cause physical discomfort
— practical joker *noun*
prac·ti·cal·ly \ˈprak-ti-k(ə-)lē\ *adverb* (1623)
1 : in a practical manner ⟨look *practically* at the problem⟩
2 : ALMOST, NEARLY ⟨*practically* everyone went to the party⟩
practical nurse *noun* (1921)
: a nurse who cares for the sick professionally without having the training or experience required of a registered nurse; *especially* **:** LICENSED PRACTICAL NURSE
practical theology *noun* (circa 1909)
: the study of the institutional activities of religion (as preaching, church administration, pastoral care, and liturgics)
¹prac·tice *or* **prac·tise** \ˈprak-təs\ *verb* **prac·ticed** *or* **prac·tised; prac·tic·ing** *or* **prac·tis·ing** [Middle English *practisen,* from Middle French *practiser,* from *practique, pratique* practice, noun, from Late Latin *practice,* from Greek *praktikē,* from feminine of *praktikos*] (14th century)
transitive verb
1 a : CARRY OUT, APPLY ⟨*practice* what you preach⟩ **b :** to do or perform often, customarily, or habitually ⟨*practice* politeness⟩ **c :** to be professionally engaged in ⟨*practice* medicine⟩
2 a : to perform or work at repeatedly so as to become proficient ⟨*practice* the act⟩ **b :** to train by repeated exercises ⟨*practice* pupils in penmanship⟩
3 *obsolete* **:** PLOT
intransitive verb
1 : to do repeated exercises for proficiency
2 : to pursue a profession actively
3 *archaic* **:** INTRIGUE
4 : to do something customarily
5 : to take advantage of someone ⟨he *practised* on their credulity with huge success —*Times Literary Supplement*⟩
— prac·tic·er *noun*
²practice *also* **practise** *noun* (15th century)
1 a : actual performance or application ⟨ready to carry out in *practice* what they advocated in principle⟩ **b :** a repeated or customary action ⟨had this irritating *practice*⟩ **c :** the usual way of doing something ⟨local *practices*⟩ **d :** the form, manner, and order of conducting legal suits and prosecutions
2 a : systematic exercise for proficiency ⟨*practice* makes perfect⟩ **b :** the condition of being proficient through systematic exercise ⟨get in *practice*⟩
3 a : the continuous exercise of a profession **b :** a professional business; *especially* **:** one constituting an incorporeal property
synonym see HABIT
prac·ticed *or* **prac·tised** \ˈprak-təst\ *adjective* (1568)
1 : EXPERIENCED, SKILLED
2 : learned by practice
prac·tice–teach \ˈprak-təs-ˈtēch\ *intransitive verb* **-taught** \-ˈtȯt\; **-teach·ing** [back-formation from *practice teaching*] (1952)
: to engage in practice teaching
— practice teacher *noun*
practice teaching *noun* (circa 1913)
: teaching by a student under the supervision of an experienced teacher
practicing *or* **practising** *adjective* (1625)
: actively engaged in a specified career or way of life ⟨a *practicing* physician⟩
prac·ti·cum \ˈprak-ti-kəm\ *noun* [German *Praktikum,* from Late Latin *practicum,* neuter of *practicus* practical] (circa 1909)
: a course of study designed especially for the preparation of teachers and clinicians that involves the supervised practical application of previously studied theory
prac·ti·tion·er \prak-ˈti-sh(ə-)nər\ *noun* [alteration of earlier *practician,* from Middle English (Scots) *pratician,* from Middle French *practicien,* from *pratique*] (1535)
1 : one who practices; *especially* **:** one who practices a profession
2 *Christian Science* **:** an authorized healer
prae·ci·pe \ˈpre-sə-ˌpē, ˈprē-\ *noun* [Middle English *presepe,* from Medieval Latin *praecipe,* from Latin, imperative of *praecipere* to instruct — more at PRECEPT] (15th century)
1 : any of various legal writs commanding a person to do something or to appear and show cause why he or she should not
2 : a written order requesting a clerk or prothonotary of a court to issue a writ and specifying the contents of the writ
prae·di·al *variant of* PREDIAL

prae·mu·ni·re \ˌprē-myu̇-ˈnīr-ē\ *noun* [Middle English *praemunire facias,* from Medieval Latin, that you cause to warn; from prominent words in the writ] (1529)
: an offense against the English Crown punishable chiefly by forfeiture and originally committed by asserting papal legal supremacy in England

prae·no·men \prē-ˈnō-mən\ *noun, plural* **-nomens** *or* **-no·mi·na** \-ˈnä-mə-nə, -ˈnō-\ [Latin, from *prae-* pre- + *nomen* name — more at NAME] (1706)
: the first of the usual three names of an ancient Roman male

prae·sid·i·um *variant of* PRESIDIUM

prae·tor \ˈprē-tər\ *noun* [Middle English *pretor,* from Latin *praetor*] (15th century)
: an ancient Roman magistrate ranking below a consul and having chiefly judicial functions
— **prae·to·ri·al** \prē-ˈtōr-ē-əl, -ˈtȯr-\ *adjective*
— **prae·tor·ship** \ˈprē-tər-ˌship\ *noun*

prae·to·ri·an \prē-ˈtōr-ē-ən, -ˈtȯr-\ *adjective* (15th century)
1 *often capitalized* **:** of, forming, or resembling the Roman imperial bodyguard
2 : of or relating to a praetor
— **praetorian** *noun, often capitalized*

prag·mat·ic \prag-ˈma-tik\ *also* **prag·mat·i·cal** \-ti-kəl\ *adjective* [Latin *pragmaticus* skilled in law or business, from Greek *pragmatikos,* from *pragmat-, pragma* deed, from *prassein* to do — more at PRACTICAL] (1616)
1 *archaic* **a** (1) **:** BUSY (2) **:** OFFICIOUS **b :** OPINIONATED
2 : relating to matters of fact or practical affairs often to the exclusion of intellectual or artistic matters **:** practical as opposed to idealistic ⟨*pragmatic* men of power have had no time or inclination to deal with . . . social morality —K. B. Clark⟩
3 : relating to or being in accordance with philosophical pragmatism
— **pragmatic** *noun*
— **prag·mat·i·cal·ly** \-ti-k(ə-)lē\ *adverb*

prag·mat·i·cism \prag-ˈma-tə-ˌsi-zəm\ *noun* (1905)
: the philosophic doctrine of C. S. Peirce
— **prag·mat·i·cist** \-sist\ *noun*

prag·mat·ics \prag-ˈma-tiks\ *noun plural but singular or plural in construction* (1937)
1 : a branch of semiotic that deals with the relation between signs or linguistic expressions and their users
2 : linguistics concerned with the relationship of sentences to the environment in which they occur

pragmatic sanction *noun* (1643)
: a solemn decree of a sovereign on a matter of primary importance and with the force of fundamental law

prag·ma·tism \ˈprag-mə-ˌti-zəm\ *noun* (circa 1864)
1 : a practical approach to problems and affairs ⟨tried to strike a balance between principles and *pragmatism*⟩
2 : an American movement in philosophy founded by C. S. Peirce and William James and marked by the doctrines that the meaning of conceptions is to be sought in their practical bearings, that the function of thought is to guide action, and that truth is preeminently to be tested by the practical consequences of belief
— **prag·ma·tist** \-mə-tist\ *adjective or noun*
— **prag·ma·tis·tic** \ˌprag-mə-ˈtis-tik\ *adjective*

prai·rie \ˈprer-ē\ *noun, often attributive* [French, from (assumed) Vulgar Latin *prataria,* from Latin *pratum* meadow] (circa 1682)
1 : land in or predominantly in grass
2 : a tract of grassland: as **a :** a large area of level or rolling land in the Mississippi valley that in its natural uncultivated state usually has deep fertile soil, a cover of tall coarse grasses, and few trees **b :** one of the dry treeless plateaus east of the Rocky Mountains that merge on their east side with the prairies proper and are characterized by shorter grasses and drier less fertile soil

prairie chicken *noun* (1691)
: a grouse (*Tympanuchus cupido*) chiefly of the prairies of the central U.S.; *also* **:** a closely related smaller grouse (*T. pallidicinctus*)

prairie chicken

prairie dog *noun* (1774)
: any of a genus (*Cynomys*) of gregarious burrowing rodents of the squirrel family chiefly of central and western U.S. plains; *especially* **:** a black-tailed rodent (*C. ludovicianus*) that usually lives in extensive colonial burrows

prairie dog

prairie schooner *noun* (1841)
: a covered wagon used by pioneers in cross-country travel — called also *prairie wagon*

prairie soil *noun* (1817)
: any of a zonal group of soils developed in a temperate relatively humid climate under tall grass

prairie wolf *noun* (1804)
: COYOTE

[1]**praise** \ˈprāz\ *verb* **praised; prais·ing** [Middle English, from Old French *preisier* to prize, praise, from Late Latin *pretiare* to prize, from Latin *pretium* price — more at PRICE] (13th century)
transitive verb
1 : to express a favorable judgment of **:** COMMEND
2 : to glorify (a god or saint) especially by the attribution of perfections
intransitive verb
: to express praise
— **prais·er** *noun*

[2]**praise** *noun* (14th century)
1 a : an expression of approval **:** COMMENDATION **b :** WORSHIP
2 a : VALUE, MERIT **b** *archaic* **:** one that is praised

praise·wor·thy \ˈprāz-ˌwər-t͟hē\ *adjective* (15th century)
: LAUDABLE
— **praise·wor·thi·ly** \-t͟hə-lē\ *adverb*
— **praise·wor·thi·ness** \-t͟hē-nəs\ *noun*

Pra·krit \ˈprä-ˌkrit, -krət\ *noun* [Sanskrit *prākṛta,* from *prākṛta* natural, vulgar] (1766)
: any or all of the ancient Indo-Aryan languages or dialects other than Sanskrit — see INDO-EUROPEAN LANGUAGES table

pra·line \ˈpra-ˌlēn, ˈprā-, ˈprō-\ *noun* [French, from Count Plessis-*Praslin* (died 1675) French soldier] (1723)
: a confection of nuts and sugar: **a :** almonds cooked in boiling sugar until brown and crisp **b :** a patty of creamy brown sugar and pecan meats

prall·tril·ler \ˈpräl-ˌtri-lər\ *noun* [German, from *prallen* to rebound + *Triller* trill] (circa 1841)
: a musical ornament made by a quick alternation of a principal tone with the tone above

[1]**pram** \ˈpräm, ˈpram\ *noun* [Middle Dutch *praem* & Middle Low German *prām*] (1548)
: a small lightweight nearly flat-bottomed boat with a broad transom and usually squared-off bow

[2]**pram** \ˈpram\ *noun* [by shortening & alteration from *perambulator*] (1884)
chiefly British **:** BABY CARRIAGE

[1]**prance** \ˈpran(t)s\ *verb* **pranced; pranc·ing** [Middle English *prauncen*] (14th century)
intransitive verb
1 : to spring from the hind legs or move by so doing
2 : to ride on a prancing horse
3 : to walk or move in a spirited manner **:** STRUT; *also* **:** to dance about
transitive verb
: to cause (a horse) to prance
— **pranc·er** \ˈpran(t)-sər\ *noun*

[2]**prance** *noun* (1751)
: an act or instance of prancing; *specifically* **:** a prancing movement

pran·di·al \ˈpran-dē-əl\ *adjective* [Latin *prandium* late breakfast, luncheon] (1820)
: of or relating to a meal

prang \ˈpraŋ\ *transitive verb* [origin unknown] (1941)
chiefly British **:** to have an accident with **:** cause to crash
— **prang** *noun, chiefly British*

[1]**prank** \ˈpraŋk\ *noun* [obsolete *prank* to play tricks] (circa 1529)
: TRICK: **a** *obsolete* **:** a malicious act **b :** a mildly mischievous act **c :** a ludicrous act

[2]**prank** *verb* [probably from Dutch *pronken* to strut; akin to Middle High German ge*brunkel* glitter of metal] (15th century)
intransitive verb
: to show oneself off
transitive verb
: to dress or adorn gaily or showily

prank·ish \ˈpraŋ-kish\ *adjective* (1827)
1 : full of pranks ⟨a *prankish* child⟩
2 : having the nature of a prank ⟨*prankish* acts⟩
— **prank·ish·ly** *adverb*
— **prank·ish·ness** *noun*

prank·ster \ˈpraŋ(k)-stər\ *noun* (1927)
: a person who plays pranks

prase \ˈprāz, ˈprās\ *noun* [French, from Latin *prasius,* from Greek *prasios,* from *prasios,* adjective, of the color of a leek from *prason* leek; akin to Latin *porrum* leek] (1788)
: a chalcedony that is translucent and yellowish green

pra·seo·dym·i·um \ˌprā-zē-ō-ˈdi-mē-əm, ˌprā-sē-\ *noun* [New Latin, alteration of *praseodidymium,* irregular from Greek *prasios,* adjective + New Latin *didymium* didymium] (1885)
: a yellowish white trivalent metallic element of the rare-earth group used chiefly in the form of its salts in coloring glass greenish yellow — see ELEMENT table

prate \ˈprāt\ *intransitive verb* **prat·ed; prat·ing** [Middle English, from Middle Dutch; akin to Middle Low German *pratten* to pout] (15th century)
: to talk long and idly **:** CHATTER
— **prate** *noun*
— **prat·er** *noun*
— **prat·ing·ly** \ˈprā-tiŋ-lē\ *adverb*

prat·fall \ˈprat-ˌfȯl\ *noun* [*prat* buttocks + *fall*] (1938)
1 : a fall on the buttocks
2 : a humiliating mishap or blunder

pra·tin·cole \ˈpra-tᵊn-ˌkōl, ˈprā-, -tiŋ-\ *noun* [ultimately from Latin *pratum* meadow + *incola* inhabitant, from *in-* + *colere* to cultivate — more at WHEEL] (1773)
: any of several Old World shore-inhabiting birds (genera *Glareolus* and *Stiltia* of the family Glareolidae) with a short bill and a forked tail

pra·tique \pra-ˈtēk\ *noun* [French, literally, practice — more at PRACTICE] (1609)
: clearance given an incoming ship by the health authority of a port

[1]**prat·tle** \ˈpra-tᵊl\ *verb* **prat·tled; prat·tling** \ˈprat-liŋ, ˈpra-tᵊl-iŋ\ [Low German *pratelen;* akin to Middle Dutch *praten* to prate] (1532)
intransitive verb
1 : PRATE
2 : to utter or make meaningless sounds suggestive of the chatter of children **:** BABBLE
transitive verb
: to say in an unaffected or childish manner
— prat·tler \ˈprat-lər, ˈpra-tᵊl-ər\ *noun*
— prat·tling·ly \-liŋ-lē, -tᵊl-iŋ-\ *adverb*

[2]**prattle** *noun* (1555)
1 : trifling or empty talk
2 : a sound that is meaningless, repetitive, and suggestive of the chatter of children

prau \ˈprau̇, ˈprä-ˌü\ *noun* [Malay *pĕrahu*] (1582)
: any of various Indonesian boats usually without a deck that are propelled especially by sails or paddles

[1]**prawn** \ˈpro̊n, ˈprän\ *noun* [Middle English *prane*] (15th century)
: any of numerous widely distributed edible decapod crustaceans (as of the genera *Pandalus* and *Peneus*) that resemble shrimps and have large compressed abdomens; *also* **:** SHRIMP

[2]**prawn** *intransitive verb* (1886)
: to fish for or with prawns
— prawn·er *noun*

prax·e·ol·o·gy \ˌprak-sē-ˈä-lə-jē\ *noun* [alteration of earlier *praxiology,* from *praxis* + *-o-* + *-logy*] (1904)
: the study of human action and conduct
— prax·e·o·log·i·cal \-sē-ə-ˈlä-ji-kəl\ *adjective*

prax·is \ˈprak-səs\ *noun, plural* **prax·es** \-ˌsēz\ [Medieval Latin, from Greek, doing, action, from *prassein* to do, practice — more at PRACTICAL] (1581)
: ACTION, PRACTICE: as **a :** exercise or practice of an art, science, or skill **b :** customary practice or conduct

pray \ˈprā\ *verb* [Middle English, from Old French *preier,* from Latin *precari,* from *prec-, prex* request, prayer; akin to Old High German *frāgēn* to ask, Sanskrit *pṛcchati* he asks] (13th century)
transitive verb
1 : ENTREAT, IMPLORE — often used as a function word in introducing a question, request, or plea ⟨*pray* be careful⟩
2 : to get or bring by praying
intransitive verb
1 : to make a request in a humble manner
2 : to address God or a god with adoration, confession, supplication, or thanksgiving

[1]**prayer** \ˈprar, ˈprer\ *noun, often attributive* [Middle English, from Middle French *preiere,* from Medieval Latin *precaria,* from Latin, feminine of *precarius* obtained by entreaty, from *prec-, prex*] (14th century)
1 a (1) **:** an address (as a petition) to God or a god in word or thought ⟨said a *prayer* for the success of the voyage⟩ (2) **:** a set order of words used in praying **b :** an earnest request or wish
2 : the act or practice of praying to God or a god ⟨kneeling in *prayer*⟩
3 : a religious service consisting chiefly of prayers — often used in plural
4 : something prayed for
5 : a slight chance ⟨haven't got a *prayer*⟩

[2]**pray·er** \ˈprā-ər, ˈpre(-ə)r\ *noun* [Middle English *prayere,* from *prayen* to pray + [2]*-er*] (14th century)
: one that prays **:** SUPPLIANT

prayer beads \ˈprar-, ˈprer-\ *noun plural* (1630)
: a string of beads by which prayers are counted; *specifically* **:** ROSARY

prayer book *noun* (circa 1597)
: a book containing prayers and often other forms and directions for worship

prayer·ful \ˈprar-fəl, ˈprer-\ *adjective* (1626)
1 : DEVOUT
2 : EARNEST, SINCERE
— prayer·ful·ly \-fə-lē\ *adverb*
— prayer·ful·ness *noun*

prayer meeting *noun* (1780)
: a usually informal gathering for worship and prayer; *especially* **:** a Protestant worship service usually held on a week night — called also *prayer service*

prayer rug *noun* (circa 1890)
: a small Oriental rug used by Muslims to kneel on when praying

prayer shawl *noun* (1905)
: TALLITH

prayer wheel *noun* (1814)
: a cylinder of wood or metal that revolves on an axis and contains written prayers and that is used in praying by Tibetan Buddhists

praying mantis *noun* (circa 1890)
: MANTIS; *especially* **:** a European mantis (*Mantis religiosa*) that has been introduced into the U.S. — called also *praying mantid*

pre- *prefix* [Middle English, from Old French & Latin; Old French, from Latin *prae-,* from *prae* in front of, before — more at FOR]
1 a (1) **:** earlier than **:** prior to **:** before ⟨*Pre*cambrian⟩ ⟨*pre*historic⟩ (2) **:** preparatory or prerequisite to ⟨*pre*medical⟩ **b :** in advance **:** beforehand ⟨*pre*cancel⟩ ⟨*pre*pay⟩
2 : in front of **:** anterior to ⟨*pre*axial⟩ ⟨*pre*molar⟩

pre·ad·mis·sion
pre·adult
pre·ag·ri·cul·tur·al
pre·an·es·thet·ic
pre·an·nounce
pre·ap·prove
pre·ar·range
pre·ar·range·ment
pre·as·sem·bled
pre·as·sign
pre·bake
pre·bat·tle
pre·bib·li·cal
pre·book
pre·break·fast
pre·cap·i·tal·ist
pre·chill
pre·Christ·mas
pre·clear
pre·clear·ance
pre·code
pre·co·i·tal
pre·col·lege
pre·col·le·giate
pre·co·lo·nial
pre·com·bus·tion
pre·com·mit·ment
pre·com·pute
pre·com·put·er
pre·con·cert
pre·con·quest
pre·con·so·nan·tal
pre·con·struct·ed
pre·con·ven·tion
pre·con·vic·tion
pre·cool
pre·cop·u·la·to·ry
pre·crash
pre·crease
pre·cri·sis
pre·cut
pre·dawn
pre·de·fine
pre·de·liv·ery
pre·de·par·ture
pre·des·ig·nate
pre·de·val·u·a·tion
pre·de·vel·op·ment
pre·din·ner
pre·dis·charge
pre·dis·cov·ery
pre·dive
pre·drill
pre·dy·nas·tic
pre·elec·tion
pre·elec·tric
pre·em·bar·go
pre·em·ploy·ment
pre·en·roll·ment
pre·erect
pre·es·tab·lish
pre·eth·i·cal
pre·ex·per·i·ment
pre·fade
pre·fas·cist
pre·feu·dal
pre·fight
pre·file
pre·filled
pre·fi·nance
pre·fire
pre·flame
pre·flight
pre·for·mat
pre·for·mu·late
pre·fresh·man
pre·fro·zen
pre·game
pre·gen·i·tal
pre·har·vest
pre·head·ache
pre·hir·ing
pre·hol·i·day
pre·hu·man
pre·in·au·gu·ral
pre·in·cor·po·ra·tion
pre·in·duc·tion
pre·in·dus·tri·al
pre·in·ter·view
pre·in·va·sion
pre·kin·der·gar·ten
pre·launch
pre·life
pre·lit·er·ary
pre·log·i·cal
pre·lunch
pre·lun·cheon
pre·made
pre·man·u·fac·ture
pre·mar·i·tal
pre·mar·i·tal·ly
pre·mar·ket
pre·mar·ket·ing
pre·mar·riage
pre·meal
pre·mea·sure
pre·me·di·e·val
pre·meet
pre·mei·ot·ic
pre·men·o·paus·al
pre·merg·er
pre·mi·gra·tion
pre·mod·ern
pre·mod·i·fi·ca·tion
pre·mod·i·fy
pre·moist·en
pre·mold
pre·molt
pre·mor·al
pre·my·cot·ic
pre·noon
pre·no·ti·fi·ca·tion
pre·no·ti·fy
pre·num·ber
pre·open·ing
pre·op·er·a·tion·al
pre·or·der
pre·paste
pre·per·for·mance
pre·pill
pre·plan
pre·por·tion
pre·pre·pared
pre·pres·i·den·tial
pre·price
pre·pri·ma·ry
pre·pro·duc·tion
pre·pro·gram
pre·psy·che·del·ic
pre·pub·li·ca·tion
pre·punch
pre·pu·pal
pre·pur·chase
pre·qual·i·fi·ca·tion
pre·qual·i·fy
pre·race
pre·re·ces·sion
pre·re·cord·ed
pre·re·hears·al
pre·re·lease
pre·re·quire
pre·re·tire·ment
pre·re·turn
pre·re·view
pre·re·vi·sion·ist
pre·rev·o·lu·tion
pre·rev·o·lu·tion·ary
pre·rinse
pre·ri·ot
pre·rock
pre·ro·man·tic
pre·sale
pre·sched·ule
pre·screen
pre·sea·son
pre·sen·tence
pre·sen·tenc·ing
pre·ser·vice
pre·set
pre·set·tle·ment
pre·show
pre·slaugh·ter
pre·sleep
pre·slice
pre·song
pre·spec·i·fy
pre·split
pre·stamp
pre·ster·il·ize
pre·stor·age
pre·strike
pre·struc·ture
pre·sum·mit
pre·sur·gery
pre·sweet·en
pre·symp·tom·at·ic
pre·tape
pre·tax
pre·tech·no·log·i·cal
pre·tele·vi·sion
pre·ter·mi·na·tion
pre·the·a·ter
pre·tour·na·ment
pre·train
pre·trav·el
pre·treat
pre·treat·ment
pre·trial
pre·trimmed
pre·type
pre·uni·fi·ca·tion
pre·uni·ver·si·ty
pre·vi·a·ble
pre·war
pre·warn
pre·wash
pre·wean·ing
pre·work
pre·wrap

preach \ˈprēch\ *verb* [Middle English *prechen,* from Old French *prechier,* from Late Latin *praedicare,* from Latin, to proclaim publicly, from *prae-* pre- + *dicare* to proclaim — more at DICTION] (13th century)
intransitive verb
1 : to deliver a sermon
2 : to urge acceptance or abandonment of an idea or course of action; *specifically* **:** to exhort in an officious or tiresome manner
transitive verb
1 : to set forth in a sermon ⟨*preach* the gospel⟩
2 : to advocate earnestly ⟨*preached* revolution⟩
3 : to deliver (as a sermon) publicly
4 : to bring, put, or affect by preaching ⟨*preached* the . . . church out of debt —*American Guide Series: Virginia*⟩
— preach·er *noun*
— preach·ing·ly \ˈprē-chiŋ-lē\ *adverb*

preach·ify \ˈprē-chə-ˌfī\ *intransitive verb* **-ified; -ify·ing** (1775)
: to preach ineptly or tediously

preach·ment \ˈprēch-mənt\ *noun* (14th century)
1 : the act or practice of preaching
2 : SERMON, EXHORTATION; *specifically* **:** a tedious or unwelcome one

preachy \ˈprē-chē\ *adjective* **preach·i·er; -est** (1819)

: marked by obvious moral exhortation : DIDACTIC
— **preach·i·ly** \-chə-lē\ *adverb*
— **preach·i·ness** \-chē-nəs\ *noun*

pre·ad·ap·ta·tion \ˌprē-ˌa-ˌdap-ˈtā-shən\ *noun* (1886)
1 : the possession by an organism or taxonomic group of characters that are not adapted to the ancestral environment but favor survival in some other environment
2 : a preadaptive character

pre·adapt·ed \ˌprē-ə-ˈdap-təd\ *adjective* (1915)
: characterized by preadaptation

pre·adap·tive \-ˈdap-tiv\ *adjective* (1915)
: of, relating to, or characterized by preadaptation

pre·ad·o·les·cence \ˌprē-ˌa-dᵊl-ˈe-sᵊn(t)s\ *noun* (1930)
: the period of human development just preceding adolescence; *specifically* **:** the period between the approximate ages of 9 and 12
— **pre·ad·o·les·cent** \-sᵊnt\ *adjective or noun*

pre·am·ble \ˈprē-ˌam-bəl, prē-ˈ\ *noun* [Middle English, from Middle French *preambule,* from Medieval Latin *preambulum,* from Late Latin, neuter of *praeambulus* walking in front of, from Latin *prae-* + *ambulare* to walk] (14th century)
1 : an introductory statement; *especially* **:** the introductory part of a constitution or statute that usually states the reasons for and intent of the law
2 : an introductory fact or circumstance; *especially* **:** one indicating what is to follow

pre·amp \ˈprē-ˌamp\ *noun* (1949)
: PREAMPLIFIER

pre·am·pli·fi·er \(ˌ)prē-ˈam-plə-ˌfī(-ə)r\ *noun* (1935)
: an amplifier designed to amplify extremely weak electrical signals before they are fed to additional amplifier circuits

pre·atom·ic \ˌprē-ə-ˈtä-mik\ *adjective* (1914)
: of or relating to a time before the use of the atomic bomb and atomic energy

pre·ax·i·al \(ˌ)prē-ˈak-sē-əl\ *adjective* (1872)
: situated in front of an axis of the body

preb·end \ˈpre-bənd\ *noun* [Middle English *prebende,* from Middle French, from Medieval Latin *praebenda,* from Late Latin, subsistence allowance granted by the state, from Latin, feminine of *praebendus,* gerundive of *praebēre* to offer, from *prae-* + *habēre* to hold — more at GIVE] (15th century)
1 : a stipend furnished by a cathedral or collegiate church to a clergyman (as a canon) in its chapter
2 : PREBENDARY
— **pre·ben·dal** \pri-ˈben-dᵊl, ˈpre-bən-\ *adjective*

preb·en·dary \ˈpre-bən-ˌder-ē\ *noun, plural* **-dar·ies** (15th century)
1 : a clergyman receiving a prebend for officiating and serving in the church
2 : an honorary canon in a cathedral chapter

pre·bi·o·log·i·cal \ˌprē-ˌbī-ə-ˈlä-ji-kəl\ *also* **pre·bi·o·log·ic** \-jik\ *adjective* (1953)
: of, relating to, or being chemical or environmental precursors of the origin of life ⟨*prebiological* molecules⟩

pre·bi·ot·ic \ˌprē-bī-ˈä-tik\ *adjective* (1958)
: PREBIOLOGICAL

pre·cal·cu·lus \(ˌ)prē-ˈkal-kyə-ləs\ *adjective* (1964)
: relating to or being mathematical prerequisites for the study of calculus
— **precalculus** *noun*

Pre·cam·bri·an \(ˌ)prē-ˈkam-brē-ən, -ˈkām-\ *adjective* (1864)
: of, relating to, or being the earliest era of geological history or the corresponding system of rocks characterized especially by the appearance of single-celled organisms and is equivalent to the Archean and Proterozoic eons — see GEOLOGIC TIME table
— **Precambrian** *noun*

¹pre·can·cel \(ˌ)prē-ˈkan(t)-səl\ *transitive verb* (1921)
: to cancel (a postage stamp) in advance of use
— **pre·can·cel·la·tion** \ˌprē-ˌkan(t)-sə-ˈlā-shən\ *noun*

²precancel *noun* (1929)
: a precanceled postage stamp

pre·can·cer·ous \(ˌ)prē-ˈkan(t)s-rəs, -ˈkan(t)-sə-\ *adjective* [International Scientific Vocabulary] (1882)
: tending to become cancerous ⟨a *precancerous* lesion⟩

pre·car·i·ous \pri-ˈkar-ē-əs, -ˈker-\ *adjective* [Latin *precarius* obtained by entreaty, uncertain — more at PRAYER] (1646)
1 : depending on the will or pleasure of another
2 : dependent on uncertain premises **:** DUBIOUS ⟨*precarious* generalizations⟩
3 a : dependent on chance circumstances, unknown conditions, or uncertain developments **b :** characterized by a lack of security or stability that threatens with danger
synonym see DANGEROUS
— **pre·car·i·ous·ly** *adverb*
— **pre·car·i·ous·ness** *noun*

pre·cast \ˌprē-ˈkast, ˈprē-ˌ\ *adjective* (1914)
: being concrete that is cast in the form of a structural element (as a panel or beam) before being placed in final position

prec·a·to·ry \ˈpre-kə-ˌtōr-ē, -ˌtȯr-\ *adjective* [Late Latin *precatorius,* from Latin *precari* to pray — more at PRAY] (1636)
: expressing a wish

pre·cau·tion \pri-ˈkȯ-shən\ *noun* [French *précaution,* from Late Latin *praecaution-, praecautio,* from Latin *praecavēre* to guard against, from *prae-* + *cavēre* to be on one's guard — more at HEAR] (1603)
1 : care taken in advance **:** FORESIGHT ⟨warned of the need for *precaution*⟩
2 : a measure taken beforehand to prevent harm or secure good **:** SAFEGUARD
— **pre·cau·tion·ary** \-shə-ˌner-ē\ *adjective*

pre·cede \pri-ˈsēd\ *verb* **pre·ced·ed; pre·ced·ing** [Middle English, from Middle French *preceder,* from Latin *praecedere,* from *prae-* pre- + *cedere* to go] (15th century)
transitive verb
1 : to surpass in rank, dignity, or importance
2 : to be, go, or come ahead or in front of
3 : to be earlier than
4 : to cause to be preceded **:** PREFACE
intransitive verb
: to go or come before

pre·ce·dence \ˈpre-sə-dən(t)s, pri-ˈsē-dᵊn(t)s\ *noun* (1588)
1 a *obsolete* **:** ANTECEDENT **b :** the fact of preceding in time
2 a : the right to superior honor on a ceremonial or formal occasion **b :** the order of ceremonial or formal preference **c :** priority of importance ⟨your safety takes *precedence*⟩

pre·ce·den·cy \-dən(t)-sē, -dᵊn(t)-sē\ *noun* (1612)
: PRECEDENCE

¹pre·ce·dent \pri-ˈsē-dᵊnt, ˈpre-sə-dənt\ *adjective* [Middle English, from Middle French, from Latin *praecedent-, praecedens,* present participle of *praecedere*] (15th century)
: prior in time, order, arrangement, or significance

²prec·e·dent \ˈpre-sə-dənt\ *noun* (15th century)
1 : an earlier occurrence of something similar
2 a : something done or said that may serve as an example or rule to authorize or justify a subsequent act of the same or an analogous kind ⟨a verdict that had no *precedent*⟩ **b :** the convention established by such a precedent or by long practice
3 : a person or thing that serves as a model

pre·ced·ing \pri-ˈsē-diŋ\ *adjective* (15th century)
: that immediately precedes in time or place ⟨the *preceding* day⟩ ⟨*preceding* paragraphs⟩ ☆

pre·cen·sor \(ˌ)prē-ˈsen(t)-sər\ *transitive verb* (1942)
: to censor (a publication or film) before its release to the public

pre·cen·tor \pri-ˈsen-tər\ *noun* [Late Latin *praecentor,* from Latin *praecinere* to sing before, from *prae-* + *canere* to sing — more at CHANT] (1613)
: a leader of the singing of a choir or congregation
— **pre·cen·to·ri·al** \ˌprē-ˌsen-ˈtōr-ē-əl, -ˈtȯr-\ *adjective*
— **pre·cen·tor·ship** \pri-ˈsen-tər-ˌship\ *noun*

pre·cept \ˈprē-ˌsept\ *noun* [Middle English, from Latin *praeceptum,* from neuter of *praeceptus,* past participle of *praecipere* to take beforehand, instruct, from *prae-* + *capere* to take — more at HEAVE] (14th century)
1 : a command or principle intended especially as a general rule of action
2 : an order issued by legally constituted authority to a subordinate official
synonym see LAW

pre·cep·tive \pri-ˈsep-tiv\ *adjective* (15th century)
: giving precepts **:** DIDACTIC

pre·cep·tor \pri-ˈsep-tər, ˈprē-ˌ\ *noun* (15th century)
1 a : TEACHER, TUTOR **b :** the headmaster or principal of a school
2 : the head of a preceptory of Knights Templars
— **pre·cep·tor·ship** \-ˌship\ *noun*

¹pre·cep·to·ri·al \pri-ˌsep-ˈtōr-ē-əl, ˌprē-, -ˈtȯr-\ *adjective* (circa 1741)
: of, relating to, or making use of preceptors

²preceptorial *noun* (circa 1952)
: a college course that emphasizes independent reading, discussion in small groups, and individual conferences with the teacher

pre·cep·to·ry \pri-ˈsep-t(ə-)rē, ˈprē-ˌ\ *noun, plural* **-ries** (1540)
1 : a subordinate house or community of the Knights Templars; *broadly* **:** COMMANDERY 1
2 : COMMANDERY 2

pre·cess \prē-ˈses, ˈprē-ˌ\ *verb* [back-formation from *precession*] (1892)
intransitive verb
: to progress with a movement of precession
transitive verb
: to cause to precess

☆ SYNONYMS
Preceding, antecedent, foregoing, previous, prior, former, anterior mean being before. PRECEDING usually implies being immediately before in time or in place ⟨the *preceding* sentence⟩. ANTECEDENT applies to order in time and may suggest a causal relation ⟨conditions *antecedent* to the revolution⟩. FOREGOING applies chiefly to statements ⟨the *foregoing* remarks⟩. PREVIOUS and PRIOR imply existing or occurring earlier, but PRIOR often adds an implication of greater importance ⟨a child from a *previous* marriage⟩ ⟨a *prior* obligation⟩. FORMER implies always a definite comparison or contrast with something that is latter ⟨the *former* name of the company⟩. ANTERIOR applies to position before or ahead of usually in space, sometimes in time or order ⟨the *anterior* lobe of the brain⟩.

pre·ces·sion \prē-ˈse-shən\ *noun* [New Latin *praecession-, praecessio,* from Medieval Latin, act of preceding, from Latin *praecedere* to precede] (1879)
: a comparatively slow gyration of the rotation axis of a spinning body about another line intersecting it so as to describe a cone
— **pre·ces·sion·al** \-ˈsesh-nəl, -ˈse-shə-nᵊl\ *adjective*

precession of the equinoxes (1621)
: a slow westward motion of the equinoxes along the ecliptic caused by the gravitational action of sun and moon upon the protuberant matter about the earth's equator

pre–Chel·le·an \(ˌ)prē-ˈshe-lē-ən\ *adjective* (1916)
: of or relating to a Lower Paleolithic culture preceding the Abbevillian and characterized by crudely flaked stone hand axes

pre–Chris·tian \-ˈkris-chən, -ˈkrish-\ *adjective* (1828)
: of, relating to, or being a time before the beginning of the Christian era

pré·cieux \prā-syœ\ *or* **pré·cieuse** \-syœz\ *adjective* [French *précieux,* masculine, & *précieuse,* feminine, literally, precious, from Middle French *precios*] (1727)
: PRECIOUS 3

pre·cinct \ˈprē-ˌsiŋ(k)t\ *noun* [Middle English, from Medieval Latin *praecinctum,* from Latin, neuter of *praecinctus,* past participle of *praecingere* to gird about, from *prae-* pre- + *cingere* to gird — more at CINCTURE] (15th century)
1 : a part of a territory with definite bounds or functions often established for administrative purposes **:** DISTRICT: as **a :** a subdivision of a county, town, city, or ward for election purposes **b :** a division of a city for police control
2 a : an enclosure bounded by the walls of a building — often used in plural **b :** a sphere of thought, action, or influence — often used in plural
3 a *plural* **:** the region immediately surrounding a place **:** ENVIRONS **b :** PLACE, LOCALE
4 : BOUNDARY — often used in plural ⟨a ruined tower within the *precincts* of the squire's grounds —T. L. Peacock⟩

pre·ci·os·i·ty \ˌpre-shē-ˈä-sə-tē, -sē-\ *noun, plural* **-ties** (1866)
1 : fastidious refinement
2 : an instance of preciosity

1pre·cious \ˈpre-shəs\ *adjective* [Middle English, from Middle French *precios,* from Latin *pretiosus,* from *pretium* price — more at PRICE] (13th century)
1 : of great value or high price
2 : highly esteemed or cherished
3 : excessively refined **:** AFFECTED
4 : GREAT, THOROUGHGOING ⟨*precious* scoundrel⟩
— **pre·cious·ness** *noun*

2precious *adverb* (1595)
: VERY, EXTREMELY ⟨has *precious* little to say⟩

pre·cious·ly *adverb* (14th century)
1 : in a precious manner
2 : PRECIOUS

pre·ci·pe *variant of* PRAECIPE

prec·i·pice \ˈpre-s(ə-)pəs\ *noun* [Middle French, from Latin *praecipitium,* from *praecipit-, praeceps* headlong, from *prae-* + *caput* head — more at HEAD] (1613)
1 : a very steep or overhanging place
2 : a hazardous situation; *broadly* **:** BRINK

pre·cip·i·ta·ble \pri-ˈsi-pə-tə-bəl\ *adjective* (1670)
: capable of being precipitated

pre·cip·i·tance \pri-ˈsi-pə-tən(t)s\ *noun* (1667)
: PRECIPITANCY

pre·cip·i·tan·cy \-tən(t)-sē\ *noun* (1646)
: undue hastiness or suddenness

1pre·cip·i·tant \-tənt\ *adjective* (1671)
: PRECIPITATE
— **pre·cip·i·tant·ly** *adverb*
— **pre·cip·i·tant·ness** *noun*

2precipitant *noun* (circa 1685)
: a precipitating agent; *especially* **:** one that causes the formation of a precipitate

1pre·cip·i·tate \pri-ˈsi-pə-ˌtāt\ *verb* **-tat·ed; -tat·ing** [Latin *praecipitatus,* past participle of *praecipitare,* from *praecipit-, praeceps*] (1528)
transitive verb
1 a : to throw violently **:** HURL ⟨the quandaries into which the release of nuclear energy has *precipitated* mankind —A. B. Arons⟩ **b :** to throw down
2 : to bring about especially abruptly ⟨*precipitate* a scandal that would end with his expulsion —John Cheever⟩
3 a : to cause to separate from solution or suspension **b :** to cause (vapor) to condense and fall or deposit
intransitive verb
1 a : to fall headlong **b :** to fall or come suddenly into some condition
2 : to move or act precipitately
3 a : to separate from solution or suspension **b :** to condense from a vapor and fall as rain or snow
— **pre·cip·i·ta·tive** \-ˌtā-tiv\ *adjective*
— **pre·cip·i·ta·tor** \-ˌtā-tər\ *noun*

2pre·cip·i·tate \pri-ˈsi-pə-tət, -ˌtāt\ *noun* [New Latin *praecipitatum,* from Latin, neuter of *praecipitatus*] (1594)
1 : a substance separated from a solution or suspension by chemical or physical change usually as an insoluble amorphous or crystalline solid
2 : a product, result, or outcome of some process or action

3pre·cip·i·tate \pri-ˈsi-pə-tət\ *adjective* (1615)
1 a : falling, flowing, or rushing with steep descent **b :** PRECIPITOUS, STEEP
2 : exhibiting violent or unwise speed ☆
— **pre·cip·i·tate·ly** *adverb*
— **pre·cip·i·tate·ness** *noun*

pre·cip·i·ta·tion \pri-ˌsi-pə-ˈtā-shən\ *noun* (1502)
1 : the quality or state of being precipitate **:** HASTINESS
2 : an act, process, or instance of precipating; *especially* **:** the process of forming a precipitate
3 : something precipitated: as **a :** a deposit on the earth of hail, mist, rain, sleet, or snow; *also* **:** the quantity of water deposited **b :** PRECIPITATE 1

pre·cip·i·tin \pri-ˈsi-pə-tən\ *noun* [International Scientific Vocabulary, from *precipitate*] (1900)
: an antibody that forms a precipitate when it unites with its antigen

pre·cip·i·tin·o·gen \pri-ˌsi-pə-ˈti-nə-jən\ *noun* (1904)
: an antigen that stimulates the production of a specific precipitin

pre·cip·i·tous \pri-ˈsi-pə-təs\ *adjective* [French *précipiteux,* from Middle French, from Latin *precipitium* precipice] (1613)
1 : PRECIPITATE 2
2 a : very steep, perpendicular, or overhanging in rise or fall ⟨a *precipitous* slope⟩ **b :** having precipitous sides ⟨a *precipitous* gorge⟩ **c :** having a very steep ascent ⟨a *precipitous* street⟩
synonym see STEEP
— **pre·cip·i·tous·ly** *adverb*
— **pre·cip·i·tous·ness** *noun*

pré·cis \prā-ˈsē, ˈprā-(ˌ)sē\ *noun, plural* **pré·cis** \-ˈsēz, -(ˌ)sēz\ [French, from *précis* precise] (1760)
: a concise summary of essential points, statements, or facts

pre·cise \pri-ˈsīs\ *adjective* [Middle English, from Middle French *precis,* from Latin *praecisus,* past participle of *praecidere* to cut off, from *prae-* + *caedere* to cut] (15th century)
1 : exactly or sharply defined or stated
2 : minutely exact
3 : strictly conforming to a pattern, standard, or convention
4 : distinguished from every other ⟨at just that *precise* moment⟩
synonym see CORRECT
— **pre·cise·ness** *noun*

pre·cise·ly *adverb* (14th century)
: EXACTLY ⟨*precisely* two o'clock⟩ — sometimes used as an intensive ⟨was popular *precisely* because he was so kind⟩

pre·ci·sian \pri-ˈsi-zhən\ *noun* (1571)
1 : a person who stresses or practices scrupulous adherence to a strict standard especially of religious observance or morality
2 : PURITAN 1

1pre·ci·sion \pri-ˈsi-zhən\ *noun* (1740)
1 : the quality or state of being precise **:** EXACTNESS
2 a : the degree of refinement with which an operation is performed or a measurement stated — compare ACCURACY 2b **b :** the accuracy (as in binary or decimal places) with which a number can be represented usually expressed in terms of the number of computer words available for representation ⟨double *precision* arithmetic permits the representation of an expression by two computer words⟩
3 : RELEVANCE 2
— **pre·ci·sion·ist** \-ˈsi-zhə-nist, -ˈsizh-nist\ *noun*

2precision *adjective* (1875)
1 : adapted for extremely accurate measurement or operation
2 : held to low tolerance in manufacture
3 : marked by precision of execution

pre·clin·i·cal \(ˌ)prē-ˈkli-ni-kəl\ *adjective* (1926)
1 : of, relating to, or concerned with the period preceding clinical manifestations
2 : of, relating to, or being the period in medical or dental education preceding the clinical study of medicine or dentistry

pre·clude \pri-ˈklüd\ *transitive verb* **pre·clud·ed; pre·clud·ing** [Latin *praecludere,* from *prae-* + *claudere* to close — more at CLOSE] (1629)
1 *archaic* **:** CLOSE
2 : to make impossible by necessary consequence **:** rule out in advance
— **pre·clu·sion** \-ˈklü-zhən\ *noun*
— **pre·clu·sive** \-ˈklü-siv, -ziv\ *adjective*
— **pre·clu·sive·ly** *adverb*

pre·co·cial \pri-ˈkō-shəl\ *adjective* [New Latin *praecoces* precocial birds, from Latin, plural of *praecoc-, precox*] (circa 1872)
: capable of a high degree of independent activity from birth ⟨ducklings are *precocial*⟩ — compare ALTRICIAL

pre·co·cious \pri-ˈkō-shəs\ *adjective* [Latin *praecoc-, praecox* early ripening, precocious, from *prae-* + *coquere* to cook — more at COOK] (1650)
1 : exceptionally early in development or occurrence
2 : exhibiting mature qualities at an unusually early age
— **pre·co·cious·ly** *adverb*
— **pre·co·cious·ness** *noun*

☆ SYNONYMS
Precipitate, headlong, abrupt, impetuous, sudden mean showing undue haste or unexpectedness. PRECIPITATE stresses lack of due deliberation and implies prematureness of action ⟨the army's *precipitate* withdrawal⟩. HEADLONG stresses rashness and lack of forethought ⟨a *headlong* flight from arrest⟩. ABRUPT stresses curtness and a lack of warning or ceremony ⟨an *abrupt* refusal⟩. IMPETUOUS stresses extreme impatience or impulsiveness ⟨an *impetuous* lover proposing marriage⟩. SUDDEN stresses unexpectedness and sharpness or violence of action ⟨flew into a *sudden* rage⟩.

— **pre·coc·i·ty** \pri-ˈkä-sə-tē\ *noun*
pre·cog·ni·tion \ˌprē-(ˌ)käg-ˈni-shən\ *noun* [Late Latin *praecognition-, praecognitio,* from Latin *praecognoscere* to know beforehand, from *prae-* + *cognoscere* to know — more at COGNITION] (circa 1611)
: clairvoyance relating to an event or state not yet experienced
— **pre·cog·ni·tive** \(ˌ)prē-ˈkäg-nə-tiv\ *adjective*
pre–Co·lum·bi·an \ˌprē-kə-ˈləm-bē-ən\ *adjective* (1888)
: preceding or belonging to the time before the arrival of Columbus in America
pre·con·ceive \ˌprē-kən-ˈsēv\ *transitive verb* (1558)
: to form (as an opinion) prior to actual knowledge or experience ⟨*preconceived* notions⟩
pre·con·cep·tion \-kən-ˈsep-shən\ *noun* (1625)
1 : a preconceived idea
2 : PREJUDICE
pre·con·cert \ˌprē-kən-ˈsərt\ *transitive verb* (1748)
: to settle by prior agreement
¹pre·con·di·tion \-kən-ˈdi-shən\ *noun* (1825)
: PREREQUISITE
²precondition *transitive verb* (1922)
: to put in a proper or desired condition or frame of mind especially in preparation
¹pre·con·scious \(ˌ)prē-ˈkän(t)-shəs\ *adjective* (1860)
: not present in consciousness but capable of being recalled without encountering any inner resistance or repression
— **pre·con·scious·ly** *adverb*
²preconscious *noun* (circa 1922)
: the preconscious part of the psyche especially in psychoanalysis
pre·con·tact \ˌprē-ˈkän-ˌtakt\ *adjective* (circa 1909)
: of or relating to the period before contact of an indigenous people with an outside culture
pre·cook \(ˌ)prē-ˈku̇k\ *transitive verb* (1926)
: to cook partially or entirely before final cooking or reheating
pre·crit·i·cal \-ˈkri-ti-kəl\ *adjective* (1881)
: prior to the development of critical capacity
pre·cur·sor \pri-ˈkər-sər, ˈprē-ˌ\ *noun* [Middle English *precursoure,* from Latin *praecursor,* from *praecurrere* to run before, from *prae-* pre- + *currere* to run — more at CURRENT] (15th century)
1 a : one that precedes and indicates the approach of another **b :** PREDECESSOR
2 : a substance, cell, or cellular component from which another substance, cell, or cellular component is formed
synonym see FORERUNNER
— **pre·cur·so·ry** \-ˈkərs-rē, -ˈkər-sə-\ *adjective*
pre·da·ceous *or* **pre·da·cious** \pri-ˈdā-shəs\ *adjective* [Latin *praedari* to prey upon (from *praeda* prey) + English *-aceous* or *-acious* (as in *rapacious*) — more at PREY] (1713)
1 : living by preying on other animals **:** PREDATORY
2 *usually predacious* **:** tending to devour or despoil **:** RAPACIOUS
— **pre·da·ceous·ness** *noun*
— **pre·dac·i·ty** \-ˈda-sə-tē\ *noun*
pre·date \(ˌ)prē-ˈdāt\ *transitive verb* (circa 1864)
: ANTEDATE
pre·da·tion \pri-ˈdā-shən\ *noun* [Middle English *predacion,* from Latin *praedation-, praedatio,* from *praedari*] (15th century)
1 : the act of preying or plundering **:** DEPREDATION
2 : a mode of life in which food is primarily obtained by the killing and consuming of animals
predation pressure *noun* (1942)
: the effects of predation on a natural community especially with respect to the survival of species preyed upon
pred·a·tor \ˈpre-də-tər, -ˌtȯr\ *noun* (1912)
1 : one that preys, destroys, or devours
2 : an animal that lives by predation
pred·a·to·ry \ˈpre-də-ˌtōr-ē, -ˌtȯr-\ *adjective* (1589)
1 a : of, relating to, or practicing plunder, pillage, or rapine **b :** inclined or intended to injure or exploit others for personal gain or profit ⟨*predatory* pricing practices⟩
2 : living by predation **:** PREDACEOUS; *also* **:** adapted to predation
pre·de·cease \ˌprē-di-ˈsēs\ *verb* **-ceased; -ceas·ing** (1593)
transitive verb
: to die before (another person)
intransitive verb
: to die first
— **predecease** *noun*
pre·de·ces·sor \ˈpre-də-ˌse-sər, ˈprē-; ˌpre-də-ˈ, ˌprē-\ *noun* [Middle English *predecessour,* from Middle French *predecesseur,* from Late Latin *praedecessor,* from Latin *prae-* pre- + *decessor* retiring governor, from *decedere* to depart, retire from office — more at DECEASE] (14th century)
1 : one that precedes; *especially* **:** a person who has previously occupied a position or office to which another has succeeded
2 *archaic* **:** ANCESTOR
pre·des·ti·nar·i·an \(ˌ)prē-ˌdes-tə-ˈner-ē-ən, ˌprē-des-\ *noun* [*predestination* + *-arian*] (1667)
: one who believes in predestination
— **predestinarian** *adjective*
— **pre·des·ti·nar·i·an·ism** \-ē-ə-ˌni-zəm\ *noun*
¹pre·des·ti·nate \prē-ˈdes-tə-nət, -ˌnāt\ *adjective* [Middle English, from Latin *praedestinatus,* past participle of *praedestinare*] (14th century)
: destined, fated, or determined beforehand
²pre·des·ti·nate \-ˌnāt\ *transitive verb* **-nated; -nat·ing** [Middle English, from Latin *praedestinatus,* past participle] (15th century)
1 : to foreordain to an earthly or eternal lot or destiny by divine decree
2 *archaic* **:** PREDETERMINE
pre·des·ti·na·tion \(ˌ)prē-ˌdes-tə-ˈnā-shən, ˌprē-des-\ *noun* (14th century)
1 : the act of predestinating **:** the state of being predestinated
2 : the doctrine that God in consequence of his foreknowledge of all events infallibly guides those who are destined for salvation
pre·des·ti·na·tor \prē-ˈdes-tə-ˌnā-tər\ *noun* (1579)
1 *archaic* **:** PREDESTINARIAN
2 : one that predestinates
pre·des·tine \(ˌ)prē-ˈdes-tən\ *transitive verb* [Middle English, from Middle French or Latin; Middle French *predestiner,* from Latin *praedestinare,* from *prae-* + *destinare* to determine — more at DESTINE] (14th century)
: to destine, decree, determine, appoint, or settle beforehand; *especially* **:** PREDESTINATE 1
pre·de·ter·mi·na·tion \ˌprē-di-ˌtər-mə-ˈnā-shən\ *noun* (1647)
1 : the act of predetermining **:** the state of being predetermined: as **a :** the ordaining of events beforehand **b :** a fixing or settling in advance
2 : a purpose formed beforehand
pre·de·ter·mine \-di-ˈtər-mən\ *transitive verb* [Late Latin *praedeterminare,* from Latin *prae-* + *determinare* to determine] (1625)
1 a : FOREORDAIN, PREDESTINE **b :** to determine beforehand
2 : to impose a direction or tendency on beforehand
pre·de·ter·min·er \-di-ˈtər-mə-nər\ *noun* (1959)
: a limiting noun modifier (as *both* or *all*) characterized by occurrence before the determiner in a noun phrase
pre·di·a·be·tes \ˌprē-ˌdī-ə-ˈbē-tēz, -təs\ *noun* (1935)
: an inapparent abnormal state that precedes the development of clinically evident diabetes
— **pre·di·a·bet·ic** \-ˈbe-tik\ *adjective or noun*
pre·di·al \ˈprē-dē-əl\ *adjective* [Middle English *prediall,* from Medieval Latin *praedialis,* from Latin *praedium* landed property, from *praed-, praes* bondsman, from *prae-* + *vad-, vas* surety — more at WED] (15th century)
: of or relating to land or its products
¹pred·i·ca·ble \ˈpre-di-kə-bəl\ *noun* [Medieval Latin *praedicabile,* from neuter of *praedicabilis*] (1551)
: something that may be predicated; *especially* **:** one of the five most general kinds of attribution in traditional logic that include genus, species, difference, property, and accident
²predicable *adjective* [Medieval Latin *praedicabilis,* from Late Latin *praedicare* to predicate] (circa 1598)
: capable of being asserted
pre·dic·a·ment \pri-ˈdi-kə-mənt, *1 is usually* ˈpre-di-kə-\ *noun* [Middle English, from Late Latin *praedicamentum,* from *praedicare*] (14th century)
1 : the character, status, or classification assigned by a predication; *specifically* **:** CATEGORY 1
2 : CONDITION, STATE; *especially* **:** a difficult, perplexing, or trying situation
¹pred·i·cate \ˈpre-di-kət\ *noun* [Middle English, from Late Latin *praedicatum,* from neuter of *praedicatus*] (15th century)
1 a : something that is affirmed or denied of the subject in a proposition in logic **b :** a term designating a property or relation
2 : the part of a sentence or clause that expresses what is said of the subject and that usually consists of a verb with or without objects, complements, or adverbial modifiers
— **pred·i·ca·tive** \-kə-tiv, -ˌkā-\ *adjective*
— **pred·i·ca·tive·ly** *adverb*
²pred·i·cate \ˈpre-də-ˌkāt\ *transitive verb* **-cat·ed; -cat·ing** [Late Latin *praedicatus,* past participle of *praedicare* to assert, predicate logically, preach, from Latin, to proclaim publicly, assert — more at PREACH] (1552)
1 a : AFFIRM, DECLARE **b** *archaic* **:** PREACH
2 a : to assert to be a quality, attribute, or property — used with following *of* ⟨*predicates* intelligence of humans⟩ **b :** to make (a term) the predicate in a proposition
3 : FOUND, BASE — usually used with *on* ⟨the theory is *predicated* on recent findings⟩
4 : IMPLY
³pred·i·cate \ˈpre-di-kət\ *adjective* (1887)
: completing the meaning of a copula ⟨*predicate* adjective⟩ ⟨*predicate* noun⟩
predicate calculus *noun* (1950)
: the branch of symbolic logic that uses symbols for quantifiers and for arguments and predicates of propositions as well as for unanalyzed propositions and logical connectives — called also *functional calculus;* compare PROPOSITIONAL CALCULUS
predicate nominative *noun* (1887)
: a noun or pronoun in the nominative or common case completing the meaning of a copula
pred·i·ca·tion \ˌpre-də-ˈkā-shən\ *noun* [Middle English *predicacion,* from Middle French *predication,* from Latin *praedication-, praedicatio,* from *praedicare*] (14th century)
1 *archaic* **a :** an act of proclaiming or preaching **b :** SERMON

2 : an act or instance of predicating: as **a :** the expression of action, state, or quality by a grammatical predicate **b :** the logical affirmation of something about another; *especially* **:** assignment of something to a class

pred·i·ca·to·ry \'pre-di-kə-ˌtōr-ē, -ˌtȯr-\ *adjective* [Late Latin *praedicatorius,* from *praedicare* to preach] (1611)
: of or relating to preaching

pre·dict \pri-'dikt\ *verb* [Latin *praedictus,* past participle of *praedicere,* from *prae-* pre- + *dicere* to say — more at DICTION] (circa 1632)
transitive verb
: to declare or indicate in advance; *especially* **:** foretell on the basis of observation, experience, or scientific reason
intransitive verb
: to make a prediction
synonym see FORETELL
— pre·dict·abil·i·ty \-ˌdik-tə-'bi-lə-tē\ *noun*
— pre·dict·able \-'dik-tə-bəl\ *adjective*
— pre·dic·tive \-'dik-tiv\ *adjective*
— pre·dic·tive·ly \-lē\ *adverb*
— pre·dic·tor \-'dik-tər\ *noun*

pre·dict·ably \-'dik-tə-blē\ *adverb* (1914)
1 : in a manner that can be predicted ⟨works quickly and *predictably*⟩
2 : as one could predict **:** as one would expect ⟨*predictably,* the politicians howled⟩

pre·dic·tion \pri-'dik-shən\ *noun* (1561)
1 : an act of predicting
2 : something that is predicted **:** FORECAST

pre·di·gest \ˌprē-dī-'jest, -də-\ *transitive verb* (1663)
1 : to subject to predigestion
2 : to simplify for easy use ⟨*predigested* classics for children⟩

pre·di·ges·tion \-'jes-chən, -'jesh-\ *noun* (circa 1612)
: artificial or natural partial digestion of food ⟨enzymatic *predigestion*⟩ ⟨microbial *predigestion*⟩

pre·di·lec·tion \ˌpre-dᵊl-'ek-shən, ˌprē-\ *noun* [French *prédilection,* from Medieval Latin *praediligere* to love more, prefer, from Latin *prae-* + *diligere* to love — more at DILIGENT] (1742)
: an established preference for something ☆

pre·dis·pose \ˌprē-di-'spōz\ (1646)
transitive verb
1 : to dispose in advance ⟨a good teacher *predisposes* children to learn⟩
2 : to make susceptible ⟨malnutrition *predisposes* one to disease⟩
intransitive verb
: to bring about susceptibility
synonym see INCLINE
— pre·dis·po·si·tion \ˌprē-ˌdis-pə-'zi-shən\ *noun*

pred·nis·o·lone \pred-'ni-sə-ˌlōn\ *noun* [blend of *prednisone* and [1]*-ol*] (1955)
: a glucocorticoid $C_{21}H_{28}O_5$ that is a dehydrogenated analogue of hydrocortisone and is used especially as an anti-inflammatory drug

pred·ni·sone \'pred-nə-ˌsōn *also* -ˌzōn\ *noun* [probably from *pregnane* ($C_{21}H_{36}$) + *diene* (compound containing two double bonds) + *cortisone*] (1955)
: a glucocorticoid $C_{21}H_{26}O_5$ that is a dehydrogenated analogue of cortisone and is used as an anti-inflammatory agent especially in the treatment of arthritis, as an antineoplastic agent, and as an immunosuppressant

pre·doc·tor·al \(ˌ)prē-'däk-t(ə-)rəl\ *adjective* (1937)
: of, relating to, or engaged in academic study leading to the doctoral degree

pre·dom·i·nance \pri-'dä-mə-nən(t)s, -'däm-nən(t)s\ *noun* (1602)
: the quality or state of being predominant

pre·dom·i·nan·cy \-nən(t)-sē\ *noun* (1598)
: PREDOMINANCE

pre·dom·i·nant \-nənt\ *adjective* [Middle French, from Medieval Latin *praedominant-, praedominans,* present participle of *praedominari* to predominate, from Latin *prae-* + *dominari* to rule, govern — more at DOMINATE] (1576)
1 : having superior strength, influence, or authority **:** PREVAILING
2 : being most frequent or common
synonym see DOMINANT
usage see PREDOMINATE

pre·dom·i·nant·ly \-nənt-lē\ *adverb* (1681)
: for the most part **:** MAINLY
usage see PREDOMINATE

[1]**pre·dom·i·nate** \-nət\ *adjective* [alteration of *predominant*] (1591)
: PREDOMINANT □
— pre·dom·i·nate·ly *adverb*

[2]**pre·dom·i·nate** \pri-'dä-mə-ˌnāt\ *verb* [Medieval Latin *praedominatus,* past participle of *praedominari*] (1594)
intransitive verb
1 : to hold advantage in numbers or quantity
2 : to exert controlling power or influence **:** PREVAIL
transitive verb
: to exert control over **:** DOMINATE
— pre·dom·i·na·tion \-ˌdä-mə-'nā-shən\ *noun*

pree \'prē\ *transitive verb* **preed; pree·ing** [alteration of *preve* to prove, test, from Middle English *preven,* from Old French *preuv-,* stem of *prover* to prove] (1680)
Scottish **:** to taste tentatively **:** SAMPLE

pre·eclamp·sia \ˌprē-i-'klam(p)-sē-ə\ *noun* [New Latin] (1923)
: a toxic condition developing in late pregnancy that is characterized by a sudden rise in blood pressure, excessive weight gain, generalized edema, albuminuria, severe headache, and visual disturbances
— pre·eclamp·tic \-'klam(p)-tik\ *adjective*

pre·emer·gence \ˌprē-ə-'mər-jən(t)s\ *adjective* (1935)
: used or occurring before emergence of seedlings above the ground ⟨*preemergence* herbicides⟩

pre·emer·gent \-jənt\ *adjective* (1959)
: PREEMERGENCE

pree·mie \'prē-mē\ *noun* [*pre*mature + *-ie*] (1927)
: a premature baby

pre·em·i·nence \prē-'e-mə-nən(t)s\ *noun* (13th century)
: the quality or state of being preeminent **:** SUPERIORITY

pre·em·i·nent \-nənt\ *adjective* [Middle English, from Late Latin *praeeminent-, praeeminens,* from Latin, present participle of *praeeminēre* to be outstanding, from *prae-* + *eminēre* to stand out — more at EMINENT] (15th century)
: having paramount rank, dignity, or importance **:** OUTSTANDING
— pre·em·i·nent·ly *adverb*

pre·empt \prē-'em(p)t\ *verb* [back-formation from *preemption*] (1850)
transitive verb
1 : to acquire (as land) by preemption
2 : to seize upon to the exclusion of others **:** take for oneself ⟨the movement was then *preempted* by a lunatic fringe⟩
3 : to replace with something considered to be of greater value or priority **:** take precedence over ⟨the program did not appear, having been *preempted* by a baseball game —Robert MacNeil⟩
4 : to gain a commanding or preeminent place in
5 : to prevent from happening or taking place **:** FORESTALL, PRECLUDE
intransitive verb
: to make a preemptive bid in bridge
— pre·emp·tor \-'em(p)-tər\ *noun*

pre·emp·tion \-'em(p)-shən\ *noun* [Medieval Latin *praeemption-, praeemptio* previous purchase, from *praeemere* to buy before, from Latin *prae-* pre- + *emere* to buy — more at REDEEM] (1602)
1 a : the right of purchasing before others; *especially* **:** one given by the government to the actual settler upon a tract of public land **b :** the purchase of something under this right
2 : a prior seizure or appropriation **:** a taking possession before others

pre·emp·tive \-'em(p)-tiv\ *adjective* (1855)
1 a : of or relating to preemption **b :** having power to preempt
2 *of a bid in bridge* **:** higher than necessary and intended to shut out bids by the opponents
3 : giving a stockholder first option to purchase new stock in an amount proportionate to his existing holdings
4 : marked by the seizing of the initiative **:** initiated by oneself ⟨a *preemptive* attack⟩
— pre·emp·tive·ly *adverb*

[1]**preen** \'prēn\ *noun* [Middle English *prene,* from Old English *prēon;* akin to Middle High German *pfrieme* awl] (before 12th century)
1 *dialect chiefly British* **:** PIN
2 *dialect chiefly British* **:** BROOCH

[2]**preen** *transitive verb* (1572)
chiefly Scottish **:** PIN

[3]**preen** *verb* [Middle English *preinen*] (14th century)
transitive verb
1 : to dress or smooth (oneself) up **:** PRIMP
2 *of a bird* **:** to trim or dress with a bill
3 : to pride or congratulate (oneself) for achievement
intransitive verb
1 : to make oneself sleek
2 : GLOAT, SWELL
— preen·er *noun*

pre–en·gi·neered \ˌprē-ˌen-jə-'nird\ *adjective* (1951)
: constructed of or employing prefabricated modules ⟨a *pre-engineered* building⟩

pre·ex·il·ic \ˌprē-eg-'zi-lik\ *adjective* (1880)
: previous to the exile of the Jews to Babylon in about 600 B.C.

pre·ex·ist \ˌprē-ig-'zist\ (1599)
intransitive verb

☆ SYNONYMS
Predilection, prepossession, prejudice, bias mean an attitude of mind that predisposes one to favor something. PREDILECTION implies a strong liking deriving from one's temperament or experience ⟨a *predilection* for horror movies⟩. PREPOSSESSION suggests a fixed conception likely to preclude objective judgment of anything counter to it ⟨a *prepossession* against technology⟩. PREJUDICE usually implies an unfavorable prepossession and connotes a feeling rooted in suspicion, fear, or intolerence ⟨a mindless *prejudice* against the unfamiliar⟩. BIAS implies an unreasoned and unfair distortion of judgment in favor of or against a person or thing ⟨the common *bias* against overweight people⟩.

□ USAGE
predominate *Predominant* and *predominate* are synonymous adjectives. *Predominant* is the older and more common form. During this century a number of handbooks and commentators have held *predominate* to be a mistake—a few insisting that the word is only a verb. But they are wrong. As an adjective *predominate* is somewhat more likely to turn up in technical writing than in general writing. The adverbs *predominantly* and *predominately* are a more even match in frequency than their base adjectives are. *Predominately* seems to be more widely used than *predominate* and hence to be more often found in general writing.

: to exist earlier or before
transitive verb
: ANTEDATE

pre·ex·is·tence \-ig-ˈzis-tən(t)s\ *noun* (circa 1652)
: existence in a former state or previous to something else; *specifically* **:** existence of the soul before its union with the body
— pre·ex·is·tent \-tənt\ *adjective*

pre·fab \(ˌ)prē-ˈfab, ˈprē-ˌ\ *noun* (1937)
: a prefabricated structure
— prefab *adjective*

pre·fab·ri·cate \(ˌ)prē-ˈfa-bri-ˌkāt\ *transitive verb* (1932)
1 : to fabricate the parts of at a factory so that construction consists mainly of assembling and uniting standardized parts
2 : to produce artificially
— pre·fab·ri·ca·tion \ˌprē-ˌfa-bri-ˈkā-shən\ *noun*

¹pref·ace \ˈpre-fəs\ *noun* [Middle English, from Middle French, from Medieval Latin *prephatia,* alteration of Latin *praefation-, praefatio* foreword, from *praefari* to say beforehand, from *prae-* pre- + *fari* to say — more at BAN] (14th century)
1 *often capitalized* **:** a variable doxology beginning with the Sursum Corda and ending with the Sanctus in traditional eucharistic liturgies
2 : the introductory remarks of a speaker or author
3 : APPROACH, PRELIMINARY

²preface *verb* **pref·aced; pref·ac·ing** (1619)
intransitive verb
: to make introductory remarks
transitive verb
1 : to say or write as preface ⟨a note *prefaced* to the manuscript⟩
2 : PRECEDE, HERALD
3 : to introduce by or begin with a preface
4 : to stand in front of ⟨a porch *prefaces* the entrance⟩
5 : to be a preliminary to
— pref·ac·er *noun*

pref·a·to·ry \ˈpre-fə-ˌtōr-ē, -ˌtȯr-\ *adjective* [Latin *praefari*] (1675)
1 : of, relating to, or constituting a preface
2 : located in front

pre·fect \ˈprē-ˌfekt\ *noun* [Middle English, from Middle French, from Latin *praefectus,* from past participle of *praeficere* to place at the head of, from *prae-* + *facere* to make — more at DO] (14th century)
1 : any of various high officials or magistrates of differing functions and ranks in ancient Rome
2 : a chief officer or chief magistrate
3 : a student monitor in a private school

prefect apostolic *noun* (circa 1888)
: a Roman Catholic clergyman and usually a priest with quasi-episcopal jurisdiction over a district of a missionary territory

pre·fec·ture \ˈprē-ˌfek-chər\ *noun* (15th century)
1 : the office or term of office of a prefect
2 : the official residence of a prefect
3 : the district governed by a prefect
— pre·fec·tur·al \prē-ˈfek-chə-rəl, pri-ˈ\ *adjective*

prefecture apostolic *noun* (1911)
: the district under a prefect apostolic

pre·fer \pri-ˈfər\ *transitive verb* **pre·ferred; pre·fer·ring** [Middle English *preferren,* from Middle French *preferer,* from Latin *praeferre* to put before, prefer, from *prae-* + *ferre* to carry — more at BEAR] (14th century)
1 : to promote or advance to a rank or position
2 : to like better or best ⟨*prefers* sports to reading⟩ ⟨*prefers* to watch TV⟩
3 : to give (a creditor) priority
4 *archaic* **:** to put or set forward or before someone **:** RECOMMEND
5 : to bring or lay against someone ⟨won't *prefer* charges⟩
6 : to bring forward or lay before one for consideration
— pre·fer·rer *noun*

pref·er·a·ble \ˈpre-f(ə-)rə-bəl, ˈpre-fər-bəl *also* pri-ˈfər-ə-bəl\ *adjective* (1666)
: having greater value or desirability **:** being preferred
— pref·er·a·bil·i·ty \ˌpre-f(ə-)rə-ˈbi-lə-tē\ *noun*
— pref·er·a·bly \-blē\ *adverb*

pref·er·ence \ˈpre-fərn(t)s, ˈpre-f(ə-)rən(t)s\ *noun* [Middle English *preferraunce,* from Middle French *preferance,* from Medieval Latin *praeferentia,* from Latin *praeferent-, praeferens,* present participle of *praeferre*] (15th century)
1 a : the act of preferring **:** the state of being preferred **b :** the power or opportunity of choosing
2 : one that is preferred
3 : the act, fact, or principle of giving advantages to some over others
4 : priority in the right to demand and receive satisfaction of an obligation
5 : ORIENTATION 2 ⟨sexual *preference*⟩
synonym see CHOICE

pref·er·en·tial \ˌpre-fə-ˈren(t)-shəl\ *adjective* (1849)
1 : showing preference
2 : employing or creating a preference in trade relations
3 : designed to permit expression of preference among candidates ⟨a *preferential* primary⟩
4 : giving preference especially in hiring to union members ⟨a *preferential* shop⟩
— pref·er·en·tial·ly \-ˈren(t)-sh(ə-)lē\ *adverb*

pre·fer·ment \pri-ˈfər-mənt\ *noun* (15th century)
1 a : advancement or promotion in dignity, office, or station **b :** a position or office of honor or profit
2 : priority or seniority in right especially to receive payment or to purchase property on equal terms with others
3 : the act of bringing forward (as charges)

preferred stock *noun* (circa 1859)
: stock guaranteed priority by a corporation's charter over common stock in the payment of dividends and usually in the distribution of assets

pre·fig·u·ra·tion \(ˌ)prē-ˌfi-gyə-ˈrā-shən, -gə-\ *noun* (14th century)
1 : the act of prefiguring **:** the state of being prefigured
2 : something that prefigures

pre·fig·u·ra·tive \ˌprē-ˈfi-gyə-rə-tiv, -gə-\ *adjective* (1504)
: of, relating to, or showing by prefiguration
— pre·fig·u·ra·tive·ly *adverb*
— pre·fig·u·ra·tive·ness *noun*

pre·fig·ure \ˌprē-ˈfi-gyər, *especially British* -ˈfi-gər\ *transitive verb* [Middle English, from Late Latin *praefigurare,* from Latin *prae-* pre- + *figurare* to shape, picture, from *figura* figure] (15th century)
1 : to show, suggest, or announce by an antecedent type, image, or likeness
2 : to picture or imagine beforehand **:** FORESEE
— pre·fig·ure·ment \-mənt\ *noun*

¹pre·fix *transitive verb* [Middle English, from Middle French *prefixer,* from *pre-* + *fixer* to fix, from *fix* fixed, from Latin *fixus* — more at FIX] (15th century)
1 \(ˌ)prē-ˈfiks\ **:** to fix or appoint beforehand
2 \ˈprē-ˌ, prē-ˈ\ [partly from ²*prefix*] **:** to place in front; *especially* **:** to add as a prefix ⟨*prefix* a syllable to a word⟩

²pre·fix \ˈprē-ˌfiks\ *noun* [New Latin *praefixum,* from Latin, neuter of *praefixus,* past participle of *praefigere* to fasten before, from *prae-* + *figere* to fasten — more at FIX] (1646)
1 : an affix attached to the beginning of a word, base, or phrase and serving to produce a derivative word or an inflectional form — compare SUFFIX
2 : a title used before a person's name
— pre·fix·al \ˈprē-ˌfik-səl, prē-ˈ\ *adjective*

³pre·fix *same as* ²\ *adjective* (1971)
: characterized by placement of an operator before its operand or before its two operands if it is a binary operator — compare INFIX, POSTFIX

pre·fo·cus \(ˌ)prē-ˈfō-kəs\ *transitive verb* (1948)
: to focus beforehand (as automotive headlights before installation)

pre·form \ˈprē-ˌfȯrm, (ˌ)prē-ˈ\ *transitive verb* [Latin *praeformare,* from *prae-* + *formare* to form, from *forma* form] (1601)
1 : to form or shape beforehand
2 : to bring to approximate preliminary shape and size
— pre·form \ˈprē-ˌfȯrm\ *noun*

pre·for·ma·tion \ˌprē-fȯr-ˈmā-shən\ *noun* (1732)
1 : previous formation
2 : the now discredited theory that every germ cell contains the organism of its kind fully formed and that development involves merely an increase in size — compare EPIGENESIS 1, HOMUNCULUS 2
— pre·for·ma·tion·ist \-sh(ə-)nist\ *noun or adjective*

¹pre·fron·tal \(ˌ)prē-ˈfrən-tᵊl\ *adjective* (1854)
: anterior to or involving the anterior part of a frontal structure ⟨a *prefrontal* bone⟩

²prefrontal *noun* (1854)
: a prefrontal part (as a bone)

pre·gan·gli·on·ic \ˌprē-ˌgaŋ-glē-ˈä-nik\ *adjective* (1895)
: proximal to a ganglion; *specifically* **:** of, relating to, or being a usually medullated axon arising from a cell body in the central nervous system and terminating in an autonomic ganglion — compare POSTGANGLIONIC

preg·na·ble \ˈpreg-nə-bəl\ *adjective* [alteration of Middle English *prenable,* from Middle French — more at IMPREGNABLE] (14th century)
: vulnerable to capture ⟨a *pregnable* fort⟩
— preg·na·bil·i·ty \ˌpreg-nə-ˈbi-lə-tē\ *noun*

preg·nan·cy \ˈpreg-nən(t)-sē\ *noun, plural* **-cies** (15th century)
1 : the quality of being pregnant (as in meaning)
2 : the condition of being pregnant **:** GESTATION
3 : an instance of being pregnant

¹preg·nant \ˈpreg-nənt\ *adjective* [Middle English *preignant,* from Middle French, from present participle of *preindre* to press, from Latin *premere* — more at PRESS] (14th century)
archaic **:** COGENT

²pregnant *adjective* [Middle English, from Latin *praegnant-, praegnans,* alteration of *praegnas,* from *prae-* pre- + *-gnas* (akin to *gignere* to give birth to) — more at KIN] (15th century)
1 : abounding in fancy, wit, or resourcefulness **:** INVENTIVE ⟨all this has been said . . . by great and *pregnant* artists —*Times Literary Supplement*⟩
2 : rich in significance or implication **:** MEANINGFUL, PROFOUND ⟨the *pregnant* phrases of the Bible —Edmund Wilson⟩ ⟨a *pregnant* pause⟩
3 : containing unborn young within the body **:** GRAVID
4 : having possibilities of development or consequence **:** involving important issues **:** MOMENTOUS ⟨draw inspiration from the heroic achievements of that *pregnant* age —Kemp Malone⟩

5 *obsolete* **:** INCLINED, DISPOSED ⟨your own most *pregnant* and vouchsafed ear —Shakespeare⟩
6 : FULL, TEEMING
— **preg·nant·ly** *adverb*

preg·nen·o·lone \preg-ˈne-nᵊl-ˌōn\ *noun* [International Scientific Vocabulary *pregnene* ($C_{21}H_{34}$) + ¹*-ol* + *-one*] (1936)
: an unsaturated hydroxy steroid ketone $C_{21}H_{32}O_2$ that is formed by the oxidation of steroids (as cholesterol) and yields progesterone on dehydrogenation

pre·heat \(ˌ)prē-ˈhēt\ *transitive verb* (1898)
: to heat beforehand; *especially* **:** to heat (an oven) to a designated temperature before using for cooking
— **pre·heat·er** *noun*

pre·hen·sile \prē-ˈhen(t)-səl, -ˈhen-ˌsīl\ *adjective* [French *préhensile,* from Latin *prehensus,* past participle of *prehendere* to seize — more at GET] (circa 1785)
1 : adapted for seizing or grasping especially by wrapping around ⟨*prehensile* tail⟩
2 : gifted with mental grasp or moral or aesthetic perception
— **pre·hen·sil·i·ty** \(ˌ)prē-ˌhen-ˈsi-lə-tē\ *noun*

pre·hen·sion \prē-ˈhen(t)-shən\ *noun* (circa 1828)
1 : the act of taking hold, seizing, or grasping
2 a : mental understanding **:** COMPREHENSION **b :** apprehension by the senses

pre–His·pan·ic \ˌprē-(h)i-ˈspa-nik\ *adjective* (1919)
: of, relating to, or being the time prior to Spanish conquests in the western hemisphere

pre·his·to·ri·an \ˌprē-(h)is-ˈtōr-ē-ən, -ˈtȯr-\ *noun* (1893)
: an archaeologist who specializes in prehistory

pre·his·tor·ic \ˌprē-(h)is-ˈtȯr-ik, -ˈtär-\ *also* **pre·his·tor·i·cal** \-i-kəl\ *adjective* (1851)
1 : of, relating to, or existing in times antedating written history
2 : of or relating to a language in a period of its development from which contemporary records of its sounds and forms have not been preserved
— **pre·his·tor·i·cal·ly** \-i-k(ə-)lē\ *adverb*

pre·his·to·ry \ˌprē-ˈhis-t(ə-)rē\ *noun* (1871)
1 : the study of prehistoric humankind
2 : a history of the antecedents of an event, situation, or thing
3 : the prehistoric period of human evolution

pre·hom·i·nid \-ˈhä-mə-nəd\ *noun* [ultimately from Latin *pre-* + *homin-, homo* human being — more at HOMAGE] (1939)
: any of various extinct primates resembling or ancestral to humans
— **prehominid** *adjective*

pre·ig·ni·tion \ˌprē-ig-ˈni-shən\ *noun* (1898)
: ignition in an internal combustion engine while the inlet valve is open or before compression is completed

pre·im·plan·ta·tion \ˌprē-ˌim-ˌplan-ˈtā-shən\ *adjective* (1945)
: of, involving, or being an embryo before uterine implantation

pre·judge \(ˌ)prē-ˈjəj\ *transitive verb* [Middle French *prejuger,* from Latin *praejudicare,* from *prae-* + *judicare* to judge — more at JUDGE] (1579)
: to judge before hearing or before full and sufficient examination
— **pre·judg·er** *noun*
— **pre·judg·ment** \-ˈjəj-mənt\ *noun*

¹**prej·u·dice** \ˈpre-jə-dəs\ *noun* [Middle English, from Old French, from Latin *praejudicium* previous judgment, damage, from *prae-* + *judicium* judgment — more at JUDICIAL] (13th century)
1 : injury or damage resulting from some judgment or action of another in disregard of one's rights; *especially* **:** detriment to one's legal rights or claims
2 a (1) **:** preconceived judgment or opinion (2) **:** an adverse opinion or leaning formed without just grounds or before sufficient knowledge **b :** an instance of such judgment or opinion **c :** an irrational attitude of hostility directed against an individual, a group, a race, or their supposed characteristics
synonym see PREDILECTION

²**prejudice** *transitive verb* **-diced; -dic·ing** (15th century)
1 : to injure or damage by some judgment or action (as in a case of law)
2 : to cause to have prejudice

prej·u·diced \-dəst\ *adjective* (1579)
: resulting from or having a prejudice or bias for or especially against

prej·u·di·cial \ˌpre-jə-ˈdi-shəl\ *adjective* (15th century)
1 : tending to injure or impair **:** DETRIMENTAL
2 : leading to premature judgment or unwarranted opinion
— **prej·u·di·cial·ly** \-ˈdi-sh(ə-)lē\ *adverb*
— **prej·u·di·cial·ness** \-shəl-nəs\ *noun*

prel·a·cy \ˈpre-lə-sē\ *noun, plural* **-cies** (14th century)
1 : the office or dignity of a prelate
2 : episcopal church government

pre·lap·sar·i·an \ˌprē-ˌlap-ˈser-ē-ən\ *adjective* [*pre-* + Latin *lapsus* slip, fall — more at LAPSE] (1879)
: characteristic of or belonging to the time or state before the fall of mankind

prel·ate \ˈpre-lət *also* ˈprē-ˌlāt\ *noun* [Middle English *prelat,* from Old French, from Medieval Latin *praelatus,* literally, one receiving preferment, from Latin (past participle of *praeferre* to prefer), from *prae-* + *latus,* past participle of *ferre* to carry — more at TOLERATE, BEAR] (13th century)
: an ecclesiastic (as a bishop or abbot) of superior rank

prel·a·ture \ˈpre-lə-ˌchu̇r, -chər, -ˌtyu̇r, -ˌtu̇r\ *noun* (1607)
1 : PRELACY 1
2 : a body of prelates

pre·lect \pri-ˈlekt\ *intransitive verb* [Latin *praelectus,* past participle of *praelegere,* from *prae-* + *legere* to read — more at LEGEND] (1785)
: to discourse publicly **:** LECTURE
— **pre·lec·tion** \-ˈlek-shən\ *noun*

pre·li·ba·tion \ˌprē-lī-ˈbā-shən\ *noun* [Latin *praelibation-, praelibatio,* from *praelibare* to taste beforehand, from *prae-* + *libare* to pour as an offering, taste — more at LIBATION] (1526)
: FORETASTE

pre·lim \ˈprē-ˌlim, pri-ˈ\ *noun or adjective* (1891)
: PRELIMINARY

¹**pre·lim·i·nary** \pri-ˈli-mə-ˌner-ē\ *noun, plural* **-nar·ies** [French *préliminaires,* plural, from Medieval Latin *praeliminaris,* adjective, preliminary, from Latin *prae-* pre- + *limin-, limen* threshold] (1656)
: something that precedes or is introductory or preparatory: as **a :** a preliminary scholastic examination **b** *plural, British* **:** FRONT MATTER **c :** a preliminary heat or trial (as of a race) **d :** a minor match preceding the main event (as of a boxing card)

²**preliminary** *adjective* (circa 1667)
: coming before and usually forming a necessary prelude to something else
— **pre·lim·i·nar·i·ly** \-ˌli-mə-ˈner-ə-lē\ *adverb*

pre·lit·er·ate \ˌprē-ˈli-t(ə-)rət\ *adjective* (1925)
1 a : not yet employing writing as a cultural medium **b :** lacking the use of writing
2 : antedating the use of writing
— **preliterate** *noun*

¹**pre·lude** \ˈprel-ˌyüd, ˈprāl-; ˈpre-ˌlüd, ˈprā-; *sense 1 also* ˈprē-ˌlüd\ *noun* [Middle French, from Medieval Latin *praeludium,* from Latin *praeludere* to play beforehand, from *prae-* + *ludere* to play — more at LUDICROUS] (1561)
1 : an introductory performance, action, or event preceding and preparing for the principal or a more important matter
2 a : a musical section or movement introducing the theme or chief subject (as of a fugue or suite) or serving as an introduction to an opera or oratorio **b :** an opening voluntary **c :** a separate concert piece usually for piano or orchestra and based entirely on a short motif

²**prelude** *verb* **pre·lud·ed; pre·lud·ing** (1655)
transitive verb
1 : to serve as a prelude to
2 : to play as a prelude
intransitive verb
: to give or serve as a prelude; *especially* **:** to play a musical introduction
— **pre·lud·er** *noun*

pre·lu·sion \pri-ˈlü-zhən\ *noun* [Latin *praelusion-, praelusio,* from *praeludere*] (1597)
: PRELUDE, INTRODUCTION

pre·lu·sive \-ˈlü-siv, -ziv\ *adjective* (1605)
: constituting or having the form of a prelude **:** INTRODUCTORY
— **pre·lu·sive·ly** *adverb*

pre·ma·lig·nant \ˌprē-mə-ˈlig-nənt\ *adjective* (circa 1897)
: PRECANCEROUS

pre·man \(ˌ)prē-ˈman, ˈprē-ˌman\ *noun* (1921)
: any of several extinct primates ancestral to humans and especially recent humans

pre·ma·ture \ˌprē-mə-ˈtyu̇r, -ˈtu̇r, -ˈchu̇r *also* ˌpre-\ *adjective* [Latin *praematurus* too early, from *prae-* + *maturus* ripe, mature] (circa 1529)
: happening, arriving, existing, or performed before the proper, usual, or intended time; *especially* **:** born after a gestation period of less than 37 weeks ⟨*premature* babies⟩
— **premature** *noun*
— **pre·ma·ture·ly** *adverb*
— **pre·ma·ture·ness** *noun*
— **pre·ma·tu·ri·ty** \-ˈtyu̇r-ə-tē, -ˈtu̇r-, -ˈchu̇r-\ *noun*

pre·max·il·la \ˌprē-mak-ˈsi-lə\ *noun* [New Latin] (1866)
: either of a pair of bones of the upper jaw of vertebrates between and in front of the maxillae
— **pre·max·il·lary** \(ˌ)prē-ˈmak-sə-ˌler-ē, *chiefly British* ˌprē-mak-ˈsi-lə-rē\ *adjective or noun*

¹**pre·med** \(ˌ)prē-ˈmed, ˈprē-ˌmed\ *noun* (circa 1928)
: a premedical student or course of study

²**premed** *adjective* (1950)
: PREMEDICAL

pre·med·i·cal \(ˌ)prē-ˈme-di-kəl\ *adjective* (1904)
: preceding and preparing for the professional study of medicine

pre·med·i·tate \(ˌ)prē-ˈme-də-ˌtāt\ *verb* [Latin *praemeditatus,* past participle of *praemeditari,* from *prae-* + *meditari* to meditate] (circa 1548)
transitive verb
: to think about and revolve in the mind beforehand
intransitive verb
: to think, consider, or deliberate beforehand
— **pre·med·i·ta·tor** \-ˌtā-tər\ *noun*

pre·med·i·tat·ed *adjective* (1590)
: characterized by fully conscious willful intent and a measure of forethought and planning ⟨*premeditated* murder⟩
— **pre·med·i·tat·ed·ly** *adverb*

pre·med·i·ta·tion \(ˌ)prē-ˌme-də-ˈtā-shən\ *noun* (15th century)
: an act or instance of premeditating; *specifically* **:** consideration or planning of an act beforehand that shows intent to commit that act

pre·med·i·ta·tive \(ˌ)prē-ˈme-də-ˌtā-tiv\ *adjective* (1858)
: given to or characterized by premeditation

pre·men·stru·al \(ˌ)prē-ˈmen(t)-strə-wəl, -strəl\ *adjective* (1885)
: of, relating to, occurring in, or being the period just preceding menstruation ⟨*premenstrual* tension⟩ ⟨*premenstrual* women⟩
— pre·men·stru·al·ly *adverb*

premenstrual syndrome *noun* (1982)
: a varying group of symptoms manifested by some women prior to menstruation that may include emotional instability, irritability, insomnia, fatigue, anxiety, depression, headache, edema, and abdominal pain — abbreviation *PMS*

pre·mie *variant of* PREEMIE

[1]pre·mier \pri-ˈmir, -ˈmyir, -ˈmē-ər; ˈprē-ˌ, ˈpre-ˌ\ *adjective* [Middle English *primier,* from Middle French *premier* first, chief, from Latin *primarius* of the first rank — more at PRIMARY] (15th century)
1 : first in position, rank, or importance
2 : first in time **:** EARLIEST

[2]premier *noun* [French, from *premier,* adjective] (1711)
: PRIME MINISTER
— pre·mier·ship \-ˌship\ *noun*

pre·mier dan·seur \prə-myā-dä^{n}-sœr\ *noun* [French] (1828)
: the principal male dancer in a ballet company

[1]pre·miere \pri-ˈmyer, -ˈmir, -ˈmē-ər; ˌpri-mē-ˈer\ *adjective* [alteration of [1]*premier*] (1768)
: PREMIER

[2]premiere *also* **pre·mière** *same as* [1]PREMIERE\ *noun* [French *première,* from feminine of *premier* first] (1889)
1 : a first performance or exhibition ⟨the *premiere* of a play⟩
2 : the chief actress of a theatrical cast

[3]premiere *also* **pre·mière** *or* **pre·mier** *same as* [1]PREMIERE\ *verb* **premiered** *also* **pre·mièred; pre·mier·ing** *also* **pre·mièr·ing** (1933)
transitive verb
: to give a first public performance of
intransitive verb
1 : to have a first public performance
2 : to appear for the first time as a star performer

pre·mière dan·seuse \prə-myer-dä^{n}-sœz\ *noun* [French] (1828)
: PRIMA BALLERINA

pre·mil·le·nar·i·an·ism \ˌprē-ˌmi-lə-ˈner-ē-ə-ˌni-zəm\ *noun* (1844)
: PREMILLENNIALISM
— pre·mil·le·nar·i·an \-ē-ən\ *adjective or noun*

pre·mil·len·ni·al \ˌprē-mə-ˈle-nē-əl\ *adjective* (1846)
1 : coming before a millennium
2 : holding or relating to premillennialism
— pre·mil·len·ni·al·ly \-nē-ə-lē\ *adverb*

pre·mil·len·ni·al·ism \-nē-ə-ˌli-zəm\ *noun* (circa 1883)
: the view that Christ's return will usher in a future millennium of Messianic rule mentioned in Revelation
— pre·mil·len·ni·al·ist \-nē-ə-list\ *noun*

[1]prem·ise *also* **prem·iss** \ˈpre-məs\ *noun* [in sense 1, from Middle English *premisse,* from Middle French, from Medieval Latin *praemissa,* from Latin, feminine of *praemissus,* past participle of *praemittere* to place ahead, from *prae-* pre- + *mittere* to send; in other senses, from Middle English *premisses,* from Medieval Latin *praemissa,* from Latin, neuter plural of *praemissus*] (14th century)
1 a : a proposition antecedently supposed or proved as a basis of argument or inference; *specifically* **:** either of the first two propositions of a syllogism from which the conclusion is drawn **b :** something assumed or taken for granted **:** PRESUPPOSITION
2 *plural* **:** matters previously stated; *specifically* **:** the preliminary and explanatory part of a deed or of a bill in equity
3 *plural* [from its being identified in the premises of the deed] **a :** a tract of land with the buildings thereon **b :** a building or part of a building usually with its appurtenances (as grounds)

[2]pre·mise \ˈpre-məs *also* pri-ˈmīz\ *transitive verb* **pre·mised; pre·mis·ing** (1526)
1 a : to set forth beforehand as an introduction or a postulate **b :** to offer as a premise in an argument
2 : POSTULATE
3 : to base on certain assumptions

[1]pre·mi·um \ˈprē-mē-əm\ *noun* [Latin *praemium* booty, profit, reward, from *prae-* + *emere* to take, buy — more at REDEEM] (1601)
1 a : a reward or recompense for a particular act **b :** a sum over and above a regular price paid chiefly as an inducement or incentive **c :** a sum in advance of or in addition to the nominal value of something ⟨bonds callable at a *premium* of six percent⟩ **d :** something given free or at a reduced price with the purchase of a product or service
2 : the consideration paid for a contract of insurance
3 : a high value or a value in excess of that normally or usually expected ⟨put a *premium* on accuracy⟩

[2]premium *adjective* (1844)
: of exceptional quality or amount; *also* **:** higher-priced

[1]pre·mix \(ˌ)prē-ˈmiks, ˈprē-ˌ\ *transitive verb* (1927)
: to mix before use

[2]pre·mix \ˈprē-ˌmiks\ *noun* (1937)
: a mixture of ingredients designed to be mixed with other ingredients before use

pre·mo·lar \(ˌ)prē-ˈmō-lər\ *adjective* (circa 1859)
: situated in front of or preceding the molar teeth; *especially* **:** being or relating to those teeth of a mammal in front of the true molars and behind the canines when the latter are present
— premolar *noun*

pre·mon·ish \-ˈmä-nish\ (1526)
transitive verb
archaic **:** FOREWARN
intransitive verb
archaic **:** to give warning in advance

pre·mo·ni·tion \ˌprē-mə-ˈni-shən, ˌpre-\ *noun* [Middle English, from Middle French, from Late Latin *praemonition-, praemonitio,* from Latin *praemonēre* to warn in advance, from *prae-* + *monēre* to warn — more at MIND] (15th century)
1 : previous notice or warning **:** FOREWARNING
2 : anticipation of an event without conscious reason **:** PRESENTIMENT

pre·mon·i·to·ry \pri-ˈmä-nə-ˌtōr-ē, -ˌtȯr-\ *adjective* (1647)
: giving warning ⟨a *premonitory* symptom⟩
— pre·mon·i·to·ri·ly \-ˌmä-nə-ˈtōr-ə-lē, -ˈtȯr-\ *adverb*

Pre·mon·stra·ten·sian \ˌprē-ˌmän(t)-strə-ˈten(t)-shən\ *noun* [Medieval Latin *praemonstratensis,* from *praemonstratensis* of Prémontré, from *Praemonstratus* Prémontré] (1695)
: a member of an order of canons regular founded by Saint Norbert at Prémontré near Laon, France, in 1120

pre·mune \(ˌ)prē-ˈmyün\ *adjective* [back-formation from *premunition*] (1948)
: exhibiting premunition

pre·mu·ni·tion \ˌprē-myu̇-ˈni-shən\ *noun* [Latin *praemunition-, praemunitio* advance fortification, from *praemunire* to fortify in advance, from *prae-* + *munire* to fortify — more at MUNITION] (1607)
1 *archaic* **:** an advance provision of protection
2 a : resistance to a disease due to the existence of its causative agent in a state of physiological equilibrium in the host **b :** immunity to a particular infection due to previous presence of the causative agent

pre·name \ˈprē-ˌnām\ *noun* (1894)
: FORENAME

pre·na·tal \(ˌ)prē-ˈnā-t^{ə}l\ *adjective* (1826)
1 : occurring, existing, or performed before birth ⟨*prenatal* care⟩ ⟨the *prenatal* period⟩
2 : providing or receiving prenatal medical care ⟨a *prenatal* clinic⟩ ⟨*prenatal* patients⟩
— pre·na·tal·ly \-t^{ə}l-ē\ *adverb*

[1]pre·nom·i·nate \(ˌ)prē-ˈnä-mə-nət\ *adjective* [Late Latin *praenominatus,* past participle of *praenominare* to name before, from Latin *prae-* + *nominare* to name — more at NOMINATE] (1513)
obsolete **:** previously mentioned

[2]pre·nom·i·nate \-ˌnāt\ *transitive verb* (1547)
obsolete **:** to mention previously
— pre·nom·i·na·tion \(ˌ)prē-ˌnä-mə-ˈnā-shən\ *noun, obsolete*

pre·no·tion \(ˌ)prē-ˈnō-shən, ˈprē-ˌ\ *noun* [Latin *praenotion-, praenotio* preconception, from *prae-* + *notio* idea, conception — more at NOTION] (1588)
1 : PRESENTIMENT, PREMONITION
2 : PRECONCEPTION

[1]pren·tice \ˈpren-təs\ *noun* [Middle English *prentis,* short for *apprentis*] (14th century)
: APPRENTICE 1, LEARNER
— prentice *adjective*

[2]prentice *transitive verb* **pren·ticed; pren·tic·ing** (1598)
: APPRENTICE

pre·nup·tial \(ˌ)prē-ˈnəp-shəl, -chəl, ÷-chə-wəl\ *adjective* (1869)
: made or occurring before marriage ⟨a *prenuptial* agreement⟩

pre·oc·cu·pan·cy \(ˌ)prē-ˈä-kyə-pən(t)-sē\ *noun* (circa 1755)
1 : an act or the right of taking possession before another
2 : the condition of being completely busied or preoccupied

pre·oc·cu·pa·tion \(ˌ)prē-ˌä-kyə-ˈpā-shən\ *noun* (1603)
1 : an act of preoccupying **:** the state of being preoccupied
2 a : extreme or excessive concern with something **b :** something that preoccupies one

pre·oc·cu·pied \(ˌ)prē-ˈä-kyə-ˌpīd\ *adjective* (1842)
1 : previously applied to another group and unavailable for use in a new sense — used of a biological generic or specific name
2 a : lost in thought; *also* **:** absorbed in some preoccupation **b :** already occupied

pre·oc·cu·py \-ˌpī\ *transitive verb* [Latin *praeoccupare,* literally, to seize in advance, from *prae-* + *occupare* to seize, occupy] (1567)
1 : to engage or engross the interest or attention of beforehand or preferentially
2 : to take possession of or fill beforehand or before another

pre·op·er·a·tive \(ˌ)prē-ˈä-p(ə-)rə-tiv, -pə-ˌrā-\ *adjective* (1904)
: occurring before a surgical operation
— pre·op·er·a·tive·ly *adverb*

pre·or·dain \ˌprē-ȯr-ˈdān\ *transitive verb* (1533)
: to decree or ordain in advance **:** FOREORDAIN
— pre·or·dain·ment \-mənt\ *noun*
— pre·or·di·na·tion \(ˌ)prē-ˌȯr-d^{ə}n-ˈā-shən\ *noun*

pre·ovu·la·to·ry \(ˌ)prē-ˈäv-yə-lə-ˌtōr-ē, -ˌtȯr-, -ˈōv-\ *adjective* (1935)
: occurring or existing in or typical of the period immediately preceding ovulation ⟨*preovulatory* oocytes⟩ ⟨a *preovulatory* surge of luteinizing hormone⟩

pre–owned \(ˌ)prē-ˈōnd, ˈprē-ˌ\ *adjective* (1964)

: SECONDHAND

[1]**prep** \'prep\ *noun* (1862)
1 : PREPARATION
2 : PREPARATORY SCHOOL
3 : a preliminary trial for a racehorse

[2]**prep** *verb* **prepped; prep·ping** (1915)
intransitive verb
1 : to attend preparatory school
2 [short for *prepare*] **:** to get ready
transitive verb
: PREPARE; *especially* **:** to prepare for operation or examination

pre·pack·age \(,)prē-'pa-kij\ *transitive verb* (1945)
: to package (as food or a manufactured article) before offering for sale to the consumer

prep·a·ra·tion \,pre-pə-'rā-shən\ *noun* [Middle English *preparacion,* from Middle French *preparation,* from Latin *praeparation-, praeparatio,* from *praeparare* to prepare] (14th century)
1 : the action or process of making something ready for use or service or of getting ready for some occasion, test, or duty
2 : a state of being prepared **:** READINESS
3 : a preparatory act or measure
4 : something that is prepared; *specifically* **:** a medicinal substance made ready for use ⟨a *preparation* for colds⟩

[1]**pre·par·a·tive** \pri-'par-ə-tiv\ *noun* (14th century)
: something that prepares the way for or serves as a preliminary to something else **:** PREPARATION

[2]**preparative** *adjective* (circa 1530)
: PREPARATORY
— pre·par·a·tive·ly *adverb*

pre·par·a·tor \pri-'par-ə-tər\ *noun* (1762)
: one that prepares; *specifically* **:** a person who prepares scientific specimens or museum displays

pre·pa·ra·to·ry \pri-'par-ə-,tōr-ē, -,tȯr- *also* 'pre-p(ə-)rə-\ *adjective* (15th century)
: preparing or serving to prepare for something **:** INTRODUCTORY
— pre·pa·ra·to·ri·ly \pri-,par-ə-'tōr-ə-lē, -'tȯr- *also* ,pre-p(ə-)rə-\ *adverb*

preparatory school *noun* (1822)
1 : a usually private school preparing students primarily for college
2 *British* **:** a private elementary school preparing students primarily for British public schools

preparatory to *preposition* (1649)
: in preparation for

pre·pare \pri-'par, -'per\ *verb* **pre·pared; pre·par·ing** [Middle English, from Middle French *preparer,* from Latin *praeparare,* from *prae-* pre- + *parare* to procure, prepare — more at PARE] (15th century)
transitive verb
1 a : to make ready beforehand for some purpose, use, or activity ⟨*prepare* food for dinner⟩ **b :** to put in a proper state of mind ⟨is *prepared* to listen⟩
2 : to work out the details of **:** plan in advance ⟨*preparing* strategy for the coming campaign⟩
3 a : to put together **:** COMPOUND ⟨*prepare* a prescription⟩ **b :** to put into written form ⟨*prepare* a report⟩
intransitive verb
: to get ready ⟨*preparing* for a career⟩
— pre·par·er *noun*

prepared *adjective* (1663)
: subjected to a special process or treatment
— pre·pared·ly \-'pard-lē, -'perd-; -'par-əd-, -'per-\ *adverb*

pre·pared·ness \pri-'par-əd-nəs, -'per- *also* -'pard-nəs *or* -'perd-nəs\ *noun* (1590)
: the quality or state of being prepared; *especially* **:** a state of adequate preparation in case of war

pre·pay \(,)prē-'pā\ *transitive verb* **-paid** \-'pād\; **-pay·ing** (1839)
: to pay or pay the charge on in advance
— pre·pay·ment \-'pā-mənt\ *noun*

pre·pense \pri-'pen(t)s\ *adjective* [by shortening & alteration from earlier *purpensed,* from Middle English, past participle of *purpensen* to deliberate, premeditate, from Middle French *purpenser,* from Old French, from *pur-* for + *penser* to think — more at PURCHASE, PENSIVE] (1702)
: planned beforehand **:** PREMEDITATED — usually used postpositively ⟨malice *prepense*⟩
— pre·pense·ly *adverb*

pre·plant \,prē-'plant, 'prē-,\ *also* **pre·plant·ing** \-'plan-tiŋ\ *adjective* (1961)
: occurring or used before planting a crop ⟨*preplant* soil fertilization⟩

pre·pon·der·ance \pri-'pän-d(ə-)rən(t)s\ *noun* (1681)
1 : a superiority in weight, power, importance, or strength
2 a : a superiority or excess in number or quantity **b :** MAJORITY

pre·pon·der·an·cy \-d(ə-)rən(t)-sē\ *noun* (1646)
: PREPONDERANCE

pre·pon·der·ant \pri-'pän-d(ə-)rənt\ *adjective* (15th century)
1 : having superior weight, force, or influence
2 : having greater prevalence
synonym see DOMINANT
— pre·pon·der·ant·ly *adverb*

[1]**pre·pon·der·ate** \pri-'pän-də-,rāt\ *verb* **-at·ed; -at·ing** [Latin *praeponderatus,* past participle of *praeponderare,* from *prae-* + *ponder-, pondus* weight — more at PENDANT] (1623)
intransitive verb
1 : to exceed in weight
2 : to exceed in influence, power, or importance
3 : to exceed in numbers
transitive verb
1 *archaic* **:** OUTWEIGH
2 *archaic* **:** to weigh down
— pre·pon·der·a·tion \-,pän-də-'rā-shən, ,prē-\ *noun*

[2]**pre·pon·der·ate** \-'pän-də-rət\ *adjective* (1802)
: PREPONDERANT
— pre·pon·der·ate·ly *adverb*

prep·o·si·tion \,pre-pə-'zi-shən\ *noun* [Middle English *preposicioun,* from Latin *praeposition-, praepositio,* from *praeponere* to put in front, from *prae-* pre- + *ponere* to put — more at POSITION] (14th century)
: a function word that typically combines with a noun phrase to form a phrase which usually expresses a modification or predication ■
— prep·o·si·tion·al \-'zish-nəl, -'zi-shə-nᵊl\ *adjective*
— prep·o·si·tion·al·ly *adverb*

pre·pos·i·tive \pri-'pä-zə-tiv, -'päz-tiv\ *adjective* [Late Latin *praepositivus,* from Latin *praepositus,* past participle of *praeponere*] (1583)
: put before **:** PREFIXED
— pre·pos·i·tive·ly *adverb*

pre·pos·sess \,prē-pə-'zes *also* -'ses\ *transitive verb* (1614)
1 *obsolete* **:** to take previous possession of
2 : to cause to be preoccupied
3 : to influence beforehand especially favorably

pre·pos·sess·ing *adjective* (1642)
1 *archaic* **:** creating prejudice
2 : tending to create a favorable impression **:** ATTRACTIVE

pre·pos·ses·sion \,prē-pə-'ze-shən *also* -'se-\ *noun* (1648)
1 *archaic* **:** prior possession
2 : an attitude, belief, or impression formed beforehand **:** PREJUDICE
3 : an exclusive concern with one idea or object **:** PREOCCUPATION
synonym see PREDILECTION

pre·pos·ter·ous \pri-'päs-t(ə-)rəs\ *adjective* [Latin *praeposterus,* literally, in the wrong order, from *prae-* + *posterus* hinder, following — more at POSTERIOR] (1542)
: contrary to nature, reason, or common sense **:** ABSURD
— pre·pos·ter·ous·ly *adverb*
— pre·pos·ter·ous·ness *noun*

pre·po·ten·cy \(,)prē-'pō-tᵊn(t)-sē\ *noun* (1646)
1 : the quality or state of being prepotent **:** PREDOMINANCE
2 : unusual ability of an individual or strain to transmit its characters to offspring because of homozygosity for numerous dominant genes

pre·po·tent \-tᵊnt\ *adjective* [Middle English, from Latin *praepotent-, praepotens,* from *prae-* + *potens* powerful — more at POTENT] (15th century)
1 a : having exceptional power, authority, or influence **b :** exceeding others in power
2 : exhibiting genetic prepotency
— pre·po·tent·ly *adverb*

[1]**prep·py** *or* **prep·pie** \'pre-pē\ *noun, plural* **prep·pies** [[1]*preposition*] (1967)
1 : a student at or a graduate of a preparatory school
2 : a person deemed to dress or behave like a preppy

[2]**preppy** *or* **preppie** *adjective* (1967)
1 : relating to, characteristic of, or being a preppy
2 : relating to or being a style of dress characterized especially by classic clothing and neat appearance
— prep·pi·ly \'pre-pə-lē\ *adverb*
— prep·pi·ness \'pre-pē-nəs\ *noun*

pre·pran·di·al \(,)prē-'pran-dē-əl\ *adjective* (1822)
: of, relating to, or suitable for the time just before dinner ⟨a *preprandial* drink⟩

pre·preg \,prē-'preg, 'prē-,\ *noun* [*pre-* + *impregnated*] (1954)
: a reinforcing or molding material (as paper or glass cloth) already impregnated with a synthetic resin

[1]**pre·print** \'prē-,print, ,prē-'print\ *noun* (1889)
1 : an issue of a technical paper often in preliminary form before its publication in a journal
2 : something (as an advertisement) printed before the rest of the publication in which it is to appear

☐ USAGE

preposition Is it all right to end a sentence with a preposition? The answer is yes, even though many people cherish the notion that it is not. The notion of the inappropriateness of the deferred preposition goes back to the middle of the 17th century and is probably based on Latin grammar. But modern commentators know that the construction is standard. It is even required by these constructions: a restrictive clause introduced by *that* ⟨the magazine that I most wanted to have a story published *in* —Truman Capote⟩, a restrictive clause from which *that* has been omitted ⟨as soon as I finish the short story I'm working *on* —F. Scott Fitzgerald⟩, a clause introduced by *what* ⟨we can only speculate what they might have been *about* —Elizabeth Drew⟩, an infinitive clause ⟨both bought new suits to be married *in* —Russell Baker⟩, a participial clause ⟨relatively new and therefore worth keeping track *of* —Kenneth G. Wilson⟩, and a passive construction ⟨people deficient in learning were looked down *on* —Cynthia Ozick⟩. Questions, too, require the deferred preposition ⟨has the car been paid *for*?⟩ ⟨what does she look *like*?⟩. The preposition at the end is so natural that consciously avoiding it can cause mistakes ⟨a young man *with* whom most Americans could identify *with* —Herbert Warren Wind⟩.

2pre·print \(ˌ)prē-ˈprint\ *transitive verb* (1926)
: to print in advance for later use

pre·pro·cess \(ˌ)prē-ˈprä-ˌses, -ˈprō-, -səs\ *transitive verb* (1942)
: to do preliminary processing of (as data)
— **pre·pro·ces·sor** \-ˌse-sər; -sə-sər, -ˌsȯr\ *noun*

pre·pro·fes·sion·al \ˌprē-prə-ˈfesh-nəl, -ˈfe-shə-nᵊl\ *adjective* (1926)
: of or relating to the period preceding specific study for or practice of a profession

prep school *noun* (1895)
: PREPARATORY SCHOOL

pre·pu·ber·al \(ˌ)prē-ˈpyü-b(ə-)rəl\ *adjective* (circa 1935)
: PREPUBERTAL

pre·pu·ber·tal \-bər-tᵊl\ *adjective* (1859)
: of or relating to prepuberty

pre·pu·ber·ty \-bər-tē\ *noun* (1922)
: the period immediately preceding puberty

pre·pu·bes·cence \ˌprē-pyü-ˈbe-sᵊn(t)s\ *noun* (1916)
: PREPUBERTY

pre·pu·bes·cent \-sᵊnt\ *adjective* (1904)
: PREPUBERTAL
— **prepubescent** *noun*

pre·puce \ˈprē-ˌpyüs\ *noun* [Middle English, from Middle French, from Latin *praeputium*] (15th century)
: FORESKIN; *also* **:** a similar fold investing the clitoris
— **pre·pu·tial** \prē-ˈpyü-shəl\ *adjective*

pre·quel \ˈprē-kwəl\ *noun* [*pre-* + *-quel* (as in *sequel*)] (1972)
: a literary or dramatic work whose story precedes that of an earlier work

Pre–Ra·pha·el·ite \(ˌ)prē-ˈra-fē-ə-ˌlīt, -ˈrā-, -ˈrä-\ *noun* (1850)
1 a : a member of a brotherhood of artists formed in England in 1848 to restore the artistic principles and practices regarded as characteristic of Italian art before Raphael **b :** an artist or writer influenced by this brotherhood
2 : a modern artist dedicated to restoring early Renaissance ideals or methods
— **Pre–Raphaelite** *adjective*
— **Pre–Ra·pha·el·it·ism** \-ˌlī-ˌti-zəm\ *noun*

pre·reg·is·tra·tion \ˌprē-ˌre-jə-ˈstrā-shən\ *noun* (1967)
: a special registration (as for returning students) prior to an official registration period
— **pre·reg·is·ter** \(ˌ)prē-ˈre-jə-stər\ *intransitive verb*

pre·req·ui·site \(ˌ)prē-ˈre-kwə-zət\ *noun* (1633)
: something that is necessary to an end or to the carrying out of a function
— **prerequisite** *adjective*

pre·rog·a·tive \pri-ˈrä-gə-tiv\ *noun* [Middle English, from Middle French & Latin; Middle French, from Latin *praerogativa,* Roman century voting first in the comitia, privilege, from feminine of *praerogativus* voting first, from *praerogatus,* past participle of *praerogare* to ask for an opinion before another, from *prae-* + *rogare* to ask — more at RIGHT] (15th century)
1 a : an exclusive or special right, power, or privilege: as (1) **:** one belonging to an office or an official body (2) **:** one belonging to a person, group, or class of individuals (3) **:** one possessed by a nation as an attribute of sovereignty **b :** the discretionary power inhering in the British Crown
2 : a distinctive excellence
— **pre·rog·a·tived** \-tivd\ *adjective*

1pres·age \ˈpre-sij, *also* pri-ˈsāj\ *noun* [Middle English, from Latin *praesagium,* from *praesagus* having a foreboding, from *prae-* + *sagus* prophetic — more at SEEK] (14th century)
1 : something that foreshadows or portends a future event **:** OMEN
2 : an intuition or feeling of what is going to happen in the future
3 *archaic* **:** PROGNOSTICATION
4 : warning or indication of the future
— **pre·sage·ful** \pri-ˈsāj-fəl\ *adjective*

2pre·sage \ˈpre-sij, pri-ˈsāj\ *verb* **pre·saged; pre·sag·ing** (1562)
transitive verb
1 : to give an omen or warning of **:** FORESHADOW
2 : FORETELL, PREDICT
intransitive verb
: to make or utter a prediction
— **pre·sag·er** *noun, obsolete*

pre·sanc·ti·fied \(ˌ)prē-ˈsaŋ(k)-ti-ˌfīd\ *adjective* (1758)
: consecrated at a previous service — used of eucharistic elements

pres·by·ope \ˈprez-bē-ˌōp; ˈpres-bē-, -pē-\ *noun* [probably from French, from Greek *presbys* old man + *ōps* eye — more at EYE] (circa 1857)
: one affected with presbyopia

pres·by·o·pia \ˌprez-bē-ˈō-pē-ə, ˌpres-\ *noun* [New Latin] (1793)
: a visual condition which becomes apparent especially in middle age and in which loss of elasticity of the lens of the eye causes defective accommodation and inability to focus sharply for near vision
— **pres·by·o·pic** \-ˈō-pik, -ˈä-\ *adjective or noun*

pres·by·ter \ˈprez-bə-tər, ˈpres-\ *noun* [Late Latin, elder, priest, from Greek *presbyteros,* comparative of *presbys* old man, elder; akin to Greek *pro* before and Greek *bainein* to go — more at FOR, COME] (1597)
1 : a member of the governing body of an early Christian church
2 : a member of the order of priests in churches having episcopal hierarchies that include bishops, priests, and deacons
3 : ELDER 4b
— **pres·byt·er·ate** \prez-ˈbi-tə-rət, pres-, -ˌrāt\ *noun*

1pres·by·te·ri·al \ˌprez-bə-ˈtir-ē-əl, ˌpres-\ *adjective* (circa 1600)
: of or relating to presbyters or a presbytery
— **pres·by·te·ri·al·ly** \-ē-ə-lē\ *adverb*

2presbyterial *noun, often capitalized* (1928)
: an organization of Presbyterian women associated with a presbytery

1Pres·by·te·ri·an \-ē-ən\ *noun* (1640)
: a member of a Presbyterian church

2Presbyterian *adjective* (1641)
1 *often not capitalized* **:** characterized by a graded system of representative ecclesiastical bodies (as presbyteries) exercising legislative and judicial powers
2 : of, relating to, or constituting a Protestant Christian church that is presbyterian in government and traditionally Calvinistic in doctrine
— **Pres·by·te·ri·an·ism** \-ē-ə-ˌni-zəm\ *noun*

pres·by·tery \ˈprez-bə-ˌter-ē, ˈpres-, -bə-trē\ *noun, plural* **-ter·ies** [Middle English & Late Latin; Middle English *presbytory* part of church reserved for clergy, from Late Latin *presbyterium* group of presbyters, part of church reserved for clergy, from Greek *presbyterion* group of presbyters, from *presbyteros* elder, priest] (15th century)
1 : the part of a church reserved for the officiating clergy
2 : a ruling body in presbyterian churches consisting of the ministers and representative elders from congregations within a district
3 : the jurisdiction of a presbytery
4 : the house of a Roman Catholic parish priest

1pre·school \ˈprē-ˌskül, (ˌ)prē-ˈ\ *adjective* (1914)
: of, relating to, or constituting the period in a child's life from infancy to the age of five or six that ordinarily precedes attendance at elementary school

2pre·school \ˈprē-ˌskül\ *noun* (circa 1925)
: NURSERY SCHOOL, KINDERGARTEN

pre·school·er \-ˌskü-lər\ *noun* (1946)
1 : a child not yet old enough for school
2 : a child attending a preschool

pre·science \ˈpre-sh(ē-)ən(t)s, ˈprē-, -s(ē-)ən(t)s\ *noun* [Middle English, from Late Latin *praescientia,* from Latin *praescient-, praesciens,* present participle of *praescire* to know beforehand, from *prae-* + *scire* to know — more at SCIENCE] (14th century)
: foreknowledge of events: **a :** divine omniscience **b :** human anticipation of the course of events **:** FORESIGHT
— **pre·scient** \-sh(ē-)ənt, -s(ē-)ənt\ *adjective*
— **pre·scient·ly** *adverb*

pre·sci·en·tif·ic \ˌprē-ˌsī-ən-ˈti-fik\ *adjective* (1858)
: of, relating to, or having the characteristics of a period before the rise of modern science or a state prior to the application of the scientific method

pre·scind \pri-ˈsind\ *verb* [Latin *praescindere* to cut off in front, from *prae-* + *scindere* to cut — more at SHED] (1650)
intransitive verb
: to withdraw one's attention
transitive verb
: to detach for purposes of thought

pre·score \(ˌ)prē-ˈskōr, -ˈskȯr\ *transitive verb* (1937)
: to record (as sound) in advance for use when the corresponding scenes are photographed in making movies

pre·scribe \pri-ˈskrīb\ *verb* **pre·scribed; pre·scrib·ing** [Middle English, from Latin *praescribere* to write at the beginning, dictate, order, from *prae-* + *scribere* to write — more at SCRIBE] (15th century)
intransitive verb
1 : to lay down a rule **:** DICTATE
2 [Middle English, from Medieval Latin *praescribere,* from Latin, to write at the beginning] **:** to claim a title to something by right of prescription
3 : to write or give medical prescriptions
4 : to become by prescription invalid or unenforceable
transitive verb
1 a : to lay down as a guide, direction, or rule of action **:** ORDAIN **b :** to specify with authority
2 : to designate or order the use of as a remedy
— **pre·scrib·er** *noun*

pre·script \ˈprē-ˌskript, pri-ˈ\ *adjective* [Middle English, from Latin *praescriptus,* past participle] (circa 1540)
: prescribed as a rule
— **pre·script** \ˈprē-ˌskript\ *noun*

pre·scrip·tion \pri-ˈskrip-shən\ *noun* [partly from Middle English *prescripcion* establishment of a claim, from Middle French *prescription,* from Late Latin *praescription-, praescriptio,* from Latin, act of writing at the beginning, order, limitation of subject matter, from *praescribere;* partly from Latin *praescription-, praescriptio* order] (14th century)
1 a : the establishment of a claim of title to something under common law usually by use and enjoyment for a period fixed by statute **b :** the right or title acquired under common law by such possession
2 : the process of making claim to something by long use and enjoyment
3 : the action of laying down authoritative rules or directions
4 a : a written direction for a therapeutic or corrective agent; *specifically* **:** one for the

preparation and use of a medicine **b :** a prescribed medicine **c :** something like a doctor's prescription ⟨*prescriptions* for economic recovery⟩
5 a : ancient or long continued custom **b :** a claim founded upon ancient custom or long continued use
6 : something prescribed as a rule

prescription drug *noun* (1951)
: a drug that can be obtained only by means of a physician's prescription

pre·scrip·tive \pri-ˈskrip-tiv\ *adjective* (1748)
1 : serving to prescribe ⟨*prescriptive* rules of usage⟩
2 : acquired by, founded on, or determined by prescription or by long-standing custom
— pre·scrip·tive·ly *adverb*

pre·se·lect \ˌprē-sə-ˈlekt\ *transitive verb* (circa 1859)
: to choose in advance usually on the basis of a particular criterion
— pre·se·lec·tion \-ˈlek-shən\ *noun*

pre·sell \(ˌ)prē-ˈsel\ *transitive verb* **-sold** \-ˈsōld\; **-sell·ing** (1947)
1 : to precondition (as a customer) for subsequent purchase or create advance demand for (as a product) especially through marketing strategies
2 : to sell in advance ⟨raised money to publish the book by *preselling* film rights⟩

pres·ence \ˈpre-zᵊn(t)s\ *noun* (14th century)
1 : the fact or condition of being present
2 a : the part of space within one's immediate vicinity **b :** the neighborhood of one of superior or especially royal rank
3 *archaic* **:** COMPANY 2a
4 : one that is present: as **a :** the actual person or thing that is present **b :** something present of a visible or concrete nature
5 a : the bearing, carriage, or air of a person; *especially* **:** stately or distinguished bearing **b :** a quality of poise and effectiveness that enables a performer to achieve a close relationship with his audience
6 : something (as a spirit) felt or believed to be present

presence of mind (1665)
: self-control so maintained in an emergency or in an embarrassing situation that one can say or do the right thing

[1]pres·ent \ˈpre-zᵊnt\ *noun* [Middle English, from Old French, from *presenter*] (13th century)
: something presented **:** GIFT

[2]pre·sent \pri-ˈzent\ *verb* [Middle English, from Old French *presenter,* from Latin *praesentare,* from *praesent-, praesens,* adjective] (14th century)
transitive verb
1 a (1) **:** to bring or introduce into the presence of someone especially of superior rank or status (2) **:** to introduce socially **b :** to bring (as a play) before the public
2 : to make a gift to
3 : to give or bestow formally
4 a : to lay (as a charge) before a court as an object of inquiry **b :** to bring a formal public charge, indictment, or presentment against
5 : to nominate to a benefice
6 a : to offer to view **:** SHOW **b :** to bring to one's attention ⟨this *presents* a problem⟩
7 : to act the part of **:** PERFORM
8 : to aim, point, or direct (as a weapon) so as to face something or in a particular direction
intransitive verb
1 : to present a weapon
2 : to become manifest
3 : to come forward as a patient
4 : to make a presentation
synonym see GIVE
— pre·sent·er *noun*

[3]pres·ent \ˈpre-zᵊnt\ *adjective* [Middle English, from Old French, from Latin *praesent-, praesens,* from present participle of *praeesse* to be before one, from *prae-* pre- + *esse* to be — more at IS] (14th century)
1 : now existing or in progress
2 a : being in view or at hand **b :** existing in something mentioned or under consideration
3 : constituting the one actually involved, at hand, or being considered
4 : of, relating to, or constituting a verb tense that is expressive of present time or the time of speaking
5 *obsolete* **:** ATTENTIVE
6 *archaic* **:** INSTANT, IMMEDIATE
— pres·ent·ness *noun*

[4]pres·ent \ˈpre-zᵊnt\ *noun* (14th century)
1 a *obsolete* **:** present occasion or affair **b** *plural* **:** the present words or statements; *specifically* **:** the legal instrument or other writing in which these words are used
2 a : the present tense of a language **b :** a verb form in the present tense
3 : the present time
— at present : at or during this time **:** NOW

pre·sent·able \pri-ˈzen-tə-bəl\ *adjective* (circa 1626)
1 : capable of being presented
2 : being in condition to be seen or inspected especially by the critical
— pre·sent·abil·i·ty \-ˌzen-tə-ˈbi-lə-tē\ *noun*
— pre·sent·able·ness \-ˈzen-tə-bəl-nəs\ *noun*
— pre·sent·ably \-blē\ *adverb*

pre·sent arms \pri-ˈzent-\ *noun* [from the command *present arms!*] (circa 1884)
1 : a position in the manual of arms in which the rifle is held vertically in front of the body
2 : a command to assume the position of present arms or to give a hand salute

pre·sen·ta·tion \ˌprē-ˌzen-ˈtā-shən, ˌpre-zᵊn-, ˌprē-zᵊn-\ *noun* (15th century)
1 a : the act of presenting **b :** the act, power, or privilege especially of a patron of applying to the bishop or ordinary for instituting someone into a benefice
2 : something presented: as **a :** a symbol or image that represents something **b :** something offered or given **:** GIFT **c :** something set forth for the attention of the mind **d :** a descriptive or persuasive account (as by a salesman of a product)
3 : the position in which the fetus lies in the uterus in labor with respect to the mouth of the uterus
4 : an immediate object of perception, cognition, or memory
5 *often capitalized* **:** a church feast on November 21 celebrating the presentation of the Virgin Mary in the temple
6 : the method by which radio, navigation, or radar information is given to the operator (as the pilot of an airplane)
— pre·sen·ta·tion·al \-shnəl, -shə-nᵊl\ *adjective*

pre·sen·ta·tive \pri-ˈzen-tə-tiv, ˈpre-zᵊn-ˌtā-\ *adjective* (circa 1842)
: known, knowing, or capable of being known directly rather than through cogitation

pres·ent–day \ˈpre-zᵊnt-ˈdā\ *adjective* (1887)
: now existing or occurring

pre·sen·tee \ˌpre-zᵊn-ˈtē, pri-ˌzen-\ *noun* (15th century)
: one who is presented or to whom something is presented

pre·sen·tient \pri-ˈsen(t)-sh(ē-)ənt, ˈprē-; pri-ˈzen(t)-\ *adjective* [Latin *praesentient-, praesentiens,* present participle of *praesentire*] (1814)
: having a presentiment

pre·sen·ti·ment \pri-ˈzen-tə-mənt\ *noun* [French *pressentiment,* from Middle French, from *pressentir* to have a presentiment, from Latin *praesentire* to feel beforehand, from *prae-* + *sentire* to feel — more at SENSE] (1714)
: a feeling that something will or is about to happen **:** PREMONITION
— pre·sen·ti·men·tal \-ˌzen-tə-ˈmen-tᵊl\ *adjective*

pres·ent·ism \ˈpre-zᵊn-ˌti-zəm\ *noun* [[3]present] (1923)
: an outlook dominated by present-day attitudes and experiences
— pres·ent·ist \-zᵊn-tist\ *adjective*

pres·ent·ly \ˈpre-zᵊnt-lē\ *adverb* (14th century)
1 a *archaic* **:** at once **b :** before long **:** without undue delay
2 : at the present time **:** NOW ■

pre·sent·ment \pri-ˈzent-mənt\ *noun* (14th century)
1 : the act of presenting to an authority a formal statement of a matter to be dealt with; *specifically* **:** the notice taken or statement made by a grand jury of an offense from their own knowledge without a bill of indictment laid before them
2 : the act of offering at the proper time and place a document (as a bill of exchange) that calls for acceptance or payment by another
3 a : the act of presenting to view or consciousness **b :** something set forth, presented, or exhibited **c :** the aspect in which something is presented

present participle *noun* (1864)
: a participle that typically expresses present action in relation to the time expressed by the finite verb in its clause and that in English is formed with the suffix *-ing* and is used in the formation of the progressive tenses

present perfect *adjective* (1887)
: of, relating to, or constituting a verb tense that is formed in English with *have* and that expresses action or state completed at the time of speaking
— present perfect *noun*

present tense *noun* (14th century)
: the tense of a verb that expresses action or state in the present time and is used of what occurs or is true at the time of speaking and of what is habitual or characteristic or is always or necessarily true, that is sometimes used to refer to action in the past, and that is sometimes used for future events

present value *noun* (1831)
: the sum of money which if invested now at a given rate of compound interest will accumulate exactly to a specified amount at a specified future date

pres·er·va·tion·ist \ˌpre-zər-ˈvā-sh(ə-)nist\ *noun* (1927)
: one who advocates preservation (as of a biological species or a historical landmark)

[1]pre·ser·va·tive \pri-ˈzər-və-tiv\ *adjective* (14th century)
: having the power of preserving

[2]preservative *noun* (15th century)
: something that preserves or has the power of preserving; *specifically* **:** an additive used to protect against decay, discoloration, or spoilage

□ USAGE
presently Both senses 1b and 2 are flourishing in current English, but many commentators have objected to sense 2. Since this sense has been in continuous use since the 15th century, it is not clear why it is objectionable. Perhaps a note in the *Oxford English Dictionary* (1909) that the sense has been obsolete since the 17th century in literary English is to blame, but the note goes on to observe that the sense is in regular use in most English dialects. The last citation in that dictionary is from a 1901 Leeds newspaper, written in Standard English. Sense 2 is most common in contexts relating to business and politics ⟨the fastest-rising welfare cost is Medicaid, *presently* paid by the states and cities —William Safire⟩

[1]pre•serve \pri-ˈzərv\ *verb* **pre•served; pre•serv•ing** [Middle English, from Middle French *preserver,* from Medieval Latin *praeservare,* from Late Latin, to observe beforehand, from Latin *prae-* + *servare* to keep, guard, observe — more at CONSERVE] (14th century)
transitive verb
1 : to keep safe from injury, harm, or destruction **:** PROTECT
2 a : to keep alive, intact, or free from decay **b :** MAINTAIN
3 a : to keep or save from decomposition **b :** to can, pickle, or similarly prepare for future use
4 : to keep up and reserve for personal or special use
intransitive verb
1 : to make preserves
2 : to raise and protect game for purposes of sport
3 : to be able to be preserved (as by canning)
— pre•serv•abil•i•ty \-ˌzər-və-ˈbi-lə-tē\ *noun*
— pre•serv•able \-ˈzər-və-bəl\ *adjective*
— pres•er•va•tion \ˌpre-zər-ˈvā-shən\ *noun*
— pre•serv•er \pri-ˈzər-vər\ *noun*

[2]preserve *noun* (1600)
1 : fruit canned or made into jams or jellies or cooked whole or in large pieces in a syrup so as to keep its shape — often used in plural
2 : an area restricted for the protection and preservation of natural resources (as animals or trees); *especially* **:** one used primarily for regulated hunting or fishing
3 : something regarded as reserved for certain persons

pre•shrink \(ˌ)prē-ˈshriŋk, *especially Southern* -ˈsriŋk\ *transitive verb* **-shrank** \-ˈshraŋk, -ˈsraŋk\; **-shrunk** \-ˈshrəŋk, -ˈsrəŋk\ (1926)
: to shrink (as a fabric) before making into a garment so that it will not shrink much when washed

pre•side \pri-ˈzīd\ *intransitive verb* **pre•sid•ed; pre•sid•ing** [Latin *praesidēre* to guard, preside over, from *prae-* + *sedēre* to sit — more at SIT] (1608)
1 : to exercise guidance, direction, or control
2 a : to occupy the place of authority **:** act as president, chairman, or moderator **b :** to occupy a position similar to that of a president or chairman
3 : to occupy a position of featured instrumental performer — usually used with *at* ⟨*presided* at the organ⟩
— pre•sid•er *noun*

pres•i•den•cy \ˈpre-zə-dən(t)-sē, ˈprez-dən(t)- *also* ˈpre-zə-ˌden(t)-sē\ *noun, plural* **-cies** (1591)
1 a : the office of president **b** (1) **:** the office of president of the U.S. (2) **:** the American governmental institution comprising the office of president and various associated administrative and policy-making agencies
2 : the term during which a president holds office
3 : the action or function of one that presides **:** SUPERINTENDENCE
4 : a Mormon executive council of the church or a stake consisting of a president and two counselors

pres•i•dent \ˈpre-zə-dənt, ˈprez-dənt, ˈpre-zə-ˌdent *in rapid speech* ˈpre-zᵊnt\ *noun* [Middle English, from Middle French, from Latin *praesident-, praesidens,* from present participle of *praesidēre*] (14th century)
1 : an official chosen to preside over a meeting or assembly
2 : an appointed governor of a subordinate political unit
3 : the chief officer of an organization (as a corporation or institution) usually entrusted with the direction and administration of its policies
4 : the presiding officer of a governmental body
5 a : an elected official serving as both chief of state and chief political executive in a republic having a presidential government **b :** an elected official having the position of chief of state but usually only minimal political powers in a republic having a parliamentary government
— pres•i•den•tial \ˌpre-zə-ˈden(t)-shəl, ˌprez-ˈden(t)-\ *adjective*
— pres•i•den•tial•ly \-sh(ə-)lē\ *adverb*
— pres•i•dent•ship \ˈpre-zə-dənt-ˌship, ˈprez-dənt-, ˈpre-zə-ˌdent-\ *noun*

presidential government *noun* (1902)
: a system of government in which the president is constitutionally independent of the legislature

Presidents' Day *noun* (1952)
: WASHINGTON'S BIRTHDAY 2

pre•sid•i•al \pri-ˈsi-dē-əl, prī-, -ˈzi-\ *adjective* [Late Latin *praesidialis,* from Latin *praesidium* garrison, from *praesid-, praeses* guard, governor, from *praesidēre*] (1611)
1 [French *présidial,* from Middle French, alteration of *presidal,* from Late Latin *praesidalis* of a provincial governor, from Latin *praesid-, praeses*] **:** PROVINCIAL 1
2 : of, having, or constituting a garrison
3 : of or relating to a president **:** PRESIDENTIAL

pre•sid•i•ary \-dē-ˌer-ē\ *adjective* (1599)
: PRESIDIAL 2

pre•si•dio \pri-ˈsē-dē-ˌō, -ˈsi-, -ˈzē-, -ˈzi-\ *noun, plural* **-di•os** [Spanish, from Latin *praesidium*] (1763)
: a garrisoned place; *especially* **:** a military post or fortified settlement in areas currently or originally under Spanish control

pre•sid•i•um \pri-ˈsi-dē-əm, prī-, -ˈzi-\ *noun, plural* **-ia** \-dē-ə\ *or* **-iums** [Russian *prezidium,* from Latin *praesidium* garrison] (1920)
1 : a permanent executive committee selected especially in Communist countries to act for a larger body
2 : a nongovernmental executive committee

pre•sig•ni•fy \(ˌ)prē-ˈsig-nə-ˌfī\ *transitive verb* [Latin *praesignificare,* from *prae-* + *significare* to signify] (1586)
: to intimate or signify beforehand **:** PRESAGE

[1]pre•soak \(ˌ)prē-ˈsōk\ *transitive verb* (1919)
: to soak beforehand

[2]pre•soak \ˈprē-ˌsōk\ *noun* (1919)
1 : an instance of presoaking
2 : a preparation used in presoaking clothes

pre–So•crat•ic \ˌprē-sə-ˈkra-tik, -sō-\ *adjective* (1871)
: of or relating to Greek philosophers before Socrates
— pre–Socratic *noun*

pre•sort \(ˌ)prē-ˈsȯrt\ *transitive verb* (1951)
: to sort (outgoing mail) by zip code usually before delivery to a post office

[1]press \ˈpres\ *noun* [Middle English *presse,* from Old French, from *presser* to press] (13th century)
1 a : a crowd or crowded condition **:** THRONG **b :** a thronging or crowding forward or together
2 a : an apparatus or machine by which a substance is cut or shaped, an impression of a body is taken, a material is compressed, pressure is applied to a body, liquid is expressed, or a cutting tool is fed into the work by pressure **b :** a building containing presses or a business using presses
3 : CLOSET, CUPBOARD
4 a : an action of pressing or pushing **:** PRESSURE **b :** an aggressive pressuring defense employed in basketball often over the entire court area
5 : the properly smoothed and creased condition of a freshly pressed garment ⟨out of *press*⟩
6 a : PRINTING PRESS **b :** the act or the process of printing **c :** a printing or publishing establishment
7 a : the gathering and publishing or broadcasting of news **:** JOURNALISM **b :** newspapers, periodicals, and often radio and television news broadcasting **c :** news reporters, publishers, and broadcasters **d :** comment or notice in newspapers and periodicals ⟨is getting a good *press*⟩
8 : any of various pressure devices (as one for keeping sporting gear from warping when not in use)
9 : a lift in weight lifting in which the weight is raised to shoulder height and then smoothly extended overhead without assist from the legs — compare CLEAN AND JERK, SNATCH

[2]press *verb* [Middle English, from Middle French *presser,* from Latin *pressare,* frequentative of *premere* to press; probably akin to Russian na*peret'* to press] (14th century)
transitive verb
1 : to act upon through steady pushing or thrusting force exerted in contact **:** SQUEEZE
2 a : ASSAIL, HARASS **b :** AFFLICT, OPPRESS
3 a : to squeeze out the juice or contents of **b :** to squeeze with apparatus or instruments to a desired density, smoothness, or shape
4 a : to exert influence on **:** CONSTRAIN **b :** to try hard to persuade **:** BESEECH, ENTREAT
5 : to move by means of pressure
6 a : to lay stress or emphasis on **b :** to insist on or request urgently
7 : to follow through (a course of action)
8 : to clasp in affection or courtesy
9 : to make (a phonograph record) from a matrix
intransitive verb
1 : to crowd closely **:** MASS
2 : to force or push one's way
3 : to seek urgently **:** CONTEND
4 : to require haste or speed in action
5 : to exert pressure
6 : to take or hold a press
7 : to employ a press in basketball
— press•er *noun*
— press the flesh : to greet and shake hands with people especially while campaigning for political office

[3]press *verb* [alteration of obsolete *prest* to enlist by giving pay in advance] (1578)
transitive verb
1 : to force into service especially in an army or navy **:** IMPRESS
2 a : to take by authority especially for public use **:** COMMANDEER **b :** to take and force into any usually emergency service
intransitive verb
: to impress men as soldiers or sailors

[4]press *noun* (1599)
1 : impressment into service especially in a navy
2 *obsolete* **:** a warrant for impressing recruits

press agent *noun* [[1]*press*] (1883)
: an agent employed to establish and maintain good public relations through publicity
— press–agent *verb*
— press–agent•ry \-ˈā-jən-trē\ *noun*

press•board \ˈpres-ˌbōrd, -ˌbȯrd\ *noun* (1849)
1 : IRONING BOARD; *especially* **:** a small one for sleeves
2 : a strong highly glazed composition board resembling vulcanized fiber

press box *noun* (1889)
: a space reserved for reporters (as at a stadium)

press cloth *noun* (1899)
: a cloth used between an iron and a garment

press conference *noun* (1937)
: an interview or announcement given by a public figure to the press by appointment

press–gang \ˈpres-ˌgaŋ\ *noun* [[4]*press*] (1693)
: a detachment of men under command of an officer empowered to force men into military or naval service

— **press–gang** *transitive verb*
press·ing *adjective* (1616)
1 : urgently important **:** CRITICAL
2 : EARNEST, WARM
— **press·ing·ly** \'pre-siŋ-lē\ *adverb*
press kit *noun* (1968)
: a collection of promotional materials for distribution to the press
press·man \'pres-mən, -ˌman\ *noun* (1598)
1 : an operator of a press; *especially* **:** the operator of a printing press
2 *British* **:** NEWSPAPERMAN
press·mark \-ˌmärk\ *noun* [[1]*press* (closet)] (1802)
chiefly British **:** a mark or number assigned to a book to indicate its location in a library
press of sail (1794)
: the fullest amount of sail that a ship can crowd on — called also *press of canvas*
pres·sor \'pre-ˌsȯr, -sər\ *adjective* [Late Latin, one that presses, from Latin *premere* to press — more at PRESS] (circa 1890)
: raising or tending to raise blood pressure; *also* **:** involving vasoconstriction
press·room \'pres-ˌrüm, -ˌru̇m\ *noun* (1683)
1 : a room in a printing plant containing the printing presses
2 : a room (as at the White House) for the use of members of the press
press·run \-ˌrən\ *noun* (1945)
: a continuous operation of a printing press producing a specified number of copies; *also* **:** the number of copies printed
press secretary *noun* (1951)
: a person officially in charge of press relations for a usually prominent public figure ⟨the President's *press secretary*⟩
press–up \'pres-(ˌ)əp\ (1936)
British **:** PUSH-UP
[1]**pres·sure** \'pre-shər\ *noun* [Middle English, from Late Latin *pressura,* from Latin, action of pressing, pressure, from *pressus,* past participle of *premere*] (14th century)
1 a : the burden of physical or mental distress **b :** the constraint of circumstance **:** the weight of social or economic imposition
2 : the application of force to something by something else in direct contact with it **:** COMPRESSION
3 *archaic* **:** IMPRESSION, STAMP
4 a : the action of a force against an opposing force **b :** the force or thrust exerted over a surface divided by its area **c :** ELECTROMOTIVE FORCE
5 : the stress or urgency of matters demanding attention **:** EXIGENCY ⟨people who work well under *pressure*⟩
6 : the force of selection that results from one or more agents and tends to reduce a population of organisms ⟨population *pressure*⟩ ⟨predation *pressure*⟩
7 : atmospheric pressure
8 : a sensation aroused by moderate compression of a body part or surface
— **pres·sure·less** *adjective*
[2]**pressure** *transitive verb* **pres·sured; pres·sur·ing** \'pre-sh(ə-)riŋ\ (1938)
1 : to apply pressure to
2 : PRESSURIZE
3 : to cook in a pressure cooker
pressure cabin *noun* (1935)
: a pressurized cabin
pressure cooker *noun* (1915)
1 : an airtight utensil for quick cooking or preserving of foods by means of superheated steam under pressure
2 : a situation or environment that is fraught with emotional or social pressures
— **pressure–cook** *verb*
pressure gauge *noun* (1862)
: a gauge for indicating fluid pressure
pressure group *noun* (1928)
: an interest group organized to influence public and especially government policy but not to elect candidates to office
pressure point *noun* (1909)
: a point where a blood vessel runs near a bone and can be compressed (as to check bleeding) by the application of pressure against the bone
pressure suit *noun* (1936)
: an inflatable suit for high-altitude or space flight to protect the body from low pressure
pressure wave *noun* (1942)
: a wave (as a sound wave) in which the propagated disturbance is a variation of pressure in a material medium — called also *P-wave*
pres·sur·ise *British variant of* PRESSURIZE
pres·sur·ize \'pre-shə-ˌrīz\ *transitive verb* **-ized; -iz·ing** (1938)
1 : to confine the contents of under a pressure greater than that of the outside atmosphere; *especially* **:** to maintain near-normal atmospheric pressure in during high-altitude or spaceflight (as by means of a supercharger)
2 : to apply pressure to
3 : to design to withstand pressure
— **pres·sur·i·za·tion** \ˌpre-sh(ə-)rə-'zā-shən\ *noun*
— **pres·sur·iz·er** *noun*
press·work \'pres-ˌwərk\ *noun* (1771)
: the operation, management, or product of a printing press; *especially* **:** the branch of printing concerned with the actual transfer of ink from form or plates to paper
prest \'prest\ *adjective* [Middle English, from Old French, from Latin *praestus* — more at PRESTO] (14th century)
obsolete **:** READY
pres·ti·dig·i·ta·tion \ˌpres-tə-ˌdi-jə-'tā-shən\ *noun* [French, from *prestidigitateur* prestidigitator, from *preste* nimble, quick (from Italian *presto*) + Latin *digitus* finger — more at DIGIT] (1859)
: SLEIGHT OF HAND, LEGERDEMAIN
— **pres·ti·dig·i·ta·tor** \-'di-jə-ˌtā-tər\ *noun*
pres·tige \pre-'stēzh, -'stēj\ *noun, often attributive* [French, from Middle French, conjuror's trick, illusion, from Latin *praestigiae,* plural, conjuror's tricks, from *praestringere* to graze, blunt, constrict, from *prae-* + *stringere* to bind tight — more at STRAIN] (1829)
1 : standing or estimation in the eyes of people **:** weight or credit in general opinion
2 : commanding position in people's minds ◆
synonym see INFLUENCE
— **pres·tige·ful** \-fəl\ *adjective*
pres·ti·gious \pre-'sti-jəs, -'stē- *also* prə-\ *adjective* [Latin *praestigiosus,* from *praestigiae*] (1546)
1 *archaic* **:** of, relating to, or marked by illusion, conjuring, or trickery
2 : having prestige **:** HONORED
— **pres·ti·gious·ly** *adverb*
— **pres·ti·gious·ness** *noun*
pres·tis·si·mo \pre-'sti-sə-ˌmō\ *adverb or adjective* [Italian, from *presto* + *-issimo,* suffix denoting a high degree] (circa 1724)
: faster than presto — used as a direction in music
[1]**pres·to** \'pres-(ˌ)tō\ *adverb or adjective* [Italian, quick, quickly, from Latin *praestus* ready, from *praesto,* adverb, on hand; akin to Latin *prae* before — more at FOR] (circa 1599)
1 : suddenly as if by magic **:** IMMEDIATELY
2 : at a rapid tempo — used as a direction in music
[2]**presto** *noun, plural* **prestos** (1869)
: a presto musical passage or movement
[1]**pre·stress** \(ˌ)prē-'stres\ *transitive verb* (1934)
: to introduce internal stresses into (as a structural beam) to counteract the stresses that will result from applied load (as in incorporating cables under tension in concrete)
[2]**pre·stress** \'prē-ˌstres, ˌprē-'\ *noun* (1934)
1 : the stresses introduced in prestressing
2 : the process of prestressing
3 : the condition of being prestressed
pre·sum·able \pri-'zü-mə-bəl\ *adjective* (1692)
: capable of being presumed **:** acceptable as an assumption
pre·sum·ably \-blē\ *adverb* (1846)
: by reasonable assumption
pre·sume \pri-'züm\ *verb* **pre·sumed; pre·sum·ing** [Middle English, from Late Latin & Middle French; Late Latin *praesumere* to dare, from Latin, to anticipate, assume, from *prae-* + *sumere* to take; Middle French *presumer* to assume, from Latin *praesumere* — more at CONSUME] (14th century)
transitive verb
1 : to undertake without leave or clear justification **:** DARE
2 : to expect or assume especially with confidence
3 : to suppose to be true without proof ⟨*presumed* innocent until proved guilty⟩
4 : to take for granted **:** IMPLY
intransitive verb
1 : to act or proceed presumptuously or on a presumption
2 : to go beyond what is right or proper
— **pre·sumed·ly** \-'zü-məd-lē, -'zümd-lē\ *adverb*
— **pre·sum·er** *noun*
presuming *adjective* (15th century)
: PRESUMPTUOUS
— **pre·sum·ing·ly** \-'zü-miŋ-lē\ *adverb*
pre·sump·tion \pri-'zəm(p)-shən\ *noun* [Middle English *presumpcioun,* from Old French *presumption,* from Late Latin & Latin; Late Latin *praesumption-, praesumptio* presumptuous attitude, from Latin, assumption, from *praesumere*] (13th century)
1 : presumptuous attitude or conduct **:** AUDACITY
2 a : an attitude or belief dictated by probability **:** ASSUMPTION **b :** the ground, reason, or evidence lending probability to a belief
3 : a legal inference as to the existence or truth of a fact not certainly known that is drawn from the known or proved existence of some other fact
pre·sump·tive \-'zəm(p)-tiv\ *adjective* (15th century)
1 : based on probability or presumption
2 : giving grounds for reasonable opinion or belief
3 : being an embryonic precursor with the potential for forming a particular structure or tissue in the normal course of development ⟨*presumptive* retina⟩
— **pre·sump·tive·ly** *adverb*
pre·sump·tu·ous \pri-'zəm(p)-chə-wəs, -chəs, -shəs\ *adjective* [Middle English, from Middle French *presumptueux,* from Late Latin *praesumptuosus,* irregular from *praesumptio*] (14th century)

◇ WORD HISTORY
prestige The word *prestige* has witnessed a curious change in sense over the course of its history. French *prestige,* the immediate source of the English word, is a learned borrowing from Latin *praestigiae* "conjurer's tricks." Consequently its original meaning was "illusion produced by magic" or "enchantment, charm," hence by extension "charm of a dazzling personality, glamour," and then more generally "ability to create an impression." English borrowed the original French meaning "illusion" in the 17th century and the sense "ability to create an impression" in the 19th century, but the English and French words are no longer accurate translations for each other. English *prestige* now usually means "standing or estimation in the eyes of people," and a person or thing can be of either low or high prestige.

: overstepping due bounds (as of propriety or courtesy) : taking liberties
— **pre·sump·tu·ous·ly** *adverb*
— **pre·sump·tu·ous·ness** *noun*
pre·sup·pose \ˌprē-sə-ˈpōz\ *transitive verb* [Middle English, from Middle French *presupposer,* from Medieval Latin *praesupponere* (perfect indicative *praesupposui*), from Latin *prae-* + Medieval Latin *supponere* to suppose — more at SUPPOSE] (15th century)
1 : to suppose beforehand
2 : to require as an antecedent in logic or fact
— **pre·sup·po·si·tion** \(ˌ)prē-ˌsə-pə-ˈzi-shən\ *noun*
— **pre·sup·po·si·tion·al** \-ˈzish-nəl, -ˈzi-shə-nᵊl\ *adjective*
pre·syn·ap·tic \ˌprē-sə-ˈnap-tik\ *adjective* (1909)
: situated or occurring just before a nerve synapse
— **pre·syn·ap·ti·cal·ly** \-ti-k(ə-)lē\ *adverb*
pret–a–por·ter *or* **prêt–à–por·ter** \ˌpret-ä-pȯr-ˈtā\ *noun* [French, ready to wear] (1959)
: ready-to-wear clothes
¹pre·teen \ˈprē-ˈtēn, -ˌtēn\ *noun* (1952)
: a boy or girl not yet 13 years old
²preteen *adjective* (1954)
1 : relating to or produced for children especially in the 9 to 12 year-old age group ⟨*preteen* fashions⟩
2 : being younger than 13
pre·teen–ag·er \(ˌ)prē-ˈtē-nā-jər\ *noun* (1965)
: PRETEEN
¹pre·tend \pri-ˈtend\ *verb* [Middle English, from Latin *praetendere* to allege as an excuse, literally, to stretch out, from *prae-* pre- + *tendere* to stretch — more at THIN] (15th century)
transitive verb
1 : to give a false appearance of being, possessing, or performing ⟨does not *pretend* to be a psychiatrist⟩
2 a : to make believe **:** FEIGN ⟨he *pretended* deafness⟩ **b :** to claim, represent, or assert falsely ⟨*pretending* an emotion he could not really feel⟩
3 *archaic* **:** VENTURE, UNDERTAKE
intransitive verb
1 : to feign an action, part, or role in play
2 : to put in a claim
synonym see ASSUME
²pretend *adjective* (1911)
: IMAGINARY, MAKE-BELIEVE
pre·tend·ed *adjective* (15th century)
: professed or avowed but not genuine ⟨*pretended* affection⟩
— **pre·tend·ed·ly** *adverb*
pre·tend·er \pri-ˈten-dər\ *noun* (1609)
: one that pretends: as **a :** one who lays claim to something; *specifically* **:** a claimant to a throne who is held to have no just title **b :** one who makes a false or hypocritical show
pre·tense *or* **pre·tence** \ˈprē-ˌten(t)s, pri-ˈ\ *noun* [Middle English, from Middle French *pretensse,* from (assumed) Medieval Latin *praetensa,* from Late Latin, feminine of *praetensus,* past participle of Latin *praetendere*] (15th century)
1 : a claim made or implied; *especially* **:** one not supported by fact
2 a : mere ostentation **:** PRETENTIOUSNESS ⟨confuse dignity with pomposity and *pretense* —Bennett Cerf⟩ **b :** a pretentious act or assertion
3 : an inadequate or insincere attempt to attain a certain condition or quality
4 : professed rather than real intention or purpose **:** PRETEXT ⟨was there under false *pretenses*⟩
5 : MAKE-BELIEVE, FICTION
6 : false show **:** SIMULATION ⟨saw through his *pretense* of indifference⟩

¹pre·ten·sion \pri-ˈten(t)-shən\ *noun* (15th century)
1 : an allegation of doubtful value **:** PRETEXT
2 : a claim or an effort to establish a claim
3 : a claim or right to attention or honor because of merit
4 : an aspiration or intention that may or may not reach fulfillment ⟨has serious literary *pretensions*⟩
5 : VANITY, PRETENTIOUSNESS
synonym see AMBITION
— **pre·ten·sion·less** \-ləs\ *adjective*
²pre·ten·sion \ˌprē-ˈten(t)-shən\ *transitive verb* [*pre-* + ²*tension*] (1937)
: PRESTRESS
pre·ten·tious \pri-ˈten(t)-shəs\ *adjective* [French *prétentieux,* from *prétention* pretension, from Medieval Latin *praetention-, praetentio,* from Latin *praetendere*] (1837)
1 : characterized by pretension: as **a :** making usually unjustified or excessive claims (as of value or standing) ⟨the *pretentious* fraud who assumes a love of culture that is alien to him —Richard Watts⟩ **b :** expressive of affected, unwarranted, or exaggerated importance, worth, or stature ⟨*pretentious* language⟩ ⟨*pretentious* houses⟩
2 : making demands on one's skill, ability, or means **:** AMBITIOUS ⟨the *pretentious* daring of the Green Mountain Boys in crossing the lake —*American Guide Series: Vermont*⟩
synonym see SHOWY
— **pre·ten·tious·ly** *adverb*
— **pre·ten·tious·ness** *noun*
¹pret·er·it *or* **pret·er·ite** \ˈpre-tə-rət\ *adjective* [Middle English *preterit,* from Middle French, from Latin *praeteritus,* from past participle of *praeterire* to go by, pass, from *praeter* beyond, past, by (from comparative of *prae* before) + *ire* to go — more at FOR, ISSUE] (14th century)
archaic **:** BYGONE, FORMER
²preterit *or* **preterite** *noun* (14th century)
: PAST TENSE
pre·term \(ˌ)prē-ˈtərm, ˈprē-ˌ\ *adjective* (1928)
: of, relating to, being, or brought forth by premature birth ⟨*preterm* infant⟩ ⟨*preterm* labor⟩
pre·ter·mi·nal \(ˌ)prē-ˈtərm-nəl, -ˈtər-mə-\ *adjective* (1947)
: occurring before death ⟨*preterminal* cancer⟩
pre·ter·mis·sion \ˌprē-tər-ˈmi-shən\ *noun* [Latin *praetermission-, praetermissio,* from *praetermittere*] (1583)
: the act or an instance of pretermitting **:** OMISSION
pre·ter·mit \-ˈmit\ *transitive verb* **-mit·ted; -mit·ting** [Latin *praetermittere,* from *praeter* by, past + *mittere* to let go, send] (1538)
1 : to let pass without mention or notice **:** OMIT
2 : to leave undone **:** NEGLECT
3 : SUSPEND, BREAK OFF
pre·ter·nat·u·ral \ˌprē-tər-ˈna-chə-rəl, -ˈnach-rəl\ *adjective* [Medieval Latin *praeternaturalis,* from Latin *praeter naturam* beyond nature] (1580)
1 : existing outside of nature
2 : exceeding what is natural or regular **:** EXTRAORDINARY ⟨wits trained to *preternatural* acuteness by the debates —G. L. Dickinson⟩
3 : inexplicable by ordinary means; *especially* **:** PSYCHIC ⟨*preternatural* phenomena⟩
— **pre·ter·nat·u·ral·ly** \-ˈna-chə-rə-lē, ˈnach-rə-, ˈna-chər-\ *adverb*
— **pre·ter·nat·u·ral·ness** \-ˈna-chə-rəl-nəs, -ˈnach-rəl-\ *noun*
pre·test \ˈprē-ˌtest\ *noun* (1926)
: a preliminary test: as **a :** a test of the effectiveness or safety of a product prior to its sale **b :** a test to evaluate the preparedness of students for further studies
— **pre·test** \(ˌ)prē-ˈtest\ *transitive verb*
pre·text \ˈprē-ˌtekst\ *noun* [Latin *praetextus,* from *praetexere* to assign as a pretext, screen, extend in front, from *prae-* + *texere* to weave — more at TECHNICAL] (1513)
: a purpose or motive alleged or an appearance assumed in order to cloak the real intention or state of affairs
synonym see APOLOGY
pre·tor, pre·to·ri·an *variant of* PRAETOR, PRAETORIAN
pret·ti·fy \ˈpri-ti-ˌfī, ˈpu̇r-, ˈpru̇-\ *transitive verb* **-fied; -fy·ing** (1850)
: to make pretty
— **pret·ti·fi·ca·tion** \ˌpri-ti-fə-ˈkā-shən, ˌpu̇r-, ˌpru̇-\ *noun*
— **pret·ti·fi·er** \ˈpri-ti-ˌfī(-ə)r, ˈpu̇r-, ˈpru̇-\ *noun*
pret·ti·ness \ˈpri-tē-nəs, ˈpu̇r-, ˈpru̇-\ *noun* (1649)
1 : the quality or state of being pretty
2 : something pretty
¹pret·ty \ˈpri-tē, ˈpu̇r- *also* ˈpru̇-\ *adjective* **pret·ti·er; -est** [Middle English *praty, prety,* from Old English *prættig* tricky, from *prætt* trick; akin to Old Norse *prettr* trick] (15th century)
1 a : ARTFUL, CLEVER **b :** PAT, APT
2 a : pleasing by delicacy or grace **b :** having conventionally accepted elements of beauty **c :** appearing or sounding pleasant or nice but lacking strength, force, manliness, purpose, or intensity ⟨*pretty* words that make no sense —Elizabeth B. Browning⟩
3 a : MISERABLE, TERRIBLE ⟨a *pretty* mess you've gotten us into⟩ **b** *chiefly Scottish* **:** STOUT
4 : moderately large **:** CONSIDERABLE ⟨a very *pretty* profit⟩ ⟨cost a *pretty* penny⟩
synonym see BEAUTIFUL
— **pret·ti·ly** \-tᵊl-ē\ *adverb*
— **pret·ty·ish** \-tē-ish\ *adjective*
²pret·ty \ˈpu̇r-tē, ˈpər-tē, ˈpri-tē *also* ˈpru̇-tē; *before "near(ly)" often* ˈpərt *or* ˈprit *or* ˈpru̇t\ *adverb* (1565)
1 : in some degree **:** MODERATELY ⟨*pretty* cold weather⟩
2 : in a pretty manner **:** PRETTILY ⟨pop vocalists who can sing *pretty* —Gerald Levitch⟩ ■
³pret·ty \ˈpri-tē, ˈpu̇r-tē *also* ˈpru̇-tē\ *noun, plural* **pretties** (1736)
1 *plural* **:** dainty clothes; *especially* **:** LINGERIE
2 : a pretty person or thing
⁴pret·ty *same as* ³\ *transitive verb* **pret·tied; pret·ty·ing** (1909)
: to make pretty — usually used with *up* ⟨curtains to *pretty* up the room⟩
pret·zel \ˈpret-səl\ *noun* [German *Brezel,* ultimately from Latin *brachiatus* having branches like arms, from *brachium* arm — more at BRACE] (circa 1838)

□ USAGE
pretty Some handbooks complain that *pretty* is overworked and recommend the selection of a more specific word or restrict *pretty* to informal or colloquial contexts. *Pretty* is used to tone down a statement and is in wide use across the whole spectrum of English. It is common in informal speech and writing but is neither rare nor wrong in serious discourse ⟨he may, if he be *pretty* well off or clever, qualify himself as a doctor —G. B. Shaw⟩ ⟨a return to those traditions of American foreign policy which worked *pretty* well for over a century —H. S. Commager⟩ ⟨the arguments for buying expensive books have to be *pretty* cogent —*Times Literary Supplement*⟩.

: a brittle or chewy glazed usually salted slender bread often shaped like a loose knot ◆

pre•vail \pri-ˈvā(ə)l\ *intransitive verb* [Middle English, from Latin *praevalēre,* from *prae-* pre- + *valēre* to be strong — more at WIELD] (15th century)
1 : to gain ascendancy through strength or superiority **:** TRIUMPH
2 : to be or become effective or effectual
3 : to use persuasion successfully ⟨*prevailed* on him to sing⟩
4 : to be frequent **:** PREDOMINATE ⟨the west winds that *prevail* in the mountains⟩
5 : to be or continue in use or fashion **:** PERSIST ⟨a custom that still *prevails*⟩

prev•a•lence \ˈpre-və-lən(t)s, ˈprev-lən(t)s\ *noun* (1713)
1 : the quality or state of being prevalent
2 : the degree to which something is prevalent; *especially* **:** the percentage of a population that is affected with a particular disease at a given time

prev•a•lent \-lənt\ *adjective* [Latin *praevalent-, praevalens* very powerful, from present participle of *praevalēre*] (1576)
1 *archaic* **:** POWERFUL
2 : being in ascendancy **:** DOMINANT
3 : generally or widely accepted, practiced, or favored **:** WIDESPREAD
— prevalent *noun*
— prev•a•lent•ly *adverb*

pre•var•i•cate \pri-ˈvar-ə-ˌkāt\ *intransitive verb* **-cat•ed; -cat•ing** [Latin *praevaricatus,* past participle of *praevaricari* to act in collusion, literally, to straddle, from *prae-* + *varicare* to straddle, from *varus* bowlegged] (circa 1631)
: to deviate from the truth **:** EQUIVOCATE
synonym see LIE
— pre•var•i•ca•tion \-ˌvar-ə-ˈkā-shən\ *noun*
— pre•var•i•ca•tor \-ˈvar-ə-ˌkā-tər\ *noun*

pre•ve•nient \pri-ˈvēn-yənt\ *adjective* [Latin *praevenient-, praeveniens,* present participle of *praevenire*] (circa 1656)
: ANTECEDENT, ANTICIPATORY
— pre•ve•nient•ly *adverb*

pre•vent \pri-ˈvent\ *verb* [Middle English, to anticipate, from Latin *praeventus,* past participle of *praevenire* to come before, anticipate, forestall, from *prae-* + *venire* to come — more at COME] (15th century)
transitive verb
1 *archaic* **a :** to be in readiness for (as an occasion) **b :** to meet or satisfy in advance **c :** to act ahead of **d :** to go or arrive before
2 : to deprive of power or hope of acting or succeeding
3 : to keep from happening or existing ⟨steps to *prevent* war⟩
4 : to hold or keep back **:** HINDER, STOP — often used with *from*
intransitive verb
: to interpose an obstacle ☆
— pre•vent•abil•i•ty \-ˌven-tə-ˈbi-lə-tē\ *noun*
— pre•vent•able *also* **pre•vent•ible** \-ˈven-tə-bəl\ *adjective*
— pre•vent•er *noun*

pre•ven•ta•tive \-ˈven-tə-tiv\ *adjective or noun* (circa 1666)
: PREVENTIVE

pre•ven•tion \pri-ˈven(t)-shən\ *noun* (1582)
: the act of preventing or hindering

¹pre•ven•tive \-ˈven-tiv\ *noun* (circa 1639)
: something that prevents; *especially* **:** something used to prevent disease

²preventive *adjective* (1639)
1 : devoted to or concerned with prevention **:** PRECAUTIONARY ⟨*preventive* steps against soil erosion⟩
2 : undertaken to forestall anticipated hostile action ⟨a *preventive* coup⟩
— pre•ven•tive•ly *adverb*
— pre•ven•tive•ness *noun*

pre•ver•bal \(ˌ)prē-ˈvər-bəl\ *adjective* (1921)
1 : occurring before the verb
2 : having not yet acquired the faculty of speech ⟨a *preverbal* child⟩

¹pre•view \ˈprē-ˌvyü\ *transitive verb* (1607)
1 : to see beforehand; *specifically* **:** to view or to show in advance of public presentation
2 : to give a preliminary survey of
— pre•view•er \-ˌvyü-ər\ *noun*

²preview *noun* (1922)
1 : an advance showing or performance (as of a motion picture or play)
2 *also* **pre•vue** \-ˌvyü\ **:** a showing of snatches from a motion picture advertised for appearance in the near future — called also *trailer*
3 : an advance statement, sample, or survey

pre•vi•ous \ˈprē-vē-əs\ *adjective* [Latin *praevius* leading the way, from *prae-* pre- + *via* way — more at WAY] (1625)
1 : going before in time or order **:** PRIOR
2 : acting too soon **:** PREMATURE
synonym see PRECEDING
— pre•vi•ous•ly *adverb*
— pre•vi•ous•ness *noun*

previous question *noun* (circa 1715)
: a parliamentary motion to put the pending question to an immediate vote without further debate or amendment that if defeated has the effect of permitting resumption of debate

previous to *preposition* (1702)
: PRIOR TO, BEFORE

¹pre•vi•sion \prē-ˈvi-zhən\ *noun* [Middle English *previsioun,* from Old French *prevision,* from Late Latin *praevision-, praevisio,* from Latin *praevidēre* to foresee, from *prae-* + *vidēre* to see — more at WIT] (15th century)
1 : FORESIGHT, PRESCIENCE
2 : FORECAST, PROGNOSTICATION
— pre•vi•sion•al \-ˈvizh-nəl, -ˈvi-zhə-nᵊl\ *adjective*
— pre•vi•sion•ary \-ˈvi-zhə-ˌner-ē\ *adjective*

²prevision *transitive verb* **pre•vi•sioned; pre•vi•sion•ing** \-ˈvi-zhə-niŋ, -ˈvizh-niŋ\ (1891)
: FORESEE

pre•vo•cal•ic \ˌprē-vō-ˈka-lik, -və-\ *adjective* [International Scientific Vocabulary] (1899)
: immediately preceding a vowel

pre•vo•ca•tion•al \ˌprē-vō-ˈkā-shnəl, -shə-nᵊl\ *adjective* (1914)
: given or required before admission to a vocational school

pre•writ•ing \ˈprē-ˌrī-tiŋ\ *noun* (1968)
: the formulation and organization of ideas preparatory to writing

prexy \ˈprek-sē\ *also* **prex** \ˈpreks\ *noun, plural* **prex•ies** *also* **prex•es** [*prexy* from *prex,* by shortening & alteration from *president*] (1871)
slang **:** PRESIDENT — used chiefly of a college president

¹prey \ˈprā\ *noun, plural* **prey** *also* **preys** [Middle English *preie,* from Old French, from Latin *praeda;* akin to Latin *prehendere* to grasp, seize — more at GET] (13th century)
1 *archaic* **:** SPOIL, BOOTY
2 a : an animal taken by a predator as food **b :** one that is helpless or unable to resist attack **:** VICTIM ⟨was *prey* to his own appetites⟩
3 : the act or habit of preying

²prey *intransitive verb* **preyed; prey•ing** [Middle English, from Old French *preier,* from Latin *praedari,* from *praeda*] (14th century)
1 : to make raids for the sake of booty
2 a : to seize and devour prey **b :** to commit violence or robbery or fraud
3 : to have an injurious, destructive, or wasting effect
— prey•er *noun*

prez \ˈprez\ *noun, plural* **prez•es** \ˈpre-zəz\ [by shortening & alteration] (1892)
slang **:** PRESIDENT

Pri•am \ˈprī-əm, -ˌam\ *noun* [Latin *Priamus,* from Greek *Priamos*]
: the father of Hector and Paris and king of Troy during the Trojan War

pri•a•pic \prī-ˈā-pik, -ˈa-\ *adjective* [Latin *priapus* lecher, from *Priapus*] (1786)
1 : PHALLIC
2 : relating to or preoccupied with virility

Pri•a•pus \prī-ˈā-pəs\ *noun* [Latin, from Greek *Priapos*]
: a Greek and Roman god of gardens and male generative power

¹price \ˈprīs\ *noun* [Middle English *pris,* from Old French, from Latin *pretium* price, money; probably akin to Sanskrit *prati-* against, in return — more at PROS-] (13th century)
1 *archaic* **:** VALUE, WORTH
2 a : the quantity of one thing that is exchanged or demanded in barter or sale for another **b :** the amount of money given or set as consideration for the sale of a specified thing
3 : the terms for the sake of which something is done or undertaken: as **a :** an amount sufficient to bribe one ⟨believed every man had his *price*⟩ **b :** a reward for the apprehension or death of a person ⟨an outlaw with a *price* on his head⟩
4 : the cost at which something is obtained ⟨the *price* of freedom is restraint —J. Irwin Miller⟩

²price *transitive verb* **priced; pric•ing** (15th century)
1 : to set a price on
2 : to find out the price of
3 : to drive by raising prices excessively ⟨*priced* themselves out of the market⟩
— pric•er *noun*

price–cut•ter \ˈprīs-ˌkə-tər\ *noun* (1901)
: one that reduces prices especially to a level designed to cripple competition
— price–cut•ting \-tiŋ\ *noun*

priced \ˈprīst\ *adjective* (1722)
: having a specified price — used in combination ⟨low-*priced* merchandise⟩

☆ SYNONYMS
Prevent, anticipate, forestall mean to deal with beforehand. PREVENT implies taking advance measures against something possible or probable ⟨measures taken to *prevent* leaks⟩. ANTICIPATE may imply merely getting ahead of another by being a precursor or forerunner or it may imply checking another's intention by acting first ⟨*anticipated* the question by making a statement⟩. FORESTALL implies a getting ahead so as to stop or interrupt something in its course ⟨hoped to *forestall* the sale⟩.

◇ WORD HISTORY
pretzel Pretzels were apparently known to Americans by 1856, when the word *pretzel,* borrowed from German *Prezel* or *Brezel,* appears in a Sacramento, California, newspaper among a list of items claimed to be typical of German cuisine, such as wurst and lager beer. This familiar hard, knot-shaped bread has been baked, at least in Germanic countries, for centuries, as is evidenced by its appearance in a painting by the 16th century Flemish artist Pieter Brueghel. The origins of the pretzel, however, are even earlier, for the word is found in Old High German glosses as *brecila* or *brezitela.* These words were borrowed from the same vernacular Latin source that yielded Medieval Latin *bracidellus* and 15th century Italian *bracciatello.* These were names for a sweet, ring-shaped pastry—all ultimately from derivatives of Latin *brachiatus* "branched, having branches like arms." Twisted pastries were apparently so called because of the similarity between their knotted shape and a pair of folded arms.

price–earnings ratio *noun* (1961)
: a measure of the value of a common stock determined as the ratio of its market price to its earnings per share and usually expressed as a simple numeral

price–fix·ing \'prīs-ˌfik-siŋ\ *noun* (1920)
: the setting of prices artificially (as by producers or government) contrary to free market operations

price index *noun* (1886)
: an index number expressing the level of a group of commodity prices relative to the level of the prices of the same commodities during an arbitrarily chosen base period and used to indicate changes in the level of prices from one period to another

price·less \'prīs-ləs\ *adjective* (1593)
1 a : having a value beyond any price **:** INVALUABLE **b :** costly because of rarity or quality **:** PRECIOUS
2 : having worth in terms of other than market value
3 : delightfully amusing, odd, or absurd
— price·less·ly *adverb*

price support *noun* (1945)
: artificial maintenance of prices (as of a raw material) at some predetermined level usually through government action

price tag *noun* (1881)
1 : a tag on merchandise showing the price at which it is offered for sale
2 : PRICE, COST

price war *noun* (1925)
: commercial competition characterized by the repeated cutting of prices below those of competitors

pric·ey *also* **pricy** \'prī-sē\ *adjective* **pric·i·er; -est** (1932)
: EXPENSIVE

[1]prick \'prik\ *noun* [Middle English *prikke,* from Old English *prica;* akin to Middle Dutch *pric* prick] (before 12th century)
1 : a mark or shallow hole made by a pointed instrument
2 a : a pointed instrument or weapon **b :** a sharp projecting organ or part
3 : an instance of pricking or the sensation of being pricked: as **a :** a nagging or sharp feeling of remorse, regret, or sorrow **b :** a slight sharply localized discomfort ⟨the *prick* of a needle⟩
4 : PENIS — usually considered vulgar
5 : a spiteful or contemptible man often having some authority — usually considered vulgar

[2]prick (before 12th century)
transitive verb
1 : to pierce slightly with a sharp point
2 : to affect with anguish, grief, or remorse ⟨doubt began to *prick* him —Philip Hale⟩
3 : to ride, guide, or urge on with or as if with spurs **:** GOAD
4 : to mark, distinguish, or note by means of a small mark
5 : to trace or outline with punctures
6 : to remove (a young seedling) from the seedbed to another suitable for further growth — usually used with *out*
7 : to cause to be or stand erect ⟨a dog *pricking* its ears⟩
intransitive verb
1 a : to prick something or cause a pricking sensation **b :** to feel discomfort as if from being pricked
2 a : to urge a horse with the spur **b :** to ride fast
3 : THRUST
4 : to become directed upward **:** POINT
— prick up one's ears : to listen intently

prick·er \'pri-kər\ *noun* (14th century)
1 : one that pricks: as **a :** a rider of horses **b :** a military light horseman
2 a : THORN, PRICKLE **b :** BRIER

prick·et \'pri-kət\ *noun* [Middle English *priket,* from *prikke*] (15th century)
1 a : a spike on which a candle is stuck **b :** a candlestick with such a point
2 : a buck in the second year of life

[1]prick·le \'pri-kəl\ *noun* [Middle English *prikle,* from Old English *pricle;* akin to Old English *prica* prick] (15th century)
1 : a fine sharp process or projection; *especially* **:** a sharp pointed emergence arising from the epidermis or bark of a plant
2 : a prickling or tingling sensation

[2]prickle *verb* **prick·led; prick·ling** \-k(ə-)liŋ\ (1513)
transitive verb
1 : to prick slightly
2 : to produce prickles in
intransitive verb
: to cause or feel a prickling, tingling, or stinging sensation

prick·ly \'pri-k(ə-)lē\ *adjective* **prick·li·er; -est** (1578)
1 : full of or covered with prickles; *especially* **:** distinguished from related kinds by the presence of prickles
2 : marked by prickling **:** STINGING ⟨a *prickly* sensation⟩
3 a : TROUBLESOME, VEXATIOUS ⟨*prickly* issues⟩ **b :** easily irritated ⟨had a *prickly* disposition⟩
— prick·li·ness *noun*

prickly ash *noun* (1709)
: a prickly aromatic shrub or small tree (*Zanthoxylum americanum*) of the rue family with yellowish flowers

prickly heat *noun* (1736)
: a noncontagious cutaneous eruption of red pimples with intense itching and tingling caused by inflammation around the sweat ducts

prickly pear *noun* (1612)
1 : OPUNTIA; *especially* **:** any of those with flat spiny joints — called also *prickly pear cactus;* compare CHOLLA
2 : the pulpy pear-shaped edible fruit of various prickly pears (as *Opuntia ficus-indica*)

prickly pear 1

prickly poppy *noun* (1724)
: any of a genus (*Argemone*) of plants of the poppy family with white or yellow flowers and prickly leaves and fruits

[1]pride \'prīd\ *noun* [Middle English, from Old English *prȳde,* from *prūd* proud — more at PROUD] (before 12th century)
1 : the quality or state of being proud: as **a :** inordinate self-esteem **:** CONCEIT **b :** a reasonable or justifiable self-respect **c :** delight or elation arising from some act, possession, or relationship ⟨parental *pride*⟩
2 : proud or disdainful behavior or treatment **:** DISDAIN
3 a : ostentatious display **b :** highest pitch **:** PRIME
4 : a source of pride **:** the best in a group or class
5 : a company of lions
6 : a showy or impressive group ⟨a *pride* of dancers⟩

[2]pride *transitive verb* **prid·ed; prid·ing** (13th century)
: to indulge (as oneself) in pride

pride·ful \'prīd-fəl\ *adjective* (15th century)
: full of pride: as **a :** DISDAINFUL, HAUGHTY **b :** EXULTANT, ELATED
— pride·ful·ly \-fə-lē\ *adverb*
— pride·ful·ness *noun*

pride of place (1605)
: the highest or first position

prie–dieu \(ˌ)prē-'dyə(r), prē-dyœ\ *noun, plural* **prie–dieux** \-'dyə(r)(z), -dyœ(z)\ [French, literally, pray God] (1760)
1 : a kneeling bench designed for use by a person at prayer and fitted with a raised shelf on which the elbows or a book may be rested
2 : a low armless upholstered chair with a high straight back

prie-dieu 1

pri·er \'prī(-ə)r\ *noun* (1552)
: one that pries; *especially* **:** an inquisitive person

priest \'prēst\ *noun* [Middle English *preist,* from Old English *prēost,* ultimately from Late Latin *presbyter* — more at PRESBYTER] (before 12th century)
: one authorized to perform the sacred rites of a religion especially as a mediatory agent between humans and God; *specifically* **:** an Anglican, Eastern Orthodox, or Roman Catholic clergyman ranking below a bishop and above a deacon

priest·ess \'prēs-təs\ *noun* (1693)
1 : a woman authorized to perform the sacred rites of a religion
2 : a woman regarded as a leader (as of a movement)
usage see -ESS

priest·hood \'prēst-ˌhu̇d, 'prē-ˌstu̇d\ *noun* (before 12th century)
1 : the office, dignity, or character of a priest
2 : the whole body of priests

priest·ly \'prēst-lē\ *adjective* (before 12th century)
1 : of or relating to a priest or the priesthood **:** SACERDOTAL
2 : characteristic of or befitting a priest
— priest·li·ness *noun*

priest–rid·den \'prēst-ˌri-dᵊn\ *adjective* (1653)
: controlled or oppressed by a priest

[1]prig \'prig\ *noun* [*prig* to steal] (1610)
: THIEF

[2]prig *noun* [probably from [1]*prig*] (1676)
1 *archaic* **:** FOP
2 *archaic* **:** FELLOW, PERSON
3 : one who offends or irritates by observance of proprieties (as of speech or manners) in a pointed manner or to an obnoxious degree
— prig·gery \-gə-rē\ *noun*
— prig·gish \'pri-gish\ *adjective*
— prig·gish·ly *adverb*
— prig·gish·ness *noun*

prig·gism \'pri-ˌgi-zəm\ *noun* (circa 1805)
: stilted adherence to convention

[1]prill \'pril\ *transitive verb* [origin unknown] (1944)
: to convert (as a molten solid) into spherical pellets usually by forming into drops in a spray and allowing the drops to solidify

[2]prill *noun* (1952)
: a pellet made by prilling

[1]prim \'prim\ *transitive verb* **primmed; prim·ming** [origin unknown] (1706)
1 : to give a prim or demure expression to ⟨*primming* her thin lips after every mouthful —John Buchan⟩
2 : to dress primly

[2]prim *adjective* **prim·mer; prim·mest** (1709)
1 a : stiffly formal and proper **:** DECOROUS **b :** PRUDISH
2 : NEAT, TRIM ⟨*prim* hedges⟩
— prim·ly *adverb*
— prim·ness *noun*

pri·ma ballerina \'prē-mə-\ *noun* [Italian, leading ballerina] (1870)
: the principal female dancer in a ballet company

pri·ma·cy \'prī-mə-sē\ *noun* (14th century)
1 : the state of being first (as in importance, order, or rank) **:** PREEMINENCE ⟨the *primacy* of

intellectual and esthetic over materialistic values —T. R. McConnell⟩
2 : the office, rank, or preeminence of an ecclesiastical primate
pri•ma don•na \ˌpri-mə-ˈdä-nə, ˌprē-\ *noun, plural* **prima donnas** [Italian, literally, first lady] (1782)
1 : a principal female singer in an opera or concert organization
2 : an extremely sensitive, vain, or undisciplined person
¹pri•ma fa•cie \ˌprī-mə-ˈfā-shə, -shē, -sē *also* -shē-ˌē, -sē-ˌē\ *adverb* [Middle English, from Latin] (15th century)
: at first view **:** on the first appearance
²prima facie *adjective* (1800)
1 : true, valid, or sufficient at first impression **:** APPARENT ⟨the theory . . . gives a *prima facie* solution —R. J. Butler⟩
2 : SELF-EVIDENT
3 : legally sufficient to establish a fact or a case unless disproved ⟨*prima facie* evidence⟩
pri•mal \ˈprī-məl\ *adjective* [Medieval Latin *primalis,* from Latin *primus* first — more at PRIME] (1602)
1 : ORIGINAL, PRIMITIVE ⟨village life continued in its *primal* innocence —Van Wyck Brooks⟩
2 : first in importance **:** FUNDAMENTAL ⟨our *primal* concern⟩
pri•mal•i•ty \prī-ˈma-lə-tē\ *noun* (1919)
: the property of being a prime number
primal scream therapy *noun* (1971)
: psychotherapy in which the patient recalls and reenacts a particularly disturbing past experience usually occurring early in life and expresses normally repressed anger or frustration especially through spontaneous and unrestrained screams, hysteria, or violence — called also *primal therapy*
pri•mar•i•ly \prī-ˈmer-ə-lē *also* prə-, *chiefly British* ˈprī-mər-ə-lē\ *adverb* (1601)
1 : for the most part **:** CHIEFLY ⟨has now become *primarily* a residential town —S. P. B. Mais⟩
2 : in the first place **:** ORIGINALLY
¹pri•ma•ry \ˈprī-ˌmer-ē, ˈprī-mə-rē, ˈprīm-rē\ *adjective* [Middle English, from Late Latin *primarius* basic, primary, from Latin, principal, from *primus*] (15th century)
1 a : first in order of time or development **:** PRIMITIVE ⟨the *primary* stage of civilization⟩ **b :** of or relating to formations of the Paleozoic and earlier periods
2 a : of first rank, importance, or value **:** PRINCIPAL ⟨the *primary* purpose⟩ **b :** BASIC, FUNDAMENTAL ⟨security is a *primary* need⟩ **c :** of, relating to, or constituting the principal quills of a bird's wing **d :** of or relating to agriculture, forestry, and the extractive industries or their products **e :** expressive of present or future time ⟨*primary* tense⟩ **f :** of, relating to, or constituting the strongest of the three or four degrees of stress recognized by most linguists ⟨the first syllable of *basketball* carries *primary* stress⟩
3 a : DIRECT, FIRSTHAND ⟨*primary* sources of information⟩ **b :** not derivable from other colors, odors, or tastes **c :** preparatory to something else in a continuing process ⟨*primary* instruction⟩ **d :** of or relating to a primary school ⟨*primary* education⟩ **e :** belonging to the first group or order in successive divisions, combinations, or ramifications ⟨*primary* nerves⟩ **f :** of, relating to, or constituting the inducing current or its circuit in an induction coil or transformer **g :** directly derived from ores ⟨*primary* metals⟩ **h :** of, relating to, or being the amino acid sequence in proteins ⟨*primary* protein structure⟩
4 : resulting from the substitution of one of two or more atoms or groups in a molecule; *especially* **:** being or characterized by a carbon atom having a bond to only one other carbon atom
5 : of, relating to, involving, or derived from primary meristem ⟨*primary* tissue⟩ ⟨*primary* growth⟩
6 : of, relating to, or involved in the production of organic substances by green plants ⟨*primary* productivity⟩
²primary *noun, plural* **-ries** (circa 1772)
1 : something that stands first in rank, importance, or value **:** FUNDAMENTAL — usually used in plural
2 : the celestial body around which one or more other celestial bodies revolve; *especially* **:** the more massive usually brighter component of a binary star system
3 : one of the usually 9 or 10 strong quills on the distal joint of a bird's wing — see WING illustration
4 a : PRIMARY COLOR **b :** a primary-color sensation
5 a : CAUCUS **b :** an election in which qualified voters nominate or express a preference for a particular candidate or group of candidates for political office, choose party officials, or select delegates for a party convention
6 : the coil that is connected to the source of electricity in an induction coil or transformer — called also *primary coil*
primary atypical pneumonia *noun* (circa 1944)
: any of a group of pneumonias (as Q fever and psittacosis) caused especially by viruses, mycoplasmas, rickettsias, and chlamydias
primary cell *noun* (1902)
: a cell that converts chemical energy into electrical energy by irreversible chemical reactions
primary color *noun* (1612)
: any of a set of colors from which all other colors may be derived
primary meristem *noun* (1875)
: meristem (as procambium) derived from the apical meristem
primary root *noun* (circa 1890)
: the root of a plant that develops first and originates from the radicle
primary school *noun* (1802)
1 : a school usually including the first three grades of elementary school but sometimes also including kindergarten
2 : ELEMENTARY SCHOOL
primary syphilis *noun* (circa 1890)
: the first stage of syphilis that is marked by the development of a chancre and the spread of the causative spirochete in the tissues of the body
primary tooth *noun* (circa 1898)
: MILK TOOTH
primary wall *noun* (circa 1933)
: the first-formed wall of a plant cell that is produced around the protoplast and usually has plasmodesmata
pri•mate \ˈprī-ˌmāt *or especially for 1* -mət\ *noun* [Middle English *primat,* from Old French, from Medieval Latin *primat-, primas* archbishop, from Latin, leader, from *primus*] (13th century)
1 *often capitalized* **:** a bishop who has precedence in a province, group of provinces, or a nation
2 *archaic* **:** one first in authority or rank **:** LEADER
3 [New Latin *Primates,* from Latin, plural of *primat-, primas*] **:** any of an order (Primates) of mammals comprising humans, apes, monkeys, and related forms (as lemurs and tarsiers)
— pri•mate•ship \-ˌship\ *noun*
— pri•ma•tial \prī-ˈmā-shəl\ *adjective*
pri•ma•tol•o•gy \ˌprī-mə-ˈtä-lə-jē\ *noun* (1926)
: the study of primates especially other than recent humans (*Homo sapiens*)
— pri•ma•to•log•i•cal \-mə-tᵊl-ˈä-ji-kəl\ *adjective*
— pri•ma•tol•o•gist \-mə-ˈtä-lə-jist\ *noun*
¹prime \ˈprīm\ *noun* [Middle English, from Old English *prīm,* from Latin *prima hora* first hour] (before 12th century)
1 a *often capitalized* **:** the second of the canonical hours **b :** the first hour of the day usually considered either as 6 a.m. or the hour of sunrise
2 a : the earliest stage **b :** SPRING **c :** YOUTH
3 : the most active, thriving, or successful stage or period ⟨in the *prime* of his life⟩
4 : the chief or best individual or part **:** PICK ⟨*prime* of the flock, and choicest of the stall —Alexander Pope⟩
5 : PRIME NUMBER
6 a : the first note or tone of a musical scale **:** TONIC **b :** the interval between two notes on the same staff degree
7 : the symbol ′
8 : PRIME RATE
²prime *adjective* [Middle English, from Middle French, feminine of *prin* first, from Latin *primus;* akin to Latin *prior*] (14th century)
1 : first in time **:** ORIGINAL
2 a : of, relating to, or being a prime number — compare RELATIVELY PRIME **b :** having no polynomial factors other than itself and no monomial factors other than 1 ⟨a *prime* polynomial⟩ **c :** expressed as a product of prime factors (as prime numbers and prime polynomials) ⟨a *prime* factorization⟩
3 a : first in rank, authority, or significance **:** PRINCIPAL **b :** having the highest quality or value ⟨*prime* farmland⟩ **c :** of the highest grade regularly marketed — used of meat and especially beef
4 : not deriving from something else **:** PRIMARY
— prime•ly *adverb*
— prime•ness *noun*
³prime *verb* **primed; prim•ing** [probably from ¹*prime*] (1513)
transitive verb
1 : FILL, LOAD
2 a : to prepare for firing by supplying with priming **b :** to insert a primer into (a cartridge case)
3 : to apply the first color, coating, or preparation to ⟨*prime* a wall⟩
4 : to put into working order by filling or charging with something ⟨*prime* a pump with water⟩
5 : to instruct beforehand **:** COACH ⟨*primed* the witness⟩
6 : STIMULATE
intransitive verb
: to become prime
— prime the pump : to take steps to encourage the growth or functioning of something
prime cost *noun* (1718)
: the combined total of raw material and direct labor costs incurred in production; *broadly* **:** cost less vendor's or agent's commission for charges
prime meridian *noun* (circa 1859)
: the meridian of 0 degrees longitude which runs through the original site of the Royal Observatory at Greenwich, England, and from which other longitudes are reckoned
prime minister *noun* (1655)
1 : the chief minister of a ruler or state
2 : the official head of a cabinet or ministry; *especially* **:** the chief executive of a parliamentary government
— prime ministerial *adjective*
— prime ministership *noun*
— prime ministry *noun*
prime mover *noun* [translation of Medieval Latin *primus motor*] (1809)
1 a : an initial source of motive power (as a windmill, waterwheel, turbine, or internal combustion engine) designed to receive and modify force and motion as supplied by some

natural source and apply them to drive machinery **b :** a powerful tractor or truck usually with all-wheel drive
2 : the self-moved being that is the source of all motion
3 : the original or most effective force in an undertaking or work ⟨education is . . . a *prime mover* of cultural and societal change —R. C. Buck⟩

prime number *noun* (1570)
: any integer other than 0 or ± 1 that is not divisible without remainder by any other integers except ± 1 and ± the integer itself

[1]prim·er \'pri-mər, *chiefly British* 'prī-mər\ *noun* [Middle English, from Medieval Latin *primarium,* from Late Latin, neuter of *primarius* primary] (14th century)
1 : a small book for teaching children to read
2 : a small introductory book on a subject

[2]prim·er \'prī-mər\ *noun* (1819)
1 : a device for priming; *especially* **:** a cap, tube, or wafer containing percussion powder or compound used to ignite an explosive charge
2 : material used in priming a surface — called also *prime coat*
3 : a molecule (as a short strand of RNA or DNA) whose presence is required for formation of another molecule (as a longer chain of DNA)

prime rate *noun* (1958)
: an interest rate formally announced by a bank to be the lowest available at a particular time to its most credit-worthy customers — called also *prime interest rate*

pri·me·ro \pri-'mer-(,)ō, -'mir-\ *noun* [modification of Spanish *primera,* from feminine of *primer* first, from Latin *primarius*] (1533)
: a card game popular in the 16th and 17th centuries

prime time *noun* (1958)
1 : the time period when the television or radio audience is the largest; *also* **:** prime-time television
2 : the choicest or busiest time
— prime–time *adjective*

pri·me·val \prī-'mē-vəl\ *adjective* [Latin *primaevus,* from *primus* first + *aevum* age — more at AYE] (1662)
1 : of or relating to the earliest ages (as of the world or human history) **:** ANCIENT, PRIMITIVE ⟨100 acres of *primeval* forest which has never felt an ax —Mary R. Zimmer⟩
2 : PRIMORDIAL 1b
— pri·me·val·ly \-və-lē\ *adverb*

priming *noun* (15th century)
1 : the act of one that primes
2 : the explosive used in priming a charge
3 : [2]PRIMER 2

pri·mip·a·ra \prī-'mi-pə-rə\ *noun, plural* **-ras** *or* **-rae** \-,rē, -,rī\ [Latin, from *primus* first + *-para* -para] (circa 1842)
1 : an individual bearing a first offspring
2 : an individual that has borne only one offspring
— pri·mip·a·rous \prī-'mi-pə-rəs\ *adjective*

[1]prim·i·tive \'pri-mə-tiv\ *adjective* [Middle English *primitif,* from Latin *primitivus* first formed, from *primitiae* first fruits, from *primus* first — more at PRIME] (14th century)
1 a : not derived **:** ORIGINAL, PRIMARY **b :** assumed as a basis; *especially* **:** AXIOMATIC ⟨*primitive* concepts⟩
2 a : of or relating to the earliest age or period **:** PRIMEVAL ⟨the *primitive* church⟩ **b :** closely approximating an early ancestral type **:** little evolved ⟨*primitive* mammals⟩ **c :** belonging to or characteristic of an early stage of development **:** CRUDE, RUDIMENTARY ⟨*primitive* technology⟩ **d :** of, relating to, or constituting the assumed parent speech of related languages ⟨*primitive* Germanic⟩
3 a : ELEMENTAL, NATURAL ⟨our *primitive* feelings of vengeance —John Mackwood⟩ **b :** of, relating to, or produced by a people or culture that is nonindustrial and often nonliterate and tribal ⟨*primitive* art⟩ **c :** NAIVE **d** (1) **:** SELF-TAUGHT, UNTUTORED ⟨*primitive* craftsmen⟩ (2) **:** produced by a self-taught artist ⟨a *primitive* painting⟩
— prim·i·tive·ly *adverb*
— prim·i·tive·ness *noun*
— prim·i·tiv·i·ty \,pri-mə-'ti-və-tē\ *noun*

[2]primitive *noun* (15th century)
1 a : something primitive; *specifically* **:** a primitive idea, term, or proposition **b :** a root word
2 a (1) **:** an artist of an early period of a culture or artistic movement (2) **:** a later imitator or follower of such an artist **b** (1) **:** a self-taught artist (2) **:** an artist whose work is marked by directness and naïveté **c :** a work of art produced by a primitive artist **d :** a typically rough or simple usually handmade and antique home accessory or furnishing
3 a : a member of a primitive people **b :** an unsophisticated person

prim·i·tiv·ism \'pri-mə-ti-,vi-zəm\ *noun* (1861)
1 a : belief in the superiority of a simple way of life close to nature **b :** belief in the superiority of nonindustrial society to that of the present
2 : the style of art of primitive peoples or primitive artists
— prim·i·tiv·ist \-vist\ *noun or adjective*
— prim·i·tiv·is·tic \,pri-mə-ti-'vis-tik\ *adjective*

[1]pri·mo \'prē-(,)mō\ *noun, plural* **primos** [Italian, from *primo* first, from Latin *primus*] (1792)
: the first or leading part (as in a duet or trio)

[2]pri·mo \'prē-(,)mō, 'prī-\ *adverb* [Latin, from *primus*] (circa 1901)
: in the first place

[3]pri·mo \'prē-(,)mō\ *adjective* [probably from Italian, chief, first] (1972)
slang **:** of the finest quality **:** EXCELLENT

pri·mo·gen·i·tor \,prī-mō-'je-nə-tər\ *noun* [Late Latin, from Latin *primus* + *genitor* begetter, from *gignere* to beget — more at KIN] (1654)
: ANCESTOR, FOREFATHER

pri·mo·gen·i·ture \-,chùr, -chər, -,tyùr, -,tùr\ *noun* [Late Latin *primogenitura,* from Latin *primus* + *genitura* birth, from *genitus,* past participle of *gignere*] (1602)
1 : the state of being the firstborn of the children of the same parents
2 : an exclusive right of inheritance belonging to the eldest son

pri·mor·di·al \prī-'mòr-dē-əl\ *adjective* [Middle English, from Late Latin *primordialis,* from Latin *primordium* origin, from *primus* first + *ordiri* to begin — more at PRIME, ORDER] (14th century)
1 a : first created or developed **:** PRIMEVAL 1 **b :** existing in or persisting from the beginning (as of a solar system or universe) ⟨a *primordial* gas cloud⟩ **c :** earliest formed in the growth of an individual or organ **:** PRIMITIVE
2 : FUNDAMENTAL, PRIMARY ⟨*primordial* human joys —Sir Winston Churchill⟩
— pri·mor·di·al·ly \-dē-ə-lē\ *adverb*

pri·mor·di·um \-dē-əm\ *noun, plural* **-dia** \-dē-ə\ [New Latin, from Latin] (1671)
: the rudiment or commencement of a part or organ

primp \'primp\ *verb* [perhaps alteration of [1]*prim*] (1801)
transitive verb
: to dress, adorn, or arrange in a careful or finicky manner
intransitive verb
: to dress or groom oneself carefully ⟨*primps* for hours before a date⟩

prim·rose \'prim-,rōz\ *noun* [Middle English *primerose,* from Middle French, from *prime* (feminine of *prin* prime) + *rose* rose — more at PRIME, ROSE] (14th century)
: any of a genus (*Primula* of the family Primulaceae, the primrose family) of perennial herbs with large tufted basal leaves and showy variously colored flowers — compare EVENING PRIMROSE

primrose path *noun* (1602)
1 : a path of ease or pleasure and especially sensual pleasure ⟨himself the *primrose path* of dalliance treads —Shakespeare⟩
2 : a path of least resistance

primrose yellow *noun* (1882)
1 : a light to moderate greenish yellow
2 : a light to moderate yellow

prim·u·la \'prim-yə-lə\ *noun* [Medieval Latin, from *primula veris,* literally, first fruit of spring] (1753)
: PRIMROSE

pri·mum mo·bi·le \'prī-məm-'mō-bə-lē, 'prē-\ *noun, plural* **primum mobiles** [Middle English, from Medieval Latin, literally, first moving thing] (15th century)
: the outermost concentric sphere conceived in medieval astronomy as carrying the spheres of the fixed stars and the planets in its daily revolution

pri·mus \'prī-məs\ *noun, often capitalized* [Medieval Latin, one who is first, magnate, from Latin, first — more at PRIME] (1724)
: the presiding bishop of the Scottish Episcopal Church

pri·mus in·ter pa·res \'prī-məs-,in-tər-'par-ēz, 'prē-məs-\ *noun* [Latin] (1813)
: first among equals

prince \'prin(t)s\ *noun* [Middle English, from Old French, from Latin *princip-, princeps* leader, initiator, from *primus* first + *capere* to take — more at HEAVE] (13th century)
1 a : MONARCH, KING **b :** the ruler of a principality or state
2 : a male member of a royal family; *especially* **:** a son of the sovereign
3 : a nobleman of varying rank and status
4 : one likened to a prince; *especially* **:** a man of high rank or of high standing in his class or profession
— prince·ship \'prin(t)s-,ship\ *noun*

Prince Al·bert \-'al-bərt\ *noun* [*Prince Albert* Edward (later Edward VII king of England)] (1884)
: a double-breasted frock coat with the upper part fitted to the body

Prince Charming *noun* [*Prince Charming,* hero of the fairy tale *Cinderella* by Charles Perrault] (1862)
: a suitor who fulfills the dreams of his beloved; *also* **:** a man of often specious charm toward women

prince consort *noun, plural* **princes consort** (1861)
: the husband of a reigning queen

Prince Albert

prince·dom \'prin(t)s-dəm, -təm\ *noun* (1560)
1 : the jurisdiction, sovereignty, rank, or estate of a prince
2 : PRINCIPALITY 3 — usually used in plural

prince·let \'prin(t)s-lət\ *noun* (1682)
: PRINCELING

prince·li·ness \-lē-nəs\ *noun* (1571)
1 : princely conduct or character
2 : LUXURY, MAGNIFICENCE

prince·ling \'prin(t)s-liŋ\ *noun* (1794)
: a petty or insignificant prince

prince·ly \'prin(t)s-lē\ *adjective* **prince·li·er; -est** (15th century)
1 : of or relating to a prince **:** ROYAL
2 : befitting a prince **:** NOBLE, MAGNIFICENT ⟨*princely* manners⟩ ⟨a *princely* sum⟩
— princely *adverb*

\ə\ **abut** \ᵊ\ kitten \ər\ **further** \a\ ash \ā\ ace \ä\ mop, mar \aù\ **out** \ch\ **chin** \e\ bet \ē\ easy \g\ go \i\ hit \ī\ ice \j\ job \ŋ\ sing \ō\ go \ò\ law \òi\ boy \th\ thin \th\ the \ü\ loot \ù\ foot \y\ yet \zh\ vision *see also* Guide to Pronunciation

Prince of Wales \-'wā(ə)lz\ (15th century)
: the male heir apparent to the British throne — used as a title only after it has been specifically conferred by the sovereign

prince's–feath·er \'prin(t)-səz-,fe-thər\ *noun* (1629)
: a showy widely cultivated annual plant (*Amaranthus hybridus erythrostachys*) of the amaranth family having dense usually red spikes of flowers

¹prin·cess \'prin(t)-səs, 'prin-,ses, (*British usually*) prin-'ses\ *noun* (14th century)
1 *archaic* **:** a woman having sovereign power
2 : a female member of a royal family; *especially* **:** a daughter or granddaughter of a sovereign
3 : the consort of a prince
4 : one likened to a princess; *especially* **:** a woman of high rank or of high standing in her class or profession ⟨a pop music *princess*⟩

²princess *same as* ¹\ *or* **prin·cesse** \prin-'ses\ *adjective* [French *princesse* princess, from *prince*] (1867)
: close-fitting and usually with gores from neck to flaring hemline ⟨a *princess* gown⟩

Princess Royal *noun, plural* **Princesses Royal** (circa 1649)
: the eldest daughter of a British sovereign — a title granted for life and used only after it has been specifically conferred by the sovereign

¹prin·ci·pal \'prin(t)-s(ə-)pəl, -sə-bəl\ *adjective* [Middle English, from Old French, from Latin *principalis,* from *princip-, princeps*] (14th century)
1 : most important, consequential, or influential **:** CHIEF
2 : of, relating to, or constituting principal or a principal
usage see PRINCIPLE
— prin·ci·pal·ly \-sə-p(ə-)lē, -sə-bə-lē, -splē\ *adverb*

²principal *noun* (14th century)
1 : a person who has controlling authority or is in a leading position: as **a :** a chief or head man or woman **b :** the chief executive officer of an educational institution **c :** one who employs another to act as agent subject to the employer's general control and instruction; *specifically* **:** the person from whom an agent's authority derives **d :** the chief or an actual participant in a crime **e :** the person primarily or ultimately liable on a legal obligation **f :** a leading performer **:** STAR
2 : a matter or thing of primary importance: as **a** (1) **:** a capital sum placed at interest, due as a debt, or used as a fund (2) **:** the corpus of an estate, portion, devise, or bequest **b :** the construction that gives shape and strength to a roof and is usually one of several trusses; *broadly* **:** the most important member of a piece of framing
usage see PRINCIPLE
— prin·ci·pal·ship \'prin(t)-s(ə-)pəl-,ship, -sə-bəl-\ *noun*

principal diagonal *noun* (1965)
: the diagonal in a square matrix that runs from upper left to lower right

prin·ci·pal·i·ty \,prin(t)-sə-'pa-lə-tē\ *noun, plural* **-ties** (14th century)
1 a : the state, office, or authority of a prince **b :** the position or responsibilities of a principal (as of a school)
2 : the territory or jurisdiction of a prince **:** the country that gives title to a prince
3 *plural* **:** an order of angels — see CELESTIAL HIERARCHY

principal parts *noun plural* (1870)
: a series of verb forms from which all the other forms of a verb can be derived including in English the infinitive, the past tense, and the present and past participles

prin·cip·i·um \prin-'si-pē-əm, prin-'ki-\ *noun, plural* **-ia** \-pē-ə\ [Latin, beginning, basis] (1600)
: a fundamental principle

prin·ci·ple \'prin(t)-s(ə-)pəl, -sə-bəl\ *noun* [Middle English, modification of Middle French *principe,* from Latin *principium* beginning, from *princip-, princeps* initiator — more at PRINCE] (14th century)
1 a : a comprehensive and fundamental law, doctrine, or assumption **b** (1) **:** a rule or code of conduct (2) **:** habitual devotion to right principles ⟨a man of *principle*⟩ **c :** the laws or facts of nature underlying the working of an artificial device
2 : a primary source **:** ORIGIN
3 a : an underlying faculty or endowment ⟨such *principles* of human nature as greed and curiosity⟩ **b :** an ingredient (as a chemical) that exhibits or imparts a characteristic quality
4 *capitalized, Christian Science* **:** a divine principle **:** GOD ■
— in principle : with respect to fundamentals ⟨prepared to accept the proposition *in principle*⟩

prin·ci·pled \-s(ə-)pəld, -sə-bəld\ *adjective* (1642)
: exhibiting, based on, or characterized by principle — often used in combination

prin·cox \'prin-,käks, 'priŋ-\ *noun* [origin unknown] (1540)
archaic **:** a pert youth **:** COXCOMB

prink \'priŋk\ *verb* [probably alteration of ²*prank*] (1576)
: PRIMP
— prink·er *noun*

¹print \'print\ *noun* [Middle English *preinte,* from Middle French, from *preint,* past participle of *preindre* to press, from Latin *premere* — more at PRESS] (14th century)
1 a : a mark made by pressure **:** IMPRESSION **b :** something impressed with a print or formed in a mold
2 a : printed state or form **b :** the printing industry
3 a : PRINTED MATTER **b** *plural* **:** printed publications
4 : printed letters **:** TYPE
5 a (1) **:** a copy made by printing (2) **:** a reproduction of an original work of art (as a painting) made by a photomechanical process (3) **:** an original work of art (as a woodcut, etching, or lithograph) intended for graphic reproduction and produced by or under the supervision of the artist who designed it **b :** cloth with a pattern or figured design applied by printing; *also* **:** an article of such cloth **c :** a photographic or motion-picture copy; *especially* **:** one made from a negative
— in print : procurable from the publisher
— out of print : not procurable from the publisher

²print (14th century)
transitive verb
1 a : to impress something in or on **b :** to stamp (as a mark) in or on something
2 a : to make a copy of by impressing paper against an inked printing surface **b** (1) **:** to impress (as wallpaper) with a design or pattern (2) **:** to impress (a pattern or design) on something **c :** to publish in print **d :** PRINT OUT; *also* **:** to display on a surface (as a computer screen) for viewing
3 : to write in letters shaped like those of ordinary roman text type
4 : to make (a positive picture) on a sensitized photographic surface from a negative or a positive
intransitive verb
1 a : to work as a printer **b :** to produce printed matter
2 : to produce something in printed form
3 : to write or hand-letter in imitation of unjoined printed characters

³print *adjective* (1953)
: of, relating to, or writing for printed publications ⟨*print* journalists⟩

print·able \'prin-tə-bəl\ *adjective* (1837)
1 : capable of being printed or of being printed from
2 : considered fit to publish
— print·abil·i·ty \,prin-tə-'bi-lə-tē\ *noun*

printed circuit *noun* (1946)
: a circuit for electronic apparatus made by depositing conductive material in continuous paths from terminal to terminal on an insulating surface

printed matter *noun* (1876)
: matter printed by any of various mechanical processes that is eligible for mailing at a special rate

print·er \'prin-tər\ *noun* (1504)
: one that prints: as **a :** a person engaged in printing **b :** a device used for printing; *especially* **:** a machine for printing from photographic negatives **c :** a device (as a dot matrix printer) that produces printout

printer's devil *noun* (1763)
: an apprentice in a printing office

print·ery \'prin-tə-rē\ *noun, plural* **-er·ies** (1638)
: PRINTING OFFICE

print·head \'print-,hed\ *noun* (1968)
: a usually movable part of a computer printer that contains the printing elements

print·ing *noun* (14th century)
1 : the act or product of one that prints
2 : reproduction in printed form
3 : the art, practice, or business of a printer
4 : IMPRESSION 4c
5 *plural* **:** paper to be printed on

printing office *noun* (1733)
: an establishment where printing is done

printing press *noun* (1588)
: a machine that produces printed copies

print·less \'print-ləs\ *adjective* (1610)
: making, bearing, or taking no imprint

print·mak·ing \-,mā-kiŋ\ *noun* (1928)
: the design and production of prints by an artist
— print·mak·er \-kər\ *noun*

print·out \'print-,au̇t\ *noun* (1953)
: a printed record produced automatically (as by a computer)

print out *transitive verb* (1953)
: to make a printout of

pri·on \'prē-,än\ *noun* [*pro*tinaceous + *in*fectious + ²*-on*] (1982)
: a protein particle that lacks nucleic acid and is sometimes held to be the cause of various infectious diseases of the nervous system (as scrapie and Creutzfeldt-Jakob disease)

¹pri·or \'prī(-ə)r\ *noun* [Middle English, from Old English & Old French; both from Medieval Latin, from Late Latin, administrator, from Latin, former, superior] (before 12th century)
1 : the superior ranking next to the abbot of a monastery
2 : the superior of a house or group of houses of any of various religious communities
— pri·or·ate \'prī-ə-rət\ *noun*
— pri·or·ship \'prī(-ə)r-,ship\ *noun*

²prior *adjective* [Latin, former, superior, comparative of Old Latin *pri* before; akin to Latin *priscus* ancient, *prae* before — more at FOR] (1714)
1 : earlier in time or order **:** PREVIOUS ⟨by *prior* agreement⟩
2 : taking precedence (as in importance)
synonym see PRECEDING
— pri·or·ly *adverb*

pri·or·ess \'prī-ə-rəs\ *noun* (14th century)
: a nun corresponding in rank to a prior

□ USAGE
principle Although nearly every handbook and many dictionaries warn against confusing *principle* and *principal,* many people still do. *Principle* is only a noun; *principal* is both adjective and noun. If you are unsure which noun you want, read the definitions in this dictionary.

pri·or·i·tize \prī-ˈȯr-ə-ˌtīz, -ˈär-; ˈprī-ə-rə-\ *transitive verb* **-tized; -tiz·ing** (1964)
: to list or rate (as projects or goals) in order of priority
usage see -IZE
— **pri·or·i·ti·za·tion** \prī-ˌȯr-ə-tə-ˈzā-shən, -är-; ˌprī-ə-rə-\ *noun*

pri·or·i·ty \prī-ˈȯr-ə-tē, -ˈär-\ *noun, plural* **-ties** (14th century)
1 a (1) **:** the quality or state of being prior (2) **:** precedence in date or position of publication — used of taxa **b** (1) **:** superiority in rank, position, or privilege (2) **:** legal precedence in exercise of rights over the same subject matter
2 : a preferential rating; *especially* **:** one that allocates rights to goods and services usually in limited supply ⟨that project has top *priority*⟩
3 : something given or meriting attention before competing alternatives

prior restraint *noun* (1951)
: governmental prohibition imposed on expression before the expression actually takes place

prior to *preposition* (1714)
: in advance of **:** BEFORE ■

pri·o·ry \ˈprī-(ə-)rē\ *noun, plural* **-ries** [Middle English *priorie,* from Anglo-French, from Medieval Latin *prioria,* from *prior*] (13th century)
: a religious house under a prior or prioress

prise \ˈprīz\ *chiefly British variant of* 5PRIZE

prism \ˈpri-zəm\ *noun* [Late Latin *prismat-, prisma,* from Greek, literally, anything sawn, from *priein* to saw] (1570)
1 : a polyhedron with two polygonal faces lying in parallel planes and with the other faces parallelograms
2 a : a transparent body that is bounded in part by two nonparallel plane faces and is used to refract or disperse a beam of light **b :** a prism-shaped decorative glass luster
3 : a crystal form whose faces are parallel to one axis; *especially* **:** one whose faces are parallel to the vertical axis
4 : a medium that distorts, slants, or colors whatever is viewed through it

pris·mat·ic \priz-ˈma-tik\ *adjective* (1709)
1 : relating to, resembling, or constituting a prism
2 a : formed by a prism **b :** resembling the colors formed by refraction of light through a prism ⟨*prismatic* effects⟩
3 : highly colored **:** BRILLIANT ⟨*prismatic* lyrics⟩
4 : having such symmetry that a general form with faces cutting all axes at unspecified intercepts is a prism ⟨*prismatic* crystals⟩
— **pris·mat·i·cal·ly** \-ti-k(ə-)lē\ *adverb*

pris·ma·toid \ˈpriz-mə-ˌtȯid\ *noun* [Late Latin *prismat-, prisma* prism] (circa 1890)
: a polyhedron that has all of its vertices in two parallel planes

pris·moid \ˈpriz-ˌmȯid\ *noun* (circa 1704)
: a prismatoid whose parallel bases have the same number of sides
— **pris·moi·dal** \priz-ˈmȯi-dᵊl\ *adjective*

1**pris·on** \ˈpri-zᵊn\ *noun* [Middle English, from Old French, from Latin *prehension-, prehensio* act of seizing, from *prehendere* to seize — more at GET] (12th century)
1 : a state of confinement or captivity
2 : a place of confinement especially for lawbreakers; *specifically* **:** an institution (as one under state jurisdiction) for confinement of persons convicted of serious crimes — compare JAIL

2**prison** *transitive verb* (14th century)
: IMPRISON, CONFINE

prison camp *noun* (circa 1908)
1 : a camp for the confinement of reasonably trustworthy prisoners usually employed on government projects
2 : a camp for prisoners of war or political prisoners

pris·on·er \ˈpriz-nər, ˈpri-zᵊn-ər\ *noun* (14th century)
1 : a person deprived of liberty and kept under involuntary restraint, confinement, or custody; *especially* **:** one on trial or in prison
2 : someone restrained as if in prison ⟨a *prisoner* of her own conscience⟩

prisoner of war (1678)
: a person captured in war; *especially* **:** a member of the armed forces of a nation who is taken by the enemy during combat

prisoner's base *noun* (circa 1773)
: a game in which players on each of two teams seek to tag and imprison players of the other team who have ventured out of their home territory

pris·sy \ˈpri-sē\ *adjective* **pris·si·er; -est** [probably blend of *prim* and *sissy*] (1895)
: being prim and precise **:** FINICKY
— **pris·si·ly** \ˈpri-sə-lē\ *adverb*
— **pris·si·ness** \ˈpri-sē-nəs\ *noun*

pris·tane \ˈpris-ˌtān\ *noun* [Latin *pristis* shark, sawfish; from its occurrence in the liver oils of sharks] (1923)
: an isoprenoid hydrocarbon $C_{19}H_{40}$ that usually accompanies phytane

pris·tine \ˈpris-ˌtēn, pri-ˈstēn, *especially British* ˈpris-ˌtīn\ *adjective* [Latin *pristinus;* akin to Latin *prior*] (1534)
1 : belonging to the earliest period or state **:** ORIGINAL ⟨the hypothetical *pristine* lunar atmosphere⟩
2 a : not spoiled, corrupted, or polluted (as by civilization) **:** PURE ⟨a *pristine* forest⟩ **b :** fresh and clean as or as if new ⟨*pristine* hard-backs in uniform editions to fill our built-in bookcases —Michiko Kakutani⟩
— **pris·tine·ly** *adverb*

prith·ee \ˈpri-t͟hē, -thē\ *interjection* [alteration of (*I*) *pray thee*] (circa 1591)
archaic — used to express a wish or request

pri·va·cy \ˈprī-və-sē, *especially British* ˈpri-\ *noun, plural* **-cies** (15th century)
1 a : the quality or state of being apart from company or observation **:** SECLUSION **b :** freedom from unauthorized intrusion ⟨one's right to *privacy*⟩
2 *archaic* **:** a place of seclusion
3 a : SECRECY **b :** a private matter **:** SECRET

pri·vat·do·cent *or* **pri·vat·do·zent** \pri-ˈvät-(ˌ)dō(t)-ˌsent\ *noun* [German *Privatdozent,* from *privat* private + *Dozent* teacher, from Latin *docent-, docens,* present participle of *docēre* to teach — more at DOCILE] (1854)
: an unsalaried university lecturer or teacher in German-speaking countries remunerated directly by students' fees

1**pri·vate** \ˈprī-vət\ *adjective* [Middle English *privat,* from Latin *privatus,* from past participle of *privare* to deprive, release, from *privus* private, individual; probably akin to Latin *pro* for, in front of — more at FOR] (14th century)
1 a : intended for or restricted to the use of a particular person, group, or class ⟨a *private* park⟩ **b :** belonging to or concerning an individual person, company, or interest ⟨a *private* house⟩ **c** (1) **:** restricted to the individual or arising independently of others ⟨*private* opinion⟩ (2) **:** carried on by the individual independently of the usual institutions ⟨*private* study⟩; *also* **:** being educated by independent study or a tutor or in a private school ⟨*private* students⟩ **d :** not general in effect ⟨a *private* statute⟩ **e :** of, relating to, or receiving hospital service in which the patient has more privileges than a semiprivate or ward patient
2 a (1) **:** not holding public office or employment ⟨a *private* citizen⟩ (2) **:** not related to one's official position **:** PERSONAL ⟨*private* correspondence⟩ **b :** being a private ⟨a *private* soldier⟩
3 a : withdrawn from company or observation **:** SEQUESTERED ⟨a *private* retreat⟩ **b :** not known or intended to be known publicly **:** SECRET **c :** preferring to keep personal affairs to oneself **:** valuing privacy highly **d :** unsuitable for public use or display
4 : not having shares that can be freely traded on the open market ⟨a *private* company⟩
— **pri·vate·ly** *adverb*
— **pri·vate·ness** *noun*

2**private** *noun* (15th century)
1 *archaic* **:** one not in public office
2 *obsolete* **:** PRIVACY
3 *plural* **:** PRIVATE PARTS
4 a : a person of low rank in various organizations (as a police or fire department) **b :** an enlisted man of the lowest rank in the marine corps or of one of the two lowest ranks in the army
— **in private :** not openly or in public

private detective *noun* (1868)
: PRIVATE INVESTIGATOR

private enterprise *noun* (1844)
: FREE ENTERPRISE

pri·va·teer \ˌprī-və-ˈtir\ *noun* (1664)
: an armed private ship licensed to attack enemy shipping; *also* **:** a sailor on such a ship
— **privateer** *intransitive verb*

private eye *noun* (1938)
: PRIVATE INVESTIGATOR

private first class *noun* (1918)
: an enlisted man ranking in the army above a private and below a corporal and in the marine corps above a private and below a lance corporal

private investigator *noun* (1940)
: a person not a member of a police force who is licensed to do detective work (as investigation of suspected wrongdoing or searching for missing persons)

private law *noun* (1773)
: a branch of law concerned with private persons, property, and relationships — compare PUBLIC LAW

private parts *noun plural* (1785)
: the external genital and excretory organs

private school *noun* (1857)
: a school that is established, conducted, and primarily supported by a nongovernmental agency

private treaty *noun* (1858)
: a sale of property on terms determined by conference of the seller and buyer — compare AUCTION

pri·va·tion \prī-ˈvā-shən\ *noun* [Middle English *privacion,* from Middle French *privation,* from Latin *privation-, privatio,* from *privare* to deprive] (14th century)
1 : an act or instance of depriving **:** DEPRIVATION
2 : the state of being deprived; *especially* **:** lack of what is needed for existence

pri·vat·ise *British variant of* PRIVATIZE

pri·vat·ism \ˈprī-və-ˌti-zəm\ *noun* [*private*] (1950)
: the attitude of being uncommitted to or avoiding involvement in anything beyond one's immediate interests

□ USAGE
prior to Sometimes termed pompous or affected, *prior to* is a synonym of *before* that most often appears in rather formal contexts, such as the annual reports of corporations. It may occasionally emphasize the notion of anticipation ⟨if page makeup decisions are verified and approved *prior to* typesetting, proofreading of pages afterward becomes unnecessary —*Publishers Weekly*⟩. It is also used sometimes in less formal contexts ⟨they have taught her how to walk (which I thought she did rather well *prior to* the expenditure of the 200 bucks) —Groucho Marx (letter)⟩.

¹priv·a·tive \'pri-və-tiv\ *adjective* (14th century)
: constituting or predicating privation or absence of a quality (as the prefixes *a-, un-, non-*)
— priv·a·tive·ly *adverb*

²priv·a·tive *noun* (1588)
: a privative term, expression, or proposition; *also* **:** a privative prefix or suffix

pri·vat·ize \'prī-və-ˌtīz\ *transitive verb* (1948)
: to make private; *especially* **:** to change (as a business or industry) from public to private control or ownership
— pri·vat·iza·tion \ˌprī-və-tə-'zā-shən\ *noun*

priv·et \'pri-vət\ *noun* [origin unknown] (1542)
: a European deciduous shrub (*Ligustrum vulgare*) of the olive family with semievergreen leaves and small white flowers that is widely used for hedges; *broadly* **:** any of various congeneric shrubs

¹priv·i·lege \'priv-lij, 'pri-və-\ *noun* [Middle English, from Old French, from Latin *privilegium* law for or against a private person, from *privus* private + *leg-, lex* law] (12th century)
: a right or immunity granted as a peculiar benefit, advantage, or favor **:** PREROGATIVE; *especially* **:** such a right or immunity attached specifically to a position or an office

²privilege *transitive verb* **-leged; -leg·ing** (14th century)
: to grant a privilege to

priv·i·leged \-lijd\ *adjective* (14th century)
1 : having or enjoying one or more privileges ⟨*privileged* classes⟩
2 : not subject to the usual rules or penalties because of some special circumstance; *especially* **:** not subject to disclosure in a court of law ⟨a *privileged* communication⟩

priv·i·ty \'pri-və-tē\ *noun, plural* **-ties** [Middle English *privite* privacy, secret, from Old French *privité,* from Medieval Latin *privitat-, privitas,* from Latin *privus* private — more at PRIVATE] (1523)
1 a : a relationship between persons who successively have a legal interest in the same right or property **b :** an interest in a transaction, contract, or legal action to which one is not a party arising out of a relationship to one of the parties
2 : private or joint knowledge of a private matter; *especially* **:** cognizance implying concurrence

¹privy \'pri-vē\ *adjective* [Middle English *prive,* from Old French *privé,* from Latin *privatus* private] (14th century)
1 a : PRIVATE, WITHDRAWN **b :** SECRET
2 : belonging or relating to a person in one's individual rather than official capacity
3 : admitted as one sharing in a secret ⟨*privy* to the conspiracy⟩
— priv·i·ly \-və-lē\ *adverb*

²privy *noun, plural* **priv·ies** (14th century)
1 a : a small building having a bench with holes through which the user may defecate or urinate **b :** TOILET 3b
2 : a person having a legal interest of privity

privy council *noun* (14th century)
1 *archaic* **:** a secret or private council
2 *P&C capitalized* **:** a body of officials and dignitaries chosen by the British monarch as an advisory council to the Crown usually functioning through its committees
3 : a usually appointive advisory council to an executive
— privy councillor *noun*

privy purse *noun* (1765)
: an allowance for the private expenses of the British sovereign

prix fixe \'prē-'fēks, -'fiks\ *noun* [French, fixed price] (1883)
: a complete meal offered at a fixed price; *also* **:** the price charged

¹prize \'prīz\ *noun* [Middle English *pris* prize, price — more at PRICE] (14th century)
1 : something offered or striven for in competition or in contests of chance; *also* **:** PREMIUM 1d
2 : something exceptionally desirable
3 *archaic* **:** a contest for a reward **:** COMPETITION

²prize *adjective* (1803)
1 a : awarded or worthy of a prize **b :** awarded as a prize **c :** entered for the sake of a prize ⟨a *prize* drawing⟩
2 : outstanding of a kind ⟨raised *prize* hogs⟩

³prize *transitive verb* **prized; priz·ing** [Middle English *prisen,* from Middle French *prisier,* from Late Latin *pretiare,* from Latin *pretium* price, value — more at PRICE] (14th century)
1 : to estimate the value of **:** RATE
2 : to value highly **:** ESTEEM
synonym see APPRECIATE

⁴prize *noun* [Middle English *prise,* from Middle French, act of taking, from *prendre* to take, from Latin *prehendere* — more at GET] (14th century)
1 : something taken by force, stratagem, or threat; *especially* **:** property lawfully captured at sea in time of war
2 : an act of capturing or taking; *especially* **:** the wartime capture of a ship and its cargo at sea
synonym see SPOIL

⁵prize *transitive verb* **prized; priz·ing** [*prize* lever] (1686)
: to press, force, or move with a lever **:** PRY

prize·fight \'prīz-ˌfīt\ *noun* (1824)
: a professional boxing match
— prize·fight·er \-ˌfī-tər\ *noun*

prize·fight·ing \-ˌfī-tiŋ\ *noun* (1706)
: professional boxing

prize money *noun* (1748)
1 : a part of the proceeds of a captured ship formerly divided among the officers and men making the capture
2 : money offered in prizes

priz·er \'prī-zər\ *noun* (1599)
archaic **:** one that contends for a prize

prize·win·ner \'prīz-ˌwi-nər\ *noun* (1893)
: a winner of a prize

prize·win·ning \-ˌwi-niŋ\ *adjective* (1919)
: having won or of a quality to win a prize ⟨a *prizewinning* design⟩

¹pro \'prō\ *noun, plural* **pros** [Middle English, from Latin, preposition, for — more at FOR] (15th century)
1 : an argument or evidence in affirmation ⟨an appraisal of the *pros* and cons⟩
2 : the affirmative side or one holding it

²pro *adverb* [*pro-*] (15th century)
: on the affirmative side **:** in affirmation ⟨much has been written *pro* and con⟩

³pro *preposition* [Latin] (1837)
: in favor of **:** FOR

⁴pro *noun or adjective* (1866)
: PROFESSIONAL

¹pro- *prefix* [Middle English, from Old French, from Latin, from Greek, before, forward, forth, for, from *pro* — more at FOR]
1 a : earlier than **:** prior to **:** before ⟨*pro*thalamion⟩ **b :** rudimentary **:** PROT- ⟨*pro*nucleus⟩ **c :** precursory ⟨*pro*insulin⟩
2 a : located in front of or at the front of **:** anterior to ⟨*pro*cephalic⟩ ⟨*pro*ventriculus⟩ **b :** front **:** anterior ⟨*pro*thorax⟩
3 : projecting ⟨*pro*gnathous⟩

²pro- *prefix* [Latin *pro* in front of, before, for, forward — more at FOR]
1 : taking the place of **:** substituting for ⟨*pro*cathedral⟩ ⟨*pro*caine⟩
2 : favoring **:** supporting **:** championing ⟨*pro*-American⟩

proa \'prō-ə\ *variant of* PRAU

pro-abor·tion \ˌprō-ə-'bȯr-shən\ *adjective* (1972)
: favoring the legalization of abortion
— pro-abor·tion·ist \-sh(ə-)nist\ *noun*

pro·ac·tive \(ˌ)prō-'ak-tiv\ *adjective* (1933)
1 [¹*pro-*] **:** relating to, caused by, or being interference between previous learning and the recall or performance of later learning ⟨*proactive* inhibition of memory⟩
2 [²*pro-* + re*active*] **:** acting in anticipation of future problems, needs, or changes

prob·a·bi·lism \'prä-bə-bə-ˌli-zəm\ *noun* [French *probabilisme,* from Latin *probabilis* probable] (circa 1843)
1 : a theory that in disputed moral questions any solidly probable course may be followed even though an opposed course is or appears more probable
2 : a theory that certainty is impossible especially in the sciences and that probability suffices to govern belief and action
— prob·a·bi·list \-list\ *adjective or noun*

prob·a·bi·lis·tic \ˌprä-bə-bə-'lis-tik\ *adjective* (1864)
1 : of or relating to probabilism
2 : of, relating to, or based on probability
— prob·a·bi·lis·ti·cal·ly *adverb*

prob·a·bil·i·ty \ˌprä-bə-'bi-lə-tē\ *noun, plural* **-ties** (15th century)
1 : the quality or state of being probable
2 : something (as an event or circumstance) that is probable
3 a (1) **:** the ratio of the number of outcomes in an exhaustive set of equally likely outcomes that produce a given event to the total number of possible outcomes (2) **:** the chance that a given event will occur **b :** a branch of mathematics concerned with the study of probabilities
4 : a logical relation between statements such that evidence confirming one confirms the other to some degree

probability density *noun* (1939)
: PROBABILITY DENSITY FUNCTION; *also* **:** a particular value of a probability density function

probability density function *noun* (1957)
1 : PROBABILITY FUNCTION
2 : a function of a continuous random variable whose integral over an interval gives the probability that its value will fall within the interval

probability distribution *noun* (1937)
: PROBABILITY FUNCTION; *also* **:** PROBABILITY DENSITY FUNCTION 2

probability function *noun* (1906)
: a function of a discrete random variable that gives the probability that a specified value will occur

¹prob·a·ble \'prä-bə-bəl, 'prä(b)-bəl\ *adjective* [Middle English, provable, from Middle French, from Latin *probabilis* commendable, probable, from *probare* to test, approve, prove — more at PROVE] (1606)
1 : supported by evidence strong enough to establish presumption but not proof ⟨a *probable* hypothesis⟩
2 : establishing a probability ⟨*probable* evidence⟩
3 : likely to be or become true or real ⟨*probable* events⟩

²probable *noun* (1647)
: one that is probable

probable cause *noun* (circa 1676)
: a reasonable ground for supposing that a charge is well-founded

prob·a·bly \'prä-bə-blē, 'prä(b)-blē\ *adverb* (1613)
: insofar as seems reasonably true, factual, or to be expected **:** without much doubt ⟨is *probably* happy⟩ ⟨it will *probably* rain⟩

pro·band \'prō-ˌband, prō-'\ *noun* [Latin *probandus,* gerundive of *probare*] (circa 1929)
: SUBJECT 3c(2)

pro·bang \'prō-ˌbaŋ\ *noun* [origin unknown] (1657)
: a slender flexible rod with a sponge on one end used especially for removing obstructions from the esophagus

¹pro·bate \'prō-ˌbāt, *British also* -bit\ *noun* [Middle English *probat,* from Latin *probatum,*

neuter of *probatus,* past participle of *probare*] (15th century)
1 a : the action or process of proving before a competent judicial authority that a document offered for official recognition and registration as the last will and testament of a deceased person is genuine **b :** the judicial determination of the validity of a will
2 : the officially authenticated copy of a probated will

²pro·bate \-ˌbāt\ *transitive verb* **pro·bat·ed; pro·bat·ing** (1570)
1 : to establish (a will) by probate as genuine and valid
2 : to put (a convicted offender) on probation

probate court *noun* (circa 1847)
: a court that has jurisdiction chiefly over the probate of wills and administration of deceased persons' estates

pro·ba·tion \prō-ˈbā-shən\ *noun* [Middle English *probacioun,* from Middle French *probation,* from Latin *probation-, probatio,* from *probare*] (15th century)
1 : critical examination and evaluation or subjection to such examination and evaluation
2 a : subjection of an individual to a period of testing and trial to ascertain fitness (as for a job or school) **b :** the action of suspending the sentence of a convicted offender and giving the offender freedom during good behavior under the supervision of a probation officer **c :** the state or a period of being subject to probation
— **pro·ba·tion·al** \-shnəl, -shə-nᵊl\ *adjective*
— **pro·ba·tion·al·ly** *adverb*
— **pro·ba·tion·ary** \-shə-ˌner-ē\ *adjective*

pro·ba·tion·er \-sh(ə-)nər\ *noun* (1603)
1 : a person (as a newly admitted student nurse) whose fitness is being tested during a trial period
2 : a convicted offender on probation

probation officer *noun* (1880)
: an officer appointed to investigate, report on, and supervise the conduct of convicted offenders on probation

pro·ba·tive \ˈprō-bə-tiv\ *adjective* (15th century)
1 : serving to test or try **:** EXPLORATORY
2 : serving to prove **:** SUBSTANTIATING

pro·ba·to·ry \ˈprō-bə-ˌtōr-ē, -ˌtȯr-\ *adjective* (1625)
: PROBATIVE

¹probe \ˈprōb\ *noun* [Medieval Latin *proba* examination, from Latin *probare*] (1580)
1 : a slender medical instrument used especially for exploration (as of a wound or bodily cavity)
2 a : any of various testing devices or substances: as (1) **:** a pointed metal tip for making electrical contact with a circuit element being checked (2) **:** a usually small object that is inserted into something so as to test conditions at a given point (3) **:** a device used to penetrate or send back information especially from outer space or a celestial body (4) **:** a device (as an ultrasound generator) or a substance (as DNA in genetic research) used to obtain specific information for diagnostic or experimental purposes **b :** a pipe on the receiving airplane thrust into the drogue of the delivering airplane in air refueling
3 a : the action of probing **b :** a penetrating or critical investigation **c :** a tentative exploratory advance or survey

²probe *verb* **probed; prob·ing** (1649)
transitive verb
1 : to search into and explore with great thoroughness **:** subject to a penetrating investigation
2 : to examine with a probe ⟨unmanned vehicles *probed* space⟩
intransitive verb
: to make a searching exploratory investigation
synonym see ENTER
— **prob·er** *noun*

pro·ben·e·cid \prō-ˈbe-nə-səd\ *noun* [irregular from *propyl* + *benzoic acid*] (1950)
: a drug $C_{13}H_{19}NO_4S$ that acts on renal tubular function and is used to increase the concentration of some drugs (as penicillin) in the blood by inhibiting their excretion and to increase the excretion of urates in gout

prob·it \ˈprä-bət\ *noun* [*prob*ability un*it*] (1934)
: a unit of measurement of statistical probability based on deviations from the mean of a normal distribution

pro·bi·ty \ˈprō-bə-tē\ *noun* [Middle English *probite,* from Middle French *probité,* from Latin *probitat-, probitas,* from *probus* honest — more at PROVE] (15th century)
: adherence to the highest principles and ideals **:** UPRIGHTNESS
synonym see HONESTY

¹prob·lem \ˈprä-bləm, -bᵊm, -ˌblem\ *noun* [Middle English *probleme,* from Middle French, from Latin *problema,* from Greek *problēma,* literally, obstacle, from *proballein* to throw forward, from *pro-* forward + *ballein* to throw — more at PRO-, DEVIL] (14th century)
1 a : a question raised for inquiry, consideration, or solution **b :** a proposition in mathematics or physics stating something to be done
2 a : an intricate unsettled question **b :** a source of perplexity, distress, or vexation **c :** difficulty in understanding or accepting ⟨I have a *problem* with your saying that⟩
synonym see MYSTERY

²problem *adjective* (1894)
1 : dealing with a problem of conduct or social relationship ⟨a *problem* play⟩
2 : difficult to deal with ⟨a *problem* child⟩

¹prob·lem·at·ic \ˌprä-blə-ˈma-tik\ *or* **prob·lem·at·i·cal** \-ti-kəl\ *adjective* (1609)
1 a : posing a problem **:** difficult to solve or decide **b :** not definite or settled **:** UNCERTAIN ⟨their future remains *problematic*⟩ **c :** open to question or debate **:** QUESTIONABLE
2 : expressing or supporting a possibility
synonym see DOUBTFUL
— **prob·lem·at·i·cal·ly** \-ti-k(ə-)lē\ *adverb*

²problematic *noun* (1957)
: something that is problematic **:** a problematic aspect or concern

pro bono \ˌprō-ˈbō-(ˌ)nō\ *adjective* [Latin *pro bono publico* for the public good] (1970)
: being, involving, or doing legal work donated especially for the public good ⟨*pro bono* work⟩

pro·bos·ci·de·an \prə-ˌbä-sə-ˈdē-ən\ *or* **pro·bos·cid·i·an** \prə-ˌbä-ˈsi-dē-ən, (ˌ)prō-\ *noun* [ultimately from Latin *proboscid-, proboscis*] (circa 1859)
: any of an order (Proboscidea) of large mammals comprising the elephants and extinct related forms
— **proboscidean** *adjective*

pro·bos·cis \prə-ˈbä-səs, -ˈbäs-kəs\ *noun, plural* **-bos·cis·es** *also* **-bos·ci·des** \-ˈbä-sə-ˌdēz\ [Latin, from Greek *proboskis,* from *pro-* + *boskein* to feed] (1576)
1 a : the trunk of an elephant; *also* **:** any long flexible snout **b :** the human nose especially when prominent
2 : any of various elongated or extensible tubular processes (as the sucking organ of a butterfly) of the oral region of an invertebrate

pro·caine \ˈprō-ˌkān\ *noun* [International Scientific Vocabulary ²*pro-* + *-caine*] (1918)
: a basic ester $C_{13}H_{20}N_2O_2$ of para-aminobenzoic acid; *also* **:** its crystalline hydrochloride used as a local anesthetic

pro·cam·bi·um \(ˌ)prō-ˈkam-bē-əm\ *noun* [New Latin] (1875)
: the part of the primary meristem of a plant that forms cambium and primary vascular tissues
— **pro·cam·bi·al** \-bē-əl\ *adjective*

pro·car·ba·zine \prō-ˈkär-bə-ˌzēn, -zən\ *noun* [*pro*pyl + *carb*amic acid + *azine*] (1965)
: an antineoplastic drug $C_{12}H_{19}N_3O$ that is a monoamine oxidase inhibitor used in the form of its hydrochloride especially in the palliative treatment of Hodgkin's disease

pro·cary·ote *variant of* PROKARYOTE

pro·ca·the·dral \ˌprō-kə-ˈthē-drəl\ *noun* (1868)
: a parish church used as a cathedral

¹pro·ce·dur·al \prə-ˈsē-jə-rəl, -ˈsēj-rəl\ *adjective* (1889)
: of or relating to procedure; *especially* **:** of or relating to the procedure used by courts or other bodies administering substantive law
— **pro·ce·dur·al·ly** *adverb*

²procedural *noun* (1974)
: a realist crime novel with a specific focus ⟨a courtroom *procedural*⟩; *especially* **:** POLICE PROCEDURAL

procedural due process *noun* (1938)
: DUE PROCESS 1

pro·ce·dure \prə-ˈsē-jər\ *noun* [French *procédure,* from Middle French, from *proceder*] (circa 1611)
1 a : a particular way of accomplishing something or of acting **b :** a step in a procedure
2 a : a series of steps followed in a regular definite order ⟨legal *procedure*⟩ ⟨a surgical *procedure*⟩ **b :** a series of instructions for a computer that has a name by which it can be called into action
3 a : a traditional or established way of doing things **b :** PROTOCOL 3a

pro·ceed \prō-ˈsēd, prə-\ *intransitive verb* [Middle English *proceden,* from Middle French *proceder,* from Latin *procedere,* from *pro-* forward + *cedere* to go — more at PRO-] (14th century)
1 : to come forth from a source **:** ISSUE
2 a : to continue after a pause or interruption **b :** to go on in an orderly regulated way
3 a : to begin and carry on an action, process, or movement **b :** to be in the process of being accomplished
4 : to move along a course **:** ADVANCE
synonym see SPRING

pro·ceed·ing *noun* (15th century)
1 : legal action ⟨a divorce *proceeding*⟩
2 : PROCEDURE
3 *plural* **:** EVENTS, HAPPENINGS
4 : TRANSACTION
5 *plural* **:** an official record of things said or done

pro·ceeds \ˈprō-ˌsēdz\ *noun plural* (1665)
1 : the total amount brought in ⟨the *proceeds* of a sale⟩
2 : the net amount received (as for a check or from an insurance settlement) after deduction of any discount or charges

pro·ce·phal·ic \ˌprō-sə-ˈfa-lik\ *adjective* (1874)
: relating to, forming, or situated on or near the front of the head

pro·cer·coid \(ˌ)prō-ˈsər-ˌkȯid\ *noun* [*pro-* + Greek *kerkos* tail] (1926)
: the solid first parasitic larva of some tapeworms that develops usually in the body cavity of a copepod

¹pro·cess \ˈprä-ˌses, ˈprō-, -səs\ *noun, plural* **pro·cess·es** \-ˌse-səz, -sə-, -ˌsēz\ [Middle English *proces,* from Middle French, from Latin *processus,* from *procedere*] (14th century)
1 a : PROGRESS, ADVANCE ⟨in the *process* of time⟩ **b :** something going on **:** PROCEEDING
2 a (1) **:** a natural phenomenon marked by gradual changes that lead toward a particular

result ⟨the *process* of growth⟩ (2) **:** a natural continuing activity or function ⟨such life *processes* as breathing⟩ **b :** a series of actions or operations conducing to an end; *especially* **:** a continuous operation or treatment especially in manufacture
3 a : the whole course of proceedings in a legal action **b :** the summons, mandate, or writ used by a court to compel the appearance of the defendant in a legal action or compliance with its orders
4 : a prominent or projecting part of an organism or organic structure ⟨a bone *process*⟩
5 : [5]CONK

[2]process *transitive verb* (1532)
1 a : to proceed against by law **:** PROSECUTE **b** (1) **:** to take out a summons against (2) **:** to serve a summons on
2 a : to subject to a special process or treatment (as in the course of manufacture) **b** (1) **:** to subject to or handle through an established usually routine set of procedures ⟨*process* insurance claims⟩ (2) **:** to subject to examination or analysis ⟨computers *process* data⟩ **c :** to work (hair) into a conk

[3]process *adjective* (1888)
1 : treated or made by a special process especially when involving synthesis or artificial modification
2 : made by or used in a mechanical or photomechanical duplicating process
3 : of or involving illusory effects usually introduced during processing of the film

[4]pro·cess \prə-ˈses\ *intransitive verb* [back-formation from [1]*procession*] (1814)
chiefly British **:** to move in a procession

process cheese *noun* (1926)
: a cheese made by blending several lots of cheese — called also *processed cheese*

pro·cess·ible *or* **pro·cess·able** \ˈprä-ˌse-sə-bəl, ˈprō-\ *adjective* (1954)
: suitable for processing **:** capable of being processed
— **pro·cess·ibil·i·ty** *or* **pro·cess·abil·i·ty** \ˌprä-ˌse-sə-ˈbi-lə-tē, ˌprō-\ *noun*

[1]pro·ces·sion \prə-ˈse-shən\ *noun* [Middle English *processioun,* from Old French *procession,* from Late Latin & Latin; Late Latin *procession-, processio* religious procession, from Latin, act of proceeding, from *procedere*] (12th century)
1 a : a group of individuals moving along in an orderly often ceremonial way **b :** SUCCESSION, SEQUENCE
2 a : continuous forward movement **:** PROGRESSION **b :** EMANATION ⟨the Holy Spirit's *procession* from the Father⟩

[2]procession *intransitive verb* (1691)
archaic **:** to go in procession

[1]pro·ces·sion·al \prə-ˈsesh-nəl, -ˈse-shə-nᵊl\ *noun* (15th century)
1 : a book containing material for a procession
2 : a musical composition (as a hymn) designed for a procession
3 : a ceremonial procession

[2]processional *adjective* (circa 1611)
: of, relating to, or moving in a procession
— **pro·ces·sion·al·ly** *adverb*

pro·ces·sor \ˈprä-ˌse-sər, ˈprō-\ *noun* (1909)
1 : one that processes ⟨scrap *processor*⟩ ⟨agricultural *processor*⟩
2 a (1) **:** COMPUTER (2) **:** the part of a computer system that operates on data — called also *central processing unit* **b :** a computer program (as a compiler) that puts another program into a form acceptable to the computer
3 : FOOD PROCESSOR

pro·cès–ver·bal \prō-ˌsā-vər-ˈbäl, -(ˌ)ver-\ *noun, plural* **pro·cès–ver·baux** \-ˈbō\ [French, literally, verbal trial] (1635)
: an official written record

pro–choice \(ˌ)prō-ˈchȯis\ *adjective* (1975)
: favoring the legalization of abortion
— **pro–choic·er** \-ˈchȯi-sər\ *noun*

pro·claim \prō-ˈklām, prə-\ *transitive verb* [Middle English *proclamen,* from Middle French or Latin; Middle French *proclamer,* from Latin *proclamare,* from *pro-* before + *clamare* to cry out — more at PRO-, CLAIM] (14th century)
1 a : to declare publicly, typically insistently, proudly, or defiantly and in either speech or writing **:** ANNOUNCE **b :** to give outward indication of **:** SHOW
2 : to declare or declare to be solemnly, officially, or formally ⟨*proclaim* an amnesty⟩ ⟨*proclaim* the country a republic⟩
3 : to praise or glorify openly or publicly **:** EXTOL
synonym see DECLARE
— **pro·claim·er** *noun*

proc·la·ma·tion \ˌprä-klə-ˈmā-shən\ *noun* [Middle English *proclamacion,* from Middle French *proclamation,* from Latin *proclamation-, proclamatio,* from *proclamare*] (14th century)
1 : the action of proclaiming **:** the state of being proclaimed
2 : something proclaimed; *specifically* **:** an official formal public announcement

pro·clit·ic \(ˌ)prō-ˈkli-tik\ *noun* [New Latin *procliticus,* from Greek *pro-* + Late Latin *-cliticus* (as in *encliticus* enclitic)] (circa 1864)
: a clitic that is associated with a following word
— **proclitic** *adjective*

pro·cliv·i·ty \prō-ˈkli-və-tē\ *noun, plural* **-ties** [Latin *proclivitas,* from *proclivis* sloping, prone, from *pro-* forward + *clivus* slope — more at PRO-, DECLIVITY] (circa 1591)
: an inclination or predisposition toward something; *especially* **:** a strong inherent inclination toward something objectionable
synonym see LEANING

Proc·ne \ˈpräk-nē\ *noun* [Latin, from Greek *Proknē*]
: the wife of Tereus who is changed into a swallow while fleeing from him

pro·con·sul \(ˌ)prō-ˈkän(t)-səl\ *noun* [Middle English, from Latin, from *pro consule* for a consul] (14th century)
1 : a governor or military commander of an ancient Roman province
2 : an administrator in a modern colony, dependency, or occupied area usually with wide powers
— **pro·con·su·lar** \-s(ə-)lər\ *adjective*
— **pro·con·su·late** \-s(ə-)lət\ *noun*
— **pro·con·sul·ship** \-səl-ˌship\ *noun*

pro·cras·ti·nate \prə-ˈkras-tə-ˌnāt, prō-\ *verb* **-nat·ed; -nat·ing** [Latin *procrastinatus,* past participle of *procrastinare,* from *pro-* forward + *crastinus* of tomorrow, from *cras* tomorrow] (1588)
transitive verb
: to put off intentionally and habitually
intransitive verb
: to put off intentionally the doing of something that should be done
synonym see DELAY
— **pro·cras·ti·na·tion** \-ˌkras-tə-ˈnā-shən\ *noun*
— **pro·cras·ti·na·tor** \-ˈkras-tə-ˌnā-tər\ *noun*

pro·cre·ant \ˈprō-krē-ənt\ *adjective* (1588)
1 : producing offspring
2 *archaic* **:** of or relating to procreation

pro·cre·ate \-ˌāt\ *verb* **-at·ed; -at·ing** [Latin *procreatus,* past participle of *procreare,* from *pro-* forth + *creare* to create — more at PRO-, CREATE] (1536)
transitive verb
: to beget or bring forth (offspring) **:** PROPAGATE
intransitive verb
: to beget or bring forth offspring **:** REPRODUCE
— **pro·cre·ation** \ˌprō-krē-ˈā-shən\ *noun*
— **pro·cre·ative** \ˈprō-krē-ˌā-tiv\ *adjective*
— **pro·cre·ator** \-ˌā-tər\ *noun*

pro·crus·te·an \prə-ˈkrəs-tē-ən, prō-\ *adjective, often capitalized* (circa 1846)
1 : of, relating to, or typical of Procrustes
2 : marked by arbitrary often ruthless disregard of individual differences or special circumstances

procrustean bed *noun, often P capitalized* (1844)
: a scheme or pattern into which someone or something is arbitrarily forced

Pro·crus·tes \prə-ˈkrəs-(ˌ)tēz, prō-\ *noun* [Latin, from Greek *Prokroustēs*]
: a villainous son of Poseidon in Greek mythology who forces travelers to fit into his bed by stretching their bodies or cutting off their legs

pro·cryp·tic \(ˌ)prō-ˈkrip-tik\ *adjective* [*pro-* (as in *protect*) + *cryptic*] (1891)
: of, relating to, or being a concealing pattern or shade of coloring especially in insects

proc·to·dae·um \ˌpräk-tə-ˈdē-əm\ *noun, plural* **-daea** \-ˈdē-ə\ *or* **-dae·ums** [New Latin, from Greek *prōktos* anus + *hodos* way] (1878)
: the posterior ectodermal part of the alimentary canal formed in the embryo by invagination of the outer body wall

proc·tol·o·gy \präk-ˈtä-lə-jē\ *noun* [Greek *prōktos* + English *-logy*] (1899)
: a branch of medicine dealing with the structure and diseases of the anus, rectum, and sigmoid colon
— **proc·to·log·ic** \ˌpräk-tə-ˈlä-jik\ *or* **proc·to·log·i·cal** \-ji-kəl\ *adjective*
— **proc·tol·o·gist** \präk-ˈtä-lə-jist\ *noun*

proc·tor \ˈpräk-tər\ *noun* [Middle English *procutour* procurator, proctor, alteration of *procuratour*] (14th century)
: SUPERVISOR, MONITOR; *specifically* **:** one appointed to supervise students (as at an examination)
— **proctor** *verb*
— **proc·to·ri·al** \präk-ˈtōr-ē-əl, -ˈtȯr-\ *adjective*
— **proc·tor·ship** \ˈpräk-tər-ˌship\ *noun*

pro·cum·bent \prō-ˈkəm-bənt\ *adjective* [Latin *procumbent-, procumbens,* present participle of *procumbere* to fall or lean forward, from *pro-* forward + *-cumbere* to lie down] (1668)
1 : being or having stems that trail along the ground without rooting
2 : lying face down

proc·u·ra·tion \ˌprä-kyə-ˈrā-shən\ *noun* [Middle English *procuratioun,* from Middle French *procuration,* from Latin *procuration-, procuratio,* from *procurare*] (15th century)
1 a : the act of appointing another as one's agent or attorney **b :** the authority vested in one so appointed
2 : the action of obtaining something (as supplies) **:** PROCUREMENT

proc·u·ra·tor \ˈprä-kyə-ˌrā-tər\ *noun* (14th century)
1 : one that manages another's affairs **:** AGENT
2 : an officer of the Roman empire entrusted with management of the financial affairs of a province and often having administrative powers as agent of the emperor
— **proc·u·ra·to·ri·al** \ˌprä-kyə-rə-ˈtōr-ē-əl, -ˈtȯr-\ *adjective*

pro·cure \prə-ˈkyu̇r, prō-\ *verb* **pro·cured; pro·cur·ing** [Middle English, from Late Latin *procurare,* from Latin, to take care of, from *pro-* for + *cura* care] (14th century)
transitive verb
1 a : to get possession of **:** obtain by particular care and effort **b :** to get and make available for promiscuous sexual intercourse
2 : BRING ABOUT, ACHIEVE
intransitive verb
: to procure women
— **pro·cur·able** \-ˈkyu̇r-ə-bəl\ *adjective*
— **pro·cure·ment** \-ˈkyu̇r-mənt\ *noun*

pro·cur·er \-ˈkyu̇r-ər\ *noun* (1538)
: one that procures; *especially* **:** PANDER

Pro·cy·on \ˈprō-sē-ˌän, ˈprä-, -ən\ *noun* [Latin, from Greek *Prokyōn,* literally, fore-dog; from its rising before Sirius]

: the brightest star in the constellation Canis Minor

[1]**prod** \'präd\ *verb* **prod·ded; prod·ding** [origin unknown] (1535)
transitive verb
1 a : to thrust a pointed instrument into **:** PRICK
b : to incite to action **:** STIR
2 : to poke or stir as if with a prod
intransitive verb
: to urge someone on
— **prod·der** *noun*

[2]**prod** *noun* (circa 1787)
1 : a pointed instrument used to prod
2 : an incitement to act

[1]**prod·i·gal** \'prä-di-gəl\ *adjective* [Latin *prodigus,* from *prodigere* to drive away, squander, from *pro-, prod-* forth + *agere* to drive — more at PRO-, AGENT] (circa 1520)
1 : recklessly extravagant
2 : characterized by wasteful expenditure **:** LAVISH
3 : yielding abundantly **:** LUXURIANT — often used with *of* ⟨nature has been so *prodigal* of her bounty —H. T. Buckle⟩
synonym see PROFUSE
— **prod·i·gal·i·ty** \ˌprä-də-'ga-lə-tē\ *noun*
— **prod·i·gal·ly** \'prä-di-g(ə-)lē\ *adverb*

[2]**prodigal** *noun* (1596)
: one who spends or gives lavishly and foolishly

pro·di·gious \prə-'di-jəs\ *adjective* (15th century)
1 a *obsolete* **:** being an omen **:** PORTENTOUS **b** *archaic* **:** resembling or befitting a prodigy **:** STRANGE, UNUSUAL
2 : exciting amazement or wonder
3 : extraordinary in bulk, quantity, or degree **:** ENORMOUS
synonym see MONSTROUS
— **pro·di·gious·ly** *adverb*
— **pro·di·gious·ness** *noun*

prod·i·gy \'prä-də-jē\ *noun, plural* **-gies** [Middle English, from Latin *prodigium* omen, monster, from *pro-, prod-* + *-igium* (akin to *aio* I say) — more at ADAGE] (15th century)
1 a : a portentous event **:** OMEN **b :** something extraordinary or inexplicable
2 a : an extraordinary, marvelous, or unusual accomplishment, deed, or event **b :** a highly talented child or youth

pro·dro·mal \(ˌ)prō-'drō-məl\ *adjective* (1716)
: PRECURSORY; *especially* **:** marked by prodromes

pro·drome \'prō-ˌdrōm\ *noun* [French, literally, precursor, from Greek *prodromos,* from *pro-* before + *dromos* act of running, racecourse — more at PRO-, DROMEDARY] (circa 1834)
: a premonitory symptom of disease

[1]**pro·duce** \prə-'düs, prō-, -'dyüs\ *verb* **pro·duced; pro·duc·ing** [Middle English (Scots), from Latin *producere,* from *pro-* forward + *ducere* to lead — more at TOW] (15th century)
transitive verb
1 : to offer to view or notice
2 : to give birth or rise to **:** YIELD
3 : to extend in length, area, or volume ⟨*produce* a side of a triangle⟩
4 : to present to the public on the stage or screen or over radio or television
5 a : to cause to have existence or to happen **:** BRING ABOUT **b :** to give being, form, or shape to **:** MAKE; *especially* **:** MANUFACTURE
6 : to compose, create, or bring out by intellectual or physical effort
7 : to cause to accrue
intransitive verb
: to bear, make, or yield something
— **pro·duc·ible** \-'dü-sə-bəl, -'dyü-\ *adjective*

[2]**pro·duce** \'prä-(ˌ)düs, 'prō- *also* -(ˌ)dyüs\ *noun* (1695)
1 a : something produced **b :** the amount produced **:** YIELD
2 : agricultural products and especially fresh fruits and vegetables as distinguished from grain and other staple crops
3 : the progeny usually of a female animal

pro·duc·er \prə-'dü-sər, prō-, -'dyü-\ *noun* (1513)
1 : one that produces; *especially* **:** one that grows agricultural products or manufactures crude materials into articles of use
2 : a furnace or apparatus that produces combustible gas to be used for fuel by circulating air or a mixture of air and steam through a layer of incandescent fuel
3 : a person who supervises or finances the production of a stage or screen production or radio or television program
4 : any of various organisms (as a green plant) which produce their own organic compounds from simple precursors (as carbon dioxide and inorganic nitrogen) and many of which are food sources for other organisms — compare CONSUMER

producer gas *noun* (1895)
: a fuel gas made in a producer and consisting chiefly of carbon monoxide, hydrogen, and nitrogen

producer goods *noun plural* (1948)
: goods (as tools and raw materials) used to produce other goods and satisfy human wants only indirectly

prod·uct \'prä-(ˌ)dəkt\ *noun* [in sense 1, from Middle English, from Medieval Latin *productum,* from Latin, something produced, from neuter of *productus,* past participle of *producere;* in other senses, from Latin *productum*] (15th century)
1 : the number or expression resulting from the multiplication together of two or more numbers or expressions
2 a : something produced **b :** something resulting from or necessarily following from a set of conditions ⟨a *product* of his environment⟩
3 : the amount, quantity, or total produced
4 : CONJUNCTION 5

pro·duc·tion \prə-'dək-shən, prō-\ *noun* (15th century)
1 a : something produced **:** PRODUCT **b** (1) **:** a literary or artistic work (2) **:** a work presented on the stage or screen or over the air **c :** something exaggerated out of proportion to its importance
2 a : the act or process of producing **b :** the creation of utility; *especially* **:** the making of goods available for use
3 : total output especially of a commodity or an industry
— **pro·duc·tion·al** \-shnəl, -shə-nᵊl\ *adjective*

production control *noun* (1929)
: systematic planning, coordinating, and directing of all manufacturing activities and influences to insure having goods made on time, of adequate quality, and at reasonable cost

production line *noun* (1935)
: LINE 6j

pro·duc·tive \prə-'dək-tiv, prō-\ *adjective* (1612)
1 : having the quality or power of producing especially in abundance ⟨*productive* fishing waters⟩
2 : effective in bringing about ⟨investigating committees have been *productive* of much good —R. K. Carr⟩
3 a : yielding results, benefits, or profits **b :** yielding or devoted to the satisfaction of wants or the creation of utilities
4 : continuing to be used in the formation of new words or constructions ⟨*un-* is a *productive* prefix⟩
5 : raising mucus or sputum (as from the bronchi) ⟨a *productive* cough⟩
— **pro·duc·tive·ly** *adverb*
— **pro·duc·tive·ness** *noun*

pro·duc·tiv·i·ty \ˌprō-dək-'ti-və-tē, ˌprä-, prə-ˌdək-\ *noun* (circa 1810)
1 : the quality or state of being productive
2 : rate of production especially of food by the utilization of solar energy by producer organisms

pro·em \'prō-ˌem, -əm\ *noun* [Middle English *proheme,* from Middle French, from Latin *prooemium,* from Greek *prooimion,* from *pro-* + *oimē* song; probably akin to Hittite *išamai-* song, Sanskrit *syati* he binds — more at SINEW] (14th century)
1 : preliminary comment **:** PREFACE
2 : PRELUDE
— **pro·emi·al** \prō-'ē-mē-əl, -'e-\ *adjective*

pro·en·zyme \(ˌ)prō-'en-ˌzīm\ *noun* [International Scientific Vocabulary] (circa 1900)
: ZYMOGEN

pro·es·trus \(ˌ)prō-'es-trəs\ *noun* [New Latin] (1923)
: a period immediately preceding estrus characterized by preparatory physiological changes

prof \'präf\ *noun* (1838)
: PROFESSOR

pro–fam·i·ly \ˌprō-'fam-lē, -'fa-mə-\ *adjective* (1926)
1 : favoring or encouraging traditional family structures and values
2 : opposing abortion and often birth control

prof·a·na·tion \ˌprä-fə-'nā-shən, ˌprō-\ *noun* (1552)
: the act or an instance of profaning

pro·fa·na·to·ry \prō-'fa-nə-ˌtōr-ē, prə-, -'fā-, -ˌtȯr-\ *adjective* (1853)
: tending to profane **:** DESECRATING

[1]**pro·fane** \prō-'fān, prə-\ *transitive verb* **pro·faned; pro·fan·ing** [Middle English *prophanen,* from Latin *profanare,* from *profanus*] (14th century)
1 : to treat (something sacred) with abuse, irreverence, or contempt **:** DESECRATE
2 : to debase by a wrong, unworthy, or vulgar use
— **pro·fan·er** *noun*

[2]**profane** *adjective* [Middle English *prophane,* from Middle French, from Latin *profanus,* from *pro-* before + *fanum* temple — more at PRO-, FEAST] (15th century)
1 : not concerned with religion or religious purposes **:** SECULAR
2 : not holy because unconsecrated, impure, or defiled **:** UNSANCTIFIED
3 : serving to debase or defile what is holy **:** IRREVERENT
4 a : not being among the initiated **b :** not possessing esoteric or expert knowledge
— **pro·fane·ly** *adverb*
— **pro·fane·ness** \-'fān-nəs\ *noun*

pro·fan·i·ty \prō-'fa-nə-tē, prə-\ *noun, plural* **-ties** (1607)
1 a : the quality or state of being profane **b :** the use of profane language
2 a : profane language **b :** an utterance of profane language

pro·fess \prə-'fes, prō-\ *verb* [in sense 1, from Middle English, from *profes,* adjective, having professed one's vows, from Middle French, from Late Latin *professus,* from Latin, past participle of *profitēri* to profess, confess, from *pro-* before + *fatēri* to acknowledge; in other senses, from Latin *professus,* past participle — more at CONFESS] (14th century)
transitive verb
1 : to receive formally into a religious community following a novitiate by acceptance of the required vows
2 a : to declare or admit openly or freely **:** AFFIRM **b :** to declare in words or appearances only **:** PRETEND, CLAIM
3 : to confess one's faith in or allegiance to

4 a : to practice or claim to be versed in (a calling or profession) **b :** to teach as a professor
intransitive verb
1 : to make a profession or avowal
2 *obsolete* **:** to profess friendship

pro·fessed \-ˈfest\ *adjective* (circa 1569)
1 : openly and freely declared or acknowledged **:** AFFIRMED
2 : professing to be qualified; *also* **:** EXPERT

pro·fess·ed·ly \prə-ˈfe-səd-lē, -ˈfest-lē\ *adverb* (1570)
1 : by profession or declaration **:** AVOWEDLY
2 : with pretense **:** ALLEGEDLY

pro·fes·sion \prə-ˈfe-shən\ *noun* [Middle English *professioun,* from Old French *profession,* from Late Latin & Latin; Late Latin *profession-, professio,* from Latin, public declaration, from *profitēri*] (13th century)
1 : the act of taking the vows of a religious community
2 : an act of openly declaring or publicly claiming a belief, faith, or opinion **:** PROTESTATION
3 : an avowed religious faith
4 a : a calling requiring specialized knowledge and often long and intensive academic preparation **b :** a principal calling, vocation, or employment **c :** the whole body of persons engaged in a calling

¹pro·fes·sion·al \prə-ˈfesh-nəl, -ˈfe-shə-nᵊl\ *adjective* (circa 1748)
1 a : of, relating to, or characteristic of a profession **b :** engaged in one of the learned professions **c** (1) **:** characterized by or conforming to the technical or ethical standards of a profession (2) **:** exhibiting a courteous, conscientious, and generally businesslike manner in the workplace
2 a : participating for gain or livelihood in an activity or field of endeavor often engaged in by amateurs ⟨a *professional* golfer⟩ **b :** having a particular profession as a permanent career ⟨a *professional* soldier⟩ **c :** engaged in by persons receiving financial return ⟨*professional* football⟩
3 : following a line of conduct as though it were a profession ⟨a *professional* patriot⟩
— pro·fes·sion·al·ly *adverb*

²professional *noun* (1811)
: one that is professional; *especially* **:** one that engages in a pursuit or activity professionally

professional corporation *noun* (1970)
: a corporation organized by one or more licensed individuals (as a doctor or lawyer) especially for the purpose of providing professional services and obtaining tax advantages

pro·fes·sion·al·ism \-ˈfesh-nə-ˌli-zəm, -ˈfe-shə-nᵊl-ˌi-\ *noun* (1856)
1 : the conduct, aims, or qualities that characterize or mark a profession or a professional person
2 : the following of a profession (as athletics) for gain or livelihood

pro·fes·sion·al·ize \-ˌlīz, -ˌīz\ *transitive verb* **-ized; -iz·ing** (1856)
: to give a professional character to
— pro·fes·sion·al·i·za·tion \-ˌfesh-nə-lə-ˈzā-shən, -ˌfe-shə-nᵊl-ə-\ *noun*

pro·fes·sor \prə-ˈfe-sər\ *noun* (14th century)
1 : one that professes, avows, or declares
2 a : a faculty member of the highest academic rank at an institution of higher education **b :** a teacher at a university, college, or sometimes secondary school **c :** one that teaches or professes special knowledge of an art, sport, or occupation requiring skill
— pro·fes·so·ri·al \ˌprō-fə-ˈsōr-ē-əl, ˌprä-, -ˈsȯr-\ *adjective*
— pro·fes·so·ri·al·ly \-ē-ə-lē\ *adverb*
— pro·fes·sor·ship \prə-ˈfe-sər-ˌship\ *noun*

pro·fes·sor·ate \prə-ˈfe-sə-rət\ *noun* (1860)
: the office, term of office, or position of a professor

pro·fes·so·ri·at \ˌprō-fə-ˈsōr-ē-ət, ˌprä-, -ˈsȯr-, -ē-ˌat\ *or* **pro·fes·so·ri·ate** \-ət, -ˌāt\ *noun* [modification of French *professorat,* from *professeur* professor, from Latin *professor,* from *profitēri*] (1858)
1 : the body of college and university teachers at an institution or in society
2 : the office, duties, or position of a professor

¹prof·fer \ˈprä-fər\ *transitive verb* **prof·fered; prof·fer·ing** \-f(ə-)riŋ\ [Middle English *profren,* from Anglo-French *profrer,* from Old French *poroffrir,* from *por-* forth (from Latin *pro-*) + *offrir* to offer — more at PRO-] (14th century)
: to present for acceptance **:** TENDER, OFFER

²proffer *noun* (14th century)
: OFFER, SUGGESTION

pro·fi·cien·cy \prə-ˈfi-shən(t)-sē\ *noun* (1544)
1 : advancement in knowledge or skill **:** PROGRESS
2 : the quality or state of being proficient

pro·fi·cient \-shənt\ *adjective* [Latin *proficient-, proficiens,* present participle of *proficere* to go forward, accomplish, from *pro-* forward + *facere* to make — more at PRO-, DO] (circa 1590)
: well advanced in an art, occupation, or branch of knowledge ☆
— proficient *noun*
— pro·fi·cient·ly *adverb*

¹pro·file \ˈprō-ˌfīl\ *noun* [Italian *profilo,* from *profilare* to draw in outline, from *pro-* forward (from Latin) + *filare* to spin, from Late Latin — more at FILE] (circa 1656)
1 : a representation of something in outline; *especially* **:** a human head or face represented or seen in a side view
2 : an outline seen or represented in sharp relief **:** CONTOUR
3 : a side or sectional elevation: as **a :** a drawing showing a vertical section of the ground **b :** a vertical section of a soil from the ground surface to the underlying unweathered material
4 : a set of data often in graphic form portraying the significant features of something ⟨a corporation's earnings *profile*⟩; *especially* **:** a graph representing the extent to which an individual exhibits traits or abilities as determined by tests or ratings
5 : a concise biographical sketch
6 : degree or level of public exposure ⟨trying to keep a low *profile*⟩ ⟨a job with a high *profile*⟩
synonym see OUTLINE

²profile *transitive verb* **pro·filed; pro·fil·ing** (1715)
1 : to represent in profile or by a profile **:** produce (as by drawing, writing, or graphing) a profile of
2 : to shape the outline of by passing a cutter around
— pro·fil·er *noun*

¹prof·it \ˈprä-fət\ *noun, often attributive* [Middle English, from Middle French, from Latin *profectus* advance, profit, from *proficere*] (14th century)
1 : a valuable return **:** GAIN
2 : the excess of returns over expenditure in a transaction or series of transactions; *especially* **:** the excess of the selling price of goods over their cost
3 : net income usually for a given period of time
4 : the ratio of profit for a given year to the amount of capital invested or to the value of sales
5 : the compensation accruing to entrepreneurs for the assumption of risk in business enterprise as distinguished from wages or rent
— prof·it·less \-ləs\ *adjective*
— prof·it·wise \-ˌwīz\ *adverb*

²profit (14th century)
intransitive verb
1 : to be of service or advantage **:** AVAIL
2 : to derive benefit **:** GAIN
3 : to make a profit
transitive verb
: to be of service to **:** BENEFIT

prof·it·able \ˈprä-fə-tə-bəl, ˈpräf-tə-bəl\ *adjective* (14th century)
: affording profits **:** yielding advantageous returns or results
— prof·it·abil·i·ty \ˌprä-fə-tə-ˈbi-lə-tē\ *noun*
— prof·it·able·ness \ˈprä-fə-tə-bəl-nəs\ *noun*
— prof·it·ably \-blē\ *adverb*

profit and loss *noun* (1588)
: a summary account used at the end of an accounting period to collect the balances of the nominal accounts so that the net profit or loss may be shown

prof·i·teer \ˌprä-fə-ˈtir\ *noun* (1912)
: one who makes what is considered an unreasonable profit especially on the sale of essential goods during times of emergency
— profiteer *intransitive verb*

pro·fit·er·ole \prə-ˈfi-tə-ˌrōl\ *noun* [French, perhaps from *profit* profit] (1884)
: a miniature cream puff with a sweet or savory filling

profit sharing *noun* (1881)
: a system or process under which employees receive a part of the profits of an industrial or commercial enterprise

profit system *noun* (1945)
: FREE ENTERPRISE

prof·li·ga·cy \ˈprä-fli-gə-sē\ *noun* (1738)
: the quality or state of being profligate

¹prof·li·gate \ˈprä-fli-gət, -ˌgāt\ *adjective* [Latin *profligatus,* from past participle of *profligare* to strike down, from *pro-* forward, down + *-fligare* (akin to *fligere* to strike); akin to Greek *phlibein* to squeeze] (1647)
1 : completely given up to dissipation and licentiousness
2 : wildly extravagant **:** PRODIGAL
— prof·li·gate·ly *adverb*

²profligate *noun* (1709)
: a person given to wildly extravagant and usually grossly self-indulgent expenditure

pro·flu·ent \ˈprä-ˌflü-ənt, ˈprō-; prō-ˈflü-\ *adjective* [Middle English, from Latin *profluent-, profluens,* present participle of *profluere* to flow forth, from *pro-* forth + *fluere* to flow — more at PRO-, FLUID] (15th century)
: flowing copiously or smoothly

pro for·ma \(ˌ)prō-ˈfȯr-mə\ *adjective* [Latin, for form] (circa 1580)
1 : made or carried out in a perfunctory manner or as a formality
2 : provided in advance to prescribe form or describe items ⟨*pro forma* invoice⟩

¹pro·found \prə-ˈfau̇nd, prō-\ *adjective* [Middle English, from Middle French *profond* deep, from Latin *profundus,* from *pro-* before + *fundus* bottom — more at PRO-, BOTTOM] (14th century)
1 a : having intellectual depth and insight **b :** difficult to fathom or understand
2 a : extending far below the surface **b :** coming from, reaching to, or situated at a depth **:** DEEP-SEATED ⟨a *profound* sigh⟩

> ☆ SYNONYMS
> **Proficient, adept, skilled, skillful, expert** mean having great knowledge and experience in a trade or profession. PROFICIENT implies a thorough competence derived from training and practice ⟨*proficient* in translating foreign languages⟩. ADEPT implies special aptitude as well as proficiency ⟨*adept* at doing long division⟩. SKILLED stresses mastery of technique ⟨a *skilled* surgeon⟩. SKILLFUL implies individual dexterity in execution or performance ⟨*skillful* drivers⟩. EXPERT implies extraordinary proficiency and often connotes knowledge as well as technical skill ⟨*expert* in the evaluation of wines⟩.

3 a : characterized by intensity of feeling or quality **b :** all encompassing **:** COMPLETE ⟨*profound* sleep⟩
— **pro·found·ly** \-ˈfau̇n(d)-lē\ *adverb*
— **pro·found·ness** \-ˈfau̇n(d)-nəs\ *noun*

²profound *noun* (1621)
archaic **:** something that is very deep; *specifically* **:** the depths of the sea

pro·fun·di·ty \prə-ˈfən-də-tē\ *noun, plural* **-ties** [Middle English *profundite,* from Middle French *profundité,* from Latin *profunditat-, profunditas* depth, from *profundus*] (15th century)
1 a : intellectual depth **b :** something profound or abstruse
2 : the quality or state of being profound or deep

pro·fuse \prə-ˈfyüs, prō-\ *adjective* [Middle English, from Latin *profusus,* past participle of *profundere* to pour forth, from *pro-* forth + *fundere* to pour — more at FOUND] (15th century)
1 : pouring forth liberally **:** EXTRAVAGANT ⟨*profuse* in their thanks⟩
2 : exhibiting great abundance **:** BOUNTIFUL ⟨a *profuse* harvest⟩ ☆
— **pro·fuse·ly** *adverb*
— **pro·fuse·ness** *noun*

pro·fu·sion \-ˈfyü-zhən\ *noun* (1545)
1 : lavish expenditure **:** EXTRAVAGANCE
2 : the quality or state of being profuse
3 : great quantity **:** lavish display or supply ⟨snow falling in *profusion*⟩

¹prog \ˈpräg\ *intransitive verb* **progged; prog·ging** [origin unknown] (1624)
chiefly dialect **:** to search about; *especially* **:** FORAGE

²prog *noun* (1655)
chiefly dialect **:** FOOD, VICTUALS

pro·gen·i·tor \prō-ˈje-nə-tər, prə-\ *noun* [Middle English, from Middle French *progeniteur,* from Latin *progenitor,* from *progignere* to beget, from *pro-* forth + *gignere* to beget — more at KIN] (14th century)
1 a : an ancestor in the direct line **:** FOREFATHER **b :** a biologically ancestral form
2 : PRECURSOR, ORIGINATOR ⟨*progenitors* of socialist ideas —*Times Literary Supplement*⟩

prog·e·ny \ˈprä-jᵊn-ē\ *noun, plural* **-nies** [Middle English *progenie,* from Middle French, from Latin *progenies,* from *progignere*] (14th century)
1 a : DESCENDANTS, CHILDREN **b :** offspring of animals or plants
2 : OUTCOME, PRODUCT
3 : a body of followers, disciples, or successors

pro·ges·ta·tion·al \ˌprō-ˌjes-ˈtā-shnəl, -shə-nᵊl\ *adjective* (1923)
: preceding pregnancy or gestation; *especially* **:** of, relating to, inducing, or constituting the modifications of the female mammalian system associated especially with ovulation and corpus luteum formation ⟨*progestational* hormones⟩

pro·ges·ter·one \prō-ˈjes-tə-ˌrōn\ *noun* [*progestin* + *-sterone*] (1935)
: a female steroid sex hormone $C_{21}H_{30}O_2$ that is secreted by the corpus luteum to prepare the endometrium for implantation and later by the placenta during pregnancy to prevent rejection of the developing embryo or fetus

pro·ges·tin \-ˈjes-tən\ *noun* [*pro-* + *gestation* + ¹*-in*] (1930)
: a progestational hormone; *especially* **:** PROGESTERONE

pro·ges·to·gen \-ˈjes-tə-jən\ *noun* [*progestational* + *-ogen* (as in *estrogen*)] (1942)
: any of several progestational steroids (as progesterone)
— **pro·ges·to·gen·ic** \-ˌjes-tə-ˈje-nik\ *adjective*

pro·glot·tid \(ˌ)prō-ˈglä-təd\ *noun* [New Latin *proglottis*] (1878)
: a segment of a tapeworm containing both male and female reproductive organs

pro·glot·tis \(ˌ)prō-ˈglä-təs\ *noun, plural* **-glot·ti·des** \-ˈglä-tə-ˌdēz\ [New Latin *proglottid-, proglottis,* from Greek *proglōttis* tip of the tongue, from *pro-* + *glōtta* tongue — more at GLOSS] (1855)
: PROGLOTTID

prog·na·thism \ˈpräg-nə-ˌthi-zəm, präg-ˈnā-\ *noun* (circa 1864)
: prognathous condition

prog·na·thous \-thəs\ *adjective* (1836)
: having the jaws projecting beyond the upper part of the face

prog·no·sis \präg-ˈnō-səs\ *noun, plural* **-no·ses** \-ˌsēz\ [Late Latin, from Greek *prognōsis,* literally, foreknowledge, from *progignōskein* to know before, from *pro-* + *gignōskein* to know — more at KNOW] (1655)
1 : the prospect of recovery as anticipated from the usual course of disease or peculiarities of the case
2 : FORECAST, PROGNOSTICATION

¹prog·nos·tic \präg-ˈnäs-tik\ *noun* [Middle English *pronostique,* from Middle French, from Latin *prognosticum,* from Greek *prognōstikon,* from neuter of *prognōstikos* foretelling, from *progignōskein*] (14th century)
1 : something that foretells **:** PORTENT
2 : PROGNOSTICATION, PROPHECY

²prognostic *adjective* (1603)
: of, relating to, or serving as ground for prognostication or a prognosis ⟨*prognostic* weather charts⟩ ⟨favorable *prognostic* signs⟩

prog·nos·ti·cate \präg-ˈnäs-tə-ˌkāt\ *transitive verb* **-cat·ed; -cat·ing** (15th century)
1 : to foretell from signs or symptoms **:** PREDICT
2 : PRESAGE
synonym see FORETELL
— **prog·nos·ti·ca·tive** \-ˌkā-tiv\ *adjective*
— **prog·nos·ti·ca·tor** \-ˌkā-tər\ *noun*

prog·nos·ti·ca·tion \(ˌ)präg-ˌnäs-tə-ˈkā-shən\ *noun* (15th century)
1 : an indication in advance **:** FORETOKEN
2 a : an act, the fact, or the power of prognosticating **:** FORECAST **b :** FOREBODING

pro·grade \ˈprō-ˌgrād\ *adjective* [Latin *pro-* forward + English *-grade* (as in *retrograde*)] (1967)
: having or being a direction of rotation or revolution that is counterclockwise as viewed from the north pole of the sky or a planet

¹pro·gram \ˈprō-ˌgram, -grəm\ *noun* [French *programme* agenda, public notice, from Greek *programma,* from *prographein* to write before, from *pro-* before + *graphein* to write — more at CARVE] (1633)
1 [Late Latin *programma,* from Greek] **:** a public notice
2 a : a brief usually printed outline of the order to be followed, of the features to be presented, and the persons participating (as in a public exercise or performance) **b :** the performance of a program; *especially* **:** a performance broadcast on radio or television
3 : a plan or system under which action may be taken toward a goal
4 : CURRICULUM
5 : PROSPECTUS, SYLLABUS
6 a : a plan for the programming of a mechanism (as a computer) **b :** a sequence of coded instructions that can be inserted into a mechanism (as a computer); *also* **:** such a sequence that is part of an organism's genotype or behavioral repertoire

²program *also* **programme** *transitive verb* **-grammed** *or* **-gramed; -gram·ming** *or* **-gram·ing** (1896)
1 a : to arrange or furnish a program of or for **:** BILL **b :** to enter in a program
2 : to work out a sequence of operations to be performed by (a mechanism) **:** provide with a program
3 a : to insert a program for (a particular action) into or as if into a mechanism **b :** to control by or as if by a program **c** (1) **:** to code in an organism's program (2) **:** to provide with a biological program ⟨cells *programmed* to synthesize hemoglobin⟩
4 : to predetermine the thinking, behavior, or operations of as if by computer programming ⟨children are *programmed* into violence —Lisa A. Richette⟩
— **pro·gram·ma·bil·i·ty** \(ˌ)prō-ˌgra-mə-ˈbi-lə-tē\ *noun*
— **pro·gram·ma·ble** \ˈprō-ˌgra-mə-bəl\ *adjective or noun*

program director *noun* (1953)
: one in charge of planning and scheduling program material for a radio or television station or network

pro·gram·mat·ic \ˌprō-grə-ˈma-tik\ *adjective* (1896)
1 : relating to program music
2 : of, relating to, resembling, or having a program
— **pro·gram·mat·i·cal·ly** \-ti-k(ə-)lē\ *adverb*

programme *chiefly British variant of* PROGRAM

programmed instruction *noun* (1962)
: instruction through information given in small steps with each requiring a correct response by the learner before going on to the next step

pro·gram·mer *also* **pro·gram·er** \ˈprō-ˌgra-mər, -grə-\ *noun* (circa 1890)
: one that programs: as **a :** a person who prepares and tests programs for devices (as computers) **b :** one that programs a mechanism **c :** one that prepares instructional or educational programs

pro·gram·ming *also* **pro·gram·ing** \-miŋ\ *noun* (1940)
1 : the planning, scheduling, or performing of a program
2 a : the process of instructing or learning by means of an instructional program **b :** the process of preparing an instructional program

program music *noun* (1879)
: music intended to suggest a sequence of images or incidents

program trading *noun* (1985)
: computerized trading of large blocks of stocks in one market against stock index futures in another

¹prog·ress \ˈprä-grəs, -ˌgres, *US also & British usually* ˈprō-ˌgres\ *noun* [Middle English, from Latin *progressus* advance, from *progredi* to go forth, from *pro-* forward + *gradi* to go — more at PRO-, GRADE] (15th century)

☆ SYNONYMS
Profuse, lavish, prodigal, luxuriant, lush, exuberant mean giving or given out in great abundance. PROFUSE implies pouring forth without restraint ⟨uttered *profuse* apologies⟩. LAVISH suggests an unstinted or unmeasured profusion ⟨a *lavish* wedding reception⟩. PRODIGAL implies reckless or wasteful lavishness threatening to lead to early exhaustion of resources ⟨*prodigal* spending⟩. LUXURIANT suggests a rich and splendid abundance ⟨*luxuriant* sidewhiskers⟩. LUSH suggests rich, soft luxuriance ⟨a *lush* green lawn⟩. EXUBERANT implies marked vitality or vigor in what produces abundantly ⟨an *exuberant* imagination⟩.

1 a (1) **:** a royal journey marked by pomp and pageant (2) **:** a state procession **b :** a tour or circuit made by an official (as a judge) **c :** an expedition, journey, or march through a region
2 : a forward or onward movement (as to an objective or to a goal) **:** ADVANCE
3 : gradual betterment; *especially* **:** the progressive development of mankind
— in progress : going on **:** OCCURRING

²pro·gress \prə-ˈgres\ *intransitive verb* (1539)
1 : to move forward **:** PROCEED
2 : to develop to a higher, better, or more advanced stage

pro·gres·sion \prə-ˈgre-shən\ *noun* (15th century)
1 : a sequence of numbers in which each term is related to its predecessor by a uniform law
2 a : the action or process of progressing **:** ADVANCE **b :** a continuous and connected series **:** SEQUENCE
3 a : succession of musical tones or chords **b :** the movement of musical parts in harmony **c :** SEQUENCE 2c
— pro·gres·sion·al \-ˈgresh-nəl, -ˈgre-shən-ᵊl\ *adjective*

¹pro·gres·sive \prə-ˈgre-siv\ *adjective* (circa 1612)
1 a : of, relating to, or characterized by progress **b :** making use of or interested in new ideas, findings, or opportunities **c :** of, relating to, or constituting an educational theory marked by emphasis on the individual child, informality of classroom procedure, and encouragement of self-expression
2 : of, relating to, or characterized by progression
3 : moving forward or onward **:** ADVANCING
4 a : increasing in extent or severity ⟨a *progressive* disease⟩ **b :** increasing in rate as the base increases ⟨a *progressive* tax⟩
5 *often capitalized* **:** of or relating to political Progressives
6 : of, relating to, or constituting a verb form that expresses action or state in progress at the time of speaking or a time spoken of
— pro·gres·sive·ly *adverb*
— pro·gres·sive·ness *noun*

²progressive *noun* (1846)
1 a : one that is progressive **b :** one believing in moderate political change and especially social improvement by governmental action
2 *capitalized* **:** a member of any of various U.S. political parties: as **a :** a member of a predominantly agrarian minor party that around 1912 split off from the Republicans; *specifically* **:** BULL MOOSE **b :** a follower of Robert M. La Follette in the presidential campaign of 1924 **c :** a follower of Henry A. Wallace in the presidential campaign of 1948

Progressive Conservative *adjective* (1944)
: of or relating to a major political party in Canada traditionally advocating economic nationalism and close ties with the United Kingdom and the Commonwealth
— Progressive Conservative *noun*

pro·gres·siv·ism \prə-ˈgre-si-ˌvi-zəm\ *noun* (1892)
1 : the principles, beliefs, or practices of progressives
2 *capitalized* **:** the political and economic doctrines advocated by the Progressives
3 : the theories of progressive education
— pro·gres·siv·ist \-vist\ *noun or adjective*
— pro·gres·siv·is·tic \-ˌgre-si-ˈvis-tik\ *adjective*

pro·gres·siv·i·ty \ˌprō-(ˌ)gre-ˈsi-və-tē\ *noun* (1883)
: the quality or state of being a progressive tax

pro·hib·it \prō-ˈhi-bət, prə-\ *transitive verb* [Middle English, from Latin *prohibitus,* past participle of *prohibēre* to keep off, from *pro-* forward + *habēre* to hold — more at PRO-, GIVE] (15th century)
1 : to forbid by authority **:** ENJOIN
2 a : to prevent from doing something **b :** PRECLUDE
synonym see FORBID

pro·hi·bi·tion \ˌprō-ə-ˈbi-shən *also* ˌprō-hə-\ *noun* (14th century)
1 : the act of prohibiting by authority
2 : an order to restrain or stop
3 *often capitalized* **:** the forbidding by law of the manufacture, transportation, and sale of alcoholic liquors except for medicinal and sacramental purposes

pro·hi·bi·tion·ist \-sh(ə-)nist\ *noun* (circa 1846)
: one who favors prohibition; *especially, capitalized* **:** a member of a minor U.S. political party advocating prohibition

pro·hib·i·tive \prō-ˈhi-bə-tiv, prə-\ *adjective* (15th century)
1 : tending to prohibit or restrain
2 : tending to preclude use or purchase ⟨*prohibitive* costs⟩
3 : almost certain to perform as predicted ⟨a *prohibitive* favorite⟩
— pro·hib·i·tive·ly *adverb*
— pro·hib·i·tive·ness *noun*

pro·hib·i·to·ry \-ˈhi-bə-ˌtōr-ē, -ˌtȯr-\ *adjective* (circa 1591)
: PROHIBITIVE

pro·insulin \(ˌ)prō-ˈin(t)-s(ə-)lən\ *noun* (1916)
: a single-chain pancreatic polypeptide precursor of insulin that gives rise to the double chain of insulin by loss of the middle part of the molecule

¹proj·ect \ˈprä-ˌjekt, -jikt *also* ˈprō-\ *noun* [Middle English *proiecte,* from Medieval Latin *projectum,* from Latin, neuter of *projectus,* past participle of *proicere* to throw forward, from *pro-* + *jacere* to throw — more at JET] (15th century)
1 : a specific plan or design **:** SCHEME
2 *obsolete* **:** IDEA
3 : a planned undertaking: as **a :** a definitely formulated piece of research **b :** a large usually government-supported undertaking **c :** a task or problem engaged in usually by a group of students to supplement and apply classroom studies
4 : a usually public housing development consisting of houses or apartments built and arranged according to a single plan
synonym see PLAN

²pro·ject \prə-ˈjekt\ *verb* [partly modification of Middle French *projeter,* from Old French *porjeter* to throw forward, from *por-* (from Latin *porro* forward; akin to Greek *pro* forward) + *jeter* to throw; partly from Latin *projectus,* past participle — more at FOR, JET] (15th century)
transitive verb
1 a : to devise in the mind **:** DESIGN **b :** to plan, figure, or estimate for the future ⟨*project* expenditures for the coming year⟩
2 : to throw or cast forward **:** THRUST
3 : to put or set forth **:** present for consideration
4 : to cause to protrude
5 : to cause (light or shadow) to fall into space or (an image) to fall on a surface ⟨*project* a beam of light⟩
6 : to reproduce (as a point, line, or area) on a surface by motion in a prescribed direction
7 : to display outwardly especially to an audience ⟨*project* an image⟩ ⟨an actress who could *project* amorality —*Current Biography*⟩
8 : to attribute (one's own ideas, feelings, or characteristics) to other people or to objects ⟨a nation is an entity on which one can *project* many of the worst of one's instincts —*Times Literary Supplement*⟩
intransitive verb
1 : to jut out **:** PROTRUDE
2 a : to come across vividly **:** give an impression **b :** to make oneself heard clearly
— pro·ject·able \-ˈjek-tə-bəl\ *adjective*

¹pro·jec·tile \prə-ˈjek-tᵊl, -ˌtīl, *chiefly British* ˈprä-jik-ˌtīl\ *noun* (1665)
1 : a body projected by external force and continuing in motion by its own inertia; *especially* **:** a missile for a weapon (as a firearm)
2 : a self-propelling weapon (as a rocket)

²projectile *adjective* (1715)
1 : projecting or impelling forward ⟨a *projectile* force⟩
2 : capable of being thrust forward

pro·jec·tion \prə-ˈjek-shən\ *noun* (1557)
1 a : a systematic presentation of intersecting coordinate lines on a flat surface upon which features from the curved surface of the earth or the celestial sphere may be mapped **b :** the process or technique of reproducing a spatial object upon a plane or curved surface or a line by projecting its points; *also* **:** a graph or figure so formed
2 : a transforming change
3 : the act of throwing or thrusting forward
4 : the forming of a plan **:** SCHEMING
5 a (1) **:** a jutting out (2) **:** a part that juts out **b :** a view of a building or architectural element
6 a : the act of perceiving a mental object as spatially and sensibly objective; *also* **:** something so perceived **b :** the attribution of one's own ideas, feelings, or attitudes to other people or to objects; *especially* **:** the externalization of blame, guilt, or responsibility as a defense against anxiety
7 : the display of motion pictures by projecting an image from them upon a screen
8 a : the act of projecting especially to an audience **b :** control of the volume, clarity, and distinctness of a voice to gain greater audibility
9 : an estimate of future possibilities based on a current trend ☆
— pro·jec·tion·al \-shnəl, -shə-nᵊl\ *adjective*

projection booth *noun* (circa 1928)
: a booth in a theater or hall for housing and operating a projector and especially a motion-picture projector

pro·jec·tion·ist \prə-ˈjek-sh(ə-)nist\ *noun* (1922)
: one that makes projections: as **a :** CARTOGRAPHER **b :** a person who operates a motion-picture projector or television equipment

pro·jec·tive \prə-ˈjek-tiv\ *adjective* (1682)
1 : relating to, produced by, or involving geometric projection
2 : of or relating to something that indicates the psychodynamic constitution of an individual ⟨*projective* tests⟩
— pro·jec·tive·ly *adverb*

projective geometry *noun* (1885)
: a branch of geometry that deals with the properties of configurations that are unaltered by projection

pro·jec·tor \prə-ˈjek-tər\ *noun* (1596)
1 : one that plans a project; *specifically* **:** PROMOTER
2 : one that projects: as **a :** a device for projecting a beam of light **b :** an optical instrument for projecting an image upon a surface **c :** a machine for projecting motion pictures on a screen
3 : an imagined line from an object to a surface along which projection takes place

☆ SYNONYMS
Projection, protrusion, protuberance, bulge mean an extension beyond the normal line or surface. PROJECTION implies a jutting out especially at a sharp angle ⟨those *projections* along the wall are safety hazards⟩. PROTRUSION suggests a thrusting out so that the extension seems a deformity ⟨the bizarre *protrusions* of a coral reef⟩. PROTUBERANCE implies a growing or swelling out in rounded form ⟨a skin disease marked by warty *protuberances*⟩. BULGE suggests an expansion caused by internal pressure ⟨*bulges* in the tile floor⟩.

pro·jet \prō-ˈzhā, ˈprō-ˌ\ *noun, plural* **projets** \-ˈzhā(z), -ˌzhā(z)\ [French, from Middle French *pourget,* from *pourjeter, projeter*] (1808)
1 : PLAN; *especially* **:** a draft of a proposed measure or treaty
2 : a projected or proposed design

pro·kary·ote \(ˌ)prō-ˈkar-ē-ˌōt\ *noun* [New Latin *Prokaryotes,* proposed subdivision of protists, from [1]*pro-* + *kary-* + *-otes,* plural noun suffix, from Greek *-ōtos* — more at -OTIC] (1963)
: a cellular organism (as a bacterium or a blue-green alga) that does not have a distinct nucleus — compare EUKARYOTE
— pro·kary·ot·ic \-ˌkar-ē-ˈä-tik\ *adjective*

pro·lac·tin \prō-ˈlak-tən\ *noun* [[2]*pro-* + *lact-* + [1]*-in*] (1932)
: a protein hormone of the anterior lobe of the pituitary that induces lactation

pro·la·min *or* **pro·la·mine** \ˈprō-lə-mən, -ˌmēn\ *noun* [International Scientific Vocabulary *proline* + *ammonia* + [1]*-in,* [2]*-ine*] (1908)
: any of various simple proteins (as zein) that are found especially in seeds and are insoluble in absolute alcohol or water

pro·lan \ˈprō-ˌlan\ *noun* [German, from Latin *proles* progeny — more at PROLETARIAN] (1931)
: either of two gonadotrophic hormones: **a :** FOLLICLE-STIMULATING HORMONE **b :** LUTEINIZING HORMONE

[1]**pro·lapse** \prō-ˈlaps, ˈprō-ˌ\ *noun* [New Latin *prolapsus,* from Late Latin, fall, from Latin *prolabi* to fall or slide forward, from *pro-* forward + *labi* to slide — more at PRO-, SLEEP] (circa 1834)
: the falling down or slipping of a body part from its usual position or relations

[2]**pro·lapse** \prō-ˈlaps\ *intransitive verb* **pro·lapsed; pro·laps·ing** (1876)
: to undergo prolapse

pro·late \ˈprō-ˌlāt\ *adjective* [Latin *prolatus* (past participle of *proferre* to bring forward, extend) from *pro-* forward + *latus,* past participle of *ferre* to carry — more at BEAR, TOLERATE] (1694)
: EXTENDED; *especially* **:** elongated in the direction of a line joining the poles ⟨a *prolate* spheroid⟩

prole \ˈprōl\ *noun or adjective* (1887)
: PROLETARIAN

pro·leg \ˈprō-ˌleg, -ˌlāg\ *noun* (1816)
: a fleshy leg that occurs on an abdominal segment of some insect larvae but not in the adult

pro·le·gom·e·non \ˌprō-li-ˈgä-mə-ˌnän, -nən\ *noun, plural* **-e·na** \-nə\ [Greek, neuter present passive participle of *prolegein* to say beforehand, from *pro-* before + *legein* to say — more at LEGEND] (circa 1652)
: prefatory remarks; *specifically* **:** a formal essay or critical discussion serving to introduce and interpret an extended work
— pro·le·gom·e·nous \-nəs\ *adjective*

pro·lep·sis \prō-ˈlep-səs\ *noun, plural* **-lep·ses** \-ˌsēz\ [Greek *prolēpsis,* from *prolambanein* to take beforehand, from *pro* before + *lambanein* to take — more at LATCH] (1578)
: ANTICIPATION: as **a :** the representation or assumption of a future act or development as if presently existing or accomplished **b :** the application of an adjective to a noun in anticipation of the result of the action of the verb (as in "while yon slow oxen turn the *furrowed* plain")
— pro·lep·tic \-ˈlep-tik\ *adjective*
— pro·lep·ti·cal·ly \-ti-k(ə-)lē\ *adverb*

pro·le·tar·i·an \ˌprō-lə-ˈter-ē-ən\ *noun* [Latin *proletarius,* from *proles* progeny, from *pro-* forth + *-oles* (akin to *alere* to nourish) — more at OLD] (1658)
: a member of the proletariat
— proletarian *adjective*

pro·le·tar·i·an·ise *British variant of* PROLETARIANIZE

pro·le·tar·i·an·ize \-ˈter-ē-ə-ˌnīz\ *transitive verb* **-ized; -iz·ing** (1887)
: to reduce to a proletarian status or level
— pro·le·tar·i·an·i·za·tion \-ˌter-ē-ə-nə-ˈzā-shən\ *noun*

pro·le·tar·i·at \ˌprō-lə-ˈter-ē-ət, -ˈtar-, -ē-ˌat\ *noun* [French *prolétariat,* from Latin *proletarius*] (1853)
1 : the lowest social or economic class of a community
2 : the laboring class; *especially* **:** the class of industrial workers who lack their own means of production and hence sell their labor to live ◆

pro–life \(ˌ)prō-ˈlīf\ *adjective* (1961)
: ANTIABORTION
— pro–lif·er \-ˈlī-fər\ *noun*

pro·lif·er·ate \prə-ˈli-fə-ˌrāt\ *verb* **-at·ed; -at·ing** [back-formation from *proliferation,* from French *prolifération,* from *proliférer* to proliferate, from *prolifère* reproducing freely, from Latin *proles* + *-fer* -ferous] (1873)
intransitive verb
1 : to grow by rapid production of new parts, cells, buds, or offspring
2 : to increase in number as if by proliferating **:** MULTIPLY
transitive verb
: to cause to grow by proliferating
— pro·lif·er·a·tion \-ˌli-fə-ˈrā-shən\ *noun*
— pro·lif·er·a·tive \-ˈli-fə-ˌrā-tiv, -f(ə-)rə-tiv\ *adjective*

pro·lif·ic \prə-ˈli-fik\ *adjective* [French *prolifique,* from Latin *proles*] (1650)
1 : producing young or fruit especially freely **:** FRUITFUL
2 *archaic* **:** causing abundant growth, generation, or reproduction
3 : marked by abundant inventiveness or productivity ⟨a *prolific* composer⟩
synonym see FERTILE
— pro·lif·i·ca·cy \-ˈli-fi-kə-sē\ *noun*
— pro·lif·i·cal·ly \-fi-k(ə-)lē\ *adverb*
— pro·lif·ic·ness \-fik-nəs\ *noun*

pro·li·fic·i·ty \ˌprō-lə-ˈfi-sə-tē\ *noun* (1725)
: prolific power or character

pro·line \ˈprō-ˌlēn\ *noun* [German *Prolin*] (1904)
: an amino acid $C_5H_9NO_2$ that can be synthesized by animals from glutamate

pro·lix \prō-ˈliks, ˈprō-(ˌ)\ *adjective* [Middle English, from Middle French & Latin; Middle French *prolixe,* from Latin *prolixus* extended, from *pro-* forward + *liquēre* to be fluid — more at LIQUID] (15th century)
1 : unduly prolonged or drawn out **:** too long
2 : marked by or using an excess of words
synonym see WORDY
— pro·lix·i·ty \prō-ˈlik-sə-tē\ *noun*
— pro·lix·ly *adverb*

pro·loc·u·tor \prō-ˈlä-kyə-tər\ *noun* [Middle English, from Latin, from *pro-* for + *locutor* speaker, from *loqui* to speak] (15th century)
1 : one who speaks for another **:** SPOKESMAN
2 : presiding officer **:** CHAIRMAN

pro·log·ize \ˈprō-ˌlȯ-ˌgīz, -ˌlä-; -lə-ˌjīz\ *or* **pro·logu·ize** \-ˌlȯ-ˌgīz, -ˌlä-\ *intransitive verb* **-log·ized** *or* **-logu·ized; -log·iz·ing** *or* **-logu·iz·ing** (1608)
: to write or speak a prologue

pro·logue *also* **pro·log** \ˈprō-ˌlȯg, -ˌläg\ *noun* [Middle English *prolog,* from Middle French *prologue,* from Latin *prologus* preface to a play, from Greek *prologos* part of a Greek play preceding the entry of the chorus, from *pro-* before + *legein* to speak — more at PRO-, LEGEND] (14th century)
1 : the preface or introduction to a literary work
2 a : a speech often in verse addressed to the audience by an actor at the beginning of a play **b :** the actor speaking such a prologue
3 : an introductory or preceding event or development

pro·long \prə-ˈlȯŋ\ *transitive verb* [Middle English, from Middle French *prolonguer,* from Late Latin *prolongare,* from Latin *pro-* forward + *longus* long] (15th century)
1 : to lengthen in time **:** CONTINUE
2 : to lengthen in extent, scope, or range
synonym see EXTEND
— pro·long·er \-ˈlȯŋ-ər\ *noun*

pro·lon·ga·tion \(ˌ)prō-ˌlȯŋ-ˈgā-shən, prə-\ *noun* (15th century)
1 : an extension or lengthening in time or duration
2 : an expansion or continuation in extent, scope, or range

pro·lu·sion \prō-ˈlü-zhən\ *noun* [Latin *prolusion-, prolusio,* from *proludere* to play beforehand, from *pro-* before + *ludere* to play — more at LUDICROUS] (1601)
1 : a preliminary trial or exercise **:** PRELUDE
2 : an introductory and often tentative discourse
— pro·lu·so·ry \-ˈlü-sə-rē, -zə-; -ˈlüz-rē, -ˈlüz-\ *adjective*

prom \ˈpräm\ *noun* [short for *promenade*] (1894)
1 : a formal dance given by a high school or college class
2 *British* **:** PROMENADE 2

[1]**prom·e·nade** \ˌprä-mə-ˈnād, -ˈnäd\ *verb* **-nad·ed; -nad·ing** [[2]*promenade*] (1588)
intransitive verb
1 : to take or go on a promenade
2 : to perform a promenade in a dance
transitive verb
: to walk about in or on
— prom·e·nad·er *noun*

[2]**promenade** *noun* [French, from *promener* to take for a walk, from Latin *prominare* to drive forward, from *pro-* forward + *minare* to drive — more at AMENABLE] (1648)
1 : a place for strolling
2 : a leisurely walk or ride especially in a public place for pleasure or display
3 a : a ceremonious opening of a formal ball consisting of a grand march of all the guests **b :** a figure in a square dance in which couples move counterclockwise in a circle

◇ WORD HISTORY

proletariat The Latin adjective *proletarius* "pertaining to the lowest class of citizens" was a derivative of *proles* "offspring, progeny," the Roman conception apparently being that the humblest citizens had nothing to contribute to the state but their offspring. *Proletarius* was borrowed into French as *prolétaire,* which in the writing of early French socialists such as Saint-Simon was applied to the working class, that is, those people who did not own property and lived solely off their wages. In 1832 the socialist philosopher Pierre Leroux first used the derivative noun *prolétariat* to refer to this class. The concept of the proletariat was refined by Karl Marx, whose "scientific socialism" ousted French utopianism, and French *prolétariat* was borrowed into nearly every language of Europe, including English. The phrase *dictature du prolétariat* "dictatorship of the proletariat," which apparently first surfaced in Paris in 1850, in a manifesto of a revolutionary group inspired by the agitator Louis-Auguste Blanqui, was taken up by Marx in the same year and became a staple of communism. Under Lenin it was synonymous with the dictatorship of the party and became a license for ruthless suppression of dissent in the name of the proletariat.

promenade deck *noun* (1829)
: an upper deck or an area on a deck of a passenger ship where passengers stroll

Pro·me·the·an \prə-ˈmē-thē-ən\ *adjective* (1588)
: of, relating to, or resembling Prometheus, his experiences, or his art; *especially* : daringly original or creative

Pro·me·theus \-thē-əs, -ˌthyüs\ *noun* [Latin, from Greek *Promētheus*]
: a Titan who is chained and tortured by Zeus for stealing fire from heaven and giving it to mankind

pro·me·thi·um \-thē-əm\ *noun* [New Latin, from Latin *Prometheus*] (1948)
: a radioactive metallic element of the rare-earth group obtained as a fission product of uranium or from neutron-irradiated neodymium — see ELEMENT table

prom·i·nence \ˈprä-mə-nən(t)s, ˈpräm-nən(t)s\ *noun* (1598)
1 : something prominent : PROJECTION ⟨a rocky *prominence*⟩
2 : the quality, state, or fact of being prominent or conspicuous
3 : a mass of gas resembling a cloud that arises from the chromosphere of the sun

prom·i·nent \-nənt\ *adjective* [Middle English *promynent,* from Latin *prominent-, prominens,* from present participle of *prominēre* to jut forward, from *pro-* forward + *-minēre* (akin to *mont-, mons* mountain) — more at MOUNT] (15th century)
1 : standing out or projecting beyond a surface or line : PROTUBERANT
2 a : readily noticeable : CONSPICUOUS **b** : widely and popularly known : LEADING
synonym see NOTICEABLE
— **prom·i·nent·ly** *adverb*

pro·mis·cu·i·ty \ˌprä-mə-ˈskyü-ə-tē, ˌprō-\ *noun, plural* **-ties** (circa 1849)
1 : a miscellaneous mixture or mingling of persons or things
2 : promiscuous sexual behavior

pro·mis·cu·ous \prə-ˈmis-kyə-wəs\ *adjective* [Latin *promiscuus,* from *pro-* forth + *miscēre* to mix — more at PRO-, MIX] (1603)
1 : composed of all sorts of persons or things
2 : not restricted to one class, sort, or person : INDISCRIMINATE ⟨education . . . cheapened through the *promiscuous* distribution of diplomas —Norman Cousins⟩
3 : not restricted to one sexual partner
4 : CASUAL, IRREGULAR ⟨*promiscuous* eating habits⟩
— **pro·mis·cu·ous·ly** *adverb*
— **pro·mis·cu·ous·ness** *noun*

¹prom·ise \ˈprä-məs\ *noun* [Middle English *promis,* from Latin *promissum,* from neuter of *promissus,* past participle of *promittere* to send forth, promise, from *pro-* forth + *mittere* to send] (15th century)
1 a : a declaration that one will do or refrain from doing something specified **b** : a legally binding declaration that gives the person to whom it is made a right to expect or to claim the performance or forbearance of a specified act
2 : reason to expect something ⟨little *promise* of relief⟩; *especially* : ground for expectation of success, improvement, or excellence ⟨shows considerable *promise*⟩
3 : something that is promised

²promise *verb* **prom·ised; prom·is·ing** (15th century)
transitive verb
1 : to pledge to do, bring about, or provide ⟨*promise* aid⟩
2 *archaic* : WARRANT, ASSURE
3 *chiefly dialect* : BETROTH
4 : to suggest beforehand : give promise of ⟨dark clouds *promise* rain⟩
intransitive verb
1 : to make a promise
2 : to give ground for expectation : be imminent
— **prom·is·ee** \ˌprä-mə-ˈsē\ *noun*
— **prom·i·sor** \-ˈsȯr\ *also* **prom·is·er** \ˈprä-mə-sər\ *noun*

promised land *noun* (1667)
: something and especially a place or condition believed to promise final satisfaction or realization of hopes

prom·is·ing \ˈprä-mə-siŋ\ *adjective* (1601)
: full of promise : likely to succeed or to yield good results
— **prom·is·ing·ly** \-siŋ-lē\ *adverb*

prom·is·so·ry \ˈprä-mə-ˌsōr-ē, -ˌsȯr-\ *adjective* [Middle English *promissorye,* from Medieval Latin *promissorius,* from Latin *promittere*] (15th century)
: containing or conveying a promise or assurance

promissory note *noun* (1710)
: a written promise to pay at a fixed or determinable future time a sum of money to a specified individual or to bearer

pro·mo \ˈprō-(ˌ)mō\ *noun, plural* **promos** *often attributive* [short for *promotional*] (1946)
: a promotional announcement, blurb, or appearance

prom·on·to·ry \ˈprä-mən-ˌtōr-ē, -ˌtȯr-\ *noun, plural* **-ries** [Latin *promunturium, promontorium;* probably akin to *prominēre* to jut forth — more at PROMINENT] (1548)
1 a : a high point of land or rock projecting into a body of water **b** : a prominent mass of land overlooking or projecting into a lowland
2 : a bodily prominence

pro·mote \prə-ˈmōt\ *transitive verb* **pro·mot·ed; pro·mot·ing** [Middle English, from Latin *promotus,* past participle of *promovēre,* literally, to move forward, from *pro-* forward + *movēre* to move] (14th century)
1 a : to advance in station, rank, or honor : RAISE **b** : to change (a pawn) into a piece in chess by moving to the eighth rank **c** : to advance (a student) from one grade to the next higher grade
2 a : to contribute to the growth or prosperity of : FURTHER ⟨*promote* international understanding⟩ **b** : to help bring (as an enterprise) into being : LAUNCH **c** : to present (merchandise) for buyer acceptance through advertising, publicity, or discounting
3 *slang* : to get possession of by doubtful means or by ingenuity
synonym see ADVANCE
— **pro·mot·abil·i·ty** \-ˌmō-tə-ˈbi-lə-tē\ *noun*
— **pro·mot·able** \-ˈmō-tə-bəl\ *adjective*

pro·mot·er \-ˈmō-tər\ *noun* (14th century)
1 : one that promotes; *especially* : one who assumes the financial responsibilities of a sporting event (as a boxing match) including contracting with the principals, renting the site, and collecting gate receipts
2 *obsolete* : PROSECUTOR
3 : a substance that in very small amounts is able to increase the activity of a catalyst
4 : a binding site in a DNA chain at which RNA polymerase binds to initiate transcription of messenger RNA by one or more nearby structural genes

pro·mo·tion \prə-ˈmō-shən\ *noun* (15th century)
1 : the act or fact of being raised in position or rank : PREFERMENT
2 : the act of furthering the growth or development of something; *especially* : the furtherance of the acceptance and sale of merchandise through advertising, publicity, or discounting
— **pro·mo·tion·al** \-shnəl, -shə-nᵊl\ *adjective*

pro·mo·tive \-ˈmō-tiv\ *adjective* (1644)
: tending or serving to promote
— **pro·mo·tive·ness** *noun*

¹prompt \ˈpräm(p)t\ *transitive verb* [Middle English, from Medieval Latin *promptare,* from Latin *promptus* prompt] (14th century)
1 : to move to action : INCITE
2 : to assist (one acting or reciting) by suggesting or saying the next words of something forgotten or imperfectly learned : CUE
3 : to serve as the inciting cause of
— **prompt·er** *noun*

²prompt *adjective* (1784)
: of or relating to prompting actors

³prompt *adjective* [Middle English, from Middle French or Latin; Middle French, from Latin *promptus* ready, prompt, from past participle of *promere* to bring forth, from *pro-* forth + *emere* to take — more at REDEEM] (15th century)
1 : being ready and quick to act as occasion demands
2 : performed readily or immediately ⟨*prompt* assistance⟩
synonym see QUICK
— **prompt·ly** \ˈpräm(p)t-lē, ˈpräm-plē\ *adverb*
— **prompt·ness** \ˈpräm(p)t-nəs, ˈprämp-nəs\ *noun*

⁴prompt *noun, plural* **prompts** \ˈpräm(p)ts, ˈprämps\ (1597)
1 [¹*prompt*] : something that prompts : REMINDER
2 [³*prompt*] : a limit of time given for payment of an account for goods purchased; *also* : the contract by which this time is fixed

prompt·book \ˈpräm(p)t-ˌbu̇k, ˈprämp-ˌbu̇k\ *noun* (1809)
: a copy of a play with directions for performance used by a theater prompter

promp·ti·tude \ˈpräm(p)-tə-ˌtüd, -ˌtyüd\ *noun* [Middle English, from Middle French or Late Latin; Middle French, from Late Latin *promptitudo,* from Latin *promptus*] (15th century)
: the quality or habit of being prompt : PROMPTNESS

prompt side *noun* (1824)
1 : the side of the stage adjacent to the prompter's corner
2 : the side of the stage to the right of an actor facing the audience

pro·mul·gate \ˈprä-məl-ˌgāt; prō-ˈməl-, prə-ˈ, ˈprō-(ˌ)\ *transitive verb* **-gat·ed; -gat·ing** [Latin *promulgatus,* past participle of *promulgare,* from *pro-* forward + *-mulgare* (probably akin to *mulgēre* to milk, extract) — more at EMULSION] (1530)
1 : to make known by open declaration : PROCLAIM
2 a : to make known or public the terms of (a proposed law) **b** : to put (a law) into action or force
synonym see DECLARE
— **pro·mul·ga·tion** \ˌprä-məl-ˈgā-shən; ˌprō-(ˌ)məl-, (ˌ)prō-ˌ\ *noun*
— **pro·mul·ga·tor** \ˈprä-məl-ˌgā-tər; prō-ˈməl-, prə-ˈ, ˈprō-(ˌ)\ *noun*

pro·na·tion \prō-ˈnā-shən\ *noun* [*pronate,* from Late Latin *pronatus,* past participle of *pronare* to bend forward, from Latin *pronus*] (1666)
1 : rotation of the hand and forearm so that the palm faces backwards or downwards
2 : rotation of the medial bones in the midtarsal region of the foot inward and downward so that in walking the foot tends to come down on its inner margin
— **pro·nate** \ˈprō-ˌnāt\ *verb*

pro·na·tor \ˈprō-ˌnā-tər\ *noun* (circa 1741)
: a muscle that produces pronation

prone \ˈprōn\ *adjective* [Middle English, from Latin *pronus* bent forward, tending; akin to Latin *pro* forward — more at FOR] (14th century)
1 : having a tendency or inclination : being likely ⟨*prone* to forget names⟩ ⟨accident-*prone*⟩

2 a : having the front or ventral surface downward **b :** lying flat or prostrate ☆ ▣
— **prone** *adverb*
— **prone·ly** *adverb*
— **prone·ness** \ˈprōn-nəs\ *noun*

pro·neph·ros \(ˌ)prō-ˈne-frəs, -ˌfräs\ *noun* [New Latin, from Greek *pro-* + *nephros* kidney — more at NEPHRITIS] (1881)
: either member of the first and most anterior pair of the three successive paired vertebrate renal organs that functions in the adults of amphioxus and some lampreys, functions temporarily in larval fishes and amphibians, and is present but nonfunctional in embryos of reptiles, birds, and mammals — compare MESONEPHROS, METANEPHROS
— **pro·neph·ric** \-frik\ *adjective*

¹prong \ˈprȯŋ, ˈpräŋ\ *noun* [Middle English *pronge*] (15th century)
1 : FORK
2 : a tine of a fork
3 : a slender pointed or projecting part: as **a :** a fang of a tooth **b :** a point of an antler
4 : something resembling a prong

²prong *transitive verb* (1848)
: to stab, pierce, or break up with a pronged device

pronged \ˈprȯŋd, ˈpräŋd\ *adjective* (1767)
1 : having a usually specified number of prongs — usually used in combination ⟨a three-*pronged* fork⟩
2 : having a usually specified number of parts or approaches ⟨a two-*pronged* strategy⟩

prong·horn \ˈprȯŋ-ˌhȯrn, ˈpräŋ-\ *noun, plural* **pronghorn** *or* **pronghorns** (1823)
: a ruminant mammal (*Antilocapra americana*) of treeless parts of western North America that resembles an antelope — called also *pronghorn antelope*

pronghorn

pro·nom·i·nal \prō-ˈnä-mə-nᵊl, -ˈnam-nəl\ *adjective* [Late Latin *pronominalis,* from Latin *pronomin-, pronomen*] (1680)
1 : of, relating to, or constituting a pronoun
2 : resembling a pronoun in identifying or specifying without describing ⟨the *pronominal* adjective *this* in *this* dog⟩
— **pro·nom·i·nal·ly** *adverb*

pro·noun \ˈprō-ˌnau̇n\ *noun* [Middle English *pronom,* from Latin *pronomin-, pronomen,* from *pro-* for + *nomin-, nomen* name — more at PRO-, NAME] (1530)
: any of a small set of words in a language that are used as substitutes for nouns or noun phrases and whose referents are named or understood in the context

pro·nounce \prə-ˈnau̇n(t)s\ *verb* **pro·nounced; pro·nounc·ing** [Middle English, from Middle French *prononcier,* from Latin *pronuntiare,* from *pro-* forth + *nuntiare* to report, from *nuntius* messenger — more at PRO-] (14th century)
transitive verb
1 : to declare officially or ceremoniously ⟨the minister *pronounced* them husband and wife⟩
2 : to declare authoritatively or as an opinion ⟨doctors *pronounced* him fit to resume duties⟩
3 a : to employ the organs of speech to produce ⟨*pronounce* these words⟩; *especially* **:** to say correctly ⟨I can't *pronounce* his name⟩ **b :** to represent in printed characters the spoken counterpart of (an orthographic representation) ⟨both dictionaries *pronounce clique* the same⟩
4 : RECITE ⟨speak the speech, I pray you, as I *pronounced* it to you —Shakespeare⟩
intransitive verb
1 : to pass judgment
2 : to produce the components of spoken language
— **pro·nounce·abil·i·ty** \-ˌnau̇n(t)-sə-ˈbi-lə-tē\ *noun*
— **pro·nounce·able** \-ˈnau̇n(t)-sə-bəl\ *adjective*
— **pro·nounc·er** *noun*

pro·nounced \-ˈnau̇n(t)st\ *adjective* (circa 1741)
: strongly marked **:** DECIDED
— **pro·nounced·ly** \-ˈnau̇n(t)-səd-lē, -ˈnau̇n(t)st-lē\ *adverb*

pro·nounce·ment \prə-ˈnau̇n(t)s-mənt\ *noun* (1593)
1 : a usually formal declaration of opinion
2 : an authoritative announcement

pronouncing *adjective* (1764)
: relating to or indicating pronunciation ⟨a *pronouncing* dictionary⟩

pron·to \ˈprän-ˌtō\ *adverb* [Spanish, from Latin *promptus* prompt] (circa 1740)
: without delay

¹pro·nu·clear \(ˌ)prō-ˈnü-klē-ər, -ˈnyü-, ÷-kyə-lər\ *adjective* [*pronucleus*] (circa 1890)
: of, relating to, or resembling a pronucleus

²pronuclear *adjective* [²*pro-* + *nuclear*] (1971)
: advocating the use of nuclear-powered generating stations

pro·nu·cle·us \(ˌ)prō-ˈnü-klē-əs, -ˈnyü-\ *noun* [New Latin] (1880)
: the haploid nucleus of a male or female gamete (as an egg or sperm) up to the time of fusion with that of another gamete in fertilization

pro·nun·ci·a·men·to \prō-ˌnən(t)-sē-ə-ˈmen-(ˌ)tō\ *noun, plural* **-tos** *or* **-toes** [modification of Spanish *pronunciamiento,* from *pronunciar* to pronounce, from Latin *pronuntiare*] (1835)
: PROCLAMATION, PRONOUNCEMENT

pro·nun·ci·a·tion \prə-ˌnən(t)-sē-ˈā-shən *also* ÷-ˌnau̇n(t)-\ *noun* [Middle English *pronunciacion,* from Middle French *prononciation,* from Latin *pronuntiation-, pronuntiatio,* from *pronuntiare*] (15th century)
: the act or manner of pronouncing something ▣
— **pro·nun·ci·a·tion·al** \-shnəl, -shə-nᵊl\ *adjective*

¹proof \ˈprüf\ *noun* [Middle English, alteration of *preove,* from Old French *preuve,* from Late Latin *proba,* from Latin *probare* to prove — more at PROVE] (13th century)
1 a : the cogency of evidence that compels acceptance by the mind of a truth or a fact **b :** the process or an instance of establishing the validity of a statement especially by derivation from other statements in accordance with principles of reasoning
2 *obsolete* **:** EXPERIENCE
3 : something that induces certainty or establishes validity
4 *archaic* **:** the quality or state of having been tested or tried; *especially* **:** unyielding hardness
5 : evidence operating to determine the finding or judgment of a tribunal
6 a *plural* **proofs** *or* **proof :** a copy (as of typeset text) made for examination or correction **b :** a test impression of an engraving, etching, or lithograph **c :** a coin that is struck from a highly-polished die on a polished planchet, is not intended for circulation, and sometimes differs in metallic content from coins of identical design struck for circulation **d :** a test photographic print made from a negative
7 : a test applied to articles or substances to determine whether they are of standard or satisfactory quality
8 a : the minimum alcoholic strength of proof spirit **b :** strength with reference to the standard for proof spirit; *specifically* **:** alcoholic strength indicated by a number that is twice the percent by volume of alcohol present ⟨whiskey of 90 *proof* is 45% alcohol⟩

²proof *adjective* (1592)
1 : able to resist or repel ⟨boots that were . . . *proof* against cold and wet —Robertson Davies⟩ — often used in combination ⟨wind*proof*⟩
2 : used in proving or testing or as a standard of comparison

☆ SYNONYMS
Prone, supine, prostrate, recumbent mean lying down. PRONE implies a position with the front of the body turned toward the supporting surface ⟨push-ups require a *prone* position⟩. SUPINE implies lying on one's back and suggests inertness or abjectness ⟨lying *supine* on the couch⟩. PROSTRATE implies lying full-length as in submission, defeat, or physical collapse ⟨a runner fell *prostrate* at the finish line⟩. RECUMBENT implies the posture of one sleeping or resting ⟨a patient comfortably *recumbent* in a hospital bed⟩. See in addition LIABLE.

☐ USAGE
prone Most commentators insist on the distinction between *prone* and *supine* spelled out in the synonymy paragraph above. The distinction is indeed observed by those writing on physiology and anatomy, and *prone* always means lying on one's belly to those who shoot guns and write about it. But *prone* has wider application: it is used of inanimate objects to mean simply "lying flat" ⟨the *prone* golden autumn harvest of leaves —Michael P. O'Connor⟩ and is also used of the body when its orientation is uncertain or unimportant ⟨I caught sight of the large *prone* figure in bed —D. H. Lawrence⟩ ⟨I too have been *prone* on my couch this week, a victim of the common cold —Flannery O'Connor⟩. Sometimes it is used where *supine* could be used ⟨he lies *prone,* his face to the sky —James Joyce⟩; such use may come from a conscious avoidance of the pejorative overtones of sense 2 of *supine. Supine* is a relatively rare word, so apparently most writers find a way around it. Both *prone* and *supine* are more commonly used in their figurative senses.

pronunciation As if to prove the inherently controversial nature of English pronunciation, the very word *pronunciation* has given rise to not one, but two disputes. The first has to do with the sound of the letter *c* before *i*: should it be articulated as an \s\ as in *précis* or as an \sh\ as in *specialty*? The *ci* sequence in *species, uncial, preciosity,* and *sociologist* is heard both ways, though in many familiar words of Latin derivation the *c* before *i* is pronounced only as \sh\: *special, precious, superficial, judicious.* No fast rule obtains for the Latinate *ci,* though, and thus common usage has dictated the \s\ sound for *pronunciation.* The second issue is that some substitute the syllable \ˌnau̇n(t)\ for \ˌnən(t)\, perhaps wanting unconsciously to preserve the relatedness of *pronounce* and *pronunciation.* The same speakers do not, however, extend the analogy to the pronunciation of *annunciation* or *denunciation.* The \-ˌnau̇n(t)-\ variant has never established itself as an accepted form. Again, common usage overwhelmingly dictates the *pronunciation* \prə-ˌnən(t)-sē-ˈā-shən\.

PROOFREADERS' MARKS

or or	delete; take it out
	close up; print as one word
	delete and close up
^ or > or	caret; insert here (something
#	insert a space
eq#	space evenly where indicated
stet	let marked text stand as set
tr	transpose; change order the
/	used to separate two or more marks and often as a concluding stroke at the end of an insertion
[	set farther to the left
]	set farther to the right
	set ae or fl as ligatures æ or ﬂ
=	straighten alignment
‖	straighten or align
x	imperfect or broken character
□	indent or insert em quad space
¶	begin a new paragraph
(sp)	spell out (set 5 lbs. as five pounds)
cap	set in capitals (CAPITALS)
sm cap or s.c.	set in small capitals (SMALL CAPITALS)
lc	set in Lowercase (lowercase)
ital	set in italic (*italic*)
rom	set in roman (roman)
bf	set in boldface (**boldface**)
= or -/ or or /H/	hyphen
1/N or en or /N/	en dash (1965–72)
1/M or em or /M/	em — or long — dash
∨	superscript or superior (as in πr^2)
∧	subscript or inferior (as in H_2O)
or	centered (for a centered dot in $p \cdot q$)
	comma
	apostrophe
⊙	period
; or ;/	semicolon
: or (:)	colon
or	quotation marks
(/)	parentheses
[/]	brackets
OK/?	query to author: has this been set as intended?
⊥ or ⊥ [1]	push down a work-up
9 [1]	turn over an inverted letter
wf [1]	wrong font; a character of the wrong size or esp. style

[1] The last three symbols are unlikely to be needed in marking proofs of photocomposed matter.

3 : of standard strength or quality or alcoholic content

³proof *transitive verb* (1745)
1 a : to make or take a proof of **b :** PROOFREAD
2 : to give a resistant quality to
3 : to activate (yeast) by mixing with water and sometimes sugar or milk
— proof·er *noun*

proof·read \'prüf-ˌrēd\ *transitive verb* **-read** \-ˌred\; **-read·ing** [back-formation from *proofreader*] (1920)
: to read and mark corrections in (as a proof)

proof·read·er \-ˌrē-dər\ *noun* (1832)
: a person who proofreads

proof·room \'prüf-ˌrüm, -ˌru̇m\ *noun* (1903)
: a room for proofreading

proof spirit *noun* (1790)
: an alcoholic liquor or mixture of ethanol and water that contains 50% ethanol by volume at 60°F (16°C)

¹prop \'präp\ *noun* [Middle English *proppe,* from Middle Dutch, stopper; akin to Middle Low German *proppe* stopper] (15th century)
: something that props or sustains **:** SUPPORT

²prop *transitive verb* **propped; prop·ping** (1538)
1 a : to support by placing something under or against — often used with *up* **b :** to support by placing against something
2 : SUSTAIN, STRENGTHEN

³prop *noun* (1841)
: PROPERTY 3

⁴prop *noun* (1914)
: PROPELLER

prop- *combining form* [International Scientific Vocabulary, from *propionic (acid)*]
: related to propionic acid ⟨propane⟩ ⟨*propyl*⟩

pro·pae·deu·tic \ˌprō-pi-'dü-tik, -'dyü-\ *noun* [Greek *propaideuein* to teach beforehand, from *pro-* before + *paideuein* to teach, from *paid-, pais* child — more at PRO-, FEW] (1798)
: preparatory study or instruction
— propaedeutic *adjective*

pro·pa·gan·da \ˌprä-pə-'gan-də, ˌprō-\ *noun* [New Latin, from *Congregatio de propaganda fide* Congregation for propagating the faith, organization established by Pope Gregory XV (died 1623)] (1718)
1 *capitalized* **:** a congregation of the Roman curia having jurisdiction over missionary territories and related institutions
2 : the spreading of ideas, information, or rumor for the purpose of helping or injuring an institution, a cause, or a person
3 : ideas, facts, or allegations spread deliberately to further one's cause or to damage an opposing cause; *also* **:** a public action having such an effect ◆
— pro·pa·gan·dist \-dist\ *noun or adjective*
— pro·pa·gan·dis·tic \-ˌgan-'dis-tik\ *adjective*
— pro·pa·gan·dis·ti·cal·ly \-ti-k(ə-)lē\ *adverb*

pro·pa·gan·dize \-'gan-ˌdīz\ *verb* **-dized; -diz·ing** (1844)
transitive verb
: to subject to propaganda; *also* **:** to carry on propaganda for
intransitive verb
: to carry on propaganda
— pro·pa·gan·diz·er \-ˌdī-zər\ *noun*

prop·a·gate \'prä-pə-ˌgāt\ *verb* **-gat·ed; -gat·ing** [Latin *propagatus,* past participle of *propagare* to set slips, propagate, from *propages* slip, offspring, from *pro-* before + *pangere* to fasten — more at PRO-, PACT] (circa 1570)
transitive verb
1 : to cause to continue or increase by sexual or asexual reproduction
2 : to pass along to offspring
3 a : to cause to spread out and affect a greater number or greater area **:** EXTEND **b :** to foster

◇ WORD HISTORY

propaganda In 1622, in an effort to centralize the missionary activities of the Roman Catholic Church overseas, Pope Gregory XV issued a bull by which he established the *Congregatio de propaganda fide* ("Congregation for the Propagation of the Faith"), a committee of cardinals charged with the administration of ecclesiastical affairs in non-Catholic countries. The name of the office was informally shortened to *Propaganda,* a word that became current in several European vernaculars, including English. During the French Revolution, the French form *propagande* was generalized to mean any group whose purpose was to spread views or principles, especially of a political nature, and then to mean the activity of spreading such views. The latter meaning of the word was taken up by English, German, Russian, and other languages. The association of *propaganda* in various languages with the promulgation of subversive political or economic ideas by revolutionary groups gave it a negative connotation in the 19th and 20th centuries. In extremist circles of both the left and right, though, it simply came to mean "inculcation of correct political views," and German *Propaganda* was a positive term to both the Nazis and the Communist regime that succeeded the Nazis in eastern Germany.

growing knowledge of, familiarity with, or acceptance of (as an idea or belief) **:** PUBLICIZE **c :** to transmit (as sound or light) through a medium
intransitive verb
1 : to multiply sexually or asexually
2 : INCREASE, EXTEND
3 : to travel through space or a material — used of wave energy (as light, sound, or radio waves)
— **prop·a·ga·ble** \'prä-pə-gə-bəl\ *adjective*
— **prop·a·ga·tive** \-,gā-tiv\ *adjective*
— **prop·a·ga·tor** \-,gā-tər\ *noun*

prop·a·ga·tion \,prä-pə-'gā-shən\ *noun* (15th century)
: the act or action of propagating: as **a :** increase (as of a kind of organism) in numbers **b :** the spreading of something (as a belief) abroad or into new regions **c :** enlargement or extension (as of a crack) in a solid body

prop·a·gule \'prä-pə-,gyü(ə)l\ *noun* [New Latin *propagulum,* from Latin *propages* slip] (1858)
: a structure (as a cutting, a seed, or a spore) that propagates a plant

pro·pane \'prō-,pān\ *noun* [International Scientific Vocabulary *prop-* + *-ane*] (1866)
: a heavy flammable gaseous alkane C_3H_8 found in crude petroleum and natural gas and used especially as fuel and in chemical synthesis

pro·pel \prə-'pel\ *transitive verb* **pro·pelled; pro·pel·ling** [Middle English *propellen,* from Latin *propellere,* from *pro-* before + *pellere* to drive — more at FELT] (15th century)
: to drive forward or onward by or as if by means of a force that imparts motion

[1]**pro·pel·lant** *also* **pro·pel·lent** \-'pe-lənt\ *adjective* (1644)
: capable of propelling

[2]**propellant** *also* **propellent** *noun* (1814)
: something that propels: as **a :** an explosive for propelling projectiles **b :** fuel plus oxidizer used by a rocket engine **c :** a gas kept under pressure in a bottle or can for expelling the contents when the pressure is released

pro·pel·ler *also* **pro·pel·lor** \prə-'pe-lər\ *noun* (1780)
: one that propels; *especially* **:** a device that consists of a central hub with radiating blades placed and twisted so that each forms part of a helical surface and that is used to propel a vehicle (as a ship or airplane)

pro·pend \prō-'pend\ *intransitive verb* [Latin *propendēre,* from *pro-* before + *pendēre* to hang — more at PENDANT] (1545)
obsolete **:** INCLINE

pro·pense \prō-'pen(t)s\ *adjective* [Latin *propensus,* past participle of *propendēre*] (1528)
archaic **:** leaning or inclining toward **:** DISPOSED

pro·pen·si·ty \prə-'pen(t)-sə-tē\ *noun, plural* **-ties** (1570)
: an often intense natural inclination or preference
synonym see LEANING

[1]**prop·er** \'prä-pər\ *adjective* [Middle English *propre* proper, own, from Middle French, from Latin *proprius* own] (14th century)
1 a : referring to one individual only **b :** belonging to one **:** OWN **c :** appointed for the liturgy of a particular day **d :** represented heraldically in natural color
2 : belonging characteristically to a species or individual **:** PECULIAR
3 *chiefly dialect* **:** GOOD-LOOKING, HANDSOME
4 : very good **:** EXCELLENT
5 *chiefly British* **:** UTTER, ABSOLUTE
6 : strictly limited to a specified thing, place, or idea ⟨the city *proper*⟩
7 a : strictly accurate **:** CORRECT **b** *archaic* **:** VIRTUOUS, RESPECTABLE **c :** strictly decorous **:** GENTEEL
8 : marked by suitability, rightness, or appropriateness **:** FIT
9 : being a mathematical subset (as a subgroup) that does not contain all the elements of the inclusive set from which it is derived
synonym see FIT
— **prop·er·ly** *adverb*
— **prop·er·ness** *noun*

[2]**proper** *noun* (15th century)
1 : the parts of the Mass that vary according to the liturgical calendar
2 : the part of a missal or breviary containing the proper of the Mass and the offices proper to the holy days of the liturgical year

[3]**proper** *adverb* (15th century)
chiefly dialect **:** in a thorough manner **:** COMPLETELY

proper adjective *noun* (1905)
: an adjective that is formed from a proper noun and that is usually capitalized in English

pro·per·din \prō-'pər-dᵊn\ *noun* [probably from [1]*pro-* + Latin *perdere* to destroy + English [1]*-in* — more at PERDITION] (1954)
: a serum protein that participates in destruction of bacteria, neutralization of viruses, and lysis of red blood cells

proper fraction *noun* (1674)
: a fraction in which the numerator is less or of lower degree than the denominator

proper noun *noun* (circa 1890)
: a noun that designates a particular being or thing, does not take a limiting modifier, and is usually capitalized in English — called also *proper name*

prop·er·tied \'prä-pər-ted\ *adjective* (circa 1772)
: possessing property

prop·er·ty \'prä-pər-tē\ *noun, plural* **-ties** [Middle English *proprete,* from Middle French *proprété,* from Latin *proprietat-, proprietas,* from *proprius* own] (14th century)
1 a : a quality or trait belonging and especially peculiar to an individual or thing **b :** an effect that an object has on another object or on the senses **c :** VIRTUE 3 **d :** an attribute common to all members of a class
2 a : something owned or possessed; *specifically* **:** a piece of real estate **b :** the exclusive right to possess, enjoy, and dispose of a thing **:** OWNERSHIP **c :** something to which a person or business has a legal title **d :** one (as a performer) under contract whose work is especially valuable
3 : an article or object used in a play or motion picture except painted scenery and costumes
synonym see QUALITY
— **prop·er·ty·less** \-ləs\ *adjective*
— **prop·er·ty·less·ness** \-nəs\ *noun*

property damage insurance *noun* (circa 1946)
: insurance protecting against all or part of an individual's legal liability for damage done (as by his or her automobile) to the property of another

property right *noun* (1903)
: a legal right or interest in or against specific property

property tax *noun* (1808)
: a tax levied on real or personal property

pro·phage \'prō-,fāj, -,fäzh\ *noun* (1951)
: an intracellular form of a bacteriophage in which it is harmless to the host, is usually integrated into the hereditary material of the host, and reproduces when the host does

pro·phase \-,fāz\ *noun* [International Scientific Vocabulary] (1884)
1 : the initial stage of mitosis and of the mitotic division of meiosis characterized by the condensation of chromosomes consisting of two chromatids, disappearance of the nucleolus and nuclear membrane, and formation of mitotic spindle
2 : the initial stage of the first division of meiosis in which the chromosomes become visible, homologous pairs of chromosomes undergo synapsis and crossing over, chiasmata appear, chromosomes condense with homologues visible as tetrads, and the nuclear membrane and nucleolus disappear — compare DIAKINESIS, DIPLOTENE, LEPTOTENE, PACHYTENE, ZYGOTENE
— **pro·pha·sic** \(,)prō-'fā-zik\ *adjective*

proph·e·cy *also* **proph·e·sy** \'prä-fə-sē\ *noun, plural* **-cies** *also* **-sies** [Middle English *prophecie,* from Old French, from Late Latin *prophetia,* from Greek *prophēteia,* from *prophētēs* prophet] (13th century)
1 : an inspired utterance of a prophet
2 : the function or vocation of a prophet; *specifically* **:** the inspired declaration of divine will and purpose
3 : a prediction of something to come

proph·e·sy \'prä-fə-,sī\ *verb* **-sied; -sy·ing** [Middle English *prophesien,* from Middle French *prophesier,* from Old French, from *prophecie*] (14th century)
transitive verb
1 : to utter by or as if by divine inspiration
2 : to predict with assurance or on the basis of mystic knowledge
3 : PREFIGURE
intransitive verb
1 : to speak as if divinely inspired
2 : to give instruction in religious matters **:** PREACH
3 : to make a prediction
synonym see FORETELL
— **proph·e·si·er** \-,sī(-ə)r\ *noun*

proph·et \'prä-fət\ *noun* [Middle English *prophete,* from Old French, from Latin *propheta,* from Greek *prophētēs,* from *pro* for + *phanai* to speak — more at FOR, BAN] (12th century)
1 : one who utters divinely inspired revelations; *specifically, often capitalized* **:** the writer of one of the prophetic books of the Old Testament
2 : one gifted with more than ordinary spiritual and moral insight; *especially* **:** an inspired poet
3 : one who foretells future events **:** PREDICTOR
4 : an effective or leading spokesman for a cause, doctrine, or group
5 *Christian Science* **a :** a spiritual seer **b :** disappearance of material sense before the conscious facts of spiritual Truth
— **proph·et·hood** \-,hu̇d\ *noun*

proph·et·ess \'prä-fə-təs\ *noun* (14th century)
: a woman who is a prophet

pro·phet·ic \prə-'fe-tik\ *also* **pro·phet·i·cal** \-'fe-ti-kəl\ *adjective* (15th century)
1 : of, relating to, or characteristic of a prophet or prophecy
2 : foretelling events **:** PREDICTIVE
— **pro·phet·i·cal·ly** \-ti-k(ə-)lē\ *adverb*

Proph·ets \'prä-fəts\ *noun plural*
: the second part of the Jewish scriptures — see BIBLE table

[1]**pro·phy·lac·tic** \,prō-fə-'lak-tik *also* ,prä-\ *adjective* [Greek *prophylaktikos,* from *prophylassein* to be on guard, from *pro-* before + *phylassein* to guard, from *phylak-, phylax* guard] (1574)
1 : guarding from or preventing disease
2 : tending to prevent or ward off **:** PREVENTIVE
— **pro·phy·lac·ti·cal·ly** \-ti-k(ə-)lē\ *adverb*

[2]**prophylactic** *noun* (1642)
: something prophylactic; *especially* **:** a device and especially a condom for preventing venereal infection or conception

pro·phy·lax·is \-'lak-səs\ *noun, plural* **-lax·es** \-'lak-,sēz\ [New Latin, from Greek *prophylaktikos*] (circa 1842)

: measures designed to preserve health (as of an individual or of society) and prevent the spread of disease

¹pro·pine \prə-'pēn, -'pīn\ *transitive verb* **pro·pined; pro·pin·ing** [Middle English, from Middle French *propiner,* from Latin *propinare* to present, drink to someone's health, from Greek *propinein* literally, to drink first, from *pro-* + *pinein* to drink — more at POTABLE] (15th century)
chiefly Scottish : to present or give especially as a token of friendship

²propine *noun* (15th century)
Scottish : a gift in return for a favor

pro·pin·qui·ty \prə-'piŋ-kwə-tē\ *noun* [Middle English *propinquite,* from Latin *propinquitat-, propinquitas* kinship, proximity, from *propinquus* near, akin, from *prope* near — more at APPROACH] (14th century)
1 : nearness of blood : KINSHIP
2 : nearness in place or time : PROXIMITY

pro·pi·o·nate \'prō-pē-ə-ˌnāt\ *noun* [International Scientific Vocabulary] (1862)
: a salt or ester of propionic acid

pro·pi·on·ic acid \ˌprō-pē-'ä-nik-\ *noun* [International Scientific Vocabulary ¹*pro-* + Greek *piōn* fat; akin to Sanskrit *pīvan* swelling, fat] (1850)
: a liquid sharp-odored fatty acid $C_3H_6O_2$ found in milk and distillates of wood, coal, and petroleum

pro·pi·ti·ate \prō-'pi-shē-ˌāt\ *transitive verb* **-at·ed; -at·ing** [Latin *propitiatus,* past participle of *propitiare,* from *propitius* propitious] (1583)
: to gain or regain the favor or goodwill of : APPEASE, CONCILIATE
synonym see PACIFY
— **pro·pi·ti·a·tor** \-ˌā-tər\ *noun*

pro·pi·ti·a·tion \prō-ˌpi-shē-'ā-shən\ *noun* (14th century)
1 : the act of propitiating
2 : something that propitiates; *specifically* : an atoning sacrifice

pro·pi·ti·a·to·ry \prō-'pi-sh(ē-)ə-ˌtōr-ē, -ˌtȯr-\ *adjective* (1551)
1 : intended to propitiate : EXPIATORY
2 : of or relating to propitiation

pro·pi·tious \prə-'pi-shəs\ *adjective* [Middle English *propicious,* from Latin *propitius,* probably from *pro-* for + *petere* to seek — more at PRO-, FEATHER] (15th century)
1 : favorably disposed : BENEVOLENT
2 : being of good omen : AUSPICIOUS ⟨*propitious* sign⟩
3 : tending to favor : ADVANTAGEOUS
synonym see FAVORABLE
— **pro·pi·tious·ly** *adverb*
— **pro·pi·tious·ness** *noun*

pro·plas·tid \(ˌ)prō-'plas-təd\ *noun* [International Scientific Vocabulary] (1922)
: a minute cytoplasmic body from which a plastid is formed

prop·man \'präp-ˌman\ *noun* (circa 1937)
: a man in charge of stage properties

prop·o·lis \'prä-pə-ləs\ *noun* [Middle English *propoleos,* from Medieval Latin, alteration of Latin *propolis,* from Greek, from *pro-* for + *polis* city — more at PRO-, POLICE] (15th century)
: a brownish resinous material of waxy consistency collected by bees from the buds of trees and used as a cement

pro·pone \prə-'pōn\ *transitive verb* **pro·poned; pro·pon·ing** [Middle English (Scots), from Latin *proponere* — more at PROPOUND] (14th century)
1 *Scottish* : PROPOSE, PROPOUND
2 *Scottish* : to put forward (a defense)

pro·po·nent \prə-'pō-nənt, 'prō-ˌ\ *noun* [Latin *proponent-, proponens,* present participle of *proponere*] (1588)
: one who argues in favor of something : ADVOCATE

¹pro·por·tion \prə-'pōr-shən, -'pȯr-\ *noun* [Middle English *proporcion,* from Middle French *proportion,* from Latin *proportion-, proportio,* from *pro* for + *portion-, portio* portion — more at FOR] (14th century)
1 : harmonious relation of parts to each other or to the whole : BALANCE, SYMMETRY
2 a : proper or equal share ⟨each did her *proportion* of the work⟩ **b :** QUOTA, PERCENTAGE
3 : the relation of one part to another or to the whole with respect to magnitude, quantity, or degree : RATIO
4 : SIZE, DIMENSION
5 : a statement of equality between two ratios in which the first of the four terms divided by the second equals the third divided by the fourth (as in 4/2=10/5) — compare EXTREME 1b, MEAN 1c
— **in proportion :** PROPORTIONAL 1

²proportion *transitive verb* **pro·por·tioned; pro·por·tion·ing** \-sh(ə-)niŋ\ (14th century)
1 : to adjust (a part or thing) in size relative to other parts or things
2 : to make the parts of harmonious or symmetrical
3 : APPORTION, ALLOT

pro·por·tion·able \-sh(ə-)nə-bəl\ *adjective* (14th century)
: PROPORTIONAL, PROPORTIONATE
— **pro·por·tion·ably** \-blē\ *adverb, archaic*

¹pro·por·tion·al \prə-'pōr-shnəl, -'pȯr-, -shə-nᵊl\ *noun* (14th century)
: a number or quantity in a proportion

²proportional *adjective* (15th century)
1 a : corresponding in size, degree, or intensity **b :** having the same or a constant ratio ⟨corresponding sides of similar triangles are *proportional*⟩
2 : regulated or determined in size or degree with reference to proportions ⟨a *proportional* system of immigration quotas⟩
— **pro·por·tion·al·i·ty** \-ˌpōr-shə-'na-lə-tē, -ˌpȯr-\ *noun*
— **pro·por·tion·al·ly** \-'pōr-shnə-lē, -'pȯr-, -shə-nᵊl-ē\ *adverb*

proportional parts *noun plural* (circa 1890)
: fractional parts of the difference between successive entries in a table for use in linear interpolation

proportional representation *noun* (1870)
: an electoral system designed to represent in a legislative body each political group or party in proportion to its actual voting strength in the electorate

proportional tax *noun* (circa 1943)
: a tax in which the tax rate remains constant regardless of the amount of the tax base

¹pro·por·tion·ate \prə-'pōr-sh(ə-)nət, -'pȯr-\ *adjective* (14th century)
: PROPORTIONAL 1
— **pro·por·tion·ate·ly** *adverb*

²pro·por·tion·ate \-shə-ˌnāt\ *transitive verb* **-at·ed; -at·ing** (1570)
: to make proportionate : PROPORTION

pro·pos·al \prə-'pō-zəl\ *noun* (1653)
1 : an act of putting forward or stating something for consideration
2 a : something proposed : SUGGESTION **b :** OFFER; *specifically* : an offer of marriage

pro·pose \prə-'pōz\ *verb* **pro·posed; pro·pos·ing** [Middle English, from Middle French *proposer,* from Latin *proponere* (perfect indicative *proposui*) — more at PROPOUND] (14th century)
intransitive verb
1 : to form or put forward a plan or intention ⟨man *proposes,* but God disposes⟩
2 *obsolete* : to engage in talk or discussion
3 : to make an offer of marriage
transitive verb
1 a : to set before the mind (as for discussion, imitation, or action) ⟨*propose* a plan for settling the dispute⟩ **b :** to set before someone and especially oneself as an aim or intent ⟨*proposed* to spend the summer in Italy⟩
2 a : to set forth for acceptance or rejection ⟨*propose* terms for peace⟩ ⟨*propose* a topic for debate⟩ **b :** to recommend to fill a place or vacancy : NOMINATE ⟨*propose* them for membership⟩ **c :** to offer as a toast ⟨*propose* the happiness of the couple⟩
— **pro·pos·er** *noun*

¹prop·o·si·tion \ˌprä-pə-'zi-shən\ *noun* (14th century)
1 a (1) **:** something offered for consideration or acceptance : PROPOSAL (2) **:** a request for sexual intercourse **b :** the point to be discussed or maintained in argument usually stated in sentence form near the outset **c :** a theorem or problem to be demonstrated or performed
2 a : an expression in language or signs of something that can be believed, doubted, or denied or is either true or false **b :** the objective meaning of a proposition
3 : something of an indicated kind ⟨getting there is a tough *proposition*⟩ ⟨the farm was never a paying *proposition*⟩
— **prop·o·si·tion·al** \-'zish-nəl, -'zi-shə-nᵊl\ *adjective*

²proposition *transitive verb* **-si·tioned; -si·tion·ing** \-'zi-sh(ə-)niŋ\ (1924)
: to make a proposal to; *especially* : to suggest sexual intercourse to

propositional calculus *noun* (1903)
: the branch of symbolic logic that uses symbols for unanalyzed propositions and logical connectives only — called also *sentential calculus*; compare PREDICATE CALCULUS

propositional function *noun* (1903)
1 : SENTENTIAL FUNCTION
2 : something that is designated or expressed by a sentential function

pro·pos·i·tus \prō-'pä-zə-təs\ *noun, plural* **-i·ti** \-ˌtī\ [New Latin, from Latin, past participle of *proponere*] (1899)
: the person immediately concerned : SUBJECT

pro·pound \prə-'pau̇nd\ *transitive verb* [alteration of earlier *propone,* from Middle English (Scots) *proponen,* from Latin *proponere* to display, propound, from *pro-* before + *ponere* to put, place — more at PRO-, POSITION] (1537)
: to offer for discussion or consideration
— **pro·pound·er** *noun*

pro·poxy·phene \prō-'päk-sə-ˌfēn\ *noun* [*prop-* + *oxy* + *-phene* (alteration of *phenyl*)] (1955)
: an analgesic $C_{22}H_{29}NO_2$ structurally related to methadone but less addicting that is administered in the form of its hydrochloride

pro·prae·tor *or* **pro·pre·tor** \(ˌ)prō-'prē-tər\ *noun* [Latin *propraetor,* from *pro-* (as in *proconsul*) + *praetor*] (circa 1580)
: a praetor of ancient Rome sent out to govern a province

pro·pran·o·lol \prō-'pra-nə-ˌlȯl, -ˌlōl\ *noun* [*propyl* + *propanol* + ¹*-ol*] (1964)
: a beta-adrenergic blocking agent $C_{16}H_{21}NO_2$ used in the form of its hydrochloride in the treatment of abnormal heart rhythms and angina pectoris

¹pro·pri·e·tary \prə-'prī-ə-ˌter-ē\ *noun, plural* **-tar·ies** (15th century)
1 : one that possesses, owns, or holds exclusive right to something; *specifically* : PROPRIETOR 1
2 : something that is used, produced, or marketed under exclusive legal right of the inventor or maker; *specifically* : a drug (as a patent medicine) that is protected by secrecy, patent, or copyright against free competition as to name, product, composition, or process of manufacture
3 : a business secretly owned by and run as a cover for an intelligence organization

²proprietary *adjective* [Late Latin *proprietarius,* from Latin *proprietas* property — more at PROPERTY] (1589)
1 : of, relating to, or characteristic of a proprietor ⟨*proprietary* rights⟩
2 : used, made, or marketed by one having the exclusive legal right ⟨a *proprietary* process⟩

3 : privately owned and managed and run as a profit-making organization ⟨a *proprietary* clinic⟩

pro•pri•e•tor \prə-ˈprī-ə-tər\ *noun* [alteration of [1]*proprietary*] (1637)
1 : one granted ownership of a colony (as one of the original American colonies) and full prerogatives of establishing a government and distributing land
2 a : one who has the legal right or exclusive title to something **:** OWNER **b :** one having an interest (as control or present use) less than absolute and exclusive right
— **pro•pri•e•tor•ship** \-ˌship\ *noun*

pro•pri•e•to•ri•al \prə-ˌprī-ə-ˈtōr-ē-əl, prō-, -ˈtȯr-\ *adjective* (1851)
: PROPRIETARY 1

pro•pri•e•tress \-ˈprī-ə-trəs\ *noun* (1692)
: a woman who is a proprietor

pro•pri•e•ty \prə-ˈprī-ə-tē\ *noun, plural* **-ties** [Middle English *propriete,* from Middle French *proprieté, proprete* property, quality of a person or thing — more at PROPERTY] (14th century)
1 *obsolete* **:** true nature
2 *obsolete* **:** a special characteristic **:** PECULIARITY
3 : the quality or state of being proper **:** APPROPRIATENESS
4 a : conformity to what is socially acceptable in conduct or speech **b :** fear of offending against conventional rules of behavior especially between the sexes **c** *plural* **:** the customs and manners of polite society

pro•pri•o•cep•tion \ˌprō-prē-ō-ˈsep-shən\ *noun* [*proprioceptive* + *-ion*] (1906)
: the reception of stimuli produced within the organism

pro•pri•o•cep•tive \-ˈsep-tiv\ *adjective* [Latin *proprius* own + English *-ceptive* (as in *receptive*)] (1906)
: of, relating to, or being stimuli arising within the organism

pro•pri•o•cep•tor \-tər\ *noun* [New Latin, from Latin *proprius* + New Latin *-ceptor* (as in *receptor*)] (1906)
: a sensory receptor (as a muscle spindle) excited by proprioceptive stimuli

prop root *noun* (1905)
: a root that serves as a prop or support to the plant

prop•to•sis \präp-ˈtō-səs, prō-ˈtō-\ *noun* [New Latin, from Late Latin, falling forward, from Greek *proptōsis,* from *propiptein* to fall forward, from *pro-* forward + *piptein* to fall — more at PRO-, FEATHER] (1676)
: forward projection or displacement especially of the eyeball

pro•pul•sion \prə-ˈpəl-shən\ *noun* [Latin *propellere* to propel] (1626)
1 : the action or process of propelling
2 : something that propels

pro•pul•sive \-ˈpəl-siv\ *adjective* [Latin *propulsus*] (1758)
: tending or having power to propel

pro•pyl \ˈprō-pəl\ *noun, often attributive* [International Scientific Vocabulary *prop-* + *-yl*] (1850)
: either of two isomeric alkyl groups C_3H_7 derived from propane — often used in combination

pro•py•lae•um \ˌprä-pə-ˈlē-əm, ˌprō-\ *noun, plural* **-laea** \-ˈlē-ə\ [Latin, from Greek *propylaion,* from *pro-* before + *pylē* gate — more at PRO-] (circa 1706)
: a vestibule or entrance of architectural importance before a building or enclosure — often used in plural

pro•pyl•ene \ˈprō-pə-ˌlēn\ *noun* (1850)
: a flammable gaseous hydrocarbon C_3H_6 obtained by cracking petroleum hydrocarbons and used chiefly in organic synthesis

propylene glycol *noun* (1885)
: a sweet hygroscopic viscous liquid $C_3H_8O_2$ made especially from propylene and used especially as an antifreeze and solvent and in brake fluids

pro ra•ta \(ˌ)prō-ˈrā-tə, -ˈrä-, -ˈra-\ *adverb* [Latin] (1575)
: proportionately according to an exactly calculable factor (as share or liability)
— **pro rata** *adjective*

pro•rate \(ˌ)prō-ˈrāt, ˈprō-ˌ\ *verb* **pro•rat•ed; pro•rat•ing** [*pro rata*] (1860)
transitive verb
: to divide, distribute, or assess proportionately
intransitive verb
: to make a pro rata distribution
— **pro•ra•tion** \prō-ˈrā-shən\ *noun*

pro•ro•gate \ˈprō-rə-ˌgāt\ *transitive verb* **-gat•ed; -gat•ing** (1534)
: PROROGUE
— **pro•ro•ga•tion** \ˌprō-rə-ˈgā-shən\ *noun*

pro•rogue \prə-ˈrōg\ *verb* **pro•rogued; pro•rogu•ing** [Middle English *prorogen,* from Middle French *proroguer,* from Latin *prorogare,* from *pro-* before + *rogare* to ask — more at PRO-, RIGHT] (15th century)
transitive verb
1 : DEFER, POSTPONE
2 : to terminate a session of (as a British parliament) by royal prerogative
intransitive verb
: to suspend or end a legislative session

pros *plural of* PRO

pros- *prefix* [Late Latin, from Greek, from *proti, pros* face to face with, toward, in addition to, near; akin to Sanskrit *prati-* near, toward, against, in return, Greek *pro* before — more at FOR]
: in front ⟨*pros*encephalon⟩

pro•sa•ic \prō-ˈzā-ik\ *adjective* [Late Latin *prosaicus,* from Latin *prosa* prose] (circa 1656)
1 a : characteristic of prose as distinguished from poetry **:** FACTUAL **b :** DULL, UNIMAGINATIVE
2 : EVERYDAY, ORDINARY
— **pro•sa•i•cal•ly** \-ˈzā-ə-k(ə-)lē\ *adverb*

pro•sa•ism \ˈprō-(ˌ)zā-ˌi-zəm\ *noun* (1787)
1 : a prosaic manner, style, or quality
2 : a prosaic expression

pro•sa•ist *noun* [Latin *prosa* prose] (1803)
1 \ˈprō-(ˌ)zā-ist, -zā-ˌist\ **:** a prose writer
2 \prō-ˈzā-ist\ **:** a prosaic person

pro•sa•teur \ˌprō-zə-ˈtər\ *noun* [French, from Italian *prosatore,* from Medieval Latin *prosator,* from Latin *prosa*] (1880)
: a writer of prose

pro•sau•ro•pod \prō-ˈsȯr-ə-ˌpäd\ *noun* [New Latin *Prosauropoda,* from [1]*pro-* + *Sauropoda* — more at SAUROPOD] (1941)
: any of a group (Prosauropoda) of chiefly herbivorous Triassic dinosaurs that are probably ancestral to sauropods

pro•sce•ni•um \prō-ˈsē-nē-əm\ *noun* [Latin, from Greek *proskēnion* front of the building forming the background for a dramatic performance, stage, from *pro-* + *skēnē* building forming the background for a dramatic performance — more at SCENE] (1606)
1 a : the stage of an ancient Greek or Roman theater **b :** the part of a modern stage in front of the curtain **c :** the wall that separates the stage from the auditorium and provides the arch that frames it
2 : FOREGROUND

pro•sciut•to \prō-ˈshü-(ˌ)tō\ *noun, plural* **-ti** \-(ˌ)tē\ *or* **-tos** [Italian, alteration of *presciutto,* from *pre-* (from Latin *prae-* pre-) + *asciutto* dried out, from Latin *exsuctus,* from past participle of *exsugere* to suck out, from *ex-* + *sugere* to suck — more at SUCK] (circa 1929)
: dry-cured spiced Italian ham usually sliced thin

pro•scribe \prō-ˈskrīb\ *transitive verb* **pro•scribed; pro•scrib•ing** [Latin *proscribere* to publish, proscribe, from *pro-* before + *scribere* to write — more at SCRIBE] (1560)
1 : to condemn or forbid as harmful or unlawful **:** PROHIBIT
2 : to publish the name of as condemned to death with the property of the condemned forfeited to the state
— **pro•scrib•er** *noun*

pro•scrip•tion \prō-ˈskrip-shən\ *noun* [Middle English *proscripcion,* from Latin *proscription-, proscriptio,* from *proscribere*] (14th century)
1 : the act of proscribing **:** the state of being proscribed
2 : an imposed restraint or restriction **:** PROHIBITION
— **pro•scrip•tive** \-ˈskrip-tiv\ *adjective*
— **pro•scrip•tive•ly** *adverb*

[1]**prose** \ˈprōz\ *noun* [Middle English, from Middle French, from Latin *prosa,* from feminine of *prorsus, prosus,* straightforward, being in prose, contraction of *proversus,* past participle of *provertere* to turn forward, from *pro-* forward + *vertere* to turn — more at PRO-, WORTH] (14th century)
1 a : the ordinary language people use in speaking or writing **b :** a literary medium distinguished from poetry especially by its greater irregularity and variety of rhythm and its closer correspondence to the patterns of everyday speech
2 : a prosaic style, quality, or condition

[2]**prose** *adjective* (14th century)
1 : of, relating to, or written in prose
2 : PROSAIC

[3]**prose** *intransitive verb* **prosed; pros•ing** (1642)
1 : to write prose
2 : to write or speak in a prosaic manner

pro•sec•tor \prō-ˈsek-tər\ *noun* [probably from French *prosecteur,* from Late Latin *prosector* anatomist, from Latin *prosecare* to cut away, from *pro-* forth + *secare* to cut — more at PRO-, SAW] (circa 1857)
: a person who makes dissections for anatomic demonstrations

pros•e•cute \ˈprä-si-ˌkyüt\ *verb* **-cut•ed; -cut•ing** [Middle English, from Latin *prosecutus,* past participle of *prosequi* to pursue — more at PURSUE] (15th century)
transitive verb
1 : to follow to the end **:** pursue until finished ⟨was . . . ordered to *prosecute* the war with . . . vigor —Marjory S. Douglas⟩
2 : to engage in **:** PERFORM
3 a : to bring legal action against for redress or punishment of a crime or violation of law ⟨*prosecuted* them for fraud⟩ **b :** to institute legal proceedings with reference to ⟨*prosecute* a claim⟩
intransitive verb
: to institute and carry on a legal suit or prosecution
— **pros•e•cut•able** \ˌprä-sə-ˈkyü-tə-bəl\ *adjective*

prosecuting attorney *noun* (1832)
: an attorney who conducts proceedings in a court on behalf of the government **:** DISTRICT ATTORNEY

pros•e•cu•tion \ˌprä-si-ˈkyü-shən\ *noun* (1567)
1 : the act or process of prosecuting; *specifically* **:** the institution and continuance of a criminal suit involving the process of pursuing formal charges against an offender to final judgment
2 : the party by whom criminal proceedings are instituted or conducted
3 *obsolete* **:** PURSUIT

pros•e•cu•tor \'prä-si-ˌkyü-tər\ *noun* (circa 1670)
1 : a person who institutes an official prosecution before a court
2 : PROSECUTING ATTORNEY

pros•e•cu•to•ri•al \ˌprä-si-kyü-'tōr-ē-əl, -'tȯr-\ *adjective* (1968)
: of, relating to, or being a prosecutor or prosecution

[1]**pros•e•lyte** \'prä-sə-ˌlīt\ *noun* [Middle English *proselite,* from Late Latin *proselytus* proselyte, alien resident, from Greek *prosēlytos,* from *pros* near + *-ēlytos* (akin to *ēlythe* he went) — more at PROS-, ELASTIC] (14th century)
: a new convert; *specifically* **:** a convert to Judaism

[2]**proselyte** *verb* **-lyt•ed; -lyt•ing** (1624)
: PROSELYTIZE

pros•e•ly•tise *British variant of* PROSELYTIZE

pros•e•ly•tism \'prä-sə-ˌlī-ˌti-zəm, 'prä-s(ə-)lə-\ *noun* (circa 1660)
1 : the act of becoming or condition of being a proselyte **:** CONVERSION
2 : the act or process of proselytizing

pros•e•ly•tize \'prä-s(ə-)lə-ˌtīz\ *verb* **-tized; -tiz•ing** (1679)
intransitive verb
1 : to induce someone to convert to one's faith
2 : to recruit someone to join one's party, institution, or cause
transitive verb
: to recruit or convert especially to a new faith, institution, or cause
— pros•e•ly•ti•za•tion \ˌprä-s(ə-)lə-tə-'zā-shən, ˌprä-sə-ˌlī-tə-\ *noun*
— pros•e•ly•tiz•er \'prä-s(ə-)lə-ˌtī-zər\ *noun*

pro•sem•i•nar \(ˌ)prō-'se-mə-ˌnär\ *noun* (circa 1922)
: a directed course of study like a graduate seminar but often open to advanced undergraduates

pros•en•ceph•a•lon \ˌpräs-ˌen-'se-fə-ˌlän, -lən\ *noun* [New Latin] (1846)
: FOREBRAIN
— pros•en•ce•phal•ic \-sə-'fa-lik\ *adjective*

prose poem *noun* (1842)
: a composition in prose that has some of the qualities of a poem
— prose poet *noun*

pros•er \'prō-zər\ *noun* (1627)
1 : a writer of prose
2 : one who talks or writes tediously

Pro•ser•pi•na \prə-'sər-pə-nə\ *or* **Pros•er•pine** \'prä-sər-ˌpīn\ *noun* [Latin]
: PERSEPHONE

pro•sim•i•an \prō-'si-mē-ən\ *noun* [New Latin *Prosimii,* from [1]*pro-* + Latin *simia* ape — more at SIMIAN] (circa 1890)
: any of a suborder (Prosimii) of lower primates (as lemurs)
— prosimian *adjective*

pro•sit \'prō-zət, -sət\ *or* **prost** \'prōst\ *interjection* [German, from Latin *prosit* may it be beneficial, from *prodesse* to be useful — more at PROUD] (1846)
— used to wish good health especially before drinking

pro•so \'prō-(ˌ)sō\ *noun* [Russian] (1917)
: MILLET 1a

pros•o•branch \'prä-sə-ˌbraŋk\ *noun, plural* **-branchs** [New Latin *Prosobranchia,* from *proso-* in front (from Greek *prosō* forward) + *branchia* gills (from Greek)] (1851)
: any of a subclass (Prosobranchia) of gastropod mollusks that have the loop of visceral nerves twisted into a figure eight, the sexes usually separate, and usually an operculum
— prosobranch *adjective*

pro•sod•ic \prə-'sä-dik *also* -'zä-\ *or* **pro•sod•i•cal** \-di-kəl\ *adjective* (1774)
: of or relating to prosody
— pro•sod•i•cal•ly \-di-k(ə-)lē\ *adverb*

pros•o•dy \'prä-sə-dē, -zə-\ *noun, plural* **-dies** [Middle English, from Latin *prosodia* accent of a syllable, from Greek *prosōidia* song sung to instrumental music, accent, from *pros* in addition to + *ōidē* song — more at PROS-, ODE] (15th century)
1 : the study of versification; *especially* **:** the systematic study of metrical structure
2 : a particular system, theory, or style of versification
3 : the rhythmic and intonational aspect of language
— pros•o•dist \-dist\ *noun*

pro•so•ma \(ˌ)prō-'sō-mə\ *noun* [New Latin, from Greek *pro-* + *sōma* body] (1872)
: the anterior region of the body of an invertebrate when not readily analyzable into its primitive segmentation; *especially* **:** CEPHALOTHORAX

pros•o•pog•ra•phy \ˌprä-sə-'pä-grə-fē\ *noun* [New Latin *prosopographia,* from Greek *prosōpon* person + *-graphia* -graphy] (1929)
: a study that identifies and relates a group of persons or characters within a particular historical or literary context
— pros•o•po•graph•i•cal \-pə-'gra-fi-kəl\ *adjective*

pro•so•po•poe•ia \prə-ˌsō-pə-'pē-ə, ˌprä-sə-pə-\ *noun* [Latin, from Greek *prosōpopoiia,* from *prosōpon* mask, person (from *pros-* + *ōps* face) + *poiein* to make — more at EYE, POET] (circa 1555)
1 : a figure of speech in which an imaginary or absent person is represented as speaking or acting
2 : PERSONIFICATION

[1]**pros•pect** \'prä-ˌspekt\ *noun* [Middle English, from Latin *prospectus* view, prospect, from *prospicere* to look forward, exercise foresight, from *pro-* forward + *specere* to look — more at PRO-, SPY] (15th century)
1 : EXPOSURE 3b
2 a (1) **:** an extensive view (2) **:** a mental consideration **:** SURVEY **b :** a place that commands an extensive view **:** LOOKOUT **c :** something extended to the view **:** SCENE **d** *archaic* **:** a sketch or picture of a scene
3 *obsolete* **:** ASPECT
4 a : the act of looking forward **:** ANTICIPATION **b :** a mental picture of something to come **:** VISION **c :** something that is awaited or expected **:** POSSIBILITY **d** *plural* (1) **:** financial expectations (2) **:** CHANCES
5 : a place showing signs of containing a mineral deposit
6 a : a potential buyer or customer **b :** a likely candidate ☆
— in prospect : possible or likely for the future

[2]**pros•pect** \'prä-ˌspekt, *chiefly British* prə-'\ (1841)
intransitive verb
: to explore an area especially for mineral deposits
transitive verb
: to inspect (a region) for mineral deposits; *broadly* **:** EXPLORE
— pros•pec•tor \-ˌspek-tər, -'spek-\ *noun*

pro•spec•tive \prə-'spek-tiv *also* 'prä-ˌ, prō-', prä-'\ *adjective* (circa 1699)
1 : relating to or effective in the future
2 a : likely to come about **:** EXPECTED ⟨the *prospective* benefits of this law⟩ **b :** likely to be or become ⟨a *prospective* mother⟩
— pro•spec•tive•ly *adverb*

pro•spec•tus \prə-'spek-təs, prä-\ *noun, plural* **-tus•es** [Latin, prospect] (1765)
1 : a preliminary printed statement that describes an enterprise (as a business or publication) and that is distributed to prospective buyers, investors, or participants
2 : something (as a statement or situation) that forecasts the course or nature of something

pros•per \'präs-pər\ *verb* **pros•pered; pros•per•ing** \-p(ə-)riŋ\ [Middle English, from Middle French *prosperer,* from Latin *prosperare* to cause to succeed, from *prosperus* favorable] (14th century)
intransitive verb
1 : to succeed in an enterprise or activity; *especially* **:** to achieve economic success
2 : to become strong and flourishing
transitive verb
: to cause to succeed or thrive

pros•per•i•ty \prä-'sper-ə-tē\ *noun* (13th century)
: the condition of being successful or thriving; *especially* **:** economic well-being

Pros•pero \'präs-pə-ˌrō\ *noun*
: the rightful duke of Milan in Shakespeare's *The Tempest*

pros•per•ous \'präs-p(ə-)rəs\ *adjective* [Middle English, from Middle French *prospereus,* from Latin *prosperus*] (15th century)
1 : AUSPICIOUS, FAVORABLE
2 a : marked by success or economic well-being **b :** enjoying vigorous and healthy growth **:** FLOURISHING
— pros•per•ous•ly *adverb*
— pros•per•ous•ness *noun*

pross \'präs\ *or* **pros•sie** \'prä-sē\ *or* **pros•tie** \'präs-tē\ *noun* (circa 1902)
slang **:** PROSTITUTE 1a

pros•ta•cy•clin \ˌpräs-tə-'sī-klən\ *noun* [*prosta-* (as in *prostaglandin*) + *cycl-* + [1]*-in*] (1976)
: a prostaglandin that is a metabolite of arachidonic acid, inhibits platelet aggregation, and dilates blood vessels

pros•ta•glan•din \ˌpräs-tə-'glan-dən\ *noun* [*prostate gland* + [1]*-in;* from its occurrence in the seminal fluid of animals] (1936)
: any of various oxygenated unsaturated cyclic fatty acids of animals that perform a variety of hormonelike actions (as in controlling blood pressure or smooth muscle contraction)

pros•tate \'präs-ˌtāt\ *noun* [New Latin *prostata* prostate gland, from Greek *prostatēs,* from *proïstanai* to put in front, from *pro-* before + *histanai* to cause to stand — more at PRO-, STAND] (1646)
: PROSTATE GLAND
— pros•tat•ic \prä-'sta-tik\ *adjective*

pros•ta•tec•to•my \ˌpräs-tə-'tek-tə-mē\ *noun, plural* **-mies** (circa 1890)
: surgical removal or resection of the prostate gland

prostate gland *noun* (1828)
: a firm partly muscular partly glandular body that is situated about the base of the mammalian male urethra and secretes an alkaline viscid fluid which is a major constituent of the ejaculatory fluid

pros•ta•tism \'präs-tə-ˌti-zəm\ *noun* (circa 1900)
: disease of the prostate; *especially* **:** a disorder resulting from obstruction of the bladder neck by an enlarged prostate

pros•ta•ti•tis \ˌpräs-tə-'tī-təs\ *noun* [New Latin] (circa 1844)
: inflammation of the prostate gland

pros•the•sis \präs-'thē-səs, 'präs-thə-\ *noun, plural* **-the•ses** \-ˌsēz\ [New Latin, from Greek, addition, from *prostithenai* to add to,

☆ SYNONYMS
Prospect, outlook, anticipation, foretaste mean an advance realization of something to come. PROSPECT implies expectation of a particular event, condition, or development of definite interest or concern ⟨the *prospect* of a quiet weekend⟩. OUTLOOK suggests a forecasting of the future ⟨a favorable *outlook* for the economy⟩. ANTICIPATION implies a prospect or outlook that involves advance suffering or enjoyment of what is foreseen ⟨the *anticipation* of her arrival⟩. FORETASTE implies an actual though brief or partial experience of something forthcoming ⟨the frost was a *foretaste* of winter⟩.

from *pros-* in addition to + *tithenai* to put — more at PROS-, DO] (circa 1900)
: an artificial device to replace a missing part of the body

pros·thet·ic \präs-ˈthe-tik\ *adjective* (circa 1890)
1 : of or relating to a prosthesis or prosthetics
2 : of, relating to, or constituting a nonprotein group of a conjugated protein
— pros·thet·i·cal·ly \-ti-k(ə-)lē\ *adverb*

pros·thet·ics \-tiks\ *noun plural but singular or plural in construction* (circa 1894)
: the surgical or dental specialty concerned with the design, construction, and fitting of prostheses

pros·thet·ist \ˈpräs-thə-tist\ *noun* (1902)
: a specialist in prosthetics

pros·tho·don·tics \ˌpräs-thə-ˈdän-tiks\ *noun plural but singular or plural in construction* [New Latin *prosthodontia,* from *prosthesis* + *-odontia*] (1947)
: prosthetic dentistry

pros·tho·don·tist \-ˈdän-tist\ *noun* (1917)
: a specialist in prosthodontics

[1]**pros·ti·tute** \ˈpräs-tə-ˌtüt, -ˌtyüt\ *transitive verb* **-tut·ed; -tut·ing** [Latin *prostitutus,* past participle of *prostituere,* from *pro-* before + *statuere* to station — more at PRO-, STATUTE] (1530)
1 : to offer indiscriminately for sexual intercourse especially for money
2 : to devote to corrupt or unworthy purposes **:** DEBASE ⟨*prostitute* one's talents⟩
— pros·ti·tu·tor \-ˌtü-tər, -ˌtyü-\ *noun*

[2]**prostitute** *adjective* (1563)
: devoted to corrupt purposes **:** PROSTITUTED

[3]**prostitute** *noun* (1613)
1 a : a woman who engages in promiscuous sexual intercourse especially for money **:** WHORE **b :** a male who engages in sexual and especially homosexual practices for money
2 : a person (as a writer or painter) who deliberately debases his or her talents (as for money)

pros·ti·tu·tion \ˌpräs-tə-ˈtü-shən, -ˈtyü-\ *noun* (1553)
1 : the act or practice of indulging in promiscuous sexual relations especially for money
2 : the state of being prostituted **:** DEBASEMENT

pro·sto·mi·um \prō-ˈstō-mē-əm\ *noun, plural* **-mia** \-mē-ə\ [New Latin, from Greek *pro-* + *stoma* mouth — more at STOMACH] (1870)
: the portion of the head of various worms that is situated in front of the mouth
— pro·sto·mi·al \-mē-əl\ *adjective*

[1]**pros·trate** \ˈprä-ˌstrāt\ *adjective* [Middle English *prostrat,* from Latin *prostratus,* past participle of *prosternere,* from *pro-* before + *sternere* to spread out, throw down — more at STREW] (14th century)
1 : stretched out with face on the ground in adoration or submission; *also* **:** lying flat
2 : completely overcome and lacking vitality, will, or power to rise ⟨was *prostrate* from the heat⟩
3 : trailing on the ground **:** PROCUMBENT ⟨*prostrate* shrub⟩
synonym see PRONE

[2]**pros·trate** \ˈprä-ˌstrāt, *especially British* prä-ˈ\ *transitive verb* **pros·trat·ed; pros·trat·ing** (15th century)
1 : to throw or put into a prostrate position
2 : to put (oneself) in a humble and submissive posture or state ⟨the whole town had to *prostrate* itself in official apology —Claudia Cassidy⟩
3 : to reduce to submission, helplessness, or exhaustion ⟨was *prostrated* with grief⟩

pros·tra·tion \prä-ˈstrā-shən\ *noun* (14th century)
1 a : the act of assuming a prostrate position **b :** the state of being in a prostrate position **:** ABASEMENT
2 a : complete physical or mental exhaustion **:** COLLAPSE **b :** the process of being made powerless or the condition of powerlessness ⟨the country suffered economic *prostration* after the war⟩

prosy \ˈprō-zē\ *adjective* **pros·i·er; -est** [[1]*prose*] (1814)
: lacking in qualities that seize the attention or strike the imagination **:** COMMONPLACE; *especially* **:** tediously dull in speech or manner
— pros·i·ly \-zə-lē\ *adverb*
— pros·i·ness \-zē-nəs\ *noun*

prot- *or* **proto-** *combining form* [Middle English *protho-,* from Middle French, from Late Latin *proto-,* from Greek *prōt-, prōto-,* from *prōtos;* akin to Greek *pro* before — more at FOR]
1 a : first in time ⟨*proto*history⟩ **b :** beginning **:** giving rise to ⟨*proto*planet⟩
2 : parent substance of a (specified) substance ⟨*prot*actinium⟩
3 : first formed **:** primary ⟨*proto*xylem⟩
4 *capitalized* **:** relating to or constituting the recorded or assumed language that is ancestral to a language or to a group of related languages or dialects ⟨*Proto*-Indo-European⟩

prot·ac·tin·i·um \ˌprō-ˌtak-ˈti-nē-əm\ *noun* [New Latin] (1918)
: a shiny radioactive metallic element of relatively short life — see ELEMENT table

pro·tag·o·nist \prō-ˈta-gə-nist\ *noun* [Greek *prōtagōnistēs,* from *prōt-* prot- + *agōnistēs* competitor at games, actor, from *agōnizesthai* to compete, from *agōn* contest, competition at games — more at AGONY] (1671)
1 a : the principal character in a literary work (as a drama or story) **b :** a leading actor, character, or participant in a literary work or real event
2 : a leader, proponent, or supporter of a cause **:** CHAMPION
3 : a muscle that by its contraction actually causes a particular movement

prot·amine \ˈprō-tə-ˌmēn\ *noun* [International Scientific Vocabulary *prot-* + *amine*] (1874)
: any of various strongly basic proteins of relatively low molecular weight that are rich in arginine and are found associated especially with DNA in place of histone in the sperm cells of various animals (as fish)

prot·a·sis \ˈprä-tə-səs\ *noun, plural* **-a·ses** \-ˌsēz\ [Late Latin, from Greek, premise of a syllogism, conditional clause, from *proteinein* to stretch out before, put forward, from *pro-* + *teinein* to stretch — more at THIN] (circa 1568)
1 : the introductory part of a play or narrative poem
2 : the subordinate clause of a conditional sentence — compare APODOSIS
— pro·tat·ic \prä-ˈta-tik, prō-\ *adjective*

prote- *or* **proteo-** *combining form* [International Scientific Vocabulary, from French *protéine*]
: protein ⟨*proteo*lysis⟩ ⟨*prote*ose⟩

pro·tea \ˈprō-tē-ə\ *noun* [New Latin, from Latin *Proteus* Proteus] (1825)
: any of a genus (*Protea* of the family Proteaceae, the protea family) of evergreen shrubs often grown for their showy bracts and dense flower heads

protea

pro·te·an \ˈprō-tē-ən, prō-ˈtē-\ *adjective* (1598)
1 : of or resembling Proteus in having a varied nature or ability to assume different forms
2 : displaying great diversity or variety **:** VERSATILE

pro·te·ase \ˈprō-tē-ˌās, -ˌāz\ *noun* [International Scientific Vocabulary] (1903)
: any of numerous enzymes that hydrolyze proteins and are classified according to the most prominent functional group (as serine or cysteine) at the active site — called also *proteinase*

pro·tect \prə-ˈtekt\ *transitive verb* [Middle English, from Latin *protectus,* past participle of *protegere,* from *pro-* in front + *tegere* to cover — more at PRO-, THATCH] (15th century)
1 : to cover or shield from exposure, injury, or destruction **:** GUARD
2 : to maintain the status or integrity of especially through financial or legal guarantees: as **a :** to save from contingent financial loss **b :** to foster or shield from infringement or restriction ⟨salesmen with *protected* territories⟩; *specifically* **:** to restrict competition for (as domestic industries) by means of tariffs or trade controls
synonym see DEFEND
— pro·tec·tive \-ˈtek-tiv\ *adjective*
— pro·tec·tive·ly *adverb*
— pro·tec·tive·ness *noun*

pro·tec·tant \prə-ˈtek-tənt\ *noun* (1935)
: a protecting agent

pro·tec·tion \prə-ˈtek-shən\ *noun* (14th century)
1 : the act of protecting **:** the state of being protected
2 a : one that protects **b :** supervision or support of one that is smaller and weaker
3 : the freeing of the producers of a country from foreign competition in their home market by restrictions (as high duties) on foreign competitive goods
4 a : immunity from prosecution purchased by criminals through bribery **b :** money extorted by racketeers posing as a protective association
5 : COVERAGE 1a

pro·tec·tion·ist \-sh(ə-)nist\ *noun* (1844)
: an advocate of government economic protection for domestic producers through restrictions on foreign competitors
— pro·tec·tion·ism \-shə-ˌni-zəm\ *noun*
— protectionist *adjective*

protective tariff *noun* (1838)
: a tariff intended primarily to protect domestic producers rather than to yield revenue

pro·tec·tor \prə-ˈtek-tər\ *noun* (14th century)
1 a : one that protects **:** GUARDIAN **b :** a device used to prevent injury **:** GUARD
2 a : one having the care of a kingdom during the king's minority **:** REGENT **b :** the executive head of the Commonwealth of England, Scotland, and Ireland from 1653 to 1659 — called also *Lord Protector of the Commonwealth*
— pro·tec·tor·ship \-ˌship\ *noun*

pro·tec·tor·al \-ˈtek-t(ə-)rəl\ *adjective* (1657)
: of or relating to a protector or protectorate

pro·tec·tor·ate \-ˈtek-t(ə-)rət\ *noun* (1692)
1 a : government by a protector **b** *capitalized* **:** the government of England (1653–59) under the Cromwells **c :** the rank, office, or period of rule of a protector
2 a : the relationship of superior authority assumed by one power or state over a dependent one **b :** the dependent political unit in such a relationship

pro·tec·to·ry \-ˈtek-t(ə-)rē\ *noun, plural* **-ries** (1885)
: an institution for the protection and care usually of homeless or delinquent children

pro·tec·tress \-ˈtek-trəs\ *noun* (1570)
: a woman who is a protector

pro·té·gé \ˈprō-tə-ˌzhā, ˌprō-tə-ˈ\ *noun* [French, from past participle of *protéger* to protect, from Latin *protegere*] (1787)

\ə\ **abut** \ᵊ\ **kitten** \ər\ **further** \a\ **ash** \ā\ **ace** \ä\ **mop, mar** \au̇\ **out** \ch\ **chin** \e\ **bet** \ē\ **easy** \g\ **go** \i\ **hit** \ī\ **ice** \j\ **job** \ŋ\ **sing** \ō\ **go** \ȯ\ **law** \ȯi\ **boy** \th\ **thin** \t͟h\ **the** \ü\ **loot** \u̇\ **foot** \y\ **yet** \zh\ **vision** *see also* Guide to Pronunciation

: one who is protected or trained or whose career is furthered by a person of experience, prominence, or influence
pro·té·gée \'prō-tə-,zhā, ,prō-tə-'\ *noun* [French, feminine of *protégé*] (1778)
: a female protégé
pro·tein \'prō-,tēn *also* 'prō-tē-ən\ *noun, often attributive* [French *protéine,* from Late Greek *prōteios* primary, from Greek *prōtos* first — more at PROT-] (circa 1844)
1 : any of numerous naturally occurring extremely complex substances that consist of amino-acid residues joined by peptide bonds, contain the elements carbon, hydrogen, nitrogen, oxygen, usually sulfur, and occasionally other elements (as phosphorus or iron), and include many essential biological compounds (as enzymes, hormones, or immunoglobulins)
2 : the total nitrogenous material in plant or animal substances
pro·tein·a·ceous \,prō-tᵊn-'ā-shəs; ,prō-,tē(-ə)-'nā-shəs\ *adjective* (circa 1844)
: of, relating to, resembling, or being protein
pro·tein·ase \'prō-tᵊn-,ās, -,āz; 'prō-,tē(-ə)-,nās, -,nāz\ *noun* [International Scientific Vocabulary] (circa 1929)
: PROTEASE
pro·tein·uria \,prō-tᵊn-'ůr-ē-ə, -'yůr-; ,prō-tē(-ə)-'nůr-ē-ə, -'nyůr-\ *noun* [New Latin, from International Scientific Vocabulary *protein* + New Latin *-uria*] (1911)
: the presence of excess protein in the urine
pro tem \(,)prō-'tem\ *adverb* (1828)
: PRO TEMPORE
pro tem·po·re \prō-'tem-pə-rē\ *adverb* [Middle English, from Latin] (15th century)
: for the time being
pro·tend \prō-'tend\ *verb* [Middle English, from Latin *protendere,* from *pro-* + *tendere* to stretch — more at THIN] (15th century)
transitive verb
1 *archaic* **:** to stretch forth
2 *archaic* **:** EXTEND
intransitive verb
archaic **:** STICK OUT, PROTRUDE
pro·ten·sive \-'ten(t)-siv\ *adjective* [Latin *protensus,* past participle of *protendere*] (1671)
1 *archaic* **:** having continuance in time
2 *archaic* **:** having lengthwise extent or extensiveness
— pro·ten·sive·ly *adverb*
pro·teo·gly·can \,prō-tē-ə-'glī-,kan\ [International Scientific Vocabulary] (1968)
: any of a class of glycoproteins of high molecular weight that are found especially in the extracellular matrix of connective tissue
pro·te·ol·y·sis \,prō-tē-'ä-lə-səs\ *noun* [New Latin] (1880)
: the hydrolysis of proteins or peptides with formation of simpler and soluble products
pro·teo·lyt·ic \,prō-tē-ə-'li-tik\ *adjective* (1877)
: of, relating to, or producing proteolysis
— pro·teo·lyt·i·cal·ly \-ti-k(ə-)lē\ *adverb*
pro·te·ose \'prō-tē-,ōs, -,ōz\ *noun* [International Scientific Vocabulary] (circa 1890)
: any of various water-soluble protein derivatives formed by partial hydrolysis of proteins
Pro·te·ro·zo·ic \,prä-tə-rə-'zō-ik, ,prō-\ *adjective* [Greek *proteros* former, earlier (from *pro* before) + International Scientific Vocabulary *-zoic* — more at FOR] (1899)
: of, relating to, or being the eon of geologic time or the corresponding system of rocks that includes the interval between the Archean and Phanerozoic eons, perhaps exceeds in length all of subsequent geological time, and is marked by rocks that contain fossils indicating the first appearance of eukaryotic organisms (as algae) — see GEOLOGIC TIME table
— Proterozoic *noun*

¹pro·test \'prō-,test\ *noun* [Middle English, from Middle French, from *protester*] (15th century)
1 : a solemn declaration of opinion and usually of dissent: as **a :** a sworn declaration that payment of a note or bill has been refused and that all responsible signers or debtors are liable for resulting loss or damage **b :** a declaration made especially before or while paying that a tax is illegal and that payment is not voluntary
2 : the act of objecting or a gesture of disapproval ⟨resigned in *protest*⟩; *especially* **:** a usually organized public demonstration of disapproval
3 : a complaint, objection, or display of unwillingness usually to an idea or a course of action ⟨went under *protest*⟩
4 : an objection made to an official or a governing body of a sport
²pro·test \prə-'test, 'prō-,, prō-'\ *verb* [Middle English, from Middle French *protester,* from Latin *protestari,* from *pro-* forth + *testari* to call to witness — more at PRO-, TESTAMENT] (15th century)
transitive verb
1 : to make solemn declaration or affirmation of ⟨*protest* my innocence⟩
2 : to execute or have executed a formal protest against (as a bill or note)
3 : to make a statement or gesture in objection to ⟨*protested* the abuses of human rights⟩
intransitive verb
1 : to make a protestation
2 : to make or enter a protest
synonym see ASSERT
— pro·test·er *or* **pro·tes·tor** \-'tes-tər, -,tes-\ *noun*
¹prot·es·tant \'prä-təs-tənt, *2 is also* prə-'tes-\ *noun* [Middle French, from Latin *protestant-, protestans,* present participle of *protestari*] (1539)
1 *capitalized* **a :** any of a group of German princes and cities presenting a defense of freedom of conscience against an edict of the Diet of Spires in 1529 intended to suppress the Lutheran movement **b :** a member of any of several church denominations denying the universal authority of the Pope and affirming the Reformation principles of justification by faith alone, the priesthood of all believers, and the primacy of the Bible as the only source of revealed truth; *broadly* **:** a Christian not of a Catholic or Eastern church
2 : one who makes or enters a protest
— Prot·es·tant·ism \'prä-təs-tən-,ti-zəm\ *noun*
²protestant *adjective* (1539)
1 *capitalized* **:** of or relating to Protestants, their churches, or their religion
2 : making or sounding a protest ⟨the two *protestant* ladies up and marched out —*Time*⟩
Protestant ethic *noun* (1926)
: an ethic that stresses the virtue of hard work, thrift, and self-discipline
pro·tes·ta·tion \,prä-təs-'tā-shən, ,prō-, -,tes-\ *noun* (14th century)
: the act of protesting **:** a solemn declaration or avowal
pro·te·us \'prō-tē-əs\ *noun, plural* **-tei** \-tē-,ī\ [New Latin, from Latin, Proteus] (1896)
: any of a genus (*Proteus*) of aerobic usually motile enterobacteria that include saprophytes in decaying organic matter and forms associated with urinary tract infections
Pro·teus \'prō-,tyüs, -tē-əs, -,tüs\ *noun* [Latin, from Greek *Prōteus*]
: a Greek sea god capable of assuming different forms
pro·tha·la·mi·on \,prō-thə-'lā-mē-ən, -,än\ *or* **pro·tha·la·mi·um** \-mē-əm\ *noun, plural* **-mia** \-mē-ə\ [New Latin, from Greek *pro-* + *-thalamion* (as in *epithalamion*)] (1597)
: a song in celebration of a marriage

pro·thal·li·um \prō-'tha-lē-əm\ *noun, plural* **-thal·lia** \-lē-ə\ [New Latin, from *pro-* + *thallus*] (1858)
1 : the gametophyte of a pteridophyte (as a fern) that is typically a small flat green thallus attached to the soil by rhizoids
2 : a greatly reduced structure of a seed plant corresponding to the pteridophyte prothallium
pro·thal·lus \(,)prō-'tha-ləs\ *noun* [New Latin] (1854)
: PROTHALLIUM
proth·e·sis \'prä-thə-səs\ *noun, plural* **-e·ses** \-,sēz\ [Late Latin, alteration of *prosthesis,* from Greek, literally, addition — more at PROSTHESIS] (circa 1550)
: the addition of a sound to the beginning of a word (as in Old French *estat*—whence English *estate*—from Latin *status*)
— pro·thet·ic \prä-'the-tik\ *adjective*
pro·tho·no·ta·ry \prō-'thä-nə-,ter-ē, ,prō-thə-'nō-tə-rē\ *or* **pro·to·no·ta·ry** \prō-'tä-nə-,ter-ē, ,prō-tə-'nō-tə-rē\ *noun, plural* **-ries** [Middle English *prothonotarie,* from Late Latin *protonotarius,* from *prot-* + Latin *notarius* stenographer — more at NOTARY PUBLIC] (15th century)
: a chief clerk of any of various courts of law
— pro·tho·no·tar·i·al \prō-,thä-nə-'ter-ē-əl, ,pro-thə-nō-'ter-ē-əl\ *adjective*
pro·tho·rac·ic \,prō-thə-'ra-sik\ *adjective* (1826)
: of or relating to the prothorax
prothoracic gland *noun* (1887)
: one of a pair of thoracic endocrine organs in some insects that control molting
pro·tho·rax \(,)prō-'thōr-,aks, -'thȯr-\ *noun* [New Latin *prothorac-, prothorax,* from ¹*pro-* + *thorax*] (1826)
: the anterior segment of the thorax of an insect — see INSECT illustration
pro·throm·bin \(,)prō-'thräm-bən\ *noun* [International Scientific Vocabulary] (1898)
: a plasma protein produced in the liver in the presence of vitamin K and converted into thrombin in the clotting of blood
pro·tist \'prō-(,)tist\ *noun* [New Latin *Protista,* from Greek, neuter plural of *prōtistos* very first, primal, from superlative of *prōtos* first — more at PROT-] (1889)
: any of a taxonomic group and especially a kingdom (Protista) of unicellular, colonial, or multicellular organisms usually including the protozoans and most algae and in various classifications prokaryotes, some or all fungi, and the sponges
— pro·tis·tan \prō-'tis-tən\ *adjective or noun*
pro·ti·um \'prō-tē-əm, 'prō-shē-\ *noun* [New Latin, from Greek *prōtos* first] (1933)
: the ordinary light hydrogen isotope of atomic mass 1
proto- — see PROT-
pro·to·col \'prō-tə-,kȯl, -,kōl, -,käl, -kəl\ *noun* [Middle French *prothocole,* from Medieval Latin *protocollum,* from Late Greek *prōtokollon* first sheet of a papyrus roll bearing data of manufacture, from Greek *prōt-* prot- + *kollan* to glue together, from *kolla* glue; perhaps akin to Middle Dutch *helen* to glue] (1541)
1 : an original draft, minute, or record of a document or transaction
2 a : a preliminary memorandum often formulated and signed by diplomatic negotiators as a basis for a final convention or treaty **b :** the records or minutes of a diplomatic conference or congress that show officially the agreements arrived at by the negotiators
3 a : a code prescribing strict adherence to correct etiquette and precedence (as in diplomatic exchange and in the military services) **b :** a set of conventions governing the treatment and especially the formatting of data in an electronic communications system

4 : a detailed plan of a scientific or medical experiment, treatment, or procedure ◆

pro·to·derm \'prō-tə-ˌdərm\ *noun* [International Scientific Vocabulary] (circa 1932)
: the outer primary meristem of a plant or plant part

pro·to·gal·axy \ˌprō-tō-'ga-lək-sē\ *noun* (1950)
: a cloud of gas believed to be the precursor to a galaxy

pro·to·his·to·ry \-'his-t(ə-)rē\ *noun* [International Scientific Vocabulary] (1903)
: the study of human beings in the times that immediately antedate recorded history
— **pro·to·his·to·ri·an** \-(h)is-'tōr-ē-ən, -'tȯr-\ *noun*
— **pro·to·his·tor·ic** \-'tȯr-ik, -'tär-\ *adjective*

pro·to·hu·man \-'hyü-mən, -'yü-\ *adjective* (circa 1909)
: of, relating to, or resembling an early hominid (as an australopithecine)
— **protohuman** *noun*

pro·to·lan·guage \'prō-tō-ˌlaŋ-gwij\ *noun* (1948)
: an assumed or recorded ancestral language

pro·to·mar·tyr \'prō-tō-ˌmär-tər\ *noun* [Middle English *prothomartir,* from Middle French, from Late Latin *protomartyr,* from Late Greek *prōtomartyr-, prōtomartys,* from Greek *prōt-* + *martyr-, martys* martyr] (15th century)
: the first martyr in a cause or region

pro·ton \'prō-ˌtän\ *noun* [Greek *prōton,* neuter of *prōtos* first — more at PROT-] (1920)
: an elementary particle that is identical with the nucleus of the hydrogen atom, that along with neutrons is a constituent of all other atomic nuclei, that carries a positive charge numerically equal to the charge of an electron, and that has a mass of 1.673×10^{-24} gram
— **pro·ton·ic** \prō-'tä-nik\ *adjective*

pro·ton·ate \'prō-tə-ˌnāt\ *verb* **-at·ed; -at·ing** (1945)
transitive verb
: to add a proton to
intransitive verb
: to acquire an additional proton
— **pro·ton·ation** \ˌprō-tə-'nā-shən\ *noun*

pro·to·ne·ma \ˌprō-tə-'nē-mə\ *noun, plural* **-ne·ma·ta** \-'nē-mə-tə, -'ne-\ [New Latin *protonemat-, protonema,* from *prot-* + Greek *nēma* thread — more at NEMAT-] (1857)
: the primary usually filamentous thalloid stage of the gametophyte in mosses and in some liverworts comparable to the prothallium in ferns
— **pro·to·ne·mal** \-'nē-məl\ *adjective*
— **pro·to·ne·ma·tal** \-'nē-mə-tᵊl, -'ne-\ *adjective*

protonotary apostolic *or* **prothonotary apostolic** *noun, plural* **protonotaries apostolic** *or* **prothonotaries apostolic** (1682)
: a priest of the chief college of the papal curia who keeps records of consistories and canonizations and signs papal bulls; *also* **:** an honorary member of this college

proton synchrotron *noun* (1947)
: a synchrotron in which protons are accelerated by means of frequency modulation of the radio-frequency accelerating voltage so that they have energies of billions of electron volts

pro·to·path·ic \ˌprō-tə-'pa-thik\ *adjective* [International Scientific Vocabulary, from Middle Greek *prōtopathēs* affected first, from Greek *prōt-* prot- + *pathos* experience, suffering — more at PATHOS] (1905)
: of, relating to, or being cutaneous sensory reception responsive only to rather gross stimuli

pro·to·phlo·em \-'flō-ˌem\ *noun* (1884)
: the first-formed phloem that develops from procambium, consists of narrow thin-walled cells capable of a limited amount of stretching, and is usually associated with a region of rapid growth

pro·to·plan·et \'prō-tō-ˌpla-nət\ *noun* (1949)
: a hypothetical whirling gaseous mass within a giant cloud of gas and dust that rotates around a sun and is believed to give rise to a planet
— **pro·to·plan·e·tary** \ˌprō-tō-'pla-nə-ˌter-ē\ *adjective*

pro·to·plasm \'prō-tə-ˌpla-zəm\ *noun* [German *Protoplasma,* from *prot-* + New Latin *plasma*] (1848)
1 : the organized colloidal complex of organic and inorganic substances (as proteins and water) that constitutes the living nucleus, cytoplasm, plastids, and mitochondria of the cell
2 : CYTOPLASM
— **pro·to·plas·mic** \ˌprō-tə-'plaz-mik\ *adjective*

pro·to·plast \'prō-tə-ˌplast\ *noun* [Middle French *protoplaste,* from Late Latin *protoplastus* first human, from Greek *prōtoplastos* first formed, from *prōt-* prot- + *plastos* formed, from *plassein* to mold — more at PLASTER] (1532)
1 : one that is formed first **:** PROTOTYPE
2 : a plant cell that has had its cell wall removed; *also* **:** the nucleus, cytoplasm, and plasma membrane of a cell as distinguished from inert walls and inclusions

pro·to·por·phy·rin \ˌprō-tō-'pȯr-f(ə-)rən\ *noun* [International Scientific Vocabulary] (1925)
: a purple porphyrin acid $C_{34}H_{34}N_4O_4$ obtained from hemin or heme by removal of bound iron

pro·to·star \'prō-tō-ˌstär\ *noun* (1947)
: a cloud of gas and dust in space believed to develop into a star

pro·to·stele \'prō-tə-ˌstēl, ˌprō-tə-'stē-lē\ *noun* (1901)
: a stele forming a solid rod with the phloem surrounding the xylem
— **pro·to·ste·lic** \ˌprō-tə-'stē-lik\ *adjective*

pro·to·stome \'prō-tə-ˌstōm\ *noun* [New Latin *Protostomia,* from *prot-* + Greek *stoma* mouth — more at STOMACH] (circa 1961)
: any of a major group (Protostomia) of bilateral metazoan animals (as mollusks, annelids, and arthropods) characterized in typical forms by determinate and spiral cleavage, formation of a mouth and anus directly from the blastopore, and formation of the coelom by splitting of the embryonic mesoderm — compare DEUTEROSTOME

pro·to·troph \'prō-tə-ˌtrōf, -ˌträf\ *noun* [back-formation from *prototrophic*] (1946)
: a prototrophic individual

pro·to·tro·phic \ˌprō-tə-'trō-fik\ *adjective* [International Scientific Vocabulary] (1900)
: having the nutritional requirements of the normal or wild type
— **pro·to·tro·phy** \prō-'tä-trə-fē\ *noun*

pro·to·typ·al \ˌprō-tə-'tī-pəl\ *adjective* (circa 1693)
: PROTOTYPICAL

pro·to·type \'prō-tə-ˌtīp\ *noun* [French, from Greek *prōtotypon,* from neuter of *prōtotypos* archetypal, from *prōt-* + *typos* type] (1552)
1 : an original model on which something is patterned **:** ARCHETYPE
2 : an individual that exhibits the essential features of a later type
3 : a standard or typical example
4 : a first full-scale and usually functional form of a new type or design of a construction (as an airplane)

pro·to·typ·i·cal \ˌprō-tə-'ti-pi-kəl\ *also* **pro·to·typ·ic** \-pik\ *adjective* (1650)
: of, relating to, or being a prototype
— **pro·to·typ·i·cal·ly** \-pi-k(ə-)lē\ *adverb*

pro·to·xy·lem \ˌprō-tə-'zī-ləm, -ˌlem\ *noun* (1887)
: the first-formed xylem developing from procambium and consisting of narrow cells with annular, spiral, or scalariform wall thickenings

pro·to·zo·al \ˌprō-tə-'zō-əl\ *adjective* (1890)
: of or relating to protozoans

pro·to·zo·an \-'zō-ən\ *noun* [New Latin *Protozoa,* from *prot-* + *-zoa*] (circa 1864)
: any of a phylum or subkingdom (Protozoa) of chiefly motile and heterotrophic unicellular protists (as amoebas, trypanosomes, sporozoans, and paramecia) that are represented in almost every kind of habitat and include some pathogenic parasites of humans and domestic animals
— **protozoan** *adjective*

pro·to·zo·ol·o·gy \-zō-'ä-lə-jē, -zə-'wä-\ *noun* [New Latin *Protozoa* + International Scientific Vocabulary *-logy*] (1904)
: a branch of zoology dealing with protozoans
— **pro·to·zo·ol·o·gist** \-zō-'ä-lə-jist, -zə-'wä-\ *noun*

pro·to·zo·on \-'zō-ˌän\ *noun, plural* **-zoa** \-'zō-ə\ [New Latin, from singular of *Protozoa*] (circa 1853)
: PROTOZOAN

pro·tract \prō-'trakt, prə-\ *transitive verb* [Latin *protractus,* past participle of *protrahere,* literally, to draw forward, from *pro-* forward + *trahere* to draw — more at PRO-] (1540)
1 *archaic* **:** DELAY, DEFER
2 : to prolong in time or space **:** CONTINUE
3 : to extend forward or outward — compare RETRACT 1
synonym see EXTEND
— **pro·trac·tive** \-'trak-tiv\ *adjective*

protracted meeting *noun* (1832)
: a protracted revival meeting

pro·trac·tile \-'trak-tᵊl, -ˌtīl\ *adjective* [Latin *protractus*] (1828)
: capable of being thrust out ⟨*protractile* jaws⟩

pro·trac·tion \-'trak-shən\ *noun* [Late Latin *protraction-, protractio* act of drawing out, from *protrahere*] (1535)
1 : the act of protracting **:** the state of being protracted
2 : the drawing to scale of an area of land

pro·trac·tor \-'trak-tər\ *noun* (circa 1611)
1 a : one that protracts **b :** a muscle that extends a part
2 : an instrument for laying down and measuring angles in drawing and plotting

◇ WORD HISTORY

protocol In ancient Greece a book consisted of papyrus sheets glued together along their edges to form a roll. The first sheet in the roll was known in Late Greek as the *protokollon,* from *proto-,* "first, beginning," and *-kollon,* a derivative of Greek *kolla* "glue." From *protokollon* came Medieval Latin *protocollum* and hence Middle French *prothocole.* By the time this ancient Greek word made its way into English via French in the 16th century as *protocol,* it had come to denote a "first" document of a different kind: the original draft, minute, or record of a transaction. A century later the word referred specifically to the first draft or record of a diplomatic document, such as a treaty or declaration. The diplomatic connection was reinforced in the 19th century when the official records or minutes of a diplomatic conference or congress also came under the rubric of *protocol.* Not long afterward, the French used *protocole* in the name of a department within the Ministry of Foreign Affairs whose province was the etiquette to be observed in official ceremonies and diplomatic relations. This use by the French is responsible for what is now the most familiar meaning of *protocol* in English.

pro·trep·tic \prō-ˈtrep-tik\ *noun* [Late Latin *protrepticus* hortatory, encouraging, from Greek *protreptikos,* from *protrepein* to turn forward, urge on, from *pro-* + *trepein* to turn] (circa 1656)
: an utterance (as a speech) designed to instruct and persuade
— **protreptic** *adjective*

pro·trude \prō-ˈtrüd\ *verb* **pro·trud·ed; pro·trud·ing** [Latin *protrudere,* from *pro-* + *trudere* to thrust — more at THREAT] (1620)
transitive verb
1 *archaic* **:** to thrust forward
2 : to cause to project
intransitive verb
: to jut out from the surrounding surface or context ⟨a handkerchief *protruding* from his breast pocket⟩
— **pro·tru·si·ble** \-ˈtrü-sə-bəl, -zə-\ *adjective*

pro·tru·sion \prō-ˈtrü-zhən\ *noun* [Latin *protrudere*] (1646)
1 : the act of protruding **:** the state of being protruded
2 : something (as a part or excrescence) that protrudes
synonym see PROJECTION

pro·tru·sive \-ˈtrü-siv, -ziv\ *adjective* (1676)
1 *archaic* **:** thrusting forward
2 : PROMINENT, PROTUBERANT ⟨a *protrusive* jaw⟩
3 : OBTRUSIVE, PUSHING ⟨a coarse *protrusive* manner⟩
— **pro·tru·sive·ly** *adverb*
— **pro·tru·sive·ness** *noun*

pro·tu·ber·ance \prō-ˈtü-b(ə-)rən(t)s, -ˈtyü-\ *noun* (1646)
1 : something that is protuberant
2 : the quality or state of being protuberant
synonym see PROJECTION

pro·tu·ber·ant \-b(ə-)rənt\ *adjective* [Late Latin *protuberant-, protuberans,* present participle of *protuberare* to bulge out, from Latin *pro-* forward + *tuber* excrescence, swelling; perhaps akin to Latin *tumēre* to swell — more at THUMB] (1646)
1 : thrusting out from a surrounding or adjacent surface often as a rounded mass **:** PROMINENT
2 : forcing itself into consciousness **:** OBTRUSIVE
— **pro·tu·ber·ant·ly** *adverb*

proud \ˈprau̇d\ *adjective* [Middle English, from Old English *prūd,* probably from Old French *prod, prud, prou* capable, good, valiant, from Late Latin *prode* advantage, advantageous, back-formation from Latin *prodesse* to be advantageous, from *pro-, prod-* for, in favor + *esse* to be — more at PRO-, IS] (before 12th century)
1 : feeling or showing pride: as **a :** having or displaying excessive self-esteem **b :** much pleased **:** EXULTANT **c :** having proper self-respect
2 a : marked by stateliness **:** MAGNIFICENT **b :** giving reason for pride **:** GLORIOUS ⟨the *proudest* moment in her life⟩
3 : VIGOROUS, SPIRITED ⟨a *proud* steed⟩
4 *chiefly British* **:** raised above a surrounding area ⟨a *proud* design on a stamp⟩ ☆
— **proud·ly** *adverb*

proud flesh *noun* (14th century)
: an excessive growth of granulation tissue (as in an ulcer)

proud·ful \ˈprau̇d-fəl\ *adjective* (14th century)
chiefly dialect **:** marked by or full of pride

proud·heart·ed \-ˈhär-təd\ *adjective* (14th century)
: proud in spirit **:** HAUGHTY

proust·ite \ˈprü-ˌstīt\ *noun* [French, from Joseph L. *Proust* (died 1826) French chemist] (1835)
: a mineral that consists of a red silver arsenic sulfide and occurs in crystals or massively

pro·vas·cu·lar \(ˌ)prō-ˈvas-kyə-lər\ *adjective* (circa 1948)
: of, relating to, or being procambium

prove \ˈprüv\ *verb* **proved; proved** *or* **prov·en** \ˈprü-vən, *British also* ˈprō-\; **prov·ing** \ˈprü-viŋ\ [Middle English, from Old French *prover,* from Latin *probare* to test, approve, prove, from *probus* good, honest, from *pro-* for, in favor + *-bus* (akin to Old English *bēon* to be) — more at PRO-, BE] (13th century)
transitive verb
1 *archaic* **:** to learn or find out by experience
2 a : to test the truth, validity, or genuineness of ⟨the exception *proves* the rule⟩ ⟨*prove* a will at probate⟩ **b :** to test the worth or quality of; *specifically* **:** to compare against a standard — sometimes used with *up* or *out* **c :** to check the correctness of (as an arithmetic result)
3 a : to establish the existence, truth, or validity of (as by evidence or logic) ⟨*prove* a theorem⟩ ⟨the charges were never *proved* in court⟩ **b :** to demonstrate as having a particular quality or worth ⟨the vaccine has been *proven* effective after years of tests⟩ ⟨*proved* herself a great actress⟩
4 : to show (oneself) to be worthy or capable ⟨eager to *prove* myself in the new job⟩
intransitive verb
: to turn out especially after trial or test ⟨the new drug *proved* effective⟩ ■
— **prov·able** \ˈprü-və-bəl\ *adjective*
— **prov·able·ness** *noun*
— **prov·ably** \-blē\ *adverb*
— **prov·er** \ˈprü-vər\ *noun*

prov·e·nance \ˈpräv-nən(t)s, ˈprä-və-ˌnän(t)s\ *noun* [French, from *provenir* to come forth, originate, from Latin *provenire,* from *pro-* forth + *venire* to come — more at PRO-, COME] (1785)
1 : ORIGIN, SOURCE
2 : the history of ownership of a valued object or work of art or literature

¹Pro·ven·çal \ˌprō-ˌvän-ˈsäl, ˌprä-vən-, *sense 1 also* prə-ˈven(t)-səl\ *adjective* [Middle French, from *Provence* Provence] (1589)
1 : of, relating to, or characteristic of Provence or the people of Provence
2 *or* **Pro·ven·çale :** cooked with garlic, onion, mushrooms, tomato, olive oil, and herbs ⟨scallops *Provençal*⟩

²Pro·ven·çal \ˌprō-ˌvän-ˈsäl, ˌprä-vən-\ *noun* (1600)
1 : a native or inhabitant of Provence
2 : a Romance language spoken in southern France; *especially* **:** the dialect of this language spoken in Provence

prov·en·der \ˈprä-vən-dər\ *noun* [Middle English, from Middle French *provende, provendre,* from Medieval Latin *provenda,* alteration of *praebenda* prebend] (14th century)
1 : dry food for domestic animals **:** FEED
2 : FOOD, VICTUALS

pro·ve·nience \prə-ˈvē-nyən(t)s, -nē-ən(t)s\ *noun* [alteration of *provenance*] (1882)
: ORIGIN, SOURCE

prov·en·ly \ˈprü-vən-lē, *British also* ˈprō-\ *adverb* (1887)
: demonstrably as stated **:** without doubt or uncertainty

pro·ven·tric·u·lus \ˌprō-ven-ˈtri-kyə-ləs\ *noun, plural* **-li** \-ˌlī, -ˌlē\ [New Latin] (circa 1836)
1 : the glandular or true stomach of a bird that is situated between the crop and gizzard
2 : a muscular dilatation of the foregut in most mandibulate insects that is armed internally with chitinous teeth or plates for triturating food
3 : the thin-walled sac in front of the gizzard of an earthworm

prove out *intransitive verb* (1941)
: to turn out to be satisfactory or as expected

¹prov·erb \ˈprä-ˌvərb\ *noun* [Middle English *proverbe,* from Middle French, from Latin *proverbium,* from *pro-* + *verbum* word — more at WORD] (14th century)
1 : a brief popular epigram or maxim **:** ADAGE
2 : BYWORD 4

²proverb *transitive verb* (14th century)
1 : to speak of proverbially
2 *obsolete* **:** to provide with a proverb

pro–verb \ˈprō-ˌvərb, -ˈvərb\ *noun* (1907)
: a form of the verb *do* used to avoid repetition of a verb (as *do* in "act as I do")

pro·ver·bi·al \prə-ˈvər-bē-əl\ *adjective* (1548)
1 : of, relating to, or resembling a proverb
2 : that has become a proverb or byword **:** commonly spoken of ⟨the *proverbial* smoking gun⟩
— **pro·ver·bi·al·ly** \-ə-lē\ *adverb*

Prov·erbs \ˈprä-ˌvərbz\ *noun plural but singular in construction*
: a collection of moral sayings and counsels forming a book of canonical Jewish and Christian Scripture — see BIBLE table

pro·vide \prə-ˈvīd\ *verb* **pro·vid·ed; pro·vid·ing** [Middle English, from Latin *providēre,* literally, to see ahead, from *pro-* forward + *vidēre* to see — more at PRO-, WIT] (15th century)
intransitive verb
1 : to take precautionary measures ⟨*provide* for the common defense —*U.S. Constitution*⟩
2 : to make a proviso or stipulation ⟨the constitution . . . *provides* for an elected two-chamber legislature —*Current Biography*⟩
3 : to make preparation to meet a need ⟨*provide* for entertainment⟩; *especially* **:** to supply something for sustenance or support ⟨*provides* for the poor⟩
transitive verb
1 *archaic* **:** to prepare or get ready in advance

☆ SYNONYMS
Proud, arrogant, haughty, lordly, insolent, overbearing, supercilious, disdainful mean showing scorn for inferiors. PROUD may suggest an assumed superiority or loftiness ⟨too *proud* to take charity⟩. ARROGANT implies a claiming for oneself of more consideration or importance than is warranted ⟨a conceited and *arrogant* executive⟩. HAUGHTY suggests a consciousness of superior birth or position ⟨a *haughty* aristocrat⟩. LORDLY implies pomposity or an arrogant display of power ⟨a *lordly* condescension⟩. INSOLENT implies contemptuous haughtiness ⟨ignored by an *insolent* waiter⟩. OVERBEARING suggests a tyrannical manner or an intolerable insolence ⟨an *overbearing* supervisor⟩. SUPERCILIOUS implies a cool, patronizing haughtiness ⟨an aloof and *supercilious* manner⟩. DISDAINFUL suggests a more active and openly scornful superciliousness ⟨*disdainful* of their social inferiors⟩.

□ USAGE
prove The past participle *proven,* originally the past participle of *preve,* a Middle English variant of *prove* that survived in Scotland, has gradually worked its way into standard English over the past three and a half centuries. It seems to have first become established in legal use and to have come only slowly into literary use. Tennyson was one of its earliest frequent users, probably for metrical reasons. It was disapproved by 19th century grammarians, one of whom included it in a list of "words that are not words." Surveys made some 30 or 40 years ago indicated that *proved* was about four times as frequent as *proven.* But our evidence from the last 10 or 15 years shows this no longer to be the case. As a past participle *proven* is now about as frequent as *proved* in all contexts. As an attributive adjective ⟨*proved* or *proven* gas reserves⟩ *proven* is much more common than *proved.*

2 a : to supply or make available (something wanted or needed) ⟨*provided* new uniforms for the band⟩; *also* **:** AFFORD ⟨curtains *provide* privacy⟩ **b :** to make something available to ⟨*provide* the children with free balloons⟩
3 : to have as a condition **:** STIPULATE ⟨the contract *provides* that certain deadlines will be met⟩

provided *conjunction* [Middle English, past participle of *providen* to provide] (15th century)
: on condition that **:** with the understanding **:** IF
usage see PROVIDING

prov·i·dence \'prä-və-dən(t)s, -,den(t)s\ *noun* [Middle English, from Middle French, from Latin *providentia,* from *provident-, providens*] (14th century)
1 a *often capitalized* **:** divine guidance or care **b** *capitalized* **:** God conceived as the power sustaining and guiding human destiny
2 : the quality or state of being provident

prov·i·dent \-dənt, -,dent\ *adjective* [Middle English, from Latin *provident-, providens,* from present participle of *providēre*] (15th century)
1 : making provision for the future **:** PRUDENT
2 : FRUGAL, SAVING
— prov·i·dent·ly *adverb*

prov·i·den·tial \,prä-və-'den(t)-shəl\ *adjective* (1648)
1 : of, relating to, or determined by Providence
2 *archaic* **:** marked by foresight **:** PRUDENT
3 : occurring by or as if by an intervention of Providence ⟨a *providential* escape⟩
synonym see LUCKY
— prov·i·den·tial·ly \-'den(t)-sh(ə-)lē\ *adverb*

pro·vid·er \prə-'vī-dər\ *noun* (1523)
: one that provides; *especially* **:** BREADWINNER

providing *conjunction* [Middle English, present participle of *providen*] (15th century)
: on condition that **:** in case ■

prov·ince \'prä-vən(t)s\ *noun* [Middle English, from Middle French, from Latin *provincia*] (14th century)
1 a : a country or region brought under the control of the ancient Roman government **b :** an administrative district or division of a country **c** *plural* **:** all of a country except the metropolises
2 a : a division of a country forming the jurisdiction of an archbishop or metropolitan **b :** a territorial unit of a religious order
3 a : a biogeographic division of less rank than a region **b :** an area that exhibits essential continuity of geological history; *also* **:** one characterized by particular structural or petrological features
4 a : proper or appropriate function or scope **:** SPHERE **b :** a department of knowledge or activity
synonym see FUNCTION

[1]**pro·vin·cial** \prə-'vin(t)-shəl\ *noun* [in sense 1, from Middle English, from Middle French or Medieval Latin; Middle French, from Medieval Latin *provincialis,* from *provincia* ecclesiastical province; in other senses, from Latin *provincialis,* from *provincia* province] (14th century)
1 : the superior of a province of a Roman Catholic religious order
2 : one living in or coming from a province
3 a : a person of local or restricted interests or outlook **b :** a person lacking urban polish or refinement

[2]**provincial** *adjective* (14th century)
1 : of, relating to, or coming from a province
2 a : limited in outlook **:** NARROW **b :** lacking the polish of urban society **:** UNSOPHISTICATED
3 : of or relating to a decorative style (as in furniture) marked by simplicity, informality, and relative plainness; *especially* **:** FRENCH PROVINCIAL
— pro·vin·cial·ly \-'vin(t)-sh(ə-)lē\ *adverb*

pro·vin·cial·ism \-shə-,li-zəm\ *noun* (1770)
1 : a dialectal or local word, phrase, or idiom
2 : the quality or state of being provincial

pro·vin·cial·ist \-'vin(t)-sh(ə-)list\ *noun* (1656)
: a native or inhabitant of a province

pro·vin·ci·al·i·ty \prə-,vin(t)-shē-'a-lə-tē\ *noun, plural* **-ties** (1782)
1 : PROVINCIALISM 2
2 : an act or instance of provincialism

pro·vin·cial·ize \-'vin(t)-shə-,līz\ *transitive verb* **-ized; -iz·ing** (1829)
: to make provincial
— pro·vin·cial·i·za·tion \-,vin(t)-sh(ə-)lə-'zā-shən\ *noun*

proving ground *noun* (circa 1890)
1 : a place for scientific experimentation or testing (as of vehicles or weapons)
2 : a place where something is developed or tried out

pro·vi·rus \(,)prō-'vī-rəs\ *noun* [New Latin] (1949)
: a form of a virus that is integrated into the genetic material of a host cell and by replicating with it can be transmitted from one cell generation to the next without causing lysis
— pro·vi·ral \-rəl\ *adjective*

[1]**pro·vi·sion** \prə-'vi-zhən\ *noun* [Middle English, from Middle French, from Late Latin & Latin; Late Latin *provision-, provisio* act of providing, from Latin, foresight, from *providēre* to see ahead — more at PROVIDE] (14th century)
1 a : the act or process of providing **b :** the fact or state of being prepared beforehand **c :** a measure taken beforehand to deal with a need or contingency **:** PREPARATION ⟨made *provision* for replacements⟩
2 : a stock of needed materials or supplies; *especially* **:** a stock of food — usually used in plural
3 : PROVISO, STIPULATION

[2]**provision** *transitive verb* **pro·vi·sioned; pro·vi·sion·ing** \-'vi-zhə-niŋ, -'vizh-niŋ\ (1809)
: to supply with provisions

[1]**pro·vi·sion·al** \prə-'vizh-nəl, -'vi-zhə-n^{ə}l\ *adjective* (1601)
: serving for the time being **:** TEMPORARY
— pro·vi·sion·al·ly *adverb*

[2]**provisional** *noun* (1886)
: a postage stamp for use until a regular issue appears — compare DEFINITIVE

pro·vi·sion·ary \prə-'vi-zhə-,ner-ē\ *adjective* (1617)
: PROVISIONAL

pro·vi·sion·er \-'vi-zhə-nər, -'vizh-nər\ *noun* (1866)
: a furnisher of provisions

pro·vi·so \prə-'vī-(,)zō\ *noun, plural* **-sos** *or* **-soes** [Middle English, from Medieval Latin *proviso quod* provided that] (15th century)
1 : an article or clause (as in a contract) that introduces a condition
2 : a conditional stipulation

pro·vi·so·ry \-'vī-zə-rē, -'vīz-rē\ *adjective* (circa 1611)
1 : containing or subject to a proviso **:** CONDITIONAL
2 : PROVISIONAL

pro·vi·ta·min \(,)prō-'vī-tə-mən\ *noun* (1927)
: a precursor of a vitamin convertible into the vitamin in an organism

Pro·vo \'prō-(,)vō\ *noun, plural* **Provos** [*Pro*visional I.R.A., name of the faction + [1]*-o*] (1971)
: a member of the extremist faction of the Irish Republican Army

pro·vo·ca·teur \prō-,vä-kə-'tər\ *noun* (1919)
: AGENT PROVOCATEUR

prov·o·ca·tion \,prä-və-'kā-shən\ *noun* [Middle English *provocacioun,* from Middle French *provocation,* from Latin *provocation-, provocatio,* from *provocare*] (14th century)
1 : the act of provoking **:** INCITEMENT
2 : something that provokes, arouses, or stimulates

pro·voc·a·tive \prə-'vä-kə-tiv\ *adjective* (15th century)
: serving or tending to provoke, excite, or stimulate
— provocative *noun*
— pro·voc·a·tive·ly *adverb*
— pro·voc·a·tive·ness *noun*

pro·voke \prə-'vōk\ *transitive verb* **pro·voked; pro·vok·ing** [Middle English, from Middle French *provoquer,* from Latin *provocare,* from *pro-* forth + *vocare* to call, from *voc-, vox* voice — more at PRO-, VOICE] (14th century)
1 a *archaic* **:** to arouse to a feeling or action **b :** to incite to anger
2 a : to call forth (as a feeling or action) **:** EVOKE ⟨*provoke* laughter⟩ **b :** to stir up purposely ⟨*provoke* a fight⟩ **c :** to provide the needed stimulus for ⟨will *provoke* a lot of discussion⟩ ☆
— pro·vok·er *noun*

pro·vok·ing \-'vō-kiŋ\ *adjective* (1642)
: causing mild anger **:** ANNOYING
— pro·vok·ing·ly \-kiŋ-lē\ *adverb*

pro·vo·lo·ne \,prō-və-'lō-nē\ *noun* [Italian, augmentative of *provola,* a kind of cheese] (1912)
: a usually firm pliant often smoked cheese of Italian origin

pro·vost \'prō-,vōst, 'prä-vəst, 'prō-vəst, *especially attributive* ,prō-(,)vō\ *noun* [Middle English, from Old English *profost* & Old French *provost,* from Medieval Latin *propositus,* alteration of *praepositus,* from Latin, one in charge, director, from past participle of *praeponere* to place at the head — more at PREPOSITION] (before 12th century)
1 : the chief dignitary of a collegiate or cathedral chapter
2 : the chief magistrate of a Scottish burgh
3 : the keeper of a prison
4 : a high-ranking university administrative officer

provost court *noun* (1864)
: a military court usually for the trial of minor offenses within an occupied hostile territory

provost guard *noun* (1862)
: a police detail of soldiers under the authority of the provost marshal

provost marshal *noun* (1535)
: an officer who supervises the military police of a command

☆ SYNONYMS
Provoke, excite, stimulate, pique, quicken mean to arouse as if by pricking. PROVOKE directs attention to the response called forth ⟨my stories usually *provoke* laughter⟩. EXCITE implies a stirring up or moving profoundly ⟨news that *excited* anger and frustration⟩. STIMULATE suggests a rousing out of lethargy, quiescence, or indifference ⟨*stimulating* conversation⟩. PIQUE suggests stimulating by mild irritation or challenge ⟨that remark *piqued* my interest⟩. QUICKEN implies beneficially stimulating and making active or lively ⟨the high salary *quickened* her desire to have the job⟩. See in addition IRRITATE.

□ USAGE
providing Although occasionally still disapproved, *providing* is as well established as a conjunction as *provided* is. *Provided* is more common.

[1]prow \ˈprau̇\ *adjective* [Middle English, from Middle French *prou* — more at PROUD] (14th century)
archaic **:** VALIANT, GALLANT

[2]prow \ˈprau̇, *archaic* ˈprō\ *noun* [Middle French *proue,* probably from Old Italian dialect *prua,* from Latin *prora,* from Greek *prōira*] (1555)
1 : the bow of a ship **:** STEM
2 : a pointed projecting front part

prow·ess \ˈprau̇-əs *also* ˈprō-\ *noun* [Middle English *prouesse,* from Old French *proesse,* from *prou*] (13th century)
1 : distinguished bravery; *especially* **:** military valor and skill
2 : extraordinary ability ⟨his *prowess* on the football field⟩

[1]prowl \ˈprau̇(ə)l\ *verb* [Middle English *prollen*] (14th century)
intransitive verb
: to move about or wander stealthily in or as if in search of prey
transitive verb
: to roam over in a predatory manner
— prowl·er \ˈprau̇-lər\ *noun*

[2]prowl *noun* (1803)
: an act or instance of prowling
— on the prowl : in the act of prowling; *also* **:** in search of something ⟨his fourth wife had just left him, and he was *on the prowl* again —Mary McCarthy⟩

prowl car *noun* (1937)
: SQUAD CAR

prox·e·mics \präk-ˈsē-miks\ *noun plural but singular or plural in construction* [*prox*imity + *-emics* (as in *phonemics*)] (1963)
: the study of the nature, degree, and effect of the spatial separation individuals naturally maintain (as in various social and interpersonal situations) and of how this separation relates to environmental and cultural factors
— prox·e·mic \-mik\ *adjective*

prox·i·mal \ˈpräk-sə-məl\ *adjective* [Latin *proximus*] (1727)
1 : situated close to **:** PROXIMATE
2 : next to or nearest the point of attachment or origin, a central point, or the point of view; *especially* **:** located toward the center of the body — compare DISTAL
3 : of, relating to, or being the mesial and distal surfaces of a tooth
— prox·i·mal·ly \-mə-lē\ *adverb*

proximal convoluted tubule *noun* (circa 1899)
: the convoluted portion of the vertebrate nephron that lies between Bowman's capsule and the loop of Henle and functions especially in the resorption of sugar, sodium and chloride ions, and water from the glomerular filtrate — called also *proximal tubule*

prox·i·mate \ˈpräk-sə-mət\ *adjective* [Latin *proximatus,* past participle of *proximare* to approach, from *proximus* nearest, next, superlative of *prope* near — more at APPROACH] (1661)
1 : immediately preceding or following (as in a chain of events, causes, or effects) ⟨*proximate,* rather than ultimate, goals —Reinhold Niebuhr⟩
2 a : very near **:** CLOSE **b :** soon forthcoming **:** IMMINENT
— prox·i·mate·ly *adverb*
— prox·i·mate·ness *noun*

prox·im·i·ty \präk-ˈsi-mə-tē\ *noun* [Middle French *proximité,* from Latin *proximitat-, proximitas,* from *proximus*] (15th century)
: the quality or state of being proximate **:** CLOSENESS

proximity fuze *noun* (1945)
: a fuze for a projectile that uses the principle of radar to detect the presence of a target within the projectile's effective range

prox·i·mo \ˈpräk-sə-ˌmō\ *adjective* [Latin *proximo mense* in the next month] (1855)
: of or occurring in the next month after the present

proxy \ˈpräk-sē\ *noun, plural* **prox·ies** [Middle English *procucie,* contraction of *procuracie,* from Anglo-French, from Medieval Latin *procuratia,* alteration of Latin *procuratio* procuration] (15th century)
1 : the agency, function, or office of a deputy who acts as a substitute for another
2 a : authority or power to act for another **b :** a document giving such authority; *specifically* **:** a power of attorney authorizing a specified person to vote corporate stock
3 : a person authorized to act for another **:** PROCURATOR
— proxy *adjective*

proxy marriage *noun* (1900)
: a marriage celebrated in the absence of one of the contracting parties who is represented at the ceremony by a proxy

prude \ˈprüd\ *noun* [French, good woman, prudish woman, short for *prudefemme* good woman, from Old French *prode femme*] (1704)
: a person who is excessively or priggishly attentive to propriety or decorum; *especially* **:** a woman who shows or affects extreme modesty

pru·dence \ˈprü-dᵊn(t)s\ *noun* [Middle English, from Middle French, from Latin *prudentia,* alteration of *providentia* — more at PROVIDENCE] (14th century)
1 : the ability to govern and discipline oneself by the use of reason
2 : sagacity or shrewdness in the management of affairs
3 : skill and good judgment in the use of resources
4 : caution or circumspection as to danger or risk

pru·dent \-dᵊnt\ *adjective* [Middle English, from Middle French, from Latin *prudent-, prudens,* contraction of *provident-, providens* — more at PROVIDENT] (14th century)
: characterized by, arising from, or showing prudence: as **a :** marked by wisdom or judiciousness **b :** shrewd in the management of practical affairs **c :** marked by circumspection **:** DISCREET **d :** PROVIDENT, FRUGAL
synonym see WISE
— pru·dent·ly *adverb*

pru·den·tial \prü-ˈden(t)-shəl\ *adjective* (15th century)
1 : of, relating to, or proceeding from prudence
2 : exercising prudence especially in business matters
— pru·den·tial·ly \-ˈden(t)-sh(ə-)lē\ *adverb*

prud·ery \ˈprü-d(ə-)rē\ *noun, plural* **-er·ies** (1709)
1 : the characteristic quality or state of a prude
2 : a prudish act or remark

prud·ish \ˈprü-dish\ *adjective* (1717)
: marked by prudery **:** PRIGGISH
— prud·ish·ly *adverb*
— prud·ish·ness *noun*

pru·i·nose \ˈprü-ə-ˌnōs\ *adjective* [Latin *pruinosus* covered with hoarfrost, from *pruina* hoarfrost — more at FREEZE] (circa 1826)
: covered with whitish dust or bloom ⟨*pruinose* stems⟩

[1]prune \ˈprün\ *noun* [Middle English, from Middle French, plum, from Latin *prunum* — more at PLUM] (14th century)
: a plum dried or capable of drying without fermentation

[2]prune *verb* **pruned; prun·ing** [Middle English *prouynen,* from Middle French *proignier,* probably alteration of *provigner* to layer, from *provain* layer, from Latin *propagin-, propago,* from *pro-* forward + *pangere* to fix — more at PRO-, PACT] (15th century)
transitive verb
1 a : to reduce especially by eliminating superfluous matter ⟨*pruned* the text⟩ ⟨*prune* the budget⟩ **b :** to remove as superfluous ⟨*prune* away all ornamentation⟩
2 : to cut off or cut back parts of for better shape or more fruitful growth
intransitive verb
: to cut away what is unwanted or superfluous
— prun·er *noun*

pru·nel·la \prü-ˈne-lə\ *also* **pru·nelle** \-ˈnel\ *noun* [French *prunelle,* literally, sloe, from diminutive of *prune* plum] (1670)
1 : a twilled woolen dress fabric
2 : a heavy woolen fabric used for the uppers of shoes

pruning hook *noun* (1611)
: a pole bearing a curved blade for pruning plants

pru·nus \ˈprü-nəs\ *noun* [New Latin, from Latin, plum tree, from Greek *proumnē*] (1901)
: any of a genus (*Prunus*) of drupaceous trees or shrubs of the rose family that have showy clusters of usually white or pink flowers first appearing in the spring often before the leaves including many grown for ornament or for their fruit (as the plum, cherry, or apricot)

pru·ri·ence \ˈpru̇r-ē-ən(t)s\ *noun* (1781)
: the quality or state of being prurient

pru·ri·en·cy \-ən(t)-sē\ *noun* (1795)
: PRURIENCE

pru·ri·ent \-ənt\ *adjective* [Latin *prurient-, pruriens,* present participle of *prurire* to itch, crave; akin to Latin *pruna* glowing coal, Sanskrit *ploṣati* he singes, and probably to Latin *pruina* hoarfrost — more at FREEZE] (1592)
: marked by or arousing an immoderate or unwholesome interest or desire; *especially* **:** marked by, arousing, or appealing to unusual sexual desire
— pru·ri·ent·ly *adverb*

pru·ri·go \prü-ˈrī-(ˌ)gō, -ˈrē-\ *noun* [New Latin, from Latin, itch, from *prurire*] (circa 1646)
: a chronic inflammatory skin disease marked by itching papules

pru·rit·ic \-ˈri-tik\ *adjective* (1899)
: of, relating to, or marked by itching

pru·ri·tus \-ˈrī-təs, -ˈrē-\ *noun* [Latin, from *prurire*] (1653)
: ITCH 1a

Prus·sian blue \ˈprə-shən-\ *noun* [*Prussia,* Germany] (1724)
1 : any of numerous blue iron pigments formerly regarded as ferric ferrocyanide
2 : a dark blue crystalline hydrated ferric ferrocyanide $Fe_4[Fe(CN)_6]_3{\cdot}xH_2O$ used as a test for ferric iron
3 : a greenish blue

prus·sian·ise *British variant of* PRUSSIANIZE

Prus·sian·ism \ˈprə-shə-ˌni-zəm\ *noun* (1856)
: the practices or policies (as the advocacy of militarism) held to be typically Prussian

prus·sian·ize \-ˌnīz\ *transitive verb* **-ized; -iz·ing** *often capitalized* (1861)
: to make Prussian in character or principle (as in authoritarian control or rigid discipline)
— prus·sian·i·za·tion \ˌprə-shə-nə-ˈzā-shən\ *noun*

pru·tah *or* **pru·ta** \prü-ˈtä\ *noun, plural* **pru·toth** \-ˈtōt, -ˈtōth, -ˈtōs\ *or* **pru·tot** \-ˈtōt, -ˈtōs\ [New Hebrew *pĕrūṭāh,* from Late Hebrew, a small coin] (1949)
1 : a former monetary unit of Israel equivalent to 1/1000 pound
2 : a coin representing one prutah

[1]pry \ˈprī\ *intransitive verb* **pried; pry·ing** [Middle English *prien*] (14th century)
: to look closely or inquisitively; *also* **:** to make a nosy or presumptuous inquiry

[2]pry *transitive verb* **pried; pry·ing** [probably back-formation from [5]*prize*] (circa 1806)
1 : to raise, move, or pull apart with a lever **:** PRIZE
2 : to extract, detach, or open with difficulty ⟨*pried* the secret out of my sister⟩

[3]pry *noun* (1823)
1 : a tool for prying
2 : LEVERAGE

pry·er *variant of* PRIER

prying *adjective* (1552)

: impertinently or officiously inquisitive or interrogatory
synonym see CURIOUS
— pry·ing·ly \-iŋ-lē\ *adverb*

Prze·wal·ski's horse \pshə-'väl-skēz-, shə-, ˌpər-zhə-'väl-\ *noun* [Nikolaĭ M. *Przhevalskiĭ* (died 1888) Russian soldier & explorer] (1881)
: a small stocky bay- or dun-colored wild horse (*Equus caballus przewalskii* synonym *E. przewalskii*) of central Asia having a large head and short erect mane — called also *Prze·wal·ski horse* \-skē-\

psalm \'säm, 'sälm, 'sȯm, 'sȯlm, *New England also* 'såm\ *noun, often capitalized* [Middle English, from Old English *psealm,* from Late Latin *psalmus,* from Greek *psalmos,* literally, twanging of a harp, from *psallein* to pluck, play a stringed instrument] (before 12th century)
: a sacred song or poem used in worship; *especially* **:** one of the biblical hymns collected in the Book of Psalms

psalm·book \-ˌbu̇k\ *noun* (12th century)
archaic **:** PSALTER

psalm·ist \'sä-mist, 'säl-, 'sȯ-, 'sȯl-, *New England also* 'så-mist\ *noun* (15th century)
: a writer or composer of especially biblical psalms

psalm·o·dy \'sä-mə-dē, 'säl-, 'sȯ-, 'sȯl, *New England also* 'så-mə-dē\ *noun* [Middle English *psalmodie,* from Late Latin *psalmodia,* from Late Greek *psalmōidia,* literally, singing to the harp, from Greek *psalmos* + *aidein* to sing — more at ODE] (14th century)
1 : the act, practice, or art of singing psalms in worship
2 : a collection of psalms

Psalms \'sämz, 'sälmz, 'sȯmz, 'sȯlmz, *New England also* 'såmz\ *noun plural but singular in construction*
: a collection of sacred poems forming a book of canonical Jewish and Christian Scripture — see BIBLE table

Psal·ter \'säl-tər, 'sȯl-\ *noun* [Middle English, from Old English *psalter* & Old French *psaltier,* from Late Latin *psalterium,* from Late Greek *psaltērion,* from Greek, psaltery] (before 12th century)
: the Book of Psalms; *also* **:** a collection of Psalms for liturgical or devotional use

psal·te·ri·um \säl-'tir-ē-əm, sȯl-\ *noun, plural* **-ria** \-ē-ə\ [New Latin, from Late Latin, psalter; from the resemblance of the folds to the pages of a book] (circa 1846)
: OMASUM

psal·tery *also* **psal·try** \'säl-t(ə-)rē, 'sȯl-\ *noun, plural* **-ter·ies** *also* **-tries** [Middle English *psalterie,* from Middle French, from Latin *psalterium,* from Greek *psaltērion,* from *psallein* to play on a stringed instrument] (14th century)
: an ancient musical instrument resembling the zither

p's and q's \ˌpēz-^{ə}n-'kyüz\ *noun plural* [from the phrase *mind one's p's and q's,* alluding to the difficulty a child learning to write has in distinguishing between *p* and *q*] (1779)
1 : something (as one's manners) that one should be mindful of ⟨better watch his *p's and q's* when I get a six-gun of my own —Jean Stafford⟩
2 : best behavior ⟨being on her *p's and q's* for two solid days was too much —Guy McCrone⟩

pse·phol·o·gy \sē-'fä-lə-jē\ *noun* [Greek *psēphos* pebble, ballot, vote; from the use of pebbles by the ancient Greeks in voting] (1952)
: the scientific study of elections
— pse·pho·log·i·cal \ˌsē-fə-'lä-ji-kəl\ *adjective*
— pse·phol·o·gist \sē-'fä-lə-jist\ *noun*

pseud \'süd\ *noun* [short for *pseudo-intellectual*] (1964)
British **:** a person who pretends to be an intellectual

pseud- *or* **pseudo-** *combining form* [Middle English, from Late Latin, from Greek, from *pseudēs,* from *pseudesthai* to lie; akin to Armenian *sut* lie and probably to Greek *psychein* to breathe — more at PSYCH-]
: false **:** spurious ⟨*pseudo*classic⟩ ⟨*pseudo*podium⟩

pseud·ep·i·graph \sü-'de-pə-ˌgraf\ *noun* (1884)
: PSEUDEPIGRAPHON 2

pseud·epig·ra·phon \ˌsü-di-'pi-grə-ˌfän\ *noun, plural* **-pha** \-fə\ [New Latin, singular of *pseudepigrapha,* from Greek, neuter plural of *pseudepigraphos* falsely inscribed, from *pseud-* + *epigraphein* to inscribe — more at EPIGRAM] (1692)
1 *plural* **:** APOCRYPHA
2 : any of various pseudonymous or anonymous Jewish religious writings of the period 200 B.C. to 200 A.D.; *especially* **:** one of such writings (as the Psalms of Solomon) not included in any canon of biblical Scripture — usually used in plural

pseud·epig·ra·phy \-fē\ *noun* [Greek *pseudepigraphos*] (circa 1842)
: the ascription of false names of authors to works

pseu·do \'sü-(ˌ)dō\ *adjective* [Middle English, from *pseudo-*] (15th century)
: being apparently rather than actually as stated **:** SHAM, SPURIOUS ⟨distinction between true and *pseudo* humanism —K. F. Reinhardt⟩

pseu·do·al·lele \ˌsü-dō-ə-'lē(ə)l\ *noun* (1948)
: any of two or more closely linked genes that act usually as if a single member of an allelic pair but occasionally undergo crossing-over and recombination

pseu·do·cho·lin·es·ter·ase \'sü-dō-ˌkō-lə-'nes-tə-ˌrās, -ˌrāz\ *noun* (1943)
: CHOLINESTERASE 2

pseu·do·clas·sic \ˌsü-dō-'kla-sik\ *adjective* (1899)
: pretending to be or erroneously regarded as classic
— pseudoclassic *noun*

pseu·do·clas·si·cism \-'kla-sə-ˌsi-zəm\ *noun* (1871)
: imitative representation of classicism in literature and art

pseu·do·coel \'sü-də-ˌsēl\ *noun* (1887)
: a body cavity that is not a product of gastrulation and is not lined with a well-defined mesodermal membrane

pseu·do·coe·lom·ate \ˌsü-dō-'sē-lə-ˌmāt\ *noun* (1940)
: an invertebrate (as a nematode or rotifer) having a body cavity that is a pseudocoel
— pseudocoelomate *adjective*

pseu·do·cy·e·sis \-sī-'ē-səs\ *noun* [New Latin, from *pseud-* + *cyesis* pregnancy, from Greek *kyēsis,* from *kyein* to be pregnant — more at CYME] (circa 1817)
: a psychosomatic state that occurs without conception and is marked by some of the physical symptoms and changes in hormonal balance of pregnancy

pseu·do·mo·nad \ˌsü-də-'mō-ˌnad, -nəd\ *noun* [New Latin *Pseudomonad-, Pseudomonas*] (1921)
: any of a genus (*Pseudomonas*) of gram-negative rod-shaped motile bacteria including some that produce a greenish fluorescent water-soluble pigment and some that are saprophytes or plant or animal pathogens

pseu·do·mo·nas \-nəs\ *noun, plural* **-mo·na·des** \-'mō-nə-ˌdēz, -'mä-\ [New Latin, from *pseud-* + *monad-, monas* monad] (1903)
: PSEUDOMONAD

pseu·do·morph \'sü-də-ˌmȯrf\ *noun* [probably from French *pseudomorphe,* from *pseud-* + *-morphe* -morph] (1849)
1 : a mineral having the characteristic outward form of another species
2 : a deceptive or irregular form
— pseu·do·mor·phic \ˌsü-də-'mȯr-fik\ *adjective*
— pseu·do·mor·phism \-ˌfi-zəm\ *noun*
— pseu·do·mor·phous \-fəs\ *adjective*

pseu·do·nym \'sü-d^{ə}n-ˌim\ *noun* [French *pseudonyme,* from Greek *pseudōnymos* bearing a false name, from *pseud-* + *onyma* name — more at NAME] (1833)
: a fictitious name; *especially* **:** PEN NAME

pseu·do·nym·i·ty \ˌsü-d^{ə}n-'i-mə-tē\ *noun* (1877)
: the use of a pseudonym; *also* **:** the fact or state of being signed with a pseudonym

pseu·don·y·mous \sü-'dä-nə-məs\ *adjective* [Greek *pseudōnymos*] (circa 1706)
: bearing or using a fictitious name ⟨a *pseudonymous* report⟩; *also* **:** being a pseudonym
— pseu·don·y·mous·ly *adverb*
— pseu·don·y·mous·ness *noun*

pseu·do·pa·ren·chy·ma \ˌsü-dō-pə-'reŋ-kə-mə\ *noun* [New Latin] (1875)
: compactly interwoven short-celled filaments in a thallophyte that resemble parenchyma of higher plants
— pseu·do·par·en·chy·ma·tous \-ˌpar-ən-'ki-mə-təs, -'kī-\ *adjective*

pseu·do·pod \'sü-də-ˌpäd\ *noun* [New Latin *pseudopodium*] (1874)
: PSEUDOPODIUM
— pseu·dop·o·dal \sü-'dä-pə-d^{ə}l\ *or* **pseu·do·po·di·al** \ˌsü-də-'pō-dē-əl\ *adjective*

pseu·do·po·di·um \ˌsü-də-'pō-dē-əm\ *noun, plural* **-po·dia** \-dē-ə\ [New Latin] (1854)
1 : a temporary protrusion or retractile process of the cytoplasm of a cell that functions (as in an amoeba) especially in a locomotor or food gathering capacity — see AMOEBA illustration
2 : a slender leafless branch of the gametophyte in various mosses that often bears gemmae

pseu·do·preg·nan·cy \ˌsü-dō-'preg-nən(t)-sē\ *noun* (1860)
1 : PSEUDOCYESIS
2 : an anestrous state resembling pregnancy that occurs in various mammals usually after an infertile copulation
— pseu·do·preg·nant \-nənt\ *adjective*

pseu·do·ran·dom \-'ran-dəm\ *adjective* (1949)
: being or involving entities (as numbers) that are selected by a definite computational process but that satisfy one or more standard tests for statistical randomness

pseu·do·sci·ence \ˌsü-dō-'sī-ən(t)s\ *noun* (1844)
: a system of theories, assumptions, and methods erroneously regarded as scientific
— pseu·do·sci·en·tif·ic \-ˌsī-ən-'ti-fik\ *adjective*
— pseu·do·sci·en·tist \-'sī-ən-tist\ *noun*

pseu·do·scor·pi·on \-'skȯr-pē-ən\ *noun* [New Latin *Pseudoscorpiones,* from *pseud-* + Latin *scorpion-, scorpio* scorpion] (1835)
: any of a widely distributed order (Pseudoscorpionida synonym Pseudoscorpiones) of tiny arachnids that have no caudal stinger and feed on tiny invertebrates (as insects and mites)

pseu·do·so·phis·ti·ca·tion \'sü-dō-sə-ˌfis-tə-'kā-shən\ *noun* (1903)
: false or feigned sophistication
— pseu·do·so·phis·ti·cat·ed \-sə-'fis-ti-ˌkā-təd\ *adjective*

pseu·do·tu·ber·cu·lo·sis \-tu̇-ˌbər-kyə-'lō-səs, -tyu̇-\ *noun* [New Latin] (1900)
: any of several diseases that are characterized by the formation of granulomas resembling tubercular nodules but are not caused by the tubercle bacillus

pshaw \'shȯ\ *interjection* (1673)

\ə\ **abut** \$^{ə}$\ **kitten** \ər\ **further** \a\ **ash** \ā\ **ace** \ä\ **mop, mar** \au̇\ **out** \ch\ **chin** \e\ **bet** \ē\ **easy** \g\ **go** \i\ **hit** \ī\ **ice** \j\ **job** \ŋ\ **sing** \ō\ **go** \ȯ\ **law** \ȯi\ **boy** \th\ **thin** \t͟h\ **the** \ü\ **loot** \u̇\ **foot** \y\ **yet** \zh\ **vision** *see also* Guide to Pronunciation

— used to express irritation, disapproval, contempt, or disbelief

[1]psi \'sī, 'psī\ *noun* [Middle English, from Medieval Latin, from Late Greek, from Greek *psei*] (15th century)
: the 23d letter of the Greek alphabet — see ALPHABET table

[2]psi \'sī\ *noun* [probably by shortening & alteration from *psychic*] (circa 1946)
: parapsychological psychic phenomena or powers

psi·lo·cy·bin \ˌsī-lə-'sī-bən\ *noun* [New Latin *Psilocybe*, fungus genus + [1]*-in*] (1958)
: a hallucinogenic indole $C_{12}H_{17}N_2O_4P$ obtained from a fungus (*Psilocybe mexicana*)

psi·lo·phyte \'sī-lə-ˌfīt\ *noun* [New Latin *Psilophyton*, genus of plants, from Greek *psilos* bare, mere (probably akin to Greek *psēn* to rub) + *phyton* plant — more at PHYT-] (circa 1911)
: any of an order (Psilophytales) of extinct simple dichotomously branched plants from the Paleozoic of Europe and eastern Canada that include the oldest known land plants with vascular structure
— **psi·lo·phyt·ic** \ˌsī-lə-'fi-tik\ *adjective*

psi particle \'sī-, 'psī-\ *noun* (1974)
: J/PSI PARTICLE

psit·ta·cine \'si-tə-ˌsīn\ *adjective* [Latin *psittacinus*, from *psittacus* parrot, from Greek *psittakos*] (1874)
: of or relating to the parrots
— **psittacine** *noun*

psit·ta·co·sis \ˌsi-tə-'kō-səs\ *noun* [New Latin, from Latin *psittacus*] (1897)
: an infectious disease of birds caused by a bacterium (*Chlamydia psittaci*), marked by diarrhea and wasting, and transmissible to humans in whom it usually occurs as an atypical pneumonia accompanied by high fever
— **psit·ta·co·tic** \-'kä-tik, -'kō-\ *adjective*

pso·cid \'sō-səd\ *noun* [ultimately from New Latin *Psocus*, genus of lice] (1891)
: any of an order (Psocoptera synonym Corrodentia) of minute usually winged primitive insects (as a book louse)

pso·ri·a·sis \sə-'rī-ə-səs\ *noun* [New Latin, from Greek *psōriasis*, from *psōrian* to have the itch, from *psōra* itch; akin to Greek *psēn* to rub] (circa 1684)
: a chronic skin disease characterized by circumscribed red patches covered with white scales
— **pso·ri·at·ic** \ˌsōr-ē-'a-tik, ˌsór-\ *adjective or noun*

psych *also* **psyche** \'sīk\ *transitive verb* **psyched; psych·ing** [by shortening] (1917)
1 : PSYCHOANALYZE
2 a : to anticipate correctly the intentions or actions of **:** OUTGUESS **b :** to analyze or figure out (as a problem or course of action) ⟨I *psyched* it all out by myself and decided —David Hulburd⟩
3 a : to make psychologically uneasy **:** INTIMIDATE, SCARE ⟨pressure doesn't *psych* me —Jerry Quarry⟩ — often used with *out* **b :** to make (as oneself) psychologically ready especially for performance — often used with *up* ⟨*psyched* herself up for the race⟩

psych- *or* **psycho-** *combining form* [Greek, from *psychē* breath, principle of life, life, soul, from *psychein* to breathe; akin to Sanskrit *babhasti* he blows]
1 : mind **:** mental processes and activities ⟨*psycho*dynamic⟩ ⟨*psycho*logy⟩
2 : psychological methods ⟨*psycho*analysis⟩ ⟨*psycho*therapy⟩
3 : brain ⟨*psycho*surgery⟩
4 : mental and ⟨*psycho*somatic⟩

psych·as·the·nia \ˌsī-kəs-'thē-nē-ə\ *noun* [New Latin] (1900)
: a neurotic state characterized especially by phobias, obsessions, or compulsions that one knows are irrational
— **psych·as·then·ic** \-'the-nik\ *adjective or noun*

Psy·che \'sī-kē\ *noun* [Latin, from Greek *psychē* soul]
1 : a princess loved by Cupid
2 *not capitalized* [Greek *psychē*] **a :** SOUL, SELF **b :** MIND

psy·che·de·lia \ˌsī-kə-'dēl-yə\ *noun* [New Latin, from English *psychedelic*] (1967)
: the world of people, phenomena, or items associated with psychedelic drugs

[1]psy·che·del·ic \ˌsī-kə-'de-lik\ *noun* [irregular from *psych-* + Greek *dēloun* to show, from *dēlos* evident; akin to Sanskrit *dīdeti* it shines, Latin *dies* day — more at DEITY] (1956)
: a psychedelic drug (as LSD)

[2]psychedelic *adjective* (1957)
1 a : of, relating to, or being drugs (as LSD) capable of producing abnormal psychic effects (as hallucinations) and sometimes psychic states resembling mental illness **b :** produced by or associated with the use of psychedelic drugs ⟨a *psychedelic* experience⟩
2 : imitating, suggestive of, or reproducing effects (as distorted or bizarre images or sounds) resembling those produced by psychedelic drugs ⟨*psychedelic* color schemes⟩
— **psy·che·del·i·cal·ly** \-'de-li-k(ə-)lē\ *adverb*

psy·chi·a·try \sə-'kī-ə-trē, sī-\ *noun* [probably from French *psychiatrie*, from *psychiatre* psychiatrist, from *psych-* psych- + Greek *iatros* physician — more at -IATRY] (circa 1846)
: a branch of medicine that deals with mental, emotional, or behavioral disorders
— **psy·chi·at·ric** \ˌsī-kē-'a-trik\ *adjective*
— **psy·chi·at·ri·cal·ly** \-tri-k(ə-)lē\ *adverb*
— **psy·chi·a·trist** \sə-'kī-ə-trist, sī-\ *noun*

[1]psy·chic \'sī-kik\ *also* **psy·chi·cal** \-ki-kəl\ *adjective* [Greek *psychikos* of the soul, from *psychē* soul] (1642)
1 : of or relating to the psyche **:** PSYCHOGENIC
2 : lying outside the sphere of physical science or knowledge **:** immaterial, moral, or spiritual in origin or force
3 : sensitive to nonphysical or supernatural forces and influences **:** marked by extraordinary or mysterious sensitivity, perception, or understanding
— **psy·chi·cal·ly** \-ki-k(ə-)lē\ *adverb*

[2]psychic *noun* (1871)
1 a : a person apparently sensitive to nonphysical forces **b :** MEDIUM 2d
2 : psychic phenomena

psy·cho \'sī-(ˌ)kō\ *noun, plural* **psychos** [short for *psychopath*] (1942)
: a deranged or psychopathic person — not used technically
— **psycho** *adjective*

psy·cho·acous·tics \ˌsī-kō-ə-'kü-stiks\ *noun plural but singular in construction* (1948)
: a branch of science dealing with hearing, the sensations produced by sounds, and the problems of communication
— **psy·cho·acous·tic** \-stik\ *adjective*

psy·cho·ac·tive \ˌsī-kō-'ak-tiv\ *adjective* (1961)
: affecting the mind or behavior ⟨*psychoactive* drugs⟩

psy·cho·anal·y·sis \ˌsī-kō-ə-'na-lə-səs\ *noun* [New Latin] (1906)
: a method of analyzing psychic phenomena and treating emotional disorders that involves treatment sessions during which the patient is encouraged to talk freely about personal experiences and especially about early childhood and dreams
— **psy·cho·an·a·lyst** \-'a-nᵊl-ist\ *noun*

psy·cho·an·a·lyt·ic \-ˌa-nᵊl-'i-tik\ *also* **psy·cho·an·a·lyt·i·cal** \-ti-kəl\ *adjective* (1906)
: of, relating to, or employing psychoanalysis or its principles and techniques
— **psy·cho·an·a·lyt·i·cal·ly** \-ti-k(ə-)lē\ *adverb*

psy·cho·an·a·lyze \-'a-nᵊl-ˌīz\ *transitive verb* (1911)
: to treat by means of psychoanalysis

psy·cho·bab·ble \'sī-kō-ˌba-bəl\ *noun* (1975)
: a predominantly metaphorical language for expressing one's feelings; *also* **:** psychological jargon
— **psy·cho·bab·bler** \-ˌba-blər\ *noun*

psy·cho·bi·og·ra·phy \ˌsī-kō-bī-'ä-grə-fē, -bē-\ *noun* (1931)
: a biography written from a psychodynamic or psychoanalytic point of view; *also* **:** the application of such a point of view when writing a biography
— **psy·cho·bi·og·ra·pher** \-fər\ *noun*
— **psy·cho·bio·graph·i·cal** \-ˌbī-ə-'gra-fi-kəl\ *adjective*

psy·cho·bi·ol·o·gy \-bī-'ä-lə-jē\ *noun* [International Scientific Vocabulary] (1902)
: the study of mental functioning and behavior in relation to other biological processes
— **psy·cho·bi·o·log·i·cal** \-ˌbī-ə-'lä-ji-kəl\ *also* **psy·cho·bi·o·log·ic** \-jik\ *adjective*
— **psy·cho·bi·ol·o·gist** \-bī-'ä-lə-jist\ *noun*

psy·cho·chem·i·cal \-'ke-mi-kəl\ *noun* (1956)
: a psychoactive chemical
— **psychochemical** *adjective*

psy·cho·dra·ma \-'drä-mə, -'dra-\ *noun* (1937)
1 : an extemporized dramatization designed to afford catharsis and social relearning for one or more of the participants from whose life history the plot is abstracted
2 : a dramatic narrative or event characterized by psychological overtones
— **psy·cho·dra·mat·ic** \-drə-'ma-tik\ *adjective*

psy·cho·dy·nam·ics \-dī-'na-miks, -də-\ *noun plural but singular or plural in construction* (1874)
1 : the psychology of mental or emotional forces or processes developing especially in early childhood and their effects on behavior and mental states
2 : explanation or interpretation (as of behavior or mental states) in terms of mental or emotional forces or processes
3 : motivational forces acting especially at the unconscious level
— **psy·cho·dy·nam·ic** \-mik\ *adjective*
— **psy·cho·dy·nam·i·cal·ly** \-mi-k(ə-)lē\ *adverb*

psy·cho·gen·e·sis \-'je-nə-səs\ *noun* [New Latin] (1838)
1 : the origin and development of mental functions, traits, or states
2 : development from mental as distinguished from physical origins
— **psy·cho·ge·net·ic** \-jə-'ne-tik\ *adjective*

psy·cho·gen·ic \-'je-nik\ *adjective* (1902)
: originating in the mind or in mental or emotional conflict
— **psy·cho·gen·i·cal·ly** \-ni-k(ə-)lē\ *adverb*

psy·cho·graph \'sī-kə-ˌgraf\ *noun* (1916)
: PSYCHOBIOGRAPHY

psy·cho·his·to·ry \'sī-kō-ˌhis-t(ə-)rē\ *noun* (1934)
: historical analysis or interpretation employing psychological and psychoanalytic methods; *also* **:** a work of history employing such methods
— **psy·cho·his·to·ri·an** \ˌsī-kō-(h)is-'tȯr-ē-ən, -'tȯr-, -'tär-\ *noun*
— **psy·cho·his·to·ri·cal** \-'tȯr-i-kəl, -'tär-\ *adjective*

psy·cho·ki·ne·sis \ˌsī-kō-kə-'nē-səs, -kī-\ *noun* [New Latin] (1914)
: movement of physical objects by the mind without use of physical means — compare PRECOGNITION, TELEKINESIS
— **psy·cho·ki·net·ic** \-'ne-tik\ *adjective*

psy·cho·lin·guis·tics \ˌsī-kō-liŋ-'gwis-tiks\ *noun plural but singular in construction* (1936)

: the study of the mental faculties involved in the perception, production, and acquisition of language
— **psy·cho·lin·guist** \-ˈliŋ-gwist\ *noun*
— **psy·cho·lin·guis·tic** \-tik\ *adjective*

psy·cho·log·i·cal \ˌsī-kə-ˈlä-ji-kəl\ *also* **psy·cho·log·ic** \-jik\ *adjective* (circa 1688)
1 a : of or relating to psychology **b :** MENTAL
2 : directed toward the will or toward the mind specifically in its conative function ⟨*psychological* warfare⟩
— **psy·cho·log·i·cal·ly** \-ji-k(ə-)lē\ *adverb*

psychological moment *noun* (1871)
: the occasion when the mental atmosphere is most certain to be favorable to the full effect of an action or event

psy·chol·o·gise *British variant of* PSYCHOLOGIZE

psy·chol·o·gism \sī-ˈkä-lə-ˌji-zəm\ *noun* (1858)
: a theory that applies psychological conceptions to the interpretation of historical events or logical thought

psy·chol·o·gize \-ˌjīz\ *verb* **-gized; -giz·ing** (1810)
intransitive verb
: to speculate in psychological terms or on psychological motivations
transitive verb
: to explain or interpret in psychological terms

psy·chol·o·gy \-jē\ *noun, plural* **-gies** [New Latin *psychologia,* from *psych-* + *-logia* -logy] (1653)
1 : the science of mind and behavior
2 a : the mental or behavioral characteristics of an individual or group **b :** the study of mind and behavior in relation to a particular field of knowledge or activity
3 : a treatise on psychology
— **psy·chol·o·gist** \-jist\ *noun*

psy·cho·met·ric \ˌsī-kə-ˈme-trik\ *adjective* (1854)
: of or relating to psychometrics or psychometry
— **psy·cho·met·ri·cal·ly** \-tri-k(ə-)lē\ *adverb*

psy·cho·me·tri·cian \-mə-ˈtri-shən\ *noun* (circa 1939)
1 : a person (as a clinical psychologist) who is skilled in the administration and interpretation of objective psychological tests
2 : a psychologist who devises, constructs, and standardizes psychometric tests

psy·cho·met·rics \-ˈme-triks\ *noun plural but singular in construction* (circa 1924)
: the psychological theory or technique of mental measurement

psy·chom·e·try \sī-ˈkä-mə-trē\ *noun* (circa 1842)
1 : divination of facts concerning an object or its owner through contact with or proximity to the object
2 : PSYCHOMETRICS

psy·cho·mo·tor \ˌsī-kə-ˈmō-tər\ *adjective* [International Scientific Vocabulary] (1878)
: of or relating to motor action directly proceeding from mental activity

psy·cho·neu·ro·sis \ˌsī-kō-nu̇-ˈrō-səs, -nyu̇-\ *noun* [New Latin] (1883)
: NEUROSIS; *especially* **:** a neurosis based on emotional conflict in which an impulse that has been blocked seeks expression in a disguised response or symptom
— **psy·cho·neu·rot·ic** \-ˈrä-tik\ *adjective or noun*

psy·cho·path \ˈsī-kə-ˌpath\ *noun* [International Scientific Vocabulary] (1885)
: a mentally ill or unstable person; *especially* **:** a person having a psychopathic personality

¹psy·cho·path·ic \ˌsī-kə-ˈpa-thik\ *adjective* (1847)
: of, relating to, or characterized by psychopathy
— **psy·cho·path·i·cal·ly** \-thi-k(ə-)lē\ *adverb*

²psychopathic *noun* (circa 1890)
: PSYCHOPATH

psychopathic personality *noun* (circa 1923)
1 : an emotionally and behaviorally disordered state characterized by clear perception of reality except for the individual's social and moral obligations and often by the pursuit of immediate personal gratification in criminal acts, drug addiction, or sexual perversion
2 : an individual having a psychopathic personality

psy·cho·pa·thol·o·gy \ˌsī-kō-pə-ˈthä-lə-jē, -pa-\ *noun* [International Scientific Vocabulary] (1847)
: the study of psychological and behavioral dysfunction occurring in mental disorder or in social disorganization; *also* **:** such dysfunction
— **psy·cho·path·o·log·ic** \-ˌpa-thə-ˈlä-jik\ *or* **psy·cho·path·o·log·i·cal** \-ji-kəl\ *adjective*
— **psy·cho·path·o·log·i·cal·ly** \-ji-k(ə-)lē\ *adverb*
— **psy·cho·pa·thol·o·gist** \-pə-ˈthä-lə-jist, -pa-\ *noun*

psy·chop·a·thy \sī-ˈkä-pə-thē\ *noun* [International Scientific Vocabulary] (1847)
: mental disorder; *especially* **:** extreme mental disorder marked usually by egocentric and antisocial activity

psy·cho·phar·ma·col·o·gy \ˌsī-kō-ˌfär-mə-ˈkä-lə-jē\ *noun* (1920)
: the study of the effect of drugs on the mind and behavior
— **psy·cho·phar·ma·co·log·i·cal** \-mə-kə-ˈlä-ji-kəl\ *or* **psy·cho·phar·ma·co·log·ic** \-jik\ *adjective*
— **psy·cho·phar·ma·col·o·gist** \-ˈkä-lə-jist\ *noun*

psy·cho·phys·i·cal \ˌsī-kō-ˈfi-zi-kəl\ *adjective* (1847)
: of or relating to psychophysics; *also* **:** sharing mental and physical qualities
— **psy·cho·phys·i·cal·ly** \-k(ə-)lē\ *adverb*

psychophysical parallelism *noun* (1894)
: PARALLELISM 4

psy·cho·phys·ics \ˌsī-kō-ˈfi-ziks\ *noun plural but singular in construction* [International Scientific Vocabulary] (1878)
: a branch of psychology concerned with the effect of physical processes (as intensity of stimulation) on the mental processes of an organism
— **psy·cho·phys·i·cist** \-ˈfi-zə-sist, -ˈfiz-sist\ *noun*

psy·cho·phys·i·o·log·i·cal \ˌsī-kō-ˌfi-zē-ə-ˈlä-ji-kəl\ *also* **psy·cho·phys·i·o·log·ic** \-jik\ *adjective* (1839)
1 : of or relating to physiological psychology
2 : combining or involving mental and bodily processes
— **psy·cho·phys·i·o·log·i·cal·ly** \-ji-k(ə-)lē\ *adverb*

psy·cho·phys·i·ol·o·gy \-ˌfi-zē-ˈä-lə-jē\ *noun* [International Scientific Vocabulary] (1839)
: PHYSIOLOGICAL PSYCHOLOGY
— **psy·cho·phys·i·ol·o·gist** \-jist\ *noun*

psy·cho·sex·u·al \-ˈsek-shə-wəl, -shwəl, -shəl\ *adjective* (1897)
1 : of or relating to the mental, emotional, and behavioral aspects of sexual development
2 : of or relating to mental or emotional attitudes concerning sexual activity
3 : of or relating to the physiological psychology of sex
— **psy·cho·sex·u·al·ly** *adverb*

psy·cho·sex·u·al·i·ty \-ˌsek-shə-ˈwa-lə-tē\ *noun* (1910)
: the psychic factors of sex

psy·cho·sis \sī-ˈkō-səs\ *noun, plural* **-cho·ses** \-ˌsēz\ [New Latin] (1847)
: fundamental mental derangement (as schizophrenia) characterized by defective or lost contact with reality

psy·cho·so·cial \ˌsī-kō-ˈsō-shəl\ *adjective* (1899)
1 : involving both psychological and social aspects ⟨*psychosocial* adjustment in marriage⟩
2 : relating social conditions to mental health ⟨*psychosocial* medicine⟩
— **psy·cho·so·cial·ly** \-ˈsō-sh(ə-)lē\ *adverb*

psy·cho·so·mat·ic \-sə-ˈma-tik\ *adjective* [International Scientific Vocabulary] (1863)
1 : of, relating to, concerned with, or involving both mind and body ⟨the *psychosomatic* nature of man —Herbert Ratner⟩
2 : of, relating to, involving, or concerned with bodily symptoms caused by mental or emotional disturbance ⟨*psychosomatic* illness⟩ ⟨*psychosomatic* medicine⟩
— **psy·cho·so·mat·i·cal·ly** \-ti-k(ə-)lē\ *adverb*

psy·cho·so·mat·ics \-sə-ˈma-tiks\ *noun plural but singular in construction* (1938)
: a branch of medical science dealing with interrelationships between the mind or emotions and the body and especially with the relation of psychic conflict to somatic symptomatology

psy·cho·sur·gery \-ˈsər-jə-rē, -ˈsərj-rē\ *noun* (1936)
: cerebral surgery employed in treating psychic symptoms
— **psy·cho·sur·geon** \-ˈsər-jən\ *noun*
— **psy·cho·sur·gi·cal** \-ˈsər-ji-kəl\ *adjective*

psy·cho·syn·the·sis \-ˈsin(t)-thə-səs\ *noun* (1919)
: a form of psychotherapy combining psychoanalytic techniques with meditation and exercise

psy·cho·ther·a·peu·tic \-ˌther-ə-ˈpyü-tik\ *adjective* [International Scientific Vocabulary] (circa 1888)
: of, relating to, or used in psychotherapy
— **psy·cho·ther·a·peu·ti·cal·ly** \-ti-k(ə-)lē\ *adverb*

psy·cho·ther·a·py \-ˈther-ə-pē\ *noun* [International Scientific Vocabulary] (circa 1890)
: treatment of mental or emotional disorder or of related bodily ills by psychological means
— **psy·cho·ther·a·pist** \-pist\ *noun*

psy·chot·ic \sī-ˈkä-tik\ *adjective* (circa 1890)
: of, relating to, marked by, or affected with psychosis ⟨a *psychotic* patient⟩ ⟨*psychotic* behavior⟩
— **psychotic** *noun*
— **psy·chot·i·cal·ly** \-ti-k(ə-)lē\ *adverb*

psy·choto·mi·met·ic \sī-ˌkä-tō-mə-ˈme-tik, -mī-\ *adjective* [*psychotic* + *-o-* + *mimetic*] (1957)
: of, relating to, involving, or inducing psychotic alteration of behavior and personality ⟨*psychotomimetic* drugs⟩
— **psychotomimetic** *noun*
— **psy·choto·mi·met·i·cal·ly** \-ti-k(ə-)lē\ *adverb*

psy·cho·tro·pic \ˌsī-kə-ˈtrō-pik\ *adjective* (1948)
: acting on the mind ⟨*psychotropic* drugs⟩
— **psychotropic** *noun*

psych–out \ˈsīk-ˌau̇t\ *noun* (1971)
: an act or an instance of psyching someone out

psychro- *combining form* [Greek, from *psychros,* from *psychein* to cool]
: cold ⟨*psychrometer*⟩

psy·chrom·e·ter \sī-ˈkrä-mə-tər\ *noun* [International Scientific Vocabulary] (1838)
: a hygrometer consisting essentially of two similar thermometers with the bulb of one being kept wet so that the cooling that results from evaporation makes it register a lower

temperature than the dry one and with the difference between the readings constituting a measure of the dryness of the atmosphere
— **psy·chro·met·ric** \ˌsī-krə-ˈme-trik\ *adjective*
— **psy·chrom·e·try** \sī-ˈkrä-mə-trē\ *noun*

psy·chro·phil·ic \ˌsī-krō-ˈfi-lik\ *adjective* (circa 1903)
: thriving at a relatively low temperature ⟨*psychrophilic* bacteria⟩

psyl·la \ˈsi-lə\ *noun* [New Latin, genus name, from Greek, flea; akin to Latin *pulex* flea, Sanskrit *pluṣi*] (1852)
: any of various plant lice (family Psyllidae) including economically important plant pests — compare PEAR PSYLLA

psyl·lid \ˈsi-ləd\ *noun* [ultimately from New Latin *Psylla*] (1899)
: PSYLLA
— **psyllid** *adjective*

psyl·li·um seed \ˈsi-lē-əm-\ *noun* [New Latin *psyllium,* from Greek *psyllion* fleawort, from *psylla*] (1897)
: the seed of a fleawort (especially *Plantago psyllium*) that has the property of swelling and becoming gelatinous when moist and is used as a mild laxative — called also *psyllium*

psy·war \ˈsī-ˌwȯr\ *noun* [by shortening] (1951)
: psychological warfare

ptar·mi·gan \ˈtär-mi-gən\ *noun, plural* **-gan** *or* **-gans** [modification of Scottish Gaelic *tarmachan*] (1599)
: any of various grouses (genus *Lagopus*) of northern regions with completely feathered feet

ptarmigan: *A* summer plumage; *B* winter plumage

PT boat \ˌpē-ˈtē-, ˈpē-ˌ\ *noun* [patrol torpedo] (1941)
: a small fast patrol craft usually armed with torpedoes, machine guns, and depth charges — called also *PT*

PTC \ˌpē-(ˌ)tē-ˈsē\ *noun* (1932)
: PHENYLTHIOCARBAMIDE

pter·an·o·don \tə-ˈra-nə-ˌdän, -ˈrä-\ *noun* [New Latin, from Greek *pteron* wing + *anodōn* toothless from *an-* + *odōn, odous* tooth — more at FEATHER, TOOTH] (1897)
: any of a genus (*Pteranodon*) of Cretaceous pterosaurs having a backwardly directed bony crest on the skull and a wingspan of about 25 feet (7.7 meters)

pterid- *or* **pterido-** *combining form* [Greek *pterid-, pteris;* akin to Greek *pteron* wing, feather]
: fern ⟨*pterido*logy⟩

pter·i·dine \ˈter-ə-ˌdēn\ *noun* [International Scientific Vocabulary *pter*in + *-idine*] (circa 1943)
: a yellow crystalline bicyclic base $C_6H_4N_4$ that is a structural constituent especially of various animal pigments

pter·i·dol·o·gy \ˌter-ə-ˈdä-lə-jē\ *noun* (1855)
: the study of ferns
— **pter·i·do·log·i·cal** \ˌter-ə-də-ˈlä-ji-kəl\ *adjective*
— **pter·i·dol·o·gist** \-ˈdä-lə-jist\ *noun*

pte·ri·do·phyte \tə-ˈri-də-ˌfīt, ˈter-ə-dō-\ *noun* [New Latin *Pteridophyta,* from Greek *pterid-, pteris* fern + *phyton* plant — more at PHYT-] (1880)
: any of a division (Pteridophyta) of vascular plants (as a fern) that have roots, stems, and leaves but lack flowers or seeds

pte·ri·do·sperm \tə-ˈri-də-ˌspərm, ˈter-ə-dō-\ *noun* [International Scientific Vocabulary] (1904)
: SEED FERN

pter·in \ˈter-ən\ *noun* [International Scientific Vocabulary *pter-* (from Greek *pteron* wing) + [1]*-in;* from its being a factor in the pigments of butterfly wings] (1934)
: any of various compounds that contain the bicyclic ring system characteristic of pteridine

ptero·dac·tyl \ˌter-ə-ˈdak-tᵊl\ *noun* [New Latin *Pterodactylus,* genus of reptiles, from Greek *pteron* wing + *daktylos* finger — more at FEATHER] (1830)
: any of various pterosaurs (suborder Pterodactyloidea) of the Late Jurassic and Cretaceous having a rudimentary tail and a beak with reduced dentition; *broadly* : PTEROSAUR

ptero·pod \ˈter-ə-ˌpäd\ *noun* [New Latin *Pteropoda,* group name, from Greek *pteron* wing + New Latin *-poda*] (1835)
: any of the opisthobranch mollusks comprising two orders (Thecosomata and Gymnosomata) and having the anterior lobes of the foot expanded into broad thin winglike swimming organs

ptero·saur \ˈter-ə-ˌsȯr\ *noun* [New Latin *Pterosauria,* from Greek *pteron* wing + *sauros* lizard] (1862)
: any of an order (Pterosauria) of extinct flying reptiles existing from the Late Triassic throughout the Jurassic and most of the Cretaceous and having a featherless wing membrane extending from the side of the body along the arm to the end of the greatly elongated fourth digit

pter·o·yl·glu·tam·ic acid \ˌter-ə-ˌwil-glü-ˈta-mik-\ *noun* [International Scientific Vocabulary *pteroyl* (the radical $(C_{13}H_{11}N_6O)CO$) + *glutamic acid*] (1943)
: FOLIC ACID

[1]**pter·y·goid** \ˈter-ə-ˌgȯid\ *adjective* [New Latin *pterygoides,* from Greek *pterygoeidēs,* literally, shaped like a wing, from *pteryg-, pteryx* wing; akin to Greek *pteron* wing — more at FEATHER] (1722)
: of, relating to, or lying in the region of the inferior part of the sphenoid bone of the vertebrate skull

[2]**pterygoid** *noun* (1831)
: a pterygoid part (as a bone, muscle, or nerve)

pterygoid process *noun* (1741)
: a process extending downward from each side of the sphenoid bone in humans and other mammals

pter·y·la \ˈter-ə-lə\ *noun, plural* **-lae** \-ˌlē, -ˌlī\ [New Latin, from Greek *pteron* + *hylē* wood, forest] (1867)
: one of the definite areas of the skin of a bird on which feathers grow

Ptol·e·ma·ic \ˌtä-lə-ˈmā-ik\ *adjective* [Greek *Ptolemaikos,* from *Ptolemaios* Ptolemy] (1674)
1 : of or relating to Ptolemy the geographer and astronomer who flourished at Alexandria about A.D. 130
2 : of or relating to the Greco-Egyptian Ptolemies ruling Egypt from 323 to 30 B.C.

Ptolemaic system *noun* (circa 1771)
: the system of planetary motions according to which the earth is at the center with the sun, moon, and planets revolving around it

pto·maine \ˈtō-ˌmān, tō-ˈ\ *noun* [Italian *ptomaina,* from Greek *ptōma* fall, fallen body, corpse, from *piptein* to fall — more at FEATHER] (1880)
: any of various organic bases which are formed by the action of putrefactive bacteria on nitrogenous matter and some of which are poisonous

ptomaine poisoning *noun* (1893)
: food poisoning caused by bacteria or bacterial products

pto·sis \ˈtō-səs\ *noun, plural* **pto·ses** \-ˌsēz\ [New Latin, from Greek *ptōsis* act of falling, from *piptein*] (1743)
: a sagging or prolapse of an organ or part; *especially* : a drooping of the upper eyelid

pty·a·lin \ˈtī-ə-lən\ *noun* [Greek *ptyalon* saliva, from *ptyein* to spit — more at SPEW] (1845)
: an amylase found in the saliva of many animals that converts starch into sugar

pty·a·lism \-ˌli-zəm\ *noun* [New Latin *ptyalismus,* from Greek *ptyalismos,* from *ptyalizein* to salivate, from *ptyalon*] (1676)
: an excessive flow of saliva

p–type \ˈpē-ˌtīp\ *adjective* [*p*ositive *type*] (1946)
: relating to or being a semiconductor in which charge is carried by holes — compare N-TYPE

pub \ˈpəb\ *noun* (circa 1859)
1 *chiefly British* : PUBLIC HOUSE 2
2 : an establishment where alcoholic beverages are sold and consumed

pub crawler *noun* (1910)
: one who goes from bar to bar
— **pub–crawl** *intransitive verb*
— **pub crawl** *noun*

pu·ber·tal \ˈpyü-bər-tᵊl\ *or* **pu·ber·al** \ˈpyü-bə-rəl\ *adjective* [*pubertal* from *puberty; puberal* from Medieval Latin *puberalis,* from Latin *puber*] (circa 1837)
: of or relating to puberty

pu·ber·ty \ˈpyü-bər-tē\ *noun* [Middle English *puberte,* from Latin *pubertas,* from *puber* pubescent] (14th century)
1 : the condition of being or the period of becoming first capable of reproducing sexually marked by maturing of the genital organs, development of secondary sex characteristics, and in the human and in higher primates by the first occurrence of menstruation in the female
2 : the age at which puberty occurs often construed legally as 14 in boys and 12 in girls

pu·ber·u·lent \pyü-ˈber-ə-lənt, -yə-lənt\ *adjective* [Latin *puber* pubescent + English *-ulent* (as in *pulverulent*)] (circa 1859)
: covered with fine pubescence

pu·bes \ˈpyü-(ˌ)bēz\ *noun, plural* **pubes** [New Latin, from Latin, manhood, body hair, pubic region; akin to Latin *puber* pubescent] (circa 1570)
1 : the hair that appears on the lower part of the hypogastric region at puberty
2 : the pubic region

pu·bes·cence \pyü-ˈbe-sᵊn(t)s\ *noun* (15th century)
1 : the quality or state of being pubescent
2 : a pubescent covering or surface

pu·bes·cent \-sᵊnt\ *adjective* [Latin *pubescent-, pubescens,* present participle of *pubescere* to reach puberty, become covered as with hair, from *pubes*] (1646)
1 a : arriving at or having reached puberty **b :** of or relating to puberty
2 : covered with fine soft short hairs — compare VILLOUS

pu·bic \ˈpyü-bik\ *adjective* (1831)
: of, relating to, or situated in or near the region of the pubes or the pubis

pu·bis \ˈpyü-bəs\ *noun, plural* **pu·bes** \-(ˌ)bēz\ [New Latin *os pubis,* literally, bone of the pubic region] (1597)
: the ventral and anterior of the three principal bones composing either half of the pelvis — called also *pubic bone*

[1]**pub·lic** \ˈpə-blik\ *adjective* [Middle English *publique,* from Middle French, from Latin *publicus;* akin to Latin *populus* the people] (14th century)
1 a : exposed to general view : OPEN **b :** WELL-KNOWN, PROMINENT **c :** PERCEPTIBLE, MATERIAL
2 a : of, relating to, or affecting all the people or the whole area of a nation or state ⟨*public* law⟩ **b :** of or relating to a government **c :** of, relating to, or being in the service of the community or nation
3 a : of or relating to people in general : UNIVERSAL **b :** GENERAL, POPULAR
4 : of or relating to business or community interests as opposed to private affairs : SOCIAL

5 : devoted to the general or national welfare **:** HUMANITARIAN
6 a : accessible to or shared by all members of the community **b :** capitalized in shares that can be freely traded on the open market — often used with *go*
— pub·lic·ness *noun*
²public *noun* (15th century)
1 : a place accessible or visible to the public — usually used in the phrase *in public*
2 : the people as a whole **:** POPULACE
3 : a group of people having common interests or characteristics; *specifically* **:** the group at which a particular activity or enterprise aims
public–address system *noun* (1923)
: an apparatus including a microphone and loudspeakers used for broadcasting to a large audience in an auditorium or outdoors
pub·li·can \'pə-bli-kən\ *noun* [Middle English, from Old French, from Latin *publicanus* tax farmer, from *publicum* public revenue, from neuter of *publicus*] (13th century)
1 a : a Jewish tax collector for the ancient Romans **b :** a collector of taxes or tribute
2 *chiefly British* **:** the licensee of a public house
public assistance *noun* (1884)
: government aid to needy, blind, aged, or disabled persons and to dependent children
pub·li·ca·tion \ˌpə-blə-'kā-shən\ *noun* [Middle English *publicacioun,* from Middle French *publication,* from Latin *publication-, publicatio,* from *publicare,* from *publicus* public] (14th century)
1 : the act or process of publishing
2 : a published work
public defender *noun* (1918)
: a lawyer usually holding public office whose duty is to defend accused persons unable to pay for legal assistance
public domain *noun* (1832)
1 : land owned directly by the government
2 : the realm embracing property rights that belong to the community at large, are unprotected by copyright or patent, and are subject to appropriation by anyone
public health *noun* (1617)
: the art and science dealing with the protection and improvement of community health by organized community effort and including preventive medicine and sanitary and social science
public house *noun* (1658)
1 : INN, HOSTELRY
2 *chiefly British* **:** a licensed saloon or bar
pub·li·cise *British variant of* PUBLICIZE
pub·li·cist \'pə-blə-sist\ *noun* (1792)
1 a : an expert in international law **b :** an expert or commentator on public affairs
2 : one that publicizes; *specifically* **:** PRESS AGENT
pub·lic·i·ty \(ˌ)pə-'bli-sə-tē, -'blis-tē\ *noun* (1791)
1 : the quality or state of being public
2 a : an act or device designed to attract public interest; *specifically* **:** information with news value issued as a means of gaining public attention or support **b :** the dissemination of information or promotional material **c :** paid advertising **d :** public attention or acclaim
pub·li·cize \'pə-blə-ˌsīz\ *transitive verb* **-cized; -ciz·ing** (1925)
: to bring to the attention of the public **:** ADVERTISE
public land *noun* (1789)
: land owned by a government; *specifically* **:** that part of the U.S. public domain subject to sale or disposal under the homestead laws
public law *noun* (1773)
1 : a legislative enactment affecting the public at large
2 : a branch of law concerned with regulating the relations of individuals with the government and the organization and conduct of the government itself — compare PRIVATE LAW
pub·lic·ly \'pə-bli-klē\ *also* **pub·li·cal·ly** \-li-k(ə-)lē\ *adverb* (1567)
1 : in a manner observable by or in a place accessible to the public **:** OPENLY
2 a : by the people generally **b :** by a government
public officer *noun* (1925)
: a person who has been legally elected or appointed to office and who exercises governmental functions
public relations *noun plural but usually singular in construction, often attributive* (1807)
: the business of inducing the public to have understanding for and goodwill toward a person, firm, or institution; *also* **:** the degree of understanding and goodwill achieved
public sale *noun* (1678)
: AUCTION 1
public school *noun* (1580)
1 : an endowed secondary boarding school in Great Britain offering a classical curriculum and preparation for the universities or public service
2 : a free tax-supported school controlled by a local governmental authority
public servant *noun* (1676)
: a government official or employee
public service *noun* (circa 1576)
1 : the business of supplying a commodity (as electricity or gas) or service (as transportation) to any or all members of a community
2 : a service rendered in the public interest
3 : governmental employment; *especially* **:** CIVIL SERVICE
public–service corporation *noun* (1904)
: a quasi-public corporation
public speaking *noun* (1762)
1 : the act or process of making speeches in public
2 : the art of effective oral communication with an audience
public–spirited *adjective* (1677)
: motivated by devotion to the general welfare
— pub·lic–spir·it·ed·ness *noun*
public television *noun* (1965)
: television supported by public funds and private contributions rather than by commercials
public utility *noun* (1903)
: a business organization (as an electric company) performing a public service and subject to special governmental regulation
public works *noun plural* (1676)
: works (as schools, highways, docks) constructed for public use or enjoyment especially when financed and owned by the government
pub·lish \'pə-blish\ *verb* [Middle English, modification of Middle French *publier,* from Latin *publicare,* from *publicus* public] (14th century)
transitive verb
1 a : to make generally known **b :** to make public announcement of
2 a : to disseminate to the public **b :** to produce or release for distribution; *specifically* **:** PRINT 2c **c :** to issue the work of (an author)
intransitive verb
1 : to put out an edition
2 : to have one's work accepted for publication
— pub·lish·able \-bli-shə-bəl\ *adjective*
pub·lish·er \'pə-bli-shər\ *noun* (15th century)
: one that publishes something; *especially* **:** a person or corporation whose business is publishing
pub·lish·ing \-shiŋ\ *noun* (1580)
: the business or profession of the commercial production and issuance of literature, information, musical scores or sometimes recordings, or art ⟨newspaper *publishing*⟩ ⟨software *publishing*⟩
puc·coon \(ˌ)pə-'kün\ *noun* [Virginia Algonquian *poughkone*] (1612)
1 : any of several American plants (as bloodroot) yielding a red or yellow pigment
2 : a pigment from a puccoon
puce \'pyüs\ *noun* [French, literally, flea, from Latin *pulic-, pulex* — more at PSYLLA] (1882)
: a dark red
¹puck \'pək\ *noun* [Middle English *puke,* from Old English *pūca;* akin to Old Norse *pūki* devil] (before 12th century)
1 *archaic* **:** an evil spirit **:** DEMON
2 : a mischievous sprite **:** HOBGOBLIN; *specifically, capitalized* **:** ROBIN GOODFELLOW
²puck *noun* [English dialect *puck* to poke, hit, probably from Irish *poc* butt, stroke in hurling, literally, buck (male deer)] (1891)
: a vulcanized rubber disk used in ice hockey
pucka *variant of* PUKKA
¹puck·er \'pə-kər\ *verb* **puck·ered; puck·er·ing** \-k(ə-)riŋ\ [probably irregular from ¹*poke*] (1598)
intransitive verb
: to become wrinkled or constricted
transitive verb
: to contract into folds or wrinkles
²pucker *noun* (circa 1750)
: a fold or wrinkle in a normally even surface
puck·ery \'pə-k(ə-)rē\ *adjective* (1830)
: that puckers or causes puckering
puck·ish \'pə-kish\ *adjective* [¹*puck*] (1874)
: IMPISH, WHIMSICAL
— puck·ish·ly *adverb*
— puck·ish·ness *noun*
pud \'pu̇d\ *noun* (1706)
British **:** PUDDING
pud·ding \'pu̇-diŋ\ *noun* [Middle English] (13th century)
1 : BLOOD SAUSAGE
2 a (1) **:** a boiled or baked soft food usually with a cereal base ⟨corn *pudding*⟩ (2) **:** a dessert of a soft, spongy, or thick creamy consistency ⟨chocolate *pudding*⟩ (3) *British* **:** DESSERT 1 **b :** a dish often containing suet or having a suet crust and originally boiled in a bag ⟨steak and kidney *pudding*⟩
pudding stone *noun* (1753)
: CONGLOMERATE
¹pud·dle \'pə-dᵊl\ *noun* [Middle English *podel;* akin to Low German *pudel* puddle, Old English *pudd* ditch] (14th century)
1 : a very small pool of usually dirty or muddy water
2 a : an earthy mixture (as of clay, sand, and gravel) worked while wet into a compact mass that becomes impervious to water when dry **b :** a thin mixture of soil and water for puddling plants
²puddle *verb* **pud·dled; pud·dling** \'pəd-liŋ, 'pə-dᵊl-iŋ\ (15th century)
intransitive verb
: to dabble or wade around in a puddle
transitive verb
1 : to make muddy or turbid **:** MUDDLE
2 a : to work (a wet mixture of earth or concrete) into a dense impervious mass **b :** to subject (iron) to the process of puddling
3 a : to strew with puddles **b :** to compact (soil) especially by working when too wet **c :** to dip the roots of (a plant) in a thin mud before transplanting
— pud·dler \'pəd-lər, 'pə-dᵊl-ər\ *noun*
puddle duck *noun* (1877)
: DABBLER b
puddle jumper *noun* (1942)
slang **:** LIGHTPLANE
pud·dling \'pəd-liŋ, 'pə-dᵊl-iŋ\ *noun* (1839)
: the process of converting pig iron into wrought iron or rarely steel by subjecting it to heat and frequent stirring in a furnace in the presence of oxidizing substances
pu·den·cy \'pyü-dᵊn(t)-sē\ *noun* [Latin *pudentia,* from *pudent-, pudens,* present participle of *pudēre* to be ashamed, make ashamed] (1611)
: MODESTY

pu·den·dum \pyu̇-ˈden-dəm\ *noun, plural* **-da** \-də\ [New Latin, singular of Latin *pudenda,* from neuter plural of *pudendus,* gerundive of *pudēre* to be ashamed] (1634)
: the external genital organs of a human being and especially of a woman — usually used in plural
— pu·den·dal \-dᵊl\ *adjective*

pudgy \ˈpə-jē\ *adjective* **pudg·i·er; -est** [origin unknown] (1836)
: being short and plump **:** CHUBBY
— pudg·i·ness *noun*

pu·di·bund \ˈpyü-də-ˌbənd\ *adjective* [Latin *pudibundus,* from *pudēre* to be ashamed + *-bundus* (as in *moribundus* moribund)] (circa 1656)
: PRUDISH

pueb·lo \ˈpwe-(ˌ)blō; pü-ˈe-, pyü-\ *noun, plural* **-los** [Spanish, village, literally, people, from Latin *populus*] (1808)
1 a : the communal dwelling of an Indian village of Arizona, New Mexico, and adjacent areas consisting of contiguous flat-roofed stone or adobe houses in groups sometimes several stories high **b :** an Indian village of the southwestern U.S.
2 *capitalized* **:** a member of a group of Indian peoples of the southwestern U.S.

pu·er·ile \ˈpyu̇(-ə)r-əl, -ˌīl\ *adjective* [French or Latin; French *puéril,* from Latin *puerilis,* from *puer* boy, child; akin to Sanskrit *putra* son, child and perhaps to Greek *pais* boy, child — more at FEW] (1661)
1 : JUVENILE
2 : CHILDISH, SILLY ⟨*puerile* remarks⟩
— pu·er·ile·ly \-ə(l)-lē, -ˌīl-lē\ *adverb*
— pu·er·il·i·ty \ˌpyu̇(-ə)r-ˈi-lə-tē\ *noun*

pu·er·il·ism \ˈpyu̇(-ə)r-ə-ˌli-zəm, ˈpyu̇(-ə)r-ˌī-\ *noun* (1924)
: childish behavior especially as a symptom of mental disorder

pu·er·per·al \pyü-ˈər-p(ə-)rəl\ *adjective* [Latin *puerpera* woman in childbirth, from *puer* child + *parere* to give birth to — more at PARE] (1768)
: of, relating to, or occurring during childbirth or the period immediately following ⟨*puerperal* infection⟩ ⟨*puerperal* depression⟩

puerperal fever *noun* (1768)
: an abnormal condition that results from infection of the placental site following delivery or abortion and is characterized in mild form by fever but in serious cases the infection may spread through the uterine wall or pass into the bloodstream — called also *childbed fever, puerperal sepsis*

pu·er·pe·ri·um \ˌpyü-ər-ˈpir-ē-əm\ *noun, plural* **-ria** \-ē-ə\ [Latin, from *puerpera*] (circa 1890)
: the period between childbirth and the return of the uterus to its normal size

[1]puff \ˈpəf\ *verb* [Middle English, from Old English *pyffan,* of imitative origin] (before 12th century)
intransitive verb
1 a (1) **:** to blow in short gusts (2) **:** to exhale forcibly **b :** to breathe hard **:** PANT **c :** to emit small whiffs or clouds (as of smoke) often as an accompaniment to vigorous action ⟨*puff* at a pipe⟩
2 : to speak or act in a scornful, conceited, or exaggerated manner
3 a : to become distended **:** SWELL — usually used with *up* **b :** to open or appear in or as if in a puff
4 : to form a chromosomal puff
transitive verb
1 a : to emit, propel, blow, or expel by or as if by puffs **:** WAFT **b :** to draw on (as a cigar, cigarette, or pipe) with intermittent exhalations of smoke
2 a : to distend with or as if with air or gas **:** INFLATE **b :** to make proud or conceited **:** ELATE **c** (1) **:** to praise extravagantly and usually with exaggeration (2) **:** ADVERTISE

[2]puff *noun* (13th century)
1 a : an act or instance of puffing **:** WHIFF **b :** a slight explosive sound accompanying a puff **c :** a perceptible cloud or aura emitted in a puff **d :** DRAW 1a
2 : a light round hollow pastry
3 a : a slight swelling **:** PROTUBERANCE **b :** a fluffy mass: as (1) **:** POUF 2 (2) **:** a small fluffy pad for applying cosmetic powder (3) **:** a soft loose roll of hair (4) **:** a quilted bed covering
4 : a commendatory or promotional notice or review
5 : an enlarged region of a chromosome that is associated with intensely active genes involved in RNA synthesis
— puff·i·ness \ˈpə-fē-nəs\ *noun*
— puffy \ˈpə-fē\ *adjective*

puff adder *noun* (1789)
1 : a thick-bodied extremely venomous African viper (*Bitis arietans*)
2 : HOGNOSE SNAKE

puff·ball \ˈpəf-ˌbȯl\ *noun* (1649)
: any of various globose and often edible fungi (especially family Lycoperdaceae) that discharge ripe spores in a smokelike cloud when pressed or struck

puff·er \ˈpə-fər\ *noun* (1629)
1 : one that puffs
2 a : any of a family (Tetraodontidae) of chiefly tropical marine bony fishes which can distend themselves to a globular form and most of which are highly poisonous — called also *blowfish, globefish* **b :** any of various fish of the same order (Tetraodontiformes synonym Plectognathi) as the puffers

puff·ery \ˈpə-f(ə-)rē\ *noun* (1782)
: exaggerated commendation especially for promotional purposes **:** HYPE

puf·fin \ˈpə-fən\ *noun* [Middle English *pophyn*] (14th century)
: any of several seabirds (genera *Fratercula* and *Lunda*) having a short neck and a deep grooved parti-colored laterally compressed bill

puffin

puff pastry *noun* (1611)
: a pastry dough containing many alternating layers of butter and dough or the light flaky pastry made from it — called also *puff paste*

[1]pug \ˈpəg\ *noun* [obsolete *pug* hobgoblin, monkey] (1789)
1 : a small sturdy compact dog of a breed of Asian origin with a close coat, tightly curled tail, and broad wrinkled face
2 a : PUG NOSE **b :** a close knot or coil of hair **:** BUN

pug 1

[2]pug *transitive verb* **pugged; pug·ging** [origin unknown] (1843)
: to work and mix (as clay) when wet especially to make more homogeneous and easier to handle (as in throwing or molding wares)

[3]pug *noun* [by shortening & alteration from *pugilist*] (1858)
: [1]BOXER 1

[4]pug *noun* [Hindi *pag* foot] (1865)
: FOOTPRINT; *especially* **:** a print of a wild mammal

pug·ga·ree *or* **pug·a·ree** *or* **pug·gree** \ˈpə-g(ə-)rē\ *noun* [Hindi *pagṛī* turban] (1665)
: a light scarf wrapped around a sun helmet or used as a hatband

pu·gi·lism \ˈpyü-jə-ˌli-zəm\ *noun* [Latin *pugil* boxer; akin to Latin *pugnus* fist — more at PUNGENT] (1791)
: [2]BOXING
— pu·gi·lis·tic \ˌpyü-jə-ˈlis-tik\ *adjective*

pu·gi·list \ˈpyü-jə-list\ *noun* (1790)
: FIGHTER; *especially* **:** a professional boxer

pug·mark \ˈpəg-ˌmärk\ *noun* (1922)
: [4]PUG

pug mill *noun* [[2]*pug*] (1824)
: a machine in which materials (as clay and water) are mixed, blended, or kneaded into a desired consistency

pug·na·cious \ˌpəg-ˈnā-shəs\ *adjective* [Latin *pugnac-, pugnax,* from *pugnare* to fight — more at PUNGENT] (1642)
: having a quarrelsome or combative nature **:** TRUCULENT
synonym see BELLIGERENT
— pug·na·cious·ly *adverb*
— pug·na·cious·ness *noun*
— pug·nac·i·ty \-ˈna-sə-tē\ *noun*

pug nose *noun* [[1]*pug*] (1778)
: a nose having a slightly concave bridge and flattened nostrils
— pug–nosed \ˈpəg-ˌnōzd\ *adjective*

puis·ne \ˈpyü-nē\ *adjective* [Middle French *puisné* younger — more at PUNY] (1688)
chiefly British **:** inferior in rank ⟨*puisne* judge⟩
— puisne *noun*

puis·sance \ˈpwi-sᵊn(t)s, ˈpyü-ə-sən(t)s, pyü-ˈi-sᵊn(t)s\ *noun* [Middle English, from Middle French, from Old French, from *puissant* powerful, from *poeir* to be able, be powerful — more at POWER] (15th century)
: STRENGTH, POWER

puis·sant \-sᵊnt, -sənt\ *adjective* (15th century)
: having puissance **:** POWERFUL

puke \ˈpyük\ *verb* **puked; puk·ing** [origin unknown] (1600)
: VOMIT
— puke *noun*

puk·ka \ˈpə-kə\ *adjective* [Hindi *pakkā* cooked, ripe, solid, from Sanskrit *pakva;* akin to Greek *pessein* to cook — more at COOK] (1698)
: GENUINE, AUTHENTIC; *also* **:** FIRST-CLASS

pul \ˈpül\ *noun, plural* **puls** \ˈpülz\ *or* **pul** [Persian *pūl*] (1927)
— see *afghani* at MONEY table

pu·la \ˈpü-lə, ˈpyü-\ *noun, plural* **pula** [Tswana, literally, rain (used as a greeting)] (1976)
— see MONEY table

Pu·las·ki \pə-ˈlas-kē, pyü-\ *noun* [Edward C. *Pulaski,* 20th century American forest ranger] (1924)
: a single-bit ax with an adze-shaped hoe extending from the back

pul·chri·tude \ˈpəl-krə-ˌtüd, -ˌtyüd\ *noun* [Middle English, from Latin *pulchritudin-, pulchritudo,* from *pulchr-, pulcher* beautiful] (15th century)
: physical comeliness
— pul·chri·tu·di·nous \ˌpəl-krə-ˈtüd-nəs, -ˈtyüd-; -ˈtü-dᵊn-əs, -ˈtyü-\ *adjective*

pule \ˈpyü(ə)l\ *intransitive verb* **puled; pul·ing** [probably imitative] (1534)
: WHINE, WHIMPER

pu·li \ˈpu̇-lē, ˈpyü-\ *noun, plural* **pu·lik** \-lik\ *or* **pulis** \-lēz\ [Hungarian] (1936)
: any of a breed of medium-sized Hungarian sheepdogs with a thick woolly coat hanging in long thin cords

Pu·lit·zer prize \ˈpu̇-lət-sər-, ˈpyü-\ *noun* (1918)
: any of various annual prizes (as for outstanding literary or journalistic achievement) established by the will of Joseph Pulitzer — called also *Pulitzer*

[1]pull \ˈpu̇l *also* ˈpəl\ *verb* [Middle English, from Old English *pullian;* akin to Middle Low German *pulen* to shell, cull] (before 12th century)
transitive verb
1 a : to exert force upon so as to cause or tend to cause motion toward the force **b :** to stretch (cooling candy) repeatedly ⟨*pull* taffy⟩ **c :** to strain abnormally ⟨*pull* a tendon⟩ **d :** to hold back (a racehorse) from winning **e :** to work (an oar) by drawing back strongly

2 a : to draw out from the skin ⟨*pull* feathers from a rooster's tail⟩ **b :** to pluck from a plant or by the roots ⟨*pull* flowers⟩ ⟨*pull* turnips⟩ **c :** EXTRACT ⟨*pull* a tooth⟩
3 : to hit (a ball) toward the left from a right-handed swing or toward the right from a left-handed swing — compare PUSH
4 : to draw apart **:** REND, TEAR
5 : to print (as a proof) by impression
6 : to remove from a place or situation ⟨*pull* the engine⟩ ⟨*pulled* the pitcher in the third inning⟩ ⟨*pulled* the show⟩
7 : to bring (a weapon) into the open ⟨*pulled* a knife⟩
8 a : COMMIT, PERPETRATE ⟨*pull* a robbery⟩ ⟨*pull* a prank⟩ **b :** to carry out as an assignment or duty
9 : PUT ON, ASSUME ⟨*pull* a grin⟩
10 a : to draw the support or attention of **:** ATTRACT ⟨*pull* votes⟩ — often used with *in* **b :** OBTAIN, SECURE ⟨*pulled* a B in the course⟩
11 : to demand or obtain an advantage over someone by the assertion of ⟨*pull* rank⟩
intransitive verb
1 a : to use force in drawing, dragging, or tugging **b :** to move especially through the exercise of mechanical energy ⟨the car *pulled* clear of the rut⟩ **c** (1) **:** to take a drink (2) **:** to draw hard in smoking ⟨*pulled* at a pipe⟩ **d :** to strain against the bit
2 : to draw a gun
3 : to admit of being pulled
4 : to feel or express strong sympathy **:** ROOT ⟨*pulling* for my team to win⟩
5 *of an offensive lineman in football* **:** to move back from the line of scrimmage and toward one flank to provide blocking for a ballcarrier
— **pull·er** *noun*
— **pull a fast one :** to perpetrate a trick or fraud
— **pull punches** *also* **pull a punch :** to refrain from using all the force at one's disposal
— **pull oneself together :** to regain one's composure
— **pull one's leg :** to deceive someone playfully **:** HOAX
— **pull one's weight :** to do one's full share of the work
— **pull out all the stops :** to use all one's resources without restraint
— **pull stakes** *or* **pull up stakes :** to move out **:** LEAVE
— **pull strings** *also* **pull wires :** to exert hidden influence or control
— **pull the plug 1 :** to disconnect a medical life-support system **2 :** to withdraw essential and especially financial support
— **pull the rug from under :** to weaken or unsettle especially by removing support or assistance from
— **pull the string :** to throw a change-up
— **pull the wool over one's eyes :** to blind to the true situation **:** HOODWINK
— **pull together :** to work in harmony **:** COOPERATE

²pull *noun, often attributive* (14th century)
1 a : the act or an instance of pulling **b** (1) **:** a draft of liquid (2) **:** an inhalation of smoke **c :** the effort expended in moving ⟨a long *pull* uphill⟩ **d :** force required to overcome resistance to pulling ⟨trigger *pull*⟩
2 a : ADVANTAGE **b :** special influence
3 : PROOF 6a
4 : a device for pulling something or for operating by pulling ⟨drawer *pull*⟩
5 : a force that attracts, compels, or influences **:** ATTRACTION
6 : an injury resulting from abnormal straining or stretching ⟨a muscle *pull*⟩ ⟨a groin *pull*⟩

pull away *intransitive verb* (circa 1934)
1 : to draw oneself back or away **:** WITHDRAW
2 : to move off or ahead

pull·back \ˈpu̇l-ˌbak\ *noun* (1668)
: a pulling back; *especially* **:** an orderly withdrawal of troops from a position or area

pull down *transitive verb* (15th century)
1 a : DEMOLISH, DESTROY **b :** to hunt down **:** OVERCOME
2 a : to bring to a lower level **:** REDUCE **b :** to depress in health, strength, or spirits
3 : to draw as wages or salary

pul·let \ˈpu̇-lət\ *noun* [Middle English *polet* young fowl, from Middle French *poulet,* from Old French, diminutive of *poul* cock, from Late Latin *pullus,* from Latin, young of an animal, chicken, sprout — more at FOAL] (14th century)
: a young hen; *specifically* **:** a hen of the domestic chicken less than a year old

pul·ley \ˈpu̇-lē\ *noun, plural* **pulleys** [Middle English *pouley,* from Middle French *poulie,* probably ultimately from Greek *polos* axis, pole — more at POLE] (14th century)
1 : a sheave or small wheel with a grooved rim and with or without the block in which it runs used singly with a rope or chain to change the direction and point of application of a pulling force and in various combinations to increase the applied force especially for lifting weights
2 : a pulley or pulleys with ropes to form a tackle that constitutes one of the simple machines
3 : a wheel used to transmit power by means of a band, belt, cord, rope, or chain passing over its rim

pull in (1605)
transitive verb
1 : CHECK, RESTRAIN
2 : ARREST
intransitive verb
: to arrive at a destination or come to a stop

Pull·man \ˈpu̇l-mən\ *noun* [George M. *Pullman*] (1867)
1 : a railroad passenger car with specially comfortable furnishings for day or especially for night travel
2 : a large suitcase

pull off *transitive verb* (1883)
: to carry out despite difficulties **:** accomplish successfully against odds

pul·lo·rum disease \pə-ˈlōr-əm-, -ˈlȯr-\ *noun* [New Latin *pullorum* (specific epithet of *Salmonella pullorum*), from Latin, of chickens (genitive plural of *pullus*)] (1929)
: a destructive typically diarrheal salmonellosis especially of young domestic chickens that is caused by a bacterium (*Salmonella pullorum*)

pull·out \ˈpu̇l-ˌau̇t\ *noun* (1825)
1 : the act or an instance of pulling out: as **a :** the action in which an airplane goes from a dive to horizontal flight **b :** PULLBACK
2 : something that can be pulled out

pull out *intransitive verb* (1855)
1 : LEAVE, DEPART
2 : WITHDRAW

¹pull·over \ˈpu̇l-ˌō-vər\ *noun* (1899)
: a pullover garment (as a sweater)

²pullover *adjective* (1907)
: put on by being pulled over the head

pull over (1930)
intransitive verb
: to steer one's vehicle to the side of the road
transitive verb
: to cause to pull over ⟨*pulled* him *over* for speeding⟩

pull round (1891)
intransitive verb
chiefly British **:** to regain one's health
transitive verb
chiefly British **:** to restore to good health

pull tab *noun* (1963)
: a metal tab (as on a can) pulled to open the container

pull through (1852)
intransitive verb
: to survive a dangerous or difficult situation
transitive verb
: to help survive a dangerous or difficult situation

pul·lu·late \ˈpəl-yə-ˌlāt\ *intransitive verb* **-lat·ed; -lat·ing** [Latin *pullulatus,* past participle of *pullulare,* from *pullulus,* diminutive of *pullus* chicken, sprout — more at FOAL] (1619)
1 a : GERMINATE, SPROUT **b :** to breed or produce freely
2 : SWARM, TEEM
— **pul·lu·la·tion** \ˌpəl-yə-ˈlā-shən\ *noun*

pull–up \ˈpu̇l-ˌəp\ *noun* (1938)
: CHIN-UP

pull up (1623)
transitive verb
1 : to bring to a stop **:** HALT
2 : CHECK, REBUKE
intransitive verb
1 a : to check oneself **b :** to come to an often abrupt halt **:** STOP
2 : to draw even with others in a race

pul·mo·nary \ˈpu̇l-mə-ˌner-ē, ˈpəl-\ *adjective* [Latin *pulmonarius,* from *pulmon-, pulmo* lung; akin to Greek *pleumōn* lung, Sanskrit *kloman* right lung] (1704)
1 : relating to, functioning like, or associated with the lungs
2 : PULMONATE
3 : carried on by the lungs

pulmonary artery *noun* (1704)
: an artery that conveys venous blood from the heart to the lungs — see HEART illustration

pulmonary circulation *noun* (circa 1890)
: the passage of blood from the right side of the heart through arteries to the lungs where it picks up oxygen and is returned to the left side of the heart by veins

pulmonary vein *noun* (1704)
: a valveless vein that returns oxygenated blood from the lungs to the heart

¹pul·mo·nate \ˈpu̇l-mə-ˌnāt, ˈpəl-\ *adjective* [Latin *pulmon-, pulmo* lung] (circa 1859)
1 : having lungs or organs resembling lungs
2 : of or relating to a subclass (Pulmonata) of gastropod mollusks having a respiratory sac and comprising most land snails and slugs and many freshwater snails

²pulmonate *noun* (1883)
: a pulmonate gastropod

pul·mon·ic \pu̇l-ˈmä-nik, ˌpəl-\ *adjective* [Latin *pulmon-, pulmo*] (1661)
: PULMONARY

pul·mo·tor \ˈpu̇l-ˌmō-tər, ˈpəl-\ *noun* [from *Pulmotor,* a trademark] (1911)
: a respiratory apparatus for pumping oxygen or air into and out of the lungs (as of an asphyxiated person)

¹pulp \ˈpəlp\ *noun* [Middle English *pulpe,* from Middle French *poulpe,* from Latin *pulpa* flesh, pulp] (14th century)
1 a (1) **:** the soft, succulent part of a fruit usually composed of mesocarp (2) **:** stem pith when soft and spongy **b :** a soft mass of vegetable matter (as of apples) from which most of the water has been extracted by pressure **c :** the soft sensitive tissue that fills the central cavity of a tooth — see TOOTH illustration **d :** a material prepared by chemical or mechanical means from various materials (as wood or rags) for use in making paper and cellulose products
2 : pulverized ore mixed with water
3 a : pulpy condition or character **b :** something in such a condition or having such a character
4 : a magazine or book printed on cheap paper (as newsprint) and often dealing with sensational material
— **pulp·i·ness** \ˈpəl-pē-nəs\ *noun*
— **pulpy** \ˈpəl-pē\ *adjective*

²pulp (1683)
transitive verb
1 : to reduce to pulp **:** cause to appear pulpy

2 : to deprive of the pulp
3 : to produce or reproduce (written matter) in pulp form
intransitive verb
: to become pulp or pulpy
— pulp·er *noun*

pulp·al \'pəl-pəl\ *adjective* (1903)
: of or relating to pulp especially of a tooth ⟨a *pulpal* abscess⟩
— pulp·al·ly \'pəl-pə-lē\ *adverb*

pul·pit \'pu̇l-ˌpit *also* 'pəl-, -pət\ *noun* [Middle English, from Late Latin *pulpitum*, from Latin, staging, platform] (14th century)
1 : an elevated platform or high reading desk used in preaching or conducting a worship service
2 a : the preaching profession **b :** a preaching position

pulp·wood \'pəlp-ˌwu̇d\ *noun* (1885)
: a wood (as of aspen, hemlock, pine, or spruce) used in making pulp for paper

pul·que \'pül-ˌkā; 'pül-kē, 'pu̇l-\ *noun* [Mexican Spanish] (1693)
: a Mexican alcoholic beverage made from the fermented sap of various agaves (as *Agave atrovirens*)

pul·sant \'pəl-sənt\ *adjective* (1709)
: pulsating with activity

pul·sar \'pəl-ˌsär\ *noun* [*pulse* + *-ar* (as in *quasar*)] (1968)
: a celestial source of pulsating electromagnetic radiation (as radio waves) characterized by a short relatively constant interval (as .033 second) between pulses that is held to be a rotating neutron star

pul·sate \'pəl-ˌsāt *also* ˌpəl-'\ *intransitive verb* **pul·sat·ed; pul·sat·ing** [Latin *pulsatus*, past participle of *pulsare*, frequentative of *pellere*] (1794)
1 : to exhibit a pulse or pulsation **:** BEAT
2 : to throb or move rhythmically **:** VIBRATE

pul·sa·tile \'pəl-sə-tᵊl, -ˌtīl\ *adjective* (1541)
: of or marked by pulsation

pul·sa·tion \ˌpəl-'sā-shən\ *noun* (1541)
1 : rhythmical throbbing or vibrating (as of an artery); *also* **:** a single beat or throb
2 : a periodically recurring alternate increase and decrease of a quantity (as pressure, volume, or voltage)

pul·sa·tor \'pəl-ˌsā-tər, ˌpəl-'\ *noun* (1890)
: something that beats or throbs in working

¹pulse \'pəls\ *noun* [Middle English *puls*, from Old French *pouls* porridge, from Latin *pult-, puls*, probably from Greek *poltos*] (13th century)
: the edible seeds of various leguminous crops (as peas, beans, or lentils); *also* **:** a plant yielding pulse

²pulse *noun* [Middle English *puls*, from Middle French *pouls*, from Latin *pulsus*, literally, beating, from *pellere* to drive, push, beat — more at FELT] (14th century)
1 a : a regular throbbing caused in the arteries by the contractions of the heart **b :** the palpable beat resulting from such pulse as detected in a superficial artery; *also* **:** the number of individual beats in a specified time period (as one minute) ⟨a resting *pulse* of 70⟩
2 a : underlying sentiment or opinion or an indication of it **b :** VITALITY
3 a : rhythmical beating, vibrating, or sounding **b :** BEAT, THROB
4 a : a transient variation of a quantity (as electrical current or voltage) whose value is normally constant **b** (1) **:** an electromagnetic wave or modulation thereof of brief duration (2) **:** a brief disturbance of pressure in a medium; *especially* **:** a sound wave or short train of sound waves
5 : a dose of a substance especially when applied over a short period of time ⟨*pulse*-labeled DNA⟩

³pulse *verb* **pulsed; puls·ing** (15th century)
intransitive verb
: to exhibit a pulse or pulsation **:** THROB
transitive verb
1 : to drive by or as if by a pulsation
2 : to cause to pulsate
3 a : to produce or modulate (as electromagnetic waves) in the form of pulses ⟨*pulsed* waves⟩ **b :** to cause (an apparatus) to produce pulses
— puls·er *noun*

pulse–jet engine \'pəls-'jet-\ *noun* (1949)
: a jet engine designed to produce a pulsating thrust by the intermittent flow of hot gases

pul·ver·a·ble \'pəl-və-rə-bəl, 'pəlv-rə-\ *adjective* (circa 1617)
: capable of being pulverized

pul·ver·ise *British variant of* PULVERIZE

pul·ver·ize \'pəl-və-ˌrīz\ *verb* **-ized; -iz·ing** [Middle English, from Middle French *pulveriser*, from Late Latin *pulverizare*, from Latin *pulver-, pulvis* dust, powder — more at POWDER] (15th century)
transitive verb
1 : to reduce (as by crushing, beating, or grinding) to very small particles **:** ATOMIZE
2 : ANNIHILATE, DEMOLISH
intransitive verb
: to become pulverized
— pul·ver·iz·able \ˌpəl-və-'rī-zə-bəl\ *adjective*
— pul·ver·i·za·tion \ˌpəl-və-rə-'zā-shən, ˌpəlv-rə-\ *noun*
— pul·ver·iz·er \'pəl-və-ˌrī-zər\ *noun*

pul·ver·u·lent \ˌpəl-'ver-yə-lənt, -'ver-ə-\ *adjective* [Latin *pulverulentus* dusty, from *pulver-, pulvis*] (circa 1656)
1 : consisting of or reducible to fine powder
2 : being or looking dusty **:** CRUMBLY

pul·vil·lus \ˌpəl-'vi-ləs\ *noun, plural* **-vil·li** \-'vi-ˌlī, -(ˌ)lē\ [New Latin, from Latin, diminutive of *pulvinus*] (circa 1826)
: one of the lobed hairy adhesive organs that terminate the feet of dipteran flies

pul·vi·nus \ˌpəl-'vī-nəs, -'vē-\ *noun, plural* **-vi·ni** \-'vī-ˌnī, -'vē-(ˌ)nē\ [New Latin, from Latin, cushion] (1857)
: a mass of large thin-walled cells surrounding a vascular strand at the base of a petiole or petiolule and functioning in turgor movements of leaves or leaflets

pu·ma \'pü-mə, 'pyü-\ *noun, plural* **pumas** *also* **puma** [Spanish, from Quechua] (1777)
: COUGAR; *also* **:** the fur or pelt of a cougar

pum·ice \'pə-məs\ *noun* [Middle English *pomis*, from Middle French, from Latin *pumic-, pumex* — more at FOAM] (15th century)
: a volcanic glass full of cavities and very light in weight used especially in powder form for smoothing and polishing
— pu·mi·ceous \pyü-'mi-shəs, ˌpə-\ *adjective*

pum·ic·ite \'pə-mə-ˌsīt\ *noun* (1916)
: PUMICE

pum·mel \'pə-məl\ *verb* **-meled** *also* **-melled; -mel·ing** *also* **-mel·ling** \'pə-mə-liŋ, 'pəm-liŋ\ [alteration of *pommel*] (1548)
: POUND, BEAT

pum·me·lo *variant of* POMELO 2

¹pump \'pəmp\ *noun* [Middle English *pumpe, pompe*, from Middle Low German *pumpe* or Middle Dutch *pompe*, perhaps from Spanish *bomba*, of imitative origin] (15th century)
1 : a device that raises, transfers, or compresses fluids or that attenuates gases especially by suction or pressure or both
2 : HEART
3 : an act or the process of pumping
4 : an energy source (as light) for pumping atoms or molecules
5 : a mechanism (as the sodium pump) for pumping atoms, ions, or molecules

²pump (1508)
intransitive verb
1 : to work a pump **:** raise or move a fluid with a pump
2 : to exert oneself to pump or as if to pump something
3 : to move in a manner that resembles the action of a pump handle
transitive verb
1 a : to raise (as water) with a pump **b :** to draw fluid from with a pump
2 : to pour forth, deliver, or draw with or as if with a pump ⟨*pumped* money into the economy⟩ ⟨*pump* new life into the classroom⟩
3 a : to question persistently **b :** to elicit by persistent questioning
4 a : to operate by manipulating a lever **b :** to manipulate as if operating a pump handle ⟨*pumped* my hand warmly⟩ **c :** to cause to move with an action resembling that of a pump handle ⟨a runner *pumping* her arms⟩
5 : to transport (as ions) against a concentration gradient by the expenditure of energy
6 a : to excite (as atoms or molecules) especially so as to cause emission of coherent monochromatic electromagnetic radiation (as in a laser) **b :** to energize (as a laser) by pumping
— pump iron : to lift weights

³pump *noun* [origin unknown] (1555)
: a shoe that grips the foot chiefly at the toe and heel; *especially* **:** a close-fitting woman's dress shoe with a moderate to high heel

pumped storage *noun* (1927)
: a hydroelectric system in which electricity is generated during periods of high demand by the use of water that has been pumped into a reservoir at a higher altitude during periods of low demand

pump·er \'pəm-pər\ *noun* (1660)
: one that pumps; *especially* **:** a fire truck equipped with a pump

pum·per·nick·el \'pəm-pər-ˌni-kəl\ *noun* [German, from *pumpern* to break wind + *Nickel* goblin; from its reputed indigestibility] (1756)
: a dark coarse sourdough bread made of unbolted rye flour

pump·kin \'pəm(p)-kən, ÷'pəŋ-kən\ *noun, often attributive* [alteration of earlier *pumpion*, modification of French *popon, pompon* melon, pumpkin, from Latin *pepon-, pepo*, from Greek *pepōn*, from *pepōn* ripened; akin to Greek *pessein* to cook, ripen — more at COOK] (1654)
1 a : the usually round orange fruit of a vine (*Cucurbita pepo*) of the gourd family widely cultivated as food **b :** WINTER CROOKNECK **c** *British* **:** any of various large-fruited winter squashes (*C. maxima*)
2 : a usually hairy prickly vine that produces pumpkins

pump·kin·seed \-ˌsēd\ *noun* (1814)
: a brilliantly colored North American freshwater sunfish (*Lepomis gibbosus*) with a reddish spot on the operculum

pump priming *noun* (1936)
: government investment expenditures designed to induce a self-sustaining expansion of economic activity

pump up *transitive verb* (1791)
1 a : to fill with enthusiasm or excitement **b :** to fill with or as if with air **:** INFLATE
2 : INCREASE 1

¹pun \'pən\ *noun* [perhaps from Italian *puntiglio* fine point, quibble — more at PUNCTILIO] (1662)
: the usually humorous use of a word in such a way as to suggest two or more of its meanings or the meaning of another word similar in sound

²pun *intransitive verb* **punned; pun·ning** (1670)
: to make puns

pu·na \'pü-nə\ *noun* [American Spanish, from Quechua] (1613)
: a treeless windswept tableland or basin in the higher Andes

¹punch \'pənch\ *verb* [Middle English, from Middle French *poinçonner* to prick, stamp, from *poinçon* puncheon] (14th century)
transitive verb

1 a : PROD, POKE **b :** DRIVE, HERD ⟨*punching* cattle⟩
2 a : to strike with a forward thrust especially of the fist **b :** to drive or push forcibly by or as if by a punch **c :** to hit (a ball) with less than a full swing
3 : to emboss, cut, perforate, or make with or as if with a punch
4 a : to push down so as to produce a desired result ⟨*punch* buttons on a jukebox⟩ **b :** to hit or press down the operating mechanism of ⟨*punch* a time clock⟩ ⟨*punch* a typewriter⟩ **c :** to produce by or as if by punching keys ⟨*punch* out a tune on the piano⟩ **d :** to enter (as data) by punching keys
5 : to give emphasis to
intransitive verb
: to perform the action of punching something
— punch·er *noun*

[2]punch *noun* (14th century)
1 : the action of punching
2 : a quick blow with or as if with the fist
3 : effective energy or forcefulness ⟨a story that packs a *punch*⟩ ⟨political *punch*⟩
— punch·less \ˈpənch-ləs\ *adjective*
— to the punch : to the first blow or to decisive action — usually used with *beat*

[3]punch *noun* [probably short for *puncheon*] (14th century)
1 a : a tool usually in the form of a short rod of steel that is variously shaped at one end for different operations (as forming, perforating, embossing, or cutting) **b :** a short tapering steel rod for driving the heads of nails below a surface **c :** a steel die faced with a letter in relief that is forced into a softer metal to form an intaglio matrix from which foundry type is cast **d :** a device or machine for cutting holes or notches (as in paper or cardboard)
2 : a hole or notch from a perforating operation

[4]punch *noun* [perhaps from Hindi *pā̃c* five, from Sanskrit *pañca;* akin to Greek *pente* five; from its originally having five ingredients — more at FIVE] (1632)
: a hot or cold drink that is usually a combination of hard liquor, wine, or beer and nonalcoholic beverages; *also* **:** a drink that is a mixture of nonalcoholic beverages

Punch–and–Judy show \ˌpənch-ən-ˈjü-dē-\ *noun* (1876)
: a traditional puppet show in which the little hook-nosed humpback Punch fights comically with his wife Judy

punch·ball \ˈpənch-ˌbȯl\ *noun* (1932)
: baseball adapted for small areas in which a rubber ball is hit with a fist instead of a bat

punch·board \-ˌbōrd, -ˌbȯrd\ *noun* (1912)
: a small board that has many holes each filled with a rolled-up printed slip to be punched out on payment of a nominal sum in an effort to obtain a slip that entitles the player to a designated prize

punch bowl *noun* (1692)
: a large bowl from which a beverage (as punch) is served

punch–drunk \ˈpənch-ˌdrəŋk\ *adjective* [[2]*punch*] (1918)
1 : suffering cerebral injury from many minute brain hemorrhages as a result of repeated head blows received in boxing
2 : behaving as if punch-drunk **:** DAZED, CONFUSED

punched card \ˈpəncht-\ *noun* (1919)
: a card in which holes are punched in designated positions to represent data — called also *Hollerith card, punch card*

[1]pun·cheon \ˈpən-chən\ *noun* [Middle English *ponson,* from Middle French *poinçon* pointed tool, king post, from (assumed) Vulgar Latin *punction-, punctio* pointed tool, from Latin, action of pricking, from *pungere* to prick — more at PUNGENT] (14th century)
1 : a pointed tool for piercing or for working on stone
2 a : a short upright framing timber **b :** a split log or heavy slab with the face smoothed

[2]puncheon *noun* [Middle English *poncion,* from Middle French *ponchon, poinçon,* of unknown origin] (15th century)
: a large cask of varying capacity

punch in *intransitive verb* (1926)
: to record the time of one's arrival or beginning work by punching a time clock

pun·chi·nel·lo \ˌpən-chə-ˈne-(ˌ)lō\ *noun* [modification of Italian dialect *polecenella*] (1666)
1 *capitalized* **:** a fat short humpbacked clown or buffoon in Italian puppet shows
2 *plural* **-los :** a squat grotesque person

Punchinello

punching bag *noun* (1886)
: a stuffed or inflated bag that is usually suspended for free movement and that is punched for exercise or for training in boxing

punch line *noun* (1921)
: the sentence, statement, or phrase (as in a joke) that makes the point

punch–out \ˈpənch-ˌau̇t\ *noun* (1928)
: FISTFIGHT

punch out (1973)
intransitive verb
1 : to record the time of one's stopping work or departure by punching a time clock
2 : to bail out of an aircraft using an ejection seat
transitive verb
1 : to beat up
2 : STRIKE OUT

punch press *noun* (1911)
: a press equipped with cutting, shaping, or combination dies for working on material (as metal)

punch–up \ˈpənch-ˌəp\ *noun* (1958)
chiefly British **:** FISTFIGHT

punch up *transitive verb* (circa 1959)
: to give energy or forcefulness to ⟨jokes added to *punch up* a speech⟩

punchy \ˈpən-chē\ *adjective* **punch·i·er; -est** (1917)
1 : having punch **:** FORCEFUL, SPIRITED
2 : PUNCH-DRUNK

punc·tate \ˈpəŋk-ˌtāt\ *adjective* [New Latin *punctatus,* from Latin *punctum* point — more at POINT] (circa 1760)
1 : marked with minute spots or depressions ⟨a *punctate* leaf⟩
2 : characterized by dots or points ⟨*punctate* skin lesions⟩
— punc·ta·tion \ˌpəŋk-ˈtā-shən\ *noun*

punc·til·io \ˌpəŋk-ˈti-lē-ˌō\ *noun, plural* **-i·os** [Italian & Spanish; Italian *puntiglio* point of honor, scruple, from Spanish *puntillo,* from diminutive of *punto* point, from Latin *punctum*] (1596)
1 : a minute detail of conduct in a ceremony or in observance of a code
2 : careful observance of forms (as in social conduct)

punc·til·i·ous \-lē-əs\ *adjective* (1634)
: marked by or concerned about precise accordance with the details of codes or conventions
synonym see CAREFUL
— punc·til·i·ous·ly *adverb*
— punc·til·i·ous·ness *noun*

punc·tu·al \ˈpəŋk-chə-wəl, -chəl\ *adjective* [Middle English, having a sharp point, from Medieval Latin *punctualis* of a point, from Latin *punctus* pricking, point, from *pungere* to prick — more at PUNGENT] (1675)
: being on time **:** PROMPT
— punc·tu·al·i·ty \ˌpəŋk-chə-ˈwa-lə-tē\ *noun*
— punc·tu·al·ly \ˈpəŋk-chə-wə-lē, -chə-lē\ *adverb*

punc·tu·ate \ˈpəŋk-chə-ˌwāt\ *verb* **-at·ed; -at·ing** [Medieval Latin *punctuatus,* past participle of *punctuare* to point, provide with punctuation marks, from Latin *punctus* point] (circa 1818)
transitive verb
1 : to mark or divide (written matter) with punctuation marks
2 : to break into or interrupt at intervals ⟨the steady click of her needles *punctuated* the silence —Edith Wharton⟩
3 : ACCENTUATE, EMPHASIZE
intransitive verb
: to use punctuation marks
— punc·tu·a·tor \-ˌwā-tər\ *noun*

punctuated equilibrium *noun* (1978)
: evolution that is characterized by long periods of stability in the characteristics of an organism and short periods of rapid change during which new forms appear especially from small subpopulations of the ancestral form in restricted parts of its geographic range; *also* **:** a theory or model of evolution emphasizing this — compare GRADUALISM 2

punc·tu·a·tion \ˌpəŋk-chə-ˈwā-shən\ *noun* (circa 1539)
1 : the act of punctuating **:** the state of being punctuated
2 : the act or practice of inserting standardized marks or signs in written matter to clarify the meaning and separate structural units; *also* **:** a system of punctuation
3 : something that contrasts or accentuates

punctuation mark *noun* (1860)
: any of various standardized marks or signs used in punctuation

PUNCTUATION MARKS

.	period (*or* full stop)
,	comma
;	semicolon
:	colon
’	apostrophe
‘ ’	quotation marks, single (*or chiefly British* inverted commas)
“ ”	quotation marks, double (*or chiefly British* inverted commas)
« »	guillemets
?	question mark (*or* interrogation point)
¿ ?	question marks, Spanish
!	exclamation point
¡ !	exclamation points, Spanish
‽	interrobang
/	slash (*or* diagonal *or* slant *or* solidus *or* virgule)
. . . *or*	ellipsis (*or* suspension points)
-	hyphen
⸗	double hyphen
–	dash (*or* en dash)
—	dash (*or* em dash)
~	swung dash
( )	parentheses (*or* brackets)
[ ]	brackets, square (*or* braces)
⟨ ⟩	brackets, angle
{ }	braces (*or* curly brackets)

¹punc·ture \ˈpəŋk-chər\ *noun* [Middle English, from Latin *punctura,* from *punctus,* past participle of *pungere*] (14th century)
1 : an act of puncturing
2 : a hole, wound, or perforation made by puncturing
3 : a minute depression

²puncture *verb* **punc·tured; punc·tur·ing** \ˈpəŋk-chə-riŋ, ˈpəŋk-shriŋ\ (1699)
transitive verb
1 : to pierce with or as if with a pointed instrument or object
2 : to make useless or ineffective as if by a puncture **:** DEFLATE
intransitive verb
: to become punctured

puncture vine *noun* (1911)
: an Old World annual prostrate herb (*Tribulus terrestris*) of the caltrop family that has hard spiny pods and is a troublesome weed especially in the western U.S. — called also *caltrop, puncture-weed*

pun·dit \ˈpən-dət\ *noun* [Hindi *paṇḍit,* from Sanskrit *paṇḍita,* from *paṇḍita* learned] (1672)
1 : PANDIT
2 : a learned man **:** TEACHER
3 : one who gives opinions in an authoritative manner **:** CRITIC
— pun·dit·ry \-də-trē\ *noun*

pung \ˈpəŋ\ *noun* [short for earlier *tow-pong,* of Algonquian origin; akin to Micmac *tobâgun* drag made with skin] (1825)
New England **:** a sleigh with a box-shaped body

pun·gen·cy \ˈpən-jən(t)-sē\ *noun* (1649)
: the quality or state of being pungent

pun·gent \-jənt\ *adjective* [Latin *pungent-, pungens,* present participle of *pungere* to prick, sting; akin to Latin *pugnus* fist, *pugnare* to fight, Greek *pygmē* fist] (1597)
1 : sharply painful
2 : having a stiff and sharp point ⟨*pungent* leaves⟩
3 a : marked by a sharp incisive quality **:** CAUSTIC ⟨a *pungent* critic⟩ ⟨*pungent* language⟩ **b :** being sharp and to the point
4 : causing a sharp or irritating sensation; *especially* **:** ACRID ☆
— pun·gent·ly *adverb*

pun·gle \ˈpəŋ-gəl\ *verb* **pun·gled; pun·gling** \ˈpəŋ-g(ə-)liŋ\ [Spanish *póngale* put it down] (1851)
transitive verb
: to make a payment or contribution of (money) — usually used with *up*
intransitive verb
: PAY, CONTRIBUTE — usually used with *up*

¹Pu·nic \ˈpyü-nik\ *adjective* [Latin *punicus,* from *Poenus* inhabitant of Carthage; akin to Greek *Phoinix* Phoenician] (1533)
1 : of or relating to Carthage or the Carthaginians
2 : FAITHLESS, TREACHEROUS

²Punic *noun* (1673)
: the Phoenician dialect of ancient Carthage

pun·ish \ˈpə-nish\ *verb* [Middle English *punisshen,* from Middle French *puniss-,* stem of *punir,* from Latin *punire,* from *poena* penalty — more at PAIN] (14th century)
transitive verb
1 a : to impose a penalty on for a fault, offense, or violation **b :** to inflict a penalty for the commission of (an offense) in retribution or retaliation
2 a : to deal with roughly or harshly **b :** to inflict injury on **:** HURT
intransitive verb
: to inflict punishment ☆
— pun·ish·abil·i·ty \ˌpə-nish-ə-ˈbi-lə-tē\ *noun*
— pun·ish·able \ˈpə-nish-ə-bəl\ *adjective*
— pun·ish·er *noun*

pun·ish·ment \ˈpə-nish-mənt\ *noun* (15th century)
1 : the act of punishing
2 a : suffering, pain, or loss that serves as retribution **b :** a penalty inflicted on an offender through judicial procedure
3 : severe, rough, or disastrous treatment

pu·ni·tion \pyü-ˈni-shən\ *noun* [Middle English *punicion,* from Middle French *punition,* from Latin *punition-, punitio,* from *punire*] (15th century)
: PUNISHMENT

pu·ni·tive \ˈpyü-nə-tiv\ *adjective* [French *punitif,* from Medieval Latin *punitivus,* from Latin *punitus,* past participle of *punire*] (1624)
: inflicting, involving, or aiming at punishment
— pu·ni·tive·ly *adverb*
— pu·ni·tive·ness *noun*

punitive damages *noun plural* (circa 1890)
: damages awarded in excess of normal compensation to the plaintiff to punish a defendant for a serious wrong

Pun·jabi \ˌpən-ˈjä-bē, -ˈja-\ *noun* [Hindi *pañjābī,* from *pañjābī* of Punjab, from Persian *panjābī,* from *Panjāb* Punjab] (1846)
1 : PANJABI 1
2 : a native or inhabitant of the Punjab region of the northwestern Indian subcontinent
— Punjabi *adjective*

¹punk \ˈpəŋk\ *noun* [origin unknown] (1596)
1 *archaic* **:** PROSTITUTE
2 [probably partly from ³*punk*] **:** NONSENSE, FOOLISHNESS
3 a : a young inexperienced person **:** BEGINNER, NOVICE; *especially* **:** a young man **b :** a usually petty gangster, hoodlum, or ruffian **c :** a youth used as a homosexual partner
4 a : PUNK ROCK **b :** a punk rock musician **c :** one who affects punk styles

²punk *adjective* (1896)
1 : very poor **:** INFERIOR ⟨played a *punk* game⟩
2 : being in poor health ⟨said that she was feeling *punk*⟩
3 a : of or relating to punk rock **b :** relating to or being a style (as of dress or hair) inspired by punk rock
— punk·ish \ˈpəŋ-kish\ *adjective*

³punk *noun* [perhaps alteration of *spunk*] (1687)
1 : wood so decayed as to be dry, crumbly, and useful for tinder
2 : a dry spongy substance prepared from fungi (genus *Fomes*) and used to ignite fuses especially of fireworks

pun·kah \ˈpəŋ-kə\ *noun* [Hindi *pākhā*] (1787)
: a fan used especially in India that consists of a canvas-covered frame suspended from the ceiling and that is operated by a cord

punk·er \ˈpəŋ-kər\ *noun* (1977)
: ¹PUNK 4b, c

pun·kie *also* **pun·ky** \ˈpəŋ-kē\ *noun, plural* **punkies** [New York Dutch *punki,* modification of Delaware (Munsee) *pónkwəs*] (1769)
: BITING MIDGE

pun·kin \ˈpəŋ-kən\ *variant of* PUMPKIN

punk rock *noun* (1971)
: rock music marked by extreme and often deliberately offensive expressions of alienation and social discontent
— punk rocker *noun*

¹punky \ˈpəŋ-kē\ *adjective* **punk·i·er; -est** [³*punk*] (1872)
: resembling punk in being soft or rotted
— punk·i·ness *noun*

²punky *adjective* [¹*punk*] (1972)
: resembling or typical of a punk

pun·net \ˈpə-nət\ *noun* [origin unknown] (circa 1822)
British **:** a small basket for fruits or vegetables

Pun·nett square \ˈpə-nət-\ *noun* [Reginald C. Punnett (died 1967) English geneticist] (1942)
: an $n \times n$ square used in genetics to calculate the frequencies of the different genotypes and phenotypes among the offspring of a cross

pun·ny \ˈpə-nē\ *adjective* **pun·ni·er; -est** (1947)
: constituting or involving a pun

pun·ster \ˈpən(t)-stər\ *noun* (1700)
: one who is given to punning

¹punt \ˈpənt\ *noun* [(assumed) Middle English, from Old English, from Latin *ponton-, ponto*] (before 12th century)
: a long narrow flat-bottomed boat with square ends usually propelled with a pole

²punt *transitive verb* (1816)
: to propel (as a punt) with a pole

³punt *intransitive verb* [French *ponter,* from *ponte* point in some games, play against the banker, from Spanish *punto* point, from Latin *punctum* — more at POINT] (1712)
1 : to play at a gambling game against the banker
2 *British* **:** GAMBLE

⁴punt *verb* [origin unknown] (1845)
transitive verb
: to kick (a football or soccer ball) with the top of the foot before the ball which is dropped from the hands hits the ground
intransitive verb
: to punt a ball

⁵punt *noun* (1845)
: the act or an instance of punting a ball

⁶punt \ˈpu̇nt\ *noun* [Irish, pound, from English *pound*] (1975)
: the monetary pound of Ireland

punt·er \ˈpən-tər\ *noun* (circa 1706)
: one that punts: as **a** *chiefly British* **:** a person who gambles; *especially* **:** one who bets against a bookmaker **b :** one who uses a punt in boating **c :** one who punts a ball

punt formation *noun* (1949)
: an offensive football formation in which a back making a punt stands approximately 10 yards behind the line and the other backs are in blocking position close to the line

pun·ty \ˈpən-tē\ *noun, plural* **punties** [French *pontil*] (1662)
: a metal rod used for fashioning hot glass

pu·ny \ˈpyü-nē\ *adjective* **pu·ni·er; -est** [Middle French *puisné* younger, literally, born afterward, from *puis* afterward + *né* born] (1593)

☆ SYNONYMS

Pungent, piquant, poignant, racy mean sharp and stimulating to the mind or the senses. PUNGENT implies a sharp, stinging, or biting quality especially of odors ⟨a cheese with a *pungent* odor⟩. PIQUANT suggests a power to whet the appetite or interest through tartness or mild pungency ⟨a *piquant* sauce⟩. POIGNANT suggests something is sharply or piercingly effective in stirring one's consciousness or emotions ⟨felt a *poignant* sense of loss⟩. RACY implies having a strongly characteristic natural quality fresh and unimpaired ⟨spontaneous, *racy* prose⟩.

Punish, chastise, castigate, chasten, discipline, correct mean to inflict a penalty on in requital for wrongdoing. PUNISH implies subjecting to a penalty for wrongdoing ⟨*punished* for stealing⟩. CHASTISE may apply to either the infliction of corporal punishment or to verbal censure or denunciation ⟨*chastised* his son for neglecting his studies⟩. CASTIGATE usually implies a severe, typically public censure ⟨an editorial *castigating* the entire city council⟩. CHASTEN suggests any affliction or trial that leaves one humbled or subdued ⟨*chastened* by a landslide election defeat⟩. DISCIPLINE implies a punishing or chastening in order to bring under control ⟨parents must *discipline* their children⟩. CORRECT implies punishing aimed at reforming an offender ⟨the function of prison is to *correct* the wrongdoer⟩.

: slight or inferior in power, size, or importance **:** WEAK ◇
— **pu·ni·ly** \'pyü-nᵊl-ē\ *adverb*
— **pu·ni·ness** \'pyü-nē-nəs\ *noun*

[1]pup \'pəp\ *noun* [short for *puppy*] (1773)
: a young dog; *also* **:** one of the young of various animals (as a seal or rat)

[2]pup *intransitive verb* **pupped; pup·ping** (1787)
: to give birth to pups

pu·pa \'pyü-pə\ *noun, plural* **pu·pae** \-(ˌ)pē *also* -ˌpī\ *or* **pupas** [New Latin, from Latin *pupa* doll] (1815)
: an intermediate usually quiescent stage of a metamorphic insect (as a bee, moth, or beetle) that occurs between the larva and the imago, is usually enclosed in a cocoon or protective covering, and undergoes internal changes by which larval structures are replaced by those typical of the imago
word history see LARVA
— **pu·pal** \'pyü-pəl\ *adjective*

pu·par·i·um \pyü-'par-ē-əm, -'per-\ *noun, plural* **pu·par·ia** \-ē-ə\ [New Latin, from *pupa*] (1815)
: a rigid outer shell formed from the larval skin that covers some pupae (as of a dipteran fly)

pu·pate \'pyü-ˌpāt\ *intransitive verb* **pu·pat·ed; pu·pat·ing** (circa 1879)
: to become a pupa **:** pass through a pupal stage
— **pu·pa·tion** \pyü-'pā-shən\ *noun*

pup·fish \'pəp-ˌfish\ *noun* (1949)
: any of several killifishes (genus *Cyprinodon* of the family Cyprinodontidae) especially of warm streams and springs of the western U.S.

[1]pu·pil \'pyü-pəl\ *noun* [Middle English *pupille* minor ward, from Middle French, from Latin *pupillus* male ward (from diminutive of *pupus* boy) & *pupilla* female ward, from diminutive of *pupa* girl, doll] (1536)
1 : a child or young person in school or in the charge of a tutor or instructor **:** STUDENT
2 : one who has been taught or influenced by a famous or distinguished person ◇

[2]pupil *noun* [Middle French *pupille,* from Latin *pupilla,* from diminutive of *pupa* doll; from the tiny image of oneself seen reflected in another's eye] (1567)
: the contractile aperture in the iris of the eye ◇
— **pu·pil·lary** \'pyü-pə-ˌler-ē\ *adjective*

pu·pil·age *or* **pu·pil·lage** \'pyü-pə-lij\ *noun* (circa 1599)
: the state or period of being a pupil

pup·pet \'pə-pət\ *noun, often attributive* [Middle English *popet,* from Middle French *poupette,* diminutive of (assumed) *poupe* doll, from Latin *pupa*] (1538)
1 a : a small-scale figure (as of a person or animal) usually with a cloth body and hollow head that fits over and is moved by the hand **b :** MARIONETTE
2 : DOLL 1
3 : one whose acts are controlled by an outside force or influence
— **pup·pet·like** \-ˌlīk\ *adjective*

pup·pe·teer \ˌpə-pə-'tir\ *noun* (circa 1923)
: one who manipulates puppets

pup·pet·ry \'pə-pə-trē\ *noun, plural* **-ries** (1528)
1 : the production or creation of puppets or puppet shows
2 : the art of manipulating puppets

pup·py \'pə-pē\ *noun, plural* **puppies** [Middle English *popi,* from Middle French *poupée* doll, toy, from (assumed) *poupe* doll] (1591)
: a young domestic dog; *specifically* **:** one less than a year old
— **pup·py·hood** \-ˌhu̇d\ *noun*
— **pup·py·ish** \-ish\ *adjective*
— **pup·py·like** \-ˌlīk\ *adjective*

puppy dog *noun* (1595)
: a domestic dog; *especially* **:** one having the lovable attributes of a puppy

puppy love *noun* (1834)
: transitory affection felt by a boy or girl for one of the opposite sex

pup tent *noun* (1863)
: a low small tent for two persons usually consisting of two halves fastened together

Pu·ra·na \pu̇-'rä-nə\ *noun, often capitalized* [Sanskrit *purāṇa,* from *purāṇa* ancient, from *purā* formerly; akin to Sanskrit *pura* before, Greek *para* beside, *pro* before — more at FOR] (1696)
: one of a class of Hindu sacred writings chiefly from A.D. 300 to A.D. 750 comprising popular myths and legends and other traditional lore
— **Pu·ra·nic** \-nik\ *adjective*

pur·blind \'pər-ˌblīnd\ *adjective* [Middle English *pur blind,* from *pur* purely, wholly, from *pur* pure] (14th century)
1 a *obsolete* **:** wholly blind **b :** partly blind
2 : lacking in vision, insight, or understanding **:** OBTUSE
— **pur·blind·ly** \-ˌblīn(d)-lē\ *adverb*
— **pur·blind·ness** \-ˌblīn(d)-nəs\ *noun*

[1]pur·chase \'pər-chəs\ *verb* **pur·chased; pur·chas·ing** [Middle English *purchacen,* from Old French *purchacier* to seek to obtain, from *por-, pur-* for, forward (modification of Latin *pro-*) + *chacier* to pursue, chase — more at PRO-] (14th century)
transitive verb
1 a *archaic* **:** GAIN, ACQUIRE **b :** to acquire (real estate) by means other than descent or inheritance **c :** to obtain by paying money or its equivalent **:** BUY **d :** to obtain by labor, danger, or sacrifice
2 : to apply a device for obtaining a mechanical advantage to (as something to be moved); *also* **:** to move by a purchase
3 : to constitute the means for buying ⟨our dollars *purchase* less each year⟩
intransitive verb
: to purchase something
— **pur·chas·able** \-chə-sə-bəl\ *adjective*
— **pur·chas·er** *noun*

[2]purchase *noun* (14th century)
1 : an act or instance of purchasing
2 : something obtained especially for a price in money or its equivalent
3 a (1) **:** a mechanical hold or advantage applied to the raising or moving of heavy bodies (2) **:** an apparatus or device by which advantage is gained **b** (1) **:** an advantage (as a firm hold or position) used in applying one's power ⟨clutching the steering wheel for more *purchase* —Barry Crump⟩ (2) **:** a means of exerting power

pur·dah \'pər-də\ *noun* [Hindi *parda,* literally, screen, veil] (1865)
1 : seclusion of women from public observation among Muslims and some Hindus especially in India
2 : a state of seclusion or concealment

pure \'pyu̇r\ *adjective* **pur·er; pur·est** [Middle English *pur,* from Old French, from Latin *purus;* akin to Old High German *fowen* to sift, Sanskrit *punāti* he cleanses, Middle Irish *ūr* fresh, new] (14th century)
1 a (1) **:** unmixed with any other matter ⟨*pure* gold⟩ (2) **:** free from dust, dirt, or taint ⟨*pure* food⟩ (3) **:** SPOTLESS, STAINLESS **b :** free from harshness or roughness and being in tune — used of a musical tone **c** *of a vowel* **:** characterized by no appreciable alteration of articulation during utterance
2 a : being thus and no other **:** SHEER, UNMITIGATED ⟨*pure* folly⟩ **b** (1) **:** ABSTRACT, THEORETICAL (2) **:** A PRIORI ⟨*pure* mechanics⟩ **c :** not directed toward exposition of reality or solution of practical problems ⟨*pure* literature⟩ **d :** being nonobjective and to be appraised on formal and technical qualities only ⟨*pure* form⟩
3 a (1) **:** free from what vitiates, weakens, or pollutes (2) **:** containing nothing that does not properly belong **b :** free from moral fault or guilt **c :** marked by chastity **:** CONTINENT **d** (1) **:** of pure blood and unmixed ancestry (2) **:** homozygous in and breeding true for one or more characters **e :** ritually clean
synonym see CHASTE
— **pure·ness** *noun*

pure–blood·ed \'pyu̇r-ˌblə-dəd\ *or* **pure–blood** \-ˌbləd\ *adjective* (1821)
: FULL-BLOODED 1
— **pure·blood** \-ˌbləd\ *noun*

pure·bred \-'bred, -ˌbred\ *adjective* (1868)
: bred from members of a recognized breed, strain, or kind without admixture of other blood over many generations
— **pure·bred** \-ˌbred\ *noun*

pure democracy *noun* (circa 1910)
: democracy in which the power is exercised directly by the people rather than through representatives

[1]pu·ree *or* **pu·rée** \pyu̇-'rā, -'rē\ *noun* [French *purée,* from Middle French, from feminine of *puré,* past participle of *purer* to purify, strain, from Latin *purare* to purify, from *purus*] (1707)
1 : a paste or thick liquid suspension usually made from cooked food ground finely
2 : a thick soup made of pureed vegetables

[2]puree *transitive verb* **pu·reed; pu·ree·ing** (1928)
: to make a puree of

pure imaginary *noun* (1947)

◇ WORD HISTORY

puny In medieval French *puisné,* literally "born afterward," was used to mean "younger" in reference to the relationship between two people. When borrowed into English, *puisne* and the phonetic spelling *puny* came to be used of anyone in a lesser position relative to others. From this use a specific legal sense developed. A *puisne* (or *puny*) judge became the designation for a junior or subordinate judge in the superior courts. By Shakespeare's time *puny* had largely lost its implication of relative rank and become nearly synonymous with "weak" or "feeble," a sense the word retains.

pupil If you look into another person's eyes, you see reflected within the iris a tiny image of your own face. The Romans, comparing this image of a miniature human to a doll, called the opening in the iris that seems to hold the image *pupilla,* a diminutive of the word *pupa,* meaning "doll." This Latin word, via medieval French *pupille,* was borrowed into English as *pupil.* Another sense of Latin *pupa* was "girl," and a young girl, typically an orphan, who was under the care of a guardian was called by the same diminutive form *pupilla* as the pupil of the eye. The masculine counterpart of *pupilla* was *pupillus,* a diminutive of *pupus,* meaning "boy." In medieval French and later in Middle English, the word *pupille* served for both sexes. It had about the same meaning as the Latin word, an orphan in a guardian's care, though in the 15th century *pupille* began to be used in French for any child being instructed by a tutor. English *pupil* adopted this meaning in the 16th century, but it has gradually expanded in scope to include the children under instruction in an elementary school (for which the Modern French counterpart is *élève.*)

: a complex number that is the product of a real number other than zero and the imaginary unit
— **pure imaginary** *adjective*

pure•ly \'pyůr-lē\ *adverb* (14th century)
1 : WHOLLY, COMPLETELY ⟨a selection based *purely* on merit⟩
2 : without admixture of anything injurious or foreign
3 : SIMPLY, MERELY ⟨read *purely* for relaxation⟩
4 : in a chaste or innocent manner

pur•fle \'pər-fəl\ *transitive verb* **pur•fled; pur•fling** \-f(ə-)liŋ\ [Middle English *purfilen,* from Middle French *porfiler,* from (assumed) Vulgar Latin *profilare,* from Latin *pro-* forward + Late Latin *filare* to spin — more at PRO-, FILE] (14th century)
: to ornament the border or edges of
— **purfle** *noun*

pur•ga•tion \,pər-'gā-shən\ *noun* (14th century)
: the act or result of purging

¹pur•ga•tive \'pər-gə-tiv\ *adjective* [Middle English *purgatif,* from Middle French, from Late Latin *purgativus,* from Latin *purgatus,* past participle] (15th century)
: purging or tending to purge

²purgative *noun* (1626)
: a purging medicine **:** CATHARTIC

pur•ga•to•ri•al \,pər-gə-'tōr-ē-əl, -'tȯr-\ *adjective* (15th century)
1 : of, relating to, or suggestive of purgatory
2 : cleansing of sin **:** EXPIATORY

pur•ga•to•ry \'pər-gə-,tōr-ē, -,tȯr-\ *noun, plural* **-ries** [Middle English, from Anglo-French or Medieval Latin; Anglo-French *purgatorie,* from Medieval Latin *purgatorium,* from Late Latin, neuter of *purgatorius* purging, from Latin *purgare*] (13th century)
1 : an intermediate state after death for expiatory purification; *specifically* **:** a place or state of punishment wherein according to Roman Catholic doctrine the souls of those who die in God's grace may make satisfaction for past sins and so become fit for heaven
2 : a place or state of temporary suffering or misery

¹purge \'pərj\ *verb* **purged; purg•ing** [Middle English, from Middle French *purgier,* from Latin *purigare, purgare* to purify, purge, from *purus* pure + *-igare* (akin to *agere* to drive, do) — more at ACT] (14th century)
transitive verb
1 a : to clear of guilt **b :** to free from moral or ceremonial defilement
2 a : to cause evacuation from (as the bowels) **b** (1) **:** to make free of something unwanted ⟨*purge* a manhole of gas⟩ ⟨*purge* yourself of fear⟩ (2) **:** to free (as a boiler) of sediment or relieve (as a steam pipe) of trapped air by bleeding **c** (1) **:** to rid (as a nation or party) by a purge (2) **:** to get rid of ⟨the leaders had been *purged*⟩ ⟨*purge* money-losing operations⟩
intransitive verb
1 : to become purged
2 : to have or produce frequent evacuations
3 : to cause purgation
— **purg•er** *noun*

²purge *noun* (1563)
1 : something that purges; *especially* **:** PURGATIVE
2 a : an act or instance of purging **b :** the removal of elements or members regarded as undesirable and especially as treacherous or disloyal

pu•ri \'pů-rē\ *noun, plural* **puri** *or* **puris** [Hindi *pūrī,* from Sanskrit *pūra*] (circa 1885)
: a puffy fried wheat cake of India

pu•ri•fi•ca•tion \,pyůr-ə-fə-'kā-shən\ *noun* (14th century)
: the act or an instance of purifying or of being purified

pu•ri•fi•ca•tor \'pyůr-ə-fə-,kā-tər\ *noun* (1853)
1 : a linen cloth used to wipe the chalice after celebration of the Eucharist
2 : one that purifies

pu•ri•fi•ca•to•ry \pyů-'ri-fi-kə-,tōr-ē, 'pyůr-(ə-)fə-kə-, -,tȯr-\ *adjective* (1610)
: serving, tending, or intended to purify

pu•ri•fy \'pyůr-ə-,fī\ *verb* **-fied; -fy•ing** [Middle English *purifien,* from Middle French *purifier,* from Latin *purificare,* from Latin *purus* + *-ificare* -ify] (14th century)
transitive verb
: to make pure: as **a :** to clear from material defilement or imperfection **b :** to free from guilt or moral or ceremonial blemish **c :** to free from undesirable elements
intransitive verb
: to grow or become pure or clean
— **pu•ri•fi•er** \-,fī(-ə)r\ *noun*

Pu•rim \'pů-rim, 'pyů-, -,ēm; pů-'rim, pyů-, -'rēm\ *noun* [Hebrew *pūrīm,* literally, lots; from the casting of lots by Haman (Esther 9:24–26)] (1535)
: a Jewish holiday celebrated on the 14th of Adar in commemoration of the deliverance of the Jews from the massacre plotted by Haman

pu•rine \'pyůr-,ēn\ *noun* [German *Purin,* from Latin *purus* pure + New Latin *uricus* uric (from English *uric*) + German *-in* ²-ine] (1898)
1 : a crystalline base $C_5H_4N_4$ that is the parent of compounds of the uric-acid group
2 : a derivative of purine; *especially* **:** a base (as adenine or guanine) that is a constituent of DNA or RNA

pur•ism \'pyůr-,i-zəm\ *noun* (1803)
1 : an example of rigid adherence to or insistence on purity or nicety especially in use of words; *especially* **:** a word, phrase, or sense used chiefly by purists
2 : the quality or practice of adherence to purity especially in language

pur•ist \'pyůr-ist\ *noun* (circa 1706)
: one who adheres strictly and often excessively to a tradition; *especially* **:** one preoccupied with the purity of a language and its protection from the use of foreign or altered forms
— **pu•ris•tic** \pyů-'ris-tik\ *adjective*
— **pu•ris•ti•cal•ly** \-ti-k(ə-)lē\ *adverb*

¹pu•ri•tan \'pyůr-ə-tᵊn\ *noun* [probably from Late Latin *puritas* purity] (1572)
1 *capitalized* **:** a member of a 16th and 17th century Protestant group in England and New England opposing as unscriptural the ceremonial worship and the prelacy of the Church of England
2 : one who practices or preaches a more rigorous or professedly purer moral code than that which prevails

²puritan *adjective, often capitalized* (1589)
: of or relating to puritans, the Puritans, or puritanism

pu•ri•tan•i•cal \,pyůr-ə-'ta-ni-kəl\ *adjective* (1607)
1 : PURITAN
2 : of, relating to, or characterized by a rigid morality
— **pu•ri•tan•i•cal•ly** \-k(ə-)lē\ *adverb*

pu•ri•tan•ism \'pyůr-ə-tᵊn-,i-zəm\ *noun* (1573)
1 *capitalized* **:** the beliefs and practices characteristic of the Puritans
2 : strictness and austerity especially in matters of religion or conduct

pu•ri•ty \'pyůr-ə-tē\ *noun* [Middle English *purete,* from Old French *pureté,* from Late Latin *puritat-, puritas,* from Latin *purus* pure] (13th century)
1 : the quality or state of being pure
2 : SATURATION 4a

Pur•kin•je cell \(,)pər-'kin-jē-\ *noun* [Jan *Purkinje*] (circa 1890)
: any of numerous nerve cells that occupy the middle layer of the cerebellar cortex and are characterized by a large globose body with massive dendrites directed outward and a single slender axon directed inward

Purkinje fiber *noun* (circa 1890)
: any of the modified cardiac muscle fibers that have few nuclei, granulated central cytoplasm, and sparse peripheral striations and make up a network of conducting tissue in the myocardium

¹purl \'pər(-ə)l\ *noun* [Middle English] (14th century)
1 : gold or silver thread or wire for embroidering or edging
2 : the intertwisting of thread that knots a stitch usually along an edge
3 : PURL STITCH

²purl (1526)
transitive verb
1 a : to embroider with gold or silver thread **b :** to edge or border with gold or silver embroidery
2 : to knit in purl stitch
intransitive verb
: to do knitting in purl stitch

³purl *noun* [perhaps of Scandinavian origin; akin to Norwegian *purla* to ripple] (circa 1522)
1 : a purling or swirling stream or rill
2 : a gentle murmur or movement (as of purling water)

⁴purl *intransitive verb* (1591)
1 : EDDY, SWIRL
2 : to make a soft murmuring sound like that of a purling stream

pur•lieu \'pərl-(,)yü, 'pər-(,)lü\ *noun* [Middle English *purlewe* land severed from an English royal forest by perambulation, from Anglo-French *puralé* perambulation, from Old French *puraler* to go through, from *pur-* for, through + *aler* to go — more at PURCHASE] (15th century)
1 a : an outlying or adjacent district **b** *plural* **:** ENVIRONS, NEIGHBORHOOD
2 a : a frequently visited place **:** HAUNT **b** *plural* **:** CONFINES, BOUNDS

pur•lin \'pər-lən\ *noun* [origin unknown] (15th century)
: a horizontal member in a roof

pur•loin \(,)pər-'lȯin, 'pər-,\ *transitive verb* [Middle English, to put away, misappropriate, from Anglo-French *purloigner,* from Old French *porloigner* to put off, delay, from *por-* forward + *loing* at a distance, from Latin *longe,* from *longus* long — more at PURCHASE, LONG] (15th century)
: to appropriate wrongfully and often by a breach of trust
synonym see STEAL
— **pur•loin•er** *noun*

purl stitch *noun* [¹*purl*] (1885)
: a knitting stitch usually made with the yarn at the front of the work by inserting the right needle into the front of a loop on the left needle from the right, catching the yarn with the right needle, and bringing it through to form a new loop — compare KNIT STITCH

pu•ro•my•cin \,pyůr-ə-'mī-sᵊn\ *noun* [*pur*ine + *-o-* + *-mycin*] (1953)
: an antibiotic $C_{22}H_{29}N_7O_5$ that is obtained from an actinomycete (*Streptomyces alboniger*) and is used especially as a potent inhibitor of protein synthesis

¹pur•ple \'pər-pəl\ *adjective* **pur•pler** \-p(ə-)lər\; **pur•plest** \-p(ə-)ləst\ [Middle English *purpel,* alteration of *purper,* from Old English *purpuran* of purple, genitive of *purpure* purple color, from Latin *purpura,* from Greek *porphyra*] (before 12th century)
1 : REGAL, IMPERIAL
2 : of the color purple
3 a : highly rhetorical **:** ORNATE **b :** marked by profanity

²purple *noun* (15th century)
1 a (1) **:** cloth dyed purple (2) **:** a garment of such color; *especially* **:** a purple robe worn as an emblem of rank or authority **b** (1) **:** TYRIAN PURPLE (2) **:** any of various colors that fall about midway between red and blue in hue **c** (1) **:** a mollusk (as of the genus *Purpura*) yielding a purple dye and especially the Tyrian purple of ancient times (2) **:** a pigment or dye that colors purple

2 a : imperial or regal rank or power **b :** high rank or station

³purple *verb* **pur·pled; pur·pling** \-p(ə-)liŋ\ (15th century)
transitive verb
: to make purple
intransitive verb
: to become purple

pur·ple·heart \ˈpər-pəl-ˌhärt\ *noun* (1796)
: a strong durable purplish wood that is obtained from various leguminous trees (genus *Peltogyne*); *also* **:** a tree producing such wood

Purple Heart *noun* (1932)
: a U.S. military decoration awarded to any member of the armed forces wounded or killed in action

purple loosestrife *noun* (1548)
: an Old World marsh herb (*Lythrum salicaria*) of the loosestrife family that is naturalized in the eastern U.S. and has long spikes of purple flowers

purple martin *noun* (1743)
: a large swallow (*Progne subis*) of North America the males of which have bluish black plumage

purple passage *noun* [translation of Latin *pannus purpureus* purple patch; from the traditional splendor of purple cloth as contrasted with plainer materials] (1895)
1 : a passage conspicuous for brilliance or effectiveness in a work that is dull, commonplace, or uninspired
2 *chiefly British* **:** a piece of obtrusively ornate writing — called also *purple patch*

purple scale *noun* (circa 1909)
: a brownish or purplish armored scale (*Lepidosaphes beckii*) that is destructive to citrus fruit

pur·plish \ˈpər-p(ə-)lish\ *adjective* (1562)
: somewhat purple

pur·ply \ˈpər-p(ə-)lē\ *adjective* (1725)
: PURPLISH

¹pur·port \ˈpər-ˌpōrt, -ˌpȯrt\ *noun* [Middle English, from Anglo-French, content, tenor, from *purporter* to contain, from Old French *porporter* to convey, from *por-* forward + *porter* to carry — more at PURCHASE, PORT] (15th century)
: meaning conveyed, professed, or implied **:** IMPORT; *also* **:** SUBSTANCE, GIST

²pur·port \(ˌ)pər-ˈpōrt, -ˈpȯrt\ *transitive verb* (1528)
1 : to have the often specious appearance of being, intending, or claiming (something implied or inferred) ⟨a book that *purports* to be an objective analysis⟩; *also* **:** CLAIM ⟨foreign novels which he *purports* to have translated —Mary McCarthy⟩
2 : INTEND, PURPOSE

pur·port·ed \-ˈpōr-təd, -ˈpȯr-\ *adjective* (1894)
: REPUTED, ALLEGED ⟨took gullible tourists to *purported* ancient sites⟩

pur·port·ed·ly \-lē\ *adverb* (1942)
: it is purported **:** OSTENSIBLY, ALLEGEDLY

¹pur·pose \ˈpər-pəs\ *noun* [Middle English *purpos,* from Old French, from *purposer* to purpose, from Latin *proponere* (perfect indicative *proposui*) to propose — more at PROPOUND] (14th century)
1 a : something set up as an object or end to be attained **:** INTENTION **b :** RESOLUTION, DETERMINATION
2 : a subject under discussion or an action in course of execution
synonym see INTENTION
— on purpose : by intent **:** INTENTIONALLY

²purpose *transitive verb* **pur·posed; pur·pos·ing** (14th century)
: to propose as an aim to oneself

pur·pose–built \ˌpər-pəs-ˈbilt\ *adjective* (1954)
chiefly British **:** built for a particular purpose

pur·pose·ful \ˈpər-pəs-fəl\ *adjective* (1853)
1 : having a purpose: as **a :** MEANINGFUL ⟨*purposeful* activities⟩ **b :** INTENTIONAL ⟨*purposeful* ambiguity⟩
2 : full of determination ⟨a *purposeful* man⟩
— pur·pose·ful·ly \-fə-lē\ *adverb*
— pur·pose·ful·ness *noun*

pur·pose·less \-ləs\ *adjective* (circa 1552)
: having no purpose **:** AIMLESS, MEANINGLESS
— pur·pose·less·ly *adverb*
— pur·pose·less·ness *noun*

pur·pose·ly \-lē\ *adverb* (15th century)
: with a deliberate or express purpose

pur·po·sive \ˈpər-pə-siv, (ˌ)pər-ˈpō-\ *adjective* (1855)
1 : serving or effecting a useful function though not as a result of planning or design
2 : having or tending to fulfill a conscious purpose or design **:** PURPOSEFUL
— pur·po·sive·ly *adverb*
— pur·po·sive·ness *noun*

pur·pu·ra \ˈpər-pyə-rə, -pə-rə\ *noun* [New Latin, from Latin, purple color] (1753)
: any of several hemorrhagic states characterized by patches of purplish discoloration resulting from extravasation of blood into the skin and mucous membranes
— pur·pu·ric \ˌpər-ˈpyu̇r-ik\ *adjective*

pur·pure \ˈpər-pyər\ *noun* [Middle English, from Old English, purple] (1535)
: the heraldic color purple

¹purr \ˈpər\ *noun* [imitative] (1601)
: a low vibratory murmur typical of an apparently contented or pleased cat

²purr *intransitive verb* (1620)
1 : to make a purr or a sound like a purr ⟨cars *purring* along the highway⟩
2 a : to speak in a manner that resembles a purr **b :** to speak in a malicious catty manner
— purr·ing·ly \-iŋ-lē\ *adverb*

¹purse \ˈpərs\ *noun* [Middle English *purs,* from Old English, modification of Medieval Latin *bursa,* from Late Latin, ox hide, from Greek *byrsa*] (before 12th century)
1 a (1) **:** a small bag for money (2) **:** a receptacle (as a pocketbook) for carrying money and often other small objects **b :** a receptacle (as a pouch) shaped like a purse
2 a : RESOURCES, FUNDS **b :** a sum of money offered as a prize or present; *also* **:** the total amount of money offered in prizes for a given event
— purse·like \-ˌlīk\ *adjective*

²purse *transitive verb* **pursed; purs·ing** (14th century)
1 : to put into a purse
2 : PUCKER, KNIT

purse–proud \ˈpərs-ˌprau̇d\ *adjective* (1681)
: proud because of one's wealth especially in the absence of other distinctions

purs·er \ˈpər-sər\ *noun* [Middle English, from *purs* purse] (15th century)
1 : an official on a ship responsible for papers and accounts and on a passenger ship also for the comfort and welfare of passengers
2 : a steward on an airliner

purse seine *noun* (1870)
: a large seine designed to be set by two boats around a school of fish and so arranged that after the ends have been brought together the bottom can be closed

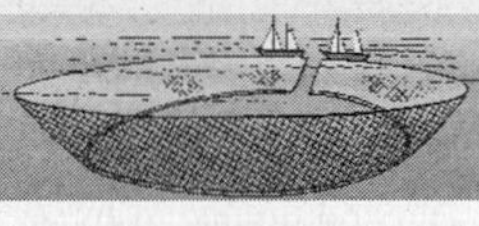
purse seine

— purse seiner *noun*
— purse seining *noun*

purse strings *noun plural* (15th century)
: financial resources; *also* **:** control over these resources

purs·lane \ˈpər-slən, -ˌslān\ *noun* [Middle English, from Middle French *porcelaine,* from Late Latin *porcillagin-, porcillago,* alteration of Latin *porcillaca,* alteration of *portulaca*] (14th century)
: any of a family (Portulacaceae, the purslane family) of cosmopolitan usually succulent herbs; *especially* **:** a fleshy-leaved trailing plant (*Portulaca oleracea*) with tiny yellow flowers that is a common troublesome weed but is sometimes eaten as a potherb or in salads

pur·su·ance \pər-ˈsü-ən(t)s\ *noun* (1605)
: the act of pursuing; *especially* **:** a carrying out or into effect **:** PROSECUTION ⟨in *pursuance* of his duties⟩

pur·su·ant to \-ənt-\ *preposition* (1648)
: in carrying out **:** in conformity with **:** ACCORDING TO

pur·sue \pər-ˈsü, -ˈsyü\ *verb* **pur·sued; pur·su·ing** [Middle English, from Anglo-French *pursuer,* from Old French *poursuir,* from Latin *prosequi,* from *pro-* forward + *sequi* to follow — more at PRO-, SUE] (14th century)
transitive verb
1 : to follow in order to overtake, capture, kill, or defeat
2 : to find or employ measures to obtain or accomplish **:** SEEK ⟨*pursue* a goal⟩
3 : to proceed along ⟨*pursues* a northern course⟩
4 a : to engage in ⟨*pursue* a hobby⟩ **b :** to follow up or proceed with ⟨*pursue* an argument⟩
5 : to continue to afflict **:** HAUNT ⟨was *pursued* by horrible memories⟩
6 : ²CHASE 1c ⟨*pursued* by dozens of fans⟩
intransitive verb
: to go in pursuit
synonym see CHASE
— pur·su·er *noun*

pur·suit \pər-ˈsüt, -ˈsyüt\ *noun* [Middle English, from Middle French *poursuite,* from *poursuir*] (14th century)
1 : the act of pursuing
2 : an activity that one engages in as a vocation, profession, or avocation **:** OCCUPATION
synonym see WORK

pursuit plane *noun* (circa 1918)
: a fighter plane especially of the period before World War II

pur·sui·vant \ˈpər-si-vənt, -swi-\ *noun* [Middle English *pursevant* attendant of a herald, from Middle French *poursuivant,* literally, follower, from present participle of *poursuir, poursuivre* to pursue] (14th century)
1 : an officer of arms ranking below a herald but having similar duties
2 : FOLLOWER, ATTENDANT

¹pur·sy \ˈpə-sē, ˈpər-sē\ *adjective* **pur·si·er; -est** [Middle English, from Anglo-French *pursif,* alteration of Middle French *polsif,* from *poulser, polser* to beat, push, pant — more at PUSH] (15th century)
1 : short-winded especially because of corpulence
2 : FAT
— pur·si·ness *noun*

²pursy \ˈpər-sē\ *adjective* **purs·i·er; -est** [¹*purse*] (1552)
1 : having a puckered appearance
2 : PURSE-PROUD

pur·te·nance \ˈpərt-nən(t)s, ˈpər-tᵊn-ən(t)s\ *noun* [Middle English, literally, appendage, modification of Middle French *partenance,* from *partenir* to pertain — more at PERTAIN] (15th century)
: ENTRAILS, PLUCK

pu·ru·lence \ˈpyu̇r-ə-lən(t)s, ˈpyu̇r-yə-\ *noun* (1597)
: the quality or state of being purulent; *also* **:** PUS

pu·ru·lent \-lənt\ *adjective* [Latin *purulentus,* from *pur-, pus* pus] (1597)
1 : containing, consisting of, or being pus ⟨a *purulent* discharge⟩
2 : accompanied by suppuration

pur·vey \(ˌ)pər-ˈvā, ˈpər-ˌ\ *transitive verb* **pur·veyed; pur·vey·ing** [Middle English *purveien,* from Middle French *porveeir,* from Latin *providēre* to provide] (14th century)
1 : to supply (as provisions) usually as a matter of business
2 : PEDDLE 2

pur·vey·ance \-ən(t)s\ *noun* (14th century)
: the act or process of purveying or procuring

pur·vey·or \-ər\ *noun* (14th century)
1 : one that purveys
2 : VICTUALLER, CATERER

pur·view \ˈpər-ˌvyü\ *noun* [Middle English *purveu,* from Anglo-French *purveu est* it is provided (opening phrase of a statute)] (15th century)
1 a : the body or enacting part of a statute **b :** the limit, purpose, or scope of a statute
2 : the range or limit of authority, competence, responsibility, concern, or intention
3 : range of vision, understanding, or cognizance

pus \ˈpəs\ *noun* [Latin *pur-, pus* — more at FOUL] (1541)
: thick opaque usually yellowish white fluid matter formed by suppuration and composed of exudate containing white blood cells, tissue debris, and microorganisms

Pu·sey·ism \ˈpyü-zē-ˌi-zəm, -sē-\ *noun* [Edward Bouverie *Pusey*] (1838)
: TRACTARIANISM
— Pu·sey·ite \-ˌīt\ *noun*

¹push \ˈpu̇sh\ *verb* [Middle English *pusshen,* from Middle French *poulser* to beat, push, from Old French, from Latin *pulsare,* frequentative of *pellere* to drive, strike — more at FELT] (13th century)
transitive verb
1 a : to press against with force in order to drive or impel **b :** to move or endeavor to move away or ahead by steady pressure without striking
2 a : to thrust forward, downward, or outward **b :** to cause to increase **:** RAISE ⟨*push* prices to record levels⟩ **c :** to hit (a ball) toward the right from a right-handed swing or toward the left from a left-handed swing — compare PULL
3 a : to press or urge forward to completion **b :** to urge or press the advancement, adoption, or practice of ⟨*pushed* a bill in the legislature⟩; *especially* **:** to make aggressive efforts to sell ⟨we're *pushing* ham this week⟩ **c :** to engage in the illicit sale of (narcotics)
4 : to bear hard upon so as to involve in difficulty ⟨poverty *pushed* them to the breaking point⟩
5 : to approach in age or number ⟨grandmother must be *pushing* 75⟩
intransitive verb
1 : to press against something with steady force in or as if in order to impel
2 : to press forward energetically against opposition
3 : to exert oneself continuously, vigorously, or obtrusively to gain an end ⟨*pushing* for higher wages⟩
— push one's luck : to take an increasing risk

²push *noun* (1563)
1 : a vigorous effort to attain an end **:** DRIVE: **a :** a military assault or offensive **b :** an advance that overcomes obstacles **c :** a campaign to promote a product
2 : a time for action **:** EMERGENCY
3 a : an act of pushing **:** SHOVE **b** (1) **:** a physical force steadily applied in a direction away from the body exerting it (2) **:** a nonphysical pressure **:** INFLUENCE, URGE **c :** vigorous enterprise or energy
4 a : an exertion of influence to promote another's interests **b :** stimulation to activity **:** IMPETUS
— push comes to shove : a decisive moment comes ⟨backed down when *push came to shove*⟩

push around *transitive verb* (1923)
: to impose on contemptuously

push·ball \ˈpu̇sh-ˌbȯl\ *noun* (1896)
: a game in which each of two sides endeavors to push an inflated originally leather-covered ball six feet (1.8 meters) in diameter across its opponents' goal; *also* **:** the ball used

push–bike \-ˌbīk\ *noun* (1913)
British **:** BICYCLE — called also *push bicycle*

push broom *noun* (1926)
: a long-handled wide brush that is designed to be pushed and is used for sweeping

push–button *adjective* (1916)
1 : operated or done by means of push buttons ⟨a *push-button* phone⟩
2 : using or dependent on complex and more or less self-operating mechanisms that are put in operation by a simple act comparable to pushing a button ⟨*push-button* warfare⟩

push button *noun* (1878)
: a small button or knob that when pushed operates something especially by closing an electric circuit

push·cart \ˈpu̇sh-ˌkärt\ *noun* (1893)
: a cart or barrow pushed by hand

push·chair \-ˌcher, -ˌchar\ *noun* (1921)
chiefly British **:** STROLLER

push·down \-ˌdau̇n\ *noun* (1961)
: a store of data (as in a computer) from which the most recently stored item must be the first retrieved — called also *pushdown list, pushdown stack*

push·er \ˈpu̇-shər\ *noun* (1591)
: one that pushes; *especially* **:** one that pushes illegal drugs

push·ful \-fəl\ *adjective* (1896)
chiefly British **:** PUSHING
— push·ful·ness *noun, chiefly British*

push·ing *adjective* (1692)
1 : marked by ambition, energy, enterprise, and initiative
2 : marked by tactless forwardness or officious intrusiveness

push off *intransitive verb* (1925)
: SET OUT ⟨we *pushed off* for home⟩

push on *intransitive verb* (1718)
: to continue on one's way **:** PROCEED

push·over \ˈpu̇sh-ˌō-vər\ *noun* (1906)
1 : something accomplished without difficulty **:** SNAP
2 : an opponent who is easy to defeat or a victim who is capable of no effective resistance
3 : someone unable to resist an attraction or appeal **:** SUCKER

push·pin \-ˌpin\ *noun* (1907)
: a pin that has a roughly cylindrical head and that is easily inserted and withdrawn (as from a bulletin board)

push–pull \-ˈpu̇l\ *adjective* (1922)
: relating to or being an arrangement of two electronic circuit elements (as transistors) such that an alternating input causes them to send current through a load alternately ⟨a *push-pull* circuit⟩
— push–pull *noun*

Push·tu \ˈpəsh-(ˌ)tü\ *variant of* PASHTO

Push·tun *variant of* PASHTUN

push–up \ˈpu̇sh-ˌəp\ *noun* (1942)
: a conditioning exercise performed in a prone position by raising and lowering the body with the straightening and bending of the arms while keeping the back straight and supporting the body on the hands and toes

pushy \ˈpu̇-shē\ *adjective* **push·i·er; -est** (1936)
: aggressive often to an objectionable degree **:** FORWARD
— push·i·ly \ˈpu̇-shə-lē\ *adverb*
— push·i·ness \ˈpu̇-shē-nəs\ *noun*

pu·sil·la·nim·i·ty \ˌpyü-sə-lə-ˈni-mə-tē *also* ˌpyü-zə-\ *noun* (14th century)
: the quality or state of being pusillanimous **:** COWARDLINESS

pu·sil·lan·i·mous \-ˈla-nə-məs\ *adjective* [Late Latin *pusillanimis,* from Latin *pusillus* very small (diminutive of *pusus* boy) + *animus* spirit; perhaps akin to Latin *puer* child — more at PUERILE, ANIMATE] (1586)
: lacking courage and resolution **:** marked by contemptible timidity
synonym see COWARDLY
— pu·sil·lan·i·mous·ly *adverb*

¹puss \ˈpu̇s\ *noun* [origin unknown] (circa 1530)
1 : CAT
2 : GIRL

²puss *noun* [Irish *pus* mouth] (circa 1890)
slang **:** FACE

puss·ley \ˈpəs-lē\ *noun* [by alteration] (1833)
: PURSLANE

¹pussy \ˈpu̇-sē\ *noun, plural* **puss·ies** [¹*puss*] (1726)
1 : CAT
2 : a catkin of the pussy willow

²pus·sy \ˈpu̇-sē\ *noun, plural* **pussies** [perhaps of Low German or Scandinavian origin; akin to Old Norse *pūss* pocket, pouch, Low German *pūse* vulva, Old English *pusa* bag] (circa 1879)
1 : VULVA — usually considered vulgar
2 a : SEXUAL INTERCOURSE — usually considered vulgar **b :** the female partner in sexual intercourse — usually considered vulgar

³pus·sy \ˈpə-sē\ *adjective* **pus·si·er; -est** (circa 1890)
: full of or resembling pus

⁴pus·sy \ˈpə-sē\ *variant of* ¹PURSY

pussy·cat \ˈpu̇-sē-ˌkat\ *noun* (1805)
1 : CAT
2 : one that is weak, compliant, or amiable **:** SOFTY

pussy·foot \ˈpu̇-sē-ˌfu̇t\ *intransitive verb* (1903)
1 : to tread or move warily or stealthily
2 : to refrain from committing oneself
— pussy·foot·er *noun*

pussy·toes \ˈpu̇-sē-ˌtōz\ *or* **puss·y's–toes** \-sēz-\ *noun plural but singular or plural in construction* (1892)
: any of a genus (*Antennaria*) of woolly or hoary chiefly temperate composite herbs that have small usually whitish discoid flower heads and a pappus formed of club-shaped bristles

pussy willow \ˈpu̇-sē-\ *noun* (1869)
: a willow (as the American *Salix discolor*) having large cylindrical silky aments

pussy willow

¹pus·tu·lant \ˈpəs-chə-lənt; ˈpəs-tyə-, -tə-\ *noun* (1871)
: an agent (as a chemical) that induces pustule formation

²pustulant *adjective* (circa 1890)
: producing pustules

pus·tu·lar \-lər\ *adjective* (1739)
1 : of, relating to, or resembling pustules
2 : covered with pustular prominences **:** PUSTULATED

pus·tu·lat·ed \-ˌlā-təd\ *adjective* (1732)
: covered with pustules

pus·tu·la·tion \ˌpəs-chə-ˈlā-shən, ˌpəs-tyə-, -tə-\ *noun* (circa 1860)
1 : the act of producing pustules **:** the state of having pustules
2 : PUSTULE

pus·tule \ˈpəs-(ˌ)chü(ə)l, -(ˌ)tyü(ə)l, -(ˌ)tü(ə)l\ *noun* [Middle English, from Latin *pustula;* akin to Lithuanian *pusti* to blow, Greek *physa* breath] (14th century)
1 : a small circumscribed elevation of the skin containing pus and having an inflamed base
2 : a small often distinctively colored elevation or spot resembling a blister or pimple

¹put \ˈpu̇t\ *verb* **put; put·ting** [Middle English *putten;* akin to Old English *putung* instigation, Middle Dutch *poten* to plant] (12th century)
transitive verb

1 a : to place in a specified position or relationship **:** LAY ⟨*put* the book on the table⟩ **b :** to move in a specified direction **c** (1) **:** to send (as a weapon or missile) into or through something **:** THRUST (2) **:** to throw with an overhand pushing motion ⟨*put* the shot⟩ **d :** to bring into a specified state or condition ⟨a reapportionment . . . that was *put* into effect at the September primaries —*Current Biography*⟩
2 a : to cause to endure or suffer something **:** SUBJECT ⟨*put* traitors to death⟩ **b :** IMPOSE, INFLICT ⟨*put* a special tax on luxuries⟩
3 a : to set before one for judgment or decision ⟨*put* the question⟩ **b :** to call for a formal vote on ⟨*put* the motion⟩
4 a (1) **:** to convey into another form ⟨want to *put* my feelings into words⟩ (2) **:** to translate into another language or style ⟨*put* the poem into English⟩ (3) **:** ADAPT ⟨lyrics *put* to music⟩ **b :** EXPRESS, STATE ⟨*putting* it mildly⟩
5 a : to devote (oneself) to an activity or end ⟨*put* himself to winning back their confidence⟩ **b :** APPLY ⟨*put* her mind to the problem⟩ **c :** ASSIGN ⟨*put* them to work⟩ **d :** to cause to perform an action **:** URGE ⟨*put* the horse over the fence⟩ **e :** IMPEL, INCITE ⟨*put* them into a frenzy⟩
6 a : REPOSE, REST ⟨*puts* his faith in reason⟩ **b :** INVEST ⟨*put* her money in the company⟩
7 a : to give as an estimate ⟨*put* the time as about eleven⟩ **b :** ATTACH, ATTRIBUTE ⟨*puts* a high value on their friendship⟩ **c :** IMPUTE ⟨*put* the blame on the partners⟩
8 : BET, WAGER ⟨*put* $2 on the favorite⟩
intransitive verb
1 : to start in motion **:** GO; *especially* **:** to leave in a hurry
2 *of a ship* **:** to take a specified course ⟨*put* down the river⟩
— put forth 1 a : ASSERT, PROPOSE **b :** to make public **:** ISSUE **2 :** to bring into action **:** EXERT **3 :** to produce or send out by growth ⟨*put forth* leaves⟩ **4 :** to start out
— put forward : PROPOSE ⟨*put forward* a theory⟩
— put in mind : REMIND
— put one's finger on : IDENTIFY ⟨*put his finger on* the cause of the trouble⟩
— put one's foot down : to take a firm stand
— put one's foot in one's mouth : to make a tactless or embarrassing blunder
— put paid to *British* **:** to finish off **:** WIPE OUT
— put the arm on *or* **put the bite on :** to ask for money
— put the finger on : to inform on ⟨*put the finger on* . . . heroin pushers —Barrie Zwicker⟩
— put the make on : to make sexual advances toward
— put to bed : to make the final preparations for printing (as a newspaper)
— put together 1 : to create as a unified whole **:** CONSTRUCT **2 :** ADD, COMBINE
— put to it : to give difficulty to **:** press hard ⟨had been *put to it* to keep up⟩

²put *noun* (14th century)
1 : a throw made with an overhand pushing motion; *specifically* **:** the act or an instance of putting the shot
2 : an option to sell a specified amount of a security (as a stock) or commodity (as wheat) at a fixed price at or within a specified time — compare CALL 3d

³put *adjective* (1848)
: being in place **:** FIXED, SET ⟨stay *put* until I call⟩

put about (1748)
intransitive verb
of a ship **:** to change direction **:** go on another tack
transitive verb
: to cause to change course or direction

put across *transitive verb* (1919)
1 : PUT OVER 3
2 : to convey effectively or forcefully

put–and–take \ˌpu̇t-ᵊn-ˈtāk\ *noun* (1922)
: any of various games of chance played with a teetotum or with dice in which players contribute to a pool and take from it according to the instructions on the top or dice

pu·ta·tive \ˈpyü-tə-tiv\ *adjective* [Middle English, from Late Latin *putativus,* from Latin *putatus,* past participle of *putare* to think] (15th century)
1 : commonly accepted or supposed
2 : assumed to exist or to have existed
— pu·ta·tive·ly *adverb*

put away *transitive verb* (14th century)
1 a : DISCARD, RENOUNCE ⟨to *put* grief *away* is disloyal to the memory of the departed —H. A. Overstreet⟩ **b :** DIVORCE
2 : to eat or drink up **:** CONSUME
3 a : to confine especially in a mental institution **b :** BURY **c :** KILL

put by *transitive verb* (15th century)
1 *archaic* **:** REJECT
2 : to lay aside **:** SAVE

put–down \ˈpu̇t-ˌdau̇n\ *noun* (1962)
: an act or instance of putting down; *especially* **:** a humiliating remark **:** SQUELCH

put down *transitive verb* (14th century)
1 : to bring to an end **:** STOP ⟨*put down* a riot⟩
2 a : DEPOSE, DEGRADE **b :** DISPARAGE, BELITTLE ⟨mentioned his poetry only to *put* it *down*⟩ **c :** DISAPPROVE, CRITICIZE ⟨was *put down* for the way she dressed⟩ **d :** HUMILIATE, SQUELCH ⟨*put* him *down* with a sharp retort⟩
3 : to make ineffective **:** CHECK ⟨*put down* the gossip⟩
4 : to do away with (as an injured, sick, or aged animal) **:** DESTROY
5 a : to put in writing ⟨*put* it *down* truthfully⟩ **b :** to enter in a list ⟨*put* me *down* for a donation⟩
6 a : to place in a category ⟨I *put* him *down* as a hypochondriac —O. S. J. Gogarty⟩ **b :** ATTRIBUTE ⟨*put* it *down* to inexperience⟩
7 : to pack or preserve for future use
8 : CONSUME ⟨*putting down* helping after helping —Carson McCullers⟩
— put down roots : to establish a permanent residence

put in (15th century)
transitive verb
1 : to make a formal offer or declaration of ⟨*put in* a plea of guilty⟩
2 : to come in with **:** INTERPOSE ⟨*put in* a word for his brother⟩
3 : to spend (time) especially at some occupation or job ⟨*put in* six hours at the office⟩
4 : PLANT ⟨*put in* a crop⟩
intransitive verb
1 : to call at or enter a place; *especially* **:** to enter a harbor or port
2 : to make an application, request, or offer — often used with *for* ⟨had to retire and *put in* for a pension —Seymour Nagan⟩

put·log \ˈpu̇t-ˌlȯg, ˈpət-, -ˌläg\ *noun* [probably alteration of earlier *putlock,* perhaps from ³*put* + ²*lock*] (1645)
: one of the short timbers that support the flooring of a scaffold

put off *transitive verb* (14th century)
1 a : DISCONCERT **b :** REPEL
2 a : to hold back to a later time **b :** to induce to wait ⟨*put* the bill collector *off*⟩
3 : to rid oneself of **:** TAKE OFF
4 : to sell or pass fraudulently

¹put–on \ˈpu̇t-ˈȯn, -ˈän\ *adjective* (1621)
: PRETENDED, ASSUMED

²put–on \ˈpu̇t-ˌȯn, -ˌän\ *noun* (circa 1927)
1 : an instance of putting someone on ⟨conversational *put-ons* are related to old-fashioned joshing —Jacob Brackman⟩
2 : PARODY, SPOOF ⟨a kind of *put-on* of every pretentious film ever made —C. A. Ridley⟩

put on *transitive verb* (15th century)
1 a : to dress oneself in **:** DON **b :** to make part of one's appearance or behavior **c :** FEIGN ⟨*put* a saintly manner *on*⟩
2 : to cause to act or operate **:** APPLY ⟨*put on* more speed⟩
3 a : ADD ⟨*put on* weight⟩ **b :** EXAGGERATE, OVERSTATE ⟨he's *putting* it *on* when he makes such claims⟩
4 : PERFORM, PRODUCE ⟨*put on* a play⟩
5 a : to mislead deliberately especially for amusement ⟨the interviewer . . . must be put down — or possibly, *put on* —Melvin Maddocks⟩ **b :** KID ⟨you're *putting* me *on*⟩

put·out \ˈpu̇t-ˌau̇t\ *noun* (1885)
: the retiring of a base runner or batter by a defensive player in baseball

put out (14th century)
transitive verb
1 : EXTINGUISH ⟨*put* the fire *out*⟩
2 : EXERT, USE ⟨*put out* considerable effort⟩
3 : PUBLISH, ISSUE
4 : to produce for sale
5 a : DISCONCERT, EMBARRASS **b :** ANNOY, IRRITATE **c :** INCONVENIENCE ⟨don't *put* yourself *out* for us⟩
6 : to cause to be out (as in baseball or cricket)
intransitive verb
1 : to set out from shore
2 : to make an effort
3 : to engage in sexual intercourse

put over *transitive verb* (1528)
1 : POSTPONE, DELAY
2 : PUT ACROSS 2
3 : to achieve or carry through by deceit or trickery ⟨*put* one *over* on me⟩

pu·tre·fac·tion \ˌpyü-trə-ˈfak-shən\ *noun* [Middle English *putrefaccion,* from Late Latin *putrefaction-, putrefactio,* from Latin *putrefacere*] (14th century)
1 : the decomposition of organic matter; *especially* **:** the typically anaerobic splitting of proteins by bacteria and fungi with the formation of foul-smelling incompletely oxidized products
2 : the state of being putrefied **:** CORRUPTION
— pu·tre·fac·tive \-ˈfak-tiv\ *adjective*

pu·tre·fy \ˈpyü-trə-ˌfī\ *verb* **-fied; -fy·ing** [Middle English *putrefien,* from Middle French & Latin; Middle French *putrefier,* from Latin *putrefacere,* from *putrēre* to be rotten + *facere* to make — more at DO] (14th century)
transitive verb
: to make putrid
intransitive verb
: to undergo putrefaction
synonym see DECAY

pu·tres·cence \pyü-ˈtre-sᵊn(t)s\ *noun* (1646)
: the state of being putrescent

pu·tres·cent \-sᵊnt\ *adjective* [Latin *putrescent-, putrescens,* present participle of *putrescere* to grow rotten, inchoative of *putrēre*] (1732)
1 : undergoing putrefaction **:** becoming putrid
2 : of or relating to putrefaction

pu·tres·ci·ble \-ˈtre-sə-bəl\ *adjective* (1797)
: liable to become putrid

pu·tres·cine \-ˈtre-ˌsēn\ *noun* [International Scientific Vocabulary, from Latin *putrescere*] (1887)
: a crystalline slightly poisonous ptomaine $C_4H_{12}N_2$ that occurs in small amounts in virtually all living things

pu·trid \ˈpyü-trəd\ *adjective* [Latin *putridus,* from *putrēre* to be rotten, from *puter, putris* rotten; akin to Latin *putēre* to stink — more at FOUL] (1598)
1 a : being in a state of putrefaction **:** ROTTEN **b :** of, relating to, or characteristic of putrefaction **:** FOUL ⟨a *putrid* odor⟩
2 a : morally corrupt **b :** totally objectionable
synonym see MALODOROUS
— pu·trid·i·ty \pyü-ˈtri-də-tē\ *noun*

— **pu·trid·ly** \'pyü-trəd-lē\ *adverb*

putsch \'pu̇ch\ *noun* [German] (1920)
: a secretly plotted and suddenly executed attempt to overthrow a government

putsch·ist \'pu̇-chist\ *noun* (1898)
: one who takes part in a putsch

putt \'pət\ *noun* [Scots, literally, shove, gentle push, from *putt, put* to put] (1743)
: a golf stroke made on a putting green to cause the ball to roll into or near the hole
— **putt** *verb*

put·tee \ˌpə-'tē, pu̇-; 'pə-tē\ *noun* [Hindi *paṭṭī* strip of cloth, from Sanskrit *paṭṭikā*] (1886)
1 : a cloth strip wrapped around the leg from ankle to knee
2 : a usually leather legging secured by a strap or catch or by laces

puttee

¹put·ter \'pu̇-tər\ *noun* (14th century)
: one that puts ⟨a *putter* of questions⟩

²putt·er \'pə-tər\ *noun* (1743)
1 : a golf club used in putting
2 : one that putts

³put·ter \'pə-tər\ *intransitive verb* [alteration of *potter*] (circa 1877)
1 : to move or act aimlessly or idly
2 : to work at random **:** TINKER
— **put·ter·er** \-tər-ər\ *noun*

put through *transitive verb* (1852)
1 : to carry to a successful conclusion ⟨*put through* a number of reforms⟩
2 a : to make a telephone connection for **b :** to obtain a connection for (a telephone call)

putt·ing green \'pə-tiŋ-\ *noun* (1841)
: a smooth grassy area at the end of a golf fairway containing the hole; *also* **:** a similar area usually with many holes that is used for practice

put·to \'pü-(ˌ)tō\ *noun, plural* **put·ti** \-(ˌ)tē\ [Italian, literally, boy, from (assumed) Vulgar Latin *puttus,* alteration of Latin *putus;* akin to Latin *puer* boy — more at PUERILE] (1644)
: a figure of an infant boy especially in European art of the Renaissance — usually used in plural

¹put·ty \'pə-tē\ *noun, plural* **putties** [French *potée* potter's glaze, literally, potful, from Old French, from *pot* pot — more at POTTAGE] (circa 1706)
1 a : a doughlike material typically made of whiting and linseed oil that is used especially to fasten glass in window frames and to fill crevices in woodwork **b :** any of various substances resembling putty in appearance, consistency, or use
2 : a light brownish gray to light grayish brown color
3 : one who is easily manipulated ⟨is *putty* in her hands⟩
— **put·ty·less** *adjective*
— **put·ty·like** *adjective*

²putty *transitive verb* **put·tied; put·ty·ing** (1734)
: to use putty on or apply putty to

putty knife *noun* (1858)
: an implement with a broad flat metal blade used especially for applying putty and for scraping

put·ty·root \'pə-tē-ˌrüt, -ˌru̇t\ *noun* (1817)
: a North American orchid (*Aplectrum hyemale*) having a slender naked rootstock and producing a solitary leaf and a scape bearing a raceme of brownish flowers

put–up \'pu̇t-'əp\ *adjective* (1810)
: arranged secretly beforehand

put up (14th century)
transitive verb
1 a : to place in a container or receptacle ⟨*put* his lunch *up* in a bag⟩ **b :** to put away (a sword) in a scabbard **:** SHEATHE **c :** to prepare so as to preserve for later use **:** CAN **d :** to put in storage
2 : to start (game) from cover
3 : to nominate for election
4 : to offer up (as a prayer)
5 : SET 16
6 : to offer for public sale ⟨*put* their possessions *up* for auction⟩
7 : to give food and shelter to **:** ACCOMMODATE
8 : to arrange (as a plot or scheme) with others ⟨*put up* a job to steal the jewels⟩
9 : BUILD, ERECT
10 a : to make a display of ⟨*put up* a bluff⟩ **b :** to engage in ⟨*put up* a struggle against odds⟩
11 a : CONTRIBUTE, PAY **b :** to offer as a prize or stake
12 *chiefly British* **:** to increase the amount of **:** RAISE
intransitive verb
: LODGE
— **put up to :** INCITE, INSTIGATE ⟨they *put* him *up to* playing the prank⟩
— **put up with :** to endure or tolerate without complaint or attempt at reprisal

put–up·on \'pu̇t-ə-ˌpȯn, -ˌpän\ *adjective* (1920)
: imposed upon **:** taken advantage of

¹puz·zle \'pə-zəl\ *verb* **puz·zled; puz·zling** \'pə-zə-liŋ, 'pəz-liŋ\ [origin unknown] (1602)
transitive verb
1 : to offer or represent to (as a person) a problem difficult to solve or a situation difficult to resolve **:** challenge mentally; *also* **:** to exert (as oneself) over such a problem or situation ⟨they *puzzled* their wits to find a solution⟩
2 *archaic* **:** COMPLICATE, ENTANGLE
3 : to solve with difficulty or ingenuity ⟨*puzzle* out an answer to a riddle⟩
intransitive verb
1 : to be uncertain as to action or choice
2 : to attempt a solution of a puzzle by guesswork or experiment ☆
— **puz·zler** \'pə-zə-lər, 'pəz-lər\ *noun*

²puzzle *noun* (circa 1612)
1 : the state of being puzzled **:** PERPLEXITY
2 a : something that puzzles **b :** a question, problem, or contrivance designed for testing ingenuity
synonym see MYSTERY

puz·zle·head·ed \'pə-zəl-ˌhe-dəd\ *adjective* (circa 1784)
: having or based on confused attitudes or ideas
— **puz·zle·head·ed·ness** *noun*

puz·zle·ment \'pə-zəl-mənt\ *noun* (1822)
1 : the state of being puzzled **:** PERPLEXITY
2 : PUZZLE

puzzling *adjective* (1666)
: difficult to understand or solve
— **puz·zling·ly** *adverb*

P–wave \'pē-ˌwāv\ *noun* [by shortening] (1936)
: PRESSURE WAVE

py- *or* **pyo-** *combining form* [Greek, from *pyon* pus — more at FOUL]
: pus ⟨*pyemia*⟩ ⟨*pyorrhea*⟩

pya \'pyä, pē-'ä\ *noun* [Burmese] (1952)
— see *kyat* at MONEY table

pyc·nid·i·um \pik-'ni-dē-əm\ *noun, plural* **-ia** \-dē-ə\ [New Latin, from Greek *pyknos* dense] (1857)
: a flask-shaped fruiting body bearing conidiophores and conidia on the interior and occurring in various imperfect fungi and ascomycetes
— **pyc·nid·i·al** \-dē-əl\ *adjective*

pyc·no·go·nid \pik-'nä-gə-nəd, ˌpik-nə-'gä-nəd\ *noun* [ultimately from Greek *pyknos* + *gony* knee — more at KNEE] (1881)
: SEA SPIDER

pyc·nom·e·ter \pik-'nä-mə-tər\ *noun* [Greek *pyknos* + International Scientific Vocabulary *-meter*] (1858)
: a standard vessel often provided with a thermometer for measuring and comparing the densities of liquids or solids

pye–dog \'pī-ˌdȯg\ *noun* [perhaps from Hindi *pāhī* outsider] (1864)
: a half-wild dog common about Asian villages

pyel- *or* **pyelo-** *combining form* [New Latin, pelvis, from Greek *pyelos* basin; akin to Greek *plynein* to wash, *plein* to sail — more at FLOW]
: renal pelvis ⟨*pyelitis*⟩

py·eli·tis \ˌpī-ə-'lī-təs\ *noun* [New Latin] (circa 1842)
: inflammation of the lining of the renal pelvis

py·elo·ne·phri·tis \ˌpī-(ə-)lō-ni-'frī-təs\ *noun* [New Latin] (1866)
: inflammation of both the lining of the pelvis and the parenchyma of the kidney
— **py·elo·ne·phrit·ic** \-'fri-tik\ *adjective*

py·emia \pī-'ē-mē-ə\ *noun* [New Latin] (circa 1857)
: septicemia caused by pus-forming bacteria and accompanied by multiple abscesses

py·gid·i·um \pī-'ji-dē-əm\ *noun, plural* **-ia** \-dē-ə\ [New Latin, from Greek *pygidion,* diminutive of *pygē* rump] (circa 1849)
: a caudal structure or the terminal body region of various invertebrates
— **py·gid·i·al** \-dē-əl\ *adjective*

pyg·mae·an *or* **pyg·me·an** \pig-'mē-ən, 'pig-mē-\ *adjective* [Latin *pygmaeus*] (1667)
: PYGMY

Pyg·ma·lion \pig-'māl-yən, -'mā-lē-ən\ *noun* [Latin, from Greek *Pygmaliōn*]
: a king of Cyprus who makes a female figure of ivory that is brought to life for him by Aphrodite

pyg·moid \'pig-ˌmȯid\ *adjective* (circa 1930)
: resembling or having the characteristics of the Pygmies

pyg·my \'pig-mē\ *noun, plural* **pygmies** [Middle English *pigmei,* from Latin *pygmaeus* of a pygmy, dwarfish, from Greek *pygmaios,* from *pygmē* fist, measure of length — more at PUNGENT] (14th century)
1 *often capitalized* **:** any of a race of dwarfs described by ancient Greek authors
2 *capitalized* **:** any of a small people of equatorial Africa ranging under five feet (1.5 meters) in height
3 a : a short insignificant person **:** DWARF **b :** something very small of its kind
— **pygmy** *adjective*

pygmy chimpanzee *noun* (1962)
: an anthropoid ape (*Pan paniscus*) of Zaire that has a lighter build than the related common chimpanzee (*P. troglodytes*) — called also *pygmy chimp*

py·ja·mas \pə-'jä-məz\ *chiefly British variant of* PAJAMAS

pyk·nic \'pik-nik\ *adjective* [International Scientific Vocabulary, from Greek *pyknos* dense, stocky] (1925)

☆ SYNONYMS
Puzzle, perplex, bewilder, distract, nonplus, confound, dumbfound mean to baffle and disturb mentally. PUZZLE implies existence of a problem difficult to solve ⟨the persistent fever *puzzled* the doctor⟩. PERPLEX adds a suggestion of worry and uncertainty especially about making a necessary decision ⟨a behavior that *perplexed* her friends⟩. BEWILDER stresses a confusion of mind that hampers clear and decisive thinking ⟨a *bewildering* number of possibilities⟩. DISTRACT implies agitation or uncertainty induced by conflicting preoccupations or interests ⟨*distracted* by personal problems⟩. NONPLUS implies a bafflement that makes orderly planning or deciding impossible ⟨the remark left us utterly *nonplussed*⟩. CONFOUND implies temporary mental paralysis caused by astonishment or profound abasement ⟨the tragic news *confounded* us all⟩. DUMBFOUND suggests intense but momentary confounding; often the idea of astonishment is so stressed that it becomes a near synonym of *astound* ⟨was at first too *dumbfounded* to reply⟩.

: characterized by shortness of stature, broadness of girth, and powerful muscularity : ENDOMORPHIC 2
— **pyknic** *noun*

py·lon \'pī-ˌlän, -lən\ *noun* [Greek *pylōn,* from *pylē* gate] (1850)
1 a : a usually massive gateway **b :** an ancient Egyptian gateway building in a truncated pyramidal form **c :** a monumental mass flanking an entranceway or an approach to a bridge
2 a *chiefly British* **:** a tower for supporting either end of usually a number of wires over a long span **b :** any of various towerlike structures
3 a : a post or tower marking a prescribed course of flight for an airplane **b :** TRAFFIC CONE
4 : a rigid structure on the outside of an aircraft for supporting something (as an engine or missile) — see AIRPLANE illustration

py·lo·ric \pī-'lȯr-ik, pə-, -'lōr-\ *adjective* (1807)
: of or relating to the pylorus; *also* : of, relating to, or situated in or near the posterior part of the stomach

py·lo·rus \-əs\ *noun, plural* **py·lo·ri** \-'lōr-ˌī, -(ˌ)ē\ [Late Latin, from Greek *pylōros,* literally, gatekeeper, from *pylē*] (1615)
: the opening from the vertebrate stomach into the intestine

pyo·der·ma \ˌpī-ə-'dər-mə\ *noun* [New Latin] (1930)
: a bacterial skin inflammation marked by pus-filled lesions

pyo·gen·ic \-'je-nik\ *adjective* [International Scientific Vocabulary] (circa 1847)
: producing pus ⟨*pyogenic* bacteria⟩; *also* : marked by pus production ⟨*pyogenic* meningitis⟩

py·or·rhea \ˌpī-ə-'rē-ə\ *noun* [New Latin] (1878)
: purulent inflammation of the sockets of the teeth leading usually to loosening of the teeth

pyr- *or* **pyro-** *combining form* [Middle English, from Middle French, from Late Latin, from Greek, from *pyr* — more at FIRE]
1 : fire **:** heat ⟨*pyro*meter⟩ ⟨*pyr*heliometer⟩
2 a : produced by or as if by the action of heat ⟨*pyro*electricity⟩ **b :** derived from a corresponding ortho acid by loss usually of one molecule of water from two molecules of acid ⟨*pyro*phosphoric acid⟩
3 : fever ⟨*pyro*genic⟩

pyr·acan·tha \ˌpī-rə-'kan(t)-thə\ *noun* [New Latin, from Greek *pyrakantha,* a tree, from *pyr-* + *akantha* thorn] (1705)
: any of a small genus (*Pyracantha*) of Eurasian thorny evergreen or semievergreen shrubs of the rose family with alternate leaves, corymbs of white flowers, and small red or orange pomes

py·ral·id \pī-'ra-ləd\ *noun* [ultimately from Latin *pyralis,* fly fabled as living in fire, from Greek, from *pyr* fire] (circa 1890)
: any of a very large heterogeneous family (Pyralidae) of mostly small slender long-legged moths
— **pyralid** *adjective*

¹pyr·a·mid \'pir-ə-ˌmid\ *noun* [Latin *pyramid-, pyramis,* from Greek] (1549)
1 a : an ancient massive structure found especially in Egypt having typically a square ground plan, outside walls in the form of four triangles that meet in a point at the top, and inner sepulchral chambers **b :** a structure or object of similar form

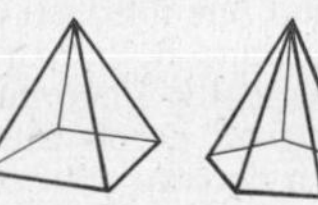

pyramid 2

2 : a polyhedron having for its base a polygon and for faces triangles with a common vertex — see VOLUME table
3 : a crystalline form each face of which intersects the vertical axis and either two lateral axes or in the tetragonal system one lateral axis
4 : an anatomical structure resembling a pyramid: as **a :** any of the conical masses that project from the renal medulla into the kidney pelvis **b :** either of two large bundles of motor fibers from the cerebral cortex that reach the medulla oblongata and are continuous with the pyramidal tracts of the spinal cord
5 : an immaterial structure built on a broad supporting base and narrowing gradually to an apex ⟨the socioeconomic *pyramid*⟩
— **py·ra·mi·dal** \pə-'ra-mə-dᵊl, ˌpir-ə-'mi-\ *adjective*
— **py·ra·mi·dal·ly** *adverb*
— **pyr·a·mid·i·cal** \ˌpir-ə-'mi-di-kəl\ *adjective*

²pyramid (circa 1900)
intransitive verb
1 : to speculate (as on a security or commodity exchange) by using paper profits as margin for additional transactions
2 : to increase rapidly and progressively step by step on a broad base
transitive verb
1 : to arrange or build up as if on the base of a pyramid
2 : to use (as profits) in speculative pyramiding
3 : to increase the impact of (as a tax assessed at the production level) on the ultimate consumer by treating as a cost subject to markup

pyramidal tract *noun* (circa 1890)
: any of four columns of motor fibers that run in pairs on each side of the spinal cord and are continuations of the pyramids of the medulla oblongata

Pyr·a·mus \'pir-ə-məs\ *noun* [Latin, from Greek *Pyramos*]
: a legendary youth of Babylon who dies for love of Thisbe

py·ran \'pīr-ˌan\ *noun* [International Scientific Vocabulary] (1904)
: either of two cyclic compounds C_5H_6O that contain five carbon atoms and one oxygen atom in the ring

py·ra·nose \'pī-rə-ˌnōs, -ˌnōz\ *noun* [International Scientific Vocabulary] (1927)
: a monosaccharide in the form of a cyclic hemiacetal containing a pyran ring

py·ran·o·side \pī-'ra-nə-ˌsīd\ *noun* (1932)
: a glycoside containing the pyran ring

pyr·ar·gy·rite \pī-'rär-jə-ˌrīt\ *noun* [German *Pyrargyrit,* from Greek *pyr-* + *argyros* silver — more at ARGENT] (1849)
: a mineral consisting of silver antimony sulfide that occurs in rhombohedral crystals or in massive form and has a dark red or black color with a metallic luster

pyre \'pīr\ *noun* [Latin *pyra,* from Greek, from *pyr* fire — more at FIRE] (1658)
: a combustible heap for burning a dead body as a funeral rite; *broadly* : a pile of material to be burned ⟨a *pyre* of dead leaves⟩

py·re·noid \pī-'rē-ˌnȯid, 'pī-rə-\ *noun* [International Scientific Vocabulary, from New Latin *pyrena* stone of a fruit, from Greek *pyrēnoun;* akin to Greek *pyros* wheat grain, wheat — more at FURZE] (circa 1875)
: one of the protein bodies in the chromatophores of various lower organisms (as some algae) that act as centers for starch deposition

py·re·thrin \pī-'rē-thrən, -'re-\ *noun* [International Scientific Vocabulary, from Latin *pyrethrum*] (1924)
: either of two oily liquid esters $C_{21}H_{28}O_3$ and $C_{22}H_{28}O_5$ having insecticidal properties and occurring especially in the flowers of pyrethrum

py·re·throid \-'rē-ˌthrȯid, -'re-\ *noun* [*pyrethrin* + *-oid*] (1949)
: any of various synthetic compounds that are related to the pyrethrins and resemble them in insecticidal properties
— **pyrethroid** *adjective*

py·re·thrum \pī-'rē-thrəm, -'re-\ *noun* [Latin, pellitory, from Greek *pyrethron,* from *pyr* fire] (circa 1543)
1 : any of several chrysanthemums with finely divided often aromatic leaves including ornamentals as well as important sources of insecticides
2 : an insecticide made from the dried heads of any of several Old World chrysanthemums

py·ret·ic \pī-'re-tik\ *adjective* [New Latin *pyreticus,* from Greek *pyretikos,* from *pyretos* fever, from *pyr*] (circa 1858)
: of or relating to fever : FEBRILE

Py·rex \'pīr-ˌeks\ *trademark*
— used for borosilicate glass and glassware resistant to heat, chemicals, and electricity

py·rex·ia \pī-'rek-sē-ə\ *noun* [New Latin, from Greek *pyressein* to be feverish, from *pyretos*] (1769)
: abnormal elevation of body temperature : FEVER
— **py·rex·i·al** \-sē-əl\ *adjective*
— **py·rex·ic** \-sik\ *adjective*

pyr·he·li·om·e·ter \ˌpīr-ˌhē-lē-'ä-mə-tər, ˌpir-\ *noun* [International Scientific Vocabulary] (1863)
: an instrument for measuring the sun's radiant energy as received at the earth
— **pyr·he·lio·met·ric** \-lē-ə-'me-trik\ *adjective*

py·ric \'pī-rik, 'pir-ik\ *adjective* [French *pyrique,* from Greek *pyr*] (1946)
: resulting from, induced by, or associated with burning

pyr·i·dine \'pir-ə-ˌdēn\ *noun* [*pyr-* + *-ide* + ²*-ine*] (1851)
: a toxic water-soluble flammable liquid base C_5H_5N of pungent odor that is the parent of many naturally occurring organic compounds and is used as a solvent and as a denaturant for alcohol and in the manufacture of pharmaceuticals and waterproofing agents

pyr·i·dox·al \ˌpir-ə-'däk-ˌsal\ *noun* [International Scientific Vocabulary, from *pyridoxine*] (1944)
: a crystalline aldehyde $C_8H_9NO_3$ of the vitamin B_6 group that occurs as a phosphate and is active as a coenzyme

pyr·i·dox·amine \ˌpir-ə-'däk-sə-ˌmēn\ *noun* [International Scientific Vocabulary *pyridox*ine + *amine*] (1944)
: a crystalline amine $C_8H_{12}N_2O_2$ of the vitamin B_6 group that occurs as a phosphate and is active as a coenzyme

pyr·i·dox·ine \ˌpir-ə-'däk-ˌsēn, -sən\ *noun* [*pyridine* + *ox-* + ²*-ine*] (1939)
: a crystalline phenolic alcohol $C_8H_{11}NO_3$ of the vitamin B_6 group found especially in cereals and convertible in the organism into pyridoxal and pyridoxamine

pyr·i·form \'pir-ə-ˌfȯrm\ *adjective* [New Latin *pyriformis,* from Medieval Latin *pyrum* pear (alteration of Latin *pirum*) + Latin *-iformis* -iform] (1741)
: having the form of a pear

py·ri·meth·amine \ˌpī-rə-'me-thə-ˌmēn\ *noun* [*pyrim*idine + *eth*yl + *amine*] (1952)
: a folic acid antagonist $C_{12}H_{13}ClN_4$ used in the treatment of malaria and of toxoplasmosis

py·rim·i·dine \pī-'ri-mə-ˌdēn, pə-\ *noun* [International Scientific Vocabulary, alteration of *pyridine*] (1885)
1 : a feeble organic base $C_4H_4N_2$ of penetrating odor
2 : a derivative of pyrimidine; *especially* : a base (as cytosine, thymine, or uracil) that is a constituent of DNA or RNA

py·rite \'pī-ˌrīt\ *noun* [Latin *pyrites*] (1588)

: a common mineral that consists of iron disulfide, has a pale brass-yellow color and metallic luster, and is burned in making sulfur dioxide and sulfuric acid

py·rites \pə-ˈrī-tēz, pī-; ˈpī-ˌrīts\ *noun, plural* **pyrites** [Latin, flint, from Greek *pyritēs* of or in fire, from *pyr* fire] (1543)
: any of various metallic-looking sulfides of which pyrite is the commonest
— py·rit·ic \-ˈri-tik\ *adjective*

py·ro·cat·e·chol \ˌpī-rō-ˈka-tə-ˌkȯl, -ˌkōl\ *noun* [International Scientific Vocabulary] (1890)
: a crystalline phenol $C_6H_6O_2$ obtained by pyrolysis of various natural substances (as resins and lignins) but usually made synthetically and used especially as a photographic developer and in organic synthesis

py·ro·clas·tic \-ˈklas-tik\ *adjective* (1887)
: formed by or involving fragmentation as a result of volcanic or igneous action

py·ro·elec·tric·i·ty \ˈpī-ˌrō-ə-ˌlek-ˈtri-sə-tē, -ˈtris-tē\ *noun* [International Scientific Vocabulary] (circa 1834)
: a state of electrical polarization produced (as in a crystal) by a change of temperature
— py·ro·elec·tric \-ˈlek-trik\ *adjective*

py·ro·gal·lol \ˌpī-rō-ˈga-ˌlȯl, -ˌlōl; -ˈgȯ-\ *noun* [International Scientific Vocabulary *pyro-* + *gallic* (acid) + [1]*-ol*] (1876)
: a poisonous bitter crystalline phenol $C_6H_6O_3$ with weak acid properties that is obtained usually by pyrolysis of gallic acid and used especially as a mild reducing agent (as in photographic developing)

py·ro·gen \ˈpī-rə-jən\ *noun* [International Scientific Vocabulary] (circa 1890)
: a fever-producing substance

py·ro·gen·ic \ˌpī-rō-ˈje-nik\ *adjective* [International Scientific Vocabulary] (1853)
1 : producing or produced by heat or fever
2 : of or relating to igneous origin
— py·ro·ge·nic·i·ty \ˌpī-rō-jə-ˈni-sə-tē\ *noun*

py·ro·la \pī-ˈrō-lə\ *noun* [New Latin, probably from Latin *pirum* pear] (1578)
: WINTERGREEN 1

py·ro·lig·ne·ous acid \ˌpī-rō-ˈlig-nē-əs-\ *noun* [French *pyroligneux,* from *pyr-* + *ligneux* woody, from Latin *lignosus,* from *lignum* wood — more at LIGNEOUS] (circa 1790)
: an acid reddish brown aqueous liquid containing chiefly acetic acid, methanol, wood oils, and tars that is obtained by destructive distillation of wood

py·ro·lu·site \ˌpī-rō-ˈlü-ˌsīt\ *noun* [German *Pyrolusit,* from Greek *pyr-* + *lousis* washing, from *louein* to wash — more at LYE] (1828)
: a mineral consisting of manganese dioxide that is of an iron-black or dark steel-gray color and metallic luster, is usually soft, and is the most important ore of manganese

py·rol·y·sate \pī-ˈrä-lə-ˌzāt, -ˌsāt\ *or* **py·rol·y·zate** \-ˌzāt\ *noun* (1944)
: a product of pyrolysis

py·rol·y·sis \pī-ˈrä-lə-səs\ *noun* [New Latin] (circa 1890)
: chemical change brought about by the action of heat
— py·ro·lyt·ic \ˌpī-rə-ˈli-tik\ *adjective*
— py·ro·lyt·i·cal·ly \-ti-k(ə-)lē\ *adverb*

py·ro·lyze *also* **py·ro·lize** \ˈpī-rə-ˌlīz\ *transitive verb* **-lyzed** *also* **-lized; -lyz·ing** *also* **-liz·ing** (1932)
: to subject to pyrolysis
— py·ro·lyz·able \ˌpī-rə-ˈlī-zə-bəl\ *adjective*
— py·ro·lyz·er *noun*

py·ro·man·cy \-ˌman(t)-sē\ *noun* [Middle English *pyromancie,* from Middle French, from Late Latin *pyromantia,* from Greek *pyromanteia,* from *pyr* fire + *manteia* divination — more at -MANCY] (14th century)
: divination by means of fire or flames

py·ro·ma·nia \ˌpī-rō-ˈmā-nē-ə, -nyə\ *noun* [New Latin] (circa 1842)
: an irresistible impulse to start fires
— py·ro·ma·ni·ac \-nē-ˌak\ *noun*
— py·ro·ma·ni·a·cal \-mə-ˈnī-ə-kəl\ *adjective*

py·ro·met·al·lur·gy \-ˈme-tᵊl-ˌər-jē, *especially British* -mə-ˈta-lər-\ *noun* [International Scientific Vocabulary] (1908)
: chemical metallurgy depending on heat action (as roasting and smelting)
— py·ro·met·al·lur·gi·cal \-ˌme-tᵊl-ˈər-ji-kəl\ *adjective*

py·rom·e·ter \pī-ˈrä-mə-tər\ *noun* [International Scientific Vocabulary] (1796)
: an instrument for measuring temperatures especially when beyond the range of mercurial thermometers
— py·ro·met·ric \ˌpī-rə-ˈme-trik\ *adjective*
— py·ro·met·ri·cal·ly \-tri-k(ə-)lē\ *adverb*
— py·rom·e·try \pī-ˈrä-mə-trē\ *noun*

py·ro·mor·phite \ˌpī-rə-ˈmȯr-ˌfīt\ *noun* [German *Pyromorphit,* from Greek *pyr-* + *morphē* form] (circa 1814)
: a mineral consisting essentially of a lead chloride and phosphate

py·ro·nine \ˈpī-rə-ˌnēn\ *noun* [International Scientific Vocabulary, irregular from *pyr-* + [2]*-ine*] (1895)
: any of several basic xanthene dyes used chiefly as biological stains

py·ro·nin·o·phil·ic \ˌpī-rə-ˌnē-nə-ˈfi-lik\ *adjective* (1946)
: staining selectively with pyronines ⟨*pyroninophilic* cells⟩

py·rope \ˈpīr-ˌōp\ *noun* [Middle English *pirope,* a red gem, from Middle French, from Latin *pyropus,* a red bronze, from Greek *pyrōpos,* literally, fiery-eyed, from *pyr-* + *ōp-, ōps* eye — more at EYE] (1804)
: a magnesium-aluminum garnet that is deep red in color and is frequently used as a gem

py·ro·phor·ic \ˌpī-rə-ˈfȯr-ik, -ˈfär-\ *adjective* [New Latin *pyrophorus,* from Greek *pyrophoros* fire-bearing, from *pyr-* + *-phoros* carrying — more at -PHORE] (1836)
1 : igniting spontaneously
2 : emitting sparks when scratched or struck especially with steel

py·ro·phos·phate \-ˈfäs-ˌfāt\ *noun* (1836)
: a salt or ester of pyrophosphoric acid

py·ro·phos·pho·ric acid \-fäs-ˈfȯr-ik-, -ˈfär-; -ˈfäs-f(ə-)rik-\ *noun* [International Scientific Vocabulary] (1832)
: a crystalline acid $H_4P_2O_7$ formed when orthophosphoric acid is heated or prepared in the form of salts by heating acid salts of orthophosphoric acid

py·ro·phyl·lite \ˌpī-rō-ˈfi-ˌlīt, pī-ˈrä-fə-ˌlīt\ *noun* [German *Pyrophyllit,* from Greek *pyr-* + *phyllon* leaf — more at BLADE] (1830)
: a white or greenish mineral that is a hydrous aluminum silicate, resembles talc, occurs in a foliated form or in compact masses, and is used especially in ceramic wares

py·ro·sis \pī-ˈrō-səs\ *noun* [New Latin, from Greek *pyrōsis* burning, from *pyroun* to burn, from *pyr* fire — more at FIRE] (1789)
: HEARTBURN

[1]py·ro·tech·nic \ˌpī-rə-ˈtek-nik\ *also* **py·ro·tech·ni·cal** \-ni-kəl\ *adjective* [French *pyrotechnique,* from Greek *pyr* fire + *technē* art — more at TECHNICAL] (1825)
: of or relating to pyrotechnics
— py·ro·tech·ni·cal·ly \-ni-k(ə-)lē\ *adverb*

[2]pyrotechnic *noun* (1840)
1 a : FIREWORK **b :** any of various similar devices (as for igniting a rocket or producing an explosion)
2 : a combustible substance used in a firework

py·ro·tech·nics \ˌpī-rə-ˈtek-niks\ *noun plural* (1729)
1 *singular or plural in construction* **:** the art of making or the manufacture and use of fireworks
2 a : a display of fireworks **b :** a spectacular display (as of extreme virtuosity) ⟨verbal *pyrotechnics*⟩ ⟨keyboard *pyrotechnics*⟩
— py·ro·tech·nist \-ˈtek-nist\ *noun*

py·rox·ene \pī-ˈräk-ˌsēn, pə-\ *noun* [French *pyroxène,* from Greek *pyr-* + *xenos* stranger] (1800)
: any of a group of igneous-rock-forming silicate minerals that contain calcium, sodium, magnesium, iron, or aluminum, usually occur in short prismatic crystals or massive form, are often laminated, and vary in color from white to dark green or black
— py·rox·e·nic \ˌpī-ˌräk-ˈsē-nik, pə-, -ˈse-\ *adjective*
— py·rox·e·noid \pī-ˈräk-sə-ˌnȯid, pə-\ *adjective or noun*

py·rox·e·nite \pī-ˈräk-sə-ˌnīt, pə-\ *noun* (circa 1862)
: an igneous rock that is free from olivine and is composed essentially of pyroxene
— py·rox·e·nit·ic \-ˌräk-sə-ˈni-tik\ *adjective*

py·rox·y·lin \pī-ˈräk-sə-lən, pə-\ *noun* [International Scientific Vocabulary *pyr-* + Greek *xylon* wood] (circa 1847)
1 : a flammable mixture of nitrocelluloses used especially in making plastics and water-repellent coatings (as lacquers)
2 : a pyroxylin product

Pyr·rha \ˈpir-ə\ *noun* [Latin, from Greek]
: the wife of Deucalion

pyr·rhic \ˈpir-ik\ *noun* [Latin *pyrrhichius,* from Greek (*pous*) *pyrrhichios,* from *pyrrhichē,* a kind of dance] (1626)
: a metrical foot consisting of two short or unaccented syllables

Pyr·rhic \ˈpir-ik\ *adjective* [*Pyrrhus,* king of Epirus who sustained heavy losses in defeating the Romans] (1885)
: achieved at excessive cost ⟨a *Pyrrhic* victory⟩; *also* **:** costly to the point of negating or outweighing expected benefits ⟨a great but *Pyrrhic* act of ingenuity⟩

Pyr·rho·nism \ˈpir-ə-ˌni-zəm\ *noun* [French *pyrrhonisme,* from *Pyrrhon* Pyrrho, 4th century B.C. Greek philosopher, from Greek *Pyrrhōn*] (circa 1670)
1 : the doctrines of a school of ancient extreme skeptics who suspended judgment on every proposition — compare ACADEMICISM
2 : total or radical skepticism
— Pyr·rho·nist \-nist\ *noun*

pyr·rho·tite \ˈpir-ə-ˌtīt\ *noun* [modification of German *Pyrrhotin,* from Greek *pyrrhotēs* redness, from *pyrrhos* red, from *pyr* fire — more at FIRE] (1868)
: a bronze-colored mineral of metallic luster that consists of ferrous sulfide and is attracted by a magnet

Pyr·rhus \ˈpir-əs\ *noun* [Latin, from Greek *Pyrrhos*]
: a son of Achilles and slayer of Priam at the taking of Troy

pyr·role \ˈpir-ˌōl\ *noun* [Greek *pyrrhos*] (1835)
: a toxic liquid heterocyclic compound C_4H_5N that has a ring consisting of four carbon atoms and one nitrogen atom, polymerizes readily in air, and is the parent compound of many biologically important substances (as bile pigments, porphyrins, and chlorophyll); *broadly* **:** a derivative of pyrrole
— pyr·ro·lic \pi-ˈrō-lik\ *adjective*

py·ru·vate \pī-ˈrü-ˌvāt\ *noun* (1855)
: a salt or ester of pyruvic acid

py·ru·vic acid \pī-ˈrü-vik-\ *noun* [International Scientific Vocabulary *pyr-* + Latin *uva* grapes; from its importance in fermentation — more at UVULA] (1838)
: a 3-carbon keto acid $C_3H_4O_3$ that in carbo-

hydrate metabolism is an important intermediate product formed as pyruvate by glycolysis

¹Py·thag·o·re·an \pə-ˌtha-gə-ˈrē-ən, (ˌ)pī-\ *noun* (1550)
: any of a group professing to be followers of the Greek philosopher Pythagoras

²Pythagorean *adjective* (circa 1580)
: of, relating to, or associated with the Greek philosopher Pythagoras, his philosophy, or the Pythagoreans

Py·thag·o·re·an·ism \-ˈrē-ə-ˌni-zəm\ *noun* (circa 1727)
: the doctrines and theories of Pythagoras and the Pythagoreans who developed some basic principles of mathematics and astronomy, originated the doctrine of the harmony of the spheres, and believed in metempsychosis, the eternal recurrence of things, and the mystical significance of numbers

Pythagorean theorem *noun* (circa 1909)
: a theorem in geometry: the square of the length of the hypotenuse of a right triangle equals the sum of the squares of the lengths of the other two sides

Pyth·i·ad \ˈpi-thē-ˌad, -əd\ *noun* [Greek *Pythia,* the Pythian games, from neuter plural of *pythios*] (1842)
: the 4-year period between celebrations of the Pythian games in ancient Greece

¹Pyth·i·an \ˈpi-thē-ən\ *adjective* [Latin *pythius* of Delphi, from Greek *pythios,* from *Pythō* Pytho, name for Delphi, Greece] (1603)
1 : of or relating to games celebrated at Delphi every four years
2 : of or relating to Delphi or its oracle of Apollo

²Pythian *noun* (1903)
: KNIGHT OF PYTHIAS

Pyth·i·as \ˈpi-thē-əs\ *noun* [Greek]
: a friend of Damon condemned to death by Dionysius of Syracuse

py·thon \ˈpī-ˌthän, -thən\ *noun* [Latin, monstrous serpent killed by Apollo, from Greek *Pythōn,* from *Pythō* Delphi] (1836)
: any of various large constricting snakes (as a boa); *especially* **:** any of the large oviparous snakes (subfamily Pythoninae of the family Boidae) of Africa, Asia, Australia, and adjacent islands that include some of the largest existing snakes

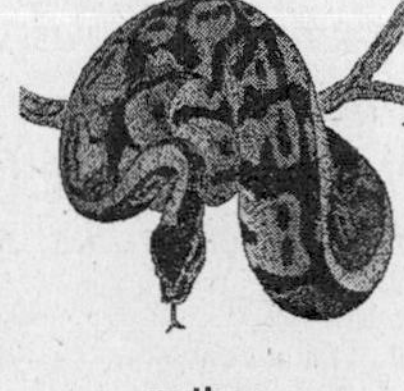
python

py·tho·ness \ˈpī-thə-nəs, ˈpi-\ *noun* [Middle English *Phitonesse,* from Middle French *pithonisse,* from Late Latin *pythonissa,* from Greek *Pythōn,* spirit of divination, perhaps from *Pythō,* seat of the Delphic oracle] (14th century)
1 : a woman who practices divination
2 : a prophetic priestess of Apollo
— py·thon·ic \pī-ˈthä-nik\ *adjective*

py·uria \pī-ˈyu̇r-ē-ə\ *noun* [New Latin] (circa 1811)
: pus in the urine; *also* **:** a condition characterized by pus in the urine

pyx \ˈpiks\ *noun* [Middle English, from Medieval Latin *pyxis,* from Latin, box, from Greek, from *pyxos* box (shrub)] (15th century)
1 : a container for the reserved host; *especially* **:** a small round metal receptacle used to carry the Eucharist to the sick
2 : a box used in a mint for deposit of sample coins reserved for testing weight and fineness

pyx·ie \ˈpik-sē\ *noun* [by shortening & alteration from New Latin *Pyxidanthera*] (1882)
: a creeping evergreen dicotyledonous shrub (*Pyxidanthera barbulata* of the family Diapensiaceae) of the sandy pine barrens of the Atlantic coast of the U.S. that has white or pink pentamerous flowers

pyx·is \ˈpik-səs\ *noun, plural* **pyx·i·des** \-sə-ˌdēz\ [New Latin, from Latin, box] (1845)
: a capsular fruit that dehisces so that the upper part falls off like a cap

Q is the seventeenth letter of the English alphabet. It comes from Phoenician qōph, *representing a guttural* k *sound, which survived in a western Greek alphabet as* koppa *and passed through Etruscan and Latin into English as* Q. *In English, as in Latin, this sign is normally used in combination with* u *to represent the sound of* kw, *as in* quiet. *In some English words of French origin, however,* qu *has the sound of plain* k, *as in* antique. Q *seldom appears without* u, *except in the transliteration of the Semitic letter* koph *(Arabic* qāf*), as in* Iraq. Q *is rarely found in Old English, where the usual spelling for the* kw- *sound was* cw *or* cu *(as in* cwēn *"queen"). Qu became the usual form in Middle English, under the influence of French and Latin. The small form of* q *developed in the late Roman period when scribes began forming the capital letter with one pen stroke, making its tail extend vertically along the right side of the letter.*

q \ˈkyü\ *noun, plural* **q's** *or* **qs** \ˈkyüz\ *often capitalized, often attributive* (before 12th century)
1 a : the 17th letter of the English alphabet **b :** a graphic representation of this letter **c :** a speech counterpart of orthographic *q*
2 : a graphic device for reproducing the letter *q*
3 : one designated *q* especially as the 17th in order or class
4 : something shaped like the letter Q

Q–boat \ˈkyü-ˌbōt\ *noun* (1918)
: Q-SHIP

Q fever \ˈkyü-\ *noun* [query] (1937)
: a disease that is characterized by high fever, chills, and muscular pains, is caused by a rickettsial bacterium (*Coxiella burnetii*), and is transmitted by raw milk, by droplet infection, or by ticks

qin·tar \kyin-ˈtär, kin-\ *noun, plural* **qin·dar·ka** \-ˈdär-kə\ *also* **qintars** *or* **qin·dars** \-ˈdärz\ [Albanian] (circa 1929)
— see *lek* at MONEY table

qi·vi·ut \ˈkē-vē-ət, -vē-ˌüt\ *noun* [Inuit] (1958)
: the wool of the undercoat of the musk ox

qoph \ˈkōf\ *noun* [Hebrew *qōph*] (circa 1823)
: the 19th letter of the Hebrew alphabet — see ALPHABET table

Q–ship \ˈkyü-ˌship\ *noun* (1919)
: an armed ship disguised as a merchant or fishing ship to decoy enemy submarines into gun range

q.t. \ˌkyü-ˈtē\ *noun, often Q&T capitalized* [abbreviation] (circa 1887)
: QUIET — usually used in the phrase *on the q.t.*

qua \ˈkwä *also* ˈkwā\ *preposition* [Latin, which way, as, from ablative singular feminine of *qui* who — more at WHO] (1647)
: in the capacity or character of **:** AS ⟨confusion of the role of scientist *qua* scientist with that of scientist as citizen —Philip Handler⟩

quaa·lude \ˈkwā-ˌlüd\ *noun* [from *Quaalude,* a trademark] (1966)
: a tablet or capsule of methaqualone

[1]quack \ˈkwak\ *intransitive verb* [alteration of *queck* to quack, from Middle English *queken,* from *queke,* interjection, of imitative origin] (14th century)
: to make the characteristic cry of a duck

[2]quack *noun* (1839)
: a noise made by quacking

[3]quack *intransitive verb* [[4]*quack*] (1628)
: to act like a quack

[4]quack *noun* [short for *quacksalver*] (1638)
1 : CHARLATAN 2
2 : a pretender to medical skill
— quack·ish \ˈkwa-kish\ *adjective*

[5]quack *adjective* (1653)
: of, relating to, or characteristic of a quack; *especially* **:** pretending to cure diseases

quack·ery \ˈkwa-k(ə-)rē\ *noun* (circa 1711)
: the practices or pretensions of a quack

quack grass \ˈkwak-\ *noun* [alteration of *quick* (grass), alteration of *quitch* (grass)] (circa 1818)
: a European grass (*Agropyron repens*) that is naturalized throughout North America as a weed and spreads by creeping rhizomes — called also *couch grass, quitch, twitch, witchgrass*

quack·sal·ver \ˈkwak-ˌsal-vər\ *noun* [obsolete Dutch (now *kwakzalver*)] (1579)
: CHARLATAN, QUACK

[1]quad \ˈkwäd\ *noun* (1820)
: QUADRANGLE

[2]quad *noun* [short for *quadrat*] (circa 1879)
: a type-metal space that is 1 en or more in width

[3]quad *transitive verb* **quad·ded; quad·ding** (1888)
: to fill out (as a typeset line) with blank space

[4]quad *noun* (1896)
: QUADRUPLET

[5]quad *adjective* (1970)
: QUADRAPHONIC

[6]quad *noun* (1971)
: quadraphonic sound

[7]quad *noun* [short for *quadrillion*] (1974)
: a unit of energy equal to one quadrillion British thermal units

quad·ran·gle \ˈkwä-ˌdraŋ-gəl\ *noun* [Middle English, from Middle French, from Late Latin *quadriangulum,* from Latin, neuter of *quadriangulus* quadrangular, from *quadri-* + *angulus* angle] (15th century)
1 : QUADRILATERAL
2 a : a 4-sided enclosure especially when surrounded by buildings **b :** the buildings enclosing a quadrangle
3 : a tract of country represented by one of a series of map sheets (as published by the U.S. Geological Survey)
— qua·dran·gu·lar \kwä-ˈdraŋ-gyə-lər\ *adjective*

quad·rant \ˈkwä-drənt\ *noun* [Middle English, from Latin *quadrant-, quadrans* fourth part; akin to Latin *quattuor* four — more at FOUR] (15th century)
1 a : an instrument for measuring altitudes consisting commonly of a graduated arc of 90 degrees with an index or vernier and usually having a plumb line or spirit level for fixing the vertical or horizontal direction **b :** a device or mechanical part shaped like or suggestive of the quadrant of a circle

quadrant 2

2 a : an arc of 90 degrees that is one quarter of a circle **b :** the area bounded by a quadrant and two radii
3 a : any of the four parts into which a plane is divided by rectangular coordinate axes lying in that plane **b :** any of the four quarters into which something is divided by two real or imaginary lines that intersect each other at right angles
— qua·dran·tal \kwä-ˈdran-tᵊl\ *adjective*

Qua·dran·tid \kwä-ˈdran-təd\ *noun* [New Latin *Quadrant-, Quadrans* (*Muralis*) mural quadrant, a group of stars in the constellation Draco from which the shower appears to radiate]
: any of a group of meteors that appear annually about January 3

quad·ra·phon·ic *also* **quad·ri·phon·ic** \ˌkwä-drə-ˈfä-nik\ *adjective* [irregular from *quadri-* + *-phonic* (as in *stereophonic*)] (circa 1970)
: of, relating to, or using four channels for the transmission, recording, or reproduction of sound
— quad·ra·phon·ics *or* **quad·ri·phon·ics** \-niks\ *noun plural but singular in construction*

quad·rat \ˈkwä-drət, -ˌdrat\ *noun* [alteration of [2]*quadrate*] (1683)
1 : [2]QUAD
2 : a usually rectangular plot used for ecological or population studies

[1]quad·rate \ˈkwä-ˌdrāt, -drət\ *adjective* [Middle English, from Latin *quadratus,* past participle of *quadrare* to make square, fit, from *quadrum* square; akin to Latin *quattuor* four] (14th century)
1 : being square or approximately square
2 *of a heraldic cross* **:** expanded into a square at the junction of the arms — see CROSS illustration
3 : of, relating to, or constituting a bony or cartilaginous element of each side of the skull to which the lower jaw is articulated in most vertebrates below mammals

[2]quadrate *noun* (15th century)
1 : an approximately square or cubical area, space, or body
2 : a quadrate bone

qua·drat·ic \kwä-ˈdra-tik\ *adjective* (1668)
: involving terms of the second degree at most ⟨*quadratic* function⟩ ⟨*quadratic* equations⟩
— quadratic *noun*
— qua·drat·i·cal·ly \-ti-k(ə-)lē\ *adverb*

quadratic form *noun* (1859)
: a homogeneous polynomial (as $x^2 + 5xy + y^2$) of the second degree

quad·ra·ture \ˈkwä-drə-ˌchu̇r, -chər, -ˌtyu̇r, -ˌtu̇r\ *noun* (1591)
1 : a configuration in which two celestial bodies (as the moon and the sun) have an angular separation of 90 degrees as seen from the earth
2 : the process of finding a square equal in area to a given area

qua·dren·ni·al \kwä-ˈdre-nē-əl\ *adjective* (circa 1656)
1 : consisting of or lasting for four years
2 : occurring or being done every four years
— quadrennial *noun*
— qua·dren·ni·al·ly \-nē-ə-lē\ *adverb*

qua·dren·ni·um \-nē-əm\ *noun, plural* **-ni·ums** *or* **-nia** \-nē-ə\ [Latin *quadriennium,* from *quadri-* + *annus* year — more at ANNUAL] (1754)
: a period of four years

quadri- *or* **quadr-** *or* **quadru-** *combining form* [Middle English, from Latin; akin to Latin *quattuor* four]
1 a : four ⟨*quadri*lateral⟩ ⟨*quadru*manous⟩ **b :** square ⟨*quadric*⟩
2 : fourth ⟨*quadri*centennial⟩

quad·ric \ˈkwä-drik\ *adjective* [International Scientific Vocabulary] (1858)

: QUADRATIC ⟨*quadric* surface⟩ — used where there are more than two variables
— **quadric** *noun*

quad·ri·cen·ten·ni·al \ˌkwä-drə-sen-ˈte-nē-əl\ *noun* (1882)
: a 400th anniversary or its celebration

quad·ri·ceps \ˈkwä-drə-ˌseps\ *noun* [New Latin *quadricipit-, quadriceps,* from *quadri-* + *-cipit-, -ceps* (as in *bicipit-, biceps* biceps)] (1840)
: the greater extensor muscle of the front of the thigh that is divided into four parts

qua·dri·ga \kwä-ˈdrē-gə\ *noun, plural* **-gae** \-ˌgī\ [Latin, singular of *quadrigae* team of four, contraction of *quadrijugae,* feminine plural of *quadrijugus* yoked four abreast, from *quadri-* + *jungere* to yoke, join — more at YOKE] (circa 1741)
: a chariot drawn by four horses abreast

1**quad·ri·lat·er·al** \ˌkwä-drə-ˈla-t(ə-)rəl\ *noun* [Latin *quadrilaterus* four-sided, from *quadri-* + *later-, latus* side] (1650)
: a polygon of four sides

2**quadrilateral** *adjective* (1656)
: having four sides

1**qua·drille** \kwä-ˈdril, kwə-, kə-\ *noun* [French, group of knights engaged in a carousel, from Spanish *cuadrilla* troop, from diminutive of *cuadra* square, from Latin *quadra, quadrum*] (1726)
1 : a four-handed variant of ombre popular especially in the 18th century
2 : a square dance for four couples made up of five or six figures chiefly in 6/8 and 2/4 time; *also* **:** music for this dance

2**quadrille** *adjective* [French *quadrillé*] (circa 1885)
: marked with squares or rectangles

qua·dril·lion \kwä-ˈdril-yən\ *noun* [French, from Middle French, from *quadri-* + *-illion* (as in *million*)] (1674)
— see NUMBER table
— **quadrillion** *adjective*
— **qua·dril·lionth** \-yən(t)th\ *adjective or noun*

quad·ri·par·tite \ˌkwä-drə-ˈpär-ˌtīt\ *adjective* [Middle English, from Latin *quadripartitus,* from *quadri-* + *partitus,* past participle of *partire* to divide, from *part-, pars* part] (15th century)
1 : consisting of or divided into four parts
2 : shared or participated in by four parties or persons ⟨a *quadripartite* agreement⟩

quad·ri·ple·gic \ˌkwä-drə-ˈplē-jik\ *noun* [*quadriplegia,* from New Latin] (1921)
: one affected with paralysis of both arms and both legs
— **quad·ri·ple·gia** \-j(ē-)ə\ *noun*

1**quad·ri·va·lent** \ˌkwä-drə-ˈvā-lənt, *in sense 2* kwä-ˈdri-və-lənt\ *adjective* [International Scientific Vocabulary] (1865)
1 : TETRAVALENT
2 : composed of four homologous chromosomes synapsed in meiotic prophase

2**quadrivalent** *noun* (1923)
: a quadrivalent chromosomal group

qua·driv·i·al \kwä-ˈdri-vē-əl\ *adjective* (15th century)
1 : of or relating to the quadrivium
2 : having four ways or roads meeting in a point

qua·driv·i·um \-vē-əm\ *noun* [Late Latin, from Latin, crossroads, from *quadri-* + *via* way — more at WAY] (1804)
: a group of studies consisting of arithmetic, music, geometry, and astronomy and forming the upper division of the seven liberal arts in medieval universities — compare TRIVIUM

qua·droon \kwä-ˈdrün\ *noun* [modification of Spanish *cuarterón,* from *cuarto* fourth, from Latin *quartus* — more at QUART] (1707)
: a person of one-quarter black ancestry

qua·dru·ma·nous \kwä-ˈdrü-mə-nəs\ *adjective* [ultimately from Latin *quadri-* + *manus* hand — more at MANUAL] (1819)
: of, relating to, or being the primates excluding humans which are distinguished by hand-shaped feet

qua·drum·vir \kwä-ˈdrəm-vər\ *noun* [back-formation from *quadrumvirate*] (1790)
: a member of a quadrumvirate

qua·drum·vi·rate \-və-rət\ *noun* [*quadri-* + *-umvirate* (as in *triumvirate*)] (1752)
: a group or association of four

quad·ru·ped \ˈkwä-drə-ˌped\ *noun* [Latin *quadruped-, quadrupes,* from *quadruped-, quadrupes,* adjective, having four feet, from *quadri-* + *ped-, pes* foot — more at FOOT] (1646)
: an animal having four feet
— **quadruped** *adjective*
— **qua·dru·pe·dal** \kwä-ˈdrü-pə-dᵊl, ˌkwä-drə-ˈpe-\ *adjective*

1**qua·dru·ple** \kwä-ˈdrü-pəl, -ˈdrə-; ˈkwä-drə-\ *verb* **qua·dru·pled; qua·dru·pling** \-p(ə-)liŋ\ (14th century)
transitive verb
: to make four times as great or as many
intransitive verb
: to become four times as great or as numerous

2**quadruple** *noun* (15th century)
: a sum four times as great as another

3**quadruple** *adjective* [Middle French or Latin; Middle French, from Latin *quadruplus,* from *quadri-* + *-plus* multiplied by — more at -FOLD] (1557)
1 : having four units or members
2 : being four times as great or as many
3 : marked by four beats per measure ⟨*quadruple* meter⟩
— **qua·dru·ply** \-ˈdrü-plē, -ˈdrə-, -drə-\ *adverb*
— **qua·dru·plic·i·ty** \ˌkwä-drü-ˈpli-sə-tē\ *noun*

qua·dru·plet \kwä-ˈdrü-plət, -ˈdrə-; ˈkwä-drə-plət\ *noun* (1787)
1 : one of four offspring born at one birth
2 : a combination of four of a kind
3 : a group of four musical notes to be performed in the time ordinarily given to three of the same kind

1**qua·dru·pli·cate** \kwä-ˈdrü-pli-kət\ *adjective* [Latin *quadruplicatus,* past participle of *quadruplicare* to quadruple, from *quadruplic-, quadruplex* fourfold, from *quadri-* + *-plic-, -plex* fold — more at -FOLD] (1657)
1 : consisting of or existing in four corresponding or identical parts or examples ⟨*quadruplicate* invoices⟩
2 : being the fourth of four things exactly alike

2**qua·dru·pli·cate** \-plə-ˌkāt\ *transitive verb* **-cat·ed; -cat·ing** (circa 1661)
1 : to make quadruple or fourfold
2 : to prepare in quadruplicate
— **qua·dru·pli·ca·tion** \-ˌdrü-plə-ˈkā-shən\ *noun*

3**qua·dru·pli·cate** \kwä-ˈdrü-pli-kət\ *noun* (1790)
1 : four copies all alike — used with *in* ⟨typed in *quadruplicate*⟩
2 : one of four things exactly alike; *specifically* **:** one of four identical copies

quad·ru·pole \ˈkwä-drə-ˌpōl\ *noun* [International Scientific Vocabulary *quadri-* + *pole*] (1922)
: a system composed of two dipoles of equal but oppositely directed moment

quae·re \ˈkwir-ē, ˈkwer-\ *noun* [Latin, imperative of *quaerere* to seek, question] (1589)
archaic **:** QUERY

quaes·tor \ˈkwe-stər, ˈkwē-\ *noun* [Middle English *questor,* from Latin *quaestor,* from *quaerere*] (14th century)
: one of numerous ancient Roman officials concerned chiefly with financial administration

quaff \ˈkwäf, ˈkwaf\ *verb* [origin unknown] (1523)
intransitive verb
: to drink deeply
transitive verb
: to drink (a beverage) deeply
— **quaff** *noun*
— **quaff·er** *noun*

quag \ˈkwag, ˈkwäg\ *noun* [origin unknown] (1589)
: MARSH, BOG

quag·ga \ˈkwa-gə, ˈkwä-\ *noun* [obsolete Afrikaans (now *kwagga*), from Khoikhoi *quácha*] (1785)
: an extinct mammal (*Equus quagga*) of southern Africa related to the zebras

quag·gy \ˈkwa-gē, ˈkwä-\ *adjective* (1610)
1 : MARSHY
2 : FLABBY

quag·mire \ˈkwag-ˌmīr, ˈkwäg-\ *noun* (circa 1580)
1 : soft miry land that shakes or yields under the foot
2 : a difficult, precarious, or entrapping position **:** PREDICAMENT

qua·hog *also* **qua·haug** \ˈkō-ˌhȯg, ˈkwȯ-, ˈkwō-, -ˌhäg\ *noun* [Narraganset *poquaûhock*] (1753)
: a thick-shelled edible clam (*Mercenaria mercenaria*) of the U.S.

quai \ˈkā\ *noun* [French, from Middle French *cai* — more at QUAY] (1862)
: QUAY

quaich *or* **quaigh** \ˈkwāk\ *noun* [Scottish Gaelic *cuach*] (1546)
chiefly Scottish **:** a small shallow drinking vessel with ears for use as handles

1**quail** \ˈkwā(ə)l\ *noun, plural* **quail** *or* **quails** [Middle English *quaille,* from Middle French, from Medieval Latin *quaccula,* of imitative origin] (14th century)
: any of numerous small gallinaceous birds: as **a :** an Old World migratory game bird (*Coturnix coturnix*) **b :** BOBWHITE

quail

2**quail** *verb* [Middle English, from Middle Dutch *quelen*] (15th century)
intransitive verb
1 a *chiefly dialect* **:** WITHER, DECLINE **b :** to give way **:** FALTER ⟨his courage never *quailed*⟩
2 : to recoil in dread or terror **:** COWER ⟨the strongest *quail* before financial ruin —Samuel Butler (died 1902)⟩
transitive verb
archaic **:** to make fearful
synonym see RECOIL

quaint \ˈkwānt\ *adjective* [Middle English *cointe,* from Old French, from Latin *cognitus,* past participle of *cognoscere* to know — more at COGNITION] (13th century)
1 *obsolete* **:** EXPERT, SKILLED
2 a : marked by skillful design ⟨*quaint* with many a device in India ink —Herman Melville⟩ **b :** marked by beauty or elegance
3 a : unusual or different in character or appearance **:** ODD ⟨figures of fun, *quaint* people —Herman Wouk⟩ **b :** pleasingly or strikingly old-fashioned or unfamiliar
synonym see STRANGE
— **quaint·ly** *adverb*
— **quaint·ness** *noun*

1**quake** \ˈkwāk\ *intransitive verb* **quaked; quak·ing** [Middle English, from Old English *cwacian*] (before 12th century)
1 : to shake or vibrate usually from shock or instability
2 : to tremble or shudder usually from cold or fear

2**quake** *noun* (14th century)

: an instance of shaking or trembling (as of the earth or moon); *especially* : EARTHQUAKE

quak·er \ˈkwā-kər\ *noun* (1597)
1 : one that quakes
2 *capitalized* : FRIEND 5
— **Quak·er·ish** \ˈkwā-k(ə-)rish\ *adjective*
— **Quak·er·ism** \-kə-ˌri-zəm\ *noun*
— **Quak·er·ly** \-kər-lē\ *adjective*

Quaker gun *noun* [from opposition to war as a basic Quaker tenet] (1809)
: a dummy piece of artillery usually made of wood

quak·er–la·dies \ˌkwā-kər-ˈlā-dēz\ *noun plural* (1871)
: BLUETS

quaking aspen *noun* (1843)
: an aspen (*Populus tremuloides*) chiefly of the U.S. and Canada that has small nearly circular leaves with flattened petioles and finely serrate margins

qua·le \ˈkwä-lē, -ˌlā\ *noun, plural* **qua·lia** \ˈkwä-lē-ə\ [Latin, neuter of *qualis* of what kind] (1675)
1 : a property (as redness) considered apart from things having the property : UNIVERSAL
2 : a property as it is experienced as distinct from any source it might have in a physical object

qual·i·fi·able \ˌkwä-lə-ˈfī-ə-bəl\ *adjective* (1611)
: capable of qualifying or being qualified

qual·i·fi·ca·tion \ˌkwä-lə-fə-ˈkā-shən\ *noun* (circa 1544)
1 : a restriction in meaning or application : a limiting modification ⟨this statement stands without *qualification*⟩
2 a *obsolete* : NATURE **b** *archaic* : CHARACTERISTIC
3 a : a quality or skill that fits a person (as for an office) ⟨the applicant with the best *qualifications*⟩ **b** : a condition or standard that must be complied with (as for the attainment of a privilege) ⟨a *qualification* for membership⟩

qual·i·fied \ˈkwä-lə-ˌfīd\ *adjective* (1558)
1 a : fitted (as by training or experience) for a given purpose : COMPETENT **b** : having complied with the specific requirements or precedent conditions (as for an office or employment) : ELIGIBLE
2 : limited or modified in some way ⟨*qualified* approval⟩
— **qual·i·fied·ly** \-ˌfī(-ə)d-lē\ *adverb*

qual·i·fi·er \-ˌfī(-ə)r\ *noun* (1561)
: one that qualifies: as **a** : one that satisfies requirements or meets a specified standard **b** : a word (as an adjective) or word group that limits or modifies the meaning of another word (as a noun) or word group

qual·i·fy \ˈkwä-lə-ˌfī\ *verb* **-fied; -fy·ing** [Middle French *qualifier,* from Medieval Latin *qualificare,* from Latin *qualis*] (1533)
transitive verb
1 a : to reduce from a general to a particular or restricted form : MODIFY **b** : to make less harsh or strict : MODERATE **c** : to alter the strength or flavor of **d** : to limit or modify the meaning of (as a noun)
2 : to characterize by naming an attribute : DESCRIBE ⟨cannot *qualify* it as . . . either glad or sorry —T. S. Eliot⟩
3 a : to fit by training, skill, or ability for a special purpose **b** (1) : to declare competent or adequate : CERTIFY (2) : to invest with legal capacity : LICENSE
intransitive verb
1 : to be or become fit (as for an office) : meet the required standard
2 : to acquire legal or competent power or capacity ⟨has just *qualified* as a lawyer⟩
3 a : to exhibit a required degree of ability in a preliminary contest ⟨*qualified* for the finals⟩ **b** : to shoot well enough to earn a marksmanship badge

qual·i·ta·tive \ˈkwä-lə-ˌtā-tiv\ *adjective* (1607)
: of, relating to, or involving quality or kind
— **qual·i·ta·tive·ly** *adverb*

qualitative analysis *noun* (1842)
: chemical analysis designed to identify the components of a substance or mixture

¹qual·i·ty \ˈkwä-lə-tē\ *noun, plural* **-ties** [Middle English *qualite,* from Old French *qualité,* from Latin *qualitat-, qualitas,* from *qualis* of what kind; akin to Latin *qui* who — more at WHO] (14th century)
1 a : peculiar and essential character : NATURE ⟨her ethereal *quality* —Gay Talese⟩ **b** : an inherent feature : PROPERTY ⟨had a *quality* of stridence, dissonance —Roald Dahl⟩ **c** : CAPACITY, ROLE ⟨in the *quality* of reader and companion —Joseph Conrad⟩
2 a : degree of excellence : GRADE ⟨the *quality* of competing air service —*Current Biography*⟩ **b** : superiority in kind ⟨merchandise of *quality*⟩
3 a : social status : RANK **b** : ARISTOCRACY
4 a : a distinguishing attribute : CHARACTERISTIC ⟨possesses many fine *qualities*⟩ **b** *archaic* : an acquired skill : ACCOMPLISHMENT
5 : the character in a logical proposition of being affirmative or negative
6 : vividness of hue
7 a : TIMBRE **b** : the identifying character of a vowel sound determined chiefly by the resonance of the vocal chambers in uttering it
8 : the attribute of an elementary sensation that makes it fundamentally unlike any other sensation ☆

²quality *adjective* (1701)
: being of high quality

quality assurance *noun* (1982)
: a program for the systematic monitoring and evaluation of the various aspects of a project, service, or facility to ensure that standards of quality are being met

quality circle *noun* (1980)
: a group of employees who volunteer to meet regularly to discuss and propose solutions to problems (as of quality or productivity) in the workplace

quality control *noun* (1935)
: an aggregate of activities (as design analysis and inspection for defects) designed to ensure adequate quality especially in manufactured products
— **quality controller** *noun*

quality point *noun* (1948)
: GRADE POINT

quality point average *noun* (circa 1972)
: GRADE POINT AVERAGE

qualm \ˈkwäm *also* ˈkwȯm *or* ˈkwälm\ *noun* [origin unknown] (circa 1530)
1 : a sudden attack of illness, faintness, or nausea
2 : a sudden access of usually disturbing emotion (as doubt or fear)
3 : a feeling of uneasiness about a point especially of conscience or propriety ☆
— **qualmy** *adjective*

qualm·ish \ˈkwä-mish *also* ˈkwȯ- *or* ˈkwäl-\ *adjective* (1548)
1 a : feeling qualms : NAUSEATED **b** : overly scrupulous : SQUEAMISH
2 : of, relating to, or producing qualms
— **qualm·ish·ly** *adverb*
— **qualm·ish·ness** *noun*

qua·mash \ˈkwä-mish\ *variant of* CAMAS

quan·da·ry \ˈkwän-d(ə-)rē\ *noun, plural* **-ries** [origin unknown] (1579)
: a state of perplexity or doubt

quan·tal \ˈkwän-tᵊl\ *adjective* (1933)
1 [Latin *quanti* how many, plural of *quantus*] : of, relating to, or having only two experimental alternatives (as dead or alive, all or none)
2 [*quantum*] : of or relating to a quantum

quan·ti·fi·ca·tion \ˌkwän-tə-fə-ˈkā-shən\ *noun* (circa 1840)
: the operation of quantifying
— **quan·ti·fi·ca·tion·al** \-shnəl, -shə-nᵊl\ *adjective*
— **quan·ti·fi·ca·tion·al·ly** *adverb*

quan·ti·fi·er \ˈkwän-tə-ˌfī(-ə)r\ *noun* (1876)
: one that quantifies: as **a** : a prefixed operator that binds the variables in a logical formula by specifying their quantity **b** : a limiting noun modifier (as *five* in "the five young men") expressive of quantity and characterized by occurrence before the descriptive adjectives in a noun phrase

quan·ti·fy \-ˌfī\ *transitive verb* **-fied; -fy·ing** [Medieval Latin *quantificare,* from Latin *quantus* how much] (circa 1840)
1 a (1) : to limit by a quantifier (2) : to bind by prefixing a quantifier **b** : to make explicit the logical quantity of
2 : to determine, express, or measure the quantity of
— **quan·ti·fi·able** \ˌkwän-tə-ˈfī-ə-bəl\ *adjective*

quan·ti·tate \ˈkwän-tə-ˌtāt\ *transitive verb* **-tat·ed; -tat·ing** [back-formation from *quantitative*] (1927)
1 : to measure or estimate the quantity of; *especially* : to measure or determine precisely
2 : to express in quantitative terms
— **quan·ti·ta·tion** \ˌkwän-tə-ˈtā-shən\ *noun*

quan·ti·ta·tive \ˈkwän-tə-ˌtā-tiv\ *adjective* [Medieval Latin *quantitativus,* from Latin *quantitat-, quantitas* quantity] (1581)
1 : of, relating to, or expressible in terms of quantity
2 : of, relating to, or involving the measurement of quantity or amount
3 : based on quantity; *specifically, of classical verse* : based on temporal quantity or duration of sounds
— **quan·ti·ta·tive·ly** *adverb*
— **quan·ti·ta·tive·ness** *noun*

quantitative analysis *noun* (circa 1847)
: chemical analysis designed to determine the amounts or proportions of the components of a substance

quantitative inheritance *noun* (circa 1929)
: genic inheritance of a character (as human skin color) controlled by polygenes

quan·ti·ty \ˈkwän-tə-tē\ *noun, plural* **-ties** [Middle English *quantite,* from Middle French *quantité,* from Latin *quantitat-, quantitas,* from *quantus* how much, how large; akin to Latin *quam* how, as, *quando* when, *qui* who — more at WHO] (14th century)

☆ SYNONYMS

Quality, property, character, attribute mean an intelligible feature by which a thing may be identified. QUALITY is a general term applicable to any trait or characteristic whether individual or generic ⟨material with a silky *quality*⟩. PROPERTY implies a characteristic that belongs to a thing's essential nature and may be used to describe a type or species ⟨the *property* of not conducting heat⟩. CHARACTER applies to a peculiar and distinctive quality of a thing or a class ⟨remarks of an unseemly *character*⟩. ATTRIBUTE implies a quality ascribed to a thing or a being ⟨the traditional *attributes* of a military hero⟩.

Qualm, scruple, compunction, demur mean a misgiving about what one is doing or going to do. QUALM implies an uneasy fear that one is not following one's conscience or better judgment ⟨no *qualms* about plagiarizing⟩. SCRUPLE implies doubt of the rightness of an act on grounds of principle ⟨no *scruples* against buying stolen goods⟩. COMPUNCTION implies a spontaneous feeling of responsibility or compassion for a potential victim ⟨had *compunctions* about lying⟩. DEMUR implies hesitation caused by objection to an outside suggestion or influence ⟨accepted her decision without *demur*⟩.

1 **a :** an indefinite amount or number **b :** a determinate or estimated amount **c :** total amount or number **d :** a considerable amount or number — often used in plural ⟨generous *quantities* of luck —H. E. Putsch⟩
2 a : the aspect in which a thing is measurable in terms of greater, less, or equal or of increasing or decreasing magnitude **b :** the subject of a mathematical operation **c :** an individual considered with respect to a given situation ⟨an unknown *quantity* . . . as attorney general —Tom Wicker⟩
3 a : duration and intensity of speech sounds as distinct from their individual quality or phonemic character; *specifically* **:** the relative length or brevity of a prosodic syllable in some languages (as Greek and Latin) **b :** the relative duration or time length of a speech sound or sound sequence
4 : the character of a logical proposition as being universal, particular, or singular

quantity theory *noun* (1888)
: a theory in economics: changes in the price level tend to vary directly with the amount of money in circulation and the rate of its circulation

quan·tize \ˈkwän-ˌtīz\ *transitive verb* **quan·tized; quan·tiz·ing** [*quantum*] (1922)
1 : to subdivide (as energy) into small but measurable increments
2 : to calculate or express in terms of quantum mechanics
— **quan·ti·za·tion** \ˌkwän-tə-ˈzā-shən\ *noun*
— **quan·tiz·er** \ˈkwän-ˌtī-zər\ *noun*

¹quan·tum \ˈkwän-təm\ *noun, plural* **quan·ta** \ˈkwän-tə\ [Latin, neuter of *quantus* how much] (1567)
1 a : QUANTITY, AMOUNT **b :** PORTION, PART **c :** gross quantity **:** BULK
2 a : any of the very small increments or parcels into which many forms of energy are subdivided **b :** any of the small subdivisions of a quantized physical magnitude (as magnetic moment)

²quantum *adjective* (1942)
: LARGE, SIGNIFICANT ⟨a *quantum* improvement⟩

quantum chromodynamics *noun plural but singular in construction* (1975)
: a theory of fundamental particles based on the assumption that quarks are distinguished by differences in color and are held together (as in hadrons) by an exchange of gluons

quantum electrodynamics *noun plural but usually singular in construction* (1927)
: quantum mechanics applied to electrical interactions (as between nuclear particles)

quantum field theory *noun* (1948)
: a theory in physics: the interaction of two separate physical systems (as particles) is attributed to a field that extends from one to the other and is manifested in a particle exchange between the two systems

quantum jump *noun* (1926)
1 : an abrupt transition (as of an electron, an atom, or a molecule) from one discrete energy state to another
2 : QUANTUM LEAP

quantum leap *noun* (1956)
: an abrupt change, sudden increase, or dramatic advance

quantum mechanics *noun plural but singular or plural in construction* (1922)
: a theory of matter that is based on the concept of the possession of wave properties by elementary particles, that affords a mathematical interpretation of the structure and interactions of matter on the basis of these properties, and that incorporates within it quantum theory and the uncertainty principle — called also *wave mechanics*
— **quantum mechanical** *adjective*
— **quantum mechanically** *adverb*

quantum number *noun* (1902)
: any of a set of integers or half odd integers that indicate the magnitude of various discrete quantities (as electric charge) of a particle or system and that serve to define its state

quantum theory *noun* (1912)
: a theory in physics based on the concept of the subdivision of radiant energy into finite quanta and applied to numerous processes involving transference or transformation of energy in an atomic or molecular scale

¹quar·an·tine \ˈkwȯr-ən-ˌtēn, ˈkwär-\ *noun* [partly modification of French *quarantaine,* from Old French, from *quarante* forty, from Latin *quadraginta,* from *quadra-* (akin to *quattuor* four) + *-ginta* (akin to *viginti* twenty); partly modification of Italian *quarantena* quarantine of a ship, from *quaranta* forty, from Latin *quadraginta* — more at FOUR, VIGESIMAL] (1609)
1 : a period of 40 days
2 a : a term during which a ship arriving in port and suspected of carrying contagious disease is held in isolation from the shore **b :** a regulation placing a ship in quarantine **c :** a place where a ship is detained during quarantine
3 a : a restraint upon the activities or communication of persons or the transport of goods designed to prevent the spread of disease or pests **b :** a place in which those under quarantine are kept
4 : a state of enforced isolation

²quarantine *verb* **-tined; -tin·ing** (1804)
transitive verb
1 : to detain in or exclude by quarantine
2 : to isolate from normal relations or communication ⟨*quarantine* an aggressor⟩
intransitive verb
: to establish or declare a quarantine

quare \ˈkwar, ˈkwer, ˈkwär\ *dialect variant of* ¹QUEER

quark \ˈkwȯrk, ˈkwärk\ *noun* [coined by Murray Gell-Mann] (1964)
: any of several elementary particles that are postulated to come in pairs (as in the up and down varieties) of similar mass with one member having a charge of $+\frac{2}{3}$ and the other a charge of $-\frac{1}{3}$ and are held to make up hadrons ◆

¹quar·rel \ˈkwȯr(-ə)l, ˈkwär(-ə)l\ *noun* [Middle English, from Old French, square-headed arrow, building stone, from (assumed) Vulgar Latin *quadrellum,* diminutive of Latin *quadrum* square — more at QUADRATE] (13th century)
: a square-headed bolt or arrow especially for a crossbow

²quarrel *noun* [Middle English *querele,* from Middle French, complaint, from Latin *querela,* from *queri* to complain] (14th century)
1 : a ground of dispute or complaint ⟨have no *quarrel* with a different approach⟩
2 : a usually verbal conflict between antagonists **:** ALTERCATION

³quarrel *intransitive verb* **-reled** *or* **-relled; -rel·ing** *or* **-rel·ling** (14th century)
1 : to find fault ⟨many people *quarrel* with the idea —*Johns Hopkins Magazine*⟩
2 : to contend or dispute actively ⟨*quarreled* frequently with his superiors —*London Calling*⟩
— **quar·rel·er** *or* **quar·rel·ler** *noun*

quar·rel·some \ˈkwȯr(-ə)l-səm, ˈkwär(-ə)l-\ *adjective* (1596)
: apt or disposed to quarrel in an often petty manner **:** CONTENTIOUS
synonym see BELLIGERENT
— **quar·rel·some·ly** *adverb*
— **quar·rel·some·ness** *noun*

quar·ri·er \ˈkwȯr-ē-ər, ˈkwär-\ *noun* (14th century)
: a worker in a stone quarry

¹quar·ry \ˈkwȯr-ē, ˈkwär-\ *noun, plural* **quarries** [Middle English *querre* entrails of game given to the hounds, from Middle French *cuiree,* from *cuir* skin, hide (on which the entrails were placed), from Latin *corium* — more at CUIRASS] (14th century)
1 *obsolete* **:** a heap of the game killed in a hunt
2 : GAME; *specifically* **:** game hunted with hawks
3 : one that is sought or pursued **:** PREY ◆

²quarry *noun, plural* **quarries** [Middle English *quarey,* alteration of *quarrere,* from Middle French *quarriere,* from (assumed) Old French *quarre* squared stone, from Latin *quadrum* square] (14th century)
1 : an open excavation usually for obtaining building stone, slate, or limestone
2 : a rich source

³quarry *verb* **quar·ried; quar·ry·ing** (1774)
transitive verb
1 : to dig or take from or as if from a quarry ⟨*quarry* marble⟩
2 : to make a quarry in ⟨*quarry* a hill⟩
intransitive verb
: to delve in or as if in a quarry

⁴quarry *noun, plural* **quarries** [alteration of ¹*quarrel*] (1555)
: a diamond-shaped pane of glass, stone, or tile

quarrying *noun* (circa 1828)
: the business, occupation, or act of extracting useful material (as building stone) from quarries

quar·ry·man \ˈkwȯr-ē-mən, ˈkwär-\ *noun* (15th century)
: QUARRIER

quart \ˈkwȯrt\ *noun* [Middle English, one fourth of a gallon, from Middle French *quarte,*

◇ WORD HISTORY

quark As a name for an elementary particle, *quark* was coined by the Nobel-Prize-winning American physicist Murray Gell-Mann in 1963. According to his own account, he was fond of using whimsical names for peculiar objects, and "quork" (rhyming with *pork*) spontaneously came to mind as he started working with hypothetical, fractionally-charged constituents of baryons and mesons. Some months later he ran across the line "Three quarks for Muster Mark" in James Joyce's *Finnegans Wake,* a work that he had perused for a number of years. Joyce's word struck the physicist as an especially apt name for the hypothetical particles, so he adopted Joyce's spelling but retained his original pronunciation, claiming *quart* as a model. Joyce himself apparently took the word from German, where it refers to a cheese that in an early stage of its manufacture resembles cottage cheese. It is also used like the German word *Quatsch* to mean "trivial nonsense."

quarry *Quarry* in the meaning "game" or "prey" can be traced to a hunting ritual that dates back to medieval times. At the end of a successful hunt the hounds used in the pursuit were rewarded with a part of the slain animal's entrails. Traditionally the entrails were spread out on the animal's hide. The word for the hounds' feast in Middle French was *cuiree,* a derivative of *cuir,* meaning "skin, hide." *Cuiree* was borrowed into Middle English as *querre* or *quirre.* Over time the meaning of *querre* shifted from the slain animal's entrails to the live animal itself, seen as the object of a hunt. Now a *quarry* may be anything that is pursued.

from Old French, from feminine of *quart,* adjective, fourth, from Latin *quartus;* akin to Latin *quattuor* four — more at FOUR] (14th century)
1 — see WEIGHT table
2 : a vessel or measure having a capacity of one quart

¹quar·tan \'kwȯr-tᵊn\ *adjective* [Middle English *quarteyne,* from Middle French *(fievre) quartaine* quartan fever, from Latin *(febris) quartana,* from *quartanus* of the fourth, from *quartus*] (14th century)
: occurring every fourth day reckoning inclusively; *specifically* **:** recurring at approximately 72-hour intervals

²quartan *noun* (14th century)
: an intermittent fever that recurs at approximately 72-hour intervals; *especially* **:** a quartan malaria

¹quar·ter \'kwȯ(r)-tər *also* 'kȯ(r)-\ *noun* [Middle English, from Old French *quartier,* from Latin *quartarius,* from *quartus* fourth] (14th century)
1 : one of four equal parts into which something is divisible **:** a fourth part ⟨in the top *quarter* of his class⟩
2 : any of various units of capacity or weight equal to or derived from one fourth of some larger unit
3 : any of various units of length or area equal to one fourth of some larger unit
4 : the fourth part of a measure of time: as **a :** one of a set of four 3-month divisions of a year ⟨business was up during the third *quarter*⟩ **b :** a school term of about 12 weeks **c :** QUARTER HOUR ⟨a *quarter* after three⟩
5 a : a coin worth a quarter of a dollar **b :** the sum of 25 cents
6 a : one limb of a quadruped with the adjacent parts; *especially* **:** one fourth part of the carcass of a slaughtered animal including a leg **b** *plural, British* **:** HINDQUARTER 2
7 a : the region or direction lying under any of the four divisions of the horizon **b :** one of the four parts into which the horizon is divided or the cardinal point corresponding to it **c :** a compass point or direction other than the cardinal points **d** (1) **:** an unspecified person or group ⟨financial help from many *quarters* —*Current Biography*⟩ (2) **:** a point, direction, or place not definitely identified ⟨the view to the rear *quarter* —*Consumer Reports*⟩
8 a : a division or district of a town or city ⟨he describes the immigrant *quarter* —Alfred Kazin⟩ **b :** the inhabitants of such a quarter
9 a : an assigned station or post **b** *plural* **:** an assembly of a ship's company for ceremony, drill, or emergency **c** *plural* **:** living accommodations **:** LODGINGS ⟨show you to your *quarters*⟩
10 : merciful consideration of an opponent; *specifically* **:** the clemency of not killing a defeated enemy
11 : a fourth part of the moon's period
12 : the side of a horse's hoof between the toe and the heel — see HOOF illustration
13 a : any of the four parts into which a heraldic field is divided **b :** a bearing or charge occupying the first fourth part of a heraldic field
14 : the state of two machine parts that are exactly at right angles to one another or are spaced about a circle so as to subtend a right angle at the center of the circle
15 a : the stern area of a ship's side **b :** the part of the yardarm outside the slings
16 : one side of the upper of a shoe or boot from heel to vamp
17 : one of the four equal periods into which the playing time of some games is divided

²quarter (14th century)
transitive verb
1 a : to cut or divide into four equal or nearly equal parts ⟨*quarter* an apple⟩ ⟨condemned to be hanged, drawn, and *quartered*⟩ **b** *archaic* **:** DIVIDE
2 : to provide with lodging or shelter
3 : to crisscross (an area) in many directions
4 a : to arrange or bear (as different coats of arms) quarterly on one escutcheon **b :** to add (a coat of arms) to others on one escutcheon **c :** to divide (a shield) into distinct sections (as by stripes)
5 : to adjust or locate (as cranks) at right angles in a machine
intransitive verb
1 : LODGE, DWELL
2 : to crisscross a district
3 : to change from one quarter to another ⟨the moon *quarters*⟩
4 : to strike on a ship's quarter ⟨the wind was *quartering*⟩

³quarter *adjective* (14th century)
: consisting of or equal to a quarter

quar·ter·age \'kwȯ(r)-tə-rij\ *noun* (14th century)
: a quarterly payment, tax, wage, or allowance

¹quar·ter·back \'kwȯ(r)-tər-ˌbak\ *noun* (1879)
1 : an offensive back in football who usually lines up behind the center, calls the signals, and directs the offensive play of the team
2 : one who gives direction and leadership

²quarterback (1944)
transitive verb
1 : to direct the offensive play of (as a football team)
2 : to give executive direction to **:** BOSS ⟨*quarterbacked* the original buying syndicate —*Time*⟩
intransitive verb
: to play quarterback

quarterback sneak *noun* (circa 1923)
: a usually quick run with the ball by a quarterback into the middle of the offensive line

quar·ter–bound \'kwȯ(r)-tər-'baund\ *adjective* (circa 1888)
of a book **:** bound in material of two qualities with the material of better quality on the spine only
— quarter binding *noun*

quarter day *noun* (15th century)
chiefly British **:** the day which begins a quarter of the year and on which a quarterly payment often falls due

quar·ter·deck \'kwȯ(r)-tər-ˌdek\ *noun* (1627)
1 : the stern area of a ship's upper deck
2 : a part of a deck on a naval vessel set aside by the captain for ceremonial and official use

¹quar·ter·fi·nal \ˌkwȯ(r)-tər-'fī-nᵊl\ *noun* (1927)
1 *plural* **:** a quarterfinal round
2 : a quarterfinal match
— quar·ter·fi·nal·ist \-nᵊl-ist\ *noun*

²quarterfinal *adjective* (circa 1934)
1 : immediately preceding the semifinal in an elimination tournament
2 : of or participating in a quarterfinal

quarter horse *noun* [from its high speed for distances up to a quarter of a mile] (1834)
: any of a breed of compact muscular saddle horses characterized by great endurance and by high speed for short distances

quarter horse

quarter hour *noun* (1883)
1 : any of the quarter points of an hour
2 : fifteen minutes
3 : a unit of academic credit representing an hour of class (as lecture class) or three hours of laboratory work each week for an academic quarter

¹quar·ter·ing \'kwȯ(r)-tə-riŋ\ *noun* (15th century)
1 a : the division of an escutcheon containing different coats of arms into four or more compartments **b :** a quarter of an escutcheon or the coat of arms on it
2 : a line of usually noble or distinguished ancestry

²quartering *adjective* (circa 1692)
1 : coming from a point well abaft the beam of a ship but not directly astern ⟨*quartering* waves⟩
2 : lying at right angles

¹quar·ter·ly \'kwȯ(r)-tər-lē\ *adverb* (14th century)
1 : in heraldic quarters or quarterings
2 : at 3-month intervals

²quarterly *adjective* (15th century)
1 : computed for or payable at 3-month intervals ⟨*quarterly* premium⟩
2 : recurring, issued, or spaced at 3-month intervals
3 : divided into heraldic quarters or compartments

³quarterly *noun, plural* **-lies** (1830)
: a periodical published four times a year

Quarterly Meeting *noun* (1675)
: an organizational unit of the Society of Friends usually composed of several Monthly Meetings

quar·ter·mas·ter \'kwȯ(r)-tər-ˌmas-tər\ *noun* (15th century)
1 : a petty officer who attends to a ship's helm, binnacle, and signals
2 : an army officer who provides clothing and subsistence for a body of troops

quar·tern \'kwȯ(r)-tərn\ *noun* [Middle English *quarteron,* from Old French, quarter of a pound, quarter of a hundred, from *quartier* quarter] (14th century)
: a fourth part (as of a unit of measurement)

quarter note *noun* (1763)
: a musical note with the time value of ¼ of a whole note — see NOTE illustration

quarter rest *noun* (circa 1890)
: a musical rest corresponding in time value to a quarter note — see REST illustration

quar·ter·sawn \'kwȯ(r)-tər-'sȯn\ *also* **quar·ter·sawed** \-'sȯd\ *adjective* (circa 1890)
: sawed from quartered logs so that the annual rings are nearly at right angles to the wide face — used of boards and planks

quarter section *noun* (1804)
: a tract of land that is half a mile square and contains 160 acres in the U.S. government system of land surveying

quarter sessions *noun plural* (1577)
: a former English local court with limited original and appellate criminal and sometimes civil jurisdiction and often administrative functions held quarterly usually by two justices of the peace in a county or by a recorder in a borough

quar·ter·staff \'kwȯ(r)-tər-ˌstaf\ *noun, plural* **-staves** \-ˌstavz, -ˌstāvz\ (1550)
: a long stout staff formerly used as a weapon and wielded with one hand in the middle and the other between the middle and the end

quarter tone *noun* (circa 1776)
1 : a musical interval of one half a semitone
2 : a tone at an interval of one quarter

quar·tet *also* **quar·tette** \kwȯr-'tet\ *noun* [Italian *quartetto,* from *quarto* fourth, from Latin *quartus* — more at QUART] (1773)
1 : a musical composition for four instruments or voices
2 : a group or set of four; *especially* **:** the performers of a quartet

quar·tic \'kwȯr-tik\ *adjective* [Latin *quartus* fourth] (circa 1890)
: of the fourth degree ⟨*quartic* equation⟩
— quartic *noun*

quar·tile \'kwȯr-ˌtīl, -tᵊl\ *noun* [International Scientific Vocabulary, from Latin *quartus*] (1879)
: any of the three values that divide the items of a frequency distribution into four classes with each containing one fourth of the total population; *also* **:** any one of the four classes

quar·to \'kwȯr-(,)tō\ *noun, plural* **quartos** [Latin, ablative of *quartus* fourth] (1589)
1 : the size of a piece of paper cut four from a sheet; *also* **:** paper or a page of this size
2 : a book printed on quarto pages

quartz \'kwȯrts\ *noun* [German *Quarz*] (circa 1631)
1 : a mineral consisting of silicon dioxide occurring in colorless and transparent or colored hexagonal crystals or in crystalline masses
2 : a quartz crystal that when placed in an electric field oscillates at a constant frequency and is used to control devices which require precise regulation ⟨a *quartz* watch⟩
— quartz·ose \'kwȯrt-,sōs\ *adjective*

quartz glass *noun* (1903)
: vitreous silica prepared from pure quartz and noted for its transparency to ultraviolet radiation

quartz heater *noun* (1980)
: a portable electric radiant heater that has heating elements sealed in quartz-glass tubes producing infrared radiation in front of a reflective backing

quartz–iodine lamp *noun* (circa 1964)
: an incandescent lamp that has a quartz bulb and a tungsten filament with the bulb containing iodine which reacts with the vaporized tungsten to prevent excessive blackening of the bulb

quartz·ite \'kwȯrt-,sīt\ *noun* [International Scientific Vocabulary] (circa 1847)
: a compact granular rock composed of quartz and derived from sandstone by metamorphism
— quartz·it·ic \kwȯrt-'si-tik\ *adjective*

qua·sar \'kwā-,zär *also* -,sär\ *noun* [*quasi*-stell*ar*] (1964)
: any of a class of celestial objects that resemble stars but whose large redshift and apparent brightness imply extreme distance and huge energy output

¹quash \'kwäsh, 'kwȯsh\ *transitive verb* [Middle English *quashen* to smash, from Middle French *quasser, casser,* from Latin *quassare* to shake violently, shatter, frequentative of *quatere* to shake] (13th century)
: to suppress or extinguish summarily and completely ⟨*quash* a rebellion⟩

²quash *transitive verb* [Middle English *quassen,* from Middle French *casser, quasser* to annul, from Late Latin *cassare,* from Latin *cassus* void] (14th century)
: to nullify especially by judicial action ⟨*quash* an indictment⟩

qua·si \'kwā-,zī, -,sī; 'kwä-zē, -sē\ *adjective* [*quasi*-] (1642)
1 : having some resemblance usually by possession of certain attributes ⟨a *quasi* corporation⟩
2 : having a legal status only by operation or construction of law and without reference to intent ⟨a *quasi* contract⟩

quasi- *combining form* [Latin *quasi* as if, as it were, approximately, from *quam* as + *si* if — more at QUANTITY, SO]
1 : in some sense or degree ⟨*quasi*periodic⟩
2 : resembling in some degree ⟨*quasi*particle⟩

qua·si–ju·di·cial \,kwā-,zī-jü-'di-shəl, -,sī-, ,kwä-zē-, -sē-\ *adjective* (1836)
1 : having a partly judicial character by possession of the right to hold hearings on and conduct investigations into disputed claims and alleged infractions of rules and regulations and to make decisions in the general manner of courts ⟨*quasi-judicial* bodies⟩
2 : essentially judicial in character but not within the judicial power or function especially as constitutionally defined ⟨*quasi-judicial* review⟩
— qua·si–ju·di·cial·ly \-'di-sh(ə-)lē\ *adverb*

qua·si–leg·is·la·tive \-'le-jəs-,lā-tiv\ *adjective* (circa 1934)
1 : having a partly legislative character by possession of the right to make rules and regulations having the force of law ⟨a *quasi-legislative* agency⟩
2 : essentially legislative in character but not within the legislative power or function especially as constitutionally defined ⟨*quasi-legislative* powers⟩

Qua·si·mo·do \,kwä-si-'mō-(,)dō, ,kwä-zi-\ *noun* [Medieval Latin *quasi modo geniti infantes* as newborn babes (words of the introit for Low Sunday)] (circa 1847)
: LOW SUNDAY

qua·si·par·ti·cle \,kwā-,zī-'pär-ti-kəl, -,sī-, ,kwä-zē-, -sē-\ *noun* (1957)
: a composite entity (as a vibration in a solid) that is analogous in its behavior to a single particle

qua·si·pe·ri·od·ic \-,pir-ē-'ä-dik\ *adjective* (circa 1890)
: almost but not quite periodic; *especially* **:** periodic on a small scale but unpredictable at some larger scale
— qua·si·pe·ri·od·ic·i·ty \-,pir-ē-ə-'di-sə-tē\ *noun*

qua·si–pub·lic \-'pə-blik\ *adjective* (1888)
: essentially public (as in services rendered) although under private ownership or control

qua·si–stel·lar object \-'ste-lər-\ *noun* (1964)
: QUASAR

quas·sia \'kwä-shə\ *noun* [New Latin, genus name of a South American tree, from *Quassi* 18th century Surinam slave who discovered the medicinal value of quassia] (1770)
: a drug from the heartwood of various tropical trees of the ailanthus family used especially as a bitter tonic and remedy for roundworms in children and as an insecticide

qua·ter·cen·te·na·ry \,kwä-tər-sen-'te-nə-rē, -'sen-tᵊn-,er-ē, -sen-'tē-nə-rē\ *noun* [Latin *quater* four times + English *centenary* — more at QUATERNION] (1883)
: a year marking a 400th anniversary

¹qua·ter·na·ry \'kwä-tə(r)-,ner-ē, kwə-'tər-nə-rē\ *adjective* [Latin *quaternarius,* from *quaterni* four each] (1605)
1 a : of, relating to, or consisting of four units or members **b :** of, relating to, or being a number system with a base of four
2 *capitalized* **:** of, relating to, or being the geological period from the end of the Tertiary to the present time or the corresponding system of rocks — see GEOLOGIC TIME table
3 : consisting of, containing, or being an atom bonded to four other atoms

²quaternary *noun, plural* **-ries** (1880)
1 *capitalized* **:** the Quaternary period or system of rocks
2 : a member of a group fourth in order or rank

quaternary ammonium compound *noun* (circa 1934)
: any of numerous strong bases and their salts derived from ammonium by replacement of the hydrogen atoms with organic radicals and important especially as surface-active agents, disinfectants, and drugs

qua·ter·ni·on \kwə-'tər-nē-ən, kwä-\ *noun* [Middle English *quaternyoun,* from Late Latin *quaternion-, quaternio,* from Latin *quaterni* four each, from *quater* four times; akin to Latin *quattuor* four — more at FOUR] (14th century)
1 : a set of four parts, things, or persons
2 a : a generalized complex number that is composed of a real number and a vector and that depends on one real and three imaginary units **b** *plural* **:** the calculus of quaternions

qua·ter·ni·ty \kwə-'tər-nə-tē, kwä-\ *noun, plural* **-ties** [Late Latin *quaternitas,* from Latin *quaterni* four each] (1529)
: a union of a group or set of four

qua·train \'kwä-,trān, kwä-'\ *noun* [Middle French, from *quatre* four, from Latin *quattuor*] (1582)
: a unit or group of four lines of verse

qua·tre·foil \'ka-tər-,fȯil, 'ka-trə-\ *noun* [Middle English *quaterfoil* set of four leaves, from Middle French *quatre* + Middle English *-foil* (as in *trefoil*)] (15th century)
1 : a conventionalized representation of a flower with four petals or of a leaf with four leaflets
2 : a 4-lobed foliation in architecture

quat·tro·cen·to \,kwä-trō-'chen-(,)tō\ *noun, often capitalized* [Italian, literally, four hundred, from *quattro* four (from Latin *quattuor*) + *cento* hundred — more at CINQUECENTO] (circa 1854)
: the 15th century especially with reference to Italian literature and art

quat·tu·or·de·cil·lion \,kwä-tə-,wȯr-di-'sil-yən\ *noun, often attributive* [Latin *quattuordecim* fourteen (from *quattuor* four + *decem* ten) + English *-illion* (as in *million*) — more at TEN] (circa 1903)
— see NUMBER table

¹qua·ver \'kwā-vər\ *verb* **qua·vered; qua·ver·ing** \'kwā-və-riŋ, 'kwāv-riŋ\ [Middle English, frequentative of *quaven* to tremble] (15th century)
intransitive verb
1 : TREMBLE
2 : TRILL
3 : to utter sound in tremulous tones
transitive verb
: to utter quaveringly
— qua·ver·ing·ly *adverb*
— qua·very \'kwā-və-rē, 'kwāv-rē\ *adjective*

²quaver *noun* (1570)
1 : EIGHTH NOTE
2 : TRILL 1
3 : a tremulous sound

quay \'kē, 'kā, 'kwā\ *noun* [alteration of earlier *key,* from Middle English, from Middle French dialect *cai,* probably of Celtic origin; akin to Breton *kae* hedge, enclosure; akin to Old English *hecg* hedge] (1696)
: a structure built parallel to the bank of a waterway for use as a landing place

quay·age \-ij\ *noun* (circa 1756)
1 : a charge for use of a quay
2 : room on or for quays
3 : a system of quays

quay·side \-,sīd\ *noun* (1903)
: land bordering a quay

quean \'kwēn, 'kwān\ *noun* [Middle English *quene,* from Old English *cwene;* akin to Old English *cwēn* woman, queen] (before 12th century)
1 : a disreputable woman; *specifically* **:** PROSTITUTE
2 *chiefly Scottish* **:** WOMAN; *especially* **:** one that is young or unmarried

quea·sy *also* **quea·zy** \'kwē-zē\ *adjective* **quea·si·er; -est** [Middle English *coysy, qwesye*] (15th century)
1 a : causing nausea ⟨*queasy* motion⟩ **b :** suffering from nausea **:** NAUSEATED
2 : full of doubt **:** HAZARDOUS
3 a : causing uneasiness **b** (1) **:** DELICATE, SQUEAMISH (2) **:** ill at ease
— quea·si·ly \-zə-lē\ *adverb*
— quea·si·ness \-zē-nəs\ *noun*

Que·bec \kwi-'bek *also* ki-\ (1952)
— a communications code word for the letter *q*

Que·be·cois *or* **Qué·bé·cois** *or* **Qué·be·cois** \,kā-bə-'kwä, -,be-\ *noun, plural* **Quebecois** *or* **Québécois** *or* **Québecois** \-'kwä(z)\ [French *québecois, québécois,* from *Québec* Quebec] (1873)
: a native or inhabitant of Quebec; *specifically* **:** a French-speaking native or inhabitant of Quebec

— **Quebecois** *or* **Québécois** *or* **Québe·cois** *adjective*

que·bra·cho \kā-ˈbrä-(ˌ)chō, ki-\ *noun* [American Spanish, alteration of *quiebracha,* from Spanish *quiebra* it breaks + *hacha* ax] (circa 1881)
1 : any of several trees of southern South America with hard wood: as **a :** a tree (*Aspidosperma quebracho*) of the dogbane family which occurs in Argentina and Chile and whose dried bark is used as a respiratory sedative in dyspnea and in asthma **b :** a chiefly Argentine tree (*Schinopsis lorentzii*) of the cashew family with dense wood rich in tannins
2 a : the wood of a quebracho **b :** a tannin-rich extract of the Argentine quebracho used in tanning leather

Que·chua \ˈke-chə-wə, ˈkech-wə\ *noun, plural* **Quechua** *or* **Quechuas** [Spanish, probably from Southern Peruvian Quechua *qheswa* (*simi*), literally, valley speech] (1840)
1 : a family of languages spoken by Indian peoples of Peru, Bolivia, Ecuador, Chile, and Argentina
2 a : a member of an Indian people of central Peru **b :** a group of peoples forming the dominant element of the Inca Empire
— **Que·chu·an** \-wən\ *adjective or noun*

¹queen \ˈkwēn\ *noun* [Middle English *quene,* from Old English *cwēn* woman, wife, queen; akin to Gothic *qens* wife, Greek *gynē* woman, Sanskrit *jani*] (before 12th century)
1 a : the wife or widow of a king **b :** the wife or widow of a tribal chief
2 a : a female monarch **b :** a female chieftain
3 a : a woman eminent in rank, power, or attractions ⟨a movie *queen*⟩ **b :** a goddess or a thing personified as female and having supremacy in a specified realm **c :** an attractive girl or woman; *especially* **:** a beauty contest winner
4 : the most privileged piece of each color in a set of chessmen having the power to move in any direction across any number of unoccupied squares
5 : a playing card marked with a stylized figure of a queen
6 : the fertile fully developed female of social bees, ants, and termites whose function is to lay eggs
7 : a mature female cat kept especially for breeding
8 : a male homosexual; *especially* **:** an effeminate one — often used disparagingly

²queen (1611)
intransitive verb
1 : to act like a queen; *especially* **:** to put on airs — usually used with *it* ⟨*queens* it over her friends⟩
2 : to become a queen in chess
transitive verb
: to promote (a pawn) to a queen in chess

Queen Anne \-ˈan\ *adjective* [*Queen Anne* of England] (1863)
1 : of, relating to, or having the characteristics of a style of furniture originating in England under Dutch influence especially during the first half of the 18th century that is marked by extensive use of upholstery, marquetry, and oriental fabrics
2 : of, relating to, or having the characteristics of a style of English building of the early 18th century characterized by modified classic ornament and the use of red brickwork in which even relief ornament is carved

Queen Anne's lace *noun* (1895)
: a widely naturalized Eurasian biennial herb (*Daucus carota*) which has a whitish acrid taproot and from which the cultivated carrot originated — called also *wild carrot*

queen consort *noun, plural* **queens consort** (1765)
: the wife of a reigning king

queen·ly \ˈkwēn-lē\ *adjective* **queen·li·er; -est** (15th century)
1 : of, relating to, or befitting a queen
2 : having royal rank
3 : MONARCHICAL
— **queen·li·ness** *noun*
— **queenly** *adverb*

queen mother *noun* (1577)
: a queen dowager who is mother of the reigning sovereign

queen post *noun* (1823)
: one of two vertical tie post in a truss (as of a roof)

d g e f

de gf **queen posts**

queen regnant *noun, plural* **queens regnant** (circa 1639)
: a queen reigning in her own right

Queen's Bench *noun* (circa 1809)
: a division of the English superior courts system that hears civil and criminal court cases — used during the reign of a queen

Queen's Counsel *noun* (circa 1860)
: a barrister selected to serve as counsel to the British crown — used during the reign of a queen

queen·ship \ˈkwēn-ˌship\ *noun* (1536)
1 : the rank, dignity, or state of being a queen
2 : a regal quality like that of a queen

queen·side \-ˌsīd\ *noun* (1897)
: the side of a chessboard containing the file on which the queen sits at the beginning of the game

queen–size *adjective* (1959)
1 : having dimensions of approximately 60 inches by 80 inches (about 1.5 by 2.0 meters) — used of a bed; compare FULL-SIZE, KING-SIZE, TWIN-SIZE
2 : of a size that fits a queen-size bed ⟨a *queen-size* sheet⟩

queen substance *noun* (1954)
: a pheromone secreted by queen bees that is consumed by worker bees and inhibits ovary development

¹queer \ˈkwir\ *adjective* [origin unknown] (1508)
1 a : WORTHLESS, COUNTERFEIT ⟨*queer* money⟩ **b :** QUESTIONABLE, SUSPICIOUS
2 a : differing in some odd way from what is usual or normal **b** (1) **:** ECCENTRIC, UNCONVENTIONAL (2) **:** mildly insane **:** TOUCHED **c :** absorbed or interested to an extreme or unreasonable degree **:** OBSESSED **d :** HOMOSEXUAL — usually used disparagingly
3 : not quite well
synonym see STRANGE
— **queer·ish** \-ish\ *adjective*
— **queer·ly** *adverb*
— **queer·ness** *noun*

²queer *transitive verb* (circa 1812)
1 : to spoil the effect or success of ⟨*queer* one's plans⟩
2 : to put or get into an embarrassing or disadvantageous situation

³queer *noun* (circa 1812)
: one that is queer; *especially* **:** HOMOSEXUAL — usually used disparagingly

¹quell \ˈkwel\ *transitive verb* [Middle English, to kill, quell, from Old English *cwellan* to kill; akin to Old High German *quellen* to torture, kill, *quāla* torment, Lithuanian *gelti* to hurt] (13th century)
1 : to thoroughly overwhelm and reduce to submission or passivity ⟨*quell* a riot⟩
2 : QUIET, PACIFY ⟨*quell* fears⟩
— **quell·er** *noun*

²quell *noun* [Middle English, from *quellen* to kill] (15th century)
1 *obsolete* **:** SLAUGHTER
2 *archaic* **:** the power of quelling

quench \ˈkwench\ *verb* [Middle English, from Old English *-cwencan;* akin to Old English *-cwincan* to vanish, Old Frisian *quinka*] (12th century)
transitive verb
1 a : PUT OUT, EXTINGUISH **b :** to put out the light or fire of ⟨*quench* glowing coals with water⟩ **c :** to cool (as heated metal) suddenly by immersion (as in oil or water) **d :** to cause to lose heat or warmth ⟨you have *quenched* the warmth of France toward you —Alfred Tennyson⟩
2 a : to bring (something immaterial) to an end typically by satisfying, damping, cooling, or decreasing ⟨a rational understanding of the laws of nature can *quench* impossible desires —Lucius Garvin⟩ ⟨the praise that *quenches* all desire to read the book —T. S. Eliot⟩ **b :** to terminate by or as if by destroying **:** ELIMINATE ⟨the Commonwealth party *quenched* a whole generation of play-acting —Margery Bailey⟩ ⟨*quench* a rebellion⟩ **c :** to relieve or satisfy with liquid ⟨*quenched* his thirst at a wayside spring⟩
intransitive verb
1 : to become extinguished **:** COOL
2 : to become calm **:** SUBSIDE
— **quench·able** \ˈkwen-chə-bəl\ *adjective*
— **quench·er** *noun*
— **quench·less** \ˈkwench-ləs\ *adjective*

que·nelle \kə-ˈnel\ *noun* [French, from German *Knödel* dumpling, from Middle High German; akin to Old High German *knoto* knot — more at KNOT] (1845)
: a poached oval dumpling of pureed forcemeat (as of pike) often served in a cream sauce

quer·ce·tin \ˈkwər-sə-tən\ *noun* [International Scientific Vocabulary, from Latin *quercetum* oak forest, from *quercus* oak — more at FIR] (1857)
: a yellow crystalline pigment $C_{15}H_{10}O_7$ occurring usually in the form of glycosides in various plants

quer·cit·ron \ˈkwər-ˌsi-trən, ˌkwər-ˈ\ *noun* [blend of New Latin *Quercus* and International Scientific Vocabulary *citron*] (1794)
1 : a large timber oak (*Quercus velutina*) of the eastern and central U.S.
2 : the bark of the quercitron that is rich in tannin and a dye containing quercetin; *also* **:** the dye

que·rist \ˈkwir-əst, ˈkwer-\ *noun* [Latin *quaerere* to ask] (1633)
: one who inquires

quern \ˈkwərn\ *noun* [Middle English, from Old English *cweorn;* akin to Old High German *quirn* hand mill, Old Church Slavonic *žrŭny*] (before 12th century)
: a primitive hand mill for grinding grain

quer·u·lous \ˈkwer-yə-ləs, -ə-ləs *also* ˈkwir-\ *adjective* [Middle English *querelose,* from Latin *querulus,* from *queri* to complain] (15th century)
1 : habitually complaining
2 : FRETFUL, WHINING ⟨a *querulous* voice⟩
— **quer·u·lous·ly** *adverb*
— **quer·u·lous·ness** *noun*

¹que·ry \ˈkwir-ē, ˈkwer-\ *noun, plural* **queries** [alteration of earlier *quere,* from Latin *quaere,* imperative of *quaerere* to ask] (circa 1635)
1 : QUESTION, INQUIRY
2 : a question in the mind **:** DOUBT
3 : QUESTION MARK

²query *transitive verb* **que·ried; que·ry·ing** (1654)
1 : to ask questions of especially with a desire for authoritative information
2 : to ask questions about especially in order to resolve a doubt
3 : to put as a question
4 : to mark with a query
synonym see ASK
— **que·ri·er** *noun*

que·sa·dil·la \ˌkā-sə-ˈdē-ə *also* -ˈthē- *or* -ˈthēl-yə\ *noun* [Mexican Spanish, from Spanish, cheese pastry, diminutive of *quesada,* from *queso* cheese, from Latin *caseus*] (1944)

: a wheat tortilla filled with a savory mixture, folded, fried in deep fat, and topped with cheese

[1]**quest** \ˈkwest\ *noun* [Middle English, search, pursuit, investigation, inquest, from Middle French *queste* search, pursuit, from (assumed) Vulgar Latin *quaesta,* from Latin, feminine of *quaestus,* past participle of *quaerere*] (14th century)
1 a : a jury of inquest **b :** INVESTIGATION
2 : an act or instance of seeking: **a :** PURSUIT, SEARCH **b :** a chivalrous enterprise in medieval romance usually involving an adventurous journey
3 *obsolete* **:** a person or group of persons who search or make inquiry

[2]**quest** (14th century)
intransitive verb
1 *of a dog* **a :** to search a trail **b :** BAY
2 : to go on a quest
transitive verb
1 : to search for
2 : to ask for
— quest·er *noun*

[1]**ques·tion** \ˈkwes-chən, ˈkwesh-\ *noun* [Middle English, from Middle French, from Latin *quaestion-, quaestio,* from *quaerere* to seek, ask] (14th century)
1 a (1) **:** an interrogative expression often used to test knowledge (2) **:** an interrogative sentence or clause **b :** a subject or aspect in dispute or open for discussion **:** ISSUE; *broadly* **:** PROBLEM, MATTER **c** (1) **:** a subject or point of debate or a proposition to be voted on in a meeting (2) **:** the bringing of such to a vote **d :** the specific point at issue
2 a : an act or instance of asking **:** INQUIRY **b :** INTERROGATION; *also* **:** a judicial or official investigation **c :** torture as part of an examination **d** (1) **:** OBJECTION, DISPUTE ⟨true beyond *question*⟩ (2) **:** room for doubt or objection ⟨little *question* of his skill⟩ (3) **:** CHANCE, POSSIBILITY ⟨no *question* of escape⟩

[2]**question** (15th century)
transitive verb
1 : to ask a question of or about
2 : to interrogate intensively **:** CROSS-EXAMINE
3 a : DOUBT, DISPUTE **b :** to subject to analysis **:** EXAMINE
intransitive verb
: to ask questions **:** INQUIRE
synonym see ASK
— ques·tion·er *noun*

ques·tion·able \ˈkwes-chə-nə-bəl, ˈkwesh-, *in rapid speech* ˈkwesh-nə-\ *adjective* (1590)
1 *obsolete* **:** inviting inquiry
2 *obsolete* **:** liable to judicial inquiry or action
3 : affording reason for being doubted, questioned, or challenged **:** not certain or exact **:** PROBLEMATIC ⟨milk of *questionable* purity⟩ ⟨a *questionable* decision⟩
4 : attended by well-grounded suspicions of being immoral, crude, false, or unsound **:** DUBIOUS ⟨*questionable* motives⟩
synonym see DOUBTFUL
— ques·tion·able·ness *noun*
— ques·tion·ably \-blē\ *adverb*

ques·tion·ary \ˈkwes-chə-ˌner-ē, ˈkwesh-\ *noun, plural* **-ar·ies** (1887)
: QUESTIONNAIRE

ques·tion·less \ˈkwes-chən-ləs, ˈkwesh-\ *adjective* (1532)
1 : INDUBITABLE, UNQUESTIONABLE
2 : UNQUESTIONING

question mark *noun* (1869)
1 : something unknown, unknowable, or uncertain
2 : a mark ? used in writing and printing at the conclusion of a sentence to indicate a direct question

ques·tion·naire \ˌkwes-chə-ˈnar, -ˈner, ˌkwesh-\ *noun* [French, from *questionner* to question, from Middle French, from *question,* noun] (1899)
1 : a set of questions for obtaining statistically useful or personal information from individuals
2 : a written or printed questionnaire often with spaces for answers
3 : a survey made by the use of a questionnaire

question time *noun* (1885)
: a period in a session of a British parliamentary body during which members may put questions to ministers on matters concerning their departments

ques·tor \ˈkwes-tər\ *variant of* QUAESTOR

quet·zal \ket-ˈsäl, -ˈsal\ *noun, plural* **quet·zals** *or* **quet·za·les** \-ˈsä-(ˌ)lās, -ˈsa-\ [American Spanish, from Nahuatl *quetzalli* tail coverts of the quetzal] (1827)
1 : a Central American trogon (*Pharomachrus mocinno*) that has brilliant green plumage above, a red breast, and in the male long upper tail coverts
2 *plural quetzales* — see MONEY table

quetzal 1

Quet·zal·co·a·tl \ˌkwet-səl-kə-ˈwä-tᵊl, ˌket-, -səl-ˈkwä-; ket-ˌsäl-, -ˌsal-\ *noun* [Nahuatl *Quetzalcōātl*]
: a chief Toltec and Aztec god identified with the wind and air and represented by a feathered serpent

[1]**queue** \ˈkyü\ *noun* [French, literally, tail, from Latin *cauda, coda*] (1748)
1 : a braid of hair usually worn hanging at the back of the head
2 : a waiting line especially of persons or vehicles
3 a : a sequence of messages or jobs held in auxiliary storage awaiting transmission or processing **b :** a data structure that consists of a list of records such that records are added at one end and removed from the other

[2]**queue** *verb* **queued; queu·ing** *or* **queue·ing** (1777)
transitive verb
: to arrange or form in a queue
intransitive verb
: to line up or wait in a queue — often used with *up*
— queu·er *noun*

[1]**quib·ble** \ˈkwi-bəl\ *verb* **quib·bled; quib·bling** \-b(ə-)liŋ\ (1656)
intransitive verb
1 : to evade the point of an argument by caviling about words
2 a : CAVIL, CARP **b :** BICKER
transitive verb
: to subject to quibbles
— quib·bler \-b(ə-)lər\ *noun*

[2]**quibble** *noun* [probably diminutive of obsolete *quib* quibble] (1670)
1 : an evasion of or shift from the point
2 : a minor objection or criticism

quiche \ˈkēsh\ *noun* [French, from French dialect (Lorraine)] (1941)
: an unsweetened custard pie usually having a savory filling (as spinach, mushrooms, or ham)

quiche lor·raine \-lə-ˈrān, -lō-\ *noun, often L capitalized* [French, quiche of Lorraine] (1941)
: a quiche containing cheese and bacon bits

[1]**quick** \ˈkwik\ *adjective* [Middle English *quik,* from Old English *cwic;* akin to Old Norse *kvikr* living, Latin *vivus* living, *vivere* to live, Greek *bios, zōē* life] (before 12th century)
1 : not dead **:** LIVING, ALIVE
2 : acting or capable of acting with speed: as **a** (1) **:** fast in understanding, thinking, or learning **:** mentally agile ⟨a *quick* wit⟩ ⟨*quick* thinking⟩ (2) **:** reacting to stimuli with speed and keen sensitivity (3) **:** aroused immediately and intensely ⟨*quick* tempers⟩ **b** (1) **:** fast in development or occurrence ⟨a *quick* succession of events⟩ (2) **:** done or taking place with rapidity ⟨gave them a *quick* look⟩ **c :** marked by speed, readiness, or promptness of physical movement ⟨walked with *quick* steps⟩ **d :** inclined to hastiness (as in action or response) ⟨*quick* to criticize⟩ **e :** capable of being easily and speedily prepared ⟨a *quick* and tasty dinner⟩
3 *archaic* **a :** not stagnant **:** RUNNING, FLOWING **b :** MOVING, SHIFTING ⟨*quick* mud⟩
4 *archaic* **:** FIERY, GLOWING
5 *obsolete* **a :** PUNGENT **b :** CAUSTIC
6 *archaic* **:** PREGNANT
7 : having a sharp angle ⟨a *quick* turn in the road⟩ ☆
— quick·ly *adverb*
— quick·ness *noun*

[2]**quick** *noun* (before 12th century)
1 quick *plural* **:** living beings
2 [probably of Scandinavian origin; akin to Old Norse *kvika* sensitive flesh, from *kvikr* living] **a :** a painfully sensitive spot or area of flesh (as that underlying a fingernail or toenail) **b :** the inmost sensibilities ⟨hurt to the *quick* by the remark⟩ **c :** the very center of something **:** HEART
3 *archaic* **:** LIFE 1l

[3]**quick** *adverb* (14th century)
: in a quick manner

quick assets *noun plural* (1891)
: cash, accounts receivable, and other current assets excluding inventories

quick bread *noun* (1918)
: bread made with a leavening agent (as baking powder or baking soda) that permits immediate baking of the dough or batter mixture

quick·en \ˈkwi-kən\ *verb* **quick·ened; quick·en·ing** \ˈkwi-kə-niŋ, ˈkwik-niŋ\ (14th century)
transitive verb
1 a : to make alive **:** REVIVE **b :** to cause to be enlivened **:** STIMULATE
2 *archaic* **a :** KINDLE **b :** to cause to burn more intensely
3 : to make more rapid **:** HASTEN, ACCELERATE ⟨*quickened* her steps⟩
4 a : to make (a curve) sharper **b :** to make (a slope) steeper
intransitive verb
1 : to quicken something
2 : to come to life; *especially* **:** to enter into a phase of active growth and development ⟨seeds *quickening* in the soil⟩
3 : to reach the stage of gestation at which fetal motion is felt

☆ SYNONYMS
Quick, prompt, ready, apt mean able to respond without delay or hesitation or indicative of such ability. QUICK stresses instancy of response and is likely to connote native rather than acquired power ⟨*quick* reflexes⟩ ⟨a keen *quick* mind⟩. PROMPT is more likely to connote training and discipline that fits one for instant response ⟨*prompt* emergency medical care⟩. READY suggests facility or fluency in response ⟨backed by a pair of *ready* assistants⟩. APT stresses the possession of qualities (as intelligence, a particular talent, or a strong bent) that makes quick effective response possible ⟨an *apt* student⟩ ⟨her answer was *apt* and to the point⟩. See in addition FAST.

4 : to shine more brightly ⟨watched the dawn *quickening* in the east⟩
5 : to become more rapid ⟨her pulse *quickened* at the sight⟩ ☆
— quick·en·er \ˈkwi-kə-nər, ˈkwik-nər\ *noun*

quick fix *noun* (1970)
: an expedient usually temporary or inadequate solution to a problem

quick–freeze \ˈkwik-ˈfrēz\ *transitive verb* **-froze** \-ˈfrōz\; **-fro·zen** \-ˈfrō-zᵊn\; **-freez·ing** (1930)
: to freeze (food) for preservation so rapidly that ice crystals formed are too small to rupture the cells and the natural juices and flavor are preserved

quick·ie \ˈkwi-kē\ *noun, often attributive* (circa 1926)
: something done or made in a hurry: as **a :** a quickly and usually cheaply produced work (as a motion picture or book) **b :** a hastily performed act of sexual intercourse

quick kick *noun* (circa 1940)
: a punt in football on first, second, or third down made from a running or passing formation and designed to take the opposing team by surprise

quick·lime \ˈkwik-ˌlīm\ *noun* (14th century)
: ¹LIME 2a

quick·sand \ˈkwik-ˌsand\ *noun* (14th century)
1 : sand readily yielding to pressure; *especially* **:** a deep mass of loose sand mixed with water into which heavy objects readily sink
2 : something that entraps or frustrates

quick·set \-ˌset\ *noun* (15th century)
chiefly British **:** plant cuttings set in the ground to grow especially in a hedgerow; *also* **:** a hedge or thicket especially of hawthorn grown from quickset

¹quick·sil·ver \-ˌsil-vər\ *noun* (before 12th century)
: MERCURY 2a ◆

²quicksilver *adjective* (1655)
: resembling or suggestive of quicksilver; *especially* **:** MERCURIAL 3

quick·step \-ˌstep\ *noun* (circa 1811)
: a spirited march tune usually accompanying a march in quick time

quick–tem·pered \-ˈtem-pərd\ *adjective* (1830)
: easily angered **:** IRASCIBLE

quick time *noun* (circa 1802)
: a rate of marching in which 120 steps each 30 inches in length are taken in one minute

quick–wit·ted \ˈkwik-ˈwi-təd\ *adjective* (1530)
: quick in perception and understanding **:** mentally alert
synonym see INTELLIGENT
— quick–wit·ted·ly *adverb*
— quick–wit·ted·ness *noun*

¹quid \ˈkwid\ *noun, plural* **quid** *also* **quids** [origin unknown] (1688)
British **:** a pound sterling

²quid *noun* [English dialect, cud, from Middle English *quide,* from Old English *cwidu, cwudu* — more at CUD] (circa 1727)
: a cut or wad of something chewable

quid·di·ty \ˈkwi-də-tē\ *noun, plural* **-ties** [Middle English *quidite,* from Medieval Latin *quidditat-, quidditas* essence, from Latin *quid* what, neuter of *quis* who — more at WHO] (14th century)
1 : whatever makes something the type that it is **:** ESSENCE
2 a : a trifling point **:** QUIBBLE **b :** CROTCHET, ECCENTRICITY

quid·nunc \ˈkwid-ˌnəŋk\ *noun* [Latin *quid nunc* what now?] (1709)
: a person who seeks to know all the latest news or gossip **:** BUSYBODY

quid pro quo \ˌkwid-ˌprō-ˈkwō\ *noun* [New Latin, something for something] (1591)
: something given or received for something else; *also* **:** a deal arranging a quid pro quo

qui·es·cence \kwī-ˈe-sᵊn(t)s, kwē-\ *noun* (circa 1631)
: the quality or state of being quiescent

qui·es·cent \-sᵊnt\ *adjective* [Latin *quiescent-, quiescens,* present participle of *quiescere* to become quiet, rest, from *quies*] (1605)
1 : marked by inactivity or repose **:** tranquilly at rest
2 : causing no trouble or symptoms ⟨*quiescent* gallstones⟩
synonym see LATENT
— qui·es·cent·ly *adverb*

¹qui·et \ˈkwī-ət\ *noun* [Middle English, from Latin *quiet-, quies* rest, quiet — more at WHILE] (14th century)
: the quality or state of being quiet **:** TRANQUILLITY
— on the quiet : in a secretive manner **:** in secret

²quiet *adjective* [Middle English, from Middle French, from Latin *quietus,* from past participle of *quiescere*] (14th century)
1 a : marked by little or no motion or activity **:** CALM ⟨a *quiet* sea⟩ **b :** GENTLE, EASYGOING ⟨a *quiet* temperament⟩ **c :** not interfered with ⟨*quiet* reading⟩ **d :** enjoyed in peace and relaxation ⟨a *quiet* cup of tea⟩
2 a : free from noise or uproar **:** STILL **b :** UNOBTRUSIVE, CONSERVATIVE ⟨*quiet* clothes⟩
3 : SECLUDED ⟨a *quiet* nook⟩
— qui·et·ly *adverb*
— qui·et·ness *noun*

³quiet *adverb* (1573)
: in a quiet manner ⟨a *quiet*-running engine⟩

⁴quiet *verb* [Middle English, from Late Latin *quietare* to set free, to calm, from Latin *quietus*] (14th century)
transitive verb
1 : to cause to be quiet **:** CALM
2 : to make secure by freeing from dispute or question ⟨*quiet* title to a property⟩
intransitive verb
: to become quiet — usually used with *down*
— qui·et·er *noun*

qui·et·en \ˈkwī-ə-tᵊn\ *verb* **qui·et·ened; qui·et·en·ing** \ˈkwī-ət-niŋ, ˈkwī-ə-tᵊn-iŋ\ (circa 1828)
chiefly British **:** QUIET

qui·et·ism \ˈkwī-ə-ˌti-zəm\ *noun* (1687)
1 a : a system of religious mysticism teaching that perfection and spiritual peace are attained by annihilation of the will and passive absorption in contemplation of God and divine things **b :** a passive withdrawn attitude or policy toward the world or worldly affairs
2 : a state of calmness or passivity
— qui·et·ist \-tist\ *adjective or noun*
— qui·et·ist·ic \ˌkwī-ə-ˈtis-tik\ *adjective*

qui·etude \ˈkwī-ə-ˌtüd, -ˌtyüd\ *noun* [Middle French, from Late Latin *quietudo,* from Latin *quietus*] (1597)
: a quiet state **:** REPOSE

qui·etus \kwī-ˈē-təs, -ˈā-\ *noun* [Middle English *quietus est,* from Medieval Latin, he is quit, formula of discharge from obligation] (1540)
1 : final settlement (as of a debt)
2 : removal from activity; *especially* **:** DEATH
3 : something that quiets or represses

quiff \ˈkwif\ *noun* [origin unknown] (circa 1890)
British **:** a prominent forelock

¹quill \ˈkwil\ *noun* [Middle English *quil* hollow reed, bobbin; akin to Middle High German *kil* large feather] (15th century)
1 a (1) **:** a bobbin, spool, or spindle on which filling yarn is wound (2) **:** a hollow shaft often surrounding another shaft and used in various mechanical devices **b :** a roll of dried bark ⟨cinnamon *quills*⟩
2 a (1) **:** the hollow horny shaft of a feather — see FEATHER illustration (2) **:** FEATHER; *especially* **:** one of the large stiff feathers of the wing or tail **b :** one of the hollow sharp spines of a porcupine or hedgehog **c :** ³PEN 3
3 : something made from or resembling the quill of a feather; *especially* **:** a pen for writing
4 : a float for a fishing line

²quill *transitive verb* (1783)
1 : to pierce with quills
2 a : to wind (thread or yarn) on a quill **b :** to make a series of small rounded ridges in (cloth)

quill·back \ˈkwil-ˌbak\ *noun, plural* **quillback** *or* **quillbacks** (1882)
: any of several suckers; *especially* **:** a small fish (*Carpiodes cyprinus*) of central and eastern North America that has the first ray of the dorsal fin much elongated

quill·work \-ˌwərk\ *noun* (1843)
: ornamental work in porcupine or bird quills

¹quilt \ˈkwilt\ *noun* [Middle English *quilte* mattress, quilt, from Old French *cuilte,* from Latin *culcita* mattress] (14th century)
1 : a bed coverlet of two layers of cloth filled with padding (as down or batting) held in place by ties or stitched designs
2 : something that is quilted or resembles a quilt

²quilt (1555)
transitive verb
1 a : to fill, pad, or line like a quilt **b** (1) **:** to stitch, sew, or cover with lines or patterns like those used in quilts (2) **:** to stitch (designs) through layers of cloth **c :** to fasten between two pieces of material
2 : to stitch or sew in layers with padding in between
intransitive verb
1 : to make quilts
2 : to do quilted work
— quilt·er *noun*

quilt·ing *noun* (1609)
1 : material that is quilted or used for making quilts
2 : the process of quilting

quin·a·crine \ˈkwi-nə-ˌkrēn\ *noun* [*quinine* + *acridine*] (circa 1934)
: an antimalarial drug derived from acridine and used especially in the form of its dihydrochloride $C_{23}H_{30}ClN_3O \cdot 2HCl \cdot 2H_2O$

quince \ˈkwin(t)s\ *noun* [Middle English *quynce* quinces, plural of *coyn, quyn* quince,

☆ SYNONYMS
Quicken, animate, enliven, vivify mean to make alive or lively. QUICKEN stresses a sudden renewal of life or activity especially in something inert ⟨the arrival of spring *quickens* the earth⟩. ANIMATE emphasizes the imparting of motion or vitality to what is or might be mechanical or artificial ⟨happiness *animated* his conversation⟩. ENLIVEN suggests a stimulus that arouses from dullness or torpidity ⟨*enlivened* her lectures with humorous anecdotes⟩. VIVIFY implies a freshening or energizing through renewal of vitality ⟨new blood needed to *vivify* the dying club⟩. See in addition PROVOKE.

◇ WORD HISTORY
quicksilver The silver-colored metal known as mercury is, unlike other metallic elements, liquid at room temperature. Its ready fluidity has long suggested a semblance of life. Fittingly, the ancient Romans called the metal *argentum vivum* "living silver." Our English word is a loan translation of the Latin phrase. First attested in the Old English form *cwicseolfor, quicksilver* represents a use of *quick* in its original sense "alive, living." The chemical symbol for mercury, Hg, and the less common name *hydrargyrum,* of which the symbol is an abbreviation, are ultimately from Greek *hydrargyros,* literally, "watery silver."

from Middle French *coin,* from Latin *cotoneum,* alteration *cydonium,* from Greek *kydōnion*] (14th century)
1 : the fruit of a central Asian tree (*Cydonia oblonga*) of the rose family that resembles a hard-fleshed yellow apple and is used especially in preserves
2 : a tree that bears quinces

quince 1

quin·cen·te·na·ry \ˌkwin-sen-ˈte-nə-rē, -ˈsent°n-ˌer-ē, *especially British* -sen-ˈtē-nə-rē\ *noun* [Latin *quin*que five + English *centenary*] (1879)
: a 500th anniversary or its celebration
— quincentenary *adjective*

quin·centen·ni·al \-sen-ˈte-nē-əl\ *noun* [Latin *quin*que five + English *centennial*] (1884)
: QUINCENTENARY
— quincentennial *adjective*

quin·cunx \ˈkwin-ˌkəŋ(k)s\ *noun* [Latin *quincunc-, quincunx,* literally, five twelfths, from *quinque* five + *uncia* twelfth part — more at FIVE, OUNCE] (1658)
: an arrangement of five things in a square or rectangle with one at each corner and one in the middle
— quin·cun·cial \kwin-ˈkən(t)-shəl\ *or* **quin·cunx·ial** \-ˈkəŋk-sē-əl\ *adjective*

quin·de·cil·lion \ˌkwin-di-ˈsil-yən\ *noun, often attributive* [Latin *quindecim* fifteen (from *quinque* five + *decem* ten) + English *-illion* (as in *million*) — more at TEN] (1903)
— see NUMBER table

quin·i·dine \ˈkwi-nə-ˌdēn\ *noun* [International Scientific Vocabulary, from *quinine*] (1836)
: an alkaloid $C_{20}H_{24}N_2O_2$ that is stereoisomeric with quinine and is used in treating cardiac rhythm irregularities

qui·nie·la \kwin-ˈye-lə\ *or* **qui·nel·la** \kwi-ˈne-lə\ *noun* [American Spanish *quiniela,* a game of chance resembling a lottery] (1905)
: a bet in which the bettor picks the first and second place finishers but need not designate their order of finish in order to win — compare PERFECTA

qui·nine \ˈkwī-ˌnīn *also* ˈkwi- *or* kwi-ˈnīn *or* ki-ˈnēn *or* kwi-ˈnēn\ *noun* [Spanish *quina* cinchona, from Quechua *kina* bark] (1826)
1 : a bitter crystalline alkaloid $C_{20}H_{24}N_2O_2$ from cinchona bark used in medicine
2 : a salt of quinine used especially as an antipyretic, antimalarial, and bitter tonic

quinine water *noun* (1953)
: TONIC WATER

qui·noa \ˈkēn-ˌwä\ *noun* [Spanish, from Quechua *kinua*] (1625)
: a pigweed (*Chenopodium quinoa*) of the high Andes whose seeds are ground and widely used as food in Peru

quin·o·line \ˈkwi-n°l-ˌēn\ *noun* [International Scientific Vocabulary *quinine* + [3]*-ol* + [2]*-ine*] (1845)
1 : a pungent oily nitrogenous base C_9H_7N obtained usually by distillation of coal tar or by synthesis from aniline that is the parent compound of many alkaloids, drugs, and dyes
2 : a derivative of quinoline

qui·none \kwi-ˈnōn, ˈkwi-ˌ\ *noun* [International Scientific Vocabulary *quinine* + *-one*] (1853)
1 : either of two isomeric cyclic crystalline compounds $C_6H_4O_2$ that are di-keto derivatives of dihydro-benzene
2 : any of various usually yellow, orange, or red quinonoid compounds including several that are biologically important as coenzymes, hydrogen acceptors, or vitamins

qui·no·noid \kwi-ˈnō-ˌnȯid, ˈkwi-nə-\ *or* **quin·oid** \ˈkwi-ˌnȯid\ *adjective* (1878)
: resembling quinone especially in having a benzene nucleus containing two double bonds within the nucleus

quin·quen·ni·al \kwin-ˈkwe-nē-əl, kwiŋ-\ *adjective* (15th century)
1 : consisting of or lasting for five years
2 : occurring or being done every five years
— quinquennial *noun*
— quin·quen·ni·al·ly \-nē-ə-lē\ *adverb*

quin·quen·ni·um \-nē-əm\ *noun, plural* **-ni·ums** *or* **-nia** \-nē-ə\ [Latin, from *quinque* five + *annus* year — more at FIVE, ANNUAL] (1621)
: a period of five years

quin·sy \ˈkwin-zē\ *noun* [Middle English *quinesie,* from Middle French *quinancie,* from Late Latin *cynanche,* from Greek *kynanchē,* from *kyn-, kyōn* dog + *anchein* to strangle — more at HOUND, ANGER] (14th century)
: an abscess in the tissue around a tonsil usually resulting from bacterial infection and often accompanied by pain and fever

quint \ˈkwint\ *noun* (1935)
: QUINTUPLET

quin·ta \ˈkin-tə, ˈkēn-\ *noun* [Spanish & Portuguese, quinta, farm rented at one fifth of its income, from Latin, feminine of *quintus* fifth] (1754)
: a country villa or estate especially in Portugal or Latin America

quin·tain \ˈkwin-t°n\ *noun* [Middle English *quintaine,* from Middle French, perhaps from Latin *quintana* street in a Roman camp separating the fifth maniple from the sixth where a market was held, from feminine of *quintanus* fifth in rank, from *quintus* fifth] (15th century)
: an object to be tilted at; *especially* **:** a post with a revolving crosspiece that has a target at one end and a sandbag at the other end

quin·tal \ˈkwin-t°l, ˈkan-\ *noun* [Middle English, from Middle French, from Medieval Latin *quintale,* from Arabic *qinṭār,* from Late Greek *kentēnarion,* from Late Latin *centenarium,* from Latin, neuter of *centenarius* consisting of a hundred — more at CENTENARY] (15th century)
1 : HUNDREDWEIGHT
2 : a unit of weight equal to 100 kilograms (about 220 pounds)

quin·tes·sence \kwin-ˈte-s°n(t)s\ *noun* [Middle English, from Middle French *quinte essence,* from Medieval Latin *quinta essentia,* literally, fifth essence] (15th century)
1 : the fifth and highest element in ancient and medieval philosophy that permeates all nature and is the substance composing the celestial bodies
2 : the essence of a thing in its purest and most concentrated form
3 : the most typical example or representative ◆
— quin·tes·sen·tial \ˌkwin-tə-ˈsen(t)-shəl\ *adjective*
— quin·tes·sen·tial·ly *adverb*

quin·tet *also* **quin·tette** \kwin-ˈtet\ *noun* [*quintet* from Italian *quintetto,* from *quinto* fifth, from Latin *quintus; quintette* from French, from Italian *quintetto*] (1811)
1 : a musical composition or movement for five instruments or voices
2 : a group or set of five: as **a :** the performers of a quintet **b :** a basketball team

[1]**quin·tic** \ˈkwin-tik\ *adjective* [Latin *quintus* fifth] (1853)
: of the fifth degree

[2]**quintic** *noun* (1856)
: a polynomial or a polynomial equation of the fifth degree

quin·tile \ˈkwin-ˌtīl\ *noun* [Latin *quintus* + English [2]*-ile*] (circa 1928)
: any of the four values that divide the items of a frequency distribution into five classes with each containing one fifth of the total population; *also* **:** any one of the five classes

quin·til·lion \kwin-ˈtil-yən\ *noun* [Latin *quintus* + English *-illion* (as in *million*)] (1674)
— see NUMBER table
— quintillion *adjective*
— quin·til·lionth \-yən(t)th\ *adjective or noun*

[1]**quin·tu·ple** \kwin-ˈtü-pəl, -ˈtyü-, -ˈtə-; ˈkwin-tə-\ *adjective* [Middle French, from Medieval Latin *quintuplus,* from Latin *quintus* fifth + *-plus* -fold; akin to Latin *quinque* five — more at FIVE, -FOLD] (1570)
1 : being five times as great or as many
2 : having five units or members
3 : marked by five beats per measure ⟨*quintuple* meter⟩
— quintuple *noun*

[2]**quintuple** *verb* **quin·tu·pled; quin·tu·pling** \-p(ə-)liŋ\ (1639)
transitive verb
: to make five times as great or as many
intransitive verb
: to become five times as much or as numerous

quin·tu·plet \kwin-ˈtə-plət, -ˈtü-, -ˈtyü-; ˈkwin-tə-\ *noun* (1873)
1 : a combination of five of a kind
2 : one of five offspring born at one birth

[1]**quin·tu·pli·cate** \kwin-ˈtü-pli-kət, -ˈtyü-\ *adjective* [Medieval Latin *quintuplicatus,* past participle of *quintuplicare* to quintuple, from *quintuplus* quintuple] (1656)
1 : consisting of or existing in five corresponding or identical parts or examples ⟨*quintuplicate* invoices⟩
2 : being the fifth of five things exactly alike ⟨file the *quintuplicate* copy⟩

[2]**quintuplicate** *noun* (1851)
1 : one of five things exactly alike; *specifically* **:** one of five identical copies
2 : five copies all alike — used with *in* ⟨typed in *quintuplicate*⟩

[3]**quin·tu·pli·cate** \-plə-ˌkāt\ *transitive verb* **-cat·ed; -cat·ing** (circa 1890)
1 : to make quintuple or fivefold
2 : to prepare in quintuplicate

[1]**quip** \ˈkwip\ *noun* [earlier *quippy,* perhaps from Latin *quippe* indeed, to be sure (often ironic), from *quid* what — more at QUIDDITY] (1532)
1 a : a clever usually taunting remark **:** GIBE **b :** a witty or funny observation or response usually made on the spur of the moment

◇ WORD HISTORY
quintessence Before the advent of modern science it was thought by those who speculated about nature's secrets that there were four elements: earth, air, fire, and water. These combined in varying proportions to form all the different substances found on Earth. The stars and planets, however, seemed to be composed of an altogether different, less substantial element. This fifth substance or *quintessence* (from Medieval Latin *quinta essentia,* itself a loan translation of Greek *pemptē ousia* "fifth substance" or "fifth essence") was believed to permeate all space beyond the earth's atmosphere. The quintessence was conceived by some as being composed of pure and invisible light or fire. Alchemists believed that the quintessence was innate in all matter, and so they labored in vain to extract it from other materials. The hope was that the quintessence might provide a panacea for disease or even for mortality itself. Though discarded long ago, this alchemistic belief gave rise to the customary meaning of the word today: "the essence of a thing in its purest and most concentrated form."

2 : QUIBBLE, EQUIVOCATION
3 : something strange, droll, curious, or eccentric **:** ODDITY

²quip *verb* **quipped; quip·ping** (1579)
intransitive verb
: to make quips **:** GIBE
transitive verb
: to jest or gibe at
— quip·per \ˈkwi-pər\ *noun*

quip·ster \ˈkwip-stər\ *noun* (1876)
: one who is given to quipping

qui·pu \ˈkē-(ˌ)pü\ *noun* [Spanish *quipo,* from Quechua *khipu*] (1704)
: a device made of a main cord with smaller varicolored cords attached and knotted and used by the ancient Peruvians (as for calculating)

¹quire \ˈkwīr\ *noun* [Middle English *quair* four sheets of paper folded once, collection of sheets, from Middle French *quaer,* from (assumed) Vulgar Latin *quaternum* set of four, from Latin *quaterni* four each, set of four — more at QUATERNION] (15th century)
: a collection of 24 or sometimes 25 sheets of paper of the same size and quality **:** one twentieth of a ream

²quire *variant of* CHOIR

Qui·ri·nus \kwə-ˈrī-nəs, -ˈrē-\ *noun* [Latin]
: an early state god of the Romans later identified with Romulus

¹quirk \ˈkwərk\ *noun* [origin unknown] (1565)
1 a : an abrupt twist or curve **b :** a peculiar trait **:** IDIOSYNCRASY **c :** ACCIDENT, VAGARY ⟨a *quirk* of fate⟩
2 : a groove separating a bead or other molding from adjoining members
— quirk·i·ly \ˈkwər-kə-lē\ *adverb*
— quirk·i·ness \-kē-nəs\ *noun*
— quirk·ish \ˈkwər-kish\ *adjective*
— quirky \-kē\ *adjective*

²quirk *verb* (1596)
: CURVE, TWIST ⟨*quirked* his eyebrows⟩

¹quirt \ˈkwərt\ *noun* [Mexican Spanish *cuarta*] (1845)
: a riding whip with a short handle and a rawhide lash

²quirt *transitive verb* (1887)
: to strike or drive with a quirt

quis·ling \ˈkwiz-liŋ\ *noun, often attributive* [Vidkun *Quisling* (died 1945) Norwegian politician who collaborated with the Nazis] (1940)
: TRAITOR 2, COLLABORATOR
— quis·ling·ism \-liŋ-ˌi-zəm\ *noun*

¹quit \ˈkwit\ *adjective* [Middle English *quite, quit,* from Old French *quite*] (13th century)
: released from obligation, charge, or penalty; *especially* **:** FREE

²quit *verb* **quit** *also* **quit·ted; quit·ting** [Middle English *quiten, quitten,* from Middle French *quiter, quitter,* from Old French, from *quite* free of, released, literally, at rest, from Latin *quietus* quiet, at rest] (13th century)
transitive verb
1 : to make full payment of **:** PAY UP ⟨*quit* a debt⟩
2 : to set free **:** RELIEVE, RELEASE ⟨*quit* oneself of fear⟩
3 : CONDUCT, ACQUIT ⟨the youths *quit* themselves like men⟩
4 a : to depart from or out of **b :** to leave the company of **c :** GIVE UP 1, 2 ⟨*quit* a job⟩ ⟨*quit* smoking⟩
intransitive verb
1 : to cease normal, expected, or necessary action
2 : to give up employment
3 : to admit defeat **:** GIVE UP
synonym see STOP

³quit *noun* (circa 1923)
: the act or an instance of quitting a job

quitch \ˈkwich\ *noun* [(assumed) Middle English *quicche,* from Old English *cwice;* akin to Old High German *quecca* couch grass] (before 12th century)
: QUACK GRASS

quit·claim \ˈkwit-ˌklām\ *transitive verb* (14th century)
: to release or relinquish a legal claim to; *especially* **:** to release a claim to or convey by a quitclaim deed
— quitclaim *noun*

quitclaim deed *noun* (1756)
: a legal instrument used to release one person's right, title, or interest to another without providing a guarantee or warranty of title

quite \ˈkwīt\ *adverb* [Middle English, from *quite,* adjective, quit] (14th century)
1 : WHOLLY, COMPLETELY ⟨not *quite* finished⟩
2 : to an extreme **:** POSITIVELY ⟨*quite* sure⟩ — often used as an intensifier with *a* ⟨*quite* a swell guy⟩ ⟨*quite* a beauty⟩
3 : to a considerable extent **:** RATHER ⟨*quite* near⟩
usage see PLENTY
— quite a bit : a considerable amount
— quite a few : MANY

quit·rent \ˈkwit-ˌrent\ *noun* (15th century)
: a fixed rent payable to a feudal superior in commutation of services; *specifically* **:** a fixed rent due from a socage tenant

quits \ˈkwits\ *adjective* [Middle English, quit, probably from Medieval Latin *quittus,* alteration of Latin *quietus* at rest] (15th century)
: being on even terms by repayment or requital

quit·tance \ˈkwi-tᵊn(t)s\ *noun* (14th century)
1 a : discharge from a debt or an obligation **b :** a document evidencing quittance
2 : RECOMPENSE, REQUITAL

quit·ter \ˈkwi-tər\ *noun* (1611)
: one that quits; *especially* **:** one that gives up too easily **:** DEFEATIST

quit·tor \ˈkwi-tər\ *noun* [Middle English *quiture* pus, probably from Old French, act of boiling, from Latin *coctura,* from *coctus,* past participle of *coquere* to cook — more at COOK] (1703)
: a purulent inflammation of the feet especially of horses and donkeys

¹quiv·er \ˈkwi-vər\ *noun* [Middle English, from Middle French *quivre,* of Germanic origin; akin to Old English *cocer* quiver, Old High German *kohhari*] (14th century)
1 : a case for carrying or holding arrows
2 : the arrows in a quiver

²quiver *intransitive verb* **quiv·ered; quiv·er·ing** \ˈkwi-və-riŋ, ˈkwiv-riŋ\ [Middle English, probably from *quiver* agile, quick; akin to Old English *cwiferlice* zealously] (15th century)
: to shake or move with a slight trembling motion
— quiv·er·ing·ly *adverb*

³quiver *noun* (1786)
: the act or action of quivering **:** TREMOR

qui vive \kē-ˈvēv\ *noun* [French *qui-vive,* from *qui vive?* long live who?, challenge of a French sentry] (1726)
: ALERT, LOOKOUT — used in the phrase *on the qui vive*

qui·xote \ˈkwik-sət, kē-ˈhō-tē, -ˈō-\ *noun, often capitalized* [Don *Quixote,* hero of the novel *Don Quixote de la Mancha* (1605, 1615) by Cervantes] (1648)
: a quixotic person
— quix·o·tism \ˈkwik-sə-ˌti-zəm\ *noun*
— quix·o·try \-sə-trē\ *noun*

quix·ot·ic \kwik-ˈsä-tik\ *adjective* [Don *Quixote*] (1815)
1 : foolishly impractical especially in the pursuit of ideals; *especially* **:** marked by rash lofty romantic ideas or extravagantly chivalrous action
2 : CAPRICIOUS, UNPREDICTABLE ◆
synonym see IMAGINARY
— quix·ot·i·cal \-ti-kəl\ *adjective*
— quix·ot·i·cal·ly \-ti-k(ə-)lē\ *adverb*

¹quiz \ˈkwiz\ *noun, plural* **quiz·zes** [origin unknown] (1749)
1 : an eccentric person
2 : PRACTICAL JOKE
3 : the act or action of quizzing; *specifically* **:** a short oral or written test

²quiz *transitive verb* **quizzed; quiz·zing** (1794)
1 : to make fun of **:** MOCK
2 : to look at inquisitively
3 : to question closely
— quiz·zer *noun*

quiz·mas·ter \ˈkwiz-ˌmas-tər\ *noun* (1943)
: one who puts the questions to contestants in a quiz show

quiz show *noun* (1944)
: an entertainment program (as on radio or television) in which contestants answer questions — called also *quiz program*

quiz·zi·cal \ˈkwi-zi-kəl\ *adjective* (1800)
1 : comically quaint
2 : mildly teasing or mocking ⟨a *quizzical* remark⟩
3 : expressive of puzzlement, curiosity, or disbelief ⟨raised a *quizzical* eyebrow⟩
— quiz·zi·cal·i·ty \ˌkwi-zə-ˈka-lə-tē\ *noun*
— quiz·zi·cal·ly \ˈkwi-zi-k(ə-)lē\ *adverb*

quod \ˈkwäd\ *noun* [origin unknown] (circa 1700)
slang British **:** PRISON

quod·li·bet \ˈkwäd-lə-ˌbet\ *noun* [Middle English, from Medieval Latin *quodlibetum,* from Latin *quodlibet,* neuter of *quilibet* any whatever, from *qui* who, what + *libet* it pleases, from *libēre* to please — more at WHO, LOVE] (14th century)
1 : a philosophical or theological point proposed for disputation; *also* **:** a disputation on such a point
2 : a whimsical combination of familiar melodies or texts

¹quoin \ˈkȯin, ˈkwȯin\ *noun* [alteration of ¹*coin*] (1532)
1 a : a solid exterior angle (as of a building) **b :** one of the members (as a block) forming a quoin and usually differentiated from the adjoining walls by material, texture, color, size, or projection

quoin 1b

2 : the keystone or a voussoir of an arch
3 : a wooden or expandable metal block used by printers to lock up a form within a chase

²quoin *transitive verb* (1683)
1 : to equip (a type form) with quoins
2 : to provide with quoins ⟨*quoined* walls⟩

¹quoit \ˈkȯit, ˈkwȯit, ˈkwāt\ *noun* [Middle English *coite*] (15th century)
1 : a flattened ring of iron or circle of rope used in a throwing game

◇ WORD HISTORY

quixotic The word *quixotic* reflects the lasting place in world literature held by Miguel de Cervantes Saavedra's novel *Don Quixote de la Mancha* (1605 and 1615). The story concerns Alonso Quijano, who, inspired by the romantic ideals of the days of chivalry, sets out determined to have his own chivalric adventures and in the process undo the wrongs of the world. With the title of Don Quixote and the viewpoint of a romantic idealist, he undergoes a series of adventures that turn into comic misadventures. The would-be knight returns to his home in La Mancha a tired and disillusioned old man. *Don Quixote* was translated into English not long after its Spanish publication and within a few decades gave rise to *Quixote* as a synonym for an impractical idealist. The 19th century derivative *quixotic* now evokes the unrealizable schemes of modern Don Quixotes.

2 *plural but singular in construction* **:** a game in which the quoits are thrown at an upright pin in an attempt to ring the pin or come as near to it as possible

²quoit *transitive verb* (1597)
: to throw like a quoit

quon·dam \ˈkwän-dəm, -ˌdam\ *adjective* [Latin, at one time, formerly, from *quom, cum* when; akin to Latin *qui* who — more at WHO] (1539)
: FORMER, SOMETIME ⟨a *quondam* friend⟩

Quon·set \ˈkwän(t)-sət, ˈkwän-zət\ *trademark* — used for a prefabricated shelter set on a foundation of bolted steel trusses and built of a semicircular arching roof of corrugated metal insulated with wood fiber

quo·rum \ˈkwōr-əm, ˈkwȯr-\ *noun* [Middle English, quorum of justices of the peace, from Latin, of whom, genitive plural of *qui* who; from the wording of the commission formerly issued to justices of the peace] (1602)
1 : a select group
2 : the number (as a majority) of officers or members of a body that when duly assembled is legally competent to transact business
3 : a Mormon body comprising those in the same grade of priesthood

quo·ta \ˈkwō-tə\ *noun* [Medieval Latin, from Latin *quota pars* how great a part] (1618)
1 : a proportional part or share; *especially* **:** the share or proportion assigned to each in a division or to each member of a body
2 : the number or amount constituting a proportional share

quot·able \ˈkwō-tə-bəl *also* ˈkō-\ *adjective* (1811)
: fit for or worth quoting
— quot·abil·i·ty \ˌkwō-tə-ˈbi-lə-tē *also* ˌkō-\ *noun*

quo·ta·tion \kwō-ˈtā-shən *also* kō-\ *noun* (1646)
1 a : the act or process of quoting **b** (1) **:** the naming or publishing of current bids and offers or prices of securities or commodities (2) **:** the bids, offers, or prices so named or published; *especially* **:** the highest bid and lowest offer for a particular security in a given market at a given time
2 : something that is quoted; *especially* **:** a passage referred to, repeated, or adduced

quotation mark *noun* (circa 1859)
: one of a pair of punctuation marks " " or ' ' used chiefly to indicate the beginning and the end of a quotation in which the exact phraseology of another or of a text is directly cited

¹quote \ˈkwōt *also* ˈkōt\ *verb* **quot·ed; quot·ing** [Medieval Latin *quotare* to mark the number of, number references, from Latin *quotus* of what number or quantity, from *quot* how many, (as) many as; akin to Latin *qui* who — more at WHO] (1582)
transitive verb
1 a : to speak or write (a passage) from another usually with credit acknowledgment **b :** to repeat a passage from especially in substantiation or illustration
2 : to cite in illustration ⟨*quote* cases⟩
3 a : to state (the current price or bid-offer spread) for a commodity, stock, or bond **b :** to give exact information on
4 : to set off by quotation marks
intransitive verb
: to inform a hearer or reader that matter following is quoted
— quot·er *noun*

²quote *noun* (1888)
1 : QUOTATION
2 : QUOTATION MARK — often used orally to indicate the beginning of a direct quotation

quoth \ˈkwōth *also* ˈkōth\ *verb past* [Middle English, past of *quethen* to say, from Old English *cwethan;* akin to Old High German *quedan* to say] (before 12th century)
archaic **:** SAID — used chiefly in the first and third persons with a postpositive subject

quotha \ˈkwō-thə\ *interjection* [alteration of *quoth he*] (1519)
archaic — used especially to express surprise or contempt

quo·tid·i·an \kwō-ˈti-dē-ən\ *adjective* [Middle English *cotidian,* from Middle French, from Latin *quotidianus, cotidianus,* from *quotidie* every day, from *quot* (as) many as + *dies* day — more at DEITY] (14th century)
1 : occurring every day ⟨*quotidian* fever⟩
2 a : belonging to each day **:** EVERYDAY ⟨*quotidian* routine⟩ **b :** COMMONPLACE, ORDINARY ⟨*quotidian* drabness⟩
— quotidian *noun*

quo·tient \ˈkwō-shənt\ *noun* [Middle English *quocient,* modification of Latin *quotiens* how many times, from *quot* how many] (15th century)
1 : the number resulting from the division of one number by another
2 : the numerical ratio usually multiplied by 100 between a test score and a measurement on which that score might be expected largely to depend
3 : QUOTA, SHARE

quotient group *noun* (1893)
: a group whose elements are the cosets of a normal subgroup of a given group — called also *factor group*

quotient ring *noun* (circa 1958)
: a ring whose elements are the cosets of an ideal in a given ring

quo war·ran·to \ˌkwō-wə-ˈrän-(ˌ)tō, -ˈran-; -ˈwȯr-ən-ˌtō, -ˈwär-\ *noun* [Middle English *quo waranto,* from Medieval Latin *quo warranto* by what warrant; from the wording of the writ] (15th century)
1 a : an English writ formerly requiring a person to show by what authority he exercises a public office, franchise, or liberty **b :** a legal proceeding for a like purpose begun by an information
2 : the legal action begun by a quo warranto

Qur·'an *also* **Qur·an** \kə-ˈran, -ˈrän; ku̇-ˈran, -ˈrän\ *variant of* KORAN

QWER·TY \ˈkwər-tē, ˈkwer-\ *noun, often not capitalized* [from the first six letters in the second row of the keyboard] (1929)
: a standard typewriter keyboard — called also *QWERTY keyboard*

R *is the eighteenth letter of the English alphabet. It came through Latin by way of Etruscan from Greek* rhō, *which originated as a borrowing of Phoenician* rês. *In English* r *usually denotes the consonant shared by* run, dry, *and* very. *In words derived from Greek the letter* h *that is usually written after* r *to represent* rhō *with a rough breathing mark over it (as in* rhapsody *and* rhetoric) *does not affect the pronunciation of the English word. In many regional dialects of the United States (as those of eastern New England, metropolitan New York, and the South) and in England, an* r *sound at the end of a syllable is "dropped," that is, either it is not pronounced and the preceding vowel is lengthened slightly or it is pronounced as the neutral schwa sound with little or no* r *quality in it. The small* r *form developed in the late Roman period through the gradual degeneration of the loop and tail of the capital form.*

r \'är\ *noun, plural* **r's** *or* **rs** \'ärz\ *often capitalized, often attributive* (before 12th century)
1 a : the 18th letter of the English alphabet **b :** a graphic representation of this letter **c :** a speech counterpart of orthographic *r*
2 : a graphic device for reproducing the letter *r*
3 : one designated *r* especially as the 18th in order or class
4 : something shaped like the letter R

R *certification mark*
— used to certify that a motion picture is of such a nature that admission is restricted to persons over a specified age (as 17) unless accompanied by a parent or guardian; compare G, NC-17, PG, PG-13

Ra \'rä\ *noun* [Egyptian *r'*]
: the Egyptian sun-god and chief deity

ra·ba·to \rə-'bä-(,)tō\ *noun, plural* **-tos** [modification of Middle French *rabat,* literally, act of turning down, from *rabattre*] (1591)
: a wide lace-edged collar of the early 17th century often stiffened to stand high at the back

¹rab·bet \'ra-bət\ *noun* [Middle English *rabet,* from Middle French *rabat* act of beating down, from Old French *rabattre* to beat down, reduce — more at REBATE] (14th century)
: a channel, groove, or recess cut out of the edge or face of a surface; *especially* **:** one that is intended to receive another member (as a panel)

²rabbet (15th century)
transitive verb
1 : to unite the edges of in a rabbet joint
2 : to cut a rabbet in
intransitive verb
: to become joined by a rabbet

rabbet joint *noun* (circa 1828)
: a joint formed by fitting together rabbeted boards or timbers

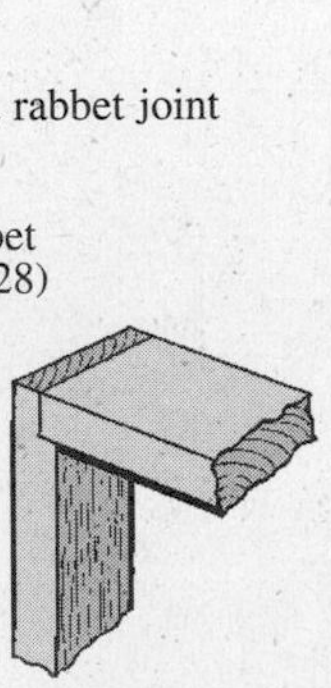

rabbet joint

rab·bi \'ra-,bī\ *noun* [Middle English, from Old English, from Late Latin, from Greek *rhabbi,* from Hebrew *rabbī* my master, from *rabh* master + *-ī* my] (before 12th century)
1 : MASTER, TEACHER — used by Jews as a term of address
2 : a Jew qualified to expound and apply the halakah and other Jewish law
3 : a Jew trained and ordained for professional religious leadership; *specifically* **:** the official leader of a Jewish congregation

rab·bin \'ra-bən\ *noun* [French] (1579)
: RABBI

rab·bin·ate \'ra-bə-nət, -,nāt\ *noun* (1702)
1 : the office or tenure of a rabbi
2 : the whole body of rabbis

rab·bin·ic \rə-'bi-nik, ra-\ *or* **rab·bin·i·cal** \-ni-kəl\ *adjective* (1612)
1 : of or relating to rabbis or their writings
2 : of or preparing for the rabbinate
3 : comprising or belonging to any of several sets of Hebrew characters simpler than the square Hebrew letters
— rab·bin·i·cal·ly \-ni-k(ə-)lē\ *adverb*

Rabbinic Hebrew *noun* (circa 1909)
: the Hebrew used especially by medieval rabbis

rab·bin·ism \'ra-bə-,ni-zəm\ *noun* (1652)
: rabbinic teachings and traditions

¹rab·bit \'ra-bət\ *noun, plural* **rabbit** *or* **rabbits** *often attributive* [Middle English *rabet,* probably from Middle French dialect (Walloon) *robett,* from Middle Dutch *robe*] (14th century)
1 : any of a family (Leporidae) of long-eared short-tailed lagomorph mammals with long hind legs: **a :** any of various lagomorphs that are born naked, blind, and helpless, that are sometimes gregarious, and that include especially the cottontails of the New World and a small Old World mammal (*Oryctolagus cuniculus*) that is the source of various domestic breeds **b :** HARE
2 : the pelt of a rabbit
3 : WELSH RABBIT
4 a : a figure of a rabbit sped mechanically along the edge of a dog track as an object of pursuit **b :** a runner in a long-distance race who sets a fast pace for the field in the first part of the race
— rab·bity \-bə-tē\ *adjective*

²rabbit *intransitive verb* (1852)
: to hunt rabbits
— rab·bit·er *noun*

rab·bit·brush \'ra-bət-,brəsh\ *noun* (circa 1890)
: any of several low branching composite shrubs (genus *Chrysothamnus* and especially *C. nauseosus*) of the alkali plains of western North America that are characterized by linear entire leaves and clusters of golden yellow or white flowers

rabbit ears *noun plural* (1952)
: an indoor dipole television antenna consisting of two usually extensible rods connected to a base to form a V shape

rabbit fever *noun* (1925)
: TULAREMIA

rabbit punch *noun* (1915)
: a short chopping blow delivered to the back of the neck or the base of the skull
— rabbit–punch *transitive verb*

rab·bit·ry \'ra-bə-trē\ *noun, plural* **-ries** (1838)
: a place where domestic rabbits are kept; *also* **:** a rabbit-raising enterprise

¹rab·ble \'ra-bəl\ *noun* [Middle English *rabel* pack of animals] (14th century)
1 : a disorganized or confused collection of things
2 a : a disorganized or disorderly crowd of people **:** MOB **b :** the lowest class of people

²rabble *transitive verb* **rab·bled; rab·bling** \-b(ə-)liŋ\ (1644)
: to insult or assault by or as a mob

rab·ble·ment \'ra-bəl-mənt\ *noun* (1548)
1 : RABBLE
2 : DISTURBANCE

rab·ble–rous·er \'ra-bəl-,rau̇-zər\ *noun* (1843)
: one that stirs up (as to hatred or violence) the masses of the people **:** DEMAGOGUE
— rab·ble–rous·ing \-ziŋ\ *noun or adjective*

Ra·be·lai·sian \,ra-bə-'lā-zhən, -zē-ən\ *adjective* (1817)
1 : of, relating to, or characteristic of Rabelais or his works
2 : marked by gross robust humor, extravagance of caricature, or bold naturalism

Ra·bi \'rə-bē\ *noun* [Arabic *rabī'*] (circa 1769)
: either of two months of the Islamic year: **a :** the 3d month **b :** the 4th month — see MONTH table

ra·bic \'rā-bik\ *adjective* (1885)
: of or relating to rabies

ra·bid \'ra-bəd *also* 'rā-\ *adjective* [Latin *rabidus* mad, from *rabere*] (1611)
1 a : extremely violent **:** FURIOUS **b :** going to extreme lengths in expressing or pursuing a feeling, interest, or opinion
2 : affected with rabies
— ra·bid·i·ty \rə-'bi-də-tē, ra-, rā-\ *noun*
— ra·bid·ly \'ra-bəd-lē *also* 'rā-\ *adverb*
— ra·bid·ness *noun*

ra·bies \'rā-bēz\ *noun, plural* **rabies** [New Latin, from Latin, madness, from *rabere* to rave — more at RAGE] (circa 1598)
: an acute virus disease of the nervous system of warm-blooded animals usually transmitted through the bite of a rabid animal and typically characterized by increased salivation, abnormal behavior, and eventual paralysis and death

rac·coon \ra-'kün *also* rə-\ *noun, plural* **raccoon** *or* **raccoons** [Virginia Algonquian *raugroughcun, arocoun*] (1608)
1 a : a small nocturnal carnivore (*Procyon lotor*) of North America that is chiefly gray, has a black mask and bushy ringed tail, lives chiefly in trees, and has a varied diet including small animals, fruits, and nuts **b :** the pelt of this animal

raccoon 1a

2 : any of several animals resembling or related to the raccoon ◆

raccoon dog *noun* (1868)
: a small omnivorous canid (*Nyctereutes procyonoides*) of eastern Asia having a long yellowish brown coat and facial markings resembling that of a raccoon — called also *tanuki*

¹race \'rās\ *noun* [Middle English *ras,* from Old Norse *rās;* akin to Old English *rǣs* rush] (14th century)
1 *chiefly Scottish* **:** the act of running
2 a : a strong or rapid current of water flowing through a narrow channel **b :** a watercourse used industrially **c :** the current flowing in such a course

3 a : a set course or duration of time **b :** the course of life
4 a : a contest of speed **b** *plural* **:** a meeting in which several races (as for horses) are run **c :** a contest or rivalry involving progress toward a goal ⟨pennant *race*⟩
5 : a track or channel in which something rolls or slides; *specifically* **:** a groove (as for the balls) in a bearing — see ROLLER BEARING illustration

²race *verb* **raced; rac·ing** (15th century)
intransitive verb
1 : to compete in a race
2 : to go or move at top speed or out of control
3 : to revolve too fast under a diminished load
transitive verb
1 : to engage in a race with
2 a : to enter in a race **b :** to drive or ride at high speed **c :** to transport or propel at maximum speed
3 : to speed (as an engine) without a working load or with the transmission disengaged

³race *noun* [Middle French, generation, from Old Italian *razza*] (1580)
1 : a breeding stock of animals
2 a : a family, tribe, people, or nation belonging to the same stock **b :** a class or kind of people unified by community of interests, habits, or characteristics ⟨the English *race*⟩
3 a : an actually or potentially interbreeding group within a species; *also* **:** a taxonomic category (as a subspecies) representing such a group **b :** BREED **c :** a division of mankind possessing traits that are transmissible by descent and sufficient to characterize it as a distinct human type
4 *obsolete* **:** inherited temperament or disposition
5 : distinctive flavor, taste, or strength

race·course \'rās-ˌkōrs, -ˌkȯrs\ *noun* (1764)
1 : a course for racing
2 : RACEWAY 1

race·horse \-ˌhȯrs\ *noun* (circa 1626)
: a horse bred or kept for racing

ra·ce·mate \rā-'sē-ˌmāt, rə-; 'ra-sə-\ *noun* (1907)
: a racemic compound or mixture

ra·ceme \rā-'sēm, rə-\ *noun* [Latin *racemus* bunch of grapes; probably akin to Greek *rhag-, rhax* grape] (1785)
: a simple inflorescence (as in the lily-of-the-valley) in which the flowers are borne on short stalks of about equal length at equal distances along an elongated axis and open in succession toward the apex — see INFLORESCENCE illustration

ra·ce·mic \-'sē-mik\ *adjective* (1892)
: of, relating to, or constituting a compound or mixture that is composed of equal amounts of dextrorotatory and levorotatory forms of the same compound and is not optically active

ra·ce·mi·za·tion \rā-ˌsē-mə-'zā-shən, rə-; ˌra-sə-mə-\ *noun* (1895)
: the action or process of changing from an optically active compound into a racemic compound or mixture
— ra·ce·mize \rā-'sē-ˌmīz, rə-; 'ra-sə-\ *verb*

ra·ce·mose \'ra-sə-ˌmōs; rā-'sē-, rə-\ *adjective* [Latin *racemosus* full of clusters, from *racemus*] (1698)
: having or growing in the form of a raceme

rac·er \'rā-sər\ *noun* (1649)
1 : one that races or is used for racing
2 : any of various active American colubrid snakes (genus *Coluber* and *Mastigophis*); *especially* **:** BLACK RACER

race riot *noun* (1890)
: a riot caused by racial dissensions or hatreds

race runner *noun* (1915)
: a North American lizard (*Cnemidophorus sexlineatus*) that moves swiftly

race·track \'rās-ˌtrak\ *noun* (1859)
: a usually oval course for racing

race·track·er \-ˌtra-kər\ *noun* (1953)
: one who frequents a racetrack

race·walk·ing \-ˌwȯ-kiŋ\ *noun* (1962)
: the competitive sport of racing at a fast walk while maintaining continuous foot contact with the ground and keeping the supporting leg straight
— race·walk·er \-kər\ *noun*

race·way \-ˌwā\ *noun* (1828)
1 : a canal for a current of water
2 : a channel for loosely holding electrical wires in buildings
3 : ¹RACE 5
4 : a course for racing; *especially* **:** a track for harness racing

rach·et \'ra-chət\ *variant of* RATCHET

ra·chis \'rā-kəs, 'ra-\ *noun, plural* **ra·chis·es** *also* **ra·chi·des** \'ra-kə-ˌdēz, 'rā-\ [New Latin *rachid-, rachis,* from Greek *rhachis;* akin to Greek *rhachos* thorn, Lithuanian *ražas* dry twig, tine] (1842)
1 : SPINAL COLUMN
2 : an axial structure: as **a** (1) **:** the elongated axis of an inflorescence (2) **:** an extension of the petiole of a compound leaf that bears the leaflets **b :** the distal part of the shaft of a feather that bears the web

ra·chit·ic \rə-'ki-tik\ *adjective* [New Latin *rachitis* rickets, from Greek *rhachitis* spinal disease, from *rhachis*] (1797)
: RICKETY

ra·cial \'rā-shəl\ *adjective* (1862)
1 : of, relating to, or based on a race
2 : existing or occurring between races
— ra·cial·ly \-shə-lē\ *adverb*

ra·cial·ism \'rā-shə-ˌli-zəm\ *noun* (1907)
: RACISM
— ra·cial·ist \ list\ *noun or adjective*
— ra·cial·is·tic \ˌrā-shə-'lis-tik\ *adjective*

rac·ing \'rā-siŋ\ *noun* (1680)
: the sport or profession of engaging in or holding races

racing form *noun* (1946)
: an information sheet giving details of past performance (as for racehorses) for use by bettors

rac·ism \'rā-ˌsi-zəm *also* -ˌshi-\ *noun* (1936)
1 : a belief that race is the primary determinant of human traits and capacities and that racial differences produce an inherent superiority of a particular race
2 : racial prejudice or discrimination
— rac·ist \-sist *also* -shist\ *noun or adjective*

¹rack \'rak\ *noun* [Middle English *rak,* probably of Scandinavian origin; akin to Swedish dialect *rak* wreck; akin to Old English *wrecan* to drive — more at WREAK] (14th century)
: a wind-driven mass of high often broken clouds

²rack *intransitive verb* (1590)
: to fly or scud in high wind

³rack *noun* [Middle English, probably from Middle Dutch *rec* framework; akin to Old English *reccan* to stretch, Greek *oregein* — more at RIGHT] (14th century)
1 : a framework for holding fodder for livestock
2 : an instrument of torture on which a body is stretched
3 a (1) **:** a cause of anguish or pain (2) **:** acute suffering **b :** the action of straining or wrenching
4 : a framework, stand, or grating on or in which articles are placed
5 : a frame placed in a stream to stop fish and floating or suspended matter
6 a : a bar with teeth on one face for gearing with a pinion or worm gear to transform rotary motion to linear motion or vice versa (as in an automobile steering mechanism or microscope drawtube) **b :** a notched bar used as a ratchet to engage with a pawl, click, or detent
7 : a pair of antlers
8 : a triangular frame used to set up the balls in a pool game; *also* **:** the balls as set up
— rack·ful \-ˌfu̇l\ *noun*
— on the rack : under great mental or emotional stress

⁴rack (15th century)
transitive verb
1 : to torture on the rack
2 : to cause to suffer torture, pain, or anguish
3 a : to stretch or strain violently ⟨*racked* his brains⟩ **b :** to raise (rents) oppressively **c :** to harass or oppress with high rents or extortions
4 : to work or treat (material) on a rack
5 : to work by a rack and pinion or worm so as to extend or contract ⟨*rack* a camera⟩
6 : to seize (as parallel ropes of a tackle) together
7 : to place (as pool balls) in a rack
intransitive verb
: to become forced out of shape or out of plumb
synonym see AFFLICT
— rack·er *noun*
— rack·ing·ly \'ra-kiŋ-lē\ *adverb*

⁵rack *transitive verb* [Middle English *rakken,* from Old Provençal *arraca,* from *raca* stems and husks of pressed grapes] (15th century)
: to draw off (as wine) from the lees

⁶rack *intransitive verb* [probably alteration of ¹*rock*] (1530)
of a horse **:** to go at a rack

⁷rack *noun* (1580)
: either of two gaits of a horse: **a :** PACE 4b **b :** a fast showy 4-beat gait

⁸rack *noun* [perhaps from ³*rack*] (1570)
1 : the neck and spine of a forequarter of veal, pork, or especially mutton
2 : the rib section of a foresaddle of lamb used for chops or as a roast — see LAMB illustration

⁹rack *noun* [alteration of *wrack*] (1599)
: DESTRUCTION ⟨*rack* and ruin⟩

¹rack·et *also* **rac·quet** \'ra-kət\ *noun* [Mid-

◇ WORD HISTORY
raccoon The Algonquian-speaking Indians of tidewater Virginia contributed to American history one of its oldest stock images, the young princess Pocahontas beseeching her father Powhatan to spare the life of the English colonist John Smith—a tale related by Smith himself and the subject of much fictional elaboration later. Descendants of Powhatan's people still live in parts of Virginia, though, like the tribes of southern New England who were also very early victims of white colonization, they maintain few traces of their original culture, and their language has long been extinct. Before its demise, however, Virginia Algonquian contributed to English some of the oldest and best known loanwords from Native American languages, including *hickory, hominy, moccasin, opossum,* and *tomahawk.* Another familiar word from Virginia Algonquian is *raccoon,* which John Smith first recorded as *aroughcun, rarowcun,* and *raugroughcun.* Like the opossum, the raccoon was unknown to Europeans, who borrowed the indigenous name for it, as is clear from John Smith's remark in describing the animal to his readers: "There is a beast they call *Aroughcun,* much like a badger."

dle French *raquette,* ultimately from Arabic *rāḥah* palm of the hand] (circa 1520)
1 : a lightweight implement that consists of a netting (as of nylon) stretched in a usually oval open frame with a handle attached and that is used for striking the ball or shuttlecock in various games (as tennis, racquets, or badminton)

racket 1: *A* tennis, *B* racquetball, *C* badminton

2 *usually* **racquets** *plural but singular in construction* **:** a game for two or four players with ball and racket on a 4-walled court
²racket *noun* [origin unknown] (1565)
1 : confused clattering noise **:** CLAMOR
2 a : social whirl or excitement **b :** the strain of exciting or trying experiences
3 a : a fraudulent scheme, enterprise, or activity **b :** a usually illegitimate enterprise made workable by bribery or intimidation **c :** an easy and lucrative means of livelihood **d** *slang* **:** OCCUPATION, BUSINESS
³racket *intransitive verb* (1609)
1 : to engage in active social life
2 : to move with or make a racket
¹rack·e·teer \ˌra-kə-ˈtir\ *noun* (1928)
: one who obtains money by an illegal enterprise usually involving intimidation
²racketeer (1928)
intransitive verb
: to carry on a racket
transitive verb
: to practice extortion on
rack·ety \ˈra-kə-tē\ *adjective* (1773)
1 : NOISY
2 : FLASHY, ROWDY
3 : RICKETY
rack·le \ˈra-kəl\ *adjective* [Middle English *rakel*] (14th century)
chiefly Scottish **:** IMPETUOUS
rack railway *noun* (1884)
: a railway having between its rails a rack that meshes with a gear wheel or pinion of the locomotive for traction on steep grades
rack–rent *transitive verb* (1748)
: to subject to rack rent
rack rent *noun* [⁴*rack*] (1607)
1 : an excessive or unreasonably high rent
2 *British* **:** the highest rent that can be earned on a property
rack–rent·er \ˈrak-ˌren-tər\ *noun* (1680)
: one that pays or exacts rack rent
rack up *transitive verb* (1949)
: ACCUMULATE, GAIN ⟨*racked up* their tenth victory⟩
ra·clette \ra-ˈklet, rä-\ *noun* [French, literally, scraper, from *racler* to scrape, from Middle French, from Old Provençal *rasclar,* from (assumed) Vulgar Latin *rasiculare,* from Latin *rasus,* past participle of *radere* to scrape — more at RODENT] (circa 1949)
: a Swiss dish consisting of cheese melted over a fire and then scraped onto bread or boiled potatoes; *also* **:** the cheese used in this dish
ra·con \ˈrā-ˌkän\ *noun* [*ra*dar bea*con*] (1945)
: RADAR BEACON
ra·con·teur \ˌra-ˌkän-ˈtər, -kən-\ *noun* [French, from Middle French, from *raconter* to tell, from Old French, from *re-* + *aconter, acompter* to tell, count — more at ACCOUNT] (1828)
: a person who excels in telling anecdotes
ra·coon *variant of* RACCOON
rac·quet·ball \ˈra-kət-ˌbȯl\ *noun* (1968)
: a game similar to handball that is played on a 4-walled court with a short-handled racket and a larger ball
¹racy \ˈrā-sē\ *adjective* **rac·i·er; -est** [³*race*] (circa 1650)
1 a : full of zest or vigor **b :** having a strongly marked quality **:** PIQUANT ⟨a *racy* flavor⟩ **c :** RISQUÉ, SUGGESTIVE
2 : having the distinctive quality of something in its original or most characteristic form
synonym see PUNGENT
— rac·i·ly \ˈrā-sə-lē\ *adverb*
— rac·i·ness \-sē-nəs\ *noun*
²racy *adjective* **rac·i·er; -est** [¹*race*] (1841)
: having a body fitted for racing **:** long-bodied and lean
¹rad \ˈrad\ *noun* [*r*adiation *a*bsorbed *d*ose] (1918)
: a unit of absorbed dose of ionizing radiation equal to an energy of 100 ergs per gram of irradiated material
²rad *adjective* (1982)
slang **:** COOL 7, RADICAL
ra·dar \ˈrā-ˌdär\ *noun, often attributive* [*ra*dio *d*etecting *a*nd *r*anging] (1941)
: a device or system consisting usually of a synchronized radio transmitter and receiver that emits radio waves and processes their reflections for display and is used especially for detecting and locating objects (as aircraft) or surface features (as of a planet)
radar astronomy *noun* (1959)
: astronomy in which celestial bodies in the solar system are studied by analyzing the return of radio waves directed at them
radar beacon *noun* (1945)
: a radar transmitter that upon receiving a radar signal emits a signal which reinforces the normal reflected signal or which introduces a code into the reflected signal especially for identification purposes
ra·dar·scope \ˈrā-ˌdär-ˌskōp\ *noun* [*radar* + oscillo*scope*] (1945)
: the oscilloscope or screen serving as the visual indicator in a radar receiver
¹rad·dle \ˈra-dᵊl\ *noun* [Middle English *radel,* from diminutive of *rad-, red* red] (14th century)
: RED OCHER
²raddle *transitive verb* **rad·dled; rad·dling** \ˈrad-liŋ, ˈra-dᵊl-iŋ\ (1631)
: to mark or paint with raddle
³raddle *transitive verb* **rad·dled; rad·dling** \ˈrad-liŋ, ˈra-dᵊl-iŋ\ [English dialect *raddle* supple stick interwoven with others as in making a fence] (1671)
: to twist together **:** INTERWEAVE
rad·dled \ˈra-dᵊld\ *adjective* [origin unknown] (1694)
1 : being in a state of confusion **:** lacking composure
2 : BROKEN-DOWN, WORN
radi- *or* **radio-** *combining form* [French, from Latin *radius* ray]
1 : radiant energy **:** radiation ⟨*radio*active⟩ ⟨*radio*paque⟩
2 : radioactive ⟨*radio*element⟩
3 : radium **:** X rays ⟨*radio*therapy⟩
4 : radioactive isotopes especially as produced artificially ⟨*radio*carbon⟩
5 : radio ⟨*radio*telegraphy⟩
¹ra·di·al \ˈrā-dē-əl\ *adjective* [Medieval Latin *radialis,* from Latin *radius* ray] (1570)
1 : arranged or having parts arranged like rays
2 a : relating to, placed like, or moving along a radius **b :** characterized by divergence from a center
3 : of, relating to, or adjacent to a bodily radius
4 : developing uniformly around a central axis
— ra·di·al·ly \-ə-lē\ *adverb*
²radial *noun* (1872)
1 a : a radial part **b :** RAY
2 : a body part (as an artery) lying near or following the course of the radius
3 : a pneumatic tire in which the ply cords that extend to the beads are laid at approximately 90 degrees to the centerline of the tread — called also *radial-ply tire, radial tire*
radial cleavage *noun* (1973)
: holoblastic cleavage that is typical of deuterostomes and that is characterized by arrangement of the blastomeres of each upper tier directly over those of the next lower tier resulting in radial symmetry around the pole to pole axis of the embryo — compare SPIRAL CLEAVAGE
radial engine *noun* (1909)
: a usually internal combustion engine with cylinders arranged radially like the spokes of a wheel
radial ker·a·tot·o·my \-ˌker-ə-ˈtä-tə-mē\ *noun* (1980)
: multiple incision of the cornea in a radial pattern that is performed to correct myopia
radial symmetry *noun* (circa 1890)
: the condition of having similar parts regularly arranged around a central axis
— radially symmetrical *adjective*
ra·di·an \ˈrā-dē-ən\ *noun* (1879)
: a unit of plane angular measurement that is equal to the angle at the center of a circle subtended by an arc equal in length to the radius
ra·di·ance \ˈrā-dē-ən(t)s\ *noun* (1601)
1 : the quality or state of being radiant
2 : a deep pink
3 : the flux density of radiant energy per unit solid angle and per unit projected area of radiating surface
ra·di·an·cy \-ən(t)-sē\ *noun* (1646)
: RADIANCE
¹ra·di·ant \ˈrā-dē-ənt\ *adjective* (15th century)
1 a : radiating rays or reflecting beams of light **b :** vividly bright and shining **:** GLOWING
2 : marked by or expressive of love, confidence, or happiness ⟨a *radiant* smile⟩
3 a : emitted or transmitted by radiation **b :** emitting or relating to radiant heat
synonym see BRIGHT
— ra·di·ant·ly *adverb*
²radiant *noun* (circa 1741)
: something that radiates: as **a :** a point in the heavens at which the visible parallel paths of meteors appear to meet when traced backward **b :** the part of a gas or electric heater that becomes incandescent
radiant energy *noun* (circa 1890)
: energy traveling as electromagnetic waves
radiant flux *noun* (1917)
: the rate of emission or transmission of radiant energy
radiant heat *noun* (1794)
: heat transmitted by radiation as contrasted with that transmitted by conduction or convection
¹ra·di·ate \ˈrā-dē-ˌāt\ *verb* **-at·ed; -at·ing** [Latin *radiatus,* past participle of *radiare,* from *radius* ray] (circa 1619)
intransitive verb
1 : to proceed in a direct line from or toward a center
2 : to send out rays **:** shine brightly
3 a : to issue in or as if in rays **b :** to evolve by adaptive radiation
transitive verb
1 : to send out in or as if in rays
2 : IRRADIATE, ILLUMINATE
3 : to spread abroad or around as if from a center
²ra·di·ate \ˈrā-dē-ət, -ˌāt\ *adjective* (1668)
: having rays or radial parts: as **a :** having ray flowers **b :** characterized by radial symmetry **:** radially symmetrical
— ra·di·ate·ly *adverb*
ra·di·a·tion \ˌrā-dē-ˈā-shən\ *noun* (15th century)
1 a : the action or process of radiating **b :** the process of emitting radiant energy in the form

of waves or particles **c** (1) **:** the combined processes of emission, transmission, and absorption of radiant energy (2) **:** the transfer of heat by radiation — compare CONDUCTION, CONVECTION
2 a : something that is radiated **b :** energy radiated in the form of waves or particles
3 : radial arrangement
4 : ADAPTIVE RADIATION
5 : RADIATOR
— **ra·di·a·tion·al** \-shnəl, -shə-nᵊl\ *adjective*
— **ra·di·a·tion·less** \-shən-ləs\ *adjective*
— **ra·di·a·tive** \'rā-dē-ˌā-tiv\ *adjective*

radiation sickness *noun* (1924)
: sickness that results from exposure to radiation and is commonly marked by fatigue, nausea, vomiting, loss of teeth and hair, and in more severe cases by damage to blood-forming tissue with decrease in red and white blood cells and with bleeding

ra·di·a·tor \'rā-dē-ˌā-tər, *dialect* 'ra-\ *noun* (1836)
: one that radiates: as **a :** any of various devices (as a nest of pipes or tubes) for transferring heat from a fluid within to an area or object outside **b :** a transmitting antenna

[1]**rad·i·cal** \'ra-di-kəl\ *adjective* [Middle English, from Late Latin *radicalis,* from Latin *radic-, radix* root — more at ROOT] (14th century)
1 : of, relating to, or proceeding from a root: as **a** (1) **:** of or growing from the root of a plant ⟨*radical* tubers⟩ (2) **:** growing from the base of a stem, from a rootlike stem, or from a stem that does not rise above the ground ⟨*radical* leaves⟩ **b :** of, relating to, or constituting a linguistic root **c :** of or relating to a mathematical root **d :** designed to remove the root of a disease or all diseased tissue ⟨*radical* surgery⟩
2 : of or relating to the origin **:** FUNDAMENTAL
3 a : marked by a considerable departure from the usual or traditional **:** EXTREME **b :** tending or disposed to make extreme changes in existing views, habits, conditions, or institutions **c :** of, relating to, or constituting a political group associated with views, practices, and policies of extreme change **d :** advocating extreme measures to retain or restore a political state of affairs ⟨the *radical* right⟩
4 *slang* **:** EXCELLENT, COOL ◆
— **rad·i·cal·ness** *noun*

[2]**radical** *noun* (1641)
1 a : a root part **b :** a basic principle **:** FOUNDATION
2 a : ROOT 6 **b :** a sound or letter belonging to a radical
3 : one who is radical
4 : FREE RADICAL; *also* **:** a group of atoms bonded together that is considered an entity in various kinds of reactions
5 a : a mathematical expression indicating a root by means of a radical sign **b :** RADICAL SIGN

radical chic *noun* (1970)
: a fashionable practice among socially prominent people of associating with radicals or members of minority groups

rad·i·cal·ise *British variant of* RADICALIZE

rad·i·cal·ism \'ra-di-kə-ˌli-zəm\ *noun* (1820)
1 : the quality or state of being radical
2 : the doctrines or principles of radicals

rad·i·cal·ize \-kə-ˌlīz\ *transitive verb* **-ized; -iz·ing** (1830)
: to make radical especially in politics
— **rad·i·cal·i·za·tion** \ˌra-di-kə-lə-'zā-shən\ *noun*

rad·i·cal·ly \'ra-di-k(ə-)lē\ *adverb* (15th century)
1 : in origin or essence
2 : in a radical or extreme manner

radical sign *noun* (1668)
: the sign $\sqrt{}$ or $\surd$ placed before an expression to denote that the square root is to be extracted or that the root marked by an index (as in $\sqrt[3]{}$ or $\sqrt[3]{}$ for the cube root) is to be extracted

rad·i·cand \ˌra-də-'kand\ *noun* [Latin *radicandum,* neuter of *radicandus,* gerundive of *radicari*] (circa 1890)
: the quantity under a radical sign

ra·dic·chio \ra-'di-kē-ō\ *noun* [Italian, chicory, from (assumed) Vulgar Latin *radiculus,* alteration of Latin *radicula*] (1968)
: a chicory of a red variety with variegated leaves that is used as a salad green

radices *plural of* RADIX

rad·i·cle \'ra-di-kəl\ *noun* [Latin *radicula,* diminutive of *radic-, radix*] (1671)
1 : the lower part of the axis of a plant embryo or seedling: **a :** the embryonic root of a seedling **b :** HYPOCOTYL **c :** the hypocotyl and the root together
2 : RADICAL

ra·dic·u·lar \rə-'di-kyə-lər, ra-\ *adjective* (1830)
1 : of or relating to a plant radicle
2 : of, relating to, or involving a nerve root ⟨*radicular* pain⟩

radii *plural of* RADIUS

[1]**ra·dio** \'rā-dē-ˌō\ *adjective* [[2]*radio* or *radio-*] (1887)
1 : of, relating to, or operated by radiant energy
2 : of or relating to electric currents or phenomena (as electromagnetic radiation) of frequencies between about 15,000 and 10^{11} hertz
3 a : of, relating to, or used in radio or a radio set **b :** specializing in radio or associated with the radio industry **c** (1) **:** transmitted by radio (2) **:** making or participating in radio broadcasts **d :** controlled or directed by radio

[2]**radio** *noun, plural* **ra·di·os** [short for *radiotelegraphy*] (1903)
1 a : the wireless transmission and reception of electric impulses or signals by means of electromagnetic waves **b :** the use of these waves for the wireless transmission of electric impulses into which sound is converted
2 : a radio message
3 : a radio receiving set
4 a : a radio transmitting station **b :** a radio broadcasting organization **c :** the radio broadcasting industry **d :** communication by radio

[3]**radio** (1913)
transitive verb
1 : to send or communicate by radio
2 : to send a radio message to
intransitive verb
: to send or communicate something by radio

radio- — see RADI-

ra·dio·ac·tive \ˌrā-dē-ō-'ak-tiv\ *adjective* [International Scientific Vocabulary] (1898)
: of, caused by, or exhibiting radioactivity
— **ra·dio·ac·tive·ly** *adverb*

ra·dio·ac·tiv·i·ty \-ak-'ti-və-tē\ *noun* [International Scientific Vocabulary] (1899)
: the property possessed by some elements (as uranium) or isotopes (as carbon 14) of spontaneously emitting energetic particles (as electrons or alpha particles) by the disintegration of their atomic nuclei; *also* **:** the rays emitted

ra·dio·al·ler·go·sor·bent \ˌrā-dē-ō-ə-ˌlər-gō-'sȯr-bənt\ *adjective* [*radi-* + *allergen* + *-o-* + *sorbent*] (1967)
: relating to or being a blood analysis that tests for allergen-specific antibodies and is used to detect allergic reactions

radio astronomy *noun* (1948)
: astronomy dealing with radio waves received from outside the earth's atmosphere
— **radio astronomer** *noun*
— **radio astronomical** *adjective*

ra·dio·au·to·graph \ˌrā-dē-ō-'ȯ-tə-ˌgraf\ *noun* (1941)
: AUTORADIOGRAPH
— **ra·dio·au·to·graph·ic** \-ˌȯ-tə-'gra-fik\ *adjective*
— **ra·dio·au·tog·ra·phy** \-ȯ-'tä-grə-fē\ *noun*

radio beacon *noun* (1919)
: a radio transmitting station that transmits special radio signals for use (as on a landing field) in determining the direction or position of those receiving them

ra·dio·bi·ol·o·gy \ˌrā-dē-ō-bī-'ä-lə-jē\ *noun* (1919)
: a branch of biology dealing with the effects of radiation or radioactive materials on biological systems
— **ra·dio·bi·o·log·i·cal** \-ˌbī-ə-'lä-ji-kəl\ *also* **ra·dio·bi·o·log·ic** \-jik\ *adjective*
— **ra·dio·bi·o·log·i·cal·ly** \-ji-k(ə-)lē\ *adverb*
— **ra·dio·bi·ol·o·gist** \-bī-'ä-lə-jist\ *noun*

radio car *noun* (1925)
: an automobile equipped with radio communication

ra·dio·car·bon \ˌrā-dē-ō-'kär-bən\ *noun, often attributive* [International Scientific Vocabulary] (1939)
: radioactive carbon; *especially* **:** CARBON 14

radiocarbon dating *noun* (1951)
: CARBON DATING
— **radiocarbon–date** \-'dāt\ *transitive verb*

ra·dio·chem·is·try \ˌrā-dē-ō-'ke-mə-strē\ *noun* (1904)
: a branch of chemistry dealing with radioactive substances and phenomena including tracer studies
— **ra·dio·chem·i·cal** \-'ke-mi-kəl\ *adjective*
— **ra·dio·chem·i·cal·ly** \-k(ə-)lē\ *adverb*
— **ra·dio·chem·ist** \-'ke-mist\ *noun*

ra·dio·chro·mato·gram \-krō-'ma-tə-ˌgram, -krə-\ *noun* (1951)
: a chromatogram revealing one or more radioactive substances

radio compass *noun* (1918)
: a direction finder used in navigation

ra·dio·ecol·o·gy \ˌrā-dē-ō-i-'kä-lə-jē\ *noun* (1956)
: the study of the effects of radiation and radioactive substances on ecological communities

ra·dio·el·e·ment \-'e-lə-mənt\ *noun* [International Scientific Vocabulary] (1903)
: a radioactive element

radio frequency *noun* (1915)
: any of the electromagnetic wave frequencies that lie in the range extending from below 3 kilohertz to about 300 gigahertz and that in-

◇ WORD HISTORY
radical The Late Latin adjective *radicalis* "rooted, having roots," a derivative of Latin *radix* "root," was borrowed into later Middle English as *radical.* Aside from its literal meaning "of a root," *radical* in the 16th century meant "primary" or "fundamental," that is, effecting the "root" of something. In the 17th and 18th centuries *radical* increasingly began to modify nouns such as *change* or *cure* and gradually shifted from the sense "fundamental" to "thorough" and then "extreme." This shift was complete by about the year 1800, when *radical reform* became a political catchphrase, and an advocate of radical reform was called a *radical.* In the 20th century *radical,* when applied to opinions or to advocates of a particular cause, is practically synonymous with *extremist.*

RADIO FREQUENCIES

CLASS	ABBREVIATION	RANGE
extremely low frequency	ELF	below 3 kilohertz
very low frequency	VLF	3 to 30 kilohertz
low frequency	LF	30 to 300 kilohertz
medium frequency	MF	300 to 3000 kilohertz
high frequency	HF	3 to 30 megahertz
very high frequency	VHF	30 to 300 megahertz
ultrahigh frequency	UHF	300 to 3000 megahertz
superhigh frequency	SHF	3 to 30 gigahertz
extremely high frequency	EHF	30 to 300 gigahertz

clude the frequencies used for radio and television transmission

radio galaxy *noun* (1960)
: a galaxy that is a powerful source of radio waves

ra·dio·gen·ic \ˌrā-dē-ō-ˈje-nik\ *adjective* (1935)
: produced by or determined from radioactivity ⟨*radiogenic* isotopes⟩ ⟨*radiogenic* tumors⟩

ra·dio·gram \ˈrā-dē-ō-ˌgram\ *noun* (1896)
1 : RADIOGRAPH
2 : a message transmitted by radiotelegraphy
3 [short for *radiogramophone*] *British* **:** a combined radio receiver and record player

¹ra·dio·graph \-ˌgraf\ *noun* (1896)
: a picture produced on a sensitive surface by a form of radiation other than visible light; *specifically* **:** an X-ray or gamma ray photograph
— ra·dio·graph·ic \ˌrā-dē-ō-ˈgra-fik\ *adjective*
— ra·dio·graph·i·cal·ly \-fi-k(ə-)lē\ *adverb*

²radiograph *transitive verb* (1896)
: to make a radiograph of

ra·di·og·ra·phy \ˌrā-dē-ˈä-grə-fē\ *noun* [International Scientific Vocabulary] (1896)
: the art, act, or process of making radiographs

ra·dio·im·mu·no·as·say \ˌrā-dē-ˌō-i-myə-nō-ˈa-ˌsā, -i-ˌmyü-, -a-ˈsā\ *noun* (1961)
: immunoassay of a substance that has been radioactively labeled
— ra·dio·im·mu·no·as·say·able \-ə-bəl\ *adjective*

ra·dio·iso·tope \ˌrā-dē-ō-ˈī-sə-ˌtōp\ *noun* [International Scientific Vocabulary] (1946)
: a radioactive isotope
— ra·dio·iso·to·pic \-ˌī-sə-ˈtä-pik, -ˈtō-\ *adjective*
— ra·dio·iso·to·pi·cal·ly \-pi-k(ə-)lē\ *adverb*

ra·dio·la·bel \-ˈlā-bəl\ *transitive verb* (1953)
: to label with a radioactive atom or substance

ra·di·o·lar·i·an \ˌrā-dē-ō-ˈlar-ē-ən, -ˈler-\ *noun* [ultimately from Late Latin *radiolus* small sunbeam, from diminutive of Latin *radius* ray — more at RAY] (1877)
: any of three classes (Acantharia, Polycystina, and Phaeodaria) of usually spherical marine protozoans having radiating threadlike pseudopodia and often a siliceous skeleton of spicules
— radiolarian *adjective*

ra·dio·log·i·cal \ˌrā-dē-ə-ˈlä-ji-kəl\ *or* **ra·dio·log·ic** \-jik\ *adjective* (1909)
1 : of or relating to radiology
2 : of or relating to nuclear radiation
— ra·dio·log·i·cal·ly \-k(ə-)lē\ *adverb*

ra·di·ol·o·gist \ˌrā-dē-ˈä-lə-jist\ *noun* (1906)
: a physician specializing in the use of radiant energy for diagnostic and therapeutic purposes

ra·di·ol·o·gy \-jē\ *noun* (1900)
1 : a branch of medicine concerned with the use of radiant energy (as X rays and radium) in the diagnosis and treatment of disease
2 : the science of radioactive substances and high-energy radiations

ra·dio·lu·cent \ˌrā-dē-ō-ˈlü-sᵊnt\ *adjective* (1917)
: partly or wholly permeable to radiation ⟨*radiolucent* tissues⟩
— ra·dio·lu·cen·cy \-sᵊn(t)-sē\ *noun*

ra·di·ol·y·sis \ˌrā-dē-ˈä-lə-səs\ *noun* [New Latin] (1948)
: chemical decomposition by the action of radiation
— ra·dio·lyt·ic \-dē-ə-ˈli-tik\ *adjective*

ra·dio·man \ˈrā-dē-ō-ˌman\ *noun* (1921)
: a radio operator or technician

ra·di·om·e·ter \ˌrā-dē-ˈä-mə-tər\ *noun* (1875)
: an instrument for measuring the intensity of radiant energy by the torsional twist of suspended vanes that are blackened on one side and exposed to a source of radiant energy; *also* **:** an instrument for measuring electromagnetic or acoustic radiation
— ra·di·om·e·try \-mə-trē\ *noun*

radiometer

ra·dio·met·ric \ˌrā-dē-ō-ˈme-trik\ *adjective* [International Scientific Vocabulary] (1877)
1 : relating to, using, or measured by a radiometer
2 : of or relating to the measurement of geologic time by means of the rate of disintegration of radioactive elements
— ra·dio·met·ri·cal·ly \-tri-k(ə-)lē\ *adverb*

ra·dio·mi·met·ic \-mə-ˈme-tik, -mī-\ *adjective* [International Scientific Vocabulary] (1947)
: producing effects similar to those of radiation

ra·dio·nu·clide \ˌrā-dē-ō-ˈnü-ˌklīd, -ˈnyü-\ *noun* (1947)
: a radioactive nuclide

ra·di·opaque \ˌrā-dē-ō-ˈpāk\ *adjective* (circa 1923)
: being opaque to various forms of radiation (as X rays)

ra·dio·phar·ma·ceu·ti·cal \ˌrā-dē-ō-ˌfär-mə-ˈsü-ti-kəl\ *noun* (1952)
: a radioactive drug used for diagnostic or therapeutic purposes
— radiopharmaceutical *adjective*

ra·dio·phone \ˈrā-dē-ə-ˌfōn\ *noun* (1919)
: RADIOTELEPHONE

ra·dio·pho·to \ˌrā-dē-ō-ˈfō-(ˌ)tō\ *noun* (1929)
: a picture transmitted by radio

ra·dio·pro·tec·tive \-prə-ˈtek-tiv\ *adjective* (1956)
: serving to protect or aiding in protecting against the injurious effect of radiations ⟨*radioprotective* drugs⟩
— ra·dio·pro·tec·tion \-ˈtek-shən\ *noun*

radio range *noun* (1929)
: a radio facility for aircraft navigation

ra·dio·sen·si·tive \ˌrā-dē-ō-ˈsen(t)-sə-tiv, -ˈsen(t)-stiv\ *adjective* (1920)
: sensitive to the effects of radiant energy ⟨*radiosensitive* cancer cells⟩
— ra·dio·sen·si·tiv·i·ty \-ˌsen(t)-sə-ˈti-və-tē\ *noun*

ra·dio·sonde \ˈrā-dē-ō-ˌsänd\ *noun* [International Scientific Vocabulary] (1937)
: a miniature radio transmitter that is carried aloft (as by an unmanned balloon) with instruments for broadcasting the humidity, temperature, and pressure

radio spectrum *noun* (1929)
: the region of the electromagnetic spectrum usually including frequencies below 30,000 megahertz in which radio or radar transmission and detection techniques may be used

radio star *noun* (1948)
: a cosmic radio source; *especially* **:** a point source of radio emissions

ra·dio·stron·tium \ˌrā-dē-ō-ˈsträn(t)-sh(ē-)əm, -ˈsträn-tē-əm\ *noun* [New Latin] (1941)
: radioactive strontium; *especially* **:** STRONTIUM 90

ra·dio·tele·graph \-ˈte-lə-ˌgraf\ *noun* [International Scientific Vocabulary] (1903)
: WIRELESS TELEGRAPHY
— ra·dio·te·leg·ra·phy \-tə-ˈle-grə-fē\ *noun*

ra·dio·te·lem·e·try \-tə-ˈle-mə-trē\ *noun* (1951)
1 : TELEMETRY
2 : BIOTELEMETRY
— ra·dio·tele·met·ric \-ˌte-lə-ˈme-trik\ *adjective*

ra·dio·tele·phone \-ˈte-lə-ˌfōn\ *noun* [International Scientific Vocabulary] (1904)
: an apparatus for carrying on wireless telephony by radio waves
— ra·dio·te·le·pho·ny \-tə-ˈle-fə-nē, -ˈte-lə-ˌfō-nē\ *noun*

radio telescope *noun* (1929)
: a radio receiver-antenna combination used for observation in radio astronomy

ra·dio·ther·a·py \ˌrā-dē-ō-ˈther-ə-pē\ *noun* [International Scientific Vocabulary] (1903)
: the treatment of disease with radiation (as X rays)
— ra·dio·ther·a·pist \-pist\ *noun*

ra·dio·tho·ri·um \-ˈthōr-ē-əm, -ˈthȯr-\ *noun* [New Latin] (1905)
: a radioactive isotope of thorium with the mass number 228

ra·dio·trac·er \ˈrā-dē-ō-ˌtrā-sər\ *noun* (1946)
: a radioactive tracer

ra·dio–ul·na \ˌrā-dē-ō-ˈəl-nə\ *noun* [New Latin] (1960)
: a single bone in the forelimb of an amphibian (as a frog) that represents fusion of the separate radius and ulna of higher vertebrate forms

radio wave *noun* (1916)
: an electromagnetic wave with radio frequency

rad·ish \ˈra-dish *also* ˈre-\ *noun* [Middle English, alteration of Old English *rædic,* from Latin *radic-, radix* root, radish — more at ROOT] (15th century)
: the pungent fleshy root of a widely cultivated Eurasian plant (*Raphanus sativus*) of the mustard family usually eaten raw; *also* **:** a plant that produces radishes

ra·di·um \ˈrā-dē-əm\ *noun, often attributive* [New Latin, from Latin *radius* ray] (1899)
: an intensely radioactive brilliant white metallic element that resembles barium chemically, occurs in combination in minute quantities in minerals (as pitchblende or carnotite), emits alpha particles and gamma rays to form radon, and is used chiefly in luminous materials and in the treatment of cancer — see ELEMENT table

radium therapy *noun* (1904)
: RADIOTHERAPY

ra·di·us \ˈrā-dē-əs\ *noun, plural* **ra·dii** \-dē-ˌī\ *also* **ra·di·us·es** [Latin, ray, radius] (circa 1611)
1 : a line segment extending from the center of a circle or sphere to the circumference or bounding surface
2 a : the bone on the thumb side of the human forearm; *also* **:** a corresponding part of vertebrates above fishes **b :** the third and usually largest vein of an insect's wing
3 a : the length of a radius ⟨a truck with a short turning *radius*⟩ **b :** the circular area de-

fined by a stated radius **c :** a bounded or circumscribed area
4 : a radial part
5 : the distance from a center line or point to an axis of rotation

radius of curvature *noun* (circa 1753)
: the reciprocal of the curvature of a curve

radius vector *noun* (circa 1753)
1 a : the line segment or its length from a fixed point to a variable point **b :** the linear polar coordinate of a variable point
2 : a straight line joining the center of an attracting body (as the sun) with that of a body (as a planet) in orbit around it

ra·dix \'rā-diks\ *noun, plural* **ra·di·ces** \'rā-də-ˌsēz, 'ra-\ *or* **ra·dix·es** \'rā-dik-səz\ [Latin, root — more at ROOT] (1798)
1 : the base of a number system or of logarithms
2 : the primary source

ra·dome \'rā-ˌdōm\ *noun* [*radar dome*] (circa 1944)
: a plastic housing sheltering the antenna assembly of a radar set especially on an airplane

ra·don \'rā-ˌdän\ *noun* [International Scientific Vocabulary, from *radium*] (1918)
: a heavy radioactive gaseous element formed by the decay of radium — see ELEMENT table

rad·u·la \'ra-jə-lə\ *noun, plural* **-lae** \-ˌlē, -ˌlī\ *also* **-las** [New Latin, from Latin, scraper, from *radere* to scrape — more at RODENT] (circa 1859)
: a horny band or ribbon in mollusks other than bivalves that bears minute teeth on its dorsal surface and tears up food and draws it into the mouth
— rad·u·lar \-lər\ *adjective*

rad·waste \'rad-ˌwāst\ *noun, often attributive* [by shortening] (1973)
: radioactive waste

raff \'raf\ *noun* [Middle English *raf* rubbish] (14th century)
: RIFFRAFF

raf·fia \'ra-fē-ə\ *noun* [Malagasy *rafia*] (1882)
: the fiber of the raffia palm used especially for tying plants and making baskets and hats

raffia palm *noun* (1897)
: a pinnate-leaved palm (*Raphia farinifera* synonym *R. ruffia*) of Madagascar that is valued for the fiber from its leafstalks

raf·fi·nose \'ra-fə-ˌnōs, -ˌnōz\ *noun* [French, from *raffiner* to refine, from *re-* + *affiner* to make fine, from *a-* ad- (from Latin *ad-*) + *fin* fine] (1876)
: a crystalline slightly sweet sugar $C_{18}H_{32}O_{16}$ obtained commercially from cottonseed meal and present in many plant products

raff·ish \'ra-fish\ *adjective* (1801)
1 : marked by or suggestive of flashy vulgarity or crudeness
2 : marked by a careless unconventionality
: RAKISH
— raff·ish·ly *adverb*
— raff·ish·ness *noun*

[1]**raffle** \'ra-fəl\ *verb* **raf·fled; raf·fling** \'ra-f(ə-)liŋ\ (circa 1680)
intransitive verb
: to engage in a raffle
transitive verb
: to dispose of by means of a raffle ⟨*raffle* off a turkey⟩

[2]**raffle** *noun* [Middle English *rafle,* a dice game, from Middle French, dice game in which all the stakes can be won in a throw, literally, rake for a fire, from Middle High German *raffel* rake for a fire, from *raffen* to snatch, gather] (1766)
: a lottery in which the prize is won by one of numerous persons buying chances

[3]**raffle** *noun* [probably from French *rafle* act of snatching, sweeping, from Middle French, rake for a fire] (1881)
: RUBBISH; *especially* **:** a jumble or tangle of nautical equipment

raf·fle·sia \rə-'flē-zh(ē-)ə, ra-\ *noun* [New Latin, from Sir Stamford *Raffles* (died 1826) English colonial administrator] (1830)
: any of a genus (*Rafflesia* of the family Rafflesiaceae) of Malaysian dicotyledonous plants that are parasitic in other plants and have fleshy usually foul-smelling apetalous flowers emerging from the host, imbricated scales in place of leaves, and no stems

[1]**raft** \'raft\ *noun* [Middle English *rafte* rafter, raft, from Old Norse *raptr* rafter] (15th century)
1 a : a collection of logs or timber fastened together for conveyance by water **b :** a flat structure for support or transportation on water
2 : a floating cohesive mass
3 : an aggregation of animals (as waterfowl) resting on the water

[2]**raft** (1706)
transitive verb
1 : to transport in the form of or by means of a raft; *also* **:** to convey (as pebbles) in floating ice or masses of organic material
2 : to make into a raft
intransitive verb
: to travel by raft

[3]**raft** *noun* [alteration of *raff* jumble] (1830)
: a large collection or number

[1]**raf·ter** \'raf-tər\ *noun* [Middle English, from Old English *ræfter;* akin to Old Norse *raptr* rafter] (before 12th century)
: any of the parallel beams that support a roof — see RIDGEPOLE illustration
— raf·tered \-tərd\ *adjective*

[2]**raft·er** \'raf-tər\ *noun* [[2]*raft*] (1809)
1 : one who maneuvers logs into position and binds them into rafts
2 : one who travels by raft

rafts·man \'raf(t)s-mən\ *noun* (1776)
: a man engaged in rafting

[1]**rag** \'rag\ *noun* [Middle English *ragge,* from (assumed) Old English *ragg,* from Old Norse *rǫgg* tuft, shagginess] (14th century)
1 a : a waste piece of cloth **b** *plural* **:** clothes usually in poor or ragged condition **c :** CLOTHING ⟨the *rag* trade⟩
2 : something resembling a rag
3 : NEWSPAPER

[2]**rag** *noun* [Middle English *ragge*] (14th century)
1 : any of various hard rocks
2 : a large roofing slate that is rough on one side

[3]**rag** *transitive verb* **ragged** \'ragd\; **rag·ging** [origin unknown] (1739)
1 : to rail at **:** SCOLD
2 : TORMENT, TEASE

[4]**rag** *noun* (1864)
chiefly British **:** an outburst of boisterous fun; *also* **:** PRANK

[5]**rag** *noun* [short for *ragtime*] (1897)
: a composition in ragtime

ra·ga \'rä-gə\ *noun* [Sanskrit *rāga,* literally, color, tone; akin to Sanskrit *rajyati* it reddens, Greek *rhezein* to dye] (1788)
1 : one of the ancient traditional melodic patterns or modes in Indian music
2 : an improvisation based on a traditional raga — compare TALA

rag·a·muf·fin \'ra-gə-ˌmə-fən\ *noun* [Middle English *Ragamuffyn,* name for a ragged, oafish person] (1581)
: a ragged often disreputable person; *especially* **:** a poorly clothed often dirty child

rag·bag \'rag-ˌbag\ *noun* (1820)
1 : a bag for scraps
2 : a miscellaneous collection

rag doll *noun* (1853)
: a stuffed usually painted cloth doll

[1]**rage** \'rāj\ *noun* [Middle English, from Old French, from Late Latin *rabia,* from Latin *rabies* rage, madness, from *rabere* to be mad; akin to Sanskrit *rabhas* violence] (14th century)
1 a : violent and uncontrolled anger **b :** a fit of violent wrath **c** *archaic* **:** INSANITY
2 : violent action (as of wind or sea)
3 : an intense feeling **:** PASSION
4 : a fad pursued with intense enthusiasm ⟨was all the *rage*⟩
synonym see ANGER, FASHION

[2]**rage** *intransitive verb* **raged; rag·ing** (14th century)
1 : to be in a rage
2 : to be in tumult
3 : to prevail uncontrollably

rag·ged \'ra-gəd\ *adjective* (14th century)
1 : roughly unkempt
2 : having an irregular edge or outline
3 a : torn or worn to tatters **b :** worn-out from stress and strain ⟨ran herself *ragged*⟩
4 : wearing tattered clothes
5 a : STRAGGLY **b :** executed in an irregular or uneven manner **c** *of a sound* **:** HARSH, DISSONANT
— rag·ged·ly *adverb*
— rag·ged·ness *noun*

ragged robin *noun* (1741)
: a perennial herb (*Lychnis flos-cuculi*) cultivated for its pink flowers with narrow-lobed petals

rag·gedy \'ra-gə-dē\ *adjective* (1890)
: RAGGED

rag·gle–tag·gle \'ra-gəl-ˌta-gəl\ *adjective* [irregular from *ragtag*] (1904)
: MOTLEY

ra·gi \'ra-gē, 'rä-\ *noun* [perhaps from Deccan Hindi *rāgī*] (1792)
: an Old World cereal grass (*Eleusine coracana*) yielding a staple food crop especially in India and Africa; *also* **:** the seeds of ragi used for food

raging *adjective* (15th century)
1 : causing great pain or distress
2 : VIOLENT, WILD
3 : EXTRAORDINARY, TREMENDOUS ⟨a *raging* success⟩

rag·lan \'ra-glən\ *noun* [F.J.H. Somerset, Baron *Raglan* (died 1855) British field marshal] (circa 1859)
: a loose overcoat with raglan sleeves

raglan sleeve *noun* (circa 1924)
: a sleeve that extends to the neckline with slanted seams from the underarm to the neck

rag·man \'rag-ˌman\ *noun* (1586)
: a man who collects or deals in rags

Rag·na·rok \'rag-nə-ˌräk, -rək\ *noun* [Old Norse *Ragnarǫk,* literally, fate of the gods, from *ragna,* genitive plural of *regin* gods + *rǫk* fate, course (later rendered as *Ragnarøkkr,* literally, twilight of the gods)]
: the final destruction of the world in the conflict between the Aesir and the powers of Hel led by Loki — called also *Twilight of the Gods*

ra·gout \ra-'gü\ *noun* [French *ragoût,* from *ragoûter* to revive the taste, from Middle French *ragouster,* from *re-* + *a-* ad- (from Latin *ad-*) + *goust* taste, from Latin *gustus;* akin to Latin *gustare* to taste — more at CHOOSE] (circa 1657)
1 : well-seasoned meat and vegetables cooked in a thick sauce
2 : MIXTURE, MÉLANGE

rag·pick·er \'rag-ˌpi-kər\ *noun* (1860)
: one who collects rags and refuse for a living

rag·tag \'rag-ˌtag\ *adjective* [*ragtag and bobtail*] (1882)
1 : RAGGED, UNKEMPT
2 : MOTLEY 2 ⟨a *ragtag* bunch of misfits⟩

ragtag and bobtail *noun* [[1]*rag* + [1]*tag*] (1820)
: RABBLE

rag·time \'rag-ˌtīm\ *noun* [probably from *ragged* + *time*] (1897)
1 : rhythm characterized by strong syncopation in the melody with a regularly accented accompaniment

\ə\ abut \ᵊ\ kitten \ər\ further \a\ ash \ā\ ace \ä\ mop, mar \au̇\ out \ch\ chin \e\ bet \ē\ easy \g\ go \i\ hit \ī\ ice \j\ job \ŋ\ sing \ō\ go \ȯ\ law \ȯi\ boy \th\ thin \t͟h\ the \ü\ loot \u̇\ foot \y\ yet \zh\ vision *see also* Guide to Pronunciation

2 : music having ragtime rhythm
rag·top \-ˌtäp\ *noun* (1953)
: a convertible automobile
rag·weed \-ˌwēd\ *noun* (1790)
: any of various chiefly North American weedy composite herbs (genus *Ambrosia*) that produce highly allergenic pollen
rag·wort \-ˌwərt, -ˌwȯrt\ *noun* (14th century)
: any of several senecios; *especially* **:** TANSY RAGWORT
rah \ˈrä, ˈrȯ\ *interjection* (1870)
: HURRAH — used especially to cheer on a team
rah–rah \ˈrä-(ˌ)rä, ˈrȯ-(ˌ)rȯ\ *adjective* [reduplication of *rah*] (1911)
: marked by the enthusiastic expression of college spirit
¹raid \ˈrād\ *noun* [Middle English (Scots) *rade,* from Old English *rād* ride, raid — more at ROAD] (15th century)
1 a : a hostile or predatory incursion **b :** a surprise attack by a small force
2 a : a brief foray outside one's usual sphere **b :** a sudden invasion by officers of the law **c :** a daring operation against a competitor **d :** the recruiting of personnel (as faculty, executives, or athletes) from competing organizations
3 : the act of mulcting public money
4 : an attempt by professional operators to depress stock prices by concerted selling
²raid (1865)
intransitive verb
: to conduct or take part in a raid
transitive verb
: to make a raid on
raid·er \ˈrā-dər\ *noun* (1863)
: one that raids: as **a :** a fast lightly armed ship operating against merchant shipping **b :** a soldier specially trained for close-range fighting **c :** one that attempts a usually hostile takeover of a business corporation ⟨corporate *raiders*⟩
¹rail \ˈrā(ə)l\ *noun* [Middle English *raile,* from Middle French *reille* ruler, bar, from Latin *regula* ruler, from *regere* to keep straight, direct, rule — more at RIGHT] (14th century)
1 a : a bar extending from one post or support to another and serving as a guard or barrier **b :** a structural member or support
2 a : RAILING 1 **b :** a light structure serving as a guard at the outer edge of a ship's deck **c :** a fence bounding a racetrack
3 a : a bar of rolled steel forming a track for wheeled vehicles **b :** TRACK **c :** RAILROAD
²rail *transitive verb* (14th century)
: to provide with a railing **:** FENCE
³rail *noun, plural* **rail** *or* **rails** [Middle English *raile,* from Middle French *raale*] (15th century)
: any of numerous wading birds (family Rallidae, the rail family) that are of small or medium size and have short rounded wings, a short tail, and usually very long toes which enable them to run on the soft mud of marshes

rail

⁴rail *intransitive verb* [Middle English, from Middle French *railler* to mock, from Old Provençal *ralhar* to babble, joke, from (assumed) Vulgar Latin *ragulare* to bray, from Late Latin *ragere* to neigh] (15th century)
: to revile or scold in harsh, insolent, or abusive language
synonym see SCOLD
— rail·er *noun*
rail·bird \ˈrā(ə)l-ˌbərd\ *noun* (1892)
: a racing enthusiast who sits on or near the track rail to watch a race or workout
rail·bus \-ˌbəs\ *noun* (1933)
: a passenger car with an automotive engine for operation on rails
rail·car \-ˌkär\ *noun* (1834)
1 : a railroad car
2 : a self-propelled railroad car
rail·head \ˈrā(ə)l-ˌhed\ *noun* (1896)
: a point on a railroad at which traffic may originate or terminate
rail·ing \ˈrā-liŋ\ *noun* (15th century)
1 : a barrier consisting of a rail and supports
2 : RAILS; *also* **:** material for making rails
rail·lery \ˈrā-lə-rē\ *noun, plural* **-ler·ies** [French *raillerie,* from Middle French, from *railler* to mock] (1653)
1 : good-natured ridicule **:** BANTER
2 : JEST
¹rail·road \ˈrā(ə)l-ˌrōd, ˈre(ə)l-; ˈre-ˌrōd\ *noun* (1825)
: a permanent road having a line of rails fixed to ties and laid on a roadbed and providing a track for cars or equipment drawn by locomotives or propelled by self-contained motors; *also* **:** such a road and its assets constituting a single property
²railroad (1877)
transitive verb
1 a : to convict with undue haste and by means of false charges or insufficient evidence **b :** to push through hastily or without due consideration
2 : to transport by railroad
intransitive verb
: to work for a railroad company
— rail·road·er *noun*
railroad flat *noun* (1947)
: an apartment having a series of narrow rooms arranged in line
rail·road·ing *noun* (1870)
: construction or operation of a railroad
railroad worm *noun* (1909)
1 [probably from its dissemination by railroad] **:** APPLE MAGGOT
2 [from the rows of luminescent spots along its sides making it resemble a lighted train] **:** the larva or wingless female of any of several South American beetles (genus *Phrixothrix* of the family Cantharidae)
rail–split·ter \ˈrā(ə)l-ˌspli-tər\ *noun* (1860)
: one that makes logs into fence rails
rail·way \-ˌwā\ *noun* (1812)
: RAILROAD; *especially* **:** a railroad operating with light equipment or within a small area
rai·ment \ˈrā-mənt\ *noun* [Middle English *rayment,* short for *arrayment,* from *arrayen* to array] (15th century)
: CLOTHING, GARMENTS
¹rain \ˈrān\ *noun, often attributive* [Middle English *reyn,* from Old English *regn, rēn;* akin to Old High German *regan* rain] (before 12th century)
1 a : water falling in drops condensed from vapor in the atmosphere **b :** the descent of this water **c :** water that has fallen as rain **:** RAINWATER
2 a : a fall of rain **:** RAINSTORM **b** *plural* **:** the rainy season
3 : rainy weather
4 : a heavy fall of particles or bodies
²rain (before 12th century)
intransitive verb
1 : to send down rain
2 : to fall as water in drops from the clouds
3 : to fall like rain
transitive verb
1 : to pour down
2 : to bestow abundantly
— rain cats and dogs : to rain heavily
rain·bird \ˈrān-ˌbərd\ *noun* (1555)
: any of numerous birds (especially of the family Cuculidae) whose cries are popularly believed to augur rain
rain·bow \-ˌbō\ *noun* (before 12th century)
1 : an arc or circle that exhibits in concentric bands the colors of the spectrum and that is formed opposite the sun by the refraction and reflection of the sun's rays in raindrops, spray, or mist
2 a : a multicolored array **b :** a wide assortment or range
3 [from the impossibility of reaching the rainbow, at whose foot a pot of gold is said to be buried] **:** an illusory goal or hope
4 : RAINBOW TROUT
— rain·bow·like *adjective*
rainbow fish *noun* (1888)
: any of numerous brilliantly colored fishes (as a wrasse, parrot fish, or guppy)
rainbow runner *noun* (1940)
: a large brilliantly marked blue and yellow carangid food and sport fish (*Elagatis bipinnulata*) common in warm seas
rainbow trout *noun* (1882)
: a large stout-bodied salmonid fish (*Oncorhynchus mykiss* synonym *Salmo gairdneri*) of western North America that is related to the Pacific salmon and is typically greenish above and white on the belly with a pink, red, or lavender stripe along each side of the body and with profuse black dots — compare STEELHEAD

rainbow trout

rain check *noun* (1884)
1 : a ticket stub good for a later performance when the scheduled one is rained out
2 : an assurance of a deferred extension of an offer; *especially* **:** an assurance that a customer can take advantage of a sale later if the item or service offered is not available (as by being sold out)
rain·coat \ˈrān-ˌkōt\ *noun* (1830)
: a waterproof or water-resistant coat
rain·drop \-ˌdräp\ *noun* (before 12th century)
: a drop of rain
rain·fall \-ˌfȯl\ *noun* (1854)
1 : the amount of precipitation usually measured by the depth in inches
2 : RAIN 2a
rain forest *noun* (1903)
1 : a tropical woodland with an annual rainfall of at least 100 inches (254 centimeters) and marked by lofty broad-leaved evergreen trees forming a continuous canopy — called also *tropical rain forest*
2 : TEMPERATE RAIN FOREST
rain gauge *noun* (1769)
: an instrument for measuring the quantity of precipitation
rain·mak·er \ˈrān-ˌmā-kər\ *noun* (1775)
1 : a person who produces or attempts to produce rain by artificial means
2 : a person (as a partner in a law firm) who brings in new business; *also* **:** a person whose influence can initiate progress or ensure success
— rain·mak·ing \-ˌmā-ˈkiŋ\ *noun*
rain out *transitive verb* (1928)
: to interrupt or prevent by rain
rain·proof \ˈrān-ˌprüf\ *adjective* (1831)
: impervious to rain
rain·spout \-ˌspau̇t\ *noun* (1922)
: GUTTER 1a; *also* **:** DOWNSPOUT
rain·squall \-ˌskwȯl\ *noun* (1849)
: a squall accompanied by rain
rain·storm \-ˌstȯrm\ *noun* (1816)
: a storm of or with rain
rain tree *noun* (circa 1890)
: MONKEYPOD
rain·wash \ˈrān-ˌwȯsh, -ˌwäsh\ *noun* (1876)
: the washing away of material by rain; *also* **:** the material so washed away
rain·wa·ter \-ˌwȯ-tər, -ˌwä-\ *noun* (before 12th century)
: water fallen as rain that has not collected soluble matter from the soil and is therefore soft
rain·wear \-ˌwar, -ˌwer\ *noun* (1939)
: waterproof or water-resistant clothing

rainy \ˈrā-nē\ *adjective* **rain·i·er; -est** (before 12th century)
: marked by, abounding with, or bringing rain

rainy day *noun* (circa 1580)
: a period of want or need

[1]**raise** \ˈrāz\ *verb* **raised; rais·ing** [Middle English, from Old Norse *reisa* — more at REAR] (13th century)
transitive verb
1 : to cause or help to rise to a standing position
2 a : AWAKEN, AROUSE **b :** to stir up **:** INCITE ⟨*raise* a rebellion⟩ **c :** to flush (game) from cover **d :** to recall from or as if from death **e :** to establish radio communication with
3 a : to set upright by lifting or building **b :** to lift higher **c :** to place higher in rank or dignity **:** ELEVATE **d :** HEIGHTEN, INVIGORATE ⟨*raise* the spirits⟩ **e :** to end or suspend the operation or validity of ⟨*raise* a siege⟩
4 : to get together for a purpose **:** COLLECT ⟨*raise* funds⟩
5 a : to breed and bring (an animal) to maturity **b :** GROW, CULTIVATE ⟨*raise* cotton⟩ **c :** to bring up (a child) **:** REAR
6 a : to give rise to **:** PROVOKE ⟨*raise* a commotion⟩ **b :** to give voice to ⟨*raise* a cheer⟩
7 : to bring up for consideration or debate ⟨*raise* an issue⟩
8 a : to increase the strength, intensity, or pitch of **b :** to increase the degree of **c :** to cause to rise in level or amount ⟨*raise* the rent⟩ **d** (1) **:** to increase the amount of (a poker bet) (2) **:** to bet more than (a previous bettor) **e** (1) **:** to make a higher bridge bid in (a partner's suit) (2) **:** to increase the bid of (one's partner)
9 : to make light and porous ⟨*raise* dough⟩
10 : to cause to ascend
11 : to multiply (a quantity) by itself a specified number of times
12 : to bring in sight on the horizon by approaching ⟨*raise* land⟩
13 a : to bring up the nap of (cloth) **b :** to cause (as a blister) to form on the skin
14 : to increase the nominal value of fraudulently ⟨*raise* a check⟩
15 : to articulate (a sound) with the tongue in a higher position
intransitive verb
1 *dialect* **:** RISE
2 : to increase a bet or bid ■
synonym see LIFT
— **rais·er** *noun*
— **raise Cain** *or* **raise hell 1 :** to act wildly **:** create a disturbance **2 :** to scold or upbraid someone especially loudly
— **raise eyebrows :** to cause surprise or astonishment

[2]**raise** *noun* (1538)
1 : an act of raising or lifting
2 : a rising stretch of road **:** an upward grade **:** RISE
3 : an increase in amount: as **a :** an increase of a bet or bid **b :** an increase in wages or salary
4 : a vertical or inclined opening or passageway connecting one mine working area with another at a higher level

raised *adjective* (1599)
1 a : done in relief **b :** having a nap
2 : leavened with yeast rather than with baking powder or baking soda

rai·sin \ˈrā-zᵊn\ *noun* [Middle English, from Middle French, grape, from Latin *racemus* cluster of grapes or berries — more at RACEME] (14th century)
: a grape of any of several varieties that has been dried in the sun or by artificial heat

rai·son d'être *also* **rai·son d'etre** \ˌrā-ˌzō^{n}-ˈdetrᵊ\ *noun* [French] (1864)
: reason or justification for existence

raj \ˈräj\ *noun* [Hindi *rāj,* from Sanskrit *rājya;* akin to Sanskrit *rājan* king] (1800)
: RULE; *especially, often capitalized* **:** the former British rule of the Indian subcontinent

ra·ja *or* **ra·jah** \ˈrä-jə, -(ˌ)jä, -zhə, -(ˌ)zhä\ *noun* [Hindi *rājā,* from Sanskrit *rājan* king — more at ROYAL] (1555)
1 : an Indian or Malay prince or chief
2 : the bearer of a title of nobility among the Hindus

Ra·jab \rə-ˈjab\ *noun* [Arabic] (circa 1771)
: the 7th month of the Islamic year — see MONTH table

Ra·jas·tha·ni \ˌrä-jə-ˈstä-nē, ˌrä-zhə-\ *noun* [Hindi *Rājasthānī,* from *Rājasthān* Rajasthan] (1901)
: the Indo-Aryan dialects of Rajasthan

Raj·put *or* **Raj·poot** \ˈräj-ˌpu̇t, ˈräzh-\ *noun* [Hindi *rājpūt,* from Sanskrit *rājaputra* king's son, from *rājan* king + *putra* son — more at FEW] (1598)
: a member of a dominant military caste of northern India

[1]**rake** \ˈrāk\ *noun* [Middle English, from Old English *racu;* akin to Old High German *rehho* rake] (before 12th century)
1 a : an implement equipped with projecting prongs to gather material (as leaves) or for loosening or smoothing the surface of the ground **b :** a machine for gathering hay
2 : an implement like a rake

[2]**rake** *transitive verb* **raked; rak·ing** (13th century)
1 : to gather, loosen, or smooth with or as if with a rake
2 : to gain rapidly or in abundance ⟨*rake* in a fortune⟩
3 a : to touch in passing over lightly **b :** SCRATCH
4 : to censure severely
5 : to search through **:** RANSACK
6 : to sweep the length of especially with gunfire **:** ENFILADE
7 : to glance over rapidly
— **rak·er** *noun*

[3]**rake** *noun* [origin unknown] (1626)
1 : inclination from the perpendicular; *especially* **:** the overhang of a ship's bow or stern
2 : inclination from the horizontal **:** SLOPE
3 : the angle between the top cutting surface of a tool and a plane perpendicular to the surface of the work
4 : the angle between a wing-tip edge that is sensibly straight in planform and the plane of symmetry of an airplane

[4]**rake** *intransitive verb* **raked; rak·ing** (1691)
: to incline from the perpendicular

[5]**rake** *noun* [short for *rakehell*] (1653)
: a dissolute person **:** LIBERTINE

rake·hell \ˈrāk-ˌhel\ *noun* (1554)
: LIBERTINE 2
— **rakehell** *or* **rake·helly** \-ˌhe-lē\ *adjective*

rake–off \ˈrāk-ˌȯf\ *noun* [*rake off,* verb; from the use of a rake by a croupier to collect the operator's profits in a gambling casino] (1888)
: a percentage or cut taken (as by an operator)

rake up *transitive verb* (1581)
: to make known or public **:** UNCOVER ⟨*rake up* a scandal⟩

ra·ki \rə-ˈkē; ˈra-kē, ˈrä-\ *noun* [Turkish, from Arabic *ʽaraqī,* literally, of liquor, from *ʽaraq* liquor, arrack] (1675)
: a Turkish liqueur flavored with aniseed

[1]**rak·ish** \ˈrā-kish\ *adjective* [[5]*rake*] (1706)
: of, relating to, or characteristic of a rake **:** DISSOLUTE

[2]**rakish** *adjective* [probably from [4]*rake;* from the raking masts of pirate ships] (1824)
1 : having a trim or streamlined appearance suggestive of speed ⟨a *rakish* ship⟩
2 : dashingly or carelessly unconventional **:** JAUNTY ⟨*rakish* clothes⟩

rak·ish·ly *adverb* (1838)
: in a rakish manner

rak·ish·ness *noun* (circa 1828)
: the quality or state of being rakish

rale \ˈral, ˈräl\ *noun* [French *râle,* from *râler* to make a rattling sound in the throat] (1828)
: an abnormal sound heard accompanying the normal respiratory sounds on auscultation of the chest

ral·len·tan·do \ˌrä-lən-ˈtän-(ˌ)dō\ *adverb or adjective* [Italian, literally, slowing down, verbal of *rallentare* to slow down again, from *re-* + *allentare* to slow down, from Late Latin, from Latin *al-* ad- + *lentus* slow, pliant — more at LITHE] (circa 1811)
: RITARDANDO

[1]**ral·ly** \ˈra-lē\ *verb* **ral·lied; ral·ly·ing** [French *rallier,* from Old French *ralier,* from *re-* + *alier* to unite — more at ALLY] (1603)
transitive verb
1 a : to muster for a common purpose **b :** to recall to order
2 a : to arouse for action **b :** to rouse from depression or weakness
intransitive verb
1 : to come together again to renew an effort
2 : to join in a common cause
3 : RECOVER, REBOUND
4 : to engage in a rally

[2]**rally** *noun, plural* **rallies** (1651)
1 a : a mustering of scattered forces to renew an effort **b :** a summoning up of strength or courage after weakness or dejection **c :** a recovery of price after a decline **d :** a renewed offensive
2 : a mass meeting intended to arouse group enthusiasm

□ USAGE
raise Some critics still offer the prescription "raise plants and animals and rear children." What is wrong with raising children? Actually, nothing. Sense 5c of *raise* dropped out of fashionable use in England around 1800 but continued in American use. Throughout the 19th century, English reviewers, who commonly found much to fault in the writing of American authors, seem to have included this use of *raise* among other objectionable usages. American literati appear to have been somewhat sensitive to the criticism, and *raise* came to be viewed as something less than elegant. In the opinion of some New Englanders, including Noah Webster, this use of *raise* was a Southernism. The two leading 19th century American dictionaries, Webster's and Worcester's, were both edited by New Englanders, and both labeled *raise* as a Southern usage. So when the question was picked up by handbooks, the dictionaries could be cited as evidence the usage was not standard, but regional. We now know that the description of *raise* as a Southernism was not quite accurate—John Adams had used it. The usage was simply American ⟨grew up, and married, and *raised* a large family —Mark Twain⟩ ⟨the small town in which I was *raised* —Sherwood Anderson⟩ ⟨a very great advantage to have been *raised* in an atmosphere of evangelical piety — Irving Howe⟩. We now know too that *raise* was used of humans in the English of Scotland, Ireland, and some northern and eastern counties of England. And it seems not to have died entirely in the mainstream even in England ⟨she had been *raised* in virtual seclusion —Lady Antonia Fraser⟩. This sense is standard and can be used without a second thought.

3 : a series of shots interchanged between players (as in tennis) before a point is won
4 *also* **ral·lye** [French *rallye,* from English [1]*rally*] **:** an automobile competition using public roads and ordinary traffic rules with the object of maintaining a specified average speed between checkpoints over a route unknown to the participants until the start of the event

[3]rally *transitive verb* **ral·lied; ral·ly·ing** [French *railler* to mock, rally — more at RAIL] (1668)
: to attack with raillery **:** BANTER

ral·ly·ing \'ra-lē-iŋ\ *noun* (1957)
: the sport of driving in automobile rallies

[1]ram \'ram\ *noun* [Middle English, from Old English *ramm;* akin to Old High German *ram*] (before 12th century)
1 a : a male sheep **b** *capitalized* **:** ARIES
2 a : BATTERING RAM **b :** a warship with a heavy beak at the prow for piercing an enemy ship
3 : any of various guided pieces for exerting pressure or for driving or forcing something by impact: as **a :** the plunger of a hydrostatic press or force pump **b :** the weight that strikes the blow in a pile driver

[2]ram *verb* **rammed; ram·ming** [Middle English *rammen,* probably from *ram,* noun] (14th century)
intransitive verb
1 : to strike with violence **:** CRASH
2 : to move with extreme rapidity
transitive verb
1 : to force in by or as if by driving
2 a : to make compact (as by pounding) **b :** CRAM, CROWD
3 : to force passage or acceptance of ⟨*ram* home an idea⟩
4 : to strike against violently
— ram·mer *noun*

RAM \'ram\ *noun* (1957)
: RANDOM-ACCESS MEMORY

Ra·ma \'rä-mə\ *noun* [Sanskrit *Rāma*]
: a deity or deified hero of later Hinduism worshiped as an avatar of Vishnu

Ram·a·dan \'rä-mə-ˌdän, ˌrä-mə-'\ *noun* [Arabic *Ramaḍān*] (circa 1595)
: the 9th month of the Islamic year observed as sacred with fasting practiced daily from dawn to sunset — see MONTH table

ra·mate \'rā-ˌmāt\ *adjective* [Latin *ramus* branch — more at RAMIFY] (1897)
: RAMOSE

[1]ram·ble \'ram-bəl\ *verb* **ram·bled; ram·bling** \-b(ə-)liŋ\ [perhaps from Middle English *romblen,* frequentative of *romen* to roam] (1620)
intransitive verb
1 a : to move aimlessly from place to place **b :** to explore idly
2 : to talk or write in a desultory or long-winded wandering fashion
3 : to grow or extend irregularly
transitive verb
: to wander over **:** ROAM
synonym see WANDER
— ram·bling·ly \-b(ə-)liŋ-lē\ *adverb*

[2]ramble *noun* (1654)
: a leisurely excursion for pleasure; *especially* **:** an aimless walk

ram·bler \'ram-blər\ *noun* (1624)
1 : one that rambles
2 : any of various climbing roses with long flexible canes and rather small often double flowers in large clusters
3 : RANCH HOUSE

ram·bouil·let \ˌram-bə-'lā, -bü-'yā\ *noun, often capitalized* [*Rambouillet,* France] (1906)
: any of a breed of large sturdy sheep developed in France

ram·bunc·tious \ram-'bəŋk-shəs\ *adjective* [probably alteration of *rumbustious*] (1830)
: marked by uncontrollable exuberance **:** UNRULY
— ram·bunc·tious·ly *adverb*
— ram·bunc·tious·ness *noun*

ram·bu·tan \ram-'bü-tᵊn\ *noun* [Malay] (1707)
: a bright red spiny Malayan fruit closely related to the litchi; *also* **:** a tree (*Nephelium lappaceum*) of the soapberry family that bears this fruit

ram·e·kin *also* **ram·e·quin** \'ram-kən, 'ra-mi-\ *noun* [French *ramequin,* from Low German *ramken,* diminutive of *ram* cream] (circa 1706)
1 : a preparation of cheese especially with bread crumbs or eggs baked in a mold or shell
2 : an individual baking dish

ra·met \'rā-ˌmet\ *noun* [Latin *ramus* branch] (1929)
: an independent member of a clone

ra·mie \'rā-mē, 'ra-\ *noun* [Malay *rami*] (1832)
1 : an Asian perennial plant (*Boehmeria nivea*) of the nettle family
2 a : the strong lustrous bast fiber of ramie capable of being spun or woven **b :** fabric made of ramie often resembling linen or silk

ram·i·fi·ca·tion \ˌra-mə-fə-'kā-shən\ *noun* (1665)
1 a : BRANCH, OFFSHOOT **b :** a branched structure
2 a : the act or process of branching **b :** arrangement of branches (as on a plant)
3 : CONSEQUENCE, OUTGROWTH ⟨the *ramifications* of a problem⟩

ram·i·fy \'ra-mə-ˌfī\ *verb* **-fied; -fy·ing** [Middle English *ramifien,* from Middle French *ramifier,* from Medieval Latin *ramificare,* from Latin *ramus* branch; akin to Latin *radix* root — more at ROOT] (15th century)
intransitive verb
1 : to split up into branches or constituent parts
2 : to send forth branches or extensions
transitive verb
1 : to cause to branch
2 : to separate into divisions

Ra·mism \'rā-ˌmi-zəm\ *noun* [Petrus *Ramus* (died 1572) French philosopher] (1710)
: the doctrines of Ramus based on opposition to Aristotelianism and advocacy of a new logic blended with rhetoric
— Ra·mist \-mist\ *noun or adjective*

ram·jet \'ram-ˌjet\ *noun* (1942)
: a jet engine that consists essentially of a hollow tube without mechanical components and depends on the aircraft's speed of flight for the air compression necessary to produce the thrust obtained from the expansion of gases caused by the combustion of fuel

ra·mose \'rā-ˌmōs\ *adjective* [Latin *ramosus,* from *ramus* branch] (1689)
: consisting of or having branches ⟨a *ramose* sponge⟩

[1]ramp \'ramp\ *verb* [Middle English, from Middle French *ramper* to crawl, rear, of Germanic origin; akin to Old High German *rimpfan* to wrinkle — more at RUMPLE] (14th century)
intransitive verb
1 a : to stand or advance menacingly with forelegs or with arms raised **b :** to move or act furiously
2 : to creep up — used especially of plants
transitive verb
[[4]*ramp* (electrical waveform)] **:** to increase or decrease especially at a constant rate — usually used with *up* or *down* ⟨*ramp* up production⟩

[2]ramp *noun* (1671)
: the act or an instance of ramping

[3]ramp *noun* [back-formation from *ramps,* alteration of *rams,* from Middle English, from Old English *hramsa;* akin to Old High German *ramusia* ramp, Greek *krommyon* onion] (1598)
: any of various alliums used for food

[4]ramp *noun* [French *rampe,* from *ramper,* from Middle French] (1778)
1 : a short bend, slope, or curve usually in the vertical plane where a handrail or coping changes its direction
2 : a sloping way: as **a :** a sloping floor, walk, or roadway leading from one level to another **b :** a stairway for entering or leaving an airplane **c :** a slope for launching boats
3 : APRON 2h

[1]ram·page \'ram-ˌpāj, (ˌ)ram-'\ *intransitive verb* **ram·paged; ram·pag·ing** [Scots] (1808)
: to rush wildly about

[2]ram·page \'ram-ˌpāj\ *noun* (1861)
: a course of violent, riotous, or reckless action or behavior
— ram·pa·geous \ram-'pā-jəs\ *adjective*
— ram·pa·geous·ly *adverb*
— ram·pa·geous·ness *noun*

ram·pan·cy \'ram-pən(t)-sē\ *noun* (1664)
: the quality or state of being rampant

ram·pant \'ram-pənt *also* -ˌpant\ *adjective* [Middle English, from Middle French, present participle of *ramper*] (14th century)
1 a : rearing upon the hind legs with forelegs extended **b :** standing on one hind foot with one foreleg raised above the other and the head in profile — used of a heraldic animal
2 a : marked by a menacing wildness, extravagance, or absence of restraint **b :** WIDESPREAD
3 : having one impost or abutment higher than the other ⟨a *rampant* arch⟩
— ram·pant·ly *adverb*

ram·part \'ram-ˌpärt, -pərt\ *noun* [Middle French, from *ramparer* to fortify, from *re-* + *emparer* to defend, from Old Provençal *antparar,* from (assumed) Vulgar Latin *anteparare,* from Latin *ante* before + *parare* to prepare — more at ANTE-, PARE] (1536)
1 : a protective barrier **:** BULWARK
2 : a broad embankment raised as a fortification and usually surmounted by a parapet
3 : a wall-like ridge (as of rock fragments, earth, or debris)

ram·pike \-ˌpīk\ *noun* [origin unknown] (1853)
: an erect broken or dead tree

[1]ram·rod \'ram-ˌräd\ *noun* (1757)
1 : a rod for ramming home the charge in a muzzle-loading firearm
2 : a cleaning rod for small arms
3 : BOSS, OVERSEER

[2]ramrod *adjective* (1905)
: marked by rigidity, severity, or stiffness

[3]ramrod *transitive verb* (circa 1940)
: to direct, supervise, and control

ram·shack·le \'ram-ˌsha-kəl\ *adjective* [alteration of earlier *ransackled,* from past participle of obsolete *ransackle,* frequentative of *ransack*] (1830)
1 : appearing ready to collapse **:** RICKETY
2 : carelessly or loosely constructed

rams·horn \'ramz-ˌhȯrn\ *noun* (1901)
: any of various snails (as genera *Planorbis, Helisoma,* and *Planorbarius*) often used as aquarium scavengers

ra·mus \'rā-məs\ *noun, plural* **ra·mi** \-ˌmī\ [New Latin, from Latin, branch — more at RAMIFY] (1803)
: a projecting part, elongated process, or branch: as **a :** the posterior more or less vertical part on each side of the lower jaw that articulates with the skull **b :** a branch of a nerve

ran *past of* RUN

[1]ranch \'ranch\ *noun* [Mexican Spanish *rancho* small ranch, from Spanish, camp, hut & Spanish dialect, small farm, from Old Spanish *ranchearse* to take up quarters, from Middle French *se ranger* to take up a position, from *ranger* to set in a row — more at RANGE] (1831)
1 : a large farm for raising horses, beef cattle, or sheep
2 : a farm or area devoted to a particular specialty

3 : RANCH HOUSE ◆

²ranch (1866)
intransitive verb
: to live or work on a ranch
transitive verb
1 : to work as a rancher on
2 : to raise on a ranch

ranch dressing *noun* (1984)
: a creamy salad dressing usually containing milk or buttermilk and mayonnaise

ranch·er \'ran-chər\ *noun* (1836)
: one who owns or works on a ranch

ran·che·ro \ran-'cher-(,)ō, rän-\ *noun, plural* **-ros** [Mexican Spanish, from *rancho*] (1826)
: RANCHER; *also* **:** RANCH 1

ranch house *noun* (1862)
1 : the main dwelling house on a ranch
2 : a one-story house typically with a low-pitched roof and an open plan
word history see RANCH

ranch·man \'ranch-mən\ *noun* (1856)
: RANCHER

ran·cho \'ran-(,)chō, 'rän-\ *noun, plural* **ran·chos** [Mexican Spanish, small ranch] (1808)
: RANCH 1
word history see RANCH

ran·cid \'ran(t)-səd\ *adjective* [Latin *rancidus,* from *rancēre* to be rancid] (1646)
1 : having a rank smell or taste
2 : OFFENSIVE
— ran·cid·i·ty \ran-'si-də-tē\ *noun*
— ran·cid·ness \'ran(t)-səd-nəs\ *noun*

ran·cor \'raŋ-kər, -,kȯr\ *noun* [Middle English *rancour,* from Middle French *ranceur,* from Late Latin *rancor* rancidity, rancor, from Latin *rancēre*] (14th century)
: bitter deep-seated ill will
synonym see ENMITY

ran·cor·ous \'raŋ-k(ə-)rəs\ *adjective* (1590)
: marked by rancor
— ran·cor·ous·ly *adverb*

ran·cour *British variant of* RANCOR

rand \'rand, 'ränd, 'ränt\ *noun, plural* **rand** [the *Rand,* South Africa] (circa 1932)
1 — see MONEY table
2 : a former monetary unit of Botswana, Lesotho, and Swaziland

R and D *noun* (1966)
: research and development

¹ran·dom \'ran-dəm\ *noun* [Middle English, impetuosity, from Middle French *randon,* from Old French, from *randir* to run, of Germanic origin; akin to Old High German *rinnan* to run — more at RUN] (1561)
: a haphazard course
— at random : without definite aim, direction, rule, or method

²random *adjective* (1565)
1 a : lacking a definite plan, purpose, or pattern **b :** made, done, or chosen at random ⟨read *random* passages from the book⟩
2 a : relating to, having, or being elements or events with definite probability of occurrence ⟨*random* processes⟩ **b :** being or relating to a set or to an element of a set each of whose elements has equal probability of occurrence ⟨a *random* sample⟩; *also* **:** characterized by procedures designed to obtain such sets or elements ⟨*random* sampling⟩ ☆
— ran·dom·ly *adverb*
— ran·dom·ness *noun*

³random *adverb* (1618)
: in a random manner

random–access *adjective* (1953)
: permitting access to stored data in any order the user desires

random–access memory *noun* (1955)
: a computer memory that provides the main internal storage available to the user for programs and data — called also *RAM;* compare READ-ONLY MEMORY

ran·dom·ize \'ran-də-,mīz\ *transitive verb* **-ized; -iz·ing** (1926)
: to select, assign, or arrange in a random way
— ran·dom·i·za·tion \,ran-də-mə-'zā-shən\ *noun*
— ran·dom·iz·er *noun*

randomized block *noun* (circa 1942)
: an experimental design (as in horticulture) in which different treatments are distributed in random order in a block or plot — called also *randomized block design*

random variable *noun* (1949)
: a variable that is itself a function of the result of a statistical experiment in which each outcome has a definite probability of occurrence — called also *variate*

random walk *noun* (1941)
: a process (as Brownian motion or genetic drift) consisting of a sequence of steps (as movements or changes in gene frequency) each of whose characteristics (as magnitude and direction) is determined by chance

¹randy \'ran-dē\ *adjective* [probably from obsolete *rand* to rant] (1698)
1 *chiefly Scottish* **:** having a coarse manner
2 : LUSTFUL, LECHEROUS

²randy *noun, plural* **rand·ies** (1762)
chiefly Scottish **:** a scolding or dissolute woman

rang *past of* RING

¹range \'rānj\ *noun, often attributive* [Middle English, row of persons, from Middle French *renge,* from Old French *rengier* to range] (14th century)
1 a (1) **:** a series of things in a line **:** ROW (2) **:** a series of mountains (3) **:** one of the north-south rows of townships in a U.S. public-land survey that are numbered east and west from the principal meridian of the survey **b :** an aggregate of individuals in one order **c :** a direction line
2 : a cooking stove that has an oven and a flat top with burners or heating elements
3 a : a place that may be ranged over **b :** an open region over which animals (as livestock) may roam and feed **c :** the region throughout which a kind of organism or ecological community naturally lives or occurs
4 : the act of ranging about
5 a (1) **:** the horizontal distance to which a projectile can be propelled (2) **:** the horizontal distance between a weapon and target **b :** the maximum distance a vehicle or craft can travel without refueling **c** (1) **:** a place where shooting is practiced (2) **:** DRIVING RANGE
6 a : the space or extent included, covered, or used **:** SCOPE **b :** the extent of pitch covered by a melody or lying within the capacity of a voice or instrument
7 a : a sequence, series, or scale between limits ⟨a wide *range* of patterns⟩ **b :** the limits of a series **:** the distance or extent between possible extremes **c :** the difference between the least and greatest values of an attribute or of the variable of a frequency distribution
8 a : the set of values a function may take on **b :** the class of admissible values of a variable
9 : LINE 11 ☆

²range *verb* **ranged; rang·ing** [Middle English, from Middle French *ranger,* from Old French *rengier,* from *renc, reng* line, place, row — more at RANK] (14th century)
transitive verb
1 a : to set in a row or in the proper order **b :** to place among others in a position or situation **c :** to assign to a category **:** CLASSIFY
2 a : to rove over or through **b :** to sail or pass along
3 : to arrange (an anchor cable) on deck
4 : to graze (livestock) on a range
5 : to determine or give the elevation necessary for (a gun) to propel a projectile to a given distance
intransitive verb
1 a : to roam at large or freely **b :** to move over an area so as to explore it
2 : to take a position
3 a : to correspond in direction or line **:** ALIGN **b :** to extend in a particular direction
4 : to have range
5 : to change or differ within limits
6 *of an organism* **:** to live or occur in or be native to a region
7 : to obtain the range of an object by instrument (as radar or laser)
word history see RANCH

range finder *noun* (1872)
1 : an instrument used in gunnery to determine the distance of a target

☆ SYNONYMS

Random, haphazard, casual mean determined by accident rather than design. RANDOM stresses lack of definite aim, fixed goal, or regular procedure ⟨a *random* selection of books⟩. HAPHAZARD applies to what is done without regard for regularity or fitness or ultimate consequence ⟨a *haphazard* collection of rocks⟩. CASUAL suggests working or acting without deliberation, intention, or purpose ⟨a *casual* collector⟩.

Range, gamut, compass, sweep, scope, orbit mean the extent that lies within the powers of something (as to cover or control). RANGE is a general term indicating the extent of one's perception or the extent of powers, capacities, or possibilities ⟨the entire *range* of human experience⟩. GAMUT suggests a graduated series running from one possible extreme to another ⟨a performance that ran the *gamut* of emotions⟩. COMPASS implies a sometimes limited extent of perception, knowledge, or activity ⟨your concerns lie beyond the narrow *compass* of this study⟩. SWEEP suggests extent, often circular or arc-shaped, of motion or activity ⟨the book covers the entire *sweep* of criminal activity⟩. SCOPE is applicable to an area of activity, predetermined and limited, but somewhat flexible ⟨as time went on, the *scope* of the investigation widened⟩. ORBIT suggests an often circumscribed range of activity or influence within which forces work toward accommodation ⟨within that restricted *orbit* they tried to effect social change⟩.

◇ WORD HISTORY

ranch Old French *renc* or *ranc* "row, line" (the source of English *rank*) gave rise to a verb *rengier* "to set in rows, place in proper order" (the source of English *range*). The reflexive form of this verb in Middle French, *se ranger,* meant "to line up, take a position." By the 16th century, *se ranger* had found its way into Spanish as *rancharse,* which meant in military contexts "to take up quarters, be billeted." The conquistadores who took this verb to the Americas were also familiar with a derivative noun *rancho,* denoting originally a camp or temporary dwelling such as a hut. The word *rancho* remains widely used in American Spanish, with much local variation in meaning. Both the general sense "habitation" and the specifically Mexican Spanish sense "small farm" were adopted into American English in the 19th century directly as *rancho* and in the Anglicized form *ranch.* By the late 19th century, *ranch* most often denoted not a small farm but a large tract of land devoted to livestock-raising—a meaning the word still maintains. The one-story, rambling dwelling typical of ranches was the prototype of that staple of American domestic architecture, the *ranch house.*

2 : a surveying instrument (as a transit) for determining quickly the distances, bearings, and elevations of distant objects
3 : a usually built-in adjustable optical device for focusing a camera that automatically indicates the correct focus (as when two parts of a split image are brought together)

range·land \'rānj-ˌland\ *noun* (1935)
: land used or suitable for range

rang·er \'rān-jər\ *noun* (14th century)
1 a : the keeper of a British royal park or forest **b :** FOREST RANGER
2 : one that ranges
3 a : one of a body of organized armed men who range over a region especially to enforce the law **b :** a soldier specially trained in close-range fighting and in raiding tactics

rangy \'rān-jē\ *adjective* **rang·i·er; -est** (1868)
1 : able to range for considerable distances
2 a : long-limbed and long-bodied ⟨*rangy* cattle⟩ **b :** being tall and slender
3 : having room for ranging
4 : having great scope
— rang·i·ness *noun*

ra·ni *or* **ra·nee** \rä-'nē, 'rä-ˌnē\ *noun* [Hindi *rānī,* from Sanskrit *rājñī,* feminine of *rājan* king — more at ROYAL] (1673)
: a Hindu queen **:** a rajah's wife

ra·nid \'ra-nəd, 'rā-\ *noun* [ultimately from Latin *rana* frog] (circa 1934)
: any of a large family (Ranidae) of frogs distinguished by slightly dilated transverse sacral processes

¹rank \'raŋk\ *adjective* [Middle English, from Old English *ranc* overbearing, strong; akin to Old Norse *rakkr* erect and perhaps to Old English *riht* right — more at RIGHT] (13th century)
1 : luxuriantly or excessively vigorous in growth
2 : offensively gross or coarse **:** FOUL
3 *obsolete* **:** grown too large
4 a : shockingly conspicuous ⟨must lecture him on his *rank* disloyalty —David Walden⟩ **b :** OUTRIGHT — used as an intensive ⟨*rank* beginners⟩
5 *archaic* **:** LUSTFUL, RUTTISH
6 : offensive in odor or flavor; *especially* **:** RANCID
7 : PUTRID, FESTERING
8 : high in amount or degree **:** FRAUGHT
synonym see MALODOROUS, FLAGRANT
— rank·ly *adverb*
— rank·ness *noun*

²rank *noun* [Middle English, from Middle French *renc, reng,* of Germanic origin; akin to Old High German *hring* ring — more at RING] (14th century)
1 a : ROW, SERIES **b :** a row of people **c** (1) **:** a line of soldiers ranged side by side in close order (2) *plural* **:** ARMED FORCES (3) *plural* **:** the body of enlisted personnel **d :** any of the rows of squares that extend across a chessboard perpendicular to the files **e** *British* **:** STAND 6
2 a : relative standing or position **b :** a degree or position of dignity, eminence, or excellence **:** DISTINCTION ⟨soon took *rank* as a leading attorney —J. D. Hicks⟩ **c :** high social position ⟨the privileges of *rank*⟩ **d :** a grade of official standing in a hierarchy
3 : an orderly arrangement **:** FORMATION
4 : an aggregate of individuals classed together — usually used in plural
5 : the order according to some statistical characteristic (as the score on a test)
6 : any of a series of classes of coal based on increasing alteration of the parent vegetable matter, increasing carbon content, and increasing fuel value
7 : the number of linearly independent rows or columns in a matrix
word history see RANCH

³rank (1573)
transitive verb
1 : to arrange in lines or in a regular formation
2 : to determine the relative position of **:** RATE
3 : to take precedence of
intransitive verb
1 : to form or move in ranks
2 : to take or have a position in relation to others

rank and file *noun* (1598)
1 : the enlisted personnel of an armed force
2 : the individuals who constitute the body of an organization, society, or nation as distinguished from the leaders
— rank–and–file \ˌraŋk-ᵊn-'fīl\ *adjective*
— rank and fil·er \-'fī-lər\ *noun*

rank correlation *noun* (1907)
: a measure of correlation depending on rank

rank·er \'raŋ-kər\ *noun* (1878)
: one who serves or has served in the ranks; *especially* **:** a commissioned officer promoted from the ranks

Ran·kine \'raŋ-kən\ *adjective* [William J. M. *Rankine* (died 1872) Scottish engineer & physicist] (circa 1926)
: being, according to, or relating to an absolute-temperature scale on which the unit of measurement equals a Fahrenheit degree and on which the freezing point of water is 491.67° and the boiling point 671.67°

rank·ing *adjective* (1862)
: having a high position: as **a :** FOREMOST ⟨*ranking* poet⟩ **b :** being next to the chairman in seniority ⟨*ranking* committee member⟩

ran·kle \'raŋ-kəl\ *verb* **ran·kled; ran·kling** \-k(ə-)liŋ\ [Middle English *ranclen* to fester, from Middle French *rancler,* from Old French *draoncler, raoncler,* from *draoncle, raoncle* festering sore, from Medieval Latin *dracunculus,* from Latin, diminutive of *draco* serpent — more at DRAGON] (1606)
intransitive verb
1 : to cause anger, irritation, or deep bitterness
2 : to feel anger and irritation
transitive verb
: to cause irritation or bitterness in ◆

ran·sack \'ran-ˌsak, (ˌ)ran-'\ *transitive verb* [Middle English *ransaken,* from Old Norse *rannsaka,* from *rann* house + *-saka* (akin to Old English *sēcan* to seek) — more at SEEK] (13th century)
1 a : to search thoroughly **b :** to examine closely and carefully
2 : to search through to commit robbery **:** PLUNDER
— ran·sack·er *noun*

¹ran·som \'ran(t)-səm\ *noun* [Middle English *ransoun,* from Old French *rançon,* from Latin *redemption-, redemptio* — more at REDEMPTION] (13th century)
1 : a consideration paid or demanded for the release of someone or something from captivity
2 : the act of ransoming

²ransom *transitive verb* (14th century)
1 : to deliver especially from sin or its penalty
2 : to free from captivity or punishment by paying a price
synonym see RESCUE
— ran·som·er *noun*

¹rant \'rant\ *verb* [obsolete Dutch *ranten, randen*] (1602)
intransitive verb
1 : to talk in a noisy, excited, or declamatory manner
2 : to scold vehemently
transitive verb
: to utter in a bombastic declamatory fashion
— rant·er *noun*
— rant·ing·ly \'ran-tiŋ-lē\ *adverb*

²rant *noun* (1649)
1 a : a bombastic extravagant speech **b :** bombastic extravagant language
2 *dialect British* **:** a rousing good time

ran·u·la \'ran-yə-lə\ *noun* [New Latin, from Latin, swelling on the tongue of cattle, from diminutive of *rana* frog] (15th century)
: a cyst formed under the tongue by obstruction of a gland duct

ra·nun·cu·lus \rə-'nəŋ-kyə-ləs\ *noun, plural* **-lus·es** *or* **-li** \-ˌlī, -ˌlē\ [New Latin, from Latin, from diminutive of *rana* frog] (1543)
: BUTTERCUP

¹rap \'rap\ *noun* [Middle English *rappe*] (14th century)
1 : a sharp blow or knock
2 a : a sharp rebuke or criticism **b :** a negative and often undeserved reputation or charge — usually used with *bum* or *bad* ⟨given a bum *rap* by the press⟩
3 a : the responsibility for or adverse consequences of an action ⟨refused to take the *rap*⟩ **b :** a criminal charge **c :** a prison sentence

²rap *verb* **rapped; rap·ping** (14th century)
transitive verb
1 : to strike with a sharp blow
2 : to utter suddenly and forcibly
3 : to cause to be or come by raps ⟨*rap* the meeting to order⟩
4 : to criticize sharply
intransitive verb
1 : to strike a quick sharp blow
2 : to make a short sharp sound

³rap *transitive verb* **rapped** *also* **rapt** \'rapt\; **rap·ping** [back-formation from *rapt*] (1528)
1 : to snatch away or upward
2 : ENRAPTURE

⁴rap *noun* [perhaps from ¹*rap*] (1834)
: a minimum amount or degree (as of care or consideration) **:** the least bit ⟨doesn't care a *rap*⟩

⁵rap *intransitive verb* **rapped; rap·ping** [perhaps from ¹*rap*] (1929)
1 : to talk freely and frankly
2 : to perform rap music

⁶rap *noun* (1967)
1 : TALK, CONVERSATION; *also* **:** a line of talk **:** PATTER
2 a : a rhythmic chanting often in unison of usually rhymed couplets to a musical accompaniment **b :** a piece so performed

ra·pa·cious \rə-'pā-shəs\ *adjective* [Latin *rapac-, rapax,* from *rapere* to seize — more at RAPID] (1651)
1 : excessively grasping or covetous
2 : living on prey
3 : RAVENOUS
synonym see VORACIOUS
word history see RAPT
— ra·pa·cious·ly *adverb*
— ra·pa·cious·ness *noun*

ra·pac·i·ty \rə-'pa-sə-tē\ *noun* (1543)
: the quality of being rapacious

¹rape \'rāp\ *noun* [Middle English, from Latin *rapa, rapum* turnip, rape; akin to Old High German *rāba* turnip, rape, Lithuanian *ropė*] (14th century)
: a European herb (*Brassica napus*) of the mustard family grown as a forage crop for sheep and hogs and for its seeds which yield rapeseed oil and are a bird food — compare CANOLA

◇ WORD HISTORY

rankle The modern meaning of the verb *rankle* "to cause anger, irritation, or bitterness" is a figurative extension of the earlier meaning "to fester, swell," said of an ulcer or festering sore. The Middle English predecessor of this verb, *ranclen* or *ranklen,* was a loanword from Middle French *rancler,* contracted from Old French *draoncler* or *raoncler,* a derivative of *draoncle* "festering sore." *Draoncle* is descended from Latin *dracunculus,* which in classical Latin is a diminutive of *draco* "snake," though in spoken Latin, to judge by its offspring in Romance languages, it must also have had the meaning "festering sore." The comparison of something both painful and destructive of body tissue to a biting animal is not surprising; another example is the use of *cancer* "crab" in Latin to denote gangrene or a malignant tumor.

word history see RAPT

[2]rape *transitive verb* **raped; rap·ing** [Middle English, from Latin *rapere*] (14th century)
1 a *archaic* **:** to seize and take away by force **b :** DESPOIL
2 : to commit rape on
— rap·er *noun*
— rap·ist \ˈrā-pist\ *noun*

[3]rape *noun* (14th century)
1 : an act or instance of robbing or despoiling or carrying away a person by force
2 a : sexual intercourse with a woman by a man without her consent and chiefly by force or deception — compare STATUTORY RAPE **b :** unlawful sexual intercourse by force or threat other than by a man with a woman
3 : an outrageous violation

[4]rape *noun* [French *râpe* grape stalk] (1657)
: grape pomace

rape·seed \ˈrāp-ˌsēd\ *noun* (1577)
: the seed of the rape plant

rapeseed oil *noun* (1816)
: a nondrying or semidrying oil obtained from rapeseed and turnip seed and used chiefly as a lubricant, illuminant, and food — called also *rape oil;* compare CANOLA OIL

Ra·pha·el \ˈra-fē-əl, ˈrā-, -ˌel\ *noun* [Late Latin, from Greek *Rhaphaēl,* from Hebrew *Rĕphāʼēl*]
: one of the four archangels named in Hebrew tradition

ra·phe \ˈrā-(ˌ)fē\ *noun* [New Latin, from Greek *rhaphē* seam, from *rhaptein* to sew] (circa 1753)
1 : the seamlike union of the two lateral halves of a part or organ (as the tongue) having externally a ridge or furrow
2 a : the part of the stalk of an anatropous ovary that is united in growth to the outside covering and forms a ridge along the body of the ovule **b :** the median line or slit of a diatom's valve

ra·phia \ˈrā-fē-ə, ˈra-\ *noun* [New Latin, genus of palms, from Malagasy *rafia* raffia] (circa 1866)
: RAFFIA

raph·ide \ˈra-ˌfīd\ *noun, plural* **raph·ides** \ˈra-ˌfīdz, ˈra-fə-ˌdēz\ [French & New Latin; French *raphide,* from New Latin *raphides,* plural, from Greek *rhaphides,* plural of *rhaphid-, rhaphis* needle, from *rhaptein*] (circa 1842)
: any of the needle-shaped crystals usually of calcium oxalate that develop as metabolic by-products in plant cells

[1]rap·id \ˈra-pəd\ *adjective* [Latin *rapidus* seizing, sweeping, rapid, from *rapere* to seize, sweep away; akin to Lithuanian *aprėpti* to embrace] (1634)
: marked by a fast rate of motion, activity, succession, or occurrence
synonym see FAST
— rap·id·ly *adverb*
— rap·id·ness *noun*

[2]rapid *noun* (1765)
: a part of a river where the current is fast and the surface is usually broken by obstructions — usually used in plural but singular or plural in construction

rapid eye movement *noun* (1916)
: a rapid conjugate movement of the eyes associated especially with REM sleep

rapid eye movement sleep *noun* (1965)
: REM SLEEP

rap·id–fire \ˌra-pəd-ˈfīr\ *adjective* (1890)
1 : firing or adapted for firing shots in rapid succession
2 : marked by rapidity, liveliness, or sharpness

ra·pid·i·ty \rə-ˈpi-də-tē, ra-\ *noun* (1654)
: the quality or state of being rapid

rapid transit *noun* (1873)
: fast passenger transportation (as by subway) in urban areas

ra·pi·er \ˈrā-pē-ər\ *noun* [Middle French (*espee*) *rapiere*] (1553)
: a straight 2-edged sword with a narrow pointed blade

rap·ine \ˈra-pən, -ˌpīn\ *noun* [Middle English *rapyne,* from Latin *rapina,* from *rapere* to seize, rob] (15th century)
: PILLAGE, PLUNDER

ra·pi·ni *also* **rap·pi·ni** \ra-ˈpē-nē\ *noun* [Italian *rapini,* plural of *rapino,* diminutive of *rapo* turnip, from Latin *rapum* — more at RAPE] (1942)
: BROCCOLI RABE

rap·pa·ree \ˌra-pə-ˈrē\ *noun* [Irish *rapaire, ropaire,* literally, thruster, stabber, from *rop* thrust, stab] (1690)
1 : an Irish irregular soldier or bandit
2 : VAGABOND, PLUNDERER

rap·pee \ra-ˈpā\ *noun* [French (*tabac*) *râpé,* literally, grated tobacco] (circa 1740)
: a pungent snuff made from dark tobacco leaves

rap·pel \rə-ˈpel, ra-\ *intransitive verb* **-pelled** *also* **-peled; -pel·ling** *also* **-pel·ing** [French, literally, recall, from Old French *rapel,* from *rapeler* to recall, from *re-* + *apeler* to appeal, call — more at APPEAL] (1944)
: to descend (as from a cliff) by sliding down a rope passed under one thigh, across the body, and over the opposite shoulder or through a special friction device
— rappel *noun*

rap·pen \ˈräp-ən\ *noun, plural* **rappen** [German, from German dialect, literally, raven] (1838)
: the centime of Switzerland

rap·per \ˈra-pər\ *noun* (1640)
: one that raps or is used for rapping: as **a :** a door knocker **b :** a performer of rap music

rap·port \ra-ˈpōr, rə-, -ˈpȯr\ *noun* [French, from *rapporter* to bring back, refer, from Old French *raporter* to bring back, from *re-* + *aporter* to bring, from Latin *apportare,* from *ad-* ad- + *portare* to carry — more at FARE] (circa 1661)
: RELATION; *especially* **:** relation marked by harmony, conformity, accord, or affinity

rap·por·teur \ˌra-ˌpōr-ˈtər, -ˌpȯr-\ *noun* [Middle French, from *rapporter* to bring back, report] (circa 1500)
: a person who gives reports (as at a meeting of a learned society)

rap·proche·ment \ˌra-ˌprōsh-ˈmä^{n}, -ˌprȯsh-; ra-ˈprōsh-ˌ\ *noun* [French, from *rapprocher* to bring together, from Middle French, from *re-* + *approcher* to approach, from Old French *aprochier*] (1809)
: establishment of or state of having cordial relations

rap·scal·lion \rap-ˈskal-yən\ *noun* [alteration of earlier *rascallion,* irregular from *rascal*] (1699)
: RASCAL, NE'ER-DO-WELL

rap sheet *noun* (1960)
: a police arrest record especially for an individual

rapt \ˈrapt\ *adjective* [Middle English, from Latin *raptus,* past participle of *rapere* to seize — more at RAPID] (14th century)
1 : lifted up and carried away
2 : transported with emotion **:** ENRAPTURED
3 : wholly absorbed **:** ENGROSSED ◆
— rapt·ly \ˈrap(t)-lē\ *adverb*
— rapt·ness \ˈrap(t)-nəs\ *noun*

rap·tor \ˈrap-tər, -ˌtȯr\ *noun* [New Latin *Raptores,* former order name, from Latin, plural of *raptor* plunderer, from *rapere*] (1873)
: BIRD OF PREY
word history see RAPT

rap·to·ri·al \rap-ˈtōr-ē-əl, -ˈtȯr-\ *adjective* (1823)
1 : PREDACEOUS 1
2 : adapted to seize prey
3 : of, relating to, or being a bird of prey

[1]rap·ture \ˈrap-chər\ *noun* [Latin *raptus*] (1629)
1 a : a state or experience of being carried away by overwhelming emotion **b :** a mystical experience in which the spirit is exalted to a knowledge of divine things
2 : an expression or manifestation of ecstasy or passion
synonym see ECSTASY
word history see RAPT
— rap·tur·ous \ˈrap-chə-rəs, ˈrap-shrəs\ *adjective*
— rap·tur·ous·ly *adverb*
— rap·tur·ous·ness *noun*

[2]rapture *transitive verb* **rap·tured; rap·tur·ing** (1637)
: ENRAPTURE

rapture of the deep (1953)
: NITROGEN NARCOSIS

ra·ra avis \ˌrar-ə-ˈā-vəs, ˌrer-; ˌrär-ə-ˈä-wəs\ *noun, plural* **ra·ra avis·es** \-ˈā-və-səz\ *or* **ra·rae aves** \ˌrär-ˌī-ˈä-ˌwās\ [Latin, rare bird] (1607)
: RARITY 2

[1]rare \ˈrar, ˈrer\ *adjective* **rar·er; rar·est** [Middle English, from Latin *rarus*] (14th century)
1 : marked by wide separation of component particles **:** THIN ⟨*rare* air⟩
2 a : marked by unusual quality, merit, or appeal **:** DISTINCTIVE **b :** superlative or extreme of its kind
3 : seldom occurring or found **:** UNCOMMON
synonym see CHOICE, INFREQUENT
— rare·ness *noun*

[2]rare *adjective* **rar·er; rar·est** [alteration of earlier *rere,* from Middle English, from Old English *hrēre* boiled lightly; akin to Old English *hrēran* to stir, Old High German *hruoren*] (1784)
: cooked so that the inside is still red ⟨*rare* roast beef⟩

rare bird *noun* (1890)
: RARITY 2, RARA AVIS

rare·bit \ˈrar-bət, ˈrer-\ *noun* [(*Welsh*) *rarebit*] (circa 1785)
: WELSH RABBIT

rare earth *noun* (1875)
1 : any of a group of similar oxides of metals or a mixture of such oxides occurring together in widely distributed but relatively scarce minerals

◇ WORD HISTORY

rapt Current usage of the adjective *rapt* reflects its history only dimly. It is borrowed from Latin *raptus,* the past participle of *rapere* "to seize, carry off." This Latin verb is the ultimate source of a number of other English words, including *rapacious* and *raptor,* which still convey something of the idea of seizing, as well as *rapture,* now more usual in the figurative sense of being carried away with emotional ecstasy. In Middle English *rapt* simply functioned as the past participle of *rapen,* which meant both "to rape" or "to abduct (a woman) with the intent of raping her" and, with a wholly different connotation, "to carry off" or "to transport" by divine agency. Fifteenth century authors use the latter sense when they speak of a saint being "rapt in to paradys" or of someone whose body "was rapte up fro the erthe." In early Modern English *rape,* with its long vowel, and *rapt,* with its short vowel, went their own separate ways. While *rape* maintained its brutal literal sense, *rapt* came to mean solely "carried away in spirit" and hence "transported by emotion" or "wholly absorbed or engrossed," which are its usual modern senses.

2 : RARE EARTH ELEMENT

rare earth element *noun* (1942)
: any of a series of metallic elements of which the oxides are classed as rare earths and which include the elements of the lanthanide series and sometimes yttrium and scandium — called also *rare earth metal*

rar·ee–show \'rar-ē-ˌshō, 'rer-\ *noun* [alteration of *rare show*] (1684)
: a small display or scene viewed in a box **:** PEEP SHOW; *broadly* **:** an unusual or amazing show or spectacle

rar·e·fac·tion \ˌrar-ə-'fak-shən, ˌrer-\ *noun* [French or Medieval Latin; French *raréfaction,* from Medieval Latin *rarefaction-, rarefactio,* from Latin *rarefacere* to rarefy] (1603)
1 : the action or process of rarefying
2 : the quality or state of being rarefied
3 : a state or region of minimum pressure in a medium traversed by compressional waves (as sound waves)
— rar·e·fac·tion·al \-shnəl, -shə-nᵊl\ *adjective*

rar·e·fied *also* **rar·i·fied** \'rar-ə-ˌfīd, 'rer-\ *adjective* (1941)
1 : of, relating to, or interesting to a select group **:** ESOTERIC
2 : very high

rar·e·fy *also* **rar·i·fy** \-ə-ˌfī\ *verb* **-fied; -fy·ing** [Middle English *rarefien, rarifien,* from Middle French *rarefier,* modification of Latin *rarefacere,* from *rarus* rare + *facere* to make — more at DO] (14th century)
transitive verb
1 : to make rare, thin, porous, or less dense **:** to expand without the addition of matter
2 : to make more spiritual, refined, or abstruse
intransitive verb
: to become less dense

rare·ly \'rar-lē, 'rer-\ *adverb* (1552)
1 : not often **:** SELDOM
2 : with rare skill **:** EXCELLENTLY
3 : in an extreme or exceptional manner

rare·ripe \'rar-ˌrīp, 'rer-\ *noun* [English dialect *rare* early + English *ripe*] (1722)
1 : an early ripening fruit or vegetable
2 *dialect* **:** GREEN ONION

rar·ing \'rer-ən, 'rer-, -iŋ\ *adjective* [from present participle of English dialect *rare* to rear, alteration of English *rear*] (1909)
: full of enthusiasm or eagerness ⟨ready and *raring* to go⟩

rar·i·ty \'rar-ə-tē, 'rer-\ *noun, plural* **-ties** (1542)
1 : the quality, state, or fact of being rare
2 : one that is rare

ras·bo·ra \raz-'bōr-ə, -'bȯr-\ *noun* [New Latin] (1931)
: any of a genus (*Rasbora*) of tiny brilliantly colored cyprinid freshwater fishes often kept in tropical aquariums

ras·cal \'ras-kəl\ *noun* [Middle English *rascaile* rabble, one of the rabble] (15th century)
1 : a mean, unprincipled, or dishonest person
2 : a mischievous person or animal
— rascal *adjective*

ras·cal·i·ty \ra-'ska-lə-tē\ *noun, plural* **-ties** (circa 1577)
1 : RABBLE
2 a : the character or actions of a rascal **:** KNAVERY **b :** a rascally act

ras·cal·ly \'ras-kə-lē\ *adjective* (1594)
: of or characteristic of a rascal
— rascally *adverb*

rase \'rāz\ *transitive verb* **rased; ras·ing** [Middle English, from Middle French *raser,* from (assumed) Vulgar Latin *rasare,* frequentative of Latin *radere* to scrape, shave — more at RODENT] (14th century)
1 *archaic* **:** ERASE
2 *archaic* **:** RAZE 2

¹rash \'rash\ *adverb* [Middle English (northern dialect) *rasch* quickly; akin to Old High German *rasc* fast] (15th century)
archaic **:** in a rash manner

²rash *adjective* (1509)
1 : marked by or proceeding from undue haste or lack of deliberation or caution
2 *obsolete* **:** quickly effective
synonym see ADVENTUROUS
— rash·ly *adverb*
— rash·ness *noun*

³rash *noun* [obsolete French *rache* scurf, from (assumed) Vulgar Latin *rasica,* from *rasicare* to scratch, from Latin *rasus,* past participle of *radere*] (1709)
1 : an eruption on the body
2 : a large number of instances in a short period ⟨a *rash* of complaints⟩

rash·er \'ra-shər\ *noun* [perhaps from obsolete *rash* to cut, from Middle English *rashen*] (1592)
: a thin slice of bacon or ham broiled or fried; *also* **:** a portion consisting of several such slices

¹rasp \'rasp\ *verb* [Middle English, from (assumed) Middle French *rasper,* of Germanic origin; akin to Old High German *raspōn* to scrape together] (14th century)
transitive verb
1 : to rub with something rough; *specifically* **:** to abrade with a rasp
2 : to grate upon **:** IRRITATE
3 : to utter in a raspy tone
intransitive verb
1 : SCRAPE
2 : to produce a grating sound
— rasp·er *noun*
— rasp·ing·ly \'ras-piŋ-lē\ *adverb*

²rasp *noun* (circa 1512)
1 : a coarse file with cutting points instead of lines
2 : something used for rasping
3 a : an act of rasping **b :** a rasping sound, sensation, or effect

rasp·ber·ry \'raz-ˌber-ē, -b(ə-)rē\ *noun* [English dialect *rasp* raspberry + English *berry*] (circa 1616)
1 a : any of various usually black or red edible berries that are aggregate fruits consisting of numerous small drupes on a fleshy receptacle and that are usually rounder and smaller than the closely related blackberries **b :** a plant (genus *Rubus*) of the rose family that bears raspberries
2 [short for *raspberry tart,* rhyming slang for *fart*] **:** a sound of contempt made by protruding the tongue between the lips and expelling air forcibly to produce a vibration; *broadly* **:** an expression of disapproval or contempt

raspy \'ras-pē\ *adjective* (1838)
1 : HARSH, GRATING
2 : IRRITABLE

Ras·ta \'ras-tə, 'räs-\ *noun* (1955)
: RASTAFARIAN
— Rasta *adjective*

Ras·ta·far·i·an \ˌras-tə-'far-ē-ən, ˌräs-tə-'fär-\ *noun* [*Ras Tafari,* precoronation name of Haile Selassie] (1955)
: an adherent of Rastafarianism
— Rastafarian *adjective*

Ras·ta·far·i·an·ism \-ē-ə-ˌni-zəm\ *noun* (1968)
: a religious cult among black Jamaicans that teaches the eventual redemption of blacks and their return to Africa, employs the ritualistic use of marijuana, forbids the cutting of hair, and venerates Haile Selassie as a god

ras·ter \'ras-tər\ *noun* [German, from Latin *raster, rastrum* rake, from *radere* to scrape] (1934)
: a scan pattern (as of the electron beam in a cathode-ray tube) in which an area is scanned from side to side in lines from top to bottom; *also* **:** a pattern of closely spaced rows of dots that form the image on a cathode-ray tube (as of a television or computer display)

ra·sure \'rā-shər, -zhər\ *noun* [Middle French, from Latin *rasura,* from *rasus,* past participle of *radere*] (1508)
: ERASURE, OBLITERATION

¹rat \'rat\ *noun* [Middle English, from Old English *ræt;* akin to Old High German *ratta* rat and perhaps to Latin *rodere* to gnaw — more at RODENT] (before 12th century)
1 a : any of numerous rodents (*Rattus* and related genera) differing from the related mice by considerably larger size and by structural details (as of the teeth) **b :** any of various similar rodents
2 : a contemptible person: as **a :** one who betrays or deserts friends or associates **b :** SCAB 3b **c :** INFORMER 2
3 : a pad over which a woman's hair is arranged
4 : a person who spends much time especially idly in a specified place ⟨a mall *rat*⟩
— rat·like \-ˌlīk\ *adjective*

²rat *verb* **rat·ted; rat·ting** (1812)
intransitive verb
1 : to betray, desert, or inform on one's associates — usually used with *on*
2 : to catch or hunt rats
3 : to work as a scab
transitive verb
: to give (hair) the effect of greater quantity by use of a rat

rat·able *or* **rate·able** \'rā-tə-bəl\ *adjective* (1503)
: capable of being rated, estimated, or apportioned
— rat·ably \-blē\ *adverb*

rat·a·fia \ˌra-tə-'fē-ə\ *noun* [French] (1699)
1 : a liqueur made from an infusion of macerated fruit or fruit juice in a liquor (as brandy) and often flavored with almonds
2 : a sweet biscuit made of almond paste

rat·a·plan \'ra-tə-ˌplan\ *noun* [French, of imitative origin] (circa 1848)
: the iterative sound of beating ⟨a rolling *rataplan* of drums —*Time*⟩

rat–a–tat \'ra-tə-ˌtat\ *or* **rat–a–tat–tat** \ˌra-tə-ˌta(t)-'tat\ *noun* [imitative] (1681)
: a rapid succession of knocking, tapping, or cracking sounds

ra·ta·tou·ille \ˌra-ˌta-'twē, ˌrä-ˌtä-, -'tü-ē\ *noun* [French, from blend of *ratouiller* to disturb, shake and *tatouiller* to stir] (circa 1877)
: a seasoned stew made of eggplant, tomatoes, green peppers, squash, and sometimes meat

rat·bag \'rat-ˌbag\ *noun* (1890)
chiefly Australian **:** a stupid, eccentric, or disagreeable person

rat–bite fever *noun* (1910)
: either of two febrile bacterial diseases of humans usually transmitted by the bite of a rat

rat cheese *noun* (1939)
: CHEDDAR

¹ratch·et \'ra-chət\ *noun* [alteration of earlier *rochet,* from French, alteration of Middle French *rocquet* ratchet, bobbin, of Germanic origin; akin to Old High German *rocko* distaff — more at ROCK] (1654)
1 : a mechanism that consists of a bar or wheel having inclined teeth into which a pawl drops so that motion can be imparted to the wheel or bar, governed, or prevented and that is used in a hand tool (as a wrench or screwdriver) to allow effective motion in one direction only
2 : a pawl or detent for holding or propelling a ratchet wheel

²ratchet (1973)
transitive verb
: to cause to move by steps or degrees — usually used with *up* or *down* ⟨tried to *ratchet* down the debt⟩
intransitive verb
: to proceed by steps or degrees

ratchet wheel *noun* (1777)
: a toothed wheel held in position or turned by an engaging pawl

ratchet wheel: *1* **wheel,** *2* **reciprocating lever,** *3* **pawl for communicating motion,** *4* **pawl for preventing backward motion**

[1]**rate** \ˈrāt\ *verb* **rat·ed; rat·ing** [Middle English] (14th century)
transitive verb
1 : to rebuke angrily or violently
2 *obsolete* **:** to drive away by scolding
intransitive verb
: to voice angry reprimands

[2]**rate** *noun* [Middle English, from Middle French, from Medieval Latin *rata,* from Latin (*pro*) *rata* (*parte*) according to a fixed proportion] (15th century)
1 a : reckoned value **:** VALUATION **b** *obsolete* **:** ESTIMATION
2 *obsolete* **:** a fixed quantity
3 a : a fixed ratio between two things **b :** a charge, payment, or price fixed according to a ratio, scale, or standard: as (1) **:** a charge per unit of a public-service commodity (2) **:** a charge per unit of freight or passenger service (3) **:** a unit charge or ratio used by a government for assessing property taxes (4) *British* **:** a local tax
4 a : a quantity, amount, or degree of something measured per unit of something else **b :** an amount of payment or charge based on another amount; *specifically* **:** the amount of premium per unit of insurance
5 : relative condition or quality **:** CLASS
— at any rate : in any case **:** ANYWAY

[3]**rate** *verb* **rat·ed; rat·ing** (15th century)
transitive verb
1 *obsolete* **:** ALLOT
2 a : to set an estimate on **:** VALUE, ESTEEM ⟨black is *rated* very high this season⟩ **b :** to determine or assign the relative rank or class of **:** GRADE ⟨*rate* a seaman⟩ **c :** to estimate the normal capacity or power of
3 : CONSIDER, REGARD ⟨was *rated* an excellent pianist⟩
4 : to fix the amount of premium to be charged per unit of insurance on
5 : to have a right to **:** DESERVE ⟨she *rated* special privileges⟩
intransitive verb
: to enjoy a status of special privilege ⟨really *rates* with the boss⟩
synonym see ESTIMATE

ra·tel \ˈrā tᵊl, ˈrä \ *noun* [Afrikaans, literally, rattle, from Dutch, from Middle Dutch — more at RATTLE] (1777)
: an African or Asian nocturnal carnivorous mammal (*Mellivora capensis*) of the weasel family that resembles a badger

rate·me·ter \ˈrāt-ˌmē-tər\ *noun* (1949)
: an instrument that indicates the counting rate of an electronic counter

rate of change (circa 1939)
: a value that results from dividing the change in a function of a variable by the change in the variable ⟨velocity is the *rate of change* in distance with respect to time⟩

rate of exchange (circa 1741)
: the amount of one currency that will buy a given amount of another

rate of interest (1785)
: the percentage usually on an annual basis that is paid for the use of money borrowed from another

rate·pay·er \ˈrāt-ˌpā-ər\ *noun* (1845)
1 *British* **:** TAXPAYER
2 : one who pays for the consumption of electricity according to established rates

rat·er \ˈrā-tər\ *noun* (1611)
1 : one that rates; *specifically* **:** a person who estimates or determines a rating
2 : one having a specified rating or class — usually used in combination ⟨first-*rater*⟩

rat fink *noun* (1964)
: FINK, INFORMER

rat·fish \ˈrat-ˌfish\ *noun* (1882)
: CHIMAERA; *especially* **:** a silvery iridescent white-spotted chimaera (*Hydrolagus colliei*) of cold deep waters of the Pacific coast of North America

rathe \ˈrāth, ˈrath\ *adjective* [Middle English, quick, from Old English *hræth,* alteration of *hræd;* akin to Old High German *hrad* quick] (14th century)
archaic **:** EARLY ⟨bring the *rathe* primrose that forsaken dies —John Milton⟩

rath·er \ˈra-thər, ˈrä-, ˈrə- *also* ˈre-; *interjectionally* ˈra-ˈthər, ˈrä-, ˈrə-\ *adverb* [Middle English, from Old English *hrathor,* comparative of *hrathe* quickly; akin to Old High German *rado* quickly, Old English *hræd* quick] (before 12th century)
1 : with better reason or more propriety **:** more properly ⟨this you should pity *rather* than despise —Shakespeare⟩
2 : more readily or willingly **:** PREFERABLY ⟨I'd *rather* not go⟩ ⟨would *rather* read than watch television⟩ — often used interjectionally to express affirmation
3 : more correctly speaking ⟨my father, or *rather* my stepfather⟩
4 : to the contrary **:** INSTEAD ⟨was no better but *rather* grew worse —Mark 5:26 (Revised Standard Version)⟩
5 : in some degree **:** SOMEWHAT ⟨it's *rather* warm⟩ — often used as a mild intensive ⟨spent *rather* a lot of money⟩
— the rather *archaic* **:** the more quickly or readily

rather than *conjunction* (14th century)
: and not ⟨obscures *rather than* resolves the problem⟩ ⟨why do one thing *rather than* another?⟩

raths·kel·ler \ˈrät-ˌske-lər, ˈrat-, ˈrath-\ *noun* [obsolete German (now *Ratskeller*), city-hall basement restaurant, from *Rat* council + *Keller* cellar] (1900)
: a usually basement tavern or restaurant

rat·i·cide \ˈra-tə-ˌsīd\ *noun* (1908)
: a substance for killing rats

rat·i·fy \ˈra-tə-ˌfī\ *transitive verb* **-fied; -fy·ing** [Middle English *ratifien,* from Middle French *ratifier,* from Medieval Latin *ratificare,* from Latin *ratus* determined, from past participle of *reri* to calculate — more at REASON] (14th century)
: to approve and sanction formally **:** CONFIRM ⟨*ratify* a treaty⟩
— rat·i·fi·ca·tion \ˌra-tə-fə-ˈkā-shən\ *noun*
— rat·i·fi·er \ˈra-tə-ˌfī(-ə)r\ *noun*

ra·ti·né \ˌra-tə-ˈnā\ *or* **ra·tine** \ˌra-tə-ˈnā, ra-ˈtēn\ *noun* [French *ratiné*] (circa 1914)
1 : a nubby ply yarn of various fibers made by twisting under tension a thick and a thin yarn
2 : a rough bulky fabric usually woven loosely in plain weave from ratiné yarns

rat·ing \ˈrā-tiŋ\ *noun* (1702)
1 : a classification according to grade; *specifically* **:** a military or naval specialist classification
2 *chiefly British* **:** a naval enlisted man
3 a : relative estimate or evaluation **:** STANDING ⟨the school has a good academic *rating*⟩ **b :** an estimate of an individual's or business's credit and responsibility **c :** an estimate of the percentage of the public listening to or viewing a particular radio or television program
4 : a stated operating limit of a machine expressible in power units (as kilowatts of a direct-current generator) or in characteristics (as voltage)

ra·tio \ˈrā-(ˌ)shō, -shē-ˌō\ *noun, plural* **ra·tios** [Latin, computation, reason — more at REASON] (1660)
1 a : the indicated quotient of two mathematical expressions **b :** the relationship in quantity, amount, or size between two or more things **:** PROPORTION
2 : the expression of the relative values of gold and silver as determined by a country's currency laws

ra·ti·o·ci·nate \ˌra-tē-ˈō-sᵊn-ˌāt, ˌra-shē-, -ˈä-\ *intransitive verb* **-nat·ed; -nat·ing** [Latin *ratiocinatus,* past participle of *ratiocinari* to reckon, from *ratio* + *-cinari* (as in *vaticinari* to prophesy) — more at VATICINATE] (1643)
: REASON
— ra·ti·o·ci·na·tor \-ˌā-tər\ *noun*

ra·ti·o·ci·na·tion \-ˌō-sᵊn-ˈā-shən, -ˌä-\ *noun* (circa 1530)
1 : the process of exact thinking **:** REASONING
2 : a reasoned train of thought
— ra·ti·o·ci·na·tive \-ˈō-sᵊn-ˌā-tiv, -ˈä-\ *adjective*

[1]**ra·tion** \ˈra-shən, ˈrā-\ *noun* [French, from Latin *ration-, ratio* computation, reason] (circa 1711)
1 a : a food allowance for one day **b** *plural* **:** FOOD, PROVISIONS
2 : a share especially as determined by supply

[2]**ration** *transitive verb* **ra·tioned; ra·tion·ing** \ˈra-sh(ə-)niŋ, ˈrā-\ (1859)
1 : to supply with or put on rations
2 a : to distribute as rations — often used with *out* **b :** to distribute equitably **c :** to use sparingly

[1]**ra·tio·nal** \ˈrash-nəl, ˈra-shə-nᵊl\ *adjective* [Middle English *racional,* from Latin *rationalis,* from *ration-, ratio*] (14th century)
1 a : having reason or understanding **b :** relating to, based on, or agreeable to reason **:** REASONABLE ⟨a *rational* explanation⟩ ⟨*rational* behavior⟩
2 : involving only multiplication, division, addition, and subtraction and only a finite number of times
3 : relating to, consisting of, or being one or more rational numbers ⟨a *rational* root of an equation⟩
— ra·tio·nal·ly *adverb*
— ra·tio·nal·ness *noun*

[2]**rational** *noun* (1606)
: something rational; *specifically* **:** RATIONAL NUMBER

ra·tio·nale \ˌra-shə-ˈnal\ *noun* [Latin, neuter of *rationalis*] (1657)
1 : an explanation of controlling principles of opinion, belief, practice, or phenomena
2 : an underlying reason **:** BASIS

rational function *noun* (1885)
: a function that is the quotient of two polynomials; *also* **:** POLYNOMIAL

ra·tio·nal·ise *British variant of* RATIONALIZE

ra·tio·nal·ism \ˈrash-nə-ˌli-zəm, ˈra-shə-nᵊl-ˌi-\ *noun* (1827)
1 : reliance on reason as the basis for establishment of religious truth
2 a : a theory that reason is in itself a source of knowledge superior to and independent of sense perceptions **b :** a view that reason and experience rather than the nonrational are the fundamental criteria in the solution of problems
3 : FUNCTIONALISM
— ra·tio·nal·ist \-nə-list, -nᵊl-ist\ *noun*
— rationalist *or* **ra·tio·nal·is·tic** \ˌrash-nə-ˈlis-tik, ˌra-shə-nᵊl-ˈis-\ *adjective*
— ra·tio·nal·is·ti·cal·ly \-ti-k(ə-)lē\ *adverb*

ra·tio·nal·i·ty \ˌra-shə-ˈna-lə-tē\ *noun, plural* **-ties** (1628)
1 : the quality or state of being rational
2 : the quality or state of being agreeable to reason **:** REASONABLENESS
3 : a rational opinion, belief, or practice — usually used in plural

ra·tio·nal·ize \ˈrash-nə-ˌlīz, ˈra-shə-nᵊl-ˌīz\ *verb* **-ized; -iz·ing** (1803)
transitive verb
1 : to free (a mathematical expression) from irrational parts ⟨*rationalize* a denominator⟩
2 : to bring into accord with reason or cause something to seem reasonable: as **a :** to substitute a natural for a supernatural explanation of

⟨*rationalize* a myth⟩ **b :** to attribute (one's actions) to rational and creditable motives without analysis of true and especially unconscious motives ⟨*rationalized* his dislike of his brother⟩
3 : to apply the principles of scientific management to (as an industry or its operations) for a desired result (as increased efficiency)
intransitive verb
: to provide plausible but untrue reasons for conduct
— ra·tio·nal·iz·able \ˌrash-nə-ˈlī-zə-bəl, ˌra-shə-nᵊl-ˈī-\ *adjective*
— ra·tio·nal·i·za·tion \ˌrash-nə-lə-ˈzā-shən, ˌra-shə-nᵊl-ə-\ *noun*
— ra·tio·nal·iz·er \ˈrash-nə-ˌlī-zər, ˈra-shə-nᵊl-ˌī-\ *noun*

rational number *noun* (1904)
: an integer or the quotient of an integer divided by a nonzero integer

rat·ite \ˈra-ˌtīt\ *noun* [ultimately from Latin *ratitus* marked with the figure of a raft, from *ratis* raft] (circa 1890)
: a bird with a flat breastbone; *especially* **:** any of various mostly flightless birds (as an ostrich, rhea, emu, moa, or kiwi) with small or rudimentary wings and no keel on the sternum that are probably of polyphyletic origin and are assigned to a number of different orders
— ratite *adjective*

rat·line \ˈrat-lən\ *noun* [Middle English *radelyng*] (15th century)
: any of the small transverse ropes attached to the shrouds of a ship so as to form the steps of a rope ladder

ratline

[1]ra·toon \ra-ˈtün\ *noun* [Spanish *retoño,* from *retoñar* to sprout, from *re-* (from Latin) + *otoñar* to grow in autumn, from *otoño* autumn, from Latin *autumnus*] (1631)
1 : a shoot of a perennial plant (as sugarcane)
2 : a crop (as of bananas) produced on ratoons

[2]ratoon (1756)
intransitive verb
: to sprout or spring up from the root
transitive verb
: to grow or produce (a crop) from or on ratoons

rat race *noun* (1939)
: strenuous, wearisome, and usually competitive activity or rush

rats \ˈrats\ *interjection* (1890)
— used to express disappointment, frustration, or disgust

rat snake *noun* (1860)
: any of numerous large harmless rat-eating colubrid snakes (especially genus *Elaphe*) — called also *chicken snake*

rat·tail \ˈrat-ˌtāl\ *noun* (1705)
1 : a horse's tail with little or no hair
2 : GRENADIER 2

rattail cactus *noun* (1900)
: a commonly cultivated tropical American cactus (*Aporocactus flagelliformis*) with showy crimson flowers

rat–tail file *noun* (1744)
: a round slender tapered file

rat·tan \ra-ˈtan, rə-\ *noun* [Malay *rotan*] (1660)
1 : a rattan cane or switch
2 a : a climbing palm (especially of the genera *Calamus* and *Daemonorops*) with very long tough stems **b :** a part of the stem of a rattan used especially for walking sticks and wickerwork

rat·teen \ra-ˈtēn\ *noun* [French *ratine*] (1685)
archaic **:** a coarse woolen fabric

rat·ter \ˈra-tər\ *noun* (1857)
: one that catches rats; *specifically* **:** a rat-catching dog or cat

[1]rat·tle \ˈra-tᵊl\ *verb* **rat·tled; rat·tling** \ˈrat-liŋ, ˈra-tᵊl-iŋ\ [Middle English *ratelen;* akin to Middle Dutch *ratel* rattle] (14th century)
intransitive verb
1 : to make a rapid succession of short sharp noises ⟨the windows *rattled* in the wind⟩
2 : to chatter incessantly and aimlessly
3 : to move with a clatter or rattle; *also* **:** to be or move about in a place or station too large or grand ⟨*rattled* around the big old house⟩
transitive verb
1 : to say, perform, or affect in a brisk lively fashion ⟨*rattled* off four magnificent backhands —Kim Chapin⟩
2 : to cause to make a rattling sound
3 : ROUSE; *specifically* **:** to beat (a cover) for game
4 : to upset to the point of loss of poise and composure
synonym see EMBARRASS

[2]rattle *noun* (1519)
1 a : a device that produces a rattle; *specifically* **:** a case containing pellets used as a baby's toy **b :** the sound-producing organ on a rattlesnake's tail
2 a : a rapid succession of sharp clattering sounds **b :** NOISE, RACKET
3 : a throat noise caused by air passing through mucus and heard especially at the approach of death

[3]rattle *transitive verb* **rat·tled; rat·tling** \ˈrat-liŋ, ˈra-tᵊl-iŋ\ [irregular from *ratline*] (1729)
: to furnish with ratlines

rat·tle·brain \ˈra-tᵊl-ˌbrān\ *noun* (1709)
: a flighty or thoughtless person
— rat·tle·brained \-ˌbrānd\ *adjective*

rat·tler \ˈrat-lər, ˈra-tᵊl-ər\ *noun* (15th century)
1 : one that rattles
2 : RATTLESNAKE

rat·tle·snake \ˈra-tᵊl-ˌsnāk\ *noun* (1630)
: any of the American pit vipers comprising two genera (*Crotalus* and *Sistrurus*) and having horny interlocking joints at the end of the tail that make a sharp rattling sound when shaken

rattlesnake

rattlesnake plantain *noun* (1778)
: an orchid (genus *Goodyera*) with checked or mottled leaves

rattlesnake root *noun* (1682)
: any of various plants formerly believed to be distasteful to rattlesnakes or effective against their venom: as **a :** any of a genus (*Prenanthes* and especially *P. altissima*) of composite plants that have lobed or pinnatifid leaves and small heads of drooping ligulate flowers **b :** SENECA SNAKEROOT

rattlesnake weed *noun* (1760)
: a hawkweed (*Hieracium venosum*) with purple-veined leaves

rat·tle·trap \ˈra-tᵊl-ˌtrap\ *noun* (1822)
: something rattly or rickety; *especially* **:** an old car
— rattletrap *adjective*

[1]rat·tling \ˈrat-liŋ\ *adjective* (1560)
1 : LIVELY, BRISK ⟨moved at a *rattling* pace⟩
2 : extraordinarily good **:** SPLENDID
— rat·tling·ly \-liŋ-lē\ *adverb*

[2]rattling *adverb* (1829)
: to an extreme degree **:** VERY ⟨a *rattling* good argument —E. A. Betts⟩

rat·tly \ˈrat-lē, ˈra-tᵊl-ē\ *adjective* (1881)
: likely to rattle **:** making a rattle

rat·ton \ˈra-tᵊn, ˈrä-\ *noun* [Middle English *ratoun,* from Middle French *raton,* diminutive of *rat,* probably of Germanic origin; akin to Old English *ræt* rat] (14th century)
chiefly dialect **:** RAT

rat·trap \ˈra(t)-ˌtrap\ *noun* (15th century)
1 : a trap for rats
2 : a dirty dilapidated structure
3 : a hopeless situation

rat trap cheese *noun* (1927)
: CHEDDAR

rat·ty \ˈra-tē\ *adjective* **rat·ti·er; -est** (1865)
1 a : infested with rats **b :** of, relating to, or suggestive of a rat
2 : SHABBY, UNKEMPT ⟨a *ratty* brown overcoat —John Lardner⟩
3 a : DESPICABLE, TREACHEROUS **b :** IRRITABLE ⟨feeling *ratty* as hell —Richard Bissell⟩

rau·cous \ˈrȯ-kəs\ *adjective* [Latin *raucus* hoarse; akin to Latin *ravis* hoarseness] (1769)
1 : disagreeably harsh or strident **:** HOARSE ⟨*raucous* voices⟩
2 : boisterously disorderly ⟨a . . . *raucous* frontier town —Truman Capote⟩
synonym see LOUD
— rau·cous·ly *adverb*
— rau·cous·ness *noun*

raunch \ˈrȯnch, ˈränch\ *noun* [back-formation from *raunchy*] (1964)
: VULGARITY, LEWDNESS

raun·chy \ˈrȯn-chē, ˈrän-\ *adjective* **raun·chi·er; -est** [origin unknown] (1939)
1 : SLOVENLY, DIRTY ⟨a *raunchy* panhandler⟩; *also* **:** very smelly ⟨*raunchy* sneakers⟩
2 : OBSCENE, SMUTTY ⟨*raunchy* jokes⟩
— raun·chi·ly \ˈrȯn-chə-lē, ˈrän-\ *adverb*
— raun·chi·ness \-chē-nəs\ *noun*

rau·wol·fia \raů-ˈwůl-fē-ə, rȯ-\ *noun* [New Latin, from Leonhard *Rauwolf* (died 1596) German botanist] (1752)
1 : any of a large pantropical genus (*Rauvolfia* synonym *Rauwolfia*) of somewhat poisonous trees and shrubs of the dogbane family that yield emetic and purgative substances
2 : a medicinal extract from the root of an Asian rauwolfia (*Rauvolfia serpentina*) used in the treatment of hypertension and mental disorders

[1]rav·age \ˈra-vij\ *noun* [French, from Middle French, from *ravir* to ravish — more at RAVISH] (circa 1611)
1 : an act or practice of ravaging
2 : damage resulting from ravaging **:** violently destructive effect ⟨the *ravages* of time⟩

[2]ravage *verb* **rav·aged; rav·ag·ing** (circa 1611)
transitive verb
: to wreak havoc on **:** affect destructively
intransitive verb
: to commit destructive actions ☆
— rav·age·ment \-vij-mənt\ *noun*
— rav·ag·er *noun*

[1]rave \ˈrāv\ *verb* **raved; rav·ing** [Middle English] (14th century)
intransitive verb

☆ SYNONYMS
Ravage, devastate, waste, sack, pillage, despoil mean to lay waste by plundering or destroying. RAVAGE implies violent often cumulative depredation and destruction ⟨a hurricane *ravaged* the coast⟩. DEVASTATE implies the complete ruin and desolation of a wide area ⟨an earthquake *devastated* the city⟩. WASTE may imply producing the same result by a slow process rather than sudden and violent action ⟨years of drought had *wasted* the area⟩. SACK implies carrying off all valuable possessions from a place ⟨barbarians *sacked* ancient Rome⟩. PILLAGE implies ruthless plundering at will but without the completeness suggested by SACK ⟨settlements *pillaged* by Vikings⟩. DESPOIL applies to looting or robbing of a place or person without suggesting accompanying destruction ⟨the Nazis *despoiled* the art museums⟩.

1 a : to talk irrationally in or as if in delirium **b :** to speak out wildly **c :** to talk with extreme enthusiasm ⟨*raved* about its beauty⟩
2 : to move or advance violently **:** STORM ⟨the iced gusts still *rave* and beat —John Keats⟩
transitive verb
: to utter in madness or frenzy
— rav·er *noun*

[2]rave *noun* (1598)
1 : an act or instance of raving
2 : an extravagantly favorable criticism ⟨the play received the critics' *raves*⟩

[1]rav·el \'ra-vəl\ *verb* **-eled** *or* **-elled; -el·ing** *or* **-el·ling** \'rav-liŋ, 'ra-və-\ [Dutch *rafelen,* from *rafel* loose thread] (1582)
transitive verb
1 a : to separate or undo the texture of **:** UNRAVEL **b :** to undo the intricacies of **:** DISENTANGLE
2 : ENTANGLE, CONFUSE
intransitive verb
1 *obsolete* **:** to become entangled or confused
2 : to become unwoven, untwisted, or unwound **:** FRAY
3 : BREAK UP, CRUMBLE
— rav·el·er \'rav-lər, 'ra-və-\ *noun*
— rav·el·ment \'ra-vəl-mənt\ *noun*

[2]ravel *noun* (1634)
: an act or result of raveling: as **a :** something tangled **b :** something raveled out; *specifically* **:** a loose thread

raveling *or* **ravelling** *noun* (1658)
: RAVEL b

[1]ra·ven \'rā-vən\ *noun* [Middle English, from Old English *hræfn;* akin to Old High German *hraban* raven, Latin *corvus,* Greek *korax*] (before 12th century)
: a large glossy black corvine bird (*Corvus corax*) of Europe, Asia, northern Africa, and America

raven

[2]raven *adjective* (1634)
: shiny and black like a raven's feathers ⟨*raven* hair⟩

[3]rav·en \'ra-vən\ *verb* **rav·ened; rav·en·ing** \'ra-və-niŋ, 'rav-niŋ\ [Middle French *raviner* to rush, take by force, from *ravine* rapine] (1530)
transitive verb
1 : to devour greedily
2 : DESPOIL ⟨men . . . *raven* the earth, destroying its resources —*New Yorker*⟩
intransitive verb
1 : to feed greedily
2 : to prowl for food **:** PREY
3 : PLUNDER
— rav·en·er \'ra-və-nər, 'rav-nər\ *noun*

rav·en·ous \'ra-və-nəs, 'rav-nəs\ *adjective* (15th century)
1 : RAPACIOUS ⟨*ravenous* wolves⟩
2 : very eager or greedy for food, satisfaction, or gratification ⟨a *ravenous* appetite⟩
synonym see VORACIOUS
— rav·en·ous·ly *adverb*
— rav·en·ous·ness *noun*

rav·in \'ra-vən\ *noun* [Middle English, from Middle French *ravine*] (14th century)
1 : PLUNDER, PILLAGE
2 a : an act or habit of preying **b :** something seized as prey

ra·vine \rə-'vēn\ *noun* [French, from Middle French, rapine, rush, from Latin *rapina* rapine] (1760)
: a small narrow steep-sided valley that is larger than a gully and smaller than a canyon and that is usually worn by running water

rav·ined \'ra-vənd\ *adjective* (1606)
obsolete **:** RAPACIOUS, RAVENOUS

[1]rav·ing \'rā-viŋ\ *noun* (14th century)
: irrational, incoherent, wild, or extravagant utterance or declamation — usually used in plural

[2]raving *adjective* (15th century)
1 : talking wildly or irrationally ⟨a *raving* lunatic⟩
2 : RAVISHING ⟨a *raving* beauty⟩

rav·i·o·li \ˌra-vē-'ō-lē, ˌrä-\ *noun, plural* **ravioli** *also* **rav·i·o·lis** \-lēz\ [Italian, from Italian dialect, plural of *raviolo,* literally, little turnip, diminutive of *rava* turnip, from Latin *rapa* — more at RAPE] (circa 1611)
: pasta in the form of little cases of dough containing a savory filling (as of meat or cheese)

rav·ish \'ra-vish\ *transitive verb* [Middle English *ravisshen,* from Middle French *raviss-,* stem of *ravir,* from (assumed) Vulgar Latin *rapire,* alteration of Latin *rapere* to seize, rob — more at RAPID] (14th century)
1 a : to seize and take away by violence **b :** to overcome with emotion (as joy or delight) ⟨*ravished* by the scenic beauty⟩ **c :** RAPE 2
2 : PLUNDER, ROB
— rav·ish·er *noun*
— rav·ish·ment \-mənt\ *noun*

rav·ish·ing \'ra-vi-shiŋ\ *adjective* (14th century)
: unusually attractive, pleasing, or striking
— rav·ish·ing·ly \-shiŋ-lē\ *adverb*

[1]raw \'rȯ\ *adjective* **raw·er** \'rȯ(-ə)r\; **raw·est** \'rȯ-əst\ [Middle English, from Old English *hrēaw;* akin to Old High German *hrō* raw, Latin *crudus* raw, *cruor* blood, Greek *kreas* flesh] (before 12th century)
1 : not cooked
2 a (1) **:** being in or nearly in the natural state **:** not processed or purified ⟨*raw* fibers⟩ ⟨*raw* sewage⟩ (2) **:** not diluted or blended ⟨*raw* spirits⟩ **b :** unprepared or imperfectly prepared for use **c :** not being in polished, finished, or processed form ⟨*raw* data⟩ ⟨a *raw* draft of a thesis⟩
3 a (1) **:** having the surface abraded or chafed (2) **:** very irritated ⟨a *raw* sore throat⟩ **b :** lacking covering **:** NAKED **c :** not protected **:** susceptible to hurt ⟨*raw* emotions⟩
4 a : lacking experience or understanding **:** GREEN ⟨a *raw* recruit⟩ **b** (1) **:** marked by absence of refinements (2) **:** VULGAR, COARSE **c :** not tempered **:** UNBRIDLED ⟨*raw* power⟩
5 : disagreeably damp or cold
synonym see RUDE
— raw·ly *adverb*
— raw·ness *noun*

[2]raw *noun* (1823)
: a raw place or state
— in the raw 1 : in a natural, unrefined, or crude state **2 :** NAKED ⟨slept *in the raw*⟩

raw bar *noun* (1943)
: a counter (as in a restaurant) that serves raw shellfish

raw·boned \'rȯ-ˌbōnd\ *adjective* (1591)
: relatively thin with prominent bone structure; *also* **:** heavy-framed and rugged but not attractively built
synonym see LEAN

raw deal *noun* (circa 1912)
: an instance of unfair treatment

[1]raw·hide \'rȯ-ˌhīd\ *noun* (1829)
1 : a whip of untanned hide
2 : untanned cattle skin

[2]rawhide *transitive verb* **raw·hid·ed; raw·hid·ing** (1858)
1 : to whip or drive with or as if with a rawhide
2 : CHASTISE 2

ra·win·sonde \'rā-wən-ˌsänd\ *noun* [*ra*dar + *win*d + radio*sonde*] (1946)
: a radiosonde tracked by a radio direction-finding device to determine the velocity of winds aloft

raw material *noun* (1796)
: crude or processed material that can be converted by manufacture, processing, or combination into a new and useful product ⟨wheat . . . is *raw material* for the flour mill —C. A. Koepke⟩; *broadly* **:** something with a potential for improvement, development, or elaboration ⟨perplexities are often the *raw material* of discoveries —Agnes M. Clerke⟩

raw score *noun* (1920)
: an individual's actual achievement score (as on a test) before being adjusted for relative position in the test group

rax \'raks\ *verb* [Middle English (northern dialect) *raxen,* from Old English *raxan;* akin to Old English *reccan* to stretch — more at RACK] (before 12th century)
chiefly Scottish **:** STRETCH

[1]ray \'rā\ *noun* [Middle English *raye,* from Middle French *raie,* from Latin *raia*] (14th century)
: any of an order (Rajiformes) of cartilaginous fishes (as stingrays and skates) having the body flattened dorsoventrally, the eyes on the upper surface, and enlarged pectoral fins fused with the head

[2]ray *noun* [Middle English, from Middle French *rai,* from Latin *radius* rod, ray] (14th century)
1 a : any of the lines of light that appear to radiate from a bright object **b :** a beam of radiant energy (as light) of small cross section **c** (1) **:** a stream of material particles traveling in the same line (as in radioactive phenomena) (2) **:** a single particle of such a stream
2 a : light cast by rays **:** RADIANCE **b :** a moral or intellectual light
3 : a thin line suggesting a ray: as **a :** any of a group of lines diverging from a common center **b :** HALF LINE
4 a : one of the bony rods that extend and support the membrane in the fin of a fish **b :** one of the radiating divisions of the body of a radiate animal (as a starfish)
5 a : a branch or flower stalk of an umbel **b** (1) **:** MEDULLARY RAY (2) **:** VASCULAR RAY **c :** RAY FLOWER 1
6 : PARTICLE, TRACE ⟨*ray* of hope⟩
— rayed \'rād\ *adjective*

[3]ray (1598)
intransitive verb
1 a : to shine in or as if in rays **b :** to issue as rays
2 : to extend like the radii of a circle **:** RADIATE
transitive verb
1 : to emit in rays
2 : to furnish or mark with rays

ray flower *noun* (1842)
1 : one of the marginal flowers of the head in a composite plant (as the aster) that also has disk flowers
2 : the entire head in a plant (as chicory) that lacks disk flowers

Ray·leigh scattering \'rā-lē-\ *noun* [John W. S. *Rayleigh*] (1937)
: scattering of light by particles small enough to render the effect selective so that different colors are deflected through different angles

ray·less \'rā-ləs\ *adjective* (1747)
: having, admitting, or emitting no rays; *especially* **:** DARK
— ray·less·ness *noun*

rayless goldenrod *noun* (1923)
: a shrubby to herbaceous composite plant (*Haplopappus heterophyllus* synonym *Isocoma wrightii*) especially of open saline ground from Texas to Arizona and northern Mexico that lacks ray flowers and causes trembles in cattle

ray·on \'rā-ˌän\ *noun* [irregular from [2]*ray*] (1924)
1 : any of a group of smooth textile fibers made from regenerated cellulose by extrusion through minute holes
2 : a rayon yarn, thread, or fabric

raze \'rāz\ *transitive verb* **razed; raz·ing** [alteration of *rase*] (1536)
1 a *archaic* **:** ERASE **b :** to scrape, cut, or shave off
2 : to destroy to the ground **:** DEMOLISH
— raz·er *noun*

ra·zee \rā-'zē\ *noun* [French (*vaisseau*) *rasé,* literally, cut-off ship] (1794)
: a wooden warship with the upper deck cut away

ra·zor \'rā-zər\ *noun* [Middle English *rasour,* from Old French *raseor,* from *raser* to raze, shave — more at RASE] (14th century)
: a keen-edged cutting instrument for shaving or cutting hair

ra·zor·back \'rā-zər-ˌbak\ *noun* (1849)
: a thin-bodied long-legged feral hog chiefly of the southeastern U.S.

ra·zor–backed \-ˌbakt\ *or* **ra·zor·back** \-ˌbak\ *adjective* (1829)
: having a sharp narrow back ⟨a *razor-backed* horse⟩

ra·zor·bill \-ˌbil\ *noun* (1674)
: a North Atlantic auk (*Alca torda*) with the plumage black above and white below and a compressed sharp-edged bill — called also *razor-billed auk*

razor clam *noun* (1882)
: any of a family (Solenidae) of marine bivalve mollusks having a long narrow thin shell

¹razz \'raz\ *noun* [short for *razzberry* sound of contempt, alteration of *raspberry*] (circa 1919)
: RASPBERRY 2

²razz *transitive verb* (1921)
: HECKLE, DERIDE ⟨the fans *razzed* the visiting players⟩

razz·a·ma·tazz \ˌra-zə-mə-'taz\ *chiefly British variant of* RAZZMATAZZ

raz·zle–daz·zle \ˌra-zəl-'da-zəl\ *noun* [reduplication of *dazzle*] (1889)
1 : a state of confusion or hilarity
2 : a complex maneuver (as in sports) designed to confuse an opponent
3 : a confusing or colorful often gaudy action or display
— razzle–dazzle *adjective*

razz·ma·tazz \ˌraz-mə-'taz\ *noun* [probably alteration of *razzle-dazzle*] (1942)
1 : RAZZLE-DAZZLE 3
2 : DOUBLE-TALK 2
3 : VIM, ZING

RBI \ˌär-(ˌ)bē-'ī, 'ri-bē\ *noun, plural* **RBIs** *or* **RBI** [run batted in] (1948)
: a run in baseball that is driven in by a batter; *also* **:** official credit to a batter for driving in a run ⟨led the league in *RBIs*⟩

r color *noun* (1937)
: an acoustic effect of a simultaneously articulated \r\ imparted to a vowel by retroflexion or constriction of the tongue
— r–col·ored \'är-ˌkə-lərd\ *adjective*

-rd *symbol*
— used after the figure 3 to indicate the ordinal number *third* ⟨3rd⟩ ⟨83rd⟩

¹re \'rā\ *noun* [Medieval Latin, from the syllable sung to this note in a medieval hymn to Saint John the Baptist] (14th century)
: the 2d tone of the diatonic scale in solmization

²re \'rā, 'rē\ *preposition* [Latin, ablative of *res* thing — more at REAL] (1707)
: with regard to **:** IN RE

re- *prefix* [Middle English, from Old French, from Latin *re-, red-* back, again, against]
1 : again **:** anew ⟨*retell*⟩
2 : back **:** backward ⟨*recall*⟩

re·ac·cel·er·ate
re·ac·cept
re·ac·ces·sion
re·ac·cli·ma·tize
re·ac·cred·it
re·ac·cred·i·ta·tion
re·ac·quaint
re·ac·quire
re·ac·qui·si·tion
re·ac·ti·vate
re·ac·ti·va·tion
re·ad·dress
re·ad·just
re·ad·just·ment
re·ad·mis·sion
re·ad·mit
re·adopt
re·af·firm
re·af·fir·ma·tion
re·af·fix
re·al·lo·cate
re·al·lo·ca·tion
re·anal·y·sis
re·an·a·lyze
re·an·i·mate
re·an·i·ma·tion
re·an·nex
re·an·nex·a·tion
re·ap·pear
re·ap·pear·ance
re·ap·pli·ca·tion
re·ap·ply
re·ap·point
re·ap·point·ment
re·ap·prais·al
re·ap·praise
re·ap·pro·pri·ate
re·ap·prove
re·ar·gue
re·ar·gu·ment
re·arous·al
re·arouse
re·ar·range
re·ar·range·ment
re·ar·rest
re·ar·tic·u·late
re·as·cend
re·as·cent
re·as·sem·blage
re·as·sem·ble
re·as·sem·bly
re·as·sert
re·as·ser·tion
re·as·sess
re·as·sess·ment
re·as·sign
re·as·sign·ment
re·as·sume
re·at·tach
re·at·tach·ment
re·at·tain
re·at·tempt
re·at·tri·bute
re·at·tri·bu·tion
re·au·tho·ri·za·tion
re·au·tho·rize
re·awak·en
re·bait
re·bal·ance
re·bap·tism
re·bap·tize
re·be·gin
re·bid
re·bind
re·blend
re·bloom
re·board
re·boil
re·book
re·boot
re·bore
re·bot·tle
re·breed
re·buri·al
re·bury
re·buy
re·cal·cu·late
re·cal·cu·la·tion
re·cal·i·brate
re·cal·i·bra·tion
re·cen·tral·i·za·tion
re·cen·tri·fuge
re·cer·ti·fi·ca·tion
re·cer·ti·fy
re·chal·lenge
re·chan·nel
re·char·ter
re·check
re·cho·reo·graph
re·chris·ten
re–Chris·tian·ize
re·chro·mato·graph
re·chro·ma·tog·ra·phy
re·cir·cu·late
re·cir·cu·la·tion
re·clad
re·clas·si·fi·ca·tion
re·clas·si·fy
re·clothe
re·cock
re·cod·i·fi·ca·tion
re·cod·i·fy
re·col·o·ni·za·tion
re·col·o·nize
re·col·or
re·com·bine
re·com·mence
re·com·mence·ment
re·com·mis·sion
re·com·pi·la·tion
re·com·pile
re·com·pu·ta·tion
re·com·pute
re·con·ceive
re·con·cen·trate
re·con·cen·tra·tion
re·con·cep·tion
re·con·cep·tu·al·i·za·tion
re·con·cep·tu·al·ize
re·con·dense
re·con·fig·u·ra·tion
re·con·fig·ure
re·con·nect
re·con·nec·tion
re·con·quer
re·con·quest
re·con·se·crate
re·con·se·cra·tion
re·con·sol·i·date
re·con·tact
re·con·tam·i·nate
re·con·tam·i·na·tion
re·con·tex·tu·al·ize
re·con·tour
re·con·vene
re·con·ver·sion
re·con·vert
re·con·vict
re·con·vic·tion
re·con·vince
re·copy
re·cork
re·cross
re·cul·ti·vate
re·date
re·ded·i·cate
re·ded·i·ca·tion
re·de·fect
re·de·liv·er
re·de·liv·ery
re·de·pos·it
re·de·ter·mi·na·tion
re·de·ter·mine
re·di·ges·tion
re·dis·cov·er
re·dis·cov·ery
re·dis·cuss
re·dis·play
re·dis·pose
re·dis·po·si·tion
re·dis·solve
re·dis·till
re·dis·til·la·tion
re·di·vide
re·di·vi·sion
re·don
re·draft
re·draw
re·dream
re·drill
re·dub
re·el·i·gi·bil·i·ty
re·el·i·gi·ble
re·emerge
re·emer·gence
re·emis·sion
re·emit
re·em·pha·sis
re·em·pha·size
re·em·ploy
re·em·ploy·ment
re·en·coun·ter
re·en·dow
re·en·er·gize
re·en·gage
re·en·gage·ment
re·en·gi·neer
re·en·grave
re·en·list
re·en·list·ment
re·en·roll
re·en·throne
re·equip
re·equip·ment
re·erect
re·es·ca·late
re·es·ca·la·tion
re·es·tab·lish
re·es·tab·lish·ment
re·es·ti·mate
re·eval·u·ate
re·eval·u·a·tion
re·ex·am·i·na·tion
re·ex·am·ine
re·ex·pe·ri·ence
re·ex·plore
re·ex·port
re·ex·por·ta·tion
re·ex·pose
re·ex·po·sure
re·ex·press
re·face
re·feed
re·feel
re·fence
re·fight
re·fig·ure
re·file
re·find
re·fire
re·fix
re·float
re·fold
re·forge
re·for·mat
re·for·mu·late
re·for·mu·la·tion
re·for·ti·fi·ca·tion
re·for·ti·fy
re·found
re·foun·da·tion
re·frame
re·freeze
re·fry
re·fur·nish
re·gain
re·gath·er
re·gear
re·gild
re·give
re·glaze
re·grade
re·graft
re·grant
re·green
re·grind
re·groom
re·groove
re·growth
re·han·dle
re·hang
re·heat
re·hinge
re·hire
re·hos·pi·tal·i·za·tion
re·hos·pi·tal·ize
re·hu·man·ize
re·hyp·no·tize
re·iden·ti·fy
re·ig·nite
re·ig·ni·tion
re·im·age
re·im·merse
re·im·plant
re·im·plan·ta·tion
re·im·port
re·im·por·ta·tion
re·im·pose
re·im·po·si·tion
re·in·cor·po·rate
re·in·cor·po·ra·tion
re·in·dict
re·in·dict·ment
re·in·fes·ta·tion
re·in·flate
re·in·fla·tion
re·in·hab·it
re·ini·tiate
re·in·ject
re·in·jec·tion
re·in·jure
re·in·ju·ry
re·ink
re·in·ner·vate
re·in·ner·va·tion
re·in·oc·u·late
re·in·oc·u·la·tion
re·in·sert
re·in·ser·tion
re·in·spect
re·in·spec·tion
re·in·spire
re·in·stall
re·in·stal·la·tion
re·in·sti·tute
re·in·sti·tu·tion·al·i·za·tion
re·in·ter
re·in·ter·view
re·in·tro·duce
re·in·tro·duc·tion
re·in·vade
re·in·va·sion
re·in·ves·ti·gate
re·in·ves·ti·ga·tion
re·in·vig·o·rate
re·in·vig·o·ra·tion
re·in·vig·o·ra·tor
re·jack·et
re·judge
re·jug·gle
re·key
re·key·board
re·kin·dle
re·knit
re·la·bel
re·lac·quer
re·land·scape
re·launch
re·learn
re·lend
re·li·cense
re·li·cen·sure
re·light
re·link
re·liq·ue·fy
re·load
re·lock
re·look
re·lu·bri·cate
re·lu·bri·ca·tion
re·mar·ket
re·mar·riage
re·mar·ry
re·mate
re·ma·te·ri·al·ize
re·mea·sure
re·mea·sure·ment
re·meet
re·melt
re·merge
re·mi·gra·tion
re·mil·i·ta·ri·za·tion
re·mil·i·ta·rize
re·mix
re·mo·bi·li·za·tion
re·mo·bi·lize
re·moist·en
re·mold
re·mon·e·ti·za·tion

re•mon•e•tize
re•mo•ti•vate
re•mo•ti•va•tion
re•my•thol•o•gize
re•nail
re•name
re•na•tion•al•i•za•tion
re•na•tion•al•ize
re•nest
re•num•ber
re•ob•serve
re•oc•cu•pa•tion
re•oc•cu•py
re•oc•cur
re•oc•cur•rence
re•oil
re•op•er•ate
re•op•er•a•tion
re•or•ches•trate
re•or•ches•tra•tion
re•ori•ent
re•ori•en•tate
re•ori•en•ta•tion
re•out•fit
re•ox•i•da•tion
re•ox•i•dize
re•pack
re•paint
re•park
re•patch
re•pat•tern
re•pave
re•peg
re•peo•ple
re•pho•to•graph
re•phrase
re•plan
re•plas•ter
re•plate
re•pledge
re•plot
re•plumb
re•pol•ish
re•poll
re•pop•u•lar•ize
re•pop•u•late
re•pop•u•la•tion
re•pot
re•pres•sur•ize
re•price
re•pri•vat•i•za•tion
re•pri•vat•ize
re•pro•vi•sion
re•pump
re•punc•tu•a•tion
re•pur•chase
re•pu•ri•fy
re•rack
re•raise
re•read
re•read•ing
re•re•cord
re•reg•is•ter
re•reg•is•tra•tion
re•reg•u•late
re•reg•u•la•tion
re•re•lease
re•re•mind
re•re•peat
re•re•view
re•rig
re•roof
re•route
re•sail
re•sam•ple
re•saw
re•school
re•score
re•screen
re•sculpt
re•seal
re•seal•able
re•sea•son
re•seat
re•se•cure
re•see
re•seg•re•gate
re•seg•re•ga•tion
re•sell
re•sell•er
re•sen•si•tize
re•sen•tence
re•ser•vice
re•set•tle
re•set•tle•ment
re•sew
re•shin•gle
re•shoe
re•shoot
re•show
re•sight
re•sil•ver
re•site
re•size
re•slate
re•soak
re•so•cial•i•za•tion
re•so•cial•ize
re•sod
re•sol•der
re•sole
re•so•lid•i•fi•ca•tion
re•so•lid•i•fy
re•sow
re•spir•i•tu•al•ize
re•spot
re•spray
re•sprout
re•sta•bi•lize
re•stack
re•stage
re•stamp
re•stim•u•late
re•stim•u•la•tion
re•stock
re•stoke
re•strength•en
re•stress
re•study
re•stuff
re•style
re•sub•mis•sion
re•sub•mit
re•sum•mon
re•sup•ply
re•sur•vey
re•syn•the•sis
re•syn•the•size
re•sys•tem•a•tize
re•tack•le
re•tag
re•tar•get
re•taste
re•teach
re•team
re•test
re•tex•ture
re•thread
re•tie
re•tight•en
re•tile
re•time
re•trace
re•trans•fer
re•trans•form
re•trans•for•ma•tion
re•trans•mis•sion
re•trans•mit
re•try
re•tune
re•type
re•uni•fy
re•uni•fi•ca•tion
re•up•hol•ster
re•uti•li•za•tion
re•uti•lize
re•vac•ci•nate
re•vac•ci•na•tion
re•val•i•date
re•val•i•da•tion
re•val•o•ri•za•tion
re•val•o•rize
re•vict•ual
re•vi•su•al•i•za•tion
re•vote
re•warm
re•wash
re•weave
re•weigh
re•wet
re•wire
re•wrap
re•zone

're \(ə)r\ *verb* (1591)
: ARE ⟨you*'re* right⟩

re•ab•sorb \ˌrē-əb-ˈsȯrb, -ˈzȯrb\ *transitive verb* (circa 1774)
: to take up (something previously secreted or emitted) ⟨sugars *reabsorbed* in the kidney⟩; *also* : RESORB 2

[1]reach \ˈrēch\ *verb* [Middle English *rechen,* from Old English *rǣcan;* akin to Old High German *reichen* to reach, Lithuanian *raižytis* to stretch oneself] (before 12th century)
transitive verb
1 a : to stretch out : EXTEND **b :** THRUST
2 a : to touch or grasp by extending a part of the body (as a hand) or an object ⟨couldn't *reach* the apple⟩ **b :** to pick up and draw toward one : TAKE **c** (1) **:** to extend to ⟨the shadow *reached* the wall⟩ (2) **:** to get up to or as far as : come to ⟨your letter *reached* me yesterday⟩ ⟨his voice *reached* the last rows⟩ ⟨they hoped to *reach* an agreement⟩ **d** (1) **:** ENCOMPASS (2) **:** to make an impression on (3) **:** to communicate with
3 : to hand over : PASS
intransitive verb
1 a : to make a stretch with or as if with one's hand **b :** to strain after something
2 a : PROJECT, EXTEND ⟨his land *reaches* to the river⟩ **b :** to arrive at or come to something ⟨as far as the eye could *reach*⟩
3 : to sail on a reach
— **reach•able** \ˈrē-chə-bəl\ *adjective*
— **reach•er** *noun*

[2]reach *noun* (1536)
1 : a continuous stretch or expanse; *especially* : a straight portion of a stream or river
2 a (1) **:** the action or an act of reaching (2) **:** an individual part of a progression or journey **b** (1) **:** a reachable distance ⟨within *reach*⟩ (2) **:** ability to reach ⟨had a long *reach*⟩ **c :** an extent or range especially of knowledge or comprehension
3 : a bearing shaft or coupling pole; *especially* : the rod joining the hind axle to the forward bolster of a wagon
4 : the tack sailed by a ship with the wind coming just forward of the beam or with the wind directly abeam or abaft the beam
5 : ECHELON, LEVEL — usually used in plural ⟨the higher *reaches* of academic life⟩

reach–me–down \ˈrēch-mē-ˌdau̇n\ *adjective or noun* (1862)
chiefly British : HAND-ME-DOWN

re•act \rē-ˈakt\ *verb* [New Latin *reactus,* past participle of *reagere,* from Latin *re-* + *agere* to act — more at AGENT] (1644)
intransitive verb
1 : to exert a reciprocal or counteracting force or influence — often used with *on* or *upon*
2 : to respond to a stimulus
3 : to act in opposition to a force or influence — usually used with *against*
4 : to move or tend in a reverse direction
5 : to undergo chemical reaction
transitive verb
: to cause to react

re•ac•tance \rē-ˈak-tən(t)s\ *noun* (1893)
: the part of the impedance of an alternating-current circuit that is due to capacitance or inductance or both and that is expressed in ohms

re•ac•tant \-tənt\ *noun* (circa 1920)
: a substance that enters into and is altered in the course of a chemical reaction

re•ac•tion \rē-ˈak-shən\ *noun* (circa 1611)
1 a : the act or process or an instance of reacting **b :** resistance or opposition to a force, influence, or movement; *especially* **:** tendency toward a former and usually outmoded political or social order or policy
2 : a response to some treatment, situation, or stimulus ⟨her stunned *reaction* to the news⟩; *also* : such a response expressed verbally ⟨critical *reaction* to the play⟩
3 : bodily response to or activity aroused by a stimulus: **a :** an action induced by vital resistance to another action; *especially* : the response of tissues to a foreign substance (as an antigen or infective agent) **b :** depression or exhaustion due to excessive exertion or stimulation **c :** heightened activity and overaction succeeding depression or shock **d :** a mental or emotional disorder forming an individual's response to his or her life situation
4 : the force that a body subjected to the action of a force from another body exerts in the opposite direction
5 a (1) **:** chemical transformation or change : the interaction of chemical entities (2) **:** the state resulting from such a reaction **b :** a process involving change in atomic nuclei

re•ac•tion•ary \rē-ˈak-shə-ˌner-ē\ *adjective* (1840)
: relating to, marked by, or favoring reaction; *especially* : ultraconservative in politics
— **reactionary** *noun*
— **re•ac•tion•ary•ism** \-ˌi-zəm\ *noun*

re•ac•tive \rē-ˈak-tiv\ *adjective* (1794)
1 : of, relating to, or marked by reaction or reactance
2 a : readily responsive to a stimulus **b :** occurring as a result of stress or emotional upset ⟨*reactive* depression⟩
— **re•ac•tive•ly** *adverb*
— **re•ac•tive•ness** *noun*
— **re•ac•tiv•i•ty** \(ˌ)rē-ˌak-ˈti-və-tē\ *noun*

re•ac•tor \rē-ˈak-tər\ *noun* (circa 1904)
1 : one that reacts
2 : a device (as a coil, winding, or conductor of small resistance) used to introduce reactance into an alternating-current circuit
3 a : a vat for an industrial chemical reaction **b :** a device for the controlled release of nuclear energy (as for producing heat)

[1]read \ˈrēd\ *verb* **read** \ˈred\; **read•ing** \ˈrē-diŋ\ [Middle English *reden* to advise, interpret, read, from Old English *rǣdan;* akin to Old High German *rātan* to advise, Sanskrit *rādhnoti* he achieves, prepares] (before 12th century)
transitive verb
1 a (1) **:** to receive or take in the sense of (as letters or symbols) especially by sight or touch (2) **:** to study the movements of (as lips) with mental formulation of the communication expressed (3) **:** to utter aloud the printed or written words of ⟨*read* them a story⟩ **b :** to learn from what one has seen or found in writing or printing **c :** to deliver aloud by or as if by reading; *specifically* : to utter interpretively **d** (1) **:** to become acquainted with or look over the contents of (as a book) (2) **:** to make a study of ⟨*read* law⟩ (3) **:** to read the works of **e :** to check (as copy or proof) for errors **f** (1) **:** to receive and understand (a voice message) by radio (2) **:** UNDERSTAND, COMPREHEND
2 a : to interpret the meaning or significance of ⟨*read* palms⟩ **b :** FORETELL, PREDICT ⟨able to *read* his fortune⟩
3 : to recognize or interpret as if by reading: as **a :** to learn the nature of by observing outward expression or signs ⟨*reads* him like a book⟩ **b :** to note the action or characteristics of in order to anticipate what will happen ⟨a good canoeist *reads* the rapids⟩ ⟨a golfer *reading* a green⟩; *also* : to predict the movement of (a putt) by reading a green **c :** to anticipate by observation of an opponent's position or movement ⟨*read* a blitz⟩

4 a : to attribute a meaning to (as something read) **:** INTERPRET ⟨how do you *read* this passage⟩ **b :** to attribute (a meaning) to something read or considered ⟨*read* a nonexistent meaning into her words⟩
5 : to use as a substitute for or in preference to another word or phrase in a particular passage, text, or version ⟨*read hurry* for *harry*⟩ — often used to introduce a clarifying substitute for a euphemistic or misleading word or phrase ⟨a friendly, *read* nosy, coworker⟩
6 : INDICATE ⟨the thermometer *reads* zero⟩
7 : to interpret (a musical work) in performance
8 a : to acquire (information) from storage; *especially* **:** to sense the meaning of (data) in recorded and coded form — used of a computer or data processor **b :** to read the coded information on (as a floppy disk)
intransitive verb
1 a : to perform the act of reading words **:** read something **b** (1) **:** to learn something by reading (2) **:** to pursue a course of study
2 a : to yield a particular meaning or impression when read **b :** to be readable or read in a particular manner or to a particular degree ⟨this book *reads* smoothly⟩
3 : to consist of specific words, phrases, or other similar elements ⟨a passage that *reads* differently in older versions⟩
— read between the lines : to understand more than is directly stated
— read the riot act 1 : to order a mob to disperse **2 a :** to order or warn to cease something **b :** to protest vehemently **c :** to reprimand severely

[2]read \ˈred\ *adjective* (1586)
: instructed by or informed through reading

[3]read \ˈrēd\ *noun* (1825)
1 *chiefly British* **:** a period of reading ⟨it was a night . . . for a *read* and a long sleep —William Sansom⟩
2 : something (as a book) that is read ⟨a novel that's a good *read*⟩
3 : the action or an instance of reading

read·able \ˈrē-də-bəl\ *adjective* (15th century)
: able to be read easily: as **a :** LEGIBLE **b :** interesting to read
— read·abil·i·ty \ˌrē-də-ˈbi-lə-tē\ *noun*
— read·able·ness \ˈrē-də-bəl-nəs\ *noun*
— read·ably \-blē\ *adverb*

read·er \ˈrē-dər\ *noun* (before 12th century)
1 a : one that reads **b :** one appointed to read to others: as (1) **:** LECTOR (2) **:** one chosen to read aloud selected material in a Christian Science church or society **c** (1) **:** PROOFREADER (2) **:** one who evaluates manuscripts (3) **:** one who reads periodical literature to discover items of special interest or value **d :** an employee who reads and records the indications of meters **e :** a teacher's assistant who reads and marks student papers
2 *British* **:** one who reads lectures or expounds subjects to students
3 a : a device for projecting a readable image of a transparency **b :** a unit that scans material recorded (as on punched cards) for storage or computation
4 a : a book for instruction and practice especially in reading **b :** ANTHOLOGY

read·er·ly \-lē\ *adjective* (1959)
: of, relating to, or typical of a reader

read·er·ship \-ˌship\ *noun* (1719)
1 a : the quality or state of being a reader **b :** the office or position of a reader
2 : the mass or a particular group of readers ⟨a magazine's *readership*⟩

read·i·ly \ˈre-dᵊl-ē\ *adverb* (14th century)
: in a ready manner: as **a :** without hesitating **:** WILLINGLY ⟨*readily* accepted advice⟩ **b :** without much difficulty **:** EASILY ⟨for reasons that anyone could *readily* understand⟩

read·ing \ˈrē-diŋ\ *noun* (before 12th century)
1 : the act of reading
2 a : material read or for reading **b :** extent of material read
3 a : a particular version **b :** data indicated by an instrument
4 a : a particular interpretation of something (as a law) **b :** a particular performance of something (as a musical work)
5 : an indication of a certain state of affairs ⟨a study to get some *reading* of shoppers' preferences⟩

reading desk *noun* (1703)
: LECTERN

reading frame *noun* (1965)
: any of the three possible ways of reading a sequence of nucleotides as a series of triplets

read–only memory *noun* (1961)
: a usually small computer memory that contains special-purpose information (as a program) which cannot be altered — called also *ROM*; compare RANDOM-ACCESS MEMORY

read·out \ˈrēd-ˌau̇t\ *noun* (1652)
1 : the process of reading
2 a : the process of removing information from an automatic device (as an electronic computer) and displaying it in an understandable form **b :** the information removed from such a device and displayed or recorded (as by magnetic tape or printing device) **c :** an electronic device that presents information in visual form
3 : the radio transmission of data or pictures from a space vehicle

read out *transitive verb* (1600)
1 a : to read aloud **b :** to produce a readout of
2 : to expel from an organization or group

[1]ready \ˈre-dē\ *adjective* **read·i·er; -est** [Middle English *redy;* akin to Old English *gerǣde* ready, Gothic *garaiths* arranged] (13th century)
1 a : prepared mentally or physically for some experience or action **b :** prepared for immediate use ⟨dinner is *ready*⟩
2 a : willingly disposed **:** INCLINED ⟨*ready* to agree to his proposal⟩ **b :** likely to do something indicated ⟨a house that looks *ready* to collapse⟩
3 : displayed readily and spontaneously ⟨a *ready* wit⟩
4 : immediately available ⟨had little *ready* cash⟩
synonym see QUICK
— read·i·ness *noun*
— at the ready : ready for immediate use ⟨kept their guns *at the ready*⟩

[2]ready *transitive verb* **read·ied; ready·ing** (14th century)
: to make ready

ready box *noun* (1942)
: a box placed near a gun (as on a ship) to hold ammunition kept ready for immediate use

[1]ready–made \ˌred-ē-ˈmād\ *adjective* (15th century)
1 : made beforehand especially for general sale ⟨*ready-made* suits⟩
2 : lacking originality or individuality
3 : readily available ⟨her illness provided a *ready-made* excuse⟩

[2]ready–made *noun* (1882)
1 : something (as a garment) that is ready-made
2 *usually* **ready·made** [French *ready-made,* from English] **:** a commonplace artifact (as a comb or a pair of ice tongs) selected and displayed as a work of art

ready room *noun* (1941)
: a room in which pilots or astronauts are briefed and await orders

ready–to–wear *adjective* (1895)
1 *of clothing* **:** READY-MADE
2 : dealing in ready-made clothes ⟨*ready-to-wear* stores⟩
— ready–to–wear *noun*

ready–wit·ted \ˌre-dē-ˈwi-təd\ *adjective* (1581)
: QUICK-WITTED

re·af·for·es·ta·tion \ˌrē-ə-ˌfȯr-ə-ˈstā-shən, -ˌfär-\ *noun* (1884)
chiefly British **:** REFORESTATION
— re·af·for·est \-ˈfȯr-əst, -ˈfär-\ *transitive verb, chiefly British*

re·agent \rē-ˈā-jənt\ *noun* [New Latin *reagent-, reagens,* present participle of *reagere* to react — more at REACT] (1797)
: a substance used (as in detecting or measuring a component, in preparing a product, or in developing photographs) because of its chemical or biological activity

re·ag·gre·gate \(ˌ)rē-ˈa-gri-ˌgāt\ *transitive verb* (1849)
: to cause to re-form into an aggregate or a whole
— re·ag·gre·gate \-gət\ *noun*
— re·ag·gre·ga·tion \-ˌa-gri-ˈgā-shən\ *noun*

re·agin \rē-ˈā-jən, -gən\ *noun* [International Scientific Vocabulary, from *reagent*] (circa 1911)
1 : a substance in the blood of persons with syphilis responsible for positive serological reactions for syphilis
2 : an antibody in the blood of individuals with some forms of allergy possessing the power of passively sensitizing the skin of normal individuals
— re·agin·ic \ˌrē-ə-ˈji-nik, -ˈgi-\ *adjective*

[1]re·al \ˈrē(-ə)l, ˈri(-ə)l\ *adjective* [Middle English, real, relating to things (in law), from Middle French, from Medieval Latin & Late Latin; Medieval Latin *realis* relating to things (in law), from Late Latin, real, from Latin *res* thing, fact; akin to Sanskrit *rayi* property] (14th century)
1 : of or relating to fixed, permanent, or immovable things (as lands or tenements)
2 a : not artificial, fraudulent, illusory, or apparent **:** GENUINE ⟨*real* gold⟩; *also* **:** being precisely what the name implies ⟨a *real* professional⟩ **b** (1) **:** occurring in fact ⟨a story of *real* life⟩ (2) **:** of or relating to practical or everyday concerns or activities ⟨left school to live in the *real* world⟩ (3) **:** existing as a physical entity and having properties that deviate from an ideal, law, or standard ⟨a *real* gas⟩ — compare IDEAL 3b **c :** having objective independent existence ⟨unable to believe that what he saw was *real*⟩ **d :** FUNDAMENTAL, ESSENTIAL **e** (1) **:** belonging to or having elements or components that belong to the set of real numbers ⟨the *real* roots of an equation⟩ ⟨a *real* matrix⟩ (2) **:** concerned with or containing real numbers ⟨*real* analysis⟩ (3) **:** REAL-VALUED ⟨*real* variable⟩ **f :** measured by purchasing power ⟨*real* income⟩ ⟨*real* dollars⟩ **g :** COMPLETE, UTTER ⟨a *real* fiasco⟩
3 *of a particle* **:** capable of being detected — compare VIRTUAL 3
— re·al·ness *noun*
— for real 1 : in earnest **:** SERIOUSLY ⟨fighting *for real*⟩ **2 :** GENUINE ⟨couldn't believe the threats were *for real*⟩

[2]real *noun* (circa 1626)
: a real thing; *especially* **:** a mathematically real quantity

[3]real *adverb* (1718)
: VERY ⟨he was *real* cool —H. M. McLuhan⟩ ■

□ USAGE
real Most handbooks consider the adverb *real* to be informal and more suitable to speech than writing. Our evidence shows these observations to be true in the main, but *real* is becoming increasingly common in writing of an informal, conversational style. It is used as an intensifier only and is not interchangeable with *really* except in that use.

[4]**re·al** \rā-ˈäl\ *noun, plural* **reals** *or* **re·ales** \-ˈä-(ˌ)läs\ [Spanish, from *real* royal, from Latin *regalis* — more at ROYAL] (1555)
: a former monetary unit and coin of Spain and its possessions

[5]**re·al** \rā-ˈäl\ *noun, plural* **reals** *or* **reis** \ˈrāsh, ˈrās, ˈrāzh, ˈrāz\ [Portuguese, from *real* royal, from Latin *regalis*] (1951)
1 : a former monetary unit and coin of Portugal
2 — see MONEY table

real estate *noun* (1666)
: property in buildings and land

real focus *noun* (1909)
: a point at which rays (as of light) converge or from which they diverge

re·al·gar \rē-ˈal-ˌgär, -gər\ *noun* [Middle English, from Medieval Latin, from Catalan, from Arabic *rahj al-ghār* powder of the mine] (15th century)
: an orange-red mineral consisting of arsenic sulfide and having a resinous luster

re·a·lia \rē-ˈa-lē-ə, -ˈā-\ *noun plural* [Late Latin, neuter plural of *realis* real] (1937)
: objects or activities used to relate classroom teaching to the real life especially of peoples studied

re·align \ˌrē-ə-ˈlīn\ *transitive verb* (1899)
: to align again; *especially* **:** to reorganize or make new groupings of
— **re·align·ment** \-mənt\ *noun*

real image *noun* (1899)
: an optical image formed of real foci

re·al·ism \ˈrē-ə-ˌli-zəm, ˈri-ə-\ *noun* (1817)
1 : concern for fact or reality and rejection of the impractical and visionary
2 a : a doctrine that universals exist outside the mind; *specifically* **:** the conception that an abstract term names an independent and unitary reality **b :** the conception that objects of sense perception or cognition exist independently of the mind — compare NOMINALISM
3 : fidelity in art and literature to nature or to real life and to accurate representation without idealization
— **re·al·ist** \-list\ *adjective or noun*
— **re·al·is·tic** \ˌrē-ə-ˈlis-tik, ˌri-ə-\ *adjective*
— **re·al·is·ti·cal·ly** \-ti-k(ə-)lē\ *adverb*

re·al·i·ty \rē-ˈa-lə-tē\ *noun, plural* **-ties** (1550)
1 : the quality or state of being real
2 a (1) **:** a real event, entity, or state of affairs ⟨his dream became a *reality*⟩ (2) **:** the totality of real things and events ⟨trying to escape from *reality*⟩ **b :** something that is neither derivative nor dependent but exists necessarily
— **in reality :** in actual fact

reality check *noun* (1986)
: something that clarifies or serves as a reminder of reality often by correcting a misconception

re·al·i·za·tion \ˌrē-ə-lə-ˈzā-shən, ˌri-ə-\ *noun* (circa 1611)
1 : the action of realizing **:** the state of being realized
2 : something realized

re·al·ize \ˈrē-ə-ˌlīz, ˈri-ə-\ *transitive verb* **-ized; -iz·ing** [French *réaliser,* from Middle French *realiser,* from *real* real] (circa 1611)
1 a : to bring into concrete existence **:** ACCOMPLISH ⟨finally *realized* her goal⟩ **b :** to cause to seem real **:** make appear real ⟨a book in which the characters are carefully *realized*⟩
2 a : to convert into actual money ⟨*realized* assets⟩ **b :** to bring or get by sale, investment, or effort **:** GAIN
3 : to conceive vividly as real **:** be fully aware of ⟨did not *realize* the risk she was taking⟩
synonym see THINK
— **re·al·iz·able** \ˌrē-ə-ˈlī-zə-bəl, ˌri-\ *adjective*
— **re·al·iz·er** *noun*

real–life *adjective* (1938)
: existing or occurring in reality **:** drawn from or drawing on actual events or situations ⟨*real-life* problems⟩ ⟨*real-life* drama⟩

re·al·ly \ˈrē-(ə-)lē, ˈri-\ *adverb* (15th century)
1 a : in reality **:** ACTUALLY ⟨things as they *really* are⟩ ⟨there was nothing peculiar about her doing this, *really* —Peter Taylor⟩ **b :** TRULY, UNQUESTIONABLY — used as an intensifier ⟨a *really* beautiful day⟩
2 — used to emphasize an assertion ⟨you *really* should read Yeats⟩ ⟨*really,* you're being ridiculous⟩

realm \ˈrelm\ *noun* [Middle English *realme,* from Old French *reialme,* alteration of *reiame,* from Latin *regimen* rule — more at REGIMEN] (13th century)
1 : KINGDOM 2
2 : SPHERE, DOMAIN ⟨within the *realm* of possibility⟩
3 : a primary marine or terrestrial biogeographic division of the earth's surface

real number *noun* (1909)
: one of the numbers that have no imaginary parts and comprise the rationals and the irrationals

real part *noun* (1949)
: the term in a complex number (as 2 in $2 + 3i$) that does not contain the imaginary unit as a factor

re·al·po·li·tik \rā-ˈäl-ˌpō-li-ˌtēk\ *noun, often capitalized* [German, from *real* actual + *Politik* politics] (1914)
: politics based on practical and material factors rather than on theoretical or ethical objectives

real presence *noun, often R&P capitalized* (1559)
: the doctrine that Christ is actually present in the Eucharist

real time *noun* (1953)
: the actual time during which something takes place ⟨the computer may partly analyze the data in *real time* (as it comes in) —R. H. March⟩
— **real–time** *adjective*

Re·al·tor \ˈrē(-ə)l-tər, -ˌtȯr, ÷ˈrē-lə-tər *also* rē-ˈal-tər\ *collective mark*
— used for a real estate agent who is a member of the National Association of Realtors

re·al·ty \ˈrē(-ə)l-tē\ *noun* [*real* + *-ty* (as in *property*)] (1670)
: REAL ESTATE

real–valued *adjective* (1965)
: taking on only real numbers for values

real–world *adjective* (1963)
: REAL-LIFE

[1]**ream** \ˈrēm\ *noun* [Middle English *reme,* from Middle French *raime,* from Arabic *rizmah,* literally, bundle] (14th century)
1 : a quantity of paper being 20 quires or variously 480, 500, or 516 sheets
2 : a great amount — usually used in plural

[2]**ream** *transitive verb* [perhaps from (assumed) Middle English *remen* to open up, from Old English *rēman;* akin to Old English *rȳman* to open up, *rūm* space — more at ROOM] (1815)
1 a : to widen the opening of (a hole) **:** COUNTERSINK **b** (1) **:** to enlarge or dress out (a hole) with a reamer (2) **:** to enlarge the bore of (as a gun) in this way **c :** to remove by reaming
2 a : to press out with a reamer **b :** to press out the juice of (as an orange) with a reamer
3 : CHEAT, VICTIMIZE
4 : REPRIMAND

ream·er \ˈrē-mər\ *noun* (1825)
: one that reams: as **a :** a rotating finishing tool with cutting edges used to enlarge or shape a hole **b :** a fruit juice extractor with a ridged and pointed center rising from a shallow dish

reap \ˈrēp\ *verb* [Middle English *repen,* from Old English *reopan*] (before 12th century)
transitive verb
1 a (1) **:** to cut with a sickle, scythe, or reaping machine (2) **:** to clear of a crop by reaping **b :** to gather by reaping **:** HARVEST
2 : OBTAIN, WIN
intransitive verb
: to reap something

reap·er \ˈrē-pər\ *noun* (before 12th century)
: one that reaps; *especially* **:** any of various machines for reaping grain

reap·hook \ˈrēp-ˌhu̇k\ *noun* (1591)
: a hand implement with a hook-shaped blade used in reaping

re·ap·por·tion \ˌrē-ə-ˈpōr-shən, -ˈpȯr-\ (circa 1828)
transitive verb
: to apportion (as a house of representatives) anew
intransitive verb
: to make a new apportionment
— **re·ap·por·tion·ment** \-shən-mənt\ *noun*

[1]**rear** \ˈrir, *transitive verb 4 & intransitive verb 2 are also* ˈrar *or* ˈrer\ *verb* [Middle English *reren,* from Old English *rǣran;* akin to Old Norse *reisa* to raise, Old English *rīsan* to rise] (before 12th century)
transitive verb
1 : to erect by building **:** CONSTRUCT
2 : to raise upright
3 a (1) **:** to breed and raise (an animal) for use or market (2) **:** BRING UP 1 **b :** to cause (as plants) to grow
4 : to cause (a horse) to rise up on the hind legs
intransitive verb
1 : to rise high
2 *of a horse* **:** to rise up on the hind legs
synonym see LIFT
— **rear·er** *noun*

[2]**rear** \ˈrir\ *noun* [Middle English *rere,* short for *rerewarde* rearward] (14th century)
1 : the back part of something: as **a :** the unit (as of an army) or area farthest from the enemy **b :** the part of something located opposite its front ⟨the *rear* of a house⟩ **c :** BUTTOCKS
2 : the space or position at the back ⟨moved to the *rear*⟩

[3]**rear** \ˈrir\ *adjective* [Middle English *rere,* from Middle French *rere* backward, behind, from Latin *retro-* — more at RETRO] (14th century)
: being at the back

[4]**rear** \ˈrir\ *adverb* (1855)
: toward or from the rear — usually used in combination ⟨a *rear*-driven car⟩

rear admiral *noun* (1589)
: a commissioned officer in the navy who ranks above a commodore and in the coast guard who ranks above a captain and whose insignia is two stars

rear echelon *noun* (circa 1934)
: an element of a military headquarters or unit located at a considerable distance from the front and concerned especially with administrative and supply duties

rear–end \ˈrir-ˈend, -ˌend\ *transitive verb* (1957)
: to crash into the back of (as an automobile)

rear end *noun* (circa 1930)
: BUTTOCKS

rear·guard \ˈrir-ˌgärd\ *adjective* (1898)
: of or relating to resistance especially to sweeping social forces ⟨fought a *rearguard* action against automation⟩

rear guard \-ˈgärd, -ˌgärd\ *noun* [Middle French *reregarde,* from Old French, from *rere* + *garde* guard] (1659)
: a military detachment detailed to bring up and protect the rear of a main body or force

re·arm \(ˌ)rē-'ärm\ (1871)
transitive verb
: to arm (as a nation or a military force) again with new or better weapons
intransitive verb
: to become armed again
— re·ar·ma·ment \-'är-mə-mənt\ *noun*

rear·most \'rir-ˌmōst\ *adjective* (1718)
: farthest in the rear **:** LAST

rear·view mirror \'rir-ˌvyü-\ *noun* (1926)
: a mirror (as in an automobile) that gives a view of the area behind a vehicle

[1]rear·ward \'rir-ˌwȯrd\ *noun* [Middle English *rerewarde,* from Anglo-French; akin to Old French *reregarde* rear guard] (14th century)
: REAR; *especially* **:** the rear division (as of an army)

[2]rear·ward \-wərd\ *adjective* [[2]*rear* + *-ward*] (1598)
1 : located at, near, or toward the rear
2 : directed toward the rear

[3]rear·ward \-wərd\ *also* **rear·wards** \-wərdz\ *adverb* (1625)
: at, near, or toward the rear **:** BACKWARD

[1]rea·son \'rē-zᵊn\ *noun* [Middle English *resoun,* from Old French *raison,* from Latin *ration-, ratio* reason, computation, from *reri* to calculate, think; probably akin to Gothic *rathjo* account, explanation] (13th century)
1 a : a statement offered in explanation or justification ⟨gave *reasons* that were quite satisfactory⟩ **b :** a rational ground or motive ⟨a good *reason* to act soon⟩ **c :** a sufficient ground of explanation or of logical defense; *especially* **:** something (as a principle or law) that supports a conclusion or explains a fact ⟨the *reasons* behind her client's action⟩ **d :** the thing that makes some fact intelligible **:** CAUSE ⟨the *reason* for earthquakes⟩ ⟨the real *reason* why he wanted me to stay —Graham Greene⟩
2 a (1) **:** the power of comprehending, inferring, or thinking especially in orderly rational ways **:** INTELLIGENCE (2) **:** proper exercise of the mind (3) **:** SANITY **b :** the sum of the intellectual powers
3 *archaic* **:** treatment that affords satisfaction
■
— in reason : RIGHTLY, JUSTIFIABLY
— within reason : within reasonable limits
— with reason : with good cause

[2]reason *verb* **rea·soned; rea·son·ing** \'rēz-niŋ, 'rē-zᵊn-iŋ\ (15th century)
intransitive verb
1 a *obsolete* **:** to take part in conversation, discussion, or argument **b :** to talk with another so as to influence his actions or opinions ⟨can't *reason* with her⟩
2 : to use the faculty of reason so as to arrive at conclusions
transitive verb
1 *archaic* **:** to justify or support with reasons
2 : to persuade or influence by the use of reason
3 : to discover, formulate, or conclude by the use of reason ⟨a carefully *reasoned* analysis⟩
synonym see THINK
— rea·son·er \'rēz-nər, 'rē-zᵊn-ər\ *noun*

rea·son·able \'rēz-nə-bəl, 'rē-zᵊn-ə-bəl\ *adjective* (14th century)
1 a : being in accordance with reason ⟨a *reasonable* theory⟩ **b :** not extreme or excessive ⟨*reasonable* requests⟩ **c :** MODERATE, FAIR ⟨a *reasonable* chance⟩ ⟨a *reasonable* price⟩ **d :** INEXPENSIVE
2 a : having the faculty of reason **b :** possessing sound judgment
— rea·son·abil·i·ty \ˌrēz-nə-'bi-lə-tē, ˌrē-zᵊn-ə-\ *noun*
— rea·son·able·ness \'rēz-nə-bəl-nəs, 'rē-zᵊn-ə-\ *noun*
— rea·son·ably \-blē\ *adverb*

reasoning *noun* (14th century)
1 : the use of reason; *especially* **:** the drawing of inferences or conclusions through the use of reason
2 : an instance of the use of reason **:** ARGUMENT

rea·son·less \'rē-zᵊn-ləs\ *adjective* (14th century)
1 : not having the faculty of reason ⟨a *reasonless* brute⟩
2 : not reasoned **:** SENSELESS ⟨*reasonless* hostility⟩
3 : not based on or supported by reasons ⟨a *reasonless* accusation⟩
— rea·son·less·ly *adverb*

re·as·sur·ance \ˌrē-ə-'shu̇r-ən(t)s\ *noun* (circa 1611)
1 : the action of reassuring **:** the state of being reassured
2 : REINSURANCE

re·as·sure \ˌrē-ə-'shu̇r\ *transitive verb* (1598)
1 : to assure anew ⟨*reassured* him that the work was on schedule⟩
2 : to restore to confidence ⟨felt *reassured* by their earnest promise to do better⟩
3 : REINSURE
— re·as·sur·ing·ly \-'shu̇r-iŋ-lē\ *adverb*

re·ata \rē-'a-tə, -'ä-\ *noun* [American Spanish — more at LARIAT] (1846)
: LARIAT

Re·au·mur \ˌrā-ō-'myu̇r, 'rā-ō-ˌ\ *adjective* [René Antoine Ferchault de *Réaumur*] (1814)
: relating to or conforming to a thermometric scale on which the boiling point of water is at 80° above the zero of the scale and the freezing point is at zero

[1]reave \'rēv\ *verb* **reaved** *or* **reft** \'reft\; **reav·ing** [Middle English *reven,* from Old English *rēafian;* akin to Old High German *roubōn* to rob, Latin *rumpere* to break] (before 12th century)
intransitive verb
: PLUNDER, ROB
transitive verb
1 *archaic* **a** (1) **:** ROB, DESPOIL (2) **:** to deprive one of **b :** SEIZE
2 *archaic* **:** to carry or tear away
— reav·er *noun*

[2]reave *transitive verb* **reaved** *or* **reft** \'reft\; **reav·ing** [Middle English *reven,* probably modification of Old Norse *rīfa* to rive] (13th century)
archaic **:** BURST

reb \'reb\ *noun* [short for *rebel*] (1862)
: JOHNNY REB

Reb \'reb\ *noun* [Yiddish, from Hebrew *rabbī* my master, rabbi] (1882)
: RABBI, MISTER — used as a title

re·bar \'rē-ˌbär\ *noun* [*re*inforcing *bar*] (1953)
: a steel rod with ridges for use in reinforced concrete

re·bar·ba·tive \ri-'bär-bə-tiv\ *adjective* [French *rébarbatif,* from Middle French, from *rebarber* to be repellent, from *re-* + *barbe* beard, from Latin *barba* — more at BEARD] (1892)
: REPELLENT, IRRITATING
— re·bar·ba·tive·ly *adverb*

[1]re·bate \'rē-ˌbāt, ri-'\ *verb* **re·bat·ed; re·bat·ing** [Middle English, from Middle French *rabattre* to beat down again, from Old French, from *re-* + *abattre* to beat down, from *a-* (from Latin *ad-*) + *battre* to beat, from Latin *battuere*] (14th century)
transitive verb
1 : to reduce the force or activity of **:** DIMINISH
2 : to reduce the sharpness of **:** BLUNT
3 a : to make a rebate of **b :** to give a rebate to
intransitive verb
: to give rebates
— re·bat·er *noun*

[2]re·bate \'rē-ˌbāt\ *noun* (1656)
: a return of a part of a payment

[3]re·bate \'ra-bət, 'rē-ˌbāt\ *chiefly British variant of* RABBET

re·ba·to \ri-'bä-(ˌ)tō\ *variant of* RABATO

reb·be \'re-bə\ *noun* [Yiddish *rebe,* from Hebrew *rabbī* rabbi] (1881)
: a Jewish spiritual leader or teacher **:** RABBI

re·bec *or* **re·beck** \'rē-ˌbek, 're-(ˌ)bek\ *noun* [Middle English *rebecke,* from Middle French *rebec,* alteration of Old French *rebebe,* from Old Provençal *rebeb,* from Arabic *rabāb*] (15th century)
: an ancient bowed usually 3-stringed musical instrument with a pear-shaped body and slender neck

rebec

Re·bek·ah \ri-'be-kə\ *noun* [Hebrew *Ribhqāh*]
: the wife of Isaac

[1]reb·el \'re-bəl\ *adjective* [Middle English, from Old French *rebelle,* from Latin *rebellis,* from *re-* + *bellum* war, from Old Latin *duellum*] (14th century)
1 a : opposing or taking arms against a government or ruler **b :** of or relating to rebels ⟨the *rebel* camp⟩
2 : DISOBEDIENT, REBELLIOUS

[2]rebel *noun* (14th century)
: one who rebels or participates in a rebellion

[3]re·bel \ri-'bel\ *intransitive verb* **re·belled; re·bel·ling** (14th century)
1 a : to oppose or disobey one in authority or control **b :** to renounce and resist by force the authority of one's government
2 a : to act in or show opposition or disobedience ⟨*rebelled* against the conventions of polite society⟩ **b :** to feel or exhibit anger or revulsion ⟨*rebelled* at the injustice of life⟩

re·bel·lion \ri-'bel-yən\ *noun* (14th century)
1 : opposition to one in authority or dominance
2 a : open, armed, and usually unsuccessful defiance of or resistance to an established gov-

□ USAGE
reason *The reason is because* is a short way of referring to a number of locutions in which a primary clause with *reason* as its subject is followed by a subordinate clause introduced by *because* ⟨the *reason* I didn't answer the other letter was because mine was already on the way to you —Ernest Hemingway⟩. This locution has been used in standard English since the 16th century at least, but it has been attacked since a commentator named Robert Baker decided it made no sense in 1770. Modern objections are that the expression is redundant or illogical (which follows Baker) or that it is ungrammatical (which is incorrect). The universal recommendation is to replace *the reason is because* with *the reason is that.* Our evidence shows that *the reason is that* is more often used in edited prose than *the reason is because,* the latter is the more common in speech. We find *because* quite common in informal prose ⟨the *reason* the story has never been made into a film is because I won't sign a contract —E. B. White⟩ ⟨the *reason* I had not seen the nomination in the Tazewell paper was . . . because I did not then, as I do not now, take that paper —Abraham Lincoln⟩ ⟨the only *reason* my appearances are rare . . . is because nobody asks me oftener —Groucho Marx⟩ ⟨now the *reason* hair falls off is because it hangs down —Lewis Carroll⟩. Although this locution is perfectly ordinary, perfectly understandable, and perfectly standard, the handbooks continue to advise against it. They omit, however, Baker's other observation, that "there are scarce any even of our greatest authors, that avoid this way of speaking."

ernment **b :** an instance of such defiance or resistance ☆

re·bel·lious \-yəs\ *adjective* (15th century)
1 a : given to or engaged in rebellion ⟨*rebellious* troops⟩ **b :** of, relating to, or characteristic of a rebel or rebellion ⟨a *rebellious* speech⟩
2 : resisting treatment or management **:** REFRACTORY
— re·bel·lious·ly *adverb*
— re·bel·lious·ness *noun*

rebel yell *noun* (1863)
: a prolonged high-pitched yell often uttered by Confederate soldiers in the U.S. Civil War

re·birth \(ˌ)rē-ˈbərth, ˈrē-ˌ\ *noun* (1837)
1 a : a new or second birth **:** METEMPSYCHOSIS **b :** spiritual regeneration
2 : RENAISSANCE, REVIVAL ⟨a *rebirth* of nationalism⟩

Re·blo·chon \rə-blō-ˈshō^{n}\ *noun* [French *reblochon,* from French dialect (Savoy)] (1908)
: a semisoft creamy mild-flavored French cheese

reb·o·ant \ˈre-bə-wənt\ *adjective* [Latin *reboant-, reboans,* present participle of *reboare* to resound, from *re-* + *boare* to cry aloud, roar, from Greek *boan,* of imitative origin] (1830)
: marked by reverberation

re·born \(ˌ)rē-ˈbȯrn\ *adjective* (1598)
: born again **:** REGENERATED, REVIVED

1re·bound \ˈrē-ˌbau̇nd, ri-ˈ\ *verb* [Middle English, from Middle French *rebondir,* from Old French, from *re-* + *bondir* to bound — more at BOUND] (14th century)
intransitive verb
1 a : to spring back on or as if on collision or impact with another body **b :** to recover from setback or frustration
2 : REECHO
3 : to gain possession of a rebound in basketball
transitive verb
: to cause to rebound
— re·bound·er \ˈrē-ˌbau̇n-dər, ri-ˈ\ *noun*

2re·bound \ˈre-ˌbau̇nd, ri-ˈ\ *noun* (1530)
1 a : the action of rebounding **:** RECOIL **b :** an upward leap or movement **:** RECOVERY ⟨a sharp *rebound* in prices⟩
2 a : a basketball or hockey puck that rebounds **b :** the act or an instance of gaining possession of a basketball rebound ⟨leads the league in *rebounds*⟩
3 : a reaction to setback, frustration, or crisis ⟨on the *rebound* from an unhappy love affair⟩

re·bo·zo \ri-ˈbō-(ˌ)zō, -(ˌ)sō\ *noun, plural* **-zos** [Spanish, shawl, from *rebozar* to muffle, alteration of *embozar* to muffle, probably from (assumed) Vulgar Latin *imbucciare,* from Latin *in-* + *bucca* cheek] (1807)
: a long scarf worn chiefly by Mexican women

re·branch \(ˌ)rē-ˈbranch\ *intransitive verb* (1888)
: to form secondary branches

re·broad·cast \(ˌ)rē-ˈbrȯd-ˌkast\ *transitive verb* **-cast; -cast·ing** (1923)
1 : to broadcast again (a radio or television program being simultaneously received from another source)
2 : to repeat (a broadcast) at a later time
— rebroadcast *noun*

re·buff \ri-ˈbəf\ *transitive verb* [Middle French *rebuffer,* from Old Italian *ribuffare* to reprimand, from *ribuffo* reprimand] (circa 1586)
: to reject or criticize sharply **:** SNUB
— rebuff *noun*

re·build \(ˌ)rē-ˈbild\ *verb* **-built** \-ˈbilt\; **-build·ing** (1537)
transitive verb
1 a : to make extensive repairs to **:** RECONSTRUCT ⟨*rebuild* a war-torn city⟩ **b :** to restore to a previous state ⟨*rebuild* inventories⟩
2 : to make extensive changes in **:** REMODEL ⟨*rebuild* society⟩
intransitive verb
: to build again ⟨planned to *rebuild* after the fire⟩
synonym see MEND

1re·buke \ri-ˈbyük\ *transitive verb* **re·buked; re·buk·ing** [Middle English, from Old North French *rebuker*] (14th century)
1 a : to criticize sharply **:** REPRIMAND **b :** to serve as a rebuke to
2 : to turn back or keep down **:** CHECK
synonym see REPROVE
— re·buk·er *noun*

2rebuke *noun* (15th century)
: an expression of strong disapproval **:** REPRIMAND

re·bus \ˈrē-bəs\ *noun* [Latin, by things, ablative plural of *res* thing — more at REAL] (1605)
: a representation of words or syllables by pictures of objects or by symbols whose names resemble the intended words or syllables in sound; *also* **:** a riddle made up of such pictures or symbols

rebus

re·but \ri-ˈbət\ *verb* **re·but·ted; re·but·ting** [Middle English, from Middle French *reboter,* from *re-* + *boter* to butt — more at BUTT] (14th century)
transitive verb
1 : to drive or beat back **:** REPEL
2 a : to contradict or oppose by formal legal argument, plea, or countervailing proof **b :** to expose the falsity of **:** REFUTE
intransitive verb
: to make or furnish an answer or counter proof
— re·but·ta·ble \-ˈbə-tə-bəl\ *adjective*

re·but·tal \ri-ˈbə-t^{ə}l\ *noun* (1830)
: the act of rebutting especially in a legal suit; *also* **:** argument or proof that rebuts

1re·but·ter \-ˈbə-tər\ *noun* [Anglo-French *rebuter,* from Old French *reboter* to rebut] (1540)
: the answer of a defendant in matter of fact to a plaintiff's surrejoinder

2rebutter *noun* (1794)
: one that rebuts

re·cal·ci·trance \ri-ˈkal-sə-trən(t)s\ *noun* (1856)
: the state of being recalcitrant

re·cal·ci·tran·cy \-trən(t)-sē\ *noun* (1869)
: RECALCITRANCE

re·cal·ci·trant \-trənt\ *adjective* [Late Latin *recalcitrant-, recalcitrans,* present participle of *recalcitrare* to be stubbornly disobedient, from Latin, to kick back, from *re-* + *calcitrare* to kick, from *calc-, calx* heel] (1843)
1 : obstinately defiant of authority or restraint
2 a : difficult to manage or operate **b :** not responsive to treatment **c :** RESISTANT ⟨this subject is *recalcitrant* both to observation and to experiment —G. G. Simpson⟩ ◆
synonym see UNRULY
— recalcitrant *noun*

1re·call \ri-ˈkȯl\ *transitive verb* (1582)
1 a : to call back ⟨was *recalled* to active duty⟩ **b :** to bring back to mind ⟨*recall* those early years⟩ **c :** to remind one of **:** RESEMBLE ⟨a playwright who *recalls* the Elizabethan dramatists⟩
2 : CANCEL, REVOKE
3 : RESTORE, REVIVE
synonym see REMEMBER
— re·call·abil·i·ty \-ˌkȯ-lə-ˈbi-lə-tē\ *noun*
— re·call·able \-ˈkȯ-lə-bəl\ *adjective*
— re·call·er *noun*

2re·call \ri-ˈkȯl, ˈrē-ˌ\ *noun* (1611)
1 : a call to return ⟨a *recall* of workers after a layoff⟩
2 : the right or procedure by which an official may be removed by vote of the people
3 : remembrance of what has been learned or experienced
4 : the act of revoking
5 : a public call by a manufacturer for the return of a product that may be defective or contaminated

re·ca·mier \ˌrā-käm-ˈyā\ *noun* [from its appearance in a portrait of Mme. Récamier by Jacques-Louis David] (1924)
: a sometimes backless couch with a high curved headrest and low footrest

recamier

re·can·a·li·za·tion \(ˌ)rē-ˌka-n^{ə}l-ə-ˈzā-shən\ *noun* (1953)
: the process of restoring flow to or reuniting an interrupted channel of a bodily tube (as a blood vessel or vas deferens)
— re·can·a·lize \-kə-ˈna-ˌlīz, -ˈka-n^{ə}l-ˌīz\ *transitive verb*

re·cant \ri-ˈkant\ *verb* [Latin *recantare,* from *re-* + *cantare* to sing — more at CHANT] (1535)
transitive verb
1 : to withdraw or repudiate (a statement or belief) formally and publicly **:** RENOUNCE
2 : REVOKE
intransitive verb
: to make an open confession of error
synonym see ABJURE
— re·can·ta·tion \ˌrē-ˌkan-ˈtā-shən\ *noun*

1re·cap \ˈrē-ˌkap\ *noun* [by shortening] (circa 1926)

☆ SYNONYMS
Rebellion, revolution, uprising, revolt, insurrection, mutiny mean an outbreak against authority. REBELLION implies an open formidable resistance that is often unsuccessful ⟨open *rebellion* against the officers⟩. REVOLUTION applies to a successful rebellion resulting in a major change (as in government) ⟨a political *revolution* that toppled the monarchy⟩. UPRISING implies a brief, limited and often immediately ineffective rebellion ⟨quickly put down the *uprising*⟩. REVOLT and INSURRECTION imply an armed uprising that quickly fails or succeeds ⟨a *revolt* by the Young Turks that surprised party leaders⟩ ⟨an *insurrection* of oppressed laborers⟩. MUTINY applies to group insubordination or insurrection especially against naval authority ⟨a *mutiny* led by the ship's cook⟩.

◇ WORD HISTORY
recalcitrant Underlying the word *recalcitrant* is the image of a stubborn beast of burden kicking back at its handlers. The rather uncommon Latin word *calcitrare* "to kick with the heels" was a derivative of an unattested noun *calcitrum,* which presumably meant "kick" and was itself formed from the well-attested *calx* "heel." The verb *recalcitrare* "to kick back," formed from *calcitrare* with the addition of the prefix *re-* "back," is known only once from classical Latin. But in the Latin of the church fathers the word occurs more frequently and in the figurative usage "to be disobedient, rebel," in part as a loan translation of the Greek verb *apolaktizein.* The now rare English word *recalcitrate* "to show opposition to" was borrowed from Latin *recalcitrare* in the 17th century, but much more common now is *recalcitrant* (based on the present participle of *recalcitrare*), which is not attested until the 19th century.

: RECAPITULATION

²re·cap \'rē-,kap, ri-'\ *verb* **re·capped; re·cap·ping** (1945)

: RECAPITULATE

³re·cap \'rē-,kap\ *noun* [⁴*recap*] (1940)

: RETREAD 1

⁴re·cap \(,)rē-'kap\ *transitive verb* **re·capped; re·cap·ping** [*re-* + ¹*cap*] (1941)

: RETREAD

— **re·cap·pa·ble** \-'ka-pə-bəl\ *adjective*

re·cap·i·tal·i·za·tion \(,)rē-,ka-pə-tᵊl-ə-'zā-shən, -,kap-tᵊl-\ *noun* (1920)

: a revision of the capital structure of a corporation

re·cap·i·tal·ize \(,)rē-'ka-pə-tᵊl-,īz, -'kap-tᵊl-\ *transitive verb* (1904)

: to change the capital structure of

re·ca·pit·u·late \,rē-kə-'pi-chə-,lāt\ *verb* **-lat·ed; -lat·ing** [Late Latin *recapitulatus,* past participle of *recapitulare* to restate by heads, sum up, from Latin *re-* + *capitulum* division of a book — more at CHAPTER] (1570)

transitive verb

: to repeat the principal points or stages of : SUMMARIZE

intransitive verb

: SUM UP

re·ca·pit·u·la·tion \-,pi-chə-'lā-shən\ *noun* (14th century)

1 : a concise summary

2 : the hypothetical occurrence in an individual organism's development of successive stages resembling the series of ancestral types from which it has descended so that the ontogeny of the individual is a recapitulation of the phylogeny of its group

3 : the third section of a sonata form

¹re·cap·ture \(,)rē-'kap-chər\ *noun* (1752)

1 a : the act of retaking **b :** an instance of being retaken

2 : the retaking of a prize or goods under international law

3 : a government seizure under law of earnings or profits beyond a fixed amount

²recapture *transitive verb* (1799)

1 a : to capture again **b :** to experience again ⟨by no effort of the imagination could she *recapture* the ecstasy —Ellen Glasgow⟩

2 : to take (as a portion of earnings or profits above a fixed amount) by law or through negotiations under law

re·cast \(,)rē-'kast\ *transitive verb* **-cast; -cast·ing** (1603)

: to cast again ⟨*recast* a gun⟩ ⟨*recast* a play⟩; *also* : REMODEL, REFASHION ⟨*recasts* his political image to fit the times⟩

— **re·cast** \'rē-,kast, (,)rē-'\ *noun*

rec·ce \'re-kē\ *noun, often attributive* [by shortening & alteration] (1941)

: RECONNAISSANCE

¹re·cede \ri-'sēd\ *intransitive verb* **re·ced·ed; re·ced·ing** [Middle English, from Latin *recedere* to go back, from *re-* + *cedere* to go] (15th century)

1 a : to move back or away : WITHDRAW **b :** to slant backward

2 : to grow less or smaller : DIMINISH, DECREASE ☆

²re·cede \(,)rē-'sēd\ *transitive verb* [*re-* + *cede*] (1771)

: to cede back to a former possessor

¹re·ceipt \ri-'sēt\ *noun* [Middle English *receite,* from Old North French, from Medieval Latin *recepta,* probably from Latin, neuter plural of *receptus,* past participle of *recipere* to receive] (14th century)

1 : RECIPE

2 a *obsolete* : RECEPTACLE **b** *archaic* : a revenue office

3 : the act or process of receiving

4 : something received — usually used in plural

5 : a writing acknowledging the receiving of goods or money

²receipt *transitive verb* (1787)

1 : to give a receipt for or acknowledge the receipt of

2 : to mark as paid

re·ceiv·able \ri-'sē-və-bəl\ *adjective* (14th century)

1 : capable of being received

2 : subject to call for payment ⟨notes *receivable*⟩

re·ceiv·ables \-bəlz\ *noun plural* (1863)

: amounts of money receivable

re·ceive \ri-'sēv\ *verb* **re·ceived; re·ceiv·ing** [Middle English, from Old North French *receivre,* from Latin *recipere,* from *re-* + *capere* to take — more at HEAVE] (14th century)

transitive verb

1 : to come into possession of : ACQUIRE ⟨*receive* a gift⟩

2 a : to act as a receptacle or container for ⟨the cistern *receives* water from the roof⟩ **b :** to assimilate through the mind or senses ⟨*receive* new ideas⟩

3 a : to permit to enter : ADMIT **b :** WELCOME, GREET **c :** to react to in a specified manner

4 : to accept as authoritative, true, or accurate : BELIEVE

5 a : to support the weight or pressure of : BEAR **b :** to take (a mark or impression) from the weight of something ⟨some clay *receives* clear impressions⟩ **c :** ACQUIRE, EXPERIENCE ⟨*received* his early schooling at home⟩ **d :** to suffer the hurt or injury of ⟨*received* a broken nose⟩

intransitive verb

1 : to be a recipient

2 : to be at home to visitors ⟨*receives* on Tuesdays⟩

3 : to convert incoming radio waves into perceptible signals

4 : to prepare to take possession of the ball from a kick in football

received *adjective* (15th century)

: generally accepted : COMMON ⟨a healthy skepticism about *received* explanations —B. K. Lewalski⟩

Received Pronunciation *noun* (1869)

: the pronunciation of Received Standard

Received Standard *noun* (1913)

: a traditionally prestigious form of English spoken at the English public schools, at the universities of Oxford and Cambridge, and by many educated British people elsewhere

re·ceiv·er \ri-'sē-vər\ *noun* (14th century)

: one that receives: as **a :** TREASURER **b** (1) **:** a person appointed to hold in trust and administer property under litigation (2) **:** a person appointed to settle the affairs of a business involving a public interest or to manage a corporation during reorganization **c :** one that receives stolen goods : FENCE **d :** a device for converting signals (as electromagnetic waves) into audio or visual form: as (1) **:** a device in a telephone for converting electric impulses or varying current into sound (2) **:** a radio receiver with a tuner and amplifier on one chassis **e** (1) **:** CATCHER (2) **:** a member of the offensive team in football eligible to catch a forward pass

receiver general *noun, plural* **receivers general** (15th century)

: a public officer in charge of the treasury (as of Massachusetts)

re·ceiv·er·ship \ri-'sē-vər-,ship\ *noun* (15th century)

1 : the office or function of a receiver

2 : the state of being in the hands of a receiver

receiving blanket *noun* (1926)

: a small lightweight blanket used to wrap an infant (as after bathing)

receiving end *noun* (1937)

: the position of being a recipient or especially a victim — usually used in the phrase *on the receiving end*

receiving line *noun* (1933)

: a group of people who stand in a line and individually welcome guests (as at a wedding reception)

re·cen·cy \'rē-sᵊn(t)-sē\ *noun* (1612)

: the quality or state of being recent

re·cen·sion \ri-'sen(t)-shən\ *noun* [Latin *recension-, recensio* enumeration, from *recensēre* to review, from *re-* + *censēre* to assess, tax — more at CENSOR] (circa 1828)

1 : a critical revision of a text

2 : a text established by critical revision

re·cent \'rē-sᵊnt\ *adjective* [Middle English, from Middle French or Latin; Middle French, from Latin *recent-, recens;* perhaps akin to Greek *kainos* new] (15th century)

1 a : having lately come into existence : NEW, FRESH **b :** of or relating to a time not long past

2 *capitalized* : HOLOCENE

— **re·cent·ness** *noun*

re·cent·ly *adverb* (1533)

: during a recent period of time : LATELY

re·cep·ta·cle \ri-'sep-ti-kəl\ *noun* [Middle English, from Latin *receptaculum,* from *receptare* to receive, frequentative of *recipere* to receive] (15th century)

1 : one that receives and contains something : CONTAINER

2 [New Latin *receptaculum,* from Latin] **a :** the end of the flower stalk upon which the floral organs are borne **b :** a modified branch bearing sporangia in a cryptogamous plant

3 : a mounted female electrical fitting that contains the live parts of the circuit

re·cep·tion \ri-'sep-shən\ *noun* [Middle English *recepcion,* from Middle French or Latin; Middle French *reception,* from Latin *reception-, receptio,* from *recipere*] (15th century)

1 : the act or action or an instance of receiving: as **a :** RECEIPT ⟨the *reception* and distribution of funds⟩ **b :** ADMISSION ⟨*reception* into the church⟩ **c :** RESPONSE, REACTION ⟨the play met with a mixed *reception*⟩ **d :** the receiving of a radio or television broadcast **e :** the catching of a forward pass by a receiver

2 : a social gathering often for the purpose of extending a formal welcome

re·cep·tion·ist \-sh(ə-)nist\ *noun* (1901)

: a person employed to greet telephone callers, visitors, patients, or clients

re·cep·tive \ri-'sep-tiv\ *adjective* (15th century)

1 : able or inclined to receive; *especially* : open and responsive to ideas, impressions, or suggestions

2 a *of a sensory end organ* : fit to receive and transmit stimuli **b :** SENSORY

— **re·cep·tive·ly** *adverb*

— **re·cep·tive·ness** *noun*

— **re·cep·tiv·i·ty** \,rē-,sep-'ti-və-tē, ri-\ *noun*

re·cep·tor \ri-'sep-tər\ *noun* (1898)

: RECEIVER: as **a :** a cell or group of cells that receives stimuli : SENSE ORGAN **b :** a chemical group or molecule (as a protein) on the cell surface or in the cell interior that has an affinity for a specific chemical group, molecule, or virus

¹re·cess \'rē-,ses, ri-'\ *noun* [Latin *recessus,* from *recedere* to recede] (1531)

1 : the action of receding : RECESSION

2 : a hidden, secret, or secluded place or part

3 a : INDENTATION, CLEFT ⟨a deep *recess* in the hill⟩ **b :** ALCOVE ⟨a *recess* lined with books⟩

☆ SYNONYMS

Recede, retreat, retract, back mean to move backward. RECEDE implies a gradual withdrawing from a forward or high fixed point in time or space ⟨the flood waters gradually *receded*⟩. RETREAT implies withdrawal from a point or position reached ⟨*retreating* soldiers⟩. RETRACT implies drawing back from an extended position ⟨a cat *retracting* its claws⟩. BACK is used with *up, down, out,* or *off* to refer to any retrograde motion ⟨*backed* off on the throttle⟩.

4 : a suspension of business or procedure often for rest or relaxation ⟨children playing at *recess*⟩

²recess (1809)
transitive verb
1 : to put into a recess ⟨*recessed* lighting⟩
2 : to make a recess in
3 : to interrupt for a recess
intransitive verb
: to take a recess

¹re·ces·sion \ri-ˈse-shən\ *noun* (circa 1652)
1 : the act or action of receding **:** WITHDRAWAL
2 : a departing procession (as of clergy and choir at the end of a church service)
3 : a period of reduced economic activity
— re·ces·sion·ary \-shə-ˌner-ē\ *adjective*

²re·ces·sion \(ˌ)rē-ˈse-shən\ *noun* [*re-* + *cession*] (1828)
: the act of ceding back to a former possessor

¹re·ces·sion·al \ri-ˈsesh-nəl, -ˈse-shə-nᵊl\ *adjective* (1867)
: of or relating to a withdrawal

²recessional *noun* (1867)
1 : a hymn or musical piece at the conclusion of a service or program
2 : ¹RECESSION 2

¹re·ces·sive \ri-ˈse-siv\ *adjective* (circa 1673)
1 a : tending to recede **b :** WITHDRAWN 2
2 a : producing little or no phenotypic effect when occurring in heterozygous condition with a contrasting allele ⟨*recessive* genes⟩ **b :** expressed only when the determining gene is in the homozygous condition ⟨*recessive* traits⟩
— re·ces·sive·ly *adverb*
— re·ces·sive·ness *noun*

²recessive *noun* (1900)
1 : an organism possessing one or more recessive characters
2 : a recessive character or gene

re·charge \(ˌ)rē-ˈchärj\ (1598)
intransitive verb
1 : to make a new attack
2 : to regain energy or spirit
transitive verb
1 : to charge again; *especially* **:** to restore anew the active materials in (a storage battery)
2 : to inspire or invigorate afresh **:** RENEW
— re·charge \(ˌ)rē-ˈchärj, ˈrē-ˌ\ *noun*
— re·charge·able \(ˌ)rē-ˈchär-jə-bəl\ *adjective*
— re·charg·er \-jər\ *noun*

ré·chauf·fé \ˌrā-shō-ˈfā, -ˈshō-ˌ\ *noun* [French, from *réchauffé* warmed-over, from past participle of *réchauffer* to warm over, from *ré-* re- + *chauffer* to warm, from Middle French *chaufer* — more at CHAFE] (1805)
1 : REHASH
2 : a warmed-over dish of food

re·cheat \ri-ˈchēt\ *noun* [Middle English *rechate,* from *rechaten* to blow the recheat, from Middle French *rachater* to assemble, rally, from *re-* + *achater* to acquire, from (assumed) Vulgar Latin *accaptare* — more at CATE] (15th century)
: a hunting call sounded on a horn to assemble the hounds

re·cher·ché \rə-ˌsher-ˈshā, -ˈsher-ˌ\ *adjective* [French, from past participle of *rechercher* to seek out, from Middle French *recherchier* — more at RESEARCH] (1722)
1 a : EXQUISITE, CHOICE **b :** EXOTIC, RARE
2 : excessively refined **:** AFFECTED
3 : PRETENTIOUS, OVERBLOWN

re·cid·i·vism \ri-ˈsi-də-ˌvi-zəm\ *noun* (1886)
: a tendency to relapse into a previous condition or mode of behavior; *especially* **:** relapse into criminal behavior

re·cid·i·vist \-vist\ *noun* [French *récidiviste,* from *récidiver* to relapse, from Medieval Latin *recidivare,* from Latin *recidivus* recurring, from *recidere* to fall back, from *re-* + *cadere* to fall — more at CHANCE] (1880)
: one who relapses; *specifically* **:** an habitual criminal
— recidivist *adjective*
— re·cid·i·vis·tic \-ˌsi-də-ˈvis-tik\ *adjective*

rec·i·pe \ˈre-sə-(ˌ)pē\ *noun* [Latin, take, imperative of *recipere* to take, receive — more at RECEIVE] (1584)
1 : PRESCRIPTION 4a
2 : a set of instructions for making something from various ingredients
3 : a formula or procedure for doing or attaining something ⟨a *recipe* for success⟩

re·cip·i·ent \ri-ˈsi-pē-ənt\ *noun* [Latin *recipient-, recipiens,* present participle of *recipere*] (1558)
: one that receives **:** RECEIVER
— recipient *adjective*

¹re·cip·ro·cal \ri-ˈsi-prə-kəl\ *adjective* [Latin *reciprocus* returning the same way, alternating] (1570)
1 a : inversely related **:** OPPOSITE **b :** of, constituting, or resulting from paired crosses in which the kind that supplies the male parent of the first cross supplies the female parent of the second cross and vice versa
2 : shared, felt, or shown by both sides
3 : serving to reciprocate **:** consisting of or functioning as a return in kind ⟨the *reciprocal* devastation of nuclear war⟩
4 a : mutually corresponding ⟨agreed to extend *reciprocal* privileges to each other's citizens⟩ **b :** marked by or based on reciprocity ⟨*reciprocal* trade agreements⟩
— re·cip·ro·cal·ly \-k(ə-)lē\ *adverb*

²reciprocal *noun* (1570)
1 : something in a reciprocal relationship to another
2 : either of a pair of numbers (as ⅔ and 3/2 or 9 and 1/9) whose product is one; *broadly* **:** MULTIPLICATIVE INVERSE

reciprocal pronoun *noun* (1844)
: a pronoun (as *each other*) used when its referents are predicated to bear the same relationship to one another

re·cip·ro·cate \ri-ˈsi-prə-ˌkāt\ *verb* **-cat·ed; -cat·ing** (1607)
transitive verb
1 : to give and take mutually
2 : to return in kind or degree ⟨*reciprocate* a compliment gracefully⟩
intransitive verb
1 : to make a return for something ⟨we hope to *reciprocate* for your kindness⟩
2 : to move forward and backward alternately ⟨a *reciprocating* valve⟩ ☆
— re·cip·ro·ca·tor \-ˌkā-tər\ *noun*

reciprocating engine *noun* (1822)
: an engine in which the to-and-fro motion of one or more pistons is transformed into the rotary motion of a crankshaft

re·cip·ro·ca·tion \ri-ˌsi-prə-ˈkā-shən\ *noun* (1561)
1 a : a mutual exchange **b :** a return in kind or of like value
2 : an alternating motion
— re·cip·ro·ca·tive \-ˈsi-prə-ˌkā-tiv, -kə-\ *adjective*

rec·i·proc·i·ty \ˌre-sə-ˈprä-s(ə-)tē\ *noun, plural* **-ties** (1766)
1 : the quality or state of being reciprocal **:** mutual dependence, action, or influence
2 : a mutual exchange of privileges; *specifically* **:** a recognition by one of two countries or institutions of the validity of licenses or privileges granted by the other

re·ci·sion \ri-ˈsi-zhən\ *noun* [Middle French, alteration of *rescision,* from Late Latin *rescission-, rescissio* rescission] (1611)
: an act of rescinding **:** CANCELLATION

re·cit·al \ri-ˈsī-tᵊl\ *noun* (1536)
1 a : a detailed account **:** ENUMERATION ⟨a *recital* of names and dates⟩ **b :** the act or process or an instance of reciting **c :** DISCOURSE, NARRATION ⟨a colorful *recital* of a night on the town⟩
2 a : a concert given by an individual musician or dancer or by a dance troupe **b :** a public exhibition of skill given by music or dance pupils
— re·cit·al·ist \-tᵊl-ist\ *noun*

rec·i·ta·tion \ˌre-sə-ˈtā-shən\ *noun* (15th century)
1 : the act of enumerating ⟨a *recitation* of relevant details⟩
2 : the act or an instance of reading or repeating aloud especially publicly
3 a : a student's oral reply to questions **b :** a class period especially in association with and for review of a lecture

rec·i·ta·tive \ˌre-sə-tə-ˈtēv, ˌres-tə-\ *noun* [Italian *recitativo,* from *recitare* to recite, from Latin] (1656)
1 : a rhythmically free vocal style that imitates the natural inflections of speech and that is used for dialogue and narrative in operas and oratorios; *also* **:** a passage to be delivered in this style
2 : RECITATION 2
— recitative *adjective*

rec·i·ta·ti·vo \-ˈtē-(ˌ)vō\ *noun, plural* **-vi** \-(ˌ)vē\ *or* **-vos** [Italian] (1645)
: RECITATIVE 1

re·cite \ri-ˈsīt\ *verb* **re·cit·ed; re·cit·ing** [Middle English, to state formally, from Middle French or Latin; Middle French *reciter* to recite, from Latin *recitare,* from *re-* + *citare* to summon — more at CITE] (15th century)
transitive verb
1 : to repeat from memory or read aloud publicly
2 a : to relate in full ⟨*recites* dull anecdotes⟩ **b :** to give a recital of **:** DETAIL ⟨*recited* a catalog of offenses⟩
3 : to repeat or answer questions about (a lesson)
intransitive verb
1 : to repeat or read aloud something memorized or prepared
2 : to reply to a teacher's question on a lesson
— re·cit·er *noun*

reck \ˈrek\ *verb* [Middle English, to take heed, from Old English *reccan;* akin to Old High German *ruohhen* to take heed] (before 12th century)
intransitive verb
1 : WORRY, CARE
2 *archaic* **:** to be of account or interest **:** MATTER
transitive verb
1 *archaic* **:** to care for **:** REGARD
2 *archaic* **:** to matter to **:** CONCERN

reck·less \ˈre-kləs\ *adjective* (before 12th century)

> ☆ SYNONYMS
> **Reciprocate, retaliate, requite, return** mean to give back usually in kind or in quantity. RECIPROCATE implies a mutual or equivalent exchange or a paying back of what one has received ⟨*reciprocated* their hospitality by inviting them for a visit⟩. RETALIATE usually implies a paying back of injury in exact kind, often vengefully ⟨the enemy *retaliated* by executing their prisoners⟩. REQUITE implies a paying back according to one's preference and often not equivalently ⟨*requited* her love with cold indifference⟩. RETURN implies a paying or giving back ⟨*returned* their call⟩ ⟨*return* good for evil⟩.

1 : marked by lack of proper caution **:** careless of consequences
2 : IRRESPONSIBLE ⟨*reckless* charges⟩
synonym see ADVENTUROUS
— reck·less·ly *adverb*
— reck·less·ness *noun*

reck·on \ˈre-kən\ *verb* **reck·oned; reck·on·ing** \ˈre-kə-niŋ, ˈrek-niŋ\ [Middle English *rekenen,* from Old English *-recenian* (as in *gerecenian* to narrate); akin to Old English *reccan*] (13th century)
transitive verb
1 a : COUNT ⟨*reckon* the days till Christmas⟩ **b :** ESTIMATE, COMPUTE ⟨*reckon* the height of a building⟩ **c :** to determine by reference to a fixed basis ⟨the existence of the U.S. is *reckoned* from the Declaration of Independence⟩
2 : to regard or think of as **:** CONSIDER
3 *chiefly dialect* **:** THINK, SUPPOSE ⟨I *reckon* I've outlived my time —Ellen Glasgow⟩
intransitive verb
1 : to settle accounts
2 : to make a calculation
3 a : JUDGE **b** *chiefly dialect* **:** SUPPOSE, THINK
4 : to accept something as certain **:** place reliance ⟨I *reckon* on your promise to help⟩
— reckon with : to take into consideration
— reckon without : to fail to consider **:** IGNORE

reckoning *noun* (14th century)
1 : the act or an instance of reckoning: as **a :** ACCOUNT, BILL **b :** COMPUTATION **c :** calculation of a ship's position
2 : a settling of accounts ⟨day of *reckoning*⟩
3 : a summing up

re·claim \ri-ˈklām\ *transitive verb* [Middle English *reclamen,* from Middle French *reclamer* to call back, from Latin *reclamare* to cry out against, from *re-* + *clamare* to cry out — more at CLAIM] (14th century)
1 a : to recall from wrong or improper conduct **:** REFORM **b :** TAME, SUBDUE
2 a : to rescue from an undesirable state **b :** to make available for human use by changing natural conditions ⟨*reclaim* swampland⟩
3 : to obtain from a waste product or by-product **:** RECOVER
4 a : to demand or obtain the return of **b :** to regain possession of
synonym see RESCUE
— re·claim·able \-ˈklā-mə-bəl\ *adjective*

rec·la·ma·tion \ˌre-klə-ˈmā-shən\ *noun* [French *réclamation,* from Latin *reclamation-, reclamatio,* from *reclamare*] (1633)
: the act or process of reclaiming: as **a :** REFORMATION, REHABILITATION **b :** restoration to use **:** RECOVERY

ré·clame \rā-ˈkläm\ *noun* [French, advertising, from *réclamer* to appeal, from Middle French *reclamer*] (1883)
1 : a gift for dramatization or publicity **:** SHOWMANSHIP
2 : public acclaim **:** VOGUE

re·cline \ri-ˈklīn\ *verb* **re·clined; re·clin·ing** [Middle English, from Middle French or Latin; Middle French *recliner,* from Latin *reclinare,* from *re-* + *clinare* to bend — more at LEAN] (15th century)
transitive verb
: to cause or permit to incline backwards
intransitive verb
1 : to lean or incline backwards
2 : REPOSE, LIE

re·clin·er \-ˈklī-nər\ *noun* (1928)
: a chair with an adjustable back and footrest

re·clos·able \(ˌ)rē-ˈklō-zə-bəl\ *adjective* (1965)
: capable of being tightly closed again after opening ⟨*reclosable* packages of bacon⟩

¹re·cluse \ˈre-ˌklüs, ri-ˈklüs, ˈre-ˌklüz\ *adjective* [Middle English, from Old French *reclus,* literally, shut up, from Late Latin *reclusus,* past participle of *recludere* to shut up, from Latin *re-* + *claudere* to close — more at CLOSE] (13th century)
: marked by withdrawal from society **:** SOLITARY
— re·clu·sive \ri-ˈklü-siv, -ziv\ *adjective*
— re·clu·sive·ly *adverb*
— re·clu·sive·ness *noun*

²recluse *noun* (13th century)
: a person who leads a secluded or solitary life

re·clu·sion \ri-ˈklü-zhən\ *noun* (15th century)
: the state of being recluse

rec·og·nise *chiefly British variant of* RECOGNIZE

rec·og·ni·tion \ˌre-kig-ˈni-shən, -kəg-\ *noun* [Middle English *recognicion,* from Latin *recognition-, recognitio,* from *recognoscere*] (15th century)
1 : the action of recognizing **:** the state of being recognized: as **a :** ACKNOWLEDGMENT; *especially* **:** formal acknowledgment of the political existence of a government or nation **b :** knowledge or feeling that someone or something present has been encountered before
2 : special notice or attention
3 : the sensing and encoding of printed or written data by a machine ⟨optical character *recognition*⟩ ⟨magnetic ink character *recognition*⟩

re·cog·ni·zance \ri-ˈkäg-nə-zən(t)s, -ˈkä-nə-\ *noun* [Middle English, alteration of *reconissaunce,* from Middle French *reconoissance* recognition, from *reconoistre* to recognize] (14th century)
1 a : an obligation of record entered into before a court or magistrate requiring the performance of an act (as appearance in court) usually under penalty of a money forfeiture ⟨released on his own *recognizance*⟩ **b :** the sum liable to forfeiture upon such an obligation
2 *archaic* **:** TOKEN, PLEDGE

rec·og·nize \ˈre-kig-ˌnīz, -kəg-\ *transitive verb* **-nized; -niz·ing** [modification of Middle French *reconoiss-,* stem of *reconoistre,* from Latin *recognoscere,* from *re-* + *cognoscere* to know — more at COGNITION] (circa 1532)
1 : to acknowledge formally: as **a :** to admit as being lord or sovereign **b :** to admit as being of a particular status **c :** to admit as being one entitled to be heard **:** give the floor to **d :** to acknowledge the de facto existence or the independence of
2 : to acknowledge or take notice of in some definite way: as **a :** to acknowledge with a show of appreciation ⟨*recognize* an act of bravery with the award of a medal⟩ **b :** to acknowledge acquaintance with ⟨*recognize* a neighbor with a nod⟩
3 a : to perceive to be something or someone previously known ⟨*recognized* the word⟩ **b :** to perceive clearly **:** REALIZE
— rec·og·niz·abil·i·ty \ˌre-kig-ˌnī-zə-ˈbi-lə-tē, -kəg-\ *noun*
— rec·og·niz·able \ˈre-kəg-ˌnī-zə-bəl, -kig-\ *adjective*
— rec·og·niz·ably \-blē\ *adverb*
— rec·og·niz·er *noun*

¹re·coil \ri-ˈkȯi(ə)l\ *intransitive verb* [Middle English *reculen,* from Middle French *reculer,* from *re-* + *cul* backside — more at CULET] (14th century)
1 a : to fall back under pressure **b :** to shrink back physically or emotionally
2 : to spring back to or as if to a starting point **:** REBOUND
3 *obsolete* **:** DEGENERATE ☆

²re·coil \ˈrē-ˌkȯil, ri-ˈkȯi(ə)l\ *noun* (14th century)
1 : the act or action of recoiling; *especially* **:** the kickback of a gun upon firing
2 : REACTION ⟨the *recoil* from the rigors of Calvinism —Edmund Wilson⟩

re·coil·less \-ˌkȯil-ləs, -ˈkȯi(ə)l-\ *adjective* (1943)
: venting expanding propellant gas before recoil is produced ⟨*recoilless* rifle⟩ ⟨*recoilless* airgun⟩

recoil–operated *adjective* (1942)
of a firearm **:** utilizing the movement of parts in recoil to operate the action

re·coin \(ˌ)rē-ˈkȯin\ *transitive verb* (1685)
: to coin again or anew; *especially* **:** REMINT
— re·coin·age \-ˈkȯi-nij\ *noun*

rec·ol·lect \ˌre-kə-ˈlekt\ *verb* [Medieval Latin *recollectus,* past participle of *recolligere,* from Latin, to gather again] (1559)
transitive verb
1 : to bring back to the level of conscious awareness **:** REMEMBER ⟨trying to *recollect* the name⟩
2 : to remind (oneself) of something temporarily forgotten
intransitive verb
: to call something to mind
synonym see REMEMBER

re–col·lect \ˌrē-kə-ˈlekt\ *transitive verb* [partly from Latin *recollectus,* past participle of *recolligere,* from *re-* + *colligere* to collect; partly from *re-* + *collect*] (1604)
: to collect again; *especially* **:** RALLY, RECOVER

re·col·lect·ed \ˌrē-kə-ˈlek-təd\ *adjective* (1627)
: COMPOSED, CALM

rec·ol·lec·tion \ˌre-kə-ˈlek-shən\ *noun* (1624)
1 a : tranquillity of mind **b :** religious contemplation
2 a : the action or power of recalling to mind **b :** something recalled to the mind
synonym see MEMORY

re·com·bi·nant \(ˌ)rē-ˈkäm-bə-nənt\ *adjective* (1942)
1 : relating to or exhibiting genetic recombination ⟨*recombinant* progeny⟩
2 : relating to or containing recombinant DNA; *also* **:** produced by recombinant DNA technology
— recombinant *noun*

recombinant DNA *noun* (1975)
: genetically engineered DNA prepared in vitro by cutting up DNA molecules and splicing together specific DNA fragments usually from more than one species of organism

re·com·bi·na·tion \ˌrē-ˌkäm-bə-ˈnā-shən\ *noun* (1903)
: the formation by the processes of crossing-over and independent assortment of new combinations of genes in progeny that did not occur in the parents
— re·com·bi·na·tion·al \-shnəl, -shə-nᵊl\ *adjective*

rec·om·mend \ˌre-kə-ˈmend\ *transitive verb* [Middle English, to praise, from Medieval Latin *recommendare,* from Latin *re-* + *commendare* to commend] (14th century)
1 a : to present as worthy of acceptance or trial ⟨*recommended* the medicine⟩ **b :** to endorse as fit, worthy, or competent ⟨*recommends* her for the position⟩
2 : ENTRUST, COMMIT ⟨*recommended* his soul to God⟩

☆ SYNONYMS
Recoil, shrink, flinch, wince, blench, quail mean to draw back in fear or distaste. RECOIL implies a start or movement away through shock, fear, or disgust ⟨*recoiled* at the suggestion of stealing⟩. SHRINK suggests an instinctive recoil through sensitiveness, scrupulousness, or cowardice ⟨*shrank* from the unpleasant truth⟩. FLINCH implies a failure to endure pain or face something dangerous or frightening with resolution ⟨faced her accusers without *flinching*⟩. WINCE suggests a slight involuntary physical reaction (as a start or recoiling) ⟨*winced* in pain⟩. BLENCH implies fainthearted flinching ⟨stood their ground without *blenching*⟩. QUAIL suggests shrinking and cowering in fear ⟨*quailed* before the apparition⟩.

3 : to make acceptable ⟨has other points to *recommend* it⟩
4 : ADVISE ⟨*recommend* that the matter be dropped⟩
— rec·om·mend·able \-ˈmen-də-bəl\ *adjective*
— rec·om·men·da·to·ry \-də-ˌtōr-ē, -ˌtȯr-\ *adjective*

rec·om·men·da·tion \ˌre-kə-mən-ˈdā-shən, -ˌmen-\ *noun* (15th century)
1 a : the act of recommending **b :** something (as a procedure) recommended
2 : something that recommends or expresses commendation

re·com·mit \ˌrē-kə-ˈmit\ *transitive verb* (1621)
1 : to refer (as a bill) back to a committee
2 : to entrust or consign again
— re·com·mit·ment \-mənt\ *noun*
— re·com·mit·tal \-ˈmi-tᵊl\ *noun*

[1]rec·om·pense \ˈre-kəm-ˌpen(t)s\ *transitive verb* **-pensed; -pens·ing** [Middle English, from Middle French *recompenser,* from Late Latin *recompensare,* from Latin *re-* + *compensare* to compensate] (15th century)
1 a : to give something to by way of compensation (as for a service rendered or damage incurred) **b :** to pay for
2 : to return in kind **:** REQUITE
synonym see PAY

[2]recompense *noun* (15th century)
: an equivalent or a return for something done, suffered, or given **:** COMPENSATION ⟨offered in *recompense* for injuries⟩

re·com·pose \ˌrē-kəm-ˈpōz\ *transitive verb* (1611)
1 : to compose again **:** REARRANGE
2 : to restore to composure
— re·com·po·si·tion \(ˌ)rē-ˌkäm-pə-ˈzi-shən\ *noun*

re·con \ri-ˈkän, ˈrē-ˌkän\ *noun* (1918)
: RECONNAISSANCE

rec·on·cile \ˈre-kən-ˌsīl\ *verb* **-ciled; -cil·ing** [Middle English, from Middle French or Latin; Middle French *reconcilier,* from Latin *reconciliare,* from *re-* + *conciliare* to conciliate] (14th century)
transitive verb
1 a : to restore to friendship or harmony ⟨*reconciled* the factions⟩ **b :** SETTLE, RESOLVE ⟨*reconcile* differences⟩
2 : to make consistent or congruous ⟨*reconcile* an ideal with reality⟩
3 : to cause to submit to or accept something unpleasant ⟨was *reconciled* to hardship⟩
4 a : to check (a financial account) against another for accuracy **b :** to account for
intransitive verb
: to become reconciled
synonym see ADAPT
— rec·on·cil·abil·i·ty \ˌre-kən-ˌsī-lə-ˈbi-lə-tē\ *noun*
— rec·on·cil·able \ˌre-kən-ˈsī-lə-bəl, ˈre-kən-ˌ\ *adjective*
— rec·on·cile·ment \ˈre-kən-ˌsīl-mənt\ *noun*
— rec·on·cil·er *noun*

rec·on·cil·i·a·tion \ˌre-kən-ˌsi-lē-ˈā-shən\ *noun* [Middle English, from Latin *reconciliation-, reconciliatio,* from *reconciliare*] (14th century)
1 : the action of reconciling **:** the state of being reconciled
2 : the Roman Catholic sacrament of penance
— rec·on·cil·ia·to·ry \-ˈsil-yə-ˌtōr-ē, -ˈsi-lē-ə-, -ˌtȯr-\ *adjective*

re·con·dite \ˈre-kən-ˌdīt, ri-ˈkän-\ *adjective* [Latin *reconditus,* past participle of *recondere* to conceal, from *re-* + *condere* to store up, from *com-* + *-dere* to put — more at COM-, DO] (1649)
1 : hidden from sight **:** CONCEALED
2 : difficult or impossible for one of ordinary understanding or knowledge to comprehend **:** DEEP ⟨a *recondite* subject⟩
3 : of, relating to, or dealing with something little known or obscure ⟨*recondite* fact about the origin of the holiday —Floyd Dell⟩
— re·con·dite·ly *adverb*
— re·con·dite·ness *noun*

re·con·di·tion \ˌrē-kən-ˈdi-shən\ *transitive verb* (1920)
1 : to restore to good condition (as by replacing parts)
2 : to condition (as a person or a person's attitudes) anew; *also* **:** to reinstate (a response) in an organism

re·con·firm \ˌrē-kən-ˈfərm\ *transitive verb* (1611)
: to confirm again; *also* **:** to establish more strongly
— re·con·fir·ma·tion \(ˌ)rē-ˌkän-fər-ˈmā-shən\ *noun*

re·con·nais·sance \ri-ˈkä-nə-zən(t)s, -sən(t)s\ *noun* [French, literally, recognition, from Middle French *reconoissance* — more at RECOGNIZANCE] (1810)
: a preliminary survey to gain information; *especially* **:** an exploratory military survey of enemy territory

re·con·noi·ter *or* **re·con·noi·tre** \ˌrē-kə-ˈnȯi-tər, ˌre-kə-\ *verb* **-noi·tered** *or* **-noi·tred; -noi·ter·ing** *or* **-noi·tring** \-ˈnȯi-tə-riŋ, -ˈnȯi-triŋ\ [obsolete French *reconnoître,* literally, to recognize, from Middle French *reconoistre* — more at RECOGNIZE] (1707)
transitive verb
: to make a reconnaissance of
intransitive verb
: to engage in reconnaissance

re·con·sid·er \ˌrē-kən-ˈsi-dər\ (1571)
transitive verb
: to consider again especially with a view to changing or reversing
intransitive verb
: to consider something again
— re·con·sid·er·a·tion \-ˌsi-də-ˈrā-shən\ *noun*

re·con·sti·tute \(ˌ)rē-ˈkän(t)-stə-ˌtüt, -ˌtyüt\ *transitive verb* (1812)
: to constitute again or anew; *especially* **:** to restore to a former condition by adding water
— re·con·sti·tu·tion \(ˌ)rē-ˌkän(t)-stə-ˈtü-shən, -ˈtyü-\ *noun*

re·con·struct \ˌrē-kən-ˈstrəkt\ *transitive verb* (1768)
: to construct again: as **a :** to establish or assemble again **b :** to build up again mentally **:** RE-CREATE ⟨*reconstructing* a lost civilization⟩
— re·con·struct·ible \-ˈstrək-tə-bəl\ *adjective*
— re·con·struc·tive \-tiv\ *adjective*
— re·con·struc·tor \-tər\ *noun*

re·con·struc·tion \ˌrē-kən-ˈstrək-shən\ *noun* (1791)
1 a : the action of reconstructing **:** the state of being reconstructed **b** *often capitalized* **:** the reorganization and reestablishment of the seceded states in the Union after the American Civil War
2 : something reconstructed

re·con·struc·tion·ism \-shə-ˌni-zəm\ *noun, often capitalized* (1942)
1 : a movement in 20th century American Judaism that advocates a creative adjustment to contemporary conditions through the cultivation of traditions and folkways shared by all Jews
2 : advocacy of post-Civil War reconstruction
— re·con·struc·tion·ist \-sh(ə-)nist\ *adjective or noun,* often capitalized

reconstructive surgery *noun* (1943)
: surgery to restore function or normal appearance by remaking defective organs or parts

re·con·vey \ˌrē-kən-ˈvā\ *transitive verb* (1506)
: to convey back to a previous position or owner
— re·con·vey·ance \-ˈvā-ən(t)s\ *noun*

[1]re·cord \ri-ˈkȯrd\ *verb* [Middle English, literally, to recall, from Old French *recorder,* from Latin *recordari,* from *re-* + *cord-, cor* heart — more at HEART] (14th century)
transitive verb
1 a (1) **:** to set down in writing **:** furnish written evidence of (2) **:** to deposit an authentic official copy of ⟨*record* a deed⟩ **b :** to state for or as if for the record ⟨voted in favor but *recorded* certain reservations⟩ **c** (1) **:** to register permanently by mechanical means ⟨earthquake shocks *recorded* by a seismograph⟩ (2) **:** INDICATE, READ ⟨the thermometer *recorded* 90°⟩
2 : to give evidence of
3 : to cause (as sound, visual images, or data) to be registered on something (as a disc or magnetic tape) in reproducible form
intransitive verb
: to record something
— re·cord·able \-ˈkȯr-də-bəl\ *adjective*

[2]rec·ord \ˈre-kərd *also* -ˌkȯrd\ *noun* (14th century)
1 : the state or fact of being recorded
2 : something that records: as **a :** something that recalls or relates past events **b :** an official document that records the acts of a public body or officer **c :** an authentic official copy of a document deposited with a legally designated officer **d :** the official copy of the papers used in a law case
3 a (1) **:** a body of known or recorded facts about something or someone especially with reference to a particular sphere of activity that often forms a discernible pattern ⟨a good academic *record*⟩ ⟨a liberal voting *record*⟩ (2) **:** a collection of related items of information (as in a database) treated as a unit **b** (1) **:** an attested top performance (2) **:** an unsurpassed statistic
4 : something on which sound or visual images have been recorded; *specifically* **:** a disc with a spiral groove carrying recorded sound for phonograph reproduction
— for the record : for public knowledge **:** on the record
— off the record : not for publication ⟨spoke *off the record*⟩ ⟨remarks that were *off the record*⟩
— of record : being documented or attested ⟨a partner *of record* in several firms⟩
— on record 1 : in the position of having publicly declared oneself ⟨went *on record* as opposed to higher taxes⟩ **2 :** being known, published, or documented ⟨the judge's opinion is *on record*⟩
— on the record : for publication

[3]record *same as* [2]\ *adjective* (1893)
: of, relating to, or being one that is extraordinary among or surpasses others of its kind

re·cor·da·tion \ˌre-ˌkȯr-ˈdā-shən; ˌrē-, ri-\ *noun* (circa 1812)
: the action or process of recording

record changer *noun* (1931)
: a phonograph with a device that automatically positions and plays successively each of a stack of records; *also* **:** the automatic device on a record changer

re·cord·er \ri-ˈkȯr-dər\ *noun* (15th century)
1 a : the chief judicial magistrate of some British cities and boroughs **b :** a municipal judge with criminal jurisdiction of first instance and sometimes limited civil jurisdiction
2 : one that records
3 : any of a group of wind instruments ranging from soprano to bass that are characterized by

a conical tube, a whistle mouthpiece, and eight finger holes ◈

recorder 3

re·cord·ing \ri-ˈkȯr-diŋ\ *noun* (1932)
: RECORD 4

re·cord·ist \ri-ˈkȯr-dist\ *noun* (circa 1930)
: one who records sound (as on magnetic tape)

¹re·count \ri-ˈkau̇nt\ *transitive verb* [Middle English, from Middle French *reconter,* from *re-* + *conter* to count, relate — more at COUNT] (15th century)
: to relate in detail : NARRATE
— **re·count·er** *noun*

²re·count \(ˌ)rē-ˈkau̇nt\ *transitive verb* [*re-* + *count*] (1764)
: to count again

³re·count \ˈrē-ˌkau̇nt, (ˌ)rē-ˈ\ *noun* (1884)
: a second or fresh count

re·coup \ri-ˈküp\ *verb* [French *recouper* to cut back, from Old French, from *re-* + *couper* to cut — more at COPE] (1628)
transitive verb
1 a : to get an equivalent for (as losses) : make up for **b :** REIMBURSE, COMPENSATE ⟨*recoup* a person for losses⟩
2 : REGAIN ⟨an attempt to *recoup* his fortune⟩
intransitive verb
: to make good or make up for something lost; *also* : RECUPERATE
— **re·coup·able** \-ˈkü-pə-bəl\ *adjective*
— **re·coup·ment** \-ˈküp-mənt\ *noun*

re·course \ˈrē-ˌkōrs, -ˌkȯrs, ri-ˈ\ *noun* [Middle English *recours,* from Middle French, from Late Latin *recursus,* from Latin, act of running back, from *recurrere* to run back — more at RECUR] (14th century)
1 a : a turning to someone or something for help or protection **b :** a source of help or strength : RESORT
2 : the right to demand payment from the maker or endorser of a negotiable instrument (as a check)

re·cov·er \ri-ˈkə-vər\ *verb* **re·cov·ered; re·cov·er·ing** \-ˈkə-və-riŋ, -ˈkəv-riŋ\ [Middle English, from Middle French *recoverer,* from Latin *recuperare,* from *re-* + (assumed) Latin *caperare,* from Latin *capere* to take — more at HEAVE] (14th century)
transitive verb
1 : to get back : REGAIN
2 a : to bring back to normal position or condition ⟨stumbled, then *recovered* himself⟩ **b** *archaic* : RESCUE
3 a : to make up for ⟨*recover* increased costs through higher prices⟩ **b :** to gain by legal process
4 *archaic* : REACH
5 : to find or identify again ⟨*recover* a comet⟩
6 a : to obtain from an ore, a waste product, or a by-product **b :** to save from loss and restore to usefulness : RECLAIM
intransitive verb
1 : to regain a normal position or condition (as of health) ⟨*recovering* from a cold⟩
2 : to obtain a final legal judgment in one's favor
— **re·cov·er·abil·i·ty** \-ˌkə-və-rə-ˈbi-lə-tē, -ˌkəv-rə-\ *noun*
— **re·cov·er·able** \-ˈkə-və-rə-bəl, -ˈkəv-rə-\ *adjective*
— **re·cov·er·er** \-ˈkə-vər-ər\ *noun*

re–cov·er \(ˌ)rē-ˈkə-vər\ *transitive verb* (15th century)
: to cover again or anew

re·cov·ery \ri-ˈkə-və-rē, -ˈkəv-rē\ *noun, plural* **-er·ies** (15th century)
: the act, process, or an instance of recovering; *especially* : an economic upturn (as after a depression)

recovery room *noun* (1916)
: a hospital room equipped for meeting postoperative emergencies

¹rec·re·ant \ˈre-krē-ənt\ *adjective* [Middle English, from Middle French, from present participle of *recroire* to renounce one's cause in a trial by battle, from *re-* + *croire* to believe, from Latin *credere* — more at CREED] (14th century)
1 : crying for mercy : COWARDLY
2 : unfaithful to duty or allegiance

²recreant *noun* (15th century)
1 : COWARD
2 : APOSTATE, DESERTER

rec·re·ate \ˈre-krē-ˌāt\ *verb* **-at·ed; -at·ing** [Latin *recreatus,* past participle of *recreare*] (1530)
transitive verb
: to give new life or freshness to : REFRESH
intransitive verb
: to take recreation
— **rec·re·a·tive** \-ˌā-tiv\ *adjective*

re–cre·ate \ˌrē-krē-ˈāt\ *transitive verb* (1587)
: to create again; *especially* : to form anew in the imagination
— **re–cre·at·able** \-ˈā-tə-bəl\ *adjective*
— **re–cre·a·tion** \-ˈā-shən\ *noun*
— **re–cre·a·tive** \-ˈā-tiv\ *adjective*
— **re–cre·a·tor** \-ˈā-tər\ *noun*

rec·re·a·tion \ˌre-krē-ˈā-shən\ *noun* [Middle English *recreacion,* from Middle French *recreation,* from Latin *recreation-, recreatio* restoration to health, from *recreare* to create anew, restore, refresh, from *re-* + *creare* to create] (15th century)
: refreshment of strength and spirits after work; *also* : a means of refreshment or diversion : HOBBY
— **rec·re·a·tion·al** \-shnəl, -shə-nᵊl\ *adjective*

recreational vehicle *noun* (1966)
: a vehicle designed for recreational use (as in camping); *especially* : MOTOR HOME

rec·re·a·tion·ist \-sh(ə-)nist\ *noun* (1904)
: a person who seeks recreation especially in the outdoors

recreation room *noun* (1854)
1 : a room (as a rumpus room) used for recreation and relaxation — called also *rec room*
2 : a public room (as in a hospital) for recreation and social activities

re·crim·i·na·tion \ri-ˌkri-mə-ˈnā-shən\ *noun* [Medieval Latin *recrimination-, recriminatio,* from *recriminare* to make a retaliatory charge, from Latin *re-* + *criminari* to accuse — more at CRIMINATE] (circa 1611)
: a retaliatory accusation; *also* : the making of such accusations ⟨endless *recrimination*⟩
— **re·crim·i·nate** \-ˈkri-mə-ˌnāt\ *intransitive verb*
— **re·crim·i·na·tive** \-ˌnā-tiv\ *adjective*
— **re·crim·i·na·to·ry** \-ˈkri-mə-nə-ˌtōr-ē, -ˈkrim-nə-, -ˌtȯr-\ *adjective*

re·cru·desce \ˌrē-krü-ˈdes\ *intransitive verb* **-desced; -desc·ing** [Latin *recrudescere* to become raw again, from *re-* + *crudescere* to become raw, from *crudus* raw — more at RAW] (1884)
: to break out or become active again

re·cru·des·cence \-ˈde-sᵊn(t)s\ *noun* (circa 1721)
: a new outbreak after a period of abatement or inactivity : RENEWAL

re·cru·des·cent \-sᵊnt\ *adjective* (circa 1727)
: breaking out again : RENEWING

¹re·cruit \ri-ˈkrüt\ (1643)
transitive verb
1 a (1) **:** to fill up the number of (as an army) with new members : REINFORCE (2) **:** to enlist as a member of an armed service **b :** to increase or maintain the number of ⟨America *recruited* her population from Europe⟩ **c :** to secure the services of : ENGAGE, HIRE **d :** to seek to enroll ⟨*recruit* students⟩
2 : REPLENISH
3 : to restore or increase the health, vigor, or intensity of
intransitive verb
: to enlist new members
— **re·cruit·er** *noun*

²recruit *noun* [French *recrute, recrue* fresh growth, new levy of soldiers, from Middle French, from *recroistre* to grow up again, from Latin *recrescere,* from *re-* + *crescere* to grow — more at CRESCENT] (circa 1645)
1 : a fresh or additional supply
2 : a newcomer to a field or activity; *specifically* : a newly enlisted or drafted member of the armed forces
3 : a former enlisted man of the lowest rank in the army

re·cruit·ment \ri-ˈkrüt-mənt\ *noun* (circa 1828)
1 : the action or process of recruiting
2 : the process of adding new individuals to a population or subpopulation (as of breeding individuals) by growth, reproduction, immigration, and stocking; *also* : a measure (as in numbers or biomass) of recruitment

re·crys·tal·lize \(ˌ)rē-ˈkris-tə-ˌlīz\ *verb* (1797)
: to crystallize again or repeatedly
— **re·crys·tal·li·za·tion** \(ˌ)rē-ˌkris-tə-lə-ˈzā-shən\ *noun*

rec·tal \ˈrek-tᵊl\ *adjective* (circa 1859)
: relating to, affecting, or being near the rectum
— **rec·tal·ly** *adverb*

rect·an·gle \ˈrek-ˌtaŋ-gəl\ *noun* [Medieval Latin *rectangulus* having a right angle, from Latin *rectus* right + *angulus* angle — more at RIGHT, ANGLE] (1571)
: a parallelogram all of whose angles are right angles; *especially* : one with adjacent sides of unequal length

rect·an·gu·lar \rek-ˈtaŋ-gyə-lər\ *adjective* (1624)
1 : shaped like a rectangle ⟨a *rectangular* area⟩
2 a : crossing, lying, or meeting at a right angle ⟨*rectangular* axes⟩ **b :** having edges, surfaces, or faces that meet at right angles : having faces or surfaces shaped like rectangles ⟨*rectangular* parallelepipeds⟩ ⟨a *rectangular* solid⟩
— **rect·an·gu·lar·i·ty** \(ˌ)rek-ˌtaŋ-gyə-ˈlar-ə-tē\ *noun*
— **rect·an·gu·lar·ly** \rek-ˈtaŋ-gyə-lər-lē\ *adverb*

rectangular coordinate *noun* (circa 1864)
: a Cartesian coordinate of a Cartesian coordinate system whose straight-line axes or coordinate planes are perpendicular

◇ WORD HISTORY

recorder There is little obvious connection between the sweet-sounding wind instrument called the recorder and modern electronic devices that copy sound and image. In actuality, the origin of the woodwind's name is rather mysterious, and it has no relation to current senses of the verb *record.* The customary, though not totally satisfactory, solution is to see it as a derivative of Middle English *recorden* "to repeat, rehearse," one of several early senses that Modern English *record* has not retained. In Elizabethan English *record* could mean "to play (a tune)," but this more appropriate sense may come from the instrument's name. The oldest reference to the name appears to be a Latin notation in the household accounts of Henry, Earl of Derby, dated 1388, that mentions a *fistula nomine Ricordo* ("pipe called Ricordo"). If this is a hint of the word's origin, it sends us off in a quite different direction, for "Ricordo" appears to be Italian *ricordo* "remembrance, memento." Other languages offer no illumination, since their names for the instrument simply consist of the word for "flute" with some modifier, such as German *Blockflöte* ("fipple flute") or French *flûte à neuf trous* ("flute with nine holes").

rec·ti·fi·able \'rek-tə-ˌfī-ə-bəl\ *adjective* [*rectify* (to determine the length of an arc)] (1816)
: having finite length ⟨a *rectifiable* curve⟩
— rec·ti·fi·abil·i·ty \ˌrek-tə-ˌfī-ə-'bi-lə-tē\ *noun*

rec·ti·fi·er \'rek-tə-ˌfī(-ə)r\ *noun* (1611)
: one that rectifies; *specifically* **:** a device for converting alternating current into direct current

rec·ti·fy \'rek-tə-ˌfī\ *transitive verb* **-fied; -fy·ing** [Middle English *rectifien,* from Middle French *rectifier,* from Medieval Latin *rectificare,* from Latin *rectus* right — more at RIGHT] (14th century)
1 : to set right **:** REMEDY
2 : to purify (as alcohol) especially by repeated or fractional distillation
3 : to correct by removing errors **:** ADJUST ⟨*rectify* the calendar⟩
4 : to make (an alternating current) unidirectional
synonym see CORRECT
— rec·ti·fi·ca·tion \ˌrek-tə-fə-'kā-shən\ *noun*

rec·ti·lin·e·ar \ˌrek-tə-'li-nē-ər\ *adjective* [Late Latin *rectilineus,* from Latin *rectus* + *linea* line] (1659)
1 : moving in or forming a straight line ⟨*rectilinear* motion⟩
2 : characterized by straight lines
3 : PERPENDICULAR 3
— rec·ti·lin·e·ar·ly *adverb*

rec·ti·tude \'rek-tə-ˌtüd, -ˌtyüd\ *noun* [Middle English, from Middle French, from Late Latin *rectitudo,* from Latin *rectus* straight, right] (15th century)
1 : the quality or state of being straight
2 : moral integrity **:** RIGHTEOUSNESS
3 : the quality or state of being correct in judgment or procedure

rec·ti·tu·di·nous \ˌrek-tə-'tüd-nəs, -'tyüd-; 'tü dᵊn-əs, -'tyü-\ *adjective* [Late Latin *rectitudin-, rectitudo* rectitude] (1897)
1 : characterized by rectitude
2 : piously self-righteous

rec·to \'rek-(ˌ)tō\ *noun, plural* **rectos** [New Latin *recto* (*folio*) on the right-hand leaf] (1824)
1 : the side of a leaf (as of a manuscript) that is to be read first
2 : a right-hand page — compare VERSO

rec·tor \'rek-tər\ *noun* [Middle English, from Latin, from *regere* to direct — more at RIGHT] (14th century)
1 : one that directs **:** LEADER
2 a : a clergyman (as of the Protestant Episcopal Church) in charge of a parish **b :** an incumbent of a Church of England benefice in full possession of its rights **c :** a Roman Catholic priest directing a church with no pastor or one whose pastor has other duties
3 : the head of a university or school
— rec·tor·ate \-t(ə-)rət\ *noun*
— rec·to·ri·al \rek-'tōr-ē-əl, -'tór-\ *adjective*
— rec·tor·ship \'rek-tər-ˌship\ *noun*

rec·to·ry \'rek-t(ə-)rē\ *noun, plural* **-ries** (1594)
1 : a benefice held by a rector
2 : a residence of a rector or a parish priest

rec·trix \'rek-triks\ *noun, plural* **rec·tri·ces** \'rek-trə-ˌsēz, rek-'trī-(ˌ)sēz\ [New Latin, from Latin, feminine of *rector* one that directs] (1768)
: any of the quill feathers of a bird's tail that are important in controlling flight direction — see BIRD illustration

rec·tum \'rek-təm\ *noun, plural* **rectums** *or* **rec·ta** \-tə\ [Middle English, from Medieval Latin, from *rectum intestinum,* literally, straight intestine] (15th century)
: the terminal part of the intestine from the sigmoid flexure to the anus

rec·tus \'rek-təs\ *noun, plural* **rec·ti** \-ˌtī, -ˌtē\ [New Latin, from *rectus musculus* straight muscle] (circa 1704)
: any of several straight muscles (as of the abdomen)

re·cum·ben·cy \ri-'kəm-bən(t)-sē\ *noun, plural* **-cies** (1646)
: the state of leaning, resting, or reclining **:** REPOSE; *also* **:** a recumbent position

re·cum·bent \-bənt\ *adjective* [Latin *recumbent-, recumbens,* present participle of *recumbere* to lie down, from *re-* + *-cumbere* to lie down; akin to Latin *cubare* to lie] (1705)
1 a : suggestive of repose **:** LEANING, RESTING **b :** lying down
2 : representing a person lying down ⟨a *recumbent* statue⟩
synonym see PRONE

re·cu·per·ate \ri-'kü-pə-ˌrāt, -'kyü-\ *verb* **-at·ed; -at·ing** [Latin *recuperatus,* past participle of *recuperare* — more at RECOVER] (1542)
transitive verb
: to get back **:** REGAIN
intransitive verb
: to regain a former state or condition; *especially* **:** to recover health or strength
— re·cu·per·a·tion \-ˌkü-pə-'rā-shən, -ˌkyü-\ *noun*

re·cu·per·a·tive \-'kü-pə-ˌrā-tiv, -'kyü-, -p(ə-)rə-tiv\ *adjective* (circa 1828)
1 : of or relating to recuperation ⟨*recuperative* powers⟩
2 : aiding in recuperation **:** RESTORATIVE

re·cur \ri-'kər\ *intransitive verb* **re·curred; re·cur·ring** [Middle English *recurren* to return, from Latin *recurrere,* literally, to run back, from *re-* + *currere* to run — more at CAR] (1529)
1 : to have recourse **:** RESORT
2 : to go back in thought or discourse
3 a : to come up again for consideration **b :** to come again to mind
4 : to occur again after an interval **:** occur time after time
— re·cur·rence \-'kər-ən(t)s, -'kə-rən(t)s\ *noun*

re·cur·rent \-'kər-ənt, -'kə-rənt\ *adjective* [Latin *recurrent-, recurrens,* present participle of *recurrere*] (1611)
1 : running or turning back in a direction opposite to a former course — used of various nerves and branches of vessels in the arms and legs
2 : returning or happening time after time ⟨*recurrent* complaints⟩
— re·cur·rent·ly *adverb*

recurring decimal *noun* (1801)
: REPEATING DECIMAL

re·cur·sion \ri-'kər-zhən\ *noun* [Late Latin *recursion-, recursio,* from *recurrere*] (1616)
1 : RETURN
2 : the determination of a succession of elements (as numbers or functions) by operation on one or more preceding elements according to a rule or formula involving a finite number of steps
3 : a computer programming technique involving the use of a procedure, subroutine, function, or algorithm that calls itself in a step having a termination condition so that successive repetitions are processed up to the critical step until the condition is met at which time the rest of each repetition is processed from the last one called to the first — compare ITERATION

re·cur·sive \ri-'kər-siv\ *adjective* (1934)
1 : of, relating to, or involving recursion ⟨a *recursive* function in a computer program⟩
2 : of, relating to, or constituting a procedure that can repeat itself indefinitely ⟨a *recursive* rule in a grammar⟩
— re·cur·sive·ly *adverb*
— re·cur·sive·ness *noun*

re·curved \(ˌ)rē-'kərvd\ *adjective* (1597)
: curved backward or inward

re·cu·san·cy \'re-kyə-zən(t)-sē, ri-'kyü-\ *noun* (circa 1600)
: the act or state of being a recusant

re·cu·sant \-zənt\ *noun* [Latin *recusant-, recusans,* present participle of *recusare* to refuse, from *re-* + *causari* to give a reason, from *causa* cause, reason] (circa 1553)
1 : an English Roman Catholic of the time from about 1570 to 1791 who refused to attend services of the Church of England and thereby committed a statutory offense
2 : one who refuses to accept or obey established authority
— recusant *adjective*

re·cuse \ri-'kyüz\ *transitive verb* **re·cused; re·cus·ing** [Middle English, to refuse, from Middle French *recuser,* from Latin *recusare*] (1949)
: to disqualify (oneself) as judge in a particular case; *broadly* **:** to remove (oneself) from participation to avoid a conflict of interest
— re·cus·al \-'kyü-zəl\ *noun*

re·cut \(ˌ)rē-'kət, 'rē-ˌ\ *transitive verb* (1664)
1 : to cut again
2 : to edit (as a film) anew

[1]re·cy·cle \(ˌ)rē-'sī-kəl\ (1926)
transitive verb
1 : to pass again through a series of changes or treatments: as **a :** to process (as liquid body waste, glass, or cans) in order to regain material for human use **b :** RECOVER 6
2 : to adapt to a new use **:** ALTER
3 : to bring back **:** REUSE ⟨*recycles* a number of good anecdotes —Larry McMurtry⟩
4 : to make ready for reuse ⟨a plan to *recycle* vacant tenements⟩
intransitive verb
1 : to return to an earlier point in a countdown
2 : to return to an original condition so that operation can begin again — used of an electronic device
— re·cy·cla·ble \-k(ə-)lə-bəl\ *adjective or noun*
— re·cy·cler \-k(ə-)lər\ *noun*

[2]recycle *noun* (1926)
: the process of recycling

[1]red \'red\ *adjective* **red·der; red·dest** [Middle English, from Old English *rēad;* akin to Old High German *rōt* red, Latin *ruber* & *rufus,* Greek *erythros*] (before 12th century)
1 a : of the color red **b :** having red as a distinguishing color
2 a (1) **:** flushed especially with anger or embarrassment (2) **:** RUDDY, FLORID (3) **:** being or having skin of a coppery hue **b :** BLOODSHOT ⟨eyes *red* from crying⟩ **c :** in the color range between a moderate orange and russet or bay **d :** tinged with red **:** REDDISH
3 : heated to redness **:** GLOWING
4 a : inciting or endorsing radical social or political change especially by force **b** *often capitalized* **:** COMMUNIST **c** *often capitalized* **:** of or relating to a communist country and especially to the U.S.S.R.

[2]red *noun* (before 12th century)
1 : a color whose hue resembles that of blood or of the ruby or is that of the long-wave extreme of the visible spectrum
2 : red clothing ⟨the lady in *red*⟩
3 : one that is of a red or reddish color: as **a :** RED WINE **b :** an animal with a red or reddish coat
4 a : a pigment or dye that colors red **b :** a shade or tint of red
5 a : one who advocates the violent overthrow of an existing social or political order **b** *capitalized* **:** COMMUNIST
6 [from the bookkeeping practice of entering debit items in red ink] **:** the condition of showing a loss — usually used with *the* ⟨in the *red*⟩; compare BLACK

re·dact \ri-ˈdakt\ *transitive verb* [Middle English, from Latin *redactus,* past participle of *redigere*] (15th century)
1 : to put in writing **:** FRAME
2 : to select or adapt for publication **:** EDIT

re·dac·tion \-ˈdak-shən\ *noun* [French *rédaction,* from Late Latin *redaction-, redactio* act of reducing, compressing, from Latin *redigere* to bring back, reduce, from *re-, red-* re- + *agere* to lead — more at AGENT] (1785)
1 : an act or instance of redacting something
2 : a work that has been redacted **:** EDITION, VERSION
— re·dac·tion·al \-shnəl, -shə-nᵊl\ *adjective*

re·dac·tor \-ˈdak-tər\ *noun* (1816)
: one who redacts a work; *especially* **:** EDITOR

red admiral *noun* (1840)
: a nymphalid butterfly (*Vanessa atalanta*) that is common in both Europe and America, has broad orange-red bands on the forewings, and feeds chiefly on nettles in the larval stage

red admiral

red alert *noun* (circa 1951)
: the final stage of alert in which enemy attack appears imminent; *broadly* **:** a state of alert brought on by impending danger

red alga *noun* (1852)
: any of a division (Rhodophyta) of chiefly marine algae that have predominantly red pigmentation

red ant *noun* (1667)
: any of various reddish ants (as the pharaoh ant)

red·ar·gue \ri-ˈdär-(ˌ)gyü\ *transitive verb* **-gued; -gu·ing** [Middle English, from Latin *redarguere,* from *red-* + *arguere* to demonstrate, prove — more at ARGUE] (1627)
archaic **:** CONFUTE, DISPROVE

red–bait \ˈred-ˌbāt\ *verb, often R capitalized* (1940)
transitive verb
: to subject (as a person or group) to red-baiting
intransitive verb
: to engage in red-baiting
— red–bait·er *noun, often R capitalized*

red–bait·ing *noun, often R capitalized* (1928)
: the act of attacking or persecuting as a Communist or as communistic

red bay *noun* (circa 1730)
: a small tree (*Persea borbonia*) of the southern U.S. that has dark red heartwood

red·bel·ly dace \ˈred-ˈbe-lē-\ *noun* (1884)
: either of two small brightly marked North American cyprinid fishes (*Phoxinus eos* and *P. erythrogaster*) — called also *red-bel·lied dace* \-lēd-\

red·bird \ˈred-ˌbərd\ *noun* (1669)
: any of several birds (as a cardinal, several tanagers, or the bullfinch) with predominantly red plumage

red blood cell *noun* (1910)
: any of the hemoglobin-containing cells that carry oxygen to the tissues and are responsible for the red color of vertebrate blood — called also *erythrocyte, red blood corpuscle, red cell, red corpuscle*

red–blood·ed \ˈred-ˈblə-dəd\ *adjective* (1881)
: VIGOROUS, LUSTY

red·bone \ˈred-ˌbōn\ *noun* (1916)
: any of a breed of moderate-sized speedy dark red or red and tan coonhounds of American origin

red·breast \ˈred-ˌbrest\ *noun* (15th century)
1 : a bird (as a robin) with a reddish breast
2 : a reddish-bellied sunfish (*Lepomis auritus*) of the eastern U.S. — called also *red-breasted bream*

red·brick \-ˌbrik\ *adjective* (1943)
1 : built of red brick
2 *often capitalized* [from the common use of red brick in constructing the buildings of recently founded universities] **:** of, relating to, or being the British Universities founded in the 19th or early 20th century — compare OXBRIDGE, PLATEGLASS

red·bud \-ˌbəd\ *noun* (1705)
: any of several leguminous trees (genus *Cercis*) with usually pale rosy pink flowers

red bug *noun* (1804)
Southern & Midland **:** CHIGGER 2

red·cap \ˈred-ˌkap\ *noun* (1918)
: a baggage porter (as at a railroad station) — compare SKYCAP

red–carpet *adjective* [from the traditional laying down of a red carpet for important guests to walk on] (1952)
: marked by ceremonial courtesy ⟨*red-carpet* treatment⟩

red carpet *noun* (1951)
: a greeting or reception marked by ceremonial courtesy — usually used in the phrase *roll out the red carpet*

red cedar *noun* (1682)
1 : a common juniper (*Juniperus virginiana*) chiefly of the eastern U.S. that has dark green closely imbricated scalelike leaves; *also* **:** a related tree (*J. silicicola*) of the southeastern U.S.
2 : an arborvitae (*Thuja plicata*) of the Pacific Northwest
3 : the red or reddish brown wood of a red cedar

red cent *noun* (circa 1839)
: PENNY — used for emphasis in negative constructions

red clover *noun* (before 12th century)
: a Eurasian clover (*Trifolium pratense*) that has globose heads of reddish purple flowers and is widely cultivated as a hay, forage, and cover crop

red·coat \ˈred-ˌkōt\ *noun* (1520)
: a British soldier especially in America during the Revolutionary War

red coral *noun* (circa 1864)
: a gorgonian (*Corallium nobile*) of the Mediterranean and adjacent parts of the Atlantic having a hard stony skeleton of a delicate red or pink color used for ornaments and jewelry

Red Cross *noun* (1863)
: a red Greek cross on a white background used as the emblem of the International Red Cross

red currant *noun* (1622)
: either of two currants (*Ribes sativum* and *R. rubrum*) often cultivated for their fruit; *also* **:** the fruit

¹redd \ˈred\ *verb* **redd·ed** *or* **redd; redd·ing** [Middle English (Scots), to clear, perhaps alteration of *ridden* — more at RID] (circa 1520)
transitive verb
chiefly dialect **:** to set in order — usually used with *up* or *out*
intransitive verb
chiefly dialect **:** to make things tidy — usually used with *up*

²redd *noun* [origin unknown] (1808)
: the spawning ground or nest of various fishes

red deer *noun* (15th century)
: ELK 1b — used for one of the Old World elk

red·den \ˈre-dᵊn\ *verb* **red·dened; red·den·ing** \ˈred-niŋ, ˈre-dᵊn-iŋ\ (circa 1611)
transitive verb
: to make red or reddish
intransitive verb
: to become red; *especially* **:** BLUSH

red·dish \ˈre-dish\ *adjective* (14th century)
: tinged with red
— red·dish·ness *noun*

red dog *noun* (1953)
: BLITZ 2b
— red dog *verb*

red drum *noun* (1709)
: CHANNEL BASS

¹rede \ˈrēd\ *transitive verb* [Middle English — more at READ] (before 12th century)
1 *dialect* **:** to give counsel to **:** ADVISE
2 *dialect* **:** INTERPRET, EXPLAIN

²rede *noun* (before 12th century)
1 *chiefly dialect* **:** COUNSEL, ADVICE
2 *archaic* **:** ACCOUNT, STORY

red·ear \ˈred-ˌir\ *noun* (circa 1948)
: a common sunfish (*Lepomis microlophus*) of the southern and eastern U.S. resembling the bluegill but having the back part of the operculum bright orange-red — called also *redear sunfish, shellcracker*

re·dec·o·rate \(ˌ)rē-ˈde-kə-ˌrāt\ (circa 1611)
transitive verb
: to freshen or change in appearance **:** REFURBISH
intransitive verb
: to freshen or change a decorative scheme
— re·dec·o·ra·tion \(ˌ)rē-ˌde-kə-ˈrā-shən\ *noun*
— re·dec·o·ra·tor \(ˌ)rē-ˈde-kə-ˌrā-tər\ *noun*

re·deem \ri-ˈdēm\ *transitive verb* [Middle English *redemen,* modification of Middle French *redimer,* from Latin *redimere,* from *re-, red-* re- + *emere* to take, buy; akin to Lithuanian *imti* to take] (15th century)
1 a : to buy back **:** REPURCHASE **b :** to get or win back
2 : to free from what distresses or harms: as **a :** to free from captivity by payment of ransom **b :** to extricate from or help to overcome something detrimental **c :** to release from blame or debt **:** CLEAR **d :** to free from the consequences of sin
3 : to change for the better **:** REFORM
4 : REPAIR, RESTORE
5 a : to free from a lien by payment of an amount secured thereby **b** (1) **:** to remove the obligation of by payment ⟨the U.S. Treasury *redeems* savings bonds on demand⟩ (2) **:** to exchange for something of value ⟨*redeem* trading stamps⟩ **c :** to make good **:** FULFILL
6 a : to atone for **:** EXPIATE **b** (1) **:** to offset the bad effect of (2) **:** to make worthwhile **:** RETRIEVE
synonym see RESCUE
— re·deem·able \-ˈdē-mə-bəl\ *adjective*

re·deem·er \-ˈdē-mər\ *noun* (15th century)
: a person who redeems; *especially, capitalized* **:** JESUS

re·deem·ing \-ˈdē-miŋ\ *adjective* (1631)
: serving to offset or compensate for a defect ⟨her performance is the film's *redeeming* feature⟩

re·de·fine \ˌrē-di-ˈfīn\ *transitive verb* (1872)
1 : to define (as a concept) again **:** REFORMULATE ⟨had to *redefine* their terms⟩
2 a : to reexamine or reevaluate especially with a view to change **b :** TRANSFORM 1c
— re·def·i·ni·tion \(ˌ)rē-ˌde-fə-ˈni-shən\ *noun*

re·demp·tion \ri-ˈdem(p)-shən\ *noun* [Middle English *redempcioun,* from Middle French *redemption,* from Latin *redemption-, redemptio,* from *redimere* to redeem] (14th century)
: the act, process, or an instance of redeeming

re·demp·tion·er \-sh(ə-)nər\ *noun* (1771)
: an immigrant to America in the 18th and 19th centuries who obtained passage by becoming an indentured servant

re·demp·tive \-ˈdem(p)-tiv\ *adjective* (15th century)
: of, relating to, or bringing about redemption

Re·demp·tor·ist \ri-ˈdem(p)-t(ə-)rist\ *noun* [French *rédemptoriste,* from Late Latin *redemptor* redeemer, from Latin, contractor, from *redimere*] (circa 1842)
: a member of the Congregation of the Most Holy Redeemer founded by Saint Alphonsus Liguori in Scala, Italy, in 1732 and devoted to preaching

re·demp·to·ry \ri-ˈdem(p)-t(ə-)rē\ *adjective* (1602)
: serving to redeem

re·de·ploy \ˌrē-di-ˈplȯi\ (1945) *transitive verb* **:** to transfer from one area or activity to another *intransitive verb* **:** to relocate men or equipment — **re·de·ploy·ment** \-mənt\ *noun*

re·de·scribe \ˌrē-di-ˈskrīb\ *transitive verb* (1871) **:** to describe anew or again; *especially* **:** to give a new and more complete description to (a biological taxon)

re·de·scrip·tion \-ˈskrip-shən\ *noun* (1884) **:** a new and more complete description especially of a biological taxon

re·de·sign \ˌrē-di-ˈzīn\ *transitive verb* (1891) **:** to revise in appearance, function, or content — **redesign** *noun*

re·de·vel·op \ˌrē-di-ˈve-ləp\ *transitive verb* (1882) **:** to develop again; *especially* **:** REDESIGN, REBUILD — **re·de·vel·op·er** *noun*

re·de·vel·op·ment \-mənt\ *noun* (1873) **:** the act or process of redeveloping; *especially* **:** renovation of a blighted area

red–eye \ˈred-ˌī\ *noun* (1819) **1 :** cheap whiskey **2 :** a late night or overnight flight

red·eye gravy \ˈred-ˌī-\ *noun* (1947) **:** gravy made from the juices of ham and often flavored with coffee

red fescue *noun* (1900) **:** a perennial pasture and turf grass (*Festuca rubra*) of Eurasia and North America with creeping rootstocks, erect culms, and reddish spikelets

red·fish \ˈred-ˌfish\ *noun* (15th century) **:** any of various reddish fishes: as **a** (1) **:** a marine scorpaenid food fish (*Sebastes marinus*) of the northern coasts of Europe and North America that is usually bright rose-red when mature — called also *rosefish* (2) **:** a fish (*Sebastes mentella*) related to the redfish **b :** CHANNEL BASS

red flag *noun* (1777) **1 :** a warning signal **2 :** something that attracts usually irritated attention

red flannel hash *noun* (circa 1907) **:** hash made especially from beef, potatoes, and beets

red fox *noun* (1778) **:** a usually orange-red to reddish brown Holarctic fox (*Vulpes vulpes*) that has a white-tipped tail

red fox

red giant *noun* (1916) **:** a star that has low surface temperature and a diameter that is large relative to the sun

red–green blindness *noun* (1888) **:** dichromatism in which the spectrum is seen in tones of yellow and blue — called also *red-green color blindness*

Red Guard *noun* (1966) **:** a member of a paramilitary youth organization in China in the 1960s

red gum *noun* (1788) **1 a :** any of several Australian eucalyptus trees (especially *Eucalyptus camaldulensis, E. amygdalina,* and *E. calophylla*) **b :** eucalyptus gum **2 :** SWEET GUM

red–hand·ed \ˈred-ˈhan-dəd\ *adverb or adjective* (1819) **:** in the act of committing a crime or misdeed ⟨caught *red-handed*⟩

red·head \ˈred-ˌhed\ *noun* (1664) **1 :** a person having red hair **2 :** a North American duck (*Aythya americana*) resembling the related canvasback but having a shorter bill with a black tip and in the male a brighter reddish head

red·head·ed \-ˌhe-dəd\ *adjective* (1565) **:** having red hair or a red head

red heat *noun* (1686) **:** the state of being red-hot; *also* **:** the temperature at which a substance is red-hot

red herring *noun* (15th century) **1 :** a herring cured by salting and slow smoking to a dark brown color **2** [from the practice of drawing a red herring across a trail to confuse hunting dogs] **:** something that distracts attention from the real issue

red·horse \ˈred-ˌhȯrs\ *noun* (1796) **:** any of numerous large suckers (genus *Moxostoma*) of North American rivers and lakes with the males having red fins especially in the breeding season

red–hot \ˈred-ˈhät\ *adjective* (14th century) **:** extremely hot: as **a :** glowing with heat **b :** exhibiting or marked by intense emotion, enthusiasm, or violence ⟨a *red-hot* campaign⟩ **c :** FRESH, NEW ⟨*red-hot* news⟩ **d :** extremely popular

red hot *noun* (1835) **1 :** one who shows intense emotion or partisanship **2 :** HOT DOG 1 **3 :** a small red candy strongly flavored with cinnamon

re·dia \ˈrē-dē-ə\ *noun, plural* **re·di·ae** \-dē-ˌē\ *also* **re·di·as** [New Latin, from Francesco *Redi* (died 1698?) Italian naturalist] (1877) **:** a larva produced within the sporocyst of many trematodes that produces another generation of rediae or develops into a cercaria — **re·di·al** \-dē-əl\ *adjective*

re·dial \ˈrē-ˌdī(-ə)l, rē-ˈ\ *noun* (1980) **:** a function on a telephone that automatically repeats the dialing of the last number called; *also* **:** a button that invokes this function — **re·dial** *verb*

Red Indian *noun* (1831) *chiefly British* **:** AMERICAN INDIAN

red·in·gote \ˈre-diŋ-ˌgōt\ *noun* [French, modification of English *riding coat*] (1793) **:** a fitted outer garment: as **a :** a double-breasted coat with wide flat cuffs and collar worn by men in the 18th century **b :** a woman's lightweight coat open at the front **c :** a dress with a front gore of contrasting material

redingote a

red ink *noun* [from the use of red ink in financial statements to indicate a loss] (1926) **1 :** a business loss **:** DEFICIT **2 :** the condition of showing a business loss

red·in·te·grate \ri-ˈdin-tə-ˌgrāt, re-\ *transitive verb* [Middle English, from Latin *redintegratus,* past participle of *redintegrare,* from *re-, red-* re- + *integrare* to make complete — more at INTEGRATE] (15th century) *archaic* **:** to restore to a former and especially sound state

red·in·te·gra·tion \ri-ˌdin-tə-ˈgrā-shən, re-\ *noun* (15th century) **1** *archaic* **:** restoration to a former state **2 a :** revival of the whole of a previous mental state when a phase of it recurs **b :** arousal of any response by a part of the complex of stimuli that originally aroused that response — **red·in·te·gra·tive** \-ˈdin-tə-ˌgrā-tiv\ *adjective*

re·di·rect \ˌrē-də-ˈrekt, ˌrē-(ˌ)dī-\ *transitive verb* (1844) **:** to change the course or direction of — **re·di·rec·tion** \-ˈrek-shən\ *noun*

[1]re·dis·count \(ˌ)rē-ˈdis-ˌkau̇nt, ˌrē-dis-ˈ\ *transitive verb* (1866) **:** to discount again (as commercial paper) — **re·dis·count·able** \-ˌkau̇n-tə-bəl, -ˈkau̇n-\ *adjective*

[2]re·dis·count \(ˌ)rē-ˈdis-ˌkau̇nt\ *noun* (1892) **1 :** the act or process of rediscounting **2 :** negotiable paper that is rediscounted

re·dis·trib·ute \ˌrē-də-ˈstri-byət\ *transitive verb* (1611) **1 :** to alter the distribution of **:** REALLOCATE **2 :** to spread to other areas — **re·dis·tri·bu·tion** \(ˌ)rē-ˌdis-trə-ˈbyü-shən\ *noun* — **re·dis·tri·bu·tion·al** \-shnəl, -shə-nᵊl\ *adjective* — **re·dis·trib·u·tive** \ˌrē-də-ˈstri-byə-tiv\ *adjective*

re·dis·tri·bu·tion·ist \(ˌ)rē-ˌdis-trə-ˈbyü-sh(ə-)nist\ *noun* (1979) **:** one that believes in or advocates a welfare state

re·dis·trict \(ˌ)rē-ˈdis-(ˌ)trikt\ (1850) *transitive verb* **:** to divide anew into districts; *specifically* **:** to revise the legislative districts of *intransitive verb* **:** to revise legislative districts

red·i·vi·vus \ˌre-də-ˈvī-vəs, -ˈvē-\ *adjective* [Late Latin, from Latin, reused] (1675) **:** brought back to life **:** REBORN — used postpositively

red lead *noun* (15th century) **:** an orange-red to brick-red lead oxide Pb_3O_4 used in storage-battery plates, in glass and ceramics, and as a paint pigment

red leaf *noun* (1909) **:** any of several plant diseases characterized by reddening of the foliage

red·leg \ˈred-ˌleg, -ˌlāg\ *noun* (1900) **:** ARTILLERYMAN

red–legged grasshopper *noun* (1867) **:** a widely distributed and sometimes highly destructive small North American grasshopper (*Melanoplus femur-rubrum*) with red hind legs — called also *red-legged locust*

red–let·ter \ˈred-ˌle-tər\ *adjective* [from the practice of marking holy days in red letters in church calendars] (1704) **:** of special significance

word history see RUBRIC

red light *noun* (1849) **:** a warning signal; *especially* **:** a red traffic signal

red–light district *noun* (1900) **:** a district in which houses of prostitution are numerous

[1]red·line \ˈred-ˈlīn\ *noun* (1952) **:** a recommended safety limit **:** the fastest, farthest, or highest point or degree considered safe; *also* **:** the red line which marks this point on a gauge

[2]red·line \ˈred-ˌlīn, -ˈlīn\ (1968) *intransitive verb* **:** to withhold home-loan funds or insurance from neighborhoods considered poor economic risks *transitive verb* **:** to discriminate against in housing or insurance

red·ly \ˈred-lē\ *adverb* (1611) **:** in a red manner **:** with red color

red man *noun* (1725) **1 :** AMERICAN INDIAN **2** *R&M capitalized* [Improved Order of *Red Men*] **:** a member of a major benevolent and fraternal order

red maple *noun* (1770) **:** a common tree (*Acer rubrum*) of the eastern and central U.S. that grows chiefly on moist soils, has reddish twigs and flowers, and yields a lighter and softer wood than the sugar maple

red marrow *noun* (1900)

: reddish bone marrow that is the seat of blood-cell production

red mass *noun, often R&M capitalized* (1889)
: a votive mass of the Holy Spirit celebrated in red vestments especially at the opening of courts and congresses

red mite *noun* (1894)
: any of several mites having a red color: as **a** : EUROPEAN RED MITE **b** : CITRUS RED MITE

red mulberry *noun* (1717)
: a North American forest tree (*Morus rubra*) with toothed leaves and soft durable wood; *also* : its edible usually purple fruit

red mullet *noun* (1762)
: GOATFISH

red·neck \'red-ˌnek\ *noun* (1830)
1 : a white member of the Southern rural laboring class — sometimes used disparagingly
2 : a person whose behavior and opinions are similar to those attributed to rednecks — often used disparagingly
— **redneck** *also* **red·necked** \-ˌnekt\ *adjective*

red·ness \-nəs\ *noun* (before 12th century)
: the quality or state of being red or red-hot

re·do \(ˌ)rē-'dü\ *transitive verb* **-did** \-'did\; **-done** \-'dən\; **-do·ing** \-'dü-iŋ\; **-does** \-'dəz\ (1597)
1 : to do over or again
2 : REDECORATE
— **re·do** \'rē-ˌdü, ˌrē-'dü\ *noun*

red oak *noun* (1634)
1 : any of numerous American oaks (as *Quercus rubra* and *Quercus falcata*) that have four stamens in each floret, acorns with the inner surface of the shell lined with woolly hairs, the acorn cap covered with thin scales, and leaf veins that usually run beyond the margin of the leaf to form bristles
2 : the wood of red oak

red ocher *noun* (1572)
: a red earthy hematite used as a pigment

red·o·lence \'re-dᵊl-ən(t)s\ *noun* (15th century)
1 : an often pungent or agreeable odor
2 : the quality or state of being redolent
synonym see FRAGRANCE

red·o·lent \-ənt\ *adjective* [Middle English, from Middle French, from Latin *redolent-, redolens,* present participle of *redolēre* to emit a scent, from *re-, red-* + *olēre* to smell — more at ODOR] (15th century)
1 : exuding fragrance : AROMATIC
2 a : full of a specified fragrance : SCENTED ⟨air *redolent* of seaweed⟩ **b** : EVOCATIVE, SUGGESTIVE ⟨a city *redolent* of antiquity⟩
synonym see ODOROUS
— **red·o·lent·ly** *adverb*

red osier *noun* (1807)
: a common shrubby North American dogwood (*Cornus sericea* synonym *C. stolonifera*) with reddish purple twigs, white flowers, and globose blue or whitish fruit

re·dou·ble \(ˌ)rē-'də-bəl\ (15th century)
transitive verb
1 : to make twice as great in size or amount; *broadly* : INTENSIFY, STRENGTHEN
2 a *obsolete* : to echo back **b** *archaic* : REPEAT
intransitive verb
1 : to become redoubled
2 *archaic* : RESOUND
3 : to double an opponent's double in bridge
— **redouble** *noun*

re·doubt \ri-'dau̇t\ *noun* [French *redoute,* from Italian *ridotto,* from Medieval Latin *reductus* secret place, from Latin, withdrawn, from past participle of *reducere* to lead back — more at REDUCE] (circa 1608)
1 a : a small usually temporary enclosed defensive work **b** : a defended position : protective barrier
2 : a secure retreat : STRONGHOLD

re·doubt·able \ri-'dau̇-tə-bəl\ *adjective* [Middle English *redoutable,* from Middle French, from *redouter* to dread, from *re-* + *douter* to doubt] (15th century)
1 : causing fear or alarm : FORMIDABLE
2 : ILLUSTRIOUS, EMINENT; *broadly* : worthy of respect
— **re·doubt·ably** \-blē\ *adverb*

re·dound \ri-'dau̇nd\ *intransitive verb* [Middle English, from Middle French *redonder,* from Latin *redundare,* from *re-, red-* re- + *unda* wave — more at WATER] (14th century)
1 *archaic* : to become swollen : OVERFLOW
2 : to have an effect for good or ill ⟨new power alignments which may or may not *redound* to the faculty's benefit —G. W. Bonham⟩
3 : to become transferred or added : ACCRUE
4 : REBOUND, REFLECT

red·out \'red-ˌau̇t\ *noun* (1942)
: a condition in which centripetal acceleration (as that created when an aircraft abruptly enters a dive) drives blood to the head and causes reddening of the visual field and headache

re·dox \'rē-ˌdäks\ *adjective* [*red*uction + *ox*idation] (1928)
: of or relating to oxidation-reduction

red panda *noun* (1955)
: PANDA 1

red–pen·cil \'red-ˌpen(t)-səl\ *transitive verb* (1946)
1 : CENSOR
2 : CORRECT, REVISE

red pepper *noun* (circa 1591)
: CAYENNE PEPPER

red pine *noun* (1809)
1 : a North American pine (*Pinus resinosa*) that has reddish bark and two long needles in each cluster
2 : the relatively hard wood of the red pine that consists chiefly of sapwood

red·poll \'red-ˌpōl\ *noun* (1738)
: either of two small finches (genus *Carduelis* synonym *Acanthis*) having brownish streaked plumage and a red or rosy crown; *especially* : one (*C. flammea*) found in northern regions of both of the New and Old World

red poll *noun, often R&P capitalized* [alteration of *red polled*] (1891)
: any of a breed of large hornless red beef cattle of English origin

[1]**re·dress** \ri-'dres\ *transitive verb* [Middle English, from Middle French *redresser,* from Old French *redrecier,* from *re-* + *drecier* to make straight — more at DRESS] (14th century)
1 a (1) : to set right : REMEDY (2) : to make up for : COMPENSATE **b** : to remove the cause of (a grievance or complaint) **c** : to exact reparation for : AVENGE
2 *archaic* **a** : to requite (a person) for a wrong or loss **b** : HEAL
synonym see CORRECT
— **re·dress·er** *noun*

[2]**re·dress** \ri-'dres, 'rē-ˌ\ *noun* (14th century)
1 a : relief from distress **b** : means or possibility of seeking a remedy ⟨without *redress*⟩
2 : compensation for wrong or loss : REPARATION
3 a : an act or instance of redressing **b** : RETRIBUTION, CORRECTION

red ribbon *noun* (1927)
: a red ribbon usually with appropriate words or markings awarded the second-place winner in a competition

red·root \'red-ˌrüt, -ˌru̇t\ *noun* (1709)
1 : a perennial herb (*Lachnanthes caroliniana* synonym *L. tinctoria*) of the bloodwort family of the eastern U.S. whose red root is the source of a dye
2 : NEW JERSEY TEA
3 : BLOODROOT
4 : a pigweed (*Amaranthus retroflexus*) that bears greenish flowers in dense spikes with bracts almost twice as long as the sepals

red rust *noun* (1846)
1 : the uredinial stage of a rust
2 : the diseased condition produced by red rust

red salmon *noun* (1881)
: SOCKEYE

red–shaft·ed flicker \'red-'shaf-təd-\ *noun* (1846)
: a flicker of western North America with light red on the underside of the tail and wings, a gray nape with no red, and in the male red on each cheek

red·shank \'red-ˌshaŋk\ *noun* (1525)
: a common Old World sandpiper (*Tringa totanus*) with pale red legs and feet

red·shift \'red-'shift\ *noun* (1923)
: a displacement of the spectrum of a celestial body toward longer wavelengths that is a consequence of the Doppler effect or the gravitational field of the source
— **red·shift·ed** *adjective*

red·shirt \'red-ˌshərt\ *noun* [from the red jersey commonly worn by such a player in practice scrimmages against the regulars] (1955)
: a college athlete who is kept out of varsity competition for a year in order to extend eligibility
— **redshirt** *verb*

red–shoul·dered hawk \-ˌshōl-dərd-\ *noun* (1812)
: a common North American hawk (*Buteo lineatus*) that has a banded tail and a light spot on the underside of the wings toward the tips

red sin·dhi \-'sin-dē\ *noun* [[1]*red* + *sindhi* one belonging to Sind, Pakistan] (1946)
: any of a breed of humped rather small red dairy cattle developed in southwestern Asia and extensively used for crossbreeding with European stock in tropical areas

red siskin *noun* (1948)
: a finch (*Carduelis cucullata*) of northern South America that is scarlet with black head, wings, and tail

red·skin \'red-ˌskin\ *noun* (1699)
: AMERICAN INDIAN — usually taken to be offensive

red snapper *noun* (1755)
: any of various reddish fishes (as of the genera *Lutjanus* and *Sebastes*) including several food fishes

red snow *noun* (1678)
: snow colored by various airborne dusts or by a growth of algae (as of the genus *Chlamydomonas*) that contain red pigment and live in the upper layer of snow; *also* : an alga causing red snow

red soil *noun* (1889)
: any of a group of zonal soils that develop in a warm temperate moist climate under deciduous or mixed forests and that have thin organic and organic-mineral layers overlying a yellowish brown leached layer resting on an illuvial red horizon — called also *red podzolic soil*

red spider *noun* (1646)
: SPIDER MITE

red spruce *noun* (1777)
: a spruce (*Picea rubens*) of eastern North America that has pubescent twigs and yellowish green needles and is an important source of lumber and pulpwood

red squill *noun* (1738)
1 : a red-bulbed form of squill (*Urginea maritima*)
2 : a rat poison derived from the bulb of red squill

red squirrel *noun* (1682)
: a common and widely distributed North American squirrel (*Tamiasciurus hudsonicus*) that has reddish upper parts and is smaller than the gray squirrel

red star *noun* (1903)
: a star having a very low surface temperature and a red color

red·start \'red-ˌstärt\ *noun* [*red* + obsolete *start* handle, tail] (circa 1570)
1 : a small Old World songbird (*Phoenicurus phoenicurus* of the family Turdidae) with the male having a white brow, black throat, and chestnut breast and tail

2 : an American warbler (*Setophaga ruticilla* of the family Parulidae) with a black and orange male

red–tailed hawk \'red-ˌtāld-\ *noun* (1805)
: a widely distributed chiefly rodent-eating New World hawk (*Buteo jamaicensis*) that is usually mottled dusky above and white streaked dusky and tinged with buff below and has a rather short typically reddish tail — called also *redtail*

red-tailed hawk

red tape *noun* [from the red tape formerly used to bind legal documents in England] (1736)
: official routine or procedure marked by excessive complexity which results in delay or inaction

red tide *noun* (1904)
: seawater discolored by the presence of large numbers of dinoflagellates (especially of the genera *Gonyaulax* and *Gymnodinium*) which produce a toxin poisonous especially to many forms of marine vertebrate life and to humans who consume contaminated shellfish — compare SAXITOXIN

red•top \'red-ˌtäp\ *noun* (1790)
: any of various grasses (genus *Agrostis*) with usually reddish panicles; *especially* **:** an important forage and lawn grass (*A. alba* synonym *A. gigantea*) of eastern North America

re•duce \ri-'düs, -'dyüs\ *verb* **re•duced; re•duc•ing** [Middle English, to lead back, from Latin *reducere,* from *re-* + *ducere* to lead — more at TOW] (14th century)
transitive verb
1 a : to draw together or cause to converge **:** CONSOLIDATE ⟨*reduce* all the questions to one⟩ **b** (1) **:** to diminish in size, amount, extent, or number ⟨*reduce* taxes⟩ ⟨*reduce* the likelihood of war⟩ (2) **:** to decrease the volume and concentrate the flavor of by boiling ⟨add the wine and *reduce* the sauce for two minutes⟩ **c :** to narrow down **:** RESTRICT ⟨the Indians were *reduced* to small reservations⟩ **d :** to make short er **:** ABRIDGE
2 *archaic* **:** to restore to righteousness **:** SAVE
3 : to bring to a specified state or condition ⟨the impact of the movie *reduced* them to tears⟩
4 a : to force to capitulate **b :** FORCE, COMPEL
5 a : to bring to a systematic form or character ⟨*reduce* natural events to laws⟩ **b :** to put down in written or printed form ⟨*reduce* an agreement to writing⟩
6 : to correct (as a fracture) by bringing displaced or broken parts back into their normal positions
7 a : to lower in grade or rank **:** DEMOTE **b :** to lower in condition or status **:** DOWNGRADE
8 a : to diminish in strength or density **b :** to diminish in value
9 a (1) **:** to change the denominations or form of without changing the value (2) **:** to construct a geometrical figure similar to but smaller than (a given figure) **b :** to transpose from one form into another **:** CONVERT **c :** to change (an expression) to an equivalent but more fundamental expression ⟨*reduce* a fraction⟩
10 : to break down (as by crushing or grinding) **:** PULVERIZE
11 a : to bring to the metallic state by removal of nonmetallic elements ⟨*reduce* an ore by heat⟩ **b :** DEOXIDIZE **c :** to combine with or subject to the action of hydrogen **d** (1) **:** to change (an element or ion) from a higher to a lower oxidation state (2) **:** to add one or more electrons to (an atom or ion or molecule)
12 : to change (a stressed vowel) to an unstressed vowel
intransitive verb
1 a (1) **:** to become diminished or lessened; *especially* **:** to lose weight by dieting (2) **:** to become reduced ⟨ferrous iron *reduces* to ferric iron⟩ **b :** to become concentrated or consolidated **c :** to undergo meiosis
2 : to become converted or equated
synonym see DECREASE, CONQUER
— re•duc•er *noun*
— re•duc•ibil•i•ty \-ˌdü-sə-'bi-lə-tē, -ˌdyü-\ *noun*
— re•duc•ible \-'dü-sə-bəl, -'dyü-\ *adjective*
— re•duc•ibly \-blē\ *adverb*

reducing agent *noun* (1885)
: a substance that reduces a chemical compound usually by donating electrons

re•duc•tant \ri-'dək-tənt\ *noun* (1925)
: REDUCING AGENT

re•duc•tase \-ˌtās, -ˌtāz\ *noun* (1902)
: an enzyme that catalyzes reduction

re•duc•tio ad ab•sur•dum \ri-'dək-tē-ˌō-ˌad-əb-'sər-dəm, -'dək-sē-ō-, -shē-, -'zər-\ *noun* [Late Latin, literally, reduction to the absurd] (1741)
1 : disproof of a proposition by showing an absurdity to which it leads when carried to its logical conclusion
2 : the carrying of something to an absurd extreme

re•duc•tion \ri-'dək-shən\ *noun* [Middle English *reduccion* restoration, from Middle French *reduction,* from Late Latin & Latin; Late Latin *reduction-, reductio* reduction (in a syllogism), from Latin, restoration, from *reducere*] (1546)
1 : the act or process of reducing **:** the state of being reduced
2 a : something made by reducing **b :** the amount by which something is reduced
3 [Spanish *reducción,* from Latin *reduction-, reductio*] **:** a South American Indian settlement directed by Jesuit missionaries
4 : MEIOSIS 2; *specifically* **:** production of the gametic chromosome number in the first meiotic division
— re•duc•tion•al \-shnəl, -shə-nᵊl\ *adjective*

reduction division *noun* (1891)
: the usually first division of meiosis in which chromosome reduction occurs; *also* **:** MEIOSIS 2

reduction gear *noun* (1896)
: a combination of gears used to reduce the input speed (as of a marine turbine) to a lower output speed (as of a ship's propeller)

re•duc•tion•ism \ri-'dək-shə-ˌni-zəm\ *noun* (1943)
1 : the attempt to explain all biological processes by the same explanations (as by physical laws) that chemists and physicists use to interpret inanimate matter; *also* **:** the theory that complete reductionism is possible
2 : a procedure or theory that reduces complex data or phenomena to simple terms; *especially* **:** OVERSIMPLIFICATION
— re•duc•tion•ist \-sh(ə-)nist\ *noun or adjective*
— re•duc•tion•is•tic \-ˌdək-shə-'nis-tik\ *adjective*

re•duc•tive \ri-'dək tiv\ *adjective* (1633)
1 : of, relating to, causing, or involving reduction
2 : of or relating to reductionism **:** REDUCTIONISTIC
— re•duc•tive•ly *adverb*
— re•duc•tive•ness *noun*

re•dun•dan•cy \ri-'dən-dən(t)-sē\ *noun, plural* **-cies** (circa 1602)
1 a : the quality or state of being redundant **:** SUPERFLUITY **b :** the use of redundant components; *also* **:** such components **c** *chiefly British* **:** dismissal from a job especially by layoff
2 : PROFUSION, ABUNDANCE
3 a : superfluous repetition **:** PROLIXITY **b :** an act or instance of needless repetition
4 : the part of a message that can be eliminated without loss of essential information

re•dun•dant \-dənt\ *adjective* [Latin *redundant-, redundans,* present participle of *redundare* to overflow — more at REDOUND] (1594)
1 a : exceeding what is necessary or normal **:** SUPERFLUOUS **b :** characterized by or containing an excess; *specifically* **:** using more words than necessary **c :** characterized by similarity or repetition ⟨a group of particularly *redundant* brick buildings⟩ **d** *chiefly British* **:** no longer needed for a job and hence laid off
2 : PROFUSE, LAVISH
3 : serving as a duplicate for preventing failure of an entire system (as a spacecraft) upon failure of a single component
— re•dun•dant•ly *adverb*

re•du•pli•cate \ri-'dü-pli-ˌkāt, 'rē-, -'dyü-\ *transitive verb* [Late Latin *reduplicatus,* past participle of *reduplicare,* from Latin *re-* + *duplicare* to double — more at DUPLICATE] (circa 1570)
1 : to make or perform again **:** COPY, REPEAT
2 : to form (a word) by reduplication
— re•du•pli•cate \-kət\ *adjective*

re•du•pli•ca•tion \ri-ˌdü-pli-'kā-shən, ˌrē-, -ˌdyü-\ *noun* (1555)
1 : an act or instance of doubling or reiterating
2 a : an often grammatically functional repetition of a radical element or a part of it occurring usually at the beginning of a word and often accompanied by change of the radical vowel **b** (1) **:** a word or form produced by reduplication (2) **:** the repeated element in such a word or form
3 : ANADIPLOSIS
— re•du•pli•ca•tive \ri-'dü-pli-ˌkā-tiv, 'rē-, -'dyü-\ *adjective*
— re•du•pli•ca•tive•ly *adverb*

re•du•vi•id \ri-'dü-vē-əd, -'dyü-\ *noun* [ultimately from Latin *reduvia* hangnail] (1888)
: ASSASSIN BUG
— reduviid *adjective*

re•dux \(ˌ)rē-'dəks, 'rē-ˌ\ *adjective* [Latin, returning, from *reducere* to lead back] (1873)
: brought back — used postpositively

red•ware \'red-ˌwar, -ˌwer\ *noun* (circa 1797)
: earthenware pottery made of clay containing considerable iron oxide

red water *noun* (1594)
: any of several cattle diseases characterized by hematuria

red wheat *noun* (1523)
: a wheat that has red grains

red wine *noun* (circa 1754)
: a wine with a predominantly red color derived during fermentation from the natural pigment in the skins of dark-colored grapes

red•wing \'red-ˌwiŋ\ *noun* (1657)
1 : a European thrush (*Turdus iliacus* synonym *T. musicus*) having the underwing coverts red
2 : RED-WINGED BLACKBIRD

red–winged blackbird \'red-ˌwiŋd-\ *noun* (1797)
: a North American blackbird (*Agelaius phoeniceus*) of which the adult male is black with a patch of bright scarlet at the bend of the wings bordered behind with yellow or buff — called also *redwing blackbird*

red wolf *noun* (1840)
: a wolf (*Canis rufus* synonym *C. niger*) originally of the southeastern U.S.

red•wood \'red-ˌwu̇d\ *noun* (1634)
1 : any of various woods yielding a red dye
2 : a tree that yields a red dyewood or produces red or reddish wood
3 a : a commercially important coniferous timber tree (*Sequoia sempervirens*) of the bald cypress family that grows chiefly in coastal California and sometimes reaches a height of

360 feet (110 meters) — called also *coast redwood* **b :** its brownish red durable wood

red worm *noun* (1935)
: BLOODWORM

re·echo \(,)rē-ˈe-(,)kō\ (1590)
intransitive verb
: to repeat or return an echo **:** echo again or repeatedly **:** REVERBERATE
transitive verb
: to echo back **:** REPEAT

¹reed \ˈrēd\ *noun* [Middle English *rede,* from Old English *hrēod;* akin to Old High German *hriot* reed] (before 12th century)
1 a : any of various tall grasses with slender often prominently jointed stems that grow especially in wet areas **b :** a stem of a reed **c :** a person or thing too weak to rely on **:** one easily swayed or overcome

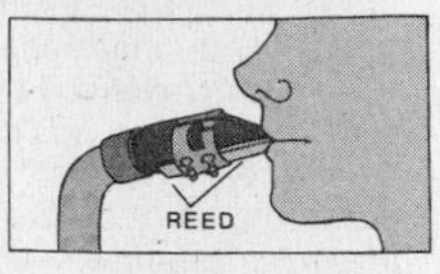

reed 6a

2 : a growth or mass of reeds; *specifically* **:** reeds for thatching
3 : ARROW
4 : a wind instrument made from the hollow joint of a plant
5 : an ancient Hebrew unit of length equal to 6 cubits
6 a : a thin elastic tongue (as of cane, wood, metal, or plastic) fastened at one end over an air opening in a wind instrument (as a clarinet, organ pipe, or accordion) and set in vibration by an air current **b :** a woodwind instrument that produces sound by the vibrating of a reed against the mouthpiece ⟨the *reeds* of an orchestra⟩
7 : a device on a loom resembling a comb and used to space warp yarns evenly
8 : REEDING 1a
— reed·like \-,līk\ *adjective*

²reed *transitive verb* (1951)
: MILL 2

reed·buck \ˈrēd-,bək\ *noun, plural* **reedbuck** *also* **reedbucks** (1834)
: any of a genus (*Redunca*) of fawn-colored African antelopes in which the females are hornless

reed·ed \ˈrē-dəd\ *adjective* (1829)
: decorated with reeds or reeding ⟨a bed with *reeded* posts⟩

re·ed·i·fy \(,)rē-ˈe-də-,fī\ *transitive verb* **-fied; -fy·ing** [Middle English *reedifien,* from Middle French *reedifier,* from Late Latin *reaedificare,* from Latin *re-* + *aedificare* to build — more at EDIFY] (15th century)
British **:** REBUILD

reed·ing \ˈrē-diŋ\ *noun* (1815)
1 a : a small convex molding — see MOLDING illustration **b :** decoration by series of reedings
2 : MILLING

re·edit \(,)rē-ˈe-dət\ *transitive verb* (1797)
: to edit again **:** make a new edition of
— re·edi·tion \,rē-ə-ˈdi-shən\ *noun*

reed·man \ˈrēd-,man\ *noun* (1938)
: one who plays a reed instrument

reed organ *noun* (1851)
: a keyboard wind instrument in which the wind acts on a set of free reeds

reed pipe *noun* (circa 1741)
: a pipe-organ pipe producing its tone by vibration of a beating reed in a current of air

re·ed·u·cate \(,)rē-ˈe-jə-,kāt\ *transitive verb* (1808)
: to train again; *especially* **:** to rehabilitate through education
— re·ed·u·ca·tion \(,)rē-,e-jə-ˈkā-shən\ *noun*
— re·ed·u·ca·tive \(,)rē-ˈe-jə-,kā-tiv\ *adjective*

reedy \ˈrē-dē\ *adjective* **reed·i·er; -est** (14th century)
1 : abounding in or covered with reeds
2 : made of or resembling reeds; *especially* **:** SLENDER, FRAIL
3 : having the tone quality of a reed instrument
— reed·i·ness \-nəs\ *noun*

¹reef \ˈrēf\ *noun* [Middle English *riff,* from Old Norse *rif;* akin to Old English *ribb* rib] (14th century)
1 : a part of a sail taken in or let out in regulating size
2 : reduction in sail area by reefing

²reef (1667)
transitive verb
1 : to reduce the area of (a sail) by rolling or folding a portion
2 : to lower or bring inboard (a spar) wholly or partially
intransitive verb
: to reduce a sail by taking in a reef
— reef·able \ˈrē-fə-bəl\ *adjective*

³reef *noun* [Dutch *rif,* probably of Scandinavian origin; akin to Old Norse *rif* reef of a sail] (1584)
1 a : a chain of rocks or coral or a ridge of sand at or near the surface of water **b :** a hazardous obstruction
2 : LODE, VEIN
— reefy \ˈrē-fē\ *adjective*

¹reef·er \ˈrē-fər\ *noun* (1818)
1 : one that reefs
2 : a close-fitting usually double-breasted jacket or coat of thick cloth

²ree·fer \ˈrē-fər\ *noun* [by shortening & alteration] (1914)
1 : REFRIGERATOR
2 : a refrigerator car, truck, trailer, or ship

³ree·fer *noun* [probably modification of Mexican Spanish *grifa*] (1931)
: a marijuana cigarette; *also* **:** MARIJUANA 2

reef knot *noun* (1841)
: a square knot used in reefing a sail

¹reek \ˈrēk\ *noun* [Middle English *rek,* from Old English *rēc;* akin to Old High German *rouh* smoke] (before 12th century)
1 *chiefly dialect* **:** SMOKE
2 : VAPOR, FOG
3 : a strong or disagreeable fume or odor

²reek (before 12th century)
intransitive verb
1 : to emit smoke or vapor
2 a : to give off or become permeated with a strong or offensive odor **b :** to give a strong impression of some constituent quality or feature ⟨a neighborhood that *reeks* of poverty⟩
3 : EMANATE
transitive verb
1 : to subject to the action of smoke or vapor
2 : EXUDE, GIVE OFF ⟨a politician who *reeks* charm⟩
— reek·er *noun*
— reeky \ˈrē-kē\ *adjective*

¹reel \ˈrē(ə)l\ *noun* [Middle English, from Old English *hrēol;* akin to Old Norse *hræll* weaver's reed, Greek *krekein* to weave] (before 12th century)
1 : a revolvable device on which something flexible is wound: as **a :** a small windlass at the butt of a fishing rod for the line **b** *chiefly British* **:** a spool or bobbin for sewing thread **c :** a flanged spool for photographic film; *especially* **:** one for motion pictures
2 : a quantity of something wound on a reel

²reel (14th century)
transitive verb
1 : to wind on or as if on a reel
2 : to draw by reeling a line ⟨*reel* a fish in⟩
intransitive verb
: to turn a reel
— reel·able \ˈrē-lə-bəl\ *adjective*

³reel *verb* [Middle English *relen,* probably from *reel,* noun] (14th century)
intransitive verb
1 a : to turn or move round and round **b :** to be in a whirl
2 : to behave in a violent disorderly manner
3 : to waver or fall back (as from a blow)
4 : to walk or move unsteadily
transitive verb
: to cause to reel

⁴reel *noun* (1572)
: a reeling motion

⁵reel *noun* [probably from ⁴*reel*] (circa 1585)
1 : a lively Scottish-Highland dance; *also* **:** the music for this dance
2 : VIRGINIA REEL

re·elect \,rē-ə-ˈlekt\ *transitive verb* (1601)
: to elect for another term in office
— re·elec·tion \-ˈlek-shən\ *noun*

reel·er \ˈrē-lər\ *noun* (circa 1598)
1 : one that reels
2 : a motion picture having a specified number of reels ⟨a two-*reeler*⟩

reel off *transitive verb* (1837)
1 : to tell or recite readily and usually at length ⟨*reel off* a few jokes to break the ice⟩
2 : to chalk up usually as a series

reel–to–reel *adjective* (1961)
: of, relating to, or utilizing magnetic tape that requires threading on a take-up reel ⟨a *reel-to-reel* tape recorder⟩

re·em·broi·der \,rē-əm-ˈbrȯi-dər\ *transitive verb* (1927)
: to outline a design (as on lace) with embroidery stitching

re·en·act \,rē-ə-ˈnakt\ *transitive verb* (circa 1676)
1 : to enact (as a law) again
2 : to act or perform again
3 : to repeat the actions of (an earlier event or incident)
— re·en·act·ment \-ˈnak(t)-mənt\ *noun*

re·en·force \,rē-ən-ˈfōrs, -ˈfȯrs\ *variant of* REINFORCE

re·en·ter \(,)rē-ˈen-tər\ (15th century)
transitive verb
1 : to enter (something) again
2 : to return to and enter
intransitive verb
: to enter again

re·en·trance \(,)rē-ˈen-trən(t)s\ *noun* (1594)
: REENTRY

¹re·en·trant \-trənt\ *adjective* (1781)
: directed inward

²reentrant *noun* (1899)
1 : one that reenters
2 : one that is reentrant
3 : an indentation in a landform

re·en·try \(,)rē-ˈen-trē\ *noun* (15th century)
1 : a retaking possession; *especially* **:** entry by a lessor on leased premises on the tenant's failure to perform the conditions of the lease
2 : a second or new entry
3 : a playing card that will enable a player to regain the lead
4 : the action of reentering the earth's atmosphere after travel in space

reest \ˈrēst\ *intransitive verb* [probably short for Scots *arreest* to arrest, from Middle English (Scots) *arreisten,* from Middle French *arester* — more at ARREST] (1786)
chiefly Scottish **:** BALK

¹reeve \ˈrēv\ *noun* [Middle English *reve,* from Old English *gerēfa,* from *ge-* (associative prefix) + *-rēfa* (akin to Old English *-rōf* number, Old High German *ruova*) — more at CO-] (before 12th century)
1 : a local administrative agent of an Anglo-Saxon king
2 : a medieval English manor officer responsible chiefly for overseeing the discharge of feudal obligations
3 a : the council president in some Canadian municipalities **b :** a local official charged with enforcement of specific regulations ⟨deer *reeve*⟩

²reeve *verb* **rove** \ˈrōv\ *or* **reeved; reev·ing** [origin unknown] (1627)
transitive verb
1 : to pass (as a rope) through a hole or opening
2 : to fasten by passing through a hole or around something

3 : to pass a rope through
intransitive verb
of a rope **:** to pass through a block or similar device

³reeve *noun* [origin unknown] (1634)
: the female of the ruff (sandpiper)

ref \'ref\ *noun* (1899)
: a referee in a game or sport

re·fash·ion \(,)rē-'fa-shən\ *transitive verb* (1803)
: REMAKE, ALTER

re·fect \ri-'fekt\ *transitive verb* [Latin *refectus,* past participle of *reficere*] (14th century)
archaic **:** to refresh with food or drink

re·fec·tion \ri-'fek-shən\ *noun* [Middle English *refeccioun,* from Middle French *refection,* from Latin *refection-, refectio,* from *reficere* to restore, from *re-* + *facere* to make — more at DO] (14th century)
1 : refreshment of mind, spirit, or body; *especially* **:** NOURISHMENT
2 a : the taking of refreshment **b :** food and drink together **:** REPAST

re·fec·to·ry \ri-'fek-t(ə-)rē\ *noun, plural* **-ries** [Late Latin *refectorium,* from Latin *reficere*] (15th century)
: a dining hall (as in a monastery or college)

refectory table *noun* (1923)
: a long table with heavy legs

re·fel \ri-'fel\ *transitive verb* **re·felled; re·fel·ling** [Latin *refellere* to prove false, refute, from *re-* + *fallere* to deceive] (1530)
obsolete **:** REJECT, REPULSE

re·fer \ri-'fər\ *verb* **re·ferred; re·fer·ring** [Middle English *referren,* from Latin *referre* to bring back, report, refer, from *re-* + *ferre* to carry — more at BEAR] (14th century)
transitive verb
1 a (1) **:** to think of, regard, or classify within a general category or group (2) **:** to explain in terms of a general cause **b :** to allot to a particular place, stage, or period **c :** to regard as coming from or located in a specific area
2 a : to send or direct for treatment, aid, information, or decision ⟨*refer* a patient to a specialist⟩ ⟨*refer* a bill back to a committee⟩ **b :** to direct for testimony or guaranty as to character or ability
intransitive verb
1 a : to have relation or connection **:** RELATE **b :** to direct attention usually by clear and specific mention ⟨no one *referred* to yesterday's quarrel⟩
2 : to have recourse **:** glance briefly ⟨*referred* frequently to his notes while speaking⟩
— re·fer·able \'re-f(ə-)rə-bəl, ri-'fər-ə-\ *adjective*
— re·fer·rer \ri-'fər-ər\ *noun*

¹ref·er·ee \,re-fə-'rē\ *noun* (1621)
1 : one to whom a thing is referred: as **a :** a person to whom a legal matter if referred for investigation and report or for settlement **b :** a person who reviews a paper and especially a technical paper and recommends that it should or should not be published **c** *chiefly British* **:** REFERENCE 4a
2 : a sports official usually having final authority in administering a game

²referee *verb* **-eed; -ee·ing** (1889)
transitive verb
1 : to conduct (as a match or game) as referee
2 a : to arbitrate (as a legal matter) as a judge or third party **b :** to review (as a technical paper) before publication
intransitive verb
: to act as a referee

¹ref·er·ence \'re-fərn(t)s, 're-f(ə-)rən(t)s\ *noun* (1589)
1 : the act of referring or consulting
2 : a bearing on a matter **:** RELATION ⟨in *reference* to your recent letter⟩
3 : something that refers: as **a :** ALLUSION, MENTION **b :** something (as a sign or indication) that refers a reader or consulter to another source of information (as a book or passage) **c :** consultation of sources of information
4 : one referred to or consulted: as **a :** a person to whom inquiries as to character or ability can be made **b :** a statement of the qualifications of a person seeking employment or appointment given by someone familiar with the person **c** (1) **:** a source of information (as a book or passage) to which a reader or consulter is referred (2) **:** a work (as a dictionary or encyclopedia) containing useful facts or information **d :** DENOTATION, MEANING

²reference *adjective* (1856)
: used or usable for reference; *especially* **:** constituting a standard for measuring or constructing

³reference *transitive verb* **-enced; -enc·ing** (1891)
1 a : to supply with references **b :** to cite in or as a reference
2 : to put in a form (as a table) adapted to easy reference

reference mark *noun* (1856)
: a conventional mark (as *, †, or ‡) placed in written or printed text to direct the reader's attention especially to a footnote

ref·er·en·dum \,re-fə-'ren-dəm\ *noun, plural* **-da** \-də\ *or* **-dums** [New Latin, from Latin, neuter of *referendus,* gerundive of *referre* to refer] (1847)
1 a : the principle or practice of submitting to popular vote a measure passed on or proposed by a legislative body or by popular initiative **b :** a vote on a measure so submitted
2 : a diplomatic agent's note asking for government instructions

ref·er·ent \'re-f(ə-)rənt\ *noun* [Latin *referent-, referens,* present participle of *referre*] (1844)
: one that refers or is referred to; *especially* **:** the thing that a symbol (as a word or sign) stands for
— referent *adjective*

ref·er·en·tial \,re-fə-'ren(t)-shəl\ *adjective* (1660)
: of, containing, or constituting a reference
— ref·er·en·tial·i·ty \-,ren(t)-shē-'a-lə-tē\ *noun*
— ref·er·en·tial·ly \-'ren(t)-sh(ə-)lē\ *adverb*

re·fer·ral \ri-'fər-əl\ *noun* (1927)
1 : the act, action, or an instance of referring ⟨gave the patient a *referral* to a specialist⟩
2 : one that is referred

¹re·fill \(,)rē-'fil\ (1681)
transitive verb
: to fill again **:** REPLENISH
intransitive verb
: to become filled again
— re·fill·able \-'fi-lə-bəl\ *adjective*

²re·fill \'rē-,fil\ *noun* (1886)
1 : a product or a container and a product used to refill the exhausted supply of a device
2 : something provided again; *especially* **:** a second or later filling of a medical prescription

re·fi·nance \,rē-fə-'nan(t)s, (,)rē-'fī-,, ,rē-(,)fī-'\ (1908)
transitive verb
: to renew or reorganize the financing of
intransitive verb
: to finance something anew

re·fine \ri-'fīn\ *verb* **re·fined; re·fin·ing** (1582)
transitive verb
1 : to free (as metal, sugar, or oil) from impurities or unwanted material
2 : to free from moral imperfection **:** ELEVATE
3 : to improve or perfect by pruning or polishing ⟨*refine* a poetic style⟩
4 : to reduce in vigor or intensity
5 : to free from what is coarse, vulgar, or uncouth
intransitive verb
1 : to become pure or perfected
2 : to make improvement by introducing subtleties or distinctions
— re·fin·er *noun*

re·fined \ri-'fīnd\ *adjective* (1588)
1 : FASTIDIOUS, CULTIVATED
2 : free from impurities
3 : PRECISE, EXACT ⟨a *refined* test for radioactivity⟩

re·fine·ment \ri-'fīn-mənt\ *noun* (circa 1611)
1 : the action or process of refining
2 : the quality or state of being refined **:** CULTIVATION
3 a : a refined feature or method **b :** a highly refined distinction **:** SUBTLETY **c :** a contrivance or device intended to improve or perfect

re·fin·ery \ri-'fī-nə-rē, -'fīn-rē\ *noun, plural* **-er·ies** (circa 1741)
: a building and equipment for refining or processing (as oil or sugar)

re·fin·ish \(,)rē-'fi-nish\ (1931)
transitive verb
: to give (as furniture) a new surface
intransitive verb
: to refinish furniture
— re·fin·ish·er *noun*

¹re·fit \(,)rē-'fit\ (1666)
transitive verb
: to fit out or supply again
intransitive verb
: to obtain repairs or fresh supplies or equipment

²re·fit \'rē-,fit, (,)rē-'\ *noun* (1799)
: the action of refitting; *especially* **:** a refitting and renovating of a ship

re·fla·tion \(,)rē-'flā-shən\ *noun* [*re-* + *-flation* (as in *deflation*)] (1932)
: restoration of deflated prices to a desirable level
— re·flate \(,)rē-'flāt\ *verb*
— re·fla·tion·ary \-shə-,ner-ē\ *adjective*

re·flect \ri-'flekt\ *verb* [Middle English, from Latin *reflectere* to bend back, from *re-* + *flectere* to bend] (15th century)
transitive verb
1 *archaic* **:** to turn into or away from a course **:** DEFLECT
2 : to prevent passage of and cause to change direction ⟨a mirror *reflects* light⟩
3 : to bend or fold back
4 : to give back or exhibit as an image, likeness, or outline **:** MIRROR ⟨the clouds were *reflected* in the water⟩
5 : to bring or cast as a result ⟨his attitude *reflects* little credit on his judgment⟩
6 : to make manifest or apparent **:** SHOW ⟨the pulse *reflects* the condition of the heart⟩
7 : REALIZE, CONSIDER
intransitive verb
1 : to throw back light or sound
2 a : to think quietly and calmly **b :** to express a thought or opinion resulting from reflection
3 a : to tend to bring reproach or discredit ⟨an investigation that *reflects* on all the members of the department⟩ **b :** to bring about a specified appearance or characterization ⟨an act which *reflects* well on him⟩ **c :** to have a bearing or influence
synonym see THINK

re·flec·tance \ri-'flek-tən(t)s\ *noun* (1926)
: the fraction of the total radiant flux incident upon a surface that is reflected and that varies according to the wavelength distribution of the incident radiation — called also *re·flec·tiv·i·ty* \,rē-,flek-'ti-və-tē, ri-\

reflecting telescope *noun* (circa 1704)
: REFLECTOR 2

re·flec·tion \ri-'flek-shən\ *noun* [Middle English, alteration of *reflexion,* from Late Latin *reflexion-, reflexio* act of bending back, from Latin *reflectere*] (14th century)
1 : an instance of reflecting; *especially* **:** the return of light or sound waves from a surface

2 : the production of an image by or as if by a mirror
3 a : the action of bending or folding back **b :** a reflected part **:** FOLD
4 : something produced by reflecting: as **a :** an image given back by a reflecting surface **b :** an effect produced by an influence ⟨the high crime rate is a *reflection* of our violent society⟩
5 : an often obscure or indirect criticism **:** REPROACH ⟨a *reflection* on his character⟩
6 : a thought, idea, or opinion formed or a remark made as a result of meditation
7 : consideration of some subject matter, idea, or purpose
8 *obsolete* **:** turning back **:** RETURN
9 a : a transformation of a figure in which each point is replaced by a point symmetric with respect to a line or plane **b :** a transformation that involves reflection in more than one axis of a rectangular coordinate system
— **re•flec•tion•al** \-shnəl, -shə-nᵊl\ *adjective*

re•flec•tive \ri-ˈflek-tiv\ *adjective* (1627)
1 : capable of reflecting light, images, or sound waves
2 : marked by reflection **:** THOUGHTFUL, DELIBERATIVE
3 : of, relating to, or caused by reflection ⟨*reflective* glare⟩
4 : REFLEXIVE 3
— **re•flec•tive•ly** *adverb*
— **re•flec•tive•ness** *noun*

re•flec•tom•e•ter \ˌrē-ˌflek-ˈtä-mə-tər, ri-\ *noun* (1891)
: a device for measuring the reflectance of radiant energy (as light)
— **re•flec•tom•e•try** \-mə-trē\ *noun*

re•flec•tor \ri-ˈflek-tər\ *noun* (1665)
1 : one that reflects; *especially* **:** a polished surface for reflecting light or other radiation
2 : a telescope in which the principal focusing element is a mirror

re•flec•tor•ize \-tə-ˌrīz\ *transitive verb* **-ized; -iz•ing** (1940)
1 : to make reflecting
2 : to provide with reflectors

¹re•flex \ˈrē-ˌfleks\ *noun* [Latin *reflexus,* past participle of *reflectere* to reflect] (1508)
1 a : reflected heat, light, or color **b :** a mirrored image **c :** a copy exact in essential or peculiar features
2 a : an automatic and often inborn response to a stimulus that involves a nerve impulse passing inward from a receptor to a nerve center and thence outward to an effector (as a muscle or gland) without reaching the level of consciousness — compare HABIT **b :** the process that culminates in a reflex and comprises reception, transmission, and reaction — called also *reflex action* **c** *plural* **:** the power of acting or responding with adequate speed **d :** a way of thinking or behaving
3 : a linguistic element (as a word or sound) or system (as writing) that is derived from a prior and especially an older element or system ⟨*boat* is the *reflex* of Old English *bāt*⟩

²reflex *adjective* [Latin *reflexus*] (1649)
1 : directed back on the mind or its operations **:** INTROSPECTIVE
2 : bent, turned, or directed back **:** REFLECTED ⟨a stem with *reflex* leaves⟩
3 : produced or carried out in reaction, resistance, or return
4 *of an angle* **:** being between 180° and 360°
5 : of, relating to, or produced by a reflex without intervention of consciousness
— **re•flex•ly** *adverb*

reflex arc *noun* (1882)
: the complete nervous path involved in a reflex

re•flexed \ˈrē-ˌflekst, ri-ˈ\ *adjective* [Latin *reflexus* + English ¹*-ed*] (1733)
: bent or curved backward or downward ⟨*reflexed* petals⟩ ⟨*reflexed* leaves⟩

re•flex•ion *chiefly British variant of* REFLECTION

¹re•flex•ive \ri-ˈflek-siv\ *adjective* [Medieval Latin *reflexivus,* from Latin *reflexus*] (1640)
1 a : directed or turned back on itself **b :** marked by or capable of reflection **:** REFLECTIVE
2 : of, relating to, characterized by, or being a relation that exists between an entity and itself ⟨the relation "is equal to" is *reflexive* but the relation "is the father of" is not⟩
3 : of, relating to, or constituting an action (as in "he perjured himself") directed back on the agent or the grammatical subject
4 : characterized by habitual and unthinking behavior
— **re•flex•ive•ly** *adverb*
— **re•flex•ive•ness** *noun*
— **re•flex•iv•i•ty** \ˌrē-ˌflek-ˈsi-və-tē, ri-\ *noun*

²reflexive *noun* (1866)
: REFLEXIVE PRONOUN

reflexive pronoun *noun* (1867)
: a pronoun referring to the subject of the sentence, clause, or verbal phrase in which it stands; *specifically* **:** a personal pronoun compounded with *-self*

re•flex•ol•o•gy \ˌrē-ˌflek-ˈsä-lə-jē\ *noun* [International Scientific Vocabulary] (1923)
1 : the study and interpretation of behavior in terms of simple and complex reflexes
2 : massage of the hands or feet based on the belief that pressure applied to specific points on these extremities benefits other parts of the body

re•flow \(ˌ)rē-ˈflō\ *intransitive verb* (14th century)
1 : to flow back **:** EBB
2 : to flow in again
— **re•flow** \ˈrē-ˌflō\ *noun*

ref•lu•ence \ˈre-ˌflü-ən(t)s, re-ˈflü-\ *noun* (15th century)
archaic **:** REFLUX 1

re•flu•ent \-ənt\ *adjective* [Middle English, from Latin *refluent-, refluens,* present participle of *refluere* to flow back, from *re-* + *fluere* to flow — more at FLUID] (15th century)
: flowing back

¹re•flux \ˈrē-ˌfləks\ *noun* [Middle English, from Medieval Latin *refluxus,* from Latin *re-* + *fluxus* flow — more at FLUX] (15th century)
1 : a flowing back **:** EBB
2 : a process of refluxing or condition of being refluxed

²re•flux \ri-ˈfləks, ˈrē-ˌ\ *transitive verb* (1926)
: to cause to flow back or return; *especially* **:** to heat so that the vapors formed condense and return to be heated again

re•fo•cus \(ˌ)rē-ˈfō-kəs\ (circa 1865)
transitive verb
1 : to focus again
2 : to change the emphasis or direction of ⟨had *refocused* his life⟩
intransitive verb
1 : to focus something again
2 : to change emphasis or direction

re•for•es•ta•tion \(ˌ)rē-ˌfȯr-ə-ˈstā-shən, -ˌfär-\ *noun* (1887)
: the action of renewing forest cover by planting seeds or young trees
— **re•for•est** \(ˌ)rē-ˈfȯr-əst, -ˈfär-\ *transitive verb*

¹re•form \ri-ˈfȯrm\ *verb* [Middle English, from Middle French *reformer,* from Latin *reformare,* from *re-* + *formare* to form, from *forma* form] (14th century)
transitive verb
1 a : to put or change into an improved form or condition **b :** to amend or improve by change of form or removal of faults or abuses
2 : to put an end to (an evil) by enforcing or introducing a better method or course of action
3 : to induce or cause to abandon evil ways ⟨*reform* a drunkard⟩
4 a : to subject (hydrocarbons) to cracking **b :** to produce (as gasoline or gas) by cracking
intransitive verb
: to become changed for the better
synonym see CORRECT
— **re•form•abil•i•ty** \-ˌfȯr-mə-ˈbi-lə-tē\ *noun*
— **re•form•able** \-ˈfȯr-mə-bəl\ *adjective*

²reform *noun* (1663)
1 : amendment of what is defective, vicious, corrupt, or depraved
2 : a removal or correction of an abuse, a wrong, or errors
3 *capitalized* **:** REFORM JUDAISM

³reform *adjective* (1819)
: relating to or favoring reform

re–form \(ˌ)rē-ˈfȯrm\ (14th century)
transitive verb
: to form again
intransitive verb
: to take form again ⟨the ice *re-formed* on the lake⟩

re•for•mate \ri-ˈfȯr-ˌmāt, -mət\ *noun* (1949)
: a product of hydrocarbon reforming

ref•or•ma•tion \ˌre-fər-ˈmā-shən\ *noun* (15th century)
1 : the act of reforming **:** the state of being reformed
2 *capitalized* **:** a 16th century religious movement marked ultimately by rejection or modification of some Roman Catholic doctrine and practice and establishment of the Protestant churches
— **ref•or•ma•tion•al** \-shnəl, -shə-nᵊl\ *adjective*

re•for•ma•tive \ri-ˈfȯr-mə-tiv\ *adjective* (1593)
: intended or tending to reform

¹re•for•ma•to•ry \ri-ˈfȯr-mə-ˌtȯr-ē, -ˌtȯr-\ *adjective* (1589)
: REFORMATIVE

²reformatory *noun, plural* **-ries** (1834)
: a penal institution to which especially young or first offenders are committed for training and reformation

re•formed *adjective* (1563)
1 : changed for the better
2 *capitalized* **:** PROTESTANT; *specifically* **:** of or relating to the chiefly Calvinist Protestant churches formed in various continental European countries

reformed spelling *noun* (circa 1934)
: any of several methods of spelling English words that use letters with more phonetic consistency than conventional spelling and that usually discard some silent letters (as in *pedagog* for *pedagogue*)

re•form•er \ri-ˈfȯr-mər\ *noun* (1548)
1 : one that works for or urges reform
2 *capitalized* **:** a leader of the Protestant Reformation
3 : an apparatus for cracking oils or gases to form specialized products

re•form•ism \ri-ˈfȯr-ˌmi-zəm\ *noun* (1904)
: a doctrine, policy, or movement of reform
— **re•form•ist** \-mist\ *noun or adjective*

Reform Judaism *noun* (circa 1905)
: Judaism marked by a liberal approach in nonobservance of much legal tradition regarded as irrelevant to the present and in shortening and simplification of traditional ritual — compare CONSERVATIVE JUDAISM, ORTHODOX JUDAISM

reform school *noun* (circa 1859)
: a reformatory for boys or girls

re•fract \ri-ˈfrakt\ *transitive verb* [Latin *refractus,* past participle of *refringere* to break open, break up, from *re-* + *frangere* to break — more at BREAK] (1612)
1 a : to subject (as a ray of light) to refraction **b :** to alter or distort as if by refraction ⟨to *refract* that familiar world through the mind and heart of a romantic . . . woman —Anton Myrer⟩
2 : to determine the refracting power of

re·frac·tile \-'frak-tᵊl, -ˌtīl\ *adjective* (circa 1849)
: capable of refracting **:** REFRACTIVE

refracting telescope *noun* (1764)
: REFRACTOR

re·frac·tion \ri-'frak-shən\ *noun* (1603)
1 : deflection from a straight path undergone by a light ray or energy wave in passing obliquely from one medium (as air) into another (as glass) in which its velocity is different
2 : the change in the apparent position of a celestial body due to bending of the light rays emanating from it as they pass through the atmosphere; *also* **:** the correction to be applied to the apparent position of a body because of this bending
3 : the action of distorting an image by viewing through a medium; *also* **:** an instance of this

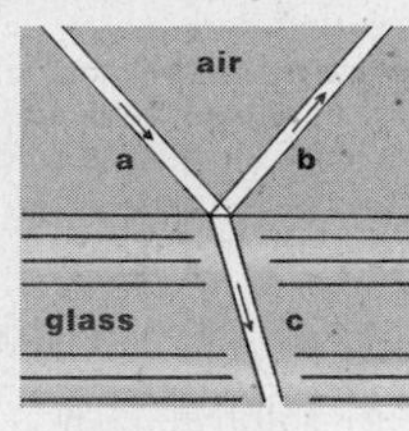

refraction 1: *a* light ray, *b* reflected ray, *c* refracted ray

re·frac·tive \ri-'frak-tiv\ *adjective* (1673)
1 : having power to refract
2 : relating or due to refraction
— re·frac·tive·ly *adverb*
— re·frac·tive·ness *noun*
— re·frac·tiv·i·ty \ˌrē-ˌfrak-'ti-və-tē, ri-\ *noun*

refractive index *noun* (1839)
: INDEX OF REFRACTION

re·frac·tom·e·ter \ˌrē-ˌfrak-'tä-mə-tər, ri-\ *noun* [International Scientific Vocabulary] (circa 1859)
: an instrument for measuring indices of refraction
— re·frac·to·met·ric \ri-ˌfrak-tə-'me-trik\ *adjective*
— re·frac·tom·e·try \ˌrē-ˌfrak-'tä-mə-trē, ri-\ *noun*

re·frac·tor \ri-'frak-tər\ *noun* (1769)
: a telescope whose principal focusing element is a lens

[1]re·frac·to·ry \ri-'frak-t(ə-)rē\ *adjective* [alteration of *refractary,* from Latin *refractarius,* irregular from *refragari* to oppose, from *re-* + *-fragari* (as in *suffragari* to support with one's vote)] (1606)
1 : resisting control or authority **:** STUBBORN, UNMANAGEABLE
2 a : resistant to treatment or cure ⟨a *refractory* lesion⟩ **b :** unresponsive to stimulus **c :** IMMUNE, INSUSCEPTIBLE ⟨after recovery they were *refractory* to infection⟩
3 : difficult to fuse, corrode, or draw out; *especially* **:** capable of enduring high temperature
synonym see UNRULY
— re·frac·to·ri·ly \-t(ə-)rə-lē; ˌrē-ˌfrak-'tōr-ə-lē, ri-, -'tȯr-\ *adverb*
— re·frac·to·ri·ness \ri-'frak-t(ə-)rē-nəs\ *noun*

[2]refractory *noun, plural* **-ries** (1627)
: a refractory person or thing; *especially* **:** a heat-resisting ceramic material

refractory period *noun* (circa 1880)
: the brief period immediately following the response especially of a muscle or nerve before it recovers the capacity to make a second response — called also *refractory phase*

[1]re·frain \ri-'frān\ *verb* [Middle English *refreynen,* from Middle French *refreiner, refrener,* from Latin *refrenare,* from *re-* + *frenum* bridle — more at FRENUM] (14th century)
transitive verb
archaic **:** CURB, RESTRAIN
intransitive verb
: to keep oneself from doing, feeling, or indulging in something and especially from following a passing impulse
— re·frain·ment \-mənt\ *noun*

[2]refrain *noun* [Middle English *refreyn,* from Middle French *refrain,* from *refraindre, refreindre* to resound, from Latin *refringere* to break up] (14th century)
: a regularly recurring phrase or verse especially at the end of each stanza or division of a poem or song **:** CHORUS; *also* **:** the musical setting of a refrain

re·fran·gi·ble \ri-'fran-jə-bəl\ *adjective* [irregular from Latin *refringere* to break up] (1673)
: capable of being refracted
— re·fran·gi·bil·i·ty \-ˌfran-jə-'bi-lə-tē\ *noun*
— re·fran·gi·ble·ness \-'fran-jə-bəl-nəs\ *noun*

re·fresh \ri-'fresh\ *verb* [Middle English *refresshen,* from Middle French *refreschir,* from Old French, from *re-* + *freis* fresh — more at FRESH] (14th century)
transitive verb
1 : to restore strength and animation to **:** REVIVE
2 : to freshen up **:** RENOVATE
3 a : to restore or maintain by renewing supply **:** REPLENISH **b :** AROUSE, STIMULATE ⟨let me *refresh* your memory⟩
4 : to run water over or restore water to
intransitive verb
1 : to become refreshed
2 : to take refreshment
3 : to lay in fresh provisions
synonym see RENEW

re·fresh·en \ri-'fre-shən, ˌrē-\ *transitive verb* [*re-* + *freshen*] (1782)
: REFRESH

re·fresh·er \ri-'fre-shər\ *noun* (15th century)
1 : something (as a drink) that refreshes
2 : REMINDER
3 : review or instruction designed especially to keep one abreast of professional developments ⟨*refresher* course⟩

re·fresh·ing \-shiŋ\ *adjective* (circa 1580)
: serving to refresh; *especially* **:** agreeably stimulating because of freshness or newness
— re·fresh·ing·ly \-shiŋ-lē\ *adverb*

re·fresh·ment \ri-'fresh-mənt\ *noun* (14th century)
1 : the act of refreshing **:** the state of being refreshed
2 a : something (as food or drink) that refreshes **b** *plural* (1) **:** a light meal (2) **:** assorted light foods

re·fried beans \'rē-ˌfrīd-\ *noun plural* (1957)
: beans cooked with seasonings, fried, then mashed and fried again

[1]re·frig·er·ant \ri-'fri-jə-rənt, -'frij-rənt\ *adjective* (1599)
: allaying heat or fever

[2]refrigerant *noun* (1676)
: a refrigerant agent or agency: as **a :** a medication for reducing body heat **b :** a substance used in refrigeration

re·frig·er·ate \ri-'fri-jə-ˌrāt\ *transitive verb* **-at·ed; -at·ing** [Latin *refrigeratus,* past participle of *refrigerare,* from *re-* + *frigerare* to cool, from *frigor-, frigus* cold — more at FRIGID] (1534)
: to make or keep cold or cool; *specifically* **:** to freeze or chill (as food) for preservation
— re·frig·er·a·tion \-ˌfri-jə-'rā-shən\ *noun*

re·frig·er·a·tor \ri-'fri-jə-ˌrā-tər\ *noun* (1803)
: something that refrigerates; *especially* **:** a room or appliance for keeping food or other items cool

reft *past of* REAVE

re·fu·el \(ˌ)rē-'fyü(-ə)l\ (1811)
transitive verb
: to provide with additional fuel
intransitive verb
: to take on additional fuel

[1]ref·uge \'re-(ˌ)fyüj *also* -(ˌ)fyüzh\ *noun* [Middle English, from Middle French, from Latin *refugium,* from *refugere* to escape, from *re-* + *fugere* to flee — more at FUGITIVE] (14th century)
1 : shelter or protection from danger or distress
2 : a place that provides shelter or protection
3 : something to which one has recourse in difficulty

[2]refuge *verb* **ref·uged; ref·ug·ing** (1594)
transitive verb
: to give refuge to
intransitive verb
: to seek or take refuge

ref·u·gee \ˌre-fyu̇-'jē, 're-fyu̇-ˌ\ *noun* [French *réfugié,* past participle of *(se) réfugier* to take refuge, from Latin *refugium*] (1685)
: one that flees; *especially* **:** a person who flees to a foreign country or power to escape danger or persecution
— ref·u·gee·ism \-ˌi-zəm\ *noun*

re·fu·gi·um \ri-'fyü-jē-əm\ *noun, plural* **-gia** \-jē-ə\ [New Latin, from Latin, refuge] (1943)
: an area of relatively unaltered climate that is inhabited by plants and animals during a period of continental climatic change (as a glaciation) and remains as a center of relict forms from which a new dispersion and speciation may take place after climatic readjustment

re·ful·gence \ri-'fu̇l-jən(t)s, -'fəl-\ *noun* [Latin *refulgentia,* from *refulgent-, refulgens,* present participle of *refulgēre* to shine brightly, from *re-* + *fulgēre* to shine — more at FULGENT] (1634)
: a radiant or resplendent quality or state **:** BRILLIANCE
— re·ful·gent \-jənt\ *adjective*

[1]re·fund \ri-'fənd, 'rē-ˌ\ *transitive verb* [Middle English, from Middle French & Latin; Middle French *refonder,* from Latin *refundere,* literally, to pour back, from *re-* + *fundere* to pour — more at FOUND] (15th century)
1 : to give or put back
2 : to return (money) in restitution, repayment, or balancing of accounts
— re·fund·abil·i·ty \ri-ˌfən-də-'bi-lə-tē, (ˌ)rē-\ *noun*
— re·fund·able \-'fən-də-bəl\ *adjective*

[2]re·fund \'rē-ˌfənd\ *noun* (1866)
1 : the act of refunding
2 : a sum refunded

[3]re·fund \(ˌ)rē-'fənd\ *transitive verb* [*re-* + [2]*fund*] (circa 1860)
: to fund again

re·fur·bish \ri-'fər-bish\ *transitive verb* (1611)
: to brighten or freshen up **:** RENOVATE
— re·fur·bish·er *noun*
— re·fur·bish·ment \-bish-mənt\ *noun*

re·fus·al \ri-'fyü-zəl\ *noun* (15th century)
1 : the act of refusing or denying
2 : the opportunity or right of refusing or taking before others

[1]re·fuse \ri-'fyüz\ *verb* **re·fused; re·fus·ing** [Middle English, from Middle French *refuser,* from (assumed) Vulgar Latin *refusare,* perhaps blend of Latin *refutare* to refute and *recusare* to demur — more at RECUSE] (14th century)
transitive verb
1 : to express oneself as unwilling to accept ⟨*refuse* a gift⟩ ⟨*refuse* a promotion⟩
2 a : to show or express unwillingness to do or comply with ⟨*refused* to answer the question⟩ **b :** DENY ⟨they were *refused* admittance to the game⟩
3 *obsolete* **:** GIVE UP, RENOUNCE
4 *of a horse* **:** to decline to jump or leap over
intransitive verb
: to withhold acceptance, compliance, or permission
synonym see DECLINE
— re·fus·er *noun*

²re·fuse \'re-ˌfyüs, -ˌfyüz\ *noun* [Middle English, from Middle French *refus* rejection, from Old French, from *refuser*] (14th century)
1 : the worthless or useless part of something **:** LEAVINGS
2 : TRASH, GARBAGE

³ref·use \'re-ˌfyüs, -ˌfyüz\ *adjective* (15th century)
: thrown aside or left as worthless

re·fuse·nik *also* **re·fus·nik** \ri-'fyüz-(ˌ)nik\ *noun* [part translation of Russian *otkaznik,* from *otkaz* refusal] (1974)
: a Soviet citizen and especially a Jew refused permission to emigrate

ref·u·ta·tion \ˌre-fyu̇-'tā-shən\ *noun* (circa 1548)
: the act or process of refuting

re·fute \ri-'fyüt\ *transitive verb* **re·fut·ed; re·fut·ing** [Latin *refutare* to check, suppress, refute] (1597)
1 : to prove wrong by argument or evidence **:** show to be false or erroneous
2 : to deny the truth or accuracy of ⟨*refuted* the allegations⟩
— re·fut·able \-'fyü-tə-bəl\ *adjective*
— re·fut·ably \-blē\ *adverb*
— re·fut·er *noun*

reg \'reg\ *noun* [by shortening] (circa 1925)
: REGULATION ⟨federal *regs*⟩

re·gal \'rē-gəl\ *adjective* [Middle English, from Middle French or Latin; Middle French, from Latin *regalis* — more at ROYAL] (14th century)
1 : of, relating to, or suitable for a king
2 : of notable excellence or magnificence **:** SPLENDID
— re·gal·i·ty \ri-'ga-lə-tē\ *noun*
— re·gal·ly \'rē-gə-lē\ *adverb*

¹re·gale \ri-'gā(ə)l\ *verb* **re·galed; re·gal·ing** [French *régaler,* from Middle French, from *regale,* noun] (circa 1656)
transitive verb
1 : to entertain sumptuously **:** feast with delicacies
2 : to give pleasure or amusement to ⟨*regaled* us with tall tales⟩
intransitive verb
: to feast oneself **:** FEED

²regale *noun* [French *régal,* from Middle French *regale,* from *re-* + *galer* to have a good time — more at GALLANT] (1670)
1 : a sumptuous feast
2 : a choice piece especially of food

re·ga·lia \ri-'gāl-yə\ *noun plural* [Medieval Latin, from Latin, neuter plural of *regalis*] (circa 1540)
1 : royal rights or prerogatives
2 a : the emblems, symbols, or paraphernalia indicative of royalty **b :** decorations or insignia indicative of an office or membership
3 : special dress; *especially* **:** FINERY

¹re·gard \ri-'gärd\ *noun* [Middle English, from Middle French, from Old French, from *regarder*] (14th century)
1 *archaic* **:** APPEARANCE
2 a : ATTENTION, CONSIDERATION ⟨due *regard* should be given to all facets of the question⟩ **b :** a protective interest **:** CARE ⟨ought to have more *regard* for his health⟩
3 : LOOK, GAZE
4 a : the worth or estimation in which something or someone is held ⟨a man of small *regard*⟩ **b** (1) **:** a feeling of respect and affection **:** ESTEEM ⟨his hard work won him the *regard* of his colleagues⟩ (2) *plural* **:** friendly greetings implying such feeling ⟨give him my *regards*⟩
5 : a basis of action or opinion **:** MOTIVE
6 : an aspect to be taken into consideration **:** RESPECT ⟨is a small school, and is fortunate in this *regard*⟩
7 *obsolete* **:** INTENTION
— in regard to : with respect to **:** CONCERNING
— with regard to : in regard to

²regard *verb* [Middle English, from Middle French *regarder* to look back at, regard, from Old French, from *re-* + *garder* to guard, look at — more at GUARD] (14th century)
transitive verb
1 : to consider and appraise usually from a particular point of view ⟨is highly *regarded* as a mechanic⟩
2 : to pay attention to **:** take into consideration or account
3 a : to show respect or consideration for **b :** to hold in high esteem
4 : to look at
5 *archaic* **:** to relate to
intransitive verb
1 : to look attentively **:** GAZE
2 : to pay attention **:** HEED ☆
— as regards : with respect to **:** CONCERNING

re·gar·dant \ri-'gär-dᵊnt\ *adjective* [Middle English *regardand,* from Middle French *regardant,* present participle of *regarder*] (15th century)
: looking backward over the shoulder — used of a heraldic animal

re·gard·ful \ri-'gärd-fəl\ *adjective* (circa 1586)
1 : HEEDFUL, OBSERVANT
2 : full or expressive of regard or respect **:** RESPECTFUL
— re·gard·ful·ly \-fə-lē\ *adverb*
— re·gard·ful·ness *noun*

re·gard·ing *preposition* (1866)
: with respect to **:** CONCERNING

¹re·gard·less \ri-'gärd-ləs\ *adjective* (1591)
: HEEDLESS, CARELESS
— re·gard·less·ly *adverb*
— re·gard·less·ness *noun*

²regardless *adverb* (1872)
: despite everything ⟨went ahead with their plans *regardless*⟩
usage see IRREGARDLESS

regardless of *preposition* (1784)
: without taking into account ⟨accepts all *regardless of* age⟩; *also* **:** in spite of ⟨*regardless of* our mistakes⟩

re·gat·ta \ri-'gä-tə, -'ga-\ *noun* [Italian *regata*] (1652)
: a rowing, speedboat, or sailing race or a series of such races

re·gen·cy \'rē-jən(t)-sē\ *noun, plural* **-cies** (15th century)
1 : the office, jurisdiction, or government of a regent or body of regents
2 : a body of regents
3 : the period of rule of a regent or body of regents

Regency *adjective* (1880)
: of, relating to, or characteristic of the styles of George IV's regency as Prince of Wales during the period 1811–20

re·gen·er·a·cy \ri-'je-nə-rə-sē, -'jen-rə-\ *noun* (1626)
: the state of being regenerated

¹re·gen·er·ate \-rət\ *adjective* [Middle English *regenerat,* from Latin *regeneratus,* past participle of *regenerare* to regenerate, from *re-* + *generare* to beget — more at GENERATE] (15th century)
1 : formed or created again
2 : spiritually reborn or converted
3 : restored to a better, higher, or more worthy state
— re·gen·er·ate·ly *adverb*
— re·gen·er·ate·ness *noun*

²re·gen·er·ate \ri-'je-nə-ˌrāt\ (1541)
intransitive verb
1 : to become formed again
2 : to become regenerate **:** REFORM
3 : to undergo regeneration
transitive verb
1 a : to subject to spiritual regeneration **b :** to change radically and for the better
2 a : to generate or produce anew; *especially* **:** to replace (a body part) by a new growth of tissue **b :** to produce again chemically sometimes in a physically changed form
3 : to restore to original strength or properties
— re·gen·er·a·ble \-'je-nə-rə-bəl, -'jen-rə-\ *adjective*

³re·gen·er·ate *same as* ¹\ *noun* (circa 1569)
: one that is regenerated: as **a :** an individual who is spiritually reborn **b** (1) **:** an organism that has undergone regeneration (2) **:** a regenerated body part

regenerated cellulose *noun* (1904)
: cellulose obtained in a changed form by chemical treatment (as of a cellulose solution or derivative)

re·gen·er·a·tion \ri-ˌje-nə-'rā-shən, ˌrē-\ *noun* (14th century)
1 : an act or the process of regenerating **:** the state of being regenerated
2 : spiritual renewal or revival
3 : renewal or restoration of a body or bodily part after injury or as a normal process
4 : utilization by special devices of heat or other products that would ordinarily be lost

re·gen·er·a·tive \ri-'je-nə-ˌrā-tiv, -'je-nə-rə-, -'jen-rə-\ *adjective* (14th century)
1 : of, relating to, or marked by regeneration
2 : tending to regenerate

re·gen·er·a·tor \ri-'je-nə-ˌrā-tər\ *noun* (circa 1550)
1 : one that regenerates
2 : a device used especially with hot-air engines or gas furnaces in which incoming air or gas is heated by contact with masses (as of brick) previously heated by outgoing hot air or gas

re·gent \'rē-jənt\ *noun* [Middle English, from Middle French or Medieval Latin; Middle French, from Medieval Latin *regent-, regens,* from Latin, present participle of *regere* to rule — more at RIGHT] (15th century)
1 : one who governs a kingdom in the minority, absence, or disability of the sovereign
2 : one who rules or reigns **:** GOVERNOR
3 : a member of a governing board (as of a state university)
— regent *adjective*
— re·gent·al \-jən-tᵊl\ *adjective*

reg·gae \'re-(ˌ)gā, 'rā-\ *noun* [origin unknown] (1968)
: popular music of Jamaican origin that combines native styles with elements of rock and soul music and is performed at moderate tempos with the accent on the offbeat

reg·i·cide \'re-jə-ˌsīd\ *noun* [Latin *reg-, rex* king + English *-cide* — more at ROYAL] (circa 1548)
1 : one who kills a king
2 : the killing of a king
— reg·i·ci·dal \ˌre-jə-'sī-dᵊl\ *adjective*

re·gime *also* **ré·gime** \rā-'zhēm, ri- *also* ri-'jēm\ *noun* [French *régime,* from Latin *regimin-, regimen*] (1776)
1 a : REGIMEN 1 **b :** a regular pattern of occurrence or action (as of seasonal rainfall) **c :** the characteristic behavior or orderly procedure of a natural phenomenon or process
2 a : mode of rule or management **b :** a form of government ⟨a socialist *regime*⟩ **c :** a government in power ⟨predicted that the new *regime* would fall⟩ **d :** a period of rule

☆ SYNONYMS
Regard, respect, esteem, admire mean to recognize the worth of a person or thing. REGARD is a general term that is usually qualified ⟨he is highly *regarded* in the profession⟩. RESPECT implies a considered evaluation or estimation ⟨after many years they came to *respect* her views⟩. ESTEEM implies greater warmth of feeling accompanying a high valuation ⟨no citizen of the town was more highly *esteemed*⟩. ADMIRE suggests usually enthusiastic appreciation and often deep affection ⟨a friend that I truly *admire*⟩.

reg·i·men \'re-jə-mən *also* 're-zhə-\ *noun* [Middle English, from Latin *regimin-, regimen* rule, from *regere* to rule] (14th century)
1 a : a systematic plan (as of diet, therapy, or medication) especially when designed to improve and maintain the health of a patient **b :** a regular course of action and especially of strenuous training ⟨the daily *regimen* of a top ballet dancer⟩
2 : GOVERNMENT, RULE
3 : REGIME 1c

1reg·i·ment \'re-jə-mənt, 'rej-mənt\ *noun* [Middle English, from Middle French, from Late Latin *regimentum,* from Latin *regere*] (14th century)
1 *archaic* **:** governmental rule
2 : a military unit consisting usually of a number of battalions

2reg·i·ment \'re-jə-ˌment\ *transitive verb* (1617)
1 : to form into or assign to a regiment
2 a : to organize rigidly especially for the sake of regulation or control ⟨*regiment* an entire country⟩ **b :** to subject to order or uniformity
— reg·i·men·ta·tion \ˌre-jə-mən-'tā-shən, -ˌmen-\ *noun*

reg·i·men·tal \ˌre-jə-'men-tᵊl\ *adjective* (1659)
1 : of or relating to a regiment
2 : AUTHORITATIVE, DICTATORIAL

reg·i·men·tals \-tᵊlz\ *noun plural* (1742)
1 : a regimental uniform
2 : military dress

re·gion \'rē-jən\ *noun* [Middle English, from Middle French, from Latin *region-, regio,* from *regere* to rule] (14th century)
1 : an administrative area, division, or district; *especially* **:** the basic administrative unit for local government in Scotland
2 a : an indefinite area of the world or universe ⟨few unknown *regions* left on earth⟩ **b :** a broad geographical area distinguished by similar features ⟨the Appalachian *region*⟩ **c** (1) **:** a major world area that supports a characteristic fauna (2) **:** an area characterized by the prevalence of one or more vegetational climax types
3 a : any of the major subdivisions into which the body or one of its parts is divisible **b :** an indefinite area surrounding a specified body part ⟨a pain in the *region* of the heart⟩
4 : a sphere of activity or interest **:** FIELD
5 : any of the zones into which the atmosphere is divided according to height or the sea according to depth
6 : an open connected set together with none, some, or all of the points on its boundary ⟨a simple closed curve divides a plane into two *regions*⟩

1re·gion·al \'rēj-nəl, 'rē-jə-nᵊl\ *adjective* (15th century)
1 : affecting a particular region **:** LOCALIZED
2 : of, relating to, characteristic of, or serving a region ⟨a *regional* high school⟩
3 : marked by regionalism ⟨*regional* art⟩

2regional *noun* (1936)
: something (as a branch of an organization or an edition of a magazine) that serves a region

re·gion·al·ism \'rēj-nə-ˌli-zəm, 'rē-jə-nᵊl-ˌi-\ *noun* (1881)
1 a : consciousness of and loyalty to a distinct region with a homogeneous population **b :** development of a political or social system based on one or more such areas
2 : emphasis on regional locale and characteristics in art or literature
3 : a characteristic feature (as of speech) of a geographic area
— re·gion·al·ist \-list, -ist\ *noun or adjective*
— re·gion·al·is·tic \ˌrēj-nə-'lis-tik, ˌrē-jə-nᵊl-'is-\ *adjective*

re·gion·al·ize \'rēj-nə-ˌlīz, 'rē-jə-nᵊl-ˌīz\ *transitive verb* **-ized; -iz·ing** (1921)
: to divide into regions or administrative districts **:** arrange regionally
— re·gion·al·i·za·tion \ˌrēj-nə-lə-'zā-shən, ˌrē-jə-nᵊl-ə-\ *noun*

re·gion·al·ly \'rēj-nə-lē, 'rē-jə-nᵊl-ē\ *adverb* (1879)
: on a regional basis

re·gis·seur *or* **ré·gis·seur** \ˌrā-zhi-'sər\ *noun* [French *régisseur,* from *régir* to direct, from Latin *regere* to rule] (1828)
: a director responsible for staging a theatrical work (as a ballet)

1reg·is·ter \'re-jə-stər\ *noun* [Middle English *registre,* from Middle French, from Medieval Latin *registrum,* alteration of Late Latin *regesta,* plural, register, from Latin, neuter plural of *regestus,* past participle of *regerere* to bring back, pile up, collect, from *re-* + *gerere* to bear] (14th century)
1 : a written record containing regular entries of items or details
2 a : a book or system of public records **b :** a roster of qualified or available individuals ⟨a civil service *register*⟩
3 : an entry in a register
4 a : a set of organ pipes of like quality **:** STOP **b** (1) **:** the range of a human voice or a musical instrument (2) **:** a portion of such a range similarly produced or of the same quality **c :** any of the varieties of a language that a speaker uses in a particular social context
5 a : a device regulating admission of air to fuel **b :** a grille often with shutters for admitting heated air or for ventilation
6 : REGISTRATION, REGISTRY
7 a : an automatic device registering a number or a quantity **b :** a number or quantity so registered
8 : a condition of correct alignment or proper relative position
9 : a device (as in a computer) for storing small amounts of data; *especially* **:** one in which data can be both stored and operated on

2register *verb* **reg·is·tered; reg·is·ter·ing** \-st(ə-)riŋ\ (14th century)
transitive verb
1 a : to make or secure official entry of in a register **b :** to enroll formally especially as a voter or student **c :** to record automatically **:** INDICATE **d :** to make a record of **:** NOTE **e :** PERCEIVE; *also* **:** COMPREHEND
2 : to make or adjust so as to correspond exactly
3 : to secure special protection for (a piece of mail) by prepayment of a fee
4 : to convey an impression of **:** EXPRESS
5 : ACHIEVE ⟨*registered* an impressive victory⟩
intransitive verb
1 a : to enroll one's name in a register ⟨*registered* at the hotel⟩ **b :** to enroll one's name officially as a prerequisite for voting **c :** to enroll formally as a student
2 a : to correspond exactly **b :** to be in correct alignment or register
3 : to make or convey an impression

3register *noun* [probably alteration of Middle English *registrer*] (circa 1532)
: REGISTRAR

registered *adjective* (1861)
1 a : having the owner's name entered in a register ⟨*registered* security⟩ **b :** recorded as the owner of a security
2 : recorded on the basis of pedigree or breed characteristics in the studbook of a breed association
3 : qualified formally or officially

registered mail *noun* (1886)
: mail recorded in the post office of mailing and at each successive point of transmission and guaranteed special care in delivery

registered nurse *noun* (1896)
: a graduate trained nurse who has been licensed by a state authority after qualifying for registration

register ton *noun* (circa 1909)
: TON 1a

reg·is·tra·ble \'re-jə-strə-bəl\ *also* **reg·is·ter·able** \-st(ə-)rə-bəl\ *adjective* (1765)
: capable of being registered

reg·is·trant \'re-jə-strənt\ *noun* (circa 1890)
: one that registers or is registered

reg·is·trar \'re-jə-ˌsträr\ *noun* [alteration of Middle English *registrer,* from Middle French *registreur,* from *registrer* to register, from Medieval Latin *registrare,* from *registrum*] (1675)
: an official recorder or keeper of records: as **a :** an officer of an educational institution responsible for registering students, keeping academic records, and corresponding with applicants and evaluating their credentials **b :** an admitting officer at a hospital **c** *chiefly British* **:** RESIDENT 3

reg·is·tra·tion \ˌre-jə-'strā-shən\ *noun* (circa 1566)
1 : the act of registering
2 : an entry in a register
3 : the number of individuals registered **:** ENROLLMENT
4 a : the art or act of selecting and adjusting pipe organ stops **b :** the combination of stops selected for performing a particular organ work
5 : a document certifying an act of registering

reg·is·try \'re-jə-strē\ *noun, plural* **-tries** (1589)
1 : REGISTRATION, ENROLLMENT
2 : the nationality of a ship according to its entry in a register **:** FLAG
3 : a place of registration
4 a : an official record book **b :** an entry in a registry

re·gius professor \'rē-j(ē-)əs-\ *noun* [New Latin, royal professor] (1621)
: a holder of a professorship founded by royal subsidy at a British university

reg·let \'re-glət\ *noun* [French *réglet,* from Middle French *reglet* straightedge, from *regle* rule, from Latin *regula* — more at RULE] (1664)
1 : a flat narrow architectural molding
2 : a strip of wood used like a lead between lines of type

reg·nal \'reg-nᵊl\ *adjective* [Medieval Latin *regnalis,* from Latin *regnum* reign — more at REIGN] (1612)
: of or relating to a king or his reign; *specifically* **:** calculated from a monarch's accession to the throne ⟨in his eighth *regnal* year⟩

reg·nant \'reg-nənt\ *adjective* [Latin *regnant-, regnans,* present participle of *regnare* to reign, from *regnum*] (1600)
1 : exercising rule **:** REIGNING
2 a : having the chief power **:** DOMINANT **b :** of common or widespread occurrence

reg·num \'reg-nəm\ *noun, plural* **reg·na** \-nə\ [Latin] (circa 1890)
: KINGDOM

reg·o·lith \'re-gə-ˌlith\ *noun* [Greek *rhēgos* blanket + English *-lith;* akin to Greek *rhezein* to dye — more at RAGA] (1897)
: unconsolidated residual or transported material that overlies the solid rock on the earth, moon, or a planet

reg·o·sol \'re-gə-ˌsäl, -ˌsȯl\ *noun* [*rego-* (as in *regolith*) + Latin *solum* soil — more at SOLE] (1949)
: any of a group of azonal soils consisting chiefly of imperfectly consolidated material and having no clear-cut and specific morphology

re·greet \(ˌ)rē-'grēt\ *transitive verb* (1593)
archaic **:** to greet in return

regreets *noun plural* (1596)
obsolete **:** GREETINGS

1re·gress \'rē-ˌgres\ *noun* [Middle English, from Latin *regressus,* from *regredi* to go back, from *re-* + *gradi* to go — more at GRADE] (14th century)

1 a : an act or the privilege of going or coming back **b :** REENTRY 1
2 : movement backward to a previous and especially worse or more primitive state or condition
3 : the act of reasoning backward

²re·gress \ri-ˈgres\ (1552)
intransitive verb
1 a : to make or undergo regress **:** RETROGRADE **b :** to be subject to or exhibit regression
2 : to tend to approach or revert to a mean
transitive verb
: to induce a state of psychological regression in
— **re·gres·sor** \-ˈgre-sər\ *noun*

re·gres·sion \ri-ˈgre-shən\ *noun* (1597)
1 : the act or an instance of regressing
2 : a trend or shift toward a lower or less perfect state: as **a :** progressive decline of a manifestation of disease **b** (1) **:** gradual loss of differentiation and function by a body part especially as a physiological change accompanying aging (2) **:** gradual loss of memories and acquired skills **c :** reversion to an earlier mental or behavioral level **d :** a functional relationship between two or more correlated variables that is often empirically determined from data and is used especially to predict values of one variable when given values of the others ⟨the *regression* of *y* on *x* is linear⟩; *specifically* **:** a function that yields the mean value of a random variable under the condition that one or more independent variables have specified values
3 : retrograde motion

re·gres·sive \ri-ˈgre-siv\ *adjective* (1634)
1 : tending to regress or produce regression
2 : being, characterized by, or developing in the course of an evolutionary process involving increasing simplification of bodily structure
3 : decreasing in rate as the base increases ⟨a *regressive* tax⟩
— **re·gres·sive·ly** *adverb*
— **re·gres·sive·ness** *noun*
— **re·gres·siv·i·ty** \ˌrē-ˌgre-ˈsi-və-tē\ *noun*

¹re·gret \ri-ˈgret\ *verb* **re·gret·ted; re·gret·ting** [Middle English *regretten,* from Middle French *regreter,* from Old French, from *re-* + *-greter* (perhaps of Germanic origin; akin to Old Norse *grāta* to weep) — more at GREET] (14th century)
transitive verb
1 a : to mourn the loss or death of **b :** to miss very much
2 : to be very sorry for ⟨*regrets* his mistakes⟩
intransitive verb
: to experience regret
— **re·gret·ter** *noun*

²regret *noun* (1590)
1 : sorrow aroused by circumstances beyond one's control or power to repair
2 a : an expression of distressing emotion (as sorrow or disappointment) **b** *plural* **:** a note politely declining an invitation
synonym see SORROW
— **re·gret·ful** \-ˈgret-fəl\ *adjective*
— **re·gret·ful·ness** *noun*

re·gret·ful·ly \ri-ˈgret-fə-lē\ *adverb* (1682)
1 : with regret
2 : it is to be regretted

re·gret·ta·ble \ri-ˈgre-tə-bəl\ *adjective* (1603)
: deserving regret

re·gret·ta·bly \-blē\ *adverb* (1866)
1 : to a regrettable extent ⟨a *regrettably* steep decline in wages⟩
2 : it is to be regretted ⟨*regrettably,* they could not attend⟩

re·group \(ˌ)rē-ˈgrüp\ (1885)
transitive verb
: to form into a new grouping ⟨in order to subtract 129 from 531 *regroup* 531 into 5 hundreds, 2 tens, and 11 ones⟩ ⟨*regroup* military forces⟩
intransitive verb
1 : to reorganize (as after a setback) for renewed activity
2 : to alter the tactical formation of a military force

re·grow \(ˌ)rē-ˈgrō\ *verb* **-grew** \-ˈgrü\; **-grown** \-ˈgrōn\; **-grow·ing** (1872)
transitive verb
: to grow (as a missing part) anew
intransitive verb
: to continue growth after interruption or injury

¹reg·u·lar \ˈre-gyə-lər, ˈre-g(ə-)lər\ *adjective* [Middle English *reguler,* from Middle French, from Late Latin *regularis* regular, from Latin, of a bar, from *regula* rule — more at RULE] (14th century)
1 : belonging to a religious order
2 a : formed, built, arranged, or ordered according to some established rule, law, principle, or type **b** (1) **:** both equilateral and equiangular ⟨a *regular* polygon⟩ (2) **:** having faces that are congruent regular polygons and all the polyhedral angles congruent ⟨a *regular* polyhedron⟩ **c** *of a flower* **:** having the arrangement of floral parts exhibiting radial symmetry with members of the same whorl similar in form
3 a : ORDERLY, METHODICAL ⟨*regular* habits⟩ **b :** recurring, attending, or functioning at fixed or uniform intervals ⟨a *regular* income⟩ ⟨a *regular* churchgoer⟩
4 a : constituted, conducted, or done in conformity with established or prescribed usages, rules, or discipline **b :** NORMAL, STANDARD: as (1) **:** ABSOLUTE, COMPLETE ⟨a *regular* fool⟩ ⟨the office seemed like a *regular* madhouse⟩ (2) **:** thinking or behaving in an acceptable, normal, or agreeable manner ⟨was a *regular* guy⟩ **c** (1) **:** conforming to the normal or usual manner of inflection (2) **:** WEAK 7 **d** *of a postage stamp* **:** issued in large numbers over a long period for general use in prepayment of postage
5 : of, relating to, or constituting the permanent standing military force of a state ⟨*regular* army⟩ ⟨*regular* soldiers⟩ ☆

²regular *noun* (15th century)
1 : one who is regular: as **a :** one of the regular clergy **b :** a soldier in a regular army **c :** one who can be trusted or depended on ⟨a party *regular*⟩ **d :** a player on an athletic team who usually starts every game **e :** one who is usually present or participating; *especially* **:** a long-standing regular customer
2 : something of average or medium size; *especially* **:** a clothing size designed to fit a person of average height

reg·u·lar·i·ty \ˌre-gyə-ˈlar-ə-tē\ *noun, plural* **-ties** (1603)
1 : the quality or state of being regular
2 : something that is regular

reg·u·lar·ize \ˈre-gyə-lə-ˌrīz\ *transitive verb* **-ized; -iz·ing** (1833)
: to make regular by conformance to law, rules, or custom
— **reg·u·lar·i·za·tion** \ˌre-gyə-lə-rə-ˈzā-shən\ *noun*

reg·u·lar·ly \ˈre-gyə-lər-lē, ˈre-gyə(r)-lē\ *adverb* (14th century)
1 : in a regular manner
2 : on a regular basis **:** at regular intervals

regular solid *noun* (1841)
: any of the five possible regular polyhedrons that include the regular forms of the tetrahedron, hexahedron, octahedron, dodecahedron, and icosahedron

reg·u·late \ˈre-gyə-ˌlāt\ *transitive verb* **-lat·ed; -lat·ing** [Middle English, from Late Latin *regulatus,* past participle of *regulare,* from Latin *regula* rule] (15th century)
1 a : to govern or direct according to rule **b** (1) **:** to bring under the control of law or constituted authority (2) **:** to make regulations for or concerning ⟨*regulate* the industries of a country⟩
2 : to bring order, method, or uniformity to ⟨*regulate* one's habits⟩
3 : to fix or adjust the time, amount, degree, or rate of ⟨*regulate* the pressure of a tire⟩
— **reg·u·la·tive** \-ˌlā-tiv\ *adjective*
— **reg·u·la·to·ry** \-lə-ˌtōr-ē, -ˌtȯr-\ *adjective*

¹reg·u·la·tion \ˌre-gyə-ˈlā-shən, ˌre-gə-\ *noun* (1665)
1 : the act of regulating **:** the state of being regulated
2 a : an authoritative rule dealing with details or procedure ⟨safety *regulations*⟩ **b :** a rule or order issued by an executive authority or regulatory agency of a government and having the force of law
3 a : the process of redistributing material (as in an embryo) to restore a damaged or lost part independent of new tissue growth **b :** the mechanism by which an early embryo maintains normal development
synonym see LAW

²regulation *adjective* (circa 1839)
: conforming to regulations **:** OFFICIAL

reg·u·la·tor \ˈre-gyə-ˌlā-tər\ *noun* (1655)
1 : one that regulates
2 : REGULATORY GENE

regulatory gene *or* **regulator gene** *noun* (1961)
: a gene that regulates the expression of one or more structural genes by controlling the production of a protein (as a genetic repressor) which regulates their rate of transcription

reg·u·lus \ˈre-gyə-ləs\ *noun* [New Latin, from Latin, petty king, from *reg-, rex* king — more at ROYAL]
1 *capitalized* **:** a first-magnitude star in the constellation Leo
2 [Medieval Latin, metallic antimony, from Latin] **:** the more or less impure mass of metal formed beneath the slag in smelting and reducing ores

re·gur·gi·tate \(ˌ)rē-ˈgər-jə-ˌtāt\ *verb* **-tat·ed; -tat·ing** [Medieval Latin *regurgitatus,* past participle of *regurgitare,* from Latin *re-* + Late Latin *gurgitare* to engulf, from Latin *gurgit-, gurges* whirlpool — more at VORACIOUS] (1653)
intransitive verb
: to become thrown or poured back
transitive verb
: to throw or pour back or out from or as if from a cavity ⟨*regurgitate* food⟩ ⟨memorized facts to *regurgitate* on the exam⟩

re·gur·gi·ta·tion \(ˌ)rē-ˌgər-jə-ˈtā-shən\ *noun* (1601)
: an act of regurgitating: as **a :** the casting up of incompletely digested food (as by some birds in feeding their young) **b :** the backward flow of blood through a defective heart valve

re·hab \ˈrē-ˌhab\ *noun, often attributive* [short for *rehabilitation* or *rehabilitate*] (1941)
1 : the action or process of rehabilitating **:** REHABILITATION
2 : a rehabilitated building or dwelling
— **rehab** *transitive verb*
— **re·hab·ber** \-ˌha-bər\ *noun*

> ☆ SYNONYMS
> **Regular, normal, typical, natural** mean being of the sort or kind that is expected as usual, ordinary, or average. REGULAR stresses conformity to a rule, standard, or pattern ⟨the club's *regular* monthly meeting⟩. NORMAL implies lack of deviation from what has been discovered or established as the most usual or expected ⟨*normal* behavior for a two-year-old⟩. TYPICAL implies showing all important traits of a type, class, or group and may suggest lack of strong individuality ⟨a *typical* small town⟩. NATURAL applies to what conforms to a thing's essential nature, function, or mode of being ⟨the *natural* love of a mother for her child⟩.

re•ha•bil•i•tant \,rē-ə-'bi-lə-tənt, ,rē-hə-\ *noun* (1961)
: a disabled person undergoing rehabilitation

re•ha•bil•i•tate \,rē-ə-'bi-lə-,tāt, ,rē-hə-\ *transitive verb* **-tat•ed; -tat•ing** [Medieval Latin *rehabilitatus,* past participle of *rehabilitare,* from Latin *re-* + Late Latin *habilitare* to habilitate] (circa 1581)
1 a : to restore to a former capacity **:** REINSTATE **b :** to restore to good repute **:** reestablish the good name of
2 a : to restore to a former state (as of efficiency, good management, or solvency) ⟨*rehabilitate* slum areas⟩ **b :** to restore or bring to a condition of health or useful and constructive activity
— re•ha•bil•i•ta•tion \-,bi-lə-'tā-shən\ *noun*
— re•ha•bil•i•ta•tive \-'bi-lə-,tā-tiv\ *adjective*
— re•ha•bil•i•ta•tor \-,tā-tər\ *noun*

¹re•hash \(,)rē-'hash\ *transitive verb* (circa 1822)
1 : to talk over or discuss again
2 : to present or use again in another form without substantial change or improvement

²re•hash \'rē-,hash\ *noun* (1849)
1 : something that is rehashed
2 : the action or process of rehashing

re•hear \(,)rē-'hir\ *transitive verb* **-heard** \-'hərd\; **-hear•ing** \-'hir-iŋ\ (1756)
: to hear again or anew especially judicially

rehearing *noun* (1686)
: a second or new hearing by the same tribunal

re•hears•al \ri-'hər-səl\ *noun* (14th century)
1 : something recounted or told again **:** RECITAL
2 a : a private performance or practice session preparatory to a public appearance **b :** a practice exercise **:** TRIAL

re•hearse \ri-'hərs\ *verb* **re•hearsed; re•hears•ing** [Middle English *rehersen,* from Middle French *rehercier,* literally, to harrow again, from *re-* + *hercier* to harrow, from *herce* harrow — more at HEARSE] (14th century)
transitive verb
1 a : to say again **:** REPEAT **b :** to recite aloud in a formal manner
2 : to present an account of **:** RELATE ⟨*rehearse* a familiar story⟩
3 : to recount in order **:** ENUMERATE ⟨*rehearsed* their demands⟩
4 a : to give a rehearsal of **b :** to train or make proficient by rehearsal
5 : to perform or practice as if in a rehearsal
intransitive verb
: to engage in a rehearsal
word history see HEARSE
— re•hears•er *noun*

re•house \(,)rē-'hau̇z\ *transitive verb* (1820)
: to house again or anew; *especially* **:** to establish in a new or different housing unit of a better quality

re•hy•drate \(,)rē-'hī-,drāt\ *transitive verb* (1943)
: to restore fluid to (something dehydrated)
— re•hy•drat•able \-,drā-tə-bəl\ *adjective*
— re•hy•dra•tion \,rē-,hī-'drā-shən\ *noun*

reichs•mark \'rīks-,märk\ *noun, plural* **reichsmarks** *also* **reichsmark** [German, from *Reich* empire, kingdom + *Mark* mark] (1924)
: the German mark from 1925 to 1948

re•ifi•ca•tion \,rā-ə-fə-'kā-shən, ,rē-\ *noun* (1846)
: the process or result of reifying

re•ify \'rā-ə-,fī, 'rē-\ *transitive verb* **re•ified; re•ify•ing** [Latin *res* thing — more at REAL] (1854)
: to regard (something abstract) as a material or concrete thing

¹reign \'rān\ *noun* [Middle English *regne,* from Old French, from Latin *regnum,* from *reg-, rex* king — more at ROYAL] (13th century)
1 a : royal authority **:** SOVEREIGNTY ⟨under the *reign* of the Stuart kings⟩ **b :** the dominion, sway, or influence of one resembling a monarch ⟨the *reign* of the Puritan ministers⟩
2 : the time during which one (as a sovereign) reigns

²reign *intransitive verb* (14th century)
1 a : to possess or exercise sovereign power **:** RULE **b :** to hold office as chief of state although possessing little governing power ⟨in England the sovereign *reigns* but does not rule⟩
2 : to exercise authority in the manner of a monarch
3 : to be predominant or prevalent ⟨chaos *reigned* in the classroom⟩

reign of terror [*Reign of Terror,* a period of the French Revolution that was conspicuous for mass executions of political suspects] (1801)
: a state or a period of time marked by violence often committed by those in power that produces widespread terror

re•imag•ine \,rē-i-'ma-jən\ *transitive verb* (circa 1934)
: to imagine again or anew; *especially* **:** to form a new conception of **:** RE-CREATE

re•im•burse \,rē-əm-'bərs\ *transitive verb* **-bursed; -burs•ing** [*re-* + obsolete English *imburse* to put in the pocket, pay, from Medieval Latin *imbursare,* from Latin *in-* in- + Medieval Latin *bursa* purse — more at PURSE] (1611)
1 : to pay back to someone **:** REPAY ⟨*reimburse* travel expenses⟩
2 : to make restoration or payment of an equivalent to ⟨*reimburse* him for his traveling expenses⟩
synonym see PAY
— re•im•burs•able \-'bər-sə-bəl\ *adjective*
— re•im•burse•ment \-'bərs-mənt\ *noun*

re•im•pres•sion \,rē-əm-'pre-shən\ *noun* (1684)
: REPRINT a

¹rein \'rān\ *noun* [Middle English *reine,* from Middle French *rene,* from (assumed) Vulgar Latin *retina,* from Latin *retinēre* to restrain more at RETAIN] (14th century)
1 : a strap fastened to a bit by which a rider or driver controls an animal — usually used in plural
2 a : a restraining influence **:** CHECK ⟨kept a tight *rein* on the proceedings⟩ **b :** controlling or guiding power — usually used in plural ⟨the *reins* of government⟩
3 : opportunity for unhampered activity or use ⟨gave full *rein* to her imagination⟩

²rein (15th century)
transitive verb
1 : to control or direct with or as if with reins
2 : to check or stop by or as if by a pull at the reins ⟨*reined* in her horse⟩ ⟨couldn't *rein* his impatience⟩
intransitive verb
1 *archaic* **:** to submit to the use of reins
2 : to stop or slow up one's horse or oneself by or as if by pulling the reins

re•in•car•nate \,rē-ən-'kär-,nāt, (,)rē-'in-,\ *transitive verb* (1858)
: to incarnate again

re•in•car•na•tion \,rē-(,)in-(,)kär-'nā-shən\ *noun* (1858)
1 a : the action of reincarnating **:** the state of being reincarnated **b :** rebirth in new bodies or forms of life; *especially* **:** a rebirth of a soul in a new human body
2 : a fresh embodiment

rein•deer \'rān-,dir\ *noun* [Middle English *reindere,* from Old Norse *hreinn* reindeer + Middle English *deer* animal, deer] (14th century)
: CARIBOU — used especially for the Old World caribou

reindeer moss *noun* (circa 1753)
: a gray, erect, tufted, and much-branched lichen (*Cladonia rangiferina*) that forms extensive patches in arctic and north-temperate regions, constitutes a large part of the food of caribou, and is sometimes eaten by humans — called also *reindeer lichen*

re•in•dus•tri•al•i•za•tion \,rē-in-,dəs-trē-ə-lə-'zā-shən\ *noun* (1979)
: a policy of stimulating economic growth especially through government aid to revitalize and modernize aging industries and encourage growth of new ones
— re•in•dus•tri•al•ize \-'dəs-trē-ə-,līz\ *verb*

re•in•fec•tion \,rē-ən-'fek-shən\ *noun* (1882)
: infection following recovery from or superimposed on infection of the same type

re•in•force \,rē-ən-'fōrs, -'fȯrs\ *verb* [*re-* + *inforce,* alteration of *enforce*] (1600)
transitive verb
1 : to strengthen or increase by fresh additions ⟨*reinforce* our troops⟩ ⟨were *reinforcing* their pitching staff⟩
2 : to strengthen by additional assistance, material, or support **:** make stronger or more pronounced ⟨*reinforce* levees⟩ ⟨*reinforce* the elbows of a jacket⟩ ⟨*reinforce* ideas⟩
3 : to stimulate (as an experimental animal or a student) with a reinforcer; *also* **:** to encourage (a response) with a reinforcer
intransitive verb
: to seek or get reinforcements
— re•in•force•able \-'fōr-sə-bəl, -'fȯr-\ *adjective*

reinforced concrete *noun* (1902)
: concrete in which metal (as steel) is embedded so that the two materials act together in resisting forces

re•in•force•ment \,rē-ən-'fōrs-mənt, -'fȯrs-\ *noun* (1617)
1 : the action of reinforcing **:** the state of being reinforced
2 : something that reinforces

re•in•forc•er \-'fōr-sər, -'fȯr-\ *noun* (1955)
: a stimulus (as a reward or the removal of an electric shock) that increases the probability of a desired response in operant conditioning by being applied or effected following the desired response

reins \'rānz\ *noun plural* [Middle English, from Middle French & Latin; Middle French, from Latin *renes*] (14th century)
1 a : KIDNEYS **b :** the region of the kidneys **:** LOINS
2 : the seat of the feelings or passions

reins•man \'rānz-mən\ *noun* (1855)
: a skilled driver or rider of horses

re•in•state \,rē-ən-'stāt\ *transitive verb* **-stat•ed; -stat•ing** (1628)
1 : to place again (as in possession or in a former position)
2 : to restore to a previous effective state
— re•in•state•ment \-'stāt-mənt\ *noun*

re•in•sur•ance \,rē-ən-'shu̇r-ən(t)s *also* (,)rē-'in-,\ *noun* (1755)
: insurance by another insurer of all or a part of a risk previously assumed by an insurance company

re•in•sure \,rē-ən-'shu̇r\ (1755)
transitive verb
1 : to insure again by transferring to another insurance company all or a part of a liability assumed
2 : to insure again by assuming all or a part of the liability of an insurance company already covering a risk
intransitive verb
: to provide increased insurance
— re•in•sur•er *noun*

re•in•te•grate \(,)rē-'in-tə-,grāt\ *transitive verb* [Medieval Latin *reintegratus,* past participle of *reintegrare* to renew, reinstate, from Latin *re-* + *integrare* to integrate] (1626)

: to integrate again into an entity : restore to unity
— **re·in·te·gra·tion** \(ˌ)rē-ˌin-tə-ˈgrā-shən\ *noun*
— **re·in·te·gra·tive** \(ˌ)rē-ˈin-tə-ˌgrā-tiv\ *adjective*

re·in·ter·pret \ˌrē-ən-ˈtər-prət, -pət\ *transitive verb* (1611)
: to interpret again; *specifically* : to give a new or different interpretation to
— **re·in·ter·pre·ta·tion** \-ˌtər-prə-ˈtā-shən, -pə-\ *noun*

re·in·vent \ˌrē-ən-ˈvent\ *transitive verb* (1686)
1 : to make as if for the first time something already invented ⟨*reinvent* the wheel⟩
2 : to remake or redo completely
3 : to bring into use again
— **re·in·ven·tion** \-ˈven(t)-shən\ *noun*

re·in·vest \ˌrē-ən-ˈvest\ *transitive verb* (1611)
1 : to invest again or anew
2 a : to invest (as income from investments) in additional securities **b** : to invest (as earnings) in a business rather than distribute as dividends or profits

re·in·vest·ment \-ˈves(t)-mənt\ *noun* (1611)
1 : the action of reinvesting : the state of being reinvested
2 : a second or repeated investment

reis *plural of* REAL

re·is·sue \(ˌ)rē-ˈi-(ˌ)shü, *chiefly British* -ˈis-(ˌ)yü\ (circa 1618)
intransitive verb
: to come forth again
transitive verb
: to issue again; *especially* : to cause to become available again
— **reissue** *noun*

re·it·er·ate \rē-ˈi-tə-ˌrāt\ *transitive verb* **-at·ed; -at·ing** [Middle English, from Latin *reiteratus,* past participle of *reiterare* to repeat, from *re-* + *iterare* to iterate] (15th century)
: to state or do over again or repeatedly sometimes with wearying effect
— **re·it·er·a·tion** \(ˌ)rē-ˌi-tə-ˈrā-shən\ *noun*
— **re·it·er·a·tive** \rē-ˈi-tə-ˌrā-tiv, -t(ə-)rə-tiv\ *adjective*
— **re·it·er·a·tive·ly** *adverb*

Rei·ter's syndrome \ˈrī-tərz-\ *noun* [Hans *Reiter* (died 1969) German physician] (circa 1947)
: a disease that is usually initiated by infection in genetically predisposed individuals and is characterized usually by recurrence of arthritis, conjunctivitis, and urethritis — called also *Reiter's disease*

reive \ˈrēv\ *verb* **reived; reiv·ing** [Middle English (Scots) *reifen,* from Old English *rēafian* to rob — more at REAVE] (before 12th century)
Scottish : RAID
— **reiv·er** *noun, Scottish*

¹re·ject \ri-ˈjekt\ *transitive verb* [Middle English, from Latin *rejectus,* past participle of *reicere,* from *re-* + *jacere* to throw — more at JET] (15th century)
1 a : to refuse to accept, consider, submit to, take for some purpose, or use ⟨*rejected* the suggestion⟩ ⟨*reject* a manuscript⟩ **b** : to refuse to hear, receive, or admit : REBUFF, REPEL ⟨parents who *reject* their children⟩ **c** : to refuse as lover or spouse
2 *obsolete* : to cast off
3 : THROW BACK, REPULSE
4 : to spew out
5 : to subject to immunological rejection
synonym see DECLINE
— **re·ject·er** *or* **re·jec·tor** \-ˈjek-tər\ *noun*
— **re·ject·ing·ly** \-tiŋ-lē\ *adverb*
— **re·jec·tive** \-ˈjek-tiv\ *adjective*

²re·ject \ˈrē-ˌjekt\ *noun* (circa 1555)
: a rejected person or thing; *especially* : one rejected as not wanted, unsatisfactory, or not fulfilling standard requirements

re·ject·ee \ri-ˌjek-ˈtē, ˌrē-\ *noun* (1941)
: one that is rejected; *especially* : a person rejected as unfit for military service

re·jec·tion \ri-ˈjek-shən\ *noun* (circa 1552)
1 a : the action of rejecting : the state of being rejected **b** : the immunological process of sloughing off foreign tissue or an organ (as a transplant) by the recipient organism
2 : something rejected

rejection slip *noun* (1906)
: a printed slip enclosed with a rejected manuscript returned by an editor to an author

re·jig·ger \(ˌ)rē-ˈji-gər\ *transitive verb* [*re-* + ³*jigger*] (1942)
: ALTER, REARRANGE

re·joice \ri-ˈjȯis\ *verb* **re·joiced; re·joic·ing** [Middle English, from Middle French *rejoiss-,* stem of *rejoir,* from *re-* + *joir* to rejoice, from Latin *gaudēre* — more at JOY] (14th century)
transitive verb
: to give joy to : GLADDEN
intransitive verb
: to feel joy or great delight
— **re·joic·er** *noun*
— **re·joic·ing·ly** \-ˈjȯi-siŋ-lē\ *adverb*
— **rejoice in** : HAVE, POSSESS

rejoicing *noun* (14th century)
1 : the action of one that rejoices
2 : an instance, occasion, or expression of joy : FESTIVITY

re·join \ri-ˈjȯin, *transitive verb 1 is* (ˌ)rē-\ *verb* [Middle English, from Middle French *rejoin-,* stem of *rejoindre,* from *re-* + Old French *joindre* to join — more at JOIN] (15th century)
intransitive verb
: to answer the replication of the plaintiff
transitive verb
1 : to join again
2 : to say often sharply or critically in response especially as a reply to a reply
synonym see ANSWER

re·join·der \ri-ˈjȯin-dər\ *noun* [Middle English *rejoiner,* from Middle French *rejoindre* to rejoin] (15th century)
1 : the defendant's answer to the plaintiff's replication
2 : REPLY; *specifically* : an answer to a reply

re·ju·ve·nate \ri-ˈjü-və-ˌnāt\ *verb* **-nat·ed; -nat·ing** [*re-* + Latin *juvenis* young — more at YOUNG] (1807)
transitive verb
1 a : to make young or youthful again : give new vigor to **b** : to restore to an original or new state ⟨*rejuvenate* old cars⟩
2 a : to stimulate (a stream) to renewed erosive activity especially by uplift **b** : to develop youthful features of topography in
intransitive verb
: to cause or undergo rejuvenescence
synonym see RENEW
— **re·ju·ve·na·tion** \ri-ˌjü-və-ˈnā-shən, ˌrē-\ *noun*
— **re·ju·ve·na·tor** \ri-ˈjü-və-ˌnā-tər\ *noun*

re·ju·ve·nes·cence \ri-ˌjü-və-ˈne-sən(t)s, ˌrē-\ *noun* [Medieval Latin *rejuvenescere* to become young again, from Latin *re-* + *juvenescere* to become young, from *juvenis*] (circa 1631)
: a renewal of youthfulness or vigor : REJUVENATION
— **re·ju·ve·nes·cent** \-sᵊnt\ *adjective*

¹re·lapse \ri-ˈlaps, ˈrē-ˌ\ *noun* [Middle English, from Medieval Latin *relapsus,* from Latin *relabi* to slide back, from *re-* + *labi* to slide — more at SLEEP] (15th century)
1 : the act or an instance of backsliding, worsening, or subsiding
2 : a recurrence of symptoms of a disease after a period of improvement

²re·lapse \ri-ˈlaps\ *intransitive verb* **re·lapsed; re·laps·ing** (1568)
1 : to slip or fall back into a former worse state
2 : SINK, SUBSIDE ⟨*relapse* into deep thought⟩
— **re·laps·er** *noun*

relapsing fever *noun* (1849)
: a variable acute epidemic disease that is marked by recurring high fever usually lasting 3 to 7 days and is caused by a spirochete (genus *Borrelia*) transmitted by the bites of lice and ticks

re·late \ri-ˈlāt\ *verb* **re·lat·ed; re·lat·ing** [Latin *relatus* (past participle of *referre* to carry back), from *re-* + *latus,* past participle of *ferre* to carry — more at TOLERATE, BEAR] (1530)
transitive verb
1 : to give an account of : TELL
2 : to show or establish logical or causal connection between ⟨seeks to *relate* crime to poverty⟩
intransitive verb
1 : to apply or take effect retroactively
2 : to have relationship or connection ⟨the readings *relate* to his lectures⟩
3 : to have or establish a relationship : INTERACT ⟨the way a child *relates* to a teacher⟩
4 : to respond especially favorably ⟨can't *relate* to that kind of music⟩
— **re·lat·able** \-ˈlā-tə-bəl\ *adjective*
— **re·lat·er** *or* **re·la·tor** \-ˈlā-tər\ *noun*

related *adjective* (circa 1663)
1 : connected by reason of an established or discoverable relation
2 : connected by common ancestry or sometimes by marriage
3 : having close harmonic connection — used of tones, chords, or tonalities
— **re·lat·ed·ly** *adverb*
— **re·lat·ed·ness** *noun*

re·la·tion \ri-ˈlā-shən\ *noun* [Middle English *relacioun,* from Middle French *relation,* from Latin *relation-, relatio,* from *referre* (past participle *relatus*) to carry back] (14th century)
1 : the act of telling or recounting : ACCOUNT
2 : an aspect or quality (as resemblance) that connects two or more things or parts as being or belonging or working together or as being of the same kind ⟨the *relation* of time and space⟩; *specifically* : a property (as one expressed by *is equal to, is less than,* or *is the brother of*) that holds between an ordered pair of objects
3 : the referring by a legal fiction of an act to a prior date as the time of its taking effect
4 a (1) : a person connected by consanguinity or affinity : RELATIVE (2) : a person legally entitled to a share of the property of an intestate **b** : relationship by consanguinity or affinity : KINSHIP
5 : REFERENCE, RESPECT ⟨in *relation* to⟩
6 : the attitude or stance which two or more persons or groups assume toward one another ⟨race *relations*⟩
7 a : the state of being mutually or reciprocally interested (as in social or commercial matters) **b** *plural* (1) : DEALINGS, AFFAIRS ⟨foreign *relations*⟩ (2) : INTERCOURSE (3) : SEXUAL INTERCOURSE

re·la·tion·al \-shnəl, -shə-nᵊl\ *adjective* (1662)
1 : of or relating to kinship
2 : characterized or constituted by relations
3 : having the function chiefly of indicating a relation of syntax ⟨*has* is notional in *he has luck, relational* in *he has gone*⟩
4 : relating to, using, or being a method of organizing data in a database so that it is perceived by the user as a set of tables
— **re·la·tion·al·ly** *adverb*

relational grammar *noun* (1982)
: a grammar based on a theory in which grammatical relations (as subject or object) are primitives in terms of which syntactic operations are defined

re·la·tion·ship \-shən-ˌship\ *noun* (circa 1744)
1 : the state of being related or interrelated ⟨studied the *relationship* between the variables⟩
2 : the relation connecting or binding participants in a relationship: as **a** : KINSHIP **b** : a specific instance or type of kinship

3 a : a state of affairs existing between those having relations or dealings ⟨had a good *relationship* with his family⟩ **b :** a romantic or passionate attachment

[1]**rel·a·tive** \'re-lə-tiv\ *noun* (14th century)
1 : a word referring grammatically to an antecedent
2 : a thing having a relation to or connection with or necessary dependence on another thing
3 a : a person connected with another by blood or affinity **b :** an animal or plant related to another by common descent
4 : a relative term

[2]**relative** *adjective* (15th century)
1 : introducing a subordinate clause qualifying an expressed or implied antecedent ⟨*relative* pronoun⟩; *also* **:** introduced by such a connective ⟨*relative* clause⟩
2 : RELEVANT, PERTINENT ⟨matters *relative* to world peace⟩
3 : not absolute or independent **:** COMPARATIVE ⟨the *relative* isolation of life in the country⟩
4 : having the same key signature — used of major and minor keys and scales
5 : expressed as the ratio of the specified quantity (as an error in measuring) to the total magnitude (as the value of a measured quantity) or to the mean of all the quantities involved

relative humidity *noun* (1820)
: the ratio of the amount of water vapor actually present in the air to the greatest amount possible at the same temperature

rel·a·tive·ly *adverb* (1561)
: to a relative degree or extent **:** SOMEWHAT

relatively prime *adjective* (circa 1890)
of integers **:** having no common factors except +1 and −1 ⟨12 and 25 are *relatively prime*⟩

relative to *preposition* (1660)
: with regard to **:** in connection with

relative wind *noun* (circa 1918)
: the motion of the air relative to a body in it

rel·a·tiv·ism \'re-lə-ti-ˌvi-zəm\ *noun* (1865)
1 a : a theory that knowledge is relative to the limited nature of the mind and the conditions of knowing **b :** a view that ethical truths depend on the individuals and groups holding them
2 : RELATIVITY
— **rel·a·tiv·ist** \-vist\ *noun*

rel·a·tiv·is·tic \ˌre-lə-ti-'vis-tik\ *adjective* (1886)
1 : of, relating to, or characterized by relativity or relativism
2 : moving at a velocity such that there is a significant change in properties (as mass) in accordance with the theory of relativity ⟨a *relativistic* electron⟩
— **rel·a·tiv·is·ti·cal·ly** \-'vis-ti-k(ə-)lē\ *adverb*

rel·a·tiv·i·ty \ˌre-lə-'ti-və-tē\ *noun, plural* **-ties** (circa 1834)
1 a : the quality or state of being relative **b :** something that is relative
2 : the state of being dependent for existence on or determined in nature, value, or quality by relation to something else
3 a : a theory which is based on the two postulates (1) that the speed of light in a vacuum is constant and independent of the source or observer and (2) that the mathematical forms of the laws of physics are invariant in all inertial systems and which leads to the assertion of the equivalence of mass and energy and of change in mass, dimension, and time with increased velocity — called also *special relativity, special theory of relativity* **b :** an extension of the theory to include gravitation and related acceleration phenomena — called also *general relativity, general theory of relativity*
4 : RELATIVISM 1b

rel·a·tiv·ize \'re-lə-tə-ˌvīz\ *transitive verb* **-ized; -iz·ing** (1935)
: to treat or describe as relative

re·lax \ri-'laks\ *verb* [Middle English, from Latin *relaxare,* from *re-* + *laxare* to loosen, from *laxus* loose — more at SLACK] (15th century)
transitive verb
1 : to make less tense or rigid **:** SLACKEN ⟨*relaxed* his grip⟩
2 : to make less severe or stringent **:** MODIFY ⟨*relax* immigration laws⟩
3 : to make soft or enervated
4 : to relieve from nervous tension
intransitive verb
1 : to become lax, weak, or loose **:** REST
2 : to become less intense or severe ⟨hoped the committee would *relax* in its opposition⟩
3 *of a muscle or muscle fiber* **:** to become inactive and lengthen
4 : to cast off social restraint, nervous tension, or anxiety ⟨couldn't *relax* in crowds⟩
5 : to seek rest or recreation ⟨*relax* at the seashore⟩
6 : to relieve constipation
7 : to attain equilibrium following the abrupt removal of some influence (as light, high temperature, or stress)
— **re·lax·er** *noun*

[1]**re·lax·ant** \ri-'lak-sənt\ *adjective* (1771)
: of, relating to, or producing relaxation ⟨an anesthetic and *relaxant* agent⟩

[2]**relaxant** *noun* (circa 1847)
: a substance (as a drug) that relaxes; *specifically* **:** one that relieves muscular tension

re·lax·a·tion \ˌrē-ˌlak-'sā-shən, ri-ˌlak-, *especially British* ˌre-lək-\ *noun* (1548)
1 : the act of relaxing or state of being relaxed
2 : a relaxing or recreative state, activity, or pastime **:** DIVERSION
3 : the lengthening that characterizes inactive muscle fibers or muscles

re·laxed \ri-'lakst\ *adjective* (1623)
1 : freed from or lacking in precision or stringency
2 : set or being at rest or at ease
3 : easy of manner **:** INFORMAL
— **re·laxed·ly** \-'lak-səd-lē, -'lakst-le\ *adverb*
— **re·laxed·ness** \-'lak-səd-nəs, -'laks(t)-nəs\ *noun*

re·lax·in \ri-'lak-sən\ *noun* (1930)
: a sex hormone of the corpus luteum that facilitates birth by causing relaxation of the pelvic ligaments

[1]**re·lay** \'rē-ˌlā\ *noun* (1659)
1 a : a supply (as of horses) arranged beforehand for successive relief **b :** a number of persons who relieve others in some work ⟨worked in *relays* around the clock⟩
2 a : a race between teams in which each team member successively covers a specified portion of the course **b :** one of the divisions of a relay
3 : an electromagnetic device for remote or automatic control that is actuated by variation in conditions of an electric circuit and that operates in turn other devices (as switches) in the same or a different circuit
4 : SERVOMOTOR
5 : the act of passing along (as a message or ball) by stages; *also* **:** one of such stages

[2]**re·lay** \'rē-ˌlā, ri-'lā\ *transitive verb* **re·layed; re·lay·ing** [Middle English, to hunt with relays, from Middle French *relaier,* from Old French, from *re-* + *laier* to leave — more at DELAY] (1788)
1 a : to place or dispose in relays **b :** to provide with relays
2 : to pass along by relays ⟨news was *relayed* to distant points⟩
3 : to control or operate by a relay

[3]**re·lay** \(ˌ)rē-'lā\ *transitive verb* **-laid** \-'lād\; **-lay·ing** [*re-* + [1]*lay*] (1757)
: to lay again ⟨*relay* track⟩

[1]**re·lease** \ri-'lēs\ *transitive verb* **re·leased; re·leas·ing** [Middle English *relesen,* from Middle French *relessier,* from Latin *relaxare* to relax] (14th century)
1 : to set free from restraint, confinement, or servitude ⟨*release* hostages⟩ ⟨*release* pent-up emotions⟩ ⟨*release* the brakes⟩; *also* **:** to let go **:** DISMISS ⟨*released* from her job⟩
2 : to relieve from something that confines, burdens, or oppresses ⟨was *released* from her promise⟩
3 : to give up in favor of another **:** RELINQUISH ⟨*release* a claim to property⟩
4 : to give permission for publication, performance, exhibition, or sale of; *also* **:** to make available to the public ⟨the commission *released* its findings⟩ ⟨*release* a new movie⟩
synonym see FREE
— **re·leas·able** \-'lē-sə-bəl\ *adjective*

[2]**release** *noun* (14th century)
1 : relief or deliverance from sorrow, suffering, or trouble
2 a : discharge from obligation or responsibility **b** (1) **:** relinquishment of a right or claim (2) **:** an act by which a legal right is discharged; *specifically* **:** a conveyance of a right in lands or tenements to another having an estate in possession
3 a : the act or an instance of liberating or freeing (as from restraint) **b :** the act or manner of concluding a musical tone or phrase **c :** the act or manner of ending a sound **:** the movement of one or more vocal organs in quitting the position for a speech sound
4 : an instrument effecting a legal release
5 : the state of being freed
6 : a device adapted to hold or release a mechanism as required
7 a : the act of permitting performance or publication; *also* **:** PERFORMANCE, PUBLICATION ⟨became a best-seller on its *release*⟩ **b :** the matter released; *especially* **:** a statement prepared for the press

re–lease \(ˌ)rē-'lēs\ *transitive verb* (1828)
: to lease again

released time *noun* (1941)
: time off from regularly scheduled activities (as school) given to take part in some other specified activity

release print *noun* (1937)
: a motion-picture film released for public showing

re·leas·er \ri-'lē-sər\ *noun* (15th century)
: one that releases; *specifically* **:** a stimulus that serves as the initiator of complex reflex behavior

rel·e·gate \'re-lə-ˌgāt\ *transitive verb* **-gat·ed; -gat·ing** [Latin *relegatus,* past participle of *relegare,* from *re-* + *legare* to send with a commission — more at LEGATE] (1599)
1 : to send into exile **:** BANISH
2 : ASSIGN: as **a :** to assign to a place of insignificance or of oblivion **:** put out of sight or mind **b :** to assign to an appropriate place or situation on the basis of classification or appraisal **c :** to submit to someone or something for appropriate action **:** DELEGATE
synonym see COMMIT
— **rel·e·ga·tion** \ˌre-lə-'gā-shən\ *noun*

re·lent \ri-'lent\ *verb* [Middle English, to melt, soften, from Anglo-French *relenter,* from *re-* + Latin *lentare* to bend, from *lentus* soft, pliant, slow — more at LITHE] (1526)
intransitive verb
1 a : to become less severe, harsh, or strict usually from reasons of humanity **b :** to cease resistance **:** GIVE IN
2 : LET UP, SLACKEN
transitive verb
obsolete **:** SOFTEN, MOLLIFY
synonym see YIELD

re·lent·less \-ləs\ *adjective* (1592)

: showing or promising no abatement of severity, intensity, strength, or pace **:** UNRELENTING ⟨*relentless* pressure⟩ ⟨a *relentless* campaign⟩
— **re·lent·less·ly** *adverb*
— **re·lent·less·ness** *noun*

rel·e·vance \ˈre-lə-vən(t)s\ *noun* (1733)
1 a : relation to the matter at hand **b :** practical and especially social applicability **:** PERTINENCE ⟨giving *relevance* to college courses⟩
2 : the ability (as of an information retrieval system) to retrieve material that satisfies the needs of the user

rel·e·van·cy \-vən(t)-sē\ *noun, plural* **-cies** (1561)
: RELEVANCE; *also* **:** something relevant

rel·e·vant \ˈre-lə-vənt\ *adjective* [Medieval Latin *relevant-, relevans,* from Latin, present participle of *relevare* to raise up — more at RELIEVE] (1560)
1 a : having significant and demonstrable bearing on the matter at hand **b :** affording evidence tending to prove or disprove the matter at issue or under discussion ⟨*relevant* testimony⟩ **c :** having social relevance
2 : PROPORTIONAL, RELATIVE ☆
— **rel·e·vant·ly** *adverb*

re·li·abil·i·ty \ri-ˌlī-ə-ˈbi-lə-tē\ *noun* (1816)
1 : the quality or state of being reliable
2 : the extent to which an experiment, test, or measuring procedure yields the same results on repeated trials

[1]re·li·able \ri-ˈlī-ə-bəl\ *adjective* (1569)
1 : suitable or fit to be relied on **:** DEPENDABLE
2 : giving the same result on successive trials
— **re·li·able·ness** *noun*
— **re·li·ably** \-blē\ *adverb*

[2]reliable *noun* (1890)
: one that is reliable

re·li·ance \ri-ˈlī-ən(t)s\ *noun* (1607)
1 : the act of relying **:** the state of being reliant
2 : something or someone relied on

re·li·ant \-ənt\ *adjective* (1859)
: having reliance on something or someone **:** DEPENDENT
— **re·li·ant·ly** *adverb*

rel·ic \ˈre-lik\ *noun* [Middle English *relik,* from Old French *relique,* from Medieval Latin *reliquia,* from Late Latin *reliquiae,* plural, remains of a martyr, from Latin, remains, from *relinquere* to leave behind — more at RELINQUISH] (13th century)
1 a : an object esteemed and venerated because of association with a saint or martyr **b :** SOUVENIR, MEMENTO
2 *plural* **:** REMAINS, CORPSE
3 : a survivor or remnant left after decay, disintegration, or disappearance
4 : a trace of some past or outmoded practice, custom, or belief

[1]rel·ict \ˈre-likt\ *noun* [in sense 1, from Middle English *relicte,* from Late Latin *relicta,* from Latin, feminine of *relictus,* past participle of *relinquere;* in senses 2 & 3, from *relict* residual, adjective, from Latin *relictus*] (15th century)
1 : WIDOW
2 : a persistent remnant of an otherwise extinct flora or fauna or kind of organism
3 a : a relief feature or rock remaining after other parts have disappeared **b :** something left unchanged

[2]relict *adjective* (15th century)
: of, relating to, or being a relict ⟨*relict* populations⟩

re·lic·tion \ri-ˈlik-shən\ *noun* [Latin *reliction-, relictio* act of leaving behind, from *relinquere*] (circa 1676)
1 : the gradual recession of water leaving land permanently uncovered
2 : land uncovered by reliction

[1]re·lief \ri-ˈlēf\ *noun* [Middle English, from Middle French, from Old French, from *relever* to relieve] (14th century)
1 : a payment made by a feudal tenant to a lord on succeeding to an inherited estate
2 a : removal or lightening of something oppressive, painful, or distressing **b :** WELFARE 2a **c :** military assistance to an endangered post or force **d :** means of breaking or avoiding monotony or boredom **:** DIVERSION
3 : release from a post or from the performance of duty
4 : one that takes the place of another on duty
5 : legal remedy or redress
6 [French] **a :** a mode of sculpture in which forms and figures are distinguished from a surrounding plane surface **b :** sculpture or a sculptural form executed in this mode **c :** projecting detail, ornament, or figures
7 : sharpness of outline due to contrast ⟨a roof in bold *relief* against the sky⟩
8 : the elevations or inequalities of a land surface
9 : the pitching done by a relief pitcher ⟨two innings of hitless *relief*⟩

relief 6b

[2]relief *adjective* (1838)
1 : providing relief
2 : characterized by surface inequalities
3 : of or used in letterpress

relief map *noun* (1876)
: a map representing topographic relief

relief pitcher *noun* (circa 1949)
: a baseball pitcher who takes over for another during a game

relief printing *noun* (1875)
: LETTERPRESS 1

re·lieve \ri-ˈlēv\ *verb* **re·lieved; re·liev·ing** [Middle English *releven,* from Middle French *relever* to raise, relieve, from Latin *relevare,* from *re-* + *levare* to raise — more at LEVER] (14th century)
transitive verb
1 a : to free from a burden **:** give aid or help to **b :** to set free from an obligation, condition, or restriction **c :** to ease of a burden, wrong, or oppression by judicial or legislative interposition
2 a : to bring about the removal or alleviation of **:** MITIGATE **b :** ROB, DEPRIVE
3 a : to release from a post, station, or duty **b :** to take the place of
4 : to remove or lessen the monotony of
5 a : to set off by contrast **b :** to raise in relief
6 : to discharge the bladder or bowels of (oneself)
intransitive verb
1 : to bring or give relief
2 : to stand out in relief
3 : to serve as a relief pitcher ☆
— **re·liev·able** \-ˈlē-və-bəl\ *adjective*

re·lieved \ri-ˈlēvd\ *adjective* (1869)
: experiencing or showing relief especially from anxiety or pent-up emotions
— **re·liev·ed·ly** \-ˈlē-vəd-lē\ *adverb*

re·liev·er \ri-ˈlē-vər\ *noun* (15th century)
: one that relieves; *especially* **:** RELIEF PITCHER

re·lie·vo \ri-ˈlē-(ˌ)vō, rēl-ˈyā-(ˌ)vō\ *noun, plural* **-vos** [Italian *rilievo,* from *rilevare* to raise, from Latin *relevare*] (1625)
: RELIEF 6

re·li·gio- \ri-ˈli-j(ē-)ō\ *combining form*
: religion and ⟨*religio*-political⟩

re·li·gion \ri-ˈli-jən\ *noun* [Middle English *religioun,* from Latin *religion-, religio* supernatural constraint, sanction, religious practice, perhaps from *religare* to restrain, tie back — more at RELY] (13th century)
1 a : the state of a religious ⟨a nun in her 20th year of *religion*⟩ **b** (1) **:** the service and worship of God or the supernatural (2) **:** commitment or devotion to religious faith or observance
2 : a personal set or institutionalized system of religious attitudes, beliefs, and practices
3 *archaic* **:** scrupulous conformity **:** CONSCIENTIOUSNESS
4 : a cause, principle, or system of beliefs held to with ardor and faith
— **re·li·gion·less** *adjective*

re·li·gion·ist \-ˈli-jə-nist, -ˈlij-nist\ *noun* (1653)
: a person adhering to a religion; *especially* **:** a religious zealot

re·li·gi·ose \ri-ˈli-jē-ˌōs\ *adjective* [*religion* + [1]*-ose*] (1853)
: RELIGIOUS; *especially* **:** excessively, obtrusively, or sentimentally religious
— **re·li·gi·os·i·ty** \-ˌli-jē-ˈä-sə-tē\ *noun*

[1]re·li·gious \ri-ˈli-jəs\ *adjective* [Middle English, from Old French *religieus,* from Latin *religiosus,* from *religio*] (13th century)
1 : relating to or manifesting faithful devotion to an acknowledged ultimate reality or deity ⟨a *religious* person⟩ ⟨*religious* attitudes⟩
2 : of, relating to, or devoted to religious beliefs or observances
3 a : scrupulously and conscientiously faithful **b :** FERVENT, ZEALOUS
— **re·li·gious·ly** *adverb*
— **re·li·gious·ness** *noun*

[2]religious *noun, plural* **religious** [Middle English, from Old French *religieus,* from *religieus,* adjective] (13th century)
: a member of a religious order under monastic vows

re·line \(ˌ)rē-ˈlīn\ *transitive verb* (1851)
: to put new lines on or a new lining in

re·lin·quish \ri-ˈliŋ-kwish, -ˈlin-\ *transitive verb* [Middle English *relinquisshen,* from Middle French *relinquiss-,* stem of *relinquir,* from Latin *relinquere* to leave behind, from *re-* + *linquere* to leave — more at LOAN] (15th century)
1 : to withdraw or retreat from **:** leave behind
2 : GIVE UP ⟨*relinquish* a title⟩

☆ SYNONYMS

Relevant, germane, material, pertinent, apposite, applicable, apropos mean relating to or bearing upon the matter in hand. RELEVANT implies a traceable, significant, logical connection ⟨found material *relevant* to her case⟩. GERMANE may additionally imply a fitness for or appropriateness to the situation or occasion ⟨a point not *germane* to the discussion⟩. MATERIAL implies so close a relationship that it cannot be dispensed with without serious alteration of the case ⟨facts *material* to the investigation⟩. PERTINENT stresses a clear and decisive relevance ⟨a *pertinent* observation⟩. APPOSITE suggests a felicitous relevance ⟨add an *apposite* quotation to the definition⟩. APPLICABLE suggests the fitness of bringing a general rule or principle to bear upon a particular case ⟨the rule is not *applicable* in this case⟩. APROPOS suggests being both relevant and opportune ⟨the quip was *apropos*⟩.

Relieve, alleviate, lighten, assuage, mitigate, allay mean to make something less grievous. RELIEVE implies a lifting of enough of a burden to make it tolerable ⟨took an aspirin to *relieve* the pain⟩. ALLEVIATE implies temporary or partial lessening of pain or distress ⟨the lotion *alleviated* the itching⟩. LIGHTEN implies reducing a burdensome or depressing weight ⟨good news would *lighten* our worries⟩. ASSUAGE implies softening or sweetening what is harsh or disagreeable ⟨ocean breezes *assuaged* the intense heat⟩. MITIGATE suggests a moderating or countering of the effect of something violent or painful ⟨the need to *mitigate* barbaric laws⟩. ALLAY implies an effective calming or soothing of fears or alarms ⟨*allayed* their fears⟩.

3 a : to stop holding physically **:** RELEASE ⟨slowly *relinquished* his grip on the bar⟩ **b :** to give over possession or control of **:** YIELD ⟨few leaders willingly *relinquish* power⟩ ☆
— re·lin·quish·ment \-mənt\ *noun*

rel·i·quary \'re-lə-ˌkwer-ē\ *noun, plural* **-quar·ies** [French *reliquaire,* from Medieval Latin *reliquiarium,* from *reliquia* relic — more at RELIC] (circa 1656)
: a container or shrine in which sacred relics are kept

re·lique \ri-'lēk, 're-lik\ *archaic variant of* RELIC

re·liq·ui·ae \ri-'li-kwē-ˌī, -kwē-ˌē\ *noun plural* [Latin — more at RELIC] (1654)
: remains of the dead **:** RELICS

[1]**rel·ish** \'re-lish\ *noun* [alteration of Middle English *reles* taste, from Old French, something left behind, release, from *relessier* to release] (1530)
1 : characteristic flavor; *especially* **:** pleasing or zestful flavor
2 : a quantity just sufficient to flavor or characterize **:** TRACE
3 a : enjoyment of or delight in something that satisfies one's tastes, inclinations, or desires ⟨eat with great *relish*⟩ **b :** a strong liking **:** INCLINATION ⟨has little *relish* for sports⟩
4 a : something adding a zestful flavor; *especially* **:** a condiment (as of pickles or green tomatoes) eaten with other food to add flavor **b :** APPETIZER, HORS D'OEUVRE

[2]**relish** (1586)
transitive verb
1 : to add relish to
2 : to be pleased or gratified by **:** ENJOY
3 : to eat or drink with pleasure
4 : to appreciate with taste and discernment
intransitive verb
: to have a characteristic or pleasing taste
— rel·ish·able \'re-li-shə-bəl\ *adjective*

re·live \(ˌ)rē-'liv\ (1548)
intransitive verb
: to live again
transitive verb
: to live over again; *especially* **:** to experience again in the imagination

re·lo·cate \(ˌ)rē-'lō-ˌkat, ˌrē-lō-'\ (1834)
transitive verb
: to locate again **:** establish or lay out in a new place
intransitive verb
: to move to a new location
— re·lo·cat·able \-'lō-ˌkā-tə-bəl, -ˌlō-'kā-\ *adjective*
— re·lo·ca·tion \ˌrē-lō-'kā-shən\ *noun*

re·lo·cat·ee \ˌrē-lə-ˌkā-'tē, ˌrē-ˌlō-kə-'tē\ *noun* (1954)
: one who moves to a new location **:** one that is relocated

re·lu·cent \ri-'lü-sᵊnt\ *adjective* [Latin *relucent-, relucens,* past participle of *relucēre* to shine back, from *re-* + *lucēre* to shine — more at LIGHT] (15th century)
: reflecting light **:** SHINING

re·luct \ri-'ləkt\ *intransitive verb* [Latin *reluctari*] (1547)
: to show reluctance

re·luc·tance \ri-'lək-tən(t)s\ *noun* (1710)
1 : the quality or state of being reluctant
2 : the opposition offered in a magnetic circuit to magnetic flux; *specifically* **:** the ratio of the magnetic potential difference to the corresponding flux

re·luc·tan·cy \-tən(t)-sē\ *noun* (1634)
: RELUCTANCE

re·luc·tant \ri-'lək-tənt\ *adjective* [Latin *reluctant-, reluctans,* present participle of *reluctari* to struggle against, from *re-* + *luctari* to struggle] (1667)
: feeling or showing aversion, hesitation, or unwillingness ⟨*reluctant* to get involved⟩; *also* **:** having or assuming a specified role unwillingly ⟨a *reluctant* hero⟩
synonym see DISINCLINED
— re·luc·tant·ly *adverb*

re·luc·tate \ri-'lək-ˌtāt\ *intransitive verb* **-tat·ed; -tat·ing** (1643)
: RELUCT
— re·luc·ta·tion \ri-ˌlək-'tā-shən, ˌrē-\ *noun*

re·lume \(ˌ)rē-'lüm\ *transitive verb* **re·lumed; re·lum·ing** [irregular from Late Latin *reluminare,* from Latin *re-* + *luminare* to light up — more at ILLUMINATE] (1604)
archaic **:** to light or light up again **:** REKINDLE

re·ly \ri-'lī\ *intransitive verb* **re·lied; re·ly·ing** [Middle English *relien* to rally, from Middle French *relier* to connect, rally, from Latin *religare* to tie back, from *re-* + *ligare* to tie — more at LIGATURE] (1574)
1 : to be dependent ⟨the system on which we *rely* for water⟩
2 : to have confidence based on experience ⟨someone you can *rely* on⟩
— re·li·er \-'lī(-ə)r\ *noun*

rem \'rem\ *noun* [*r*oentgen *e*quivalent *m*an] (1947)
: the dosage of an ionizing radiation that will cause the same biological effect as one roentgen of X-ray or gamma-ray exposure

REM \'rem\ *noun* (1957)
: RAPID EYE MOVEMENT

[1]**re·main** \ri-'mān\ *intransitive verb* [Middle English, from Middle French *remaindre,* from Latin *remanēre,* from *re-* + *manēre* to remain — more at MANSION] (14th century)
1 a : to be a part not destroyed, taken, or used up ⟨only a few ruins *remain*⟩ **b :** to be something yet to be shown, done, or treated ⟨it *remains* to be seen⟩
2 : to stay in the same place or with the same person or group; *especially* **:** to stay behind
3 : to continue unchanged ⟨the fact *remains* that we can't go⟩

[2]**remain** *noun* (15th century)
1 *obsolete* **:** STAY
2 : a remaining part or trace — usually used in plural
3 *plural* **:** a dead body

[1]**re·main·der** \ri-'mān-dər\ *noun* [Middle English, from Anglo-French, from Middle French *remaindre*] (14th century)
1 : an interest or estate in property that follows and is dependent on the termination of a prior intervening possessory estate created at the same time by the same instrument
2 a : a remaining group, part, or trace **b** (1) **:** the number left after a subtraction (2) **:** the final undivided part after division that is less or of lower degree than the divisor
3 : a book sold at a reduced price by the publisher after sales have slowed

[2]**remainder** *adjective* (1567)
: LEFTOVER, REMAINING

[3]**remainder** *transitive verb* **-dered; -der·ing** \-d(ə-)riŋ\ (1904)
: to dispose of as remainders

remainder theorem *noun* (1886)
: a theorem in algebra: if $f(x)$ is a polynomial in x then the remainder on dividing $f(x)$ by $x - a$ is $f(a)$

[1]**re·make** \(ˌ)rē-'māk\ *transitive verb* **-made** \-'mād\; **-mak·ing** (circa 1635)
: to make anew or in a different form
— re·mak·er \-'ma-kər\ *noun*

[2]**re·make** \'rē-ˌmāk\ *noun* (1936)
: one that is remade; *especially* **:** a new version of a motion picture

re·man \(ˌ)rē-'man\ *transitive verb* (1666)
1 : to man again or anew
2 : to imbue with courage again

re·mand \ri-'mand\ *transitive verb* [Middle English *remaunden,* from Middle French *remander,* from Late Latin *remandare* to send back word, from Latin *re-* + *mandare* to order — more at MANDATE] (15th century)
: to order back: as **a :** to send back (a case) to another court or agency for further action **b :** to return to custody pending trial or for further detention
— remand *noun*

rem·a·nence \'re-mə-nən(t)s, ri-'mā-\ *noun* (circa 1880)
: the magnetic induction remaining in a magnetized substance no longer under external magnetic influence

rem·a·nent \-nənt\ *adjective* [Middle English, from Latin *remanent-, remanens,* present participle of *remanēre* to remain] (15th century)
1 : RESIDUAL, REMAINING
2 : of, relating to, or characterized by remanence

re·man·u·fac·ture \(ˌ)rē-ˌma-nyə-'fak-chər, -ˌma-nə-\ *transitive verb* (1851)
: to manufacture into a new product
— remanufacture *noun*
— re·man·u·fac·tur·er \-chər-ər\ *noun*

re·map \(ˌ)rē-'map\ *transitive verb* (1931)
: to map again; *also* **:** to lay out in a new pattern

[1]**re·mark** \ri-'märk\ *noun* [French *remarque,* from Middle French, from *remarquer* to remark, from *re-* re- + *marquer* to mark — more at MARQUE] (1660)
1 : the act of remarking **:** NOTICE
2 : an expression of opinion or judgment
3 : mention of that which deserves attention or notice

[2]**remark** (1675)
transitive verb
1 : to take notice of **:** OBSERVE
2 : to express as an observation or comment **:** SAY
intransitive verb
: to notice something and comment thereon — used with *on* or *upon*

re·mark·able \ri-'mär-kə-bəl\ *adjective* (circa 1604)
: worthy of being or likely to be noticed especially as being uncommon or extraordinary
synonym see NOTICEABLE
— re·mark·able·ness *noun*

re·mark·ably \-blē\ *adverb* (1638)
1 : in a remarkable manner ⟨*remarkably* talented⟩
2 : as is remarkable ⟨*remarkably,* no one was hurt⟩

re·marque \ri-'märk\ *noun* [French *remarque* remark, note, from Middle French, from *remarquer*] (1882)
1 : a drawn, etched, or incised scribble or sketch done on the margin of a plate or stone and removed before the regular printing
2 : a proof taken before remarques have been removed

re·mas·ter \(ˌ)rē-'mas-tər\ *transitive verb* (1964)
: to create a new master of especially by altering or enhancing the sound quality of an older recording

re·match \(ˌ)rē-'mach, 'rē-ˌ\ *noun* (1941)

☆ SYNONYMS
Relinquish, yield, resign, surrender, abandon, waive mean to give up completely. RELINQUISH usually does not imply strong feeling but may suggest some regret, reluctance, or weakness ⟨*relinquished* her crown⟩. YIELD implies concession or compliance or submission to force ⟨the troops *yielded* ground grudgingly⟩. RESIGN emphasizes voluntary relinquishment or sacrifice without struggle ⟨*resigned* her position⟩. SURRENDER implies a giving up after a struggle to retain or resist ⟨*surrendered* their claims⟩. ABANDON stresses finality and completeness in giving up ⟨*abandoned* all hope⟩. WAIVE implies conceding or forgoing with little or no compulsion ⟨*waived* the right to a trial by jury⟩.

: a second match between the same contestants or teams

re·me·di·a·ble \ri-ˈmē-dē-ə-bəl\ *adjective* (15th century)
: capable of being remedied
— **re·me·di·abil·i·ty** \ri-ˌmē-dē-ə-ˈbi-lə-tē\ *noun*

re·me·di·al \ri-ˈmē-dē-əl\ *adjective* (1651)
1 : intended as a remedy
2 : concerned with the correction of faulty study habits and the raising of a pupil's general competence ⟨*remedial* reading courses⟩; *also* **:** receiving or requiring remedial instruction ⟨*remedial* students⟩
— **re·me·di·al·ly** \-ə-lē\ *adverb*

re·me·di·ate \ri-ˈmē-dē-ət\ *adjective* (1605)
archaic **:** REMEDIAL

re·me·di·a·tion \ri-ˌmē-dē-ˈā-shən\ *noun* (1818)
: the act or process of remedying ⟨*remediation* of reading problems⟩
— **re·me·di·ate** \-ˈmē-dē-ˌāt\ *transitive verb*

¹rem·e·dy \ˈre-mə-dē\ *noun, plural* **-dies** [Middle English *remedie,* from Anglo-French, from Latin *remedium,* from *re-* + *mederi* to heal — more at MEDICAL] (13th century)
1 : a medicine, application, or treatment that relieves or cures a disease
2 : something that corrects or counteracts
3 : the legal means to recover a right or to prevent or obtain redress for a wrong
— **rem·e·di·less** *adjective*

²remedy *transitive verb* **-died; -dy·ing** (15th century)
: to provide or serve as a remedy for **:** RELIEVE ⟨*remedy* a social evil⟩
synonym see CORRECT

re·mem·ber \ri-ˈmem-bər\ *verb* **-bered; -ber·ing** \-b(ə-)riŋ\ [Middle English *remembren,* from Middle French *remembrer,* from Late Latin *rememorari,* from Latin *re-* + Late Latin *memorari* to be mindful of, from Latin *memor* mindful — more at MEMORY] (14th century)
transitive verb
1 : to bring to mind or think of again ⟨*remembers* the old days⟩
2 *archaic* **a :** BETHINK 1b **b :** REMIND
3 a : to keep in mind for attention or consideration ⟨*remembers* friends at Christmas⟩ **b :** REWARD ⟨was *remembered* in the will⟩
4 : to retain in the memory ⟨*remember* the facts until the test is over⟩
5 : to convey greetings from ⟨*remember* me to her⟩
6 : RECORD, COMMEMORATE
intransitive verb
1 : to exercise or have the power of memory
2 : to have a recollection or remembrance ☆
— **re·mem·ber·abil·i·ty** \-ˌmem-b(ə-)rə-ˈbi-lə-tē\ *noun*
— **re·mem·ber·able** \-ˈmem-b(ə-)rə-bəl\ *adjective*
— **re·mem·ber·er** \-bər-ər\ *noun*

re·mem·brance \ri-ˈmem-brən(t)s *also* -bə-rən(t)s\ *noun* (14th century)
1 : the state of bearing in mind
2 a : the ability to remember **:** MEMORY **b :** the period over which one's memory extends
3 : an act of recalling to mind
4 : a memory of a person, thing, or event
5 a : something that serves to keep in or bring to mind **:** REMINDER **b :** COMMEMORATION, MEMORIAL **c :** a greeting or gift recalling or expressing friendship or affection
synonym see MEMORY

Remembrance Day *noun* (1918)
: November 11 set aside in commemoration of the end of hostilities in 1918 and 1945 and observed as a legal holiday in Canada; *also* **:** REMEMBRANCE SUNDAY

re·mem·branc·er \ri-ˈmem-brən(t)-sər\ *noun* (15th century)
1 : any of several English officials
2 : one that reminds

Remembrance Sunday *noun* (1942)
: a Sunday that is usually closest to November 11 and that in Great Britain is set aside in commemoration of the end of hostilities in 1918 and 1945

re·mex \ˈrē-ˌmeks\ *noun, plural* **rem·i·ges** \ˈre-mə-ˌjēz\ [New Latin *remig-, remex,* from Latin, oarsman, from *remus* oar + *agere* to drive — more at ROW, AGENT] (1767)
: a primary or secondary quill feather of the wing of a bird

re·mind \ri-ˈmīnd\ *transitive verb* (1660)
: to put in mind of something **:** cause to remember
synonym see REMEMBER
— **re·mind·er** *noun*

re·mind·ful \-ˈmīn(d)-fəl\ *adjective* (1810)
1 : MINDFUL
2 : tending to remind **:** SUGGESTIVE, EVOCATIVE

rem·i·nisce \ˌre-mə-ˈnis\ *intransitive verb* **-nisced; -nisc·ing** [back-formation from *reminiscence*] (1829)
: to indulge in reminiscence
synonym see REMEMBER
— **rem·i·nis·cer** \-ˈni-sər\ *noun*

rem·i·nis·cence \-ˈni-sᵊn(t)s\ *noun* (1589)
1 : apprehension of a Platonic idea as if it had been known in a previous existence
2 a : recall to mind of a long-forgotten experience or fact **b :** the process or practice of thinking or telling about past experiences
3 a : a remembered experience **b :** an account of a memorable experience — often used in plural
4 : something so like another as to be regarded as an unconscious repetition, imitation, or survival
synonym see MEMORY

rem·i·nis·cent \-sᵊnt\ *adjective* [Latin *reminiscent-, reminiscens,* present participle of *reminisci* to remember, from *re-* + *-minisci* (akin to Latin *ment-, mens* mind) — more at MIND] (1765)
1 : of the character of or relating to reminiscence
2 : marked by or given to reminiscence
3 : tending to remind **:** SUGGESTIVE
— **rem·i·nis·cent·ly** *adverb*

rem·i·nis·cen·tial \ˌre-mə-(ˌ)ni-ˈsen(t)-shəl\ *adjective* (1646)
: REMINISCENT

re·mint \(ˌ)rē-ˈmint\ *transitive verb* (1823)
: to melt down (old or worn coin) and make into new coin

re·mise \ri-ˈmīz\ *transitive verb* **re·mised; re·mis·ing** [Middle English, from Middle French *remis,* past participle of *remettre* to put back, from Latin *remittere* to send back] (15th century)
: to give, grant, or release a claim to **:** DEED

re·miss \ri-ˈmis\ *adjective* [Middle English, from Latin *remissus,* from past participle of *remittere* to send back, relax] (15th century)
1 : negligent in the performance of work or duty **:** CARELESS
2 : showing neglect or inattention **:** LAX
synonym see NEGLIGENT
— **re·miss·ly** *adverb*
— **re·miss·ness** *noun*

re·mis·si·ble \ri-ˈmi-sə-bəl\ *adjective* (1577)
: capable of being forgiven ⟨*remissible* sins⟩
— **re·mis·si·bly** \-blē\ *adverb*

re·mis·sion \ri-ˈmi-shən\ *noun* (13th century)
1 : the act or process of remitting
2 : a state or period during which something is remitted

¹re·mit \ri-ˈmit\ *verb* **re·mit·ted; re·mit·ting** [Middle English *remitten,* from Latin *remittere* to send back, from *re-* + *mittere* to send] (14th century)
transitive verb
1 a : to lay aside (a mood or disposition) partly or wholly **b :** to desist from (an activity) **c :** to let (as attention or diligence) slacken **:** RELAX
2 a : to release from the guilt or penalty of ⟨*remit* sins⟩ **b :** to refrain from exacting ⟨*remit* a tax⟩ **c :** to cancel or refrain from inflicting ⟨*remit* the penalty⟩ **d :** to give relief from (suffering)
3 : to submit or refer for consideration, judgment, decision, or action; *specifically* **:** REMAND
4 : to restore or consign to a former status or condition
5 : POSTPONE, DEFER
6 : to send (money) to a person or place especially in payment of a demand, account, or draft
intransitive verb
1 a : to abate in force or intensity **:** MODERATE **b** *of a disease or abnormality* **:** to abate symptoms for a period
2 : to send money (as in payment)
— **re·mit·ment** \-ˈmit-mənt\ *noun*
— **re·mit·ta·ble** \-ˈmi-tə-bəl\ *adjective*
— **re·mit·ter** *noun*

²re·mit \ri-ˈmit, ˈrē-ˌ\ *noun* (15th century)
1 : an act of remitting
2 : something remitted to another person or authority

re·mit·tal \ri-ˈmi-tᵊl\ *noun* (1596)
: REMISSION

re·mit·tance \ri-ˈmi-tᵊn(t)s\ *noun* (1705)
1 a : a sum of money remitted **b :** an instrument by which money is remitted
2 : transmittal of money (as to a distant place)

remittance man *noun* (1886)
: one living abroad on remittances from home

re·mit·tent \ri-ˈmi-tᵊnt\ *adjective* [Latin *remittent-, remittens,* present participle of *remittere*] (1693)
of a disease **:** marked by alternating periods of abatement and increase of symptoms

¹rem·nant \ˈrem-nənt\ *noun* [Middle English, contraction of *remenant,* from Middle French, from present participle of *remenoir* to remain, from Latin *remanēre* — more at REMAIN] (14th century)
1 a : a usually small part, member, or trace remaining **b :** a small surviving group — often used in plural
2 : an unsold or unused end of piece goods

²remnant *adjective* (1550)
: still remaining

re·mod·el \(ˌ)rē-ˈmä-dᵊl\ *transitive verb* (1789)
: to alter the structure of **:** REMAKE

re·mon·strance \ri-ˈmän(t)-strən(t)s\ *noun* (1585)
1 : an earnest presentation of reasons for opposition or grievance; *especially* **:** a document formally stating such points
2 : an act or instance of remonstrating

re·mon·strant \-strənt\ *adjective* (1641)
: vigorously objecting or opposing
— **remonstrant** *noun*
— **re·mon·strant·ly** *adverb*

re·mon·strate \ˈre-mən-ˌstrāt, ri-ˈmän-\ *verb* **-strat·ed; -strat·ing** [Medieval Latin *remonstratus,* past participle of *remonstrare* to

☆ SYNONYMS
Remember, recollect, recall, remind, reminisce mean to bring an image or idea from the past into the mind. REMEMBER implies a keeping in memory that may be effortless or unwilled ⟨*remembers* that day as though it were yesterday⟩. RECOLLECT implies a bringing back to mind what is lost or scattered ⟨as near as I can *recollect*⟩. RECALL suggests an effort to bring back to mind and often to re-create in speech ⟨can't *recall* the words of the song⟩. REMIND suggests a jogging of one's memory by an association or similarity ⟨that *reminds* me of a story⟩. REMINISCE implies a casual often nostalgic recalling of experiences long past and gone ⟨old college friends like to *reminisce*⟩.

demonstrate, from Latin *re-* + *monstrare* to show — more at MUSTER] (1695)
intransitive verb
: to present and urge reasons in opposition : EXPOSTULATE — usually used with *with*
transitive verb
: to say or plead in protest, reproof, or opposition
— **re·mon·stra·tion** \ˌre-mən-ˈstrā-shən, ri-ˈmän-\ *noun*
— **re·mon·stra·tive** \ri-ˈmän(t)-strə-tiv\ *adjective*
— **re·mon·stra·tive·ly** *adverb*
— **re·mon·stra·tor** \ˈre-mən-ˌstrā-tər, ri-ˈmän-\ *noun*

rem·o·ra \ˈre-mə-rə *also* ri-ˈmōr-ə *or* -ˈmȯr-\ *noun* [Late Latin, from Latin, delay, from *remorari* to delay, from *re-* + *morari* to delay — more at MORATORIUM] (1567)
1 : any of a family (Echeneididae) of marine bony fishes that have the anterior dorsal fin modified into a suctorial disk on the head by means of which they cling especially to other fishes
2 : HINDRANCE, DRAG

remoras on tiger shark

re·morse \ri-ˈmȯrs\ *noun* [Middle English, from Middle French *remors,* from Medieval Latin *remorsus,* from Late Latin, act of biting again, from Latin *remordēre* to bite again, from *re-* + *mordēre* to bite — more at MORDANT] (14th century)
1 : a gnawing distress arising from a sense of guilt for past wrongs : SELF-REPROACH
2 *obsolete* **:** COMPASSION ◆
synonym see PENITENCE

re·morse·ful \-ˈmȯrs-fəl\ *adjective* (1592)
: motivated or marked by remorse
— **re·morse·ful·ly** \-fə-lē\ *adverb*
— **re·morse·ful·ness** *noun*

re·morse·less \-ləs\ *adjective* (1593)
1 : having no remorse : MERCILESS
2 : RELENTLESS
— **re·morse·less·ly** *adverb*
— **re·morse·less·ness** *noun*

¹re·mote \ri-ˈmōt\ *adjective* **re·mot·er; -est** [Middle English, from Latin *remotus,* from past participle of *removēre* to remove] (15th century)
1 : separated by an interval or space greater than usual ⟨an involucre *remote* from the flower⟩
2 : far removed in space, time, or relation : DIVERGENT ⟨the *remote* past⟩ ⟨comments *remote* from the truth⟩
3 : OUT-OF-THE-WAY, SECLUDED ⟨a *remote* cabin in the hills⟩
4 : acting, acted on, or controlled indirectly or from a distance ⟨*remote* computer operation⟩; *also* **:** relating to the acquisition of information about a distant object (as by radar or photography) without coming into physical contact with it ⟨*remote* sensing⟩
5 : not arising from a primary or proximate action
6 : small in degree : SLIGHT ⟨a *remote* possibility⟩
7 : distant in manner : ALOOF
— **re·mote·ly** *adverb*
— **re·mote·ness** *noun*

²remote *noun* (1937)
1 : a radio or television program or a portion of a program originating outside the studio
2 : REMOTE CONTROL 2

remote control *noun* (1904)
1 : control (as by radio signal) of operation from a point at some distance removed
2 : a device or mechanism for controlling something from a distance

re·mo·tion \ri-ˈmō-shən\ *noun* (15th century)
1 : the quality or state of being remote
2 : the act of removing : REMOVAL
3 *obsolete* **:** DEPARTURE

¹re·mount \(ˌ)rē-ˈmau̇nt\ *verb* [Middle English, partly from *re-* + *mounten* to mount, partly from Middle French *remonter,* from *re-* + *monter* to mount] (15th century)
transitive verb
1 : to mount (something) again ⟨*remount* a picture⟩
2 : to furnish remounts to
intransitive verb
1 : to mount again
2 : REVERT

²re·mount \ˈrē-ˌmau̇nt, (ˌ)rē-ˈ\ *noun* (1781)
: a fresh horse to replace one no longer available

re·mov·al \ri-ˈmü-vəl\ *noun* (1597)
: the act or process of removing : the fact of being removed

¹re·move \ri-ˈmüv\ *verb* **re·moved; re·mov·ing** [Middle English, from Old French *removoir,* from Latin *removēre,* from *re-* + *movēre* to move] (14th century)
transitive verb
1 a : to change the location, position, station, or residence of ⟨*remove* soldiers to the front⟩ **b :** to transfer (a legal proceeding) from one court to another
2 : to move by lifting, pushing aside, or taking away or off ⟨*remove* your hat⟩
3 : to dismiss from office
4 : to get rid of : ELIMINATE ⟨*remove* a tumor surgically⟩
intransitive verb
1 : to change location, station, or residence ⟨*removing* from the city to the suburbs⟩
2 : to go away
3 : to be capable of being removed
— **re·mov·abil·i·ty** \-ˌmü-və-ˈbi-lə-tē\ *noun*
— **re·mov·able** *also* **re·move·able** \ri-ˈmü-və-bəl\ *adjective*
— **re·mov·able·ness** \-ˈmü-və-bəl-nəs\ *noun*
— **re·mov·ably** \-blē\ *adverb*
— **re·mov·er** *noun*

²remove *noun* (1553)
1 : REMOVAL; *specifically* **:** MOVE 2c
2 a : a distance or interval separating one person or thing from another **b :** a degree or stage of separation

removed *adjective* (circa 1548)
1 a : distant in degree of relationship **b :** of a younger or older generation ⟨a second cousin's child is a second cousin once *removed*⟩
2 : separate or remote in space, time, or character

REM sleep *noun* (1965)
: a state of sleep that recurs cyclically several times during a normal period of sleep and that is characterized by increased neuronal activity of the forebrain and midbrain, by depressed muscle tone, and especially in humans by dreaming, rapid eye movements, and vascular congestion of the sex organs — called also *paradoxical sleep, rapid eye movement sleep*

re·mu·da \ri-ˈmü-də, -ˈmyü-\ *noun* [American Spanish, relay of horses, from Spanish, exchange, from *remudar* to exchange, from *re-* + *mudar* to change, from Latin *mutare* — more at MUTABLE] (circa 1892)
: the herd of horses from which those to be used for the day are chosen

re·mu·ner·ate \ri-ˈmyü-nə-ˌrāt\ *transitive verb* **-at·ed; -at·ing** [Latin *remuneratus,* past participle of *remunerare* to recompense, from *re-* + *munerare* to give, from *muner-, munus* gift — more at MEAN] (1523)
1 : to pay an equivalent for ⟨their services were generously *remunerated*⟩
2 : to pay an equivalent to for a service, loss, or expense : RECOMPENSE
synonym see PAY
— **re·mu·ner·a·tor** \-ˌrā-tər\ *noun*
— **re·mu·ner·a·to·ry** \-rə-ˌtōr-ē, -ˌtȯr-\ *adjective*

re·mu·ner·a·tion \ri-ˌmyü-nə-ˈrā-shən\ *noun* (15th century)
1 : something that remunerates : RECOMPENSE, PAY
2 : an act or fact of remunerating

re·mu·ner·a·tive \ri-ˈmyü-nə-rə-tiv, -ˌrā-\ *adjective* (circa 1677)
1 : serving to remunerate
2 : providing remuneration : PROFITABLE
— **re·mu·ner·a·tive·ly** *adverb*
— **re·mu·ner·a·tive·ness** *noun*

Re·mus \ˈrē-məs\ *noun* [Latin]
: a son of Mars slain by his twin brother Romulus

re·nais·sance \ˌre-nə-ˈsän(t)s, -ˈzän(t)s, -ˈsäⁿs, -ˈzäⁿs, ˈre-nə-ˌ, *chiefly British* ri-ˈnā-sᵊn(t)s\ *noun, often attributive* [French, from Middle French, rebirth, from *renaistre* to be born again, from Latin *renasci,* from *re-* + *nasci* to be born — more at NATION] (1845)
1 *capitalized* **a :** the transitional movement in Europe between medieval and modern times beginning in the 14th century in Italy, lasting into the 17th century, and marked by a humanistic revival of classical influence expressed in a flowering of the arts and literature and by the beginnings of modern science **b :** the period of the Renaissance **c :** the neoclassic style of architecture prevailing during the Renaissance
2 *often capitalized* **:** a movement or period of vigorous artistic and intellectual activity
3 : REBIRTH, REVIVAL

Renaissance man *noun* (1906)
: a person who has wide interests and is expert in several areas

re·nal \ˈrē-nᵊl\ *adjective* [French or Late Latin; French *rénal,* from Late Latin *renalis,* from Latin *renes* kidneys] (circa 1656)
: relating to, involving, or located in the region of the kidneys : NEPHRITIC

renal clearance *noun* (1948)
: CLEARANCE 3

re·na·scence \ri-ˈna-sᵊn(t)s, -ˈnā-\ *noun, often capitalized* (1727)
: RENAISSANCE

re·na·scent \-sᵊnt\ *adjective* [Latin *renascent-, renascens,* present participle of *renasci*] (circa 1727)

◇ WORD HISTORY

remorse *Remorse* is a gnawing distress that arises from a sense of guilt. In the light of its etymology, "gnawing" is a description of special relevance, for *remorse* is borrowed from Medieval Latin *remorsus,* a noun derivative of Latin *remordere* "to bite again, vex persistently." When a monk known to us as Dan Michel of Northgate, in the year 1340, finished his English translation of a French manual on sin and penitence, he called it *Ayenbite of Inwyt*—which we would render in Modern English as "Remorse of Conscience." The word *ayenbite* was Dan Michel's piece-by-piece translation of Latin *remorsus,* with *ayen-* ("again" in Modern English) rendering *re-* and *bite* rendering *morsus.* Similar loan translations of Latin words with *ayen-* for *re-* were quite popular in Middle English, though they have not survived. In many cases we have simply borrowed the Latin word, as has happened with *remorse.*

: rising again into being or vigor
re·na·ture \(ˌ)rē-ˈnā-chər\ *transitive verb* **re·na·tured; re·na·tur·ing** \-ˈnā-chə-riŋ, -ˈnāch-riŋ\ [*re-* + de*nature*] (1926)
: to restore (as a denatured protein) to an original or normal condition
— **re·na·tur·ation** \(ˌ)rē-ˌnā-chə-ˈrā-shən\ *noun*
ren·con·tre \räⁿ-kōⁿtrᵊ, ren-ˈkän-tər\ *or* **ren·coun·ter** \ren-ˈkau̇n-tər\ *noun* [*rencounter* from Middle French *rencontre,* from *rencontrer; rencontre* from French] (1523)
1 : a hostile meeting or a contest between forces or individuals **:** COMBAT
2 : a casual meeting
ren·coun·ter \ren-ˈkau̇n-tər\ *transitive verb* [Middle French *rencontrer* to meet by chance or in hostility, from *re-* + *encontrer* to encounter] (1549)
archaic **:** to meet casually
rend \ˈrend\ *verb* **rent** \ˈrent\ *also* **rend·ed** \ˈren-dəd\; **rend·ing** [Middle English, from Old English *rendan;* akin to Old Frisian *renda* to tear and perhaps to Sanskrit *randhra* hole] (before 12th century)
transitive verb
1 : to remove from place by violence **:** WREST
2 : to split or tear apart or in pieces by violence
3 : to tear (the hair or clothing) as a sign of anger, grief, or despair
4 a : to lacerate mentally or emotionally **b :** to pierce with sound **c :** to divide (as a nation) into contesting factions
intransitive verb
1 : to perform an act of tearing or splitting
2 : to become torn or split
synonym see TEAR
¹ren·der \ˈren-dər\ *verb* **ren·dered; ren·der·ing** \-d(ə-)riŋ\ [Middle English *rendren,* from Middle French *rendre* to give back, yield, from (assumed) Vulgar Latin *rendere,* alteration of Latin *reddere,* partly from *re-* + *dare* to give & partly from *re-* + *-dere* to put — more at DATE, DO] (14th century)
transitive verb
1 a : to melt down ⟨*render* suet⟩; *also* **:** to extract by melting ⟨*render* lard⟩ **b :** to treat so as to convert into industrial fats and oils or fertilizer
2 a : to transmit to another **:** DELIVER **b :** GIVE UP, YIELD **c :** to furnish for consideration, approval, or information: as (1) **:** to hand down (a legal judgment) (2) **:** to agree on and report (a verdict)
3 a : to give in return or retribution **b** (1) **:** GIVE BACK, RESTORE (2) **:** REFLECT, ECHO **c :** to give in acknowledgment of dependence or obligation **:** PAY **d :** to do (a service) for another
4 a (1) **:** to cause to be or become **:** MAKE ⟨enough rainfall . . . to *render* irrigation unnecessary —P. E. James⟩ ⟨*rendered* him helpless⟩ (2) **:** IMPART **b** (1) **:** to reproduce or represent by artistic or verbal means **:** DEPICT (2) **:** to give a performance of (3) **:** to produce a copy or version of ⟨the documents are *rendered* in the original French⟩ (4) **:** to execute the motions of ⟨*render* a salute⟩ **c :** TRANSLATE
5 : to direct the execution of **:** ADMINISTER ⟨*render* justice⟩
6 : to apply a coat of plaster or cement directly to
intransitive verb
: to give recompense
— **ren·der·able** \-d(ə-)rə-bəl\ *adjective*
— **ren·der·er** \-dər-ər\ *noun*
²render *noun* (1647)
: a return especially in goods or services due from a feudal tenant to his lord
¹ren·dez·vous \ˈrän-di-ˌvü, -dā-\ *noun, plural* **ren·dez·vous** \-ˌvüz\ [Middle French, from *rendez vous* present yourselves] (1591)
1 a : a place appointed for assembling or meeting **b :** a place of popular resort **:** HAUNT
2 : a meeting at an appointed place and time
3 : the process of bringing two spacecraft together
²rendezvous *verb* **-voused** \-ˌvüd\; **-vous·ing** \-ˌvü-iŋ\; **-vouses** \-ˌvüz\ (circa 1645)
intransitive verb
: to come together at a rendezvous
transitive verb
1 : to bring together at a rendezvous
2 : to meet at a rendezvous
ren·di·tion \ren-ˈdi-shən\ *noun* [obsolete French, from Middle French, alteration of *reddition,* from Late Latin *reddition-, redditio,* from Latin *reddere* to return] (1601)
: the act or result of rendering: as **a :** SURRENDER **b :** TRANSLATION **c :** PERFORMANCE, INTERPRETATION
ren·dzi·na \ren-ˈjē-nə\ *noun* [Polish *rędzina* rich limy soil] (1922)
: any of a group of dark grayish brown intrazonal soils developed in grassy regions of high to moderate humidity from soft calcareous marl or chalk
¹ren·e·gade \ˈre-ni-ˌgād\ *noun* [Spanish *renegado,* from Medieval Latin *renegatus,* from past participle of *renegare* to deny, from Latin *re-* + *negare* to deny — more at NEGATE] (1583)
1 : a deserter from one faith, cause, or allegiance to another
2 : an individual who rejects lawful or conventional behavior
²renegade *intransitive verb* **-gad·ed; -gad·ing** (circa 1611)
: to become a renegade
³renegade *adjective* (1705)
1 : having deserted a faith, cause, or religion for a hostile one
2 : having rejected tradition **:** UNCONVENTIONAL
re·nege \ri-ˈnig, -ˈneg, -ˈnēg, -ˈnāg\ *verb* **re·neged; re·neg·ing** [Medieval Latin *renegare*] (1548)
transitive verb
: DENY, RENOUNCE
intransitive verb
1 *obsolete* **:** to make a denial
2 : REVOKE
3 : to go back on a promise or commitment
— **re·neg·er** *noun*
re·ne·go·tia·ble \ˌrē-ni-ˈgō-sh(ē-)ə-bəl\ *adjective* (1943)
: subject to renegotiation ⟨*renegotiable* mortgages⟩ ⟨*renegotiable* rates⟩
re·ne·go·ti·ate \ˌrē-ni-ˈgō-shē-ˌāt\ *transitive verb* (circa 1934)
: to negotiate again (as to adjust interest rates or repayments or to get more money) ⟨*renegotiate* a loan⟩ ⟨*renegotiate* a contract⟩
— **re·ne·go·ti·a·tion** \ˌrē-ni-ˌgō-shē-ˈā-shən, -sē-ˈā-\ *noun*
re·new \ri-ˈnü, -ˈnyü\ (14th century)
transitive verb
1 : to make like new **:** restore to freshness, vigor, or perfection ⟨as we *renew* our strength in sleep⟩
2 : to make new spiritually **:** REGENERATE
3 a : to restore to existence **:** REVIVE **b :** to make extensive changes in **:** REBUILD
4 : to do again **:** REPEAT
5 : to begin again **:** RESUME
6 : REPLACE, REPLENISH ⟨*renew* water in a tank⟩
7 a : to grant or obtain an extension of or on ⟨*renew* a license⟩ **b :** to grant or obtain an extension on the loan of ⟨*renew* a library book⟩
intransitive verb
1 : to become new or as new
2 : to begin again **:** RESUME
3 : to make a renewal (as of a lease) ☆
— **re·new·er** *noun*
re·new·able \-ˈnü-ə-bəl, -ˈnyü-\ *adjective* (1727)
1 : capable of being renewed ⟨*renewable* contracts⟩
2 : capable of being replaced by natural ecological cycles or sound management practices ⟨*renewable* resources⟩
— **re·new·abil·i·ty** \-ˌnü-ə-ˈbi-lə-tē, -ˌnyü-\ *noun*
— **re·new·ably** \-ˈnü-ə-blē, -ˈnyü-\ *adverb*
re·new·al \ri-ˈnü-əl, -ˈnyü-\ *noun* (circa 1686)
1 : the act or process of renewing **:** REPETITION
2 : the quality or state of being renewed
3 : something (as a subscription to a magazine) renewed
4 : something used for renewing; *specifically* **:** an expenditure that betters existing fixed assets
5 : the rebuilding of a large area (as of a city) by a public authority
reni- *or* **reno-** *combining form* [Latin *renes* kidneys]
: kidney ⟨*reni*form⟩
re·ni·form \ˈrē-nə-ˌfȯrm, ˈre-\ *adjective* [New Latin *reniformis,* from *reni-* + *-formis* -form] (circa 1753)
: suggesting a kidney in outline
re·nin \ˈrē-nən, ˈre-\ *noun* [International Scientific Vocabulary, from Latin *renes*] (1906)
: a proteolytic enzyme of the kidney that plays a major role in the release of angiotensin
re·ni·ten·cy \ˈre-nə-tən(t)-sē, ri-ˈnī-tᵊn(t)-\ *noun* (1613)
: RESISTANCE, OPPOSITION
re·ni·tent \ˈre-nə-tənt, ri-ˈnī-tᵊnt\ *adjective* [French or Latin; French *rénitent,* from Latin *renitent-, renitens,* present participle of *reniti* to resist, from *re-* + *niti* to strive — more at NISUS] (1701)
1 : resisting physical pressure
2 : resisting constraint or compulsion **:** RECALCITRANT
ren·min·bi \ˈren-ˈmin-ˈbē\ *noun plural* [Chinese (Beijing) *rénmínbì,* from *rénmín* people + *bì* currency] (1957)
: the currency of the People's Republic of China consisting of yuan
ren·net \ˈre-nət\ *noun* [Middle English, from (assumed) Middle English *rennen* to cause to coagulate, from Old English *gerennan,* from *ge-* together + (assumed) Old English *rennan* to cause to run; akin to Old High German *rennen* to cause to run, Old English *rinnan* to run — more at CO-, RUN] (15th century)
1 a : the contents of the stomach of an unweaned animal and especially a calf **b :** the lining membrane of a stomach or one of its compartments (as the fourth of a ruminant) used for curdling milk; *also* **:** a preparation of the stomach of animals used for this purpose
2 a : RENNIN **b :** a substitute for rennin
ren·nin \ˈre-nən\ *noun* [rennet + ¹*-in*] (1897)
: an enzyme that coagulates milk and is used in making cheese and junkets; *especially* **:** one from the mucous membrane of the stomach of a calf
re·no·gram \ˈrē-nə-ˌgram\ *noun* (1952)
: a photographic depiction of the course of renal excretion of a radioactively labeled substance
— **re·no·graph·ic** \ˌrē-nə-ˈgra-fik\ *adjective*
— **re·nog·ra·phy** \rē-ˈnä-grə-fē\ *noun*

☆ SYNONYMS
Renew, restore, refresh, renovate, rejuvenate mean to make like new. RENEW implies a restoration of what had become faded or disintegrated so that it seems like new ⟨efforts to *renew* the splendor of the old castle⟩. RESTORE implies a return to an original state after depletion or loss ⟨*restored* a fine piece of furniture⟩. REFRESH implies the supplying of something necessary to restore lost strength, animation, or power ⟨a *refreshing* drink⟩. RENOVATE suggests a renewing by cleansing, repairing, or rebuilding ⟨the apartment has been entirely *renovated*⟩. REJUVENATE suggests the restoration of youthful vigor, powers, or appearance ⟨the change in jobs *rejuvenated* her spirits⟩.

re•nom•i•nate \(ˌ)rē-ˈnä-mə-ˌnāt\ *transitive verb* (1864)
: to nominate again especially for a succeeding term
— **re•nom•i•na•tion** \(ˌ)rē-ˌnä-mə-ˈnā-shən\ *noun*

re•nounce \ri-ˈnau̇n(t)s\ *verb* **re•nounced; re•nounc•ing** [Middle English, from Middle French *renoncer,* from Latin *renuntiare,* from *re-* + *nuntiare* to report, from *nuntius* messenger] (14th century)
transitive verb
1 : to give up, refuse, or resign usually by formal declaration ⟨*renounce* his errors⟩
2 : to refuse to follow, obey, or recognize any further **:** REPUDIATE ⟨*renounce* the authority of the church⟩
intransitive verb
1 : to make a renunciation
2 : to fail to follow suit in a card game
synonym see ABDICATE, ABJURE
— **re•nounce•ment** \-ˈnau̇n(t)s-mənt\ *noun*
— **re•nounc•er** *noun*

re•no•vas•cu•lar \ˌrē-nō-ˈvas-kyə-lər\ *adjective* (1961)
: of, relating to, or involving the blood vessels of the kidneys ⟨*renovascular* hypertension⟩

ren•o•vate \ˈre-nə-ˌvāt\ *transitive verb* **-vat•ed; -vat•ing** [Latin *renovatus,* past participle of *renovare,* from *re-* + *novare* to make new, from *novus* new — more at NEW] (circa 1522)
1 : to restore to a former better state (as by cleaning, repairing, or rebuilding)
2 : to restore to life, vigor, or activity **:** REVIVE ⟨the church was *renovated* by a new ecumenical spirit⟩
synonym see RENEW
— **ren•o•va•tion** \ˌre-nə-ˈvā-shən\ *noun*
— **ren•o•va•tive** \ˈre-nə-ˌvā-tiv\ *adjective*
— **ren•o•va•tor** \-ˌvā-tər\ *noun*

^{1}**re•nown** \ri-ˈnau̇n\ *noun* [Middle English, from Middle French *renon,* from Old French, from *renomer* to celebrate, from *re-* + *nomer* to name, from Latin *nominare,* from *nomin-, nomen* name — more at NAME] (14th century)
1 : a state of being widely acclaimed and highly honored **:** FAME
2 *obsolete* **:** REPORT, RUMOR

^{2}**renown** *transitive verb* (15th century)
: to give renown to

re•nowned *adjective* (14th century)
: having renown **:** CELEBRATED
synonym see FAMOUS

^{1}**rent** \ˈrent\ *noun* [Middle English *rente,* from Old French, income from a property, from (assumed) Vulgar Latin *rendita,* from feminine of *renditus,* past participle of *rendere* to yield — more at RENDER] (12th century)
1 : property (as a house) rented or for rent
2 a : a usually fixed periodical return made by a tenant or occupant of property to the owner for the possession and use thereof; *especially* **:** an agreed sum paid at fixed intervals by a tenant to the landlord **b :** the amount paid by a hirer of personal property to the owner for the use thereof
3 a : the portion of the income of an economy (as of a nation) attributable to land as a factor of production in addition to capital and labor **b :** ECONOMIC RENT
— **for rent :** available for use or service in return for payment

^{2}**rent** (15th century)
transitive verb
1 : to grant the possession and enjoyment of in exchange for rent
2 : to take and hold under an agreement to pay rent
intransitive verb
1 : to be for rent
2 a : to obtain the possession and use of a place or article in exchange for rent **b :** to allow the possession and use of property in exchange for rent
synonym see HIRE
— **rent•abil•i•ty** \ˌren-tə-ˈbi-lə-tē\ *noun*
— **rent•able** \ˈren-tə-bəl\ *adjective*

^{3}**rent** *past and past participle of* REND

^{4}**rent** *noun* [English dialect *rent* to rend, from Middle English, alteration of *renden* — more at REND] (1535)
1 : an opening made by or as if by rending
2 : a split in a party or organized group **:** SCHISM
3 : an act or instance of rending

rent–a–car \ˈrent-ə-ˌkär\ *noun* (1935)
: a rented car

^{1}**rent•al** \ˈren-t^{ə}l\ *noun* (14th century)
1 : an amount paid or collected as rent
2 : something that is rented
3 : an act of renting
4 : a business that rents something

^{2}**rental** *adjective* (15th century)
1 a : of or relating to rent **b :** available for rent
2 : dealing in rental property ⟨a *rental* agency⟩

rental library *noun* (1928)
: a commercially operated library (as in a store) that lends books at a fixed charge per book per day — called also *lending library*

rent control *noun* (1931)
: government regulation of the amount charged as rent for housing and often also of eviction
— **rent–controlled** *adjective*

rente \ˈrä^{n}(n)t\ *noun* [French] (1873)
: a government security (as in France) paying interest; *also* **:** the interest paid

rent•er \ˈren-tər\ *noun* (1655)
: one that rents; *specifically* **:** the lessee or tenant of property

ren•tier \rä^{n}-tyā\ *noun* [French, from Old French, from *rente*] (circa 1847)
: a person who lives on income from property or securities

rent strike *noun* (1964)
: a refusal by a group of tenants to pay rent (as in protest against high rates)

re•nun•ci•a•tion \ri-ˌnən(t)-sē-ˈā-shən\ *noun* [Middle English, from Latin *renuntiation-, renuntiatio,* from *renuntiare* to renounce] (14th century)
: the act or practice of renouncing **:** REPUDIATION; *specifically* **:** ascetic self-denial
— **re•nun•ci•a•tive** \ri-ˈnən(t)-sē-ˌā-tiv\ *adjective*
— **re•nun•ci•a•to•ry** \-sē-ə-ˌtōr-ē, -ˌtȯr-\ *adjective*

re•of•fer \(ˌ)rē-ˈȯ-fər, -ˈä-\ *transitive verb* (1920)
: to offer (a security issue) for public sale

re•open \(ˌ)rē-ˈō-pən, -ˈō-p^{ə}m\ (1733)
transitive verb
1 : to open again
2 a : to take up again **:** RESUME ⟨*reopen* discussion⟩ **b :** to resume discussion or consideration of ⟨*reopen* a contract⟩
3 : to begin again
intransitive verb
: to open again ⟨school *reopens* in September⟩

^{1}**re•or•der** \(ˌ)rē-ˈȯr-dər\ (1656)
transitive verb
1 : to arrange in a different way
2 : to give a reorder for
intransitive verb
: to place a reorder

^{2}**reorder** *noun* (1901)
: an order like a previous order placed with the same supplier

re•or•ga•ni•za•tion \(ˌ)rē-ˌȯr-gə-nə-ˈzā-shən, -ˌȯrg-nə-\ *noun* (1813)
: the act or process of reorganizing **:** the state of being reorganized; *especially* **:** the financial reconstruction of a business concern
— **re•or•ga•ni•za•tion•al** \-shnəl, -shə-n^{ə}l\ *adjective*

re•or•ga•nize \(ˌ)rē-ˈȯr-gə-ˌnīz\ (circa 1686)
transitive verb
: to organize again or anew
intransitive verb
: to reorganize something
— **re•or•ga•niz•er** *noun*

reo•vi•rus \ˌrē-ō-ˈvī-rəs\ *noun* [*r*espiratory *e*nteric *o*rphan (i.e., unidentified) *virus*] (1959)
: any of a group of double-stranded RNA viruses lacking a lipoprotein envelope that includes many pathogens of plants or animals

^{1}**rep** \ˈrep\ *noun* (circa 1705)
slang **:** REPUTATION; *especially* **:** status in a group (as a gang)

^{2}**rep** *noun* (1848)
: REPRESENTATIVE ⟨sales *reps*⟩

^{3}**rep** *or* **repp** \ˈrep\ *noun* [French *reps,* modification of English *ribs,* plural of *rib*] (1860)
: a plain-weave fabric with prominent rounded crosswise ribs

^{4}**rep** *noun* (1925)
: REPERTORY 2b, 3

^{5}**rep** *noun* [*r*oentgen *e*quivalent *p*hysical] (1947)
: the dosage of an ionizing radiation that will develop the same amount of energy upon absorption in human tissue as one roentgen of X-ray or gamma-ray exposure

^{6}**rep** *noun* [short for *repetition*] (1978)
: a repetition of a specified movement or exercise (as in weight lifting)

re•pack•age \(ˌ)rē-ˈpa-kij\ *transitive verb* (1946)
: to package again or anew; *specifically* **:** to put into a more efficient or attractive form
— **re•pack•ag•er** *noun*

^{1}**re•pair** \ri-ˈpar, -ˈper\ *intransitive verb* [Middle English, from Middle French *repairier* to go back to one's country, from Late Latin *repatriare,* from Latin *re-* + *patria* native country — more at EXPATRIATE] (14th century)
1 a : to betake oneself **:** GO ⟨*repaired* to the judge's chambers⟩ **b :** to come together **:** RALLY
2 *obsolete* **:** RETURN

^{2}**repair** *noun* (14th century)
1 : the act of repairing **:** RESORT
2 : a popular gathering place

^{3}**repair** *verb* [Middle English, from Middle French *reparer,* from Latin *reparare,* from *re-* + *parare* to prepare — more at PARE] (14th century)
transitive verb
1 a : to restore by replacing a part or putting together what is torn or broken **:** FIX ⟨*repair* a shoe⟩ **b :** to restore to a sound or healthy state **:** RENEW ⟨*repair* his strength⟩
2 : to make good **:** compensate for **:** REMEDY ⟨*repair* a gap in my reading⟩
intransitive verb
: to make repairs
synonym see MEND
— **re•pair•abil•i•ty** \-ˌpar-ə-ˈbi-lə-tē, -ˌper-\ *noun*
— **re•pair•able** \-ˈpar-ə-bəl, -ˈper-\ *adjective*
— **re•pair•er** \-ˈpar-ər, -ˈper-\ *noun*

^{4}**repair** *noun* (15th century)
1 a : an instance or result of repairing **b :** the act or process of repairing **c :** the replacement of destroyed cells or tissues by new formations
2 a : relative condition with respect to soundness or need of repairing **b :** the state of being in good or sound condition

re•pair•man \ri-ˈpar-ˌman, -ˈper-, -mən\ *noun* (1871)
: one who repairs; *specifically* **:** one whose occupation is to make repairs in a mechanism

re•pand \ri-ˈpand\ *adjective* [Latin *repandus* spread out, from *repandere* to open wide, from *re-* + *pandere* to spread — more at FATHOM] (circa 1760)
: having a slightly undulating margin ⟨a *repand* leaf⟩ ⟨a *repand* colony of bacteria⟩

rep•a•ra•ble \ˈre-p(ə-)rə-bəl\ *adjective* (1570)
: capable of being repaired

rep•a•ra•tion \ˌre-pə-ˈrā-shən\ *noun* [Middle English, from Middle French, from Late Latin *reparation-, reparatio,* from Latin *reparare*] (14th century)
1 a : a repairing or keeping in repair **b** *plural* **:** REPAIRS
2 a : the act of making amends, offering expiation, or giving satisfaction for a wrong or injury **b :** something done or given as amends or satisfaction
3 : the payment of damages **:** INDEMNIFICATION; *specifically* **:** compensation in money or materials payable by a defeated nation for damages to or expenditures sustained by another nation as a result of hostilities with the defeated nation — usually used in plural

re•par•a•tive \ri-ˈpar-ə-tiv\ *adjective* (1656)
1 : of, relating to, or effecting repair
2 : serving to make amends

rep•ar•tee \ˌre-pər-ˈtē, -ˌpär-, -ˈtā\ *noun* [French *repartie,* from *repartir* to retort, from Middle French, from *re-* + *partir* to divide — more at PART] (circa 1645)
1 a : a quick and witty reply **b :** a succession or interchange of clever retorts **:** amusing and usually light sparring with words
2 : adroitness and cleverness in reply **:** skill in repartee
synonym see WIT

¹re•par•ti•tion \ˌre-ˌpär-ˈti-shən, ˌrē-\ *noun* [probably from Spanish *repartición,* from *repartir* to distribute, from *re-* + *partir* to divide, from Latin *partire* — more at PART] (1555)
: DISTRIBUTION

²re•par•ti•tion \ˌrē-ˌpär-ˈti-shən\ *noun* [*re-* + *partition*] (1835)
: a second or additional dividing or distribution

re•pass \(ˌ)rē-ˈpas\ *verb* [Middle English, from Middle French *repasser,* from Old French, from *re-* + *passer* to pass] (15th century)
intransitive verb
: to pass again especially in the opposite direction **:** RETURN
transitive verb
1 : to pass through, over, or by again ⟨*repass* the house⟩
2 : to cause to pass again
3 : to adopt again ⟨*repassed* the resolution⟩
— re•pas•sage \-ˈpa-sij\ *noun*

¹re•past \ri-ˈpast, ˈrē-ˌ\ *noun* [Middle English, from Middle French, from Old French, from *repaistre* to feed, from *re-* + *paistre* to feed, from Latin *pascere* — more at FOOD] (14th century)
1 : something taken as food **:** MEAL
2 : the act or time of taking food

²re•past \ri-ˈpast\ (15th century)
transitive verb
obsolete **:** FEED
intransitive verb
: to take food **:** FEAST

re•pa•tri•ate \(ˌ)rē-ˈpā-trē-ˌāt, -ˈpa-\ *transitive verb* **-at•ed; -at•ing** [Late Latin *repatriatus,* past participle of *repatriare* to go back to one's country — more at REPAIR] (1611)
: to restore or return to the country of origin, allegiance, or citizenship ⟨*repatriate* prisoners of war⟩
— re•pa•tri•ate \-trē-ət, -trē-ˌāt\ *noun*
— re•pa•tri•a•tion \(ˌ)rē-ˌpā-trē-ˈā-shən, -ˌpa-\ *noun*

re•pay \(ˌ)rē-ˈpā\ *verb* **-paid** \-ˈpād\; **-pay•ing** (15th century)
transitive verb
1 a : to pay back ⟨*repay* a loan⟩ **b :** to give or inflict in return or requital ⟨*repay* evil for evil⟩
2 : to make a return payment to **:** COMPENSATE, REQUITE
3 : to make requital for **:** RECOMPENSE ⟨the success that *repays* hard work⟩
intransitive verb
: to make return payment or requital
synonym see PAY
— re•pay•able \-ˈpā-ə-bəl\ *adjective*
— re•pay•ment \-ˈpā-mənt\ *noun*

re•peal \ri-ˈpē(ə)l\ *transitive verb* [Middle English *repelen,* from Middle French *repeler,* from Old French, from *re-* + *apeler* to appeal, call] (14th century)
1 : to rescind or annul by authoritative act; *especially* **:** to revoke or abrogate by legislative enactment
2 : ABANDON, RENOUNCE
3 *obsolete* **:** to summon to return **:** RECALL
— repeal *noun*
— re•peal•able \-ˈpē-lə-bəl\ *adjective*

re•peal•er \ri-ˈpē-lər\ *noun* (1765)
: one that repeals; *specifically* **:** a legislative act that abrogates an earlier act

¹re•peat \ri-ˈpēt\ *verb* [Middle English *repeten,* from Middle French *repeter,* from Latin *repetere* to return to, repeat, from *re-* + *petere* to go to, seek — more at FEATHER] (14th century)
transitive verb
1 a : to say or state again **b :** to say over from memory **:** RECITE **c :** to say after another
2 a : to make, do, or perform again ⟨*repeat* an experiment⟩ **b :** to make appear again ⟨the curtains *repeat* the wallpaper pattern⟩ ⟨will *repeat* the program tomorrow⟩ **c :** to go through or experience again ⟨had to *repeat* third grade⟩
3 : to express or present (oneself) again in the same words, terms, or form
intransitive verb
: to say, do, or accomplish something again; *especially* **:** to win (as a sports championship) another time in succession
— re•peat•abil•i•ty \-ˌpē-tə-ˈbi-lə-tē\ *noun*
— re•peat•able \-ˈpē-tə-bəl\ *adjective*

²re•peat \ri-ˈpēt, ˈrē-ˌ\ *noun* (15th century)
1 a : something repeated **:** REPETITION **b :** a musical passage to be repeated in performance; *also* **:** a sign placed before and after such a passage **c :** a usually transcribed repetition of a radio or television program **d :** a genetic duplication in which the duplicated parts are adjacent to each other along the chromosome
2 : the act of repeating

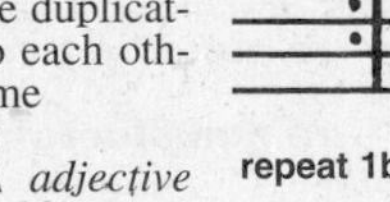
repeat 1b

re•peat•ed \ri-ˈpē-təd\ *adjective* (1611)
1 : renewed or recurring again and again ⟨*repeated* changes of plan⟩
2 : said, done, or presented again

re•peat•ed•ly *adverb* (circa 1718)
: AGAIN AND AGAIN

re•peat•er \ri-ˈpē-tər\ *noun* (1598)
: one that repeats: as **a :** one who relates or recites **b :** a watch or clock with a striking mechanism that upon pressure of a spring will indicate the time in hours or quarters and sometimes minutes **c :** a firearm having a magazine that holds a number of cartridges loaded one at a time into the chamber by the action of the piece **d :** an habitual violator of the laws **e :** one who votes illegally by casting more than one ballot in an election **f :** a student enrolled in a class or course for a second or subsequent time **g :** a device for receiving electronic communication signals and delivering corresponding amplified ones

re•peat•ing *adjective* (1824)
of a firearm **:** designed to load cartridges from a magazine

repeating decimal *noun* (1773)
: a decimal in which after a certain point a particular digit or sequence of digits repeats itself indefinitely — compare TERMINATING DECIMAL

re•pe•chage \ˌre-pə-ˈshäzh, ˈre-pə-ˌ\ *noun* [French *repêchage* second chance, reexamination for a candidate who has failed, from *repêcher* to fish out, rescue, from *re-* + *pêcher* to fish, from Latin *piscari* — more at PISCATORY] (1928)
: a trial heat (as in rowing) in which first-round losers are given another chance to qualify for the semifinals

re•pel \ri-ˈpel\ *verb* **re•pelled; re•pel•ling** [Middle English *repellen,* from Latin *repellere,* from *re-* + *pellere* to drive — more at FELT] (15th century)
transitive verb
1 a : to drive back **:** REPULSE **b :** to fight against **:** RESIST
2 : TURN AWAY, REJECT ⟨*repelled* the insinuation⟩
3 a : to drive away **:** DISCOURAGE ⟨foul words and frowns must not *repel* a lover —Shakespeare⟩ **b :** to be incapable of adhering to, mixing with, taking up, or holding **c :** to force away or apart or tend to do so by mutual action at a distance
4 : to cause aversion in **:** DISGUST
intransitive verb
: to cause aversion
— re•pel•ler *noun*

re•pel•len•cy \ri-ˈpe-lən(t)-sē\ *noun* (1747)
: the quality or capacity of repelling

¹re•pel•lent *also* **re•pel•lant** \ri-ˈpe-lənt\ *adjective* [Latin *repellent-, repellens,* present participle of *repellere*] (1643)
1 : serving or tending to drive away or ward off — often used in combination ⟨a mosquito-*repellent* spray⟩
2 : arousing aversion or disgust **:** REPULSIVE
— re•pel•lent•ly *adverb*

²repellent *also* **repellant** *noun* (1661)
: something that repels; *especially* **:** a substance that repels insects

¹re•pent \ri-ˈpent\ *verb* [Middle English, from Old French *repentir,* from *re-* + *pentir* to be sorry, from Latin *paenitēre* — more at PENITENT] (14th century)
intransitive verb
1 : to turn from sin and dedicate oneself to the amendment of one's life
2 a : to feel regret or contrition **b :** to change one's mind
transitive verb
1 : to cause to feel regret or contrition
2 : to feel sorrow, regret, or contrition for
— re•pent•er *noun*

²re•pent \ˈrē-pənt\ *adjective* [Latin *repent-, repens,* present participle of *repere* to creep — more at REPTILE] (1669)
: CREEPING, PROSTRATE ⟨*repent* stems⟩

re•pen•tance \ri-ˈpen-tᵊn(t)s\ *noun* (14th century)
: the action or process of repenting especially for misdeeds or moral shortcomings
synonym see PENITENCE

re•pen•tant \-tᵊnt\ *adjective* (13th century)
1 : experiencing repentance **:** PENITENT
2 : expressive of repentance
— re•pen•tant•ly *adverb*

re•per•cus•sion \ˌrē-pər-ˈkə-shən, ˌre-\ *noun* [Latin *repercussion-, repercussio,* from *repercutere* to drive back, from *re-* + *percutere* to beat — more at PERCUSSION] (1536)
1 : REFLECTION, REVERBERATION
2 a : an action or effect given or exerted in return **:** a reciprocal action or effect **b :** a widespread, indirect, or unforeseen effect of an act, action, or event — usually used in plural
— re•per•cus•sive \-ˈkə-siv\ *adjective*

rep•er•toire \ˈre-pə(r)-ˌtwär\ *noun* [French *répertoire,* from Late Latin *repertorium*] (1847)
1 a : a list or supply of dramas, operas, pieces, or parts that a company or person is prepared to perform **b :** a supply of skills, devices, or expedients ⟨part of the *repertoire* of a quarterback⟩; *broadly* **:** AMOUNT, SUPPLY ⟨an endless

repertoire of summer clothes⟩ **c :** a list or supply of capabilities ⟨the instruction *repertoire* of a computer⟩
2 a : the complete list or supply of dramas, operas, or musical works available for performance ⟨our modern orchestral *repertoire*⟩ **b :** the complete list or supply of skills, devices, or ingredients used in a particular field, occupation, or practice ⟨the *repertoire* of literary criticism⟩

rep•er•to•ry \'re-pə(r)-ˌtōr-ē, -ˌtȯr-\ *noun, plural* **-ries** [Late Latin *repertorium* list, from Latin *reperire* to find, from *re-* + *parere* to produce — more at PARE] (1593)
1 : a place where something may be found **:** REPOSITORY
2 a : REPERTOIRE **b :** a company that presents several different plays, operas, or pieces usually alternately in the course of a season at one theater **c :** a theater housing such a company
3 : the production and presentation of plays by a repertory company ⟨acting in *repertory*⟩

rep•e•tend \'re-pə-ˌtend\ *noun* [Latin *repetendus* to be repeated, gerundive of *repetere* to repeat] (1874)
: a repeated sound, word, or phrase; *specifically* **:** REFRAIN

rep•e•ti•tion \ˌre-pə-'ti-shən\ *noun* [Middle English *repeticioun,* from Latin *repetition-, repetitio,* from *repetere* to repeat] (15th century)
1 : the act or an instance of repeating or being repeated
2 : MENTION, RECITAL
— rep•e•ti•tion•al \-'tish-nəl, -'ti-shə-nᵊl\ *adjective*

rep•e•ti•tious \-'ti-shəs\ *adjective* (1675)
: characterized or marked by repetition; *especially* **:** tediously repeating
— rep•e•ti•tious•ly *adverb*
— rep•e•ti•tious•ness *noun*

re•pet•i•tive \ri-'pe-tə-tiv\ *adjective* (1839)
1 : containing repetition
2 : REPETITIOUS
— re•pet•i•tive•ly *adverb*
— re•pet•i•tive•ness *noun*

re•pine \ri-'pīn\ *intransitive verb* (circa 1530)
1 : to feel or express dejection or discontent
2 : to long for something
— re•pin•er *noun*

re•place \ri-'plās\ *transitive verb* (1595)
1 : to restore to a former place or position ⟨*replace* cards in a file⟩
2 : to take the place of especially as a substitute or successor
3 : to put something new in the place of ⟨*replace* a worn carpet⟩ ☆
— re•place•able \-'plā-sə-bəl\ *adjective*
— re•plac•er *noun*

re•place•ment \ri-'plās-mənt\ *noun* (circa 1790)
1 : the action or process of replacing **:** the state of being replaced
2 : one that replaces another especially in a job or function

replacement set *noun* (1959)
: a set of elements any one of which may be used to replace a given variable or placeholder in a mathematical sentence or expression (as an equation)

re•plant \(ˌ)rē-'plant\ *transitive verb* (1575)
1 : to plant again or anew
2 : to provide with new plants
3 : to subject to replantation

re•plan•ta•tion \ˌrē-(ˌ)plan-'tā-shən\ *noun* (1870)
: reattachment or reinsertion of a bodily part (as a limb or tooth) after separation from the body

¹re•play \(ˌ)rē-'plā\ *transitive verb* (1884)
: to play again or over

²re•play \'rē-ˌplā\ *noun* (1895)
1 a : an act or instance of replaying **b :** the playing of a tape (as a videotape)
2 : REPETITION, REENACTMENT ⟨don't want a *replay* of our old mistakes⟩

re•plead•er \(ˌ)rē-'plē-dər\ *noun* [*replead* to plead again + *-er* (as in *misnomer*)] (1607)
1 : a second legal pleading
2 : the right of pleading again granted usually when the issue raised is immaterial or insufficient

re•plen•ish \ri-'ple-nish\ *verb* [Middle English *replenisshen,* from Middle French *repleniss-,* stem of *replenir* to fill, from Old French, from *re-* + *plein* full, from Latin *plenus* — more at FULL] (14th century)
transitive verb
1 a : to fill with persons or animals **:** STOCK **b** *archaic* **:** to supply fully **:** PERFECT **c :** to fill with inspiration or power **:** NOURISH
2 a : to fill or build up again ⟨*replenished* his glass⟩ **b :** to make good **:** REPLACE
intransitive verb
: to become full **:** fill up again
— re•plen•ish•able \-ni-shə-bəl\ *adjective*
— re•plen•ish•er *noun*
— re•plen•ish•ment \-nish-mənt\ *noun*

re•plete \ri-'plēt\ *adjective* [Middle English, from Middle French & Latin; Middle French *replet,* from Latin *repletus,* past participle of *replēre* to fill up, from *re-* + *plēre* to fill — more at FULL] (14th century)
1 : fully or abundantly provided or filled ⟨a book *replete* with . . . delicious details —William Safire⟩
2 a : abundantly fed **b :** FAT, STOUT
3 : COMPLETE
synonym see FULL
— re•plete•ness *noun*

re•ple•tion \ri-'plē-shən\ *noun* (14th century)
1 : the act of eating to excess **:** the state of being fed to excess **:** SURFEIT
2 : the condition of being filled up or overcrowded
3 : fulfillment of a need or desire **:** SATISFACTION

¹re•plev•in \ri-'ple-vən\ *noun* [Middle English, from Anglo-French *replevine,* from *replevir* to give security, from Old French, from *re-* + *plevir* to pledge, from (assumed) Late Latin *plebere* — more at PLEDGE] (15th century)
1 : the recovery by a person of goods or chattels claimed to be wrongfully taken or detained upon the person's giving security to try the matter in court and return the goods if defeated in the action
2 : the writ or the common-law action whereby goods and chattels are replevied

²replevin *transitive verb* (1678)
: REPLEVY

¹re•plevy \ri-'ple-vē\ *noun, plural* **re•plev•ies** [Middle English, from Anglo-French *replevir,* verb] (15th century)
: REPLEVIN

²replevy *transitive verb* **re•plev•ied; re•plevy•ing** (1596)
: to take or get back by a writ for replevin
— re•plevi•able \-vē-ə-bəl\ *adjective*

rep•li•ca \'re-pli-kə\ *noun* [Italian, repetition, from *replicare* to repeat, from Late Latin, from Latin, to fold back — more at REPLY] (1824)
1 : an exact reproduction (as of a painting) executed by the original artist ⟨a *replica* of this was painted . . . this year —Constance Strachey⟩
2 : a copy exact in all details ⟨DNA makes a *replica* of itself⟩ ⟨sailed a *replica* of the Viking ship⟩; *broadly* **:** COPY ⟨this faithful, pathetic *replica* of a Midwestern suburb —G. F. Kennan⟩ ☐
synonym see REPRODUCTION

rep•li•ca•ble \'re-plə-kə-bəl\ *adjective* (1950)
: capable of replication ⟨*replicable* experimental results⟩
— rep•li•ca•bil•i•ty \ˌre-plə-kə-'bi-lə-tē\ *noun*

rep•li•case \'re-pli-ˌkās, -ˌkāz\ *noun* [*replica*tion + *-ase*] (1963)
: a polymerase that promotes synthesis of a particular RNA in the presence of a template of RNA

¹rep•li•cate \'re-plə-ˌkāt\ *verb* **-cat•ed; -cat•ing** [Late Latin *replicatus,* past participle of *replicare*] (15th century)
transitive verb
: DUPLICATE, REPEAT ⟨*replicate* a statistical experiment⟩
intransitive verb
: to undergo replication **:** produce a replica of itself ⟨virus particles *replicating* in cells⟩

²rep•li•cate \-kət\ *adjective* (1922)
: MANIFOLD, REPEATED

³rep•li•cate \-kət\ *noun* (1929)
: one of several identical experiments, procedures, or samples

rep•li•ca•tion \ˌre-plə-'kā-shən\ *noun* (14th century)
1 a : ANSWER, REPLY **b** (1) **:** an answer to a reply **:** REJOINDER (2) **:** a plaintiff's reply to a defendant's plea, answer, or counterclaim
2 : ECHO, REVERBERATION
3 a : COPY, REPRODUCTION **b :** the action or process of reproducing
4 : performance of an experiment or procedure more than once; *especially* **:** systematic or random repetition of agricultural test rows or plats to reduce error

rep•li•ca•tive \'re-pli-ˌkā-tiv\ *adjective* (circa 1890)
: of, relating to, involved in, or characterized by replication ⟨the *replicative* form of tobacco mosaic virus⟩

rep•li•con \'re-pli-ˌkän\ *noun* [*replic*ate + ²*-on*] (1963)
: a linear or circular section of DNA or RNA which replicates sequentially as a unit

¹re•ply \ri-'plī\ *verb* **re•plied; re•ply•ing** [Middle English *replien,* from Middle French *replier* to fold again, from Latin *replicare* to fold back, from *re-* + *plicare* to fold — more at PLY] (14th century)
intransitive verb
1 a : to respond in words or writing **b :** ECHO, RESOUND **c :** to make a legal replication

☆ SYNONYMS
Replace, displace, supplant, supersede mean to put out of a usual or proper place or into the place of another. REPLACE implies a filling of a place once occupied by something lost, destroyed, or no longer usable or adequate ⟨*replaced* the broken window⟩. DISPLACE implies an ousting or dislodging ⟨war had *displaced* thousands⟩. SUPPLANT implies either a dispossessing or usurping of another's place, possessions, or privileges or an uprooting of something and its replacement with something else ⟨was abruptly *supplanted* in her affections by another⟩. SUPERSEDE implies replacing a person or thing that has become superannuated, obsolete, or otherwise inferior ⟨the new edition *supersedes* all previous ones⟩.

☐ USAGE
replica A number of language commentators would have us believe that sense 1 is the only proper meaning of the word. But their battle was lost more than a century ago; the word has been applied to various kinds of copies since the second half of the 19th century. Sense 1 is, in fact, the least common sense in current usage.

2 : to do something in response; *specifically* **:** to return gunfire or an attack
transitive verb
: to give as an answer
synonym see ANSWER
— re·pli·er \-ˈplī(-ə)r\ *noun*

[2]reply *noun, plural* **replies** (1560)
1 : something said, written, or done in answer or response
2 : REPLICATION 1b(2)

re·po \ˈrē-ˌpō\ *noun, plural* **repos** [by shortening & alteration] (1963)
: REPURCHASE AGREEMENT

re·po·lar·i·za·tion \ˌrē-ˌpō-lə-rə-ˈzā-shən\ *noun* (1922)
: polarization of a muscle fiber, cell, or membrane following depolarization
— re·po·lar·ize \(ˌ)rē-ˈpō-lə-ˌrīz\ *verb*

[1]re·port \ri-ˈpōrt, -ˈpȯrt\ *noun* [Middle English, from Middle French, from Old French, from *reporter* to report, from Latin *reportare,* from *re-* + *portare* to carry — more at FARE] (14th century)
1 a : common talk or an account spread by common talk **:** RUMOR **b :** quality of reputation ⟨a witness of good *report*⟩
2 a : a usually detailed account or statement ⟨a news *report*⟩ **b :** an account or statement of a judicial opinion or decision **c :** a usually formal record of the proceedings of a meeting or session
3 : an explosive noise
— on report : subject to disciplinary action

[2]report (14th century)
transitive verb
1 a : to give an account of **:** RELATE **b :** to describe as being in a specified state ⟨*reported* him much improved⟩
2 a : to serve as carrier of (a message) **b :** to relate the words or sense of (something said) **c :** to make a written record or summary of **d** (1) **:** to watch for and write about the newsworthy aspects or developments of **:** COVER (2) **:** to prepare or present an account of for broadcast
3 a (1) **:** to give a formal or official account or statement of ⟨the treasurer *reported* a balance of ten dollars⟩ (2) **:** to return or present (a matter referred for consideration) with conclusions or recommendations **b :** to announce or relate as the result of investigation ⟨*reported* no sign of disease⟩ **c :** to announce the presence, arrival, or sighting of **d :** to make known to the proper authorities ⟨*report* a fire⟩ **e :** to make a charge of misconduct against
intransitive verb
1 a : to give an account **:** TELL **b :** to present oneself **c :** to account for oneself ⟨*reported* sick on Friday⟩
2 : to make, issue, or submit a report
3 : to act in the capacity of a reporter

re·port·able \ri-ˈpōr-tə-bəl, -ˈpȯr-\ *adjective* (1858)
1 : worth reporting ⟨*reportable* news⟩
2 : required by law to be reported ⟨*reportable* income⟩ ⟨*reportable* diseases⟩

re·port·age \ri-ˈpōr-tij, -ˈpȯr-, *especially for 2* ˌre-pər-ˈtäzh, ˌre-ˌpȯr-ˈ\ *noun* [French, from *reporter* to report] (circa 1864)
1 a : the act or process of reporting news **b :** something (as news) that is reported
2 : writing intended to give an account of observed or documented events

report card *noun* (1920)
1 : a report on a student that is periodically submitted by a school to the student's parents or guardian
2 : an evaluation of performance

re·port·ed·ly \ri-ˈpōr-təd-lē, -ˈpȯr-\ *adverb* (1901)
: according to report

re·port·er \ri-ˈpōr-tər, -ˈpȯr-\ *noun* (14th century)
: one that reports: as **a :** one who makes authorized statements of law decisions or legislative proceedings **b :** one who makes a shorthand record of a speech or proceeding **c** (1) **:** one employed by a newspaper, magazine, or television company to gather and report news (2) **:** one who broadcasts news
— re·por·to·ri·al \ˌre-pə(r)-ˈtōr-ē-əl, ˌrē-, -ˈtȯr-\ *adjective*
— re·por·to·ri·al·ly \-ē-ə-lē\ *adverb*

report out *transitive verb* (1907)
: to return after consideration and often with revisions to a legislative body for action ⟨after much debate the committee *reported* the bill *out*⟩

report stage *noun* (circa 1906)
: the stage in the British legislative process preceding the third reading and concerned especially with amendments and details

re·pos·al \ri-ˈpō-zəl\ *noun* (1605)
obsolete **:** the act of reposing

[1]re·pose \ri-ˈpōz\ *verb* **re·posed; re·pos·ing** [Middle English, from Middle French *reposer,* from Old French, from Late Latin *repausare,* from Latin *re-* + Late Latin *pausare* to stop, from Latin *pausa* pause] (15th century)
transitive verb
: to lay at rest
intransitive verb
1 a : to lie at rest **b :** to lie dead ⟨*reposing* in state⟩ **c :** to remain still or concealed
2 : to take a rest
3 *archaic* **:** RELY
4 : to rest for support **:** LIE

[2]repose *noun* (1509)
1 a : a state of resting after exertion or strain; *especially* **:** rest in sleep **b :** eternal or heavenly rest ⟨pray for the *repose* of a soul⟩
2 a : a place of rest **b :** PEACE, TRANQUILLITY ⟨the *repose* of the bayous⟩ **c :** a harmony in the arrangement of parts and colors that is restful to the eye
3 a : lack of activity **:** QUIESCENCE **b :** cessation or absence of activity, movement, or animation ⟨the face in *repose* is grave and thoughtful⟩
4 : composure of manner **:** POISE

[3]re·pose *transitive verb* **re·posed; re·pos·ing** [Middle English, to replace, from Latin *reponere* (perfect indicative *reposui*)] (15th century)
1 *archaic* **:** to put away or set down **:** DEPOSIT
2 a : to place (as confidence or trust) in someone or something **b :** to place for control, management, or use

re·pose·ful \ri-ˈpōz-fəl\ *adjective* (1852)
: of a kind to induce ease and relaxation
— re·pose·ful·ly \-fə-lē\ *adverb*
— re·pose·ful·ness *noun*

re·pos·it *transitive verb* **re·pos·it·ed** \-ˈpä-zə-təd, -ˈpäz-təd\; **re·pos·it·ing** \-ˈpä-zə-tiŋ, -ˈpäz-tiŋ\ [Latin *repositus,* past participle of *reponere* to replace, from *re-* + *ponere* to place — more at POSITION] (circa 1641)
1 \ri-ˈpä-zət\ **:** DEPOSIT, STORE
2 \(ˌ)rē-\ **:** to put back in place **:** REPLACE

[1]re·po·si·tion \ˌrē-pə-ˈzi-shən, ˌre-\ *noun* (1588)
: the act of repositing **:** the state of being reposited

[2]re·po·si·tion \ˌrē-pə-ˈzi-shən\ *transitive verb* (circa 1859)
1 : to change the position of
2 : to revise the marketing strategy of (a product or a company) so as to increase sales

[1]re·pos·i·to·ry \ri-ˈpä-zə-ˌtōr-ē, -ˌtȯr-\ *noun, plural* **-ries** (15th century)
1 : a place, room, or container where something is deposited or stored **:** DEPOSITORY
2 : a side altar in a Roman Catholic church where the consecrated Host is reserved from Maundy Thursday until Good Friday
3 : one that contains or stores something nonmaterial ⟨considered the book a *repository* of knowledge⟩
4 : a place or region richly supplied with a natural resource
5 : a person to whom something is confided or entrusted

[2]repository *adjective* (1950)
of a drug **:** designed to act over a prolonged period ⟨*repository* penicillin⟩

re·pos·sess \ˌrē-pə-ˈzes *also* -ˈses\ *transitive verb* (15th century)
1 a : to regain possession of **b :** to resume possession of in default of the payment of installments due
2 : to restore to possession
— re·pos·ses·sion \-ˈze-shən *also* -ˈse-\ *noun*
— re·pos·sess·or \-ˈze-sər *also* -ˈse-\ *noun*

[1]re·pous·sé \rə-ˌpü-ˈsā, -ˈpü-ˌ\ *adjective* [French, literally, pushed back] (1858)
1 : shaped or ornamented with patterns in relief made by hammering or pressing on the reverse side — used especially of metal
2 : formed in relief

[2]repoussé *noun* (1858)
1 : repoussé work
2 : repoussé decoration

re·pow·er \(ˌ)rē-ˈpau̇(-ə)r\ *transitive verb* (1954)
: to provide again or anew with power; *especially* **:** to provide (as a boat) with a new engine

repp *variant of* REP

rep·re·hend \ˌre-pri-ˈhend\ *transitive verb* [Middle English, from Latin *reprehendere,* literally, to hold back, from *re-* + *prehendere* to grasp — more at GET] (14th century)
: to voice disapproval of **:** CENSURE
synonym see CRITICIZE

rep·re·hen·si·ble \ˌre-pri-ˈhen(t)-sə-bəl\ *adjective* (14th century)
: worthy of or deserving reprehension **:** CULPABLE
— rep·re·hen·si·bil·i·ty \-ˌhen(t)-sə-ˈbi-lə-tē\ *noun*
— rep·re·hen·si·ble·ness \-ˈhen(t)-sə-bəl-nəs\ *noun*
— rep·re·hen·si·bly \-blē\ *adverb*

rep·re·hen·sion \-ˈhen(t)-shən\ *noun* [Middle English *reprehensioun,* from Middle French or Latin; Middle French *reprehension,* from Latin *reprehension-, reprehensio,* from *reprehendere*] (14th century)
: the act of reprehending **:** CENSURE

rep·re·hen·sive \-ˈhen(t)-siv\ *adjective* (1589)
: serving to reprehend **:** conveying reprehension or reproof

rep·re·sent \ˌre-pri-ˈzent\ *verb* [Middle English, from Middle French *representer,* from Latin *repraesentare,* from *re-* + *praesentare* to present] (14th century)
transitive verb
1 : to bring clearly before the mind **:** PRESENT ⟨a book which *represents* the character of early America⟩
2 : to serve as a sign or symbol of ⟨the flag *represents* our country⟩
3 : to portray or exhibit in art **:** DEPICT
4 : to serve as the counterpart or image of **:** TYPIFY ⟨a movie hero who *represents* the ideals of the culture⟩
5 a : to produce on the stage **b :** to act the part or role of
6 a (1) **:** to take the place of in some respect (2) **:** to act in the place of or for usually by legal right **b :** to serve especially in a legislative body by delegated authority usually resulting from election
7 : to describe as having a specified character or quality ⟨*represents* himself as a friend⟩
8 a : to give one's impression and judgment of **:** state in a manner intended to affect action or judgment **b :** to point out in protest or remonstrance
9 : to serve as a specimen, example, or instance of

10 a : to form an image or representation of in the mind **b** (1) **:** to apprehend (an object) by means of an idea (2) **:** to recall in memory
11 : to correspond to in essence **:** CONSTITUTE
intransitive verb
: to make representations against something **:** PROTEST
— rep·re·sent·able \-'zen-tə-bəl\ *adjective*
— rep·re·sent·er *noun*

re-pre·sent \ˌrē-pri-'zent\ *transitive verb* (1564)
: to present again or anew
— re-pre·sen·ta·tion \ˌrē-ˌprē-ˌzen-'tā-shən, -ˌprē-zᵊn-, -ˌprē-zᵊn-\ *noun*

rep·re·sen·ta·tion \ˌre-pri-ˌzen-'tā-shən, -zən-\ *noun* (15th century)
1 : one that represents: as **a :** an artistic likeness or image **b** (1) **:** a statement or account made to influence opinion or action (2) **:** an incidental or collateral statement of fact on the faith of which a contract is entered into **c :** a dramatic production or performance **d** (1) **:** a usually formal statement made against something or to effect a change (2) **:** a usually formal protest
2 : the act or action of representing **:** the state of being represented: as **a :** REPRESENTATIONALISM 2 **b** (1) **:** the action or fact of one person standing for another so as to have the rights and obligations of the person represented (2) **:** the substitution of an individual or class in place of a person (as a child for a deceased parent) **c :** the action of representing or the fact of being represented especially in a legislative body
3 : the body of persons representing a constituency
— rep·re·sen·ta·tion·al \-shnəl, -shə-nᵊl\ *adjective*
— rep·re·sen·ta·tion·al·ly *adverb*

rep·re·sen·ta·tion·al·ism \-shnə-ˌli-zəm, -shə-nᵊl-ˌi-\ *noun* (1899)
1 : the doctrine that the immediate object of knowledge is an idea in the mind distinct from the external object which is the occasion of perception
2 : the theory or practice of realistic representation in art
— rep·re·sen·ta·tion·al·ist \-list, -ist\ *noun*

[1]rep·re·sen·ta·tive \ˌre-pri-'zen-tə-tiv\ *adjective* (1532)
1 : serving to represent
2 a : standing or acting for another especially through delegated authority **b :** of, based on, or constituting a government in which the many are represented by persons chosen from among them usually by election
3 : serving as a typical or characteristic example ⟨a *representative* moviegoer⟩
4 : of or relating to representation or representationalism
— rep·re·sen·ta·tive·ly *adverb*
— rep·re·sen·ta·tive·ness *noun*
— rep·re·sen·ta·tiv·i·ty \-ˌzen-tə-'ti-və-tē\ *noun*

[2]representative *noun* (1647)
1 : a typical example of a group, class, or quality **:** SPECIMEN
2 : one that represents another or others: as **a** (1) **:** one that represents a constituency as a member of a legislative body (2) **:** a member of the house of representatives of the U.S. Congress or a state legislature **b :** one that represents another as agent, deputy, substitute, or delegate usually being invested with the authority of the principal **c :** one that represents a business organization **d :** one that represents another as successor or heir

re·press \ri-'pres\ *verb* [Middle English, from Latin *repressus,* past participle of *reprimere* to check, from *re-* + *premere* to press — more at PRESS] (14th century)
transitive verb
1 a : to check by or as if by pressure **:** CURB ⟨injustice was *repressed*⟩ **b :** to put down by force **:** SUBDUE ⟨*repress* a disturbance⟩
2 a : to hold in by self-control ⟨*repressed* a laugh⟩ **b :** to prevent the natural or normal expression, activity, or development of ⟨*repressed* her anger⟩
3 : to exclude from consciousness
4 : to inactivate (a gene or formation of a gene product) by allosteric combination at a DNA binding site
intransitive verb
: to take repressive action
— re·press·ibil·i·ty \-ˌpre-sə-'bi-lə-tē\ *noun*
— re·press·ible \-'pre-sə-bəl\ *adjective*
— re·pres·sive \-'pre-siv\ *adjective*
— re·pres·sive·ly *adverb*
— re·pres·sive·ness *noun*

re-press \(ˌ)rē-'pres\ *transitive verb* (14th century)
: to press again ⟨*re-press* a record⟩

re·pressed \ri-'prest\ *adjective* (1665)
1 : subjected to or marked by repression
2 : characterized by restraint

re·pres·sion \ri-'pre-shən\ *noun* (1533)
1 a : the action or process of repressing **:** the state of being repressed ⟨*repression* of unpopular opinions⟩ **b :** an instance of repressing ⟨racial *repressions*⟩
2 a : a process by which unacceptable desires or impulses are excluded from consciousness and left to operate in the unconscious **b :** an item so excluded
— re·pres·sion·ist \-sh(ə-)nist\ *adjective*

re·pres·sor \ri-'pre-sər\ *noun* [New Latin] (1611)
: one that represses; *especially* **:** a protein that is determined by a regulatory gene, binds to a genetic operator, and inhibits the initiation of transcription of messenger RNA

re·priev·al \ri-'prē-vəl\ *noun* (circa 1586)
archaic **:** REPRIEVE

[1]re·prieve \ri-'prēv\ *transitive verb* **re·prieved; re·priev·ing** [alteration of earlier *repry,* perhaps from Middle French *repris,* past participle of *reprendre* to take back] (1596)
1 : to delay the punishment of (as a condemned prisoner)
2 : to give relief or deliverance to for a time

[2]reprieve *noun* (1592)
1 a : the act of reprieving **:** the state of being reprieved **b :** a formal temporary suspension of the execution of a sentence especially of death
2 : an order or warrant for a reprieve
3 : a temporary respite (as from pain or trouble)

[1]rep·ri·mand \'re-prə-ˌmand\ *noun* [French *réprimande,* from Latin *reprimenda,* feminine of *reprimendus,* gerundive of *reprimere* to check — more at REPRESS] (1636)
: a severe or formal reproof

[2]reprimand *transitive verb* (1681)
: to reprove sharply or censure formally usually from a position of authority
synonym see REPROVE

[1]re·print \(ˌ)rē-'print\ *transitive verb* (1551)
: to print again **:** make a reprint of

[2]re·print \'rē-ˌprint, (ˌ)rē-'\ *noun* (1611)
: a reproduction of printed matter: as **a :** a subsequent printing of a book already published that preserves the identical text of the previous printing **b :** OFFPRINT **c :** matter (as an article) that has appeared in print before

re·print·er \(ˌ)rē-'prin-tər\ *noun* (1689)
: one that publishes a reprint

re·pri·sal \ri-'prī-zəl\ *noun* [Middle English *reprisail,* from Middle French *reprisaille,* from Old Italian *ripresaglia,* from *ripreso,* past participle of *riprendere* to take back, from *ri-* re- (from Latin *re-*) + *prendere* to take, from Latin *prehendere* — more at GET] (15th century)
1 a : the act or practice in international law of resorting to force short of war in retaliation for damage or loss suffered **b :** an instance of such action
2 *obsolete* **:** PRIZE
3 : the regaining of something (as by recapture)
4 : something (as a sum of money) given or paid in restitution — usually used in plural
5 : a retaliatory act

[1]re·prise \ri-'prēz, *1 is also* -'prīz\ *noun* [Middle English, from Middle French, literally, action of taking back, from Old French, from *reprendre* to take back, from *re-* + *prendre* to take, from Latin *prehendere*] (15th century)
1 : a deduction or charge made yearly out of a manor or estate — usually used in plural
2 : a recurrence, renewal, or resumption of an action
3 a : a musical repetition: (1) **:** the repetition of the exposition preceding the development (2) **:** RECAPITULATION **b :** a repeated performance **:** REPETITION

[2]re·prise \ri-'prīz, *3 is* -'prēz\ *transitive verb* **re·prised; re·pris·ing** [Middle French *reprise* action of taking back] (15th century)
1 *archaic* **:** TAKE BACK; *especially* **:** to recover by force
2 *archaic* **:** COMPENSATE
3 a : to repeat the performance of **b :** RECAPITULATE

re·pris·ti·nate \(ˌ)rē-'pris-tə-ˌnāt\ *transitive verb* **-nat·ed; -nat·ing** [*re-* + *pristine* + [4]*-ate*] (1659)
: to restore to an original state or condition
— re·pris·ti·na·tion \(ˌ)rē-ˌpris-tə-'nā-shən\ *noun*

re·pro \'rē-(ˌ)prō\ *noun, plural* **repros** [short for *reproduction*] (1946)
1 : a clear sharp proof made especially from a letterpress printing surface to serve as photographic copy for a printing plate
2 : REPRODUCTION 2

[1]re·proach \ri-'prōch\ *noun* [Middle English *reproche,* from Middle French, from Old French, from *reprochier* to reproach, from (assumed) Vulgar Latin *repropiare,* from Latin *re-* + *prope* near — more at APPROACH] (14th century)
1 : an expression of rebuke or disapproval
2 : the act or action of reproaching or disapproving ⟨was beyond *reproach*⟩
3 a : a cause or occasion of blame, discredit, or disgrace **b :** DISCREDIT, DISGRACE
4 *obsolete* **:** one subjected to censure or scorn
— re·proach·ful \-fəl\ *adjective*
— re·proach·ful·ly \-fə-lē\ *adverb*
— re·proach·ful·ness *noun*

[2]reproach *transitive verb* (14th century)
1 : to express disappointment in or displeasure with (a person) for conduct that is blameworthy or in need of amendment
2 : to make (something) a matter of reproach
3 : to bring into discredit
synonym see REPROVE
— re·proach·able \-'prō-chə-bəl\ *adjective*
— re·proach·er *noun*
— re·proach·ing·ly \-'prō-chiŋ-lē\ *adverb*

rep·ro·bance \'re-prə-bən(t)s\ *noun* (1604)
archaic **:** REPROBATION

[1]rep·ro·bate \'re-prə-ˌbāt\ *transitive verb* **-bat·ed; -bat·ing** [Middle English, from Late Latin *reprobatus,* past participle of *reprobare* — more at REPROVE] (15th century)
1 : to condemn strongly as unworthy, unacceptable, or evil ⟨*reprobating* the laxity of the age⟩
2 : to foreordain to damnation
3 : to refuse to accept **:** REJECT
synonym see CRITICIZE
— rep·ro·ba·tive \'re-prə-ˌbā-tiv\ *adjective*

— **rep•ro•ba•to•ry** \-bə-ˌtōr-ē, -ˌtȯr-\ *adjective*

²reprobate *adjective* (15th century)
1 *archaic* **:** rejected as worthless or not standing a test **:** CONDEMNED
2 a : foreordained to damnation **b :** morally corrupt **:** DEPRAVED
3 : expressing or involving reprobation
4 : of, relating to, or characteristic of a reprobate

³reprobate *noun* (1545)
: a reprobate person

rep•ro•ba•tion \ˌre-prə-ˈbā-shən\ *noun* (14th century)
: the act of reprobating or the state of being reprobated

re•pro•cess \(ˌ)rē-ˈprä-ˌses, -ˈprō-, -səs\ *transitive verb* (1921)
: to subject to a special process or treatment in preparation for reuse; *especially* **:** to extract uranium and plutonium from (the spent fuel rods of a nuclear reactor) for use again as fuel

re•pro•duce \ˌrē-prə-ˈdüs, -ˈdyüs\ (1611)
transitive verb
: to produce again: as **a :** to produce (new individuals of the same kind) by a sexual or asexual process **b :** to cause to exist again or anew ⟨*reproduce* water from steam⟩ **c :** to imitate closely ⟨sound-effects can *reproduce* the sound of thunder⟩ **d :** to present again **e :** to make a representation (as an image or copy) of ⟨*reproduce* a face on canvas⟩ **f :** to revive mentally **:** RECALL **g :** to translate (a recording) into sound
intransitive verb
1 : to undergo reproduction
2 : to produce offspring
— **re•pro•duc•er** *noun*
— **re•pro•duc•ibil•i•ty** \-ˌdü-sə-ˈbi-lə-tē, -ˌdyü-\ *noun*
— **re•pro•duc•ible** \-ˈdü-sə-bəl, -ˈdyü-\ *adjective or noun*
— **re•pro•duc•ibly** \-blē\ *adverb*

re•pro•duc•tion \ˌrē-prə-ˈdək-shən\ *noun* (1659)
1 : the act or process of reproducing; *specifically* **:** the process by which plants and animals give rise to offspring and which fundamentally consists of the segregation of a portion of the parental body by a sexual or an asexual process and its subsequent growth and differentiation into a new individual
2 : something reproduced **:** COPY
3 : young seedling trees in a forest ☆

reproduction proof *noun* (1945)
: REPRO 1

¹re•pro•duc•tive \ˌrē-prə-ˈdək-tiv\ *adjective* (1753)
: of, relating to, or capable of reproduction
— **re•pro•duc•tive•ly** *adverb*

²reproductive *noun* (1934)
: an actual or potential parent; *specifically* **:** a sexually functional social insect

re•pro•gram \(ˌ)rē-ˈprō-ˌgram, -grəm\ (1959)
transitive verb
: to program anew; *especially* **:** to write new programs for (as a computer)
intransitive verb
: to rewrite or revise a program especially of a computer
— **re•pro•gram•ma•ble** \-ˈprō-ˌgra-mə-bəl, -ˌprō-ˈgra-\ *adjective*

re•prog•ra•phy \ri-ˈprä-grə-fē\ *noun* [*repro*duction + *-graphy*] (1956)
: facsimile reproduction (as by photocopying) of graphic matter
— **re•prog•ra•pher** \-grə-fər\ *noun*
— **re•pro•graph•ic** \ˌrē-prə-ˈgra-fik, ˌre-\ *adjective*
— **re•pro•graph•ics** \-fiks\ *noun plural*

re•proof \ri-ˈprüf\ *noun* [Middle English *reprof,* from Middle French *reprove,* from Old French, from *reprover*] (14th century)
: criticism for a fault **:** REBUKE

re•prove \ri-ˈprüv\ *verb* **re•proved; re•prov•ing** [Middle English, from Middle French *reprover,* from Late Latin *reprobare* to disapprove, condemn, from Latin *re-* + *probare* to test, approve — more at PROVE] (14th century)
transitive verb
1 : to scold or correct usually gently or with kindly intent
2 : to express disapproval of **:** CENSURE ⟨it is not for me to *reprove* popular taste —D. W. Brogan⟩
3 *obsolete* **:** DISPROVE, REFUTE
4 *obsolete* **:** CONVINCE, CONVICT
intransitive verb
: to express rebuke or reproof ☆
— **re•prov•er** *noun*
— **re•prov•ing•ly** \-ˈprü-viŋ-lē\ *adverb*

¹rep•tile \ˈrep-ˌtīl, -tᵊl\ *noun* [Middle English *reptil,* from Middle French or Late Latin; Middle French *reptile* (feminine), from Late Latin *reptile* (neuter), from neuter of *reptilis* creeping, from Latin *reptus,* past participle of *repere* to crawl; akin to Lithuanian *rėplioti* to crawl] (14th century)
1 : an animal that crawls or moves on its belly (as a snake) or on small short legs (as a lizard)
2 : any of a class (Reptilia) of air-breathing vertebrates that include the alligators and crocodiles, lizards, snakes, turtles, and extinct related forms (as dinosaurs and pterosaurs) and are characterized by a completely ossified skeleton with a single occipital condyle, a distinct quadrate bone usually immovably articulated with the skull, ribs attached to the sternum, and a body usually covered with scales or bony plates
3 : a groveling or despised person

²reptile *adjective* (1607)
: characteristic of a reptile **:** REPTILIAN

¹rep•til•ian \rep-ˈti-lē-ən, -ˈtil-yən\ *adjective* (circa 1846)
1 : resembling or having the characteristics of the reptiles
2 : of or relating to the reptiles
3 : cold-bloodedly treacherous ⟨a *reptilian* villain —Theodore Dreiser⟩

²reptilian *noun* (circa 1847)
: REPTILE 2

re•pub•lic \ri-ˈpə-blik\ *noun* [French *république,* from Middle French *republique,* from Latin *respublica,* from *res* thing, wealth + *publica,* feminine of *publicus* public — more at REAL, PUBLIC] (1604)
1 a (1) **:** a government having a chief of state who is not a monarch and who in modern times is usually a president (2) **:** a political unit (as a nation) having such a form of government **b** (1) **:** a government in which supreme power resides in a body of citizens entitled to vote and is exercised by elected officers and representatives responsible to them and governing according to law (2) **:** a political unit (as a nation) having such a form of government **c :** a usually specified republican government of a political unit ⟨the French Fourth *Republic*⟩
2 : a body of persons freely engaged in a specified activity ⟨the *republic* of letters⟩
3 : a constituent political and territorial unit of the former nations of Czechoslovakia, the U.S.S.R., or Yugoslavia

¹re•pub•li•can \ri-ˈpə-bli-kən\ *noun* (1697)
1 : one that favors or supports a republican form of government
2 *capitalized* **a :** a member of a political party advocating republicanism **b :** a member of the Democratic-Republican party or of the Republican party of the U.S.

²republican *adjective* (1712)
1 a : of, relating to, or having the characteristics of a republic **b :** favoring, supporting, or advocating a republic **c :** belonging or appropriate to one living in or supporting a republic ⟨*republican* simplicity⟩
2 *capitalized* **a :** DEMOCRATIC-REPUBLICAN **b :** of, relating to, or constituting the one of the two major political parties evolving in the U.S. in the mid-19th century that is usually primarily associated with business, financial, and some agricultural interests and is held to favor a restricted governmental role in economic life

re•pub•li•can•ism \ri-ˈpə-bli-kə-ˌni-zəm\ *noun* (1689)
1 : adherence to or sympathy for a republican form of government
2 : the principles or theory of republican government
3 *capitalized* **a :** the principles, policy, or practices of the Republican party of the U.S. **b :** the Republican party or its members

re•pub•li•can•ize \-kə-ˌnīz\ *transitive verb* **-ized; -iz•ing** (1797)
: to make republican in character, form, or principle

re•pub•li•ca•tion \(ˌ)rē-ˌpə-blə-ˈkā-shən\ *noun* (1730)
1 : the act or action of republishing **:** the state of being republished
2 : something that has been republished

re•pub•lish \(ˌ)rē-ˈpə-blish\ *transitive verb* (1592)
1 : to publish again or anew
2 : to execute (a will) anew
— **re•pub•lish•er** *noun*

re•pu•di•ate \ri-ˈpyü-dē-ˌāt\ *transitive verb* **-at•ed; -at•ing** [Latin *repudiatus,* past participle of *repudiare,* from *repudium* rejection of a prospective spouse, divorce, probably from *re-* + *pudēre* to shame] (1545)
1 : to divorce or separate formally from (a woman)
2 : to refuse to have anything to do with **:** DISOWN
3 a : to refuse to accept; *especially* **:** to reject as unauthorized or as having no binding force **b :** to reject as untrue or unjust ⟨*repudiate* a charge⟩
4 : to refuse to acknowledge or pay
synonym see DECLINE
— **re•pu•di•a•tor** \-ˌā-tər\ *noun*

☆ SYNONYMS

Reproduction, duplicate, copy, facsimile, replica mean a thing made to closely resemble another. REPRODUCTION implies an exact or close imitation of an existing thing ⟨*reproductions* from the museum's furniture collection⟩. DUPLICATE implies a double or counterpart exactly corresponding to another thing ⟨a *duplicate* of a house key⟩. COPY applies especially to one of a number of things reproduced mechanically ⟨printed a thousand *copies* of the lithograph⟩. FACSIMILE suggests a close reproduction often of graphic matter that may differ in scale ⟨a *facsimile* of a rare book⟩. REPLICA implies the exact reproduction of a particular item in all details ⟨a *replica* of the Mayflower⟩ but not always in the same scale ⟨miniature *replicas* of classic cars⟩.

Reprove, rebuke, reprimand, admonish, reproach, chide mean to criticize adversely. REPROVE implies an often kindly intent to correct a fault ⟨gently *reproved* my table manners⟩. REBUKE suggests a sharp or stern reproof ⟨the papal letter *rebuked* dissenting clerics⟩. REPRIMAND implies a severe, formal, often public or official rebuke ⟨*reprimanded* by the ethics committee⟩. ADMONISH suggests earnest or friendly warning and counsel ⟨*admonished* by my parents to control expenses⟩. REPROACH and CHIDE suggest displeasure or disappointment expressed in mild reproof or scolding ⟨*reproached* him for tardiness⟩ ⟨*chided* by their mother for untidiness⟩.

re·pu·di·a·tion \ri-ˌpyü-dē-ˈā-shən\ *noun* (1545)
: the act of repudiating **:** the state of being repudiated; *especially* **:** the refusal of public authorities to acknowledge or pay a debt
— re·pu·di·a·tion·ist \-sh(ə-)nist\ *noun*

re·pugn \ri-ˈpyün\ *verb* [Middle English, from Middle French & Latin; Middle French *repugner,* from Latin *repugnare*] (14th century)
intransitive verb
archaic **:** to offer opposition, objection, or resistance
transitive verb
: to contend against **:** OPPOSE

re·pug·nance \ri-ˈpəg-nən(t)s\ *noun* (15th century)
1 a : the quality or fact of being contradictory or inconsistent **b :** an instance of such contradiction or inconsistency
2 : strong dislike, distaste, or antagonism

re·pug·nan·cy \-nən(t)-sē\ *noun, plural* **-cies** (15th century)
: REPUGNANCE

re·pug·nant \-nənt\ *adjective* [Middle English, opposed, contradictory, incompatible, from Middle French, from Latin *repugnant-, repugnans,* present participle of *repugnare* to fight against, from *re-* + *pugnare* to fight — more at PUNGENT] (15th century)
1 : INCOMPATIBLE, INCONSISTENT
2 *archaic* **:** HOSTILE
3 : exciting distaste or aversion
— re·pug·nant·ly *adverb*

¹re·pulse \ri-ˈpəls\ *transitive verb* **re·pulsed; re·puls·ing** [Latin *repulsus,* past participle of *repellere* to repel] (15th century)
1 : to drive or beat back **:** REPEL
2 : to repel by discourtesy, coldness, or denial
3 : to cause repulsion in

²repulse *noun* (1533)
1 : REBUFF, REJECTION
2 : the action of repelling an attacker **:** the fact of being repelled

re·pul·sion \ri-ˈpəl-shən\ *noun* (15th century)
1 : the action of repulsing **:** the state of being repulsed
2 : the action of repelling **:** the force with which bodies, particles, or like forces repel one another
3 : a feeling of aversion **:** REPUGNANCE

re·pul·sive \-siv\ *adjective* (1598)
1 : tending to repel or reject **:** COLD, FORBIDDING
2 : serving or able to repulse
3 : arousing aversion or disgust
— re·pul·sive·ly *adverb*
— re·pul·sive·ness *noun*

re·pur·chase agreement \(ˌ)rē-ˈpər-chəs-\ *noun* (1952)
: a contract giving the seller of securities (as treasury bills) the right to repurchase after a stated period and the buyer the right to retain interest earnings

rep·u·ta·ble \ˈre-pyə-tə-bəl\ *adjective* (1674)
1 : enjoying good repute **:** held in esteem
2 : employed widely or sanctioned by good writers
— rep·u·ta·bil·i·ty \ˌre-pyə-tə-ˈbi-lə-tē\ *noun*
— rep·u·ta·bly \ˈre-pyə-tə-blē\ *adverb*

rep·u·ta·tion \ˌre-pyə-ˈtā-shən\ *noun* [Middle English *reputacioun,* from Latin *reputation-, reputatio* consideration, from *reputare*] (14th century)
1 a : overall quality or character as seen or judged by people in general **b :** recognition by other people of some characteristic or ability ⟨has the *reputation* of being clever⟩
2 : a place in public esteem or regard **:** good name
— rep·u·ta·tion·al \-shnəl, -shə-nᵊl\ *adjective*

¹re·pute \ri-ˈpyüt\ *transitive verb* **re·put·ed; re·put·ing** [Middle English, from Middle French *reputer,* from Latin *reputare* to reckon up, think over, from *re-* + *putare* to reckon] (15th century)
: BELIEVE, CONSIDER

²repute *noun* (1551)
1 : the character or status commonly ascribed to one **:** REPUTATION
2 : the state of being favorably known, spoken of, or esteemed

reputed *adjective* (1549)
1 : having a good repute **:** REPUTABLE
2 : being such according to reputation or popular belief

re·put·ed·ly *adverb* (1687)
: according to reputation or general belief

¹re·quest \ri-ˈkwest\ *noun* [Middle English *requeste,* from Middle French, from (assumed) Vulgar Latin *requaesta,* from feminine of *requaestus,* past participle of *requaerere* to require] (14th century)
1 : the act or an instance of asking for something
2 : something asked for
3 : the condition or fact of being requested ⟨available on *request*⟩
4 : the state of being sought after **:** DEMAND

²request *transitive verb* (1533)
1 : to make a request to or of ⟨*requested* her to write a paper⟩
2 : to ask as a favor or privilege ⟨*requests* to be excused⟩
3 *obsolete* **:** to ask (a person) to come or go to a thing or place
4 : to ask for ⟨*requested* a brief delay⟩
synonym see ASK
— re·quest·er *or* **re·quest·or** \-ˈkwes-tər\ *noun*

re·qui·em \ˈre-kwē-əm *also* ˈrā- *or* ˈrē-\ *noun* [Middle English, from Latin (first word of the introit of the requiem mass), accusative of *requies* rest, from *re-* + *quies* quiet, rest — more at WHILE] (14th century)
1 : a mass for the dead
2 a : a solemn chant (as a dirge) for the repose of the dead **b :** something that resembles such a solemn chant
3 a : a musical setting of the mass for the dead **b :** a musical composition in honor of the dead

requiem shark *noun* [obsolete French *requiem* shark, alteration of French *requin*] (1900)
: any of a family (Carcharhinidae) of sharks (as the tiger shark) that includes some dangerous to humans

re·qui·es·cat \ˌre-kwē-ˈes-ˌkät, -ˌkat; ˌrā-kwē-ˈes-ˌkät\ *noun* [Latin, may he (or she) rest, from *requiescere* to rest, from *re-* + *quiescere* to be quiet, from *quies*] (1824)
: a prayer for the repose of a dead person

re·quire \ri-ˈkwīr\ *verb* **re·quired; re·quir·ing** [Middle English *requeren,* from Middle French *requerre,* from (assumed) Vulgar Latin *requaerere* to seek for, need, require, alteration of Latin *requirere,* from *re-* + *quaerere* to seek, ask] (14th century)
transitive verb
1 a : to claim or ask for by right and authority **b** *archaic* **:** REQUEST
2 a : to call for as suitable or appropriate ⟨the occasion *requires* formal dress⟩ **b :** to demand as necessary or essential **:** have a compelling need for ⟨all living beings *require* food⟩
3 : to impose a compulsion or command on **:** COMPEL
4 *chiefly British* **:** to feel or be obliged — used with a following infinitive ⟨one does not *require* to be a specialist —Elizabeth Bowen⟩
intransitive verb
archaic **:** ASK
synonym see DEMAND

re·quire·ment \-ˈkwīr-mənt\ *noun* (1662)
: something required: **a :** something wanted or needed **:** NECESSITY ⟨production was not sufficient to satisfy military *requirements*⟩ **b :** something essential to the existence or occurrence of something else **:** CONDITION ⟨failed to meet the school's *requirements* for graduation⟩

req·ui·site \ˈre-kwə-zət\ *adjective* [Middle English, from Latin *requisitus,* past participle of *requirere*] (15th century)
: ESSENTIAL, NECESSARY
— requisite *noun*
— req·ui·site·ness *noun*

req·ui·si·tion \ˌre-kwə-ˈzi-shən\ *noun* [Middle English *requisicion,* from Middle French or Medieval Latin; Middle French *requisition,* from Medieval Latin *requisition-, requisitio,* from Latin, act of searching, from *requirere*] (15th century)
1 a : the act of formally requiring or calling upon someone to perform an action **b :** a formal demand made by one nation upon another for the surrender or extradition of a fugitive from justice
2 a : the act of requiring something to be furnished **b :** a demand or application made usually with authority: as (1) **:** a demand made by military authorities upon civilians for supplies or other needs (2) **:** a written request for something authorized but not made available automatically
3 : the state of being in demand or use
— requisition *transitive verb*

re·quit·al \ri-ˈkwī-tᵊl\ *noun* (1582)
1 : something given in return, compensation, or retaliation
2 : the act or action of requiting **:** the state of being requited

re·quite \ri-ˈkwīt\ *transitive verb* **re·quit·ed; re·quit·ing** [*re-* + obsolete *quite* to quit, pay, from Middle English *quiten* — more at QUIT] (1529)
1 a : to make return for **:** REPAY **b :** to make retaliation for **:** AVENGE
2 : to make suitable return to for a benefit or service or for an injury
synonym see RECIPROCATE
— re·quit·er *noun*

re·ra·di·ate \(ˌ)rē-ˈrā-dē-ˌāt\ *transitive verb* (1913)
: to radiate again or anew; *especially* **:** to emit (energy) in the form of radiation after absorbing incident radiation
— re·ra·di·a·tion \(ˌ)rē-ˌrā-dē-ˈā-shən\ *noun*

rere·dos \ˈrer-ə-ˌdäs *also* ˈrir-ə-ˌdäs *or* ˈrir-ˌdäs\ *noun* [Middle English, from Anglo-French *areredos,* from Middle French *arrere* behind + *dos* back, from Latin *dorsum* — more at ARREAR] (14th century)
: a usually ornamental wood or stone screen or partition wall behind an altar

rereward *noun* [Middle English *rerewarde,* from Anglo-French, from Old French *rere* behind + Old North French *warde* guard; akin to Old French *garde* guard — more at REAR GUARD] (14th century)
obsolete **:** REAR GUARD

¹re·run \(ˌ)rē-ˈrən\ *transitive verb* **-ran** \-ˈran\; **-run; -run·ning** (1804)
: to run again or anew

²re·run \ˈrē-ˌrən, (ˌ)rē-ˈ\ *noun* (circa 1934)
: the act or action or an instance of rerunning **:** REPETITION; *especially* **:** a movie or television show that is rerun

res \ˈrās, ˈrēz\ *noun, plural* **res** [Latin — more at REAL] (1851)
: a particular thing **:** MATTER — used especially in legal phrases

res ad·ju·di·ca·ta \ˈrēz-ə-ˌjü-di-ˈkä-tə\ *noun* [Late Latin] (1902)
: RES JUDICATA

re·sal·able \(ˌ)rē-ˈsā-lə-bəl\ *adjective* (1866)
: fit for resale

re·sale \ˈrē-ˌsāl, (ˌ)rē-ˈsā(ə)l\ *noun* (1625)
1 : the act of selling again usually to a new party

2 a : a secondhand sale **b :** an additional sale to the same buyer

re·scale \(ˌ)rē-ˈskā(ə)l\ *transitive verb* (1944)
: to plan, establish, or formulate on a new and usually smaller scale

re·sched·ule \(ˌ)rē-ˈske-(ˌ)jü(ə)l, -jəl, *Canadian also* -ˈshe-, *British usually* -ˈshe-(ˌ)dyü(ə)l\ *transitive verb* (1965)
: to schedule or plan again according to a different timetable; *especially* **:** to defer required payment of (a debt or loan)

re·scind \ri-ˈsind\ *transitive verb* [Latin *rescindere* to annul, from *re-* + *scindere* to cut — more at SHED] (1643)
1 : to take away **:** REMOVE
2 a : TAKE BACK, CANCEL ⟨refused to *rescind* the order⟩ **b :** to abrogate (a contract) and restore the parties to the positions they would have occupied had there been no contract
3 : to make void (as an act) by action of the enacting authority or a superior authority **:** REPEAL
— re·scind·er *noun*
— re·scind·ment \-ˈsin(d)-mənt\ *noun*

re·scis·sion \ri-ˈsi-zhən\ *noun* [Late Latin *rescission-, rescissio,* from Latin *rescindere*] (1611)
: an act of rescinding

re·scis·so·ry \-ˈsi-zə-rē, -ˈsi-sə-\ *adjective* (1605)
: relating to or tending to or having the effect of rescission

re·script \ˈrē-ˌskript\ *noun* [Middle English *rescripte,* from Latin *rescriptum,* from neuter of *rescriptus,* past participle of *rescribere* to write in reply, from *re-* + *scribere* to write — more at SCRIBE] (15th century)
1 : a written answer of a Roman emperor or of a pope to a legal inquiry or petition
2 : an official or authoritative order, decree, edict, or announcement
3 : an act or instance of rewriting

res·cue \ˈres-(ˌ)kyü\ *transitive verb* **res·cued; res·cu·ing** [Middle English, from Middle French *rescourre,* from Old French, from *re-* + *escourre* to shake out, from Latin *excutere,* from *ex-* + *quatere* to shake] (14th century)
: to free from confinement, danger, or evil **:** SAVE, DELIVER: as **a :** to take (as a prisoner) forcibly from custody **b :** to recover (as a prize) by force **c :** to deliver (as a place under siege) by armed force ☆
— res·cu·able \-ə-bəl\ *adjective*
— rescue *noun*
— res·cu·er *noun*

rescue mission *noun* (1902)
: a city religious mission seeking to convert and rehabilitate the down-and-out

¹re·search \ri-ˈsərch, ˈrē-ˌ\ *noun* [Middle French *recerche,* from *recerchier* to investigate thoroughly, from Old French, from *re-* + *cerchier* to search — more at SEARCH] (1577)
1 : careful or diligent search
2 : studious inquiry or examination; *especially* **:** investigation or experimentation aimed at the discovery and interpretation of facts, revision of accepted theories or laws in the light of new facts, or practical application of such new or revised theories or laws
3 : the collecting of information about a particular subject

²research (1593)
transitive verb
1 : to search or investigate exhaustively ⟨*research* a problem⟩
2 : to do research for ⟨*research* a book⟩
intransitive verb
: to engage in research
— re·search·able \ri-ˈsər-chə-bəl, ˈrē-ˌ\ *adjective*
— re·search·er *noun*

re·search·ist \ri-ˈsər-chist, ˈrē-ˌsər-\ *noun* (1921)
: one engaged in research

re·seau \rā-ˈzō, ri-\ *noun, plural* **re·seaux** \-ˈzōz\ [French *réseau,* from Old French *resel,* diminutive of *rais* net, from Latin *retis, rete*] (1578)
1 : a net ground or foundation in lace
2 : a grid photographed by a separate exposure onto a plate containing star images to facilitate astronomical measurements

re·sect \ri-ˈsekt\ *transitive verb* [Latin *resectus,* past participle of *resecare* to cut off, from *re-* + *secare* to cut — more at SAW] (1846)
: to perform resection on
— re·sect·abil·i·ty \-ˌsek-tə-ˈbi-lə-tē\ *noun*
— re·sect·able \-ˈsek-tə-bəl\ *adjective*

re·sec·tion \ri-ˈsek-shən\ *noun* (1775)
: the surgical removal of part of an organ or structure

re·se·da \ˈrā-zə-ˌdä\ *noun* [French *réséda,* from *réséda,* a mignonette] (1873)
: a grayish green color

re·seed \(ˌ)rē-ˈsēd\ (1888)
transitive verb
1 : to sow seed on again or anew
2 : to maintain (itself) by self-sown seed
intransitive verb
: to maintain itself by self-sown seed

re·sem·blance \ri-ˈzem-blən(t)s\ *noun* (14th century)
1 a : the quality or state of resembling; *especially* **:** correspondence in appearance or superficial qualities **b :** a point of likeness **:** SIMILARITY
2 : REPRESENTATION, IMAGE
3 *archaic* **:** characteristic appearance
4 *obsolete* **:** PROBABILITY
synonym see LIKENESS

re·sem·blant \-blənt\ *adjective* (14th century)
: marked by or showing resemblance

re·sem·ble \ri-ˈzem-bəl\ *transitive verb* **re·sem·bled; re·sem·bling** \-b(ə-)liŋ\ [Middle English, from Middle French *resembler, ressembler,* from Old French, from *re-* + *sembler* to be like, seem, from Latin *similare* to copy, from *similis* like — more at SAME] (14th century)
1 : to be like or similar to
2 *archaic* **:** to represent as like

re·send \(ˈ)rē-ˈsend\ *transitive verb* **-sent** \-ˈsent\; **-send·ing** (1554)
: to send again or back

re·sent \ri-ˈzent\ *transitive verb* [French *ressentir* to be emotionally sensible of, from Old French, from *re-* + *sentir* to feel, from Latin *sentire* — more at SENSE] (1628)
: to feel or express annoyance or ill will at

re·sent·ful \-fəl\ *adjective* (1656)
1 : full of resentment **:** inclined to resent
2 : caused or marked by resentment
— re·sent·ful·ly \-fə-lē\ *adverb*
— re·sent·ful·ness *noun*

re·sent·ment \ri-ˈzent-mənt\ *noun* (1619)
: a feeling of indignant displeasure or persistent ill will at something regarded as a wrong, insult, or injury
synonym see OFFENSE

re·ser·pine \ri-ˈsər-ˌpēn, -pən\ *noun* [German *Reserpin,* probably irregular from New Latin *Rauwolfia serpentina,* a species of rauwolfia] (1952)
: an alkaloid $C_{33}H_{40}N_2O_9$ extracted especially from the root of rauwolfias and used in the treatment of hypertension, mental disorders, and tension states

res·er·va·tion \ˌre-zər-ˈvā-shən\ *noun* (15th century)
1 : an act of reserving something: as **a** (1) **:** the act or fact of a grantor's reserving some newly created thing out of the thing granted (2) **:** the right or interest so reserved **b :** the setting of limiting conditions or withholding from complete exposition ⟨answered without *reservation*⟩ **c :** an arrangement to have something (as a hotel room) held for one's use; *also* **:** a promise, guarantee, or record of such engagement
2 a : a limiting condition ⟨agreed, but with *reservations*⟩ **b :** DOUBT, MISGIVING ⟨had serious *reservations* about marriage⟩
3 : something reserved: as **a :** a tract of public land set aside (as for use by American Indians) **b :** an area in which hunting is not permitted; *especially* **:** one set aside as a secure breeding place
— res·er·va·tion·ist \-sh(ə-)nist\ *noun*

¹re·serve \ri-ˈzərv\ *transitive verb* **re·served; re·serv·ing** [Middle English, from Middle French *reserver,* from Latin *reservare,* literally, to keep back, from *re-* + *servare* to keep — more at CONSERVE] (14th century)
1 a : to hold in reserve **:** keep back ⟨*reserve* grain for seed⟩ **b :** to set aside (part of the consecrated elements) at the Eucharist for future use **c :** to retain or hold over to a future time or place **:** DEFER ⟨*reserve* one's judgment on a plan⟩ **d :** to make legal reservation of
2 : to set or have set aside or apart ⟨*reserve* a hotel room⟩
synonym see KEEP
— re·serv·able \-ˈzər-və-bəl\ *adjective*

²reserve *noun, often attributive* (1648)
1 : something reserved or set aside for a particular purpose, use, or reason: as **a** (1) **:** a military force withheld from action for later decisive use — usually used in plural (2) **:** forces not in the field but available (3) **:** the military forces of a country not part of the regular services; *also* **:** RESERVIST **b :** a tract (as of public land) set apart **:** RESERVATION
2 : something stored or kept available for future use or need **:** STOCK
3 : an act of reserving **:** QUALIFICATION
4 a : restraint, closeness, or caution in one's words and actions **b :** forbearance from making a full explanation, complete disclosure, or free expression of one's mind
5 *archaic* **:** SECRET
6 a : money or its equivalent kept in hand or set apart usually to meet liabilities **b :** the liquid resources of a nation for meeting international payments
7 : SUBSTITUTE
8 : RESERVE PRICE
9 : a wine made from select grapes, bottled on the maker's premises, and aged specially and often longer
— in reserve : held back for future or special use

reserve bank *noun* (1905)
: a central bank holding reserves of other banks

reserve clause *noun* (1944)
: a clause formerly placed in a professional athlete's contract that reserved for the club the exclusive right automatically to renew the

☆ SYNONYMS
Rescue, deliver, redeem, ransom, reclaim, save mean to set free from confinement or danger. RESCUE implies freeing from imminent danger by prompt or vigorous action ⟨*rescued* the crew of a sinking ship⟩. DELIVER implies release usually of a person from confinement, temptation, slavery, or suffering ⟨*delivered* his people from bondage⟩. REDEEM implies releasing from bondage or penalties by giving what is demanded or necessary ⟨job training designed to *redeem* school dropouts from chronic unemployment⟩. RANSOM specifically applies to buying out of captivity ⟨tried to *ransom* the kidnap victim⟩. RECLAIM suggests a bringing back to a former state or condition of someone or something abandoned or debased ⟨*reclaimed* long-abandoned farms⟩. SAVE may replace any of the foregoing terms; it may further imply a preserving or maintaining for usefulness or continued existence ⟨an operation that *saved* my life⟩.

contract and that bound the athlete to the club until retirement or until the athlete was traded or released

re·served \ri-ˈzərvd\ *adjective* (1601)
1 : restrained in words and actions
2 : kept or set apart or aside for future or special use
synonym see SILENT
— re·serv·ed·ly \-ˈzər-vəd-lē\ *adverb*
— re·serv·ed·ness \-ˈzər-vəd-nəs\ *noun*

reserved power *noun* (1838)
: a political power reserved by a constitution to the exclusive jurisdiction of a specified political authority

reserve price *noun* (1919)
: a price announced at an auction as the lowest that will be considered

re·serv·ist \ri-ˈzər-vist\ *noun* (1876)
: a member of a military reserve

res·er·voir \ˈre-zə-ˌvwär, -zər-, -ˌvwȯr *also* -ˌvȯi\ *noun* [French *réservoir,* from Middle French, from *reserver*] (1690)
1 : a place where something is kept in store: as **a :** an artificial lake where water is collected and kept in quantity for use **b :** a part of an apparatus in which a liquid is held **c :** SUPPLY, STORE ⟨a large *reservoir* of educated people⟩
2 : an extra supply **:** RESERVE
3 : an organism in which a parasite that is pathogenic for some other species lives and multiplies without damaging its host; *also* **:** a noneconomic organism within which a pathogen of economic or medical importance flourishes

re·set \(ˌ)rē-ˈset\ *transitive verb* **-set; -set·ting** (1655)
1 : to set again or anew ⟨*reset* type⟩ ⟨*reset* a diamond⟩ ⟨*reset* a circuit breaker⟩
2 : to change the reading of often to zero ⟨*reset* an odometer⟩
— re·set \ˈrē ˌset\ *noun*
— re·set·table \-ˈse-tə-bəl\ *adjective*

res ges·tae \ˈrās-ˈges-ˌtī, ˈrēz-ˈjes-(ˌ)tē\ *noun plural* [Latin] (1616)
: things done; *especially* **:** the facts that form the environment of a litigated issue and are admissible in evidence

resh \ˈrāsh\ *noun* [Hebrew *rēsh*] (circa 1823)
: the 20th letter of the Hebrew alphabet — see ALPHABET table

re·shape \(ˌ)rē-ˈshāp\ *transitive verb* (1827)
: to give a new form or orientation to **:** REORGANIZE
— re·shap·er *noun*

re·shuf·fle \(ˌ)rē-ˈshə-fəl\ *transitive verb* (1830)
1 : to shuffle (as cards) again
2 : to reorganize usually by the redistribution of existing elements ⟨the cabinet was *reshuffled* by the prime minister⟩
— reshuffle *noun*

re·sid \ri-ˈzid\ *noun* (1967)
: RESIDUAL OIL

re·side \ri-ˈzīd\ *intransitive verb* **re·sid·ed; re·sid·ing** [Middle English, from Middle French or Latin; Middle French *resider,* from Latin *residēre* to sit back, remain, abide, from *re-* + *sedēre* to sit — more at SIT] (15th century)
1 a : to be in residence as the incumbent of a benefice or office **b :** to dwell permanently or continuously **:** occupy a place as one's legal domicile
2 a : to be present as an element or quality **b :** to be vested as a right
— re·sid·er *noun*

res·i·dence \ˈre-zə-dən(t)s, ˈrez-dən(t)s, ˈre-zə-ˌden(t)s\ *noun* (14th century)
1 a : the act or fact of dwelling in a place for some time **b :** the act or fact of living or regularly staying at or in some place for the discharge of a duty or the enjoyment of a benefit
2 a (1) **:** the place where one actually lives as distinguished from one's domicile or a place of temporary sojourn (2) **:** DOMICILE 2a **b :** the place where a corporation is actually or officially established **c :** the status of a legal resident
3 a : a building used as a home **:** DWELLING **b :** housing or a unit of housing provided for students
4 a : the period or duration of abode in a place **b :** a period of active and especially full-time study, research, or teaching at a college or university
— in residence : engaged to live and work at a particular place often for a specified time ⟨poet *in residence* at a university⟩

residence time *noun* (1954)
: the duration of persistence of a mass or substance in a medium or place

res·i·den·cy \ˈre-zə-dən(t)-sē, ˈrez-dən(t)-, ˈre-zə-ˌden(t)-\ *noun, plural* **-cies** (1579)
1 : a usually official place of residence
2 : a territory in a protected state in which the powers of the protecting state are executed by a resident agent
3 : a period of advanced training in a medical specialty after graduation from medical school and licensing to practice medicine

¹res·i·dent \ˈre-zə-dənt, ˈrez-dənt, ˈre-zə-ˌdent\ *adjective* [Middle English, from Latin *resident-, residens,* present participle of *residēre*] (14th century)
1 a : living in a place for some length of time **:** RESIDING **b :** serving in a regular or full-time capacity ⟨the *resident* engineer for a highway department⟩; *also* **:** being in residence
2 : PRESENT, INHERENT
3 : not migratory

²resident *noun* (15th century)
1 : one who resides in a place
2 : a diplomatic agent residing at a foreign court or seat of government; *especially* **:** one exercising authority in a protected state as representative of the protecting power
3 : a physician serving a residency

resident commissioner *noun* (1902)
1 : a nonvoting representative of a dependency in the U.S. House of Representatives
2 : a resident administrator in a British colony or possession

res·i·den·tial \ˌre-zə-ˈden(t)-shəl, ˌrez-ˈden(t)-\ *adjective* (1654)
1 a : used as a residence or by residents **b :** providing living accommodations for students ⟨a *residential* college⟩
2 : restricted to or occupied by residences ⟨a *residential* neighborhood⟩
3 : of or relating to residence or residences
— res·i·den·tial·ly \-ˈden(t)-sh(ə-)lē\ *adverb*

¹re·sid·u·al \ri-ˈzi-jə-wəl, -jəl; -ˈzij-wəl\ *noun* [Latin *residuum* residue] (1557)
1 : REMAINDER, RESIDUUM: as **a :** the difference between results obtained by observation and by computation from a formula or between the mean of several observations and any one of them **b :** a residual product or substance **c :** an internal aftereffect of experience or activity that influences later behavior; *especially* **:** a disability remaining from a disease or operation
2 : a payment (as to an actor or writer) for each rerun after an initial showing (as of a TV show)

²residual *adjective* (1570)
1 : of, relating to, or constituting a residue
2 : leaving a residue that remains effective for some time
— re·sid·u·al·ly *adverb*

residual oil *noun* (circa 1948)
: fuel oil that remains after the removal of valuable distillates (as gasoline) from petroleum and that is used especially by industry — called also *resid*

residual power *noun* (1919)
: power held to remain at the disposal of a governmental authority after an enumeration or delegation of specified powers to other authorities

re·sid·u·ary \ri-ˈzi-jə-ˌwer-ē\ *adjective* (1726)
: of, relating to, or constituting a residue ⟨*residuary* estate⟩

res·i·due \ˈre-zə-ˌdü, -ˌdyü\ *noun* [Middle English, from Middle French *residu,* from Latin *residuum,* from neuter of *residuus* left over, from *residēre* to remain] (14th century)
: something that remains after a part is taken, separated, or designated **:** REMNANT, REMAINDER: as **a :** the part of a testator's estate remaining after the satisfaction of all debts, charges, allowances, and previous devises and bequests **b :** the remainder after subtracting a multiple of a modulus from an integer or a power of the integer that can appear as the second of the two terms in an appropriate congruence ⟨2 and 7 are *residues* of 12 modulo 5⟩ **c :** a constituent structural unit (as a group or monomer) of a usually complex molecule ⟨amino acid *residues* from hydrolysis of protein⟩

residue class *noun* (1948)
: the set of elements (as integers) that leave the same remainder when divided by a given modulus

re·sid·u·um \ri-ˈzi-jə-wəm\ *noun, plural* **re·sid·ua** \-wə\ [Latin] (1672)
: something residual: as **a :** RESIDUE a **b :** a residual product (as from the distillation of petroleum)

re·sign \ri-ˈzīn\ *verb* [Middle English, from Middle French *resigner,* from Latin *resignare,* literally, to unseal, cancel, from *re-* + *signare* to sign, seal — more at SIGN] (14th century)
transitive verb
1 : RELEGATE, CONSIGN; *especially* **:** to give (oneself) over without resistance ⟨*resigned* herself to her fate⟩
2 : to give up deliberately; *especially* **:** to renounce (as a right or position) by a formal act
intransitive verb
1 : to give up one's office or position **:** QUIT
2 : to accept something as inevitable **:** SUBMIT
synonym see RELINQUISH, ABDICATE
— re·sign·ed·ly \-ˈzī-nəd-lē\ *adverb*
— re·sign·ed·ness \-ˈzī-nəd-nəs\ *noun*
— re·sign·er \-ˈzī-nər\ *noun*

re–sign \(ˌ)rē-ˈsīn\ (1805)
transitive verb
: to sign again
intransitive verb
: to sign up again

res·ig·na·tion \ˌre-zig-ˈnā-shən\ *noun* (14th century)
1 a : an act or instance of resigning something **:** SURRENDER **b :** a formal notification of resigning
2 : the quality or state of being resigned **:** SUBMISSIVENESS

re·sile \ri-ˈzī(ə)l\ *intransitive verb* **re·siled; re·sil·ing** [Late Latin & Latin; Late Latin *resilire* to withdraw, from Latin, to recoil] (1529)
: RECOIL, RETRACT; *especially* **:** to return to a prior position

re·sil·ience \ri-ˈzil-yən(t)s\ *noun* (1824)
1 : the capability of a strained body to recover its size and shape after deformation caused especially by compressive stress
2 : an ability to recover from or adjust easily to misfortune or change

re·sil·ien·cy \-yən(t)-sē\ *noun* (circa 1836)
: RESILIENCE

re·sil·ient \-yənt\ *adjective* [Latin *resilient-, resiliens,* present participle of *resilire* to jump back, recoil, from *re-* + *salire* to leap — more at SALLY] (1674)

: characterized or marked by resilience: as **a :** capable of withstanding shock without permanent deformation or rupture **b :** tending to recover from or adjust easily to misfortune or change
synonym see ELASTIC
— re·sil·ient·ly *adverb*
1res·in \'re-zᵊn\ *noun* [Middle English, from Middle French *resine,* from Latin *resina;* akin to Greek *rhētinē* pine resin] (14th century)
1 a : any of various solid or semisolid amorphous fusible flammable natural organic substances that are usually transparent or translucent and yellowish to brown, are formed especially in plant secretions, are soluble in organic solvents (as ether) but not in water, are electrical nonconductors, and are used chiefly in varnishes, printing inks, plastics, and sizes and in medicine **b :** ROSIN
2 a : any of a large class of synthetic products that have some of the physical properties of natural resins but are different chemically and are used chiefly in plastics **b :** any various products made from a natural resin or a natural polymer
— res·in·ous \'re-zᵊn-əs, 'rez-nəs\ *adjective*
2resin *transitive verb* **res·ined; res·in·ing** \'re-zᵊn-iŋ, 'rez-niŋ\ (1865)
: to treat with resin
res·in·ate \'re-zᵊn-,āt\ *transitive verb* **-at·ed; -at·ing** (circa 1890)
: to impregnate or flavor with resin
resin canal *noun* (1884)
: a tubular intercellular space in gymnosperms and some angiosperms that is lined with epithelial cells which secrete resin — called also *resin duct*
res·in·oid \'re-zᵊn-,ȯid\ *noun* (1880)
: GUM RESIN
1re·sist \ri-'zist\ *verb* [Middle English, from Middle French or Latin; Middle French *resister,* from Latin *resistere,* from *re-* + *sistere* to take a stand; akin to Latin *stare* to stand — more at STAND] (14th century)
intransitive verb
: to exert force in opposition
transitive verb
1 : to exert oneself so as to counteract or defeat
2 : to withstand the force or effect of
synonym see OPPOSE
2resist *noun* (1836)
: something (as a coating) that protects against a chemical, electrical, or physical action
re·sis·tance \ri-'zis-tən(t)s\ *noun* (14th century)
1 a : an act or instance of resisting **:** OPPOSITION **b :** a means of resisting
2 : the ability to resist; *especially* **:** the inherent capacity of a living being to resist untoward circumstances (as disease, malnutrition, or toxic agents)
3 : an opposing or retarding force
4 a : the opposition offered by a body or substance to the passage through it of a steady electric current **b :** a source of resistance
5 *often capitalized* **:** an underground organization of a conquered or nearly conquered country engaging in sabotage and secret operations against occupation forces and collaborators
1re·sis·tant \-tənt\ *adjective* (15th century)
: giving or capable of resistance — often used in combination ⟨wrinkle-*resistant* clothes⟩
2resistant *noun* (1600)
: one that resists **:** RESISTER
re·sist·er \ri-'zis-tər\ *noun* (14th century)
: one that resists; *especially* **:** one who actively opposes the policies of a government
re·sist·ibil·i·ty \ri-,zis-tə-'bi-lə-tē\ *noun* (1617)
1 : the quality or state of being resistible
2 : ability to resist
re·sist·ible \ri-'zis-tə-bəl\ *adjective* (1608)
: capable of being resisted
re·sis·tive \ri-'zis-tiv\ *adjective* (1603)
: marked by resistance — often used in combination ⟨fire-*resistive* material⟩
— re·sis·tive·ly *adverb*
— re·sis·tive·ness *noun*
re·sis·tiv·i·ty \ri-,zis-'ti-və-tē, ,rē-\ *noun, plural* **-ties** (1885)
1 : the longitudinal electrical resistance of a uniform rod of unit length and unit cross-sectional area **:** the reciprocal of conductivity
2 : capacity for resisting **:** RESISTANCE
re·sist·less \ri-'zist-ləs\ *adjective* (1586)
1 : IRRESISTIBLE
2 : offering no resistance
— re·sist·less·ly *adverb*
— re·sist·less·ness *noun*
re·sis·tor \ri-'zis-tər\ *noun* (1905)
: a device that has electrical resistance and that is used in an electric circuit for protection, operation, or current control
re·sit·ting \(,)rē-'si-tiŋ\ *noun* (1661)
: a sitting (as of a legislature) for a second time **:** another sitting
res ju·di·ca·ta \'rēz-,jü-di-'kä-tə\ *noun* [Latin, judged matter] (1693)
: a matter finally decided on its merits by a court having competent jurisdiction and not subject to litigation again between the same parties
re·sol·u·ble \ri-'zäl-yə-bəl\ *adjective* [Late Latin *resolubilis,* from Latin *resolvere* to resolve, unloose] (1602)
: capable of being resolved
1res·o·lute \'re-zə-,lüt, -lət\ *adjective* [Latin *resolutus,* past participle of *resolvere*] (1533)
1 : marked by firm determination **:** RESOLVED
2 : BOLD, STEADY
synonym see FAITHFUL
— res·o·lute·ly \-,lüt-lē, -lət-; ,re-zə-'lüt-\ *adverb*
— res·o·lute·ness \-,lüt-nəs, -lət-, -'lüt-\ *noun*
2resolute *noun* (1602)
: a resolute person
res·o·lu·tion \,re-zə-'lü-shən\ *noun* [Middle English, from Middle French or Latin; Middle French *resolution,* from Latin *resolution-, resolutio,* from *resolvere*] (14th century)
1 : the act or process of reducing to simpler form: as **a :** the act of analyzing a complex notion into simpler ones **b :** the act of answering **:** SOLVING **c :** the act of determining **d :** the passing of a voice part from a dissonant to a consonant tone or the progression of a chord from dissonance to consonance **e :** the separating of a chemical compound or mixture into its constituents **f** (1) **:** the division of a prosodic element into its component parts (2) **:** the substitution in Greek or Latin prosody of two short syllables for a long syllable **g :** the analysis of a vector into two or more vectors of which it is the sum **h :** the process or capability of making distinguishable the individual parts of an object, closely adjacent optical images, or sources of light
2 : the subsidence of a pathological state (as inflammation)
3 a : something that is resolved ⟨made a *resolution* to mend my ways⟩ **b :** firmness of resolve
4 : a formal expression of opinion, will, or intent voted by an official body or assembled group
5 : the point in a literary work at which the chief dramatic complication is worked out
synonym see COURAGE
1re·solve \ri-'zälv, -'zȯlv *also* -'zäv *or* -'zȯv\ *verb* **re·solved; re·solv·ing** [Middle English, from Latin *resolvere* to unloose, dissolve, from *re-* + *solvere* to loosen, release — more at SOLVE] (14th century)
transitive verb
1 *obsolete* **:** DISSOLVE, MELT
2 a : BREAK UP, SEPARATE ⟨the prism *resolved* the light into a play of color⟩; *also* **:** to change by disintegration **b :** to reduce by analysis ⟨*resolve* the problem into simple elements⟩ **c :** to distinguish between or make independently visible adjacent parts of **d :** to separate (a racemic compound or mixture) into the two components
3 : to cause resolution of (a pathological state)
4 a : to deal with successfully **:** clear up ⟨*resolve* doubts⟩ ⟨*resolve* a dispute⟩ **b :** to find an answer to **c :** to make clear or understandable **d :** to find a mathematical solution of **e :** to split up (as a vector) into two or more components especially in assigned directions
5 : to reach a firm decision about ⟨*resolve* to get more sleep⟩ ⟨*resolve* disputed points in a text⟩
6 a : to declare or decide by a formal resolution and vote **b :** to change by resolution or formal vote ⟨the house *resolved* itself into a committee⟩
7 : to make (as voice parts) progress from dissonance to consonance
8 : to work out the resolution of (as a play)
intransitive verb
1 : to become separated into component parts; *also* **:** to become reduced by dissolving or analysis
2 : to form a resolution **:** DETERMINE
3 : CONSULT, DELIBERATE
4 : to progress from dissonance to consonance
synonym see DECIDE
— re·solv·able \-'zäl-və-bəl, -'zȯl- *also* -'zä-və- *or* -'zȯ-və-\ *adjective*
— re·solv·er *noun*
2resolve *noun* (1591)
1 : fixity of purpose **:** RESOLUTENESS
2 : something that is resolved
3 : a legal or official determination; *especially* **:** a formal resolution
re·sol·vent \ri-'zäl-vənt, -'zȯl-\ *noun* (1851)
: a means of solving something (as an equation)
— resolvent *adjective*
resolving power *noun* (1879)
1 : the ability of an optical system to form distinguishable images of objects separated by small angular distances
2 : the ability of a photographic film or plate to reproduce the fine detail of an optical image
res·o·nance \'re-zᵊn-ən(t)s, 'rez-nən(t)s\ *noun* (15th century)
1 a : the quality or state of being resonant **b** (1) **:** a vibration of large amplitude in a mechanical or electrical system caused by a relatively small periodic stimulus of the same or nearly the same period as the natural vibration period of the system (2) **:** the state of adjustment that produces resonance in a mechanical or electrical system
2 a : the intensification and enriching of a musical tone by supplementary vibration **b :** a quality imparted to voiced sounds by vibration in anatomical resonating chambers or cavities (as the mouth or the nasal cavity) **c :** a quality of richness or variety **d :** a quality of evoking response ⟨how much *resonance* the scandal seems to be having —*U.S. News & World Report*⟩
3 : the sound elicited on percussion of the chest
4 : the conceptual alternation of a chemical species (as a molecule or ion) between two or more equivalent allowed structural representations differing only in the placement of electrons that aids in understanding the actual state of the species as an amalgamation of its possible structures and the usually higher-than-expected stability of the species
5 a : the enhancement of an atomic, nuclear, or particle reaction or a scattering event by excitation of internal motion in the system **b :** MAGNETIC RESONANCE
6 : an extremely short-lived elementary particle
7 : a synchronous gravitational relationship of two celestial bodies (as moons) that orbit a

third (as a planet) which can be expressed as a simple ratio of their orbital periods

res·o·nant \'re-zᵊn-ənt, 'rez-nənt\ *adjective* (1592)

1 : continuing to sound **:** ECHOING

2 a : capable of inducing resonance **b :** relating to or exhibiting resonance

3 a : intensified and enriched by or as if by resonance **b :** marked by grandiloquence

— resonant *noun*

— res·o·nant·ly *adverb*

res·o·nate \'re-zᵊn-,āt\ *verb* **-nat·ed; -nat·ing** [Latin *resonatus,* past participle of *resonare* to resound — more at RESOUND] (1873)

intransitive verb

1 : to produce or exhibit resonance

2 : to respond as if by resonance ⟨*resonate* to the music⟩; *also* **:** to have a repetitive pattern that resembles resonance

3 : to relate harmoniously

transitive verb

: to subject to resonating

res·o·na·tor \-,ā-tər\ *noun* (circa 1869)

: something that resounds or resonates: as **a :** a hollow metallic container for producing microwaves or a piezoelectric crystal put into oscillation by the oscillations of an outside source **b :** a device for increasing the resonance of a musical instrument

re·sorb \(,)rē-'sȯrb, -'zȯrb\ *verb* [Latin *resorbēre,* from *re-* + *sorbēre* to suck up — more at ABSORB] (1640)

transitive verb

1 : to swallow or suck in again

2 : to break down and assimilate (something previously differentiated)

intransitive verb

: to undergo resorption

res·or·cin \rə-'zȯr-sᵊn\ *noun* [International Scientific Vocabulary *res-* (from Latin *resina* resin) + *orcin,* a phenol ($C_7H_8O_2$)] (circa 1868)

: RESORCINOL

res·or·cin·ol \-,ȯl, -,ōl\ *noun* (1881)

: a crystalline phenol $C_6H_6O_2$ obtained from various resins or artificially and used especially in making dyes, pharmaceuticals, and resins

re·sorp·tion \(,)rē-'sȯrp-shən, -'zȯrp-\ *noun* [Latin *resorbēre*] (circa 1820)

: the action or process of resorbing something

— re·sorp·tive \-tiv\ *adjective*

¹re·sort \ri-'zȯrt\ *noun* [Middle English, from Middle French, resource, recourse, from *resortir* to rebound, resort, from Old French, from *re-* + *sortir* to escape, sally] (14th century)

1 a : one that affords aid or refuge **:** RESOURCE ⟨went to them as a last *resort*⟩ **b :** RECOURSE

2 a : frequent, habitual, or general visiting ⟨a place of popular *resort*⟩ **b :** persons who frequent a place **:** THRONG **c** (1) **:** a frequently visited place **:** HAUNT (2) **:** a place providing recreation and entertainment especially to vacationers

synonym see RESOURCE

²resort *intransitive verb* (15th century)

1 : to go especially frequently or habitually **:** REPAIR

2 : to have recourse ⟨*resort* to force⟩

re–sort \(,)rē-'sȯrt\ *transitive verb* (1889)

: to sort again

re·sort·er \ri-'zȯr-tər\ *noun* (1917)

: a frequenter of resorts

re·sound \ri-'zau̇nd *also* -'sau̇nd\ *verb* [Middle English *resounen,* from Middle French *resoner,* from Latin *resonare,* from *re-* + *sonare* to sound — more at SOUND] (14th century)

intransitive verb

1 : to become filled with sound **:** REVERBERATE

2 a : to sound loudly **b :** to produce a sonorous or echoing sound

3 : to become renowned

transitive verb

1 : to extol loudly or widely **:** CELEBRATE

2 : ECHO, REVERBERATE

3 : to sound or utter in full resonant tones

re·sound·ing *adjective* (15th century)

1 : producing or characterized by resonant sound **:** RESONATING

2 a : impressively sonorous **b :** EMPHATIC, UNEQUIVOCAL ⟨a *resounding* success⟩

— re·sound·ing·ly \-'zau̇n-diŋ-lē *also* -'sau̇n-\ *adverb*

re·source \'rē-,sōrs, -,sȯrs, -,zōrs, -,zȯrs, ri-'\ *noun* [French *ressource,* from Old French *ressourse* relief, resource, from *resourdre* to relieve, literally, to rise again, from Latin *resurgere* — more at RESURRECTION] (1611)

1 a : a source of supply or support **:** an available means — usually used in plural **b :** a natural source of wealth or revenue — often used in plural **c :** computable wealth — usually used in plural **d :** a source of information or expertise

2 : something to which one has recourse in difficulty **:** EXPEDIENT

3 : a possibility of relief or recovery

4 : a means of spending one's leisure time

5 : an ability to meet and handle a situation **:** RESOURCEFULNESS ☆

re·source·ful \ri-'sōrs-fəl, -'sȯrs-, -'zōrs-, -'zȯrs-\ *adjective* (1851)

: able to meet situations **:** capable of devising ways and means

— re·source·ful·ly \-fə-lē\ *adverb*

— re·source·ful·ness *noun*

¹re·spect \ri-'spekt\ *noun* [Middle English, from Latin *respectus,* literally, act of looking back, from *respicere* to look back, regard, from *re-* + *specere* to look — more at SPY] (14th century)

1 : a relation or reference to a particular thing or situation ⟨remarks having *respect* to an earlier plan⟩

2 : an act of giving particular attention **:** CONSIDERATION

3 a : high or special regard **:** ESTEEM **b :** the quality or state of being esteemed **c** *plural* **:** expressions of respect or deference ⟨paid our *respects*⟩

4 : PARTICULAR, DETAIL ⟨a good plan in some *respects*⟩

— in respect of *chiefly British* **:** with respect to **:** CONCERNING

— in respect to : with respect to **:** CONCERNING

— with respect to : with reference to **:** in relation to

²respect *transitive verb* (1560)

1 a : to consider worthy of high regard **:** ESTEEM **b :** to refrain from interfering with

2 : to have reference to **:** CONCERN

synonym see REGARD

— re·spect·er *noun*

¹re·spect·able \ri-'spek-tə-bəl\ *adjective* (1599)

1 : worthy of respect **:** ESTIMABLE

2 : decent or correct in character or behavior **:** PROPER

3 a : fair in size or quantity ⟨*respectable* amount⟩ **b :** moderately good **:** TOLERABLE

4 : fit to be seen **:** PRESENTABLE ⟨*respectable* clothes⟩

— re·spect·abil·i·ty \-,spek-tə-'bi-lə-tē\ *noun*

— re·spect·able·ness \-'spek-tə-bəl-nəs\ *noun*

— re·spect·ably \-blē\ *adverb*

²respectable *noun* (1814)

: a respectable person

re·spect·ful \ri-'spekt-fəl\ *adjective* (1687)

: marked by or showing respect or deference

— re·spect·ful·ly \-fə-lē\ *adverb*

— re·spect·ful·ness *noun*

re·spect·ing *preposition* (1611)

1 : in view of **:** CONSIDERING

2 : with respect to **:** CONCERNING

re·spec·tive \ri-'spek-tiv\ *adjective* (1592)

1 *obsolete* **:** PARTIAL, DISCRIMINATIVE

2 : PARTICULAR, SEPARATE ⟨their *respective* homes⟩

— re·spec·tive·ness *noun*

re·spec·tive·ly *adverb* (1605)

1 : in particular **:** SEPARATELY ⟨could not recognize the solutions as salty or sour, *respectively*⟩

2 : in the order given ⟨Mary and Anne were *respectively* 12 and 16 years old⟩

re·spell \(,)rē-'spel\ *transitive verb* (1806)

: to spell again or in another way; *especially* **:** to spell out according to a phonetic system

— re·spell·ing \-'spe-liŋ\ *noun*

re·spi·ra·ble \'res-p(ə-)rə-bəl, ri-'spī-rə-\ *adjective* (1779)

: fit for breathing; *also* **:** capable of being taken in by breathing ⟨*respirable* particles of ash⟩

res·pi·ra·tion \,res-pə-'rā-shən\ *noun* [Middle English *respiracioun,* from Latin *respiration-, respiratio,* from *respirare*] (15th century)

1 a : the placing of air or dissolved gases in intimate contact with the circulating medium of a multicellular organism (as by breathing) **b :** a single complete act of breathing

2 : the physical and chemical processes by which an organism supplies its cells and tissues with the oxygen needed for metabolism and relieves them of the carbon dioxide formed in energy-producing reactions

3 : any of various energy-yielding oxidative reactions in living matter

— re·spi·ra·to·ry \'res-p(ə-)rə-,tōr-ē, ri-'spī-rə-, -,tȯr-\ *adjective*

res·pi·ra·tor \'res-pə-,rā-tər\ *noun* (1836)

1 : a device worn over the mouth or nose for protecting the respiratory tract

2 : a device for maintaining artificial respiration

respiratory pigment *noun* (1896)

: any of various permanently or intermittently colored conjugated proteins and especially hemoglobin that function in the transfer of oxygen in cellular respiration

respiratory quotient *noun* (circa 1890)

: a ratio indicating the relation of the volume of carbon dioxide given off in respiration to that of the oxygen consumed

respiratory system *noun* (1940)

: a system of organs subserving the function of respiration and in air-breathing vertebrates consisting typically of the lungs and their nervous and circulatory supply and the channels by which these are continuous with the outer air

re·spire \ri-'spīr\ *verb* **re·spired; re·spir·ing** [Middle English, from Latin *respirare,* from *re-* + *spirare* to blow, breathe] (15th century)

intransitive verb

☆ SYNONYMS

Resource, resort, expedient, shift, makeshift, stopgap mean something one turns to in the absence of the usual means or source of supply. RESOURCE and RESORT apply to anything one falls back upon ⟨exhausted all of their *resources*⟩ ⟨a last *resort*⟩. EXPEDIENT may apply to any device or contrivance used when the usual one is not at hand or not possible ⟨a flimsy *expedient*⟩. SHIFT implies a tentative or temporary imperfect expedient ⟨desperate *shifts* to stave off foreclosure⟩. MAKESHIFT implies an inferior expedient adopted because of urgent need or countenanced through indifference ⟨old equipment employed as a *makeshift*⟩. STOPGAP applies to something used temporarily as an emergency measure ⟨a new law intended only as a *stopgap*⟩.

1 : BREATHE; *specifically* **:** to inhale and exhale air successively
2 *of a cell or tissue* **:** to take up oxygen and produce carbon dioxide through oxidation
transitive verb
: BREATHE

res·pi·rom·e·ter \ˌres-pə-ˈrä-mə-tər\ *noun* (circa 1883)
: an instrument for studying the character and extent of respiration
— res·pi·ro·met·ric \-rō-ˈme-trik\ *adjective*
— res·pi·rom·e·try \-ˈrä-mə-trē\ *noun*

[1]re·spite \ˈres-pət *also* ri-ˈspīt, *British usually* ˈres-ˌpīt\ *noun* [Middle English *respit,* from Old French, from Medieval Latin *respectus,* from Latin, act of looking back — more at RESPECT] (13th century)
1 : a period of temporary delay; *especially* **:** REPRIEVE 1b
2 : an interval of rest or relief

[2]respite *transitive verb* **re·spit·ed; re·spit·ing** (14th century)
1 : to grant a respite to
2 : PUT OFF, DELAY

re·splen·dence \ri-ˈsplen-dən(t)s\ *noun* (15th century)
: the quality or state of being resplendent **:** SPLENDOR

re·splen·den·cy \-dən(t)-sē\ *noun* (1611)
: RESPLENDENCE

re·splen·dent \-dənt\ *adjective* [Latin *resplendent-, resplendens,* present participle of *resplendēre* to shine back, from *re-* + *splendēre* to shine — more at SPLENDID] (15th century)
: shining brilliantly **:** characterized by a glowing splendor ⟨meadows *resplendent* with wildflowers —*Outdoor World*⟩
synonym see SPLENDID
— re·splen·dent·ly *adverb*

[1]re·spond \ri-ˈspänd\ *noun* (15th century)
: an engaged pillar supporting an arch or closing a colonnade or arcade

[2]respond *verb* [Middle French *respondre,* from Latin *respondēre* to promise in return, answer, from *re-* + *spondēre* to promise — more at SPOUSE] (1719)
intransitive verb
1 : to say something in return **:** make an answer ⟨*respond* to criticism⟩
2 a : to react in response ⟨*responded* to a call for help⟩ **b :** to show favorable reaction ⟨*respond* to surgery⟩
3 : to be answerable ⟨*respond* in damages⟩
transitive verb
: REPLY
synonym see ANSWER
— re·spond·er \-ˈspän-dər\ *noun*

[1]re·spon·dent \ri-ˈspän-dənt\ *noun* [Latin *respondent-, respondens,* present participle of *respondēre*] (1528)
1 : one who responds: as **a :** one who maintains a thesis in reply **b** (1) **:** one who answers in various legal proceedings (as in equity cases) (2) **:** the prevailing party in the lower court **c :** a person who responds to a poll
2 : a reflex that occurs in response to a specific external stimulus — compare OPERANT

[2]respondent *adjective* (1726)
1 : making response **:** RESPONSIVE; *especially* **:** being a respondent at law
2 : relating to or being behavior or responses to a stimulus that are followed by a reward ⟨*respondent* conditioning⟩ — compare OPERANT 3

re·sponse \ri-ˈspän(t)s\ *noun* [Middle English & Latin; Middle English *respounse,* from Middle French *respons,* from Latin *responsum* reply, from neuter of *responsus,* past participle of *respondēre*] (14th century)
1 : an act of responding
2 : something constituting a reply or a reaction: as **a :** a verse, phrase, or word sung or said by the people or choir after or in reply to the officiant in a liturgical service **b :** the activity or inhibition of previous activity of an organism or any of its parts resulting from stimulation **c :** the output of a transducer or detecting device resulting from a given input

re·spon·si·bil·i·ty \ri-ˌspän(t)-sə-ˈbi-lə-tē\ *noun, plural* **-ties** (1787)
1 : the quality or state of being responsible: as **a :** moral, legal, or mental accountability **b :** RELIABILITY, TRUSTWORTHINESS
2 : something for which one is responsible **:** BURDEN

re·spon·si·ble \ri-ˈspän(t)-sə-bəl\ *adjective* (1643)
1 a : liable to be called on to answer **b** (1) **:** liable to be called to account as the primary cause, motive, or agent ⟨a committee *responsible* for the job⟩ (2) **:** being the cause or explanation ⟨mechanical defects were *responsible* for the accident⟩ **c :** liable to legal review or in case of fault to penalties
2 a : able to answer for one's conduct and obligations **:** TRUSTWORTHY **b :** able to choose for oneself between right and wrong
3 : marked by or involving responsibility or accountability ⟨*responsible* financial policies⟩ ⟨a *responsible* job⟩
4 : politically answerable; *especially* **:** required to submit to the electorate if defeated by the legislature — used especially of the British cabinet ☆
— re·spon·si·ble·ness *noun*
— re·spon·si·bly \-blē\ *adverb*

re·spon·sions \ri-ˈspän(t)-shənz\ *noun plural* [Middle English *responcioun* response, sum to be paid, from Middle French or Medieval Latin; Middle French *responsion,* from Medieval Latin *responsion-, responsio,* from Latin, answer, from *respondēre*] (1813)
: an examination required for matriculation as an undergraduate at Oxford

re·spon·sive \ri-ˈspän(t)-siv\ *adjective* (15th century)
1 : giving response **:** constituting a response **:** ANSWERING ⟨a *responsive* glance⟩ ⟨*responsive* aggression⟩
2 : quick to respond or react appropriately or sympathetically **:** SENSITIVE
3 : using responses ⟨*responsive* worship⟩
— re·spon·sive·ly *adverb*
— re·spon·sive·ness *noun*

re·spon·so·ry \-ˈspän(t)s-(ə-)rē\ *noun, plural* **-ries** [Middle English, from Medieval Latin *responsorius,* from Latin *respondēre*] (15th century)
: a set of versicles and responses sung or said after or during a lection

re·spon·sum \ri-ˈspän(t)-səm\ *noun, plural* **-sa** \-sə\ [New Latin, from Latin, reply, response] (1896)
: a written decision from a rabbinic authority in response to a submitted question or problem

res pu·bli·ca \ˈrās-ˈpü-bli-ˌkä\ *noun* [Latin — more at REPUBLIC] (circa 1898)
1 : COMMONWEALTH, STATE, REPUBLIC
2 : COMMONWEAL

res·sen·ti·ment \rə-ˌsä^{n}-tē-ˈmä^{n}\ *noun* [French, resentment, from *ressentir* to resent — more at RESENT] (1941)
: deep-seated resentment, frustration, and hostility accompanied by a sense of being powerless to express these feelings directly

[1]rest \ˈrest\ *noun* [Middle English, from Old English; akin to Old High German *rasta* rest and perhaps to Old High German *ruowa* calm] (before 12th century)
1 : REPOSE, SLEEP; *specifically* **:** a bodily state characterized by minimal functional and metabolic activities
2 a : freedom from activity or labor **b :** a state of motionlessness or inactivity **c :** the repose of death
3 : a place for resting or lodging
4 : peace of mind or spirit
5 a (1) **:** a rhythmic silence in music (2) **:** a character representing such a silence **b :** a brief pause in reading
6 : something used for support
— at rest 1 : resting or reposing especially in sleep or death **2 :** QUIESCENT, MOTIONLESS **3 :** free of anxieties

rest 5a(2): ***1*** **whole,** ***2*** **half,** ***3*** **quarter,** ***4*** **eighth,** ***5*** **sixteenth**

[2]rest (before 12th century)
intransitive verb
1 a : to get rest by lying down; *especially* **:** SLEEP **b :** to lie dead
2 : to cease from action or motion **:** refrain from labor or exertion
3 : to be free from anxiety or disturbance
4 : to sit or lie fixed or supported ⟨a column *rests* on its pedestal⟩
5 a : to remain confident **:** TRUST ⟨cannot *rest* on that assumption⟩ **b :** to be based or founded ⟨the verdict *rested* on several sound precedents⟩
6 : to remain for action or accomplishment ⟨the answer *rests* with you⟩
7 *of farmland* **:** to remain idle or uncropped
8 : to bring to an end voluntarily the introduction of evidence in a law case
transitive verb
1 : to give rest to
2 : to set at rest
3 : to place on or against a support
4 : to cause to be firmly fixed ⟨*rested* all hope in his child⟩
5 : to stop voluntarily from presenting evidence pertinent to (a case at law)
— rest·er *noun*

[3]rest *noun* [Middle English *reste,* literally, stoppage, short for *areste,* from Middle French, from Old French, from *arester* to arrest] (14th century)
: a projection or attachment on the side of the breastplate of medieval armor for supporting the butt of a lance

[4]rest *noun* [Middle English, from Middle French *reste,* from *rester* to remain, from Latin *restare,* from *re-* + *stare* to stand — more at STAND] (15th century)
: something that remains over **:** REMAINDER ⟨ate the *rest* of the candy⟩
— for the rest : with regard to remaining issues or needs

re·start \(ˌ)rē-ˈstärt\ (1845)
transitive verb
1 : to start anew
2 : to resume (as an activity) after interruption
intransitive verb
: to resume operation
— re·start \ˈrē-ˌstärt, (ˌ)re-ˈ\ *noun*

☆ SYNONYMS
Responsible, answerable, accountable, amenable, liable mean subject to being held to account. RESPONSIBLE implies holding a specific office, duty, or trust ⟨the bureau *responsible* for revenue collection⟩. ANSWERABLE suggests a relation between one having a moral or legal obligation and a court or other authority charged with oversight of its observance ⟨an intelligence agency *answerable* to Congress⟩. ACCOUNTABLE suggests imminence of retribution for unfulfilled trust or violated obligation ⟨elected officials are *accountable* to the voters⟩. AMENABLE and LIABLE stress the fact of subjection to review, censure, or control by a designated authority under certain conditions ⟨laws are *amenable* to judicial review⟩ ⟨not *liable* for the debts of the former spouse⟩.

— **re•start•able** \-'stär-tə-bəl\ *adjective*

re•state \(,)rē-'stāt\ *transitive verb* (circa 1713)
: to state again or in another way

re•state•ment \-mənt\ *noun* (1803)
1 : something that is restated
2 : the act of restating

res•tau•rant \'res-t(ə-)rənt, -tə-,ränt, -,tränt, -tərnt\ *noun* [French, from present participle of *restaurer* to restore, from Latin *restaurare*] (1827)
: a business establishment where meals or refreshments may be procured ◆

res•tau•ra•teur \,res-tə-rə-'tər\ *also* **res•tau•ran•teur** \-,rän-\ *noun* [French *restaurateur,* from Late Latin *restaurator* restorer, from Latin *restaurare*] (1796)
: the operator or proprietor of a restaurant

rest•ful \'rest-fəl\ *adjective* (14th century)
1 : marked by, affording, or suggesting rest and repose ⟨a *restful* color scheme⟩
2 : being at rest **:** QUIET
synonym see COMFORTABLE
— **rest•ful•ly** \-fə-lē\ *adverb*
— **rest•ful•ness** *noun*

rest home *noun* (1926)
: an establishment that provides housing and general care for the aged or the convalescent

rest house *noun* (1807)
: a building used for shelter by travelers

rest•ing *adjective* (14th century)
1 : being or characterized by dormancy **:** QUIESCENT ⟨a *resting* spore⟩ ⟨bulbs in the *resting* state⟩
2 : not undergoing or marked by division **:** VEGETATIVE ⟨a *resting* nucleus⟩

res•ti•tute \'res-tə-,tüt, -,tyüt\ *verb* **-tut•ed; -tut•ing** [Latin *restitutus,* past participle of *restituere*] (circa 1500)
transitive verb
1 : to restore to a former state or position
2 : GIVE BACK; *especially* **:** REFUND
intransitive verb
: to undergo restitution

res•ti•tu•tion \,res-tə-'tü-shən, -'tyü-\ *noun* [Middle English, from Middle French, from Latin *restitution-, restitutio,* from *restituere* to restore, from *re-* + *statuere* to set up — more at STATUTE] (14th century)
1 : an act of restoring or a condition of being restored: as **a :** a restoration of something to its rightful owner **b :** a making good of or giving an equivalent for some injury
2 : a legal action serving to cause restoration of a previous state

res•tive \'res-tiv\ *adjective* [Middle English *restyf,* from Middle French *restif,* from *rester* to stop behind, remain] (15th century)
1 : stubbornly resisting control **:** BALKY
2 : marked by impatience **:** FIDGETY
synonym see CONTRARY
— **res•tive•ly** *adverb*
— **res•tive•ness** *noun*

rest•less \'rest-ləs\ *adjective* (before 12th century)
1 : lacking or denying rest **:** UNEASY ⟨a *restless* night⟩
2 : continuously moving **:** UNQUIET ⟨the *restless* sea⟩
3 : characterized by or manifesting unrest especially of mind ⟨*restless* pacing⟩; *also* **:** CHANGEFUL, DISCONTENTED
— **rest•less•ly** *adverb*
— **rest•less•ness** *noun*

rest mass *noun* (1914)
: the mass of a body exclusive of additional mass the body acquires by its motion according to the theory of relativity

re•stor•able \ri-'stōr-ə-bəl, -'stȯr-\ *adjective* (1611)
: fit for restoring or reclaiming

re•stor•al \-əl\ *noun* (1611)
: RESTORATION

res•to•ra•tion \,res-tə-'rā-shən\ *noun* (14th century)
1 : an act of restoring or the condition of being restored: as **a :** a bringing back to a former position or condition **:** REINSTATEMENT ⟨the *restoration* of peace⟩ **b :** RESTITUTION **c :** a restoring to an unimpaired or improved condition ⟨the *restoration* of a painting⟩ **d :** the replacing of missing teeth or crowns
2 : something that is restored; *especially* **:** a representation or reconstruction of the original form (as of a fossil or a building)
3 *capitalized* **a :** the reestablishing of the monarchy in England in 1660 under Charles II **b :** the period in English history usually held to coincide with the reign of Charles II but sometimes to extend through the reign of James II

¹re•stor•ative \ri-'stōr-ə-tiv, -'stȯr-\ *adjective* (14th century)
: of or relating to restoration; *especially* **:** having power to restore

²restorative *noun* (15th century)
: something that serves to restore to consciousness, vigor, or health

re•store \ri-'stōr, -'stȯr\ *transitive verb* **re•stored; re•stor•ing** [Middle English, from Old French *restorer,* from Latin *restaurare* to renew, rebuild, alteration of *instaurare* to renew] (14th century)
1 : GIVE BACK, RETURN
2 : to put or bring back into existence or use
3 : to bring back to or put back into a former or original state **:** RENEW
4 : to put again in possession of something
synonym see RENEW
— **re•stor•er** *noun*

re•strain \ri-'strān\ *transitive verb* [Middle English *restraynen,* from Middle French *restraindre,* from Latin *restringere* to restrain, restrict, from *re-* + *stringere* to bind tight — more at STRAIN] (14th century)
1 a : to prevent from doing, exhibiting, or expressing something ⟨*restrained* the child from jumping⟩ **b :** to limit, restrict, or keep under control ⟨try to *restrain* your anger⟩
2 : to moderate or limit the force, effect, development, or full exercise of ⟨*restrain* trade⟩
3 : to deprive of liberty; *especially* **:** to place under arrest or restraint ☆
— **re•strain•able** \-'strā-nə-bəl\ *adjective*
— **re•strain•er** *noun*

re•strained \ri-'strānd\ *adjective* (14th century)
: marked by restraint **:** being without excess or extravagance
— **re•strain•ed•ly** \-'strā-nəd-lē\ *adverb*

restraining order *noun* (circa 1876)
1 : a preliminary legal order sometimes issued to keep a situation unchanged pending decision upon an application for an injunction
2 : a legal order issued against an individual to restrict or prohibit access or proximity to another specified individual

re•straint \ri-'strānt\ *noun* [Middle English, from Middle French *restrainte,* from *restraindre*] (15th century)
1 a : an act of restraining **:** the state of being restrained **b** (1) **:** a means of restraining **:** a restraining force or influence (2) **:** a device that restricts movement ⟨a *restraint* for children riding in cars⟩
2 : a control over the expression of one's emotions or thoughts

re•strict \ri-'strikt\ *transitive verb* [Latin *restrictus,* past participle of *restringere*] (1535)
1 : to confine within bounds **:** RESTRAIN
2 : to place under restrictions as to use or distribution
synonym see LIMIT

re•strict•ed *adjective* (circa 1828)
: subject or subjected to restriction: as **a :** not general **:** LIMITED ⟨the decision had a *restricted* effect⟩ **b :** available to the use of particular groups or specifically excluding others ⟨a *restricted* country club⟩ **c :** not intended for general circulation or release ⟨a *restricted* document⟩
— **re•strict•ed•ly** *adverb*

re•stric•tion \ri-'strik-shən\ *noun* [Middle English *restriccioun,* from Late Latin *restriction-, restrictio,* from Latin *restringere*] (15th century)
1 : something that restricts: as **a :** a regulation that restricts or restrains ⟨*restrictions* for hunters⟩ **b :** a limitation on the use or enjoyment of property or a facility
2 : an act of restricting **:** the condition of being restricted

restriction enzyme *noun* (1965)
: any of various enzymes that break DNA into fragments at specific sites in the interior of the molecule — called also *restriction endonuclease*

re•stric•tion•ism \-shə-,ni-zəm\ *noun* (1937)
: a policy or philosophy favoring restriction (as of trade or immigration)
— **re•stric•tion•ist** \-sh(ə-)nist\ *adjective or noun*

re•stric•tive \ri-'strik-tiv\ *adjective* (1579)
1 a : of or relating to restriction **b :** serving or tending to restrict ⟨*restrictive* regulations⟩
2 : limiting the reference of a modified word or phrase
3 : prohibiting further negotiation
— **restrictive** *noun*
— **re•stric•tive•ly** *adverb*
— **re•stric•tive•ness** *noun*

☆ SYNONYMS

Restrain, check, curb, bridle mean to hold back from or control in doing something. RESTRAIN suggests holding back by force or persuasion from acting or from going to extremes ⟨*restrained* themselves from laughing⟩. CHECK implies restraining or impeding a progress, activity, or impetus ⟨trying to *check* government spending⟩. CURB suggests an abrupt or drastic checking ⟨learn to *curb* your appetite⟩. BRIDLE implies keeping under control by subduing or holding in ⟨*bridle* an impulse to throw the book down⟩.

◇ WORD HISTORY

restaurant The *restaurant*—both word and institution—allegedly originated in Paris in the second half of the 18th century. According to an oft-repeated account first published in 1853, the first *restaurant* was opened in 1765 by a Parisian named Boulanger, though the word is not definitely attested in French before 1803. Boulanger's establishment offered to its customers *bouillons restaurants* ("restorative broths," from which the name *restaurant* supposedly derived), fowl, and eggs under the Latin motto *Venite ad me omnes qui stomacho laboratis et ego vos restaurabo* ("Come to me all who suffer from pain of the stomach and I will restore you"), which punningly alludes to both *restaurant* and the invitation of Matthew 11:28 (Revised Standard Version) "Come to me all who labor and are heavy-laden, and I will give you rest." Similar but more elaborate establishments soon opened, though cooked meat dishes (*ragoûts*) briefly remained the domain of someone called a *traiteur* "treater," a sort of caterer who had preceded the *restaurateur* in providing food to Frenchmen who did not have their own kitchens. (*Traiteur* was borrowed into Italian as *trattore,* the base of *trattoria,* a common Italian word for "restaurant" now familiar to many English-speaking diners.)

restrictive clause *noun* (circa 1895)
: a descriptive clause that is essential to the definiteness of the word it modifies (as *that you ordered* in "the book that you ordered is out of print")

re·strike \(ˌ)rē-ˈstrīk, ˈrē-ˌ\ *noun* (1899)
: a coin or medal struck from an original die at some time after the original issue

rest room *noun* (1899)
: a room or suite of rooms providing toilets and lavatories

re·struc·ture \(ˌ)rē-ˈstrək-chər\ (1942)
transitive verb
: to change the makeup, organization, or pattern of
intransitive verb
: to restructure something

¹re·sult \ri-ˈzəlt\ *intransitive verb* [Middle English, from Medieval Latin *resultare,* from Latin, to rebound, from *re-* + *saltare* to leap — more at SALTATION] (15th century)
1 a : to proceed or arise as a consequence, effect, or conclusion ⟨death *resulted* from the disease⟩ **b :** to have an issue or result ⟨the disease *resulted* in death⟩
2 : REVERT 2

²result *noun* (1647)
1 : something that results as a consequence, issue, or conclusion; *also* **:** beneficial or tangible effect **:** FRUIT
2 : something obtained by calculation or investigation
— re·sult·ful \-fəl\ *adjective*
— re·sult·less \-ləs\ *adjective*

¹re·sul·tant \ri-ˈzəl-tᵊnt\ *adjective* (1639)
: derived from or resulting from something else
— re·sul·tant·ly *adverb*

²resultant *noun* (1815)
: something that results **:** OUTCOME; *specifically* **:** the single vector that is the sum of a given set of vectors

re·sume \ri-ˈzüm\ *verb* **re·sumed; re·sum·ing** [Middle English, from Middle French or Latin; Middle French *resumer,* from Latin *resumere,* from *re-* + *sumere* to take up, take — more at CONSUME] (15th century)
transitive verb
1 : to assume or take again **:** REOCCUPY ⟨*resumed* his seat by the fire —Thomas Hardy⟩
2 : to return to or begin again after interruption ⟨*resumed* her work⟩
3 : to take back to oneself
4 : to pick up again
5 : REITERATE, SUMMARIZE
intransitive verb
: to begin again something interrupted

ré·su·mé *or* **re·su·me** *or* **re·su·mé** \ˈre-zə-ˌmā, ˌre-zə-ˈ *also* ˈrā- *or* ˌrā-\ *noun* [French *résumé,* from past participle of *résumer* to resume, summarize, from Middle French *resumer*] (1804)
1 : SUMMARY
2 : CURRICULUM VITAE

re·sump·tion \ri-ˈzəm(p)-shən\ *noun* [Middle English, from Middle French or Late Latin; Middle French *resomption,* from Late Latin *resumption-, resumptio,* from Latin *resumere*] (15th century)
1 : an act or instance of resuming **:** RECOMMENCEMENT
2 : a return to payment in specie

re·su·pi·nate \ri-ˈsü-pə-ˌnāt\ *adjective* [Latin *resupinatus,* past participle of *resupinare* to bend back to a supine position, from *re-* + *supinus* supine] (circa 1776)
1 : inverted in position ⟨*resupinate* orchid flowers⟩
2 : having or being a fruiting body lying flat on the substrate with the hymenium at the periphery or over the whole surface ⟨*resupinate* fungi⟩ ⟨*resupinate* sporophores⟩

re·sur·face \(ˌ)rē-ˈsər-fəs\ (1894)
transitive verb
: to provide with a new or fresh surface
intransitive verb
: to come again to the surface (as of the water); *broadly* **:** to appear or show up again
— re·sur·fac·er \-fə-sər\ *noun*

re·surge \ri-ˈsərj\ *intransitive verb* **re·surged; re·surg·ing** [Latin *resurgere*] (1575)
: to undergo a resurgence

re·sur·gence \ri-ˈsər-jən(t)s\ *noun* (circa 1834)
: a rising again into life, activity, or prominence **:** RENASCENCE

re·sur·gent \-jənt\ *adjective* [Latin *resurgent-, resurgens,* present participle of *resurgere*] (1808)
: undergoing or tending to produce resurgence

res·ur·rect \ˌre-zə-ˈrekt\ *transitive verb* [back-formation from *resurrection*] (1772)
1 : to raise from the dead
2 : to bring to view, attention, or use again

res·ur·rec·tion \ˌre-zə-ˈrek-shən\ *noun* [Middle English, from Late Latin *resurrection-, resurrectio* act of rising from the dead, from *resurgere* to rise from the dead, from Latin, to rise again, from *re-* + *surgere* to rise — more at SURGE] (14th century)
1 a *capitalized* **:** the rising of Christ from the dead **b** *often capitalized* **:** the rising again to life of all the human dead before the final judgment **c :** the state of one risen from the dead
2 : RESURGENCE, REVIVAL
3 *Christian Science* **:** a spiritualization of thought **:** material belief that yields to spiritual understanding
— res·ur·rec·tion·al \-shnəl, -shə-nᵊl\ *adjective*

res·ur·rec·tion·ist \-sh(ə-)nist\ *noun* (1776)
1 : BODY SNATCHER
2 : one who resurrects

re·sus·ci·tate \ri-ˈsə-sə-ˌtāt\ *verb* **-tat·ed; -tat·ing** [Latin *resuscitatus,* past participle of *resuscitare* to reawaken, from *re-* + *suscitare* to rouse, from *sub-, sus-* up + *citare* to put in motion, stir — more at SUB-, CITE] (1532)
transitive verb
: to revive from apparent death or from unconsciousness; *also* **:** REVITALIZE
intransitive verb
: COME TO, REVIVE
— re·sus·ci·ta·tion \ri-ˌsə-sə-ˈtā-shən, ˌrē-\ *noun*
— re·sus·ci·ta·tive \ri-ˈsə-sə-ˌtā-tiv\ *adjective*

re·sus·ci·ta·tor \ri-ˈsə-sə-ˌtā-tər\ *noun* (circa 1843)
: one that resuscitates; *specifically* **:** an apparatus used to restore respiration (as to a partially asphyxiated person)

ret \ˈret\ *verb* **ret·ted; ret·ting** [Middle English, from Middle Dutch] (14th century)
transitive verb
: to soak (as flax) to loosen the fiber from the woody tissue
intransitive verb
: to become retted

re·ta·ble \ˈrē-ˌtā-bəl, rē-ˈ\ *noun* [French, from Middle French, modification of Old Provençal *retaule,* alteration of *reretaule,* ultimately from Latin *retro-* + *tabula* board, tablet] (circa 1823)
: a raised shelf above an altar for the altar cross, the altar lights, and flowers

¹re·tail \ˈrē-ˌtāl, *especially for 2 also* ri-ˈtā(ə)l\ *verb* [Middle English, from Middle French *retaillier* to cut back, divide into pieces, from Old French, from *re-* + *taillier* to cut — more at TAILOR] (15th century)
transitive verb
1 : to sell in small quantities directly to the ultimate consumer
2 : TELL, RETELL
intransitive verb
: to sell at retail
— re·tail·er *noun*

²re·tail \ˈrē-ˌtāl\ *noun* (15th century)
: the sale of commodities or goods in small quantities to ultimate consumers; *also* **:** the industry of such selling
— at retail 1 : at a retailer's price **2 :** ⁴RETAIL

³re·tail \ˈrē-ˌtāl\ *adjective* (1601)
: of, relating to, or engaged in the sale of commodities at retail ⟨*retail* trade⟩

⁴re·tail \ˈrē-ˌtāl\ *adverb* (1784)
: in small quantities **:** from a retailer

re·tail·ing \ˈrē-ˌtā-liŋ\ *noun* (14th century)
: the activities involved in the selling of goods to ultimate consumers for personal or household consumption

re·tain \ri-ˈtān\ *transitive verb* [Middle English *reteinen, retainen,* from Middle French *retenir,* from Latin *retinēre* to hold back, keep, restrain, from *re-* + *tenēre* to hold — more at THIN] (15th century)
1 a : to keep in possession or use **b :** to keep in one's pay or service; *specifically* **:** to employ by paying a retainer **c :** to keep in mind or memory **:** REMEMBER
2 : to hold secure or intact
synonym see KEEP

retained object *noun* (circa 1904)
: an object of a verb in the predicate of a passive construction (as *me* in "a book was given me" and *book* in "I was given a book")

¹re·tain·er \ri-ˈtā-nər\ *noun* (1540)
1 a : a person attached or owing service to a household; *especially* **:** SERVANT **b :** EMPLOYEE
2 : one that retains
3 : any of various devices used for holding something

²retainer *noun* [Middle English *reteiner* act of withholding, from *reteinen* + Anglo-French *-er* (as in *weyver* waiver)] (1775)
1 : the act of a client by which the services of a lawyer, counselor, or adviser are engaged
2 : a fee paid to a lawyer or professional adviser for advice or services or for a claim on services when needed

¹re·take \(ˌ)rē-ˈtāk\ *transitive verb* **-took** \-ˈtu̇k\; **-tak·en** \-ˈtā-kən\; **-tak·ing** (15th century)
1 : to take or receive again
2 : RECAPTURE
3 : to photograph again

²re·take \ˈrē-ˌtāk\ *noun* (1916)
: a subsequent filming, photographing, or recording undertaken to improve upon the first; *also* **:** an instance of this

re·tal·i·ate \ri-ˈta-lē-ˌāt\ *verb* **-at·ed; -at·ing** [Late Latin *retaliatus,* past participle of *retaliare,* from Latin *re-* + *talio* legal retaliation] (1611)
transitive verb
: to repay (as an injury) in kind
intransitive verb
: to return like for like; *especially* **:** to get revenge
synonym see RECIPROCATE
— re·tal·i·a·tion \ri-ˌta-lē-ˈā-shən, ˌrē-\ *noun*
— re·tal·i·a·tive \ri-ˈta-lē-ˌā-tiv\ *adjective*
— re·tal·ia·to·ry \-ˈtal-yə-ˌtōr-ē, -ˈta-lē-ə-, -ˌtȯr-\ *adjective*

¹re·tard \ri-ˈtärd\ *verb* [Middle English, from Middle French or Latin; Middle French *retarder,* from Latin *retardare,* from *re-* + *tardus* slow] (15th century)
transitive verb
1 : to slow up especially by preventing or hindering advance or accomplishment **:** IMPEDE
2 : to delay academic progress by failure to promote
intransitive verb
: to undergo retardation
synonym see DELAY
— re·tard·er *noun*

²re·tard *noun* (1788)
1 \ri-ˈtärd\ **:** a holding back or slowing down **:** RETARDATION

2 \ˈrē-ˌtärd\ **:** a retarded person; *also* **:** a person held to resemble a retarded person in behavior — often taken to be offensive

re·tar·dant \ri-ˈtär-dᵊnt\ *adjective* (1642)
: serving or tending to retard ⟨a growth-*retardant* substance⟩
— retardant *noun*

re·tar·date \-ˈtär-ˌdāt, -dət\ *noun* (1915)
: a mentally retarded person

re·tar·da·tion \ˌrē-ˌtär-ˈdā-shən, ri-\ *noun* (15th century)
1 : an act or instance of retarding
2 : the extent to which something is retarded
3 : a musical suspension; *specifically* **:** one that resolves upward
4 a : an abnormal slowness of thought or action; *also* **:** less than normal intellectual competence usually characterized by an IQ of less than 70 **b :** slowness in development or progress

re·tard·ed \ri-ˈtär-dəd\ *adjective* (1895)
: slow or limited in intellectual or emotional development or academic progress

retch \ˈrech, *especially British* ˈrēch\ *verb* [(assumed) Middle English *rechen* to spit, retch, from Old English *hrǣcan* to spit, hawk; akin to Old Norse *hrækja* to spit] (circa 1798)
transitive verb
: VOMIT 1
intransitive verb
: to make an effort to vomit; *also* **:** VOMIT
— retch *noun*

re·te \ˈrē-tē, ˈrā-\ *noun, plural* **re·tia** \ˈrē-tē-ə, ˈrā-\ [New Latin, from Latin, net] (1541)
1 : a network especially of blood vessels or nerves **:** PLEXUS
2 : an anatomical part resembling or including a network

re·tell \(ˌ)rē-ˈtel\ *transitive verb* **-told** \-ˈtōld\; **-tell·ing** (1593)
1 : to tell again or in another form
2 : to count again

re·tell·ing *noun* (1883)
: a new version of a story ⟨a *retelling* of a Greek legend⟩

re·ten·tion \ri-ˈten(t)-shən\ *noun* [Middle English *retencioun,* from Latin *retention-, retentio,* from *retinēre* to retain — more at RETAIN] (14th century)
1 a : the act of retaining **:** the state of being retained **b :** abnormal retaining of a fluid or secretion in a body cavity
2 a : power of retaining **:** RETENTIVENESS **b :** an ability to retain things in mind; *specifically* **:** a preservation of the aftereffects of experience and learning that makes recall or recognition possible
3 : something retained

re·ten·tive \-ˈten-tiv\ *adjective* [Middle English *retentif,* from Middle French & Medieval Latin; Middle French, from Medieval Latin *retentivus,* from Latin *retentus,* past participle of *retinēre*] (14th century)
: having the power, property, or capacity of retaining ⟨soils *retentive* of moisture⟩; *especially* **:** retaining knowledge easily
— re·ten·tive·ly *adverb*
— re·ten·tive·ness *noun*

re·ten·tiv·i·ty \ˌrē-ˌten-ˈti-və-tē, ri-\ *noun* (1881)
: the power of retaining; *specifically* **:** the capacity for retaining magnetism after the action of the magnetizing force has ceased

re·think \(ˌ)rē-ˈthiŋk\ *verb* **-thought** \-ˈthȯt\; **-think·ing** (1700)
transitive verb
: to think about again **:** RECONSIDER
intransitive verb
: to engage in reconsideration
— re·think \ˈrē-ˌthiŋk, -ˈthiŋk\ *noun*
— re·think·er *noun*

ret·i·cence \ˈre-tə-sən(t)s\ *noun* (1603)
1 : the quality or state of being reticent **:** RESERVE, RESTRAINT
2 : an instance of being reticent
3 : RELUCTANCE 1

ret·i·cen·cy \-sən(t)-sē\ *noun, plural* **-cies** (1617)
: RETICENCE

ret·i·cent \-sənt\ *adjective* [Latin *reticent-, reticens,* present participle of *reticēre* to keep silent, from *re-* + *tacēre* to be silent — more at TACIT] (circa 1834)
1 : inclined to be silent or uncommunicative in speech **:** RESERVED
2 : restrained in expression, presentation, or appearance ⟨the room has an aspect of *reticent* dignity —A. N. Whitehead⟩
3 : RELUCTANT
synonym see SILENT
— ret·i·cent·ly *adverb*

ret·i·cle \ˈre-ti-kəl\ *noun* [Latin *reticulum* network] (circa 1731)
: a scale on transparent material (as in an optical instrument) used especially for measuring or aiming

re·tic·u·lar \ri-ˈti-kyə-lər\ *adjective* (1597)
1 a : RETICULATE **b :** of, relating to, or forming a reticulum
2 : INTRICATE

reticular formation *noun* (1887)
: a mass of nerve cells and fibers situated primarily in the brain stem and functioning upon stimulation especially in arousal of the organism

[1]re·tic·u·late \-lət, -ˌlāt\ *adjective* [Latin *reticulatus,* from *reticulum*] (1658)
1 : resembling a net; *especially* **:** having veins, fibers, or lines crossing ⟨a *reticulate* leaf⟩
2 : being or involving evolutionary change dependent on genetic recombination involving diverse interbreeding populations
— re·tic·u·late·ly *adverb*

[2]re·tic·u·late \-ˌlāt\ *verb* **-lat·ed; -lat·ing** [back-formation from *reticulated,* adjective, reticulate] (1787)
transitive verb
: to divide, mark, or construct so as to form a network
intransitive verb
: to become reticulated

re·tic·u·la·tion \ri-ˌti-kyə-ˈlā-shən\ *noun* (1671)
: a reticulated formation **:** NETWORK; *also* **:** something reticulated

ret·i·cule \ˈre-ti-ˌkyü(ə)l\ *noun* [French *réticule,* from Latin *reticulum* network, network bag, from diminutive of *rete* net] (circa 1738)
1 : RETICLE
2 : a woman's drawstring bag used especially as a carryall

re·tic·u·lo·cyte \ri-ˈti-kyə-lō-ˌsīt\ *noun* [New Latin *reticulum* + International Scientific Vocabulary *-cyte*] (1922)
: an immature red blood cell that appears especially during regeneration of lost blood and has a fine basophilic reticulum formed of ribosomal remains

re·tic·u·lo·en·do·the·li·al \ri-ˌti-kyə-lō-ˌen-də-ˈthē-lē-əl\ *adjective* [New Latin *reticulum* + *endothelium*] (circa 1923)
: of, relating to, or being the reticuloendothelial system

reticuloendothelial system *noun* (circa 1923)
: a diffuse system of cells arising from mesenchyme and comprising all the phagocytic cells of the body except the circulating white blood cells

re·tic·u·lum \ri-ˈti-kyə-ləm\ *noun* [New Latin, from Latin, network] (circa 1658)
1 : the second compartment of the stomach of a ruminant in which folds of the mucous membrane form hexagonal cells — compare ABOMASUM, OMASUM, RUMEN
2 : a reticular structure **:** NETWORK; *especially* **:** interstitial tissue composed of reticulum cells

reticulum cell *noun* (1912)
: one of the branched anastomosing reticuloendothelial cells that form an intricate interstitial network ramifying through other tissues and organs

retin- *or* **retino-** *combining form* [*retina*]
1 : retina ⟨*retin*itis⟩ ⟨*retino*scopy⟩
2 : retinol ⟨*retin*oid⟩

ret·i·na \ˈre-tᵊn-ə, ˈret-nə\ *noun, plural* **retinas** *or* **ret·i·nae** \-tᵊn-ˌē, -ˌī\ [Middle English *rethina,* from Medieval Latin *retina,* probably from Latin *rete* net] (14th century)
: the sensory membrane that lines the eye, is composed of several layers including one containing the rods and cones, and functions as the immediate instrument of vision by receiving the image formed by the lens and converting it into chemical and nervous signals which reach the brain by way of the optic nerve — see EYE illustration

ret·i·nac·u·lum \ˌre-tᵊn-ˈa-kyə-ləm\ *noun, plural* **-la** \-lə\ [New Latin, from Latin, halter, cable, from *retinēre* to hold back — more at RETAIN] (circa 1825)
: a small structure on the forewings of many lepidopterous insects (as a noctuid moth) that catches and holds the frenulum

[1]ret·i·nal \ˈre-tᵊn-əl, ˈret-nəl\ *adjective* (1838)
: of, relating to, involving, or being a retina

[2]ret·i·nal \ˈre-tᵊn-ˌal, -ˌȯl\ *noun* [*retin-* + [3]*-al*] (1944)
: a yellowish to orange aldehyde $C_{20}H_{28}O$ derived from vitamin A that in combination with proteins forms the visual pigments of the retinal rods and cones

ret·i·nene \ˈre-tᵊn-ˌēn\ *noun* (1934)
: RETINAL

ret·i·ni·tis \ˌre-tᵊn-ˈī-təs\ *noun* [New Latin] (1861)
: inflammation of the retina

retinitis pig·men·to·sa \-ˌpig-mən-ˈtō-sə, -(ˌ)men-, -zə\ *noun* [New Latin, pigmented retinitis] (1861)
: any of several hereditary progressive degenerative diseases of the eye marked by night blindness in the early stages, atrophy and pigment changes in the retina, constriction of the visual field, and eventual blindness

ret·i·no·blas·to·ma \ˌre-tᵊn-ō-ˌblas-ˈtō-mə\ *noun* [*retin-* + *blast-* + *-oma*] (1924)
: a malignant tumor of the retina that develops during childhood, is derived from retinal germ cells, and is associated with a chromosomal abnormality

ret·i·no·ic acid \ˌre-tᵊn-ˈō-ik-\ *noun* (1965)
: an acid $C_{20}H_{28}O_2$ derived from vitamin A and used especially in the treatment of acne

ret·i·noid \ˈre-tᵊn-ˌȯid\ *noun* (1976)
: any of various synthetic or naturally occurring analogues of vitamin A

ret·i·nol \ˈre-tᵊn-ˌȯl, -ˌōl\ *noun* [*retin-* + [1]*-ol;* from its being the source of retinal] (1960)
: the chief and typical vitamin A

ret·i·nop·a·thy \ˌre-tᵊn-ˈä-pə-thē\ *noun* (1932)
: any of various noninflammatory disorders of the retina including some that cause blindness

ret·i·nos·co·py \ˌre-tᵊn-ˈäs-kə-pē\ *noun* (1884)
: observation of the retina of the eye especially to determine the state of refraction

ret·i·no·tec·tal \ˌre-tᵊn-ō-ˈtek-təl\ *adjective* [*retin-* + *tect*um + [1]*-al*] (1951)
: of, relating to, or being the nerve fibers connecting the retina and the tectum of the midbrain ⟨*retinotectal* pathways⟩

ret·i·nue \ˈre-tᵊn-ˌü, -ˌyü\ *noun* [Middle English *retenue,* from Middle French, from feminine of *retenu,* past participle of *retenir* to retain] (14th century)
: a group of retainers or attendants

re·tin·u·la \re-ˈtin-yə-lə\ *noun, plural* **-lae** \-ˌlē, -ˌlī\ *also* **-las** [New Latin, diminutive of Medieval Latin *retina*] (1878)
: the neural receptor of a single facet of an arthropod compound eye

— **re·tin·u·lar** \-lər\ *adjective*

re·tir·ant \ri-ˈtī-rənt\ *noun* (1948)
: RETIREE

re·tire \ri-ˈtīr\ *verb* **re·tired; re·tir·ing** [Middle French *retirer,* from *re-* + *tirer* to draw] (1533)
intransitive verb
1 : to withdraw from action or danger **:** RETREAT
2 : to withdraw especially for privacy
3 : to move back **:** RECEDE
4 : to withdraw from one's position or occupation **:** conclude one's working or professional career
5 : to go to bed
transitive verb
1 : WITHDRAW: as **a :** to march (a military force) away from the enemy **b :** to withdraw from circulation or from the market **:** RECALL **c :** to withdraw from usual use or service
2 : to cause to retire from one's position or occupation
3 a : to put out (a batter or batsman) in baseball or cricket **b :** to cause (a side) to end a turn at bat in baseball
4 : to win permanent possession of (as a trophy)

re·tired \ri-ˈtīrd\ *adjective* (1590)
1 : SECLUDED ⟨*retired* village⟩
2 : withdrawn from one's position or occupation **:** having concluded one's working or professional career
3 : received by or due to one in retirement
— **re·tired·ly** \-ˈtī-rəd-lē, -ˈtīrd-\ *adverb*
— **re·tired·ness** \-ˈtīrd-nəs\ *noun*

re·tir·ee \ri-ˌtī-ˈrē\ *noun* (1945)
: a person who has retired from a working or professional career

¹re·tire·ment \ri-ˈtīr-mənt\ *noun* (1596)
1 a : an act of retiring **:** the state of being retired **b :** withdrawal from one's position or occupation or from active working life **c :** the age at which one normally retires ⟨reaches *retirement* in May⟩
2 : a place of seclusion or privacy

²retirement *adjective* (1919)
: of, relating to, or designed for retired persons

re·tir·ing \ri-ˈtīr-iŋ\ *adjective* (1766)
: RESERVED, SHY
— **re·tir·ing·ly** \-iŋ-lē\ *adverb*
— **re·tir·ing·ness** *noun*

re·tool \(ˌ)rē-ˈtül\ *transitive verb* (1927)
1 : to reequip with tools
2 : REORGANIZE

¹re·tort \ri-ˈtȯrt\ *verb* [Latin *retortus,* past participle of *retorquēre,* literally, to twist back, hurl back, from *re-* + *torquēre* to twist — more at TORTURE] (circa 1557)
transitive verb
1 : to pay or hurl back **:** RETURN ⟨*retort* an insult⟩
2 a : to make a reply to **b :** to say in reply
3 : to answer (as an argument) by a counter argument
intransitive verb
1 : to answer back usually sharply
2 : to return an argument or charge
3 : RETALIATE
synonym see ANSWER

²retort *noun* (1600)
: a quick, witty, or cutting reply; *especially* **:** one that turns back or counters the first speaker's words

³re·tort \ri-ˈtȯrt, ˈrē-ˌ\ *noun* [Middle French *retorte,* from Medieval Latin *retorta,* from Latin, feminine of *retortus;* from its shape] (1605)
: a vessel or chamber in which substances are distilled or decomposed by heat

retort

⁴re·tort \ri-ˈtȯrt, ˈrē-ˌ\ *transitive verb* (1850)
: to treat (as oil shale) by heating in a retort

re·tort pouch \ri-ˈtȯrt-, ˈrē-ˌ\ *noun* (1977)
: a flexible package in which prepared food is hermetically sealed for long-term unrefrigerated storage

¹re·touch \(ˌ)rē-ˈtəch\ *verb* [French *retoucher,* from Middle French, from *re-* + *toucher* to touch] (1685)
transitive verb
1 : to rework in order to improve **:** TOUCH UP
2 : to alter (as a photographic negative) to produce a more desirable appearance
3 : to color (new growth of hair) to match previously dyed, tinted, or bleached hair
intransitive verb
: to make or give retouches
— **re·touch·er** *noun*

²re·touch \ˈrē-ˌtəch, (ˌ)rē-ˈ\ *noun* (1703)
: the act, process, or an instance of retouching; *especially* **:** the retouching of a new growth of hair

re·tract \ri-ˈtrakt\ *verb* [Middle English, from Latin *retractus,* past participle of *retrahere* — more at RETREAT] (15th century)
transitive verb
1 : to draw back or in ⟨cats *retract* their claws⟩
2 a : TAKE BACK, WITHDRAW ⟨*retract* a confession⟩ **b :** DISAVOW
intransitive verb
1 : to draw or pull back
2 : to recant or disavow something
synonym see ABJURE, RECEDE
— **re·tract·able** \-ˈtrak-tə-bəl\ *adjective*

re·trac·tile \ri-ˈtrak-tᵊl, -ˌtīl\ *adjective* (1777)
: capable of being drawn back or in ⟨*retractile* claws⟩
— **re·trac·til·i·ty** \ˌrē-ˌtrak-ˈti-lə-tē, ri-\ *noun*

re·trac·tion \ri-ˈtrak-shən\ *noun* (14th century)
1 : an act of recanting; *specifically* **:** a statement made by one retracting
2 : an act of retracting **:** the state of being retracted
3 : the ability to retract

re·trac·tor \ri-ˈtrak-tər\ *noun* (1837)
: one that retracts: as **a :** a surgical instrument for holding open the edges of a wound **b :** a muscle that draws in an organ or part

re·train \(ˌ)rē-ˈtrān\ (1918)
transitive verb
: to train again or anew
intransitive verb
: to become trained again
— **re·train·able** \-ˈtrā-nə-bəl\ *adjective*

re·tral \ˈrē-trəl, ˈre-\ *adjective* [Latin *retro* back — more at RETRO-] (1875)
1 : situated at or toward the back **:** POSTERIOR
2 : BACKWARD, RETROGRADE
— **re·tral·ly** \-trə-lē\ *adverb*

re·trans·late \ˌrē-tran(t)s-ˈlāt, -tranz-\ (1860)
transitive verb
: to translate (a translation) into another language; *also* **:** to give a new form to
intransitive verb
: to retranslate something
— **re·trans·la·tion** \-ˈlā-shən\ *noun*

¹re·tread \(ˌ)rē-ˈtred\ *transitive verb* **re·tread·ed; re·tread·ing** (1907)
1 : to bond or vulcanize a new tread to the prepared surface of (a worn tire)
2 : to make over as if new ⟨*retread* an old plot⟩

²re·tread \ˈrē-ˌtred\ *noun* (1914)
1 : a retreaded tire
2 : something made or done again especially in slightly revised form **:** REMAKE
3 a : one (as a retired person) who is retrained for work **b :** one (as an athlete) who has previously held the same or a similar position

re–tread \(ˌ)rē-ˈtred\ *transitive verb* **-trod** \-ˈträd\; **-trod·den** \-ˈträ-dᵊn\ *or* **-trod; -tread·ing** (1598)
: to tread again

¹re·treat \ri-ˈtrēt\ *noun* [Middle English *retret,* from Middle French *retrait,* from past participle of *retraire* to withdraw, from Latin *retrahere,* from *re-* + *trahere* to draw] (14th century)
1 a (1) **:** an act or process of withdrawing especially from what is difficult, dangerous, or disagreeable (2) **:** the process of receding from a position or state attained ⟨the *retreat* of a glacier⟩ ⟨the slow *retreat* of an epidemic⟩ **b** (1) **:** the usually forced withdrawal of troops from an enemy or from an advanced position (2) **:** a signal for retreating **c** (1) **:** a signal given by bugle at the beginning of a military flag-lowering ceremony (2) **:** a military flag-lowering ceremony
2 : a place of privacy or safety **:** REFUGE
3 : a period of group withdrawal for prayer, meditation, study, and instruction under a director

²retreat (15th century)
intransitive verb
1 : to make a retreat **:** WITHDRAW
2 : to slope backward
transitive verb
: to draw or lead back **:** REMOVE; *specifically* **:** to move (a piece) back in chess
synonym see RECEDE
— **re·treat·er** *noun*

re·treat·ant \-ˈtrē-tᵊnt\ *noun* (1880)
: a person on a religious retreat

re·trench \ri-ˈtrench\ *verb* [obsolete French *retrencher* (now *retrancher*), from Middle French *retrenchier,* from *re-* + *trenchier* to cut] (1625)
transitive verb
1 a : CUT DOWN, REDUCE **b :** to cut out **:** EXCISE
2 : to pare away **:** REMOVE
intransitive verb
: to make retrenchments; *specifically* **:** ECONOMIZE
synonym see SHORTEN

re·trench·ment \-mənt\ *noun* (circa 1600)
: REDUCTION, CURTAILMENT; *specifically* **:** a cutting of expenses

re·tri·al \(ˌ)rē-ˈtrī(-ə)l\ *noun* (1875)
: a second trial, experiment, or test; *specifically* **:** a second judicial trial

ret·ri·bu·tion \ˌre-trə-ˈbyü-shən\ *noun* [Middle English *retribucioun,* from Middle French *retribution,* from Late Latin *retribution-, retributio,* from Latin *retribuere* to pay back, from *re-* + *tribuere* to pay — more at TRIBUTE] (14th century)
1 : RECOMPENSE, REWARD
2 : the dispensing or receiving of reward or punishment especially in the hereafter
3 : something given or exacted in recompense; *especially* **:** PUNISHMENT

re·trib·u·tive \ri-ˈtri-byə-tiv\ *adjective* (1678)
: of, relating to, or marked by retribution
— **re·trib·u·tive·ly** *adverb*

re·trib·u·to·ry \-byə-ˌtōr-ē, -ˌtȯr-\ *adjective* (circa 1615)
: RETRIBUTIVE

re·triev·al \ri-ˈtrē-vəl\ *noun* (circa 1643)
1 : an act or process of retrieving
2 : possibility of being retrieved or of recovering ⟨beyond *retrieval*⟩

¹re·trieve \ri-ˈtrēv\ *verb* **re·trieved; re·triev·ing** [Middle English *retreven,* modification of Middle French *retrouver* to find again, from *re-* + *trouver* to find, from (assumed) Vulgar Latin *tropare* to find, compose — more at TROUBADOUR] (15th century)
transitive verb
1 : to discover and bring in (killed or wounded game)
2 : to call to mind again
3 : to get back again **:** REGAIN
4 a : RESCUE, SALVAGE **b :** to return (as a ball or shuttlecock that is difficult to reach) successfully
5 : RESTORE, REVIVE ⟨his writing *retrieves* the past⟩
6 : to remedy the evil consequences of **:** CORRECT

7 : to get and bring back; *especially* **:** to recover (as information) from storage
intransitive verb
: to bring in game ⟨a dog that *retrieves* well⟩; *also* **:** to bring back an object thrown by a person
— **re·triev·abil·i·ty** \-ˌtrē-və-ˈbi-lə-tē\ *noun*
— **re·triev·able** \-ˈtrē-və-bəl\ *adjective*

[2]retrieve *noun* (1575)
1 : RETRIEVAL
2 : the successful return of a ball that is difficult to reach or control (as in tennis)

re·triev·er \ri-ˈtrē-vər\ *noun* (15th century)
: one that retrieves; *especially* **:** a dog of any of several breeds (as a golden retriever) having a heavy water-resistant coat and used especially for retrieving game

ret·ro \ˈre-(ˌ)trō\ *adjective* [French *rétro,* short for *rétrospectif* retrospective] (1974)
: relating to, reviving, or being the styles or especially the fashions of the past **:** fashionably nostalgic or old-fashioned ⟨a *retro* look⟩

retro- *prefix* [Middle English, from Latin, from *retro,* from *re-* + *-tro* (as in *intro* within) — more at INTRO-]
1 : backward **:** back ⟨*retro*-rocket⟩
2 : situated behind ⟨*retro*peritoneal⟩

ret·ro·ac·tion \ˌre-trō-ˈak-shən\ *noun* (circa 1738)
1 [*retroactive*] **:** retroactive operation (as of a law or tax)
2 [*retro-* + *action*] **:** a reciprocal action **:** REACTION

ret·ro·ac·tive \-ˈak-tiv\ *adjective* [French *rétroactif,* from Latin *retroactus,* past participle of *retroagere* to drive back, reverse, from *retro-* + *agere* to drive — more at AGENT] (1611)
: extending in scope or effect to a prior time or to conditions that existed or originated in the past; *especially* **:** made effective as of a date prior to enactment, promulgation, or imposition ⟨*retroactive* tax⟩
— **ret·ro·ac·tive·ly** *adverb*
— **ret·ro·ac·tiv·i·ty** \-ˌak-ˈti-və-tē\ *noun*

ret·ro·cede \ˌre-trō-ˈsēd\ *verb* **-ced·ed; -ced·ing** [Latin *retrocedere,* from *retro-* + *cedere* to go, cede] (1654)
intransitive verb
: to go back **:** RECEDE
transitive verb
[French *rétrocéder,* from Medieval Latin *retrocedere,* from Latin *retro-* + *cedere* to cede]
: to cede back (as a territory)
— **ret·ro·ces·sion** \-ˈse-shən\ *noun*

ret·ro·dict \ˌre-trə-ˈdikt\ *transitive verb* [*retro-* + *predict*] (1949)
: to utilize present information or ideas to infer or explain (a past event or state of affairs) ⟨*retrodict* past eclipses⟩
— **ret·ro·dic·tion** \-ˈdik-shən\ *noun*
— **ret·ro·dic·tive** \-ˈdik-tiv\ *adjective*

ret·ro–engine \ˈre-trō-ˌen-jən\ *noun* (1965)
: RETRO-ROCKET

ret·ro·fire \-ˌfīr\ (1961)
transitive verb
: to cause (a retro-rocket) to become ignited
intransitive verb
of a retro-rocket **:** to become ignited
— **retrofire** *noun*

ret·ro·fit \ˌre-trō-ˈfit\ *transitive verb* (1953)
1 : to furnish (as a computer, airplane, or building) with new or modified parts or equipment not available or considered necessary at the time of manufacture
2 : to install (new or modified parts or equipment) in something previously manufactured or constructed
— **retrofit** *noun*

ret·ro·flex \ˈre-trə-ˌfleks\ *adjective* [New Latin *retroflexus,* from Latin *retro-* + *flexus,* past participle of *flectere* to bend] (1776)
1 : turned or bent abruptly backward
2 : articulated with the tongue tip turned up or curled back just under the hard palate ⟨*retroflex* vowel⟩

ret·ro·flex·ion *or* **ret·ro·flec·tion** \ˌre-trə-ˈflek-shən\ *noun* (1845)
1 : the state of being bent back; *especially* **:** the bending back of an organ (as a uterus) upon itself
2 : the act or process of bending back
3 : retroflex articulation

ret·ro·gra·da·tion \ˌre-trō-grā-ˈdā-shən, -grə-\ *noun* (circa 1545)
: the action or process of retrograding

[1]ret·ro·grade \ˈre-trə-ˌgrād\ *adjective* [Middle English, from Latin *retrogradus,* from *retrogradi*] (14th century)
1 a (1) *of a celestial body* **:** having a direction contrary to that of the general motion of similar bodies (2) **:** having or being a direction of rotation or revolution that is clockwise as viewed from the north pole of the sky or a planet **b :** moving, occurring, or performed in a backward direction or opposite to the usual direction ⟨a *retrograde* step⟩ ⟨*retrograde* peristalsis⟩ **c :** contrary to the normal order **:** INVERSE
2 : tending toward or resulting in a worse or previous state
3 *archaic* **:** CONTRADICTORY, OPPOSED
4 : characterized by retrogression
5 : affecting a period prior to a precipitating cause ⟨*retrograde* amnesia⟩
6 : RETRO
— **ret·ro·grade·ly** *adverb*

[2]retrograde *adverb* (circa 1619)
: BACKWARD, REVERSELY

[3]retrograde *verb* [Latin *retrogradi,* from *retro-* + *gradi* to go — more at GRADE] (1582)
transitive verb
archaic **:** to turn back **:** REVERSE
intransitive verb
1 a : to go back **:** RETREAT **b :** to go back over or recapitulate something
2 : to decline to a worse condition

ret·ro·gress \ˌre-trə-ˈgres\ *intransitive verb* [Latin *retrogressus,* past participle of *retrogradi*] (1819)
: to move backward **:** REVERT

ret·ro·gres·sion \-ˈgre-shən\ *noun* (1646)
1 : REGRESSION 3
2 : return to a former and less complex level of development or organization

ret·ro·gres·sive \-ˈgre-siv\ *adjective* (1802)
: characterized by retrogression: as **a :** going or directed backward **b :** declining from a better to a worse state **c :** passing from a higher to a lower level of organization ⟨*retrogressive* evolution⟩
— **ret·ro·gres·sive·ly** *adverb*

ret·ro·pack \ˈre-trō-ˌpak\ *noun* (1962)
: a system of auxiliary rockets on a spacecraft that produces thrust in the direction opposite to the motion of the spacecraft and that is used to reduce speed

ret·ro·per·i·to·ne·al \ˌre-trō-ˌper-ə-tᵊn-ˈē-əl\ *adjective* (1874)
: situated behind the peritoneum
— **ret·ro·per·i·to·ne·al·ly** \-ə-lē\ *adverb*

ret·ro·re·flec·tion \ˌre-trō-ri-ˈflek-shən\ *noun* (circa 1965)
: the action or use of a retroreflector
— **ret·ro·re·flec·tive** \-ˈflek-tiv\ *adjective*

ret·ro·re·flec·tor \-ˈflek-tər\ *noun* (1946)
: a device that reflects radiation (as light) so that the paths of the rays are parallel to those of the incident rays

ret·ro–rock·et \ˈre-trō-ˌrä-kət\ *noun* (1957)
: an auxiliary rocket engine (as on a spacecraft) used in decelerating

re·trorse \ˈrē-ˌtrȯrs\ *adjective* [Latin *retrorsus,* contraction of *retroversus*] (circa 1825)
: bent backward or downward

[1]ret·ro·spect \ˈre-trə-ˌspekt\ *noun* [probably from *retro-* + *prospect*] (1602)
1 *archaic* **:** reference to or regard of a precedent or authority
2 : a review of or meditation on past events
— **in retrospect :** in considering the past or a past event

[2]retrospect *adjective* (1709)
: RETROSPECTIVE

[3]retrospect (1659)
intransitive verb
1 : to engage in retrospection
2 : to refer back **:** REFLECT
transitive verb
: to go back over in thought

ret·ro·spec·tion \ˌre-trə-ˈspek-shən\ *noun* (1674)
: the act or process or an instance of surveying the past

[1]ret·ro·spec·tive \-ˈspek-tiv\ *adjective* (1664)
1 a (1) **:** of, relating to, or given to retrospection (2) **:** based on memory ⟨a *retrospective* report⟩ **b :** being a retrospective ⟨a *retrospective* exhibition⟩
2 : affecting things past **:** RETROACTIVE
— **ret·ro·spec·tive·ly** *adverb*

[2]retrospective *noun* (1932)
: a generally comprehensive exhibition or performance of the work of an artist over a span of years

re·trous·sé \rə-ˌtrü-ˈsā, rə-ˈtrü-ˌ, ˌre-trü̇-ˈ\ *adjective* [French, from past participle of *retrousser* to tuck up, from Middle French, from *re-* + *trousser* to truss, tuck up — more at TRUSS] (1802)
: turned up ⟨*retroussé* nose⟩

ret·ro·ver·sion \ˌre-trō-ˈvər-zhən *also* -shən\ *noun* [Latin *retroversus* turned backward, from *retro-* + *versus,* past participle of *vertere* to turn — more at WORTH] (1776)
1 : the bending backward of the uterus and cervix
2 : the act or process of turning back or regressing

ret·ro·vi·rus \ˈre-trō-ˌvī-rəs\ *noun* (1975)
: any of a group of RNA-containing viruses (as HIV) that produce reverse transcriptase by means of which DNA is produced using their RNA as a template and incorporated into the genome of infected cells and that include numerous tumorigenic viruses
— **ret·ro·vi·ral** \-rəl\ *adjective*

ret·si·na \ret-ˈsē-nə\ *noun* [New Greek, perhaps from Italian *resina* resin, from Latin — more at RESIN] (1940)
: a resin-flavored Greek wine

[1]re·turn \ri-ˈtərn\ *verb* [Middle English *retournen,* from Middle French *retourner,* from *re-* + *tourner* to turn — more at TURN] (14th century)
intransitive verb
1 a : to go back or come back again ⟨*return* home⟩ **b :** to go back in thought or practice **:** REVERT ⟨soon *returned* to her old habit⟩
2 : to pass back to an earlier possessor
3 : REPLY, RETORT
transitive verb
1 a : to give (as an official account) to a superior **b** *British* **:** to elect (a candidate) as attested by official report or returns **c :** to bring back (as a writ or verdict) to an office or tribunal
2 a : to bring, send, or put back to a former or proper place ⟨*return* the gun to its holster⟩ **b :** to restore to a former or to a normal state
3 a : to send back **:** VISIT — usually used with *on* or *upon* **b** *obsolete* **:** RETORT
4 : to bring in (as profit) **:** YIELD
5 a : to give or perform in return **:** REPAY ⟨*return* a compliment⟩ **b :** to give back to the owner **c :** REFLECT ⟨*return* an echo⟩

\ə\ **abut** \ᵊ\ **kitten** \ər\ **further** \a\ **ash** \ā\ **ace** \ä\ **mop, mar** \au̇\ **out** \ch\ **chin** \e\ **bet** \ē\ **easy** \g\ **go** \i\ **hit** \ī\ **ice** \j\ **job** \ŋ\ **sing** \ō\ **go** \ȯ\ **law** \ȯi\ **boy** \th\ **thin** \th\ **the** \ü\ **loot** \u̇\ **foot** \y\ **yet** \zh\ **vision** *see also* Guide to Pronunciation

6 : to cause (as a wall) to continue in a different direction (as at a right angle)
7 : to lead (a specified suit or specified card of a suit) in response to a partner's earlier lead
8 a : to hit back (a ball or shuttlecock) **b :** to run with (a football) after a change of possession (as by a punt or a fumble)
synonym see RECIPROCATE
— re•turn•er *noun*

²return *noun* (14th century)
1 a : the act of coming back to or from a place or condition **b :** a regular or frequent returning **:** RECURRENCE
2 a (1) **:** the delivery of a legal order (as a writ) to the proper officer or court (2) **:** an endorsed certificate stating an official's action in the execution of such an order (3) **:** the sending back of a commission with the certificate of the commissioners **b :** an account or formal report **c** (1) **:** a report of the results of balloting — usually used in plural ⟨election *returns*⟩ (2) **:** an official declaration of the election of a candidate (3) *chiefly British* **:** ELECTION **d** (1) **:** a formal statement on a required legal form showing taxable income, allowable deductions and exemptions, and the computation of the tax due (2) **:** a list of taxable property
3 a : the continuation usually at a right angle of the face or of a member of a building or of a molding or group of moldings **b :** a turn, bend, or winding back (as in a rod, stream, or trench) **c :** a means for conveying something (as water) back to its starting point
4 a : a quantity of goods, consignment, or cargo coming back in exchange for goods sent out as a mercantile venture **b :** the value of or profit from such venture **c** (1) **:** the profit from labor, investment, or business **:** YIELD (2) *plural* **:** RESULTS **d :** the rate of profit in a process of production per unit of cost
5 a : the act of returning something to a former place, condition, or ownership **:** RESTITUTION **b :** something returned; *especially, plural* **:** unsold publications returned to the publisher for cash or credit
6 a : something given in repayment or reciprocation **b :** ANSWER, RETORT
7 : an answering play: as **a :** a lead in a suit previously led by one's partner in a card game **b :** the action or an instance of returning a ball (as in football or tennis)
— in return : in compensation or repayment

³return *adjective* (1676)
1 a : having or formed by a change of direction ⟨a *return* facade⟩ **b :** doubled on itself ⟨a *return* flue⟩
2 : played, delivered, or given in return **:** taking place for the second time ⟨a *return* meeting for the two champions⟩
3 : used or taken on returning ⟨the *return* road⟩
4 : returning or permitting return ⟨a *return* valve⟩
5 : of, relating to, or causing a return to a place or condition

¹re•turn•able \ri-ˈtər-nə-bəl\ *adjective* (15th century)
1 : legally required to be returned, delivered, or argued at a specified time or place ⟨a writ *returnable* on the date indicated⟩
2 a : capable of returning or of being returned (as for reuse) **b :** permitted to be returned

²returnable *noun* (1963)
: something designed to be returned (as for recycling); *especially* **:** a returnable beverage container

re•turn•ee \ri-ˌtər-ˈnē\ *noun* (1944)
: one who returns; *especially* **:** one returning to the U.S. after military service overseas

re•tuse \ri-ˈtüs, -ˈtyüs\ *adjective* [Latin *retusus* blunted, from past participle of *retundere* to pound back, blunt, from *re-* + *tundere* to beat, pound — more at CONTUSION] (circa 1753)
: having the apex rounded or obtuse with a slight notch

¹Reu•ben \ˈrü-bən\ *noun* [Hebrew *Rĕʼūbhēn*]
: a son of Jacob and the traditional eponymous ancestor of one of the tribes of Israel

²Reuben *noun* [probably from *Reuben* Kulakofsky (died 1960) American grocer] (1956)
: a grilled sandwich of corned beef, Swiss cheese, and sauerkraut usually on rye bread ◆

³Reuben *noun* [*Reuben* L. Goldberg (died 1970) American cartoonist] (1958)
: a statuette awarded annually by a professional organization for notable achievement in cartoon artistry

re•union \(ˌ)rē-ˈyün-yən\ *noun* (1610)
1 : an act of reuniting **:** the state of being reunited
2 : a reuniting of persons after separation

re•union•ist \-yə-nist\ *noun* (1866)
: an advocate of reunion (as of sects or parties)
— re•union•is•tic \(ˌ)rē-ˌyün-yə-ˈnis-tik\ *adjective*

re•unite \ˌrē-yü-ˈnīt\ *verb* [Medieval Latin *reunitus,* past participle of *reunire,* from Latin *re-* + Late Latin *unire* to unite — more at UNITE] (15th century)
transitive verb
: to bring together again
intransitive verb
: to come together again **:** REJOIN

re–up \(ˌ)rē-ˈəp\ *intransitive verb* [*re-* + sign *up*] (circa 1906)
: to enlist again

re•us•able \(ˌ)rē-ˈyü-zə-bəl\ *adjective* (1943)
: capable of being used again or repeatedly
— re•us•abil•i•ty \(ˌ)rē-ˌyü-zə-ˈbi-lə-tē\ *noun*

¹re•use \(ˌ)rē-ˈyüz\ *transitive verb* (1843)
: to use again especially after reclaiming or reprocessing ⟨the need to *reuse* scarce resources⟩

²reuse \-ˈyüs\ *noun* (1866)
: further or repeated use

¹rev \ˈrev\ *noun* [short for *revolution*] (circa 1890)
: a revolution of a motor

²rev *verb* **revved; rev•ving** (1920)
transitive verb
1 a : to step up the number of revolutions per minute of — often used with *up* ⟨*rev* up the engine⟩ **b :** INCREASE — used with *up* ⟨*rev* up production⟩
2 : to drive or operate especially at high speed — often used with *up*
3 : to make more active or effective — used with *up*
4 : to stir up **:** EXCITE — usually used with *up*
intransitive verb
1 : to operate at an increased speed of revolution — usually used with *up*
2 : to increase in amount or activity — used with *up*

re•val•u•ate \(ˌ)rē-ˈval-yə-ˌwāt\ *transitive verb* [back-formation from *revaluation*] (1921)
: REVALUE; *specifically* **:** to increase the value of (as currency)
— re•val•u•a•tion \(ˌ)rē-ˌval-yə-ˈwā-shən\ *noun*

re•val•ue \(ˌ)rē-ˈval-(ˌ)yü\ *transitive verb* (1592)
1 : to value (as currency) anew
2 : to make a new valuation of **:** REAPPRAISE

re•vamp \(ˌ)rē-ˈvamp\ *transitive verb* (1850)
1 : REMAKE, REVISE
2 : RENOVATE, RECONSTRUCT
— re•vamp \ˈrē-ˌ, ri-ˈ\ *noun*

re•vanche \rə-ˈvänsh\ *noun* [French, from Middle French, alteration of *revenche* — more at REVENGE] (1882)
: REVENGE; *especially* **:** a usually political policy designed to recover lost territory or status
— re•vanch•ism \-ˈvän-ˌshi-zəm\ *noun*

¹re•vanch•ist \-ˈvän-shist\ *noun* (1926)
: one who advocates a policy of revanche

²revanchist *adjective* (1948)
: of or relating to a policy of revanche

re•vas•cu•lar•i•za•tion \ˌrē-ˌvas-kyə-lə-rə-ˈzā-shən\ *noun* (1951)
: a surgical procedure for the provision of a new, additional, or augmented blood supply to a body part or organ

¹re•veal \ri-ˈvē(ə)l\ *transitive verb* [Middle English *revelen,* from Middle French *reveler,* from Latin *revelare* to uncover, reveal, from *re-* + *velare* to cover, veil, from *velum* veil] (14th century)
1 : to make known through divine inspiration
2 : to make (something secret or hidden) publicly or generally known ⟨*reveal* a secret⟩
3 : to open up to view **:** DISPLAY ⟨the uncurtained window *revealed* a cluttered room⟩ ☆
— re•veal•able \-ˈvē-lə-bəl\ *adjective*
— re•veal•er *noun*

²reveal *noun* [alteration of earlier *revale,* from Middle English *revalen* to lower, from Middle French *revaler,* from *re-* + *val* valley — more at VALE] (1688)
: the side of an opening (as for a window) between a frame and the outer surface of a wall; *also* **:** JAMB

re•veal•ing *adjective* (circa 1925)
: allowing a look at or an understanding of something inner or hidden **:** INSIGHTFUL
— re•veal•ing•ly *adverb*

re•veal•ment \ri-ˈvē(ə)l-mənt\ *noun* (1584)

☆ SYNONYMS
Reveal, disclose, divulge, tell, betray mean to make known what has been or should be concealed. REVEAL may apply to supernatural or inspired revelation of truths beyond the range of ordinary human vision or reason ⟨divine will as *revealed* in sacred writings⟩. DISCLOSE may imply a discovering but more often an imparting of information previously kept secret ⟨candidates must *disclose* their financial assets⟩. DIVULGE implies a disclosure involving some impropriety or breach of confidence ⟨refused to *divulge* an anonymous source⟩. TELL implies an imparting of necessary or useful information ⟨*told* them what he had overheard⟩. BETRAY implies a divulging that represents a breach of faith or an involuntary or unconscious disclosure ⟨a blush that *betrayed* her embarrassment⟩.

◇ WORD HISTORY
Reuben The history of American culinary items can be difficult to trace once we confront people's sometimes unconsciously embellished recollections with documentary evidence. One difficult item is the Reuben, winner of a national sandwich contest in 1956 after being submitted by Fern Snider, a waitress from Omaha, Nebraska. Ms. Snider had acquired the recipe from a former employer, the Schimmel family, who had for a number of years managed the Blackstone Hotel in downtown Omaha. According to local tradition, the combination of rye bread, corned beef, Swiss cheese, and sauerkraut had been dreamed up in the 1920's to feed participants in a late-night poker game at the hotel by Reuben Kulakofsky, an Omaha grocer. Charles Schimmel, the hotel's owner, was so taken with the sandwich that he then put it on their restaurant menu designated by its inventor's name. Is this tale fact or folklore? We can say for certain that there was a Reuben Kulakofsky—his 1960 obituary appeared in the *Omaha World-Herald*—and a sandwich called the "rueben" (not further described, hence presumably familiar to habitués) appears on a Blackstone Hotel menu datable to about 1944. The argument for the Reuben's Nebraskan origin is hence fairly convincing.

: an act of revealing

re·veg·e·tate \(ˌ)rē-ˈve-jə-ˌtāt\ *transitive verb* (1804)
: to provide (barren or denuded land) with a new vegetative cover
— **re·veg·e·ta·tion** \(ˌ)rē-ˌve-jə-ˈtā-shən\ *noun*

rev·eil·le \ˈre-və-lē, *British* ri-ˈva-li *or* -ˈve-\ *noun* [modification of French *réveillez,* imperative plural of *réveiller* to awaken, from Middle French *reveiller,* from *re-* + *eveiller* to awaken, from (assumed) Vulgar Latin *exvigilare,* from Latin *ex-* + *vigilare* to keep watch, stay awake — more at VIGILANT] (1644)
1 : a signal to get up mornings
2 : a bugle call at about sunrise signaling the first military formation of the day; *also* **:** the formation so signaled

^{1}**rev·el** \ˈre-vəl\ *intransitive verb* **-eled** *or* **-elled; -el·ing** *or* **-el·ling** \ˈre-və-liŋ, ˈrev-liŋ\ [Middle English, from Middle French *reveler,* literally, to rebel, from Latin *rebellare*] (14th century)
1 : to take part in a revel **:** CAROUSE
2 : to take intense pleasure or satisfaction

^{2}**revel** *noun* (14th century)
: a usually wild party or celebration

rev·e·la·tion \ˌre-və-ˈlā-shən\ *noun* [Middle English, from Middle French, from Late Latin *revelation-, revelatio,* from Latin *revelare* to reveal] (14th century)
1 a : an act of revealing or communicating divine truth **b :** something that is revealed by God to humans
2 a : an act of revealing to view or making known **b :** something that is revealed; *especially* **:** an enlightening or astonishing disclosure ⟨shocking *revelations*⟩ **c :** a pleasant often enlightening surprise ⟨her talent was a *revelation*⟩
3 *capitalized* **:** an apocalyptic writing addressed to early Christians of Asia Minor and included as a book in the New Testament — called also *Apocalypse*; see BIBLE table

Rev·e·la·tions \-shənz\ *noun plural but singular in construction*
: REVELATION 3

rev·e·la·tor \ˈre-və-ˌlā-tər\ *noun* (1801)
: one that reveals; *especially* **:** one that reveals the will of God

rev·e·la·to·ry \ˈre-və-lə-ˌtōr-ē, -ˌtȯr-, ri-ˈve-lə-\ *adjective* (1882)
: of or relating to revelation **:** serving to reveal something

rev·el·er *or* **rev·el·ler** \ˈre-və-lər, ˈrev-lər\ *noun* (14th century)
: one who engages in revelry

rev·el·ry \ˈre-vəl-rē\ *noun* (15th century)
: noisy partying or merrymaking

rev·e·nant \ˈre-və-ˌnäⁿ, -nənt\ *noun* [French, from present participle of *revenir* to return] (1827)
: one that returns after death or a long absence
— **revenant** *adjective*

^{1}**re·venge** \ri-ˈvenj\ *transitive verb* **re·venged; re·veng·ing** [Middle English, from Middle French *revengier,* from Old French, from *re-* + *vengier* to avenge — more at VENGEANCE] (14th century)
1 : to avenge (as oneself) usually by retaliating in kind or degree
2 : to inflict injury in return for ⟨*revenge* an insult⟩
— **re·veng·er** *noun*

^{2}**revenge** *noun* [Middle French *revenge, revenche,* from *revengier, revenchier* to revenge] (circa 1547)
1 : a desire for revenge
2 : an act or instance of retaliating in order to get even
3 : an opportunity for getting satisfaction

re·venge·ful \ri-ˈvenj-fəl\ *adjective* (circa 1586)
: full of or prone to revenge **:** determined to get even
— **re·venge·ful·ly** \-fə-lē\ *adverb*
— **re·venge·ful·ness** *noun*

rev·e·nue \ˈre-və-ˌnü, -ˌnyü\ *noun, often attributive* [Middle English, from Middle French, from *revenir* to return, from Latin *revenire,* from *re-* + *venire* to come — more at COME] (15th century)
1 : the total income produced by a given source ⟨a property expected to yield a large annual *revenue*⟩
2 : the gross income returned by an investment
3 : the yield of sources of income (as taxes) that a political unit (as a nation or state) collects and receives into the treasury for public use
4 : a government department concerned with the collection of the national revenue

revenue bond *noun* (1856)
: a bond issued by a public agency authorized to build, acquire, or improve a revenue-producing property (as a toll road) and payable out of revenue derived from such property

rev·e·nu·er \ˈre-və-ˌnü-ər, -ˌnyü-\ *noun* (1880)
: a revenue officer or boat

revenue stamp *noun* (1862)
: a stamp (as on a cigar box) for use as evidence of payment of a tax

revenue tariff *noun* (1820)
: a tariff intended wholly or primarily to produce public revenue — compare PROTECTIVE TARIFF

re·verb \ri-ˈvərb, ˈrē-ˌ\ *noun* [short for *reverberation*] (1953)
: an electronically produced echo effect in recorded music; *also* **:** a device for producing reverb

re·ver·ber·ant \ri-ˈvər-b(ə-)rənt\ *adjective* (circa 1798)
1 : tending to reverberate
2 : marked by reverberation **:** RESONANT
— **re·ver·ber·ant·ly** *adverb*

^{1}**re·ver·ber·ate** \-bə-ˌrāt\ *verb* **-at·ed; -at·ing** [Latin *reverberatus,* past participle of *reverberare,* from *re-* + *verberare* to lash, from *verber* rod — more at VERVAIN] (15th century)
transitive verb
1 : REFLECT
2 : REPEL
3 : ECHO
intransitive verb
1 a : to become driven back **b :** to become reflected
2 : to continue in or as if in a series of echoes **:** RESOUND

^{2}**re·ver·ber·ate** \-b(ə-)rət\ *adjective* (1603)
: REVERBERANT

re·ver·ber·a·tion \ri-ˌvər-bə-ˈrā-shən\ *noun* (14th century)
1 : an act of reverberating **:** the state of being reverberated
2 a : something that is reverberated **b :** an effect or impact that resembles an echo

re·ver·ber·a·tive \ri-ˈvər-bə-ˌrā-tiv, -b(ə-)rə-\ *adjective* (1716)
1 : constituting reverberation
2 : tending to reverberate **:** REVERBERANT

re·ver·ber·a·to·ry \ri-ˈvər-b(ə-)rə-ˌtōr-ē, -bə-ˌtōr-, -ˌtȯr-\ *adjective* (1605)
: acting by reverberation

reverberatory furnace *noun* (1672)
: a furnace in which heat is radiated from the roof onto the material treated

^{1}**re·vere** \ri-ˈvir\ *transitive verb* **re·vered; re·ver·ing** [Latin *revereri,* from *re-* + *vereri* to fear, respect — more at WARY] (circa 1661)
: to show devoted deferential honor to **:** regard as worthy of great honor ⟨*revere* the aged⟩ ⟨*revere* tradition⟩ ☆

^{2}**revere** *noun* [by alteration] (circa 1934)
: REVERS

^{1}**rev·er·ence** \ˈrev-rən(t)s, ˈre-və-; ˈre-vərn(t)s\ *noun* (14th century)
1 : honor or respect felt or shown **:** DEFERENCE; *especially* **:** profound adoring awed respect
2 : a gesture of respect (as a bow)
3 : the state of being revered
4 : one held in reverence — used as a title for a clergyman
synonym see HONOR

^{2}**reverence** *transitive verb* **-enced; -enc·ing** (14th century)
: to regard or treat with reverence
synonym see REVERE
— **rev·er·enc·er** *noun*

^{1}**rev·er·end** \ˈrev-rənd, ˈre-və-; ˈre-vərnd\ *adjective* [Middle English, from Middle French, from Latin *reverendus,* gerundive of *revereri*] (15th century)
1 : worthy of reverence **:** REVERED
2 a : of or relating to the clergy **b :** being a member of the clergy — used as a title ⟨the *Reverend* Mr. Doe⟩ ⟨the *Reverend* John Doe⟩ ⟨the *Reverend* Mrs. Jane Doe⟩

^{2}**reverend** *noun* (1608)
: a member of the clergy — sometimes used in plural as a title

rev·er·ent \ˈrev-rənt, ˈre-və-; ˈre-vərnt\ *adjective* [Middle English, from Latin *reverent-, reverens,* present participle of *revereri*] (14th century)
: expressing or characterized by reverence **:** WORSHIPFUL
— **rev·er·ent·ly** *adverb*

rev·er·en·tial \ˌre-və-ˈren(t)-shəl\ *adjective* (circa 1555)
1 : expressing or having a quality of reverence ⟨*reverential* awe⟩
2 : inspiring reverence
— **rev·er·en·tial·ly** \-ˈren(t)-sh(ə-)lē\ *adverb*

rev·er·ie *also* **rev·ery** \ˈre-və-rē, ˈrev-rē\ *noun, plural* **rev·er·ies** [French *rêverie,* from Middle French, delirium, from *resver, rever* to wander, be delirious] (1657)
1 : DAYDREAM
2 : the condition of being lost in thought

re·vers \ri-ˈvir, -ˈver\ *noun, plural* **re·vers** \-ˈvirz, -ˈverz\ [French, literally, reverse, from Middle French, from *revers,* adjective] (1869)
: a lapel especially on a woman's garment

re·ver·sal \ri-ˈvər-səl\ *noun* (15th century)
1 : an act or the process of reversing
2 : a conversion of a photographic positive into a negative or vice versa
3 : a change of fortune usually for the worse

^{1}**re·verse** \ri-ˈvərs\ *adjective* [Middle English *revers,* from Middle French, from Latin *reversus,* past participle of *revertere* to turn back — more at REVERT] (14th century)

☆ SYNONYMS
Revere, reverence, venerate, worship, adore mean to honor and admire profoundly and respectfully. REVERE stresses deference and tenderness of feeling ⟨a professor *revered* by generations of students⟩. REVERENCE presupposes an intrinsic merit and inviolability in the one honored and a corresponding depth of feeling in the one honoring ⟨*reverenced* the academy's code of honor⟩. VENERATE implies a holding as holy or sacrosanct because of character, association, or age ⟨heroes still *venerated*⟩. WORSHIP implies homage usually expressed in words or ceremony ⟨*worships* their memory⟩. ADORE implies love and stresses the notion of an individual and personal attachment ⟨a doctor *adored* by her patients⟩.

1 a : opposite or contrary to a previous or normal condition ⟨*reverse* order⟩ **b :** having the back presented to the observer or opponent
2 : coming from the rear of a military force
3 : acting, operating, or arranged in a manner contrary to the usual
4 : effecting reverse movement ⟨*reverse* gear⟩
5 : so made that the part which normally prints in color appears white against a colored background
— re·verse·ly *adverb*

[2]reverse *verb* **re·versed; re·vers·ing** (14th century)
transitive verb
1 a : to turn completely about in position or direction **b :** to turn upside down **:** INVERT
2 : ANNUL: as **a :** to overthrow, set aside, or make void (a legal decision) by a contrary decision **b :** to cause to take an opposite point of view **c :** to change to the contrary ⟨*reverse* a policy⟩
3 : to cause to go in the opposite direction; *especially* **:** to cause (as an engine) to perform its action in the opposite direction
intransitive verb
1 : to turn or move in the opposite direction
2 : to put a mechanism (as an engine) in reverse ☆
— re·vers·er *noun*
— reverse field : to make a sudden reversal in direction or opinion

[3]reverse *noun* (14th century)
1 : something directly contrary to something else **:** OPPOSITE
2 : an act or instance of reversing; *especially* **:** DEFEAT, SETBACK
3 : the back part of something
4 a (1) **:** a gear that reverses something; *also* **:** the whole mechanism brought into play when such a gear is used (2) **:** movement in reverse **b :** an offensive play in football in which a back moving in one direction gives the ball to a player moving in the opposite direction
— in reverse : in an opposite manner or direction

reverse discrimination *noun* (1969)
: discrimination against whites or males (as in employment or education)

reverse osmosis *noun* (1955)
: the movement of fresh water through a semipermeable membrane when pressure is applied to a solution (as seawater) on one side of it

reverse tran·scrip·tase \-ˌtran-'skrip-(ˌ)tās, -(ˌ)tāz\ *noun* (1971)
: a polymerase that catalyzes the formation of DNA using RNA as a template and that is found especially in retroviruses

reverse transcription *noun* (1971)
: the process of synthesizing double-stranded DNA using RNA as a template and reverse transcriptase as a catalyst

[1]re·vers·ible \ri-'vər-sə-bəl\ *adjective* (1648)
: capable of being reversed or of reversing: as **a :** capable of going through a series of actions (as changes) either backward or forward ⟨a *reversible* chemical reaction⟩ **b :** having two finished usable sides ⟨*reversible* fabric⟩ **c :** wearable with either side out ⟨a *reversible* coat⟩
— re·vers·ibil·i·ty \-ˌvər-sə-'bi-lə-tē\ *noun*
— re·vers·ibly \-'vər-sə-blē\ *adverb*

[2]reversible *noun* (1863)
: a reversible cloth or article of clothing

re·ver·sion \ri-'vər-zhən, -shən\ *noun* [Middle English, from Middle French, from Latin *reversion-, reversio* act of returning, from *revertere*] (15th century)
1 a : the part of a simple estate remaining in the control of its owner after the owner has granted therefrom a lesser particular estate **b :** a future interest in property left in the control of a grantor or the grantor's successor
2 : the right of succession or future possession or enjoyment
3 a : an act or the process of returning (as to a former condition) **b :** a return toward an ancestral type or condition **:** reappearance of an ancestral character
4 : an act or instance of turning the opposite way **:** the state of being so turned
5 : a product of reversion; *specifically* **:** an organism with an atavistic character **:** THROWBACK

re·ver·sion·al \-'vərzh-nəl, -'vərsh-; -'vər-zhə-n^{ə}l, -shə-\ *adjective* (1675)
: REVERSIONARY

re·ver·sion·ary \-'vər-zhə-ˌner-ē, -shə-\ *adjective* (1720)
: of, relating to, constituting, or involving especially a legal reversion

re·ver·sion·er \-'vərzh-nər, -'vərsh-; -'vər-zhə-nər, -shə-\ *noun* (1614)
: one that has or is entitled to a reversion; *broadly* **:** one having a vested right to a future estate

re·vert \ri-'vərt\ *intransitive verb* [Middle English, from Middle French *revertir,* from Latin *revertere,* transitive verb, to turn back & *reverti,* intransitive verb, to return, come back, from *re-* + *vertere, verti* to turn — more at WORTH] (15th century)
1 : to come or go back (as to a former condition, period, or subject)
2 : to return to the proprietor or his heirs at the end of a reversion
3 : to return to an ancestral type
— re·vert·er *noun*
— re·vert·ible \-'vər-tə-bəl\ *adjective*

re·ver·tant \ri-'vər-t^{ə}nt\ *noun* (1955)
: a mutant gene, individual, or strain that regains a former capability (as the production of a particular protein) by undergoing further mutation ⟨yeast *revertants*⟩
— revertant *adjective*

re·vest \(ˌ)rē-'vest\ *transitive verb* (1561)
: REINSTATE, REINVEST

re·vet \ri-'vet\ *transitive verb* **re·vet·ted; re·vet·ting** [French *revêtir,* literally, to clothe again, dress up, from Latin *revestire,* from *re-* + *vestire* to clothe — more at VEST] (1812)
: to face (as an embankment) with a revetment

re·vet·ment \-'vet-mənt\ *noun* (1779)
1 : a facing (as of stone or concrete) to sustain an embankment
2 : EMBANKMENT; *especially* **:** a barricade to provide shelter (as against bomb fragments or strafing)

[1]re·view \ri-'vyü\ *noun* [Middle English, from Middle French *reveue,* from *revoir* to look over, from *re-* + *voir* to see — more at VIEW] (15th century)
1 a : a formal military inspection **b :** a military ceremony honoring a person or an event
2 : REVISION 1a
3 : a general survey (as of the events of a period)
4 : an act or the process of reviewing
5 : judicial reexamination (as of the proceedings of a lower tribunal by a higher)
6 a : a critical evaluation (as of a book or play) **b :** a magazine devoted chiefly to reviews and essays
7 a : a retrospective view or survey (as of one's life) **b** (1) **:** renewed study of material previously studied (2) **:** an exercise facilitating such study
8 : REVUE

[2]re·view \ri-'vyü, *1 is also* 'rē-\ *verb* [in senses 1 & 2, from *re-* + *view;* in other senses, from [1]*review*] (1576)
transitive verb
1 : to view or see again
2 : to examine or study again; *especially* **:** to reexamine judicially
3 : to look back on **:** take a retrospective view of
4 a : to go over or examine critically or deliberately ⟨*reviewed* the results of the study⟩ **b :** to give a critical evaluation of ⟨*review* a novel⟩
5 : to hold a review of ⟨*review* troops⟩
intransitive verb
1 : to study material again **:** make a review ⟨*review* for a test⟩
2 : to write reviews
— re·view·able \ri-'vyü-ə-bəl\ *adjective*

re·view·er \ri-'vyü-ər\ *noun* (1651)
: one that reviews; *especially* **:** a writer of critical reviews

re·vile \ri-'vī(ə)l\ *verb* **re·viled; re·vil·ing** [Middle English, from Middle French *reviler* to despise, from *re-* + *vil* vile] (14th century)
transitive verb
: to subject to verbal abuse **:** VITUPERATE
intransitive verb
: to use abusive language **:** RAIL
synonym see SCOLD
— re·vile·ment \-'vī(ə)l-mənt\ *noun*
— re·vil·er *noun*

re·vis·al \ri-'vī-zəl\ *noun* (1612)
: an act of revising **:** REVISION

[1]re·vise \'rē-ˌvīz, ri-'\ *noun* (1591)
1 : an act of revising **:** REVISION
2 : a printing proof that incorporates changes marked in a previous proof

[2]re·vise \ri-'vīz\ *verb* **re·vised; re·vis·ing** [Middle French *reviser,* from Latin *revisere* to look at again, frequentative of *revidēre* to see again, from *re-* + *vidēre* to see — more at WIT] (1596)
transitive verb
1 a : to look over again in order to correct or improve ⟨*revise* a manuscript⟩ **b** *British* **:** to study again **:** REVIEW
2 a : to make a new, amended, improved, or up-to-date version of ⟨*revise* a dictionary⟩ **b :** to provide with a new taxonomic arrangement ⟨*revising* the alpine ferns⟩
intransitive verb
British **:** REVIEW 1
synonym see CORRECT
— re·vis·able \-'vī-zə-bəl\ *adjective*
— re·vis·er *or* **re·vi·sor** \-'vī-zər\ *noun*

Revised Standard Version *noun* (1946)
: a revision of the American Standard Version of the Bible published in 1946 and 1952

Revised Version *noun* (1880)
: a British revision of the Authorized Version of the Bible published in 1881 and 1885

re·vi·sion \ri-'vi-zhən\ *noun* (1611)
1 a : an act of revising **b :** a result of revising **:** ALTERATION
2 : a revised version
— re·vi·sion·ary \-zhə-ˌner-ē\ *adjective*

re·vi·sion·ism \ri-'vi-zhə-ˌni-zəm\ *noun* (1903)
1 : a movement in revolutionary Marxian socialism favoring an evolutionary rather than a revolutionary spirit
2 : advocacy of revision (as of a doctrine or policy or in historical analysis)
— re·vi·sion·ist \-nist\ *noun or adjective*

[1]re·vis·it \(ˌ)rē-'vi-zət\ *transitive verb* (15th century)
: to visit again **:** return to ⟨*revisit* the old neighborhood⟩; *also* **:** to consider or take up again ⟨reluctant to *revisit* past disputes⟩

[2]revisit *noun* (1623)
: a second or subsequent visit

re·vi·so·ry \ri-'vī-zə-rē, -'vīz-rē\ *adjective* (circa 1841)

☆ SYNONYMS
Reverse, transpose, invert mean to change to the opposite position. REVERSE is the most general term and may imply change in order, side, direction, meaning ⟨*reversed* his position on the trade agreement⟩. TRANSPOSE implies a change in order or relative position of units often through exchange of position ⟨*transposed* the letters to form an anagram⟩. INVERT applies chiefly to turning upside down or inside out ⟨a stamp with an *inverted* picture of an airplane⟩.

: having the power or purpose to revise ⟨a *revisory* committee⟩ ⟨a *revisory* function⟩

re·vi·tal·ise *British variant of* REVITALIZE

re·vi·tal·ize \(ˌ)rē-ˈvī-tᵊl-ˌīz\ *transitive verb* **-ized; -iz·ing** (1869)
: to give new life or vigor to
— re·vi·tal·i·za·tion \(ˌ)rē-ˌvī-tᵊl-ə-ˈzā-shən\ *noun*

re·viv·al \ri-ˈvī-vəl\ *noun* (1651)
1 : an act or instance of reviving **:** the state of being revived: as **a :** renewed attention to or interest in something **b :** a new presentation or publication of something old **c** (1) **:** a period of renewed religious interest (2) **:** an often highly emotional evangelistic meeting or series of meetings
2 : restoration of force, validity, or effect (as to a contract)

re·viv·al·ism \-ˈvī-və-ˌli-zəm\ *noun* (1815)
1 : the spirit or methods characteristic of religious revivals
2 : a tendency or desire to revive or restore

re·viv·al·ist \-ˈvī-və-list, -ˈvīv-list\ *noun* (1820)
1 : one who conducts religious revivals; *specifically* **:** a clergyman who travels about to conduct revivals
2 : one who revives or restores something disused
— revivalist *adjective*
— re·viv·al·is·tic \-ˌvī-və-ˈlis-tik\ *adjective*

re·vive \ri-ˈvīv\ *verb* **re·vived; re·viv·ing** [Middle English, from Middle French *revivre,* from Latin *revivere* to live again, from *re-* + *vivere* to live — more at QUICK] (15th century)
intransitive verb
: to return to consciousness or life **:** become active or flourishing again
transitive verb
1 : to restore to consciousness or life
2 : to restore from a depressed, inactive, or unused state **:** bring back
3 : to renew in the mind or memory
— re·viv·able \-ˈvī-və-bəl\ *adjective*
— re·viv·er *noun*

re·viv·i·fy \rē-ˈvi-və-ˌfī\ *transitive verb* [French *révivifier,* from Late Latin *revivificare,* from Latin *re-* + Late Latin *vivificare* to vivify] (1675)
: to give new life to **:** REVIVE
— re·viv·i·fi·ca·tion \-ˌvi-və-fə-ˈkā-shən\ *noun*

re·vi·vis·cence \ˌrē-ˌvī-ˈvi-sᵊn(t)s, ri-\ *noun* [Latin *reviviscere* to come to life again, from *re-* + *viviscere* to come to life, from *vivus* alive, living — more at QUICK] (1626)
: an act of reviving **:** the state of being revived
— re·vi·vis·cent \-sᵊnt\ *adjective*

rev·o·ca·ble \ˈre-və-kə-bəl *also* ri-ˈvō-\ *also* **re·vok·able** \ri-ˈvō-kə-bəl\ *adjective* [Middle English, from Middle French, from Latin *revocabilis,* from *revocare*] (15th century)
: capable of being revoked

rev·o·ca·tion \ˌre-və-ˈkā-shən; ri-ˌvō-, ˌrē-\ *noun* [Middle English, from Middle French, from Latin *revocation-, revocatio,* from *revocare*] (15th century)
: an act or instance of revoking

¹re·voke \ri-ˈvōk\ *verb* **re·voked; re·vok·ing** [Middle English, from Middle French *revoquer,* from Latin *revocare,* from *re-* + *vocare* to call, from *voc-, vox* voice — more at VOICE] (14th century)
transitive verb
1 : to annul by recalling or taking back **:** RESCIND ⟨*revoke* a will⟩
2 : to bring or call back
intransitive verb
: to fail to follow suit when able in a card game in violation of the rules
— re·vok·er *noun*

²revoke *noun* (1709)
: an act or instance of revoking in a card game

¹re·volt \ri-ˈvōlt *also* -ˈvȯlt\ *verb* [Middle French *revolter,* from Old Italian *rivoltare* to overthrow, from (assumed) Vulgar Latin *revolvitare,* frequentative of Latin *revolvere* to revolve, roll back] (1539)
intransitive verb
1 : to renounce allegiance or subjection (as to a government) **:** REBEL
2 a : to experience disgust or shock **b :** to turn away with disgust
transitive verb
: to cause to turn away or shrink with disgust or abhorrence
— re·volt·er *noun*

²revolt *noun* (1560)
1 : a renouncing of allegiance (as to a government or party); *especially* **:** a determined armed uprising
2 : a movement or expression of vigorous dissent
synonym see REBELLION

re·volt·ing *adjective* (1806)
: extremely offensive
— re·volt·ing·ly *adverb*

rev·o·lute \ˈre-və-ˌlüt\ *adjective* [Latin *revolutus,* past participle of *revolvere*] (circa 1753)
: rolled backward or downward ⟨a leaf with *revolute* margins⟩

rev·o·lu·tion \ˌre-və-ˈlü-shən\ *noun* [Middle English *revolucioun,* from Middle French *revolution,* from Late Latin *revolution-, revolutio,* from Latin *revolvere* to revolve] (14th century)
1 a (1) **:** the action by a celestial body of going round in an orbit or elliptical course; *also* **:** apparent movement of such a body round the earth (2) **:** the time taken by a celestial body to make a complete round in its orbit (3) **:** the rotation of a celestial body on its axis **b :** completion of a course (as of years); *also* **:** the period made by the regular succession of a measure of time or by a succession of similar events **c** (1) **:** a progressive motion of a body round an axis so that any line of the body parallel to the axis returns to its initial position while remaining parallel to the axis in transit and usually at a constant distance from it (2) **:** motion of any figure about a center or axis ⟨*revolution* of a right triangle about one of its legs generates a cone⟩ (3) **:** ROTATION 1b
2 a : a sudden, radical, or complete change **b :** a fundamental change in political organization; *especially* **:** the overthrow or renunciation of one government or ruler and the substitution of another by the governed **c :** activity or movement designed to effect fundamental changes in the socioeconomic situation **d :** a fundamental change in the way of thinking about or visualizing something **:** a change of paradigm ⟨the Copernican *revolution*⟩ **e :** a changeover in use or preference especially in technology ⟨the computer *revolution*⟩ ⟨the foreign car *revolution*⟩
synonym see REBELLION

¹rev·o·lu·tion·ary \-shə-ˌner-ē\ *adjective* (1774)
1 a : of, relating to, or constituting a revolution ⟨*revolutionary* war⟩ **b :** tending to or promoting revolution ⟨a *revolutionary* party⟩ **c :** constituting or bringing about a major or fundamental change ⟨*revolutionary* styling⟩ ⟨a *revolutionary* new product⟩
2 *capitalized* **:** of or relating to the American Revolution or to the period in which it occurred
— rev·o·lu·tion·ar·i·ly \-ˌlü-shə-ˈner-ə-lē\ *adverb*
— rev·o·lu·tion·ar·i·ness \-ˈlü-shə-ˌner-ē-nəs\ *noun*

²revolutionary *noun, plural* **-ar·ies** (1850)
1 : one engaged in a revolution
2 : an advocate or adherent of revolutionary doctrines

rev·o·lu·tion·ise *British variant of* REVOLUTIONIZE

rev·o·lu·tion·ist \ˌre-və-ˈlü-sh(ə-)nist\ *noun* (1710)
: REVOLUTIONARY
— revolutionist *adjective*

rev·o·lu·tion·ize \-shə-ˌnīz\ *verb* **-ized; -iz·ing** (1797)
transitive verb
1 : to overthrow the established government of
2 : to imbue with revolutionary doctrines
3 : to change fundamentally or completely
intransitive verb
: to engage in revolution
— rev·o·lu·tion·iz·er *noun*

re·volve \ri-ˈvälv, -ˈvȯlv *also* -ˈväv *or* -ˈvȯv\ *verb* **re·volved; re·volv·ing** [Middle English, from Latin *revolvere* to roll back, cause to return, from *re-* + *volvere* to roll — more at VOLUBLE] (15th century)
transitive verb
1 : to turn over at length in the mind **:** PONDER ⟨*revolve* a scheme⟩
2 a *obsolete* **:** to cause to go round in an orbit **b :** ROTATE 1
intransitive verb
1 : RECUR
2 a : to ponder something **b :** to remain under consideration ⟨ideas *revolved* in his mind⟩
3 a : to move in a curved path round a center or axis **b :** to turn or roll round on an axis
4 : to come to a main or central point **:** CENTER ⟨the dispute *revolved* around wages⟩
— re·volv·able \-ˈväl-və-bəl, -ˈvȯl- *also* -ˈvä-və- *or* -ˈvȯ-və-\ *adjective*

re·volv·er \ri-ˈväl-vər, -ˈvȯl- *also* -ˈvä-vər *or* -ˈvȯ-vər\ *noun* (circa 1835)
1 : one that revolves
2 : a handgun with a cylinder of several chambers brought successively into line with the barrel and discharged with the same hammer

revolving *adjective* (1599)
1 : tending to revolve or recur; *especially* **:** recurrently available
2 : turning around on or as if on an axis ⟨a *revolving* platform⟩

revolving charge account *noun* (1967)
: a charge account under which payment is made in monthly installments and includes a carrying charge

revolving credit *noun* (1919)
: a credit which may be used repeatedly up to the limit specified after partial or total repayments have been made

revolving–door *adjective* (1973)
: characterized by a frequently repeated cycle of leaving and returning ⟨*revolving-door* governments⟩

revolving fund *noun* (1920)
: a fund set up for specified purposes with the proviso that repayments to the fund may be used again for these purposes

re·vue \ri-ˈvyü\ *noun* [French, from Middle French *reveue* review — more at REVIEW] (1872)
: a theatrical production consisting typically of brief loosely connected often satirical skits, songs, and dances

re·vulsed \ri-ˈvəlst\ *adjective* [Latin *revulsus,* past participle of *revellere* + English *-ed*] (circa 1934)
: affected with or having undergone revulsion

re·vul·sion \ri-ˈvəl-shən\ *noun* [Latin *revulsion-, revulsio* act of tearing away, from *revellere* to pluck away, from *re-* + *vellere* to pluck — more at VULNERABLE] (1609)
1 : a strong pulling or drawing away **:** WITHDRAWAL
2 a : a sudden or strong reaction or change **b :** a sense of utter distaste or repugnance
— re·vul·sive \-ˈvəl-siv\ *adjective*

revved *past and past participle of* REV

revving *present participle of* REV

re·wake \(ˌ)rē-ˈwāk\ *verb* **-waked** *or* **-woke** \-ˈwōk\; **-waked** *or* **-wo·ken** \-ˈwō-kən\ *or* **-woke; -wak·ing** (1593)
transitive verb
: to waken again or anew
intransitive verb
: to become awake again

re·wak·en \(ˌ)rē-ˈwā-kən\ *verb* (1638)
: REWAKE

¹re·ward \ri-ˈwȯrd\ *transitive verb* [Middle English, from Old North French *rewarder* to regard, reward, from *re-* + *warder* to watch, guard, of Germanic origin; akin to Old High German *wartēn* to watch — more at WARD] (14th century)
1 : to give a reward to or for
2 : RECOMPENSE
— re·ward·able \-ˈwȯr-də-bəl\ *adjective*
— re·ward·er *noun*

²reward *noun* (14th century)
1 : something that is given in return for good or evil done or received and especially that is offered or given for some service or attainment
2 : a stimulus administered to an organism following a correct or desired response that increases the probability of occurrence of the response

re·ward·ing *adjective* (1697)
1 : yielding or likely to yield a reward **:** VALUABLE, SATISFYING ⟨a *rewarding* experience⟩
2 : serving as a reward ⟨a *rewarding* smile of thanks⟩
— re·ward·ing·ly *adverb*

¹re·wind \(ˌ)rē-ˈwīnd\ *transitive verb* **-wound** \-ˈwau̇nd\; **-wind·ing** (1717)
: to wind again; *especially* **:** to reverse the winding of (as film)

²re·wind \ˈrē-ˌwīnd, (ˌ)rē-ˈ\ *noun* (1926)
1 : something that rewinds or is rewound
2 : an act of rewinding

re·word \(ˌ)rē-ˈwərd\ *transitive verb* (1602)
1 : to repeat in the same words
2 : to alter the wording of; *also* **:** to restate in other words

re·work \(ˌ)rē-ˈwərk\ *transitive verb* (1842)
: to work again or anew: as **a :** REVISE **b :** to reprocess (as used material) for further use

¹re·write \(ˌ)rē-ˈrīt\ *verb* **-wrote** \-ˈrōt\; **-writ·ten** \-ˈri-tᵊn\; **-writ·ing** \-ˈrī-tiŋ\ (1567)
transitive verb
1 : to write in reply
2 : to make a revision of (as a story) **:** cause to be revised: as **a :** to put (contributed material) into form for publication **b :** to alter (previously published material) for use in another publication
intransitive verb
: to revise something previously written
— re·writ·er *noun*

²re·write \ˈrē-ˌrīt\ *noun* (1914)
1 : a piece of writing (as a news story) constructed by rewriting
2 : an act or instance of rewriting

re·write man \ˈrē-ˌrīt-ˌman\ *noun* (1901)
: a newspaperman who specializes in rewriting

re·write rule \ˈrē-ˌrīt-\ *noun* (1961)
: a rule in a grammar which specifies the constituents of a single symbol

rex \ˈreks\ *noun, plural* **rex·es** *or* **rex** [modification of French *castorrex, castorex,* a variety of rabbit, perhaps from Latin *castor* beaver + *rex* king — more at CASTOR, ROYAL] (1920)
: an animal (as a domestic rabbit) showing a genetic recessive variation in which the guard hairs are very short or entirely lacking; *especially* **:** any of the slender domestic cats having a short curly undercoat and no guard hairs and belonging to either of two breeds

rex

Reye's syndrome \ˈrīz-, ˈrāz-\ *noun* [R.D.K. *Reye* (died 1977) Australian pathologist] (1965)
: an often fatal encephalopathy especially of childhood characterized by fever, vomiting, fatty infiltration of the liver, and swelling of the kidneys and brain — called also *Reye syndrome*

rey·nard \ˈrā-nərd, ˈre-, -ˌnär(d)\ *noun, often capitalized* [Middle English *Renard,* name of the fox who is hero of the French beast epic *Roman de Renart,* from Middle French *Renart, Renard*] (14th century)
: FOX

Rey·nolds number \ˈre-nᵊldz-\ *noun* [Osborne *Reynolds* (died 1912) English physicist] (1910)
: a number characteristic of the flow of a fluid in a pipe or past an obstruction

R factor \ˈär-\ *noun* [*r*esistance] (1962)
: a group of genes present in some bacteria that provide a basis for resistance to antibiotics and can be transferred from cell to cell by conjugation

Rh \ˌär-ˈāch\ *adjective* (1940)
: of, relating to, or being an Rh factor ⟨*Rh* antigens⟩ ⟨*Rh* sensitization in pregnancy⟩

rhabdo- *combining form* [Late Greek, from Greek *rhabdos* rod — more at VERVAIN]
: rodlike structure ⟨*rhabdo*virus⟩

rhab·do·coele \ˈrab-də-ˌsēl\ *noun* [New Latin *Rhabdocoela,* from *rhabdo-* + New Latin *-coela* -coele] (circa 1909)
: a turbellarian worm (order Rhabdocoela) with an unbranched intestine

rhab·dom \ˈrab-ˌdäm, -dəm\ *or* **rhab·dome** \-ˌdōm\ *noun* [New Latin, from Late Greek *rhabdōma* bundle of rods, from Greek *rhabdos* rod] (1878)
: one of the minute rodlike structures in the retinulae in the compound eyes of arthropods

rhab·do·man·cy \ˈrab-də-ˌman(t)-sē\ *noun* [Late Greek *rhabdomanteia,* from Greek *rhabdos* rod + *-manteia* -mancy] (1646)
: divination by rods or wands
— rhab·do·man·cer \-ˌman(t)-sər\ *noun*

rhab·do·mere \-ˌmir\ *noun* (circa 1884)
: a division of a rhabdom

rhab·do·myo·sar·co·ma \ˈrab-(ˌ)dō-ˌmī-ə-sär-ˈkō-mə\ *noun, plural* **-mas** *or* **-ma·ta** \-mə-tə\ [New Latin, from *rhabdo-* + *my-* + *sarcoma*] (1898)
: a malignant tumor composed of striated muscle fibers

rhab·do·vi·rus \-ˌvī-rəs\ *noun* [New Latin] (1966)
: any of a group of RNA-containing rod- or bullet-shaped viruses found in plants and animals and including the causative agents of rabies and vesicular stomatitis

rhad·a·man·thine \ˌra-də-ˈman(t)-thən, -ˈman-ˌthīn\ *adjective, often capitalized* [*Rhadamanthus*] (1840)
: rigorously strict or just

Rhad·a·man·thus \ˌra-də-ˈman(t)-thəs\ *noun*
: a judge of the underworld in Greek mythology

Rhae·to–Ro·mance \ˌrē-tō-rō-ˈman(t)s, -tō-rə-; -ˈrō-ˌman(t)s\ *also* **Rhae·to–Ro·man·ic** \-rō-ˈma-nik\ *noun* [Latin *Rhaetus* of Rhaetia, ancient Roman province] (1867)
: a group of Romance languages spoken in eastern Switzerland and northeastern Italy

rham·nose \ˈram-ˌnōs, -ˌnōz\ *noun* [International Scientific Vocabulary, from New Latin *Rhamnus,* genus of the buckthorn; from its being produced from a plant of this genus] (1888)
: a crystalline sugar $C_6H_{12}O_5$ that occurs usually in the form of a glycoside in many plants and is obtained in the common dextrorotatory L form

rhap·sode \ˈrap-ˌsōd\ *noun* [French, from Greek *rhapsōidos*] (1834)
: RHAPSODIST

rhap·sod·ic \rap-ˈsä-dik\ *also* **rhap·sod·i·cal** \-di-kəl\ *adjective* (1782)
1 : extravagantly emotional **:** RAPTUROUS
2 : resembling or characteristic of a rhapsody
— rhap·sod·i·cal·ly \-di-k(ə-)lē\ *adverb*

rhap·so·dist \ˈrap-sə-dist\ *noun* (circa 1656)
1 : a professional reciter of epic poems
2 : one who writes or speaks rhapsodically

rhap·so·dize \-sə-ˌdīz\ *intransitive verb* **-dized; -diz·ing** (1806)
: to speak or write in a rhapsodic manner ⟨*rhapsodize* about a new book⟩

rhap·so·dy \ˈrap-sə-dē\ *noun, plural* **-dies** [Latin *rhapsodia,* from Greek *rhapsōidia* recitation of selections from epic poetry, rhapsody, from *rhapsōidos* rhapsodist, from *rhaptein* to sew, stitch together + *aidein* to sing — more at ODE] (1542)
1 : a portion of an epic poem adapted for recitation
2 *archaic* **:** a miscellaneous collection
3 a (1) **:** a highly emotional utterance (2) **:** a highly emotional literary work (3) **:** effusively rapturous or extravagant discourse **b :** RAPTURE, ECSTASY
4 : a musical composition of irregular form having an improvisatory character

rhat·a·ny \ˈra-tᵊn-ē\ *noun* [Spanish *ratania* & Portuguese *ratânhia*] (1808)
1 : the dried root of either of two South American shrubs (*Krameria triandra* and *K. argentea* of the family Krameriaceae) used as an astringent
2 : a plant yielding rhatany

rhea \ˈrē-ə\ *noun* [New Latin, genus of birds, probably from Latin *Rhea,* mother of Zeus, from Greek] (1797)
: either of two South American ratite birds (*Rhea americana* and *Pterocnemia pennata* of the family Rheidae) that resemble but are smaller than the African ostrich and that have three toes, a fully feathered head and neck, an undeveloped tail, and pale gray to brownish feathers that droop over the rump and back

rhea

rhe·bok \ˈrē-ˌbäk\ *noun* [Afrikaans *reebok,* from Dutch, male roe deer, from *ree* roe + *boc* buck] (1834)
: a brownish gray antelope (*Pelea capreolus*) of southern Africa

rhe·ni·um \ˈrē-nē-əm\ *noun* [New Latin, from Latin *Rhenus* Rhine River] (1925)
: a rare heavy metallic element that resembles manganese, is obtained either as a powder or as a silver-white hard metal, and is used in catalysts and thermocouples — see ELEMENT table

rheo- *combining form* [Greek *rhein* to flow — more at STREAM]
: flow **:** current ⟨*rheo*stat⟩

rhe·ol·o·gy \rē-ˈä-lə-jē\ *noun* [International Scientific Vocabulary] (1929)
: a science dealing with the deformation and flow of matter; *also* **:** the ability to flow or be deformed
— rhe·o·log·i·cal \ˌrē-ə-ˈlä-ji-kəl\ *adjective*
— rhe·o·log·i·cal·ly \-k(ə-)lē\ *adverb*
— rhe·ol·o·gist \rē-ˈä-lə-jist\ *noun*

rhe·om·e·ter \rē-ˈä-mə-tər\ *noun* [International Scientific Vocabulary] (circa 1859)
: an instrument for measuring flow (as of viscous substances)

rheo·stat \ˈrē-ə-ˌstat\ *noun* (1843)
: a resistor for regulating a current by means of variable resistances
— rheo·stat·ic \ˌrē-ə-ˈsta-tik\ *adjective*

rhe·sus monkey \'rē-səs-\ *noun* [New Latin *Rhesus,* genus of monkeys, from Latin, a mythical king of Thrace, from Greek *Rhēsos*] (1841)
: a pale brown Asian macaque (*Macaca mulatta*) often used in medical research

rhe·tor \'rē-ˌtȯr, 're-; 'rē-tər, 're-\ *noun* [Middle English *rethor,* from Latin *rhetor,* from Greek *rhētōr*] (14th century)
: RHETORICIAN 1

rhet·o·ric \'re-tə-rik\ *noun* [Middle English *rethorik,* from Middle French *rethorique,* from Latin *rhetorica,* from Greek *rhētorikē,* literally, art of oratory, from feminine of *rhētorikos* of an orator, from *rhētōr* orator, rhetorician, from *eirein* to say, speak — more at WORD] (14th century)
1 : the art of speaking or writing effectively: as **a :** the study of principles and rules of composition formulated by critics of ancient times **b :** the study of writing or speaking as a means of communication or persuasion
2 a : skill in the effective use of speech **b :** a type or mode of language or speech; *also* **:** insincere or grandiloquent language
3 : verbal communication **:** DISCOURSE

rhe·tor·i·cal \ri-'tȯr-i-kəl, -'tär-\ *also* **rhe·tor·ic** \ri-'tȯr-ik, -'tär-\ *adjective* (15th century)
1 a : of, relating to, or concerned with rhetoric **b :** employed for rhetorical effect; *especially* **:** asked merely for effect with no answer expected ⟨a *rhetorical* question⟩
2 a : given to rhetoric **:** GRANDILOQUENT **b :** VERBAL
— rhe·tor·i·cal·ly \-i-k(ə-)lē\ *adverb*

rhet·o·ri·cian \ˌre-tə-'ri-shən\ *noun* (15th century)
1 a : a master or teacher of rhetoric **b :** ORATOR
2 : an eloquent or grandiloquent writer or speaker

rheum \'rüm\ *noun* [Middle English *reume,* from Middle French, from Latin *rheuma,* from Greek, literally, flow, flux, from *rhein* to flow — more at STREAM] (14th century)
1 : a watery discharge from the mucous membranes especially of the eyes or nose
2 *archaic* **:** TEARS
— rheumy \'rü-mē\ *adjective*

[1]**rheu·mat·ic** \rü-'ma-tik\ *adjective* [Middle English *rewmatik* subject to rheum, from Latin *rheumaticus,* from Greek *rheumatikos,* from *rheumat-, rheuma*] (1711)
: of, relating to, characteristic of, or affected with rheumatism
— rheu·mat·i·cal·ly \-ti-k(ə-)lē\ *adverb*

[2]**rheumatic** *noun* (1884)
: one affected with rheumatism

rheumatic fever *noun* (1782)
: an acute disease that occurs chiefly in children and young adults and is characterized by fever, by inflammation and pain in and around the joints, and by inflammatory involvement of the pericardium and heart valves

rheu·ma·tism \'rü-mə-ˌti-zəm, 'rü-\ *noun* [Latin *rheumatismus* flux, rheum, from Greek *rheumatismos,* from *rheumatizesthai* to suffer from a flux, from *rheumat-, rheuma* flux] (1677)
1 : any of various conditions characterized by inflammation or pain in muscles, joints, or fibrous tissue ⟨muscular *rheumatism*⟩
2 : RHEUMATOID ARTHRITIS

rheu·ma·tiz \-ˌtiz\ *noun* (1760)
chiefly dialect **:** RHEUMATISM

rheu·ma·toid \-ˌtȯid\ *adjective* [International Scientific Vocabulary, from *rheumatism*] (1871)
: characteristic of or affected with rheumatoid arthritis

rheumatoid arthritis *noun* (1859)
: a usually chronic disease of unknown cause that is characterized especially by pain, stiffness, inflammation, swelling, and sometimes destruction of joints

rheumatoid factor *noun* (1960)
: an autoantibody of high molecular weight that is usually present in rheumatoid arthritis

rheu·ma·tol·o·gy \ˌrü-mə-'tä-lə-jē, ˌrü-\ *noun* (circa 1941)
: a medical science dealing with rheumatic diseases
— rheu·ma·tol·o·gist \-jist\ *noun*

Rh factor \ˌär-'āch-\ *noun* [*rh*esus monkey (in which it was first detected)] (1942)
: any of one or more genetically determined antigens present in the red blood cells of most persons and of higher animals and capable of inducing intense immunogenic reactions

rhin- *or* **rhino-** *combining form* [New Latin, from Greek, from *rhin-, rhis*]
: nose ⟨*rhino*plasty⟩

rhi·nal \'rī-nᵊl\ *adjective* (circa 1859)
: of or relating to the nose **:** NASAL

rhin·en·ceph·a·lon \ˌrī-(ˌ)nen-'se-fə-ˌlän, -lən\ *noun* [New Latin] (1846)
: the chiefly olfactory part of the forebrain
— rhin·en·ce·phal·ic \ˌrī-ˌnen-sə-'fa-lik\ *adjective*

rhine·stone \'rīn-ˌstōn\ *noun* [*Rhine* River] (circa 1888)
: a colorless imitation stone of high luster made of glass, paste, or gem quartz
— rhine·stoned \-ˌstōnd\ *adjective*

Rhine wine \'rīn-\ *noun* (1843)
1 : a usually white wine produced in the Rhine valley
2 : a wine similar to Rhine wine produced elsewhere

rhi·ni·tis \rī-'nī-təs\ *noun* [New Latin] (circa 1884)
: inflammation of the mucous membrane of the nose

[1]**rhi·no** \'rī-(ˌ)nō\ *noun* [origin unknown] (1670)
: MONEY, CASH

[2]**rhino** *noun, plural* **rhino** *or* **rhinos** (1884)
: RHINOCEROS

rhi·noc·er·os \rī-'näs-rəs, rə-, -'nä-sə-\ *noun, plural* **-noc·er·os·es** *or* **-noc·er·os** *or* **-noc·eri** \-'nä-sə-ˌrī\ [Middle English *rinoceros,* from Latin *rhinocerot-, rhinoceros,* from Greek *rhinokerōt-, rhinokerōs,* from *rhin-* + *keras* horn — more at HORN] (14th century)
: any of a family (Rhinocerotidae) of large heavyset herbivorous perissodactyl mammals of Africa and Asia that have one or two upright keratinous horns on the snout and thick gray to brown skin with little hair
word history see UNICORN

rhinoceros beetle *noun* (1681)
: any of various large chiefly tropical scarab beetles (subfamily Dynastinae) having projecting horns on thorax and head

rhinoceros beetle

rhi·no·plas·ty \'rī-nō-ˌplas-tē\ *noun, plural* **-ties** (1842)
: plastic surgery on the nose usually for cosmetic purposes

rhi·nos·co·py \rī-'näs-kə-pē\ *noun* [International Scientific Vocabulary] (1861)
: examination of the nasal passages

rhi·no·vi·rus \ˌrī-nō-'vī-rəs\ *noun* [New Latin] (1961)
: any of a group of picornaviruses that are related to the enteroviruses and are associated with disorders of the upper respiratory tract (as the common cold)

rhiz- *or* **rhizo-** *combining form* [New Latin, from Greek, from *rhiza* — more at ROOT]
: root ⟨*rhizo*plane⟩

-rhiza *or* **-rrhiza** *noun combining form, plural* **-zae** *or* **-zas** [New Latin, from Greek *rhiza*]
: root **:** part resembling or connected with a root ⟨coleo*rhiza*⟩ ⟨mycor*rhiza*⟩

rhi·zo·bi·um \rī-'zō-bē-əm\ *noun, plural* **-bia** \-bē-ə\ [New Latin, from *rhiz-* + Greek *bios* life — more at QUICK] (1921)
: any of a genus (*Rhizobium*) of small heterotrophic soil bacteria capable of forming symbiotic nodules on the roots of leguminous plants and of there becoming bacteroids that fix atmospheric nitrogen
— rhi·zo·bi·al \-bē-əl\ *adjective*

rhi·zoc·to·nia \ˌrī-ˌzäk-'tō-nē-ə\ *noun* [New Latin, from *rhiz-* + Greek *-ktonos* killing, from *kteinein* to kill; akin to Sanskrit *kṣaṇoti* he wounds] (1897)
: any of a form genus (*Rhizoctonia*) of imperfect fungi that includes major plant pathogens

rhi·zoid \'rī-ˌzȯid\ *noun* (1875)
: a rootlike structure
— rhi·zoi·dal \rī-'zȯi-dᵊl\ *adjective*

rhi·zo·ma·tous \rī-'zō-mə-təs\ *adjective* [International Scientific Vocabulary, from New Latin *rhizomat-, rhizoma*] (1847)
: having, resembling, or being a rhizome ⟨*rhizomatous* plants⟩

rhi·zome \'rī-ˌzōm\ *noun* [New Latin *rhizomat-, rhizoma,* from Greek *rhizōmat-, rhizōma* mass of roots, from *rhizoun* to cause to take root, from *rhiza* root — more at ROOT] (1845)
: a somewhat elongate usually horizontal subterranean plant stem that is often thickened by deposits of reserve food material, produces shoots above and roots below, and is distinguished from a true root in possessing buds, nodes, and usually scalelike leaves
— rhi·zo·mic \rī-'zō-mik, -'zä-\ *adjective*

rhi·zo·plane \'rī-zə-ˌplān\ *noun* (1949)
: the external surface of roots together with closely adhering soil particles and debris

rhi·zo·pod \'rī-zə-ˌpäd\ *noun* [New Latin *Rhizopoda,* from *rhiz-* + *-poda* -pod] (1851)
: any of a superclass (Rhizopoda) of usually creeping protozoans (as an amoeba or a foraminifer) having lobate or rootlike pseudopodia

rhi·zo·pus \'rī-zə-pəs, -ˌpu̇s\ *noun* [New Latin, from *rhiz-* + Greek *pous* foot — more at FOOT] (1887)
: any of a genus (*Rhizopus*) of mold fungi including some economically valuable forms and some plant or animal pathogens (as a bread mold)

rhi·zo·sphere \-ˌsfir\ *noun* [International Scientific Vocabulary] (1929)
: soil that surrounds and is influenced by the roots of a plant

rhi·zot·o·my \rī-'zä-tə-mē\ *noun, plural* **-mies** [International Scientific Vocabulary] (1911)
: the operation of cutting the anterior or posterior spinal nerve roots

Rh–neg·a·tive \ˌär-ˌāch-'ne-gə-tiv\ *adjective* (1945)
: lacking Rh factor in the blood

rho \'rō\ *noun* [Greek *rhō,* of Semitic origin; akin to Hebrew *rēsh* resh] (15th century)
: the 17th letter of the Greek alphabet — see ALPHABET table

rhod- *or* **rhodo-** *combining form* [New Latin, from Latin, from Greek, from *rhodon* rose — more at ROSE]
: rose **:** red ⟨*rhodo*lite⟩

rho·da·mine \'rō-də-ˌmēn\ *noun, often capitalized* [International Scientific Vocabulary] (1888)
: any of a group of yellowish red to blue fluorescent dyes; *especially* **:** a brilliant bluish red dye made by fusing an amino derivative of phenol with phthalic anhydride and used especially in coloring paper and as a biological stain

Rhode Island Red *noun* [*Rhode Island,* U.S. state] (1896)
: any of an American breed of general-purpose domestic fowls having a long heavy body, smooth yellow or reddish legs, and rich brownish red plumage

Rhode Island White *noun* (circa 1923)
: any of an American breed of domestic fowls resembling Rhode Island Reds but having pure white plumage

Rhodes grass \ˈrōdz-\ *noun* [Cecil J. *Rhodes*] (1915)
: an African perennial grass (*Chloris gayana*) widely cultivated as a forage grass especially in dry regions

Rho·de·sian man \rō-ˈdē-zh(ē-)ən-\ *noun* [Northern *Rhodesia,* Africa] (1921)
: an extinct African hominid (*Homo sapiens rhodesiensis*) having long bones of modern type, a skull with prominent brow ridges and large face but human palate and dentition, and a simple but relatively large brain

Rhodesian Ridge·back \-ˈrij-ˌbak\ *noun* (1925)
: any of an African breed of powerful long-bodied hunting dogs having a dense harsh short tan coat with a characteristic crest of reversed hair along the spine

Rhodes scholar \ˈrōd(z)-\ *noun* (1902)
: a holder of one of numerous scholarships founded under the will of Cecil J. Rhodes that can be used at Oxford University for two or three years and are open to candidates from the Commonwealth and the U.S.

rho·di·um \ˈrō-dē-əm\ *noun* [New Latin, from Greek *rhodon* rose] (1804)
: a white hard ductile metallic element that is resistant to acids, occurs in platinum ores, and is used in alloys with platinum — see ELEMENT table

rho·do·chro·site \ˌrō-də-ˈkrō-ˌsīt\ *noun* [German *Rhodocrosit,* from Greek *rhodochrōs* rose-colored, from *rhod-* + *chrōs* color — more at CHROMATIC] (1836)
: a rose red mineral consisting essentially of manganese carbonate

rho·do·den·dron \ˌrō-də-ˈden-drən\ *noun* [New Latin, from Latin, oleander, from Greek, from *rhod-* + *dendron* tree — more at DENDR-] (1664)
: any of a genus (*Rhododendron*) of the heath family of widely cultivated shrubs and trees with alternate leaves and showy flowers; *especially* **:** one with leathery evergreen leaves as distinguished from a deciduous azalea

rho·do·lite \ˈrō-dᵊl-ˌīt\ *noun* (1897)
: a pink or purple garnet used as a gem

rho·do·mon·tade *variant of* RODOMONTADE

rho·do·nite \ˈrō-dᵊn-ˌīt\ *noun* [German *Rhodonit,* from Greek *rhodon* rose] (1823)
: a pale red triclinic mineral that consists essentially of manganese silicate and is used as an ornamental stone

rho·dop·sin \rō-ˈdäp-sən\ *noun* [International Scientific Vocabulary *rhod-* + Greek *opsis* sight, vision + International Scientific Vocabulary ¹*-in* — more at OPTIC] (1886)
: a red photosensitive pigment in the retinal rods of marine fishes and most higher vertebrates that is important in vision in dim light — called also *visual purple*

rho·do·ra \rō-ˈdōr-ə, -ˈdȯr-\ *noun* [New Latin, alteration of Latin *rodarum,* a plant] (circa 1731)
: an azalea (*Rhododendron canadense*) of northeastern North America that has spring-flowering pink blossoms

rhomb \ˈräm(b)\ *noun, plural* **rhombs** \ˈrämz\ [Middle French *rhombe,* from Latin *rhombus*] (circa 1578)
1 : RHOMBUS
2 : RHOMBOHEDRON

rhomb- *or* **rhombo-** *combining form* [New Latin, from Greek *rhombos*]
: rhombus ⟨*rhomb*encephalon⟩

rhomb·en·ceph·a·lon \ˌräm-(ˌ)ben-ˈse-fə-ˌlän, -lən\ *noun* [New Latin] (1897)
: HINDBRAIN 1

rhom·bic \ˈräm-bik\ *adjective* (1701)
1 : having the form of a rhombus
2 : ORTHORHOMBIC

rhom·bo·he·dron \ˌräm-bō-ˈhē-drən\ *noun, plural* **-drons** *or* **-dra** \-drə\ [New Latin] (1836)
: a parallelepiped whose faces are rhombuses
— rhom·bo·he·dral \-drəl\ *adjective*

¹**rhom·boid** \ˈräm-ˌbȯid\ *noun* [Middle French *rhomboïde,* from Latin *rhomboides,* from Greek *rhomboeidēs* resembling a rhombus, from *rhombos*] (1570)
: a parallelogram with no right angles and with adjacent sides of unequal length

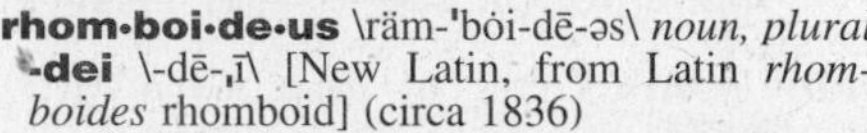
rhomboid

²**rhom·boid** \ˈräm-ˌbȯid\ *or* **rhom·boi·dal** \räm-ˈbȯi-dᵊl\ *adjective* (1693)
: shaped somewhat like a rhombus or rhomboid

rhom·boi·de·us \räm-ˈbȯi-dē-əs\ *noun, plural* **-dei** \-dē-ˌī\ [New Latin, from Latin *rhomboides* rhomboid] (circa 1836)
: either of two muscles that lie beneath the trapezius muscle and connect the spinous processes of various vertebrae with the medial border of the scapula

rhom·bus \ˈräm-bəs\ *noun, plural* **rhom·bus·es** *or* **rhom·bi** \-ˌbī, -ˌbē\ [Latin, from Greek *rhombos* piece of wood whirled on a string, lozenge, from *rhembein* to whirl] (circa 1567)
: a parallelogram with four equal sides and sometimes one with no right angles

rhon·chus \ˈräŋ-kəs\ *noun, plural* **rhon·chi** \ˈräŋ-ˌkī\ [Late Greek, from *rhenchein* to snore, wheeze; probably akin to Old Irish *sreinnid* he snores] (1829)
: a whistling or snoring sound heard on auscultation of the chest when the air channels are partly obstructed

Rh–pos·i·tive \ˌär-ˌāch-ˈpä-zə-tiv, -ˈpäz-tiv\ *adjective* (1942)
: containing Rh factor in the red blood cells

rhu·barb \ˈrü-ˌbärb\ *noun* [Middle English *rubarbe,* from Middle French *reubarbe,* from Medieval Latin *reubarbarum,* alteration of *rha barbarum,* literally, barbarian rhubarb] (15th century)
1 : any of a genus (*Rheum*) of Asian plants of the buckwheat family having large leaves with thick succulent petioles often used as food
2 : the dried rhizome and roots of any of several rhubarbs grown in China and Tibet and used as a purgative and stomachic
3 : a heated dispute or controversy

rhumb \ˈrəm(b)\ *noun, plural* **rhumbs** \ˈrəmz\ [Spanish *rumbo* rhumb, rhumb line] (1578)
1 : a line or course on a single bearing
2 : any of the points of the mariner's compass

rhumba *variant of* RUMBA

rhumb line *noun* [Spanish *rumbo*] (circa 1795)
: a line on the surface of the earth that follows a single compass bearing and makes equal oblique angles with all meridians — called also *loxodrome*

rhus \ˈrüs\ *noun, plural* **rhus·es** *or* **rhus** [New Latin, from Latin, sumac, from Greek *rhous*] (circa 1611)
: any of a genus (*Rhus*) of shrubs and trees (as sumac or poison ivy) of the cashew family that are native to temperate and warm regions, have compound trifoliolate or pinnate leaves, and sometimes produce substances causing dermatitis

¹**rhyme** \ˈrīm\ *noun* [Middle English *rime,* from Old French] (13th century)
1 a (1) **:** rhyming verse (2) **:** POETRY **b :** a composition in verse that rhymes
2 a : correspondence in terminal sounds of units of composition or utterance (as two or more words or lines of verse) **b :** one of two or more words thus corresponding in sound **c :** correspondence of other than terminal word sounds: as (1) **:** ALLITERATION (2) **:** INTERNAL RHYME
3 : RHYTHM, MEASURE
— rhyme·less *adjective*

²**rhyme** *verb* **rhymed; rhym·ing** (14th century)
transitive verb
1 : to relate or praise in rhyming verse
2 a : to put into rhyme **b :** to compose (verse) in rhyme **c :** to cause to rhyme **:** use as rhyme
intransitive verb
1 : to make rhymes; *also* **:** to compose rhyming verse
2 *of a word or verse* **:** to end in syllables that are rhymes
3 : to be in accord **:** HARMONIZE
— rhym·er *noun*

rhyme or reason *noun* (15th century)
: good sense or reason

rhyme royal \-ˈrȯi(-ə)l\ *noun* (circa 1841)
: a stanza of seven lines in iambic pentameter with a rhyme scheme of *ababbcc*

rhyme scheme *noun* (1917)
: the arrangement of rhymes in a stanza or a poem

rhyme·ster \ˈrīm(p)-stər\ *noun* (1589)
: an inferior poet

rhyming slang *noun* (1859)
: slang in which the word intended is replaced by a word or phrase that rhymes with it (as *loaf of bread* for *head*) or the first part of the phrase (as *loaf* for *head*)

rhyn·cho·ce·pha·lian \ˌriŋ-kō-sə-ˈfāl-yən\ *noun* [ultimately from Greek *rhynchos* beak, snout + *kephalē* head — more at CEPHALIC] (1886)
: any of an order (Rhynchocephalia) of reptiles resembling lizards that includes the tuatara as the only living member
— rhynchocephalian *adjective*

rhy·o·lite \ˈrī-ə-ˌlīt\ *noun* [German *Rhyolith,* from Greek *rhyax* stream, stream of lava (from *rhein*) + German *-lith* -lite] (1868)
: a very acid volcanic rock that is the lava form of granite
— rhy·o·lit·ic \ˌrī-ə-ˈli-tik\ *adjective*

rhythm \ˈri-t͟həm\ *noun* [Middle French & Latin; Middle French *rhythme,* from Latin *rhythmus,* from Greek *rhythmos,* probably from *rhein* to flow — more at STREAM] (1560)
1 a : an ordered recurrent alternation of strong and weak elements in the flow of sound and silence in speech **b :** a particular example or form of rhythm ⟨iambic *rhythm*⟩
2 a : the aspect of music comprising all the elements (as accent, meter, and tempo) that relate to forward movement **b :** a characteristic rhythmic pattern ⟨rumba *rhythm*⟩; *also* **:** ¹METER 2 **c :** the group of instruments in a band supplying the rhythm — called also *rhythm section*
3 a : movement or fluctuation marked by the regular recurrence or natural flow of related elements **b :** the repetition in a literary work of phrase, incident, character type, or symbol
4 : a regularly recurrent quantitative change in a variable biological process ⟨a circadian *rhythm*⟩ — compare BIORHYTHM
5 : the effect created by the elements in a play, movie, or novel that relate to the temporal development of the action
6 : RHYTHM METHOD

rhythm and blues *noun* (1949)
: popular music typically including elements of blues and black folk music and marked by a strong beat and simple chord structure

rhythm band *noun* (circa 1943)
: a band usually composed of schoolchildren who play simple percussion instruments (as rhythm sticks, sleigh bells, or tambourines) to learn fundamentals of coordination and music

rhyth·mic \ˈrit͟h-mik\ *or* **rhyth·mi·cal** \-mi-kəl\ *adjective* (1589)

1 : marked by or moving in pronounced rhythm
2 : of, relating to, or involving rhythm
— rhyth•mi•cal•ly \-mi-k(ə-)lē\ *adverb*

rhyth•mic•i•ty \rith-'mi-sə-tē\ *noun* (1901)
: the state of being rhythmic or of responding rhythmically

rhyth•mics \'rith-miks\ *noun plural but singular or plural in construction* (circa 1859)
: the science or theory of rhythms

rhyth•mist \'ri-thə-mist, 'rith-mist\ *noun* (1864)
: one who studies or has a feeling for rhythm

rhyth•mize \-,mīz\ *transitive verb* **-mized; -miz•ing** (1885)
: to order or compose rhythmically
— rhyth•mi•za•tion \,ri-thə-mə-'zā-shən, ,rith-mə-\ *noun*

rhythm method *noun* (1940)
: a method of birth control involving continence during the period in which ovulation is most likely to occur

rhythm stick *noun* (1952)
: one of a pair of plain or notched wood sticks that are struck or rubbed together to produce various percussive sounds and are used especially by young children in rhythm bands

rhyt•i•dome \'ri-tə-,dōm, 'rī-\ *noun* [Greek *rhytidōma* wrinkle, from *rhytidoun* to wrinkle, from *rhytid-, rhytis* wrinkle] (1881)
: the bark external to the last formed periderm

rhy•ton \'rī-,tän\ *noun* [Greek, neuter of *rhytos* flowing, from *rhein* to flow — more at STREAM] (1850)
: any of various ornate drinking vessels of ancient times typically shaped in part like an animal or animal's head

[1]ri•al \rē-'ȯl, -'äl\ *noun* [Persian, from Arabic *riyāl* riyal] (1932)
— see MONEY table

[2]rial *variant of* RIYAL

ri•al•to \rē-'al-(,)tō\ *noun, plural* **-tos** [*Rialto,* island and district in Venice] (1549)
1 : EXCHANGE, MARKETPLACE
2 : a theater district

ri•ant \'rī-ənt, 'rē-; rē-'äⁿ\ *adjective* [Middle French, present participle of *rire* to laugh, from Latin *ridere*] (1567)
: GAY, MIRTHFUL
— ri•ant•ly \'rī-ənt-lē, 'rē-\ *adverb*

ri•a•ta \rē-'a-tə, -'ä-\ *noun* [modification of American Spanish *reata*] (1846)
: LARIAT

[1]rib \'rib\ *noun* [Middle English, from Old English *ribb;* akin to Old High German *rippi* rib, Old Church Slavonic *rebro,* and probably to Greek *erephein* to roof over] (before 12th century)
1 a : any of the paired curved bony or partly cartilaginous rods that stiffen the walls of the body of most vertebrates and protect the viscera **b :** a cut of meat including a rib — see BEEF illustration **c** [from the account of Eve's creation from Adam's rib in Genesis 2:21–22] **:** WIFE
2 : something resembling a rib in shape or function: as **a** (1) **:** a traverse member of the frame of a ship that runs from keel to deck (2) **:** a light fore-and-aft member in an airplane's wing **b :** one of the stiff strips supporting an umbrella's fabric **c :** one of the arches in Romanesque and Gothic vaulting meeting and crossing one another and dividing the whole vaulted space into triangles
3 : an elongated ridge: as **a** (1) **:** a vein of an insect's wing (2) **:** one of the primary veins of a leaf **b :** one of the ridges in a knitted or woven fabric

rib 1a

[2]rib *transitive verb* **ribbed; rib•bing** (circa 1547)
1 : to furnish or enclose with ribs
2 : to form vertical ridges in in knitting
— rib•ber *noun*

[3]rib *noun* [[4]*rib*] (1929)
1 : JOKE
2 : PARODY

[4]rib *transitive verb* **ribbed; rib•bing** [probably from [1]*rib;* from the tickling of the ribs to cause laughter] (1930)
: to poke fun at **:** KID
— rib•ber *noun*

[1]rib•ald \'ri-bəld *also* 'ri-,bȯld, 'rī-,bȯld\ *noun* [Middle English, from Middle French *ribaut, ribauld* wanton, rascal, from *riber* to be wanton, of Germanic origin; akin to Old High German *rīban* to be wanton, literally, to rub] (13th century)
: a ribald person

[2]ribald *adjective* (1508)
1 : CRUDE, OFFENSIVE ⟨*ribald* language⟩
2 : characterized by or using coarse indecent humor
synonym see COARSE

rib•ald•ry \'ri-bəl-drē *also* 'rī-\ *noun, plural* **-ries** (14th century)
1 : a ribald quality or element
2 a : ribald language or humor **b :** an instance of ribald language or humor

rib•and \'ri bənd\ *noun* [Middle English, alteration of *riban*] (15th century)
: a ribbon used especially as a decoration

ri•ba•vi•rin \,rī-bə-'vī-rən\ *noun* [perhaps from *ribo*nucleic *a*cid + *vir*us + [1]*-in*] (1976)
: a synthetic broad-spectrum antiviral nucleoside $C_8H_{12}N_4O_5$

rib•band \'ri(b)-,band, 'ri-bən(d)\ *noun* [[1]*rib* + [1]*band*] (1711)
: a long narrow strip or bar used in shipbuilding; *especially* **:** one bent and bolted longitudinally to the frames to hold them in position during construction

rib•bing \'ri-biŋ\ *noun* (1564)
: an arrangement of ribs

[1]rib•bon \'ri-bən\ *noun* [Middle English *riban,* from Middle French *riban, ruban*] (14th century)
1 a : a flat or tubular narrow closely woven fabric (as of silk or rayon) used for trimmings or knitting **b :** a narrow fabric used for tying packages **c :** a piece of usually multicolored ribbon worn as a military decoration or in place of a medal **d :** a strip of colored satin given for winning a place in a competition
2 : a strip of inked fabric (as in a typewriter)
3 *plural* **:** reins for controlling an animal
4 : TATTER, SHRED — usually used in plural
5 : RIBBAND
— rib•bon•like \-,līk\ *adjective*

[2]ribbon *transitive verb* (1716)
1 a : to adorn with ribbons **b :** to divide into ribbons **c :** to cover with or as if with ribbons
2 : to rip to shreds

ribbon development *noun* (1927)
: a system of buildings built side by side along a road

rib•bon•fish \'ri-bən-,fish\ *noun* (circa 1798)
: any of a family (Trachipteridae) of elongate greatly compressed marine bony fishes (as a dealfish)

ribbon worm *noun* (1855)
: NEMERTEAN

rib•by \'ri-bē\ *adjective* (1849)
: showing or marked by ribs

rib cage *noun* (1909)
: the bony enclosing wall of the chest consisting chiefly of the ribs and the structures connecting them

ri•bes \'rī-(,)bēz\ *noun, plural* **ribes** [New Latin, from Medieval Latin, currant, from Arabic *rībās* rhubarb] (1543)
: any of a genus (*Ribes*) of shrubs (as a currant or a gooseberry) of the saxifrage family that have small racemose variously colored flowers and pulpy two-seeded to many-seeded berries

rib eye *noun* (1926)
: the large piece of meat that lies along the outer side of the rib (as of a steer)

rib•grass \'rib-,gras\ *noun* (circa 1500)
: [1]PLANTAIN; *especially* **:** an Old World plantain (*Plantago lanceolata*) with long narrow ribbed leaves

rib•let \'ri-blət\ *noun* (1943)
: one of the rib ends in the strip of breast of lamb or veal — see LAMB illustration

ribo- *combining form* [*ribose*]
1 : ribose ⟨*ribo*flavin⟩
2 : ribonucleic acid ⟨*ribo*some⟩

ri•bo•fla•vin \,rī-bə-'flā-vən, 'rī-bə-,\ *noun* [International Scientific Vocabulary *ribo-* + Latin *flavus* yellow — more at BLUE] (1935)
: a yellow crystalline compound $C_{17}H_{20}N_4O_6$ that is a growth-promoting member of the vitamin B complex and occurs both free (as in milk) and combined (as in liver) — called also *vitamin B_2, vitamin G*

ri•bo•nu•cle•ase \,rī-bō-'nü-klē-,ās, -'nyü-, -,āz\ *noun* (1942)
: an enzyme that catalyzes the hydrolysis of RNA

ri•bo•nu•cle•ic acid \,rī-bō-nu̇-'klē-ik-, -nyu̇-, -,klā-\ *noun* (1931)
: RNA

ri•bo•nu•cleo•pro•tein \-,nü-klē-ō-'prō-,tēn, -'prō-tē-ən\ *noun* (1940)
: a nucleoprotein that contains RNA

ri•bo•nu•cle•o•side \-'nü-klē-ə-,sīd, -'nyü-\ *noun* (1931)
: a nucleoside that contains ribose

ri•bo•nu•cle•o•tide \-,tīd\ *noun* (1929)
: a nucleotide that contains ribose and occurs especially as a constituent of RNA

ri•bose \'rī-,bōs, -,bōz\ *noun* [International Scientific Vocabulary, from *ribonic acid* (an acid $C_5H_{10}O_6$ obtained by oxidation of ribose)] (1892)
: a pentose $C_5H_{10}O_5$ found especially in the dextrorotatory form and obtained especially from RNA

ribosomal RNA *noun* (1961)
: RNA that is a fundamental structural element of ribosomes

ri•bo•some \'rī-bə-,sōm\ *noun* (1958)
: any of the RNA-rich cytoplasmic granules that are sites of protein synthesis — see CELL illustration
— ri•bo•som•al \,rī-bə-'sō-məl\ *adjective*

rib roast *noun* (circa 1890)
: a cut of meat containing the large piece that lies along the outer side of the rib — see BEEF illustration

rib•wort \'rib-,wərt, -,wȯrt\ *noun* (14th century)
: RIBGRASS

rice \'rīs\ *noun* [Middle English *rys,* from Old French *ris,* from Old Italian *riso,* from Greek *oryza, oryzon,* of Iranian origin; akin to Pashto *wriže* rice; akin to Sanskrit *vrīhi* rice] (13th century)
: the starchy seeds of an annual cereal grass (*Oryza sativa*) that are cooked and used for food; *also* **:** this cereal grass that occurs in southeastern Asia and is widely cultivated in warm climates for its seeds and by-products

rice•bird \'rīs-,bərd\ *noun* (1731)
: any of several small birds common in rice fields; *especially* **:** BOBOLINK

rice paper *noun* [from its resemblance to paper made from rice straw] (1822)

\ə\ abut \ᵊ\ kitten \ər\ further \a\ ash \ā\ ace \ä\ mop, mar \au̇\ out \ch\ chin \e\ bet \ē\ easy \g\ go \i\ hit \ī\ ice \j\ job \ŋ\ sing \ō\ go \ȯ\ law \ȯi\ boy \th\ thin \t͟h\ the \ü\ loot \u̇\ foot \y\ yet \zh\ vision *see also* Guide to Pronunciation

: a thin papery material made from the pith of a small Asian tree or shrub (*Tetrapanax papyriferum*) of the ginseng family

rice polishings *noun plural* (circa 1934)
: the inner bran layer of rice rubbed off in milling

ric•er \'rī-sər\ *noun* (1896)
: a kitchen utensil in which soft foods are pressed through a perforated container to produce strings

ri•cer•car \,rē-(,)chər-'kär\ *or* **ri•cer•ca•re** \-'kä-(,)rā\ *noun, plural* **ricercars** *or* **ri•cer•ca•ri** \-'kä-(,)rē\ [Italian, from *ricercare* to seek again, seek out, from *ri-* re- (from Latin *re-*) + *cercare* to seek, from Late Latin *circare* to go about — more at SEARCH] (1789)
: any of various usually keyboard musical forms especially of the 16th and 17th centuries in either quasi-improvisatory toccata style or strict polyphonic fugal style

rich \'rich\ *adjective* [Middle English *riche,* from Old English *rīce;* akin to Old High German *rīhhi* rich, Old English *rīce* kingdom, Old High German *rīhhi,* noun; all from prehistoric Germanic words borrowed from Celtic words akin to Old Irish *rí* (genitive *ríg*) king — more at ROYAL] (before 12th century)
1 : having abundant possessions and especially material wealth
2 a : having high value or quality **b :** well supplied ⟨a city *rich* in traditions⟩
3 : magnificently impressive **:** SUMPTUOUS
4 a : vivid and deep in color ⟨a *rich* red⟩ **b :** full and mellow in tone and quality ⟨a *rich* voice⟩ **c :** having a strong fragrance ⟨*rich* perfumes⟩
5 : highly productive or remunerative ⟨a *rich* mine⟩
6 a : having abundant plant nutrients ⟨*rich* soil⟩ **b :** highly seasoned, fatty, oily, or sweet ⟨*rich* foods⟩ **c :** high in the combustible component ⟨a *rich* fuel mixture⟩ **d :** high in some component ⟨cholesterol-*rich* foods⟩
7 a : ENTERTAINING; *also* **:** LAUGHABLE **b :** MEANINGFUL, SIGNIFICANT ⟨*rich* allusions⟩ **c :** LUSH ⟨*rich* meadows⟩
8 : pure or nearly pure ⟨*rich* lime⟩ ☆
— rich•ness *noun*

Rich•ard Roe \,rich-ərd-'rō\ *noun* (1768)
: a party to legal proceedings whose true name is unknown — compare JOHN DOE

rich•en \'ri-chən\ *transitive verb* **rich•ened; rich•en•ing** \'ri-chə-niŋ, 'rich-niŋ\ (1878)
: to make rich or richer

rich•es \'ri-chəz\ *noun plural* [Middle English, singular or plural, from *richesse,* literally, richness, from Old French, from *riche* rich, of Germanic origin; akin to Old English *rīce* rich] (13th century)
: things that make one rich **:** WEALTH

rich•ly \'rich-lē\ *adverb* [Middle English *richely,* from Old English *rīclīce,* from *rīce* rich] (before 12th century)
1 : in a rich manner
2 : in full measure **:** AMPLY ⟨praise *richly* deserved⟩

Rich•ter scale \'rik-tər-\ *noun* [Charles F. *Richter*] (1938)
: an open-ended logarithmic scale for expressing the magnitude of a seismic disturbance (as an earthquake) in terms of the energy dissipated in it with 1.5 indicating the smallest earthquake that can be felt, 4.5 an earthquake causing slight damage, and 8.5 a very devastating earthquake

ri•cin \'rī-sᵊn, 'ri-\ *noun* [Latin *ricinus* castor-oil plant] (1896)
: a poisonous protein in the castor bean

ri•cin•o•le•ic acid \,rī-sᵊn-ō-'lē-ik-, ,ri-, -'lā-\ *noun* [Latin *ricinus* + English *oleic acid*] (1848)
: an oily unsaturated hydroxy fatty acid $C_{18}H_{34}O_3$ that occurs in castor oil as a glyceride and yields esters important as plasticizers

¹rick \'rik\ *noun* [Middle English *reek,* from Old English *hrēac;* akin to Old Norse *hraukr* rick] (before 12th century)
1 : a stack (as of hay) in the open air
2 : a pile of material (as cordwood) split from short logs

²rick *transitive verb* (1623)
: to pile (as hay) in ricks

³rick *transitive verb* [perhaps from Middle English *wrikken* to move unsteadily] (1798)
chiefly British **:** WRENCH, SPRAIN

rick•ets \'ri-kəts\ *noun plural but singular in construction* [origin unknown] (1634)
: a deficiency disease that affects the young during the period of skeletal growth, is characterized especially by soft and deformed bones, and is caused by failure to assimilate and use calcium and phosphorus normally due to inadequate sunlight or vitamin D

rick•ett•sia \ri-'ket-sē-ə\ *noun, plural* **-si•as** *or* **-si•ae** \-sē-,ē, -,ī\ *also* **-sia** [New Latin, genus name, from Howard T. *Ricketts* (died 1910) American pathologist] (1919)
: any of a family (Rickettsiaceae) of rod-shaped, coccoid, or diplococcus-shaped, often pleomorphic bacteria that cause various diseases (as typhus)
— rick•ett•si•al \-sē-əl\ *adjective*

rick•ety \'ri-kə-tē\ *adjective* (1683)
1 : affected with rickets
2 a : lacking stability or firmness **:** SHAKY 2a ⟨a *rickety* coalition⟩ **b :** in unsound physical condition ⟨*rickety* veterans⟩ ⟨*rickety* stairs⟩

rick•ey \'ri-kē\ *noun, plural* **rickeys** [probably from the name *Rickey*] (1895)
: a drink containing liquor, lime juice, sugar, and soda water; *also* **:** a similar drink without liquor

rick•rack *or* **ric•rac** \'rik-,rak\ *noun* [reduplication of ⁴*rack*] (1884)
: a flat braid woven to form zigzags and used especially as trimming on clothing

rick•sha *or* **rick•shaw** \'rik-,shȯ\ *noun* [alteration of *jinrikisha*] (1887)
: a small covered 2-wheeled vehicle usually for one passenger that is pulled by one man and that was used originally in Japan

¹ric•o•chet \'ri-kə-,shā, *British also* -,shet\ *noun* [French] (1769)
: a glancing rebound (as of a projectile off a flat surface); *also* **:** an object that ricochets

²ricochet *intransitive verb* **-cheted** \-,shād\ *or* **-chet•ted** \-,she-təd\; **-chet•ing** \-,shā-iŋ\ *or* **-chet•ting** \-,she-tiŋ\ (1828)
: to skip with or as if with glancing rebounds

ri•cot•ta \ri-'kä-tə, -'kȯ-\ *noun* [Italian, from feminine of past participle of *ricuocere* to cook again, from Latin *recoquere,* from *re-* + *coquere* to cook — more at COOK] (1877)
: a white unripened whey cheese of Italy that resembles cottage cheese; *also* **:** a similar cheese made in the U.S. from whole or skim milk

ric•tal \'rik-tᵊl\ *adjective* (1825)
: of or relating to the rictus

ric•tus \'rik-təs\ *noun* [New Latin, from Latin, open mouth, from *ringi* to open the mouth; akin to Old Church Slavonic *rǫgŭ* mockery] (1827)
1 : the gape of a bird's mouth
2 a : the mouth orifice **b :** a gaping grin or grimace

rid \'rid\ *transitive verb* **rid** *also* **rid•ded; rid•ding** [Middle English *ridden* to clear, from Old Norse *rythja;* akin to Old High German *riutan* to clear land] (13th century)
1 *archaic* **:** SAVE, RESCUE
2 : to make free **:** RELIEVE, DISENCUMBER ⟨*rid* the language of impropriety⟩ ⟨be *rid* of worries⟩ ⟨get *rid* of that junk⟩

rid•dance \'ri-dᵊn(t)s\ *noun* (1533)
1 : an act of ridding
2 : DELIVERANCE, RELIEF — often used in the phrase *good riddance*

rid•den \'ri-dᵊn\ *adjective* (1653)
1 : harassed, oppressed, or obsessed by — usually used in combination ⟨guilt-*ridden*⟩ ⟨debt-*ridden*⟩
2 : excessively full of or supplied with — usually used in combination ⟨slum-*ridden*⟩

¹rid•dle \'ri-dᵊl\ *noun* [Middle English *redels, ridel,* from Old English *rǣdelse* opinion, conjecture, riddle; akin to Old English *rǣdan* to interpret — more at READ] (before 12th century)
1 : a mystifying, misleading, or puzzling question posed as a problem to be solved or guessed **:** CONUNDRUM, ENIGMA
2 : something or someone difficult to understand
synonym see MYSTERY

²riddle *verb* **rid•dled; rid•dling** \'rid-liŋ, 'ri-dᵊl-iŋ\ (1571)
intransitive verb
: to speak in or propound riddles
transitive verb
1 : to find the solution of **:** EXPLAIN
2 : to set a riddle for **:** PUZZLE
— rid•dler \'rid-lər, 'ri-dᵊl-ər\ *noun*

³riddle *noun* [Middle English *riddil,* from Old English *hriddel;* akin to Latin *cribrum* sieve, *cernere* to sift — more at CERTAIN] (before 12th century)
: a coarse sieve

⁴riddle *transitive verb* **rid•dled; rid•dling** \'rid-liŋ, 'ri-dᵊl-iŋ\ (13th century)
1 : to separate (as grain from chaff) with a riddle **:** SCREEN
2 : to pierce with many holes ⟨*riddled* the car with bullets⟩
3 : to spread through **:** PERMEATE ⟨a book *riddled* with errors⟩

rid•dling \'rid-liŋ, 'ri-dᵊl-iŋ\ *adjective* (1591)
: containing or presenting riddles

¹ride \'rīd\ *verb* **rode** \'rōd\ *or chiefly dialect* **rid** \'rid\; **rid•den** \'ri-dᵊn\ *or chiefly dialect* **rid** *or* **rode; rid•ing** \'rī-diŋ\ [Middle English, from Old English *rīdan;* akin to Old High German *rītan* to ride, Middle Irish *réidid* he rides] (before 12th century)
intransitive verb
1 a : to sit and travel on the back of an animal that one directs **b :** to travel in or on a conveyance
2 : to travel as if on a conveyance **:** be borne ⟨*rode* on a wave of popularity⟩
3 a : to lie moored or anchored ⟨a ship *rides* at anchor⟩ **b :** SAIL **c :** to move like a floating object ⟨the moon *rode* in the sky⟩
4 : to become supported on a point or surface
5 a : to travel over a surface ⟨the car *rides* well⟩ **b :** to move on the body ⟨shorts that *ride* up⟩
6 : to continue without interference ⟨let it *ride*⟩
7 a : to be contingent **:** DEPEND ⟨plans on which the future *rides*⟩ **b :** to become bet ⟨a lot of money *riding* on the favorite⟩
transitive verb
1 a : to travel on ⟨*ride* a bike⟩ ⟨*ride* the bus⟩ **b :** to move with like a rider ⟨*ride* the waves⟩

☆ SYNONYMS
Rich, wealthy, affluent, opulent mean having goods, property, and money in abundance. RICH implies having more than enough to gratify normal needs or desires ⟨became *rich* through shrewd investing⟩. WEALTHY stresses the possession of property and intrinsically valuable things ⟨*wealthy* landowners⟩. AFFLUENT suggests prosperity and an increasing wealth ⟨an *affluent* society⟩. OPULENT suggests lavish expenditure and display of great wealth, more often applying to things than people ⟨an *opulent* mansion⟩.

2 a : to traverse by conveyance ⟨*rode* 500 miles⟩ **b :** to ride a horse in ⟨*ride* a race⟩
3 : SURVIVE, OUTLAST — usually used with *out* ⟨*rode* out the gale⟩
4 : to traverse on horseback to inspect or maintain ⟨*ride* fence⟩
5 : to mount in copulation
6 a : OBSESS, OPPRESS ⟨*ridden* by anxiety⟩ **b :** to harass persistently **:** NAG **c :** TEASE, RIB
7 : CARRY, CONVEY
8 : to project over **:** OVERLAP
9 : to give with (a punch) to soften the impact
10 : to keep in partial engagement by resting a foot continuously on the pedal ⟨*ride* the brakes⟩
— rid·able *or* **ride·able** \ˈrī-də-bəl\ *adjective*
— ride circuit : to hold court in the various towns of a judicial circuit
— ride for a fall : to court disaster
— ride herd on : to keep a check on
— ride high : to experience success
— ride shotgun : to ride in the front passenger seat of a motor vehicle

²ride *noun* (1759)
1 : an act of riding; *especially* **:** a trip on horseback or by vehicle
2 : a way (as a road or path) suitable for riding
3 : any of various mechanical devices (as at an amusement park) for riding on
4 a : a trip on which gangsters take a victim to murder him **b :** something likened to such a trip ⟨take the taxpayers for a *ride*⟩
5 : a means of transportation
6 : the qualities of travel comfort in a vehicle

rid·er \ˈrī-dər\ *noun* (14th century)
1 : one that rides
2 a : an addition to a document often attached on a separate piece of paper **b :** a clause appended to a legislative bill to secure a usually distinct object
3 : something used to overlie another or to move along on another piece
— rid·er·less \-ləs\ *adjective*

rid·er·ship \ˈrī-dər-ˌship\ *noun* (1968)
: the number of persons who ride a system of public transportation

¹ridge \ˈrij\ *noun* [Middle English *rigge,* from Old English *hrycg;* akin to Old High German *hrukki* ridge, back] (before 12th century)
1 : an elevated body part (as along the backbone)
2 a : a range of hills or mountains **b :** an elongate elevation on an ocean bottom
3 : an elongate crest or a linear series of crests
4 : a raised strip (as of plowed ground)
5 : the line of intersection at the top between the opposite slopes or sides of a roof
— ridged \ˈrijd\ *adjective*

²ridge *verb* **ridged; ridg·ing** (1523)
transitive verb
: to form into a ridge
intransitive verb
: to extend in ridges

ridge·line \ˈrij-ˌlīn\ *noun* (1833)
: a line marking or following a ridge top

ridge·ling *or* **ridg·ling** \ˈrij-liŋ\ *noun* [perhaps from ¹*ridge;* from the supposition that the undescended testis remains near the animal's back] (1555)
1 : a partially castrated male animal
2 : a male animal having one or both testes retained in the inguinal canal

ridge·pole \ˈrij-ˌpōl\ *noun* (1788)
1 : the horizontal pole at the top of a tent
2 : the highest horizontal timber in a roof against which the upper ends of the rafters are fixed

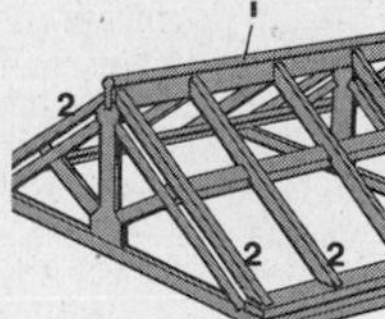

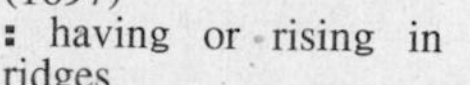
1 ridgepole, 2 rafters

ridgy \ˈri-jē\ *adjective* (1697)
: having or rising in ridges

¹rid·i·cule \ˈri-də-ˌkyü(ə)l\ *noun* [French or Latin; French, from Latin *ridiculum* jest] (1690)
: the act of exposing to laughter **:** DERISION, MOCKERY

²ridicule *transitive verb* **-culed; -cul·ing** (circa 1700)
: to make fun of ☆
— rid·i·cul·er *noun*

ri·dic·u·lous \rə-ˈdi-kyə-ləs\ *adjective* [Latin *ridiculosus* (from *ridiculum* jest, from neuter of *ridiculus*) or *ridiculus,* literally, laughable, from *ridēre* to laugh] (1550)
: arousing or deserving ridicule **:** ABSURD, PREPOSTEROUS
synonym see LAUGHABLE
— ri·dic·u·lous·ly *adverb*
— ri·dic·u·lous·ness *noun*

¹rid·ing \ˈrī-diŋ\ *noun* (14th century)
: the action or state of one that rides

²rid·ing *adjective* (15th century)
1 : used for or when riding ⟨a *riding* horse⟩ ⟨*riding* boots⟩
2 : operated by a rider ⟨a *riding* mower⟩

³ri·ding \ˈrī-diŋ\ *noun* [Middle English *-redying* or *trithing,* alteration of (assumed) Old English *thriding,* from Old Norse *thrithjungr* third part, from *thrithi* third; akin to Old English *thridda* third — more at THIRD] (15th century)
1 : one of the three administrative jurisdictions into which Yorkshire, England, was formerly divided
2 : an administrative jurisdiction or electoral district in a British dominion (as Canada)

rid·ley \ˈrid-lē\ *noun* [origin unknown] (1926)
: either of two sea turtles: **a :** one (*Lepidochelys kempii*) of the Gulf of Mexico and the Atlantic coast of the U.S. — called also *Kemp's ridley* **b :** one (*L. olivacea*) of the Pacific, Atlantic, and Indian oceans

ri·dot·to \ri-ˈdä-(ˌ)tō\ *noun, plural* **-tos** [Italian, retreat, place of entertainment, redoubt — more at REDOUBT] (1722)
: a public entertainment consisting of music and dancing often in masquerade popular in 18th century England

ri·el \rē-ˈel\ *noun* [origin unknown] (1956)
— see MONEY table

Rie·mann·ian geometry \rē-ˈmä-nē-ən-\ *noun* [G. F. B. *Riemann*] (1904)
: a non-Euclidean geometry in which straight lines are geodesics and in which the parallel postulate is replaced by the postulate that every pair of straight lines intersects

Rie·mann integral \ˈrē-ˌmän-, -mən-\ *noun* (1914)
: a definite integral defined as the limit of sums found by partitioning the interval comprising the domain of definition into subintervals, by finding the sum of products each of which consists of the width of a subinterval multiplied by the value of the function at some point in it, and by letting the maximum width of the subintervals approach zero

Ries·ling \ˈrēz-liŋ, ˈrēs-\ *noun* [German] (1833)
: a white wine that ranges from dry to very sweet and is made from a single variety of grape originally grown in Germany; *also* **:** the grape

ri·fam·pin \rī-ˈfam-pən\ *or* **ri·fam·pi·cin** \rī-ˈfam-pə-sən\ *noun* [International Scientific Vocabulary, alteration of *rifamycin,* antibiotic derived from *Streptomyces mediterranei*] (1966)
: a semisynthetic antibiotic $C_{43}H_{58}N_4O_{12}$ that acts against some viruses and bacteria especially by inhibiting RNA synthesis

rife \ˈrīf\ *adjective* [Middle English *ryfe,* from Old English *rȳfe;* akin to Old Norse *rīfr* abundant] (12th century)
1 : prevalent especially to an increasing degree ⟨fear was *rife* in the people⟩
2 : ABUNDANT, COMMON
3 : copiously supplied **:** ABOUNDING — usually used with *with* ⟨*rife* with rumors⟩
— rife *adverb*
— rife·ly *adverb*

¹riff \ˈrif\ *noun* [probably by shortening & alteration from *refrain*] (1935)
1 : an ostinato phrase (as in jazz) typically supporting a solo improvisation; *also* **:** a piece based on such a phrase
2 : a short succinct usually witty comment; *also* **:** BIT, ROUTINE

²riff *intransitive verb* (1950)
: to perform a riff

³riff *verb* [short for *riffle*] (1952)
: RIFFLE, SKIM ⟨*riff* pages⟩

Riff \ˈrif\ *noun, plural* **Riffs** *or* **Riffi** \ˈri-fē\ *or* **Riff** (1903)
: a Berber of the Rif Mountains in northern Morocco

Riff·ian \ˈri-fē-ən\ *noun* (1867)
: RIFF

¹rif·fle \ˈri-fəl\ *verb* **rif·fled; rif·fling** \ˈri-f(ə-)liŋ\ [²*riffle*] (1752)
intransitive verb
1 : to form, flow over, or move in riffles
2 : to flip cursorily **:** THUMB ⟨*riffle* through the catalog⟩
transitive verb
1 : to ruffle slightly **:** RIPPLE
2 a : to leaf through hastily; *specifically* **:** to leaf (as a stack of paper) by sliding a thumb along the edge of the leaves **b :** to shuffle (playing cards) by separating the deck into two parts and riffling with the thumbs so the cards intermix
3 : to manipulate (small objects) idly between the fingers

²riffle *noun* [perhaps alteration of *ruffle*] (1785)
1 a : a shallow extending across a streambed and causing broken water **b :** a stretch of water flowing over a riffle
2 : a small wave or succession of small waves **:** RIPPLE
3 a : any of various contrivances (as blocks or rails) laid on the bottom of a sluice or launder to make a series of grooves or interstices to catch and retain a mineral (as gold) **b :** a groove or interstice so formed
4 : a cleat or bar fastened to an inclined surface in a gold-washing apparatus to catch and hold mineral grains
5 [¹*riffle*] **a :** the act or process of shuffling (as cards) **b :** the sound made while doing this

rif·fler \ˈri-flər\ *noun* [French *rifloir,* from *rifler* to file, rifle] (circa 1797)
: a small filing or scraping tool

riff·raff \ˈrif-ˌraf\ *noun* [Middle English *ryffe raffe,* from *rif and raf* every single one, from Middle French *rif et raf* completely, from *rifler* to scratch, plunder + *raffe* act of sweeping] (15th century)
1 a : disreputable persons **b :** RABBLE **c :** one of the riffraff

☆ SYNONYMS
Ridicule, deride, mock, taunt mean to make an object of laughter of. RIDICULE implies a deliberate often malicious belittling ⟨consistently *ridiculed* everything she said⟩. DERIDE suggests contemptuous and often bitter ridicule ⟨*derided* their efforts to start their own business⟩. MOCK implies scorn often ironically expressed as by mimicry or sham deference ⟨youngsters began to *mock* the helpless wino⟩. TAUNT suggests jeeringly provoking insult or challenge ⟨hometown fans *taunted* the visiting team⟩.

2 : REFUSE, RUBBISH ◆
— **riffraff** *adjective*

¹ri·fle \ˈrī-fəl\ *verb* **ri·fled; ri·fling** \-f(ə-)liŋ\ [Middle English, from Middle French *rifler* to scratch, file, plunder, of Germanic origin; akin to Old High German *riffilōn* to saw, obsolete Dutch *rijffelen* to scrape] (14th century)
transitive verb
1 : to ransack especially with the intent to steal
2 : to steal and carry away
intransitive verb
: to engage in ransacking and stealing
— **ri·fler** \-f(ə-)lər\ *noun*

²rifle *transitive verb* **ri·fled; ri·fling** \-f(ə-)liŋ\ [perhaps from French *rifler* to scratch, file] (1635)
: to cut spiral grooves into the bore of ⟨*rifled* arms⟩ ⟨*rifled* pipe⟩

³rifle *noun* (1770)
1 a : a shoulder weapon with a rifled bore **b :** a rifled artillery piece
2 *plural* **:** soldiers armed with rifles
word history see MUSKET

⁴rifle *transitive verb* **ri·fled; ri·fling** \-f(ə-)liŋ\ [³*rifle*] (1937)
: to propel (as a ball) with great force or speed

ri·fle·bird \ˈrī-fəl-ˌbərd\ *noun* (1831)
: any of several birds of paradise (genus *Ptiloris*)

ri·fle·man \-mən\ *noun* (1775)
1 : a soldier armed with a rifle
2 : one skilled in shooting with a rifle

ri·fle·ry \ˈrī-fəl-rē\ *noun* (1935)
: the practice of shooting at targets with a rifle

ri·fling \ˈrī-f(ə-)liŋ\ *noun* (1797)
1 : the act or process of making spiral grooves
2 : a system of spiral grooves in the surface of the bore of a gun causing a projectile when fired to rotate about its longer axis

¹rift \ˈrift\ *noun* [Middle English, of Scandinavian origin; akin to Danish & Norwegian *rift* fissure, Old Norse *rīfa* to rive — more at RIVE] (14th century)
1 a : FISSURE, CREVASSE **b :** FAULT 5
2 : a clear space or interval
3 : BREACH, ESTRANGEMENT

²rift (14th century)
intransitive verb
: to burst open
transitive verb
1 : CLEAVE, DIVIDE ⟨hills were *rifted* by the earthquake⟩
2 : PENETRATE

rift valley *noun* (1894)
: an elongated valley formed by the depression of a block of the earth's crust between two faults or groups of faults of approximately parallel strike

¹rig \ˈrig\ *transitive verb* **rigged; rig·ging** [Middle English *riggen*] (15th century)
1 : to fit out (as a ship) with rigging
2 : CLOTHE, DRESS — usually used with *out*
3 : to furnish with special gear **:** EQUIP
4 a : to put in condition or position for use **:** ADJUST, ARRANGE ⟨a car *rigged* for manual control⟩ **b :** CONSTRUCT ⟨*rig* up a temporary shelter⟩

²rig *noun* (1822)
1 : the distinctive shape, number, and arrangement of sails and masts of a ship
2 : EQUIPAGE; *especially* **:** a carriage with its horse
3 : CLOTHING, DRESS
4 : tackle, equipment, or machinery fitted for a specified purpose ⟨an oil-drilling *rig*⟩
5 : a tractor-trailer combination

³rig *transitive verb* **rigged; rig·ging** [*rig,* noun, a swindle] (1851)
1 : to manipulate or control usually by deceptive or dishonest means ⟨*rig* an election⟩
2 : to fix in advance for a desired result ⟨*rig* a quiz program⟩

rig·a·doon \ˌri-gə-ˈdün\ *or* **ri·gau·don** \rē-gō-dō^{n}\ *noun* [French *rigaudon*] (1691)
: a lively dance of the 17th and 18th centuries; *also* **:** the music for a rigadoon

rig·a·ma·role *variant of* RIGMAROLE

rig·a·to·ni \ˌri-gə-ˈtō-nē\ *noun* [Italian, plural, from *rigato* furrowed, fluted, from past participle of *rigare* to furrow, flute, from *riga* line, of Germanic origin; akin to Old High German *rīga* line — more at ROW] (circa 1923)
: pasta made in short wide fluted tubes

Ri·gel \ˈrī-jəl, -gəl; ˈri-jəl\ *noun* [Arabic *Rijl,* literally, foot]
: a first-magnitude star in the left foot of the constellation Orion

rig·ger \ˈri-gər\ *noun* (1611)
1 : one that rigs
2 : a long slender pointed sable paintbrush

rig·ging \ˈri-giŋ, -gən\ *noun* (1594)
1 a : lines and chains used aboard a ship especially in working sail and supporting masts and spars **b :** a similar network (as in theater scenery) used for support and manipulation
2 : CLOTHING

¹right \ˈrīt\ *adjective* [Middle English, from Old English *riht;* akin to Old High German *reht* right, Latin *rectus* straight, right, *regere* to lead straight, direct, rule, *rogare* to ask, Greek *oregein* to stretch out] (before 12th century)
1 : RIGHTEOUS, UPRIGHT
2 : being in accordance with what is just, good, or proper ⟨*right* conduct⟩
3 a : agreeable to a standard **b :** conforming to facts or truth **:** CORRECT ⟨the *right* answer⟩
4 : SUITABLE, APPROPRIATE ⟨the *right* man for the job⟩
5 : STRAIGHT ⟨a *right* line⟩
6 : GENUINE, REAL
7 a : of, relating to, situated on, or being the side of the body which is away from the heart and on which the hand is stronger in most people **b :** located nearer to the right hand than to the left **c :** located to the right of an observer facing the object specified or directed as the right arm would point when raised out to the side **d** (1) **:** located on the right of an observer facing in the same direction as the object specified ⟨stage *right*⟩ (2) **:** located on the right when facing downstream ⟨the *right* bank of a river⟩
8 : having the axis perpendicular to the base ⟨*right* cone⟩
9 : of, relating to, or constituting the principal or more prominent side of an object ⟨made sure the socks were *right* side out⟩
10 : acting or judging in accordance with truth or fact ⟨time proved her *right*⟩
11 a : being in good physical or mental health or order ⟨not in his *right* mind⟩ **b :** being in a correct or proper state ⟨put things *right*⟩
12 : most favorable or desired **:** PREFERABLE; *also* **:** socially acceptable ⟨knew all the *right* people⟩
13 *often capitalized* **:** of, adhering to, or constituted by the Right especially in politics
synonym see CORRECT
— **right·ness** *noun*

²right *noun* [Middle English, from Old English *riht,* from *riht,* adjective] (before 12th century)
1 : qualities (as adherence to duty or obedience to lawful authority) that together constitute the ideal of moral propriety or merit moral approval
2 : something to which one has a just claim: as **a :** the power or privilege to which one is justly entitled **b** (1) **:** the interest that one has in a piece of property — often used in plural ⟨mineral *rights*⟩ (2) *plural* **:** the property interest possessed under law or custom and agreement in an intangible thing especially of a literary and artistic nature ⟨film *rights* of the novel⟩
3 : something that one may properly claim as due
4 : the cause of truth or justice
5 a : RIGHT HAND 1a; *also* **:** a blow struck with this hand ⟨gave him a hard *right* on the jaw⟩ **b :** the location or direction of the right side ⟨woods on his *right*⟩ **c :** the part on the right side **d :** RIGHT FIELD
6 a : the true account or correct interpretation **b :** the quality or state of being factually correct
7 *often capitalized* **a :** the part of a legislative chamber located to the right of the presiding officer **b :** the members of a continental European legislative body occupying the right as a result of holding more conservative political views than other members
8 a *often capitalized* **:** individuals sometimes professing opposition to change in the established order and favoring traditional attitudes and practices and sometimes advocating the forced establishment of an authoritarian order (as in government) **b** *often capitalized* **:** a conservative position
9 a : a privilege given stockholders to subscribe pro rata to a new issue of securities generally below market price **b :** the negotiable certificate evidencing such privilege — usually used in plural
— **right·most** \-ˌmōst\ *adjective*
— **by rights :** with reason or justice **:** PROPERLY
— **in one's own right :** by virtue of one's own qualifications or properties
— **of right 1 :** as an absolute right **2 :** legally or morally exactable
— **to rights :** into proper order

³right *adverb* (before 12th century)
1 : according to right ⟨live *right*⟩
2 : in the exact location, position, or moment **:** PRECISELY ⟨*right* at his fingertips⟩ ⟨quit *right* then and there⟩
3 : in a suitable, proper, or desired manner ⟨knew he wasn't doing it *right*⟩
4 : in a direct line, course, or manner **:** DIRECTLY, STRAIGHT ⟨go *right* home⟩ ⟨came *right* out and said it⟩
5 : according to fact or truth **:** TRULY ⟨guessed *right*⟩
6 a : all the way ⟨windows *right* to the floor⟩ **b :** in a complete manner ⟨felt *right* at home⟩
7 : without delay **:** IMMEDIATELY ⟨*right* after lunch⟩
8 : to a great degree **:** VERY ⟨a *right* pleasant day⟩
9 : on or to the right ⟨looked left and *right*⟩

⁴right (before 12th century)
transitive verb
1 a : to do justice to **:** redress the injuries of ⟨so just is God to *right* the innocent —Shakespeare⟩ **b :** JUSTIFY, VINDICATE ⟨felt the need to *right* himself in court⟩
2 : AVENGE ⟨vows to *right* the injustice done to his family⟩

◇ WORD HISTORY

riffraff The word *riffraff* seems to represent a triumph of sound over sense, for it is a compound of two quite different words that have little semantic connection. In Middle French the verb *rifler* meant "to scratch" or "to plunder" and the noun *rafle* or *raffe* meant "the act of sweeping or snatching." From these words about the late 13th century the French contrived the phrase *rif et raf,* used to mean "completely," "altogether." Later variants such as *rifle ou* [or] *rafle* and *rif ny* [nor] *raf* show that the phrase most likely also meant "anything" or with a negative adverb "nothing." Medieval French *rif et raf* was thus borrowed into Middle English as *rif and raf,* in the sense "one and all, every single one." No doubt it was the lack of exclusivity implied by "every single one" that led to *ryffe raffe* being used in early Modern English as a collective noun for the multitude of people of the lowest or most disreputable sort.

3 a : to adjust or restore to the proper state or condition ⟨*right* the economy⟩ **b :** to bring or restore to an upright position ⟨*right* a capsized boat⟩
intransitive verb
: to become upright
— **right·er** *noun*

right and left *adverb* (1829)
: on both or all sides **:** in every direction **:** EVERYWHERE

right angle *noun* (15th century)
: the angle bounded by two lines perpendicular to each other **:** an angle of 90° or ½ π radians
— **right–an·gled** \'rīt-'aŋ-gəld\ *or* **right–an·gle** \-gəl\ *adjective*

right ascension *noun* (15th century)
: the arc of the celestial equator between the vernal equinox and the point where the hour circle through the given body intersects the equator reckoned eastward commonly in terms of the corresponding interval of sidereal time in hours, minutes, and seconds

right away *adverb* (1818)
: without delay or hesitation **:** IMMEDIATELY

right circular cone *noun* (circa 1889)
: CONE 1a

right circular cylinder *noun* (1877)
: a cylinder with the bases circular and with the axis joining the two centers of the bases perpendicular to the planes of the two bases

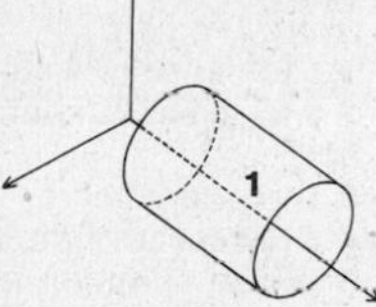

right circular cylinder: *1* **axis**

righ·teous \'rī-chəs\ *adjective* [alteration of earlier *rightuous,* alteration of Middle English *rightwise, rightwos,* from Old English *rihtwīs,* from *riht,* noun, right + *wīs* wise] (1535)
1 : acting in accord with divine or moral law **:** free from guilt or sin
2 a : morally right or justifiable ⟨a *righteous* decision⟩ **b :** arising from an outraged sense of justice or morality ⟨*righteous* indignation⟩
3 *slang* **:** GENUINE, GOOD
synonym see MORAL
— **righ·teous·ly** *adverb*
— **righ·teous·ness** *noun*

right field *noun* (1857)
1 : the part of the baseball outfield to the right looking out from home plate
2 : the position of the player defending right field
— **right fielder** *noun*

right·ful \'rīt-fəl\ *adjective* (14th century)
1 : JUST, EQUITABLE
2 a : having a just or legally established claim **:** LEGITIMATE ⟨the *rightful* owner⟩ **b :** held by right or just claim **:** LEGAL ⟨*rightful* authority⟩
3 : PROPER, FITTING ⟨assured of his *rightful* place in history —Brian Duff⟩
— **right·ful·ly** \-fə-lē\ *adverb*
— **right·ful·ness** *noun*

right–hand \'rīt-ˌhand\ *adjective* (1592)
1 : situated on the right
2 : RIGHT-HANDED
3 : chiefly relied on ⟨*right-hand* man⟩

right hand *noun* (before 12th century)
1 a : the hand on a person's right side **b :** an indispensable person
2 a : the right side **b :** a place of honor

right–hand·ed \-'han-dəd\ *adjective* (14th century)
1 : using the right hand habitually or more easily than the left; *also* **:** swinging from right to left ⟨a *right-handed* batter⟩
2 : relating to, designed for, or done with the right hand
3 a : having the same direction or course as the movement of the hands of a watch viewed from in front **:** CLOCKWISE **b :** having a spiral structure or form that ascends or advances to the right ⟨a *right-handed* screw⟩ ⟨a *right-handed* double helix of DNA⟩
4 *of a door* **:** opening to the right away from one
— **right–handed** *adverb*
— **right–hand·ed·ly** *adverb*
— **right–hand·ed·ness** *noun*

right–hand·er \-'han-dər\ *noun* (1857)
1 : a blow struck with the right hand
2 : a right-handed person; *especially* **:** a right-handed pitcher

right·ism \'rī-ˌti-zəm\ *noun, often capitalized* (1939)
1 : the principles and views of the Right
2 : advocacy of or adherence to the doctrines of the Right
— **right·ist** \'rī-tist\ *noun or adjective, often capitalized*

right·ly \'rīt-lē\ *adverb* (before 12th century)
1 : in accordance with right conduct **:** FAIRLY, JUSTLY
2 : in the right or proper manner **:** PROPERLY, FITTINGLY
3 : according to truth or fact **:** CORRECTLY, EXACTLY

right–mind·ed \-'mīn-dəd\ *adjective* (circa 1586)
: having a right or honest mind ⟨a *right-minded* citizen⟩
— **right–mind·ed·ness** *noun*

right now *adverb* (14th century)
1 : RIGHT AWAY
2 : at present

righto \ˌrīt-'ō, 'rī-(ˌ)tō\ *interjection* (1896)
— used to express cheerful concurrence, assent, or understanding

right off *adverb* (1790)
: RIGHT AWAY **:** at once
— **right off the bat :** RIGHT OFF

right–of–way \ˌrīt-ə(v)-'wā\ *noun, plural* **rights–of–way** *also* **right–of–ways** (1768)
1 : a legal right of passage over another person's ground
2 a : the area over which a right-of-way exists **b :** the strip of land over which is built a public road **c :** the land occupied by a railroad especially for its main line **d :** the land used by a public utility (as for a transmission line)
3 a : a precedence in passing accorded to one vehicle over another by custom, decision, or statute **b :** the right of traffic to take precedence **c :** the right to take precedence over others ⟨gave the bill the *right-of-way* in the Senate⟩

right on *adjective* (1925)
1 : exactly correct — often used interjectionally to express agreement
2 *usually* **right–on :** attuned to the spirit of the times

Right Reverend (15th century)
— used as a title for high ecclesiastical officials

right shoulder arms *noun* (1902)
: a position in the manual of arms in which the butt of the rifle is held in the right hand with the barrel resting on the right shoulder; *also* **:** a command to assume this position

right–to–life \'rīt-tə-'līf\ *adjective* (1973)
: opposed to abortion
— **right–to–lif·er** \-'lī-fər\ *noun*

right–to–work *adjective* (1949)
: opposing or banning the closed shop and the union shop

right triangle *noun* (1924)
: a triangle having a right angle — see TRIANGLE illustration

right·ward \'rīt-wərd\ *adjective* (1825)
: being toward or on the right

right whale *noun* (1725)
: any of a family (Balaenidae) of baleen whales having very long baleen, a large head on a stocky body, a smooth throat, and short broad rounded flippers

right wing *noun* (1905)
1 : the rightist division of a group or party
2 : RIGHT 8
— **right–wing** \'rīt-'wiŋ, -ˌwiŋ\ *adjective*
— **right–wing·er** \'rīt-'wiŋ-ər, ˌrīt-\ *noun*

right whale

righty \'rī-tē\ *noun, plural* **right·ies** (1949)
: RIGHT-HANDER 2

rig·id \'ri-jəd\ *adjective* [Middle English *rigide,* from Middle French or Latin; Middle French, from Latin *rigidus,* from *rigēre* to be stiff] (15th century)
1 a : deficient in or devoid of flexibility ⟨*rigid* price controls⟩ ⟨a *rigid* bar of metal⟩ **b :** appearing stiff and unyielding ⟨his face *rigid* with pain⟩
2 a : inflexibly set in opinion **b :** strictly observed ⟨adheres to a *rigid* schedule⟩
3 : firmly inflexible rather than lax or indulgent ⟨a *rigid* disciplinarian⟩
4 : precise and accurate in procedure ⟨*rigid* control of the manufacturing process⟩
5 *of an airship* **:** having the outer shape maintained by a fixed framework ☆
— **rig·id·ly** *adverb*
— **rig·id·ness** *noun*

ri·gid·i·fy \rə-'ji-də-ˌfī\ *verb* **-fied; -fy·ing** (1842)
transitive verb
: to make rigid
intransitive verb
: to become rigid
— **ri·gid·i·fi·ca·tion** \-ˌji-də-fə-'kā-shən\ *noun*

ri·gid·i·ty \rə-'ji-də-tē\ *noun, plural* **-ties** (1624)
1 : the quality or state of being rigid
2 : one that is rigid (as in form or conduct)

rig·ma·role \'ri-gə-mə-ˌrōl, 'rig-mə-\ *noun* [alteration of obsolete *ragman roll* long list, catalog] (circa 1736)
1 : confused or meaningless talk
2 : a complex and ritualistic procedure

rig·or \'ri-gər\ *noun* [Middle English *rigour,* from Middle French *rigueur,* from Latin *rigor,* literally, stiffness, from *rigēre* to be stiff] (14th century)
1 a (1) **:** harsh inflexibility in opinion, temper, or judgment **:** SEVERITY (2) **:** the quality of being unyielding or inflexible **:** STRICTNESS (3) **:** severity of life **:** AUSTERITY **b :** an act or instance of strictness, severity, or cruelty
2 : a tremor caused by a chill
3 : a condition that makes life difficult, challenging, or uncomfortable; *especially* **:** extremity of cold
4 : strict precision **:** EXACTNESS ⟨logical *rigor*⟩
5 a *obsolete* **:** RIGIDITY, STIFFNESS **b :** rigidness or torpor of organs or tissue that prevents response to stimuli

rig·or·ism \'ri-gə-ˌri-zəm\ *noun* (1704)
: rigidity in principle or practice

☆ SYNONYMS
Rigid, rigorous, strict, stringent mean extremely severe or stern. RIGID implies uncompromising inflexibility ⟨*rigid* rules of conduct⟩. RIGOROUS implies the imposition of hardship and difficulty ⟨the *rigorous* training of recruits⟩. STRICT emphasizes undeviating conformity to rules, standards, or requirements ⟨*strict* enforcement of the law⟩. STRINGENT suggests severe, tight restriction or limitation ⟨*stringent* standards of admission⟩. See in addition STIFF.

— **rig·or·ist** \-rist\ *noun or adjective*
— **rig·or·is·tic** \ˌri-gə-ˈris-tik\ *adjective*
rig·or mor·tis \ˌri-gər-ˈmȯr-təs *also chiefly British* ˌrī-ˌgȯr-\ *noun* [New Latin, stiffness of death] (circa 1847)
: temporary rigidity of muscles occurring after death
rig·or·ous \ˈri-g(ə-)rəs\ *adjective* (15th century)
1 : manifesting, exercising, or favoring rigor **:** very strict
2 a : marked by extremes of temperature or climate **b :** HARSH, SEVERE
3 : scrupulously accurate **:** PRECISE
synonym see RIGID
— **rig·or·ous·ly** *adverb*
— **rig·or·ous·ness** *noun*
rig·our *chiefly British variant of* RIGOR
rijst·ta·fel \ˈrīs-ˌtä-fəl\ *noun* [Dutch, from *rijst* rice + *tafel* table] (1889)
: an Indonesian meal consisting of rice and a variety of accompanying dishes (as meat, seafood, and vegetables)
Riks·mål *or* **Riks·maal** \ˈriks-ˌmȯl, ˈrēks-\ *noun* [Norwegian, from *rik* kingdom + *mål* speech] (1913)
: BOKMÅL
rile \ˈrī(ə)l\ *transitive verb* **riled; ril·ing** [variant of *roil*] (1825)
1 : to make agitated and angry **:** UPSET
2 : ROIL 1 ◆
synonym see IRRITATE
ril·ey \ˈrī-lē\ *adjective* (1805)
1 : TURBID
2 : ANGRY
¹rill \ˈril\ *noun* [Dutch *ril* or Low German *rille;* akin to Old English *rīth* rivulet] (1538)
: a very small brook
²rill *intransitive verb* (1610)
: to flow like a rill
³rill \ˈril\ *or* **rille** \ˈril, ˈri-lə\ *noun* [German *Rille,* literally, channel made by a small stream, from Low German, rill] (1868)
: any of several long narrow valleys on the moon's surface
rill·et \ˈri-lət\ *noun* (1538)
: a little rill
ril·lettes \ri-ˈlets, -ˈyet\ *noun plural* [French, plural, diminutive of *rille,* singular, piece of pork, from Middle French, dialect variant of *reille* board, lath, from Latin *regula* straightedge — more at RULE] (1889)
: cooked shredded meat (as pork or duck) or fish preserved in fat
¹rim \ˈrim\ *noun* [Middle English, from Old English *rima;* akin to Old Norse *rimi* strip of land] (13th century)
1 a : BRINK **b :** the outer often curved or circular edge or border of something
2 a : the outer part of a wheel joined to the hub usually by spokes **b :** a removable outer metal band on an automobile wheel to which the tire is attached
3 : FRAME 4c(1)
— **rim·less** \-ləs\ *adjective*
²rim *verb* **rimmed; rim·ming** (1794)
transitive verb
1 : to serve as a rim for **:** BORDER ⟨cliffs *rimming* the camp⟩
2 : to run around the rim of ⟨putts that *rim* the cup⟩
intransitive verb
: to form or show a rim
¹rime \ˈrīm\ *noun* [Middle English *rim,* from Old English *hrīm;* akin to Old Norse *hrīm* frost] (before 12th century)
1 : FROST 1b
2 : an accumulation of granular ice tufts on the windward sides of exposed objects that is formed from supercooled fog or cloud and built out directly against the wind
3 : CRUST, INCRUSTATION ⟨a *rime* of snow⟩
²rime *transitive verb* **rimed; rim·ing** (circa 1755)
: to cover with or as if with rime
³rime, rime·ster *variant of* RHYME, RHYMESTER
rim·fire \ˈrim-ˌfīr\ *adjective* (1868)
of a cartridge **:** having the priming distributed in the rim of the shell
— **rimfire** *noun*
rim·land \ˈrim-ˌland\ *noun* (1944)
: a region on the periphery of the heartland
rimmed \ˈrimd\ *adjective* (1729)
: having a rim — usually used in combination ⟨dark-*rimmed* glasses⟩ ⟨red-*rimmed* eyes⟩
rim·rock \ˈrim-ˌräk\ *noun* (1860)
1 : a top stratum or overlying strata of resistant rock of a plateau that outcrops to form a vertical face
2 : the edge or face of a rimrock outcrop
rimy \ˈrī-mē\ *adjective* **rim·i·er; -est** [Old English *hrīmig,* from *hrīm*] (before 12th century)
: covered with rime **:** FROSTY
rind \ˈrīnd, *dialect* ˈrīn\ *noun* [Middle English, from Old English; akin to Old High German *rinda* bark, and probably to Old English *rendan* to rend] (before 12th century)
1 : the bark of a tree
2 : a usually hard or tough outer layer **:** PEEL, CRUST ⟨grated lemon *rind*⟩
— **rind·ed** \ˈrīn-dəd\ *adjective*
rin·der·pest \ˈrin-dər-ˌpest\ *noun* [German, from *Rinder,* plural, cattle + *Pest* pestilence] (1865)
: an acute infectious febrile disease of ruminant mammals (as cattle) that is caused by a virus and is marked by diarrhea and inflammation of mucous membranes
¹ring \ˈriŋ\ *noun* [Middle English, from Old English *hring;* akin to Old High German *hring* ring, Old Church Slavonic *krǫgŭ* circle] (before 12th century)
1 : a circular band for holding, connecting, hanging, pulling, packing, or sealing ⟨a key *ring*⟩ ⟨a towel *ring*⟩
2 : a circlet usually of precious metal worn on the finger
3 a : a circular line, figure, or object ⟨smoke *ring*⟩ **b :** an encircling arrangement ⟨a *ring* of suburbs⟩ **c :** a circular or spiral course — often used in plural in the phrase *run rings around*
4 a (1) **:** an often circular space especially for exhibitions or competitions; *especially* **:** such a space at a circus (2) **:** a structure containing such a ring **b :** a square enclosure in which boxers or wrestlers contest
5 : a band believed to be composed of rocky fragments revolving around a planet (as Saturn)
6 : ANNUAL RING
7 a : an exclusive combination of persons for a selfish and often corrupt purpose (as to control a market) ⟨a wheat *ring*⟩ **b :** GANG
8 : the field of a political contest **:** RACE
9 : food in the shape of a circle
10 : an arrangement of atoms represented in formulas or models in a cyclic manner — called also *cycle*
11 : a set of mathematical elements that is closed under two binary operations of which the first forms a commutative group with the set and the second is associative over the set and is distributive with respect to the first operation
12 *plural* **a :** a pair of usually rubber-covered metal rings suspended from a ceiling or crossbar to a height of approximately eight feet above the floor and used for hanging, swinging, and balancing feats in gymnastics **b :** an event in gymnastics competition in which the rings are used
13 : ²BOXING ⟨ended his *ring* career⟩
— **ring·like** \ˈriŋ-ˌlīk\ *adjective*
²ring *verb* **ringed; ring·ing** \ˈriŋ-iŋ\ (14th century)
transitive verb
1 : to provide with a ring
2 : to place or form a ring around **:** ENCIRCLE ⟨police *ringed* the building⟩
3 : GIRDLE 2
4 : to throw a ringer over (the peg) in a game (as horseshoes or quoits)
intransitive verb
1 a : to move in a ring **b :** to rise in the air spirally
2 : to form or take the shape of a ring
³ring *verb* **rang** \ˈraŋ\; **rung** \ˈrəŋ\; **ring·ing** \ˈriŋ-iŋ\ [Middle English, from Old English *hringan;* akin to Old Norse *hringja* to ring] (before 12th century)
intransitive verb
1 : to sound resonantly or sonorously ⟨the doorbell *rang*⟩ ⟨cheers *rang* out⟩
2 a : to be filled with a reverberating sound **:** RESOUND ⟨the halls *rang* with laughter⟩ **b :** to have the sensation of being filled with a humming sound ⟨his ears *rang*⟩
3 : to cause something to ring ⟨*ring* for the butler⟩
4 a : to be filled with talk or report ⟨the whole land *rang* with her fame⟩ **b :** to have great renown **c :** to sound repetitiously ⟨their praise *rang* in his ears⟩
5 : to have a sound or character expressive of some quality ⟨a story that *rings* true⟩
6 *chiefly British* **:** to make a telephone call — usually used with *up*
transitive verb
1 : to cause to sound especially by striking
2 : to make (a sound) by or as if by ringing a bell
3 : to announce by or as if by ringing
4 : to repeat often, loudly, or earnestly
5 a : to summon especially by bell **b** *chiefly British* **:** TELEPHONE — usually used with *up*
— **ring a bell :** to arouse a response ⟨that name *rings a bell*⟩
— **ring down the curtain :** to conclude a performance or an action
— **ring off the hook :** to ring frequently or constantly with incoming calls ⟨the telephone was *ringing off the hook*⟩
— **ring the changes** *or* **ring changes :** to run through the range of possible variations
— **ring up the curtain :** to begin a performance or an action
⁴ring *noun* (1549)
1 : a set of bells
2 : a clear resonant sound made by or resembling that made by vibrating metal
3 : resonant tone **:** SONORITY

◇ WORD HISTORY

rile English words that now have the sound \ȯi\ such as *boil* and *voice* fall into two major groups of distinct historical origin. The diphthong (vowel combination) of the *voice* group had a sound in early Modern English nearly the same as present-day \ȯi\; the *boil* group, on the other hand, had a sound close to \əi\. Because the \əi\ sound of the *boil* words was virtually identical to the sound that *i* had come to acquire in words such as *price* and *write,* the two merged, so that *boil* rhymed with *bile* and *toil* with *tile.* This pronunciation remained standard into the 18th century, though literate people, under the influence of English spelling, had begun to pronounce *boil* and *toil* with the diphthong of *toy* and *choice.* By the 19th century, standard pronunciation had shifted completely, and spellings such as *pizen* or *pison* for *poison* were used by Mark Twain and other American authors to characterize uneducated rural speech. The word *roil* underwent a unique development: in the sense "to make angry" it was disassociated from the spelling *roil* and written *rile,* becoming thereby a different word. Originally a colloquialism, it has since become standard written English.

4 : a loud sound continued, repeated, or reverberated
5 : a sound or character expressive of some particular quality ⟨the story had a familiar *ring*⟩
6 a : the act or an instance of ringing **b :** a telephone call ⟨give me a *ring* in the morning⟩

ring–a–le·vio \ˌriŋ-ə-ˈlē-vē-ˌō\ *or* **ring–a–lie·vo** \-ˈlē-(ˌ)vō\ *noun* [alteration of earlier *ring relievo,* from [1]*ring* + *relieve*] (circa 1901)
: a game in which players on one team are given time to hide and are then sought out by members of the other team who try to capture them, keep them in a place of confinement, and keep them from being released by their teammates

ring–around–a–rosy \ˌriŋ-ə-ˌrau̇n-də-ˈrō-zē\ *or* **ring–around–the–rosy** \-ˌrau̇n(d)-t͟hə-\ *noun* (circa 1886)
: a children's singing game in which players dance around in a circle and at a given signal squat — called also *ring-a-rosy* \ˌriŋ-ə-ˈrō-zē\

ring·bark \ˈriŋ-ˌbärk\ *transitive verb* (1892)
: GIRDLE 2

ring binder *noun* (1929)
: a loose-leaf binder in which split metal rings attached to a metal back hold the perforated sheets of paper

ring·bolt \ˈriŋ-ˌbōlt\ *noun* (1599)
: an eyebolt with a ring through its eye

ring·bone \-ˌbōn\ *noun* (1523)
: a bony outgrowth on the pastern bones of the horse usually producing lameness

ring dance *noun* (1600)
: ROUND DANCE 1

ring·dove \ˈriŋ-ˌdəv\ *noun* (1538)
1 : a chiefly European pigeon (*Columba palumbus*) with a whitish patch on each side of the neck and wings edged with white
2 : a small dove (*Streptopelia risoria*) of southeastern Europe and Asia — called also *ringed turtle dove*

ringed \ˈriŋd\ *adjective* (14th century)
1 : encircled or marked with or as if with rings
2 : composed or formed of rings

[1]**ring·er** \ˈriŋ-ər\ *noun* (15th century)
1 : one that sounds especially by ringing
2 a (1) **:** one that enters a competition under false representations (2) **:** IMPOSTER, FAKE **b :** one that strongly resembles another — often used with *dead* ⟨he's a dead *ringer* for the senator⟩

[2]**ringer** *noun* (1863)
: one that encircles or puts a ring around (as a quoit or horseshoe that lodges so as to surround the peg)

Ring·er's solution \ˈriŋ-ərz-\ *noun* [Sidney *Ringer* (died 1910) English physician] (1893)
: a balanced aqueous solution that contains chloride, sodium, potassium, calcium, bicarbonate, and phosphate ions and that is used in physiological experiments to provide a medium essentially isotonic to many animal tissues — called also *Ringer solution*

ring finger *noun* (before 12th century)
: the third finger especially of the left hand counting the forefinger as the first

ring·git \ˈriŋ-git\ *noun, plural* **ringgit** *or* **ringgits** [Malay, literally, toothed, serrated] (1967)
— see MONEY table

ring·ing \ˈriŋ-iŋ\ *adjective* (14th century)
1 : clear and full in tone **:** RESOUNDING ⟨a *ringing* baritone⟩
2 : vigorously unequivocal **:** DECISIVE ⟨a *ringing* condemnation of immorality⟩
— **ring·ing·ly** \-iŋ-lē\ *adverb*

ring·lead·er \ˈriŋ-ˌlē-dər\ *noun* (1503)
: a leader of a ring of individuals engaged especially in improper or unlawful activities

ring·let \ˈriŋ-lət\ *noun* (1555)
1 : a small ring or circle
2 : CURL; *especially* **:** a long curl of hair

ring·mas·ter \ˈriŋ-ˌmas-tər\ *noun* (1873)
: one in charge of performances in a ring (as of a circus); *broadly* **:** a supervisor or moderator especially of a performance or presentation

ring·neck \-ˌnek\ *noun* (1791)
: a ring-necked animal

ring–necked \ˈriŋ-ˌnekt\ *or* **ring–neck** \ˈriŋ-ˌnek\ *adjective* (1817)
: having a ring of color about the neck

ring–necked duck *noun* (1831)
: an American scaup duck (*Aythya collaris*) that has a white ring around the bill and in the male a faint narrow chestnut ring encircling the neck

ring–necked pheasant *noun* (1834)
: a Eurasian pheasant (*Phasianus colchicus*) that has been widely introduced as a game bird in North America and that has a white neck ring and an iridescent green and purplish head with red wattles around the eyes

ring-necked pheasant

ring off *intransitive verb* (1882)
chiefly British **:** HANG UP 1

ring–po·rous \ˈriŋ-ˌpōr-əs, -ˌpȯr-\ *adjective* (1902)
: having vessels more numerous and usually larger in cross section in the springwood with a resulting more or less distinct line between the springwood and the last season's wood — compare DIFFUSE-POROUS

ring road *noun* (1928)
chiefly British **:** a highway skirting an urban area

[1]**ring·side** \ˈriŋ-ˌsīd\ *noun* (1866)
1 : the area just outside a ring especially in which a contest occurs
2 : a place from which one may have a close view

[2]**ringside** *adjective* (1896)
: being at the ringside ⟨a *ringside* seat⟩

ring spot *noun* (1923)
1 : a lesion of plant tissue consisting of yellowish, purplish, or necrotic, often concentric rings
2 : a viral disease of plants that is characterized by ring spots

ring stand *noun* (circa 1865)
: a metal stand consisting of a long upright rod attached to a heavy rectangular base that is used with rings and clamps for supporting laboratory apparatus

ring·straked \ˈriŋ-ˌstrākt\ *adjective* (1611)
archaic **:** marked with circular stripes

ring·tail \-ˌtāl\ *noun* (1844)
1 : RACCOON
2 : a carnivore (*Bassariscus astutus*) of the western U.S. and Mexico that is related to and resembles the raccoon — called also *cacomistle, civet cat, ringtail cat, ringtailed cat*
3 : CAPUCHIN 3

ring–tailed \-ˌtā(ə)ld\ *adjective* (1729)
1 : having a tail marked with rings of differing colors
2 : having a tail carried in a form approximating a circle ⟨a *ring-tailed* Afghan hound⟩

ring·taw \-ˌtȯ\ *noun* (1828)
: a game of marbles in which marbles are placed in a circle on the ground and shot at from the edge of the circle with the object being to knock them out of the circle

ring·toss \-ˌtȯs, -ˌtäs\ *noun* (1871)
: a game in which the object is to toss a ring so that it will fall over an upright stick

ring up *transitive verb* (1937)
1 : to total and record especially by means of a cash register
2 : ACHIEVE ⟨*rang up* many social triumphs⟩

ring·worm \ˈriŋ-ˌwərm\ *noun* (15th century)
: any of several contagious fungal diseases of the skin, hair, or nails of humans and domestic animals that are characterized by ring-shaped discolored skin patches covered with vesicles and scales

rink \ˈriŋk\ *noun* [Middle English (Scots) *rinc* area in which a contest takes place, from Middle French *renc* place, row — more at RANK] (1787)
1 a : a smooth extent of ice marked off for curling or ice hockey **b :** a surface of ice for ice-skating; *also* **:** a building containing such a rink **c :** an enclosure for roller-skating
2 : an alley for lawn bowling
3 : a team in bowls or curling

rinky–dink \ˈriŋ-kē-ˌdiŋk\ *adjective* [origin unknown] (1913)
1 : SMALL-TIME
2 : OLD-FASHIONED

[1]**rinse** \ˈrin(t)s, *especially dialect* ˈrench\ *transitive verb* **rinsed; rins·ing** [Middle English *rincen,* from Middle French *rincer,* from (assumed) Vulgar Latin *recentiare,* from Latin *recent-, recens* fresh, recent] (14th century)
1 : to cleanse by flushing with liquid (as water) — often used with *out* ⟨*rinse* out the mouth⟩
2 a : to cleanse (as of soap used in washing) by clear water **b :** to treat (hair) with a rinse
3 : to remove (dirt or impurities) by washing lightly or in water only
— **rins·er** *noun*

[2]**rinse** *noun* (1837)
1 : the act or process of rinsing
2 a : liquid used for rinsing **b :** a solution that temporarily tints hair

rins·ing *noun* (1818)
1 : DREGS, RESIDUE — usually used in plural
2 : water that has been used for rinsing — usually used in plural

rio·ja \rē-ˈō-(ˌ)hä\ *noun, often capitalized* (1907)
: a wine from the Rioja region of Spain; *especially* **:** a dry red wine from this region

[1]**ri·ot** \ˈrī-ət\ *noun* [Middle English, from Old French, dispute, from *rioter* to quarrel] (13th century)
1 *archaic* **a :** profligate behavior **:** DEBAUCHERY **b :** unrestrained revelry **c :** noise, uproar, or disturbance made by revelers
2 a : public violence, tumult, or disorder **b :** a violent public disorder; *specifically* **:** a tumultuous disturbance of the public peace by three or more persons assembled together and acting with a common intent
3 : a random or disorderly profusion ⟨the woods were a *riot* of color⟩
4 : one that is wildly amusing ⟨the new comedy is a *riot*⟩

[2]**riot** *intransitive verb* (14th century)
1 : to indulge in revelry or wantonness
2 : to create or engage in a riot
— **ri·ot·er** *noun*

riot act *noun* [the *Riot Act,* English law of 1715 providing for the dispersal of riots upon command of legal authority] (1819)
: a vigorous reprimand or warning — used in the phrase *read the riot act*

riot gun *noun* (1916)
: a small arm used to disperse rioters rather than to inflict serious injury or death; *especially* **:** a short-barreled shotgun

ri·ot·ous \ˈrī-ə-təs\ *adjective* (15th century)
1 a : of the nature of a riot **:** TURBULENT **b :** participating in riot
2 : ABUNDANT, EXUBERANT ⟨the garden was *riotous* with flowers⟩
— **ri·ot·ous·ly** *adverb*
— **ri·ot·ous·ness** *noun*

[1]**rip** \ˈrip\ *verb* **ripped; rip·ping** [probably from Flemish *rippen* to strip off roughly] (15th century)

transitive verb
1 a : to tear or split apart or open **b :** to saw or split (wood) with the grain
2 : to slash or slit with or as if with a sharp blade
3 : to hit sharply ⟨*ripped* a double to left field⟩
4 : to utter violently **:** spit out ⟨*ripped* out an oath⟩
5 : CRITICIZE, PUT DOWN
intransitive verb
1 : to become ripped **:** REND
2 : to rush headlong ⟨*ripped* past second base⟩
synonym see TEAR
— rip into : to tear into **:** ATTACK

²rip *noun* (1711)
1 : a rent made by ripping **:** TEAR
2 : CUT 5b

³rip *noun* [perhaps from ²*rip*] (1775)
1 : a body of water made rough by the meeting of opposing tides, currents, or winds
2 : a current of water roughened by passing over an irregular bottom
3 : RIP CURRENT

⁴rip *noun* [perhaps by shortening & alteration from *reprobate*] (1797)
: a dissolute person **:** LIBERTINE

ri·par·i·an \rə-ˈper-ē-ən, rī-\ *adjective* [Latin *riparius* — more at RIVER] (circa 1841)
: relating to or living or located on the bank of a natural watercourse (as a river) or sometimes of a lake or a tidewater

riparian right *noun* (circa 1860)
: a right (as access to or use of the shore, bed, and water) of one owning riparian land

rip cord *noun* (1907)
1 : a cord by which the gasbag of a balloon may be ripped open for a limited distance to release the gas quickly and so cause immediate descent
2 : a cord or wire pulled in making a descent to release the pilot parachute which lifts the main parachute out of its container

rip current *noun* (1936)
: a strong usually narrow surface current flowing outward from a shore that results from the return flow of waves and wind-driven water

ripe \ˈrīp\ *adjective* **rip·er; rip·est** [Middle English, from Old English *rīpe;* akin to Old English *rīpan, reopan* to reap] (before 12th century)
1 : fully grown and developed **:** MATURE
2 : having mature knowledge, understanding, or judgment
3 : of advanced years **:** LATE ⟨lived to a *ripe* old age⟩
4 a : SUITABLE, APPROPRIATE ⟨the time seemed *ripe* for the experiment⟩ **b :** fully prepared **:** READY ⟨the colonies were *ripe* for revolution⟩
5 a : brought by aging to full flavor or the best state **:** MELLOW ⟨*ripe* cheese⟩ **b :** SMELLY, STINKING
6 : ruddy, plump, or full like ripened fruit
7 : INDECENT ⟨*ripe* language⟩
— ripe·ly *adverb*
— ripe·ness *noun*

rip·en \ˈrī-pən, ˈrī-pᵊm\ *verb* **rip·ened; rip·en·ing** \ˈrī-pə-niŋ, ˈrīp-niŋ\ (1561)
intransitive verb
: to grow or become ripe
transitive verb
1 : to make ripe
2 a : to bring to completeness or perfection **b :** to age or cure (cheese) to develop characteristic flavor, odor, body, texture, and color **c :** to improve flavor and tenderness of (beef or game) by aging under refrigeration
— rip·en·er \ˈrī-pə-nər, ˈrīp-nər\ *noun*

ri·pie·no \ri-ˈpyā-(ˌ)nō, -ˈpye-\ *noun, plural* **-ni** \-(ˌ)nē\ *or* **-nos** [Italian, literally, filled up] (circa 1930)
: TUTTI

rip–off \ˈrip-ˌȯf\ *noun* (1969)
1 : an act or instance of stealing **:** THEFT; *also* **:** a financial exploitation
2 : a usually cheap exploitive imitation

rip off *transitive verb* (1967)
1 a : ROB; *also* **:** CHEAT, DEFRAUD **b :** STEAL
2 : to copy or imitate blatantly or unscrupulously
3 : to perform, achieve, or score quickly or easily ⟨*ripped off* 10 straight points⟩

ri·poste \ri-ˈpōst\ *noun* [French, modification of Italian *risposta,* literally, answer, from *rispondere* to respond, from Latin *respondēre*] (1707)
1 : a fencer's quick return thrust following a parry
2 : a retaliatory verbal sally **:** RETORT
3 : a retaliatory maneuver or measure
— riposte *verb*

ripped \ˈript\ *adjective* (1970)
slang **:** being under the influence of alcohol or drugs **:** HIGH, STONED

rip·per \ˈri-pər\ *noun* (1611)
1 : one that rips; *especially* **:** a machine used to break up solid material (as rock or ore)
2 : an excellent example or instance of its kind

rip·ping \ˈri-piŋ\ *adjective* [probably from present participle of ¹*rip*] (1846)
chiefly British **:** EXCELLENT, DELIGHTFUL ⟨wrote me some *ripping* letters⟩ ⟨had a *ripping* time⟩

¹rip·ple \ˈri-pəl\ *verb* **rip·pled; rip·pling** \-p(ə-)liŋ\ [perhaps frequentative of ¹*rip*] (circa 1671)
intransitive verb
1 a : to become lightly ruffled or covered with small waves **b :** to flow in small waves **c :** to fall in soft undulating folds ⟨the scarf *rippled* to the floor⟩
2 : to flow with a light rise and fall of sound or inflection ⟨laughter *rippled* over the audience⟩
3 : to move with an undulating motion or so as to cause ripples ⟨the canoe *rippled* through the water⟩
4 : to have or produce a ripple effect **:** SPREAD ⟨the news *rippled* outwards⟩
transitive verb
1 : to stir up small waves on
2 : to impart a wavy motion or appearance to ⟨*rippling* his arm muscles⟩
3 : to utter or play with a slight rise and fall of sound
— rip·pler \-p(ə-)lər\ *noun*

²ripple *noun* (1755)
1 a : a shallow stretch of rough water in a stream **b** (1) **:** the ruffling of the surface of water (2) **:** a small wave
2 a : RIPPLE MARK **b :** a sound like that of rippling water ⟨a *ripple* of laughter⟩ **c :** a usually slight noticeable effect or reaction
— rip·ply \ˈri-p(ə-)lē\ *adjective*

ripple effect *noun* (1966)
: a spreading, pervasive, and usually unintentional effect or influence ⟨the automotive industry has a *ripple effect* on many other industries⟩ — compare DOMINO EFFECT

ripple mark *noun* (1833)
1 : one of a series of small ridges produced especially on sand by the action of wind, a current of water, or waves
2 : a striation across the grain of wood especially on the tangential surface
— rip·ple–marked \ˈri-pəl-ˌmärkt\ *adjective*

¹rip·rap \ˈrip-ˌrap\ *noun* [obsolete *riprap* sound of rapping] (1833)
1 : a foundation or sustaining wall of stones or chunks of concrete thrown together without order (as in deep water); *also* **:** a layer of this or similar material on an embankment slope to prevent erosion
2 : material used for riprap

²riprap *transitive verb* (1848)
1 : to form a riprap in or upon
2 : to strengthen or support with a riprap

rip–roar·ing \ˈrip-ˈrōr-iŋ, -ˈrȯr-\ *adjective* (1834)
: noisily excited or exciting

rip·saw \ˈrip-ˌsȯ\ *noun* (1846)
: a coarse-toothed saw used to cut wood in the direction of the grain — compare CROSSCUT SAW

rip·snort·er \ˈrip-ˈsnȯr-tər\ *noun* (1840)
: something extraordinary **:** HUMDINGER ⟨the finale was a *ripsnorter*⟩
— rip·snort·ing \-tiŋ\ *adjective*

rip·stop \ˈrip-ˌstäp\ *adjective* (1949)
: of, relating to, or being a fabric woven with a double thread at regular intervals so that small tears do not spread ⟨*ripstop* nylon⟩
— ripstop *noun*

rip·tide \ˈrip-ˌtīd\ *noun* (1862)
: RIP CURRENT

Rip·u·ar·i·an \ˌri-pyə-ˈwer-ē-ən\ *adjective* [Medieval Latin *Ripuarius*] (1781)
: of, relating to, or constituting a group of Franks settling in the 4th century on the Rhine near Cologne

Rip van Win·kle \ˌrip-(ˌ)van-ˈwiŋ-kəl, -vən-\ *noun*
: a ne'er-do-well in a story in Washington Irving's *Sketch Book* who sleeps for 20 years

¹rise \ˈrīz\ *intransitive verb* **rose** \ˈrōz\; **ris·en** \ˈri-zᵊn\; **ris·ing** \ˈrī-ziŋ\ [Middle English, from Old English *rīsan;* akin to Old High German *rīsan* to rise] (before 12th century)
1 a : to assume an upright position especially from lying, kneeling, or sitting **b :** to get up from sleep or from one's bed
2 : to return from death
3 : to take up arms ⟨*rise* in rebellion⟩
4 : to respond warmly **:** APPLAUD — usually used with *to* ⟨the audience *rose* to her verve and wit⟩
5 *chiefly British* **:** to end a session **:** ADJOURN
6 : to appear above the horizon ⟨the sun *rises* at six⟩
7 a : to move upward **:** ASCEND **b :** to increase in height, size, or volume ⟨the river *rose* after the heavy rains⟩
8 : to extend above other objects ⟨mountain peaks *rose* to the west⟩
9 a : to become heartened or elated ⟨his spirits *rose*⟩ **b :** to increase in fervor or intensity ⟨my anger *rose* as I thought about the insult⟩
10 a : to attain a higher level or rank ⟨officers who *rose* from the ranks⟩ **b :** to increase in quantity or number
11 a : to take place **:** HAPPEN **b :** to come into being **:** ORIGINATE
12 : to follow as a consequence **:** RESULT
13 : to exert oneself to meet a challenge ⟨*rise* to the occasion⟩
synonym see SPRING

²rise \ˈrīz *also* ˈrīs\ *noun* (15th century)
1 a : a spot higher than surrounding ground **:** HILLTOP **b :** an upward slope ⟨a *rise* in the road⟩
2 : an act of rising or a state of being risen: as **a :** a movement upward **:** ASCENT **b :** emergence (as of the sun) above the horizon **c :** the upward movement of a fish to seize food or bait
3 : BEGINNING, ORIGIN ⟨the river had its *rise* in the mountain⟩
4 : the distance or elevation of one point above another
5 a : an increase especially in amount, number, or volume **b** *chiefly British* **:** RAISE 3b **c :** an increase in price, value, rate, or sum ⟨a *rise* in the cost of living⟩
6 : an angry reaction ⟨got a *rise* out of him⟩
7 : the distance from the crotch to the waistline on pants

ris·er \ˈrī-zər\ *noun* (15th century)
1 : one that rises (as from sleep)
2 : the upright member between two stair treads
3 : a stage platform on which performers are placed for greater visibility
4 : a vertical pipe (as for water or gas) or a vertical portion of an electric wiring system
5 : one of the straps that connects a parachutist's harness with the shroud lines

ris·i·bil·i·ty \ˌri-zə-ˈbi-lə-tē\ *noun, plural* **-ties** (1620)
1 : LAUGHTER
2 : the ability or inclination to laugh — often used in plural ⟨our *risibilities* support us as we skim over the surface of a deep issue —J. A. Pike⟩

ris·i·ble \ˈri-zə-bəl\ *adjective* [Late Latin *risibilis,* from Latin *risus,* past participle of *ridēre* to laugh] (1557)
1 a : capable of laughing **b :** disposed to laugh
2 : arousing or provoking laughter; *especially* **:** LAUGHABLE
3 : associated with, relating to, or used in laughter ⟨*risible* muscles⟩

ris·i·bles \-bəlz\ *noun plural* (1785)
: sense of the ridiculous **:** sense of humor

¹ris·ing \ˈrī-ziŋ\ *noun* (14th century)
: INSURRECTION, UPRISING

²rising *adverb* (circa 1772)
: approaching a stated age **:** NEARLY ⟨a red cow *rising* four years old —*Lancaster (Pennsylvania) Journal*⟩

rising diphthong *noun* (1888)
: a diphthong in which the second element is more sonorous than the first (as \wi\ in \ˈkwit\ *quit*)

rising rhythm *noun* (1881)
: rhythm with stress occurring regularly on the last syllable of each foot — compare FALLING RHYTHM

¹risk \ˈrisk\ *noun* [French *risque,* from Italian *risco*] (circa 1661)
1 : possibility of loss or injury **:** PERIL
2 : someone or something that creates or suggests a hazard
3 a : the chance of loss or the perils to the subject matter of an insurance contract; *also* **:** the degree of probability of such loss **b :** a person or thing that is a specified hazard to an insurer ⟨a poor *risk* for insurance⟩ **c :** an insurance hazard from a specified cause or source ⟨war *risk*⟩
— risk·less \ˈris-kləs\ *adjective*
— at risk : exposed to a usually specified danger or loss ⟨patients *at risk* of infection⟩

²risk *transitive verb* (circa 1687)
1 : to expose to hazard or danger ⟨*risked* her life⟩
2 : to incur the risk or danger of ⟨*risked* breaking his neck⟩
— risk·er *noun*

risk capital *noun* (1944)
: VENTURE CAPITAL

risky \ˈris-kē\ *adjective* **risk·i·er; -est** (1827)
: attended with risk or danger **:** HAZARDOUS
synonym see DANGEROUS
— risk·i·ness *noun*

ri·sor·gi·men·to \(ˌ)rē-ˌzȯr-ji-ˈmen-(ˌ)tō, -ˌsȯr-\ *noun, plural* **-tos** [Italian, literally, rising again, from *risorgere* to rise again, from Latin *resurgere* — more at RESURRECTION] (1902)
: a time of renewal or renaissance **:** REVIVAL; *specifically* **:** the 19th century movement for Italian political unity

ri·sot·to \ri-ˈsȯ-(ˌ)tō, -ˈzȯ-\ *noun, plural* **-tos** [Italian, from *riso* rice — more at RICE] (1855)
: rice cooked in meat stock and seasoned (as with Parmesan cheese or saffron)

ris·qué \ri-ˈskā\ *adjective* [French, from past participle of *risquer* to risk, from *risque*] (1867)
: verging on impropriety or indecency **:** OFF-COLOR

Rit·a·lin \ˈri-tə-lən\ *trademark*
— used for methylphenidate

ri·tard \ri-ˈtärd\ *noun* (circa 1890)
: RITARDANDO

¹ri·tar·dan·do \ri-ˌtär-ˈdän-(ˌ)dō, ˌrē-\ *adverb or adjective* [Italian, from Latin *retardandum,* gerund of *retardare* to retard] (circa 1811)
: with a gradual slackening in tempo — used as a direction in music

²ritardando *noun, plural* **-dos** (1889)
: a ritardando passage

rite \ˈrīt\ *noun* [Middle English, from Latin *ritus;* akin to Greek *arithmos* number — more at ARITHMETIC] (14th century)
1 a : a prescribed form or manner governing the words or actions for a ceremony **b :** the liturgy of a church or group of churches
2 : a ceremonial act or action ⟨initiation *rites*⟩
3 : a division of the Christian church using a distinctive liturgy

rite de pas·sage \ˌrēt-də-pa-ˈsäzh, -pä-\ *noun, plural* **rites de passage** \ˌrēt(s)-də-\ [French] (1911)
: RITE OF PASSAGE

rite of passage (1909)
: a ritual associated with a crisis or a change of status (as marriage, illness, or death) for an individual

ri·tor·nel·lo \ˌri-tər-ˈne-(ˌ)lō, ˌri-ˌtȯr-\ *noun, plural* **-nel·li** \-ˈne-(ˌ)lē\ *or* **-nellos** [Italian, diminutive of *ritorno* return, from *ritornare* to return, from *ri-* re- + *tornare* to turn, from Latin, to turn on a lathe — more at TURN] (1675)
1 a : a short recurrent instrumental passage in a vocal composition **b :** an instrumental interlude in early opera
2 : a tutti passage in a concerto or rondo refrain

¹rit·u·al \ˈri-chə-wəl, -chəl; ˈrich-wəl\ *adjective* [Latin *ritualis,* from *ritus* rite] (1570)
1 : of or relating to rites or a ritual **:** CEREMONIAL ⟨a *ritual* dance⟩
2 : according to religious law or social custom ⟨*ritual* purity⟩
— rit·u·al·ly *adverb*

²ritual *noun* (1649)
1 : the established form for a ceremony; *specifically* **:** the order of words prescribed for a religious ceremony
2 a : ritual observance; *specifically* **:** a system of rites **b :** a ceremonial act or action **c :** a customarily repeated often formal act or series of acts

rit·u·al·ism \ˈri-chə wə-ˌli-zəm, -chə-ˌli-; ˈrich-wə-\ *noun* (1843)
1 : the use of ritual
2 : excessive devotion to ritual
— rit·u·al·ist \-list\ *noun*
— rit·u·al·is·tic \ˌri-chə-wə-ˈlis-tik, -chə-ˈlis-; ˌrich-wə-\ *adjective*
— rit·u·al·is·ti·cal·ly \-ti-k(ə-)lē\ *adverb*

rit·u·al·ize \-ˌlīz\ *verb* **-ized; -iz·ing** (1842)
intransitive verb
: to practice ritualism
transitive verb
1 : to make a ritual of
2 : to impose a ritual on
— rit·u·al·i·za·tion \ˌri-chə-wə-lə-ˈzā-shən, -chə-lə-; ˌrich-wə-\ *noun*

ritzy \ˈrit-sē\ *adjective* **ritz·i·er; -est** [*Ritz* hotels, noted for their opulence] (1920)
1 : SNOBBISH
2 : ostentatiously smart **:** FASHIONABLE, POSH
— ritz·i·ness *noun*

¹ri·val \ˈrī-vəl\ *noun* [Middle French or Latin; Middle French, from Latin *rivalis* one using the same stream as another, rival in love, from *rivalis* of a stream, from *rivus* stream — more at RUN] (1577)
1 a : one of two or more striving to reach or obtain something that only one can possess **b :** one striving for competitive advantage
2 *obsolete* **:** COMPANION, ASSOCIATE
3 : one that equals another in desired qualities **:** PEER ◆

²rival *adjective* (1590)
: having the same pretensions or claims **:** COMPETING

³rival *verb* **ri·valed** *or* **ri·valled; ri·val·ing** *or* **ri·val·ling** \ˈrīv-(ə-)liŋ\ (1605)
intransitive verb
: to act as a rival **:** COMPETE
transitive verb
1 : to be in competition with
2 : to strive to equal or excel **:** EMULATE
3 : to possess qualities or aptitudes that approach or equal (those of another)

ri·val·rous \ˈrī-vəl-rəs\ *adjective* (1812)
: given to rivalry **:** COMPETITIVE

ri·val·ry \ˈrī-vəl-rē\ *noun, plural* **-ries** (1598)
: the act of rivaling **:** the state of being a rival **:** COMPETITION

rive \ˈrīv\ *verb* **rived** \ˈrīvd\; **riv·en** \ˈri-vən\ *also* **rived; riv·ing** \ˈrī-viŋ\ [Middle English, from Old Norse *rīfa;* akin to Greek *ereipein* to tear down] (14th century)
transitive verb
1 a : to wrench open or tear apart or to pieces **:** REND **b :** to split with force or violence ⟨lightning *rived* the tree⟩
2 a : to divide into pieces **:** SHATTER **b :** FRACTURE
intransitive verb
: to become split **:** CRACK
synonym see TEAR

riv·er \ˈri-vər\ *noun, often attributive* [Middle English *rivere,* from Old French, from (assumed) Vulgar Latin *riparia,* from Latin, feminine of *riparius* riparian, from *ripa* bank, shore; perhaps akin to Greek *ereipein* to tear down] (14th century)
1 a : a natural stream of water of usually considerable volume **b :** WATERCOURSE
2 a : something resembling a river ⟨a *river* of lava⟩ **b** *plural* **:** large or overwhelming quantities ⟨drank *rivers* of coffee⟩
word history see RIVAL
— up the river : to or in prison ⟨takes the rap and goes *up the river* —Nigel Balchin⟩

riv·er·bank \ˈri-vər-ˌbaŋk\ *noun* (1565)
: the bank of a river

riv·er·bed \-ˌbed\ *noun* (1833)
: the channel occupied by a river

river blindness *noun* (1953)
: ONCHOCERCIASIS

riv·er·boat \-ˌbōt\ *noun* (1565)
: a boat for use on a river

river duck *noun* (1837)
: DABBLER b

riv·er·front \-ˌfrənt\ *noun* (1855)
: the land or area along a river

river horse *noun* (1601)
: HIPPOPOTAMUS

riv·er·ine \ˈri-və-ˌrīn, -ˌrēn\ *adjective* (1860)
1 : relating to, formed by, or resembling a river
2 : living or situated on the banks of a river

riv·er·side \ˈri-vər-ˌsīd\ *noun* (14th century)

◇ WORD HISTORY

rival *Rival* is borrowed, directly or indirectly through French, from Latin *rivalis,* which, as an adjective derivative of *rivus* "brook or stream," meant "of a brook or stream." As a noun, *rivalis* occurs only in its plural forms in Latin texts, referring literally to those who used the same stream as a source of water. Just as neighbors are likely to dispute each other's rights to a common source of water, so too contention is inevitable when two or more persons strive to obtain something that only one can possess. Thus Latin *rivalis* also developed a sense relating to competition in other areas, especially love, and in this sense it came into French and English. Despite its aquatic background, *rival* is unrelated to *river,* which descends ultimately not from Latin *rivus,* but rather from an unrelated word *ripa* "bank, shore." The Old French word *rivere* or *riviere* (from an unattested spoken Latin word *riparia*) meant "riverbank" or "land along a river" as well as "watercourse, stream."

: the side or bank of a river

riv•er•ward \-wərd\ *or* **riv•er•wards** \-wərdz\ *adverb or adjective* (1833)
: toward a river

[1]**riv•et** \ˈri-vət\ *noun* [Middle English *ryvette,* from Middle French *rivet,* from *river* to attach] (15th century)
: a headed pin or bolt of metal used for uniting two or more pieces by passing the shank through a hole in each piece and then beating or pressing down the plain end so as to make a second head

[2]**rivet** *transitive verb* (15th century)
1 : to fasten with or as if with rivets
2 : to upset the end or point of (as a metallic pin, rod, or bolt) by beating or pressing so as to form a head
3 : to fasten firmly ⟨they *rivet* these feelings . . . tightly together —Michael Novak⟩
4 : to attract and hold (as the attention) completely
— riv•et•er *noun*

riv•et•ing \ˈri-və-tiŋ\ *adjective* (1677)
: having the power to fix the attention **:** ENGROSSING, FASCINATING
— riv•et•ing•ly \-lē\ *adverb*

ri•vi•era \ˌri-vē-ˈer-ə, ri-ˈvyer-\ *noun, often capitalized* [from the *Riviera,* region in southeastern France and northwestern Italy] (1766)
: a coastal region frequented as a resort area and usually marked by a mild climate

ri•vi•ère \ˌri-vē-ˈer, ri-ˈvyer\ *noun* [French, literally, river, from Old French *rivere*] (1879)
: a necklace of precious stones (as diamonds)

riv•u•let \ˈri-vyə-lət, -və-\ *noun* [Italian *rivoletto,* diminutive of *rivolo,* from Latin *rivulus,* diminutive of *rivus* stream — more at RUN] (1587)
: a small stream

[1]**ri•yal** \rē-ˈyäl, -ˈyal\ *noun* [Arabic *riyāl,* from Spanish *real* real] (1928)
— see MONEY table

[2]**riyal** *variant of* RIAL

RN \ˌär-ˈen\ *noun* (1903)
: REGISTERED NURSE

RNA \ˌär-(ˌ)en-ˈā\ *noun* [*ribo*nucleic *a*cid] (1948)
: any of various nucleic acids that contain ribose and uracil as structural components and are associated with the control of cellular chemical activities — compare MESSENGER RNA, RIBOSOMAL RNA, TRANSFER RNA

RNA polymerase *noun* (circa 1962)
: any of a group of enzymes that promote the synthesis of RNA using DNA or RNA as a template

RN•ase \ˌär-ˈen-ˌās, -ˌāz\ *or* **RNA•ase** \ˌär-(ˌ)en-ˈā-ˌās, -ˈā-ˌāz\ *noun* [*RNA* + *-ase*] (1957)
: RIBONUCLEASE

[1]**roach** \ˈrōch\ *noun, plural* **roach** *also* **roach•es** [Middle English *roche,* from Middle French] (13th century)
1 : a silver-green European freshwater cyprinid fish (*Rutilus rutilus*); *also* **:** any of various related fishes (as some shiners)
2 : any of several American freshwater sunfishes (family Centrarchidae)

[2]**roach** *noun* [origin unknown] (1794)
1 : a curved cut in the edge of a sail to prevent chafing or to secure a better fit
2 : a roll of hair brushed straight back from the forehead or side of the head

[3]**roach** *transitive verb* (1818)
1 : to cut (as a horse's mane) so that the remainder stands upright
2 : to cause to arch; *specifically* **:** to brush (the hair) in a roach — often used with *up*

[4]**roach** *noun* [by shortening] (circa 1848)
1 : COCKROACH
2 : the butt of a marijuana cigarette

roach back *noun* (1874)
: an arched back (as of a dog)

roach clip *noun* (1968)
: a metal clip that resembles tweezers and is used by marijuana smokers to hold a roach — called also *roach holder*

road \ˈrōd\ *noun* [Middle English *rode,* from Old English *rād* ride, journey; akin to Old English *rīdan* to ride] (14th century)
1 : ROADSTEAD — often used in plural
2 a : an open way for vehicles, persons, and animals; *especially* **:** one lying outside of an urban district **:** HIGHWAY **b :** ROADBED 2b
3 : ROUTE, PATH
4 : RAILWAY
5 : a series of scheduled visits or appearances (as games or performances) in several locations or the travel necessary to make these visits ⟨the team is on the *road*⟩ ⟨on tour with the musical's *road* company⟩
— road•less \ˈrōd-ləs\ *adjective*
— down the road : in or into the future

roach back

road•abil•i•ty \ˌrō-də-ˈbi-lə-tē\ *noun* (circa 1914)
: the qualities (as steadiness and balance) desirable in an automobile on the road

road agent *noun* (1863)
: a highwayman who formerly operated especially on stage routes in unsettled districts

road•bed \ˈrōd-ˌbed\ *noun* (circa 1840)
1 a : the bed on which the ties, rails, and ballast of a railroad rest **b :** the ballast or the upper surface of the ballast on which the ties rest
2 a : the earth foundation of a road prepared for surfacing **b :** the part of the surface of a road traveled by vehicles

road•block \-ˌbläk\ *noun* (1940)
1 a : a barricade often with traps or mines for holding up an enemy at a point on a road covered by fire **b :** a road barricade set up especially by law enforcement officers
2 : an obstruction in a road
3 : something (as a fact, condition, or countermeasure) that blocks progress or prevents accomplishment of an objective
— roadblock *transitive verb*

road hog *noun* (1891)
: a driver of an automotive vehicle who obstructs others especially by occupying part of another's traffic lane

road•hold•ing \ˈrōd-ˌhōl-diŋ\ *noun* (1932)
chiefly British **:** the qualities of an automobile that tend to make it respond precisely to the driver's steering

road•house \ˈrōd-ˌhau̇s\ *noun* (1857)
: an inn or tavern usually outside city limits providing liquor and usually meals, dancing, and often gambling

road•ie \ˈrō-dē\ *noun* [*road* + *-ie*] (1969)
: one who works (as by moving heavy equipment) for traveling entertainers

road•kill \ˈrōd-ˌkil\ *noun* (1979)
: an animal that has been killed on a road by a motor vehicle
— road–killed \-ˌkild\ *adjective*

road map *noun* (1883)
1 : a map showing roads especially for automobile travel
2 a : a detailed plan to guide progress toward a goal **b :** a detailed explanation

road metal *noun* (1818)
: broken stone or cinders used in making and repairing roads or ballasting railroads

road racing *noun* (1828)
: racing over public roads; *especially* **:** automobile racing over roads or over a closed course designed to simulate public roads
— road race *noun*

road roller *noun* (circa 1876)
: one that rolls roadways; *specifically* **:** a machine with heavy wide smooth rollers for compacting roadbeds

road•run•ner \ˈrōd-ˌrə-nər\ *noun* (1856)
: a largely terrestrial bird (*Geococcyx californianus*) of the cuckoo family that has a long tail and a crest, is a speedy runner, and ranges from the southwestern U.S. to Mexico; *also* **:** a closely related bird (*G. velox*) of Mexico and Central America

roadrunner

road show *noun* (1908)
1 : a theatrical performance given by a troupe on tour
2 : a special engagement of a new motion picture usually at increased prices
3 : a promotional presentation or meeting conducted in a series of locations

road•side \ˈrōd-ˌsīd\ *noun* (1744)
: the strip of land along a road **:** the side of a road
— roadside *adjective*

road•stead \ˈrōd-ˌsted\ *noun* (1556)
: a place less enclosed than a harbor where ships may ride at anchor

road•ster \ˈrōd-stər\ *noun* (1818)
1 a : a horse for riding or driving on roads **b :** a utility saddle horse of the hackney type
2 a : a light carriage **:** BUGGY **b :** an automobile with an open body that seats two and has a folding fabric top and a luggage compartment in the rear

road test *noun* (1906)
1 : a test of a vehicle under practical operating conditions on the road
2 : a test on the road of a person's driving ability as a requirement for a driver's license
— road test *transitive verb*

road•way \ˈrōd-ˌwā\ *noun* (1600)
1 a : the strip of land over which a road passes **b :** ROAD; *specifically* **:** ROADBED 2b
2 : the part of a bridge used by vehicles

road•work \-ˌwərk\ *noun* (1869)
1 : work done in constructing or repairing roads
2 : conditioning for an athletic contest (as a boxing match) consisting mainly of long runs

road•wor•thy \-ˌwər-t͟hē\ *adjective* (1819)
: fit for use on the road
— road•wor•thi•ness *noun*

roam \ˈrōm\ *verb* [Middle English *romen*] (14th century)
intransitive verb
1 : to go from place to place without purpose or direction **:** WANDER
2 : to travel purposefully unhindered through a wide area ⟨cattle *roaming* in search of water⟩
transitive verb
: to range or wander over
synonym see WANDER
— roam *noun*
— roam•er *noun*

[1]**roan** \ˈrōn *also* ˈrō-ən\ *adjective* [Middle French, from Old Spanish *roano*] (1530)
: having the base color (as red, black, or brown) muted and lightened by admixture of white hairs ⟨a *roan* horse⟩ ⟨a *roan* calf⟩

[2]**roan** *noun* (1580)
1 : an animal (as a horse) with a roan coat — usually used of a red roan when unqualified
2 : the color of a roan horse — used especially when the base color is red

[3]**roan** *noun* [origin unknown] (1818)
: sheepskin tanned with sumac and colored and finished to imitate morocco

[1]**roar** \ˈrōr, ˈrȯr\ *verb* [Middle English *roren,* from Old English *rārian;* akin to Old High German *rērēn* to bleat] (before 12th century)
intransitive verb
1 a : to utter or emit a full loud prolonged sound **b :** to sing or shout with full force
2 a : to make or emit a loud confused sound (as background reverberation or rumbling) **b :** to laugh loudly

3 a : to be boisterous or disorderly **b :** to proceed or rush with great noise or commotion
4 : to make a loud noise during inhalation (as that of a horse affected with roaring)
transitive verb
1 : to utter or proclaim with a roar
2 : to cause to roar

[2]roar *noun* (14th century)
1 : the deep cry of a wild animal
2 : a loud deep cry (as of pain or anger)
3 : a loud continuous confused sound ⟨the *roar* of the crowd⟩
4 : a boisterous outcry

roar·er \ˈrōr-ər, ˈrȯr-\ *noun* (14th century)
1 : one that roars
2 : a horse subject to roaring

[1]roar·ing \ˈrōr-iŋ, ˈrȯr-\ *adjective* (14th century)
1 : making or characterized by a sound resembling a roar **:** LOUD ⟨*roaring* applause⟩
2 : marked by prosperity especially of a temporary nature **:** BOOMING
3 : INTENSE, UTTER ⟨in the *roaring* heat⟩ ⟨a *roaring* success⟩
— roar·ing·ly *adverb*

[2]roaring *adverb* (1697)
: EXTREMELY ⟨was *roaring* hungry —Herman Wouk⟩

[3]roaring *noun* (1823)
: noisy inhalation in a horse caused by nerve paralysis and muscular atrophy and constituting an unsoundness

roaring boy *noun* (circa 1590)
: a noisy street bully especially of Elizabethan and Jacobean England who intimidated passersby

[1]roast \ˈrōst\ *verb* [Middle English *rosten,* from Old French *rostir,* of Germanic origin; akin to Old High German *rōsten* to roast] (13th century)
transitive verb
1 a : to cook by exposing to dry heat (as in an oven or before a fire) or by surrounding with hot embers, sand, or stones ⟨*roast* a potato in ashes⟩ **b :** to dry and parch by exposure to heat ⟨*roast* coffee⟩
2 : to heat (inorganic material) with access of air and without fusing to effect change (as expulsion of volatile matter, oxidation, or removal of sulfur from sulfide ores)
3 : to heat to excess ⟨*roasted* by the summer sun⟩
4 : to subject to severe criticism or ridicule ⟨films have been *roasted* by most critics —H. J. Seldes⟩
5 : to honor (a person) at a roast
intransitive verb
1 : to cook food by heat
2 : to undergo being roasted

[2]roast *noun* (14th century)
1 : a piece of meat suitable for roasting
2 : a gathering at which food is roasted before an open fire or in hot ashes or sand
3 : an act or process of roasting; *specifically* **:** severe banter or criticism
4 : a banquet honoring a person (as a celebrity) who is subjected to humorous tongue-in-cheek ridicule by friends

[3]roast *adjective* (14th century)
: that has been roasted ⟨*roast* beef⟩

roast·er \ˈrōs-tər\ *noun* (15th century)
1 : one that roasts
2 : a device for roasting
3 : something adapted to roasting: as **a :** a suckling pig **b :** a bird fit for roasting; *especially* **:** a young chicken of more than four pounds (1.8 kilograms) dressed weight

roast·ing ear \ˈrōs-tiŋ-ˌir, *sense 2 usually* ˈrō-sᵊn-ˌir *or* ˈrōs-ˌnir\ *noun* (1650)
1 : an ear of young corn roasted or suitable for roasting usually in the husk
2 *chiefly Southern & Midland* **:** an ear of corn suitable for boiling or steaming

rob \ˈräb\ *verb* **robbed; rob·bing** [Middle English *robben,* from Old French *rober,* of Germanic origin; akin to Old High German *roubōn* to rob — more at REAVE] (13th century)
transitive verb
1 a (1) **:** to take something away from by force **:** steal from (2) **:** to take personal property from by violence or threat **b** (1) **:** to remove valuables without right from (a place) (2) **:** to take the contents of (a receptacle) **c :** to take away as loot **:** STEAL ⟨*rob* jewelry⟩
2 a : to deprive of something due, expected, or desired **b :** to withhold unjustly or injuriously
intransitive verb
: to commit robbery ☐
— rob·ber *noun*

ro·ba·lo \rō-ˈbä-(ˌ)lō\ *noun, plural* **-los** *or* **-lo** [Spanish] (circa 1890)
: SNOOK 1

ro·band \ˈrō-ˌband, -bənd\ *noun* [probably from Middle Dutch *rabant*] (1762)
: a piece of spun yarn or marline used to fasten the head of a sail to a spar

robber baron *noun* (1878)
: an American capitalist of the latter part of the 19th century who became wealthy through exploitation (as of natural resources, governmental influence, or low wage scales)

robber fly *noun* (1871)
: any of a family (Asilidae) of predaceous flies including some closely resembling bumblebees

rob·bery \ˈrä-b(ə-)rē\ *noun, plural* **-ber·ies** (13th century)
: the act or practice of robbing; *specifically* **:** larceny from the person or presence of another by violence or threat

[1]robe \ˈrōb\ *noun* [Middle English, from Old French, robe, booty, of Germanic origin; akin to Old High German *roubōn* to rob] (13th century)
1 a : a long flowing outer garment; *especially* **:** one used for ceremonial occasions or as a symbol of office or profession **b :** a loose garment (as a bathrobe) for informal wear especially at home
2 : COVERING, MANTLE ⟨peaks on the axis of the range in their *robes* of snow and light —John Muir (died 1914)⟩
3 : a covering of pelts or fabric for the lower body used while driving or at outdoor events

[2]robe *verb* **robed; rob·ing** (14th century)
transitive verb
: to clothe or cover with or as if with a robe
intransitive verb
1 : to put on a robe
2 : DRESS

robe de cham·bre \ˌrōb-də-ˈshäⁿbrᵊ, -ˈshäm-brə\ *noun, plural* **robes de chambre** \ˌrōb(z)-\ [French] (1731)
: DRESSING GOWN

rob·in \ˈrä-bən\ *noun* [short for *robin redbreast*] (1549)
1 a : a small chiefly European thrush (*Erithacus rubecula*) resembling a warbler and having a brownish olive back and orangish face and breast **b :** any of various Old World songbirds that are related to or resemble the European robin
2 : a large North American thrush (*Turdus migratorius*) with olivaceous to slate gray upperparts, blackish head and tail, black and whitish streaked throat, and dull reddish breast and underparts

Rob·in Good·fel·low \ˈrä-bən-ˈgu̇d-ˌfe-(ˌ)lō\ *noun*
: a mischievous sprite in English folklore

Robin Hood \-ˌhu̇d\ *noun* [*Robin Hood,* legendary English outlaw who gave to the poor what he stole from the rich] (1597)
: a person or group likened to a heroic outlaw; *especially* **:** one that robs the rich and gives to the poor

robin red·breast \-ˈred-ˌbrest\ *noun* [Middle English, from *Robin,* nickname for *Robert*] (15th century)
: ROBIN

Rob·in·son Cru·soe \ˈrä-bə(n)-sən-ˈkrü-(ˌ)sō\ *noun*
: a shipwrecked sailor in Defoe's *Robinson Crusoe* who lives for many years on a desert island

Rob·in·son projection \ˈrä-bən-sən-\ *noun* [Arthur H. *Robinson* (born 1915) American geographer] (1978)
: a compromise map projection showing the poles as lines rather than points and more accurately portraying high latitutude lands and water to land ratio

ro·ble \ˈrō-(ˌ)blā\ *noun* [American Spanish, from Spanish, oak, from Latin *robur*] (1864)
: any of several oaks of California and Mexico

ro·bot \ˈrō-ˌbät, -bət\ *noun* [Czech, from *robota* compulsory labor; akin to Old High German *arabeit* trouble, Latin *orbus* orphaned — more at ORPHAN] (1923)
1 a : a machine that looks like a human being and performs various complex acts (as walking or talking) of a human being; *also* **:** a similar but fictional machine whose lack of capacity for human emotions is often emphasized **b :** an efficient insensitive person who functions automatically
2 : a device that automatically performs complicated often repetitive tasks
3 : a mechanism guided by automatic controls ◆
— ro·bot·ism \ˈrō-ˌbä-ˌti-zəm, -bə-\ *noun*

ro·bot·ic \rō-ˈbä-tik, rə-\ *adjective* (1941)
1 : of or relating to mechanical robots
2 : having the characteristics of a robot ⟨performs with *robotic* consistency⟩
— ro·bot·i·cal·ly \-ti-k(ə-)lē\ *adverb*

☐ USAGE

rob Sense *transitive verb* 1c, in which the direct object is the thing stolen, is sometimes considered to be wrong, or perhaps archaic. The sense has been in use since the 13th century and is found in earlier literature ⟨contrive to *rob* the honey and subvert the hive —John Dryden⟩. It is still in use though not as common as other senses ⟨then *robbed* $100 after the clerk fled —*Springfield (Massachusetts) Morning Union*⟩.

◇ WORD HISTORY

robot In 1923 the play *R.U.R.*, by Czech author Karel Čapek, opened in London and New York. The title in the Czech version as well as its English translation stands for "Rossum's Universal Robots." In Čapek's drama this is the name of a company formed by Rossum, an English scientist, to manufacture "robots," humanlike machines that do manual labor. Another scientist decides to make the machines more human, which he does by gradually adding such traits as the capacity to feel pain. Eventually the robots, created to serve humans, come to dominate them completely. Čapek based his coinage *robot* on the Czech word *robota* "compulsory labor." Highly successful in its day, his play caught the imagination of the public. *Robot* was soon used generically for a humanoid machine but then also for a human who has become machinelike and stolid as a result of performing mindless tasks. Appropriately, given the origin of *robot,* the word *robotics* was also first used in a work of fiction, by the American writer Isaac Asimov.

ro·bot·ics \rō-ˈbä-tiks\ *noun plural but singular in construction* (1941)
: technology dealing with the design, construction, and operation of robots in automation
word history see ROBOT

ro·bot·i·za·tion \ˌrō-ˌbä-tə-ˈzā-shən, -bə-\ *noun* (circa 1927)
1 : AUTOMATION
2 : the process of turning a human being into a robot

ro·bot·ize \ˈrō-ˌbä-ˌtīz, -bə-\ *transitive verb* **-ized; -iz·ing** (1927)
1 : to make automatic **:** equip with robots
2 : to turn (a human being) into a robot

Rob Roy \ˈräb-ˈrȯi\ *noun* [*Rob Roy,* nickname of Robert McGregor (died 1734) Scottish freebooter] (1919)
: a manhattan made with Scotch whisky

ro·bust \rō-ˈbəst, ˈrō-(ˌ)bəst\ *adjective* [Latin *robustus* oaken, strong, from *robor-, robur* oak, strength] (1549)
1 a : having or exhibiting strength or vigorous health **b :** having or showing vigor, strength, or firmness ⟨a *robust* debate⟩ ⟨a *robust* faith⟩ **c :** strongly formed or constructed **:** STURDY ⟨a *robust* plastic⟩
2 : ROUGH, RUDE ⟨stories . . . laden with *robust,* down-home imagery —*Playboy*⟩
3 : requiring strength or vigor ⟨*robust* work⟩
4 : FULL-BODIED ⟨*robust* coffee⟩; *also* **:** HEARTY ⟨a *robust* dinner⟩
5 : relating to, resembling, or being any of the primitive, relatively large, heavyset hominids (genus *Australopithecus* and especially *A. robustus* and *A. boisei*) characterized especially by heavy molars and small incisors adapted to a vegetarian diet — compare GRACILE 3
synonym see HEALTHY
— ro·bust·ly *adverb*
— ro·bust·ness \-ˈbəs(t)-nəs, -(ˌ)bəs(t)-\ *noun*

ro·bus·ta \rō-ˈbəs-tə\ *noun, often attributive* [New Latin *robusta,* specific epithet of *Coffea robusta,* synonym of *Coffea canephora*] (1909)
1 : a coffee (*Coffea canephora*) that is indigenous to central Africa but has been introduced elsewhere (as in Java)
2 a : the seed of robusta coffee **b :** coffee brewed from the seed of robusta coffee

ro·bus·tious \rō-ˈbəs-chəs\ *adjective* (circa 1548)
1 : ROBUST
2 : vigorous in a rough or unrefined way **:** BOISTEROUS
— ro·bus·tious·ly *adverb*
— ro·bus·tious·ness *noun*

roc \ˈräk\ *noun* [Arabic *rukhkh*]
: a legendary bird of great size and strength believed to inhabit the Indian Ocean area

ro·caille \rō-ˈkī, rä-\ *noun* [French, literally, stone debris, from Middle French *roquailles,* plural, rocky terrain, from *roc* rock, alteration of *roche,* from (assumed) Vulgar Latin *rocca*] (1856)
1 : a style of ornament developed in the 18th century and characterized by sinuous foliate forms
2 : ROCOCO

Roche limit \ˈrōsh-, ˈrōsh-\ *noun* [E. A. *Roche* (died 1883) French mathematician] (1889)
: the distance between a planet's center and its satellite within which the satellite cannot approach without becoming disrupted

Ro·chelle salt \rō-ˈshel-\ *noun* [La *Rochelle,* France] (1753)
: a crystalline salt $KNaC_4H_4O_6{\cdot}4H_2O$ that is a mild purgative

roche mou·ton·née \ˈrōsh-ˌmü-tᵊn-ˈā, ˈrōsh-\ *noun, plural* **roches mou·ton·nées** *same or* -ˈāz\ [French, literally, fleecy rock] (1843)
: an elongate rounded ice-sculptured hillock of bedrock

roch·et \ˈrä-chət\ *noun* [Middle English, from Middle French, from Old French, from (assumed) Old French *roc* coat, of Germanic origin; akin to Old High German *roc* coat] (13th century)
: a white linen vestment resembling a surplice with close-fitting sleeves worn especially by bishops and privileged prelates

¹rock \ˈräk\ *verb* [Middle English *rokken,* from Old English *roccian;* akin to Old High German *rucken* to cause to move] (12th century)
transitive verb
1 a : to move back and forth in or as if in a cradle **b :** to wash (placer gravel) in a cradle
2 a : to cause to sway back and forth ⟨a boat *rocked* by the waves⟩ **b** (1) **:** to cause to shake violently (2) **:** to daze with or as if with a vigorous blow ⟨a hard right *rocked* the contender⟩ (3) **:** to astonish or disturb greatly ⟨the scandal *rocked* the community⟩
intransitive verb
1 : to become moved backward and forward under often violent impact; *also* **:** to move gently back and forth
2 : to move forward at a steady pace; *also* **:** to move forward at a high speed ⟨the train *rocked* through the countryside⟩
3 : to sing, dance to, or play rock music
synonym see SHAKE
— rock the boat : to do something that disturbs the equilibrium of a situation

²rock *noun, often attributive* (1823)
1 : a rocking movement
2 : popular music usually played on electronically amplified instruments and characterized by a persistent heavily accented beat, much repetition of simple phrases, and often country, folk, and blues elements

³rock *noun* [Middle English *roc,* from Middle Dutch *rocke;* akin to Old High German *rocko* distaff] (14th century)
1 : DISTAFF
2 : the wool or flax on a distaff

⁴rock *noun* [Middle English *rokke,* from Old North French *roque,* from (assumed) Vulgar Latin *rocca*] (14th century)
1 : a large mass of stone forming a cliff, promontory, or peak
2 : a concreted mass of stony material; *also* **:** broken pieces of such masses
3 : consolidated or unconsolidated solid mineral matter; *also* **:** a particular mass of it
4 a : something like a rock in firmness: (1) **:** FOUNDATION, SUPPORT (2) **:** REFUGE ⟨a *rock* of independent thought . . . in an ocean of parochialism —Thomas Molnar⟩ **b :** something that threatens or causes disaster — often used in plural
5 a : a flavored stick candy with color running through **b :** ROCK CANDY 1
6 *slang* **a :** GEM **b :** DIAMOND
7 a : a small crystallized mass of crack cocaine **b :** CRACK 9
— rock *adjective*
— rock·like \ˈräk-ˌlīk\ *adjective*
— between a rock and a hard place *also* **between the rock and the hard place :** in a difficult or uncomfortable position with no attractive way out
— on the rocks 1 : in or into a state of destruction or wreckage ⟨their marriage went *on the rocks*⟩ **2 :** on ice cubes ⟨bourbon *on the rocks*⟩

rock·a·bil·ly \ˈrä-kə-ˌbi-lē\ *noun* [²*rock* + *-a-* (as in *rock-a-bye,* phrase used to put a child to sleep) + hill*billy*] (1956)
: pop music marked by features of rock and country music

rock and roll *noun* (1954)
: ²ROCK 2

rock and roller *noun* (1956)
: ROCKER 3

rock·a·way \ˈrä-kə-ˌwā\ *noun* [perhaps from *Rockaway,* New Jersey] (1846)
: a light low four-wheel carriage with a fixed top and open sides

rock bass *noun* (1811)
: a brown spotted sunfish (*Ambloplites rupestris*) found especially in the upper Mississippi valley and Great Lakes region

rock–bottom *adjective* (1884)
: being the very lowest ⟨*rock-bottom* off-season rates⟩; *also* **:** FUNDAMENTAL ⟨the *rock-bottom* question⟩

rock bottom *noun* (1890)
: the lowest or most fundamental part or level

rock·bound \ˈräk-ˌbau̇nd\ *adjective* (1840)
: fringed, surrounded, or covered with rocks **:** ROCKY

rock brake *noun* (circa 1850)
: any of several ferns that grow chiefly on or among rocks

rock candy *noun* (1723)
1 : boiled sugar crystallized in large masses on string
2 : ⁴ROCK 5a

rock climbing *noun* (1892)
: mountain climbing on rocky cliffs
— rock climb *intransitive verb*
— rock climber *noun*

Rock Cornish *noun* (1956)
: a crossbred domestic fowl produced by interbreeding Cornish and white Plymouth Rock fowls and used especially for small roasters

rock crystal *noun* (1666)
: CRYSTAL 1

rock dove *noun* (1655)
: a bluish gray dove (*Columba livia*) of Europe and Asia that is the ancestor of many domesticated pigeons and of the feral pigeons found in cities and towns throughout most of the world — called also *rock pigeon*

rock·er \ˈrä-kər\ *noun* (1760)
1 a : either of two curving pieces of wood or metal on which an object (as a cradle) rocks **b :** any of various objects (as a rocking chair or an infant's toy having a seat placed between side pieces) that rock on rockers **c :** any of various objects in the form of a rocker or with parts resembling a rocker (as a skate with a curved blade) **d** (1) **:** any of the curved stripes at the lower part of a chevron worn by a noncommissioned officer above the rank of sergeant (2) **:** the curved stripe at the upper part of a chevron worn by a chief petty officer
2 : any of various devices that work with a rocking motion
3 : a rock performer, song, or enthusiast
— off one's rocker : in a state of extreme confusion or insanity ⟨went *off her rocker,* and had to be put away —Mervyn Wall⟩

rocker arm *noun* (1860)
: a center-pivoted lever to push an automotive engine valve down

rock·ery \ˈrä-k(ə-)rē\ *noun, plural* **-er·ies** [⁴*rock* + *-ery*] (1845)
chiefly British **:** ROCK GARDEN

¹rock·et \ˈrä-kət, rä-ˈket\ *noun* [Middle French *roquette,* from Old Italian *rochetta,* diminutive of *ruca* arugula, from Latin *eruca*] (1530)
: any of several plants of the mustard family: as **a :** ARUGULA **b :** DAME'S ROCKET

²rock·et \ˈrä-kət\ *noun, often attributive* [Italian *rocchetta,* literally, small distaff, from diminutive of *rocca* distaff, of Germanic origin; akin to Old High German *rocko* distaff] (1611)
1 a : a firework consisting of a case partly filled with a combustible composition fastened to a guiding stick and propelled through the air by the rearward discharge of the gases liberated by combustion **b :** a similar device used as an incendiary weapon or as a propelling unit (as for a lifesaving line)
2 : a jet engine that operates on the same principle as the firework rocket, consists essentially of a combustion chamber and an exhaust nozzle, carries either liquid or solid propellants which provide the fuel and oxygen needed for combustion and thus make the engine independent of the oxygen of the air, and is used especially for the propulsion of a missile (as a bomb or shell) or a vehicle (as an airplane)

3 : a rocket-propelled bomb, missile, or projectile

³rock·et \'rä-kət\ (1837)
transitive verb
: to convey or propel by means of or as if by a rocket
intransitive verb
1 : to rise up swiftly, spectacularly, and with force
2 : to travel rapidly in or as if in a rocket

rock·e·teer \ˌrä-kə-'tir\ *noun* (1832)
1 : one who fires, pilots, or rides in a rocket
2 : a scientist who specializes in rocketry

rocket plane *noun* (1928)
: an airplane propelled by rockets

rock·et·ry \'rä-kə-trē\ *noun* (1930)
: the study of, experimentation with, or use of rockets

rocket ship *noun* (1927)
: a rocket-propelled spaceship

rocket sled *noun* (1954)
: a rocket-propelled vehicle that runs usually on a single rail and that is used especially in aeronautical experimentation

rock·fall \'räk-ˌfȯl\ *noun* (1924)
: a mass of falling or fallen rocks

rock·fish \-ˌfish\ *noun* (1598)
: any of various important market fishes that live among rocks or on rocky bottoms: as **a :** any of numerous scorpaenid fishes (especially genus *Sebastes*) **b :** STRIPED BASS **c :** any of several groupers

rock garden *noun* (1836)
: a garden laid out among rocks or decorated with rocks and adapted for the growth of particular kinds of plants (as alpines)

rock hind *noun* (circa 1867)
: a red-spotted tan to olive brown grouper (*Epinephelus adscensionis*) of the Atlantic especially from Massachusetts to southeastern Brazil

rock·hop·per \'räk-ˌhä-pər\ *noun* (1875)
: a small penguin (*Eudyptes crestatus*) with a short thick bill and a yellow crest

rock hound *noun* (1915)
1 : a specialist in geology
2 : an amateur rock and mineral collector
— rock·hound·ing \'räk-ˌhau̇n-diŋ\ *noun*

rock·i·ness \'rä-kē-nəs\ *noun* (1611)
: the quality or state of being rocky

rocking chair *noun* (1766)
: a chair mounted on rockers

rocking horse *noun* (1724)
: a toy horse mounted on rockers — called also *hobbyhorse*

rock·ling \'rä-kliŋ\ *noun* (1602)
: any of several small rather elongate marine bony fishes (especially genera *Enchelyopus* and *Gaidropsarus*) of the cod family

rock lobster *noun* (1884)
1 : SPINY LOBSTER
2 : the flesh of a spiny lobster especially when canned or frozen for use as food

rock maple *noun* (1775)
: SUGAR MAPLE 1

rock 'n' roll, rock 'n' roller *variant of* ROCK AND ROLL, ROCK AND ROLLER

rock oil *noun* (1668)
: PETROLEUM

rock pigeon *noun* (1611)
: ROCK DOVE

rock rabbit *noun* (1840)
1 : HYRAX
2 : PIKA

rock–ribbed \'räk-'ribd\ *adjective* (1776)
1 : ¹ROCKY 1
2 : firm and inflexible in doctrine or integrity ⟨a *rock-ribbed* conservative community —John Hale⟩

rock·rose \'räk-ˌrōz\ *noun* (1731)
: any of various shrubs or woody herbs (family Cistaceae, the rockrose family) with simple entire leaves and a capsular fruit

rock salt *noun* (1707)
: common salt occurring in solid form as a mineral; *also* **:** salt artificially prepared in large crystals or masses

rock·shaft \'räk-ˌshaft\ *noun* (circa 1864)
: a shaft that oscillates on its journals instead of revolving

rock tripe *noun* (1854)
: any of various dark leathery umbilicate foliose lichens (as of the genus *Umbilicaria*) that are widely distributed on rocks in boreal and alpine areas and are sometimes used as food

rock wallaby *noun* (1841)
: any of various medium-sized kangaroos (genus *Petrogale*)

rock·weed \'räk-ˌwēd\ *noun* (1626)
: any of various coarse brown algae (family Fucaceae, especially genera *Fucus* and *Ascophyllum*) growing in marine environments attached to rocks

rock wool *noun* (circa 1909)
: mineral wool made by blowing a jet of steam through molten rock (as limestone or siliceous rock) or through slag and used chiefly for heat and sound insulation

¹rocky \'rä-kē\ *adjective* **rock·i·er; -est** [Middle English *rokky,* from *rokke* rock] (15th century)
1 : abounding in or consisting of rocks
2 : difficult to impress or affect **:** INSENSITIVE
3 : firmly held **:** STEADFAST

²rocky *adjective* **rock·i·er; -est** [¹*rock*] (1737)
1 : UNSTABLE, WOBBLY
2 : physically upset or mentally confused (as from drinking excessively)
3 : marked by obstacles **:** DIFFICULT ⟨a financially *rocky* year —Michael Murray⟩

Rocky Mountain sheep *noun* [*Rocky Mountains,* North America] (1817)
: BIGHORN

Rocky Mountain spotted fever *noun* (1905)
: an acute disease that is characterized by chills, fever, prostration, pains in muscles and joints, and a red to purple eruption and that is caused by a rickettsia (*Rickettsia rickettsii*) usually transmitted by an ixodid tick and especially either the American dog tick or a wood tick (*Dermacentor andersoni*)

¹ro·co·co \rə-'kō-(ˌ)kō, rō-kə-'kō\ *noun* (1840)
: rococo work or style

²rococo *adjective* [French, irregular from *rocaille* rocaille] (1841)
1 a : of or relating to an artistic style especially of the 18th century characterized by fanciful curved asymmetrical forms and elaborate ornamentation **b :** of or relating to an 18th century musical style marked by light gay ornamentation and departure from thorough-bass and polyphony
2 : excessively ornate or intricate

rod \'räd\ *noun* [Middle English, from Old English *rodd;* akin to Old Norse *rudda* club] (before 12th century)
1 a (1) **:** a straight slender stick growing on or cut from a tree or bush (2) **:** OSIER (3) **:** a stick or bundle of twigs used to punish; *also* **:** PUNISHMENT (4) **:** a shepherd's cudgel (5) **:** a pole with a line and usually a reel attached for fishing **b** (1) **:** a slender bar (as of wood or metal) (2) **:** a bar or staff for measuring (3) **:** SCEPTER; *also* **:** a wand or staff carried as a badge of office (as of marshal)
2 a : a unit of length — see WEIGHT table **b :** a square rod
3 : any of the long rod-shaped photosensitive receptors in the retina responsive to faint light — compare CONE 3a
4 : a rod-shaped bacterium
5 *slang* **:** HANDGUN
— rod·less \-ləs\ *adjective*
— rod·like \-ˌlīk\ *adjective*

rode *past and chiefly dialect past participle of* RIDE

ro·dent \'rō-dᵊnt\ *noun* [ultimately from Latin *rodent-, rodens,* present participle of *rodere* to gnaw; akin to Latin *radere* to scrape, scratch, Sanskrit *radati* he gnaws] (1859)
1 : any of an order (Rodentia) of relatively small gnawing mammals (as a mouse, a squirrel, or a beaver) that have in both jaws a single pair of incisors with a chisel-shaped edge
2 : a small mammal (as a rabbit or a shrew) other than a true rodent
— rodent *adjective*

ro·den·ti·cide \rō-'den-tə-ˌsīd\ *noun* (circa 1935)
: an agent that kills, repels, or controls rodents

rodent ulcer *noun* [Latin *rodent-, rodens* gnawing] (1853)
: a chronic persisting ulcer of the exposed skin and especially of the face that is destructive locally, spreads slowly, and is usually a carcinoma derived from basal cells — called also *rodent cancer*

¹ro·deo \'rō-dē-ˌō, rə-'dā-(ˌ)ō\ *noun, plural* **ro·de·os** [Spanish, from *rodear* to surround, from *rueda* wheel, from Latin *rota* — more at ROLL] (1834)
1 : ROUNDUP
2 a : a public performance featuring bronco riding, calf roping, steer wrestling, and Brahma bull riding **b :** a contest resembling a rodeo

²rodeo *intransitive verb* (1951)
: to participate in a rodeo

rod·man \'räd-mən, -ˌman\ *noun* (1853)
: a surveyor's assistant who holds the leveling rod

ro·do·mon·tade \ˌrä-də-mən-'tād, ˌrō-, -'täd\ *noun* [Middle French, from Italian *Rodomonte,* character in *Orlando Innamorato* by Matteo M. Boiardo] (1612)
1 : a bragging speech
2 : vain boasting or bluster **:** RANT
— rodomontade *adjective*

¹roe \'rō\ *noun, plural* **roe** *or* **roes** [Middle English *ro,* from Old English *rā;* akin to Old High German *rēh* roe] (before 12th century)
: DOE

²roe *noun* [Middle English *roof, roughe, row;* akin to Old Norse *hrogn* roe and probably to Lithuanian *kurkulai* frog's eggs] (15th century)
1 : the eggs of a fish especially when still enclosed in the ovarian membrane
2 : the eggs or ovaries of an invertebrate (as the coral of a lobster)

roe·buck \'rō-ˌbək\ *noun, plural* **roebuck** *or* **roebucks** (14th century)
: ROE DEER; *especially* **:** the male roe deer

roe deer *noun* (1575)
: either of two small European or Asian deer (*Capreolus capreolus* and *C. pygarus*) that have erect cylindrical antlers forked at the summit, are reddish brown in summer and grayish in winter, have a white rump patch, and are noted for their nimbleness and grace

roe deer

¹roent·gen \'rent-gən, 'rənt-, -jən, -shən\ *adjective* [International Scientific Vocabulary, from Wilhelm *Röntgen*] (1896)
: of or relating to X rays ⟨*roentgen* examinations⟩

²roentgen *noun* (1922)
: the international unit of x-radiation or gamma radiation equal to the amount of radiation that produces in one cubic centimeter of dry air at 0°C and standard atmospheric pressure

ionization of either sign equal to one electrostatic unit of charge

roent·gen·o·gram \-gə-nə-ˌgram, -jə-, -shə-\ *noun* [International Scientific Vocabulary] (circa 1904)
: a photograph made with X rays

roent·gen·og·ra·phy \ˌrent-gə-ˈnä-grə-fē, ˌrənt-, -jə-, -shə-\ *noun* [International Scientific Vocabulary] (1905)
: photography by means of X rays
— roent·gen·o·graph·ic \-nə-ˈgra-fik\ *adjective*
— roent·gen·o·graph·i·cal·ly \-fi-k(ə-)lē\ *adverb*

roent·gen·ol·o·gy \-ˈnä-lə-jē\ *noun* [International Scientific Vocabulary] (1905)
: a branch of radiology that deals with the use of X rays for diagnosis or treatment of disease
— roent·gen·o·log·ic \-nə-ˈlä-jik\ *or* **roent·gen·o·log·i·cal** \-ji-kəl\ *adjective*
— roent·gen·o·log·i·cal·ly \-ji-k(ə-)lē\ *adverb*
— roent·gen·ol·o·gist \-ˈnä-lə-jist\ *noun*

roentgen ray *noun, often 1st R capitalized* (circa 1890)
: X RAY

ro·ga·tion \rō-ˈgā-shən\ *noun* [Middle English *rogacion,* from Late Latin *rogation-, rogatio,* from Latin, questioning, from *rogare* to ask — more at RIGHT] (14th century)
1 : LITANY, SUPPLICATION
2 : the religious observance of the Rogation Days — often used in plural

Rogation Day *noun* (15th century)
: any of the days of prayer especially for the harvest observed on the three days before Ascension Day and by Roman Catholics also on April 25

rog·er \ˈrä-jər\ *interjection* [from *Roger,* former communications code word for the letter *r*] (circa 1941)
— used especially in radio and signaling to indicate that a message has been received and understood

[1]rogue \ˈrōg\ *noun* [origin unknown] (1561)
1 : VAGRANT, TRAMP
2 : a dishonest or worthless person **:** SCOUNDREL
3 : a mischievous person **:** SCAMP
4 : a horse inclined to shirk or misbehave
5 : an individual exhibiting a chance and usually inferior biological variation
— rogu·ish \ˈrō-gish\ *adjective*
— rogu·ish·ly *adverb*
— rogu·ish·ness *noun*

[2]rogue *intransitive verb* **rogued; rogu·ing** *or* **rogue·ing** (1766)
: to weed out inferior, diseased, or nontypical individuals from a crop plant or a field

[3]rogue *adjective* (1872)
1 *of an animal* **:** being vicious and destructive
2 : resembling or suggesting a rogue elephant especially in being isolated and dangerous or uncontrollable ⟨capsized by a *rogue* wave⟩

rogue elephant *noun* (1859)
: a vicious elephant that separates from the herd and roams alone

rogu·ery \ˈrō-g(ə-)rē\ *noun, plural* **-er·ies** (1592)
1 : an act or behavior characteristic of a rogue
2 : mischievous play

rogues' gallery *noun* (1859)
: a collection of pictures of persons arrested as criminals; *also* **:** a collection or list likened to a rogues' gallery ⟨a *rogues' gallery* of infectious diseases⟩

roil \ˈrȯi(ə)l, *verb transitive 2 is also* ˈrī(ə)l\ *verb* [origin unknown] (1590)
transitive verb
1 a : to make turbid by stirring up the sediment or dregs of **b :** to stir up **:** DISTURB, DISORDER
2 : RILE 1
intransitive verb
: to move turbulently **:** be in a state of turbulence or agitation
word history see RILE

roily \ˈrȯi-lē\ *adjective* (1823)
1 : full of sediment or dregs **:** MUDDY
2 : TURBULENT ⟨*roily* waters⟩

[1]rois·ter \ˈrȯi-stər\ *noun* [Middle French *rustre* lout, alteration of *ruste,* from *ruste,* adjective, rude, rough, from Latin *rusticus* rural — more at RUSTIC] (1551)
archaic **:** one that roisters **:** ROISTERER

[2]roister *intransitive verb* **rois·tered; rois·ter·ing** \-st(ə-)riŋ\ (1582)
: to engage in noisy revelry **:** CAROUSE
— rois·ter·er \-stər-ər\ *noun*
— rois·ter·ous \-st(ə-)rəs\ *adjective*
— rois·ter·ous·ly *adverb*

Ro·land \ˈrō-lənd\ *noun* [French]
: a stalwart defender of the Christians against the Saracens in the Charlemagne legends who is killed at Roncesvalles

role *also* **rôle** \ˈrōl\ *noun* [French *rôle,* literally, roll, from Old French *rolle*] (1606)
1 a (1) **:** a character assigned or assumed (2) **:** a socially expected behavior pattern usually determined by an individual's status in a particular society **b :** a part played by an actor or singer
2 : a function or part performed especially in a particular operation or process ⟨played a major *role* in the negotiations⟩
3 : an identifier attached to an index term to show functional relationships between terms

role model *noun* (1957)
: a person whose behavior in a particular role is imitated by others

role–play \ˈrōl-ˌplā, -ˈplā\ (1949)
transitive verb
: ACT OUT ⟨students were asked to *role-play* the thoughts and feelings of each character —R. G. Lambert⟩
intransitive verb
: to play a role

rolf \ˈrȯlf *also* ˈrȯf\ *transitive verb, often capitalized* (1970)
: to practice Rolfing on
— rolf·er \ˈrȯl-fər *also* ˈrȯ-\ *noun, often capitalized*

Rolf·ing \ˈrȯl-fiŋ *also* ˈrȯ-\ *service mark*
— used for a system of muscle massage intended to serve as both physical and emotional therapy

[1]roll \ˈrōl\ *noun* [Middle English *rolle,* from Old French, from Latin *rotulus,* diminutive of *rota* wheel; akin to Old High German *rad* wheel, Welsh *rhod,* Sanskrit *ratha* wagon] (13th century)
1 a (1) **:** a written document that may be rolled up **:** SCROLL; *specifically* **:** a document containing an official or formal record ⟨the *rolls* of parliament⟩ (2) **:** a manuscript book **b :** a list of names or related items **:** CATALOG **c :** an official list: as (1) **:** MUSTER ROLL (2) **:** a list of members of a school or class or of members of a legislative body
2 : something that is rolled up into a cylinder or ball or rounded as if rolled: as **a :** a quantity (as of fabric or paper) rolled up to form a single package **b :** a hairdo in which some or all of the hair is rolled or curled up or under ⟨a pageboy *roll*⟩ **c :** any of various food preparations rolled up for cooking or serving; *especially* **:** a small piece of baked yeast dough **d :** a cylindrical twist of tobacco **e :** a flexible case (as of leather) in which articles may be rolled and fastened by straps or clasps **f** (1) **:** paper money folded or rolled into a wad (2) *slang* **:** BANKROLL
3 : something that performs a rolling action or movement **:** ROLLER: as **a :** a wheel for making decorative lines on book covers; *also* **:** a design impressed by such a tool **b :** a typewriter platen

[2]roll *verb* [Middle English, from Middle French *roller,* from (assumed) Vulgar Latin *rotulare,* from Latin *rotulus*] (14th century)
transitive verb
1 a : to impel forward by causing to turn over and over on a surface **b :** to cause to revolve by turning over and over on or as if on an axis **c :** to cause to move in a circular manner **d :** to form into a mass by turning over and over **e :** to impel forward with an easy continuous motion
2 a : to put a wrapping around **:** ENFOLD, ENVELOP **b :** to wrap round on itself **:** shape into a ball or roll; *also* **:** to produce by such shaping ⟨*rolled* his own cigarettes⟩
3 a : to press, spread, or level with a roller **:** make smooth, even, or compact ⟨hulled and *rolled* oats⟩ **b :** to spread out **:** EXTEND ⟨*roll* out the red carpet⟩
4 a : to move on rollers or wheels **b :** to cause to begin operating or moving ⟨*roll* the cameras⟩
5 a : to sound with a full reverberating tone ⟨*rolled* out the words⟩ **b :** to make a continuous beating sound upon **:** sound a roll upon ⟨*rolled* their drums⟩ **c :** to utter with a trill ⟨*rolled* his *r*'s⟩ **d :** to play (a chord) in arpeggio style
6 : to rob (a drunk, sleeping, or unconscious person) usually by going through the pockets
intransitive verb
1 a : to move along a surface by rotation without sliding **b** (1) **:** to turn over and over ⟨the children *rolled* in the grass⟩ (2) **:** to luxuriate in an abundant supply **:** WALLOW ⟨fairly *rolling* in money⟩
2 a : to move onward or around as if by completing a revolution **:** ELAPSE, PASS ⟨the months *roll* on⟩ **b :** to shift the gaze continually ⟨eyes *rolling* in terror⟩ **c :** to revolve on an axis
3 : to move about **:** ROAM, WANDER
4 a : to go forward in an easy, gentle, or undulating manner ⟨the waves *rolled* in⟩ **b :** to flow in a continuous stream **:** POUR ⟨money was *rolling* in⟩ **c :** to flow as part of a stream of words **d :** to have an undulating contour ⟨*rolling* prairie⟩ **e :** to lie extended **:** STRETCH
5 a : to travel in a vehicle **b :** to become carried on a stream **c :** to move on wheels
6 a : to make a deep reverberating sound ⟨the thunder *rolls*⟩ **b :** TRILL
7 a : to swing from side to side ⟨the ship heaved and *rolled*⟩ **b :** to walk with a swinging gait **:** SWAY **c :** to move so as to lessen the impact of a blow — used with *with* ⟨*rolled* with the punch⟩
8 a : to take the form of a cylinder or ball **b :** to respond to rolling in a specified way
9 a : to get under way **:** begin to move or operate **b :** to move forward **:** develop and maintain impetus; *especially* **:** to proceed or progress with notable ease or success ⟨the team was *rolling*⟩
10 a : BOWL **b :** to execute a somersault
11 *of a football quarterback* **:** to run toward one flank usually parallel to the line of scrimmage especially before throwing a pass — often used with *out*
— roll the bones : to shoot craps

[3]roll *noun* (1688)
1 a : a sound produced by rapid strokes on a drum **b :** a sonorous and often rhythmical flow of speech **c :** a heavy reverberatory sound ⟨the *roll* of cannon⟩ **d :** a chord in arpeggio style **e :** a trill of some birds (as a canary)
2 : a rolling movement or an action or process involving such movement ⟨a *roll* of the dice⟩ ⟨an airplane's takeoff *roll*⟩: as **a :** a swaying movement of the body **b :** a side-to-side movement (as of a ship or train) **c** (1) **:** a flight maneuver in which a complete revolution about the longitudinal axis of an airplane is made with the horizontal direction of flight being approximately maintained (2) **:** the motion of an aircraft or spacecraft about its longitudinal axis **d :** SOMERSAULT **e :** the movement of a curling stone after impact with another stone

— **on a roll :** in the midst of a series of successes **:** on a hot streak — sometimes used with a modifier ⟨has been *on a* brilliant *roll*⟩
roll·back \ˈrōl-ˌbak\ *noun* (1942)
: the act or an instance of rolling back ⟨a government-ordered *rollback* of gasoline prices⟩
roll back *transitive verb* (1942)
1 : to reduce (as a commodity price) to or toward a previous level on a national scale
2 : to cause to retreat or withdraw **:** push back
3 : RESCIND ⟨attempted to *roll back* antipollution standards⟩
roll bar *noun* (circa 1952)
: an overhead metal bar on an automobile that is designed to protect the occupant in case of a rollover
roll cage *noun* (1966)
: a protective framework of metal bars encasing the driver of a vehicle (as a racing car)
roll call *noun* (1775)
1 : the act or an instance of calling off a list of names (as for checking attendance); *also* **:** a time for a roll call
2 : [6]LIST 1
[1]**roll·er** \ˈrō-lər\ *noun* (13th century)
1 a : a revolving cylinder over or on which something is moved or which is used to press, shape, spread, or smooth something **b :** a cylinder or rod on which something (as a shade) is rolled up
2 a : a long heavy ocean wave **b :** a tumbler pigeon
3 : one that rolls or performs a rolling operation
4 : a slowly rolling ground ball
[2]**rol·ler** \ˈrō-lər\ *noun* [German, from *rollen* to roll, reverberate, from Middle French *roller* — more at ROLL] (1678)
1 : any of numerous mostly brightly colored nonpasserine Old World birds (family Coraciidae) that perform rolling aerial dives during courtship displays
2 : a canary having a song in which the notes are soft and run together
roller bearing *noun* (1857)
: a bearing in which the journal rotates in peripheral contact with a number of rollers usually contained in a cage

roller bearing: *1* outer race, *2* cage, *3* roller, *4* inner race

roll·er–coast·er \ˈrō-lər-ˌkō-stər, ˈrō-lə-ˌkō-\ *adjective* (1940)
: marked by numerous ups and downs ⟨an entertainer's *roller-coaster* career⟩
roll·er coast·er \ˈrō-lər-ˌkō-stər, ˈrō-lē-ˌkō-\ *noun* (1888)
1 : an elevated railway (as in an amusement park) constructed with sharp curves and steep inclines on which cars roll
2 : something resembling a roller coaster; *especially* **:** behavior, events, or experiences characterized by sudden and extreme changes ⟨an emotional *roller coaster*⟩
Roller Derby *service mark*
— used for an entertainment involving two roller-skating teams on an oval track in which each team attempts to maneuver a skater into position to score points by circling the track and lapping opponents within a given time period
roller rink *noun* (1885)
: RINK 1c
roller skate *noun* (1863)
: a shoe with a set of wheels attached for skating over a flat surface; *also* **:** a metal frame with wheels attached that can be fitted to the sole of a shoe
— **roller–skate** *intransitive verb*
— **roller skater** *noun*
roller towel *noun* (1845)
: an endless towel hung from a roller

Rolle's theorem \ˈrȯlz-, ˈrōlz-\ *noun* [Michel *Rolle* (died 1719) French mathematician] (circa 1891)
: a theorem in mathematics: if a curve is continuous, crosses the x-axis at two points, and has a tangent at every point between the two intercepts, its tangent is parallel to the x-axis at some point between the intercepts
roll film *noun* (1895)
: a strip of film for still camera use wound on a spool
rol·lick \ˈrä-lik\ *intransitive verb* [origin unknown] (1826)
: to move or behave in a carefree joyous manner **:** FROLIC
— **rollick** *noun*
rol·lick·ing \ˈrä-li-kiŋ\ *adjective* (1811)
: boisterously carefree, joyful, or high-spirited ⟨a *rollicking* adventure film⟩
rolling hitch *noun* (circa 1769)
: a hitch for fastening a line to a spar or to the standing part of another line that will not slip when the pull is parallel to the spar or line — see KNOT illustration
rolling mill *noun* (1787)
: an establishment where metal is rolled into plates and bars
rolling pin *noun* (circa 1589)
: a long cylinder for rolling out dough
rolling stock *noun* (1853)
: the wheeled vehicles owned and used by a railroad or motor carrier
roll–neck \ˈrōl-ˌnek\ *noun, often attributive* (1943)
British **:** TURTLENECK
— **roll–necked** \-ˌnekt\ *adjective*
roll–off \ˈrō-ˌlȯf\ *noun* (1947)
: a play-off match in bowling
roll·out \ˈrō-ˌlau̇t\ *noun* (1952)
1 : the public introduction of a new aircraft; *broadly* **:** the widespread public introduction of a new product
2 : a football play in which the quarterback rolls to the left or right
roll out (1884)
intransitive verb
: to get out of bed
transitive verb
: to introduce (as a new product) especially for widespread sale to the public
roll·over \ˈrō-ˌlō-vər\ *noun* (1945)
1 : the act or process of rolling over
2 : a motor vehicle accident in which the vehicle overturns
roll over *transitive verb* (1949)
1 a : to defer payment of (an obligation) **b :** to renegotiate the terms of (a financial agreement)
2 : to place (invested funds) in a new investment of the same kind **:** REINVEST ⟨*roll over* IRA funds⟩
roll–over arm *noun* (circa 1925)
: a fully upholstered chair or sofa arm curving outward from the seat
roll·top desk \ˈrōl-ˈtäp-\ *noun* (1887)
: a writing desk with a sliding cover often of parallel slats fastened to a flexible backing

rolltop desk

roll up (1859)
transitive verb
: to increase or acquire by successive accumulations **:** ACCUMULATE ⟨*rolled up* a large majority⟩
intransitive verb
1 : to become larger by successive accumulations
2 : to arrive in a vehicle
[1]**ro·ly–po·ly** \ˌrō-lē-ˈpō-lē\ *adjective* [reduplication of *roly,* from [2]*roll*] (1820)
: being short and pudgy **:** ROTUND
[2]**roly–poly** *noun, plural* **-lies** (1836)
1 : a roly-poly person or thing
2 *British* **:** a sweet dough spread with a filling, rolled, and baked or steamed — called also *roly-poly pudding*
Rom \ˈrōm\ *noun, plural* **Rom** *also* **Roma** \ˈrō-mə\ [Romany, married man, husband, male Gypsy, from Sanskrit *ḍomba, ḍoma* low caste male musician] (1841)
: GYPSY 1
ROM \ˈräm\ *noun* (1966)
: READ-ONLY MEMORY
Ro·ma·ic \rō-ˈmā-ik\ *noun* [New Greek *Rhōmaiikos,* from Greek *Rhōmaïkos* Roman, from *Rhōmē* Rome] (1810)
: the modern Greek vernacular
— **Romaic** *adjective*
ro·maine \rō-ˈmān, ˈrō-ˌ\ *noun* [French, from feminine of *romain* Roman, from Latin *Romanus*] (1907)
: a lettuce that belongs to a cultivar of garden lettuce (*Lactuca sativa*) and has long crisp leaves and columnar heads — called also *cos lettuce*
ro·man \rō-ˈmä^{n}\ *noun* [French, from Old French *romans* romance] (1765)
: a metrical romance
[1]**Ro·man** \ˈrō-mən\ *noun* [partly from Middle English, from Old English, from Latin *Romanus,* adjective & noun, from *Roma* Rome; partly from Middle English *Romain,* from Old French, from Latin *Romanus*] (before 12th century)
1 : a native or resident of Rome
2 : ROMAN CATHOLIC — often taken to be offensive
3 *not capitalized* **:** roman letters or type
[2]**Roman** *adjective* (14th century)
1 : of or relating to Rome or the people of Rome; *specifically* **:** characteristic of the ancient Romans ⟨*Roman* fortitude⟩
2 a : LATIN 1a **b :** of or relating to the Latin alphabet
3 *not capitalized* **:** of or relating to a type style with upright characters — compare ITALIC
4 : of or relating to the see of Rome or the Roman Catholic Church
5 : having a semicircular intrados ⟨*Roman* arch⟩
6 : having a prominent slightly aquiline bridge ⟨*Roman* nose⟩
ro·man à clef \rō-ˌmä(ⁿ)n-(ˌ)ä-ˈklā\ *noun, plural* **romans à clef** \-ˌmä^{n}-(ˌ)zä-\ [French, literally, novel with a key] (1893)
: a novel in which real persons or actual events figure under disguise
Roman candle *noun* (1834)
: a cylindrical firework that discharges at intervals balls or stars of fire
[1]**Roman Catholic** *noun* (1605)
: a member of the Roman Catholic Church
[2]**Roman Catholic** *adjective* (1614)
: of, relating to, or being a Christian church having a hierarchy of priests and bishops under the pope, a liturgy centered in the Mass, veneration of the Virgin Mary and saints, clerical celibacy, and a body of dogma including transubstantiation and papal infallibility
Roman Catholicism *noun* (circa 1823)
: the faith, doctrine, or polity of the Roman Catholic Church
[1]**ro·mance** \rō-ˈman(t)s, rə-; ˈrō-ˌ\ *noun* [Middle English *romauns,* from Old French *romans* French, something written in French, from Latin *romanice* in the Roman manner, from *romanicus* Roman, from *Romanus*] (14th century)
1 a (1) **:** a medieval tale based on legend, chivalric love and adventure, or the supernatural (2) **:** a prose narrative treating imaginary characters involved in events remote in time or place and usually heroic, adventurous, or

mysterious (3) **:** a love story **b :** a class of such literature
2 : something (as an extravagant story or account) that lacks basis in fact
3 : an emotional attraction or aura belonging to an especially heroic era, adventure, or activity
4 : LOVE AFFAIR
5 *capitalized* **:** the Romance languages ◆

[2]romance *verb* **ro·manced; ro·manc·ing** (1671)
intransitive verb
1 : to exaggerate or invent detail or incident
2 : to entertain romantic thoughts or ideas
transitive verb
1 : to try to influence or curry favor with especially by lavishing personal attention, gifts, or flattery
2 : to carry on a love affair with

[3]romance *noun* (circa 1854)
: a short instrumental piece in ballad style

Ro·mance \rō-ˈman(t)s, rə-; ˈrō-ˌ\ *adjective* (1690)
: of, relating to, or being any of several languages developed from Latin (as Italian, French, and Spanish)

ro·manc·er \rō-ˈman(t)-sər, rə-; ˈrō-ˌ\ *noun* (1654)
1 : a writer of romance
2 : one that romances

Roman collar *noun* (circa 1890)
: CLERICAL COLLAR

Ro·man·esque \ˌrō-mə-ˈnesk\ *adjective* (1819)
: of or relating to a style of architecture developed in Italy and western Europe between the Roman and the Gothic styles and characterized in its development after 1000 by the use of the round arch and vault, substitution of piers for columns, decorative use of arcades, and profuse ornament
— Romanesque *noun*

ro·man–fleuve \rō-ˌmä^{n}-ˈflœv, -ˈflə(r)v\ *noun, plural* **ro·mans–fleuves** \-ˌmä^{n}-ˈflœv, ˈflə(r)v(z)\ [French, literally, river novel] (1935)
: a novel in the form of a long usually easygoing chronicle of a social group (as a family or a community)

Roman holiday *noun* (1886)
1 : a time of debauchery or of sadistic enjoyment
2 : a destructive or tumultuous disturbance **:** RIOT

Ro·ma·nian \rů-ˈmā-nē-ən, rō-, -nyən\ *noun* (1868)
1 : a native or inhabitant of Romania
2 : the Romance language of the Romanians

Ro·man·ic \rō-ˈma-nik\ *adjective* (1708)
: ROMANCE
— Romanic *noun*

ro·man·ise *British variant of* ROMANIZE

Ro·man·ism \ˈrō-mə-ˌni-zəm\ *noun* (1674)
: ROMAN CATHOLICISM — often taken to be offensive

Ro·man·ist \-nist\ *noun* (1523)
1 : ROMAN CATHOLIC — often taken to be offensive
2 : a specialist in the language, culture, or law of ancient Rome
— Romanist *or* **Ro·man·is·tic** \ˌrō-mə-ˈnis-tik\ *adjective*

ro·man·ize \ˈrō-mə-ˌnīz\ *transitive verb* **-ized; -iz·ing** (1607)
1 *often capitalized* **:** to make Roman in character
2 : to write or print (as a language) in the Latin alphabet ⟨*romanize* Chinese⟩
3 *capitalized* **a :** to convert to Roman Catholicism **b :** to give a Roman Catholic character to
— ro·man·i·za·tion \ˌrō-mə-nə-ˈzā-shən\ *noun, often capitalized*

roman law *noun, often R capitalized* (1660)
: the legal system of the ancient Romans that includes written and unwritten law, is based on the traditional law and the legislation of the city of Rome, and in form comprises legislation of the assemblies, resolves of the senate, enactments of the emperors, edicts of the praetors, writings of the jurisconsults, and the codes of the later emperors

Roman numeral *noun* (1735)
: a numeral in a system of notation that is based on the ancient Roman system — see NUMBER table

Ro·ma·no \rə-ˈmä-(ˌ)nō, rō-\ *noun* [Italian, Roman, from Latin *Romanus*] (1908)
: a hard sharp cheese of Italian origin that is often served grated

Ro·mans \ˈrō-mənz\ *noun plural but singular in construction*
: a letter on doctrine written by Saint Paul to the Christians of Rome and included as a book in the New Testament — see BIBLE table

Ro·mansh *or* **Ro·mansch** \rō-ˈmänch, -ˈmanch\ *noun* [Romansh *romonsch*] (1663)
: the Rhaeto-Romance dialects spoken in the Grisons, Switzerland

[1]ro·man·tic \rō-ˈman-tik, rə-\ *adjective* [French *romantique,* from obsolete *romant* romance, from Old French *romans*] (1650)
1 : consisting of or resembling a romance
2 : having no basis in fact **:** IMAGINARY
3 : impractical in conception or plan **:** VISIONARY
4 a : marked by the imaginative or emotional appeal of what is heroic, adventurous, remote, mysterious, or idealized **b** *often capitalized* **:** of, relating to, or having the characteristics of romanticism **c :** of or relating to music of the 19th century characterized by an emphasis on subjective emotional qualities and freedom of form; *also* **:** of or relating to a composer of this music
5 a : having an inclination for romance **:** responsive to the appeal of what is idealized, heroic, or adventurous **b :** marked by expressions of love or affection **c :** conducive to or suitable for lovemaking
6 : of, relating to, or constituting the part of the hero especially in a light comedy
— ro·man·ti·cal·ly \-ti-k(ə-)lē\ *adverb*

[2]romantic *noun* (1679)
1 : a romantic person, trait, or component
2 *capitalized* **:** a romantic writer, artist, or composer

ro·man·ti·cise *British variant of* ROMANTICIZE

ro·man·ti·cism \rō-ˈman-tə-ˌsi-zəm, rə-\ *noun* (1823)
1 *often capitalized* **a** (1) **:** a literary, artistic, and philosophical movement originating in the 18th century, characterized chiefly by a reaction against neoclassicism and an emphasis on the imagination and emotions, and marked especially in English literature by sensibility and the use of autobiographical material, an exaltation of the primitive and the common man, an appreciation of external nature, an interest in the remote, a predilection for melancholy, and the use in poetry of older verse forms (2) **:** an aspect of romanticism **b :** adherence to a romantic attitude or style
2 : the quality or state of being romantic
— ro·man·ti·cist \-sist\ *noun, often capitalized*

ro·man·ti·cize \-ˈman-tə-ˌsīz\ *verb* **-cized; -ciz·ing** (1818)
transitive verb
: to make romantic **:** treat as idealized or heroic
intransitive verb
1 : to hold romantic ideas
2 : to present details, incidents, or people in a romantic way
— ro·man·ti·ci·za·tion \-ˌman-tə-sə-ˈzā-shən\ *noun*

Ro·ma·ny \ˈrä-mə-nē, ˈrō-\ *noun, plural* **Ro·manies** [Romany *romani,* feminine of *romano,* adjective, Gypsy, from *rom* Gypsy man — more at ROM] (circa 1812)
1 : GYPSY 1
2 : the Indo-Aryan language of the Gypsies
— Romany *adjective*

ro·maunt \rō-ˈmȯnt, -ˈmänt\ *noun* [Middle English, from Middle French *romant*] (1530)
archaic **:** ROMANCE 1a(1)

rom·el·dale \ˈrä-məl-ˌdāl\ *noun, often capitalized* [blend of *Romney* (Marsh), *Rambouillet,* and *Corriedale*] (circa 1948)
: any of an American breed of utility sheep yielding a heavy fleece of fine wool and producing a quickly maturing high-grade market lamb

[1]Ro·meo \ˈrō-mē-ˌō, *in Shakespeare also* ˈrōm-(ˌ)yō\ *noun, plural* **Ro·me·os**
1 : the hero of Shakespeare's *Romeo and Juliet* who dies for love of Juliet
2 : a male lover

[2]Romeo (1952)
— a communications code word for the letter *r*

Rom·ish \ˈrō-mish\ *adjective* (1531)
: ROMAN CATHOLIC — usually used disparagingly
— Rom·ish·ly *adverb*
— Rom·ish·ness *noun*

Rom·ney Marsh \ˈräm-nē-, ˈrəm-\ *noun* [*Romney Marsh,* pasture tract in England] (1832)
: any of a British breed of hardy long-wooled mutton-type sheep especially adapted to damp or marshy regions — called also *Romney*

[1]romp \ˈrämp, ˈrȯmp\ *noun* [partly alteration of [2]*ramp;* partly alteration of *ramp* bold woman] (1706)
1 : one that romps; *especially* **:** a romping girl or woman
2 a : high-spirited, carefree, and boisterous play **b :** something suggestive of such play: as (1) **:** a light fast-paced narrative, dramatic, or musical work usually in a comic mood (2) **:** an episode of lovemaking
3 : an easy winning pace; *also* **:** RUNAWAY

[2]romp *intransitive verb* [alteration of [1]*ramp*] (1709)
1 : to run or play in a lively, carefree, or boisterous manner
2 : to move or proceed in a brisk, easy, or playful manner
3 : to win a contest easily

romp·er \ˈräm-pər, ˈrȯm-\ *noun* (1842)
1 : one that romps
2 : a one-piece garment especially for children with the lower part shaped like bloomers — usually used in plural

◇ WORD HISTORY

romance The Latin word *Romanus* "of Rome, Roman" was used by the inhabitants of Gaul in the declining years of the Roman Empire to refer to their own Latin-derived speech, Gallo-Romance, as opposed to the language of the Franks and other Germanic peoples who had overwhelmed Gaul by the end of the 5th century. In the Middle Ages, the Latin adverb *Romanice,* a derivative of *Romanus,* meant "in the vernacular," alluding to the languages that had developed out of Gallo-Romance (namely Old French in the north of what had been Gaul and Old Occitan in the south), as opposed to Latin, the language of the church and of most formal writing. The Old French outcome of *Romanice* was *romans* or *romanz,* which was used as a noun referring not only to Old French, but to works composed in it, and then specifically to a verse narrative in the vernacular, typically one dealing with chivalric love and adventure. Middle English *romaunce* was borrowed from Old French with all these senses. The close association of the word with tales of love has led to the use of Modern English *romance* for both love stories in fiction and real love affairs.

Rom·u·lus \ˈräm-yə-ləs\ *noun* [Latin]
: a son of Mars and legendary founder of Rome

ron·deau \ˈrän-(ˌ)dō, rän-ˈdō\ *noun, plural* **ron·deaux** \-(ˌ)dōz, -ˈdōz\ [Middle French *rondel, rondeau*] (1525)
1 a : a fixed form of verse based on two rhyme sounds and consisting usually of 13 lines in three stanzas with the opening words of the first line of the first stanza used as an independent refrain after the second and third stanzas **b :** a poem in this form
2 : a monophonic trouvère song with a 2-part refrain

ron·del \ˈrän-dᵊl, rän-ˈdel\ *or* **ron·delle** \rän-ˈdel\ *noun* [Middle English, from Old French, literally, small circle — more at ROUNDEL] (14th century)
1 *usually rondelle* **:** a circular object; *especially* **:** a circular jewel or jeweled ring
2 a *usually rondel* **:** a fixed form of verse based on two rhyme sounds and consisting usually of 14 lines in three stanzas in which the first two lines of the first stanza are repeated as the refrain of the second and third stanzas **b :** a poem in this form **c :** RONDEAU 1

ron·de·let \ˌrän-də-ˈlet, -ˈlā\ *noun* (15th century)
: a modified rondeau consisting usually of seven lines in which the first line of four syllables is repeated as the third line and as the final line or refrain and the remaining lines are made up of eight syllables each

ron·do \ˈrän-(ˌ)dō, rän-ˈdō\ *noun, plural* **ron·dos** [Italian *rondò,* from Middle French *rondeau*] (1797)
1 : an instrumental composition typically with a refrain recurring four times in the tonic and with three couplets in contrasting keys
2 : the musical form of a rondo used especially for a movement in a concerto or sonata

ron·dure \ˈrän-jər, -(ˌ)dyu̇r, -(ˌ)du̇r\ *noun* [French *rondeur* roundness, from Middle French, from *rond* round, from Old French *roont* — more at ROUND] (circa 1600)
1 : ROUND 1a
2 : gracefully rounded curvature

ron·yon \ˈrən-yən, ˈrän-\ *noun* [perhaps modification of Middle French *rogne* scab] (1598)
obsolete **:** a mangy or scabby creature

rood \ˈrüd\ *noun* [Middle English, from Old English *rōd* rod, rood; akin to Old High German *ruota* rod and perhaps to Old Russian *ratište* lance] (before 12th century)
1 : a cross or crucifix symbolizing the cross on which Jesus Christ died; *specifically* **:** a large crucifix on a beam or screen at the entrance of the chancel of a medieval church
2 a : any of various units of land area; *especially* **:** a British unit equal to ¼ acre **b :** any of various units of length; *especially* **:** a British unit equal to seven or eight yards or sometimes a rod

¹roof \ˈrüf, ˈru̇f\ *noun, plural* **roofs** \ˈrüfs, ˈru̇fs *also* ˈrüvz, ˈru̇vz\ [Middle English, from Old English *hrōf;* akin to Old Norse *hrōf* roof of a boathouse and perhaps to Old Church Slavonic *stropŭ* roof] (before 12th century)
1 a (1) **:** the cover of a building (2) **:** material used for a roof **:** ROOFING **b :** the roof of a dwelling conventionally designating the home itself ⟨didn't have a *roof* over my head⟩ ⟨they share the same *roof*⟩
2 a : the highest point **:** SUMMIT **b :** an upper limit **:** CEILING
3 a : the vaulted upper boundary of the mouth **b :** a covering structure of any of various parts of the body ⟨*roof* of the skull⟩
4 : something suggesting a roof: as **a :** a canopy of leaves and branches **b :** the top over the passenger section of a vehicle
— roofed \ˈrüft, ˈru̇ft\ *adjective*
— roof·less \ˈrü-fləs, ˈru̇-\ *adjective*
— roof·like \-ˌlīk\ *adjective*

²roof *transitive verb* (15th century)

roof 1a(1): *1* gambrel, *2* mansard, *3* hip, *4* lean-to

1 a : to cover with or as if with a roof **b :** to provide with a particular kind of roof or roofing — often used in combination ⟨slate-roofed houses⟩
2 : to constitute a roof over
— roof·er *noun*

roof garden *noun* (1893)
: a restaurant or nightclub at the top of a building often in connection with or decorated to suggest an outdoor garden

roof·ing *noun* (15th century)
: material for a roof

roof·line \ˈrüf-ˌlīn, ˈru̇f-\ *noun* (1857)
: the profile of a roof (as of a house)

¹roof·top \-ˌtäp\ *noun* (1611)
: ROOF; *especially* **:** the outer surface of a usually flat roof ⟨sunning themselves on the *rooftop*⟩

²rooftop *adjective* (1935)
: situated or taking place on a rooftop

roof·tree \ˈrüf-ˌtrē, ˈru̇f-\ *noun* (14th century)
: RIDGEPOLE

¹rook \ˈru̇k\ *noun* [Middle English, from Old English *hrōc;* akin to Old High German *hruoh* rook] (before 12th century)
: a common Old World gregarious bird (*Corvus frugilegus*) about the size and color of the related American crow but having a bare patch of skin at the base of the bill

rook

²rook *transitive verb* (circa 1590)
: to defraud by cheating or swindling

³rook *noun* [Middle English *rok,* from Middle French *roc,* from Arabic *rukhkh,* from Persian *rukh*] (14th century)
: either of two pieces of the same color in a set of chessmen having the power to move along the ranks or files across any number of unoccupied squares — called also *castle*

rook·ery \ˈru̇-kə-rē\ *noun, plural* **-er·ies** (1725)
1 a : the nests or breeding place of a colony of rooks; *also* **:** a colony of rooks **b :** a breeding ground or haunt especially of gregarious birds or mammals; *also* **:** a colony of such birds or mammals
2 : a crowded dilapidated tenement or group of dwellings
3 : a place teeming with like individuals

rook·ie \ˈru̇-kē\ *noun* [perhaps alteration of *recruit*] (1892)
1 : RECRUIT; *also* **:** NOVICE
2 : a first-year participant in a major professional sport

rooky \ˈru̇-kē\ *adjective* (1605)
: full of or containing rooks

¹room \ˈrüm, ˈru̇m\ *noun* [Middle English, from Old English *rūm;* akin to Old High German *rūm* room, Latin *rur-, rus* open land] (before 12th century)
1 : an extent of space occupied by or sufficient or available for something ⟨in the country where there is *room* to run and play⟩
2 a *obsolete* **:** an appropriate or designated position, post, or station **b :** PLACE, STEAD ⟨in whose *room* I am now assuming the pen —Sir Walter Scott⟩
3 a : a partitioned part of the inside of a building; *especially* **:** such a part used as a lodging **b :** the people in a room
4 : a suitable or fit occasion or opportunity **:** CHANCE ⟨left no *room* for doubt⟩
— roomed \ˈrümd, ˈru̇md\ *adjective*

²room (1817)
intransitive verb
: to occupy a room especially as a lodger
transitive verb
: to accommodate with lodgings

room and board *noun* (1955)
: lodging and food usually furnished for a set price or as part of wages

room·er \ˈrü-mər, ˈru̇-\ *noun* (circa 1871)
: one who occupies a rented room in another's house

room·ette \rü-ˈmet, ru̇-\ *noun* (1937)
: a small private single room on a railroad sleeping car

room·ful \ˈrüm-ˌfu̇l, ˈru̇m-\ *noun* (1710)
: as much or as many as a room will hold; *also* **:** the persons or objects in a room

rooming house *noun* (1893)
: a house where lodgings are provided for rent

rooming–in \ˈrü-miŋ-ˈin, ˈru̇-\ *noun* (1943)
: an arrangement in a hospital whereby a newborn infant is kept in a crib at the mother's bedside instead of in a nursery

room·mate \ˈrüm-ˌmāt, ˈru̇m-\ *noun* (1789)
: one of two or more persons sharing the same room or living quarters — called also *room·ie* \ˈrü-mē, ˈru̇-\

room service *noun* (1930)
: service provided to hotel guests in their rooms; *also* **:** the hotel department responsible for such service ⟨ordered a meal from *room service*⟩

roomy \ˈrü-mē, ˈru̇-\ *adjective* **room·i·er; -est** (1627)
1 : having ample room **:** SPACIOUS
2 *of a female mammal* **:** having a large or well-proportioned body suited for breeding
— room·i·ness *noun*

roor·back \ˈru̇r-ˌbak\ *noun* [from an attack on James K. Polk in 1844 purporting to quote from an invented book by a Baron von *Roorback*] (1855)
: a defamatory falsehood published for political effect

roose \ˈrüz\ *transitive verb* [Middle English *rusen,* from Old Norse *hrōsa*] (14th century)
chiefly dialect **:** PRAISE

¹roost \ˈrüst\ *noun* [Middle English, from Old English *hrōst;* akin to Old Saxon *hrōst* attic] (before 12th century)
1 a : a support on which birds rest **b :** a place where winged animals and especially birds customarily roost
2 : a group of birds (as fowl) roosting together

²roost (1530)
intransitive verb
1 : to settle down for rest or sleep **:** PERCH
2 : to settle oneself as if on a roost
transitive verb
: to supply a roost for or put to roost

roost·er \ˈrüs-tər *also* ˈru̇s-\ *noun* (1772)
1 a : an adult male domestic fowl **:** COCK **b :** an adult male of various birds other than the domestic fowl
2 : a cocky or vain man

rooster tail *noun* (1946)
: a high arching spray of water thrown up behind a fast-moving motorboat

¹root \ˈrüt, ˈru̇t\ *noun, often attributive* [Middle English, from Old English *rōt,* from Old

\ə\ **abut** \ᵊ\ kitten \ər\ **further** \a\ ash \ā\ **ace** \ä\ mop, mar \au̇\ **out** \ch\ **chin** \e\ bet \ē\ **easy** \g\ go \i\ hit \ī\ ice \j\ job \ŋ\ **sing** \ō\ **go** \ȯ\ law \ȯi\ **boy** \th\ **thin** \t͟h\ **the** \ü\ **loot** \u̇\ **foot** \y\ yet \zh\ vision *see also* Guide to Pronunciation

Norse; akin to Old English *wyrt* root, Latin *radix,* Greek *rhiza*] (12th century)
1 a : the usually underground part of a seed plant body that originates usually from the hypocotyl, functions as an organ of absorption, aeration, and food storage or as a means of anchorage and support, and differs from a stem especially in lacking nodes, buds, and leaves **b :** any subterranean plant part (as a true root or a bulb, tuber, rootstock, or other modified stem) especially when fleshy and edible
2 a : the part of a tooth within the socket; *also* **:** any of the processes into which this part is often divided — see TOOTH illustration **b :** the enlarged basal part of a hair within the skin **c :** the proximal end of a nerve **d :** the part of an organ or physical structure by which it is attached to the body ⟨the *root* of the tongue⟩
3 a : something that is an origin or source (as of a condition or quality) ⟨the love of money is the *root* of all evil —1 Timothy 6:10 (Authorized Version)⟩ **b :** one or more progenitors of a group of descendants — usually used in plural **c :** an underlying support **:** BASIS **d :** the essential core **:** HEART — often used in the phrase *at root* **e :** close relationship with an environment **:** TIE — usually used in plural
4 a : a quantity taken an indicated number of times as an equal factor ⟨2 is a fourth *root* of 16⟩ **b :** a number that reduces an equation to an identity when it is substituted for one variable
5 a : the lower part **:** BASE **b :** the part by which an object is attached to something else
6 : the simple element inferred as the basis from which a word is derived by phonetic change or by extension (as composition or the addition of an affix or inflectional ending)
7 : the tone from whose overtones a chord is composed **:** the lowest tone of a chord in normal position
synonym see ORIGIN
— root·ed \'rü-təd, 'ru̇-\ *adjective*
— root·less \'rüt-ləs, 'ru̇t-\ *adjective*
— root·less·ness *noun*
— root·like \-ˌlīk\ *adjective*

²root (14th century)
transitive verb
1 a : to furnish with or enable to develop roots **b :** to fix or implant by or as if by roots
2 : to remove altogether by or as if by pulling out by the roots — usually used with *out* ⟨*root* out dissenters⟩
intransitive verb
1 : to grow roots or take root
2 : to have an origin or base

³root *verb* [alteration of *wroot,* from Middle English *wroten,* from Old English *wrōtan;* akin to Old High German *ruozzan* to root] (1532)
intransitive verb
1 : to turn up or dig in the earth with the snout **:** GRUB
2 : to poke or dig about
transitive verb
: to turn over, dig up, or discover and bring to light — usually used with *out* ⟨*root* out the cause of the problem⟩

⁴root \'rüt *also* 'ru̇t\ *intransitive verb* [perhaps alteration of ²*rout*] (1889)
1 : to noisily applaud or encourage a contestant or team **:** CHEER
2 : to wish the success of or lend support to someone or something
— root·er *noun*

root·age \'rü-tij, 'ru̇-\ *noun* (circa 1895)
1 : a developed system of roots
2 : ROOT 3a

root beer *noun* (1843)
: a sweetened carbonated beverage flavored with extracts of roots (as sarsaparilla) and herbs

root canal *noun* (1893)
: the part of the pulp cavity lying in the root of a tooth; *also* **:** a dental operation to save a tooth by removing the contents of its root canal and filling the cavity with a protective substance

root cap *noun* (1875)
: a protective cap of parenchyma cells that covers the terminal meristem in most root tips

root cellar *noun* (1822)
: a pit used for the storage especially of root crops

root crop *noun* (1834)
: a crop (as turnips) grown for its enlarged roots

root·ed·ness \'rü-təd-nəs, 'ru̇-\ *noun* (1642)
: the quality or state of having roots

root hair *noun* (1857)
: a filamentous extension of an epidermal cell near the tip of a rootlet that functions in absorption of water and minerals

root·hold \'rüt-ˌhōld, 'ru̇t-\ *noun* (1864)
1 : the anchorage of a plant to soil through the growing and spreading of roots
2 : a place where plants may obtain a roothold

root knot *noun* (1889)
: a plant disease caused by nematodes that produce characteristic enlargements on the roots and stunt the growth of the plant

root–knot nematode *noun* (1922)
: any of several small plant-parasitic nematodes (genus *Meloidogyne*) that cause root knot

root·let \'rüt-lət, 'ru̇t-\ *noun* (circa 1793)
: a small root

root–mean–square *noun* (1895)
: the square root of the arithmetic mean of the squares of a set of numbers

root pressure *noun* (1875)
: the chiefly osmotic pressure by which water rises into the stems of plants from the roots

root rot *noun* (1883)
: any of various plant diseases characterized by decay of the roots and caused especially by fungi

root·stock \'rüt-ˌstäk, 'ru̇t-\ *noun* (1832)
1 : a rhizomatous underground part of a plant
2 : a stock for grafting consisting of a root or a piece of root; *broadly* **:** STOCK 3b

rooty \'rü-tē, 'ru̇-\ *adjective* (15th century)
: full or consisting of roots ⟨*rooty* soil⟩

¹rope \'rōp\ *noun* [Middle English, from Old English *rāp;* akin to Old High German *reif* hoop] (before 12th century)
1 a : a large stout cord of strands of fibers or wire twisted or braided together **b :** a long slender strip of material used as rope ⟨rawhide *rope*⟩ **c :** a hangman's noose **d :** LARIAT
2 : a row or string consisting of things united by or as if by braiding, twining, or threading
3 *plural* **:** special or basic techniques or procedures ⟨show him the *ropes*⟩
4 : LINE DRIVE
— rope·like \-ˌlīk\ *adjective*
— on the ropes : in a defensive and often helpless position

²rope *verb* **roped; rop·ing** (14th century)
transitive verb
1 a : to bind, fasten, or tie with a rope or cord **b :** to partition, separate, or divide by a rope ⟨*rope* off the street⟩ **c :** LASSO
2 : to draw as if with a rope **:** LURE
intransitive verb
: to take the form of or twist in the manner of rope
— rop·er *noun*

rope·danc·er \'rōp-ˌdan(t)-sər\ *noun* (1648)
: one that dances, walks, or performs acrobatic feats on a rope high in the air
— rope·danc·ing \-siŋ\ *noun*

rop·ery \'rō-p(ə-)rē\ *noun* [probably from the thought that the perpetrator deserved the gallows] (1592)
archaic **:** roguish tricks or banter

rope tow *noun* (1948)
: SKI TOW 1

rope·walk \'rōp-ˌwȯk\ *noun* (1672)
: a long covered walk, building, or room where ropes are manufactured

rope·walk·er \-ˌwȯ-kər\ *noun* (1615)
: an acrobat who walks on a rope high in the air

rope·way \-ˌwā\ *noun* (1889)
1 : an endless aerial cable moved by a stationary engine and used to transport freight (as logs and ore)
2 : a fixed cable or a pair of fixed cables between supporting towers serving as a track for suspended passenger or freight carriers

ropy *also* **rop·ey** \'rō-pē\ *adjective* **rop·i·er; -est** (15th century)
1 a : capable of being drawn into a thread **:** VISCOUS; *also* **:** tending to adhere in stringy masses **b :** having a gelatinous or slimy quality from bacterial or fungal contamination ⟨*ropy* milk⟩ ⟨*ropy* flour⟩
2 a : resembling rope **b :** MUSCULAR, SINEWY
3 *usually ropey, slang* **:** extremely unsatisfactory **:** LOUSY
— rop·i·ness *noun*

roque \'rōk\ *noun* [alteration of *croquet*] (1899)
: croquet played on a hard-surfaced court with a raised border

Roque·fort \'rōk-fərt\ *trademark*
— used for a pungent French blue cheese made from sheep's milk

ro·que·laure \ˌrō-kə-'lōr, ˌrä-, -'lȯr\ *noun* [French, from the Duc de *Roquelaure* (died 1738) French marshal] (1716)
: a knee-length cloak worn especially in the 18th and 19th centuries

ror·qual \'rȯr-kwəl, -ˌkwȯl\ *noun* [French, from Norwegian *røyrhval,* from Old Norse *reytharhvalr,* from *reythr* rorqual + *hvalr* whale] (1827)
: any of a family (Balaenopteridae) of large baleen whales (as a blue whale or humpback whale) having the skin of the throat marked with deep longitudinal furrows

Ror·schach \'rȯr-ˌshäk\ *adjective* (1927)
: of, relating to, used in connection with, or resulting from the Rorschach test

Rorschach test *noun* [Hermann *Rorschach* (died 1922) Swiss psychiatrist] (1927)
: a personality and intelligence test in which a subject interprets inkblot designs in terms that reveal intellectual and emotional factors — called also *Rorschach, Rorschach inkblot test*

ro·sa·ceous \rō-'zā-shəs\ *adjective* [ultimately from Latin *rosa*] (1731)
1 : of or relating to the rose family
2 : of, relating to, or resembling a rose especially in having a 5-petaled regular corolla

ro·sar·i·an \rō-'zar-ē-ən, -'zer-\ *noun* (1864)
: a cultivator of roses

ro·sa·ry \'rō-zə-rē, 'rōz-rē\ *noun, plural* **-ries** [Medieval Latin *rosarium,* from Latin, rose garden, from neuter of *rosarius* of roses, from *rosa* rose] (1547)
1 *often capitalized* **:** a Roman Catholic devotion consisting of meditation on usually five sacred mysteries during recitation of five decades of Hail Marys of which each begins with an Our Father and ends with a Gloria Patri
2 : a string of beads used in counting prayers especially of the Roman Catholic rosary

rosary pea *noun* (circa 1866)
1 : a tropical leguminous twining herb (*Abrus precatorius*) that bears jequirity beans and has a root used as a substitute for licorice — called also *Indian licorice, jequirity bean*
2 : JEQUIRITY BEAN 1

ros·coe \'räs-(ˌ)kō\ *noun* [probably from the name *Roscoe*] (circa 1914)
slang **:** HANDGUN

¹rose *past of* RISE

²rose \'rōz\ *noun* [Middle English, from Old English, from Latin *rosa;* akin to Greek *rhodon* rose, Persian *gul*] (before 12th century)
1 a : any of a genus (*Rosa* of the family Rosaceae, the rose family) of usually prickly shrubs with pinnate leaves and showy flowers

having five petals in the wild state but being often double or partly double under cultivation **b :** the flower of a rose
2 : something resembling a rose in form: as **a** (1) **:** COMPASS CARD (2) **:** a circular card with radiating lines used in other instruments **b :** a rosette especially on a shoe **c :** ROSE CUT
3 *plural* **:** a comfortable situation or an easily accomplished task ⟨it was not all sunshine and *roses* —Anthony Lewis⟩
4 : a moderate purplish red
5 : a plane curve which consists of three or more loops meeting at the origin and whose equation in polar coordinates is of the form $\rho = a \sin n\theta$ or $\rho = a \cos n\theta$ where n is an integer greater than zero
— **rose·like** \-ˌlīk\ *adjective*
— **under the rose :** SUB ROSA

[3]rose *adjective* (14th century)
1 a : containing or used for roses **b :** of or relating to a rose **c :** flavored, scented, or colored with or like roses
2 : of the color rose

ro·sé \rō-ˈzā\ *noun* [French] (1897)
: a light pink table wine made from red grapes by removing the skins after fermentation has begun

ro·se·ate \ˈrō-zē-ət, -zē-ˌāt\ *adjective* [Latin *roseus* rosy, from *rosa*] (1589)
1 : resembling a rose especially in color
2 : overly optimistic **:** viewed favorably
— **ro·se·ate·ly** *adverb*

roseate spoonbill *noun* (circa 1785)
: a spoonbill (*Ajaia ajaja*) that is found from the southern U.S. to Patagonia and has chiefly pink plumage

rose·bay \ˈrōz-ˌbā\ *noun* (1760)
1 : RHODODENDRON; *especially* **:** GREAT LAUREL
2 : FIREWEED b

rosebay rhododendron *noun* (circa 1949)
: GREAT LAUREL

rose–breast·ed grosbeak \ˈrōz-ˌbres-təd-\ *noun* (1810)
: a grosbeak (*Pheucticus ludovicianus*) chiefly of eastern North America that in the male is chiefly black and white with a rose-red breast and in the female is grayish brown with a streaked breast

rose·bud \ˈrōz-ˌbəd\ *noun* (15th century)
: the bud of a rose

rose·bush \-ˌbu̇sh\ *noun* (1587)
: a shrub that produces roses

rose chafer *noun* (1704)
: a common North American beetle (*Macrodactylus subspinosus*) that feeds on plant roots as a larva and on leaves and flowers (as of rose or grapevines) as an adult — called also *rose bug*

rose–col·ored \ˈrōz-ˌkə-lərd\ *adjective* (1526)
1 : having a rose color
2 : seeing or seen in a promising light **:** OPTIMISTIC

rose–colored glasses *noun plural* (1950)
: favorably disposed opinions **:** optimistic eyes ⟨views the world through *rose-colored glasses*⟩

rose cut *noun* (circa 1842)
: a form in which gems (as diamonds) are cut that usually has a flat circular base and facets in two ranges rising to a point
— **rose–cut** *adjective*

rose fever *noun* (1851)
: hay fever occurring in the spring or early summer — called also *rose cold*

rose·fish \ˈrōz-ˌfish\ *noun* (1731)
: REDFISH

rose geranium *noun* (1832)
: any of several pelargoniums grown for their fragrant 3- to 5-lobed leaves and small pink flowers

rose hip *noun* (1857)
: the ripened accessory fruit of a rose that consists of a fleshy receptacle enclosing numerous achenes

ro·se·ma·ling \ˈrō-zə-ˌmä-liŋ, -sə-\ *noun* [Norwegian, from *rose* rose + *maling* painting] (1942)
: painted or sometimes carved decoration (as on furniture, walls, or wooden dinnerware) in Scandinavian peasant style that consists especially of floral designs and inscriptions

rose mallow *noun* (1857)
: any of several hibiscuses with large rose-colored flowers; *especially* **:** a showy plant (*Hibiscus moscheutos*) of the salt marshes of the eastern U.S.

rose·mary \ˈrōz-ˌmer-ē\ *noun, plural* **-mar·ies** [Middle English *rosmarine,* from Latin *rosmarinus,* from *ror-, ros* dew + *marinus* of the sea; akin to Sanskrit *rasa* sap, juice — more at MARINE] (14th century)
: a fragrant shrubby mint (*Rosmarinus officinalis*) of southern Europe and Asia Minor; *also* **:** its leaves used as a seasoning

rose of Jer·i·cho \-ˈjer-i-ˌkō\ [Middle English, from *Jericho,* ancient city in Palestine] (15th century)
: an Asian plant (*Anastatica hierochuntica*) of the mustard family that rolls up when dry and expands when moistened

rose of Shar·on \-ˈshar-ən, -ˈsher-\ [Plain of *Sharon,* Palestine] (circa 1847)
: a commonly cultivated Asian shrub or small tree (*Hibiscus syriacus*) having showy bell-shaped rose, purple, or white flowers

rose oil *noun* (1552)
: a fragrant essential oil obtained from roses and used chiefly in perfumery and in flavoring

ro·se·o·la \ˌrō-zē-ˈō-lə, rō-ˈzē-ə-lə\ *noun* [New Latin, from Latin *roseus* rosy, from *rosa* rose] (circa 1818)
: a rose-colored eruption in spots or a disease marked by such an eruption; *especially* **:** ROSEOLA INFANTUM
— **ro·se·o·lar** \-lər\ *adjective*

roseola in·fan·tum \-in-ˈfan-təm\ *noun* [Latin, infant roseola] (circa 1935)
: a mild disease of infants and children characterized by fever lasting usually three days followed by an eruption of rose-colored spots

rose pink *noun* (circa 1859)
: a moderate pink

rose slug *noun* (1877)
: the slimy green larva of either of two sawflies (*Claudius isomerus* and *Endelomyia aethiops*) that feed on the parenchyma of and skeletonize the leaves of roses

ro·set \ˈrō-zət\ *noun* [alteration of Middle English *rosin*] (14th century)
chiefly Scottish **:** RESIN

Ro·set·ta stone \rō-ˈze-tə-\ *noun* [*Rosetta,* Egypt] (circa 1859)
1 : a black basalt stone found in 1799 that bears an inscription in hieroglyphics, demotic characters, and Greek and is celebrated for having given the first clue to the decipherment of Egyptian hieroglyphics
2 : one that gives a clue to understanding

ro·sette \rō-ˈzet\ *noun* [French, literally, small rose, from Old French, from *rose,* from Latin *rosa*] (1790)
1 : an ornament usually made of material gathered or pleated so as to resemble a rose and worn as a badge of office, as evidence of having won a decoration (as the Medal of Honor), or as trimming
2 : a disk of foliage or a floral design usually in relief used as a decorative motif

rosette 2

3 : a structure or color marking on an animal suggestive of a rosette; *especially* **:** one of the groups of spots on a leopard
4 : a cluster of leaves in crowded circles or spirals arising basally from a crown (as in the dandelion) or apically from an axis with greatly shortened internodes (as in many tropical palms)
5 : a food decoration or garnish in the shape of a rose ⟨icing *rosettes*⟩ ⟨carrot *rosettes*⟩

rose·wa·ter \ˈrōz-ˌwȯ-tər, -ˌwä-\ *adjective* (1840)
1 : affectedly nice or delicate
2 : having the odor of rose water

rose water *noun* (14th century)
: a watery solution of the odoriferous constituents of the rose used as a perfume or a flavoring

rose window *noun* (1773)
: a circular window filled with tracery

rose window

rose·wood \ˈrōz-ˌwu̇d\ *noun* (1660)
: any of various tropical trees (especially genus *Dalbergia*) yielding valuable cabinet woods of a dark red or purplish color streaked and variegated with black; *also* **:** the wood

Rosh Ha·sha·nah \ˌräsh-(h)ə-ˈshä-nə, ˌrōsh-, -ˈshō-\ *noun* [Late Hebrew *rōsh hashshānāh,* literally, beginning of the year] (1846)
: the Jewish New Year observed on the first and by Orthodox and Conservative Jews also on the second of Tishri

Ro·si·cru·cian \ˌrō-zə-ˈkrü-shən, ˌrä-\ *noun* [Christian *Rosenkreutz* (New Latin *Rosae Crucis*) reputed 15th century founder of the movement] (1624)
1 : an adherent of a 17th and 18th century movement devoted to esoteric wisdom with emphasis on psychic and spiritual enlightenment
2 : a member of one of several organizations held to be descended from the Rosicrucians
— **Rosicrucian** *adjective*
— **Ro·si·cru·cian·ism** \-shə-ˌni-zəm\ *noun*

ros·i·ly \ˈrō-zə lē\ *adverb* (1809)
1 : in an optimistic manner
2 : with a rosy color or tinge

[1]ros·in \ˈrä-zᵊn, ˈrȯ-, *dialect* ˈrȯ-zəm\ *noun* [Middle English, modification of Middle French *resine* resin] (13th century)
: a translucent amber-colored to almost black brittle friable resin that is obtained by chemical means from the oleoresin or deadwood of pine trees or from tall oil and used especially in making varnish, paper size, soap, and soldering flux and in rosining violin bows

[2]rosin *transitive verb* **ros·ined; ros·in·ing** \ˈräz-niŋ, ˈrȯz-; ˈrä-zᵊn-iŋ, ˈrȯ-\ (15th century)
: to rub or treat (as the bow of a violin) with rosin

ros·in·weed \ˈrä-zᵊn-ˌwēd, ˈrȯ-\ *noun* (1831)
: any of several American plants (as the compass plant) having resinous foliage or a resinous odor

ros·tel·lum \rä-ˈste-ləm\ *noun* [New Latin, from Latin, diminutive of *rostrum* beak] (circa 1826)
: a small process resembling a beak **:** a diminutive rostrum: as **a :** an extension of the stigma of an orchid flower **b :** an anterior prolongation of the head of a tapeworm bearing hooks
— **ros·tel·lar** \rä-ˈste-lər\ *adjective*

ros·ter \ˈräs-tər *also* ˈrȯs- *or* ˈrōs-\ *noun* [Dutch *rooster,* literally, gridiron; from the parallel lines] (1727)

1 a : a roll or list of personnel **b :** such a list giving the order in which a duty is to be performed ⟨duty *roster*⟩ **c :** the persons listed on a roster
2 : an itemized list

ros·tral \ˈräs-trəl *also* ˈrȯs-\ *adjective* [New Latin *rostralis,* from Latin *rostrum*] (1709)
1 : of or relating to a rostrum
2 : situated toward the oral or nasal region: as **a** *of a part of the spinal cord* **:** SUPERIOR 6a **b** *of a part of the brain* **:** ANTERIOR, VENTRAL
— ros·tral·ly \-trə-lē\ *adverb*

ros·trate \ˈräs-ˌtrāt, -trət *also* ˈrȯs-\ *adjective* (circa 1819)
: having a rostrum

ros·trum \ˈräs-trəm *also* ˈrȯs-\ *noun, plural* **ros·tra** \-trə\ *or* **rostrums** [Latin, beak, ship's beak, from *rodere* to gnaw — more at RODENT] (1542)
1 [Latin *Rostra,* plural, a platform for speakers in the Roman Forum decorated with the beaks of captured ships, from plural of *rostrum*] **a :** an ancient Roman platform for public orators **b :** a stage for public speaking **c :** a raised platform on a stage
2 : the curved end of a ship's prow; *especially* **:** the beak of a war galley
3 : a bodily part or process suggesting a bird's bill: as **a :** the beak, snout, or proboscis of any of various insects or arachnids **b :** the often spinelike anterior median prolongation of the carapace of a crustacean (as a crayfish or lobster) ◆

rosy \ˈrō-zē\ *adjective* **ros·i·er; -est** (14th century)
1 a : of the color rose **b :** having a pinkish usually healthy-looking complexion **:** BLOOMING **c :** marked by blushes
2 : characterized by or tending to promote optimism
— ros·i·ness *noun*

¹rot \ˈrät\ *verb* **rot·ted; rot·ting** [Middle English *roten,* from Old English *rotian;* akin to Old High German *rōzzēn* to rot] (before 12th century)
intransitive verb
1 a : to undergo decomposition from the action of bacteria or fungi **b :** to become unsound or weak (as from use or chemical action)
2 a : to go to ruin **:** DETERIORATE **b :** to become morally corrupt **:** DEGENERATE
transitive verb
: to cause to decompose or deteriorate with or as if with rot
synonym see DECAY

²rot *noun* (14th century)
1 a : the process of rotting **:** the state of being rotten **:** DECAY **b :** something rotten or rotting
2 a *archaic* **:** a wasting putrescent disease **b :** any of several parasitic diseases especially of sheep marked by necrosis and wasting **c :** plant disease marked by breakdown of tissues and caused especially by fungi or bacteria
3 : NONSENSE — often used interjectionally

ro·ta \ˈrō-tə\ *noun* [Latin, wheel — more at ROLL] (1673)
1 *chiefly British* **a :** a fixed order of rotation (as of persons or duties) **b :** a roll or list of persons **:** ROSTER
2 *capitalized* [Medieval Latin, from Latin] **:** a tribunal of the papal curia exercising jurisdiction especially in matrimonial cases appealed from diocesan courts

ro·ta·me·ter \ˈrō-tə-ˌmē-tər, rō-ˈta-mə-tər\ *noun* [Latin *rota* + English *-meter*] (1907)
: a gauge that consists of a graduated glass tube containing a free float for measuring the flow of a fluid

Ro·tar·i·an \rō-ˈter-ē-ən\ *noun* [*Rotary (club)*] (1911)
: a member of a major national and international service club

¹ro·ta·ry \ˈrō-tə-rē\ *adjective* [Medieval Latin *rotarius,* from Latin *rota* wheel] (circa 1731)
1 a : turning on an axis like a wheel **b :** taking place about an axis ⟨*rotary* motion⟩
2 : having an important part that turns on an axis ⟨*rotary* cutter⟩
3 : characterized by rotation
4 : of, relating to, or being a press in which paper is printed by rotation in contact with a curved printing surface attached to a cylinder

²rotary *noun, plural* **-ries** (circa 1888)
1 : a rotary machine
2 : a road junction formed around a central circle about which traffic moves in one direction only — called also *circle, traffic circle*

rotary cultivator *noun* (1926)
: ROTOTILLER

rotary engine *noun* (1837)
1 : any of various engines (as a turbine) in which power is applied to vanes or similar parts constrained to move in a circular path
2 : a radial engine in which the cylinders revolve about a stationary crankshaft

rotary–wing aircraft *noun* (1935)
: ROTORCRAFT; *specifically* **:** HELICOPTER — usually used in plural

¹ro·tate \ˈrō-ˌtāt\ *adjective* [Latin *rota*] (1785)
: having the parts flat and spreading or radiating like the spokes of a wheel ⟨*rotate* blue flowers⟩

²ro·tate \ˈrō-ˌtāt, *especially British* rō-ˈ\ *verb* **ro·tat·ed; ro·tat·ing** [Latin *rotatus,* past participle of *rotare,* from *rota* wheel — more at ROLL] (1808)
intransitive verb
1 : to turn about an axis or a center **:** REVOLVE; *especially* **:** to move in such a way that all particles follow circles with a common angular velocity about a common axis
2 a : to perform an act, function, or operation in turn **b :** to pass or alternate in a series
transitive verb
1 a : to cause to turn or move about an axis or a center **b :** to cause (a plane region or line) to sweep out a volume or surface by moving around an axis so that each of its points remains at a constant distance from the axis ⟨generate a torus by *rotating* a circle about an external line⟩
2 : to cause to grow in rotation ⟨*rotate* crops⟩
3 : to cause to pass or act in a series **:** ALTERNATE
4 : to exchange (individuals or units) with others
— ro·tat·able \ˈrō-ˌtā-tə-bəl *also* rō-ˈ\ *adjective*

ro·ta·tion \rō-ˈtā-shən\ *noun* (1555)
1 a (1) **:** the action or process of rotating on or as if on an axis or center (2) **:** the act or an instance of rotating something **b :** one complete turn **:** the angular displacement required to return a rotating body or figure to its original orientation
2 a : return or succession in a series ⟨*rotation* of the seasons⟩ **b :** the growing of different crops in succession in one field usually in a regular sequence
3 : the turning of a body part about its long axis as if on a pivot
4 : a game of pool in which all 15 object balls are shot in numerical order
5 : the series of pitchers on a baseball team who regularly start successive games in turn
— ro·ta·tion·al \-shnəl, -shə-nᵊl\ *adjective*

ro·ta·tive \ˈrō-ˌtā-tiv *also* rō-ˈ\ *adjective* (1778)
1 : turning like a wheel **:** ROTARY
2 : relating to, occurring in, or characterized by rotation
— ro·ta·tive·ly *adverb*

ro·ta·tor \ˈrō-ˌtā-tər *also* rō-ˈ\ *noun* (1676)
: one that rotates or causes rotation; *especially, plural* **-tors** *or* **-tor·es** \ˌrō-tə-ˈtōr-ˌēz, -ˈtȯr-\ **:** a muscle that partially rotates a part on its axis

rotator cuff *noun* (1961)
: a supporting and strengthening structure of the shoulder joint that is made up of the capsule of the shoulder joint blended with tendons and muscles as they pass to the capsule or across it to insert on the head of the humerus

ro·ta·to·ry \ˈrō-tə-ˌtōr-ē, -ˌtȯr-, *British* -t(ə-)ri *also* rō-ˈtā-tə-ri\ *adjective* (1755)
1 : of, relating to, or producing rotation
2 : occurring in rotation

ro·ta·vi·rus \ˈrō-tə-ˌvī-rəs\ *noun* [New Latin, from Latin *rota* wheel + New Latin *virus*] (1974)
: a reovirus that has a double-layered capsid and that causes diarrhea especially in infants

¹rote \ˈrōt\ *noun* [Middle English, from Middle French, of Germanic origin; akin to Old High German *hruozza* crowd] (14th century)
: ³CROWD 1

²rote *noun* [Middle English] (14th century)
1 : the use of memory usually with little intelligence ⟨learn by *rote*⟩
2 : routine or repetition carried out mechanically or unthinkingly ⟨a joyless sense of order, *rote,* and commercial hustle —L. L. King⟩

³rote *adjective* (1641)
1 : learned or memorized by rote
2 : MECHANICAL 3a

⁴rote *noun* [perhaps of Scandinavian origin; akin to Old Norse *rauta* to roar — more at ROUT] (1610)
: the noise of surf on the shore

ro·te·none \ˈrō-tᵊn-ˌōn\ *noun* [International Scientific Vocabulary, from Japanese *roten* derris plant] (1924)
: a crystalline insecticide $C_{23}H_{22}O_6$ that is obtained from the roots of several tropical plants (as a derris) and is of low toxicity for warm-blooded animals and is used especially in home gardens

rot·gut \ˈrät-ˌgət\ *noun* (1633)
: cheap or inferior liquor

ro·ti \ˈrō-tē\ *noun* [Hindi *roṭī* bread; akin to Sanskrit *roṭika,* kind of bread] (1920)
: a round soft flat unleavened bread; *also* **:** such a bread wrapped around a filling and eaten as a sandwich

ro·ti·fer \ˈrō-tə-fər\ *noun* [ultimately from Latin *rota* + *-fer*] (1793)
: any of a class (Rotifera of the phylum Aschelminthes) of minute usually microscopic but many-celled chiefly freshwater aquatic invertebrates having the anterior end modified into a retractile disk bearing circles of strong cilia that often give the appearance of rapidly revolving wheels

ro·tis·ser·ie \rō-ˈtis-rē, -ˈti-sə-\ *noun* [French *rôtisserie,* from Middle French *rostisserie,* from *rostir* to roast — more at ROAST] (circa 1920)
1 : a restaurant specializing in broiled and barbecued meats
2 : an appliance fitted with a spit on which food is rotated before or over a source of heat

ro·to \ˈrō-(ˌ)tō\ *noun, plural* **rotos** (1926)
: ROTOGRAVURE

◇ WORD HISTORY

rostrum In Latin the primary meaning of *rostrum* was "beak," the noun having been derived from the verb *rodere,* meaning "to gnaw." Eventually *rostrum* came to be used metaphorically for a metal-pointed beam projecting from a galley's bow and used for ramming enemy ships. (Using the same metaphor, English calls this beam a *beak.*) In 338 B.C. the beaks of ships captured from the people of Antium (now called Anzio) were used to decorate the orators' platform in the Roman Forum. According to tradition this event gave to the platform the name *Rostra* (the plural of *rostrum*). Later *rostra* was applied to other platforms from which a speaker addressed an assembly. It was in this general sense that the singular form *rostrum* was first used in English in the 18th century.

ro·to·gra·vure \ˌrō-tə-grə-ˈvyu̇r\ *noun* [German *Rotogravur,* blend of Latin *rota* wheel and German *Photogravur* photogravure] (1913)
1 : PHOTOGRAVURE
2 : a section of a newspaper devoted to rotogravure pictures

ro·tor \ˈrō-tər\ *noun* [contraction of *rotator*] (1903)
1 : a part that revolves in a stationary part; *especially* **:** the rotating member of an electrical machine
2 : a complete system of horizontal rotating blades that supplies lift for a rotorcraft

ro·tor·craft \-ˌkraft\ *noun, plural* **rotorcraft** (1940)
: an aircraft (as a helicopter) whose lift is derived principally from rotating airfoils

ro·to·till \ˈrō-tə-ˌtil\ *transitive verb* [back-formation from *rototiller*] (1939)
: to till or plow (soil) with a rototiller

ro·to·till·er \-ˌti-lər\ *noun* [from *Rototiller,* a trademark] (1923)
: a landscaping implement with engine-powered rotating blades used to lift and turn over soil

[1]**rot·ten** \ˈrä-t^{ə}n\ *adjective* [Middle English *roten,* from Old Norse *rotinn;* akin to Old English *rotian* to rot] (13th century)
1 : having rotted **:** PUTRID
2 : morally corrupt
3 : extremely unpleasant or inferior ⟨a *rotten* day⟩ ⟨a *rotten* job⟩
4 : very uncomfortable ⟨feeling *rotten*⟩
5 : of very poor quality **:** LOUSY, ABOMINABLE ⟨a *rotten* show⟩ ⟨what *rotten* luck⟩
— rot·ten·ly *adverb*
— rot·ten·ness \-t^{ə}n-(n)əs\ *noun*

[2]**rotten** *adverb* (1880)
: to an extreme degree ⟨spoiled *rotten*⟩

rotten borough *noun* (1812)
: an election district that has many fewer inhabitants than other election districts with the same voting power

rot·ten·stone \ˈrä-t^{ə}n-ˌstōn\ *noun* (1677)
: a decomposed siliceous limestone used for polishing

rot·ter \ˈrä-tər\ *noun* (1894)
: a thoroughly objectionable person

rott·wei·ler \ˈrät-ˌwī-lər, ˈrȯt-ˌvī-\ *noun, often capitalized* [German, from *Rottweil,* city in Germany] (1907)
: any of a breed of tall powerful black-and-tan short-haired dogs of German origin that are commonly used as guard dogs

rottweiler

ro·tund \rō-ˈtənd, ˈrō-ˌ\ *adjective* [Latin *rotundus,* probably alteration of (assumed) Old Latin *retundus;* akin to Latin *rota* wheel — more at ROLL] (1705)
1 : marked by roundness **:** ROUNDED
2 : marked by fullness of sound or cadence **:** OROTUND, SONOROUS ⟨a master of *rotund* phrase⟩
3 : notably plump **:** CHUBBY
— ro·tun·di·ty \rō-ˈtən-də-tē\ *noun*
— ro·tund·ly \-ˈtənd-lē, ˈrō-ˌ\ *adverb*
— ro·tund·ness \rō-ˈtən(d)-nəs, ˈrō-ˌ\ *noun*

ro·tun·da \rō-ˈtən-də\ *noun* [Italian *rotonda,* from Latin *rotunda,* feminine of *rotundus*] (circa 1700)
1 : a round building; *especially* **:** one covered by a dome
2 a : a large round room **b :** a large central area (as in a hotel)

ro·tu·ri·er \rō-ˈtu̇r-ē-ˌā, -ˈtyu̇r-\ *noun* [Middle French] (1586)
: a person not of noble birth

rou·ble *variant of* RUBLE

roué \ru̇-ˈā\ *noun* [French, literally, broken on the wheel, from past participle of *rouer* to break on the wheel, from Medieval Latin *rotare,* from Latin, to rotate; from the feeling that such a person deserves this punishment] (1800)
: a man devoted to a life of sensual pleasure **:** RAKE ◆

[1]**rouge** \ˈrüzh, *especially Southern* ˈrüj\ *noun* [French, from Middle French, from *rouge* red, from Latin *rubeus* reddish — more at RUBY] (1753)
1 : any of various cosmetics for coloring the cheeks or lips red
2 : a red powder consisting essentially of ferric oxide used in polishing glass, metal, or gems and as a pigment

[2]**rouge** *verb* **rouged; roug·ing** (1777)
transitive verb
1 : to apply rouge to
2 : to cause to redden
intransitive verb
: to use rouge

[1]**rough** \ˈrəf\ *adjective* **rough·er; rough·est** [Middle English, from Old English *rūh;* akin to Old High German *rūh* rough, Lithuanian *raukas* wrinkle] (before 12th century)
1 a : marked by inequalities, ridges, or projections on the surface **:** COARSE **b :** covered with or made up of coarse and often shaggy hair ⟨*rough*-coated collie⟩ — compare SMOOTH, WIREHAIRED **c** (1) **:** having a broken, uneven, or bumpy surface ⟨*rough* terrain⟩ (2) **:** difficult to travel through or penetrate **:** WILD ⟨into the *rough* woods —P. B. Shelley⟩
2 a : TURBULENT, TEMPESTUOUS ⟨*rough* seas⟩ **b** (1) **:** characterized by harshness, violence, or force (2) **:** presenting a challenge **:** DIFFICULT ⟨*rough* to deal with —R. M. McAlmon⟩
3 : coarse or rugged in character or appearance: as **a :** harsh to the ear **b :** crude in style or expression **c :** INDELICATE **d :** marked by a lack of refinement or grace **:** UNCOUTH
4 a : CRUDE, UNFINISHED ⟨*rough* carpentry⟩ **b :** executed or ventured hastily, tentatively, or imperfectly ⟨a *rough* draft⟩ ⟨*rough* estimate⟩; *also* **:** APPROXIMATE ⟨this gives a *rough* idea⟩ ☆
— rough·ish \ˈrə-fish\ *adjective*
— rough·ness \ˈrəf-nəs\ *noun*

[2]**rough** *adverb* (14th century)
: ROUGHLY 1

[3]**rough** *noun* (15th century)
1 : uneven ground covered with high grass, brush, and stones; *specifically* **:** such ground bordering a golf fairway
2 : the rugged or disagreeable side or aspect ⟨hiking-camping admirers of nature in the *rough* —Eleanor Stirling⟩
3 a : something in a crude, unfinished, or preliminary state **b :** broad outline **:** general terms ⟨the question . . . has been discussed in *rough* —*Manchester Guardian Weekly*⟩ **c :** a hasty preliminary drawing or layout
4 : ROWDY

[4]**rough** *transitive verb* (1763)
1 : ROUGHEN
2 a : to subject to abuse **:** MANHANDLE, BEAT — usually used with *up* **b :** to subject to unnecessary and intentional violence in a sport
3 : to calk or otherwise roughen (a horse's shoes) to prevent slipping
4 a : to shape, make, or dress in a rough or preliminary way **b :** to indicate the chief lines of ⟨*rough* out the structure of a building⟩
— rough·er *noun*
— rough it : to live under harsh or primitive conditions

rough·age \ˈrə-fij\ *noun* (circa 1900)
: FIBER 1d; *also* **:** food containing much indigestible material acting as fiber

rough–and–ready \ˌrə-fən-ˈre-dē\ *adjective* (1810)
: crude in nature, method, or manner but effective in action or use

[1]**rough–and–tum·ble** \-ˈtəm-bəl\ *noun* (1792)
: rough disorderly unrestrained fighting or struggling; *also* **:** INFIGHTING

[2]**rough–and–tumble** *adjective* (1832)
1 : marked by rough-and-tumble ⟨grew up in a *rough-and-tumble* atmosphere —E. J. Kahn⟩; *also* **:** ROUGH-AND-READY
2 : put together haphazardly **:** MAKESHIFT ⟨a *rough-and-tumble* fence⟩

rough bluegrass *noun* (circa 1925)
: a forage grass (*Poa trivialis*) of Eurasia and northern Africa that is naturalized in North America

rough breathing *noun* (1899)
1 : a mark ʽ used in Greek over some initial vowels to show that they are aspirated or over ρ to show that it is voiceless
2 : the sound indicated by a mark ʽ over a Greek vowel or ρ

[1]**rough·cast** \ˈrəf-ˌkast, *for 2 also* -ˈkast\ *transitive verb* **-cast; -cast·ing** (1565)
1 : to plaster (as a wall) with roughcast
2 : to shape or form roughly

[2]**rough·cast** \-ˌkast\ *noun* (1579)
1 : a rough model
2 : a plaster of lime mixed with shells or pebbles used for covering buildings
3 : a rough surface finish (as of a plaster wall)

rough cut *noun* (1937)

☆ SYNONYMS
Rough, harsh, uneven, rugged, scabrous mean not smooth or even. ROUGH implies points, bristles, ridges, or projections on the surface ⟨a *rough* wooden board⟩. HARSH implies a surface or texture distinctly unpleasant to the touch ⟨a *harsh* fabric that chafes the skin⟩. UNEVEN implies a lack of uniformity in height, breadth, or quality ⟨an old house with *uneven* floors⟩. RUGGED implies irregularity or roughness of land surface and connotes difficulty of travel ⟨a *rugged* landscape⟩. SCABROUS implies scaliness or prickliness of surface ⟨a *scabrous* leaf⟩. See in addition RUDE.

◇ WORD HISTORY
roué In ancient and medieval times wheels were used as instruments of torture and execution, employed in devices that in one way or another disjointed or mutilated the victim. In 16th century France the use of the wheel in a somewhat more symbolic way was introduced as a means of capital punishment: a criminal would have his limbs and chest crushed by iron bars, after which he would be tied onto a carriage wheel placed upright on a post and exposed until he died. The French verb denoting this means of execution was *rouer,* a derivative of *roue* "wheel," corresponding to Medieval Latin *rotare* "to break on the wheel," a derivative of Latin *rota* "wheel." About the year 1720 Philippe II the Duke of Orléans and Regent of France, who was himself a libertine, allegedly called his debauched companions *roués,* meaning literally "those executed on the wheel." The customary interpretation is that this epithet was a wry acknowledgment, or even boast, that they all deserved to be so killed for their wicked ways. *Roué* became a generic word in French for someone devoted to a life of sensual pleasure and by 1800 had been borrowed into English.

: a print of an incompletely edited motion picture

[1]**rough–dry** \ˈrəf-ˈdrī\ *transitive verb* (1837)
: to dry (laundry) without smoothing or ironing

[2]**rough–dry** *adjective* (1856)
: being dry after laundering but not ironed or smoothed over ⟨*rough-dry* clothes⟩

rough·en \ˈrə-fən\ *verb* **rough·ened; rough·en·ing** \ˈrə-fə-niŋ, ˈrəf-niŋ\ (1582)
transitive verb
: to make rough or rougher ⟨her hands were *roughened* by work —Ellen Glasgow⟩
intransitive verb
: to become rough

rough fish *noun* (1843)
: a fish that is neither a sport fish nor an important food for sport fishes

rough–hew \ˈrəf-ˈhyü\ *transitive verb* **-hewed; -hewn** \-ˈhyün\; **-hew·ing** (1530)
1 : to hew (as timber) coarsely without smoothing or finishing
2 : to form crudely

rough–hewn \-ˈhyün\ *adjective* (1530)
1 : being in a rough, unsmoothed, or unfinished state **:** crudely formed ⟨*rough-hewn* beams⟩
2 : lacking refinement ⟨he was rather attractive, in a *rough-hewn* kind of way —Jan Speas⟩

[1]**rough·house** \ˈrəf-ˌhau̇s\ *noun* (1887)
: violence or rough boisterous play

[2]**rough·house** \-ˌhau̇s, -ˌhau̇z\ *verb* **rough·housed; rough·hous·ing** (1902)
transitive verb
: to treat in a boisterously rough manner
intransitive verb
: to engage in roughhouse

rough·leg \ˈrəf-ˌleg, -ˌlāg\ *noun* (1895)
: ROUGH-LEGGED HAWK

rough–legged hawk *noun* (1811)
: a large circumpolar arctic hawk (*Buteo lagopus*) that winters southward and typically has a white tail with a wide black band or bands at the tip

rough lemon *noun* (1900)
1 : a hybrid lemon that forms a large spreading thorny tree, bears rough-skinned nearly globular acid fruit, and is important chiefly as a rootstock for other citrus trees
2 : the fruit of a rough lemon

rough·ly \ˈrə-flē\ *adverb* (14th century)
1 : in a rough manner: as **a :** with harshness or violence ⟨treated the prisoner *roughly*⟩ **b :** in crude fashion **:** IMPERFECTLY ⟨*roughly* dressed lumber⟩
2 : without completeness or exactness **:** APPROXIMATELY ⟨*roughly* 20 percent⟩

[1]**rough·neck** \ˈrəf-ˌnek\ *noun* (1836)
1 a : a rough or uncouth person **b :** ROWDY, TOUGH
2 : a worker of an oil-well-drilling crew other than the driller

[2]**roughneck** *adjective* (1916)
: having the characteristics of or suitable for a roughneck

rough·ri·der \ˈrəf-ˈrī-dər\ *noun* (1733)
1 : one who is accustomed to riding unbroken or little-trained horses
2 *usually* **Rough Rider :** a member of the 1st U.S. Volunteer Cavalry regiment in the Spanish-American War commanded by Theodore Roosevelt

[1]**rough·shod** \-ˌshäd\ *adjective* (circa 1688)
1 : shod with calked shoes
2 : marked by tyrannical force ⟨*roughshod* rule⟩

[2]**roughshod** *adverb* (1813)
: in a roughshod manner ⟨rode *roughshod* over the opposition⟩

rough trade *noun* (circa 1935)
: male homosexuals who are or affect to be rugged and potentially violent; *also* **:** such a homosexual

rouille \ˈrü-ē, rüy\ *noun* [French, literally, rust; from its color] (1951)
: a peppery garlic sauce

rou·lade \rü-ˈläd\ *noun* [French, literally, act of rolling] (circa 1706)
1 : a florid vocal embellishment sung to one syllable
2 : a slice of usually stuffed meat that is rolled, browned, and steamed or braised

rou·leau \rü-ˈlō\ *noun, plural* **rou·leaux** \-ˈlōz\ [French] (1693)
: a little roll; *especially* **:** a roll of coins put up in paper

[1]**rou·lette** \rü-ˈlet\ *noun* [French, literally, small wheel, from Old French *roelete,* diminutive of *roele* small wheel, from Late Latin *rotella,* diminutive of Latin *rota* wheel — more at ROLL] (1745)
1 : a gambling game in which players bet on which compartment of a revolving wheel a small ball will come to rest in
2 a : any of various toothed wheels or disks (as for producing rows of dots on engraved plates or for making short consecutive incisions in paper to facilitate subsequent division) **b :** tiny slits between rows of stamps in a sheet that are made by a roulette and serve as an aid in separation — compare PERFORATION

[2]**roulette** *transitive verb* **rou·lett·ed; rou·lett·ing** (1867)
: to make roulettes in

Rou·ma·nian \ru̇-ˈmā-nē-ən, -nyən\ *variant of* RUMANIAN

[1]**round** \ˈrau̇nd\ *adjective* [Middle English, from Old French *roont,* from Latin *rotundus* — more at ROTUND] (14th century)
1 a (1) **:** having every part of the surface or circumference equidistant from the center (2) **:** CYLINDRICAL ⟨a *round* peg⟩ **b :** approximately round ⟨a *round* face⟩
2 : well filled out **:** PLUMP, SHAPELY
3 a : COMPLETE, FULL ⟨a *round* dozen⟩ ⟨a *round* ton⟩ **b :** approximately correct; *especially* **:** exact only to a specific decimal or place ⟨use the *round* number 1400 for the exact figure 1411⟩ **c :** substantial in amount **:** AMPLE ⟨a good *round* price —T. B. Costain⟩
4 : direct in utterance **:** OUTSPOKEN ⟨a *round* oath⟩
5 : moving in or forming a circle
6 a : brought to completion or perfection **:** FINISHED **b :** presented with lifelike fullness or vividness
7 : delivered with a swing of the arm ⟨a *round* blow⟩
8 a : having full or unimpeded resonance or tone **:** SONOROUS **b :** pronounced with rounded lips **:** LABIALIZED
9 : of or relating to handwriting predominantly curved rather than angular
— round·ness \ˈrau̇n(d)-nəs\ *noun*

[2]**round** *adverb* (14th century)
: AROUND 1, 2b, 3, 4, 5

[3]**round** *noun* (14th century)
1 a : something (as a circle, globe, or ring) that is round **b** (1) **:** a knot of people (2) **:** a circle of things
2 : ROUND DANCE 1
3 : a musical canon sung in unison in which each part is continuously repeated
4 a : a rung of a ladder or a chair **b :** a rounded molding
5 a : a circling or circuitous path or course **b :** motion in a circle or a curving path
6 a : a route or circuit habitually covered (as by a security guard or police officer) **b :** a series of similar or customary calls or stops ⟨making the *rounds* of his friends —*Current Biography*⟩; *especially* **:** a series of regularly scheduled professional calls on hospital patients made by a doctor or nurse — usually used in plural
7 : a drink of liquor apiece served at one time to each person in a group ⟨I'll buy the next *round*⟩
8 : a sequence of recurring routine or repetitive actions or events ⟨went about my *round* of chores⟩ ⟨the newest *round* of talks⟩
9 : a period of time that recurs in a fixed pattern ⟨the daily *round*⟩
10 a : one shot fired by a weapon or by each man in a military unit **b :** a unit of ammunition consisting of the parts necessary to fire one shot
11 a : a unit of action in a contest or game which comprises a stated period, covers a prescribed distance, includes a specified number of plays, or gives each player one turn **b :** a division of a tournament in which each contestant plays an opponent
12 : a prolonged burst (as of applause)
13 a : a cut of meat (as beef) especially between the rump and the lower leg — see BEEF illustration **b :** a slice of food ⟨a *round* of bread⟩
14 : a rounded or curved part
— in the round 1 : in full sculptured form unattached to a background **2 :** with an inclusive or comprehensive view or representation **3 :** with a center stage surrounded by an audience ⟨theater *in the round*⟩

[4]**round** (14th century)
transitive verb
1 a : to make round **b** (1) **:** to make (the lips) round and protruded (as in the pronunciation of \ü\) (2) **:** to pronounce with lip rounding **:** LABIALIZE
2 a : GO AROUND **b :** to pass part of the way around
3 : ENCIRCLE, ENCOMPASS
4 : to bring to completion or perfection — often used with *off* or *out*
5 : to express as a round number — often used with *off* ⟨11.3572 *rounded* off to two decimal places becomes 11.36⟩
intransitive verb
1 a : to become round, plump, or shapely **b :** to reach fullness or completion
2 : to follow a winding course **:** BEND
— round on : to turn against **:** ASSAIL

[5]**round** *preposition* (1602)
1 : AROUND
2 : all during **:** THROUGHOUT ⟨*round* the year⟩

[6]**round** *transitive verb* [alteration of Middle English *rounen,* from Old English *rūnian;* akin to Old English *rūn* mystery — more at RUNE] (circa 1529)
1 *archaic* **:** WHISPER
2 *archaic* **:** to speak to in a whisper

[1]**round·about** \ˈrau̇n-də-ˌbau̇t\ *adjective* (1608)
: CIRCUITOUS, INDIRECT ⟨had to take a *roundabout* course⟩
— round·about·ness *noun*

[2]**roundabout** *noun* (1755)
1 : a circuitous route **:** DETOUR
2 *British* **:** MERRY-GO-ROUND
3 : a short close-fitting jacket worn by men and boys especially in the 19th century
4 *British* **:** ROTARY 2

round angle *noun* (circa 1934)
: an angle of 360° or 2π radians

round clam *noun* (circa 1843)
: QUAHOG

round dance *noun* (1683)
1 : a folk dance in which participants form a ring and move in a prescribed direction
2 : a ballroom dance in which couples progress around the room
3 : a series of movements performed by a bee to indicate that a source of food is nearby

round·ed \ˈrau̇n-dəd\ *adjective* (1712)
1 : made round **:** flowing rather than jagged or angular
2 : fully developed
— round·ed·ness *noun*

roun·del \ˈrau̇n-dᵊl\ *noun* [Middle English, from Old French *rondel,* from *roont* round — more at ROUND] (14th century)
1 : a round figure or object (as a circular panel, window, or niche)
2 a : RONDEL 2a **b :** an English modified rondeau

roun·de·lay \'rau̇n-də-ˌlā\ *noun* [Middle English, modification of Middle French *rondelet,* diminutive of *rondel*] (15th century)
1 : a simple song with a refrain
2 : a poem with a refrain recurring frequently or at fixed intervals as in a rondel

round·er \'rau̇n-dər\ *noun* (1828)
1 *plural but singular in construction* **:** a game of English origin that is played with ball and bat and that somewhat resembles baseball
2 : a dissolute person **:** WASTREL
3 : one that rounds by hand or by machine
4 : a boxing match lasting a specified number of rounds — usually used in combination ⟨a 10-*rounder*⟩

Round·head \'rau̇nd-ˌhed\ *noun* [from the Puritans' cropping their hair short in contrast to the Cavaliers] (1642)
1 : a member of the parliamentary party in England at the time of Charles I and Oliver Cromwell
2 : PURITAN 1

round·head·ed \-'he-dəd\ *adjective* (1729)
: having a round head; *specifically* **:** BRACHYCEPHALIC
— round·head·ed·ness *noun*

round·house \'rau̇nd-ˌhau̇s\ *noun* (1589)
1 *archaic* **:** LOCKUP
2 : a circular building for housing and repairing locomotives
3 : a blow delivered with a wide swing

round·ish \'rau̇n-dish\ *adjective* (1545)
: somewhat round

round lot *noun* (circa 1902)
: the standard unit of trading in a security market usually amounting to 100 shares of stock

round·ly \'rau̇n(d)-lē\ *adverb* (15th century)
1 a : in a complete or thorough manner **:** THOROUGHLY ⟨*roundly* disliked⟩ ⟨*roundly* satisfying⟩ **b :** by nearly everyone **:** WIDELY ⟨*roundly* praised⟩
2 : in a plainspoken manner **:** BLUNTLY ⟨told them *roundly* they would get no help⟩
3 : with vigor or asperity ⟨*roundly* attacked the plan⟩

round–rob·in \'rau̇nd-ˌrä-bən\ *noun* [from the name *Robin*] (1730)
1 a : a written petition, memorial, or protest to which the signatures are affixed in a circle so as not to indicate who signed first **b :** a statement signed by several persons **c :** something (as a letter) sent in turn to the members of a group each of whom signs and forwards it sometimes after adding comment
2 : ROUND TABLE 2
3 : a tournament in which every contestant meets every other contestant in turn
4 : SERIES, ROUND

round–shoul·dered \'rau̇n(d)-ˌshōl-dərd\ *adjective* (1586)
: having the shoulders stooping or rounded

rounds·man \'rau̇n(d)z-mən\ *noun* (1795)
1 : one that makes rounds
2 : a supervisory police officer of the grade of sergeant or just below

round steak *noun* (1876)
: a steak cut from the round of beef — see BEEF illustration

round ta·ble \'rau̇n(d)-ˌtā-bəl\ *noun* (14th century)
1 a *R&T capitalized* **:** the large circular table of King Arthur and his knights **b :** the knights of King Arthur
2 *usually* **round·ta·ble :** a conference for discussion or deliberation by several participants; *also* **:** the participants in such a conference

round–the–clock *adjective* (1937)
: AROUND-THE-CLOCK

round–trip \'rau̇n(d)-ˌtrip\ *noun, often attributive* (1860)
: a trip to a place and back usually over the same route

round·up \'rau̇n-ˌdəp\ *noun* (1873)
1 a (1) **:** the act or process of collecting animals (as cattle) by riding around them and driving them in (2) **:** the cowboys and ranch personnel engaged in a cattle roundup **b :** a gathering in of scattered persons or things ⟨a *roundup* of all suspects⟩
2 : a summary of information ⟨a *roundup* of the news⟩

round up *transitive verb* (1844)
1 : to collect (as cattle) by means of a roundup
2 : to gather in or bring together from various quarters

round window *noun* (circa 1903)
: FENESTRA 1b

round·wood \'rau̇nd-ˌwu̇d\ *noun* (1910)
: timber used (as for poles) without being squared by sawing or hewing

round·worm \'rau̇nd-ˌwərm\ *noun* (1565)
: NEMATODE; *also* **:** a related round-bodied unsegmented worm (as a spiny-headed worm) as distinguished from a flatworm

roup \'rüp, 'rau̇p\ *noun* [origin unknown] (circa 1808)
: TRICHOMONIASIS c

[1]rouse \'rau̇z\ *verb* **roused; rous·ing** [Middle English, to shake the feathers] (1531)
transitive verb
1 *archaic* **:** to cause to break from cover
2 a : to stir up **:** EXCITE ⟨was *roused* to fury⟩ **b :** to arouse from or as if from sleep or repose **:** AWAKEN
intransitive verb
1 : to become aroused **:** AWAKEN
2 : to become stirred
— rouse·ment \'rau̇z-mənt\ *noun*
— rous·er *noun*

[2]rouse *noun* (circa 1802)
: an act or instance of rousing; *especially* **:** an excited stir

[3]rouse *noun* [alteration (from misdivision of *to drink carouse*) of *carouse*] (1602)
1 *obsolete* **:** DRINK, TOAST
2 *archaic* **:** CAROUSAL

rouse·about \'rau̇-zə-ˌbau̇t\ *noun* (1861)
Australian **:** an unskilled worker

rous·ing \'rau̇ ziŋ\ *adjective* (1641)
1 a : giving rise to excitement **:** STIRRING **b :** BRISK, LIVELY
2 : EXCEPTIONAL, SUPERLATIVE
— rous·ing·ly *adverb*

Rous sarcoma \'rau̇s-\ *noun* [F. Peyton *Rous* (died 1970) American physician] (circa 1925)
: a readily transplantable malignant fibrosarcoma of chickens that is caused by a specific carcinogenic virus

Rous·seau·ism \rü-'sō-ˌi-zəm\ *noun* (1865)
1 : the philosophical, educational, and political doctrines of Jean Jacques Rousseau
2 : the return to or glorification of a simpler and more primitive way of life
— Rous·seau·ist \-ist\ *noun*
— Rous·seau·is·tic \ˌrü-ˌsō-'is-tik, ru̇-\ *adjective*

roust \'rau̇st\ *transitive verb* [alteration of [1]*rouse*] (1658)
: to drive (as from bed) roughly or unceremoniously

roust·about \'rau̇s-tə-ˌbau̇t\ *noun* (1868)
1 a : DECKHAND **b :** LONGSHOREMAN
2 : an unskilled or semiskilled laborer especially in an oil field or refinery
3 : a circus worker who erects and dismantles tents, cares for the grounds, and handles animals and equipment

roust·er \'rau̇s-tər\ *noun* (1883)
1 : DECKHAND
2 : LONGSHOREMAN

[1]rout \'rau̇t\ *noun* [Middle English *route,* from Middle French, troop, defeat, from (assumed) Vulgar Latin *rupta,* from Latin, feminine of *ruptus,* past participle of *rumpere* to break — more at REAVE] (13th century)
1 : a crowd of people **:** THRONG; *specifically* **:** RABBLE 2b
2 a : DISTURBANCE **b** *archaic* **:** FUSS
3 : a fashionable gathering

[2]rout \'rōt, 'rüt\ *intransitive verb* [Middle English *rowten,* from Old Norse *rauta;* akin to Old English *rēotan* to weep, Latin *rudere* to roar] (14th century)
dialect chiefly British **:** to low loudly **:** BELLOW — used of cattle

[3]rout \'rau̇t\ *verb* [alteration of [3]*root*] (circa 1564)
intransitive verb
1 : to poke around with the snout **:** ROOT ⟨pigs *routing* in the earth⟩
2 : to search haphazardly
transitive verb
1 a *archaic* **:** to dig up with the snout **b :** to gouge out or make a furrow in (as wood or metal)
2 a : to force out as if by digging — usually used with *out* **b :** to cause to emerge especially from bed
3 : to come up with **:** UNCOVER

[4]rout \'rau̇t\ *noun* [Middle French *route* troop, defeat] (1598)
1 : a state of wild confusion or disorderly retreat
2 a : a disastrous defeat **:** DEBACLE **b :** a precipitate flight

[5]rout \'rau̇t\ *transitive verb* (1600)
1 a : to disorganize completely **:** DEMORALIZE **b :** to put to precipitate flight **c :** to defeat decisively or disastrously ⟨the discomfiture of seeing their party *routed* at the polls —A. N. Holcombe⟩
2 : to drive out **:** DISPEL

[1]route \'rüt, 'rau̇t\ *noun* [Middle English, from Old French, from (assumed) Vulgar Latin *rupta (via),* literally, broken way, from Latin *rupta,* feminine of *ruptus,* past participle] (13th century)
1 a : a traveled way **:** HIGHWAY ⟨the main *route* north⟩ **b :** a means of access **:** CHANNEL ⟨the *route* to social mobility —T. F. O'Dea⟩
2 : a line of travel **:** COURSE
3 a : an established or selected course of travel or action **b :** an assigned territory to be systematically covered ⟨a newspaper *route*⟩

[2]route *transitive verb* **rout·ed; rout·ing** (1832)
1 : to send by a selected route **:** DIRECT ⟨was *routed* along the scenic shore road⟩
2 : to divert in a specified direction

route·man \'rüt-mən, 'rau̇t-ˌman\ *noun* (1918)
: one who is responsible for making sales or deliveries on an assigned route

[1]rout·er \'rau̇-tər\ *noun* (1846)
: one that routs: as **a :** a routing plane **b :** a machine with a revolving vertical spindle and cutter for milling out the surface of wood or metal

[2]rout·er \'rü-tər, 'rau̇-\ *noun* (1903)
: one that routes

[3]rout·er \'rü-tər, 'rau̇-\ *noun* [*route* (race of a mile or more)] (circa 1951)
: a horse trained for distance races

route step *noun* (1867)
: a style of marching in which troops maintain prescribed intervals but are not required to keep in step or to maintain silence — called also *route march*

route·way \'rüt-ˌwā, 'rau̇t-\ *noun* (1946)
chiefly British **:** ROUTE 3a

routh \'rau̇th, 'rüth\ *noun* [origin unknown] (circa 1689)
chiefly Scottish **:** PLENTY

[1]rou·tine \rü-'tēn\ *noun* [French, from Middle French, from *route* traveled way] (1676)
1 a : a regular course of procedure ⟨if resort to legal action becomes a campus *routine* —J. A. Perkins⟩ **b :** habitual or mechanical performance of an established procedure ⟨the *routine* of factory work⟩
2 : a reiterated speech or formula ⟨the old "After you" *routine* —Ray Russell⟩

\ə\ **abut** \ᵊ\ **kitten** \ər\ **further** \a\ **ash** \ā\ **ace** \ä\ **mop, mar** \au̇\ **out** \ch\ **chin** \e\ **bet** \ē\ **easy** \g\ **go** \i\ **hit** \ī\ **ice** \j\ **job** \ŋ\ **sing** \ō\ **go** \ȯ\ **law** \ȯi\ **boy** \th\ **thin** \t͟h\ **the** \ü\ **loot** \u̇\ **foot** \y\ **yet** \zh\ **vision** *see also* Guide to Pronunciation

3 : a worked-out part (as of an entertainment or sports contest) that may be often repeated ⟨a dance *routine*⟩ ⟨a gymnastic *routine*⟩; *especially* **:** a theatrical number
4 : a sequence of computer instructions for performing a particular task

[2]rou·tine \rü-ˈtēn, ˈrü-ˌ\ *adjective* (1817)
1 : of a commonplace or repetitious character **:** ORDINARY
2 : of, relating to, or being in accordance with established procedure ⟨*routine* business⟩
— rou·tine·ly *adverb*

rou·tin·ize \rü-ˈtē-ˌnīz, ˈrü-tᵊn-ˌīz\ *transitive verb* **-ized; -iz·ing** (1921)
: to discipline in or reduce to a routine
— rou·tin·i·za·tion \(ˌ)rü-ˌtē-nə-ˈzā-shən, ˌrü-tᵊn-ə-\ *noun*

roux \ˈrü\ *noun, plural* **roux** \ˈrüz\ [French, from *beurre roux* brown butter] (1813)
: a cooked mixture of flour and fat used as a thickening agent in a soup or a sauce

[1]rove \ˈrōv\ *verb* **roved; rov·ing** [Middle English *roven* to shoot at random, wander] (1536)
intransitive verb
: to move aimlessly **:** ROAM
transitive verb
: to wander through or over
synonym see WANDER

[2]rove *noun* (1606)
: an act or instance of wandering

[3]rove *past and past participle of* REEVE

[4]rove *transitive verb* **roved; rov·ing** [origin unknown] (1789)
: to join (textile fibers) with a slight twist and draw out into roving

[5]rove *noun* (1789)
: ROVING

rove beetle *noun* [perhaps from [1]*rove*] (circa 1771)
: any of a family (Staphylinidae) of often predatory active beetles having a long body and very short wing covers beneath which the wings are folded transversely — called also *staphylinid*

[1]ro·ver \ˈrō-vər\ *noun* [Middle English, from Middle Dutch, from *roven* to rob; akin to Old English *rēafian* to reave — more at REAVE] (14th century)
: PIRATE

[2]rov·er \ˈrō-vər\ *noun* [Middle English, from *roven*] (15th century)
1 : a random or long-distance mark in archery — usually used in plural
2 : WANDERER, ROAMER
3 : a player who is not assigned to a specific position on a team and who plays wherever needed
4 : a vehicle for exploring the surface of an extraterrestrial body (as the moon or Mars)

[1]rov·ing \ˈrō-viŋ\ *adjective* [[1]*rove*] (1596)
1 a : capable of being shifted from place to place **:** MOBILE **b :** not restricted as to location or area of concern
2 : inclined to ramble or stray ⟨a *roving* fancy⟩

[2]roving *noun* [[4]*rove*] (1802)
: a slightly twisted roll or strand of usually textile fibers

[1]row \ˈrō\ *verb* [Middle English, from Old English *rōwan;* akin to Middle High German *rüejen* to row, Latin *remus* oar] (before 12th century)
intransitive verb
1 : to propel a boat by means of oars
2 : to move by or as if by the propulsion of oars
transitive verb
1 a : to propel with or as if with oars **b :** to be equipped with (a specified number of oars) **c** (1) **:** to participate in (a rowing match) (2) **:** to compete against in rowing (3) **:** to pull (an oar) in a crew
2 : to transport in an oar-propelled boat
— row·er \ˈrō(-ə)r\ *noun*

[2]row *noun* (1832)
: an act or instance of rowing

[3]row *noun* [Middle English *rawe;* akin to Old English *rǣw* row, Old High German *rīga* line, and perhaps to Sanskrit *rikhati* he scratches] (13th century)
1 : a number of objects arranged in a usually straight line ⟨a *row* of bottles⟩; *also* **:** the line along which such objects are arranged ⟨planted the corn in parallel *rows*⟩
2 a : WAY, STREET **b :** a street or area dominated by a specific kind of enterprise or occupancy ⟨doctors' *row*⟩
3 : TWELVE-TONE ROW
4 a : a continuous strip usually running horizontally or parallel to a base line **b :** a horizontal arrangement of items
— in a row : one after another **:** SUCCESSIVELY

[4]row *transitive verb* (1657)
: to form into rows

[5]row \ˈrau̇\ *noun* [origin unknown] (1746)
: a noisy disturbance or quarrel

[6]row \ˈrau̇\ *intransitive verb* (1797)
: to engage in a row **:** have a quarrel

row·an \ˈrau̇-ən, ˈrō-ən\ *noun* [of Scandinavian origin; akin to Old Norse *reynir* rowan; akin to Old English *rēad* red — more at RED] (1804)
1 a : a Eurasian mountain ash (*Sorbus aucuparia*) with flat corymbs of white flowers followed by small red pomes **b :** an American mountain ash (*Sorbus americana*) with flat corymbs of white flowers followed by small orange red pomes
2 : the fruit of a rowan

rowan 1b: leaves and fruit

row·an·ber·ry \-ˌber-ē\ *noun* (1814)
: ROWAN 2

row·boat \ˈrō-ˌbōt\ *noun* (1538)
: a small boat designed to be rowed

[1]row·dy \ˈrau̇-dē\ *adjective* **row·di·er; -est** [perhaps irregular from [5]*row*] (1819)
: coarse or boisterous in behavior **:** ROUGH; *also* **:** characterized by such behavior ⟨*rowdy* local bars⟩
— row·di·ly \ˈrau̇-dᵊl-ē\ *adverb*
— row·di·ness \ˈrau̇-dē-nəs\ *noun*
— row·dy·ish \-ish\ *adjective*

[2]rowdy *noun, plural* **rowdies** (1819)
: a rowdy person **:** TOUGH

row·dy·ism \ˈrau̇-dē-ˌi-zəm\ *noun* (1842)
: rowdy character or behavior

[1]row·el \ˈrau̇(-ə)l\ *noun* [Middle English *rowelle,* from Middle French *rouelle* small wheel, from Old French *roele* — more at ROULETTE] (15th century)
: a revolving disk with sharp marginal points at the end of a spur

[2]rowel *transitive verb* **-eled** *or* **-elled; -el·ing** *or* **-el·ling** (1599)
1 : to goad with or as if with a rowel
2 : VEX, TROUBLE

row·en \ˈrau̇-ən\ *noun* [Middle English *rowein,* from (assumed) Old North French *rewain;* akin to Old French *regaïn* aftermath, from *re-* + *gaïn* aftermath, of Germanic origin; akin to Old High German *weida* pasture, *weidanōn* to hunt for food — more at GAIN] (15th century)
: AFTERMATH 1

row house \ˈrō-\ *noun* (1936)
: one of a series of houses connected by common sidewalls and forming a continuous group

row·ing \ˈrō-iŋ\ *noun* (before 12th century)
1 : the propelling of a boat by means of oars **:** the action of one that rows
2 : the sport of racing in shells

rowing boat *noun* (1820)
chiefly British **:** ROWBOAT

rowing machine *noun* (1848)
: an exercise machine that simulates the action of rowing

row·lock \ˈrä-lək, ˈrə-; ˈrō-ˌläk\ *noun* [probably by alteration] (circa 1750)
chiefly British **:** OARLOCK

[1]roy·al \ˈrȯi(-ə)l\ *adjective* [Middle English *roial,* from Middle French, from Latin *regalis,* from *reg-, rex* king; akin to Old Irish *rī* (genitive *rīg*) king, Sanskrit *rājan,* Latin *regere* to rule — more at RIGHT] (14th century)
1 a : of kingly ancestry ⟨the *royal* family⟩ **b :** of, relating to, or subject to the crown ⟨the *royal* estates⟩ **c :** being in the crown's service ⟨*Royal* Air Force⟩
2 a : suitable for royalty **:** MAGNIFICENT **b :** requiring no exertion **:** EASY ⟨there is no *royal* road to logic —Justus Buchler⟩
3 a : of superior size, magnitude, or quality ⟨a patronage of *royal* dimensions —J. H. Plumb⟩ — often used as an intensive ⟨a *royal* pain⟩ **b :** established or chartered by the crown
4 : of, relating to, or being a part (as a mast, sail, or yard) next above the topgallant
— roy·al·ly \ˈrȯi-ə-lē\ *adverb*

[2]royal *noun* (14th century)
1 : a person of royal blood
2 : a small sail on the royal mast immediately above the topgallant sail
3 : a stag of 8 years or more having antlers with at least 12 points

royal blue *noun* (1789)
: a vivid purplish blue

royal flush *noun* (circa 1868)
: a straight flush having an ace as the highest card — see POKER illustration

roy·al·ism \ˈrȯi-ə-ˌli-zəm\ *noun* (1793)
: MONARCHISM

roy·al·ist \-list\ *noun* (1643)
1 *often capitalized* **:** an adherent of a king or of monarchical government: as **a :** CAVALIER 3 **b :** TORY 4
2 : a reactionary business tycoon
— royalist *adjective*

royal jelly *noun* (1817)
: a highly nutritious secretion of the pharyngeal glands of the honeybee that is fed to the very young larvae in a colony and to all queen larvae

royal palm *noun* (1861)
: any of a genus (*Roystonea*) of palms chiefly of the Caribbean region; *especially* **:** a tall graceful pinnate-leaved palm (*R. regia*) native to Cuba that is widely planted for ornament

royal poinciana *noun* (circa 1900)
: a showy tropical tree (*Delonix regia* synonym *Poinciana regia*) widely planted for its immense racemes of scarlet and orange flowers — called also *flamboyant, peacock flower*

royal purple *noun* (1661)
: a dark reddish purple

roy·al·ty \ˈrȯi(-ə)l-tē\ *noun, plural* **-ties** [Middle English *roialte,* from Middle French *roialté,* from Old French, from *roial*] (14th century)
1 a : royal status or power **:** SOVEREIGNTY **b :** a right or perquisite of a sovereign (as a percentage paid to the crown of gold or silver taken from mines)
2 : regal character or bearing **:** NOBILITY
3 a : persons of royal lineage **b :** a person of royal rank ⟨how to address *royalties* —George Santayana⟩ **c :** a privileged class
4 : a right of jurisdiction granted to an individual or corporation by a sovereign
5 a : a share of the product or profit reserved by the grantor especially of an oil or mining lease **b :** a payment made to an author or composer for each copy of a work sold or to an inventor for each article sold under a patent

royster *variant of* ROISTER

roz·zer \ˈrä-zər\ *noun* [origin unknown] (1893)
slang British **:** POLICE OFFICER

RPG \ˌär-(ˌ)pē-ˈjē\ *noun* [report program generator] (1966)

: a computer language that generates programs from the user's specifications especially to produce business reports

RPV \ˌär-(ˌ)pē-ˈvē\ *noun* [*r*emotely *p*iloted *v*ehicle] (1970)
: an unmanned aircraft flown by remote control and used especially for reconnaissance

-rrhagia *noun combining form* [New Latin, from Greek, from *rhēgnynai* to break, burst; probably akin to Lithuanian *rėžti* to cut]
: abnormal or excessive discharge or flow ⟨metro*rrhagia*⟩

-rrhea *noun combining form* [Middle English *-ria,* from Late Latin *-rrhoea,* from Greek *-rrhoia,* from *rhoia,* from *rhein* to flow — more at STREAM]
: flow **:** discharge ⟨logo*rrhea*⟩ ⟨leuko*rrhea*⟩

-rrhiza — see -RHIZA

-rrhoea *chiefly British variant of* -RRHEA

rRNA \ˌär-(ˌ)är-(ˌ)en-ˈā\ *noun* (circa 1965)
: RIBOSOMAL RNA

ru•a•na \rü-ˈä-nə\ *noun* [American Spanish, from Spanish, woolen fabric] (1903)
: a woolen covering resembling a poncho

¹rub \ˈrəb\ *verb* **rubbed; rub•bing** [Middle English *rubben;* akin to Icelandic *rubba* to scrape] (14th century)
intransitive verb
1 a : to move along the surface of a body with pressure **:** GRATE **b** (1) **:** to fret or chafe with or as if with friction (2) **:** to cause discontent, irritation, or anger
2 : to continue in a situation usually with slight difficulty ⟨in spite of financial difficulties, he is *rubbing* along⟩
3 : to admit of being rubbed (as for erasure or obliteration)
transitive verb
1 a : to subject to or as if to the action of something moving especially back and forth with pressure and friction **b** (1) **:** to cause (a body) to move with pressure and friction along a surface (2) **:** to treat in any of various ways by rubbing **c :** to bring into reciprocal back-and-forth or rotary contact
2 : ANNOY, IRRITATE
— **rub elbows** *or* **rub shoulders :** to associate closely **:** MINGLE
— **rub one's nose in :** to bring forcefully or repeatedly to one's attention
— **rub the wrong way :** to arouse the antagonism or displeasure of **:** IRRITATE

²rub *noun* (1586)
1 a : an unevenness of surface (as of the ground in lawn bowling) **b :** OBSTRUCTION, DIFFICULTY ⟨the *rub* is that so few of the scholars have any sense of this truth themselves —Benjamin Farrington⟩ **c :** something grating to the feelings (as a gibe or harsh criticism) **d :** something that mars serenity
2 : the application of friction with pressure ⟨an alcohol *rub*⟩

Ru•bai•yat stanza \ˈrü-bē-ˌät-, -ˌbī-, -ˌat-\ *noun* [*The Rubáiyát of Omar Khayyám,* quatrains translated by Edward FitzGerald (1859)] (1940)
: an iambic pentameter quatrain with a rhyme scheme *aaba*

ru•basse \rü-ˈbas, ˈrü-ˌ\ *noun* [French *rubace,* irregular from *rubis* ruby — more at RUBY] (circa 1890)
: a quartz stained a ruby red

ru•ba•to \rü-ˈbä-(ˌ)tō\ *noun, plural* **-tos** [Italian, literally, robbed] (circa 1883)
: a fluctuation of tempo within a musical phrase often against a rhythmically steady accompaniment

¹rub•ber \ˈrə-bər\ *noun* (1536)
1 a : one that rubs **b :** an instrument or object (as a rubber eraser) used in rubbing, polishing, scraping, or cleaning **c :** something that prevents rubbing or chafing
2 [from its use in erasers] **a :** an elastic substance that is obtained by coagulating the milky juice of any of various tropical plants (as of the genera *Hevea* and *Ficus*), is essentially a polymer of isoprene, and is prepared as sheets and then dried — called also *caoutchouc, india rubber* **b :** any of various synthetic rubberlike substances **c :** natural or synthetic rubber modified by chemical treatment to increase its useful properties (as toughness and resistance to wear) and used especially in tires, electrical insulation, and waterproof materials
3 : something made of or resembling rubber: as **a :** a rubber overshoe **b** (1) **:** a rubber tire (2) **:** the set of tires on a vehicle **c :** a rectangular slab of white rubber in the middle of a baseball infield on which a pitcher stands while pitching **d :** CONDOM
— **rubber** *adjective*

²rubber *noun* [origin unknown] (1599)
1 : a contest consisting of an odd number of games won by the side that takes a majority (as two out of three)
2 : an odd game played to determine the winner of a tie

rubber band *noun* (1886)
: a continuous band of rubber used in various ways (as for holding together a sheaf of papers)

rubber bridge *noun* (1936)
: a form of contract bridge in which settlement is made at the end of each rubber

rubber cement *noun* (1886)
: an adhesive consisting typically of a dispersion of vulcanized rubber in an organic solvent

rubber check *noun* [from its coming back like a bouncing rubber ball] (1926)
: a check returned by a bank because of insufficient funds in the payer's account

rub•ber•ized \ˈrə-bə-ˌrīzd\ *adjective* (1908)
: coated or saturated with rubber or a rubber solution

rub•ber•like \ˈrə-bər-ˌlīk\ *adjective* (1922)
: resembling rubber especially in physical properties (as elasticity and toughness)

¹rub•ber•neck \-ˌnek\ *noun* (circa 1896)
1 : an overly inquisitive person
2 : TOURIST; *especially* **:** one on a guided tour

²rubberneck *intransitive verb* (1896)
1 : to look about, stare, or listen with exaggerated curiosity
2 : to go on a tour **:** SIGHTSEE
— **rub•ber•neck•er** \-ˌne-kər\ *noun*

rubber plant *noun* (1888)
: a plant that yields rubber; *especially* **:** a tall tropical Asian tree (*Ficus elastica*) of the mulberry family that is frequently dwarfed as an ornamental

rubber–stamp *transitive verb* (1918)
1 : to approve, endorse, or dispose of as a matter of routine or at the command of another
2 : to mark with a rubber stamp

rubber stamp *noun* (1881)
1 : a stamp of rubber for making imprints
2 a : a person who echoes or imitates others **b :** a body or person that approves or endorses a program or policy with little or no dissent or discussion
3 a : a stereotyped copy or expression ⟨the usual *rubber stamps* of criticism —H. L. Mencken⟩ **b :** a routine endorsement or approval
— **rubber–stamp** *adjective*

rubber tree *noun* (1847)
: a tree that yields rubber; *especially* **:** a South American tree (*Hevea brasiliensis*) of the spurge family that is cultivated in plantations and is a chief source of rubber

rubber tree

rub•bery \ˈrə-b(ə-)rē\ *adjective* (1907)
: resembling rubber (as in elasticity, consistency, or texture) ⟨*rubbery* legs⟩

rub•bing \ˈrə-biŋ\ *noun* (1845)
: an image of a raised, incised, or textured surface obtained by placing paper over it and rubbing the paper with a colored substance

rubbing alcohol *noun* (circa 1931)
: a cooling and soothing liquid for external application that contains approximately 70 percent denatured ethyl alcohol or isopropanol

rub•bish \ˈrə-bish, *dialect* -bij\ *noun* [Middle English *robys*] (15th century)
1 : useless waste or rejected matter **:** TRASH
2 : something that is worthless or nonsensical ⟨few real masterpieces are forgotten and not much *rubbish* survives —William Bridges-Adams⟩
— **rub•bishy** \-bə-shē\ *adjective*

¹rub•ble \ˈrə-bəl\ *noun* [Middle English *robyl*] (14th century)
1 a : broken fragments (as of rock) resulting from the decay or destruction of a building ⟨fortifications knocked into *rubble* —C. S. Forester⟩ **b :** a miscellaneous confused mass or group of usually broken or worthless things ⟨lay in a pile of *rubble,* only this time there was more of it, additional gear having hit the deck —K. M. Dodson⟩
2 : waterworn or rough broken stones or bricks used in coarse masonry or in filling courses of walls
3 : rough stone as it comes from the quarry

²rubble *transitive verb* **rub•bled; rub•bling** \-b(ə-)liŋ\ (1926)
: to reduce to rubble

rub•down \ˈrəb-ˌdau̇n\ *noun* (1896)
: a brisk rubbing of the body

rube \ˈrüb\ *noun* [*Rube,* nickname for *Reuben*] (1896)
: an awkward unsophisticated person **:** RUSTIC

¹ru•be•fa•cient \ˌrü-bə-ˈfā-shənt\ *adjective* [Latin *rubefacient-, rubefaciens,* present participle of *rubefacere* to make red, from *rubeus* reddish + *facere* to make — more at RUBY, DO] (1804)
: causing redness (as of the skin)

²rubefacient *noun* (1805)
: a substance for external application that produces redness of the skin

Rube Gold•berg \ˈrüb-ˈgōl(d)-ˌbərg\ *also* **Rube Gold•berg•i•an** \-ˌbər-gē-ən, -ˌbərg-yən\ *adjective* [Reuben (*Rube*) L. *Goldberg* (died 1970) American cartoonist] (1931)
: accomplishing by complex means what seemingly could be done simply ⟨a kind of *Rube Goldberg* contraption . . . with five hundred moving parts —L. T. Grant⟩; *also* **:** characterized by such complex means

ru•bel•la \rü-ˈbe-lə\ *noun* [New Latin, from Latin, feminine of *rubellus* reddish, from *ruber* red — more at RED] (1883)
: GERMAN MEASLES

ru•bel•lite \rü-ˈbe-ˌlīt, ˈrü-bə-ˌlīt\ *noun* [Latin *rubellus*] (circa 1796)
: a red tourmaline used as a gem

Ru•ben•esque \ˌrü-bə-ˈnesk\ *adjective* (1925)
: of, relating to, or suggestive of the painter Rubens or his works; *especially* **:** plump or rounded usually in a pleasing or attractive way ⟨a *Rubenesque* figure⟩

ru•be•o•la \ˌrü-bē-ˈō-lə, rü-ˈbē-ə-lə\ *noun* [New Latin, from neuter plural of (assumed) New Latin *rubeolus* reddish, from Latin *rubeus*] (1803)
: MEASLES

Ru•bi•con \ˈrü-bi-ˌkän\ *noun* [Latin *Rubicon-, Rubico,* river of northern Italy forming part of the boundary between Cisalpine Gaul and Italy whose crossing by Julius Caesar in 49 B.C. was regarded by the Senate as an act of war] (1626)

: a bounding or limiting line; *especially* **:** one that when crossed commits a person irrevocably

ru·bi·cund \'rü-bi-(,)kənd\ *adjective* [Middle English *rubicunde,* from Latin *rubicundus,* from *rubēre* to be red; akin to Latin *rubeus*] (15th century)
: RUDDY
— ru·bi·cun·di·ty \,rü-bi-'kən-də-tē\ *noun*

ru·bid·i·um \rü-'bi-dē-əm\ *noun* [New Latin, from Latin *rubidus* red, from *rubēre*] (1861)
: a soft silvery metallic element of the alkali metal group that reacts violently with water and bursts into flame spontaneously in air — see ELEMENT table

rub in *transitive verb* (1851)
: to harp on (as something unpleasant) **:** EMPHASIZE

ru·bi·ous \'rü-bē-əs\ *adjective* (1601)
: RED, RUBY

ru·ble \'rü-bəl\ *noun* [Russian *rubl'*] (1554)
— see MONEY table

rub off *intransitive verb* (1950)
: to become transferred ⟨bad habits *rubbed off* on them⟩ ⟨carbon *rubs off* on your hands⟩
— rub–off \'rəb-,ȯf\ *noun*

rub out *transitive verb* (14th century)
1 : to obliterate or extinguish by or as if by rubbing
2 : to destroy completely; *specifically* **:** KILL, MURDER ⟨somebody *rubbed* him *out* . . . with a twenty-two —Raymond Chandler⟩
— rub·out \'rəb-,au̇t\ *noun*

ru·bric \'rü-brik, -,brik\ *noun* [Middle English *rubrike* red ocher, heading in red letters of part of a book, from Middle French *rubrique,* from Latin *rubrica,* from *rubr-, ruber* red] (14th century)
1 a : an authoritative rule; *especially* **:** a rule for conduct of a liturgical service **b** (1) **:** NAME, TITLE; *specifically* **:** the title of a statute (2) **:** something under which a thing is classed **:** CATEGORY ⟨the sensations falling under the general *rubric,* "pressure" —F. A. Geldard⟩ **c :** an explanatory or introductory commentary **:** GLOSS; *specifically* **:** an editorial interpolation
2 : a heading of a part of a book or manuscript done or underlined in a color (as red) different from the rest
3 : an established rule, tradition, or custom ◇
— rubric *or* **ru·bri·cal** \-bri-kəl\ *adjective*
— ru·bri·cal·ly \-bri-k(ə-)lē\ *adverb*

ru·bri·cate \'rü-bri-,kāt\ *transitive verb* **-cat·ed; -cat·ing** (1570)
1 : to write or print as a rubric
2 : to provide with a rubric
— ru·bri·ca·tion \,rü-bri-'kā-shən\ *noun*
— ru·bri·ca·tor \'rü-bri-,kā-tər\ *noun*

rub up *transitive verb* (1572)
1 : to revive or refresh knowledge of **:** RECALL
2 : to improve the keenness of (a mental faculty)

ru·bus \'rü-bəs\ *noun, plural* **rubus** [New Latin, from Latin, blackberry] (circa 1921)
: any of a genus (*Rubus*) of plants (as a blackberry or a raspberry) of the rose family having 3- to 7-foliolate or simple lobed leaves, white or pink flowers, and a mass of carpels ripening into an aggregate fruit composed of many drupelets

¹ru·by \'rü-bē\ *noun, plural* **rubies** [Middle English, from Middle French *rubis, rubi,* alteration of *robin,* from Medieval Latin *rubinus,* from Latin *rubeus* reddish; akin to Latin *ruber* red — more at RED] (14th century)
1 a : a precious stone that is a red corundum **b :** something (as a watch bearing) made of ruby
2 a : the dark red color of the ruby **b :** something resembling a ruby in color

²ruby *adjective* (1508)
: of the color ruby

ruby glass *noun* (1797)
: glass of a deep red color containing selenium, an oxide of copper, or a chloride of gold

ruby spinel *noun* (1839)
: a usually red spinel used as a gem

ru·by·throat \'rü-bē-,thrōt\ *noun* (circa 1783)
: RUBY-THROATED HUMMINGBIRD

ru·by–throat·ed hummingbird \'rü-bē-,thrō-təd-\ *noun* (circa 1782)
: a hummingbird (*Archilochus colubris*) of eastern North America having a bright green back, whitish underparts, and in the adult male a red throat with metallic reflections

ruche \'rüsh\ *or* **ruch·ing** \'rü-shiŋ\ *noun* [French *ruche* literally, beehive, from Medieval Latin *rusca* bark] (1827)
: a pleated, fluted, or gathered strip of fabric used for trimming
— ruched \'rüsht\ *adjective*

¹ruck \'rək\ *noun* [Middle English, of Scandinavian origin; akin to Old Norse *hraukr* rick — more at RICK] (15th century)
1 a : the usual run of persons or things **:** GENERALITY ⟨trying to rise above the *ruck* —Richard Holt⟩ **b :** an indistinguishable gathering **:** JUMBLE
2 : the persons or things following the vanguard ⟨finished the race in the *ruck*⟩

²ruck *verb* [*ruck,* noun, wrinkle] (1812)
: PUCKER, WRINKLE

ruck·sack \'rək-,sak, 'ru̇k-\ *noun* [German, from German dialect, from *Rucken* back + *Sack* sack] (1866)
: KNAPSACK

ruck·us \'rə-kəs *also* 'rü- *or* 'ru̇-\ *noun* [probably blend of *ruction* and *rumpus*] (circa 1890)
: ROW, DISTURBANCE ⟨raise a *ruckus*⟩

ruc·tion \'rək-shən\ *noun* [perhaps by shortening & alteration from *insurrection*] (circa 1825)
1 : a noisy fight
2 : DISTURBANCE, UPROAR

rud·beck·ia \,rəd-'be-kē-ə, rüd-\ *noun* [New Latin, from Olof *Rudbeck* (died 1702) Swedish scientist] (circa 1759)
: any of a genus (*Rudbeckia*) of North American perennial composite herbs having showy flower heads with mostly yellow ray flowers and a conical chaffy receptacle

rudd \'rəd, 'ru̇d\ *noun* [probably from *rud* redness, red ocher, from Middle English *rude,* from Old English *rudu* — more at RUDDY] (1526)
: a freshwater European cyprinid fish (*Scardinius erythrophthalmus*) resembling the roach

rud·der \'rə-dər\ *noun* [Middle English *rother,* from Old English *rōther* paddle; akin to Old English *rōwan* to row] (14th century)
1 : a flat piece or structure of wood or metal attached upright to the stern of a boat or ship so that it can be turned causing the vessel's head to turn in the same direction
2 : a movable auxiliary airfoil on an airplane usually attached at the rear end that serves to control direction of flight in the horizontal plane — see AIRPLANE illustration
3 : a guiding force or strategy
— rud·der·less \-ləs\ *adjective*

rud·der·post \-,pōst\ *noun* (1691)
1 : the shaft of a rudder
2 : an additional sternpost in a ship with a single screw propeller to which the rudder is attached

¹rud·dle \'rə-dᵊl\ *noun* [diminutive of *rud* red ocher] (1538)
: RED OCHER

²ruddle *transitive verb* **rud·dled; rud·dling** \'rəd-liŋ, 'rə-dᵊl-iŋ\ (1718)
: to color with or as if with red ocher **:** REDDEN

rud·dock \'rə-dək, 'ru̇-\ *noun* [Middle English *ruddok,* from Old English *rudduc;* akin to Old English *rudu*] (before 12th century)
archaic **:** ROBIN 1a

rud·dy \'rə-dē\ *adjective* **rud·di·er; -est** [Middle English *rudi,* from Old English *rudig,* from *rudu* redness; akin to Old English *rēad* red — more at RED] (before 12th century)
1 : having a healthy reddish color
2 : RED, REDDISH
3 *British* — used as an intensive ⟨bellowed like a *ruddy* bull when she wanted food —Doreen Tovey⟩
— rud·di·ly \'rə-dᵊl-ē\ *adverb*
— rud·di·ness \'rə-dē-nəs\ *noun*

ruddy duck *noun* (1814)
: an American duck (*Oxyura jamaicensis*) which has a wedge-shaped tail of stiff sharp feathers, a broad bill, and white cheeks and the adult male of which has rich brownish red on the upper side and a blue bill during breeding season

rude \'rüd\ *adjective* **rud·er; rud·est** [Middle English, from Middle French, from Latin *rudis;* probably akin to Latin *rudus* rubble] (14th century)
1 a : being in a rough or unfinished state **:** CRUDE ⟨*rude* line illustrations⟩ **b :** NATURAL, RAW ⟨*rude* cotton⟩ **c :** PRIMITIVE, UNDEVELOPED ⟨peasants use *rude* wooden plows —Jack Raymond⟩ **d :** SIMPLE, ELEMENTAL ⟨landscape done in *rude* whites, blacks, deep browns —Richard Harris⟩
2 : lacking refinement or delicacy: **a :** IGNORANT, UNLEARNED **b :** INELEGANT, UNCOUTH **c :** offensive in manner or action **:** DISCOURTEOUS **d :** UNCIVILIZED, SAVAGE **e :** COARSE, VULGAR
3 : marked by or suggestive of lack of training or skill **:** INEXPERIENCED ⟨*rude* workmanship⟩
4 : ROBUST, STURDY ⟨in *rude* health⟩
5 : occurring abruptly and disconcertingly ⟨a *rude* awakening⟩ ☆
— rude·ly *adverb*

rude·ness *noun* (14th century)
1 : the quality or state of being rude
2 : a rude action

¹ru·der·al \'rü-də-rəl\ *adjective* [New Latin *ruderalis,* from Latin *ruder-, rudus* rubble] (circa 1858)
: growing where the natural vegetational cover has been disturbed by humans ⟨*ruderal* weeds of old fields and roadsides⟩

²ruderal *noun* (circa 1928)
: a weedy and commonly introduced plant growing where the vegetational cover has been interrupted

ru·di·ment \'rü-də-mənt\ *noun* [Latin *rudimentum* beginning, from *rudis* raw, rude] (1548)

☆ SYNONYMS

Rude, rough, crude, raw mean lacking in social refinement. RUDE implies ignorance of or indifference to good form; it may suggest intentional discourtesy ⟨*rude* behavior⟩. ROUGH is likely to stress lack of polish and gentleness ⟨*rough* manners⟩. CRUDE may apply to thought or behavior limited to the gross, the obvious, or the primitive and ignorant of civilized amenities ⟨a *crude* joke⟩. RAW suggests being untested, inexperienced, or unfinished ⟨turning *raw* youths into polished performers⟩.

◇ WORD HISTORY

rubric Descended ultimately from Latin *rubrica,* a derivative of *ruber* "red," the word *rubric* was used in Middle English to designate red ocher, a red, earthy hematite used as a pigment. Yet it also meant "heading for part of a treatise," the predecessor of present-day English *rubric* in the sense "category, something under which a thing is classed." This semantic transformation derives from the ancient practice of putting chapter headings or explanatory notes in a manuscript in red ink to contrast with the black ink of the text. Similarly, red ink was used to enter saints' names and holy days in calendars, and we still speak of a day with special significance as a "red-letter day."

1 : a basic principle or element or a fundamental skill — usually used in plural ⟨teaching themselves the *rudiments* of rational government —G. B. Galanti⟩
2 a : something unformed or undeveloped **:** BEGINNING — usually used in plural ⟨the *rudiments* of a plan⟩ **b** (1) **:** a body part so deficient in size or structure as to be entirely unable to perform its normal function (2) **:** an organ just beginning to develop **:** ANLAGE
— **ru·di·men·tal** \ˌrü-də-ˈmen-tᵊl\ *adjective*

ru·di·men·ta·ry \ˌrü-də-ˈmen-tə-rē, -ˈmen-trē\ *adjective* (1839)
1 : consisting in first principles **:** FUNDAMENTAL ⟨had only a *rudimentary* formal education —D. J. Boorstin⟩
2 : of a primitive kind ⟨the equipment of these past empire-builders was *rudimentary* —A. J. Toynbee⟩
3 : very imperfectly developed or represented only by a vestige ⟨the *rudimentary* tail of a hyrax⟩
— **ru·di·men·tar·i·ly** \-ˌmen-ˈter-ə-lē, -ˈmen-trə-lē\ *adverb*
— **ru·di·men·ta·ri·ness** \-ˈmen-tə-rē-nəs, -ˈmen-trē-\ *noun*

[1]rue \ˈrü\ *noun* [Middle English *rewe,* from Old English *hrēow;* akin to Old High German *hriuwa* sorrow] (before 12th century)
: REGRET, SORROW

[2]rue *verb* **rued; ru·ing** (12th century)
transitive verb
: to feel penitence, remorse, or regret for
intransitive verb
: to feel sorrow, remorse, or regret

[3]rue *noun* [Middle English, from Middle French, from Latin *ruta,* from Greek *rhytē*] (13th century)
: a European strong-scented perennial woody herb (*Ruta graveolens* of the family Rutaceae, the rue family) that has bitter leaves used medicinally

rue anemone *noun* (circa 1818)
: a delicate vernal North American herb (*Anemonella thalictroides*) of the buttercup family with white flowers resembling those of the wood anemone

rue·ful \ˈrü-fəl\ *adjective* (13th century)
1 : exciting pity or sympathy **:** PITIABLE ⟨*rueful* squalid poverty . . . by every wayside —John Morley⟩
2 : MOURNFUL, REGRETFUL ⟨troubled her with a *rueful* disquiet —W. M. Thackeray⟩
— **rue·ful·ly** \-fə-lē\ *adverb*
— **rue·ful·ness** *noun*

ru·fes·cent \rü-ˈfe-sᵊnt\ *adjective* [Latin *rufescent-, rufescens,* present participle of *rufescere* to become reddish, from *rufus* red — more at RED] (1817)
: REDDISH

[1]ruff *also* **ruffe** \ˈrəf\ *noun* [Middle English *ruf*] (15th century)
: a small freshwater European perch (*Acerina cernua*)

[2]ruff *noun* [probably back-formation from *ruffle*] (1555)
1 : a large round collar of pleated muslin or linen worn by men and women of the late 16th and early 17th centuries
2 : a fringe or frill of long hairs or feathers growing around or on the neck
3 : a common Eurasian sandpiper (*Philomachus pugnax*) whose male during the breeding season has a large ruff of erectile feathers on the neck
— **ruffed** \ˈrəft\ *adjective*

ruff 1

[3]ruff *verb* [Middle French *roffler*] (1598)
intransitive verb
: to take a trick with a trump
transitive verb
: to play a trump on (a card previously led or played)

[4]ruff *noun* (circa 1828)
: the act of trumping

ruffed grouse *noun* (circa 1782)
: a North American grouse (*Bonasa umbellus*) of heavy forests that has a ruff of erectile black feathers in the male that is lighter and less conspicuous in the female

ruf·fi·an \ˈrə-fē-ən\ *noun* [Middle French *rufian*] (1531)
: a brutal person **:** BULLY
— **ruffian** *adjective*
— **ruf·fi·an·ism** \-ə-ˌni-zəm\ *noun*
— **ruf·fi·an·ly** *adjective*

[1]ruf·fle \ˈrə-fəl\ *verb* **ruf·fled; ruf·fling** \-f(ə-)liŋ\ [Middle English *ruffelen;* akin to Low German *ruffelen* to crumple] (14th century)
transitive verb
1 a : ROUGHEN, ABRADE **b :** TROUBLE, VEX ⟨is not *ruffled* by such barbs —Bruce Anderson⟩
2 : to erect (as feathers) in or like a ruff
3 a : to flip through (as pages) **b :** SHUFFLE
4 : to make into a ruffle
intransitive verb
: to become ruffled ⟨their dispositions *ruffle* perceptibly —*Life*⟩

[2]ruffle *noun* (1534)
1 : COMMOTION, BRAWL
2 : a state or cause of irritation
3 a : a strip of fabric gathered or pleated on one edge **b :** [2]RUFF 2
4 : an unevenness or disturbance of surface **:** RIPPLE
— **ruf·fly** \ˈrə-f(ə-)lē\ *adjective*

[3]ruffle *noun* [*ruff* a drumbeat] (circa 1802)
: a low vibrating drumbeat less loud than a roll

ru·fi·yaa \ˈrü-fē-ˌyä\ *noun, plural* **rufiyaa** [probably from Divehi (Indo-Aryan language of the Maldive Islands), from Hindi *rupīyā, rūpaiyā* rupee] (1982)
— see MONEY table

RU 486 \ˈär-ˌyü-ˌfȯr-ˌā-tē-ˈsiks, -ˌfȯr-\ *noun* [*R*oussel-*U*CLAF, the drug's French manufacturer + *486,* laboratory serial number] (1983)
: a drug $C_{29}H_{35}NO_2$ taken orally to induce abortion especially early in pregnancy by blocking the body's use of progesterone

ru·fous \ˈrü-fəs\ *adjective* [Latin *rufus* red — more at RED] (1782)
: REDDISH

rug \ˈrəg\ *noun* [(assumed) Middle English, rag, tuft, of Scandinavian origin; akin to Old Norse *rǫgg* tuft] (1591)
1 : LAP ROBE
2 : a piece of thick heavy fabric that usually has a nap or pile and is used as a floor covering
3 : a floor mat of an animal pelt ⟨a bearskin *rug*⟩
4 *slang* **:** TOUPEE 2
5 *British* **:** a blanket for an animal (as a horse or dog)

ru·ga \ˈrü-gə\ *noun, plural* **ru·gae** \-ˌgī, -ˌgē, -ˌjē\ [New Latin, from Latin, wrinkle — more at CORRUGATE] (1773)
: an anatomical fold or wrinkle especially of the viscera — usually used in plural

rug·by \ˈrəg-bē\ *noun, often capitalized* [*Rugby* School, Rugby, England] (1864)
: a football game in which play is continuous without time-outs or substitutions, interference and forward passing are not permitted, and kicking, dribbling, lateral passing, and tackling are featured

rug·ged \ˈrə-gəd\ *adjective* [Middle English, from (assumed) Middle English *rug*] (14th century)
1 *obsolete* **:** SHAGGY, HAIRY
2 : having a rough uneven surface **:** JAGGED ⟨*rugged* mountains⟩
3 : TURBULENT, STORMY ⟨*rugged* weather⟩
4 a : seamed with wrinkles and furrows **:** WEATHERED — used of a human face **b :** showing facial signs of strength ⟨*rugged* good looks⟩
5 a : AUSTERE, STERN **b :** COARSE, RUDE **c :** rough and strong in character
6 a : presenting a severe test of ability, stamina, or resolution **b :** strongly built or constituted **:** ROBUST ⟨those that survive are stalwart, *rugged* men —L. D. Stamp⟩
synonym see ROUGH
— **rug·ged·ly** *adverb*
— **rug·ged·ness** *noun*

rug·ged·ize \ˈrə-gə-ˌdīz\ *transitive verb* **-ized; -iz·ing** (1950)
: to strengthen (as a machine) for better resistance to wear, stress, and abuse ⟨a *ruggedized* camera⟩
— **rug·ged·i·za·tion** \ˌrə-gə-də-ˈzā-shən\ *noun*

rug·ger \ˈrə-gər\ *noun* [by alteration] (1893)
British **:** RUGBY; *also* **:** a rugby player

ru·go·la \ˈrü-gə-lə\ *noun* [probably from Italian dialect; akin to Italian dialect *ruga* arugula, Italian *ruca* — more at ROCKET] (1973)
: ARUGULA

ru·go·sa rose \rü-ˈgō-sə-, -zə-\ *noun* [New Latin *rugosa,* specific epithet of *Rosa rugosa* rugose rose] (1892)
: any of various garden roses descended from a rose (*Rosa rugosa*) introduced from China and Japan — called also *rugosa*

ru·gose \ˈrü-ˌgōs\ *adjective* [Latin *rugosus,* from *ruga*] (1676)
1 : full of wrinkles ⟨*rugose* cheeks⟩
2 : having the veinlets sunken and the spaces between elevated ⟨*rugose* leaves of the sage⟩
— **ru·gos·i·ty** \rü-ˈgä-sə-tē\ *noun*

ru·gu·lose \ˈrü-gyə-ˌlōs\ *adjective* [New Latin *rugula,* diminutive of Latin *ruga*] (circa 1819)
: having small rugae **:** finely wrinkled

[1]ru·in \ˈrü-ən, -ˌin; ˈrün\ *noun* [Middle English *ruine,* from Middle French, from Latin *ruina,* from *ruere* to rush headlong, fall, collapse] (12th century)
1 a *archaic* **:** a falling down **:** COLLAPSE ⟨from age to age . . . the crash of *ruin* fitfully resounds —William Wordsworth⟩ **b :** physical, moral, economic, or social collapse
2 a : the state of being ruined — archaic except in plural ⟨the city lay in *ruins*⟩ **b :** the remains of something destroyed — usually used in plural ⟨the *ruins* of an ancient temple⟩ ⟨the *ruins* of his life⟩
3 : a cause of destruction
4 a : the action of destroying, laying waste, or wrecking **b :** DAMAGE, INJURY
5 : a ruined building, person, or object
— **ru·in·ate** \ˈrü-ə-ˌnāt, -nət\ *adjective*
— **ruinate** \-ˌnāt\ *transitive verb*

[2]ruin (1585)
transitive verb
1 : to reduce to ruins **:** DEVASTATE
2 a : to damage irreparably **b :** BANKRUPT, IMPOVERISH ⟨*ruined* by stock speculation⟩
3 : to subject to frustration, failure, or disaster ⟨will *ruin* your chances of promotion⟩
intransitive verb
: to become ruined
— **ru·in·er** *noun*

ru·in·a·tion \ˌrü-ə-ˈnā-shən\ *noun* (1664)
: RUIN, DESTRUCTION

ru·in·ous \ˈrü-ə-nəs\ *adjective* (14th century)
1 : DILAPIDATED, RUINED
2 : causing or tending to cause ruin
— **ru·in·ous·ly** *adverb*
— **ru·in·ous·ness** *noun*

[1]rule \ˈrül\ *noun* [Middle English *reule,* from Old French, from Latin *regula* straightedge,

rule, from *regere* to direct — more at RIGHT] (13th century)
1 a : a prescribed guide for conduct or action **b :** the laws or regulations prescribed by the founder of a religious order for observance by its members **c :** an accepted procedure, custom, or habit **d** (1) **:** a usually written order or direction made by a court regulating court practice or the action of parties (2) **:** a legal precept or doctrine **e :** a regulation or bylaw governing procedure or controlling conduct
2 a (1) **:** a usually valid generalization (2) **:** a generally prevailing quality, state, or mode ⟨fair weather was the *rule* yesterday —*N.Y. Times*⟩ **b :** a standard of judgment **:** CRITERION **c :** a regulating principle **d :** a determinate method for performing a mathematical operation and obtaining a certain result
3 a : the exercise of authority or control **:** DOMINION **b :** a period during which a specified ruler or government exercises control
4 a : a strip of material marked off in units used especially for measuring **:** RULER 3, TAPE MEASURE **b :** a metal strip with a type-high face that prints a linear design; *also* **:** a linear design produced by or as if by such a strip
synonym see LAW
— as a rule : for the most part **:** GENERALLY

²rule *verb* **ruled; rul·ing** (13th century)
transitive verb
1 a : to exert control, direction, or influence on ⟨the passions that *rule* our minds⟩ **b :** to exercise control over especially by curbing or restraining ⟨*rule* a fractious horse⟩ ⟨*ruled* his appetites firmly⟩
2 a : to exercise authority or power over often harshly or arbitrarily ⟨the speaker *ruled* the legislature with an iron hand⟩ **b :** to be preeminent in **:** DOMINATE
3 : to determine and declare authoritatively; *especially* **:** to command or determine judicially
4 a (1) **:** to mark with lines drawn along or as if along the straight edge of a ruler (2) **:** to mark (a line) on a paper with a ruler **b :** to arrange in a line
intransitive verb
1 a : to exercise supreme authority **b :** to be first in importance or prominence **:** PREDOMINATE ⟨the physical did not *rule* in her nature —Sherwood Anderson⟩
2 : to exist in a specified state or condition
3 : to lay down a legal rule
synonym see DECIDE

ruled surface *noun* (1862)
: a surface generated by a moving straight line with the result that through every point on the surface a line can be drawn lying wholly in the surface

rule·less \ˈrül-ləs\ *adjective* (15th century)
: not restrained or regulated by law

rule of the road (1871)
: a customary practice (as driving always on a particular side of the road or yielding the right of way) developed in the interest of safety and often subsequently reinforced by law; *especially* **:** any of the rules making up a code governing ships in matters relating to mutual safety

rule of thumb (1692)
1 : a method of procedure based on experience and common sense
2 : a general principle regarded as roughly correct but not intended to be scientifically accurate

rule out *transitive verb* (1869)
1 : EXCLUDE, ELIMINATE
2 : to make impossible **:** PREVENT ⟨heavy rain *ruled out* the picnic⟩

rul·er \ˈrü-lər\ *noun* (14th century)
1 : one that rules; *specifically* **:** SOVEREIGN
2 : a worker or a machine that rules paper
3 : a smooth-edged strip (as of wood or metal) that is usually marked off in units (as inches) and is used as a straightedge or for measuring
— rul·er·ship \-ˌship\ *noun*

¹rul·ing \ˈrü-liŋ\ *noun* (15th century)
: an official or authoritative decision, decree, statement, or interpretation (as by a judge on a point of law)

²ruling *adjective* (1593)
1 a : exerting power or authority ⟨the *ruling* party⟩ **b :** CHIEF, PREDOMINATING ⟨a *ruling* passion⟩
2 : generally prevailing

ruly \ˈrü-lē\ *adjective* [back-formation from *unruly*] (15th century)
: OBEDIENT, ORDERLY ⟨a *ruly* crowd⟩

¹rum \ˈrəm\ *noun* [probably short for obsolete *rumbullion* rum] (1654)
1 : an alcoholic beverage distilled from a fermented cane product (as molasses)
2 : alcoholic liquor ⟨the demon *rum*⟩

²rum *adjective* **rum·mer; rum·mest** [origin unknown] (1752)
1 *chiefly British* **:** QUEER, ODD ⟨writing is a *rum* trade . . . and what is all right one day is all wrong the next —Angela Thirkell⟩
2 *chiefly British* **:** DIFFICULT, DANGEROUS

ru·ma·ki \rə-ˈmä-kē\ *noun* [perhaps modification of Japanese *harumaki,* translation of Chinese (Beijing) *chūnjuǎn* spring roll] (1961)
: a cooked appetizer made of pieces of usually marinated chicken liver wrapped together with sliced water chestnuts in bacon slices

Ru·ma·nian \ru̇-ˈmā-nē-ən, -nyən\ *noun* (1878)
: ROMANIAN
— Rumanian *adjective*

rum·ba \ˈrəm-bə, ˈrüm-, ˈrüm-\ *noun* [American Spanish] (1922)
: a ballroom dance of Cuban origin in 2/4 or 4/4 time with a basic pattern of step-close-step and marked by a delayed transfer of weight and pronounced hip movements; *also* **:** the music for this dance

¹rum·ble \ˈrəm-bəl\ *verb* **rum·bled; rum·bling** \-b(ə-)liŋ\ [Middle English; akin to Middle High German *rummeln* to rumble] (14th century)
intransitive verb
1 : to make a low heavy rolling sound ⟨thunder *rumbling* in the distance⟩
2 : to travel with a low reverberating sound ⟨wagons *rumbled* into town⟩
3 : to speak in a low rolling tone
4 : to engage in a rumble
transitive verb
1 : to utter or emit in a low rolling voice
2 *British* **:** to reveal or discover the true character of
— rum·bler \-b(ə-)lər\ *noun*

²rumble *noun* (14th century)
1 a : a low heavy continuous reverberating often muffled sound (as of thunder) **b :** low frequency noise in phonographic playback caused by the transmission of mechanical vibrations by the turntable to the pickup
2 : a seat for servants behind the body of a carriage
3 a : widespread expression of dissatisfaction or unrest **b :** a street fight especially among gangs

rumble seat *noun* (1912)
: a folding seat in the back of an automobile (as a coupe or roadster) not covered by the top

rum·bling \ˈrəm-bliŋ\ *noun* (14th century)
1 : RUMBLE
2 : general but unofficial talk or opinion often of dissatisfaction — usually used in plural ⟨*rumblings* of political trouble —Anthony Burgess⟩

rum·bly \ˈrəm-b(ə-)lē\ *adjective* (1874)
: tending to rumble or rattle

rum·bus·tious \ˌrəm-ˈbəs-chəs\ *adjective* [alteration of *robustious*] (1778)
chiefly British **:** RAMBUNCTIOUS
— rum·bus·tious·ly *adverb, chiefly British*
— rum·bus·tious·ness *noun, chiefly British*

ru·men \ˈrü-mən\ *noun, plural* **ru·mi·na** \-mə-nə\ *or* **rumens** [New Latin *rumin-, rumen,* from Latin] (circa 1728)
: the large first compartment of the stomach of a ruminant in which cellulose is broken down by the action of symbiotic microorganisms — compare ABOMASUM, OMASUM, RETICULUM
— ru·mi·nal \-mə-nᵊl\ *adjective*

¹ru·mi·nant \ˈrü-mə-nənt\ *noun* (1661)
: a ruminant mammal

²ruminant *adjective* (1691)
1 a (1) **:** chewing the cud (2) **:** characterized by chewing again what has been swallowed **b :** of or relating to a suborder (Ruminantia) of even-toed hoofed mammals (as sheep, giraffes, deer, and camels) that chew the cud and have a complex usually 4-chambered stomach
2 : given to or engaged in contemplation **:** MEDITATIVE ⟨stood there with her hands clasped in this attitude of *ruminant* relish —Thomas Wolfe⟩
— ru·mi·nant·ly *adverb*

ru·mi·nate \ˈrü-mə-ˌnāt\ *verb* **-nat·ed; -nat·ing** [Latin *ruminatus,* past participle of *ruminari* to chew the cud, muse upon, from *rumin-, rumen* rumen; perhaps akin to Sanskrit *romantha* act of chewing the cud] (1533)
transitive verb
1 : to go over in the mind repeatedly and often casually or slowly
2 : to chew repeatedly for an extended period
intransitive verb
1 : to a chew again what has been chewed slightly and swallowed **:** chew the cud
2 : to engage in contemplation **:** REFLECT ◆
synonym see PONDER
— ru·mi·na·tion \ˌrü-mə-ˈnā-shən\ *noun*
— ru·mi·na·tive \ˈrü-mə-ˌnā-tiv\ *adjective*
— ru·mi·na·tive·ly *adverb*
— ru·mi·na·tor \-ˌnā-tər\ *noun*

¹rum·mage \ˈrə-mij\ *verb* **rum·maged; rum·mag·ing** [²*rummage*] (circa 1595)
intransitive verb
1 : to make a thorough search or investigation
2 : to engage in an undirected or haphazard search
transitive verb
1 : to make a thorough search through **:** RANSACK ⟨*rummaged* the attic⟩
2 : to examine minutely and completely
3 : to discover by searching
— rum·mag·er *noun*

²rummage *noun* [obsolete English *rummage* act of packing cargo, modification of Middle French *arrimage*] (1598)
1 a : a confused miscellaneous collection **b :** items for sale at a rummage sale
2 : a thorough search especially among a confusion of objects

rummage sale *noun* (circa 1858)

◇ WORD HISTORY

ruminate Hoofed animals such as deer, sheep, and domestic cattle digest their food in stages, first swallowing the vegetation that they crop into a special stomach, where it is partially digested, then bringing it back up for some further leisurely chewing. This process is known as chewing the cud, and the animals that engage in it are called *ruminants,* from the present participle of the Latin verb *ruminari* "to chew the cud," a derivative of *rumen,* the Latin name for a ruminant's first stomach. The similarity between recalling food to the mouth for further chewing and recalling an idea to the mind for further thought—as well, perhaps, as the contemplative expression characteristic of a cow in mid-chew—led the Romans to use *ruminari* in an extended sense "to think over, meditate on." The English verb *ruminate,* from the past participle of *ruminari,* retains both the literal and figurative applications of its Latin source.

: a usually informal sale of miscellaneous goods; *especially* **:** a sale of donated articles conducted by a nonprofit organization (as a church or charity) to help support its programs

rum·mer \ˈrə-mər\ *noun* [German or Dutch; German *Römer,* from Dutch *roemer*] (1654)
: a large-bowled footed drinking glass often elaborately etched or engraved

[1]**rum·my** \ˈrə-mē\ *adjective* **rum·mi·er; -est** [[2]*rum*] (1828)
: QUEER, ODD ⟨were still feeling a little *rummy* from our trip up the escalator —*New Yorker*⟩

[2]**rummy** *noun, plural* **rummies** [[1]*rum*] (1851)
: DRUNKARD

[3]**rummy** *noun* [perhaps from [1]*rummy*] (1915)
: any of several card games for two or more players in which each player tries to assemble groups of three or more cards of the same rank or suit and to be the first to meld them all

[1]**ru·mor** \ˈrü-mər\ *noun* [Middle English *rumour,* from Middle French, from Latin *rumor* clamor, gossip; akin to Old English *rēon* to lament, Sanskrit *rauti* he roars] (14th century)
1 : talk or opinion widely disseminated with no discernible source
2 : a statement or report current without known authority for its truth
3 *archaic* **:** talk or report of a notable person or event
4 : a soft low indistinct sound **:** MURMUR

[2]**rumor** *transitive verb* **ru·mored; ru·mor·ing** (1594)
: to tell or spread by rumor

ru·mor·mon·ger \-ˌməŋ-gər, -ˌmäŋ-\ *noun* (1884)
: a person who spreads rumors
— ru·mor·mon·ger·ing \-gər-iŋ\ *noun*

ru·mour \ˈrü-mər\ *chiefly British variant of* RUMOR

rump \ˈrəmp\ *noun* [Middle English, of Scandinavian origin; akin to Icelandic *rumpr* rump; akin to Middle High German *rumph* torso] (15th century)
1 a : the upper rounded part of the hindquarters of a quadruped mammal **b :** BUTTOCKS **c :** the sacral or dorsal part of the posterior end of a bird
2 : a cut of meat (as beef) between the loin and round — see BEEF illustration
3 : a small or inferior remnant or offshoot; *especially* **:** a group (as a parliament) carrying on in the name of the original body after the departure or expulsion of a large number of its members

[1]**rum·ple** \ˈrəm-pəl\ *noun* (circa 1520)
: FOLD, WRINKLE

[2]**rumple** *verb* **rum·pled; rum·pling** \-p(ə-)liŋ\ [Dutch *rompelen;* akin to Old High German *rimpfan* to wrinkle] (1603)
transitive verb
1 : WRINKLE, CRUMPLE
2 : to make unkempt **:** TOUSLE
intransitive verb
: to become rumpled

rum·ply \ˈrəm-p(ə-)lē\ *adjective* **rum·pli·er; -est** (1833)
: having rumples

rum·pus \ˈrəm pəs\ *noun* [origin unknown] (1764)
: a usually noisy commotion

rumpus room *noun* (1939)
: a room usually in the basement of a home that is used for games, parties, and recreation

rum·run·ner \ˈrəm-ˌrə-nər\ *noun* (1920)
: a person or ship engaged in bringing prohibited liquor ashore or across a border
— rum–run·ning \-ˌrə-niŋ\ *adjective or noun*

[1]**run** \ˈrən\ *verb* **ran** \ˈran\ *also chiefly dialect* **run; run; run·ning** [Middle English *ronnen,* alteration of *rinnen,* intransitive verb (from Old English *iernan, rinnan* & Old Norse *rinna*) & of *rennen,* transitive verb, from Old Norse *renna;* akin to Old High German *rinnan,* intransitive verb, to run, Sanskrit *riṇāti* he causes to flow, and probably to Latin *rivus* stream] (before 12th century)
intransitive verb
1 a : to go faster than a walk; *specifically* **:** to go steadily by springing steps so that both feet leave the ground for an instant in each step **b** *of a horse* **:** to move at a fast gallop **c :** FLEE, RETREAT, ESCAPE ⟨dropped the gun and *ran*⟩ **d :** to utilize a running play on offense — used of a football team
2 a : to go without restraint **:** move freely about at will ⟨let chickens *run* loose⟩ **b :** to keep company **:** CONSORT ⟨a ram *running* with ewes⟩ ⟨*ran* with a wild crowd when he was young⟩ **c :** to sail before the wind in distinction from reaching or sailing close-hauled **d :** ROAM, ROVE ⟨*running* about with no overcoat⟩
3 a : to go rapidly or hurriedly **:** HASTEN ⟨*run* and fetch the doctor⟩ **b :** to go in urgency or distress **:** RESORT ⟨*runs* to mother at every little difficulty⟩ **c :** to make a quick, easy, or casual trip or visit ⟨*ran* over to borrow some sugar⟩
4 a : to contend in a race **b :** to enter into an election contest
5 a : to move on or as if on wheels **:** GLIDE ⟨file drawers *running* on ball bearings⟩ **b :** to roll forward rapidly or freely **c :** to pass or slide freely ⟨a rope *runs* through the pulley⟩ **d :** to ravel lengthwise ⟨stockings guaranteed not to *run*⟩
6 : to sing or play a musical passage quickly ⟨*run* up the scale⟩
7 a : to go back and forth **:** PLY ⟨the train *runs* between New York and Washington⟩ **b** *of fish* **:** to migrate or move in considerable numbers; *especially* **:** to move up or down a river to spawn
8 a : TURN, ROTATE ⟨a swiftly *running* grindstone⟩ **b :** FUNCTION, OPERATE ⟨the engine *runs* on gasoline⟩
9 a (1) **:** to continue in force, operation, or production ⟨the contract has two more years to *run*⟩ ⟨the play *ran* for six months⟩ (2) **:** to have a specified duration, extent, or length ⟨the manuscript *runs* nearly 500 pages⟩ **b :** to accompany as a valid obligation or right ⟨a right-of-way that *runs* with the land⟩ **c :** to continue to accrue or become payable ⟨interest on the loan *runs* from July 1st⟩
10 : to pass from one state to another ⟨*run* into debt⟩
11 a : to flow rapidly or under pressure **b :** MELT, FUSE **c :** SPREAD, DISSOLVE ⟨colors guaranteed not to *run*⟩ **d :** to discharge pus or serum ⟨a *running* sore⟩
12 a : to develop rapidly in some specific direction; *especially* **:** to throw out an elongated shoot of growth **b :** to tend to produce or develop a specified quality or feature ⟨they *run* to big noses in that family⟩
13 a : to lie in or take a certain direction ⟨the boundary line *runs* east⟩ **b :** to lie or extend in relation to something **c :** to go back **:** REACH **d** (1) **:** to be in a certain form or expression ⟨the letter *runs* as follows⟩ (2) **:** to be in a certain order of succession
14 a : to occur persistently ⟨musical talent *runs* in the family⟩ **b** (1) **:** to remain of a specified size, amount, character, or quality ⟨profits were *running* high⟩ (2) **:** to have or maintain a relative position or condition (as in a race) ⟨*ran* third⟩ ⟨*running* late⟩ **c :** to exist or occur in a continuous range of variation ⟨shades *run* from white to dark gray⟩
15 a : to spread or pass quickly from point to point ⟨chills *ran* up her spine⟩ **b :** to be current **:** CIRCULATE ⟨speculation *ran* rife⟩
transitive verb
1 a : to cause (an animal) to go rapidly **:** ride or drive fast **b :** to bring to a specified condition by or as if by running ⟨*ran* himself to death⟩ **c :** to go in pursuit of **:** HUNT, CHASE ⟨dogs that *run* deer⟩ **d :** to follow the trail of backward **:** TRACE ⟨*ran* the rumor to its source⟩ **e :** to enter, register, or enroll as a contestant in a race **f :** to put forward as a candidate for office
2 a : to drive (livestock) especially to a grazing place **b :** to provide pasturage for (livestock) **c :** to keep or maintain (livestock) on or as if on pasturage
3 a : to pass over or traverse with speed **b :** to accomplish or perform by or as if by running ⟨*ran* a great race⟩ ⟨*running* errands for a bank⟩ **c :** to slip or go through or past ⟨*run* a blockade⟩ ⟨*run* a red light⟩
4 a : to cause to penetrate or enter **:** THRUST ⟨*ran* a splinter into her toe⟩ **b :** STITCH **c :** to cause to pass **:** LEAD ⟨*run* a wire in from the antenna⟩ **d :** to cause to collide ⟨*ran* his head into a post⟩ **e :** SMUGGLE ⟨*run* guns⟩
5 : to cause to pass lightly or quickly over, along, or into something ⟨*ran* her eye down the list⟩
6 a : to cause or allow (as a vehicle or a vessel) to go in a specified manner or direction ⟨*ran* the car off the road⟩ **b :** OPERATE ⟨*run* a lathe⟩ **c :** to direct the business or activities of **:** MANAGE, CONDUCT ⟨*run* a factory⟩
7 a : to be full of or drenched with ⟨streets *ran* blood⟩ **b :** CONTAIN, ASSAY
8 a : to cause to move or flow in a specified way or into a specified position ⟨*run* cards into a file⟩ **b :** to cause to produce a flow (as of water) ⟨*run* the faucet⟩; *also* **:** to prepare by running a faucet ⟨*run* a hot bath⟩
9 a : to melt and cast in a mold ⟨*run* bullets⟩ **b :** TREAT, PROCESS, REFINE ⟨*run* oil in a still⟩ ⟨*run* a problem through a computer⟩
10 : to make oneself liable to **:** INCUR ⟨*ran* the risk of discovery⟩
11 : to mark out **:** DRAW ⟨*run* a contour line on a map⟩
12 a : to permit (as charges) to accumulate before settling ⟨*run* an account at the grocery⟩ — often used with *up* ⟨*ran* up a large phone bill⟩ **b :** COST 1 ⟨rooms that *run* $50 a night⟩
13 a : to produce by or as if by printing — usually used with *off* ⟨*ran* off 10,000 copies of the first edition⟩ **b :** to carry in a printed medium **:** PRINT ⟨every newspaper *ran* the story⟩
14 a : to make (a series of counts) without a miss ⟨*run* 19 in an inning in billiards⟩ **b :** to lead winning cards of (a suit) successively **c :** to alter by addition ⟨*ran* his record to six wins and four losses⟩
15 : to make (a golf ball) roll forward after alighting ■
— run across : to meet with or discover by chance
— run a fever *or* **run a temperature :** to have a fever
— run after 1 : PURSUE, CHASE; *especially* **:** to seek the company of **2 :** to take up with **:** FOLLOW ⟨*run after* new theories⟩
— run against 1 : to meet suddenly or unexpectedly **2 :** to work or take effect unfavorably to **:** DISFAVOR, OPPOSE
— run foul of 1 : to collide with ⟨*ran foul of* a hidden reef⟩ **2 :** to come into conflict with ⟨*run foul of* the law⟩

□ USAGE
run The past tense *run* still survives in speech in southern England and in the speech especially of older people in some parts of the U.S. It was formerly used in literature, and was a standard variant in our dictionaries from 1828 until 1934. Grammarians have generally opposed it, and many people consider it nonstandard. Just about everybody uses *ran* in writing now.

— **run into 1 a :** to change or transform into **:** BECOME **b :** to merge with **c :** to mount up to ⟨their yearly income often *runs into* six figures⟩ **2 a :** to collide with **b :** to meet by chance ⟨*ran into* an old classmate the other day⟩
— **run low on :** to approach running out of ⟨*running low on* options⟩
— **run rings around :** to show marked superiority over **:** defeat decisively or overwhelmingly
— **run riot 1 :** to act wildly or without restraint **2 :** to occur in profusion ⟨daffodils *running riot*⟩
— **run short :** to become insufficient
— **run short of :** to use up **:** run low on
— **run to :** to mount up to ⟨the book *runs to* 500 pages⟩
— **run upon :** to run across **:** meet with

²run *noun* (14th century)
1 a : an act or the action of running **:** continued rapid movement **b :** a quickened gallop **c :** a migration of fish (as up or down a river) especially to spawn; *also* **:** such fish in the process of migration **d :** a running race ⟨a mile *run*⟩ **e :** a score made in baseball by a runner reaching home plate safely **f :** strength or ability to run **g :** a gain of a usually specified distance made on a running play in football ⟨scored on a 25-yard *run*⟩; *also* **:** a running play **h :** a sustained usually aggressive effort (as to win or obtain something) ⟨making a *run* at the championship⟩
2 a *chiefly Midland* **:** CREEK 2 **b :** something that flows in the course of an operation or during a particular time ⟨the first *run* of sap in sugar maples⟩
3 a : the stern of the underwater body of a ship from where it begins to curve or slope upward and inward **b :** the direction in which a vein of ore lies **c :** a direction of secondary or minor cleavage **:** GRAIN ⟨the *run* of a mass of granite⟩ **d :** a horizontal distance (as that covered by a flight of steps) **e :** general tendency or direction
4 : a continuous period or series especially of things of identical or similar sort ⟨a *run* of bad luck⟩: as **a :** a rapid passage up or down a scale in vocal or instrumental music **b :** a number of rapid small dance steps executed in even tempo **c :** the act of making successively a number of successful shots or strokes; *also* **:** the score thus made ⟨a *run* of 20 in billiards⟩ **d :** an unbroken course of performances or showings **e :** a set of consecutive measurements, readings, or observations **f :** persistent and heavy demands from depositors, creditors, or customers ⟨a *run* on a bank⟩ **g :** SEQUENCE 2b
5 : the quantity of work turned out in a continuous operation ⟨a press *run* of 10,000 copies⟩
6 : the usual or normal kind, character, type, or group ⟨the average *run* of students⟩
7 a : the distance covered in a period of continuous traveling or sailing **b :** a course or trip especially if mapped out and traveled with regularity **c :** a news reporter's regular territory **:** BEAT **d :** freedom of movement in or access to a place or area ⟨has the *run* of the house⟩
8 a : the period during which a machine or plant is in continuous operation **b :** the use of machinery for a single set of processing procedures ⟨a computer *run*⟩
9 a : a way, track, or path frequented by animals **b :** an enclosure for domestic animals where they may feed or exercise **c** *Australian* (1) **:** a large area of land used for grazing ⟨a sheep *run*⟩ (2) **:** RANCH, STATION ⟨*run*-holder⟩ **d :** an inclined passageway
10 a : an inclined course (as for skiing or bobsledding) **b :** a support (as a track, pipe, or trough) on which something runs
11 a : a ravel in a knitted fabric (as in hosiery) caused by the breaking of stitches **b :** a paint defect caused by excessive flow
12 *plural but singular or plural in construction* **:** DIARRHEA — used with *the*
— **on the run 1 :** in haste **:** without pausing ⟨ate lunch *on the run*⟩ **2 :** in retreat **:** in flight (as from the law) ⟨an escaped convict *on the run*⟩

³run *adjective* (1774)
1 a : being in a melted state ⟨*run* butter⟩ **b :** made from molten material **:** cast in a mold ⟨*run* metal⟩
2 *of fish* **:** having made a migration or spawning run ⟨a fresh *run* salmon⟩
3 : exhausted or winded from running

run·about \'rə-nə-ˌbau̇t\ *noun* (1549)
1 : one who wanders about **:** STRAY
2 : a light usually open wagon, car, or motorboat

run·a·gate \'rə-nə-ˌgāt\ *noun* [alteration of *renegate,* from Medieval Latin *renegatus* — more at RENEGADE] (1547)
1 : VAGABOND
2 : FUGITIVE, RUNAWAY

run along *intransitive verb* (1902)
: to go away **:** be on one's way **:** DEPART

run·around \'rə-nə-ˌrau̇nd\ *noun* (1915)
1 : deceptive or delaying action especially in response to a request ⟨tired of getting the *runaround*⟩
2 : matter typeset in shortened measure to run around something (as a cut)

¹run·away \'rə-nə-ˌwā\ *noun* (1547)
1 : one that runs away from danger, duty, or restraint **:** FUGITIVE
2 : the act of running away out of control; *also* **:** something (as a horse) that is running out of control
3 : a one-sided or overwhelming victory

²runaway *adjective* (1548)
1 a : running away **:** FUGITIVE **b :** leaving to gain special advantages (as lower wages) or avoid disadvantages (as governmental or union restrictions) ⟨*runaway* shipping firms⟩ ⟨a *runaway* shop⟩
2 : accomplished by elopement or during flight
3 : won by or having a long lead ⟨a *runaway* success⟩; *also* **:** extremely successful ⟨a *runaway* best-seller⟩
4 : subject to uncontrolled changes ⟨*runaway* inflation⟩
5 : being or operating out of control ⟨a *runaway* oil well⟩ ⟨a *runaway* nuclear reactor⟩

run away *intransitive verb* (13th century)
1 a : to leave quickly in order to avoid or escape something **b :** to leave home; *especially* **:** ELOPE
2 : to run out of control **:** STAMPEDE, BOLT
3 : to gain a substantial lead **:** win by a large margin
— **run away with 1 :** to take away in haste or secretly; *especially* **:** STEAL **2 :** to outshine the others in (a theatrical performance) **3 :** to carry or drive beyond prudent or reasonable limits ⟨your imagination *ran away with* you⟩

run·back \'rən-ˌbak\ *noun* (1929)
: a run made in football after catching an opponent's kick or intercepting a pass

run·ci·ble spoon \'rən(t)-sə-bəl-\ *noun* [coined with an obscure meaning by Edward Lear] (1871)
: a sharp-edged fork with three broad curved prongs

run·ci·nate \'rən(t)-sə-ˌnāt\ *adjective* [Latin *runcinatus,* past participle of *runcinare* to plane off, from *runcina* plane] (1776)
: pinnately cut with the lobes pointing downward ⟨*runcinate* leaves of the dandelion⟩ — see LEAF illustration

run·dle \'rən-dᵊl\ *noun* [Middle English *roundel* circle — more at ROUNDEL] (1565)
1 : a step of a ladder **:** RUNG
2 : the drum of a windlass or capstan

rund·let *or* **run·let** \'rən(d)-lət\ *noun* [Middle English *roundelet,* from Middle French *rondelet* — more at ROUNDELAY] (14th century)
: a small barrel **:** KEG

run·down \'rən-ˌdau̇n\ *noun* (1908)
1 : a play in baseball in which a base runner who is caught off base is chased by two or more opposing players who throw the ball from one to another in an attempt to tag the runner out
2 : a item-by-item report or review **:** SUMMARY

run–down \'rən-'dau̇n\ *adjective* (1866)
1 : WORN-OUT, EXHAUSTED
2 : completely unwound
3 : being in poor repair **:** DILAPIDATED

run down (circa 1578)
transitive verb
1 a : to collide with and knock down **b :** to run against and cause to sink
2 a : to chase to exhaustion or until captured **b :** to trace the source of **c :** to tag out (a base runner) between bases on a rundown
3 : DISPARAGE
intransitive verb
1 : to cease to operate because of the exhaustion of motive power ⟨the clock *ran down*⟩
2 : to decline in physical condition or vigor

rune \'rün\ *noun* [Old Norse & Old English *rūn* mystery, runic character, writing; akin to Old High German *rūna* secret discussion, Old Irish *rún* mystery] (1690)
1 : any of the characters of any of several alphabets used by the Germanic peoples from about the 3d to the 13th centuries
2 : MYSTERY, MAGIC
3 [Finnish *runo,* of Germanic origin; akin to Old Norse *rūn*] **a :** a Finnish or Old Norse poem **b :** POEM, SONG
— **ru·nic** \'rü-nik\ *adjective*

f u th o r k g
w h n i j ch p eo s t
b e m l ng oe d a ae y ea

rune 1: Anglo-Saxon runic alphabet

¹rung *past participle of* RING

²rung \'rəŋ\ *noun* [Middle English, from Old English *hrung* crossbar; akin to Gothic *hrunga* staff and perhaps to Old English *hring* ring — more at RING] (14th century)
1 a : a rounded part placed as a crosspiece between the legs of a chair **b :** one of the crosspieces of a ladder
2 *Scottish* **:** a heavy staff or cudgel
3 : a spoke of a wheel
4 : a level in a hierarchy ⟨rise a few *rungs* on the social scale —H. W. Van Loon⟩

run–in \'rən-ˌin\ *noun* (1857)
1 *British* **:** the final part of a race or racetrack
2 : ALTERCATION, QUARREL
3 : something inserted as a substantial addition in copy or typeset matter

run in (1817)
transitive verb
1 a : to insert as additional matter **b :** to make (typeset matter) continuous without a paragraph or other break
2 : to arrest for a usually minor offense
3 *chiefly British* **:** to operate (a new machine) carefully until there is efficient running
intransitive verb
: to pay a casual visit

run·less \'rən-ləs\ *adjective* (1921)
: scoring or allowing no runs

run·let \'rən-lət\ *noun* (circa 1755)
: RIVULET, STREAMLET

run·nel \'rə-nᵊl\ *noun* [alteration of Middle English *rinel,* from Old English *rynel;* akin to Old English *rinnan* to run — more at RUN] (before 12th century)
: RIVULET, STREAMLET

run·ner \'rə-nər\ *noun* (14th century)
1 a : one that runs **:** RACER **b :** BASE RUNNER **c :** BALLCARRIER
2 a : MESSENGER **b :** one that smuggles or distributes illicit or contraband goods (as drugs, liquor, or guns)
3 : any of several large active carangid fishes

4 a : either of the longitudinal pieces on which a sled or sleigh slides **b :** the part of a skate that slides on the ice **:** BLADE **c :** the support of a drawer or a sliding door
5 a : a growth produced by a plant in running; *especially* **:** STOLON 1a **b :** a plant that forms or spreads by means of runners **c :** a twining vine (as a scarlet runner)
6 a : a long narrow carpet for a hall or staircase **b :** a narrow decorative cloth cover for a table or dresser top

runner bean *noun* (1882)
chiefly British **:** SCARLET RUNNER

run·ner–up \'rə-nə-ˌrəp, ˌrə-nə-'\ *noun, plural* **run·ners–up** \-nər-ˌzəp, -'zəp\ *also* **runner–ups** (1842)
: the competitor that does not win first place in a contest; *especially* **:** one that finishes in second place

[1]**run·ning** \'rə-niŋ\ *noun* (before 12th century)
1 a : the action of running **b :** RACE
2 : physical condition for running
3 : MANAGEMENT, CARE
— in the running 1 : competing in a contest **2 :** having a chance to win a contest
— out of the running 1 : not competing in a contest **2 :** having no chance of winning a contest

[2]**running** *adjective* (14th century)
1 : CURSIVE, FLOWING
2 : FLUID, RUNNY
3 a : INCESSANT, CONTINUOUS ⟨a *running* battle⟩ **b :** made during the course of a process or activity ⟨a *running* commentary on the game⟩
4 : measured in a straight line ⟨cost of lumber per *running* foot⟩
5 a : initiated or performed while running or with a running start ⟨*running* catch⟩ **b :** of, relating to, or being a football play in which the ball is advanced by running rather than by passing ⟨their *running* game was off⟩ **c :** designed for use by runners ⟨a *running* track⟩ ⟨*running* shoes⟩
6 : fitted or trained for running rather than walking, trotting, or jumping ⟨a *running* horse⟩

[3]**running** *adverb* (1719)
: in succession **:** CONSECUTIVELY ⟨three days *running*⟩

running back *noun* (1924)
: a football back (as a halfback or fullback) who carries the ball on running plays

running board *noun* (1817)
: a footboard especially at the side of an automobile

running board

running dog *noun* (1927)
: one who does someone else's bidding **:** LACKEY

running gear *noun* (1662)
1 : the working and carrying parts of a machine (as a locomotive)
2 : the parts of an automobile chassis not used in developing, transmitting, and controlling power

running hand *noun* (1648)
: handwriting in which the letters are usually slanted and the words formed without lifting the pen

running head *noun* (1839)
: a headline repeated on consecutive pages (as of a book) — called also *running headline*

running knot *noun* (1648)
: a knot that slips along the rope or line round which it is tied; *especially* **:** an overhand slipknot

running light *noun* (1881)
: any of the lights carried by a vehicle (as a ship) under way at night that indicate size, position, and course

running mate *noun* (1868)
1 : a horse entered in a race to set the pace for a horse of the same owner or stable
2 : a candidate running for a subordinate place on a ticket; *especially* **:** the candidate for vice-president
3 : COMPANION

running start *noun* (1926)
: FLYING START 1

running stitch *noun* (1848)
: a small even stitch run in and out in cloth

running title *noun* (1668)
: the title or short title of a volume printed at the top of left-hand text pages or sometimes of all text pages

running water *noun* (1912)
: water distributed through pipes and fixtures ⟨a cabin with hot and cold *running water*⟩

run·ny \'rə-nē\ *adjective* (1817)
: having a tendency to run: as **a :** extremely or excessively soft and liquid ⟨a *runny* dough⟩ **b :** secreting a thin flow of mucus ⟨a *runny* nose⟩

run·off \'rən-ˌȯf\ *noun* (1873)
1 : a final race, contest, or election to decide an earlier one that has not resulted in a decision in favor of any one competitor
2 : the portion of precipitation on land that ultimately reaches streams often with dissolved or suspended material

run off (1683)
transitive verb
1 a : to recite, compose, or produce rapidly **b :** to cause to be run or played to a finish **c :** to decide (as a race) by a runoff **d :** CARRY OUT
2 : to drain off **:** DRAW OFF
3 a : to drive off (as trespassers) **b :** to steal (as cattle) by driving away
intransitive verb
: RUN AWAY 1
— run off with : to carry off **:** STEAL

run–of–paper \ˌrən-əv-'pā-pər\ *adjective* (circa 1923)
: to be placed anywhere in a newspaper at the option of the editor ⟨*run-of-paper* advertisement⟩

run–of–the–mill \ˌrən-ə(v)-t͟hə-'mil\ *adjective* (1930)
: not outstanding in quality or rarity **:** AVERAGE, ORDINARY

run–of–the–mine \-'mīn\ *or* **run–of–mine** \-əv-'mīn\ *adjective* (1903)
1 : not graded ⟨*run-of-the-mine* coal⟩
2 : RUN-OF-THE-MILL

[1]**run–on** \'rən-'ȯn, -'än\ *adjective* (1877)
: continuing without rhetorical pause from one line of verse into another

[2]**run–on** \-ˌȯn, -ˌän\ *noun* (circa 1909)
: something (as a dictionary entry) that is run on

run on (15th century)
intransitive verb
1 : to talk or narrate at length
2 : to keep going **:** CONTINUE
transitive verb
1 : to continue (matter in type) without a break or a new paragraph **:** RUN IN
2 : to place or add (as an entry in a dictionary) at the end of a paragraphed item

run–on sentence *noun* (1914)
: a sentence containing a comma splice

run out (14th century)
intransitive verb
1 a : to come to an end **:** EXPIRE ⟨time *ran out*⟩ **b :** to become exhausted or used up ⟨the gasoline *ran out*⟩
2 : to jut out
transitive verb
1 : to finish out (as a course, series, or contest) **:** COMPLETE
2 a : to fill out (a typeset line) with quads, leaders, or ornaments **b :** to set (as the first line of a paragraph) with a hanging indention
3 : to exhaust (oneself) in running
4 : to cause to leave by force or coercion **:** EXPEL
— run out of : to use up the available supply of
— run out on : DESERT

run·over \'rən-ˌō-vər\ *noun* (1927)
: matter for publication that exceeds the space allotted

run–over \'rən-'ō-vər\ *adjective* (circa 1934)
: extending beyond the allotted space

run over (15th century)
intransitive verb
1 : to exceed a limit
2 : OVERFLOW
transitive verb
1 : to go over, examine, repeat, or rehearse quickly
2 : to collide with, knock down, and often drive over ⟨*ran over* a dog⟩

runt \'rənt\ *noun* [origin unknown] (1501)
1 *chiefly Scottish* **:** a hardened stalk or stem of a plant
2 : an animal unusually small of its kind; *especially* **:** the smallest of a litter of pigs
3 : a person of small stature
— runt·i·ness \'rən-tē-nəs\ *noun*
— runt·ish \-tish\ *adjective*
— runty \-tē\ *adjective*

run–through \'rən-ˌthrü\ *noun* (1923)
: a usually cursory reading, summary, or rehearsal

run through *transitive verb* (15th century)
1 : PIERCE
2 : to spend or consume wastefully and rapidly
3 : to read or rehearse without pausing
4 a : CARRY OUT, DO **b :** to subject to a process

run–up \'rən-ˌəp\ *noun* (1834)
1 : the act of running up something
2 : a usually sudden increase in volume or price
3 *chiefly British* **:** a period immediately preceding an action or event

run up (1664)
intransitive verb
: to grow rapidly **:** shoot up
transitive verb
1 : to increase by bidding **:** bid up
2 : to stitch together quickly
3 : to erect hastily
4 : to achieve by accumulating ⟨*ran up* a big lead⟩

run·way \'rən-ˌwā\ *noun* (1833)
1 a : a beaten path made by animals **b :** a passageway for animals
2 : a paved strip of ground on a landing field for the landing and takeoff of airplanes
3 a : a narrow platform from a stage into an auditorium **b :** a platform along which models walk in a fashion show
4 : RUN 10b
5 : the area or path along which a jumper, pole vaulter, or javelin thrower runs

ru·pee \rü-'pē, 'rü-ˌpē\ *noun* [Hindi *rūpaiyā,* from Sanskrit *rūpya* coined silver] (1610)
— see MONEY table

ru·pi·ah \rü-'pē-ə\ *noun, plural* **rupiah** *or* **rupiahs** [Malay, from Hindi *rūpaiyā* rupee] (1947)
— see MONEY table

[1]**rup·ture** \'rəp(t)-shər\ *noun* [Middle English *ruptur,* from Middle French or Latin; Middle French *rupture,* from Latin *ruptura* fracture, from *ruptus,* past participle of *rumpere* to break — more at REAVE] (15th century)
1 : breach of peace or concord; *specifically* **:** open hostility or war between nations
2 a : the tearing apart of a tissue ⟨*rupture* of the heart muscle⟩ ⟨*rupture* of an intervertebral disk⟩ **b :** HERNIA
3 : a breaking apart or the state of being broken apart

[2]**rupture** *verb* **rup·tured; rup·tur·ing** \-sh(ə-)riŋ\ (1739)
transitive verb

1 a : to part by violence **:** BREAK, BURST **b :** to create or induce a breach of
2 : to produce a rupture in
intransitive verb
: to have or undergo a rupture

ru·ral \'rür-əl\ *adjective* [Middle English, from Middle French, from Latin *ruralis,* from *rur-, rus* open land — more at ROOM] (15th century)
: of or relating to the country, country people or life, or agriculture
— ru·ral·i·ty \rü-'ra-lə-tē\ *noun*
— ru·ral·ly \'rür-ə-lē\ *adverb*

rural dean *noun* (circa 1628)
: DEAN 1b

rural free delivery *noun* (1892)
: free delivery of mail to a rural area — called also *rural delivery*

ru·ral·ist \'rür-ə-list\ *noun* (1739)
: one who lives in a rural area

rural route *noun* (1898)
: a mail-delivery route in a rural free delivery area

rur·ban \'rər-bən, 'rür-\ *adjective* [blend of *rural* and *urban*] (1918)
: of, relating to, or constituting an area which is chiefly residential but where some farming is carried on

Ru·ri·tan \'rür-ə-tən\ *noun* [*Ruritan National,* a service club] (1968)
: a member of a major national service club

Ru·ri·ta·ni·an \'rür-ə-,tā-nē-ən, ,rür-ə-'\ *adjective* [*Ruritania,* fictional kingdom in the novel *Prisoner of Zenda* (1894) by Anthony Hope] (1896)
: of, relating to, or having the characteristics of an imaginary place of high romance

ruse \'rüs, 'rüz\ *noun* [French, from Middle French, from *ruser* to dodge, deceive] (1625)
: a wily subterfuge
synonym see TRICK

¹rush \'rəsh\ *noun* [Middle English, from Old English *rysc;* akin to Middle High German *rusch* rush, Lithuanian *regzti* to knit] (before 12th century)
: any of various monocotyledonous often tufted marsh plants (as of the genera *Juncus* and *Scirpus* of the family Juncaceae, the rush family) with cylindrical often hollow stems which are used in bottoming chairs and plaiting mats
— rushy \'rə-shē\ *adjective*

²rush *verb* [Middle English *russhen,* from Middle French *ruser* to put to flight, repel, deceive, from Latin *recusare* to refuse — more at RECUSANT] (14th century)
intransitive verb
1 : to move forward, progress, or act with haste or eagerness or without preparation
2 : to advance a football by running plays ⟨*rushed* for a total of 150 yards⟩
transitive verb
1 : to push or impel on or forward with speed, impetuosity, or violence
2 : to perform in a short time or at high speed
3 : to urge to an unnatural or extreme speed ⟨don't *rush* me⟩
4 : to run toward or against in attack **:** CHARGE
5 a : to carry (a ball) forward in a running play **b :** to move in quickly on (a kicker or passer) to hinder, prevent, or block a kick or pass — used especially of defensive linemen
6 a : to lavish attention on **:** COURT **b :** to try to secure a pledge of membership (as in a fraternity) from

³rush *noun* (14th century)
1 a : a violent forward motion **b :** ATTACK, ONSET **c :** a surging of emotion
2 a : a burst of activity, productivity, or speed **b :** a sudden insistent demand
3 : a thronging of people usually to a new place in search of wealth ⟨gold *rush*⟩
4 a : the act of carrying a football during a game **:** running play **b :** the action or an instance of rushing a passer or kicker in football ⟨pass *rush*⟩
5 a : a round of attention usually involving extensive social activity **b :** a drive by a fraternity or sorority to recruit new members
6 : a print of a motion-picture scene processed directly after the shooting for review by the director or producer — usually used in plural
7 a : the immediate pleasurable feeling produced by a drug (as heroin or amphetamine) — called also *flash* **b :** a sudden feeling of intense pleasure or euphoria **:** THRILL

⁴rush *adjective* (1887)
: requiring or marked by special speed or urgency ⟨*rush* orders⟩ ⟨the *rush* season⟩ ⟨a *rush* job⟩

rush candle *noun* (1591)
: RUSHLIGHT

rush·ee \,rə-'shē\ *noun* (circa 1916)
: a college or university student who is being rushed by a fraternity or sorority

rush·er \'rə-shər\ *noun* (1654)
: one that rushes; *especially* **:** BALLCARRIER

rush hour *noun* (1898)
: a period of the day when the demands especially of traffic or business are at a peak

rush·ing *noun* (1883)
: the act of advancing a football by running plays **:** the use of running plays; *also* **:** yardage gained by running plays

rush·light \'rəsh-,līt\ *noun* (1710)
: a candle that consists of the pith of a rush dipped in grease

rusk \'rəsk\ *noun* [modification of Spanish & Portuguese *rosca* coil, twisted roll] (1595)
1 : hard crisp bread originally used as ship's stores
2 : a sweet or plain bread baked, sliced, and baked again until dry and crisp

Russ \'rəs, 'rüs, 'rus\ *noun, plural* **Russian** *or* **Russ·es** [ultimately from Old Russian *Rus'* the East Slavic-speaking lands] (1537)
: RUSSIAN
— Russ *adjective*

¹rus·set \'rə-sət\ *noun* [Middle English, from Old French *rousset,* from *rousset,* adjective, russet, from *rous* russet, from Latin *russus* red; akin to Latin *ruber* red — more at RED] (13th century)
1 : coarse homespun usually reddish brown cloth
2 : a strong brown
3 : any of various winter apples having russet rough skins
4 : IDAHO

²russet *adjective* (15th century)
: of the color russet

rus·set·ing *also* **rus·set·ting** \'rə-sə-tiŋ\ *noun* (1912)
: a brownish roughened area on the skin of fruit (as apples) caused by injury

Rus·sia leather \'rə-shə-\ *noun* [*Russia,* Europe] (1658)
: leather made by tanning various skins with willow, birch, or oak and then rubbing the flesh side with a phenolic oil distilled from a European birch — called also *Russia calf*

Rus·sian \'rə-shən\ *noun* (1538)
1 a : a native or inhabitant of Russia **b :** a member of the dominant Slavic-speaking ethnic group of Russia **c :** a person of Russian descent
2 : a Slavic language of the Russian people spoken as a second language by many non-Russian ethnic groups of the Soviet Union and its successor states
— Russian *adjective*
— Rus·sian·ness \-nəs\ *noun*

Russian blue *noun, often B capitalized* (1889)
: any of a breed of slender long-bodied large-eared domestic cats with short silky bluish gray fur

Russian dressing *noun* (1922)
: a dressing (as of mayonnaise or oil and vinegar) with added chili sauce, chopped pickles, or pimientos

Rus·sian·ize \'rə-shə-,nīz\ *transitive verb* **-ized; -iz·ing** (1831)
: to make Russian
— Rus·sian·i·za·tion \,rə-shə-nə-'zā-shən\ *noun*

Russian olive *noun* (1913)
: a chiefly silvery Eurasian large shrub or small tree (*Elaeagnus angustifolia* of the family Elaeagnaceae) cultivated especially as a shelterbelt plant

Russian roulette *noun* (1937)
1 : an act of bravado consisting of spinning the cylinder of a revolver loaded with one cartridge, pointing the muzzle at one's own head, and pulling the trigger
2 : something resembling Russian roulette in its potential for disaster ⟨taking cocaine is playing *Russian roulette* —Jonathan Nicholas⟩

Russian thistle *noun* (1894)
: a prickly European saltwort (*Salsola kali*) that is a serious pest in North America — called also *Russian tumbleweed*

Russian wolfhound *noun* (1872)
: BORZOI

Rus·si·fy \'rə-sə-,fī\ *transitive verb* **-fied; -fy·ing** (1865)
: RUSSIANIZE
— Rus·si·fi·ca·tion \,rə-sə-fə-'kā-shən\ *noun*

Russ·ki *or* **Russ·ky** *or* **Russ·kie** \'rəs-kē, 'rus-\ *noun, plural* **Russkies** *or* **Russkis** [Russian *russkiĭ,* adjective & noun, Russian, from Old Russian, from *Rus'*] (1858)
: RUSSIAN 1

Rus·so- \'rə-(,)sō, ,rə-, -(,)shō\ *combining form* [*Russia & Russian*]
: Russian and ⟨the *Russo*-Japanese war⟩

¹rust \'rəst\ *noun* [Middle English, from Old English *rūst;* akin to Old English *rēad* red — more at RED] (before 12th century)
1 a : the reddish brittle coating formed on iron especially when chemically attacked by moist air and composed essentially of hydrated ferric oxide **b :** a comparable coating produced on a metal other than iron by corrosion **c :** something resembling rust **:** ACCRETION
2 : corrosive or injurious influence or effect
3 : any of numerous destructive diseases of plants produced by fungi (order Uredinales) and characterized by usually reddish brown pustular lesions; *also* **:** a fungus causing this
4 : a strong reddish brown

²rust (13th century)
intransitive verb
1 : to form rust **:** become oxidized ⟨iron *rusts*⟩
2 : to degenerate especially from inaction, lack of use, or passage of time ⟨most men would . . . have allowed their faculties to *rust* —T. B. Macaulay⟩
3 : to become reddish brown as if with rust ⟨the leaves slowly *rusted*⟩
4 : to be affected with a rust fungus
transitive verb
1 : to cause (a metal) to form rust ⟨keep up your bright swords, for the dew will *rust* them —Shakespeare⟩
2 : to impair or corrode by or as if by time, inactivity, or deleterious use
3 : to cause to become reddish brown **:** turn the color of rust

rust belt *noun, often R&B capitalized* (1983)
: the northeastern and midwestern states of the U.S. in which heavy industry has declined

rust bucket *noun* (1945)
: an old and dilapidated ship

¹rus·tic \'rəs-tik\ *also* **rus·ti·cal** \-ti-kəl\ *adjective* [Middle English *rustik,* from Middle French *rustique,* from Latin *rusticus,* from *rus* open land — more at ROOM] (15th century)
1 : of, relating to, or suitable for the country **:** RURAL
2 a : made of the rough limbs of trees ⟨*rustic* furniture⟩ **b :** finished by rusticating ⟨a *rustic* joint in masonry⟩

3 a : characteristic of or resembling country people **b :** lacking in social graces or polish
4 : appropriate to the country (as in plainness or sturdiness) ⟨heavy *rustic* boots⟩
— **rus·ti·cal·ly** \-ti-k(ə-)lē\ *adverb*
— **rus·tic·i·ty** \ˌrəs-ˈti-sə-tē\ *noun*

²rustic *noun* (circa 1550)
1 : an inhabitant of a rural area
2 a : an awkward coarse person **b :** an unsophisticated rural person

rus·ti·cate \ˈrəs-ti-ˌkāt\ *verb* **-cat·ed; -cat·ing** (1660)
intransitive verb
: to go into or reside in the country **:** follow a rustic life
transitive verb
1 *chiefly British* **:** to suspend from school or college
2 : to build or face with usually rough-surfaced masonry blocks having beveled or rebated edges producing pronounced joints ⟨a *rusticated* facade⟩
3 a : to compel to reside in the country **b :** to cause to become rustic **:** implant rustic mannerisms in
— **rus·ti·ca·tion** \ˌrəs-ti-ˈkā-shən\ *noun*
— **rus·ti·ca·tor** \ˈrəs-ti-ˌkā-tər\ *noun*

¹rus·tle \ˈrə-səl\ *verb* **rus·tled; rus·tling** \ˈrə-s(ə-)liŋ\ [Middle English *rustelen*] (14th century)
intransitive verb
1 : to make or cause a rustle
2 a : to act or move with energy or speed **b :** to forage food
3 : to steal cattle
transitive verb
1 : to cause to rustle
2 a : to obtain by one's own exertions — often used with *up* ⟨able to *rustle* up $5,000 bail —Jack McCallum⟩ **b :** FORAGE
3 : to steal (as livestock) especially from a farm or ranch
— **rus·tler** \-s(ə-)lər\ *noun*

²rustle *noun* (1759)
: a quick succession or confusion of small sounds

rust mite *noun* (1884)
: any of various small gall mites that burrow in the surface of leaves or fruits usually producing brown or reddish patches

rust·proof \ˈrəst-ˌprüf\ *adjective* (1691)
: incapable of rusting

¹rusty \ˈrəs-tē\ *adjective* **rust·i·er; -est** (before 12th century)
1 : affected by or as if by rust; *especially* **:** stiff with or as if with rust
2 : inept and slow through lack of practice or old age
3 a : of the color rust **b :** dulled in color or appearance by age and use ⟨a *rusty* old suit of clothes⟩
4 : OUTMODED
5 : HOARSE, GRATING
— **rust·i·ly** \-tə-lē\ *adverb*
— **rust·i·ness** \-tē-nəs\ *noun*

²rus·ty \ˈrəs-tē\ *adjective* **rus·ti·er; -est** [alteration of *restive*] (1625)
chiefly dialect **:** ILL-NATURED, SURLY

¹rut \ˈrət\ *noun* [Middle English *rutte,* from Middle French *rut* roar, from Late Latin *rugitus,* from Latin *rugire* to roar; akin to Middle Irish *rucht* roar, Old Church Slavonic *ružati* to neigh] (15th century)
1 : an annually recurrent state of sexual excitement in the male deer; *broadly* **:** sexual excitement in a mammal (as estrus in the female) especially when periodic
2 : the period during which rut normally occurs — often used with *the*

²rut *intransitive verb* **rut·ted; rut·ting** (1625)
: to be in or enter into a state of rut

³rut *noun* [perhaps modification of Middle French *route* way, route] (1580)
1 a : a track worn by a wheel or by habitual passage **b :** a groove in which something runs **c :** CHANNEL, FURROW
2 : a usual or fixed practice; *especially* **:** a monotonous routine ⟨fall easily into a conversational *rut*⟩

⁴rut *transitive verb* **rut·ted; rut·ting** (1607)
: to make a rut in **:** FURROW

ru·ta·ba·ga \ˌrü-tə-ˈbā-gə, ˌrů-, -ˈbe-; ˈrü-tə-ˌ, ˈrů-\ *noun* [Swedish dialect *rotabagge,* from *rot* root + *bagge* bag] (1799)
: a turnip (*Brassica napus napobrassica*) that usually produces a large yellowish root

ruth \ˈrüth\ *noun* [Middle English *ruthe,* from *ruen* to rue] (13th century)
1 : compassion for the misery of another
2 : sorrow for one's own faults **:** REMORSE

Ruth \ˈrüth\ *noun* [Hebrew *Rūth*]
1 : a Moabite woman who accompanied Naomi to Bethlehem and became the ancestress of David
2 : a short narrative book of canonical Jewish and Christian Scriptures — see BIBLE table

ru·the·ni·um \rü-ˈthē-nē-əm\ *noun* [New Latin, from Medieval Latin *Ruthenia* Ruthenia] (1848)
: a hard brittle grayish polyvalent rare metallic element occurring in platinum ores and used in hardening platinum alloys — see ELEMENT table

ruth·er·ford·ium \ˌrə-t͟hə(r)-ˈfȯr-dē-əm\ *noun* [New Latin, from Ernest *Rutherford*] (1969)
: UNNILQUADIUM

ruth·ful \ˈrüth-fəl\ *adjective* (13th century)
1 : full of ruth **:** TENDER
2 : full of sorrow **:** WOEFUL
3 : causing sorrow
— **ruth·ful·ly** \-fə-lē\ *adverb*
— **ruth·ful·ness** *noun*

ruth·less \ˈrüth-ləs *also* ˈrůth-\ *adjective* (14th century)
: having no ruth **:** MERCILESS, CRUEL
— **ruth·less·ly** *adverb*
— **ruth·less·ness** *noun*

ru·ti·lant \ˈrü-tᵊl-ənt\ *adjective* [Middle English *rutilaunt,* from Latin *rutilant-, rutilans,* past participle of *rutilare* to glow reddish, from *rutilus* ruddy; probably akin to Latin *ruber* red — more at RED] (15th century)
: having a reddish glow

ru·tile \ˈrü-ˌtēl\ *noun* [German *Rutil,* from Latin *rutilus*] (1803)
: a mineral that consists of titanium dioxide usually with a little iron, is typically of a reddish brown color but sometimes deep red or black, and has a brilliant metallic or adamantine luster

rut·tish \ˈrə-tish\ *adjective* (1601)
: inclined to rut **:** LUSTFUL
— **rut·tish·ly** *adverb*
— **rut·tish·ness** *noun*

rut·ty \ˈrə-tē\ *adjective* **rut·ti·er; -est** (1596)
: full of ruts

RV \ˌär-ˈvē\ *noun* (1967)
: RECREATIONAL VEHICLE

R–value \ˈär-ˌval-(ˌ)yü\ *noun* [probably from thermal *resistance*] (1948)
: a measure of resistance to the flow of heat through a given thickness of a material (as insulation) with higher numbers indicating better insulating properties — compare U-VALUE

Rx \ˌär-ˈeks\ *noun* [alteration of ℞, symbol used at the beginning of a prescription, abbreviation for Latin *recipe,* literally, take — more at RECIPE] (1926)
: a medical prescription

-ry *noun suffix* [Middle English *-rie,* from Old French, short for *-erie* -ery]
: -ERY ⟨wizard*ry*⟩ ⟨citizen*ry*⟩ ⟨ancient*ry*⟩

rya \ˈrē-ə\ *noun* [*Rya,* village in southwest Sweden] (1945)
: a Scandinavian handwoven rug with a deep resilient comparatively flat pile; *also* **:** the weave typical of this rug

¹rye \ˈrī\ *noun* [Middle English, from Old English *ryge;* akin to Old High German *rocko* rye, Lithuanian *rugys*] (before 12th century)
1 : a hardy annual grass (*Secale cereale*) that is widely grown for grain and as a cover crop
2 : the seeds of rye
3 : RYE BREAD
4 : RYE WHISKEY

²rye *noun* [Romany *rai* gentleman, master, from Sanskrit *rājan* king — more at ROYAL] (1851)
: a male Gypsy

rye bread *noun* (1579)
: bread made wholly or in part of rye flour; *especially* **:** a light bread often with caraway seeds

rye·grass \ˈrī-ˌgras\ *noun* (1747)
: any of several grasses (genus *Lolium*); *especially* **:** either of two grasses (*Latin perenne* and *Latin multiflorum*) that are used especially for pasture, a cover crop, or lawn grass

rye whiskey *noun* (1785)
: a whiskey distilled from rye or from rye and malt

S is the nineteenth letter of the English alphabet. It came through Latin via Etruscan from Greek sigma, *which borrowed the character from the Phoenician* šîn, *pronounced "sh." In English,* s *usually represents the sibilant sound in* soon, hates, *and* hats. *In certain positions, however,* s *may represent the sound of* z *(as in* deserve; loves; heads*), and it sometimes has the value of* sh *or* zh, *as in* sure *and* measure. *In a few words it is silent (as* isle *and* debris*). With the letter* h *it forms the digraph* sh, *as in* ship. *In related languages, English* s *corresponds to German* s, *Latin* s *or* r, *Greek* s *or* h *(rough breathing), and Sanskrit* s *or* ṣ. *For example, English* sit *is historically related to German* sitzen, *Latin* sedere, *Greek* hezesthai, *and Sanskrit* sadas *("a seat"). The small form of* s *began as a simple reduction in size of the capital form. In printing, the "long* s*" (ſ), formerly much used, now appears only rarely, usually joined with a following consonant such as* s *or* t.

s \'es\ *noun, plural* **s's** *or* **ss** \'e-səz\ *often capitalized, often attributive* (before 12th century)
1 a : the 19th letter of the English alphabet **b :** a graphic representation of this letter **c :** a speech counterpart of orthographic *s*
2 : a graphic device for reproducing the letter *s*
3 : one designated *s* especially as the 19th in order or class
4 [abbreviation for *satisfactory*] **a :** a grade rating a student's work as satisfactory **b :** one graded or rated with an S
5 : something shaped like the letter S

[1]**-s** \s *after a voiceless consonant,* z *after a voiced consonant or a vowel*\ *noun plural suffix* [Middle English *-es, -s,* from Old English *-as,* nominative & accusative plural ending of some masculine nouns; akin to Old Saxon *-os*]
— used to form the plural of most nouns that do not end in *s, z, sh, ch,* or postconsonantal *y* ⟨head*s*⟩ ⟨book*s*⟩ ⟨boy*s*⟩ ⟨belief*s*⟩, to form the plural of proper nouns that end in postconsonantal *y* ⟨Mary*s*⟩, and with or without a preceding apostrophe to form the plural of abbreviations, numbers, letters, and symbols used as nouns ⟨MC*s*⟩ ⟨PhD'*s*⟩ ⟨4*s*⟩ ⟨the 1940'*s*⟩ ⟨$*s*⟩ ⟨B'*s*⟩; compare [1]-ES

[2]**-s** *adverb suffix* [Middle English *-es, -s,* plural ending of nouns, from *-es,* genitive singular ending of nouns (functioning adverbially), from Old English *-es*]
— used to form adverbs denoting usual or repeated action or state ⟨always at home Sunday*s*⟩ ⟨morning*s* he stops by the newsstand⟩

[3]**-s** *verb suffix* [Middle English (Northern & North Midland dialect) *-es,* from Old English (Northumbrian dialect) *-es, -as,* probably from Old English *-es, -as,* 2d singular present indicative ending — more at -EST]
— used to form the third person singular present of most verbs that do not end in *s, z, sh, ch,* or postconsonantal *y* ⟨fall*s*⟩ ⟨take*s*⟩ ⟨play*s*⟩; compare [2]-ES

[1]**'s** *same as* -'S\ *verb* [contraction of *is, has, does*] (1584)
1 a : IS ⟨she'*s* here⟩ **b :** WAS ⟨when'*s* the last time you ate?⟩
2 : HAS ⟨he'*s* seen them⟩
3 : DOES ⟨what'*s* he want?⟩

[2]**'s** \s\ *pronoun* [by contraction] (1588)
: US — used with *let* ⟨let'*s*⟩

-'s \s *after voiceless consonants other than* s, sh, ch; z *after vowels and voiced consonants other than* z, zh, j; əz *after* s, sh, ch, z, zh, j\ *noun suffix or pronoun suffix* [Middle English *-es, -s,* genitive singular ending, from Old English *-es;* akin to Old High German *-es,* genitive singular ending, Greek *-oio, -ou,* Sanskrit *-asya*]
— used to form the possessive of singular nouns ⟨boy'*s*⟩, of plural nouns not ending in *s* ⟨children'*s*⟩, of some pronouns ⟨anyone'*s*⟩, and of word groups functioning as nouns ⟨the man in the corner'*s* hat⟩ or pronouns ⟨someone else'*s*⟩

Saa·nen \'sä-nən, 'zä-\ *noun* [*Saanen,* locality in southwest Switzerland] (1906)
: any of a Swiss breed of usually white and hornless short-haired dairy goats

sab·a·dil·la \ˌsa-bə-'di-lə, -'dē-yə\ *noun* [Spanish *cebadilla*] (1812)
: a Mexican plant (*Schoenocaulon officinalis*) of the lily family; *also* **:** its seeds that are used as a source of veratrine and in insecticides

sa·ba·yon \sȧ-bȧ-yōⁿ\ *noun* [French, modification of Italian *zabaione*] (1906)
: ZABAGLIONE

sab·bat \'sa-bət\ *noun, often capitalized* [French, literally, sabbath, from Latin *sabbatum*] (1652)
: a midnight assembly of diabolists (as witches and sorcerers) held especially in medieval and Renaissance times to renew allegiance to the devil through mystic rites and orgies

[1]**Sab·ba·tar·i·an** \ˌsa-bə-'ter-ē-ən\ *noun* [Latin *sabbatarius,* from *sabbatum* sabbath] (1613)
1 : one who observes the Sabbath on Saturday in conformity with the letter of the fourth commandment
2 : an adherent of Sabbatarianism

[2]**Sabbatarian** *adjective* (circa 1631)
1 : of or relating to the Sabbath
2 : of or relating to the Sabbatarians or Sabbatarianism

Sab·ba·tar·i·an·ism \ˌsa-bə-'ter-ē-ə-ˌni-zəm\ *noun* (circa 1674)
: strict and often rigorous observance of the Sabbath

Sab·bath \'sa-bəth\ *noun* [Middle English *sabat,* from Old French & Old English, from Latin *sabbatum,* from Greek *sabbaton,* from Hebrew *shabbāth,* literally, rest] (before 12th century)
1 a : the seventh day of the week observed from Friday evening to Saturday evening as a day of rest and worship by Jews and some Christians **b :** Sunday observed among Christians as a day of rest and worship
2 : a time of rest

[1]**sab·bat·i·cal** \sə-'ba-ti-kəl\ *or* **sab·bat·ic** \-tik\ *adjective* [Late Latin *sabbaticus,* from Greek *sabbatikos,* from *sabbaton*] (1645)
1 : of or relating to the sabbath ⟨*sabbatical* laws⟩
2 : of or relating to a sabbatical year

[2]**sabbatical** *noun* (1903)
1 : SABBATICAL YEAR 2
2 : LEAVE 1b
3 : a break or change from a normal routine (as of employment)

sabbatical year *noun* (1599)
1 *often S capitalized* **:** a year of rest for the land observed every seventh year in ancient Judea
2 : a leave often with pay granted usually every seventh year (as to a college professor) for rest, travel, or research — called also *sabbatical leave*

Sa·bel·li·an \sə-'be-lē-ən\ *noun* [Latin *Sabellus* Sabine] (1601)
1 : a member of one of a group of early Italian peoples including Sabines and Samnites
2 : one or all of several little known languages or dialects of ancient Italy presumably closely related to Oscan and Umbrian
— **Sabellian** *adjective*

[1]**sa·ber** *or* **sa·bre** \'sā-bər\ *noun* [French *sabre,* modification of German dialect *Sabel,* from Middle High German, probably of Slavic origin; akin to Russian *sablya* saber] (1680)
1 : a cavalry sword with a curved blade, thick back, and guard
2 a : a light fencing or dueling sword having an arched guard that covers the back of the hand and a tapering flexible blade with a full cutting edge along one side and a partial cutting edge on the back at the tip — compare ÉPÉE, FOIL **b :** the sport of fencing with the saber

saber 1

[2]**saber** *or* **sabre** *transitive verb* **sa·bered** *or* **sa·bred; sa·ber·ing** *or* **sa·bring** \-b(ə-)riŋ\ (1790)
: to strike, cut, or kill with a saber

sa·ber·met·rics \ˌsā-bər-'me-triks\ *noun plural but singular in construction* [*saber-* (from Society for American Baseball Research) + *-metrics* (as in *econometrics*)] (1982)
: the statistical analysis of baseball data
— **sa·ber·me·tri·cian** \ˌsā-bər-mə-'tri-shən\ *noun*

saber rattling *noun* (1922)
: ostentatious display of military power

saber saw *noun* (1953)
: a light portable electric saw with a pointed reciprocating blade

sa·ber–toothed \'sā-bər-ˌtütht\ *adjective* (1849)
: having long sharp canine teeth

saber–toothed tiger *noun* (1849)
: any of numerous extinct cats (as genus *Smilodon*) widely distributed from the Oligocene through the Pleistocene and characterized by long curving upper canines — called also *saber-toothed cat*

sa·bin \'sā-bən\ *noun* [Wallace C. W. *Sabine* (died 1919) American physicist] (1934)
: a unit of acoustic absorption equivalent to the absorption by one square foot of a perfect absorber

Sa·bine \'sā-ˌbīn, *especially British* 'sa-\ *noun* [Middle English *Sabin,* from Latin *Sabinus*] (14th century)
1 : a member of an ancient people of the Apennines northeast of Latium
2 : the Italic language of the Sabine people
— **Sabine** *adjective*

[1]**sa·ble** \'sā-bəl\ *noun, plural* **sables** [Middle English, sable or its fur, the heraldic color black, black, from Middle French, sable or its fur, the heraldic color black, from Middle Low German *sabel* sable or its fur, from Middle High German *zobel,* of Slavic origin; akin to Russian *sobol'* sable or its fur] (14th century)
1 a : the color black **b :** black clothing worn in mourning — usually used in plural
2 a *or plural* **sable** (1) **:** a carnivorous mammal (*Martes zibellina*) of the weasel family that occurs chiefly in northern Asia (2) **:** any of various animals related to the sable **b :** the fur or pelt of a sable

3 a : the usually dark brown color of the fur of the sable **b :** a grayish yellowish brown
²sable *adjective* (15th century)
1 : of the color black
2 : DARK, GLOOMY
sa·ble·fish \'sā-bəl-ˌfish\ *noun* (1917)
: a large spiny-finned gray to blackish bony fish (*Anoplopoma fimbria* of the family Anoplomatidae) of the Pacific coast that is an important food fish and has a liver rich in vitamins
sa·bot \sa-'bō, 'sa-(ˌ)bō, *for 1b also* 'sa-bət\ *noun* [French] (1607)
1 a : a wooden shoe worn in various European countries **b** (1) **:** a strap across the instep in a shoe especially of the sandal type (2) **:** a shoe having a sabot strap
2 : a thrust-transmitting carrier that positions a missile in a gun barrel or launching tube and that prevents the escape of gas ahead of the missile
3 : SHOE 6
¹sab·o·tage \'sa-bə-ˌtäzh\ *noun* [French, from *saboter* to clatter with sabots, botch, sabotage, from *sabot*] (1910)
1 : destruction of an employer's property (as tools or materials) or the hindering of manufacturing by discontented workers
2 : destructive or obstructive action carried on by a civilian or enemy agent to hinder a nation's war effort
3 a : an act or process tending to hamper or hurt **b :** deliberate subversion ◆
²sabotage *transitive verb* **-taged; -tag·ing** (1913)
: to practice sabotage on
sab·o·teur \ˌsa-bə-'tər, -'tu̇r, -'tyu̇r\ *noun* [French, from *saboter*] (1921)
: one that practices sabotage
sa·bra \'sä-brə\ *noun, often capitalized* [New Hebrew *ṣabhār,* literally, prickly pear] (1945)
: a native-born Israeli
sac \'sak\ *noun* [French, literally, bag, from Latin *saccus* — more at SACK] (1741)
: a pouch within an animal or plant often containing a fluid ⟨a synovial *sac*⟩
— sac·like \-ˌlīk\ *adjective*
Sac \'sak, 'sȯk\ *variant of* SAUK
sa·ca·huis·te \ˌsa-kə-'wis-tə, ˌsä-, -tē\ *or* **sa·ca·huis·ta** \-tə\ *noun* [Mexican Spanish *zacahuiscle,* from (assumed) Nahuatl *zacahuitztli,* from Nahuatl *zacatl* grass, hay + *huitztli* thorn] (1896)
: a bear grass (*Nolina texana*) that may cause poisoning in livestock
sac·a·ton \'sa-kə-ˌtōn\ *noun* [American Spanish *zacatón,* from *zacate* coarse grass, from Nahuatl *zacatl*] (1846)
: a coarse perennial grass (*Sporobolus wrightii*) of the southwestern U.S. that is used for hay in alkaline regions
sac·cade \sa-'käd\ *noun* [French, twitch, jerk, from Middle French, from *saquer* to pull, draw] (1938)
: a small rapid jerky movement of the eye especially as it jumps from fixation on one point to another (as in reading)
— sac·cad·ic \-'kä-dik\ *adjective*
sac·cate \'sa-ˌkāt\ *adjective* [New Latin *saccatus,* from Latin *saccus*] (1830)
: having the form of a sac or pouch ⟨*saccate* pollen grains⟩
sacchar- *or* **sacchari-** *or* **saccharo-** *combining form* [Latin *saccharum,* from Greek *sakcharon,* from Prakrit *sakkharā,* from Sanskrit *śarkarā* gravel, sugar]
: sugar ⟨*saccharify*⟩ ⟨*saccharometer*⟩
sac·cha·rase \'sa-kə-ˌrās, -ˌrāz\ *noun* [International Scientific Vocabulary] (1920)
: INVERTASE
sac·cha·ride \'sa-kə-ˌrīd\ *noun* (1895)
: a simple sugar, combination of sugars, or polymerized sugar **:** CARBOHYDRATE
sac·char·i·fi·ca·tion \sə-ˌkar-ə-fə-'kā-shən\ *noun* (1839)
: the process of breaking a complex carbohydrate (as starch or cellulose) into simple sugars
— sac·char·i·fy \sə-'kar-ə-ˌfī, sa-\ *transitive verb*
sac·cha·rim·e·ter \ˌsa-kə-'ri-mə-tər\ *noun* [International Scientific Vocabulary] (1874)
: a device for measuring the amount of sugar in a solution; *specifically* **:** a polarimeter so used
sac·cha·rin \'sa-k(ə-)rən\ *noun* [International Scientific Vocabulary] (1885)
: a crystalline compound $C_7H_5NO_3S$ that is unrelated to the carbohydrates, is several hundred times sweeter than cane sugar, and is used as a calorie-free sweetener
sac·cha·rine \'sa-k(ə-)rən, -kə-ˌrēn, -kə-ˌrīn\ *adjective* [Latin *saccharum*] (circa 1674)
1 a : of, relating to, or resembling that of sugar ⟨*saccharine* taste⟩ **b :** yielding or containing sugar ⟨*saccharine* vegetables⟩
2 : overly or sickishly sweet ⟨*saccharine* flavor⟩
3 : ingratiatingly or affectedly agreeable or friendly
4 : overly sentimental **:** MAWKISH
— sac·cha·rin·i·ty \ˌsa-kə-'ri-nə-tē\ *noun*
sac·cha·roi·dal \ˌsa-kə-'rȯi-dᵊl\ *adjective* (1838)
: having or being a fine granular texture like that of loaf sugar ⟨*saccharoidal* marble⟩
sac·cha·rom·e·ter \-'rä-mə-tər\ *noun* (1784)
: SACCHARIMETER; *especially* **:** a hydrometer with a special scale
sac·cha·ro·my·ces \-rō-'mī-(ˌ)sēz\ *noun* [New Latin, from *sacchar-* + *-myces* fungus, from Greek *mykēs* — more at MYC-] (1873)
: any of a genus (*Saccharomyces* of the family Saccharomycetaceae) of usually unicellular yeasts (as a brewer's yeast) that are distinguished by their sparse or absent mycelium and by their facility in reproducing asexually by budding
sac·cu·lar \'sa-kyə-lər\ *adjective* (circa 1859)
: resembling a sac ⟨a *saccular* aneurysm⟩
sac·cu·lat·ed \-ˌlā-təd\ *also* **sac·cu·late** \-ˌlāt, -lət\ *adjective* (circa 1836)
: having or formed of a series of saccular expansions
— sac·cu·la·tion \ˌsa-kyə-'lā-shən\ *noun*
sac·cule \'sa-(ˌ)kyü(ə)l\ *noun* [New Latin *sacculus,* from Latin, diminutive of *saccus* bag — more at SACK] (circa 1839)
: a little sac; *specifically* **:** the smaller chamber of the membranous labyrinth of the ear
sac·cu·lus \'sa-kyə-ləs\ *noun, plural* **-li** \-ˌlī, -ˌlē\ [New Latin] (1748)
: SACCULE
sac·er·do·tal \ˌsa-sər-'dō-tᵊl, ˌsa-kər-\ *adjective* [Middle English, from Middle French, from Latin *sacerdotalis,* from *sacerdot-, sacerdos* priest, from *sacer* sacred + *-dot-, -dos* (akin to *facere* to make) — more at SACRED, DO] (15th century)
1 : of or relating to priests or a priesthood **:** PRIESTLY
2 : of, relating to, or suggesting sacerdotalism
— sac·er·do·tal·ly \-tᵊl-ē\ *adverb*
sac·er·do·tal·ism \-tᵊl-ˌi-zəm\ *noun* (1856)
: religious belief emphasizing the powers of priests as essential mediators between God and mankind
— sac·er·do·tal·ist \-tᵊl-ist\ *noun*
sac fungus *noun* (circa 1929)
: ASCOMYCETE
sa·chem \'sā-chəm, 'sa-\ *noun* [Narraganset *sâchim*] (1622)
1 : a North American Indian chief; *especially* **:** the chief of a confederation of the Algonquian tribes of the North Atlantic coast
2 : a Tammany leader
— sa·chem·ic \sā-'che-mik, sa-\ *adjective*
Sa·cher torte \'sä-kər-, 'zä-\ *noun* [German *Sachertorte,* from *Sacher* (name of a family of 19th and 20th century Austrian restaurant proprietors) + German *Torte* torte] (1906)
: a rich chocolate torte with an apricot jam filling
sa·chet \sa-'shā\ *noun* [Middle French, from Old French, diminutive of *sac* bag — more at SAC] (15th century)
1 : a small bag or packet
2 : a small bag containing a perfumed powder or potpourri used to scent clothes and linens
— sa·cheted \-'shād\ *adjective*
¹sack \'sak\ *noun* [Middle English *sak* bag, sackcloth, from Old English *sacc,* from Latin *saccus* bag & Late Latin *saccus* sackcloth, both from Greek *sakkos* bag, sackcloth, of Semitic origin; akin to Hebrew *śaq* bag, sackcloth] (before 12th century)
1 : a usually rectangular-shaped bag (as of paper, burlap, or canvas)
2 : the amount contained in a sack; *especially* **:** a fixed amount of a commodity used as a unit of measure
3 a : a woman's loose-fitting dress **b :** a short usually loose-fitting coat for women and children **c :** SACQUE 2
4 : DISMISSAL
5 a : HAMMOCK, BUNK **b :** BED
6 : a base in baseball
7 : an instance of sacking the quarterback in football
— sack·ful \-ˌfu̇l\ *noun*
²sack *transitive verb* (14th century)
1 : to put in or as if in a sack
2 : to dismiss especially summarily
3 : to tackle (the quarterback) behind the line of scrimmage in football
— sack·er *noun*
³sack *noun* [modification of Middle French *sec* dry, from Latin *siccus;* probably akin to Old High German *sīhan* to filter, Sanskrit *siñcati* he pours] (circa 1532)
: any of several white wines imported to England from Spain and the Canary Islands during the 16th and 17th centuries
⁴sack *transitive verb* [⁵*sack*] (circa 1547)
1 : to plunder (as a town) especially after capture
2 : to strip of valuables **:** LOOT
synonym see RAVAGE
— sack·er *noun*

◇ WORD HISTORY
sabotage Because the word *sabotage* appears to have some relation to French *sabot* "wooden shoe," the suggestion has often been made that *sabotage* in France originally alluded to disgruntled industrial workers throwing their sabots into machinery in order to damage it. In fact, there is no evidence to support such an etymology. The meaning "to damage an employer's property" of the French verb *saboter,* attested in the first decade of our century, is clearly built on an older sense "to perform haphazardly, botch, bungle," first attested in 1808. This meaning is in turn usually explained as proceeding from a yet older sense, "to make a clattering noise with sabots," on the premise that walking with wooden shoes suggests clumsy performance. Though the hypothesis is not implausible, *saboter* is attested in the 17th century with the sense "to torment," and as early as the 14th century in the meaning "to shake, knock." It is difficult to determine if these diverse meanings constitute a single line of development, or if the many associations evoked in Frenchmen by this piece of peasant footwear caused repeated episodes of verb creation from the same noun.

[5]sack *noun* [Middle French *sac,* from Old Italian *sacco,* literally, bag, from Latin *saccus*] (1549)
: the plundering of a captured town
sack·but \'sak-(,)bət\ *noun* [Middle French *saqueboute,* literally, hooked lance, from Old French, from *saquer* to pull + *boter* to push — more at BUTT] (1533)
: the medieval and Renaissance trombone
sack·cloth \'sa(k)-,klȯth\ *noun* [[1]*sack*] (13th century)
1 : a coarse cloth of goat or camel's hair or of flax, hemp, or cotton
2 : a garment of sackcloth worn as a sign of mourning or penitence
sack coat *noun* (1847)
: a man's jacket with a straight back
sack·ing \'sa-kiŋ\ *noun* (1707)
: material for sacks; *especially* **:** a coarse fabric (as burlap)
sack out *intransitive verb* [[1]*sack*] (1946)
: to go to bed **:** go to sleep
sack race *noun* (1859)
: a jumping race in which each contestant's legs are enclosed in a sack
sacque \'sak\ *noun* [alteration of [1]*sack*] (1846)
1 : SACK 3a, b
2 : an infant's usually short jacket that fastens at the neck
[1]sa·cral \'sa-krəl, 'sā-\ *adjective* (1767)
: of, relating to, or lying near the sacrum
[2]sa·cral \'sā-krəl, 'sa-\ *adjective* [Latin *sacr-, sacer* — more at SACRED] (1882)
: HOLY, SACRED
sac·ra·ment \'sa-krə-mənt\ *noun* [Middle English *sacrement, sacrament,* from Old French & Late Latin; Old French, from Late Latin *sacramentum,* from Latin, oath of allegiance, obligation, from *sacrare* to consecrate] (13th century)
1 a : a Christian rite (as baptism or the Eucharist) that is believed to have been ordained by Christ and that is held to be a means of divine grace or to be a sign or symbol of a spiritual reality **b :** a religious rite or observance comparable to a Christian sacrament
2 *capitalized* **a :** COMMUNION 2a **b :** BLESSED SACRAMENT
3 : something likened to a religious sacrament ⟨saw voting as a *sacrament* of democracy⟩
[1]sac·ra·men·tal \,sa-krə-'men-tᵊl\ *adjective* (15th century)
1 : of, relating to, or having the character of a sacrament
2 : suggesting a sacrament (as in sacredness)
— sac·ra·men·tal·ly \-tᵊl-ē\ *adverb*
[2]sacramental *noun* (15th century)
: an action or object (as the rosary) of ecclesiastical origin that serves to express or increase devotion
sac·ra·men·tal·ism \-tᵊl-,i-zəm\ *noun* (1861)
: belief in or use of sacramental rites, acts, or objects; *specifically* **:** belief that the sacraments are inherently efficacious and necessary for salvation
sac·ra·men·tal·ist \-tᵊl-ist\ *noun* (1840)
1 : SACRAMENTARIAN
2 : an adherent of sacramentalism
Sac·ra·men·tar·i·an \,sa-krə-,men-'ter-ē-ən, -mən-\ *noun* (1535)
1 : one who interprets sacraments as merely visible symbols
2 : SACRAMENTALIST
— Sacramentarian *adjective*
— Sac·ra·men·tar·i·an·ism \-ē-ə-,ni-zəm\ *noun*
sa·crar·i·um \sə-'krer-ē-əm, sa-, sā-\ *noun, plural* **-ia** \-ē-ə\ [Medieval Latin, from Latin, shrine, from *sacr-, sacer* sacred] (1727)
1 a : SANCTUARY 1b **b :** SACRISTY **c :** PISCINA
2 : an ancient Roman shrine or sanctuary in a temple or a home holding sacred objects
sa·cred \'sā-krəd\ *adjective* [Middle English, from past participle of *sacren* to consecrate, from Old French *sacrer,* from Latin *sacrare,* from *sacr-, sacer* sacred; akin to Latin *sancire* to make sacred, Hittite *šaklāi-* rite] (14th century)
1 a : dedicated or set apart for the service or worship of a deity ⟨a tree *sacred* to the gods⟩ **b :** devoted exclusively to one service or use (as of a person or purpose) ⟨a fund *sacred* to charity⟩
2 a : worthy of religious veneration **:** HOLY **b :** entitled to reverence and respect
3 : of or relating to religion **:** not secular or profane ⟨*sacred* music⟩
4 *archaic* **:** ACCURSED
5 a : UNASSAILABLE, INVIOLABLE **b :** highly valued and important ⟨a *sacred* responsibility⟩
— sa·cred·ly *adverb*
— sa·cred·ness *noun*
sacred baboon *noun* [from its veneration by the ancient Egyptians] (circa 1890)
: HAMADRYAS BABOON
sacred cow *noun* [from the veneration of the cow by Hindus] (1910)
: one that is often unreasonably immune from criticism or opposition
sacred mushroom *noun* (1930)
1 : any of various New World hallucinogenic fungi (genus *Psilocybe*) used especially in some Indian ceremonies
2 : MESCAL BUTTON
[1]sac·ri·fice \'sa-krə-,fīs, *also* -fəs *or* -,fīz\ *noun* [Middle English, from Old French, from Latin *sacrificium,* from *sacr-, sacer* + *facere* to make — more at DO] (13th century)
1 : an act of offering to a deity something precious; *especially* **:** the killing of a victim on an altar
2 : something offered in sacrifice
3 a : destruction or surrender of something for the sake of something else **b :** something given up or lost ⟨the *sacrifices* made by parents⟩
4 : LOSS ⟨goods sold at a *sacrifice*⟩
5 : SACRIFICE HIT
[2]sacrifice *verb* **-ficed; -fic·ing** (14th century)
transitive verb
1 : to offer as a sacrifice
2 : to suffer loss of, give up, renounce, injure, or destroy especially for an ideal, belief, or end
3 : to sell at a loss
intransitive verb
1 : to make or perform the rites of a sacrifice
2 : to make a sacrifice hit in baseball
— sac·ri·fic·er *noun*
sacrifice fly *noun* (1944)
: an outfield fly in baseball caught by a fielder after which a runner scores
sacrifice hit *noun* (1880)
: a bunt in baseball that allows a runner to advance one base while the batter is put out
sac·ri·fi·cial \,sa-krə-'fi-shəl\ *adjective* (1607)
1 : of, relating to, of the nature of, or involving sacrifice
2 : of or relating to a metal that serves as the anode and is electrolytically consumed instead of another metal that is present
— sac·ri·fi·cial·ly \-shə-lē\ *adverb*
sac·ri·lege \'sa-krə-lij\ *noun* [Middle English, from Middle French, from Latin *sacrilegium,* from *sacrilegus* one who steals sacred things, from *sacr-, sacer* + *legere* to gather, steal — more at LEGEND] (14th century)
1 : a technical and not necessarily intrinsically outrageous violation (as improper reception of a sacrament) of what is sacred because consecrated to God
2 : gross irreverence toward a hallowed person, place, or thing
— sac·ri·le·gious \÷,sa-krə-'li-jəs *also* -'lē-\ *adjective*
— sac·ri·le·gious·ly *adverb*
— sac·ri·le·gious·ness *noun*
sac·ris·tan \'sa-krə-stən\ *noun* [Middle English, from Medieval Latin *sacristanus,* from Latin *sacr-, sacer*] (14th century)
: a person in charge of the sacristy and ceremonial equipment; *also* **:** SEXTON
sac·ris·ty \'sa-krə-stē\ *noun, plural* **-ties** [Middle English *sacristie,* from Medieval Latin *sacristia,* from *sacrista* sacristan, from Latin *sacr-, sacer*] (15th century)
: a room in a church where sacred vessels and vestments are kept and where the clergy vests
[1]sa·cro·il·i·ac \,sa-krō-'i-lē-,ak, ,sā-\ *adjective* [probably from French *sacro-iliaque,* from New Latin *sacrum* + French *iliaque* iliac] (1831)
: of, relating to, or being the region of juncture of the sacrum and ilium
[2]sacroiliac *noun* (1936)
: the sacroiliac region; *also* **:** its firm fibrous cartilage
sac·ro·sanct \'sa-krō-,saŋ(k)t\ *adjective* [Latin *sacrosanctus,* probably from *sacro sanctus* hallowed by a sacred rite] (1601)
1 : most sacred or holy **:** INVIOLABLE
2 : treated as if holy **:** immune from criticism or violation ⟨politically *sacrosanct* programs⟩
— sac·ro·sanc·ti·ty \,sa-krō-'saŋ(k)-tə-tē\ *noun*
sa·crum \'sa-krəm, 'sā-\ *noun, plural* **sa·cra** \'sa-krə, 'sā-\ [New Latin, from Late Latin *os sacrum* last bone of the spine, literally, holy bone, translation of Greek *hieron osteon*] (1753)
: the part of the vertebral column that is directly connected with or forms a part of the pelvis and in humans consists of five fused vertebrae
sad \'sad\ *adjective* **sad·der; sad·dest** [Middle English, from Old English *sæd* sated; akin to Old High German *sat* sated, Latin *satis* enough] (13th century)
1 a : affected with or expressive of grief or unhappiness **:** DOWNCAST **b** (1) **:** causing or associated with grief or unhappiness **:** DEPRESSING ⟨*sad* news⟩ (2) **:** REGRETTABLE, DEPLORABLE ⟨a *sad* relaxation of morals —C. W. Cunnington⟩ **c :** of little worth
2 : of a dull somber color ◆
— sad·ly *adverb*
— sad·ness *noun*
sad·den \'sa-dᵊn\ *verb* **sad·dened; sad·den·ing** \'sad-niŋ, 'sa-dᵊn-iŋ\ (1628)
transitive verb
: to make sad
intransitive verb
: to become sad

◇ WORD HISTORY
sad The word *sad* goes back to the remote Germanic past of the English language, though the ordinary modern meanings of the word, such as "downcast," "depressing," or "deplorable," give little hint of its history. Its Old English ancestor *sæd* meant "filled, full, satiated," a meaning matched by cognate words in other early Germanic languages, and continued by Modern German *satt* "full, having had enough to eat." In Middle English, though *sad* continued to have the Old English sense "full, sated," it underwent a great burst of semantic diversification. Common senses were "firmly established, fixed," "solid, weighty," "sober, serious," "powerful, strong," "hard, stout (of a blow)," "true, real," "deep, intense (of a color)." The meaning "sorrowful" was in use fairly early, by about 1300, though peculiarly it alone has been retained, as virtually all the other senses developed in Middle English gradually fell from use in early Modern English. The sense "sober, serious" occurs notably in Samuel Taylor Coleridge's poem "The Rime of the Ancient Mariner" ("A sadder and a wiser man,/He rose the following morn"), though Coleridge may have already been employing it as a conscious archaism.

1sad·dle \ˈsa-dᵊl\ *noun, often attributive* [Middle English *sadel,* from Old English *sadol;* akin to Old High German *satul* saddle] (before 12th century)
1 a (1) **:** a girthed usually padded and leather-covered seat for the rider of an animal (as a horse) (2) **:** a part of a driving harness comparable to a saddle that is used to keep the breeching in place **b :** a seat to be straddled by the rider of a vehicle (as a bicycle)

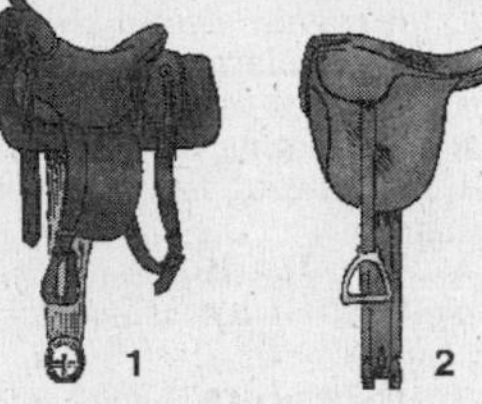

saddle 1a(1): *1* western, *2* English

2 : a device mounted as a support and often shaped to fit the object held
3 a : a ridge connecting two higher elevations **b :** a pass in a mountain range
4 a : both sides of the unsplit back of a carcass including both loins **b :** a colored marking on the back of an animal **c :** the rear part of a male fowl's back extending to the tail — see DUCK illustration
5 : the central part of the backbone of the binding of a book
6 : a piece of leather across the instep of a shoe
— sad·dle·less \-dᵊl-(l)əs\ *adjective*
— in the saddle : in control

2saddle *verb* **sad·dled; sad·dling** \ˈsad-liŋ, ˈsa-dᵊl-iŋ\ (before 12th century)
transitive verb
1 : to put a saddle on
2 a : to place under a burden or encumbrance **b :** to place (an onerous responsibility) on a person or group
intransitive verb
: to mount a saddled horse

sad·dle·bag \ˈsa-dᵊl-ˌbag\ *noun* (1773)
: one of a pair of covered pouches laid across the back of a horse behind the saddle or hanging over the rear wheel of a bicycle or motorcycle

saddle blanket *noun* (1737)
: a folded blanket or pad under a saddle to prevent galling the horse

sad·dle·bow \ˈsa-dᵊl-ˌbō\ *noun* (before 12th century)
: the arch in or the pieces forming the front of a saddle

sad·dle·bred \-ˌbred\ *noun* (1948)
: AMERICAN SADDLEBRED

sad·dle·cloth \-ˌklȯth\ *noun* (15th century)
: a cloth placed under or over a saddle

sad·dled prominent \ˈsa-dᵊld-\ *noun* [from the hump or prominence on the back of the larva] (1910)
: a moth (*Heterocampa guttivitta*) whose larva is a serious defoliator of hardwood trees in the eastern and midwestern U.S.

saddle horn *noun* (1856)
: a hornlike prolongation of the pommel of a stock saddle

saddle horse *noun* (1662)
: a horse suited for or trained for riding

saddle leather *noun* (1832)
: leather made of cowhide that is vegetable tanned and used especially for saddlery; *also* **:** smooth polished leather simulating this

sad·dler \ˈsad-lər\ *noun* (14th century)
: one that makes, repairs, or sells saddles and other furnishings for horses

sad·dlery \ˈsad-lə-rē, ˈsa-dᵊl-rē\ *noun, plural* **-dler·ies** (15th century)
: the trade, articles of trade, or shop of a saddler

saddle seat *noun* (1895)
: a slightly concave chair seat (as of a Windsor chair) with sometimes a thickened ridge at the center front

saddle shoe *noun* (1939)
: an oxford-style shoe having a saddle of contrasting color or leather — called also *saddle oxford*

saddle soap *noun* (1889)
: a mild soap used for cleansing and conditioning leather

saddle sore *noun* (1946)
1 : a gall or open sore developing on the back of a horse at points of pressure from an ill-fitting or ill-adjusted saddle
2 : an irritation or sore on parts of the rider chafed by the saddle

sad·dle·tree \ˈsa-dᵊl-ˌtrē\ *noun* (15th century)
: the frame of a saddle

Sad·du·ce·an \ˌsa-jə-ˈsē-ən, ˌsa-dyə-\ *adjective* (1593)
: of or relating to the Sadducees

Sad·du·cee \ˈsa-jə-ˌsē, ˈsa-dyə-\ *noun* [Middle English *saducee,* from Old English *sadduce,* from Late Latin *sadducaeus,* from Greek *saddoukaios,* from Late Hebrew *ṣāddūqī*] (before 12th century)
: a member of a Jewish party of the intertestamental period consisting of a traditional ruling class of priests and rejecting doctrines not in the Law (as resurrection, retribution in a future life, and the existence of angels)
— Sad·du·cee·ism \-ˌi-zəm\ *noun*

sa·dhe \ˈ(t)sä-də, -dē\ *noun* [Hebrew *ṣādhē*] (circa 1899)
: the 18th letter of the Hebrew alphabet — see ALPHABET table

sa·dhu *also* **sad·dhu** \ˈsä-(ˌ)dü\ *noun* [Sanskrit *sādhu*] (1845)
: a usually Hindu mendicant ascetic

sad·iron \ˈsad-ˌī(-ə)rn\ *noun* [*sad* (compact, heavy) + *iron*] (1761)
: a flatiron pointed at both ends and having a removable handle

sa·dism \ˈsā-ˌdi-zəm, ˈsa-\ *noun* [International Scientific Vocabulary, from Marquis de *Sade*] (1888)
1 : a sexual perversion in which gratification is obtained by the infliction of physical or mental pain on others (as on a love object) — compare MASOCHISM
2 a : delight in cruelty **b :** excessive cruelty
— sa·dist \ˈsā-dist, ˈsa-\ *noun*
— sa·dis·tic \sə-ˈdis-tik *also* sā- *or* sa-\ *adjective*
— sa·dis·ti·cal·ly \-ti-k(ə-)lē\ *adverb*

sa·do·mas·och·ism \ˌsā-(ˌ)dō-ˈma-sə-ˌki-zəm, ˌsa-, -ˈma-zə-\ *noun* [International Scientific Vocabulary *sadi*sm + *-o-* + *masochism*] (1922)
: the derivation of pleasure from the infliction of physical or mental pain either on others or on oneself
— sa·do·mas·och·ist \-kist\ *noun or adjective*
— sa·do·mas·och·is·tic \-ˌma-sə-ˈkis-tik, -ˌma-zə-\ *adjective*

sad sack *noun* (1943)
: an inept person; *especially* **:** an inept soldier
— sad–sack *adjective*

Sa·far \sə-ˈfär\ *noun* [Arabic *ṣafar*] (circa 1771)
: the 2d month of the Islamic year — see MONTH table

sa·fa·ri \sə-ˈfär-ē, -ˈfar-\ *noun* [Arabic *safarīy* of a trip] (1896)
1 : the caravan and equipment of a hunting expedition especially in eastern Africa; *also* **:** such a hunting expedition
2 : JOURNEY, EXPEDITION ⟨an arctic *safari*⟩
— safari *intransitive verb*

safari jacket *noun* (1951)
: a usually belted shirt jacket with pleated expansible pockets

safari suit *noun* (1967)
: a safari jacket with matching pants

1safe \ˈsāf\ *adjective* **saf·er; saf·est** [Middle English *sauf,* from Old French, from Latin *salvus* safe, healthy; akin to Latin *solidus* solid, Greek *holos* whole, safe, Sanskrit *sarva* entire] (14th century)
1 : free from harm or risk **:** UNHURT
2 a : secure from threat of danger, harm, or loss **b :** successful at getting to a base in baseball without being put out
3 : affording safety or security from danger, risk, or difficulty
4 *obsolete, of mental or moral faculties* **:** HEALTHY, SOUND
5 a : not threatening danger **:** HARMLESS **b :** unlikely to produce controversy or contradiction
6 a : not likely to take risks **:** CAUTIOUS **b :** TRUSTWORTHY, RELIABLE
— safe *or* **safe·ly** *adverb*
— safe·ness *noun*

2safe *noun* (15th century)
1 : a place or receptacle to keep articles (as valuables) safe
2 : CONDOM

safe–con·duct \ˈsāf-ˈkän-(ˌ)dəkt\ *noun* [Middle English *sauf conduit,* from Old French, safe conduct] (14th century)
1 : protection given a person passing through a military zone or occupied area
2 : a document authorizing safe-conduct

safe·crack·er \ˈsāf-ˌkra-kər\ *noun* (circa 1825)
: one that breaks open safes to steal
— safe·crack·ing \-kiŋ\ *noun*

safe–deposit box *noun* (1882)
: a box (as in the vault of a bank) for safe storage of valuables — called also *safety-deposit box*

1safe·guard \ˈsāf-ˌgärd\ *noun* [Middle English *saufgarde,* from Middle French *saufegarde,* from Old French, from *sauve* safe + *garde* guard] (14th century)
1 a : PASS, SAFE-CONDUCT **b :** CONVOY, ESCORT
2 a : a precautionary measure, stipulation, or device **b :** a technical contrivance to prevent accident

2safeguard *transitive verb* (15th century)
1 : to provide a safeguard for
2 : to make safe **:** PROTECT
synonym see DEFEND

safe house *noun* (1946)
: a place where one may engage in secret activities or take refuge

safe·keep·ing \ˈsāf-ˈkē-piŋ\ *noun* (15th century)
1 : the act or process of preserving in safety
2 : the state of being preserved in safety

safe·light \ˈsāf-ˌlīt\ *noun* (1903)
: a darkroom lamp with a filter to screen out rays that are harmful to sensitive film or paper

safe sex *noun* (1985)
: sexual activity and especially sexual intercourse in which various measures (as the use of latex condoms or the practice of monogamy) are taken to avoid disease (as AIDS) transmitted by sexual contact

1safe·ty \ˈsāf-tē\ *noun, plural* **safeties** [Middle English *saufte,* from Middle French *sauveté,* from Old French, from *sauve,* feminine of *sauf* safe] (14th century)
1 : the condition of being safe from undergoing or causing hurt, injury, or loss
2 : a device (as on a weapon or a machine) designed to prevent inadvertent or hazardous operation
3 a (1) **:** a situation in football in which a member of the offensive team is tackled behind its own goal line that counts two points for the defensive team — compare TOUCHBACK (2) **:** a member of a defensive backfield in football who occupies the deepest position in order to receive a kick, defend against a forward pass, or stop a ballcarrier **b :** a billiard shot made with no attempt to score or so as to

leave the balls in an unfavorable position for the opponent **c :** BASE HIT

²safety *transitive verb* **safe·tied; safe·ty·ing** (1940)
: to protect against failure, breakage, or accident ⟨*safety* a rifle⟩

safety belt *noun* (circa 1858)
: a belt fastening a person to an object (as a car seat) to prevent falling or injury

safety glass *noun* (1919)
: transparent material that is made by laminating a sheet of transparent plastic between sheets of clear glass and is used especially for windows (as of automobiles) likely to be subjected to shock or impact

safety lamp *noun* (1816)
: a miner's lamp constructed to avoid explosion in an atmosphere containing flammable gas usually by enclosing the flame in fine wire gauze

safe·ty·man \ˈsāf-tē-ˌman\ *noun* (1927)
: SAFETY 3a(2)

safety match *noun* (1863)
: a match capable of being struck and ignited only on a specially prepared friction surface

safety net *noun* (1950)
: something that provides security against misfortune or difficulty

safety pin *noun* (1857)
: a pin in the form of a clasp with a guard covering its point when fastened

safety razor *noun* (circa 1875)
: a razor provided with a guard for the blade to prevent deep cuts in the skin

safety valve *noun* (1813)
1 : an automatic escape or relief valve (as for a steam boiler)
2 : an outlet for pent-up energy or emotion
3 : something that relieves the pressure of overcrowding

saf·flow·er \ˈsa-ˌflau̇(-ə)r\ *noun* [Middle French *saffleur,* from Old Italian *saffiore,* from Arabic *aṣfar,* a yellow plant] (1642)
: a widely cultivated Old World composite herb (*Carthamus tinctorius*) with large orange or red flower heads and seeds rich in oil; *also* **:** a red dyestuff prepared from the flower heads

safflower oil *noun* (circa 1857)
: an edible drying oil obtained from the seeds of the safflower

saf·fron \ˈsa-frən\ *noun* [Middle English, from Old French *safran,* from Medieval Latin *safranum,* from Arabic *zaʽfarān*] (13th century)
1 a : the deep orange aromatic pungent dried stigmas of a purple-flowered crocus (*Crocus sativus*) used to color and flavor foods and formerly as a dyestuff and in medicine **b :** the crocus supplying saffron
2 : a moderate orange to orange yellow

saf·ra·nine \ˈsa-frə-ˌnēn, -nən\ *or* **saf·ra·nin** \-nən\ *noun* [International Scientific Vocabulary, from French or German *safran* saffron] (1868)
1 : any of various usually red synthetic dyes that are amino derivatives of bases
2 : any of various mixtures of safranine salts used in dyeing and as biological stains

saf·role \ˈsaf-ˌrōl\ *noun* [International Scientific Vocabulary, from sas*safras* + *-ole*] (1869)
: a poisonous oily cyclic ether $C_{10}H_{10}O_2$ that is the principal component of sassafras oil and is used chiefly in perfumery

¹sag \ˈsag\ *verb* **sagged; sag·ging** [Middle English *saggen,* probably of Scandinavian origin; akin to Swedish *sacka* to sag] (14th century)
intransitive verb
1 : to droop, sink, or settle from or as if from pressure or loss of tautness
2 a : to lose firmness, resiliency, or vigor ⟨spirits *sagging* from overwork⟩ **b :** to decline especially from a thriving state
3 : DRIFT
4 : to fail to stimulate or retain interest
transitive verb
: to cause to sag **:** leave slack in

²sag *noun* (1580)
1 : a tendency to drift (as of a ship to leeward)
2 a : a sagging part ⟨the *sag* in a rope⟩ **b :** a drop or depression below the surrounding area **c :** an instance or amount of sagging
3 : a temporary decline (as in the price of a commodity)

sa·ga \ˈsä-gə *also* ˈsa-\ *noun* [Old Norse — more at SAW] (1709)
1 : a prose narrative recorded in Iceland in the 12th and 13th centuries of historic or legendary figures and events of the heroic age of Norway and Iceland
2 : a modern heroic narrative resembling the Icelandic saga
3 : a long detailed account ⟨a *saga* of the Old South⟩

sa·ga·cious \sə-ˈgā-shəs, si-\ *adjective* [Latin *sagac-, sagax,* from *sagire* to perceive keenly; akin to Latin *sagus* prophetic — more at SEEK] (1607)
1 *obsolete* **:** keen in sense perception
2 a : of keen and farsighted penetration and judgment **:** DISCERNING ⟨*sagacious* judge of character⟩ **b :** caused by or indicating acute discernment ⟨*sagacious* purchase of stock⟩
synonym see SHREWD
— sa·ga·cious·ly *adverb*
— sa·ga·cious·ness *noun*

sa·gac·i·ty \sə-ˈga-sə-tē, si-\ *noun* (15th century)
: the quality of being sagacious

sag·a·more \ˈsa-gə-ˌmōr, -ˌmȯr\ *noun* [Eastern Abenaki *sàkəmα*] (1613)
1 : a subordinate chief of the Algonquian Indians of the North Atlantic coast
2 : SACHEM 1

saga novel *noun* (circa 1938)
: ROMAN-FLEUVE

¹sage \ˈsāj\ *adjective* **sag·er; sag·est** [Middle English, from Old French, from (assumed) Vulgar Latin *sapius,* from Latin *sapere* to taste, have good taste, be wise; akin to Oscan *sipus* knowing, Old Saxon an*sebbian* to perceive] (14th century)
1 a : wise through reflection and experience **b** *archaic* **:** GRAVE, SOLEMN
2 : proceeding from or characterized by wisdom, prudence, and good judgment ⟨*sage* advice⟩
synonym see WISE
— sage·ly *adverb*
— sage·ness *noun*

²sage *noun* (14th century)
1 : one (as a profound philosopher) distinguished for wisdom
2 : a mature or venerable man of sound judgment

³sage *noun* [Middle English, from Middle French *sauge,* from Latin *salvia,* from *salvus* healthy; from its use as a medicinal herb — more at SAFE] (14th century)
1 : a mint (*Salvia officinalis*) with grayish green aromatic leaves used especially in flavoring meats; *broadly* **:** SALVIA
2 : SAGEBRUSH

sage·brush \ˈsāj-ˌbrəsh\ *noun* (1850)
: any of several North American hoary composite subshrubs (genus *Artemisia*); *especially* **:** one (*A. tridentata*) having a bitter juice and an odor resembling sage and often covering vast tracts of alkaline plains in the western U.S.

sage cheese *noun* (1699)
: a cheese similar to mild cheddar flecked with green and flavored with sage

sage grouse *noun* (1876)
: a large grouse (*Centrocercus urophasianus*) of the dry sagebrush plains of western North America that has mottled gray and buff plumage with a contrasting black belly

sag·ger *or* **sag·gar** \ˈsa-gər\ *noun* [probably alteration of *safeguard*] (1752)
: a box made of fireclay in which delicate ceramic pieces are fired

sag·it·tal \ˈsa-jə-tᵊl\ *adjective* [Middle English *sagittale,* from Medieval Latin *sagittalis,* from Latin *sagitta* arrow] (14th century)
1 : of or relating to the suture between the parietal bones of the skull
2 : of, relating to, situated in, or being the median plane of the body or any plane parallel thereto
— sag·it·tal·ly \-tᵊl-ē\ *adverb*

Sag·it·tar·i·an \ˌsa-jə-ˈter-ē-ən\ *noun* (1911)
: SAGITTARIUS 2b

Sag·it·tar·i·us \-ē-əs\ *noun* [Latin (genitive *Sagittarii*), literally, archer, from *sagitta*]
1 : a southern zodiacal constellation pictured as a centaur shooting an arrow and containing the point in the sky where the center of the Milky Way galaxy is located
2 a : the 9th sign of the zodiac in astrology — see ZODIAC table **b :** one born under the sign of Sagittarius

sag·it·tate \ˈsa-jə-ˌtāt\ *adjective* [Latin *sagitta*] (1760)
: shaped like an arrowhead; *specifically* **:** elongated, triangular, and having the two basal lobes prolonged downward ⟨*sagittate* leaf⟩ — see LEAF illustration

sa·go \ˈsā-(ˌ)gō\ *noun, plural* **sagos** [Malay *sagu* sago palm] (circa 1580)
: a dry granulated or powdered starch prepared from the pith of a sago palm and used in foods and as textile stiffening

sago palm *noun* (1769)
: a plant that yields sago; *especially* **:** any of various lofty pinnate-leaved Indian and Malaysian palms (genus *Metroxylon*)

sa·gua·ro \sə-ˈwär-ə, -ˈgwär-, -ō\ *noun, plural* **-ros** [Mexican Spanish] (1856)
: an arborescent cactus (*Carnegiea gigantea*) of desert regions of the southwestern U.S. and Mexico that has a tall columnar simple or sparsely branched trunk of up to 60 feet (18 meters) and bears white flowers and edible fruit

saguaro

Sa·hap·ti·an \sə-ˈhap-tē-ən\ *variant of* SHAHAPTIAN

sa·hib \ˈsä-ˌ(h)ib, -ˌ(h)ēb, ˌsä-ˈ\ *noun* [Hindi *sāheb,* from Arabic *ṣāḥib*] (1673)
: SIR, MASTER — used especially among the native inhabitants of colonial India when addressing or speaking of a European of some social or official status

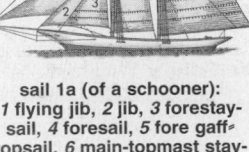

¹said *past and past participle of* SAY

²said \ˈsed\ *adjective* [past participle of *say*] (14th century)
: AFOREMENTIONED

¹sail \ˈsā(ə)l, *as last element in compounds often* səl\ *noun* [Middle English, from Old English *segl;* akin to Old High German *segal* sail] (before 12th century)
1 a (1) **:** an extent of fabric (as canvas) by means of which wind is used to propel a ship through water (2) **:** the sails of a ship **b** *plural usually* **sail :** a ship equipped with sails
2 : an extent of fabric used in propelling a wind-driven vehicle (as an iceboat)

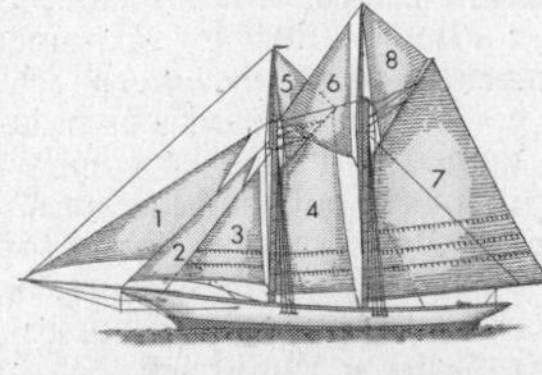

sail 1a (of a schooner): *1* **flying jib,** *2* **jib,** *3* **forestaysail,** *4* **foresail,** *5* **fore gaff-topsail,** *6* **main-topmast staysail,** *7* **mainsail,** *8* **main gaff-topsail**

3 : something that resembles a sail; *especially* **:** a streamlined conning tower on a submarine
4 : a passage by a sailing craft **:** CRUISE
— sailed \ˈsā(ə)ld\ *adjective*
— under sail : in motion with sails set

²sail (before 12th century)
intransitive verb
1 a : to travel on water in a ship **b :** YACHT
2 a : to travel on water by the action of wind upon sails or by other means **b :** to move or proceed easily, gracefully, nonchalantly, or without resistance ⟨*sails* through all sorts of contradictions —Vicki Hearne⟩ ⟨the bill *sailed* through the legislature⟩
3 : to begin a water voyage ⟨*sail* with the tide⟩
transitive verb
1 a : to travel on (water) by means of motive power (as sail) **b :** to glide through
2 : to direct or manage the motion of (as a ship)
— sail·able \ˈsā-lə-bəl\ *adjective*
— sail into : to attack vigorously or sharply ⟨*sailed into* me for being late⟩

sail·board \ˈsā(ə)l-ˌbōrd, -ˌbȯrd\ *noun* (1962)
: a modified surfboard having a mast mounted on a universal joint and sailed by one person standing up
— sail·board·ing *noun*

sail·boat \ˈsā(ə)l-ˌbōt\ *noun* (1798)
: a boat usually propelled by sail
— sail·boat·er \-ˌbō-tər\ *noun*
— sail·boat·ing *noun*

sail·cloth \-ˌklȯth\ *noun* (13th century)
: a heavy canvas used for sails, tents, or upholstery; *also* **:** a lightweight canvas used for clothing

sail·er \ˈsā-lər\ *noun* (15th century)
: a ship or boat especially having specified sailing qualities

sail·fish \ˈsā(ə)l-ˌfish\ *noun* (1879)
: any of a genus (*Istiophorus* and especially *I. platypterus*) of billfishes having a very large dorsal fin

sail·ing \ˈsā-liŋ\ *noun* (before 12th century)
1 a : the technical skill of managing a ship **:** NAVIGATION **b :** the method of determining the course to be followed to reach a given point
2 a : the sport of handling or riding in a sailboat **b :** a departure from a port

sail·or \ˈsā-lər\ *noun* [alteration of *sailer*] (circa 1642)
1 a : one that sails; *especially* **:** MARINER **b** (1) **:** a member of a ship's crew (2) **:** SEAMAN 2b
2 : a traveler by water
3 : a stiff straw hat with a low flat crown and straight circular brim

sailor collar *noun* (1895)
: a broad collar having a square flap across the back and tapering to a V in the front

sail·or's–choice \ˌsā-lərz-ˈchȯis\ *noun* (1850)
: any of several small grunts of the Western Atlantic: as **a :** PINFISH **b :** PIGFISH

sail·plane \ˈsā(ə)l-ˌplān\ *noun* (1922)
: a glider of such design that it is able to rise in an upward air current
— sailplane *intransitive verb*
— sail·plan·er *noun*

sai·min \ˈsī-ˈmin\ *noun* [probably from Chinese (Guangdong) *sai mihn* fine noodles] (1949)
: a Hawaiian noodle soup

sain \ˈsān\ *transitive verb* [Middle English, from Old English *segnian,* from Late Latin *signare,* from Latin, to mark — more at SIGN] (before 12th century)
1 *dialect British* **:** to make the sign of the cross on (oneself)
2 *dialect British* **:** BLESS

sain·foin \ˈsān-ˌfȯin, ˈsan-\ *noun* [French, from Middle French, from *sain* healthy (from Latin *sanus*) + *foin* hay, from Latin *fenum*] (1626)
: a pink-flowered Eurasian perennial leguminous herb (*Onobrychis viciaefolia* synonym *O. viciifolia*) grown for forage

¹saint \ˈsānt, *before a name* (ˌ)sānt *or* sənt\ *noun* [Middle English, from Old French, from Late Latin *sanctus,* from Latin, sacred, from past participle of *sancire* to make sacred — more at SACRED] (13th century)
1 : one officially recognized especially through canonization as preeminent for holiness
2 a : one of the spirits of the departed in heaven **b :** ANGEL 1a
3 a : one of God's chosen and usually Christian people **b** *capitalized* **:** a member of any of various Christian bodies; *specifically* **:** LATTER-DAY SAINT
4 : one eminent for piety or virtue
5 : an illustrious predecessor
— saint·dom \ˈsānt-dəm\ *noun*
— saint·like \ˈsānt-ˌlīk\ *adjective*

²saint \ˈsānt\ *transitive verb* (13th century)
: to recognize or designate as a saint; *specifically* **:** CANONIZE

Saint Ag·nes' Eve \-ˈag-nə-səz-, -ˈag-nəs-\ *noun* [*Saint Agnes*] (1820)
: the night of January 20 when a woman is traditionally held to have a revelation of her future husband

Saint An·drew's cross \-ˈan-ˌdrüz-\ *noun* [*Saint Andrew* (died about A.D. 60), apostle who, according to tradition, was crucified on a cross of this type] (1615)
: a figure of a cross that has the form of two intersecting oblique bars — see CROSS illustration

Saint An·tho·ny's cross \-ˈan(t)-thə-nēz-, *chiefly British* -ˈan-tə-\ *noun* [*Saint Anthony*] (1885)
: TAU CROSS

Saint Anthony's fire *noun* (14th century)
: any of several inflammations or gangrenous conditions (as erysipelas or ergotism) of the skin

saint au·gus·tine grass \-ˈȯ-gə-ˌstēn-\ *noun, often S&A capitalized* [probably from *Saint Augustine,* Fla.] (1900)
: a perennial much-branched creeping grass (*Stenotaphrum secundatum*) of the southern U.S. that is valuable as a sand binder and as sod grass

Saint Ber·nard \-bər-ˈnärd\ *noun* [the hospice of Grand *Saint Bernard,* where such dogs were first bred] (1839)
: any of a Swiss alpine breed of tall powerful working dogs used especially formerly in aiding lost travelers

Saint Bernard

saint·ed \ˈsān-təd\ *adjective* (1598)
1 : befitting or relating to a saint
2 : SAINTLY, PIOUS
3 : entered into heaven **:** DEAD
4 : much admired **:** IDOLIZED

Saint El·mo's fire \ˈsānt-ˈel-(ˌ)mōz-\ *noun* [*Saint Elmo* (*Erasmus*) (died 303) Italian bishop & patron saint of sailors] (1814)
: a flaming phenomenon sometimes seen in stormy weather at prominent points on an airplane or ship and on land that is of the nature of a brush discharge of electricity — called also *Saint Elmo's light*

Saint Emi·lion \ˌsaⁿ-tā-mēl-ˈyōⁿ\ *noun* [*Saint-Émilion,* village in SW France] (1833)
: a red Bordeaux wine

saint·hood \ˈsānt-ˌhu̇d\ *noun* (1550)
1 : the quality or state of being a saint
2 : saints as a group

Saint John's bread \-ˈjänz-ˌbred\ *noun* (1591)
: CAROB 2

Saint–John's–wort \-ˌwərt, -ˌwȯrt\ *noun* [*Saint John* the Baptist] (15th century)
: any of a genus (*Hypericum* of the family Guttiferae, the Saint-John's-wort family) of herbs and shrubs with showy pentamerous yellow flowers

Saint Lou·is encephalitis \-ˈlü-əs-\ *noun* [*Saint Louis,* Mo.] (1934)
: a North American viral encephalitis that is transmitted by several culex mosquitoes

saint·ly \ˈsānt-lē\ *adjective* **saint·li·er; -est** (1534)
: relating to, resembling, or befitting a saint **:** HOLY
— saint·li·ness *noun*

Saint Mar·tin's summer \-ˈmär-tᵊnz-\ *noun* [*Saint Martin's* Day, November 11] (1591)
: Indian summer when occurring in November

Saint Pat·rick's Day \-ˈpa-triks-\ *noun* (1844)
: March 17 observed by the Roman Catholic Church in honor of Saint Patrick and celebrated in Ireland in commemoration of his death

saint's day *noun* (15th century)
: a day in a church calendar on which a saint is commemorated

saint·ship \ˈsānt-ˌship\ *noun* (1631)
: SAINTHOOD 1

Saint Val·en·tine's Day \-ˈva-lən-ˌtīnz-\ *noun* [*Saint Valentine* (died about 270) Italian priest] (14th century)
: February 14 observed in honor of Saint Valentine and as a time for sending valentines

Saint Vi·tus' dance \-ˈvī-təs-, -ˈvī-tə-səz-\ *noun* [*Saint Vitus,* 3d century Christian child martyr] (1621)
: CHOREA — called also *Saint Vitus's dance* *same*\

saith \ˈseth, ˈsā-əth\ *archaic present 3d singular of* SAY

saithe \ˈsāth, ˈsāt͟h\ *noun, plural* **saithe** [of Scandinavian origin; akin to Old Norse *seithr* coalfish] (1632)
: POLLACK

Sai·va \ˈsī-və, ˈshī-\ *noun* [Sanskrit *Śaiva,* from *Śiva* Siva] (1810)
: a member of a major Hindu sect devoted to the cult of Siva
— Sai·vism \-ˌvi-zəm\ *noun*

¹sake \ˈsāk\ *noun* [Middle English, dispute, guilt, purpose, from Old English *sacu* guilt, action at law; akin to Old High German *sahha* action at law, cause, Old English *sēcan* to seek — more at SEEK] (13th century)
1 : END, PURPOSE ⟨for the *sake* of argument⟩
2 a : the good, advantage, or enhancement of some entity (as an ideal) ⟨free to pursue learning for its own *sake* —M. S. Eisenhower⟩ **b :** personal or social welfare, safety, or benefit

²sa·ke *or* **sa·ki** \ˈsä-kē\ *noun* [Japanese *sake*] (1687)
: a Japanese alcoholic beverage of fermented rice usually served hot

sa·ker \ˈsā-kər\ *noun* [Middle English *sagre,* from Middle French *sacre,* from Arabic *ṣaqr*] (15th century)
: an Old World falcon (*Falco cherrug*) used in falconry

Sak·ti \ˈsäk-tē, ˈshäk-\, **Sak·tism** *variant of* SHAKTI, SHAKTISM

sal \ˈsal\ *noun* [Middle English, from Latin — more at SALT] (14th century)
: SALT

¹sa·laam \sə-ˈläm\ *noun* [Arabic *salām,* literally, peace] (1613)
1 : a salutation or ceremonial greeting in the East
2 : an obeisance performed by bowing very low and placing the right palm on the forehead

²salaam (1693)
transitive verb
: to greet or pay homage to with a salaam

intransitive verb
: to perform a salaam

sal·able *or* **sale·able** \ˈsā-lə-bəl\ *adjective* (1530)
: capable of being or fit to be sold **:** MARKETABLE
— sal·abil·i·ty \ˌsā-lə-ˈbi-lə-tē\ *noun*

sa·la·cious \sə-ˈlā-shəs\ *adjective* [Latin *salac-, salax,* from *salire* to move spasmodically, leap — more at SALLY] (circa 1645)
1 : arousing or appealing to sexual desire or imagination **:** LASCIVIOUS
2 : LECHEROUS, LUSTFUL
— sa·la·cious·ly *adverb*
— sa·la·cious·ness *noun*

sal·ad \ˈsa-ləd\ *noun* [Middle English *salade,* from Middle French, from Old Provençal *salada,* from *salar* to salt, from *sal* salt, from Latin] (14th century)
1 a : green vegetables (as lettuce, endive, or romaine) and often tomatoes, cucumbers, or radishes served with dressing **b :** a dish of meat, fish, shellfish, eggs, fruits, or vegetables singly or in combination usually served cold with a dressing
2 : a green vegetable or herb grown for salad; *especially* **:** LETTUCE
3 : a usually incongruous mixture **:** HODGEPODGE

salad bar *noun* (1973)
: a self-service counter (as in a restaurant) featuring an array of salad makings and dressings

salad days *noun plural* (1606)
: time of youthful inexperience or indiscretion ⟨my *salad days* when I was green in judgment —Shakespeare⟩; *also* **:** an early flourishing period **:** HEYDAY

salad dressing *noun* (circa 1839)
: a dressing either uncooked (as French dressing) or cooked (as a boiled dressing) that is used for salad

salad oil *noun* (1537)
: an edible vegetable oil suitable for use in salad dressings

sa·lal \sə-ˈlal, sa-\ *noun* [Chinook Jargon, from Lower Chinook *sálal*] (1825)
: a small shrub (*Gaultheria shallon*) of the heath family found on the Pacific coast of North America and bearing edible grape-sized dark purple berries

sal·a·man·der \ˈsa-lə-ˌman-dər *also* ˌsa-lə-ˈ\ *noun* [Middle English *salamandre,* from Middle French, from Latin *salamandra,* from Greek] (14th century)
1 : a mythical animal having the power to endure fire without harm
2 : an elemental being in the theory of Paracelsus inhabiting fire
3 : any of numerous amphibians (order Caudata) superficially resembling lizards but scaleless and covered with a soft moist skin and breathing by gills in the larval stage
4 : an article used in connection with fire: as **a :** a cooking utensil for browning a food (as pastry or pudding) **b :** a portable stove **c :** a cooking device with an overhead heat source like a broiler
— sal·a·man·drine \ˌsa-lə-ˈman-drən\ *adjective*

sa·la·mi \sə-ˈlä-mē\ *noun* [Italian, plural of *salame* salami, from *salare* to salt, from *sale* salt, from Latin *sal* — more at SALT] (1852)
: a highly seasoned sausage of pork and beef either dried or fresh

sal ammoniac *noun* [Middle English *sal armoniak,* from Latin *sal ammoniacus,* literally, salt of Ammon] (14th century)
: AMMONIUM CHLORIDE

sa·lar·i·at \sə-ˈlar-ē-ət, -ˈler-\ *noun* [French, from *salaire* salary (from Latin *salarium*) + *-ariat* (as in *prolétariat* proletariat)] (1917)
: the class or body of salaried persons usually as distinguished from wage earners

sal·a·ry \ˈsal-rē, ˈsa-lə-\ *noun, plural* **-ries** [Middle English *salarie,* from Latin *salarium* pension, salary, from neuter of *salarius* of salt, from *sal* salt — more at SALT] (13th century)
: fixed compensation paid regularly for services ◇
— sal·a·ried \-rēd\ *adjective*

sal·a·ry·man \-ˌman\ *noun* [Japanese *sararīman,* from English *salary* + *man*] (1962)
: a Japanese white-collar businessman

sal·chow \ˈsal-ˌkau̇, -ˌkȯv, -(ˌ)kō\ *noun* [Ulrich *Salchow* (died 1949) Swedish figure skater] (1940)
: a figure-skating jump with a takeoff from the back inside edge of one skate followed by a full turn in the air and a landing on the back outside edge of the opposite skate

sale \ˈsā(ə)l\ *noun* [Middle English, from Old English *sala,* from Old Norse — more at SELL] (before 12th century)
1 : the act of selling; *specifically* **:** the transfer of ownership of and title to property from one person to another for a price
2 a : opportunity of selling or being sold **:** DEMAND **b :** distribution by selling
3 : public disposal to the highest bidder **:** AUCTION
4 : a selling of goods at bargain prices
5 *plural* **a :** operations and activities involved in promoting and selling goods or services ⟨vice-president in charge of *sales*⟩ **b :** gross receipts
— for sale : available for purchase
— on sale 1 : for sale **2 :** available for purchase at a reduced price

sa·lep \ˈsa-ləp, sə-ˈlep\ *noun* [French or Spanish, both from Arabic dialect *saḥlab,* perhaps alteration of Arabic (*khusy ath-*)*thaʽlab,* literally, testicles of the fox] (1736)
: the starchy or mucilaginous dried tubers of various Old World orchids (especially genus *Orchis*) used for food or in medicine

sal·e·ra·tus \ˌsa-lə-ˈrā-təs\ *noun* [New Latin *sal aeratus* aerated salt] (1837)
: a leavening agent consisting of potassium or sodium bicarbonate

sale·room \ˈsā(ə)l-ˌrüm, -ˌru̇m\ *chiefly British variant of* SALESROOM

sales \ˈsā(ə)lz\ *adjective* (1913)
: of, relating to, or used in selling

sales·clerk \ˈsā(ə)lz-ˌklərk\ *noun* (1926)
: a salesperson in a store

sales·girl \-ˌgər(-ə)l\ *noun* (1887)
: SALESWOMAN

Sa·le·sian \sə-ˈlē-zhən, sā-\ *noun* (1884)
: a member of the Society of Saint Francis de Sales founded by Saint John Bosco in Turin, Italy in the 19th century and devoted chiefly to education

sales·la·dy \ˈsā(ə)lz-ˌlā-dē\ *noun* (1856)
: SALESWOMAN

sales·man \ˈsā(ə)lz-mən\ *noun* (1523)
: one who sells in a given territory, in a store, or by telephone

sales·man·ship \-ˌship\ *noun* (1880)
1 : the skill or art of selling
2 : ability or effectiveness in selling or in presenting persuasively ⟨political *salesmanship*⟩

sales·peo·ple \-ˌpē-pəl\ *noun plural* (1876)
: persons employed to sell goods or services

sales·per·son \-ˌpər-sᵊn\ *noun* (1901)
: a salesman or saleswoman

sales·room \ˈsā(ə)lz-ˌrüm, -ˌru̇m\ *noun* (1840)
: a place where goods are displayed for sale; *especially* **:** an auction room

sales slip *noun* (1926)
: a receipt for a purchase

sales tax *noun* (1921)
: a tax levied on the sale of goods and services that is usually calculated as a percentage of the purchase price and collected by the seller

sales·wom·an \ˈsā(ə)lz-ˌwu̇-mən\ *noun* (1704)
: a woman who sells in a given territory, in a store, or by telephone

sal·ic \ˈsa-lik\ *adjective* [by alteration] (1902)
: SIALIC

Sa·lic \ˈsā-lik, ˈsa-\ *adjective* [Middle French or Medieval Latin; Middle French *salique,* from Medieval Latin *Salicus,* from Late Latin *Salii* Salic Franks] (circa 1548)
: of, relating to, or being a Frankish people that settled on the IJssel river early in the 4th century

sal·i·cin \ˈsa-lə-sən\ *noun* [French *salicine,* from Latin *salic-, salix* willow — more at SALLOW] (1830)
: a bitter white crystalline glucoside $C_{13}H_{18}O_7$ found in the bark and leaves of several willows and poplars and used in medicine like salicylic acid

Salic law *noun* (1599)
1 : a rule held to derive from the legal code of the Salic Franks excluding females from the line of succession to a throne
2 : the legal code of the Salic Franks

sa·lic·y·late \sə-ˈli-sə-ˌlāt\ *noun* (1842)
: a salt or ester of salicylic acid

sal·i·cyl·ic acid \ˌsa-lə-ˈsi-lik-\ *noun* [International Scientific Vocabulary, from *salicyl* (the group HOC_6H_4CO)] (1840)
: a crystalline phenolic acid $C_7H_6O_3$ used especially in the form of salts and other derivatives as an analgesic and antipyretic

sa·lience \ˈsā-lyən(t)s, -lē-ən(t)s\ *noun* (1836)
1 : the quality or state of being salient
2 : a striking point or feature **:** HIGHLIGHT

sa·lien·cy \-lyən(t)-sē, -lē-ən(t)-\ *noun, plural* **-cies** (1664)
: SALIENCE

¹sa·lient \ˈsā-lyənt, -lē-ənt\ *adjective* [Latin *salient-, saliens,* present participle of *salire* to leap — more at SALLY] (1646)
1 : moving by leaps or springs **:** JUMPING
2 : jetting upward ⟨a *salient* fountain⟩
3 a : projecting beyond a line, surface, or level **b :** standing out conspicuously **:** PROMINENT; *especially* **:** of notable significance ⟨similar to . . . Prohibition, but there are a couple of *salient* differences —Tony Gibbs⟩
synonym see NOTICEABLE
— sa·lient·ly *adverb*

²salient *noun* (1828)
: something (as a promontory) that projects outward or upward from its surroundings; *especially* **:** an outwardly projecting part of a fortification, trench system, or line of defense

¹sa·line \ˈsā-ˌlēn, -ˌlīn\ *adjective* [Middle English, from Latin *salinus,* from *sal* salt — more at SALT] (15th century)
1 : consisting of or containing salt ⟨a *saline* solution⟩
2 : of, relating to, or resembling salt **:** SALTY ⟨a *saline* taste⟩

◇ WORD HISTORY

salary The word *salary* is a loan from Latin *salarium,* a derivative of *sal* "salt," and perhaps originally short for *salarium argentum* "salt money." According to a customary explanation, it was originally money paid to Roman soldiers with which they were supposed to buy salt, but nothing in the attested history of the Latin word supports this. From the evidence of extant documents and inscriptions, the *salarium* was a fixed payment, introduced under the rule of Augustus, that was made to officials of a certain rank. The word was also applied to fees given to scholars and physicians by the Roman state or a community, as well as a payment to army officers. Presumably *salarium* was a kind of euphemism, since the sums involved in such payments were in excess of what would have been needed just to buy salt. Our modern idea of providing public office holders with a fixed income was not well developed in ancient Rome, where personal relations rather than bureaucratic rules determined much behavior and the lines between bribe, gift, and payment were often fuzzy.

3 : consisting of or relating to the salts of the alkali metals or of magnesium ⟨a *saline* cathartic⟩
— **sa•lin•i•ty** \sā-ˈli-nə-tē, sə-\ *noun*

²saline *noun* (1662)
1 : a metallic salt; *especially* **:** a salt of potassium, sodium, or magnesium with a cathartic action
2 : a saline solution; *especially* **:** one isotonic with body fluids

sa•li•nize \ˈsa-lə-ˌnīz *also* ˈsā-\ *transitive verb* **-nized; -niz•ing** (1926)
: to treat or impregnate with salt
— **sa•li•ni•za•tion** \ˌsa-lə-nə-ˈzā-shən *also* ˌsā-\ *noun*

sa•li•nom•e•ter \ˌsa-lə-ˈnä-mə-tər, ˌsā-\ *noun* [International Scientific Vocabulary *saline* + *-o-* + *-meter*] (1844)
: an instrument (as a hydrometer) for measuring the amount of salt in a solution

Sa•lique \ˈsā-lik, ˈsa-; sa-ˈlēk, sā-\ *variant of* SALIC

Salis•bury steak \ˈsȯlz-ˌber-ē-, ˈsalz-, -b(ə-)rē-\ *noun* [J. H. *Salisbury,* 19th century English physician] (1897)
: ground beef mixed with egg, milk, bread crumbs, and seasonings and formed into a large patty and cooked

Sa•lish \ˈsā-lish\ *noun* (1831)
1 : a group of American Indian peoples of British Columbia and the northwestern U.S.
2 : the family of languages spoken by the Salish peoples
— **Sa•lish•an** \-li-shən\ *adjective or noun*

sa•li•va \sə-ˈlī-və\ *noun* [Latin] (15th century)
: a slightly alkaline secretion of water, mucin, protein, salts, and often a starch-splitting enzyme (as ptyalin) that is secreted into the mouth by salivary glands, lubricates ingested food, and often begins the breakdown of starches

sal•i•vary \ˈsa-lə-ˌver-ē\ *adjective* (1709)
: of or relating to saliva or the glands that secrete it; *especially* **:** producing or carrying saliva

sal•i•vate \ˈsa-lə-ˌvāt\ *intransitive verb* **-vat•ed; -vat•ing** (circa 1706)
1 : to have a flow of saliva especially in excess
2 : DROOL 2
— **sal•i•va•tion** \ˌsa-lə-ˈvā-shən\ *noun*
— **sal•i•va•tor** \ˈsa-lə-ˌvā-tər\ *noun*

Salk vaccine \ˈsȯk-, ˈsȯlk-\ *noun* [Jonas *Salk*] (1954)
: a vaccine consisting of poliomyelitis virus inactivated with formaldehyde

sal•let \ˈsa-lət\ *noun* [Middle English, from Middle French *sallade*] (15th century)
: a light 15th century helmet with or without a visor and with a projection over the neck

¹sal•low \ˈsa-(ˌ)lō\ *noun* [Middle English, from Old English *sealh;* akin to Old High German *salha* sallow, Latin *salix* willow] (before 12th century)
: any of several Old World broad-leaved willows (as *Salix caprea*) including important sources of charcoal and tanbark

²sallow *adjective* [Middle English *salowe,* from Old English *salu;* akin to Old High German *salo* murky, Russian *solovyĭ* yellowish gray] (before 12th century)
: of a grayish greenish yellow color
— **sal•low•ish** \ˈsa-lə-wish\ *adjective*
— **sal•low•ness** \ˈsa-lō-nəs, -lə-\ *noun*

¹sal•ly \ˈsa-lē\ *noun, plural* **sallies** [Middle French *saillie,* from Old French, from *saillir* to rush forward, from Latin *salire* to leap; akin to Greek *hallesthai* to leap] (1560)
1 : an action of rushing or bursting forth; *especially* **:** a sortie of troops from a defensive position to attack the enemy
2 a : a brief outbreak **:** OUTBURST **b :** a witty or imaginative saying **:** QUIP
3 : a venture or excursion usually off the beaten track **:** JAUNT

²sally *intransitive verb* **sal•lied; sal•ly•ing** (1560)
1 : to leap out or burst forth suddenly
2 : SET OUT, DEPART — often used with *forth*

Sal•ly Lunn \ˌsa-lē-ˈlən\ *noun* [*Sally Lunn,* 18th century English baker] (1780)
: a slightly sweetened yeast-leavened bread

sally port *noun* (1649)
: a gate or passage in a fortified place for use by troops making a sortie

sal•ma•gun•di \ˌsal-mə-ˈgən-dē\ *noun* [French *salmigondis*] (circa 1674)
1 : a salad plate of chopped meats, anchovies, eggs, and vegetables arranged in rows for contrast and dressed with a salad dressing
2 : a heterogeneous mixture **:** POTPOURRI

sal•mi \ˈsal-mē\ *noun* [French *salmis,* short for *salmigondis*] (1759)
: a ragout of partly roasted game stewed in a rich sauce

salm•on \ˈsa-mən\ *noun, plural* **salmon** *also* **salmons** [Middle English *samon,* from Middle French, from Latin *salmon-, salmo*] (13th century)
1 a : a large anadromous salmonid fish (*Salmo salar*) of the North Atlantic noted as a game and food fish — called also *Atlantic salmon* **b :** any of various anadromous salmonid fishes other than the salmon; *especially* **:** PACIFIC SALMON **c :** a fish (as a barramunda) resembling a salmon
2 : the variable color of salmon's flesh averaging a strong yellowish pink

salm•on•ber•ry \-ˌber-ē\ *noun* (1844)
: a showy red-flowered raspberry (*Rubus spectabilis*) of the Pacific coast; *also* **:** its edible salmon-colored fruit

sal•mo•nel•la \ˌsal-mə-ˈne-lə\ *noun, plural* **-nel•lae** \-ˈne-(ˌ)lē, -ˌlī\ *or* **-nellas** *or* **-nella** [New Latin, from Daniel E. *Salmon* (died 1914) American veterinarian] (1913)
: any of a genus (*Salmonella*) of usually motile enterobacteria that are pathogenic for humans and other warm-blooded animals and cause food poisoning, gastrointestinal inflammation, typhoid fever, or septicemia

sal•mo•nel•lo•sis \ˌsal-mə-(ˌ)ne-ˈlō-səs\ *noun, plural* **-lo•ses** \-ˌsēz\ [New Latin] (circa 1913)
: infection with or disease caused by salmonellae

sal•mo•nid \ˈsa-mə-nid, ˈsal-\ *noun* [New Latin *Salmonidae,* from *Salmon-, Salmo,* genus name, from Latin *salmo* salmon] (1868)
: any of a family (Salmonidae) of elongate bony fishes (as a salmon or trout) that have the last three vertebrae upturned
— **salmonid** *adjective*

salm•on•oid \ˈsa-mə-ˌnȯid\ *noun* (circa 1842)
: SALMONID; *also* **:** a related fish
— **salmonoid** *adjective*

salmon pink *noun* (1882)
: a strong yellowish pink

Sa•lo•me \sə-ˈlō-mē, ˈsa-lə-(ˌ)mā\ *noun* [Late Latin, from Greek *Salōmē*]
: a niece of Herod Antipas given the head of John the Baptist as a reward for her dancing

sa•lom•e•ter \sā-ˈlä-mə-tər, sə-\ *noun* [Latin *sal* salt + English *-o-* + *-meter*] (1860)
: a hydrometer for indicating the percentage of salt in a solution

sa•lon \sə-ˈlän, ˈsa-ˌlän, sa-ˈlōⁿ\ *noun* [French] (1699)
1 : an elegant apartment or living room (as in a fashionable home)
2 : a fashionable assemblage of notables (as literary figures, artists, or statesmen) held by custom at the home of a prominent person
3 a : a hall for exhibition of art **b** *capitalized* **:** an annual exhibition of works of art
4 : a stylish business establishment or shop ⟨a beauty *salon*⟩

sa•loon \sə-ˈlün\ *noun* [French *salon,* from Italian *salone,* augmentative of *sala* hall, of Germanic origin; akin to Old High German *sal* hall; akin to Lithuanian *sala* village] (1728)
1 a *chiefly British* **:** SALON 1 **b** *chiefly British* **:** an often elaborately decorated public hall **c** (1) **:** a usually large public cabin on a ship (as for dining) (2) **:** the living area on a yacht **d** *chiefly British* **:** SALON 4 **e :** BARROOM
2 : SALON 2
3 *British* **a :** PARLOR CAR **b :** SEDAN 2a — called also *saloon car*

salp \ˈsalp\ *also* **sal•pa** \ˈsal-pə\ *noun* [New Latin, from Latin, a kind of deep-sea fish, from Greek *salpē*] (1835)
: any of various transparent barrel-shaped or fusiform free-swimming tunicates (class Thaliacea) abundant in warm seas

sal•pi•glos•sis \ˌsal-pə-ˈglä-səs\ *noun* [New Latin, irregular from Greek *salpinx* trumpet + *glōssa* tongue — more at GLOSS] (1827)
: any of a small genus (*Salpiglossis*) of Chilean herbs of the nightshade family with large funnel-shaped varicolored flowers often strikingly marked

sal•pin•gi•tis \ˌsal-pən-ˈjī-təs\ *noun* [New Latin, from *salping-, salpinx* fallopian or eustachian tube, from Greek, trumpet] (1860)
: inflammation of a fallopian or eustachian tube

sal•sa \ˈsȯl-sə, ˈsäl-\ *noun* [Spanish, literally, sauce, from Latin, feminine of *salsus* salted — more at SAUCE] (circa 1962)
1 : a spicy sauce of tomatoes, onions, and hot peppers
2 : popular music of Latin American origin that has absorbed characteristics of rhythm and blues, jazz, and rock

sal•si•fy \ˈsal-sə-fē, -ˌfī\ *noun* [French *salsifis,* from Italian *salsefica, sassefrica*] (circa 1706)
: a European biennial composite herb (*Tragopogon porrifolius*) with a long fusiform edible root — called also *oyster plant, vegetable oyster*

sal soda *noun* (15th century)
: WASHING SODA

¹salt \ˈsȯlt\ *noun* [Middle English, from Old English *sealt;* akin to Old High German *salz* salt, Lithuanian *saldus* sweet, Latin *sal* salt, Greek *hals* salt, sea] (before 12th century)
1 a : a crystalline compound NaCl that consists of sodium chloride, is abundant in nature, and is used especially to season or preserve food or in industry — called also *common salt* **b :** a substance (as washing soda) resembling common salt **c** *plural* (1) **:** a mineral or saline mixture (as Epsom salts) used as an aperient or cathartic (2) **:** SMELLING SALTS **d :** any of numerous compounds that result from replacement of part or all of the acid hydrogen of an acid by a metal or a group acting like a metal **:** an ionic crystalline compound
2 : a container for salt at table — often used in the phrases *above the salt* and *below the salt* alluding to the former custom of seating persons of higher rank above and those of lower rank below a saltcellar placed in the middle of a long table
3 a : an ingredient that gives savor, piquancy, or zest **:** FLAVOR ⟨a people . . . full of life, vigor, and the *salt* of personality —Clifton Fadiman⟩ **b :** sharpness of wit **:** PUNGENCY **c :** COMMON SENSE **d :** RESERVE, SKEPTICISM — usually used in the phrases *with a grain of salt* and *with a pinch of salt* **e :** a dependable steadfast person or group of people — usually used in the phrase *salt of the earth*
4 : SAILOR ⟨a tale worthy of an old *salt*⟩
5 : KEEP 3 — usually used in the phrase *worth one's salt*
— **salt•like** \-ˌlīk\ *adjective*

²salt *transitive verb* (before 12th century)

1 a : to treat, provide, or season with common salt **b :** to preserve (food) with salt or in brine **c :** to supply (as an animal) with salt
2 : to give flavor or piquancy to (as a story)
3 : to enrich (as a mine) artificially by secretly placing valuable mineral in some of the working places
4 : to sprinkle with or as if with a salt
— **salt·er** \ˈsȯl-tər\ *noun*

[3]salt *adjective* (before 12th century)
1 a : SALINE, SALTY **b :** being or inducing the one of the four basic taste sensations that is suggestive of seawater — compare BITTER, SOUR, SWEET
2 : cured or seasoned with salt **:** SALTED
3 : overflowed with salt water ⟨a *salt* pond⟩
4 : SHARP, PUNGENT
— **salt·ness** *noun*

[4]salt *adjective* [by shortening & alteration from *assaut,* from Middle English *a sawt,* from Middle French *a saut,* literally, on the jump] (1598)
obsolete **:** LUSTFUL, LASCIVIOUS

salt–and–pepper *adjective* (1915)
: having black-and-white or dark and light color intermingled in small flecks ⟨a *salt-and-pepper* suit⟩

sal·ta·rel·lo \ˌsal-tə-ˈre-(ˌ)lō, ˌsäl-\ *noun, plural* **-los** [Italian, from *saltare* to jump, from Latin] (circa 1724)
: an Italian dance with a lively hop step beginning each measure

sal·ta·tion \sal-ˈtā-shən, sȯl-\ *noun* [Latin *saltation-, saltatio,* from *saltare* to leap, dance, frequentative of *salire* to leap — more at SALLY] (1646)
1 a : the action or process of leaping or jumping **b :** DANCE
2 a : the origin of a new species or a higher taxon in essentially a single evolutionary step that in some especially former theories is held to be due to a major mutation or to unknown causes — compare DARWINISM, NEO-DARWINISM, PUNCTUATED EQUILIBRIUM **b :** MUTATION — used especially of bacteria and fungi

sal·ta·to·ri·al \ˌsal-tə-ˈtōr-ē-əl, ˌsȯl-, -ˈtȯr-\ *adjective* (1789)
: relating to, marked by, or adapted for leaping ⟨*saltatorial* legs of a grasshopper⟩

sal·ta·to·ry \ˈsal-tə-ˌtōr-ē, ˈsȯl-, -ˌtȯr-\ *adjective* (1656)
1 *archaic* **:** of or relating to dancing
2 : proceeding by leaps rather than by gradual transitions **:** DISCONTINUOUS

salt away *transitive verb* (circa 1890)
: to lay away (as money) safely **:** SAVE

salt·box \ˈsȯlt-ˌbäks\ *noun* (1876)
: a frame dwelling with two stories in front and one behind and a roof with a long rear slope

saltbox

salt·bush \-ˌbu̇sh\ *noun* (1863)
: any of various shrubby plants of the goosefoot family that thrive in dry alkaline soil; *especially* **:** any of numerous oraches that are important browse plants in dry regions

salt·cel·lar \ˈsȯlt-ˌse-lər\ *noun* [Middle English *salt saler,* from *salt* + *saler* saltcellar, from Middle French, from Latin *salarius* of salt — more at SALARY] (14th century)
: a small container for holding salt at the table

salt dome *noun* (1908)
: a domical anticline in sedimentary rock that has a mass of rock salt as its core

salt flat *noun* (1816)
: a salt-encrusted flat area resulting from evaporation of a former body of water

salt gland *noun* (1950)
: a gland (as of a marine bird) capable of excreting a concentrated salt solution

salt grass *noun* (1704)
: a grass (especially *Distichlis spicata*) native to an alkaline habitat (as a salt marsh)

sal·tim·boc·ca \ˌsȯl-təm-ˈbä-kə\ *noun* [Italian, from *saltare* to jump + *in* in + *bocca* mouth] (1937)
: scallops of veal prepared with sage, slices of ham, and sometimes cheese and served with a wine sauce

sal·tine \sȯl-ˈtēn\ *noun* (1907)
: a thin crisp cracker usually sprinkled with salt

salt·ing \ˈsȯl-tiŋ\ *noun* (1712)
chiefly British **:** land flooded regularly by tides — usually used in plural

sal·tire \ˈsȯl-ˌtīr, ˈsal-\ *noun* [Middle English *sautire,* from Middle French *saultoir* X-shaped animal barricade that can be jumped over by people, saltire, from *saulter* to jump, from Latin *saltare* — more at SALTATION] (14th century)
: a heraldic charge consisting of a cross formed by a bend and a bend sinister crossing in the center

salt lake *noun* (1763)
: a landlocked body of water that has become salty through evaporation

salt·less \ˈsȯlt-ləs\ *adjective* (14th century)
1 : having no salt
2 : INSIPID

salt lick *noun* (1751)
: LICK 3

salt marsh *noun* (before 12th century)
: flat land subject to overflow by salt water

salt out (1939)
transitive verb
: to precipitate, coagulate, or separate (as a dissolved substance or lyophilic sol) especially from a solution by the addition of salt
intransitive verb
: to become salted out

salt pan *noun* (15th century)
: an undrained natural depression in which water gathers and leaves a deposit of salt on evaporation

salt·pe·ter \ˈsȯlt-ˈpē-tər\ *noun* [Middle English *salt petre,* alteration of *salpetre,* from Middle French, from Medieval Latin *sal petrae,* literally, salt of the rock] (14th century)
1 : POTASSIUM NITRATE
2 : SODIUM NITRATE

salt pork *noun* (1723)
: fat pork cured in salt or brine

salt·shak·er \ˈsȯlt-ˌshā-kər\ *noun* (1895)
: a container with a perforated top for sprinkling salt

salt·wa·ter \ˈsȯlt-ˌwȯ-tər, -ˌwä-\ *adjective* (before 12th century)
: relating to, living in, located near, or consisting of salt water

salt·works \ˈsȯlt-ˌwərks\ *noun plural but singular or plural in construction* (1565)
: a plant where salt is prepared commercially

salt·wort \-ˌwərt, -ˌwȯrt\ *noun* (1568)
1 : any of a genus (*Salsola*) of plants (as the Russian thistle) of the goosefoot family of which some have been used in making soda ash
2 : a low-growing strong-smelling coastal shrub (*Batis maritima* of the family Bataceae) of warm parts of the New World

salty \ˈsȯl-tē\ *adjective* **salt·i·er; -est** (15th century)
1 : of, seasoned with, or containing salt
2 : smacking of the sea or nautical life
3 a : PIQUANT **b :** EARTHY, CRUDE ⟨*salty* language⟩
— **salt·i·ly** \-tə-lē\ *adverb*
— **salt·i·ness** \-tē-nəs\ *noun*

sa·lu·bri·ous \sə-ˈlü-brē-əs\ *adjective* [Latin *salubris;* akin to *salvus* safe, healthy — more at SAFE] (1547)
: favorable to or promoting health or well-being
synonym see HEALTHFUL
— **sa·lu·bri·ous·ly** *adverb*
— **sa·lu·bri·ous·ness** *noun*
— **sa·lu·bri·ty** \-brə-tē\ *noun*

sa·lu·ki \sə-ˈlü-kē\ *noun* [Arabic *salūqīy* of Saluq, from *Salūq* Saluq, ancient city in Arabia] (1809)
: any of an ancient northern African and Asian breed of tall swift slender hunting dogs having long narrow heads, long silky ears, and a smooth silky coat

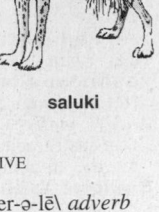

saluki

sal·u·tary \ˈsal-yə-ˌter-ē\ *adjective* [Middle French *salutaire,* from Latin *salutaris,* from *salut-, salus* health] (15th century)
1 : producing a beneficial effect **:** REMEDIAL ⟨*salutary* influences⟩
2 : promoting health **:** CURATIVE
synonym see HEALTHFUL
— **sal·u·tar·i·ly** \ˌsal-yə-ˈter-ə-lē\ *adverb*
— **sal·u·tar·i·ness** \ˈsal-yə-ˌter-ē-nəs\ *noun*

sal·u·ta·tion \ˌsal-yə-ˈtā-shən\ *noun* (14th century)
1 a : an expression of greeting, goodwill, or courtesy by word, gesture, or ceremony **b** *plural* **:** REGARDS
2 : the word or phrase of greeting (as *Gentlemen* or *Dear Sir* or *Madam*) that conventionally comes immediately before the body of a letter
— **sal·u·ta·tion·al** \-shnəl, -shə-nᵊl\ *adjective*

sa·lu·ta·to·ri·an \sə-ˌlü-tə-ˈtōr-ē-ən, -ˈtȯr-\ *noun* (circa 1847)
: the student usually having the second highest rank in a graduating class who delivers the salutatory address at the commencement exercises

[1]sa·lu·ta·to·ry \sə-ˈlü-tə-ˌtōr-ē, -ˌtȯr-\ *adjective* (1702)
: of or relating to a salutation **:** expressing or containing a welcome or greeting

[2]salutatory *noun, plural* **-ries** (1779)
: an address or statement of welcome or greeting

[1]sa·lute \sə-ˈlüt\ *verb* **sa·lut·ed; sa·lut·ing** [Middle English, from Latin *salutare,* from *salut-, salus* health, safety, greeting, from *salvus* safe, healthy — more at SAFE] (14th century)
transitive verb
1 a : to address with expressions of kind wishes, courtesy, or honor **b :** to give a sign of respect, courtesy, or goodwill to **:** GREET
2 : to become apparent to (one of the senses)
3 a : to honor (as a person, nation, or event) by a conventional military or naval ceremony **b :** to show respect and recognition to (a military superior) by assuming a prescribed position **c :** to express commendation of **:** PRAISE
intransitive verb
: to make a salute
— **sa·lut·er** *noun*

[2]salute *noun* (14th century)
1 : GREETING, SALUTATION
2 a : a sign, token, or ceremony expressing goodwill, compliment, or respect ⟨the festival was a *salute* to the arts⟩ **b :** the position (as of the hand) or the entire attitude of a person saluting a superior
3 : FIRECRACKER

sal·u·tif·er·ous \ˌsal-yə-ˈti-f(ə-)rəs\ *adjective* [Latin *salutifer,* from *salut-, salus* + *-i-* + *-fer* -ferous] (circa 1540)
: SALUTARY

salv·able \ˈsal-və-bəl\ *adjective* [Late Latin *salvare* to save — more at SAVE] (1667)
: capable of being saved or salvaged

[1]sal·vage \ˈsal-vij\ *noun* [French, from Middle French, from *salver* to save — more at SAVE] (1645)
1 a : compensation paid for saving a ship or its cargo from the perils of the sea or for the

lives and property rescued in a wreck **b :** the act of saving or rescuing a ship or its cargo **c :** the act of saving or rescuing property in danger (as from fire)
2 a : property saved from destruction in a calamity (as a wreck or fire) **b :** something extracted (as from rubbish) as valuable or useful

²salvage *transitive verb* **sal·vaged; sal·vag·ing** (1889)
: to rescue or save especially from wreckage or ruin
— sal·vage·abil·i·ty \ˌsal-vi-jə-ˈbi-lə-tē\ *noun*
— sal·vage·able \ˈsal-vi-jə-bəl\ *adjective*
— sal·vag·er *noun*

sal·var·san \ˈsal-vər-ˌsan\ *noun* [from *Salvarsan,* a trademark] (1909)
: ARSPHENAMINE

sal·va·tion \sal-ˈvā-shən\ *noun* [Middle English, from Old French, from Late Latin *salvation-, salvatio,* from *salvare* to save — more at SAVE] (13th century)
1 a : deliverance from the power and effects of sin **b :** the agent or means that effects salvation **c** *Christian Science* **:** the realization of the supremacy of infinite Mind over all bringing with it the destruction of the illusion of sin, sickness, and death
2 : liberation from ignorance or illusion
3 a : preservation from destruction or failure **b :** deliverance from danger or difficulty
— sal·va·tion·al \-shnəl, -shə-nᵊl\ *adjective*

Salvation Army *noun* (1878)
: an international religious and charitable group organized on military lines and founded in 1865 by William Booth for evangelizing and social betterment (as of the poor)

sal·va·tion·ism \sal-ˈvā-shə-ˌni-zəm\ *noun* (1883)
: religious teaching emphasizing the saving of the soul

Sal·va·tion·ist \-sh(ə-)nist\ *noun* (1882)
1 : a soldier or officer of the Salvation Army
2 *often not capitalized* **:** EVANGELIST
— salvationist *adjective, often capitalized*

¹salve \ˈsav, ˈsäv, ˈsȧv, ˈsalv, ˈsälv\ *noun* [Middle English, from Old English *sealf;* akin to Old High German *salba* salve, Greek *olpē* oil flask] (before 12th century)
1 : an unctuous adhesive substance for application to wounds or sores
2 : a remedial or soothing influence or agency ⟨a *salve* to their hurt feelings⟩

²salve *transitive verb* **salved; salv·ing** (before 12th century)
1 : to remedy (as disease) with or as if with a salve
2 : QUIET, ASSUAGE ⟨give him a raise in salary to *salve* his feelings —Upton Sinclair⟩

³salve \ˈsalv\ *transitive verb* **salved; salv·ing** [back-formation from *salvage*] (circa 1706)
: SALVAGE
— sal·vor \ˈsal-vər, -ˌvȯr\ *noun*

sal·ver \ˈsal-vər\ *noun* [modification of French *salve,* from Spanish *salva* sampling of food to detect poison, tray, from *salvar* to save, sample food to detect poison, from Late Latin *salvare* to save — more at SAVE] (circa 1661)
: a tray especially for serving food or beverages

sal·ver·form \ˈsal-vər-ˌfȯrm\ *adjective* (1821)
: tubular with a spreading limb — used of a gamopetalous corolla

sal·via \ˈsal-vē-ə\ *noun* [New Latin, from Latin, sage — more at SAGE] (1601)
: any of a large and widely distributed genus (*Salvia*) of herbs and shrubs of the mint family having a 2-lipped open calyx and two anthers; *especially* **:** one (*S. splendens*) with scarlet flowers

sal·vif·ic \sal-ˈvi-fik\ *adjective* [Late Latin *salvificus,* from Latin *salvus* safe + *-ficus* -fic] (1591)
: having the intent or power to save or redeem ⟨the *salvific* life and death of Christ —E. A. Walsh⟩

¹sal·vo \ˈsal-(ˌ)vō\ *noun, plural* **salvos** *or* **salvoes** [Italian *salva,* from French *salve,* from Latin, hail!, from *salvus* healthy — more at SAFE] (1591)
1 a : a simultaneous discharge of two or more guns in military action or as a salute **b :** the release all at one time of a rack of bombs or rockets (as from an airplane) **c :** a series of shots by an artillery battery with each gun firing one round in turn after a prescribed interval **d :** the bombs or projectiles released in a salvo
2 : something suggestive of a salvo: as **a :** a sudden burst ⟨a *salvo* of cheers⟩ **b :** a spirited attack ⟨the first *salvo* of a political campaign⟩

²salvo (1839)
transitive verb
: to release a salvo of
intransitive verb
: to fire a salvo

³salvo *noun, plural* **salvos** [Medieval Latin *salvo jure* with the right reserved] (1621)
1 : a mental reservation **:** PROVISO
2 : a means of safeguarding one's name or honor or allaying one's conscience **:** SALVE

sal vo·la·ti·le \ˌsal-və-ˈla-tᵊl-ē\ *noun* [New Latin, literally, volatile salt] (1654)
: SMELLING SALTS

SAM \ˈsam, ˌes-(ˌ)ā-ˈem\ *noun* (1950)
: a surface-to-air missile

sa·ma·ra \ˈsa-mə-rə; sə-ˈmar-ə, -ˈmär-\ *noun* [New Latin, from Latin, seed of the elm] (1577)
: a dry indehiscent usually one-seeded winged fruit (as of an ash or elm tree) — called also *key*

Sa·mar·i·tan \sə-ˈmar-ə-tᵊn, -ˈmer-\ *noun* [Middle English, from Old English, from Late Latin *samaritanus,* noun & adjective, from Greek *samaritēs* inhabitant of Samaria, from *Samaria*] (before 12th century)
1 : a native or inhabitant of Samaria
2 [from the parable of the good Samaritan in Luke 10:30–37] **:** one ready and generous in helping those in distress
— samaritan *adjective, often capitalized*

sa·mar·i·um \sə-ˈmer-ē-əm, -ˈmar-\ *noun* [New Latin, from French *samarskite*] (1879)
: a pale gray lustrous metallic element used especially in alloys that form permanent magnets — see ELEMENT table

sa·mar·skite \sə-ˈmär-ˌskīt, ˈsa-mər-\ *noun* [German *Samarskit,* from V. E. *Samarskiĭ-Bykhovets* (died 1870) Russian mining engineer] (1849)
: a black or brownish black orthorhombic mineral that is a complex oxide of rare earths, uranium, iron, lead, thorium, niobium, tantalum, titanium, and tin

sam·ba \ˈsam-bə, ˈsäm-\ *noun* [Portuguese] (1885)
: a Brazilian dance of African origin with a basic pattern of step-close-step-close and characterized by a dip and spring upward at each beat of the music; *also* **:** the music for this dance
— samba *intransitive verb*

sam·bar *or* **sam·bur** \ˈsäm-bər, ˈsam-\ *noun* [Hindi *sā̃bar,* from Sanskrit *śambara*] (1698)
: a large Asian deer (*Cervus unicolor*) with the male having strong 3-pointed antlers and long coarse hair on the throat

sam·bo \ˈsam-(ˌ)bō, ˈsäm-\ *noun* [Russian, from *samozashchita bez oruzhiya* self-defense without weapons] (1972)
: an international style of wrestling employing judo techniques

Sam Browne \ˈsam-ˈbrau̇n\ *noun* [Sir *Sam*uel James *Browne* (died 1901) British army officer] (1915)
: a leather belt for a dress uniform supported by a light strap passing over the right shoulder

¹same \ˈsām\ *adjective* [Middle English, from Old Norse *samr;* akin to Old High German *sama* same, Latin *simulis* like, *simul* together, at the same time, *similis* like, *sem-* one, Greek *homos* same, *hama* together, *hen-, heis* one] (13th century)
1 a : resembling in every relevant respect **b :** conforming in every respect — used with *as*
2 a : being one without addition, change, or discontinuance **:** IDENTICAL **b :** being the one under discussion or already referred to
3 : corresponding so closely as to be indistinguishable
4 : equal in size, shape, value, or importance — usually used with *the* or a demonstrative (as *that, those*) in all senses ☆

²same *pronoun* (14th century)
1 : something identical with or similar to another
2 : something or someone previously mentioned or described — often used with *the* or a demonstrative (as *that, those*) in both senses
— all the same *or* **just the same :** despite everything **:** NEVERTHELESS

³same *adverb* (1766)
: in the same manner — used with *the* or a demonstrative (as *that, those*)

sa·mekh \ˈsä-ˌmek\ *noun* [Hebrew *sāmekh*] (1823)
: the 15th letter of the Hebrew alphabet — see ALPHABET table

same·ness \ˈsām-nəs\ *noun* (1581)
1 : the quality or state of being the same **:** IDENTITY, SIMILARITY
2 : MONOTONY, UNIFORMITY

sam·i·sen \ˈsa-mə-ˌsen\ *noun* [Japanese] (1864)
: a 3-stringed Japanese musical instrument resembling a banjo

sa·mite \ˈsa-ˌmīt, ˈsā-\ *noun* [Middle English *samit,* from Middle French, from Medieval Latin *examitum, samitum,* from Middle Greek *hexamiton,* from Greek, neuter of *hexamitos* of six threads, from *hexa-* + *mitos* thread of the warp] (13th century)
: a rich medieval silk fabric interwoven with gold or silver

sa·miz·dat \ˈsä-mez-ˌdät\ *noun* [Russian, from *sam-* self- + *izdat*el'stvo publishing house] (1967)
: a system in the U.S.S.R. and countries within its orbit by which government-suppressed literature was clandestinely printed and distributed; *also* **:** such literature

sam·let \ˈsam-lət\ *noun* [irregular from *salmon* + *-let*] (1655)
: PARR

☆ SYNONYMS
Same, selfsame, very, identical, equivalent, equal mean not different or not differing from one another. SAME may imply and SELFSAME always implies that the things under consideration are one thing and not two or more things ⟨took the *same* route⟩ ⟨derived from the *selfsame* source⟩. VERY, like SELFSAME, may imply identity, or, like SAME, may imply likeness in kind ⟨the *very* point I was trying to make⟩. IDENTICAL may imply selfsameness or suggest absolute agreement in all details ⟨their test answers were *identical*⟩. EQUIVALENT implies amounting to the same thing in worth or significance ⟨two houses *equivalent* in market value⟩. EQUAL implies being identical in value, magnitude, or some specified quality ⟨*equal* shares in the business⟩.

Sam·mar·i·nese \(ˌ)sa(m)-ˌmar-ə-ˈnēz, -ˈnēs\ *noun, plural* **-ne·si** \-ˈnā-zē\ [Italian *sammarinese,* from *San Marino*] (1938)
: a native or inhabitant of San Marino

Sam·nite \ˈsam-ˌnīt\ *noun* [*Samnium,* Italy] (14th century)
: a member of an ancient people of central Italy

Sa·mo·an \sə-ˈmō-ən\ *noun* (1846)
1 : the Polynesian language of the Samoans
2 : a native or inhabitant of Samoa
— Samoan *adjective*

Samoa time *noun* (1983)
: the time of the 11th time zone west of Greenwich that includes American Samoa

sam·o·var \ˈsa-mə-ˌvär\ *noun* [Russian, from *samo-* self + *varit'* to boil] (1830)
1 : an urn with a spigot at its base used especially in Russia to boil water for tea
2 : an urn similar to a Russian samovar with a device for heating the contents

samovar 1

Sam·o·yed *also* **Sam·o·yede** \ˈsa-mə-ˌyed, -ˌmȯi-\ *noun* [Russian *samoed*] (1589)
1 : a member of any of a group of peoples inhabiting the far north of European Russia and parts of northwestern Siberia
2 : the family of Uralic languages spoken by the Samoyed people
3 : any of a Siberian breed of medium-sized white or cream-colored sled dogs
— Samoyed *adjective*
— Sam·o·yed·ic \ˌsa-mə-ˈye-dik, -ˌmȯi-\ *adjective*

samp \ˈsamp\ *noun* [modification of Narraganset *nasàump* corn mush] (1643)
: coarse hominy or a boiled cereal made from it

sam·pan \ˈsam-ˌpan\ *noun* [Chinese (Guangdong) *sàambáan,* from *sàam* three + *báan* board, plank] (1620)
: a flat-bottomed Chinese skiff usually propelled by two short oars

sam·phire \ˈsam-ˌfīr\ *noun* [alteration of earlier *sampiere,* from Middle French (*herbe de*) *Saint Pierre,* literally, Saint Peter's herb] (1545)
1 : a fleshy European seacoast plant (*Crithmum maritimum*) of the carrot family that is sometimes pickled
2 : a common glasswort (*Salicornia europaea*) that is sometimes pickled

[1]**sam·ple** \ˈsam-pəl\ *noun* [Middle English, from Middle French *essample,* from Latin *exemplum* — more at EXAMPLE] (15th century)
1 : a representative part or a single item from a larger whole or group especially when presented for inspection or shown as evidence of quality **:** SPECIMEN
2 : a finite part of a statistical population whose properties are studied to gain information about the whole
synonym see INSTANCE

[2]**sample** *transitive verb* **sam·pled; sam·pling** \-p(ə-)liŋ\ (1767)
: to take a sample of or from; *especially* **:** to judge the quality of by a sample **:** TEST ⟨*sampled* his output for defects⟩ ⟨*sample* a wine⟩

[3]**sample** *adjective* (1820)
: serving as an illustration or example ⟨*sample* questions⟩

[1]**sam·pler** \ˈsam-plər\ *noun* (1523)
: a decorative piece of needlework typically having letters or verses embroidered on it in various stitches as an example of skill

[2]**sam·pler** \-p(ə-)lər\ *noun* (1778)
1 : one that collects, prepares, or examines samples
2 : something containing representative specimens or selections ⟨a *sampler* of nineteen poets —K. E. Judd⟩; *also* **:** ASSORTMENT

sample space *noun* (1951)
: a set in which all of the possible outcomes of a statistical experiment are represented as points

sam·pling \ˈsam-pliŋ, *for 1 & 3* -p(ə-)liŋ\ *noun* (1778)
1 : the act, process, or technique of selecting a suitable sample; *specifically* **:** the act, process, or technique of selecting a representative part of a population for the purpose of determining parameters or characteristics of the whole population
2 : a small part selected as a sample for inspection or analysis ⟨ask a *sampling* of people which candidate they favor⟩
3 : the introduction or promotion of a product by distributing trial packages of it

sam·sa·ra \səm-ˈsär-ə\ *noun* [Sanskrit *saṁsāra,* literally, passing through] (1886)
: the indefinitely repeated cycles of birth, misery, and death caused by karma

Sam·son \ˈsam(p)-sən\ *noun* [Late Latin, from Greek *Sampsōn,* from Hebrew *Shimshōn*]
: a Hebrew hero who wreaked havoc among the Philistines by means of his great strength

Sam·so·ni·an \sam(p)-ˈsō-nē-ən\ *adjective* [*Samson*] (circa 1623)
: of heroic strength or proportions **:** MIGHTY

Sam·u·el \ˈsam-yə-wəl, -yəl\ *noun* [Late Latin, from Greek *Samouel,* from Hebrew *Shĕmūʼēl*]
1 : the early Hebrew judge who successively anointed Saul and David king
2 : either of two narrative and historical books of canonical Jewish and Christian Scriptures — see BIBLE table

sam·u·rai \ˈsa-mə-ˌrī, ˈsam-yə-\ *noun, plural* **samurai** [Japanese] (1727)
1 : a military retainer of a Japanese daimyo practicing the code of conduct of Bushido
2 : the warrior aristocracy of Japan

San \ˈsän\ *noun* (1876)
1 *plural in construction* **:** BUSHMEN
2 : BUSHMAN 2

san·a·tive \ˈsa-nə-tiv\ *adjective* [Middle English *sanatif,* from Middle French, from Late Latin *sanativus,* from Latin *sanatus,* past participle of *sanare* to cure, from *sanus* healthy] (15th century)
: having the power to cure or heal **:** CURATIVE, RESTORATIVE

san·a·to·ri·um \ˌsa-nə-ˈtōr-ē-əm, -ˈtȯr-\ *noun, plural* **-riums** *or* **-ria** \-ē-ə\ [New Latin, from Late Latin, neuter of *sanatorius* curative, from *sanare*] (1839)
1 : an establishment that provides therapy combined with a regimen (as of diet and exercise) for treatment or rehabilitation
2 a : an institution for rest and recuperation (as of convalescents) **b :** an establishment for the treatment of the chronically ill

san·be·ni·to \ˌsan-bə-ˈnē-(ˌ)tō, ˌsam-\ *noun, plural* **-tos** [Spanish *sambenito,* from *San Benito* Saint Benedict of Nursia] (circa 1560)
1 : a sackcloth coat worn by penitents on being reconciled to the church
2 : a Spanish Inquisition garment resembling a scapular and being either yellow with red crosses for the penitent or black with painted devils and flames for the impenitent condemned to an auto-da-fé

San·cerre \sä^{n}-ser\ *noun* [*Sancerre,* village in France] (circa 1946)
: a dry white wine from the Loire valley of France

San·cho Pan·za \ˌsan-chō-ˈpan-zə, ˌsän-chō-ˈpän-\ *noun* [Spanish]
: the squire of Don Quixote in Cervantes' *Don Quixote*

sanc·ti·fi·ca·tion \ˌsaŋ(k)-tə-fə-ˈkā-shən\ *noun* (14th century)
1 : an act of sanctifying
2 a : the state of being sanctified **b :** the state of growing in divine grace as a result of Christian commitment after baptism or conversion

sanc·ti·fi·er \ˈsaŋ(k)-tə-ˌfī(-ə)r\ *noun* (1548)
: one that sanctifies; *specifically, capitalized* **:** HOLY SPIRIT

sanc·ti·fy \-ˌfī\ *transitive verb* **-fied; -fy·ing** [Middle English *sanctifien,* from Middle French *sanctifier,* from Late Latin *sanctificare,* from Latin *sanctus* sacred — more at SAINT] (14th century)
1 : to set apart to a sacred purpose or to religious use **:** CONSECRATE
2 : to free from sin **:** PURIFY
3 a : to impart or impute sacredness, inviolability, or respect to **b :** to give moral or social sanction to
4 : to make productive of holiness or piety ⟨observe the day of the sabbath, to *sanctify* it —Deuteronomy 5:12 (Douay Version)⟩

sanc·ti·mo·nious \ˌsaŋ(k)-tə-ˈmō-nē-əs, -nyəs\ *adjective* (1603)
1 : affecting piousness **:** hypocritically devout; *also* **:** indicative of affected piousness ⟨the king's *sanctimonious* rebuke —G. B. Shaw⟩
2 *obsolete* **:** possessing sanctity **:** HOLY
— sanc·ti·mo·nious·ly *adverb*
— sanc·ti·mo·nious·ness *noun*

sanc·ti·mo·ny \ˈsaŋ(k)-tə-ˌmō-nē\ *noun, plural* **-nies** [Middle French *sanctimonie,* from Latin *sanctimonia,* from *sanctus*] (circa 1541)
1 *obsolete* **:** HOLINESS
2 : affected or hypocritical holiness

[1]**sanc·tion** \ˈsaŋ(k)-shən\ *noun* [Middle French or Latin; Middle French, from Latin *sanction-, sanctio,* from *sancire* to make holy — more at SACRED] (15th century)
1 : a formal decree; *especially* **:** an ecclesiastical decree
2 a *obsolete* **:** a solemn agreement **:** OATH **b :** something that makes an oath binding
3 : the detriment, loss of reward, or coercive intervention annexed to a violation of a law as a means of enforcing the law
4 a : a consideration, principle, or influence (as of conscience) that impels to moral action or determines moral judgment **b :** a mechanism of social control for enforcing a society's standards **c :** explicit or official approval, permission, or ratification **:** APPROBATION
5 : an economic or military coercive measure adopted usually by several nations in concert for forcing a nation violating international law to desist or yield to adjudication

[2]**sanction** *transitive verb* **sanc·tioned; sanc·tion·ing** \-sh(ə-)niŋ\ (1778)
1 : to make valid or binding usually by a formal procedure (as ratification)
2 : to give effective or authoritative approval or consent to
synonym see APPROVE
— sanc·tion·able \-sh(ə-)nə-bəl\ *adjective*

sanc·ti·ty \ˈsaŋ(k)-tə-tē\ *noun, plural* **-ties** [Middle English *sauncite,* from Middle French *saincteté,* from Latin *sanctitat-, sanctitas,* from *sanctus* sacred] (14th century)
1 : holiness of life and character **:** GODLINESS
2 a : the quality or state of being holy or sacred **:** INVIOLABILITY **b** *plural* **:** sacred objects, obligations, or rights

sanc·tu·ary \ˈsaŋ(k)-chə-ˌwer-ē\ *noun, plural* **-ar·ies** [Middle English *sanctuarie,* from Middle French *sainctuarie,* from Late Latin *sanctuarium,* from Latin *sanctus*] (14th century)
1 : a consecrated place: as **a :** the ancient Hebrew temple at Jerusalem or its holy of holies **b** (1) **:** the most sacred part of a religious building (as the part of a Christian church in which the altar is placed) (2) **:** the room in which general worship services are held (3) **:** a place (as a church or a temple) for worship
2 a (1) **:** a place of refuge and protection (2) **:** a refuge for wildlife where predators are controlled and hunting is illegal **b :** the immunity from law attached to a sanctuary

sanc·tum \ˈsaŋ(k)-təm\ *noun, plural* **sanc·tums** *also* **sanc·ta** \-tə\ [Late Latin, from Latin, neuter of *sanctus* sacred] (1577)
1 : a sacred place

2 : a place where one is free from intrusion ⟨an editor's *sanctum*⟩ ⟨the inner *sanctums* of research⟩

sanc·tum sanc·to·rum \ˌsaŋ(k)-təm-ˌsaŋ(k)-ˈtōr-əm, -ˈtȯr-\ *noun* [Late Latin] (1558)
1 : HOLY OF HOLIES
2 : SANCTUM 2

Sanc·tus \ˈsaŋ(k)-təs; ˈsäŋ(k)-təs, -ˌtüs\ *noun* [Middle English, from Late Latin *Sanctus, sanctus, sanctus* Holy, holy, holy, opening of a hymn sung by the angels in Isaiah 6:3] (15th century)
: an ancient Christian hymn of adoration sung or said immediately before the prayer of consecration in traditional liturgies

Sanctus bell *noun* (15th century)
: a bell rung by the server at several points (as at the Sanctus) during the mass

[1]sand \ˈsand\ *noun* [Middle English, from Old English; akin to Old High German *sant* sand, Latin *sabulum,* Greek *psammos*] (before 12th century)
1 a : a loose granular material that results from the disintegration of rocks, consists of particles smaller than gravel but coarser than silt, and is used in mortar, glass, abrasives, and foundry molds **b :** soil containing 85 percent or more of sand and a maximum of 10 percent of clay; *broadly* **:** sandy soil
2 a : a tract of sand **:** BEACH **b :** a sandbank or sandbar
3 : the sand in an hourglass; *also* **:** the moments of a lifetime — usually used in plural ⟨the *sands* of this government run out very rapidly —H. J. Laski⟩
4 : an oil-producing formation of sandstone or unconsolidated sand
5 : firm resolution
6 : a yellowish gray color

[2]sand *transitive verb* (14th century)
1 : to sprinkle or dust with or as if with sand
2 : to cover or fill with sand
3 : to smooth or dress by grinding or rubbing with an abrasive (as sandpaper)

san·dal \ˈsan-dᵊl\ *noun* [Middle English *sandalie,* from Latin *sandalium,* from Greek *sandalion,* diminutive of *sandalon* sandal] (14th century)
1 : a shoe consisting of a sole strapped to the foot
2 : a low-cut shoe that fastens by an ankle strap
3 : a strap to hold on a slipper or low shoe
4 : a rubber overshoe cut very low
— san·daled \ˈsan-dᵊld\ *adjective*

san·dal·wood \-ˌwu̇d\ *noun* [*sandal* sandalwood (from Middle English, from Middle French, from Medieval Latin *sandalum,* from Late Greek *santalon,* ultimately from Sanskrit *candana,* of Dravidian origin; akin to Tamil *cāntu* sandalwood tree) + [2]*wood*] (circa 1511)
1 : the compact close-grained fragrant yellowish heartwood of a parasitic tree (*Santalum album* of the family Santalaceae, the sandalwood family) of southern Asia much used in ornamental carving and cabinetwork; *also* **:** the tree that yields this wood
2 : any of various trees other than the sandalwood some of which yield dyewoods; *also* **:** the fragrant wood of such a tree

sandalwood oil *noun* (1851)
: an essential oil obtained from sandalwood: as **a :** a pale yellow somewhat viscous aromatic liquid obtained from a sandalwood (*Santalum album*) and used chiefly in perfumes and soaps **b :** an oil obtained from a sandalwood (*Eucarya spicata*) of Australia

san·da·rac \ˈsan-də-ˌrak\ *noun* [Latin *sandaraca* red coloring, from Greek *sandarakē* realgar, red pigment from realgar] (1543)
: a brittle faintly aromatic translucent resin obtained from a northern African tree (*Tetraclinis articulata*) of the cypress family and used chiefly in making varnish and as incense; *also* **:** a similar resin obtained from any of several Australian trees (genus *Callitris*) of the same family

[1]sand·bag \ˈsan(d)-ˌbag\ *noun* (1590)
: a bag filled with sand and used in fortifications, as ballast, or as a weapon

[2]sandbag *transitive verb* (1860)
1 : to bank, stop up, or weight with sandbags
2 a : to hit or stun with or as if with a sandbag **b :** to treat unfairly or harshly **c :** to coerce by crude means ⟨are raiding the Treasury and *sandbagging* the government —C. W. Ferguson⟩ **d :** to conceal or misrepresent one's true position, potential, or intent especially in order to take advantage of
— sand·bag·ger *noun*

sand·bank \ˈsan(d)-ˌbaŋk\ *noun* (15th century)
: a large deposit of sand forming a mound, hillside, bar, or shoal

sand·bar \-ˌbär\ *noun* (1766)
: a ridge of sand built up by currents especially in a river or in coastal waters

[1]sand·blast \-ˌblast\ *noun* (1871)
: a stream of sand projected by compressed air (as for engraving, cutting, or cleaning glass or stone)

[2]sandblast *transitive verb* (1888)
: to affect or treat with or as if with a sandblast
— sand·blast·er *noun*

sand–blind \ˈsan(d)-ˌblīnd\ *adjective* [Middle English, probably from (assumed) Middle English *samblind,* from Old English *sam-* half (akin to Old High German *sāmi-* half) + *blind* — more at SEMI-] (15th century)
: having poor eyesight **:** PURBLIND

sand bluestem *noun* (circa 1946)
: a tall rhizomatous grass (*Andropogon hallii*) of the western U.S. used for forage and as a soil binder

sand·box \ˈsan(d)-ˌbäks\ *noun* (1688)
: a box or receptacle containing loose sand: as **a :** a shaker for sprinkling sand on wet ink **b :** a box that contains sand for children to play in

sand·bur \ˈsan(d)-ˌbər\ *noun* (1830)
: any of a genus (*Cenchrus*) of grasses producing spikelets enclosed in ovoid spiny involucres that form burs; *also* **:** one of these burs

sand–cast \-ˌkast\ *transitive verb* **-cast; -cast·ing** (1928)
: to make (a casting) by pouring metal in a sand mold

sand casting *noun* (1926)
: a casting made in a mold of sand

sand crack *noun* (1754)
: a fissure in the wall of a horse's hoof often causing lameness

sand dab *noun* (1946)
: any of several Pacific flounders (genus *Citharichthys* of the family Bothidae); *especially* **:** a common food fish (*C. sordidus*)

sand dollar *noun* (1884)
: any of numerous flat circular sea urchins (order Clypeasteroida) that live chiefly in shallow water on sandy bottoms

sand·er \ˈsan-dər\ *noun* (1627)
: one that sands: as **a :** a device for spreading sand on newly surfaced or icy roads; *also* **:** the device together with the truck that bears it **b :** a machine or device that smooths, polishes, or scours by means of abrasive material usually in the form of a disk or belt — called also *sanding machine*

sand·er·ling \ˈsan-dər-liŋ\ *noun* [*sand* + *-erling,* perhaps from Old English *yrthling,* kind of bird found in fields, literally, plowman, from *yrth, earth* plowing, from *erian* to plow — more at ARABLE] (1602)
: a small widely distributed sandpiper (*Calidris alba*) with pale gray and white plumage in winter

sand flea *noun* (1796)
1 : a flea (as a chigoe) found in sandy places
2 : BEACH FLEA

sand fly *noun* (1736)
: any of various small biting dipteran flies (families Psychodidae, Simuliidae, and Ceratopogonidae)

sand·fly fever \ˈsan(d)-ˌflī-\ *noun* (1910)
: a virus disease of brief duration that is characterized by fever, headache, pain in the eyes, malaise, and leukopenia and is transmitted by the bite of a sand fly (*Phlebotomus papatasii*) — called also *phlebotomus fever*

sand·glass \ˈsan(d)-ˌglas\ *noun* (1556)
: an instrument (as an hourglass) for measuring time by the running of sand

sand·grouse \-ˌgrau̇s\ *noun* (1783)
: any of numerous birds (family Pteroclidae) of arid parts of southern Europe, Asia, and Africa that are closely related to the pigeons but have precocial downy young

san·dhi \ˈsan-dē, ˈsän-, ˈsən-\ *noun* [Sanskrit *saṁdhi,* literally, placing together] (1806)
: modification of the sound of a morpheme (as a word or affix) conditioned by syntactic context in which it is uttered (as pronunciation of *-ed* as \d\ in *glazed* and as \t\ in *paced* or occurrence of *a* in *a cow* and of *an* in *an old cow*)

sand·hill crane \ˈsand-ˌhil-\ *noun* (1805)
: a crane (*Grus canadensis*) of North America and Siberia that has a red crown and is chiefly bluish gray tinged with a sandy yellow

sand·hog \ˈsand-ˌhȯg, -ˌhäg\ *noun* (1903)
: a laborer who works in a caisson in driving underwater tunnels

San·di·nis·ta \ˌsan-də-ˈnēs-tə, ˌsän-\ *noun* [American Spanish, from Augusto César *Sandino* (died 1933) Nicaraguan rebel leader] (1974)
: a member of a military and political coalition taking power in Nicaragua in 1979

S and L *noun* (1951)
: SAVINGS AND LOAN ASSOCIATION

sand lance *noun* (1776)
: any of several small elongate marine bony fishes (genus *Ammodytes* of the family Ammodytidae) that associate in large schools and remain buried in sandy beaches at ebb tide — called also *sand eel, sand launce*

sand lily *noun* (circa 1900)
: a western North American spring herb (*Leucocrinum montanum*) of the lily family with narrow linear leaves and fragrant salverform flowers

sand·lot \ˈsan(d)-ˌlät\ *noun* (1878)
: a vacant lot especially when used (as by children) for usually unorganized sports
— sandlot *adjective*
— sand·lot·ter \-ˌlä-tər\ *noun*

sand·man \ˈsan(d)-ˌman\ *noun*
: a genie in folklore who makes children sleepy by sprinkling sand in their eyes

sand myrtle *noun* (1814)
: a variable low-branching evergreen upland shrub (*Leiophyllum buxifolium*) of the heath family found in the southeastern U.S.

sand·paint·ing \ˈsan(d)-ˌpān-tiŋ\ *noun* (1900)
: a Navajo and Pueblo Indian ceremonial design made of various materials (as colored sands) on a flat surface of sand or buckskin

[1]sand·pa·per \-ˌpā-pər\ *noun* (1825)
: paper covered on one side with abrasive material (as sand) glued fast and used for smoothing and polishing
— sand·pa·pery \-p(ə-)rē\ *adjective*

[2]sandpaper *transitive verb* (1846)
: to rub with or as if with sandpaper

sand·pile \ˈsan(d)-ˌpīl\ *noun* (1901)
: a pile of sand; *especially* **:** sand for children to play in

sand·pip·er \-ˌpī-pər\ *noun* (1674)
: any of numerous small shorebirds (family Scolopacidae) distinguished from the related plovers chiefly by the longer and soft-tipped bill

sandpiper

sand·pit \-ˌpit\ *noun* (1898)
British : SANDBOX b

sand·shoe \-ˌshü\ *noun* (1855)
chiefly Australian & New Zealand : TENNIS SHOE

sand·soap \ˈsan(d)-ˌsōp\ *noun* (1854)
: a gritty soap for all-purpose cleaning

sand·spur \-ˌspər\ *noun* (circa 1898)
: SANDBUR

sand·stone \-ˌstōn\ *noun* (1668)
: a sedimentary rock consisting of usually quartz sand united by some cement (as silica or calcium carbonate)

sand·storm \-ˌstȯrm\ *noun* (1774)
: a windstorm (as in a desert) driving clouds of sand before it

sand table *noun* (1812)
1 : a table holding sand for children to mold
2 : a table bearing a relief model of a terrain built to scale for study or demonstration especially of military tactics

sand trap *noun* (1922)
: an artificial hazard on a golf course consisting of a depression containing sand

sand verbena *noun* (1898)
: any of several western North American herbs (genus *Abronia*) of the four-o'clock family with flowers like the verbena; *especially* : either of two plants (*A. latifolia* and *A. umbellata*) of the Pacific coast

¹sand·wich \ˈsan(d)-ˌwich, ˈsam-\ *noun* [John Montagu, 4th Earl of *Sandwich* (died 1792) English diplomat] (1762)
1 a : two or more slices of bread or a split roll having a filling in between **b :** one slice of bread covered with food
2 : something resembling a sandwich; *especially* : composite structural material consisting of layers often of high-strength facings bonded to a low strength central core ◆

²sandwich *transitive verb* (1861)
1 : to make into or as if into a sandwich; *especially* : to insert or enclose between usually two things of another quality or character
2 : to make a place for — often used with *in* or *between*

sandwich board *noun* (1897)
: two usually hinged boards designed for hanging from the shoulders with one board before and one behind and used especially for advertising or picketing

sandwich coin *noun* (1965)
: a clad coin

sandwich man *noun* (1864)
: one who wears a sandwich board

sand·worm \ˈsan(d)-ˌwərm\ *noun* (1776)
: any of various sand-dwelling polychaete worms: as **a :** any of several large burrowing worms (especially genus *Nereis*) often used as bait **b :** LUGWORM

sand·wort \ˈsan(d)-ˌwərt, -ˌwȯrt\ *noun* (1597)
: any of a genus (*Arenaria*) of low tufted herbs of the pink family growing usually in dry sandy regions

sandy \ˈsan-dē\ *adjective* **sand·i·er; -est** (before 12th century)
1 a : consisting of or containing sand : full of sand **b :** sprinkled with sand
2 : of the color sand
— sand·i·ness *noun*

sane \ˈsān\ *adjective* **san·er; san·est** [Latin *sanus* healthy, sane] (1628)
1 : proceeding from a sound mind : RATIONAL
2 : mentally sound; *especially* : able to anticipate and appraise the effect of one's actions
3 : healthy in body
synonym see WISE
— sane·ly *adverb*
— sane·ness \ˈsān-nəs\ *noun*

San·for·ized \ˈsan-fə-ˌrīzd\ *trademark*
— used for fabrics that are shrunk by a mechanical process before being manufactured into articles (as clothing)

sang *past of* SING

san·ga·ree \ˌsaŋ-gə-ˈrē\ *noun* [Spanish *sangría,* literally, act of bleeding, from *sangre* blood, from Latin *sanguin-, sanguis*] (1736)
1 : a sweetened iced drink of wine or sometimes of ale, beer, or liquor garnished with nutmeg
2 : SANGRIA

sang·froid \ˈsäⁿ-ˈf(r)wä\ *noun* [French *sang-froid,* literally, cold blood] (1750)
: self-possession or imperturbability especially under strain
synonym see EQUANIMITY

San·greal \ˈsan-ˈgrā(ə)l, ˈsaŋ-\ *noun* [Middle English *Sangrayll,* from Middle French *Saint Graal* Holy Grail]
: GRAIL

san·gria \saŋ-ˈgrē-ə, san-\ *noun* [Spanish] (1736)
: a usually iced punch made of red wine, fruit juice, and soda water

san·gui·nar·ia \ˌsaŋ-gwə-ˈner-ē-ə, -ˈnar-\ *noun* [New Latin, from Latin, an herb that stanches blood, from feminine of *sanguinarius* sanguinary] (1808)
1 : BLOODROOT
2 : the rhizome and roots of a bloodroot used as an expectorant and emetic

san·gui·nary \ˈsaŋ-gwə-ˌner-ē\ *adjective* [Latin *sanguinarius,* from *sanguin-, sanguis* blood] (1623)
1 : BLOODTHIRSTY, MURDEROUS ⟨*sanguinary* hatred⟩
2 : attended by bloodshed : BLOODY ⟨this bitter and *sanguinary* war —T. H. D. Mahoney⟩
3 : consisting of blood ⟨a *sanguinary* stream⟩
synonym see BLOODY
— san·gui·nar·i·ly \ˌsaŋ-gwə-ˈner-ə-lē\ *adverb*

¹san·guine \ˈsaŋ-gwən\ *adjective* [Middle English *sanguin,* from Middle French, from Latin *sanguineus,* from *sanguin-, sanguis*] (14th century)
1 : BLOODRED
2 a : consisting of or relating to blood **b :** BLOODTHIRSTY, SANGUINARY **c** *of the complexion* : RUDDY
3 : having blood as the predominating bodily humor; *also* : having the bodily conformation and temperament held characteristic of such predominance and marked by sturdiness, high color, and cheerfulness
4 : CONFIDENT, OPTIMISTIC
— san·guine·ly *adverb*
— san·guine·ness \-gwən-nəs\ *noun*
— san·guin·i·ty \saŋ-ˈgwi-nə-tē, san-\ *noun*

²sanguine *noun* (1500)
: a moderate to strong red

san·guin·e·ous \san-ˈgwi-nē-əs, saŋ-\ *adjective* [Latin *sanguineus*] (circa 1520)
1 : BLOODRED
2 : of, relating to, or involving bloodshed : BLOODTHIRSTY
3 : of, relating to, or containing blood

San·he·drin \san-ˈhe-drən, sän-; san-ˈhē-, ˈsa-nə-\ *noun* [Late Hebrew *sanhedhrīn* (*gĕdhōlāh*) (great) Sanhedrin, from Greek *synedrion* council, from *synedros* sitting in council, from *syn-* + *hedra* seat — more at SIT] (1588)
: the supreme council and tribunal of the Jews during postexilic times headed by a High Priest and having religious, civil, and criminal jurisdiction

san·i·cle \ˈsa-ni-kəl\ *noun* [Middle English, from Middle French, from Medieval Latin *sanicula*] (14th century)
: any of several plants sometimes held to have healing powers; *especially* : a plant (genus *Sanicula*) of the carrot family with a root used in folk medicine as an anodyne or astringent

san·i·tar·i·an \ˌsa-nə-ˈter-ē-ən\ *noun* (1859)
: a specialist in sanitary science and public health ⟨milk *sanitarian*⟩

san·i·tar·i·um \ˌsa-nə-ˈter-ē-əm\ *noun, plural* **-i·ums** *or* **-ia** \-ē-ə\ [New Latin, from Latin *sanitat-, sanitas* health] (1851)
: SANATORIUM

san·i·tary \ˈsa-nə-ˌter-ē\ *adjective* [French *sanitaire,* from Latin *sanitas*] (1842)
1 : of or relating to health ⟨*sanitary* measures⟩
2 : of, relating to, or used in the disposal especially of domestic waterborne waste ⟨*sanitary* sewage⟩
3 : characterized by or readily kept in cleanliness ⟨*sanitary* packages⟩
— san·i·tar·i·ly \ˌsa-nə-ˈter-ə-lē\ *adverb*

sanitary landfill *noun* (1968)
: LANDFILL

sanitary napkin *noun* (1917)
: a disposable absorbent pad used postpartum or during menstruation to absorb the uterine flow

sanitary ware *noun* (1872)
: ceramic plumbing fixtures (as sinks, lavatories, or toilet bowls)

san·i·tate \ˈsa-nə-ˌtāt\ *transitive verb* **-tat·ed; -tat·ing** [back-formation from *sanitation*] (1882)
: to make sanitary especially by providing with sanitary appliances or facilities

san·i·ta·tion \ˌsa-nə-ˈtā-shən\ *noun* (1848)
1 : the act or process of making sanitary
2 : the promotion of hygiene and prevention of disease by maintenance of sanitary conditions

san·i·tize \ˈsa-nə-ˌtīz\ *transitive verb* **-tized; -tiz·ing** [Latin *sanitas*] (1836)
1 : to make sanitary (as by cleaning or sterilizing)
2 : to make more acceptable by removing unpleasant or undesired features ⟨*sanitize* a document⟩
— san·i·ti·za·tion \ˌsa-nə-tə-ˈzā-shən\ *noun*

san·i·to·ri·um \ˌsa-nə-ˈtōr-ē-əm, -ˈtȯr-\ *noun, plural* **-ri·ums** *or* **-ria** \-ē-ə\ [by alteration (influenced by *sanitarium*)] (1917)
: SANATORIUM

san·i·ty \ˈsa-nə-tē\ *noun* [Middle English *sanite,* from Latin *sanitat-, sanitas* health, sanity, from *sanus* healthy, sane] (15th century)

◇ WORD HISTORY

sandwich John Montagu, the 4th Earl of Sandwich (1718–1792), was not the most distinguished figure in 18th century British politics, and his maladministration of the Royal Navy as First Lord of the Admiralty had a role in the American colonies winning their war for independence from Britain. Among his few positive deeds was the funding of Captain James Cook's third and last expedition to the Pacific, as a result of which Cook named a newly discovered Polynesian archipelago the Sandwich Islands. Unfortunately for the perpetuation of the Earl's memory, the archipelago has since become much better known as the Hawaiian Islands. The Earl's chief claim to fame lies elsewhere, however. According to the French historian and memoirist Pierre-Jean Grosley, who lived in London in the 1760s, the Earl once spent 24 hours at the gaming tables without eating anything but slices of cold beef between pieces of toast. Thus, allegedly, was born the *sandwich,* the archetypal convenience food. The word as well as the item have since been borrowed into most languages of Europe, from Portuguese *sanduíche* to Serbo-Croatian *sandvič.*

: the quality or state of being sane; *especially* **:** soundness or health of mind

San Ja·cin·to Day \ˌsan-jə-ˈsin-tə-, -hə-ˌsin-\ *noun* (1907)
: April 21 observed as a legal holiday in Texas in commemoration of the battle of San Jacinto in 1836 by which independence from Mexico was won

San Jo·se scale \ˌsa-nə-ˈzā-, ˌsan-(h)ō-\ *noun* [*San Jose,* Calif.] (1887)
: a scale insect (*Quadraspidiotus perniciosus*) probably of Asian origin that is naturalized in the U.S. and is destructive to fruit trees

sank *past of* SINK

San·khya \ˈsäŋ-kyə\ *noun* [Sanskrit *sāṁkhya,* literally, based on calculation] (1788)
: an orthodox Hindu philosophy teaching salvation through knowledge of the dualism of matter and souls

sann hemp \ˈsən-, ˈsän-\ *noun* [Hindi *san*] (1939)
: SUNN

san·nup \ˈsa-nəp\ *noun* [Eastern Abenaki *sénαpe* man, male human] (1628)
: a married male American Indian

sann·ya·si \(ˌ)sən-ˈyä-sē\ *or* **sann·ya·sin** \-ˈyä-sᵊn\ *noun* [Hindi *sannyāsī,* from Sanskrit *sannyāsin*] (1613)
: a Hindu mendicant ascetic

[1]sans \ˈsanz\ *preposition* [Middle English *saun, sans,* from Middle French *san, sans,* modification of Latin *sine* without — more at SUNDER] (14th century)
: WITHOUT ⟨my love to thee is sound, *sans* crack or flaw —Shakespeare⟩

[2]sans \ˈsanz\ *noun, plural* **sans** (circa 1909)
: SANS SERIF

sans·cu·lotte \ˌsanz-kü-ˈlät, -kyü-\ *noun* [French *sans-culotte,* literally, without breeches] (1790)
1 : an extreme radical republican in France at the time of the Revolution
2 : a radical or violent extremist in politics
— sans·cu·lott·ic \-ˈlä-tik\ *adjective*
— sans·cu·lott·ish \-tish\ *adjective*
— sans·cu·lott·ism \-ˌti-zəm\ *noun*

san·sei \ˌsän-ˈsā, ˈsän-ˌ\ *noun, plural* **sansei** *often capitalized* [Japanese *san* third + *sei* generation] (1940)
: a son or daughter of nisei parents who is born and educated in America and especially in the U.S.

san·se·vie·ria \ˌsan(t)-sə-ˈvir-ē-ə\ *noun* [New Latin, from Raimondo di Sangro, prince of *San Severo* (died 1774) Italian scholar] (1804)
: any of a genus (*Sansevieria*) of tropical herbs of the agave family with showy mottled sword-shaped leaves usually yielding a strong fiber

sansevieria

San·skrit \ˈsan-ˌskrit, ˈsan(t)-skrət\ *noun* [Sanskrit *saṁskṛta,* literally, perfected, from *sam* together + *karoti* he makes] (1696)
1 : an ancient Indo-Aryan language that is the classical language of India and of Hinduism
2 : classical Sanskrit together with the older Vedic and various later modifications of classical Sanskrit — see INDO-EUROPEAN LANGUAGES table
— Sanskrit *adjective*
— San·skrit·ic \san-ˈskri-tik\ *adjective*
— San·skrit·ist \-ˌskri-tist\ *noun*

sans ser·if *or* **san·ser·if** \san-ˈser-əf, ˈsanz-\ *noun* [probably from *sans* + modification of Dutch *schreef* stroke — more at SERIF] (1830)
: a letter or typeface with no serifs

San·ta Ana \ˌsan-tə-ˈa-nə\ *noun* [*Santa Ana* Mountains in southern Calif.] (1880)
: a strong hot dry foehn wind from the north, northeast, or east in southern California

San·ta Claus \ˈsan-tə-ˌklȯz *also* ˈsan-tē-\ *noun* [modification of Dutch *Sinterklaas,* alteration of *Sint Nikolaas* Saint Nicholas]
: a plump white-bearded and red-suited old man in modern folklore who delivers presents to good children at Christmastime — called also *Santa*

San·ta Ger·tru·dis \ˌsan-tə-(ˌ)gər-ˈtrü-dəs\ *noun* [*Santa Gertrudis,* section of the King Ranch, Kingsville, Texas] (1942)
: any of a breed of red beef cattle developed from a Brahman-shorthorn cross and valued for their hardiness in hot climates

Santa Gertrudis

San·te·ria *also* **San·te·ría** \ˌsan-tə-ˈrē-ə, ˌsän-\ *noun* [American Spanish *santería,* from *santero* practitioner of Santeria, from *santo* Yoruba deity, literally, saint's image, saint, from Spanish] (1950)
: a religion practiced originally in Cuba in which Yoruba deities are identified with Roman Catholic saints

san·tir \san-ˈtir\ *or* **san·tour** \-ˈtu̇r\ *noun* [Arabic *santīr, santūr,* from Greek *psaltērion* psaltery] (1853)
: a Persian dulcimer

san·to \ˈsän-(ˌ)tō\ *noun, plural* **santos** [Spanish, literally, saint, from Late Latin *sanctus* — more at SAINT] (1834)
: a painted or carved wooden image of a saint common especially in Mexico and the southwestern U.S.

san·to·li·na \ˌsan-tᵊl-ˈē-nə\ *noun* [New Latin, alteration of Latin *santonica,* an herb, feminine of *santonicus* of the Santoni, from *Santoni,* a people of Aquitania] (1578)
: any of a genus (*Santolina*) of Mediterranean composite subshrubs that have dissected leaves and clustered flower heads lacking ray flowers

san·to·nin \ˈsan-tᵊn-ən, san-ˈtä-nən\ *noun* [International Scientific Vocabulary, from New Latin *santonica,* from Latin] (1838)
: a poisonous slightly bitter crystalline compound $C_{15}H_{18}O_3$ found especially in the unopened flower heads of several artemisias (especially *Artemisia maritima*) and used as an anthelmintic

[1]sap \ˈsap\ *noun* [Middle English, from Old English *sæp;* akin to Old High German *saf* sap] (before 12th century)
1 a : the fluid part of a plant; *specifically* **:** a watery solution that circulates through a plant's vascular system **b** (1) **:** a body fluid (as blood) essential to life, health, or vigor (2) **:** bodily health and vigor
2 : a foolish gullible person
3 : BLACKJACK, BLUDGEON
— sap·less \ˈsa-pləs\ *adjective*
— sap·less·ness \-nəs\ *noun*

[2]sap *transitive verb* **sapped; sap·ping** (1725)
1 : to drain or deprive of sap
2 : to knock out with a sap

[3]sap *verb* **sapped; sap·ping** [Middle French *sapper,* from Old Italian *zappare,* from *zappa* hoe] (1598)
intransitive verb
: to proceed by digging a sap
transitive verb
1 : to subvert by digging or eroding the substratum or foundation **:** UNDERMINE
2 a : to gradually diminish the supply or intensity of ⟨*sapped* her strength⟩ **b :** to weaken or exhaust the energy or vitality of
3 : to operate against or pierce by a sap
synonym see WEAKEN

[4]sap *noun* [Middle French *sape,* from *saper*] (1642)
: the extension of a trench to a point beneath an enemy's fortifications

sap green *noun* (1578)
: a strong yellow green

sap·head \ˈsap-ˌhed\ *noun* (1798)
: a weak-minded stupid person **:** SAP
— sap·head·ed \-ˈhe-dəd\ *adjective*

sa·phe·nous \sə-ˈfē-nəs, ˈsa-fə-nəs\ *adjective* [*saphena* saphenous vein, from Middle English, from Medieval Latin, from Arabic *ṣāfin*] (1840)
: of, relating to, associated with, or being either of the two chief superficial veins of the leg ⟨*saphenous* nerve⟩

sap·id \ˈsa-pəd\ *adjective* [Latin *sapidus* tasty, from *sapere* to taste — more at SAGE] (1623)
1 : having flavor **:** FLAVORFUL
2 *archaic* **:** agreeable to the mind
— sa·pid·i·ty \sa-ˈpi-də-tē\ *noun, archaic*

sa·pi·ence \ˈsā-pē-ən(t)s, ˈsa-\ *noun* [Middle English, from Middle French, from Latin *sapientia,* from *sapient-, sapiens,* present participle] (14th century)
: WISDOM, SAGACITY

sa·pi·ens \ˈsa-pē-ənz, ˈsā-, -ˌenz\ *adjective* [New Latin (specific epithet of *Homo sapiens*), from Latin, present participle of *sapere*] (1939)
: of, relating to, or being recent humans (*Homo sapiens*) as distinguished from various fossil hominids

sa·pi·ent \ˈsā-pē-ənt, ˈsa-\ *adjective* [Middle English, from Middle French, from Latin *sapient-, sapiens,* from present participle of *sapere* to taste, be wise — more at SAGE] (15th century)
: possessing or expressing great sagacity
synonym see WISE
— sa·pi·ent·ly *adverb*

sap·ling \ˈsa-pliŋ, -plən\ *noun* (14th century)
1 : a young tree; *specifically* **:** one not over four inches in diameter at breast height
2 : YOUTH 2a

sap·o·dil·la \ˌsa-pə-ˈdi-lə, -ˈdē-yə\ *noun* [Spanish *zapotillo,* diminutive of *zapote* sapodilla, from Nahuatl *tzapotl*] (1697)
: a tropical evergreen tree (*Manilkara zapota* synonym *Achras zapota* of the family Sapotaceae, the sapodilla family) with hard reddish wood, a latex that yields chicle, and a rough-skinned brownish edible fruit; *also* **:** its fruit

sa·po·ge·nin \ˌsa-pə-ˈje-nən, sə-ˈpä-jə-nən\ *noun* [International Scientific Vocabulary *saponin* + *-genin* (compound formed from another compound)] (circa 1862)
: a nonsugar portion of a saponin that is typically obtained by hydrolysis, has either a complex terpenoid or a steroidal structure, and in the latter case forms a practicable starting point in the synthesis of steroid hormones

sap·o·na·ceous \ˌsa-pə-ˈnā-shəs\ *adjective* [New Latin *saponaceus,* from Latin *sapon-, sapo* soap, of Germanic origin; akin to Old English *sāpe* soap] (1710)
: resembling or having the qualities of soap
— sap·o·na·ceous·ness *noun*

sa·pon·i·fy \sə-ˈpä-nə-ˌfī\ *verb* **-fied; -fy·ing** [French *saponifier,* from Latin *sapon-, sapo*] (1821)
transitive verb
: to convert (as fat) into soap; *specifically* **:** to hydrolyze (a fat) with alkali to form a soap and glycerol
intransitive verb
: to undergo saponifying
— sa·pon·i·fi·able \-ˌfī-ə-bəl\ *adjective*
— sa·pon·i·fi·ca·tion \-ˌpä-nə-fə-ˈkā-shən\ *noun*

\ə\ abut \ᵊ\ kitten \ər\ further \a\ ash \ā\ ace \ä\ mop, mar \au̇\ out \ch\ chin \e\ bet \ē\ easy \g\ go \i\ hit \ī\ ice \j\ job \ŋ\ sing \ō\ go \ȯ\ law \ȯi\ boy \th\ thin \t͟h\ the \ü\ loot \u̇\ foot \y\ yet \zh\ vision *see also* Guide to Pronunciation

— **sa·pon·i·fi·er** \-ˈpä-nə-ˌfī(-ə)r\ *noun*

sa·po·nin \ˈsa-pə-nən, sə-ˈpō-\ *noun* [French *saponine,* from Latin *sapon-, sapo*] (1831)
: any of various mostly toxic glucosides that occur in plants (as soapwort or soapbark) and are characterized by the property of producing a soapy lather; *especially* **:** a hygroscopic amorphous saponin mixture used especially as a foaming and emulsifying agent and detergent

sap·o·nite \ˈsa-pə-ˌnīt\ *noun* [Swedish *saponit,* from Latin *sapon-, sapo* soap] (circa 1849)
: a hydrous magnesium aluminum silicate occurring in soft soapy amorphous masses and filling veins and cavities (as in serpentine)

sap·per \ˈsa-pər\ *noun* (1626)
1 : a military specialist in field fortification work (as sapping)
2 : a military demolitions specialist

[1]**sap·phic** \ˈsa-fik\ *adjective* (1501)
1 *capitalized* **:** of or relating to the Greek lyric poet Sappho
2 : of, relating to, or consisting of a 4-line strophe made up of chiefly trochaic and dactylic feet
3 : LESBIAN 2

[2]**sapphic** *noun* (1586)
1 : a verse having the metrical pattern of one of the first three lines of a sapphic strophe
2 : a sapphic strophe

sap·phire \ˈsa-ˌfīr\ *noun* [Middle English *safir,* from Old French, from Latin *sapphirus,* from Greek *sappheiros,* perhaps of Semitic origin; akin to Hebrew *sappīr* sapphire] (13th century)
1 a : a gem variety of corundum in transparent or translucent crystals of a color other than red; *especially* **:** one of a transparent rich blue **b :** a gem of such corundum
2 : a deep purplish blue color
— **sapphire** *adjective*

sap·phi·rine \ˈsa-fə-ˌrīn, ˈsa-ˌfīr-ˌēn, sa-ˈfī-rən\ *adjective* (15th century)
1 : made of sapphire
2 : resembling sapphire especially in color

sap·phism \ˈsa-ˌfi-zəm\ *noun* [*Sappho* + *-ism;* from the belief that Sappho was homosexual] (circa 1890)
: LESBIANISM

sap·pi·ness \ˈsa-pē-nəs\ *noun* (1552)
1 : the state of being full of or smelling of sap
2 : the quality or state of being sappy **:** FOOLISHNESS

sap·py \ˈsa-pē\ *adjective* **sap·pi·er; -est** (12th century)
1 : abounding with sap
2 : resembling or consisting largely of sapwood
3 a : overly sweet or sentimental **b :** lacking in good sense **:** SILLY

sapr- *or* **sapro-** *combining form* [New Latin, from Greek *sapros* rotten]
1 : dead or decaying organic matter ⟨*sapro*phyte⟩
2 : decay **:** putrefaction ⟨*sapro*genic⟩

sap·ro·gen·ic \ˌsa-prə-ˈje-nik\ *adjective* (1876)
: of, causing, or resulting from putrefaction
— **sap·ro·ge·nic·i·ty** \-jə-ˈni-sə-tē\ *noun*

sap·ro·lite \ˈsa-prə-ˌlīt\ *noun* (1894)
: disintegrated rock that lies in its original place

sa·proph·a·gous \sa-ˈprä-fə-gəs\ *adjective* [New Latin *saprophagus,* from *sapr-* + *-phagus* -phagous] (1819)
: feeding on decaying matter

sap·ro·phyte \ˈsa-prə-ˌfīt\ *noun* [International Scientific Vocabulary] (1875)
: a saprophytic organism; *especially* **:** a plant living on dead or decaying organic matter

sap·ro·phyt·ic \ˌsa-prə-ˈfi-tik\ *adjective* (1882)
: obtaining food by absorbing dissolved organic material; *especially* **:** obtaining nourishment osmotically from the products of organic breakdown and decay
— **sap·ro·phyt·i·cal·ly** \-ti-k(ə-)lē\ *adverb*

sap·ro·zo·ic \ˌsa-prə-ˈzō-ik\ *adjective* (circa 1920)
: SAPROPHYTIC — used of animals (as protozoans)

sap·sa·go \sap-ˈsā-(ˌ)gō, ˈsap-sə-ˌgō\ *noun* [modification of German *Schabziger*] (circa 1846)
: a very hard green skim-milk cheese flavored with the powdered leaves of an aromatic legume (*Trigonella coerulea*)

sap·suck·er \ˈsap-ˌsə-kər\ *noun* (1805)
: any of a genus (*Sphyrapicus*) of small North American woodpeckers that drill holes in trees in order to obtain sap and insects for food

sap·wood \-ˌwu̇d\ *noun* (1791)
: the younger softer living or physiologically active outer portion of wood that lies between the cambium and the heartwood and is more permeable, less durable, and usually lighter in color than the heartwood

sar·a·band *or* **sar·a·bande** \ˈsar-ə-ˌband\ *noun* [French *sarabande,* from Spanish *zarabanda*] (1616)
1 : a stately court dance of the 17th and 18th centuries resembling the minuet
2 : the music for the saraband in slow triple time with accent on the second beat

Sar·a·cen \ˈsar-ə-sən\ *noun* [Middle English, from Old English, from Late Latin *Saracenus,* from Late Greek *Sarakēnos*] (before 12th century)
: a member of a nomadic people of the deserts between Syria and Arabia; *broadly* **:** ARAB
— **Saracen** *adjective*
— **Sar·a·cen·ic** \ˌsar-ə-ˈse-nik\ *adjective*

Sa·rah \ˈser-ə, ˈsar-ə, ˈsā-rə\ *noun* [Hebrew *Śārāh*]
1 : the wife of Abraham and mother of Isaac
2 : a kinswoman of Tobias married to him

sa·ran \sə-ˈran\ *noun* [from *Saran,* a trademark] (1940)
: a tough flexible thermoplastic resin

sarape \sə-ˈrä-pē\ *variant of* SERAPE

Sar·a·to·ga trunk \ˌsar-ə-ˈtō-gə-\ *noun* [*Saratoga* Springs, N.Y.] (1858)
: a large traveling trunk usually with a rounded top

sarc- *or* **sarco-** *combining form* [Greek *sark-, sarko-,* from *sark-, sarx*]
1 : flesh ⟨*sarc*oid⟩
2 : striated muscle ⟨*sarco*lemma⟩

sar·casm \ˈsär-ˌka-zəm\ *noun* [French or Late Latin; French *sarcasme,* from Late Latin *sarcasmos,* from Greek *sarkasmos,* from *sarkazein* to tear flesh, bite the lips in rage, sneer, from *sark-, sarx* flesh; probably akin to Avestan *thwarəs-* to cut] (1550)
1 : a sharp and often satirical or ironic utterance designed to cut or give pain ⟨tired of continual *sarcasms*⟩
2 a : a mode of satirical wit depending for its effect on bitter, caustic, and often ironic language that is usually directed against an individual **b :** the use or language of sarcasm ⟨this is no time to indulge in *sarcasm*⟩
synonym see WIT

sar·cas·tic \sär-ˈkas-tik\ *adjective* (1695)
1 : having the character of sarcasm ⟨*sarcastic* criticism⟩
2 : given to the use of sarcasm **:** CAUSTIC ⟨a *sarcastic* critic⟩ ☆
— **sar·cas·ti·cal·ly** \-ti-k(ə-)lē\ *adverb*

[1]**sarce·net** \ˈsär-snət\ *noun* [Middle English, from Anglo-French *sarzinett*] (15th century)
: a soft thin silk in plain or twill weaves; *also* **:** a garment made of this

[2]**sarcenet** *adjective* (1521)
1 *archaic* **:** made of sarcenet
2 *archaic* **:** soft like sarcenet

sar·coid \ˈsär-ˌkȯid\ *noun* (1899)
1 : any of various diseases characterized especially by the formation of nodules in the skin
2 : a nodule characteristic of sarcoid or of sarcoidosis

sar·coid·o·sis \ˌsär-ˌkȯi-ˈdō-səs\ *noun, plural* **-o·ses** \-ˌsēz\ [New Latin] (1936)
: a chronic disease of unknown cause that is characterized by the formation of nodules especially in the lymph nodes, lungs, bones, and skin

sar·co·lem·ma \ˌsär-kə-ˈle-mə\ *noun* [New Latin, from *sarc-* + Greek *lemma* husk — more at LEMMA] (1840)
: the thin transparent homogeneous sheath enclosing a striated muscle fiber
— **sar·co·lem·mal** \-məl\ *adjective*

sar·co·ma \sär-ˈkō-mə\ *noun, plural* **-mas** *also* **-ma·ta** \-mə-tə\ [New Latin, from Greek *sarkōmat-, sarkōma* fleshy growth, from *sarkoun* to grow flesh, from *sark-, sarx*] (1804)
: a malignant neoplasm arising in tissue of mesodermal origin (as connective tissue, bone, cartilage, or striated muscle)
— **sar·co·ma·tous** \sär-ˈkō-mə-təs\ *adjective*

sar·co·ma·to·sis \(ˌ)sär-ˌkō-mə-ˈtō-səs\ *noun, plural* **-to·ses** \-ˌsēz\ [New Latin] (circa 1890)
: a disease characterized by the presence and spread of sarcomas

sar·co·mere \ˈsär-kə-ˌmir\ *noun* (1891)
: any of the repeating structural units of striated muscle fibrils

sar·coph·a·gus \sär-ˈkä-fə-gəs\ *noun, plural* **-gi** \-ˌgī, -ˌjī, -ˌgē\ *also* **-gus·es** [Latin *sarcophagus* (*lapis*) limestone used for coffins, from Greek (*lithos*) *sarkophagos,* literally, flesh-eating stone, from *sark-* sarc- + *phagein* to eat — more at BAKSHEESH] (1619)
: a stone coffin; *broadly* **:** COFFIN

sar·co·plasm \ˈsär-kə-ˌpla-zəm\ *noun* [New Latin *sarcoplasma*] (1899)
: the cytoplasm of a striated muscle fiber
— **sar·co·plas·mic** \-ˈplaz-mik\ *adjective*

sarcoplasmic reticulum *noun* (1953)
: the specialized endoplasmic reticulum of cardiac muscle and skeletal striated muscle that functions especially as a storage and release area for calcium

sar·cop·tic mange \(ˌ)sär-ˈkäp-tik-\ *noun* [New Latin *Sarcoptes,* from *sarc-* + Greek *koptein* to cut — more at CAPON] (1886)
: mange caused by mites (genus *Sarcoptes*) burrowing in the skin especially of the head and face

sar·co·some \ˈsär-kə-ˌsōm\ *noun* [New Latin *sarcosoma,* from *sarc-* + *-soma* -some] (1899)
: a mitochondrion of a striated muscle fiber
— **sar·co·som·al** \ˌsär-kə-ˈsō-məl\ *adjective*

sard \ˈsärd\ *noun* [French *sarde,* from Latin *sarda*] (14th century)
: a reddish brown variety of chalcedony sometimes classified as a variety of carnelian

sardar *variant of* SIRDAR

sar·dine \sär-ˈdēn\ *noun, plural* **sardines** *also* **sardine** [Middle English *sardeine,* from

☆ SYNONYMS
Sarcastic, satiric, ironic, sardonic mean marked by bitterness and a power or will to cut or sting. SARCASTIC implies an intentional inflicting of pain by deriding, taunting, or ridiculing ⟨a critic famous mainly for his *sarcastic* remarks⟩. SATIRIC implies that the intent of the ridiculing is censure and reprobation ⟨a *satiric* look at contemporary sexual mores⟩. IRONIC implies an attempt to be amusing or provocative by saying usually the opposite of what is meant ⟨made the *ironic* observation that the government could always be trusted⟩. SARDONIC implies scorn, mockery, or derision that is manifested by either verbal or facial expression ⟨surveyed the scene with a *sardonic* smile⟩.

Middle French *sardine,* from Latin *sardina*] (14th century)
1 : any of several small or immature clupeid fishes; *especially* **:** the young of the European pilchard (*Sardina pilchardus*) when of a size suitable for preserving for food
2 : any of various small fishes (as an anchovy) resembling the true sardines or similarly preserved for food

Sar·din·ian \sär-ˈdi-nē-ən, -ˈdin-yən\ *noun* (1598)
1 : a native or inhabitant of Sardinia
2 : the Romance language of central and southern Sardinia
— Sardinian *adjective*

sar·don·ic \sär-ˈdä-nik\ *adjective* [French *sardonique,* from Greek *sardonios*] (1638)
: disdainfully or skeptically humorous **:** derisively mocking ⟨a *sardonic* comment⟩ ⟨his *sardonic* expression⟩
synonym see SARCASTIC
— sar·don·i·cal·ly \-ni-k(ə-)lē\ *adverb*

sar·don·i·cism \sär-ˈdä-nə-ˌsi-zəm\ *noun* (1926)
: sardonic quality or humor

sar·don·yx \sär-ˈdä-niks *also* ˈsär-dᵊn-iks\ *noun* [Middle English *sardonix,* from Latin *sardonyx,* from Greek] (14th century)
: an onyx having parallel layers of sard

sar·gas·so \sär-ˈga-(ˌ)sō\ *noun, plural* **-sos** [Portuguese *sargaço*] (1598)
1 : GULFWEED, SARGASSUM
2 : a mass of floating vegetation and especially sargassums

sar·gas·sum \sär-ˈga-səm\ *noun* [New Latin, genus name, from International Scientific Vocabulary *sargasso*] (circa 1890)
: any of a genus (*Sargassum*) of brown algae that have a branching thallus with lateral outgrowths differentiated as leafy segments, air bladders, or spore-bearing structures **:** GULFWEED

sarge \ˈsärj\ *noun* [by shortening & alteration] (1867)
: SERGEANT

sa·ri *also* **sa·ree** \ˈsär-ē\ *noun* [Hindi *sāṛī,* from Sanskrit *śāṭī* strip of cloth] (1785)
: a garment of southern Asian women that consists of several yards of lightweight cloth draped so that one end forms a skirt and the other a head or shoulder covering

sari

sa·rin \ˈsär-ən, zä-ˈrēn\ *noun* [German] (1951)
: an extremely toxic chemical warfare agent $C_4H_{10}FO_2P$ that is a powerful cholinesterase inhibitor

sark \ˈsärk\ *noun* [Middle English (Scots) *serk,* from Old English *serc;* akin to Old Norse *serkr* shirt] (before 12th century)
dialect chiefly British **:** SHIRT

sa·rod *also* **sa·rode** \sə-ˈrōd\ *noun* [Hindi *sarod,* from Persian] (1865)
: a lute of northern India
— sa·rod·ist \-ˈrō-dist\ *noun*

sa·rong \sə-ˈrȯŋ, -ˈräŋ\ *noun* [Malay] (1830)
: a loose garment made of a long strip of cloth wrapped around the body that is worn as a skirt or dress by men and women chiefly of the Malay Archipelago and the Pacific islands

Sar·pe·don \sär-ˈpē-dᵊn\ *noun* [Latin, from Greek *Sarpēdōn*]
: a son of Zeus and Europa and king of Lycia killed in the Trojan War

sar·ra·ce·nia \ˌsar-ə-ˈsē-nē-ə, -ˈse-\ *noun* [New Latin, from Michel *Sarrazin* (died 1734) French physician & naturalist] (1884)
: any of a genus (*Sarracenia* of the family Sarraceniaceae) that includes the insectivorous bog pitcher plants of eastern North America

sar·sa·pa·ril·la \ˌsas-pə-ˈri-lə, ˌsärs-, -ˈre-; ˌsa-sə-, ˌsär-sə-\ *noun* [Spanish *zarzaparrilla,* from *zarza* bush + *parrilla,* diminutive of *parra* vine] (1577)
1 a : any of various tropical American greenbriers **b :** the dried roots of a sarsaparilla used especially as a flavoring
2 : any of various plants (as wild sarsaparilla) that resemble or are used as a substitute for sarsaparilla
3 : a sweetened carbonated beverage flavored with sassafras and oil distilled from a European birch

sarsenet *variant of* SARCENET

sar·to·ri·al \sär-ˈtōr-ē-əl, sə(r)-, -ˈtȯr-\ *adjective* [Medieval Latin *sartor*] (1823)
: of or relating to a tailor or tailored clothes; *broadly* **:** of or relating to clothes
— sar·to·ri·al·ly \-ē-ə-lē\ *adverb*

sar·to·ri·us \sär-ˈtōr-ē-əs, -ˈtȯr-\ *noun, plural* **-rii** \-ē-ˌī, -ē-ˌē\ [New Latin, from Medieval Latin *sartor* tailor, from Latin *sarcire* to mend] (1704)
: a muscle that crosses the front of the thigh obliquely, assists in rotating the leg to the cross-legged position in which the knees are spread wide apart, and in humans is the longest muscle

Sar·um \ˈsar-əm, ˈser-\ *adjective* [*Sarum,* old borough near Salisbury, England] (1570)
: of or relating to the Roman rite as modified in Salisbury and used in England, Wales, and Ireland before the Reformation

¹sash \ˈsash\ *noun* [Arabic *shāsh* muslin] (1681)
: a band worn about the waist or over one shoulder and used as a dress accessory or the emblem of an honorary or military order
— sashed \ˈsasht\ *adjective*

²sash *noun, plural* **sash** *also* **sash·es** [probably modification of French *châssis* chassis (taken as plural)] (1681)
: the framework in which panes of glass are set in a window or door; *also* **:** such a framework together with its panes forming a usually movable part of a window

¹sa·shay \sa-ˈshā *also* sī-\ *intransitive verb* [alteration of *chassé*] (1836)
1 : to make a chassé
2 a : WALK, GLIDE, GO **b :** to strut or move about in an ostentatious or conspicuous manner **c :** to proceed or move in a diagonal or sideways manner

²sashay *noun* [by alteration] (1900)
1 : TRIP, EXCURSION
2 : a square-dance figure in which partners sidestep in a circle around each other with the man moving behind the woman
3 : CHASSÉ

sa·shi·mi \ˈsä-shə-mē\ *noun* [Japanese] (1880)
: a Japanese dish consisting of thinly sliced raw fish served with a sauce for dipping

sas·ka·toon \ˌsas-kə-ˈtün\ *noun* [modification of Cree *misa·skwato·min* serviceberry fruit] (1810)
: SERVICEBERRY

Sas·quatch \ˈsas-ˌkwach, -ˌkwäch\ *noun* [Halkomelem (Salishan language of southwestern British Columbia) *sésq̓əc*] (1929)
: a hairy creature like a human being reported to exist in the northwestern U.S. and western Canada and said to be a primate between 6 and 15 feet (1.8 and 4.6 meters) tall — called also *bigfoot*

¹sass \ˈsas\ *noun* [alteration of ¹*sauce*] (1835)
: impudent speech ◆

²sass *transitive verb* (1856)
: to talk impudently or disrespectfully to

sas·sa·fras \ˈsa-sə-ˌfras\ *noun* [Spanish *sasafrás*] (1577)
1 : an eastern North American tree (*Sassafras albidum*) of the laurel family having both ovate and lobed aromatic leaves
2 : the carcinogenic dried root bark of the sassafras used formerly as a diaphoretic or flavoring agent

¹Sas·sa·ni·an *or* **Sa·sa·ni·an** \sə-ˈsā-nē-ən, sa-ˈsā-\ *adjective* (1788)
: of, relating to, or having the characteristics of the Sassanid dynasty of ancient Persia or its art or architecture

²Sassanian *or* **Sasanian** *noun* (1855)
: SASSANID

Sas·sa·nid *or* **Sa·sa·nid** \sə-ˈsä-nəd, -ˈsa-; ˈsa-sᵊn-əd\ *noun* [New Latin *Sassanidae* Sassanids, from *Sassan,* founder of the dynasty] (1776)
: a member of a dynasty of Persian kings of the 3d to 7th centuries
— Sassanid *or* **Sasanid** *adjective*

sassy \ˈsa-sē\ *adjective* **sass·i·er; -est** [alteration of *saucy*] (1833)
1 : IMPUDENT, SAUCY
2 : VIGOROUS, LIVELY
3 : distinctively smart and stylish ⟨a *sassy* black-and-white bow tie —Jean Stafford⟩

sat *past and past participle of* SIT

Sa·tan \ˈsā-tᵊn\ *noun* [Middle English, from Old English, from Late Latin, from Greek, from Hebrew *śāṭān* adversary] (before 12th century)
: the adversary of God and lord of evil in Judaism and Christianity

sa·tang \sə-ˈtäŋ\ *noun, plural* **satang** *or* **satangs** [Thai *sataŋ*] (circa 1915)
— see *baht* at MONEY table

sa·tan·ic \sə-ˈta-nik, sā-\ *adjective* (1667)
1 : of, relating to, or characteristic of Satan or satanism ⟨*satanic* pride⟩ ⟨*satanic* rites⟩
2 : characterized by extreme cruelty or viciousness
— sa·tan·i·cal·ly \-ni-k(ə-)lē\ *adverb*

sa·tan·ism \ˈsā-tᵊn-ˌi-zəm\ *noun, often capitalized* (1565)
1 : innate wickedness **:** DIABOLISM
2 : obsession with or affinity for evil; *specifically* **:** the worship of Satan marked by the travesty of Christian rites
— sa·tan·ist \-ist\ *noun, often capitalized*

satch·el \ˈsa-chəl\ *noun* [Middle English *sachel,* from Middle French, from Late Latin *saccellum,* diminutive of Latin *sacculus,* diminutive of *saccus* bag — more at SACK] (14th century)
: a small bag often with a shoulder strap
— satch·el·ful \-ˌfu̇l\ *noun*

¹sate \ˈsāt, ˈsat\ *archaic past of* SIT

◇ WORD HISTORY

sass The pronunciation of the *au* in *sauce, saucer,* and *sausage* with \a\ rather than the \ȯ\ that most Americans and Britons have in these words is the result of a change that took place in English toward the end of the Middle Ages, when the diphthong (vowel combination) \au̇\ that *au* then represented lost its final element and changed to a short *a* sound. By the 17th century, some Englishmen were pronouncing *sauce* to rhyme with *pass.* This pronunciation was brought to the thirteen American colonies, where the spelling *sas* appears for *sauce* by 1775. This word differed not only in pronunciation but in meaning from what we now think of as sauce. It usually denoted a side dish of vegetables or stewed fruit rather than a liquid garnish. In the 19th century United States there was *long sass* (carrots, parsnips, and beets) and *short sass* (potatoes, onions, and turnips). Aside from *applesauce* and *cranberry sauce* these usages are now passing away, but *sass* in the sense "impudent speech," along with its derivative *sassy,* still survives, and many people no longer associate the two words *sass* and *sauce.*

²sate \ˈsāt\ *transitive verb* **sat·ed; sat·ing** [probably by shortening & alteration from *satiate*] (1602)
1 : to cloy with overabundance **:** GLUT
2 : to appease (as a thirst) by indulging to the full
synonym see SATIATE

sa·teen \sa-ˈtēn, sə-\ *noun* [alteration of *satin*] (circa 1878)
: a smooth durable lustrous fabric usually made of cotton in satin weave

sat·el·lite \ˈsa-tᵊl-ˌīt\ *noun* [Middle French, from Latin *satellit-, satelles* attendant] (circa 1548)
1 : a hired agent or obsequious follower **:** MINION, SYCOPHANT
2 a : a celestial body orbiting another of larger size **b :** a manufactured object or vehicle intended to orbit the earth, the moon, or another celestial body
3 : someone or something attendant, subordinate, or dependent; *especially* **:** a country politically and economically dominated or controlled by another more powerful country
4 : a usually independent urban community situated near but not immediately adjacent to a large city
— satellite *adjective*

satellite DNA *noun* (1969)
: a fraction of a eukaryotic organism's DNA that differs in density from most of its DNA as determined by centrifugation, that apparently consists of short repetitive nucleotide sequences, that does not undergo transcription, and that in some organisms (as the mouse) is found especially in centromeric regions

sa·tem \ˈsä-təm\ *adjective* [Avestan *satəm* hundred; from the fact that its initial sound (derived from an alveolar fricative) is the representative of an Indo-European palatal stop — more at HUNDRED] (1901)
: of, relating to, or constituting an Indo-European language group in which the palatal stops became in prehistoric times palatal or alveolar fricatives — compare CENTUM

sa·ti \(ˌ)sə-ˈtē, ˈsə-ˌtē\ *variant of* SUTTEE

sa·tia·ble \ˈsā-shə-bəl\ *adjective* (1570)
: capable of being appeased or satisfied

¹sa·tiate \ˈsā-sh(ē-)ət\ *adjective* (15th century)
: filled to satiety

²sa·ti·ate \ˈsā-shē-ˌāt\ *transitive verb* **-at·ed; -at·ing** [Latin *satiatus,* past participle of *satiare,* from *satis* enough — more at SAD] (15th century)
: to satisfy (as a need or desire) fully or to excess ☆
— sa·ti·a·tion \ˌsā-shē-ˈā-shən, ˌsā-sē-\ *noun*

sa·ti·ety \sə-ˈtī-ə-tē *also* ˈsā-sh(ē-)ə-\ *noun* [Middle French *satieté,* from Latin *satietat-, satietas,* from *satis*] (1533)
1 : the quality or state of being fed or gratified to or beyond capacity **:** SURFEIT, FULLNESS
2 : the revulsion or disgust caused by overindulgence or excess

¹sat·in \ˈsa-tᵊn\ *noun* [Middle English, from Middle French, probably from Arabic *zaytūnī,* literally, of Zaytūn, seaport in China during the Middle Ages] (14th century)
: a fabric (as of silk) in satin weave with lustrous face and dull back

²satin *adjective* (1521)
1 : made of or covered with satin
2 : suggestive of satin especially in smooth lustrous appearance or sleekness to touch

sat·in·et \ˌsa-tᵊn-ˈet\ *noun* (1703)
1 : a thin silk satin or imitation satin
2 : a variation of satin weave used in making satinet

satin stitch *noun* (1684)
: an embroidery stitch nearly alike on both sides and worked so closely as to resemble satin

satin weave *noun* (circa 1883)
: a weave in which warp threads interlace with filling threads to produce a smooth-faced fabric

satin stitch

sat·in·wood \ˈsa-tᵊn-ˌwu̇d\ *noun* (1792)
1 a : an East Indian tree (*Chloroxylon swietenia*) of the rue family that yields a lustrous yellowish brown wood **b :** a tree (as *Zanthoxylum flavum* of the rue family) with wood resembling true satinwood
2 : the wood of a satinwood

sat·iny \ˈsat-nē, ˈsa-tᵊn-ē\ *adjective* (1786)
: having or resembling the soft usually lustrous smoothness of satin

sat·ire \ˈsa-ˌtīr\ *noun* [Middle French or Latin; Middle French, from Latin *satura, satira,* perhaps from (*lanx*) *satura* dish of mixed ingredients, from feminine of *satur* well-fed; akin to Latin *satis* enough — more at SAD] (1501)
1 : a literary work holding up human vices and follies to ridicule or scorn
2 : trenchant wit, irony, or sarcasm used to expose and discredit vice or folly
synonym see WIT

sa·tir·ic \sə-ˈtir-ik\ *or* **sa·tir·i·cal** \-i-kəl\ *adjective* (1509)
1 : of, relating to, or constituting satire ⟨*satiric* writers⟩
2 : manifesting or given to satire
synonym see SARCASTIC
— sa·tir·i·cal·ly \-i-k(ə-)lē\ *adverb*

sat·i·rise *British variant of* SATIRIZE

sat·i·rist \ˈsa-tə-rist\ *noun* (1589)
: one that satirizes; *especially* **:** a writer of satire

sat·i·rize \-ˌrīz\ *verb* **-rized; -riz·ing** (1601)
intransitive verb
: to utter or write satire
transitive verb
: to censure or ridicule by means of satire
— sat·i·ri·za·ble \ˌsa-tə-ˈrī-zə-bəl\ *adjective*

sat·is·fac·tion \ˌsa-təs-ˈfak-shən\ *noun* [Middle English, from Middle French, from Late Latin *satisfaction-, satisfactio,* from Latin, reparation, amends, from *satisfacere* to satisfy] (14th century)
1 a : the payment through penance of the temporal punishment incurred by a sin **b :** reparation for sin that meets the demands of divine justice
2 a : fulfillment of a need or want **b :** the quality or state of being satisfied **:** CONTENTMENT **c :** a source or means of enjoyment **:** GRATIFICATION
3 a : compensation for a loss or injury **:** ATONEMENT, RESTITUTION **b :** the discharge of a legal obligation or claim **c :** VINDICATION
4 : convinced assurance or certainty ⟨proved to the *satisfaction* of the court⟩

sat·is·fac·to·ry \ˌsa-təs-ˈfak-t(ə-)rē\ *adjective* (15th century)
: giving satisfaction **:** ADEQUATE
— sat·is·fac·to·ri·ly \-t(ə-)rə-lē\ *adverb*
— sat·is·fac·to·ri·ness *noun*

sat·is·fi·able \ˈsa-təs-ˌfī-ə-bəl\ *adjective* (1609)
: capable of being satisfied

sat·is·fy \ˈsa-təs-ˌfī\ *verb* **-fied; -fy·ing** [Middle English *satisfien,* from Middle French *satisfier,* modification of Latin *satisfacere,* from *satis* enough + *facere* to do, make — more at SAD, DO] (15th century)
transitive verb
1 a : to carry out the terms of (as a contract) **:** DISCHARGE **b :** to meet a financial obligation to
2 : to make reparation to (an injured party) **:** INDEMNIFY
3 a : to make happy **:** PLEASE **b :** to gratify to the full **:** APPEASE
4 a : CONVINCE **b :** to put an end to (doubt or uncertainty) **:** DISPEL
5 a : to conform to (as specifications) **:** be adequate to (an end in view) **b :** to make true by fulfilling a condition ⟨values that *satisfy* an equation⟩ ⟨*satisfy* a hypothesis⟩
intransitive verb
: to be adequate **:** SUFFICE; *also* **:** PLEASE
synonym see PAY
— sat·is·fy·ing·ly \-iŋ-lē\ *adverb*

sa·to·ri \sə-ˈtōr-ē, sä-, -ˈtȯr-\ *noun* [Japanese] (1727)
: a state of intuitive illumination sought in Zen Buddhism

sa·trap \ˈsā-ˌtrap *also* ˈsa-ˌtrap *or* ˈsa-trəp\ *noun* [Middle English, from Latin *satrapes,* from Greek *satrapēs,* from Old Persian *khshathrapāvan,* literally, protector of the dominion] (14th century)
1 : the governor of a province in ancient Persia
2 a : RULER **b :** a subordinate official **:** HENCHMAN

sa·tra·py \ˈsā-trə-pē, ˈsa-, -ˌtra-pē\ *noun, plural* **-pies** (1603)
: the territory or jurisdiction of a satrap

sat·su·ma \sat-ˈsü-mə, ˈsat-sə-\ *noun* [*Satsuma,* former province in Kyushu, Japan] (1882)
1 : any of several cultivated mandarin trees that bear medium-sized largely seedless fruits with thin smooth skin
2 : the fruit of a satsuma

sat·u·ra·ble \ˈsach-rə-bəl, ˈsa-chə-\ *adjective* (1570)
: capable of being saturated

sat·u·rant \-rənt\ *noun* (circa 1775)
: something that saturates

¹sat·u·rate \ˈsa-chə-ˌrāt\ *transitive verb* **-rat·ed; -rat·ing** [Latin *saturatus,* past participle of *saturare,* from *satur* well-fed — more at SATIRE] (1538)
1 : to satisfy fully **:** SATIATE
2 : to treat, furnish, or charge with something to the point where no more can be absorbed, dissolved, or retained ⟨water *saturated* with salt⟩
3 a : to fill completely with something that permeates or pervades ⟨book is *saturated* with Hollywood, old and new —Newgate Callendar⟩ **b :** to load to capacity
4 : to cause to combine till there is no further tendency to combine
synonym see SOAK
— sat·u·ra·tor \-ˌrā-tər\ *noun*

²sat·u·rate \ˈsach-rət, ˈsa-chə-\ *adjective* (1782)
: SATURATED

sat·u·rat·ed \ˈsa-chə-ˌrā-təd\ *adjective* (1728)
1 : full of moisture **:** made thoroughly wet
2 a : being a solution that is unable to absorb or dissolve any more of a solute at a given temperature and pressure **b :** being an organic compound having no double or triple bonds between carbon atoms
3 *of a color* **:** having high saturation **:** PURE

☆ SYNONYMS
Satiate, sate, surfeit, cloy, pall, glut, gorge mean to fill to repletion. SATIATE and SATE may sometimes imply only complete satisfaction but more often suggest repletion that has destroyed interest or desire ⟨years of globe-trotting had *satiated* their interest in travel⟩ ⟨readers were *sated* with sensationalistic stories⟩. SURFEIT implies a nauseating repletion ⟨*surfeited* themselves with junk food⟩. CLOY stresses the disgust or boredom resulting from such surfeiting ⟨sentimental pictures that *cloy* after a while⟩. PALL emphasizes the loss of ability to stimulate interest or appetite ⟨a life of leisure eventually begins to *pall*⟩. GLUT implies excess in feeding or supplying ⟨a market *glutted* with diet books⟩. GORGE suggests glutting to the point of bursting or choking ⟨*gorged* themselves with chocolate⟩.

sat·u·ra·tion \ˌsa-chə-ˈrā-shən\ *noun* (circa 1554)
1 a : the act of saturating **:** the state of being saturated **b :** SATIETY, SURFEIT
2 : conversion of an unsaturated to a saturated chemical compound (as by hydrogenation)
3 : a state of maximum impregnation: as **a :** complete infiltration **:** PERMEATION **b :** the presence in air of the most water possible under existent pressure and temperature **c :** magnetization to the point beyond which a further increase in the intensity of the magnetizing force will produce no further magnetization
4 a : chromatic purity **:** freedom from dilution with white **b** (1) **:** degree of difference from the gray having the same lightness — used of an object color (2) **:** degree of difference from the achromatic light-source color of the same brightness — used of a light-source color; compare HUE 2c
5 : the supplying of a market with as much of a product as it will absorb
6 : an overwhelming concentration of military forces or firepower

Sat·ur·day \ˈsa-tər-dē, -(ˌ)dā\ *noun* [Middle English *saterday,* from Old English *sæterndæg* (akin to Old Frisian *sāterdei*), from Latin *Saturnus* Saturn + Old English *dæg* day] (before 12th century)
: the seventh day of the week
— Sat·ur·days \-dēz, -(ˌ)dāz\ *adverb*

Saturday night special *noun* (1968)
: a cheap easily concealed handgun

Sat·urn \ˈsa-tərn\ *noun* [Latin *Saturnus*]
1 : a Roman god of agriculture and father by Ops of Jupiter
2 : the planet 6th in order from the sun — see PLANET table

sat·ur·na·lia \ˌsa-tər-ˈnāl-yə, -ˈnā-lē-ə\ *noun plural but singular or plural in construction* [Latin, from neuter plural of *saturnalis* of Saturn, from *Saturnus*] (1591)
1 *capitalized* **:** the festival of Saturn in ancient Rome beginning on Dec. 17
2 *singular, plural* **saturnalias** *also* **saturnalia a :** an unrestrained often licentious celebration **:** ORGY **b :** EXCESS, EXTRAVAGANCE
— sat·ur·na·lian \-ˈnāl-yən, -ˈnā-lē-ən\ *adjective*
— sat·ur·na·lian·ly *adverb*

Sa·tur·ni·an \sa-ˈtər-nē-ən, sə-\ *adjective* (1557)
1 : of, relating to, or influenced by the planet Saturn
2 *archaic* **:** of or relating to the god Saturn or the golden age of his reign

sa·tur·ni·id \-nē-əd\ *noun* [New Latin *Saturniidae,* from *Saturnia,* genus of moths, from Latin, daughter of the god Saturn] (circa 1909)
: any of a family (Saturniidae) of usually large stout strong-winged moths (as a luna moth or a cecropia moth) with hairy bodies
— saturniid *adjective*

sat·ur·nine \ˈsa-tər-ˌnīn\ *adjective* (15th century)
1 : born under or influenced astrologically by the planet Saturn
2 a : cold and steady in mood **:** slow to act or change **b :** of a gloomy or surly disposition **c :** having a sardonic aspect ⟨a *saturnine* smile⟩
synonym see SULLEN

sat·urn·ism \ˈsa-tər-ˌni-zəm\ *noun* [*saturn* lead] (1855)
: LEAD POISONING

sa·tya·gra·ha \(ˌ)sə-ˈtyä-grə-hə, ˈsə-tyə-\ *noun* [New Sanskrit *satyāgraha,* from Sanskrit *satya* truth + *āgraha* persistence] (1920)
: pressure for social and political reform through friendly passive resistance practiced by M. K. Gandhi and his followers in India

sa·tyr \ˈsā-tər, *chiefly British* ˈsa-\ *noun* [Middle English, from Latin *satyrus,* from Greek *satyros*]
1 *often capitalized* **:** a sylvan deity in Greek mythology having certain characteristics of a horse or goat and fond of Dionysian revelry
2 a : a lecherous man **b :** one having satyriasis
3 : any of various usually brown and gray satyrid butterflies
— sa·tyr·ic \sā-ˈtir-ik, sə-, sa-\ *adjective*

sa·ty·ri·a·sis \ˌsā-tə-ˈrī-ə-səs, ˌsa-\ *noun* [Late Latin, from Greek, from *satyros*] (15th century)
: excessive or abnormal sexual craving in the male

sa·ty·rid \sə-ˈtī-rəd\ *noun* [New Latin *Satyridae,* ultimately from Greek *satyros*] (1901)
: any of a family (Satyridae) of usually brownish butterflies that feed on grasses as larvae and have one or more forewing veins swollen basally
— satyrid *adjective*

satyr play *noun* (1929)
: a comic play of ancient Greece burlesquing a mythological subject and having a chorus representing satyrs

[1]**sauce** \ˈsȯs, *usually* ˈsas *for 4*\ *noun* [Middle English, from Middle French, from Latin *salsa,* feminine of *salsus* salted, from past participle of *sallere* to salt, from *sal* salt — more at SALT] (14th century)
1 : a condiment or relish for food; *especially* **:** a fluid dressing or topping
2 : something that adds zest or piquancy
3 : stewed fruit eaten with other food or as a dessert
4 : pert or impudent language or actions
5 *slang* **:** LIQUOR — used with *the*

[2]**sauce** \ˈsȯs, *usually* ˈsas *for 3*\ *transitive verb* **sauced; sauc·ing** (15th century)
1 a : to dress with relish or seasoning **b :** to cover or serve with a sauce
2 a *archaic* **:** to modify the harsh or unpleasant characteristics of **b :** to give zest or piquancy to
3 : to be rude or impudent to

sauce·boat \ˈsȯs-ˌbōt\ *noun* (1747)
: a low boat-shaped pitcher for serving sauces and gravies

sauce·box \ˈsȯs-ˌbäks, ˈsas-\ *noun* (1588)
: a saucy impudent person

sauce·pan \ˈsȯs-ˌpan, *especially British* -pən\ *noun* (1686)
: a small deep cooking pan with a handle

sau·cer \ˈsȯ-sər\ *noun* [Middle English, plate containing sauce, from Middle French *saussier,* from *sausse, sauce*] (1607)
1 : a small shallow dish in which a cup is set at table
2 : something resembling a saucer especially in shape; *especially* **:** FLYING SAUCER
— sau·cer·like \-ˌlīk\ *adjective*

saucy \ˈsȯ-sē, ˈsa-\ *adjective* **sauc·i·er; -est** (1508)
1 : served with or having the consistency of sauce
2 a : impertinently bold and impudent **b :** amusingly forward and flippant **:** IRREPRESSIBLE
3 : SMART, TRIM ⟨a *saucy* little hat⟩
— sauc·i·ly \-sə-lē\ *adverb*
— sauc·i·ness \-sē-nəs\ *noun*

sau·er·bra·ten \ˈsau̇(-ə)r-ˌbrä-tᵊn\ *noun* [German, from *sauer* sour + *Braten* roast meat] (1889)
: oven-roasted or pot-roasted beef marinated before cooking in vinegar with peppercorns, garlic, onions, and bay leaves

sau·er·kraut \ˈsau̇(-ə)r-ˌkrau̇t\ *noun* [German, from *sauer* sour + *Kraut* greens] (1617)
: cabbage cut fine and fermented in a brine made of its own juice with salt

sau·ger \ˈsȯ-gər\ *noun* [origin unknown] (1882)
: a pike perch (*Stizostedion canadense*) similar to but smaller than the walleye

saugh *or* **sauch** \ˈsäk, ˈsȯk\ *noun* [Middle English (Scots) *sauch,* from Old English *salh,* alteration of *sealh* — more at SALLOW] (before 12th century)
chiefly Scottish **:** SALLOW

Sauk \ˈsȯk\ *noun, plural* **Sauk** *or* **Sauks** (1762)
: a member of an American Indian people formerly living in what is now Wisconsin

Saul \ˈsȯl\ *noun* [Late Latin *Saulus,* from Greek *Saulos,* from Hebrew *Shā'ūl*]
1 : the first king of Israel
2 : the apostle Paul — called also *Saul of Tarsus*

sau·na \ˈsȯ-nə, ˈsau̇-nə\ *noun* [Finnish] (1891)
1 : a Finnish steam bath in which the steam is provided by water thrown on hot stones; *also* **:** a bathhouse or room used for such a bath
2 : a dry heat bath; *also* **:** a room or cabinet used for such a bath

saun·ter \ˈsȯn-tər, ˈsän-\ *intransitive verb* [probably from Middle English *santren* to muse] (circa 1667)
: to walk about in an idle or leisurely manner **:** STROLL
— saunter *noun*
— saun·ter·er \-tər-ər\ *noun*

sau·rel \sȯ-ˈrel\ *noun* [French, from Late Latin *saurus* horse mackerel, from Greek *sauros* horse mackerel, lizard] (1882)
: JACK MACKEREL

sau·ri·an \ˈsȯr-ē-ən\ *noun* [New Latin *Sauria,* from New Latin *saurus* lizard, from Greek *sauros*] (circa 1829)
: any of a suborder (Sauria) of reptiles including the lizards and in older classifications the crocodiles and various extinct forms (as the dinosaurs and ichthyosaurs) that resemble lizards
— saurian *adjective*

saur·is·chi·an \(ˌ)sȯ-ˈris-kē-ən\ *noun* [New Latin *Saurischia,* from Greek *sauros* lizard + New Latin *ischium* ischium] (circa 1891)
: any of an order (Saurischia) of herbivorous or carnivorous dinosaurs (as a brontosaurus) that have a triradiate pelvis with the pubis typically pointed forward — compare ORNITHISCHIAN
— saurischian *adjective*

sau·ro·pod \ˈsȯr-ə-ˌpäd\ *noun* [New Latin *Sauropoda,* from Greek *sauros* lizard + New Latin *-poda*] (circa 1891)
: any of a suborder (Sauropoda) of quadrupedal saurischian dinosaurs comprising herbivorous forms with a long neck and tail, small head, and more or less plantigrade 5-toed limbs
— sauropod *adjective*

sau·ry \ˈsȯr-ē\ *noun, plural* **sauries** [New Latin *saurus* lizard] (circa 1771)
1 : a slender long-beaked fish (*Scombresox saurus*) related to the needlefishes and found in temperate parts of the Atlantic
2 : a widely distributed fish (*Cololabris saira*) of the Pacific similar to the related Atlantic saury

sau·sage \ˈsȯ-sij\ *noun* [Middle English *sausige,* from Old North French *saussiche,* from Late Latin *salsicia,* from Latin *salsus* salted — more at SAUCE] (15th century)
: a highly seasoned minced meat (as pork) usually stuffed in casings of prepared animal intestine; *also* **:** a link or patty of sausage

[1]**sau·té** *also* **sau·te** \sō-ˈtā, sȯ-\ *noun* [French, from past participle of *sauter* to jump, from Latin *saltare* — more at SALTATION] (1813)
: a sautéed dish
— sauté *adjective*

[2]**sauté** *also* **saute** *transitive verb* **sau·téed** *or* **sau·téd; sau·té·ing** (1859)
: to fry in a small amount of fat

sau·ternes \sō-ˈtərn, sȯ-, -ˈtern\ *noun, often capitalized* [French, from *Sauternes,* commune in France] (1711)

\ə\ **abut** \ᵊ\ **kitten** \ər\ **further** \a\ **ash** \ā\ **ace** \ä\ **mop, mar** \au̇\ **out** \ch\ **chin** \e\ **bet** \ē\ **easy** \g\ **go** \i\ **hit** \ī\ **ice** \j\ **job** \ŋ\ **sing** \ō\ **go** \ȯ\ **law** \ȯi\ **boy** \th\ **thin** \t͟h\ **the** \ü\ **loot** \u̇\ **foot** \y\ **yet** \zh\ **vision** *see also* **Guide to Pronunciation**

1 : a full-bodied sweet white wine from the Bordeaux region of France
2 *usually* **sauterne :** a semidry to semisweet American white wine that is a blend of several grapes

sau·vi·gnon blanc \ˌsō-vēn-ˈyōⁿ-ˈbläⁿ\ *noun* [French, white sauvignon (variety of grape)] (1941)
: a dry white wine made from a grape originally grown in Bordeaux and the Loire valley

¹sav·age \ˈsa-vij\ *adjective* [Middle English *sauvage,* from Middle French, from Medieval Latin *salvaticus,* alteration of Latin *silvaticus* of the woods, wild, from *silva* wood, forest] (13th century)
1 a : not domesticated or under human control **:** UNTAMED ⟨*savage* beasts⟩ **b :** lacking the restraints normal to civilized human beings **:** FIERCE, FEROCIOUS
2 : WILD, UNCULTIVATED ⟨seldom have I seen such *savage* scenery —Douglas Carruthers⟩
3 a : BOORISH, RUDE ⟨the *savage* bad manners of most motorists —M. P. O'Connor⟩ **b :** MALICIOUS
4 : lacking complex or advanced culture **:** UNCIVILIZED ◆
synonym see FIERCE
— sav·age·ly *adverb*
— sav·age·ness *noun*

²savage *noun* (15th century)
1 : a person belonging to a primitive society
2 : a brutal person
3 : a rude or unmannerly person

³savage *transitive verb* **sav·aged; sav·ag·ing** (1880)
: to attack or treat brutally

sav·age·ry \ˈsa-vij-rē, -vi-jə-\ *noun, plural* **-ries** (1595)
1 a : the quality of being savage **b :** an act of cruelty or violence
2 : an uncivilized state

sav·ag·ism \ˈsa-vi-ˌji-zəm\ *noun* (1796)
: SAVAGERY

sa·van·na *also* **sa·van·nah** \sə-ˈva-nə\ *noun* [Spanish *zavana,* from Taino *zabana*] (1555)
1 : a treeless plain especially in Florida
2 : a tropical or subtropical grassland containing scattered trees and drought-resistant undergrowth

sa·vant \sa-ˈvänt, sə-, -ˈväⁿ; sə-ˈvant, ˈsa-vənt\ *noun* [French, from Middle French, from present participle of *savoir* to know, from Latin *sapere* to be wise — more at SAGE] (1719)
1 : a person of learning; *especially* **:** one with detailed knowledge in some specialized field (as of science or literature)
2 : IDIOT SAVANT 1

sav·a·rin \ˈsa-və-rən\ *noun* [French, from Anthelme Brillat-*Savarin* (died 1826) French politician, writer, and gourmet] (1877)
: a rich yeast cake baked in a ring mold and soaked in a rum or kirsch syrup

sa·vate \sə-ˈvät, sa-, -ˈvat\ *noun* [French, literally, old shoe] (1862)
: a form of boxing in which blows are delivered with either the hands or the feet

¹save \ˈsāv\ *verb* **saved; sav·ing** [Middle English, from Old French *salver,* from Late Latin *salvare,* from Latin *salvus* safe — more at SAFE] (13th century)
transitive verb
1 a : to deliver from sin **b :** to rescue or deliver from danger or harm **c :** to preserve or guard from injury, destruction, or loss
2 a : to put aside as a store or reserve **:** ACCUMULATE **b :** to spend less by ⟨*save* 25%⟩
3 a : to make unnecessary **:** AVOID ⟨it *saves* an hour's driving⟩ **b** (1) **:** to keep from being lost to an opponent (2) **:** to prevent an opponent from scoring or winning
4 : MAINTAIN, PRESERVE ⟨*save* appearances⟩
intransitive verb
1 : to rescue or deliver someone
2 a : to put aside money **b :** to avoid unnecessary waste or expense **:** ECONOMIZE **c :** to spend less money ⟨buy now and *save*⟩
3 : to make a save
synonym see RESCUE
— sav·able *or* **save·able** \ˈsā-və-bəl\ *adjective*
— sav·er *noun*

²save *noun* (1890)
1 : a play that prevents an opponent from scoring or winning
2 : the action of a relief pitcher in baseball in successfully protecting a team's lead

³save *preposition* [Middle English *sauf,* from Middle French, from *sauf,* adjective, safe — more at SAFE] (14th century)
: other than **:** BUT, EXCEPT ⟨no hope *save* one⟩

⁴save *conjunction* (14th century)
1 : except for the fact that **:** ONLY — used with *that* ⟨of his earlier years little is known, *save* that he studied violin —J. N. Burk⟩
2 : BUT, EXCEPT — used before a word often taken to be the subject of a clause ⟨no one knows about it *save* she⟩

save–all \ˈsā-ˌvȯl\ *noun* (circa 1645)
: something that prevents waste, loss, or damage (as a receptacle for catching waste products for further utilization)

sav·e·loy \ˈsa-və-ˌlȯi\ *noun* [modification of French *cervelas,* from Middle French, from Old Italian *cervellata,* literally, pig's brains, from *cervello* brain, from Latin *cerebellum* — more at CEREBELLUM] (1837)
British **:** a ready-cooked highly seasoned dry sausage

sav·in \ˈsa-vən\ *noun* [Middle English, from Middle French *savine,* from Latin (*herba*) *sabina,* literally, Sabine plant] (14th century)
1 : a Eurasian juniper (*Juniperus sabina*) with dark foliage and small yellowish green berries
2 : RED CEDAR 1

¹sav·ing \ˈsā-viŋ\ *noun* [Middle English, from gerund of *saven* to save] (14th century)
1 : preservation from danger or destruction **:** DELIVERANCE
2 : the act or an instance of economizing
3 a *plural* **:** money put by **b :** the excess of income over consumption expenditures — often used in plural **c :** a usually specified lower cost — often used in plural ⟨a *savings* of 50%⟩

²saving *preposition* [Middle English, from present participle of *saven*] (14th century)
1 : EXCEPT, SAVE
2 : without disrespect to

³saving *conjunction* (15th century)
: EXCEPT, SAVE

saving grace *noun* (1597)
: a redeeming quality or factor

savings account *noun* (1911)
: an account (as in a bank) on which interest is usually paid and from which withdrawals can be made usually only by presentation of a passbook or by written authorization on a prescribed form

savings and loan association *noun* (circa 1924)
: a cooperative association organized to hold savings of members in the form of dividend-bearing shares and to invest chiefly in home mortgage loans

savings bank *noun* (1817)
: a bank organized to hold funds of individual depositors in interest-bearing accounts and to make long-term investments (as in home mortgage loans)

savings bond *noun* (1948)
: a nontransferable registered U.S. bond issued in denominations of $50 to $10,000

sav·ior *or* **sav·iour** \ˈsāv-yər *also* -ˌyȯr\ *noun* [Middle English *saveour,* from Middle French, from Late Latin *salvator,* from *salvare* to save] (14th century)
1 : one that saves from danger or destruction
2 : one who brings salvation; *specifically, capitalized* **:** JESUS 1

sa·voir faire \ˌsav-ˌwär-ˈfar, -ˈfer\ *noun* [French *savoir-faire,* literally, knowing how to do] (1815)
: capacity for appropriate action; *especially* **:** a polished sureness in social behavior
synonym see TACT

¹sa·vor *also* **sa·vour** \ˈsā-vər\ *noun* [Middle English, from Old French, from Latin *sapor,* from *sapere* to taste — more at SAGE] (13th century)
1 : the taste or smell of something
2 : a particular flavor or smell
3 : a distinctive quality
— sa·vor·less \-ləs\ *adjective*
— sa·vor·ous \ˈsā-vər-əs, ˈsāv-rəs\ *adjective*

²savor *also* **savour** *verb* **sa·vored; sa·vor·ing** \ˈsā-vər-iŋ, ˈsāv-riŋ\ (14th century)
intransitive verb
: to have a specified smell or quality **:** SMACK
transitive verb
1 : to give flavor to **:** SEASON
2 a : to have experience of **:** TASTE **b :** to taste or smell with pleasure **:** RELISH **c :** to delight in **:** ENJOY
— sa·vor·er \ˈsā-vər-ər\ *noun*

¹sa·vory *also* **sa·voury** \ˈsā-və-rē, ˈsāv-rē\ *adjective* (13th century)
: having savor: as **a :** piquantly pleasant to the mind ⟨a *savory* collection of essays⟩ **b :** morally exemplary **:** EDIFYING ⟨his reputation was anything but *savory*⟩ **c :** pleasing to the sense of taste especially by reason of effective seasoning
synonym see PALATABLE
— sa·vor·i·ly \-rə-lē\ *adverb*
— sa·vor·i·ness \-rē-nəs\ *noun*

²savory *also* **savoury** *noun, plural* **sa·vor·ies** (1661)
British **:** a dish of stimulating flavor served usually at the end of dinner but sometimes as an appetizer

³sa·vo·ry \ˈsā-və-rē, ˈsāv-rē\ *noun, plural* **-ries** [Middle English *saverey*] (14th century)
: either of two aromatic mints: **a :** SUMMER SAVORY **b :** WINTER SAVORY

Sa·voy·ard \sə-ˈvȯi-ˌärd, ˌsa-ˌvȯi-ˈärd, ˌsav-ˌwä-ˈyär(d)\ *noun* [*Savoy* Theater, London, built for the presentation of Gilbert and Sullivan operas] (1890)
: a devotee, performer, or producer of the comic operas of W. S. Gilbert and A. S. Sullivan

sa·voy cabbage \sə-ˈvȯi-, ˈsa-ˌvȯi-\ *noun, often S capitalized* [translation of French *chou de Savoie* cabbage of Savoy] (1707)
: a cabbage with compact heads of wrinkled and curled leaves

¹sav·vy \ˈsa-vē\ *verb* **sav·vied; sav·vy·ing** [alteration of *sabi* know (in English-based creoles and pidgins), from Portuguese *sabe* he knows, from *saber* to know, from Latin *sapere* to be wise — more at SAGE] (1785)

◇ WORD HISTORY

savage In Latin the adjective *silvaticus,* a derivative of *silva* "forest," meant, when applied to animals or plants, "growing or living in the forest." Because forest life is wild rather than domesticated, the adjective easily acquired the meaning "wild, uncultivated" in literary Late Latin as well as the spoken Latin of the declining Roman Empire. Old French inherited Late Latin *silvaticus,* altered to *salvaticus,* as *salvage.* In later medieval French this became *sauvage,* which was borrowed into Middle English in the 13th century. The Old French adjective retained the source meanings "wild, uncultivated (of fruit)" and "untamed (of animals)." But *sauvage* could also be applied to humans, in which case its meanings could range from "lacking civilization, barbarous" to "fierce, cruel." English *savage* has had all these meanings at some point in its history, though the "fierce" sense is the one we now perhaps associate most closely with the word.

: UNDERSTAND
²savvy *noun* (circa 1785)
: practical know-how ⟨political *savvy*⟩
— **savvy** *adjective*

¹saw *past of* SEE

²saw \ˈsȯ\ *noun* [Middle English *sawe,* from Old English *sagu;* akin to Old High German *sega* saw, Latin *secare* to cut] (before 12th century)
: a hand or power tool or a machine used to cut hard material (as wood, metal, or bone) and equipped usually with a toothed blade or disk
— **saw·like** \-ˌlīk\ *adjective*

³saw *verb* **sawed** \ˈsȯd\; **sawed** *or* **sawn** \ˈsȯn\; **saw·ing** \ˈsȯ(-)iŋ\ (13th century)
transitive verb
1 : to cut with a saw
2 : to produce or form by cutting with a saw
3 : to slash as though with a saw
intransitive verb
1 a : to use a saw **b :** to cut with or as if with a saw
2 : to undergo cutting with a saw
3 : to make motions as though using a saw ⟨*sawed* at the reins⟩
— **saw·er** \ˈsȯ(-ə)r\ *noun*

⁴saw *noun* [Middle English *sawe,* from Old English *sagu* discourse; akin to Old High German & Old Norse *saga* tale, Old English *secgan* to say — more at SAY] (before 12th century)
: MAXIM, PROVERB

saw·bones \ˈsȯ-ˌbōnz\ *noun, plural* **saw·bones** *or* **saw·bones·es** (1837)
slang **:** PHYSICIAN, SURGEON

saw·buck \ˈsȯ-ˌbək\ *noun* (1850)
1 *slang* **:** a 10-dollar bill
2 : SAWHORSE; *especially* **:** one with X-shaped ends

saw·dust \ˈsȯ-(ˌ)dəst\ *noun* (1530)
: fine particles (as of wood) made by a saw in cutting

sawed–off \ˈsȯ-ˌdȯf\ *adjective* (1869)
1 : having an end sawed off ⟨a *sawed-off* shot gun⟩
2 : of less than average height

saw·fish \ˈsȯ-ˌfish\ *noun* (1664)
: any of a family (Pristidae) of large elongate rays that resemble sharks but have a long flattened snout with a row of serrate structures along each edge and that live in tropical and subtropical shallow seas and in or near the mouths of rivers

saw·fly \-ˌflī\ *noun* (1773)
: any of numerous hymenopterous insects (superfamily Tenthredinoidea) with the female usually having a sawlike ovipositor and with the larva resembling a plant-feeding caterpillar

saw grass *noun* (1822)
: any of various sedges (as of the genus *Cladium*) having the edges of the leaves set with minute sharp teeth

saw·horse \ˈsȯ-ˌhȯrs\ *noun* (1778)
: a frame on which wood is laid for sawing by hand; *especially* **:** HORSE 2b
word history see EASEL

sawhorses

saw·log \-ˌlȯg, -ˌläg\ *noun* (1756)
: a log of suitable size for sawing into lumber

saw·mill \-ˌmil\ *noun* (1553)
: a mill or machine for sawing logs

saw·ney \ˈsȯ-nē\ *noun* [perhaps from *sawney* Scotsman, from Scots, alteration of *Sandy,* short for *Alexander*] (circa 1700)
chiefly British **:** FOOL, SIMPLETON
— **sawney** *adjective*

saw palmetto *noun* (1797)
: any of several shrubby palms chiefly of the southern U.S. and West Indies that have spiny-toothed petioles; *especially* **:** a common palm (*Serenoa repens*) of the southeastern U.S. with a usually creeping stem

saw set *noun* (1846)
: an instrument used to set the teeth of saws

saw·tim·ber \ˈsȯ-ˌtim-bər\ *noun* (1901)
: timber suitable for sawing into lumber

saw·tooth \-ˌtüth\ *adjective* (circa 1859)
: having serrations **:** arranged or having parts arranged like the teeth of a saw ⟨a *sawtooth* roof⟩

saw–toothed \-ˈtütht\ *adjective* (circa 1857)
1 : having teeth like those of a saw ⟨a *saw-toothed* shark⟩
2 : SAWTOOTH

saw–whet owl \ˈsȯ-ˌhwet-, -ˌwet-\ *noun* [from the supposed resemblance of its cry to the sound made in filing a saw] (1834)
: a very small harsh-voiced North American owl (*Aegolius acadicus*) that is largely dark brown above and chestnut streaked with white beneath — called also *saw-whet*

saw·yer \ˈsȯ-yər, ˈsȯi-ər\ *noun* (13th century)
1 : one that saws
2 : any of several large longicorn beetles whose larvae bore large holes in timber or dead wood
3 : a tree fast in the bed of a stream with its branches projecting to the surface

sax \ˈsaks\ *noun* (circa 1923)
: SAXOPHONE

sax·horn \ˈsaks-ˌhȯrn\ *noun* [Antoine *Sax* (died 1894) Belgian instrument maker + English *horn*] (1844)
: any of a group of valved brass instruments ranging from soprano to bass and characterized by a conical tube, oval shape, and cup-shaped mouthpiece

sax·ic·o·lous \sak-ˈsi-kə-ləs\ *adjective* [Latin *saxum* rock (akin to Latin *secare* to cut) + English *-colous* — more at SAW] (1856)
: inhabiting or growing among rocks ⟨*saxicolous* lichens⟩

sax·i·frage \ˈsak-sə-frij, -ˌfrāj\ *noun* [Middle English, from Middle French, from Late Latin *saxifraga,* from Latin, feminine of *saxifragus* breaking rocks, from *saxum* rock + *frangere* to break — more at BREAK] (14th century)
: any of a genus (*Saxifraga* of the family Saxifragaceae, the saxifrage family) of chiefly perennial herbs with showy pentamerous flowers and often with basal tufted leaves

sax·i·tox·in \ˌsak-sə-ˈtäk-sən\ *noun* [*saxi-* (from New Latin *Saxidomus giganteus,* species of butter clam from which it is isolated) + *toxin*] (1962)
: a potent nonprotein neurotoxin $C_{10}H_{17}N_7O_4 \cdot 2HCl$ that originates in dinoflagellates (genus *Gonyaulax*) found in red tides and that sometimes occurs in and renders toxic normally edible mollusks which feed on them

Sax·on \ˈsak-sən\ *noun* [Middle English, from Late Latin *Saxones* Saxons, of Germanic origin; akin to Old English *Seaxan* Saxons] (13th century)
1 a (1) **:** a member of a Germanic people that entered and conquered England with the Angles and Jutes in the 5th century A.D. and merged with them to form the Anglo Saxon people (2) **:** an Englishman or lowlander as distinguished from a Welshman, Irishman, or Highlander **b :** a native or inhabitant of Saxony
2 a : the Germanic language or dialect of any of the Saxon peoples **b :** the Germanic element in the English language especially as distinguished from the French and Latin
— **Saxon** *adjective*

sax·o·ny \ˈsak-s(ə-)nē\ *noun, plural* **-nies** *often capitalized* [*Saxony,* Germany] (1842)
1 a : a fine soft woolen fabric **b :** a fine closely twisted knitting yarn
2 : a Wilton jacquard carpet

sax·o·phone \ˈsak-sə-ˌfōn\ *noun* [French, from Antoine *Sax* (died 1894) Belgian instrument maker + French *-phone*] (1851)
: one of a group of single-reed woodwind instruments ranging from soprano to bass and characterized by a conical metal tube and finger keys
— **sax·o·phon·ic** \ˌsak-sə-ˈfō-nik, -ˈfä-\ *adjective*
— **sax·o·phon·ist** \ˈsak-sə-ˌfō-nist, *especially British* sak-ˈsä-fə-\ *noun*

saxophone

sax·tu·ba \ˈsaks-ˌtü-bə, -ˌtyü-\ *noun* [Antoine *Sax* + English *tuba*] (1856)
: a bass saxhorn

¹say \ˈsā, *Southern also* ˈse\ *verb* **said** \ˈsed, *especially when subject follows* səd\; **say·ing** \ˈsā-iŋ\; **says** \ˈsez, *sometimes* ˈsāz, *especially when subject follows* səz\ [Middle English, from Old English *secgan;* akin to Old High German *sagēn* to say, Lithuanian *sakyti,* Greek *ennepein* to speak, tell] (before 12th century)
transitive verb
1 a : to express in words **:** STATE **b :** to state as opinion or belief **:** DECLARE
2 a : UTTER, PRONOUNCE **b :** RECITE, REPEAT ⟨*say* your prayers⟩
3 a : INDICATE, SHOW ⟨the clock *says* five minutes after twelve⟩ **b :** to give expression to **:** COMMUNICATE ⟨a glance that *said* all that was necessary⟩
intransitive verb
: to express oneself **:** SPEAK
— **say·er** \ˈsā-ər, ˈse(-ə)r\ *noun*
— **say uncle :** to admit defeat
— **that is to say :** in other words **:** in effect

²say *noun, plural* **says** \ˈsāz, *Southern also* ˈsez\ (1571)
1 *archaic* **:** something that is said **:** STATEMENT
2 : an expression of opinion ⟨had my *say*⟩
3 : a right or power to influence action or decision; *especially* **:** the authority to make final decisions

³say *adverb* [from imperative of ¹*say*] (1596)
1 : ABOUT, APPROXIMATELY ⟨the property is worth, *say,* four million dollars⟩
2 : for example **:** AS ⟨if we compress any gas, *say* oxygen⟩

say·able \ˈsā-ə-bəl, ˈse-\ *adjective* (1856)
1 : capable of being said
2 : capable of being spoken effectively or easily ⟨readings in *sayable* Chinese —*Linguistic Reporter*⟩

say·est \ˈsā-əst\ *archaic 2d person singular of* SAY

say·ing \ˈsā-iŋ, ˈse-\ *noun* (14th century)
: something said; *especially* **:** ADAGE

say–so \ˈsā-(ˌ)sō, ˈse-\ *noun* (1637)
1 a : one's unsupported assertion or assurance **b :** an authoritative pronouncement ⟨left the hospital on the *say-so* of his doctor⟩
2 : a right of final decision ⟨has the ultimate *say-so* on what will be taught⟩

say·yid \ˈsī-yəd, ˈsā-; ˈsīd, ˈsād\ *noun* [Arabic] (1788)
1 : an Islamic chief or leader
2 : LORD, SIR — used as a courtesy title for a Muslim of rank or lineage

Saz·e·rac \ˈsa-zə-ˌrak\ *trademark*
— used for a cocktail with a whiskey base

¹scab \ˈskab\ *noun* [Middle English, of Scandinavian origin; akin to Old Swedish *skabbr* scab; akin to Old English *sceabb* scab, Latin *scabere* to scratch — more at SHAVE] (13th century)
1 : scabies of domestic animals
2 : a crust of hardened blood and serum over a wound

3 a : a contemptible person **b** (1) **:** a worker who refuses to join a labor union (2) **:** a union member who refuses to strike or returns to work before a strike has ended (3) **:** a worker who accepts employment or replaces a union worker during a strike (4) **:** one who works for less than union wages or on nonunion terms
4 : any of various bacterial or fungus diseases of plants characterized by crustaceous spots; *also* **:** one of the spots

²scab *intransitive verb* **scabbed; scab·bing** (1683)
1 : to become covered with a scab
2 : to act as a scab

scab·bard \'ska-bərd\ *noun* [Middle English *scaubert,* from Anglo-French *escaubers*] (13th century)
: a sheath for a sword, dagger, or bayonet
— scabbard *transitive verb*

scab·by \'ska-bē\ *adjective* **scab·bi·er; -est** (15th century)
1 a : covered with or full of scabs ⟨*scabby* skin⟩ **b :** diseased with scab ⟨a *scabby* animal⟩ ⟨*scabby* potatoes⟩
2 : MEAN, CONTEMPTIBLE ⟨a *scabby* trick⟩

sca·bies \'skā-bēz\ *noun, plural* **scabies** [Latin, from *scabere* to scratch] (1814)
: contagious itch or mange especially with exudative crusts that is caused by parasitic mites (especially *Sarcoptes scabiei*)
— sca·bi·et·ic \ˌskā-bē-'e-tik\ *adjective*

¹sca·bi·ous \'skā-bē-əs, 'ska-\ *noun* [Middle English *scabiose,* from Medieval Latin *scabiosa,* from Latin, feminine of *scabiosus,* adjective] (14th century)
: any of a genus (*Scabiosa*) of Old World herbs of the teasel family with terminal flower heads subtended by a leafy involucre

²scabious *adjective* [Latin *scabiosus,* from *scabies*] (1603)
1 : SCABBY
2 : of, relating to, or resembling scabies ⟨*scabious* eruptions⟩

scab·land \'skab-ˌland\ *noun* (1904)
: a region characterized by elevated tracts of rocky land with little or no soil cover and traversed or isolated by postglacial dry stream channels — usually used in plural

sca·brous \'ska-brəs *also* 'skā-\ *adjective* [Latin *scabr-, scaber* rough, scurfy; akin to Latin *scabere* to scratch — more at SCAB] (1646)
1 : DIFFICULT, KNOTTY ⟨a *scabrous* problem⟩
2 : rough to the touch: as **a :** having small raised dots, scales, or points ⟨a *scabrous* leaf⟩ **b :** covered with raised, roughened, or unwholesome patches ⟨*scabrous* paint⟩ ⟨yellowed *scabrous* skin⟩
3 : dealing with suggestive, indecent, or scandalous themes **:** SALACIOUS; *also* **:** SQUALID
synonym see ROUGH
— sca·brous·ly *adverb*
— sca·brous·ness *noun*

¹scad \'skad\ *noun, plural* **scad** *also* **scads** [origin unknown] (1602)
: any of several carangid fishes (especially of the genus *Decapterus*)

²scad *noun* [probably alteration of English dialect *scald* a multitude] (1869)
: a large number or quantity — usually used in plural ⟨*scads* of money⟩

scaf·fold \'ska-fəld *also* -ˌfōld\ *noun* [Middle English, from Old North French *escafaut,* modification of (assumed) Vulgar Latin *catafalicum,* from Greek *kata-* cata- + Latin *fala* siege tower] (14th century)
1 a : a temporary or movable platform for workers (as bricklayers, painters, or miners) to stand or sit on when working at a height above the floor or ground **b :** a platform on which a criminal is executed (as by hanging or beheading) **c :** a platform at a height above ground or floor level
2 : a supporting framework

scaf·fold·ing \-fəl-diŋ, -ˌfōl-\ *noun* (14th century)
: a system of scaffolds; *also* **:** material for scaffolds

scag \'skag\ *noun* [origin unknown] (1967)
slang **:** HEROIN

sca·glio·la \skal-'yō-lə, -'yȯ-\ *noun, often attributive* [Italian, literally, little chip] (1747)
: an imitation marble used for floors, columns, and ornamental interior work

scal·able \'skā-lə-bəl\ *adjective* (circa 1580)
: capable of being scaled

sca·lade \skə-'lād, -'läd\ *or* **sca·la·do** \-'lā-(ˌ)dō, -'lä-\ *noun, plural* **-lades** *or* **-la·dos** [obsolete Italian *scalada,* from *scalare* to scale, from *scala* ladder, staircase, from Late Latin — more at SCALE] (1591)
archaic **:** ESCALADE

¹sca·lar \'skā-lər, -ˌlär\ *adjective* [Latin *scalaris,* from *scalae* stairs, ladder — more at SCALE] (circa 1656)
1 : having an uninterrupted series of steps **:** GRADUATED ⟨*scalar* chain of authority⟩ ⟨*scalar* cells⟩
2 a : capable of being represented by a point on a scale ⟨*scalar* quantity⟩ **b :** of or relating to a scalar or scalar product ⟨*scalar* multiplication⟩

²scalar *noun* (1846)
1 : a real number rather than a vector
2 : a quantity (as mass or time) that has a magnitude describable by a real number and no direction

sca·la·re \skə-'lar-ē, -'ler-, -'lär-\ *noun* [New Latin, specific epithet, from Latin, neuter of *scalaris;* from the barred pattern on its body] (1928)
: a black and silver laterally compressed South American cichlid fish (*Pterophyllum scalare*) popular in aquariums

sca·lar·i·form \skə-'lar-ə-ˌfȯrm\ *adjective* [New Latin *scalariformis,* from Latin *scalaris* + *-iformis* -iform] (1836)
: resembling a ladder especially in having transverse bars or markings like the rungs of a ladder ⟨*scalariform* cells in plants⟩
— sca·lar·i·form·ly *adverb*

scalar product *noun* (1878)
: a real number that is the product of the lengths of two vectors and the cosine of the angle between them — called also *dot product, inner product*

scal·a·wag \'ska-li-ˌwag\ *noun* [origin unknown] (circa 1848)
1 : SCAMP, REPROBATE
2 : a white Southerner acting in support of the reconstruction governments after the American Civil War often for private gain

¹scald \'skȯld\ *verb* [Middle English, from Old North French *escalder,* from Late Latin *excaldare* to wash in warm water, from Latin *ex-* + *calida, calda* warm water, from feminine of *calidus* warm, from *calēre* to be warm — more at LEE] (13th century)
transitive verb
1 : to burn with or as if with hot liquid or steam
2 a : to subject to the action of boiling water or steam **b :** to bring to a temperature just below the boiling point ⟨*scald* milk⟩
3 : SCORCH
intransitive verb
1 : to scald something
2 : to become scalded

²scald *noun* (1601)
1 : an injury to the body caused by scalding
2 : an act or process of scalding
3 : any of various conditions or diseases of plants or fruits marked especially by a usually brownish discoloration of tissue

³scald *adjective* [*scall* + *¹-ed*] (1561)
1 *archaic* **:** SCABBY, SCURFY
2 *archaic* **:** SHABBY, CONTEMPTIBLE

⁴scald \'skȯld, 'skäld\ *variant of* SKALD

⁵scald \'skȯld\ *adjective* [alteration of *scalded*] (1791)
: subjected to scalding ⟨coffee . . . with *scald* cream —Charles Kingsley⟩

scald·ing \'skȯl-diŋ\ *adjective* (13th century)
1 : hot enough to scald ⟨*scalding* water⟩
2 a : having or producing the feeling of being burned ⟨*scalding* sun⟩ ⟨*scalding* sand⟩ **b :** SCATHING ⟨*scalding* criticism⟩

¹scale \'skā(ə)l\ *noun* [Middle English, bowl, scale of a balance, from Old Norse *skāl;* akin to Old Norse *skel* shell — more at SHELL] (14th century)
1 a : either pan or tray of a balance **b :** a beam that is supported freely in the center and has two pans of equal weight suspended from its ends — usually used in plural
2 : an instrument or machine for weighing

²scale *verb* **scaled; scal·ing** (1603)
transitive verb
: to weigh in scales
intransitive verb
: to have a specified weight on scales

³scale *noun* [Middle English, from Middle French *escale,* of Germanic origin; akin to Old English *scealu* shell, husk — more at SHELL] (14th century)
1 a : a small, flattened, rigid, and definitely circumscribed plate forming part of the external body covering especially of a fish **b :** a small thin plate suggesting a fish scale ⟨*scales* of mica⟩ ⟨the *scales* on a moth's wing⟩ **c :** the scaly covering of a scaled animal
2 : a small thin dry lamina shed (as in many skin diseases) from the skin
3 : a thin coating, layer, or incrustation: **a :** a usually black scaly coating of oxide forming on the surface of a metal (as iron) when it is heated for processing **b :** a hard incrustation usually rich in sulfate of calcium that is deposited on the inside of a vessel (as a boiler) in which water is heated
4 a : a modified leaf protecting a seed plant bud before expansion **b :** a thin, membranous, chaffy, or woody bract
5 a : any of the small overlapping usually metal pieces forming the outer surface of scale armor **b :** SCALE ARMOR
6 a : SCALE INSECT **b :** infestation with or disease caused by scale insects
— scaled \'skā(ə)ld\ *adjective*
— scale·less \'skā(ə)l-ləs\ *adjective*

⁴scale *verb* **scaled; scal·ing** (15th century)
transitive verb
1 : to remove the scale or scales from (as by scraping) ⟨*scale* a fish⟩
2 : to take off in thin layers or scales
3 : to form scale on ⟨hard water *scales* a boiler⟩
4 : to throw (as a thin flat stone) so that the edge cuts the air or so that it skips on water **:** SKIM
intransitive verb
1 : to separate and come off in scales **:** FLAKE
2 : to shed scales ⟨*scaling* skin⟩
3 : to become encrusted with scale

⁵scale *verb* **scaled; scal·ing** [Middle English, from *⁶scale*] (14th century)
transitive verb
1 a : to attack with or take by means of scaling ladders ⟨*scale* a castle wall⟩ **b :** to climb up or reach by means of a ladder **c :** to reach the highest point of **:** SURMOUNT
2 a : to arrange in a graduated series ⟨*scale* a test⟩ **b** (1) **:** to measure by or as if by a scale (2) **:** to measure or estimate the sound content of (as logs) **c :** to pattern, make, regulate, set, or estimate according to some rate or standard **:** ADJUST ⟨a production schedule *scaled* to actual need⟩ — often used with *down* or *up* ⟨*scale* down imports⟩
intransitive verb
1 : to climb by or as if by a ladder
2 : to rise in a graduated series
3 : MEASURE

⁶scale *noun* [Middle English, from Late Latin *scala* ladder, staircase, from Latin *scalae,* plural, stairs, rungs, ladder; akin to Latin *scandere* to climb — more at SCAN] (15th century)

1 a *obsolete* **:** LADDER **b** *archaic* **:** a means of ascent
2 : a graduated series of musical tones ascending or descending in order of pitch according to a specified scheme of their intervals
3 : something graduated especially when used as a measure or rule: as **a :** a series of marks or points at known intervals used to measure distances (as the height of the mercury in a thermometer) **b :** an indication of the relationship between the distances on a map and the corresponding actual distances **c :** RULER 3
4 a : a graduated series or scheme of rank or order ⟨a *scale* of taxation⟩ **b :** MINIMUM WAGE 2
5 a : a proportion between two sets of dimensions (as between those of a drawing and its original) **b :** a distinctive relative size, extent, or degree ⟨projects done on a large *scale*⟩
6 : a graded series of tests or of performances used in rating individual intelligence or achievement
— **scale** *adjective*
— **to scale :** according to the proportions of an established scale of measurement ⟨floor plans drawn *to scale*⟩

[7]scale *noun* [[5]*scale*] (circa 1587)
1 *obsolete* **:** ESCALADE
2 : an estimate of the amount of sound lumber in logs or standing timber

scale armor *noun* (1842)
: armor of small metallic scales on leather or cloth

scale–down \ˈskā(ə)l-ˌdau̇n\ *noun* (1931)
: a reduction according to a fixed ratio ⟨a *scale-down* of debts⟩

scale insect *noun* (1840)
: any of numerous small but very prolific homopterous insects (superfamily Coccoidea) that have winged males, wingless scale-covered females attached to the host plant, and young that suck the juices of plants and some of which are economic pests — compare LAC

scale•like \ˈskā(ə)l-ˌlīk\ *adjective* (1883)
: resembling a scale ⟨*scalelike* design⟩; *specifically* **:** reduced to a minute appressed element resembling a scale

sca•lene \ˈskā-ˌlēn, skā-ˈ\ *adjective* [Late Latin *scalenus,* from Greek *skalēnos,* literally, uneven; perhaps akin to Greek *skolios* crooked, *skelos* leg — more at ISOSCELES] (1734)
of a triangle **:** having the three sides of unequal length — see TRIANGLE illustration

scal•er \ˈskā-lər\ *noun* (1568)
1 : one that scales
2 : a dental instrument for removing tartar from teeth
3 : an electronic device that operates a recorder or produces an output pulse after a specified number of input impulses

scale–up \ˈskā-ˌləp\ *noun* (1945)
: an increase according to a fixed ratio

scall \ˈskȯl\ *noun* [Middle English, from Old Norse *skalli* bald head; akin to Swedish *skulle* skull] (14th century)
: a scurf or scabby disorder (as of the scalp)

scal•lion \ˈskal-yən\ *noun* [Middle English *scaloun,* from Anglo-French *scalun,* from (assumed) Vulgar Latin *escalonia,* from Latin *ascalonia* (*caepa*) onion of Ascalon, from feminine of *ascalonius* of Ascalon, from *Ascalon-, Ascalo* Ascalon, seaport in southern Palestine] (14th century)
1 : SHALLOT
2 : LEEK
3 : an onion forming a thick basal portion without a bulb; *also* **:** GREEN ONION

[1]scal•lop \ˈskä-ləp, ˈska-, ˈskȯ-\ *noun* [Middle English *scalop,* from Middle French *escalope* shell, of Germanic origin; akin to Middle Dutch *schelpe* shell] (15th century)
1 a : any of numerous marine bivalve lamellibranch mollusks (family Pectinidae) that have a radially ribbed shell with the edge undulated and that swim by opening and closing the valves **b :** the adductor muscle of a scallop as an article of food
2 a : a valve or shell of a scallop **b :** a baking dish shaped like a valve of a scallop
3 : one of a continuous series of circle segments or angular projections forming a border
4 : PATTYPAN
5 [French *escalope,* probably from Middle French, shell] **:** a thin slice of boneless meat (as veal)

[2]scallop (1737)
transitive verb
1 [from the use of a scallop shell as a baking dish] **:** to bake in a sauce usually covered with seasoned bread or cracker crumbs ⟨*scalloped* potatoes⟩
2 a : to shape, cut, or finish in scallops **b :** to form scallops in
intransitive verb
: to gather or dredge scallops

scal•lop•er \-lə-pər\ *noun* (circa 1881)
1 : a person who dredges for or gathers scallops
2 : a boat equipped and used to dredge for scallops

scal•lo•pi•ni *or* **sca•lop•pi•ne** \ˌskä-lə-ˈpē-nē, ˌska-\ *noun* [Italian *scaloppine,* ultimately from French *escalope* thin slice of meat, probably from Middle French, shell] (1946)
: thin slices of meat (as veal) sautéed or coated with flour and fried

scal•ly•wag *variant of* SCALAWAG

sca•lo•gram \ˈskā-lə-ˌgram\ *noun* [[6]*scale* + *-o-* + *-gram*] (1944)
: an arrangement of items (as problems on a test or features of speech) in ascending order so that the presence or accomplishment of an item at one level implies the presence of or the capability to accomplish items at all lower levels

[1]scalp \ˈskalp\ *noun* [Middle English, of Scandinavian origin; akin to Old Norse *skālpr* sheath; akin to Middle Dutch *schelpe* shell] (14th century)
1 a : the part of the integument of the human head usually covered with hair in both sexes **b :** the part of an animal (as a wolf or fox) corresponding to the human scalp
2 a : a part of the human scalp with attached hair cut or torn from an enemy as a token of victory (as by Indian warriors of North America) **b :** a trophy of victory or accomplishment
3 *chiefly Scottish* **:** a projecting mass of bare ground or rock

[2]scalp (1676)
transitive verb
1 a : to deprive of the scalp **b :** to remove an upper part from
2 : to remove a desired constituent from and discard the rest
3 : to buy and sell so as to make small quick profits ⟨*scalp* stocks⟩ ⟨*scalp* grain⟩; *especially* **:** to resell at greatly increased prices ⟨*scalp* theater tickets⟩
intransitive verb
1 : to take scalps
2 : to profit by slight market fluctuations
— **scalp•er** *noun*

scal•pel \ˈskal-pəl *also* skal-ˈpel\ *noun* [Latin *scalpellus, scalpellum,* diminutive of *scalper, scalprum* chisel, knife, from *scalpere* to scratch, carve] (1742)
: a small straight thin-bladed knife used especially in surgery

scalp lock *noun* (1826)
: a long tuft of hair on the crown of the otherwise shaved head especially of a warrior of some American Indian tribes

scaly \ˈskā-lē\ *adjective* **scal•i•er; -est** (14th century)
1 a : covered with, composed of, or rich in scale or scales **b :** FLAKY
2 : of or relating to scaly animals
3 : DESPICABLE, POOR
4 : infested with scale insects ⟨*scaly* fruit⟩
— **scal•i•ness** *noun*

scaly anteater *noun* (1840)
: PANGOLIN

scam \ˈskam\ *noun* [origin unknown] (1963)
: a fraudulent or deceptive act or operation ⟨an insurance *scam*⟩

scam•mo•ny \ˈska-mə-nē\ *noun, plural* **-nies** [Middle English *scamonie,* from Old English *scammoniam,* from Latin *scammonia,* from Greek *skammōnia*] (before 12th century)
1 : a twining convolvulus (*Convolvulus scammonia*) of Asia Minor with a large thick root
2 a : the dried root of scammony **b :** a cathartic resin obtained from scammony

[1]scamp \ˈskamp\ *noun* [obsolete *scamp* to roam about idly] (1808)
1 : RASCAL, ROGUE
2 : an impish or playful young person
— **scamp•ish** \ˈskam-pish\ *adjective*

[2]scamp *transitive verb* [origin unknown] (1837)
: to perform or deal with in a hasty, neglectful, or imperfect manner

[1]scam•per \ˈskam-pər\ *intransitive verb* **scam•pered; scam•per•ing** \-p(ə-)riŋ\ [probably from obsolete Dutch *schampen* to flee, from Middle French *escamper,* from Italian *scampare,* from (assumed) Vulgar Latin *excampare* to decamp, from Latin *ex-* + *campus* field] (1691)
: to run nimbly and usually playfully about

[2]scamper *noun* (1697)
: a playful or hurried run or movement

scam•pi \ˈskam-pē, ˈskäm-\ *noun, plural* **scampi** [Italian, plural of *scampo,* a European lobster] (1925)
: SHRIMP; *especially* **:** large shrimp prepared with a garlic-flavored sauce

[1]scan \ˈskan\ *verb* **scanned; scan•ning** [Middle English *scannen,* from Late Latin *scandere,* from Latin, to climb; akin to Middle Irish *sceinnid* he springs, Sanskrit *skandati* he leaps] (14th century)
transitive verb
1 : to read or mark so as to show metrical structure
2 : to examine by point-by-point observation or checking: **a :** to investigate thoroughly by checking point by point and often repeatedly ⟨a fire lookout *scanning* the hills with binoculars⟩ **b :** to glance from point to point of often hastily, casually, or in search of a particular item ⟨*scan* the want ads looking for a job⟩
3 a : to examine especially systematically with a sensing device (as a photometer or a beam of radiation) usually to obtain information **b :** to pass an electron beam over and convert (an image) into variations of electrical properties (as voltage) that convey information electronically **c :** to pass over in the formation of an image ⟨the electron beam *scans* the picture tube⟩
intransitive verb
1 : to scan verse
2 : to conform to a metrical pattern
synonym see SCRUTINIZE
— **scan•na•ble** \ˈska-nə-bəl\ *adjective*

[2]scan *noun* (1706)
1 : the act or process of scanning
2 : a radar or television trace
3 a : a depiction (as a photograph) of the distribution of a radioactive material in something (as a bodily organ) **b :** an image of a bodily part produced (as by computer) by combining radiographic data obtained from several angles or sections

[1]scan•dal \ˈskan-dᵊl\ *noun* [Middle English, from Late Latin *scandalum* stumbling block, offense, from Greek *skandalon* trap, stumbling block, offense; akin to Latin *scandere* to climb] (13th century)

1 a : discredit brought upon religion by unseemly conduct in a religious person **b :** conduct that causes or encourages a lapse of faith or of religious obedience in another
2 : loss of or damage to reputation caused by actual or apparent violation of morality or propriety **:** DISGRACE
3 a : a circumstance or action that offends propriety or established moral conceptions or disgraces those associated with it **b :** a person whose conduct offends propriety or morality
4 : malicious or defamatory gossip
5 : indignation, chagrin, or bewilderment brought about by a flagrant violation of morality, propriety, or religious opinion
synonym see OFFENSE

²scandal *transitive verb* (1592)
1 *obsolete* **:** DISGRACE
2 *chiefly dialect* **:** DEFAME, SLANDER

scan·dal·ise *chiefly British variant of* SCANDALIZE

scan·dal·ize \'skan-də-ˌlīz\ *transitive verb* **-ized; -iz·ing** (1566)
1 : to speak falsely or maliciously of
2 *archaic* **:** to bring into reproach
3 : to offend the moral sense of **:** SHOCK

scan·dal·mon·ger \'skan-dᵊl-ˌməŋ-gər, -ˌmäŋ-\ *noun* (1721)
: a person who circulates scandal
— scan·dal·mon·ger·ing \-g(ə-)riŋ\ *noun*

scan·dal·ous \'skan-dᵊl-əs\ *adjective* (1603)
1 : LIBELOUS, DEFAMATORY
2 : offensive to propriety or morality **:** SHOCKING
— scan·dal·ous·ly *adverb*
— scan·dal·ous·ness *noun*

scandal sheet *noun* (1904)
: a newspaper or periodical dealing to a large extent in scandal and gossip

scan·dent \'skan-dənt\ *adjective* [Latin *scandent-, scandens,* present participle of *scandere* to climb — more at SCAN] (circa 1682)
: characterized by a climbing mode of growth ⟨*scandent* stems⟩ ⟨*scandent* vines⟩

Scan·di·an \'skan-dē-ən\ *adjective* [Latin *Scandia*] (1668)
1 : SCANDINAVIAN
2 : of or relating to the languages of Scandinavia
— Scandian *noun*

Scan·di·na·vian \ˌskan-də-'nā-vē-ən, -vyən\ *noun* (1766)
1 : the North Germanic languages
2 a : a native or inhabitant of Scandinavia **b :** a person of Scandinavian descent
— Scandinavian *adjective*

scan·di·um \'skan-dē-əm\ *noun* [New Latin, from Latin *Scandia,* ancient name of southern Scandinavian peninsula] (1879)
: a white metallic element found in association with rare earth elements — see ELEMENT table

scan·ner \'ska-nər\ *noun* (1557)
: one that scans: as **a :** a device for sensing recorded data **b :** a device (as a CAT scanner) used for scanning something (as the human body)

scanning electron microscope *noun* (1953)
: an electron microscope in which a beam of focused electrons moves across the object with the secondary electrons produced by the object and the electrons scattered by the object being collected to form a three-dimensional image on a cathode-ray tube — called also *scanning microscope*
— scanning electron microscopy *noun*

scanning tunneling microscope *noun* (1983)
: a microscope that makes use of the phenomenon of electron tunneling to map the positions of individual atoms in a surface or to move atoms around on a surface
— scanning tunneling microscopy *noun*

scan·sion \'skan(t)-shən\ *noun* [Late Latin *scansion-, scansio,* from Latin, act of climbing, from *scandere*] (1671)
: the analysis of verse to show its meter

¹scant \'skant\ *adjective* [Middle English, from Old Norse *skamt,* neuter of *skammr* short] (14th century)
1 *dialect* **a :** excessively frugal **b :** not prodigal **:** CHARY
2 a : barely or scarcely sufficient; *especially* **:** not quite coming up to a stated measure **b :** lacking in amplitude or quantity
3 : having a small or insufficient supply ⟨he's fat, and *scant* of breath —Shakespeare⟩
synonym see MEAGER
— scant·ly *adverb*
— scant·ness *noun*

²scant *adverb* (15th century)
dialect **:** SCARCELY, HARDLY

³scant *transitive verb* (circa 1580)
1 : to provide an incomplete supply of
2 : to make small, narrow, or meager
3 : to give scant attention to **:** SLIGHT
4 : to provide with a meager or inadequate portion or supply **:** STINT

scant·ies \'skan-tēz\ *noun plural* [blend of *¹scant* and *panties*] (1929)
: abbreviated panties for women

scant·ling \'skant-liŋ, -lən\ *noun* [alteration of Middle English *scantilon,* literally, mason's or carpenter's gauge, from Old North French *escantillon*] (1555)
1 a : the dimensions of timber and stone used in building **b :** the dimensions of a frame or strake used in shipbuilding
2 : a small quantity, amount, or proportion **:** MODICUM
3 : a small piece of lumber (as an upright piece in house framing)

scanty \'skan-tē\ *adjective* **scant·i·er; -est** [English dialect *scant* scanty supply, from Middle English, from Old Norse *skamt,* from neuter of *skammr* short] (1660)
: limited or less than sufficient in degree, quantity, or extent
synonym see MEAGER
— scant·i·ly \'skan-tᵊl-ē\ *adverb*
— scant·i·ness \'skan-tē-nəs\ *noun*

¹scape \'skāp\ *verb* **scaped; scap·ing** [Middle English, short for *escapen*] (13th century)
: ESCAPE

²scape *noun* [Latin *scapus* shaft, stalk — more at SHAFT] (1601)
1 : a peduncle arising at or beneath the surface of the ground in an acaulescent plant (as the tulip); *broadly* **:** a flower stalk
2 : the shaft of an animal part (as an antenna or feather)

³scape *noun* [*landscape*] (1773)
: a view or picture of a scene — usually used in combination ⟨city*scape*⟩
word history see LANDSCAPE

¹scape·goat \'skāp-ˌgōt\ *noun* [*¹scape;* intended as translation of Hebrew *'azāzēl* (probably name of a demon), as if *'ēz 'ōzēl* goat that departs—Leviticus 16:8 (Authorized Version)] (1530)
1 : a goat upon whose head are symbolically placed the sins of the people after which he is sent into the wilderness in the biblical ceremony for Yom Kippur
2 a : one that bears the blame for others **b :** one that is the object of irrational hostility ◆

²scapegoat *transitive verb* (1943)
: to make a scapegoat of
— scape·goat·ism \-ˌgō-ˌti-zəm\ *noun*

scape·grace \'skāp-ˌgrās\ *noun* [*¹scape*] (1809)
: an incorrigible rascal

scaph·oid \'ska-ˌfȯid\ *noun* [New Latin *scaphoides,* from Greek *skaphoeidēs,* from *skaphos* boat] (1846)
: the bone of the thumb side of the carpus that is the largest in the proximal row; *also* **:** the navicular bone of the tarsus
— scaphoid *adjective*

scap·o·lite \'ska-pə-ˌlīt\ *noun* [French, from Latin *scapus* shaft + French *-o-* + *-lite;* from the prismatic shape of its crystals] (1802)
: any of a group of minerals that are essentially complex silicates of aluminum, calcium, and sodium and that include some used as semiprecious stones

sca·pose \'skā-ˌpōs\ *adjective* (circa 1903)
: bearing, resembling, or consisting of a scape ⟨the terminal *scapose* flowers of a pyrola⟩

scap·u·la \'ska-pyə-lə\ *noun, plural* **-lae** \-ˌlē, -ˌlī\ *or* **-las** [New Latin, from Latin, shoulder blade, shoulder] (1578)
: either of a pair of large triangular bones lying one in each dorsal lateral part of the thorax, being the principal bone of the corresponding half of the shoulder girdle, and articulating with the corresponding clavicle or coracoid — called also *shoulder blade*

¹scap·u·lar \-lər\ *noun* [Middle English *scapulare,* from Late Latin, from Latin *scapula* shoulder] (15th century)
1 a : a long wide band of cloth with an opening for the head worn front and back over the shoulders as part of a monastic habit **b :** a pair of small cloth squares joined by shoulder tapes and worn under the clothing on the breast and back as a sacramental and often also as a badge of a third order or confraternity
2 a : SCAPULA **b :** one of the feathers covering the base of a bird's wing — see BIRD illustration

²scapular *adjective* [New Latin *scapularis,* from *scapula*] (1713)
: of or relating to the shoulder, the scapula, or scapulars

scapular medal *noun* (1912)
: a medal worn in place of a sacramental scapular

¹scar \'skär\ *noun* [Middle English *skere,* from Old Norse *sker* skerry; probably akin to Old Norse *skera* to cut — more at SHEAR] (14th century)
1 : an isolated or protruding rock
2 : a steep rocky eminence **:** a bare place on the side of a mountain

²scar *noun* [Middle English *escare, scar,* from Middle French *escare* scab, from Late Latin *eschara,* from Greek, hearth, brazier, scab] (14th century)
1 : a mark left (as in the skin) by the healing of injured tissue

◇ WORD HISTORY

scapegoat An ancient Hebrew custom called for the high priest on Yom Kippur, the Jewish Day of Atonement, to take two goats, one to be sacrificed to the Lord, the other to carry the sins of the people away into the desert. After the Lord's goat had been sacrificed, the priest confessed the sins of the people over the head of the live goat, which was then released "to Azazel." Modern commentators have usually interpreted Azazel as the name of a demon living in the desert, but the name was interpreted by ancient Biblical translators as referring to the goat itself, and hence being a form of the Hebrew phrase *'ez 'ōzēl,* which means "goat that departs." In the Greek translation of the passage referring to Azazel's goat, the goat became *tragos apopompaios* "goat sent out," and in the Latin Vulgate the animal became *caper emissarius* "emissary goat." Perpetuating the tradition, William Tyndale in his 1530 English translation of the Pentateuch rendered the Latin as *scapegoat,* that is, "escape goat." Although the goat for Azazel did in fact escape sacrifice, he bore the heavy burden of the people's sins. It is for this reason that we call anyone who bears the blame for others' transgressions or mistakes a *scapegoat.*

2 a : a mark left on a stem or branch by a fallen leaf or harvested fruit **b :** CICATRIX 2
3 : a mark or indentation resulting from damage or wear
4 : a lasting moral or emotional injury ⟨one of his men had been killed . . . in a manner that left a *scar* upon his mind —H. G. Wells⟩
— scar·less \-ləs\ *adjective*

[3]**scar** *verb* **scarred; scar·ring** (1555)
transitive verb
1 : to mark with a scar
2 : to do lasting injury to
intransitive verb
1 : to form a scar
2 : to become scarred

scar·ab \ˈskar-əb\ *noun* [Middle French *scarabee,* from Latin *scarabaeus*] (1579)
1 : any of a family (Scarabaeidae) of stout-bodied beetles (as a dung beetle) with lamellate or flabellate antennae
2 : a stone or faience beetle used in ancient Egypt as a talisman, ornament, and a symbol of resurrection

scar·a·bae·us \ˌskar-ə-ˈbē-əs\ *noun* [Latin] (1664)
: SCARAB 2

scar·a·mouch *or* **scar·a·mouche** \ˈskar-ə-ˌmüsh, ˌmüch, -ˌmau̇ch\ *noun* [French *Scaramouche,* from Italian *Scaramuccia,* from *scaramuccia* skirmish] (1662)
1 *capitalized* **:** a stock character in the Italian commedia dell'arte that burlesques the Spanish don and is characterized by boastfulness and cowardliness
2 a : a cowardly buffoon **b :** RASCAL, SCAMP

[1]**scarce** \ˈskers, ˈskars\ *adjective* **scarc·er; scarc·est** [Middle English *scars,* from Old North French *escars,* from (assumed) Vulgar Latin *excarpsus,* literally, plucked out, past participle of Latin *excerpere* to pluck out — more at EXCERPT] (14th century)
1 : deficient in quantity or number compared with the demand **:** not plentiful or abundant
2 : intentionally absent ⟨made himself *scarce* at inspection time⟩
synonym see INFREQUENT
— scarce·ness *noun*

[2]**scarce** *adverb* (15th century)
: SCARCELY, HARDLY

scarce·ly *adverb* (14th century)
1 a : by a narrow margin **:** only just ⟨had *scarcely* rung the bell when the door flew open —Agnes S. Turnbull⟩ **b :** almost not ⟨could *scarcely* see for the fog⟩
2 a : certainly not ⟨could *scarcely* interfere⟩ **b :** probably not ⟨there could *scarcely* have been found a leader better equipped —V. L. Parrington⟩

scar·ci·ty \ˈsker-sə-tē, ˈskar-, -stē\ *noun, plural* **-ties** (14th century)
: the quality or state of being scarce; *especially* **:** want of provisions for the support of life

[1]**scare** \ˈsker, ˈskar\ *verb* **scared; scar·ing** [Middle English *skerren,* from Old Norse *skirra,* from *skjarr* shy, timid] (13th century)
transitive verb
: to frighten especially suddenly **:** ALARM
intransitive verb
: to become scared
— scar·er *noun*

[2]**scare** *noun* (circa 1548)
1 : a sudden fright
2 : a widespread state of alarm **:** PANIC
— scare *adjective*

scare·crow \ˈsker-ˌkrō, ˈskar-\ *noun* (1573)
1 a : an object usually suggesting a human figure that is set up to frighten birds away from crops **b :** something frightening but harmless
2 : a skinny or ragged person

scared *adjective* (1590)
: thrown into or being in a state of fear, fright, or panic ⟨*scared* of snakes⟩ ⟨*scared* to go out⟩

scaredy–cat \ˈsker-dē-ˌkat, ˈskar-\ *noun* [*scared* (past participle of *scare*) + [1]-y + *cat*] (1948)
: an unduly fearful person

scare·head \ˈsker-ˌhed, ˈskar-\ *noun* (1887)
: a big, sensational, or alarming newspaper headline

scare·mon·ger \-ˌməŋ-gər, -ˌmäŋ-\ *noun* (1888)
: one inclined to raise or excite alarms especially needlessly

scare up *transitive verb* (1841)
: to find or get together with considerable labor or difficulty **:** scrape up ⟨managed to *scare up* the money⟩

[1]**scarf** \ˈskärf\ *noun, plural* **scarfs** [Middle English *skarf,* probably of Scandinavian origin; akin to Old Norse *skarfr* scarf] (15th century)
1 : either of the chamfered or cutaway ends that fit together to form a scarf joint
2 : a joint made by chamfering, halving, or notching two pieces to correspond and lapping and bolting them — called also *scarf joint*

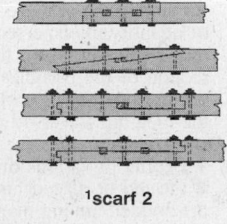

[1]**scarf 2**

[2]**scarf** *also* **scarph** \ˈskärf\ *transitive verb* (1627)
1 : to unite by a scarf joint
2 : to form a scarf on

[3]**scarf** *noun, plural* **scarves** \ˈskärvz\ *or* **scarfs** [Old North French *escarpe* sash, sling] (1555)
1 a : a military or official sash usually indicative of rank **b** *archaic* **:** TIPPET 3
2 : a broad band of cloth worn about the shoulders, around the neck, or over the head
3 : RUNNER 6b

[4]**scarf** *transitive verb* (1598)
1 : to wrap, cover, or adorn with or as if with a scarf
2 : to wrap or throw on (a scarf or mantle) loosely

[5]**scarf** *transitive verb* [by alteration] (circa 1960)
: [3]SCOFF 1 ⟨*scarfed* down my sandwich⟩

scarf·pin \ˈskärf-ˌpin\ *noun* (1859)
: TIEPIN

scarf·skin \ˈskärf-ˌskin\ *noun* [[3]*scarf*] (1615)
: EPIDERMIS; *especially* **:** that forming the cuticle of a nail

scar·i·fi·ca·tion \ˌskar-ə-fə-ˈkā-shən, ˌsker-\ *noun* (14th century)
1 : the act or process of scarifying
2 : a mark or marks made by scarifying

[1]**scar·i·fy** \ˈskar-ə-ˌfī, ˈsker-\ *transitive verb* **-fied; -fy·ing** [Middle English *scarifien,* from Middle French *scarifier,* from Late Latin *scarificare,* alteration of Latin *scarifare,* from Greek *skariphasthai* to scratch an outline, sketch — more at SCRIBE] (14th century)
1 : to make scratches or small cuts in (as the skin) ⟨*scarify* an area for vaccination⟩
2 : to lacerate the feelings of
3 : to break up and loosen the surface of (as a field or road)
4 : to cut or soften the wall of (a hard seed) to hasten germination
— scar·i·fi·er \-ˌfī(-ə)r\ *noun*

[2]**scarify** *transitive verb* **-fied; -fy·ing** (1794)
: SCARE, FRIGHTEN
— scar·i·fy·ing·ly \-ˌfī-iŋ-lē\ *adverb*

scar·i·ous \ˈsker-ē-əs, ˈskar-\ *adjective* [New Latin *scariosus*] (circa 1806)
: dry and membranous in texture ⟨a *scarious* bract⟩

scar·la·ti·na \ˌskär-lə-ˈtē-nə\ *noun* [New Latin, from Medieval Latin *scarlata* scarlet] (1803)
: SCARLET FEVER
— scar·la·ti·nal \-ˈtē-nᵊl\ *adjective*

[1]**scar·let** \ˈskär-lət\ *noun* [Middle English *scarlat, scarlet,* from Old French or Medieval Latin; Old French *escarlate,* from Medieval Latin *scarlata,* from Persian *saqalāt,* a kind of rich cloth] (13th century)
1 : scarlet cloth or clothes
2 : any of various bright reds

[2]**scarlet** *adjective* (14th century)
1 : of the color scarlet
2 a : grossly and glaringly offensive ⟨sinning in flagrant and *scarlet* fashion —G. W. Johnson⟩ **b** [from the use of the word in Isaiah 1:18 & Revelation 17:1–6 (Authorized Version)] **:** of, characterized by, or associated with sexual immorality ⟨a *scarlet* woman⟩

scarlet fever *noun* (1676)
: an acute contagious febrile disease caused by hemolytic streptococci (especially various strains of *Streptococcus pyogenes*) and characterized by inflammation of the nose, throat, and mouth, generalized toxemia, and a red rash

scarlet letter *noun* [from such a letter in the novel *The Scarlet Letter* (1850) by Nathaniel Hawthorne] (1850)
: a scarlet A worn as a punitive mark of adultery

scarlet pimpernel *noun* (1855)
1 : a common pimpernel (*Anagallis arvensis*) having scarlet, white, or purplish flowers that close in cloudy weather
2 [*Scarlet Pimpernel,* assumed name of the hero of *The Scarlet Pimpernel* (1905), novel by Baroness Orczy] **:** a person who rescues others from mortal danger by smuggling them across a border

scarlet runner *noun* (1806)
: a tropical American high-climbing bean (*Phaseolus coccineus*) that has large bright red flowers and red-and-black seeds and is grown widely as an ornamental and as a food bean — called also *scarlet runner bean*

scarlet sage *noun* (circa 1890)
: a garden salvia (*Salvia splendens*) of Brazil with long racemes of intense typically scarlet flowers

scarlet tanager *noun* (1810)
: a common American tanager (*Piranga olivacea*) with the male having scarlet plumage and black wings during the breeding season and the female having chiefly olive plumage

scarp \ˈskärp\ *noun* [Italian *scarpa*] (1589)
1 : the inner side of a ditch below the parapet of a fortification
2 a : a line of cliffs produced by faulting or erosion — see FAULT illustration **b :** a low steep slope along a beach caused by wave erosion
— scarped *adjective*

scar·per \ˈskär-pər\ *intransitive verb* [probably ultimately from Italian *scappare,* from (assumed) Vulgar Latin *excappare* — more at ESCAPE] (circa 1846)
British **:** FLEE, RUN AWAY; *broadly* **:** LEAVE, DEPART

scar·ry \ˈskär-ē\ *adjective* [[2]*scar*] (1653)
: bearing marks of wounds **:** SCARRED

[1]**scart** \ˈskärt\ *verb* [Middle English, alteration of *scratten* to scratch] (14th century)
chiefly Scottish **:** SCRATCH, SCRAPE

[2]**scart** *noun* (circa 1585)
chiefly Scottish **:** SCRATCH, MARK; *especially* **:** one made in writing

scar tissue *noun* (1875)
: the connective tissue forming a scar and composed chiefly of fibroblasts in recent scars and largely of dense collagenous fibers in old scars

scary \ˈsker-ē, ˈskar-\ *adjective* **scar·i·er; -est** (1582)
1 : causing fright **:** ALARMING ⟨a *scary* story⟩
2 : easily scared **:** TIMID
3 : feeling alarm or fright **:** FRIGHTENED

— **scar·i·ly** \'sker-ə-lē, 'skar-\ *adverb*

[1]scat \'skat\ *intransitive verb* **scat·ted; scat·ting** [*scat,* interjection used to drive away a cat] (1838)
1 : to go away quickly
2 : to move fast **:** SCOOT

[2]scat *noun* [perhaps from Greek *skat-, skōr* excrement — more at SCATOLOGY] (1927)
: an animal fecal dropping

[3]scat *noun* [origin unknown] (1929)
: jazz singing with nonsense syllables

[4]scat *intransitive verb* **scat·ted; scat·ting** (1935)
: to improvise nonsense syllables usually to an instrumental accompaniment **:** sing scat

scat·back \'skat-,bak\ *noun* [[1]*scat* + *back*] (1945)
: an offensive back in football who is an especially fast and elusive ballcarrier

[1]scathe \'skāth, 'skāth\ *noun* [Middle English *skathe,* from Old Norse *skathi;* akin to Old English *sceatha* injury, Greek *askēthēs* unharmed] (13th century)
: HARM, INJURY
— **scathe·less** \-ləs\ *adjective*

[2]scathe \'skāth\ *transitive verb* **scathed; scath·ing** (13th century)
1 : to do harm to; *specifically* **:** SCORCH, SEAR
2 : to assail with withering denunciation

scath·ing \'skā-thiŋ\ *adjective* (1794)
: bitterly severe ⟨a *scathing* condemnation⟩
synonym see CAUSTIC
— **scath·ing·ly** \-thiŋ-lē\ *adverb*

sca·tol·o·gy \ska-'tä-lə-jē, skə-\ *noun* [Greek *skat-, skōr* excrement; akin to Old English *scearn* dung, Latin mu*scerdae* mouse droppings] (1876)
1 : interest in or treatment of obscene matters especially in literature
2 : the biologically oriented study of excrement (as for taxonomic purposes or for the determination of diet)
— **scat·o·log·i·cal** \,ska-tᵊl-'ä-ji-kəl\ *adjective*

scatt \'skat\ *noun* [Middle English *scat,* from Old Norse *skattr;* akin to Old English *sceat* property, money, a small coin, Old Church Slavonic *skotŭ* domestic animal] (13th century)
archaic **:** TAX, TRIBUTE

[1]scat·ter \'ska-tər\ *verb* [Middle English *scateren*] (14th century)
transitive verb
1 a : to cause to separate widely **b :** to cause to vanish
2 *archaic* **:** to fling away heedlessly **:** SQUANDER
3 : to distribute irregularly
4 : to sow by casting in all directions **:** STREW
5 a : to reflect irregularly and diffusely **b :** to cause (a beam of radiation) to diffuse or disperse
6 : to divide into ineffectual small portions
intransitive verb
1 : to separate and go in various directions **:** DISPERSE
2 : to occur or fall irregularly or at random ☆
— **scat·ter·er** \-tər-ər\ *noun*

[2]scatter *noun* (1642)
1 : the act of scattering
2 : a small quantity or number irregularly distributed or strewn about **:** SCATTERING
3 : the state or extent of being scattered; *especially* **:** DISPERSION

scat·ter·ation \,ska-tə-'rā-shən\ *noun* (1776)
1 : the act or process of scattering **:** the state of being scattered
2 : the movement of people and industry away from the city; *also* **:** the resulting regional urbanization
3 : a policy of distributing funds and energies in too many ineffectively small units

scat·ter·brain \'ska-tər-,brān\ *noun* (1790)
: a giddy heedless person

scat·ter·brained \-,brānd\ *adjective* (1747)
: having the characteristics of a scatterbrain

scatter diagram *noun* (1925)
: a two-dimensional graph in rectangular coordinates consisting of points whose coordinates represent values of two variables under study

scat·ter·good \'ska-tər-,gu̇d\ *noun* (1577)
: a wasteful person **:** SPENDTHRIFT

scat·ter·gram \-,gram\ *noun* (1938)
: SCATTER DIAGRAM

scat·ter·gun \-,gən\ *noun or adjective* (1836)
: SHOTGUN

[1]scat·ter·ing *noun* (14th century)
1 : an act or process in which something scatters or is scattered
2 : something scattered: as **a :** a small number or quantity interspersed here and there ⟨a *scattering* of visitors⟩ **b :** the random change in direction of the particles constituting a beam or wave front due to collision with particles of the medium traversed

[2]scattering *adjective* (15th century)
1 : going in various directions
2 : found or placed far apart and in no order
3 : divided among many or several ⟨*scattering* votes⟩
— **scat·ter·ing·ly** \'ska-tə-riŋ-lē\ *adverb*

scatter rug *noun* (1926)
: a rug of such a size that several can be used in a room

scat·ter·shot \'ska-tər-,shät\ *adjective* (1951)
: broadly and often randomly inclusive **:** SHOTGUN

scat·ty \'ska-tē\ *adjective* **scat·ti·er; -est** [probably from *scatt*erbrain + [1]-y] (1911)
chiefly British **:** CRAZY

scaup \'sko̊p\ *noun, plural* **scaup** *or* **scaups** [short for *scaup duck; scaup* probably alteration of *scalp* bed of shellfish] (1797)
: either of two diving ducks (*Aythya affinis* or *A. marila*) with the male having a glossy purplish or greenish head and a black breast and tail

scav·enge \'ska-vənj, -vinj\ *verb* **scav·enged; scav·eng·ing** [back-formation from *scavenger*] (circa 1644)
transitive verb
1 a (1) **:** to remove (as dirt or refuse) from an area (2) **:** to clean away dirt or refuse from **:** CLEANSE ⟨*scavenge* a street⟩ **b :** to feed on (carrion or refuse)
2 a : to remove (burned gases) from the cylinder of an internal combustion engine after a working stroke **b :** to remove (as an undesirable constituent) from a substance or region by chemical or physical means **c :** to clean and purify (molten metal) by taking up foreign elements in chemical union
3 : to salvage from discarded or refuse material; *also* **:** to salvage usable material from
intransitive verb
: to work or act as a scavenger

scav·en·ger \'ska-vən-jər\ *noun* [alteration of earlier *scavager,* from Middle English *skawager* customs collector, from *skawage* customs, from Old North French *escauwage* inspection, from *escauwer* to inspect, of Germanic origin; akin to Old English *scēawian* to look at — more at SHOW] (1530)
1 *chiefly British* **:** a person employed to remove dirt and refuse from streets
2 : one that scavenges: as **a :** a garbage collector **b :** a junk collector **c :** a chemically active substance acting to make innocuous or remove an undesirable substance
3 : an organism that feeds habitually on refuse or carrion

scavenger hunt *noun* (1936)
: a game in which players try to acquire without buying specified items within a time limit

sce·na \'shā-(,)nä\ *noun* [Italian, literally, scene, from Latin] (1819)
: an elaborate solo vocal composition that consists of a recitative usually followed by one or more aria sections

sce·nar·io \sə-'nar-ē-,ō, -'ner-\ *noun, plural* **-i·os** [Italian, from Latin *scaenarium* place for erecting stages, from *scaena* stage] (1878)
1 a : an outline or synopsis of a play; *especially* **:** a plot outline used by actors of the commedia dell'arte **b :** the libretto of an opera
2 a : SCREENPLAY **b :** SHOOTING SCRIPT
3 : a sequence of events especially when imagined; *especially* **:** an account or synopsis of a possible course of action or events ⟨his *scenario* for a settlement envisages . . . reunification —Selig Harrison⟩

sce·nar·ist \-'nar-ist, -'ner-\ *noun* (1920)
: a writer of scenarios

scend \'send\ *noun* [perhaps short for *ascend*] (circa 1883)
1 : the lift of a wave **:** SEND
2 : the upward movement of a pitching ship

scene \'sēn\ *noun* [Middle French, stage, from Latin *scena, scaena* stage, scene, probably from Etruscan, from Greek *skēnē* shelter, tent, building forming the background for a dramatic performance, stage; perhaps akin to Greek *skia* shadow — more at SHINE] (1540)
1 : one of the subdivisions of a play: as **a :** a division of an act presenting continuous action in one place **b :** a single situation or unit of dialogue in a play ⟨the love *scene*⟩ **c :** a motion-picture or television episode or sequence
2 a : a stage setting **b :** a real or imaginary prospect suggesting a stage setting ⟨a sylvan *scene*⟩
3 : the place of an occurrence or action **:** LOCALE ⟨*scene* of the crime⟩
4 : an exhibition of anger or indecorous behavior ⟨make a *scene*⟩
5 a : sphere of activity ⟨the drug *scene*⟩ **b :** SITUATION ⟨a bad *scene*⟩
— **behind the scenes 1 :** out of public view; *also* **:** in secret **2 :** in a position to see the hidden workings ⟨taken *behind the scenes* and told just how in fact the actual government . . . has operated —William Clark⟩

scen·ery \'sē-nə-rē, 'sēn-rē\ *noun, plural* **-er·ies** (1770)
1 : the painted scenes or hangings and accessories used on a theater stage
2 : a picturesque view or landscape

scene·shift·er \'sēn-,shif-tər\ *noun* (1752)
: a worker who moves the scenes in a theater

scene–steal·er \-,stē-lər\ *noun* (1949)
: an actor who attracts attention when another is intended to be the center of attention

sce·nic \'sē-nik *also* 'se-\ *also* **sce·ni·cal** \-ni-kəl\ *adjective* (1623)
1 : of or relating to the stage, a stage setting, or stage representation
2 : of or relating to natural scenery ⟨a *scenic* view⟩
3 : representing graphically an action, event, or episode ⟨a *scenic* bas-relief⟩
— **sce·ni·cal·ly** \-ni-k(ə-)lē\ *adverb*

scenic railway *noun* (1894)
chiefly British **:** a miniature railway (as in an amusement park) with artificial scenery along the way

☆ SYNONYMS
Scatter, disperse, dissipate, dispel mean to cause to separate or break up. SCATTER implies a force that drives parts or units irregularly in many directions ⟨the bowling ball *scattered* the pins⟩. DISPERSE implies a wider separation and a complete breaking up of a mass or group ⟨police *dispersed* the crowd⟩. DISSIPATE stresses complete disintegration or dissolution and final disappearance ⟨the fog was *dissipated* by the morning sun⟩. DISPEL stresses a driving away or getting rid of as if by scattering ⟨an authoritative statement that *dispelled* all doubt⟩.

sce·nog·ra·phy \sē-ˈnä-grə-fē\ *noun* [Greek *skēnographia* painting of scenery, from *skēnē* + *-graphia* -graphy] (1645)
: the art of perspective representation especially as applied to the painting of stage scenery (as by the ancient Greeks)
— sce·nog·ra·pher \-grə-fər\ *noun*
— sce·no·graph·ic \ˌsē-nə-ˈgra-fik\ *adjective*

[1]scent \ˈsent\ *noun* [Middle English *sent,* from *senten*] (14th century)
1 : effluvia from a substance that affect the sense of smell: as **a :** an odor left by an animal on a surface passed over **b :** a characteristic or particular odor; *especially* **:** one that is agreeable
2 a : power of smelling **:** sense of smell ⟨a keen *scent*⟩ **b :** power of detection **:** NOSE ⟨a *scent* for heresy⟩
3 : a course of pursuit or discovery ⟨throw one off the *scent*⟩
4 : INKLING, INTIMATION ⟨a *scent* of trouble⟩
5 : PERFUME 2
6 : bits of paper dropped in the game of hare and hounds
7 : a mixture prepared for use as a lure in hunting or fishing
synonym see FRAGRANCE, SMELL
— scent·less \ˈsent-ləs\ *adjective*

[2]scent *verb* [Middle English *senten,* from Middle French *sentir* to feel, smell, from Latin *sentire* to perceive, feel — more at SENSE] (15th century)
transitive verb
1 a : to perceive by the olfactory organs **:** SMELL **b :** to get or have an inkling of ⟨*scent* trouble⟩
2 : to imbue or fill with odor ⟨*scented* the air with perfume⟩
intransitive verb
1 : to yield an odor of some specified kind ⟨this *scents* of sulfur⟩; *also* **:** to bear indication or suggestions
2 : to use the nose in seeking or tracking prey

scent·ed *adjective* (1666)
: having scent: as **a :** having the sense of smell **b :** having a perfumed smell **c :** having or exhaling an odor

[1]scep·ter \ˈsep-tər\ *noun* [Middle English *sceptre,* from Middle French *ceptre,* from Latin *sceptrum,* from Greek *skēptron* staff, scepter, from *skēptesthai* to prop oneself — more at SHAFT] (14th century)
1 : a staff or baton borne by a sovereign as an emblem of authority
2 : royal or imperial authority **:** SOVEREIGNTY

[2]scepter *transitive verb* **scep·tered; scep·ter·ing** \-t(ə-)riŋ\ (1526)
: to invest with the scepter in token of royal authority

scep·tered \ˈsep-tərd\ *adjective* (1513)
1 : invested with a scepter or sovereign authority
2 : of or relating to a sovereign or to royalty

scep·tic, scep·ti·cal, scep·ti·cism *variant of* SKEPTIC, SKEPTICAL, SKEPTICISM

sceptre *British variant of* SCEPTER

scha·den·freu·de \ˈshä-d^{ə}n-ˌfrȯi-də\ *noun, often capitalized* [German, from *Schaden* damage + *Freude* joy] (1895)
: enjoyment obtained from the troubles of others

[1]sched·ule \ˈske-(ˌ)jü(ə)l, -jəl, *Canadian also* ˈshe-, *British usually* ˈshe-(ˌ)dyü(ə)l\ *noun* [alteration of Middle English *cedule,* from Middle French, slip of paper, note, from Late Latin *schedula* slip of paper, diminutive of (assumed) Latin *scheda* strip of papyrus, probably back-formation from Latin *schedium* impromptu speech, from Greek *schedion,* from neuter of *schedios* casual; akin to Greek *schedon* near at hand, *echein* to seize, have] (1560)
1 a *obsolete* **:** a written document **b :** a statement of supplementary details appended to a legal or legislative document
2 : a written or printed list, catalog, or inventory; *also* **:** TIMETABLE 1
3 : PROGRAM; *especially* **:** a procedural plan that indicates the time and sequence of each operation ⟨finished on *schedule*⟩
4 : a body of items to be dealt with **:** AGENDA
5 : a governmental list of drugs all subject to the same legal restrictions and controls □

[2]schedule *transitive verb* **sched·uled; sched·ul·ing** (1862)
1 a : to place in a schedule **b :** to make a schedule of
2 : to appoint, assign, or designate for a fixed time
— sched·ul·er *noun*

schee·lite \ˈshā-ˌlīt\ *noun* [German *Scheelit,* from Karl W. *Scheele* (died 1786) Swedish chemist] (circa 1837)
: a mineral consisting of the tungstate of calcium that is a source of tungsten and its compounds

Sche·her·a·zade \shə-ˌher-ə-ˈzäd; -ˈzä-də, -dē\ *noun* [German *Scheherezade,* from Persian *Shīrazād*]
: the fictional wife of an oriental king and the narrator of the tales in the *Arabian Nights' Entertainments*

sche·ma \ˈskē-mə\ *noun, plural* **sche·ma·ta** \-mə-tə\ *also* **schemas** [Greek *schēmat-, schēma*] (circa 1890)
1 : a diagrammatic presentation; *broadly* **:** a structured framework or plan **:** OUTLINE
2 : a mental codification of experience that includes a particular organized way of perceiving cognitively and responding to a complex situation or set of stimuli

[1]sche·mat·ic \ski-ˈma-tik\ *adjective* [New Latin *schematicus,* from Greek *schēmat-, schēma*] (1701)
: of or relating to a scheme or schema
— sche·mat·i·cal·ly \-ti-k(ə-)lē\ *adverb*

[2]schematic *noun* (1929)
: a schematic drawing or diagram

sche·ma·tism \ˈskē-mə-ˌti-zəm\ *noun* (1660)
: the disposition of constituents in a pattern or according to a scheme **:** DESIGN; *also* **:** a particular systematic disposition of parts

sche·ma·tize \ˈskē-mə-ˌtīz\ *transitive verb* **-tized; -tiz·ing** [Greek *schēmatizein,* from *schēmat-, schēma*] (1828)
1 : to form or to form into a scheme or systematic arrangement
2 : to express or depict schematically
— sche·ma·ti·za·tion \ˌskē-mə-tə-ˈzā-shən\ *noun*

[1]scheme \ˈskēm\ *noun* [Latin *schemat-, schema* arrangement, figure, from Greek *schēmat-, schēma,* from *echein* to have, hold, be in (such) a condition; akin to Old English *sige* victory, Sanskrit *sahate* he prevails] (1610)
1 a *archaic* (1) **:** a mathematical or astronomical diagram (2) **:** a representation of the astrological aspects of the planets at a particular time **b :** a graphic sketch or outline
2 : a concise statement or table **:** EPITOME
3 : a plan or program of action; *especially* **:** a crafty or secret one
4 : a systematic or organized framework **:** DESIGN
synonym see PLAN

[2]scheme *verb* **schemed; schem·ing** (1767)
transitive verb
: to form a scheme for
intransitive verb
: to form plans; *also* **:** PLOT, INTRIGUE
— schem·er *noun*

schem·ing *adjective* (circa 1828)
: given to forming schemes; *especially* **:** shrewdly devious and intriguing

[1]scher·zan·do \skert-ˈsän-(ˌ)dō\ *adverb or adjective* [Italian, from verbal of *scherzare* to joke, of Germanic origin; akin to Middle High German *scherzen* to leap for joy, joke; perhaps akin to Greek *skairein* to gambol] (circa 1811)
: in sportive manner **:** PLAYFULLY — used as a direction in music indicating style and tempo ⟨allegretto *scherzando*⟩

[2]scherzando *noun, plural* **-dos** (circa 1876)
: a passage or movement in scherzando style

scher·zo \ˈskert-(ˌ)sō\ *noun, plural* **scher·zos** *or* **scher·zi** \-(ˌ)sē\ [Italian, literally, joke, from *scherzare*] (1852)
: a sprightly humorous instrumental musical composition or movement commonly in quick triple time

Schick test \ˈshik-\ *noun* [Béla *Schick*] (1916)
: a serological test for susceptibility to diphtheria by cutaneous injection of a diluted diphtheria toxin that causes an area of reddening and induration in susceptible individuals

Schiff's reagent \ˈshifs-\ *noun* [Hugo *Schiff* (died 1915) German chemist] (1897)
: a solution of fuchsine decolorized by treatment with sulfur dioxide that gives a useful test for aldehydes because they restore the dye's color — called also *Schiff reagent* \ˈshif-\; compare FEULGEN REACTION

schil·ler \ˈshi-lər\ *noun* [German] (1885)
: a bronzy iridescent luster (as of a mineral)

schil·ling \ˈshi-liŋ\ *noun* [German, from Old High German *skilling,* a gold coin — more at SHILLING] (1753)
— see MONEY table

schip·per·ke \ˈski-pər-kē, ˈshi-, -kə; -pərk\ *noun* [Flemish, diminutive of *schipper* skipper; from its use as a watchdog on boats — more at SKIPPER] (1887)
: any of a Belgian breed of small stocky black dogs with foxy head and heavy coat

schipperke

schism \ˈsi-zəm, ˈski- *also* ˈshi-; *among clergy usually* ˈsi-\ *noun* [Middle English *scisme,* from Middle French *cisme,* from Late Latin *schismat-, schisma,* from Greek, cleft, division, from *schizein* to split — more at SHED] (14th century)
1 : DIVISION, SEPARATION; *also* **:** DISCORD, DISHARMONY

□ USAGE
schedule British pronunciation maintains an aura of prestige in American ears, and thus some critics insist that the American habit of saying *sch* like \sk\ and not \sh\ in *schedule* is incorrect. In Middle English the word began with a simple \s\, spelled *c.* Both the \sk\ and \sh\ sounds in *schedule* are innovations. In eighteenth century Britain the *sch* was pronounced either as \s\ or \sk\; the \sh\ variant is a later development modeled along the lines of French pronunciation. Noah Webster advised Americans to pronounced the *sch* as \sk\ in his 1828 dictionary. The \sk\ variant is closer in spirit to the Latin and Greek ancestors of *schedule,* and though it does not have the allure of the distinctive but quirky Received Pronunciation, it is universal in educated American speech.

2 a : formal division in or separation from a church or religious body **b :** the offense of promoting schism ■

¹schis·mat·ic \siz-ˈma-tik, ski-\ *noun* (14th century)
: one who creates or takes part in schism

²schismatic *also* **schis·mat·i·cal** \-ti-kəl\ *adjective* (15th century)
: of, relating to, or guilty of schism
— **schis·mat·i·cal·ly** \-ti-k(ə-)lē\ *adverb*

schis·ma·tize \ˈsiz-mə-ˌtīz, ˈskiz-\ *verb* **-tized; -tiz·ing** (1601)
intransitive verb
: to take part in schism; *especially* **:** to make a breach of union (as in the church)
transitive verb
: to induce into schism

schist \ˈshist\ *noun* [French *schiste,* from Latin *schistos (lapis),* literally, fissile stone, from Greek *schistos* that may be split, from *schizein*] (circa 1782)
: a metamorphic crystalline rock that has a closely foliated structure and can be split along approximately parallel planes

schis·tose \ˈshis-ˌtōs\ *adjective* (1794)
: of or relating to schist **:** having the character or structure of a schist
— **schis·tos·i·ty** \shis-ˈtä-sə-tē\ *noun*

schis·to·some \ˈshis-tə-ˌsōm\ *noun* [New Latin *Schistosoma,* from Greek *schistos* + *sōma* body] (1905)
: any of a genus (*Schistosoma*) of elongated trematode worms with the sexes separate that parasitize the blood vessels of birds and mammals and cause a destructive human schistosomiasis; *broadly* **:** a worm of the family (Schistosomatidae) that includes this genus
— **schis·to·som·al** \ˌshis-tə-ˈsō-məl\ *adjective*
— **schistosome** *adjective*

schis·to·so·mi·a·sis \ˌshis-tə-sō-ˈmī-ə-səs\ *noun, plural* **-a·ses** \-ˌsēz\ [New Latin, from *Schistosoma*] (1906)
: infestation with or disease caused by schistosomes; *specifically* **:** a severe endemic disease of humans in much of Asia, Africa, and South America marked especially by blood loss and tissue damage

schiz- *or* **schizo-** *combining form* [New Latin, from Greek *schizo-,* from *schizein* to split — more at SHED]
1 : split **:** cleft ⟨*schizo*carp⟩
2 : characterized by or involving cleavage ⟨*schizo*gony⟩
3 : schizophrenia ⟨*schizo*id⟩

schizo \ˈskit-(ˌ)sō\ *noun, plural* **schiz·os** (1945)
: a schizophrenic individual

schizo·carp \ˈski-zə-ˌkärp, ˈskit-sə-\ *noun* [International Scientific Vocabulary] (1870)
: a dry compound fruit that splits at maturity into several indehiscent one-seeded carpels

schi·zog·o·ny \ski-ˈzä-gə-nē, skit-ˈsä-\ *noun* [New Latin *schizogonia,* from *schiz-* + Latin *-gonia* -gony] (1887)
: asexual reproduction by multiple segmentation characteristic of sporozoans (as the malaria parasite)
— **schizo·gon·ic** \ˌski-zə-ˈgä-nik, ˌskit-sə-\ *or* **schi·zog·o·nous** \ski-ˈzä-gə-nəs, skit-ˈsä-gə-\ *adjective*

schiz·oid \ˈskit-ˌsȯid\ *adjective* [International Scientific Vocabulary] (1924)
: characterized by, resulting from, tending toward, or suggestive of schizophrenia
— **schizoid** *noun*

schiz·ont \ˈski-ˌzänt, ˈskit-ˌsänt\ *noun* [International Scientific Vocabulary] (1900)
: a multinucleate sporozoan that reproduces by schizogony

schizo·phrene \ˈskit-sə-ˌfrēn\ *noun* [International Scientific Vocabulary, probably back-formation from New Latin *schizophrenia*] (1925)
: one affected with schizophrenia **:** SCHIZOPHRENIC

schizo·phre·nia \ˌskit-sə-ˈfrē-nē-ə\ *noun* [New Latin] (1912)
1 : a psychotic disorder characterized by loss of contact with the environment, by noticeable deterioration in the level of functioning in everyday life, and by disintegration of personality expressed as disorder of feeling, thought (as in hallucinations and delusions), and conduct — called also *dementia praecox*
2 : contradictory or antagonistic qualities or attitudes ⟨both parties . . . have exhibited *schizophrenia* over the desired outcome —Elizabeth Drew⟩
— **schizo·phren·ic** \-ˈfre-nik\ *adjective or noun*
— **schizo·phren·i·cal·ly** \-ni-k(ə-)lē\ *adverb*

schizy *or* **schiz·zy** \ˈskit-sē\ *adjective* [by shortening & alteration] (1927)
: SCHIZOID

schle·miel \shlə-ˈmē(ə)l\ *noun* [Yiddish *shlemil*] (1892)
: an unlucky bungler **:** CHUMP

schlepp *or* **schlep** \ˈshlep\ *verb* [Yiddish *shlepn,* from Middle High German *sleppen,* from Middle Low German *slēpen*] (1922)
transitive verb
: DRAG, HAUL
intransitive verb
: to proceed or move slowly, tediously, or awkwardly

schlie·ren \ˈshlir-ən\ *noun plural* [German] (1898)
1 : small masses or streaks in an igneous rock that differ in composition from the main body
2 : regions of varying refraction in a transparent medium often caused by pressure or temperature differences and detectable especially by photographing the passage of a beam of light
— **schlie·ric** \ˈshlir-ik\ *adjective*

schlock \ˈshläk\ *also* **schlocky** \ˈshlä-kē\ *adjective* [perhaps from Yiddish *shlak* evil, nuisance, literally, blow, from Middle High German *slag, slac,* from Old High German *slag,* from *slahan* to strike — more at SLAY] (1915)
: of low quality or value
— **schlock** *noun*

schm- *or* **shm-** \shm\ *prefix* [Yiddish *shm-*]
— used to form a rhyming term of derision by replacing the initial consonant or consonant cluster of a word or by preceding the initial vowel ⟨fancy, *schm*ancy, I prefer plain⟩ ⟨Godfather-*shm*odfather—enough already —Judith Crist⟩

schmaltz *also* **schmalz** \ˈshmȯlts, ˈshmälts\ *noun* [Yiddish *shmalts,* literally, rendered fat, from Middle High German *smalz;* akin to Old High German *smelzan* to melt — more at SMELT] (1935)
1 : sentimental or florid music or art
2 : SENTIMENTALITY
— **schmaltzy** \ˈshmȯlt-sē, ˈshmält-\ *adjective*

schmear *or* **schmeer** \ˈshmir\ *noun* [Yiddish *shmir* smear; akin to Old High German *smero* grease — more at SMEAR] (1965)
: an aggregate of related things ⟨the whole *schmear*⟩

Schmidt camera \ˈshmit-\ *noun* [B. *Schmidt* (died 1935) German optical scientist] (1936)
: a photographic telescope with specialized optics that correct for spherical aberration and coma — called also *Schmidt telescope*

schmo *or* **schmoe** \ˈshmō\ *noun, plural* **schmoes** [origin unknown] (1947)
slang **:** JERK 4

schmooze *or* **shmooze** \ˈshmüz\ *intransitive verb* [Yiddish *shmuesn,* from *schmues* talk, from Hebrew *shĕmu'ōth* news, rumor] (1897)
: to converse informally **:** CHAT

schmuck \ˈshmək\ *noun* [Yiddish *shmok,* literally, penis] (1892)
slang **:** JERK 4

schnapps \ˈshnaps\ *noun, plural* **schnapps** [German *Schnaps,* literally, dram of liquor, from Low German *snaps* dram, mouthful, from *snappen* to snap] (1818)
: any of various liquors of high alcoholic content; *especially* **:** strong Holland gin

schnau·zer \ˈshnau̇t-sər, ˈshnau̇-zər\ *noun* [German, from *Schnauze* snout — more at SNOUT] (1923)
: a dog of any of three breeds that originated in Germany and are characterized by a wiry coat, long head, small ears, heavy eyebrows, and long hair on the muzzle: **a :** STANDARD SCHNAUZER **b :** GIANT SCHNAUZER **c :** MINIATURE SCHNAUZER

schnit·zel \ˈshnit-səl\ *noun* [German, literally, shaving, chip, diminutive of *Schnitz* slice, from Middle High German *snitz;* akin to Old High German *snīdan* to cut, Old English *snīthan,* and perhaps to Czech *snět* bough] (1854)
: a seasoned and garnished veal cutlet

schnook \ˈshnu̇k\ *noun* [origin unknown] (1940)
slang **:** a stupid or unimportant person **:** DOLT

schnor·kel \ˈshnȯr-kəl\ *variant of* SNORKEL

schnor·rer \ˈshnȯr-ər, ˈshnōr-\ *noun* [Yiddish *shnorer*] (1892)
: BEGGAR; *especially* **:** one who wheedles others into supplying his wants

schnoz·zle \ˈshnä-zəl\ *noun* [probably modification of Yiddish *shnoitsl,* diminutive of *shnoits* snout; akin to German *Schnauze* snout, muzzle — more at SNOUT] (1937)
slang **:** NOSE

scho·la can·to·rum \ˌskō-lə-kan-ˈtōr-əm, -ˈtȯr-\ *noun, plural* **scho·lae can·torum** \-ˌlē-, -ˌlā-, -ˌlī-\ [Medieval Latin, school of singers] (1782)
1 : a singing school especially for church choristers; *specifically* **:** the choir or choir school of a monastery or of a cathedral
2 : an enclosure designed for a choir and located in the center of the nave in early church buildings

schol·ar \ˈskä-lər\ *noun* [Middle English *scoler,* from Old English *scolere* & Old French *escoler,* from Medieval Latin *scholaris,* from Late Latin, of a school, from Latin *schola* school] (before 12th century)
1 : one who attends a school or studies under a teacher **:** PUPIL
2 a : one who has done advanced study in a special field **b :** a learned person
3 : a holder of a scholarship

schol·ar·ly \-lē\ *adjective* (1638)
: of, characteristic of, or suitable to learned persons **:** LEARNED, ACADEMIC

□ USAGE

schism With the rebirth of classical learning in the Renaissance came a "silent" revolution in English spelling, whereby words formerly spelled as they were pronounced (like *onur, douten, indite,* and *receite*) were given extra letters to match their ancestral forms in Latin and Greek (yielding *honor, doubt, indict,* and *receipt*). The modern spelling of Middle English *scisme* is a product of this reform, as is the modern confusion over its pronunciation. The \ˈski-zəm\ variant is a spelling pronunciation, the \ˈshi-zəm\ variant is another spelling pronunciation following the conventions of modern French or German, and the \ˈsi-zəm\ variant retains the original Middle English and Middle French sound. Some insist the original \ˈsi-zəm\ alone is correct, but these critics are not similarly exercised by the modern spelling pronunciations of *hermit* and *hostage,* which like *schism* originally had no *h* in spelling or pronunciation. Many educated speakers today use the fully acceptable spelling pronunciation \ˈski-zəm\.

schol·ar·ship \-ˌship\ *noun* (circa 1536)
1 **:** a grant-in-aid to a student (as by a college or foundation)
2 **:** the character, qualities, activity, or attainments of a scholar **:** LEARNING
3 **:** a fund of knowledge and learning ⟨drawing on the *scholarship* of the ancients⟩
synonym see KNOWLEDGE

Scholarship level *noun* (1947)
: S LEVEL

¹scho·las·tic \skə-ˈlas-tik\ *adjective* [Medieval Latin & Latin; Medieval Latin *scholasticus* of the schoolmen, from Latin, of a school, from Greek *scholastikos,* from *scholazein* to keep a school, from *scholē* school] (1596)
1 a *often capitalized* **:** of or relating to Scholasticism ⟨*scholastic* theology⟩ ⟨*scholastic* philosophy⟩ **b :** suggestive or characteristic of a scholastic especially in subtlety or aridity **:** PEDANTIC ⟨dull *scholastic* reports⟩
2 **:** of or relating to schools or scholars; *especially* **:** of or relating to high school or secondary school
— scho·las·ti·cal·ly \-ti-k(ə-)lē\ *adverb*

²scholastic *noun* (1644)
1 a *capitalized* **:** a Scholastic philosopher **b :** PEDANT, FORMALIST
2 [New Latin *scholasticus,* from Latin *scholasticus,* adjective] **:** a student in a scholasticate
3 **:** one who adopts academic or traditional methods in art

scho·las·ti·cate \skə-ˈlas-tə-ˌkāt, -ti-kət\ *noun* [New Latin *scholasticatus,* from *scholasticus* student in a scholasticate] (1875)
: a college-level school of general study for those preparing for membership in a Roman Catholic religious order

scho·las·ti·cism \skə-ˈlas-tə-ˌsi-zəm\ *noun* (circa 1782)
1 *capitalized* **a :** a philosophical movement dominant in western Christian civilization from the 9th until the 17th century and combining religious dogma with the mystical and intuitional tradition of patristic philosophy especially of Saint Augustine and later with Aristotelianism **b :** NEO SCHOLASTICISM
2 a **:** close adherence to the traditional teachings or methods of a school or sect **b :** pedantic adherence to scholarly methods

scho·li·ast \ˈskō-lē-ˌast, -lē-əst\ *noun* [Middle Greek *scholiastēs,* from *scholiazein* to write scholia on, from Greek *scholion*] (1583)
: a maker of scholia **:** COMMENTATOR, ANNOTATOR
— scho·li·as·tic \ˌskō-lē-ˈas-tik\ *adjective*

scho·li·um \ˈskō-lē-əm\ *noun, plural* **-lia** \-lē-ə\ *or* **-li·ums** [New Latin, from Greek *scholion* comment, scholium, from diminutive of *scholē* lecture] (1535)
1 **:** a marginal annotation or comment (as on the text of a classic by an early grammarian)
2 **:** a remark or observation subjoined but not essential to a demonstration or a train of reasoning

¹school \ˈskül\ *noun* [Middle English *scole,* from Old English *scōl,* from Latin *schola,* from Greek *scholē* leisure, discussion, lecture, school; perhaps akin to Greek *echein* to hold — more at SCHEME] (before 12th century)
1 **:** an organization that provides instruction: as **a :** an institution for the teaching of children **b :** COLLEGE, UNIVERSITY **c** (1) **:** a group of scholars and teachers pursuing knowledge together that with similar groups constituted a medieval university (2) **:** one of the four faculties of a medieval university (3) **:** an institution for specialized higher education often associated with a university ⟨the *school* of engineering⟩ **d :** an establishment offering specialized instruction ⟨a secretarial *school*⟩ ⟨driving *schools*⟩
2 a (1) **:** the process of teaching or learning especially at a school (2) **:** attendance at a school (3) **:** a session of a school **b :** a school building **c :** the students attending a school; *also* **:** its teachers and students
3 **:** a source of knowledge ⟨experience was his *school*⟩
4 a **:** persons who hold a common doctrine or follow the same teacher (as in philosophy, theology, or medicine) ⟨the Aristotelian *school*⟩ **b :** a group of artists under a common influence **c :** persons of similar opinions or behavior ⟨other *schools* of thought⟩
5 **:** the regulations governing military drill of individuals or units; *also* **:** the exercises carried out ⟨the *school* of the soldier⟩ ◆

²school *transitive verb* (15th century)
1 a **:** to teach or drill in a specific knowledge or skill ⟨well *schooled* in languages⟩ **b :** to discipline or habituate to something ⟨*school* oneself in patience⟩
2 **:** to educate in an institution of learning
synonym see TEACH

³school *noun* [Middle English *scole,* from Middle Dutch *schole;* akin to Old English *scolu* multitude and probably to Old English *scylian* to separate — more at SKILL] (15th century)
: a large number of fish or aquatic animals of one kind swimming together

⁴school *intransitive verb* (1597)
: to swim or feed in a school ⟨bluefish are *schooling*⟩

school–age *adjective* (1741)
: old enough to go to school ⟨*school-age* children⟩

school·bag \ˈskül-ˌbag\ *noun* (1895)
: a bag for carrying schoolbooks and school supplies

school board *noun* (1836)
: a board in charge of local public schools

school·book \-ˌbu̇k\ *noun* (1745)
: a school textbook

¹school·boy \-ˌbȯi\ *noun* (1588)
: a boy attending school

²schoolboy *adjective* (1687)
1 **:** of, relating to, or characteristic of a schoolboy ⟨*schoolboy* pranks⟩
2 **:** of, relating to, or being a sport for high school or prep school boys; *also* **:** being one who participates in such sports ⟨a former *schoolboy* and collegiate goalie —D. J. Barr⟩

school·boy·ish \-ˌbȯi-ish\ *adjective* (1831)
: SCHOOLBOY 1

school bus *noun* (1908)
: a vehicle used for transporting children to or from school or on activities connected with school

school·child \ˈskül-ˌchīld\ *noun* (1840)
: a child attending school

school committee *noun* (1787)
: SCHOOL BOARD

school district *noun* (1809)
: a unit for administration of a public-school system often comprising several towns within a state

school·fel·low \ˈskül-ˌfe-(ˌ)lō\ *noun* (15th century)
: SCHOOLMATE

school·girl \-ˌgər(-ə)l\ *noun* (1777)
: a girl attending school

school·house \-ˌhau̇s\ *noun* (14th century)
: a building used as a school and especially as an elementary school

school·ing *noun* (15th century)
1 a **:** instruction in school **:** EDUCATION **b :** training, guidance, or discipline derived from experience
2 *archaic* **:** REPROOF
3 **:** the cost of instruction and maintenance at school
4 **:** the training of a horse to service; *especially* **:** the teaching and exercising of horse and rider in the formal techniques of equitation

school·kid \ˈskül-ˌkid\ *noun* (1938)
: a child or teenager attending school

school–leav·er \ˈskül-ˌlē-vər\ *noun* (1925)
British **:** one who has left school usually after completing a course of study

school·man \ˈskül-mən, -ˌman\ *noun* (circa 1540)
1 a **:** one skilled in academic disputation **b** *capitalized* **:** SCHOLASTIC 1a
2 **:** EDUCATOR 1, 2b

school·marm \-ˌmä(r)m\ *or* **school·ma'am** \-ˌmäm, -ˌmam\ *noun* [*school* + *marm,* alteration of *ma'am*] (1831)
1 **:** a woman schoolteacher especially in a rural or small-town school
2 **:** a person who exhibits characteristics attributed to schoolteachers (as strict adherence to arbitrary rules)
— school·marm·ish \-ˌmä(r)-mish\ *adjective*

school·mas·ter \-ˌmas-tər\ *noun* (13th century)
1 **:** a man who teaches school
2 **:** one that disciplines or directs
3 **:** a reddish brown edible snapper (*Lutjanus apodus*) of the tropical Atlantic and the Gulf of Mexico
— school·mas·ter·ish \-tə-rish\ *adjective*
— school·mas·ter·ly *adjective*

school·mate \-ˌmāt\ *noun* (1563)
: a companion at school

school·mis·tress \-ˌmis-trəs\ *noun* (15th century)
: a woman who teaches school

school·room \-ˌrüm, -ˌru̇m\ *noun* (1773)
: CLASSROOM

school·teach·er \-ˌtē-chər\ *noun* (circa 1847)
: one who teaches school

school·time \-ˌtīm\ *noun* (1740)
1 **:** the time for beginning a session of school or during which school is held
2 **:** the period of life spent in school or in study

school·work \-ˌwərk\ *noun* (1857)
: lessons done in class or assigned to be done at home

◇ WORD HISTORY

school The work we associate with elementary education today may make it hard to believe that the Greek ancestor of the word *school* meant "leisure," but such is the case. Ancient Greek *scholē* "rest, leisure" came to be applied to the learned discussions and philosophical disputations in which the cream of Greek society spent their free time (of which they had much). Its use was then extended to groups who listened to the lectures of a particular philosopher, and finally, in Hellenistic and later Greek, to the set of beliefs held by a particular philosophical group. When Latin *schola* was borrowed from Greek, the emphasis fell more on the place where a particular teacher expounded his views. The Romans later generalized *schola* to "place of instruction" and applied it to an establishment where a scholar tutored young students in the basics of writing and rhetoric. (The earlier Latin word for "place of instruction" was usually *ludus,* which had originally meant, ironically, "play, sport.") *Schola* also was applied in Latin to assembly places for fraternities of priests or other groups; a distant reflection of this sense, filtered through centuries of Jewish history, is the use of *shul* in Yiddish to mean "synagogue."

schoo·ner \'skü-nər\ *noun* [origin unknown] (1716)
1 : a typically 2-masted fore-and-aft rigged vessel with a foremast and a mainmast stepped nearly amidships
2 : a larger-than-usual drinking glass (as for beer)
3 : PRAIRIE SCHOONER

schooner 1

schooner rig *noun* (1866)
: FORE-AND-AFT RIG
— schoo·ner–rigged \'skü-nə(r)-'rigd\ *adjective*

schorl \'shȯr(ə)l\ *noun* [German *Schörl*] (1779)
: TOURMALINE; *especially* **:** tourmaline of the black variety

schot·tische \'shä-tish, shä-'tēsh\ *noun* [German, from *schottisch* Scottish, from *Schotte* Scotsman; akin to Old English *Scottas* Scots] (1849)
1 : a round dance resembling a slow polka
2 : music for the schottische

schrod *variant of* SCROD

Schrö·ding·er equation \'shrā-diŋ-ər, 'shrœ̄-, 'shrə(r)-\ *noun* [Erwin *Schrödinger*] (1936)
: an equation that describes the wave nature of elementary particles and is fundamental to the description of the properties of all matter

schtick *variant of* SHTICK

schuss \'shu̇s, 'shüs\ *intransitive verb* [*schuss,* noun, from German *Schuss,* literally, shot, from Old High German *scuz* — more at SHOT] (1940)
: to ski directly down a slope at high speed
— schuss *noun*

schuss·boom·er \-,bü-mər\ *noun* (1953)
: a skier who schusses

schwa \'shwä\ *noun* [German, from Hebrew *schĕwā'*] (1895)
1 : an unstressed mid-central vowel (as the usual sound of the first and last vowels of the English word *America*)
2 : the symbol ə used for the schwa sound and less widely for a similarly articulated stressed vowel (as in *cut*)

Schwann cell \'shwän-\ *noun* [Theodor *Schwann* (died 1882) German naturalist] (circa 1909)
: the myelin-secreting cell surrounding a myelinated nerve fiber between two nodes of Ranvier

schwar·me·rei \,shver-mə-'rī\ *noun* [German *Schwärmerei,* from *schwärmen* to be enthusiastic, literally, to swarm] (1845)
: excessive or unwholesome sentiment

sci·at·ic \sī-'a-tik\ *adjective* [Middle French *sciatique,* from Late Latin *sciaticus,* alteration of Latin *ischiadicus* of sciatica, from Greek *ischiadikos,* from *ischiad-, ischias* sciatica, from *ischion* ischium] (1586)
1 : of, relating to, or situated near the hip
2 : of, relating to, or caused by sciatica ⟨*sciatic* pains⟩

sci·at·i·ca \sī-'a-ti-kə\ *noun* [Middle English, from Medieval Latin, from Late Latin, feminine of *sciaticus*] (14th century)
: pain along the course of a sciatic nerve especially in the back of the thigh; *broadly* **:** pain in the lower back, buttocks, hips, or adjacent parts

sciatic nerve *noun* (1741)
: either of the pair of largest nerves in the body that arise one on each side from the nerve plexus supplying the posterior limb and pelvic region and that pass out of the pelvis and down the back of the thigh

sci·ence \'sī-ən(t)s\ *noun* [Middle English, from Middle French, from Latin *scientia,* from *scient-, sciens* having knowledge, from present participle of *scire* to know; probably akin to Sanskrit *chyati* he cuts off, Latin *scindere* to split — more at SHED] (14th century)
1 : the state of knowing **:** knowledge as distinguished from ignorance or misunderstanding
2 a : a department of systematized knowledge as an object of study ⟨the *science* of theology⟩ **b :** something (as a sport or technique) that may be studied or learned like systematized knowledge ⟨have it down to a *science*⟩
3 a : knowledge or a system of knowledge covering general truths or the operation of general laws especially as obtained and tested through scientific method **b :** such knowledge or such a system of knowledge concerned with the physical world and its phenomena **:** NATURAL SCIENCE
4 : a system or method reconciling practical ends with scientific laws ⟨culinary *science*⟩
5 *capitalized* **:** CHRISTIAN SCIENCE

science fiction *noun* (1851)
: fiction dealing principally with the impact of actual or imagined science on society or individuals or having a scientific factor as an essential orienting component

sci·en·tial \sī-'en(t)-shəl\ *adjective* (15th century)
1 : relating to or producing knowledge or science
2 : having efficient knowledge **:** CAPABLE

sci·en·tif·ic \,sī-ən-'ti-fik\ *adjective* [Medieval Latin *scientificus* producing knowledge, from Latin *scient-, sciens* + *-i-* + *-ficus* -fic] (1589)
: of, relating to, or exhibiting the methods or principles of science
— sci·en·tif·i·cal·ly \-fi-k(ə-)lē\ *adverb*

scientific method *noun* (1854)
: principles and procedures for the systematic pursuit of knowledge involving the recognition and formulation of a problem, the collection of data through observation and experiment, and the formulation and testing of hypotheses

scientific notation *noun* (circa 1934)
: a widely used floating-point system in which numbers are expressed as products consisting of a number between 1 and 10 multiplied by an appropriate power of 10

sci·en·tism \'sī-ən-,ti-zəm\ *noun* (1877)
1 : methods and attitudes typical of or attributed to the natural scientist
2 : an exaggerated trust in the efficacy of the methods of natural science applied to all areas of investigation (as in philosophy, the social sciences, and the humanities)

sci·en·tist \'sī-ən-tist\ *noun* [Latin *scientia*] (1834)
1 : a person learned in science and especially natural science **:** a scientific investigator
2 *capitalized* **:** CHRISTIAN SCIENTIST

sci·en·tize \'sī-ən-,tīz\ *transitive verb* **-tized; -tiz·ing** (1917)
: to treat with a scientific approach ⟨the attempt to *scientize* reality, to name it and classify it —John Fowles⟩

sci–fi \'sī-'fī\ *adjective* [*sci*ence *fi*ction] (1955)
: of, relating to, or being science fiction ⟨a *sci-fi* film⟩
— sci–fi *noun*

sci·li·cet \'skē-li-,ket; 'sī-lə-,set, 'si-\ *adverb* [Middle English, from Latin, surely, to wit, from *scire* to know + *licet* it is permitted, from *licēre* to be permitted — more at LICENSE] (14th century)
: TO WIT, NAMELY

scil·la \'si-lə, 'ski-\ *noun* [New Latin, from Latin, squill — more at SQUILL] (1824)
: any of a genus (*Scilla*) of Old World bulbous herbs of the lily family with narrow basal leaves and pink, blue, or white racemose flowers

scim·i·tar \'si-mə-tər, -,tär\ *noun* [Italian *scimitarra*] (circa 1548)
: a saber having a curved blade with the edge on the convex side and used chiefly by Arabs and Turks

scin·tig·ra·phy \sin-'ti-grə-fē\ *noun* [*scinti*llation + *-graphy;* from the scintillation counter used to record radiation on the picture] (1958)
: a diagnostic technique in which a two-dimensional picture of a bodily radiation source is obtained by the use of radioisotopes
— scin·ti·graph·ic \,sin-tə-'gra-fik\ *adjective*

scin·til·la \sin-'ti-lə\ *noun* [Latin] (1692)
: SPARK, TRACE

scin·til·lant \'sin-t^{ə}l-ənt\ *adjective* (1737)
: that scintillates **:** SPARKLING
— scin·til·lant·ly *adverb*

scin·til·late \'sin-t^{ə}l-,āt\ *verb* **-lat·ed; -lat·ing** [Latin *scintillatus,* past participle of *scintillare* to sparkle, from *scintilla* spark] (circa 1623)
intransitive verb
1 : to emit sparks **:** SPARK
2 : to emit quick flashes as if throwing off sparks **:** SPARKLE
transitive verb
: to throw off as a spark or as sparkling flashes ⟨*scintillate* witticisms⟩
— scin·til·la·tor \-,ā-tər\ *noun*

scin·til·lat·ing *adjective* (1883)
: brilliantly lively, stimulating, or witty ⟨*scintillating* conversation⟩

scin·til·la·tion \,sin-t^{ə}l-'ā-shən\ *noun* (circa 1623)
1 : an act or instance of scintillating; *especially* **:** rapid changes in the brightness of a celestial body
2 a : a spark or flash emitted in scintillating **b :** a flash of light produced in a phosphor by an ionizing event
3 : a brilliant outburst (as of wit)
4 : a flash of the eye

scintillation counter *noun* (1948)
: a device for detecting and registering individual scintillations (as in radioactive emission)

scin·til·lom·e·ter \,sin-t^{ə}l-'ä-mə-tər\ *noun* [Latin *scintilla* + International Scientific Vocabulary *-o-* + *-meter*] (1877)
: SCINTILLATION COUNTER

sci·o·lism \'sī-ə-,li-zəm\ *noun* [Late Latin *sciolus* smatterer, from diminutive of Latin *scius* knowing, from *scire* to know — more at SCIENCE] (1816)
: a superficial show of learning
— sci·o·list \-list\ *noun*
— sci·o·lis·tic \,sī-ə-'lis-tik\ *adjective*

sci·on \'sī-ən\ *noun* [Middle English, from Middle French *cion,* of Germanic origin; akin to Old High German *chīnan* to sprout, split open, Old English *cīnan* to gape] (13th century)
1 : a detached living portion of a plant joined to a stock in grafting and usually supplying solely aerial parts to a graft
2 : DESCENDANT, CHILD

sci·re fa·cias \,sī-rē-'fā-sh(ē-)əs\ *noun* [Middle English, from Medieval Latin, you should cause to know] (15th century)
1 : a judicial writ founded on some matter of record and requiring the party proceeded against to show cause why the record should not be enforced, annulled, or vacated
2 : a legal proceeding instituted by a scire facias

sci·roc·co \shi-'rä-(,)kō, sə-\ *variant of* SIROCCO

scir·rhous \'sir-əs, 'skir-\ *adjective* [New Latin *scirrhosus,* from *scirrhus* scirrhous tumor, from Greek *skiros, skirrhos* overgrown land, hardened tumor] (1563)
: of, relating to, or being a hard slow-growing malignant tumor having a preponderance of fibrous tissue

scis·sile \'si-səl, -,sīl\ *adjective* [French, from Latin *scissilis,* from *scissus,* past participle of *scindere* to split — more at SHED] (1621)

: capable of being cut smoothly or split easily ⟨a *scissile* peptide bond⟩

scis·sion \ˈsi-zhən\ *noun* [Middle English *scissione,* from Middle French *scission,* from Late Latin *scission-, scissio,* from Latin *scindere*] (15th century)
1 : a division or split in a group or union **:** SCHISM
2 : an action or process of cutting, dividing, or splitting **:** the state of being cut, divided, or split

[1]scis·sor \ˈsi-zər\ *noun* [Middle English *sisoure,* from Middle French *cisoire,* from Late Latin *cisorium* cutting instrument, irregular from Latin *caedere* to cut] (15th century)
: SCISSORS

[2]scissor *transitive verb* **scis·sored; scis·sor·ing** \ˈsi-zə-riŋ, ˈsiz-riŋ\ (1612)
: to cut, cut up, or cut off with scissors or shears

scis·sors \ˈsi-zərz\ *noun plural but singular or plural in construction* (14th century)
1 : a cutting instrument having two blades whose cutting edges slide past each other
2 a : a gymnastic feat in which the leg movements suggest the opening and closing of scissors **b :** SCISSORS HOLD

scissors–and–paste *adjective* (1902)
: being a compilation rather than an effort of original and independent investigation

scissors hold *noun* (1909)
: a wrestling hold in which the legs are locked around the head or body of an opponent

scissors kick *noun* (circa 1930)
: a swimming kick used especially in sidestrokes in which the legs move like scissors

scis·sor–tailed flycatcher \ˈsi-zər-ˌtāld-\ *noun* (circa 1909)
: a flycatcher (*Muscivora forficata*) of the southern U.S., Mexico, and Central America that has a deeply forked tail — called also *scissortail*

sclaff \ˈsklaf\ *intransitive verb* [Scots, from *sclaff,* noun, literally, blow with the palm; probably of imitative origin] (1893)
: to scrape the ground instead of hitting the ball cleanly on a golf stroke
— sclaff *noun*
— sclaff·er *noun*

scler- *or* **sclero-** *combining form* [New Latin, from Greek *sklēr-, sklēro-,* from *sklēros* — more at SKELETON]
1 : hard ⟨*scler*ite⟩ ⟨*sclero*derma⟩
2 : hardness ⟨*sclero*meter⟩

sclera \ˈskler-ə\ *noun* [New Latin, from Greek *sklēros*] (1888)
: the dense fibrous opaque white outer coat enclosing the eyeball except the part covered by the cornea — see EYE illustration
— scler·al \-əl\ *adjective*

scler·e·id \ˈskler-ē-əd\ *noun* [International Scientific Vocabulary, irregular from Greek *sklēros*] (circa 1900)
: a variably shaped sclerenchymatous cell of a higher plant that is often nearly isodiametric

scle·ren·chy·ma \sklə-ˈreŋ-kə-mə\ *noun* [New Latin] (1875)
: a protective or supporting tissue in higher plants composed of cells with walls thickened and often lignified
— scler·en·chy·ma·tous \ˌskler-ən-ˈki-mə-təs, -ˈkī-\ *adjective*

scler·ite \ˈskler-ˌīt\ *noun* [International Scientific Vocabulary] (1861)
: a hard chitinous or calcareous plate, piece, or spicule (as of the arthropod integument)

sclero·der·ma \ˌskler-ə-ˈdər-mə\ *noun* [New Latin] (circa 1860)
: a usually slowly progressive disease marked by the deposition of fibrous connective tissue in the skin and often in internal organs

scle·rom·e·ter \sklə-ˈrä-mə-tər\ *noun* [International Scientific Vocabulary] (circa 1879)
: an instrument for determining the relative hardnesses of materials

sclero·pro·tein \ˌskler-ō-ˈprō-ˌtēn, -ˈprō-tē-ən\ *noun* [International Scientific Vocabulary] (1907)
: any of various fibrous proteins especially from connective and skeletal tissues

scle·ros·ing \sklə-ˈrō-siŋ, -ziŋ\ *adjective* [New Latin *scleros*is + English [1]*-ing*] (1885)
: causing or characterized by sclerosis ⟨*sclerosing* agents⟩

scle·ro·sis \sklə-ˈrō-səs\ *noun* [Middle English *sclirosis* tumor, from Medieval Latin, from Greek *sklērōsis* hardening, from *sklēroun* to harden, from *sklēros*] (1846)
: pathological hardening of tissue especially from overgrowth of fibrous tissue or increase in interstitial tissue; *also* **:** a disease characterized by sclerosis

[1]scle·rot·ic \sklə-ˈrä-tik\ *adjective* (1543)
1 : being or relating to the sclera
2 : of, relating to, or affected with sclerosis
3 : grown rigid or unresponsive especially with age ⟨*sclerotic* institutions⟩

[2]sclerotic *noun* [Medieval Latin *sclerotica,* from (assumed) Greek *sklērōtos,* verbal of Greek *sklēroun* to harden] (1690)
: SCLERA

sclerotic coat *noun* (1741)
: SCLERA

scler·o·tin \ˈskler-ə-tən, sklə-ˈrō-tᵊn\ *noun* [probably *scler-* + *-tin* (as in *chitin*)] (1940)
: an insoluble tanned protein permeating and stiffening the chitin of the cuticle of arthropods

scle·ro·tium \sklə-ˈrō-sh(ē-)əm\ *noun, plural* **-tia** \-sh(ē-)ə\ [New Latin, from (assumed) Greek *sklērōtos*] (1871)
: a compact mass of hardened mycelium stored with reserve food material that in some higher fungi becomes detached and remains dormant until a favorable opportunity for growth occurs
— scle·ro·tial \-ˈrō-shəl\ *adjective*

scler·o·tized \ˈskler-ə-ˌtīzd\ *adjective* [[1]*sclerotic* + *-ize* + [1]*-ed*] (circa 1890)
: hardened especially by the formation of sclerotin ⟨*sclerotized* insect cuticle⟩
— scler·o·ti·za·tion \ˌskler-e-tə-ˈzā-shən\ *noun*

[1]scoff \ˈskäf, ˈskȯf\ *noun* [Middle English *scof,* probably of Scandinavian origin; akin to obsolete Danish *skof* jest; akin to Old Frisian *skof* mockery] (14th century)
1 : an expression of scorn, derision, or contempt **:** GIBE
2 : an object of scorn, mockery, or derision

[2]scoff (14th century)
intransitive verb
: to show contempt by derisive acts or language
transitive verb
: to treat or address with derision **:** MOCK ☆
— scoff·er *noun*

[3]scoff *verb* [alteration of dialect *scaff* to eat greedily] (1846)
transitive verb
1 : to eat greedily
2 : SEIZE — often used with *up*
intransitive verb
: to eat something greedily

scoff·law \-ˌlȯ\ *noun* (1924)
: a contemptuous law violator

[1]scold \ˈskōld\ *noun* [Middle English *scald, scold,* perhaps of Scandinavian origin; akin to Old Norse *skāld* poet, skald, Icelandic *skālda* to make scurrilous verse] (12th century)
1 a : one who scolds habitually or persistently **b :** a woman who disturbs the public peace by noisy and quarrelsome or abusive behavior
2 : SCOLDING

[2]scold (14th century)
intransitive verb
1 *obsolete* **:** to quarrel noisily
2 : to find fault noisily or angrily
transitive verb
: to censure severely or angrily **:** REBUKE ☆
— scold·er *noun*

scold·ing *noun* (1547)
1 : the action of one who scolds
2 : a harsh reproof

sco·le·cite \ˈskä-lə-ˌsīt, ˈskō-\ *noun* [German *Skolezit,* from Greek *skōlēk-, skōlēx* worm; from the motion of some forms when heated] (circa 1823)
: a zeolite mineral that is a hydrous calcium aluminum silicate and occurs in radiating groups of crystals, in fibrous masses, and in nodules

sco·lex \ˈskō-ˌleks\ *noun, plural* **sco·li·ces** \-lə-ˌsēz\ *also* **sco·le·ces** \ˈskō-lə-ˌsēz, ˈskä-\ [New Latin *scolic-, scolex,* from Greek *skōlēk-, skōlēx* worm; akin to Greek *skolios* crooked, *skelos* leg — more at ISOSCELES] (1855)
: the head of a tapeworm either in the larva or adult stage

sco·li·o·sis \ˌskō-lē-ˈō-səs\ *noun, plural* **-o·ses** \-ˌsēz\ [New Latin, from Greek *skoliōsis* crookedness of a bodily part, from *skolios*] (circa 1706)
: a lateral curvature of the spine
— sco·li·ot·ic \-ˈä-tik\ *adjective*

scol·lop \ˈskä-ləp, ˈskȯ-\ *variant of* SCALLOP

scol·o·pen·dra \ˈskä-lə-ˈpen-drə\ *noun* [New Latin, genus of centipedes, from Latin, a kind of millipede, from Greek *skolopendra*] (1608)
: CENTIPEDE

scom·broid \ˈskäm-ˌbrȯid\ *noun* [ultimately from Greek *skombros* mackerel] (circa 1842)
: any of a suborder (Scombroidei) of marine bony fishes (as mackerels, tunas, albacores, bonitos, and swordfishes) of great economic importance as food fishes
— scombroid *adjective*

[1]sconce \ˈskän(t)s\ *noun* [Middle English, from Middle French *esconse* screened lantern, from Old French, from feminine of *escons,*

☆ SYNONYMS

Scoff, jeer, gibe, fleer, sneer, flout mean to show one's contempt in derision or mockery. SCOFF stresses insolence, disrespect, or incredulity as motivating the derision ⟨*scoffed* at their concerns⟩. JEER suggests a coarser more undiscriminating derision ⟨the crowd *jeered* at the prisoners⟩. GIBE implies taunting either good-naturedly or in sarcastic derision ⟨hooted and *gibed* at the umpire⟩. FLEER suggests grinning or grimacing derisively ⟨the saucy jackanapes *fleered* at my credulity⟩. SNEER stresses insulting by contemptuous facial expression, phrasing, or tone of voice ⟨*sneered* at anything romantic⟩. FLOUT stresses contempt shown by refusal to heed ⟨*flouted* the conventions of polite society⟩.

Scold, upbraid, berate, rail, revile, vituperate mean to reproach angrily and abusively. SCOLD implies rebuking in irritation or ill temper justly or unjustly ⟨angrily *scolding* the children⟩. UPBRAID implies censuring on definite and usually justifiable grounds ⟨*upbraided* her assistants for poor research⟩. BERATE suggests prolonged and often abusive scolding ⟨*berated* continually by an overbearing boss⟩. RAIL (*at* or *against*) stresses an unrestrained berating ⟨*railed* loudly at their insolence⟩. REVILE implies a scurrilous, abusive attack prompted by anger or hatred ⟨an alleged killer *reviled* in the press⟩. VITUPERATE suggests a violent reviling ⟨was *vituperated* for betraying his friends⟩.

past participle of *escondre* to hide, from Latin *abscondere* — more at ABSCOND] (15th century)
1 : a bracket candlestick or group of candlesticks; *also* **:** an electric light fixture patterned on a candle sconce
2 : HEAD, SKULL
²sconce *noun* [Dutch *schans,* from German *Schanze*] (1571)
: a detached defensive work
scone \'skōn, 'skän\ *noun* [perhaps from Dutch *schoonbrood* fine white bread, from *schoon* pure, clean + *brood* bread] (1513)
: a rich quick bread cut into usually triangular shapes and cooked on a griddle or baked on a sheet
¹scoop \'sküp\ *noun* [Middle English *scope,* from Middle Dutch *schope;* akin to Old High German *skepfen* to shape — more at SHAPE] (14th century)
1 a : a large ladle **b :** a deep shovel or similar implement for digging, dipping, or shoveling **c :** a usually hemispherical utensil for dipping food **d :** a small spoon-shaped utensil or instrument for cutting or gouging
2 : the action of scooping
3 a : a hollow place **:** CAVITY **b :** a part forming or surrounding an opening for channeling a fluid (as air) into a desired path
4 a : information especially of immediate interest **b :** BEAT 5b
5 : a rounded and usually low-cut neckline on a woman's garment — called also *scoop neck*
— scoop·ful \-ˌfu̇l\ *noun*
²scoop *transitive verb* (14th century)
1 a : to take out or up with or as if with a scoop **:** DIP **b :** to pick up quickly or surreptitiously with or as if with a sweep of the hand — often used with *up*
2 : to empty by ladling out the contents
3 : to make hollow **:** DIG OUT
4 : BEAT 5a(2)
— scoop·er *noun*
scoot \'sküt\ *intransitive verb* [perhaps of Scandinavian origin; akin to Old Norse *skjōta* to shoot — more at SHOOT] (1758)
1 : to move swiftly
2 : to slide especially while seated ⟨*scoot* over and let me sit down⟩
— scoot *noun*
scoot·er \'skü-tər\ *noun* (1916)
1 : a child's foot-operated vehicle consisting of a narrow footboard mounted between two wheels tandem with an upright steering handle attached to the front wheel
2 : MOTOR SCOOTER
scop \'shōp, 'skōp, 'skäp\ *noun* [Old English; akin to Old High German *schof* poet] (before 12th century)
: an Old English bard or poet
¹scope \'skōp\ *noun* [Italian *scopo* purpose, goal, from Greek *skopos;* akin to Greek *skeptesthai* to watch, look at — more at SPY] (circa 1555)
1 : INTENTION, OBJECT
2 : space or opportunity for unhampered motion, activity, or thought
3 : extent of treatment, activity, or influence
4 : range of operation: as **a :** the range of a logical operator **:** a string in predicate calculus that is governed by a quantifier **b :** a grammatical constituent that determines the interpretation of a predicate or quantifier
synonym see RANGE
²scope *noun* [*-scope*] (1872)
1 : an instrument (as a telescope or radarscope) for viewing
2 : HOROSCOPE
³scope *transitive verb* **scoped; scop·ing** [perhaps from ²*scope*] (1974)
: to look at especially for the purpose of evaluation — often used with *out* ⟨*scoped* her out from across the room —Tim Allis⟩

-scope *noun combining form* [New Latin *-scopium,* from Greek *-skopion;* akin to Greek *skeptesthai*]
: means (as an instrument) for viewing or observing ⟨endo*scope*⟩ ⟨spectro*scope*⟩
sco·pol·amine \skō-'pä-lə-ˌmēn, -mən\ *noun* [German *Scopolamin,* from New Latin *Scopolia,* genus of plants + German *Amin* amine] (1892)
: a poisonous alkaloid $C_{17}H_{21}NO_4$ found in various plants (as jimsonweed) of the nightshade family and used especially as a truth serum or usually with morphine as a sedative in surgery and obstetrics — called also *hyoscine*
-scopy *noun combining form* [Greek *-skopia,* from *skeptesthai*]
: viewing **:** observation ⟨spectro*scopy*⟩
scor·bu·tic \skȯr-'byü-tik\ *adjective* [New Latin *scorbuticus,* from *scorbutus* scurvy, probably of Germanic origin; akin to Old English *scurf* scurf] (1655)
: of, relating to, producing, or affected with scurvy
¹scorch \'skȯrch\ *verb* [Middle English; akin to Middle English *scorcnen,* probably of Scandinavian origin; akin to Old Norse *skorpna* to shrivel up — more at SHRIMP] (14th century)
transitive verb
1 : to burn a surface of so as to change its color and texture
2 a : to dry or shrivel with or as if with intense heat **:** PARCH **b :** to afflict painfully with censure or sarcasm
3 : DEVASTATE; *especially* **:** to destroy (as property of possible use to an advancing enemy) before abandoning — used in the phrase *scorched earth*
intransitive verb
1 : to become scorched
2 : to travel at great and usually excessive speed
3 : to cause intense heat or mental anguish ⟨*scorching* sun⟩ ⟨*scorching* fury⟩
— scorch·ing·ly \'skȯr-chiŋ-lē\ *adverb*
²scorch *noun* (15th century)
1 : a result of scorching
2 : a browning of plant tissues usually from disease or heat
³scorch *transitive verb* [Middle English, perhaps blend of *scoren* to score and *scocchen* to scotch] (14th century)
dialect British **:** CUT, SLASH
scorched *adjective* (1590)
: parched or discolored by scorching
scorch·er \'skȯr-chər\ *noun* (1842)
: one that scorches; *especially* **:** a very hot day
¹score \'skōr, 'skȯr\ *noun, plural* **scores** [Middle English *scor,* from Old Norse *skor* notch, tally, twenty; akin to Old English *sceran* to cut — more at SHEAR] (14th century)
1 *or plural* **score a :** TWENTY **b :** a group of 20 things — often used in combination with a cardinal number ⟨four*score*⟩ **c :** an indefinitely large number
2 a : a line (as a scratch or incision) made with or as if with a sharp instrument **b** (1) **:** a mark used as a starting point or goal (2) **:** a mark used for keeping account
3 a : an account or reckoning originally kept by making marks on a tally **b :** amount due **:** INDEBTEDNESS
4 : GRUDGE ⟨a *score* to settle⟩
5 a : REASON, GROUND **b :** SUBJECT, TOPIC
6 a : the copy of a musical composition in written or printed notation **b :** a musical composition; *specifically* **:** the music for a movie or theatrical production **c :** a complete description of a dance composition in choreographic notation
7 a : a number that expresses accomplishment (as in a game or test) or excellence (as in quality) either absolutely in points gained or by comparison to a standard **b :** an act (as a goal, run, or touchdown) in any of various games or contests that gains points

8 : success in obtaining something (as money or drugs) especially through illegal or irregular means
9 : the stark inescapable facts of a situation ⟨knows the *score*⟩
²score *verb* **scored; scor·ing** (14th century)
transitive verb
1 a : to keep a record or account of by or as if by notches on a tally **:** RECORD **b :** to enter in a record **c :** to mark with significant lines or notches (as in keeping account)
2 : to mark with lines, grooves, scratches, or notches
3 : BERATE, SCOLD; *also* **:** DENOUNCE
4 a (1) **:** to make (a score) in a game or contest ⟨*scored* a touchdown⟩ ⟨*scored* three points⟩ (2) **:** to enable (a base runner) to make a score (3) **:** to have as a value in a game or contest **:** COUNT ⟨a touchdown *scores* six points⟩ **b** (1) **:** ACHIEVE, WIN ⟨*scored* a dazzling success⟩ (2) **:** ACQUIRE ⟨help a traveler *score* local drugs —Poitor Koper⟩
5 : to determine the merit of **:** GRADE
6 a : to write or arrange (music) for a specific performance medium **b :** to make an orchestration of **c :** to compose a score for (a movie)
intransitive verb
1 : to keep score in a game or contest
2 : to make a score in a game or contest
3 a : to gain or have the advantage **b :** to be successful: as (1) **:** to succeed in having sexual intercourse (2) **:** to manage to obtain illicit drugs **c :** ³RATE
— scor·er *noun*
— score points : to gain favor, status, or advantage
score·board \'skōr-ˌbōrd, 'skȯr-ˌbȯrd\ *noun* (1826)
: a large board for displaying the score of a game or match
score·card \-ˌkärd\ *noun* (circa 1877)
: a card for recording the score of a game
score·keep·er \-ˌkē-pər\ *noun* (1880)
: one that keeps score; *specifically* **:** an official who records the score during a game or contest
score·less \-ləs\ *adjective* (1885)
: having no score
sco·ria \'skōr-ē-ə, 'skȯr-\ *noun, plural* **-ri·ae** \-ē-ˌē, -ē-ˌī\ [Middle English, from Latin, from Greek *skōria,* from *skōr* excrement — more at SCATOLOGY] (14th century)
1 : the refuse from melting of metals or reduction of ores **:** SLAG
2 : rough vesicular cindery lava
— sco·ri·a·ceous \ˌskōr-ē-'ā-shəs, ˌskȯr-\ *adjective*
¹scorn \'skȯrn\ *noun* [Middle English, from Old French *escarn,* of Germanic origin; akin to Old High German *scern* jest] (13th century)
1 : open dislike and disrespect or derision often mixed with indignation
2 : an expression of contempt or derision
3 : an object of extreme disdain, contempt, or derision **:** something contemptible
²scorn (13th century)
transitive verb
: to treat with scorn **:** reject or dismiss as contemptible or unworthy ⟨*scorned* the traditions of their ancestors⟩ ⟨*scorned* to reply to the charge⟩
intransitive verb
: to show disdain or derision **:** SCOFF
synonym see DESPISE
— scorn·er *noun*
scorn·ful \'skȯrn-fəl\ *adjective* (14th century)
: full of scorn **:** CONTEMPTUOUS
— scorn·ful·ly \-fə-lē\ *adverb*
— scorn·ful·ness *noun*
scor·pae·nid \skȯr-'pē-nəd\ *noun* [ultimately from Greek *skorpaina,* a kind of fish] (1885)
: any of a family (Scorpaenidae) of marine spiny-finned fishes that includes the scorpion fishes

— **scorpaenid** *adjective*

Scor·pio \'skȯr-pē-,ō\ *noun* [Latin (genitive *Scorpionis*), from Greek *Skorpios*, literally, scorpion]
1 : SCORPIUS
2 a : the 8th sign of the zodiac in astrology — see ZODIAC table **b :** one born under this sign

scor·pi·on \'skȯr-pē-ən\ *noun* [Middle English, from Old French, from Latin *scorpion-, scorpio*, from Greek *skorpios*] (12th century)
1 a : any of an order (Scorpionida) of nocturnal arachnids that have an elongated body and a narrow segmented tail bearing a venomous stinger at the tip **b** *capitalized* **:** SCORPIO
2 : a scourge probably studded with metal
3 : something that incites to action like the sting of an insect

scorpion 1a

scorpion fish *noun* (1661)
: any of various scorpaenid fishes; *especially* **:** one (as genus *Scorpaena*) with a venomous spine on the dorsal fin

scorpion fly *noun* (1668)
: any of a family (Panorpidae) of insects that have cylindrical bodies, a long beak with biting mouthparts, and the male genitalia enlarged into a swollen bulb; *broadly* **:** an insect of the order (Mecoptera) that includes this family

Scor·pi·us \'skȯr-pē-əs\ *noun* [Latin (genitive *Scorpii*), from Greek *Skorpios*, literally, scorpion]
: a southern zodiacal constellation partly in the Milky Way and between Libra and Sagittarius

scot \'skät\ *noun* [Middle English, from Old Norse *skot* shot, contribution — more at SHOT] (14th century)
: money assessed or paid

Scot \'skät\ *noun* [Middle English *Scottes* Scots, from Old English *Scottas* from Late Latin *Scotus*] (before 12th century)
1 : a member of a Celtic people of northern Ireland settling in Scotland about A.D. 500
2 a : a native or inhabitant of Scotland **b :** a person of Scottish descent

scot and lot *noun* (15th century)
1 : a parish assessment formerly laid on subjects in Great Britain according to their ability to pay
2 : obligations of all kinds taken as a whole

[1]**scotch** \'skäch\ *transitive verb* [Middle English *scocchen* to gash] (15th century)
1 *archaic* **:** CUT, GASH, SCORE; *also* **:** WOUND ⟨we have *scotched* the snake, not killed it —Shakespeare⟩
2 : to put an end to ⟨*scotched* rumors of a military takeover⟩

[2]**scotch** *noun* (15th century)
: a superficial cut **:** SCORE

[3]**scotch** *noun* [origin unknown] (1639)
: a chock to prevent rolling or slipping

[4]**scotch** *transitive verb* (1642)
1 : to block with a chock
2 : HINDER, THWART

[1]**Scotch** \'skäch\ *adjective* [contraction of *Scottish*] (1591)
1 : SCOTTISH
2 : inclined to frugality ▪

[2]**Scotch** *noun* (circa 1700)
1 : SCOTS
2 *plural in construction* **:** the people of Scotland
3 *often not capitalized* **:** whiskey distilled in Scotland especially from malted barley — called also *Scotch whisky*

[3]**Scotch** *trademark*
— used for any of numerous adhesive tapes

Scotch broom *noun* (circa 1818)
: a deciduous broom (*Cytisus scoparius*) of western Europe that is widely cultivated for its bright yellow or partly red flowers and that has become a pest in some areas (as California)

Scotch broth *noun* (1834)
: a soup made from beef or mutton and vegetables and thickened with barley

Scotch egg *noun* (1809)
: a hard-boiled egg wrapped in sausage meat, covered with bread crumbs, and fried

Scotch–Irish *adjective* (1744)
: of, relating to, or descended from Scottish settlers in northern Ireland

Scotch·man \'skäch-mən\ *noun* (15th century)
: SCOTSMAN

Scotch pine *noun* (1731)
: a pine (*Pinus sylvestris*) of northern Europe and Asia with spreading or pendulous branches, short rigid twisted needles, and hard yellow wood that provides valuable timber

Scotch terrier *noun* (1810)
: SCOTTISH TERRIER

Scotch verdict *noun* (1912)
1 : a verdict of not proven that is allowed by Scottish criminal law in some cases instead of a verdict of not guilty
2 : an inconclusive decision or pronouncement

Scotch·wom·an \'skäch-,wu̇-mən\ *noun* (1818)
: SCOTSWOMAN

Scotch woodcock *noun* (1879)
: buttered toast spread with anchovy paste and scrambled egg

sco·ter \'skō-tər\ *noun, plural* **scoters** *or* **scoter** [origin unknown] (circa 1674)
: any of a genus (*Melanitta*) of sea ducks especially of northern coasts of Europe and North America that have males with chiefly black plumage

scot–free \'skät-'frē\ *adjective* [*scot*] (1528)
: completely free from obligation, harm, or penalty

sco·tia \'skō-sh(ē-)ə, -tē-ə\ *noun* [Latin, from Greek *skotia*, from feminine of *skotios* dark, shadowy, from *skotos* darkness — more at SHADE] (1563)
: a concave molding used especially in classical architecture in the bases of columns — see BASE illustration

Scot·ic \'skä-tik\ *adjective* (1796)
: of or relating to the ancient Scots

Sco·tism \'skō-,ti-zəm\ *noun* (circa 1871)
: the doctrines of Duns Scotus
— **Sco·tist** \'skō-tist\ *noun*

Scot·land Yard \,skät-lən(d)-'yärd\ *noun* [*Scotland Yard*, street in London, formerly the headquarters of the metropolitan police] (1864)
: the detective department of the London metropolitan police

sco·to·ma \skə-'tō-mə\ *noun, plural* **-mas** *or* **-ma·ta** \-mə-tə\ [New Latin, from Medieval Latin, dimness of vision, from Greek *skotōmat-, skotōma*, from *skotoun* to darken, from *skotos*] (1875)
: a blind or dark spot in the visual field

sco·to·pic \skə-'tō-pik, -'tä-\ *adjective* [New Latin *scotopia* scotopic vision, from Greek *skotos* darkness + New Latin *-opia*] (1915)
: relating to or being vision in dim light with dark-adapted eyes which involves only the retinal rods as light receptors

[1]**Scots** \'skäts\ *adjective* [Middle English *Scottis*, alteration of *Scottish*] (14th century)
: SCOTTISH — used especially of the people and language and in legal context
usage see SCOTCH

[2]**Scots** *noun* (1542)
: the English language of Scotland

Scots·man \'skäts-mən\ *noun* (14th century)
: a native or inhabitant of Scotland

Scots pine *noun* (1797)
chiefly British **:** SCOTCH PINE

Scots·wom·an \'skäts-,wu̇-mən\ *noun* (1820)
: a woman who is a native or inhabitant of Scotland

Scot·ti·cism \'skä-tə-,si-zəm\ *noun* [Late Latin *scotticus* of the ancient Scots, from *Scotus* Scot] (1706)
: a word, phrase, or expression characteristic of Scottish English

Scot·tie \'skä-tē\ *noun* (circa 1896)
1 : SCOTSMAN
2 : SCOTTISH TERRIER

[1]**Scot·tish** \'skä-tish\ *adjective* [Middle English, from *Scottes* Scotsmen] (13th century)
: of, relating to, or characteristic of Scotland, Scots, or the Scots
usage see SCOTCH
— **Scot·tish·ness** \-nəs\ *noun*

[2]**Scottish** *noun* (1759)
: SCOTS

Scottish deerhound *noun* (1891)
: any of an old breed of dogs of Scottish origin that have the general form of a greyhound but are larger and taller with a rough coat

Scottish Gaelic *noun* (1956)
: the Gaelic language of Scotland

Scottish rite *noun* (1903)
1 : a ceremonial observed by one of the Masonic systems
2 : a system or organization that observes the Scottish rite and confers the 4th through the 33d degrees

Scottish terrier *noun* (1837)
: any of an old Scottish breed of terrier that has short legs, a large head with small erect ears and a powerful muzzle, a broad deep chest, and a very hard coat of wiry hair

Scottish terrier

scoun·drel \'skau̇n-drəl\ *noun* [origin unknown] (1589)
: a disreputable person **:** RASCAL
— **scoundrel** *adjective*
— **scoun·drel·ly** \-drə-lē\ *adjective*

[1]**scour** \'skau̇(-ə)r\ *verb* [Middle English, probably from Middle Dutch *schuren*, from Old French *escurer*, from Late Latin *excurare* to clean off, from Latin, to take good care of, from *ex-* + *curare* to care for, from *cura* care] (13th century)
transitive verb
1 a : to rub hard especially with a rough material for cleansing **b :** to remove by rubbing hard and washing
2 *archaic* **:** to clear (a region) of enemies or outlaws
3 : to clean by purging **:** PURGE
4 : to remove dirt and debris from (as a pipe or ditch)
5 : to free from foreign matter or impurities by or as if by washing ⟨*scour* wool⟩

□ USAGE
Scotch In Scotland *Scottish* is the preferred adjective, *Scotch* is usually limited to familiar compounds like *Scotch pine* and *Scotch whiskey,* and *Scots* refers to language and law. In North America *Scotch* is commonly used in all adjectival applications, partly because *Scotch* was the usual adjective in the 18th and 19th centuries when many Scots immigrated here. A rule of thumb which works well for Americans who do not want to offend but do not know the finer points of Scottish practice is to use *Scottish* for people and *Scotch* for things.

6 : to clear, dig, or remove by or as if by a powerful current of water
intransitive verb
1 : to perform a process of scouring
2 : to suffer from diarrhea or dysentery
3 : to become clean and bright by rubbing
— **scour·er** *noun*

²scour *noun* (1681)
1 : a place scoured by running water
2 : scouring action (as of a glacier)
3 : DIARRHEA, DYSENTERY — usually used in plural but singular or plural in construction
4 : SCOURING 1; *also* **:** damage done by scouring action

³scour *verb* [Middle English *scuren,* probably of Scandinavian origin; akin to Swedish *skura* to rush] (14th century)
intransitive verb
: to move about quickly especially in search
transitive verb
: to go through or range over in or as if in a search

¹scourge \'skərj *also* 'skōrj, 'skȯrj, 'skůrj\ *noun* [Middle English, from Anglo-French *escorge,* from (assumed) Old French *escorgier* to whip, from Old French *es-* ex- + Latin *corrigia* whip] (13th century)
1 : WHIP; *especially* **:** one used to inflict pain or punishment
2 : an instrument of punishment or criticism
3 : a cause of widespread or great affliction

²scourge *transitive verb* **scourged; scourg·ing** (13th century)
1 : FLOG, WHIP
2 a : to punish severely **b :** AFFLICT **c :** to drive as if by blows of a whip **d :** CHASTISE
— **scourg·er** *noun*

scour·ing \'skaůr-iŋ\ *noun* (1588)
1 : material removed by scouring or cleaning
2 : the lowest rank of society — usually used in plural

scouring rush *noun* (circa 1818)
: EQUISETUM; *especially* **:** one (*Equisetum hyemale*) with strongly siliceous stems formerly used for scouring

scouse \'skaůs\ *noun* (1840)
1 : LOBSCOUSE
2 *capitalized* **a :** SCOUSER **b :** a dialect of English spoken in Liverpool

Scous·er \'skaů-sər\ *noun* (1959)
: a native or inhabitant of Liverpool, England

¹scout \'skaůt\ *verb* [Middle English, from Middle French *escouter* to listen, from Latin *auscultare* — more at AUSCULTATION] (14th century)
intransitive verb
1 : to explore an area to obtain information (as about an enemy)
2 a : to make a search **b :** to work as a talent scout
transitive verb
1 : to observe in order to obtain information or evaluate
2 : to explore in order to obtain information
3 : to find by making a search — often used with *up*

²scout *noun* (1534)
1 a : one sent to obtain information; *especially* **:** a soldier, ship, or plane sent out in war to reconnoiter **b :** WATCHMAN, LOOKOUT **c :** TALENT SCOUT
2 a : the act of scouting **b :** a scouting expedition **:** RECONNAISSANCE
3 *often capitalized* **:** a member of any of various scouting movements: as **a :** BOY SCOUT **b :** GIRL SCOUT
4 : INDIVIDUAL, PERSON — used chiefly in the phrase *good scout*

³scout *verb* [probably of Scandinavian origin; akin to Old Norse *skūti* taunt; akin to Old English *scēotan* to shoot — more at SHOOT] (1605)
transitive verb
1 : MOCK
2 : to reject scornfully
intransitive verb
: SCOFF

scout car *noun* (1933)
1 : a military reconnaissance vehicle
2 : SQUAD CAR

scout·craft \'skaůt-ˌkraft\ *noun* (1908)
: the craft, skill, or practice of a scout

scout·er \'skaů-tər\ *noun* (1642)
1 : one that scouts
2 *often capitalized* **:** an adult leader in the Boy Scouts of America

scouth \'sküth, 'skaůth\ *noun* [origin unknown] (1591)
Scottish **:** PLENTY

scout·ing \'skaů-tiŋ\ *noun* (1644)
1 : the action of one that scouts
2 *often capitalized* **:** the activities of various national and worldwide organizations for youth directed to developing character, citizenship, and individual skills

scout·mas·ter \'skaůt-ˌmas-tər\ *noun* (1579)
: the leader of a band of scouts; *specifically* **:** the adult leader of a troop of Boy Scouts

scow \'skaů\ *noun* [Dutch *schouw;* akin to Old High German *scalta* punt pole] (1669)
: a large flat-bottomed boat with broad square ends used chiefly for transporting bulk material (as ore, sand, or refuse)

¹scowl \'skaů(ə)l\ *verb* [Middle English *skoulen,* probably of Scandinavian origin; akin to Danish *skule* to scowl] (14th century)
intransitive verb
1 : to contract the brow in an expression of displeasure
2 : to exhibit a threatening aspect
transitive verb
: to express with a scowl
— **scowl·er** *noun*
— **scowl·ing·ly** \'skaů-liŋ-lē\ *adverb*

²scowl *noun* (1500)
: a facial expression of displeasure **:** FROWN

¹scrab·ble \'skra-bəl\ *verb* **scrab·bled; scrab·bling** \-b(ə-)liŋ\ [Dutch *schrabbelen* to scratch] (1537)
intransitive verb
1 : SCRAWL, SCRIBBLE
2 : to scratch, claw, or grope about clumsily or frantically
3 a : SCRAMBLE, CLAMBER **b :** to struggle by or as if by scraping or scratching ⟨*scrabble* for survival⟩
transitive verb
1 : SCRAMBLE
2 : SCRIBBLE
— **scrab·bler** \-b(ə-)lər\ *noun*

²scrabble *noun* (1842)
1 : SCRIBBLE
2 : a repeated scratching or clawing
3 : SCRAMBLE

scrab·bly \'skra-b(ə-)lē\ *adjective* (1945)
1 : SCRATCHY, RASPY
2 : SPARSE, SCRUBBY

¹scrag \'skrag\ *noun* [perhaps alteration of ²*crag*] (1542)
1 : a rawboned or scrawny person or animal
2 : the lean end of a neck of mutton or veal; *broadly* **:** NECK — called also *scrag end*

²scrag *transitive verb* **scragged; scrag·ging** (1756)
1 a : to execute by hanging or garroting **b :** to wring the neck of
2 a : CHOKE **b :** MANHANDLE 1 **c :** KILL, MURDER

scrag·gly \'skra-g(ə-)lē\ *adjective* (1869)
: irregular in form or growth ⟨*scraggly* hills⟩ ⟨a *scraggly* beard⟩; *also* **:** UNKEMPT

¹scrag·gy \'skra-gē\ *adjective* **scrag·gi·er; -est** [Middle English *scraggi;* akin to English dialect *scrag* tree stump, uneven ground, Middle English *scrogge* bush] (13th century)
: ROUGH, JAGGED; *also* **:** SCRAGGLY

²scraggy *adjective* **scrag·gi·er; -est** [¹*scrag*] (1611)
: being lean and long **:** SCRAWNY

¹scram \'skram\ *intransitive verb* **scrammed; scram·ming** [short for *scramble*] (circa 1928)
: to go away at once ⟨*scram,* you're not wanted⟩

²scram *noun* (1953)
: a rapid emergency shutdown of a nuclear reactor

¹scram·ble \'skram-bəl\ *verb* **scram·bled; scram·bling** \-b(ə-)liŋ\ [perhaps alteration of ¹*scrabble*] (circa 1586)
intransitive verb
1 a : to move or climb hastily on all fours **b :** to move with urgency or panic
2 a : to struggle eagerly or unceremoniously for possession of something ⟨*scramble* for front seats⟩ **b :** to get or gather something with difficulty or in irregular ways ⟨*scramble* for a living⟩
3 : to spread or grow irregularly **:** SPRAWL, STRAGGLE
4 : to take off quickly in response to an alert
5 *of a football quarterback* **:** to run with the ball after the pass protection breaks down
transitive verb
1 : to collect by scrambling
2 a : to toss or mix together **:** JUMBLE **b :** to prepare (eggs) by stirring during frying
3 : to cause or order (a fighter-interceptor group) to scramble
4 : to disarrange the elements of a transmission (as a telephone or television signal) in order to make unintelligible to interception
— **scram·bler** \-b(ə-)lər\ *noun*

²scramble *noun* (1674)
1 : the act or an instance of scrambling
2 : a disordered mass **:** JUMBLE ⟨a . . . *scramble* of patterns and textures —*Vogue*⟩
3 : a rapid emergency takeoff of fighter-interceptor planes

scran·nel \'skra-nᵊl\ *adjective* [origin unknown] (1637)
: HARSH, UNMELODIOUS

¹scrap \'skrap\ *noun, often attributive* [Middle English, from Old Norse *skrap* scraps; akin to Old Norse *skrapa* to scrape] (14th century)
1 *plural* **:** fragments of discarded or leftover food
2 a : a small detached piece ⟨a *scrap* of paper⟩ **b :** a fragment of something written, printed, or spoken ⟨*scraps* of conversation⟩ **c :** the least bit ⟨not a *scrap* of evidence⟩
3 *plural* **:** CRACKLINGS
4 a : fragments of stock removed in manufacturing **b :** manufactured articles or parts rejected or discarded and useful only as material for reprocessing; *especially* **:** waste and discarded metal

²scrap *transitive verb* **scrapped; scrap·ping** (circa 1891)
1 : to convert into scrap
2 : to abandon or get rid of as no longer of enough worth or effectiveness to retain ⟨*scrap* outworn methods⟩
synonym see DISCARD

³scrap *noun* [origin unknown] (circa 1889)
: FIGHT

⁴scrap *intransitive verb* **scrapped; scrap·ping** (circa 1895)
: QUARREL, FIGHT

scrap·book \'skrap-ˌbůk\ *noun* (1825)
: a blank book in which miscellaneous items (as newspaper clippings or pictures) are collected and preserved

¹scrape \'skrāp\ *verb* **scraped; scrap·ing** [Middle English, from Old Norse *skrapa;* akin to Old English *scrapian* to scrape, Latin *scrobis* ditch, Russian *skresti* to scrape] (14th century)
transitive verb
1 a : to remove from a surface by usually repeated strokes of an edged instrument **b :** to make (a surface) smooth or clean with strokes of an edged instrument or an abrasive
2 a : to grate harshly over or against **b :** to damage or injure the surface of by contact with a rough surface **c :** to draw roughly or noisily over a surface

3 : to collect by or as if by scraping — often used with *up* or *together* ⟨*scrape* up the price of a ticket⟩
intransitive verb
1 : to move in sliding contact with a rough surface
2 : to accumulate money by small economies
3 : to draw back the foot along the ground in making a bow
4 : to make one's way with difficulty or succeed by a narrow margin ⟨just *scraped* by at school⟩
— scrap·er *noun*
— scrape a leg : to make a low bow

²scrape *noun* (15th century)
1 a : the act or process of scraping **b :** a sound made by scraping **c :** damage or injury caused by scraping **:** ABRASION ⟨bumps and *scrapes*⟩
2 : a bow made with a drawing back of the foot along the ground
3 a : a distressing encounter ⟨a *scrape* with death⟩ **b :** QUARREL, FIGHT

scrap heap *noun* (circa 1902)
1 : a pile of discarded metal
2 : the place where useless things are discarded

scra·pie \'skrā-pē\ *noun* [¹*scrape*] (1910)
: a usually fatal disease of the nervous system especially of sheep that is characterized by twitching, intense itching, excessive thirst, emaciation, weakness, and finally paralysis

scrap·page \'skra-pij\ *noun* (circa 1909)
1 : the scrapping of discarded objects (as automobiles)
2 : the rate at which objects are scrapped

scrap·per \'skra-pər\ *noun* (1874)
: FIGHTER, QUARRELER

scrap·pi·ly \'skra-pə-lē\ *adverb* (1886)
: in a scrappy manner

scrap·pi·ness \'skra-pē-nəs\ *noun* (1867)
: the quality or state of being scrappy

scrap·ple \'skra-pəl\ *noun* [diminutive of ¹*scrap*] (1855)
: a seasoned mixture of ground meat (as pork) and cornmeal set in a mold and served sliced and fried

¹scrap·py \'skra-pē\ *adjective* **scrap·pi·er; -est** [¹*scrap*] (1837)
: consisting of scraps ⟨*scrappy* meals⟩

²scrappy *adjective* **scrap·pi·er; -est** [³*scrap*] (circa 1896)
1 : QUARRELSOME
2 : having an aggressive and determined spirit **:** FEISTY

¹scratch \'skrach\ *verb* [Middle English *scracchen,* probably blend of *scratten* to scratch and *cracchen* to scratch] (15th century)
transitive verb
1 : to scrape or dig with the claws or nails
2 : to rub and tear or mark the surface of with something sharp or jagged
3 : to scrape or rub lightly (as to relieve itching)
4 : to scrape together **:** collect with difficulty or by effort
5 : to write or draw on a surface
6 a : to cancel or erase by or as if by drawing a line through **b :** to withdraw (an entry) from competition
7 : SCRIBBLE, SCRAWL
8 : to scrape along a rough surface ⟨*scratch* a match⟩
intransitive verb
1 : to use the claws or nails in digging, tearing, or wounding
2 : to scrape or rub oneself lightly (as to relieve itching)
3 : to gather money or get a living by hard work and especially through irregular means and sacrifice
4 : to make a thin grating sound
5 : to withdraw from a contest or engagement
6 : to make a scratch in billiards or pool
— scratch·er *noun*
— scratch one's back : to accommodate with a favor especially in expectation of like return
— scratch the surface : to make a superficial effort or modest start

²scratch *noun* (circa 1586)
1 : a mark or injury produced by scratching; *also* **:** a slight wound
2 : SCRAWL, SCRIBBLE
3 : the sound made by scratching
4 a : the starting line in a race **b :** a point at the beginning of a project at which nothing has been done ahead of time ⟨build a school system from *scratch*⟩
5 a : a test of courage **b :** satisfactory condition, level, or performance ⟨not up to *scratch*⟩
6 : a contestant whose name is withdrawn
7 : poultry feed (as mixed grains) scattered on the litter or ground especially to induce birds to exercise — called also *scratch feed*
8 a : a shot in billiards or pool that ends a player's turn; *specifically* **:** a shot in pool in which the cue ball falls into the pocket **b :** a shot that scores by chance **:** FLUKE
9 *slang* **:** MONEY, FUNDS

³scratch *adjective* (1897)
1 : made as or used for a tentative effort
2 : made or done by chance and not as intended ⟨a *scratch* shot⟩
3 : arranged or put together with little selection **:** HAPHAZARD ⟨a *scratch* team⟩
4 : having no handicap or allowance ⟨a *scratch* golfer⟩
5 : made from scratch **:** made with basic ingredients ⟨a *scratch* cake⟩

scratch·board \'skrach-ˌbōrd, -ˌbȯrd\ *noun* (circa 1908)
: a black-surfaced cardboard having an undercoat of white clay on which an effect resembling engraving is achieved by scratching away portions of the surface to produce white lines

scratch hit *noun* (1903)
: a batted ball not solidly hit yet credited to the batter as a base hit

scratch pad *noun* (1895)
: a pad of scratch paper

scratch paper *noun* (1899)
: paper that may be used for casual writing

scratch sheet *noun* (1939)
: a racing publication listing competitors scratched from races and giving odds

scratch test *noun* (1937)
: a test for allergic susceptibility made by rubbing an extract of an allergy-producing substance into small breaks or scratches in the skin

scratchy \'skra-chē\ *adjective* **scratch·i·er; -est** (1827)
1 : marked or made with scratches ⟨*scratchy* drawing⟩ ⟨*scratchy* handwriting⟩
2 : likely to scratch **:** PRICKLY ⟨*scratchy* undergrowth⟩
3 : making a scratching noise
4 : uneven in quality **:** RAGGED
5 : causing tingling or itching **:** IRRITATING ⟨*scratchy* wool⟩
— scratch·i·ly \-chə-lē\ *adverb*
— scratch·i·ness \-chē-nəs\ *noun*

scrawl \'skrȯl\ *verb* [origin unknown] (1612)
transitive verb
: to write or draw awkwardly, hastily, or carelessly
intransitive verb
: to write awkwardly or carelessly
— scrawl *noun*
— scrawl·er *noun*
— scrawly \'skrȯ-lē\ *adjective*

scraw·ny \'skrȯ-nē\ *adjective* **scraw·ni·er; -est** [origin unknown] (1833)
: exceptionally thin and slight or meager in body ⟨*scrawny* scrub cattle⟩
synonym see LEAN
— scraw·ni·ness *noun*

screak \'skrēk\ *intransitive verb* [of Scandinavian origin; akin to Old Norse *skrækja* to screech] (circa 1500)
: to make a harsh shrill noise **:** SCREECH
— screak *noun*
— screaky \'skrē-kē\ *adjective*

¹scream \'skrēm\ *verb* [Middle English *scremen;* akin to Middle Dutch *schreem* scream] (12th century)
intransitive verb
1 a (1) **:** to voice a sudden sharp loud cry (2) **:** to produce harsh high tones **b :** to make a noise resembling a scream ⟨a *screaming* siren⟩ **c :** to move with great rapidity
2 a : to speak or write with intense or hysterical emotion **b :** to protest or complain vehemently **c :** to laugh hysterically
3 : to produce a vivid startling effect
transitive verb
: to utter with or as if with a scream

²scream *noun* (1605)
1 : a loud sharp penetrating cry or noise
2 : one that is very funny

scream·er \'skrē-mər\ *noun* (1712)
1 : one that screams
2 : any of a small family (Anhimidae) of South American wetland birds having a large body, long legs, and spurred wings
3 : a sensationally startling headline

scream·ing *adjective* (1848)
1 : so striking or conspicuous as to attract notice as if by screaming ⟨*screaming* headlines⟩ ⟨a *screaming* need for reform⟩ ⟨dressed in *screaming* red⟩
2 : so funny as to provoke screams of laughter ⟨a *screaming* farce⟩
— scream·ing·ly *adverb*

screaming mee·mies \-'mē-mēz\ *noun plural but singular in construction* [origin unknown] (1942)
: nervous hysteria **:** JITTERS

scree \'skrē\ *noun* [of Scandinavian origin; akin to Old Norse *skritha* landslide, from *skrītha* to creep; akin to Old High German *scrītan* to go, Lithuanian *skriesti* to turn] (circa 1781)
: an accumulation of loose stones or rocky debris lying on a slope or at the base of a hill or cliff **:** TALUS

¹screech \'skrēch\ *noun* (1560)
1 : a high shrill piercing cry usually expressing pain or terror
2 : a sound resembling a screech

²screech *verb* [alteration of earlier *scritch,* from Middle English *scrichen;* akin to Old Norse *skrækja* to screech] (1577)
intransitive verb
1 : to utter a high shrill piercing cry **:** make an outcry usually in terror or pain
2 : to make a shrill high-pitched sound resembling a screech; *also* **:** to move with such a sound ⟨the car *screeched* to a stop⟩
transitive verb
: to utter with or as if with a screech
— screech·er *noun*

screech owl *noun* (1593)
1 : BARN OWL
2 : any of numerous New World owls (genus *Otus*); *especially* **:** a small North American owl (*O. asio*) with a pair of tufts of lengthened feathers on the head resembling ears

screechy \'skrē-chē\ *adjective* (circa 1830)
: producing a screech

screed \'skrēd\ *noun* [Middle English *screde* fragment, alteration of Old English *scrēade* — more at SHRED] (circa 1789)
1 a : a lengthy discourse **b :** an informal piece of writing
2 : a strip (as of plaster of the thickness planned for the coat) laid on as a guide

3 : a leveling device drawn over freshly poured concrete

¹screen \'skrēn\ *noun* [Middle English *screne,* from Middle French *escren,* from Middle Dutch *scherm;* akin to Old High German *skirm* shield; probably akin to Sanskrit *carman* skin, *kṛnāti* he injures — more at SHEAR] (14th century)

1 : a protective or ornamental device (as a movable partition) shielding an area from heat or drafts or from view

2 : something that shelters, protects, or hides: as **a :** a growth or stand of trees, shrubs, or plants **b :** a protective formation of troops, ships, or planes **c :** something that covers or disguises the true nature (as of an activity or feeling) ⟨greets strangers with a *screen* of excessive friendliness —Tony Schwartz⟩ **d** (1) **:** a maneuver in various sports (as basketball or ice hockey) whereby an opponent is legally impeded or his view of the play is momentarily blocked (2) **:** SCREEN PASS

3 a : a perforated plate or cylinder or a meshed wire or cloth fabric usually mounted and used to separate coarser from finer parts **b :** a system for examining and separating into different groups **c :** a piece of apparatus designed to prevent agencies in one part from affecting other parts ⟨an optical *screen*⟩ ⟨an electric *screen*⟩ ⟨a magnetic *screen*⟩ **d :** a frame holding a usually metallic netting used especially in a window or door to exclude pests (as insects)

4 a : a flat surface on which a picture or series of pictures is projected or reflected **b :** the surface on which the image appears in an electronic display (as in a television set, radar receiver, or computer terminal)

5 : a glass plate ruled with crossing opaque lines through which an image is photographed in making a halftone

6 : the motion-picture medium or industry

²screen (15th century)

transitive verb

1 : to guard from injury or danger

2 a : to give shelter or protection to with or as if with a screen **b :** to separate with or as if with a screen; *also* **:** to shield (an opponent) from a play or from view of a play

3 a : to pass (as coal, gravel, or ashes) through a screen to separate the fine part from the coarse; *also* **:** to remove by a screen **b** (1) **:** to examine usually methodically in order to make a separation into different groups (2) **:** to select or eliminate by a screening process

4 : to provide with a screen to keep out pests (as insects)

5 a (1) **:** to present (as a motion picture) for viewing on a screen (2) **:** to view the presentation of (as a motion picture) **b :** to present in a motion picture

intransitive verb

1 : to appear on a motion-picture screen

2 : to provide a screen in a game or sport

synonym see HIDE

— screen·able \'skrē-nə-bəl\ *adjective*

— screen·er *noun*

screen·ing \'skrē-niŋ\ *noun* (1725)

1 : the act or process of one that screens

2 *plural but singular or plural in construction* **:** material (as waste or fine coal) separated out by means of a screen

3 : metal or plastic mesh (as for window screens)

4 : a showing of a motion picture

screen·land \'skrēn-ˌland\ *noun* (1925)

: FILMDOM

screen memory *noun* (1923)

: a recollection of early childhood that may be falsely recalled or magnified in importance and that masks another memory of deep emotional significance

screen pass *noun* (circa 1949)

: a forward pass in football to a receiver at or behind the line of scrimmage who is protected by a screen of blockers

screen·play \'skrēn-ˌplā\ *noun* (1916)

: the script and often shooting directions of a story prepared for motion-picture production

screen test *noun* (1927)

: a short film sequence for assessing the ability or suitability of a person for a motion-picture role

— screen–test *transitive verb*

screen·writ·er \'skrēn-ˌrī-tər\ *noun* (1921)

: a writer of screenplays

¹screw \'skrü\ *noun* [Middle English, from Middle French *escroe* female screw, nut, from Medieval Latin *scrofa,* from Latin, sow] (15th century)

1 a : a simple machine of the inclined plane type consisting of a spirally grooved solid cylinder and a correspondingly grooved hollow cylinder into which it fits **b :** a nail-shaped or rod-shaped piece with a spiral groove and a slotted or recessed head designed to be inserted into material by rotating (as with a screwdriver) and used for fastening pieces of solid material together

2 a : a screwlike form **:** SPIRAL **b :** a turn of a screw; *also* **:** a twist like the turn of a screw **c :** a screwlike device (as a corkscrew)

3 : a worn-out horse

4 *chiefly British* **:** a small packet (as of tobacco)

5 : a prison guard

6 : one who bargains shrewdly; *also* **:** SKINFLINT

7 : a propeller especially of a ship

8 a : THUMBSCREW 2 **b :** pressure or punitive measures intended to coerce — used chiefly in the phrase *put the screws on* or *put the screws to*

9 a : an act of sexual intercourse — usually considered vulgar **b :** a partner in sexual intercourse — usually considered vulgar

— screw·like \-ˌlīk\ *adjective*

— have a screw loose : to be mentally unbalanced

²screw (1605)

transitive verb

1 a (1) **:** to attach, fasten, or close by means of a screw (2) **:** to unite or separate by means of a screw or a twisting motion ⟨*screw* the two pieces together⟩ (3) **:** to press tightly in a device (as a vise) operated by a screw (4) **:** to operate, tighten, or adjust by means of a screw (5) **:** to torture by means of a thumbscrew **b :** to cause to rotate spirally about an axis

2 a (1) **:** to twist into strained configurations **:** CONTORT ⟨*screwed* up his face⟩ (2) **:** SQUINT (3) **:** CRUMPLE **b :** to furnish with a spiral groove or ridge **:** THREAD

3 : to increase the intensity, quantity, or capability of ⟨trying to *screw* up courage to confess —Will Scott⟩

4 a (1) **:** to mistreat or exploit through extortion, trickery, or unfair actions; *especially* **:** to deprive of or cheat out of something due or expected ⟨*screwed* out of a job⟩ (2) **:** to treat so as to bring about injury or loss (as to a person's reputation) ⟨use the available Federal machinery to *screw* our political enemies —J. W. Dean III⟩ — often used as a generalized curse ⟨*screw* you!⟩ **b :** to extract by pressure or threat

5 : to copulate with — usually considered vulgar

intransitive verb

1 : to rotate like or as a screw

2 : to turn or move with a twisting or writhing motion

3 : COPULATE — usually considered vulgar

— screw·er *noun*

screw around *intransitive verb* (1939)

1 : to waste time with unproductive activity **:** DALLY

2 : to have sexual relations with someone outside of a marriage or steady relationship **:** be sexually promiscuous

¹screw·ball \'skrü-ˌbȯl\ *noun* (1928)

1 : a baseball pitch that spins and breaks in the opposite direction to a curve

2 : a whimsical, eccentric, or crazy person **:** ZANY

²screwball *adjective* (circa 1936)

: crazily eccentric or whimsical **:** ZANY

screw·bean \'skrü-ˌbēn\ *noun* (1866)

1 : a leguminous shrub or small tree (*Prosopis pubescens*) of the southwestern U.S. and northern Mexico — called also *screwbean mesquite*

2 : a spirally twisted sweet pod that is the fruit of the screwbean

screw·driv·er \'skrü-ˌdrī-vər\ *noun* (1779)

1 : a tool for turning screws

2 : vodka and orange juice served with ice

screw eye *noun* (1873)

: a wood screw with a head in the form of a loop

screw jack *noun* (1719)

: a screw-operated jack for lifting, exerting pressure, or adjusting position (as of a machine part)

screw pine *noun* (1836)

: any of a genus (*Pandanus* of the family Pandanaceae, the screw-pine family) of tropical monocotyledonous Old World plants with slender palmlike stems, often huge prop roots, and terminal crowns of swordlike leaves

screw propeller *noun* (1839)

: PROPELLER

screw thread *noun* (circa 1812)

1 : the projecting helical rib of a screw

2 : one complete turn of a screw thread

screw-up \'skrü-ˌəp\ *noun* (circa 1960)

1 : one who screws up

2 : BOTCH, BLUNDER

screw up (1680)

transitive verb

1 : to tighten, fasten, or lock by or as if by a screw

2 a : BUNGLE, BOTCH **b :** to cause to act or function in a crazy or confused way **:** CONFOUND, DISTURB

intransitive verb

: to botch an activity or undertaking

screw·worm \'skrü-ˌwərm\ *noun* (1879)

: a blowfly (*Cochliomyia hominivorax*) of the warmer parts of America whose larva develops in sores or wounds or in the nostrils of mammals including humans with serious or sometimes fatal results; *especially* **:** its larva

2 : any of several flies other than the screwworm and especially their larvae which parasitize the flesh of mammals

screwy \'skrü-ē\ *adjective* **screw·i·er; -est** (1887)

1 : crazily absurd, eccentric, or unusual

2 : CRAZY, INSANE

— screw·i·ness *noun*

scrib·al \'skrī-bəl\ *adjective* (1857)

: of, relating to, or due to a scribe

scrib·ble \'skri-bəl\ *verb* **scrib·bled; scrib·bling** \-b(ə-)liŋ\ [Middle English *scriblen,* from Medieval Latin *scribillare,* from Latin *scribere* to write] (15th century)

transitive verb

1 : to write hastily or carelessly without regard to legibility or form

2 : to cover with careless or worthless writings or drawings

intransitive verb

: to write or draw hastily and carelessly

— scribble *noun*

scrib·bler \'skri-b(ə-)lər\ *noun* (circa 1553)

1 : one that scribbles

2 : a minor or insignificant author

¹scribe \'skrīb\ *noun* [Middle English, from Latin *scriba* official writer, from *scribere* to write; akin to Greek *skariphasthai* to scratch an outline] (14th century)

1 : one of a learned class in ancient Israel through New Testament times studying the Scriptures and serving as copyists, editors, teachers, and jurists

2 a : an official or public secretary or clerk **b :** a copier of manuscripts
3 : WRITER; *specifically* **:** JOURNALIST

[2]scribe *intransitive verb* **scribed; scrib·ing** (1782)
: to work as a scribe **:** WRITE

[3]scribe *transitive verb* **scribed; scrib·ing** [probably short for *describe*] (1678)
1 : to mark a line on by cutting or scratching with a pointed instrument
2 : to make by cutting or scratching

[4]scribe *noun* (1812)
: SCRIBER

scrib·er \'skrī-bər\ *noun* (circa 1836)
: a sharp-pointed tool for making marks and especially for marking off material (as wood or metal) to be cut

scrieve \'skrēv\ *intransitive verb* [of Scandinavian origin; akin to Old Norse *skrefa* to stride] (1785)
Scottish **:** to move along swiftly and smoothly

scrim \'skrim\ *noun* [origin unknown] (1792)
1 : a durable plain-woven usually cotton fabric for use in clothing, curtains, building, and industry
2 : a theater drop that appears opaque when a scene in front is lighted and transparent or translucent when a scene in back is lighted
3 : something likened to a theater scrim

[1]scrim·mage \'skri-mij\ *noun* [Middle English *scrymmage,* alteration of *skyrmissh* skirmish] (15th century)
1 a : a minor battle **:** SKIRMISH **b :** a confused fight **:** SCUFFLE
2 a : the interplay between two football teams that begins with the snap of the ball and continues until the ball is dead **b :** practice play (as in football or basketball) between two squads

[2]scrimmage *verb* **scrim·maged; scrim·mag·ing** (circa 1825)
intransitive verb
: to take part in a scrimmage
transitive verb
: to play a scrimmage against
— scrim·mag·er *noun*

scrimmage line *noun* (circa 1909)
: LINE OF SCRIMMAGE

scrimp \'skrimp\ *verb* [perhaps of Scandinavian origin; akin to Swedish *skrympa* to shrink, Middle Low German *schrempen* to contract — more at SHRIMP] (circa 1774)
transitive verb
1 : to be stingy in providing for
2 : to make too small, short, or scanty
intransitive verb
: to be frugal or stingy
— scrimpy \'skrim-pē\ *adjective*

scrim·shan·der \'skrim-,shan-dər\ *noun* [origin unknown] (1851)
: a person who creates scrimshaw

[1]scrim·shaw \'skrim-,shȯ\ *verb* [origin unknown] (circa 1826)
transitive verb
: to carve or engrave into scrimshaw
intransitive verb
: to produce scrimshaw

[2]scrimshaw *noun* (circa 1864)
1 : any of various carved or engraved articles made originally by American whalers usually from baleen or whale ivory
2 : scrimshawed work
3 : the art, practice, or technique of producing scrimshaw

[1]scrip \'skrip\ *noun* [Middle English *scrippe,* from Medieval Latin *scrippum* pilgrim's knapsack] (13th century)
archaic **:** a small bag or wallet

[2]scrip *noun* [alteration of *script*] (1590)
1 : a short writing (as a certificate, schedule, or list)
2 : a small piece
3 a : any of various documents used as evidence that the holder or bearer is entitled to receive something (as a fractional share of stock or an allotment of land) **b :** paper currency or a token issued for temporary use in an emergency

[1]script \'skript\ *noun* [Middle English, from Latin *scriptum* thing written, from neuter of *scriptus,* past participle of *scribere* to write — more at SCRIBE] (14th century)
1 a : something written **:** TEXT **b :** an original or principal instrument or document **c** (1) **:** MANUSCRIPT 1 (2) **:** the written text of a stage play, screenplay, or broadcast; *specifically* **:** the one used in production or performance
2 a : a style of printed letters that resembles handwriting **b :** written characters **:** HANDWRITING **c :** ALPHABET
3 : a plan of action

[2]script *transitive verb* (1935)
1 : to prepare a script for or from
2 : to provide carefully considered details for (as a plan of action)

script·er \'skrip-tər\ *noun* (1939)
: SCRIPTWRITER

scrip·to·ri·um \skrip-'tōr-ē-əm, -'tȯr-\ *noun, plural* **-ria** \-ē-ə\ [Medieval Latin, from Latin *scribere*] (1774)
: a copying room for the scribes in a medieval monastery

scrip·tur·al \'skrip(t)-sh(ə-)rəl\ *adjective* (1641)
: of, relating to, contained in, or according to a sacred writing; *especially* **:** BIBLICAL
— scrip·tur·al·ly *adverb*

scrip·ture \'skrip(t)-shər\ *noun* [Middle English, from Late Latin *scriptura,* from Latin, act or product of writing, from *scriptus*] (14th century)
1 a (1) *capitalized* **:** the books of the Bible — often used in plural (2) *often capitalized* **:** a passage from the Bible **b :** a body of writings considered sacred or authoritative
2 : something written ⟨the primitive man's awe for any *scripture* —George Santayana⟩

script·writ·er \'skrip(t)-,rī-tər\ *noun* (1935)
: a person who writes scripts

scriv·en·er \'skriv-nər, 'skri-və-\ *noun* [Middle English *scriveiner,* alteration of *scrivein,* from Middle French *escrivein,* from (assumed) Vulgar Latin *scriban-, scriba,* alteration of Latin *scriba* scribe] (14th century)
1 : a professional or public copyist or writer **:** SCRIBE
2 : NOTARY PUBLIC

scrod \'skräd\ *noun* [origin unknown] (1841)
: a young fish (as a cod or haddock); *especially* **:** one split and boned for cooking

scrof·u·la \'skrȯ-fyə-lə, 'skrä-\ *noun* [New Latin, back-formation from Late Latin *scrofulae,* plural, swellings of the lymph nodes of the neck, from plural of *scrofula,* diminutive of Latin *scrofa* breeding sow] (1791)
: tuberculosis of lymph nodes especially in the neck

scrof·u·lous \-ləs\ *adjective* (1612)
1 : of, relating to, or affected with scrofula
2 a : having a diseased run-down appearance **b :** morally contaminated

[1]scroll \'skrōl\ *noun* [Middle English *scrowle,* blend of *rolle* roll and *scrowe* scrap, scroll (from Middle French *escroue,* of Germanic origin; akin to Old English *screade* shred)] (15th century)
1 a : a roll (as of papyrus, leather, or parchment) for writing a document **b** *archaic* **:** a written message **c :** ROSTER, LIST **d :** a riband with rolled ends often inscribed with a motto
2 a : something resembling a scroll in shape; *especially* **:** a spiral or convoluted form in ornamental design derived from the curves of a loosely or partly rolled parchment scroll **b :** the curved head of a bowed stringed musical instrument — see VIOLIN illustration

scroll 1

[2]scroll (1973)
intransitive verb
: to move text or graphics up or down or across a display screen as if by unrolling a scroll
transitive verb
: to cause (text or graphics on a display screen) to move in scrolling

scroll saw *noun* (1851)
1 : FRETSAW
2 : JIGSAW 1

scroll·work \'skrōl-,wərk\ *noun* (1739)
: ornamentation characterized by scrolls; *especially* **:** fancy designs in wood often made with a scroll saw

scrooge \'skrüj\ *noun, often capitalized* [Ebenezer *Scrooge,* character in the story *A Christmas Carol* (1843) by Charles Dickens] (1899)
: a miserly person

scro·tum \'skrō-təm\ *noun, plural* **scro·ta** \-tə\ *or* **scrotums** [Latin; akin to Latin *scrautum* quiver] (1597)
: the external pouch that in most mammals contains the testes
— scro·tal \'skrō-tᵊl\ *adjective*

scrouge \'skrau̇j, 'skrüj\ *verb* **scrouged; scroug·ing** [alteration of English dialect *scruze* to squeeze] (1755)
chiefly dialect **:** CROWD, PRESS

scrounge \'skrau̇nj\ *verb* **scrounged; scroung·ing** [alteration of English dialect *scrunge* to wander about idly] (circa 1909)
transitive verb
1 : STEAL, SWIPE
2 a : to get as needed by or as if by foraging, scavenging, or borrowing **b :** FINAGLE, WHEEDLE — often used with *up*
intransitive verb
: to search about and turn up something needed from whatever source is available; *also* **:** to actively seek money, work, or sustenance from any available source
— scroung·er *noun*

scroungy \'skrau̇n-jē\ *adjective* **scroung·i·er; -est** (circa 1960)
: being shabby, dirty, or unkempt

[1]scrub \'skrəb\ *noun, often attributive* [Middle English, alteration of *schrobbe* shrub — more at SHRUB] (14th century)
1 a : a stunted tree or shrub **b :** vegetation consisting chiefly of scrubs **c :** a tract covered with scrub
2 : a domestic animal of mixed or unknown parentage and usually inferior conformation **:** MONGREL
3 : a person of insignificant size or standing
4 : a player not belonging to the first string

[2]scrub *verb* **scrubbed; scrub·bing** [of Low German or Scandinavian origin; akin to Middle Low German & Middle Dutch *schrubben* to scrub, Swedish *skrubba*] (circa 1595)
transitive verb
1 a (1) **:** to clean with hard rubbing **:** SCOUR (2) **:** to remove by scrubbing **b :** to subject to friction **:** RUB
2 : WASH 6c(2)
3 : CANCEL, ELIMINATE
intransitive verb
1 : to use hard rubbing in cleaning
2 : to prepare for surgery by scrubbing oneself
— scrub·ba·ble \'skrə-bə-bəl\ *adjective*

[3]scrub *noun* (1621)
1 : an act or instance of scrubbing; *especially* **:** CANCELLATION
2 : one that scrubs

[1]scrubbed \'skrə-bəd\ *adjective* [[1]*scrub*] (1596)
archaic **:** SCRUBBY 1

²scrubbed \'skrəbd\ *adjective* [from past participle of ²*scrub*] (1948)
: giving the impression of being clean or wholesome as if from scrubbing ⟨days when studios manufactured *scrubbed* public images for their stars —Sally Helgesen⟩
scrub·ber \'skrə-bər\ *noun* (1839)
: one that scrubs; *especially* **:** an apparatus for removing impurities especially from gases
scrub brush *noun* (1897)
: a brush with hard bristles for heavy cleaning — called also *scrubbing brush*
scrub·by \'skrə-bē\ *adjective* **scrub·bi·er; -est** [¹*scrub*] (1591)
1 : inferior in size or quality **:** STUNTED ⟨*scrubby* cattle⟩
2 : covered with or consisting of scrub
3 : SHABBY, PALTRY
scrub·land \'skrəb-ˌland\ *noun* (1779)
: land covered with scrub
scrub nurse *noun* (1905)
: a nurse who assists the surgeon in an operating room
scrub oak *noun* (1766)
: any of various chiefly American oaks (as *Quercus ilicifolia* of the northeastern U.S.) of small size and usually shrubby habit
scrub pine *noun* (1791)
: a pine of dwarf, straggly, or scrubby growth usually by reason of environmental conditions; *specifically* **:** a pine tree unsuitable for lumber by reason of inferior or defective growth
scrub typhus *noun* (1929)
: TSUTSUGAMUSHI DISEASE
scrub·wom·an \'skrəb-ˌwu̇-mən\ *noun* (1873)
: CHARWOMAN
scruff \'skrəf\ *noun* [alteration of earlier *scuff,* of unknown origin] (1790)
: the back of the neck **:** NAPE
scruffy \'skrə-fē\ *adjective* **scruff·i·er; -est** [English dialect *scruff* something worthless] (1871)
: UNKEMPT, SLOVENLY, SHAGGY
— scruff·i·ly \-fə-lē\ *adverb*
— scruff·i·ness \-fē-nəs\ *noun*
scrum \'skrəm\ *noun* [short for *scrummage,* alteration of *scrimmage*] (1857)
1 *or* **scrum·mage** \'skrə-mij\ **:** a rugby play in which the forwards of each side come together in a tight formation and struggle to gain possession of the ball when it is tossed in among them
2 *British* **:** MADHOUSE 2
— scrummage *intransitive verb*
scrump·tious \'skrəm(p)-shəs\ *adjective* [perhaps alteration of *sumptuous*] (1830)
: DELIGHTFUL, EXCELLENT; *especially* **:** DELICIOUS
— scrump·tious·ly *adverb*
¹scrunch \'skrənch, 'skrünch\ *verb* [alteration of ¹*crunch*] (circa 1790)
transitive verb
1 : CRUNCH, CRUSH
2 a : to draw or squeeze together tightly **b :** CRUMPLE — often used with *up* **c :** to cause (as one's features) to draw together — usually used with *up*
intransitive verb
1 : to move with or make a crunching sound
2 : CROUCH, HUNCH; *also* **:** SQUEEZE
²scrunch *noun* (1857)
: a crunching sound
¹scru·ple \'skrü-pəl\ *noun* [Middle English *scriple,* from Latin *scrupulus* a unit of weight, diminutive of *scrupus* sharp stone] (14th century)
1 — see WEIGHT table
2 : a minute part or quantity **:** IOTA
²scruple *noun* [Middle English *scrupul,* from Middle French *scrupule,* from Latin *scrupulus,* diminutive of *scrupus* source of uneasiness, literally, sharp stone] (15th century)
1 : an ethical consideration or principle that inhibits action
2 : the quality or state of being scrupulous
3 : mental reservation ◆
synonym see QUALM
³scruple *intransitive verb* **scru·pled; scru·pling** \-p(ə-)liŋ\ (1627)
1 : to have scruples
2 : to show reluctance on grounds of conscience **:** HESITATE
scru·pu·los·i·ty \ˌskrü-pyə-'lä-sə-tē\ *noun* (1526)
1 : the quality or state of being scrupulous
2 : ²SCRUPLE 1
scru·pu·lous \'skrü-pyə-ləs\ *adjective* [Middle English, from Latin *scrupulosus,* from *scrupulus*] (15th century)
1 : having moral integrity **:** acting in strict regard for what is considered right or proper
2 : punctiliously exact **:** PAINSTAKING ⟨working with *scrupulous* care⟩
synonym see UPRIGHT, CAREFUL
word history see SCRUPLE
— scru·pu·lous·ly *adverb*
— scru·pu·lous·ness *noun*
scru·ta·ble \'skrü-tə-bəl\ *adjective* [Late Latin *scrutabilis* searchable, from Latin *scrutari*] (circa 1600)
: capable of being deciphered **:** COMPREHENSIBLE
scru·ti·neer \ˌskrü-tᵊn-'ir\ *noun* (1557)
1 : one that examines
2 *British* **:** one who takes or counts votes
scru·ti·nise *British variant of* SCRUTINIZE
scru·ti·nize \'skrü-tᵊn-ˌīz\ *verb* **-nized; -niz·ing** (1671)
transitive verb
: to examine closely and minutely
intransitive verb
: to make a scrutiny ☆
— scru·ti·niz·er *noun*
scru·ti·ny \'skrü-tᵊn-ē, 'skrüt-nē\ *noun, plural* **-nies** [Latin *scrutinium,* from *scrutari* to search, examine, probably from *scruta* trash] (1604)
1 : a searching study, inquiry, or inspection **:** EXAMINATION
2 : a searching look
3 : close watch **:** SURVEILLANCE
scu·ba \'skü-bə\ *noun, often attributive* [*s*elf-*c*ontained *u*nderwater *b*reathing *a*pparatus] (1952)
: an apparatus utilizing a portable supply of compressed gas (as air) supplied at a regulated pressure and used for breathing while swimming underwater
scuba diver *noun* (1958)
: one who swims underwater with the aid of scuba gear
— scuba dive *intransitive verb*
¹scud \'skəd\ *intransitive verb* **scud·ded; scud·ding** [probably of Scandinavian origin; akin to Norwegian *skudda* to push] (1532)
1 : to move or run swiftly especially as if driven forward
2 : to run before a gale
²scud *noun* (1609)
1 : the action of scudding **:** RUSH
2 a : loose vapory clouds driven swiftly by the wind **b** (1) **:** a slight sudden shower (2) **:** mist, rain, snow, or spray driven by the wind **c :** a gust of wind
scu·do \'skü-(ˌ)dō\ *noun, plural* **scu·di** \-(ˌ)dē\ [Italian, literally, shield, from Latin *scutum* — more at ESQUIRE] (1644)
1 : a gold or silver coin formerly used in Italy
2 : a unit of value equivalent to a scudo
¹scuff \'skəf\ *verb* [probably of Scandinavian origin; akin to Swedish *skuffa* to push] (1768)
intransitive verb
1 a : to walk without lifting the feet **:** SHUFFLE **b :** to poke or shuffle a foot in exploration or embarrassment
2 : to become scratched, chipped, or roughened by wear
transitive verb
1 : ³CUFF
2 a : to scrape (the feet) along a surface while walking or back and forth while standing **b :** to poke at with the toe
3 : to scratch, gouge, or wear away the surface of
²scuff *noun* (1899)
1 a : a noise of or as if of scuffing **b :** the act or an instance of scuffing **c :** a mark or injury caused by scuffing
2 : a flat-soled slipper without quarter or heel strap — compare MULE
scuf·fle \'skə-fəl\ *intransitive verb* **scuf·fled; scuf·fling** \-f(ə-)liŋ\ [probably of Scandinavian origin; akin to Swedish *skuffa* to push] (1590)
1 a : to struggle at close quarters with disorder and confusion **b :** to struggle (as by working odd jobs) to get by
2 a : to move with a quick shuffling gait **:** SCURRY **b :** SHUFFLE
— scuffle *noun*
scuffle hoe *noun* (1856)
: a garden hoe that has both edges sharpened and can be pushed forward or drawn back
¹scull \'skəl\ *noun* [Middle English *sculle*] (14th century)
1 a : an oar used at the stern of a boat to propel it forward with a thwartwise motion **b :** either of a pair of oars usually less than 10 feet (3 meters) in length and operated by one person
2 : a racing shell propelled by one or two persons using sculls
²scull (1624)
transitive verb
: to propel (a boat) by sculls or by a large oar worked thwartwise
intransitive verb
: to scull a boat
— scull·er *noun*
scul·lery \'skə-lə-rē, 'skəl-rē\ *noun, plural* **-ler·ies** [Middle English, department of household in charge of dishes, from Middle French *escuelerie,* from *escuelle* bowl, from Latin *scutella* drinking bowl — more at SCUTTLE] (15th century)

☆ SYNONYMS
Scrutinize, scan, inspect, examine mean to look at or over. SCRUTINIZE stresses close attention to minute detail ⟨*scrutinized* the hospital bill⟩. SCAN implies a surveying from point to point often suggesting a cursory overall observation ⟨*scanned* the wine list⟩. INSPECT implies scrutinizing for errors or defects ⟨*inspected* my credentials⟩. EXAMINE suggests a scrutiny in order to determine the nature, condition, or quality of a thing ⟨*examined* the specimens⟩.

◇ WORD HISTORY
scruple In Latin *scrupus* meant literally "sharp stone" and metaphorically "source of uneasiness," alluding to the way a sharp pebble that has worked its way into one's shoe becomes a nagging cause of discomfort. The much more commonly used diminutive form *scrupulus* had the same metaphorical meaning "source of anxiety." The derived adjective *scrupulosus* had likewise both a concrete meaning, "full of sharp rocks, jagged," and a metaphorical one—not "full of sources of anxiety," as one might expect, but "marked by careful attention to detail, taking precautions." Thus it denoted the attitude of someone following a rocky path and trying to avoid a spill. Both *scrupulus* and *scrupulosus* found their way into English in their metaphorical senses, respectively as *scruple* "an ethical consideration" and *scrupulous.*

: a room for cleaning and storing dishes and cooking utensils and for doing messy kitchen work

scul·lion \'skəl-yən\ *noun* [Middle English *sculion,* from Middle French *escouillon* dishcloth, alteration of *escouvillon,* from *escouve* broom, from Latin *scopae,* literally, twigs bound together] (15th century)
: a kitchen helper

scul·pin \'skəl-pən\ *noun, plural* **sculpins** *also* **sculpin** [origin unknown] (1672)
1 : any of a family (Cottidae) of spiny large-headed broad-mouthed often scaleless bony fishes
2 : a scorpion fish (*Scorpaena guttata*) of the southern California coast caught for food and sport

sculpin 1

sculpt \'skəlpt\ *verb* [French *sculpter,* alteration of obsolete *sculper,* from Latin *sculpere*] (1864)
: CARVE, SCULPTURE

sculp·tor \'skəlp-tər\ *noun* [Latin, from *sculpere*] (1634)
: an artist who makes sculptures

sculp·tress \-trəs\ *noun* (1662)
: a woman artist who makes sculptures

sculp·tur·al \'skəlp-chə-rəl, 'skəlp-shrəl\ *adjective* (1819)
1 : of or relating to sculpture
2 : resembling sculpture **:** SCULPTURESQUE
— **sculp·tur·al·ly** *adverb*

¹sculp·ture \'skəlp-chər\ *noun* [Middle English, from Latin *sculptura,* from *sculptus,* past participle of *sculpere* to carve, alteration of *scalpere* to scratch, carve] (14th century)
1 a : the action or art of processing (as by carving, modeling, or welding) plastic or hard materials into works of art **b** (1) **:** work produced by sculpture (2) **:** a three-dimensional work of art (as a statue)
2 : impressed or raised markings or a pattern of such especially on a plant or animal part

²sculpture *verb* **sculp·tured; sculp·tur·ing** \'skəlp-chə-riŋ, 'skəlp-shriŋ\ (1645)
transitive verb
1 a : to form an image or representation of from solid material (as wood or stone) **b :** to form into a three-dimensional work of art
2 : to change (the form of the earth's surface) by natural processes (as erosion and deposition)
3 : to shape by or as if by carving or molding
intransitive verb
: to work as a sculptor

sculp·tur·esque \ˌskəlp-chə-'resk\ *adjective* (1835)
: done in the manner of or resembling sculpture
— **sculp·tur·esque·ly** *adverb*

¹scum \'skəm\ *noun* [Middle English, from Middle Dutch *schum;* akin to Old High German *scūm* foam] (14th century)
1 a : extraneous matter or impurities risen to or formed on the surface of a liquid often as a foul filmy covering **b :** the scoria of metals in a molten state **:** DROSS **c :** a slimy film on a solid or gelatinous object
2 a : REFUSE **b :** a low, vile, or worthless person or group of people
— **scum·my** \'skə-mē\ *adjective*

²scum *intransitive verb* **scummed; scum·ming** (1661)
: to become covered with or as if with scum

scum·bag \'skəm-ˌbag *also* -ˌbäg\ *noun* (1967)
slang **:** a dirty or despicable person

¹scum·ble \'skəm-bəl\ *transitive verb* **scum·bled; scum·bling** \-b(ə-)liŋ\ [perhaps frequentative of ²*scum*] (1798)
1 a : to make (as color or a painting) less brilliant by covering with a thin coat of opaque or semiopaque color **b :** to apply (a color) in this manner
2 : to soften the lines or colors of (a drawing) by rubbing lightly

²scumble *noun* (1834)
1 : the act or effect of scumbling
2 : a material used for scumbling

scun·gil·li \skün-'jē-lē, -'gē-, -'ji-\ *noun* [Italian dialect (Neapolitan) *scuncigli,* plural of *scunciglio* conch] (1945)
: conch used as food

¹scun·ner \'skə-nər\ *intransitive verb* [Middle English (Scots) *skunniren*] (14th century)
chiefly Scottish **:** to be in a state of disgusted irritation

²scunner *noun* (circa 1520)
: an unreasonable or extreme dislike or prejudice

scup \'skəp\ *noun, plural* **scup** *also* **scups** [short for *scuppaug,* modification of Narraganset *mishcùppaûog*] (circa 1848)
: a porgy (*Stenotomus chrysops*) occurring along the Atlantic coast of the U.S. chiefly from North Carolina to Maine and used as a panfish

¹scup·per \'skə-pər\ *noun* [Middle English *skopper*] (15th century)
1 : an opening cut through the bulwarks of a ship so that water falling on deck may flow overboard
2 : an opening in the wall of a building through which water can drain from a floor or flat roof

²scupper *transitive verb* [origin unknown] (1899)
British **:** to defeat or put an end to **:** DO IN 1a

scup·per·nong \-ˌnȯŋ, -ˌnäŋ\ *noun* [*Scuppernong,* river and lake in North Carolina] (1811)
1 : MUSCADINE; *especially* **:** a cultivated muscadine with yellowish green plum-flavored fruits
2 : a sweet aromatic amber-colored wine made from scuppernongs

scurf \'skərf\ *noun* [Middle English, from Old English, of Scandinavian origin; akin to Icelandic *skurfa* scurf; akin to Old High German *scorf* scurf, Old English *sceorfan* to scarify] (before 12th century)
1 : thin dry scales detached from the epidermis especially in an abnormal skin condition; *specifically* **:** DANDRUFF
2 a : something like flakes or scales adhering to a surface **b :** the foul remains of something adherent
3 a : a scaly deposit or covering on some plant parts; *also* **:** a localized or general darkening and roughening of a plant surface usually more pronounced than russeting **b :** a plant disease characterized by scurf
— **scurfy** \'skər-fē\ *adjective*

scur·rile *or* **scur·ril** \'skər-əl, 'skə-rəl\ *adjective* [Middle French *scurrile,* from Latin *scurrilis,* from *scurra* buffoon] (1567)
: SCURRILOUS

scur·ril·i·ty \skə-'ri-lə-tē\ *noun, plural* **-ties** (1508)
1 : the quality or state of being scurrilous
2 a : scurrilous or abusive language **b :** an offensively rude or abusive remark

scur·ri·lous \'skər-ə-ləs, 'skə-rə-\ *adjective* (1576)
1 a : using or given to coarse language **b :** being vulgar and evil ⟨*scurrilous* imposters who used a religious exterior to rob poor people —Edwin Benson⟩
2 : containing obscenities, abuse, or slander ⟨a . . . campaign filled with *scurrilous* charges and countercharges —A. D. Graeff⟩
— **scur·ri·lous·ly** *adverb*
— **scur·ri·lous·ness** *noun*

scur·ry \'skər-ē, 'skə-rē\ *intransitive verb* **scur·ried; scur·ry·ing** [short for *hurry-scurry,* reduplication of *hurry*] (1810)
1 : to move in or as if in a brisk pace **:** SCAMPER
2 : to move around in an agitated, confused, or fluttering manner
— **scurry** *noun*

¹scur·vy \'skər-vē\ *noun* [²*scurvy*] (circa 1565)
: a disease marked by spongy gums, loosening of the teeth, and a bleeding into the skin and mucous membranes and caused by a lack of vitamin C

²scurvy *adjective* [*scurf*] (1579)
: disgustingly mean or contemptible **:** DESPICABLE ⟨a *scurvy* trick⟩
synonym see CONTEMPTIBLE
— **scur·vi·ly** \-və-lē\ *adverb*
— **scur·vi·ness** \-vē-nəs\ *noun*

scurvy grass *noun* (1597)
: a cruciferous herb (as *Cochlearia officinalis*) formerly believed useful in preventing or treating scurvy

scut \'skət\ *noun* [origin unknown] (circa 1530)
: a short erect tail (as of a hare)

scu·tage \'skü-tij, 'skyü-\ *noun* [Middle English, from Medieval Latin *scutagium,* from Latin *scutum* shield — more at ESQUIRE] (15th century)
: a tax levied on a vassal or a knight in lieu of military service

¹scutch \'skəch\ *transitive verb* [obsolete French *escoucher,* from (assumed) Vulgar Latin *excuticare* to beat out, from Latin *excutere,* from *ex-* + *quatere* to shake, strike] (1733)
: to separate the woody fiber from (flax or hemp) by beating

²scutch *noun* (circa 1791)
1 : SCUTCHER
2 : a bricklayer's hammer for cutting, trimming, and dressing bricks

scutch·eon \'skə-chən\ *noun* [Middle English *scochon,* from Middle French *escuchon*] (14th century)
: ESCUTCHEON

scutch·er \'skə-chər\ *noun* (1776)
: an implement or machine for scutching flax or cotton

scute \'sküt, 'skyüt\ *noun* [New Latin *scutum,* from Latin, shield — more at ESQUIRE] (1898)
: an external bony or horny plate or large scale

scu·tel·late \skü-'te-lət, skyü-; 'sk(y)ü-tᵊl-ˌāt\ *or* **scu·tel·lat·ed** \'sk(y)ü-tᵊl-ˌā-təd\ *adjective* (1785)
: having or covered with scutella

scu·tel·lum \skü-'te-ləm, skyü-\ *noun, plural* **-la** \-lə\ [New Latin, diminutive of Latin *scutum* shield] (circa 1760)
1 : a hard plate or scale (as on the thorax of an insect or the tarsus of a bird)
2 : the shield-shaped cotyledon of a monocotyledon (as a grass)
— **scu·tel·lar** \-lər\ *adjective*

scut·ter \'skə-tər\ *intransitive verb* [alteration of ⁵*scuttle*] (1781)
: SCURRY, SCAMPER

¹scut·tle \'skə-tᵊl\ *noun* [Middle English *scutel,* from Latin *scutella* drinking bowl, tray, diminutive of *scutra* platter] (15th century)
1 : a shallow open basket for carrying something (as grain or garden produce)
2 : a metal pail that usually has a bail and a sloped lip and is used especially for carrying coal

²scuttle *noun* [Middle English *skottell*] (15th century)
1 : a small opening in a wall or roof furnished with a lid: as **a :** a small opening or hatchway in the deck of a ship large enough to admit a person and with a lid for covering it **b :** a small hole in the side or bottom of a ship fitted with a covering or glazed
2 : a covering that closes a scuttle

³scuttle *transitive verb* **scut·tled; scut·tling** \'skət-liŋ, 'skə-tᵊl-iŋ\ (1642)

1 : to cut a hole through the bottom, deck, or side of (a ship); *specifically* **:** to sink or attempt to sink by making holes through the bottom
2 : DESTROY, WRECK; *also* **:** SCRAP 2

[4]scuttle *noun* [perhaps blend of *scud* and *shuttle*] (1623)
1 : a quick shuffling pace
2 : a short swift run

[5]scuttle *intransitive verb* **scut·tled; scut·tling** \'skət-liŋ, 'skə-tᵊl-iŋ\ (1657)
: SCURRY

scut·tle·butt \'skə-tᵊl-ˌbət\ *noun* [[2]*scuttle* + [3]*butt*] (1805)
1 a : a cask on shipboard to contain fresh water for a day's use **b :** a drinking fountain on a ship or at a naval or marine installation
2 : RUMOR, GOSSIP

scu·tum \'skü-təm, 'skyü-\ *noun, plural* **scu·ta** \-tə\ [New Latin, from Latin, shield — more at ESQUIRE] (1771)
: a bony, horny, or chitinous plate **:** SCUTE

scut work \'skət-\ *noun* [probably from medical argot *scut* junior intern] (circa 1962)
: routine and often menial labor

scuz·zy \'skə-zē\ *adjective* **scuz·zi·er; -est** [origin unknown] (1969)
slang **:** dirty, shabby, or foul in condition or character

Scyl·la \'si-lə\ *noun* [Latin, from Greek *Skyllē*]
: a nymph changed into a monster in Greek mythology who terrorizes mariners in the Strait of Messina
— between Scylla and Charybdis
: between two equally hazardous alternatives

scy·phis·to·ma \sī-'fis-tə-mə\ *noun, plural* **-mae** \-(ˌ)mē\ *also* **-mas** [New Latin, from Latin *scyphus* cup + Greek *stoma* mouth — more at STOMACH] (1878)
: a sexually produced scyphozoan larva that ultimately repeatedly constricts transversely to form free-swimming medusae

scy·pho·zo·an \ˌsī-fə-'zō-ən\ *noun* [New Latin *Scyphozoa,* from Latin *scyphus* + New Latin *-zoa*] (circa 1909)
: any of a class (Scyphozoa) of coelenterates that comprise jellyfishes lacking a true polyp and usually a velum
— scyphozoan *adjective*

[1]scythe \'sīth, 'sī\ *noun* [Middle English *sithe,* from Old English *sīthe;* akin to Old English *sagu* saw — more at SAW] (before 12th century)
: an implement used for mowing (as grass) and composed of a long curving blade fastened at an angle to a long handle

scythe

[2]scythe *verb* **scythed; scyth·ing** (circa 1580)
intransitive verb
: to use a scythe
transitive verb
: to cut with or as if with a scythe

Scyth·i·an \'si-thē-ən, -thē-\ *noun* [Latin *Scytha,* from Greek *Skythēs*] (15th century)
1 : a member of an ancient nomadic people inhabiting Scythia
2 : the Iranian language of the Scythians
— Scythian *adjective*

sea \'sē\ *noun* [Middle English *see,* from Old English *sǣ;* akin to Old High German *sē* sea, Gothic *saiws*] (before 12th century)
1 a : a great body of salty water that covers much of the earth; *broadly* **:** the waters of the earth as distinguished from the land and air **b :** a body of salt water of second rank more or less landlocked ⟨the Mediterranean *sea*⟩ **c :** OCEAN **d :** an inland body of water — used especially for names of such bodies ⟨the Caspian *Sea*⟩ ⟨the *Sea* of Galilee⟩
2 a : surface motion on a large body of water or its direction; *also* **:** a large swell or wave — often used in plural ⟨heavy *seas*⟩ **b :** the disturbance of the ocean or other body of water due to the wind
3 : something likened to the sea especially in vastness ⟨the crowd was a *sea* of faces⟩
4 : the seafaring life
5 : [3]MARE
— sea *adjective*
— at sea 1 : on the sea; *specifically* **:** on a sea voyage **2 :** LOST, BEWILDERED
— to sea : to or on the open waters of the sea

sea anchor *noun* (1769)
: a drag typically of canvas thrown overboard to retard the drifting of a ship or seaplane and to keep its head to the wind

sea anemone *noun* (1742)
: any of numerous usually solitary anthozoan polyps (order Actiniaria) whose form, bright and varied colors, and cluster of tentacles superficially resemble a flower

sea·bag \'sē-ˌbag\ *noun* (1919)
: a cylindrical canvas bag used especially by a sailor for clothes and other gear

sea bass *noun* (1765)
1 : any of numerous marine bony fishes (family Serranidae) that are usually smaller and more active than the groupers; *especially* **:** a food and sport fish (*Centropristis striata*) of the Atlantic coast of the U.S.
2 : any of numerous croakers or drums including noted sport and food fishes

sea·bed \-ˌbed\ *noun* (1838)
: the floor of a sea or ocean

Sea·bee \'sē-(ˌ)bē\ *noun* [alteration of *cee* + *bee;* from the initials of *construction battalion*] (1942)
: a member of one of the U.S. Navy construction battalions for building naval shore facilities in combat zones

sea·bird \'sē-ˌbərd\ *noun* (1589)
: a bird (as a gull or albatross) frequenting the open ocean

sea biscuit *noun* (circa 1690)
: HARDTACK 1

sea·board \'sē-ˌbōrd, -ˌbȯrd\ *noun* (1788)
: SEACOAST; *also* **:** the country bordering a seacoast
— seaboard *adjective*

sea·boot \-ˌbüt\ *noun* (1851)
: a very high waterproof boot used especially by sailors and fishermen

sea·borg·i·um \sē-'bȯr-gē-əm\ *noun* [New Latin, from Glenn T. *Seaborg*] (1994)
: UNNILHEXIUM

sea·borne \-ˌbōrn, -ˌbȯrn\ *adjective* (1823)
1 : borne over or on the sea ⟨a *seaborne* invasion⟩
2 : carried on by oversea shipping ⟨*seaborne* trade⟩

sea bream *noun* (circa 1530)
: any of numerous marine bony fishes (as of the family Sparidae)

sea breeze *noun* (1697)
: a cooling breeze blowing generally in the daytime inland from the sea

sea captain *noun* (1612)
: the master especially of a merchant vessel

sea change *noun* (1610)
1 *archaic* **:** a change brought about by the sea
2 : a marked change **:** TRANSFORMATION

sea chest *noun* (1669)
: a sailor's storage chest for personal property

sea·coast \'sē-ˌkōst\ *noun* (14th century)
: the shore or border of the land adjacent to the sea

sea cow *noun* (1613)
: SIRENIAN

sea·craft \'sē-ˌkraft\ *noun* (1727)
1 : skill in navigation
2 : seagoing ships

sea crayfish *noun* (1601)
: SPINY LOBSTER

sea cucumber *noun* (1601)
: any of a class (Holothurioidea) of echinoderms having a tough muscular elongate body with tentacles surrounding the mouth — called also *holothurian*

sea devil *noun* (1634)
: DEVILFISH 1

sea dog *noun* (1823)
: a veteran sailor

sea duck *noun* (1753)
: a diving duck (as a scoter, merganser, or eider) that frequents the sea

sea duty *noun* (1946)
: duty in the U.S. Navy performed with a deployable unit (as a ship or aircraft squadron)

sea eagle *noun* (1668)
: any of various fish-eating eagles (especially genus *Haliaeetus*)

Sea Explorer *noun* (1948)
: an Explorer in a scouting program that teaches seamanship

sea fan *noun* (1633)
: a gorgonian with a fan-shaped skeleton; *especially* **:** one (*Gorgonia flabellum*) of Florida and the West Indies

sea·far·er \'sē-ˌfar-ər, -ˌfer-\ *noun* [*sea* + [1]*fare* + [2]*-er*] (1513)
: MARINER

sea·far·ing \-ˌfar-iŋ, -ˌfer-\ *noun* (1592)
: the use of the sea for travel or transportation
— seafaring *adjective*

sea fire *noun* (1814)
: marine bioluminescence

sea·floor \'sē-ˌflōr, -ˌflȯr\ *noun* (1855)
: SEABED

sea·food \-ˌfüd\ *noun* (1836)
: edible marine fish and shellfish

sea·fowl \-ˌfau̇l\ *noun* (14th century)
: SEABIRD

sea·front \-ˌfrənt\ *noun* (1879)
: the waterfront of a seaside place

sea·girt \'sē-ˌgərt\ *adjective* (1616)
: surrounded by the sea

sea·go·ing \-ˌgō-iŋ, -ˌgȯ(-)iŋ\ *adjective* (1828)
: OCEANGOING

sea grape *noun* (1806)
: a tree (*Coccoloba uvifera*) of the buckwheat family that inhabits sandy shores from Florida to South America, has rounded leaves, and bears clusters of purple to whitish edible berries

sea grass *noun* (1578)
: any of various grasslike plants that inhabit coastal areas; *especially* **:** EELGRASS 1

sea green *noun* (1598)
1 : a moderate green or bluish green
2 : a moderate yellow green

sea·gull \'sē-ˌgəl\ *noun* (1542)
: a gull frequenting the sea; *broadly* **:** GULL

sea hare *noun* (1593)
: any of various large opisthobranch mollusks (especially genus *Aplysia*) that have an arched back and two anterior tentacles and have the shell much reduced or missing

sea holly *noun* (1548)
: a European coastal herb (*Eryngium maritimum*) of the carrot family with spiny leaves and pale blue flowers

sea horse *noun* (circa 1500)
1 : WALRUS
2 : a mythical creature half horse and half fish
3 : any of a genus (*Hippocampus* of the family Syngnathidae) of small bony fishes that have the head angled downward toward the body which is carried vertically and are equipped with a prehensile tail

sea horse 3

sea island cotton *noun, often S&I capitalized* [Sea Islands, chain of islands off the southeastern U.S. coast] (1805)
: a cotton (*Gossypium barbadense*) with especially long silky fiber — called also *sea island*

sea kale *noun* (1699)
: a succulent Eurasian perennial herb (*Crambe maritima*) of the mustard family used as a potherb

sea king *noun* (1819)
: a Norse pirate chief

[1]seal \ˈsē(ə)l\ *noun, plural* **seals** *also* **seal** [Middle English *sele,* from Old English *seolh;* akin to Old High German *selah* seal] (before 12th century)
1 : any of numerous carnivorous marine mammals (families Phocidae and Otariidae) that live chiefly in cold regions and have limbs modified into webbed flippers adapted primarily to swimming; *especially* **:** a fur seal or hair seal as opposed to a sea lion
2 a : the pelt of a fur seal **b :** leather made from the skin of a seal
3 : a dark brown

[2]seal *intransitive verb* (1828)
: to hunt seals

[3]seal *noun* [Middle English *seel,* from Old French, from Latin *sigillum* seal, from diminutive of *signum* sign, seal — more at SIGN] (13th century)
1 a : something that confirms, ratifies, or makes secure **:** GUARANTEE, ASSURANCE **b** (1) **:** a device with a cut or raised emblem, symbol, or word used especially to certify a signature or authenticate a document (2) **:** a medallion or ring face bearing such a device incised so that it can be impressed on wax or moist clay; *also* **:** a piece of wax or a wafer bearing such an impression **c :** an impression, device, or mark given the effect of a common-law seal by statute law or by American local custom recognized by judicial decision **d :** a usually ornamental adhesive stamp that may be used to close a letter or package; *especially* **:** one given in a fund-raising campaign
2 a : something that secures (as a wax seal on a document) **b :** a closure that must be broken to be opened and that thus reveals tampering **c** (1) **:** a tight and perfect closure (as against the passage of gas or water) (2) **:** a device to prevent the passage or return of gas or air into a pipe or container
3 : a seal that is a symbol or mark of office
— under seal : with an authenticating seal affixed

[4]seal *transitive verb* (14th century)
1 a : to confirm or make secure by or as if by a seal **b :** to solemnize for eternity (as a marriage) by a Mormon rite
2 a : to set or affix an authenticating seal to; *also* **:** AUTHENTICATE, RATIFY **b :** to mark with a stamp or seal usually as an evidence of standard exactness, legal size, weight, or capacity, or merchantable quality
3 a : to fasten with or as if with a seal to prevent tampering **b :** to close or make secure against access, leakage, or passage by a fastening or coating **c :** to fix in position or close breaks in with a filling (as of plaster)
4 : to determine irrevocably or indisputably ⟨that answer *sealed* our fate⟩

sea lamprey *noun* (1879)
: a large anadromous lamprey (*Petromyzon marinus*) that has a mottled upper surface, is an ectoparasite of fish, and is sometimes used as food

sea–lane \ˈsē-ˌlān\ *noun* (1927)
: an established sea route

seal•ant \ˈsē-lənt\ *noun* (1944)
: a sealing agent ⟨radiator *sealant*⟩

sea lavender *noun* (1597)
: any of a genus (*Limonium*) of chiefly perennial herbs of the plumbago family with small flowers and basal leaves

sea lawyer *noun* (1848)
: an argumentative captious sailor

sealed–beam \ˈsē(ə)l(d)-ˈbēm\ *adjective* (1939)
: being an electric lamp with a prefocused reflector and lens sealed in the lamp vacuum

sea legs *noun plural* (1712)
: bodily adjustment to the motion of a ship indicated especially by ability to walk steadily and by freedom from seasickness

[1]seal•er \ˈsē-lər\ *noun* (15th century)
1 : an official who attests or certifies conformity to a standard of correctness
2 : a coat (as of size) applied to prevent subsequent coats of paint or varnish from sinking in

[2]sealer *noun* (1842)
: a person or a ship engaged in hunting seals

sea lettuce *noun* (1668)
: any of a genus (*Ulva*) of seaweeds with green fronds sometimes eaten as salad

sea level *noun* (1806)
: the level of the surface of the sea especially at its mean position midway between mean high and low water

sea lily *noun* (1876)
: CRINOID; *especially* **:** a stalked crinoid

sealing wax *noun* (14th century)
: a resinous composition that is plastic when warm and is used for sealing (as letters, dry cells, or cans)

sea lion *noun* (1697)
: any of several Pacific eared seals (as genera *Eumetopius* and *Zalophus*) that are usually larger than the related fur seals and lack a thick underfur

seal off *transitive verb* (1931)
: to close tightly

seal point *noun* [[1]*seal* (the color)] (1939)
: a coat color of cats characterized by a cream or fawn body with dark brown points; *also* **:** a Siamese cat with such coloring

seal ring *noun* (1608)
: a finger ring engraved with a seal **:** SIGNET RING

seal•skin \ˈsē(ə)l-ˌskin\ *noun* (14th century)
1 : the fur or pelt of a fur seal
2 : a garment (as a jacket, coat, or cape) of sealskin
— sealskin *adjective*

Sea•ly•ham terrier \ˈsē-lē-ˌham-, *especially British* -lē-əm-\ *noun* [*Sealyham,* Pembrokeshire, Wales] (1907)
: any of a breed of short-legged long-headed strong-jawed heavy-boned chiefly white terriers developed in Wales

[1]seam \ˈsēm\ *noun* [Middle English *seem,* from Old English *sēam;* akin to Old English *sīwian* to sew — more at SEW] (before 12th century)
1 a : the joining of two pieces (as of cloth or leather) by sewing usually near the edge **b :** the stitching used in such a joining
2 : the space between adjacent planks or strakes of a ship
3 a : a line, groove, or ridge formed by the abutment of edges **b :** a thin layer or stratum (as of rock) between distinctive layers; *also* **:** a bed of valuable mineral and especially coal irrespective of thickness **c :** a line left by a cut or wound; *also* **:** WRINKLE
4 : a weak or vulnerable area or gap ⟨found a *seam* in the zone defense⟩
— seam•like \-ˌlīk\ *adjective*
— at the seams : ENTIRELY, COMPLETELY ⟨falling apart *at the seams*⟩

[2]seam (1582)
transitive verb
1 a : to join by sewing **b :** to join as if by sewing (as by welding, riveting, or heat-sealing)
2 : to mark with lines suggesting seams
intransitive verb
: to become fissured or ridgy
— seam•er *noun*

sea–maid \ˈsē-ˌmād\ *or* **sea–maid•en** \-ˌmā-dᵊn\ *noun* (1590)
: MERMAID; *also* **:** a goddess or nymph of the sea

sea•man \ˈsē-mən\ *noun* (before 12th century)
1 : SAILOR, MARINER
2 a : any of the three ranks below petty officer in the navy or coast guard **b :** an enlisted man in the navy or coast guard ranking above a seaman apprentice and below a petty officer

seaman apprentice *noun* (1947)
: an enlisted man in the navy or coast guard ranking above a seaman recruit and below a seaman

sea•man•like \ˈsē-mən-ˌlīk\ *adjective* (1796)
: characteristic of or befitting a competent seaman

sea•man•ly \-lē\ *adjective* (1798)
: SEAMANLIKE

seaman recruit *noun* (1947)
: an enlisted man of the lowest rank in the navy or coast guard

sea•man•ship \ˈsē-mən-ˌship\ *noun* (1766)
: the art or skill of handling, working, and navigating a ship

sea•mark \-ˌmärk\ *noun* (15th century)
1 : a line on a coast marking the tidal limit
2 : an elevated object serving as a beacon to mariners

sea mew *noun* (15th century)
: SEAGULL; *especially* **:** a common gull (*Larus canus*) chiefly of Europe and northwestern North America

sea mile *noun* (1796)
: NAUTICAL MILE

seam•less \ˈsēm-ləs\ *adjective* (15th century)
1 : having no seams
2 : having no awkward transitions or indications of disparity **:** perfectly smooth ⟨a *seamless* fusion of beauty and intelligence —Jack Kroll *et al.*⟩
— seam•less•ly \-lē\ *adverb*
— seam•less•ness \-nəs\ *noun*

sea•mount \ˈsē-ˌmau̇nt\ *noun* (1941)
: a submarine mountain rising above the deep-sea floor

sea mouse *noun* (circa 1520)
: any of various large broad marine polychaete worms (especially genus *Aphrodite*) covered with hairlike setae

seam•ster \ˈsēm(p)-stər *also* ˈsem(p)-\ *noun* [Middle English *semester, semster,* from Old English *sēamestre* seamstress, tailor, from *sēam* seam] (before 12th century)
: a person employed at sewing; *especially* **:** TAILOR

seam•stress \-strəs\ *noun* (1644)
: a woman whose occupation is sewing

seamy \ˈsē-mē\ *adjective* **seam•i•er; -est** (1604)
1 *archaic* **:** having the rough side of the seam showing
2 a : UNPLEASANT **b :** DEGRADED, SORDID
— seam•i•ness *noun*

sé•ance \ˈsā-ˌän(t)s, -ˌäⁿs, sā-ˈ\ *noun* [French, from *seoir* to sit, from Latin *sedēre* — more at SIT] (1803)
1 : SESSION, SITTING
2 : a spiritualist meeting to receive spirit communications

sea nettle *noun* (1601)
: a stinging jellyfish; *especially* **:** one (*Chrysaora quinquecirrha*) occurring especially in Atlantic estuaries from Cape Cod to the West Indies

sea oats *noun plural but singular or plural in construction* (1894)
: a tall grass (*Uniola panicolata*) that has panicles resembling those of the oat, grows chiefly on the coast of the southern U.S., and is useful as a sand binder

sea onion *noun* (14th century)
: SQUILL 1a

sea otter *noun* (1664)
: a rare marine otter (*Enhydra lutris*) of the northern Pacific coasts that may attain a length of six feet (two meters), is chiefly brown but with lighter coloration on the back of the head and neck, and feeds largely on shellfish

sea pen *noun* (1763)

\ə\ **abut** \ᵊ\ **kitten** \ər\ **further** \a\ **ash** \ā\ **ace** \ä\ **mop, mar** \au̇\ **out** \ch\ **chin** \e\ **bet** \ē\ **easy** \g\ **go** \i\ **hit** \ī\ **ice** \j\ **job** \ŋ\ **sing** \ō\ **go** \ȯ\ **law** \ȯi\ **boy** \th\ **thin** \t͟h\ **the** \ü\ **loot** \u̇\ **foot** \y\ **yet** \zh\ **vision** *see also* Guide to Pronunciation

: any of numerous anthozoans (order Pennatulacea) growing in colonies with a feathery form

sea·piece \'sē-,pēs\ *noun* (1656)
: SEASCAPE 2

sea pink (1731)
1 : THRIFT 4
2 : any of a genus (*Sabatia*) of smooth slender North American herbs of the gentian family typically having pink or white cymose flowers

sea·plane \-,plān\ *noun* (1913)
: an airplane designed to take off from and land on the water

sea·port \'sē-,pōrt, -,pȯrt\ *noun* (1596)
: a port, harbor, or town accessible to seagoing ships

sea power *noun* (1849)
1 : a nation having formidable naval strength
2 : naval strength

sea puss \-,pu̇s\ *noun* [alteration of dialect *seapoose* tidal stream, from Unquachog (Algonquian language of Long Island) *seépus* river] (circa 1891)
: a swirling or along shore undertow

sea·quake \'sē-,kwāk\ *noun* [*sea* + *earthquake*] (1680)
: a submarine earthquake

[1]sear *variant of* SERE

[2]sear \'sir\ *verb* [Middle English *seren,* from Old English *sēarian* to become sere, from *sēar* sere] (before 12th century)
intransitive verb
: to cause withering or drying
transitive verb
1 : to make withered and dry **:** PARCH
2 : to burn, scorch, or injure with or as if with sudden application of intense heat
— sear·ing·ly \-iŋ-lē\ *adverb*

[3]sear *noun* (1874)
: a mark or scar left by searing

[4]sear *noun* [probably from Middle French *serre* grasp, from *serrer* to press, grasp, from Late Latin *serare* to bolt, latch, from Latin *sera* bar for fastening a door] (1596)
: the catch that holds the hammer of a gun's lock at cock or half cock

[1]search \'sərch\ *verb* [Middle English *cerchen,* from Middle French *cerchier* to go about, survey, search, from Late Latin *circare* to go about, from Latin *circum* round about — more at CIRCUM-] (14th century)
transitive verb
1 : to look into or over carefully or thoroughly in an effort to find or discover something: as **a :** to examine in seeking something ⟨*searched* the north field⟩ **b :** to look through or explore by inspecting possible places of concealment or investigating suspicious circumstances **c :** to read thoroughly **:** CHECK; *especially* **:** to examine a public record or register for information about ⟨*search* land titles⟩ **d :** to examine for articles concealed on the person **e :** to look at as if to discover or penetrate intention or nature
2 : to uncover, find, or come to know by inquiry or scrutiny — usually used with *out*
intransitive verb
1 : to look or inquire carefully ⟨*searched* for the papers⟩
2 : to make painstaking investigation or examination
— search·able \'sər-chə-bəl\ *adjective*
— search·er *noun*
— search·ing·ly \-chiŋ-lē\ *adverb*

[2]search *noun* (15th century)
1 a : an act of searching ⟨a *search* for food⟩ ⟨go in *search* of help⟩ **b :** an act of boarding and inspecting a ship on the high seas in exercise of right of search
2 *obsolete* **:** a party that searches
3 : power or range of penetrating; *also* **:** a penetrating effect

search·less \'sərch-ləs\ *adjective* (1605)
: INSCRUTABLE, IMPENETRABLE

search·light \-,līt\ *noun* (1883)
: an apparatus for projecting a powerful beam of light; *also* **:** a beam of light projected by it

search warrant *noun* (1818)
: a warrant authorizing a search (as of a house) for stolen goods or unlawful possessions

sea robin *noun* (1814)
: any of a family (Triglidae) of marine bony fishes typically having a spiny armored head and the bottom three rays of the pectoral fin on each side free of membrane and modified for use as feelers or in crawling — called also *gurnard*

sea room *noun* (circa 1554)
: room for maneuver at sea

sea rover *noun* (circa 1580)
: one that roves the sea; *specifically* **:** PIRATE

sea–run \'sē-'rən\ *adjective* (1885)
: ANADROMOUS ⟨a *sea-run* salmon⟩

sea·scape \'sē-,skāp\ *noun* (1799)
1 : a view of the sea
2 : a picture representing a scene at sea
word history see LANDSCAPE

sea scorpion *noun* (1896)
1 : SCULPIN 1
2 : EURYPTERID

Sea Scout *noun* (1911)
: SEA EXPLORER

sea serpent *noun* (1774)
: a large marine animal resembling a serpent often reported to have been seen but never proved to exist

sea·shell \'sē-,shel\ *noun* (before 12th century)
: the shell of a marine animal and especially a mollusk

sea·shore \-,shōr, -,shȯr\ *noun* (1526)
1 a : land adjacent to the sea **:** SEACOAST **b :** NATIONAL SEASHORE
2 : all the ground between the ordinary highwater and low-water marks **:** FORESHORE

sea·sick \-,sik\ *adjective* (circa 1566)
: affected with or suggestive of seasickness

sea·sick·ness \-nəs\ *noun* (1625)
: motion sickness experienced on the water

sea·side \'sē-,sīd\ *noun* (13th century)
: the district or land bordering the sea **:** country adjacent to the sea **:** SEASHORE
— seaside *adjective*

sea slug *noun* (1779)
1 : HOLOTHURIAN
2 : a naked marine gastropod; *specifically* **:** NUDIBRANCH

sea snake *noun* (1755)
1 : SEA SERPENT
2 : any of numerous venomous aquatic chiefly viviparous elapid snakes of warm seas

[1]sea·son \'sē-zᵊn\ *noun* [Middle English, from Middle French *saison,* from Old French, from Latin *sation-, satio* action of sowing, from *serere* to sow — more at SOW] (14th century)
1 a : a time characterized by a particular circumstance or feature ⟨in a *season* of religious awakening —F. A. Christie⟩ **b :** a suitable or natural time or occasion ⟨when my *season* comes to sit on David's throne —John Milton⟩ **c :** an indefinite period of time **:** WHILE ⟨sent home again to her father for a *season* —Francis Hackett⟩
2 a : a period of the year characterized by or associated with a particular activity or phenomenon ⟨hay fever *season*⟩: as (1) **:** a period associated with some phase or activity of agriculture (as growth or harvesting) (2) **:** a period in which an animal engages in some activity (as migrating or mating); *also* **:** ESTRUS, HEAT (3) **:** the period normally characterized by a particular kind of weather ⟨a long rainy *season*⟩ (4) **:** a period marked by special activity especially in some field ⟨the theatrical *season*⟩ ⟨tourist *season*⟩ (5) **:** a period in which a place is most frequented **b :** one of the four quarters into which the year is commonly divided **c :** the time of a major holiday
3 : YEAR ⟨a boy of seven *seasons*⟩
4 [Middle English *sesoun,* from *sesounen* to season] **:** SEASONING
5 : the schedule of official games played or to be played by a sports team during a playing season ⟨try to get through the *season* undefeated⟩
6 : OFF-SEASON ⟨closed for the *season*⟩
— in season 1 : at the right time **2 :** at the stage of greatest fitness (as for eating) ⟨peaches are *in season*⟩ **3 :** legally available to be hunted or caught
— out of season : not in season

[2]season *verb* **sea·soned; sea·son·ing** \'sēz-niŋ, 'sē-zᵊn-iŋ\ [Middle English *sesounen,* from Middle French *assaisoner* to ripen, season, from Old French, from *a-* (from Latin *ad-*) + *saison* season] (14th century)
transitive verb
1 a : to give (food) more flavor or zest by adding seasoning or savory ingredients **b :** to give a distinctive quality to as if by seasoning; *especially* **:** to make more agreeable ⟨advice *seasoned* with wit⟩ **c** *archaic* **:** to qualify by admixture **:** TEMPER
2 a : to treat (as wood or a skillet) so as to prepare for use **b :** to make fit by experience ⟨a *seasoned* veteran⟩
intransitive verb
: to become seasoned

sea·son·able \'sēz-nə-bəl, 'sē-zᵊn-ə-bəl\ *adjective* (14th century)
1 : suitable to the season or circumstances **:** TIMELY ⟨a *seasonable* frost⟩
2 : occurring in good or proper time **:** OPPORTUNE ⟨a *seasonable* time for discussion⟩
— sea·son·able·ness *noun*
— sea·son·ably \-blē\ *adverb*

sea·son·al \'sēz-nəl, 'sē-zᵊn-əl\ *adjective* (1838)
1 : of, relating to, or varying in occurrence according to the season ⟨*seasonal* storms⟩ ⟨*seasonal* fruits⟩
2 : affected or caused by seasonal need or availability ⟨*seasonal* unemployment⟩ ⟨*seasonal* industries⟩
— sea·son·al·i·ty \,sē-zᵊn-'a-lə-tē\ *noun*
— sea·son·al·ly \'sēz-nə-lē, 'sē-zᵊn-ə-lē\ *adverb*

seasonal affective disorder *noun* (1985)
: depression that tends to recur as the days grow shorter during the fall and winter

sea·son·er \'sēz-nər, 'sē-zᵊn-ər\ *noun* (1598)
: one that seasons: as **a :** a user of seasonings ⟨a heavy *seasoner*⟩ **b :** SEASONING

sea·son·ing \'sēz-niŋ, 'sē-zᵊn-iŋ\ *noun* (1580)
: something that serves to season; *especially* **:** an ingredient (as a condiment, spice, or herb) added to food primarily for the savor that it imparts

sea·son·less \'sē-zᵊn-ləs\ *adjective* (1816)
1 : exhibiting no seasonal changes
2 : not restricted to a particular season; *especially* **:** suitable for wearing in any season ⟨*seasonless* fabrics⟩

season ticket *noun* (1820)
: a ticket (as to all of a club's home games or for specified daily transportation) valid during a specified time

sea spider *noun* (1666)
: any of various small long-legged marine arthropods (class Pycnogonida) that superficially resemble spiders

sea squirt *noun* (1850)
: ASCIDIAN

sea star *noun* (1569)
: STARFISH

sea stores *noun plural* (1659)
: supplies (as of foodstuffs) laid in before starting on a sea voyage

sea·strand \'sē-,strand\ *noun* (before 12th century)
: SEASHORE

[1]seat \'sēt\ *noun* [Middle English *sete,* from Old Norse *sæti;* akin to Old English *gesete* seat, *sittan* to sit] (13th century)

1 a : a special chair of one in eminence; *also* **:** the status represented by it **b :** a chair, stool, or bench intended to be sat in or on **c :** the particular part of something on which one rests in sitting ⟨the *seat* of a chair⟩ ⟨trouser *seat*⟩ **d :** BUTTOCKS
2 a : a seating accommodation ⟨a *seat* for the game⟩ ⟨a 200-*seat* restaurant⟩ **b :** a right of sitting ⟨lost his *seat* in Congress⟩ **c :** membership on an exchange
3 a : a place where something specified is prevalent **:** CENTER ⟨a *seat* of learning⟩ **b :** a place from which authority is exercised ⟨the county *seat*⟩ **c :** a bodily part in which some function or condition is centered ⟨the brain as the *seat* of the mind⟩
4 : posture in or way of sitting on horseback
5 a : a part at or forming the base of something **b :** a part (as a socket) or surface on or in which another part or surface rests
— by the seat of one's pants : using experience and intuition rather than mechanical aids or formal theory

²seat (1593)
transitive verb
1 a : to install in a seat of dignity or office **b** (1) **:** to cause to sit or assist in finding a seat (2) **:** to provide seats for ⟨a theater *seating* 1000 persons⟩ **c :** to put in a sitting position
2 : to repair the seat of or provide a new seat for
3 : to fit to or with a seat ⟨*seat* a valve⟩
intransitive verb
1 *archaic* **:** to take one's seat or place
2 : to fit correctly on a seat

seat belt *noun* (1932)
: an arrangement of straps designed to hold a person steady in a seat (as in an airplane or automobile)

seat·er \'sē-tər\ *noun* (1693)
1 : one that seats
2 : one that has a specified number of seats — used in combination ⟨a two-*seater* jet⟩

seat·ing \'sē-tiŋ\ *noun* (circa 1859)
1 : the act of providing with seats
2 a : material for covering or upholstering seats **b :** a seat on or in which something rests ⟨a valve *seating*⟩

seat·mate \'sēt-ˌmāt\ *noun* (1859)
: one with whom one shares a seat (as in a vehicle with double or paired seats)

seat–of–the–pants *adjective* (1942)
: employing or based on personal experience, judgment, and effort rather than technological aids or formal theory ⟨*seat-of-the-pants* navigation⟩ ⟨a *seat-of-the-pants* decision⟩

sea·train \'sē-ˌtrān\ *noun* (1932)
: a seagoing ship equipped for carrying a train of railroad cars

sea trout *noun* (1745)
1 : any of various trouts or chars that as adults inhabit the sea but ascend rivers to spawn
2 : any of various marine fishes resembling trouts: as **a :** WEAKFISH 1 **b :** SPOTTED SEA TROUT

sea turtle *noun* (1612)
: any of two families (Cheloniidae and Dermochelyidae) of widely distributed marine turtles with the feet modified into paddles that include the green turtle, leatherback, hawksbill, loggerhead, and ridley

sea urchin *noun* (1591)
: any of numerous echinoderms (class Echinoidea) that are usually enclosed in thin brittle globular tests covered with movable spines

sea·wall \'sē-ˌwȯl\ *noun* (15th century)
: a wall or embankment to protect the shore from erosion or to act as a breakwater

¹sea·ward \'sē-wərd\ *noun* (14th century)
: the direction or side away from land and toward the open sea

²seaward *also* **sea·wards** \-wərdz\ *adverb* (1517)
: toward the sea

³seaward *adjective* (circa 1621)
1 : directed or situated toward the sea
2 : coming from the sea ⟨a *seaward* wind⟩

sea wasp *noun* (1910)
: any of various scyphozoan jellyfishes (order or suborder Cubomedusae) that sting virulently and sometimes fatally

sea·wa·ter \'sē-ˌwȯ-tər, -ˌwä-\ *noun* (before 12th century)
: water in or from the sea

sea·way \-ˌwā\ *noun* (before 12th century)
1 : the sea as a route for travel; *also* **:** an ocean traffic lane
2 : a moderate or rough sea
3 : a deep inland waterway that admits ocean shipping

sea·weed \-ˌwēd\ *noun* (1577)
1 : a mass or growth of marine plants
2 : a plant growing in the sea; *especially* **:** a marine alga (as a kelp)

sea whip *noun* (1775)
: any of various gorgonian corals with elongated flexible unbranched or little-branched skeletons

sea·wor·thy \'sē-ˌwər-t͟hē\ *adjective* (1807)
: fit or safe for a sea voyage ⟨a *seaworthy* ship⟩
— sea·wor·thi·ness \-t͟hē-nəs\ *noun*

sea wrack *noun* (1551)
: SEAWEED; *especially* **:** seaweed cast ashore in masses

se·ba·ceous \si-'bā-shəs\ *adjective* [Latin *sebaceus* made of tallow, from *sebum* tallow] (1728)
1 : secreting sebum ⟨*sebaceous* glands⟩
2 : of, relating to, or being fatty material **:** FATTY ⟨a *sebaceous* exudate⟩

se·ba·cic acid \si-'ba-sik-, ˌsē-, -'bā-\ *noun* [International Scientific Vocabulary, from Latin *sebaceus*] (1790)
: a crystalline dicarboxylic acid $C_{10}H_{18}O_4$ used especially in the manufacture of synthetic resins

seb·or·rhea \ˌse-bə-'rē-ə\ *noun* [New Latin, from Latin *sebum* + New Latin *-rrhea*] (circa 1860)
: abnormally increased secretion and discharge of sebum
— seb·or·rhe·ic \-'rē-ik\ *adjective*

se·bum \'sē-bəm\ *noun* [Latin, tallow, grease] (circa 1860)
: fatty lubricant matter secreted by sebaceous glands of the skin

sec \'sek\ *adjective* [French, literally, dry — more at SACK] (1863)
of champagne **:** moderately dry

se·cant \'sē-ˌkant, -kənt\ *noun* [New Latin *secant-, secans,* from Latin, present participle of *secare* to cut — more at SAW] (1593)
1 : a straight line cutting a curve at two or more points — see CIRCLE illustration
2 : a straight line drawn from the center of a circle through one end of a circular arc to a tangent drawn from the other end of the arc
3 a : a trigonometric function that for an acute angle is the ratio of the hypotenuse of a right triangle of which the angle is considered part and the leg adjacent to the angle **b :** a trigonometric function $sec\ \theta$ that is the reciprocal of the cosine for all real numbers θ for which the cosine is not zero and that is exactly equal to the secant of an angle of measure θ in radians

sec·a·teur \ˌse-kə-'tər, 'se-kə-ˌ\ *noun* [French *sécateur,* from Latin *secare* to cut] (1881)
chiefly British **:** pruning shears — usually used in plural

¹sec·co \'se-(ˌ)kō\ *noun* [Italian, from *secco* dry, from Latin *siccus* — more at SACK] (1852)
: the art of painting on dry plaster

²secco *adjective or adverb* [Italian, literally, dry] (circa 1854)
1 : short and very staccato — used as a direction in music
2 *of a recitative* **:** accompanied only by the instruments playing the continuo

se·cede \si-'sēd\ *intransitive verb* **se·ced·ed; se·ced·ing** [Latin *secedere,* from *sed-, se-* apart (from *sed, se* without) + *cedere* to go — more at SUICIDE] (circa 1755)
: to withdraw from an organization (as a religious communion or political party or federation)
— se·ced·er *noun*

se·cern \si-'sərn\ *transitive verb* [Latin *secernere* to separate — more at SECRET] (circa 1656)
: to discriminate in thought **:** DISTINGUISH

se·ces·sion \si-'se-shən\ *noun* [Latin *secession-, secessio,* from *secedere*] (1604)
1 : withdrawal into privacy or solitude **:** RETIREMENT
2 : formal withdrawal from an organization

se·ces·sion·ist \-'se-sh(ə-)nist\ *noun* (circa 1860)
: one who joins in a secession or maintains that secession is a right
— se·ces·sion·ism \-shə-ˌni-zəm\ *noun*
— secessionist *adjective*

se·clude \si-'klüd\ *transitive verb* **se·clud·ed; se·clud·ing** [Middle English, to keep away, from Latin *secludere* to separate, seclude, from *se-* apart + *claudere* to close — more at SECEDE, CLOSE] (circa 1533)
1 *obsolete* **:** to exclude from a privilege, rank, or dignity **:** DEBAR
2 : to remove or separate from intercourse or outside influence **:** ISOLATE
3 : SHUT OFF, SCREEN

secluded *adjective* (1604)
1 : screened or hidden from view **:** SEQUESTERED ⟨a *secluded* valley⟩
2 : living in seclusion **:** SOLITARY ⟨*secluded* monks⟩
— se·clud·ed·ly *adverb*
— se·clud·ed·ness *noun*

se·clu·sion \si-'klü-zhən\ *noun* [Medieval Latin *seclusion-, seclusio,* from Latin *secludere*] (circa 1616)
1 : the act of secluding **:** the condition of being secluded
2 : a secluded or isolated place
synonym see SOLITUDE
— se·clu·sive \-'klü-siv, -ziv\ *adjective*
— se·clu·sive·ly *adverb*
— se·clu·sive·ness *noun*

seco·bar·bi·tal \ˌse-kō-'bär-bə-ˌtȯl\ *noun* [from *Seco*nal, a trademark + *barbital*] (1951)
: a barbiturate $C_{12}H_{18}N_2O_3$ that is used chiefly in the form of its bitter hygroscopic sodium salt as a hypnotic and sedative

Sec·o·nal \'se-kə-ˌnȯl, -ˌnal, -nᵊl\ *trademark*
— used for a preparation of secobarbital

¹sec·ond \'se-kənd *also* -kənt, *especially before a consonant* -kən, -kᵊŋ\ *adjective* [Middle English, from Old French, from Latin *secundus* second, following, favorable, from *sequi* to follow — more at SUE] (13th century)
1 a : next to the first in place or time ⟨was *second* in line⟩ **b** (1) **:** next to the first in value, excellence, or degree ⟨his *second* choice of schools⟩ (2) **:** INFERIOR, SUBORDINATE ⟨was *second* to none⟩ **c :** ranking next below the top of a grade or degree in authority or precedence ⟨*second* mate⟩ **d :** ALTERNATE, OTHER ⟨elects a mayor every *second* year⟩ **e :** resembling or suggesting a prototype **:** ANOTHER ⟨a *second* Thoreau⟩ **f :** ingrained by discipline, training, or effort **:** ACQUIRED ⟨*second* nature⟩ **g :** being the forward gear or speed next higher than first in a motor vehicle
2 : relating to or having a part typically subordinate to and lower in pitch than the first part in concerted or ensemble music
— second *or* **sec·ond·ly** *adverb*

²second *noun* (14th century)

1 a — see NUMBER table **b :** one that is next after the first in rank, position, authority, or precedence ⟨the *second* in line⟩
2 : one that assists or supports another; *especially* **:** the assistant of a duelist or boxer
3 a : the musical interval embracing two diatonic degrees **b :** a tone at this interval; *specifically* **:** SUPERTONIC **c :** the harmonic combination of two tones a second apart
4 a *plural* **:** merchandise that is usually slightly flawed and does not meet the manufacturer's standard for firsts or irregulars **b :** an article of such merchandise
5 : the act or declaration by which a parliamentary motion is seconded
6 : a place next below the first in a competition, examination, or contest
7 : SECOND BASE
8 : the second forward gear or speed of a motor vehicle
9 *plural* **:** a second helping of food

[3]**second** *noun* [Middle English *secunde,* from Medieval Latin *secunda,* from Latin, feminine of *secundus* second; from its being the second sexagesimal division of a unit, as a minute is the first] (14th century)
1 a : the 60th part of a minute of angular measure **b :** the 60th part of a minute of time **:** 1/86,400 part of the mean solar day; *specifically* **:** the base unit of time in the International System of Units that is equal to the duration of 9,192,631,770 periods of the radiation corresponding to the transition between the two hyperfine levels of the ground state of the cesium-133 atom
2 : an instant of time **:** MOMENT

[4]**second** *transitive verb* [Latin *secundare,* from *secundus* second, favorable] (circa 1586)
1 a : to give support or encouragement to **:** ASSIST **b :** to support (a fighting person or group) in combat **:** bring up reinforcements for
2 a : to support or assist in contention or debate **b :** to endorse (a motion or a nomination) so that debate or voting may begin
3 \si-ˈkänd\ *chiefly British* **:** to release (as a military officer) from a regularly assigned position for temporary duty with another unit or organization
— **sec·ond·er** *noun*

[1]**sec·ond·ary** \ˈse-kən-ˌder-ē\ *adjective* (14th century)
1 a : of second rank, importance, or value **b :** of, relating to, or constituting the second strongest of the three or four degrees of stress recognized by most linguists ⟨the fourth syllable of *basketball team* carries *secondary* stress⟩ **c** *of a tense* **:** expressive of past time
2 a : immediately derived from something original, primary, or basic **b :** of, relating to, or being the induced current or its circuit in an induction coil or transformer ⟨*secondary* voltage⟩ **c :** characterized by or resulting from the substitution of two atoms or groups in a molecule ⟨a *secondary* salt⟩; *especially* **:** being, characterized by, or attached to a carbon atom having bonds to two other carbon atoms **d** (1) **:** not first in order of occurrence or development (2) **:** produced by activity of formative tissue and especially cambium other than that at a growing point ⟨*secondary* growth⟩ ⟨*secondary* phloem⟩
3 a : of, relating to, or being the second order or stage in a series **b :** of, relating to, or being the second segment of the wing of a bird or the quills of this segment **c :** of or relating to a secondary school ⟨*secondary* education⟩
— **sec·ond·ar·i·ly** \ˌse-kən-ˈder-ə-lē\ *adverb*
— **sec·ond·ar·i·ness** \ˈse-kən-ˌder-ē-nəs\ *noun*

[2]**secondary** *noun, plural* **-ar·ies** (15th century)
1 : one occupying a subordinate or auxiliary position rather than that of a principal
2 : a defensive football backfield
3 : the coil through which the secondary current passes in an induction coil or transformer — called also *secondary coil*
4 : any of the quill feathers of the forearm of a bird — see WING illustration

secondary cell *noun* (circa 1909)
: STORAGE BATTERY

secondary color *noun* (1831)
: a color formed by mixing primary colors in equal or equivalent quantities

secondary emission *noun* (1931)
: the emission of electrons from a surface that is bombarded by particles (as electrons or ions) from a primary source

secondary radiation *noun* (1900)
: radiation emitted by molecules or atoms after bombardment by a primary radiation

secondary road *noun* (1903)
1 : a road not of primary importance
2 : a feeder road

secondary root *noun* (1861)
: one of the branches of a primary root

secondary school *noun* (1835)
: a school intermediate between elementary school and college and usually offering general, technical, vocational, or college-preparatory courses

secondary sex characteristic *noun* (1927)
: a physical characteristic (as the breasts of a female mammal or the breeding plumage of a male bird) that appears in members of one sex at puberty or in seasonal breeders at the breeding season and is not directly concerned with reproduction — called also *secondary sexual characteristic*

secondary syphilis *noun* (1885)
: the second stage of syphilis that appears from 2 to 6 months after primary infection, that is marked by lesions especially in the skin but also in organs and tissues, and that lasts from 3 to 12 weeks

second banana *noun* (1953)
: a comedian who plays a supporting role to a top banana; *broadly* **:** a person in a subservient position

second base *noun* (1845)
1 : the base that must be touched second by a base runner in baseball
2 : the player position for defending the area on the first-base side of second base
— **second baseman** *noun*

sec·ond–best \ˌse-kən(d)-ˈbest, -kᵊŋ-\ *adjective* (14th century)
: next to the best

[1]**second best** *noun* (1708)
: one that is below or after the best

[2]**second best** *adverb* (1777)
: in second place

second blessing *noun* (1891)
: sanctification as a second gift of the Holy Spirit that follows an initial experience of conversion

second childhood *noun* (1641)
: DOTAGE

second–class *adjective* (circa 1838)
1 : of or relating to a second class
2 : MEDIOCRE; *also* **:** socially, politically, or economically deprived

second class *noun* (1810)
1 : the second and usually next to highest group in a classification
2 : CABIN CLASS
3 : a class of U.S. or Canadian mail comprising periodicals sent to regular subscribers

Second Coming *noun* (1644)
: the coming of Christ as judge on the last day

second–degree burn *noun* (1937)
: a burn marked by pain, blistering, and superficial destruction of dermis with edema and hyperemia of the tissues beneath the burn

Second Empire *adjective* (1873)
: of, relating to, or characteristic of a style (as of furniture) developed in France under Napoleon III and marked by heavy ornate modification of Empire styles

second estate *noun, often S&E capitalized* (circa 1935)
: the second of the traditional political classes; *specifically* **:** NOBILITY

second fiddle *noun* (1809)
: one that plays a supporting or subservient role

second growth *noun* (1829)
: forest trees that come up naturally after removal of the first growth by cutting or by fire

sec·ond–guess \ˌse-kᵊŋ-ˈges, -kən(d)-\ *transitive verb* (1941)
1 : to criticize or question actions or decisions of (someone) often after the results of those actions or decisions are known ⟨meet almost every morning and, over coffee, *second-guess* the local coach —Bruce Newman⟩; *also* **:** to engage in such criticism of (an action or decision) ⟨*second-guess* the general's strategy⟩
2 : to seek to anticipate or predict ⟨lived royally by his ability to *second-guess* the stock market —*Time*⟩
— **sec·ond–guess·er** *noun*

[1]**sec·ond·hand** \ˈse-kən(d)-ˈhand\ *adjective* (1654)
1 a : received from or through an intermediary **:** BORROWED **b :** DERIVATIVE ⟨*secondhand* ideas⟩
2 a : acquired after being used by another **:** not new ⟨*secondhand* books⟩ **b :** dealing in secondhand merchandise ⟨a *secondhand* bookstore⟩

[2]**secondhand** *adverb* (1849)
: at second hand **:** INDIRECTLY

[1]**second hand** \ˌse-kən(d)-ˈhand\ *noun* (1588)
: an intermediate person or means **:** INTERMEDIARY — usually used in the phrase *at second hand*

[2]**second hand** \ˈsek-ən(d)-ˌ\ *noun* (1759)
: the hand marking seconds on a timepiece

secondhand smoke *noun* (1976)
: tobacco smoke that is exhaled by smokers and inhaled by persons nearby

second lieutenant *noun* (1702)
: a commissioned officer of the lowest rank in the army, air force, or marine corps

second mortgage *noun* (1912)
: a mortgage the lien of which is subordinate to that of a first mortgage

se·con·do \si-ˈkōn-(ˌ)dō, -ˈkän-\ *noun, plural* **-di** \-(ˌ)dē\ [Italian, from *secondo,* adjective, second, from Latin *secundus*] (1792)
: the second part in a concerted piece; *especially* **:** the lower part (as in a piano duet)

second person *noun* (1672)
1 a : a set of linguistic forms (as verb forms, pronouns, and inflectional affixes) referring to the person or thing addressed in the utterance in which they occur **b :** a linguistic form belonging to such a set
2 : reference of a linguistic form to the person or thing addressed in the utterance in which it occurs

sec·ond–rate \ˌse-kən(d)-ˈrāt\ *adjective* (1669)
: of second or inferior quality or value **:** MEDIOCRE
— **sec·ond–rate·ness** *noun*
— **sec·ond–rat·er** \-ˈrā-tər\ *noun*

Second Reader *noun* (1895)
: a member of a Christian Science church or society chosen for a term of office to assist the First Reader in conducting services by reading aloud selections from the Bible

second reading *noun* (1647)
1 : the stage in the British legislative process following the first reading and usually providing for debate on the principal features of a bill before its submission to a committee for consideration of details
2 : the stage in the U.S. legislative process that occurs when a bill has been reported back from committee and that provides an opportunity for full debate and amendment before a vote is taken on the question of a third reading

second sight *noun* (1616)

: the capacity to see remote or future objects or events : CLAIRVOYANCE, PRECOGNITION

second–story man *noun* (1886)
: a burglar who enters a house by an upstairs window

sec·ond–string \ˌse-kən(d)-ˈstriŋ, ˌse-kᵊŋ-\ *adjective* [from the reserve bowstring carried by an archer in case the first breaks] (1922)
: being a substitute as distinguished from a regular (as on a ball team)

second thought *noun* (1633)
: reconsideration or a revised opinion of a previous often hurried decision ⟨began to have *second thoughts*⟩

second wind *noun* (1824)
: renewed energy or endurance

second world *noun, often S&W capitalized* [after *third world*] (1973)
: the Communist nations as a political and economic bloc

se·cre·cy \ˈsē-krə-sē\ *noun, plural* **-cies** [alteration of earlier *secretie,* from Middle English *secretee,* from *secre* secret, from Middle French *secré,* from Latin *secretus*] (1575)
1 : the condition of being hidden or concealed
2 : the habit or practice of keeping secrets or maintaining privacy or concealment

[1]**se·cret** \ˈsē-krət\ *adjective* [Middle English, from Middle French, from Latin *secretus,* from past participle of *secernere* to separate, distinguish, from *se-* apart + *cernere* to sift — more at SECEDE, CERTAIN] (14th century)
1 a : kept from knowledge or view **:** HIDDEN **b :** marked by the habit of discretion **:** CLOSEMOUTHED **c :** working with hidden aims or methods **:** UNDERCOVER ⟨a *secret* agent⟩ **d :** not acknowledged **:** UNAVOWED ⟨a *secret* bride⟩ **e :** conducted in secret ⟨a *secret* trial⟩
2 : remote from human frequentation or notice **:** SECLUDED
3 : revealed only to the initiated **:** ESOTERIC
4 : constructed so as to elude observation or detection ⟨a *secret* panel⟩
5 : containing information whose unauthorized disclosure could endanger national security — compare CONFIDENTIAL, TOP SECRET ☆
— **se·cret·ly** *adverb*

[2]**secret** *noun* (14th century)
1 a : something kept hidden or unexplained **:** MYSTERY **b :** something kept from the knowledge of others or shared only confidentially with a few **c :** a method, formula, or process used in an art or a manufacturing operation and divulged only to those of one's own company or craft **d** *plural* **:** the practices or knowledge making up the shared discipline or culture of an esoteric society
2 : a prayer traditionally said inaudibly by the celebrant just before the preface of the mass
3 : something taken to be a specific or key to a desired end ⟨the *secret* of longevity⟩
— **in secret :** in a private place or manner

se·cre·ta·gogue \si-ˈkrē-tə-ˌgäg\ *noun* [*secretion* + *-agogue*] (1919)
: a substance stimulating secretion (as by the stomach or pancreas)

sec·re·tar·i·at \ˌse-krə-ˈter-ē-ət, -ē-ˌat\ *noun* [French *secrétariat,* from Medieval Latin *secretariatus,* from *secretarius*] (1811)
1 : the office of secretary
2 : a secretarial corps; *specifically* **:** the clerical staff of an organization
3 : the administrative department of a governmental organization

sec·re·tary \ˈse-krə-ˌter-ē, ˈse-kə-ˌter-, *in rapid speech also* ˈsek-ˌter-, *especially British* ˈsek(r)ə-trē\ *noun, plural* **-tar·ies** [Middle English *secretarie,* from Medieval Latin *secretarius,* confidential employee, secretary, from Latin *secretum* secret, from neuter of *secretus*] (15th century)
1 : one employed to handle correspondence and manage routine and detail work for a superior
2 a : an officer of a business concern who may keep records of directors' and stockholders' meetings and of stock ownership and transfer and help supervise the company's legal interests **b :** an officer of an organization or society responsible for its records and correspondence
3 : an officer of state who superintends a government administrative department
4 a : WRITING DESK, ESCRITOIRE **b :** a writing desk with a top section for books
— **sec·re·tar·i·al** \ˌse-krə-ˈter-ē-əl\ *adjective*
— **sec·re·tary·ship** \ˈse-krə-ˌter-ē-ˌship\ *noun*

secretary 4b

secretary bird *noun* [probably from the resemblance of its crest to a bunch of quill pens stuck behind the ear] (1824)
: a large long-legged African bird of prey (*Sagittarius serpentarius* of the family Sagittariidae) that feeds largely on reptiles

secretary bird

secretary–gen·eral *noun, plural* **secretaries–general** (1701)
: a principal administrative officer

secret ballot *noun* (1917)
: AUSTRALIAN BALLOT

[1]**se·crete** \si-ˈkrēt\ *transitive verb* **se·cret·ed; se·cret·ing** [back-formation from *secretion*] (1707)
: to form and give off (a secretion)

[2]**se·crete** \si-ˈkrēt, ˈsē-krət\ *transitive verb* **se·cret·ed; se·cret·ing** [alteration of obsolete *secret,* from [1]*secret*] (1741)
1 : to deposit or conceal in a hiding place
2 : to appropriate secretly **:** ABSTRACT
synonym see HIDE

se·cre·tin \si-ˈkre-tᵊn\ *noun* [*secretion* + [1]*-in*] (1902)
: an intestinal proteinaceous hormone capable of stimulating secretion by the pancreas and liver

se·cre·tion \si-ˈkrē-shən\ *noun* [French *sécrétion,* from Latin *secretion-, secretio* separation, from *secernere* to separate — more at SECRET] (1646)
1 a : the process of segregating, elaborating, and releasing some material either functionally specialized (as saliva) or isolated for excretion (as urine) **b :** a product of secretion formed by an animal or plant; *especially* **:** one performing a specific useful function in the organism
2 [[2]*secrete*] **:** the act of hiding something **:** CONCEALMENT
— **se·cre·tion·ary** \-shə-ˌner-ē\ *adjective*

se·cre·tive \ˈsē-krə-tiv, si-ˈkrē-\ *adjective* [back-formation from *secretiveness,* part translation of French *secrétivité*] (1853)
: disposed to secrecy **:** not open or outgoing in speech, activity, or purposes
synonym see SILENT
— **se·cre·tive·ly** *adverb*
— **se·cre·tive·ness** *noun*

se·cre·tor \si-ˈkrē-tər\ *noun* (1941)
: an individual of blood group A, B, or AB who secretes the antigens characteristic of these blood groups in bodily fluids (as saliva)

se·cre·to·ry \ˈsē-krə-ˌtȯr-ē, *especially British* si-ˈkrē-t(ə-)rē\ *adjective* (1692)
: of, relating to, or promoting secretion; *also* **:** produced by secretion

secret partner *noun* (circa 1909)
: a partner whose membership in a partnership is kept secret from the public

secret police *noun* (1823)
: a police organization operating for the most part in secrecy and especially for the political purposes of its government often with terroristic methods

secret service *noun* (1737)
1 : a governmental service of a secret nature
2 *both Ss capitalized* **:** a division of the U.S. Treasury Department charged chiefly with the suppression of counterfeiting and the protection of the president

secret society *noun* (1829)
: any of various oath-bound societies often devoted to brotherhood, moral discipline, and mutual assistance

sect \ˈsekt\ *noun* [Middle English *secte,* from Middle French & Late Latin; Middle French, group, sect, from Late Latin *secta* organized ecclesiastical body, from Latin, course of action, way of life, probably from *sectari* to pursue, frequentative of *sequi* to follow — more at SUE] (14th century)
1 a : a dissenting or schismatic religious body; *especially* **:** one regarded as extreme or heretical **b :** a religious denomination
2 *archaic* **:** SEX 1 ⟨so is all her *sect* —Shakespeare⟩
3 a : a group adhering to a distinctive doctrine or to a leader **b :** PARTY **c :** FACTION

[1]**sec·tar·i·an** \sek-ˈter-ē-ən\ *adjective* (1649)
1 : of, relating to, or characteristic of a sect or sectarian
2 : limited in character or scope **:** PAROCHIAL
— **sec·tar·i·an·ism** \-ē-ə-ˌni-zəm\ *noun*

[2]**sectarian** *noun* (1819)
1 : an adherent of a sect
2 : a narrow or bigoted person

sec·tar·i·an·ize \sek-ˈter-ē-ə-ˌnīz\ *verb* **-ized; -iz·ing** (1842)
intransitive verb
: to act as sectarians
transitive verb
: to make sectarian

sec·ta·ry \ˈsek-tə-rē\ *noun, plural* **-ries** (1556)
: a member of a sect

sec·tile \ˈsek-tᵊl, -ˌtīl\ *adjective* [Latin *sectilis,* from *sectus,* past participle of *secare*] (1805)
: capable of being severed by a knife with a smooth cut
— **sec·til·i·ty** \sek-ˈti-lə-tē\ *noun*

☆ SYNONYMS
Secret, covert, stealthy, furtive, clandestine, surreptitious, underhanded mean done without attracting observation. SECRET implies concealment on any grounds for any motive ⟨met at a *secret* location⟩. COVERT stresses the fact of not being open or declared ⟨*covert* intelligence operations⟩. STEALTHY suggests taking pains to avoid being seen or heard especially in some misdoing ⟨the *stealthy* step of a burglar⟩. FURTIVE implies a sly or cautious stealthiness ⟨lovers exchanging *furtive* glances⟩. CLANDESTINE implies secrecy usually for an evil, illicit, or unauthorized purpose and often emphasizes the fear of being discovered ⟨a *clandestine* meeting of conspirators⟩. SURREPTITIOUS applies to action or behavior done secretly often with skillful avoidance of detection and in violation of custom, law, or authority ⟨the *surreptitious* stockpiling of weapons⟩. UNDERHANDED stresses fraud or deception ⟨an *underhanded* trick⟩.

[1]sec·tion \'sek-shən\ *noun* [Latin *section-, sectio,* from *secare* to cut — more at SAW] (1559)
1 a : the action or an instance of cutting or separating by cutting **b :** a part set off by or as if by cutting
2 : a distinct part or portion of something written (as a chapter, law, or newspaper)
3 a : the profile of something as it would appear if cut through by an intersecting plane **b :** the plane figure resulting from the cutting of a solid by a plane
4 : a natural subdivision of a taxonomic group
5 : a character § used as a mark for the beginning of a section and as a reference mark
6 : a piece of land one square mile in area forming one of the 36 subdivisions of a township
7 : a distinct part of a territorial or political area, community, or group of people
8 a : a part that may be, is, or is viewed as separated ⟨chop the stalks into *sections*⟩ ⟨the northern *section* of the route⟩ **b :** one segment of a fruit **:** CARPEL
9 : a basic military unit usually having a special function
10 : a very thin slice (as of tissue) suitable for microscopic examination
11 a : one of the classes formed by dividing the students taking a course **b :** one of the discussion groups into which a conference or organization is divided
12 a : a part of a permanent railroad way under the care of a particular crew **b :** one of two or more vehicles or trains which run on the same schedule
13 : one of several component parts that may be assembled or reassembled ⟨a bookcase in *sections*⟩
14 : a division of an orchestra composed of one class of instruments
15 : SIGNATURE 3b
synonym see PART

[2]section *verb* **sec·tioned; sec·tion·ing** \-sh(ə-)niŋ\ (1819)
transitive verb
1 : to cut or separate into sections
2 : to represent in sections
intransitive verb
: to become cut or separated into parts

[1]sec·tion·al \'sek-shnəl, -shə-n^{ə}l\ *adjective* (1806)
1 a : of or relating to a section **b :** local or regional rather than general in character ⟨*sectional* interests⟩
2 : consisting of or divided into sections ⟨*sectional* furniture⟩
— sec·tion·al·ly *adverb*

[2]sectional *noun* (1901)
: a piece of furniture made up of modular units capable of use separately or in various combinations

sec·tion·al·ism \'sek-shnə-ˌli-zəm, -shə-n^{ə}l-ˌi-\ *noun* (1855)
: an exaggerated devotion to the interests of a region

Section Eight *noun* [*Section VIII,* Army Regulation 615-360, in effect from December 1922 to July 1944] (1943)
: a discharge from the U.S. Army for military inaptitude or undesirable habits or traits of character; *also* **:** a soldier receiving such a discharge

section gang *noun* (1890)
: a crew of track workers employed to maintain a railroad section

section hand *noun* (1873)
: a laborer belonging to a section gang

[1]sec·tor \'sek-tər, -ˌtȯr\ *noun* [Late Latin, from Latin, cutter, from *secare* to cut — more at SAW] (1570)
1 a : a geometric figure bounded by two radii and the included arc of a circle **b** (1) **:** a subdivision of a defensive military position (2) **:** a portion of a military front or area of operation **c :** an area or portion resembling a sector ⟨bilingual *sector* of town —David Kleinberg⟩ **d :** a sociological, economic, or political subdivision of society ⟨greater cooperation between the public and private *sectors* —Peter Chapman⟩
2 : a mathematical instrument consisting of two rulers connected at one end by a joint and marked with several scales
3 : a subdivision of a track on a computer disk

[2]sec·tor \-tər\ *transitive verb* **sec·tored; sec·tor·ing** \-t(ə-)riŋ\ (1884)
: to divide into or furnish with sectors

sec·to·ri·al \sek-'tōr-ē-əl, -'tȯr-\ *adjective* (1803)
1 : of, relating to, or having the shape of a sector of a circle
2 *of a chimera* **:** having a sector of variant growth interposed in an otherwise normal body of tissue

[1]sec·u·lar \'se-kyə-lər\ *adjective* [Middle English, from Old French *seculer,* from Late Latin *saecularis,* from *saeculum* the present world, from Latin, generation, age, century, world; akin to Welsh *hoedl* lifetime] (14th century)
1 a : of or relating to the worldly or temporal ⟨*secular* concerns⟩ **b :** not overtly or specifically religious ⟨*secular* music⟩ **c :** not ecclesiastical or clerical ⟨*secular* courts⟩ ⟨*secular* landowners⟩
2 : not bound by monastic vows or rules; *specifically* **:** of, relating to, or forming clergy not belonging to a religious order or congregation ⟨a *secular* priest⟩
3 a : occurring once in an age or a century **b :** existing or continuing through ages or centuries **c :** of or relating to a long term of indefinite duration
— sec·u·lar·i·ty \ˌse-kyə-'lar-ə-tē\ *noun*
— sec·u·lar·ly \'se-kyə-lər-lē\ *adverb*

[2]secular *noun, plural* **seculars** *or* **secular** (14th century)
1 : a secular ecclesiastic (as a diocesan priest)
2 : LAYMAN

secular humanism *noun* (1933)
: HUMANISM 3; *especially* **:** humanistic philosophy viewed as a nontheistic religion antagonistic to traditional religion
— secular humanist *noun or adjective*

sec·u·lar·ise *British variant of* SECULARIZE

sec·u·lar·ism \'se-kyə-lə-ˌri-zəm\ *noun* (1851)
: indifference to or rejection or exclusion of religion and religious considerations
— sec·u·lar·ist \-rist\ *noun*
— secularist *or* **sec·u·lar·is·tic** \ˌse-kyə-lə-'ris-tik\ *adjective*

sec·u·lar·ize \'se-kyə-lə-ˌrīz\ *transitive verb* **-ized; -iz·ing** (1611)
1 : to make secular
2 : to transfer from ecclesiastical to civil or lay use, possession, or control
3 : to convert to or imbue with secularism
— sec·u·lar·i·za·tion \ˌse-kyə-lə-rə-'zā-shən\ *noun*
— sec·u·lar·iz·er *noun*

se·cund \si-'kənd, 'sē-ˌ\ *adjective* [Latin *secundus* following — more at SECOND] (circa 1777)
: having some part or element arranged on one side only

[1]se·cure \si-'kyu̇r\ *adjective* **se·cur·er; -est** [Latin *securus* safe, secure, from *se* without + *cura* care — more at SUICIDE] (circa 1533)
1 a *archaic* **:** unwisely free from fear or distrust **:** OVERCONFIDENT **b :** easy in mind **:** CONFIDENT **c :** assured in opinion or expectation **:** having no doubt
2 a : free from danger **b :** free from risk of loss **c :** affording safety ⟨a *secure* hideaway⟩ **d :** TRUSTWORTHY, DEPENDABLE ⟨*secure* foundation⟩
3 : ASSURED 1 ⟨*secure* victory⟩
— se·cure·ly *adverb*
— se·cure·ness *noun*

[2]secure *verb* **se·cured; se·cur·ing** (1593)
transitive verb
1 a : to relieve from exposure to danger **:** act to make safe against adverse contingencies ⟨*secure* a supply line from enemy raids⟩ **b :** to put beyond hazard of losing or of not receiving **:** GUARANTEE ⟨*secure* the blessings of liberty —*U.S. Constitution*⟩ **c :** to give pledge of payment to (a creditor) or of (an obligation) ⟨*secure* a note by a pledge of collateral⟩
2 a : to take (a person) into custody **:** hold fast **:** PINION **b :** to make fast **:** SEAL ⟨*secure* a door⟩
3 a : to get secure usually lasting possession or control of ⟨*secure* a job⟩ **b :** BRING ABOUT, EFFECT
4 : to release (naval personnel) from work or duty
intransitive verb
1 *of naval personnel* **:** to stop work **:** go off duty
2 *of a ship* **:** to tie up **:** BERTH
synonym see ENSURE
— se·cur·er *noun*

se·cure·ment \si-'kyu̇r-mənt\ *noun* (1622)
1 *obsolete* **:** PROTECTION
2 : the act or process of securing

se·cu·ri·tize \si-'kyu̇r-ə-ˌtīz\ *transitive verb* **-tized; -tiz·ing** (1981)
: to consolidate (as mortgage loans) and sell to other investors for resale to the public in the form of securities
— se·cu·ri·ti·za·tion \-ˌkyu̇r-ə-tə-'zā-shən\ *noun*

se·cu·ri·ty \si-'kyu̇r-ə-tē\ *noun, plural* **-ties** (15th century)
1 : the quality or state of being secure: as **a :** freedom from danger **:** SAFETY **b :** freedom from fear or anxiety **c :** freedom from the prospect of being laid off ⟨job *security*⟩
2 a : something given, deposited, or pledged to make certain the fulfillment of an obligation **b :** SURETY
3 : an evidence of debt or of ownership (as a stock certificate or bond)
4 a : something that secures **:** PROTECTION **b** (1) **:** measures taken to guard against espionage or sabotage, crime, attack, or escape (2) **:** an organization or department whose task is security

security blanket *noun* (1968)
1 : a blanket carried by a child as a protection against anxiety
2 : a usually familiar object whose presence dispels anxiety

Security Council *noun* (1944)
: a permanent council of the United Nations with primary responsibility for maintaining peace and security

security interest *noun* (1951)
: the rights that a creditor has in the personal property of a debtor that secures an obligation **:** LIEN

security police *noun* (1920)
1 : police engaged in counterespionage
2 : AIR POLICE

se·dan \si-'dan\ *noun* [origin unknown] (1635)
1 : a portable often covered chair that is designed to carry one person and that is borne on poles by two men
2 a : a 2- or 4-door automobile seating 4 or more persons and usually having a permanent top — compare COUPE **b :** a motorboat having one passenger compartment

[1]se·date \si-'dāt\ *adjective* [Latin *sedatus,* from past participle of *sedare* to calm; akin to *sedēre* to sit — more at SIT] (1663)
: keeping a quiet steady attitude or pace **:** UNRUFFLED
synonym see SERIOUS
— se·date·ly *adverb*
— se·date·ness *noun*

[2]sedate *transitive verb* **se·dat·ed; se·dat·ing** [back-formation from *sedative*] (1945)
: to dose with sedatives

se·da·tion \si-'dā-shən\ *noun* (1543)

1 : the inducing of a relaxed easy state especially by the use of sedatives
2 : a state resulting from or as if from sedation

[1]**sed·a·tive** \'se-də-tiv\ *adjective* [Middle English, alleviating pain, from Middle French *sedatif,* from Medieval Latin *sedativus,* from Latin *sedatus*] (1795)
: tending to calm, moderate, or tranquilize nervousness or excitement

[2]**sedative** *noun* (1797)
: a sedative agent or drug

sed·en·tary \'se-dᵊn-ˌter-ē\ *adjective* [Middle French *sedentaire,* from Latin *sedentarius,* from *sedent-, sedens,* present participle of *sedēre* to sit — more at SIT] (1598)
1 : not migratory **:** SETTLED ⟨*sedentary* birds⟩
2 : doing or requiring much sitting
3 : permanently attached ⟨*sedentary* barnacles⟩

se·der \'sā-dər\ *noun, often capitalized* [Hebrew *sēdher* order] (1865)
: a Jewish home or community service including a ceremonial dinner held on the first or first and second evenings of the Passover in commemoration of the exodus from Egypt

se·de·runt \sə-'dir-ənt, -'der-\ *noun* [Latin, there (they) sat (from *sedēre* to sit), word used to introduce list of those attending a session — more at SIT] (1825)
: a prolonged sitting (as for discussion)

sedge \'sej\ *noun* [Middle English *segge,* from Old English *secg;* akin to Middle High German *segge* sedge, Old English *sagu* saw — more at SAW] (before 12th century)
: any of a family (Cyperaceae, the sedge family) of usually tufted marsh plants differing from the related grasses in having achenes and solid stems; *especially* **:** any of a cosmopolitan genus (*Carex*)
— sedgy \'se-jē\ *adjective*

se·di·lia \sə-'dēl-yə, -'dil-, *especially British* -'dīl-\ *noun plural* [Latin, plural of *sedile* seat, from *sedēre*] (1793)
: seats on the south side of the chancel for the celebrant, deacon, and subdeacon

[1]**sed·i·ment** \'se-də-mənt\ *noun* [Middle French, from Latin *sedimentum* settling, from *sedēre* to sit, sink down] (1547)
1 : the matter that settles to the bottom of a liquid
2 : material deposited by water, wind, or glaciers

[2]**sed·i·ment** \-ˌment\ (1859)
transitive verb
: to deposit as sediment
intransitive verb
1 : to settle to the bottom in a liquid
2 : to deposit sediment

sed·i·ment·able \ˌse-də-'men-tə-bəl\ *adjective* (1943)
: capable of being sedimented by centrifugation ⟨*sedimentable* ribosomal particles⟩

sed·i·men·ta·ry \ˌse-də-'men-tə-rē, -'men-trē\ *adjective* (1830)
1 : of, relating to, or containing sediment ⟨*sedimentary* deposits⟩
2 : formed by or from deposits of sediment ⟨*sedimentary* rock⟩

sed·i·men·ta·tion \ˌse-də-mən-'tā-shən, -ˌmen-\ *noun* (1874)
: the action or process of forming or depositing sediment **:** SETTLING

sed·i·men·tol·o·gy \ˌse-də-mən-'tä-lə-jē, -ˌmen-\ *noun* (1932)
: a branch of science that deals with sedimentary rocks and their inclusions
— sed·i·men·to·log·ic \-ˌmen-tᵊl-'ä-jik\ *or* **sed·i·men·to·log·i·cal** \-ji-kəl\ *adjective*
— sed·i·men·to·log·i·cal·ly \-ji-k(ə-)lē\ *adverb*
— sed·i·men·tol·o·gist \-mən-'tä-lə-jist, -ˌmen-\ *noun*

se·di·tion \si-'di-shən\ *noun* [Middle English, from Middle French, from Latin *sedition-, seditio,* literally, separation, from *se-* apart + *ition-, itio* act of going, from *ire* to go — more at SECEDE, ISSUE] (14th century)
: incitement of resistance to or insurrection against lawful authority

se·di·tious \si-'di-shəs\ *adjective* (15th century)
1 : disposed to arouse or take part in or guilty of sedition
2 : of, relating to, or tending toward sedition
— se·di·tious·ly *adverb*
— se·di·tious·ness *noun*

se·duce \si-'düs, -'dyüs\ *transitive verb* **se·duced; se·duc·ing** [Middle English, from Late Latin *seducere,* from Latin, to lead away, from *se-* apart + *ducere* to lead — more at TOW] (15th century)
1 : to persuade to disobedience or disloyalty
2 : to lead astray usually by persuasion or false promises
3 : to carry out the physical seduction of **:** entice to sexual intercourse
4 : ATTRACT
synonym see LURE
— se·duc·er *noun*

se·duce·ment \-mənt\ *noun* (1586)
1 : SEDUCTION
2 : something that serves to seduce

se·duc·tion \si-'dək-shən\ *noun* [Middle French, from Late Latin *seduction-, seductio,* from Latin, act of leading aside, from *seducere*] (1526)
1 : the act of seducing to wrong; *especially* **:** the often unlawful enticement of a female to sexual intercourse
2 : something that seduces **:** TEMPTATION
3 : something that attracts or charms

se·duc·tive \-'dək-tiv\ *adjective* (1771)
: tending to seduce **:** having alluring or tempting qualities ⟨a *seductive,* sometimes disingenuous man —Thatcher Freund⟩ ⟨a *seductive* spring morning⟩
— se·duc·tive·ly *adverb*
— se·duc·tive·ness *noun*

se·duc·tress \-'dək-trəs\ *noun* [obsolete *seductor* male seducer, from Late Latin, from *seducere* to seduce] (1803)
: a woman who seduces

se·du·li·ty \si-'dü-lə-tē, -'dyü-\ *noun* (1542)
: sedulous activity **:** DILIGENCE

sed·u·lous \'se-jə-ləs\ *adjective* [Latin *sedulus,* from *sedulo* sincerely, diligently, from *se* without + *dolus* guile — more at SUICIDE] (1540)
1 : involving or accomplished with careful perseverance ⟨*sedulous* craftsmanship⟩
2 : diligent in application or pursuit ⟨a *sedulous* student⟩
synonym see BUSY
— sed·u·lous·ly *adverb*
— sed·u·lous·ness *noun*

se·dum \'sē-dəm\ *noun* [New Latin, from Latin, houseleek] (1760)
: any of a genus (*Sedum*) of fleshy widely distributed herbs of the orpine family — compare STONECROP

[1]**see** \'sē\ *verb* **saw** \'sȯ\; **seen** \'sēn\; **see·ing** \'sē-iŋ\ [Middle English *seen,* from Old English *sēon;* akin to Old High German *sehan* to see and perhaps to Latin *sequi* to follow — more at SUE] (before 12th century)
transitive verb
1 a : to perceive by the eye **b :** to perceive or detect as if by sight
2 a : to have experience of **:** UNDERGO ⟨*see* army service⟩ **b :** to come to know **:** DISCOVER **c :** to be the setting or time of ⟨the last fifty years have *seen* a sweeping revolution in science —Barry Commoner⟩
3 a : to form a mental picture of **:** VISUALIZE ⟨can still *see* her as she was years ago⟩ **b :** to perceive the meaning or importance of **:** UNDERSTAND **c :** to be aware of **:** RECOGNIZE ⟨*sees* only our faults⟩ **d :** to imagine as a possibility **:** SUPPOSE ⟨couldn't *see* him as a crook⟩
4 a : EXAMINE, WATCH ⟨want to *see* how she handles the problem⟩ **b** (1) **:** READ (2) **:** to read of **c :** to attend as a spectator ⟨*see* a play⟩
5 a : to take care of **:** provide for ⟨had enough money to *see* us through⟩ **b :** to make sure ⟨*see* that order is kept⟩
6 a : to regard as **:** JUDGE **b :** to prefer to have ⟨I'll *see* him hanged first⟩ ⟨I'll *see* you dead before I accept your terms⟩ **c :** to find acceptable or attractive ⟨can't understand what he *sees* in her⟩
7 a : to call on **:** VISIT **b** (1) **:** to keep company with especially in courtship or dating ⟨had been *seeing* each other for a year⟩ (2) **:** to grant an interview to **:** RECEIVE ⟨the president will *see* you now⟩
8 : ACCOMPANY, ESCORT ⟨*see* the guests to the door⟩
9 : to meet (a bet) in poker or to equal the bet of (a player) **:** CALL
intransitive verb
1 a : to give or pay attention **b :** to look about
2 a : to have the power of sight **b :** to apprehend objects by sight **c :** to perceive objects as if by sight
3 : to grasp something mentally
4 : to make investigation or inquiry
— see·able \-ə-bəl\ *adjective*
— see after : to attend to **:** care for
— see eye to eye : to have a common viewpoint **:** AGREE
— see things : HALLUCINATE
— see through : to grasp the true nature of ⟨*saw through* the scheme⟩
— see to : to attend to **:** care for

[2]**see** *noun* [Middle English *se,* from Old French, from Latin *sedes* seat; akin to Latin *sedēre* to sit — more at SIT] (14th century)
1 a *archaic* **:** CATHEDRA **b :** a cathedral town **c :** a seat of a bishop's office, power, or authority
2 : the authority or jurisdiction of a bishop

[1]**seed** \'sēd\ *noun, plural* **seed** *or* **seeds** [Middle English, from Old English *sǣd;* akin to Old High German *sāt* seed, Old English *sāwan* to sow — more at SOW] (before 12th century)
1 a (1) **:** the grains or ripened ovules of plants used for sowing (2) **:** the fertilized ripened ovule of a flowering plant containing an embryo and capable normally of germination to produce a new plant; *broadly* **:** a propagative plant structure (as a spore or small dry fruit) **b :** a propagative animal structure: (1) **:** MILT, SEMEN (2) **:** a small egg (as of an insect) (3) **:** a developmental form of a lower animal suitable for transplanting; *specifically* **:** SPAT **c :** the condition or stage of bearing seed ⟨in *seed*⟩
2 : PROGENY
3 : a source of development or growth **:** GERM ⟨sowed the *seeds* of discord⟩
4 : something (as a tiny particle or a bubble in glass) that resembles a seed in shape or size
5 : a competitor who has been seeded in a tournament
— seed *adjective*
— seed·ed \'sē-dəd\ *adjective*
— seed·less \'sēd-ləs\ *adjective*
— seed·like \-ˌlīk\ *adjective*
— go to seed *or* **run to seed 1 :** to develop seed **2 :** DECAY

[2]**seed** (14th century)
intransitive verb
1 : to bear or shed seed
2 : to sow seed **:** PLANT
transitive verb

1 a : to plant seeds in **:** SOW ⟨*seed* land to grass⟩ **b :** to furnish with something that causes or stimulates growth or development **c :** INOCULATE **d :** to supply with nuclei (as of crystallization or condensation); *especially* **:** to treat (a cloud) with solid particles to convert water droplets into ice crystals in an attempt to produce precipitation **e :** to cover or permeate by or as if by scattering something ⟨*seeded* [the] sea-lanes with thousands of magnetic mines —Otto Friedrich⟩
2 : PLANT 1a
3 : to extract the seeds from (as raisins)
4 a : to schedule (tournament players or teams) so that superior ones will not meet in early rounds **b :** to rank (a contestant) relative to others in a tournament on the basis of previous record ⟨the top-*seeded* tennis star⟩
seed·bed \'sēd-,bed\ *noun* (1660)
1 : soil or a bed of soil prepared for planting seed
2 : a place or source of growth or development
seed·cake \-,kāk\ *noun* (1573)
1 : a cake or cookie containing aromatic seeds (as sesame or caraway)
2 : OIL CAKE
seed coat *noun* (1796)
: an outer protective covering of a seed
seed·eat·er \'sēd-,ē-tər\ *noun* (circa 1879)
: a bird (as a finch) whose diet consists basically of seeds
seed·er \'sē-dər\ *noun* (1868)
1 : an implement for planting or sowing seeds
2 : a device for seeding fruit
3 : one that seeds clouds
seed fern *noun* (1927)
: any of an order (Pteridospermales) of extinct cycadophytes with foliage like that of ferns and with naked seeds
seed leaf *noun* (circa 1693)
: COTYLEDON 2
seed·ling \'sēd-liŋ\ *noun* (1660)
1 : a young plant grown from seed
2 a : a young tree before it becomes a sapling **b :** a nursery plant not yet transplanted
— seedling *adjective*
seed money *noun* (1943)
: money used for setting up a new enterprise
seed oyster *noun* (1885)
: a young oyster especially of a size for transplantation
seed pearl *noun* (1553)
1 : a very small and often irregular pearl
2 : minute pearls imbedded in some binding material
seed plant *noun* (1707)
: a plant that bears seeds; *specifically* **:** SPERMATOPHYTE
seed·pod \'sēd-,päd\ *noun* (1718)
: [2]POD 1
seeds·man \'sēdz-mən\ *noun* (1601)
1 : one who sows seeds
2 : a dealer in seeds
seed stock *noun* (1926)
: a supply (as of seed) for planting; *broadly* **:** a source of new individuals ⟨a *seed stock* of trout in the streams⟩
seed tick *noun* (1705)
: the 6-legged larva of a tick
seed·time \'sēd-,tīm\ *noun* (before 12th century)
1 : the season of sowing
2 : a period of original development
seedy \'sē-dē\ *adjective* **seed·i·er; -est** (1574)
1 : containing or full of seeds ⟨a *seedy* fruit⟩
2 : inferior in condition or quality: as **a :** SHABBY, RUN-DOWN ⟨*seedy* clothes⟩ **b :** somewhat disreputable **:** SQUALID ⟨a *seedy* district⟩ ⟨*seedy* entertainment⟩ **c :** slightly unwell **:** DEBILITATED ⟨felt *seedy* and went home early⟩
— seed·i·ly \'sē-dᵊl-ē\ *adverb*
— seed·i·ness \'sē-dē-nəs\ *noun*
[1]**see·ing** \'sē-iŋ\ *conjunction* (1503)
: INASMUCH AS — often used with *as* or *that*
[2]**seeing** *noun* (1903)
: the quality of the images of celestial bodies observed telescopically
Seeing Eye *trademark*
— used for a guide dog trained to lead the blind
seek \'sēk\ *verb* **sought** \'sȯt\; **seek·ing** [Middle English *seken,* from Old English *sēcan;* akin to Old High German *suohhen* to seek, Latin *sagus* prophetic, Greek *hēgeisthai* to lead] (before 12th century)
transitive verb
1 : to resort to **:** go to
2 a : to go in search of **:** look for **b :** to try to discover
3 : to ask for **:** REQUEST ⟨*seeks* advice⟩
4 : to try to acquire or gain **:** aim at ⟨*seek* fame⟩
5 : to make an attempt **:** TRY — used with *to* and an infinitive ⟨governments . . . *seek* to keep the bulk of their people contented —D. M. Potter⟩
intransitive verb
1 : to make a search or inquiry
2 a : to be sought **b :** to be lacking ⟨in critical judgment . . . they were sadly to *seek* —*Times Literary Supplement*⟩
— seek·er *noun*
seel \'sē(ə)l\ *transitive verb* [Middle English *selen,* alteration of *silen,* from Middle French *siller,* from Medieval Latin *ciliare,* from Latin *cilium* eyelid] (15th century)
1 : to close the eyes of (as a hawk) by drawing threads through the eyelids
2 *archaic* **:** to close up (one's eyes)
see·ly \'sē-lē\ *adjective* [Middle English *sely* — more at SILLY] (14th century)
archaic **:** pitiable especially because of weak physical or mental condition **:** FRAIL
seem \'sēm\ *intransitive verb* [Middle English *semen,* of Scandinavian origin; akin to Old Norse *sœma* to honor, *sœmr* fitting, *samr* same — more at SAME] (13th century)
1 : to appear to the observation or understanding
2 : to give the impression of being
[1]**seem·ing** *noun* (15th century)
: external appearance as distinguished from true character **:** LOOK
[2]**seeming** *adjective* (circa 1557)
: having an often deceptive or delusive appearance on superficial examination ⟨their wealth gave them a *seeming* security⟩
synonym see APPARENT
— seem·ing·ly \'sē-miŋ-lē\ *adverb*
seem·ly \'sēm-lē\ *adjective* **seem·li·er; -est** [Middle English *semely,* from Old Norse *sœmiligr,* from *sœmr* fitting] (13th century)
1 a : GOOD-LOOKING, HANDSOME **b :** agreeably fashioned **:** ATTRACTIVE
2 : conventionally proper **:** DECOROUS
3 : suited to the occasion, purpose, or person **:** FIT
— seem·li·ness *noun*
— seemly *adverb*
seen *past participle of* SEE
[1]**seep** \'sēp\ *intransitive verb* [alteration of earlier *sipe,* from Middle English *sipen,* from Old English *sipian;* akin to Middle Low German *sipen* to seep] (1790)
1 : to flow or pass slowly through fine pores or small openings **:** OOZE ⟨water *seeped* in through a crack⟩
2 a : to enter or penetrate slowly ⟨fear of nuclear war had *seeped* into the national consciousness —Tip O'Neill⟩ **b :** to become diffused or spread ⟨a sadness *seeped* through his being —Agnes S. Turnbull⟩
[2]**seep** *noun* (circa 1825)
1 a : a spot where a fluid (as water, oil, or gas) contained in the ground oozes slowly to the surface and often forms a pool **b :** a small spring
2 : SEEPAGE
— seepy \'sē-pē\ *adjective*
seep·age \'sē-pij\ *noun* (circa 1825)
1 : the process of seeping **:** OOZING
2 : a quantity of fluid that has seeped (as through porous material)
[1]**seer** \'sir, 'sē-ər\ *noun* (14th century)
1 : one that sees
2 a : one that predicts events or developments **b :** a person credited with extraordinary moral and spiritual insight
3 : one that practices divination especially by concentrating on a glass or crystal globe
[2]**seer** *noun, plural* **seers** *or* **seer** [Hindi *ser*] (1618)
: an Indian unit of weight equal to 2.057 pounds (0.9330 kilograms)
seer·ess \'sir-əs, 'sē-ər-əs\ *noun* (1845)
: a woman who predicts events or developments **:** PROPHETESS
seer·suck·er \'sir-,sə-kər\ *noun* [Hindi *śīrśaker,* from Persian *shīr-o-shakar,* literally, milk and sugar] (1722)
: a light fabric of linen, cotton, or rayon usually striped and slightly puckered
[1]**see·saw** \'sē-,sȯ\ *noun* [probably from reduplication of [3]*saw*] (1704)
1 : an alternating up-and-down or backward-and-forward motion or movement; *also* **:** a contest or struggle in which now one side now the other has the lead
2 a : a pastime in which two children or groups of children ride on opposite ends of a plank balanced in the middle so that one end goes up as the other goes down **b :** the plank or apparatus so used
— seesaw *adjective*
[2]**seesaw** (1712)
intransitive verb
1 a : to move backward and forward or up and down **b :** to play at seesaw
2 : ALTERNATE
transitive verb
: to cause to move in seesaw fashion
[1]**seethe** \'sēth\ *verb* **seethed; seeth·ing** [Middle English *sethen,* from Old English *sēothan;* akin to Old High German *siodan* to seethe] (before 12th century)
transitive verb
1 *archaic* **:** BOIL, STEW
2 : to soak or saturate in a liquid
intransitive verb
1 *archaic* **:** BOIL
2 a : to be in a state of rapid agitated movement **b :** to churn or foam as if boiling
3 : to suffer violent internal excitement
[2]**seethe** *noun* (1816)
: a state of seething **:** EBULLITION
seething *adjective* (14th century)
1 : intensely hot **:** BOILING ⟨a *seething* inferno⟩
2 : constantly moving or active **:** AGITATED
see–through \'sē-,thrü\ *adjective* (1945)
: TRANSPARENT 1
[1]**seg·ment** \'seg-mənt\ *noun* [Latin *segmentum,* from *secare* to cut — more at SAW] (1570)
1 a : a separate piece of something **:** BIT, FRAGMENT ⟨chop the stalks into short *segments*⟩ **b :** one of the constituent parts into which a body, entity, or quantity is divided or marked off by or as if by natural boundaries ⟨all *segments* of the population agree⟩
2 : a portion cut off from a geometric figure by one or more points, lines, or planes: as **a :** the part of a circular area bounded by a chord and an arc of that circle or so much of the area as is cut off by the chord **b :** the part of a sphere cut off by a plane or included between two parallel planes **c :** the finite part of a line between two points in the line
synonym see PART
— seg·men·tary \-mən-,ter-ē\ *adjective*
[2]**seg·ment** \'seg-,ment\ *transitive verb* (1859)
: to separate into segments **:** give off as segments
seg·men·tal \seg-'men-tᵊl\ *adjective* (1816)
1 : of, relating to, or having the form of a segment and especially the sector of a circle ⟨*segmental* fanlight⟩

2 : of, relating to, or composed of somites or metameres **:** METAMERIC
3 a : divided into segments ⟨*segmental* knowledge⟩ **b :** PARTIAL, INCOMPLETE **c :** resulting from segmentation
— seg·men·tal·ly \-tᵊl-ē\ *adverb*

seg·men·ta·tion \ˌseg-mən-ˈtā-shən, -ˌmen-\ *noun* (1851)
: the process of dividing into segments; *especially* **:** the formation of many cells from a single cell (as in a developing egg)

segmentation cavity *noun* (1888)
: BLASTOCOEL

seg·ment·ed \ˈseg-ˌmen-təd, seg-ˈ\ *adjective* (1854)
: divided into or composed of segments or sections ⟨*segmented* worms⟩

se·gno \ˈsān-(ˌ)yō\ *noun, plural* **segnos** [Italian, sign, from Latin *signum* — more at SIGN] (1908)
: a notational sign; *specifically* **:** the sign that marks the beginning or end of a musical repeat

se·go lily \ˈsē-(ˌ)gō-\ *noun* [*sego* the bulb of the sego lily, from Southern Paiute *siɣoʔo*] (1913)
: a mariposa lily (*Calochortus nuttallii*) of western North America having mostly white or in some areas mostly yellow flowers mottled with a darker color

sego lily

seg·re·gant \ˈse-gri-gənt\ *noun* (1926)
: a genetic segregate

¹seg·re·gate \ˈse-gri-ˌgāt\ *verb* **-gat·ed; -gat·ing** [Latin *segregatus,* past participle of *segregare,* from *se-* apart + *greg-, grex* herd — more at SECEDE] (1542)
transitive verb
1 : to separate or set apart from others or from the general mass **:** ISOLATE
2 : to cause or force the separation of (as from the rest of society)
intransitive verb
1 : SEPARATE, WITHDRAW
2 : to practice or enforce a policy of segregation
3 : to undergo genetic segregation
— seg·re·ga·tive \-ˌgā-tiv\ *adjective*

²seg·re·gate \ˈse-gri-gət, -ˌgāt\ *noun* (1871)
: one that is in some respect segregated; *especially* **:** one that differs genetically from the parental line because of genetic segregation

segregated *adjective* (1652)
1 a : set apart or separated from others of the same kind or group ⟨a *segregated* account in a bank⟩ **b :** divided in facilities or administered separately for members of different groups or races ⟨*segregated* education⟩ **c :** restricted to members of one group or one race by a policy of segregation ⟨*segregated* schools⟩
2 : practicing or maintaining segregation especially of races ⟨*segregated* states⟩

seg·re·ga·tion \ˌse-gri-ˈgā-shən\ *noun* (1555)
1 : the act or process of segregating **:** the state of being segregated
2 a : the separation or isolation of a race, class, or ethnic group by enforced or voluntary residence in a restricted area, by barriers to social intercourse, by separate educational facilities, or by other discriminatory means **b :** the separation for special treatment or observation of individuals or items from a larger group ⟨*segregation* of gifted children into accelerated classes⟩
3 : the separation of allelic genes that occurs typically during meiosis

seg·re·ga·tion·ist \-sh(ə-)nist\ *noun* (1913)
: a person who believes in or practices segregation especially of races
— segregationist *adjective*

¹se·gue \ˈse-(ˌ)gwā, ˈsā-\ *imperative verb* [Italian, there follows, from *seguire* to follow, from Latin *sequi* — more at SUE] (circa 1740)
1 : proceed to what follows without pause — used as a direction in music
2 : perform the music that follows like that which has preceded — used as a direction in music

²segue *intransitive verb* **se·gued; se·gue·ing** (circa 1913)
1 : to proceed without pause from one musical number or theme to another
2 : to make a transition without interruption from one activity, topic, scene, or part to another

³segue *noun* (circa 1937)
: the act or an instance of segueing

se·gui·di·lla \ˌse-gə-ˈdē-yə, -ˈdēl-yə\ *noun* [Spanish, diminutive of *seguida,* a dance, literally, sequence, from *seguido,* past participle of *seguir* to follow, from Latin *sequi*] (1763)
1 a : a Spanish dance with many regional variations **b :** the music for such a dance
2 : a Spanish stanza of four or seven short partly assonant verses

sei·cen·to \sā-ˈchen-(ˌ)tō\ *noun* [Italian, literally, six-hundred, from *sei* six (from Latin *sex*) + *cento* hundred — more at SIX, CINQUECENTO] (circa 1902)
: the 17th century; *specifically* **:** the 17th century period in Italian literature and art

seiche \ˈsāsh, ˈsēch\ *noun* [French] (circa 1839)
: an oscillation of the surface of a landlocked body of water (as a lake) that varies in period from a few minutes to several hours

sei·del \ˈsī-dᵊl, ˈzī-\ *noun* [German, from Middle High German *sīdel,* from Latin *situla* bucket] (1908)
: a large glass for beer

Seid·litz powders \ˈsed-ləts-\ *noun plural* [*Sedlitz,* (Sedlčany), village in Bohemia; from the similarity of their effect to that of the water of the village] (1815)
: effervescing salts consisting of one powder of sodium bicarbonate and Rochelle salt and another of tartaric acid that are mixed in water and drunk as a mild cathartic

sei·gneur \sān-ˈyər\ *noun, often capitalized* [Middle French, from Medieval Latin *senior,* from Latin, adjective, elder — more at SENIOR] (1592)
1 : a man of rank or authority; *especially* **:** the feudal lord of a manor
2 : a member of the landed gentry of Canada

sei·gneur·ial \-ˈyu̇r-ē-əl, -ˈyər-\ *adjective* (1656)
: of, relating to, or befitting a seigneur

sei·gneury \ˈsān-yə-rē\ *noun, plural* **-gneur·ies** (1683)
1 a : the territory under the government of a feudal lord **b :** a landed estate held in Canada by feudal tenure until 1854
2 : the manor house of a Canadian seigneur

sei·gnior \sān-ˈyȯr, ˈsān-ˌ\ *noun* [Middle English *seigniour,* from Middle French *seigneur*] (15th century)
: SEIGNEUR 1

sei·gnior·age *or* **sei·gnor·age** \ˈsān-yə-rij\ *noun* [Middle English *seigneurage,* from Middle French, right of the lord (especially to coin money), from *seigneur*] (15th century)
: a government revenue from the manufacture of coins calculated as the difference between the face value and the metal value of the coins

sei·gniory *or* **sei·gnory** \ˈsān-yə-rē\ *noun, plural* **-gnior·ies** *or* **-gnor·ies** (14th century)
1 : LORDSHIP, DOMINION; *specifically* **:** the power or authority of a feudal lord
2 : the territory over which a lord holds jurisdiction

sei·gno·ri·al \sān-ˈyōr-ē-əl, -ˈyȯr-\ *adjective* (1818)
: of, relating to, or befitting a seignior **:** MANORIAL

¹seine \ˈsān\ *noun* [Middle English, from Old English *segne,* from Latin *sagena,* from Greek *sagēnē*] (before 12th century)
: a large net with sinkers on one edge and floats on the other that hangs vertically in the water and is used to enclose fish when its ends are pulled together or are drawn ashore

²seine *verb* **seined; sein·ing** (1836)
intransitive verb
: to fish with or catch fish with a seine
transitive verb
: to fish for or in with a seine

sein·er \ˈsā-nər\ *noun* (1602)
1 : one who fishes with a seine
2 : a boat used for seining

sei·sin *or* **sei·zin** \ˈsē-zᵊn\ *noun* [Middle English *seisine,* from Old French *saisine,* from *saisir* to seize — more at SEIZE] (14th century)
1 : the possession of land or chattels
2 : the possession of a freehold estate in land by one having title thereto

seis·mic \ˈsīz-mik, ˈsīs-\ *adjective* [Greek *seismos* shock, earthquake, from *seiein* to shake; probably akin to Avestan *thwaēshō* fear] (1858)
1 : of, subject to, or caused by an earthquake; *also* **:** of or relating to an earth vibration caused by something else (as an explosion or the impact of a meteorite)
2 : of or relating to a vibration on a celestial body (as the moon) comparable to a seismic event on earth
3 : having a strong or widespread impact **:** EARTHSHAKING ⟨*seismic* social changes⟩
— seis·mi·cal·ly \-mi-k(ə-)lē\ *adverb*

seis·mic·i·ty \sīz-ˈmi-sə-tē, sīs-\ *noun* (1902)
: the relative frequency and distribution of earthquakes

seismo- *combining form* [Greek, from *seismos*]
: earthquake **:** vibration ⟨*seismometer*⟩

seis·mo·gram \ˈsīz-mə-ˌgram, ˈsīs-\ *noun* [International Scientific Vocabulary] (circa 1891)
: the record of an earth tremor by a seismograph

seis·mo·graph \-ˌgraf\ *noun* [International Scientific Vocabulary] (1858)
: an apparatus to measure and record vibrations within the earth and of the ground
— seis·mog·ra·pher \sīz-ˈmä-grə-fər, sīs-\ *noun*
— seis·mo·graph·ic \ˌsīz-mə-ˈgra-fik, ˌsīs-\ *adjective*
— seis·mog·ra·phy \sīz-ˈmä-grə-fē, sīs-\ *noun*

seis·mol·o·gy \sīz-ˈmä-lə-jē, sīs-\ *noun* [International Scientific Vocabulary] (1858)
: a science that deals with earthquakes and with artificially produced vibrations of the earth
— seis·mo·log·i·cal \ˌsīz-mə-ˈlä-ji-kəl, ˌsīs-\ *adjective*
— seis·mol·o·gist \sīz-ˈmä-lə-jist, sīs-\ *noun*

seis·mom·e·ter \sīz-ˈmä-mə-tər, sīs-\ *noun* (1841)
: a seismograph measuring the actual movements of the ground (as on the earth or the moon)
— seis·mo·met·ric \ˌsīz-mə-ˈme-trik, ˌsīs-\ *adjective*

seis·mom·e·try \sīz-ˈmä-mə-trē, sīs-\ *noun* [International Scientific Vocabulary] (1858)
: the scientific study of earthquakes

sei whale \ˈsā-, ˈsī-\ *noun* [part translation of Norwegian *seihval,* from *sei* coalfish + *hval* whale] (1912)

: a common and widely distributed dark gray rorqual (*Balaenoptera borealis*) that has a ridge on the top of the head and grows to a length of nearly 60 feet (18 meters) — called also *sei*

seize \'sēz\ *verb* **seized; seiz·ing** [Middle English *saisen,* from Old French *saisir* to put in possession of, from Medieval Latin *sacire,* of Germanic origin; perhaps akin to Old High German *sezzen* to set — more at SET] (14th century)
transitive verb
1 a *usually* **seise** \'sēz\ **:** to vest ownership of a freehold estate in **b** *often* **seise :** to put in possession of something ⟨the biographer will be *seized* of all pertinent papers⟩
2 a : to take possession of **:** CONFISCATE **b :** to take possession of by legal process
3 a : to possess or take by force **:** CAPTURE **b :** to take prisoner **:** ARREST
4 a : to take hold of **:** CLUTCH **b :** to possess oneself of **:** GRASP **c :** to understand fully and distinctly **:** APPREHEND
5 a : to attack or overwhelm physically **:** AFFLICT ⟨suddenly *seized* with an acute illness —H. G. Armstrong⟩ **b :** to possess (as one's mind) completely or overwhelmingly ⟨*seized* the popular imagination —Basil Davenport⟩
6 : to bind or fasten together with a lashing of small stuff (as yarn, marline, or fine wire)
intransitive verb
1 : to take or lay hold suddenly or forcibly
2 a : to cohere to a relatively moving part through excessive pressure, temperature, or friction — used especially of machine parts (as bearings, brakes, or pistons) **b :** to fail to operate due to the seizing of a part — used of an engine
synonym see TAKE
— seiz·er *noun*

seizing *noun* (14th century)
1 a : the cord or lashing used in binding or fastening **b :** the fastening so made — see KNOT illustration
2 : the operation of fastening together or lashing with tarred small stuff

sei·zure \'sē-zhər\ *noun* (15th century)
1 a : the act, action, or process of seizing **:** the state of being seized **b :** the taking possession of person or property by legal process
2 : a sudden attack (as of disease) ⟨an epileptic *seizure*⟩

se·jant \'sē-jənt\ *adjective* [modification of Middle French *seant,* present participle of *seoir* to sit, from Latin *sedēre* — more at SIT] (circa 1500)
: SITTING — used of a heraldic animal

sel \'sel\ *chiefly Scottish variant of* SELF

se·la·chi·an \sə-'lā-kē-ən\ *noun* [ultimately from Greek *selachos* cartilaginous phosphorescent fish; akin to Greek *selas* brightness] (1835)
: any of a variously defined group (Selachii) of cartilaginous fishes that includes all the elasmobranchs or all elasmobranchs except the chimaeras, the existing sharks and rays or in its most restricted use the existing sharks as distinguished from the rays
— selachian *adjective*

se·lag·i·nel·la \sə-ˌla-jə-'ne-lə\ *noun* [New Latin, from Latin *selagin-, selago,* a plant resembling the savin] (1835)
: any of a genus (*Selaginella*) of mossy lower tracheophytes that are related to or grouped with the club mosses and have branching stems and scalelike leaves and produce one-celled sporangia containing both megaspores and microspores

se·lah \'sē-lə, -ˌlä\ *interjection* [Hebrew *selāh*] (1530)
— a term of uncertain meaning found in the Hebrew text of the Psalms and Habakkuk carried over untranslated into some English versions

sel·couth \'sel-ˌküth\ *adjective* [Middle English, from Old English *seldcūth,* from *seldan* seldom + *cūth* known — more at UNCOUTH] (before 12th century)
archaic **:** UNUSUAL, STRANGE

¹sel·dom \'sel-dəm\ *adverb* [Middle English, from Old English *seldan;* akin to Old High German *seltan* seldom] (before 12th century)
: in few instances **:** RARELY, INFREQUENTLY

²seldom *adjective* (13th century)
: RARE, INFREQUENT

¹se·lect \sə-'lekt\ *adjective* [Latin *selectus,* past participle of *seligere* to select, from *se-* apart (from *sed, se* without) + *legere* to gather, select — more at SUICIDE, LEGEND] (1565)
1 : chosen from a number or group by fitness or preference
2 a : of special value or excellence **:** SUPERIOR, CHOICE **b :** exclusively or fastidiously chosen often with regard to social, economic, or cultural characteristics
3 : judicious or restrictive in choice **:** DISCRIMINATING ⟨pleased with the *select* appreciation of his books —Osbert Sitwell⟩
— se·lect·able \sə-'lek-tə-bəl\ *adjective*
— se·lect·ness \sə-'lek(t)-nəs\ *noun*
— se·lec·tor \sə-'lek-tər\ *noun*

²select (1567)
transitive verb
: to choose (as by fitness or excellence) from a number or group **:** pick out
intransitive verb
: to make a choice

³select *noun* (1610)
: one that is select — often used in plural

se·lect·ed *adjective* (1590)
: SELECT; *specifically* **:** of a higher grade or quality than the ordinary

se·lect·ee \sə-ˌlek-'tē\ *noun* (1940)
1 : one inducted into military service under selective service
2 : one who is chosen from a group by fitness or preference

se·lec·tion \sə-'lek-shən\ *noun* (circa 1623)
1 : the act or process of selecting **:** the state of being selected
2 : one that is selected **:** CHOICE; *also* **:** a collection of selected things
3 : a natural or artificial process that results or tends to result in the survival and propagation of some individuals or organisms but not of others with the result that the inherited traits of the survivors are perpetuated — compare DARWINISM, NATURAL SELECTION
synonym see CHOICE

se·lec·tion·ist \-sh(ə-)nist\ *noun* (1892)
: one who considers natural selection a fundamental factor in evolution
— selectionist *adjective*

se·lec·tive \sə-'lek-tiv\ *adjective* (1625)
1 : of, relating to, or characterized by selection **:** selecting or tending to select
2 : highly specific in activity or effect ⟨*selective* pesticides⟩ ⟨*selective* absorption⟩
— se·lec·tive·ly *adverb*
— se·lec·tive·ness *noun*
— se·lec·tiv·i·ty \sə-ˌlek-'ti-və-tē, ˌsē-\ *noun*

selective service *noun* (1917)
: a system under which men are called up for military service **:** DRAFT

se·lect·man \si-'lek(t)-ˌman, -ˌlek(t)-'man, -'lek(t)-mən; 'sē-ˌlek(t)-ˌman\ *noun* (1635)
: one of a board of officials elected in towns of all New England states except Rhode Island to serve as the chief administrative authority of the town

¹selen- *or* **seleno-** *combining form* [Latin *selen-,* from Greek *selēn-,* from *selēnē* — more at SELENIUM]
: moon ⟨*selen*ium⟩ ⟨*seleno*logy⟩

²selen- *or* **seleni-** *combining form* [Swedish, from New Latin *selenium*]
: selenium ⟨*seleni*ferous⟩

sel·e·nate \'se-lə-ˌnāt\ *noun* [Swedish *selenat,* from *selen* of or containing selenium, from New Latin *selenium*] (1818)
: a salt containing the anion SeO_4^{2-}

Se·le·ne \sə-'lē-nē\ *noun*
: the Greek goddess of the moon

sel·e·nide \'se-lə-ˌnīd\ *noun* (1849)
: a binary compound of selenium with a more electropositive element or group

sel·e·nif·er·ous \ˌse-lə-'ni-f(ə-)rəs\ *adjective* (1823)
: containing or yielding selenium ⟨*seleniferous* vegetation⟩ ⟨*seleniferous* soils⟩

sel·e·nite \'se-lə-ˌnīt\ *noun* [Middle English *selinete,* from Latin *selenites,* from Greek *selēnitēs* (*lithos*), literally, stone of the moon, from *selēnē;* from the belief that it waxed and waned with the moon] (15th century)
: a variety of gypsum occurring in transparent crystals or crystalline masses

se·le·ni·um \sə-'lē-nē-əm\ *noun* [New Latin, from Greek *selēnē* moon, from *selas* brightness] (1818)
: a nonmetallic element that resembles sulfur and tellurium chemically, is obtained chiefly as a by-product in copper refining, and is a photoconductive semiconductor in its crystalline form — see ELEMENT table

selenium cell *noun* (1880)
: an insulated strip of selenium mounted with electrodes and used as a photoconductive element

se·le·no·cen·tric \sə-ˌlē-nə-'sen-trik\ *adjective* [International Scientific Vocabulary] (circa 1852)
: of or relating to the center of the moon; *also* **:** referred to or involving the moon as a center

sel·e·nol·o·gy \ˌse-lə-'nä-lə-jē\ *noun* (1821)
: a branch of astronomy that deals with the moon
— se·le·no·log·i·cal \ˌse-lə-nō-'lä-ji-kəl, sə-ˌlē-nᵊl-'ä-\ *adjective*
— sel·e·nol·o·gist \ˌse-lə-'nä-lə-jist\ *noun*

Se·leu·cid \sə-'lü-səd, səl-'yü-\ *noun* [New Latin *seleucides,* from *Seleucus I*] (1851)
: a member of a Greek dynasty ruling Syria and at various times other Asian territories from 312 B.C. to 64 B.C.
— Seleucid *adjective*

¹self \'self, *Southern also* 'sef\ *pronoun* [Middle English (intensive pronoun), from Old English; akin to Old High German *selb,* intensive pronoun, and probably to Latin *suus* one's own — more at SUICIDE] (before 12th century)
: MYSELF, HIMSELF, HERSELF ⟨check payable to *self*⟩

²self *adjective* (before 12th century)
1 *obsolete* **:** IDENTICAL, SAME
2 *obsolete* **:** belonging to oneself **:** OWN
3 a : having a single character or quality throughout; *specifically* **:** having one color only ⟨a *self* flower⟩ **b :** of the same kind (as in color, material, or pattern) as something with which it is used ⟨*self* trimming⟩

³self *noun, plural* **selves** \'selvz, *Southern also* 'sevz\ (13th century)
1 a : the entire person of an individual **b :** the realization or embodiment of an abstraction
2 a (1) **:** an individual's typical character or behavior ⟨her true *self* was revealed⟩ (2) **:** an individual's temporary behavior or character ⟨his better *self*⟩ **b :** a person in prime condition ⟨feel like my old *self* today⟩
3 : the union of elements (as body, emotions, thoughts, and sensations) that constitute the individuality and identity of a person
4 : personal interest or advantage
5 : material that is part of an individual organism ⟨ability of the immune system to distinguish *self* from nonself⟩

⁴self (1914)
transitive verb
1 : INBREED
2 : to pollinate with pollen from the same flower or plant
intransitive verb
: to undergo self-pollination

self- *combining form* [Middle English, from Old English, from *self*]

1 a : oneself or itself ⟨*self*-supporting⟩ **b :** of oneself or itself ⟨*self*-abasement⟩ **c :** by oneself or itself ⟨*self*-propelled⟩ ⟨*self*-acting⟩
2 a : to, with, for, or toward oneself or itself ⟨*self*-consistent⟩ ⟨*self*-addressed⟩ ⟨*self*-love⟩ **b :** of or in oneself or itself inherently ⟨*self*-evident⟩ **c :** from or by means of oneself or itself ⟨*self*-fertile⟩

self–abase•ment
self–ab•ne•gat•ing
self–ab•ne•ga•tion
self–ac•cel•er•at•ing
self–ac•cep•tance
self–ac•cu•sa•tion
self–ac•cu•sa•to•ry
self–ac•cus•ing
self–ac•knowl•edged
self–ad•just•ing
self–ad•min•is•ter
self–ad•min•is•tra•tion
self–ad•mit•ted
self–ad•mit•ted•ly
self–ad•u•la•to•ry
self–ad•vance•ment
self–ad•ver•tise•ment
self–ad•ver•tis•er
self–af•fir•ma•tion
self–ag•gran•dize•ment
self–ag•gran•diz•ing
self–alien•ation
self–anoint•ed
self–ap•prais•al
self–ap•pro•ba•tion
self–as•sess•ment
self–as•sign•ment
self–au•then•ti•cat•ing
self–avowed
self–bet•ter•ment
self–can•cel
self–care
self–car•i•ca•ture
self–cas•ti•gate
self–cas•ti•ga•tion
self–cen•sor•ship
self–char•ac•ter•i•za•tion
self–clas•si•fi•ca•tion
self–clean•ing
self–com•mand
self–com•mun•ing
self–com•mu•nion
self–com•pla•cen•cy
self–com•pla•cent
self–con•dem•na•tion
self–con•demned
self–con•firm•ing
self–con•se•cra•tion
self–con•sti•tut•ed
self–con•sum•ing
self–con•tempt
self–cre•at•ed
self–cre•a•tion
self–crit•i•cal
self–crit•i•cism
self–cri•tique
self–cul•ti•va•tion
self–cul•ture
self–damn•ing
self–de•base•ment
self–de•ceit
self–de•ceive
self–de•cep•tion
self–de•feat•ing
self–de•lud•ed
self–de•lud•ing
self–de•lu•sion
self–den•i•grat•ing
self–den•i•gra•tion
self–de•pen•dence
self–de•pen•dent
self–dep•re•cat•ing
self–dep•re•cat•ing•ly
self–dep•re•ca•tion
self–dep•re•ca•to•ry
self–de•pre•ci•a•tion
self–de•scribed
self–de•scrip•tion
self–de•scrip•tive
self–de•vel•op•ment
self–de•vour•ing
self–dif•fer•en•ti•a•tion
self–di•rect•ed
self–di•rect•ing
self–di•rec•tion
self–di•rec•tive
self–dis•gust
self–dis•play
self–dis•sat•is•fac•tion
self–doubt
self–ed•u•cat•ed
self–ed•u•cat•ing
self–ed•u•ca•tion
self–ef•face•ment
self–ef•fac•ing
self–eman•ci•pa•tion
self–emas•cu•la•tion
self–en•closed
self–en•grossed
self–en•hance•ment
self–eval•u•ate
self–eval•u•a•tion
self–ex•clu•sion
self–ex•cul•pa•tion
self–ex•hi•bi•tion
self–ex•is•tence
self–ex•is•tent
self–ex•plain•ing
self–ex•tinc•tion
self–fi•nance
self–formed
self–gen•er•ate
self–giv•ing
self–guid•ed
self–hate
self–hat•ing
self–ha•tred
self–heal•ing
self–help
self–hum•bling
self–hu•mil•i•a•tion
self–hyp•no•sis
self–idol•a•try
self–im•posed
self–im•prove•ment
self–im•prov•er
self–in•duced
self–in•fat•u•at•ed
self–in•fla•tion
self–in•flict•ed
self–ini•ti•at•ed
self–in•struct•ed
self–in•struc•tion
self–in•struc•tion•al
self–in•ter•view
self–in•vent•ed
self–in•ven•tion
self–iso•la•tion
self–la•beled
self–lac•er•at•ing
self–lac•er•a•tion
self–loath•ing
self–lock•ing
self–lu•bri•cat•ing
self–lu•mi•nous
self–main•te•nance
self–man•age•ment
self–mas•tery
self–med•i•ca•tion
self–mock•ery
self–mock•ing
self–mor•ti•fi•ca•tion
self–mo•ti•vat•ed
self–mu•ti•lat•ing
self–mu•ti•la•tion
self–ne•gat•ing
self–ob•sessed
self–op•er•at•ing
self–or•dained
self–ori•ent•ed
self–par•o•dist
self–par•o•dy
self–penned
self–per•pet•u•at•ing
self–per•pet•u•a•tion
self–pla•gia•rism
self–pleas•ing
self–po•lic•ing
self–praise
self–pre•oc•cu•pa•tion
self–pre•oc•cu•pied
self–pre•serv•ing
self–pro•claimed
self–pro•duced
self–pro•fessed
self–pro•mot•er
self–pro•mot•ing
self–pro•mo•tion
self–pro•tec•tion
self–pro•tec•tive
self–pro•tec•tive•ness
self–pun•ish•ing
self–pun•ish•ment
self–raised
self–re•crim•i•na•tion
self–ref•er•en•tial
self–re•fer•ring
self–re•flex•ive
self–ref•or•ma•tion
self–reg•u•la•tion
self–reg•u•la•tive
self–reg•u•la•to•ry
self–re•in•forc•ing
self–re•new•al
self–re•new•ing
self–re•nounc•ing
self–re•nun•ci•a•tion
self–re•proach
self–re•proach•ful
self–re•proof
self–re•prov•ing
self–re•straint
self–rid•i•cule
self–sat•i•riz•ing
self–se•lect•ed
self–se•lec•tion
self–set
self–sur•ren•der
self–sus•tained
self–ther•a•py
self–tor•ment
self–tor•men•tor
self–tor•ture
self–tran•scen•dence
self–trans•for•ma•tion
self–un•der•stand•ing
self–val•i•dat•ing
self–wor•ship
self–wor•ship•er

self–aban•doned \ˌself-ə-ˈban-dənd\ *adjective* (1791)
: abandoned by oneself; *especially* **:** given up to one's impulses

self–aban•don•ment \ dən-mənt\ *noun* (1818)
1 : a surrender of one's selfish interests or desires
2 : a lack of self-restraint

self–ab•sorbed \-əb-ˈsȯrbd, -ˈzȯrbd\ *adjective* (1847)
: absorbed in one's own thoughts, activities, or interests

self–ab•sorp•tion \-ˈsȯrp-shən, -ˈzȯrp-\ *noun* (1862)
: preoccupation with oneself

self–abuse \-ə-ˈbyüs\ *noun* (1605)
1 : reproach of oneself
2 : MASTURBATION
3 : abuse of one's body or health

self–act•ing \-ˈak-tiŋ\ *adjective* (circa 1680)
: acting or capable of acting of or by itself **:** AUTOMATIC

self–ac•tiv•i•ty \-ˌak-ˈti-və-tē\ *noun* (1644)
: independent and especially self-determined activity

self–ac•tu•al•ize \-ˈak-ch(ə-w)ə-ˌlīz, -sh(ə-w)ə-ˌlīz\ *intransitive verb* (1874)
: to realize fully one's potential
— **self–ac•tu•al•i•za•tion** \-ˌak-ch(ə-w)ə-lə-ˈzā-shən, -sh(ə-w)ə-lə-\ *noun*

self–ad•dressed \-ə-ˈdrest, -ˈa-ˌdrest\ *adjective* (1904)
: addressed for return to the sender ⟨a *self-addressed* envelope⟩

self–ad•he•sive \-əd-ˈhē-siv, -ziv\ *adjective* (1958)
: having a side coated with an adhesive that sticks without wetting ⟨*self-adhesive* labels⟩

self–ad•just•ment \-ə-ˈjəs(t)-mənt\ *noun* (1848)
: adjustment to oneself or one's environment

self–ad•mi•ra•tion \-ˌad-mə-ˈrā-shən\ *noun* (1661)
: SELF-CONCEIT

self–af•fect•ed \-ə-ˈfek-təd\ *adjective* (1606)
: CONCEITED, SELF-LOVING

self–anal•y•sis \-ə-ˈna-lə-səs\ *noun* (1860)
: a systematic attempt by an individual to understand his or her own personality without the aid of another person

self–an•a•lyt•i•cal \-ˌa-nᵊl-ˈi-ti-kəl\ *also* **self–an•a•lyt•ic** *adjective* (1943)
: using self-analysis

self–an•ni•hi•la•tion \-ə-ˌnī-ə-ˈlā-shən\ *noun* (1647)
: annihilation of the self (as in mystical contemplation of God)

self–ap•plaud•ing \-ə-ˈplȯ-diŋ\ *adjective* (1654)
: marked by self-applause

self–ap•plause \-ˈplȯz\ *noun* (1678)
: an expression or feeling of approval of oneself

self–ap•point•ed \-ə ˈpȯin-təd\ *adjective* (1799)
: appointed by oneself **:** SELF-STYLED

self–as•sem•bly \-ə-ˈsem-blē\ *noun* (1966)
: the process by which a complex macromolecule (as collagen) or a supramolecular system (as a virus) spontaneously assembles itself from its components
— **self–as•sem•ble** \-ə-ˈsem-bəl\ *intransitive verb*

self–as•sert•ing \-ə-ˈsər-tiŋ\ *adjective* (1837)
1 : asserting oneself or one's own rights, claims, or opinions
2 a : SELF-ASSURED, CONFIDENT **b :** ARROGANT
— **self–as•sert•ing•ly** \-tiŋ-lē\ *adverb*

self–as•ser•tion \-ə-ˈsər-shən\ *noun* (1806)
1 : the act of asserting oneself or one's own rights, claims, or opinions
2 : the act of asserting one's superiority over others

self–as•ser•tive \-ˈsər-tiv\ *adjective* (1865)
: given to or characterized by self-assertion
synonym see AGGRESSIVE
— **self–as•ser•tive•ly** *adverb*
— **self–as•ser•tive•ness** *noun*

self–as•sump•tion \-ə-ˈsəm(p)-shən\ *noun* (1606)
: SELF-CONCEIT

self–as•sur•ance \-ə-ˈshu̇r-ən(t)s\ *noun* (1594)
: SELF-CONFIDENCE

self–as•sured \-ˈshu̇rd\ *adjective* (1711)
: sure of oneself **:** SELF-CONFIDENT
— **self–as•sured•ly** \-ˈshu̇r-əd-lē, -ˈshu̇rd-\ *adverb*
— **self–as•sured•ness** \-ˈshu̇r-əd-nəs, -ˈshu̇rd-\ *noun*

self–aware \-ə-ˈwar, -ˈwer\ *adjective* (circa 1934)
: characterized by self-awareness

self–aware•ness *noun* (1880)
: an awareness of one's own personality or individuality

self–belt \ˈself-ˈbelt\ *noun* (1965)
: a belt made of the same material as the garment with which it is worn
— **self–belt•ed** \-ˈbel-təd\ *adjective*

self–be•tray•al \ˌself-bi-ˈtrā(-ə)l\ *noun* (1857)
: SELF-REVELATION

self–born \-ˈbȯrn\ *adjective* (1587)
1 : arising within the self ⟨*self-born* sorrows⟩
2 : springing from a prior self ⟨phoenix rising *self-born* from the fire⟩

self–ca•ter•ing \-ˈkā-tər-iŋ\ *adjective* (1970)
British **:** provided with lodging and cooking facilities but not meals ⟨*self-catering* holiday cottages⟩

self–cen•tered \-ˈsen-tərd\ *adjective* (circa 1764)
1 : independent of outside force or influence **:** SELF-SUFFICIENT

2 : concerned solely with one's own desires, needs, or interests
— **self–cen·tered·ly** *adverb*
— **self–cen·tered·ness** *noun*
self–clos·ing \-ˈklō-ziŋ\ *adjective* (circa 1875)
: closing or shutting automatically after being opened
self–cock·ing \-ˈkä-kiŋ\ *adjective* (1862)
of a firearm **:** cocked by the operation of some part of the action ⟨*self-cocking* on pushing the bolt forward⟩
self–col·lect·ed \-kə-ˈlek-təd\ *adjective* (circa 1711)
: SELF-POSSESSED
self–col·ored \-ˈkə-lərd\ *adjective* (1759)
: of a single color ⟨a *self-colored* flower⟩
self–com·pat·i·ble \-kəm-ˈpa-tə-bəl\ *adjective* (1922)
: capable of effective self-pollination that results in the production of seeds and fruits
— **self–com·pat·i·bil·i·ty** \-ˌpa-tə-ˈbi-lə-tē\ *noun*
self–com·posed \-kəm-ˈpōzd\ *adjective* (circa 1934)
: having control over one's emotions **:** CALM
— **self–com·pos·ed·ly** \-ˈpō-zəd-lē\ *adverb*
— **self–com·posed·ness** \-ˈpō-zəd-nəs, -ˈpōz(d)-nəs\ *noun*
self–con·ceit \-kən-ˈsēt\ *noun* (circa 1589)
: an exaggerated opinion of one's own qualities or abilities **:** VANITY
— **self–con·ceit·ed** \-ˈsē-təd\ *adjective*
self–con·cept \ˈself-ˈkän-ˌsept\ *noun* (1925)
: the mental image one has of oneself
self–con·cep·tion \ˌself-kən-ˈsep-shən\ *noun* (1950)
: SELF-CONCEPT
self–con·cern \-ˈsərn\ *noun* (1681)
: a selfish or morbid concern for oneself
— **self–con·cerned** \-ˈsərnd\ *adjective*
self–con·fessed \-ˈfest\ *adjective* (circa 1900)
: openly acknowledged by oneself **:** AVOWED
— **self–con·fess·ed·ly** \-ˈfe-səd-lē, -ˈfest-lē\ *adverb*
self–con·fes·sion \-ˈfe-shən\ *noun* (circa 1961)
: open acknowledgment **:** AVOWAL
self–con·fi·dence \-ˈkän-fə-dən(t)s, -ˌden(t)s\ *noun* (1637)
: confidence in oneself and in one's powers and abilities
— **self–con·fi·dent** \-fə-dənt, -ˌdent\ *adjective*
— **self–con·fi·dent·ly** *adverb*
self–con·fron·ta·tion \-ˌkän-(ˌ)frən-ˈtā-shən\ *noun* (1961)
: SELF-ANALYSIS
self–con·grat·u·la·tion \-kən-ˌgra-chə-ˈlā-shən, -ˌgra-jə-\ *noun* (1712)
: congratulation of oneself; *especially* **:** a complacent acknowledgment of one's own superiority or good fortune
self–con·grat·u·la·to·ry \-ˈgra-chə-lə-ˌtōr-ē, -ˌgra-jə-, -ˌtȯr-\ *adjective* (1877)
: expressive of self-congratulation ⟨*self-congratulatory* memoirs⟩
self–con·scious \-ˈkän(t)-shəs\ *adjective* (circa 1680)
1 a : conscious of one's own acts or states as belonging to or originating in oneself **:** aware of oneself as an individual **b :** intensely aware of oneself **:** CONSCIOUS ⟨a rising and *self-conscious* social class⟩; *also* **:** produced or done with such awareness ⟨*self-conscious* art⟩ **2 :** uncomfortably conscious of oneself as an object of the observation of others **:** ILL AT EASE
— **self–con·scious·ly** *adverb*
— **self–con·scious·ness** *noun*
self–con·se·quence \-ˈkän(t)-sə-ˌkwen(t)s, -si-kwən(t)s\ *noun* (1778)
: SELF-IMPORTANCE
self–con·sis·ten·cy \-kən-ˈsis-tən(t)-sē\ *noun* (1692)
: the quality or state of being self-consistent
self–con·sis·tent \-tənt\ *adjective* (1683)
: having each part logically consistent with the rest
self–con·tained \-kən-ˈtānd\ *adjective* (1591)
1 a : complete in itself **:** INDEPENDENT ⟨a *self-contained* machine⟩ ⟨a *self-contained* program of study⟩ **b :** BUILT-IN ⟨a lectern with a *self-contained* light fixture⟩ **2 a :** showing self-control **b :** formal and reserved in manner
— **self–con·tained·ly** \-ˈtā-nəd-lē, -ˈtānd-lē\ *adverb*
— **self–con·tained·ness** \-ˈtā-nəd-nəs, -ˈtān(d)-nəs\ *noun*
— **self–con·tain·ment** \-ˈtān-mənt\ *noun*
self–con·tam·i·na·tion \-kən-ˌta-mə-ˈnā-shən\ *noun* (1955)
1 : contamination by oneself
2 : contamination from within
self–con·tent \-kən-ˈtent\ *noun* (1654)
: SELF-SATISFACTION
self–con·tent·ed \-ˈten-təd\ *adjective* (1818)
: SELF-SATISFIED
— **self–con·tent·ed·ly** *adverb*
— **self–con·tent·ed·ness** *noun*
self–con·tent·ment \-ˈtent-mənt\ *noun* (1815)
: SELF-SATISFACTION
self–con·tra·dic·tion \-ˌkän-trə-ˈdik-shən\ *noun* (1658)
1 : contradiction of oneself
2 : a self-contradictory statement or proposition
self–con·tra·dic·to·ry \-ˈdik-t(ə-)rē\ *adjective* (1657)
: consisting of two contradictory members or parts
self–con·trol \-kən-ˈtrōl\ *noun* (1711)
: restraint exercised over one's own impulses, emotions, or desires
— **self–con·trolled** \-ˈtrōld\ *adjective*
self–cor·rect·ing \-kə-ˈrek-tiŋ\ *adjective* (1939)
: correcting or compensating for one's own errors or weaknesses
self–cor·rec·tive \-ˈrek-tiv\ *adjective* (circa 1925)
: SELF-CORRECTING
self–deal·ing \ˈself-ˈdē-liŋ\ *noun* (1940)
: financial dealing that is not at arm's length; *especially* **:** borrowing from or lending to a company by a controlling individual primarily to the individual's own advantage
self–de·fense \-di-ˈfen(t)s\ *noun* (1651)
1 : a plea of justification for the use of force or for homicide
2 : the act of defending oneself, one's property, or a close relative
self–de·fen·sive \-ˈfen(t)-siv\ *adjective* (1828)
: of, relating to, or given to self-defense ⟨a *self-defensive* person⟩ ⟨a *self-defensive* attitude⟩
self–def·i·ni·tion \-ˌde-fə-ˈni-shən\ *noun* (1957)
: the evaluation by oneself of one's worth as an individual in distinction from one's interpersonal or social roles
self–de·ni·al \-di-ˈnī(-ə)l\ *noun* (1642)
: a restraint or limitation of one's own desires or interests
self–de·ny·ing \-ˈnī-iŋ\ *adjective* (1632)
: showing self-denial
— **self–de·ny·ing·ly** \-iŋ-lē\ *adverb*
self–de·spair \-di-ˈspar, -ˈsper\ *noun* (1677)
: despair of oneself **:** HOPELESSNESS
self–de·stroy·er \-di-ˈstrȯi(-ə)r\ *noun* (1654)
: one who destroys oneself
self–de·stroy·ing \-ˈstrȯi-iŋ\ *adjective* (1645)
: SELF-DESTRUCTIVE
self–de·struct \-di-ˈstrəkt\ *intransitive verb* (1968)
: to destroy oneself or itself
— **self–destruct** *adjective*
self–de·struc·tion \-ˈstrək-shən\ *noun* (circa 1586)
: destruction of oneself; *especially* **:** SUICIDE
self–de·struc·tive \-ˈstrək-tiv\ *adjective* (1654)
: acting or tending to harm or destroy oneself; *also* **:** SUICIDAL
— **self–de·struc·tive·ness** *noun*
self–de·ter·mi·na·tion \-di-ˌtər-mə-ˈnā-shən\ *noun* (circa 1670)
1 : free choice of one's own acts or states without external compulsion
2 : determination by the people of a territorial unit of their own future political status
self–de·ter·mined \-ˈtər-mənd\ *adjective* (circa 1670)
: determined by oneself
self–de·ter·min·ing \-ˈtər-mə-niŋ, -ˈtərm-niŋ\ *adjective* (1662)
: capable of determining one's or its own acts
self–de·ter·min·ism \-ˈtər-mə-ˌni-zəm\ *noun* (1936)
: a doctrine that the actions of a self are determined by itself
self–de·vot·ed \-di-ˈvō-təd\ *adjective* (1713)
: characterized by total devotion of oneself (as to a cause)
— **self–de·vot·ed·ness** *noun*
self–dis·ci·pline \-ˈdi-sə-plən\ *noun* (1838)
: correction or regulation of oneself for the sake of improvement
self–dis·ci·plined \-plənd\ *adjective* (1932)
: capable of or subject to self-discipline
self–dis·cov·ery \-dis-ˈkə-v(ə-)rē\ *noun* (1924)
: the act or process of achieving self-knowledge
self–dis·trust \-ˈtrəst\ *noun* (1789)
: a lack of confidence in oneself **:** DIFFIDENCE
— **self–dis·trust·ful** \-fəl\ *adjective*
self–dra·ma·ti·za·tion \ˌself-ˌdra-mə-tə-ˈzā-shən, -ˌdrä-\ *noun* (1937)
: the act or an instance of dramatizing oneself
self–dra·ma·tiz·ing \-ˈdra-mə-ˌtī-ziŋ, -ˈdrä-\ *adjective* (1938)
: seeing and presenting oneself as an important or dramatic figure
self–drive \ˈself-ˈdrīv\ *adjective* (1929)
chiefly British **:** being a rental car
self–elect·ed \ˌself-ə-ˈlek-təd\ *adjective* (1818)
: SELF-APPOINTED
self–em·ployed \-im-ˈplȯid\ *adjective* (1946)
: earning income directly from one's own business, trade, or profession rather than as a specified salary or wages from an employer
— **self–employed** *noun*
self–em·ploy·ment \-ˈplȯi-mənt\ *noun* (1745)
: the state of being self-employed
self–en·er·giz·ing \-ˈe-nər-ˌjī-ziŋ\ *adjective* (1931)
: containing means for augmentation of power within itself ⟨a *self-energizing* brake⟩
self–en·forc·ing \-in-ˈfōr-siŋ, -ˈfȯr-\ *adjective* (1952)
: containing in itself the authority or means that provide for its enforcement
self–en·rich·ment \-in-ˈrich-mənt\ *noun* (1920)
: the act or process of increasing one's intellectual or spiritual resources
self–es·teem \-ə-ˈstēm\ *noun* (1657)
1 : a confidence and satisfaction in oneself **:** SELF-RESPECT
2 : SELF-CONCEIT
self–ev·i·dence \-ˈe-və-dən(t)s, -ˌden(t)s\ *noun* (1671)
: the quality or state of being self-evident
self–ev·i·dent \-dənt, -ˌdent\ *adjective* (1671)
: evident without proof or reasoning
— **self–ev·i·dent·ly** *adverb*
self–ex·am·i·na·tion \-ig-ˌza-mə-ˈnā-shən\ *noun* (1647)

1 : a reflective examination (as of one's beliefs or motives) **:** INTROSPECTION
2 : examination of one's body especially for evidence of disease

self–ex·cit·ed \-ik-'sī-təd\ *adjective* (circa 1896)
: excited by a current produced by the generator itself ⟨*self-excited* generators⟩

self–ex·e·cut·ing \-'ek-sə-ˌkyü-tiŋ\ *adjective* (1868)
: taking effect immediately without implementing legislation ⟨a *self-executing* treaty⟩

self–ex·iled \-'eg-ˌzīld, -'ek-ˌsīld\ *adjective* (1737)
: exiled by one's own wish or decision

self–ex·plan·a·to·ry \ik-'spla-nə-ˌtōr-ē, -ˌtȯr-\ *adjective* (1898)
: explaining itself **:** capable of being understood without explanation

self–ex·plo·ra·tion \-ˌek-splə-'rā-shən, -ˌsplȯ-\ *noun* (1959)
: the examination and analysis of one's own unrealized spiritual or intellectual capacities

self–ex·pres·sion \-ik-'spre-shən\ *noun* (1892)
: the expression of one's own personality **:** assertion of one's individual traits
— **self–ex·pres·sive** \-'spre-siv\ *adjective*

self–feed \'self-'fēd\ *transitive verb* **-fed** \-'fed\; **-feed·ing** (circa 1924)
: to provide rations to (animals) in bulk so as to permit feeding as wanted

self–feed·er \-'fē-dər\ *noun* (1924)
: a device for providing feed to livestock that is equipped with a feed hopper that automatically supplies a trough below

self–feel·ing \'self-'fē-liŋ\ *noun* (1879)
: self-centered emotion

self–fer·tile \ˌself-'fər-t^{ə}l\ *adjective* (1865)
: fertile by means of its own pollen or sperm
— **self–fer·til·i·ty** \-(ˌ)fər-'ti-lə-tē\ *noun*

self–fer·til·i·za·tion \-ˌfər-t^{ə}l-ə-'zā-shən\ *noun* (1859)
: fertilization effected by union of ova with pollen or sperm from the same individual

self–fer·til·ized \-'fər-t^{ə}l-ˌīzd\ *adjective* (1871)
: fertilized by one's own pollen or sperm

self–fer·til·iz·ing \-ˌī-ziŋ\ *adjective* (1859)
: SELF-FERTILIZED

self–flag·el·la·tion \-ˌfla-jə-'lā-shən\ *noun* (1845)
: extreme criticism of oneself

self–flat·ter·ing \-'fla-tə-riŋ\ *adjective* (circa 1586)
: given to self-flattery

self–flat·tery \-tə-rē\ *noun* (1680)
: the glossing over of one's own weaknesses or mistakes and the exaggeration of one's own good qualities and achievements

self–for·get·ful \-fər-'get-fəl\ *adjective* (1848)
: having or showing no thought of self or selfish interests
— **self–for·get·ful·ly** \-fə-lē\ *adverb*
— **self–for·get·ful·ness** *noun*

self–for·get·ting \-'ge-tiŋ\ *adjective* (1847)
: SELF-FORGETFUL
— **self–for·get·ting·ly** \-tiŋ-lē\ *adverb*

self–fruit·ful \'self-'früt-fəl\ *adjective* (1940)
: capable of setting a crop of self-pollinated fruit
— **self–fruit·ful·ness** *noun*

self–ful·fill·ing \ˌself-fu̇l-'fi-liŋ\ *adjective* (1949)
1 : marked by or achieving self-fulfillment
2 : becoming real or true by virtue of having been predicted or expected ⟨a *self-fulfilling* prophecy⟩

self–ful·fill·ment \-'fil-mənt\ *noun* (circa 1864)
: fulfillment of oneself

self–giv·en \-'gi-vən\ *adjective* (1742)
1 : derived from itself ⟨a *self-given* entity⟩
2 : given by oneself ⟨*self-given* authority⟩

self–glo·ri·fi·ca·tion \-ˌglōr-ə-fə-'kā-shən, -ˌglȯr-\ *noun* (1838)
: a feeling or expression of one's own superiority

self–glo·ri·fy·ing \-'glōr-ə-ˌfī-iŋ, -'glȯr-\ *adjective* (1860)
: given to or marked by boasting **:** BOASTFUL

self–glo·ry \-'glōr-ē, -'glȯr-ē\ *noun* (1647)
: personal vanity **:** PRIDE

self–gov·er·nance \-'gə-vər-nən(t)s\ *noun* (1964)
: SELF-GOVERNMENT 2

self–gov·erned \-'gə-vərnd\ *adjective* (1709)
1 : not influenced or controlled by others
2 : exercising self-control

self–gov·ern·ing \-'gə-vər-niŋ\ *adjective* (1845)
: having control or rule over oneself; *specifically* **:** having self-government **:** AUTONOMOUS

self–gov·ern·ment \-'gə-vər(n)-mənt, -'gə-v^{ə}m-ənt\ *noun* (1734)
1 : SELF-COMMAND, SELF-CONTROL
2 : government under the control and direction of the inhabitants of a political unit rather than by an outside authority; *broadly* **:** control of one's own affairs

self–grat·i·fi·ca·tion \-ˌgra-tə-fə-'kā-shən\ *noun* (1677)
: the act of pleasing oneself or of satisfying one's desires; *especially* **:** the satisfying of one's own sexual urges

self–grat·u·la·tion \-ˌgra-chə-'lā-shən\ *noun* (1802)
: SELF-CONGRATULATION

self–grat·u·la·to·ry \-'gra-chə-lə-ˌtōr-ē, -ˌtȯr-\ *adjective* (1859)
: SELF-CONGRATULATORY

self–heal \'self-ˌhēl\ *noun* (14th century)
: a blue-flowered Eurasian mint (*Prunella vulgaris*) naturalized throughout North America and formerly considered to have medicinal properties

self·hood \-ˌhu̇d\ *noun* (1649)
1 : INDIVIDUALITY
2 : the quality or state of being selfish

self–iden·ti·cal \ˌself-ī-'den-ti-kəl, -ə-'den-\ *adjective* (1877)
: having self-identity

self–iden·ti·fi·ca·tion \-ˌden-tə-fə-'kā-shən\ *noun* (1941)
: identification with someone or something outside oneself

self–iden·ti·ty \-'den-tə-tē, -'de-nə-tē\ *noun* (1866)
1 : sameness of a thing with itself
2 : INDIVIDUALITY ⟨self-understanding is the necessary condition of a sense of *self-identity* —J. C. Murray⟩

self–ig·nite \-ig-'nīt\ *intransitive verb* (1943)
: to become ignited without flame or spark (as under high compression)

self–ig·ni·tion \-'ni-shən\ *noun* (1903)
: ignition without flame or spark

self–im·age \'self-'i-mij\ *noun* (1939)
: one's conception of oneself or of one's role

self–im·mo·la·tion \ˌself-ˌi-mə-'lā-shən\ *noun* (1817)
: a deliberate and willing sacrifice of oneself often by fire
— **self–im·mo·late** \-'i-mə-ˌlāt\ *verb*

self–im·por·tance \-im-'pȯr-t^{ə}n(t)s, -tən(t)s\ *noun* (circa 1775)
1 : an exaggerated estimate of one's own importance **:** SELF-CONCEIT
2 : arrogant or pompous behavior

self–important \-t^{ə}nt, -tənt\ *adjective* (circa 1775)
: having or showing self-importance
— **self–im·por·tant·ly** *adverb*

self–in·clu·sive \-in-'klü-siv, -ziv\ *adjective* (circa 1909)
1 : enclosing itself
2 : complete in itself

self–in·com·pat·i·ble \-ˌin-kəm-'pa-tə-bəl\ *adjective* (1922)
: incapable of effective self-pollination
— **self–in·com·pat·i·bil·i·ty** \-ˌpa-tə-'bi-lə-tē\ *noun*

self–in·crim·i·nat·ing \-in-'kri-mə-ˌnā-tiŋ\ *adjective* (1931)
: serving or tending to incriminate oneself

self–in·crim·i·na·tion \-ˌkri-mə-'nā-shən\ *noun* (1911)
: incrimination of oneself; *specifically* **:** the giving of testimony which will likely subject one to criminal prosecution

self–in·duc·tance \-'dək-tən(t)s\ *noun* (1865)
: inductance in which an electromotive force is produced by self-induction

self–in·duc·tion \-'dək-shən\ *noun* (1873)
: induction of an electromotive force in a circuit by a varying current in the same circuit

self–in·dul·gence \-'dəl-jən(t)s\ *noun* (1753)
: excessive or unrestrained gratification of one's own appetites, desires, or whims
— **self–in·dul·gent** \-jənt\ *adjective*
— **self–in·dul·gent·ly** *adverb*

self–in·sur·ance \in-'shu̇r-ən(t)s, -'in-ˌ\ *noun* (circa 1897)
: insurance of oneself or of one's own interests by the setting aside of money at regular intervals to provide a fund to cover possible losses

self–insure \-in-'shu̇r\ *verb* (1932)
: to insure oneself; *especially* **:** to practice self-insurance
— **self–insurer** \-'shu̇r-ər\ *noun*

self–in·ter·est \-'in-t(ə-)rəst; -'in-tə-ˌrest, -ˌtrest; -'in-tərst\ *noun* (1649)
1 : a concern for one's own advantage and well-being ⟨acted out of *self-interest* and fear⟩
2 : one's own interest or advantage ⟨*self-interest* requires that we be generous in foreign aid⟩
— **self–in·ter·est·ed** *adjective*
— **self–in·ter·est·ed·ness** *noun*

self–in·volved \-in-'välvd, -'vȯlvd *also* -'vävd *or* 'vȯvd\ *adjective* (1842)
: SELF-ABSORBED

self·ish \'sel-fish\ *adjective* (1640)
1 : concerned excessively or exclusively with oneself **:** seeking or concentrating on one's own advantage, pleasure, or well-being without regard for others
2 : arising from concern with one's own welfare or advantage in disregard of others ⟨a *selfish* act⟩
— **self·ish·ly** *adverb*
— **self·ish·ness** *noun*

self–jus·ti·fi·ca·tion \ˌself-ˌjəs-tə-fə-'kā-shən\ *noun* (circa 1775)
: the act or an instance of making excuses for oneself

self–jus·ti·fy·ing \-'jəs-tə-ˌfī-iŋ\ *adjective* (1740)
: seeking to justify oneself

self–know·ing \-'nō-iŋ\ *adjective* (1667)
: having self-knowledge

self–knowl·edge \-'nä-lij\ *noun* (circa 1613)
: knowledge or understanding of one's own capabilities, character, feelings, or motivations

self·less \'sel-fləs\ *adjective* (1825)
: having no concern for self **:** UNSELFISH
— **self·less·ly** *adverb*
— **self·less·ness** *noun*

self–lim·it·ed \ˌself-'li-mə-təd\ *adjective* (1845)
: limited by one's or its own nature; *specifically* **:** running a definite and limited course ⟨a *self-limited* disease⟩

self–lim·it·ing \-tiŋ\ *adjective* (1863)
: limiting oneself or itself; *especially, of a disease* **:** SELF-LIMITED

self–liq·ui·dat·ing \-'li-kwə-ˌdā-tiŋ\ *adjective* (1915)

\ə\ **abut** \$^{ə}$\ kitten \ər\ **further** \a\ **ash** \ā\ **ace** \ä\ **mop, mar** \au̇\ **out** \ch\ **chin** \e\ **bet** \ē\ **easy** \g\ **go** \i\ **hit** \ī\ **ice** \j\ **job** \ŋ\ **sing** \ō\ **go** \ȯ\ **law** \ȯi\ **boy** \th\ **thin** \t͟h\ **the** \ü\ **loot** \u̇\ **foot** \y\ **yet** \zh\ **vision** *see also* Guide to Pronunciation

1 : of or relating to a commercial transaction in which goods are converted into cash in a short time
2 : generating funds from its own operations to repay the investment made to create it ⟨a *self-liquidating* housing project⟩
self-load·er \-ˈlō-dər\ *noun* (circa 1936)
: a semiautomatic firearm
self-load·ing \-diŋ\ *adjective* (1899)
of a firearm **:** SEMIAUTOMATIC
self-love \ˈself-ˈləv\ *noun* (1563)
: love of self: **a :** CONCEIT **b :** regard for one's own happiness or advantage
— self-lov·ing \-ˈlə-viŋ\ *adjective*
self-made \ˈself-ˈmād\ *adjective* (1615)
: made such by one's own actions; *especially* **:** having achieved success or prominence by one's own efforts ⟨a *self-made* man⟩
self-mail·er \-ˈmā-lər\ *noun* (circa 1942)
: a folder that can be sent by mail without enclosure in an envelope by use of a gummed sticker or a precanceled stamp to hold the leaves together
self-mail·ing \-liŋ\ *adjective* (circa 1948)
: capable of being mailed without being enclosed in an envelope
self-moved \ˈself-ˈmüvd\ *adjective* (circa 1670)
: moved by inherent power
self-mur·der \-ˈmər-dər\ *noun* (1563)
: SELF-DESTRUCTION, SUICIDE
self·ness \ˈself-nəs\ *noun* (circa 1586)
1 : EGOISM, SELFISHNESS
2 : PERSONALITY, SELFHOOD
self-ob·ser·va·tion \ˌself-ˌäb-sər-ˈvā-shən, -zər-\ *noun* (1832)
1 : INTROSPECTION
2 : observation of one's own appearance
self-opin·ion \-ə-ˈpin-yən\ *noun* (circa 1580)
: high or exaggerated opinion of oneself **:** SELF-CONCEIT
self-opin·ion·at·ed \-yə-ˌnā-təd\ *adjective* (1671)
1 : CONCEITED
2 : stubbornly holding to one's own opinion **:** OPINIONATED
— self-opin·ion·at·ed·ness *noun*
self-or·ga·ni·za·tion \-ˌȯr-gə-nə-ˈzā-shən, -ˌȯrg-nə-\ *noun* (1898)
: organization of oneself or itself; *specifically* **:** the act or process of forming or joining a labor union
self-paced \ˈself-ˈpāst\ *adjective* (1962)
: designed to permit learning at the student's own pace ⟨*self-paced* mathematician course⟩
self-par·tial·i·ty \ˌself-ˌpär-shē-ˈa-lə-tē, -ˌpär-ˈsha-\ *noun* (1628)
1 : an excessive estimate of oneself as compared with others
2 : a prejudice in favor of one's own claims or interests
self-per·cep·tion \-pər-ˈsep-shən\ *noun* (1678)
: perception of oneself; *especially* **:** SELF-CONCEPT
self-pity \ˈself-ˈpi-tē\ *noun* (1621)
: pity for oneself; *especially* **:** a self-indulgent dwelling on one's own sorrows or misfortunes
— self-pity·ing \-tē-iŋ\ *adjective*
— self-pity·ing·ly \-iŋ-lē\ *adverb*
self-pleased \-ˈplēzd\ *adjective* (1748)
: SELF-SATISFIED
self-poise \-ˈpȯiz\ *noun* (1854)
: the quality or state of being self-poised
self-poised \-ˈpȯizd\ *adjective* (1621)
: having poise through self-command
self-pol·li·nate \ˈself-ˈpä-lə-ˌnāt\ (1890)
transitive verb
: SELF 2
intransitive verb
: to undergo self-pollination
self-pol·li·na·tion \ˌself-ˌpä-lə-ˈnā-shən\ *noun* (1872)
: the transfer of pollen from the anther of a flower to the stigma of the same flower or sometimes to that of a genetically identical flower (as of the same plant or clone)
self-por·trait \-ˈpōr-trət, -ˈpȯr-, -ˌtrāt\ *noun* (1831)
: a portrait of oneself done by oneself
self-pos·sessed \-pə-ˈzest *also* -ˈsest\ *adjective* (1818)
: having or showing self-possession **:** composed in mind or manner **:** CALM
— self-pos·sessed·ly \-ˈze-səd-lē, -ˈse-; -ˈzest-lē, -ˈsest-\ *adverb*
self-pos·ses·sion \-pə-ˈze-shən *also* -ˈse-\ *noun* (1745)
: control of one's emotions or reactions especially when under stress **:** PRESENCE OF MIND, COMPOSURE
synonym see CONFIDENCE
self-pres·er·va·tion \-ˌpre-zər-ˈvā-shən\ *noun* (circa 1614)
1 : preservation of oneself from destruction or harm
2 : a natural or instinctive tendency to act so as to preserve one's own existence
self-pride \ˈself-ˈprīd\ *noun* (circa 1586)
: pride in oneself or in that which relates to oneself
self-pro·pelled \ˌself-prə-ˈpeld\ *adjective* (1899)
1 : containing within itself the means for its own propulsion ⟨a *self-propelled* vehicle⟩
2 : mounted on or fired from a moving vehicle ⟨a *self-propelled* gun⟩
self-pro·pel·ling \-ˈpe-liŋ\ *adjective* (1862)
: SELF-PROPELLED 1
self-pro·pul·sion \-ˈpəl-shən\ *noun* (circa 1934)
: propulsion by one's own power
self-pub·lished \-ˈpə-blisht\ *adjective* (1975)
: published by the author ⟨*self-published* book⟩
self-pu·ri·fi·ca·tion \-ˌpyu̇r-ə-fə-ˈkā-shən\ *noun* (1919)
1 : purification by natural process ⟨*self-purification* of water⟩
2 : purification of oneself
self-ques·tion \ˈself-ˈkwes-chən, -ˈkwesh-\ *noun* (1917)
: a question asked of oneself by oneself
self-ques·tion·ing \-chə-niŋ\ *noun* (1856)
: examination of one's own actions and motives
self-rat·ing \-ˈrā-tiŋ\ *noun* (1925)
: determination of one's own rating with reference to a standard scale
self-re·al·i·za·tion \-ˌrē-ə-lə-ˈzā-shən, -ˌri-ə-\ *noun* (1874)
: fulfillment by oneself of the possibilities of one's character or personality
self-re·al·i·za·tion·ism \-shə-ˌni-zəm\ *noun* (circa 1874)
: the ethical theory that the highest good for a person consists in realizing or fulfilling oneself usually on the assumption that one has certain inborn abilities constituting one's real or ideal self
— self-re·al·i·za·tion·ist \-sh(ə-)nist\ *noun*
self-rec·og·ni·tion \-ˌre-kig-ˈni-shən, -kəg-\ *noun* (1946)
1 : recognition of one's own self
2 : the process by which the immune system of an organism distinguishes between the body's own chemicals, cells, and tissues and those of foreign organisms and agents
self-re·cord·ing \-ri-ˈkȯr-diŋ\ *adjective* (1875)
: making an automatic record ⟨*self-recording* instruments⟩
self-re·flec·tion \-ri-ˈflek-shən\ *noun* (1652)
: SELF-EXAMINATION 1
self-re·flec·tive \-ri-ˈflek-tiv\ *adjective* (1875)
: marked by or engaging in self-reflection
self-re·gard \-ri-ˈgärd\ *noun* (1595)
: regard for or consideration of oneself or one's own interests
self-re·gard·ing \-ri-ˈgär-diŋ\ *adjective* (1789)
: concerned with oneself or one's own interests
self-reg·u·lat·ing \-ˈre-gyə-ˌlā-tiŋ\ *adjective* (1837)
: regulating oneself or itself; *especially* **:** AUTOMATIC ⟨a *self-regulating* mechanism⟩
self-re·li·ance \-ri-ˈlī-ənts\ *noun* (1833)
: reliance on one's own efforts and abilities
self-re·li·ant \-ənt\ *adjective* (1848)
: having confidence in and exercising one's own powers or judgment
self-rep·li·cat·ing \-ˈre-plə-ˌkā-tiŋ\ *adjective* (1946)
: reproducing itself autonomously ⟨DNA is a *self-replicating* molecule⟩
— self-rep·li·ca·tion \-ˌre-plə-ˈkā-shən\ *noun*
self-re·spect \-ri-ˈspekt\ *noun* (circa 1814)
1 : a proper respect for oneself as a human being
2 : regard for one's own standing or position
self-re·spect·ing \-ri-ˈspek-tiŋ\ *adjective* (1786)
: having or characterized by self-respect
self-re·veal·ing \-ri-ˈvē-liŋ\ *adjective* (1839)
: marked by self-revelation
self-rev·e·la·tion \-ˌre-və-ˈlā-shən\ *noun* (1852)
: revelation of one's own thoughts, feelings, and attitudes especially without deliberate intent
self-re·ward·ing \-ri-ˈwȯr-diŋ\ *adjective* (1740)
: containing or producing its own reward ⟨virtue is *self-rewarding*⟩
self-righ·teous \-ˈrī-chəs\ *adjective* (circa 1680)
: convinced of one's own righteousness especially in contrast with the actions and beliefs of others **:** narrow-mindedly moralistic
— self-righ·teous·ly *adverb*
— self-righ·teous·ness *noun*
self-ris·ing flour \ˈself-ˈrī-ziŋ-\ *noun* (1854)
: a commercially prepared mixture of flour, salt, and a leavening agent
self-rule \ˈself-ˈrül\ *noun* (circa 1855)
: SELF-GOVERNMENT
self-rul·ing \-ˈrü-liŋ\ *adjective* (circa 1680)
: SELF-GOVERNING
self-sac·ri·fice \ˈself-ˈsa-krə-ˌfīs *also* -fəs *or* -ˌfīz\ *noun* (1805)
: sacrifice of oneself or one's interest for others or for a cause or ideal
self-sac·ri·fic·er \-ˌfī-sər *also* -fə- *or* -ˌfī-\ *noun* (1668)
: one that practices self-sacrifice
self-sa·cri·fic·ing \-ˌfīs-iŋ *also* -fə- *or* -ˌfī-\ *adjective* (1817)
: sacrificing oneself for others
— self-sac·ri·fic·ing·ly \-siŋ-lē\ *adverb*
self·same \ˈself-ˌsām\ *adjective* (15th century)
: being the one mentioned or in question **:** IDENTICAL ⟨left the *selfsame* day⟩
synonym see SAME
— self·same·ness \-ˌsām-nəs, -ˈsām-\ *noun*
self-sat·is·fac·tion \ˌself-ˌsa-təs-ˈfak-shən\ *noun* (1739)
: a usually smug satisfaction with oneself or one's position or achievements
self-sat·is·fied \ˈself-ˈsa-təs-ˌfīd\ *adjective* (1734)
: feeling or showing self-satisfaction
self-scru·ti·ny \ˈself-ˈskrü-tᵊn-ē, -ˈskrüt-nē\ *noun* (circa 1711)
: SELF-EXAMINATION
self-seal·ing \ˈself-ˈsē-liŋ\ *adjective* (1924)
1 : capable of sealing itself (as after puncture) ⟨a *self-sealing* tire⟩
2 : capable of being sealed by pressure without the addition of moisture ⟨*self-sealing* envelopes⟩
self-search·ing \-ˈsər-chiŋ\ *noun* (1687)

: SELF-QUESTIONING

self–seek·er \-ˈsē-kər\ *noun* (1632)
: a self-seeking person

[1]self–seek·ing \-kiŋ\ *noun* (circa 1586)
: the act or practice of selfishly advancing one's own ends

[2]self–seeking *adjective* (circa 1628)
: seeking only to further one's own interests

self–serve \ˈself-ˈsərv\ *adjective* (1926)
: permitting self-service

self–ser·vice \ˈself-ˈsər-vəs\ *noun* (1919)
: the serving of oneself (as in a restaurant or service station) with goods or services to be paid for at a cashier's desk or by means of a coin-operated mechanism
— **self–service** *adjective*

self–serv·ing \-ˈsər-viŋ\ *adjective* (1827)
: serving one's own interests often in disregard of the truth or the interests of others
— **self–serv·ing·ly** *adverb*

self–slaugh·ter \-ˈslȯ-tər\ *noun* (1602)
: SUICIDE 1a

self–slaugh·tered \-tərd\ *adjective* (1593)
: killed by oneself

self–sow \ˈself-ˈsō\ *intransitive verb* **-sowed** \-ˈsōd\; **-sown** \-ˈsōn\ *or* **-sowed; -sow·ing** (1608)
: to sow itself by dropping seeds or by natural action (as of wind or water)

self–start·er \ˌself-ˈstär-tər\ *noun* (1894)
1 : STARTER 3a
2 : a person who has initiative

self–start·ing \-ˈstär-tiŋ\ *adjective* (1866)
: capable of starting by oneself or itself

self–ster·ile \-ˈster-əl\ *adjective* (1876)
: sterile to its own pollen or sperm
— **self–ste·ril·i·ty** \-stə-ˈri-lə-tē\ *noun*

self–stick \ˈself-ˈstik\ *adjective* (1947)
: capable of adhering to a surface by application of pressure without the addition of moisture

self–stim·u·la·tion \ˌself-ˌstim-yə-ˈlā-shən\ *noun* (1947)
: stimulation of oneself as a result of one's own activity or behavior ⟨electrical *self-stimulation* of the brain⟩; *especially* : MASTURBATION
— **self–stim·u·la·to·ry** \-ˈstim-yə-lə-ˌtōr-ē, -ˌtȯr-\ *adjective*

self–study \ˈself-ˈstə-dē\ *noun* (1683)
: study of oneself; *also* : a record of observations from such study

self–styled \-ˈstī(ə)ld\ *adjective* (1823)
: called by oneself ⟨*self-styled* experts⟩

self–sub·sis·tent \ˌself-səb-ˈsis-tənt\ *adjective* (1647)
: subsisting independently of anything external to itself
— **self–sub·sis·tence** \-tən(t)s\ *noun*

self–sub·sist·ing \-ˈsis-tiŋ\ *adjective* (1654)
: SELF-SUBSISTENT

self–suf·fi·cien·cy \-sə-ˈfi-shən(t)-sē\ *noun* (1623)
: the quality or state of being self-sufficient

self–suf·fi·cient \-ˈfi-shənt\ *adjective* (1589)
1 : able to maintain oneself or itself without outside aid : capable of providing for one's own needs
2 : having an extreme confidence in one's own ability or worth : HAUGHTY, OVERBEARING

self–suf·fic·ing \-ˈfī-siŋ *also* -ziŋ\ *adjective* (1687)
: SELF-SUFFICIENT
— **self–suf·fic·ing·ly** *adverb*
— **self–suf·fic·ing·ness** *noun*

self–sug·ges·tion \-səg-ˈjes-chən, -sə-ˈ, -ˈjesh-\ *noun* (1892)
: AUTOSUGGESTION

self–sup·port \ˌself-sə-ˈpōrt, -ˈpȯrt\ *noun* (1774)
: independent support of oneself or itself
— **self–sup·port·ed** *adjective*

self–supporting *adjective* (1829)
: characterized by self-support: as **a :** meeting one's needs by one's own efforts or output **b :** supporting itself or its own weight ⟨a *self-supporting* wall⟩

self–sus·tain·ing \-sə-ˈstā-niŋ\ *adjective* (1844)
1 : maintaining or able to maintain oneself or itself by independent effort
2 : maintaining or able to maintain itself once commenced ⟨a *self-sustaining* nuclear reaction⟩

self–taught \ˈself-ˈtȯt\ *adjective* (1725)
1 : having knowledge or skills acquired by one's own efforts without formal instruction ⟨a *self-taught* musician⟩
2 : learned by oneself ⟨*self-taught* knowledge⟩

self–tol·er·ance \-ˈtä-lə-rən(t)s, -ˈtäl-rən(t)s\ *noun* (1964)
: the physiological state that exists in an organism when its immune system has proceeded far enough in the process of self-recognition to lose the capacity to attack and destroy its own bodily constituents

self–treat·ment \-ˈtrēt-mənt\ *noun* (1886)
: medication of oneself or treatment of one's own disease or condition without medical supervision or prescription

self–trust \-ˈtrəst\ *noun* (1583)
: SELF-CONFIDENCE

self–will \-ˈwil\ *noun* (14th century)
: stubborn or willful adherence to one's own desires or ideas : OBSTINACY

self–willed \-ˈwild\ *adjective* (14th century)
: governed by one's own will : not yielding to the wishes of others : OBSTINATE
— **self–willed·ly** \-ˈwild-lē\ *adverb*
— **self–willed·ness** \-ˈwild-nəs\ *noun*

self–wind·ing \-ˈwīn-diŋ\ *adjective* (1825)
: not needing to be wound by hand ⟨a *self-winding* watch⟩

self–worth \-ˈwərth\ *noun* (1944)
: SELF-ESTEEM

Sel·juk \ˈsel-ˌjük, sel-ˈ\ *or* **Sel·ju·ki·an** \sel-ˈjü-kē-ən\ *adjective* [Turkish *Selçuk*, eponymous ancestor of the dynasties] (1834)
1 : of or relating to any of several Turkish dynasties ruling over a great part of western Asia in the 11th, 12th, and 13th centuries
2 : of, relating to, or characteristic of a Turkish people ruled over by a Seljuk dynasty
— **Seljuk** *or* **Seljukian** *noun*

[1]sell \ˈsel\ *verb* **sold** \ˈsōld\; **sell·ing** [Middle English, from Old English *sellan;* akin to Old High German *sellen* to sell, Old Norse *sala* sale, Greek *helein* to take] (before 12th century)
transitive verb
1 : to deliver or give up in violation of duty, trust, or loyalty : BETRAY — often used with *out*
2 a (1) **:** to give up (property) to another for something of value (as money) (2) **:** to offer for sale **b :** to give up in return for something else especially foolishly or dishonorably ⟨*sold* his birthright for a mess of pottage⟩ **c :** to exact a price for ⟨*sold* their lives dearly⟩
3 a : to deliver into slavery for money **b :** to give into the power of another ⟨*sold* his soul to the devil⟩ **c :** to deliver the personal services of for money
4 : to dispose of or manage for profit instead of in accordance with conscience, justice, or duty ⟨*sold* their votes⟩
5 a : to develop a belief in the truth, value, or desirability of : gain acceptance for ⟨trying to *sell* a program to the Congress⟩ **b :** to persuade or influence to a course of action or to the acceptance of something ⟨*sell* children on reading⟩
6 : to impose on : CHEAT
7 a : to cause or promote the sale of ⟨using television advertising to *sell* cereal⟩ **b :** to make or attempt to make sales to **c :** to influence or induce to make a purchase
8 : to achieve a sale of ⟨*sold* a million copies⟩
intransitive verb
1 : to dispose of something by sale
2 : to achieve a sale; *also* : to achieve satisfactory sales ⟨hoped that the new line would *sell*⟩
3 : to have a specified price
— **sell·able** \ˈse-lə-bəl\ *adjective*
— **sell down the river :** to betray the faith of
— **sell short 1 :** to make a short sale **2 :** to fail to value properly : UNDERESTIMATE

[2]sell *noun* (1838)
1 : a deliberate deception : HOAX
2 : the act or an instance of selling

[3]sell *or* **selle** \ˈsel\ *noun* [Middle English *selle,* from Middle French, from Latin *sella* — more at SETTLE] (15th century)
archaic : SADDLE

[4]sell *chiefly Scottish variant of* SELF

sell·er \ˈse-lər\ *noun* (13th century)
1 : one that offers for sale
2 : a product offered for sale and selling well, to a specified extent, or in a specified manner ⟨a million-copy *seller*⟩ ⟨a poor *seller*⟩

seller's market *noun* (1932)
: a market in which goods are scarce, buyers have a limited range of choice, and prices are high — compare BUYER'S MARKET

selling climax *noun* (circa 1949)
: a sharp decline in stock prices for a short time on very heavy trading volume followed by a rally

sell·ing–plat·er \ˈse-liŋ-ˌplā-tər\ *noun* (1886)
: a horse that runs in selling races

selling point *noun* (1923)
: an aspect or detail of something that is emphasized (as in selling or promoting)

selling race *noun* (1898)
: a claiming race in which the winning horse is put up for auction

sell–off \ˈsel-ˌȯf\ *noun* (1937)
: a usually sudden sharp decline in security prices accompanied by increased volume of trading

sell off *intransitive verb* (circa 1700)
: to suffer a drop in prices

sell·out \ˈsel-ˌau̇t\ *noun* (1859)
1 : the act or an instance of selling out
2 : something sold out; *especially* : something (as a concert or contest) for which all tickets are sold
3 : one who sells out

sell out (1796)
transitive verb
1 : to sell the goods of (a debtor) in order to satisfy creditors
2 : to sell security or commodity holdings of usually to satisfy an uncovered margin
3 a : to sell all the available tickets for **b :** to sell all of ⟨the merchandise was quickly *sold out*⟩
intransitive verb
1 : to dispose of one's goods by sale; *especially* : to sell one's business
2 : to betray one's cause or associates
3 : to be or achieve a sellout

sel·syn \ˈsel-ˌsin\ *noun* [*self-syn*chronizing] (1926)
: a system comprising a generator and a motor so connected by wire that angular rotation or position in the generator is reproduced simultaneously in the motor — called also *synchro*

selt·zer \ˈselt-sər\ *noun* [modification of German *Selterser* (*Wasser*) water of Selters, from Nieder *Selters,* Germany] (1775)
: artificially carbonated water

sel·vage *or* **sel·vedge** \ˈsel-vij\ *noun* [Middle English *selvage,* probably from Middle Dutch *selvegge, selvage,* from *selv* self + *egge* edge; akin to Old English *self* and to Old English *ecg* edge — more at EDGE] (15th century)

\ə\ **abut** \ᵊ\ **kitten** \ər\ **further** \a\ **ash** \ā\ **ace** \ä\ **mop, mar** \au̇\ **out** \ch\ **chin** \e\ **bet** \ē\ **easy** \g\ **go** \i\ **hit** \ī\ **ice** \j\ **job** \ŋ\ **sing** \ō\ **go** \ȯ\ **law** \ȯi\ **boy** \th\ **thin** \t͟h\ **the** \ü\ **loot** \u̇\ **foot** \y\ **yet** \zh\ **vision** *see also* Guide to Pronunciation

1 a : the edge on either side of a woven or flat-knitted fabric so finished as to prevent raveling; *specifically* **:** a narrow border often of different or heavier threads than the fabric and sometimes in a different weave **b :** an edge (as of fabric or paper) meant to be cut off and discarded
2 : an outer or peripheral part **:** BORDER, EDGE
— sel·vaged *or* **sel·vedged** \-vijd\ *adjective*

selves *plural of* SELF

se·man·tic \si-ˈman-tik\ *also* **se·man·ti·cal** \-ti-kəl\ *adjective* [Greek *sēmantikos* significant, from *sēmainein* to signify, mean, from *sēma* sign, token] (1894)
1 : of or relating to meaning in language
2 : of or relating to semantics
— se·man·ti·cal·ly \-ti-k(ə-)lē\ *adverb*

se·man·ti·cist \-ˈman-tə-sist\ *noun* (1902)
: a specialist in semantics

se·man·tics \si-ˈman-tiks\ *noun plural but singular or plural in construction* (1893)
1 : the study of meanings: **a :** the historical and psychological study and the classification of changes in the signification of words or forms viewed as factors in linguistic development **b** (1) **:** SEMIOTIC (2) **:** a branch of semiotic dealing with the relations between signs and what they refer to and including theories of denotation, extension, naming, and truth
2 : GENERAL SEMANTICS
3 a : the meaning or relationship of meanings of a sign or set of signs; *especially* **:** connotative meaning **b :** the language used (as in advertising or political propaganda) to achieve a desired effect on an audience especially through the use of words with novel or dual meanings

¹sem·a·phore \ˈse-mə-ˌfōr, -ˌfȯr\ *noun* [Greek *sēma* sign, signal + International Scientific Vocabulary *-phore*] (1816)
1 : an apparatus for visual signaling (as by the position of one or more movable arms)
2 : a system of visual signaling by two flags held one in each hand

semaphore 2: alphabet; 3 positions following Z: error, end of word, numerals follow; numerals 1, 2, 3, 4, 5, 6, 7, 8, 9, 0 same as A through J

²semaphore *verb* **-phored; -phor·ing** (1893)
transitive verb
: to convey (information) by or as if by semaphore
intransitive verb
: to send signals by or as if by semaphore

se·ma·si·ol·o·gy \si-ˌmā-sē-ˈä-lə-jē, -ˌmā-zē-\ *noun* [International Scientific Vocabulary, from Greek *sēmasia* meaning, from *sēmainein* to mean] (1857)
: SEMANTICS 1
— se·ma·si·o·log·i·cal \-sē-ə-ˈlä-ji-kəl, -zē-\ *adjective*

¹sem·bla·ble \ˈsem-blə-bəl\ *adjective* [Middle English, from Middle French, from Old French, from *sembler* to be like, seem] (14th century)
1 : SIMILAR
2 : SUITABLE
3 : APPARENT, SEEMING
— sem·bla·bly \-blə-blē\ *adverb*

²semblable *noun* (15th century)
1 *archaic* **:** something similar **:** LIKE
2 : one that is like oneself **:** one's fellow

sem·blance \ˈsem-blən(t)s\ *noun* [Middle English, from Middle French, from Old French *sembler* to be like, seem — more at RESEMBLE] (14th century)
1 a : outward and often specious appearance or show **:** FORM ⟨wrapped in a *semblance* of composure —Harry Hervey⟩ **b :** MODICUM ⟨has been struggling to get some *semblance* of justice for his people —Bayard Rustin⟩
2 : ASPECT, COUNTENANCE
3 a : a phantasmal form **:** APPARITION **b :** IMAGE, LIKENESS
4 : actual or apparent resemblance

seme \ˈsēm\ *noun* [Greek *sēma* sign] (circa 1866)
1 : a linguistic sign
2 : any of the basic components of the meaning of a morpheme

se·mei·ol·o·gy *variant of* SEMIOLOGY

se·mei·ot·ic *variant of* SEMIOTIC

Sem·e·le \ˈse-mə-ˌlē\ *noun* [Latin, from Greek *Semelē*]
: a daughter of Cadmus consumed by flames when visited by Zeus in his divine splendor

sem·eme \ˈse-ˌmēm\ *noun* [International Scientific Vocabulary, from Greek *sēma* + International Scientific Vocabulary *-eme*] (1913)
1 : the meaning of a morpheme
2 a : SEME 2 **b :** a class of related semes
— se·me·mic \sə-ˈmē-mik\ *adjective*

se·men \ˈsē-mən\ *noun* [Middle English, from Latin, seed, semen; akin to Old High German *sāmo* seed, Latin *serere* to sow — more at sow] (14th century)
: a viscid whitish fluid of the male reproductive tract consisting of spermatozoa suspended in secretions of accessory glands

se·mes·ter \sə-ˈmes-tər\ *noun* [German, from Latin *semestris* half-yearly, from *sex* six + *mensis* month — more at SIX, MOON] (1827)
1 : either of the two usually 18-week periods of instruction into which an academic year is often divided
2 : a period of six months
— se·mes·tral \-trəl\ *or* **se·mes·tri·al** \-trē-əl\ *adjective*

semester hour *noun* (1922)
: a unit of academic credit representing an hour of class (as lecture class) or three hours of laboratory work each week for an academic semester

¹semi \ˈse-mē *also* -ˌmī\ *noun, plural* **sem·is** [short for *semidetached*] (1912)
chiefly British **:** a semidetached house

²semi \ˈse-ˌmī *also* -mē\ *noun, plural* **sem·is** (1942)
: SEMITRAILER

³semi *same as* ²\ *noun, plural* **sem·is** (1942)
: SEMIFINAL — often used in plural

semi- \ˌse-mē, -ˌmī, -mi\ *prefix* [Middle English, from Latin; akin to Old High German *sāmi-* half, Greek *hēmi-*]
1 a : precisely half of: (1) **:** forming a bisection of ⟨*semi*diameter⟩ (2) **:** being a usually vertically bisected form of (a specified architectural feature) ⟨*semi*dome⟩ **b :** half in quantity or value **:** half of or occurring halfway through a specified period of time ⟨*semi*annual⟩ ⟨*semi*monthly⟩ — compare BI-
2 : to some extent **:** partly **:** incompletely ⟨*semi*civilized⟩ ⟨*semi*-independent⟩ ⟨*semi*dry⟩ — compare DEMI-, HEMI-
3 a : partial **:** incomplete ⟨*semi*consciousness⟩ ⟨*semi*darkness⟩ **b :** having some of the characteristics of ⟨*semi*porcelain⟩ **c :** QUASI- ⟨*semi*governmental⟩ ⟨*semi*monastic⟩

semi·ab·stract \-ab-ˈstrakt, -ˈab-ˌ\ *adjective* (1871)
: having subject matter that is easily recognizable although the form is stylized ⟨*semiabstract* art⟩
— semi·ab·strac·tion \-ab-ˈstrak-shən\ *noun*

semi·an·nu·al \-ˈan-yə(-wə)l\ *adjective* (1794)
: occurring every six months or twice a year
— semi·an·nu·al·ly *adverb*

semi–an·tique \-an-ˈtēk\ *adjective* (circa 1930)
: being approximately 50 to 100 years old ⟨a *semi-antique* carpet⟩
— semi–antique *noun*

semi·aquat·ic \-ə-ˈkwä-tik, -ˈkwa-\ *adjective* (1833)
: growing equally well in or adjacent to water; *also* **:** frequenting but not living wholly in water

semi·ar·bo·re·al \-är-ˈbōr-ē-əl, -ˈbȯr-\ *adjective* (1938)
: often inhabiting and frequenting trees but not completely arboreal

semi·ar·id \-ˈar-əd\ *adjective* (1898)
: characterized by light rainfall; *especially* **:** having from about 10 to 20 inches (25 to 51 centimeters) of annual precipitation
— semi·arid·i·ty \-ə-ˈri-də-tē, -a-ˈri-\ *noun*

semi·au·to·bio·graph·i·cal \-ˌȯ-tə-ˌbī-ə-ˈgra-fi-kəl\ *adjective* (1939)
: partly autobiographical ⟨a *semiautobiographical* comedy⟩

semi·au·to·mat·ic \-ˌȯ-tə-ˈma-tik\ *adjective* (1890)
: not fully automatic: as **a :** operated partly automatically and partly by hand **b** *of a firearm* **:** employing gas pressure or force of recoil and mechanical spring action to eject the empty cartridge case after the first shot and load the next cartridge from the magazine but requiring release and another pressure of the trigger for each successive shot
— semiautomatic *noun*
— semi·au·to·mat·i·cal·ly \-ti-k(ə-)lē\ *adverb*

semi·au·ton·o·mous \-ȯ-ˈtä-nə-məs\ *adjective* (1915)
: largely self-governing within a larger political or organizational entity

semi·breve \ˈse-mē-ˌbrēv, ˈse-ˌmī-, -mi-, -ˌbrev\ *noun* (15th century)
: WHOLE NOTE

semi·cen·ten·ni·al \-sen-ˈte-nē-əl\ *noun* (1859)
: a 50th anniversary or its celebration
— semicentennial *adjective*

semi·cir·cle \ˈse-mē-ˌsər-kəl, ˈse-ˌmī-, -mi-\ *noun* [Latin *semicirculus,* from *semi-* + *circulus* circle] (1526)
1 : a half of a circle
2 : an object or arrangement of objects in the form of a half circle
— semi·cir·cu·lar \ˌse-mē-ˈsər-kyə-lər\ *adjective*

semicircular canal *noun* (1748)
: any of the loop-shaped tubular parts of the labyrinth of the ear that together constitute a sensory organ associated with the maintenance of bodily equilibrium — see EAR illustration

semi·civ·i·lized \ˌse-mē-ˈsi-və-ˌlīzd, ˌse-ˌmī-, -mi-\ *adjective* (1836)
: partly civilized

semi·clas·sic \-ˈkla-sik\ *noun* (1843)
: a semiclassical work (as of music)

semi·clas·si·cal \-si-kəl\ *adjective* (1904)
: having some of the characteristics of the classical: as **a :** of, relating to, or being a musical composition that acts as a bridge between

classical and popular music **b :** of, relating to, or being a classical composition that has developed popular appeal

semi·co·lon \'se-mē-ˌkō-lən, 'se-ˌmī-, -mi-\ *noun* (1644)
: a punctuation mark ; used chiefly in a coordinating function between major sentence elements (as independent clauses of a compound sentence)

semi·co·lo·nial \ˌse-mē-kə-'lō-nyəl, ˌse-ˌmī-, -mi-, -nē-əl\ *adjective* (1932)
1 : nominally independent but actually under foreign domination
2 : dependent on foreign nations as suppliers of manufactured goods and as purchasers of raw materials
— semi·co·lo·nial·ism \-nyə-ˌli-zəm, -nē-ə-\ *noun*

semi·col·o·ny \-'kä-lə-nē\ *noun* (1945)
: a semicolonial state

semi·com·mer·cial \-kə-'mər-shəl\ *adjective* (1926)
: of, relating to, adapted to, or characterized by limited marketing of an experimental product

semi·con·duct·ing \ˌse-mē-kən-'dək-tiŋ, ˌse-ˌmī-, -mi-\ *adjective* (1782)
: of, relating to, or having the characteristics of a semiconductor

semi·con·duc·tor \-'dək-tər\ *noun* (1838)
: any of a class of solids (as germanium or silicon) whose electrical conductivity is between that of a conductor and that of an insulator in being nearly as great as that of a metal at high temperatures and nearly absent at low temperatures

semi·con·scious \-'kän(t)-shəs\ *adjective* (1839)
: incompletely conscious **:** imperfectly aware or responsive
— semi·con·scious·ness *noun*

semi·con·ser·va·tive \-kən-'sər-və-tiv\ *adjective* (1957)
: relating to or being genetic replication in which a double-stranded molecule of nucleic acid separates into two single strands each of which serves as a template for the formation of a complementary strand that together with the template forms a complete molecule
— semi·con·ser·va·tive·ly *adverb*

semi·crys·tal·line \-'kris-tə-lən\ *adjective* (1816)
: incompletely or imperfectly crystalline

semi·cy·lin·dri·cal \-sə-'lin-dri-kəl\ *adjective* (circa 1731)
: having the shape of a longitudinal half of a cylinder

semi·dark·ness \-'därk-nəs\ *noun* (1849)
: partial darkness

semi·de·ify \-'dē-ə-ˌfī, -'dā-\ *transitive verb* (1953)
: to regard as somewhat godlike

semi·des·ert \-'de-zərt\ *noun* (1849)
: an arid area that has some of the characteristics of a desert but has greater annual precipitation

semi·de·tached \-di-'tacht\ *adjective* (1859)
: forming one of a pair of residences joined into one building by a common sidewall

semi·di·am·e·ter \-dī-'a-mə-tər\ *noun* (14th century)
: RADIUS; *specifically* **:** the apparent radius of a generally spherical celestial body

semi·di·ur·nal \-dī-'ər-nᵊl\ *adjective* (1594)
1 : relating to or accomplished in half a day
2 : occurring twice a day
3 : occurring approximately every half day ⟨the *semidiurnal* tides⟩

semi·di·vine \-də-'vīn\ *adjective* (1600)
: more than mortal but not fully divine

semi·doc·u·men·ta·ry \-ˌdä-kyə-'men-tə-rē, -'men-trē\ *noun* (1939)
: a motion picture that uses many details taken from actual events or situations in presenting a fictional story
— semidocumentary *adjective*

semi·dome \'se-mē-ˌdōm, 'se-ˌmī-, -mi-\ *noun* (1788)
: a roof or ceiling covering a semicircular or nearly semicircular room or recess
— semi·domed \-ˌdōmd\ *adjective*

semi·do·mes·ti·ca·tion \ˌse-mē-də-ˌmes-ti-'kā-shən, ˌse-ˌmī-, -mi-\ *noun* (circa 1835)
: a captive state of a wild animal in which its living conditions and often its breeding are controlled by humans
— semi·do·mes·ti·cat·ed \-də-'mes-ti-ˌkā-təd\ *adjective*

semi·dom·i·nant \-'dä-mə-nənt, -'däm-nənt\ *adjective* (1942)
: producing an intermediate phenotype in the heterozygous condition ⟨a *semidominant* mutant gene⟩

semi·dry \-'drī\ *adjective* (1878)
: moderately dry

semi·dry·ing \-'drī-iŋ\ *adjective* (1905)
: that dries imperfectly or slowly — used of some oils (as cottonseed oil)

semi·dwarf \-'dwȯrf\ *adjective* (1959)
: of or being a plant of a variety that is undersized but larger than a dwarf ⟨*semidwarf* wheats⟩
— semidwarf *noun*

semi·em·pir·i·cal \-im-'pir-ə-kəl, -em-\ *adjective* (1935)
: partly empirical; *especially* **:** involving assumptions, approximations, or generalizations designed to simplify calculation or to yield a result in accord with observation

semi·erect \-ə-'rekt\ *adjective* (1822)
1 : incompletely upright in bodily posture ⟨*semierect* primates⟩
2 : erect for half the length ⟨*semierect* stems⟩

semi·ev·er·green \-'e-vər-ˌgrēn\ *adjective* (1901)
1 : having functional and persistent foliage during part of the winter or dry season
2 : tending to be evergreen in a mild climate but deciduous in a rigorous climate

semi·feu·dal \-'fyü-dᵊl\ *adjective* (1898)
: having some characteristics of feudalism

¹semi·fi·nal \-'fī-nᵊl\ *adjective* (1884)
1 : being next to the last in an elimination tournament
2 : of or participating in a semifinal

²semi·fi·nal \'se-mē-ˌfī-nᵊl, 'se-ˌmī-, -mi-\ *noun* (1895)
1 : a semifinal match
2 : a semifinal round
— semi·fi·nal·ist \ˌse-mē-'fī-nᵊl-ist, ˌse-ˌmī-, -mi-\ *noun*

semi·fin·ished \ˌse-mē-'fi-nisht, ˌse-ˌmī-, -mi-\ *adjective* (1902)
: partially finished or processed; *especially, of steel* **:** rolled from raw ingots into shapes (as bars, billets, or plates) suitable for further processing

semi·fit·ted \-'fi-təd\ *adjective* (circa 1950)
: conforming somewhat to the lines of the body

semi·flex·i·ble \-'flek-sə-bəl\ *adjective* (1925)
1 : somewhat flexible
2 *of a book cover* **:** having a thin board stiffener under the covering material

semi·flu·id \-'flü-əd\ *adjective* (circa 1735)
: having the qualities of both a fluid and a solid **:** VISCOUS ⟨fluid and *semifluid* lubricants⟩
— semifluid *noun*

semi·for·mal \-'fȯr-məl\ *adjective* (1906)
: being or suitable for an occasion of moderate formality ⟨a *semiformal* dinner⟩ ⟨*semiformal* gowns⟩

semi·gloss \'se-mē-ˌgläs, 'se-ˌmī-, -mi-, -ˌglȯs\ *adjective* (1937)
: having a low luster; *specifically* **:** producing a finish midway between gloss and flat

semi·gov·ern·men·tal \ˌse-mē-ˌgəv-ər(n)-'men-tᵊl, ˌse-ˌmī-, -mi-, -ˌgə-vᵊm-'en-\ *adjective* (1919)
: having some governmental functions and powers

semi·group \'se-mē-ˌgrüp, 'se-ˌmī-, -mi-\ *noun* (1904)
: a mathematical set that is closed under an associative binary operation

semi–in·de·pen·dent \-ˌin-də-'pen-dənt\ *adjective* (1860)
: partially independent; *specifically* **:** SEMIAUTONOMOUS

semi·leg·end·ary \-'le-jən-ˌder-ē\ *adjective* (1878)
: having historical foundation but elaborated in legend

semi·le·thal \-'lē-thəl\ *noun* (1919)
: a mutation that in the homozygous condition produces more than 50 percent mortality but not complete mortality
— semilethal *adjective*

semi·liq·uid \-'li-kwəd\ *adjective* (1684)
: having the qualities of both a liquid and a solid **:** SEMIFLUID ⟨*semiliquid* manure⟩
— semiliquid *noun*

semi·lit·er·ate \-'li-tə-rət *also* -'li-trət\ *adjective* (1927)
1 a : able to read and write on an elementary level **b :** able to read but unable to write
2 : having limited knowledge or understanding **:** not well-versed
— semiliterate *noun*

semi·log \-'lȯg, -'läg\ *adjective* (1921)
: SEMILOGARITHMIC

semi·log·a·rith·mic \-ˌlȯ-gə-'rith-mik, -ˌlä-\ *adjective* (1919)
: having one scale logarithmic and the other arithmetic — used of graph paper or of a graph on such paper

semi·lu·nar \-'lü-nər\ *adjective* [New Latin *semilunaris,* from Latin *semi-* + *lunaris* lunar] (1597)
: shaped like a crescent

semilunar valve *noun* (circa 1719)
: any of the crescent-shaped cusps that occur as a set of three between the heart and the aorta and another of three between the heart and the pulmonary artery, are forced apart by pressure in the ventricles during systole and pushed together by pressure in the arteries during diastole, and prevent regurgitation of blood into the ventricles; *also* **:** either set of three cusps

semi·lus·trous \-'ləs-trəs\ *adjective* (1953)
: slightly lustrous

semi·ma·jor axis \-'mā-jər\ *noun* (1899)
: one half of the major axis of an ellipse (as that formed by the orbit of a planet)

semi·matte *also* **semi·mat** *or* **semi·matt** \-'mat\ *adjective* [*semi-* + ²*matte*] (1937)
: having a slight luster

semi·met·al \-'me-tᵊl\ *noun* (1661)
: an element (as arsenic) possessing metallic properties in an inferior degree and not malleable
— semi·me·tal·lic \-mə-'ta-lik\ *adjective*

semi·mi·cro \-'mī-(ˌ)krō\ *adjective* (1935)
: of, relating to, or dealing with quantities intermediate between those treated as micro and macro ⟨*semimicro* analysis for chlorine⟩ ⟨a *semimicro* balance⟩

semi·mi·nor axis \-'mī-nər-\ *noun* (1909)
: one half of the minor axis of an ellipse (as that formed by the orbit of a planet)

semi·moist \-'mȯist\ *adjective* (1903)
: slightly moist

semi·mo·nas·tic \-mə-'nas-tik\ *adjective* (1911)
: having some features characteristic of a monastic order

¹semi·month·ly \-'mən(t)th-lē\ *noun* (1851)
: a semimonthly publication

²semimonthly *adjective* (1860)
: occurring twice a month

[3]**semimonthly** *adverb* (circa 1890)
: twice a month

semi·mys·ti·cal \-'mis-ti-kəl\ *adjective* (1890)
: having some of the qualities of mysticism

sem·i·nal \'se-mə-nᵊl\ *adjective* [Middle English, from Middle French, from Latin *seminalis*, from *semin-, semen* seed — more at SEMEN] (14th century)
1 : of, relating to, or consisting of seed or semen
2 : containing or contributing the seeds of later development : CREATIVE, ORIGINAL ⟨a *seminal* book⟩ ⟨one of the most *seminal* of the great poets⟩
— **sem·i·nal·ly** \-nᵊl-ē\ *adverb*

seminal duct *noun* (circa 1909)
: a tube or passage serving especially or exclusively as an efferent duct of the testis and in the human male being made up of the tubules of the epididymis, the vas deferens, and the ejaculatory duct

seminal fluid *noun* (circa 1929)
1 : SEMEN
2 : the part of the semen that is produced by various accessory glands : semen excepting the spermatozoa

seminal vesicle *noun* (circa 1890)
: either of a pair of glandular pouches that lie one on either side of the male reproductive tract and in the human male secrete a sugar- and protein-containing fluid into the ejaculatory duct

sem·i·nar \'se-mə-ˌnär\ *noun* [German, from Latin *seminarium* nursery] (1889)
1 : a group of advanced students studying under a professor with each doing original research and all exchanging results through reports and discussions
2 a (1) **:** a course of study pursued by a seminar (2) **:** an advanced or graduate course often featuring informality and discussion **b :** a scheduled meeting of a seminar or a room for such meetings
3 : a meeting for giving and discussing information

sem·i·nar·i·an \ˌse-mə-'ner-ē-ən\ *noun* (1794)
: a student in a seminary especially of the Roman Catholic Church

sem·i·na·rist \'se-mə-nə-rist\ *noun* (1835)
: SEMINARIAN

sem·i·nary \'se-mə-ˌner-ē\ *noun, plural* **-nar·ies** [Middle English, seedbed, nursery, from Latin *seminarium*, from *semin-, semen* seed] (1542)
1 : an environment in which something originates and from which it is propagated ⟨a *seminary* of vice and crime⟩
2 a : an institution of secondary or higher education **b :** an institution for the training of candidates for the priesthood, ministry, or rabbinate ◆

semi·nat·u·ral \ˌse-mē-'na-chə-rəl, ˌse-ˌmī-, -mi-, -'nach-rəl\ *adjective* (circa 1962)
: modified by human influence but retaining many natural features ⟨a *seminatural* park⟩

sem·i·nif·er·ous \ˌse-mə-'ni-f(ə-)rəs\ *adjective* [Latin *semin-, semen* seed + English *-iferous*] (1692)
: producing or bearing seed or semen

seminiferous tubule *noun* (1860)
: any of the coiled threadlike tubules that make up the bulk of the testis and are lined with a layer of epithelial cells from which the spermatozoa are produced

Sem·i·nole \'se-mə-ˌnōl\ *noun, plural* **Seminoles** *or* **Seminole** [Creek *simanó·li* fugitive, wild, alteration of *simaló·ni*, from American Spanish *cimarrón* wild] (1771)
: a member of any of the groups of American Indians that emigrated to Florida from Georgia and Alabama in the 18th century and that are now located in southern Florida and Oklahoma

semi·no·mad \ˌse-mē-'nō-ˌmad, ˌse-ˌmī-, -mi-\ *noun* (circa 1934)
: a member of a people living usually in portable or temporary dwellings and practicing seasonal migration but having a base camp at which some crops are cultivated
— **semi·no·mad·ic** \-nō-'ma-dik\ *adjective*

semi·nude \-'nüd, -'nyüd\ *adjective* (1849)
: partially nude
— **semi·nu·di·ty** \-'nü-də-tē, -'nyü-\ *noun*

semi·of·fi·cial \-ə-'fi-shəl\ *adjective* (1806)
: having some official authority or standing
— **semi·of·fi·cial·ly** \-'fi-sh(ə-)lē\ *adverb*

se·mi·ol·o·gy \ˌsē-mē-'ä-lə-jē, ˌse-mē-, ˌsē-ˌmī-\ *noun* [Greek *sēmeion* sign] (circa 1890)
: the study of signs; *especially* : SEMIOTIC
— **se·mi·o·log·i·cal** \(ˌ)sē-ˌmī-ə-'lä-ji-kəl, ˌse-mē-ə-\ *adjective*
— **se·mi·o·log·i·cal·ly** \-ə-'lä-ji-k(ə-)lē\ *adverb*
— **se·mi·ol·o·gist** \-'ä-lə-jist\ *noun*

semi·opaque \ˌse-mē-ō-'pāk, ˌse-ˌmī-, -mi-\ *adjective* (1691)
: nearly opaque

se·mi·o·sis \ˌsē-mē-'ō-səs, ˌse-mē-, ˌsē-ˌmī-\ *noun* [New Latin, from Greek *sēmeiōsis* observation of signs, from *sēmeioun* to observe signs, from *sēmeion*] (circa 1907)
: a process in which something functions as a sign to an organism

se·mi·ot·ic \-'ä-tik\ *or* **se·mi·ot·ics** \-tiks\ *noun, plural* **semiotics** [Greek *sēmeiōtikos* observant of signs, from *sēmeiousthai* to interpret signs, from *sēmeion* sign, from *sēma* sign] (1880)
: a general philosophical theory of signs and symbols that deals especially with their function in both artificially constructed and natural languages and comprises syntactics, semantics, and pragmatics
— **semiotic** *adjective*
— **se·mi·o·ti·cian** \-ə-'ti-shən\ *noun*
— **se·mi·ot·i·cist** \-'ä-tə-sist\ *noun*

semi·pal·mat·ed \ˌse-mē-'pal-ˌmā-təd, ˌse-ˌmī-, -mi-, -'päl-, -'pä-\ *adjective* (1785)
: having the toes joined only part way down with a web ⟨a plover with *semipalmated* feet⟩

semi·par·a·sit·ic \-ˌpar-ə-'si-tik\ *adjective* (circa 1880)
: of, relating to, or being a parasitic plant that contains some chlorophyll and is capable of photo-synthesis
— **semi·par·a·site** \-'par-ə-ˌsīt\ *noun*

semi·per·ma·nent \-'pər-mə-nənt, -'pərm-nənt\ *adjective* (circa 1890)
: lasting or intended to last for a long time but not permanent

semi·per·me·able \-'pər-mē-ə-bəl\ *adjective* (1888)
: partially but not freely or wholly permeable; *specifically* : permeable to some usually small molecules but not to other usually larger particles ⟨a *semipermeable* membrane⟩
— **semi·per·me·abil·i·ty** \-ˌpər-mē-ə-'bi-lə-tē\ *noun*

semi·po·lit·i·cal \-pə-'li-ti-kəl\ *adjective* (1857)
: of, relating to, or involving some political features or activity

semi·pop·u·lar \-'pä-pyə-lər\ *adjective* (1860)
: somewhat popular

semi·por·ce·lain \-'pōr-s(ə-)lən, -'pȯr-\ *noun* (1880)
: any of several ceramic wares resembling or imitative of porcelain; *especially* : a relatively high-fired and hard-glazed white earthenware widely used for tableware

semi·por·no·graph·ic \-ˌpȯr-nə-'gra-fik\ *adjective* (1964)
: somewhat pornographic
— **semi·por·nog·ra·phy** \-pȯr-'nä-grə-fē\ *noun*

semi·post·al \-'pōs-tᵊl\ *noun* (1927)
: a postage stamp sold at a premium over its postal value especially for a humanitarian purpose

semi·pre·cious \-'pre-shəs\ *adjective* (circa 1890)
of a gemstone : of less commercial value than a precious stone

semi·pri·vate \-'prī-vət\ *adjective* (1876)
: of, receiving, or associated with hospital service giving a patient more privileges than a ward patient but fewer than a private patient

semi·pro \'se-mē-ˌprō, 'se-ˌmī-, -mi-\ *adjective or noun* (1908)
: SEMIPROFESSIONAL

[1]**semi·pro·fes·sion·al** \ˌse-mē-prə-'fesh-nəl, ˌse-ˌmī-, -mi-, -'fe-shə-nᵊl\ *noun* (1897)
: one who engages in an activity (as a sport) semiprofessionally

[2]**semiprofessional** *adjective* (1900)
1 : engaging in an activity for pay or gain but not as a full-time occupation
2 : engaged in by semiprofessional players ⟨*semiprofessional* baseball⟩
— **semi·pro·fes·sion·al·ly** *adverb*

semi·pub·lic \-'pə-blik\ *adjective* (1804)
1 : open to some persons outside the regular constituency
2 : having some features of a public institution; *specifically* : maintained as a public service by a private nonprofit organization

semi·quan·ti·ta·tive \-'kwän-tə-ˌtā-tiv\ *adjective* (1927)
: constituting or involving less than quantitative precision
— **semi·quan·ti·ta·tive·ly** *adverb*

semi·qua·ver \'se-mē-ˌkwā-vər, 'se-ˌmī-, -mi-\ *noun* (1576)
: SIXTEENTH NOTE

semi·re·li·gious \ˌse-mē-ri-'li-jəs, ˌse-ˌmī-, -mi-\ *adjective* (1864)
: somewhat religious

semi·re·tired \-ri-'tīrd\ *adjective* (1937)
: working only part-time especially because of age or ill health

semi·re·tire·ment \-'tīr-mənt\ *noun* (1923)
: the state or condition of being semiretired

semi·rig·id \-'ri-jəd\ *adjective* (1908)
1 : rigid to some degree or in some parts
2 *of an airship* : having a flexible cylindrical gas container with an attached stiffening keel that carries the load

semi·rur·al \-'rür-əl\ *adjective* (1835)
: somewhat rural

semi·sa·cred \-'sā-krəd\ *adjective* (1898)
: SEMIRELIGIOUS

semi·se·cret \-'sē-krət\ *adjective* (1917)
: not publicly announced but widely known nevertheless

◇ WORD HISTORY

seminary The English word *seminary* and its Latin source *seminarium*, a derivative of *semen* "seed," both originally denoted a nursery for young plants. Though the classical Latin word was sometimes used figuratively, English has gone much further in metaphorical extensions of the word, while the old sense "nursery" is now long obsolete. The use of *seminary* in reference to training schools for the clergy was an outgrowth of 16th century reforms in the Roman Catholic Church: decrees of the Council of Trent (1545–63) applied the Latin word *seminarium* to newly established institutes for educating priests, which were meant to raise the educational level of the clergy as well as train missionaries to non-Catholic countries. In application to such schools, English *seminary* long retained a strong association with Roman Catholicism; now the word refers equally to Catholic, Protestant, or Jewish colleges for training priests, ministers, or rabbis. *Seminary* has also been applied to other kinds of schools since the 16th century. When they were first formed in the 19th century, women's colleges were called "female seminaries" or "seminaries for young ladies."

semi·sed·en·tary \-'se-d^{ə}n-ˌter-ē\ *adjective* (circa 1930)
: sedentary during part of the year and nomadic otherwise ⟨*semisedentary* tribes⟩

semi·shrub·by \'se-mē-ˌshrə-bē, 'se-ˌmī-, -mi-, *especially Southern* -ˌsrə-\ *adjective* (1930)
: resembling or being a subshrub

semi·skilled \ˌse-mē-'skild, ˌse-ˌmī-, -mi-\ *adjective* (1916)
: having or requiring less training than skilled labor and more than unskilled labor

semi·soft \-'sȯft\ *adjective* (circa 1903)
: moderately soft; *specifically* **:** firm but easily cut ⟨*semisoft* cheese⟩

semi·sol·id \-'sä-ləd\ *adjective* (1834)
: having the qualities of both a solid and a liquid **:** highly viscous
— semisolid *noun*

semi·sub·mers·ible \-səb-'mər-sə-bəl\ *adjective* (1962)
: being a floating deepwater drilling platform that is towed to a desired location and then partially flooded for stabilization and usually anchored
— semisubmersible *noun*

semi·sweet \-'swēt\ *adjective* (1943)
: slightly sweetened ⟨*semisweet* chocolate⟩

semi·syn·thet·ic \-sin-'the-tik\ *adjective* (1937)
1 : produced by chemical alteration of a natural starting material ⟨*semisynthetic* penicillins⟩
2 : containing both chemically identified and complex natural ingredients ⟨a *semisynthetic* diet⟩

Sem·ite \'se-ˌmīt, *especially British* 'sē-ˌmīt\ *noun* [French *sémite,* from *Semitic* Shem, from Late Latin, from Greek *Sēm,* from Hebrew *Shēm*] (1848)
1 a : a member of any of a number of peoples of ancient southwestern Asia including the Akkadians, Phoenicians, Hebrews, and Arabs **b :** a descendant of these peoples
2 : a member of a modern people speaking a Semitic language

semi·ter·res·tri·al \ˌse-mē-tə-'res-trē-əl, ˌse-ˌmī-, -mi-, -'res-chəl, -'resh-\ *adjective* (1919)
1 : growing on boggy ground
2 : frequenting but not living wholly on land

[1]Se·mit·ic \sə-'mi-tik *also* -'me-\ *adjective* [German *semitisch,* from *Semit, Semite* Semite, probably from New Latin *Semita,* from Late Latin *Semitic* Shem] (1813)
1 : of, relating to, or constituting a subfamily of the Afro-Asiatic language family that includes Hebrew, Aramaic, Arabic, and Amharic
2 : of, relating to, or characteristic of the Semites
3 : JEWISH

[2]Semitic *noun* (1875)
: any or all of the Semitic languages

Se·mit·i·cist \sə-'mi-tə-sist\ *noun* (1956)
: SEMITIST

Se·mit·ics \-'mi-tiks\ *noun plural but singular in construction* (1895)
: the study of the language, literature, and history of Semitic peoples; *specifically* **:** Semitic philology

Sem·i·tism \'se-mə-ˌti-zəm\ *noun* (1851)
1 a : Semitic character or qualities **b :** a characteristic feature of a Semitic language occurring in another language
2 : policy favorable to Jews **:** predisposition in favor of Jews

Sem·i·tist \-tist\ *noun* (1885)
1 : a scholar of the Semitic languages, cultures, or histories
2 *often not capitalized* **:** a person favoring or disposed to favor the Jews

semi·ton·al \ˌse-mē-'tō-n^{ə}l, ˌse-ˌmī-, -mi-\ *adjective* (1863)
: CHROMATIC 3a, SEMITONIC
— semi·ton·al·ly *adverb*

semi·tone \'se-mē-ˌtōn, 'se-ˌmī-, -mi-\ *noun* (15th century)
: the tone at a half step; *also* **:** HALF STEP
— semi·ton·ic \ˌse-mē-'tä-nik, ˌse-ˌmī-, -mi-\ *adjective*
— semi·ton·i·cal·ly \-ni-k(ə-)lē\ *adverb*

semi·trail·er \'se-ˌmī-ˌtrā-lər, 'se-mē-, -mi-\ *noun* (1919)
1 : a freight trailer that when attached is supported at its forward end by the fifth wheel device of the truck tractor
2 : a trucking rig made up of a tractor and a semitrailer

semi·trans·lu·cent \ˌse-mē-ˌtran(t)s-'lü-s^{ə}nt, ˌse-ˌmī-, -mi-, -ˌtranz-\ *adjective* (1832)
: somewhat translucent

semi·trans·par·ent \-ˌtran(t)s-'par-ənt, -'per-\ *adjective* (circa 1793)
: imperfectly transparent

semi·trop·i·cal \-'trä-pi-kəl\ *also* **semi·trop·ic** \-pik\ *adjective* (1853)
: SUBTROPICAL

semi·trop·ics \-piks\ *noun plural* (1908)
: SUBTROPICS

semi·vow·el \'se-mē-ˌvau̇(-ə)l, 'se-ˌmī-, -mi-\ *noun* (1530)
1 : a speech sound (as \y\, \w\, or \r\) that has the articulation of a vowel but that is shorter in duration and is treated as a consonant in syllabication
2 : a letter representing a semivowel

[1]semi·week·ly \ˌse-mē-'wē-klē, ˌse-ˌmī-, -mi-\ *adjective* (1791)
: occurring twice a week
— semiweekly *adverb*

[2]semiweekly *noun* (1833)
: a semiweekly publication

semi·works \'se-mē-ˌwərks, 'se-ˌmī-, -mi-\ *noun plural, often attributive* (1926)
: a manufacturing plant operating on a limited commercial scale to provide final tests of a new product or process

semi·year·ly \ˌse-mē-'yir-lē, ˌse-ˌmī-, -mi-\ *adjective* (circa 1928)
: occurring twice a year

sem·o·li·na \ˌse-mə-'lē-nə\ *noun* [Italian *semolino,* diminutive of *semola* bran, from Latin *simila* wheat flour] (1797)
: the purified middlings of hard wheat (as durum) used especially for pasta (as macaroni or spaghetti)

sem·per·vi·vum \ˌsem-pər-'vī-vəm\ *noun* [New Latin, from Latin, neuter of *sempervivus* ever-living, from *semper* ever + *vivus* living — more at QUICK] (1591)
: any of a large genus (*Sempervivum*) of Old World fleshy herbs of the orpine family often grown as ornamentals

sem·pi·ter·nal \ˌsem-pi-'tər-n^{ə}l\ *adjective* [Middle English, from Late Latin *sempiternalis,* from Latin *sempiternus,* from *semper* ever, always, from *sem-* one, same (akin to Old Norse *samr* same) + *per* through — more at SAME, FOR] (15th century)
: of never-ending duration **:** ETERNAL
— sem·pi·ter·nal·ly *adverb*

sem·pi·ter·ni·ty \-'tər-nə-tē\ *noun* (1599)
: ETERNITY

sem·ple \'sem-pəl\ *adjective* [alteration of *simple*] (1759)
Scottish **:** of humble birth

sem·pli·ce \'sem-pli-ˌchā\ *adjective or adverb* [Italian, from Latin *simplic-, simplex* — more at SIMPLE] (circa 1740)
: SIMPLE — used as a direction in music

sem·pre \'sem-(ˌ)prā\ *adverb* [Italian, from Latin *semper*] (circa 1801)
: ALWAYS — used in music directions ⟨*sempre* legato⟩

semp·stress \'sem(p)-strəs\ *variant of* SEAMSTRESS

[1]sen \'sen\ *noun, plural* **sen** [Japanese] (1875)
— see *yen* at MONEY table

[2]sen *noun, plural* **sen** [Malay, probably from English *century*] (1952)
— see *dollar, ringgit, rupiah* at MONEY table

[3]sen *noun, plural* **sen** [Khmer *sein,* probably from French *century,* abbreviation of *centime* centime] (1964)
— see *riel* at MONEY table

se·nar·i·us \si-'nar-ē-əs, -'ner-\ *noun, plural* **se·nar·ii** \-ē-ˌī, -ē-ˌē\ [Latin, from *senarius* consisting of six each, from *seni* six each, from *sex* six — more at SIX] (1540)
: a verse consisting of six feet especially in Latin prosody

se·na·ry \'se-nə-rē, 'sē-\ *adjective* [Latin *senarius* consisting of six] (1661)
: of, based on, or characterized by six **:** compounded of six things or six parts ⟨*senary* scale⟩ ⟨*senary* division⟩

sen·ate \'se-nət\ *noun* [Middle English *senat,* from Old French, from Latin *senatus,* from *sen-, senex* old, old man — more at SENIOR] (13th century)
1 : an assembly or council usually possessing high deliberative and legislative functions: as **a :** the supreme council of the ancient Roman republic and empire **b :** the second chamber in the bicameral legislature of a major political unit (as a nation, state, or province)
2 : the hall or chamber in which a senate meets
3 : a governing body of some universities charged with maintaining academic standards and regulations and usually composed of the principal or representative members of the faculty

sen·a·tor \'se-nə-tər\ *noun* [Middle English *senatour,* from Old French *senateur,* from Latin *senator,* from *senatus*] (13th century)
: a member of a senate
— sen·a·tor·ship \-ˌship\ *noun*

sen·a·to·ri·al \ˌse-nə-'tōr-ē-əl, -'tȯr-\ *adjective* (1740)
: of, relating to, or befitting a senator or a senate ⟨*senatorial* office⟩ ⟨*senatorial* rank⟩

senatorial courtesy *noun* (1884)
: a custom of the U.S. Senate of refusing to confirm a presidential appointment of an official in or from a state when the appointment is opposed by the senators or senior senator of the president's party from that state

senatorial district *noun* (1785)
: a territorial division from which a senator is elected — compare CONGRESSIONAL DISTRICT

sen·a·to·ri·an \ˌse-nə-'tōr-ē-ən, -'tȯr-\ *adjective* (1614)
: SENATORIAL; *specifically* **:** of or relating to the ancient Roman senate

se·na·tus con·sul·tum \sə-ˌnä-təs-kən-'səl-təm, -'su̇l-\ *noun, plural* **senatus con·sul·ta** \-tə\ [Latin, decree of the senate] (1696)
: a decree of the ancient Roman senate

[1]send \'send\ *verb* **sent** \'sent\; **send·ing** [Middle English, from Old English *sendan;* akin to Old High German *sendan* to send, Old English *sith* road, journey, Old Irish *sét* path, way] (before 12th century)
transitive verb
1 : to cause to go: as **a :** to propel or throw in a particular direction **b :** DELIVER ⟨*sent* a blow to the chin⟩ **c :** DRIVE ⟨*sent* the ball between the goalposts⟩
2 : to cause to happen ⟨whatever fate may *send*⟩
3 : to dispatch by a means of communication
4 a : to direct, order, or request to go **b :** to permit or enable to attend a term or session ⟨*send* a daughter to college⟩ **c :** to direct by advice or reference **d :** to cause or order to depart **:** DISMISS
5 a : to force to go **:** drive away **b :** to cause to assume a specified state ⟨*sent* them into a rage⟩

\ə\ **abut** \$^{ə}$\ **kitten** \ər\ **further** \a\ **ash** \ā\ **ace**
\ä\ **mop, mar** \au̇\ **out** \ch\ **chin** \e\ **bet** \ē\ **easy**
\g\ **go** \i\ **hit** \ī\ **ice** \j\ **job** \ŋ\ **sing** \ō\ **go**
\ȯ\ **law** \ȯi\ **boy** \th\ **thin** \t͟h\ **the** \ü\ **loot** \u̇\ **foot**
\y\ **yet** \zh\ **vision** *see also* Guide to Pronunciation

6 : to cause to issue: as **a :** to pour out **:** DISCHARGE ⟨clouds *sending* forth rain⟩ **b :** UTTER ⟨*send* forth a cry⟩ **c :** EMIT ⟨*sent* out waves of perfume⟩ **d :** to grow out (parts) in the course of development ⟨a plant *sending* forth shoots⟩
7 : to cause to be carried to a destination; *especially* **:** to consign to death or a place of punishment
8 : to convey or cause to be conveyed or transmitted by an agent ⟨*send* a package by mail⟩ ⟨*sent* out invitations⟩
9 : to strike or thrust so as to impel violently ⟨*sent* him sprawling⟩
10 : DELIGHT, THRILL
intransitive verb
1 a : to dispatch someone to convey a message or do an errand — often used with *out* ⟨*send* out for pizza⟩ **b :** to dispatch a request or order — often used with *away*
2 : SCEND
3 : TRANSMIT
— **send·er** *noun*
— **send for :** to request by message to come **:** SUMMON
— **send packing :** to send off or dismiss roughly or in disgrace

²send *noun* (1726)
: the lift of a wave

send down *transitive verb* (1853)
British **:** to suspend or expel from a university

send in (1715)
transitive verb
1 : to cause to be delivered ⟨*send in* a letter of complaint⟩
2 : to give (one's name or card) to a servant when making a call
3 : to send (a player) into an athletic contest

send–off \'send-,ȯf\ *noun* (1872)
: a demonstration of goodwill and enthusiasm for the beginning of a new venture (as a trip)

send–up \-,əp\ *noun* (1958)
: PARODY, TAKEOFF

send up *transitive verb* (1852)
1 : to sentence to imprisonment **:** send to jail
2 : to make fun of **:** SATIRIZE, PARODY

se·ne \'sā-(,)nā\ *noun, plural* **sene** [Samoan, from English *century*] (1967)
— see *tala* at MONEY table

Sen·e·ca \'se-ni-kə\ *noun, plural* **Seneca** *or* **Senecas** [Dutch *Sennecaas,* plural, the Iroquois living west of the Mohawks] (circa 1616)
1 : a member of an American Indian people of what is now western New York
2 : the Iroquoian language of the Seneca people

seneca snakeroot *noun* (1789)
: a milkwort (*Polygala senega*) of eastern North America having racemes of small white flowers — called also *rattlesnake root, senega root*; compare SENEGA

seneca snakeroot

se·ne·cio \si-'nē-sh(ē-,)ō\ *noun, plural* **-cios** [New Latin, from Latin, old man, groundsel (from its hoary pappus), from *sen-, senic-, senex* old man] (circa 1890)
: any of a genus (*Senecio*) of widely distributed composite plants that have alternate or basal leaves and flower heads usually with yellow ray flowers

se·nec·ti·tude \si-'nek-tə-,tüd, -,tyüd\ *noun* [Medieval Latin *senectitudo,* alteration of Latin *senectus* old age, from *sen-, senic-, senex* old, old man — more at SENIOR] (1796)
: the final stage of the normal life span

sen·e·ga \'se-ni-gə\ *noun* (1748)
: the dried root of seneca snakeroot that contains an irritating saponin and is used medicinally

senega root *noun* [alteration of *Seneca root;* from its use by the Seneca as a remedy for snakebite] (circa 1846)
1 : SENECA SNAKEROOT
2 : SENEGA

se·nes·cence \si-'ne-sᵊn(t)s\ *noun* [*senescent,* from Latin *senescent-, senescens,* present participle of *senescere* to grow old, from *sen-, senex* old] (1695)
1 : the state of being old **:** the process of becoming old
2 : the growth phase in a plant or plant part (as a leaf) from full maturity to death
— **se·nes·cent** \-sᵊnt\ *adjective*

sen·e·schal \'se-nə-shəl\ *noun* [Middle English, from Middle French, of Germanic origin; akin to Gothic *sineigs* old and to Old High German *scalc* servant — more at SENIOR] (14th century)
: an agent or steward in charge of a lord's estate in feudal times

se·nhor \si-'nyȯr, -'nyȯr\ *noun, plural* **senhors** *or* **se·nho·res** \-'nyȯr-ēsh, -'nyȯr-, -ēzh, -ēs, -ēz\ [Portuguese, from Medieval Latin *senior* superior, lord, from Latin, adjective, elder] (1795)
: a Portuguese or Brazilian man — used as a title equivalent to *Mr.*

se·nho·ra \-'nyȯr-ə, -'nyȯr-\ *noun* [Portuguese, feminine of *senhor*] (1802)
: a married Portuguese or Brazilian woman — used as a title equivalent to *Mrs.*

se·nho·ri·ta \,sē-nyə-'rē-tə\ *noun* [Portuguese, from diminutive of *senhora*] (1874)
: an unmarried Portuguese or Brazilian girl or woman — used as a title equivalent to *Miss*

se·nile \'sē-,nīl *also* 'se-\ *adjective* [Latin *senilis,* from *sen-, senex* old, old man] (1661)
1 : of, relating to, exhibiting, or characteristic of old age ⟨*senile* weakness⟩; *especially* **:** exhibiting a loss of mental faculties associated with old age
2 : approaching the end of a geological cycle of erosion
— **se·nile·ly** \-,nīl-lē\ *adverb*

senile dementia *noun* (circa 1851)
: a mental disorder of old age especially of the degenerative type associated with Alzheimer's disease

se·nil·i·ty \si-'ni-lə-tē *also* se-\ *noun* (1791)
: the quality or state of being senile; *specifically* **:** the physical and mental infirmity of old age

¹se·nior \'sē-nyər\ *noun* [Middle English, from Latin, from *senior,* adjective] (14th century)
1 : a person older than another ⟨five years my *senior*⟩
2 a : a person with higher standing or rank **b :** a senior fellow of a college at an English university **c :** a student in the year preceding graduation from a school of secondary or higher level
3 *capitalized* **:** a member of a program of the Girl Scouts for girls in the ninth through twelfth grades in school
4 : SENIOR CITIZEN

²senior *adjective* [Middle English, from Latin, older, elder, comparative of *sen-, senex* old; akin to Gothic *sineigs* old, Greek *henos*] (14th century)
1 : of prior birth, establishment, or enrollment — often used to distinguish a father with the same given name as his son
2 : higher ranking **:** SUPERIOR ⟨*senior* officers⟩
3 : of or relating to seniors ⟨the *senior* class⟩
4 : having a claim on corporate assets and income prior to other securities

senior airman *noun* (circa 1977)
: an enlisted man in the air force who ranks above an airman first class but who has not been made sergeant

senior chief petty officer *noun* (circa 1960)
: an enlisted man in the navy or coast guard ranking above a chief petty officer and below a master chief petty officer

senior citizen *noun* (1938)
: an elderly person; *especially* **:** one who has retired

senior high school *noun* (1909)
: a school usually including grades 10 to 12

se·nior·i·ty \sēn-'yȯr-ə-tē, -'yär-\ *noun* (15th century)
1 : the quality or state of being senior **:** PRIORITY
2 : a privileged status attained by length of continuous service (as in a company)

senior master sergeant *noun* (circa 1962)
: a noncommissioned officer in the air force ranking above a master sergeant and below a chief master sergeant

sen·i·ti \'se-nə-tē\ *noun, plural* **seniti** [Tongan, modification of English *century*] (1967)
— see *pa'anga* at MONEY table

sen·na \'se-nə\ *noun* [New Latin, from Arabic *sanā*] (1543)
1 : any of a genus (*Cassia*) of leguminous herbs, shrubs, and trees native to warm regions; *especially* **:** one used medicinally
2 : the dried leaflets or pods of various sennas (especially *Cassia acutifolia* and *C. angustifolia*) used as a purgative

sen·net \'se-nət\ *noun* [probably alteration of obsolete *signet* signal] (circa 1590)
: a signal call on a trumpet or cornet for entrance or exit on the stage

sen·night *also* **se'n·night** \'se-,nīt\ *noun* [Middle English, from Old English *seofon nihta* seven nights] (15th century)
archaic **:** the space of seven nights and days **:** WEEK

sen·nit \'se-nət\ *noun* [origin unknown] (circa 1769)
1 : a braided cord or fabric (as of plaited rope yarns)
2 : a straw or grass braid for hats

se·nor *or* **se·ñor** \sān-'yȯr\ *noun, plural* **senors** *or* **se·ño·res** \-'yȯr-(,)ās, -'yȯr-\ [Spanish *señor,* from Medieval Latin *senior* superior, lord, from Latin, adjective, elder] (1622)
: a Spanish or Spanish-speaking man — used as a title equivalent to *Mr.*

se·no·ra *or* **se·ño·ra** \sān-'yȯr-ə, -'yȯr-\ *noun* [Spanish *señora,* feminine of *señor*] (1579)
: a married Spanish or Spanish-speaking woman — used as a title equivalent to *Mrs.*

se·no·ri·ta *or* **se·ño·ri·ta** \,sān-yə-'rē-tə\ *noun* [Spanish *señorita,* from diminutive of *señora*] (1823)
: an unmarried Spanish or Spanish-speaking girl or woman — used as a title equivalent to *Miss*

sen·ryu \'sen-rē-(,)ü\ *noun, plural* **senryu** [Japanese] (1938)
: a 3-line unrhymed Japanese poem structurally similar to haiku but treating human nature usually in an ironic or satiric vein

sensa *plural of* SENSUM

sen·sate \'sen-,sāt\ *adjective* [Middle English *sensat,* from Medieval Latin *sensatus,* from Late Latin, endowed with sense, from Latin *sensus* sense] (15th century)
1 : relating to or apprehending or apprehended through the senses
2 : preoccupied with things that can be experienced through a sense modality
— **sen·sate·ly** *adverb*

sen·sa·tion \sen-'sā-shən, sən-\ *noun* [Medieval Latin *sensation-, sensatio,* from Late Latin, understanding, idea, from Latin *sensus*] (1615)
1 a : a mental process (as seeing, hearing, or smelling) due to immediate bodily stimulation often as distinguished from awareness of the process — compare PERCEPTION **b :** awareness (as of heat or pain) due to stimulation of a sense organ **c :** a state of consciousness of a kind usually due to physical objects or internal bodily changes ⟨a burning *sensation* in his chest⟩ **d :** an indefinite bodily feeling ⟨a *sensation* of buoyancy⟩

2 : something (as a physical object, sense-datum, pain, or afterimage) that causes or is the object of sensation
3 a : a state of excited interest or feeling ⟨their elopement caused a *sensation*⟩ **b :** a cause of such excitement ⟨the show was the musical *sensation* of the season⟩; *especially* **:** one (as a person) in some respect exceptional or outstanding ⟨the rookie hitting *sensation* of the American League⟩

sen•sa•tion•al \-shnəl, -shə-nᵊl\ *adjective* (1840)
1 : of or relating to sensation or the senses
2 : arousing or tending to arouse (as by lurid details) a quick, intense, and usually superficial interest, curiosity, or emotional reaction
3 : exceedingly or unexpectedly excellent or great
— sen•sa•tion•al•ly *adverb*

sen•sa•tion•al•ise *British variant of* SENSATIONALIZE

sen•sa•tion•al•ism \-shnə-ˌli-zəm, -shə-nᵊl-ˌi-zəm\ *noun* (1846)
1 : empiricism that limits experience as a source of knowledge to sensation or sense perceptions
2 : the use or effect of sensational subject matter or treatment
— sen•sa•tion•al•ist \-list, -ist\ *adjective or noun*
— sen•sa•tion•al•is•tic \-ˌsā-shnə-ˈlis-tik, -shən-ᵊl-ˈis-tik\ *adjective*

sen•sa•tion•al•ize \-ˌlīz, -ˌīz\ *transitive verb* **-ized; -iz•ing** (1869)
: to present in a sensational manner

¹sense \ˈsen(t)s\ *noun* [Middle English, from Middle French or Latin; Middle French *sens* sensation, feeling, mechanism of perception, meaning, from Latin *sensus,* from *sentire* to perceive, feel; perhaps akin to Old High German *sinnan* to go, strive, Old English *sith* journey — more at SEND] (14th century)
1 : a meaning conveyed or intended **:** IMPORT, SIGNIFICATION; *especially* **:** one of a set of meanings a word or phrase may bear especially as segregated in a dictionary entry
2 a : the faculty of perceiving by means of sense organs **b :** a specialized animal function or mechanism (as sight, hearing, smell, taste, or touch) basically involving a stimulus and a sense organ **c :** the sensory mechanisms constituting a unit distinct from other functions (as movement or thought)
3 : conscious awareness or rationality — usually used in plural ⟨finally came to his *senses*⟩
4 a : a particular sensation or kind or quality of sensation ⟨a good *sense* of balance⟩ **b :** a definite but often vague awareness or impression ⟨felt a *sense* of insecurity⟩ ⟨a *sense* of danger⟩ **c :** a motivating awareness ⟨a *sense* of shame⟩ **d :** a discerning awareness and appreciation ⟨her *sense* of humor⟩
5 : CONSENSUS ⟨the *sense* of the meeting⟩
6 a : capacity for effective application of the powers of the mind as a basis for action or response **:** INTELLIGENCE **b :** sound mental capacity and understanding typically marked by shrewdness and practicality; *also* **:** agreement with or satisfaction of such power ⟨this decision makes *sense*⟩
7 : one of two opposite directions especially of motion (as of a point, line, or surface) ☆

²sense *transitive verb* **sensed; sens•ing** (circa 1531)
1 a : to perceive by the senses **b :** to be or become conscious of ⟨*sense* danger⟩
2 : GRASP, COMPREHEND
3 : to detect automatically especially in response to a physical stimulus (as light or movement)

sense–datum *noun, plural* **sense–data** (1882)
: an immediate unanalyzable private object of sensation

sense•ful \ˈsen(t)s-fəl\ *adjective* (1591)
: REASONABLE, JUDICIOUS

sense•less \ˈsen(t)s-ləs\ *adjective* (1557)
: destitute of, deficient in, or contrary to sense: as **a :** UNCONSCIOUS ⟨knocked *senseless*⟩ **b :** FOOLISH, STUPID ⟨it was some *senseless* practical joke —A. Conan Doyle⟩ **c :** MEANINGLESS, PURPOSELESS ⟨a *senseless* murder⟩
— sense•less•ly *adverb*
— sense•less•ness *noun*

sense organ *noun* (1854)
: a bodily structure that receives a stimulus (as heat or sound waves) and is affected in such a manner as to initiate a wave of excitation in associated sensory nerve fibers which convey specific impulses to the central nervous system where they are interpreted as corresponding sensations **:** RECEPTOR

sen•si•bil•ia \ˌsen(t)-sə-ˈbi-lē-ə, -ˈbil-yə\ *noun plural* [Late Latin, from neuter plural of Latin *sensibilis* sensible] (1856)
: what may be sensed

sen•si•bil•i•ty \ˌsen(t)-sə-ˈbi-lə-tē\ *noun, plural* **-ties** (15th century)
1 : ability to receive sensations **:** SENSITIVENESS ⟨tactile *sensibility*⟩
2 : peculiar susceptibility to a pleasurable or painful impression (as from praise or a slight) — often used in plural
3 : awareness of and responsiveness toward something (as emotion in another)
4 : refined or excessive sensitiveness in emotion and taste with especial responsiveness to the pathetic

¹sen•si•ble \ˈsen(t)-sə-bəl\ *adjective* [Middle English, from Middle French, from Latin *sensibilis,* from *sensus,* past participle of *sentire* to feel] (14th century)
1 : of a kind to be felt or perceived: as **a :** perceptible to the senses or to reason or understanding ⟨felt a *sensible* chill⟩ ⟨her distress was *sensible* from her manner⟩ **b** *archaic* **:** perceptibly large **:** CONSIDERABLE **c** (1) **:** perceptible as real or material **:** SUBSTANTIAL ⟨the *sensible* world in which we live⟩ (2) **:** of a kind to arouse emotional response ⟨his whipping was a *sensible* expression of his father's anger⟩
2 a : capable of receiving sensory impressions ⟨*sensible* to pain⟩ **b :** receptive to external influences **:** SENSITIVE ⟨the most *sensible* reaches of the spirit⟩
3 a : perceiving through the senses or mind **:** COGNIZANT ⟨*sensible* of the increasing heat⟩; *also* **:** convinced by perceived evidence **:** SATISFIED ⟨*sensible* of my error⟩ **b :** emotionally aware and responsive ⟨we are *sensible* of your problems⟩ **c :** CONSCIOUS
4 : having, containing, or indicative of good sense or reason **:** RATIONAL, REASONABLE ⟨*sensible* people⟩ ⟨made a *sensible* answer⟩
synonym see MATERIAL, PERCEPTIBLE, AWARE, WISE
— sen•si•ble•ness *noun*
— sen•si•bly \-blē\ *adverb*

²sensible *noun* (1589)
: something that can be sensed

sen•sil•lum \sen-ˈsi-ləm\ *also* **sen•sil•la** \-ˈsi-lə\ *noun, plural* **-sil•la** \-ˈsi-lə\ *also* **-sil•lae** \-ˈsi-(ˌ)lē\ [New Latin *sensillum,* diminutive of Medieval Latin *sensus* sense organ, from Latin, sense] (1925)
: a simple epithelial sense organ of an invertebrate (as an insect) usually in the form of a spine, plate, rod, cone, or peg that is composed of one or a few cells with a nerve connection

sen•si•ti•sa•tion, sen•si•tise *British variant of* SENSITIZATION, SENSITIZE

¹sen•si•tive \ˈsen(t)-sə-tiv, ˈsen(t)s-təv\ *adjective* [Middle English, from Middle French *sensitif,* from Medieval Latin *sensitivus,* probably alteration of *sensativus,* from *sensatus* sensate] (15th century)
1 : SENSORY 2
2 a : receptive to sense impressions **b :** capable of being stimulated or excited by external agents (as light, gravity, or contact) ⟨a photographic emulsion *sensitive* to red light⟩ ⟨*sensitive* protoplasm⟩
3 : highly responsive or susceptible: as **a** (1) **:** easily hurt or damaged; *especially* **:** easily hurt emotionally (2) **:** delicately aware of the attitudes and feelings of others **b :** excessively or abnormally susceptible **:** HYPERSENSITIVE ⟨*sensitive* to egg protein⟩ **c :** readily fluctuating in price or demand ⟨*sensitive* commodities⟩ **d :** capable of indicating minute differences **:** DELICATE ⟨*sensitive* scales⟩ **e :** readily affected or changed by various agents (as light or mechanical shock) **f :** highly radiosensitive
4 a : concerned with highly classified government information or involving discretionary authority over important policy matters ⟨*sensitive* documents⟩ **b :** calling for tact, care, or caution in treatment **:** TOUCHY ⟨a *sensitive* issue like race relations⟩
synonym see LIABLE
— sen•si•tive•ly *adverb*
— sen•si•tive•ness *noun*

²sensitive *noun* (1850)
1 : a person having occult or psychical abilities
2 : a sensitive person

sensitive plant *noun* (1659)
: any of several mimosas (especially *Mimosa pudica*) with leaves that fold or droop when touched; *broadly* **:** a plant responding to touch with movement

sen•si•tiv•i•ty \ˌsen(t)-sə-ˈti-və-tē\ *noun, plural* **-ties** (1803)
: the quality or state of being sensitive: as **a :** the capacity of an organism or sense organ to respond to stimulation **:** IRRITABILITY **b :** the quality or state of being hypersensitive **c :** the degree to which a radio receiving set responds to incoming waves **d :** the capacity of being easily hurt **e :** awareness of the needs and emotions of others ⟨a book written with just the right mix of empathy and *sensitivity* —L.C. Brown⟩

sen•si•ti•za•tion \ˌsen(t)-sə-tə-ˈzā-shən, ˌsen(t)s-tə-ˈzā-\ *noun* (1887)
1 : the action or process of sensitizing
2 : the quality or state of being sensitized (as to an antigen)

sen•si•tize \ˈsen(t)-sə-ˌtīz\ *verb* **-tized; -tiz•ing** [*sensitive* + *-ize*] (circa 1859)
transitive verb
: to make sensitive or hypersensitive
intransitive verb
: to become sensitive
— sen•si•tiz•er *noun*

sen•si•tom•e•ter \ˌsen(t)-sə-ˈtä-mə-tər\ *noun* [International Scientific Vocabulary *sensitive* + *-o-* + *-meter*] (1880)
: an instrument for measuring sensitivity of photographic material
— sen•si•to•met•ric \ˌsen(t)-sə-tə-ˈme-trik\ *adjective*

☆ SYNONYMS
Sense, common sense, judgment, wisdom mean ability to reach intelligent conclusions. SENSE implies a reliable ability to judge and decide with soundness, prudence, and intelligence ⟨a choice showing good *sense*⟩. COMMON SENSE suggests an average degree of such ability without sophistication or special knowledge ⟨*common sense* tells me it's wrong⟩. JUDGMENT implies sense tempered and refined by experience, training, and maturity ⟨they relied on her *judgment* for guidance⟩. WISDOM implies sense and judgment far above average ⟨a leader of rare *wisdom*⟩.

— **sen·si·tom·e·try** \ˌsen(t)-sə-ˈtä-mə-trē\ *noun*

sen·sor \ˈsen-ˌsȯr, ˈsen(t)-sər\ *noun* [Latin *sentire* to perceive + English [1]*-or* — more at SENSE] (circa 1928)
: a device that responds to a physical stimulus (as heat, light, sound, pressure, magnetism, or a particular motion) and transmits a resulting impulse (as for measurement or operating a control); *also* **:** SENSE ORGAN

sen·so·ri·al \sen-ˈsōr-ē-əl, -ˈsȯr-\ *adjective* (1768)
: SENSORY
— **sen·so·ri·al·ly** \-ə-lē\ *adverb*

sen·so·ri·mo·tor \ˌsen(t)s-rē-ˈmō-tər, ˌsen(t)-sə-\ *adjective* [*sensory* + *motor*] (1855)
: of, relating to, or functioning in both sensory and motor aspects of bodily activity

sen·so·ri·neu·ral \-ˈnu̇r-əl, -ˈnyu̇r-\ *adjective* (circa 1977)
: of, relating to, or involving the aspects of sense perception mediated by nerves ⟨*sensorineural* hearing loss⟩

sen·so·ri·um \sen-ˈsōr-ē-əm, -ˈsȯr-\ *noun, plural* **-ri·ums** *or* **-ria** \-ē-ə\ [Late Latin, sense organ, from Latin *sentire*] (1647)
: the parts of the brain or the mind concerned with the reception and interpretation of sensory stimuli; *broadly* **:** the entire sensory apparatus

sen·so·ry \ˈsen(t)s-rē, ˈsen(t)-sə-rē\ *adjective* (1749)
1 : of or relating to sensation or to the senses
2 : conveying nerve impulses from the sense organs to the nerve centers **:** AFFERENT

sensory area *noun* (1896)
: an area of the cerebral cortex that receives afferent nerve fibers from lower sensory or motor areas

sen·su·al \ˈsen(t)-sh(ə-)wəl, -shəl\ *adjective* [Middle English, from Late Latin *sensualis,* from Latin *sensus* sense] (15th century)
1 : relating to or consisting in the gratification of the senses or the indulgence of appetite **:** FLESHLY
2 : SENSORY
3 a : devoted to or preoccupied with the senses or appetites **b :** VOLUPTUOUS **c :** deficient in moral, spiritual, or intellectual interests **:** WORLDLY; *especially* **:** IRRELIGIOUS
synonym see CARNAL, SENSUOUS
— **sen·su·al·i·ty** \ˌsen(t)-shə-ˈwa-lə-tē\ *noun*
— **sen·su·al·ly** \ˈsen(t)-sh(ə-)wə-lē, ˈsen-shə-lē\ *adverb*

sen·su·al·ism \ˈsen(t)-sh(ə-)wə-ˌli-zəm, ˈsen-shə-ˌli-\ *noun* (1813)
: persistent or excessive pursuit of sensual pleasures and interests
— **sen·su·al·ist** \-list\ *noun*
— **sen·su·al·is·tic** \ˌsen(t)-sh(ə-)wə-ˈlis-tik, ˌsen-shə-ˈlis-\ *adjective*

sen·su·al·ize \ˈsen(t)-sh(ə-)wə-ˌlīz, ˈsen-shə-ˌlīz\ *transitive verb* **-ized; -iz·ing** (circa 1687)
: to make sensual
— **sen·su·al·i·za·tion** \ˌsen(t)-sh(ə-)wə-lə-ˈzā-shən, ˌsen-shə-lə-\ *noun*

sen·sum \ˈsen(t)-səm\ *noun, plural* **sen·sa** \-sə\ [Medieval Latin, from Latin, neuter of *sensus,* past participle of *sentire* to feel — more at SENSE] (1868)
: SENSE-DATUM

sen·su·ous \ˈsen(t)-sh(ə-)wəs\ *adjective* [Latin *sensus* sense + English *-ous*] (1640)
1 a : of or relating to the senses or sensible objects **b :** producing or characterized by gratification of the senses **:** having strong sensory appeal ⟨*sensuous* pleasure⟩
2 : characterized by sense impressions or imagery aimed at the senses ⟨*sensuous* verse⟩
3 : highly susceptible to influence through the senses ☆
— **sen·su·os·i·ty** \ˌsen(t)-shə-ˈwä-sə-tē\ *noun*
— **sen·su·ous·ly** \ˈsen(t)-shə-wəs-lē\ *adverb*
— **sen·su·ous·ness** *noun*

sen·su stric·to \ˈsen-(ˌ)sü-ˈstrik-(ˌ)tō\ *adverb* [New Latin] (1902)
: in a narrow or strict sense

sent *past and past participle of* SEND

sen·te \ˈsen-tē\ *noun, plural* **li·cen·te** *or* **li·sen·te** \li-ˈsen-tē\ [Sesotho, from English *century*] (1966)
— see *loti* at MONEY table

1sen·tence \ˈsen-tᵊn(t)s, -tᵊnz\ *noun* [Middle English, from Middle French, from Latin *sententia* feeling, opinion, from (assumed) *sentent-, sentens,* irregular present participle of *sentire* to feel — more at SENSE] (14th century)
1 *obsolete* **:** OPINION; *especially* **:** a conclusion given on request or reached after deliberation
2 a : JUDGMENT 2a; *specifically* **:** one formally pronounced by a court or judge in a criminal proceeding and specifying the punishment to be inflicted upon the convict **b :** the punishment so imposed ⟨serve out a *sentence*⟩
3 *archaic* **:** MAXIM, SAW
4 a : a word, clause, or phrase or a group of clauses or phrases forming a syntactic unit which expresses an assertion, a question, a command, a wish, an exclamation, or the performance of an action, that in writing usually begins with a capital letter and concludes with appropriate end punctuation, and that in speaking is distinguished by characteristic patterns of stress, pitch, and pauses **b :** a mathematical or logical statement (as an equation or a proposition) in words or symbols
5 : PERIOD 2b

2sentence *transitive verb* **sen·tenced; sen·tenc·ing** (1592)
1 : to impose a sentence on
2 : to cause to suffer something ⟨*sentenced* these most primitive cultures to extinction —E. W. Count⟩

sentence fragment *noun* (1947)
: a word, phrase, or clause that usually has in speech the intonation of a sentence but lacks the grammatical structure usually found in the sentences of formal and especially written composition

sentence stress *noun* (1884)
: the manner in which stresses are distributed on the syllables of words assembled into sentences — called also *sentence accent*

sen·ten·tia \sen-ˈten(t)-sh(ē-)ə\ *noun, plural* **-ti·ae** \-shē-ˌē\ [Latin, literally, feeling, opinion] (1917)
: APHORISM — usually used in plural

sen·ten·tial \sen-ˈten(t)-shəl\ *adjective* (1646)
1 : of or relating to a sentence ⟨a relative clause with a *sentential* antecedent⟩
2 : of, relating to, or involving a proposition in logic ⟨*sentential* connective⟩

sentential calculus *noun* (1937)
: PROPOSITIONAL CALCULUS

sentential function *noun* (1937)
: an expression that contains one or more variables and becomes a declarative sentence when constants are substituted for the variables

sen·ten·tious \sen-ˈten(t)-shəs\ *adjective* [Middle English, full of meaning, from Latin *sententiosus,* from *sententia* sentence, maxim] (1509)
1 a : given to or abounding in aphoristic expression **b :** given to or abounding in excessive moralizing
2 : terse, aphoristic, or moralistic in expression **:** PITHY, EPIGRAMMATIC
— **sen·ten·tious·ly** *adverb*
— **sen·ten·tious·ness** *noun*

sen·tience \ˈsen(t)-sh(ē-)ən(t)s, ˈsen-tē-ən(t)s\ *noun* (1839)
1 : a sentient quality or state
2 : feeling or sensation as distinguished from perception and thought

sen·tient \ˈsen(t)-sh(ē-)ənt, ˈsen-tē-ənt\ *adjective* [Latin *sentient-, sentiens,* present participle of *sentire* to perceive, feel] (1632)
1 : responsive to or conscious of sense impressions
2 : AWARE
3 : finely sensitive in perception or feeling
— **sen·tient·ly** *adverb*

sen·ti·ment \ˈsen-tə-mənt\ *noun* [French or Medieval Latin; French, from Medieval Latin *sentimentum,* from Latin *sentire*] (1639)
1 a : an attitude, thought, or judgment prompted by feeling **:** PREDILECTION **b :** a specific view or notion **:** OPINION
2 a : EMOTION **b :** refined feeling **:** delicate sensibility especially as expressed in a work of art **c :** emotional idealism **d :** a romantic or nostalgic feeling verging on sentimentality
3 a : an idea colored by emotion **b :** the emotional significance of a passage or expression as distinguished from its verbal context
synonym see FEELING, OPINION

sen·ti·men·tal \ˌsen-tə-ˈmen-tᵊl\ *adjective* (1749)
1 a : marked or governed by feeling, sensibility, or emotional idealism **b :** resulting from feeling rather than reason or thought
2 : having an excess of sentiment or sensibility
— **sen·ti·men·tal·ly** *adverb*

sen·ti·men·tal·ise *British variant of* SENTIMENTALIZE

sen·ti·men·tal·ism \ˌsen-tə-ˈmen-tᵊl-ˌi-zəm\ *noun* (1817)
1 : the disposition to favor or indulge in sentimentality
2 : an excessively sentimental conception or statement
— **sen·ti·men·tal·ist** \-tᵊl-ist\ *noun*

sen·ti·men·tal·i·ty \ˌsen-tə-ˌmen-ˈta-lə-tē, -mən-\ *noun, plural* **-ties** (1770)
1 : the quality or state of being sentimental especially to excess or in affectation
2 : a sentimental idea or its expression

sen·ti·men·tal·ize \-ˈmen-tᵊl-ˌīz\ *verb* **-ized; -iz·ing** (1788)
intransitive verb
: to indulge in sentiment
transitive verb
: to look upon or imbue with sentiment
— **sen·ti·men·tal·i·za·tion** \-ˌmen-tᵊl-ə-ˈzā-shən\ *noun*

sen·ti·mo \sen-ˈtē-(ˌ)mō\ *noun, plural* **-mos** [Tagalog, from Spanish *céntimo*] (1968)
— see *peso* at MONEY table

1sen·ti·nel \ˈsent-nəl, ˈsen-tᵊn-əl\ *noun* [Middle French *sentinelle,* from Old Italian *sentinella,* from *sentina* vigilance, from *sentire* to perceive, from Latin] (1579)
: SENTRY

2sentinel *transitive verb* **-neled** *or* **-nelled; -nel·ing** *or* **-nel·ling** (1593)
1 : to watch over as a sentinel
2 : to furnish with a sentinel
3 : to post as sentinel

sen·try \ˈsen-trē\ *noun, plural* **sentries** [perhaps from obsolete *sentry* sanctuary, watchtower] (1632)
: GUARD, WATCH; *especially* **:** a soldier standing guard at a point of passage (as a gate)

☆ SYNONYMS
Sensuous, sensual, luxurious, voluptuous mean relating to or providing pleasure through gratification of the senses. SENSUOUS implies gratification of the senses for the sake of aesthetic pleasure ⟨the *sensuous* delights of great music⟩. SENSUAL tends to imply the gratification of the senses or the indulgence of the physical appetites as ends in themselves ⟨a life devoted to *sensual* pleasures⟩. LUXURIOUS suggests the providing of or indulgence of sensuous pleasure inducing bodily ease and languor ⟨a *luxurious* hotel⟩. VOLUPTUOUS implies more strongly an abandonment especially to sensual pleasure ⟨a *voluptuous* feast⟩.

sentry box *noun* (circa 1728)
: a shelter for a sentry on his post

se·pal \'sē-pəl, 'se-\ *noun* [New Latin *sepalum,* from *sep-* (irregular from Greek *skepē* covering) + *-alum* (as in *petalum* petal)] (1821)
: one of the modified leaves comprising a calyx — see FLOWER illustration

se·pal·oid \-pə-ˌlȯid\ *adjective* (1830)
: resembling or functioning as a sepal

sep·a·ra·ble \'se-p(ə-)rə-bəl\ *adjective* [Middle English, from Latin *separabilis,* from *separare*] (14th century)
1 : capable of being separated or dissociated
2 *obsolete* **:** causing separation
— sep·a·ra·bil·i·ty \ˌse-p(ə-)rə-'bi-lə-tē\ *noun*
— sep·a·ra·ble·ness *noun*

¹sep·a·rate \'se-p(ə-)ˌrāt\ *verb* **-rat·ed; -rat·ing** [Middle English, from Latin *separatus,* past participle of *separare,* from *se-* apart + *parare* to prepare, procure — more at SECEDE, PARE] (15th century)
transitive verb
1 a : to set or keep apart **:** DISCONNECT, SEVER **b :** to make a distinction between **:** DISCRIMINATE, DISTINGUISH ⟨*separate* religion from magic⟩ **c :** SORT ⟨*separate* mail⟩ **d :** to disperse in space or time **:** SCATTER ⟨widely *separated* homesteads⟩
2 *archaic* **:** to set aside for a special purpose **:** CHOOSE, DEDICATE
3 : to part by a legal separation: **a :** to sever conjugal ties with **b :** to sever contractual relations with **:** DISCHARGE ⟨*separated* from the army⟩
4 : to block off **:** SEGREGATE
5 a : to isolate from a mixture **:** EXTRACT ⟨*separate* cream from milk⟩ **b :** to divide into constituent parts
6 : to dislocate (as a shoulder) especially in sports
intransitive verb
1 : to become divided or detached
2 a : to sever an association **:** WITHDRAW **b :** to cease to live together as a married couple
3 : to go in different directions
4 : to become isolated from a mixture ☆

²sep·a·rate \'se-p(ə-)rət\ *adjective* (15th century)
1 a : set or kept apart **:** DETACHED **b** *archaic* **:** SOLITARY, SECLUDED **c :** IMMATERIAL, DISEMBODIED
2 a : not shared with another **:** INDIVIDUAL ⟨*separate* rooms⟩ **b** *often capitalized* **:** estranged from a parent body ⟨*separate* churches⟩
3 a : existing by itself **:** AUTONOMOUS **b :** dissimilar in nature or identity
synonym see DISTINCT
— sep·a·rate·ly \-p(ə-)rət-lē, 'se-pərt-lē\ *adverb*
— sep·a·rate·ness \-nəs\ *noun*

³sep·a·rate \'se-p(ə-)rət\ *noun* (1886)
1 : OFFPRINT
2 : an article of dress designed to be worn interchangeably with others to form various costume combinations — usually used in plural

sep·a·ra·tion \ˌse-pə-'rā-shən\ *noun* (14th century)
1 : the act or process of separating **:** the state of being separated
2 a : a point, line, or means of division **b :** an intervening space **:** GAP
3 a : cessation of cohabitation between a married couple by mutual agreement or judicial decree **b :** termination of a contractual relationship (as employment or military service)

sep·a·ra·tion·ist \-sh(ə-)nist\ *noun* (1831)
: SEPARATIST

sep·a·rat·ism \'se-p(ə-)rə-ˌti-zəm\ *noun* (1628)
: a belief in, movement for, or state of separation (as schism, secession, or segregation)

sep·a·rat·ist \'se-p(ə-)rə-tist, 'se-pə-ˌrā-\ *noun* (1608)
: one that favors separatism: as **a** *capitalized* **:** one of a group of 16th and 17th century English Protestants preferring to separate from rather than to reform the Church of England **b :** an advocate of independence or autonomy for a part of a political unit (as a nation) **c :** an advocate of racial or cultural separation
— separatist *adjective*
— sep·a·ra·tis·tic \ˌse-p(ə-)rə-'tis-tik\ *adjective*

sep·a·ra·tive \'se-pə-ˌrā-tiv, 'se-p(ə-)rə-\ *adjective* (1592)
: tending toward, causing, or expressing separation

sep·a·ra·tor \'se-p(ə-)ˌrā-tər\ *noun* (1607)
: one that separates; *especially* **:** a device for separating liquids of different specific gravities (as cream from milk) or liquids from solids

Se·phar·di \sə-'fär-dē\ *noun, plural* **Se·phar·dim** \-'fär-dəm\ [Late Hebrew *sĕphāradhī,* from *sĕphāradh* Spain, from Hebrew, region where Jews were once exiled (Obadiah 1:20)] (1851)
: a member of the occidental branch of European Jews settling in Spain and Portugal and later in the Balkans, the Levant, England, the Netherlands, and the Americas; *also* **:** one of their descendants — compare ASHKENAZI
— Se·phar·dic \-'fär-dik\ *adjective*

¹se·pia \'sē-pē-ə\ *noun* [Latin, cuttlefish, ink, from Greek *sēpia*] (1821)
1 a : the inky secretion of a cuttlefish **b :** a brown melanin-containing pigment from the ink of cuttlefishes
2 : a print or photograph of a brown color resembling sepia
3 : a brownish gray to dark olive brown color

²sepia *adjective* (1827)
1 : made of or done in sepia
2 : of the color sepia

se·pi·o·lite \'sē-pē-ə-ˌlīt\ *noun* [German *Sepiolith,* from Greek *sēpion* cuttlebone (from *sēpia*) + German *-lith* -lite] (1854)
: MEERSCHAUM 1

se·poy \'sē-ˌpȯi\ *noun* [Portuguese *sipai,* from Hindi *sipāhī,* from Persian, cavalryman] (circa 1718)
: a native of India employed as a soldier by a European power

sep·pu·ku \se-'pü-(ˌ)kü, 'se-pə-ˌkü\ *noun* [Japanese] (1871)
: HARA-KIRI 1

sep·sis \'sep-səs\ *noun, plural* **sep·ses** \'sep-ˌsēz\ [New Latin, from Greek *sēpsis* decay, from *sēpein* to putrefy] (1876)
: a toxic condition resulting from the spread of bacteria or their products from a focus of infection; *especially* **:** SEPTICEMIA

sept \'sept\ *noun* [perhaps from Latin *septum, saeptum* enclosure, fold — more at SEPTUM] (1517)
: a branch of a family; *especially* **:** CLAN

sep·tal \'sep-tᵊl\ *adjective* (circa 1847)
: of or relating to a septum

sep·tate \'sep-ˌtāt\ *adjective* (1846)
: divided by or having a septum

Sep·tem·ber \sep-'tem-bər, səp-\ *noun* [Middle English *Septembre,* from Old French & Old English, both from Latin *September* (seventh month), from *septem* seven — more at SEVEN] (before 12th century)
: the 9th month of the Gregorian calendar ◆

sep·te·nar·i·us \ˌsep-tə-'nar-ē-əs, -'ner-\ *noun, plural* **-nar·ii** \-ē-ˌī, -ē-ˌē\ [Latin, from *septenarius* of seven, from *septeni* seven each, from *septem* seven] (1819)
: a verse consisting of seven feet especially in Latin prosody

sep·ten·de·cil·lion \(ˌ)sep-ˌten-di-'sil-yən\ *noun, often attributive* [Latin *septendec*im seventeen (from *septem* seven + *decem* ten) + English *-illion* (as in *million*) — more at TEN] (circa 1938)
— see NUMBER table

sep·ten·ni·al \sep-'te-nē-əl\ *adjective* [Late Latin *septennium* period of seven years, from Latin *septem* + *-ennium* (as in *biennium*)] (1640)
1 : occurring or being done every seven years
2 : consisting of or lasting for seven years
— sep·ten·ni·al·ly \-ə-lē\ *adverb*

sep·ten·tri·on \sep-'ten-trē-ˌän, -trē-ən\ *noun* [Middle English, from Middle French, from Latin *septentriones,* plural, the seven stars of Ursa Major or Ursa Minor, from *septem* seven + *triones* plowing oxen] (14th century)
obsolete **:** the northern regions **:** NORTH

sep·ten·tri·o·nal \-trē-ə-nᵊl\ *adjective* (14th century)
: NORTHERN

sep·tet \sep-'tet\ *noun* [German, from Latin *septem*] (1828)
1 : a musical composition for seven instruments or voices
2 : a group or set of seven; *especially* **:** the performers of a septet

☆ SYNONYMS
Separate, part, divide, sever, sunder, divorce mean to become or cause to become disunited or disjointed. SEPARATE may imply any of several causes such as dispersion, removal of one from others, or presence of an intervening thing ⟨*separated* her personal life from her career⟩. PART implies the separating of things or persons in close union or association ⟨vowed never to *part*⟩. DIVIDE implies separating into pieces or sections by cutting or breaking ⟨civil war *divided* the nation⟩. SEVER implies violence especially in the removal of a part or member ⟨a *severed* limb⟩. SUNDER suggests violent rending or wrenching apart ⟨a city *sundered* by racial conflict⟩. DIVORCE implies separating two things that commonly interact and belong together ⟨cannot *divorce* scientific research from moral responsibility⟩.

◇ WORD HISTORY
September The earliest Roman calendar appears to have had just ten lunar months, March to December, with an uncounted intervening gap in the winter. The names of the last four months, in Latin *September, October, November,* and *December,* were formed, appropriately enough, from the numerals *septem* "seven," *octo* "eight," *novem* "nine," and *decem* "ten." The original names of *September, November,* and *December* were presumably compounds made up with the aid of an unattested element *mensri-,* itself derived from *mensis* "month" and a suffix *ri-.* Through a series of sound changes, *mensri-* became *membri-.* The resulting words should have been *Septemmember, Novemmember,* and *Decemmember,* but these likewise unattested forms appear to have undergone contraction by dropping one pair of the identical sounds found next to each other (a process called haplology) to give *September, November,* and *December. October* was simply formed by analogy to fit the shape of the other three words; its earlier name in Latin is unknown. The subsequent changes in the calendar that added two months, January and February, and made January rather than March the beginning of the year rendered the numerical sequence incorrect, but the Romans retained the old names, which we have inherited.

sep·tic \'sep-tik\ *adjective* [Latin *septicus,* from Greek *sēptikos,* from *sēpein* to putrefy] (1605)
1 : of, relating to, or causing putrefaction
2 : relating to, involving, or characteristic of sepsis

sep·ti·ce·mia \,sep-tə-'sē-mē-ə\ *noun* [New Latin, from Latin *septicus* + New Latin *-emia*] (circa 1860)
: invasion of the bloodstream by virulent microorganisms from a local seat of infection accompanied especially by chills, fever, and prostration — called also *blood poisoning*; compare SEPSIS
— sep·ti·ce·mic \-'sē-mik\ *adjective*

sep·ti·ci·dal \,sep-tə-'sī-dᵊl\ *adjective* [New Latin *septum* + Latin *-cidere* to cut, from *caedere*] (1819)
: dehiscent longitudinally at or along a septum ⟨a *septicidal* fruit⟩

septic sore throat *noun* (1924)
: STREP THROAT

septic tank *noun* (circa 1902)
: a tank in which the solid matter of continuously flowing sewage is disintegrated by bacteria

sep·til·lion \sep-'til-yən\ *noun, often attributive* [French, from Latin *sept*em + French *-illion* (as in *million*) — more at SEVEN] (1690)
— see NUMBER table

sep·tu·a·ge·nar·i·an \(,)sep-,tü-ə-jə-'ner-ē-ən, -,tyü-, -,chü-\ *noun* [Late Latin *septuagenarius* seventy years old, from Latin, of or containing seventy, from *septuageni* seventy each, from *septuaginta*] (1805)
: a person whose age is in the seventies
— septuagenarian *adjective*

Sep·tu·a·ge·si·ma \,sep-tə-wə-'je-sə-mə, -'jā-zə-\ *noun* [Middle English, from Late Latin, from Latin, feminine of *septuagesimus* seventieth, from *septuaginta* seventy; from its being approximately seventy days before Easter] (14th century)
: the third Sunday before Lent

Sep·tu·a·gint \sep-'tü-ə-jənt, -'tyü-; 'sep-tə-wə-,jint\ *noun* [Late Latin *Septuaginta,* from Latin, seventy, irregular from *septem* seven + *-ginta* (akin to Latin *viginti* twenty); from the approximate number of its translators — more at SEVEN, VIGESIMAL] (1633)
: a Greek version of the Jewish Scriptures redacted in the 3d and 2d centuries B.C. by Jewish scholars and adopted by Greek-speaking Christians
— Sep·tu·a·gin·tal \(,)sep-,tü-ə-'jin-tᵊl, -,tyü-; ,sep-tə-wə-\ *adjective*

sep·tum \'sep-təm\ *noun, plural* **sep·ta** \-tə\ [New Latin, from Latin *saeptum* enclosure, fence, wall, from *saepire* to fence in, from *saepes* fence, hedge] (1698)
: a dividing wall or membrane especially between bodily spaces or masses of soft tissue — compare DISSEPIMENT

se·pul·chral \sə-'pəl-krəl *also* -'pu̇l-\ *adjective* (1615)
: suited to or suggestive of a sepulchre **:** FUNEREAL, MORTUARY
— se·pul·chral·ly \-krə-lē\ *adverb*

¹sep·ul·chre *or* **sep·ul·cher** \'se-pəl-kər\ *noun* [Middle English *sepulcre,* from Old French, from Latin *sepulcrum, sepulchrum,* from *sepelire* to bury; akin to Greek *hepein* to care for, Sanskrit *saparyati* he honors] (13th century)
1 : a place of burial **:** TOMB
2 : a receptacle for religious relics especially in an altar

²sepulchre *or* **sepulcher** *transitive verb* **-chred** *or* **-chered; -chring** *or* **-chering** \-k(ə-)riŋ\ (1591)
1 *archaic* **:** to place in or as if in a sepulchre **:** BURY
2 *archaic* **:** to serve as a sepulchre for

sep·ul·ture \'se-pəl-,chu̇r\ *noun* [Middle English, from Old French, from Latin *sepultura,* from *sepultus,* past participle of *sepelire*] (14th century)
1 : BURIAL
2 : SEPULCHRE

se·qua·cious \si-'kwā-shəs\ *adjective* [Latin *sequac-, sequax* inclined to follow, from *sequi*] (1643)
1 *archaic* **:** SUBSERVIENT, TRACTABLE
2 : intellectually servile
— se·qua·cious·ly *adverb*
— se·quac·i·ty \-'kwa-sə-tē\ *noun*

se·quel \'sē-kwəl *also* -,kwel\ *noun* [Middle English, from Middle French *sequelle,* from Latin *sequela,* from *sequi* to follow — more at SUE] (15th century)
1 : CONSEQUENCE, RESULT
2 a : subsequent development **b :** the next installment (as of a speech or story); *especially* **:** a literary or cinematic work continuing the course of a story begun in a preceding one

se·quela \si-'kwe-lə\ *noun, plural* **se·quel·ae** \-'kwe-(,)lē\ [New Latin, from Latin, sequel] (circa 1793)
1 : an aftereffect of disease or injury
2 : a secondary result

¹se·quence \'sē-kwən(t)s, -,kwen(t)s\ *noun* [Middle English, from Medieval Latin *sequentia,* from Late Latin, sequel, literally, act of following, from Latin *sequent-, sequens,* present participle of *sequi*] (14th century)
1 : a hymn in irregular meter between the gradual and Gospel in masses for special occasions (as Easter)
2 : a continuous or connected series: as **a :** an extended series of poems united by a single theme ⟨a sonnet *sequence*⟩ **b :** three or more playing cards usually of the same suit in consecutive order of rank **c :** a succession of repetitions of a melodic phrase or harmonic pattern each in a new position **d :** a set of elements ordered so that they can be labeled with the positive integers **e** (1) **:** a succession of related shots or scenes developing a single subject or phase of a film story (2) **:** EPISODE
3 a : order of succession **b :** an arrangement of the tenses of successive verbs in a sentence designed to express a coherent relationship especially between main and subordinate parts
4 a : CONSEQUENCE, RESULT **b :** a subsequent development
5 : continuity of progression

²sequence *transitive verb* **se·quenced; se·quenc·ing** (1941)
1 : to arrange in a sequence
2 : to determine the sequence of chemical constituents (as amino-acid residues) in ⟨*sequenced* biological macromolecules⟩

se·quenc·er \'sē-kwən(t)-sər, -,kwen(t)-\ *noun* (1949)
: one that sequences: as **a :** a device for arranging (as events in the ignition of a rocket) in a sequence **b :** a device for determining the order of occurrence of amino acids in a protein

se·quen·cy \'sē-kwən(t)-sē\ *noun* [Late Latin *sequentia*] (1818)
: SEQUENCE 3a, 5

se·quent \'sē-kwənt\ *adjective* [Latin *sequent-, sequens,* present participle] (1601)
1 : CONSECUTIVE, SUCCEEDING
2 : CONSEQUENT, RESULTANT

se·quen·tial \si-'kwen(t)-shəl\ *adjective* (1854)
1 : of, relating to, or arranged in a sequence **:** SERIAL ⟨*sequential* file systems⟩
2 : following in sequence
3 : relating to or based on a method of testing a statistical hypothesis that involves examination of a sequence of samples for each of which the decision is made to accept or reject the hypothesis or to continue sampling
— se·quen·tial·ly \-'kwen(t)-sh(ə-)lē\ *adverb*

¹se·ques·ter \si-'kwes-tər\ *transitive verb* **-tered; -ter·ing** \-t(ə-)riŋ\ [Middle English *sequestren,* from Middle French *sequestrer,* from Latin *sequestrare* to hand over to a trustee, from *sequester* third party to whom disputed property is entrusted, agent, from *secus* beside, otherwise; akin to Latin *sequi* to follow] (14th century)
1 a : to set apart **:** SEGREGATE **b :** SECLUDE, WITHDRAW
2 a : to seize especially by a writ of sequestration **b :** to place (property) in custody especially in sequestration
3 : to hold (as a metallic ion) in solution usually by inclusion in an appropriate coordination complex

²sequester *noun* (1604)
obsolete **:** SEPARATION, ISOLATION

se·ques·trate \'sē-kwəs-,trāt, 'se-; si-'kwes-\ *transitive verb* **-trat·ed; -trat·ing** [Latin *sequestratus,* past participle of *sequestrare*] (15th century)
: SEQUESTER

se·ques·tra·tion \,sē-kwəs-'trā-shən, ,se-; (,)sē-,kwes-\ *noun* (15th century)
1 : the act of sequestering **:** the state of being sequestered
2 a : a legal writ authorizing a sheriff or commissioner to take into custody the property of a defendant who is in contempt until the orders of a court are complied with **b :** a deposit whereby a neutral depositary agrees to hold property in litigation and to restore it to the party to whom it is adjudged to belong

se·ques·trum \si-'kwes-trəm\ *noun, plural* **-trums** *also* **-tra** \-trə\ [New Latin, from Latin, legal sequestration, from *sequester*] (1831)
: a fragment of dead bone detached from adjoining sound bone

se·quin \'sē-kwən\ *noun* [French, from Italian *zecchino,* from *zecca* mint, from Arabic *sikkah* die, coin] (1617)
1 : an old gold coin of Italy and Turkey
2 : a small plate of shining metal or plastic used for ornamentation especially on clothing

se·quined *or* **se·quinned** \-kwənd\ *adjective* (1894)
: ornamented with or as if with sequins

se·qui·tur \'se-kwə-tər, -,tu̇r\ *noun* [Latin, it follows, 3d person singular present indicative of *sequi* to follow — more at SUE] (1836)
: the conclusion of an inference **:** CONSEQUENCE

se·quoia \si-'kwȯi-ə\ *noun* [New Latin, genus name, from *Sequoya* (George Guess)] (circa 1866)
: either of two huge coniferous California trees of the bald cypress family that may reach a height of over 300 feet (90 meters): **a :** BIG TREE **b :** REDWOOD 3a

sera *plural of* SERUM

se·rac \sə-'rak, sā-\ *noun* [French *sérac,* literally, a kind of white cheese, from Medieval Latin *seracium* whey, from Latin *serum* whey — more at SERUM] (1860)
: a pinnacle, sharp ridge, or block of ice among the crevasses of a glacier

se·ra·glio \sə-'ral-(,)yō, -'räl-\ *noun, plural* **-glios** [Italian *serraglio,* modification of Turkish *saray* palace] (1581)
1 : HAREM 1a
2 : a palace of a sultan

se·rai \sə-'rī\ *noun* [Turkish & Persian; Turkish *saray* mansion, palace, from Persian *sarāī* mansion, inn] (1609)
1 : CARAVANSARY
2 : SERAGLIO 2

ser·al \'sir-əl\ *adjective* (1916)
: of, relating to, or constituting an ecological sere

se·ra·pe \sə-'rä-pē, -'ra-\ *noun* [Mexican Spanish *sarape*] (1834)

: a colorful woolen shawl worn over the shoulders especially by Mexican men

serape

ser·aph \'ser-əf\ *noun, plural* **ser·a·phim** \-ə-ˌfim, -ˌfēm\ *or* **seraphs** [assumed singular of Hebrew *śĕrāphīm*] (1667)
: SERAPHIM 2

ser·a·phim \'ser-ə-ˌfim, -ˌfēm\ *noun plural* [Late Latin *seraphim,* plural, seraphs, from Hebrew *śĕrāphīm*] (12th century)
1 : an order of angels — see CELESTIAL HIERARCHY
2 *singular, plural* **seraphim** **:** one of the 6-winged angels standing in the presence of God
— **se·raph·ic** \sə-'ra-fik\ *adjective*
— **se·raph·i·cal·ly** \-fi-k(ə-)lē\ *adverb*

Se·ra·pis \sə-'rā-pəs\ *noun* [Latin, from Greek *Sarapis*]
: an Egyptian god combining attributes of Osiris and Apis and having a widespread cult in Ptolemaic Egypt and ancient Greece

Serb \'sərb\ *noun* [ultimately from Serbo-Croatian *srb*] (1860)
1 : a native or inhabitant of Serbia
2 : SERBIAN 2
— **Serb** *adjective*

Ser·bi·an \'sər-bē-ən\ *noun* (1848)
1 : SERB 1
2 a : the Serbo-Croatian language as spoken in Serbia **b :** a literary form of Serbo-Croatian using the Cyrillic alphabet
— **Serbian** *adjective*

Ser·bo–Cro·a·tian \ˌsər-(ˌ)bō-krō-'ā-shən\ *noun* (1883)
1 : a Slavic language spoken in Croatia, Bosnia and Herzegovina, Serbia, and Montenegro
2 : a person whose native language is Serbo-Croatian
— **Serbo–Croatian** *adjective*

[1]sere \'sir\ *adjective* [Middle English, from Old English *sēar* dry; akin to Old High German *sōrēn* to wither, Greek *hauos* dry, Lithuanian *sausas*] (before 12th century)
1 : being dried and withered
2 *archaic* **:** THREADBARE

[2]sere *noun* [Latin *series* series] (1916)
: a series of ecological communities formed in ecological succession

[1]ser·e·nade \ˌser-ə-'nād\ *noun* [French *sérénade,* from Italian *serenata,* from *sereno* clear, calm (of weather), from Latin *serenus* serene] (1649)
1 a : a complimentary vocal or instrumental performance; *especially* **:** one given outdoors at night for a woman being courted **b :** a work so performed
2 : an instrumental composition in several movements, written for a small ensemble, and midway between the suite and the symphony in style

[2]serenade *verb* **-nad·ed; -nad·ing** (1668)
intransitive verb
: to play a serenade
transitive verb
: to perform a serenade in honor of
— **ser·e·nad·er** *noun*

ser·e·na·ta \ˌser-ə-'nä-tə\ *noun* [Italian, serenade] (circa 1724)
: an 18th century secular cantata of a dramatic character usually composed in honor of an individual or event

ser·en·dip·i·tous \ˌser-ən-'di-pə-təs\ *adjective* (1943)
: obtained or characterized by serendipity ⟨*serendipitous* discoveries⟩
— **ser·en·dip·i·tous·ly** *adverb*

ser·en·dip·i·ty \-'di-pə-tē\ *noun* [from its possession by the heroes of the Persian fairy tale *The Three Princes of Serendip*] (1754)
: the faculty or phenomenon of finding valuable or agreeable things not sought for

[1]se·rene \sə-'rēn\ *adjective* [Latin *serenus* clear, cloudless, untroubled] (15th century)
1 a : clear and free of storms or unpleasant change ⟨*serene* skies⟩ **b :** shining bright and steady ⟨the moon, *serene* in glory —Alexander Pope⟩
2 : AUGUST — used as part of a title ⟨His *Serene* Highness⟩
3 : marked by or suggestive of utter calm and unruffled repose or quietude ⟨a *serene* smile⟩
synonym see CALM
— **se·rene·ly** *adverb*
— **se·rene·ness** \-'rēn-nəs\ *noun*

[2]serene *noun* (1644)
1 *archaic* **:** a serene condition or expanse (as of sky, sea, or light)
2 *archaic* **:** SERENITY, TRANQUILLITY

se·ren·i·ty \sə-'re-nə-tē\ *noun* (15th century)
: the quality or state of being serene

serf \'sərf\ *noun* [French, from Old French, from Latin *servus* slave] (1611)
: a member of a servile feudal class bound to the soil and subject to the will of his lord
— **serf·age** \'sər-fij\ *noun*
— **serf·dom** \'sərf-dəm, -təm\ *noun*

serge \'sərj\ *noun* [Middle English *sarge,* from Middle French, from (assumed) Vulgar Latin *sarica,* alteration of Latin *serica,* feminine of *sericus* silken — more at SERICEOUS] (14th century)
: a durable twilled fabric having a smooth clear face and a pronounced diagonal rib on the front and the back

ser·gean·cy \'sär-jən(t)-sē\ *noun* (circa 1670)
: the function, office, or rank of a sergeant

ser·geant \'sär-jənt\ *noun* [Middle English, servant, attendant, sergeant, from Middle French *sergent, serjant,* from Latin *servient-, serviens,* present participle of *servire* to serve] (13th century)
1 : SERGEANT AT ARMS
2 *obsolete* **:** an officer who enforces the judgments of a court or the commands of one in authority
3 : a noncommissioned officer ranking in the army and marine corps above a corporal and below a staff sergeant and in the air force above an airman first class or senior airman and below a staff sergeant; *broadly* **:** NONCOMMISSIONED OFFICER
4 : an officer in a police force ranking in the U.S. just below captain or sometimes lieutenant and in England just below inspector

sergeant at arms (14th century)
: an officer of an organization (as a legislative body or court of law) who preserves order and executes commands

sergeant first class *noun* (1948)
: a noncommissioned officer in the army ranking above a staff sergeant and below a master sergeant

sergeant fish *noun* (1873)
1 : COBIA
2 : SNOOK 1

sergeant major *noun, plural* **sergeants major** *or* **sergeant majors** (1802)
1 : a noncommissioned officer in the army, air force, or marine corps serving as chief administrative assistant in a headquarters
2 : a noncommissioned officer in the marine corps ranking above a first sergeant
3 : a bluish green to yellow bony fish (*Abudefduf saxatilis* of the family Pomacentridae) with black vertical stripes on the sides that is widely distributed in the western tropical Atlantic Ocean

sergeant major of the army (1966)
: the ranking noncommissioned officer of the army serving as adviser to the chief of staff

sergeant major of the marine corps (circa 1971)
: the ranking noncommissioned officer of the marine corps serving as adviser to the commandant

ser·geanty \'sär-jən-tē\ *noun, plural* **-geant·ies** [Middle English *sergeantie,* from Middle French *sergentie,* from *sergent* sergeant] (15th century)
: any of numerous feudal services of a personal nature by which an estate is held of the king or other lord distinct from military tenure and from socage tenure

serg·ing \'sər-jiŋ\ *noun* [*serge*] (circa 1909)
: the process of overcasting the raw edges of a piece of fabric (as a carpet) to prevent raveling

[1]se·ri·al \'sir-ē-əl\ *adjective* (1840)
1 : of, relating to, consisting of, or arranged in a series, rank, or row ⟨*serial* order⟩
2 : appearing in successive parts or numbers ⟨a *serial* story⟩
3 : belonging to a series maturing periodically rather than on a single date ⟨*serial* bonds⟩
4 : of, relating to, or being music based on a series of tones in a chosen pattern without regard for traditional tonality
5 a : effecting a series of similar acts over a period of time ⟨a *serial* killer⟩ **b :** occurring in such a series ⟨a *serial* murder⟩
6 : relating to or being a connection in a computer system in which the bits of a byte are transmitted sequentially over a single wire — compare PARALLEL
— **se·ri·al·ly** \-ə-lē\ *adverb*

[2]serial *noun* (1846)
1 a : a work appearing (as in a magazine or on television) in parts at intervals **b :** one part of a serial work **:** INSTALLMENT
2 : a publication (as a newspaper or journal) issued as one of a consecutively numbered and indefinitely continued series

se·ri·al·ise *British variant of* SERIALIZE

se·ri·al·ism \'sir-ē-ə-ˌli-zəm\ *noun* (1958)
: serial music; *also* **:** the theory or practice of composing serial music

se·ri·al·ist \-list\ *noun* (1846)
1 : a writer of serials
2 : a composer of serial music

se·ri·al·ize \-ˌlīz\ *transitive verb* **-ized; -iz·ing** (1857)
: to arrange or publish in serial form
— **se·ri·al·i·za·tion** \ˌsir-ē-ə-lə-'zā-shən\ *noun*

serial number *noun* (1896)
: a number indicating place in a series and used as a means of identification

[1]se·ri·ate \'sir-ē-ˌāt, -ē-ət\ *adjective* [Latin *series*] (1846)
: arranged in a series or succession
— **se·ri·ate·ly** *adverb*

[2]se·ri·ate \'sir-ē-ˌāt\ *transitive verb* **-at·ed; -at·ing** (circa 1899)
: to arrange in a series

[1]se·ri·a·tim \ˌsir-ē-'ā-təm, -'a-\ *adverb* [Medieval Latin, from Latin *series*] (1680)
: in a series

[2]seriatim *adjective* (1871)
: following seriatim

se·ri·ceous \sə-'ri-shəs\ *adjective* [Late Latin *sericeus* silken, from Latin *sericum* silk garment, silk, from neuter of *sericus* silken, from Greek *sērikos,* from *Sēres,* an eastern Asian people, probably the Chinese] (circa 1777)
: covered with fine silky hair ⟨*sericeous* leaf⟩

ser·i·cin \'ser-ə-sən\ *noun* [International Scientific Vocabulary, from Latin *sericum* silk] (circa 1868)
: a gelatinous protein that cements the two fibroin filaments in a silk fiber

seri·cul·ture \'ser-ə-ˌkəl-chər\ *noun* [Latin *sericum* silk + English *culture*] (circa 1854)
: the production of raw silk by raising silkworms
— **seri·cul·tur·al** \-'kəl-chə-rəl, -'kəlch-rəl\ *adjective*

\ə\ **a**but \ᵊ\ kitt**e**n \ər\ f**ur**th**er** \a\ **a**sh \ā\ **a**ce \ä\ m**o**p, m**ar** \au̇\ **ou**t \ch\ **ch**in \e\ b**e**t \ē\ **ea**sy \g\ **g**o \i\ h**i**t \ī\ **i**ce \j\ **j**ob \ŋ\ si**ng** \ō\ g**o** \ȯ\ l**aw** \ȯi\ b**oy** \th\ **th**in \t͟h\ **th**e \ü\ l**oo**t \u̇\ f**oo**t \y\ **y**et \zh\ vi**si**on *see also* Guide to Pronunciation

— **seri•cul•tur•ist** \-rist\ *noun*

se•ries \'sir-(,)ēz\ *noun, plural* **series** *often attributive* [Latin, from *serere* to join, link together; akin to Greek *eirein* to string together, *hormos* chain, necklace, and perhaps to Latin *sort-, sors* lot] (1611)
1 a : a number of things or events of the same class coming one after another in spatial or temporal succession ⟨a concert *series*⟩ ⟨the hall opened into a *series* of small rooms⟩ **b :** a set of regularly presented television programs each of which is complete in itself
2 : the indicated sum of a usually infinite sequence of numbers
3 a : the coins or currency of a particular country and period **b :** a group of postage stamps in different denominations
4 : a succession of volumes or issues published with related subjects or authors, similar format and price, or continuous numbering
5 : a division of rock formations that is smaller than a system and comprises rocks deposited during an epoch
6 : a group of chemical compounds related in composition and structure
7 : an arrangement of the parts of or elements in an electric circuit whereby the whole current passes through each part or element without branching — compare PARALLEL
8 : a set of vowels connected by ablaut (as *i, a, u* in *ring, rang, rung*)
9 : a number of games (as of baseball) played usually on consecutive days between two teams ⟨in town for a 3-game *series*⟩
10 : a group of successive coordinate sentence elements joined together
11 : SOIL SERIES
12 : three consecutive games in bowling
— **in series :** in a serial arrangement

ser•if \'ser-əf\ *noun* [probably from Dutch *schreef* stroke, line, from Middle Dutch, from *schriven* to write, from Latin *scribere* — more at SCRIBE] (1841)
: any of the short lines stemming from and at an angle to the upper and lower ends of the strokes of a letter
— **ser•ifed** *or* **ser•iffed** \-əft\ *adjective*

ABC

1 serif

seri•graph \'ser-ə-,graf\ *noun* [Latin *seri*cum silk + Greek *graphein* to write, draw — more at CARVE] (1940)
: an original silk-screen color print
— **se•rig•ra•pher** \sə-'ri-grə-fər\ *noun*
— **se•rig•ra•phy** \-fē\ *noun*

se•rin \sə-'ra^{n}\ *noun* [French] (1530)
: a small yellow and grayish Old World finch (*Serinus serinus*) that is related to the canary

ser•ine \'ser-,ēn\ *noun* [International Scientific Vocabulary *seri*cin + [2]*-ine*] (1880)
: a crystalline nonessential amino acid $C_3H_7NO_3$ that occurs especially as a structural part of many proteins

se•rio•com•ic \,sir-ē-ō-'kä-mik\ *adjective* [*serious* + *-o-* + *comic*] (1783)
: having a mixture of the serious and the comic ⟨a *seriocomic* novel⟩
— **se•rio•com•i•cal•ly** \-mi-k(ə-)lē\ *adverb*

se•ri•ous \'sir-ē-əs\ *adjective* [Middle English *seryows,* from Middle French or Late Latin; Middle French *serieux,* from Late Latin *seriosus,* alteration of Latin *serius* weighty, serious; probably akin to Old English *swǣr* heavy, sad] (15th century)
1 : thoughtful or subdued in appearance or manner **:** SOBER
2 a : requiring much thought or work ⟨*serious* study⟩ **b :** of or relating to a matter of importance ⟨a *serious* play⟩
3 a : not joking or trifling **:** being in earnest **b** *archaic* **:** PIOUS **c :** deeply interested **:** DEVOTED ⟨a *serious* musician⟩
4 a : not easily answered or solved ⟨*serious* objections⟩ **b :** having important or dangerous possible consequences ⟨a *serious* injury⟩
5 : excessive or impressive in quantity, extent, or degree **:** CONSIDERABLE ⟨making *serious* money⟩ ⟨*serious* drinking⟩ ☆
— **se•ri•ous•ness** *noun*

se•ri•ous•ly *adverb* (1509)
1 : in a sincere manner **:** EARNESTLY
2 : to a serious extent **:** SEVERELY

se•ri•ous–mind•ed \,sir-ē-əs-'mīn-dəd\ *adjective* (1845)
: having a serious disposition or trend of thought
— **se•ri•ous–mind•ed•ly** *adverb*
— **se•ri•ous–mind•ed•ness** *noun*

ser•jeant, ser•jeanty *variant of* SERGEANT, SERGEANTY

ser•jeant–at–law \,sär-jənt-ət-'lȯ\ *noun, plural* **ser•jeants–at–law** (1503)
: a member of a former class of barristers of the highest rank

ser•mon \'sər-mən\ *noun* [Middle English, from Old French, from Medieval Latin *sermon-, sermo,* from Latin, speech, conversation, from *serere* to link together — more at SERIES] (13th century)
1 : a religious discourse delivered in public usually by a clergyman as a part of a worship service
2 : a speech on conduct or duty
— **ser•mon•ic** \,sər-'mä-nik\ *adjective*

ser•mon•ette \,sər-mə-'net\ *noun* (1814)
: a short sermon

ser•mon•ize \'sər-mə-,nīz\ *verb* **-ized; -iz•ing** (1635)
intransitive verb
1 : to compose or deliver a sermon
2 : to speak didactically or dogmatically
transitive verb
: to preach to or on at length
— **ser•mon•iz•er** *noun*

Sermon on the Mount
: an ethical discourse delivered by Jesus and recorded in Matthew 5–7 and paralleled briefly in Luke 6:20–49

sero- *combining form* [Latin *serum*]
: serum ⟨*sero*logy⟩

se•ro•con•ver•sion \,sir-ō-kən-'vər-zhən, -shən\ *noun* (1963)
: the production of antibodies in response to an antigen

se•ro•di•ag•no•sis \-,dī-ig-'nō-səs\ *noun* [New Latin] (1896)
: diagnosis by the use of serum (as in the Wassermann test)
— **se•ro•di•ag•nos•tic** \-'näs-tik\ *adjective*

se•rol•o•gy \sə-'rä-lə-jē, si-\ *noun* [International Scientific Vocabulary] (1909)
: a science dealing with serums and especially their reactions and properties
— **se•ro•log•i•cal** \,sir-ə-'lä-ji-kəl\ *or* **se•ro•log•ic** \-jik\ *adjective*
— **se•ro•log•i•cal•ly** \-ji-k(ə-)lē\ *adverb*
— **se•rol•o•gist** \sə-'rä-lə-jist, si-\ *noun*

se•ro•neg•a•tive \,sir-ō-'ne-gə-tiv\ *adjective* (1927)
: having or being a negative serum reaction especially in a test for the presence of an antibody
— **se•ro•neg•a•tiv•i•ty** \-,ne-gə-'ti-və-tē\ *noun*

se•ro•pos•i•tive \-'pä-zə-tiv, -'päz-tiv\ *adjective* (circa 1930)
: having or being a positive serum reaction especially in a test for the presence of an antibody
— **se•ro•pos•i•tiv•i•ty** \-,pä-zə-'ti-və-tē\ *noun*

se•ro•pu•ru•lent \-'pyu̇r-ə-lənt, ,ser-, -'pyu̇r-yə-\ *adjective* (circa 1836)
: consisting of a mixture of serum and pus ⟨a *seropurulent* exudate⟩

se•ro•sa \sə-'rō-zə\ *noun* [New Latin, from feminine of *serosus* serous, from Latin *serum*] (circa 1890)
: a usually enclosing serous membrane
— **se•ro•sal** \-zəl\ *adjective*

se•ro•ti•nal \sə-'rät-nəl, -'rä-t^{ə}n-əl; ,ser-ə-'tī-n^{ə}l\ *adjective* [Latin *serotinus* coming late] (1898)
: of or relating to the latter and usually drier part of summer

se•rot•i•nous \sə-'rät-nəs, -'rä-t^{ə}n-əs; ,ser-ə-'tī-nəs\ *adjective* [Latin *serotinus* coming late, from *sero* late — more at SOIREE] (circa 1656)
: remaining closed on the tree with seed dissemination delayed or occurring gradually ⟨*serotinous* cones⟩

se•ro•to•ner•gic \,sir-ə-tə-'nər-jik\ *also* **se•ro•to•nin•er•gic** \,sir-ə-,tō-nə-'nər-jik\ *adjective* [*serotonin* + *-ergic*] (1957)
: liberating, activated by, or involving serotonin in the transmission of nerve impulses ⟨*serotonergic* pathways⟩

se•ro•to•nin \,sir-ə-'tō-nən, ,ser-\ *noun* [*sero-* + *tonic* + [1]*-in*] (1948)
: a phenolic amine neurotransmitter $C_{10}H_{12}N_2O$ that is a powerful vasoconstrictor and is found especially in the brain, blood serum, and gastric mucosa of mammals

se•ro•type \'sir-ə-,tīp, 'ser-\ *noun* (1946)
: a group of intimately related microorganisms distinguished by a common set of antigens; *also* **:** the set of antigens characteristic of such a group

se•rous \'sir-əs\ *adjective* [Middle English, from Medieval Latin *serosus,* from Latin *serum*] (15th century)
: of, relating to, or resembling serum; *especially* **:** of thin watery constitution ⟨a *serous* exudate⟩

serous membrane *noun* (circa 1852)
: a thin membrane (as the peritoneum) with cells that secrete a serous fluid; *especially* **:** SEROSA

se•row \sə-'rō\ *noun* [Lepcha *să-ro* long-haired Tibetan goat] (1847)
: any of several artiodactyl mammals (genus *Capricornis*) of eastern Asia that are usually rather dark and heavily built and some of which have distinct manes

ser•pent \'sər-pənt\ *noun* [Middle English, from Middle French, from Latin *serpent-, serpens,* from present participle of *serpere* to creep; akin to Greek *herpein* to creep, Sanskrit *sarpati* he creeps] (13th century)
1 a *archaic* **:** a noxious creature that creeps, hisses, or stings **b :** SNAKE
2 : DEVIL 1
3 : a treacherous person

[1]**ser•pen•tine** \'sər-pən-,tēn, -,tīn\ *adjective* [Middle English, from Middle French *serpentin,* from Late Latin *serpentinus,* from Latin *serpent-, serpens*] (15th century)
1 : of or resembling a serpent (as in form or movement)
2 : subtly wily or tempting

☆ SYNONYMS
Serious, grave, solemn, sedate, staid, sober, earnest mean not light or frivolous. SERIOUS implies a concern for what really matters ⟨a *serious* play about social injustice⟩. GRAVE implies both seriousness and dignity in expression or attitude ⟨read the proclamation in a *grave* voice⟩. SOLEMN suggests an impressive gravity utterly free from levity ⟨a sad and *solemn* occasion⟩. SEDATE implies a composed and decorous seriousness ⟨remained *sedate* amid the commotion⟩. STAID suggests a settled, accustomed sedateness and prim self-restraint ⟨a quiet and *staid* community⟩. SOBER stresses seriousness of purpose and absence of levity or frivolity ⟨a *sober* look at the state of our schools⟩. EARNEST suggests sincerity or often zealousness of purpose ⟨an *earnest* reformer⟩.

3 a : winding or turning one way and another **b :** having a compound curve whose central curve is convex
— **ser·pen·tine·ly** *adverb*

²serpentine *noun* (1519)
: something that winds sinuously

³ser·pen·tine \-ˌtēn\ *noun* [Middle English, from Medieval Latin *serpentina, serpentinum,* from Late Latin, feminine & neuter of *serpentinus* resembling a serpent] (15th century)
: a mineral or rock consisting essentially of a hydrous magnesium silicate usually having a dull green color and often a mottled appearance

ser·pig·i·nous \(ˌ)sər-ˈpi-jə-nəs\ *adjective* [Middle English *serpiginose,* from Medieval Latin *serpiginosus,* from *serpigin-, serpigo* creeping skin disease, from Latin *serpere* to creep] (15th century)
: CREEPING, SPREADING; *especially* **:** healing over in one portion while continuing to advance in another ⟨*serpiginous* ulcer⟩
— **ser·pig·i·nous·ly** *adverb*

ser·ra·nid \sə-ˈra-nəd, ˈser-ə-nəd\ *noun* [ultimately from Latin *serra* saw] (circa 1900)
: any of a large family (Serranidae) of carnivorous marine bony fishes which have an oblong compressed body covered with usually ctenoid scales and many of which are important food and sport fishes (as the sea basses) especially of warm seas
— **serranid** *adjective*

ser·ra·no \sə-ˈrä-(ˌ)nō, si-\ *noun, plural* **-nos** [probably from Mexican Spanish, from Spanish *serrano,* adjective, montane, highland, from *sierra* mountain range — more at SIERRA] (circa 1972)
: a small Mexican hot pepper

¹ser·rate \ˈser-ˌāt, sə-ˈrāt\ *adjective* [Latin *serratus,* from *serra* saw] (1668)
: notched or toothed on the edge; *specifically* **:** having marginal teeth pointing forward or toward the apex ⟨a *serrate* leaf⟩

²ser·rate \sə-ˈrāt, ˈser-ˌāt\ *transitive verb* **ser·rat·ed; ser·rat·ing** [Late Latin *serratus,* past participle of *serrare* to saw, from Latin *serra*] (1750)
: to mark or make with serrations ⟨a *serrated* knife⟩

ser·ra·tion \sə-ˈrā-shən, se-\ *noun* (1842)
1 : the condition of being serrate
2 : a formation resembling the toothed edge of a saw
3 : one of the teeth in a serrate margin

ser·ried \ˈser-ēd\ *adjective* (1667)
1 : crowded or pressed together **:** COMPACT ⟨the crowd collected in a *serried* mass —W. S. Maugham⟩
2 [by alteration] **:** marked by ridges **:** SERRATE ⟨the *serried* contours of the . . . mountains —*American Guide Series: Oregon*⟩
— **ser·ried·ly** *adverb*
— **ser·ried·ness** *noun*

ser·ry \ˈser-ē\ *verb* **ser·ried; ser·ry·ing** [Middle French *serré,* past participle of *serrer* to press, crowd — more at SEAR] (1581)
intransitive verb
archaic **:** to press together especially in ranks
transitive verb
: to crowd together

Ser·to·li cell \ˌsər-ˈtō-lē-\ *noun* [Enrico *Sertoli* (died 1910) Italian histologist] (1888)
: one of the elongated striated cells lining the seminiferous tubules that support and apparently nourish the spermatids

Ser·to·man \(ˌ)sər-ˈtō-mən\ *noun* [*Sertoma* (Club)] (1956)
: a member of a major international service club

se·rum \ˈsir-əm\ *noun, plural* **serums** *or* **se·ra** \-ə\ [Latin, whey, wheylike fluid; akin to Greek *oros* whey] (1665)
1 : the watery portion of an animal fluid remaining after coagulation: **a** (1) **:** BLOOD SERUM (2) **:** ANTISERUM **b :** WHEY **c :** a normal or pathological serous fluid (as in a blister)
2 : the watery part of a plant fluid

serum albumin *noun* (1879)
: a crystallizable albumin or mixture of albumins that normally constitutes more than half of the protein in blood serum and serves to maintain the osmotic pressure of the blood

serum globulin *noun* (circa 1890)
: a globulin or mixture of globulins occurring in blood serum and containing most of the antibodies of the blood

serum hepatitis *noun* (1932)
: HEPATITIS B

serum sickness *noun* (circa 1913)
: an allergic reaction to the injection of foreign serum manifested by urticaria, swelling, eruption, arthritis, and fever

ser·val \ˈsər-vəl, (ˌ)sər-ˈval\ *noun* [French, from Portuguese *lobo cerval* lynx, from Medieval Latin *lupus cervalis,* literally, deerlike wolf] (1771)
: a long-legged African wildcat (*Felis serval*) having large ears and a tawny black-spotted coat

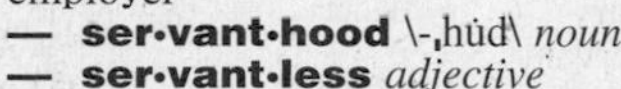
serval

ser·vant \ˈsər-vənt\ *noun* [Middle English, from Old French, from present participle of *servir*] (13th century)
: one that serves others ⟨a public *servant*⟩; *especially* **:** one that performs duties about the person or home of a master or personal employer
— **ser·vant·hood** \-ˌhu̇d\ *noun*
— **ser·vant·less** *adjective*

¹serve \ˈsərv\ *verb* **served; serv·ing** [Middle English, from Middle French *servir,* from Old French, from Latin *servire* to be a slave, serve, from *servus* slave, servant] (13th century)
intransitive verb
1 a : to be a servant **b :** to do military or naval service
2 : to assist a celebrant as server at mass
3 a : to be of use ⟨in a day when few people could write, seals *served* as signatures —Elizabeth W. King⟩ **b :** to be favorable, opportune, or convenient **c :** to be worthy of reliance or trust ⟨if memory *serves*⟩ **d :** to hold an office **:** discharge a duty or function ⟨*serve* on a jury⟩
4 : to prove adequate or satisfactory **:** SUFFICE
5 : to help persons to food: as **a :** to wait at table **b :** to set out portions of food or drink
6 : to wait on customers
7 : to put the ball or shuttlecock in play in various games (as tennis, volleyball, or badminton)
transitive verb
1 a : to be a servant to **:** ATTEND **b :** to give the service and respect due to (a superior) **c :** to comply with the commands or demands of **:** GRATIFY **d :** to give military or naval service to **e :** to perform the duties of (an office or post)
2 : to act as server at (mass)
3 *archaic* **:** to pay a lover's or suitor's court to (a lady) ⟨that gentle lady, whom I love and *serve* —Edmund Spenser⟩
4 a : to work through (a term of service) **b :** to put in (a term of imprisonment)
5 a : to wait on at table **b :** to bring (food) to a diner **c :** PRESENT, PROVIDE — usually used with *up* ⟨the novel *served* up many laughs⟩
6 a : to furnish or supply with something needed or desired **b :** to wait on (a customer) in a store **c :** to furnish professional service to
7 a : to answer the needs of **b :** to be enough for **:** SUFFICE **c :** to contribute or conduce to **:** PROMOTE
8 : to treat or act toward in a specified way ⟨he *served* me ill⟩
9 a : to bring to notice, deliver, or execute as required by law **b :** to make legal service upon (a person named in a process)
10 *of a male animal* **:** to copulate with
11 : to wind yarn or wire tightly around (a rope or stay) for protection
12 : to provide services that benefit or help
13 : to put (the ball or shuttlecock) in play (as in tennis or badminton)
— **serve one right :** to be deserved

²serve *noun* (1688)
: the act or action of putting the ball or shuttlecock in play in various games (as volleyball, badminton, or tennis); *also* **:** a turn to serve

serv·er \ˈsər-vər\ *noun* (15th century)
1 : one that serves food or drink
2 : the player who serves (as in tennis)
3 : something used in serving food or drink
4 : one that serves legal processes upon another
5 : the celebrant's assistant at low mass
6 : a computer in a network that is used to provide services (as access to files or shared peripherals or the routing of electronic mail) to other computers in the network

¹ser·vice \ˈsər-vəs\ *noun* [Middle English, from Middle French, from Latin *servitium* condition of a slave, body of slaves, from *servus* slave] (13th century)
1 a : the occupation or function of serving ⟨in active *service*⟩ **b :** employment as a servant ⟨entered his *service*⟩
2 a : the work performed by one that serves ⟨good *service*⟩ **b :** HELP, USE, BENEFIT ⟨glad to be of *service*⟩ **c :** contribution to the welfare of others **d :** disposal for use ⟨I'm entirely at your *service*⟩
3 a : a form followed in worship or in a religious ceremony ⟨the burial *service*⟩ **b :** a meeting for worship — often used in plural ⟨held evening *services*⟩
4 : the act of serving: as **a :** a helpful act ⟨did him a *service*⟩ **b :** useful labor that does not produce a tangible commodity — usually used in plural ⟨charge for professional *services*⟩ **c :** SERVE
5 : a set of articles for a particular use ⟨a silver tea *service*⟩
6 a : an administrative division (as of a government or business) ⟨the consular *service*⟩ **b :** one of a nation's military forces (as the army or navy)
7 a : a facility supplying some public demand ⟨telephone *service*⟩ ⟨bus *service*⟩ **b :** a facility providing maintenance and repair ⟨television *service*⟩
8 : the materials (as spun yarn, small lines, or canvas) used for serving a rope
9 : the act of bringing a legal writ, process, or summons to notice as prescribed by law
10 : the act of copulating with a female animal
11 : a branch of a hospital medical staff devoted to a particular specialty ⟨obstetrical *service*⟩

²service *transitive verb* **ser·viced; ser·vic·ing** (1528)
: to perform services for: as **a :** to repair or provide maintenance for **b :** to meet interest and sinking fund payments on (as government debt) **c :** to perform any of the business functions auxiliary to production or distribution of **d** *of a male animal* **:** SERVE 10
— **ser·vic·er** *noun*

³service *adjective* (1837)
1 : of or relating to the armed services
2 : used in serving or supplying ⟨delivery men use the *service* entrance⟩
3 : intended for hard or everyday use

4 a : providing services ⟨the *service* trades—from filling stations to universities —John Fischer⟩ **b :** offering repair, maintenance, or incidental services

[4]**ser·vice** \ˈsər-vəs\ *noun* [Middle English *serves,* plural of *serve* fruit of the service tree, service tree, from Old English *syrfe,* from (assumed) Vulgar Latin *sorbea,* from Latin *sorbus* service tree] (1530)
: an Old World tree (*Sorbus domestica*) resembling the related mountain ashes but having larger flowers and larger edible fruit; *also* **:** a related Old World tree (*S. torminalis*) with bitter fruits

ser·vice·able \ˈsər-və-sə-bəl\ *adjective* (14th century)
1 : HELPFUL, USEFUL
2 : fit for use ⟨her *serviceable* but not exceptional voice —Irving Kolodin⟩
— ser·vice·abil·i·ty \ˌsər-və-sə-ˈbi-lə-tē\ *noun*
— ser·vice·able·ness \ˈsər-və-sə-bəl-nəs\ *noun*
— ser·vice·ably \-blē\ *adverb*

ser·vice·ber·ry \ˈsər-vəs-ˌber-ē *also* ˈsär-\ *noun* [[4]*service*] (1784)
1 : the edible purple or red fruit of any of various North American trees or shrubs (genus *Amelanchier*) of the rose family
2 : a tree or shrub that produces serviceberries and has showy white flowers in the spring — called also *Juneberry, shadblow, shadbush*

service book *noun* (1580)
: a book setting forth forms of worship used in religious services

service box *noun* (circa 1898)
: the area in which a player stands while serving in various court games (as squash or handball)

service cap *noun* (circa 1908)
: a flat-topped visor cap worn as part of a military uniform — compare GARRISON CAP

service ceiling *noun* (1920)
: the altitude at which under standard air conditions a particular airplane can no longer rise at a rate greater than a small designated rate (as 100 feet per minute)

service charge *noun* (1917)
: a fee charged for a particular service often in addition to a standard or basic fee — called also *service fee*

service club *noun* (1926)
1 : a club of business or professional men or women organized for their common benefit and active in community service
2 : a recreation center for enlisted personnel provided by one of the armed services

service court *noun* (circa 1878)
: a part of the court into which the ball or shuttlecock must be served

service line *noun* (1875)
: a line marked on a court in various games (as handball or tennis) parallel to the front wall or to the net to mark a boundary of the service area or service court

ser·vice·man \ˈsər-vəs-ˌman, -mən\ *noun* (1899)
1 : a male member of the armed forces
2 : a man employed to repair or maintain equipment
3 : a service station attendant

service mark *noun* (1945)
: a mark or device used to identify a service (as transportation or insurance) offered to customers

service medal *noun* (circa 1934)
: a medal awarded to an individual for military service in a specified war or campaign

service module *noun* (1961)
: a space vehicle module that contains oxygen, water, fuel cells, propellant tanks, and the main rocket engine

service road *noun* (1921)
: FRONTAGE ROAD

service station *noun* (1916)
1 : a retail station for servicing motor vehicles especially with gasoline and oil
2 : a place at which some service is offered

service stripe *noun* (circa 1920)
: a stripe worn on an enlisted man's left sleeve to indicate three years of service in the army or four years in the navy

service tree *noun* (1600)
: [4]SERVICE

ser·vice·wom·an \ˈsər-vəs-ˌwu̇-mən\ *noun* (1943)
: a female member of the armed forces

ser·vi·ette \ˌsər-vē-ˈet\ *noun* [French, from Middle French, from *servir* to serve] (1818)
chiefly British **:** a table napkin

ser·vile \ˈsər-vəl, -ˌvīl\ *adjective* [Middle English, from Latin *servilis,* from *servus* slave] (15th century)
1 : of or befitting a slave or a menial position
2 : meanly or cravenly submissive **:** ABJECT
synonym see SUBSERVIENT
— ser·vile·ly \-və(l)-lē, -ˌvīl-lē\ *adverb*
— ser·vile·ness \-vəl-nəs, -ˌvīl-\ *noun*
— ser·vil·i·ty \(ˌ)sər-ˈvi-lə-tē\ *noun*

serv·ing \ˈsər-viŋ\ *noun* (1864)
: a helping of food or drink

Ser·vite \ˈsər-ˌvīt\ *noun* [Medieval Latin *Servitae,* plural, Servites, from Latin *servus*] (circa 1550)
: a member of the mendicant Order of Servants of Mary founded in Florence, Italy, in 1233
— Servite *adjective*

ser·vi·tor \ˈsər-və-tər, -ˌtȯr\ *noun* [Middle English *servitour,* from Middle French, from Late Latin *servitor,* from Latin *servire* to serve] (14th century)
: a male servant

ser·vi·tude \ˈsər-və-ˌtüd, -ˌtyüd\ *noun* [Middle English, from Middle French, from Latin *servitudo* slavery, from *servus* slave] (15th century)
1 : a condition in which one lacks liberty especially to determine one's course of action or way of life
2 : a right by which something (as a piece of land) owned by one person is subject to a specified use or enjoyment by another

ser·vo \ˈsər-(ˌ)vō\ *noun, plural* **servos** (1947)
1 : SERVOMOTOR
2 : SERVOMECHANISM

ser·vo·mech·a·nism \ˈsər-vō-ˌme-kə-ˌni-zəm\ *noun* [*servo-* (as in *servomotor*) + *mechanism*] (1926)
: an automatic device for controlling large amounts of power by means of very small amounts of power and automatically correcting the performance of a mechanism

ser·vo·mo·tor \ˈsər-vō-ˌmō-tər\ *noun* [French *servo-moteur,* from Latin *servus* slave, servant + French *-o-* + *moteur* motor, from Latin *motor* one that moves — more at MOTOR] (1889)
: a power-driven mechanism that supplements a primary control operated by a comparatively feeble force (as in a servomechanism)

-ses *plural of* -SIS

ses·a·me \ˈse-sə-mē *also* ˈse-zə-\ *noun* [alteration of earlier *sesam, sesama,* from Latin *sesamum, sesama,* from Greek *sēsamon, sēsamē,* of Semitic origin; akin to Akkadian *šamaššamu* sesame] (1682)
1 : a widely cultivated chiefly tropical or subtropical annual erect herb (*Sesamum indicum* of the family Pedaliaceae); *also* **:** its small seeds used especially as a source of oil and a flavoring agent
2 : OPEN SESAME

sesame 1

sesame oil *noun* (1870)
: a pale yellow bland semidrying fatty oil obtained from sesame seeds and used chiefly as an edible oil, as a vehicle for various pharmaceuticals, and in cosmetics and soaps

ses·a·moid \ˈse-sə-ˌmȯid\ *noun* [Greek *sēsamoeidēs,* literally, resembling sesame seed, from *sēsamon*] (circa 1696)
: a nodular mass of bone (as the patella) or cartilage in a tendon especially at a joint or bony prominence
— sesamoid *adjective*

Se·so·tho \se-ˈsō-(ˌ)thō, -(ˌ)tō\ *noun* [Sesotho] (1846)
: the Bantu language of the Basotho people

sesqui- *combining form* [Latin, one and a half, half again, literally, and a half, from *semis* half of an as, one half (probably from *semi-* + *as* as) + *-que* (enclitic) and; akin to Greek *te* and, Sanskrit *ca,* Gothic *-h, -uh*]
1 : one and a half times ⟨*sesqui*centennial⟩
2 : containing half again as many atoms ⟨*sesqui*terpene⟩
3 : intermediate **:** combination ⟨*sesqui*carbonate⟩

ses·qui·car·bon·ate \ˌses-kwi-ˈkär-bə-ˌnāt, -nət\ *noun* (1825)
: a salt that is neither a simple normal carbonate nor a simple bicarbonate but often (as $Na_2CO_3 \cdot NaHCO_3 \cdot 2H_2O$) a combination of the two

ses·qui·cen·te·na·ry \-kwi-sen-ˈte-nə-rē, -ˈsen-tᵊn-ˌer-ē, -sen-ˈtē-nə-rē\ *noun* (1954)
: SESQUICENTENNIAL

ses·qui·cen·ten·ni·al \-sen-ˈte-nē-əl\ *noun* (1880)
: a 150th anniversary or its celebration
— sesquicentennial *adjective*

ses·qui·pe·da·lian \ˌses-kwə-pə-ˈdāl-yən\ *adjective* [Latin *sesquipedalis,* literally, a foot and a half long, from *sesqui-* + *ped-, pes* foot — more at FOOT] (1656)
1 : having many syllables **:** LONG ⟨*sesquipedalian* terms⟩
2 : given to or characterized by the use of long words ⟨a *sesquipedalian* television commentator⟩

ses·qui·ter·pene \ˌses-kwə-ˈtər-ˌpēn\ *noun* (circa 1888)
: any of a class of terpenes $C_{15}H_{24}$; *also* **:** a derivative of such a terpene

ses·sile \ˈse-ˌsīl, -səl\ *adjective* [Latin *sessilis* of or fit for sitting, low, dwarf (of plants), from *sessus,* past participle of *sedēre*] (1753)
1 : attached directly by the base **:** not raised upon a stalk or peduncle ⟨a *sessile* leaf⟩ ⟨*sessile* bubbles⟩
2 : permanently attached or established **:** not free to move about ⟨*sessile* polyps⟩

ses·sion \ˈse-shən\ *noun* [Middle English, from Middle French, from Latin *session-, sessio,* literally, act of sitting, from *sedēre* to sit — more at SIT] (14th century)
1 : a meeting or series of meetings of a body (as a court or legislature) for the transaction of business ⟨morning *session*⟩
2 *plural* **a** (1) **:** a sitting of English justices of peace in execution of the powers conferred by their commissions (2) **:** an English court holding such sessions **b :** any of various courts similar to the English sessions
3 : the period between the first meeting of a legislative or judicial body and the prorogation or final adjournment
4 : the ruling body of a Presbyterian congregation consisting of the elders in active service
5 : the period during the year or day in which a school conducts classes
6 : a meeting or period devoted to a particular activity ⟨a recording *session*⟩
— ses·sion·al \ˈsesh-nəl, ˈse-shə-nᵊl\ *adjective*

session man *noun* (1958)
: a studio musician who backs up a performer at a recording session

ses·terce \ˈses-ˌtərs\ *noun* [Latin *sestertius,* from *sestertius* two and a half times as great (from its being equal originally to two and a

half asses), from *semis* half of an as, one half + *tertius* third — more at SESQUI-, THIRD] (1598)
: an ancient Roman coin equal to ¼ denarius

ses·ter·tium \se-ˈstər-sh(ē-)əm\ *noun, plural* **-tia** \-sh(ē-)ə\ [Latin, from genitive plural of *sestertius* (in the phrase *milia sestertium* thousands of sesterces)] (circa 1541)
: a unit of value in ancient Rome equal to 1000 sesterces

ses·tet \ses-ˈtet\ *noun* [Italian *sestetto,* from *sesto* sixth, from Latin *sextus* — more at SEXT] (1859)
: a stanza or a poem of six lines; *specifically* **:** the last six lines of an Italian sonnet

ses·ti·na \se-ˈstē-nə\ *noun* [Italian, from *sesto* sixth] (circa 1586)
: a lyrical fixed form consisting of six 6-line usually unrhymed stanzas in which the end words of the first stanza recur as end words of the following five stanzas in a successively rotating order and as the middle and end words of the three verses of the concluding tercet

[1]set \ˈset\ *verb* **set; set·ting** [Middle English *setten,* from Old English *settan;* akin to Old High German *sezzen* to set, Old English *sittan* to sit] (before 12th century)
transitive verb
1 : to cause to sit **:** place in or on a seat
2 a : to put (a fowl) on eggs to hatch them **b :** to put (eggs) for hatching under a fowl or into an incubator
3 : to place (oneself) in position to start running in a race
4 a : to place with care or deliberate purpose and with relative stability ⟨*set* a ladder against the wall⟩ ⟨*set* a stone on the grave⟩ **b :** TRANSPLANT 1 ⟨*set* seedlings⟩ **c** (1) **:** to make (as a trap) ready to catch prey (2) **:** to fix (a hook) firmly into the jaw of a fish **d :** to put aside (as dough containing yeast) for fermenting
5 : to direct with fixed attention ⟨*set* your mind to it⟩
6 a : to cause to assume a specified condition, relation, or occupation ⟨slaves were *set* free⟩ ⟨*set* the house on fire⟩ **b :** to cause the start of ⟨*set* a fire⟩
7 a : to appoint or assign to an office or duty **b :** POST, STATION
8 : to cause to assume a specified posture or position ⟨*set* the door ajar⟩
9 a : to fix as a distinguishing imprint, sign, or appearance ⟨the years have *set* their mark on him⟩ **b :** AFFIX **c :** APPLY ⟨*set* a match to kindling⟩
10 : to fix or decide on as a time, limit, or regulation **:** PRESCRIBE ⟨*set* a wedding day⟩ ⟨*set* the rules for the game⟩
11 a : to establish as the highest level or best performance ⟨*set* a record for the half mile⟩ **b :** to furnish as a pattern or model ⟨*set* an example of generosity⟩ **c :** to allot as a task ⟨*setting* lessons for the children to work upon at home —*Manchester Examiner*⟩
12 a : to adjust (a device and especially a measuring device) to a desired position ⟨*set* the alarm for 7:00⟩ ⟨*set* a thermostat at 68⟩; *also* **:** to adjust (as a clock) in conformity with a standard **b :** to restore to normal position or connection when dislocated or fractured ⟨*set* a broken bone⟩ **c :** to spread to the wind ⟨*set* the sails⟩
13 a : to put in order for use ⟨*set* a place for a guest⟩ **b :** to make scenically ready for a performance ⟨*set* the stage⟩ **c** (1) **:** to arrange (type) for printing ⟨*set* type by hand⟩ (2) **:** to put into type or its equivalent (as on film) ⟨*set* the first word in italic⟩
14 a : to put a fine edge on by grinding or honing ⟨*set* a razor⟩ **b :** to bend slightly the tooth points of (a saw) alternately in opposite directions **c :** to sink (the head of a nail) below the surface
15 : to fix in a desired position (as by heating or stretching)
16 : to arrange (hair) in a desired style by using implements (as curlers, rollers, or clips) and gels or lotions
17 a : to adorn with something affixed or infixed **:** STUD, DOT ⟨clear sky *set* with stars⟩ **b :** to fix (as a precious stone) in a border of metal **:** place in a setting
18 a : to hold something in regard or esteem at the rate of ⟨*sets* a great deal by daily exercise⟩ **b :** to place in a relative rank or category ⟨*set* duty before pleasure⟩ **c :** to fix at a certain amount ⟨*set* bail at $500⟩ **d :** VALUE, RATE ⟨their promises were *set* at naught⟩ **e :** to place as an estimate of worth ⟨*set* a high value on life⟩
19 : to place in relation for comparison or balance ⟨theory *set* against practice⟩
20 a : to direct to action **b :** to incite to attack or antagonism ⟨war *sets* brother against brother⟩
21 a : to place by transporting ⟨was *set* ashore on the island⟩ **b :** to put in motion **c :** to put and fix in a direction ⟨*set* our faces toward home once more⟩ **d** *of a dog* **:** to point out the position of (game) by holding a fixed attitude
22 : to defeat (an opponent or a contract) in bridge
23 a : to fix firmly **:** make immobile **:** give rigid form or condition to ⟨*set* her jaw in determination⟩ **b :** to make unyielding or obstinate
24 : to cause to become firm or solid ⟨*set* milk for cheese⟩
25 : to cause (as fruit) to develop
intransitive verb
1 *chiefly dialect* **:** SIT
2 : to be becoming **:** be suitable **:** FIT ⟨the coat *sets* well⟩
3 : to cover and warm eggs to hatch them
4 a : to become lodged or fixed ⟨the pudding *set* heavily on my stomach⟩ **b :** to place oneself in position in preparation for an action (as running)
5 *of a plant part* **:** to undergo development usually as a result of pollination
6 a : to pass below the horizon **:** go down ⟨the sun *sets*⟩ **b :** to come to an end ⟨this century *sets* with little mirth —Thomas Fuller⟩
7 : to apply oneself to some activity ⟨*set* to work⟩
8 : to have a specified direction in motion **:** FLOW, TEND ⟨the wind was *setting* from Pine Hill to the farm —Esther Forbes⟩
9 *of a dog* **:** to indicate the position of game by crouching or pointing
10 : to dance face to face with another in a square dance ⟨*set* to your partner and turn⟩
11 a : to become solid or thickened by chemical or physical alteration ⟨the cement *sets* rapidly⟩ **b** *of a dye or color* **:** to become permanent **c** *of a bone* **:** to become whole by knitting **d** *of metal* **:** to acquire a permanent twist or bend from strain ▪
— set about : to begin to do
— set apart 1 : to reserve to a particular use **2 :** to make noticeable or outstanding
— set aside 1 : to put to one side **:** DISCARD **2 :** to reserve for a purpose **:** SAVE **3 :** DISMISS **4 :** ANNUL, OVERRULE
— set at : to mount an attack on **:** ASSAIL ⟨would go although . . . devils should *set at* me —Charlotte Yonge⟩
— set eyes on : to catch sight of
— set foot in : ENTER
— set foot on : to step onto
— set forth 1 : to give an account or statement of **2 :** to start out on a journey
— set forward 1 : FURTHER **2 :** to start out on a journey
— set in motion : to give impulse to ⟨*sets* the story *in motion* vividly —Howard Thompson⟩
— set one's hand to : to become engaged in
— set one's heart on : RESOLVE *verb transitive* 5 ⟨she *set her heart on* going to medical school⟩
— set one's house in order : to organize one's affairs
— set one's sights on : to determine to pursue
— set one's teeth on edge : IRRITATE, ANNOY
— set one straight : to correct someone by providing accurate information
— set sail : to start out on a course; *especially* **:** to begin a voyage ⟨*set sail* for Bermuda⟩
— set store by *or* **set store on :** to consider valuable, trustworthy, or worthwhile
— set the stage : to provide the basis or background ⟨this trend will *set the stage* for higher earnings⟩
— set to music : to provide music or instrumental accompaniment for (a text)
— set upon : to attack usually with violence ⟨the dogs *set upon* the trespassers⟩

[2]set *noun* (14th century)
1 a : the act or action of setting **b :** the condition of being set
2 : a number of things of the same kind that belong or are used together ⟨an electric train *set*⟩
3 a : mental inclination, tendency, or habit **:** BENT ⟨a *set* toward mathematics⟩ **b :** a state of psychological preparedness usually of limited duration for action in response to an anticipated stimulus or situation ⟨the influence of mental *set* on the effect experienced with marijuana⟩
4 : direction of flow ⟨the *set* of the wind⟩
5 : form or carriage of the body or of its parts ⟨her face took on a cynical *set* —Raymond Kennedy⟩
6 : the manner of fitting or of being placed or suspended ⟨in order to give the skirt a pretty *set* —Mary J. Howell⟩
7 : amount of deflection from a straight line ⟨*set* of a saw's teeth⟩

□ USAGE
set *Set* has been used intransitively in the sense of "sit" since about the 14th century. The practice went unnoticed until Samuel Johnson observed in his 1755 dictionary that is was "commonly used in conversation" and "though undoubtedly barbarous" was "sometimes found in authors." Attempts to correct the usage have been a fixture of schoolbooks ever since. Noah Webster did not include Johnson's definition in his 1828 dictionary probably because he had earlier disapproved the use in one of his school grammars. Generations of teachers and critics have succeeded in chastening the use of most literary and learned writers; these writers have, however, used it freely to establish a character whose language use is not especially fastidious ⟨he never done nothing for three months but *set* in his back yard and learn that frog to jump —Mark Twain⟩. It turns up occasionally in general prose, but not as often as intransitive *lay* does. It still is used in familiar speech, but appears to be somewhat more marked socially than *lay* is. There is apparently a folk belief that *sit* is for people and *set* for things; such a belief may account for the "ore *setting* on the docks at Chicago" from a steel company spokesman in one of our citations. You should remember that the basic objection to intransitive *set* is in its literal meaning; it shares with *sit* some figurative meanings that draw no criticism ⟨grammatically, the first example does not *set* well —Morton S. Freeman⟩.

8 : permanent change of form (as of metal) due to repeated or excessive stress
9 : the act or result of arranging hair by curling or waving
10 *also* **sett** \'set\ **a :** a young plant or rooted cutting ready for transplanting **b :** a small bulb, corm, or tuber or a piece of tuber used for propagation ⟨onion *sets*⟩
11 *or* **sett :** the burrow of a badger
12 : the width of the body of a piece of type
13 : an artificial setting for a scene of a theatrical or film production
14 *also* **sett :** a rectangular paving stone of sandstone or granite
15 : a division of a tennis match won by the side that wins at least six games beating the opponent by two games or by winning a tie-breaker
16 : a collection of books or periodicals forming a unit
17 : a clutch of eggs
18 : the basic formation in a country-dance or square dance
19 : a session of music (as jazz or dance music) usually followed by an intermission; *also* **:** the music played at one session
20 : a group of persons associated by common interests
21 : a collection of elements and especially mathematical ones (as numbers or points) — called also *class*
22 : an apparatus of electronic components assembled so as to function as a unit ⟨a television *set*⟩
23 : a usually offensive formation in football or basketball
24 : a group of a specific number of repetitions of a particular exercise

³set *adjective* [Middle English *sett,* from Old English *gesett,* past participle of *settan*] (14th century)
1 : INTENT, DETERMINED ⟨*set* upon going⟩
2 : INTENTIONAL, PREMEDITATED ⟨did it of *set* purpose⟩
3 : fixed by authority or appointment **:** PRESCRIBED, SPECIFIED ⟨*set* hours of study⟩
4 : reluctant to change ⟨*set* in their ways⟩
5 a : IMMOVABLE, RIGID ⟨*set* frown⟩ **b :** BUILT-IN ⟨a *set* tub⟩
6 : SETTLED, PERSISTENT ⟨*set* defiance⟩
7 : being in readiness **:** PREPARED ⟨*set* for an early morning start⟩

se·ta \'sē-tə\ *noun, plural* **se·tae** \'sē-ˌtē\ [New Latin, from Latin *saeta, seta* bristle] (circa 1793)
: a slender usually rigid or bristly and springy organ or part of an animal or plant
— se·tal \'sē-tᵊl\ *adjective*

se·ta·ceous \si-'tā-shəs\ *adjective* [Latin *saeta, seta*] (1664)
1 : set with or consisting of bristles
2 : resembling a bristle in form or texture

set–aside \'set-ə-ˌsīd\ *noun* (1943)
: something (as a portion of receipts or production) that is set aside for a specified purpose

set·back \'set-ˌbak\ *noun* (1674)
1 : a checking of progress
2 : DEFEAT, REVERSE
3 : ⁴PITCH 7
4 : a placing of the face of a building on a line some distance to the rear of the building line or of the wall below; *also* **:** the rooftop area produced by a setback
5 : automatic scheduled adjustment to a lower temperature setting of a thermostat

set back *transitive verb* (1600)
1 : to slow the progress of **:** HINDER, DELAY
2 : COST ⟨a new suit will *set* you *back* $200⟩

set by *transitive verb* (1595)
: to set apart for future use

set down *transitive verb* (15th century)
1 : to cause to sit down **:** SEAT
2 : to place at rest on a surface or on the ground
3 : to suspend (a jockey) from racing
4 : to cause or allow to get off a vehicle **:** DELIVER
5 : to land (an airplane) on the ground or water
6 a : ORDAIN, ESTABLISH **b :** to put in writing
7 a : REGARD, CONSIDER ⟨*set* him *down* as a liar⟩ **b :** ATTRIBUTE

se te·nant \sə-'te-nənt, sē-; ˌse-tə-'näⁿ\ *adjective* [French, literally, holding one another] (circa 1911)
of postage stamps **:** joined together as in the original sheet but differing in design, overprint, color, or perforation

Seth \'seth\ *noun* [Hebrew *Shēth*]
: a son of Adam and Eve

¹set–in \'set-'in\ *adjective* (1534)
1 : placed, located, or built as a part of some other construction ⟨a *set-in* bookcase⟩ ⟨a *set-in* washbasin⟩
2 : cut separately and stitched in ⟨*set-in* sleeves⟩

²set–in \'set-ˌin\ *noun* (1953)
: INSERT

set in (15th century)
transitive verb
: INSERT; *especially* **:** to stitch (a small part) within a large article ⟨*set in* a sleeve of a dress⟩
intransitive verb
: to become established

set·line \'set-ˌlīn\ *noun* (1865)
: a long heavy fishing line to which several hooks are attached in series

set·off \'set-ˌȯf\ *noun* (1621)
1 : something that is set off against another thing: **a :** DECORATION, ORNAMENT **b :** COMPENSATION, COUNTERBALANCE
2 : the discharge of a debt by setting against it a distinct claim in favor of the debtor; *also* **:** the claim itself
3 : OFFSET 7a

set off (1596)
transitive verb
1 a : to put in relief **:** show up by contrast **b :** ADORN, EMBELLISH **c :** to set apart **:** make distinct or outstanding
2 a : OFFSET, COMPENSATE ⟨more variety in the Lancashire weather to *set off* its most disagreeable phases —*Geographical Journal*⟩ **b :** to make a setoff of ⟨the respective totals shall be *set off* against one another —O. R. Hobson⟩
3 a : to set in motion **:** cause to begin **b :** to cause to explode
4 : to measure off on a surface
intransitive verb
: to start out on a course or a journey ⟨*set off* for home⟩

set on (14th century)
transitive verb
1 : ATTACK
2 a *obsolete* **:** PROMOTE **b :** to urge (as a dog) to attack or pursue **c :** to incite to action **:** INSTIGATE **d :** to set to work
intransitive verb
: GO ON, ADVANCE

se·tose \'sē-ˌtōs\ *adjective* [Latin *saetosus,* from *saeta*] (1661)
: SETACEOUS, BRISTLY

set·out \'set-ˌau̇t\ *noun* (1537)
1 a (1) **:** ARRAY, DISPLAY (2) **:** ARRANGEMENT, LAYOUT **b :** BUFFET, SPREAD **c :** TURNOUT 5
2 : PARTY, ENTERTAINMENT
3 : BEGINNING, OUTSET

set out (14th century)
transitive verb
1 a : to arrange and present graphically or systematically **b :** to mark out (as a design) **:** lay out the plan of
2 : to state, describe, or recite at length ⟨distributed copies of a pamphlet *setting out* his ideas in full —S. F. Mason⟩
3 : to begin with a definite purpose **:** INTEND, UNDERTAKE
intransitive verb
: to start out on a course, a journey, or a career

set piece *noun* (circa 1909)
1 : a realistic piece of stage scenery standing by itself
2 a : a composition (as in literature, art, or music) executed in a fixed or ideal form often with studied artistry and brilliant effect **b :** a scene, depiction, speech, or event that is obviously designed to have an imposing effect
3 : a precisely planned and conducted military operation
— set–piece *adjective*

set point *noun* (1928)
: a situation (as in tennis) in which one player will win the set by winning the next point; *also* **:** the point won

set·screw \'set-ˌskrü\ *noun* (circa 1855)
1 : a screw screwed through one part tightly upon or into another part to prevent relative movement
2 : a screw for regulating a valve opening or a spring tension

set shot *noun* (1937)
: a two-handed shot in basketball taken from a stationary position

sett *variant of* SET

set·tee \se-'tē\ *noun* [alteration of *settle*] (1716)
1 : a long seat with a back
2 : a medium-sized sofa with arms and a back

set·ter \'se-tər\ *noun* (15th century)
1 : one that sets
2 : a large bird dog of a type trained to point on finding game

set theory *noun* (1936)
: a branch of mathematics or of symbolic logic that deals with the nature and relations of sets
— set theoretic *adjective*

set·ting \'se-tiŋ\ *noun* (14th century)
1 : the manner, position, or direction in which something is set
2 : the frame or bed in which a gem is set; *also* **:** style of mounting
3 a : the time, place, and circumstances in which something occurs or develops **b :** the time and place of the action of a literary, dramatic, or cinematic work **c :** the scenery used in a theatrical or film production
4 : the music composed for a text (as a poem)
5 : the articles of tableware for setting a place at table ⟨two *settings* of sterling silver⟩
6 : a batch of eggs for incubation
synonym see BACKGROUND

setting circle *noun* (circa 1891)
: a graduated scale or wheel on the mounting of an equatorial telescope for indicating right ascension or declination

setting–up exercise *noun* (circa 1900)
: any of a series of gymnastic exercises used to give an erect carriage, supple muscles, and easy control of the limbs

¹set·tle \'se-tᵊl\ *verb* **set·tled; set·tling** \'set-liŋ, 'se-tᵊl-iŋ\ [Middle English, to seat, bring to rest, come to rest, from Old English *setlan,* from *setl* seat] (1515)
transitive verb
1 : to place so as to stay
2 a : to establish in residence **b :** to furnish with inhabitants **:** COLONIZE
3 a : to cause to pack down **b :** to clarify by causing dregs or impurities to sink
4 : to make quiet or orderly
5 a : to fix or resolve conclusively ⟨*settle* the question⟩ **b :** to establish or secure permanently ⟨*settle* the order of royal succession⟩ **c :** to conclude (a lawsuit) by agreement between parties usually out of court **d :** to close (as an account) by payment often of less than is due
6 : to arrange in a desired position
7 : to make or arrange for final disposition of ⟨*settled* his affairs⟩
8 *of an animal* **:** IMPREGNATE
intransitive verb
1 : to come to rest
2 a : to sink gradually or to the bottom **b :** to become clear by the deposit of sediment or scum **c :** to become compact by sinking

3 a : to become fixed, resolved, or established ⟨a cold *settled* in his chest⟩ **b :** to establish a residence or colony ⟨*settled* in Wisconsin⟩ — often used with *down*
4 a : to become quiet or orderly **b :** to take up an ordered or stable life — often used with *down* ⟨marry and *settle* down⟩
5 a : to adjust differences or accounts **b :** to come to a decision — used with *on* or *upon* ⟨*settleed* on a new plan⟩ **c :** to conclude a lawsuit by agreement out of court
6 *of an animal* **:** CONCEIVE
synonym see DECIDE
— set·tle·able \ˈse-tᵊl-ə-bəl, ˈset-lə-bəl\ *adjective*
— settle for : to be content with
— settle one's hash : to silence or subdue someone by decisive action
— settle the stomach : to remove or relieve the distress or nausea of indigestion

²set·tle *noun* [Middle English, place for sitting, seat, chair, from Old English *setl;* akin to Old High German *sezzal* seat, Latin *sella* seat, chair, Old English *sittan* to sit] (1553)
: a wooden bench with arms, a high solid back, and an enclosed foundation which can be used as a chest

settle

set·tle·ment \ˈse-tᵊl-mənt\ *noun* (1648)
1 : the act or process of settling
2 a : an act of bestowing or giving possession under legal sanction **b :** the sum, estate, or income secured to one by such a settlement
3 a : occupation by settlers **b :** a place or region newly settled **c :** a small village
4 : SETTLEMENT HOUSE
5 : an agreement composing differences
6 : payment or adjustment of an account

settlement house *noun* (1907)
: an institution providing various community services especially to large city populations

set·tler \ˈset-lər, ˈse-tᵊl-ər\ *noun* (1696)
: one that settles (as a new region)

set·tling \ˈset-liŋ, ˈse-tᵊl-iŋ\ *noun* (1594)
: SEDIMENT, DREGS — usually used in plural

set·tlor \ˈset-ˌlȯr, ˈse-tᵊl-ˌȯr\ *noun* (1818)
: one that makes a settlement or creates a trust of property

set–to \ˈset-ˌtü\ *noun, plural* **set–tos** (1743)
: a usually brief and vigorous fight or debate

set to *intransitive verb* (circa 1525)
1 : to begin actively and earnestly
2 : to begin fighting

set·up \ˈset-ˌəp\ *noun* (1890)
1 a : carriage of the body; *especially* **:** erect and soldierly bearing **b :** CONSTITUTION, MAKEUP
2 a : the assembly and arrangement of the tools and apparatus required for the performance of an operation **b :** the preparation and adjustment of machines for an assigned task
3 a : a table setting **b :** glass, ice, and mixer served to patrons who supply their own liquor
4 a : a camera position from which a scene is filmed; *also* **:** the footage taken from one camera position **b :** the final arrangement of the scenery and properties for a scene of a theatrical or cinematic production
5 a : a position of the balls in billiards or pool from which it is easy to score **b :** a task or contest purposely made easy **c :** something easy to get or accomplish **d :** something (as a plot) that has been constructed or contrived **e :** the execution of a planned scoring play in sports
6 a : the manner in which the elements or components of a machine, apparatus, or mechanical, electrical, or hydraulic system are arranged, designed, or assembled **b :** the patterns within which political, social, or administrative forces operate **:** customary or established practice
7 : PROJECT, PLAN
8 : something done by deceit or trickery in order to compromise or frame someone

set up (13th century)
transitive verb
1 a : to raise to and place in a high position **b :** to place in view **:** POST **c :** to put forward (as a plan) for acceptance
2 a : to place upright **:** ERECT ⟨*set up* a statue⟩ **b :** to assemble the parts of and erect in position ⟨*set up* a printing press⟩ **c :** to put (a machine) in readiness or adjustment for a tooling operation
3 a : CAUSE, CREATE ⟨*set up* a clamor⟩ **b :** BRING ABOUT
4 : to place in power or in office ⟨*set up* the general as dictator⟩
5 a : to raise from depression **:** ELATE, GRATIFY **b :** to make proud or vain
6 a : to put forward or extol as a model **b :** to claim oneself to be ⟨*sets* himself *up* as an authority⟩
7 : FOUND, INAUGURATE ⟨*set up* a home for orphans⟩
8 a : to provide with means of making a living ⟨*set* him *up* in business⟩ **b :** to bring or restore to normal health **c :** to cause (one) to take on a soldierly or athletic appearance especially through drill
9 : to erect (a perpendicular or a figure) on a base in a drawing
10 a : to make taut (a stay or hawser) **b :** to tighten firmly
11 : to make carefully worked out plans for ⟨*set up* a bank robbery⟩
12 a : to pay for (drinks) **b :** to treat (someone) to something
13 a : to put in a compromising or dangerous position usually by trickery or deceit **b :** FRAME 3
14 : to execute one or more plays in preparation for scoring
intransitive verb
1 : to come into active operation or use
2 : to begin business
3 : to make pretensions ⟨has never *set up* to be a wise man —Thomas Rogers⟩
4 : to become firm or consolidated **:** HARDEN
— set up housekeeping : to establish one's living quarters
— set up shop : to establish one's business

sev·en \ˈse-vən, ˈse-bᵊm\ *noun* [Middle English, from *seven,* adjective, from Old English *seofon;* akin to Old High German *sibun* seven, Latin *septem,* Greek *hepta*] (before 12th century)
1 — see NUMBER table
2 : the seventh in a set or series ⟨the *seven* of diamonds⟩
3 : something having seven units or members
— seven *adjective*
— seven *pronoun, plural in construction*

sev·en·fold \-ˌfōld\ *adjective* (before 12th century)
1 : having seven units or members
2 : being seven times as great or as many
— sevenfold *adverb*

seven seas *noun plural* (1872)
: all the waters or oceans of the world

sev·en·teen \ˌse-vən-ˈtēn, ˌse-bᵊm-\ *noun* [*seventeen,* adjective, from Middle English *seventene,* from Old English *seofontēne;* akin to Old English *tīen* ten] (14th century)
— see NUMBER table
— seventeen *adjective*
— seventeen *pronoun, plural in construction*
— sev·en·teenth \-ˈtēn(t)th\ *adjective or noun*

seventeen–year locust *noun* (1817)
: a cicada (*Magicicada septendecim*) of the U.S. that has in the North a life of seventeen years and in the South of thirteen years of which most is spent underground as a nymph and only a few weeks as a winged adult

sev·enth \ˈse-vən(t)th, ˈse-bᵊm(t)th\ *noun, plural* **sevenths** \ˈse-vən(t)s, -vən(t)ths; ˈse-bᵊm(t)s, -bᵊm(t)ths\ (12th century)
1 — see NUMBER table
2 a : a musical interval embracing seven diatonic degrees **b :** a tone at this interval; *specifically* **:** LEADING TONE **c :** the harmonic combination of two tones a seventh apart
— seventh *adjective or adverb*

seventh chord *noun* (circa 1909)
: a chord comprising a fundamental tone with its third, fifth, and seventh

Seventh–Day *adjective* (1684)
: advocating or practicing observance of Saturday as the Sabbath

seventh heaven *noun* [from the seventh being the highest of the seven heavens of Islamic and cabalist doctrine] (1818)
: a state of extreme joy

sev·en·ty \ˈse-vən-tē, ˈse-bᵊm-, -dē\ *noun, plural* **-ties** [*seventy,* adjective, from Middle English, from Old English *seofontig,* short for *hundseofontig,* from *hundseofontig,* noun, group of seventy, from *hund* hundred + *seofon* seven + *-tig* group of ten; akin to Old English *tīen* ten] (13th century)
1 — see NUMBER table
2 *plural* **:** the numbers 70 to 79; *specifically* **:** the years 70 to 79 in a lifetime or century
3 *capitalized* **:** a Mormon elder ordained for missionary work under the apostles
— sev·en·ti·eth \-tē-əth, -dē-\ *adjective or noun*
— seventy *adjective*
— seventy *pronoun, plural in construction*

sev·en·ty–eight \ˌse-vən-tē-ˈāt, ˌse-bᵊm-, -dē-ˈāt\ *noun* (circa 1934)
1 — see NUMBER table
2 : a phonograph record designed to be played at 78 revolutions per minute — usually written 78
— seventy–eight *adjective*
— seventy–eight *pronoun, plural in construction*

sev·en–up \ˌse-və-ˈnəp, ˌse-bᵊm-ˈəp\ *noun* (1830)
: an American variety of all fours in which a total of seven points constitutes game

sev·er \ˈse-vər\ *verb* **sev·ered; sev·er·ing** \ˈsev-riŋ, ˈse-və-\ [Middle English, from Middle French *severer,* from Latin *separare* — more at SEPARATE] (14th century)
transitive verb
: to put or keep apart **:** DIVIDE; *especially* **:** to remove (as a part) by or as if by cutting
intransitive verb
: to become separated
synonym see SEPARATE

sev·er·able \ˈsev-rə-bəl, ˈse-və-\ *adjective* (1548)
: capable of being severed; *especially* **:** capable of being divided into legally independent rights or obligations
— sev·er·abil·i·ty \ˌsev-rə-ˈbi-lə-tē, ˌse-və-\ *noun*

¹sev·er·al \ˈsev-rəl, ˈse-və-\ *adjective* [Middle English, from Anglo-French, from Medieval Latin *separalis,* from Latin *separ* separate, back-formation from *separare* to separate] (15th century)
1 a : separate or distinct from one another ⟨federal union of the *several* states⟩ **b** (1) **:** individually owned or controlled **:** EXCLUSIVE ⟨a *several* fishery⟩ — compare COMMON (2) **:** of or relating separately to each individual involved ⟨a *several* judgment⟩ **c :** being separate and distinctive **:** RESPECTIVE ⟨specialists in their *several* fields⟩
2 a : more than one ⟨*several* pleas⟩ **b :** more than two but fewer than many ⟨moved *several*

inches⟩ **c** *chiefly dialect* **:** being a great many

²several *pronoun, plural in construction* (1686)
: an indefinite number more than two and fewer than many ⟨*several* of the guests⟩

sev·er·al·fold \ˌsev-rəl-ˈfōld, ˌse-və-\ *adjective* (1738)
1 : having several parts or aspects
2 : being several times as large, as great, or as many as some understood size, degree, or amount
— severalfold *adverb*

sev·er·al·ly \ˈsev-rə-lē, ˈse-və-\ *adverb* (14th century)
1 : one at a time **:** each by itself **:** SEPARATELY
2 : apart from others **:** INDEPENDENTLY

sev·er·al·ty \ˈsev-rəl-tē, ˈse-və-\ *noun* [Middle English *severalte,* from Anglo-French *several-té,* from *several*] (15th century)
1 : the quality or state of being several **:** DISTINCTNESS, SEPARATENESS
2 a : a sole, separate, and exclusive possession, dominion, or ownership **:** one's own right without a joint interest in any other person ⟨tenants in *severalty*⟩ **b :** the quality or state of being individual or particular
3 a : land owned in severalty **b :** the quality or state of being held in severalty

sev·er·ance \ˈsev-rən(t)s, ˈse-və-\ *noun* (15th century)
: the act or process of severing **:** the state of being severed

severance pay *noun* (1943)
: an allowance usually based on length of service that is payable to an employee on termination of employment

severance tax *noun* (1928)
: a tax levied by a state on the extractor of oil, gas, or minerals intended for consumption in other states — compare ROYALTY 5a

se·vere \sə-ˈvir\ *adjective* **se·ver·er; -est** [Middle French or Latin; Middle French, from Latin *severus*] (1548)
1 a : strict in judgment, discipline, or government **b :** of a strict or stern bearing or manner **:** AUSTERE
2 : rigorous in restraint, punishment, or requirement **:** STRINGENT, RESTRICTIVE
3 : strongly critical or condemnatory **:** CENSORIOUS ⟨a *severe* critic⟩
4 a : maintaining a scrupulously exacting standard of behavior or self-discipline **b :** establishing exacting standards of accuracy and integrity in intellectual processes ⟨a *severe* logician⟩
5 : sober or restrained in decoration or manner **:** PLAIN
6 a : inflicting physical discomfort or hardship **:** HARSH ⟨*severe* winters⟩ **b :** inflicting pain or distress **:** GRIEVOUS ⟨a *severe* wound⟩
7 : requiring great effort **:** ARDUOUS ⟨a *severe* test⟩
8 : of a great degree **:** SERIOUS ⟨*severe* depression⟩ ☆
— se·vere·ly *adverb*
— se·vere·ness *noun*
— se·ver·i·ty \sə-ˈver-ə-tē\ *noun*

severe combined immunodeficiency *noun* (1974)
: a rare congenital disorder of the immune system that is characterized by inability to produce a normal complement of antibodies and T cells and that usually results in early death — called also *severe combined immune deficiency*

se·vi·che \sə-ˈvē-(ˌ)chā, -chē\ *noun* [American Spanish] (1952)
: a dish of raw fish marinated in lime or lemon juice often with oil, onions, peppers, and seasonings and served especially as an appetizer

Sevres *or* **Sèvres** \ˈsev-rə, ˈsev(rᵊ)\ *noun* [*Sèvres,* France] (1786)
: an often elaborately decorated French porcelain

sev·ru·ga \sə-ˈvrü-gə, se-\ *noun* [Russian *sevryuga,* a species of sturgeon] (1591)
: a light to dark gray caviar from a sturgeon (*Acipenser sevru*) of the Caspian Sea that has very small roe; *also* **:** the fish

sew \ˈsō\ *verb* **sewed; sewn** \ˈsōn\ *or* **sewed; sew·ing** [Middle English, from Old English *sīwian;* akin to Old High German *siuwen* to sew, Latin *suere*] (before 12th century)
transitive verb
1 : to unite or fasten by stitches
2 : to close or enclose by sewing ⟨*sew* the money in a bag⟩
intransitive verb
: to practice or engage in sewing
— sew·abil·i·ty \ˌsō-ə-ˈbi-lə-tē\ *noun*
— sew·able \ˈsō-ə-bəl\ *adjective*

sew·age \ˈsü-ij\ *noun* [³*sewer*] (1834)
: refuse liquids or waste matter carried off by sewers

¹sew·er \ˈsü-ər, ˈsu̇(-ə)r\ *noun* [Middle English, from Anglo-French *asseour,* literally, seater, from Old French *asseoir* to seat — more at ASSIZE] (14th century)
: a medieval household officer often of high rank in charge of serving the dishes at table and sometimes of seating and tasting

²sew·er \ˈsō(-ə)r\ *noun* (14th century)
: one that sews

³sew·er \ˈsü-ər, ˈsu̇(-ə)r\ *noun* [Middle English, from Middle French *esseweur, seweur,* from *essewer* to drain, from (assumed) Vulgar Latin *exaquare,* from Latin *ex-* + *aqua* water — more at ISLAND] (15th century)
: an artificial usually subterranean conduit to carry off sewage and sometimes surface water (as from rainfall)

sew·er·age \ˈsü-ə-rij, ˈsu̇(-ə)r-ij\ *noun* (1834)
1 : the removal and disposal of sewage and surface water by sewers
2 : a system of sewers
3 : SEWAGE

sew·ing \ˈsō-iŋ\ *noun* (14th century)
1 : the act, method, or occupation of one that sews
2 : material that has been or is to be sewed

sew up *transitive verb* (15th century)
1 : to mend completely by sewing
2 : to get exclusive use or control of
3 : to make certain of **:** ASSURE

¹sex \ˈseks\ *noun* [Middle English, from Latin *sexus*] (14th century)
1 : either of the two major forms of individuals that occur in many species and that are distinguished respectively as female or male
2 : the sum of the structural, functional, and behavioral characteristics of living things that are involved in reproduction by two interacting parents and that distinguish males and females
3 a : sexually motivated phenomena or behavior **b :** SEXUAL INTERCOURSE
4 : GENITALIA

²sex *transitive verb* (1884)
1 : to identify the sex of ⟨*sex* chicks⟩
2 a : to increase the sexual appeal of — often used with *up* **b :** to arouse the sexual desires of

sex·a·ge·nar·i·an \ˌsek-sə-jə-ˈner-ē-ən\ *noun* [Latin *sexagenarius* of or containing sixty, sixty years old, from *sexageni* sixty each, from *sexaginta* sixty, from *sex* six + *-ginta* (akin to Latin *viginti* twenty) — more at SIX, VIGESIMAL] (1738)
: a person whose age is in the sixties
— sexagenarian *adjective*

¹sex·a·ges·i·mal \-ˈje-sə-məl\ *adjective* [Latin *sexagesimus* sixtieth, from *sexaginta* sixty] (1685)
: of, relating to, or based on the number 60

²sexagesimal *noun* (1685)
: a sexagesimal fraction

sex appeal *noun* (1924)
1 : personal appeal or physical attractiveness for members of the opposite sex
2 : general attractiveness

sex cell *noun* (1889)
: GAMETE; *also* **:** its cellular precursor

sex chromatin *noun* (1952)
: BARR BODY

sex chromosome *noun* (1906)
: a chromosome that is inherited differently in the two sexes, that is concerned directly with the inheritance of sex, and that is the seat of factors governing the inheritance of various sex-linked and sex-limited characters

sex·de·cil·lion \ˌseks-di-ˈsil-yən\ *noun, often attributive* [Latin *sedecim, sexdecim* sixteen (from *sex* six + *decem* ten) + English *-illion* (as in *million*) — more at TEN] (circa 1934)
— see NUMBER table

sexed \ˈsekst\ *adjective* (circa 1891)
1 : having sex or sexual instincts
2 : having sex appeal

sex gland *noun* (1935)
: GONAD

sex hormone *noun* (1917)
: a hormone (as from the gonads or adrenal cortex) that affects the growth or function of the reproductive organs or the development of secondary sex characteristics

sex·ism \ˈsek-ˌsi-zəm\ *noun* [¹*sex* + *-ism* (as in *racism*)] (1968)
1 : prejudice or discrimination based on sex; *especially* **:** discrimination against women
2 : behavior, conditions, or attitudes that foster stereotypes of social roles based on sex
— sex·ist \ˈsek-sist\ *adjective or noun*

sex kitten *noun* (1958)
: a young woman with conspicuous sex appeal

sex·less \ˈseks-ləs\ *adjective* (1598)
1 : lacking sex **:** NEUTER
2 : devoid of sexual interest or activity ⟨a *sexless* relationship⟩
— sex·less·ly *adverb*
— sex·less·ness *noun*

sex–lim·it·ed \ˈseks-ˈli-mə-təd\ *adjective* (1923)
: expressed in the phenotype of only one sex

sex–link·age \-ˌliŋ-kij\ *noun* (1912)
: the quality or state of being sex-linked

sex–linked \-ˌliŋ(k)t\ *adjective* (1912)
1 : located on a sex chromosome ⟨a *sex-linked* gene⟩
2 : mediated by a sex-linked gene ⟨a *sex-linked* character⟩

sex object *noun* (1911)
: a person regarded especially exclusively as an object of sexual interest

sex·ol·o·gy \sek-ˈsä-lə-jē\ *noun* (1902)
: the study of sex or of the interaction of the sexes especially among human beings
— sex·ol·o·gist \-jist\ *noun*

sex·ploi·ta·tion \ˌseks-ˌplȯi-ˈtā-shən\ *noun* [blend of *sex* and *exploitation*] (circa 1942)
: the exploitation of sex in the media and especially in film

sex·pot \ˈseks-ˌpät\ *noun* (1948)
: a conspicuously sexy woman

sex symbol *noun* (circa 1911)
: a usually renowned person (as an entertainer) noted and admired for conspicuous sex appeal

sext \ˈsekst\ *noun, often capitalized* [Middle English *sexte,* from Late Latin *sexta,* from

☆ SYNONYMS

Severe, stern, austere, ascetic mean given to or marked by strict discipline and firm restraint. SEVERE implies standards enforced without indulgence or laxity and may suggest harshness ⟨*severe* military discipline⟩. STERN stresses inflexibility and inexorability of temper or character ⟨*stern* arbiters of public morality⟩. AUSTERE stresses absence of warmth, color, or feeling and may apply to rigorous restraint, simplicity, or self-denial ⟨living an *austere* life in the country⟩. ASCETIC implies abstention from pleasure and comfort or self-indulgence as spiritual discipline ⟨the *ascetic* life of the monastic orders⟩.

Latin, sixth hour of the day, from feminine of *sextus* sixth, from *sex* six] (15th century)
: the fourth of the canonical hours

Sex·tans \'seks-ˌtanz\ *noun* [New Latin (genitive *Sextantis*), literally, sextant]
: a constellation on the equator south of Leo

sex·tant \'seks-tənt\ *noun* [New Latin *sextant-, sextans* sixth part of a circle, from Latin, sixth part, from *sextus* sixth] (1628)
: an instrument for measuring angular distances used especially in navigation to observe altitudes of celestial bodies (as in ascertaining latitude and longitude)

sextant

sex·tet \seks-'tet\ *noun* [alteration of *sestet*] (1841)
1 : a musical composition for six instruments or voices
2 : a group or set of six: as **a :** the performers of a sextet **b :** a hockey team

sex·til·lion \seks-'til-yən\ *noun, often attributive* [French, irregular from *sex-* (from Latin *sex*) + *-illion* (as in *million*)] (1690)
— see NUMBER table

sex·to \'seks-(ˌ)tō\ *noun, plural* **sextos** [Latin *sexto,* ablative of *sextus* sixth] (1847)
: SIXMO

sex·to·dec·i·mo \ˌseks-tə-'de-sə-ˌmō\ *noun, plural* **-mos** [Latin, ablative of *sextus decimus* sixteenth, from *sextus* sixth + *decimus* tenth — more at DIME] (1688)
: SIXTEENMO

sex·ton \'seks-tən\ *noun* [Middle English *secresteyn, sexteyn,* from Middle French *secrestain,* from Medieval Latin *sacristanus* — more at SACRISTAN] (14th century)
: a church officer or employee who takes care of the church property and performs related minor duties (as ringing the bell for services and digging graves)

¹sex·tu·ple \seks-'tü-pəl, -'tyü-, -'tə-; 'seks-tə-\ *adjective* [probably from Medieval Latin *sextuplus,* from Latin *sextus* sixth + *-plus* multiplied by; akin to Latin *-plex* -plex — more at -FOLD] (1626)
1 : having six units or members
2 : being six times as great or as many
3 : marked by six beats per measure of music ⟨*sextuple* time⟩
— sextuple *noun*

²sextuple *verb* **sex·tu·pled; sex·tu·pling** \-p(ə-)liŋ\ (1632)
transitive verb
: to make six times as much or as many
intransitive verb
: to become six times as much or as numerous

sex·tu·plet \seks-'tə-plət, -'tü-; -'tyü-; 'seks-t(y)ə-\ *noun* (1852)
1 : a combination of six of a kind
2 : one of six offspring born at one birth
3 : a group of six equal musical notes performed in the time ordinarily given to four of the same value

¹sex·tu·pli·cate \seks-'tü-pli-kət, -'tyü-\ *adjective* [blend of *sextuple* and *-plicate* (as in *duplicate*)] (1657)
1 : repeated six times
2 : SIXTH ⟨file the *sextuplicate* copy⟩
— sextuplicate *noun*

²sex·tu·pli·cate \-plə-ˌkāt\ *transitive verb* **-cat·ed; -cat·ing** (circa 1934)
1 : SEXTUPLE
2 : to provide in sextuplicate

sex·u·al \'sek-sh(ə-)wəl, 'sek-shəl\ *adjective* [Late Latin *sexualis,* from Latin *sexus* sex] (1651)
1 : of, relating to, or associated with sex or the sexes ⟨*sexual* differentiation⟩ ⟨*sexual* conflict⟩
2 : having or involving sex ⟨*sexual* reproduction⟩
— sex·u·al·ly \'sek-sh(ə-)wə-lē, 'sek-sh(ə-)lē\ *adverb*

sexual generation *noun* (1880)
: the generation of an organism with alternation of generations that reproduces sexually

sexual harassment *noun* (1975)
: uninvited and unwelcome verbal or physical conduct directed at an employee because of his or her sex

sexual intercourse *noun* (1799)
1 : heterosexual intercourse involving penetration of the vagina by the penis **:** COITUS
2 : intercourse involving genital contact between individuals other than penetration of the vagina by the penis

sex·u·al·i·ty \ˌsek-shə-'wa-lə-tē\ *noun* (circa 1800)
: the quality or state of being sexual: **a :** the condition of having sex **b :** sexual activity **c :** expression of sexual receptivity or interest especially when excessive

sex·u·al·ize \'sek-sh(ə-)wə-ˌlīz, 'sek-shə-ˌlīz\ *transitive verb* **-ized; -iz·ing** (1839)
: to make sexual **:** endow with a sexual character or cast

sexual relations *noun plural* (1950)
: COITUS

sexual selection *noun* (1859)
: natural selection for characters that confer success in competition for a mate as distinguished from competition with other species; *also* **:** the choice of a mate based on a preference for certain characteristics (as color or bird song)

sexy \'sek-sē\ *adjective* **sex·i·er; -est** (1925)
1 : sexually suggestive or stimulating **:** EROTIC
2 : generally attractive or interesting **:** APPEALING ⟨a *sexy* stock⟩
— sex·i·ly \-sə-lē\ *adverb*
— sex·i·ness \-sē-nəs\ *noun*

Sey·fert galaxy \'sē-fərt *also* 'sī-\ *noun* [Carl K. *Seyfert* (died 1960) American astronomer] (1959)
: any of a class of spiral galaxies that have small compact bright nuclei characterized by variability in light intensity, emission of radio waves, and spectra which indicate hot gases in rapid motion — called also *Seyfert*

sfer·ics \'sfir-iks, 'sfer-\ *noun plural* [by shortening & alteration] (1945)
: ATMOSPHERICS

¹sfor·zan·do \sfȯrt-'sän-(ˌ)dō, -'san-\ *adjective or adverb* [Italian, verbal of *sforzare* to force] (circa 1801)
: played with prominent stress or accent — used as a direction in music

²sforzando *noun, plural* **-dos** *or* **-di** \-(ˌ)dē\ (1890)
: an accented tone or chord

sfu·ma·to \sfü-'mä-(ˌ)tō\ *noun* [Italian, from past participle of *sfumare* to evaporate] (1909)
: the definition of form in painting without abrupt outline by the blending of one tone into another

sgraf·fi·to \zgra-'fē-(ˌ)tō, skra-\ *noun, plural* **-ti** \-(ˌ)tē\ [Italian, from past participle of *sgraffire* to scratch, produce sgraffito] (circa 1730)
1 : decoration by cutting away parts of a surface layer (as of plaster or clay) to expose a different colored ground — compare GRAFFITO
2 : something (as traditional Pennsylvania Dutch pottery) decorated with sgraffito

sh \sh *often prolonged*\ *interjection* (1847)
— used often in prolonged or rapidly repeated form to urge or command silence or less noise

Sha'·ban \shə-'bän\ *noun* [Arabic *sha'bān*] (circa 1771)
: the 8th month of the Islamic year — see MONTH table

Shab·bat \shə-'bät, 'shä-bəs\ *noun* [Hebrew *shabbāth*] (circa 1905)
: the Jewish Sabbath

shab·by \'sha-bē\ *adjective* **shab·bi·er; -est** [obsolete English *shab* a low fellow] (1669)
1 : clothed with worn or seedy garments ⟨a *shabby* hobo⟩
2 a : threadbare and faded from wear ⟨a *shabby* sofa⟩ **b :** ill-kept **:** DILAPIDATED ⟨a *shabby* neighborhood⟩
3 a : MEAN, DESPICABLE, CONTEMPTIBLE ⟨must feel *shabby* . . . because of his compromises —Nat Hentoff⟩ **b :** UNGENEROUS, UNFAIR ⟨laments the *shabby* way in which this country often treated a poet —Paul Engle⟩ **c :** inferior in quality ⟨*shabby* reasoning⟩
— shab·bi·ly \'sha-bə-lē\ *adverb*
— shab·bi·ness \'sha-bē-nəs\ *noun*

Sha·bu·oth \shə-'vü-ˌōt, -ˌōth, -ˌōs, -əs\ *noun* [Hebrew *shābhū'ōth,* literally, weeks] (circa 1903)
: a Jewish holiday observed on the 6th and 7th of Sivan in commemoration of the revelation of the Ten Commandments at Mount Sinai — called also *Pentecost*

shack \'shak\ *noun* [probably back-formation from English dialect *shackly* rickety] (1878)
1 : HUT, SHANTY
2 : a room or similar enclosed structure for a particular person or use ⟨a guard *shack*⟩

¹shack·le \'sha-kəl\ *noun* [Middle English *schakel,* from Old English *sceacul;* akin to Old Norse *skǫkull* pole of a cart] (before 12th century)
1 : something (as a manacle or fetter) that confines the legs or arms
2 : something that checks or prevents free action as if by fetters — usually used in plural
3 : a usually U-shaped fastening device secured by a bolt or pin through holes in the end of the two arms
4 : a length of cable or anchor chain of usually 15 fathoms

²shackle *transitive verb* **shack·led; shack·ling** \-k(ə-)liŋ\ (15th century)
1 a : to bind with shackles **:** FETTER **b :** to make fast with or as if with a shackle
2 : to deprive of freedom especially of action by means of restrictions or handicaps **:** IMPEDE
synonym see HAMPER
— shack·ler \-k(ə-)lər\ *noun*

shack·le·bone \'sha-kəl-ˌbōn, 'shä-\ *noun* (1571)
Scottish **:** WRIST

shack up *intransitive verb* (1935)
: to sleep or live together as unmarried sexual partners

shad \'shad\ *noun, plural* **shad** [(assumed) Middle English, from Old English *sceadd*] (before 12th century)
: any of several clupeid fishes (especially genus *Alosa*) that differ from the typical herrings in having a relatively deep body and in being anadromous and that include some important food fishes of Europe and North America

shad·ber·ry \-ˌber-ē\ *noun* (1847)
: SERVICEBERRY

shad·blow \'shad-ˌblō\ *noun* (1846)
: SERVICEBERRY 2

shad·bush \-ˌbu̇sh\ *noun* (circa 1818)
: SERVICEBERRY 2

shad·dock \'sha-dək\ *noun* [Captain *Shaddock,* 17th century English ship commander] (1696)
: a very large thick-rinded usually pear-shaped citrus fruit differing from the closely related grapefruit especially in its loose rind and often coarse dry pulp; *also* **:** the tree (*Citrus maxima* synonym *C. grandis*) that bears it

¹shade \'shād\ *noun* [Middle English, from Old English *sceadu;* akin to Old High German *scato* shadow, Greek *skotos* darkness] (before 12th century)

1 a : comparative darkness or obscurity owing to interception of the rays of light **b :** relative obscurity or retirement
2 a : shelter (as by foliage) from the heat and glare of sunlight **b :** a place sheltered from the sun
3 : an evanescent or unreal appearance
4 *plural* **a :** the shadows that gather as darkness comes on **b :** NETHERWORLD, HADES
5 a : a disembodied spirit **:** GHOST **b** — used to signal the similarity between a previously encountered person or situation and one at hand; usually used in plural ⟨*shades* of my childhood⟩
6 : something that intercepts or shelters from light, sun, or heat: as **a :** a device partially covering a lamp so as to reduce glare **b :** a flexible screen usually mounted on a roller for regulating the light or the view through a window **c** *plural* **:** SUNGLASSES
7 a : the reproduction of the effect of shade in painting or drawing **b :** a subdued or somber feature
8 a : a color produced by a pigment or dye mixture having some black in it **b :** a color slightly different from the one under consideration
9 a : a minute difference or variation **:** NUANCE **b :** a minute degree or quantity
10 : a facial expression of sadness or displeasure
— shade·less \-ləs\ *adjective*

²shade *verb* **shad·ed; shad·ing** (14th century)
transitive verb
1 a : to shelter or screen by intercepting radiated light or heat **b :** to cover with a shade
2 : to hide partly by or as if by a shadow
3 : to darken with or as if with a shadow
4 : to better or exceed by a shade **:** SURPASS, ECLIPSE
5 a : to represent the effect of shade or shadow on **b :** to add shading to **c :** to color so that the shades pass gradually from one to another
6 : to change by gradual transition or qualification
7 : to reduce slightly (as a price)
8 : SLANT, BIAS
intransitive verb
1 : to pass by slight changes or imperceptible degrees
2 : to undergo or exhibit minute difference or variation
— shad·er *noun*

shade–grown \'shād-ˌgrōn\ *adjective* (1922)
: grown in the shade; *specifically* **:** grown under cloth ⟨*shade-grown* tobacco⟩

shade tree *noun* (1806)
: a tree grown primarily to produce shade

shad·ing \'shā-diŋ\ *noun* (1663)
1 : the use of marking made within outlines to suggest three-dimensionality, shadow, or degrees of light and dark in a picture or drawing
2 : an interpretative effect in music gained especially by subtle changes in dynamics

sha·doof *also* **sha·duf** \shə-'düf, sha-\ *noun* [Arabic *shādūf*] (1836)
: a counterbalanced sweep used since ancient times especially in Egypt for raising water (as for irrigation)

¹shad·ow \'sha-(ˌ)dō\ *noun* [Middle English *shadwe,* from Old English *sceaduw-, sceadu* shade] (before 12th century)
1 : partial darkness or obscurity within a part of space from which rays from a source of light are cut off by an interposed opaque body
2 : a reflected image
3 : shelter from danger or observation
4 a : an imperfect and faint representation **b :** an imitation of something **:** COPY
5 : the dark figure cast upon a surface by a body intercepting the rays from a source of light
6 : PHANTOM
7 *plural* **:** DARK 1a
8 : a shaded or darker portion of a picture
9 : an attenuated form or a vestigial remnant
10 a : an inseparable companion or follower **b :** one (as a spy or detective) that shadows
11 : a small degree or portion **:** TRACE
12 : a source of gloom or unhappiness
13 a : an area near an object **:** VICINITY **b :** pervasive and dominant influence
14 : a state of ignominy or obscurity
— shad·ow·less \'sha-dō-ləs, -də-ləs\ *adjective*
— shad·ow·like \-ˌlīk\ *adjective*

²shadow (before 12th century)
transitive verb
1 *archaic* **:** SHELTER, PROTECT
2 : to cast a shadow upon **:** CLOUD
3 *obsolete* **:** to shelter from the sun
4 *obsolete* **:** CONCEAL
5 : to represent or indicate obscurely or faintly — often used with *forth* or *out*
6 : to follow especially secretly **:** TRAIL
7 *archaic* **:** SHADE 5
intransitive verb
1 : to pass gradually or by degrees
2 : to become overcast with or as if with shadows
— shad·ow·er \-dō-ər, -də-wər\ *noun*

³shadow *adjective* (1906)
1 : of, relating to, or resembling a shadow cabinet ⟨*shadow* minister of defense⟩
2 a : having an indistinct pattern ⟨*shadow* plaid⟩ **b :** having darker sections of design ⟨*shadow* lace⟩

shadow band *noun* (1900)
: one of a series of dark narrow parallel bands that appear to rush swiftly across the landscape just before or after totality in a solar eclipse

shad·ow·box \'sha-dō-ˌbäks, -də-ˌbäks\ *intransitive verb* (1919)
: to box with an imaginary opponent especially as a form of training

shadow box *noun* (circa 1909)
: a shallow enclosing case usually with a glass front in which something is set for protection and display

shadow cabinet *noun* (1906)
: a group of leaders of a parliamentary opposition who constitute the probable membership of the cabinet when their party is returned to power

shadow dance *noun* (circa 1909)
: a dance shown by throwing the shadows of dancers on a screen

shad·ow·graph \'sha-dō-ˌgraf, -də-ˌgraf\ *noun* (1888)
1 : SHADOW PLAY
2 : a photographic image resembling a shadow
— shad·ow·graphy \-ˌgra-fē\ *noun*

shadow mask *noun* (1951)
: a metal plate in a color television tube that contains minute apertures permitting passage of electron beams to specific phosphors on the screen during a scan

shadow play *noun* (1895)
: a drama exhibited by throwing shadows of puppets or actors on a screen — called also *shadow show*

shad·owy \'sha-dō-ē, -də-wē\ *adjective* (14th century)
1 a : of the nature of or resembling a shadow **:** UNSUBSTANTIAL **b :** faintly perceptible **:** INDISTINCT, VAGUE
2 : being in or obscured by shadow ⟨deep *shadowy* interiors⟩
3 : SHADY 1, 3
— shad·ow·i·ly \-wə-lē\ *adverb*
— shad·ow·i·ness \-wē-nəs\ *noun*

shady \'shā-dē\ *adjective* **shad·i·er; -est** (1579)
1 : producing or affording shade
2 : sheltered from the sun's rays
3 a : of questionable merit **:** UNCERTAIN, UNRELIABLE **b :** DISREPUTABLE
— shad·i·ly \'shā-dᵊl-ē\ *adverb*
— shad·i·ness \'shā-dē-nəs\ *noun*

¹shaft \'shaft\ *noun, plural* **shafts** \'shaf(t)s, *for 1b usually* 'shavz\ [Middle English, from Old English *sceaft;* akin to Old High German *scaft* shaft, Latin *scapus* shaft, stalk, Greek *skēptesthai* to prop oneself, lean] (before 12th century)
1 a (1) **:** the long handle of a spear or similar weapon (2) **:** SPEAR, LANCE **b** *or plural* **shaves** \'shavz\ **:** POLE; *specifically* **:** either of two long pieces of wood between which a horse is hitched to a vehicle **c** (1) **:** an arrow especially for a longbow (2) **:** the body or stem of an arrow extending from the nock to the head
2 : a sharply delineated beam of light shining through an opening
3 : something suggestive of the shaft of a spear or arrow especially in long slender cylindrical form: as **a :** the trunk of a tree **b :** the cylindrical pillar between the capital and the base **c :** the handle of a tool or instrument (as a golf club) **d :** a commonly cylindrical bar used to support rotating pieces or to transmit power or motion by rotation **e :** the stem or central axis of a feather **f :** the upright member of a cross especially below the arms **g :** a small architectural column (as at each side of a doorway) **h :** a column, obelisk, or other spire-shaped or columnar monument **i :** a vertical or inclined opening of uniform and limited cross section made for finding or mining ore, raising water, or ventilating underground workings (as in a cave) **j :** a vertical opening or passage through the floors of a building
4 a : a projectile thrown like a spear or shot like an arrow **b :** a scornful, satirical, or pithily critical remark or attack **c :** harsh or unfair treatment — usually used with *the*

²shaft *transitive verb* (1611)
1 : to fit with a shaft
2 : to treat unfairly or harshly

shaft horsepower *noun* (1908)
: horsepower transmitted by an engine shaft

shaft·ing \'shaf-tiŋ\ *noun* (1825)
: shafts or material for shafts

¹shag \'shag\ *noun* [(assumed) Middle English *shagge,* from Old English *sceacga;* akin to Old Norse *skegg* beard, *skaga* to project] (before 12th century)
1 a : a shaggy tangled mass or covering (as of hair) **b :** long coarse or matted fiber, nap, or pile
2 : tobacco cut into fine shreds
3 : CORMORANT 1

²shag *adjective* (1592)
: SHAGGY

³shag *verb* **shagged; shag·ging** (1596)
intransitive verb
: to fall or hang in shaggy masses
transitive verb
: to make rough or shaggy

⁴shag *transitive verb* **shagged; shag·ging** [origin unknown] (1904)
1 a : to chase after; *especially* **:** to chase after and return (a ball) hit usually out of play **b :** to catch (a fly) in baseball practice
2 : to chase away

⁵shag *intransitive verb* **shagged; shag·ging** [perhaps alteration of *shack* to lumber along] (1914)
1 : to move or lope along
2 : to dance the shag

⁶shag *noun* (1932)
: a dance step executed by hopping livelily on each foot in turn

shag·bark \'shag-ˌbärk\ *noun* (1777)
: SHAGBARK HICKORY

shagbark hickory *noun* (1751)
: a hickory (*Carya ovata*) with sweet edible nuts and a gray shaggy outer bark that peels off in long strips; *also* **:** its wood

shag·gy \'sha-gē\ *adjective* **shag·gi·er; -est** (circa 1590)

1 a : covered with or consisting of long, coarse, or matted hair **b :** covered with or consisting of thick, tangled, or unkempt vegetation **c :** having a rough nap, texture, or surface **d :** having hairlike processes
2 a : UNKEMPT **b :** confused or unclear in conception or thinking
— **shag·gi·ly** \ˈsha-gə-lē\ *adverb*
— **shag·gi·ness** \ˈsha-gē-nəs\ *noun*

shag·gy–dog story \ˌsha-gē-ˈdȯg-\ *noun* (1946)
: a long-drawn-out circumstantial story concerning an inconsequential happening that impresses the teller as humorous but the hearer as boring and pointless; *also* **:** a similar humorous story whose humor lies in the pointlessness or irrelevance of the punch line

shag·gy·mane \ˈsha-gē-ˌmān\ *noun* (circa 1909)
: a common edible mushroom (*Coprinus comatus*) having an elongated shaggy white pileus with deliquescing gills and black spores — called also *shaggy cap*

shaggymane

sha·green \sha-ˈgrēn, shə-\ *noun* [by folk etymology from French *chagrin,* modification of Turkish *sağrı*] (1677)
1 : an untanned leather covered with small round granulations and usually dyed green
2 : the rough skin of various sharks and rays when covered with small close-set tubercles
— **shagreen** *adjective*

shah \ˈshä, ˈshȯ\ *noun, often capitalized* [Persian *shāh* king — more at CHECK] (1566)
: a sovereign of Iran
— **shah·dom** \ˈshä-dəm, ˈshȯ-\ *noun*

Sha·hap·ti·an \shə-ˈhap-tē-ən\ *noun, plural* **Shahaptian** *or* **Shahaptians** (1836)
1 : a member of a group of American Indian peoples who formerly inhabited a large territory along the Columbia River and its tributaries
2 : the language of the Shahaptian people including Nez Percé and Yakima

shai·tan \shā-ˈtän, shī-\ *noun* [Arabic *shayṭān*] (1638)
: an evil spirit; *specifically* **:** an evil jinni

¹shake \ˈshāk\ *verb* **shook** \ˈshu̇k\; **shak·en** \ˈshā-kən\; **shak·ing** [Middle English, from Old English *sceacan;* akin to Old Norse *skaka* to shake] (before 12th century)
intransitive verb
1 : to move irregularly to and fro
2 : to vibrate especially as the result of a blow or shock
3 : to tremble as a result of physical or emotional disturbance
4 : to experience a state of instability **:** TOTTER
5 : to briskly move something to and fro or up and down especially in order to mix
6 : to clasp hands
7 : ³TRILL
transitive verb
1 : to brandish, wave, or flourish often in a threatening manner
2 : to cause to move in a usually quick jerky manner
3 : to cause to quake, quiver, or tremble
4 a : to free oneself from ⟨*shake* a habit⟩ ⟨*shake* off a cold⟩ **b :** to get away from **:** get rid of ⟨can you *shake* your friend? I want to talk to you alone —Elmer Davis⟩
5 : to lessen the stability of **:** WEAKEN ⟨*shake* one's faith⟩
6 : to bring to a specified condition by or as if by repeated quick jerky movements ⟨*shook* himself loose from the man's grasp⟩
7 : to dislodge or eject by quick jerky movements of the support or container ⟨*shook* the dust from the cloth⟩
8 : to clasp (hands) in greeting or farewell or as a sign of goodwill or agreement
9 : to stir the feelings of **:** UPSET ⟨*shook* her up⟩
10 : ³TRILL ☆
— **shak·able** *or* **shake·able** \ˈshā-kə-bəl\ *adjective*
— **shake a leg 1 :** DANCE **2 :** to hurry up

²shake *noun* (1581)
1 : an act of shaking: as **a :** an act of shaking hands **b :** an act of shaking oneself
2 a : a blow or shock that upsets the equilibrium or disturbs the balance of something **b :** EARTHQUAKE
3 *plural* **a :** a condition of trembling or nervousness; *specifically* **:** DELIRIUM TREMENS **b :** MALARIA 2a
4 : something produced by shaking: as **a :** a fissure separating annual rings of growth in timber **b** (1) **:** MILK SHAKE (2) **:** a beverage resembling a milk shake but made without milk
5 : a wavering, quivering, or alternating motion caused by a blow or shock
6 : TRILL
7 : a very brief period of time
8 *plural* **:** one that is exceptional especially in importance, ability, or merit — usually used in the phrase *no great shakes*
9 : a shingle split from a piece of log usually three or four feet (about one meter) long
10 : ³DEAL 3 ⟨a fair *shake*⟩

shake·down \ˈshāk-ˌdau̇n\ *noun* (circa 1730)
1 : an improvised bed (as one made up on the floor)
2 : a boisterous dance
3 : an act or instance of shaking someone down; *especially* **:** EXTORTION
4 : a thorough search
5 : a process or period of adjustment
6 : a testing under operating conditions of something new (as a ship) for possible faults and defects and for familiarizing the operators with it

shake down (circa 1859)
intransitive verb
1 a : to take up temporary quarters **b :** to occupy an improvised or makeshift bed
2 a : to become accustomed especially to new surroundings or duties **b :** to settle down
transitive verb
1 : to obtain money from in a deceitful or illegal manner
2 : to make a thorough search of
3 : to bring about a reduction of
4 : to give a shakedown test to

shake·out \ˈshā-ˌkau̇t\ *noun* (1895)
1 : the failure or retrenchment of a significant number of firms in the economy or a sector or an industry that usually results in a depressed market
2 : a period or process in which the relatively weak or unessential are eliminated

shak·er \ˈshā-kər\ *noun* (15th century)
1 : one that shakes: as **a :** a utensil or machine used in shaking ⟨cocktail *shaker*⟩ **b :** one that incites, promotes, or directs action ⟨a mover and *shaker*⟩
2 *capitalized* [from a dance with shaking movements performed as part of worship] **:** a member of a millenarian sect originating in England in 1747 and practicing celibacy and an ascetic communal life
— **Shaker** *adjective*
— **Shak·er·ism** \-kə-ˌri-zəm\ *noun*

¹Shake·spear·ean *or* **Shake·spear·ian** *also* **Shak·sper·ean** *or* **Shak·sper·ian** \shāk-ˈspir-ē-ən\ *adjective* (1755)
1 : of, relating to, or having the characteristics of Shakespeare or his writings
2 : evocative of a theme, setting, or event from a work of Shakespeare ⟨*Shakespearean* pageantry⟩

²Shakespearean *or* **Shakespearian** *also* **Shaksperean** *or* **Shaksperian** *noun* (1837)
: an authority on or devotee of Shakespeare

Shake·spear·eana *or* **Shake·spear·iana** \(ˌ)shāk-ˌspir-ē-ˈa-nə, -ˈä-, -ˈā-\ *noun plural* (1718)
: collected items by, about, or relating to Shakespeare

Shakespearean sonnet *noun* (1903)
: ENGLISH SONNET

shake–up \ˈshā-ˌkəp\ *noun* (1847)
: an act or instance of shaking up; *specifically* **:** an extensive and often drastic reorganization

shake up *transitive verb* (1538)
1 *obsolete* **:** CHIDE, SCOLD
2 : to jar by or as if by a physical shock ⟨the collision *shook up* both drivers⟩
3 : to effect an extensive and often drastic reorganization of

shaking palsy *noun* (1615)
: PARKINSON'S DISEASE

sha·ko \ˈsha-(ˌ)kō, ˈshā-, ˈshä-\ *noun, plural* **shakos** *or* **shakoes** [French, from Hungarian *csákó*] (1815)
: a stiff military hat with a high crown and plume

Shak·ta \ˈshäk-tə, ˈsäk-\ *noun or adjective* [Sanskrit *śākta,* from *Śakti*] (1810)
: an adherent of Shaktism

Shak·ti \-tē\ *noun* [Sanskrit *Śakti*] (1810)
: the dynamic energy of a Hindu god personified as his female consort; *broadly* **:** cosmic energy as conceived in Hindu thought

Shak·tism \-ˌti-zəm\ *noun* (1877)
: a Hindu sect worshiping Shakti under various names (as Kali or Durga) in a cult of devotion to the female principle often with magical or orgiastic rites

shaky \ˈshā-kē\ *adjective* **shak·i·er; -est** (1703)
1 : characterized by shakes ⟨*shaky* timber⟩
2 a : lacking stability **:** PRECARIOUS **b :** lacking in firmness (as of beliefs or principles) **c :** lacking in authority or reliability **:** QUESTIONABLE
3 a : somewhat unsound in health **b :** characterized by shaking
4 : likely to give way or break down
— **shak·i·ly** \-kə-lē\ *adverb*
— **shak·i·ness** \-kē-nəs\ *noun*

shale \ˈshā(ə)l\ *noun* [probably from obsolete or dialect *shale* scale, shell, from Middle English, from Old English *scealu* — more at SHELL] (1747)
: a fissile rock that is formed by the consolidation of clay, mud, or silt, has a finely stratified or laminated structure, and is composed of minerals essentially unaltered since deposition
— **shal·ey** \ˈshā-lē\ *adjective*

shale oil *noun* (1857)
: a crude dark oil obtained from oil shale by heating

shall \shəl, ˈshal\ *verb, past* **should** \shəd, ˈshu̇d\; *present singular & plural* **shall** [Middle English *shal* (1st & 3d singular present indicative), from Old English *sceal;* akin to Old High German *scal* (1st & 3d singular present indicative) ought to, must, Lithuanian *skola* debt] (before 12th century)

☆ SYNONYMS
Shake, agitate, rock, convulse mean to move up and down or to and fro with some violence. SHAKE often carries a further implication of a particular purpose ⟨*shake* well before using⟩. AGITATE suggests a violent and prolonged tossing or stirring ⟨an ocean *agitated* by storms⟩. ROCK suggests a swinging or swaying motion resulting from violent impact or upheaval ⟨the whole city was *rocked* by the explosion⟩. CONVULSE suggests a violent pulling or wrenching as of a body in a paroxysm ⟨spectators were *convulsed* with laughter⟩.

\ə\ **abut** \ᵊ\ **kitten** \ər\ **further** \a\ **ash** \ā\ **ace** \ä\ **mop, mar** \au̇\ **out** \ch\ **chin** \e\ **bet** \ē\ **easy** \g\ **go** \i\ **hit** \ī\ **ice** \j\ **job** \ŋ\ **sing** \ō\ **go** \ȯ\ **law** \ȯi\ **boy** \th\ **thin** \t͟h\ **the** \ü\ **loot** \u̇\ **foot** \y\ **yet** \zh\ **vision** *see also* Guide to Pronunciation

verbal auxiliary
1 *archaic* **a :** will have to **:** MUST **b :** will be able to **:** CAN
2 a — used to express a command or exhortation ⟨you *shall* go⟩ **b** — used in laws, regulations, or directives to express what is mandatory ⟨it *shall* be unlawful to carry firearms⟩
3 a — used to express what is inevitable or seems likely to happen in the future ⟨we *shall* have to be ready⟩ ⟨we *shall* see⟩ **b** — used to express simple futurity ⟨when *shall* we expect you⟩
4 — used to express determination ⟨they *shall* not pass⟩
intransitive verb
archaic **:** will go ⟨he to England *shall* along with you —Shakespeare⟩ ▢

shal·loon \shə-ˈlün, sha-\ *noun* [*Châlons*-sur-Marne, France] (1678)
: a lightweight twilled fabric of wool or worsted used chiefly for the linings of coats and uniforms

shal·lop \ˈsha-ləp\ *noun* [Middle French *chaloupe*] (circa 1578)
1 : a usually 2-masted ship with lugsails
2 : a small open boat propelled by oars or sails and used chiefly in shallow waters

shal·lot \shə-ˈlät *also* ˈsha-lət\ *noun* [modification of French *échalote,* from Middle French *eschalotte,* alteration of *eschaloigne,* from (assumed) Vulgar Latin *escalonia* — more at SCALLION] (1664)
1 : a bulbous perennial herb (*Allium cepa aggregatum*) that resembles an onion and produces small clustered bulbs used in seasoning
2 : GREEN ONION

¹shal·low \ˈsha-(ˌ)lō\ *adjective* [Middle English *schalowe;* probably akin to Old English *sceald* shallow — more at SKELETON] (14th century)
1 : having little depth ⟨*shallow* water⟩
2 : having little extension inward or backward ⟨office buildings have taken the form of *shallow* slabs —Lewis Mumford⟩
3 a : penetrating only the easily or quickly perceived ⟨*shallow* generalizations⟩ **b :** lacking in depth of knowledge, thought, or feeling ⟨a *shallow* demagogue⟩
4 : displacing comparatively little air **:** WEAK ⟨*shallow* breathing⟩
synonym see SUPERFICIAL
— shal·low·ly \-lō-lē, -lə-lē\ *adverb*
— shal·low·ness *noun*

²shallow (1510)
transitive verb
: to make shallow
intransitive verb
: to become shallow

³shallow *noun* (1571)
: a shallow place or area in a body of water — usually used in plural but singular or plural in construction

sha·lom \shä-ˈlōm, shə-\ *interjection* [Hebrew *shālōm* peace] (1904)
— used as a Jewish greeting and farewell

sha·lom alei·chem \ˌshò-ləm-ə-ˈlā-kəm, ˌshō-, -k̲əm\ *interjection* [Hebrew *shālōm ʽalēkhem* peace unto you] (1881)
— used as a traditional Jewish greeting

shalt \shəlt, ˈshalt\ *archaic present 2d singular of* SHALL

¹sham \ˈsham\ *noun* [perhaps from English dialect *sham* shame, alteration of English *shame*] (1677)
1 : a trick that deludes **:** HOAX
2 : cheap falseness **:** HYPOCRISY
3 : an ornamental covering for a pillow
4 : an imitation or counterfeit purporting to be genuine
5 : a person who shams
synonym see IMPOSTURE

²sham *adjective* (1681)
1 : not genuine **:** FALSE, FEIGNED
2 : having such poor quality as to seem false

³sham *verb* **shammed; sham·ming** (1755)
transitive verb
: to go through the external motions necessary to counterfeit
intransitive verb
: to act intentionally so as to give a false impression **:** FEIGN
— sham·mer \ˈsha-mər\ *noun*
synonym see ASSUME

sha·man \ˈshä-mən, ˈshā- *also* shə-ˈmän\ *noun, plural* **shamans** [ultimately from Evenki (Tungusic language of Siberia) *šamān*] (1698)
1 : a priest or priestess who uses magic for the purpose of curing the sick, divining the hidden, and controlling events
2 : one who resembles a shaman; *especially* **:** HIGH PRIEST 3
— sha·man·ic \shə-ˈma-nik, -ˈmä-\ *adjective*

sha·man·ism \-ˌni-zəm\ *noun* (1780)
: a religion practiced by indigenous peoples of far northern Europe and Siberia that is characterized by belief in an unseen world of gods, demons, and ancestral spirits responsive only to the shamans; *also* **:** any similar religion
— sha·man·ist \-nist\ *noun*
— sha·man·is·tic \ˌshä-mə-ˈnis-tik, ˌshā-mə-\ *adjective*

sham·ble \ˈsham-bəl\ *intransitive verb* **sham·bled; sham·bling** \-b(ə-)liŋ\ [*shamble* bowed, malformed] (1681)
: to walk awkwardly with dragging feet **:** SHUFFLE
— shamble *noun*

sham·bles \ˈsham-bəlz\ *noun plural but singular or plural in construction* [Middle English *shameles,* plural of *schamel* vendor's table, footstool, from Old English *sceamol* stool, from Latin *scamillum,* diminutive of *scamnum* stool, bench; perhaps akin to Sanskrit *skambha* pillar] (15th century)
1 *archaic* **:** a meat market
2 : SLAUGHTERHOUSE
3 a : a place of mass slaughter or bloodshed **b :** a scene or a state of great destruction **:** WRECKAGE **c** (1) **:** a scene or a state of great disorder or confusion (2) **:** great confusion **:** MESS

sham·bling *adjective* (1592)
: characterized by slow awkward movement

sham·bol·ic \sham-ˈbä-lik\ *adjective* [probably from *shambles*] (1970)
chiefly British **:** obviously disorganized or confused

¹shame \ˈshām\ *noun* [Middle English, from Old English *scamu;* akin to Old High German *scama* shame] (before 12th century)
1 a : a painful emotion caused by consciousness of guilt, shortcoming, or impropriety **b :** the susceptibility to such emotion
2 : a condition of humiliating disgrace or disrepute **:** IGNOMINY
3 a : something that brings censure or reproach; *also* **:** something to be regretted **:** PITY ⟨it's a *shame* you can't go⟩ **b :** a cause of feeling shame

²shame *transitive verb* **shamed; sham·ing** (13th century)
1 : to bring shame to **:** DISGRACE
2 : to put to shame by outdoing
3 : to cause to feel shame
4 : to force by causing to feel guilty ⟨*shamed* into confessing⟩

shame·faced \ˈshām-ˌfāst\ *adjective* [alteration of *shamefast*] (1593)
1 : showing modesty **:** BASHFUL
2 : showing shame **:** ASHAMED ◆
— shame·faced·ly \-ˌfā-səd-lē, -ˌfāst-lē\ *adverb*
— shame·faced·ness \-ˌfā-səd-nəs, -ˌfās(t)-nəs\ *noun*

shame·fast \ˈshām-ˌfast\ *adjective* [Middle English, from Old English *scamfæst,* from *scamu* + *fæst* fixed, fast] (before 12th century)
archaic **:** SHAMEFACED

shame·ful \ˈshām-fəl\ *adjective* (13th century)
1 a : bringing shame **:** DISGRACEFUL **b :** arousing the feeling of shame
2 *archaic* **:** full of the feeling of shame **:** ASHAMED
— shame·ful·ly \-fə-lē\ *adverb*
— shame·ful·ness *noun*

shame·less \ˈshām-ləs\ *adjective* (before 12th century)
1 : having no shame **:** insensible to disgrace
2 : showing lack of shame **:** DISGRACEFUL
— shame·less·ly *adverb*
— shame·less·ness *noun*

sham·mes \ˈshä-məs\ *noun, plural* **sham·mo·sim** \shä-ˈmò-səm\ [Yiddish *shames,* from Late Hebrew *shāmmāsh*] (1650)
1 : the sexton of a synagogue
2 : the candle or taper used to light the other candles in a Hanukkah menorah

sham·my \ˈsha-mē\ *variant of* CHAMOIS

¹sham·poo \sham-ˈpü\ *transitive verb* [Hindi *cā̃po,* imperative of *cā̃pnā* to press, shampoo] (1762)
1 *archaic* **:** MASSAGE
2 a : to wash (as the hair) with soap and water or with a special preparation **b :** to wash the hair of
— sham·poo·er *noun*

²shampoo *noun, plural* **shampoos** (1838)
1 : an act or instance of shampooing
2 : a preparation used in shampooing

sham·rock \ˈsham-ˌräk\ *noun* [Irish *seamróg,* diminutive of *seamar* clover] (1577)
: a trifoliolate plant used as a floral emblem by the Irish: as **a :** a yellow-flowered clover (*Trifolium dubium*) often regarded as the true shamrock **b :** WOOD SORREL **c :** WHITE CLOVER

sha·mus \ˈshä-məs, ˈshā-\ *noun* [perhaps from Yiddish *shames* shammes; from a jocular comparison of the duties of a sexton and those of a store detective] (1925)

▢ USAGE
shall From the reams of pronouncements written about the distinction between *shall* and *will*—dating back as far as the 17th century—it is clear that the rules laid down have never very accurately reflected actual usage. The nationalistic statements of 18th and 19th century British grammarians, who commonly cited the misuses of the Irish, the Scots, and occasionally the Americans, suggest that the traditional rules may have come closest to the usage of southern England. Some modern commentators believe that English usage is still the closest to the traditionally prescribed norms. Most modern commentators allow that *will* is more common in nearly all uses. The entries for *shall* and *will* in this dictionary show current usage.

◇ WORD HISTORY
shamefaced Old English *scamfæst* meant "bashful" or "modest" or, more literally, "held fast by shame." The second element of *shamefaced* was thus originally the same as that of *steadfast,* meaning "fixed in place." While we can be sure that the similarity of consonant sounds between *-fast* and *-faced* contributed to the alteration of *shamefast* to *shamefaced,* the belief that modesty or bashfulness is reflected in a person's face probably had some influence too. For instance, we find this sentence in a 15th century English translation from a Latin text: "Tho that haue the face sumwhate ruddy bene schamefast." (Those who have a somewhat ruddy face are shamefast.) Around the middle of the 16th century the alteration to *shamefaced* began to appear in print, and since then the folk etymology has been firmly established.

1 *slang* **:** POLICE OFFICER
2 *slang* **:** PRIVATE INVESTIGATOR

Shan \'shän, 'shan\ *noun, plural* **Shan** *or* **Shans** (1795)
1 : a member of a people living primarily in Myanmar and southern China
2 : the Thai language of the Shan

shan·dy \'shan-dē\ *noun, plural* **shandies** (1888)
1 : SHANDYGAFF
2 : a drink consisting of beer and lemonade

shan·dy·gaff \'shan-dē-,gaf\ *noun* [origin unknown] (1853)
: beer diluted with a nonalcoholic drink (as ginger beer)

Shang \'shäŋ\ *noun* [Chinese (Beijing) *Shāng*] (1669)
: a Chinese dynasty traditionally dated 1766–1122 B.C. and known especially for bronze work

shang·hai \shaŋ-'hī\ *transitive verb* **shang·haied; shang·hai·ing** [*Shanghai,* China; from the formerly widespread use of this method to secure sailors for voyages to the Orient] (1871)
1 a : to put aboard a ship by force often with the help of liquor or a drug **b :** to put by force or threat of force into or as if into a place of detention
2 : to put by trickery into an undesirable position
— shang·hai·er \-'hī(-ə)r\ *noun*

Shan·gri–la \,shaŋ-gri-'lä\ *noun* [*Shangri-La,* imaginary land depicted in the novel *Lost Horizon* (1933) by James Hilton] (1940)
1 : a remote beautiful imaginary place where life approaches perfection **:** UTOPIA
2 : a remote usually idyllic hideaway

[1]shank \'shaŋk\ *noun* [Middle English *shanke,* from Old English *scanca;* akin to Old Norse *skakkr* crooked, Greek *skazein* to limp] (before 12th century)
1 a : the part of the leg between the knee and the ankle in humans or the corresponding part in various other vertebrates **b :** LEG **c :** a cut of beef, veal, mutton, or lamb from the upper or the lower part of the leg **:** SHIN — see BEEF illustration
2 : a straight narrow usually essential part of an object: as **a :** the straight part of a nail or pin **b :** a straight part of a plant **:** STEM, STALK **c :** the part of an anchor between the ring and the crown — see ANCHOR illustration **d :** the part of a fishhook between the eye and the bend **e :** the part of a key between the handle and the bit **f :** the stem of a tobacco pipe or the part between the stem and the bowl **g :** TANG 1 **h** (1) **:** the narrow part of the sole of a shoe beneath the instep (2) **:** SHANKPIECE
3 : a part of an object by which it can be attached: as **a** (1) **:** a projection on the back of a solid button (2) **:** a short stem of thread that holds a sewn button away from the cloth **b :** the end (as of a drill bit) that is gripped in a chuck
4 a : the latter part of a period of time **b :** the early or main part of a period of time
5 *slang* **:** an often homemade knife
— shanked \'shaŋ(k)t\ *adjective*

[2]shank *transitive verb* (1927)
: to hit (a golf ball or shot) with the extreme heel of the club so that the ball goes off in an unintended direction; *also* **:** to kick (a football) in an unintended direction

shank·piece \'shaŋk-,pēs\ *noun* (1885)
: a support for the arch of the foot inserted in the shank of a shoe

shank's mare *noun* (circa 1795)
: one's own legs ⟨traveling by *shank's mare*⟩

shan't \'shant, 'shänt\ (1664)
: shall not

shan·tung \(,)shan-'təŋ\ *noun* [*Shantung* (Shandong), China] (circa 1882)
: a fabric in plain weave having a slightly irregular surface due to uneven slubbed filling yarns

[1]shan·ty \'shan-tē\ *variant of* CHANTEY

[2]shanty *noun, plural* **shanties** [probably from Canadian French *chantier* lumber camp, hut, from French, gantry, from Latin *cantherius* rafter, trellis] (1820)
: a small crudely built dwelling or shelter usually of wood

shan·ty·man \-mən, -,man\ *noun* (circa 1858)
: one who lives in a shanty

shan·ty·town \-,taùn\ *noun* (1876)
: a usually poor town or section of a town consisting mostly of shanties

shap·able *or* **shape·able** \'shā-pə-bəl\ *adjective* (1647)
1 : capable of being shaped
2 : SHAPELY

[1]shape \'shāp\ *verb* **shaped; shap·ing** [Middle English, alteration of Old English *scieppan;* akin to Old High German *skepfen* to shape] (before 12th century)
transitive verb
1 : FORM, CREATE; *especially* **:** to give a particular form or shape to
2 *obsolete* **:** ORDAIN, DECREE
3 : to adapt in shape so as to fit neatly and closely ⟨a dress *shaped* to her figure⟩
4 a : DEVISE, PLAN **b :** to embody in definite form ⟨*shaping* a folktale into an epic⟩
5 a : to make fit for (as a particular use or purpose) **:** ADAPT **b :** to determine or direct the course or character of (as life) **c :** to modify (behavior) by rewarding changes that tend toward a desired response
intransitive verb
1 : HAPPEN, BEFALL
2 : to take on or approach a mature or definite form — often used with *up*
— shap·er *noun*

[2]shape *noun* (before 12th century)
1 a : the visible makeup characteristic of a particular item or kind of item **b** (1) **:** spatial form or contour (2) **:** a standard or universally recognized spatial form
2 : the appearance of the body as distinguished from that of the face **:** FIGURE
3 a : PHANTOM, APPARITION **b :** assumed appearance **:** GUISE
4 : form of embodiment
5 : a mode of existence or form of being having identifying features
6 : something having a particular form
7 : the condition in which someone or something exists at a particular time ⟨the car was in fine *shape*⟩
— shaped \'shāpt\ *adjective*
— in shape : in an original, normal, or fit condition ⟨exercises to keep *in shape*⟩

shape·less \'shā-pləs\ *adjective* (14th century)
1 : having no definite shape
2 a : deprived of usual or normal shape **:** MISSHAPEN ⟨a *shapeless* old hat⟩ **b :** not shapely
— shape·less·ly *adverb*
— shape·less·ness *noun*

shape·ly \'shā-plē\ *adjective* **shape·li·er; -est** (14th century)
1 : having a regular or pleasing shape
2 : orderly and consistent in arrangement or plan
— shape·li·ness *noun*

shap·en \'shā-pən\ *adjective* [Middle English, from past participle of *shapen* to shape] (14th century)
: fashioned in or provided with a definite shape — usually used in combination ⟨an ill-*shapen* body⟩

shape note *noun* (1932)
: one of a system of seven notes showing the musical scale degree by the shape of the note head

shape–shift·er \'shāp-,shif-tər\ *noun* (1887)
: one that seems able to change form at will

shape–up \'shāp-,əp\ *noun* (1940)
: a system of hiring workers and especially longshoremen by the day or shift by having applicants gather for each day's selection; *also* **:** an instance of such hiring practice

shape up (circa 1920)
intransitive verb
: to improve to a good or acceptable condition or standard of behavior
transitive verb
: to bring to a good or acceptable condition or standard of behavior

shard \'shärd\ *also* **sherd** \'shərd\ *noun* [Middle English, from Old English *sceard;* akin to Old English *scieran* to cut — more at SHEAR] (before 12th century)
1 a : a piece or fragment of a brittle substance; *broadly* **:** a small piece or part **b :** SHELL, SCALE; *especially* **:** ELYTRON
2 *usually sherd* **:** fragments of pottery vessels found on sites and in refuse deposits where pottery-making peoples have lived
3 : highly angular curved glass fragments of tuffaceous sediments

[1]share \'shar, 'sher\ *noun* [Middle English *schare,* from Old English *scear;* akin to Old High German *scaro* plowshare, Old English *scieran* to cut — more at SHEAR] (before 12th century)
: PLOWSHARE

[2]share *noun* [Middle English, from Old English *scearu* cutting, tonsure; akin to Old English *scieran* to cut] (14th century)
1 a : a portion belonging to, due to, or contributed by an individual or group **b :** one's full or fair portion
2 a : the part allotted or belonging to one of a number owning together property or interest **b :** any of the equal portions into which property or invested capital is divided; *specifically* **:** any of the equal interests or rights into which the entire capital stock of a corporation is divided and ownership of which is regularly evidenced by one or more certificates **c** *plural, chiefly British* **:** STOCK 7c(1)

[3]share *verb* **shared; shar·ing** (1590)
transitive verb
1 : to divide and distribute in shares **:** APPORTION — usually used with *out* or *with*
2 a : to partake of, use, experience, occupy, or enjoy with others **b :** to have in common
3 : to grant or give a share in
intransitive verb
1 : to have a share — used with *in*
2 : to apportion and take shares of something
☆
— shar·er *noun*

share·able *or* **shar·able** \'sher-ə-bəl, 'shar-\ *adjective* (1920)
: capable of being shared
— share·abil·i·ty \,sher-ə-'bi-lə-tē, ,shar-\ *noun*

share·crop \'sher-,kräp, 'shar-\ *verb* [back-formation from *sharecropper*] (circa 1930)
intransitive verb
: to farm as a sharecropper

☆ SYNONYMS
Share, participate, partake mean to have, get, or use in common with another or others. SHARE usually implies that one as the original holder grants to another the partial use, enjoyment, or possession of a thing ⟨*shared* my toys with the others⟩. PARTICIPATE implies a having or taking part in an undertaking, activity, or discussion ⟨*participated* in sports⟩. PARTAKE implies accepting or acquiring a share especially of food or drink ⟨*partook* freely of the refreshments⟩.

transitive verb
: to farm (land) or produce (a crop) as a sharecropper

share·crop·per \-ˌkrä-pər\ *noun* (1923)
: a tenant farmer especially in the southern U.S. who is provided with credit for seed, tools, living quarters, and food, who works the land, and who receives an agreed share of the value of the crop minus charges

share·hold·er \-ˌhōl-dər\ *noun* (circa 1828)
: one that holds or owns a share in property; *especially* **:** STOCKHOLDER

share·ware \-ˌwar, -ˌwer\ *noun* (1983)
: software with usually limited capability or incomplete documentation which is available for trial use at little or no cost but which can be upgraded upon payment of a fee to the author

sha·rif \shə-ˈrēf\ *noun* [Arabic *sharīf,* literally, illustrious] (1599)
: a descendant of the prophet Muhammad through his daughter Fatima; *broadly* **:** one of noble ancestry or political preeminence in predominantly Islamic countries
— **sha·rif·ian** \-ˈrē-fē-ən\ *adjective*

[1]**shark** \ˈshärk\ *noun* [Middle English] (15th century)
: any of numerous mostly marine cartilaginous fishes of medium to large size that have a fusiform body, lateral branchial clefts, and a tough usually dull gray skin roughened by minute tubercles and are typically active predators sometimes dangerous to humans
— **shark·like** \ˈshärk-ˌlīk\ *adjective*

shark: *1* mako, *2* tiger, *3* thresher, *4* hammerhead, *5* great white

[2]**shark** *noun* [probably modification of German *Schurke* scoundrel] (1599)
1 : a rapacious crafty person who preys upon others through usury, extortion, or trickery
2 : one who excels greatly especially in a particular field

[3]**shark** (1602)
transitive verb
1 *archaic* **:** to gather hastily
2 *archaic* **:** to obtain by some irregular means
intransitive verb
1 *archaic* **:** to practice fraud or trickery
2 *archaic* **:** SNEAK

shark·skin \ˈshärk-ˌskin\ *noun* (1851)
1 : the hide of a shark or leather made from it
2 a : a smooth durable woolen or worsted suiting in twill or basket weave with small woven designs **b :** a smooth crisp fabric with a dull finish made usually of rayon in basket weave

shark sucker *noun* (circa 1850)
: REMORA 1

[1]**sharp** \ˈshärp\ *adjective* [Middle English, from Old English *scearp;* akin to Old High German *scarf* sharp and perhaps to Old English *scrapian* to scrape — more at SCRAPE] (before 12th century)
1 : adapted to cutting or piercing: as **a :** having a thin keen edge or fine point **b :** briskly or bitingly cold **:** NIPPING ⟨a *sharp* wind⟩
2 a : keen in intellect **:** QUICK-WITTED **b :** keen in perception **:** ACUTE ⟨*sharp* sight⟩ **c :** keen in attention **:** VIGILANT ⟨keep a *sharp* lookout⟩ **d :** keen in attention to one's own interest sometimes to the point of being unethical ⟨a *sharp* trader⟩
3 : keen in spirit or action: as **a :** full of activity or energy **:** BRISK ⟨*sharp* blows⟩ **b :** capable of acting or reacting strongly; *especially* **:** CAUSTIC
4 : SEVERE, HARSH: as **a :** inclined to or marked by irritability or anger ⟨a *sharp* temper⟩ **b :** causing intense mental or physical distress ⟨a *sharp* pain⟩ **c :** cutting in language or import ⟨a *sharp* rebuke⟩
5 : affecting the senses or sense organs intensely: as **a** (1) **:** having a strong odor or flavor ⟨*sharp* cheese⟩ (2) **:** ACRID **b :** having a strong piercing sound **c :** having the effect of or involving a sudden brilliant display of light ⟨a *sharp* flash⟩
6 a : terminating in a point or edge ⟨*sharp* features⟩ **b :** involving an abrupt or marked change especially in direction ⟨a *sharp* turn⟩ **c :** clear in outline or detail **:** DISTINCT ⟨a *sharp* image⟩ **d :** set forth with clarity and distinctness ⟨*sharp* contrast⟩
7 a *of a tone* **:** raised a half step in pitch **b :** higher than the proper pitch **c :** MAJOR, AUGMENTED — used of an interval in music
8 : STYLISH, DRESSY ☆
— **sharp·ly** *adverb*
— **sharp·ness** *noun*

[2]**sharp** *adverb* (before 12th century)
1 : in a sharp manner
2 : EXACTLY ⟨1:15 *sharp*⟩

[3]**sharp** *noun* (14th century)
: one that is sharp: as **a :** a sharp edge or point **b** (1) **:** a musical note or tone one half step higher than a note or tone named (2) **:** a character ♯ on a line or space of the musical staff indicating a pitch a half step higher than the degree would indicate without it **c :** a needle with a small eye for sewing by hand **d :** a real or self-styled expert; *also* **:** SHARPER

[4]**sharp** (1662)
transitive verb
: to raise (as a musical tone) in pitch; *especially* **:** to raise in pitch by a half step
intransitive verb
: to sing or play above the proper pitch

shar-pei \ˌshä-ˈpā, ˌshär-\ *noun, plural* **shar-peis** [Chinese (Guangdong) *sà* sand + *péi* fur] (1975)
: any of an ancient breed of dogs originating in China that have loose wrinkled skin especially when young, a short bristly coat, blue-black tongue, and wide blunt muzzle

sharp·en \ˈshär-pən\ *verb* **sharp·ened; sharp·en·ing** \ˈshärp-niŋ, ˈshär-pə-\ (15th century)
transitive verb
: to make sharp or sharper; *especially* **:** HONE
intransitive verb
: to become sharp or sharper
— **sharp·en·er** \ˈshärp-nər, ˈshär-pə-\ *noun*

sharp·er \ˈshär-pər\ *noun* (1681)
: CHEAT, SWINDLER; *especially* **:** a cheating gambler

sharp–eyed \ˈshärp-ˈīd\ *adjective* (1670)
: having keen sight; *also* **:** keen in observing or penetrating

sharp·ie *or* **sharpy** \ˈshär-pē\ *noun, plural* **sharp·ies** (circa 1859)
1 : a long narrow shallow-draft boat with flat or slightly V-shaped bottom and one or two masts each carrying a triangular sail
2 a : SHARPER **b :** an exceptionally keen or alert person

sharp–nosed \ˈshärp-ˈnōzd\ *adjective* (1561)
1 : keen in smelling
2 : having a pointed nose or snout

sharp practice *noun* (1845)
: the act of dealing in which advantage is taken or sought unscrupulously

sharp–set \ˈshärp-ˈset\ *adjective* (1540)
: eager in appetite or desire

sharp–shinned hawk \ˈshärp-ˈshind-\ *noun* (circa 1812)
: a common widely distributed American accipiter (*Accipiter striatus*) that is grayish above, has a chestnut breast, short rounded wings, and a tail with a notched or square tip when folded — called also *sharp-shin*

sharp·shoot·er \ˈshärp-ˌshü-tər\ *noun* (1802)
: a good marksman

sharp·shoot·ing \-ˌshü-tiŋ\ *noun* (1806)
1 : shooting with great precision
2 : accurate and usually unexpected attack (as in words)

sharp–sight·ed \-ˈsī-təd\ *adjective* (1571)
1 : having acute sight
2 : mentally keen or alert
— **sharp–sight·ed·ly** *adverb*
— **sharp–sight·ed·ness** *noun*

sharp–tongued \-ˈtəŋd\ *adjective* (1837)
: having a sharp tongue **:** harsh or bitter in speech or language

sharp–wit·ted \-ˈwi-təd\ *adjective* (circa 1586)
: having an acute mind

shash·lik *also* **shash·lick** *or* **shas·lik** \shäsh-ˈlik, ˈshäsh-lik\ *noun* [Russian *shashlyk,* probably modification of Crimean Tatar *šišlik,* from *šiš* skewer] (1926)
: KABOB

Shas·ta daisy \ˈshas-tə-\ *noun* [Mount *Shasta,* California] (circa 1893)
: a large-flowered garden daisy (*Chrysanthemum superbum* synonym *C. maximum*) that resembles the oxeye daisy

shat *past and past participle of* SHIT

[1]**shat·ter** \ˈsha-tər\ *verb* [Middle English *schateren*] (14th century)
transitive verb
1 : to cause to drop or be dispersed
2 a : to break at once into pieces **b :** to damage badly **:** RUIN
3 : to cause the disruption or annihilation of **:** DEMOLISH
intransitive verb
1 : to break apart **:** DISINTEGRATE
2 : to drop off parts (as leaves, petals, or fruit)
— **shat·ter·ing·ly** \-tə-riŋ-lē\ *adverb*

[2]**shatter** *noun* (circa 1640)
1 : FRAGMENT, SHRED — usually used in plural ⟨the broken vase lay in *shatters*⟩
2 : an act of shattering **:** the state of being shattered
3 : a result of shattering **:** SHOWER

shatter cone *noun* (1933)
: a conical fragment of rock that has striations radiating from the apex and that is formed by high pressure (as from volcanism or meteorite impact)

shat·ter·proof \ˈsha-tər-ˌprüf\ *adjective* (1930)
: proof against shattering

[1]**shave** \ˈshāv\ *verb* **shaved; shaved** *or* **shav·en** \ˈshā-vən\; **shav·ing** [Middle English, from Old English *scafan;* akin to Lithuanian *skobti* to pluck, Latin *scabere* to scratch, and perhaps to Greek *koptein* to cut — more at CAPON] (before 12th century)

☆ SYNONYMS
Sharp, keen, acute mean having or showing alert competence and clear understanding. SHARP implies quick perception, clever resourcefulness, or sometimes questionable trickiness ⟨*sharp* enough to spot a confidence game⟩. KEEN suggests quickness, enthusiasm, and a penetrating mind ⟨a *keen* observer of the political scene⟩. ACUTE implies a power to penetrate and may suggest subtlety and sharpness of discrimination ⟨an *acute* sense of style⟩.

transitive verb
1 a : to remove a thin layer from **b :** to cut off in thin layers or shreds **:** SLICE **c :** to cut off closely
2 a : to sever the hair from (the head or another part of the body) close to the roots **b :** to cut off (hair or beard) close to the skin
3 a : to discount (a note) at an exorbitant rate **b :** DEDUCT, REDUCE
4 : to come close to or touch lightly in passing
intransitive verb
1 : to cut off hair or beard close to the skin
2 : to proceed with difficulty **:** SCRAPE
[2]**shave** *noun* (before 12th century)
1 : SHAVER 3
2 : a thin slice **:** SHAVING
3 : an act or the process of shaving
shave·ling \'shāv-liŋ\ *noun* (1529)
1 : a tonsured clergyman **:** PRIEST — usually used disparagingly
2 : YOUTH, STRIPLING
shav·er \'shā-vər\ *noun* (15th century)
1 : a person who shaves
2 *archaic* **:** one who swindles
3 : a tool or machine for shaving; *especially* **:** an electric razor
4 : BOY, YOUNGSTER
shaves *plural of* SHAFT
shave·tail \'shāv-ˌtāl\ *noun* [from the practice of shaving the tails of newly broken mules to distinguish them from seasoned ones] (1846)
1 : a pack mule especially when newly broken in
2 : SECOND LIEUTENANT — usually used disparagingly
Sha·vi·an \'shā-vē-ən\ *noun* [New Latin *Shavius,* Latinized form of George Bernard *Shaw*] (1905)
: an admirer or devotee of G. B. Shaw, his writings, or his social and political theories
— Shavian *adjective*
shav·ie \'shā-vē\ *noun* [*shave* (swindle) + *-ie*] (1737)
Scottish **:** PRANK
shav·ing \'shā-viŋ\ *noun* (14th century)
1 : the act of one that shaves
2 : something shaved off ⟨wood *shavings*⟩
[1]**shaw** \'shȯ\ *noun* [Middle English, from Old English *sceaga;* akin to Old Norse *skegg* beard — more at SHAG] (before 12th century)
dialect **:** COPPICE, THICKET
[2]**shaw** *noun* [probably alteration of *show*] (1726)
chiefly British **:** the tops and stalks of a cultivated crop (as potatoes or turnips)
[1]**shawl** \'shȯl\ *noun* [Persian *shāl*] (1662)
: a square or oblong usually fabric garment or wrapper used especially as a covering for the head or shoulders
[2]**shawl** *transitive verb* (1812)
: to wrap in or as if in a shawl
shawl collar *noun* (circa 1908)
: a turned-over collar of a garment that combines with lapels forming an unbroken curving line

shawl collar

shawm \'shȯm\ *noun* [Middle English *schalme,* from Middle French *chalemie,* from Old French, alteration of *chalemel,* from Late Latin *calamellus,* diminutive of Latin *calamus* reed — more at CALAMUS] (14th century)
: an early double-reed woodwind instrument
Shaw·nee \shȯ-'nē, shä-\ *noun, plural* **Shawnee** *or* **Shawnees** [back-formation from obsolete English *Shawnese,* ultimately from Shawnee *ša·wano·ki,* literally, southerners] (1769)
1 : a member of an American Indian people originally of the central Ohio valley
2 : the Algonquian language of the Shawnee people
Shaw·wal \shə-'wäl\ *noun* [Arabic *shawwāl*] (circa 1771)
: the 10th month of the Islamic year — see MONTH table
shay \'shā\ *noun* [back-formation from *chaise,* taken as plural] (1717)
chiefly dialect **:** CHAISE 1
[1]**she** \'shē\ *pronoun* [Middle English, probably alteration of *hye,* alteration of Old English *hēo* she — more at HE] (12th century)
1 : that female one who is neither speaker nor the one addressed ⟨*she* is my wife⟩ — compare HE, HER, HERS, IT, THEY
2 — used to refer to one regarded as feminine (as by personification) ⟨*she* was a fine ship⟩
usage see HE
[2]**she** *noun* (14th century)
: a female person or animal — often used in combination ⟨*she*-cat⟩ ⟨*she*-cousin⟩
s/he \'shē-'hē; 'shē-ər-'hē; 'shē-ˌslash-'hē\ *pronoun* (1973)
: she or he — used in writing as a pronoun of common gender
shea butter \'shē-, 'shā-\ *noun* (1847)
: a pale solid fat from the seeds of the shea tree used in food, soap, and candles
sheaf \'shēf\ *noun, plural* **sheaves** \'shēvz\ [Middle English *sheef,* from Old English *scēaf;* akin to Old High German *scoub* sheaf, Russian *chub* forelock] (before 12th century)
1 : a quantity of the stalks and ears of a cereal grass or sometimes other plant material bound together
2 : something resembling a sheaf of grain ⟨a *sheaf* of papers⟩
— sheaf·like \'shēf-ˌlīk\ *adjective*
shea nut *noun* (1919)
: the seed of the shea tree
[1]**shear** \'shir\ *verb* **sheared; sheared** *or* **shorn** \'shōrn, 'shȯrn\; **shear·ing** [Middle English *sheren,* from Old English *scieran;* akin to Old Norse *skera* to cut, Latin *curtus* shortened, Greek *keirein* to cut, shear, Sanskrit *kr̥nāti* he injures] (before 12th century)
transitive verb
1 a : to cut off the hair from ⟨with crown *shorn*⟩ **b :** to cut or clip (as hair or wool) from someone or something; *also* **:** to cut something from ⟨*shear* a lawn⟩ **c** *chiefly Scottish* **:** to reap with a sickle **d :** to cut or trim with shears or a similar instrument
2 : to cut with something sharp
3 : to deprive of something as if by cutting
4 a : to subject to a shear force **b :** to cause (as a rock mass) to move along the plane of contact
intransitive verb
1 : to cut through something with or as if with a sharp instrument
2 *chiefly Scottish* **:** to reap crops with a sickle
3 : to become divided under the action of a shear ⟨the bolt may *shear* off⟩
— shear·er *noun*
[2]**shear** *noun* (before 12th century)
1 a (1) **:** a cutting implement similar or identical to a pair of scissors but typically larger — usually used in plural (2) **:** one blade of a pair of shears **b :** any of various cutting tools or machines operating by the action of opposed cutting edges of metal — usually used in plural **c** (1) **:** something resembling a shear or a pair of shears (2) **:** a hoisting apparatus consisting of two or sometimes more upright spars fastened together at their upper ends and having tackle for masting or dismasting ships or lifting heavy loads (as guns) — usually used in plural but singular or plural in construction
2 *chiefly British* **:** the action or process or an instance of shearing — used in combination to indicate the approximate age of sheep in terms of shearings undergone
3 a : internal force tangential to the section on which it acts — called also *shearing force* **b :** an action or stress resulting from applied forces that causes or tends to cause two contiguous parts of a body to slide relatively to each other in a direction parallel to their plane of contact
sheared *adjective* (1616)
: formed or finished by shearing; *especially* **:** cut to uniform length ⟨a *sheared* rug⟩
shear·ling \'shir-liŋ\ *noun* (14th century)
: skin from a recently sheared sheep or lamb that has been tanned and dressed with the wool left on
shear pin *noun* (circa 1931)
: an easily replaceable pin inserted at a critical point in a machine and designed to break when subjected to excess stress
shear·wa·ter \'shir-ˌwȯ-tər, -ˌwä-\ *noun* (circa 1671)
: any of numerous oceanic birds (especially genus *Puffinus*) that are related to the petrels and usually skim close to the waves in flight
sheath \'shēth\ *noun, plural* **sheaths** \'shēt͟hz, 'shēths\ [Middle English *shethe,* from Old English *scēath;* akin to Old High German *sceida* sheath and perhaps to Latin *scindere* to split — more at SHED] (before 12th century)
1 : a case for a blade (as of a knife)
2 : an investing cover or case of a plant or animal body or body part: as **a :** the tubular fold of skin into which the penis of many mammals is retracted **b** (1) **:** the lower part of a leaf (as of a grass) when surrounding the stem (2) **:** an ensheathing spathe
3 : any of various covering or supporting structures that are applied like or resemble in appearance or function the sheath of a blade: as **a :** SHEATHING 2 **b :** a woman's close-fitting dress usually worn without a belt **c** *British* **:** CONDOM
sheath·bill \'shēth-ˌbil\ *noun* (circa 1781)
: either of two white shorebirds (*Chionis alba* and *C. minor* of the family Chionididae) of colder parts of the southern hemisphere that have a horny sheath over the base of the upper mandible and suggest the pigeons in general appearance
sheathe \'shēt͟h\ *also* **sheath** \'shēth\ *transitive verb* **sheathed; sheath·ing** [Middle English *shethen,* from *shethe* sheath] (15th century)
1 : to put into or furnish with a sheath
2 : to plunge or bury (as a sword) in flesh
3 : to withdraw (a claw) into a sheath
4 : to case or cover with something (as sheets of metal) that protects
— sheath·er \'shē-t͟hər, -thər\ *noun*
sheath·ing \'shē-t͟hiŋ, -thiŋ\ *noun* (15th century)
1 : the action of one that sheathes something
2 : material used to sheathe something; *especially* **:** the first covering of boards or of waterproof material on the outside wall of a frame house or on a timber roof
sheath knife *noun* (1837)
: a knife having a fixed blade and designed to be carried in a sheath
shea tree \'shē-, 'shā-\ *noun* [Bambara *si*] (1799)
: a tropical African tree (*Vitellaria paradoxa* synonym *Butyrospermum parkii*) of the sapodilla family with fatty nuts that yield shea butter
[1]**sheave** \'shiv, 'shēv\ *noun* [Middle English *sheve;* akin to Old High German *scība* disk] (14th century)
: a grooved wheel or pulley (as of a pulley block)
[2]**sheave** \'shēv\ *transitive verb* **sheaved; sheav·ing** [*sheaf*] (1598)
: to gather and bind into a sheaf
she·bang \shi-'baŋ\ *noun* [origin unknown] (1869)

: everything involved in what is under consideration — usually used in the phrase *the whole shebang*

She·bat \shə-ˈbät, -ˈvät\ *noun* [Hebrew *shĕbhāṭ*] (1535)
: the 5th month of the civil year or the 11th month of the ecclesiastical year in the Jewish calendar — see MONTH table

she·been \shə-ˈbēn\ *noun* [Irish *síbín* illicit whiskey, shebeen] (circa 1787)
chiefly Irish : an unlicensed or illegally operated drinking establishment

She·chi·nah \shə-ˈk̲ē-nə, -ˈkē-nə, -ˈkī-nə\ *noun* [Hebrew *shĕkhīnāh*] (1663)
: the presence of God in the world as conceived in Jewish theology

¹shed \ˈshed\ *verb* **shed; shed·ding** [Middle English, to divide, separate, from Old English *scēadan;* akin to Old High German *skeidan* to separate, Latin *scindere* to split, cleave, Greek *schizein* to split] (before 12th century)
transitive verb
1 *chiefly dialect* : to set apart : SEGREGATE
2 : to cause to be dispersed without penetrating ⟨duck's plumage *sheds* water⟩
3 a : to cause (blood) to flow by cutting or wounding **b** : to pour forth in drops ⟨*shed* tears⟩ **c** : to give off in a stream ⟨fish *shedding* their eggs in spawning⟩ **d** : to give off or out ⟨*sheds* some light on the subject⟩
4 a (1) : to cast off (as a body covering) : MOLT (2) : to let fall (as leaves) (3) : to eject (as seed or spores) from a natural receptacle **b** : to rid oneself of temporarily or permanently as superfluous or unwanted ⟨*shed* her inhibitions⟩ ⟨the company *shed* 100 jobs⟩
intransitive verb
1 : to pour out : SPILL
2 : to become dispersed : SCATTER
3 : to cast off some natural covering (as fur or skin) ⟨the cat is *shedding*⟩
synonym see DISCARD
— **shed blood** : to cause death by violence

²shed *noun* (12th century)
1 *obsolete* : DISTINCTION, DIFFERENCE
2 : something (as the skin of a snake) that is discarded in shedding
3 : a divide of land

³shed *noun* [alteration of earlier *shadde,* probably from Middle English *shade* shade] (1557)
1 a : a slight structure built for shelter or storage; *especially* : a single-storied building with one or more sides unenclosed **b** : a building that resembles a shed
2 *archaic* : HUT
— **shed·like** \-ˌlīk\ *adjective*

⁴shed *transitive verb* **shed·ded; shed·ding** (1850)
: to put or house in a shed

she'd \ˈshēd\ (1609)
: she had : she would

shed·der \ˈshe-dər\ *noun* (14th century)
: one that sheds something: as **a** : a crab or lobster about to molt **b** : a newly molted crab

shed dormer *noun* (1948)
: a dormer with a roof sloping in the same direction as the roof from which the dormer projects

¹sheen \ˈshēn\ *adjective* [Middle English *shene,* from Old English *scīene;* akin to Old English *scēawian* to look — more at SHOW] (before 12th century)
1 *archaic* : BEAUTIFUL
2 *archaic* : SHINING, RESPLENDENT

²sheen *intransitive verb* (14th century)
: to be bright : show a sheen

³sheen *noun* (1602)
1 a : a bright or shining condition : BRIGHTNESS **b** : a subdued glitter approaching but short of optical reflection **c** : a lustrous surface imparted to textiles through finishing processes or use of shiny yarns
2 : a textile exhibiting notable sheen
— **sheeny** \ˈshē-nē\ *adjective*

sheep \ˈshēp\ *noun, plural* **sheep** *often attributive* [Middle English, from Old English *scēap;* akin to Old High German *scāf* sheep] (before 12th century)
1 : any of various ruminant mammals (genus *Ovis*) related to the goats but stockier and lacking a beard in the male; *specifically* : one (*O. aries*) long domesticated especially for its flesh and wool
2 a : a timid defenseless creature **b** : a timid docile person; *especially* : one easily influenced or led
3 : leather prepared from the skins of sheep : SHEEPSKIN

sheep·ber·ry \-ˌber-ē\ *noun* (circa 1818)
: an often shrubby North American viburnum (*Viburnum lentago*) with white flowers in flat cymes

sheep·cote \-ˌkōt, -ˌkät\ *noun* (15th century)
chiefly British : SHEEPFOLD

sheep–dip \-ˌdip\ *noun* (1865)
: a liquid preparation of usually toxic chemicals into which sheep are plunged especially to destroy parasitic arthropods

sheep·dog \-ˌdȯg\ *noun* (circa 1774)
: a dog (as a Border collie) used to tend, drive, or guard sheep

sheep fescue *noun* (1945)
: a hardy fine-foliaged perennial grass (*Festuca ovina*) widely used as a pasture grass — called also *sheep's fescue*

sheep·fold \ˈshēp-ˌfōld\ *noun* (15th century)
: a pen or shelter for sheep

sheep·herd·er \ˈshēp-ˌhər-dər\ *noun* (1871)
: a worker in charge of sheep especially on open range

sheep·herd·ing \-ˌhər-diŋ\ *noun* (1891)
: the activities of a worker engaged in tending sheep

sheep·ish \ˈshē-pish\ *adjective* (13th century)
1 : resembling a sheep in meekness, stupidity, or timidity
2 : affected by or showing embarrassment caused by consciousness of a fault ⟨a *sheepish* grin⟩
— **sheep·ish·ly** *adverb*
— **sheep·ish·ness** *noun*

sheep ked \ˈshēp-ˌked\ *noun* [*sheep* + *ked* sheep ked, of unknown origin] (1925)
: a wingless bloodsucking dipteran fly (*Melophagus ovinus*) that feeds chiefly on sheep and is a vector of sheep trypanosomiasis — called also *sheep tick*

sheep laurel *noun* (1810)
: a dwarf shrub (*Kalmia angustifolia*) of the eastern U.S. that is poisonous to young stock and resembles mountain laurel but has narrower leaves and smaller bright red flowers — called also *lambkill*

sheep's eye *noun* (circa 1529)
: a shy longing usually amorous glance — usually used in plural

sheep·shank \ˈshēp-ˌshaŋk\ *noun* (1627)
1 : a knot for shortening a line — see KNOT illustration
2 *Scottish* : something of no worth or importance

sheeps·head \ˈshēps-ˌhed\ *noun* (1643)
1 : a marine bony fish (*Archosargus probatocephalus* of the family Sparidae) of the Atlantic and Gulf coasts of the U.S. that has broad incisor teeth and is used for food
2 : FRESHWATER DRUM
3 : a largely red or rose California wrasse (*Semicossyphus pulcher*)

sheep·shear·er \ˈshēp-ˌshir-ər\ *noun* (1539)
: one that shears sheep

sheep·shear·ing \ˈshēp-ˌshir-iŋ\ *noun* (1586)
1 : the act of shearing sheep
2 : the time or season for shearing sheep

sheep·skin \-ˌskin\ *noun* (13th century)
1 a : the skin of a sheep; *also* : leather prepared from it **b** : PARCHMENT **c** : a garment made of or lined with sheepskin
2 : DIPLOMA

sheep sorrel *noun* (1806)
: a small dock (*Rumex acetosella*) of acidic soils

sheep walk *noun* (1586)
chiefly British : a pasture or range for sheep

¹sheer \ˈshir\ *verb* [perhaps alteration of ¹*shear*] (1539)
intransitive verb
: to deviate from a course : SWERVE
transitive verb
: to cause to sheer

²sheer *noun* (1670)
1 : a turn, deviation, or change in a course (as of a ship)
2 : the position of a ship riding to a single anchor and heading toward it

³sheer *adjective* [Middle English *schere* freed from guilt, probably alteration of *skere,* from Old Norse *skærr* pure; akin to Old English *scīnan* to shine] (circa 1568)
1 *obsolete* : BRIGHT, SHINING
2 : of very thin or transparent texture : DIAPHANOUS
3 a : UNQUALIFIED, UTTER ⟨*sheer* folly⟩ ⟨*sheer* ignorance⟩ **b** : being free from an adulterant : PURE, UNMIXED **c** : viewed or acting in dissociation from all else ⟨in terms of *sheer* numbers⟩
4 : marked by great and continuous steepness
synonym see STEEP
— **sheer·ly** *adverb*
— **sheer·ness** *noun*

⁴sheer *adverb* (circa 1600)
1 : in a complete manner : ALTOGETHER
2 : straight up or down without a break : PERPENDICULARLY

⁵sheer *noun* (circa 1920)
: a sheer fabric; *also* : an article of such a fabric

⁶sheer *noun* [perhaps alteration of ²*shear*] (1691)
: the fore-and-aft curvature from bow to stern of a ship's deck as shown in side elevation

sheer·legs \ˈshir-ˌlegz, -ˌlāgz\ *noun plural but singular or plural in construction* (circa 1860)
: SHEAR 1c(2)

¹sheet \ˈshēt\ *noun* [Middle English *shete,* from Old English *scȳte;* akin to Old English *scēat* edge, Old High German *scōz* flap, skirt] (before 12th century)
1 a : a broad piece of cloth; *especially* : BEDSHEET **b** : SAIL 1a(1)
2 a (1) : a usually rectangular piece of paper; *especially* : one manufactured for printing (2) : a rectangular piece of heavy paper with a plant specimen mounted on it ⟨an herbarium of 100,000 *sheets*⟩ **b** : a printed signature for a book especially before it has been folded, cut, or bound — usually used in plural **c** : a newspaper, periodical, or occasional publication ⟨a gossip *sheet*⟩ **d** : the unseparated postage stamps printed by one impression of a plate on a single piece of paper; *also* : a pane of stamps
3 : a broad stretch or surface of something ⟨a *sheet* of ice⟩
4 : a suspended or moving expanse (as of fire or rain)
5 a : a portion of something that is thin in comparison to its length and breadth **b** : a flat baking pan of tinned metal ⟨a cookie *sheet*⟩
6 : a surface or part of a surface in which it is possible to pass from any one point of it to any other without leaving the surface ⟨a hyperboloid of two *sheets*⟩
— **sheet·like** \-ˌlīk\ *adjective*

²sheet *adjective* (1582)
1 : rolled or spread out in a sheet
2 : of, relating to, or concerned with the making of sheet metal

³sheet (1606)
transitive verb
1 : to cover with a sheet : SHROUD
2 : to furnish with sheets
3 : to form into sheets

intransitive verb
: to fall, spread, or flow in a sheet ⟨the rain *sheeted* against the windows⟩
— sheet·er *noun*
— sheet home 1 : to extend (a sail) and set as flat as possible by hauling upon the sheets **2 :** to fix the responsibility for **:** bring home to one

[4]sheet *noun* [Middle English *shete,* from Old English *scēata* lower corner of a sail; akin to Old English *scӯte* sheet] (13th century)
1 : a rope or chain that regulates the angle at which a sail is set in relation to the wind
2 *plural* **:** the spaces at either end of an open boat not occupied by thwarts **:** foresheets and stern sheets together
— three sheets in the wind *or* **three sheets to the wind :** DRUNK 1a

sheet anchor *noun* (15th century)
1 : a large strong anchor formerly carried in the waist of a ship and used as a spare in an emergency
2 : something that constitutes a main support or dependence especially in danger

sheet bend *noun* (circa 1823)
: a bend or hitch used for temporarily fastening a rope to the bight of another rope or to an eye — see KNOT illustration

sheet·fed \'shēt-ˌfed\ *adjective* (1926)
: of, relating to, being, or printed by a press that prints on paper in sheet form

sheet glass *noun* (1805)
: glass made in large sheets directly from the furnace or by making a cylinder and then flattening it

sheet·ing \'shē-tiŋ\ *noun* (1711)
1 : material in the form of sheets or suitable for forming into sheets
2 : a lining (as wood or steel) used to support an embankment or the walls of an excavation

sheet lightning *noun* (1794)
: lightning in diffused or sheet form due to reflection and diffusion by the clouds and sky

sheet metal *noun* (circa 1909)
: metal in the form of a sheet

sheet music *noun* (1857)
: music printed on large unbound sheets of paper

Sheet·rock \'shēt-ˌräk\ *trademark*
— used for plasterboard

sheikh *or* **sheik** \'shēk, *also* 'shāk *for 1*\ *noun* [Arabic *shaykh*] (1577)
1 : an Arab chief
2 *usually* **sheik :** a man held to be irresistibly attractive to romantic young women

sheikh·dom *or* **sheik·dom** \-dəm, -təm\ *noun* (1860)
: a region under the rule of a sheikh

shei·la \'shē-lə\ *noun* [probably from *Sheila,* female given name] (circa 1919)
Australian & New Zealand **:** a girl or young woman

shek·el \'she-kəl\ *noun* [Hebrew *sheqel*] (15th century)
1 a : any of various ancient units of weight; *especially* **:** a Hebrew unit equal to about 252 grains troy **b :** a unit of value based on a shekel weight of gold or silver
2 : a coin weighing one shekel
3 *plural* **:** MONEY
4 — see MONEY table

Shekinah *variant of* SHECHINAH

shel·drake \'shel-ˌdrāk\ *noun* [Middle English, from *sheld-* (akin to Middle Dutch *schillede* parti-colored) + *drake*] (14th century)
1 : SHELDUCK
2 : MERGANSER

shel·duck \-ˌdək\ *noun* [*shel-* (as in *sheldrake*) + *duck*] (1707)
: any of various Old World ducks (genus *Tadorna*); *especially* **:** a common mostly black-and-white duck (*T. tadorna*) slightly larger than the mallard

shelf \'shelf\ *noun, plural* **shelves** \'shelvz\ [Middle English, probably from Old English *scylfe;* akin to Old Norse hlīth*skjalf* Odin's seat] (14th century)

shelduck

1 a : a thin flat usually long and narrow piece of material (as wood) fastened horizontally (as on a wall) at a distance from the floor to hold objects **b :** one of several similar pieces in a closet, bookcase, or similar structure **c :** the contents of a shelf
2 : something resembling a shelf in form or position: as **a :** a sandbank or ledge of rocks usually partially submerged **b :** a flat projecting layer of rock **c :** the submerged gradually sloping border of a continent or island **:** CONTINENTAL SHELF
— shelf·ful \'shelf-ˌfu̇l\ *noun*
— shelf·like \'shelf-ˌlīk\ *adjective*
— off the shelf : available from stock **:** not made to order ⟨*off the shelf* equipment⟩
— on the shelf : in a state of inactivity or uselessness

shelf fungus *noun* (circa 1903)
: BRACKET FUNGUS

shelf ice *noun* (1910)
: an extensive ice sheet originating on land but continuing out to sea beyond the depths at which it rests on the sea bottom

shelf life *noun* (1927)
: the period of time during which a material may be stored and remain suitable for use; *broadly* **:** the period of time during which something lasts or remains popular

[1]shell \'shel\ *noun* [Middle English, from Old English *sciell;* akin to Old English *scealu* shell, Old Norse *skel,* Lithuanian *skelti* to split, Greek *skallein* to hoe] (before 12th century)
1 a : a hard rigid usually largely calcareous covering or support of an animal **b :** the hard or tough often thin outer covering of an egg (as of a bird or reptile) — see EGG illustration
2 : the covering or outside part of a fruit or seed especially when hard or fibrous
3 : shell material (as of mollusks or turtles) or their substance
4 : something that resembles a shell: as **a :** a framework or exterior structure; *especially* **:** a building with an unfinished interior **b** (1) **:** an external case or outside covering ⟨the *shell* of a ship⟩ (2) **:** a thin usually spherical layer or surface enclosing a space or surrounding an object ⟨an expanding *shell* of gas around a neutron star⟩ **c :** a casing without substance ⟨mere effigies and *shells* of men —Thomas Carlyle⟩ **d :** an edible crust for holding a filling ⟨a pastry *shell*⟩ **e :** BAND SHELL **f :** a small beer glass **g :** an unlined article of outerwear
5 : a shell-bearing mollusk
6 : an impersonal attitude or manner that conceals the presence or absence of feeling ⟨he retreated into his *shell*⟩
7 : a narrow light racing boat propelled by one or more persons pulling oars or sculls
8 : any of the spaces occupied by the orbits of a group of electrons of approximately equal energy surrounding the nucleus of an atom
9 a : a projectile for cannon containing an explosive bursting charge **b :** a metal or paper case which holds the charge of powder and shot or bullet used with breech-loading small arms
10 : a plain usually sleeveless blouse or sweater
11 : a company or corporation that exists without assets or independent operations as a legal entity through which another company or corporation can conduct various dealings
— shell *adjective*

[2]shell (1562)
transitive verb
1 a : to take out of a natural enclosing cover (as a shell, husk, pod, or capsule) ⟨*shell* peanuts⟩ **b :** to separate the kernels of (as an ear of Indian corn, wheat, or oats) from the cob, ear, or husk
2 : to throw shells at, upon, or into **:** BOMBARD
3 : to score heavily against (as an opposing pitcher in baseball)
intransitive verb
1 : to fall or scale off in thin pieces
2 : to cast the shell or exterior covering **:** fall out of the pod or husk ⟨nuts which *shell* in falling⟩
3 : to gather shells (as from a beach) **:** collect shells

she'll \'shē(ə)l, shil\ (circa 1590)
: she will **:** she shall

[1]shel·lac \shə-'lak\ *noun* [[1]*shell* + *lac*] (1713)
1 : purified lac usually prepared in thin orange or yellow flakes by heating and filtering and often bleached white
2 : a preparation of lac dissolved usually in alcohol and used chiefly as a wood filler and finish
3 a : a composition containing shellac formerly used for making phonograph records **b :** an old 78 rpm phonograph record

[2]shellac *transitive verb* **shel·lacked; shel·lack·ing** (1876)
1 : to coat or otherwise treat with shellac or a shellac varnish
2 : to defeat decisively

shellacking *noun* (1931)
: a decisive defeat **:** DRUBBING

shell·back \'shel-ˌbak\ *noun* (1853)
1 : an old or veteran sailor
2 : a person who has crossed the equator and been initiated in the traditional ceremony

shell bean *noun* (1868)
1 : a bean grown primarily for its edible seeds — compare SNAP BEAN
2 : the edible seed of a bean

shell·crack·er \'shel-ˌkra kər\ *noun* (circa 1890)
: REDEAR

shelled \'sheld\ *adjective* (15th century)
1 : having a shell especially of a specified kind — often used in combination ⟨pink-*shelled*⟩ ⟨thick-*shelled*⟩
2 a : having the shell removed ⟨*shelled* oysters⟩ ⟨*shelled* nuts⟩ **b :** removed from the cob ⟨*shelled* corn⟩

shell·er \'she-lər\ *noun* (1694)
1 : one that shells ⟨a peanut *sheller*⟩
2 : a person who collects seashells

shell·fish \-ˌfish\ *noun* (before 12th century)
: an aquatic invertebrate animal with a shell; *especially* **:** an edible mollusk or crustacean

shell·fish·ery \-ˌfi-shə-rē, -ˌfish-rē\ *noun* (1885)
: a commercially exploited population of shellfish

shell game *noun* (1890)
1 : thimblerig played especially with three walnut shells
2 : FRAUD; *especially* **:** a swindle involving the substitution of something of little or no value for a valuable item

shell jacket *noun* (1840)
1 : a short tight military jacket worn buttoned up the front
2 : MESS JACKET

shell out *verb* (1801)
: PAY

shell pink *noun* (1887)
: a light yellowish pink

shell·proof \'shel-ˌprüf\ *adjective* (circa 1859)
: capable of resisting shells or bombs

shell shock *noun* (1915)

: post-traumatic stress disorder in soldiers as a result of combat experience

shell–shocked *adjective* (1918)
1 : affected with shell shock
2 : mentally confused, upset, or exhausted as a result of excessive stress

shell steak *noun* (circa 1968)
: the part of a short loin of beef that contains no tenderloin

shell•work \'shel-ˌwərk\ *noun* (1611)
: work adorned with shells or composed of a pattern of shells

shelly \'she-lē\ *adjective* **shell•i•er; -est** (1555)
1 : abounding in or covered with shells ⟨a *shelly* shore⟩
2 : of, relating to, or resembling a shell

[1]shel•ter \'shel-tər\ *noun* [origin unknown] (1585)
1 a : something that covers or affords protection ⟨a bomb *shelter*⟩ **b :** an establishment providing food and shelter (as to the homeless) **c :** an establishment that houses and feeds stray animals
2 : a position or the state of being covered and protected ⟨took *shelter*⟩
— shel•ter•less \-ləs\ *adjective*

[2]shelter *verb* **shel•tered; shel•ter•ing** \-t(ə-)riŋ\ (1590)
transitive verb
1 : to constitute or provide a shelter for **:** PROTECT ⟨has led a *sheltered* life⟩
2 : to place under shelter or protection ⟨*sheltered* himself in a mountain cave⟩
3 : to protect (income) from taxation
intransitive verb
: to take shelter
— shel•ter•er \-tər-ər\ *noun*

shel•ter•belt \'shel-tər-ˌbelt\ *noun* (1868)
: a barrier of trees and shrubs that protects (as crops) from wind and storm and lessens erosion

shelter half *noun* (1911)
: one of the halves of a shelter tent

shelter tent *noun* (1862)
: a small tent usually consisting of two interchangeable pieces **:** PUP TENT

shel•tie *or* **shel•ty** \'shel-tē\ *noun, plural* **shelties** [probably from Old Norse *Hjalti* Shetlander] (1650)
1 : SHETLAND PONY
2 : SHETLAND SHEEPDOG

shelve \'shelv\ *verb* **shelved; shelv•ing** [*shelf*] (1598)
transitive verb
1 : to furnish with shelves
2 : to place on a shelf
3 a : to remove from active service **b :** to put off or aside ⟨*shelve* a project⟩
intransitive verb
: to slope in a formation like a shelf
— shelv•er *noun*

[1]shelv•ing \'shel-viŋ\ *noun* (1687)
1 : the state or degree of sloping
2 : a sloping surface or place

[2]shelving *noun* (1817)
1 : material for shelves
2 : SHELVES

Shem \'shem\ *noun* [Hebrew *Shēm*]
: the eldest son of Noah held to be the progenitor of the Semitic peoples

She•ma \shə-'mä\ *noun* [Hebrew *shĕma'* hear, first word of Deuteronomy 6:4] (1706)
: the Jewish confession of faith made up of Deuteronomy 6:4–9 and 11:13–21 and Numbers 15:37–41

She•mi•ni Atze•reth \shə-'mē-nē-ät-'ser-ət, -əth, -əs\ *noun* [Late Hebrew *shĕmīnī 'ăṣereth*, from Hebrew *shĕmīnī* eighth + *'ăṣereth* assembly] (circa 1905)
: a Jewish festival following the seventh day of Sukkoth and marked by a special prayer for seasonal rain

Shem•ite \'she-ˌmīt\ *noun* [*Shem*] (1659)
archaic **:** SEMITE
— She•mit•ic \shə-'mi-tik\ *or* **Shem•it•ish** \'she-ˌmī-tish\ *adjective, archaic*

she•nan•i•gan \shə-'na-ni-gən\ *noun* [origin unknown] (1855)
1 : a devious trick used especially for an underhand purpose
2 a : tricky or questionable practices or conduct — usually used in plural **b :** high-spirited or mischievous activity — usually used in plural

shend \'shend\ *transitive verb* **shent** \'shent\; **shend•ing** [Middle English, from Old English *scendan;* akin to Old English *scamu* shame — more at SHAME] (before 12th century)
1 *archaic* **:** to put to shame or confusion
2 *archaic* **:** REPROVE, REVILE
3 *chiefly dialect* **a :** INJURE, MAR **b :** RUIN, DESTROY

she–oak \'shē-ˌōk\ *noun* (1792)
: any of several casuarinas

She•ol \shē-'ōl, 'shē-ˌ\ *noun* [Hebrew *Shĕ'ōl*] (1599)
: the abode of the dead in early Hebrew thought

[1]shep•herd \'she-pərd\ *noun* [Middle English *sheepherde,* from Old English *scēaphyrde,* from *scēap* sheep + *hierde* herdsman; akin to Old English *heord* herd] (before 12th century)
1 : one who tends sheep
2 : PASTOR
3 : GERMAN SHEPHERD

[2]shepherd *transitive verb* (1790)
1 : to tend as a shepherd
2 : to guide or guard in the manner of a shepherd ⟨*shepherded* the bill through Congress⟩

shepherd dog *noun* (15th century)
: SHEEPDOG

shep•herd•ess \'she-pər-dəs\ *noun* (14th century)
: a woman or girl who tends sheep; *also* **:** a rural girl or woman

shepherd's check *noun* (1896)
: a pattern of small even black-and-white checks; *also* **:** a fabric woven in this pattern — called also *shepherd's plaid*

shepherd's check

shepherd's pie *noun* (1877)
: a meat pie with a mashed potato crust

shepherd's purse *noun* (15th century)
: a white-flowered weedy annual herb (*Capsella bursa-pastoris*) of the mustard family with flat heart-shaped pods

sheq•el \'she-kəl\ *noun, plural* **sheq•a•lim** \she-'kä-lim\ *variant of* SHEKEL

Sher•a•ton \'sher-ə-tᵊn\ *adjective* [Thomas *Sheraton*] (1883)
: of, relating to, or being a style of furniture that originated in England around 1800 and is characterized by straight lines and graceful proportions

sher•bet \'shər-bət\ *also* **sher•bert** \-bərt\ *noun* [Turkish & Persian; Turkish *şerbet,* from Persian *sharbat,* from Arabic *sharbah* drink] (1603)
1 : a cold drink of sweetened and diluted fruit juice
2 : an ice with milk, egg white, or gelatin added

sherd *variant of* SHARD

sher•got•tite \'shər-gə-ˌtīt\ *noun* [*Shergotty* (Sherghati), town in India] (circa 1911)
: any of a class of rare achondritic geologically young meteorites that are apparently composed of solidified lava from celestial bodies other than earth

she•rif \shə-'rēf\ *variant of* SHARIF

sher•iff \'sher-əf\ *noun* [Middle English *shirreve,* from Old English *scīrgerēfa,* from *scīr* shire + *gerēfa* reeve — more at SHIRE, REEVE] (before 12th century)
: an important official of a shire or county charged primarily with judicial duties (as executing the processes and orders of courts and judges)
— sher•iff•dom \-əf-dəm, -təm\ *noun*

sher•lock \'shər-ˌläk, -lək\ *noun, often capitalized* [*Sherlock* Holmes, detective in stories by Sir Arthur Conan Doyle] (1903)
: DETECTIVE

Sher•pa \'sher-pə, 'shər-\ *noun* (1847)
: a member of a Tibetan people living on the high southern slopes of the Himalayas in eastern Nepal who provide support for foreign trekkers and mountain climbers

sher•ris \'sher-is\ *archaic variant of* SHERRY

sher•ry \'sher-ē\ *noun, plural* **sherries** [alteration of earlier *sherris* (taken as plural), from *Xeres* (now *Jerez*), Spain] (1597)
: a Spanish fortified wine with a distinctive nutty flavor; *also* **:** a similar wine produced elsewhere ◆

she's \'shēz\ (circa 1592)
: she is **:** she has

Shet•land \'shet-lənd\ *noun* (1836)
1 a : SHETLAND PONY **b :** SHETLAND SHEEPDOG
2 *often not capitalized* **a :** a lightweight loosely twisted yarn of Shetland wool used for knitting and weaving **b :** a fabric or a garment made from Shetland wool

Shetland pony *noun* (1801)
: any of a breed of small stocky hardy ponies that originated in the Shetland Islands

Shetland sheepdog *noun* (1909)
: any of a breed of small heavy-coated dogs developed in the Shetland Islands that resemble miniature collies

Shetland wool *noun* (1790)
: fine wool from sheep raised in the Shetland Islands; *also* **:** yarn spun from this

sheugh \'shük\ *noun* [Middle English *sogh* swamp; akin to Middle Low German *sō* gutter] (1501)
chiefly Scottish **:** DITCH, TRENCH

shew \'shō\ *British variant of* SHOW

shew•bread \'shō-ˌbred\ *noun* [translation of German *Schaubrot*] (1530)
: consecrated unleavened bread ritually placed by the Jewish priests of ancient Israel on a table in the sanctuary of the Tabernacle on the Sabbath

Shia \'shē-(ˌ)ä\ *noun* [Arabic *shī'ah* sect] (1626)
1 : the Muslims of the branch of Islam comprising sects believing in Ali and the Imams as the only rightful successors of Muhammad and in the concealment and messianic return of the last recognized Imam — compare SUNNI
2 : SHIITE
3 : the branch of Islam formed by the Shia

◇ WORD HISTORY

sherry Wines and distilled spirits are often named after the places where they are made. The region around the town of Jerez (formerly Xeres) in southern Spain produced a type of fortified wine with a rather nutty flavor that was introduced into England in the 16th century. The name *Xeres* was rendered *Sherries* or *Sherris* in English—the initial *sh* was a good approximation of the contemporary pronunciation of the Spanish sound represented by *x* (which has since changed to a \k\ or \h\ sound). Consequently, the wine from Xeres was sold under the name *sherris.* Some English speakers, however, believing that *sherris* was a plural noun, created a singular form *sherry,* which has since become standard. Other English words created by this process of back-formation from a presumed plural form are *pea* from *pease* and *cherry* from *cherise,* a northern dialectal form of Old French *cerise* "cherry."

shi•at•su *also* **shi•at•zu** \shē-ˈät-(ˌ)sü\ *noun, often capitalized* [short for Japanese *shiatsu-ryōhō,* from *shi-* finger + *-atsu* pressure + *ryōhō* treatment] (1967)
: a massage with the fingers applied to those specific areas of the body used in acupuncture

shib•bo•leth \ˈshi-bə-ləth *also* -ˌleth\ *noun* [Hebrew *shibbōleth* stream; from the use of this word in Judges 12:6 as a test to distinguish Gileadites from Ephraimites, who pronounced it *sibbōleth*] (1638)
1 a : a word or saying used by adherents of a party, sect, or belief and usually regarded by others as empty of real meaning ⟨the old *shibboleths* come rolling off their lips —Joseph Epstein⟩ **b :** a widely held belief ⟨today this book publishing *shibboleth* is a myth —L. A. Wood⟩ **c :** TRUISM, PLATITUDE ⟨some truth in the *shibboleth* that crime does not pay —Lee Rogow⟩
2 a : a use of language regarded as distinctive of a particular group ⟨accent was . . . a *shibboleth* of social class —Vivian Ducat⟩ **b :** a custom or usage regarded as distinguishing one group from others ⟨for most of the well-to-do in the town, dinner was a *shibboleth,* its hour dividing mankind —Osbert Sitwell⟩ ◆

shiel \ˈshē(ə)l\ *noun* [Middle English (northern dialect) *schele*] (13th century)
chiefly Scottish **:** SHIELING

¹shield \ˈshē(ə)ld\ *noun* [Middle English *sheld,* from Old English *scield;* akin to Old High German *scilt* shield and probably to Old English *sciell* shell] (before 12th century)
1 : a broad piece of defensive armor carried on the arm
2 : one that protects or defends **:** DEFENSE
3 : DRESS SHIELD
4 a : a device or part that serves as a protective cover or barrier **b :** a protective structure (as a carapace, scale, or plate) of some animals
5 : ESCUTCHEON; *especially* **:** one that is wide at the top and rounds to a point at the bottom
6 : the Precambrian nuclear mass of a continent that is surrounded and sometimes covered by sedimentary rocks
7 : something resembling a shield: as **a :** APOTHECIUM **b :** a police officer's badge **c :** a decorative or identifying emblem

²shield *transitive verb* (before 12th century)
1 a : to protect with or as if with a shield **:** provide with a protective cover or shelter **b :** to cut off from observation **:** HIDE
2 *obsolete* **:** FORBID
synonym see DEFEND
— shield•er *noun*

shield law *noun* (1971)
: a law that protects journalists from forced disclosure of confidential news sources

shield volcano *noun* (1911)
: a broad rounded volcano that is built up by successive outpourings of very fluid lava

shiel•ing \ˈshē-lən, -liŋ\ *noun* (1568)
1 *British* **:** a mountain hut used as a shelter by shepherds
2 *dialect British* **:** a summer pasture in the mountains

shier *comparative of* SHY

shiest *superlative of* SHY

¹shift \ˈshift\ *verb* [Middle English, from Old English *sciftan* to divide, arrange; akin to Old Norse *skipa* to arrange, assign] (13th century)
transitive verb
1 : to exchange for or replace by another **:** CHANGE
2 a : to change the place, position, or direction of **:** MOVE **b :** to make a change in (place)
3 : to change phonetically
intransitive verb
1 a : to change place or position **b :** to change direction ⟨the wind *shifted*⟩ **c :** to change gears **d :** to depress the shift key (as on a typewriter)
2 a : to assume responsibility ⟨had to *shift* for themselves⟩ **b :** to resort to expedients
3 a : to go through a change **b :** to change one's clothes **c :** to become changed phonetically
— shift•able \ˈshif-tə-bəl\ *adjective*
— shift•er *noun*
— shift gears : to make a change

²shift *noun* (1523)
1 a : a means or device for effecting an end **b** (1) **:** a deceitful or underhand scheme **:** DODGE (2) **:** an expedient tried in difficult circumstances **:** EXTREMITY
2 a *chiefly dialect* **:** a change of clothes **b** (1) *chiefly dialect* **:** SHIRT (2) **:** a woman's slip or chemise (3) **:** a usually loose-fitting or semifitted dress
3 a : a change in direction ⟨a *shift* in the wind⟩ **b :** a change in emphasis, judgment, or attitude
4 a : a group of people who work or occupy themselves in turn with other groups **b** (1) **:** a change of one group of people (as workers) for another in regular alternation (2) **:** a scheduled period of work or duty
5 : a change in place or position: as **a :** a change in the position of the hand on a fingerboard (as of a violin) **b** (1) **:** FAULT 5 (2) **:** the relative displacement of rock masses on opposite sides of a fault or fault zone **c** (1) **:** a simultaneous change of position in football by two or more players from one side of the line to the other (2) **:** a change of positions made by one or more players in baseball to provide better defense against a particular hitter **d :** a change in frequency resulting in a change in position of a spectral line or band — compare DOPPLER EFFECT **e :** a movement of bits in a computer register a specified number of places to the right or left
6 : a removal from one person or thing to another **:** TRANSFER
7 : CONSONANT SHIFT
8 : a bid in bridge in a suit other than the suit one's partner has bid — compare JUMP
9 : GEARSHIFT
synonym see RESOURCE

shift key *noun* (1893)
: a key on a keyboard (as of a typewriter) that when pressed enables an alternate set of characters to be printed

shift•less \ˈshif(t)-ləs\ *adjective* [*shift* (resourcefulness)] (1584)
1 : lacking in resourcefulness **:** INEFFICIENT
2 : lacking in ambition or incentive **:** LAZY
— shift•less•ly *adverb*
— shift•less•ness *noun*

shifty \ˈshif-tē\ *adjective* **shift•i•er; -est** (circa 1570)
1 : full of or ready with expedients **:** RESOURCEFUL
2 a : given to deception, evasion, or fraud **:** TRICKY **b :** capable of evasive movement **:** ELUSIVE ⟨a *shifty* boxer⟩
3 : indicative of a tricky nature ⟨*shifty* eyes⟩
— shift•i•ly \-tə-lē\ *adverb*
— shift•i•ness \-tē-nəs\ *noun*

shi•gel•la \shi-ˈge-lə\ *noun, plural* **-gel•lae** \-ˈge-(ˌ)lē, -(ˌ)lī\ *also* **-gellas** [New Latin, from Kiyoshi *Shiga* (died 1957) Japanese bacteriologist] (1937)
: any of a genus (*Shigella*) of nonmotile rod-shaped bacteria that cause dysenteries in animals and especially humans

shig•el•lo•sis \ˌshi-gə-ˈlō-səs\ *noun, plural* **-lo•ses** \-ˈlō-ˌsēz\ [New Latin] (1944)
: dysentery caused by shigellae

shih tzu \ˈshēd-ˈzü, ˈshēt-ˈsü\ *noun, plural* **shih tzus** *also* **shih tzu** *often S&T capitalized* [Chinese (Beijing) *shīzi* (*gǒu*), from *shīzi* lion + *gǒu* dog] (1921)
: any of a breed of small short-legged dogs of Chinese origin that have a short muzzle and a long dense coat

Shi•ism \ˈshē-ˌi-zəm\ *noun* (circa 1883)
: Islam as taught by the Shia

shii•ta•ke \shē-ˈtä-kē\ *noun* [Japanese, from *shii,* the Japanese chinquapin + *take* mushroom] (1877)
: a dark Oriental mushroom (*Lentinus edodes* of the family Agaricaceae) widely cultivated especially on woods of the beech family for its edible flavorful cap

Shi•ite \ˈshē-ˌīt\ *noun* (1728)
: a Muslim of the Shia branch of Islam

¹shi•kar \shi-ˈkär\ *noun* [Hindi *shikār,* from Persian] (circa 1610)
India **:** HUNTING

²shikar *verb* **shi•karred; shi•kar•ring** (1872)
India **:** HUNT

shi•ka•ri \shi-ˈkär-ē, -ˈkar-\ *noun* [Hindi *shikārī,* from Persian, from *shikār*] (1827)
India **:** a big game hunter; *especially* **:** a professional hunter or guide

shik•sa *or* **shik•se** \ˈshik-sə\ *noun* [Yiddish *shikse,* feminine of *sheygets* non-Jewish boy, from Hebrew *sheqeṣ* blemish, abomination] (1892)
1 : a non-Jewish girl or woman — often used disparagingly
2 : a Jewish girl or woman who does not observe Jewish precepts — used especially by Orthodox Jews

¹shill \ˈshil\ *intransitive verb* [²*shill*] (circa 1914)
1 : to act as a shill
2 : to act as a spokesperson or promoter ⟨the eminent Shakespearean producer, director, actor and star . . . is now *shilling* for a brokerage house —Andy Rooney⟩

²shill *noun* [perhaps short for *shillaber,* of unknown origin] (circa 1916)
1 : one who acts as a decoy (as for a pitchman or gambler); *also* **:** one who makes a sales pitch
2 : PITCH 8a

shil•le•lagh *also* **shil•la•lah** \shə-ˈlā-lē\ *noun* [*Shillelagh,* town in Ireland] (1772)
: CUDGEL

shil•ling \ˈshi-liŋ\ *noun* [Middle English, from Old English *scilling;* akin to Old High German *skilling,* a gold coin] (before 12th century)
1 a : a former monetary unit of the United Kingdom equal to 12 pence or 1/20 pound **b :** a former monetary unit equal to 1/20 pound of any of various countries in or formerly in the Commonwealth
2 : a coin representing one shilling
3 : any of several early American coins

◇ WORD HISTORY

shibboleth *Shibboleth* is an example of an English word whose meaning is based upon its use and not its literal sense in another language. In the 12th chapter of the biblical book of Judges there is an account of a battle between the armies of Gilead and Ephraim. The Ephraimite army was routed, and the retreating Ephraimites tried to cross the Jordan River at a ford held by the men of Gilead. Anyone wishing to pass was asked if he were an Ephraimite. If the reply was "no", he was asked to say the word *shibboleth.* In Hebrew *shibbōleth* happens to mean "stream," but the reason for choosing it as a password was that the Ephraimites could not pronounce the sound \sh\. If a man gave the password as *sibbōleth,* he betrayed his true identity and was killed. The role of *shibboleth* in this biblical episode led to its allusive use in English, and that use generated over time a number of new meanings.

4 — see MONEY table

Shil·luk \shi-ˈlük\ *noun, plural* **Shilluk** *or* **Shilluks** (1790)
1 : a member of a Nilotic people of the Sudan dwelling mainly on the west bank of the White Nile
2 : the language of the Shilluk people

[1]**shilly–shally** \ˈshi-lē-ˌsha-lē\ *adverb* [irregular reduplication of *shall I*] (1700)
: in an irresolute, undecided, or hesitating manner

[2]**shilly–shally** *adjective* (1734)
: IRRESOLUTE, VACILLATING

[3]**shilly–shally** *noun* (1755)
: INDECISION, IRRESOLUTION

[4]**shilly–shally** *intransitive verb* **shilly–shall·ied; shilly–shally·ing** (1782)
1 : to show hesitation or lack of decisiveness or resolution
2 : DAWDLE

shil·pit \ˈshil-pət\ *adjective* [origin unknown] (1812)
1 *Scottish* **:** pinched and starved in appearance
2 *Scottish* **:** WEAK, INSIPID — used of drink

[1]**shim** \ˈshim\ *noun* [origin unknown] (1860)
: a thin often tapered piece of material (as wood, metal, or stone) used to fill in space between things (as for support, leveling, or adjustment of fit)

[2]**shim** *transitive verb* **shimmed; shim·ming** (circa 1890)
: to fill out or level up by the use of a shim

[1]**shim·mer** \ˈshi-mər\ *verb* **shim·mered; shim·mer·ing** \ˈshi-mə-riŋ, ˈshim-riŋ\ [Middle English *schimeren,* from Old English *scimerian;* akin to Old English *scīnan* to shine — more at SHINE] (before 12th century)
intransitive verb
1 : to shine with a soft tremulous or fitful light **:** GLIMMER
2 : to reflect a wavering sometimes distorted visual image
transitive verb
: to cause to shimmer
synonym see FLASH

[2]**shimmer** *noun* (1821)
1 : a light that shimmers **:** subdued sparkle or sheen **:** GLIMMER
2 : a wavering sometimes distorted visual image usually resulting from heat-induced changes in atmospheric refraction
— shim·mery \ˈshi-mə-rē, ˈshim-rē\ *adjective*

[1]**shim·my** \ˈshi-mē\ *noun, plural* **shimmies** (1837)
1 [by alteration] **:** CHEMISE
2 [short for *shimmy-shake*] **:** a jazz dance characterized by a shaking of the body from the shoulders down
3 : an abnormal vibration especially in the front wheels of a motor vehicle

[2]**shimmy** *intransitive verb* **shim·mied; shim·my·ing** (1919)
1 : to shake, quiver, or tremble in or as if in dancing a shimmy
2 : to vibrate abnormally — used especially of automobiles

[1]**shin** \ˈshin\ *noun* [Middle English *shine,* from Old English *scinu;* akin to Old High German *scina* shin, Old English *scīa* shin, leg] (before 12th century)
: the front part of the vertebrate leg below the knee

[2]**shin** *verb* **shinned; shin·ning** (1829)
intransitive verb
1 : to move oneself up or down something vertical (as a pole) especially by alternately hugging it with the arms or hands and the legs
2 : to move forward rapidly on foot
transitive verb
1 : to kick or strike on the shins
2 : to climb by shinning

[3]**shin** \ˈshēn, ˈshin\ *noun* [Hebrew *shīn*] (circa 1823)
: the 22d letter of the Hebrew alphabet — see ALPHABET table

Shin \ˈshin, ˈshēn\ *noun* [Japanese, literally, truth] (1877)
: a major Japanese Buddhist sect that emphasizes salvation by faith in exclusive worship of Amida Buddha

Shi·na \ˈshē-nə\ *noun* (1854)
: an Indo-Aryan language spoken in Gilgit in northern Kashmir

shin·bone \ˈshin-ˌbōn\ *noun* (before 12th century)
: TIBIA 1

shin·dig \ˈshin-ˌdig\ *noun* [probably alteration of *shindy*] (1871)
1 a : a social gathering with dancing **b :** a usually large or lavish party
2 : SHINDY 2

shin·dy \ˈshin-dē\ *noun, plural* **shindys** *or* **shindies** [probably alteration of [1]*shinny*] (1821)
1 : SHINDIG 1
2 : FRACAS, UPROAR

[1]**shine** \ˈshīn\ *verb* **shone** \ˈshōn, *especially Canadian & British* ˈshän\ *or* **shined; shin·ing** [Middle English, from Old English *scīnan;* akin to Old High German *skīnan* to shine and perhaps to Greek *skia* shadow] (before 12th century)
intransitive verb
1 : to emit rays of light
2 : to be bright by reflection of light
3 : to be eminent, conspicuous, or distinguished ⟨*shines* in math⟩
4 : to have a bright glowing appearance ⟨his face *shone* with enthusiasm⟩
5 : to be conspicuously evident or clear
transitive verb
1 a : to cause to emit light **b :** to throw or direct the light of
2 *past & past part* **shined :** to make bright by polishing ⟨*shined* his shoes⟩

[2]**shine** *noun* (15th century)
1 : brightness caused by the emission of light
2 : brightness caused by the reflection of light **:** LUSTER
3 : BRILLIANCE, SPLENDOR
4 : fair weather **:** SUNSHINE ⟨rain or *shine*⟩
5 : TRICK, CAPER — usually used in plural
6 : LIKING, FANCY ⟨took a *shine* to him⟩
7 a : a polish or gloss given to shoes **b :** a single polishing of a pair of shoes

shin·er \ˈshī-nər\ *noun* (14th century)
1 : one that shines
2 : a silvery fish; *especially* **:** any of numerous small freshwater American cyprinid fishes (especially genus *Notropis*) — compare GOLDEN SHINER
3 : BLACK EYE 1

[1]**shin·gle** \ˈshiŋ-gəl\ *noun* [Middle English *schingel*] (13th century)
1 : a small thin piece of building material often with one end thicker than the other for laying in overlapping rows as a covering for the roof or sides of a building
2 : a small signboard especially designating a professional office — used chiefly in the phrase *hang out one's shingle*
3 : a woman's haircut with the hair trimmed short from the back of the head to the nape

[2]**shingle** *transitive verb* **shin·gled; shin·gling** \-g(ə-)liŋ\ (1562)
1 : to cover with or as if with shingles
2 : to bob and shape (the hair) in a shingle
3 : to lay out or arrange so as to overlap
— shin·gler \-g(ə-)lər\ *noun*

[3]**shingle** *noun* [probably of Scandinavian origin; akin to Norwegian *singel* coarse gravel] (15th century)
1 : coarse rounded detritus or alluvial material especially on the seashore that differs from ordinary gravel only in the larger size of the stones
2 : a place strewn with shingle
— shin·gly \-g(ə-)lē\ *adjective*

shin·gles \ˈshiŋ-gəlz\ *noun plural but singular in construction* [Middle English *schingles,* by folk etymology from Medieval Latin *cingulus,* from Latin *cingulum* girdle — more at CINGULUM] (14th century)
: HERPES ZOSTER

Shin·gon \ˈshin-ˌgän, ˈshēn-\ *noun* [Japanese] (1727)
: an esoteric Japanese Buddhist sect claiming the achievement of Buddhahood in this life through prescribed rituals

shin·ing \ˈshī-niŋ\ *adjective* (before 12th century)
1 : emitting or reflecting light
2 : bright and often splendid in appearance **:** RESPLENDENT
3 : possessing a distinguished quality **:** ILLUSTRIOUS
4 : full of sunshine

shin·leaf \ˈshin-ˌlēf\ *noun, plural* **shinleafs** (circa 1818)
: any of several wintergreens (especially *Pyrola elliptica*) with lustrous evergreen basal leaves and racemose white or pinkish flowers

shin·nery \ˈshi-nə-rē\ *noun, plural* **-ner·ies** [modification of Louisiana French *chênière,* from French *chêne* oak] (1901)
: a dense growth of small trees or an area of such growth; *especially* **:** one of scrub oak in the West and Southwest

[1]**shin·ny** *also* **shin·ney** \ˈshi-nē\ *noun* [perhaps from [1]*shin*] (1672)
: a variation of hockey played by children with a curved stick and a ball or block of wood; *also* **:** the stick used

[2]**shinny** *intransitive verb* **shin·nied; shin·ny·ing** [alteration of [2]*shin*] (1851)
: SHIN 1

shin·plas·ter \ˈshin-ˌplas-tər\ *noun* (1824)
1 : a piece of privately-issued paper currency; *especially* **:** one poorly secured and depreciated in value
2 : a piece of fractional currency

shin·splints \ˈshin-ˌsplin(t)s\ *noun plural but singular or plural in construction* (circa 1930)
: injury to and inflammation of the tibial and toe extensor muscles or their fasciae caused by repeated minimal traumas (as by running)

Shin·to \ˈshin-(ˌ)tō\ *noun* [Japanese *shintō*] (1727)
: the indigenous religion of Japan consisting chiefly in the cultic devotion to deities of natural forces and veneration of the Emperor as a descendant of the sun goddess
— Shinto *adjective*
— Shin·to·ism \-(ˌ)tō-ˌi-zəm\ *noun*
— Shin·to·ist \-ˌtō-ist\ *noun or adjective*
— Shin·to·is·tic \ˌshin-tō-ˈis-tik\ *adjective*

shiny \ˈshī-nē\ *adjective* **shin·i·er; -est** (1590)
1 a : bright with the rays of the sun **:** SUNSHINY **b :** filled with light
2 : bright in appearance **:** POLISHED ⟨*shiny* new shoes⟩
3 : rubbed or worn smooth
4 : lustrous with natural secretions ⟨a *shiny* nose⟩
— shin·i·ness *noun*

[1]**ship** \ˈship\ *noun, often attributive* [Middle English, from Old English *scip;* akin to Old High German *skif* ship] (before 12th century)
1 a : a large seagoing vessel **b :** a sailing vessel having a bowsprit and usually three masts each composed of a lower mast, a topmast, and a topgallant mast
2 : BOAT; *especially* **:** one propelled by power or sail
3 : a ship's crew
4 : FORTUNE 3 ⟨when their *ship* comes in they'll be able to live in better style⟩
5 : AIRSHIP, AIRPLANE, SPACECRAFT

[2]**ship** *verb* **shipped; ship·ping** (14th century)
transitive verb
1 a : to place or receive on board a ship for transportation by water **b :** to cause to be transported ⟨*shipped* him off to prep school⟩
2 *obsolete* **:** to provide with a ship

3 : to put in place for use ⟨*ship* the tiller⟩
4 : to take into a ship or boat ⟨*ship* the gangplank⟩
5 : to engage for service on a ship
6 : to take (as water) over the side — used of a boat or a ship
intransitive verb
1 : to embark on a ship
2 a : to go or travel by ship — often used with *out* **b :** to proceed by ship or other means under military orders — often used with *out*
3 : to engage to serve on shipboard
— **ship·pa·ble** \ˈshi-pə-bəl\ *adjective*

-ship *noun suffix* [Middle English, from Old English *-scipe;* akin to Old High German *-scaft* -ship, Old English *scieppan* to shape — more at SHAPE]
1 : state **:** condition **:** quality ⟨friend*ship*⟩
2 : office **:** dignity **:** profession ⟨clerk*ship*⟩
3 : art **:** skill ⟨horseman*ship*⟩
4 : something showing, exhibiting, or embodying a quality or state ⟨town*ship*⟩
5 : one entitled to a (specified) rank, title, or appellation ⟨his Lord*ship*⟩
6 : the body of persons participating in a specified activity ⟨reader*ship*⟩ ⟨listener*ship*⟩

ship biscuit *noun* (1799)
: HARDTACK — called also *ship bread*

[1]**ship·board** \ˈship-ˌbōrd, -ˌbȯrd\ *noun* (13th century)
1 : the side of a ship
2 : SHIP ⟨met on *shipboard*⟩

[2]**shipboard** *adjective* (1857)
: existing or taking place on board a ship

ship·borne \ˈship-ˌbōrn, -ˌbȯrn\ *adjective* (circa 1835)
: transported or designed to be transported by ship ⟨*shipborne* aircraft⟩

ship·build·er \ˈship-ˌbil-dər\ *noun* (circa 1700)
: one who designs or constructs ships
— **ship·build·ing** \-diŋ\ *noun*

ship·fit·ter \ˈship-ˌfi-tər\ *noun* (1941)
1 : one that fits together the structural members of ships and puts them into position for riveting or welding
2 : a naval enlisted man who works in sheet metal and performs the work of a plumber aboard ship

ship·lap \-ˌlap\ *noun* (1895)
: wooden sheathing in which the boards are rabbeted so that the edges of each board lap over the edges of adjacent boards to make a flush joint

ship·load \-ˈlōd, -ˌlōd\ *noun* (1639)
1 : as much or as many as will fill or load a ship
2 : an indefinitely large amount or number

ship·man \-mən\ *noun* (before 12th century)
1 : SAILOR, SEAMAN
2 : SHIPMASTER

ship·mas·ter \-ˌmas-tər\ *noun* (14th century)
: the master or commander of a ship other than a warship

ship·mate \-ˌmāt\ *noun* (1748)
: a fellow sailor

ship·ment \-mənt\ *noun* (1802)
1 : the act or process of shipping
2 : the goods shipped

ship of state *noun* (1847)
: the affairs of a state symbolized as a ship on a course

ship of the line (1706)
: a warship large enough to have a place in the line of battle

ship·own·er \ˈship-ˌō-nər\ *noun* (circa 1530)
: the owner of a ship or of a share in a ship

ship·per \ˈshi-pər\ *noun* (1755)
: one that sends goods by any form of conveyance

ship·ping \ˈshi-piŋ\ *noun* (14th century)
1 a : passage on a ship **b :** SHIPS **c :** the body of ships in one place or belonging to one port or country
2 : the act or business of one that ships

shipping clerk *noun* (circa 1858)
: one who is employed in a shipping room to assemble, pack, and send out or receive goods

ship·shape \ˌship-ˈshāp, ˈship-ˌ\ *adjective* [short for earlier *shipshapen,* from *ship* + *shapen,* archaic past participle of *shape*] (1644)
: TRIM, TIDY

ship·side \ˈship-ˌsīd\ *noun* (15th century)
: the area adjacent to a ship; *specifically* **:** a dock at which a ship loads or unloads passengers and freight

ship's papers *noun plural*
: the papers a ship is legally required to carry for due inspection to show the character of the ship and cargo

ship·way \ˈship-ˌwā\ *noun* (1834)
: the ways on which a ship is built

ship·worm \-ˌwərm\ *noun* (circa 1778)
: any of various elongated marine clams (especially family Teredinidae) that resemble worms, burrow in submerged wood, and damage wharf piles and wooden ships

[1]**ship·wreck** \-ˌrek\ *noun* [alteration of earlier *shipwrack,* from Middle English *schipwrak,* from Old English *scipwræc,* from *scip* ship + *wræc* something driven by the sea — more at WRACK] (12th century)
1 : a wrecked ship or its parts
2 : the destruction or loss of a ship
3 : an irretrievable loss or failure

[2]**shipwreck** *transitive verb* (1589)
1 a : to cause to experience shipwreck **b :** RUIN
2 : to destroy (a ship) by grounding or foundering

ship·wright \ˈship-ˌrīt\ *noun* (before 12th century)
: a carpenter skilled in ship construction and repair

ship·yard \-ˌyärd\ *noun* (circa 1700)
: a yard, place, or enclosure where ships are built or repaired

shire \ˈshīr, *in place-name compounds* ˌshir, shər\ *noun* [Middle English, from Old English *scīr* office, shire; akin to Old High German *scīra* care] (before 12th century)
1 : an administrative subdivision; *especially* **:** a county in England
2 : any of a breed of large heavy draft horses of British origin that have heavily feathered legs

shire 2

shire town *noun* (15th century)
1 *British* **:** a town that is the seat of the government of a shire
2 *New England* **:** a town where a court of superior jurisdiction (as a circuit court or a court with a jury) sits

shirk \ˈshərk\ *verb* [origin unknown] (1681)
intransitive verb
1 : to go stealthily **:** SNEAK
2 : to evade the performance of an obligation
transitive verb
: AVOID, EVADE ⟨*shirk* one's duty⟩
— **shirk·er** *noun*

shirr \ˈshər\ *transitive verb* [origin unknown] (1891)
1 : to draw (as cloth) together in a shirring
2 : to bake (eggs removed from the shell) until set

shirr·ing \ˈshər-iŋ\ *noun* (circa 1882)
: a decorative gathering (as of cloth) made by drawing up the material along two or more parallel lines of stitching

shirring

shirt \ˈshərt\ *noun* [Middle English *shirte,* from Old English *scyrte;* akin to Old Norse *skyrta* shirt, Old English *scort* short] (before 12th century)
1 : a garment for the upper part of the body: as **a :** a cloth garment usually having a collar, sleeves, a front opening, and a tail long enough to be tucked inside trousers or a skirt **b :** UNDERSHIRT
2 : all or a large part of one's money or resources ⟨lost his *shirt* on that business deal⟩
— **shirt·less** \-ləs\ *adjective*

shirt·dress \-ˌdres\ *noun* (1943)
: a tailored dress patterned after a shirt and having buttons down the front

shirt·front \-ˌfrənt\ *noun* (1838)
: the front of a shirt; *also* **:** the part of a man's shirt not covered by coat or vest

shirt·ing \ˈshər-tiŋ\ *noun* (1604)
: fabric suitable for shirts

shirt jacket *noun* (1879)
: a jacket designed in the style of a shirt — called also *shirt-jac*

shirt·mak·er \ˈshərt-ˌmā-kər\ *noun* (circa 1858)
: one that makes shirts

[1]**shirt·sleeve** \-ˌslēv\ *noun* (circa 1566)
: the sleeve of a shirt
— **in shirtsleeves :** wearing a shirt but no coat

[2]**shirtsleeve** *also* **shirt·sleeves** \-ˌslēvz\ *or* **shirt·sleeved** \-ˌslēvd\ *adjective* (1864)
1 a : being without a coat ⟨a *shirtsleeve* spectator⟩ **b :** calling for the removal of coats for the sake of comfort or efficiency ⟨*shirtsleeve* weather⟩
2 : marked by informality and directness ⟨*shirtsleeve* diplomacy⟩

[1]**shirt·tail** \ˈshərt-ˌtāl\ *adjective* (1845)
1 : very young **:** IMMATURE ⟨*shirttail* boys fishing in the creek⟩
2 : distantly and indefinitely related ⟨a *shirttail* cousin on her father's side⟩
3 : small, trivial, or short typically to the point of inadequacy ⟨has a gullied *shirttail* ranch in the hills⟩

[2]**shirttail** *noun* (1873)
1 : the part of a shirt that reaches below the waist especially in the back
2 : something small or inadequate

shirt·waist \ˈshərt-ˌwāst\ *noun* (1879)
: a woman's tailored garment (as a blouse or dress) with details copied from men's shirts

shirty \ˈshər-tē\ *adjective* (1846)
chiefly British **:** ANGRY, IRRITATED

shish ke·bab \ˈshish-kə-ˌbäb\ *noun* [Turkish *şişkebabı,* from *şiş* spit + *kebap* roast meat] (1914)
: kabob cooked on skewers

[1]**shit** \ˈshit, *interjectionally also* ˈshē-ət\ *noun* [(assumed) Middle English, from Old English *scite;* akin to Old English *-scītan* to defecate] (circa 1585)
1 : EXCREMENT — usually considered vulgar
2 : an act of defecation — usually considered vulgar
3 : NONSENSE, CRAP — usually considered vulgar
4 : any of several intoxicating or narcotic drugs; *especially* **:** HEROIN — usually considered vulgar
5 : DAMN 2 — usually considered vulgar
6 : a worthless, offensive, or detestable person — usually considered vulgar
— **shit·ty** \ˈshi-tē\ *adjective*

[2]**shit** \ˈshit\ *verb* **shit** *or* **shat** \ˈshat\; **shit·ting** [alteration of earlier *shite,* from Middle English *shiten,* from Old English *-scītan;* akin to Old High German *scīzan* to defecate and probably to Old English *scēadan* to separate — more at SHED] (circa 1720)

intransitive verb
: DEFECATE — usually considered vulgar
transitive verb
1 : to defecate in — usually considered vulgar
2 : to attempt to deceive : BULLSHIT — usually considered vulgar

shi·ta·ke *variant of* SHIITAKE

shit·tah \'shi-tə\ *noun, plural* **shittahs** *or* **shit·tim** \'shi-təm\ [Hebrew *shiṭṭāh*] (1611)
: a tree of uncertain identity but probably an acacia (as *Acacia seyal*) from the wood of which the ark and fittings of the Hebrew tabernacle were made

shit·tim·wood \'shi-təm-,wu̇d\ *noun* [Hebrew *shiṭṭīm* (plural of *shiṭṭāh*) + English *wood*] (1588)
1 : the wood of the shittah tree
2 : any of several trees (genus *Bumelia*, especially *B. lanuginosa*) of the sapodilla family of the southern U.S.; *also* **:** their hard heavy dense wood

shiv \'shiv\ *noun* [alteration of *chiv*, of unknown origin] (1674)
slang : KNIFE

Shi·va \'shi-və, 'shē-\ *variant of* SIVA

shi·vah *or* **shi·va** *also* **shi·ve** \'shi-və\ *noun* [Hebrew *shibh'āh* seven (days)] (1892)
: a traditional seven-day period of mourning the dead that is observed in Jewish homes — often used in the phrase *sit shivah*

shiv·a·ree \,shi-və-'rē, 'shi-və-,\ *noun* [modification of French *charivari* — more at CHARIVARI] (1843)
: a noisy mock serenade to a newly married couple ◆
— shivaree *transitive verb*

¹shiv·er \'shi-vər\ *noun* [Middle English; akin to Old High German *scivaro* splinter] (13th century)
: one of the small pieces into which a brittle thing is broken by sudden violence

²shiver *verb* **shiv·ered; shiv·er·ing** \'shi-və-riŋ, 'shiv-riŋ\ (13th century)
: to break into many small pieces : SHATTER

³shiver *verb* **shiv·ered; shiv·er·ing** \'shi-və-riŋ, 'shiv-riŋ\ [Middle English, alteration of *chiveren*] (15th century)
intransitive verb
1 : to undergo trembling : QUIVER
2 : to tremble in the wind as it strikes first one and then the other side (of a sail)
transitive verb
: to cause (a sail) to shiver by steering close to the wind

⁴shiver *noun* (1727)
1 : an instance of shivering : TREMBLE
2 : an intense shivery sensation especially of fear — often used in plural with *the* ⟨horror movies give him the *shivers*⟩

¹shiv·ery \'shi-və-rē, 'shiv-rē\ *adjective* (1683)
: easily broken into shivers

²shivery *adjective* (1747)
1 : characterized by shivers
2 : causing shivers

shle·miehl *variant of* SCHLEMIEL

shlep, shlepp *variant of* SCHLEPP

shlock *variant of* SCHLOCK

shm- — see SCHM-

¹shoal \'shōl\ *adjective* [alteration of Middle English *shold*, from Old English *sceald* — more at SKELETON] (1554)
: SHALLOW

²shoal *noun* (1555)
1 : SHALLOW
2 : a sandbank or sandbar that makes the water shallow; *specifically* **:** an elevation which is not rocky and on which there is a depth of water of six fathoms (11 meters) or less

³shoal (1574)
intransitive verb
: to become shallow
transitive verb
1 : to come to a shallow or less deep part of
2 : to cause to become shallow or less deep

⁴shoal *noun* [(assumed) Middle English *shole*, from Old English *scolu* multitude — more at SCHOOL] (1579)
: a large group or number : CROWD ⟨a *shoal* of fish⟩

⁵shoal (1610)
intransitive verb
: THRONG, SCHOOL

shoat \'shōt\ *noun* [Middle English *shote;* akin to Flemish *schote* shoat] (15th century)
: a young hog usually less than one year old

¹shock \'shäk\ *noun* [Middle English; akin to Middle High German *schoc* heap] (14th century)
: a pile of sheaves of grain or stalks of Indian corn set up in a field with the butt ends down

²shock *transitive verb* (15th century)
: to collect into shocks

³shock *noun, often attributive* [Middle French *choc*, from *choquer* to strike against, from Old French *choquier*, probably of Germanic origin; akin to Middle Dutch *schocken* to jolt] (1565)
1 : the impact or encounter of individuals or groups in combat
2 a : a violent shake or jar : CONCUSSION **b :** an effect of such violence
3 a (1) **:** a disturbance in the equilibrium or permanence of something (2) **:** a sudden or violent mental or emotional disturbance **b :** something that causes such disturbance **c :** a state of being so disturbed
4 : a state of profound depression of the vital processes associated with reduced blood volume and pressure and caused usually by severe especially crushing injuries, hemorrhage, or burns
5 : sudden stimulation of the nerves and convulsive contraction of the muscles caused by the discharge of electricity through the animal body
6 a : STROKE 5 **b :** CORONARY THROMBOSIS
7 : SHOCK ABSORBER

⁴shock (1576)
transitive verb
1 a : to strike with surprise, terror, horror, or disgust **b :** to cause to undergo a physical or nervous shock **c :** to subject to the action of an electrical discharge
2 : to drive by or as if by a shock
intransitive verb
1 : to meet with a shock : COLLIDE
2 : to cause surprise or shock
— shock·able \'shä-kə-bəl\ *adjective*

⁵shock *adjective* [perhaps from ¹*shock*] (1681)
: BUSHY, SHAGGY

⁶shock *noun* (1819)
: a thick bushy mass (as of hair)

shock absorber *noun* (1906)
: any of several devices for absorbing the energy of sudden impulses or shocks in machinery or structures

shock·er \'shä-kər\ *noun* (circa 1824)
: one that shocks; *especially* **:** something horrifying or offensive (as a sensational film or work of fiction)

shock front *noun* (1949)
: the advancing edge of a shock wave

shock·ing \'shä-kiŋ\ *adjective* (1703)
: extremely startling, distressing, or offensive
— shock·ing·ly \-kiŋ-lē\ *adverb*

shocking pink *noun* (1938)
: a striking, vivid, bright, or intense pink

shock·proof \'shäk-,prüf\ *adjective* (1911)
1 : incapable of being shocked
2 a : resistant to damage by shock **b :** unlikely to cause shock **:** protectively insulated ⟨a *shockproof* switch⟩

shock therapy *noun* (1917)
: the treatment of mental disorder by the artificial induction of coma or convulsions through use of drugs or electric current — called also *shock treatment*

shock troops *noun plural* (1917)
1 : troops especially suited and chosen for offensive work because of their high morale, training, and discipline
2 : a group of people militant in pressing for a cause

shock tube *noun* (1949)
: a usually enclosed tube in which experimental shock waves are produced as a result of the rupturing of a diaphragm separating two chambers containing a gas or gases at differential pressure

shock wave *noun* (1907)
1 : a compressional wave of high amplitude caused by a shock (as from an earthquake or explosion) to the medium through which the wave travels
2 : a violent often pulsating disturbance or reaction ⟨*shock waves* of rebellion⟩

shod \'shäd\ *adjective* [Middle English, from past participle of *shoen* to shoe, from Old English *scōgan*, from *scōh* shoe — more at SHOE] (13th century)
1 a : wearing footgear (as shoes) **b :** equipped with tires
2 : furnished or equipped with a shoe

¹shod·dy \'shä-dē\ *noun* [origin unknown] (1832)
1 a : a reclaimed wool from materials that are not felted that is of better quality and longer staple than mungo **b :** a fabric often of inferior quality manufactured wholly or partly from reclaimed wool
2 a : inferior, imitative, or pretentious articles or matter **b :** pretentious vulgarity

²shoddy *adjective* **shod·di·er; -est** (1847)
1 : made wholly or partly of shoddy
2 a : cheaply imitative **:** vulgarly pretentious **b :** hastily or poorly done : INFERIOR **c :** SHABBY
— shod·di·ly \'shä-dᵊl-ē\ *adverb*
— shod·di·ness \'shä-dē-nəs\ *noun*

¹shoe \'shü\ *noun* [Middle English *shoo*, from Old English *scōh;* akin to Old High German *scuoh* shoe] (before 12th century)
1 a : an outer covering for the human foot typically having a thick or stiff sole with an attached heel and an upper part of lighter material (as leather) **b :** a metal plate or rim for the hoof of an animal
2 : something resembling a shoe in function or placement
3 *plural* **:** another's place, function, or viewpoint ⟨steps from assistant stage manager into the star's *shoes* —Steven Fuller⟩

◇ WORD HISTORY

shivaree In 19th century rural America, a newly wed couple—often one whose marriage was because of discrepancy in age or some other cause considered incongruous by the local community—might have their wedding night interrupted by an unwanted disturbance. A motley band of revelers outside the windows of their home would treat them to a cacophonous serenade, delivered in derision on a battery of pots, pans, cowbells, and other ersatz musical instruments. In the eastern U.S. this custom, imported from rural England, was simply called a *serenade* or known under a variety of colorful local names. In much of the central U.S. and Canada, however, it was called a *shivaree*, a loan from French *charivari*, a word that denotes the same folk custom in France. American English *shivaree* is part of our small linguistic heritage from Mississippi Valley French, the language of French pioneers who explored and settled the Mississippi and its tributaries. The greater part of this heritage consists of place names, from the obvious (*St. Louis* and *New Orleans*) to the buried (*Ozark*, a phonetic rendering of French *Aux Arcs* "At the Arcs," *Arcs* being a French name for several tribes of Indians living in that area).

4 : a device that retards, stops, or controls the motion of an object; *especially* **:** the part of a brake that presses on the brake drum
5 a : any of various devices that are inserted in or run along a track or groove to guide a movement, provide a contact or friction grip, or protect against wear, damage, or slipping **b :** a device (as a clip or track) on a camera that permits attachment of an accessory item (as a flash unit)
6 : a dealing box designed to hold several decks of playing cards
— shoe·less *adjective*

[2]shoe *transitive verb* **shod** \'shäd\ *also* **shoed** \'shüd\; **shoe·ing** \'shü-iŋ\ (before 12th century)
1 : to furnish with a shoe
2 : to cover for protection, strength, or ornament

shoe·bill \'shü-ˌbil\ *noun* (1874)
: a large broad-billed wading bird (*Balaeniceps rex*) of the valley of the White Nile that is related to the storks and herons

shoe·black \-ˌblak\ *noun* (1778)
: BOOTBLACK

[1]shoe·horn \-ˌhȯrn\ *noun* (1589)
: a curved piece (as of horn, wood, or metal) used in putting on a shoe

[2]shoehorn *transitive verb* (1926)
: to force into a small, narrow, or insufficient space **:** SQUEEZE ⟨*shoehorn* the past, present, and future into about 500 pages —Otis Port⟩

shoe·lace \'shü-ˌlās\ *noun* (circa 1647)
: a lace or string for fastening a shoe

shoe·mak·er \-ˌmā-kər\ *noun* (14th century)
: one whose occupation is making or repairing shoes

shoe·pac *or* **shoe·pack** \'shü-ˌpak\ *noun* [by folk etymology from Delaware Jargon (Delaware-based pidgin) *seppock* shoe, from Delaware (Unami dialect) *čípahkɔ* shoes] (1755)
: a waterproof laced boot worn especially over heavy socks in cold weather

[1]shoe·string \'shü-ˌstriŋ\ *noun* (1616)
1 : SHOELACE
2 [from shoestrings being a typical item sold by itinerant vendors] **:** a small sum of money **:** capital inadequate or barely adequate to the needs of a transaction ⟨started the business on a *shoestring*⟩

[2]shoestring *adjective* (1878)
1 : narrow and long like a shoestring ⟨a *shoestring* tie⟩
2 : operating on, accomplished by, or consisting of a small amount of money or capital ⟨a *shoestring* budget⟩

shoestring catch *noun* (1926)
: a catch (as in baseball) made very close to the feet

shoe tree *noun* (1827)
: a foot-shaped device for inserting in a shoe to preserve its shape

sho·far \'shō-ˌfär, -fər\ *noun, plural* **sho·froth** \shō-'frōt, -'frōth, -'frōs\ [Hebrew *shōphār*] (1833)
: a ram's-horn trumpet blown by the ancient Hebrews in battle and high religious observances and used in synagogues before and during Rosh Hashanah and at the conclusion of Yom Kippur

shofar

[1]shog \'shäg\ *intransitive verb* **shogged; shog·ging** [Middle English *shoggen*] (15th century)
chiefly dialect **:** to move along

[2]shog *noun* (1611)
chiefly dialect **:** SHAKE, JOLT

sho·gun \'shō-gən\ *noun* [Japanese *shōgun* general] (1727)
: one of a line of military governors ruling Japan until the revolution of 1867–68
— sho·gun·al \'shō-gə-nəl\ *adjective*
— sho·gun·ate \'shō-gə-nət, -ˌnāt\ *noun*

sho·ji \'shō-(ˌ)jē\ *noun, plural* **shoji** *also* **shojis** [Japanese *shōji*] (1880)
: a paper screen serving as a wall, partition, or sliding door

sho·lom \shä-'lōm, shə-\ *variant of* SHALOM

Sho·na \'shō-nə\ *noun, plural* **Shona** *or* **Shonas** (circa 1895)
1 : a member of any of a group of Bantu peoples of Zimbabwe and southern Mozambique
2 : the group of languages spoken by the Shona

shone *past and past participle of* SHINE

[1]shoo \'shü\ *interjection* [Middle English *schowe*] (15th century)
— used especially in driving away an unwanted animal

[2]shoo *transitive verb* (circa 1798)
: to scare, drive, or send away by or as if by crying *shoo*

shoo·fly \'shü-ˌflī\ *noun* [[1]*shoo* + *fly*] (1887)
1 : a child's rocker having the seat built on or usually between supports representing an animal figure
2 : any of several plants held to repel flies

shoofly pie *noun* (1926)
: a rich pie of Pennsylvania-Dutch origin made of molasses or brown sugar sprinkled with a crumbly mixture of flour, sugar, and butter

shoo–in \'shü-ˌin\ *noun* (1937)
: one that is a certain and easy winner

[1]shook *past or chiefly dialect past participle of* SHAKE

[2]shook \'shu̇k\ *noun* [origin unknown] (1796)
1 a : a set of staves and headings for one hogshead, cask, or barrel **b :** a bundle of parts (as of boxes) ready to be put together
2 : [1]SHOCK

shook–up \ˌshu̇k-'əp\ *adjective* (1897)
: nervously upset **:** AGITATED

shoon \'shün, 'shōn\ *chiefly dialect plural of* SHOE

[1]shoot \'shüt\ *verb* **shot** \'shät\; **shoot·ing** [Middle English *sheten, shuten,* from Old English *scēotan;* akin to Old Norse *skjōta* to shoot] (before 12th century)
transitive verb
1 a (1) **:** to eject or impel or cause to be ejected or impelled by a sudden release of tension (as of a bowstring or slingshot or by a flick of a finger) ⟨*shoot* an arrow⟩ ⟨*shoot* a spitball⟩ ⟨*shoot* a marble⟩ (2) **:** to drive forth or cause to be driven forth by an explosion (as of a powder charge in a firearm or of ignited fuel in a rocket) (3) **:** to drive forth or cause to be driven forth by a sudden release of gas or air ⟨*shoot* darts from a blowgun⟩ ⟨a steam catapult *shoots* planes from a carrier⟩ (4) **:** to propel (as a ball or puck) toward a goal by striking or pushing with part of the body (as the hand or foot) or with an implement; *also* **:** to score by so doing ⟨*shoot* the winning goal⟩ ⟨*shoot* a basket⟩ (5) **:** to throw or cast off or out often with force ⟨*shoot* dice⟩ ⟨the horse *shot* his rider out of the saddle⟩ **b :** to cause (as a gun or bow) to propel a missile **c** (1) **:** to utter (as words or sounds) rapidly or suddenly or with force ⟨*shoot* out a stream of invective⟩ (2) **:** to emit (as light, flame, or fumes) suddenly and rapidly (3) **:** to send forth with suddenness or intensity ⟨*shot* a look of anger at them⟩ **d :** to discharge, dump, or empty especially by overturning, upending, or directing into a slide
2 : to affect by shooting: as **a :** to strike with a missile especially from a bow or gun; *especially* **:** to wound or kill with a missile discharged from a bow or firearm **b :** to remove or destroy by use of firearms ⟨*shot* out the light⟩; *also* **:** WRECK, EXPLODE
3 a : to push or slide (as the bolt of a door or lock) into or out of a fastening **b :** to push or thrust forward **:** stick out ⟨toads *shooting* out their tongues⟩ **c :** to put forth in growing **d :** to place, send, or bring into position abruptly
4 a : to engage in (a sport or game or a portion of a game that involves shooting) **:** PLAY ⟨*shoot* pool⟩ ⟨*shoot* a round of golf⟩ ⟨*shoot* craps⟩ **b** (1) **:** to place or offer (a bet) on the result of casting dice ⟨*shoot* $5⟩ (2) **:** to use up by or as if by betting **:** EXHAUST ⟨*shot* his whole wad on a shady deal⟩
5 a : to engage in the hunting and killing of (as game) with firearms especially as a sport ⟨*shoot* woodcock⟩ **b :** to hunt over ⟨*shoot* a tract of woodland⟩
6 a : to cause to move suddenly or swiftly forward ⟨*shot* the car onto the highway⟩ **b :** to send or carry quickly **:** DISPATCH ⟨*shoot* the letter on to me as soon as you receive it⟩
7 : to variegate as if by sprinkling color in streaks, flecks, or patches
8 : to pass swiftly by, past, or along ⟨*shooting* rapids⟩
9 : to plane (as the edge of a board) straight or true
10 a : SET OFF, DETONATE, IGNITE ⟨*shoot* a charge of dynamite⟩ **b :** to effect by blasting
11 : to determine the altitude of
12 : to take a picture or series of pictures or television images of **:** PHOTOGRAPH, FILM
13 a : to give an injection to **b :** to inject (an illicit drug) especially into the bloodstream
intransitive verb
1 a : to go or pass rapidly and precipitately ⟨sparks *shooting* all over⟩ ⟨his feet *shot* out from under him⟩ **b :** to move ahead by force of momentum **c :** to stream out suddenly **:** SPURT **d :** to dart in or as if in rays from a source of light **e :** to dart with a piercing sensation ⟨pain *shot* up my arm⟩
2 a : to cause an engine or weapon to discharge a missile **b :** to use a firearm or bow especially for sport (as in hunting)
3 : to propel a missile ⟨guns that *shoot* many miles⟩
4 : PROTRUDE, PROJECT
5 a : to grow or sprout by or as if by putting forth shoots **b :** DEVELOP, MATURE **c :** to spring or rise rapidly or suddenly — often used with *up* ⟨in a burst of growth he *shot* up to six feet tall⟩ ⟨prices *shot* up⟩
6 a : to propel an object (as a ball) in a particular way **b :** to drive the ball or puck toward a goal
7 : to cast dice
8 : to slide into or out of a fastening ⟨a bolt that *shoots* in either direction⟩
9 : to record something (as on film or videotape) with a camera
10 : to begin to speak — usually used as an imperative ⟨OK, *shoot,* what do you have to say⟩
— shoot at *or* **shoot for :** to aim at **:** strive for
— shoot from the hip : to act or speak hastily without consideration of the consequences
— shoot one's bolt : to exhaust one's capabilities and resources
— shoot oneself in the foot : to act against one's own best interests
— shoot the breeze : to converse idly **:** GOSSIP
— shoot the works 1 : to venture all one's capital on one play **2 :** to put forth all one's efforts

[2]shoot *noun* (15th century)
1 : a sending out of new growth or the growth sent out: as **a :** a stem or branch with its leaves and appendages especially when not yet mature **b :** OFFSHOOT

2 a : an act of shooting (as with a bow or a firearm): (1) **:** SHOT (2) **:** the firing of a missile especially by artillery **b** (1) **:** a hunting trip or party (2) **:** the right to shoot game in a particular area or land over which it is held **c** (1) **:** a shooting match ⟨skeet *shoot*⟩ (2) **:** a round of shots in a shooting match **d :** the action or an instance of shooting with a camera **:** a session or a series of sessions of photographing or filming ⟨a movie *shoot*⟩
3 a : a motion or movement of rapid thrusting: as (1) **:** a sudden or rapid advance (2) **:** a momentary darting sensation **:** TWINGE (3) **:** THRUST 2b (4) **:** the pace between strokes in rowing **b :** a bar of rays **:** BEAM ⟨a *shoot* of sunlight⟩
4 [probably by folk etymology from French *chute* — more at CHUTE] **a :** a rush of water down a steep or rapid **b :** a place where a stream runs or descends swiftly

³shoot *interjection* [euphemism for *shit*] (1876)
— used to express annoyance or surprise

shoot down *transitive verb* (1845)
1 : to cause to fall by shooting ⟨enemy aircraft *shot down* the helicopter⟩; *especially* **:** to kill in this way ⟨was *shot down* in cold blood⟩
2 : to put an end to **:** DEFEAT
3 : DEFLATE, RIDICULE
4 : DISCREDIT 2 ⟨*shoot down* a theory⟩

shoot–'em–up \'shüt-əm-ˌəp\ *noun* (1947)
: a movie or television show with much shooting and bloodshed

shoot•er \'shü-tər\ *noun* (13th century)
1 : one that shoots: as **a :** a person who fires a missile-discharging device (as a rifle or bow) **b :** the person who is shooting or whose turn it is to shoot
2 : something that is used in shooting: as **a :** a marble shot from the hand **b :** REVOLVER — usually used in combination ⟨six-*shooter*⟩
3 : a shot of whiskey or whiskey diluted with something (as soda); *also* **:** a shot glass containing a bit of food (as a raw oyster)

shooting gallery *noun* (1836)
1 : a usually covered range equipped with targets for practice with firearms
2 *slang* **:** a place where one can obtain narcotics and shoot up

shooting iron *noun* (1775)
: FIREARM

shooting script *noun* (1929)
1 : the final completely detailed version of a motion-picture script in which scenes are grouped in the order most convenient for shooting
2 : the final version of a television script used in the production of a program

shooting star *noun* (1593)
1 : a visual meteor appearing as a temporary streak of light in the night sky
2 : any of several North American perennial herbs (genus *Dodecatheon,* especially *D. meadia*) of the primrose family that have entire oblong leaves and showy flowers with reflexed petals

shooting stick *noun* (1926)
: a spiked stick with a top that opens into a seat

shoot–out \'shüt-ˌau̇t\ *noun* (1948)
1 : a battle fought with handguns or rifles
2 : something resembling a shoot-out; *broadly* **:** SHOWDOWN

shoot–the–chutes \ˌshüt-t͟hə-'shüts\ *noun plural but singular in construction* (1920)
: an amusement ride consisting of a steep incline down which boats slide into a pool at the bottom

shoot up (1890)
transitive verb
1 : to shoot or shoot at especially indiscriminately or recklessly ⟨cowboys *shooting up* the town⟩
2 : to inject (a narcotic drug) into a vein
intransitive verb
: to inject a narcotic into a vein
— shoot–up \'shüt-ˌəp\ *noun*

¹shop \'shäp\ *noun, often attributive* [Middle English *shoppe,* from Old English *sceoppa* booth; akin to Old High German *scopf* shed] (14th century)
1 : a handicraft establishment **:** ATELIER
2 a : a building or room stocked with merchandise for sale **:** STORE **b** *or* **shoppe** \'shäp\ **:** a small retail establishment or a department in a large one offering a specified line of goods or services ⟨a millinery *shop*⟩ ⟨a sandwich *shop*⟩
3 : a commercial establishment for the making or repair of goods or machinery ⟨machine *shop*⟩ ⟨repair *shop*⟩
4 a : a school laboratory equipped for manual training **b :** the art or science of working with tools and machinery
5 a : a business establishment **:** OFFICE **b :** SHOPTALK

²shop *verb* **shopped; shop•ping** (1764)
intransitive verb
1 a : to examine goods or services with intent to buy **b :** to hunt through a market in search of the best buy
2 : to make a search **:** HUNT
transitive verb
: to examine the stock or offerings of ⟨*shop* the stores for Christmas gift ideas⟩

shop•keep•er \'shäp-ˌkē-pər\ *noun* (1530)
: STOREKEEPER 2

shop•lift \-ˌlift\ *verb* [back-formation from *shoplifter*] (1820)
intransitive verb
: to steal displayed goods from a store
transitive verb
: to steal (displayed goods) from a store

shop•lift•er \-ˌlif-tər\ *noun* (1680)
: one who shoplifts

shop•per \'shä-pər\ *noun* (1860)
1 : one that shops
2 : one whose occupation is shopping as an agent for customers or for an employer
3 : a usually free paper carrying advertising and sometimes local news

shopping bag *noun* (1886)
: a bag (as of strong paper) that has handles and is intended for carrying purchases

shopping center *noun* (1898)
: a group of retail stores and service establishments usually with ample parking facilities and usually designed to serve a community or neighborhood — called also *shopping plaza*

shopping list *noun* (1913)
: a list of items to be purchased; *broadly* **:** a list of related items ⟨the biggest possible *shopping list* of budget cuts —Leonard Silk⟩

shopping mall *noun* (1959)
: MALL 3

shop steward *noun* (1904)
: a union member elected as the union representative of a shop or department in dealings with the management

shop•talk \'shäp-ˌtȯk\ *noun* (1881)
: the jargon or subject matter peculiar to an occupation or a special area of interest

shop•win•dow \-ˌwin-(ˌ)dō\ *noun* (15th century)
: a display window of a store

shop•worn \-ˌwōrn, -ˌwȯrn\ *adjective* (1838)
1 : faded, soiled, or otherwise impaired by remaining too long in a store
2 : stale from excessive use or familiarity ⟨*shopworn* clichés⟩
3 : WORN-OUT ⟨think of himself as a *shopworn* Hollywood cynic —A. H. Johnston⟩

¹shore \'shōr, 'shȯr\ *noun, often attributive* [Middle English, from (assumed) Old English *scor;* akin to Middle Low German *schōr* foreland and perhaps to Old English *scieran* to cut — more at SHEAR] (14th century)
1 : the land bordering a usually large body of water; *specifically* **:** COAST
2 : a boundary or the country or place that it bounds ⟨hold him accountable for difficulties beyond our *shores* that he could do nothing about —Dorothy Fosdick⟩
3 : land as distinguished from the sea ⟨shipboard and *shore* duty⟩

²shore *transitive verb* **shored; shor•ing** [Middle English; akin to Old Norse *skortha* to prop] (14th century)
1 : to support by a shore **:** PROP
2 : to give support to **:** BRACE — usually used with *up*

³shore *noun* (14th century)
: a prop for preventing sinking or sagging

shore•bird \'shōr-ˌbərd, 'shȯr-\ *noun* (circa 1672)
: any of a suborder (Charadrii) of birds (as a plover or snipe) that frequent the seashore

shore dinner *noun* (1892)
: a dinner consisting chiefly of seafoods

shore•front \-ˌfrənt\ *noun* (1919)
: land along a shore; *specifically* **:** BEACHFRONT

shore leave *noun* (1888)
: a leave of absence to go on shore granted to a sailor or naval officer

shore•line \-ˌlīn\ *noun* (1852)
1 : the line where a body of water and the shore meet
2 : the strip of land along the shoreline

shore patrol *noun* (1917)
1 : a branch of a navy that exercises guard and police functions — compare MILITARY POLICE
2 : petty officers detailed to perform police duty while a ship is in port

shore•side \-ˌsīd\ *adjective* (1883)
: situated at or near a shore

shore•ward \-wərd\ *or* **shore•wards** \-wərdz\ *adverb* (circa 1691)
: toward the shore

shor•ing \'shōr-iŋ, 'shȯr-\ *noun* (15th century)
1 : the act of supporting with or as if with a prop
2 : a system or group of shores

shorn *past participle of* SHEAR

¹short \'shȯrt\ *adjective* [Middle English, from Old English *sceort;* akin to Old High German *scurz* short, Old Norse *skortr* lack] (before 12th century)
1 a : having little length **b :** not tall or high **:** LOW
2 a : not extended in time **:** BRIEF ⟨a *short* vacation⟩ **b :** not retentive ⟨a *short* memory⟩ **c :** EXPEDITIOUS, QUICK ⟨made *short* work of the problem⟩ **d :** seeming to pass quickly ⟨made great progress in just a few *short* years⟩
3 a *of a speech sound* **:** having a relatively short duration **b :** being the member of a pair of similarly spelled vowel or vowel-containing sounds that is descended from a vowel that was short in duration but is no longer so and that does not necessarily have duration as its chief distinguishing feature ⟨*short* i in *sin*⟩ **c** *of a syllable in prosody* (1) **:** of relatively brief duration (2) **:** UNSTRESSED
4 : limited in distance ⟨a *short* trip⟩
5 a : not coming up to a measure or requirement **:** INSUFFICIENT ⟨in *short* supply⟩ **b :** not reaching far enough ⟨the throw to first was *short*⟩ **c :** enduring privation **d :** insufficiently supplied ⟨*short* of cash⟩ ⟨*short* on brains⟩
6 a : ABRUPT, CURT **b :** quickly provoked
7 : CHOPPY 2
8 : payable at an early date
9 a : containing or cooked with shortening; *also* **:** FLAKY ⟨*short* pastry⟩ **b** *of metal* **:** brittle under certain conditions
10 a : not lengthy or drawn out **b :** made briefer **:** ABBREVIATED
11 a : not having goods or property that one has sold in anticipation of a fall in prices **b :** consisting of or relating to a sale of securities or commodities that the seller does not possess or has not contracted for at the time of the sale ⟨*short* sale⟩
12 : near the end of a tour of duty
— short•ish \'shȯr-tish\ *adjective*
— in short order : with dispatch **:** QUICKLY

²short *adverb* (14th century)

1 : in a curt manner
2 : for or during a brief time ⟨*short*-lasting⟩
3 : at a disadvantage **:** UNAWARES ⟨caught *short*⟩
4 : in an abrupt manner **:** SUDDENLY ⟨the car stopped *short*⟩
5 : at some point or degree before a goal or limit aimed at or under consideration ⟨the shells fell *short*⟩ ⟨quit a month *short* of graduation⟩
6 : clean across ⟨the axle was snapped *short*⟩
7 : by or as if by a short sale
[3]**short** *noun* (circa 1586)
1 : the sum and substance **:** UPSHOT
2 a : a short syllable **b :** a short sound or signal
3 *plural* **a :** a by-product of wheat milling that includes the germ, fine bran, and some flour **b :** refuse, clippings, or trimmings discarded in various manufacturing processes
4 a : knee-length or less than knee-length trousers — usually used in plural **b** *plural* **:** short drawers **c :** a size in clothing for short men
5 a : one who operates on the short side of the market **b** *plural* **:** short-term bonds
6 *plural* **:** DEFICIENCIES
7 : SHORT CIRCUIT
8 : SHORTSTOP
9 a : SHORT SUBJECT **b :** a brief story or article (as in a newspaper)
— for short : as an abbreviation ⟨named Katherine or Kate *for short*⟩
— in short : by way of summary **:** BRIEFLY
[4]**short** *transitive verb* (1904)
1 : SHORT-CIRCUIT
2 : SHORTCHANGE, CHEAT
short account *noun* (circa 1902)
: the total of open short sales in a given subject of trade or in the market as a whole
short·age \'shȯr-tij\ *noun* (1868)
: LACK, DEFICIT
short ballot *noun* (1909)
: a ballot limiting the number of elective offices to the most important legislative and executive posts and leaving minor positions to be filled by appointment
short·bread \'shȯrt-ˌbred\ *noun* (1801)
: a thick cookie made of flour, sugar, and a large amount of shortening
short·cake \-ˌkāk\ *noun* (1594)
1 : a crisp and often unsweetened biscuit or cookie
2 a : a dessert made typically of very short baking-powder-biscuit dough spread with sweetened fruit **b :** a dish consisting of a rich biscuit split and covered with a meat mixture
short·change \-'chānj\ *transitive verb* (1903)
1 : to give less than the correct amount of change to
2 : to deprive of or give less than something due **:** CHEAT
— short·chang·er *noun*
short–cir·cuit *transitive verb* (1867)
1 : to apply a short circuit to or establish a short circuit in
2 : BYPASS
3 : FRUSTRATE, IMPEDE
short circuit *noun* (1854)
: a connection of comparatively low resistance accidentally or intentionally made between points on a circuit between which the resistance is normally much greater
short·com·ing \'shȯrt-ˌkə-miŋ, ˌshȯrt-'\ *noun* (15th century)
: DEFICIENCY, DEFECT
[1]**short·cut** \'shȯrt-ˌkət *also* -'kət\ *noun* (1637)
1 : a route more direct than the one ordinarily taken
2 : a method of doing something more directly and quickly than and often not so thoroughly as by ordinary procedure
[2]**shortcut** *verb* **-cut; -cut·ting** (1915)
transitive verb
: to shorten (as a route or procedure) by use of a shortcut; *also* **:** CIRCUMVENT
intransitive verb
: to take or use a shortcut
short–day \'shȯrt-ˌdā\ *adjective* (1920)
: responding to or relating to a short photoperiod — used especially of a plant; compare DAY-NEUTRAL, LONG-DAY
short division *noun* (circa 1890)
: mathematical division in which the successive steps are performed without writing out the remainders
short–eared owl \'shȯrt-ˌird-\ *noun* (1766)
: a medium-sized nearly cosmopolitan owl (*Asio flammeus*) that has very short ear tufts and usually nests on the ground
short·en \'shȯr-tᵊn\ *verb* **short·ened; short·en·ing** \'shȯrt-niŋ, 'shȯr-tᵊn-iŋ\ (14th century)
transitive verb
1 a : to reduce the length or duration of **b :** to cause to seem short
2 a : to reduce in power or efficiency ⟨is my hand *shortened,* that it cannot redeem —Isaiah 50:2 (Revised Standard Version)⟩ **b** *obsolete* **:** to deprive of effect
3 : to add fat to (as pastry dough) in order to make tender and flaky
intransitive verb
: to become short or shorter ☆
— short·en·er \'shȯrt-nər, 'shȯr-tᵊn-ər\ *noun*
short·en·ing \'shȯrt-niŋ, 'shȯr-tᵊn-iŋ\ *noun* (1538)
1 : the action or process of making or becoming short; *specifically* **:** the dropping of the latter part of a word so as to produce a new and shorter word of the same meaning
2 : an edible fat used to shorten baked goods
short·fall \'shȯrt-ˌfȯl\ *noun* (1895)
: a failure to come up to expectation or need; *also* **:** the amount of such failure
short–grass prairie \'shȯrt-ˌgras-\ *noun* (1844)
: PRAIRIE 2b
short·hair \-ˌhar, -ˌher\ *noun* (1903)
: a domestic cat with a short thick coat; *especially* **:** a member of any of several breeds of muscular medium- to large-sized cats with a short plushy coat
— short·haired *adjective*
short·hand \-ˌhand\ *noun* (1636)
1 : a method of writing rapidly by substituting characters, abbreviations, or symbols for letters, sounds, words, or phrases **:** STENOGRAPHY
2 : a system or instance of rapid or abbreviated communication or representation
— shorthand *adjective*
short·hand·ed \ˌshȯrt-'han-dəd\ *adjective* (1794)
: having, working with, or done with fewer than the regular or necessary number of people
short–haul \'shȯrt-ˌhȯl\ *adjective* (1895)
: traveling or involving a short distance ⟨*short-haul* flights⟩
short·horn \-ˌhȯrn\ *noun, often capitalized* (1847)
: any of a breed of red, roan, or white beef cattle originating in the north of England and including good milk-producing strains — called also *Durham*

shorthorn

short–horned grasshopper \'shȯrt-ˌhȯrn(d)-\ *noun* (circa 1890)
: any of a family (Acrididae) of grasshoppers with short antennae
short hundredweight *noun* (1924)
: HUNDREDWEIGHT 1
short·leaf pine \'shȯrt-ˌlēf-\ *noun* (1796)
: a pine (*Pinus echinata*) of the southeastern U.S. that has short flexible needles usually in clusters of two and reddish brown bark; *also* **:** its yellow wood
short line *noun* (circa 1917)
: a transportation system (as a railroad) operating over a relatively short distance
short·list \'shȯrt-ˌlist\ *noun* (1927)
: a list of candidates for final consideration (as for a position or a prize)
— short–list *transitive verb, chiefly British*
short–lived \'shȯrt-'livd *also* -'līvd\ *adjective* (1588)
: not living or lasting long ⟨*short-lived* insects⟩ ⟨*short-lived* joy⟩
usage see LONG-LIVED
short loin *noun* (circa 1923)
: a portion of the hindquarter of beef immediately behind the ribs that is usually cut into steaks — see BEEF illustration
short·ly \'shȯrt-lē\ *adverb* (before 12th century)
1 a : in a few words **:** BRIEFLY **b :** in an abrupt manner
2 a : in a short time ⟨we will be there *shortly*⟩ **b :** at a short interval ⟨*shortly* after sunset⟩
short·ness \-nəs\ *noun* (before 12th century)
: the quality or state of being short
short–nosed cattle louse \'shȯrt-ˌnōz(d)-\ *noun* (1942)
: a large bluish sucking louse (*Haematopinus eurysternus*) that attacks domestic cattle
short–order \'shȯrt-ˌȯr-dər, -'ȯr-\ *adjective* (1920)
: preparing or serving food that can be cooked quickly to a customer's order ⟨a *short-order* cook⟩
short–range \'shȯrt-'rānj\ *adjective* (1869)
1 : involving or taking into account a short period of time ⟨*short-range* plans⟩
2 : relating to or fit for short distances
short ribs *noun plural* (1912)
: a cut of beef consisting of rib ends between the rib roast and the plate — see BEEF illustration
short run *noun* (1879)
: a relatively brief period of time — often used in the phrase *in the short run*
— short–run *adjective*
short shrift *noun* (1594)
1 : barely adequate time for confession before execution
2 a : little or no attention or consideration **b :** quick work — usually used in the phrase *make short shrift of*
short sight *noun* (circa 1829)
: MYOPIA
short·sight·ed \'shȯrt-ˌsī-təd\ *adjective* (circa 1649)
1 : NEARSIGHTED
2 : lacking foresight
— short·sight·ed·ly *adverb*

☆ SYNONYMS
Shorten, curtail, abbreviate, abridge, retrench mean to reduce in extent. SHORTEN implies reduction in length or duration ⟨*shorten* a speech⟩. CURTAIL adds an implication of cutting that in some way deprives of completeness or adequacy ⟨ceremonies *curtailed* because of rain⟩. ABBREVIATE implies a making shorter usually by omitting some part ⟨using an *abbreviated* title⟩. ABRIDGE implies a reduction in compass or scope with retention of essential elements and a relative completeness in the result ⟨the *abridged* version of the novel⟩. RETRENCH suggests a reduction in extent or costs of something felt to be excessive ⟨declining business forced the company to *retrench*⟩.

\ə\ **abut** \ᵊ\ **kitten** \ər\ **further** \a\ **ash** \ā\ **ace** \ä\ **mop, mar** \au̇\ **out** \ch\ **chin** \e\ **bet** \ē\ **easy** \g\ **go** \i\ **hit** \ī\ **ice** \j\ **job** \ŋ\ **sing** \ō\ **go** \ȯ\ **law** \ȯi\ **boy** \th\ **thin** \t͟h\ **the** \ü\ **loot** \u̇\ **foot** \y\ **yet** \zh\ **vision** *see also* Guide to Pronunciation

— **short·sight·ed·ness** *noun*

short–spo·ken \-ˌspō-kən\ *adjective* (1865)
: CURT

short·stop \-ˌstäp\ *noun* (1857)
1 : the player position in baseball for defending the infield area on the third-base side of second base
2 : the player stationed in the shortstop position

short–stop \-ˌstäp\ *noun* (1936)
: STOP BATH

short story *noun* (1877)
: an invented prose narrative shorter than a novel usually dealing with a few characters and aiming at unity of effect and often concentrating on the creation of mood rather than plot

short subject *noun* (1944)
: a brief often documentary or educational film

short–tem·pered \ˌshȯrt-ˈtem-pərd\ *adjective* (1900)
: having a quick temper

short–term \ˈshȯrt-ˌtərm\ *adjective* (1901)
1 : occurring over or involving a relatively short period of time
2 a : of, relating to, or constituting a financial operation or obligation based on a brief term and especially one of less than a year **b :** generated by assets held for less than six months

short ton *noun* (1881)
— see WEIGHT table

short·wave \ˈshȯrt-ˌwāv\ *noun, often attributive* (1902)
1 : a radio wave having a wavelength between 10 and 100 meters
2 : a radio transmitter using shortwaves
3 : electromagnetic radiation having a wavelength equal to or less than that of visible light

short–weight \ˈshȯrt-ˈwāt\ *transitive verb* (1926)
: to defraud with short weight

short weight \ˈshȯrt-ˈwāt\ *noun* (1789)
: weight less than the stated weight or less than one is charged for

short–wind·ed \ˌshȯrt-ˈwin-dəd\ *adjective* (15th century)
1 : affected with or characterized by shortness of breath
2 a : BRIEF **b :** broken up into short units

shorty *or* **short·ie** \ˈshȯr-tē\ *noun, plural* **short·ies** (1888)
: one that is short

Sho·shone \shə-ˈshōn, -ˈshō-nē; ˈshō-ˌshōn\ *or* **Sho·sho·ni** \shə-ˈshō-nē\ *noun, plural* **Shoshones** *or* **Shoshoni** *also* **Shoshone** *or* **Shoshonis** (1805)
1 : a member of a group of American Indian peoples originally ranging through California, Idaho, Nevada, Utah, and Wyoming
2 : the Uto-Aztecan language of the Shoshones

[1]**shot** \ˈshät\ *noun* [Middle English, from Old English *scot;* akin to Old Norse *skot* shot, Old High German *scuz,* Old English *scēotan* to shoot — more at SHOOT] (before 12th century)
1 a : an action of shooting **b :** a directed propelling of a missile; *specifically* **:** a directed discharge of a firearm **c :** a stroke or throw in an attempt to score points in a game (as tennis, pool, or basketball); *also* **:** HOME RUN **d :** BLAST **e :** a medical or narcotics injection
2 a *plural* **shot :** something propelled by shooting; *especially* **:** small lead or steel pellets especially forming a charge for a shotgun **b :** a metal sphere of iron or brass that is heaved in the shot put
3 a : the distance that a missile is or can be thrown **b :** RANGE, REACH
4 : a charge to be paid **:** SCOT
5 : one that shoots; *especially* **:** MARKSMAN
6 a : ATTEMPT, TRY **b :** GUESS, CONJECTURE **c :** CHANCE **d :** a single appearance as an entertainer ⟨did a guest *shot* for the program⟩
7 : an effective remark
8 a : a single photographic exposure; *especially* **:** SNAPSHOT **b :** a single sequence of a motion picture or a television program shot by one camera without interruption
9 : a charge of explosives
10 a : a small measure or serving (as one ounce) of undiluted liquor **b :** a small amount applied at one time **:** DOSE
11 shot *plural* **:** SPRINKLES, JIMMIES
— **a shot :** for each one **:** APIECE
— **like a shot :** very rapidly
— **shot in the arm :** STIMULUS, BOOST
— **shot in the dark 1 :** a wild guess **2 :** an attempt that has little chance of success

[2]**shot** *past and past participle of* SHOOT

[3]**shot** *adjective* (1763)
1 a *of a fabric* **:** having contrasting and changeable color effects **:** IRIDESCENT **b :** suffused or streaked with a color ⟨hair *shot* with gray⟩ **c :** infused or permeated with a quality or element ⟨*shot* through with wit⟩
2 : having the form of pellets resembling shot
3 : reduced to a state of ruin, prostration, or uselessness ⟨his nerves are *shot*⟩

[1]**shot·gun** \ˈshät-ˌgən\ *noun* (1776)
1 : a smoothbore shoulder weapon for firing shot at short ranges
2 : an offensive football formation in which the quarterback plays a few yards behind the line of scrimmage and the other backs are scattered as flankers or slotbacks
— **shotgun** *transitive verb*
— **shot·gun·ner** \-ˌgə-nər\ *noun*

[2]**shotgun** *adjective* (1892)
1 : of, relating to, or using a shotgun
2 : involving coercion
3 : covering a wide field with hit-or-miss effectiveness

shotgun house *noun* (1940)
: a house in which all the rooms are in direct line with each other usually front to back — called also *shotgun cottage;* compare RAILROAD FLAT

shotgun marriage *noun* (1929)
1 : a marriage forced or required because of pregnancy — called also *shotgun wedding*
2 : a forced union ⟨a spate of brokerage mergers . . . hastily arranged *shotgun marriages* —John Brooks⟩

shot hole *noun* (1875)
1 : a drilled hole in which a charge of dynamite is exploded
2 : the dropping out of small rounded fragments of leaves that produces a shot-riddled appearance and is caused especially by parasitic action

shot put *noun* (1894)
: a field event in which a shot is heaved for distance
— **shot–put·ter** \ˈshät-ˌpu̇-tər\ *noun*

shot·ten \ˈshä-tᵊn\ *adjective* [Middle English *shotyn,* from past participle of *shuten* to shoot] (15th century)
: having ejected the spawn and so of inferior food value ⟨*shotten* herring⟩

should \shəd, ˈshu̇d\ [Middle English *sholde,* from Old English *sceolde* owed, was obliged to, ought to] *past of* SHALL (before 12th century)
1 — used in auxiliary function to express condition ⟨if he *should* leave his father, his father would die —Genesis 44:22 (Revised Standard Version)⟩
2 — used in auxiliary function to express obligation, propriety, or expediency ⟨'tis commanded I *should* do so —Shakespeare⟩ ⟨this is as it *should* be —H. L. Savage⟩ ⟨you *should* brush your teeth after each meal⟩
3 — used in auxiliary function to express futurity from a point of view in the past ⟨realized that she *should* have to do most of her farm work before sunrise —Ellen Glasgow⟩
4 — used in auxiliary function to express what is probable or expected ⟨with an early start, they *should* be here by noon⟩
5 — used in auxiliary function to express a request in a polite manner or to soften direct statement ⟨I *should* suggest that a guide . . . is the first essential —L. D. Reddick⟩

[1]**shoul·der** \ˈshōl-dər\ *noun* [Middle English *sholder,* from Old English *sculdor;* akin to Old High German *scultra* shoulder] (before 12th century)
1 a : the laterally projecting part of the human body formed of the bones and joints with their covering tissue by which the arm is connected with the trunk **b :** the region of the body of nonhuman vertebrates that corresponds to the shoulder but is less projecting
2 a : the two shoulders and the upper part of the back — usually used in plural **b** *plural* **:** capacity for bearing a task or blame ⟨placed the guilt squarely on his *shoulders*⟩
3 : a cut of meat including the upper joint of the foreleg and adjacent parts — see LAMB illustration
4 : the part of a garment at the wearer's shoulder
5 : an area adjacent to or along the edge of a higher, more prominent, or more important part: as **a** (1) **:** the part of a hill or mountain near the top (2) **:** a lateral protrusion or extension of a hill or mountain **b :** either edge of a roadway; *specifically* **:** the part of a roadway outside of the traveled way
6 : a rounded or sloping part (as of a stringed instrument or a bottle) where the neck joins the body
— **shoul·dered** \-dərd\ *adjective*

[2]**shoulder** *verb* **shoul·dered; shoul·der·ing** \-d(ə-)riŋ\ (14th century)
transitive verb
1 : to push or thrust with or as if with the shoulder **:** JOSTLE ⟨*shouldered* his way through the crowd⟩
2 a : to place or bear on the shoulder ⟨*shouldered* her knapsack and took off⟩ **b :** to assume the burden or responsibility of ⟨*shoulder* the blame⟩
intransitive verb
: to push with or as if with the shoulders aggressively

shoulder bag *noun* (1912)
: a handbag looped over the shoulder by a strap

shoulder belt *noun* (1967)
: an automobile safety belt worn across the torso and over the shoulder — called also *shoulder harness*

shoulder blade *noun* (14th century)
: SCAPULA

shoulder board *noun* (1945)
: one of a pair of broad pieces of stiffened cloth worn on the shoulders of a military uniform and carrying insignia

shoulder girdle *noun* (1868)
: PECTORAL GIRDLE

shoulder knot *noun* (1676)
1 : an ornamental knot of ribbon or lace worn on the shoulder in the 17th and 18th centuries
2 : a detachable ornament of braided wire cord worn on ceremonial occasions on the shoulders of a uniform by a commissioned officer

shoulder patch *noun* (1945)
: a cloth patch bearing an identifying mark and worn on one sleeve of a uniform below the shoulder

shoulder strap *noun* (1688)
: a strap that passes across the shoulder and holds up an article or garment

should·est \ˈshu̇-dəst\ *archaic past 2d singular of* SHALL

shouldn't \ˈshu̇-dᵊnt, -dᵊn, *dialect also* ˈshu̇-tᵊn(t) *or* ˈshu̇nt\ (1848)
: should not

shouldst \shədst, ˈshu̇dst, shətst, ˈshu̇tst\ *archaic past 2d singular of* SHALL

[1]**shout** \ˈshau̇t\ *verb* [Middle English] (14th century)
intransitive verb
1 : to utter a sudden loud cry

2 : to command attention as if by shouting ⟨a quality that *shouts* from good novels —John Gardner⟩
transitive verb
1 : to utter in a loud voice
2 : to cause to be, come, or stop by or as if by shouting ⟨*shouted* himself hoarse⟩ ⟨the proponents *shouted* down the opposition⟩
— shout·er *noun*

[2]shout *noun* (14th century)
: a loud cry or call

shouting distance *noun* (1930)
: a short distance **:** easy reach — usually used with *within* ⟨lived within *shouting distance* of her cousins⟩

shout song *noun* (1925)
: a rhythmic song sung at religious services especially by black Americans and characterized by responsive singing or shouting between leader and congregation

[1]shove \ˈshəv\ *verb* **shoved; shov·ing** [Middle English, from Old English *scūfan* to thrust away; akin to Old High German *scioban* to push and probably to Lithuanian *skubti* to hurry] (before 12th century)
transitive verb
1 : to push along
2 : to push or put in a rough, careless, or hasty manner **:** THRUST
3 : to force by other than physical means **:** COMPEL ⟨*shove* a bill through the legislature⟩
intransitive verb
1 : to move by forcing a way ⟨bargain hunters *shoving* up to the counter⟩
2 a : to move something by exerting force **b :** LEAVE — usually used with *off* ⟨*shoved* off for home⟩
— shov·er *noun*

[2]shove *noun* (14th century)
: an act or instance of shoving **:** a forcible push

[1]shov·el \ˈshə-vəl\ *noun* [Middle English, from Old English *scofl;* akin to Old High German *scūfla* shovel, Old English *scūfan* to thrust away] (before 12th century)
1 a : a hand implement consisting of a broad scoop or a more or less hollowed out blade with a handle used to lift and throw material **b :** something that resembles a shovel **c :** an excavating machine; *especially* **:** an hydraulic diesel-engine driven power shovel
2 : SHOVELFUL

[2]shovel *verb* **-eled** *or* **-elled; -el·ing** *or* **-el·ling** \ˈshə-və-liŋ, ˈshəv-liŋ\ (15th century)
transitive verb
1 : to take up and throw with a shovel
2 : to dig or clean out with a shovel
3 : to throw or convey roughly or in a mass as if with a shovel ⟨*shoveled* his food into his mouth⟩
intransitive verb
: to use a shovel

shov·el·er *or* **shov·el·ler** \ˈshə-və-lər, ˈshəv-lər\ *noun* (15th century)
1 : one that shovels
2 : any of several river ducks (genus *Anas*) having a large and very broad bill

shov·el·ful \ˈshə-vəl-ˌfu̇l\ *noun, plural* **shov·elfuls** \-ˌfu̇lz\ *also* **shov·els·ful** \-vəlz-ˌfu̇l\ (1533)
: as much as a shovel will hold

shovel hat *noun* (1829)
: a shallow-crowned hat with a wide brim curved up at the sides that is worn by some clergymen

shov·el·nose \-ˌnōz\ *noun* (1709)
: a shovel-nosed animal and especially a fish

shov·el–nosed \ˈshə-vəl-ˌnōzd\ *adjective* (1707)
: having a broad flat head, nose, or beak

[1]show \ˈshō\ *verb* **showed** \ˈshōd\; **shown** \ˈshōn\ *or* **showed; show·ing** [Middle English *shewen, showen,* from Old English *scēawian* to look, look at, see; akin to Old High German *scouwōn* to look, look at, and probably to Latin *cavēre* to be on one's guard] (12th century)
transitive verb
1 : to cause or permit to be seen **:** EXHIBIT ⟨*showed* pictures of the baby⟩
2 : to offer for sale ⟨stores were *showing* new spring suits⟩
3 : to present as a public spectacle **:** PERFORM
4 : to reveal by one's condition, nature, or behavior ⟨*showed* themselves to be cowards⟩
5 : to give indication or record of ⟨an anemometer *shows* wind speed⟩
6 a : to point out **:** direct attention to ⟨*showed* the view from the terrace⟩ **b :** CONDUCT, USHER ⟨*showed* me to an aisle seat⟩
7 : ACCORD, BESTOW ⟨*shows* them no mercy⟩
8 a : to set forth **:** DECLARE **b :** ALLEGE, PLEAD — used especially in law ⟨*show* cause⟩
9 a : to demonstrate or establish by argument or reasoning ⟨*show* a plan to be faulty⟩ **b :** INFORM, INSTRUCT ⟨*showed* me how to solve the problem⟩
10 : to present (an animal) for judging in a show
intransitive verb
1 a : to be or come in view ⟨3:15 *showed* on the clock⟩ **b :** to put in an appearance ⟨failed to *show*⟩
2 a : to appear in a particular way ⟨anger *showed* in their faces⟩ **b :** SEEM, APPEAR
3 a : to give a theatrical performance **b :** to be staged or presented
4 a : to appear as a contestant **b :** to present an animal in a show
5 : to finish third or at least third (as in a horse race)
6 : to exhibit one's artistic work ☆ ☆
— show·able \ˈshō-ə-bəl\ *adjective*
— show one's hand 1 : to display one's cards faceup **2 :** to declare one's intentions or reveal one's resources
— show one the door : to tell someone to get out

[2]show *noun, often attributive* (13th century)
1 : a demonstrative display ⟨a *show* of strength⟩
2 a *archaic* **:** outward appearance **b :** a false semblance **:** PRETENSE ⟨made a *show* of friendship⟩ **c :** a more or less true appearance of something **:** SIGN **d :** an impressive display ⟨his role as househusband . . . was purely for *show* —John Lahr⟩ **e :** OSTENTATION
3 : CHANCE ⟨gave him a *show* in spite of his background⟩
4 : something exhibited especially for wonder or ridicule **:** SPECTACLE
5 a : a large display or exhibition arranged to arouse interest or stimulate sales ⟨the national auto *show*⟩ **b :** a competitive exhibition of animals (as dogs) to demonstrate quality in breeding
6 a : a theatrical presentation **b :** a radio or television program **c :** ENTERTAINMENT 3a
7 : ENTERPRISE, AFFAIR ⟨they ran the whole *show*⟩
8 : an indication of metal in a mine or of gas or oil in a well
9 : third place at the finish (as of a horse race)

show–and–tell \ˈshō-ən(d)-ˈtel\ *noun* (1950)
1 : a classroom exercise in which children display an item and talk about it
2 : a public display or demonstration

show bill *noun* (1801)
: an advertising poster

show·biz \ˈshō-ˌbiz\ *noun, often attributive* [by shortening & alteration] (1945)
: SHOW BUSINESS; *also* **:** RAZZLE-DAZZLE 3
— show·biz·zy \-ˌbi-zē\ *adjective*

[1]show·boat \ˈshō-ˌbōt\ *noun* (1869)
1 : a river steamship containing a theater and carrying a troupe of actors to give plays at river communities
2 : one who tries to attract attention by conspicuous behavior

[2]showboat *intransitive verb* (1951)
: to behave in a conspicuous or ostentatious manner **:** SHOW OFF

showbread *variant of* SHEWBREAD

show business *noun* (1850)
: the arts, occupations, and businesses (as theater, motion pictures, and television) that comprise the entertainment industry

[1]show·case \ˈshō-ˌkās\ *noun* (1835)
1 : a glazed case, box, or cabinet for displaying and protecting wares in a store or articles in a museum
2 : a setting, occasion, or medium for exhibiting something or someone especially in an attractive or favorable aspect

[2]showcase *transitive verb* **show·cased; show·cas·ing** (1945)
: to exhibit especially in an attractive or favorable aspect ⟨*showcase* new talent⟩

show·down \ˈshō-ˌdau̇n\ *noun* (1884)
1 : the placing of poker hands faceup on the table to determine the winner of a pot
2 : a decisive confrontation or contest

[1]show·er \ˈshau̇(-ə)r\ *noun* [Middle English *shour,* from Old English *scūr;* akin to Old High German *scūr* shower, storm, Latin *caurus* northwest wind] (before 12th century)
1 a : a fall of rain of short duration **b :** a similar fall of sleet, hail, or snow
2 : something resembling a rain shower ⟨a *shower* of statistics⟩; *especially* **:** a fall of meteors which belong to a single group and whose trails appear to originate at the same point in space
3 : a party given by friends who bring gifts often of a particular kind
4 : a bath in which water is showered on the body; *also* **:** the apparatus that provides a shower
— show·er·less \-ləs\ *adjective*
— show·ery \ˈshau̇(-ə)-rē\ *adjective*

☆ SYNONYMS

Show, exhibit, display, expose, parade, flaunt mean to present so as to invite notice or attention. SHOW implies no more than enabling another to see or examine ⟨*showed* her snapshots to the whole group⟩. EXHIBIT stresses putting forward prominently or openly ⟨*exhibit* paintings at a gallery⟩. DISPLAY emphasizes putting in a position where others may see to advantage ⟨*display* sale items⟩. EXPOSE suggests bringing forth from concealment and displaying ⟨sought to *expose* the hypocrisy of the town fathers⟩. PARADE implies an ostentatious or arrogant displaying ⟨*parading* their piety for all to see⟩. FLAUNT suggests a shameless, boastful, often offensive parading ⟨nouveaux riches *flaunting* their wealth⟩.

Show, manifest, evidence, evince, demonstrate mean to reveal outwardly or make apparent. SHOW is the general term but sometimes implies that what is revealed must be gained by inference from acts, looks, or words ⟨careful not to *show* his true feelings⟩. MANIFEST implies a plainer, more immediate revelation ⟨*manifested* musical ability at an early age⟩. EVIDENCE suggests serving as proof of the actuality or existence of something ⟨a commitment *evidenced* by years of loyal service⟩. EVINCE implies a showing by outward marks or signs ⟨*evinced* not the slightest fear⟩. DEMONSTRATE implies showing by action or by display of feeling ⟨*demonstrated* their approval by loud applause⟩.

— **to the showers :** out of the ball game
²shower (15th century)
intransitive verb
1 : to rain or fall in or as if in a shower ⟨letters *showered* on him in praise and protest⟩
2 : to bathe in a shower
transitive verb
1 a : to wet (as with water) in a spray, fine stream, or drops **b** (1) **:** to cause to fall in a shower ⟨factory chimneys *showered* soot on the district⟩ (2) **:** to cause a shower to fall on ⟨*showered* the newlyweds with rice⟩
2 : to give in abundance ⟨*showered* the writer with honors⟩
— **show·er·er** \'shau̇(-ə)r-ər\ *noun*
³show·er \'shō(-ə)r\ *noun* (14th century)
: one that shows **:** EXHIBITOR
shower bath *noun* (1785)
: SHOWER 4
show·er·head \'shau̇(-ə)r-ˌhed\ *noun* (1925)
: a fixture for directing the spray of water in a bathroom shower
show·girl \'shō-ˌgər(-ə)l\ *noun* (1836)
: a chorus girl in a musical comedy or nightclub show
show·ing \'shō-iŋ\ *noun* (before 12th century)
1 : an act or an instance of putting something (as an artist's work) on view **:** DISPLAY
2 : PERFORMANCE, RECORD ⟨made a good *showing* in competition⟩
3 a : a statement or presentation of a case **b :** APPEARANCE, EVIDENCE
show jumping *noun* (1929)
: the competitive riding of horses one at a time over a set course of obstacles in which the winner is judged according to ability and speed
— **show jumper** *noun*
show·man \'shō-mən\ *noun* (circa 1734)
1 : the producer of a play or theatrical show
2 : an individual having a sense or knack for dramatically effective presentation
— **show·man·ship** \-ˌship\ *noun*
show–me \'shō-mē\ *adjective* (1909)
: insistent on proof or evidence
show–off \'shō-ˌȯf\ *noun, often attributive* (1843)
1 : the act of showing off
2 : one that shows off **:** EXHIBITIONIST
— **show–offy** \-ˌȯ-fē\ *adjective*
show off (circa 1793)
transitive verb
: to display proudly ⟨wanted to *show* our new car *off*⟩
intransitive verb
: to seek to attract attention by conspicuous behavior ⟨boys *showing off* for the girls⟩
show·piece \'shō-ˌpēs\ *noun* (1885)
: a prime or outstanding example used for exhibition
show·place \-ˌplās\ *noun* (1794)
: a place (as an estate or building) that is regarded as an example of beauty or excellence
show·ring \-ˌriŋ\ *noun* (1926)
: a ring (as at a cattle show) where animals are displayed
show·room \-ˌrüm, -ˌru̇m\ *noun* (1616)
: a room where merchandise is exhibited for sale or where samples are displayed
show·stop·per \-ˌstä-pər\ *noun* (1926)
1 : an act, song, or performer that wins applause so prolonged as to interrupt a performance
2 : something or someone exceptionally arresting or attractive ⟨the gold crown was the *showstopper* of the exhibition⟩
— **show·stop·ping** \-ˌstä-piŋ\ *adjective*
show trial *noun* (1937)
: a trial (as of political opponents) in which the verdict is rigged and a public confession is often extracted
show up (1826)
transitive verb
1 : to expose or discredit especially by revealing faults ⟨*showed* them *up* as frauds⟩
2 : to embarrass or cause to look bad especially by comparison ⟨trying to *show up* the boss⟩
3 : REVEAL ⟨*showed up* my ignorance⟩
intransitive verb
1 : ARRIVE, APPEAR ⟨*showed up* late for his own wedding⟩
2 : to be plainly evident
show window *noun* (1826)
1 : an outside display window in which a store exhibits merchandise
2 : a sample or setting used to exhibit or illustrate something at its best
showy \'shō-ē\ *adjective* **show·i·er; -est** (1712)
1 : making an attractive show **:** STRIKING ⟨a *showy* orchid⟩
2 : given to or marked by a flashy often tasteless display ☆
— **show·i·ly** \'shō-ə-lē\ *adverb*
— **show·i·ness** \'shō-ē-nəs\ *noun*
sho·yu \'shō-(ˌ)yü\ *noun* [Japanese *shōyu*] (1727)
: SOY 1
shrank *past of* SHRINK
shrap·nel \'shrap-nᵊl, *especially Southern* 'srap-\ *noun, plural* **shrapnel** [Henry *Shrapnel* (died 1842) English artillery officer] (1806)
1 : a projectile that consists of a case provided with a powder charge and a large number of usually lead balls and that is exploded in flight
2 : bomb, mine, or shell fragments
¹shred \'shred, *especially Southern* 'sred\ *noun* [Middle English *shrede,* from Old English *scrēade;* akin to Old High German *scrōt* piece cut off] (before 12th century)
1 a : a long narrow strip cut or torn off **b** *plural* **:** a shredded, damaged, disrupted, or ruined condition ⟨torn to *shreds* by an air attack⟩
2 : PARTICLE, SCRAP ⟨not a *shred* of evidence⟩
²shred *verb* **shred·ded; shred·ding** (before 12th century)
transitive verb
1 *archaic* **:** to cut off
2 : to cut or tear into shreds ⟨*shredded* the documents⟩
3 : DEMOLISH 2 ⟨sharp lawyers *shredding* hapless witnesses —Charles Krauthammer⟩
intransitive verb
: to come apart in or break up into shreds
— **shred·der** *noun*
shredded wheat *noun* (1898)
: a breakfast cereal made from cooked partially dried wheat that is shredded and molded into biscuits which are then oven-baked and toasted
¹shrew \'shrü, *especially Southern* 'srü\ *noun* [Middle English *shrewe* evil or scolding person, from Old English *scrēawa* shrew (animal)] (before 12th century)
1 : any of a family (Soricidae) of small chiefly nocturnal insectivores related to the moles and distinguished by a long pointed snout, very small eyes, and velvety fur
2 : an ill-tempered scolding woman
— **shrew·like** \-ˌlīk\ *adjective*
²shrew *transitive verb* (14th century)
obsolete **:** CURSE
shrewd \'shrüd, *especially Southern* 'srüd\ *adjective* [Middle English *shrewed,* from *shrewe* + ¹*-ed*] (13th century)
1 *archaic* **:** MISCHIEVOUS
2 *obsolete* **:** ABUSIVE, SHREWISH
3 *obsolete* **:** OMINOUS, DANGEROUS
4 a : SEVERE, HARD ⟨a *shrewd* knock⟩ **b :** SHARP, PIERCING ⟨a *shrewd* wind⟩
5 a : marked by clever discerning awareness and hardheaded acumen ⟨*shrewd* common sense⟩ **b :** given to wily and artful ways or dealing ⟨a *shrewd* operator⟩ ☆
— **shrewd·ly** *adverb*
— **shrewd·ness** *noun*
shrew·ish \'shrü-ish, *especially Southern* 'srü-\ *adjective* (1565)
: ILL-NATURED, INTRACTABLE
— **shrew·ish·ly** *adverb*
— **shrew·ish·ness** *noun*
shri \'shrē, 'srē\ *variant of* SRI
¹shriek \'shrēk, *especially Southern* 'srēk\ *verb* [Middle English *shreken,* probably irregular from *shriken* to shriek; akin to Old Norse *skrækja* to shriek] (15th century)
intransitive verb
1 : to utter a sharp shrill sound
2 a : to cry out in a high-pitched voice **:** SCREECH **b :** to suggest such a cry (as by vividness of expression) ⟨neon colors *shrieked* for attention —Calvin Tomkins⟩
transitive verb
1 : to utter with a shriek or sharply and shrilly ⟨*shriek* an alarm⟩
2 : to express in a manner suggestive of a shriek
²shriek *noun* (1590)
1 : a shrill usually wild or involuntary cry
2 : a sound resembling a shriek ⟨the *shriek* of chalk on the blackboard⟩
shrie·val \'shrē-vəl, *especially Southern* 'srē-\ *adjective* [obsolete *shrieve* sheriff, from Middle English *shirreve* — more at SHERIFF] (1681)
chiefly British **:** of or relating to a sheriff
shrie·val·ty \-vəl-tē\ *noun* (1502)
chiefly British **:** the office, term of office, or jurisdiction of a sheriff
shrieve \'shrēv, *especially Southern* 'srēv\ *archaic variant of* SHRIVE
shrift \'shrift, *especially Southern* 'srift\ *noun* [Middle English, from Old English *scrift,* from *scrīfan* to shrive — more at SHRIVE] (before 12th century)
1 *archaic* **a :** a remission of sins pronounced by a priest in the sacrament of reconciliation **b :** the act of shriving **:** CONFESSION
2 *obsolete* **:** CONFESSIONAL
shrike \'shrīk, *especially Southern* 'srīk\ *noun* [perhaps from (assumed) Middle English *shrik,* from Old English *scrīc* thrush; akin to Middle English *shriken* to shriek] (1544)
: any of numerous usually largely gray or brownish oscine birds (family Laniidae) that have a hooked bill, feed chiefly on insects, and often impale their prey on thorns
¹shrill \'shril, *especially Southern* 'sril\ *verb* [Middle English; probably akin to Old English *scrallettan* to resound loudly — more at SKIRL] (13th century)
transitive verb
: SCREAM

☆ SYNONYMS
Showy, pretentious, ostentatious mean given to excessive outward display. SHOWY implies an imposing or striking appearance but usually suggests cheapness or poor taste ⟨the performers' *showy* costumes⟩. PRETENTIOUS implies an appearance of importance not justified by the thing's value or the person's standing ⟨a *pretentious* parade of hard words⟩. OSTENTATIOUS stresses vainglorious display or parade ⟨the *ostentatious* summer homes of the rich⟩.

Shrewd, sagacious, perspicacious, astute mean acute in perception and sound in judgment. SHREWD stresses practical, hardheaded cleverness and judgment ⟨a *shrewd* judge of character⟩. SAGACIOUS suggests wisdom, penetration, and farsightedness ⟨*sagacious* investors got in on the ground floor⟩. PERSPICACIOUS implies unusual power to see through and understand what is puzzling or hidden ⟨a *perspicacious* counselor saw through the child's facade⟩. ASTUTE suggests shrewdness, perspicacity, and diplomatic skill ⟨an *astute* player of party politics⟩.

intransitive verb
: to utter or emit an acute piercing sound
[2]**shrill** *adjective* (14th century)
1 a : having or emitting a sharp high-pitched tone or sound **:** PIERCING **b :** accompanied by sharp high-pitched sounds or cries ⟨*shrill* gaiety⟩
2 : having a sharp or vivid effect on the senses ⟨*shrill* light⟩
3 : STRIDENT, INTEMPERATE ⟨*shrill* anger⟩ ⟨*shrill* criticism⟩
— shrill *adverb*
— shrill·ness *noun*
— shril·ly \ˈshril-lē, *especially Southern* ˈsril-\ *adverb*
[3]**shrill** *noun* (1591)
: a shrill sound ⟨the *shrill* of the ship's whistle⟩
[1]**shrimp** \ˈshrimp, *especially Southern* ˈsrimp\ *noun, plural* **shrimps** *or* **shrimp** [Middle English *shrimpe;* akin to Middle Low German *schrempen* to contract, wrinkle, Old Norse *skorpna* to shrivel up] (14th century)
1 : any of numerous mostly small and marine decapod crustaceans (suborders Dendrobranchiata and Pleocyemata) having a slender elongated body, a compressed abdomen, and a long spiny rostrum; *also* **:** a small crustacean (as an amphipod or a branchiopod) resembling the true shrimps
2 : a very small or puny person or thing
— shrimp·like \-ˌlīk\ *adjective*
— shrimpy \ˈshrim-pē, ˈsrim-\ *adjective*
[2]**shrimp** *intransitive verb* (circa 1859)
: to fish for or catch shrimps
shrimp·er \ˈshrim-pər, *especially Southern* ˈsrim-\ *noun* (1851)
1 : a shrimp fisherman
2 : a boat engaged in shrimping
shrimp pink *noun* (1882)
: a deep pink
[1]**shrine** \ˈshrīn, *especially Southern* ˈsrīn\ *noun* [Middle English, from Old English *scrīn,* from Latin *scrinium* case, chest] (before 12th century)
1 a : a case, box, or receptacle; *especially* **:** one in which sacred relics (as the bones of a saint) are deposited **b :** a place in which devotion is paid to a saint or deity **:** SANCTUARY **c :** a niche containing a religious image
2 : a receptacle (as a tomb) for the dead
3 : a place or object hallowed by its associations
[2]**shrine** *transitive verb* **shrined; shrin·ing** (14th century)
: ENSHRINE
Shrin·er \ˈshrī-nər, *especially Southern* ˈsrī-\ *noun* [Ancient Arabic Order of Nobles of the Mystic *Shrine*] (1886)
: a member of a secret fraternal society that is non-Masonic but admits only Knights Templars and 32d-degree Masons to membership
[1]**shrink** \ˈshriŋk, *especially Southern* ˈsriŋk\ *verb* **shrank** \ˈshraŋk, ˈsraŋk\ *also* **shrunk** \ˈshrəŋk, ˈsrəŋk\; **shrunk** *or* **shrunk·en** \ˈshrəŋ-kən, ˈsrəŋ-\; **shrink·ing** [Middle English, from Old English *scrincan;* akin to Middle Dutch *schrinken* to draw back] (before 12th century)
intransitive verb
1 : to contract or curl up the body or part of it **:** HUDDLE, COWER
2 a : to contract to less extent or compass **b :** to become smaller or more compacted **c :** to lose substance or weight **d :** to lessen in value **:** DWINDLE
3 : to recoil instinctively (as from something painful or horrible) ⟨*shrank* from the challenge⟩
transitive verb
: to cause to contract or shrink; *specifically* **:** to compact (cloth) by causing to contract when subjected to washing, boiling, steaming, or other processes ■
synonym see CONTRACT, RECOIL
— shrink·able \ˈshriŋ-kə-bəl, ˈsriŋ-\ *adjective*
— shrink·er *noun*
[2]**shrink** *noun* (1590)
1 : the act of shrinking
2 : SHRINKAGE
3 [short for *headshrinker*] **:** a clinical psychiatrist or psychologist ⟨regaling us with all the stories he never told his *shrink* —*Rolling Stone*⟩
shrink·age \ˈshriŋ-kij, *especially Southern* ˈsriŋ-\ *noun* (1800)
1 : the act or process of shrinking
2 : the loss in weight of livestock during shipment and in the process of preparing the meat for consumption
3 : the amount lost by shrinkage
shrinking violet *noun* (1915)
: a bashful or retiring person
shrink–wrap \ˈshriŋk-ˌrap, *especially Southern* ˈsriŋk-\ *transitive verb* (1966)
: to wrap (as a book or meat) in tough clear plastic film that is then shrunk (as by heating) to form a tightly fitting package
— shrink–wrap *noun*
shrive \ˈshrīv, *especially Southern* ˈsrīv\ *verb* **shrived** *or* **shrove** \ˈshrōv, ˈsrōv\; **shriv·en** \ˈshri-vən, ˈsri-\ *or* **shrived; shriv·ing** [Middle English, from Old English *scrīfan* to shrive, prescribe (akin to Old High German *scrīban* to write), from Latin *scribere* to write — more at SCRIBE] (before 12th century)
transitive verb
1 : to administer the sacrament of reconciliation to
2 : to free from guilt
intransitive verb
archaic **:** to confess one's sins especially to a priest
shriv·el \ˈshri-vəl, *especially Southern* ˈsri-\ *verb* **-eled** *or* **-elled; -el·ing** *or* **-el·ling** \ˈshri-və-liŋ, ˈsri-; ˈshriv-liŋ, ˈsriv-\ [origin unknown] (1588)
intransitive verb
1 : to draw into wrinkles especially with a loss of moisture
2 a : to become reduced to inanition, helplessness, or inefficiency **b :** DWINDLE
transitive verb
: to cause to shrivel
shroff \ˈshräf, ˈshrȯf, *especially Southern* ˈsräf, ˈsrȯf\ *noun* [Hindi *śarāf,* from Arabic *ṣarrāf*] (1618)
: a banker or money changer in the Far East; *especially* **:** one who tests and evaluates coin
Shrop·shire \ˈshräp-ˌshir, -shər, *especially US* -ˌshīr, *especially Southern* ˈsräp-\ *noun* [*Shropshire,* England] (1803)
: any of a breed of dark-faced hornless sheep of English origin that are raised for both mutton and wool
[1]**shroud** \ˈshrau̇d, *especially Southern* ˈsrau̇d\ *noun* [Middle English, garment, from Old English *scrūd;* akin to Old English *scrēade* shred — more at SHRED] (14th century)
1 *obsolete* **:** SHELTER, PROTECTION
2 : something that covers, screens, or guards: as **a :** one of two flanges that give peripheral support to turbine or fan bedding **b :** a usually fiberglass guard that protects a spacecraft from the heat of launching
3 : burial garment **:** WINDING-SHEET, CEREMENT
4 a : one of the ropes leading usually in pairs from a ship's mastheads to give lateral support to the masts **b :** one of the cords that suspend the harness of a parachute from the canopy

1 **shroud 4a**

[2]**shroud** (14th century)
transitive verb
1 a *archaic* **:** to cover for protection **b** *obsolete* **:** CONCEAL
2 a : to cut off from view **:** OBSCURE ⟨trees *shrouded* by a heavy fog⟩ **b :** to veil under another appearance (as by obscuring or disguising) ⟨*shrouded* the decision in a series of formalities⟩
3 : to dress for burial
intransitive verb
archaic **:** to seek shelter
Shrove·tide \ˈshrōv-ˌtīd, *especially Southern* ˈsrōv-\ *noun* [Middle English *schroftide,* from *schrof-* (from *shriven* to shrive) + *tide*] (15th century)
: the period usually of three days immediately preceding Ash Wednesday
Shrove Tuesday \ˈshrōv-, *especially Southern* ˈsrōv-\ *noun* [Middle English *schroftewesday,* from *schrof-* (as in *schroftide*) + *tewesday* Tuesday] (15th century)
: the Tuesday before Ash Wednesday
[1]**shrub** \ˈshrəb, *especially Southern* ˈsrəb\ *noun* [Middle English *schrobbe,* from Old English *scrybb* brushwood; akin to Norwegian *skrubbebær* a cornel of a dwarf species] (before 12th century)
: a low usually several-stemmed woody plant
[2]**shrub** *noun* [Arabic *sharāb* beverage] (1747)
1 : an aged blend of fruit juice, sugar, and spirits served chilled and diluted with water
2 : a beverage made by adding acidulated fruit juice to iced water
shrub·bery \ˈshrə-b(ə-)rē, *especially Southern* ˈsrə-\ *noun, plural* **-ber·ies** (1748)
: a planting or growth of shrubs
shrub·by \ˈshrə-bē, *especially Southern* ˈsrə-\ *adjective* **shrub·bi·er; -est** (1540)
1 : consisting of or covered with shrubs
2 : resembling a shrub
[1]**shrug** \ˈshrəg, *especially Southern* ˈsrəg\ *verb* **shrugged; shrug·ging** [Middle English *schruggen*] (14th century)
intransitive verb
: to raise or draw in the shoulders especially to express aloofness, indifference, or uncertainty
transitive verb
: to lift or contract (the shoulders) especially to express aloofness, indifference, or uncertainty
[2]**shrug** *noun* (1594)
1 : an act of shrugging
2 : a woman's small waist-length or shorter jacket
shrug off *transitive verb* (1904)
1 : to brush aside **:** MINIMIZE ⟨*shrugs off* the problem⟩
2 : to shake off ⟨*shrugging off* sleep⟩
3 : to remove (a garment) by wriggling out
shtetl *also* **shte·tel** \ˈshte-tᵊl, ˈshtā-\ *noun, plural* **shtet·lach** \ˈshtet-ˌläk, ˈshtāt-\ *also* **shtetels** [Yiddish *shtetl,* from Middle High German *stetel,* diminutive of *stat* place, town, city, from Old High German, place — more at STEAD] (1949)
: a small Jewish town or village formerly found in Eastern Europe

USAGE
shrink Although some people have apparently been taught that it is wrong, the past tense *shrunk* is perfectly standard if less common than *shrank* ⟨we watched fascinated, until the wolves *shrunk* to dots in the fog —Lyn Hancock⟩ ⟨the nodes *shrunk,* and for a year she was apparently free of disease —Elisabeth Rosenthal⟩ ⟨he and his chief operating officer . . . *shrunk* the headquarters staff further —*Business Week*⟩.

shtick *also* **shtik** \'shtik\ *noun* [Yiddish *shtik* pranks, literally, piece, from Middle High German *stücke,* from Old High German *stucki;* akin to Old English *stycce* piece, Old High German *stoc* stick — more at STOCK] (1959)
1 : a show-business routine, gimmick, or gag **:** BIT
2 : one's special trait, interest, or activity **:** BAG ⟨he's alive and well and now doing his *shtick* out in Hollywood —Robert Daley⟩

[1]**shuck** \'shək\ *noun* [origin unknown] (circa 1674)
1 : SHELL, HUSK: as **a :** the outer covering of a nut or of Indian corn **b :** the shell of an oyster or clam
2 : something of little value — usually used in plural often interjectionally ⟨not worth *shucks*⟩ ⟨*shucks,* it was nothing⟩

[2]**shuck** *transitive verb* (1772)
1 : to strip of shucks
2 a : to peel off (as clothing) — often used with *off* **b :** to lay aside — often used with *off* ⟨bad habits are being *shucked* off —A. W. Smith⟩
— shuck·er *noun*

[1]**shud·der** \'shə-dər\ *intransitive verb* **shud·dered; shud·der·ing** \-d(ə-)riŋ\ [Middle English *shoddren;* akin to Old High German *skutten* to shake and perhaps to Lithuanian *kutėti* to shake up] (13th century)
1 : to tremble convulsively **:** SHIVER
2 : QUIVER

[2]**shudder** *noun* (1607)
: an act of shuddering
— shud·dery \-d(ə-)rē\ *adjective*

[1]**shuf·fle** \'shə-fəl\ *verb* **shuf·fled; shuf·fling** \-f(ə-)liŋ\ [perhaps irregular from [1]*shove*] (1570)
transitive verb
1 : to mix in a mass confusedly **:** JUMBLE
2 : to put or thrust aside or under cover ⟨*shuffled* the whole matter out of his mind⟩
3 a : to rearrange (as playing cards, dominoes, or tiles) to produce a random order **b :** to move about, back and forth, or from one place to another **:** SHIFT ⟨*shuffle* funds among various accounts⟩
4 a : to move (as the feet) by sliding along or back and forth without lifting **b :** to perform (as a dance) with a dragging, sliding step
intransitive verb
1 : to work into or out of trickily ⟨*shuffled* out of the difficulty⟩
2 : to act or speak in a shifty or evasive manner
3 a : to move or walk in a sliding dragging manner without lifting the feet **b :** to dance in a lazy nonchalant manner with sliding and tapping motions of the feet **c :** to execute in a perfunctory or clumsy manner
4 : to mix playing cards or counters by shuffling
— shuf·fler \-f(ə-)lər\ *noun*

[2]**shuffle** *noun* (1628)
1 : an evasion of the issue **:** EQUIVOCATION
2 a : an act of shuffling (as of cards) **b :** a right or turn to shuffle ⟨was reminded that it was his *shuffle*⟩ **c :** JUMBLE ⟨lost in the *shuffle* of papers⟩
3 a : a dragging sliding movement; *specifically* **:** a sliding or scraping step in dancing **b :** a dance characterized by such a step

shuf·fle·board \'shə-fəl-,bōrd, -,bȯrd\ *noun* [alteration of obsolete English *shove-board*] (1836)
1 : a game in which players use long-handled cues to shove disks into scoring areas of a diagram marked on a smooth surface
2 : a diagram on which shuffleboard is played

shul \'shül\ *noun* [Yiddish, school, synagogue, from Middle High German *schuol* school] (circa 1874)
: SYNAGOGUE

shun \'shən\ *transitive verb* **shunned; shun·ning** [Middle English *shunnen,* from Old English *scunian*] (before 12th century)
: to avoid deliberately and especially habitually
synonym see ESCAPE
— shun·ner *noun*

shun·pike \'shən-,pīk\ *noun* (1862)
: a side road used to avoid the toll on or the speed and traffic of a superhighway
— shun·pik·er \-,pī-kər\ *noun*
— shun·pik·ing \-kiŋ\ *noun*

[1]**shunt** \'shənt\ *verb* [Middle English, to flinch] (13th century)
transitive verb
1 a : to turn off to one side **:** SHIFT ⟨was *shunted* aside⟩ **b :** to switch (as a train) from one track to another
2 : to provide with or divert by means of an electrical shunt
3 : to divert (blood) from one part to another by a surgical shunt
4 : SHUTTLE ⟨*shunted* the missiles from shelter to shelter⟩
intransitive verb
1 : to move to the side
2 : to travel back and forth ⟨*shunted* between the two towns⟩
— shunt·er *noun*

[2]**shunt** *noun* (circa 1859)
1 : a means or mechanism for turning or thrusting aside: as **a** *chiefly British* **:** a railroad switch **b :** a conductor joining two points in an electrical circuit so as to form a parallel or alternative path through which a portion of the current may pass (as for regulating the amount passing in the main circuit) **c :** a surgical passage created between two blood vessels to divert blood from one part to another
2 : an accident (as a collision between two cars) especially in auto racing

shush \'shəsh, 'shùsh\ *transitive verb* [imitative] (1925)
: to urge to be quiet **:** HUSH
— shush *noun*

[1]**shut** \'shət\ *verb* **shut; shut·ting** [Middle English *shutten,* from Old English *scyttan;* akin to Middle Dutch *schutten* to shut in, Old English *scēotan* to shoot — more at SHOOT] (before 12th century)
transitive verb
1 a : to move into position to close an opening ⟨*shut* the lid⟩ **b :** to prevent entrance to or passage to or from
2 : to confine by or as if by enclosure ⟨*shut* herself in her study⟩
3 : to fasten with a lock or bolt
4 : to close by bringing enclosing or covering parts together ⟨*shut* the eyes⟩
5 : to cause to cease or suspend an operation or activity — often used with *down*
intransitive verb
1 : to close itself or become closed ⟨flowers that *shut* at night⟩
2 : to cease or suspend an operation or activity — often used with *down*

[2]**shut** *adjective* (15th century)
1 : closed, fastened, or folded together
2 : RID, CLEAR, FREE — usually used with *of*

[3]**shut** *noun* (1667)
: the act of shutting

shut·down \'shət-,daùn\ *noun* (1888)
: the cessation or suspension of an operation or activity

shut down *intransitive verb* (1779)
: to settle so as to obscure vision **:** CLOSE IN ⟨the night *shut down* early⟩

shute *variant of* CHUTE

shut–eye \'shət-,ī\ *noun* (1899)
: SLEEP ⟨get some *shut-eye*⟩

[1]**shut–in** \'shət-,in\ *noun* (1903)
1 : an invalid confined to home, a room, or bed
2 : a narrow gorge-shaped part of an otherwise wide valley
3 : available oil or gas which is not being produced from an existing well

[2]**shut–in** \'shət-'in\ *adjective* (1909)
1 : confined to one's home or an institution by illness or incapacity
2 a : SECRETIVE, BROODING ⟨a bitter, *shut-in* face —Claudia Cassidy⟩ **b :** tending to avoid social contact **:** WITHDRAWN ⟨the *shut-in* personality type —S. K. Weinberg⟩

shut in *transitive verb* (14th century)
1 : CONFINE, ENCLOSE
2 : to prevent production of (oil or gas) by closing down a well

shut·off \'shət-,ȯf\ *noun* (1869)
1 : something (as a valve) that shuts off
2 : STOPPAGE, INTERRUPTION

shut off (1824)
transitive verb
1 a : to cut off (as flow or passage) **:** STOP ⟨*shuts off* the oxygen supply⟩ **b :** to stop the operation of (as a machine) ⟨*shut* the motor *off*⟩
2 : to close off **:** SEPARATE — usually used with *from* ⟨*shut off* from the rest of the world⟩
intransitive verb
: to cease operating **:** STOP ⟨*shuts off* automatically⟩

shut·out \'shət-,aùt\ *noun* (1889)
1 : a game or contest in which one side fails to score
2 : a preemptive bid in bridge

shut out *transitive verb* (14th century)
1 : EXCLUDE
2 : to prevent (an opponent) from scoring in a game or contest
3 : to forestall the bidding of (bridge opponents) by making a high or preemptive bid

[1]**shut·ter** \'shə-tər\ *noun* (1542)
1 : one that shuts
2 : a usually movable cover or screen for a window or door
3 : a mechanical device that limits the passage of light; *especially* **:** a camera attachment that exposes the film or plate by opening and closing an aperture
4 : the movable louvers in a pipe organ by which the swell box is opened
— shut·ter·less \-ləs\ *adjective*

[2]**shutter** *transitive verb* (1826)
1 : to close by or as if by shutters
2 : to furnish with shutters

shut·ter·bug \'shə-tər-,bəg\ *noun* (1940)
: a photography enthusiast

[1]**shut·tle** \'shə-tᵊl\ *noun* [Middle English *shittle,* probably from Old English *scytel* bar, bolt; akin to Old Norse *skutill* bolt, Old English *scēotan* to shoot — more at SHOOT] (14th century)
1 a : a device used in weaving for passing the thread of the weft between the threads of the warp **b :** a spindle-shaped device holding the thread in tatting, knotting, or netting **c :** a sliding thread holder for the lower thread of a sewing machine that carries the lower thread through a loop of the upper thread to make a stitch
2 : SHUTTLECOCK
3 a : a going back and forth regularly over an often short route by a vehicle **b :** an established route used in a shuttle; *also* **:** a vehicle used in a shuttle **c :** SPACE SHUTTLE
— shut·tle·less *adjective*

[2]**shuttle** *verb* **shut·tled; shut·tling** \'shət-liŋ, 'shə-tᵊl-iŋ\ (1550)
transitive verb
1 : to cause to move or travel back and forth frequently
2 : to transport in, by, or as if by a shuttle
intransitive verb
1 : to move or travel back and forth frequently
2 : to move by or as if by a shuttle

[1]**shut·tle·cock** \'shə-tᵊl-,käk\ *noun* (1522)
: a lightweight conical object with a rounded often rubber-covered nose that is used in badminton

shuttlecock

[2]**shuttlecock** *transitive verb* (1687)
: to send or toss to and fro **:** BANDY

shuttle diplomacy *noun* (1974)
: negotiations especially between nations carried on by an intermediary who shuttles back and forth between the disputants

shut up (1814)
transitive verb
: to cause (a person) to stop talking
intransitive verb
: to cease writing or speaking

[1]**shy** \'shī\ *adjective* **shi·er** *or* **shy·er** \'shī(-ə)r\; **shi·est** *or* **shy·est** \'shī-əst\ [Middle English *schey,* from Old English *scēoh;* akin to Old High German *sciuhen* to frighten off] (before 12th century)
1 : easily frightened **:** TIMID
2 : disposed to avoid a person or thing ⟨publicity *shy*⟩ ⟨book-*shy* children⟩
3 : hesitant in committing oneself **:** CIRCUMSPECT
4 : sensitively diffident or retiring **:** RESERVED ⟨a *shy* seclusive person⟩; *also* **:** expressive of such a state or nature ⟨spoke in a *shy* voice⟩
5 : SECLUDED, HIDDEN
6 : having less than the full or specified amount or number **:** SHORT ⟨just *shy* of six feet tall⟩ ⟨the stew is a little *shy* of seasoning⟩
7 : DISREPUTABLE ⟨gambling hells and *shy* saloons —*Blackwood's*⟩ ☆
— shy·ly *adverb*
— shy·ness *noun*

[2]**shy** *intransitive verb* **shied; shy·ing** (1650)
1 : to develop or show a dislike or distaste — usually used with *from* or *away from* ⟨an author who *shies* away from publicity⟩
2 : to start suddenly aside through fright or alarm

[3]**shy** *noun, plural* **shies** (1791)
: a sudden start aside (as from fright)

[4]**shy** *verb* **shied; shy·ing** [perhaps from [1]*shy*] (1787)
intransitive verb
: to make a sudden throw
transitive verb
: to throw (an object) with a jerk **:** FLING

[5]**shy** *noun, plural* **shies** (1791)
1 : the act of shying **:** TOSS, THROW
2 : a verbal fling or attack
3 : COCKSHY

[1]**shy·lock** \'shī-ˌläk\ *noun*
1 *capitalized* **:** the Jewish usurer and antagonist of Antonio in Shakespeare's *The Merchant of Venice*
2 : an extortionate creditor **:** LOAN SHARK

[2]**shylock** *intransitive verb* (circa 1934)
: to lend money at high rates of interest ⟨exposé of systematic thievery . . . *shylocking,* and murder —*Current Biography*⟩

shy·ster \'shīs-tər\ *noun* [probably from German *Scheisser,* literally, defecator] (1844)
: one who is professionally unscrupulous especially in the practice of law or politics **:** PETTIFOGGER ◆

si \'sē\ *noun* [Italian] (1728)
: [2]TI

si·al·a·gogue \sī-'a-lə-ˌgäg\ *noun* [New Latin *sialagogus* promoting the expulsion of saliva, from Greek *sialon* saliva + New Latin *-agogus* -agogue] (circa 1783)
: an agent that promotes the flow of saliva

si·al·ic \sī-'a-lik\ *adjective* [International Scientific Vocabulary *si*licon + *al*uminum] (1924)
: of, relating to, or being relatively light rock that is rich in silica and alumina and is typical of the outer layers of the earth

sialic acid *noun* [Greek *sialon* saliva] (1952)
: any of a group of reducing amido acids that are essentially carbohydrates and are found especially as components of blood glycoproteins and mucoproteins

si·a·mang \'sē-ə-ˌmaŋ, 'sī-\ *noun* [Malay] (1822)
: a black gibbon (*Hylobates syndactylus*) of Sumatra and the Malay Peninsula that is the largest of the gibbons

[1]**Si·a·mese** \ˌsī-ə-'mēz, -'mēs\ *adjective* [*Siam* (Thailand); in senses 2 & 3, from *Siamese twin*] (1693)
1 : of, relating to, or characteristic of Thailand, the Thais, or their language
2 : exhibiting great resemblance **:** very like
3 *not capitalized* **:** connecting two or more pipes or hoses so as to permit discharge in a single stream

[2]**Siamese** *noun, plural* **Siamese** (1693)
1 : THAI 2
2 : THAI 3
3 : SIAMESE CAT

Siamese cat *noun* (1871)
: any of a breed of slender blue-eyed short-haired domestic cats of oriental origin with pale fawn or gray body and darker ears, paws, tail, and face

Siamese fighting fish *noun* (1933)
: a brightly colored betta (*Betta splendens*) that has highly aggressive males and is a popular aquarium fish

Siamese fighting fish

Siamese twin *noun* [from Chang (died 1874) and Eng (died 1874) congenitally united twins born in Siam] (1829)
: one of a pair of congenitally united twins in humans or other vertebrates

[1]**sib** \'sib\ *adjective* [Middle English, from Old English *sibb,* from *sibb* kinship; akin to Old High German *sippa* kinship, family, Latin *sodalis* comrade, Greek *ēthos* custom, character, Latin *suus* one's own — more at SUICIDE] (before 12th century)
: related by blood **:** AKIN

[2]**sib** *noun* (before 12th century)
1 a : KINDRED, RELATIVES **b :** a blood relation **:** KINSMAN
2 : a brother or sister considered irrespective of sex; *broadly* **:** any plant or animal of a group sharing a degree of genetic relationship corresponding to that of human sibs
3 : a group of persons unilaterally descended from a real or supposed ancestor

Si·be·ri·an husky \sī-'bir-ē-ən-\ *noun* (1930)
: any of a breed of medium-sized thick-coated compact dogs that were developed in Siberia for use as sled dogs and that have erect ears and a bushy tail

Siberian husky

sib·i·lance \'si-bə-lən(t)s\ *noun* (1823)
: a sibilant quality or sound

[1]**sib·i·lant** \'si-bə-lənt\ *adjective* [Latin *sibilant-, sibilans,* present participle of *sibilare* to hiss, whistle, of imitative origin] (1669)
: having, containing, or producing the sound of or a sound resembling that of the *s* or the *sh* in *sash* ⟨a *sibilant* affricate⟩ ⟨a *sibilant* snake⟩
— sib·i·lant·ly *adverb*

[2]**sibilant** *noun* (1788)
: a sibilant speech sound (as English \s\, \z\, \sh\, \zh\, \ch(=tsh)\, or \j(=dzh)\)

sib·i·late \'si-bə-ˌlāt\ *verb* **-lat·ed; -lat·ing** [Latin *sibilatus,* past participle of *sibilare*] (circa 1656)
intransitive verb
1 : HISS
2 : to utter an initial sibilant **:** prefix an \s\-sound
transitive verb
1 : HISS
2 : to pronounce with an initial sibilant **:** prefix an \s\-sound to
— sib·i·la·tion \ˌsi-bə-'lā-shən\ *noun*

sib·ling \'si-bliŋ\ *noun* (before 12th century)
1 : SIB 2; *also* **:** one of two or more individuals having one common parent
2 : one of two or more things related by a common tie or characteristic

sibling species *noun* (1940)
: one of two or more species that are nearly indistinguishable morphologically

sib·yl \'si-bəl\ *noun, often capitalized* [Middle English *sibile, sybylle,* from Middle French & Latin; Middle French *sibile,* from Latin *sibylla,* from Greek] (14th century)
1 : any of several prophetesses usually accepted as 10 in number and credited to widely separate parts of the ancient world (as Babylonia, Egypt, Greece, and Italy)
2 a : a female prophet **b :** FORTUNE-TELLER
— si·byl·ic *or* **si·byl·lic** \sə-'bi-lik\ *adjective*
— sib·yl·line \'si-bə-ˌlīn, -ˌlēn\ *adjective*

[1]**sic** \'sik\ *chiefly Scottish variant of* SUCH

[2]**sic** *also* **sick** \'sik\ *transitive verb* **sicced** *also* **sicked** \'sikt\; **sic·cing** *also* **sick·ing** [alteration of *seek*] (1845)
1 : CHASE, ATTACK — usually used as a command especially to a dog ⟨*sic* 'em⟩
2 : to incite or urge to an attack, pursuit, or harassment **:** SET

[3]**sic** \'sik, 'sēk\ *adverb* [Latin, so, thus — more at SO] (circa 1859)

☆ SYNONYMS
Shy, bashful, diffident, modest, coy mean not inclined to be forward. SHY implies a timid reserve and a shrinking from familiarity or contact with others ⟨*shy* with strangers⟩. BASHFUL implies a frightened or hesitant shyness characteristic of childhood and adolescence ⟨a *bashful* boy out on his first date⟩. DIFFIDENT stresses a distrust of one's own ability or opinion that causes hesitation in acting or speaking ⟨felt *diffident* about raising an objection⟩. MODEST suggests absence of undue confidence or conceit ⟨very *modest* about her achievements⟩. COY implies an assumed or affected shyness ⟨put off by her *coy* manner⟩.

◇ WORD HISTORY
shyster The early history of *shyster* has prompted much speculation concerning its etymology, the origin of the word having been attributed to any number of languages or to any of various eponyms. The earliest known instance of *shyster* in print occurred in 1843, in the pages of *The Subterranean,* a New York City weekly that concerned itself with the goings-on in and about the Tombs (the city jail) and the local courts. Among the favorite targets of the muckraking paper's publisher and chief writer, Mike Walsh, were the unlicensed and unprincipled men who haunted the courthouse, pretending to be lawyers. After lambasting them as "pettifoggers," Walsh was approached by one of them—a Cornelius Terhune—who asked that the past masters of pettifoggery (himself included) not be lumped together with the amateurs. Terhune dismissed these lesser practitioners as *shiseters* (Walsh's spelling; he also used *shyseters*). Terhune's epithet was probably derived from the German vulgarism *Scheisser,* the literal meaning of which is "one that defecates." Walsh's repeated use of *shyster* (his final spelling) in *The Subterranean* led to the adoption of the word by other newspapers.

: intentionally so written — used after a printed word or passage to indicate that it is intended exactly as printed or to indicate that it exactly reproduces an original ⟨said he seed [*sic*] it all⟩

Sichuan *variant of* SZECHUAN

sick \'sik\ *adjective* [Middle English *sek, sik,* from Old English *sēoc;* akin to Old High German *sioh* sick] (before 12th century)
1 a (1) **:** affected with disease or ill health **:** AILING (2) **:** of, relating to, or intended for use in sickness ⟨*sick* pay⟩ ⟨a *sick* ward⟩ **b :** QUEASY, NAUSEATED ⟨*sick* to one's stomach⟩ ⟨was *sick* in the car⟩ **c :** undergoing menstruation
2 : spiritually or morally unsound or corrupt
3 a : sickened by strong emotion ⟨*sick* with fear⟩ ⟨worried *sick*⟩ **b :** having a strong distaste from surfeit **:** SATIATED ⟨*sick* of flattery⟩ **c :** filled with disgust or chagrin ⟨gossip makes me *sick*⟩ **d :** depressed and longing for something ⟨*sick* for one's home⟩
4 a : mentally or emotionally unsound or disordered **:** MORBID ⟨*sick* thoughts⟩ **b :** MACABRE, SADISTIC ⟨*sick* jokes⟩
5 : lacking vigor **:** SICKLY: as **a :** badly outclassed ⟨looked *sick* in the contest⟩ **b :** incapable of yielding a profitable crop especially because of buildup of disease organisms ⟨clover-*sick* soils⟩

sick and tired *adjective* (1783)
: thoroughly fatigued or bored; *also* **:** FED UP

sick bay *noun* (1813)
: a compartment in a ship used as a dispensary and hospital; *broadly* **:** a place for the care of the sick or injured

sick·bed \'sik-,bed\ *noun* (14th century)
: the bed on which one lies sick

sick call *noun* (1836)
: a scheduled time at which individuals (as soldiers) may report as sick to the medical officer

sick day *noun* (1968)
: a paid day of sick leave

sick·en \'si-kən\ *verb* **sick·ened; sick·en·ing** \'si-kə-niŋ, 'sik-niŋ\ (13th century)
intransitive verb
1 : to become sick
2 : to become weary or satiated
transitive verb
1 : to make sick
2 : to cause revulsion in as a result of weariness or satiety
— sick·en·er \'si-kə-nər, 'sik-nər\ *noun*

sick·en·ing \-niŋ\ *adjective* (1789)
: causing sickness or disgust ⟨a *sickening* odor⟩ ⟨a *sickening* display⟩
— sick·en·ing·ly \-niŋ-lē\ *adverb*

sick·er \'si-kər\ *adjective* [Middle English *siker,* from Old English *sicor,* from Latin *securus* secure] (before 12th century)
chiefly Scottish **:** SECURE, SAFE; *also* **:** DEPENDABLE
— sicker *adverb, chiefly Scottish*
— sick·er·ly *adverb, chiefly Scottish*

sick headache *noun* (1778)
: MIGRAINE

sick·ie \'si-kē\ *noun* (1967)
: a person who is mentally or morally sick

sick·ish \'si-kish\ *adjective* (1581)
1 *archaic* **:** somewhat ill **:** SICKLY
2 : somewhat nauseated **:** QUEASY
3 : somewhat sickening ⟨a *sickish* odor⟩
— sick·ish·ly *adverb*
— sick·ish·ness *noun*

[1]sick·le \'si-kəl\ *noun* [Middle English *sikel,* from Old English *sicol,* from Latin *secula* sickle, from *secare* to cut — more at SAW] (before 12th century)
1 : an agricultural implement consisting of a curved metal blade with a short handle fitted on a tang
2 : the cutting mechanism (as of a reaper, combine, or mower) consisting of a bar with a series of cutting elements

[1]sickle 1

[2]sickle *adjective* (1688)
: having the form of a sickle blade **:** having a curve similar to that of a sickle blade ⟨the *sickle* moon⟩

[3]sickle *verb* **sick·led; sick·ling** \'si-k(ə-)liŋ\ (1922)
transitive verb
1 : to mow or reap with a sickle
2 : to change (a red blood cell) into a sickle cell
intransitive verb
: to change into a sickle cell ⟨the ability of red blood cells to *sickle*⟩

sick leave *noun* (1840)
1 : an absence from work permitted because of illness
2 : the number of days per year for which an employer agrees to pay employees who are sick

sickle cell *noun* (1923)
: an abnormal red blood cell of crescent shape

sickle–cell anemia *noun* (1922)
: a chronic inherited anemia in which a large proportion or the majority of the red blood cells tend to sickle, which occurs primarily in individuals of African, Mediterranean, or southwest Asian ancestry, and which results from homozygosity for a semidominant gene — called also *sickle-cell disease*

sickle–cell trait *noun* (1928)
: an inherited blood condition in which some red blood cells tend to sickle but usually not enough to produce anemia, which occurs primarily in individuals of African, Mediterranean, or southwest Asian ancestry, and which results from heterozygosity for a semidominant gene

sick·le·mia \,si-kə-'lē-mē-ə\ *noun* [New Latin, from English *sickle* (cell) + New Latin *-emia*] (1932)
: SICKLE-CELL TRAIT

[1]sick·ly \'si-klē\ *adjective* (14th century)
1 : somewhat unwell; *also* **:** habitually ailing
2 : produced by or associated with sickness ⟨a *sickly* complexion⟩ ⟨a *sickly* appetite⟩
3 : producing or tending to produce disease **:** UNWHOLESOME ⟨a *sickly* climate⟩
4 a : appearing as if sick **b :** lacking in vigor **:** WEAK ⟨a *sickly* plant⟩
5 : SICKENING ⟨a *sickly* odor⟩ ⟨a *sickly* green⟩
— sick·li·ness *noun*
— sickly *adverb*

[2]sickly *transitive verb* **sick·lied; sick·ly·ing** (1763)
: to make sick or sickly

sick·ness \'sik-nəs\ *noun* (before 12th century)
1 a : ill health **:** ILLNESS **b :** a disordered, weakened, or unsound condition
2 : a specific disease
3 : NAUSEA, QUEASINESS

sicko \'si-(,)kō\ *noun, plural* **sick·os** (1963)
: SICKIE

sick–out \'sik-,au̇t\ *noun* (1951)
: an organized absence from work by workers on the pretext of sickness

sick pay *noun* (1887)
: salary or wages paid to an employee while on sick leave

sick·room \'sik-,rüm, -,ru̇m\ *noun* (1749)
: a room in which a person is confined by sickness

sic pas·sim \'sik-'pa-səm, 'sēk-'pä-sim\ *adverb* [Latin] (1921)
: so throughout — used of a word or idea to be found throughout a book or a writer's work

sid·dur \'si-dər, -,du̇r\ *noun, plural* **sid·du·rim** \sə-'du̇r-əm\ [Late Hebrew *siddūr,* literally, order, arrangement] (1864)
: a Jewish prayer book containing liturgies for daily, Sabbath, and holiday observances

[1]side \'sīd\ *noun* [Middle English, from Old English *sīde;* akin to Old High German *sīta* side, Old English *sīd* ample, wide] (before 12th century)
1 a : the right or left part of the wall or trunk of the body ⟨a pain in the *side*⟩ **b** (1) **:** one of the halves of the animal body on either side of the mesial plane (2) **:** a cut of meat including that about the ribs of one half of the body — used chiefly of smoked pork products **c :** one longitudinal half of a hide
2 : a place, space, or direction with respect to a center or to a line of division (as of an aisle, river, or street)
3 a : one of the longer bounding surfaces or lines of an object especially contrasted with the ends ⟨the *side* of a barn⟩ **b :** a line or surface forming a border or face of an object ⟨a die has six *sides*⟩ ⟨the back *side* of the moon⟩ **c :** either surface of a thin object ⟨one *side* of a record⟩ ⟨right *side* of the cloth⟩ **d :** a bounding line of a geometric figure ⟨*side* of a triangle⟩
4 a : the space beside one ⟨stood by my *side*⟩ **b :** an area next to something — usually used in combination ⟨a pool*side* interview⟩
5 : a slope (as of a hill) considered as opposed to another slope ⟨the far *side* of the hill⟩
6 a : the attitude or activity of one person or group with respect to another **:** PART ⟨there was no malice on my *side*⟩ **b :** a position that is opposite to or contrasted with another ⟨two *sides* to every question⟩ ⟨came down on the *side* of law and order⟩ **c :** a body of partisans or contestants ⟨victory for neither *side*⟩ **d :** TEAM ⟨11 players on each *side*⟩
7 : a line of descent traced through one's parent ⟨grandfather on his mother's *side*⟩
8 : an aspect or part of something contrasted with some other real or implied aspect or part ⟨the better *side* of his nature⟩ ⟨the sales *side* of the business⟩ ⟨the seasoning is a bit on the heavy *side*⟩
9 *British* **:** sideways spin imparted to a billiard or snooker ball
10 : a sheet containing the lines and cues for a single theatrical role
11 : a recording of music
12 : a side order or dish ⟨a *side* of fries⟩
— on the side 1 : in addition to the main portion **2 :** in addition to a principal occupation
— this side of : short of **:** ALMOST ⟨an attitude just *this side of* scandalous⟩

[2]side *adjective* (14th century)
1 a : situated on the side ⟨*side* window⟩ **b :** of or relating to the side
2 a : directed toward or from the side ⟨*side* thrust⟩ ⟨*side* wind⟩ **b :** INCIDENTAL, INDIRECT ⟨*side* issue⟩ ⟨*side* remark⟩ **c :** made on the side ⟨*side* payment⟩ **d :** additional to the main portion ⟨*side* order of french fries⟩

[3]side *verb* **sid·ed; sid·ing** (1591)
transitive verb
1 : to agree with **:** SUPPORT
2 : to be side by side with
3 : to set or put aside **:** clear away ⟨*side* the dishes⟩
4 : to furnish with sides or siding ⟨*side* a house⟩
intransitive verb
: to take sides **:** join or form sides ⟨*sided* with the rebels⟩

[4]side *noun* [obsolete English *side* proud, boastful] (1878)
chiefly British **:** swaggering or arrogant manner **:** PRETENTIOUSNESS

side·arm \'sīd-,ärm\ *adjective* (1908)
: of, relating to, using, or being a baseball pitching style in which the arm is not raised above the shoulder and the ball is thrown with a sideways sweep of the arm between shoulder and hip ⟨*sidearm* delivery⟩
— sidearm *adverb*

side arm *noun* (1689)

: a weapon (as a sword, revolver, or bayonet) worn at the side or in the belt

side·band \-ˌband\ *noun* (1922)
: the band of frequencies (as of radio waves) on either side of the carrier frequency produced by modulation

side·bar \-ˌbär\ *noun* (1945)
: a short news story accompanying and presenting sidelights of a major story

side bearing *noun* (circa 1894)
: the space provided at each side of a typeset letter to prevent its touching adjoining letters

side·board \ˈsīd-ˌbōrd, -ˌbȯrd\ *noun* (1671)
: a piece of dining-room furniture having compartments and shelves for holding articles of table service

sideboard

side·burns \-ˌbərnz\ *noun plural* [anagram of *burnsides*] (1887)
1 : SIDE-WHISKERS
2 : continuations of the hairline in front of the ears ◆
— side·burned \-ˌbərnd\ *adjective*

side by side *adverb* (13th century)
1 : beside one another
2 : in the same place, time, or circumstance ⟨lived peacefully *side by side* for many years⟩
— side–by–side *adjective*

side·car \ˈsīd-ˌkär\ *noun* (1904)
1 : a car attached to the side of a motorcycle for a passenger
2 : a cocktail consisting of a liqueur with lemon juice and brandy

side chain *noun* (1886)
: a shorter chain or group of atoms attached to a principal chain or to a ring in a molecule

side chair *noun* (1925)
: a chair without arms used usually in a dining room

sid·ed \ˈsī-dəd\ *adjective* (15th century)
: having sides often of a specified number or kind ⟨one-*sided*⟩ ⟨glass-*sided*⟩
— sid·ed·ness *noun*

side dish *noun* (1725)
: a food served separately along with the main course

side·dress \ˈsīd-ˌdres\ *noun* (1966)
: SIDE-DRESSING
— side–dress \ˈsīd-ˌdres\ *transitive verb*

side–dress·ing \-ˌdre-siŋ\ *noun* (1935)
1 : plant nutrients placed on or in the soil near the roots of a growing crop often by means of a cultivator having a fertilizer-distributing attachment
2 : the act or process of applying side-dressing to a crop

side drum *noun* (circa 1800)
: SNARE DRUM

side effect *noun* (1884)
: a secondary and usually adverse effect (as of a drug) ⟨toxic *side effects*⟩ — called also *side reaction*

side–glance \ˈsīd-ˌglan(t)s\ *noun* (1611)
1 : a glance directed to the side
2 : a passing allusion **:** an indirect or slight reference

side·hill \-ˌhil\ *noun* (1674)
: HILLSIDE
— sidehill *adjective*

side horse *noun* (circa 1934)
: POMMEL HORSE

side·kick \ˈsīd-ˌkik\ *noun* (1906)
: a person closely associated with another as subordinate or partner

side·light \-ˌlīt\ *noun* (1610)
1 a : light coming or produced from the side **b :** incidental light or information
2 : the red light on the port bow or the green light on the starboard bow carried by ships under way at night

¹side·line \-ˌlīn\ *noun* (1862)
1 : a line at right angles to a goal line or end line and marking a side of a court or field of play for athletic games
2 a : a line of goods sold in addition to one's principal line **b :** a business or activity pursued in addition to one's regular occupation
3 a : the space immediately outside the lines along either side of an athletic field or court **b :** a sphere of little or no participation or activity — usually used in plural

²sideline *transitive verb* (1943)
: to put out of action **:** put on the sidelines

side·lin·er \ˈsīd-ˌlī-nər\ *noun* (1947)
: one that remains on the sidelines during an activity **:** one that does not participate

¹side·ling \ˈsīd-liŋ\ *adverb* [Middle English *sidling,* from ¹*side* + ²*-ling*] (14th century)
archaic **:** in a sidelong direction **:** SIDEWAYS

²sideling *adjective* (1611)
1 *archaic* **:** directed toward one side **:** OBLIQUE
2 *archaic* **:** having an inclination **:** SLOPING ⟨*sideling* ground⟩

¹side·long \ˈsīd-ˌlȯŋ\ *adverb* [alteration of ¹*sideling*] (14th century)
1 : SIDEWAYS, OBLIQUELY
2 *archaic* **:** on the side

²sidelong *adjective* (1597)
1 : lying or inclining to one side **:** SLANTING
2 a : directed to one side ⟨*sidelong* looks⟩ **b :** indirect rather than straightforward

side·man \ˈsīd-ˌman\ *noun* (1936)
: a member of a band or orchestra and especially of a jazz or swing orchestra

side·piece \-ˌpēs\ *noun* (1802)
: a piece forming or contained in the side of something ⟨the *sidepiece* of a carriage⟩

sider- *or* **sidero-** *combining form* [Middle French, from Latin, from Greek *sidēr-, sidēro-,* from *sidēros*]
: iron ⟨hemo*sider*in⟩ ⟨*sidero*lite⟩

-sid·er \ˈsīd-ər\ *combining form*
: one placed or living in a usually specified side (as a section of the city) ⟨an east-*sider*⟩

si·de·re·al \sī-ˈdir-ē-əl, sə-\ *adjective* [Latin *sidereus,* from *sider-, sidus* star, constellation] (1647)
: of, relating to, or expressed in relation to stars or constellations **:** ASTRAL

sidereal day *noun* (1794)
: the interval between two successive transits of a point on the celestial sphere (as the vernal equinox) over the upper meridian of a place **:** 23 hours, 56 minutes, 4.1 seconds of mean time

sidereal hour *noun* (circa 1891)
: the 24th part of a sidereal day

sidereal minute *noun* (circa 1909)
: the 60th part of a sidereal hour

sidereal month *noun* (1868)
: the mean time of the moon's revolution in its orbit with reference to a star's position **:** 27 days, 7 hours, 43 minutes, 11.5 seconds of mean time

sidereal second *noun* (circa 1909)
: the 60th part of a sidereal minute

sidereal time *noun* (1812)
1 : time based on the sidereal day
2 : the hour angle of the vernal equinox at a place

sidereal year *noun* (1681)
: the time in which the earth completes one revolution in its orbit around the sun measured with respect to the fixed stars **:** 365 days, 6 hours, 9 minutes, and 9.5 seconds of mean time

¹sid·er·ite \ˈsi-də-ˌrīt\ *noun* [German *Siderit,* from Greek *sidēros* iron] (1850)
: a native ferrous carbonate $FeCO_3$ that is a valuable iron ore

²siderite *noun* (1875)
: a nickel-iron meteorite

side road *noun* (1854)
: a smaller road off a main road

si·de·ro·lite \sī-ˈdir-ə-ˌlīt, ˈsi-də-rə-\ *noun* (1863)
: a stony iron meteorite

side·sad·dle \ˈsīd-ˌsa-dᵊl\ *noun* (15th century)
: a saddle for women in which the rider sits with both legs on the same side of the horse
— sidesaddle *adverb*

side–scan sonar \-ˌskan-\ *noun* (1967)
: a sonar that scans the ocean floor to the side of a ship's track and is used especially for mapping the ocean bottom

side·show \-ˌshō\ *noun* (1846)
1 : a minor show offered in addition to a main exhibition (as of a circus)
2 : an incidental diversion or spectacle

side·slip \-ˌslip\ *intransitive verb* (1887)
1 : to skid or slide sideways
2 : to slide sideways through the air in a downward direction in an airplane along an inclined lateral axis
— sideslip *noun*

side·spin \-ˌspin\ *noun* (1926)
: a rotary motion that causes a ball to revolve horizontally

side·split·ting \-ˌspli-tiŋ\ *adjective* (1856)
: extremely funny
— side·split·ting·ly *adverb*

side·step \ˈsīd-ˌstep\ (1901)
intransitive verb
1 : to take a side step
2 : to avoid an issue or decision
transitive verb
1 : to move out of the way of **:** AVOID ⟨*sidestep* a blow⟩
2 : BYPASS, EVADE ⟨*sidestep* a question⟩
— side·step·per *noun*

side step *noun* (1789)
1 : a step aside (as in boxing to avoid a blow)
2 : a step taken sideways (as when climbing on skis)

side–strad·dle hop \ˈsīd-ˌstra-dᵊl-\ *noun* (1952)
: JUMPING JACK 2

side·stream \-ˌstrēm\ *adjective* (1951)
: relating to or being tobacco smoke that is emitted from the lighted end of a cigarette or cigar

side street *noun* (1617)
: a street joining and often terminated by a main thoroughfare

side·stroke \ˈsīd-ˌstrōk\ *noun* (1867)
: a swimming stroke which is executed on the side and in which the arms are swept in separate strokes towards the feet and downward and the legs do a scissors kick

¹side·swipe \-ˌswīp\ *transitive verb* (1904)
: to strike with a glancing blow along the side ⟨*sideswiped* a parked car⟩

²sideswipe *noun* (1917)
1 a : the action of sideswiping **b :** an instance of sideswiping **:** a glancing blow

◇ WORD HISTORY

sideburns During the American Civil War, Union general Ambrose Everett Burnside (1824–1881) cultivated long bushy side-whiskers. His appearance first struck the fancy of Washingtonians as he conducted parades and maneuvers with his regiment of Rhode Island volunteers in the early days of the war. Despite a later military career that had its ups and downs, this early popularity fostered the fashion for such whiskers in 19th century America, which originally were called *burnsides.* By the 1880s the order of the two words comprising *burnsides* was reversed to give *sideburns,* most likely by analogy with *side-whiskers.* And whereas the general's whiskers were flamboyantly bushy, today sideburns can be of any length or fullness.

2 : an incidental deprecatory remark, allusion, or reference
side table *noun* (14th century)
: a table designed to be placed against a wall
[1]**side·track** \ˈsīd-ˌtrak\ *noun* (1835)
1 : SIDING 2
2 : a position or condition of secondary importance to which one may be diverted
[2]**sidetrack** *transitive verb* (1880)
1 : to shunt aside (as to a railroad siding)
2 a : to turn aside from a purpose **:** DEFLECT **b :** to prevent action on by diversionary tactics ⟨*sidetrack* an issue⟩
side·walk \ˈsīd-ˌwȯk\ *noun* (1739)
: a usually paved walk for pedestrians at the side of a street
sidewalk superintendent *noun* (1940)
: a spectator at a building or demolition job
side·wall \ˈsīd-ˌwȯl\ *noun* (14th century)
1 : a wall forming the side of something
2 : the side of an automotive tire between the tread shoulder and the rim bead
side·ward \ˈsīd-wərd\ *or* **side·wards** \-wərdz\ *adverb* (15th century)
: toward a side
side·way \ˈsīd-ˌwā\ *adverb or adjective* (1612)
: SIDEWAYS
side·ways \-ˌwāz\ *adverb or adjective* (1577)
1 : from one side
2 : with one side forward ⟨turn *sideways*⟩
3 a : in a lateral direction or downward to one side ⟨hopped *sideways*⟩ ⟨slump *sideways*⟩ **b :** ASKANCE ⟨look *sideways* at someone⟩
side–wheel \ˈsīd-ˌhwēl, -ˌwēl\ *adjective* (1857)
: of or being a steamer having a paddle wheel on each side
— side–wheel·er \-ˌhwē-lər, -ˌwē-\ *noun*
side–whis·kers \ˈsīd-ˌhwis-kərz, -ˌwis-\ *noun plural* (1888)
: whiskers on the side of the face usually worn long
— side–whis·kered \-kərd\ *adjective*
side·wind·er \ˈsīd-ˌwīn-dər\ *noun* (1840)
1 : a heavy swinging blow from the side
2 : a small pale-colored desert rattlesnake (*Crotalus cerastes*) of the southwestern U.S. that moves by thrusting its body diagonally forward in a series of flat S-shaped curves

sidewinder 2

side·wise \ˈsīd-ˌwīz\ *adverb* (1571)
: SIDEWAYS
sid·ing \ˈsī-diŋ\ *noun* (1603)
1 *archaic* **:** the taking of sides **:** PARTISANSHIP
2 : a short railroad track connected with the main track
3 : material (as boards or metal pieces) forming the exposed surface of outside walls of frame buildings
si·dle \ˈsī-dᵊl\ *verb* **si·dled; si·dling** \ˈsīd-liŋ, ˈsī-dᵊl-iŋ\ [probably back-formation from [2]*sideling*] (1697)
intransitive verb
: to go or move with one side foremost especially in a furtive advance
transitive verb
: to cause to move or turn sideways
— sidle *noun*
siege \ˈsēj *also* ˈsēzh\ *noun* [Middle English *sege,* from Old French, seat, blockade, from *siegier* to seat, settle, from (assumed) Vulgar Latin *sedicare,* from Latin *sedēre* to sit — more at SIT] (13th century)
1 *obsolete* **:** a seat of distinction **:** THRONE
2 a : a military blockade of a city or fortified place to compel it to surrender **b :** a persistent or serious attack (as of illness)
— siege *transitive verb*
— lay siege to 1 : to besiege militarily **2 :** to pursue diligently or persistently
siege mentality *noun* (1953)
: a defensive or overly fearful attitude
Siege Perilous *noun*
: a seat at King Arthur's Round Table reserved for the knight destined to achieve the quest of the Holy Grail and fatal to any other occupying it
Sieg·fried \ˈsig-ˌfrēd, ˈsēg-\ *noun* [German]
: a hero in Germanic legend who slays a dragon guarding a gold hoard and wakes Brunhild from her enchanted sleep
Siegfried line *noun* [*Siegfried*] (1918)
: a line of German defensive fortifications facing the Maginot Line
sie·mens \ˈsē-mənz, ˈzē-\ *noun, plural* **siemens** [Werner von *Siemens* (died 1892) German electrical engineer] (circa 1933)
: a unit of conductance in the meter-kilogram-second system equivalent to one ampere per volt
si·en·na \sē-ˈe-nə\ *noun* [Italian *terra di Siena,* literally, Siena earth, from *Siena,* Italy] (1787)
: an earthy substance containing oxides of iron and usually of manganese that is brownish yellow when raw and orange red or reddish brown when burnt and is used as a pigment
si·er·o·zem \sē-ˈer-ə-ˌzem, sē-ˌer-ə-ˈzyȯm\ *noun* [Russian *serozem,* from *seryĭ* gray + *zemlya* earth; akin to Latin *humus* earth — more at HUMBLE] (1934)
: any of a group of zonal soils brownish gray at the surface and lighter below, based in a carbonate or hardpan layer, and characteristic of temperate to cool arid regions
si·er·ra \sē-ˈer-ə\ *noun* [Spanish, literally, saw, from Latin *serra*] (1600)
1 a : a range of mountains especially with a serrated or irregular outline **b :** the country about a sierra
2 : any of several large fishes (genus *Scomberomorus*) related to the mackerel
Sierra (1952)
— a communications code word for the letter *s*
si·er·ran \sē-ˈer-ən\ *adjective* (1873)
1 : of or relating to a sierra ⟨*sierran* foothills⟩
2 *capitalized* **:** of or relating to the Sierra Nevada Mountains of the western U.S.
Sierran *noun* (1906)
: a native or inhabitant of the region around the Sierra Nevada Mountains
si·es·ta \sē-ˈes-tə\ *noun* [Spanish, from Latin *sexta* (*hora*) noon, literally, sixth hour — more at SEXT] (1655)
: an afternoon nap or rest
sie·va bean \ˈsē-və-, ˈsi-vē-\ *noun* [origin unknown] (1888)
: a bean plant (*Phaseolus lunatus*) of tropical America that is closely related to and sometimes classified with the lima bean; *also* **:** its flat edible seed
[1]**sieve** \ˈsiv\ *noun* [Middle English *sive,* from Old English *sife;* akin to Old High German *sib* sieve] (before 12th century)
: a device with meshes or perforations through which finer particles of a mixture (as of ashes, flour, or sand) of various sizes may be passed to separate them from coarser ones, through which the liquid may be drained from liquid-containing material, or through which soft materials may be forced for reduction to fine particles
[2]**sieve** *verb* **sieved; siev·ing** (15th century)
: SIFT
sieve of Er·a·tos·the·nes \-ˌer-ə-ˈtäs-thə-ˌnēz\ (1803)
: a procedure for finding prime numbers that involves writing down the odd numbers from 2 up in succession and lining out every third number after 3, every fifth after 5 including those already lined out, every seventh after 7, and so on with each successive number which has not been lined out with every number that is not lined out being prime
sieve plate *noun* (1875)
: a perforated wall or part of a wall at the end of one of the individual cells making up a sieve tube
sieve tube *noun* (1875)
: a tube consisting of an end-to-end series of thin-walled living plant cells characteristic of the phloem and held to function chiefly in translocation of organic solutes
si·faka \sə-ˈfa-kə\ *noun* [Malagasy] (1845)
: any of several diurnal mostly black and white lemurs (genus *Propithecus*) with a long tail and silky fur
sift \ˈsift\ *verb* [Middle English, from Old English *siftan;* akin to Old English *sife* sieve] (before 12th century)
transitive verb
1 a : to put through a sieve ⟨*sift* flour⟩ **b :** to separate or separate out by or as if by putting through a sieve
2 : to go through especially to sort out what is useful or valuable ⟨*sifted* the evidence⟩ — often used with *through* ⟨*sift* through a pile of old letters⟩
3 : to scatter by or as if by sifting ⟨*sift* sugar on a cake⟩
intransitive verb
1 : to use a sieve
2 : to pass or fall as if through a sieve
— sift·er *noun*
sift·ing *noun* (15th century)
1 : the act or process of sifting
2 *plural* **:** sifted material
[1]**sigh** \ˈsī\ *verb* [Middle English *sihen,* alteration of *sichen,* from Old English *sīcan;* akin to Middle Dutch ver*siken* to sigh] (13th century)
intransitive verb
1 : to take a deep audible breath (as in weariness or relief)
2 : to make a sound like sighing ⟨wind *sighing* in the branches⟩
3 : GRIEVE, YEARN ⟨*sighing* for days gone by⟩
transitive verb
1 : to express by sighs
2 *archaic* **:** to utter sighs over **:** MOURN
— sigh·er \ˈsī(-ə)r\ *noun*
[2]**sigh** *noun* (14th century)
1 : an often involuntary act of sighing especially when expressing an emotion or feeling (as weariness or relief)
2 : the sound of gently moving or escaping air ⟨*sighs* of the summer breeze⟩
[1]**sight** \ˈsīt\ *noun* [Middle English, from Old English *gesiht* faculty or act of sight, thing seen; akin to Old High German *gisiht* sight, Old English *sēon* to see] (before 12th century)
1 : something that is seen **:** SPECTACLE
2 a : a thing regarded as worth seeing — usually used in plural ⟨the *sights* of the city⟩ **b :** something ludicrous or disorderly in appearance ⟨you look a *sight*⟩
3 a *chiefly dialect* **:** a great number or quantity **b :** a good deal **:** LOT ⟨a far *sight* better⟩ ⟨not by a damn *sight*⟩
4 a : the process, power, or function of seeing; *specifically* **:** the animal sense of which the end organ is the eye and by which the position, shape, and color of objects are perceived **b :** mental or spiritual perception **c :** mental view; *specifically* **:** JUDGMENT
5 a : the act of looking at or beholding **b :** INSPECTION, PERUSAL **c :** VIEW, GLIMPSE **d :** an observation to determine direction or position (as by a navigator)
6 a : a perception of an object by or as if by the eye ⟨never lost *sight* of the objective⟩ **b :** the range of vision ⟨was nowhere in *sight*⟩
7 : presentation of a note or draft to the maker or draftee **:** DEMAND
8 a : a device that aids the eye in aiming or in finding the direction of an object **b** *plural* **:** ASPIRATION ⟨set her *sights* on a medical career⟩

— **in sight :** at or within a reasonable distance or time
— **on sight :** as soon as seen ⟨ordered to shoot *on sight*⟩
— **out of sight 1 :** beyond comparison **2 :** beyond all expectation or reason **3** — used as a generalized expression of approval
— **sight for sore eyes :** one whose appearance or arrival is an occasion for joy or relief

²sight (1602)
transitive verb
1 : to get or catch sight of ⟨several whales were *sighted*⟩
2 : to look at through or as if through a sight; *especially* **:** to test for straightness
3 : to aim by means of sights
4 a : to equip with sights **b :** to adjust the sights of
intransitive verb
1 : to take aim
2 : to look carefully in a particular direction

³sight *adjective* (1801)
1 : based on recognition or comprehension without previous study ⟨*sight* translation⟩
2 : payable on presentation ⟨a *sight* draft⟩

sight·ed \ˈsī-təd\ *adjective* (1552)
: having sight ⟨clear-*sighted*⟩ ⟨a *sighted* person⟩

sight gag *noun* (1949)
: a comic bit or episode whose effect is produced by pantomime or camera shot rather than by words

sight·less \ˈsīt-ləs\ *adjective* (13th century)
1 : lacking sight **:** BLIND
2 : INVISIBLE 1
— **sight·less·ly** *adverb*
— **sight·less·ness** *noun*

sight line *noun* (1859)
: a line extending from an observer's eye to a viewed object or area (as a stage) ⟨a theater with excellent *sight lines*⟩

sight·ly \-lē\ *adjective* (1534)
1 : pleasing to the sight **:** ATTRACTIVE
2 : affording a fine view
— **sight·li·ness** *noun*
— **sightly** *adverb*

sight–read \-ˌrēd\ *verb* **-read** \-ˌred\; **-read·ing** \-ˌrē diŋ\ [back-formation from *sight reader*] (1903)
transitive verb
: to read (as a foreign language) or perform (music) without previous preparation or study
intransitive verb
: to read at sight; *especially* **:** to perform music at sight
— **sight reader** *noun*

sight rhyme *noun* (circa 1936)
: EYE RHYME

sight·see \ˈsīt-ˌsē\ *intransitive verb, past* **-saw** *present participle* **-see·ing** [back-formation from *sight-seeing*] (1824)
: to go about seeing sights of interest
— **sight·seer** \-ˌsē-ər, -ˌsir\ *noun*

sight–see·ing \ˈsīt-ˌsē-iŋ\ *adjective* (1827)
: devoted to or used for seeing sights
— **sight–seeing** *noun*

sight unseen *adverb* (1892)
: without inspection or appraisal

sig·il \ˈsi-jəl\ *noun* [Middle English *sigulle*, from Latin *sigillum* — more at SEAL] (15th century)
1 : SEAL, SIGNET
2 : a sign, word, or device held to have occult power in astrology or magic

sig·ma \ˈsig-mə\ *noun* [Greek] (1607)
1 : the 18th letter of the Greek alphabet — see ALPHABET table
2 : STANDARD DEVIATION

sig·moid \ˈsig-ˌmȯid\ *also* **sig·moi·dal** \sig-ˈmȯi-dᵊl\ *adjective* [Greek *sigmoeidēs*, from *sigma;* from a common form of sigma shaped like the Roman letter C] (1670)
1 a : curved like the letter C **b :** curved in two directions like the letter S
2 : of, relating to, or being the sigmoid flexure of the intestine
— **sig·moi·dal·ly** \sig-ˈmȯi-dᵊl-ē\ *adverb*

sigmoid flexure *noun* (1786)
: the contracted and crooked part of the colon immediately above the rectum — called also *sigmoid colon*

sig·moid·os·co·py \ˌsig-ˌmȯi-ˈdäs-kə-pē\ *noun* (circa 1900)
: the process of using a long hollow tubular instrument passed through the anus for inspection, diagnosis, treatment, and photography especially of the sigmoid flexure

¹sign \ˈsīn\ *noun* [Middle English *signe*, from Old French, from Latin *signum* mark, token, sign, image, seal; perhaps akin to Latin *secare* to cut — more at SAW] (13th century)
1 a : a motion or gesture by which a thought is expressed or a command or wish made known **b :** SIGNAL 2a **c :** a fundamental linguistic unit that designates an object or relation or has a purely syntactic function **d :** one of a set of gestures used to represent language; *also* **:** SIGN LANGUAGE
2 : a mark having a conventional meaning and used in place of words or to represent a complex notion
3 : one of the 12 divisions of the zodiac
4 a (1) **:** a character (as a flat or sharp) used in musical notation (2) **:** SEGNO **b :** a character (as ÷) indicating a mathematical operation; *also* **:** one of two characters + and − that form part of the symbol of a number and characterize it as positive or negative
5 a : a display (as a lettered board or a configuration of neon tubing) used to identify or advertise a place of business or a product **b :** a posted command, warning, or direction **c :** SIGNBOARD
6 a : something material or external that stands for or signifies something spiritual **b :** something indicating the presence or existence of something else ⟨*signs* of success⟩ ⟨a *sign* of the times⟩ **c :** PRESAGE, PORTENT ⟨*signs* of an early spring⟩ **d :** an objective evidence of plant or animal disease
7 *plural usually* **sign :** traces of a usually wild animal ⟨red fox *sign*⟩ ☆
— **signed** *adjective*

²sign *verb* [Middle English, from Middle French *signer*, from Latin *signare* to mark, sign, seal, from *signum*] (13th century)
transitive verb
1 a : CROSS 2 **b :** to place a sign on **c :** to represent or indicate by a sign
2 a : to affix a signature to **:** ratify or attest by hand or seal ⟨*sign* a bill into law⟩ ⟨the prisoner *signed* a confession⟩ **b :** to assign or convey formally ⟨*signed* over his property to his brother⟩ **c :** to write down (one's name)
3 : to communicate by making a sign or by sign language
4 : to engage or hire by securing the signature of on a contract of employment — often used with *up* or *on*
intransitive verb
1 : to write one's name in token of assent, responsibility, or obligation
2 a : to make a sign or signal **b :** to use sign language
— **sign·ee** \ˌsī-ˈnē\ *noun*
— **sign·er** \ˈsī-nər\ *noun*

sign·age \ˈsī-nij\ *noun* (1976)
: signs (as of identification, warning, or direction) or a system of such signs

¹sig·nal \ˈsig-nᵊl\ *noun* [Middle English, from Middle French, from Medieval Latin *signale*, from Late Latin, neuter of *signalis* of a sign, from Latin *signum*] (14th century)
1 : SIGN, INDICATION
2 a : an act, event, or watchword that has been agreed on as the occasion of concerted action **b :** something that incites to action
3 : something (as a sound, gesture, or object) that conveys notice or warning
4 a : an object used to transmit or convey information beyond the range of human voice **b :** the sound or image conveyed in telegraphy, telephony, radio, radar, or television **c :** a detectable physical quantity or impulse (as a voltage, current, or magnetic field strength) by which messages or information can be transmitted

²signal *verb* **sig·naled** *or* **sig·nalled; sig·nal·ing** *or* **sig·nal·ling** \-nə-liŋ\ (1805)
transitive verb
1 : to notify by a signal ⟨*signal* the fleet to turn back⟩
2 a : to communicate or indicate by or as if by signals ⟨*signaled* the end of an era⟩ **b :** to constitute a characteristic feature of (a meaningful linguistic form)
intransitive verb
: to make or send a signal
— **sig·nal·er** *or* **sig·nal·ler** *noun*

³signal *adjective* [modification of French *signalé*, past participle of *signaler* to distinguish, from Old Italian *segnalare* to signal, distinguish, from *segnale* signal, from Medieval Latin *signale*] (1641)
: distinguished from the ordinary ⟨*signal* achievement⟩

sig·nal·ise *British variant of* SIGNALIZE

sig·nal·ize \ˈsig-nə-ˌlīz\ *transitive verb* **-ized; -iz·ing** (1654)
1 : to make conspicuous **:** DISTINGUISH
2 : to point out carefully or distinctly
3 : to make signals to **:** SIGNAL; *also* **:** INDICATE
4 : to place traffic signals at or on
— **sig·nal·i·za·tion** \ˌsig-nə-lə-ˈzā-shən\ *noun*

sig·nal·ly \ˈsig-nə-lē\ *adverb* (1641)
: in a signal manner **:** NOTABLY

sig·nal·man \ˈsig-nᵊl-mən, -ˌman\ *noun* (1737)
: a person who signals or works with signals (as on a railway)

sig·nal·ment \-mənt\ *noun* [French *signalement*, from *signaler*] (1778)
: description by peculiar, appropriate, or characteristic marks; *specifically* **:** the systematic description of a person for purposes of identification

sig·na·to·ry \ˈsig-nə-ˌtōr-ē, -ˌtȯr-\ *noun, plural* **-ries** [Latin *signatorius* of sealing, from *signare*] (1866)
: a signer with another or others ⟨*signatories* to a petition⟩; *especially* **:** a government bound with others by a signed convention
— **signatory** *adjective*

sig·na·ture \ˈsig-nə-ˌchu̇r, -chər, -ˌtyu̇r, -ˌtu̇r\ *noun* [Middle French or Medieval Latin; Middle French, from Medieval Latin *signatura*, from Latin *signatus*, past participle of *signare* to sign, seal] (1536)

☆ SYNONYMS
Sign, mark, token, note, symptom mean a discernible indication of what is not itself directly perceptible. SIGN applies to any indication to be perceived by the senses or the reason ⟨encouraging *signs* for the economy⟩. MARK suggests something impressed on or inherently characteristic of a thing often in contrast to general outward appearance ⟨a *mark* of a good upbringing⟩. TOKEN applies to something that serves as a proof of something intangible ⟨this gift is a *token* of our esteem⟩. NOTE suggests a distinguishing mark or characteristic ⟨a *note* of irony in her writing⟩. SYMPTOM suggests an outward indication of an internal change or condition ⟨rampant crime is a *symptom* of that city's decay⟩.

1 a : the act of signing one's name **b :** the name of a person written with his or her own hand
2 : a feature in the appearance or qualities of a natural object formerly held to indicate its utility in medicine
3 a : a letter or figure placed usually at the bottom of the first page on each sheet of printed pages (as of a book) as a direction to the binder in arranging and gathering the sheets **b :** a folded sheet that is one unit of a book
4 a : KEY SIGNATURE **b :** TIME SIGNATURE
5 : the part of a medical prescription that contains the directions to the patient
6 : something (as a tune, style, or logo) that serves to identify; *also* **:** a characteristic mark

sign·board \'sīn-ˌbōrd, -ˌbȯrd\ *noun* (1632)
: a board bearing a notice or sign

signed \'sīnd\ *adjective* (1873)
: having a sign and especially a plus or minus sign ⟨*signed* numbers like +6 and −4⟩

¹sig·net \'sig-nət\ *noun* [Middle English, from Middle French, diminutive of *signe* sign, seal] (14th century)
1 : a seal used officially to give personal authority to a document in lieu of signature
2 : the impression made by or as if by a signet
3 : a small intaglio seal (as in a finger ring)

²signet *transitive verb* (15th century)
: to stamp or authenticate with a signet

signet ring *noun* (1681)
: a finger ring engraved with a signet, seal, or monogram **:** SEAL RING

sig·nif·i·cance \sig-'ni-fi-kən(t)s\ *noun* (13th century)
1 a : something that is conveyed as a meaning often obscurely or indirectly **b :** the quality of conveying or implying
2 a : the quality of being important **:** MOMENT **b :** the quality of being statistically significant
synonym see IMPORTANCE

significance level *noun* (1947)
: LEVEL OF SIGNIFICANCE

sig·nif·i·can·cy \sig-'ni-fi-kən(t)-sē\ *noun* (circa 1595)
: SIGNIFICANCE

sig·nif·i·cant \-kənt\ *adjective* [Latin *significant-, significans,* present participle of *significare* to signify] (1579)
1 : having meaning; *especially* **:** SUGGESTIVE ⟨a *significant* glance⟩
2 a : having or likely to have influence or effect **:** IMPORTANT ⟨a *significant* piece of legislation⟩; *also* **:** of a noticeably or measurably large amount ⟨a *significant* number of layoffs⟩ ⟨producing *significant* profits⟩ **b :** probably caused by something other than mere chance ⟨statistically *significant* correlation between vitamin deficiency and disease⟩

significant digit *noun* (1923)
: one of the digits of a number beginning with the digit farthest to the left that is not zero and ending with the last digit farthest to the right that is not zero or is a zero considered to be exact — called also *significant figure*

sig·nif·i·cant·ly \sig-'ni-fi-kənt-lē\ *adverb* (1577)
1 : in a significant manner **:** to a significant degree
2 : it is significant

significant other *noun* (1953)
: a person who is important to one's well-being; *especially* **:** a spouse or one in a similar relationship

sig·ni·fi·ca·tion \ˌsig-nə-fə-'kā-shən\ *noun* (14th century)
1 a : the act or process of signifying by signs or other symbolic means **b :** a formal notification
2 : PURPORT; *especially* **:** the meaning that a term, symbol, or character regularly conveys or is intended to convey
3 *chiefly dialect* **:** IMPORTANCE, CONSEQUENCE

sig·nif·i·ca·tive \sig-'ni-fə-ˌkā-tiv\ *adjective* (15th century)
1 : SIGNIFICANT, SUGGESTIVE
2 : INDICATIVE ⟨symptoms *significative* of malaria⟩

sig·nif·ics \sig-'ni-fiks\ *noun plural but singular or plural in construction* [*signify*] (1896)
: SEMIOTIC, SEMANTICS

signified *noun* (1939)
: a concept or meaning as distinguished from the sign through which it is communicated — compare SIGNIFIER 2

sig·ni·fi·er \'sig-nə-ˌfī(-ə)r\ *noun* (1532)
1 : one that signifies
2 : a symbol, sound, or image (as a word) that represents an underlying concept or meaning — compare SIGNIFIED

sig·ni·fy \'sig-nə-ˌfī\ *verb* **-fied; -fy·ing** [Middle English *signifien,* from Old French *signifier,* from Latin *significare* to indicate, signify, from *signum* sign] (13th century)
transitive verb
1 a : to be a sign of **:** MEAN **b :** IMPLY
2 : to show especially by a conventional token (as word, signal, or gesture)
intransitive verb
1 : to have significance **:** MATTER
2 : to engage in signifying

sig·ni·fy·ing \'sig-nə-ˌfī-iŋ\ *noun* (1959)
: a good-natured needling or goading especially among urban blacks by means of indirect gibes and clever often preposterous put-downs; *also* **:** DOZENS

sign in (1930)
intransitive verb
: to make a record of arrival by signing a register or punching a time clock
transitive verb
: to record arrival of (a person) or receipt of (an article) by signing

sign language *noun* (1847)
1 : a formal language employing a system of hand gestures for communication (as by the deaf)
2 : an unsystematic method of communicating chiefly by manual gestures used by people speaking different languages

sign of aggregation (circa 1942)
: any of various conventional devices (as braces, brackets, parentheses, or vinculums) used in mathematics to indicate that two or more terms are to be treated as one quantity

sign off *intransitive verb* (1926)
1 : to announce the end of something (as a message or broadcast)
2 : to approve or acknowledge something by or as if by a signature ⟨*sign off* on a memo⟩
— sign–off \'sī-ˌnȯf\ *noun*

sign of the cross (14th century)
: a gesture of the hand forming a cross especially on forehead, breast, and shoulders to profess Christian faith or invoke divine protection or blessing

sign on *intransitive verb* (1885)
1 : to engage oneself by or as if by a signature
2 : to announce the start of broadcasting for the day
— sign–on \'sī-ˌnȯn, -ˌnän\ *noun*

si·gnor *also* **si·gnior** \sēn-'yȯr, -'yōr\ *noun, plural* **signors** *or* **si·gno·ri** \sēn-'yōr-(ˌ)ē, -'yȯr-\ *also* **signiors** [Italian *signore, signor,* from Medieval Latin *senior* superior, lord — more at SENOR] (circa 1580)
: an Italian man usually of rank or gentility — used as a title equivalent to *Mister*

si·gno·ra \sēn-'yōr-ə, -'yȯr-\ *noun, plural* **si·gnoras** *or* **si·gno·re** \-'yōr-(ˌ)ā, -'yȯr-\ [Italian, feminine of *signore, signor*] (1763)
: a married Italian woman usually of rank or gentility — used as a title equivalent to *Mrs.*

si·gno·re \sēn-'yōr-(ˌ)ā, -'yȯr-\ *noun, plural* **si·gno·ri** \-'yōr-(ˌ)ē, -'yȯr-\ [Italian] (1594)
: SIGNOR

si·gno·ri·na \ˌsē-nyə-'rē-nə\ *noun, plural* **-nas** *or* **-ne** \-(ˌ)nā\ [Italian, from diminutive of *signora*] (1820)
: an unmarried Italian woman — used as a title equivalent to *Miss*

si·gnory *or* **si·gniory** \'sē-nyə-rē\ *noun, plural* **si·gnor·ies** *or* **si·gnior·ies** [Middle English *signorie,* from Middle French *seigneurie*] (14th century)
: SEIGNIORY

sign out (1948)
intransitive verb
: to indicate departure by signing a register
transitive verb
: to record or approve the release or departure of
— sign–out \'sī-ˌnau̇t\ *noun or adjective*

¹sign·post \'sīn-ˌpōst\ *noun* (1620)
1 : a post (as at the fork of a road) with signs on it to direct travelers
2 : GUIDE, BEACON

²signpost *transitive verb* (1895)
: to provide with signposts or guides

sign up *intransitive verb* (1903)
: to sign one's name (as to a contract) in order to obtain, do, or join something ⟨*sign up* for insurance⟩ ⟨*sign up* for classes⟩
— sign–up \'sī-ˌnəp\ *noun or adjective*

Sig·urd \'si-gu̇rd, -gərd\ *noun* [Old Norse *Sigurthr*]
: a hero in Norse mythology who slays the dragon Fafnir

sike \'sīk\ *noun* [Middle English, from Old English *sīc;* akin to Old Norse *sīk* slow stream, Old English *sicerian* to trickle] (before 12th century)
1 *dialect chiefly British* **:** a small stream; *especially* **:** one that dries up in summer
2 *dialect chiefly British* **:** DITCH

¹Sikh \'sēk\ *noun* [Hindi, literally, disciple] (1756)
: an adherent of a monotheistic religion of India founded about 1500 by Guru Nānak and marked by rejection of idolatry and caste
— Sikh·ism \'sē-ˌki-zəm\ *noun*

²Sikh *adjective* (1845)
: of or relating to Sikhs or Sikhism

si·lage \'sī-lij\ *noun* [short for *ensilage*] (1884)
: fodder converted into succulent feed for livestock through processes of anaerobic acid fermentation (as in a silo)

si·lane \'si-ˌlān, 'sī-\ *noun* [International Scientific Vocabulary *sil*icon + meth*ane*] (1916)
: any of various compounds of hydrogen and silicon that have the general formula Si_nH_{2n+2} and are analogous to alkanes

Si·las·tic \sə-'las-tik, sī-\ *trademark*
— used for a soft pliable plastic

sild \'sil(d)\ *noun, plural* **sild** *or* **silds** [Norwegian] (1921)
: a young herring other than a brisling that is canned as a sardine in Norway

¹si·lence \'sī-lən(t)s\ *noun* [Middle English, from Old French, from Latin *silentium,* from *silent-, silens*] (13th century)
1 : forbearance from speech or noise **:** MUTENESS — often used interjectionally
2 : absence of sound or noise **:** STILLNESS
3 : absence of mention: **a :** OBLIVION, OBSCURITY **b :** SECRECY ⟨weapons research was conducted in *silence*⟩

²silence *transitive verb* **si·lenced; si·lenc·ing** (1597)
1 : to compel or reduce to silence **:** STILL
2 : to restrain from expression **:** SUPPRESS
3 : to cause to cease hostile firing or criticism

si·lenc·er \'sī-lən(t)-sər\ *noun* (1600)
: one that silences: as **a** *chiefly British* **:** the muffler of an internal combustion engine **b :** a silencing device for small arms

¹si·lent \'sī-lənt\ *adjective* [Middle English *sylent,* from Latin *silent-, silens,* from present participle of *silēre* to be silent; akin to Gothic ana*silan* to cease, grow calm] (15th century)
1 a : making no utterance **:** MUTE, SPEECHLESS **b :** indisposed to speak **:** not loquacious
2 : free from sound or noise **:** STILL
3 : performed or borne without utterance **:** UNSPOKEN ⟨*silent* prayer⟩ ⟨*silent* grief⟩

4 a : making no mention ⟨history is *silent* about this person⟩ **b :** not widely or generally known or appreciated ⟨the *silent* pressures on a person in public office⟩ **c :** making no protest or outcry ⟨the *silent* majority⟩
5 : UNPRONOUNCED ⟨*silent b* in *doubt*⟩
6 : not exhibiting the usual signs or symptoms of presence ⟨a *silent* infection⟩
7 a : made without spoken dialogue ⟨*silent* movies⟩ **b :** of or relating to silent movies ☆
— si·lent·ly *adverb*
— si·lent·ness *noun*

²silent *noun* (1929)
: a motion picture made without spoken dialogue — usually used in plural

silent auction *noun* (1952)
: an auction in which sealed bids are submitted beforehand

silent butler *noun* (1937)
: a receptacle with hinged lid for collecting table crumbs and the contents of ashtrays

silent partner *noun* (1828)
1 : a partner who is known to the public but has no voice in the conduct of a firm's business
2 : SECRET PARTNER

silent service *noun* (circa 1929)
1 : NAVY — used with *the*
2 : the submarine service — used with *the*

silent treatment *noun* (1947)
: an act of completely ignoring a person or thing by resort to silence especially as a means of expressing contempt or disapproval

si·le·nus \sī-ˈlē-nəs\ *noun, plural* **-ni** \-ˌnī\ *often capitalized* [Latin, from Greek *silēnos,* from *Silēnos* foster father of Dionysus]
: a minor woodland deity and companion of Dionysus in Greek mythology with a horse's ears and tail

si·lex \ˈsī-ˌleks\ *noun* [Latin *silic-, silex* hard stone, flint] (circa 1592)
: silica or a siliceous material (as powdered tripoli) especially for use as a filler in paints or wood

¹sil·hou·ette \ˌsi-lə-ˈwet\ *noun* [French, from Étienne de *Silhouette* (died 1767) French controller general of finances; perhaps from his ephemeral tenure] (1783)
1 : a likeness cut from dark material and mounted on a light ground or one sketched in outline and solidly colored in
2 : the outline of a body viewed as circumscribing a mass ⟨the *silhouette* of a bird⟩
synonym see OUTLINE

²silhouette *transitive verb* **-ett·ed; -ett·ing** (1876)
: to represent by a silhouette; *also* **:** to project on a background like a silhouette
— sil·hou·et·tist \-ˈwe-tist\ *noun*

sil·i·ca \ˈsi-li-kə\ *noun* [New Latin, from Latin *silic-, silex* hard stone, flint] (circa 1801)
: silicon dioxide SiO_2 occurring in crystalline, amorphous, and impure forms (as in quartz, opal, and sand respectively)

silica gel *noun* (1919)
: colloidal silica resembling coarse white sand in appearance but possessing many fine pores and therefore extremely adsorbent

sil·i·cate \ˈsi-lə-ˌkāt, -kət\ *noun* [*silicic (acid)*] (1811)
: a salt or ester derived from a silicic acid; *especially* **:** any of numerous insoluble often complex metal salts that contain silicon and oxygen in the anion, constitute the largest class of minerals, and are used in building materials (as cement, bricks, and glass)

si·li·ceous *or* **si·li·cious** \sə-ˈli-shəs\ *adjective* [Latin *siliceus* of flint, from *silic-, silex* hard stone, flint] (circa 1656)
: of, relating to, or containing silica or a silicate ⟨*siliceous* limestone⟩

si·lic·ic \sə-ˈli-sik\ *adjective* [New Latin *silica* & New Latin *silicium* silicon (from *silica*)] (1817)
: of, relating to, or derived from silica or silicon

silicic acid *noun* (1817)
: any of various weakly acid substances obtained as gelatinous masses by treating silicates with acids

sil·i·cide \ˈsi-lə-ˌsīd\ *noun* [International Scientific Vocabulary *silic*on + *-ide*] (circa 1868)
: a binary compound of silicon with a more electropositive element or group

si·lic·i·fi·ca·tion \sə-ˌli-sə-fə-ˈkā-shən\ *noun* (1830)
: the action or process of silicifying **:** the state of being silicified

si·lic·i·fy \sə-ˈli-sə-ˌfī\ *verb* **-fied; -fy·ing** (circa 1828)
transitive verb
: to convert into or impregnate with silica
intransitive verb
: to become silicified

sil·i·con \ˈsi-li-kən, ˈsi-lə-ˌkän\ *noun* [New Latin *silica* + English *-on* (as in *carbon*)] (1817)
: a tetravalent nonmetallic element that occurs combined as the most abundant element next to oxygen in the earth's crust and is used especially in alloys and electronic devices — see ELEMENT table

silicon carbide *noun* (1893)
: a very hard dark crystalline compound SiC of silicon and carbon that is used as an abrasive and as a refractory and in electric resistors

sil·i·cone \ˈsi-lə-ˌkōn\ *noun* [*silic*on + *-one*] (1943)
: any of various polymeric organic silicon compounds obtained as oils, greases, or plastics and used especially for water-resistant and heat-resistant lubricants, varnishes, binders, and electric insulators

silicone rubber *noun* (1944)
: rubber made from silicone elastomers and noted for its retention of flexibility, resilience, and tensile strength over a wide temperature range

sil·i·con·ized \ˈsi-lə-kə-ˌnīzd, -ˌkō-\ *adjective* (1949)
: treated or coated with a silicone ⟨*siliconized* glassware⟩

silicon nitride *noun* (1903)
: any of several compounds of silicon and nitrogen; *specifically* **:** a compound Si_3N_4 that is a hard ceramic used in high-temperature applications and in composites

sil·i·co·sis \ˌsi-lə-ˈkō-səs\ *noun* [New Latin, from *silica* + *-osis*] (1881)
: pneumoconiosis characterized by massive fibrosis of the lungs resulting in shortness of breath and caused by prolonged inhalation of silica dusts
— sil·i·cot·ic \-ˈkä-tik\ *adjective or noun*

si·lique \sə-ˈlēk\ *noun* [French, from New Latin *siliqua,* from Latin, pod, husk] (1785)
: a narrow elongated two-valved usually many-seeded capsule that is characteristic of the mustard family, opens by sutures at either margin, and has two parietal placentas

¹silk \ˈsilk\ *noun, often attributive* [Middle English, from Old English *seolc,* probably ultimately from Greek *sērikos* silken — more at SERICEOUS] (before 12th century)
1 : a fine continuous protein fiber produced by various insect larvae usually for cocoons; *especially* **:** a lustrous tough elastic fiber produced by silkworms and used for textiles
2 : thread, yarn, or fabric made from silk filaments
3 a : a garment of silk **b** (1) **:** a distinctive silk gown worn by a King's or Queen's Counsel (2) **:** a King's or Queen's Counsel **c** *plural* **:** the colored cap and blouse of a jockey or harness horse driver made in the registered racing color of the employing stable
4 a : a filament resembling silk; *especially* **:** one produced by a spider **b :** silky material ⟨milkweed *silk*⟩; *especially* **:** the styles of an ear of Indian corn
5 : PARACHUTE
— silk·like \-ˌlīk\ *adjective*

²silk *intransitive verb* (1783)
of corn **:** to develop the silk

silk·a·line *or* **silk·o·line** \ˌsil-kə-ˈlēn\ *noun* [¹*silk* + *-oline* (as in *crinoline*)] (1896)
: a soft light cotton fabric with a smooth lustrous finish like that of silk

silk cotton *noun* (1697)
: the silky or cottony covering of seeds of various silk-cotton trees; *especially* **:** KAPOK

silk–cotton tree *noun* (1712)
: any of various tropical trees (family Bombacaceae, the silk-cotton family) with palmate leaves and large fruits with the seeds enveloped by silk cotton; *especially* **:** CEIBA 1

silk·en \ˈsil-kən\ *adjective* (before 12th century)
1 : made or consisting of silk
2 : resembling silk: as **a :** SOFT, LUSTROUS **b** (1) **:** agreeably smooth **:** HARMONIOUS (2) **:** SUAVE, INGRATIATING
3 a : dressed in silk ⟨*silken* ankles⟩ **b :** LUXURIOUS 2, 3

silk gland *noun* (1870)
: a gland that produces a viscid fluid which is extruded in filaments and hardens into silk on exposure to air: as **a :** either of a pair of greatly enlarged and modified salivary glands of an insect larva that produce a compound filament from which a larval or pupal cover (as a cocoon) is spun **b :** any of two or more abdominal glands of a spider that open through spinnerets and produce a filament used chiefly in the spinning of webs

silk hat *noun* (1834)
: a hat with a tall cylindrical crown and a silk-plush finish worn by men as a dress hat

silk oak *noun* (1866)
: any of various Australian timber trees (especially genus *Grevillea*) of the protea family that have mottled wood used in cabinetmaking and veneering — called also *silky oak*

silk screen *noun* (1930)
: a stencil process in which coloring matter is forced onto the material to be printed through the meshes of a silk or organdy screen so prepared as to have pervious printing areas and impervious nonprinting areas; *also* **:** a print made by this process
— silk–screen *transitive verb*

silk–stock·ing \ˈsilk-ˈstä-kiŋ\ *adjective* (1798)
1 : ARISTOCRATIC, WEALTHY ⟨a *silk-stocking* district⟩
2 : fashionably dressed ⟨a *silk-stocking* audience⟩
3 : of or relating to the American Federalist party

☆ SYNONYMS
Silent, taciturn, reticent, reserved, secretive mean showing restraint in speaking. SILENT implies a habit of saying no more than is needed ⟨the strong, *silent* type⟩. TACITURN implies a temperamental disinclination to speech and usually connotes unsociability ⟨*taciturn* villagers⟩. RETICENT implies a reluctance to speak out or at length, especially about one's own affairs ⟨was *reticent* about his plans⟩. RESERVED implies reticence and suggests the restraining influence of caution or formality in checking easy informal conversational exchange ⟨greetings were brief, formal, and *reserved*⟩. SECRETIVE, too, implies reticence but usually carries a suggestion of deviousness and lack of frankness or of an often ostentatious will to conceal ⟨the *secretive* research and development division⟩.

silk stocking *noun* (1891)
1 : an aristocratic or wealthy person
2 : a fashionably dressed person
3 : FEDERALIST 2

silk tree *noun* (circa 1852)
: a leguminous Asian tree (*Albizia julibrissin*) having pink flowers with long silky stamens — called also *mimosa*

silk•weed \'silk-ˌwēd\ *noun* (1784)
: MILKWEED

silk•worm \-ˌwərm\ *noun* (before 12th century)
: a moth whose larva spins a large amount of strong silk in constructing its cocoon; *especially* **:** an Asian moth (*Bombyx mori* of the family Bombycidae) whose rough wrinkled hairless yellowish caterpillar produces the silk of commerce

silky \'sil-kē\ *adjective* **silk•i•er; -est** (1611)
1 : SILKEN 1, 2
2 : having or covered with fine soft hairs, plumes, or scales
— silk•i•ly \-kə-lē\ *adverb*
— silk•i•ness \-kē-nəs\ *noun*

silky terrier *noun* (1959)
: any of a breed of low-set toy terriers of Australian origin that have a flat silky glossy coat colored blue with tan on the head, chest, and legs — called also *silky*

silky terrier

sill \'sil\ *noun* [Middle English *sille*, from Old English *syll;* akin to Old High German *swelli* beam, threshold] (before 12th century)
1 : a horizontal piece (as a timber) that forms the lowest member or one of the lowest members of a framework or supporting structure: as **a :** the horizontal member at the base of a window **b :** the threshold of a door
2 : a tabular body of igneous rock injected while molten between sedimentary or volcanic beds or along foliation planes of metamorphic rocks
3 : a submerged ridge at relatively shallow depth separating the basins of two bodies of water

sillabub *variant of* SYLLABUB

sil•li•man•ite \'si-lə-mə-ˌnīt\ *noun* [Benjamin *Silliman* (died 1864) American geologist] (circa 1830)
: a brown, grayish, or pale green mineral that consists of an aluminum silicate in orthorhombic crystals often occurring in fibrous or columnar forms

sil•ly \'si-lē\ *adjective* **sil•li•er; -est** [Middle English *sely, silly* happy, innocent, pitiable, feeble, from Old English *sǣlig*, from Old English *sǣl* happiness; akin to Old High German *sālig* happy] (14th century)
1 *archaic* **:** HELPLESS, WEAK
2 a : RUSTIC, PLAIN **b** *obsolete* **:** lowly in station **:** HUMBLE
3 a : weak in intellect **:** FOOLISH **b :** exhibiting or indicative of a lack of common sense or sound judgment ⟨a very *silly* mistake⟩ **c :** TRIFLING, FRIVOLOUS
4 : being stunned or dazed ⟨scared *silly*⟩ ⟨knocked me *silly*⟩ ◆
synonym see SIMPLE
— sil•li•ly \'si-lə-lē\ *adverb*
— sil•li•ness \'si-lē-nəs\ *noun*
— silly *noun or adverb*

silly season *noun* (1861)
1 : a period (as late summer) when newspapers often resort to trivial or frivolous matters for lack of major news stories
2 : a period marked by frivolous, outlandish, or illogical activity or behavior

si•lo \'sī-(ˌ)lō\ *noun, plural* **silos** [Spanish] (1881)
1 : a trench, pit, or especially a tall cylinder (as of wood or concrete) usually sealed to exclude air and used for making and storing silage
2 a : a deep bin for storing material (as cement or coal) **b :** an underground structure for housing a guided missile

silo 1

si•lox•ane \sə-'läk-ˌsān, sī-\ *noun* [*sil*icon + *ox*ygen + meth*ane*] (1917)
: any of various compounds containing alternate silicon and oxygen atoms in either a linear or cyclic arrangement usually with one or two organic groups attached to each silicon atom

¹silt \'silt\ *noun* [Middle English *cylte*, probably of Scandinavian origin; akin to Danish *sylt* salt marsh; akin to Old High German *sulza* salt marsh, Old English *sealt* salt] (15th century)
1 : loose sedimentary material with rock particles usually 1/20 millimeter or less in diameter; *also* **:** soil containing 80 percent or more of such silt and less than 12 percent of clay
2 : a deposit of sediment (as by a river)
— silty \'sil-tē\ *adjective*

²silt (1799)
intransitive verb
: to become choked or obstructed with silt — often used with *up* ⟨the channel *silted* up⟩
transitive verb
: to choke, fill, cover, or obstruct with silt or mud ⟨the beaver had *silted* the creek —Hugh Fosburgh⟩
— sil•ta•tion \sil-'tā-shən\ *noun*

silt•stone \'silt-ˌstōn\ *noun* (circa 1920)
: a rock composed chiefly of indurated silt

Sil•u•res \'sil-yə-ˌrēz\ *noun* [Latin] (circa 1895)
: a people of ancient Britain described by Tacitus as occupying chiefly southern Wales

Si•lu•ri•an \sī-'lu̇r-ē-ən, sə-\ *adjective* [Latin *Silures*] (1708)
1 : of or relating to the Silures or their place of habitation
2 : of, relating to, or being a period of the Paleozoic era between the Ordovician and Devonian or the corresponding system of rocks marked by numerous eurypterid crustaceans and the appearance of the first land plants — see GEOLOGIC TIME table
— Silurian *noun*

sil•va \'sil-və\ *noun* [New Latin, from Latin, wood, forest] (circa 1848)
: the forest trees of a region or country

silvan *variant of* SYLVAN

¹sil•ver \'sil-vər\ *noun* [Middle English, from Old English *seolfor;* akin to Old High German *silbar* silver, Lithuanian *sidabras*] (before 12th century)
1 : a white metallic element that is sonorous, ductile, very malleable, capable of a high degree of polish, and chiefly univalent in compounds, and that has the highest thermal and electric conductivity of any substance — see ELEMENT table
2 : silver as a commodity ⟨the value of *silver* has risen⟩
3 : coin made of silver
4 : articles (as hollowware or table flatware) made of or plated with silver; *also* **:** similar articles and especially flatware of other metals (as stainless steel)
5 : a nearly neutral slightly brownish medium gray

²silver *adjective* (before 12th century)
1 : made of silver
2 : resembling silver: as **a :** having a white lustrous sheen **b :** giving a soft resonant sound **:** dulcet in tone **c :** eloquently persuasive
3 : consisting of or yielding silver
4 : of, relating to, or characteristic of silver
5 : advocating the use of silver as a standard of currency
6 : of, relating to, or being a 25th anniversary or its celebration

³silver *transitive verb* **sil•vered; sil•ver•ing** \'sil-v(ə-)riŋ\ (14th century)
1 a : to cover with silver (as by electroplating) **b :** to coat with a substance (as a metal) resembling silver
2 a : to give a silvery luster to **b :** to make white like silver
— sil•ver•er \'sil-vər-ər\ *noun*

silver age *noun* (1565)
: an historical period of achievement secondary to that of a golden age

sil•ver•back \'sil-vər-ˌbak\ *noun* (1963)
: an older adult usually dominant male gorilla having gray or whitish hair on the back

silver bell *noun* (1785)
: any of a genus (*Halesia*) of trees and shrubs of the storax family; *especially* **:** one (*H. carolina*) of the southeastern U.S. cultivated for its bell-shaped white flowers

sil•ver•ber•ry \'sil-vər-ˌber-ē\ *noun* (1856)
: a silvery North American shrub (*Elaeagnus commutata*) of the oleaster family

silver bromide *noun* (1885)
: a compound AgBr that is extremely sensitive to light and is much used in the preparation of sensitive emulsion coatings for photographic materials

silver bullet *noun* (1935)
: something that acts as a magical weapon; *especially* **:** one that instantly solves a long-standing problem

silver certificate *noun* (1882)
: a certificate formerly issued against the deposit of silver coin as legal tender in the U.S. and its possessions

silver chloride *noun* (1885)
: a compound AgCl sensitive to light and used especially for photographic materials

silver cord *noun* [*The Silver Cord* (1926), play by Sidney Howard] (1942)
: the emotional tie between mother and child and especially son

silver fir *noun* (1707)
: any of various firs (genus *Abies*) with leaves that are white or silvery white beneath; *especially* **:** a valuable European timber tree (*A. alba*)

sil•ver•fish \'sil-vər-ˌfish\ *noun* (1703)

◇ WORD HISTORY

silly *Silly* in the now common senses "foolish" or "lacking in sound judgment" is the product of a remarkable string of sense shifts. The Anglian dialect variant *sēlig* of Old English *sǣlig* "happy, blessed, fortunate" was continued by Middle English *sely,* which developed from the Old English sense several new ones, including "innocent, harmless," then "pitiable, helpless" and finally "miserable, unfortunate"—the complete opposite of its original meaning. The "innocent, harmless" sense developed in another direction, to "simple, gullible," and finally, in early Modern English, "lacking in sound judgment." By the late 15th century, Middle English *sely* began to turn up in spellings such as *syly* and *silly,* the original vowel having turned into \i\ for most speakers. Without the change the result would have been *seely* (rhyming with *freely*), a pronunciation which some speakers seem to have retained into the 17th century. The same sound change is evident in the word *breeches* "pants," usually pronounced \'bri-chəz\ in spite of its spelling.

1 : any of various silvery fishes (as a tarpon or silversides)
2 : any of various small wingless insects (order Thysanura); *especially* **:** one (*Lepisma saccharina*) found in houses and sometimes injurious especially to sized papers or starched clothes

silver fox *noun* (circa 1792)
: a genetically determined color phase of the common red fox in which the fur is black tipped with white

silver hake *noun* (1884)
: a common hake (*Merluccius bilinearis*) of the northern Atlantic coast of the U.S. that is an important food fish

silver iodide *noun* (1885)
: a compound AgI that darkens on exposure to light and is used in photography, rainmaking, and medicine

silver lining *noun* [from the phrase "every cloud has a *silver lining*"] (1871)
: a consoling or hopeful prospect

sil·ver·ly \'sil-vər-lē\ *adverb* (1595)
: with silvery appearance or sound

silver maple *noun* (1765)
1 : a common maple (*Acer saccharinum*) of eastern North America with deeply cut leaves that are light green above and silvery white below
2 : the hard close-grained but brittle light brown wood of the silver maple

sil·vern \'sil-vərn\ *adjective* (before 12th century)
1 : made of silver
2 : resembling or characteristic of silver **:** SILVERY

silver nitrate *noun* (1885)
: an irritant compound $AgNO_3$ that in contact with organic matter turns black and is used as a chemical reagent, in photography, and in medicine especially as an antiseptic and caustic

silver paper *noun* (circa 1875)
: TIN FOIL

silver perch *noun* (1820)
: any of various somewhat silvery fishes that resemble perch: as **a :** a drum (*Bairdiella chrysoura*) that occurs especially along the more southern Atlantic coast of the U.S. — called also *mademoiselle, yellowtail* **b :** WHITE PERCH 1

silver plate *noun* (1610)
1 : domestic flatware and hollowware of silver or of a silver-plated base metal
2 : a plating of silver

sil·ver·point \'sil-vər-ˌpȯint\ *noun* (1882)
: a drawing technique utilizing a pencil of silver usually on specially prepared paper or parchment

silver protein *noun* (1928)
: any of several colloidal light-sensitive preparations of silver and protein used in aqueous solution on mucous membranes as antiseptics

silver salmon *noun* (1878)
: COHO

silver screen *noun* (1918)
1 : a motion-picture screen
2 : MOTION PICTURES

sil·ver·side \'sil-vər-ˌsīd\ *noun* (1820)
: SILVERSIDES

sil·ver·sides \'sil-vər-ˌsīdz\ *noun plural but singular or plural in construction* (1851)
: any of various small bony fishes (family Atherinidae) with a silvery stripe along each side of the body

sil·ver·smith \-ˌsmith\ *noun* (before 12th century)
: an artisan who makes articles of silverware
— sil·ver·smith·ing *noun*

silver spoon *noun* [from the phrase "born with a *silver spoon* in one's mouth" (born wealthy)] (1801)
: WEALTH; *especially* **:** inherited wealth

silver standard *noun* (1860)
: a monetary standard under which the currency unit is defined by a stated quantity of silver

Silver Star Medal *noun* (1932)
: a U.S. military decoration awarded for gallantry in action

sil·ver–tongued \'sil-vər-ˌtəŋd\ *adjective* (1592)
: marked by convincing and eloquent expression

sil·ver·ware \'sil-vər-ˌwar, -ˌwer\ *noun* (1860)
1 : SILVER PLATE 1
2 : FLATWARE

sil·ver·weed \-ˌwēd\ *noun* (1578)
: any of several cinquefoils with leaves silvery or white-tomentose beneath; *especially* **:** one (*Potentilla anserina*) with silky hairs over the entire plant

sil·very \'sil-v(ə-)rē\ *adjective* (14th century)
1 : having the luster of silver
2 : having a soft clear musical tone **:** RESONANT ⟨a *silvery* voice⟩
3 : containing or consisting of silver
— sil·ver·i·ness *noun*

sil·vex \'sil-ˌveks\ *noun* [probably from Latin *silva* wood + English *ex*terminator] (1961)
: a selective herbicide $C_9H_7Cl_3O_3$ especially effective in controlling woody plants but toxic to animals

sil·vi·cul·ture \'sil-və-ˌkəl-chər\ *noun* [French, from Latin *silva* + *cultura* culture] (1880)
: a branch of forestry dealing with the development and care of forests
— sil·vi·cul·tur·al \ˌsil-və-'kəlch-rəl, -'kəl-chə-\ *adjective*
— sil·vi·cul·tur·al·ly \-rə-lē\ *adverb*
— sil·vi·cul·tur·ist \ˌsil-və-'kəlch-rist, -'kəl-chə-\ *noun*

si·ma·zine \'sī-mə-ˌzēn\ *noun* [*sim-* (probably alteration of *sym-* symmetrical, prefix used in names of organic compounds) + tri*azine*] (1956)
: a selective herbicide $C_7H_{12}N_5Cl$ used to control weeds especially among crop plants

Sim·chas To·rah \ˌsim-k̲əs-'tōr-ə, -'tȯr-\ *noun* [Hebrew *śimḥath tōrāh* rejoicing of the Torah] (1891)
: a Jewish holiday observed on the 23d of Tishri in celebration of the completion of the annual reading of the Torah

Sim·e·on \'si-mē-ən\ *noun* [Late Latin, from Greek *Symeon,* from Hebrew *Shim'ōn*]
1 : a son of Jacob and the traditional eponymous ancestor of one of the tribes of Israel
2 : a devout man of Jerusalem held to have uttered the Nunc Dimittis on seeing the infant Jesus in the temple

¹sim·i·an \'si-mē-ən\ *adjective* [Latin *simia* ape, from *simus* snub-nosed, from Greek *simos*] (1607)
: of, relating to, or resembling monkeys or apes

²simian *noun* (1880)
: MONKEY, APE

sim·i·lar \'si-mə-lər, 'sim-lər\ *adjective* [French *similaire,* from Latin *similis* like, similar — more at SAME] (1611)
1 : having characteristics in common **:** strictly comparable
2 : alike in substance or essentials **:** CORRESPONDING ⟨no two animal habitats are exactly *similar* — W. H. Dowdeswell⟩
3 : not differing in shape but only in size or position ⟨*similar* triangles⟩ ⟨*similar* polygons⟩ ☆
— sim·i·lar·ly *adverb*

sim·i·lar·i·ty \ˌsi-mə-'lar-ə-tē\ *noun, plural* **-ties** (1664)
1 : the quality or state of being similar **:** RESEMBLANCE
2 : a comparable aspect **:** CORRESPONDENCE
synonym see LIKENESS

sim·i·le \'si-mə-(ˌ)lē\ *noun* [Middle English, from Latin, comparison, from neuter of *similis*] (14th century)
: a figure of speech comparing two unlike things that is often introduced by *like* or *as* (as in *cheeks like roses*) — compare METAPHOR

si·mil·i·tude \sə-'mi-lə-ˌtüd, -ˌtyüd\ *noun* [Middle English, from Middle French, resemblance, likeness, from Latin *similitudo,* from *similis*] (14th century)
1 a : COUNTERPART, DOUBLE **b :** a visible likeness **:** IMAGE
2 : an imaginative comparison **:** SIMILE
3 a : correspondence in kind or quality **b :** a point of comparison
synonym see LIKENESS

Sim·men·tal *also* **Sim·men·thal** \'zi-mən-ˌtäl\ *noun* [*Simmental,* valley of the Simme River in Switzerland] (1906)
: any of a breed of large buff or dull red and white cattle of Swiss origin that are used widely throughout the world for meat and milk

¹sim·mer \'si-mər\ *verb* **sim·mered; sim·mer·ing** \'si-mə-riŋ, 'sim-riŋ\ [alteration of English dialect *simper,* from Middle English *simperen,* of imitative origin] (1653)
intransitive verb
1 : to stew gently below or just at the boiling point
2 a : to be in a state of incipient development **:** FERMENT ⟨ideas *simmering* in the back of my mind⟩ **b :** to be in inward turmoil **:** SEETHE
transitive verb
: to cook slowly in a liquid just below the boiling point

²simmer *noun* (1809)
: the state of simmering

simmer down *intransitive verb* (1871)
1 : to become calm or peaceful
2 : to become reduced by or as if by simmering

sim·nel \'sim-nᵊl\ *noun* [Middle English *simenel,* from Middle French, ultimately from Latin *simila* wheat flour] (13th century)
1 : a bun or bread of fine wheat flour
2 *British* **:** a rich fruitcake sometimes coated with almond paste and baked for mid-Lent, Easter, and Christmas

si·mo·le·on \sə-'mō-lē-ən\ *noun* [origin unknown] (1896)
slang **:** DOLLAR

Si·mon \'sī-mən\ *noun* [Greek *Simōn,* from Hebrew *Shim'ōn*]
1 : PETER — called also *Simon Peter*
2 : one of the twelve disciples of Jesus — called also *Simon the Zealot*
3 : a kinsman of Jesus
4 : a Cyrenian constrained to help Jesus bear his cross to his place of crucifixion — called also *Simon the Cyrenian*
5 : SIMON MAGUS

si·mo·ni·ac \sī-'mō-nē-ˌak, sə-\ *noun* [Middle English, from Middle French or Medieval Latin; Middle French *simoniaque,* from Medieval Latin *simoniacus,* from Late Latin *simonia* simony] (14th century)
: one who practices simony
— simoniac *or* **si·mo·ni·a·cal** \ˌsī-mə-'nī-ə-kəl, ˌsi-\ *adjective*
— si·mo·ni·a·cal·ly \-k(ə-)lē\ *adverb*

☆ SYNONYMS
Similar, analogous, parallel mean closely resembling each other. SIMILAR implies the possibility of being mistaken for each other ⟨all the houses in the development are *similar*⟩. ANALOGOUS applies to things belonging in essentially different categories but nevertheless having many similarities ⟨*analogous* political systems⟩. PARALLEL suggests a marked likeness in the development of two things ⟨the *parallel* careers of two movie stars⟩.

\ə\ **abut** \ᵊ\ **kitten** \ər\ **further** \a\ **ash** \ā\ **ace** \ä\ **mop, mar** \au̇\ **out** \ch\ **chin** \e\ **bet** \ē\ **easy** \g\ **go** \i\ **hit** \ī\ **ice** \j\ **job** \ŋ\ **sing** \ō\ **go** \ȯ\ **law** \ȯi\ **boy** \th\ **thin** \t͟h\ **the** \ü\ **loot** \u̇\ **foot** \y\ **yet** \zh\ **vision** *see also* Guide to Pronunciation

si·mo·nize \'sī-mə-ˌnīz\ *transitive verb* **-nized; -niz·ing** [from *Simoniz,* a trademark] (1934)
: to polish with or as if with wax

Si·mon Le·gree \ˌsī-mən-lə-'grē\ *noun*
: a slave owner who has Tom flogged to death in Harriet B. Stowe's novel *Uncle Tom's Cabin*

Si·mon Ma·gus \-'mā-gəs\ *noun*
: a Samaritan sorcerer converted by the apostle Philip and severely rebuked by Peter for offering money for the gifts of the Holy Ghost

si·mon–pure \ˌsī-mən-'pyu̇r\ *adjective* [from *the real Simon Pure,* alluding to a character impersonated by another in the play *A Bold Stroke for a Wife* (1718) by Susannah Centlivre (died 1723) English dramatist and actress] (1840)
: of untainted purity or integrity; *also* **:** pretentiously or hypocritically pure

si·mo·ny \'sī-mə-nē, 'si-\ *noun* [Middle English *symonie,* from Late Latin *simonia,* from *Simon* Magus, Samaritan sorcerer in Acts 8: 9–24] (13th century)
: the buying or selling of a church office or ecclesiastical preferment

si·moom \sə-'müm, sī-\ *or* **si·moon** \-'mün\ *noun* [Arabic *samūm*] (1790)
: a hot dry violent dust-laden wind from Asian and African deserts

simp \'simp\ *noun* (1903)
: SIMPLETON

sim·pa·ti·co \sim-'pä-ti-ˌkō, -'pa-\ *adjective* [Italian *simpatico* & Spanish *simpático,* ultimately from Latin *sympathia* sympathy] (1864)
1 : AGREEABLE, LIKABLE
2 : being on the same wavelength **:** CONGENIAL, SYMPATHETIC

¹sim·per \'sim-pər\ *verb* **sim·pered; sim·per·ing** \-p(ə-)riŋ\ [perhaps of Scandinavian origin; akin to Danish dialect *simper* affected, coy] (circa 1563)
intransitive verb
: to smile in a silly manner
transitive verb
: to say with a simper ⟨*simpered* an apology⟩
— sim·per·er \-pər-ər\ *noun*

²simper *noun* (1599)
: a silly smile **:** SMIRK

¹sim·ple \'sim-pəl\ *adjective* **sim·pler** \-p(ə-)lər\; **sim·plest** \-p(ə-)ləst\ [Middle English, from Old French, plain, uncomplicated, artless, from Latin *simplus, simplex,* literally, single; Latin *simplus* from *sem-, sim-* one + *-plus* multiplied by; Latin *simplic-, simplex* from *sem-, sim-* + *-plic-, -plex* -fold — more at SAME, -FOLD] (13th century)
1 : free from guile **:** INNOCENT
2 a : free from vanity **:** MODEST **b :** free from ostentation or display
3 : of humble origin or modest position ⟨a *simple* farmer⟩
4 a : lacking in knowledge or expertise ⟨a *simple* amateur of the arts⟩ **b :** STUPID; *especially* **:** mentally retarded **c :** not socially or culturally sophisticated **:** NAIVE; *also* **:** CREDULOUS
5 a : SHEER, UNMIXED ⟨*simple* honesty⟩ **b :** free of secondary complications ⟨a *simple* vitamin deficiency⟩ **c** (1) **:** having only one main clause and no subordinate clauses ⟨a *simple* sentence⟩ (2) *of a subject or predicate* **:** having no modifiers, complements, or objects **d :** constituting a basic element **:** FUNDAMENTAL **e :** not made up of many like units ⟨a *simple* eye⟩
6 : free from elaboration or figuration ⟨*simple* harmony⟩
7 a (1) **:** not subdivided into branches or leaflets ⟨a *simple* stem⟩ ⟨a *simple* leaf⟩ (2) **:** consisting of a single carpel (3) **:** developing from a single ovary ⟨a *simple* fruit⟩ **b :** controlled by a single gene ⟨*simple* inherited characters⟩
8 : not limited or restricted **:** UNCONDITIONAL ⟨a *simple* obligation⟩
9 : readily understood or performed ⟨*simple* directions⟩ ⟨the adjustment was *simple* to make⟩
10 *of a statistical hypothesis* **:** specifying exact values for one or more statistical parameters — compare COMPOSITE 3 ☆
— sim·ple·ness \-pəl-nəs\ *noun*

²simple *noun* (14th century)
1 a : a person of humble birth **:** COMMONER ⟨thought very little of anybody, *simples* or gentry —Virginia Woolf⟩ **b** (1) **:** a rude or credulous person **:** IGNORAMUS (2) **:** a mentally retarded person
2 a : a medicinal plant **b :** a vegetable drug having only one ingredient
3 : one component of a complex; *specifically* **:** an unanalyzable constituent

simple closed curve *noun* (1919)
: a closed plane curve (as a circle or an ellipse) that does not intersect itself — called also *Jordan curve*

simple equation *noun* (1798)
: a linear equation

simple fraction *noun* (1910)
: a fraction having whole numbers for the numerator and denominator — compare COMPLEX FRACTION

simple fracture *noun* (1685)
: a bone fracture that does not form an open wound in the skin — compare COMPOUND FRACTURE

simple interest *noun* (1798)
: interest paid or computed on the original principal only of a loan or on the amount of an account

simple machine *noun* (1704)
: any of various elementary mechanisms formerly considered as the elements of which all machines are composed and including the lever, the wheel and axle, the pulley, the inclined plane, the wedge, and the screw

sim·ple·mind·ed \ˌsim-pəl-'mīn-dəd, 'sim-pəl-ˌ\ *adjective* (1744)
: devoid of subtlety **:** UNSOPHISTICATED; *also* **:** FOOLISH
— sim·ple·mind·ed·ly *adverb*
— sim·ple·mind·ed·ness *noun*

simple protein *noun* (circa 1909)
: a protein (as a globulin) that yields amino acids as the chief or only products of complete hydrolysis — compare CONJUGATED PROTEIN

simple sugar *noun* (1942)
: MONOSACCHARIDE

sim·ple·ton \'sim-pəl-tən\ *noun* [¹*simple* + *-ton* (as in surnames such as *Washington*)] (1650)
: a person lacking in common sense

simple vow *noun* (1759)
: a public vow taken by a religious in the Roman Catholic Church under which retention of property by the individual is permitted and marriage though illicit is valid under canon law

¹sim·plex \'sim-ˌpleks\ *adjective* [Latin *simplic-, simplex* — more at SIMPLE] (1594)
1 : SIMPLE, SINGLE
2 : allowing telecommunication in only one direction at a time ⟨*simplex* system⟩

²simplex *noun, plural* **sim·plex·es** (1892)
1 *or plural* **sim·pli·ces** \-plə-ˌsēz\ *or* **sim·pli·cia** \sim-'pli-sh(ē-)ə\ **:** a simple word
2 : a spatial configuration of *n* dimensions determined by *n* + 1 points in a space of dimension equal to or greater than *n* ⟨a triangle together with its interior determined by its three vertices is a two-dimensional *simplex* in the plane or any space of higher dimension⟩

sim·pli·cial \sim-'pli-shəl\ *adjective* (1926)
: of or relating to simplexes
— sim·pli·cial·ly \-shə-lē\ *adverb*

sim·plic·i·ty \sim-'pli-sə-tē, -'plis-tē\ *noun* [Middle English *simplicite,* from Middle French *simplicité,* from Latin *simplicitat-, simplicitas,* from *simplic-, simplex*] (14th century)
1 : the state of being simple, uncomplicated, or uncompounded
2 a : lack of subtlety or penetration **:** INNOCENCE, NAIVETÉ **b :** FOLLY, SILLINESS
3 : freedom from pretense or guile **:** CANDOR
4 a : directness of expression **:** CLARITY **b :** restraint in ornamentation **:** AUSTERITY

sim·pli·fy \'sim-plə-ˌfī\ *transitive verb* **-fied; -fy·ing** [French *simplifier,* from Medieval Latin *simplificare,* from Latin *simplus* simple] (1759)
: to make simple or simpler: as **a :** to reduce to basic essentials **b :** to diminish in scope or complexity **:** STREAMLINE ⟨was urged to *simplify* management procedures⟩ **c :** to make more intelligible **:** CLARIFY
— sim·pli·fi·ca·tion \ˌsim-plə-fə-'kā-shən\ *noun*
— sim·pli·fi·er \'sim-plə-ˌfī(-ə)r\ *noun*

sim·plism \'sim-ˌpli-zəm\ *noun* (circa 1882)
: the act or an instance of oversimplifying; *especially* **:** the reduction of a problem to a false simplicity by ignoring complicating factors

sim·plis·tic \sim-'plis-tik\ *adjective* (circa 1881)
1 : SIMPLE
2 : of, relating to, or characterized by simplism **:** OVERSIMPLE ⟨adequate, if occasionally *simplistic,* historical background —Harlow Robinson⟩
— sim·plis·ti·cal·ly \-ti-k(ə-)lē\ *adverb*

sim·ply \'sim-plē, *for 1 also* -pə-lē\ *adverb* (14th century)
1 a : without ambiguity **:** CLEARLY **b :** without embellishment **:** PLAINLY **c :** DIRECTLY, CANDIDLY
2 a : SOLELY, MERELY ⟨eats *simply* to keep alive⟩ ⟨*simply* cleaned it up and went to bed —Garrison Keillor⟩ **b :** REALLY, LITERALLY ⟨the concert was *simply* marvelous⟩ — often used as an intensive ⟨*simply* crawling with geniuses —F. Scott Fitzgerald⟩

simply connected *adjective* (1893)
: being or characterized by a surface that is divided into two separate parts by every closed curve it contains

simply ordered *adjective* (circa 1909)
: having any two elements connected by a relationship that is reflexive, antisymmetric, and transitive

Simp·son's rule \'sim(p)-sənz-\ *noun* [Thomas *Simpson* (died 1761) English mathematician] (1875)
: a method for approximating the area under a curve over a given interval that involves partitioning the interval by an odd number *n* + 1 of equally spaced ordinates and adding the areas of the *n*/2 figures formed by pairs of successive odd-numbered ordinates and the parabolas which they determine with their included even-numbered ordinates

sim·u·la·cre \'sim-yə-ˌlā-kər, -ˌla-\ *noun* [Middle English, from Middle French, from Latin *simulacrum*] (14th century)
archaic **:** SIMULACRUM

☆ SYNONYMS
Simple, foolish, silly, fatuous, asinine mean actually or apparently deficient in intelligence. SIMPLE implies a degree of intelligence inadequate to cope with anything complex or involving mental effort ⟨considered people *simple* who had trouble with computers⟩. FOOLISH implies the character of being or seeming unable to use judgment, discretion, or good sense ⟨*foolish* stunts⟩. SILLY suggests failure to act as a rational being especially by ridiculous behavior ⟨the *silly* antics of revelers⟩. FATUOUS implies foolishness, inanity, and disregard of reality ⟨*fatuous* conspiracy theories⟩. ASININE suggests utter and contemptible failure to use normal rationality or perception ⟨an *asinine* plot⟩. See in addition EASY.

sim·u·la·crum \ˌsim-yə-ˈla-krəm, -ˈlā-\ *noun, plural* **-cra** \-krə\ *also* **-crums** [Middle English, from Latin, from *simulare*] (15th century)
1 : IMAGE, REPRESENTATION ⟨a reasonable *simulacrum* of reality —Martin Mayer⟩
2 : an insubstantial form or semblance of something **:** TRACE

[1]sim·u·lar \ˈsim-yə-lər, -ˌlär\ *noun* [irregular from Latin *simulare* to simulate] (1526)
archaic **:** one that simulates **:** DISSEMBLER

[2]simular *adjective* (1611)
archaic **:** COUNTERFEIT, PRETENDED

sim·u·late \ˈsim-yə-ˌlāt\ *transitive verb* **-lat·ed; -lat·ing** [Latin *simulatus,* past participle of *simulare* to copy, represent, feign, from *similis* like — more at SAME] (1652)
1 : to give or assume the appearance or effect of often with the intent to deceive **:** IMITATE
2 : to make a simulation of (as a physical system)
synonym see ASSUME
— sim·u·la·tive \-ˌlā-tiv\ *adjective*

simulated *adjective* (1622)
: made to look genuine **:** FAKE ⟨*simulated* pearls⟩

sim·u·la·tion \ˌsim-yə-ˈlā-shən\ *noun* [Middle English *simulacion,* from Middle French, from Latin *simulation-, simulatio,* from *simulare*] (14th century)
1 : the act or process of simulating
2 : a sham object **:** COUNTERFEIT
3 a : the imitative representation of the functioning of one system or process by means of the functioning of another ⟨a computer *simulation* of an industrial process⟩ **b :** examination of a problem often not subject to direct experimentation by means of a simulating device

sim·u·la·tor \ˈsim-yə-ˌlā-tər\ *noun* (1835)
: one that simulates; *especially* **:** a device that enables the operator to reproduce or represent under test conditions phenomena likely to occur in actual performance

si·mul·cast \ˈsī məl-ˌkast *also* ˈsi-\ *verb* [*simul*taneous broad*cast*] (1948)
intransitive verb
: to broadcast simultaneously (as by radio and television)
transitive verb
: to broadcast (a program) by simulcasting
— simulcast *noun*

si·mul·ta·neous \ˌsī-məl-ˈtā-nē-əs, -nyəs *also* ˌsi-\ *adjective* [Latin *simul* at the same time + English *-taneous* (as in *instantaneous*) — more at SAME] (circa 1660)
1 : existing or occurring at the same time **:** exactly coincident
2 : satisfied by the same values of the variables ⟨*simultaneous* equations⟩
synonym see CONTEMPORARY
— si·mul·ta·ne·ity \-tə-ˈnē-ə-tē, -ˈnā-\ *noun*
— si·mul·ta·neous·ly \-ˈtā-nē-əs-lē, -nyəs-\ *adverb*
— si·mul·ta·neous·ness *noun*

[1]sin \ˈsin\ *noun* [Middle English *sinne,* from Old English *synn;* akin to Old High German *sunta* sin; probably akin to Latin *sont-, sons* guilty, *est* is — more at IS] (before 12th century)
1 a : an offense against religious or moral law **b :** an action that is or is felt to be highly reprehensible ⟨it's a *sin* to waste food⟩ **c :** an often serious shortcoming **:** FAULT
2 a : transgression of the law of God **b :** a vitiated state of human nature in which the self is estranged from God
synonym see OFFENSE

[2]sin *intransitive verb* **sinned; sin·ning** (before 12th century)
1 : to commit a sin
2 : to commit an offense or fault

[3]sin \ˈsēn, ˈsin\ *noun* [Hebrew *śin*] (circa 1823)
: the 21st letter of the Hebrew alphabet — see ALPHABET table

Sin·an·thro·pus \sī-ˈnan(t)-thrə-pəs, sə-; ˌsī-ˌnan-ˈthrō-, ˌsi-\ *noun* [New Latin, from Late Latin *Sinae,* plural, Chinese + Greek *anthrōpos* man — more at SINO-] (1928)
: PEKING MAN

Sin·bad *or* **Sind·bad** \ˈsin-ˌbad\ *noun*
: a citizen of Baghdad whose adventures at sea are told in the *Arabian Nights' Entertainments*

[1]since \ˈsin(t)s\ *adverb* [Middle English *sins,* contraction of *sithens,* from *sithen,* from Old English *siththan,* from *sīth tham* after that, from *sīth* after, late + *tham,* dative of *thæt* that; akin to Old High German *sīd* later and perhaps to Latin *setius* to a lesser degree] (before 12th century)
1 : from a definite past time until now ⟨has stayed there ever *since*⟩
2 : before the present time **:** AGO ⟨long *since* dead⟩
3 : after a time in the past **:** SUBSEQUENTLY ⟨has *since* become rich⟩

[2]since *conjunction* (15th century)
1 : at a time in the past after or later than ⟨has held two jobs *since* he graduated⟩ **:** from the time in the past when ⟨ever *since* I was a child⟩
2 *obsolete* **:** WHEN
3 : in view of the fact that **:** BECAUSE ⟨*since* it was raining she took an umbrella⟩

[3]since *preposition* (circa 1530)
: in the period after a specified time in the past **:** from a specified time in the past

sin·cere \sin-ˈsir, sən-\ **sin·cer·er; sin·cer·est** *adjective* [Middle French, from Latin *sincerus* whole, pure, genuine, probably from *sem-* one + *-cerus* (akin to Latin *crescere* to grow) — more at SAME, CRESCENT] (1533)
1 a : free of dissimulation **:** HONEST ⟨*sincere* interest⟩ **b :** free from adulteration **:** PURE ⟨a *sincere* doctrine⟩ ⟨*sincere* wine⟩
2 : marked by genuineness **:** TRUE ☆
— sin·cere·ly *adverb*
— sin·cere·ness *noun*

sin·cer·i·ty \-ˈser-ə-tē, -ˈsir-\ *noun* (15th century)
: the quality or state of being sincere **:** honesty of mind **:** freedom from hypocrisy

sin·cip·i·tal \sin-ˈsi-pə-tᵊl\ *adjective* (1653)
: of or relating to the sinciput

sin·ci·put \ˈsin(t)-sə-(ˌ)pət\ *noun, plural* **sin·ciputs** *or* **sin·cip·i·ta** \sin-ˈsi-pə-tə\ [Latin *sincipit-, sinciput,* from *semi-* + *caput* head — more at HEAD] (1578)
1 : FOREHEAD
2 : the upper half of the skull

Sind·hi \ˈsin-dē\ *noun, plural* **Sindhi** *or* **Sindhis** [Arabic *Sindī*] (1815)
1 : a member of a mostly Muslim people of Sind
2 : the Indo-Aryan language of the Sindhi people

sine \ˈsīn\ *noun* [Medieval Latin *sinus,* from Latin, curve] (1593)
1 : the trigonometric function that for an acute angle is the ratio between the leg opposite the angle when it is considered part of a right triangle and the hypotenuse
2 : a trigonometric function $\sin \theta$ that for all real numbers θ is exactly equal to the sine of an angle of measure θ in radians and that is given by the sum of the alternating series

$$\sin \theta = \theta - \frac{\theta^3}{3!} + \frac{\theta^5}{5!} - \frac{\theta^7}{7!} + \frac{\theta^9}{9!} - \cdots$$

si·ne·cure \ˈsī-ni-ˌkyu̇r, ˈsi-\ *noun* [Medieval Latin *sine cura* without cure (of souls)] (1662)
1 *archaic* **:** an ecclesiastical benefice without cure of souls
2 : an office or position that requires little or no work and that usually provides an income

sine curve *noun* (1902)
: the graph in rectangular coordinates of the equation $y=a \sin bx$ where a and b are constants

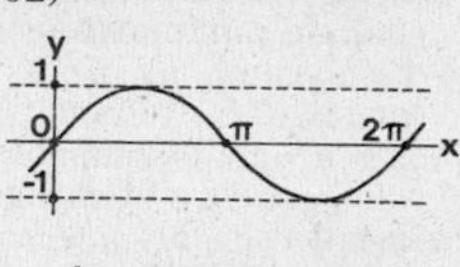

sine curve: graph of $y=a \sin bx$ where $a=1$, $b=1$

si·ne die \ˌsī-ni-ˈdī(-ˌē), ˌsi-nā-ˈdē-ˌā\ *adverb* [Latin, without day] (1607)
: without any future date being designated (as for resumption) **:** INDEFINITELY ⟨the meeting adjourned *sine die*⟩

si·ne qua non \ˌsi-ni-ˌkwä-ˈnän, -ˈnōn *also* ˌsē-; *also* ˌsī-ni-ˌkwā-ˈnän\ *noun, plural* **sine qua nons** *also* **sine qui·bus non** \-ˌkwi-(ˌ)bu̇s- *also* -ˌkwī-\ [Late Latin, without which not] (1602)
: something absolutely indispensable or essential

[1]sin·ew \ˈsin-(ˌ)yü *also* ˈsi-(ˌ)nü\ *noun* [Middle English *sinewe,* from Old English *seono;* akin to Old High German *senawa* sinew, Sanskrit *syati* he binds] (before 12th century)
1 : TENDON; *especially* **:** one dressed for use as a cord or thread
2 *obsolete* **:** NERVE
3 a : solid resilient strength **:** POWER ⟨astonishing intellectual *sinew* and clarity —Reynolds Price⟩ **b :** the chief supporting force **:** MAINSTAY — usually used in plural ⟨providing the *sinews* of better living —Sam Pollock⟩

[2]sinew *transitive verb* (circa 1614)
: to strengthen as if with sinews

sine wave *noun* (1893)
: a waveform that represents periodic oscillations in which the amplitude of displacement at each point is proportional to the sine of the phase angle of the displacement and that is visualized as a sine curve **:** SINE CURVE; *also* **:** a wave so represented

sin·ewy \ˈsin-yə-wē *also* ˈsi-nə-\ *adjective* (14th century)
1 : full of sinews; as **a :** TOUGH, STRINGY ⟨*sinewy* meat⟩ **b :** STRONG ⟨*sinewy* arms⟩
2 : marked by the strength of sinews ⟨a demanding *sinewy* intelligence —Helen Dudar⟩

sin·fo·nia \ˌsin-fə-ˈnē-ə\ *noun, plural* **-nie** \-ˈnē-ˌā\ [Italian, from Latin *symphonia* symphony] (1773)
1 : an orchestral musical composition serving as an introduction to choral works (as opera) especially in the 18th century **:** OVERTURE
2 : RITORNELLO 1, SYMPHONY 2c

sinfonia con·cer·tante \-ˌkän(t)-sər-ˈtän-tē, -shər-, -ˌtā\ *noun* [Italian, literally, symphony in concerto style] (circa 1903)
: a concerto for more than one solo instrument

☆ SYNONYMS
Sincere, wholehearted, heartfelt, hearty, unfeigned mean genuine in feeling. SINCERE stresses absence of hypocrisy, feigning, or any falsifying embellishment or exaggeration ⟨a *sincere* apology⟩. WHOLEHEARTED suggests sincerity and earnest devotion without reservation or misgiving ⟨promised our *wholehearted* support⟩. HEARTFELT suggests depth of genuine feeling outwardly expressed ⟨expresses our *heartfelt* gratitude⟩. HEARTY suggests honesty, warmth, and exuberance in displaying feeling ⟨received a *hearty* welcome⟩. UNFEIGNED stresses spontaneity and absence of pretense ⟨her *unfeigned* delight at receiving the award⟩.

sin·fo·niet·ta \\ˌsin-fən-ˈye-tə, -fōn-\\ *noun* [Italian, diminutive of *sinfonia*] (circa 1907)
1 : a symphony of less than standard length or for fewer instruments
2 : a small symphony orchestra; *especially* **:** an orchestra of strings only

sin·ful \\ˈsin-fəl\\ *adjective* (before 12th century)
1 : tainted with, marked by, or full of sin **:** WICKED
2 : such as to make one feel guilty ⟨a *sinful* chocolate cake⟩
— sin·ful·ly \\-fə-lē\\ *adverb*
— sin·ful·ness *noun*

[1]sing \\ˈsiŋ\\ *verb* **sang** \\ˈsaŋ\\ *or* **sung** \\ˈsəŋ\\; **sung; sing·ing** \\ˈsiŋ-iŋ\\ [Middle English, from Old English *singan;* akin to Old High German *singan* to sing, Greek *omphē* voice] (before 12th century)
intransitive verb
1 a : to produce musical tones by means of the voice **b :** to utter words in musical tones and with musical inflections and modulations **c :** to deliver songs as a trained or professional singer
2 : to make a shrill whining or whistling sound
3 a : to relate or celebrate something in verse **b :** to compose poetry **c :** to create in or through words a feeling or sense of song ⟨prose that *sings*⟩
4 : to produce musical or harmonious sounds
5 : BUZZ, RING
6 : to make a cry **:** CALL
7 : to give information or evidence
transitive verb
1 : to utter with musical inflections; *especially* **:** to interpret in musical tones produced by the voice
2 : to relate or celebrate in verse
3 : CHANT, INTONE
4 : to bring or accompany to a place or state by singing ⟨*sings* the child to sleep⟩
— sing·able \\ˈsiŋ-ə-bəl\\ *adjective*

[2]sing *noun* (1850)
: a session of group singing

sing–along \\ˈsiŋ-ə-ˌlȯŋ\\ *noun* (1966)
: SONGFEST; *also* **:** a song appropriate for a sing-along

[1]singe \\ˈsinj\\ *transitive verb* **singed; singe·ing** \\ˈsin-jiŋ\\ [Middle English *sengen,* from Old English *sengan;* akin to Old High German bi*sengan* to singe, Old Church Slavonic *isęknǫti* to dry up] (before 12th century)
: to burn superficially or lightly **:** SCORCH; *especially* **:** to remove the hair, down, or fuzz from usually by passing rapidly over a flame

[2]singe *noun* (1658)
: a slight burn **:** SCORCH

[1]sing·er \\ˈsiŋ-ər\\ *noun* (14th century)
: one that sings

[2]sing·er \\ˈsin-jər\\ *noun* (circa 1875)
: one that singes

singing game *noun* (1881)
: a children's game in which the players accompany their actions with the singing of a narrative song

[1]sin·gle \\ˈsiŋ-gəl\\ *adjective* [Middle English, from Middle French, from Latin *singulus* one only; akin to Latin *sem-* one — more at SAME] (14th century)
1 a : not married **b :** of or relating to celibacy
2 : unaccompanied by others **:** LONE, SOLE ⟨the *single* survivor of the disaster⟩
3 a (1) **:** consisting of or having only one part, feature, or portion ⟨*single* consonants⟩ (2) **:** consisting of one as opposed to or in contrast with many **:** UNIFORM ⟨a *single* standard for men and women⟩ (3) **:** consisting of only one in number ⟨holds to a *single* ideal⟩ **b :** having but one whorl of petals or ray flowers ⟨a *single* rose⟩
4 a : consisting of a separate unique whole **:** INDIVIDUAL ⟨every *single* citizen⟩ **b :** of, relating to, or involving only one person
5 a : FRANK, HONEST ⟨a *single* devotion⟩ **b :** exclusively attentive ⟨an eye *single* to the truth⟩
6 : UNBROKEN, UNDIVIDED
7 : having no equal or like **:** SINGULAR
8 : designed for the use of one person only ⟨a *single* room⟩ ⟨*single* bed⟩

[2]single *verb* **sin·gled; sin·gling** \\-g(ə-)liŋ\\ (1628)
transitive verb
1 : to select or distinguish from a number or group — usually used with *out*
2 a : to advance or score (a base runner) by a single **b :** to bring about the scoring of (a run) by a single
intransitive verb
: to make a single in baseball

[3]single *noun* (1646)
1 a : a separate individual person or thing **b :** an unmarried person and especially one young and socially active — usually used in plural **c :** a recording having one short tune on each side
2 : a base hit that allows the batter to reach first base
3 a *plural* **:** a tennis match or similar game with one player on each side **b :** a golf match between two players — usually used in plural
4 : a room (as in a hotel) for one guest — compare DOUBLE 7

sin·gle–ac·tion \\ˈsiŋ-gəl-ˈak-shən\\ *adjective* (1900)
of a revolver **:** that can be cocked only by manually retracting the hammer

sin·gle–blind \\ˈsiŋ-gəl-ˌblīnd\\ *adjective* (1963)
: of, relating to, or being an experimental procedure in which the experimenters but not the subjects know the makeup of the test and control groups during the actual course of the experiments — compare DOUBLE-BLIND

single bond *noun* (1903)
: a chemical bond in which one pair of electrons is shared by two atoms in a molecule especially when the atoms can share more than one pair of electrons — compare DOUBLE BOND, TRIPLE BOND

sin·gle–breast·ed \\ˈsiŋ-gəl-ˈbres-təd\\ *adjective* (1796)
: having a center closing with one row of buttons and no lap ⟨a *single-breasted* coat⟩

single combat *noun* (1610)
: combat between two persons

single cross *noun* (1940)
: a first-generation hybrid between two selected and usually inbred lines — compare DOUBLE CROSS 2

single entry *noun* (1826)
: a method of bookkeeping that recognizes only one side of a business transaction and usually consists only of a record of cash and personal accounts with debtors and creditors

single file *noun* (1670)
: [6]FILE 1
— single file *adverb*

[1]sin·gle–foot \\ˈsiŋ-gəl-ˌfu̇t\\ *noun, plural* **single–foots** (1867)
: [7]RACK b

[2]single–foot *intransitive verb* (1890)
of a horse **:** to go at a rack
— sin·gle–foot·er *noun*

[1]sin·gle–hand·ed \\ˌsiŋ-gəl-ˈhan-dəd\\ *adjective* (1709)
1 : managed or done by one person or with one on a side
2 : working alone or unassisted by others
— sin·gle–hand·ed·ly *adverb*

[2]single–handed *adverb* (1815)
: in a single-handed manner

sin·gle–hand·er \\-ˈhan-dər\\ *noun* (1946)
: a person who sails single-handed

sin·gle–heart·ed \\ˌsiŋ-gəl-ˈhär-təd\\ *adjective* (1577)
: characterized by sincerity and unity of purpose or dedication
— sin·gle–heart·ed·ly *adverb*
— sin·gle–heart·ed·ness *noun*

single knot *noun* (circa 1930)
: OVERHAND KNOT

single–lens reflex \\ˈsiŋ-gəl-ˈlenz-\\ *noun* (1940)
: a camera having a single lens that forms an image which is reflected to the viewfinder or recorded on film

sin·gle–mind·ed \\ˌsiŋ-gəl-ˈmīn-dəd, ˈsiŋ-gəl-ˌ\\ *adjective* (1860)
: having one driving purpose or resolve **:** DETERMINED, DEDICATED
— sin·gle–mind·ed·ly *adverb*
— sin·gle–mind·ed·ness *noun*

sin·gle·ness \\ˈsiŋ-gəl-nəs\\ *noun* (1560)
: the quality or state of being single

sin·gle–phase \\ˈsiŋ-gəl-ˈfāz\\ *adjective* (1900)
: of or relating to a circuit energized by a single alternating electromotive force

sin·gle–space \\-ˈspās\\ *transitive verb* (1928)
: to type or print with no blank lines between lines of text

sin·gle·stick \\-ˌstik\\ *noun* (1771)
: fighting or fencing with a wooden stick or sword held in one hand; *also* **:** the weapon used

sin·glet \\ˈsiŋ-glət\\ *noun* (circa 1746)
1 [from its having only one thickness of cloth] *chiefly British* **:** an athletic jersey **:** UNDERSHIRT
2 : an atom or molecule that has no net electronic magnetic moment; *also* **:** an excited state of an atom or molecule that is a singlet

single tax *noun* (1853)
: a tax to be levied on a single item (as real estate) as the sole source of public revenue

sin·gle·ton \\ˈsiŋ-gəl-tən\\ *noun* [French, from English *single*] (1876)
1 : a card that is the only one of its suit originally dealt to a player
2 a : an individual member or thing distinct from others grouped with it **b :** an offspring born singly ⟨*singletons* are more common than twins⟩

sin·gle–track \\ˈsiŋ-gəl-ˈtrak\\ *adjective* (1849)
1 : having only one track
2 : lacking intellectual range, receptiveness, or flexibility **:** ONE-TRACK

sin·gle·tree \\ˈsiŋ-gəl-(ˌ)trē\\ *noun* (circa 1841)
: WHIFFLETREE

sin·gle–val·ued \\ˌsiŋ-gəl-ˈval-(ˌ)yüd\\ *adjective* (1879)
: having one and only one value of the range associated with each value of the domain ⟨a *single-valued* function⟩ — compare MULTIPLE-VALUED

single wing *noun* (1945)
: an offensive football formation in which one back plays as a flanker and two backs line up four or five yards behind the line in position to receive a direct snap from center

sin·gly \\ˈsiŋ-g(ə-)lē\\ *adverb* (14th century)
1 : without the company of others **:** INDIVIDUALLY
2 : SINGLE-HANDED

[1]sing·song \\ˈsiŋ-ˌsȯŋ\\ *noun* (1609)
1 : verse with marked and regular rhythm and rhyme
2 : a voice delivery marked by a narrow range or monotonous pattern of pitch
3 *British* **:** SONGFEST
— sing·songy \\-ˌsȯŋ-ē\\ *adjective*

[2]singsong *adjective* (1734)
: having a monotonous cadence or rhythm

sing·spiel \\ˈsiŋ-ˌspēl, ˈziŋ-ˌshpēl\\ *noun* [German, from *singen* to sing + *Spiel* play] (1876)
: a musical work popular in Germany especially in the latter part of the 18th century characterized by spoken dialogue interspersed with popular or folk songs

[1]sin·gu·lar \\ˈsiŋ-gyə-lər\\ *adjective* [Middle English *singuler,* from Middle French, from Latin *singularis,* from *singulus* only one — more at SINGLE] (14th century)
1 a : of or relating to a separate person or thing **:** INDIVIDUAL **b :** of, relating to, or being a word form denoting one person, thing, or instance **c :** of or relating to a single instance or to something considered by itself

2 : distinguished by superiority **:** EXCEPTIONAL ⟨an artist of *singular* attainments⟩
3 : being out of the ordinary **:** UNUSUAL ⟨on the way home we had a *singular* adventure⟩
4 : departing from general usage or expectation **:** PECULIAR, ODD ⟨the air had a *singular* chill⟩
5 a *of a matrix* **:** having a determinant equal to zero **b** *of a linear transformation* **:** having the property that the matrix of coefficients of the new variables has a determinant equal to zero
synonym see STRANGE
— sin·gu·lar·ly *adverb*

²singular *noun* (14th century)
1 : the singular number, the inflectional form denoting it, or a word in that form
2 : a singular term

sin·gu·lar·i·ty \ˌsiŋ-gyə-ˈlar-ə-tē\ *noun, plural* **-ties** (14th century)
1 : something that is singular: as **a :** a separate unit **b :** unusual or distinctive manner or behavior **:** PECULIARITY
2 : the quality or state of being singular
3 : a point at which the derivative of a given function of a complex variable does not exist but every neighborhood of which contains points for which the derivative exists
4 : a point or region of infinite mass density at which space and time are infinitely distorted by gravitational forces and which is held to be the final state of matter falling into a black hole

sin·gu·lar·ize \ˈsiŋ-gyə-lə-ˌrīz\ *transitive verb* **-ized; -iz·ing** (1589)
: to make singular

singular point *noun* (circa 1856)
: SINGULARITY 3

Sin·ha·la \sin-ˈhä-lä, siŋ-\ *noun* [Sanskrit *Simhala* Sri Lanka] (circa 1954)
: SINHALESE 2

Sin·ha·lese *or* **Sin·gha·lese** \ˌsiŋ-gə-ˈlēz, ˌsin-(h)ə-, -ˈlēs\ *noun, plural* **Sinhalese** *or* **Singhalese** (1598)
1 : a member of a people that inhabit Sri Lanka and form a major part of its population
2 : the Indo-Aryan language of the Sinhalese people
— Sinhalese *or* **Sin·ghalese** *adjective*

si·ni·cize \ˈsī-nə-ˌsīz, ˈsi-\ *transitive verb* **-cized; -ciz·ing** *often capitalized* [Medieval Latin *sinicus* Chinese, from Late Latin *Sinae,* plural, Chinese — more at SINO-] (1889)
: to modify by Chinese influence

sin·is·ter \ˈsi-nəs-tər, *archaic* sə-ˈnis-\ *adjective* [Middle English *sinistre,* from Latin *sinistr-, sinister* on the left side, unlucky, inauspicious] (15th century)
1 *archaic* **:** UNFAVORABLE, UNLUCKY
2 *archaic* **:** FRAUDULENT
3 : singularly evil or productive of evil
4 a : of, relating to, or situated to the left or on the left side of something; *especially* **:** being or relating to the side of a heraldic shield at the left of the person bearing it **b :** of ill omen by reason of being on the left
5 : presaging ill fortune or trouble
6 : accompanied by or leading to disaster ☆ ◆
— sin·is·ter·ly *adverb*
— sin·is·ter·ness *noun*

si·nis·tral \ˈsi-nəs-trəl, sə-ˈnis-\ *adjective* (1803)
: of, relating to, or inclined to the left: as **a :** LEFT-HANDED **b** *of a gastropod shell* **:** having the whorls coiling counterclockwise down the spire when viewed with the apex toward the observer and having the aperture situated on the left of the axis when held with the spire uppermost and with the aperture opening toward the observer — compare DEXTRAL b

si·nis·trous \ˈsi-nəs-trəs, sə-ˈnis-\ *adjective* (circa 1575)
archaic **:** SINISTER

Si·nit·ic \sī-ˈni-tik, sə-\ *adjective* [Late Latin *Sinae,* plural, Chinese + English *-itic* (as in *Semitic*) — more at SINO-] (circa 1895)
: of or relating to the Chinese, their language, or their culture

¹sink \ˈsiŋk\ *verb* **sank** \ˈsaŋk\ *or* **sunk** \ˈsəŋk\; **sunk; sink·ing** [Middle English, from Old English *sincan;* akin to Old High German *sinkan* to sink] (before 12th century)
intransitive verb
1 a : to go to the bottom **:** SUBMERGE **b :** to become partly buried (as in mud) **c :** to become engulfed
2 a (1) **:** to fall or drop to a lower place or level (2) **:** to flow at a lower depth or level (3) **:** to burn with lower intensity (4) **:** to fall to a lower pitch or volume ⟨his voice *sank* to a whisper⟩ **b :** to subside gradually **:** SETTLE **c :** to disappear from view **d :** to slope gradually **:** DIP
3 a : to soak or become absorbed **:** PENETRATE **b :** to become impressively known or felt ⟨the lesson had *sunk* in⟩
4 : to become deeply absorbed ⟨*sank* into reverie⟩
5 a : to go downward in quality, state, or condition **b :** to grow less in amount or worth
6 a : to fall or drop slowly for lack of strength **b :** to become depressed **c :** to fail in health or strength
transitive verb
1 a : to cause to sink ⟨*sink* a battleship⟩ **b :** to force down especially below the earth's surface **c :** to cause (something) to penetrate
2 : to engage deeply the attention of **:** IMMERSE
3 a : to dig or bore (a well or shaft) in the earth **:** EXCAVATE **b :** to form by cutting or excising ⟨*sink* words in stone⟩
4 : to cast down or bring to a low condition or state **:** OVERWHELM, DEFEAT
5 : to lower in standing or reputation **:** ABASE
6 a : to lessen in value or amount **b :** to lower or soften (the voice) in speaking
7 : RESTRAIN, SUPPRESS ⟨*sinks* her pride and approaches the despised neighbor —Richard Harrison⟩
8 : to pay off (as a debt) **:** LIQUIDATE
9 : INVEST
10 : DROP 7c ⟨*sink* a putt⟩
— sink·able \ˈsiŋ-kə-bəl\ *adjective*

²sink *noun* (15th century)
1 a : a pool or pit for the deposit of waste or sewage **:** CESSPOOL **b :** a ditch or tunnel for carrying off sewage **:** SEWER **c :** a stationary basin connected with a drain and usually a water supply for washing and drainage
2 : a place where vice, corruption, or evil collects
3 : SUMP 3
4 a : a depression in the land surface; *especially* **:** one having a saline lake with no outlet **b :** SINKHOLE
5 : a body or process that acts as a storage device or disposal mechanism: as **a :** HEAT SINK; *broadly* **:** a device that collects or dissipates energy (as radiation) **b :** a reactant with or absorber of a substance

sink·age \ˈsiŋ-kij\ *noun* (1883)
1 : the process or degree of sinking
2 : DEPRESSION, INDENTATION
3 : the distance from the top line of a full page to the first line of sunk matter

sink·er \ˈsiŋ-kər\ *noun* (1708)
1 : one that sinks; *specifically* **:** a weight for sinking a fishing line, seine, or sounding line
2 : DOUGHNUT
3 : a fastball that sinks as it reaches the plate — called also *sinker ball*

sink·hole \ˈsiŋk-ˌhōl\ *noun* (15th century)
1 : a hollow place or depression in which drainage collects
2 : a hollow in a limestone region that communicates with a cavern or passage
3 : SINK 2

sinking fund *noun* (1724)
: a fund set up and accumulated by usually regular deposits for paying off the principal of a debt when it falls due

sin·less \ˈsin-ləs\ *adjective* (before 12th century)
: free from sin **:** IMPECCABLE
— sin·less·ly *adverb*
— sin·less·ness *noun*

sin·ner \ˈsi-nər\ *noun* (14th century)
1 : one that sins
2 : REPROBATE, SCAMP

Si·no- \ˌsī-(ˌ)nō, ˈsī-, -nə\ *combining form* [French, from Late Latin *Sinae,* plural, Chinese, from Greek *Sinai,* probably of Indo-Aryan origin; akin to Sanskrit *Cīnā,* plural, Chinese]
1 : Chinese ⟨*sinology*⟩
2 : Chinese and ⟨*Sino*-Tibetan⟩

si·no·atri·al \ˌsī-nō-ˈā-trē-əl\ *adjective* [New Latin *sinus* + *atrium*] (1913)
: of, involving, or being the sinoatrial node ⟨*sinoatrial* block⟩

sinoatrial node *noun* (1913)
: a small mass of tissue that is embedded in the musculature of the right atrium of higher vertebrates and that originates the impulses stimulating the heartbeat

si·no·logue \ˈsī-nᵊl-ˌòg, ˈsi-, -ˌäg\ *noun* [French, from Late Latin *Sinae* + French *-logue*] (1853)
: a specialist in sinology

si·nol·o·gy \sī-ˈnä-lə-jē, sə-\ *noun* [probably from French *sinologie,* from *sino-* + *-logie* -logy] (circa 1882)
: the study of the Chinese and especially their language, literature, history, and culture
— si·no·log·i·cal \ˌsī-nᵊl-ˈä-ji-kəl, ˌsi-\ *adjective*

☆ SYNONYMS
Sinister, baleful, malign mean seriously threatening evil or disaster. SINISTER suggests a general or vague feeling of fear or apprehension on the part of the observer ⟨a *sinister* aura haunts the place⟩. BALEFUL imputes perniciousness or destructiveness to something whether working openly or covertly ⟨exerting a corrupt and *baleful* influence⟩. MALIGN applies to what is inherently evil or harmful ⟨the *malign* effects of racism⟩.

◇ WORD HISTORY
sinister Words for "left" and "right" in a wide variety of languages betray an ancient prejudice toward left-handers. Because most people are right-handed and use their left hand much less deftly, the left side is typically associated with clumsiness or weakness; and in the primitive, male-dominated world of farmers, warriors, and hunters from which we all ultimately descend, an attack by a wild animal or another warrior on one's weaker left side could be fatal, so that left-handedness was also associated with evil and ill fortune. The "right" word, on the other hand, is often, as in English, associated with correctness and moral rectitude. The Latin words *sinister, scaevus,* and *laevus* all had both the literal meaning "on the left" and various negative meanings: "unlucky," "harmful," or "perverted." Idiosyncratically, Latin *sinister* could also mean "favorable" in reference to an omen, because a Roman augur, when following the Etruscan method of doing divinations, faced the south as he stood in a ritually prescribed space, and favorable omens, such as the flight of a bird, were thought to appear from the east, the augur's left side.

— **si·nol·o·gist** \sī-ˈnä-lə-jist, sə-\ *noun*
si·no·pia \sə-ˈnō-pē-ə\ *noun, plural* **-pi·as** *or* **-pie** \-pē-ˌā\ [Italian, from Latin *sinopis,* from Greek *sinōpis,* from *Sinōpē* Sinop, ancient seaport in Asia Minor] (1844)
1 : a red to reddish brown earth pigment used by the ancients that depends for its color on its content of red ferric oxide
2 : a preliminary drawing for a fresco done in sinopia
Si·no–Ti·bet·an \ˌsī-nō-tə-ˈbe-tᵊn, ˈsī-\ *noun* (1920)
: a language family comprising Tibeto-Burman and Chinese
sin·se·mil·la \ˌsin-sə-ˈmē-lə, -ˈmi-; -ˈmē-yə, -ˈmēl-\ *noun* [American Spanish, from *sin* without + *semilla* seed] (1975)
: highly potent marijuana from female plants that are specially tended and kept seedless by preventing pollination in order to induce a high resin content; *also* **:** a female hemp plant grown to produce sinsemilla
sin·syne \ˈsin-ˌsīn\ *adverb* [Middle English (Scots) *sensyne,* from *sen* since (contraction of Middle English *sithen*) + *syne* since — more at SINCE, SYNE] (14th century)
chiefly Scottish **:** since that time
sin tax *noun* (1964)
: a tax on substances or activities considered sinful or harmful (as tobacco, alcohol, or gambling)
sin·ter \ˈsin-tər\ *verb* [German *Sinter* slag, cinder, from Old High German *sintar* — more at CINDER] (1871)
transitive verb
: to cause to become a coherent mass by heating without melting
intransitive verb
: to undergo sintering
— **sinter** *noun*
— **sin·ter·abil·i·ty** \ˌsin-tə-rə-ˈbi-lə-tē\ *noun*
sin·u·ate \ˈsin-yə-wət, -ˌwāt\ *adjective* [Latin *sinuatus,* past participle of *sinuare* to bend, from *sinus* curve] (1688)
: having the margin wavy with strong indentations ⟨*sinuate* leaves⟩
sin·u·os·i·ty \ˌsin-yə-ˈwä-sə-tē\ *noun, plural* **-ties** (1598)
1 : the quality or state of being sinuous
2 : something that is sinuous
sin·u·ous \ˈsin-yə-wəs\ *adjective* [Latin *sinuosus,* from *sinus*] (1578)
1 a : of a serpentine or wavy form **:** WINDING **b :** marked by strong lithe movements
2 : INTRICATE, COMPLEX
— **sin·u·ous·ly** *adverb*
— **sin·u·ous·ness** *noun*
si·nus \ˈsī-nəs\ *noun* [Middle English, from Medieval Latin, from Latin, curve, fold, hollow] (15th century)
: CAVITY, HOLLOW: as **a :** a narrow elongated tract extending from a focus of suppuration and serving for the discharge of pus **b** (1) **:** a cavity in the substance of a bone of the skull that usually communicates with the nostrils and contains air (2) **:** a channel for venous blood (3) **:** a dilatation in a bodily canal or vessel **c :** a cleft or indentation between adjoining lobes (as of a leaf or corolla)
si·nus·i·tis \ˌsīn-yə-ˈsī-təs, ˌsī-nə-\ *noun* (1896)
: inflammation of a sinus of the skull
si·nu·soid \ˈsīn-yə-ˌsȯid, ˈsī-nə-\ *noun* [Medieval Latin *sinus* sine] (1823)
1 : SINE CURVE, SINE WAVE
2 [New Latin *sinus*] **:** a minute endothelium-lined space or passage for blood in the tissues of an organ (as the liver)
si·nu·soi·dal \ˌsīn-yə-ˈsȯi-dᵊl, ˌsī-nə-\ *adjective* (1878)
: of, relating to, shaped like, or varying according to a sine curve or sine wave ⟨*sinusoidal* motion⟩ ⟨*sinusoidal* alternating current⟩ ⟨*sinusoidal* grooves⟩
— **si·nu·soi·dal·ly** \-dᵊl-ē\ *adverb*
sinusoidal projection *noun* (1944)
: an equal-area map projection capable of showing the entire surface of the earth with all parallels as straight lines evenly spaced, the central meridian as one half the length of the equator, and all other meridians as curved lines
si·nus ve·no·sus \ˈsī-nəs-vi-ˈnō-səs\ *noun* [New Latin, venous sinus] (circa 1839)
: an enlarged pouch that adjoins the heart, is formed by the union of the large systemic veins, and is the passage through which venous blood enters the heart in lower vertebrates and in embryos of higher forms
Si·on \ˈsī-ən\ *variant of* ZION
Siou·an \ˈsü-ən\ *noun* (1889)
1 : an American Indian language family of central and southeastern North America
2 : a member of any of the peoples speaking Siouan languages
Sioux \ˈsü\ *noun, plural* **Sioux** \ˈsü, ˈsüz\ [French, short for *Nadouessioux,* from Ojibwa *na·towe·ssiw-*] (1712)
1 : DAKOTA
2 : SIOUAN
¹sip \ˈsip\ *verb* **sipped; sip·ping** [Middle English *sippen;* akin to Low German *sippen* to sip] (14th century)
intransitive verb
: to take a sip of something especially repeatedly
transitive verb
1 : to drink in small quantities
2 : to take sips from
— **sip·per** *noun*
²sip *noun* (15th century)
1 : a small draft taken with the lips
2 : the act of sipping
¹si·phon \ˈsī-fən\ *noun* [French *siphon,* from Latin *siphon-, sipho* tube, pipe, siphon, from Greek *siphōn*] (1659)
1 a : a tube bent to form two legs of unequal length by which a liquid can be transferred to a lower level over an intermediate elevation by the pressure of the atmosphere in forcing the liquid up the shorter branch of the tube immersed in it while the excess of weight of the liquid in the longer branch when once filled causes a continuous flow **b** *usually* **syphon :** a bottle for holding aerated water that is driven out through a bent tube in its neck by the pressure of the gas when a valve in the tube is opened

siphon 1a

2 : any of various tubular organs in animals and especially mollusks or arthropods that are used for drawing in or ejecting fluids — see CLAM illustration
²siphon *verb* **si·phoned; si·phon·ing** \ˈsī-fə-niŋ, ˈsīf-niŋ\ (1859)
transitive verb
: to convey, draw off, or empty by or as if by a siphon — often used with *off*
intransitive verb
: to pass by or as if by a siphon
si·pho·no·phore \sī-ˈfä-nə-ˌfōr, ˈsī-fə-nə-, -ˌfȯr\ *noun* [ultimately from Greek *siphōn* + *pherein* to carry — more at BEAR] (1883)
: any of an order (Siphonophora) of compound free-swimming or floating pelagic hydrozoans that are mostly delicate, transparent, and colored and have specialized zooids
si·pho·no·stele \sī-ˈfä-nə-ˌstēl, ˌsī-fə-nə-ˈstē-lē\ *noun* [Greek *siphōn* tube, siphon] (1902)
: a stele consisting of vascular tissue surrounding a central core of pith parenchyma
sip·pet \ˈsi-pət\ *noun* [alteration of *sop*] (1530)
chiefly British **:** a small bit of toast or fried bread especially for garnishing
sir \ˈsər\ *noun* [Middle English, from *sire*] (13th century)
1 a : a man entitled to be addressed as *sir* — used as a title before the given name of a knight or baronet and formerly sometimes before the given name of a priest **b :** a man of rank or position
2 a — used as a usually respectful form of address **b** *capitalized* — used as a conventional form of address in the salutation of a letter
Si·rach \ˈsī-rak *also* sə-ˈräk\ *noun* [Greek *Seirach*]
: a didactic book of the Roman Catholic canon of the Old Testament
sir·dar \ˈsər-ˌdär, sər-ˈ\ *noun* [Hindi *sardār,* from Persian] (1595)
1 a : a person of high rank (as an hereditary noble) especially in India **b :** the commander of the Anglo-Egyptian army
2 : one (as a foreman) holding a responsible position especially in India
¹sire \ˈsīr\ *noun* [Middle English, from Old French, from Latin *senior* older — more at SENIOR] (13th century)
1 a : FATHER **b** *archaic* **:** male ancestor **:** FOREFATHER **c :** AUTHOR, ORIGINATOR
2 a *archaic* **:** a man of rank or authority; *especially* **:** LORD — used formerly as a form of address and as a title **b** *obsolete* **:** an elderly man **:** SENIOR
3 : the male parent of an animal and especially of a domestic animal
²sire *transitive verb* **sired; sir·ing** (1611)
1 : BEGET — used especially of male domestic animals
2 : to bring into being **:** ORIGINATE
¹si·ren \ˈsī-rən, *for 3 also* sī-ˈrēn\ *noun* [Middle English, from Middle French & Latin; Middle French *sereine,* from Late Latin *sirena,* from Latin *siren,* from Greek *seirēn*] (14th century)
1 *often capitalized* **:** any of a group of female and partly human creatures in Greek mythology that lured mariners to destruction by their singing
2 a : a woman who sings with bewitching sweetness **b :** TEMPTRESS
3 a : an apparatus producing musical tones especially in acoustical studies by the rapid interruption of a current of air, steam, or fluid by a perforated rotating disk **b :** a device often electrically operated for producing a penetrating warning sound ⟨ambulance *siren*⟩ ⟨air-raid *siren*⟩
4 [New Latin, from Latin] **:** either of two North American eel-shaped amphibians that constitute a genus (*Siren*) and have small forelimbs but neither hind legs nor pelvis and have permanent external gills as well as lungs
²si·ren \ˈsī-rən\ *adjective* (1568)
: resembling that of a siren **:** ENTICING
si·re·ni·an \sī-ˈrē-nē-ən\ *noun* [New Latin *Sirenia,* from Latin *siren*] (1883)
: any of an order (Sirenia) of aquatic herbivorous mammals including the manatee, dugong, and Steller's sea cow
siren song *noun* (1568)
: an alluring utterance or appeal; *especially* **:** one that is seductive or deceptive
Sir·i·us \ˈsir-ē-əs\ *noun* [Middle English, from Latin, from Greek *Seirios*]
: a star of the constellation Canis Major that is the brightest star in the heavens — called also *Dog Star*
sir·loin \ˈsər-ˌlȯin\ *noun* [alteration of earlier *surloin,* modification of Middle French *surlonge,* from *sur* over (from Latin *super*) + *loigne, longe* loin — more at OVER, LOIN] (1554)
: a cut of meat and especially of beef from the part of the hindquarter just in front of the round — see BEEF illustration
si·roc·co \sə-ˈrä-(ˌ)kō, shə-\ *noun, plural* **-cos** [Italian *scirocco, sirocco,* from Arabic *sharq* east] (1617)

1 a : a hot dust-laden wind from the Libyan deserts that blows on the northern Mediterranean coast chiefly in Italy, Malta, and Sicily **b :** a warm moist oppressive southeast wind in the same regions
2 : a hot or warm wind of cyclonic origin from an arid or heated region

sir·rah *also* **sir·ra** \'sir-ə\ *noun* [alteration of *sir*] (1526)
obsolete — used as a form of address implying inferiority in the person addressed

sir·ree *also* **sir·ee** \(ˌ)sər-'ē\ *noun* [by alteration] (1823)
: SIR — used as an emphatic form usually after *yes* or *no*

sir–reverence *noun* [probably alteration of *save-reverence,* translation of Medieval Latin *salva reverentia* saving (your) reverence] (1575)
1 *obsolete* — used as an expression of apology before a statement that might be taken as offensive
2 *obsolete* **:** human feces; *also* **:** a lump of human feces

Sir Rog·er de Cov·er·ley \sə(r)-ˌrä-jər-di-'kə-vər-lē\ *noun* [alteration of *roger of coverley,* probably from *Roger,* male given name + *of* + *Coverley,* a fictitious place name] (1804)
: an English country-dance that resembles the Virginia reel

sirup *variant of* SYRUP

sir·vente \sir-'vänt\ *or* **sir·ven·tes** \-'ven-təs\ *noun, plural* **sir·ventes** \-'vänt, -'vän(t)s, -'ven-təs\ [French, from Provençal *sirventes,* literally, servant's song, from *sirvent* servant, from Latin *servient-, serviens,* present participle of *servire* to serve] (1819)
: a usually moral or religious song of the Provençal troubadours satirizing social vices

sis \'sis\ *noun* (1656)
: SISTER — usually used in direct address

-sis *noun suffix, plural* **-ses** [Latin, from Greek, feminine suffix of action]
: process **:** action ⟨peristal*sis*⟩

si·sal \'sī-səl, -zəl\ *noun* [*Sisal,* port in Yucatán, Mexico] (1843)
1 a : a strong durable white fiber used especially for hard fiber cordage and twine — called also *sisal hemp* **b :** a widely cultivated Mexican agave (*Agave sisalana*) whose leaves yield sisal
2 : any of several fibers similar to true sisal

sis·kin \'sis-kən\ *noun* [German dialect *Sisschen,* diminutive of Middle High German *zīse* siskin, of Slavic origin; akin to Czech *čížek* siskin] (1562)
: a small chiefly greenish and yellowish finch (*Carduelis spinus*) of Europe, Asia, and northern Africa that is related to the goldfinch — compare PINE SISKIN, RED SISKIN

sis·si·fied \'si-si-ˌfīd\ *adjective* (circa 1903)
: of, relating to, or having the characteristics of a sissy

sis·sy \'si-sē\ *noun, plural* **sissies** [*sis*] (1891)
: an effeminate man or boy; *also* **:** a timid or cowardly person
— sissy *adjective*

sis·ter \'sis-tər\ *noun* [Middle English *suster, sister,* partly from Old English *sweostor* and partly of Scandinavian origin; akin to Old Norse *systir* sister; akin to Latin *soror* sister, Sanskrit *svasṛ*] (before 12th century)
1 : a female who has one or both parents in common with another
2 *often capitalized* **a :** a member of a women's religious order (as of nuns or deaconesses); *especially* **:** one of a Roman Catholic congregation under simple vows **b :** a woman or girl who is a member of a Christian church
3 : a woman regarded as a comrade ⟨my Korean *sisters* —Alice Walker⟩
4 : one that is closely similar to or associated with another ⟨*sister* schools⟩ ⟨*sister* cities⟩
5 *chiefly British* **:** NURSE
6 a : GIRL, WOMAN; *especially* **:** a girl or woman who is black **b :** PERSON — usually used in the phrase *weak sister*
7 : a member of a sorority

sis·ter·hood \-ˌhu̇d\ *noun* (1609)
1 a : the state of being a sister **b :** sisterly relationship
2 : a community or society of sisters; *especially* **:** a society of women religious
3 : the solidarity of women based on shared conditions, experiences, or concerns

sis·ter–in–law \'sis-t(ə-)rən-ˌlȯ, -tərn-ˌlȯ\ *noun, plural* **sis·ters–in–law** \-tər-zən-\ (15th century)
1 : the sister of one's spouse
2 a : the wife of one's brother **b :** the wife of one's spouse's brother

sis·ter·ly \'sis-tər-lē\ *adjective* (circa 1570)
: of, relating to, or having the characteristics of a sister
— sisterly *adverb*

Sis·tine \'sis-ˌtēn, sis-'\ *adjective* [Italian *sistino,* from New Latin *sixtinus,* from *Sixtus,* name of some popes] (circa 1864)
1 : of or relating to any of the popes named Sixtus
2 [from Pope *Sixtus* IV (died 1484)] **:** of or relating to the Sistine chapel in the Vatican

Sis·y·phe·an \ˌsi-sə-'fē-ən\ *or* **Si·syph·i·an** \si-'si-fē-ən\ *adjective* (1635)
: of, relating to, or suggestive of the labors of Sisyphus

Sis·y·phus \'si-sə-fəs\ *noun* [Latin, from Greek *Sisyphos*]
: a legendary king of Corinth condemned eternally to repeat the cycle of rolling a heavy rock up a hill in Hades only to have it roll down again as it nears the top

¹sit \'sit\ *verb* **sat** \'sat\; **sit·ting** [Middle English *sitten,* from Old English *sittan;* akin to Old High German *sizzen* to sit, Latin *sedēre,* Greek *hezesthai* to sit, *hedra* seat] (before 12th century)
intransitive verb
1 a : to rest on the buttocks or haunches ⟨*sit* in a chair⟩ — often used with *down* **b :** PERCH, ROOST
2 : to occupy a place as a member of an official body ⟨*sit* in Congress⟩
3 : to hold a session **:** be in session for official business
4 : to cover eggs for hatching **:** BROOD
5 a : to take a position for having one's portrait painted or for being photographed **b :** to serve as a model
6 *archaic* **:** to have one's dwelling place **:** DWELL
7 a : to lie or hang relative to a wearer ⟨the collar *sits* awkwardly⟩ **b :** to affect one with or as if with weight ⟨the food *sat* heavily on his stomach⟩
8 : LIE, REST ⟨a kettle *sitting* on the stove⟩
9 a : to have a location ⟨the house *sits* well back from the road⟩ **b** *of wind* **:** to blow from a certain direction
10 : to remain inactive or quiescent ⟨the car *sits* in the garage⟩
11 : to take an examination
12 : BABY-SIT
13 : to please or agree with one — used with *with* and an adverb ⟨the decision did not *sit* well with me⟩
transitive verb
1 : to cause to be seated **:** place on or in a seat — often used with *down*
2 : to sit on (eggs)
3 : to keep one's seat on ⟨*sit* a horse⟩
4 : to provide seats or seating room for ⟨the car will *sit* six people⟩
usage see SET
— sit on 1 : to hold deliberations concerning **2 :** REPRESS, SQUELCH **3 :** to delay action or decision concerning
— sit pretty : to be in a highly favorable situation
— sit tight 1 : to maintain one's position without change **2 :** to remain quiet in or as if in hiding
— sit under : to attend religious service under the instruction or ministrations of; *also* **:** to attend the classes or lectures of

²sit *noun* (1776)
1 : the manner in which a garment fits
2 : an act or period of sitting

si·tar \si-'tär, 'si-ˌ\ *noun* [Hindi *sitār,* from Persian, a three-stringed lute, from *sih* three + *tār* string, thread] (1845)
: an Indian lute with a long neck and a varying number of strings
— si·tar·ist \-ist\ *noun*

sitar

sit·com \'sit-ˌkäm\ *noun* [*sit*uation *com*edy] (1964)
: SITUATION COMEDY

¹sit–down \'sit-'dau̇n\ *adjective* (circa 1837)
: served to seated diners ⟨a *sit-down* dinner⟩; *also* **:** of, relating to, or serving sit-down meals ⟨a *sit-down* restaurant⟩

²sit–down \'sit-ˌdau̇n\ *noun* (1936)
1 : a cessation of work by employees while maintaining continuous occupation of their place of employment as a protest and means toward forcing compliance with demands
2 : a mass obstruction of an activity by sitting down to demonstrate a grievance or to get the activity modified or halted

¹site \'sīt\ *noun* [Middle English, place, position, from Middle French or Latin; Middle French, from Latin *situs,* from *sinere* to leave, allow] (14th century)
1 a : the spatial location of an actual or planned structure or set of structures (as a building, town, or monuments) **b :** a space of ground occupied or to be occupied by a building
2 : the place, scene, or point of something

²site *transitive verb* **sit·ed; sit·ing** (15th century)
: to place on a site or in position **:** LOCATE

sith \'sith\ *or* **sith·ence** \'si-thən(t)s\ *or* **sith·ens** \'si-thənz\ *archaic variant of* SINCE

sit–in \'sit-ˌin\ *noun* (1937)
1 : SIT-DOWN 1
2 a : an act of occupying seats in a racially segregated establishment in organized protest against discrimination **b :** an act of sitting in the seats or on the floor of an establishment as a means of organized protest

sit in *intransitive verb* (1868)
1 : to take part in or be present at a session of music or discussion as a visitor
2 : to participate in a sit-in

Sit·ka spruce \'sit-kə-\ *noun* [*Sitka,* Alaska] (1895)
: a tall spruce (*Picea sitchensis*) of the northern Pacific coast that has thin reddish brown bark, flat needles, and cones with slightly toothed scales

si·tos·ter·ol \sī-'täs-tə-ˌrȯl, sə-, -ˌrōl\ *noun* [Greek *sitos* grain + English *sterol*] (1898)
: any of several sterols that are widespread especially in plant products (as wheat germ or soy bean oil) and are used as starting materials for the synthesis of steroid hormones

sit out *transitive verb* (1659)
: to refrain from participating in ⟨*sit out* the next dance⟩

sit·ter \'si-tər\ *noun* (14th century)
: one that sits; *especially* **:** one who baby-sits children

¹sit·ting \'si-tiŋ\ *noun* (13th century)
1 a : the act of one that sits **b :** a single occasion of continuous sitting (as for a portrait or meal)
2 a : a brooding over eggs for hatching **b :** SETTING 6
3 : SESSION ⟨a *sitting* of the legislature⟩

²sitting *adjective* (15th century)
1 : that is setting ⟨a *sitting* hen⟩
2 : occupying a judicial, legislative, or executive seat **:** being in office
3 : easily hit or played ⟨a *sitting* target⟩
4 a : used in or for sitting ⟨a *sitting* position⟩ **b :** performed while sitting ⟨a *sitting* shot⟩

sitting duck *noun* (1942)
: an easy or defenseless target for attack or criticism or unscrupulous dealings

sitting room *noun* (1771)
: LIVING ROOM 1

¹sit·u·ate \'si-chə-wət, -ˌwāt; 'sich-wət\ *adjective* [Middle English, from Late Latin *situatus,* from Latin *situs*] (15th century)
: having a site **:** LOCATED

²sit·u·ate \'si-chə-ˌwāt\ *transitive verb* **-at·ed; -at·ing** (circa 1532)
: to place in a site, situation, context, or category **:** LOCATE

situated *adjective* (15th century)
1 : having a site, situation, or location **:** LOCATED
2 : provided with money or possessions ⟨comfortably *situated*⟩

sit·u·a·tion \ˌsi-chə-'wā-shən\ *noun* (15th century)
1 a : the way in which something is placed in relation to its surroundings **b :** SITE **c** *archaic* **:** LOCALITY
2 *archaic* **:** state of health
3 a : position or place of employment **:** POST, JOB **b :** position in life **:** STATUS
4 : position with respect to conditions and circumstances ⟨the military *situation* remains obscure⟩
5 a : relative position or combination of circumstances at a certain moment **b :** a critical, trying, or unusual state of affairs **:** PROBLEM **c :** a particular or striking complex of affairs at a stage in the action of a narrative or drama

sit·u·a·tion·al \-shnəl, -shə-nᵊl\ *adjective* (1903)
1 : of, relating to, or appropriate to a situation
2 : of or relating to situation ethics
— sit·u·a·tion·al·ly *adverb*

situation comedy *noun* (1946)
: a radio or television comedy series that involves a continuing cast of characters in a succession of episodes

situation ethics *noun* (1955)
: a system of ethics by which acts are judged within their contexts instead of by categorical principles

sit–up \'sit-ˌəp\ *noun* (1938)
: a conditioning exercise performed from a supine position by raising the trunk to a sitting position without lifting the feet and returning to the original position

sit up *intransitive verb* (13th century)
1 a : to rise from a lying to a sitting position **b :** to sit with the back erect
2 : to show interest, alertness, or surprise ⟨*sit up* and take notice⟩
3 : to stay up after the usual time for going to bed ⟨*sat up* late to watch the movie⟩

si·tus \'sī-təs\ *noun* [Latin — more at SITE] (1701)
: the place where something exists or originates; *specifically* **:** the place where something (as a right) is held to be located in law

sitz bath \'sits-\ *noun* [part translation of German *Sitzbad,* from *Sitz* act of sitting + *Bad* bath] (1849)
: a tub in which one bathes in a sitting posture; *also* **:** a bath so taken especially therapeutically

sitz·mark \'sits-ˌmärk, 'zits-\ *noun* [part translation of German *Sitzmarke,* from *Sitz* + *Marke* mark] (1935)
: a depression left in the snow by a skier falling backward

Si·va \'si-və, 'shi-, 'sē-, 'shē-\ *noun* [Sanskrit *Śiva*] (1788)
: the god of destruction and regeneration in the Hindu sacred triad — compare BRAHMA, VISHNU

Si·van \'si-vən\ *noun* [Hebrew *Sīwān*] (1535)
: the 9th month of the civil year or the 3d month of the ecclesiastical year in the Jewish calendar — see MONTH table

Si·wash \'sī-ˌwȯsh, -ˌwäsh\ *noun* [*Siwash,* fictional college in stories by George Fitch (died 1915) American author] (1936)
: a small usually inland college that is notably provincial in outlook ⟨cheer for dear old *Siwash*⟩

six \'siks\ *noun* [Middle English, from *six,* adjective, from Old English *siex;* akin to Old High German *sehs* six, Latin *sex,* Greek *hex*] (before 12th century)
1 — see NUMBER table
2 : the sixth in a set or series ⟨the *six* of spades⟩
3 : something having six units or members: as **a :** an ice-hockey team **b :** a 6-cylinder engine or automobile
— six *adjective*
— six *pronoun, plural in construction*
— at sixes and sevens : being in disorder

six·fold \'siks-ˌfōld, -'fōld\ *adjective* (before 12th century)
1 : having six units or members
2 : being six times as great or as many
— six·fold \-'fōld\ *adverb*

six–gun \'siks-ˌgən\ *noun* (1912)
: a 6-chambered revolver

six·mo \'siks-(ˌ)mō\ *noun, plural* **sixmos** (1924)
: the size of a piece of paper cut six from a sheet; *also* **:** a book, a page, or paper of this size

six–o–six *or* **606** \ˌsiks-ˌō-'siks\ *noun* [from its having been the 606th compound tested and introduced by Paul Ehrlich (died 1915)] (1910)
: ARSPHENAMINE

six–pack \'siks-ˌpak\ *noun* (1952)
1 : six bottles or cans (as of beer) packaged and purchased as a unit
2 : the contents of a six-pack

six·pence \'siks-pən(t)s, *US also* -ˌpen(t)s\ *noun* (14th century)
1 : a former British monetary unit equal to six pennies
2 *plural* **sixpence** *or* **six·penc·es :** a coin worth sixpence

six·pen·ny \'siks-pə-nē, *US also* -ˌpe-nē\ *adjective* (15th century)
: costing or worth sixpence

sixpenny bit *noun* (circa 1887)
: SIXPENCE 2

six·pen·ny nail \'siks-ˌpe-nē-\ *noun* (15th century)
: a nail about two inches long

six–shoot·er \'sik(s)-ˌshü-tər\ *noun* (1844)
: SIX-GUN

six·teen \ˌsik-'stēn, 'siks-ˌtēn\ *noun* [Middle English *sixtene,* from Old English *sixtȳne,* adjective, from *six* six + *-tȳne* (akin to Old English *tīen* ten) — more at TEN] (before 12th century)
— see NUMBER table
— sixteen *adjective*
— sixteen *pronoun, plural in construction*
— six·teenth \-'stēn(t)th, -ˌtēn(t)th\ *adjective or noun*

six·teen·mo \sik-'stēn-(ˌ)mō\ *noun, plural* **-mos** (1847)
: the size of a piece of paper cut 16 from a sheet; *also* **:** a book, a page, or paper of this size

sixteenth note *noun* (circa 1861)
: a musical note with the time value of 1/16 of a whole note — see NOTE illustration

sixteenth rest *noun* (circa 1890)
: a musical rest corresponding in time value to a sixteenth note — see REST illustration

sixth \'siks(t)th, 'siks(t)\ *noun, plural* **sixths** \'siks(ts), 'siks(t)ths\ (12th century)
1 — see NUMBER table
2 a : a musical interval embracing six diatonic degrees **b :** a tone at this interval; *specifically* **:** SUBMEDIANT **c :** the harmonic combination of two tones a sixth apart
— sixth *adjective or adverb*
— sixth·ly \'siksth-lē, 'sikst-\ *adverb*

sixth chord *noun* (circa 1903)
: a musical chord consisting of a tone with its third and its sixth above and usually being the first inversion of a triad

sixth sense *noun* (1761)
: a power of perception like but not one of the five senses **:** a keen intuitive power

Six·tine \'sik-ˌstīn, -ˌstēn\ *variant of* SISTINE

six·ty \'siks-tē\ *noun, plural* **sixties** [Middle English, from *sixty,* adjective, from Old English *siextig,* noun, group of sixty, from *siex* six + *-tig* group of ten; akin to Old English *tīen* ten] (14th century)
1 — see NUMBER table
2 *plural* **:** the numbers 60 to 69; *specifically* **:** the years 60 to 69 in a lifetime or century
— six·ti·eth \'siks-tē-əth\ *adjective or noun*
— sixty *adjective*
— sixty *pronoun, plural in construction*
— six·ty·ish \'siks-tē-ish\ *adjective*

sixty–fourth note \ˌsiks-tē-'fōrth-, -'fȯrth-\ *noun* (circa 1890)
: a musical note with the time value of 1/64 of a whole note — see NOTE illustration

sixty–fourth rest *noun* (circa 1903)
: a musical rest corresponding in time value to a sixty-fourth note

six·ty–nine \ˌsiks-tē-'nīn\ *noun* (1924)
1 — see NUMBER table
2 : mutual cunnilingus and fellatio **:** mutual fellatio **:** mutual cunnilingus

siz·able *or* **size·able** \'sī-zə-bəl\ *adjective* (1613)
: fairly large **:** CONSIDERABLE
— siz·able·ness *noun*
— siz·ably \-blē\ *adverb*

siz·ar *also* **siz·er** \'sī-zər\ *noun* [*sizar* alteration of *sizer,* from ¹*size*] (1588)
: a student (as in the university of Cambridge) who receives an allowance toward college expenses and who originally acted as a servant to other students in return for this allowance

¹size \'sīz\ *noun* [Middle English *sise* assize, from Middle French, from Old French, short for *assise* — more at ASSIZE] (13th century)
1 *dialect British* **:** ASSIZE 2a — usually used in plural
2 *obsolete* **:** a fixed portion of food or drink
3 a : physical magnitude, extent, or bulk **:** relative or proportionate dimensions **b :** relative aggregate amount or number **c :** considerable proportions **:** BIGNESS
4 : one of a series of graduated measures especially of manufactured articles (as of clothing) conventionally identified by numbers or letters ⟨a *size* 7 hat⟩
5 : character, quality, or status of a person or thing especially with reference to importance, relative merit, or correspondence to needs ⟨try this idea on for *size*⟩
6 : actual state of affairs ⟨that's about the *size* of it⟩

²size *verb* **sized; siz·ing** (1609)
transitive verb
1 : to make a particular size **:** bring to proper or suitable size
2 : to arrange, grade, or classify according to size or bulk

intransitive verb
: to equal in size or other particular characteristic **:** COMPARE — usually used with *up* and often with *to* or *with*

[3]**size** \'sīz, ˌsīz\ *adjective* (1924)
: SIZED — usually used in combination ⟨bite-*size*⟩

[4]**size** \'sīz\ *noun* [Middle English *sise*] (15th century)
: any of various glutinous materials (as preparations of glue, flour, varnish, or resins) used for filling the pores in surfaces (as of paper, textiles, leather, or plaster) or for applying color or metal leaf (as to book edges or covers)

[5]**size** *transitive verb* **sized; siz·ing** (1667)
: to cover, stiffen, or glaze with or as if with size

sized \'sīzd, ˌsīzd\ *adjective* (1582)
1 : having a specified size or bulk — usually used in combination ⟨a small-*sized* house⟩
2 : arranged or adjusted according to size

size up *transitive verb* (1884)
: to form a judgment of

siz·ing \'sī-ziŋ\ *noun* (1825)
: [4]SIZE

[1]**siz·zle** \'si-zəl\ *verb* **siz·zled; siz·zling** \'si-zə-liŋ, 'siz-liŋ\ [perhaps frequentative of *siss* to hiss] (1603)
transitive verb
: to burn up or sear with or as if with a hissing sound
intransitive verb
1 : to make a hissing sound in or as if in burning or frying
2 : to seethe with deep anger or resentment

[2]**sizzle** *noun* (circa 1823)
: a hissing sound (as of something frying over a fire)

siz·zler \'si-zə-lər, 'siz-lər\ *noun* (1848)
: one that sizzles; *especially* **:** SCORCHER

ska \'skä\ *noun* [origin unknown] (1969)
: popular music of Jamaican origin that combines elements of traditional Caribbean rhythms and jazz

skag *variant of* SCAG

skald \'skȯld, 'skäld\ *noun* [Old Norse *skāld*] (1780)
: an ancient Scandinavian poet; *broadly* **:** BARD
— skald·ic \'skȯl dik, 'skäl-\ *adjective*

skat \'skät, 'skat\ *noun* [German, modification of Italian *scarto* discard, from *scartare* to discard, from *s-* (from Latin *ex-*) + *carta* card — more at CARD] (1864)
1 : a three-handed card game played with 32 cards in which players bid for the privilege of attempting any of several contracts
2 : a widow of two cards in skat that may be used by the winner of the bid

[1]**skate** \'skāt\ *noun, plural* **skates** *also* **skate** [Middle English *scate,* from Old Norse *skata*] (14th century)
: any of a family (Rajidae, especially genus *Raja*) of rays with the pectoral fins greatly developed giving the fish a flat diamond shape

skate

[2]**skate** *noun* [modification of Dutch *schaats* stilt, skate, from Old North French *escache* stilt, probably of Germanic origin; akin to Old English *sceacan* to shake — more at SHAKE] (1684)
1 a : a metal frame that can be fitted to the sole of a shoe and to which is attached a runner or a set of wheels for gliding over ice or a surface other than ice **b :** ROLLER SKATE **c :** ICE SKATE
2 : a period of skating

[3]**skate** *verb* **skat·ed; skat·ing** (1696)
intransitive verb
1 : to glide along on skates propelled by the alternate action of the legs
2 : to slip or glide as if on skates
3 : to proceed in a superficial or blithe manner
transitive verb
: to go along or through by skating

[4]**skate** *noun* [probably alteration of English dialect *skite* an offensive person] (1894)
1 : a thin awkward-looking or decrepit horse **:** NAG
2 : FELLOW 4c

skate·board \'skāt-ˌbōrd, -ˌbȯrd\ *noun* (1964)
: a short board mounted on small wheels that is used for coasting and often for performing athletic stunts
— skate·board·er \-ˌbōr-dər, -ˌbȯr-\ *noun*
— skate·board·ing \-diŋ\ *noun*

skat·er \'skā-tər\ *noun* (1700)
1 : one that skates
2 : WATER STRIDER

skat·ing \'skā-tiŋ\ *noun* (1723)
: the act, art, or sport of gliding on skates

ska·tole \'ska-ˌtōl, 'skā-\ *also* **ska·tol** \-ˌtȯl, -ˌtōl\ *noun* [International Scientific Vocabulary, from Greek *skat-, skōr* excrement — more at SCATOLOGY] (1879)
: a foul-smelling compound C_9H_9N found in the intestines and feces, in civet, and in several plants or made synthetically and used in perfumes as a fixative

[1]**skean** *or* **skeane** *variant of* SKEIN

[2]**skean** *or* **skene** \'skē(-ə)n\ *noun* [Middle English *skene,* from Irish *scian* & Scottish Gaelic *sgian,* from Old Irish *scían;* probably akin to Sanskrit *chyati* he cuts off — more at SCIENCE] (15th century)
: DAGGER, DIRK

ske·dad·dle \ski-'da-dᵊl\ *intransitive verb* **ske·dad·dled; ske·dad·dling** \-'dad-liŋ, -'da-dᵊl-iŋ\ [origin unknown] (1861)
: RUN AWAY, SCRAM; *especially* **:** to flee in a panic
— ske·dad·dler \ 'dad lər, -'da-dᵊl-ər\ *noun*

skeet \'skēt\ *noun* [perhaps from Norwegian *skyte* to shoot] (1926)
: trapshooting in which clay targets are thrown in such a way as to simulate the angles of flight of birds

[1]**skee·ter** \'skē-tər\ *noun* [by shortening & alteration] (1839)
1 : MOSQUITO
2 : an iceboat 16 feet (5 meters) or more in length equipped with a single sail

[2]**skeet·er** \'skē-tər\ *noun* (1926)
: a skeet shooter

skeg \'skeg\ *also* **skag** \'skag\ *noun* [Middle English *skegge,* from Old Norse *skegg* cutwater, literally, beard — more at SHAG] (13th century)
1 : the stern of the keel of a ship near the sternpost; *especially* **:** the part connecting the keel with the bottom of the rudderpost in a single-screw ship
2 : a fin situated on the rear bottom of a surfboard that is used for steering and stability

skeigh \'skēk\ *adjective* [perhaps of Scandinavian origin; akin to Swedish *skygg* shy; akin to Old English *scēoh* shy — more at SHY] (1508)
chiefly Scottish **:** proudly spirited **:** SKITTISH

[1]**skein** \'skān\ *noun* [Middle English *skeyne,* from Middle French *escaigne*] (14th century)
1 *or* **skean** *or* **skeane** \'skān\ **:** a loosely coiled length of yarn or thread wound on a reel
2 : something suggesting the twists or coils of a skein **:** TANGLE
3 : a flock of wildfowl (as geese or ducks) in flight

[2]**skein** *transitive verb* (circa 1775)
: to wind into skeins ⟨*skein* yarn⟩

skel·e·tal \'ske-lə-tᵊl\ *adjective* (1854)
: of, relating to, forming, attached to, or resembling a skeleton
— skel·e·tal·ly \-tᵊl-ē\ *adverb*

skeletal muscle *noun* (1877)
: striated muscle that is usually attached to the skeleton and is usually under voluntary control

[1]**skel·e·ton** \'ske-lə-tᵊn\ *noun* [New Latin, from Greek, neuter of *skeletos* dried up; akin to Greek *skellein* to dry up, *sklēros* hard and perhaps to Old English *sceald* shallow] (1578)
1 : a usually rigid supportive or protective structure or framework of an organism; *especially* **:** the bony or more or less cartilaginous framework supporting the soft tissues and protecting the internal organs of a vertebrate
2 : something reduced to its minimum form or essential parts
3 : an emaciated person or animal
4 a : something forming a structural framework **b :** the straight or branched chain or ring of atoms that forms the basic structure of an organic molecule
5 : something shameful and kept secret (as in a family) — often used in the phrase *skeleton in the closet*
— skel·e·ton·ic \ˌske-lə-'tä-nik\ *adjective*

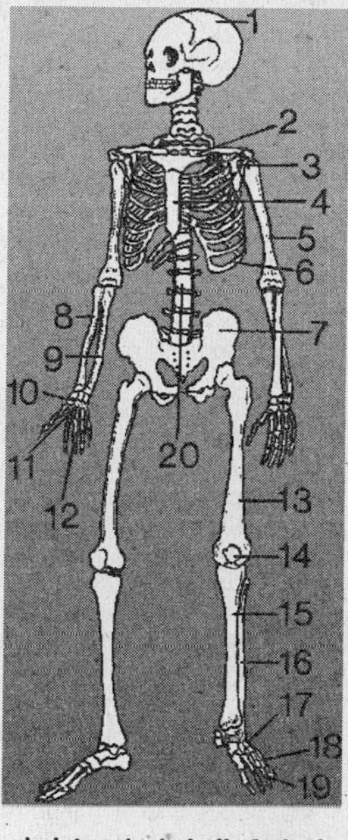

skeleton 1: ***1*** **skull,** ***2*** **clavicle,** ***3*** **scapula,** ***4*** **sternum,** ***5*** **humerus,** ***6*** **rib,** ***7*** **pelvis,** ***8*** **radius,** ***9*** **ulna,** ***10*** **carpus,** ***11*** **metacarpal bones,** ***12*** **phalanges (fingers),** ***13*** **femur,** ***14*** **patella,** ***15*** **tibia,** ***16*** **fibula,** ***17*** **tarsus,** ***18*** **metatarsal bones,** ***19*** **phalanges (toes),** ***20*** **spinal column**

[2]**skeleton** *adjective* (1778)
: of, consisting of, or resembling a skeleton

skel·e·ton·ise *British variant of* SKELETONIZE

skel·e·ton·ize \-ˌīz\ *transitive verb* **-ized; -iz·ing** (1644)
: to produce in or reduce to skeleton form ⟨*skeletonize* a leaf⟩ ⟨*skeletonize* a news story⟩ ⟨*skeletonize* a regiment⟩

skel·e·ton·iz·er \-ˌī-zər\ *noun* (circa 1891)
: any of various lepidopterous larvae that eat the parenchyma of leaves reducing them to a skeleton of veins

skeleton key *noun* (1810)
: a key with a large part of the bit filed away to enable it to open low quality locks as a master key

skel·lum \'ske-ləm\ *noun* [Dutch *schelm,* from Low German; akin to Old High German *skelmo* person deserving death] (1611)
chiefly Scottish **:** SCOUNDREL, RASCAL

[1]**skelp** \'skelp\ *verb* **skelped** \'skelpt\ *also* **skel·pit** \'skel-pət\; **skelp·ing** [Middle English] (15th century)
transitive verb
dialect British **:** STRIKE, SLAP, BEAT
intransitive verb
: to step lively **:** HUSTLE

[2]**skelp** *noun* (15th century)
dialect British **:** a smart blow **:** SLAP

skel·ter \'skel-tər\ *intransitive verb* **skel·tered; skel·ter·ing** \-t(ə-)riŋ\ [from *-skelter* (in *helter-skelter*)] (1852)

\ə\ **abut** \ᵊ\ **kitten** \ər\ **further** \a\ **ash** \ā\ **ace** \ä\ **mop, mar** \au̇\ **out** \ch\ **chin** \e\ **bet** \ē\ **easy** \g\ **go** \i\ **hit** \ī\ **ice** \j\ **job** \ŋ\ **sing** \ō\ **go** \ȯ\ **law** \ȯi\ **boy** \th\ **thin** \th\ **the** \ü\ **loot** \u̇\ **foot** \y\ **yet** \zh\ **vision** *see also* Guide to Pronunciation

: SCURRY

Skel·ton·ics \skel-'tä-niks\ *noun plural* [John *Skelton*] (1898)
: short verses of an irregular meter with two or three stresses sometimes in falling and sometimes in rising rhythm and usually with rhymed couplets

skep \'skep\ *noun* [Middle English *skeppe* basket, beehive, from Old English *sceppe* basket, from Old Norse *skeppa* bushel; akin to Old High German *sceffil* bushel, *scaf* tub] (15th century)
: BEEHIVE; *especially* : a domed hive made of twisted straw

skep·sis \'skep-səs\ *noun* [New Latin, from Greek *skepsis* examination, doubt, skeptical philosophy, from *skeptesthai*] (circa 1864)
: philosophic doubt as to the objective reality of phenomena; *broadly* : a skeptical outlook or attitude

skep·tic \'skep-tik\ *noun* [Latin or Greek; Latin *scepticus*, from Greek *skeptikos*, from *skeptikos* thoughtful, from *skeptesthai* to look, consider — more at SPY] (1587)
1 : an adherent or advocate of skepticism
2 : a person disposed to skepticism especially regarding religion or religious principles

skep·ti·cal \-ti-kəl\ *adjective* (1639)
: relating to, characteristic of, or marked by skepticism ⟨a *skeptical* listener⟩
— **skep·ti·cal·ly** \-k(ə-)lē\ *adverb*

skep·ti·cism \'skep-tə-ˌsi-zəm\ *noun* (1646)
1 : an attitude of doubt or a disposition to incredulity either in general or toward a particular object
2 a : the doctrine that true knowledge or knowledge in a particular area is uncertain **b** : the method of suspended judgment, systematic doubt, or criticism characteristic of skeptics
3 : doubt concerning basic religious principles (as immortality, providence, and revelation)
synonym see UNCERTAINTY

sker·ry \'sker-ē\ *noun, plural* **skerries** [of Scandinavian origin; akin to Old Norse *sker* skerry and to Old Norse *ey* island; akin to Latin *aqua* water — more at SCAR, ISLAND] (1612)
: a rocky isle : REEF

¹sketch \'skech\ *noun* [Dutch *schets*, from Italian *schizzo*, literally, splash, from *schizzare* to splash, of imitative origin] (1668)
1 a : a rough drawing representing the chief features of an object or scene and often made as a preliminary study **b** : a tentative draft (as for a literary work)
2 : a brief description (as of a person) or outline
3 a : a short literary composition somewhat resembling the short story and the essay but intentionally slight in treatment, discursive in style, and familiar in tone **b** : a short instrumental composition usually for piano **c** : a slight theatrical piece having a single scene; *especially* : a comic variety act

²sketch (1694)
transitive verb
: to make a sketch, rough draft, or outline of
intransitive verb
: to draw or paint a sketch
— **sketch·er** *noun*

sketch·book \'skech-ˌbu̇k\ *noun* (1820)
: a book of or for sketches

sketchy \'ske-chē\ *adjective* **sketch·i·er; -est** (1805)
1 : of the nature of a sketch : roughly outlined
2 : wanting in completeness, clearness, or substance : SLIGHT, SUPERFICIAL
— **sketch·i·ly** \'ske-chə-lē\ *adverb*
— **sketch·i·ness** \'ske-chē-nəs\ *noun*

¹skew \'skyü\ *verb* [Middle English, to escape, skew, from Old North French *escuer* to shun, of Germanic origin; akin to Old High German *sciuhen* to frighten off — more at SHY] (15th century)
intransitive verb
1 : to take an oblique course
2 : to look askance
transitive verb
1 : to make, set, or cut on the skew
2 : to distort especially from a true value or symmetrical form ⟨*skewed* statistical data⟩

²skew *adjective* (1609)
1 : set, placed, or running obliquely : SLANTING
2 : more developed on one side or in one direction than another : not symmetrical

³skew *noun* (1688)
: a deviation from a straight line : SLANT

skew·back \'skyü-ˌbak\ *noun* (1703)
: a course of masonry, a stone, or an iron plate having an inclined face against which the voussoirs of an arch abut

¹skew·bald \-ˌbȯld\ *adjective* [*skewed* (skewbald) + *bald*] (1654)
of an animal : marked with patches of white and any other color but black

²skewbald *noun* (1863)
: a skewbald horse

skew curve *noun* (circa 1889)
: a curve in three-dimensional space that does not lie in a single plane

skew distribution *noun* (circa 1931)
: an unsymmetrical frequency distribution having the mode at a different value from the mean

¹skew·er \'skyü-ər, 'skyu̇(-ə)r\ *noun* [Middle English *skeuier*] (15th century)
1 : a pin of wood or metal for fastening meat to keep it in form while roasting or to hold small pieces of meat or vegetables for broiling
2 : any of various things shaped or used like a meat skewer

²skewer *transitive verb* (1701)
1 : to fasten or pierce with or as if with a skewer
2 : to criticize or ridicule sharply and effectively

skew lines *noun plural* (1952)
: straight lines that do not intersect and are not in the same plane

skew·ness \'skyü-nəs\ *noun* (1894)
: lack of straightness or symmetry : DISTORTION; *especially* : lack of symmetry in a frequency distribution

¹ski \'skē, *British sometimes* 'shē\ *noun, plural* **skis** *also* **ski** [Norwegian, from Old Norse *skīth* stick of wood, ski; akin to Old English *scīd* board, *scēadan* to divide — more at SHED] (1755)
1 a : one of a pair of narrow strips of wood, metal, or plastic curving upward in front that are used especially for gliding over snow **b** : WATER SKI
2 : a piece of material that resembles a ski and is used as a runner on a vehicle

²ski *verb* **skied** \'skēd, 'shēd\; **ski·ing** (circa 1890)
intransitive verb
: to glide on skis in travel or as a sport
transitive verb
: to travel or pass over on skis
— **ski·able** \'skē-ə-bəl\ *adjective*
— **ski·er** *noun*

skia·gram \'skī-ə-ˌgram\ *noun* [International Scientific Vocabulary, from Greek *skia* shadow + International Scientific Vocabulary *-gram* — more at SCENE] (1801)
1 : a figure formed by shading in the outline of a shadow
2 : RADIOGRAPH

ski·bob \'skē-ˌbäb\ *noun* [¹*ski* + ⁷*bob*] (1966)
: a vehicle that resembles a bicycle with two short skis in place of wheels and that is used for gliding downhill over snow by a rider wearing miniature skis for balance
— **ski·bob·ber** \-ˌbä-bər\ *noun*
— **ski·bob·bing** \-ˌbä-biŋ\ *noun*

ski boot *noun* (1907)
: a rigid padded shoe that extends just above the ankle, is securely fastened to the foot (as with laces, buckles, or clasps), and is locked into position in a ski binding

¹skid \'skid\ *noun* [perhaps of Scandinavian origin; akin to Old Norse *skīth* stick of wood — more at SKI] (circa 1610)
1 : one of a group of objects (as planks or logs) used to support or elevate a structure or object
2 : a wooden fender hung over a ship's side to protect it in handling cargo
3 : a usually iron shoe or clog attached to a chain and placed under a wheel to prevent its turning when descending a steep hill : DRAG
4 : a timber, bar, rail, pole, or log used in pairs or sets to form a slideway (as for an incline from a truck to the sidewalk)
5 : the act of skidding : SLIP, SIDESLIP
6 : a runner used as a member of the landing gear of an airplane or helicopter
7 a *plural* : a route to defeat or downfall ⟨on the *skids*⟩ ⟨his career hit the *skids*⟩ **b** : a losing streak ⟨a five-game *skid*⟩
8 : a low platform mounted (as on wheels) on which material is set for handling and moving; *also* : PALLET 3
— **skidproof** *adjective*

²skid *verb* **skid·ded; skid·ding** (1674)
transitive verb
1 : to apply a brake or skid to : slow or halt by a skid
2 a : to haul (as logs) by dragging ⟨cutting and *skidding* firewood⟩ **b** : to haul along, slide, hoist, or store on skids
intransitive verb
1 : to slide without rotating (as a wheel held from turning while a vehicle moves onward)
2 a : to fail to grip the roadway; *especially* : to slip sideways on the road **b** *of an airplane* : to slide sideways away from the center of curvature when turning **c** : SLIDE, SLIP
3 : to fall rapidly, steeply, or far

skid·der \'ski-dər\ *noun* (1870)
1 : one that skids or uses a skid
2 : a tractor used especially for hauling logs

skid·doo *or* **ski·doo** \ski-'dü, skē-\ *intransitive verb* [probably alteration of *skedaddle*] (1903)
: to go away : DEPART

skid·dy \'ski-dē\ *adjective* **skid·di·er; -est** (1902)
: likely to skid or cause skidding ⟨a wet *skiddy* road⟩

skid road *noun* (1880)
1 : a road along which logs are skidded
2 a *West* : the part of a town frequented by loggers **b** : SKID ROW

skid row \-'rō\ *noun* [alteration of *skid road*] (circa 1931)
: a district of cheap saloons and flophouses frequented by vagrants and alcoholics

ski·ey *variant of* SKYEY

skiff \'skif\ *noun* [Middle English *skif*, from Middle French or Old Italian; Middle French *esquif*, from Old Italian *schifo*, of Germanic origin; akin to Old English *scip* ship] (15th century)
: any of various small boats; *especially* : a flat-bottomed rowboat

skif·fle \'ski-fəl\ *noun* [origin unknown] (1926)
: American jazz or folk music played entirely or in part on nonstandard instruments (as jugs, washboards, or Jew's harps); *also* : a derivative form of music formerly popular in Great Britain featuring vocals with a simple instrumental accompaniment

ski·ing *noun* (1893)
: the art or sport of sliding and jumping on skis

ski·jor·ing \'skē-ˌjōr-iŋ, -ˌjȯr-, (ˌ)skē-'\ *noun* [modification of Norwegian *skikjøring*, from *ski* + *kjøring* driving] (1910)
: a winter sport in which a person wearing skis is drawn over snow or ice (as by a horse or vehicle)

ski jump *noun* (1907)

: a jump made by a person wearing skis; *also* : a course or track especially prepared for such jumping
— **ski jump** *intransitive verb*
— **ski jumper** *noun*

skil·ful *chiefly British variant of* SKILLFUL

ski lift *noun* (1939)
: a motor-driven conveyor consisting usually of a series of bars or seats suspended from an overhead moving cable and used for transporting skiers or sightseers up a long slope

[1]skill \'skil\ *intransitive verb* [Middle English *skilen,* from Old Norse *skilja* to separate, divide; akin to Old Norse *skil* distinction] (13th century)
archaic : to make a difference : MATTER, AVAIL

[2]skill *noun* [Middle English *skil,* from Old Norse, distinction, knowledge; probably akin to Old English *scylian* to separate, *sciell* shell — more at SHELL] (13th century)
1 *obsolete* : CAUSE, REASON
2 a : the ability to use one's knowledge effectively and readily in execution or performance **b :** dexterity or coordination especially in the execution of learned physical tasks
3 : a learned power of doing something competently : a developed aptitude or ability ⟨language *skills*⟩
synonym see ART
— **skill–less** *or* **skil·less** \'skil-ləs\ *adjective*
— **skill–less·ness** *or* **skil·less·ness** *noun*

skilled \'skild\ *adjective* (1552)
1 : having acquired mastery of or skill in something (as a technique or a trade)
2 : of, relating to, or requiring workers or labor with skill and training in a particular occupation, craft, or trade
synonym see PROFICIENT

skil·let \'ski-lət\ *noun* [Middle English *skelet*] (15th century)
1 *chiefly British* **:** a small kettle or pot usually having three or four often long feet and used for cooking on the hearth
2 : FRYING PAN

skill·ful \'skil-fəl\ *adjective* (14th century)
1 : possessed of or displaying skill : EXPERT
2 : accomplished with skill
synonym see PROFICIENT
— **skill·ful·ly** \-fə-lē\ *adverb*
— **skill·ful·ness** *noun*

skil·ling \'ski-liŋ, 'shi-\ *noun* [Swedish, Norwegian, & Danish, from Old Norse *skillingr,* a gold coin; akin to Old English *scilling* shilling] (1793)
: any of various old Scandinavian units of value or the coins representing them

[1]skim \'skim\ *verb* **skimmed; skim·ming** [Middle English *skymmen*] (14th century)
transitive verb
1 a : to clear (a liquid) of scum or floating substance ⟨*skim* boiling syrup⟩ **b :** to remove (as film or scum) from the surface of a liquid **c :** to remove cream from by skimming **d :** to remove the best or most easily obtainable contents from
2 : to read, study, or examine superficially and rapidly; *especially* **:** to glance through (as a book) for the chief ideas or the plot
3 : to throw in a gliding path; *especially* **:** to throw so as to ricochet along the surface of water
4 : to cover with or as if with a film, scum, or coat
5 : to pass swiftly or lightly over
6 : to remove or conceal (as a portion of casino profits) to avoid payment of taxes
intransitive verb
1 a : to pass lightly or hastily **:** glide or skip along, above, or near a surface **b :** to give a cursory glance, consideration, or reading
2 : to become coated with a thin layer of film or scum
3 : to put on a finishing coat of plaster

[2]skim *noun* (14th century)
1 : a thin layer, coating, or film
2 : the act of skimming
3 : something skimmed; *specifically* **:** SKIM MILK

[3]skim *adjective* (1794)
1 : having the cream removed by skimming
2 : made of skim milk ⟨*skim* cheese⟩

ski mask *noun* (1966)
: a knit fabric mask that covers the head, has openings for the eyes, mouth, and sometimes the nose, and is worn especially by skiers for protection from the cold

skim·ble–skam·ble \ˌskim-bəl-'skam-bəl\ *adjective* [reduplication of English dialect *scamble* to stumble along] (1596)
: rambling and confused : SENSELESS

skim·mer \'ski-mər\ *noun* (14th century)
1 : one that skims; *specifically* **:** a flat perforated scoop or spoon used for skimming
2 : any of a small genus (*Rynchops*) of long-winged marine birds that have the lower mandible longer than the upper
3 : a usually straw flat-crowned hat with a wide straight brim
4 : a fitted sleeveless dress with a usually flaring skirt

skim milk *noun* (1596)
: milk from which the cream has been taken — called also *skimmed milk*

skimming *noun* (15th century)
: that which is skimmed from a liquid

ski·mo·bile \'skē-mō-ˌbēl\ *noun* (1944)
: SNOWMOBILE

[1]skimp \'skimp\ *adjective* [perhaps alteration of *scrimp*] (1775)
: SKIMPY

[2]skimp (circa 1879)
transitive verb
: to give insufficient or barely sufficient attention or effort to or funds for
intransitive verb
: to save by or as if by skimping

skimpy \'skim pē\ *adjective* **skimp·i·er; -est** (1842)
: deficient in supply or execution especially through skimping : SCANTY
synonym see MEAGER
— **skimp·i·ly** \-pə-lē\ *adverb*
— **skimp·i·ness** \-pē-nəs\ *noun*

[1]skin \'skin\ *noun, often attributive* [Middle English, from Old Norse *skinn;* akin to Old English *scinn* skin, Middle High German *schint* fruit peel] (13th century)
1 a (1) **:** the integument of an animal (as a fur-bearing mammal or a bird) separated from the body usually with its hair or feathers (2) **:** a usually unmounted specimen of a vertebrate (as in a museum) **b :** the hide or pelt of a game or domestic animal **c** (1) **:** the pelt of an animal prepared for use as a trimming or in a garment — compare [4]HIDE (2) **:** a sheet of parchment or vellum made from a hide (3) **:** BOTTLE 1b
2 a : the external limiting tissue layer of an animal body especially when forming a tough but flexible cover relatively impermeable from without while intact **b :** any of various outer or surface layers (as a rind, husk, or pellicle)
3 : the life or physical well-being of a person ⟨saved his own *skin*⟩
4 : a sheathing or casing forming the outside surface of a structure (as a ship or airplane)
— **skin·less** \-ləs\ *adjective*
— **by the skin of one's teeth :** by a very narrow margin
— **under one's skin :** so deeply penetrative as to irritate, stimulate, provoke thought, or otherwise excite
— **under the skin :** beneath apparent or surface differences **:** at heart

[2]skin *verb* **skinned; skin·ning** (14th century)
transitive verb
1 a : to strip, scrape, or rub off an outer covering (as the skin or rind) of **b :** to strip or peel off **c :** to cut, chip, or damage the surface of ⟨fell and *skinned* my knee⟩
2 a : to cover with or as if with skin **b :** to heal over with skin
3 a : to strip of money or property **:** FLEECE **b :** to defeat badly **c :** CENSURE, CASTIGATE
4 : to urge on and direct the course of (as a draft animal)
intransitive verb
1 : to become covered with or as if with skin
2 a : SHIN 1 **b :** to pass or get by with scant room to spare

[3]skin *adjective* (circa 1935)
: involving subjects who are nude ⟨expected to conduct *skin* searches for weapons —Diane K. Shah⟩; *especially* **:** devoted to showing nudes ⟨*skin* magazines⟩

skin–deep \'skin-'dēp\ *adjective* (1613)
1 : as deep as the skin
2 : not thorough or lasting in impression **:** SUPERFICIAL

skin diving *noun* (1938)
: the sport of swimming under water with a face mask and flippers and especially without a portable breathing device
— **skin–dive** *intransitive verb*
— **skin diver** *noun*

skin effect *noun* (1891)
: an effect characteristic of current distribution in a conductor at high frequencies by virtue of which the current density is greater near the surface of the conductor than in its interior

skin·flint \'skin-ˌflint\ *noun* (circa 1700)
: a person who would save, gain, or extort money by any means : MISER, NIGGARD

skin·ful \-ˌfu̇l\ *noun* (1788)
1 : a large or satisfying quantity especially of liquor
2 : the contents of a skin bottle

skin game *noun* (1868)
: a swindling game or trick

skin graft *noun* (1871)
: a piece of skin that is taken from a donor area to replace skin in a defective or denuded area (as one that has been burned)
— **skin grafting** *noun*

skin·head \'skin-ˌhed\ *noun* (circa 1953)
1 : a person whose hair is cut very short
2 : a usually white male belonging to any of various sometimes violent youth gangs whose members have close-shaven hair and often espouse white-supremacist beliefs

[1]skink \'skiŋk\ *transitive verb* [Middle English, from Middle Dutch *schenken;* akin to Old English *scencan* to pour out drink and probably to *scanca* shank] (15th century)
chiefly dialect **:** to draw, pour out, or serve (drink)

[2]skink *noun* [Latin *scincus,* from Greek *skinkos*] (1590)
: any of a family (Scincidae) of typically small insectivorous lizards with long tapering bodies

skink·er \'skiŋ-kər\ *noun* (1586)
: one that serves liquor : BARTENDER

skinned \'skind\ *adjective* (15th century)
: having skin especially of a specified kind usually used in combination ⟨dark-*skinned*⟩

skin·ner \'ski-nər\ *noun* (14th century)
1 a : one that deals in skins, pelts, or hides **b :** one that removes, cures, or dresses skins
2 : SHARPER
3 : a driver of draft animals : TEAMSTER

Skin·ner box \'ski-nər-ˌbäks\ *noun* [B. F. *Skinner*] (1940)
: a laboratory apparatus in which an animal is caged for experiments in operant conditioning and which typically contains a lever that must

be pressed by the animal to gain reward or avoid punishment

¹skin·ny \'ski-nē\ *adjective* **skin·ni·er; -est** (1573)
1 : resembling skin **:** MEMBRANOUS
2 a : lacking sufficient flesh **:** very thin **:** EMACIATED **b :** lacking usual or desirable bulk, quantity, qualities, or significance
synonym see LEAN
— skin·ni·ness *noun*

²skinny *noun* [perhaps from ¹*skin* + ⁴-y] (1938)
slang **:** inside information **:** DOPE ⟨the straight *skinny* on what's going on —John Geary⟩

skin·ny–dip \'ski-nē-ˌdip\ *intransitive verb* (1964)
: to swim in the nude
— skinny–dip *noun*
— skin·ny–dip·per \-ˌdi-pər\ *noun*

skin–pop \'skin-ˈpäp\ (circa 1952)
intransitive verb
: to inject a drug subcutaneously rather than into a vein
transitive verb
: to inject (a drug) by skin-popping
— skin–pop·per \-ˌpä-pər\ *noun*

skint \'skint\ *adjective* [alteration of *skinned,* past participle of ²*skin*] (circa 1925)
chiefly British **:** PENNILESS

skin test *noun* (1925)
: a test (as a scratch test) performed on the skin and used in detecting allergic hypersensitivity

skin·tight \'skin-'tīt\ *adjective* (1885)
: closely fitted to the figure

¹skip \'skip\ *verb* **skipped; skip·ping** [Middle English *skippen,* perhaps of Scandinavian origin; akin to Swedish dialect *skopa* to hop] (14th century)
intransitive verb
1 a : to move or proceed with leaps and bounds or with a skip **b :** to bound off one point after another **:** RICOCHET
2 : to leave hurriedly or secretly ⟨*skipped* out without paying their bill⟩
3 a : to pass over or omit an interval, item, or step **b :** to omit a grade in school in advancing to the next **c :** MISFIRE 1
transitive verb
1 a : to pass over without notice or mention **:** OMIT **b :** to pass by or leave out (a step in a progression or series)
2 a : to cause to skip (a grade in school) **b :** to cause to bound or skim over a surface ⟨*skip* a stone across a pond⟩
3 : to leap over lightly and nimbly
4 a : to depart from quickly and secretly ⟨*skipped* town⟩ **b :** to fail to attend or participate in ⟨*skip* the tournament⟩ ⟨*skip* the meeting⟩
— skip·pa·ble \'ski-pə-bəl\ *adjective*
— skip bail : to jump bail
— skip rope : to use a jump rope (as for exercise or a game)

²skip *noun* (15th century)
1 a : a light bounding step **b :** a gait composed of alternating hops and steps
2 : an act of omission or the thing omitted

³skip *noun* [short for ²*skipper*] (1830)
1 : the captain of a side in a game (as curling or lawn bowling) who advises the team as to the play and controls the action
2 : ²SKIPPER

⁴skip *transitive verb* **skipped; skip·ping** (1900)
: to act as skipper of

skip bomb *transitive verb* (1943)
: to attack by releasing delayed-action bombs from a low-flying airplane so that they skip along a land or water surface and strike a target

skip·jack \'skip-ˌjak\ *noun, plural* **skipjacks** *or* **skipjack** (1703)
1 : any of various fishes (as a ladyfish or bluefish) that jump above or are active at the surface of the water; *especially* **:** SKIPJACK TUNA
2 : a sailboat with vertical sides and a bottom similar to a flat V

skipjack tuna *noun* (1950)
: a relatively small scombroid food and sport fish (*Katsuwonus pelamis* synonym *Euthynnus pelamis*) that is bluish above and silvery below with oblique dark stripes on the sides and belly

ski pole *noun* (1920)
: one of a pair of lightweight poles used in skiing that have a handgrip and usually a wrist strap at one end and an encircling disk set above the point at the other end

¹skip·per \'ski-pər\ *noun* (13th century)
1 : any of various erratically active insects (as a click beetle or a water strider)
2 : one that skips
3 : SAURY
4 : any of a superfamily (Hesperioidea, especially family Hesperiidae) of stout-bodied lepidopterous insects that differ from the typical butterflies especially in wing venation and the form of the antennae

²skipper *noun* [Middle English, from Middle Dutch *schipper,* from *schip* ship; akin to Old English *scip* ship — more at SHIP] (14th century)
1 : the master of a ship; *especially* **:** the master of a fishing, small trading, or pleasure boat
2 : the captain or first pilot of an airplane
3 : a person in a position of leadership; *especially* **:** a baseball team's manager

³skip·per *transitive verb* **skip·pered; skip·per·ing** \'ski-p(ə-)riŋ\ (1893)
1 : to act as skipper of (as a boat)
2 : to act as coach of (as a team)

¹skirl \'skər(-ə)l, 'skir(-ə)l\ *verb* [Middle English (Scots) *skrillen, skirlen* to scream, shriek, of Scandinavian origin; akin to Old Swedish *skræll*a to rattle; akin to Old English *scrallettan* to resound loudly] (circa 1665)
intransitive verb
of a bagpipe **:** to emit the high shrill tone of the chanter; *also* **:** to give forth music
transitive verb
: to play (music) on the bagpipe

²skirl *noun* (1856)
: a high shrill sound produced by the chanter of a bagpipe

¹skir·mish \'skər-mish\ *noun* [Middle English *skyrmissh,* alteration of *skarmish,* from Middle French *escarmouche,* from Old Italian *scaramuccia*] (14th century)
1 : a minor fight in war usually incidental to larger movements
2 a : a brisk preliminary verbal conflict **b :** a minor dispute or contest between opposing parties

²skirmish *intransitive verb* (14th century)
1 : to engage in a skirmish
2 : to search about (as for supplies) **:** scout around
— skir·mish·er *noun*

¹skirr \'skər, 'skir\ *verb* [perhaps alteration of ³*scour*] (circa 1548)
intransitive verb
1 : to leave hastily **:** FLEE ⟨birds *skirred* off from the bushes —D. H. Lawrence⟩
2 : to run, fly, sail, or move along rapidly
transitive verb
1 : to search about in ⟨*skirr* the country round —Shakespeare⟩
2 a : to pass rapidly over **:** SKIM **b** *dialect* **:** to cause to skim

²skirr *noun* [probably imitative] (circa 1870)
: WHIR, ROAR

¹skirt \'skərt\ *noun* [Middle English, from Old Norse *skyrta* shirt, kirtle — more at SHIRT] (14th century)
1 a (1) **:** a free-hanging part of an outer garment or undergarment extending from the waist down (2) **:** a separate free-hanging outer garment or undergarment usually worn by women and girls covering some or all of the body from the waist down **b :** either of two usually leather flaps on a saddle covering the bars on which the stirrups are hung **c :** a cloth facing that hangs loosely and usually in folds or pleats from the bottom edge or across the front of a piece of furniture **d :** the lower branches of a tree when near the ground
2 a : the rim, periphery, or environs of an area **b** *plural* **:** outlying parts (as of a town or city)
3 : a part or attachment serving as a rim, border, or edging
4 : a girl or woman
— skirt·ed *adjective*

²skirt (1602)
transitive verb
1 : to form or run along the border or edge of **:** BORDER
2 a : to provide a skirt for **b :** to furnish a border or shield for
3 a : to go or pass around or about; *specifically* **:** to go around or keep away from in order to avoid danger or discovery **b :** to avoid especially because of difficulty or fear of controversy ⟨*skirted* the issue⟩ **c :** to evade or miss by a narrow margin ⟨having *skirted* disaster —Edith Wharton⟩
intransitive verb
: to be, lie, or move along an edge or border
— skirt·er *noun*

skirt·ing \'skər-tiŋ\ *noun* (1764)
1 : something that skirts: as **a :** BORDER, EDGING **b** *chiefly British* **:** BASEBOARD
2 : fabric suitable for skirts

skirt steak *noun* (circa 1909)
: a boneless strip of beef cut from the plate

ski run *noun* (1924)
: a slope or trail suitable for skiing

skit \'skit\ *noun* [origin unknown] (circa 1727)
1 : a jeering or satirical remark **:** TAUNT
2 a : a satirical or humorous story or sketch **b** (1) **:** a brief burlesque or comic sketch included in a dramatic performance (as a revue) (2) **:** a short serious dramatic piece; *especially* **:** one done by amateurs

ski touring *noun* (1935)
: cross-country skiing for pleasure

ski tow *noun* (1935)
1 : a motor-driven conveyor that is used for pulling skiers up a slope and that consists usually of an endless moving rope which a skier grasps
2 : SKI LIFT

skit·ter \'ski-tər\ *verb* [probably frequentative of English dialect *skite* to move quickly] (1845)
intransitive verb
1 : to glide or skip lightly or quickly
2 : to twitch the hook of a fishing line through or along the surface of water
transitive verb
: to cause to skitter

skit·tery \'ski-tə-rē\ *adjective* (1941)
: SKITTISH

skit·tish \'ski-tish\ *adjective* [Middle English] (15th century)
1 a : lively or frisky in action **:** CAPRICIOUS **b :** VARIABLE, FLUCTUATING
2 : easily frightened **:** RESTIVE ⟨a *skittish* horse⟩
3 a : COY, BASHFUL **b :** marked by extreme caution **:** WARY
— skit·tish·ly *adverb*
— skit·tish·ness *noun*

skit·tle \'ski-tᵊl\ *noun* [perhaps of Scandinavian origin; akin to Old Norse *skutill* bolt — more at SHUTTLE] (1634)
1 *plural but singular in construction* **:** English ninepins played with a wooden disk or ball
2 : one of the pins used in skittles

skive \'skīv\ *transitive verb* **skived; skiv·ing** [probably of Scandinavian origin; akin to Old Norse *skīfa* to slice] (circa 1825)
: to cut off (as leather or rubber) in thin layers or pieces **:** PARE

skiv·er \'skī-vər\ *noun* (1800)

1 : a thin soft leather made of the grain side of a split sheepskin, usually tanned in sumac and dyed
2 : one that skives something (as leather)

Skiv·vies \'ski-vēz\ *trademark*
— used for men's underwear

skiv·vy \'ski-vē\ *noun, plural* **skivvies** [origin unknown] (circa 1902)
British **:** a female domestic servant

ski·wear \'skē-ˌwar, -ˌwer\ *noun* (1961)
: clothing suitable for wear while skiing

sklent \'sklent\ *verb* [Middle English *sclenten* to strike obliquely, alteration of *slenten* — more at SLANT] (1805)
intransitive verb
1 *chiefly Scottish* **:** to look askance
2 *chiefly Scottish* **:** to cast aspersions
transitive verb
Scottish **:** to direct sideways **:** SLANT

skoal \'skōl\ *noun* [Danish *skaal,* literally, cup; akin to Old Norse *skāl* bowl — more at SCALE] (1600)
: TOAST, HEALTH — often used interjectionally

skosh \'skōsh\ *noun* [Japanese *sukoshi*] (1952)
: a small amount **:** BIT, SMIDGEN — used adverbially with *a* ⟨just a *skosh* bit shook —Josiah Bunting⟩

skua \'skyü-ə\ *noun* [New Latin, from Faeroese *skūgvur;* akin to Old Norse *skūfr* tassel, skua, Old English *scēaf* sheaf — more at SHEAF] (1678)
: either of two seabirds related to the jaegers: **a :** GREAT SKUA **b :** a bird (*Catharacta maccormicki*) that resembles but is slightly smaller than the great skua and that breeds in the Antarctic

skul·dug·gery *or* **skull·dug·gery** \ˌskəl-'də-g(ə-)rē, 'skəl ˌ\ *noun, plural* **-ger·ies** [origin unknown] (1867)
: a devious device or trick; *also* **:** underhanded or unscrupulous behavior

¹skulk \'skəlk\ *intransitive verb* [Middle English, of Scandinavian origin; akin to Danish *skulke* to shirk, play truant] (13th century)
1 : to move in a stealthy or furtive manner
2 a : to hide or conceal something (as oneself) often out of cowardice or fear or with sinister intent **b** *chiefly British* **:** MALINGER
synonym see LURK
— skulk·er *noun*

²skulk *noun* (14th century)
1 : one that skulks
2 : a group of foxes

skull \'skəl\ *noun* [Middle English *skulle,* of Scandinavian origin; akin to Swedish *skulle* skull] (13th century)
1 : the skeleton of the head of a vertebrate forming a bony or cartilaginous case that encloses and protects the brain and chief sense organs and supports the jaws
2 : the seat of understanding or intelligence **:** MIND
— skulled \'skəld\ *adjective*

skull and cross·bones \-'krȯs-ˌbōnz\ *noun, plural* **skulls and crossbones** (1826)
: a representation of a human skull over crossbones usually used as a warning of danger to life

skull·cap \'skəl-ˌkap\ *noun* (1682)
1 : a close-fitting cap; *especially* **:** a light cap without brim for indoor wear
2 : any of various mints (genus *Scutellaria*) having a calyx that when inverted resembles a helmet
3 : the upper portion of the skull **:** CALVARIUM

skull session *noun* (1937)
1 : a strategy class for an athletic team
2 : a meeting for consultation, discussion, or the interchange of ideas or information — called also *skull practice*

¹skunk \'skəŋk\ *noun, plural* **skunks** *also* **skunk** [of Algonquian origin; akin to Eastern Abenaki *segañk8* skunk] (1634)
1 a : any of various common omnivorous black-and-white New World mammals (especially genus *Mephitis*) of the weasel family that have a pair of perineal glands from which a secretion of pungent and offensive odor is ejected **b :** the fur of a skunk
2 : an obnoxious or disliked person

²skunk *transitive verb* (1843)
1 a : DEFEAT **b :** to shut out in a game
2 : to fail to pay; *also* **:** CHEAT

skunk cabbage *noun* (1751)
: either of two North American perennial herbs of the arum family that occur in shaded wet to swampy areas and have a fetid odor suggestive of a skunk: **a :** one (*Symplocarpus foetidus*) of eastern North America that sends up in spring a cowl-shaped brownish purple spathe **b :** one (*Lysichitum americana*) chiefly of the Pacific coast region that has a large yellow spathe

skunk cabbage

skunk works *noun plural but singular or plural in construction* [from the *Skonk Works,* illicit distillery in the comic strip *Li'l Abner* by Al Capp] (circa 1974)
: a usually small and often isolated department or facility (as for engineering research and development) that functions with minimal supervision within a company or corporation

¹sky \'skī\ *noun, plural* **skies** [Middle English, cloud, sky, from Old Norse *skȳ* cloud; akin to Old English *scēo* cloud] (13th century)
1 : the upper atmosphere or expanse of space that constitutes an apparent great vault or arch over the earth
2 : HEAVEN 2
3 a : weather in the upper atmosphere **b :** CLIMATE ⟨temperate English *skies* —G. G. Coulton⟩

²sky *transitive verb* **skied** *or* **skyed; sky·ing** (1802)
1 *chiefly British* **:** to throw or toss up **:** FLIP
2 : to hang (as a painting) above the line of vision
3 : to hit (a ball) high into the air

sky blue *noun* (1738)
: a pale to light blue color

sky·borne \'skī-ˌbōrn, -ˌbȯrn\ *adjective* (1589)
: AIRBORNE ⟨*skyborne* troops⟩

sky·box \-ˌbäks\ *noun* (1974)
: a roofed enclosure of private seats situated high in a sports stadium and typically featuring luxurious amenities

sky·cap \-ˌkap\ *noun* [¹*sky* + *-cap* (as in *redcap*)] (1941)
: one employed to carry hand luggage at an airport — compare REDCAP

sky·div·ing \-ˌdī-viŋ\ *noun* (1957)
: the sport of jumping from an airplane at a moderate altitude (as 6000 feet) and executing various body maneuvers before pulling the rip cord of a parachute
— sky diver *noun*

Skye terrier \'skī-\ *noun* [*Skye,* Scotland] (1847)
: any of a Scottish breed of terriers with a long low body and long straight coat

sky·ey \'skī-ē\ *adjective* (1603)
: of or resembling the sky **:** ETHEREAL

¹sky–high \'skī-'hī\ *adverb* (1818)
1 a : high into the air **b :** to a high or exorbitant level or degree
2 : in an enthusiastic manner
3 : to bits **:** APART ⟨blown *sky-high*⟩

²sky–high *adjective* (1945)
: excessively expensive **:** EXORBITANT

sky·hook \-ˌhu̇k\ *noun* (1915)
: a hook conceived as being suspended from the sky

sky·jack \'skī-ˌjak\ *transitive verb* [¹*sky* + *-jack* (as in *hijack*)] (1961)
: to commandeer (an airplane in flight) by the threat of violence
— sky·jack·er \-ˌja-kər\ *noun*
— sky·jack·ing \-kiŋ\ *noun*

¹sky·lark \'skī-ˌlärk\ *noun* (1686)
1 : a common largely brown Old World lark (*Alauda arvensis*) noted for its song especially as uttered in flight
2 : any of various birds resembling the skylark

²skylark *intransitive verb* (1809)
1 : to run up and down the rigging of a ship in sport
2 : FROLIC, SPORT
— sky·lark·er *noun*

sky·light \'skī-ˌlīt\ *noun* (1679)
1 : the diffused and reflected light of the sky
2 : an opening in a house roof or ship's deck that is covered with translucent or transparent material and that is designed to admit light

sky·light·ed \-ˌlī-təd\ *also* **sky·lit** \-ˌlit\ *adjective* (1849)
: having a skylight

sky·line \-ˌlīn\ *noun* (1824)
1 : the apparent juncture of earth and sky **:** HORIZON
2 : an outline (as of buildings or a mountain range) against the background of the sky

sky marshal *noun* (1968)
: an armed federal plainclothesman assigned to prevent skyjackings

sky pilot *noun* (1883)
: CLERGYMAN; *specifically* **:** CHAPLAIN

¹sky·rock·et \'skī-ˌrä-kət\ *noun* (1688)
: ²ROCKET 1a

²skyrocket (1851)
transitive verb
1 : to cause to rise or increase abruptly and rapidly
2 : CATAPULT
intransitive verb
: to shoot up abruptly ⟨prices are *skyrocketing*⟩

sky·sail \'skī-ˌsāl, -səl\ *noun* (1829)
: the sail above the royal

sky·scrap·er \-ˌskrā-pər\ *noun* (1883)
: a very tall building

sky·walk \-ˌwȯk\ *noun* (1953)
: a usually enclosed aerial walkway connecting two buildings

sky·ward \-wərd\ *adverb* (1582)
1 : toward the sky
2 : UPWARD

sky wave *noun* (1928)
: a radio wave that is propagated by means of the ionosphere

sky·way \'skī-ˌwā\ *noun* (1919)
1 : a route used by airplanes **:** AIR LANE
2 : an elevated highway
3 : SKYWALK

sky·write \-ˌrīt\ *verb* **-wrote** \-ˌrōt\; **-writ·ten** \-ˌri-tᵊn\; **-writ·ing** \-ˌrī-tiŋ\ [back-formation from *skywriting*] (1926)
transitive verb
: to letter by skywriting
intransitive verb
: to do skywriting
— sky·writ·er *noun*

sky·writ·ing \-ˌrī-tiŋ\ *noun* (1922)
: writing formed in the sky by means of a visible substance (as smoke) emitted from an airplane

¹slab \'slab\ *noun* [Middle English *slabbe*] (14th century)
1 : a thick plate or slice (as of stone, wood, or bread): as **a :** the outside piece cut from a log in squaring it **b :** concrete pavement (as of a road); *specifically* **:** a strip of concrete pavement laid as a single unjointed piece **c** (1) **:** a flat rectangular architectural element that is usually formed of a single piece or mass ⟨a

concrete foundation *slab*⟩ (2) **:** a rectangular building having little width with respect to its length and usually height
2 : something that resembles a slab (as in size) ⟨backed up by a solid *slab* of reference material —*Times Literary Supplement*⟩
— **slab·like** \-ˌlīk\ *adjective*

²slab *transitive verb* **slabbed; slab·bing** (1703)
1 a : to remove an outer slab from (as a log) **b :** to divide or form into slabs
2 : to cover or support (as a roadbed or roof) with slabs
3 : to put on thickly

³slab *adjective* [probably of Scandinavian origin; akin to obsolete Danish *slab* slippery] (1605)
dialect chiefly English **:** THICK, VISCOUS

¹slab·ber \ˈsla-bər\ *verb* **slab·bered; slab·ber·ing** \-b(ə-)riŋ\ [probably from Dutch *slabberen,* frequentative of *slabben* to slaver — more at SLAVER] (1542)
: SLOBBER, DROOL

²slabber *noun* (1718)
: SLOBBER, SLAVER

slab–sid·ed \ˈslab-ˈsī-dəd\ *adjective* (1817)
: having flat sides; *also* **:** being tall or long and lank

¹slack \ˈslak\ *adjective* [Middle English *slak,* from Old English *sleac;* akin to Old High German *slah* slack, Latin *laxus* slack, loose, *languēre* to languish, Greek *lagnos* lustful and perhaps to Greek *lēgein* to stop] (before 12th century)
1 : not using due diligence, care, or dispatch **:** NEGLIGENT
2 a : characterized by slowness, sluggishness, or lack of energy ⟨a *slack* pace⟩ **b :** moderate in some quality; *especially* **:** moderately warm ⟨a *slack* oven⟩ **c :** blowing or flowing at low speed ⟨the tide was *slack*⟩
3 a : not tight or taut ⟨a *slack* rope⟩ **b :** lacking in usual or normal firmness and steadiness **:** WEAK ⟨*slack* muscles⟩ ⟨*slack* supervision⟩
4 : wanting in activity **:** DULL ⟨a *slack* market⟩
5 : lacking in completeness, finish, or perfection ⟨a very *slack* piece of work⟩
synonym see NEGLIGENT
— **slack·ly** *adverb*
— **slack·ness** *noun*

²slack (13th century)
intransitive verb
1 : to be or become slack
2 : to shirk or evade work or duty
transitive verb
1 a : to be slack or negligent in performing or doing **b :** LESSEN, MODERATE
2 : to release tension on **:** LOOSEN
3 a : to cause to abate **b :** SLAKE 3

³slack *noun* (1756)
1 : cessation in movement or flow
2 : a part of something that hangs loose without strain ⟨take up the *slack* of a rope⟩
3 : trousers especially for casual wear — usually used in plural
4 : a dull season or period
5 a : a part that is available but not used ⟨some *slack* in the budget⟩ **b :** a part that is wanted but not supplied **:** SHORTFALL ⟨take up the *slack* in supplying oil⟩

⁴slack *noun* [Middle English *slak,* from Old Norse *slakki*] (14th century)
dialect English **:** a pass between hills

⁵slack *noun* [Middle English *sleck*] (15th century)
: the finest screenings of coal produced at a mine unusable as fuel unless cleaned

slack·en \ˈsla-kən\ *verb* **slack·ened; slack·en·ing** \ˈsla-kᵊn-iŋ\ (15th century)
transitive verb
1 : to make less active **:** slow up ⟨*slacken* speed at a crossing⟩
2 : to make slack (as by lessening tension or firmness) ⟨*slacken* sail⟩
intransitive verb
1 : to become slack or slow or negligent **:** slow down
2 : to become less active **:** SLACK
synonym see DELAY

slack·er \ˈsla-kər\ *noun* (1898)
: a person who shirks work or obligation; *especially* **:** one who evades military service in time of war

slack water *noun* (circa 1769)
: the period at the turn of the tide when there is little or no horizontal motion of tidal water — called also *slack tide*

slag \ˈslag\ *noun* [Middle Low German *slagge*] (1552)
: the dross or scoria of a metal

slain *past participle of* SLAY

slake \ˈslāk, *intransitive verb 2 & transitive verb 3 are also* ˈslak\ *verb* **slaked; slak·ing** [Middle English, from Old English *slacian,* from *sleac* slack] (14th century)
intransitive verb
1 *archaic* **:** SUBSIDE, ABATE
2 : to become slaked **:** CRUMBLE ⟨lime may *slake* spontaneously in moist air⟩
transitive verb
1 *archaic* **:** to lessen the force of **:** MODERATE
2 : SATISFY, QUENCH ⟨*slake* your thirst⟩ ⟨will *slake* your curiosity⟩
3 : to cause (as lime) to heat and crumble by treatment with water **:** HYDRATE

¹sla·lom \ˈslä-ləm\ *noun, often attributive* [Norwegian *slalåm,* literally, sloping track] (1921)
1 : skiing in a zigzag or wavy course between upright obstacles (as flags)
2 : a timed race (as on skis or in an automobile or kayak) over a winding or zigzag course past a series of flags or markers; *broadly* **:** movement over a zigzag route

²slalom *intransitive verb* (1932)
: to move over a zigzag course in or as if in a slalom

¹slam \ˈslam\ *noun* [origin unknown] (1660)
1 : GRAND SLAM
2 : LITTLE SLAM

²slam *noun* [probably of Scandinavian origin; akin to Icelandic *slæma* to slam] (1672)
1 : a heavy blow or impact
2 a : a noisy violent closing **b :** a banging noise; *especially* **:** one made by the slam of a door
3 : a cutting or violent criticism
4 : SLAMMER

³slam *verb* **slammed; slam·ming** (circa 1691)
transitive verb
1 : to strike or beat hard **:** KNOCK
2 : to shut forcibly and noisily **:** BANG
3 a : to set or slap down violently or noisily ⟨*slammed* down the phone⟩ **b :** to propel, thrust, or produce by or as if by striking hard ⟨*slam* on the brakes⟩ ⟨*slammed* the car into a wall⟩
4 : to criticize harshly
intransitive verb
1 : to make a banging noise
2 : to function (as in moving) with emphatic and usually noisy vigor ⟨the hurricane *slammed* into the coast⟩ ⟨*slammed* out of the room⟩
3 : to utter verbal abuse or harsh criticism

slam–bang \ˈslam-ˈbaŋ\ *adjective* (circa 1823)
1 : unduly loud or violent ⟨a *slam-bang* clatter⟩
2 : having fast-paced often nonstop action ⟨a *slam-bang* adventure novel⟩
3 : vigorously enthusiastic ⟨made a *slam-bang* effort to win⟩

slam dunk *noun* (1972)
: DUNK SHOT
— **slam–dunk** *verb*

slam·mer \ˈsla-mər\ *noun* (1952)
: JAIL, PRISON

¹slan·der \ˈslan-dər\ *transitive verb* **slan·dered; slan·der·ing** \-d(ə-)riŋ\ (13th century)
: to utter slander against **:** DEFAME
synonym see MALIGN
— **slan·der·er** \-dər-ər\ *noun*

²slander *noun* [Middle English *sclaundre, slaundre,* from Old French *esclandre,* from Late Latin *scandalum* stumbling block, offense — more at SCANDAL] (14th century)
1 : the utterance of false charges or misrepresentations which defame and damage another's reputation
2 : a false and defamatory oral statement about a person — compare LIBEL
— **slan·der·ous** \-d(ə-)rəs\ *adjective*
— **slan·der·ous·ly** *adverb*
— **slan·der·ous·ness** *noun*

¹slang \ˈslaŋ\ *noun* [origin unknown] (1756)
1 : language peculiar to a particular group: as **a :** ARGOT **b :** JARGON 2
2 : an informal nonstandard vocabulary composed typically of coinages, arbitrarily changed words, and extravagant, forced, or facetious figures of speech
— **slang** *adjective*
— **slang·i·ly** \ˈslaŋ-ə-lē\ *adverb*
— **slang·i·ness** \ˈslaŋ-ē-nəs\ *noun*
— **slangy** \ˈslaŋ-ē\ *adjective*

²slang (1828)
intransitive verb
: to use slang or vulgar abuse
transitive verb
: to abuse with harsh or coarse language

slanging match *noun* (1896)
chiefly British **:** a heated exchange of abuse

slan·guage \ˈslaŋ-gwij\ *noun* [blend of *slang* and *language*] (1879)
: slangy speech or writing

¹slant \ˈslant\ *noun* (1655)
1 : a slanting direction, line, or plane **:** SLOPE
2 a : something that slants **b :** DIAGONAL 3 **c :** a football running play in which the ballcarrier runs obliquely toward the line of scrimmage
3 a : a peculiar or personal point of view, attitude, or opinion **b :** a slanting view **:** GLANCE
— **slant** *adjective*
— **slant·ways** \-ˌwāz\ *adverb*
— **slant·wise** \-ˌwīz\ *adverb or adjective*
— **slanty** \ˈslan-tē\ *adjective*

²slant *verb* [Middle English *slenten* to fall obliquely, of Scandinavian origin; akin to Swedish *slinta* to slide] (1692)
intransitive verb
1 : to take a diagonal course, direction, or path
2 : to turn or incline from a right line or a level **:** SLOPE
transitive verb
1 : to give an oblique or sloping direction to
2 : to interpret or present in line with a special interest **:** ANGLE ⟨stories *slanted* toward youth⟩; *especially* **:** to maliciously or dishonestly distort or falsify
— **slant·ing·ly** \ˈslan-tiŋ-lē\ *adverb*

slant height *noun* (1798)
1 : the length of an element of a right circular cone
2 : the altitude of a side of a regular pyramid

¹slap \ˈslap\ *noun* [Middle English *slop,* from Middle Dutch; akin to Middle Dutch *slippen* to slip] (14th century)
dialect British **:** OPENING, BREACH

²slap *transitive verb* **slapped; slap·ping** [akin to Low German *slapp,* noun blow] (15th century)
1 a : to strike sharply with or as if with the open hand **b :** to cause to strike with a motion or sound like that of a blow with the open hand
2 : to put, place, or throw with careless haste or force ⟨*slapped* on a coat of paint⟩
3 : to assail verbally **:** INSULT
4 : to subject to a penalty — usually used with *with* ⟨*slapped* him with a $10 fine⟩

³slap *noun* (1648)
1 a : a blow with the open hand **b :** a quick sharp blow
2 : a noise like that of a slap

3 : REBUFF, INSULT
— **slap on the wrist :** a gentle usually ineffectual reprimand
[4]**slap** *adverb* [probably from Low German *slapp,* from *slapp,* noun] (1672)
: DIRECTLY, SMACK
slap·dash \'slap-'dash, -ˌdash\ *adjective* (circa 1792)
: HAPHAZARD, SLIPSHOD
slap down *transitive verb* (1842)
1 : to prohibit or restrain usually abruptly and with censure from acting in a specified way **:** SQUELCH
2 : to put an abrupt stop to **:** SUPPRESS
slap·hap·py \'slap-ˌha-pē\ *adjective* (1936)
1 : PUNCH-DRUNK
2 : buoyantly or recklessly carefree or foolish **:** HAPPY-GO-LUCKY
slap·jack \-ˌjak\ *noun* [[2]*slap* + *-jack* (as in *flapjack*)] (1796)
1 : PANCAKE
2 : a card game in which each player tries to be the first to slap a hand on any jack that appears faceup
slap shot *noun* (1942)
: a shot in ice hockey made with a swinging stroke
slap·stick \'slap-ˌstik\ *noun* (1896)
1 : a device made of two flat pieces of wood fastened at one end so as to make a loud noise when used by an actor to strike a person
2 : comedy stressing farce and horseplay; *also* **:** activity resembling slapstick
— **slapstick** *adjective*
slap–up \'slap-ˌəp\ *adjective* (circa 1823)
chiefly British **:** FIRST-RATE, BANG-UP
[1]**slash** \'slash\ *verb* [origin unknown] (1548)
intransitive verb
: to lash out, cut, or thrash about with or as if with an edged blade
transitive verb
1 : to cut with or as if with rough sweeping strokes
2 : CANE, LASH
3 : to cut slits in (as a garment) so as to reveal a color beneath
4 : to criticize cuttingly
5 : to reduce sharply **:** CUT
— **slash·er** *noun*
[2]**slash** *noun* (1576)
1 : the act of slashing; *also* **:** a long cut or stroke made by or as if by slashing
2 : an ornamental slit in a garment
3 a : an open tract in a forest strewn with debris (as from logging) **b :** the debris in such a tract
4 : DIAGONAL 3 — called also *slash mark*
[3]**slash** *noun* [origin unknown] (1652)
: a low swampy area often overgrown with brush
slash–and–burn *adjective* (1939)
: characterized or developed by felling and burning trees to clear land especially for temporary agriculture
[1]**slash·ing** \'sla-shiŋ\ *adjective* (1593)
1 : incisively satiric or critical
2 : DRIVING, PELTING
3 : VIVID, BRILLIANT
— **slash·ing·ly** \-shiŋ-lē\ *adverb*
[2]**slashing** *noun* (1596)
1 : the act or process of slashing
2 : an insert or layer of contrasting color revealed by a slash (as in a garment)
3 : SLASH 3
slash pine *noun* [[3]*slash*] (1882)
: a pine (*Pinus elliottii*) of the southeastern U.S. that has two or three needles in a cluster and is a source of turpentine and lumber
slash pocket *noun* (1799)
: a pocket suspended on the wrong side of a garment from a finished slit on the right side that serves as its opening
[1]**slat** \'slat\ *transitive verb* **slat·ted; slat·ting** [probably of Scandinavian origin; akin to Old Norse *sletta* to slap, throw] (1611)
1 : to hurl or throw smartly
2 : STRIKE, PUMMEL
[2]**slat** *noun* [Middle English, slate, from Middle French *esclat* splinter, from Old French, from *esclater* to burst, splinter] (1764)
1 : a thin narrow flat strip especially of wood or metal
2 *plural, slang* **:** RIBS
3 : an auxiliary airfoil at the leading edge of the wing of an airplane
— **slat** *adjective*
— **slat·ted** \'sla-təd\ *adjective*
[3]**slat** *transitive verb* **slat·ted; slat·ting** (1886)
: to make or equip with slats
[1]**slate** \'slāt\ *noun* [Middle English, from Middle French *esclat* splinter] (14th century)
1 : a piece of construction material (as laminated rock) prepared as a shingle for roofing and siding
2 : a dense fine-grained metamorphic rock produced by the compression of various sediments (as clay or shale) so as to develop a characteristic cleavage
3 : a tablet (as of slate) used for writing on
4 a : a written or unwritten record (as of deeds) ⟨started with a clean *slate*⟩ **b :** a list of candidates for nomination or election
5 a : a dark purplish gray **b :** any of various grays similar in color to common roofing slates
— **slate** *adjective*
— **slate·like** \-ˌlīk\ *adjective*
[2]**slate** *transitive verb* **slat·ed; slat·ing** (15th century)
1 : to cover with slate or a slatelike substance ⟨*slate* a roof⟩
2 : to designate for a specified purpose or action **:** SCHEDULE ⟨was *slated* to direct the play⟩
[3]**slate** *transitive verb* **slat·ed; slat·ing** [probably alteration of [1]*slat*] (1825)
1 : to thrash or pummel severely
2 *chiefly British* **:** to criticize or censure severely
slate black *noun* (circa 1890)
: a nearly neutral slightly purplish black
slate blue *noun* (1796)
: a grayish blue color
slat·er \'slā-tər\ *noun* (14th century)
1 : one that slates
2 [[1]*slate;* from its color] **a :** WOOD LOUSE **b :** any of various marine isopods
[1]**slath·er** \'sla-t͟hər\ *noun* [origin unknown] (1857)
: a great quantity — often used in plural
[2]**slather** *transitive verb* **slath·ered; slath·er·ing** \'sla-t͟hə-riŋ, 'slat͟h-riŋ\ (1866)
1 : to use or spend in a wasteful or lavish manner **:** SQUANDER
2 a : to spread thickly or lavishly **b :** to spread something thickly or lavishly on
slat·ing \'slā-tiŋ\ *noun* (15th century)
: the work of a slater
[1]**slat·tern** \'sla-tərn\ *noun* [probably from German *schlottern* to hang loosely, slouch; akin to Dutch *slodderen* to hang loosely, *slodder* slut] (circa 1639)
: an untidy slovenly woman; *also* **:** SLUT, PROSTITUTE
[2]**slattern** *adjective* (1716)
: SLATTERNLY
slat·tern·ly \'sla-tərn-lē\ *adjective* (circa 1680)
1 : untidy and dirty through habitual neglect; *also* **:** CARELESS, DISORDERLY
2 : of, relating to, or characteristic of a slut or prostitute
— **slat·tern·li·ness** *noun*
slaty *also* **slat·ey** \'slā-tē\ *adjective* (circa 1529)
: of, containing, or characteristic of slate; *also* **:** gray like slate
[1]**slaugh·ter** \'slȯ-tər\ *noun* [Middle English, of Scandinavian origin; akin to Old Norse *slātra* to slaughter; akin to Old English *sleaht* slaughter, *slēan* to slay — more at SLAY] (14th century)
1 : the act of killing; *specifically* **:** the butchering of livestock for market
2 : killing of great numbers of human beings (as in battle or a massacre) **:** CARNAGE
[2]**slaughter** *transitive verb* (1535)
1 : to kill (animals) for food **:** BUTCHER
2 a : to kill in a bloody or violent manner **:** SLAY **b :** to kill in large numbers **:** MASSACRE
3 : to discredit or demolish completely
— **slaugh·ter·er** \-tər-ər\ *noun*
slaugh·ter·house \'slȯ-tər-ˌhau̇s\ *noun* (14th century)
: an establishment where animals are butchered
slaugh·ter·ous \'slȯ-tə-rəs\ *adjective* (1582)
: of or relating to slaughter **:** MURDEROUS
— **slaugh·ter·ous·ly** *adverb*
Slav \'släv, 'slav\ *noun* [Middle English *Sclav,* from Medieval Latin *Sclavus,* from Late Greek *Sklabos,* from *Sklabēnoi* Slavs, of Slavic origin; akin to Old Russian *Slověne,* an East Slavic tribe] (14th century)
: a person whose native tongue is a Slavic language
word history see SLAVE
— **Slav** *adjective*
[1]**slave** \'slāv\ *noun* [Middle English *sclave,* from Old French or Medieval Latin; Old French *esclave,* from Medieval Latin *sclavus,* from *Sclavus* Slav; from the frequent enslavement of Slavs in central Europe] (14th century)
1 : a person held in servitude as the chattel of another
2 : one that is completely subservient to a dominating influence
3 : a device (as the printer of a computer) that is directly responsive to another
4 : DRUDGE, TOILER ◈
— **slave** *adjective*
[2]**slave** *verb* **slaved; slav·ing** (1602)
transitive verb
1 *archaic* **:** ENSLAVE
2 : to make directly responsive to another mechanism
intransitive verb
1 : to work like a slave **:** DRUDGE

◇ WORD HISTORY

slave The words *Slav* and *slave* both come ultimately from a Slavic word that various groups of early medieval Slavs used in order to distinguish themselves from their neighbors. An early form of this word was *Slověne,* applied to 9th century Slavs living outside the Greek city of Thessaloniki in South Slavic versions of the life of St. Cyril, the missionary who introduced Christianity and literacy to Slavic lands. *Slověne* is the likely source of Medieval Greek *Sklabēnoi,* from which shortened Medieval Latin *Sclavus* and hence the word for "Slav" in Western European languages derive. In Medieval Latin *Sclavus* shifted in sense from "Slav" to "Slav in servitude" to "slave" in the 10th century, when German princes in central Europe were expanding their territories eastward at the expense of West Slavic tribes too decentralized to resist invasion. Many Slavic prisoners of war were sold into slavery, often ending up in Muslim Spain, then the wealthiest part of western Europe. Captives were also taken from among Balkan Slavs by the Byzantine Greeks, especially after the fall of the first Bulgarian Empire in 1018, and were sold throughout the Mediterranean. Such traffic probably led to the adoption of Medieval Latin *sclavus* into Romance languages.

2 : to traffic in slaves
slave driver *noun* (1807)
1 : a supervisor of slaves at work
2 : a harsh taskmaster
slave·hold·er \ˈslāv-ˌhōl-dər\ *noun* (1776)
: an owner of slaves
— slave·hold·ing \-diŋ\ *adjective or noun*
slave–mak·ing ant \ˈslāv-ˌmā-kiŋ-\ *noun* (1817)
: an ant that attacks the colonies of ants of other species and carries off the larvae and pupae to be reared in its own nest as slaves
¹sla·ver \ˈsla-vər, ˈslā-, ˈslä-\ *verb* **sla·vered; sla·ver·ing** \-v(ə-)riŋ\ [Middle English, of Scandinavian origin; akin to Old Norse *slafra* to slaver; akin to Middle Dutch *slabben* to slaver] (14th century)
intransitive verb
: DROOL, SLOBBER
transitive verb
archaic **:** to smear with or as if with saliva
²slaver *noun* (14th century)
: saliva dribbling from the mouth
³slav·er \ˈslā-vər\ *noun* [¹*slave*] (1827)
1 a : a person engaged in the slave trade **b :** a ship used in the slave trade
2 : WHITE SLAVER
slav·ery \ˈslā-v(ə-)rē\ *noun* (1551)
1 : DRUDGERY, TOIL
2 : submission to a dominating influence
3 a : the state of a person who is a chattel of another **b :** the practice of slaveholding
slave state *noun* (1809)
1 : a state of the U.S. in which slavery was legal until the Civil War
2 : a nation subjected to totalitarian rule
slave trade *noun* (1734)
: traffic in slaves; *especially* **:** the buying and selling of blacks for profit prior to the American Civil War
slav·ey \ˈslā-vē\ *noun, plural* **slaveys** (circa 1812)
: DRUDGE; *especially* **:** a household servant who does general housework
¹Slav·ic \ˈslä-vik, ˈsla-\ *noun* (1812)
: a branch of the Indo-European language family containing Belorussian, Bulgarian, Czech, Polish, Serbo-Croatian, Slovene, Russian, and Ukrainian — see INDO-EUROPEAN LANGUAGES table
²Slavic *adjective* (1813)
: of, relating to, or characteristic of the Slavs or their languages
Slav·i·cist \ˈslä-və-sist, ˈsla-\ *noun* (1930)
: a specialist in the Slavic languages or literatures
slav·ish \ˈslā-vish *sometimes* ˈsla-\ *adjective* (1565)
1 a : of or characteristic of a slave; *especially* **:** basely or abjectly servile **b** *archaic* **:** DESPICABLE, LOW
2 *archaic* **:** OPPRESSIVE, TYRANNICAL
3 : copying obsequiously or without originality **:** IMITATIVE
synonym see SUBSERVIENT
— slav·ish·ly *adverb*
— slav·ish·ness *noun*
Slav·ist \ˈslä-vist, ˈsla-\ *noun* (1863)
: SLAVICIST
slav·oc·ra·cy \slā-ˈvä-krə-sē\ *noun* (1840)
: a faction of slaveholders and advocates of slavery in the South before the Civil War
¹Sla·von·ic \slə-ˈvä-nik\ *adjective* [New Latin *slavonicus,* from Medieval Latin *Sclavonia, Slavonia,* the Slavic-speaking countries, from *Sclavus* Slav] (circa 1645)
: SLAVIC
²Slavonic *noun* (1668)
1 : SLAVIC
2 : OLD CHURCH SLAVONIC
Slav·o·phile \ˈslä-və-ˌfīl, ˈsla-\ *or* **Slav·o·phil** \-ˌfil\ *noun* (1877)
: an admirer of the Slavs **:** an advocate of Slavophilism
Slav·oph·i·lism \sla-ˈvä-fə-ˌli-zəm; ˈsla-və-ˌfī-ˌli-, ˈslä-\ *noun* (1877)
: advocacy of Slavic and specifically Russian culture over western European culture especially as practiced among some members of the Russian intelligentsia in the middle 19th century
slaw \ˈslȯ\ *noun* (1861)
: COLESLAW
slay \ˈslā\ *verb* **slew** \ˈslü\ *also especially in sense 2* **slayed; slain** \ˈslān\; **slay·ing** [Middle English *slen,* from Old English *slēan* to strike, slay; akin to Old High German *slahan* to strike, Middle Irish *slachta* stricken] (before 12th century)
transitive verb
1 : to kill violently, wantonly, or in great numbers; *broadly* **:** to strike down **:** KILL
2 : to delight or amuse immensely ⟨*slayed* the audience⟩
intransitive verb
: KILL, MURDER
synonym see KILL
— slay·er *noun*
¹sleave \ˈslēv\ *noun* [²*sleave*] (1591)
archaic **:** SKEIN ⟨sleep that knits up the raveled *sleave* of care —Shakespeare⟩
²sleave *transitive verb* [(assumed) Middle English *sleven,* from Old English *-slǣfan* to cut] (circa 1628)
obsolete **:** to separate (silk thread) into filaments
sleave silk *noun* (1588)
obsolete **:** floss silk that is easily separated into filaments for embroidery
sleaze \ˈslēz *also* ˈslāz\ *noun* [back-formation from *sleazy*] (1954)
1 : sleazy quality, appearance, or behavior; *also* **:** sleazy material
2 : a sleazy person
sleaze·bag \-ˌbag *also* -ˌbāg\ *noun* (1983)
slang **:** a sleazy person
sleaze·ball \-ˌbȯl\ *noun* (1981)
slang **:** a sleazy person
sleazo \ˈslē-(ˌ)zō\ *adjective* (1972)
slang **:** SLEAZY
slea·zy \ˈslē-zē *also* ˈslā-\ *adjective* **slea·zi·er; -est** [origin unknown] (circa 1645)
1 a : lacking firmness of texture **:** FLIMSY **b :** carelessly made of inferior materials **:** SHODDY
2 a : marked by low character or quality ⟨*sleazy* tabloids⟩ **b :** SQUALID, DILAPIDATED ⟨*sleazy* bars⟩
— slea·zi·ly \-zə-lē\ *adverb*
— slea·zi·ness \-zē-nəs\ *noun*
¹sled \ˈsled\ *noun* [Middle English *sledde,* from Middle Dutch; akin to Old English *slīdan* to slide] (14th century)
1 : a vehicle on runners for transportation especially on snow or ice; *especially* **:** a small steerable one used especially by children for coasting down snow-covered hills
2 : ROCKET SLED
²sled *verb* **sled·ded; sled·ding** (1706)
transitive verb
: SLEDGE
intransitive verb
: to ride on a sled or sleigh
— sled·der *noun*
sledding *noun* (15th century)
1 a : the use of a sled **b :** the conditions under which one may use a sled
2 : GOING 4 ⟨tough *sledding*⟩
sled dog *noun* (1692)
: a dog trained to draw a sledge especially in the Arctic regions — called also *sledge dog*
¹sledge \ˈslej\ *noun* [Middle English *slegge,* from Old English *slecg;* akin to Old Norse *sleggja* sledgehammer, Old English *slēan* to strike — more at SLAY] (before 12th century)
: SLEDGEHAMMER
²sledge *verb* **sledged; sledg·ing** (1654)
: SLEDGEHAMMER
³sledge *noun* [Dutch dialect *sleedse;* akin to Middle Dutch *sledde* sled] (1617)
1 *British* **:** SLEIGH
2 : a strong heavy sled
⁴sledge *verb* **sledged; sledg·ing** (1853)
intransitive verb
1 : to travel with a sledge
2 *British* **:** to ride in a sleigh
transitive verb
: to transport on a sledge
¹sledge·ham·mer \ˈslej-ˌha-mər\ *noun* (15th century)
: a large heavy hammer that is wielded with both hands; *also* **:** something that resembles a sledgehammer in action
²sledgehammer (1834)
transitive verb
: to strike with or as if with a sledgehammer
intransitive verb
: to strike blows with or as if with a sledgehammer
³sledgehammer *adjective* (1843)
: marked by heavy-handed directness or hard-hitting force ⟨trusting in *sledgehammer* warfare —C. J. Rolo⟩
¹sleek \ˈslēk\ *verb* [Middle English *sleken,* alteration of *sliken* — more at SLICK] (15th century)
transitive verb
1 : SLICK ⟨grooms *sleeking* cooled horses —*Sunset*⟩
2 : to cover up **:** gloss over
intransitive verb
: SLICK
²sleek *adjective* [alteration of ²*slick*] (1589)
1 a : smooth and glossy as if polished ⟨*sleek* dark hair⟩ **b :** having a smooth well-groomed look ⟨*sleek* cattle⟩ **c :** healthy-looking
2 : smooth in speech or manner; *also* **:** UNCTUOUS
3 a : having a prosperous air ⟨a *sleek* apartment building with a pool on the roof⟩ **b :** having trim graceful lines ⟨a *sleek* car⟩ **c :** ELEGANT, STYLISH ⟨*sleek* restaurants⟩ ☆
— sleek·ly *adverb*
— sleek·ness *noun*
sleek·en \ˈslē-kən\ *transitive verb* **sleek·ened; sleek·en·ing** \ˈslē-kə-niŋ, ˈslēk-niŋ\ (1621)
: to make sleek
sleek·it \ˈslē-kət\ *adjective* [Scots, from past participle of ¹*sleek*] (1513)
1 *chiefly Scottish* **:** SLEEK, SMOOTH
2 *chiefly Scottish* **:** CRAFTY, DECEITFUL
¹sleep \ˈslēp\ *noun* [Middle English *slepe,* from Old English *slǣp;* akin to Old High German *slāf* sleep and perhaps to Latin *labi* to slip, slide] (before 12th century)
1 : the natural periodic suspension of consciousness during which the powers of the body are restored
2 : a state resembling sleep: as **a :** a state of torpid inactivity **b :** DEATH ⟨put a pet cat to *sleep*⟩; *also* **:** TRANCE, COMA **c :** the closing of leaves or petals especially at night **d :** a state marked by a diminution of feeling followed by tingling ⟨my foot's gone to *sleep*⟩ **e :** the state of an animal during hibernation
3 a : a period spent sleeping **b :** NIGHT **c :** a day's journey
— sleep·like \-ˌlīk\ *adjective*
²sleep *verb* **slept** \ˈslept\; **sleep·ing** (before 12th century)
intransitive verb
1 : to rest in a state of sleep
2 : to be in a state (as of quiescence or death) resembling sleep

☆ SYNONYMS
Sleek, slick, glossy mean having a smooth bright surface or appearance. SLEEK suggests a smoothness or brightness resulting from attentive grooming or physical conditioning ⟨a *sleek* racehorse⟩. SLICK suggests extreme smoothness that results in a slippery surface ⟨slipped and fell on the *slick* floor⟩. GLOSSY suggests a highly reflective surface ⟨photographs having a *glossy* finish⟩.

3 : to have sexual relations — usually used with *with*
transitive verb
1 : to be slumbering in ⟨*slept* the sleep of the dead⟩
2 : to get rid of or spend in or by sleep ⟨*sleep* away the hours⟩ ⟨*sleep* off a drunk⟩
3 : to provide sleeping accommodations for ⟨the boat *sleeps* six⟩
sleep around *intransitive verb* (1928)
: to engage in sex promiscuously
sleep·er \ˈslē-pər\ *noun* (12th century)
1 : one that sleeps
2 : a piece of timber, stone, or steel on or near the ground to support a superstructure, keep railroad rails in place, or receive floor joists **:** STRINGPIECE
3 : SLEEPING CAR
4 : someone or something unpromising or unnoticed that suddenly attains prominence or value ⟨the low-budget film became the summer's *sleeper*⟩
5 : children's pajamas usually with feet — usually used in plural
6 *chiefly British* **:** a small earring or stud worn to keep the hole of a pierced ear from closing
7 : MOLE 4
sleep–in \ˈslēp-ˈin\ *adjective* (1951)
: that lives at the place of employment ⟨a *sleep-in* maid⟩
sleep in *intransitive verb* (1827)
1 : to sleep where one is employed
2 a : OVERSLEEP **b :** to sleep late intentionally
sleeping bag *noun* (1850)
: a bag that is warmly lined or padded for sleeping outdoors or in a camp or tent
Sleeping Beauty *noun*
: a princess of a fairy tale who is wakened from an enchanted sleep by the kiss of a prince
sleeping car *noun* (1839)
: a railroad passenger car having berths for sleeping
sleeping partner *noun* (circa 1785)
chiefly British **:** SILENT PARTNER
sleeping pill *noun* (1664)
: a drug and especially a barbiturate that is taken as a tablet or capsule to induce sleep called also *sleeping tablet*
sleeping porch *noun* (1915)
: a porch or room having open sides or many windows arranged to permit sleeping in the open air
sleeping sickness *noun* (1875)
1 : a serious disease that is prevalent in much of tropical Africa, is marked by fever, protracted lethargy, tremors, and loss of weight, is caused by either of two trypanosomes (*Trypanosoma brucei gambiense* and *T. b. rhodesiense*), and is transmitted by tsetse flies
2 : any of various viral encephalitides or encephalomyelitides of which lethargy or somnolence is a prominent feature
sleep·less \ˈslēp-ləs\ *adjective* (15th century)
1 : not able to sleep ⟨lay *sleepless* with fever⟩
2 : affording no sleep ⟨*sleepless* nights⟩
3 : unceasingly active or operative ⟨*sleepless* casinos⟩
— sleep·less·ly *adverb*
— sleep·less·ness *noun*
sleep out *intransitive verb* (1908)
: to sleep outdoors
sleep·over \ˈslēp-ō-vər\ *noun* (1965)
1 : an overnight stay (as at another's home)
2 : an instance of hosting a sleepover in one's home
sleep·walk·er \ˈslēp-ˌwȯ-kər\ *noun* (1747)
: one that walks while or as if while asleep **:** SOMNAMBULIST
— sleep·walk \-ˌwȯk\ *intransitive verb or noun*
sleep·wear \-ˌwar, -ˌwer\ *noun* (1935)
: NIGHTCLOTHES
sleepy \ˈslē-pē\ *adjective* **sleep·i·er; -est** (13th century)
1 a : ready to fall asleep **b :** of, relating to, or characteristic of sleep
2 : sluggish as if from sleep **:** LETHARGIC; *also* **:** having little activity ⟨a *sleepy* coastal village⟩
3 : sleep-inducing
— sleep·i·ly \-pə-lē\ *adverb*
— sleep·i·ness \-pē-nəs\ *noun*
sleepy·head \ˈslē-pē-ˌhed\ *noun* (1577)
: a sleepy person
1sleet \ˈslēt\ *noun* [Middle English *slete;* akin to Middle High German *slōz* hailstone] (13th century)
: frozen or partly frozen rain
— sleety \ˈslē-tē\ *adjective*
2sleet *intransitive verb* (14th century)
: to shower sleet
sleeve \ˈslēv\ *noun* [Middle English *sleve,* from Old English *slīefe;* perhaps akin to Old English *slēfan* to slip (clothes) on, *slūpan* to slip, Old High German *sliofan,* Latin *lubricus* slippery] (before 12th century)
1 a : a part of a garment covering an arm **b :** SLEEVELET
2 a : a tubular part (as a hollow axle or a bushing) designed to fit over another part **b :** an open-ended flat or tubular packaging or cover; *especially* **:** JACKET 3c(2)
— sleeved \ˈslēvd\ *adjective*
— sleeve·less \ˈslēv-ləs\ *adjective*
— up one's sleeve : held secretly in reserve
sleeve·let \ˈslēv-lət\ *noun* (1889)
: a covering for the forearm to protect clothing from wear or dirt
1sleigh \ˈslā\ *noun* [Dutch *slee,* alteration of *slede;* akin to Middle Dutch *sledde* sled] (1703)
: an open usually horse-drawn vehicle with runners for use on snow or ice

sleigh

2sleigh *intransitive verb* (circa 1729)
: to drive or travel in a sleigh
sleigh bed *noun* (1902)
: a bed common especially in the first half of the 19th century having a solid headboard and footboard that roll outward at the top
sleigh bell *noun* (1772)
: any of various bells commonly attached to a sleigh or to the harness of a horse drawing a sleigh: as **a :** CASCABEL 2 **b :** a hemispherical bell with an attached clapper
sleight \ˈslīt\ *noun* [Middle English, from Old Norse *slœgth,* from *slœgr* sly — more at SLY] (14th century)
1 : deceitful craftiness; *also* **:** STRATAGEM
2 : DEXTERITY, SKILL
sleight of hand (circa 1605)
1 a : a conjuring trick requiring manual dexterity **b :** a cleverly executed trick or deception
2 a : skill and dexterity in conjuring tricks **b :** adroitness in deception
slen·der \ˈslen-dər\ *adjective* [Middle English *sclendre, slendre*] (14th century)
1 a : spare in frame or flesh; *especially* **:** gracefully slight **b :** small or narrow in circumference or width in proportion to length or height
2 : limited or inadequate in amount or scope **:** MEAGER ⟨people of *slender* means⟩
synonym see THIN
— slen·der·ly *adverb*
— slen·der·ness *noun*
slen·der·ize \-də-ˌrīz\ *transitive verb* **-ized; -iz·ing** (1923)
: to make slender
1sleuth \ˈslüth\ *noun* [short for *sleuthhound*] (1872)
: DETECTIVE ◆
2sleuth (1900)
intransitive verb
: to act as a detective **:** search for information
transitive verb
: to search for and discover
sleuth·hound \ˈslüth-ˌhau̇nd\ *noun* [Middle English *sleuth hund,* a kind of bloodhound, from *sleuth* track of an animal or person (from Old Norse *slōth*) + *hund* hound] (1856)
: DETECTIVE
S level *noun* (1951)
1 : the highest of three standardized British examinations in a secondary school subject used as a qualification for university entrance; *also* **:** successful completion of an S-level examination in a particular subject — called also *Scholarship level;* compare A LEVEL, O LEVEL
2 a : the level of education required to pass an S-level examination **b :** a course leading to an S-level examination
1slew \ˈslü\ *past of* SLAY
2slew *variant of* SLOUGH
3slew *verb* [origin unknown] (circa 1769)
transitive verb
1 : to turn (as a telescope or a ship's spar) about a fixed point that is usually the axis
2 : to cause to skid **:** VEER ⟨*slew* a car around a turn⟩
intransitive verb
1 : to turn, twist, or swing about **:** PIVOT
2 : SKID
4slew *noun* [perhaps from Irish *slua* army, host, throng, from Old Irish *slúag;* akin to Lithuanian *slaugyti* to tend] (1839)
: a large number ⟨a *slew* of books⟩
1slice \ˈslīs\ *verb* **sliced; slic·ing** [Middle English *sklicen,* from Middle French *esclicier* to splinter, from Old French, of Germanic origin; akin to Old High German *slīzan* to tear apart — more at SLIT] (1551)
transitive verb
1 : to cut with or as if with a knife
2 : to stir or spread with a slice
3 : to hit (a ball) so that a slice results
intransitive verb
1 : to slice something
2 : to move with a cutting action ⟨the ship *sliced* through the waves⟩
— slice·able \ˈslī-sə-bəl\ *adjective*
— slic·er *noun*
2slice *noun* [Middle English *sklice,* from Middle French *esclice* splinter, from Old French, from *esclicier*] (1613)
1 a : a thin flat piece cut from something **b :** a wedge-shaped piece (as of pie or cake)
2 : a spatula for spreading paint or ink

◇ WORD HISTORY
sleuth The Old Norse word *slōth* "track, trail" found its way into northern dialects of Middle English as *sloth,* with the same sense. It made a relatively fleeting impression, however, and after the 15th century the loanword does not appear anywhere. More lasting was the compound of *sloth,* in the Old Scots form *sleuth,* with the word *hund* "dog, hound." *Sleuthhund,* with many later variants in Scots such as *sloughhound, slughan, slow-hound,* and *slowan,* denoted a breed of bloodhound used especially in the Borders region of England and Scotland to track game or fugitives. Scots *sleuthhound* was borrowed into mainstream literary English in the 19th century, but only in the extended sense "keen investigator." In the U.S. the metaphoric *sleuthhound,* usually shortened to *sleuth,* acquired a more specific meaning and became an epithet for a detective.

3 : a serving knife with wedge-shaped blade ⟨a fish *slice*⟩
4 : a flight of a ball that deviates from a straight course in the direction of the dominant hand of the player propelling it; *also* **:** a ball following such a course — compare HOOK
5 : PORTION, SHARE ⟨a *slice* of the profits⟩

slice–of–life *adjective* (1895)
: of, relating to, or marked by the accurate transcription (as into drama) of a segment of actual life experience

[1]slick \'slik\ *verb* [Middle English *sliken;* akin to Old High German *slīhhan* to glide] (14th century)
transitive verb
: to make sleek or smooth
intransitive verb
: SPRUCE — usually used with *up*

[2]slick *adjective* (14th century)
1 a : having a smooth surface **:** SLIPPERY ⟨*slick* wet leaves⟩ **b :** having surface plausibility or appeal **:** GLIB, GLOSSY ⟨*slick* advertising⟩ **c :** based on stereotype **:** TRITE ⟨*slick* stories soon forgotten⟩
2 *archaic* **:** SLEEK 1
3 a : characterized by subtlety or nimble wit **:** CLEVER; *especially* **:** WILY ⟨a reputation as a *slick* operator⟩ **b :** DEFT, SKILLFUL ⟨a *slick* ballplayer⟩
4 : extremely good **:** FIRST-RATE
synonym see SLEEK, SLY
— slick *adverb*
— slick·ly *adverb*
— slick·ness *noun*

[3]slick *noun* (1849)
1 a : something that is smooth or slippery; *especially* **:** a smooth patch of water covered with a film of oil **b :** a film of oil
2 : an implement for producing a slick surface: as **a :** a flat paddle usually of steel for smoothing a sample of flour **b :** a foundry tool for smoothing the surface of a sand mold or unbaked core
3 : a shrewd untrustworthy person
4 : a popular magazine printed on coated stock and intended to appeal to sophisticated readers
5 : an automobile tire made without a tread for maximum traction (as in drag racing)

slick–ear \'slik-ˌir\ *noun* (1914)
: a range animal lacking an earmark

slick·en·side \'sli-kən-ˌsīd\ *noun* [English dialect *slicken* smooth (alteration of English [2]*slick*) + English *side*] (1822)
: a smooth often striated surface produced on rock by movement along a fault or a subsidiary fracture — usually used in plural

slick·er \'sli-kər\ *noun* (1881)
1 [[2]*slick*] **:** OILSKIN; *broadly* **:** RAINCOAT
2 [*slick* to defraud cleverly] **a :** a clever crook **:** SWINDLER **b :** a city dweller especially of natty appearance or sophisticated mannerisms

slick·rock \'slik-ˌräk\ *noun* (1925)
: smooth wind-polished rock

[1]slide \'slīd\ *verb* **slid** \'slid\; **slid·ing** \'slī-diŋ\ [Middle English, from Old English *slīdan;* akin to Middle High German *slīten* to slide] (before 12th century)
intransitive verb
1 a : to move smoothly along a surface **:** SLIP **b :** to coast over snow or ice **c** *of a base runner in baseball* **:** to fall or dive feetfirst or headfirst when approaching a base
2 a : to slip or fall by loss of footing **b :** to change position or become dislocated **:** SHIFT
3 a : to slither along the ground **:** CRAWL **b :** to stream along **:** FLOW
4 a : to move or pass smoothly or easily ⟨*slid* into the prepared speech⟩ **b :** to pass unnoticed or unremarked ⟨could have responded but let it *slide*⟩
5 a : to pass unobtrusively **:** STEAL **b :** to pass by gradations especially downward ⟨the economy *slid* from recession to depression⟩
transitive verb
1 a : to cause to glide or slip **b :** to traverse in a sliding manner
2 : to put unobtrusively or stealthily ⟨*slid* the bill into his hand⟩

[2]slide *noun* (1570)
1 a : an act or instance of sliding **b** (1) **:** a musical grace of two or more small notes (2) **:** PORTAMENTO
2 : a sliding part or mechanism: as **a** (1) **:** a U-shaped section of tube in the trombone that is pushed out and in to produce the tones between the fundamental and its harmonics (2) **:** a short U-shaped section of tube in a brass instrument that is used to adjust the pitch of the instrument or of individual valves **b** (1) **:** a moving piece (as the ram of a punch press) that is guided by a part along which it slides (2) **:** a guiding surface (as a feeding mechanism) along which something slides **c :** SLIDING SEAT
3 a : the descent of a mass of earth, rock, or snow down a hill or mountainside **b :** a dislocation in which one rock mass in a mining lode has slid on another **:** FAULT
4 a (1) **:** a slippery surface for coasting (2) **:** a chute with a slippery bed down which children slide in play **b :** a channel or track on which something is slid **c :** a sloping trough down which objects are carried by gravity ⟨a log *slide*⟩
5 a : a flat piece of glass on which an object is mounted for microscopic examination **b :** a photographic transparency on a small plate or film mounted for projection
6 : BOTTLENECK 3

slide fastener *noun* (1934)
: ZIPPER

slid·er \'slī-dər\ *noun* (1530)
1 : one that slides
2 : a fast baseball pitch that breaks slightly in the same direction as a curve

slide rule *noun* (1663)
: a manual device used for calculation that consists in its simple form of a ruler and a movable middle piece which are graduated with similar logarithmic scales

slide valve *noun* (1802)
: a valve that opens and closes a passageway by sliding over a port; *specifically* **:** such a valve often used in steam engines for admitting steam to the piston and releasing it

slide·way \'slīd-ˌwā\ *noun* (1856)
: a way along which something slides

sliding scale *noun* (1842)
1 : a wage scale geared to the selling price of the product or to the consumer price index but usually guaranteeing a minimum below which the wage will not fall
2 a : a system for raising or lowering tariffs in accord with price changes **b :** a flexible scale (as of fees or subsidies) adjusted to the needs or income of individuals ⟨the *sliding scale* of medical fees⟩

sliding seat *noun* (1874)
: a rower's seat (as in a racing shell) that slides fore and aft — called also *slide*

slier *comparative of* SLY

sliest *superlative of* SLY

[1]slight \'slīt\ *adjective* [Middle English, smooth, slight, probably from Middle Dutch *slicht;* akin to Old High German *slīhhan* to glide — more at SLICK] (14th century)
1 a : having a slim or delicate build **:** not stout or massive in body **b :** lacking in strength or substance **:** FLIMSY, FRAIL **c :** deficient in weight, solidity, or importance **:** TRIVIAL
2 : small of its kind or in amount **:** SCANTY, MEAGER
synonym see THIN
— slight·ly *adverb*
— slight·ness *noun*

[2]slight *transitive verb* (1597)
1 : to treat as slight or unimportant **:** make light of
2 : to treat with disdain or indifference
3 : to perform or attend to carelessly and inadequately
4 : [3]SLUR 3
synonym see NEGLECT

[3]slight *noun* (1701)
1 : an act or an instance of slighting
2 : an instance of being slighted **:** a humiliating discourtesy

slight·ing *adjective* (1632)
: characterized by disregard or disrespect **:** DISPARAGING ⟨a *slighting* remark⟩
— slight·ing·ly \'slī-tiŋ-lē\ *adverb*

sli·ly *variant of* SLYLY

[1]slim \'slim\ *adjective* **slim·mer; slim·mest** [Dutch, bad, inferior, from Middle Dutch *slimp* crooked, bad; akin to Middle High German *slimp* awry] (1657)
1 : of small diameter or thickness in proportion to the height or length **:** SLENDER
2 a : MEAN, WORTHLESS **b :** ADROIT, CRAFTY
3 a : inferior in quality or amount **:** SLIGHT **b :** SCANTY, SMALL ⟨a *slim* chance⟩
synonym see THIN
— slim·ly *adverb*
— slim·ness *noun*

[2]slim *verb* **slimmed; slim·ming** (1862)
transitive verb
: to make slender **:** decrease the size of
intransitive verb
: to become slender

[1]slime \'slīm\ *noun* [Middle English, from Old English *slīm;* akin to Middle High German *slīm* slime, Latin *limus* mud — more at LIME] (before 12th century)
1 : soft moist earth or clay; *especially* **:** viscous mud
2 : a viscous or glutinous substance: as **a :** a mucous or mucoid secretion of various animals (as slugs and catfishes) **b :** a product of wet crushing consisting of ore ground so fine as to pass a 200-mesh screen
3 : a repulsive or odious person

[2]slime *verb* **slimed; slim·ing** (1628)
transitive verb
1 : to smear or cover with slime
2 : to remove slime from (as fish for canning)
intransitive verb
: to become slimy

slime·ball \'slīm-ˌbȯl\ *noun* (1986)
slang **:** SLIME 3

slime mold *noun* (1880)
: any of a group (Myxomycetes or Mycetozoa) of organisms usually held to be lower fungi but sometimes considered protozoans that exist vegetatively as mobile plasmodia and reproduce by spores

slim–jim \'slim-ˈjim, -ˌjim\ *adjective* [[1]*slim* + *Jim,* nickname for *James*] (1889)
: notably slender

slim·mer \'sli-mər\ *noun* (1967)
chiefly British **:** a person dieting to lose weight **:** DIETER

slim·nas·tics \ˌslim-ˈnas-tiks\ *noun plural but singular in construction* [[1]*slim* + gym*nastics*] (1967)
: exercises designed to reduce one's weight

slim·sy *or* **slimp·sy** \'slim-zē, 'slim(p)-sē\ *adjective* [blend of *slim* and *flimsy*] (1845)
: FLIMSY, FRAIL

slimy \'slī-mē\ *adjective* **slim·i·er; -est** (14th century)
1 : of, relating to, or resembling slime **:** VISCOUS; *also* **:** covered with or yielding slime
2 : VILE, OFFENSIVE
— slim·i·ly \-mə-lē\ *adverb*
— slim·i·ness \-mē-nəs\ *noun*

[1]sling \'sliŋ\ *transitive verb* **slung** \'sləŋ\; **sling·ing** \'sliŋ-iŋ\ [Middle English, probably from Old Norse *slyngva* to hurl; akin to Old English & Old High German *slingan* to worm, twist, Lithuanian *slinkti*] (14th century)
1 : to cast with a sudden and usually sweeping or swirling motion ⟨*slung* the sack over my shoulder⟩
2 : to throw with or as if with a sling ⟨*slinging* punches⟩ ⟨*sling* mud⟩
synonym see THROW
— sling·er \'sliŋ-ər\ *noun*

[2]**sling** *noun* (14th century)
1 a : an instrument for throwing stones that usually consists of a short strap with strings fastened to its ends and is whirled round to discharge its missile by centrifugal force **b :** SLINGSHOT 1
2 a : a usually looped line (as of strap, chain, or rope) used to hoist, lower, or carry something; *especially* **:** a hanging bandage suspended from the neck to support an arm or hand **b :** a chain or rope attached to a lower yard at the middle and passing around a mast near the masthead to support a yard **c :** a chain hooked at the bow and stern of a boat for lowering or hoisting **d :** a device (as a rope net) for enclosing material to be hoisted by a tackle or crane
3 : a slinging or hurling of or as if of a missile

[3]**sling** *transitive verb* **slung** \ˈsləŋ\; **sling·ing** \ˈsliŋ-iŋ\ (1522)
1 : to place in a sling for hoisting or lowering
2 : to suspend by or as if by a sling

[4]**sling** *noun* [origin unknown] (1768)
: an alcoholic drink that is served hot or cold and that usually consists of liquor, sugar, lemon juice, and plain or carbonated water ⟨gin *sling*⟩ ⟨rum *sling*⟩

slings and arrows *noun plural* [from the phrase "the slings and arrows of outrageous fortune" in Shakespeare's *Hamlet*] (1963)
: pointed often acerbic critical attacks ⟨has suffered the *slings and arrows* of detractors —Roland Gelatt⟩

sling·shot \ˈsliŋ-ˌshät\ *noun* (1849)
1 : a forked stick with an elastic band attached for shooting small stones
2 a : a maneuver in auto racing in which a drafting car accelerates past the car in front by taking advantage of reserve power **b :** a dragster in which the driver sits behind the rear wheels

[1]**slink** \ˈsliŋk\ *verb* **slunk** \ˈsləŋk\ *also* **slinked** \ˈsliŋ(k)t\; **slink·ing** [Middle English, from Old English *slincan* to creep; akin to Old English *slingan* to worm, twist] (14th century)
intransitive verb
1 : to go or move stealthily or furtively (as in fear or shame) **:** STEAL
2 : to move in a sinuous provocative manner
transitive verb
: to give premature birth to — used especially of a domestic animal ⟨a cow that *slinks* her calf⟩
synonym see LURK

[2]**slink** *noun* (1607)
: the young of an animal (as a calf) brought forth prematurely; *also* **:** the flesh or skin of such an animal

[3]**slink** *adjective* (1750)
: born prematurely or abortively ⟨a *slink* calf⟩

slinky \ˈsliŋ-kē\ *adjective* **slink·i·er; -est** (1918)
1 : characterized by slinking **:** stealthily quiet ⟨*slinky* movements⟩
2 : sleek and sinuous in movement or outline; *especially* **:** following the lines of the figure in a gracefully flowing manner ⟨a *slinky* evening gown⟩
— **slink·i·ly** \-kə-lē\ *adverb*
— **slink·i·ness** \-kē-nəs\ *noun*

[1]**slip** \ˈslip\ *verb* **slipped; slip·ping** [Middle English *slippen*, from Middle Dutch or Middle Low German; akin to Middle High German *slipfen* to slide, Old High German *slīfan* to smooth, and perhaps to Greek *olibros* slippery] (14th century)
intransitive verb
1 a : to move with a smooth sliding motion **b :** to move quietly and cautiously **:** STEAL **c :** ELAPSE, PASS
2 a (1) **:** to escape from memory or consciousness (2) **:** to become uttered through inadvertence **b :** to pass quickly or easily away **:** become lost ⟨let an opportunity *slip*⟩
3 : to fall into error or fault **:** LAPSE
4 a : to slide out of place or away from a support or one's grasp **b :** to slide on or down a slippery surface ⟨*slip* on the stairs⟩ **c :** to flow smoothly
5 : to get speedily into or out of clothing ⟨*slipped* into his coat⟩
6 : to fall off from a standard or accustomed level by degrees **:** DECLINE
7 : SIDESLIP
transitive verb
1 : to cause to move easily and smoothly **:** SLIDE
2 a : to get away from **:** ELUDE, EVADE ⟨*slipped* his pursuers⟩ **b :** to free oneself from ⟨the dog *slipped* its collar⟩ **c :** to escape from (one's memory or notice) ⟨their names *slip* my mind⟩
3 : SHED, CAST ⟨the snake *slipped* its skin⟩
4 : to put on (a garment) quickly — usually used with *on* ⟨*slip* on a coat⟩
5 a : to let loose from a restraining leash or grasp **b :** to cause to slip open **:** RELEASE, UNDO ⟨*slip* a lock⟩ **c :** to let go of **d :** to disengage from (an anchor) instead of hauling
6 a : to insert, place, or pass quietly or secretly **b :** to give or pay on the sly
7 : SLINK, ABORT
8 : DISLOCATE ⟨*slipped* his shoulder⟩
9 : to transfer (a stitch) from one needle to another without working a stitch
10 : to avoid (a punch) by moving the body or head quickly to one side

[2]**slip** *noun* (15th century)
1 a : a sloping ramp extending out into the water to serve as a place for landing or repairing ships **b :** a ship's or boat's berth between two piers
2 : the act or an instance of departing secretly or hurriedly ⟨gave his pursuer the *slip*⟩
3 a : a mistake in judgment, policy, or procedure **b :** an unintentional and trivial mistake or fault **:** LAPSE ⟨a *slip* of the tongue⟩
4 : a leash so made that it can be quickly slipped
5 a : the act or an instance of slipping down or out of a place ⟨a *slip* on the ice⟩; *also* **:** a sudden mishap **b :** a movement dislocating parts (as of a rock or soil mass); *also* **:** the result of such movement **c :** a fall from some level or standard **:** DECLINE ⟨a *slip* in stock prices⟩
6 a : an undergarment made in dress length with shoulder straps; *also* **:** HALF-SLIP **b :** a case into which something is slipped; *specifically* **:** PILLOWCASE
7 a : the motion of the center of resistance of the float of a paddle wheel or the blade of an oar through the water horizontally **b :** retrograde movement of a belt on a pulley **c :** the amount of leakage past the piston of a pump or the impellers of a blower
8 : a disposition or tendency to slip easily
9 : the action of sideslipping **:** an instance of sideslipping
synonym see ERROR

[3]**slip** *noun* [Middle English *slippe*, probably from Middle Dutch or Middle Low German, split, slit, flap] (15th century)
1 a : a small shoot or twig cut for planting or grafting **:** SCION **b :** DESCENDANT, OFFSPRING
2 a : a long narrow strip of material **b :** a small piece of paper
3 : a young and slender person ⟨a *slip* of a girl⟩
4 : a long seat or narrow pew

[4]**slip** *transitive verb* **slipped; slip·ping** (1530)
: to take cuttings from (a plant) **:** divide into slips ⟨*slip* a geranium⟩

[5]**slip** *noun* [Middle English *slyp* slime, from Old English *slypa* slime, paste; akin to Old English *slūpan* to slip — more at SLEEVE] (1640)
: a mixture of finely divided clay and water used by potters (as for casting or decorating wares or in cementing separately formed parts)

slip·case \ˈslip-ˌkās\ *noun* (circa 1925)
: a protective container with one open end for books
— **slip·cased** \-ˌkāst\ *adjective*

slip·cov·er \ˈslip-ˌkə-vər\ *noun* (1886)
: a cover that may be slipped off and on; *specifically* **:** a removable covering for an article of furniture

slip·form \ˈslip-ˌfȯrm\ *transitive verb* (1962)
: to construct with the use of a slip form

slip form *noun* (1949)
: a form that is moved slowly as concrete is placed during construction (as of a building or pavement)

slip·knot \ˈslip-ˌnät\ *noun* (1659)
: a knot that slips along the rope or line around which it is made; *especially* **:** one made by tying an overhand knot around the standing part of a rope — see KNOT illustration

slip noose *noun* (1847)
: a noose with a slipknot

slip–on \ˈslip-ˌȯn, -ˌän\ *noun* (1815)
: an article of clothing that is easily slipped on or off: as **a :** a glove or shoe without fastenings **b :** a garment (as a girdle) that one steps into and pulls up **c :** PULLOVER

slip·over \-ˌō-vər\ *noun* (1917)
: a garment or cover that slips on and off easily; *specifically* **:** a pullover sweater

slip·page \ˈsli-pij\ *noun* (1850)
1 : an act, instance, or process of slipping
2 : a loss in transmission of power; *also* **:** the difference between theoretical and actual output (as of power)

slipped disk *noun* (1942)
: a protrusion of one of the cartilage disks between vertebrae with pressure on spinal nerves resulting in low back pain or sciatic pain

[1]**slip·per** \ˈsli-pər\ *adjective* [Middle English, from Old English *slipor*; akin to Middle Low German *slipper* slippery, *slippen* to slip] (before 12th century)
chiefly dialect **:** SLIPPERY

[2]**slipper** *noun* [Middle English, from *slippen* to slip] (15th century)
: a light low-cut shoe that is easily slipped on the foot
— **slip·pered** \-pərd\ *adjective*

slip·pery \ˈsli-p(ə-)rē\ *adjective* **slip·per·i·er; -est** [alteration of Middle English *slipper*] (circa 1500)
1 a : causing or tending to cause something to slide or fall ⟨*slippery* roads⟩ **b :** tending to slip from the grasp
2 : not firmly fixed **:** UNSTABLE
3 : not to be trusted **:** TRICKY
— **slip·per·i·ness** *noun*

slippery elm *noun* (1748)
: a large-leaved elm (*Ulmus rubra* synonym *U. fulva*) of eastern North America that has hard wood and fragrant mucilaginous inner bark; *also* **:** the bark

slip·py \ˈsli-pē\ *adjective* **slip·pi·er; -est** (1548)
: SLIPPERY

slip ring *noun* [[2]*slip*] (1898)
: one of two or more continuous conducting rings from which the brushes take or to which they deliver current in a generator or motor

slip–sheet \ˈslip-ˌshēt\ *transitive verb* (circa 1909)
: to insert slip sheets between (newly printed sheets)

slip sheet *noun* [[1]*slip*] (1903)
: a sheet of paper placed between newly printed sheets to prevent offsetting

slip·shod \ˈslip-ˈshäd\ *adjective* [[1]*slip*] (1580)
1 a : wearing loose shoes or slippers **b :** down at the heel **:** SHABBY
2 : CARELESS, SLOVENLY

slip·slop \-ˌsläp\ *noun* [reduplication of ²*slop*] (1675)
1 *archaic* **:** watery food **:** SLOPS
2 *archaic* **:** shallow talk or writing
— slip–slop *adjective*
slip·sole \-ˌsōl\ *noun* (circa 1908)
1 : a thin insole
2 : a half sole inserted between the insole or welt and the outsole of a shoe to give additional height — called also *slip tap*
slip stitch *noun* (circa 1882)
1 : a concealed stitch for sewing folded edges (as hems) made by alternately running the needle inside the fold and picking up a thread or two from the body of the article
2 : an unworked stitch; *especially* **:** a knitting stitch that is shifted from one needle to another without knitting it
¹slip·stream \ˈslip-ˌstrēm\ *noun* (1913)
1 : a stream of fluid (as air or water) driven aft by a propeller
2 : an area of reduced air pressure and forward suction immediately behind a rapidly moving vehicle
²slipstream *intransitive verb* (1957)
: to drive in the slipstream of a vehicle
slip·up \ˈslip-ˌəp\ *noun* (1854)
1 : MISTAKE
2 : MISCHANCE
slip up *intransitive verb* (1909)
: to make a mistake **:** BLUNDER
slip·ware \ˈslip-ˌwar, -ˌwer\ *noun* (1883)
: pottery coated with slip to improve or decorate the surface
slip·way \-ˌwā\ *noun* (1840)
: an inclined usually concrete surface for a ship being built or repaired
¹slit \ˈslit\ *noun* [Middle English, from *slitten*] (12th century)
: a long narrow cut or opening
— slit *adjective*
— slit·less \ˈslit-ləs\ *adjective*
²slit *transitive verb* **slit; slit·ting** [Middle English *slitten;* akin to Middle High German *slitzen* to slit, Old High German *slīzan* to tear apart, Old English *sciell* shell — more at SHELL] (14th century)
1 a : to make a slit in **b :** to cut off or away **:** SEVER **c :** to form into a slit
2 : to cut into long narrow strips
— slit·ter *noun*
slith·er \ˈsli-t͟hər\ *verb* [Middle English *slideren,* from Old English *slidrian,* frequentative of *slīdan* to slide] (before 12th century)
intransitive verb
1 : to slide on or as if on a loose gravelly surface
2 : to slip or slide like a snake
transitive verb
: to cause to slide
slith·ery \ˈsli-t͟hə-rē\ *adjective* (circa 1825)
: having a slippery surface, texture, or quality
slit trench *noun* (1942)
: a narrow trench especially for shelter in battle from bomb and shell fragments
¹sliv·er \ˈsli-vər, *2 is usually* ˈslī-\ *noun* [Middle English *slivere,* from *sliven* to slice off, from Old English *-slīfan;* akin to Old English *-slǣfan* to cut] (14th century)
1 a : a long slender piece cut or torn off **:** SPLINTER **b :** a small and narrow portion ⟨a *sliver* of land⟩ **c :** PARTICLE, SCRAP ⟨not a *sliver* of evidence⟩
2 : an untwisted strand or rope of textile fiber produced by a carding or combing machine and ready for drawing, roving, or spinning
²sliv·er \ˈsli-vər\ *verb* **sliv·ered; sliv·er·ing** \ˈsli-və-riŋ, ˈsliv-riŋ\ (1605)
transitive verb
: to cut into slivers **:** SPLINTER
intransitive verb
: to become split into slivers
sliv·o·vitz \ˈsli-və-ˌvits, ˈslē-, -ˌwits\ *noun* [Serbo-Croatian *šljivovica,* from *šljiva, sliva* plum; akin to Russian *sliva* plum — more at LIVID] (1885)
: a dry usually colorless plum brandy made especially in the Balkan countries
slob \ˈsläb\ *noun* [Irish *slab* mud, ooze, slovenly person] (1861)
: a slovenly or boorish person
— slob·bish \ˈslä-bish\ *adjective*
— slob·by \-bē\ *adjective*
¹slob·ber \ˈslä-bər\ *verb* **slob·bered; slob·ber·ing** \-b(ə-)riŋ\ [Middle English *sloberen* to eat in a slovenly manner; akin to Low German *slubberen* to sip] (1733)
intransitive verb
1 : to let saliva dribble from the mouth **:** DROOL
2 : to indulge the feelings effusively and without restraint
transitive verb
: to smear with or as if with dribbling saliva or food
— slob·ber·er \-bər-ər\ *noun*
²slobber *noun* (circa 1755)
1 : saliva drooled from the mouth
2 : driveling, sloppy, or incoherent utterance
— slob·bery \ˈslä-b(ə-)rē\ *adjective*
sloe \ˈslō\ *noun* [Middle English *slo,* from Old English *slāh;* akin to Old High German *slēha* sloe and probably to Russian *sliva* plum — more at LIVID] (before 12th century)
: the small dark globose astringent fruit of the blackthorn; *also* **:** BLACKTHORN
sloe–eyed \ˈslō-ˌīd\ *adjective* (1867)
1 : having soft dark bluish or purplish black eyes
2 : having slanted eyes
sloe gin *noun* (1895)
: a sweet reddish liqueur consisting of grain spirits flavored chiefly with sloes
¹slog \ˈsläg\ *verb* **slogged; slog·ging** [origin unknown] (1824)
transitive verb
1 : to hit hard **:** BEAT
2 : to plod (one's way) perseveringly especially against difficulty
intransitive verb
1 : to plod heavily **:** TRAMP ⟨*slogged* through the snow⟩
2 : to work hard and steadily **:** PLUG
— slog·ger *noun*
²slog *noun* (1888)
1 : hard persistent work
2 : a hard dogged march or tramp
slo·gan \ˈslō-gən\ *noun* [alteration of earlier *slogorn,* from Scottish Gaelic *sluagh-ghairm,* from *sluagh* army, host + *gairm* cry] (1513)
1 a : a war cry or rallying cry especially of a Scottish clan **b :** a word or phrase used to express a characteristic position or stand or a goal to be achieved
2 : a brief attention-getting phrase used in advertising or promotion ◆
slo·gan·eer \ˌslō-gə-ˈnir\ *noun* (1922)
: a maker or user of slogans
— sloganeer *intransitive verb*
slo·gan·ize \ˈslō-gə-ˌnīz\ *transitive verb* **-ized; -iz·ing** (1926)
: to express as a slogan
sloop \ˈslüp\ *noun* [Dutch *sloep*] (1629)
: a fore-and-aft rigged boat with one mast and a single jib
sloop of war (1704)
: a small warship with guns on only one deck

sloop

¹slop \ˈsläp\ *noun* [Middle English *sloppe,* probably from Middle Dutch *slop;* akin to Old English *oferslop* surplice] (14th century)
1 : a loose smock or overall
2 *plural* **:** short full breeches worn by men in the 16th century
3 *plural* **:** articles (as clothing) sold to sailors
²slop *noun* [Middle English *sloppe*] (15th century)
1 : soft mud **:** SLUSH
2 : thin tasteless drink or liquid food — usually used in plural
3 : liquid spilled or splashed
4 a : food waste (as garbage) fed to animals **:** SWILL 2a **b :** excreted body waste — usually used in plural
5 : sentimental effusiveness in speech or writing **:** GUSH
³slop *verb* **slopped; slop·ping** (1557)
transitive verb
1 a : to spill from a container **b :** to splash or spill liquid on **c :** to cause (a liquid) to splash
2 : to dish out messily
3 : to eat or drink greedily or noisily
4 : to feed slop to ⟨*slop* the hogs⟩
intransitive verb
1 : to tramp in mud or slush
2 : to become spilled or splashed
3 : to be effusive **:** GUSH
4 : to pass beyond or exceed a boundary or limit
slop basin *noun* (1731)
British **:** SLOP BOWL
slop bowl *noun* (1810)
: a bowl for receiving the leavings of tea or coffee cups at table
slop chest *noun* [¹*slop*] (1840)
: a store of clothing and personal requisites (as tobacco) carried on merchant ships for issue to the crew usually as a charge against their wages
¹slope \ˈslōp\ *adjective* [Middle English *slope,* adverb, obliquely] (1502)
: that slants **:** SLOPING — often used in combination ⟨*slope*-sided⟩
²slope *verb* **sloped; slop·ing** (1591)
intransitive verb
1 : to take an oblique course
2 : to lie or fall in a slant **:** INCLINE
3 : GO, TRAVEL ⟨*slopes* off into the night —Wolcott Gibbs⟩
transitive verb
: to cause to incline or slant
— slop·er *noun*
³slope *noun* (circa 1611)
1 : upward or downward slant or inclination or degree of slant
2 : ground that forms a natural or artificial incline
3 : the part of a continent draining to a particular ocean ⟨Alaska's North *Slope*⟩
4 a : the tangent of the angle made by a straight line with the x-axis **b :** the slope of the line tangent to a plane curve at a point

◇ WORD HISTORY

slogan The Scottish Gaelic word *sluagh-ghairm* "battle cry," a compound of *sluagh* "army, host" and *gairm* "cry," was borrowed into Scots, the English of lowland Scotland, as *slogorn, slughorne,* or, with loss of the *r, slogan.* A slogan in the Highlands and the Borders region was typically the name of a clan chief or rendezvous point, used originally as a summons to arms or a password. Made familiar to a wider English-speaking public in Walter Scott's poems *The Lay of the Last Minstrel* and *Marmion, slogan* was taken up outside of Scotland in the extended sense "word or phrase used to express a characteristic position." Irish and Scottish Gaelic *sluagh* (spelled *slua* in contemporary Irish) both go back to Old Irish *slóg* or *slúag,* which denoted the body of men who owed military service to their clan chief or king and took up arms at his call.

slope–intercept form *noun* (circa 1942)
: the equation of a straight line in the form $y = mx + b$ where m is the slope of the line and b is the point on the y-axis through which the line passes

slo–pitch \ˈslō-ˈpich, -ˌpich\ *noun* [alteration of *slow pitch*] (1967)
: SLOW-PITCH

slop jar *noun* (1855)
: a large pail used as a chamber pot or to receive waste water from a washbowl or the contents of chamber pots

slop pail *noun* (1864)
: a pail for toilet or household slops

slop·py \ˈslä-pē\ *adjective* **slop·pi·er; -est** (1707)
1 a : wet so as to spatter easily **:** SLUSHY ⟨a *sloppy* racetrack⟩ **b :** wet or smeared with or as if with something slopped over
2 : SLOVENLY, CARELESS ⟨a *sloppy* dresser⟩ ⟨did *sloppy* work⟩
3 : disagreeably effusive ⟨*sloppy* sentimentalism⟩
— slop·pi·ly \ˈslä-pə-lē\ *adverb*
— slop·pi·ness *noun*

sloppy joe \-ˈjō\ *noun* [probably from the name *Joe,* nickname for *Joseph*] (1961)
: ground beef cooked in a thick spicy sauce and usually served on a bun

slop·work \ˈsläp-ˌwərk\ *noun* (1849)
1 : the manufacture of cheap ready-made clothing
2 : hasty slovenly work

¹slosh \ˈsläsh, ˈslosh\ *noun* [probably blend of *slop* and *slush*] (1814)
1 : SLUSH
2 : the slap or splash of liquid

²slosh (1844)
intransitive verb
1 : to flounder or splash through water, mud, or slush
2 : to move with a splashing motion ⟨the water *sloshed* around him —Bill Alcine⟩
transitive verb
1 : to splash about in liquid
2 : to splash (a liquid) about or on something
3 : to splash with liquid

sloshed \ˈslāsht, ˈslosht\ *adjective* (circa 1946)
slang **:** DRUNK, INTOXICATED

¹slot \ˈslät\ *noun* [Middle English, the hollow running down the middle of the breast, from Middle French *esclot*] (1523)
1 a : a narrow opening or groove **:** SLIT, NOTCH ⟨a mail *slot* in a door⟩ **b :** a narrow passage or enclosure **c :** a passage through the wing of an airplane or of a missile that is located usually near the leading edge and formed between a main and an auxiliary airfoil for improving flow conditions over the wing so as to increase lift and delay stalling of the wing
2 : a place or position in an organization or sequence **:** NICHE
3 : SLOT MACHINE
4 : a gap between an end and a tackle in an offensive football line

²slot *transitive verb* **slot·ted; slot·ting** (1747)
1 : to cut a slot in
2 : to place in or assign to a slot

³slot *noun, plural* **slot** [Middle French *esclot* track] (1575)
: the track of an animal (as a deer)

slot·back \ˈslät-ˌbak\ *noun* (1959)
: an offensive football halfback who lines up just behind the slot between an offensive end and tackle

slot car *noun* (1966)
: an electric toy racing car with a pin underneath that fits into a groove on a track for guidance

sloth \ˈslòth, ˈsläth *also* ˈslōth\ *noun, plural* **sloths** *with* ths *or* t͟hz\ [Middle English *slouthe,* from *slow* slow] (12th century)
1 a : disinclination to action or labor **:** INDOLENCE **b :** spiritual apathy and inactivity ⟨the deadly sin of *sloth*⟩
2 : any of the slow-moving arboreal edentate mammals that comprise two genera (*Bradypus* and *Choloepus*), inhabit tropical forests of South and Central America, hang from the branches back downward, and feed on leaves, shoots, and fruits — compare THREE-TOED SLOTH, TWO-TOED SLOTH

sloth 2

sloth·ful \ˈslòth-fəl, ˈsläth- *also* ˈslōth-\ *adjective* (15th century)
: inclined to sloth **:** INDOLENT
synonym see LAZY
— sloth·ful·ly \-fə-lē\ *adverb*
— sloth·ful·ness *noun*

slot machine *noun* (1891)
1 : a machine whose operation is begun by dropping a coin into a slot
2 : a coin-operated gambling machine that pays off according to the matching of symbols on wheels spun by a handle — called also *one-armed bandit*

slot racing *noun* (1965)
: the racing of slot cars
— slot racer *noun*

¹slouch \ˈslau̇ch\ *noun* [origin unknown] (1515)
1 a : an awkward fellow **:** LOUT **b :** one that is unimpressive; *especially* **:** a lazy or incompetent person — used in negative constructions ⟨was no *slouch* at cooking⟩
2 : a gait or posture characterized by an ungainly stooping of the head and shoulders or excessive relaxation of body muscles

²slouch (1754)
intransitive verb
1 : to walk, stand, or sit with a slouch **:** assume a slouch
2 : DROOP
transitive verb
: to cause to droop ⟨*slouched* his shoulders⟩
— slouch·er *noun*

slouch hat *noun* (1837)
: a soft usually felt hat with a wide flexible brim

slouchy \ˈslau̇-chē\ *adjective* **slouch·i·er; -est** (circa 1693)
: lacking erectness or stiffness (as in form or posture) ⟨a *slouchy* sweater⟩ ⟨*slouchy* figures waiting in line⟩
— slouch·i·ly \-chə-lē\ *adverb*
— slouch·i·ness \-chē-nəs\ *noun*

¹slough \ˈslü, ˈslau̇; *in the US* (*except New England*) ˈslü *is usual for sense 1 with those to whom the sense is familiar; British usually* ˈslau̇ *for both senses*\ *noun* [Middle English *slogh,* from Old English *slōh;* akin to Middle High German *slouche* ditch] (before 12th century)
1 a : a place of deep mud or mire **b** (1) **:** SWAMP (2) **:** an inlet on a river; *also* **:** BACKWATER (3) **:** a creek in a marsh or tide flat
2 : a state of moral degradation or spiritual dejection
— sloughy \-ē\ *adjective*

²slough (1846)
transitive verb
: to engulf in a slough
intransitive verb
: to plod through or as if through mud **:** SLOG

³slough \ˈsləf\ *also* **sluff** *noun* [Middle English *slughe;* akin to Middle High German *slūch* snakeskin] (14th century)
1 : the cast-off skin of a snake
2 : a mass of dead tissue separating from an ulcer
3 : something that may be shed or cast off

⁴slough \ˈsləf\ *also* **sluff** (1720)
intransitive verb
1 a : to become shed or cast off **b :** to cast off one's skin **c :** to separate in the form of dead tissue from living tissue
2 : to crumble slowly and fall away
transitive verb
1 : to cast off
2 a : to get rid of or discard as irksome, objectionable, or disadvantageous — usually used with *off* **b :** to dispose of (a losing card in bridge) by discarding
synonym see DISCARD

slough of de·spond \ˌslau̇-əv-di-ˈspänd, ˌslü-\ [from the *Slough of Despond,* deep bog into which Christian falls on the way from the City of Destruction and from which Help saves him in the allegory *Pilgrim's Progress* (1678) by John Bunyan] (1776)
: a state of extreme depression

Slo·vak \ˈslō-ˌväk, -ˌvak\ *noun* [Slovak *slovák*] (1829)
1 : a member of a Slavic people of Slovakia
2 : the Slavic language of the Slovak people
— Slovak *adjective*
— Slo·va·ki·an \slō-ˈvä-kē-ən, -ˈva-\ *adjective or noun*

¹slov·en \ˈslə-vən\ *noun* [Middle English *sloveyn* rascal, perhaps from Flemish *sloovin* woman of low character] (15th century)
: one habitually negligent of neatness or cleanliness especially in personal appearance

²sloven *adjective* (1815)
: SLOVENLY

Slo·vene \ˈslō-ˌvēn\ *noun* [German *Slowene* from Slovene *Slovenec*] (1883)
1 : a member of a Slavic people living largely in Slovenia
2 : the language of the Slovenes
— Slovene *adjective*
— Slo·ve·nian \slō-ˈvē-nē-ən, -nyən\ *adjective or noun*

slov·en·ly \ˈslə-vən-lē *also* ˈslä-\ *adjective* (circa 1568)
1 a : untidy especially in personal appearance **b :** lazily slipshod ⟨*slovenly* in thought⟩
2 : characteristic of a sloven ⟨*slovenly* habits⟩
— slo·ven·li·ness *noun*
— slovenly *adverb*

¹slow \ˈslō\ *adjective* [Middle English, from Old English *slāw;* akin to Old High German *slēo* dull] (before 12th century)
1 a : mentally dull **:** STUPID ⟨a *slow* student⟩ **b :** naturally inert or sluggish
2 a : lacking in readiness, promptness, or willingness **b :** not hasty or precipitate ⟨was *slow* to anger⟩
3 a : moving, flowing, or proceeding without speed or at less than usual speed ⟨traffic was *slow*⟩ **b :** exhibiting or marked by low speed ⟨he moved with *slow* deliberation⟩ **c :** not acute ⟨a *slow* disease⟩ **d :** LOW, GENTLE ⟨*slow* fire⟩
4 : requiring a long time **:** GRADUAL ⟨a *slow* recovery⟩
5 : having qualities that hinder rapid progress or action ⟨a *slow* track⟩
6 a : registering behind or below what is correct ⟨the clock is *slow*⟩ **b :** less than the time indicated by another method of reckoning **c :** that is behind the time at a specified time or place
7 a : lacking in life, animation, or gaiety **:** BORING **b :** marked by reduced sales or patronage ⟨business was *slow*⟩
— slow·ish \ˈslō-ish\ *adjective*
— slow·ness *noun*

²slow *adverb* (15th century)

: SLOWLY ■

³slow (1557)
transitive verb
: to make slow or slower : slacken the speed of ⟨*slow* a car⟩ — often used with *down* or *up*
intransitive verb
: to go or become slower ⟨production of new cars *slowed* sharply⟩
synonym see DELAY

slow·down \'slō-ˌdau̇n\ *noun* (1897)
: a slowing down ⟨a business *slowdown*⟩

slow–foot·ed \-ˌfu̇-təd\ *adjective* (1642)
: moving at a very slow pace : PLODDING ⟨a *slow-footed* novel⟩ ⟨a *slow-footed* ship⟩
— **slow–foot·ed·ness** *noun*

slow·ly \-lē\ *adverb* (13th century)
: in a slow manner : not quickly, fast, early, rashly, or readily
usage see SLOW

slow match *noun* (circa 1802)
: a match or fuse made so as to burn slowly and evenly and used for firing (as of blasting charges)

slow–motion \'slō-'mō-shən\ *adjective* (1923)
: of, relating to, or being motion-picture or video photography in which the action that has been photographed is made to appear to occur slower than it actually occurred ⟨a *slow-motion* replay⟩; *also* : slowly moving ⟨a *slow-motion* dance⟩

slow motion *noun* (1924)
: slow-motion photography

slow–pitch \'slō-ˌpich\ *noun* (1967)
: softball which is played with 10 players on each side and in which each pitch must have an arc 3 to 10 feet high and base stealing is not allowed

slow·poke \'slō-ˌpōk\ *noun* [¹*slow* + *poke* annoyingly stupid person] (1848)
: a very slow person

slow–twitch \'slō-ˌtwich\ *adjective* (1971)
: of, relating to, or being muscle fiber that contracts slowly especially during sustained physical activity requiring endurance — compare FAST-TWITCH

slow virus *noun* (1954)
: any of various viruses with a long incubation period between infection and development of a degenerative disease (as kuru or Creutzfeldt-Jakob disease)

slow–wit·ted \-'wi-təd\ *adjective* (1571)
: mentally slow : DULL

slow·worm \-ˌwərm\ *noun* [Middle English *sloworm,* from Old English *slāwyrm,* from *slā-* (akin to Swedish *slå* earthworm) + *wyrm* worm] (before 12th century)
: a burrowing limbless European lizard (*Anguis fragilis*) with small eyes — called also *blindworm*

¹slub \'sləb\ *transitive verb* **slubbed; slub·bing** [back-formation from *slubbing*] (1834)
: to draw out and twist (as slivers of wool) slightly

²slub *noun* (1851)
: a soft thick uneven section in a yarn or thread

slub·ber \'slə-bər\ *transitive verb* **slub·bered; slub·ber·ing** \-b(ə-)riŋ\ [probably from obsolete Dutch *slubberen*] (1530)
1 *dialect chiefly English* : STAIN, SULLY
2 : to perform in a slipshod fashion

slub·bing \'slə-biŋ\ *noun* [origin unknown] (1786)
: ROVING

sludge \'sləj\ *noun* [Middle English *slugge,* perhaps alteration of *slicche* mud, slush; akin to Old High German *slīh* mire] (15th century)
1 : MUD, MIRE; *especially* : a muddy deposit (as on a riverbed) : OOZE
2 : a muddy or slushy mass, deposit, or sediment: as **a** : precipitated solid matter produced by water and sewage treatment processes **b** : muddy sediment in a steam boiler **c** : a precipitate or settling (as a mixture of impurities and acid) from a mineral oil
3 : SLUSH 5
— **sludgy** \'slə-jē\ *adjective*

¹slue \'slü\ *variant of* ¹SLOUGH

²slue *variant of* ³SLEW

³slue *noun* [²*slue*] (circa 1860)
1 : position or inclination after slewing
2 : SKID 5

¹slug \'sləg\ *noun* [Middle English *slugge,* of Scandinavian origin; akin to Norwegian dialect *slugga* to walk sluggishly] (15th century)
1 : SLUGGARD
2 : a lump, disk, or cylinder of material (as plastic or metal): as **a** (1) : a musket ball (2) : BULLET **b** : a piece of metal roughly shaped for subsequent processing **c** : a $50 gold piece **d** : a disk for insertion in a slot machine; *especially* : one used illegally instead of a coin

slug 3

3 : any of numerous chiefly terrestrial pulmonate gastropods (order Stylommatophora) that are found in most parts of the world where there is a reasonable supply of moisture and are closely related to the land snails but are long and wormlike and have only a rudimentary shell often buried in the mantle or entirely absent
4 : a smooth soft larva of a sawfly or moth that creeps like a mollusk
5 a : a quantity of liquor drunk in one swallow **b** : a detached mass of fluid (as water vapor or oil) that causes impact (as in a circulating system)
6 a : a strip of metal thicker than a printer's lead **b** : a line of type cast as one piece **c** : a usually temporary type line serving to instruct or identify
7 : the gravitational unit of mass in the foot-pound-second system to which a pound force can impart an acceleration of one foot per second per second and which is equal to the mass of an object weighing 32 pounds

²slug *transitive verb* **slugged; slug·ging** (1912)
1 : to add a printer's slug to
2 : to drink in gulps — often used with *down*

³slug *noun* [perhaps from *slug* to load with slugs] (1830)
: a heavy blow especially with the fist

⁴slug *transitive verb* **slugged; slug·ging** (circa 1861)
1 : to strike heavily with or as if with the fist or a bat
2 : FIGHT 4b — usually used in the phrase *slug it out*

slug·abed \'slə-gə-ˌbed\ *noun* (1592)
: a person who stays in bed after the usual or proper time to get up; *broadly* : SLUGGARD

slug·fest \'sləg-ˌfest\ *noun* (1916)
: a fight marked by the exchange of heavy blows; *also* : a heated dispute ⟨a vocal *slugfest*⟩

¹slug·gard \'slə-gərd\ *noun* [Middle English *sluggart*] (14th century)
: an habitually lazy person

²sluggard *adjective* (1593)
: SLUGGARDLY
— **slug·gard·ness** *noun*

slug·gard·ly \'slə-gərd-lē\ *adjective* (1865)
: lazily inactive

slug·ger \'slə-gər\ *noun* (1877)
: one that strikes hard or with heavy blows: as **a** : a prizefighter who punches hard but has usually little defensive skill **b** : a hard-hitting batter in baseball

slugging percentage *noun* (circa 1949)
: the ratio (as a rate per thousand) of the total number of bases reached on base hits to official times at bat for a baseball player — called also *slugging average*

slug·gish \'slə-gish\ *adjective* (15th century)
1 : averse to activity or exertion : INDOLENT; *also* : TORPID
2 : slow to respond (as to stimulation or treatment)
3 a : markedly slow in movement, flow, or growth **b** : economically inactive or slow
— **slug·gish·ly** *adverb*
— **slug·gish·ness** *noun*

¹sluice \'slüs\ *noun* [Middle English *sluse,* alteration of *scluse,* from Middle French *escluse,* from Late Latin *exclusa,* from Latin, feminine of *exclusus,* past participle of *excludere* to exclude] (15th century)
1 a : an artificial passage for water (as in a millstream) fitted with a valve or gate for stopping or regulating flow **b** : a body of water pent up behind a floodgate
2 : a dock gate : FLOODGATE
3 a : a stream flowing through a floodgate **b** : a channel to drain or carry off surplus water
4 : a long inclined trough usually on the ground; *especially* : such a contrivance paved usually with riffles to hold quicksilver for catching gold

²sluice *verb* **sluiced; sluic·ing** (1593)
transitive verb
1 : to draw off by or through a sluice
2 a : to wash with or in water running through or from a sluice **b** : to drench with a sudden flow : FLUSH
3 : to transport (as logs) in a sluice
intransitive verb
: to pour as if from a sluice

sluice·way \'slüs-ˌwā\ *noun* (1779)
: an artificial channel into which water is let by a sluice

sluicy \'slü-sē\ *adjective* (1697)
: falling copiously or in streams : STREAMING

¹slum \'sləm\ *noun, often attributive* [origin unknown] (1825)
: a densely populated usually urban area marked by crowding, dirty run-down housing, poverty, and social disorganization

²slum *intransitive verb* **slummed; slumming** (1884)
: to visit slums especially out of curiosity; *broadly* : to go somewhere or do something that might be considered beneath one's station — sometimes used with *it* ⟨*slumming* it in a cabin without plumbing⟩
— **slum·mer** *noun*

¹slum·ber \'sləm-bər\ *intransitive verb* **slumbered; slum·ber·ing** \-b(ə-)riŋ\ [Middle English, frequentative of *slumen* to doze, probably from *slume* slumber, from Old English *slūma;* akin to Middle High German *slumen* to slumber] (13th century)
1 a : to sleep lightly : DOZE **b** : SLEEP
2 a : to be in a torpid, slothful, or negligent state **b** : to lie dormant or latent
— **slum·ber·er** \-bər-ər\ *noun*

²slumber *noun* (14th century)
1 a : SLEEP **b** : a light sleep

□ USAGE
slow Some commentators claim that careful writers avoid the adverb *slow,* in spite of the fact that it has had over four centuries of usage ⟨have a continent forbearance till the speed of his rage goes *slower* —Shakespeare⟩. In actual practice, *slow* and *slowly* are not used in quite the same way. *Slow* is almost always used with verbs that denote movement or action, and it regularly follows the verb it modifies ⟨beans . . . are best cooked long and *slow* —Louise Prothro⟩. *Slowly* is used before the verb ⟨a sense of outrage, which *slowly* changed to shame —Paul Horgan⟩ and with participial adjectives ⟨a *slowly* dawning awareness . . . of the problem —*American Labor*⟩. *Slowly* is used after verbs where *slow* might also be used ⟨burn *slow* or *slowly*⟩ and after verbs where *slow* would be unidiomatic ⟨the leadership turned *slowly* toward bombing as a means of striking back —David Halberstam⟩.

2 : LETHARGY, TORPOR
slum·ber·ous *or* **slum·brous** \'sləm-b(ə-)rəs\ *adjective* (15th century)
1 : heavy with sleep **:** SLEEPY
2 : inducing slumber **:** SOPORIFIC
3 : marked by or suggestive of a state of sleep or lethargy ⟨a *slumberous* state of peace⟩
slumber party *noun* (1925)
: an overnight gathering especially of teenage girls usually at one of their homes
slum·bery \'sləm-b(ə-)rē\ *adjective* (14th century)
archaic **:** SLUMBEROUS
slum·gul·lion \'sləm-ˌgəl-yən, ˌsləm-'\ *noun* [perhaps from *slum* slime + English dialect *gullion* mud, cesspool] (1902)
: a meat stew
slum·lord \'sləm-ˌlȯrd\ *noun* [1*slum* + land*lord*] (1953)
: a landlord who receives unusually large profits from substandard properties
slum·my \'slə-mē\ *adjective* **slum·mi·er; -est** (1873)
: of, relating to, or suggestive of a slum ⟨*slummy* streets⟩
1slump \'sləmp\ *intransitive verb* [probably of Scandinavian origin; akin to Norwegian *slumpa* to fall] (circa 1677)
1 a : to fall or sink suddenly **b :** to drop or slide down suddenly **:** COLLAPSE ⟨*slumped* to the floor⟩
2 : to assume a drooping posture or carriage **:** SLOUCH
3 : to go into a slump ⟨sales *slumped*⟩
2slump *noun* (1887)
1 a : a marked or sustained decline especially in economic activity or prices ⟨a post-election *slump*⟩ **b :** a period of poor or losing play by a team or individual ⟨one spring I was in a batting *slump* —Ted Williams⟩
2 : a downward slide of a mass of rock or land
slump·fla·tion \ˌsləmp-'flā-shən\ *noun* [2*slump* + in*flation*] (1974)
: a state or period of combined economic decline and rising inflation
slung *past and past participle of* SLING
slung·shot \'sləŋ-ˌshät\ *noun* (1842)
: a striking weapon consisting of a small mass of metal or stone fixed on a flexible handle or strap
slunk *past and past participle of* SLINK
1slur \'slər\ *noun* [obsolete English dialect *slur* thin mud, from Middle English *sloor;* akin to Middle High German *slier* mud] (1609)
1 a : an insulting or disparaging remark or innuendo **:** ASPERSION **b :** a shaming or degrading effect **:** STAIN, STIGMA
2 : a blurred spot in printed matter **:** SMUDGE
2slur *verb* **slurred; slur·ring** (1660)
transitive verb
1 : to cast aspersions on **:** DISPARAGE
2 : to make indistinct **:** OBSCURE
intransitive verb
: to slip so as to cause a slur — used of a sheet being printed
3slur *verb* **slurred; slur·ring** [probably from Low German *slurrn* to shuffle; akin to Middle English *sloor* mud] (1660)
transitive verb
1 a : to slide or slip over without due mention, consideration, or emphasis ⟨*slurred* over certain facts⟩ **b :** to perform hurriedly **:** SKIMP ⟨let him not *slur* his lesson —R. W. Emerson⟩
2 : to perform (successive tones of different pitch) in a smooth or connected manner
3 a : to reduce, make a substitution for, or omit (sounds that would normally occur in an utterance) **b :** to utter with such reduction, substitution, or omission of sounds ⟨his speech was *slurred*⟩
intransitive verb
1 *dialect chiefly English* **:** SLIP, SLIDE
2 : DRAG, SHUFFLE
4slur *noun* (circa 1801)
1 a : a curved line connecting notes to be sung to the same syllable or performed without a break **b :** the combination of two or more slurred tones
2 : a slurring manner of speech
slurp \'slərp\ *verb* [Dutch *slurpen;* akin to Middle Low German *slorpen* to slurp] (1648)
intransitive verb
: to make a sucking noise while eating or drinking
transitive verb
: to eat or drink noisily or with a sucking sound
— slurp *noun*
1slur·ry \'slər-ē, 'slə-rē\ *noun, plural* **slur·ries** [Middle English *slory*] (15th century)
: a watery mixture of insoluble matter (as mud, lime, or plaster of paris)
2slurry *transitive verb* **slur·ried; slur·ry·ing** (1947)
: to convert into a slurry
1slush \'sləsh\ *noun* [perhaps of Scandinavian origin; akin to Norwegian *slusk* slush] (1641)
1 a : partly melted or watery snow **b :** loose ice crystals formed during the early stages of freezing of salt water
2 : soft mud **:** MIRE
3 : refuse grease and fat from cooking especially on shipboard
4 : paper pulp in water suspension
5 : trashy and usually cheaply sentimental material
6 : unsolicited writings submitted (as to a magazine) for publication
2slush (1807)
transitive verb
: to wet, splash, or paint with slush
intransitive verb
1 : to make one's way through slush
2 : to make a splashing sound
slush fund *noun* (1864)
1 : a fund raised from the sale of refuse to obtain small luxuries or pleasures for a warship's crew
2 : a fund for bribing public officials or carrying on corruptive propaganda
slushy \'slə-shē\ *adjective* **slush·i·er; -est** (1791)
: being, involving, or resembling slush: as **a :** full of or covered with slush ⟨*slushy* streets⟩ **b :** made up of or having the consistency of slush ⟨*slushy* snow⟩ ⟨a *slushy* mixture⟩ **c :** having a cheaply sentimental quality **:** TRASHY ⟨a *slushy* novel⟩
— slush·i·ness *noun*
slut \'slət\ *noun* [Middle English *slutte*] (15th century)
1 *chiefly British* **:** a slovenly woman
2 a : a promiscuous woman; *especially* **:** PROSTITUTE **b :** a saucy girl **:** MINX
— slut·tish \'slə-tish\ *adjective*
— slut·tish·ly *adverb*
— slut·tish·ness *noun*
— slut·ty \'slə-tē\ *adjective*
sly \'slī\ *adjective* **sli·er** *or* **sly·er** \'slī(-ə)r\; **sli·est** *or* **sly·est** \'slī-əst\ [Middle English *sli,* from Old Norse *slœgr;* akin to Old English *slēan* to strike — more at SLAY] (13th century)
1 *chiefly dialect* **a :** wise in practical affairs **b :** displaying cleverness **:** INGENIOUS
2 a : clever in concealing one's aims or ends **:** FURTIVE ⟨the *sly* fox⟩ **b :** lacking in straightforwardness and candor **:** DISSEMBLING ⟨a *sly* scheme⟩
3 : lightly mischievous **:** ROGUISH ⟨a *sly* jest⟩ ⟨a *sly* smile⟩ ☆
— sly·ly *adverb*
— sly·ness *noun*
— on the sly : in a manner intended to avoid notice
sly·boots \'slī-ˌbüts\ *noun plural but singular in construction* (circa 1700)
: a sly tricky person; *especially* **:** one who is cunning or mischievous in an engaging way
1smack \'smak\ *noun* [Middle English, from Old English *smæc;* akin to Old High German *smac* taste and probably to Lithuanian *smaguris* sweet tooth] (before 12th century)
1 : characteristic taste or flavor; *also* **:** a perceptible taste or tincture
2 : a small quantity
2smack *intransitive verb* (13th century)
1 : to have a taste or flavor
2 : to have a trace, vestige, or suggestion ⟨a proposal that *smacks* of treason⟩
3smack *noun* [Dutch *smak* or Low German *smack*] (1533)
: a sailing ship (as a sloop or cutter) used chiefly in coasting and fishing
4smack *verb* [akin to Middle Dutch *smacken* to strike] (1557)
transitive verb
1 : to close and open (lips) noisily and in rapid succession especially in eating
2 a : to kiss with or as if with a smack **b :** to strike so as to produce a smack
intransitive verb
: to make or give a smack
5smack *noun* (1570)
1 : a quick sharp noise made by rapidly compressing and opening the lips
2 : a loud kiss
3 : a sharp slap or blow
6smack *adverb* (1782)
: squarely and sharply **:** DIRECTLY ⟨*smack* in the middle⟩
7smack *noun* [perhaps from Yiddish *shmek* sniff, whiff, pinch (of snuff)] (circa 1960)
slang **:** HEROIN
smack–dab \'smak-'dab\ *adverb* (1892)
: EXACTLY, SQUARELY
smack·er \'sma-kər\ *noun* (1611)
1 : one that smacks
2 *slang* **:** DOLLAR
smack·ing \'sma-kiŋ\ *adjective* (1820)
: BRISK, LIVELY ⟨a *smacking* breeze⟩
1small \'smȯl\ *adjective* [Middle English *smal,* from Old English *smæl;* akin to Old High German *smal* small, Greek *mēlon* small domestic animal] (before 12th century)
1 a : having comparatively little size or slight dimensions **b :** LOWERCASE
2 a : minor in influence, power, or rank **b :** operating on a limited scale
3 : lacking in strength ⟨a *small* voice⟩
4 a : little or close to zero in an objectively measurable aspect (as quantity, amount, or value) **b :** made up of few or little units

☆ SYNONYMS
Sly, cunning, crafty, wily, tricky, foxy, artful, slick mean attaining or seeking to attain one's ends by guileful or devious means. SLY implies furtiveness, lack of candor, and skill in concealing one's aims and methods ⟨a *sly* corporate raider⟩. CUNNING suggests the inventive use of sometimes limited intelligence in overreaching or circumventing ⟨the *cunning* fox avoided the trap⟩. CRAFTY implies cleverness and subtlety of method ⟨a *crafty* lefthander⟩. WILY implies skill and deception in maneuvering ⟨the *wily* fugitive escaped the posse⟩. TRICKY is more likely to suggest shiftiness and unreliability than skill in deception and maneuvering ⟨a *tricky* political operative⟩. FOXY implies a shrewd and wary craftiness usually involving devious dealing ⟨a *foxy* publicity man planting stories⟩. ARTFUL implies indirectness in dealing and often connotes sophistication or cleverness ⟨elicited the information by *artful* questioning⟩. SLICK emphasizes smoothness and guile ⟨*slick* operators selling time-sharing⟩.

5 a : of little consequence **:** TRIVIAL, INSIGNIFICANT **b :** HUMBLE, MODEST ⟨a *small* beginning⟩
6 : limited in degree
7 a : MEAN, PETTY **b :** reduced to a humiliating position ☆
— **small·ish** \ˈsmȯ-lish\ *adjective*
— **small·ness** \ˈsmȯl-nəs\ *noun*

²small *adverb* (before 12th century)
1 : in or into small pieces
2 : without force or loudness ⟨speak as *small* as you will —Shakespeare⟩
3 : in a small manner

³small *noun* (14th century)
1 : a part smaller and especially narrower than the remainder ⟨the *small* of the back⟩
2 a *plural* **:** small-sized products **b** *plural, chiefly British* **:** SMALLCLOTHES; *especially* **:** UNDERWEAR

small arm *noun* (1689)
: a handheld firearm (as a handgun or shoulder arm) — usually used in plural

small beer *noun* (1568)
1 : weak or inferior beer
2 : something of small importance **:** TRIVIA
— **small–beer** *adjective*

small calorie *noun* (circa 1889)
: CALORIE 1a

small capital *noun* (1770)
: a letter having the form of but smaller than a capital letter (as in THESE WORDS) — called also *small cap*

small change *noun* (1819)
1 : coins of low denomination
2 : something trifling or petty

small–claims court *noun* (1925)
: a special court intended to simplify and expedite the handling of small claims on debts — called also *small-debts court*

small·clothes \ˈsmȯl-ˌklō(th)z\ *noun plural* (1796)
1 : close-fitting knee breeches worn in the 18th century
2 : small articles of clothing (as underclothing or handkerchiefs)

smaller European elm bark beetle *noun* (circa 1945)
: ELM BARK BEETLE b

small–fry \ˈsmȯl-ˌfrī\ *adjective* (1817)
1 : MINOR, UNIMPORTANT ⟨a *small-fry* politician⟩
2 : of, relating to, or intended for children **:** CHILDISH

small·hold·ing \-ˌhōl-diŋ\ *noun* (1892)
chiefly British **:** a small farm
— **small·hold·er** \-dər\ *noun*

small hours *noun plural* (circa 1837)
: the early morning hours

small intestine *noun* (1767)
: the part of the intestine that lies between the stomach and colon, consists of duodenum, jejunum, and ileum, secretes digestive enzymes, and is the chief site of the absorption of digested nutrients

small–mind·ed \ˈsmȯl-ˈmīn-dəd\ *adjective* (1847)
1 : having narrow interests, sympathies, or outlook
2 : typical of a small-minded person **:** marked by pettiness, narrowness, or meanness ⟨*small-minded* conduct⟩
— **small–mind·ed·ly** *adverb*
— **small–mind·ed·ness** *noun*

small·mouth bass \ˈsmȯl-ˌmau̇th-\ *noun* (1938)
: a black bass (*Micropterus dolomieu*) of clear rivers and lakes that is bronzy green above and lighter below and has the vertex of the angle of the jaw falling below the eye — called also *smallmouth, smallmouth black bass*

smallmouth bass

small octave *noun* (circa 1890)
: the musical octave that begins on the first C below middle C — see PITCH illustration

small potato *noun* (1831)
: one that is of trivial importance or worth — usually used in plural but singular or plural in construction

small·pox \ˈsmȯl-ˌpäks\ *noun* (1518)
: an acute contagious febrile disease caused by a poxvirus and characterized by skin eruption with pustules, sloughing, and scar formation

small–scale \-ˈskā(ə)l\ *adjective* (1852)
1 : small in scope; *especially* **:** small in output or operation
2 *of a map* **:** having a scale (as one inch to 25 miles) that permits plotting of comparatively little detail and shows mainly large features

small screen *noun* (1956)
: TELEVISION

small stuff *noun* (circa 1857)
: small rope (as spun yarn or marline) usually identified by the number of threads or yarns which it contains

small·sword \ˈsmȯl-ˌsōrd, -ˌsȯrd\ *noun* (1687)
: a light tapering sword for thrusting used chiefly in dueling and fencing

small talk *noun* (1751)
: light or casual conversation **:** CHITCHAT

small–time \ˈsmȯl-ˈtīm\ *adjective* (1910)
: insignificant in performance, scope, or standing **:** PETTY ⟨*small-time* thieves⟩
— **small–tim·er** \-ˈtī-mər\ *noun*

smalt \ˈsmȯlt\ *noun* [Middle French, from Old Italian *smalto,* of Germanic origin; akin to Old High German *smelzan* to melt — more at SMELT] (1558)
: a deep blue pigment used especially as a ceramic color and prepared by fusing together silica, potash, and oxide of cobalt and grinding to powder the resultant glass

smalt·ite \ˈsmȯl-ˌtīt\ *noun* [alteration of *smaltine,* from French, from *smalt,* from Middle French] (1868)
: a bluish white or gray isometric mineral of metallic luster that is essentially an arsenide of cobalt and nickel

smal·to \ˈsmäl-(ˌ)tō, ˈsmȯl-\ *noun, plural* **smal·ti** \-(ˌ)tē\ [Italian, smalt, smalto] (1705)
: colored glass or enamel or a piece of either used in mosaic work

sma·ragd \smə-ˈragd, ˈsmar-ˌagd\ *noun* [Middle English *smaragde,* from Latin *smaragdus*] (13th century)
: EMERALD
— **sma·rag·dine** \smə-ˈrag-dən, ˈsmar-əg-ˌdīn\ *adjective*

sma·rag·dite \smə-ˈrag-ˌdīt, ˈsmar-əg-ˌdīt\ *noun* [French, from Latin *smaragdus* emerald — more at EMERALD] (1804)
: a green foliated amphibole

smarm \ˈsmärm\ *noun* [back-formation from *smarmy*] (1937)
: smarmy language or behavior

smarmy \ˈsmär-mē\ *adjective* [*smarm* to gush, slobber] (1924)
1 : revealing or marked by a smug, ingratiating, or false earnestness ⟨a tone of *smarmy* self-satisfaction —*New Yorker*⟩
2 : of low sleazy taste or quality ⟨*smarmy* eroticism⟩
— **smarm·i·ly** \-mə-lē\ *adverb*
— **smarm·i·ness** \-mē-nəs\ *noun*

¹smart \ˈsmärt\ *adjective* (before 12th century)
1 : making one smart **:** causing a sharp stinging
2 : marked by often sharp forceful activity or vigorous strength ⟨a *smart* pull of the starter cord⟩
3 : BRISK, SPIRITED
4 a : mentally alert **:** BRIGHT **b :** KNOWLEDGEABLE **c :** SHREWD ⟨a *smart* investment⟩
5 a : WITTY, CLEVER **b :** PERT, SAUCY ⟨don't get *smart* with me⟩
6 a : NEAT, TRIM **b :** stylish or elegant in dress or appearance **c** (1) **:** SOPHISTICATED (2) **:** characteristic of or patronized by fashionable society
7 a : being a guided missile ⟨a laser-guided *smart* bomb⟩ **b :** operating by automation ⟨a *smart* machine tool⟩ **c :** INTELLIGENT 3
— **smart·ly** *adverb*
— **smart·ness** *noun*

²smart *intransitive verb* [Middle English *smerten,* from Old English *smeortan;* akin to Old High German *smerzan* to pain] (13th century)
1 : to cause or be the cause or seat of a sharp poignant pain; *also* **:** to feel or have such a pain
2 a : to feel or endure distress, remorse, or embarrassment ⟨*smarting* from wounded vanity —W. L. Shirer⟩ **b :** to pay a heavy or stinging penalty ⟨would have to *smart* for this foolishness⟩

³smart *noun* (13th century)
1 : a smarting pain; *especially* **:** a stinging local pain
2 : poignant grief or remorse ⟨was not the sort to get over *smarts* —Sir Winston Churchill⟩
3 : an affectedly witty or fashionable person
4 *plural, slang* **:** INTELLIGENCE, KNOW-HOW

⁴smart *adverb* (13th century)
: in a smart manner **:** SMARTLY

smart al·eck *also* **smart al·ec** \ˈsmärt-ˌa-lik, -ˌe-\ *noun* [*Aleck,* nickname for *Alexander*] (1865)
: an obnoxiously conceited and self-assertive person with pretensions to smartness or cleverness
— **smart–aleck** *adjective*
— **smart–al·ecky** \-ˌa-lə-kē, -ˌe-\ *adjective*

smart–ass \-ˌas\ *noun* (1964)
: SMART ALECK
— **smart–ass** *adjective*
— **smart–assed** \-ˌast\ *adjective*

smart card *noun* (1980)
: a small plastic card that has a built-in microprocessor to store and process data and records

smart·en \ˈsmär-tᵊn\ *verb* **smart·ened; smart·en·ing** \ˈsmärt-niŋ, ˈsmär-tᵊn-iŋ\ (1815)
transitive verb
: to make smart or smarter; *especially* **:** SPRUCE — usually used with *up*
intransitive verb
: to smarten oneself — used with *up*

¹smart money \ˈsmärt-ˌmə-nē\ *noun* [³*smart*] (1693)
: PUNITIVE DAMAGES

²smart money \-ˈmə-nē, -ˌmə-nē\ *noun* [¹*smart*] (1926)
1 : money ventured by one having inside information or much experience
2 : well-informed bettors or speculators

smart·weed \ˈsmärt-ˌwēd\ *noun* (circa 1787)

☆ SYNONYMS
Small, little, diminutive, minute, tiny, miniature mean noticeably below average in size. SMALL and LITTLE are often interchangeable, but SMALL applies more to relative size determined by capacity, value, number ⟨a relatively *small* backyard⟩. LITTLE is more absolute in implication often carrying the idea of petiteness, pettiness, insignificance, or immaturity ⟨your pathetic *little* smile⟩. DIMINUTIVE implies abnormal smallness ⟨*diminutive* bonsai plants⟩. MINUTE implies extreme smallness ⟨a *minute* amount of caffeine in the soda⟩. TINY is an informal equivalent to MINUTE ⟨*tiny* cracks formed in the painting⟩. MINIATURE applies to an exactly proportioned reproduction on a very small scale ⟨a dollhouse complete with *miniature* furnishings⟩.

: any of various polygonums with strong acid juice

smarty *or* **smart·ie** \ˈsmär-tē\ *noun, plural* **smart·ies** (1861)
: SMART ALECK

smarty–pants \-ˌpan(t)s\ *noun plural but singular in construction* (1941)
: SMART ALECK
— **smarty–pants** *adjective*

¹smash \ˈsmash\ *noun* [perhaps blend of ⁴*smack* and ²*mash*] (1725)
1 a : a smashing blow or attack **b :** a hard overhand stroke (as in tennis or badminton)
2 a : the action or sound of smashing; *especially* **:** a wreck due to collision **:** CRASH **b :** utter collapse **:** RUIN
3 : a striking success

²smash (1778)
transitive verb
1 : to break or crush by violence
2 a : to drive or throw violently especially with a shattering or battering effect; *also* **:** to effect in this way **b :** to hit violently **:** BATTER **c** (1) **:** to hit (as a tennis ball) with a hard overhand stroke (2) **:** to drive (a ball) with a forceful stroke
3 : to destroy utterly **:** WRECK
intransitive verb
1 : to move or become propelled with violence or crashing effect ⟨*smashed* into a tree⟩
2 : to become wrecked
3 : to go to pieces suddenly under collision or pressure
— **smash·er** *noun*

³smash *adjective* (1923)
: being a smash **:** OUTSTANDING ⟨a *smash* hit⟩

smashed \ˈsmasht\ *adjective* (circa 1959)
slang **:** DRUNK, INTOXICATED

smash·ing \ˈsma-shiŋ\ *adjective* (1833)
1 : that smashes **:** CRUSHING ⟨a *smashing* defeat⟩
2 : extraordinarily impressive or effective ⟨a *smashing* performance⟩
— **smash·ing·ly** \-shiŋ-lē\ *adverb*

smash–up \ˈsmash-ˌəp\ *noun* (1856)
1 : a collision between vehicles
2 : a complete collapse

¹smat·ter \ˈsma-tər\ *verb* [Middle English *smateren*] (15th century)
intransitive verb
: to talk superficially **:** BABBLE
transitive verb
1 : to speak with spotty or superficial knowledge ⟨*smatters* French⟩
2 : to dabble in
— **smat·ter·er** \-tər-ər\ *noun*

²smatter *noun* (1668)
: SMATTERING

smat·ter·ing \ˈsma-tə-riŋ\ *noun* (1538)
1 : superficial piecemeal knowledge ⟨a *smattering* of carpentry, house painting, bricklaying —Alva Johnston⟩
2 : a small scattered number or amount ⟨a *smattering* of spectators⟩

¹smear \ˈsmir\ *noun* [Middle English *smere,* from Old English *smeoru;* akin to Old High German *smero* grease and probably to Old Irish *smiur* marrow] (before 12th century)
1 a : a viscous or sticky substance **b :** a spot made by or as if by an unctuous or adhesive substance
2 : material smeared on a surface (as of a microscopic slide); *also* **:** a preparation made by smearing material on a surface ⟨a vaginal *smear*⟩
3 : a usually unsubstantiated charge or accusation against a person or organization

²smear *transitive verb* (before 12th century)
1 a : to overspread with something unctuous, viscous, or adhesive **:** DAUB **b :** to spread over a surface
2 a : to stain, smudge, or dirty by or as if by smearing **b :** SULLY, BESMIRCH; *specifically* **:** to vilify especially by secretly and maliciously spreading grave charges and imputations
3 : to obliterate, obscure, blur, blend, wipe out, or defeat by or as if by smearing
— **smear·er** *noun*

smear·case *also* **smier·case** \ˈsmir-ˌkās\ *noun* [modification of German *Schmierkäse,* from *schmieren* to smear + *Käse* cheese] (1829)
chiefly Midland **:** COTTAGE CHEESE

smeary \ˈsmir-ē\ *adjective* (circa 1529)
1 : marked by or covered with smears
2 : liable to cause smears ⟨*smeary* lipstick⟩

smec·tic \ˈsmek-tik\ *adjective* [Latin *smecticus* cleansing, having the properties of soap, from Greek *smēktikos,* from *smēchein* to clean] (1923)
: of, relating to, or being the phase of a liquid crystal characterized by arrangement of molecules in layers with the long molecular axes in a given layer being parallel to one another and those of other layers and perpendicular or slightly inclined to the plane of the layer — compare CHOLESTERIC, NEMATIC

smec·tite \ˈsmek-ˌtīt\ *noun* [*smectis* fuller's earth, modification of Greek *smēktris* kind of fuller's earth, from *smēchein* to clean] (1811)
: MONTMORILLONITE
— **smec·tit·ic** \(ˌ)smek-ˈti-tik\ *adjective*

smeg·ma \ˈsmeg-mə\ *noun* [New Latin, from Latin, detergent, soap, from Greek *smēgma,* from *smēchein* to wash off, clean] (circa 1819)
: the secretion of a sebaceous gland; *specifically* **:** the cheesy sebaceous matter that collects between the glans penis and the foreskin or around the clitoris and labia minora

¹smell \ˈsmel\ *verb* **smelled** \ˈsmeld\ *or* **smelt** \ˈsmelt\; **smell·ing** [Middle English] (12th century)
transitive verb
1 : to perceive the odor or scent of through stimuli affecting the olfactory nerves **:** get the odor or scent of with the nose
2 : to detect or become aware of as if by the sense of smell
3 : to emit the odor of
intransitive verb
1 : to exercise the sense of smell
2 a (1) **:** to have an odor or scent (2) **:** to have a characteristic aura or atmosphere **:** SMACK ⟨the accounts . . . seemed to me to *smell* of truth —R. S. Bourne⟩; *also* **:** SEEM, APPEAR ⟨the story didn't *smell* right⟩ **b** (1) **:** to have an offensive odor **:** STINK (2) **:** to be of bad or questionable quality ⟨all this from the moral point of view *smells* —A. F. Wills⟩
— **smell·er** *noun*
— **smell a rat :** to have a suspicion of something wrong

²smell *noun* (12th century)
1 : the property of a thing that affects the olfactory organs **:** ODOR
2 a : the process, function, or power of smelling **b :** the special sense concerned with the perception of odor
3 a : a very small amount **:** TRACE ⟨add only a *smell* of garlic⟩ **b :** a pervading or characteristic quality **:** AURA ⟨the *smell* of affluence, of power —Harry Hervey⟩
4 : an act or instance of smelling ☆

smelling salts *noun plural but singular or plural in construction* (1840)
: a usually scented aromatic preparation of ammonium carbonate and ammonia water used as a stimulant and restorative

smelly \ˈsme-lē\ *adjective* **smell·i·er; -est** (1862)
: having a smell; *especially* **:** MALODOROUS

¹smelt \ˈsmelt\ *noun, plural* **smelts** *or* **smelt** [Middle English, from Old English; akin to Norwegian *smelte* whiting] (before 12th century)
: any of a family (Osmeridae) of small salmonoid fishes that closely resemble the trouts in general structure, live along coasts and ascend rivers to spawn or are landlocked, and have delicate oily flesh with a distinctive odor and taste

²smelt *transitive verb* [Dutch or Low German *smelten;* akin to Old High German *smelzan* to melt, Old English *meltan* — more at MELT] (1543)
1 : to melt or fuse (as ore) often with an accompanying chemical change usually to separate the metal
2 : REFINE, REDUCE

smelt·er \ˈsmel-tər\ *noun* (15th century)
: one that smelts: **a :** a worker who smelts ore **b :** an owner or operator of a smeltery **c** *or* **smelt·ery** \-t(ə-)rē\ **:** an establishment for smelting

smew \ˈsmyü\ *noun* [akin to Middle High German *smiehe* smew] (1674)
: a small Eurasian merganser (*Mergus albellus*) with the male being white, gray, and black and the female chiefly gray but with a chestnut and white head

smid·gen *also* **smid·geon** *or* **smid·gin** \ˈsmi-jən\ *or* **smidge** \ˈsmij\ *noun* [probably alteration of English dialect *smitch* soiling mark] (1845)
: a small amount **:** BIT

smi·lax \ˈsmī-ˌlaks\ *noun* [Latin, bindweed, yew, from Greek] (1601)
1 : GREENBRIER
2 : a tender twining plant (*Asparagus asparagoides*) of the lily family that has ovate bright green cladophylls and is often grown in greenhouses

¹smile \ˈsmī(ə)l\ *verb* **smiled; smil·ing** [Middle English; akin to Old English *smerian* to laugh, Sanskrit *smayate* he smiles] (14th century)
intransitive verb
1 : to have, produce, or exhibit a smile
2 a : to look or regard with amusement or ridicule ⟨*smiled* at his own folly —Martin Gardner⟩ **b :** to bestow approval ⟨feeling that Heaven *smiled* on his labors —Sheila Rowlands⟩ **c :** to appear pleasant or agreeable
transitive verb
1 : to affect with or by smiling
2 : to express by a smile
— **smil·er** *noun*
— **smil·ey** \ˈsmī-lē\ *adjective*
— **smil·ing·ly** \ˈsmī-liŋ-lē\ *adverb*

²smile *noun* (15th century)
1 : a facial expression in which the eyes brighten and the corners of the mouth curve slightly upward and which expresses especially amusement, pleasure, approval, or sometimes scorn
2 : a pleasant or encouraging appearance
— **smile·less** \ˈsmī(ə)l-ləs\ *adjective*

smirch \ˈsmərch\ *transitive verb* [Middle English *smorchen*] (15th century)
1 a : to make dirty, stained, or discolored **:** SULLY **b :** to smear with something that stains or dirties
2 : to bring discredit or disgrace on
— **smirch** *noun*

☆ SYNONYMS
Smell, scent, odor, aroma mean the quality that makes a thing perceptible to the olfactory sense. SMELL implies solely the sensation without suggestion of quality or character ⟨an odd *smell* permeated the room⟩. SCENT applies to the characteristic smell given off by a substance, an animal, or a plant ⟨the *scent* of lilacs⟩. ODOR may imply a stronger or more readily distinguished scent or it may be equivalent to SMELL ⟨a cheese with a strong *odor*⟩. AROMA suggests a somewhat penetrating usually pleasant odor ⟨the *aroma* of freshly ground coffee⟩.

smirk \'smərk\ *intransitive verb* [Middle English, from Old English *smearcian* to smile; akin to Old English *smerian* to laugh] (before 12th century)
: to smile in an affected or smug manner **:** SIMPER
— smirk *noun*
smirky \'smər-kē\ *adjective* (1728)
: that smirks **:** SMIRKING
smite \'smīt\ *verb* **smote** \'smōt\; **smit·ten** \'smi-t[ə]n\ *or* **smote; smit·ing** \'smī-tiŋ\ [Middle English, from Old English *smītan* to smear, defile; akin to Old High German bi*smīzan* to defile] (12th century)
transitive verb
1 : to strike sharply or heavily especially with the hand or an implement held in the hand
2 a : to kill or severely injure by smiting **b :** to attack or afflict suddenly and injuriously ⟨*smitten* by disease⟩
3 : to cause to strike
4 : to affect as if by striking ⟨children *smitten* with the fear of hell —V. L. Parrington⟩
5 : CAPTIVATE, TAKE ⟨*smitten* with her beauty⟩
intransitive verb
: to deliver or deal a blow with or as if with the hand or something held
— smit·er \'smī-tər\ *noun*
smith \'smith\ *noun* [Middle English, from Old English; akin to Old High German *smid* smith and probably to Greek *smilē* wood-carving knife] (before 12th century)
1 : a worker in metals **:** BLACKSMITH
2 : MAKER — often used in combination ⟨gun*smith*⟩ ⟨tune*smith*⟩
smith·er·eens \ˌsmi-thə-'rēnz\ *noun plural* [perhaps from Irish *smidiríní*] (1829)
: FRAGMENTS, BITS ⟨the house was blown to *smithereens* by the explosion⟩
smith·ery \'smi-thə-rē\ *noun, plural* **-er·ies** (1625)
1 : the work, art, or trade of a smith
2 : SMITHY 1
Smith·field ham \'smith-ˌfēld-\ *noun* (1908)
: a Virginia ham produced in or near Smithfield, Va.
smith·son·ite \'smith-sə-ˌnīt\ *noun* [James *Smithson*] (1856)
: a mineral that is a carbonate of zinc and constitutes an important ore of zinc
smithy \'smi-thē *also* -thē\ *noun, plural* **smith·ies** (13th century)
1 : the workshop of a smith
2 : BLACKSMITH
[1]smock \'smäk\ *noun* [Middle English *smok,* from Old English *smoc;* akin to Old High German *smocco* adornment] (before 12th century)
1 *archaic* **:** a woman's undergarment; *especially* **:** CHEMISE
2 : a light loose garment worn especially for protection of clothing while working
[2]smock *transitive verb* (1888)
: to embroider or shirr with smocking
smock frock *noun* (circa 1800)
: a loose outer garment worn by workmen especially in Europe
smock·ing \'smä-kiŋ\ *noun* (1888)
: a decorative embroidery or shirring made by gathering cloth in regularly spaced round tucks
smog \'smäg, 'smȯg\ *noun* [*sm*oke + f*og*] (1905)
: a fog made heavier and darker by smoke and chemical fumes; *also* **:** a photochemical haze caused by the action of solar ultraviolet radiation on atmosphere polluted with hydrocarbons and oxides of nitrogen from automobile exhaust
— smog·less \'smä-gləs, 'smȯ-\ *adjective*
smog·gy \'smä-gē, 'smȯ-\ *adjective* **smog·gi·er; -est** (1905)
: characterized by or abounding in smog
smok·able *or* **smoke·able** \'smō-kə-bəl\ *adjective* (1839)
: fit for smoking

[1]smoke \'smōk\ *noun* [Middle English, from Old English *smoca;* akin to Old English *smēocan* to emit smoke, Middle High German *smouch* smoke, and probably to Greek *smychein* to smolder] (before 12th century)
1 a : the gaseous products of burning materials especially of organic origin made visible by the presence of small particles of carbon **b :** a suspension of particles in a gas
2 a : a mass or column of smoke **b :** SMUDGE
3 : fume or vapor often resulting from the action of heat on moisture
4 : something of little substance, permanence, or value
5 : something that obscures
6 a (1) **:** something (as a cigarette) to smoke (2) **:** MARIJUANA 2 **b :** an act or spell of smoking tobacco
7 a : a pale blue **b :** any of the colors of smoke
8 : pitches that are fastballs ⟨if a guy's going to hit you . . . he certainly isn't going to throw a spitter—he gives you *smoke* —Tony Conigliaro⟩
— smoke·less \'smō-kləs\ *adjective*
— smoke·like \'smōk-ˌlīk\ *adjective*
[2]smoke *verb* **smoked; smok·ing** (before 12th century)
intransitive verb
1 a : to emit or exhale smoke **b :** to emit excessive smoke
2 *archaic* **:** to undergo punishment **:** SUFFER
3 : to spread or rise like smoke
4 : to inhale and exhale the fumes of burning plant material and especially tobacco; *especially* **:** to smoke tobacco habitually
transitive verb
1 a : FUMIGATE **b :** to drive (as mosquitoes) away by smoke **c :** to blacken or discolor with smoke ⟨*smoked* glasses⟩ **d :** to cure by exposure to smoke **e :** to stupefy (as bees) by smoke
2 *archaic* **:** SUSPECT
3 : to inhale and exhale the smoke of
4 *archaic* **:** RIDICULE
smoke and mirrors *noun plural* (1982)
: something intended to disguise or draw attention away especially from an embarrassing or unpleasant issue — usually hyphenated when used attributively
smoke detector *noun* (circa 1927)
: an alarm that activates automatically when it detects smoke
smoke–filled room \'smōk-'fild-\ *noun* (1920)
: a room (as in a hotel) in which a small group of politicians carry on negotiations
smoke·house \'smōk-ˌhau̇s\ *noun* (1746)
: a building where meat or fish is cured by means of dense smoke
smoke·jack \-ˌjak\ *noun* (1675)
: a device for turning a spit by a fly or wheel moved by rising gases in a chimney
smoke jumper *noun* (1927)
: a forest firefighter who parachutes to locations otherwise difficult to reach
smokeless powder *noun* (1890)
: any of a class of explosive propellants that produce comparatively little smoke on explosion and consist mostly of gelatinized nitrocellulose
smokeless tobacco *noun* (1981)
: pulverized or shredded tobacco chewed or placed between cheek and gum
smoke out *transitive verb* (1593)
1 : to drive out by or as if by smoke
2 : to cause to be made public
smok·er \'smō-kər\ *noun* (1599)
1 : one that smokes
2 : a railroad car or compartment in which smoking is allowed
3 : an informal social gathering for men
smoke screen *noun* (1915)
1 : a screen of smoke to hinder enemy observation of a military force, area, or activity
2 : something designed to obscure, confuse, or mislead
[1]smoke·stack \'smōk-ˌstak\ *noun* (1859)
: a pipe or funnel through which smoke and gases are discharged
[2]smokestack *adjective* (1926)
: of, relating to, being, or characterized by manufacturing and especially heavy industry ⟨*smokestack* industries⟩
smoke tree *noun* (1846)
: either of two small shrubby trees (genus *Cotinus*) of the cashew family with large panicles of minute flowers that suggest a cloud of smoke: **a :** one (*C. coggygria*) of the Old World that is widely planted in the eastern U.S. **b :** one (*C. obovatus*) of the southeastern U.S. and Texas
smoking gun *noun* (1974)
: something that serves as conclusive evidence or proof especially of a crime
smoking jacket *noun* (1878)
: a loose-fitting jacket or short robe for wear at home
smoking lamp *noun* (circa 1881)
: a lamp on a ship kept lighted during the hours when smoking is allowed
smoking room *noun* (1689)
: a room (as in a hotel or club) set apart for smokers
smoky *also* **smok·ey** \'smō-kē\ *adjective* **smok·i·er; -est** (14th century)
1 : emitting smoke especially in large quantities
2 a : having the characteristics of or resembling smoke **b :** suggestive of smoke especially in flavor or odor
3 a : filled with smoke **b :** made dark or black by or as if by smoke
— smok·i·ly \-kə-lē\ *adverb*
— smok·i·ness \-kē-nəs\ *noun*
smoky quartz *noun* (1837)
: CAIRNGORM
smoky topaz *noun* (1797)
: CAIRNGORM
smol·der *or* **smoul·der** \'smōl-dər\ *intransitive verb* **smol·dered** *or* **smoul·dered; smol·der·ing** *or* **smoul·der·ing** \-d(ə-)riŋ\ [Middle English *smolderen* to smother, from *smolder* smoke, smudge; akin to Middle Dutch *smōlen* to smolder] (1529)
1 a : to burn sluggishly, without flame, and often with much smoke **b :** to be consumed by smoldering — often used with *out*
2 : to exist in a state of suppressed activity ⟨hostilities *smoldered* for years⟩
3 : to show suppressed anger, hate, or jealousy ⟨eyes *smoldering* with hate⟩
smolt \'smōlt\ *noun* [Middle English (Scots)] (15th century)
: a young salmon or sea trout about two years old that is at the stage of development when it assumes the silvery color of the adult and is ready to migrate to the sea
[1]smooch \'smüch\ *noun* (1578)
: KISS
[2]smooch *intransitive verb* [alteration of *smouch* to kiss loudly] (1588)
: KISS, PET
[3]smooch *transitive verb* [probably alteration of *smutch,* verb] (1631)
: SMUDGE, SMEAR
[4]smooch *noun* (1825)
: SMUDGE, SMEAR
— smoochy \'smü-chē\ *adjective*
[1]smooth \'smüth\ *adjective* [Middle English *smothe,* from Old English *smōth;* akin to Old Saxon *smōthi* smooth] (before 12th century)
1 a (1) **:** having a continuous even surface (2) *of a curve* **:** being the representation of a function with a continuous first derivative (3) **:** having or being a short even coat of hair ⟨a *smooth* collie⟩ — compare ROUGH, WIREHAIRED **b :** being without hair **c :** GLABROUS ⟨a *smooth* leaf⟩ **d :** causing no resistance to sliding

2 : free from difficulties or impediments ⟨the *smooth* course of his life⟩
3 : even and uninterrupted in flow or flight
4 : excessively and often artfully suave : INGRATIATING ⟨a *smooth* operator⟩
5 a : SERENE, EQUABLE ⟨a *smooth* disposition⟩ **b** : AMIABLE, COURTEOUS
6 a : not sharp or harsh ⟨a *smooth* sherry⟩ **b** : free from lumps
synonym see LEVEL, EASY, SUAVE
— **smooth** *adverb*
— **smooth·ly** *adverb*
— **smooth·ness** *noun*

²smooth *verb* **smoothed; smooth·ing; smooths** *also* **smoothes** (15th century)
transitive verb
1 : to make smooth
2 a : to free from what is harsh or disagreeable : POLISH ⟨*smoothed* out his style⟩ **b** : SOOTHE
3 : to minimize (as a fault) especially in order to allay anger or ill will ⟨his main job is to *smooth* over the friction that so often arises —Brian Crozier⟩
4 : to free from obstruction or difficulty
5 a : to press flat **b** : to remove expression from (one's face) : COMPOSE
6 : to cause to lie evenly and in order : PREEN
7 : to free (as a graph or data) from irregularities
intransitive verb
: to become smooth
— **smooth·er** *noun*

³smooth *noun* (15th century)
1 : a smooth part
2 : the act of smoothing
3 : a smoothing implement

smooth·bore \ˈsmüth-ˈbōr, -ˈbȯr\ *adjective* (1799)
of a firearm : having a barrel with an unrifled bore
— **smoothbore** \-ˌbōr, -ˌbȯr\ *noun*

smooth breathing *noun* (circa 1888)
1 : a mark ' placed over some initial vowels in Greek to show that they are not aspirated (as in ἐκεῖ pronounced \e-ˈkā\)
2 : the absence of aspiration indicated by a mark '

smooth·en \ˈsmü-thən\ *verb* **smooth·ened; smooth·en·ing** \ˈsmü-thə-niŋ, ˈsmüth-niŋ\ (1635)
transitive verb
: to make smooth
intransitive verb
: to become smooth

smooth hound *noun* [from the absence of a spine in front of the dorsal fin] (1603)
: any of several dogfishes (genus *Mustelus*) closely related to or included with the requiem sharks

smooth muscle *noun* (circa 1890)
: muscle tissue that lacks cross striations, that is made up of elongated spindle-shaped cells having a central nucleus, and that is found in vertebrate visceral structures (as the stomach and bladder) as thin sheets performing functions not subject to conscious control by the mind and in all or most of the musculature of invertebrates other than arthropods — compare STRIATED MUSCLE

smooth–tongued \ˈsmüth-ˈtəŋd\ *adjective* (1592)
: ingratiating in speech

smoothy *or* **smooth·ie** \ˈsmü-thē\ *noun, plural* **smooth·ies** (1904)
1 : a smooth-tongued person
2 a : a person with polished manners **b** : one who behaves or performs with deftness, assurance, and easy competence; *especially* : a man with an ingratiating manner toward women

smor·gas·bord \ˈsmȯr-gəs-ˌbōrd, -ˌbȯrd\ *noun* [Swedish *smörgåsbord,* from *smörgås* open sandwich + *bord* table] (1893)
1 : a luncheon or supper buffet offering a variety of foods and dishes (as hors d'oeuvres, hot and cold meats, smoked and pickled fish, cheeses, salads, and relishes)
2 : a heterogeneous mixture : MÉLANGE ◆

smote *past of* SMITE

¹smoth·er \ˈsmə-thər\ *noun* [Middle English, alteration of *smorther,* from *smoren* to smother, from Old English *smorian* to suffocate; akin to Middle Dutch *smoren* to suffocate] (13th century)
1 a : thick stifling smoke or smudge **b** : a state of being stifled or suppressed
2 : a dense cloud (as of fog, spray, or dust)
3 : a confused multitude of things : WELTER
— **smoth·ery** \ˈsmə-thə-rē, ˈsməth-rē\ *adjective*

²smother *verb* **smoth·ered; smoth·er·ing** \ˈsmə-thə-riŋ, ˈsməth-riŋ\ (circa 1520)
intransitive verb
: to be overcome or killed through or as if through lack of air
transitive verb
1 : to overcome or kill with smoke or fumes
2 a : to destroy the life of by depriving of air **b** : to overcome or discomfit through or as if through lack of air **c** : to suppress (a fire) by excluding oxygen
3 a : to cause to smolder **b** : to suppress expression or knowledge of ⟨*smothered* his rage⟩ **c** : to stop or prevent the growth or activity of ⟨*smother* a child with too much care⟩; *also* : OVERWHELM **d** : to cover thickly : BLANKET ⟨snow *smothered* the trails⟩ **e** : to overcome or vanquish quickly or decisively
4 : to cook in a covered pan or pot with little liquid over low heat

¹smudge \ˈsməj\ *verb* **smudged; smudg·ing** [Middle English *smogen*] (15th century)
transitive verb
1 a : to make a smudge on **b** : to soil as if by smudging
2 a : to rub, daub, or wipe in a smeary manner **b** : to make indistinct : BLUR
3 : to smoke or protect by means of a smudge
intransitive verb
1 : to make a smudge
2 : to become smudged

²smudge *noun* (circa 1774)
1 a : a blurry spot or streak **b** : an immaterial stain ⟨cleanse him of every last *smudge* of impropriety —Richard Hanser⟩ **c** : an indistinct mass : BLUR
2 : a smoldering mass placed on the windward side (as to protect from frost)
3 : a bid of 4 in pitch that if made wins the game
— **smudg·i·ly** \ˈsmə-jə-lē\ *adverb*
— **smudg·i·ness** \ˈsmə-jē-nəs\ *noun*
— **smudgy** \-jē\ *adjective*

smug \ˈsməg\ *adjective* **smug·ger; smug·gest** [probably modification of Low German *smuck* neat, from Middle Low German, from *smucken* to dress; akin to Old English *smoc* smock] (1551)
1 : trim or smart in dress : SPRUCE
2 : scrupulously clean, neat, or correct : TIDY
3 : highly self-satisfied
— **smug·ly** *adverb*
— **smug·ness** *noun*

smug·gle \ˈsmə-gəl\ *verb* **smug·gled; smug·gling** \-g(ə-)liŋ\ [Low German *smuggeln* & Dutch *smokkelen*] (1687)
transitive verb
1 : to import or export secretly contrary to the law and especially without paying duties imposed by law
2 : to convey or introduce surreptitiously
intransitive verb
: to import or export something in violation of the customs laws
— **smug·gler** \ˈsmə-glər\ *noun*

¹smut \ˈsmət\ *verb* **smut·ted; smut·ting** [probably alteration of earlier *smot* to stain, from Middle English *smotten;* akin to Middle High German *smutzen* to stain] (1587)
transitive verb
1 : to stain or taint with smut
2 : to affect (a crop or plant) with smut
intransitive verb
: to become affected by smut

²smut *noun* (1664)
1 : matter that soils or blackens; *specifically* : a particle of soot
2 : any of various destructive diseases especially of cereal grasses caused by parasitic fungi (order Ustilaginales) and marked by transformation of plant organs into dark masses of spores; *also* : a fungus causing a smut
3 : obscene language or matter

smutch \ˈsməch\ *noun* [akin to Middle English *smogen* to smudge] (1530)
: a dark stain : SMUDGE
— **smutch** *transitive verb*
— **smutchy** \ˈsmə-chē\ *adjective*

smut·ty \ˈsmə-tē\ *adjective* **smut·ti·er; -est** (1597)
1 : soiled or tainted with smut; *especially* : affected with smut fungus
2 : OBSCENE, INDECENT
3 : resembling smut in appearance : SOOTY
— **smut·ti·ly** \ˈsmə-tᵊl-ē\ *adverb*
— **smut·ti·ness** \ˈsmə-tē-nəs\ *noun*

¹snack \ˈsnak\ *noun* [Middle English *snak* bite, from *snaken* to bite, perhaps from Middle Dutch *snacken* to snap at — more at SNATCH] (1757)
: a light meal : food eaten between regular meals; *also* : food suitable for snacking

²snack *intransitive verb* (1807)
: to eat a snack

snack bar *noun* (1930)
: a public eating place where snacks are served usually at a counter

¹snaf·fle \ˈsna-fəl\ *noun* [origin unknown] (1533)
: a simple usually jointed bit for a bridle

²snaffle *transitive verb* **snaf·fled; snaf·fling** \ˈsna-f(ə-)liŋ\ [origin unknown] (1724)
: to obtain especially by devious or irregular means

¹sna·fu \sna-ˈfü, ˈsna-ˌfü\ *noun* [*s*ituation *no*rmal *a*ll *f*ucked *u*p (fouled up)] (circa 1941)
: CONFUSION, MUDDLE

²snafu *adjective* (1943)
: snarled or stalled in confusion : AWRY

³snafu *transitive verb* (1943)
: to bring into a state of confusion

¹snag \ˈsnag\ *noun* [of Scandinavian origin; akin to Old Norse *snagi* clothes peg] (circa 1587)

> ◇ WORD HISTORY
>
> **smorgasbord** In Swedish the word *smörgås* literally means "butter goose." *Smör* simply means "butter," and *gås* has the literal meaning "goose" and the figurative sense "a lump of butter," the result of a fancied resemblance between a buttery lump and a goose. In Swedish *smörgås* also has the extended senses "bread and butter" and "open-faced sandwich." It is in the latter sense that *smörgås* forms part of *smörgåsbord,* the name for the traditional Swedish buffet of delicacies, which are typically assembled by diners into open-faced sandwiches. (The *bord* of *smörgåsbord* is related to English *board* and simply means "table.") Swedish-style buffets first gained popularity among American diners in the early decades of the 20th century. By the 1940s *smorgasbord,* stripped of its Swedish diacritics, was sufficiently established in English to have given rise to the figurative sense "a heterogeneous mixture."

1 a : a tree or branch embedded in a lake or stream bed and constituting a hazard to navigation **b :** a standing dead tree
2 : a rough sharp or jagged projecting part **:** PROTUBERANCE: as **a :** a projecting tooth; *also* **:** a stump of a tooth **b :** one of the secondary branches of an antler
3 : a concealed or unexpected difficulty or obstacle
4 : a jagged tear made by or as if by catching on a snag ⟨a *snag* in her stocking⟩
— snag·gy \'sna-gē\ *adjective*

[2]**snag** *transitive verb* **snagged; snag·ging** (1807)
1 a : to catch and usually damage on or as if on a snag **b :** to halt or impede as if by catching on a snag
2 : to hew, trim, or cut roughly or jaggedly
3 : to clear (as a river) of snags
4 : to catch or obtain by quick action or good fortune

snag·gle·tooth \'sna-gəl-,tüth\ *noun* [English dialect *snaggle* irregularly shaped tooth + English *tooth*] (circa 1825)
: an irregular, broken, or projecting tooth
— snag·gle·toothed \-,tütht\ *adjective*

[1]**snail** \'snā(ə)l\ *noun* [Middle English, from Old English *snægl;* akin to Old High German *snecko* snail, *snahhan* to creep] (before 12th century)
1 : a gastropod mollusk especially when having an external enclosing spiral shell
2 : a slow-moving or sluggish person or thing
— snail·like \'snā(ə)l-,līk\ *adjective*

[2]**snail** *intransitive verb* (1582)
: to move, act, or go slowly or lazily

snail darter *noun* (1975)
: a darter (*Percina tanasi*) of the Tennessee River drainage system of eastern Tennessee and northern Georgia

snail fever *noun* [from the snails which serve as intermediate hosts to the schistosomes causing the disease] (1947)
: SCHISTOSOMIASIS

snail–paced \'snā(ə)l-'pāst\ *adjective* (1594)
: moving very slowly

snail's pace *noun* (15th century)
: an extremely slow pace

[1]**snake** \'snāk\ *noun* [Middle English, from Old English *snaca;* akin to Old Norse *snakr* snake, Old High German *snahhan* to crawl] (before 12th century)
1 : any of numerous limbless scaled reptiles (suborder Serpentes synonym Ophidia) with a long tapering body and with salivary glands often modified to produce venom which is injected through grooved or tubular fangs
2 : a worthless or treacherous fellow
3 : something (as a plumber's snake) resembling a snake
— snake·like \-,līk\ *adjective*

[2]**snake** *verb* **snaked; snak·ing** (1653)
transitive verb
1 : to wind (as one's way) in the manner of a snake
2 : to move (as logs) by dragging
intransitive verb
: to crawl, move, or extend silently, secretly, or sinuously

snake·bird \'snāk-,bərd\ *noun* (1791)
: ANHINGA

snake·bit \-,bit\ *or* **snake·bit·ten** \-,bi-tᵊn\ *adjective* (1957)
: UNLUCKY 1, 3

snake·bite \-,bīt\ *noun* (1839)
: the bite of a snake and especially a venomous snake

snake charmer *noun* (1836)
: an entertainer who exhibits a professed power to charm or fascinate venomous snakes

snake–dance *intransitive verb* (1922)
: to engage in a snake dance

snake dance *noun* (1772)
1 : a ceremonial dance in which snakes or their images are handled, invoked, or symbolically imitated by individual sinuous actions
2 : a group progression in a single-file serpentine path (as in celebration of an athletic victory)

snake doctor *noun* (1862)
1 : DRAGONFLY
2 : HELLGRAMMITE

snake fence *noun* (1805)
: WORM FENCE

snake in the grass (1696)
: a secretly faithless friend

snake oil *noun* (1927)
1 : any of various substances or mixtures sold (as by a traveling medicine show) as medicine usually without regard to their medical worth or properties
2 : POPPYCOCK, BUNKUM

snake pit *noun* (1946)
1 : a hospital for mental diseases
2 : a place or state of chaotic disorder and distress

snake·root \'snāk-,rüt, -,rüt\ *noun* (1635)
: any of numerous plants (as seneca snakeroot) most of which have roots sometimes believed to cure snakebites; *also* **:** the root of such a plant

snake·skin \'snāk-,skin\ *noun* (1825)
: leather prepared from the skin of a snake

snake·weed \-,wēd\ *noun* (1597)
: any of various plants popularly associated with snakes (as in appearance, habitat, or the treatment of snakebite); *especially* **:** any of an American genus (*Gutierrezia*) of composite herbs or low shrubs with clustered yellow flower heads

snaky *also* **snak·ey** \'snā-kē\ *adjective* (1567)
1 : of, formed of, or entwined with snakes ⟨the Gorgon with *snaky* hair —Joseph Addison⟩
2 : SERPENTINE, SNAKELIKE ⟨*snaky* coils⟩
3 : suggestive of a snake ⟨the oiliness and *snaky* insinuation of his demeanor —Thomas DeQuincy⟩
4 : abounding in snakes
— snak·i·ly \-kə-lē\ *adverb*

[1]**snap** \'snap\ *verb* **snapped; snap·ping** [Dutch or Low German *snappen;* akin to Middle High German *snappen* to snap] (1530)
intransitive verb
1 a : to make a sudden closing of the jaws **:** seize something sharply with the mouth ⟨fish *snapping* at the bait⟩ **b :** to grasp at something eagerly **:** make a pounce or snatch ⟨*snap* at any chance⟩
2 : to utter sharp biting words **:** bark out irritable or peevish retorts
3 a : to break suddenly with a sharp sound ⟨the twig *snapped*⟩ **b :** to give way suddenly under strain
4 : to make a sharp or crackling sound
5 : to close or fit in place with an abrupt movement or sharp sound ⟨the lock *snapped* shut⟩
6 a : to move briskly or sharply ⟨*snaps* to attention⟩ **b :** to undergo a sudden and rapid change (as from one condition to another) ⟨*snap* out of it⟩ ⟨*snapped* awake⟩
7 : SPARKLE, FLASH ⟨eyes *snapping* with fury⟩
transitive verb
1 : to seize with or as if with a snap of the jaws
2 : to take possession or advantage of suddenly or eagerly — usually used with *up* ⟨shoppers *snapping* up bargains⟩
3 a : to retort to or interrupt curtly and irritably **b :** to utter curtly or abruptly
4 : to break suddenly **:** break short or in two
5 a : to cause to make a snapping sound ⟨*snap* a whip⟩ **b :** to put into or remove from a particular position by a sudden movement or with a sharp sound ⟨*snap* the lock shut⟩
6 a : to project with a snap **b :** to put (a football) in play with a snap **c** (1) **:** to take photographically ⟨*snapping* exclusive news pictures —*Current Biography*⟩ (2) **:** to take a snapshot of

[2]**snap** *noun* (1555)
1 : an abrupt closing (as of the mouth in biting or of scissors in cutting)
2 a *archaic* **:** a share of profits or booty **b :** something that brings quick and easy profit or advantage **c :** something that is easy and presents no problems **:** CINCH
3 : a small amount **:** BIT
4 a : an act or instance of seizing abruptly **:** a sudden snatching at something **b :** a quick short movement ⟨lithe *snaps* of its body —Barbara Taylor⟩ **c :** a sudden sharp breaking
5 a : a sound made by snapping something ⟨shut the book with a *snap*⟩ **b :** a brief sharp and usually irritable speech or retort
6 : a sudden spell of weather ⟨a cold *snap*⟩
7 : a catch or fastening that closes or locks with a click ⟨the *snap* of a bracelet⟩
8 : a flat brittle cookie — compare GINGERSNAP
9 : SNAPSHOT
10 a : the condition of being vigorous in body, mind, or spirit **:** ALERTNESS, ENERGY **b :** a pleasing vigorous quality
11 : the act of a center's putting the football in play from its position on the ground by quickly passing it between his legs to a teammate (as a quarterback) standing behind him

[3]**snap** *adverb* (1583)
: with a snap

[4]**snap** *adjective* (1739)
1 : done, made, or carried through suddenly or without deliberation ⟨a *snap* judgment⟩
2 : called or taken without prior warning ⟨a *snap* election⟩
3 : fastening with a snap ⟨a *snap* lock⟩
4 : unusually easy or simple ⟨a *snap* course⟩

snap·back \'snap-,bak\ *noun* (1887)
1 : a football snap
2 : a sudden rebound or recovery ⟨a *snapback* of prices on the stock exchange⟩

snap back *intransitive verb* (1945)
: to make a quick or vigorous recovery

snap bean *noun* (1770)
: a bean grown primarily for its pods that are usually broken in pieces and cooked as a vegetable while young and tender and before the seeds have become enlarged — compare SHELL BEAN

snap–brim \'snap-,brim\ *noun, often attributive* (circa 1908)
: a usually felt hat with brim turned up in back and down in front and with a dented crown

snap·drag·on \'snap-,dra-gən\ *noun* [from the fancied resemblance of the flowers to the face of a dragon] (1593)
: any of a genus (*Antirrhinum* and especially the widely cultivated *A. majus* of the family Scrophulariaceae, the snapdragon family) of herbs having showy white, crimson, or yellow bilabiate flowers

snapdragon

snap–on \'snap-,ȯn, -,än\ *adjective* (1925)
: designed to snap into position and fit tightly ⟨*snap-on* cuffs⟩

snap pea *noun* (1980)
: a cultivated pea that has edible usually round pods easily snapped like beans and that is classified with the snow pea as a variety (*Pisum sativum macrocarpon*) — called also *sugar snap pea*

snap·per \'sna-pər\ *noun, plural* **snappers** (circa 1587)
1 : one that snaps: as **a :** something (as a remark) that gives new orientation to a situation or utterance **b** (1) **:** SNAPPING TURTLE (2) **:** CLICK BEETLE
2 *plural also* **snapper a :** any of numerous active carnivorous fishes (family Lutjanidae) of warm seas important as food and often as sport fishes **b :** any of several immature fishes (as the young of the bluefish) that resemble a snapper

snap·per–back \-,bak\ *noun* (1887)

: a football center

snapping turtle *noun* (1784)
: either of two large American aquatic turtles (family Chelydridae) with powerful jaws and a strong musky odor: **a :** one (*Chelydra serpentina*) that has the head covered with smooth skin, has large plates in a double row on the underside of the tail, and is distributed from eastern Canada to Central America **b :** ALLIGATOR SNAPPER

snapping turtle

snap·pish \'sna-pish\ *adjective* (1542)
1 a : given to curt irritable speech **b :** arising from annoyance or irascibility
2 : inclined to bite ⟨a *snappish* dog⟩
— snap·pish·ly *adverb*
— snap·pish·ness *noun*

snap·py \'sna-pē\ *adjective* **snap·pi·er; -est** (1746)
1 : SNAPPISH 1
2 a : quickly made or done **b :** marked by vigor or liveliness **c :** briskly cold **d :** STYLISH, SMART ⟨a *snappy* dresser⟩
— snap·pi·ly \'sna-pə-lē\ *adverb*
— snap·pi·ness \'sna-pē-nəs\ *noun*

snap roll *noun* (circa 1934)
: an airplane maneuver in which a rapid full revolution is completed about the plane's longitudinal axis while an approximately level line of flight is maintained

snap·shoot·er \'snap-ˌshü-tər\ *noun* (1896)
: a person who takes snapshots

snap·shot \'snap-ˌshät\ *noun* (1890)
1 : a casual photograph made typically by an amateur with a small handheld camera
2 : an impression or view of something brief or transitory

[1]snare \'snar, 'sner\ *noun* [Middle English, from Old English *sneare,* from Old Norse *snara;* akin to Old High German *snuor* cord and perhaps to Greek *narkē* numbness] (before 12th century)
1 a (1) **:** a contrivance often consisting of a noose for entangling birds or mammals (2) **:** TRAP, GIN **b** (1) **:** something by which one is entangled, involved in difficulties, or impeded (2) **:** something deceptively attractive
2 [probably from Dutch *snaar,* literally, cord; akin to Old High German *snuor*] **a :** one of the catgut strings or metal spirals of a snare drum **b :** SNARE DRUM
3 : a surgical instrument consisting usually of a wire loop constricted by a mechanism in the handle and used for removing tissue masses (as tonsils)

[2]snare *transitive verb* **snared; snar·ing** (14th century)
1 a : to capture by or as if by use of a snare **b :** to win or attain by artful or skillful maneuvers
2 : to entangle or hold as if in a snare ⟨any object that *snared* his eye —*Current Biography*⟩
synonym see CATCH
— snar·er *noun*

snare drum *noun* (circa 1859)
: a small double-headed drum with one or more snares stretched across its lower head — see DRUM illustration

snarky \'snär-kē\ *adjective* [dialect *snark* to annoy, perhaps alteration of *nark* to irritate] (1906)
: CROTCHETY, SNAPPISH

[1]snarl \'snär(-ə)l\ *verb* [[2]*snarl*] (14th century)
transitive verb
1 : to cause to become knotted and intertwined **:** TANGLE
2 : to make excessively complicated
intransitive verb
: to become snarled
— snarl·er *noun*

[2]snarl *noun* [Middle English *snarle,* probably diminutive of *snare*] (1609)
1 : a tangle especially of hairs or thread **:** KNOT
2 : a tangled situation ⟨traffic *snarls*⟩
— snarly \'snär-lē\ *adjective*

[3]snarl *verb* [frequentative of obsolete English *snar* to growl] (1589)
intransitive verb
1 : to growl with a snapping or gnashing of teeth
2 : to give vent to anger in surly language
transitive verb
: to utter or express with a snarl or by snarling
— snarl·er *noun*

[4]snarl *noun* (1613)
: a surly angry growl
— snarly \'snär-lē\ *adjective*

[1]snatch \'snach\ *verb* [Middle English *snacchen* to give a sudden snap, seize; akin to Middle Dutch *snacken* to snap at] (13th century)
intransitive verb
: to attempt to seize something suddenly
transitive verb
: to take or grasp abruptly or hastily; *also* **:** to seize or grab suddenly without permission, ceremony, or right
synonym see TAKE
— snatch·er *noun*

[2]snatch *noun* (1563)
1 a : a brief period ⟨caught *snatches* of sleep⟩ **b :** a brief, fragmentary, or hurried part **:** BIT ⟨caught *snatches* of the conversation⟩
2 a : a snatching at or of something **b** *slang* **:** an act or instance of kidnapping
3 : a lift in weight lifting in which the weight is raised from the floor directly to an overhead position in a single motion — compare CLEAN AND JERK, PRESS
4 : the female pudenda — usually considered vulgar

snatch block *noun* (circa 1625)
: a block that can be opened on one side to receive the bight of a rope

snath \'snath, 'sneth\ *or* **snathe** \'snāth, 'snāth\ *noun* [Middle English *snede,* from Old English *snǣd*] (1574)
: the handle of a scythe

snaz·zy \'sna-zē\ *adjective* **snaz·zi·er; -est** [origin unknown] (circa 1932)
: conspicuously or flashily attractive **:** FANCY

[1]sneak \'snēk\ *verb* **sneaked** \'snēkt\ *or* **snuck** \'snək\; **sneak·ing** [akin to Old English *snīcan* to sneak along, Old Norse *snīkja*] (1596)
intransitive verb
1 : to go stealthily or furtively **:** SLINK
2 : to act in or as if in a furtive manner
3 : to carry the football on a quarterback sneak
transitive verb
: to put, bring, or take in a furtive or artful manner ⟨*sneak* a smoke⟩ ■
synonym see LURK
— sneak up on : to approach or act on stealthily

[2]sneak *noun* (circa 1643)
1 : a person who acts in a stealthy, furtive, or shifty manner
2 a : a stealthy or furtive move **b :** an unobserved departure or escape
3 : SNEAKER 2
4 : QUARTERBACK SNEAK

[3]sneak *adjective* (circa 1859)
1 : carried on secretly **:** CLANDESTINE
2 : occurring without warning **:** SURPRISE ⟨a *sneak* attack⟩

sneak·er \'snē-kər\ *noun* (1598)
1 : one that sneaks
2 : a usually canvas sports shoe with a pliable rubber sole
— sneak·ered \-kərd\ *adjective*

sneak·ing \'snē-kiŋ\ *adjective* (1582)
1 : MEAN, CONTEMPTIBLE
2 : characteristic of a sneak **:** FURTIVE, UNDERHANDED
3 a : not openly expressed or acknowledged ⟨he has a *sneaking* respect for culture —H. A. Burton⟩ **b :** that is a persistent conjecture ⟨a *sneaking* suspicion⟩
— sneak·ing·ly \-kiŋ-lē\ *adverb*

sneak preview *noun* (circa 1937)
: a special advance showing of a motion picture usually announced but not named

sneak thief *noun* (circa 1859)
: a thief who steals whatever is readily available without using violence or forcibly breaking into buildings

sneaky \'snē-kē\ *adjective* **sneak·i·er; -est** (1833)
: marked by stealth, furtiveness, or shiftiness
— sneak·i·ly \-kə-lē\ *adverb*
— sneak·i·ness \-kē-nəs\ *noun*

[1]sneap \'snēp\ *transitive verb* [Middle English *snaipen,* probably of Scandinavian origin; akin to Icelandic *sneypa* to scold — more at SNUB] (14th century)
1 *archaic* **:** to blast or blight with cold **:** NIP
2 *dialect English* **:** CHIDE

[2]sneap *noun* (1597)
archaic **:** REBUKE, SNUB

sneck \'snek\ *noun* [Middle English *snekke*] (14th century)
chiefly dialect **:** LATCH

[1]sneer \'snir\ *verb* [probably akin to Middle High German *snerren* to chatter, gossip — more at SNORE] (1680)
intransitive verb
1 : to smile or laugh with facial contortions that express scorn or contempt
2 : to speak or write in a scornfully jeering manner
transitive verb
: to utter with a sneer
synonym see SCOFF
— sneer·er *noun*

[2]sneer *noun* (1707)
: the act of sneering; *also* **:** a sneering expression or remark

[1]sneeze \'snēz\ *intransitive verb* **sneezed; sneez·ing** [Middle English *snesen,* alteration of *fnesen,* from Old English *fnēosan;* akin to Middle High German *pfnūsen* to snort, sneeze, Greek *pnein* to breathe] (14th century)
: to make a sudden violent spasmodic audible expiration of breath through the nose and mouth especially as a reflex act
— sneez·er *noun*
— sneeze at : to make light of

[2]sneeze *noun* (1646)
: an act or instance of sneezing

sneeze·weed \'snēz-ˌwēd\ *noun* (circa 1837)
: any of several composite plants; *especially* **:** a North American perennial herb (*Helenium autumnale*) with yellow ray flowers and a darker globose disk

sneezy \'snē-zē\ *adjective* (1839)
: given to or causing sneezing

[1]snell \'snel\ *adjective* [Middle English, from Old English; akin to Old High German *snel* bold, agile] (before 12th century)
1 *chiefly Scottish* **:** QUICK, ACUTE

□ USAGE
sneak From its earliest appearance in print in the late 19th century as a dialectal and probably uneducated form, the past and past participle *snuck* has risen to the status of standard and to approximate equality with *sneaked.* Indications are that it is continuing to grow in frequency. It is most common in the U.S. and Canada, but has also been spotted in British and Australian English.

\ə\ **abut** \ᵊ\ **kitten** \ər\ **further** \a\ **ash** \ā\ **ace** \ä\ **mop, mar** \au̇\ **out** \ch\ **chin** \e\ **bet** \ē\ **easy** \g\ **go** \i\ **hit** \ī\ **ice** \j\ **job** \ŋ\ **sing** \ō\ **go** \ȯ\ **law** \ȯi\ **boy** \th\ **thin** \t͟h\ **the** \ü\ **loot** \u̇\ **foot** \y\ **yet** \zh\ **vision** *see also* Guide to Pronunciation

2 *chiefly Scottish* **:** KEEN, PIERCING ⟨a *snell* wind smote us —*Scotsman*⟩
3 *chiefly Scottish* **:** GRIEVOUS, SEVERE
[2]**snell** *noun* [origin unknown] (1846)
: a short line (as of gut) by which a fishhook is attached to a longer line
[1]**snick** \'snik\ *verb* [probably from obsolete *snick or snee* to engage in cut-and-thrust fighting — more at SNICKERSNEE] (circa 1700)
transitive verb
1 *archaic* **:** to cut through
2 : to cut slightly **:** NICK
intransitive verb
: to perform a light cutting action
[2]**snick** *noun* (circa 1775)
: a small cut **:** NICK
[3]**snick** *verb* [imitative] (1828)
: CLICK
[4]**snick** *noun* (circa 1886)
: a slight often metallic sound **:** CLICK
[1]**snick·er** \'sni-kər\ *intransitive verb* **snick·ered; snick·er·ing** \-k(ə-)riŋ\ [origin unknown] (1694)
: to laugh in a covert or partly suppressed manner **:** TITTER
— **snick·er·er** \-kər-ər\ *noun*
— **snick·ery** \-k(ə-)rē\ *adjective*
[2]**snicker** *noun* (1835)
: an act or sound of snickering
snick·er·snee \'sni-kə(r)-,snē\ *noun* [obsolete *snick or snee* to engage in cut-and-thrust fighting, alteration of earlier *steake or snye,* from Dutch *steken of snijden* to thrust or cut] (circa 1775)
: a large knife
snide \'snīd\ *adjective* [origin unknown] (circa 1859)
1 a : FALSE, COUNTERFEIT **b :** practicing deception **:** DISHONEST ⟨a *snide* merchant⟩
2 : unworthy of esteem **:** LOW ⟨a *snide* trick⟩
3 : slyly disparaging **:** INSINUATING ⟨*snide* remarks⟩
— **snide·ly** *adverb*
— **snide·ness** *noun*
[1]**sniff** \'snif\ *verb* [Middle English] (14th century)
intransitive verb
1 : to draw air audibly up the nose especially for smelling ⟨*sniffed* at the flowers⟩
2 : to show or express disdain or scorn
transitive verb
1 : to smell or take by inhalation through the nose
2 : to utter contemptuously
3 : to recognize or detect by or as if by smelling ⟨*sniff* out trouble⟩
[2]**sniff** *noun* (1767)
1 : an act or sound of sniffing
2 : a quantity that is sniffed
sniff·er *noun* (1864)
: one that sniffs; *especially* **:** one who takes drugs illicitly by sniffing
sniff·ish \'sni-fish\ *adjective* (1923)
: having or expressing a haughty attitude **:** DISDAINFUL, SUPERCILIOUS
— **sniff·ish·ly** *adverb*
— **sniff·ish·ness** *noun*
[1]**snif·fle** \'sni-fəl\ *intransitive verb* **snif·fled; snif·fling** \-f(ə-)liŋ\ [frequentative of *sniff*] (1632)
1 : to sniff repeatedly **:** SNUFFLE
2 : to speak with or as if with sniffling
— **snif·fler** \-f(ə-)lər\ *noun*
[2]**sniffle** *noun* (circa 1825)
1 *plural* **:** a head cold marked by nasal discharge
2 : an act or sound of sniffling
sniffy \'sni-fē\ *adjective* **sniff·i·er; -est** (1871)
: SNIFFISH, SUPERCILIOUS
— **sniff·i·ly** \'sni-fə-lē\ *adverb*
— **sniff·i·ness** \'sni-fē-nəs\ *noun*
snif·ter \'snif-tər\ *noun* [English dialect, sniff, snort, from Middle English, to sniff, snort] (1844)
1 : a small drink of distilled liquor
2 : a short-stemmed goblet with a bowl narrowing toward the top
[1]**snig·ger** \'sni-gər\ *intransitive verb* **snig·gered; snig·ger·ing** \-g(ə-)riŋ\ [by alteration] (circa 1706)
: SNICKER
— **snig·ger·er** \-gər-ər\ *noun*
[2]**snigger** *noun* (circa 1823)
: SNICKER
snig·gle \'sni-gəl\ *verb* **snig·gled; snig·gling** \-g(ə-)liŋ\ [English dialect *snig* small eel, from Middle English *snygge*] (1653)
intransitive verb
: to fish for eels by thrusting a baited hook or needle into their hiding places
transitive verb
: to catch (an eel) by sniggling
[1]**snip** \'snip\ *noun* [from or akin to Dutch & Low German *snip;* akin to Middle High German *snipfen* to snap the fingers] (1558)
1 a : a small piece that is snipped off; *also* **:** FRAGMENT, BIT **b :** a cut or notch made by snipping **c :** an act or sound of snipping
2 : a white or light mark; *especially* **:** a white spot between the nostrils of a horse
3 : a presumptuous or impertinent person; *especially* **:** an impertinent or saucy girl
4 *British* **:** BARGAIN, BUY
[2]**snip** *verb* **snipped; snip·ping** (1586)
transitive verb
: to cut or cut off with or as if with shears or scissors; *specifically* **:** to clip suddenly or by bits
intransitive verb
: to make a short quick cut with or as if with shears or scissors
— **snip·per** *noun*
[1]**snipe** \'snīp\ *noun, plural* **snipes** [Middle English, of Scandinavian origin; akin to Old Norse *snīpa* snipe; akin to Old High German *snepfa* snipe] (14th century)
1 *or plural* **snipe :** any of various usually slender-billed birds of the same family as the sandpipers; *especially* **:** any of several game birds (especially genus *Gallinago*) especially of marshy areas
2 : a contemptible person
[2]**snipe** *intransitive verb* **sniped; snip·ing** (1832)
1 : to shoot at exposed individuals (as of an enemy's forces) from a usually concealed point of vantage
2 : to aim a carping or snide attack
— **snip·er** *noun*
snip·er·scope \'snī-pər-,skōp\ *noun* (1941)
: an optical device for use especially with a rifle that allows a person to see targets better in the dark
snip·per·snap·per \'sni-pər-,sna-pər\ *noun* [origin unknown] (circa 1590)
: WHIPPERSNAPPER
snip·pet \'sni-pət\ *noun* [[1]*snip*] (1664)
: a small part, piece, or thing; *especially* **:** a brief quotable passage
snip·pety \-pə-tē\ *adjective* (1864)
1 : made up of snippets
2 [probably from [2]*snip* + *-ety* (as in *pernickety*)] **:** SNIPPY
snip·py \'sni-pē\ *adjective* **snip·pi·er; -est** [[2]*snip*] (circa 1848)
1 : SHORT-TEMPERED, SNAPPISH
2 : unduly brief or curt
3 : putting on airs **:** SNIFFY
— **snip·pi·ly** \'sni-pə-lē\ *adverb*
snips \'snips\ *noun plural but singular or plural in construction* (circa 1846)
: hand shears used especially for cutting sheet metal
snit \'snit\ *noun* [origin unknown] (1939)
: a state of agitation
[1]**snitch** \'snich\ *noun* [origin unknown] (circa 1785)
: one that snitches **:** TATTLETALE
[2]**snitch** *intransitive verb* (1801)
: INFORM, TATTLE
— **snitch·er** *noun*
[3]**snitch** *transitive verb* [probably alteration of *snatch*] (1904)
: to take by stealth **:** PILFER
[1]**sniv·el** \'sni-vəl\ *intransitive verb* **-eled** *or* **-elled; -el·ing** *or* **-el·ling** \'sni-və-liŋ, 'sniv-liŋ\ [Middle English, from (assumed) Old English *snyflan;* akin to Dutch *snuffelen* to snuffle, *snuffen* to sniff] (14th century)
1 : to run at the nose
2 : to snuff mucus up the nose audibly **:** SNUFFLE
3 : to cry or whine with snuffling
4 : to speak or act in a whining, sniffling, tearful, or weakly emotional manner
— **sniv·el·er** \'sni-və-lər, 'sniv-lər\ *noun*
[2]**snivel** *noun* (1600)
1 *plural, dialect* **:** HEAD COLD
2 : an act or instance of sniveling
snob \'snäb\ *noun* [origin unknown] (1781)
1 *British* **:** COBBLER
2 : one who blatantly imitates, fawningly admires, or vulgarly seeks association with those regarded as social superiors
3 a : one who tends to rebuff, avoid, or ignore those regarded as inferior **b :** one who has an offensive air of superiority in matters of knowledge or taste
snob appeal *noun* (1933)
: qualities in a product that appeal to the snobbery in a purchaser
snob·bery \'snä-b(ə-)rē\ *noun, plural* **-ber·ies** (1843)
1 : snobbish conduct or character **:** SNOBBISHNESS
2 : an instance of snobbery
snob·bish \'snä-bish\ *adjective* (1840)
: being, characteristic of, or befitting a snob
— **snob·bish·ly** *adverb*
— **snob·bish·ness** *noun*
snob·bism \'snä-,bi-zəm\ *noun* (1845)
: SNOBBERY
snob·by \'snä-bē\ *adjective* **snob·bi·er; -est** (1846)
: characterized by snobbery
sno–cone *variant of* SNOW CONE
snol·ly·gos·ter \'snä-lē-,gäs-tər\ *noun* [probably alteration of *snallygaster* a mythical creature that preys on poultry and children] (circa 1860)
: a shrewd unprincipled person
[1]**snood** \'snüd\ *noun* [(assumed) Middle English, from Old English *snōd*] (before 12th century)
1 a *Scottish* **:** a fillet or band for a woman's hair **b :** a net or fabric bag pinned or tied on at the back of a woman's head for holding the hair
2 : SNELL

snood 1b

[2]**snood** *transitive verb* (1714)
: to secure with a snood
[1]**snook** \'snu̇k, 'snük\ *noun, plural* **snook** *or* **snooks** [Dutch *snoek* pike, snook] (1697)
1 : a large vigorous bony fish (*Centropomus undecimalis* of the family Centropomidae) of warm seas and rivers that is a prized food and sport fish
2 : any of various marine fishes of the same family as the snook
[2]**snook** *noun* [origin unknown] (1879)
: a gesture of derision made by thumbing the nose
[1]**snook·er** \'snu̇-kər\ *noun* [origin unknown] (1889)
: a variation of pool played with 15 red balls and 6 variously colored balls
[2]**snooker** *transitive verb* (1925)
: to make a dupe of **:** HOODWINK
[1]**snoop** \'snüp\ *intransitive verb* [Dutch *snoepen* to buy or eat on the sly; akin to Dutch *snappen* to snap] (1832)
: to look or pry especially in a sneaking or meddlesome manner
— **snoop·er** *noun*

[2]**snoop** *noun* (circa 1890)
: one that snoops

snoopy \ˈsnü-pē\ *adjective* (circa 1895)
: given to snooping especially for personal information about others
— **snoop·i·ly** \-pə-lē\ *adverb*

[1]**snoot** \ˈsnüt\ *noun* [Middle English *snute*] (1861)
1 a : SNOUT **b :** NOSE
2 : a grimace expressive of contempt
3 : a snooty person **:** SNOB

[2]**snoot** *transitive verb* (1928)
: to treat with disdain **:** look down one's nose at

snooty \ˈsnü-tē\ *adjective* **snoot·i·er; -est** (1919)
1 : looking down the nose **:** showing disdain ⟨*snooty* people who won't speak to their neighbors⟩
2 : characterized by snobbery ⟨a *snooty* store⟩
— **snoot·i·ly** \ˈsnü-tᵊl-ē\ *adverb*
— **snoot·i·ness** \ˈsnü-tē-nəs\ *noun*

[1]**snooze** \ˈsnüz\ *intransitive verb* **snoozed; snooz·ing** [origin unknown] (1788)
: to take a nap **:** DOZE
— **snooz·er** *noun*

[2]**snooze** *noun* (1793)
: NAP

snoo·zle \ˈsnü-zəl\ *verb* **snoo·zled; snoo·zling** \ˈsnüz-(ə-)liŋ, ˈsnüz-liŋ\ [perhaps blend of *snooze* and *nuzzle*] (1831)
chiefly dialect **:** NUZZLE

[1]**snore** \ˈsnōr, ˈsnȯr\ *verb* **snored; snor·ing** [Middle English; akin to Middle Low German *snorren* to drone, Middle High German *snerren* to chatter] (15th century)
intransitive verb
: to breathe during sleep with a rough hoarse noise due to vibration of the soft palate
transitive verb
: to spend (time) in snoring or sleeping
— **snor·er** *noun*

[2]**snore** *noun* (1605)
1 : an act of snoring
2 : a noise of or as if of snoring

[1]**snor·kel** \ˈsnȯr-kəl\ *noun* [German *Schnorchel*] (1944)
1 : a tube housing air intake and exhaust pipes for a submarine's diesel engine that can be extended above the water's surface so that the engines can be operated while the submarine is submerged
2 : any of various devices (as for an underwater swimmer) resembling a snorkel in function

[2]**snorkel** *intransitive verb* **snor·keled; snor·kel·ing** \-k(ə-)liŋ\ (1949)
: to operate or swim submerged with only a snorkel above water
— **snor·kel·er** \-k(ə-)lər\ *noun*

[1]**snort** \ˈsnȯrt\ *verb* [Middle English] (14th century)
intransitive verb
1 a : to force air violently through the nose with a rough harsh sound **b :** to express scorn, anger, indignation, or surprise by a snort
2 : to emit explosive sounds resembling snorts
3 : to take in a drug by inhalation
transitive verb
1 : to utter with or express by a snort
2 : to expel or emit with or as if with snorts
3 : to take in (a drug) by inhalation

[2]**snort** *noun* (1808)
1 : an act or sound of snorting
2 : a drink of usually straight liquor taken in one draft

snort·er \ˈsnȯr-tər\ *noun* (1601)
1 : one that snorts
2 : something that is extraordinary or prominent **:** HUMDINGER
3 : SNORT 2

snot \ˈsnät\ *noun* [Middle English, from Old English *gesnot;* akin to Old High German *snuzza* nasal mucus] (15th century)
1 : nasal mucus
2 : a snotty person

snot·ty \ˈsnä-tē\ *adjective* **snot·ti·er; -est** (circa 1570)
1 : foul with nasal mucus
2 : annoyingly or spitefully unpleasant
— **snot·ti·ly** \ˈsnä-tᵊl-ē\ *adverb*
— **snot·ti·ness** \ˈsnä-tē-nəs\ *noun*

snout \ˈsnau̇t\ *noun* [Middle English *snute;* akin to Middle Dutch *snūt* snout, German *Schnauze*] (13th century)
1 a (1) **:** a long projecting nose (as of a swine) (2) **:** an anterior prolongation of the head of various animals (as a weevil) **:** ROSTRUM **b :** the human nose especially when large or grotesque
2 : something resembling an animal's snout in position, function, or shape: as **a :** PROW **b :** NOZZLE **c :** the terminal face of a glacier
— **snout·ed** \ˈsnau̇-təd\ *adjective*
— **snout·ish** \-tish\ *adjective*
— **snouty** \-tē\ *adjective*

snout beetle *noun* (1862)
: WEEVIL

[1]**snow** \ˈsnō\ *noun, often attributive* [Middle English, from Old English *snāw;* akin to Old High German *snēo* snow, Latin *niv-, nix,* Greek *nipha* (accusative)] (before 12th century)
1 a : precipitation in the form of small white ice crystals formed directly from the water vapor of the air at a temperature of less than 32°F (0°C) **b** (1) **:** a descent or shower of snow crystals (2) **:** a mass of fallen snow crystals
2 : something resembling snow: as **a :** a dessert made of stiffly beaten whites of eggs, sugar, and fruit pulp ⟨apple *snow*⟩ **b :** a usually white crystalline substance that condenses from a fluid phase as snow does ⟨ammonia *snow*⟩ **c** *slang* (1) **:** COCAINE (2) **:** HEROIN **d :** small transient light or dark spots on a television screen
— **snow·less** \-ləs\ *adjective*

[2]**snow** (14th century)
intransitive verb
: to fall in or as snow
transitive verb
1 : to cause to fall like or as snow
2 a : to cover, shut in, or imprison with or as if with snow **b :** to deceive, persuade, or charm glibly
3 : to whiten like snow

[1]**snow·ball** \ˈsnō-ˌbȯl\ *noun* (15th century)
1 a : a round mass of snow pressed or rolled together **b :** shaved ice molded into a ball and flavored with a syrup
2 : any of several cultivated shrubs (genus *Viburnum*) with clusters of white sterile flowers — called also *snowball bush*

[2]**snowball** (1854)
transitive verb
1 : to throw snowballs at
2 : to cause to increase or multiply at a rapidly accelerating rate
intransitive verb
1 : to engage in throwing snowballs
2 : to increase, accumulate, expand, or multiply at a rapidly accelerating rate

snow·bank \ˈsnō-ˌbaŋk\ *noun* (1779)
: a mound or slope of snow

snow belt \-ˌbelt\ *noun, often capitalized* (1874)
: a region that receives an appreciable amount of annual snowfall

snow·ber·ry \-ˌber-ē\ *noun* (1760)
: any of several white-berried shrubs (especially genus *Symphoricarpos* of the honeysuckle family); *especially* **:** a low-growing North American shrub (*S. albus*) with pink flowers in small axillary clusters

snow·bird \-ˌbərd\ *noun* (1674)
1 : any of several birds (as a junco or fieldfare) seen chiefly in winter
2 : one who travels to warm climes for the winter

snow blindness *noun* (1748)
: inflammation and photophobia caused by exposure of the eyes to ultraviolet rays reflected from snow or ice
— **snow–blind** \ˈsnō-ˌblīnd\ *or* **snow–blind·ed** \-ˌblīn-dəd\ *adjective*

snow·blow·er \ˈsnō-ˌblō(-ə)r\ *noun* (1950)
: SNOW THROWER

snow·board \-ˌbōrd, -ˌbȯrd\ *noun* (1981)
: a board like a wide ski ridden in a surfing position downhill over snow
— **snow·boarder** *noun*
— **snow·board·ing** *noun*

snow·bound \-ˈbau̇nd\ *adjective* (1814)
: shut in or blockaded by snow

snow·brush \-ˌbrəsh\ *noun* (circa 1923)
: a spreading white-flowered western North American shrub (*Ceanothus velutinus*) of the buckthorn family with scented leaves and panicles of small flowers

snow·cap \-ˌkap\ *noun* (1871)
: a covering cap of snow (as on a mountain peak)
— **snow·capped** \-ˌkapt\ *adjective*

snow cone *noun* (1964)
: SNOWBALL 1b

snow crab *noun* (1974)
: either of two crabs (*Chionoecetes opilio* and *C. bairdi*) of the northern Pacific Ocean and especially Alaska marketed for food

snow·drift \ˈsnō-ˌdrift\ *noun* (14th century)
: a bank of drifted snow

snow·drop \-ˌdräp\ *noun* (1664)
: a bulbous European herb (*Galanthus nivalis*) of the amaryllis family bearing nodding white flowers that often appear while the snow is on the ground

snow·fall \-ˌfȯl\ *noun* (1821)
: a fall of snow; *specifically* **:** the amount of snow that falls in a single storm or in a given period

snow fence *noun* (1872)
: a usually slatted fence placed across the path of prevailing winds to protect (as a building, road, or railroad track) from drifting snow by disrupting the flow of wind and causing the snow to be deposited on the lee side of the fence

snow·field \ˈsnō ˌfēld\ *noun* (1845)
: a broad level expanse of snow; *especially* **:** a mass of perennial snow at the head of a glacier

snow·flake \-ˌflāk\ *noun* (1734)
1 : a flake or crystal of snow
2 : any of a genus (*Leucojum*) of bulbous plants of the amaryllis family; *especially* **:** one (*L. vernum*) resembling the snowdrop

snow goose *noun* (1771)
: a wild goose (*Anser caerulescens* synonym *Chen caerulescens*) that has a pinkish bill and exists either as a white form with black primaries or as a grayish black form with a white head

snow job *noun* (1943)
: an intensive effort at persuasion or deception

snow leopard *noun* (1866)
: a large cat (*Panthera uncia*) of upland central Asia with long heavy grayish white fur irregularly marked with brownish black spots, rosettes, and rings

snow leopard

snow line *noun* (circa 1835)
: the lower margin of a perennial snowfield

snow·mak·er \ˈsnō-ˌmā-kər\ *noun* (1954)
: a device for making snow artificially

snow·mak·ing \-ˌmā-kiŋ\ *adjective* (1953)

: used for the production of artificial snow usually for ski slopes ⟨*snowmaking* machines⟩
snow·man \-ˌman\ *noun* (1827)
: snow shaped to resemble a human figure
snow·melt \-ˌmelt\ *noun* (circa 1946)
: runoff produced by the melting of snow
snow·mo·bile \ˈsnō-mō-ˌbēl\ *noun* (1923)
: any of various automotive vehicles for travel on snow; *specifically* : an open vehicle for usually one or two persons with steerable skis on the front and an endless belt at the rear
snow·mo·bil·ing \-ˌbē-liŋ\ *noun* (1964)
: the sport of driving a snowmobile
— **snow·mo·bil·er** \-lər\ *also* **snow·mo·bil·ist** \-list\ *noun*
snow–on–the–mountain *noun* (1873)
: a spurge (*Euphorbia marginata*) of the central and western U.S. that has showy white-bracted flower clusters and is grown as an ornamental
snow·pack \ˈsnō-ˌpak\ *noun* (circa 1946)
: a seasonal accumulation of slow-melting packed snow
snow pea *noun* (1949)
: a cultivated pea with flat edible pods that is classified with the snap pea as a variety (*Pisum sativum macrocarpon*)
snow plant *noun* (1846)
: a fleshy bright-red saprophytic chiefly California herb (*Sarcodes sanguinea*) of the wintergreen family that grows in high-altitude coniferous woods and often appears before the snow melts
¹snow·plow \ˈsnō-ˌplau̇\ *noun* (1792)
1 : any of various devices used for clearing away snow
2 : a stemming with both skis used for coming to a stop, slowing down, or descending slowly
²snowplow *intransitive verb* (1904)
: to execute a snowplow ⟨*snowplowed* to a stop⟩
snow pudding *noun* (1876)
: a pudding made very fluffy and light by the addition of whipped egg whites and gelatin
snow·scape \ˈsnō-ˌskāp\ *noun* (1886)
: a landscape covered with snow
snow·shed \-ˌshed\ *noun* (1868)
: a shelter against snowslides
¹snow·shoe \-ˌshü\ *noun* (1666)
: a light oval frame that is strengthened by two crosspieces, strung with thongs, and attached to the foot and that is used to enable a person to walk on soft snow without sinking
²snowshoe *intransitive verb* **snow·shoed; snow·shoe·ing** (1880)
: to travel on snowshoes
— **snow·sho·er** \-ˌshü-ər\ *noun*
snowshoe hare *noun* (circa 1890)
: a hare (*Lepus americanus*) of northern North America with heavy fur on the hind feet and a coat that in most populations is brown in the summer but usually white in the winter — called also *snowshoe rabbit, varying hare*

1 2
snowshoe hare

snow·slide \ˈsnō-ˌslīd\ *noun* (1841)
: an avalanche of snow
snow·storm \-ˌstȯrm\ *noun* (1771)
1 : a storm of or with snow
2 : something that resembles a snowstorm
snow·suit \-ˌsüt\ *noun* (1937)
: a one-piece or two-piece lined garment for winter wear by children
snow thrower *noun* (1954)
: a machine for removing snow (as from a driveway or sidewalk) in which a rotating spiral blade picks up and propels the snow aside
snow tire *noun* (1943)
: an automotive tire with a tread designed to give added traction on snow
snow under *transitive verb* (1880)
1 : to overwhelm especially in excess of capacity to absorb or deal with something
2 : to defeat by a large margin
snow–white \ˈsnō-ˈhwīt, -ˈwīt\ *adjective* (before 12th century)
: white as snow
snowy \ˈsnō-ē\ *adjective* **snow·i·er; -est** (before 12th century)
1 a : composed of snow or melted snow **b** : marked by or covered with snow
2 a : whitened by snow **b** : SNOW-WHITE
— **snow·i·ly** \ˈsnō-ə-lē\ *adverb*
— **snow·i·ness** \ˈsnō-ē-nəs\ *noun*
snowy egret *noun* (1869)
: a white American egret (*Egretta thula*) having a slender black bill, black legs, and yellow feet
snowy owl *noun* (1781)
: a large ground-nesting diurnal arctic owl (*Nyctea scandiaca*) that enters the chiefly northern parts of the U.S. in winter and has plumage that is sometimes nearly pure white but usually with brownish spots or bars

snowy owl

¹snub \ˈsnəb\ *transitive verb* **snubbed; snub·bing** [Middle English *snubben,* of Scandinavian origin; akin to Old Norse *snubba* to scold; akin to Icelandic *sneypa* to scold] (13th century)
1 : to check or stop with a cutting retort : REBUKE
2 a : to check (as a line) suddenly while running out especially by turning around a fixed object (as a post); *also* : to check the motion of by snubbing a line ⟨*snub* a horse to a tree⟩ **b** : to restrain the action of : SUPPRESS ⟨*snub* a vibration⟩
3 : to treat with contempt or neglect
4 : to extinguish by stubbing ⟨*snub* out a cigarette⟩
²snub *noun* (14th century)
: an act or an instance of snubbing; *especially* : SLIGHT
³snub *adjective* (1724)
1 *or* **snubbed** \ˈsnəbd\ : BLUNT, STUBBY ⟨a *snub* nose⟩
2 : used in snubbing ⟨*snub* line⟩
— **snub·ness** *noun*
snub·ber \ˈsnə-bər\ *noun* (1853)
1 : one that snubs
2 : SHOCK ABSORBER
snub·by \ˈsnə-bē\ *adjective* (1828)
1 : SNUB
2 : SNUB-NOSED
— **snub·bi·ness** *noun*
snub–nosed \ˈsnəb-ˌnōzd\ *adjective* (1725)
1 : having a stubby and usually slightly turned-up nose
2 : having a very short barrel ⟨a *snub-nosed* revolver⟩
snuck *past and past participle of* SNEAK
¹snuff \ˈsnəf\ *noun* [Middle English *snoffe*] (14th century)
1 : the charred part of a candlewick
2 a *obsolete* : UMBRAGE, OFFENSE **b** *chiefly Scottish* : HUFF
²snuff *transitive verb* (15th century)
1 : to crop the snuff of (a candle) by pinching or by the use of snuffers so as to brighten the light
2 a : to extinguish by or as if by the use of a snuffer — often used with *out* **b** : to make extinct : put an end to — usually used with *out* ⟨an accident that *snuffed* out a life⟩
³snuff *adjective* (1975)
: characterized by the sensationalistic depiction of violence; *especially* : featuring a real rather than a staged murder ⟨*snuff* movies⟩
⁴snuff *verb* [akin to Dutch *snuffen* to sniff, snuff — more at SNIVEL] (1527)
transitive verb
1 : to draw forcibly through or into the nostrils
2 : SCENT, SMELL
3 : to sniff at in order to examine — used of an animal
intransitive verb
1 : to inhale through the nose noisily and forcibly; *also* : to sniff or smell inquiringly
2 *obsolete* : to sniff loudly in or as if in disgust
3 : to take snuff
⁵snuff *noun* (1570)
: the act of snuffing : SNIFF
⁶snuff *noun* [Dutch *snuf,* short for *snuftabak,* from *snuffen* to snuff + *tabak* tobacco] (1683)
1 : a preparation of pulverized tobacco to be inhaled through the nostrils, chewed, or placed against the gums
2 : the amount of snuff taken at one time
— **up to snuff** : of sufficient quality : meeting an applicable standard
snuff·box \ˈsnəf-ˌbäks\ *noun* (circa 1687)
: a small box for holding snuff usually carried about the person
¹snuff·er \ˈsnə-fər\ *noun* (15th century)
1 : a device similar to a pair of scissors for cropping and holding the snuff of a candle — usually used in plural but singular or plural in construction
2 : a device for extinguishing candles
²snuffer *noun* (circa 1610)
: one that snuffs or sniffs
¹snuf·fle \ˈsnə-fəl\ *verb* **snuf·fled; snuf·fling** \-f(ə-)liŋ\ [akin to Dutch *snuffelen* to snuffle — more at SNIVEL] (circa 1600)
intransitive verb
1 : to snuff or sniff usually audibly and repeatedly
2 : to breathe through an obstructed nose with a sniffing sound
3 : to speak through or as if through the nose : WHINE
transitive verb
: to seek or test by or as if by repeated sniffs
— **snuf·fler** \-f(ə-)lər\ *noun*
²snuffle *noun* (circa 1764)
1 : the act or sound of snuffling
2 : a nasal twang
3 *plural* : SNIFFLES
¹snuffy \ˈsnə-fē\ *adjective* [⁴*snuff*] (1678)
1 : quick to become annoyed or take offense
2 : marked by snobbery
²snuffy *adjective* [⁶*snuff*] (1787)
1 : resembling snuff
2 a : addicted to the use of snuff **b** : having unpleasant habits
3 : soiled with snuff
¹snug \ˈsnəg\ *verb* **snugged; snug·ging** [²*snug*] (1583)
intransitive verb
: SNUGGLE
transitive verb
1 : to cause to fit closely
2 : to make snug
3 : HIDE
4 : to secure by fastening or lashing down
²snug *adjective* **snug·ger; snug·gest** [perhaps of Scandinavian origin; akin to Swedish *snygg* tidy] (circa 1595)
1 a *of a ship* : manifesting seaworthiness : TAUT **b** : TRIM, NEAT **c** : fitting closely and comfortably ⟨a *snug* coat⟩
2 a : enjoying or affording warm secure shelter or cover and opportunity for ease and contentment **b** : marked by cordiality and secure privacy
3 : affording a degree of comfort and ease
4 : offering safe concealment ⟨a *snug* hideout⟩
synonym see COMFORTABLE
— **snug** *adverb*
— **snug·ly** *adverb*
— **snug·ness** *noun*
³snug *noun* [short for *snuggery*] (1860)

British **:** a small private room or compartment in a pub

snug·gery \ˈsnə-g(ə-)rē\ *noun, plural* **-ger·ies** (1812)
chiefly British **:** a snug cozy place; *especially* **:** a small room

snug·gle \ˈsnə-gəl\ *verb* **snug·gled; snug·gling** \-g(ə-)liŋ\ [frequentative of [1]*snug*] (1687)
intransitive verb
: to curl up comfortably or cozily
transitive verb
1 : to draw close especially for comfort or in affection
2 : to make snug
— snuggle *noun*

[1]so \ˈsō, *especially before adjective or adverb followed by "that"* sə\ *adverb* [Middle English, from Old English *swā;* akin to Old High German *sō* so, Latin *sic* so, thus, *si* if, Greek *hōs* so, thus, Latin *suus* one's own — more at SUICIDE] (before 12th century)
1 a : in a manner or way indicated or suggested ⟨do you really think *so*⟩ — often used as a substitute for a preceding clause ⟨are you ready? I think *so*⟩ ⟨I didn't like it and I told her *so*⟩ **b :** in the same manner or way **:** ALSO ⟨worked hard and *so* did she⟩ **c :** THUS ⟨for *so* the Lord said —Isaiah 18:4 (Authorized Version)⟩ **d :** THEN, SUBSEQUENTLY ⟨and *so* home and to bed⟩
2 a : to an indicated or suggested extent or degree ⟨had never been *so* happy⟩ **b :** to a great extent or degree **:** VERY, EXTREMELY ⟨loves her *so*⟩ **c :** to a definite but unspecified extent or degree ⟨can only do *so* much in a day⟩ **d :** most certainly **:** INDEED ⟨you did *so* do it⟩
3 : THEREFORE, CONSEQUENTLY ⟨the witness is biased and *so* unreliable⟩ ▪

[2]so *conjunction* (before 12th century)
1 a : with the result that ⟨the acoustics are good, *so* every note is clear⟩ **b :** in order that ⟨be quiet *so* he can sleep⟩
2 *archaic* **:** provided that
3 a : for that reason **:** THEREFORE ⟨don't want to go, *so* I won't⟩ **b** (1) — used as an introductory particle ⟨*so* here we are⟩ often to belittle a point under discussion ⟨*so* what?⟩ (2) — used interjectionally to indicate awareness of a discovery ⟨*so*, that's who did it⟩ or surprised dissent ▪

[3]so *adjective* (before 12th century)
1 : conforming with actual facts **:** TRUE ⟨said things that were not *so*⟩
2 : marked by a desired order ⟨his books are always just *so*⟩
3 — used to replace a preceding adjective ⟨was witty by adult standards and of course doubly *so* by mine —Sally Kempton⟩

[4]so *pronoun* (before 12th century)
1 : such as has been specified or suggested **:** the same ⟨if you have to file a claim, do *so* as soon as possible⟩
2 — used in the phrase *or so* to indicate an estimate, approximation, or conjecture ⟨stayed a week or *so*⟩ ⟨cost $15 or *so*⟩

[5]so *variant of* SOL

[1]soak \ˈsōk\ *verb* [Middle English *soken,* from Old English *socian;* akin to Old English *sūcan* to suck] (before 12th century)
intransitive verb
1 : to lie immersed in liquid (as water) **:** become saturated by or as if by immersion
2 a : to enter or pass through something by or as if by pores or interstices **:** PERMEATE **b :** to penetrate or affect the mind or feelings — usually used with *in* or *into*
3 : to drink alcoholic beverages intemperately
transitive verb
1 : to permeate so as to wet, soften, or fill thoroughly
2 : to place in a surrounding element (as liquid) to wet or permeate thoroughly
3 : to extract by or as if by steeping ⟨*soak* the dirt out⟩
4 a : to draw in by or as if by suction or absorption ⟨*soaked* up the sunshine⟩ **b :** to intoxicate (oneself) by drinking alcoholic beverages
5 : to cause to pay an exorbitant amount ☆
— soak·er *noun*

[2]soak *noun* (15th century)
1 a : the act or process of soaking **:** the state of being soaked **b :** that (as liquid) in which something is soaked
2 : DRUNKARD
3 *slang* **:** [2]PAWN 2

[1]so-and-so \ˈsō-ən-ˌsō\ *noun, plural* **so-and-sos** *or* **so-and-so's** \-ˌsōz\ (1596)
1 : an unnamed or unspecified person, thing, or action
2 : BASTARD 3

[2]so-and-so *adverb* (1631)
1 : to an unspecified amount or degree
2 : in an unspecified manner or fashion

[1]soap \ˈsōp\ *noun* [Middle English *sope,* from Old English *sāpe;* akin to Old High German *seifa* soap] (before 12th century)
1 a : a cleansing and emulsifying agent made usually by action of alkali on fat or fatty acids and consisting essentially of sodium or potassium salts of such acids **b :** a salt of a fatty acid and a metal
2 : SOAP OPERA

[2]soap *transitive verb* (1585)
1 : to rub soap over or into
2 : FLATTER

soap·bark \ˈsōp-ˌbärk\ *noun* (1861)
: a Chilean tree (*Quillaja saponaria*) of the rose family with glossy leaves and terminal white flowers; *also* **:** its saponin-rich bark used in cleaning and in emulsifying oils

soap·ber·ry \-ˌber-ē\ *noun* (1693)
: any of a genus (*Sapindus* of the family Sapindaceae, the soapberry family) of chiefly tropical woody plants; *also* **:** the fruit of a soapberry and especially of a tree (*S. saponaria*) that is saponin-rich and used as a soap substitute

soap·box \-ˌbäks\ *noun* (1907)
: an improvised platform used by a self-appointed, spontaneous, or informal orator; *broadly* **:** something that provides an outlet for delivering opinions
— soapbox *adjective*

soap bubble *noun* (1800)
: a hollow iridescent globe formed by blowing a film of soapsuds (as from a pipe)

soap·er \ˈsō-pər\ *noun* (1946)
: SOAP OPERA

soap opera *noun* [from its sponsorship by soap manufacturers] (1939)
1 a : a serial drama performed originally on a daytime radio or television program and chiefly characterized by tangled interpersonal situations and melodramatic or sentimental treatment **b :** a series of real-life events resembling a soap opera
2 : the melodrama and sentimentality characteristic of a soap opera ⟨even cops need a little *soap opera* in their lives —Joseph Wambaugh⟩; *also* **:** something (as a novel) having such qualities

soap plant *noun* (1844)
: a plant having a part (as a root or fruit) that may be used in place of soap; *especially* **:** a California plant (*Chlorogalum pomeridianum*) of the lily family

soap·stone \ˈsōp-ˌstōn\ *noun* (circa 1681)
: a soft stone having a soapy feel and composed essentially of talc, chlorite, and often some magnetite

soap·suds \-ˌsədz\ *noun plural* (1611)
: SUDS 1

soap·wort \-ˌwərt, -ˌwȯrt\ *noun* (1548)
: BOUNCING BET

soapy \ˈsō-pē\ *adjective* **soap·i·er; -est** (1610)
1 : smeared with soap **:** LATHERED
2 : containing or combined with soap or saponin
3 a : resembling or having the qualities of soap; *especially* **:** being smooth and slippery **b :** UNCTUOUS, SUAVE
4 : of, relating to, or having the characteristics of soap opera
— soap·i·ly \-pə-lē\ *adverb*
— soap·i·ness \-pē-nəs\ *noun*

[1]soar \ˈsōr, ˈsȯr\ *intransitive verb* [Middle English *soren,* from Middle French *essorer,* from (assumed) Vulgar Latin *exaurare,* from Latin *ex-* + *aura* air — more at AURA] (14th century)
1 a : to fly aloft or about **b** (1) **:** to sail or hover in the air often at a great height **:** GLIDE (2) *of a glider* **:** to fly without engine power and without loss of altitude
2 : to rise or increase dramatically (as in position, value, or price)
3 : to ascend to a higher or more exalted level
4 : to rise to majestic stature
— soar·er *noun*

[2]soar *noun* (1596)
1 : the range, distance, or height attained in soaring
2 : the act of soaring **:** upward flight

soar·ing *noun* (15th century)
: the act or process of soaring; *specifically* **:** the act or sport of flying a heavier-than-air craft without power by utilizing ascending air currents

Soa·ve \ˈswä-(ˌ)vā\ *noun* [*Soave,* village near Verona, Italy] (1935)

☆ SYNONYMS
Soak, saturate, drench, steep, impregnate mean to permeate or be permeated with a liquid. SOAK implies usually prolonged immersion as for softening or cleansing ⟨*soak* the garment in soapy water⟩. SATURATE implies a resulting effect of complete absorption until no more liquid can be held ⟨a *saturated* sponge⟩. DRENCH implies a thorough wetting by something that pours down or is poured ⟨clothes *drenched* by a cloudburst⟩. STEEP suggests either the extraction of an essence (as of tea leaves) by the liquid or the imparting of a quality (as a color) to the thing immersed ⟨*steep* the tea for five minutes⟩. IMPREGNATE implies a thorough interpenetration of one thing by another ⟨a cake strongly *impregnated* with brandy⟩.

□ USAGE
[1]so The intensive use of the adverb *so* (sense 2b) is widely condemned in college handbooks but is nonetheless standard ⟨why is American television *so* shallow? —Anthony Lewis⟩ ⟨the cephalopod eye is an example of a remarkable evolutionary parallel because it is *so* like the eye of a vertebrate —Sarah F. Robbins⟩ ⟨the kind of sterile over-ingenuity which afflicts *so* many academic efforts —*Times Literary Supplement*⟩. There is no stigma attached to its use in negative contexts and when qualified by a dependent clause ⟨not *so* long ago⟩ ⟨was *so* good in mathematics that he began to consider engineering —*Current Biography*⟩. The denotation in these uses is, of course, slightly different (see sense 2a).

[2]so Although occasionally condemned, use of the conjunction *so* to introduce clauses of result (sense 1a) and purpose (sense 1b) is standard. In sense 1b *so that* is more common in formal contexts than *so* alone.

: a dry white Italian wine

[1]**sob** \'säb\ *verb* **sobbed; sob·bing** [Middle English *sobben*] (13th century)
intransitive verb
1 a : to catch the breath audibly in a spasmodic contraction of the throat **b :** to cry or weep with convulsive catching of the breath
2 : to make a sound like that of a sob or sobbing
transitive verb
1 : to bring (as oneself) to a specified state by sobbing ⟨*sobbed* himself to sleep⟩
2 : to utter with sobs ⟨*sobbed* out her grief⟩

[2]**sob** *noun* (14th century)
1 : an act of sobbing
2 : a sound like that of a sob

SOB \ˌes-ˌō-'bē\ *noun* [*s*on *o*f a *b*itch] (1918)
: BASTARD 3, SON OF A BITCH

[1]**so·ber** \'sō-bər\ *adjective* **so·ber·er** \-bər-ər\; **so·ber·est** \-b(ə-)rəst\ [Middle English *sobre,* from Middle French, from Latin *sobrius;* akin to Latin *ebrius* drunk] (14th century)
1 a : sparing in the use of food and drink **:** ABSTEMIOUS **b :** not addicted to intoxicating drink **c :** not drunk
2 : marked by sedate or gravely or earnestly thoughtful character or demeanor
3 : UNHURRIED, CALM
4 : marked by temperance, moderation, or seriousness
5 : subdued in tone or color
6 : showing no excessive or extreme qualities of fancy, emotion, or prejudice
synonym see SERIOUS
— so·ber·ly \-bər-lē\ *adverb*
— so·ber·ness *noun*

[2]**sober** *verb* **so·bered; so·ber·ing** \-b(ə-)riŋ\ (14th century)
transitive verb
: to make sober
intransitive verb
: to become sober — usually used with *up*

so·ber·ing *adjective* (1816)
: tending to make one thoughtful or sober

so·ber·ize \'sō-bə-ˌrīz\ *transitive verb* **-ized; -iz·ing** (1706)
archaic **:** to make sober

so·ber·sid·ed \ˌsō-bər-'sī-dəd\ *adjective* (1847)
: solemn or serious in nature or appearance
— so·ber·sid·ed·ness \-nəs\ *noun*

so·ber·sides \'sō-bər-ˌsīdz\ *noun plural but singular or plural in construction* (1705)
: a sobersided person

so·bri·ety \sə-'brī-ə-tē, sō-\ *noun* [Middle English *sobrietie,* from Middle French *sobrieté,* from Latin *sobrietat-, sobrietas,* from *sobrius*] (15th century)
: the quality or state of being sober

so·bri·quet \'sō-bri-ˌkā, -ˌket, ˌsō-bri-'\ *noun* [French] (1646)
: a descriptive name or epithet **:** NICKNAME

sob sister *noun* (1912)
1 : a journalist who specializes in writing or editing sob stories or other material of a sentimental type
2 : a sentimental and often impractical person usually engaged in good works

sob story *noun* (1913)
: a sentimental story or account intended chiefly to evoke sympathy or sadness

so·cage \'sä-kij, 'sō-\ *also* **soc·cage** \'sä-\ *noun* [Middle English, from *soc* soke] (14th century)
: a tenure of land by agricultural service fixed in amount and kind or by payment of money rent only and not burdened with any military service
— so·cag·er \-ki-jər\ *noun*

so–called \'sō-'kȯld\ *adjective* (15th century)
1 : commonly named **:** popularly so termed ⟨the *so-called* pocket veto⟩
2 : falsely or improperly so named ⟨deceived by a *so-called* friend⟩

soc·cer \'sä-kər\ *noun* [by shortening & alteration from *association football*] (1889)
: a game played on a field between two teams of 11 players each with the object to propel a round ball into the opponent's goal by kicking or by hitting it with any part of the body except the hands and arms — called also *association football* ◆

so·cia·bil·i·ty \ˌsō-shə-'bi-lə-tē\ *noun, plural* **-ties** (15th century)
: the quality or state of being sociable; *also* **:** the act or an instance of being sociable

[1]**so·cia·ble** \'sō-shə-bəl\ *adjective* [Middle French or Latin; Middle French, from Latin *sociabilis,* from *sociare* to join, associate, from *socius*] (1553)
1 : inclined by nature to companionship with others of the same species **:** SOCIAL
2 a : inclined to seek or enjoy companionship **b :** marked by or conducive to friendliness or pleasant social relations
synonym see GRACIOUS
— so·cia·ble·ness *noun*
— so·cia·bly \-blē\ *adverb*

[2]**sociable** *noun* (1750)
: an informal social gathering frequently involving a special activity or interest

[1]**so·cial** \'sō-shəl\ *adjective* [Middle English, from Latin *socialis,* from *socius* companion, ally, associate; akin to Old English *secg* man, companion, Latin *sequi* to follow — more at SUE] (14th century)
1 : involving allies or confederates ⟨the *Social* War between the Athenians and their allies⟩
2 a : marked by or passed in pleasant companionship with one's friends or associates ⟨leads a very full *social* life⟩ **b :** SOCIABLE **c :** of, relating to, or designed for sociability ⟨a *social* club⟩
3 : of or relating to human society, the interaction of the individual and the group, or the welfare of human beings as members of society ⟨*social* institutions⟩
4 a : tending to form cooperative and interdependent relationships with others of one's kind **:** GREGARIOUS **b :** living and breeding in more or less organized communities ⟨*social* insects⟩ **c** *of a plant* **:** tending to grow in groups or masses so as to form a pure stand
5 a : of, relating to, or based on rank or status in a particular society ⟨a member of our *social* set⟩ **b :** of, relating to, or characteristic of the upper classes **c :** FORMAL

[2]**social** *noun* (1870)
: SOCIABLE

social climber *noun* (1924)
: one who attempts to gain a higher social position or acceptance in fashionable society
— social climbing *noun*

social contract *noun* (circa 1850)
: an actual or hypothetical agreement among individuals forming an organized society or between the community and the ruler that defines and limits the rights and duties of each

social Darwinism *noun* (1887)
: an extension of Darwinism to social phenomena; *specifically* **:** a theory in sociology: sociocultural advance is the product of intergroup conflict and competition and the socially elite classes (as those possessing wealth and power) possess biological superiority in the struggle for existence
— social Darwinist *noun or adjective*

social democracy *noun* (1888)
: a political movement advocating a gradual and peaceful transition from capitalism to socialism by democratic means
— social democrat *noun*
— social democratic *adjective*

social disease *noun* (1891)
1 : VENEREAL DISEASE
2 : a disease (as tuberculosis) whose incidence is directly related to social and economic factors

social drinker *noun* (circa 1949)
: a person who drinks alcoholic beverages in moderation at social gatherings especially as distinguished from one who drinks habitually or to excess

social engineering *noun* (1925)
: management of human beings in accordance with their place and function in society **:** applied social science
— social engineer *noun*

social gospel *noun* (1890)
1 : the application of Christian principles to social problems
2 *S&G capitalized* **:** a movement in American Protestant Christianity especially in the first part of the 20th century to bring the social order into conformity with Christian principles

social insurance *noun* (1909)
: protection of the individual against economic hazards (as unemployment, old age, or disability) in which the government participates or enforces the participation of employers and affected individuals

so·cial·ise *British variant of* SOCIALIZE

so·cial·ism \'sō-shə-ˌli-zəm\ *noun* (1837)
1 : any of various economic and political theories advocating collective or governmental ownership and administration of the means of production and distribution of goods
2 a : a system of society or group living in which there is no private property **b :** a system or condition of society in which the means of production are owned and controlled by the state
3 : a stage of society in Marxist theory transitional between capitalism and communism and distinguished by unequal distribution of goods and pay according to work done

so·cial·ist \'sō-sh(ə-)list\ *noun* (1827)
1 : one who advocates or practices socialism
2 *capitalized* **:** a member of a party or political group advocating socialism
— socialist *adjective, often capitalized*
— so·cial·is·tic \ˌsō-shə-'lis-tik\ *adjective*
— so·cial·is·ti·cal·ly \-ti-k(ə-)lē\ *adverb*

socialist realism *noun* (1934)
: a Marxist aesthetic theory calling for the didactic use of literature, art, and music to develop social consciousness in an evolving socialist state
— socialist realist *noun or adjective*

so·cial·ite \'sō-shə-ˌlīt\ *noun* (1928)
: a socially prominent person

◇ WORD HISTORY

soccer At University College, Oxford, in 1875, it became fashionable among students to create slang versions of everyday words or phrases by clipping them to a single syllable and adding the empty suffix *-er.* Allegedly this linguistic fad originated at Rugby, one of England's exclusive private boys schools (called by tradition "public schools"). Whether this is true or not, *footer* for *football* is attested at Harrow, another private school, in 1863—the earliest of these clipped words to see print. Most examples, such as *fresher* "freshman," *brekker* "breakfast," *tosher* "student unattached to a college," and *ekker* "exercise" are unfamiliar to North Americans, and many more have proved ephemeral even in Britain. However, one *-er* coinage has become extraordinarily successful in the U.S.: the word *soccer,* clipped from *association* (or *assoc.*) *football,* the game played according to the rules of the Football Association founded in England in 1863. *Football* now denotes only the game of soccer in most of the English-speaking world and has been borrowed into many other languages around the world. But in the U.S. a different kind of kicking game claimed the word, and *soccer* has been adopted as the name for what Americans still tend to think of as an imported sport.

so·ci·al·i·ty \ˌsō-shē-ˈa-lə-tē\ *noun, plural* **-ties** (circa 1649)
1 a : SOCIABILITY **b :** an instance of social intercourse or sociability
2 : the tendency to associate in or form social groups

so·cial·ize \ˈsō-shə-ˌlīz\ *verb* **-ized; -iz·ing** (1828)
transitive verb
1 : to make social; *especially* **:** to fit or train for a social environment
2 a : to constitute on a socialistic basis ⟨*socialize* industry⟩ **b :** to adapt to social needs or uses ⟨*socialize* science⟩
3 : to organize group participation in ⟨*socialize* a recitation⟩
intransitive verb
: to participate actively in a social group
— so·cial·i·za·tion \ˌsō-sh(ə-)lə-ˈzā-shən\ *noun*
— so·cial·iz·er \ˈsō-shə-ˌlī-zər\ *noun*

socialized medicine *noun* (1938)
: medical and hospital services for the members of a class or population administered by an organized group (as a state agency) and paid for from funds obtained usually by assessments, philanthropy, or taxation

so·cial·ly \ˈsō-sh(ə-)lē\ *adverb* (circa 1763)
1 : in a social manner
2 : with respect to society
3 : by or through society

so·cial–mind·ed \ˌsō-shəl-ˈmīn-dəd\ *adjective* (1927)
: having an interest in society; *specifically* **:** actively interested in social welfare or the well-being of society as a whole

social psychology *noun* (1891)
: the study of the manner in which the personality, attitudes, motivations, and behavior of the individual influence and are influenced by social groups
— social psychologist *noun*

social science *noun* (1811)
1 : a branch of science that deals with the institutions and functioning of human society and with the interpersonal relationships of individuals as members of society
2 : a science (as economics or political science) dealing with a particular phase or aspect of human society
— social scientist *noun*

social secretary *noun* (1903)
: a personal secretary employed to handle social correspondence and appointments

social security *noun* (1908)
1 : the principle or practice or a program of public provision (as through social insurance or assistance) for the economic security and social welfare of the individual and his or her family; *especially, often both Ss capitalized* **:** a U.S. government program established in 1935 to include old-age and survivors insurance, contributions to state unemployment insurance, and old-age assistance
2 : money paid out through a social security program ⟨began collecting *social security*⟩

social service *noun* (1851)
: an activity designed to promote social well-being; *specifically* **:** organized philanthropic assistance of the sick, destitute, or unfortunate **:** WELFARE WORK

social studies *noun plural* (1926)
: a part of a school or college curriculum concerned with the study of social relationships and the functioning of society and usually made up of courses in history, government, economics, civics, sociology, geography, and anthropology

social welfare *noun* (1917)
: organized public or private social services for the assistance of disadvantaged groups; *specifically* **:** SOCIAL WORK

social work *noun* (1890)
: any of various professional activities or methods concretely concerned with providing social services and especially with the investigation, treatment, and material aid of the economically underprivileged and socially maladjusted
— social worker *noun*

so·ci·e·tal \sə-ˈsī-ə-t^{ə}l\ *adjective* (1898)
: of or relating to society **:** SOCIAL ⟨*societal* forces⟩
— so·ci·e·tal·ly \-t^{ə}l-ē\ *adverb*

[1]**so·ci·e·ty** \sə-ˈsī-ə-tē\ *noun, plural* **-ties** [Middle French *societé,* from Latin *societat-, societas,* from *socius* companion — more at SOCIAL] (1531)
1 : companionship or association with one's fellows **:** friendly or intimate intercourse **:** COMPANY
2 : a voluntary association of individuals for common ends; *especially* **:** an organized group working together or periodically meeting because of common interests, beliefs, or profession
3 a : an enduring and cooperating social group whose members have developed organized patterns of relationships through interaction with one another **b :** a community, nation, or broad grouping of people having common traditions, institutions, and collective activities and interests
4 a : a part of a community that is a unit distinguishable by particular aims or standards of living or conduct **:** a social circle or a group of social circles having a clearly marked identity ⟨move in polite *society*⟩ ⟨literary *society*⟩ **b :** a part of the community that sets itself apart as a leisure class and that regards itself as the arbiter of fashion and manners
5 a : a natural group of plants usually of a single species or habit within an association **b :** the progeny of a pair of insects when constituting a social unit (as a hive of bees); *broadly* **:** an interdependent system of organisms or biological units

[2]**society** *adjective* (1693)
: of, relating to, or characteristic of fashionable society

So·cin·i·an \sə-ˈsi-nē-ən, sō-\ *noun* [New Latin *socinianus,* from Faustus *Socinus*] (1645)
: an adherent of a 16th and 17th century theological movement professing belief in God and adherence to the Christian Scriptures but denying the divinity of Christ and consequently denying the Trinity
— Socinian *adjective*
— So·cin·i·an·ism \-nē-ə-ˌni-zəm\ *noun*

socio- *combining form* [French, from Latin *socius* companion]
1 : society **:** social ⟨*socio*gram⟩
2 : social and ⟨*socio*political⟩

so·cio·bi·ol·o·gy \ˌsō-sē-ō-bī-ˈä-lə-jē, ˌsō-shē-\ *noun* (1946)
: the comparative study of social organization in animals including humans especially with regard to its genetic basis and evolutionary history
— so·cio·bio·log·i·cal \-ˌbī-ə-ˈlä-ji-kəl\ *adjective*
— so·cio·bi·ol·o·gist \-bī-ˈä-lə-jist\ *noun*

so·cio·cul·tur·al \-ˈkəlch-rəl, -ˈkəl-chə-\ *adjective* (1928)
: of, relating to, or involving a combination of social and cultural factors
— so·cio·cul·tur·al·ly \-rə-lē\ *adverb*

so·cio·eco·nom·ic \-ˌe-kə-ˈnä-mik, -ˌē-kə-\ *adjective* (1883)
: of, relating to, or involving a combination of social and economic factors
— so·cio·eco·nom·i·cal·ly \-mi-k(ə-)lē\ *adverb*

so·cio·gram \ˈsō-sē-ə-ˌgram, ˈsō-shē-\ *noun* (1933)
: a sociometric chart plotting the structure of interpersonal relations in a group situation

so·cio·his·tor·i·cal \ˌsō-sē-ō-his-ˈtȯr-i-kəl, ˌsō-shē-, -ˈtär-\ *adjective* (1949)
: of, relating to, or involving social history or a combination of social and historical factors

so·cio·lin·guist \-ˈliŋ-gwist\ *noun* (1960)
: a linguist specializing in sociolinguistics

so·cio·lin·guis·tic \-liŋ-ˈgwis-tik\ *adjective* (1949)
1 : of or relating to the social aspects of language
2 : of or relating to sociolinguistics

so·cio·lin·guis·tics \-tiks\ *noun plural but singular in construction* (1938)
: the study of linguistic behavior as determined by sociocultural factors

so·ci·ol·o·gese \ˌsō-sē-ˌä-lə-ˈjēz, ˌsō-shē-, -ˈjēs\ *noun* [*sociology* + [2]*-ese*] (1952)
: a style of writing held to be characteristic of sociologists

so·cio·log·i·cal \ˌsō-sē-ə-ˈlä-ji-kəl, ˌsō-sh(ē-)ə-\ *also* **so·cio·log·ic** \-jik\ *adjective* (1843)
1 : of or relating to sociology or to the methodological approach of sociology
2 : oriented or directed toward social needs and problems
— so·cio·log·i·cal·ly \-ji-k(ə-)lē\ *adverb*

so·ci·ol·o·gy \ˌsō-sē-ˈä-lə-jē, ˌsō-shē-\ *noun* [French *sociologie,* from *socio-* + *-logie* -logy] (1843)
1 : the science of society, social institutions, and social relationships; *specifically* **:** the systematic study of the development, structure, interaction, and collective behavior of organized groups of human beings
2 : the scientific analysis of a social institution as a functioning whole and as it relates to the rest of society
3 : SYNECOLOGY
— so·ci·ol·o·gist \-jist\ *noun*

so·ci·om·e·try \-ˈä-mə-trē\ *noun* [International Scientific Vocabulary] (1908)
: the study and measurement of interpersonal relationships in a group of people
— so·cio·met·ric \ˌsō-sē-ə-ˈme-trik, ˌsō-shē-\ *adjective*

so·cio·path \ˈsō-sē-ə-ˌpath, ˈsō-sh(ē-)ə-\ *noun* (1930)
: PSYCHOPATH

so·cio·path·ic \ˌsō-sē-ə-ˈpa-thik, ˌsō-sh(ē-)ə-\ *adjective* (1930)
: of, relating to, or characterized by asocial or antisocial behavior or a psychopathic personality

so·cio·po·lit·i·cal \ˌsō-sē-ō-pə-ˈli-ti-kəl, ˌsō-shē-\ *adjective* (1884)
: of, relating to, or involving a combination of social and political factors

so·cio·psy·cho·log·i·cal \-ˌsī-kə-ˈlä-ji-kəl\ *adjective* (1899)
1 : of, relating to, or involving a combination of social and psychological factors
2 : of or relating to social psychology

so·cio·re·li·gious \-ri-ˈli-jəs\ *adjective* (1871)
: involving a combination of social and religious factors

so·cio·sex·u·al \-ˈsek-sh(ə-)wəl, -ˈsek-shəl\ *adjective* (1932)
: of or relating to the interpersonal aspects of sexuality

[1]**sock** \ˈsäk\ *noun, plural* **socks** [Middle English *socke,* from Old English *socc,* from Latin *soccus*] (before 12th century)
1 *archaic* **:** a low shoe or slipper
2 *also plural* **sox** \ˈsäks\ **:** a knitted or woven covering for the foot usually extending above the ankle and sometimes to the knee
3 a : a shoe worn by actors in Greek and Roman comedy **b :** comic drama
— sock·less *adjective*

[2]**sock** *verb* [origin unknown] (circa 1700)
transitive verb
: to hit, strike, or apply forcefully ⟨*sock* a home run⟩ ⟨an area *socked* by a blizzard⟩

intransitive verb
: to deliver a blow **:** HIT
— **sock it to** *slang* **:** to subject to or as if to a vigorous assault ⟨they may let you off the first time . . . but the second time they'll *sock it to* you —James Jones⟩
³sock *noun* (circa 1700)
: a vigorous or violent blow; *also* **:** ²PUNCH 3
sock away *transitive verb* [from the practice of concealing savings in the toe of a sock] (circa 1942)
: to put away (money) as savings or investment
sock·dol·a·ger *or* **sock·dol·o·ger** \säk-ˈdä-li-jər\ *noun* [origin unknown] (circa 1830)
1 : something that settles a matter **:** a decisive blow or answer **:** FINISHER
2 : something outstanding or exceptional
¹sock·et \ˈsä-kət\ *noun* [Middle English *soket,* from Anglo-French, diminutive of Old French *soc* plowshare, of Celtic origin; akin to Middle Irish *soc* plowshare, snout; akin to Old English *sugu* sow — more at SOW] (15th century)
: an opening or hollow that forms a holder for something ⟨an electric bulb *socket*⟩ ⟨the eye *socket*⟩
²socket *transitive verb* (1533)
: to provide with or support in or by a socket
socket wrench *noun* (circa 1890)
: a wrench usually in the form of a bar and removable socket made to fit a bolt or nut

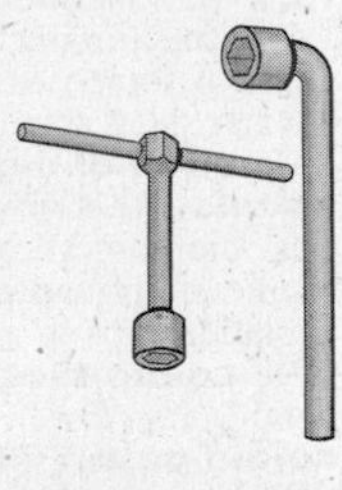

socket wrenches

sock·eye \ˈsäk-ˌī\ *noun* [by folk etymology from Northern Straits (Salishan language of southern Vancouver Island and nearby islands) *sə́qəy̓*] (1869)
: a small but commercially important Pacific salmon (*Oncorhynchus nerka*) that ascends rivers chiefly from the Columbia northward to spawn in late summer or fall — called also *red salmon*
sock in *transitive verb* [(*wind*) *sock*] (1944)
1 : to close to takeoffs or landings by aircraft ⟨an airport *socked in* by fog⟩
2 : to restrict from flying
socko \ˈsä-(ˌ)kō\ *adjective* [²*sock*] (1938)
: strikingly impressive, effective, or successful **:** OUTSTANDING
so·cle \ˈsō-kəl, ˈsä-\ *noun* [French, from Italian *zoccolo* sock, socle, from Latin *socculus,* diminutive of *soccus* sock] (circa 1704)
: a projecting usually molded member at the foot of a wall or pier or beneath the base of a column, pedestal, or superstructure
¹So·crat·ic \sə-ˈkra-tik, sō-\ *adjective* (1628)
: of or relating to Socrates, his followers, or his philosophical method of systematic doubt and questioning of another to elicit a clear expression of a truth supposed to be implicitly known by all rational beings
— **So·crat·i·cal·ly** \-ti-k(ə-)lē\ *adverb*
²Socratic *noun* (1678)
: a follower of Socrates
Socratic irony *noun* (circa 1871)
: IRONY 1
¹sod \ˈsäd\ *noun* [Middle English, from Middle Dutch or Middle Low German *sode;* akin to Old Frisian *sātha* sod] (15th century)
1 : TURF 1; *also* **:** the grass- and forb-covered surface of the ground
2 : one's native land
²sod *transitive verb* **sod·ded; sod·ding** (1653)
: to cover with sod or turfs
³sod *noun* [short for *sodomite*] (1818)
chiefly British **:** BUGGER ⟨if I ever find the *sod* I'll kill him —John Le Carré⟩ ⟨he's not a bad little *sod* taken by and large —Noel Coward⟩
⁴sod *transitive verb* (1904)
chiefly British **:** DAMN 2

so·da \ˈsō-də\ *noun* [Italian, from Arabic *suwwād,* any of several saltworts from the ashes of which sodium carbonate is obtained] (1558)
1 a : SODIUM CARBONATE **b :** SODIUM BICARBONATE **c :** SODIUM — used in combination ⟨*soda* alum⟩
2 a : SODA WATER 2a **b :** SODA POP **c :** a sweet drink consisting of soda water, flavoring, and often ice cream
3 : the faro card that shows faceup in the dealing box before play begins
soda ash *noun* (1839)
: commercial anhydrous sodium carbonate
soda biscuit *noun* (1830)
1 : a biscuit leavened with baking soda and sour milk or buttermilk
2 : SODA CRACKER
soda bread *noun* (1850)
: a quick bread made especially with buttermilk and leavened with baking soda
soda cracker *noun* (1830)
: a cracker leavened with bicarbonate of soda and cream of tartar
soda fountain *noun* (1824)
1 : an apparatus with delivery tube and faucets for drawing soda water
2 : the equipment and counter for the preparation and serving of sodas, sundaes, and ice cream
soda jerk \-ˌjərk\ *noun* (1922)
: a person who dispenses carbonated drinks and ice cream at a soda fountain — called also *soda jerker*
soda lime *noun* (1862)
: a mixture of sodium hydroxide and slaked lime used especially to absorb moisture and gases
so·da·list \ˈsō-dᵊl-ist, sō-ˈda-list\ *noun* (1794)
: a member of a sodality
so·da·lite \ˈsō-dᵊl-ˌīt\ *noun* [*soda*] (1810)
: a transparent to translucent mineral that consists of a silicate of sodium and aluminum with some chlorine, has a vitreous or greasy luster, and is found in various igneous rocks
so·dal·i·ty \sō-ˈda-lə-tē\ *noun, plural* **-ties** [Latin *sodalitat-, sodalitas* comradeship, club, from *sodalis* comrade — more at SIB] (1600)
1 : BROTHERHOOD, COMMUNITY
2 : an organized society or fellowship; *specifically* **:** a devotional or charitable association of Roman Catholic laity
soda pop *noun* (1863)
: a beverage consisting of soda water, flavoring, and a sweet syrup
soda water *noun* (1802)
1 : a weak solution of sodium bicarbonate with some acid added to cause effervescence
2 a : a beverage consisting of water highly charged with carbon dioxide **b :** SODA POP
sod·bust·er \ˈsäd-ˌbəs-tər\ *noun* (circa 1918)
: one (as a farmer or a plow) that breaks the sod
¹sod·den \ˈsä-dᵊn\ *adjective* [Middle English *soden,* from past participle of *sethen* to seethe] (1589)
1 a : dull or expressionless especially from continued indulgence in alcoholic beverages ⟨*sodden* features⟩ **b :** TORPID, SLUGGISH ⟨*sodden* minds⟩
2 a : heavy with or as if with moisture or water ⟨the *sodden* ground⟩ **b :** heavy or doughy because of imperfect cooking ⟨*sodden* biscuits⟩
— **sod·den·ly** *adverb*
— **sod·den·ness** \-dᵊn-(n)əs\ *noun*
²sodden *verb* **sod·dened; sod·den·ing** \ˈsäd-niŋ, ˈsä-dᵊn-iŋ\ (1812)
transitive verb
: to make sodden
intransitive verb
: to become soaked or saturated
so·dic \ˈsō-dik\ *adjective* (1859)
: of, relating to, or containing sodium
so·di·um \ˈsō-dē-əm\ *noun* [New Latin, from English *soda*] (1807)
: a silver white soft waxy ductile element of the alkali metal group that occurs abundantly in nature in combined form and is very active chemically — see ELEMENT table
sodium azide *noun* (circa 1937)
: a poisonous crystalline salt NaN_3 used especially to make lead azide
sodium benzoate *noun* (circa 1900)
: a crystalline or granular salt $C_7H_5O_2Na$ used chiefly as a food preservative
sodium bicarbonate *noun* (1885)
: a white crystalline weakly alkaline salt $NaHCO_3$ used especially in baking powders, fire extinguishers, and medicine — called also *baking soda, bicarbonate of soda*
sodium bo·ro·hy·dride \-ˌbōr-ə-ˈhī-ˌdrīd, -ˌbȯr-\ *noun* [*sodium* + *boron* + *hydride*] (1946)
: a crystalline compound $NaBH_4$ used in various industrial applications and as a reducing agent in organic chemistry
sodium carbonate *noun* (1868)
: a sodium salt of carbonic acid used especially in making soaps and chemicals, in water softening, in cleaning and bleaching, and in photography: as **a :** a hygroscopic crystalline anhydrous strongly alkaline salt Na_2CO_3 **b :** WASHING SODA
sodium chlorate *noun* (1885)
: a colorless crystalline salt $NaClO_3$ used especially as an oxidizing agent and weed killer
sodium chloride *noun* (1868)
: an ionic crystalline chemical compound consisting of equal numbers of sodium and chlorine atoms **:** SALT 1a
sodium citrate *noun* (1919)
: a crystalline salt $Na_3C_6H_5O_7$ used chiefly as a buffering agent, as an emulsifier, as an alkalizer and cathartic in pharmaceuticals, and as a blood anticoagulant
sodium cyanide *noun* (1885)
: a white deliquescent poisonous salt NaCN used especially in electroplating, in fumigating, and in treating steel
sodium dichromate *noun* (circa 1903)
: a red crystalline salt $Na_2Cr_2O_7$ used especially in tanning leather, in cleaning metals, and as an oxidizing agent
sodium fluoride *noun* (circa 1903)
: a poisonous crystalline salt NaF that is used in trace amounts in the fluoridation of water, in metallurgy, as a flux, and as a pesticide
sodium fluo·ro·ac·e·tate \-ˌflu̇r-ō-ˈa-sə-ˌtāt, -ˌflȯr-, -ˌflōr-\ *noun* (1945)
: a poisonous powdery compound $C_2H_2FNaO_2$ — compare 1080
sodium hydroxide *noun* (1885)
: a white brittle solid NaOH that is a strong caustic base used especially in making soap, rayon, and paper
sodium hypochlorite *noun* (1885)
: an unstable salt NaOCl produced usually in aqueous solution and used as a bleaching and disinfecting agent
sodium meta·sil·i·cate \-ˌme-tə-ˈsi-lə-ˌkāt, -ˈsi-li-kət\ *noun* (circa 1926)
: a toxic corrosive crystalline salt Na_2SiO_3 used especially as a detergent or as a substitute for phosphates in detergent formulations
sodium nitrate *noun* (1885)
: a deliquescent crystalline salt $NaNO_3$ used as a fertilizer and an oxidizing agent and in curing meat
sodium nitrite *noun* (circa 1903)
: a salt $NaNO_2$ used especially in dye manufacture and as a meat preservative
sodium pump *noun* (1951)
: a molecular mechanism by which sodium ions are actively transported across a cell membrane; *especially* **:** the one by which the appropriate internal and external concentrations of sodium and potassium ions are maintained in a nerve fiber and which involves the active transport of sodium ions outward with movement of potassium ions to the interior
sodium salicylate *noun* (circa 1904)

: a crystalline salt $NaC_7H_5O_3$ that has a sweetish saline taste and is used chiefly as an analgesic, antipyretic, and antirheumatic

sodium sulfate *noun* (1885)
: a bitter salt Na_2SO_4 used especially in detergents, in the manufacture of wood pulp and rayon, in dyeing and finishing textiles, and in its hydrated form as a cathartic — compare GLAUBER'S SALT

sodium thiosulfate *noun* (1885)
: a hygroscopic crystalline salt $Na_2S_2O_3$ used especially as a photographic fixing agent and a reducing or bleaching agent — called also *hypo*

sodium tri·poly·phos·phate \-ˌtrī-ˌpä-li-ˈfäs-ˌfāt\ *noun* (1945)
: a crystalline salt $Na_5P_3O_{10}$ that is used as a food additive and as a component in some detergents and is suspected of contributing to water pollution

sodium–vapor lamp *noun* (1936)
: an electric lamp that contains sodium vapor and electrodes between which a luminous discharge takes place and that is used especially for lighting highways

sod off *intransitive verb* [[4]*sod*] (1960)
British **:** SCRAM — usually used as a command

Sod·om \ˈsä-dəm\ *noun* [*Sodom,* ancient city destroyed by God for its wickedness in Genesis 19] (1598)
: a place notorious for vice or corruption

sod·om·ist \ˈsä-də-mist\ *noun* (1891)
: SODOMITE

sod·om·ite \-ˌmīt\ *noun* (14th century)
: one who practices sodomy

sod·om·ize \-ˌmīz\ *transitive verb* **-ized; -iz·ing** (1868)
: to perform sodomy on

sod·omy \ˈsä-də-mē\ *noun* [Middle English, from Old French *sodomie,* from Late Latin *Sodoma* Sodom; from the homosexual proclivities of the men of the city in Genesis 19:1–11] (13th century)
1 : copulation with a member of the same sex or with an animal
2 : noncoital and especially anal or oral copulation with a member of the opposite sex
— sod·om·it·ic \ˌsä-də-ˈmi-tik\ *or* **sod·om·it·i·cal** \-ti-kəl\ *adjective*

so·ev·er \sō-ˈe-vər\ *adverb* [*-soever* (as in *howsoever*)] (12th century)
1 : to any possible or known extent — used after an adjective preceded by *how* or a superlative preceded by *the* ⟨how fair *soever* she may be⟩ ⟨the most selfish *soever* in this world⟩
2 : of any or every kind that may be specified — used after a noun modified especially by *any, no,* or *what* ⟨gives no information *soever*⟩

so·fa \ˈsō-fə\ *noun* [Arabic *ṣuffah* long bench] (1717)
: a long upholstered seat usually with arms and a back and often convertible into a bed

sofa bed *noun* (1816)
: a sofa that can be made to serve as a bed by lowering its hinged upholstered back to horizontal position or by pulling out a concealed mattress

so·far \ˈsō-ˌfär\ *noun* [*so*und *f*ixing *a*nd *r*anging] (1946)
: a system for locating an underwater explosion at sea by triangulation based on the reception of the sound by three widely separated stations

so far as *conjunction* (1565)
: INSOFAR AS

sof·fit \ˈsä-fət\ *noun* [French *soffite,* from Italian *soffitto,* from (assumed) Vulgar Latin *suffictus,* past participle of Latin *suffigere* to fasten underneath — more at SUFFIX] (1592)
: the underside of a part or member of a building (as of an overhang or staircase); *especially* **:** the intrados of an arch

[1]soft \ˈsȯft\ *adjective* [Middle English, from Old English *sōfte,* alteration of *sēfte;* akin to Old High German *semfti* soft] (before 12th century)
1 a : pleasing or agreeable to the senses **:** bringing ease, comfort, or quiet ⟨the *soft* influences of home⟩ **b :** having a bland or mellow rather than a sharp or acid taste **c** (1) **:** not bright or glaring **:** SUBDUED (2) **:** having or producing little contrast or a relatively short range of tones ⟨a *soft* photographic print⟩ **d :** quiet in pitch or volume **e** *of the eyes* **:** having a liquid or gentle appearance **f :** smooth or delicate in texture, grain, or fiber ⟨*soft* cashmere⟩ ⟨*soft* fur⟩ **g** (1) **:** balmy, mild, or clement in weather or temperature (2) **:** moving or falling with slight force or impact **:** not violent ⟨*soft* breezes⟩
2 : demanding little work or effort **:** EASY ⟨a *soft* job⟩
3 a : sounding as in *ace* and *gem* respectively — used of *c* and *g* or their sound **b** *of a consonant* **:** VOICED **c :** constituting a vowel before which there is a \y\ sound or a \y\-like modification of a consonant or constituting a consonant in whose articulation there is a \y\-like modification or which is followed by a \y\ sound (as in Russian)
4 *archaic* **:** moving in a leisurely manner
5 : rising gradually ⟨a *soft* slope⟩
6 : having curved or rounded outline **:** not harsh or jagged ⟨*soft* hills against the horizon⟩
7 : marked by a gentleness, kindness, or tenderness: as **a** (1) **:** not harsh or onerous in character ⟨a policy of *soft* competition⟩ (2) **:** based on negotiation, conciliation, or flexibility rather than on force, threats, or intransigence ⟨took a *soft* line during the crisis⟩ (3) **:** tending to take a soft line — usually used with *on* ⟨*soft* on dictators⟩ ⟨*soft* on law and order⟩ **b :** tending to ingratiate or disarm **:** ENGAGING, KIND ⟨a *soft* answer turns away wrath —Proverbs 15:1 (Revised Standard Version)⟩ **c :** marked by mildness **:** UNASSUMING, LOW-KEY
8 a : emotionally suggestible or responsive **:** IMPRESSIONABLE **b :** unduly susceptible to influence **:** COMPLIANT **c :** lacking firmness or strength of character **:** FEEBLE, UNMANLY **d :** amorously attracted or emotionally involved — used with *on* ⟨has been *soft* on her for years⟩
9 a : lacking robust strength, stamina, or endurance especially because of living in ease or luxury ⟨grown *soft* and indolent⟩ **b :** weak or deficient mentally ⟨*soft* in the head⟩
10 a : yielding to physical pressure **b :** permitting someone or something to sink in — used of wet ground **c** (1) **:** of a consistency that may be shaped or molded (2) **:** capable of being spread **d :** easily magnetized and demagnetized **e :** lacking relatively or comparatively in hardness ⟨*soft* iron⟩
11 : deficient in or free from substances (as calcium and magnesium salts) that prevent lathering of soap ⟨*soft* water⟩
12 : having relatively low energy ⟨*soft* X rays⟩
13 *of news* **:** relatively less serious or significant
14 : occurring at such a speed and under such circumstances as to avoid destructive impact ⟨*soft* landing of a spacecraft on the moon⟩
15 : not protected against enemy attack ⟨a *soft* aboveground launching site⟩ ⟨*soft* targets⟩
16 : BIODEGRADABLE ⟨a *soft* detergent⟩ ⟨*soft* pesticides⟩
17 *of a drug* **:** considered less detrimental than a hard narcotic
18 : easily polarized — used of acids and bases
19 a *of currency* **:** not readily convertible **b** *of a loan* **:** not secured by collateral
20 a : being low due to sluggish market conditions ⟨*soft* prices⟩ **b :** SLUGGISH, SLOW ⟨a *soft* market⟩
21 : not firmly committed ⟨*soft* unreliable political support⟩
22 : SOFT-CORE ⟨*soft* porn⟩
23 a : being or based on interpretive or speculative data ⟨*soft* evidence⟩ **b :** utilizing or based on soft data ⟨*soft* science⟩
24 : being or using renewable sources of energy (as solar radiation, wind, or tides) ⟨*soft* technologies⟩
— soft·ish \ˈsȯf-tish\ *adjective*
— soft·ly \ˈsȯf(t)-lē\ *adverb*
— soft·ness \ˈsȯf(t)-nəs\ *noun*

[2]soft *adverb* (before 12th century)
: in a soft or gentle manner **:** SOFTLY

[3]soft *noun* (15th century)
: a soft object, material, or part ⟨the *soft* of the thumb⟩

soft·back \ˈsȯf(t)-ˌbak\ *adjective* (1958)
: SOFTCOVER
— softback *noun*

soft·ball \-ˌbȯl\ *noun* (1926)
: baseball played on a small diamond with a ball that is larger than a baseball and that is pitched underhand; *also* **:** the ball used in this game
— soft·ball·er \-ˌbȯ-lər\ *noun*

soft–boiled \-ˈbȯi(ə)ld\ *adjective* (circa 1902)
1 *of an egg* **:** boiled to a soft consistency
2 : SENTIMENTAL

soft·bound \-ˌbau̇nd\ *adjective* (1953)
: SOFTCOVER

soft chancre *noun* (1859)
: CHANCROID

soft coal *noun* (1789)
: BITUMINOUS COAL

soft–coated wheaten terrier *noun* (1948)
: any of a breed of compact medium-sized terriers developed in Ireland and having a soft abundant light fawn coat

soft–core \ˈsȯf(t)-ˈkōr, -ˈkȯr\ *adjective* [[1]*soft* + hard-*core*] (1966)
of pornography **:** containing descriptions or scenes of sex acts that are less explicit than hard-core material

soft·cov·er \-ˌkə-vər\ *adjective* (1952)
: bound in flexible covers **:** not bound in hard covers; *specifically* **:** PAPERBACK ⟨*softcover* books⟩
— softcover *noun*

soft drink *noun* (1880)
: a usually carbonated nonalcoholic beverage; *especially* **:** SODA POP

soft·en \ˈsȯ-fən\ *verb* **soft·ened; soft·en·ing** \ˈsȯ-fə-niŋ, ˈsȯf-niŋ\ (14th century)
transitive verb
1 : to make soft or softer
2 a : to weaken the military resistance or the morale of especially by harassment (as preliminary bombardment) — often used with *up* **b :** to impair the strength or resistance of — often used with *up* ⟨*soften* up a sales prospect⟩
intransitive verb
: to become soft or softer
— soft·en·er \ˈsȯ-fə-nər, ˈsȯf-nər\ *noun*

soft–fo·cus \ˈsȯf(t)-ˈfō-kəs\ *adjective* (1916)
1 *of a photographic image* **:** having unsharp outlines
2 *of a lens* **:** producing an image having unsharp outlines

soft goods *noun plural* (1894)
: goods that are not durable — used especially of textile products

soft hail *noun* (1894)
: GRAUPEL

soft·head \ˈsȯft-ˌhed\ *noun* (1650)
: a silly or feebleminded person

soft·head·ed \-ˌhe-dəd\ *adjective* (1667)
: having or indicative of a weak, unrealistic, or uncritical mind
— soft·head·ed·ly *adverb*

— **soft·head·ed·ness** *noun*
soft–heart·ed \-ˌhär-təd\ *adjective* (1593)
: emotionally responsive : SYMPATHETIC
— **soft·heart·ed·ly** *adverb*
— **soft·heart·ed·ness** *noun*
soft–land \-'land\ *verb* [back-formation from *soft landing*] (1960)
transitive verb
: to cause to make a soft landing on a celestial body (as the moon)
intransitive verb
: to make a soft landing
— **soft–lander** *noun*
soft–line \-'līn\ *adjective* (1949)
: advocating or involving a conciliatory or flexible course of action
— **soft–lin·er** \-'lī-nər\ *noun*
soft palate *noun* (circa 1811)
: the fold at the back of the hard palate that partially separates the mouth and pharynx
soft–ped·al \'sȯf(t)-'pe-d^{ə}l\ *transitive verb* (1925)
1 : PLAY DOWN, DE-EMPHASIZE ⟨*soft-pedal* the issue⟩
2 : to use the soft pedal in playing
soft pedal *noun* (1854)
1 : a foot pedal on a piano that reduces the volume of sound
2 : something that muffles, deadens, or reduces effect
soft rock *noun* (1967)
: rock music that is less driving and gentler sounding than hard rock
soft rot *noun* (1901)
: a mushy, watery, or slimy decay of plants or their parts caused by bacteria or fungi
soft sell *noun* (1954)
: the use of suggestion or gentle persuasion in selling rather than aggressive pressure — compare HARD SELL
soft·shell \'sȯf(t)-ˌshel\ *noun* (1771)
: any of a family (Trionychidae) of aquatic turtles that have sharp claws and mandibles and a flat shell covered with soft leathery skin instead of with horny plates — called also *soft-shelled turtle*

softshell

soft–shell \'sȯf(t)-ˌshel\ *or* **soft–shelled** \-'sheld\ *adjective* (1805)
: having a soft or fragile shell especially as a result of recent shedding ⟨*soft-shell* crabs⟩
soft–shell clam *noun* (1796)
: an elongated clam (*Mya arenaria*) of the east coast of North America that has a thin friable shell and long siphons and is used especially for steaming — called also *soft-shelled clam, steamer*
soft–shoe \'sȯf(t)-'shü\ *adjective* (1920)
: of or relating to tap dancing done in soft-soled shoes without metal taps
soft–soap \'sȯf(t)-'sōp\ *transitive verb* (1840)
: to soothe or persuade with flattery or blarney
— **soft–soap·er** \-ˌsō-pər\ *noun*
soft soap *noun* (1634)
1 : a semifluid soap made especially from potassium hydroxide
2 : FLATTERY
soft–spo·ken \-'spō-kən\ *adjective* (1609)
: having a mild or gentle voice; *also* : SUAVE
soft spot *noun* (1845)
1 : a sentimental weakness ⟨has a *soft spot* for him⟩
2 : a vulnerable point ⟨a *soft spot* in the defense system⟩
soft touch *noun* (1939)
: one who is easily imposed on or taken advantage of
soft·ware \'sȯft-ˌwar, -ˌwer\ *noun* (1960)
: something used or associated with and usually contrasted with hardware: as **a** : the entire set of programs, procedures, and related documentation associated with a system and especially a computer system; *specifically* : computer programs **b** : materials for use with audiovisual equipment
soft wheat *noun* (1812)
: a wheat with soft starchy kernels high in starch but usually low in gluten
[1]**soft·wood** \'sȯft-ˌwu̇d\ *noun* (1832)
1 : the wood of a coniferous tree (as a fir or pine) whether hard or soft as distinguished from that of an angiospermous tree
2 : a tree that yields softwood
[2]**softwood** *adjective* (1905)
: having or made of softwood
soft–wood·ed \'sȯft-ˌwu̇-dəd\ *adjective* (1827)
1 : having soft wood that is easy to work or finish
2 : SOFTWOOD
softy *or* **soft·ie** \'sȯf-tē\ *noun, plural* **soft·ies** [[1]*soft*] (1863)
1 : a weak or foolish person
2 : a softhearted or sentimental person
Sog·di·an \'säg-dē-ən\ *noun* [Latin *Sogdiani,* plural, from Greek *Sogdianoi,* from Old Persian *Suguda* Sogdiana] (1553)
1 : a native or inhabitant of Sogdiana
2 : an Iranian language of the Sogdians — see INDO-EUROPEAN LANGUAGES table
— **Sogdian** *adjective*
sog·gy \'sä-gē, 'sȯ-\ *adjective* **sog·gi·er; -est** [English dialect *sog* to soak] (1599)
1 : saturated or heavy with water or moisture: as **a** : WATERLOGGED, SOAKED ⟨a *soggy* lawn⟩ **b** : heavy or doughy because of imperfect cooking ⟨*soggy* bread⟩
2 : heavily dull : SPIRITLESS ⟨*soggy* prose⟩
— **sog·gi·ly** \'sä-gə-lē, 'sȯ-\ *adverb*
— **sog·gi·ness** \'sä-gē-nəs, 'sȯ-\ *noun*
soi–di·sant \ˌswä-dē-'zä^{n}\ *adjective* [French, literally, saying oneself] (1752)
: SELF-STYLED, SO-CALLED — usually used disparagingly ⟨a *soi-disant* artist⟩
soi·gné *or* **soi·gnée** \swän-'yā\ *adjective* [French, from past participle of *soigner* to take care of] (1821)
1 : WELL-GROOMED, SLEEK
2 : elegantly maintained or designed ⟨a *soigné* restaurant⟩ ⟨a *soigné* black dress⟩
[1]**soil** \'sȯi(ə)l\ *verb* [Middle English, from Old French *souillier* to wallow, soil, from *soil* abyss, pigpen, probably from Latin *solium* chair, bathtub; akin to Latin *sedēre* to sit — more at SIT] (13th century)
transitive verb
1 : to stain or defile morally : CORRUPT, POLLUTE
2 : to make unclean especially superficially : DIRTY
3 : to blacken or besmirch (as a person's reputation) by word or deed
intransitive verb
: to become soiled or dirty
[2]**soil** *noun* (1501)
1 a : SOILAGE, STAIN ⟨protect a dress from *soil*⟩ **b** : moral defilement : CORRUPTION
2 : something that spoils or pollutes: as **a** : REFUSE **b** : SEWAGE **c** : DUNG, EXCREMENT
[3]**soil** *noun* [Middle English, from Anglo-French, probably from Latin *solium* chair] (14th century)
1 : firm land : EARTH
2 a : the upper layer of earth that may be dug or plowed and in which plants grow **b** : the superficial unconsolidated and usually weathered part of the mantle of a planet and especially of the earth
3 : COUNTRY, LAND ⟨our native *soil*⟩
4 : the agricultural life or calling
5 : a medium in which something takes hold and develops
[4]**soil** *transitive verb* [origin unknown] (1605)
: to feed (livestock) in the barn or an enclosure with fresh grass or green food; *also* : to purge (livestock) by feeding on green food
[1]**soil·age** \'sȯi-lij\ *noun* [[1]*soil*] (1926)
: the act of soiling : the condition of being soiled
[2]**soilage** *noun* [[4]*soil*] (1928)
: green crops for feeding confined animals
soil bank *noun* (1955)
: acreage retired from crop cultivation and planted with soil-building plants under a plan sponsored by the U.S. government that provides subsidies to farmers for the retired land
soil·borne \'sȯil-ˌbōrn, -ˌbȯrn\ *adjective* (1944)
: transmitted by or in soil ⟨*soilborne* fungi⟩ ⟨*soilborne* diseases⟩
soil·less \'sȯi(ə)l-ləs\ *adjective* (1938)
: having, containing, or utilizing no soil ⟨*soilless* agriculture⟩
soil pipe *noun* (1833)
: a pipe for carrying off wastes from toilets
soil science *noun* (1915)
: a science dealing with soils — called also *pedology*
— **soil scientist** *noun*
soil series *noun* (1905)
: a group of soils with similar profiles developed from similar parent materials under comparable climatic and vegetational conditions
soil·ure \'sȯil-yər\ *noun* [Middle English, from Old French *soilleure,* from *souillier* to soil] (13th century)
1 : the act of soiling : the condition of being soiled
2 : STAIN, SMUDGE
soi·ree *or* **soi·rée** \swä-'rā\ *noun* [French *soirée* evening period, evening party, from Middle French, from *soir* evening, from Latin *sero* at a late hour, from *serus* late; akin to Old Irish *sír* long, lasting and perhaps to Old English *sīth* late — more at SINCE] (1802)
: a party or reception held in the evening
soi·xante–neuf \swȧ-sä^{n}t-nœf\ *noun* [French] (1888)
: SIXTY-NINE 2
[1]**so·journ** \'sō-ˌjərn, sō-'\ *noun* [Middle English *sojorn,* from Old French, from *sojorner*] (13th century)
: a temporary stay ⟨a *sojourn* in the country⟩
[2]**sojourn** *intransitive verb* [Middle English *sojornen,* from Old French *sojorner,* from (assumed) Vulgar Latin *subdiurnare,* from Latin *sub* under, during + Late Latin *diurnum* day — more at UP, JOURNEY] (14th century)
: to stay as a temporary resident : STOP ⟨*sojourned* for a month at a resort⟩
— **so·journ·er** *noun*
soke \'sōk\ *noun* [Middle English *soc, soke,* from Old English *soka,* from Medieval Latin *soca,* from Old English *sōcn* inquiry, jurisdiction; akin to Old English *sēcan* to seek] (12th century)
1 : the right in Anglo-Saxon and early English law to hold court and administer justice with the franchise to receive certain fees or fines arising from it : jurisdiction over a territory or over people
2 : the district included in a soke jurisdiction or franchise
soke·man \'sōk-mən\ *noun* (1579)
: a man who is under the soke of another
[1]**sol** \'sōl\ *also* **so** \'sō\ *noun* [Medieval Latin *sol;* from the syllable sung to this note in a medieval hymn to Saint John the Baptist] (14th century)
: the 5th tone of the diatonic scale in solmization
[2]**sol** \'säl, 'sȯl\ *noun* [Middle French — more at SOU] (1583)
: an old French coin equal to 12 deniers; *also* : a corresponding unit of value
[3]**sol** \'säl, 'sȯl\ *noun, plural* **so·les** \'sō-(ˌ)lās\ [American Spanish, from Spanish, sun, from Latin] (circa 1883)

: the basic monetary unit of Peru before 1985 and since 1990 — see MONEY table

[4]**sol** \'säl, 'sȯl\ *noun* [*-sol* (as in *hydrosol*), from *solution*] (1899)
: a fluid colloidal system; *especially* : one in which the continuous phase is a liquid

Sol \'säl\ *noun* [Middle English, from Latin]
1 : the Roman god of the sun — compare HELIOS
2 : SUN

sola *plural of* SOLUM

[1]**so·lace** \'sä-ləs *also* 'sō-\ *transitive verb* **so·laced; so·lac·ing** (13th century)
1 : to give solace to : CONSOLE
2 a : to make cheerful **b :** AMUSE
3 : ALLAY, SOOTHE ⟨*solace* grief⟩
— so·lace·ment \-mənt\ *noun*
— so·lac·er *noun*

[2]**solace** *noun* [Middle English *solas,* from Old French, from Latin *solacium,* from *solari* to console] (14th century)
1 : alleviation of grief or anxiety
2 : a source of relief or consolation

so·la·na·ceous \ˌsō-lə-'nā-shəs\ *adjective* [New Latin *Solanaceae,* family name, from *Solanum*] (1804)
: of or relating to the nightshade family of plants

so·la·nine *also* **so·la·nin** \'sō-lə-ˌnēn, -nən\ *noun* [French *solanine,* from Latin *solanum*] (1838)
: a bitter poisonous crystalline alkaloid $C_{45}H_{73}NO_{15}$ from several plants (as some potatoes or tomatoes) of the nightshade family

so·la·num \sə-'lā-nəm, -'lä-, -'la-\ *noun* [New Latin, from Latin, nightshade] (circa 1652)
: any of a genus (*Solanum*) chiefly of herbs and shrubs of the nightshade family that have often prickly-veined leaves, cymose white, purple, or yellow flowers, and a fruit that is a berry

so·lar \'sō-lər, -ˌlär\ *adjective* [Middle English, from Latin *solaris,* from *sol* sun; akin to Old English & Old Norse *sōl* sun, Lithuanian *saulė,* Greek *hēlios*] (15th century)
1 : of, derived from, or relating to the sun especially as affecting the earth
2 : measured by the earth's course in relation to the sun ⟨*solar* time⟩ ⟨*solar* year⟩; *also* : relating to or reckoned by solar time
3 a : produced or operated by the action of the sun's light or heat ⟨*solar* energy⟩ ⟨*solar* cooker⟩ **b :** utilizing the sun's rays especially to produce heat or electricity ⟨a *solar* house⟩; *also* : of or relating to such utilization ⟨*solar* design⟩ ⟨*solar* subsidies⟩

solar battery *noun* (1954)
: an array of solar cells

solar cell *noun* (1955)
: a photovoltaic cell (as one including a junction between two types of silicon semiconductors) that is able to convert light into electrical energy and is used as a power source

solar collector *noun* (1955)
: any of various devices for the absorption of solar radiation for the heating of water or buildings or the production of electricity

solar constant *noun* (1869)
: the quantity of radiant solar energy received at the outer layer of the earth's atmosphere that has a mean value of 1370 watts per square meter

solar day *noun* (1764)
: the interval between transits of the apparent or mean sun across the meridian at any place

solar eclipse *noun* (circa 1890)
: an eclipse of the sun by the moon — see ECLIPSE illustration

solar flare *noun* (1938)
: a sudden temporary outburst of energy from a small area of the sun's surface — called also *flare*

so·lar·i·um \sō-'lar-ē-əm, sə-, -'ler-\ *noun, plural* **-ia** \-ē-ə\ *also* **-ums** [Latin, porch exposed to the sun, from *sol*] (circa 1823)
: a glass-enclosed porch or room; *also* : a room (as in a hospital) used especially for sunbathing or therapeutic exposure to light

so·lar·i·za·tion \ˌsō-lə-rə-'zā-shən\ *noun* (1853)
1 : a reversal of gradation in a photographic image obtained by intense or continued exposure
2 : an act or process of solarizing

so·lar·ize \'sō-lə-ˌrīz\ *transitive verb* **-ized; -iz·ing** (1853)
1 a : to affect by the action of the sun's rays **b :** to expose to sunlight
2 : to subject (photographic materials) to solarization

solar panel *noun* (1961)
: a battery of solar cells (as in a spacecraft)

so·lar plexus \'sō-lər-\ *noun* [from the radiating nerve fibers] (1771)
1 : a nerve plexus in the abdomen that is situated behind the stomach and in front of the aorta and the crura of the diaphragm and contains several ganglia distributing nerve fibers to the viscera
2 : the pit of the stomach

solar pond *noun* (1961)
: a pool of salt water heated by the sun and used either as a direct source of heat or to provide power for an electric generator

solar sail *noun* (1958)
: a propulsive device for a spacecraft that consists of a flat material (as aluminized plastic) designed to receive thrust from solar radiation pressure

solar system *noun* (circa 1704)
: the sun together with the group of celestial bodies that are held by its attraction and revolve around it; *also* : a similar system centered on another star

solar wind *noun* (1958)
: plasma continuously ejected from the sun's surface into and through interplanetary space

sol·ation \sä-'lā-shən, sȯ-\ *noun* (1915)
: the process of changing to a sol

so·la·ti·um \sō-'lā-shē-əm\ *noun, plural* **-tia** \-shē-ə\ [Late Latin *solacium, solatium,* from Latin, solace] (1817)
: a compensation (as money) given as solace for suffering, loss, or injured feelings

sold *past and past participle of* SELL

sol·dan \'säl-dən, 'sōl-\ *noun* [Middle English, from Old French, from Arabic *sulṭān*] (14th century)
archaic : SULTAN; *especially* : the sultan of Egypt

[1]**sol·der** \'sä-dər, 'sȯ-, *British also* 'säl-dər, 'sōl-\ *noun* [Middle English *soudure,* from Middle French, from *souder* to solder, from Latin *solidare* to make solid, from *solidus* solid] (14th century)
1 : a metal or metallic alloy used when melted to join metallic surfaces; *especially* : an alloy of lead and tin so used
2 : something that unites

[2]**solder** *verb* **sol·dered; sol·der·ing** \-d(ə-)riŋ\ (15th century)
transitive verb
1 : to unite or make whole by solder
2 : to bring into or restore to firm union ⟨a friendship *soldered* by common interests⟩
intransitive verb
1 : to use solder
2 : to become united or repaired by or as if by solder
— sol·der·abil·i·ty \ˌsä-də-rə-'bi-lə-tē, ˌsȯ-\ *noun*
— sol·der·er \'sä-dər-ər, 'sȯ-\ *noun*

soldering iron *noun* (1688)
: a pointed or wedge-shaped device that is usually electrically heated and that is used for soldering

[1]**sol·dier** \'sōl-jər\ *noun* [Middle English *soudier,* from Middle French, from *soulde* pay, from Late Latin *solidus* solidus] (14th century)
1 a : one engaged in military service and especially in the army **b :** an enlisted man or woman **c :** a skilled warrior
2 : a militant leader, follower, or worker
3 a : one of a caste of wingless sterile termites usually differing from workers in larger size and head and long jaws **b :** one of a type of worker ants distinguished by exceptionally large head and jaws
4 \'sō-jər, 'sōl-\ **:** one who shirks work
— sol·dier·ly \-lē\ *adjective or adverb*
— sol·dier·ship \-ˌship\ *noun*

[2]**soldier** *intransitive verb* **sol·diered; sol·dier·ing** \'sōl-jə-riŋ, 'sōlj-riŋ\ (1647)
1 a : to serve as a soldier **b :** to behave in a soldierly manner **c :** to push doggedly forward — usually used with *on* ⟨*soldiered* on to the end⟩
2 : to make a pretense of working while really loafing

soldiering *noun* (1643)
: the life, service, or practice of one who soldiers

soldier of fortune (1661)
: one who follows a military career wherever there is promise of profit, adventure, or pleasure

soldiers' home *noun* (1860)
: an institution maintained (as by the federal or a state government) for the care and relief of military veterans

soldier's medal *noun* (circa 1930)
: a U.S. military decoration awarded for heroism not involving combat

sol·diery \'sōl-jə-rē, 'sōlj-rē\ *noun* (circa 1570)
1 a : a body of soldiers **b :** SOLDIERS, MILITARY
2 : the profession or technique of soldiering

sol·do \'sȯl-(ˌ)dō\ *noun, plural* **sol·di** \-(ˌ)dē\ [Italian, from Late Latin *solidus* solidus] (1599)
: an old Italian coin worth five centesimi

sold–out \'sōld-'au̇t\ *adjective* (1907)
: having all available tickets or accommodations sold completely and especially in advance

[1]**sole** \'sōl\ *noun* [Middle English, from Middle French, from Latin *solea* sandal, a flatfish] (13th century)
: any of various flatfishes (family Soleidae) having a small mouth, small or rudimentary fins, and small eyes placed close together and including superior food fishes (as the European Dover sole); *also* : any of various mostly market flatfishes (as lemon sole) of other families (as Pleuronectidae)

[2]**sole** *noun* [Middle English, from Middle French, from Latin *solea* sandal; akin to Latin *solum* base, ground, soil] (14th century)
1 a : the undersurface of a foot **b :** the part of an item of footwear on which the sole rests and upon which the wearer treads
2 : the usually flat or flattened bottom or lower part of something or the base on which something rests
— soled \'sōld\ *adjective*

[3]**sole** *transitive verb* **soled; sol·ing** (1570)
1 : to furnish with a sole ⟨*sole* a shoe⟩
2 : to place the sole of (a golf club) on the ground

[4]**sole** *adjective* [Middle English, alone, from Middle French *seul,* from Latin *solus*] (14th century)
1 : not married — used chiefly of women
2 *archaic* **:** having no companion : SOLITARY
3 a : having no sharer **b :** being the only one ⟨she was her mother's *sole* support⟩
4 : functioning independently and without assistance or interference ⟨let conscience be the *sole* judge⟩

\ə\ **abut** \ᵊ\ **kitten** \ər\ **further** \a\ **ash** \ā\ **ace** \ä\ **mop, mar** \au̇\ **out** \ch\ **chin** \e\ **bet** \ē\ **easy** \g\ **go** \i\ **hit** \ī\ **ice** \j\ **job** \ŋ\ **sing** \ō\ **go** \ȯ\ **law** \ȯi\ **boy** \th\ **thin** \t͟h\ **the** \ü\ **loot** \u̇\ **foot** \y\ **yet** \zh\ **vision** *see also* Guide to Pronunciation

5 : belonging exclusively or otherwise limited to one usually specified individual, unit, or group
— **sole·ness** \'sōl-nəs\ *noun*

so·le·cism \'sä-lə-ˌsi-zəm, 'sō-\ *noun* [Latin *soloecismus,* from Greek *soloikismos,* from *soloikos* speaking incorrectly, literally, inhabitant of Soloi, from *Soloi,* city in ancient Cilicia where a substandard form of Attic was spoken] (circa 1555)
1 : an ungrammatical combination of words in a sentence; *also* **:** a minor blunder in speech
2 : something deviating from the proper, normal, or accepted order
3 : a breach of etiquette or decorum
— **so·le·cis·tic** \ˌsä-lə-'sis-tik, ˌsō-\ *adjective*

sole·ly \'sō(l)-lē\ *adverb* (15th century)
1 : without another **:** SINGLY ⟨went *solely* on her way⟩
2 : to the exclusion of all else ⟨done *solely* for money⟩

sol·emn \'sä-ləm\ *adjective* [Middle English *solemne,* from Middle French, from Latin *sollemnis* regularly appointed, solemn] (14th century)
1 : marked by the invocation of a religious sanction ⟨a *solemn* oath⟩
2 : marked by the observance of established form or ceremony; *specifically* **:** celebrated with full liturgical ceremony
3 a : awe-inspiring **:** SUBLIME **b :** marked by grave sedateness and earnest sobriety **c :** SOMBER, GLOOMY
synonym see SERIOUS
— **sol·emn·ly** *adverb*
— **sol·emn·ness** *noun*

so·lem·ni·fy \sə-'lem-nə-ˌfī\ *transitive verb* **-fied; -fy·ing** (1780)
: to make solemn

so·lem·ni·ty \sə-'lem-nə-tē\ *noun, plural* **-ties** (14th century)
1 : formal or ceremonious observance of an occasion or event
2 : a solemn event or occasion
3 : a solemn condition or quality ⟨the *solemnity* of his words⟩

sol·em·nize \'sä-ləm-ˌnīz\ *verb* **-nized; -niz·ing** (14th century)
transitive verb
1 : to observe or honor with solemnity
2 : to perform with pomp or ceremony; *especially* **:** to celebrate (a marriage) with religious rites
3 : to make solemn **:** DIGNIFY
intransitive verb
: to speak or act with solemnity
— **sol·em·ni·za·tion** \ˌsä-ləm-nə-'zā-shən\ *noun*

solemn mass *noun* (15th century)
: a mass marked by the use of incense and by the presence of a deacon and a subdeacon in attendance on the celebrant

solemn vow *noun* (14th century)
: an absolute and irrevocable public vow taken by a religious in the Roman Catholic Church under which ownership of property by the individual is prohibited and marriage is invalid under canon law

so·le·noid \'sō-lə-ˌnȯid, 'sä-\ *noun* [French *solénoïde,* from Greek *sōlēnoeidēs* pipe-shaped, from Greek *sōlēn* pipe] (1827)
: a coil of wire usually in cylindrical form that when carrying a current acts like a magnet so that a movable core is drawn into the coil when a current flows and that is used especially as a switch or control for a mechanical device (as a valve)
— **so·le·noi·dal** \ˌsō-lə-'nȯi-dᵊl, ˌsä-\ *adjective*

sole·plate \'sōl-ˌplāt\ *noun* (1741)
1 : the lower plate of a studded partition on which the bases of the studs butt
2 : the undersurface of a flatiron

soles *plural of* SOL

so·le·us \'sō-lē-əs\ *noun, plural* **so·lei** \-lē-ˌī\ [New Latin, from Latin *solea* sandal — more at SOLE] (1676)
: a broad flat muscle of the calf of the leg lying immediately below the gastrocnemius

¹sol–fa \ˌsōl-'fä, 'sōl-ˌ\ (circa 1529)
intransitive verb
: to sing the sol-fa syllables
transitive verb
: to sing (as a melody) to sol-fa syllables

²sol–fa *noun* (1548)
1 : SOL-FA SYLLABLES
2 : SOLMIZATION; *also* **:** an exercise thus sung
3 : TONIC SOL-FA
— **sol–fa·ist** \-'fä(-)ist, -'fä-ˌist\ *noun*

sol–fa syllables *noun plural* (1913)
: the syllables *do, re, mi, fa, sol, la, ti,* used in singing the tones of the scale

sol·fa·ta·ra \ˌsōl-fə-'tär-ə\ *noun* [Italian, sulfur mine, from *solfo* sulfur, from Latin *sulfur*] (1777)
: a volcanic area or vent that yields only hot vapors and sulfurous gases

sol·fège \säl-'fezh\ *noun* [French, from Italian *solfeggio*] (circa 1903)
1 : the application of the sol-fa syllables to a musical scale or to a melody
2 : a singing exercise especially using sol-fa syllables; *also* **:** practice in sight-reading vocal music using the sol-fa syllables

sol·feg·gio \säl-'fe-j(ē-ˌ)ō\ *noun* [Italian, from *sol-fa*] (1774)
: SOLFÈGE

sol·gel \'säl-ˌjel, 'sȯl-\ *adjective* (1915)
: involving alternation between sol and gel states

soli *plural of* SOLO

so·lic·it \sə-'li-sət\ *verb* [Middle English, to disturb, take charge of, from Middle French *solliciter,* from Latin *sollicitare* to disturb, from *sollicitus* anxious, from *sollus* whole (from Oscan; akin to Greek *holos* whole) + *citus,* past participle of *ciēre* to move — more at SAFE, -KINESIS] (15th century)
transitive verb
1 a : to make petition to **:** ENTREAT **b :** to approach with a request or plea
2 : to urge (as one's cause) strongly
3 a : to entice or lure especially into evil **b :** to proposition (someone) especially as or in the character of a prostitute
4 : to try to obtain by usually urgent requests or pleas
intransitive verb
1 : to make solicitation **:** IMPORTUNE
2 *of a prostitute* **:** to offer to have sexual relations with someone for money
synonym see ASK

so·lic·i·tant \sə-'li-sə-tənt\ *noun* (circa 1812)
: one who solicits

so·lic·i·ta·tion \sə-ˌli-sə-'tā-shən\ *noun* (circa 1520)
1 : the practice or act or an instance of soliciting; *especially* **:** ENTREATY, IMPORTUNITY
2 : a moving or drawing force **:** INCITEMENT, ALLUREMENT

so·lic·i·tor \sə-'li-sə-tər, -'lis-tər\ *noun* (15th century)
1 : one that solicits; *especially* **:** an agent that solicits (as contributions to charity)
2 : a British lawyer who advises clients, represents them in the lower courts, and prepares cases for barristers to try in higher courts
3 : the chief law officer of a municipality, county, or government department
— **so·lic·i·tor·ship** \-ˌship\ *noun*

solicitor general *noun, plural* **solicitors general** (1647)
: a law officer appointed primarily to assist an attorney general

so·lic·i·tous \sə-'li-sə-təs, -'lis-təs\ *adjective* [Latin *sollicitus*] (1563)
1 : manifesting or expressing solicitude ⟨a *solicitous* inquiry about his health⟩
2 : full of concern or fears **:** APPREHENSIVE ⟨*solicitous* about the future⟩
3 : meticulously careful ⟨*solicitous* in matters of dress⟩
4 : full of desire **:** EAGER
— **so·lic·i·tous·ly** *adverb*
— **so·lic·i·tous·ness** *noun*

so·lic·i·tude \sə-'li-sə-ˌtüd, -ˌtyüd\ *noun* (15th century)
1 a : the state of being solicitous **:** ANXIETY **b :** attentive care and protectiveness; *also* **:** an attitude of solicitous concern or attention
2 : a cause of care or concern — usually used in plural

¹sol·id \'sä-ləd\ *adjective* [Middle English *solide,* from Middle French, from Latin *solidus;* akin to Greek *holos* whole — more at SAFE] (14th century)
1 a : being without an internal cavity ⟨a *solid* ball of rubber⟩ **b** (1) **:** printed with minimum space between lines (2) **:** joined without a hyphen ⟨a *solid* compound⟩ **c :** not interrupted by a break or opening ⟨a *solid* wall⟩
2 : having, involving, or dealing with three dimensions or with solids ⟨*solid* configuration⟩
3 a : of uniformly close and coherent texture **:** not loose or spongy **:** COMPACT **b :** possessing or characterized by the properties of a solid **:** neither gaseous nor liquid
4 : of good substantial quality or kind ⟨*solid* comfort⟩: as **a :** SOUND ⟨*solid* reasons⟩ **b :** made firmly and well ⟨*solid* furniture⟩
5 a : having no break or interruption ⟨waited three *solid* hours⟩ **b :** UNANIMOUS ⟨had the *solid* support of the party⟩ **c :** intimately friendly or associated ⟨*solid* with the boss⟩
6 a : PRUDENT; *also* **:** well-established financially **b :** serious in purpose or character
7 : of one substance or character: as **a :** entirely of one metal or containing the minimum of alloy necessary to impart hardness ⟨*solid* gold⟩ **b :** of a single color
— **sol·id·ly** *adverb*
— **sol·id·ness** *noun*

²solid *noun* (15th century)
1 : a geometrical figure or element (as a cube or sphere) having three dimensions — see VOLUME table
2 a : a substance that does not flow perceptibly under moderate stress, has a definite capacity for resisting forces (as compression or tension) which tend to deform it, and under ordinary conditions retains a definite size and shape **b :** the part of a solution or suspension that when freed from solvent or suspending medium has the qualities of a solid — usually used in plural ⟨milk *solids*⟩
3 : something that is solid: as **a :** a solid color **b :** a compound word whose members are joined together without a hyphen

³solid *adverb* (1651)
: in a solid manner; *also* **:** UNANIMOUSLY

sol·i·da·go \ˌsä-lə-'dā-(ˌ)gō, -'dä-\ *noun, plural* **-gos** [New Latin, from Medieval Latin *soldago,* an herb reputed to heal wounds, from *soldare* to make whole, from Latin *solidare,* from *solidus* solid] (circa 1771)
: any of a genus (*Solidago*) of chiefly North American composite herbs including the typical goldenrods

solid angle *noun* (circa 1704)
: the three-dimensional angular spread at the vertex of a cone measured by the area intercepted by the cone on a unit sphere whose center is the vertex of the cone

sol·i·da·rism \'sä-lə-də-ˌri-zəm\ *noun* (1906)
: SOLIDARITY
— **sol·i·da·rist** \-rist\ *noun*
— **sol·i·da·ris·tic** \ˌsä-lə-də-'ris-tik\ *adjective*

sol·i·dar·i·ty \ˌsä-lə-'dar-ə-tē\ *noun* [French *solidarité,* from *solidaire* characterized by solidarity, from Latin *solidum* whole sum, from neuter of *solidus* solid] (1841)
: unity (as of a group or class) that produces or is based on community of interests, objectives, and standards

solid geometry *noun* (1733)

: a branch of geometry that deals with figures of three-dimensional space

so·lid·i·fy \sə-ˈli-də-ˌfī\ *verb* **-fied; -fy·ing** (1799)
transitive verb
1 : to make solid, compact, or hard
2 : to make secure, substantial, or firmly fixed ⟨factors that *solidify* public opinion⟩
intransitive verb
: to become solid, compact, or hard
— so·lid·i·fi·ca·tion \-ˌli-də-fə-ˈkā-shən\ *noun*

so·lid·i·ty \sə-ˈli-də-tē\ *noun, plural* **-ties** (14th century)
1 : the quality or state of being solid
2 : something solid

sol·id–look·ing \ˌsä-ləd-ˈlu̇k-iŋ\ *adjective* (1840)
: giving an impression of solid worth or substance ⟨*solid-looking* citizens⟩

solid of revolution (1816)
: a mathematical solid conceived as formed by the revolution of a plane figure about an axis in its plane

solid–state *adjective* (circa 1951)
1 : relating to the properties, structure, or reactivity of solid material; *especially* **:** relating to the arrangement or behavior of ions, molecules, nucleons, electrons, and holes in the crystals of a substance (as a semiconductor) or to the effect of crystal imperfections on the properties of a solid substance ⟨*solid-state* physics⟩
2 a : utilizing the electric, magnetic, or optical properties of solid materials ⟨*solid-state* circuitry⟩ **b :** using semiconductor devices rather than electron tubes ⟨a *solid-state* stereo system⟩

sol·i·dus \ˈsä-lə-dəs\ *noun, plural* **-i·di** \-lə-ˌdī, -ˌdē\ [Middle English, from Late Latin, from Latin, solid] (14th century)
1 : an ancient Roman gold coin introduced by Constantine and used to the fall of the Byzantine Empire
2 [Medieval Latin, shilling, from Late Latin; from its use as a symbol for shillings] **:** DIAGONAL 3

so·li·fluc·tion \ˈsō-lə ˌflək-shən\ *noun* [Latin *solum* soil + *-i-* + *fluction-, fluctio* act of flowing, from *fluere* to flow — more at FLUID] (1906)
: the slow creeping of saturated fragmental material (as soil) down a slope that usually occurs in regions of perennial frost

so·lil·o·quist \sə-ˈli-lə-kwist\ *noun* (1804)
: one who soliloquizes

so·lil·o·quise *British variant of* SOLILOQUIZE

so·lil·o·quize \-ˌkwīz\ *intransitive verb* **-quized; -quiz·ing** (1759)
: to utter a soliloquy **:** talk to oneself
— so·lil·o·quiz·er *noun*

so·lil·o·quy \sə-ˈli-lə-kwē\ *noun, plural* **-quies** [Late Latin *soliloquium,* from Latin *solus* alone + *loqui* to speak] (circa 1613)
1 : the act of talking to oneself
2 : a dramatic monologue that gives the illusion of being a series of unspoken reflections

so·lip·sism \ˈsō-ləp-ˌsi-zəm, ˈsä-\ *noun* [Latin *solus* alone + *ipse* self] (1874)
: a theory holding that the self can know nothing but its own modifications and that the self is the only existent thing
— so·lip·sist \ˈsō-ləp-sist, ˈsä-ləp-, sə-ˈlip-\ *noun*
— so·lip·sis·tic \ˌsō-ləp-ˈsis-tik, ˌsä-\ *adjective*
— so·lip·sis·ti·cal·ly \-ti-k(ə-)lē\ *adverb*

sol·i·taire \ˈsä-lə-ˌtar, -ˌter\ *noun* [French, from *solitaire,* adjective, solitary, from Latin *solitarius*] (circa 1727)
1 : a single gem (as a diamond) set alone
2 : any of various card games that can be played by one person

[1]**sol·i·tary** \ˈsä-lə-ˌter-ē\ *adjective* [Middle English, from Latin *solitarius,* from *solitas* aloneness, from *solus* alone] (14th century)
1 a : being, living, or going alone or without companions **b :** saddened by isolation
2 : UNFREQUENTED, DESOLATE
3 a : taken, passed, or performed without companions ⟨a *solitary* ramble⟩ **b :** keeping a prisoner apart from others ⟨*solitary* confinement⟩
4 : being at once single and isolated ⟨a *solitary* example⟩
5 a : occurring singly and not as part of a group or cluster ⟨flowers terminal and *solitary*⟩ **b :** not gregarious, colonial, social, or compound ⟨*solitary* bees⟩
synonym see ALONE
— sol·i·tar·i·ly \ˌsä-lə-ˈter-ə-lē\ *adverb*
— sol·i·tar·i·ness \ˈsä-lə-ˌter-ē-nəs\ *noun*

[2]**solitary** *noun, plural* **-tar·ies** (15th century)
1 : one who lives or seeks to live a solitary life **:** RECLUSE
2 : solitary confinement in prison

sol·i·ton \ˈsä-lə-ˌtän\ *noun* [*soli*tary + [2]*-on*] (1965)
: a solitary wave (as in a gaseous plasma) that propagates with little loss of energy and retains its shape and speed after colliding with another such wave

sol·i·tude \ˈsä-lə-ˌtüd, -ˌtyüd\ *noun* [Middle English, from Middle French, from Latin *solitudin-, solitudo,* from *solus*] (14th century)
1 : the quality or state of being alone or remote from society **:** SECLUSION
2 : a lonely place (as a desert) ☆

sol·i·tu·di·nar·i·an \ˌsä-lə-ˌtü-dᵊn-ˈer-ē-ən, -ˌtyü-\ *noun* [Latin *solitudin-, solitudo* + English *-arian*] (1691)
: RECLUSE

sol·ler·et \ˌsä-lə-ˈret\ *noun* [French] (1826)
: a flexible steel shoe forming part of a medieval suit of armor — see ARMOR illustration

sol·mi·za·tion \ˌsäl-mə-ˈzā-shən\ *noun* [French *solmisation,* from *solmiser* to sol-fa, from *sol* (from Medieval Latin) + *mi* (from Medieval Latin) + *-iser* -ize] (1730)
: the act, practice, or system of using syllables to denote the tones of a musical scale

[1]**so·lo** \ˈsō-(ˌ)lō\ *noun, plural* **solos** [Italian, from *solo* alone, from Latin *solus*] (1695)
1 *or plural* **so·li** \ˈsō-(ˌ)lē\ **a :** a musical composition for a single voice or instrument with or without accompaniment **b :** the featured part of a concerto or similar work
2 : a performance in which the performer has no partner or associate
3 : any of several card games in which a player elects to play without a partner against the other players

[2]**solo** *adverb* (1712)
: without a companion **:** ALONE ⟨fly *solo*⟩

[3]**solo** *adjective* (1774)
: of, relating to, or being a solo ⟨a *solo* performance⟩

[4]**solo** *intransitive verb* **so·loed; so·lo·ing** \-(ˌ)lō-iŋ, -lə-wiŋ\ (1886)
: to perform by oneself; *especially* **:** to fly an airplane without one's instructor on board

so·lo·ist \ˈsō-lə-wist, -(ˌ)lō-ist\ *noun* (1864)
: one who performs a solo

Sol·o·mon \ˈsä-lə-mən\ *noun* [Late Latin, from Hebrew *Shĕlōmōh*]
: a son of David and 10th century B.C. king of Israel proverbial for his wisdom

Sol·o·mon·ic \ˌsä-lə-ˈmä-nik\ *adjective* (1857)
: marked by notable wisdom, reasonableness, or discretion especially under trying circumstances

Solomon's seal *noun* (1543)
1 : any of a genus (*Polygonatum*) of perennial herbs of the lily family with tubular flowers and gnarled rhizomes — called also *Solomonseal*
2 : an emblem consisting of two interlaced triangles forming a 6-pointed star and formerly used as an amulet especially against fever — compare HEXAGRAM

so·lon \ˈsō-lən, -ˌlän\ *noun* [*Solon*] (1625)
1 : a wise and skillful lawgiver
2 : a member of a legislative body

sol·on·chak \ˌsä-lən-ˈchak\ *noun* [Russian, salt marsh] (1925)
: any of a group of intrazonal strongly saline usually pale soils found especially in poorly drained arid or semiarid areas

sol·o·netz \ˌsä-lə-ˈnets\ *noun* [Russian *solonets* salt not extracted by decoction] (1924)
: any of a group of intrazonal dark hard alkaline soils evolved by leaching and alkalizing from solonchak
— sol·o·netz·ic \-ˈnet-sik\ *adjective*

so long \sō-ˈlȯŋ, sə-\ *interjection* [origin unknown] (circa 1861)
— used to express farewell

so long as *conjunction* (14th century)
1 : during and up to the end of the time that **:** WHILE
2 : provided that

sol·stice \ˈsäl-stəs, ˈsōl-, ˈsȯl-\ *noun* [Middle English, from Old French, from Latin *solstitium,* from *sol* sun + *-stit-, -stes* standing; akin to Latin *stare* to stand — more at SOLAR, STAND] (13th century)
1 : either of the two points on the ecliptic at which its distance from the celestial equator is greatest and which is reached by the sun each year about June 22d and December 22d
2 : the time of the sun's passing a solstice which occurs about June 22d to begin summer in the northern hemisphere and about December 22d to begin winter in the northern hemisphere

sol·sti·tial \säl-ˈsti-shəl, sōl-, sȯl-\ *adjective* [Middle English *solsticial,* from Old French & Latin; Old French, from Latin *solstitialis,* from *solstitium*] (14th century)
1 : of, relating to, or characteristic of a solstice and especially the summer solstice
2 : happening or appearing at or associated with a solstice

sol·u·bi·lise *British variant of* SOLUBILIZE

sol·u·bil·i·ty \ˌsäl-yə-ˈbi-lə-tē\ *noun* (1661)
1 : the quality or state of being soluble
2 : the amount of a substance that will dissolve in a given amount of another substance

sol·u·bi·lize \ˈsäl-yə-bə-ˌlīz\ *transitive verb* **-lized; -liz·ing** (circa 1926)
: to make soluble or increase the solubility of
— sol·u·bi·li·za·tion \ˌsäl-yə-bə-lə-ˈzā-shən\ *noun*

sol·u·ble \ˈsäl-yə-bəl\ *adjective* [Middle English, from Middle French, capable of being loosened or dissolved, from Late Latin *solubilis,* from Latin *solvere* to loosen, dissolve — more at SOLVE] (15th century)
1 : susceptible of being dissolved in or as if in a liquid and especially water
2 : subject to being solved or explained ⟨*soluble* questions⟩

so·lum \ˈsō-ləm\ *noun, plural* **so·la** \-lə\ *or* **solums** [New Latin, from Latin, ground, soil] (1928)

☆ SYNONYMS
Solitude, isolation, seclusion mean the state of one who is alone. SOLITUDE may imply a condition of being apart from all human beings or of being cut off by wish or circumstances from one's usual associates ⟨a few quiet hours of *solitude*⟩. ISOLATION stresses detachment from others often involuntarily ⟨the *isolation* of the village in winter⟩. SECLUSION suggests a shutting away or keeping apart from others often connoting deliberate withdrawal from the world or retirement to a quiet life ⟨lived in pastoral *seclusion*⟩.

: the altered layer of soil above the parent material that includes the A and B horizons

so·lus \'sō-ləs\ *adverb or adjective* [Latin] (1599)
: ALONE — often used in stage directions

sol·ute \'säl-ˌyüt\ *noun* [Latin *solutus,* past participle of *solvere*] (1893)
: a dissolved substance

so·lu·tion \sə-'lü-shən\ *noun* [Middle English, from Middle French, from Latin *solution-, solutio,* from *solvere* to loosen, solve] (14th century)
1 a : an action or process of solving a problem **b :** an answer to a problem **:** EXPLANATION; *specifically* **:** a set of values of the variables that satisfies an equation
2 a : an act or the process by which a solid, liquid, or gaseous substance is homogeneously mixed with a liquid or sometimes a gas or solid **b :** a homogeneous mixture formed by this process; *especially* **:** a single-phase liquid system **c :** the condition of being dissolved
3 : a bringing or coming to an end or into a state of discontinuity

solution set *noun* (1959)
: the set of values that satisfy an equation; *also* **:** TRUTH SET

So·lu·tre·an \sə-'lü-trē-ən\ *adjective* [*Solutré,* village in France] (1888)
: of or relating to an upper Paleolithic culture characterized by leaf-shaped finely flaked stone implements

solv·able \'säl-və-bəl, 'sȯl-\ *adjective* (circa 1676)
: susceptible of solution or of being solved, resolved, or explained
— solv·abil·i·ty \ˌsäl-və-'bi-lə-tē, ˌsȯl-\ *noun*

[1]sol·vate \'säl-ˌvāt, 'sȯl-\ *noun* [*sol*vent + [1]*-ate*] (1904)
: an aggregate that consists of a solute ion or molecule with one or more solvent molecules; *also* **:** a substance (as a hydrate) containing such ions

[2]solvate *transitive verb* **sol·vat·ed; sol·vat·ing** (1909)
: to make part of a solvate
— sol·va·tion \säl-'vā-shən, sȯl-\ *noun*

Sol·vay process \'säl-ˌvā-\ *noun* [Ernest *Solvay* (died 1922) Belgian chemist] (1884)
: a process for making soda from common salt by passing carbon dioxide into ammoniacal brine resulting in precipitation of sodium bicarbonate which is then calcined to carbonate

solve \'sälv, 'sȯlv\ *verb* **solved; solv·ing** [Middle English, to loosen, from Latin *solvere* to loosen, solve, dissolve, from *sed-, se-* apart + *luere* to release — more at SECEDE, LOSE] (circa 1533)
transitive verb
1 : to find a solution, explanation, or answer for ⟨*solve* a problem⟩ ⟨*solved* the crime⟩
2 : to pay (as a debt) in full
intransitive verb
: to solve something ⟨substitute the known values of the constants and *solve* for *x*⟩
— solv·er *noun*

sol·ven·cy \'säl-vən(t)-sē, 'sȯl-\ *noun* (circa 1727)
: the quality or state of being solvent

[1]sol·vent \-vənt\ *adjective* [Latin *solvent-, solvens,* present participle of *solvere* to dissolve, pay] (1630)
1 : able to pay all legal debts
2 : that dissolves or can dissolve ⟨*solvent* action of water⟩
— sol·vent·ly *adverb*

[2]solvent *noun* (1671)
1 : a usually liquid substance capable of dissolving or dispersing one or more other substances
2 : something that provides a solution
3 : something that eliminates or attenuates something especially unwanted
— sol·vent·less \-ləs\ *adjective*

sol·vol·y·sis \säl-'vä-lə-səs, sȯl-\ *noun* [New Latin, from English *sol*vent + New Latin *-o-* + *-lysis*] (1916)
: a chemical reaction (as hydrolysis) of a solvent and solute that results in the formation of new compounds
— sol·vo·lyt·ic \ˌsäl-və-'li-tik, ˌsȯl-\ *adjective*

[1]so·ma \'sō-mə\ *noun* [Sanskrit; akin to Avestan *haoma,* a Zoroastrian ritual drink, Sanskrit *sunoti* he presses out] (1827)
: an intoxicating plant juice probably from a leafless vine (*Sarcostemma intermedium*) of the milkweed family used in ancient India as an offering to the gods and as a drink of immortality by worshipers in Vedic ritual and worshiped as a Vedic god

[2]soma *noun, plural* **so·ma·ta** \'sō-mə-tə\ *or* **somas** [New Latin *somat-, soma,* from Greek *sōmat-, sōma* body] (circa 1885)
1 : the body of an organism
2 : all of an organism except the germ cells
3 : CELL BODY

So·ma·li \sō-'mä-lē, sə-\ *noun, plural* **Somali** *or* **Somalis** (1814)
1 : a member of a people of Somaliland
2 : the Cushitic language of the Somali

so many *adjective* (1533)
1 : constituting an unspecified number ⟨read *so many* chapters each night⟩
2 : constituting a group or pack ⟨behaved like *so many* animals⟩

somat- *or* **somato-** *combining form* [New Latin, from Greek *sōmat-, sōmato-,* from *sōmat-, sōma* body]
: body ⟨*somato*logy⟩

so·mat·ic \sō-'ma-tik, sə-\ *adjective* [Greek *sōmatikos,* from *sōmat-, sōma*] (circa 1775)
1 : of, relating to, or affecting the body especially as distinguished from the germ plasm or the psyche
2 : of or relating to the wall of the body **:** PARIETAL
— so·mat·i·cal·ly \-ti-k(ə-)lē\ *adverb*

somatic cell *noun* (1888)
: one of the cells of the body that compose the tissues, organs, and parts of that individual other than the germ cells

so·ma·tol·o·gy \ˌsō-mə-'tä-lə-jē\ *noun* [New Latin *somatologia,* from *somat-* + *-logia* -logy] (circa 1878)
: a branch of anthropology primarily concerned with the comparative study of human evolution, variation, and classification especially through measurement and observation
— so·ma·to·log·i·cal \ˌsō-mə-tᵊl-'ä-ji-kəl, sō-ˌma-\ *adjective*

so·mato·me·din \ˌsō-mə-tō-'mē-dᵊn, sō-ˌma-tə-\ *noun* [*somat-* + inter*med*iary + [1]*-in*] (1971)
: any of several endogenous peptides produced especially in the liver that are dependent on and probably mediate growth hormone activity (as in sulfate uptake by epiphyseal cartilage)

so·mato·pleure \sō-'ma-tə-ˌplu̇r\ *noun* [New Latin *somatopleura,* from *somat-* + Greek *pleura* side] (1874)
: a complex fold of tissue in the embryo of a craniate vertebrate consisting of an outer layer of mesoderm together with the ectoderm that sheathes it and giving rise to the amnion and chorion

so·mato·sen·so·ry \sō-ˌma-tə-'sen(t)s-rē, -'sen(t)-sə-rē\ *adjective* (1952)
: of, relating to, or being sensory activity having its origin elsewhere than in the special sense organs (as eyes and ears) and conveying information about the state of the body proper and its immediate environment

so·mato·stat·in \sō-ˌma-tə-'sta-tᵊn\ *noun* [*somat-* + Latin *stat*us (past participle of *sistere* to halt, cause to stand) + English [1]*-in;* akin to Latin *stare* to stand — more at STAND] (1973)
: a polypeptide neurohormone that is found especially in the hypothalamus and inhibits the secretion of several other hormones (as growth hormone, insulin, and gastrin)

so·mato·tro·pic hormone \-'trō-pik-\ *noun* [*somat-* + *-tropic*] (1938)
: GROWTH HORMONE 1

so·mato·tro·pin \-'trō-pən\ *also* **so·mato·tro·phin** \-fən\ *noun* [*somatotrop*ic + [1]*-in*] (1941)
: GROWTH HORMONE 1

so·mato·type \sō-'ma-tə-ˌtīp\ *noun* (1940)
: body type **:** PHYSIQUE

som·ber *or* **som·bre** \'säm-bər\ *adjective* [French *sombre*] (1760)
1 : so shaded as to be dark and gloomy
2 a : of a serious mien **:** GRAVE **b :** of a dismal or depressing character **:** MELANCHOLY **c :** conveying gloomy suggestions or ideas
3 : of a dull or heavy cast or shade **:** dark colored
— som·ber·ly *adverb*
— som·ber·ness *noun*

som·bre·ro \səm-'brer-(ˌ)ō, säm-\ *noun, plural* **-ros** [Spanish, from *sombra* shade] (1599)
: a high-crowned hat of felt or straw with a very wide brim worn especially in the Southwest and Mexico

sombrero

som·brous \'säm-brəs\ *adjective* [French *sombre*] (1730)
archaic **:** SOMBER

[1]some \'səm, *for 2 without stress*\ *adjective* [Middle English *som,* adjective & pronoun, from Old English *sum;* akin to Old High German *sum* some, Greek *hamē* somehow, *homos* same — more at SAME] (before 12th century)
1 : being an unknown, undetermined, or unspecified unit or thing ⟨*some* person knocked⟩
2 a : being one, a part, or an unspecified number of something (as a class or group) named or implied ⟨*some* gems are hard⟩ **b :** being of an unspecified amount or number ⟨give me *some* water⟩ ⟨have *some* apples⟩
3 : REMARKABLE, STRIKING ⟨that was *some* party⟩
4 : being at least one — used to indicate that a logical proposition is asserted only of a subclass or certain members of the class denoted by the term which it modifies

[2]some \'səm\ *pronoun, singular or plural in construction* (before 12th century)
1 : one indeterminate quantity, portion, or number as distinguished from the rest
2 : an indefinite additional amount ⟨ran a mile and then *some*⟩

[3]some \'səm, ˌsəm\ *adverb* (before 12th century)
1 : ABOUT ⟨*some* 80 houses⟩ ⟨twenty-*some* people⟩
2 a : in some degree **:** SOMEWHAT ⟨felt *some* better⟩ **b :** to some degree or extent **:** a little ⟨the cut bled *some*⟩ ⟨I need to work on it *some* more⟩ **c** — used as a mild intensive ⟨that's going *some*⟩ ■

[1]-some *adjective suffix* [Middle English *-som,* from Old English *-sum;* akin to Old High German *-sam* -some, Old English *sum* some]
: characterized by a (specified) thing, quality, state, or action ⟨awe*some*⟩ ⟨burden*some*⟩ ⟨cuddle*some*⟩

[2]-some *noun suffix* [Middle English (northern dialect) *-sum,* from Middle English *sum,* pronoun, one, some]
: group of (so many) members and especially persons ⟨four*some*⟩

[3]-some *noun combining form* [New Latin *-somat-, -soma,* from Greek *sōmat-, sōma*]
1 : body ⟨chromo*some*⟩
2 : chromosome ⟨mono*some*⟩

[1]some·body \'səm-(ˌ)bə-dē, -ˌbä-\ *pronoun* (14th century)

: one or some person of unspecified or indefinite identity ⟨*somebody* will come in⟩ ■

²somebody *noun* (circa 1566)
: a person of position or importance

some·day \'səm-,dā\ *adverb* (14th century)
: at some future time

some·deal \'səm-,dēl\ *adverb* (before 12th century)
archaic **:** SOMEWHAT

some·how \'səm-,hau̇\ *adverb* (1664)
: in one way or another not known or designated **:** by some means

some·one \-(,)wən\ *pronoun* (14th century)
: some person **:** SOMEBODY
usage see SOMEBODY

some·place \-,plās\ *adverb* (1880)
: SOMEWHERE

som·er·sault \'sə-mər-,sȯlt\ *noun* [Middle French *sombresaut* leap, ultimately from Latin *super* over + *saltus* leap, from *salire* to jump — more at OVER, SALLY] (circa 1530)
: a leap or roll in which a person turns forward or backward in a complete revolution bringing the feet over the head and finally landing on the feet; *also* **:** a falling or tumbling head over heels
— somersault *intransitive verb*

som·er·set \-,set\ *noun or intransitive verb* [by alteration] (1591)
: SOMERSAULT

¹some·thing \'səm(p)-thiŋ, *especially in rapid speech or for 2* 'səmp-ᵊm\ *pronoun* (before 12th century)
1 : some indeterminate or unspecified thing
2 : a person or thing of consequence
3 : one having more or less the character, qualities, or nature of something different ⟨is *something* of a bore⟩
— something else : something or someone special or extraordinary

²something *adverb* (13th century)
1 : in some degree **:** SOMEWHAT
2 — used as an intensive giving adverbial force to an adjective ⟨swears *something* awful⟩

¹some·time \'səm-,tīm\ *adverb* (14th century)
1 *archaic* **:** in the past **:** FORMERLY
2 *archaic* **:** once in a while **:** OCCASIONALLY
3 : at some time in the future ⟨I'll do it *sometime*⟩
4 : at some not specified or definitely known point of time ⟨*sometime* last night⟩

²sometime *adjective* (14th century)
1 : having been formerly **:** FORMER, LATE
2 : being so occasionally or in only some respects ⟨a *sometime* . . . father who appears and disappears —Evelyn Shelby⟩

¹some·times \'səm-,tīmz *also* (,)səm-'\ *adverb* (14th century)
: at times **:** now and then **:** OCCASIONALLY

²sometimes *adjective* (1593)
: SOMETIME

some·way \'səm-,wā\ *also* **some·ways** \-,wāz\ *adverb* (15th century)
: SOMEHOW

¹some·what \-,(h)wät, -,(h)wət, (,)səm-'\ *pronoun* (13th century)
: SOMETHING

²somewhat *adverb* (13th century)
: in some degree or measure **:** SLIGHTLY

some·when \'səm-,(h)wen\ *adverb* (1833)
: SOMETIME

¹some·where \-,(h)wer, -,(h)war, -(,)(h)wər\ *adverb* (13th century)
1 : in, at, from, or to a place unknown or unspecified ⟨mentions it *somewhere*⟩
2 : to a place symbolizing positive accomplishment or progress ⟨now we're getting *somewhere*⟩
3 : in the vicinity of **:** APPROXIMATELY ⟨*somewhere* about nine o'clock⟩

²somewhere *noun* (1647)
: an undetermined or unnamed place

some·wheres \-,(h)werz, -,(h)warz, -(,)(h)wərz\ *adverb* (1815)
: SOMEWHERE

some·whith·er \-,(h)wi-t͟hər\ *adverb* (1530)
archaic **:** to some place **:** SOMEWHERE

-somic *adjective combining form* [International Scientific Vocabulary ³*-some* + *-ic*]
: having or being a chromosome complement of which one or more but not all members exhibit (such) a degree of reduplication of chromosomes or genomes ⟨mono*somic*⟩

so·mite \'sō-,mīt\ *noun* [International Scientific Vocabulary, from Greek *sōma* body] (1869)
: one of the longitudinal series of segments into which the body of many animals (as articulate animals and vertebrates) is divided **:** METAMERE

som·me·lier \,sə-məl-'yā\ *noun, plural* **som·meliers** \-'yā(z)\ [French, from Middle French, court official charged with transportation of supplies, pack animal driver, from Old Provençal *saumalier* pack animal driver, from *sauma* pack animal, load of a pack animal, from Late Latin *sagma* packsaddle — more at SUMPTER] (1829)
: a waiter in a restaurant who has charge of wines and their service **:** a wine steward

somnambul- *combining form* [New Latin, from *somnambulus* somnambulist, from Latin *somnus* sleep + *-ambulus* (as in *funambulus* funambulist) — more at SOMNOLENT]
: somnambulism **:** somnambulist ⟨*somnambulant*⟩

som·nam·bu·lant \säm-'nam-byə-lənt\ *adjective* (1866)
: walking or addicted to walking while asleep

som·nam·bu·late \-,lāt\ *intransitive verb* **-lat·ed; -lat·ing** (1833)
: to walk when asleep
— som·nam·bu·la·tion \(,)säm-,nam-byə-'lā-shən\ *noun*

som·nam·bu·lism \säm-'nam-byə-,li-zəm\ *noun* (1797)
1 : an abnormal condition of sleep in which motor acts (as walking) are performed
2 : actions characteristic of somnambulism
— som·nam·bu·list \-list\ *noun*
— som·nam·bu·lis·tic \(,)säm-,nam-byə-'lis-tik\ *adjective*
— som·nam·bu·lis·ti·cal·ly \-ti-k(ə-)lē\ *adverb*

som·ni·fa·cient \,säm-nə-'fā-shənt\ *adjective* [Latin *somnus* sleep + English *-facient*] (circa 1890)
: HYPNOTIC 1
— somnifacient *noun*

som·nif·er·ous \säm-'ni-f(ə-)rəs\ *adjective* [Latin *somnifer* somniferous, from *somnus* + *-fer* -ferous] (1602)
: SOPORIFIC

som·no·lence \'säm-nə-lən(t)s\ *noun* (14th century)
: the quality or state of being drowsy **:** SLEEPINESS

som·no·lent \-lənt\ *adjective* [Middle English *sompnolent*, from Middle French, from Latin *somnolentus*, from *somnus* sleep; akin to Old English *swefn* sleep, Greek *hypnos*] (15th century)
1 : of a kind likely to induce sleep ⟨a *somnolent* sermon⟩
2 a : inclined to or heavy with sleep **:** DROWSY **b :** SLEEPY 2 ⟨*somnolent* rivers⟩
— som·no·lent·ly *adverb*

¹so much *adverb* (13th century)
: by the amount indicated or suggested ⟨if they lose their way, *so much* the better for us⟩

²so much *pronoun* (14th century)
1 : something (as an amount or price) unspecified or undetermined ⟨charge *so much* a mile⟩
2 : all that can be or is to be said or done ⟨*so much* for the history of the case⟩

³so much *adjective* (1557)
— used as an intensive ⟨the house burned like *so much* paper⟩ ⟨sounded like *so much* nonsense⟩

so much as *adverb* (15th century)
: EVEN 3d

son \'sən\ *noun* [Middle English *sone*, from Old English *sunu*; akin to Old High German *sun* son, Greek *hyios*] (before 12th century)
1 a : a male offspring especially of human beings **b :** a male adopted child **c :** a male descendant
2 *capitalized* **:** the second person of the Trinity
3 : a person closely associated with or deriving from a formative agent (as a nation, school, or race)
— son·hood \-,hu̇d\ *noun*

son- *or* **sono-** *combining form* [Latin *sonus* sound]
: sound ⟨*son*ic⟩ ⟨*sono*gram⟩

so·nant \'sō-nənt\ *adjective* [Latin *sonant-, sonans*, present participle of *sonare* to sound — more at SOUND] (1846)
1 : VOICED 2
2 : SYLLABIC 1a
— sonant *noun*

so·nar \'sō-,när\ *noun* [*so*und *na*vigation *r*anging] (1945)

□ USAGE

some When the adverb *some* is used to modify a number, it is almost always a round number ⟨a community of *some* 150,000 inhabitants⟩ but because *some* is slightly more emphatic than *about* or *approximately* it is occasionally used with a more exact number in an intensive function ⟨an expert parachutist, he has *some* 115 jumps to his credit —*Current Biography*⟩. When *some* is used without a number, most commentators feel that *somewhat* is to be preferred. Their advice is an oversimplification, however; only when *some* modifies an adjective, usually a comparative, will *somewhat* always substitute smoothly. When *some* modifies a verb or adverb, and especially when it follows a verb, substitution of *somewhat* may prove awkward ⟨Italy forced me to grow up *some* —E. W. Brooke⟩ ⟨I'm not a prude; I've been around *some* in my day —Roy Rogers⟩ ⟨here in Newport, both Southern Cross and Courageous practiced *some* more —W. N. Wallace⟩.

somebody The indefinite pronouns *somebody* and *someone* share with other indefinite pronouns the characteristic of taking a singular verb and, more often than not, a plural pronoun in reference. The use of the plural pronoun—*they, their, them*—has traditionally been disapproved by those who insist that agreement in number is more important than avoidance of gender-specific third-person singular pronouns. The plural pronouns are well-established ⟨*somebody* told me they thought he killed a man once —F. Scott Fitzgerald⟩ ⟨when *someone* says they want him to show up at one of their functions —Andy Rooney⟩. Use of the singular pronoun may mean that the writer has someone specific in mind ⟨it was a bit risky to bring him home as eventualities might possibly ensue (*somebody* having a temper of her own sometimes) —James Joyce⟩, but often the person is not specific ⟨if *someone* doesn't carry any money, the chances are he's loaded —Andy Rooney⟩. Usage is mixed.

There is a superstition that *someone* is preferable to *somebody*. Through the 19th century, *somebody* was the more frequent of the two, but *someone* has come on strong in the 20th. Both are impeccably standard.

: a method or device for detecting and locating objects especially underwater by means of sound waves sent out to be reflected by the objects; *also* : a device for detecting the presence of a vessel (as a submarine) by the sound it emits in water ◆

so·na·ta \sə-ˈnä-tə\ *noun* [Italian, from *sonare* to sound, from Latin] (1694)
: an instrumental musical composition typically of three or four movements in contrasting forms and keys

sonata form *noun* (1873)
: a musical form that consists basically of an exposition, a development, and a recapitulation and that is used especially for the first movement of a sonata

son·a·ti·na \ˌsä-nə-ˈtē-nə\ *noun* [Italian, diminutive of *sonata*] (circa 1801)
: a short usually simplified sonata

sonde \ˈsänd\ *noun* [French, literally, sounding line — more at SOUND] (1901)
: any of various devices for testing physical conditions (as at high altitudes, below the earth's surface, or inside the body)

sone \ˈsōn\ *noun* [International Scientific Vocabulary, from Latin *sonus* sound — more at SOUND] (1948)
: a subjective unit of loudness for an average listener equal to the loudness of a 1000-hertz sound that has an intensity 40 decibels above the listener's own threshold of hearing

son et lu·mière \ˌsō^{n}-(n)ā-lüm-ˈyer\ *noun* [French, literally, sound and light] (1957)
: an outdoor spectacle at an historic site consisting of recorded narration with light and sound effects

song \ˈsȯŋ\ *noun* [Middle English, from Old English *sang;* akin to Old English *singan* to sing] (before 12th century)
1 : the act or art of singing
2 : poetical composition
3 a : a short musical composition of words and music **b :** a collection of such compositions
4 : a distinctive or characteristic sound or series of sounds (as of a bird or insect)
5 a : a melody for a lyric poem or ballad **b :** a poem easily set to music
6 a : a habitual or characteristic manner **b :** a violent, abusive, or noisy reaction ⟨put up quite a *song*⟩
7 : a small amount ⟨sold for a *song*⟩
— song·like \-ˌlīk\ *adjective*

song and dance *noun* (1872)
1 : a theatrical performance (as a vaudeville performance) combining singing and dancing
2 : a long and often familiar statement or explanation that is usually not true or pertinent

song·bird \ˈsȯŋ-ˌbərd\ *noun* (1774)
1 a : a bird that utters a succession of musical tones **b :** an oscine bird
2 : a female singer

song·book \-ˈbu̇k\ *noun* (before 12th century)
: a collection of songs; *specifically* : a book containing vocal music (as hymns)

song cycle *noun* (1899)
: a group of related songs designed to form a musical entity

song·fest \ˈsȯŋ-fest\ *noun* (circa 1912)
: an informal session of group singing of popular or folk songs

song·ful \-fəl\ *adjective* (14th century)
: given to or suggestive of singing : MELODIOUS
— song·ful·ly \-fə-lē\ *adverb*
— song·ful·ness *noun*

song·less \-ləs\ *adjective* (circa 1805)
: lacking in, incapable of, or not given to song
— song·less·ly *adverb*

Song of Sol·o·mon \-ˈsä-lə-mən\ [from the opening verse: "The song of songs, which is Solomon's"]
: a collection of love poems forming a book in the Protestant canon of the Old Testament — see BIBLE table

Song of Songs [translation of Hebrew *shīr hashshīrīm*]
: a collection of love poems forming a book in the canonical Jewish Scriptures and in the Roman Catholic canon of the Old Testament and corresponding to the Song of Solomon in the Protestant canon of the Old Testament — see BIBLE table

song·smith \ˈsȯŋ-ˌsmith\ *noun* (1795)
: a composer of songs

song sparrow *noun* (1810)
: a common North American sparrow (*Melospiza melodia*) that is brownish above and white below with brownish streaks on the breast and that is noted for it melodious song

song·ster \ˈsȯŋ(k)-stər\ *noun* (14th century)
1 : one that sings with skill
2 : SONGBOOK

song·stress \ˈsȯŋ(k)-strəs\ *noun* (1703)
: a female singer

song thrush *noun* (1668)
: an Old World thrush (*Turdus philomelos*) that is largely brown above with brown-spotted white underparts — called also *mavis, throstle*

song·writ·er \ˈsȯŋ-ˌrī-tər\ *noun* (1821)
: a person who composes words or music or both especially for popular songs
— song·writ·ing \-ˌrī-tiŋ\ *noun*

son·ic \ˈsä-nik\ *adjective* (1923)
1 : utilizing, produced by, or relating to sound waves ⟨*sonic* altimeter⟩; *broadly* : of or involving sound ⟨*sonic* pollution⟩
2 : having a frequency within the audibility range of the human ear — used of waves and vibrations
3 : of, relating to, or being the speed of sound in air or about 761 miles per hour (1224 kilometers per hour) at sea level at 59°F (15°C)
4 : capable of uttering sounds
— son·i·cal·ly \-ni-k(ə-)lē\ *adverb*

son·i·cate \ˈsä-nə-ˌkāt\ *transitive verb* **-cat·ed; -cat·ing** [*sonic* + ^{4}*-ate*] (1960)
: to disrupt (as bacteria) by treatment with high-frequency sound waves
— son·i·ca·tion \ˌsä-nə-ˈkā-shən\ *noun*

sonic barrier *noun* (1946)
: SOUND BARRIER

sonic boom *noun* (1952)
: a sound resembling an explosion produced when a shock wave formed at the nose of an aircraft traveling at supersonic speed reaches the ground — called also *sonic bang*

son–in–law \ˈsən-ən-ˌlȯ\ *noun, plural* **sons–in–law** (14th century)
: the husband of one's daughter

son·less \ˈsən-ləs\ *adjective* (14th century)
: not possessing or never having had a son

son·ly \-lē\ *adjective* (15th century)
: FILIAL

son·net \ˈsä-nət\ *noun* [Italian *sonetto,* from Old Provençal *sonet* little song, from *son* sound, song, from Latin *sonus* sound] (1557)
: a fixed verse form of Italian origin consisting of fourteen lines that are typically five-foot iambics rhyming according to a prescribed scheme; *also* : a poem in this pattern

son·ne·teer \ˌsä-nə-ˈtir\ *noun* (1665)
1 : a composer of sonnets
2 : a minor or insignificant poet
— son·ne·teer·ing \-iŋ\ *noun*

sonnet sequence *noun* (1881)
: a series of sonnets often having a unifying theme

son·ny \ˈsə-nē\ *noun* (1850)
: a young boy — usually used in address

so·no·buoy \ˈsä-nə-ˌbȯi, -ˌbü-ē\ *noun* (1945)
: a buoy equipped for detecting underwater sounds and transmitting them by radio

son of a bitch \ˈsən-ə-və-ˌbich *also* ˈsəm-ˌbich; *as an interjection* ˌsən-ə-və-ˈbich\ *noun, plural* **sons of bitch·es** \ˌsən-zə-ˈbi-chəz\ (1671)
: BASTARD 3 — sometimes considered vulgar; sometimes used interjectionally to express surprise or disappointment

son of a gun \ˈsən-ə-və-ˌgən; *as an interjection* ˌsən-ə-və-ˈgən\ *noun, plural* **sons of guns** (1708)
— usually used as a mild or euphemistic alternative to *son of a bitch;* sometimes used interjectionally to express surprise or disappointment

son of God (14th century)
1 *often S capitalized* **:** a superhuman or divine being (as an angel)
2 *S capitalized* **:** MESSIAH 1
3 : a person established in the love of God by divine promise

son of man (14th century)
1 : a human being
2 *often S capitalized* **:** God's messiah destined to preside over the final judgment of mankind

sono·gram \ˈsä-nə-ˌgram\ *noun* (1956)
: an image produced by ultrasound

so·nog·ra·phy \sō-ˈnä-grə-fē\ *noun* (1977)
: ULTRASOUND 2

so·nor·i·ty \sə-ˈnȯr-ə-tē, -ˈnär-\ *noun, plural* **-ties** (circa 1623)
1 : the quality or state of being sonorous : RESONANCE
2 : a sonorous tone or speech

so·no·rous \sə-ˈnōr-əs, -ˈnȯr-; ˈsä-nə-rəs\ *adjective* [Latin *sonorus;* akin to Latin *sonus* sound] (1611)
1 : producing sound (as when struck)
2 : full or loud in sound
3 : imposing or impressive in effect or style
4 : having a high or an indicated degree of sonority ⟨*sonorous* sounds such as \ä\ and \ȯ\⟩
— so·no·rous·ly *adverb*
— so·no·rous·ness *noun*

son·ship \ˈsən-ˌship\ *noun* (1587)
: the relationship of son to father

sonsy *or* **sons·ie** \ˈsän(t)-sē\ *adjective* [Scots *sons* health] (1725)
chiefly dialect : BUXOM, COMELY

soon \ˈsün, *especially New England* ˈsu̇n\ *adverb* [Middle English *soone,* from Old English *sōna;* akin to Old High German *sān* immediately] (before 12th century)
1 a *obsolete* **:** at once : IMMEDIATELY **b :** before long : without undue time lapse ⟨*soon* after sunrise⟩
2 : in a prompt manner : SPEEDILY ⟨as *soon* as possible⟩ ⟨the *sooner* the better⟩
3 *archaic* **:** before the usual time

◇ WORD HISTORY

sonar The shadowy history of acronymic words such as *sonar* that have grown out of secret government projects sets many pitfalls for the etymologist. When F.V. ("Ted") Hunt, the director of the Harvard Underwater Sound Laboratory, coined *sonar,* most likely in 1942, he applied it to a specific piece of scanning equipment rather than the general technique of active echolocation, perhaps having in mind the words *so*nic, *a*zimuth, and *r*ange. Years later Hunt recalled only that he intended it as a phonetic analogue to *radar* (*ra*dio *d*etection *a*nd *r*anging), and that *sonar* had a more pleasing sound than *sodar.* The interpretation of *sonar* as *so*und *n*avigation *a*nd *r*anging was promulgated by the U.S. Navy only after the fact. *Asdic,* the British counterpart to *sonar* but developed decades earlier, also has an involved history. After Winston Churchill uttered what had long been a top secret term in the House of Commons in 1939, the British Admiralty said that *asdic* stood for *A*llied *S*ubmarine *D*etection *I*nvestigation *C*ommittee, allegedly formed during World War I. But no such committee ever existed. First noted as *asdics* in a now declassified report of 1918, the word appears to have been coined from *A*nti-*S*ubmarine *D*ivision, an Admiralty department, with *-ics* extracted from *supersonics.*

4 : in agreement with one's choice or preference **:** WILLINGLY ⟨I'd just as *soon* walk as drive⟩

soon·er \'sü-nər\ *noun* [*sooner,* comparative of *soon*] (1890)
1 : a person settling on land in the early West before its official opening to settlement in order to gain the prior claim allowed by law to the first settler after official opening
2 *capitalized* **:** a native or resident of Oklahoma — used as a nickname

sooner or later *adverb* (1577)
: at some uncertain future time **:** SOMETIME

¹soot \'sůt, 'sət, 'süt\ *noun* [Middle English, from Old English *sōt;* akin to Old Irish *suide* soot, Old English *sittan* to sit] (before 12th century)
: a black substance formed by combustion or separated from fuel during combustion, rising in fine particles, and adhering to the sides of the chimney or pipe conveying the smoke; *especially* **:** the fine powder consisting chiefly of carbon that colors smoke

²soot *transitive verb* (1602)
: to coat or cover with soot

¹sooth \'süth\ *adjective* [Middle English, from Old English *sōth;* akin to Old High German *sand* true, Latin *esse* to be] (before 12th century)
1 *archaic* **:** TRUE
2 *archaic* **:** SOFT, SWEET

²sooth *noun* (before 12th century)
1 : TRUTH, REALITY
2 *obsolete* **:** BLANDISHMENT

soothe \'sü*th*\ *verb* **soothed; sooth·ing** [Middle English *sothen* to verify, from Old English *sōthian,* from *sōth*] (1697)
transitive verb
1 : to please by or as if by attention or concern **:** PLACATE
2 : RELIEVE, ALLEVIATE
3 : to bring comfort, solace, or reassurance to
intransitive verb
: to bring peace, composure, or quietude
— sooth·er *noun*

sooth·fast \'süth-ˌfast\ *adjective* (before 12th century)
1 *archaic* **:** TRUE
2 *archaic* **:** TRUTHFUL

sooth·ing \'sü-*th*iŋ\ *adjective* (1749)
: tending to soothe; *also* **:** having a sedative effect ⟨*soothing* syrup⟩
— sooth·ing·ly \-*th*iŋ-lē\ *adverb*
— sooth·ing·ness *noun*

sooth·ly \'süth-lē\ *adverb* (before 12th century)
archaic **:** in truth **:** TRULY

sooth·say·er \-ˌsā-ər, -ˌse(-ə)r\ *noun* (14th century)
: a person who predicts the future by magical, intuitive, or more rational means **:** PROGNOSTICATOR

sooth·say·ing \-ˌsā-iŋ\ *noun* (15th century)
1 : the act of foretelling events
2 : PREDICTION, PROPHECY
— sooth·say \-ˌsā\ *intransitive verb*

sooty \'sů-tē, 'sə-, 'sü-\ *adjective* **soot·i·er; -est** (13th century)
1 a : of, relating to, or producing soot **b :** soiled with soot
2 : of the color of soot
— soot·i·ly \-tᵊl-ē\ *adverb*
— soot·i·ness \-tē-nəs\ *noun*

sooty mold *noun* (1901)
: a dark growth of fungus mycelium growing in insect honeydew on plants; *also* **:** a fungus producing such growth

sooty tern *noun* (1785)
: a widely distributed tern (*Sterna fuscata*) of tropical oceans that is blackish above and white below — called also *wideawake*

¹sop \'säp\ *noun* [Middle English *soppe,* from Old English *sopp;* akin to Old English *sūpan* to swallow — more at SUP] (before 12th century)
1 *chiefly dialect* **:** a piece of food dipped or steeped in a liquid
2 : a conciliatory or propitiatory bribe, gift, or gesture

²sop *transitive verb* **sopped; sop·ping** (circa 1529)
1 a : to steep or dip in or as if in liquid **b :** to wet thoroughly **:** SOAK
2 : MOP a
3 : to give a bribe or conciliatory gift to

so·pai·pil·la \ˌsō-pī-'pē-yə, -'pēl-yə\ *or* **so·pa·pil·la** \'sō-pə-\ *noun* [American Spanish *sopaipilla,* diminutive of Spanish *sopaipa* fritter soaked in honey, from *sopa* food soaked in milk, of Germanic origin; akin to Old English *sūpan* to swallow] (circa 1940)
: a square of deep-fried dough often sweetened and eaten as dessert

soph·ism \'sä-ˌfi-zəm\ *noun* (15th century)
1 : an argument apparently correct in form but actually invalid; *especially* **:** such an argument used to deceive
2 : SOPHISTRY 1

soph·ist \'sä-fist\ *noun* [Latin *sophista,* from Greek *sophistēs,* literally, expert, wise man, from *sophizesthai* to become wise, deceive, from *sophos* clever, wise] (1542)
1 *capitalized* **:** any of a class of ancient Greek teachers of rhetoric, philosophy, and the art of successful living prominent about the middle of the 5th century B.C. for their adroit subtle and allegedly often specious reasoning
2 : PHILOSOPHER, THINKER
3 : a captious or fallacious reasoner

so·phis·tic \sä-'fis-tik, sə-\ *or* **so·phis·ti·cal** \-ti-kəl\ *adjective* (15th century)
1 : of or relating to sophists, sophistry, or the ancient Sophists ⟨*sophistic* rhetoric⟩ ⟨*sophistic* subtleties⟩
2 : plausible but fallacious ⟨*sophistic* reasoning⟩
— so·phis·ti·cal·ly \-ti-k(ə-)lē\ *adverb*

¹so·phis·ti·cate \sə-'fis-tə-ˌkāt\ *transitive verb* **-cat·ed; -cat·ing** [Middle English, from Medieval Latin *sophisticatus,* past participle of *sophisticare,* from Latin *sophisticus* sophistic, from Greek *sophistikos,* from *sophistēs* sophist] (15th century)
1 : to alter deceptively; *especially* **:** ADULTERATE
2 : to deprive of genuineness, naturalness, or simplicity; *especially* **:** to deprive of naïveté and make worldly-wise **:** DISILLUSION
3 : to make complicated or complex

²so·phis·ti·cate \-ti-kət, -tə-ˌkāt\ *noun* (1923)
: a sophisticated person

so·phis·ti·cat·ed \-tə-ˌkā-təd\ *adjective* [Medieval Latin *sophisticatus*] (1601)
1 : not in a natural, pure, or original state **:** ADULTERATED ⟨a *sophisticated* oil⟩
2 : deprived of native or original simplicity: as **a :** highly complicated or developed **:** COMPLEX ⟨*sophisticated* electronic devices⟩ **b :** having a refined knowledge of the ways of the world cultivated especially through wide experience ⟨a *sophisticated* lady⟩
3 : devoid of grossness: as **a :** finely experienced and aware ⟨a *sophisticated* columnist⟩ **b :** intellectually appealing ⟨a *sophisticated* novel⟩ ☆
— so·phis·ti·cat·ed·ly *adverb*

so·phis·ti·ca·tion \sə-ˌfis-tə-'kā-shən\ *noun* (15th century)
1 a : the use of sophistry **:** sophistic reasoning **b :** SOPHISM, QUIBBLE
2 : the process of making impure or weak **:** ADULTERATION
3 : the process or result of becoming cultured, knowledgeable, or disillusioned; *especially* **:** CULTIVATION, URBANITY
4 : the process or result of becoming more complex, developed, or subtle

soph·ist·ry \'sä-fə-strē\ *noun* (14th century)
1 : subtly deceptive reasoning or argumentation
2 : SOPHISM 1

soph·o·more \'säf-ˌmōr, -ˌmȯr *also* 'sȯf- *or* 'sä-fə- *or* 'sȯ-fə-\ *noun* [perhaps from Greek *sophos* wise + *mōros* foolish] (1688)
: a student in the second year at college or a 4-year secondary school

soph·o·mor·ic \ˌsäf-'mōr-ik, -'mȯr-, -'mär- *also* ˌsȯf- *or* ˌsä-fə- *or* ˌsȯ-fə-\ *adjective* (1813)
1 : conceited and overconfident of knowledge but poorly informed and immature
2 : of, relating to, or characteristic of a sophomore ⟨a *sophomoric* prank⟩

So·pho·ni·as \ˌsä-fə-'nī-əs, ˌsō-\ *noun* [Late Latin, from Greek, from Hebrew *Sĕphanyāh*]
: ZEPHANIAH

so·phy \'sō-fē\ *noun* [Persian *Safī*] (1534)
archaic **:** a sovereign of Persia

-sophy *noun combining form* [Middle English *-sophie,* from Old French, from Latin *-sophia,* from Greek, from *sophia* wisdom, from *sophos*]
: knowledge **:** wisdom **:** science ⟨anthropo*sophy*⟩

so·pite \sō-'pīt\ *transitive verb* **so·pit·ed; so·pit·ing** [Latin *sopitus,* past participle of *sopire* to put to sleep; akin to Latin *sopor* deep sleep] (1542)
1 *archaic* **:** to put to sleep **:** LULL
2 *archaic* **:** to put an end to (as a claim) **:** SETTLE

sop·o·rif·er·ous \ˌsä-pə-'ri-f(ə-)rəs\ *adjective* [Latin *soporifer* soporiferous, from *sopor* + *-fer* -ferous] (1590)
: SOPORIFIC
— sop·o·rif·er·ous·ness *noun*

¹sop·o·rif·ic \-'ri-fik\ *adjective* [probably from French *soporifique,* from Latin *sopor* deep sleep; akin to Latin *somnus* sleep — more at SOMNOLENT] (1665)
1 a : causing or tending to cause sleep **b :** tending to dull awareness or alertness
2 : of, relating to, or marked by sleepiness or lethargy

²soporific *noun* (circa 1727)
: a soporific agent; *specifically* **:** HYPNOTIC 1

sop·ping \'sä-piŋ\ *adjective* (1877)
: wet through **:** SOAKING

sop·py \'sä-pē\ *adjective* **sop·pi·er; -est** (1823)
1 a : soaked through **:** SATURATED **b :** very wet
2 : SENTIMENTAL, MAWKISH
— sop·pi·ness \-nəs\ *noun*

so·pra·ni·no \ˌsō-prə-'nē-(ˌ)nō, ˌsä-\ *noun, plural* **-nos** [Italian, diminutive of *soprano*] (1905)
: a musical instrument (as a recorder or saxophone) higher in pitch than the soprano

¹so·pra·no \sə-'pra-(ˌ)nō, -'prä-\ *adjective* [Italian, adjective & noun, from *sopra* above, from Latin *supra* — more at SUPRA-] (1730)
: relating to or having the range or part of a soprano

> ☆ SYNONYMS
> **Sophisticated, worldly-wise, blasé** mean experienced in the ways of the world. SOPHISTICATED often implies refinement, urbanity, cleverness, and cultivation ⟨guests at her salon were usually rich and *sophisticated*⟩. WORLDLY-WISE suggests a close and practical knowledge of the affairs and manners of society and an inclination toward materialism ⟨a *worldly-wise* woman with a philosophy of personal independence⟩. BLASE implies a lack of responsiveness to common joys as a result of a real or affected surfeit of experience and cultivation ⟨*blasé* travelers who claimed to have been everywhere⟩.

²soprano *noun, plural* **-nos** (1738)
1 : the highest singing voice of women, boys, or castrati; *also* **:** a person having this voice
2 : the highest voice part in a 4-part chorus
3 : a member of a family of instruments having the highest range

so·ra \'sōr-ə, 'sȯr-\ *noun* [origin unknown] (1705)
: a small short-billed North American rail (*Porzana carolina*) common in marshes

sorb \'sȯrb\ *transitive verb* [back-formation from *absorb* & *adsorb*] (1909)
: to take up and hold by either adsorption or absorption
— **sorb·abil·i·ty** \ˌsȯr-bə-'bi-lə-tē\ *noun*
— **sorb·able** \'sȯr-bə-bəl\ *adjective*

Sorb \'sȯrb\ *noun* [German *Sorbe,* from Sorbian *serbje*] (1843)
1 : WEND
2 : WENDISH
— **Sor·bi·an** \'sȯr-bē-ən\ *adjective or noun*

sor·bate \'sȯr-ˌbāt, 'sȯr-bət\ *noun* (circa 1823)
: a salt or ester of sorbic acid — compare POTASSIUM SORBATE

sor·bent \'sȯr-bənt\ *noun* [Latin *sorbent-, sorbens,* present participle of *sorbēre* to suck up — more at ABSORB] (circa 1856)
: a substance that sorbs

sor·bet \sȯr-'bā *also* 'sȯr-bət\ *noun* [Middle French, from Old Italian *sorbetto,* from Turkish *şerbet* — more at SHERBET] (1864)
: a fruit-flavored ice served as a dessert or between courses as a palate refresher

sor·bic acid \'sȯr-bik-\ *noun* [*sorb* fruit of the service or related trees, from French *sorbe,* from Latin *sorbum*] (1815)
: a crystalline acid $C_6H_8O_2$ obtained from the unripe fruits of the mountain ash or synthesized and used especially as a fungicide and food preservative

sor·bi·tol \'sȯr-bə-ˌtȯl, -ˌtōl\ *noun* [*sorb* fruit of the service or related trees + *-itol*] (1895)
: a faintly sweet alcohol $C_6H_{14}O_6$ that occurs in some fruits, is made synthetically, and is used especially as a humectant and softener and in making ascorbic acid

sor·cer·er \'sȯr-sə-rər, 'sȯrs-rər\ *noun* (15th century)
: a person who practices sorcery **:** WIZARD

sor·cer·ess \-rəs\ *noun* (14th century)
: a woman who is a sorcerer

sor·cer·ous \-rəs\ *adjective* (1546)
: of or relating to sorcery **:** MAGICAL

sor·cery \-rē\ *noun* [Middle English *sorcerie,* from Middle French, from *sorcier* sorcerer, from (assumed) Vulgar Latin *sortiarius,* from Latin *sort-, sors* chance, lot — more at SERIES] (14th century)
1 : the use of power gained from the assistance or control of evil spirits especially for divining **:** NECROMANCY
2 : MAGIC 2a

sor·did \'sȯr-dəd\ *adjective* [Latin *sordidus,* from *sordes* dirt — more at SWART] (1609)
1 : marked by baseness or grossness **:** VILE ⟨*sordid* motives⟩
2 a : DIRTY, FILTHY **b :** WRETCHED, SQUALID
3 : meanly avaricious **:** COVETOUS
4 : of a dull or muddy color
synonym see MEAN
— **sor·did·ly** *adverb*
— **sor·did·ness** *noun*

sor·di·no \sȯr-'dē-(ˌ)nō\ *noun, plural* **-di·ni** \-(ˌ)nē\ [Italian, from *sordo* silent, from Latin *surdus*] (circa 1801)
: MUTE 3

¹sore \'sōr, 'sȯr\ *adjective* **sor·er; sor·est** [Middle English *sor,* from Old English *sār;* akin to Old High German *sēr* sore and probably to Old Irish *saeth* distress] (before 12th century)
1 a : causing pain or distress **b :** painfully sensitive **:** TENDER ⟨*sore* muscles⟩ **c :** hurt or inflamed so as to be or seem painful ⟨*sore* runny eyes⟩ ⟨a dog limping on a *sore* leg⟩
2 : attended by difficulties, hardship, or exertion
3 : ANGRY, IRKED ⟨a *sore* loser⟩
— **sore·ness** *noun*

²sore *noun* (before 12th century)
1 : a localized sore spot on the body; *especially* **:** one (as an ulcer) with the tissues ruptured or abraded and usually with infection
2 : a source of pain or vexation **:** AFFLICTION

³sore *adverb* (before 12th century)
: SORELY

sore·head \'sōr-ˌhed, 'sȯr-\ *noun* (1848)
: a person easily angered or disgruntled
— **sorehead** *or* **sore·head·ed** \-'he-dəd\ *adjective*

sore·ly \'sōr-lē, 'sȯr-\ *adverb* (before 12th century)
1 : in a sore manner **:** PAINFULLY
2 : VERY, EXTREMELY ⟨*sorely* needed changes⟩

sore throat *noun* (1686)
: pain in the throat due to inflammation of the fauces and pharynx

sor·ghum \'sȯr-gəm\ *noun* [New Latin, from Italian *sorgo,* from (assumed) Vulgar Latin *Syricum* (*granum*), literally, Syrian grain] (1597)
1 : any of an economically important genus (*Sorghum*) of Old World tropical grasses similar to Indian corn in habit but with the spikelets in pairs on a hairy rachis; *especially* **:** any of various cultivars (as grain sorghum or sorgo) derived from a wild form (*S. bicolor* synonym *S. vulgare*)
2 : syrup from the juice of a sorgo that resembles cane syrup
3 : something cloyingly sentimental

sorghum 1

sor·go \'sȯr-(ˌ)gō\ *noun* [Italian] (circa 1760)
: a sorghum cultivated primarily for the sweet juice in its stems from which sugar and syrup are made but also used for fodder and silage — called also *sweet sorghum*

so·ri·tes \sə-'rī-(ˌ)tēz\ *noun, plural* **sorites** [Latin, from Greek *sōritēs,* from *sōros* heap] (1551)
: an argument consisting of propositions so arranged that the predicate of any one forms the subject of the next and the conclusion unites the subject of the first proposition with the predicate of the last

So·rop·ti·mist \sə-'räp-tə-mist, sō-\ *noun* [*Soroptimist* (Club)] (1924)
: a member of a service club composed of professional women and women business executives

so·ro·ral \sə-'rōr-əl, -'rȯr-\ *adjective* [Latin *soror* sister — more at SISTER] (1858)
: of, relating to, or characteristic of a sister **:** SISTERLY

so·ro·rate \sə-'rōr-ət, -'rȯr-\ *noun* [Latin *soror*] (1910)
: the marriage of one man to two or more sisters usually successively and after the first wife has been found to be barren or after her death

so·ror·i·ty \sə-'rȯr-ə-tē, -'rär-\ *noun, plural* **-ties** [Medieval Latin *sororitas* sisterhood, from Latin *soror* sister] (1900)
: a club of women; *specifically* **:** a women's student organization (as at a college) that is formed chiefly for social purposes and has a name consisting of Greek letters

sorp·tion \'sȯrp-shən\ *noun* [back-formation from *absorption* & *adsorption*] (1909)
: the process of sorbing **:** the state of being sorbed
— **sorp·tive** \'sȯrp-tiv\ *adjective*

¹sor·rel \'sȯr-əl, 'sär-\ *noun* [Middle English *sorelle,* from Middle French *sorel,* noun & adjective, from *sor* reddish brown] (15th century)
1 : a sorrel-colored animal; *especially* **:** a light bright chestnut horse often with white mane and tail — compare ¹CHESTNUT 4, ²BAY 1
2 : a brownish orange to light brown

²sorrel *noun* [Middle English *sorel,* from Middle French *surele,* from Old French, from *sur* sour, of Germanic origin; akin to Old High German *sūr* sour — more at SOUR] (15th century)
: any of various plants or plant parts with sour juice: as **a :** any of various docks (as *Rumex acetosa* and *R. acetosella*); *also* **:** the leaves used as a potherb **b :** WOOD SORREL

sorrel tree *noun* (1687)
: SOURWOOD

¹sor·row \'sär-(ˌ)ō, 'sȯr-\ *noun* [Middle English *sorow,* from Old English *sorg;* akin to Old High German *sorga* sorrow] (before 12th century)
1 a : deep distress, sadness, or regret especially for the loss of someone or something loved **b :** resultant unhappy or unpleasant state ⟨to his *sorrow* he lost his temper⟩
2 : a cause of grief or sadness
3 : a display of grief or sadness ☆

²sorrow *intransitive verb* (before 12th century)
: to feel or express sorrow
— **sor·row·er** \'sär-ə-wər, 'sȯr-\ *noun*

sor·row·ful \-ō-fəl, -ə-fəl\ *adjective* (before 12th century)
1 : full of or marked by sorrow
2 : expressive of or inducing sorrow
— **sor·row·ful·ly** \-f(ə-)lē\ *adverb*
— **sor·row·ful·ness** \-fəl-nəs\ *noun*

sor·ry \'sär-ē, 'sȯr-\ *adjective* **sor·ri·er; -est** [Middle English *sory,* from Old English *sārig,* from *sār* sore] (before 12th century)
1 : feeling sorrow, regret, or penitence
2 : MOURNFUL, SAD
3 : inspiring sorrow, pity, scorn, or ridicule
synonym see CONTEMPTIBLE
— **sor·ri·ly** \-ə-lē\ *adverb*
— **sor·ri·ness** \-ē-nəs\ *noun*

¹sort \'sȯrt\ *noun* [Middle English, from Middle French *sorte,* probably from Medieval Latin *sort-, sors,* from Latin, chance, lot — more at SERIES] (14th century)
1 a : a group set up on the basis of any characteristic in common **:** CLASS, KIND **b :** an instance of a kind ⟨a *sort* of black Paul Bunyan, towering 6′10″ —Jack Olsen⟩ **c :** PERSON, INDIVIDUAL ⟨he's not a bad *sort*⟩
2 *archaic* **:** GROUP, COMPANY
3 a : method or manner of acting **:** WAY, MANNER **b :** CHARACTER, NATURE ⟨people of an evil *sort*⟩
4 a : a letter or character that is one element of a font **b :** a character or piece of type that is not part of a regular font
5 : an instance of sorting ⟨a numeric *sort* of a data file⟩
synonym see TYPE
— **after a sort :** in a rough or haphazard way
— **of sorts** *or* **of a sort :** of an inconsequential or mediocre quality ⟨a poet *of sorts*⟩

☆ SYNONYMS
Sorrow, grief, anguish, woe, regret mean distress of mind. SORROW implies a sense of loss or a sense of guilt and remorse ⟨a family united in *sorrow* upon the patriarch's death⟩. GRIEF implies poignant sorrow for an immediate cause ⟨the inexpressible *grief* of the bereaved parents⟩. ANGUISH suggests torturing grief or dread ⟨the *anguish* felt by the parents of the kidnapped child⟩. WOE is deep or inconsolable grief or misery ⟨cries of *woe* echoed throughout the bombed city⟩. REGRET implies pain caused by deep disappointment, fruitless longing, or unavailing remorse ⟨nagging *regret* for missed opportunities⟩.

— out of sorts 1 : somewhat ill **2 :** GROUCHY, IRRITABLE

[2]sort (14th century)
transitive verb
1 a : to put in a certain place or rank according to kind, class, or nature ⟨*sort* apples⟩ ⟨*sort* mail⟩ **b :** to arrange according to characteristics **:** CLASSIFY ⟨*sort* out colors⟩
2 *chiefly Scottish* **:** to put to rights **:** put in order
3 a : to examine in order to clarify ⟨*sorting* out his problems⟩ **b :** to free of confusion **:** CLARIFY ⟨waited until things *sorted* themselves out⟩
intransitive verb
1 : to join or associate with others especially of the same kind ⟨*sort* with thieves⟩
2 : SUIT, AGREE
— sort•able \'sȯr-tə-bəl\ *adjective*
— sort•er *noun*

sor•tie \'sȯr-tē, sȯr-'tē\ *noun* [French, from Middle French, from *sortir* to escape] (1778)
1 : a sudden issuing of troops from a defensive position against the enemy
2 : one mission or attack by a single plane
3 : FORAY, RAID
— sortie *intransitive verb*

sor•ti•lege \'sȯr-tᵊl-ij, -ˌej\ *noun* [Middle English, from Medieval Latin *sortilegium,* from Latin *sortilegus* foretelling, from *sort-, sors* lot + *-i-* + *legere* to gather — more at LEGEND] (14th century)
1 : divination by lots
2 : SORCERY

sor•ti•tion \sȯr-'ti-shən\ *noun* [Latin *sortition-, sortitio,* from *sortiri* to cast or draw lots, from *sort-, sors* lot] (1597)
: the act or an instance of casting lots

sort of \'sȯr-dəv, -dər\ *adverb* (1790)
: to a moderate degree **:** RATHER

so•rus \'sōr-əs, 'sȯr-\ *noun, plural* **so•ri** \'sōr-ˌī, 'sȯr-, -ˌē\ [New Latin, from Greek *sōros* heap] (1832)
: a cluster of plant reproductive bodies: as **a :** one of the dots on the underside of a fertile fern frond consisting of a cluster of sporangia **b :** a mass of spores bursting through the epidermis of the host plant of a parasitic fungus **c :** a cluster of gemmae on the thallus of a lichen

SOS \ˌes-(ˌ)ō-'es\ *noun* (1910)
1 : an internationally recognized signal of distress in radio code · · · – – – · · · used especially by ships calling for help
2 : a call or request for help or rescue

[1]so–so \'sō-'sō\ *adverb* (circa 1530)
: moderately well **:** TOLERABLY, PASSABLY

[2]so–so *adjective* (1542)
: neither very good nor very bad **:** MIDDLING

[1]so•ste•nu•to \ˌsōs-tə-'nü-(ˌ)tō, ˌsȯs-\ *adjective or adverb* [Italian, from past participle of *sostenere* to sustain, from Latin *sustinēre*] (circa 1724)
: sustained to or beyond the note's full value — used as a direction in music

[2]sostenuto *noun* (1757)
: a movement or passage whose notes are markedly prolonged

sot \'sät\ *noun* [Middle English, fool, from Old English *sott*] (1592)
: a habitual drunkard

so•te•ri•ol•o•gy \sō-ˌtir-ē-'ä-lə-jē\ *noun* [Greek *sōtērion* salvation (from *sōtēr* savior, preserver) + English *-logy* — more at CREOSOTE] (circa 1774)
: theology dealing with salvation especially as effected by Jesus Christ
— so•te•ri•o•log•i•cal \-ē-ə-'lä-ji-kəl\ *adjective*

so that *conjunction* (before 12th century)
: THAT 2a(1)

So•tho \'sō-(ˌ)tō\ *noun* (1928)
1 : any one of the Sotho languages and especially Sesotho
2 : a group of closely related Bantu languages of Lesotho, Botswana, and northern South Africa

so•tol \'sō-ˌtōl\ *noun* [American Spanish, from Nahuatl *zōtōlin* palm tree] (1881)
: any of several plants (genus *Dasylirion*) of the agave family of the southwestern U.S. and Mexico that resemble a yucca

sot•tish \'sä-tish\ *adjective* (1583)
: resembling a sot **:** DRUNKEN; *also* **:** DOLTISH, STUPID
— sot•tish•ly *adverb*
— sot•tish•ness *noun*

sot•to vo•ce \ˌsä-tō-'vō-chē\ *adverb or adjective* [Italian *sottovoce,* literally, under the voice] (1737)
1 : under the breath **:** in an undertone; *also* **:** in a private manner
2 : very softly — used as a direction in music

sou \'sü\ *noun, plural* **sous** \'süz\ [French, from Old French *sol,* from Late Latin *solidus* solidus] (1814)
1 : [2]SOL
2 : a 5-centime piece

sou•bise \sü-'bēz\ *noun* [French, from Charles de Rohan, Prince de *Soubise* (died 1787) French nobleman] (1822)
: a garnish or white sauce containing onions or onion purée

sou•brette \sü-'bret\ *noun* [French, from Provençal *soubreto,* feminine of *soubret* coy, from *soubra* to surmount, exceed, from Latin *superare* — more at INSUPERABLE] (1753)
1 a : a coquettish maid or frivolous young woman in comedies **b :** an actress who plays such a part
2 : a soprano who sings supporting roles in comic opera

sou•bri•quet \'sō-, ˌsō-, 'sü-, ˌsü-\ *variant of* SOBRIQUET

sou•chong \'sü-ˌchȯŋ, -ˌshȯŋ\ *noun* [of Chinese origin; akin to Chinese (Beijing) *xiǎozhǒng,* literally, small sort] (1760)
: a tea made from the larger leaves of the shoot

[1]souf•flé \sü-'flā, 'sü-ˌ\ *noun* [French, from *soufflé,* past participle of *souffler* to blow, puff up, from Latin *sufflare,* from *sub-* + *flare* to blow — more at BLOW] (1813)
: a dish that is made from a sauce, egg yolks, beaten egg whites, and a flavoring or purée (as of seafood, fruit, or vegetables) and baked until puffed up

[2]soufflé *or* **souf•fléed** \-'flād, -ˌflād\ *adjective* (1888)
: puffed up by or in cooking

sough \'sau̇, 'səf\ *intransitive verb* [Middle English *swoughen,* from Old English *swōgan;* akin to Gothic ga*swogjan* to groan, Lithuanian *svagėti* to sound] (before 12th century)
: to make a moaning or sighing sound
— sough *noun*

sought *past and past participle of* SEEK

souk \'sük\ *noun* [Arabic *sūq* market] (1826)
: a marketplace in northern Africa or the Middle East; *also* **:** a stall in such a marketplace

[1]soul \'sōl\ *noun* [Middle English *soule,* from Old English *sāwol;* akin to Old High German *sēula* soul] (before 12th century)
1 : the immaterial essence, animating principle, or actuating cause of an individual life
2 a : the spiritual principle embodied in human beings, all rational and spiritual beings, or the universe **b** *capitalized, Christian Science* **:** GOD 1b
3 : a person's total self
4 a : an active or essential part **b :** a moving spirit **:** LEADER
5 a : the moral and emotional nature of human beings **b :** the quality that arouses emotion and sentiment **c :** spiritual or moral force **:** FERVOR
6 : PERSON
7 : EXEMPLIFICATION, PERSONIFICATION ⟨she is the *soul* of integrity⟩
8 a : a strong positive feeling (as of intense sensitivity and emotional fervor) conveyed especially by black American performers **b :** NEGRITUDE **c :** SOUL MUSIC **d :** SOUL FOOD **e :** SOUL BROTHER

[2]soul *adjective* (1958)
1 : of, relating to, or characteristic of black Americans or their culture
2 : designed for or controlled by blacks ⟨*soul* radio stations⟩

soul brother *noun* (1959)
: a black male

souled \'sōld\ *adjective* (15th century)
: having a soul **:** possessing soul and feeling — usually used in combination ⟨whole-*souled* repentance⟩

soul food *noun* (1964)
: food (as chitterlings, ham hocks, and collard greens) traditionally eaten by southern black Americans

soul•ful \'sōl-fəl\ *adjective* (1860)
: full of or expressing feeling or emotion
— soul•ful•ly \-fə-lē\ *adverb*
— soul•ful•ness *noun*

soul kiss *noun* (circa 1948)
: FRENCH KISS

soul•less \'sōl-ləs\ *adjective* (15th century)
: having no soul or no greatness or warmth of mind or feeling
— soul•less•ly *adverb*
— soul•less•ness *noun*

soul mate *noun* (1822)
: a person temperamentally suited to another

soul music *noun* (1961)
: music that originated in black American gospel singing, is closely related to rhythm and blues, and is characterized by intensity of feeling and earthiness

soul–search•ing \'sōl-ˌsər-chiŋ\ *noun* (1924)
: examination of one's conscience especially with regard to motives and values

[1]sound \'sau̇nd\ *adjective* [Middle English, from Old English *gesund;* akin to Old High German *gisunt* healthy] (13th century)
1 a : free from injury or disease **:** exhibiting normal health **b :** free from flaw, defect, or decay ⟨*sound* timber⟩
2 : SOLID, FIRM; *also* **:** STABLE
3 a : free from error, fallacy, or misapprehension ⟨*sound* reasoning⟩ **b :** exhibiting or based on thorough knowledge and experience ⟨*sound* scholarship⟩ **c :** legally valid ⟨a *sound* title⟩ **d :** logically valid and having true premises **e :** agreeing with accepted views **:** ORTHODOX
4 a : THOROUGH **b :** deep and undisturbed ⟨a *sound* sleep⟩ **c :** HARD, SEVERE ⟨a *sound* whipping⟩
5 : showing good judgment or sense
synonym see HEALTHY, VALID
— sound•ly \'sau̇n(d)-lē\ *adverb*
— sound•ness \'sau̇n(d)-nəs\ *noun*

[2]sound *adverb* (14th century)
: to the full extent **:** THOROUGHLY ⟨*sound* asleep⟩

[3]sound *noun* [Middle English *soun,* from Old French *son,* from Latin *sonus,* from *sonare* to sound; akin to Old English *swinn* melody, Sanskrit *svanati* it sounds] (13th century)
1 a : a particular auditory impression **:** TONE **b :** the sensation perceived by the sense of hearing **c :** mechanical radiant energy that is transmitted by longitudinal pressure waves in a material medium (as air) and is the objective cause of hearing
2 a : a speech sound ⟨a peculiar *r-sound*⟩ **b :** value in terms of speech sounds ⟨*-cher* of *teacher* and *-ture* of *creature* have the same *sound*⟩
3 *archaic* **:** RUMOR, FAME
4 a : meaningless noise **b** *obsolete* **:** MEANING **c :** the impression conveyed **:** IMPORT
5 : hearing distance **:** EARSHOT

\ə\ **abut** \ᵊ\ **kitten** \ər\ **further** \a\ **ash** \ā\ **ace**
\ä\ **mop, mar** \au̇\ **out** \ch\ **chin** \e\ **bet** \ē\ **easy**
\g\ **go** \i\ **hit** \ī\ **ice** \j\ **job** \ŋ\ **sing** \ō\ **go**
\ȯ\ **law** \ȯi\ **boy** \th\ **thin** \t͟h\ **the** \ü\ **loot** \u̇\ **foot**
\y\ **yet** \zh\ **vision** *see also* Guide to Pronunciation

6 : recorded auditory material
7 : a particular musical style characteristic of an individual, a group, or an area ⟨the Nashville *sound*⟩
[4]**sound** (13th century)
transitive verb
1 a : to cause to sound ⟨*sound* a trumpet⟩ **b :** PRONOUNCE 3a
2 : to put into words **:** VOICE
3 a : to make known **:** PROCLAIM **b :** to order, signal, or indicate by a sound ⟨*sound* the alarm⟩
4 : to examine by causing to emit sounds ⟨*sound* the lungs⟩
5 *chiefly British* **:** to convey the impression of **:** sound like ⟨that *sounds* a logical use of resources —*Economist*⟩
intransitive verb
1 a : to make a sound **b :** RESOUND **c :** to give a summons by sound ⟨the bugle *sounds* to battle⟩
2 : to make or convey an impression especially when heard ⟨it *sounds* good to me⟩ ⟨you *sound* just like your mother⟩
— sound·able \ˈsau̇n-də-bəl\ *adjective*
[5]**sound** *noun* [Middle English, from Old English *sund* swimming, sea & Old Norse *sund* swimming, strait; akin to Old English *swimman* to swim] (14th century)
1 a : a long broad inlet of the ocean generally parallel to the coast **b :** a long passage of water connecting two larger bodies (as a sea with the ocean) or separating a mainland and an island
2 : the air bladder of a fish
[6]**sound** *verb* [Middle English, from Middle French *sonder,* from *sonde* sounding line, probably of Germanic origin; akin to Old English *sundlīne* sounding line, *sund* sea] (15th century)
transitive verb
1 : to measure the depth of **:** FATHOM
2 : to try to find out the views or intentions of **:** PROBE — often used with *out*
3 : to explore or examine (a body cavity) with a sound
intransitive verb
1 a : to ascertain the depth of water especially with a sounding line **b :** to look into or investigate the possibility ⟨sent commissioners . . . to *sound* for peace —Thomas Jefferson⟩
2 : to dive down suddenly — used of a fish or whale
[7]**sound** *noun* [French *sonde,* from Middle French, literally, sounding line] (1739)
: an elongated instrument for exploring or sounding body cavities
sound·alike \ˈsau̇n-də-ˌlīk\ *noun* (1970)
: one that sounds like another
sound–and–light show *noun* (1967)
: SON ET LUMIÈRE
sound barrier *noun* (1939)
: a sudden large increase in aerodynamic drag that occurs as the speed of an aircraft approaches the speed of sound
sound bite *noun* (1972)
: a brief recorded statement (as by a public figure) broadcast especially on a television news program; *also* **:** a brief catchy comment suitable for use as a sound bite
sound·board \ˈsau̇n(d)-ˌbōrd, -ˌbȯrd\ *noun* (1504)
1 : a thin resonant board (as the belly of a violin) so placed in an instrument as to reinforce its tones by sympathetic vibration — see VIOLIN illustration
2 : SOUNDING BOARD 1a
sound bow *noun* (1688)
: the thick part of a bell against which the clapper strikes
sound box *noun* (circa 1875)
1 : a device in an early phonograph for producing sound from a record by using the vibration of a needle to move a diaphragm
2 : a hollow chamber in a musical instrument for increasing its sonority
sound effects *noun plural* (1909)
: effects that are imitative of sounds called for in the script of a dramatic production (as a radio show) and that enhance the production's illusion of reality
sound·er \ˈsau̇n-dər\ *noun* (1575)
: one that sounds; *specifically* **:** a device for making soundings
sound hole *noun* (1611)
: an opening in the top surface of a stringed instrument (as a violin) to enhance vibration and resonance — see VIOLIN illustration
[1]**sound·ing** \ˈsau̇n-diŋ\ *adjective* (14th century)
1 : RESONANT, SONOROUS
2 a : POMPOUS **b :** IMPOSING
— sound·ing·ly \-diŋ-lē\ *adverb*
[2]**sounding** *noun* (15th century)
1 a : measurement of depth especially with a sounding line **b :** the depth so ascertained **c** *plural* **:** a place or part of a body of water where a hand sounding line will reach bottom
2 : measurement of atmospheric conditions at various heights
3 : a probe, test, or sampling of opinion or intention
sounding board *noun* (1729)
1 a : a structure behind or over a pulpit, rostrum, or platform to give distinctness and sonority to sound **b :** a device or agency that helps propagate opinions or utterances **c :** a person or group on whom one tries out an idea or opinion as a means of evaluating it
2 : SOUNDBOARD 1
sounding line *noun* (14th century)
: a line or wire weighted at one end for sounding
sounding rocket *noun* (circa 1945)
: a rocket used to obtain information concerning atmospheric conditions at various altitudes
[1]**sound·less** \ˈsau̇n(d)-ləs\ *adjective* [[6]*sound*] (1586)
: incapable of being sounded **:** UNFATHOMABLE
[2]**soundless** *adjective* [[3]*sound*] (1601)
: making no sound **:** SILENT
— sound·less·ly *adverb*
sound·man \ˈsau̇n(d)-ˌman\ *noun* (1929)
: a person who controls the volume and tone of sound picked up by a microphone (as in a recording studio or on a motion-picture set) for recording
sound off *intransitive verb* (1909)
1 : to play three chords before and after marching up and down a line of troops during a ceremonial parade or formal guard mount
2 : to count cadence while marching
3 a : to speak up in a loud voice **b :** to voice one's opinions freely and vigorously
sound pressure *noun* (1893)
: the difference between the actual pressure at any point in the field of a sound wave at any instant and the average pressure at that point
[1]**sound·proof** \ˈsau̇n(d)-ˌprüf\ *adjective* (circa 1878)
: impervious to sound
[2]**soundproof** *transitive verb* (1919)
: to insulate so as to obstruct the passage of sound
sound·stage \ˈsau̇n(d)-ˌstāj\ *noun* (1931)
: the part of a motion-picture studio in which a production is filmed
sound track *noun* (circa 1929)
1 : the area on a motion-picture film or television videotape that carries the sound record
2 : the sound recorded on a sound track; *especially* **:** the music on a sound track
sound truck *noun* (1936)
: a truck equipped with a loudspeaker
sound wave *noun* (1848)
1 : [3]SOUND 1a
2 *plural* **:** longitudinal pressure waves in any material medium regardless of whether they constitute audible sound ⟨earthquake waves and ultrasonic waves are sometimes called *sound waves*⟩
soup \ˈsüp\ *noun* [Middle English *soupe,* from Middle French, sop, soup, of Germanic origin; akin to Old Norse *soppa* soup, Old English *sopp* sop — more at SUP] (14th century)
1 : a liquid food especially with a meat, fish, or vegetable stock as a base and often containing pieces of solid food
2 : something (as a heavy fog or nitroglycerine) having or suggesting the consistency or nutrient qualities of soup
3 : an unfortunate predicament
soup·çon \süp-ˈsōⁿ, ˈsüp-ˌsän\ *noun* [French, literally, suspicion, from (assumed) Vulgar Latin *suspection-, suspectio,* from Latin *suspicere* to suspect — more at SUSPECT] (1766)
: a little bit **:** TRACE
soup du jour \ˌsüp-də-ˈzhu̇r\ *noun* [part translation of French *soupe du jour* soup of the day] (circa 1945)
: a soup that is offered by a restaurant on a particular day
soup kitchen *noun* (1839)
: an establishment dispensing minimum dietary essentials (as soup and bread) to the needy
soup·spoon \ˈsüp-ˌspün\ *noun* (1705)
: a spoon with a large or rounded bowl for eating soup
soup up *transitive verb* [*soup* (drug injected into a racehorse to improve its performance)] (circa 1933)
: to increase the power or efficiency of ⟨*soup up* an engine⟩
soupy \ˈsü-pē\ *adjective* **soup·i·er; -est** (1869)
1 : having the consistency of soup
2 : densely foggy or cloudy
3 : overly sentimental
[1]**sour** \ˈsau̇(-ə)r\ *adjective* [Middle English, from Old English *sūr;* akin to Old High German *sūr* sour, Lithuanian *sūrus* salty] (before 12th century)
1 : causing or characterized by the one of the four basic taste sensations that is produced chiefly by acids ⟨*sour* pickles⟩ — compare BITTER, SALT, SWEET
2 a (1) **:** having the acid taste or smell of or as if of fermentation **:** TURNED ⟨*sour* milk⟩ (2) **:** of or relating to fermentation **b :** smelling or tasting of decay **:** RANCID, ROTTEN ⟨*sour* breath⟩ **c** (1) **:** BAD, WRONG ⟨a project gone *sour*⟩ (2) **:** HOSTILE, DISENCHANTED ⟨went *sour* on Marxism⟩
3 a : UNPLEASANT, DISTASTEFUL **b :** CROSS, SULLEN **c :** not up to the usual, expected, or standard quality or pitch
4 : acid in reaction — used especially of soil
5 : containing malodorous sulfur compounds — used especially of petroleum products
— sour·ish \ˈsau̇(-ə)r-ish\ *adjective*
— sour·ly *adverb*
— sour·ness *noun*
[2]**sour** *noun* (before 12th century)
1 a : something sour **b :** the primary taste sensation produced by acid stimuli
2 : a cocktail consisting of a liquor (as whiskey), lemon or lime juice, sugar, and sometimes ice
[3]**sour** (14th century)
intransitive verb
: to become sour
transitive verb
: to make sour
sour ball *noun* (circa 1909)
: a spherical hard candy having a tart flavor
[1]**source** \ˈsōrs, ˈsȯrs\ *noun* [Middle English *sours,* from Middle French *sors, sourse,* from Old French, from past participle of *sourdre* to rise, spring forth, from Latin *surgere* — more at SURGE] (14th century)

1 a : a generative force **:** CAUSE **b** (1) **:** a point of origin or procurement **:** BEGINNING (2) **:** one that initiates **:** AUTHOR; *also* **:** PROTOTYPE, MODEL (3) **:** one that supplies information
2 a : the point of origin of a stream of water **:** FOUNTAINHEAD **b** *archaic* **:** SPRING, FOUNT
3 : a firsthand document or primary reference work
4 : an electrode in a field-effect transistor that supplies the charge carriers for current flow — compare DRAIN, GATE
synonym see ORIGIN
— source·less \-ləs\ *adjective*

[2]source *transitive verb* **sourced; sourc·ing** (1957)
1 : to specify the source of (as quoted material)
2 : to obtain from a source ⟨metals *sourced* from abroad⟩

[3]source *adjective* (1959)
: of, relating to, or being a computer program in its original programming language ⟨*source* listing⟩ ⟨*source* code⟩

source·book \-ˌbu̇k\ *noun* (1899)
: a fundamental document or record (as of history, literature, art, or religion) on which subsequent writings, compositions, opinions, beliefs, or practices are based; *also* **:** a collection of such documents

source language *noun* (1953)
: a language which is to be translated into another language — compare TARGET LANGUAGE

sour cherry *noun* (circa 1884)
: a widely cultivated cherry (*Prunus cerasus*) that has a round crown and bright red to almost black soft-fleshed acid fruits; *also* **:** the fruit

sour cream *noun* (1855)
: a soured cream product produced by the action of lactobacilli

sour·dough \ˈsau̇(-ə)r-ˌdō, *1 is also* -ˈdō\ *noun* (14th century)
1 : a leaven consisting of dough in which fermentation is active
2 [from the use of sourdough for making bread in prospectors' camps] **:** a veteran inhabitant and especially an old-time prospector of Alaska or northwestern Canada

sour grapes *noun plural* [from the fable ascribed to Aesop of the fox who after finding himself unable to reach some grapes he had desired disparaged them as sour] (1760)
: disparagement of something that has proven unattainable

sour gum *noun* (1785)
: BLACK GUM

sour mash *noun* (1885)
: grain mash for brewing or distilling whose initial acidity has been adjusted to optimum condition for yeast fermentation by mash from a previous run

sour orange *noun* (1748)
: a citrus tree (*Citrus aurantium*) that is used especially as a stock in grafting citrus; *also* **:** its bitter fruit

sour·puss \ˈsau̇(-ə)r-ˌpu̇s\ *noun* [[2]*puss*] (1937)
: GROUCH, KILLJOY

sour·sop \ˈsau̇(-ə)r-ˌsäp\ *noun* (1667)
: a small tropical American tree (*Annona muricata*) of the custard-apple family that has spicy odoriferous leaves; *also* **:** its large edible fruit that has fleshy spines and a slightly acid fibrous pulp

sour·wood \-ˌwu̇d\ *noun* (1709)
: a small U.S. tree (*Oxydendrum arboreum*) of the heath family with white flowers and sour-tasting leaves

sous \ˈsü\ *adjective* [French, preposition, literally, under, from Latin *subtus,* adverb, below, under; akin to Latin *sub* under — more at UP] (1687)
: being an assistant — used chiefly in titles ⟨a *sous*-chef⟩

sou·sa·phone \ˈsü-zə-ˌfōn, -sə-\ *noun* [John Philip *Sousa*] (1925)
: a large circular tuba that has a flared adjustable bell — compare HELICON

sousaphone

[1]souse \ˈsau̇s\ *verb* **soused; sous·ing** [Middle English, from Middle French *souz, souce* pickling solution, of Germanic origin; akin to Old High German *sulza* brine, Old English *sealt* salt] (14th century)
transitive verb
1 : PICKLE
2 a : to plunge in liquid **:** IMMERSE **b :** DRENCH, SATURATE
3 : to make drunk **:** INEBRIATE
intransitive verb
: to become immersed or drenched

[2]souse *noun* (14th century)
1 : something pickled; *especially* **:** seasoned and chopped pork trimmings, fish, or shellfish
2 : an act of sousing **:** WETTING
3 a : a habitual drunkard **b :** a drinking spree **:** BINGE

[3]souse *verb* **soused; sous·ing** [Middle English *souce,* noun, start of a bird's flight, alteration of *sours,* from Middle French *sourse* source — more at SOURCE] (1583)
intransitive verb
archaic **:** to swoop down **:** PLUNGE
transitive verb
archaic **:** to swoop down on

sou·tache \sü-ˈtash\ *noun* [French, from Hungarian *sujtás*] (circa 1856)
: a narrow braid with herringbone pattern used as trimming

sou·tane \sü-ˈtän, -ˈtan\ *noun* [French, from Italian *sottana,* literally, undergarment, from feminine of *sottano* being underneath, from Medieval Latin *subtanus,* from Latin *subtus* underneath — more at SOUS] (1838)
: CASSOCK

sou·ter \ˈsü-tər\ *noun* [Middle English, from Old English *sūtere,* from Latin *sutor,* from *suere* to sew — more at SEW] (before 12th century)
chiefly Scottish **:** SHOEMAKER

[1]south \ˈsau̇th\ *adverb* [Middle English, from Old English *sūth;* akin to Old High German *sund-* south and probably to Old English *sunne* sun] (before 12th century)
1 : to, toward, or in the south
2 : into a state of decline or ruin ⟨causes the sluggish economy to go *south* —G. F. Will⟩

[2]south *adjective* (12th century)
1 : situated toward or at the south ⟨the *south* entrance⟩
2 : coming from the south ⟨a *south* wind⟩

[3]south *noun* (13th century)
1 a : the direction of the south terrestrial pole **:** the direction to the right of one facing east **b :** the compass point directly opposite to north
2 *capitalized* **:** regions or countries lying to the south of a specified or implied point of orientation; *especially* **:** the southeastern part of the U.S.
3 : the right side of a church looking toward the altar from the nave
4 *often capitalized* **a :** the one of four positions at 90-degree intervals that lies to the south or at the bottom of a diagram **b :** a person (as a bridge player) occupying this position in the course of a specified activity; *specifically* **:** the declarer in bridge
5 *often capitalized* **:** the developing nations of the world **:** THIRD WORLD 3 — compare NORTH 2b

South African *noun* (1806)
: a native or inhabitant of the Republic of South Africa; *especially* **:** AFRIKANER
— South African *adjective*

south·bound \ˈsau̇th-ˌbau̇nd\ *adjective* (1885)
: traveling or heading south

south by east (circa 1771)
: a compass point that is one point east of due south **:** S11°15′E

south by west (circa 1743)
: a compass point that is one point west of due south **:** S11°15′W

South Dev·on \-ˈde-vən\ *noun* [*Devon,* England] (1897)
: any of a breed of large red dual-purpose cattle of English origin

South·down \ˈsau̇th-ˌdau̇n\ *noun* [*South Downs,* England] (1787)
: any of a breed of small medium-wooled hornless sheep of English origin

[1]south·east \sau̇-ˈthēst, *nautical* sau̇-ˈēst\ *adverb* (before 12th century)
: to, toward, or in the southeast

[2]southeast *noun* (14th century)
1 a : the general direction between south and east **b :** the point midway between the south and east compass points
2 *capitalized* **:** regions or countries lying to the southeast of a specified or implied point of orientation

[3]southeast *adjective* (14th century)
1 : coming from the southeast ⟨a *southeast* wind⟩
2 : situated toward or at the southeast ⟨the *southeast* corner⟩

southeast by east (circa 1771)
: a compass point that is one point east of due southeast **:** S56°15′E

southeast by south (1682)
: a compass point that is one point south of due southeast **:** S33°45′E

south·east·er \sau̇-ˈthē-stər, sau̇-ˈē-\ *noun* (1797)
1 : a strong southeast wind
2 : a storm with southeast winds

south·east·er·ly \-stər-lē\ *adverb or adjective* [[2]*southeast* + *-erly* (as in *easterly*)] (1708)
1 : from the southeast
2 : toward the southeast

south·east·ern \-stərn\ *adjective* [[2]*southeast* + *-ern* (as in *eastern*)] (1577)
1 *often capitalized* **:** of, relating to, or characteristic of a region conventionally designated southeast
2 : lying toward or coming from the southeast
— south·east·ern·most \-ˌmōst\ *adjective*

South·east·ern·er \-stə(r)-nər\ *noun* (1919)
: a native or inhabitant of the Southeast; *especially* **:** a native or resident of the southeastern part of the U.S.

[1]south·east·ward \sau̇-ˈthēs-twərd, sau̇-ˈēs-\ *adverb or adjective* (1528)
: toward the southeast
— south·east·wards \-twərdz\ *adverb*

[2]southeastward *noun* (1555)
: SOUTHEAST

[1]south·er·ly \ˈsə-t͟hər-lē\ *adjective or adverb* [[3]*south* + *-erly* (as in *easterly*)] (1551)
1 : situated toward or belonging to the south ⟨the *southerly* shore of the lake⟩
2 : coming from the south ⟨a *southerly* wind⟩

[2]southerly *noun, plural* **-lies** (1943)
: a wind from the south

south·ern \ˈsə-t͟hərn\ *adjective* [Middle English *southern, southren,* from Old English *sūtherne;* akin to Old High German *sundrōni* southern, Old English *sūth* south] (before 12th century)
1 *capitalized* **:** of, relating to, or characteristic of a region conventionally designated South

2 a : lying toward the south **b :** coming from the south ⟨a *southern* breeze⟩
— **south·ern·most** \-ˌmōst\ *adjective*
— **south·ern·ness** \-t͟hərn-nəs\ *noun, often capitalized*

Southern *noun* (1935)
: the dialect of English spoken in most of the Chesapeake Bay area, the coastal plain and the greater part of the upland plateau in Virginia, North Carolina, South Carolina, and Georgia, and the Gulf states at least as far west as the valley of the Brazos in Texas and sometimes taken to include the south Midland area

Southern blot *noun* [Edwin M. *Southern* 20th century British biologist] (1981)
: a blot consisting of a nitrocellulose sheet containing spots of DNA for identification by a suitable molecular probe — compare WESTERN BLOT
— **Southern blotting** *noun*

southern corn rootworm *noun* (1918)
: SPOTTED CUCUMBER BEETLE

Southern Cross *noun*
: four bright stars in the southern hemisphere, situated as if at the extremities of a Latin cross; *also* **:** the constellation of which these four stars are the brightest

Southern Crown *noun*
: CORONA AUSTRALIS

Southern English *noun* (14th century)
1 : the English spoken especially by cultivated people native to or educated in the South of England
2 : SOUTHERN

South·ern·er \ˈsə-t͟hə(r)-nər\ *noun* (1828)
: a native or inhabitant of the South; *especially* **:** a native or resident of the southern part of the U.S.

southern hemisphere *noun, often S&H capitalized* (circa 1771)
: the part of the earth that lies south of the equator

South·ern·ism \ˈsə-t͟hər-ˌni-zəm\ *noun* (1861)
1 : an attitude or trait characteristic of the South or Southerners especially in the U.S.
2 : a locution or pronunciation characteristic of the southern U.S.

southern lights *noun plural* (1775)
: AURORA AUSTRALIS

south·ern·wood \ˈsə-t͟hərn-ˌwu̇d\ *noun* (before 12th century)
: a shrubby fragrant European wormwood (*Artemisia abrotanum*) with bitter foliage

south·ing \ˈsau̇-thiŋ, -t͟hiŋ\ *noun* (1669)
1 : difference in latitude to the south from the last preceding point of reckoning
2 : southerly progress

south·land \ˈsau̇th-ˌland, -lənd\ *noun, often capitalized* (before 12th century)
: land in the south **:** the south of a country

south·paw \-ˌpȯ\ *noun* (1891)
: LEFT-HANDER; *especially* **:** a left-handed baseball pitcher ◆
— **southpaw** *adjective*

south pole *noun* (14th century)
1 a *often S&P capitalized* **:** the southernmost point of the earth **b :** the zenith of the heavens as viewed from the south terrestrial pole
2 *of a magnet* **:** the pole that points toward the south

¹South·ron \ˈsət͟h-rən\ *adjective* [Middle English (Scots)] (15th century)
chiefly Scottish **:** SOUTHERN; *specifically* **:** ENGLISH

²Southron *noun* (15th century)
: SOUTHERNER: as **a** *chiefly Scottish* **:** ENGLISHMAN **b** *chiefly Southern* **:** a native or inhabitant of the southern U.S.

south–seek·ing pole *noun* (circa 1922)
: SOUTH POLE 2

south–southeast *noun* (15th century)
: a compass point that is two points east of due south **:** S22°30′E

south–southwest *noun* (1513)
: a compass point that is two points west of due south **:** S22°30′W

¹south·ward \ˈsau̇th-wərd\ *adverb or adjective* (before 12th century)
: toward the south
— **south·wards** \-wərdz\ *adverb*

²southward *noun* (14th century)
: southward direction or part ⟨sail to the *southward*⟩

¹south·west \sau̇th-ˈwest, *nautical* sau̇-ˈwest\ *adverb* (before 12th century)
: to, toward, or in the southwest

²southwest *noun* (12th century)
1 a : the general direction between south and west **b :** the point midway between the south and west compass points
2 *capitalized* **:** regions or countries lying to the southwest of a specified or implied point of orientation

³southwest *adjective* (14th century)
1 : coming from the southwest ⟨a *southwest* wind⟩
2 : situated toward or at the southwest ⟨the *southwest* corner⟩

southwest by south (1725)
: a compass point that is one point south of due southwest **:** S33°45′W

southwest by west (circa 1771)
: a compass point that is one point west of due southwest **:** S56°15′W

south·west·er \sau̇(th)-ˈwes-tər\ *noun* (1833)
1 : a strong southwest wind
2 : a storm with southwest winds

south·west·er·ly \-tər-lē\ *adverb or adjective* [²*southwest* + *-erly* (as in *westerly*)] (1708)
1 : from the southwest
2 : toward the southwest

south·west·ern \-tərn\ *adjective* [Middle English, from Old English *sūth-westerne,* from *sūth* south + *westerne* western] (before 12th century)
1 : lying toward or coming from the southwest
2 *often capitalized* **:** of, relating to, or characteristic of a region conventionally designated Southwest
— **south·west·ern·most** \-ˌmōst\ *adjective*

southwestern corn borer *noun* (1943)
: a pyralid moth (*Diatraea grandiosella*) whose larva causes serious damage especially to corn crops by boring in the stalks

South·west·ern·er \sau̇(th)-ˈwes-tə(r)-nər\ *noun* (1860)
: a native or inhabitant of the Southwest; *especially* **:** a native or resident of the southwestern U.S.

¹south·west·ward \sau̇(th)-ˈwes-twərd\ *adverb or adjective* (1548)
: toward the southwest
— **south·west·wards** \-twərdz\ *adverb*

²southwestward *noun* (1775)
: SOUTHWEST

sou·ve·nir \ˈsü-və-ˌnir, ˌsü-və-ˈ\ *noun* [French, literally, act of remembering, from Middle French, from (*se*) *souvenir* to remember, from Latin *subvenire* to come up, come to mind — more at SUBVENTION] (1782)
: something that serves as a reminder **:** MEMENTO

souvenir sheet *noun* (1940)
: a block or set of postage stamps or a single stamp printed on a single sheet of paper often without gum or perforations and with margins containing lettering or design that identifies some notable event being commemorated

sou·vla·kia \süv-ˈlä-kē-ə\ *or* **sou·vla·ki** \-ˈlä-kē\ *noun* [New Greek *soublakia,* plural of *soublaki,* from diminutive of *soubla* skewer, from Middle Greek, from Latin *subula* awl, from *suere* to sew — more at SEW] (1950)
: SHISH KEBAB

sou'·west·er \sau̇-ˈwes-tər\ *noun* (1837)
1 : a long oilskin coat worn especially at sea during stormy weather
2 : a waterproof hat with wide slanting brim longer in back than in front

¹sov·er·eign *also* **sov·ran** \ˈsä-v(ə-)rən, -vərn *also* ˈsə-\ *noun* [Middle English *soverain,* from Old French, from *soverain,* adjective] (13th century)
1 a : one possessing or held to possess sovereignty **b :** one that exercises supreme authority within a limited sphere **c :** an acknowledged leader **:** ARBITER
2 : any of various gold coins of the United Kingdom

²sovereign *also* **sovran** *adjective* [Middle English *soverain,* from Middle French, from Old French, from (assumed) Vulgar Latin *superanus,* from Latin *super* over, above — more at OVER] (14th century)
1 a : superlative in quality **:** EXCELLENT **b :** of the most exalted kind **:** SUPREME ⟨*sovereign* virtue⟩ **c :** having generalized curative powers ⟨a *sovereign* remedy⟩ **d :** of an unqualified nature **:** UNMITIGATED ⟨*sovereign* contempt⟩ **e :** having undisputed ascendancy **:** PARAMOUNT
2 a : possessed of supreme power ⟨*sovereign* ruler⟩ **b :** unlimited in extent **:** ABSOLUTE **c :** enjoying autonomy **:** INDEPENDENT ⟨*sovereign* state⟩
3 : relating to, characteristic of, or befitting a sovereign
synonym see FREE
— **sov·er·eign·ly** *adverb*

sov·er·eign·ty *also* **sov·ran·ty** \-tē\ *noun, plural* **-ties** [Middle English *soverainte,* from Middle French *souveraineté,* from Old French, from *soverain*] (14th century)
1 *obsolete* **:** supreme excellence or an example of it
2 a : supreme power especially over a body politic **b :** freedom from external control **:** AUTONOMY **c :** controlling influence
3 : one that is sovereign; *especially* **:** an autonomous state

so·vi·et \ˈsō-vē-ˌet, ˈsä-, -vē-ət\ *noun* [Russian *sovet* council, soviet] (1917)
1 : an elected governmental council in a Communist country
2 *plural, capitalized* **a :** BOLSHEVIKS **b :** the people and especially the political and military leaders of the U.S.S.R.
— **soviet** *adjective, often capitalized*
— **so·vi·et·ism** \-vē-ə-ˌti-zəm\ *noun, often capitalized*

so·vi·et·ize \ˈsō-vē-ˌe-ˌtīz, ˈsä-, -vē-ə-\ *transitive verb* **-ized; -iz·ing** *often capitalized* (1919)
1 : to bring under Soviet control

◇ WORD HISTORY

southpaw The use of the word *southpaw* for a left-handed baseball pitcher has inspired much speculation on why southern orientation should be linked with left-handedness. The hypothesis probably repeated most often is that baseball diamonds were traditionally laid out with home plate to the west, to keep the sun out of hitters' eyes when a game continued into the late afternoon. Hence a left-handed pitcher facing the batter stood with his pitching arm to the south, inspiring the sobriquet *southpaw.* Whether or not fields have actually been so oriented, however, the origin of *southpaw* almost certainly lies outside baseball. When the word first surfaced in American English in 1848, it referred to a left-handed blow, with no allusion whatever to baseball, a game then in its early stages. More damaging yet, *south-pawed* in the sense "left-handed" has been recorded in rural dialects of northern England and Ulster, where baseball is not a traditional game. Most likely the word was brought to America by immigrants from these areas in the 18th or early 19th centuries, and we are still in the dark on why "south" should mean "left-handed."

2 : to force into conformity with Soviet cultural patterns or governmental policies
— **so·vi·et·i·za·tion** \ˌsō-vē-ˌe-tə-ˈzā-shən, -vē-ə-\ *noun, often capitalized*

So·vi·et·ol·o·gist \ˌsō-vē-ˌe-ˈtä-lə-jist, ˌsä-, -vē-ə-\ *noun* (1955)
: one who studies or is knowledgeable about the policies and practices of the Soviet government **:** KREMLINOLOGIST

sov·khoz \säf-ˈkȯz, -ˈkȯs\ *noun, plural* **sov·kho·zy** \-ˈkȯ-zē\ *or* **sov·khoz·es** [Russian, short for *sovetskoe khozyaĭstvo* soviet farm] (1921)
: a state-owned farm of the U.S.S.R. paying wages to the workers

¹sow \ˈsau̇\ *noun* [Middle English *sowe,* from Old English *sugu;* akin to Old English & Old High German *sū* sow, Latin *sus* pig, swine, hog, Greek *hys*] (before 12th century)
1 : an adult female swine; *also* **:** the adult female of various other animals (as a bear)
2 a : a channel that conducts molten metal to molds **b :** a mass of metal solidified in such a mold **:** INGOT

²sow \ˈsō\ *verb* **sowed; sown** \ˈsōn\ *or* **sowed; sow·ing** [Middle English, from Old English *sāwan;* akin to Old High German *sāwen* to sow, Latin *serere,* Lithuanian *sėti*] (before 12th century)
intransitive verb
1 : to plant seed for growth especially by scattering
2 : to set something in motion **:** begin an enterprise
transitive verb
1 a : to scatter (as seed) upon the earth for growth; *broadly* **:** PLANT 1a **b :** to strew with or as if with seed **c :** to introduce into a selected environment **:** IMPLANT
2 : to set in motion **:** FOMENT ⟨*sow* suspicion⟩
3 : to spread abroad **:** DISPERSE
— **sow·er** \ˈsō(-ə)r\ *noun*

sow·bel·ly \ˈsau̇-ˌbe-lē\ *noun* (1867)
: fat salt pork or bacon

sow bug \ˈsau̇-\ *noun* (1750)
: WOOD LOUSE

sow·ens \ˈsü-ənz, ˈsō-\ *noun plural but singular or plural in construction* [Scottish Gaelic *sùghan*] (1582)
: porridge from oat husks and siftings

sow thistle \ˈsau̇-\ *noun* (13th century)
: any of a genus (*Sonchus*) of spiny weedy European composite herbs widely naturalized in North America

sox *plural of* SOCK

soy \ˈsȯi\ *noun* [Japanese *shōyu*] (1679)
1 : a brown liquid sauce made by subjecting beans (as soybeans) to long fermentation and to digestion in brine
2 *also* **soya** \ˈsȯi-ə\ **:** SOYBEAN

soy·bean \ˈsȯi-ˌbēn\ *also* **soya bean** \ˈsȯi-ə-\ *noun* (1802)
: a hairy annual Asian legume (*Glycine max*) widely grown for its oil-rich proteinaceous seeds and for forage and soil improvement; *also* **:** its seed

soybean

soybean oil *noun* (circa 1916)
: a pale yellow drying or semidrying oil that is obtained from soybeans and is used chiefly as a food, in paints, varnishes, linoleum, printing ink, and soap, and as a source of phospholipids, fatty acids, and sterols — called also *soya oil*

soz·zled \ˈsä-zəld\ *adjective* [*sozzle* to splash, intoxicate, alteration of *sossle,* probably frequentative of British dialect *soss* to mess] (circa 1880)
: DRUNK, INTOXICATED

spa \ˈspä, ˈspȯ\ *noun* [*Spa,* watering place in Belgium] (1610)
1 a : a mineral spring **b :** a resort with mineral springs
2 : a fashionable resort or hotel
3 *New England* **:** SODA FOUNTAIN
4 : HEALTH SPA
5 : a hot tub with a whirlpool device

¹space \ˈspās\ *noun, often attributive* [Middle English, from Old French *espace,* from Latin *spatium* area, room, interval of space or time] (14th century)
1 : a period of time; *also* **:** its duration
2 a : a limited extent in one, two, or three dimensions **:** DISTANCE, AREA, VOLUME **b :** an extent set apart or available ⟨parking *space*⟩ ⟨floor *space*⟩
3 : one of the degrees between or above or below the lines of a musical staff — compare LINE
4 a : a boundless three-dimensional extent in which objects and events occur and have relative position and direction **b :** physical space independent of what occupies it — called also *absolute space*
5 : the region beyond the earth's atmosphere or beyond the solar system
6 a : a blank area separating words or lines **b :** material used to produce such blank area; *especially* **:** a piece of type less than one en in width
7 : a set of mathematical elements and especially of abstractions of all the points on a line, in a plane, or in physical space; *especially* **:** a set of mathematical entities with a set of axioms of geometric character — compare METRIC SPACE, TOPOLOGICAL SPACE, VECTOR SPACE
8 : an interval in operation during which a telegraph key is not in contact
9 a : LINAGE **b :** broadcast time available especially to advertisers
10 : accommodations on a public vehicle
11 : the opportunity to assert or experience one's identity or needs freely

²space *verb* **spaced; spac·ing** (1703)
transitive verb
: to place at intervals or arrange with space between — often used with *out*
intransitive verb
: to leave one or more blank spaces (as in a line of typing)
— **spac·er** *noun*

space–age \ˈspās-ˈāj\ *adjective* (1946)
: of, relating to, or befitting the age of space exploration; *especially* **:** MODERN ⟨*space-age* technology⟩

space·band \-ˌband\ *noun* (1904)
: a device on a linecaster that provides variable but even spacing between words in a justified line

space cadet *noun* (1979)
: a flaky, lightheaded, or forgetful person

space charge *noun* (1913)
: an electric charge distributed throughout a three-dimensional region

space·craft \ˈspās-ˌkraft\ *noun* (1930)
: a vehicle or device designed for travel or operation outside the earth's atmosphere

spaced–out \ˌspāst-ˈau̇t\ *adjective* (1937)
1 *or* **spaced** \ˈspāst\ **:** dazed or stupefied by or as if by a narcotic substance **:** HIGH
2 : of very strange character **:** WEIRD ⟨a *spaced-out* fantasy⟩

space·flight \ˈspās-ˌflīt\ *noun* (1931)
: flight beyond the earth's atmosphere

space frame *noun* (1912)
: a usually open three-dimensional framework of struts and braces (as in buildings and racing cars) which defines a structure and in which the weight of the structure is evenly distributed in all directions

space heater *noun* (1925)
: a usually portable appliance for heating a relatively small area

space heating *noun* (1934)
: heating of spaces especially for human comfort by any means (as fuel, electricity, or solar radiation) with the heater either within the space or external to it

space lattice *noun* (1895)
: LATTICE 2

space·less \ˈspās-ləs\ *adjective* (1606)
1 : having no limits **:** BOUNDLESS
2 : occupying no space

space·man \ˈspās-ˌman, -mən\ *noun* (1938)
1 : one who travels outside the earth's atmosphere
2 : a visitor to earth from outer space

space mark *noun* (circa 1890)
: the symbol #

space medicine *noun* (1949)
: a branch of medicine that deals with the physiological and biological effects on the human body of rocket or jet flight beyond the earth's atmosphere

space opera *noun* (1949)
: a futuristic melodramatic fantasy involving space travelers and extraterrestrial beings

space·port \ˈspās-ˌpōrt, -ˌpȯrt\ *noun* (1935)
: an installation for testing and launching spacecraft

space·ship \-ˌship\ *noun* (1894)
: a vehicle used for space travel

space shuttle *noun* (1969)
: a reusable spacecraft designed to transport people and cargo between earth and space

space station *noun* (1936)
: a large artificial satellite designed to be occupied for long periods and to serve as a base (as for scientific observation) — called also *space platform*

space suit *noun* (1929)
1 : a suit equipped with life supporting provisions to make life in space possible for its wearer
2 : G SUIT

space–time \ˈspās-ˈtīm, ˈspās-ˌ\ *noun* (1915)
1 : a system of one temporal and three spatial coordinates by which any physical object or event can be located — called also *space-time continuum*
2 : the whole or a portion of physical reality determinable by a usually four-dimensional coordinate system; *also* **:** the properties characteristic of such an order

space walk *noun* (1965)
: a period of activity spent outside a spacecraft by an astronaut in space
— **space·walk** \ˈspās-ˌwȯk\ *intransitive verb*
— **space·walk·er** \-ˌwȯ-kər\ *noun*

space·ward \ˈspās-wərd\ *adverb* (1958)
: toward space

spac·ey *also* **spacy** \ˈspā-sē\ *adjective* **spac·i·er; -est** (1970)
: SPACED-OUT

spa·cial *variant of* SPATIAL

spac·ing \ˈspā siŋ\ *noun* (1683)
1 a : the act of providing with spaces or placing at intervals **b :** an arrangement in space
2 a : a limited extent **:** SPACE **b :** the distance between any two objects in a usually regularly arranged series

spa·cious \ˈspā-shəs\ *adjective* [Middle English, from Middle French *spacieux,* from Latin *spatiosus,* from *spatium* space, room] (14th century)
1 : vast or ample in extent **:** ROOMY ⟨a *spacious* residence⟩

2 : large or magnificent in scale **:** EXPANSIVE ⟨a more *spacious* and stimulating existence than the farm could offer —H. L. Mencken⟩ ☆
— spa·cious·ly *adverb*
— spa·cious·ness *noun*
spack·le \ˈspa-kəl\ *transitive verb* **spack·led; spack·ling** \-k(ə-)liŋ\ [*Spackle*] (1940)
: to apply Spackle paste to
Spackle *trademark*
— used for a powder mixed with water to form a paste and used as a filler for cracks in a surface before painting
¹spade \ˈspād\ *noun* [Middle English, from Old English *spadu;* akin to Greek *spathē* blade of a sword or oar] (before 12th century)
1 : a digging implement adapted for being pushed into the ground with the foot
2 : a spade-shaped instrument
— spade·ful \-ˌfu̇l\ *noun*
— call a spade a spade 1 : to call a thing by its right name however coarse **2 :** to speak frankly
²spade *verb* **spad·ed; spad·ing** (1647)
transitive verb
: to dig up or out or shape with or as if with a spade
intransitive verb
: to use a spade
— spad·er *noun*
³spade *noun* [Italian *spada* or Spanish *espada* broadsword; both from Latin *spatha,* from Greek *spathē* blade] (1598)
1 a : a black figure that resembles a stylized spearhead on each playing card of one of the four suits; *also* **:** a card marked with this figure **b** *plural but singular or plural in construction* **:** the suit comprising cards marked spades
2 : BLACK 4 — usually taken to be offensive
— in spades : to an unusually great degree **:** in the extreme
spade beard *noun* [¹*spade*] (1598)
1 : an oblong beard with square ends
2 : a beard rounded off at the top and pointed at the bottom
— spade–beard·ed \ˈspād-ˈbir-dəd\ *adjective*
spade·fish \ˈspād-ˌfish\ *noun* (1704)
: a deep-bodied bony fish (*Chaetodipterus faber* of the family Ephippidae) that resembles the angelfishes and is found in the warmer parts of the western Atlantic
spade·foot toad \ˈspād-ˌfu̇t-\ *noun* (1867)
: any of a family (Pelobatidae) of burrowing toads having the inner bone of the tarsus edged with a strong horny sheath with which they dig
spade·work \-ˌwərk\ *noun* (1778)
1 : work done with a spade
2 : the hard plain preliminary drudgery in an undertaking
spa·dille \spə-ˈdil, -ˈdē\ *noun* [French, from Spanish *espadilla,* diminutive of *espada* broadsword, spade (in cards) — more at SPADE] (1728)
: the highest trump in various card games (as ombre)
spa·dix \ˈspā-diks\ *noun, plural* **spa·di·ces** \ˈspā-də-ˌsēz\ [New Latin *spadic-, spadix,* from Latin, frond torn from a palm tree, from Greek *spadik-, spadix,* from *span* to draw, pull] (circa 1760)
: a floral spike with a fleshy or succulent axis usually enclosed in a spathe

1 spadix

spae \ˈspā\ *transitive verb* **spaed; spae·ing** [Middle English *span,* from Old Norse *spā;* akin to Old High German *spehōn* to watch, spy — more at SPY] (14th century)
chiefly Scottish **:** FORETELL
spaetz·le \ˈshpet-slə, -sᵊl, -slē *also* ˈshpāt-\ *noun, plural* **spaetzle** *or* **spaetzles** [German *Spätzle,* from German dialect, diminutive of *Spatz* sparrow, dumpling] (1933)
: a small dumpling cooked by running batter through a colander into boiling water
spa·ghet·ti \spə-ˈge-tē\ *noun* [Italian, from plural of *spaghetto,* diminutive of *spago* cord, string, from Late Latin *spacus*] (1888)
1 : pasta made in thin solid strings
2 : insulating tubing typically of varnished cloth or of plastic for covering bare wire or holding insulated wires together
— spa·ghet·ti·like \-ˌlīk\ *adjective*
spa·ghet·ti·ni \spə-ˌge-ˈtē-nē\ *noun* [Italian, diminutive of *spaghetti*] (1923)
: a pasta thinner than spaghetti but thicker than vermicelli
spaghetti squash *noun* (1975)
: an oval winter squash with flesh that once cooked is similar in texture to spaghetti
spaghetti western *noun, often W capitalized* (1969)
: a western motion picture produced in Italy
spa·hi \ˈspä-ˌhē\ *noun* [Middle French, from Turkish *sipahi,* from Persian *sipāhī* cavalryman] (1562)
1 : one of a former corps of irregular Turkish cavalry
2 : one of a former corps of Algerian native cavalry in the French army
spake \ˈspāk\ *archaic past of* SPEAK
¹spall \ˈspȯl\ *noun* [Middle English *spalle*] (15th century)
: a small fragment or chip especially of stone
²spall (1758)
transitive verb
: to break up or reduce by or as if by chipping with a hammer
intransitive verb
1 : to break off chips, scales, or slabs **:** EXFOLIATE
2 : to undergo spallation
— spall·able \ˈspȯ-lə-bəl\ *adjective*
spall·ation \spȯ-ˈlā-shən\ *noun* (1947)
1 : a nuclear reaction in which light particles are ejected as the result of bombardment (as by high-energy protons)
2 : the process of spalling
spal·peen \spal-ˈpēn, spȯl-\ *noun* [Irish *spailpín* seasonal laborer, rascal] (1815)
chiefly Irish **:** RASCAL
¹span \ˈspan\ *archaic past of* SPIN
²span *noun* [Middle English, from Old English *spann;* akin to Old High German *spanna* span, Middle Dutch *spannen* to stretch, hitch up] (before 12th century)
1 : the distance from the end of the thumb to the end of the little finger of a spread hand; *also* **:** an English unit of length equal to 9 inches (22.9 centimeters)
2 : an extent, stretch, reach, or spread between two limits: as **a :** a limited space (as of time); *especially* **:** an individual's lifetime **b :** the spread or extent between abutments or supports (as of a bridge); *also* **:** a portion thus supported **c :** the maximum distance laterally from tip to tip of an airplane
³span *transitive verb* **spanned; span·ning** (1560)
1 a : to measure by or as if by the hand with fingers and thumb extended **b :** MEASURE
2 a : to extend across ⟨a career that *spanned* four decades⟩ **b :** to form an arch over ⟨a small bridge *spanned* the pond⟩ **c :** to place or construct a span over
3 : to be capable of expressing any element of under given operations ⟨a set of vectors that *spans* a vector space⟩
⁴span *noun* [Dutch, from Middle Dutch, from *spannen* to hitch up] (1769)
: a pair of animals (as mules) usually matched in appearance and action and driven together
spa·na·ko·pi·ta *also* **spa·no·ko·pi·ta** \ˌspä-nə-ˈkō-pē-tə, -pi-tə\ *noun* [New Greek *spanakopēta,* from *spanaki* spinach + *pēta, pita* pie] (1950)
: a traditional Greek pie of spinach, feta cheese, and seasonings baked in phyllo
span·dex \ˈspan-ˌdeks\ *noun* [anagram of *expands*] (1959)
: any of various elastic textile fibers made chiefly of polyurethane; *also* **:** clothing made of this material
span·drel *also* **span·dril** \ˈspan-drəl\ *noun* [Middle English *spandrell,* from Anglo-French *spaundre,* from Old French *espandre* to spread out — more at SPAWN] (15th century)
1 : the sometimes ornamented space between the right or left exterior curve of an arch and an enclosing right angle
2 : the triangular space beneath the string of a stair
spang \ˈspaŋ\ *adverb* [Scots *spang* to leap, cast, bang] (1843)
1 : to a complete degree
2 : in an exact or direct manner **:** SQUARELY
¹span·gle \ˈspaŋ-gəl\ *noun* [Middle English *spangel,* diminutive of *spang* shiny ornament, probably of Scandinavian origin; akin to Old Norse *spǫng* spangle; akin to Old English *spang* buckle, Middle Dutch *spannen* to stretch] (15th century)
1 : a small plate of shining metal or plastic used for ornamentation especially on clothing
2 : a small glittering object or particle
²spangle *verb* **span·gled; span·gling** \ˈspaŋ-g(ə-)liŋ\ (1598)
transitive verb
: to set or sprinkle with or as if with spangles
intransitive verb
: to glitter as if covered with spangles **:** SPARKLE
Span·glish \ˈspaŋ-glish, -lish\ *noun* [blend of *Spanish* and *English*] (1965)
: Spanish marked by numerous borrowings from English; *broadly* **:** any of various combinations of Spanish and English
Span·iard \ˈspan-yərd\ *noun* [Middle English *Spaignard,* from Middle French *Espaignart,* from *Espaigne* Spain, from Latin *Hispania*] (15th century)
: a native or inhabitant of Spain
span·iel \ˈspan-yəl *also* ˈspa-nᵊl\ *noun* [Middle English *spaniell,* from Middle French *espaignol,* literally, Spaniard, from (assumed) Vulgar Latin *Hispaniolus,* from Latin *Hispania* Spain] (14th century)
1 : a member of any of several breeds of small or medium-sized mostly short-legged dogs usually having long wavy hair, feathered legs and tail, and large drooping ears
2 : a fawning servile person
Span·ish \ˈspa-nish\ *noun* [*Spanish,* adjective, from Middle English *Spainish,* from *Spain*] (15th century)
1 : the Romance language of the largest part of Spain and of the countries colonized by Spaniards
2 *plural in construction* **:** the people of Spain
— Spanish *adjective*
— Span·ish·ness *noun*
Spanish American *noun* (1811)

☆ SYNONYMS
Spacious, commodious, capacious, ample mean larger in extent or capacity than the average. SPACIOUS implies great length and breadth ⟨a *spacious* front lawn⟩. COMMODIOUS stresses roominess and comfortableness ⟨a *commodious* and airy penthouse apartment⟩. CAPACIOUS stresses the ability to hold, contain, or retain more than the average ⟨a *capacious* suitcase⟩. AMPLE implies having a greater size, expanse, or amount than that deemed adequate ⟨*ample* closet space⟩.

1 : a resident of the U.S. whose native language is Spanish and whose culture is of Spanish origin
2 : a native or inhabitant of one of the countries of America in which Spanish is the national language
— **Spanish–American** *adjective*

Spanish bayonet *noun* (1843)
: any of several yuccas; *especially* **:** one (*Yucca aloifolia*) with a short trunk and rigid spine-tipped leaves

Spanish chestnut *noun* (1699)
: MARRON

Spanish fly *noun* (1634)
1 : a green blister beetle (*Lytta vesicatoria*) of southern Europe
2 : CANTHARIS 2

Spanish mackerel *noun* (1666)
: a large scombroid food and game fish (*Scomberomorus maculatus*) that is bluish above with oval brown spots on the sides and is found off the American Atlantic coast from Cape Ann to Brazil

Spanish moss *noun* (1823)
: an epiphytic plant (*Tillandsia usneoides*) of the pineapple family forming pendent tufts of grayish green filaments on trees from the southern U.S. to Argentina

Spanish moss

Spanish needles *noun plural but singular or plural in construction* (1743)
: any of several bur marigolds; *especially* **:** an annual (*Bidens bipinnata*) of North America and Asia having yellow flowers and dissected leaves

Spanish omelet *noun* (circa 1886)
: an omelet served with a sauce containing chopped green pepper, onion, and tomato

Spanish rice *noun* (1928)
: rice cooked with onions, green pepper, and tomatoes

[1]**spank** \'spaŋk\ *transitive verb* [imitative] (circa 1727)
: to strike especially on the buttocks with the open hand
— **spank** *noun*

[2]**spank** *intransitive verb* [back-formation from *spanking*] (circa 1810)
: to move quickly, dashingly, or spiritedly ⟨*spanking* along in his new car⟩

span·ker \'spaŋ-kər\ *noun* [origin unknown] (1794)
1 : the fore-and-aft sail on the mast nearest the stern of a square-rigged ship
2 : the sail on the sternmost mast in a schooner of four or more masts

[1]**spank·ing** \'spaŋ-kiŋ\ *adjective* [origin unknown] (1666)
1 : remarkable of its kind
2 : being fresh and strong **:** BRISK

[2]**spanking** *adverb* (1886)
: VERY ⟨a *spanking* clean floor⟩ ⟨*spanking* new⟩

span·ner \'spa-nər\ *noun* [German, instrument for winding springs, from *spannen* to stretch; akin to Middle Dutch *spannen* to stretch — more at SPAN] (circa 1790)
1 *chiefly British* **:** WRENCH
2 : a wrench that has a hole, projection, or hook at one or both ends of the head for engaging with a corresponding device on the object that is to be turned

spanner 2

span–new \'span-'nü, -'nyü\ *adjective* [Middle English, part translation of Old Norse *spānnȳr,* from *spānn* chip of wood + *nȳr* new] (14th century)
: BRAND-NEW

span·worm \'span-ˌwərm\ *noun* [[3]*span*] (1820)
: LOOPER 1

[1]**spar** \'spär\ *noun* [Middle English *sparre;* akin to Old English *spere* spear — more at SPEAR] (14th century)
1 : a stout pole
2 a : a stout rounded wood or metal piece (as a mast, boom, gaff, or yard) used to support rigging **b :** any of the main longitudinal members of the wing of an airplane that carry the ribs

[2]**spar** *intransitive verb* **sparred; spar·ring** [probably alteration of [2]*spur*] (1537)
1 a : BOX; *especially* **:** to gesture without landing a blow to draw one's opponent or create an opening **b :** to engage in a practice or exhibition bout of boxing
2 : SKIRMISH, WRANGLE
3 : to strike or fight with feet or spurs in the manner of a gamecock

[3]**spar** *noun* (1814)
1 : a movement of offense or defense in boxing
2 : a sparring match or session

[4]**spar** *noun* [Low German; akin to Old English *spærstān* gypsum, *spæren* of plaster] (1581)
: any of various nonmetallic usually cleavable and lustrous minerals

SPAR \'spär\ *noun* [*S*emper *Par*atus, motto of the U.S. Coast Guard, from New Latin, always ready] (1942)
: a member of the women's reserve of the U.S. Coast Guard

[1]**spare** \'spar, 'sper\ *verb* **spared; spar·ing** [Middle English, from Old English *sparian;* akin to Old High German *sparōn* to spare, Old English *spær,* adjective, scant] (before 12th century)
transitive verb
1 : to forbear to destroy, punish, or harm
2 : to refrain from attacking or reprimanding with necessary or salutary severity
3 : to relieve of the necessity of doing or undergoing something ⟨*spare* yourself the trouble⟩
4 : to refrain from **:** AVOID ⟨*spared* no expense⟩
5 : to use or dispense frugally — used chiefly in the negative ⟨don't *spare* the syrup⟩
6 a : to give up as not strictly needed ⟨do you have any cash to *spare*⟩ **b :** to have left over or as margin ⟨time to *spare*⟩
intransitive verb
1 : to be frugal
2 : to refrain from doing harm **:** be lenient
— **spare·able** \-ə-bəl\ *adjective*
— **spar·er** *noun*

[2]**spare** *adjective* **spar·er; spar·est** [Middle English, from Old English *spær* sparing, scant; akin to Old High German *spar* spare] (14th century)
1 : not being used; *especially* **:** held for emergency use ⟨a *spare* tire⟩
2 : being over and above what is needed **:** SUPERFLUOUS ⟨*spare* time⟩
3 : not liberal or profuse **:** SPARING ⟨a *spare* prose style⟩
4 : healthily lean
5 : not abundant or plentiful
synonym see LEAN, MEAGER
— **spare·ly** *adverb*
— **spare·ness** *noun*

[3]**spare** *noun* (1642)
1 a : a spare tire **b :** a duplicate (as a key or a machine part) kept in reserve
2 : the knocking down of all 10 pins with the first 2 balls in a frame in bowling

spare·ribs \'spar-ˌ(r)ibz, 'sper-, -əbz\ *noun plural* [by folk etymology from Low German *ribbesper* pickled pork ribs roasted on a spit, from Middle Low German, from *ribbe* rib + *sper* spear, spit] (1596)
: a cut of pork ribs separated from the bacon strip

sparge \'spärj\ *transitive verb* **sparged; sparg·ing** [probably from Middle French *espargier,* from Latin *spargere* to scatter] (1785)
1 : SPRINKLE, BESPATTER; *especially* **:** SPRAY
2 : to agitate (a liquid) by means of compressed air or gas entering through a pipe
— **sparge** *noun*
— **sparg·er** *noun*

spar·ing \'spar-iŋ, 'sper-\ *adjective* (14th century)
1 : marked by or practicing careful restraint (as in the use of resources)
2 : MEAGER, BARE ⟨the map is *sparing* of information⟩ ☆
— **spar·ing·ly** \-iŋ-lē\ *adverb*

[1]**spark** \'spärk\ *noun* [Middle English *sparke,* from Old English *spearca;* akin to Middle Dutch *sparke* spark and perhaps to Latin *spargere* to scatter] (before 12th century)
1 a : a small particle of a burning substance thrown out by a body in combustion or remaining when combustion is nearly completed **b :** a hot glowing particle struck from a larger mass; *especially* **:** one heated by friction
2 a : a luminous disruptive electrical discharge of very short duration between two conductors separated by a gas (as air) **b :** the discharge in a spark plug **c :** the mechanism controlling the discharge in a spark plug
3 : SPARKLE, FLASH
4 : something that sets off a sudden force ⟨provided the *spark* that helped the team to rally⟩
5 : a latent particle capable of growth or developing **:** GERM ⟨still retains a *spark* of decency⟩
6 *plural but singular in construction* **:** a radio operator on a ship

[2]**spark** (13th century)
intransitive verb
1 a : to throw out sparks **b :** to flash or fall like sparks
2 : to produce sparks; *specifically* **:** to have the electric ignition working
3 : to respond with enthusiasm
transitive verb
1 : to set off in a burst of activity **:** ACTIVATE ⟨the question *sparked* a lively discussion⟩ — often used with *off*
2 : to stir to activity **:** INCITE ⟨*sparked* her team to victory⟩
— **spark·er** *noun*

[3]**spark** *noun* [perhaps of Scandinavian origin; akin to Old Norse *sparkr* sprightly] (circa 1600)
1 : a foppish young man
2 : LOVER, BEAU
— **spark·ish** \'spär-kish\ *adjective*

[4]**spark** *verb* (1787)
: WOO, COURT
— **spark·er** *noun*

spark chamber *noun* (1961)
: a device usually used to detect the path of a high-energy particle that consists of a series of

☆ SYNONYMS
Sparing, frugal, thrifty, economical mean careful in the use of one's money or resources. SPARING stresses abstention and restraint ⟨*sparing* in the offering of advice⟩. FRUGAL implies absence of luxury and simplicity of lifestyle ⟨ran a *frugal* household⟩. THRIFTY stresses good management and industry ⟨*thrifty* use of nonrenewable resources⟩. ECONOMICAL stresses prudent management, lack of wastefulness, and use of things to their best advantage ⟨an *economical* health care plan⟩.

\ə\ **abut** \ᵊ\ **kitten** \ər\ **further** \a\ **ash** \ā\ **ace** \ä\ **mop, mar** \au̇\ **out** \ch\ **chin** \e\ **bet** \ē\ **easy** \g\ **go** \i\ **hit** \ī\ **ice** \j\ **job** \ŋ\ **sing** \ō\ **go** \ȯ\ **law** \ȯi\ **boy** \th\ **thin** \t͟h\ **the** \ü\ **loot** \u̇\ **foot** \y\ **yet** \zh\ **vision** *see also* Guide to Pronunciation

charged metal plates or wires separated by a gas (as neon) in which observable electric discharges follow the path of the particle

spark coil *noun* (1896)
: an induction coil for producing the spark for an internal combustion engine

spark gap *noun* (1889)
: a space between two high-potential terminals (as of an induction coil) through which pass discharges of electricity; *also* **:** a device having a spark gap

sparking plug *noun* (1902)
British **:** SPARK PLUG

[1]**spar·kle** \'spär-kəl\ *verb* **spar·kled; spar·kling** \-k(ə-)liŋ\ [Middle English, frequentative of *sparken* to spark] (13th century)
intransitive verb
1 a : to throw out sparks **b :** to give off or reflect bright moving points of light **c :** to perform brilliantly
2 : EFFERVESCE ⟨wine that *sparkles*⟩
3 : to become lively or animated ⟨the dialogue *sparkles* with wit⟩ ⟨eyes *sparkling* with anger⟩
transitive verb
: to cause to glitter or shine
synonym see FLASH
— spar·kly \-k(ə-)lē\ *adjective*

[2]**sparkle** *noun* [Middle English, diminutive of *sparke*] (14th century)
1 : a little spark **:** SCINTILLATION
2 : the quality of sparkling
3 a : ANIMATION, LIVELINESS **b :** the quality or state of being effervescent

spar·kler \'spär-klər\ *noun* (1713)
: one that sparkles: as **a :** DIAMOND **b :** a firework that throws off brilliant sparks on burning **c :** SPARKLING WINE

sparkling wine *noun* (1697)
: an effervescent table wine

spark plug *noun* (1903)
1 : a part that fits into the cylinder head of an internal combustion engine and carries two electrodes separated by an air gap across which the current from the ignition system discharges to form the spark for combustion
2 : one that initiates or gives impetus to an undertaking
— spark·plug \'spärk-ˌpləg\ *transitive verb*

sparky \'spär-kē\ *adjective* **spark·i·er; -est** (circa 1865)
: marked by animation **:** LIVELY
— spark·i·ly \-kə-lē\ *adverb*

spar·row \'spar-(ˌ)ō\ *noun* [Middle English *sparow,* from Old English *spearwa;* akin to Old High German *sparo* sparrow] (before 12th century)
1 : any of a genus (*Passer* of the family Ploceidae) of small chiefly brownish or grayish Old World oscine songbirds that include some which have been widely introduced; *especially* **:** HOUSE SPARROW
2 : any of various finches (as the song sparrow or tree sparrow) resembling the true sparrows
— spar·row·like \-ō-ˌlīk, -ə-ˌlīk\ *adjective*

sparrow hawk *noun* (15th century)
: any of various small hawks: as **a :** an Old World accipiter (*Accipiter nisus*) that is dark gray to blackish above with the female having a grayish brown barred underside and the male having a chestnut barred underside **b :** KESTREL b

sparrow hawk

sparse \'spärs\ *adjective* **spars·er; spars·est** [Latin *sparsus* spread out, from past participle of *spargere* to scatter — more at SPARK] (1753)
: of few and scattered elements; *especially* **:** not thickly grown or settled
synonym see MEAGER
— sparse·ly *adverb*
— sparse·ness *noun*
— spar·si·ty \'spär-sə-tē, -stē\ *noun*

Spar·ta·cist \'spär-tə-sist\ *noun* [German *Spartakist,* from *Spartakusbund,* literally, league of Spartakus, a revolutionary organization, from *Spartakus,* pen name of Karl Liebknecht, its cofounder] (1919)
: a member of a revolutionary political group organized in Germany in 1918 and advocating extreme socialistic doctrines

[1]**Spar·tan** \'spär-tᵊn\ *noun* (15th century)
1 : a native or inhabitant of ancient Sparta
2 : a person of great courage and self-discipline
— Spar·tan·ism \-ˌi-zəm\ *noun*

[2]**Spartan** *adjective* (1582)
1 : of or relating to Sparta in ancient Greece
2 a *often not capitalized* **:** marked by strict self-discipline or self-denial ⟨a *Spartan* athlete⟩ **b** *often not capitalized* **:** marked by simplicity, frugality, or avoidance of luxury and comfort ⟨a *Spartan* room⟩ **c :** LACONIC **d :** undaunted by pain or danger
— Spar·tan·ly *adverb*

spar·te·ine \'spär-tē-ən, 'spär-ˌtēn\ *noun* [Latin *spartum* esparto, broom + International Scientific Vocabulary *-eine* — more at ESPARTO] (1851)
: a liquid alkaloid $C_{15}H_{26}N_2$ extracted from the Scotch broom and used in medicine in the form of its sulfate

spar varnish *noun* [[1]*spar*] (circa 1909)
: an exterior waterproof varnish

spasm \'spa-zəm\ *noun* [Middle English *spasme,* from Middle French, from Latin *spasmus,* from Greek *spasmos,* from *span* to draw, pull] (14th century)
1 : an involuntary and abnormal muscular contraction
2 : a sudden violent and temporary effort or emotion ⟨a *spasm* of creativity⟩

spas·mod·ic \spaz-'mä-dik\ *adjective* [New Latin *spasmodicus,* from Greek *spasmōdēs,* from *spasmos*] (circa 1681)
1 a : relating to or affected or characterized by spasm **b :** resembling a spasm especially in sudden violence ⟨a *spasmodic* jerk⟩
2 : acting or proceeding fitfully **:** INTERMITTENT
3 : subject to outbursts of emotional excitement **:** EXCITABLE
synonym see FITFUL
— spas·mod·i·cal·ly \-di-k(ə-)lē\ *adverb*

spas·mo·lyt·ic \ˌspaz-mə-'li-tik\ *adjective* [International Scientific Vocabulary *spasmo-* (from Greek *spasmos* spasm) + *-lytic*] (circa 1935)
: tending or having the power to relieve spasms or convulsions
— spasmolytic *noun*

[1]**spas·tic** \'spas-tik\ *adjective* [Latin *spasticus,* from Greek *spastikos* drawing in, from *span*] (1753)
1 : of, relating to, characterized by, or affected with spasm ⟨a *spastic* colon⟩ ⟨a *spastic* patient⟩
2 : SPASMODIC 2 ⟨a *spastic* influx of data⟩
— spas·ti·cal·ly \-ti-k(ə-)lē\ *adverb*

[2]**spastic** *noun* (1896)
: one suffering from spastic paralysis

spas·tic·i·ty \spa-'sti-sə-tē\ *noun* (circa 1827)
: a spastic state or condition; *especially* **:** muscular hypertonicity with increased tendon reflexes

spastic paralysis *noun* (1879)
: paralysis with tonic spasm of the affected muscles and with increased tendon reflexes — compare CEREBRAL PALSY

[1]**spat** \'spat\ *past and past participle of* SPIT

[2]**spat** *noun, plural* **spat** *or* **spats** [origin unknown] (1667)
: a young bivalve (as an oyster)

[3]**spat** *noun* [short for *spatterdash* legging] (circa 1802)
: a cloth or leather gaiter covering the instep and ankle

[4]**spat** *noun* [origin unknown] (1804)
1 : a brief petty quarrel or angry outburst
2 *chiefly dialect* **:** SLAP
3 : a sound like that of rain falling in large drops

[5]**spat** *verb* **spat·ted; spat·ting** (circa 1832)
transitive verb
chiefly dialect **:** SLAP
intransitive verb
1 : to quarrel pettily or briefly
2 : to strike with a sound like that of rain falling in large drops

spate \'spāt\ *noun* [Middle English] (15th century)
1 : FRESHET, FLOOD
2 a : a large number or amount ⟨a *spate* of books on gardening⟩ **b :** a sudden or strong outburst **:** RUSH ⟨a *spate* of anger⟩

spathe \'spāth\ *noun* [New Latin *spatha,* from Latin, broadsword — more at SPADE] (1785)
: a sheathing bract or pair of bracts partly enclosing an inflorescence and especially a spadix on the same axis ⟨the *spathe* of the calla lily⟩

spath·u·late \'spath-yə-lət\ *adjective* [Late Latin *spathula, spatula* spatula] (1821)
: SPATULATE ⟨*spathulate* petals of a flower⟩

spa·tial \'spā-shəl\ *adjective* [Latin *spatium* space] (1847)
: relating to, occupying, or having the character of space
— spa·ti·al·i·ty \ˌspā-shē-'a-lə-tē\ *noun*
— spa·tial·ly \'spā-sh(ə-)lē\ *adverb*

spatial summation *noun* (1968)
: sensory summation that involves stimulation of several spatially separated neurons at the same time

spa·tio·tem·po·ral \ˌspā-shē-ō-'tem-p(ə-)rəl\ *adjective* [Latin *spatium* + *tempor-, tempus* time] (1900)
1 : having both spatial and temporal qualities
2 : of or relating to space-time
— spa·tio·tem·po·ral·ly \-p(ə-)rə-lē\ *adverb*

[1]**spat·ter** \'spa-tər\ *verb* [akin to Flemish *spetteren* to spatter] (1600)
intransitive verb
: to spurt forth in scattered drops ⟨blood *spattering* everywhere⟩
transitive verb
1 : to splash with or as if with a liquid; *also* **:** to soil in this way ⟨his coat was *spattered* with mud⟩
2 : to scatter by or as if by splashing ⟨*spatter* water⟩
3 : to cover with or as if with splashes or spots
4 : to injure by aspersion **:** DEFAME ⟨*spattered* my reputation⟩

[2]**spatter** *noun* (1797)
1 a : the act or process of spattering **:** the state of being spattered **b :** the noise of spattering
2 a : a drop or splash spattered on something or a spot or stain due to spattering **b :** a small amount or number **:** SPRINKLE ⟨a *spatter* of applause⟩

spat·ter·dock \'spa-tər-ˌdäk\ *noun* (1813)
: a common yellow North American water lily (*Nuphar advena*); *also* **:** a congeneric plant

spat·u·la \'spa-chə-lə, 'spach-lə\ *noun* [Late Latin, spoon, spatula — more at EPAULET] (1525)
: a flat thin usually metal implement used especially for spreading or mixing soft substances, scooping, or lifting

spat·u·late \'spa-chə-lət\ *adjective* (1760)
: shaped like a spatula ⟨a *spatulate* leaf⟩ ⟨*spatulate* spines of a caterpillar⟩ ⟨a *spatulate* tool⟩ — see LEAF illustration

spätz·le *variant of* SPAETZLE

spav·in \'spa-vən\ *noun* [Middle English *spavayne,* from Middle French *espavain*] (15th century)
: SWELLING; *especially* **:** a bony enlargement of the hock of a horse associated with strain

spav·ined \-vənd\ *adjective* (15th century)
1 : affected with spavin

2 : old and decrepit **:** OVER-THE-HILL

[1]**spawn** \'spȯn, 'spän\ *verb* [Middle English, from Anglo-French *espaundre,* from Old French *espandre* to spread out, expand, from Latin *expandere*] (15th century)
intransitive verb
1 : to deposit spawn
2 : to produce young especially in large numbers
transitive verb
1 a : to produce or deposit (eggs) — used of an aquatic animal **b :** to induce (fish) to spawn **c :** to plant with mushroom spawn
2 : BRING FORTH, GENERATE
— spawn·er *noun*

[2]**spawn** *noun* (15th century)
1 : the eggs of aquatic animals (as fishes or oysters) that lay many small eggs
2 : PRODUCT, OFFSPRING; *also* **:** numerous issue
3 : the seed, germ, or source of something
4 : mycelium especially prepared (as in bricks) for propagating mushrooms

spay \'spā, *nonstandard* 'spād\ *transitive verb* **spayed** \'spād, *nonstandard* 'spā-dəd\; **spay·ing** \'spā-iŋ, *nonstandard* 'spā-diŋ\ [Middle English, from Middle French *espeer* to cut with a sword, from Old French, from *espee* sword, from Latin *spatha* sword — more at SPADE] (15th century)
: to remove the ovaries of (a female animal)

spaz \'spaz\ *noun, plural* **spaz·zes** [by shortening & alteration from *spastic*] (1965)
slang **:** one who is inept **:** KLUTZ

speak \'spēk\ *verb* **spoke** \'spōk\; **spo·ken** \'spō-kən\; **speak·ing** [Middle English *speken,* from Old English *sprecan, specan;* akin to Old High German *sprehhan* to speak, Greek *spharageisthai* to crackle] (before 12th century)
intransitive verb
1 a : to utter words or articulate sounds with the ordinary voice **:** TALK **b** (1) **:** to express thoughts, opinions, or feelings orally (2) **:** to extend a greeting (3) **:** to be on speaking terms ⟨still were not *speaking* after the dispute⟩ **c** (1) **:** to express oneself before a group (2) **:** to address one's remarks ⟨*speak* to the issue⟩
2 a : to make a written statement ⟨his diaries . . . *spoke* . . . of his entrancement with death —Sy Kahn⟩ **b :** to use such an expression — often used in the phrase *so to speak* ⟨was at the enemy's gates, so to *speak* —C. S. Forester⟩ **c :** to serve as spokesman ⟨*spoke* for the whole group⟩
3 a : to express feelings by other than verbal means ⟨actions *speak* louder than words⟩ **b :** SIGNAL **c :** to be interesting or attractive **:** APPEAL ⟨great music . . . *speaks* directly to the emotions —A. N. Whitehead⟩
4 : to make a request **:** ASK ⟨*spoke* for the remaining piece of pie⟩
5 : to make a characteristic or natural sound ⟨all at once the thunder *spoke* —George Meredith⟩
6 a : TESTIFY **b :** to be indicative or suggestive ⟨his gold . . . *spoke* of riches in the land —Julian Dana⟩
transitive verb
1 a (1) **:** to utter with the speaking voice **:** PRONOUNCE (2) **:** to give a recitation of **:** DECLAIM **b :** to express orally **:** DECLARE ⟨free to *speak* their minds⟩ **c :** ADDRESS, ACCOST; *especially* **:** HAIL
2 : to make known in writing **:** STATE
3 : to use or be able to use in speaking ⟨*speaks* Spanish⟩
4 : to indicate by other than verbal means
5 *archaic* **:** DESCRIBE, DEPICT
— speak·able \'spē-kə-bəl\ *adjective*
— to speak of : worthy of mention or notice — usually used in negative constructions

-speak \-ˌspēk\ *noun combining form* [*newspeak*]
— used to form especially nonce words denoting a particular kind of jargon ⟨architect*speak*⟩ ⟨California*speak*⟩

speak·easy \'spēk-ˌē-zē\ *noun, plural* **-eas·ies** (1889)
: a place where alcoholic beverages are illegally sold; *specifically* **:** such a place during the period of prohibition in the U.S.

speak·er \'spē-kər\ *noun* (14th century)
1 a : one that speaks; *especially* **:** one who uses a language ⟨native *speakers* of French⟩ **b :** one who makes a public speech **c :** one who acts as a spokesman
2 : the presiding officer of a deliberative assembly ⟨*Speaker* of the House of Representatives⟩
3 : LOUDSPEAKER
— speak·er·ship \-ˌship\ *noun*

speak·er·phone \'spē-kər-ˌfōn\ *noun* (1955)
: a combination microphone and loudspeaker device for two-way communication by telephone lines

speak·ing \'spē-kiŋ\ *adjective* (13th century)
1 a : that speaks **:** capable of speech **b :** having a population that speaks a specified language — usually used in combination ⟨English-*speaking* countries⟩ **c :** that involves talking or giving speeches ⟨a *speaking* role⟩ ⟨a *speaking* tour⟩
2 : highly significant or expressive **:** ELOQUENT
3 : resembling a living being or a real object

speaking tube *noun* (1833)
: a pipe through which conversation may be conducted (as between different parts of a building)

speak–out \'spēk-ˌau̇t\ *noun* (1968)
: an event in which people publicly share their experiences of or views on an issue

speak out *intransitive verb* (1530)
1 : to speak loud enough to be heard
2 : to speak boldly **:** express an opinion frankly ⟨*spoke out* on the issues⟩

speak up *intransitive verb* (1705)
1 : to express an opinion freely ⟨*speak up* for truth and justice —Clive Bell⟩
2 : to speak loudly and distinctly

spean \'spēn\ *transitive verb* [Middle Dutch *spenen*] (1595)
chiefly Scottish **:** WEAN

[1]**spear** \'spir\ *noun* [Middle English *spere,* from Old English; akin to Old High German *sper* spear, Latin *sparus* hunting spear] (before 12th century)
1 : a thrusting or throwing weapon with long shaft and sharp head or blade
2 : a sharp-pointed instrument with barbs used in spearing fish
3 : SPEARMAN

[2]**spear** (15th century)
transitive verb
1 : to pierce, strike, or take with or as if with a spear ⟨*spear* salmon⟩ ⟨*speared* a chop from the platter⟩
2 : to catch (as a baseball) with a sudden thrust of the arm
intransitive verb
: to thrust at or wound something with or as if with a spear
— spear·er *noun*

[3]**spear** *adjective* (1861)
: PATERNAL, MALE ⟨the *spear* side of the family⟩ — compare DISTAFF

[4]**spear** *intransitive verb* [[5]*spear*] (1573)
of a plant **:** to thrust a spear upward

[5]**spear** *noun* [alteration of [1]*spire*] (1647)
: a usually young blade, shoot, or sprout (as of grass)

spear–car·ri·er \'spir-ˌkar-ē-ər\ *noun* (1953)
1 a : a member of an opera chorus **b :** a bit actor in a play
2 : a person whose actions are of little significance or value in an event or organization

[1]**spear·fish** \-ˌfish\ *noun* (circa 1882)
: any of several billfishes (genus *Tetrapturus*) having the anterior part of the first dorsal fin about as high as the body is deep

[2]**spearfish** *intransitive verb* (circa 1949)
: to fish with a spear

spear·gun \'spir-ˌgən\ *noun* (1951)
: a gun that shoots a spear and is used in spearfishing

[1]**spear·head** \-ˌhed\ *noun* (14th century)
1 : the sharp-pointed head of a spear
2 : a leading element, force, or influence in an undertaking or development

[2]**spearhead** *transitive verb* (1937)
: to serve as leader or leading element of

spear·man \'spir-mən\ *noun* (14th century)
: one armed with a spear

spear·mint \-ˌmint, -mənt\ *noun* (1562)
: a common mint (*Mentha spicata*) grown for flavoring and especially for its aromatic oil

spearmint

spear–throw·er \-ˌthrō(-ə)r\ *noun* (1871)
: ATLATL

spear·wort \-ˌwərt, -ˌwȯrt\ *noun* (14th century)
: any of several buttercups (especially *Ranunculus flammula*) with spear-shaped leaves

spec \'spek\ *transitive verb* **specced** *or* **spec'd** \'spekt\; **spec·cing** [[2]*specs*] (1965)
: to write specifications for

[1]**spe·cial** \'spe-shəl\ *adjective* [Middle English, from Old French or Latin; Old French *especial,* from Latin *specialis* individual, particular, from *species* species] (13th century)
1 : distinguished by some unusual quality; *especially* **:** being in some way superior ⟨our *special* blend⟩
2 : held in particular esteem ⟨a *special* friend⟩
3 a : readily distinguishable from others of the same category **:** UNIQUE ⟨they set it apart as a *special* day of thanksgiving⟩ **b :** of, relating to, or constituting a species **:** SPECIFIC
4 : being other than the usual **:** ADDITIONAL, EXTRA
5 : designed for a particular purpose or occasion ☆
— spe·cial·ness *noun*

[2]**special** *noun* (1866)
1 : one that is used for a special service or occasion ⟨caught the commuter *special* to work⟩
2 : something (as a television program) that is not part of a regular series

special assessment *noun* (1875)
: a specific tax levied on private property to meet the cost of public improvements that enhance the value of the property

special delivery *noun* (1886)
: expedited messenger delivery of mail matter for an extra fee

special district *noun* (1950)
: a political subdivision of a state established to provide a single public service (as water supply or sanitation) within a specific geographical area

☆ SYNONYMS
Special, especial, specific, particular, individual mean of or relating to one thing or class. SPECIAL stresses having a quality, character, identity, or use of its own ⟨*special* ingredients⟩. ESPECIAL may add implications of preeminence or preference ⟨a matter of *especial* importance⟩. SPECIFIC implies a quality or character distinguishing a kind or a species ⟨children with *specific* nutritional needs⟩. PARTICULAR stresses the distinctness of something as an individual ⟨a ballet step of *particular* difficulty⟩. INDIVIDUAL implies unequivocal reference to one of a class or group ⟨valued each *individual* opinion⟩.

\ə\ **abut** \ᵊ\ **kitten** \ər\ **further** \a\ **ash** \ā\ **ace** \ä\ **mop, mar** \au̇\ **out** \ch\ **chin** \e\ **bet** \ē\ **easy** \g\ **go** \i\ **hit** \ī\ **ice** \j\ **job** \ŋ\ **sing** \ō\ **go** \ȯ\ **law** \ȯi\ **boy** \th\ **thin** \t͟h\ **the** \ü\ **loot** \u̇\ **foot** \y\ **yet** \zh\ **vision** *see also* Guide to Pronunciation

special drawing rights *noun* (1967)
: a means of exchange used by governments to settle their international indebtedness

special education *noun* (1921)
: classes for students (as the handicapped) with special educational needs

special effects *noun plural* (1937)
: visual or sound effects introduced into a motion picture or a taped television production

Special Forces *noun plural* (1962)
: a branch of the army composed of soldiers specially trained in guerrilla warfare

special handling *noun* (1928)
: the handling of parcel-post or fourth-class mail as first-class but not as special-delivery matter for an extra fee

special interest *noun* (1910)
: a person or group seeking to influence legislative or government policy to further often narrowly defined interests; *especially* **:** LOBBY

spe·cial·i·sa·tion, spe·cial·ise, special·ised *British variant of* SPECIALIZATION, SPECIALIZE, SPECIALIZED

spe·cial·ism \'spe-shə-,li-zəm\ *noun* (1856)
1 : specialization in an occupation or branch of learning
2 : a field of specialization **:** SPECIALTY

spe·cial·ist \'spe-sh(ə-)list\ *noun* (1856)
1 : one who specializes in a particular occupation, practice, or branch of learning
2 : any of four enlisted ranks in the army corresponding to the grades of corporal through sergeant first class
— specialist *or* **spe·cial·is·tic** \,spe-shə-'lis-tik\ *adjective*

spe·ci·al·i·ty \,spe-shē-'a-lə-tē\ *noun, plural* **-ties** (15th century)
1 : a special mark or quality
2 : a special object or class of objects
3 a : a special aptitude or skill **b :** SPECIALTY 3

spe·cial·i·za·tion \,spe-sh(ə-)lə-'zā-shən\ *noun* (1843)
1 : a making or becoming specialized
2 a : structural adaptation of a body part to a particular function or of an organism for life in a particular environment **b :** a body part or an organism adapted by specialization

spe·cial·ize \'spe-shə-,līz\ *verb* **-ized; -iz·ing** (1613)
transitive verb
1 : to make particular mention of **:** PARTICULARIZE
2 : to apply or direct to a specific end or use ⟨*specialized* their study⟩
intransitive verb
1 : to concentrate one's efforts in a special activity, field, or practice
2 : to undergo specialization; *especially* **:** to change adaptively

specialized *adjective* (1853)
1 : characterized by or exhibiting biological specialization; *especially* **:** highly differentiated especially in a particular direction or for a particular end
2 : designed or fitted for one particular purpose or occupation ⟨*specialized* personnel⟩

spe·cial·ly \'spe-sh(ə-)lē\ *adverb* (14th century)
1 : in a special manner ⟨treated her friends *specially*⟩
2 a : for a special purpose ⟨dresses made *specially* for the occasion⟩ ⟨wines *specially* selected to match each course⟩ **b :** in particular **:** SPECIFICALLY ⟨made *specially* for you⟩
3 : ESPECIALLY 2 ⟨makes a *specially* fine curry⟩ ⟨was *specially* pleased with the gift⟩

special pleading *noun* (1684)
1 : the allegation of special or new matter to offset the effect of matter pleaded by the opposite side and admitted, as distinguished from a direct denial of the matter pleaded
2 : misleading argument that presents one point or phase as if it covered the entire question at issue

special relativity *noun* (1941)
: RELATIVITY 3a

special theory of relativity (1920)
: RELATIVITY 3a

spe·cial·ty \'spe-shəl-tē\ *noun, plural* **-ties** [Middle English *specialte,* from Middle French *especialté,* from Late Latin *specialitat-, specialitas,* from Latin *specialis* special] (15th century)
1 : a distinctive mark or quality
2 a : a special object or class of objects: as (1) **:** a legal agreement embodied in a sealed instrument (2) **:** a product of a special kind or of special excellence ⟨fried chicken is my *specialty*⟩ **b :** the state of being special, distinctive, or peculiar
3 : something in which one specializes

spe·ci·a·tion \,spē-shē-'ā-shən, -sē-\ *noun* (1906)
: the process of biological species formation
— spe·ci·ate \'spē-shē-,āt, -sē-\ *intransitive verb*
— spe·ci·a·tion·al \,spē-shē-'ā-shnəl, -sē-, -shə-nᵊl\ *adjective*

¹spe·cie \'spē-shē, -sē\ *noun* [from *in specie,* from Latin, in kind] (1617)
: money in coin
— in specie : in the same or like form or kind ⟨ready to return insult *in specie*⟩; *also* **:** in coin

²specie *noun* [back-formation from *species* (taken as a plural)] (1711)
nonstandard **:** SPECIES

¹spe·cies \'spē-(,)shēz, -(,)sēz\ *noun, plural* **species** [Middle English, from Latin, appearance, kind, species, from *specere* to look — more at SPY] (14th century)
1 a : KIND, SORT **b :** a class of individuals having common attributes and designated by a common name; *specifically* **:** a logical division of a genus or more comprehensive class **c :** the human race **:** human beings — often used with *the* ⟨survival of the *species* in the nuclear age⟩ **d** (1) **:** a category of biological classification ranking immediately below the genus or subgenus, comprising related organisms or populations potentially capable of interbreeding, and being designated by a binomial that consists of the name of a genus followed by a Latin or latinized uncapitalized noun or adjective agreeing grammatically with the genus name (2) **:** an individual or kind belonging to a biological species **e :** a particular kind of atomic nucleus, atom, molecule, or ion
2 : the consecrated eucharistic elements of the Roman Catholic or Eastern Orthodox Eucharist
3 a : a mental image; *also* **:** a sensible object **b :** an object of thought correlative with a natural object
word history see SPICE

²species *adjective* (1899)
: belonging to a biological species as distinguished from a horticultural variety ⟨a *species* rose⟩

spe·cies·ism \'spē-shē-,zi-zəm, -sē-\ *noun* (1973)
1 : prejudice or discrimination based on species; *especially* **:** discrimination against animals
2 : the assumption of human superiority on which speciesism is based

¹spe·cif·ic \spi-'si-fik\ *adjective* [Late Latin *specificus,* from Latin *species*] (circa 1631)
1 a : constituting or falling into a specifiable category **b :** sharing or being those properties of something that allow it to be referred to a particular category
2 a : restricted to a particular individual, situation, relation, or effect ⟨a disease *specific* to horses⟩ **b :** exerting a distinctive influence (as on a body part or a disease) ⟨*specific* antibodies⟩
3 : free from ambiguity **:** ACCURATE ⟨a *specific* statement of faith⟩
4 : of, relating to, or constituting a species and especially a biologic species
5 a : being any of various arbitrary physical constants and especially one relating a quantitative attribute to unit mass, volume, or area **b :** imposed at a fixed rate per unit (as of weight or count) ⟨*specific* import duties⟩ — compare AD VALOREM
synonym see SPECIAL, EXPLICIT
— spe·cif·i·cal·ly \-fi-k(ə-)lē\ *adverb*

²specific *noun* (1661)
1 a : something peculiarly adapted to a purpose or use **b :** a drug or remedy having a specific mitigating effect on a disease
2 a : a characteristic quality or trait **b :** DETAILS, PARTICULARS — usually used in plural ⟨haggling over the legal and financial *specifics* of independence —*Time*⟩ **c** *plural* **:** SPECIFICATION 2a

spec·i·fi·ca·tion \,spe-sə-fə-'kā-shən, ,spes-fə-\ *noun* (1633)
1 : the act or process of specifying
2 a : a detailed precise presentation of something or of a plan or proposal for something — usually used in plural **b :** a statement of legal particulars (as of charges or of contract terms); *also* **:** a single item of such statement **c :** a written description of an invention for which a patent is sought

specific epithet *noun* (1906)
: the Latin or latinized noun or adjective that follows the genus name in a taxonomic binomial

specific gravity *noun* (1666)
: the ratio of the density of a substance to the density of some substance (as pure water) taken as a standard when both densities are obtained by weighing in air

specific heat *noun* (1832)
: the heat in calories required to raise the temperature of one gram of a substance one degree Celsius

specific impulse *noun* (1947)
: the thrust produced per unit rate of consumption of the propellant that is usually expressed in pounds of thrust per pound of propellant used per second and that is a measure of the efficiency of a rocket engine

spec·i·fic·i·ty \,spe-sə-'fi-sə-tē\ *noun* (1876)
: the quality or condition of being specific: as **a :** the condition of being peculiar to a particular individual or group of organisms ⟨host *specificity* of a parasite⟩ **b :** the condition of participating in or catalyzing only one or a few chemical reactions ⟨the *specificity* of an enzyme⟩

specific performance *noun* (circa 1876)
1 : the performance of a legal contract strictly or substantially according to its terms
2 : an equitable remedy enjoining specific performance

spec·i·fy \'spe-sə-,fī\ *transitive verb* **-fied; -fy·ing** [Middle English *specifien,* from Middle French *specifier,* from Late Latin *specificare,* from *specificus*] (14th century)
1 : to name or state explicitly or in detail
2 : to include as an item in a specification
— spec·i·fi·able \,spe-sə-'fī-ə-bəl\ *adjective*
— spec·i·fi·er \'spe-sə-,fī(-ə)r\ *noun*

spec·i·men \'spes-mən, 'spe-sə-\ *noun* [Latin, from *specere* to look at, look — more at SPY] (1610)
1 a : an individual, item, or part considered typical of a group, class, or whole **b :** a portion or quantity of material for use in testing, examination, or study ⟨a urine *specimen*⟩
2 a : something that obviously belongs to a particular category but is noticed by reason of an individual distinguishing characteristic **b :** PERSON, INDIVIDUAL ⟨he's a tough *specimen*⟩
3 : a plant grown for exhibition or in the open to display its full development ⟨*specimen* trees⟩ ⟨*specimen* plantings⟩
synonym see INSTANCE

spe·ci·os·i·ty \,spē-shē-'ä-sə-tē\ *noun* (1608)
: the quality or state of being specious **:** SPECIOUSNESS

spe·cious \'spē-shəs\ *adjective* [Middle English, visually pleasing, from Latin *speciosus* beautiful, plausible, from *species*] (1513)
1 *obsolete* **:** SHOWY
2 : having deceptive attraction or allure
3 : having a false look of truth or genuineness **:** SOPHISTIC
— **spe·cious·ly** *adverb*
— **spe·cious·ness** *noun*

[1]**speck** \'spek\ *noun* [Middle English *specke,* from Old English *specca*] (before 12th century)
1 : a small discoloration or spot especially from stain or decay
2 : a very small amount **:** BIT
3 : something marked or marred with specks
— **specked** \'spekt\ *adjective*

[2]**speck** *transitive verb* (1580)
: to produce specks on or in

[1]**speck·le** \'spe-kəl\ *noun* [Middle English; akin to Old English *specca*] (15th century)
: a little speck (as of color)

[2]**speckle** *transitive verb* **speck·led; speck·ling** \-k(ə-)liŋ\ (circa 1570)
1 : to mark with speckles
2 : to be distributed in or on like speckles

speckled perch *noun* (1877)
: BLACK CRAPPIE

speckled trout *noun* (1805)
1 : BROOK TROUT
2 : SPOTTED SEA TROUT

speckle interferometry *noun* (1970)
: a technique for generating a clear composite image of a celestial object blurred by atmospheric turbulence in which a large number of short-exposure photographs are mathematically correlated by a computer

[1]**specs** \'speks\ *noun plural* [contraction of *spectacles*] (1807)
: GLASSES

[2]**specs** *noun plural* [by contraction] (1942)
: SPECIFICATIONS

spec·ta·cle \'spek-ti-kəl *also* -ˌti-kəl\ *noun* [Middle English, from Middle French, from Latin *spectaculum,* from *spectare* to watch, frequentative of *specere* to look, look at — more at SPY] (14th century)
1 a : something exhibited to view as unusual, notable, or entertaining; *especially* **:** an eye-catching or dramatic public display **b :** an object of curiosity or contempt ⟨made a *spectacle* of herself⟩
2 *plural* **:** GLASSES
3 : something (as natural markings on an animal) suggesting a pair of glasses

spec·ta·cled \-kəld\ *adjective* (1607)
1 : having or wearing spectacles
2 : having markings suggesting a pair of spectacles ⟨a *spectacled* alligator⟩

[1]**spec·tac·u·lar** \spek-'ta-kyə-lər, spək-\ *adjective* [Latin *spectaculum*] (1682)
: of, relating to, or being a spectacle **:** STRIKING, SENSATIONAL ⟨a *spectacular* display of fireworks⟩
— **spec·tac·u·lar·ly** *adverb*

[2]**spectacular** *noun* (1890)
: something that is spectacular; *especially* **:** an elaborate film, television, or theatrical production

spec·tate \'spek-ˌtāt\ *intransitive verb* **spec·tat·ed; spec·tat·ing** [back-formation from *spectator*] (1858)
: to be present as a spectator (as at a sports event)

spec·ta·tor \'spek-ˌtā-tər, spek-'\ *noun* [Latin, from *spectare* to watch] (circa 1586)
1 : one who looks on or watches
2 : a woman's pump usually having contrasting colors with a perforated design at the toe and sometimes heel
— **spectator** *adjective*
— **spec·ta·tor·i·al** \ˌspek-tə-'tōr-ē-əl, -'tȯr-\ *adjective*
— **spec·ta·tor·ship** \'spek-ˌtā-tər-ˌship, spek-'\ *noun*

spec·ter *or* **spec·tre** \'spek-tər\ *noun* [French *spectre,* from Latin *spectrum* appearance, specter, from *specere* to look, look at — more at SPY] (1605)
1 : a visible disembodied spirit **:** GHOST
2 : something that haunts or perturbs the mind **:** PHANTASM ⟨the *specter* of hunger⟩

spec·ti·no·my·cin \ˌspek-tə-nō-'mī-sᵊn\ *noun* [*spect-* (from New Latin *spectabilis,* specific epithet of *Streptomyces spectabilis*) + *actinomycin*] (1964)
: a white crystalline broad-spectrum antibiotic $C_{14}H_{24}N_2O_7$ produced by a bacterium (*Streptomyces spectabilis*) that is used clinically especially in the form of its hydrochloride to treat gonorrhea

spec·tral \'spek-trəl\ *adjective* (1815)
1 : of, relating to, or suggesting a specter **:** GHOSTLY
2 : of, relating to, or made by a spectrum
— **spec·tral·ly** \'spek-trə-lē\ *adverb*

spectral line *noun* (1866)
: one of a series of linear images of the narrow slit of a spectrograph or similar instrument corresponding to a component of the spectrum of the radiation emitted by a particular source

spectro- *combining form* [New Latin *spectrum*]
: spectrum ⟨*spectro*scope⟩

spec·tro·flu·o·rom·e·ter \'spek-(ˌ)trō-ˌflu̇-'rä-mə-tər, -flō-, -flō-\ *also* **spec·tro·flu·o·rim·e·ter** \-'ri-\ *noun* (1957)
: a device for measuring and recording fluorescence spectra
— **spec·tro·flu·o·ro·met·ric** \-ˌflu̇r-ə-'me-trik, -ˌflȯ-, -ˌflō-\ *adjective*
— **spec·tro·flu·o·rom·e·try** \-flu̇-'rä-mə-trē, -ˌflȯ-, -ˌflō-\ *noun*

spec·tro·gram \'spek-t(r)ə-ˌgram\ *noun* [International Scientific Vocabulary] (1892)
: a photograph or diagram of a spectrum

spec·tro·graph \-ˌgraf\ *noun* [International Scientific Vocabulary] (1884)
: an instrument for dispersing radiation (as electromagnetic radiation or sound waves) into a spectrum and photographing or mapping the spectrum
— **spec·tro·graph·ic** \ˌspek-t(r)ə-'gra-fik\ *adjective*
— **spec·tro·graph·i·cal·ly** \-fi-k(ə-)lē\ *adverb*
— **spec·trog·ra·phy** \spek-'trä-grə-fē\ *noun*

spec·tro·he·lio·gram \ˌspek-trō-'hē-lē-ə-ˌgram\ *noun* (1905)
: a photograph of the sun that is made by monochromatic light and shows the sun's faculae and prominences

spec·tro·he·lio·graph \-ˌgraf\ *noun* [International Scientific Vocabulary] (1892)
: an apparatus for making spectroheliograms
— **spec·tro·he·li·og·ra·phy** \-ˌhē-lē-'ä-grə-fē\ *noun*

spec·tro·he·lio·scope \-'hē-lē-ə-ˌskōp\ *noun* [International Scientific Vocabulary] (1906)
1 : SPECTROHELIOGRAPH
2 : an instrument similar to a spectroheliograph used for visual as distinguished from photographic observations

spec·trom·e·ter \spek-'trä-mə-tər\ *noun* [International Scientific Vocabulary] (1874)
1 : an instrument used for measuring wavelengths of light spectra
2 : any of various analytical instruments in which an emission (as of particles or radiation) is dispersed according to some property (as mass or energy) of the emission and the amount of dispersion is measured ⟨nuclear magnetic resonance *spectrometer*⟩
— **spec·tro·met·ric** \ˌspek-trə-'me-trik\ *adjective*
— **spec·trom·e·try** \spek-'trä-mə-trē\ *noun*

spec·tro·pho·tom·e·ter \ˌspek-trō-fə-'tä-mə-tər\ *noun* [International Scientific Vocabulary] (1881)
: a photometer for measuring the relative intensities of the light in different parts of a spectrum
— **spec·tro·pho·to·met·ric** \-trə-ˌfō-tə-'me-trik\ *also* **spec·tro·pho·to·met·ri·cal** \-tri-kəl\ *adjective*
— **spec·tro·pho·to·met·ri·cal·ly** \-tri-k(ə-)lē\ *adverb*
— **spec·tro·pho·tom·e·try** \ˌspek-(ˌ)trō-fə-'tä-mə-trē\ *noun*

spec·tro·scope \'spek-trə-ˌskōp\ *noun* [International Scientific Vocabulary] (1861)
: an instrument for forming and examining spectra especially in the visible region of the electromagnetic spectrum
— **spec·tro·scop·ic** \ˌspek-trə-'skä-pik\ *adjective*
— **spec·tro·scop·i·cal·ly** \-pi-k(ə-)lē\ *adverb*
— **spec·tros·co·pist** \spek-'träs-kə-pist\ *noun*

spec·tros·co·py \spek-'träs-kə-pē\ *noun* (1870)
1 : the production and investigation of spectra
2 : the process or technique of using a spectroscope or spectrometer

spec·trum \'spek-trəm\ *noun, plural* **spec·tra** \-trə\ *or* **spectrums** [New Latin, from Latin, appearance — more at SPECTER] (1671)
1 a : a continuum of color formed when a beam of white light is dispersed (as by passage through a prism) so that its component wavelengths are arranged in order **b :** any of various continua that resemble a spectrum in consisting of an ordered arrangement by a particular characteristic (as frequency or energy): as (1) **:** ELECTROMAGNETIC SPECTRUM (2) **:** RADIO SPECTRUM (3) **:** the range of frequencies of sound waves (4) **:** MASS SPECTRUM **c :** the representation (as a plot) of a spectrum
2 a : a continuous sequence or range ⟨a wide *spectrum* of interests⟩ **b :** kinds of organisms associated with a particular situation (as an environment) or susceptible to an agent (as an antibiotic)

spec·u·lar \'spe-kyə-lər\ *adjective* [Latin *specularis* of a mirror, from *speculum*] (1661)
: of, relating to, or having the qualities of a mirror
— **spec·u·lar·i·ty** \ˌspe-kyə-'lar-ə-tē\ *noun*
— **spec·u·lar·ly** \'spe-kyə-lər-lē\ *adverb*

spec·u·late \'spe-kyə-ˌlāt\ *verb* **-lat·ed; -lat·ing** [Latin *speculatus,* past participle of *speculari* to spy out, examine, from *specula* watchtower, from *specere* to look, look at — more at SPY] (1599)
intransitive verb
1 a : to meditate on or ponder a subject **:** REFLECT **b :** to review something idly or casually and often inconclusively
2 : to assume a business risk in hope of gain; *especially* **:** to buy or sell in expectation of profiting from market fluctuations
transitive verb
1 : to take to be true on the basis of insufficient evidence **:** THEORIZE ⟨*speculated* that a virus caused the disease⟩
2 : to be curious or doubtful about **:** WONDER ⟨*speculates* whether it will rain all vacation⟩
synonym see THINK
— **spec·u·la·tor** \-ˌlā-tər\ *noun*

spec·u·la·tion \ˌspe-kyə-'lā-shən\ *noun* (14th century)
: an act or instance of speculating: as **a :** assumption of unusual business risk in hopes of

obtaining commensurate gain **b :** a transaction involving such speculation

spec·u·la·tive \'spe-kyə-lə-tiv, -ˌlā-\ *adjective* (14th century)
1 : involving, based on, or constituting intellectual speculation; *also* **:** theoretical rather than demonstrable ⟨*speculative* knowledge⟩
2 : marked by questioning curiosity ⟨gave him a *speculative* glance⟩
3 : of, relating to, or being a financial speculation ⟨*speculative* stocks⟩ ⟨*speculative* venture⟩
— spec·u·la·tive·ly *adverb*

spec·u·lum \'spe-kyə-ləm\ *noun, plural* **-la** \-lə\ *also* **-lums** [Middle English, from Latin, mirror, from *specere*] (15th century)
1 : an instrument inserted into a body passage for inspection or medication
2 : a drawing or table showing the relative positions of all the planets (as in an astrological nativity)
3 : a patch of color on the secondaries of most ducks and some other birds

speech \'spēch\ *noun* [Middle English *speche,* from Old English *sprǣc, spǣc;* akin to Old English *sprecan* to speak — more at SPEAK] (before 12th century)
1 a : the communication or expression of thoughts in spoken words **b :** exchange of spoken words **:** CONVERSATION
2 a : something that is spoken **:** UTTERANCE **b :** a usually public discourse **:** ADDRESS
3 a : LANGUAGE, DIALECT **b :** an individual manner or style of speaking
4 : the power of expressing or communicating thoughts by speaking

speech community *noun* (1894)
: a group of people sharing characteristic patterns of vocabulary, grammar, and pronunciation

speech form *noun* (1863)
: LINGUISTIC FORM

speech·ify \'spē-chə-ˌfī\ *intransitive verb* **-ified; -ify·ing** (1723)
: to make a speech

speech·less \'spēch-ləs\ *adjective* (before 12th century)
1 : unable to speak **:** DUMB
2 : not speaking **:** SILENT
3 : not capable of being expressed in words
— speech·less·ly *adverb*
— speech·less·ness *noun*

speech·writ·er \-ˌrī-tər\ *noun* (1834)
: a person who writes speeches (as for a politician)

¹speed \'spēd\ *noun* [Middle English *spede,* from Old English *spēd;* akin to Old High German *spuot* prosperity, speed, Old English *spōwan* to succeed, Latin *spes* hope, Latin *spēti* to be in time] (before 12th century)
1 *archaic* **:** prosperity in an undertaking **:** SUCCESS
2 a : the act or state of moving swiftly **:** SWIFTNESS **b :** rate of motion: as (1) **:** VELOCITY 1, 3a (2) **:** the magnitude of a velocity irrespective of direction **c :** IMPETUS
3 : swiftness or rate of performance or action
4 a : the sensitivity of a photographic film, plate, or paper expressed numerically **b :** the light-gathering power of a lens or optical system **c :** the time during which a camera shutter is open
5 : a transmission gear in automotive vehicles or bicycles — usually used in combination ⟨a ten-*speed* bicycle⟩
6 : someone or something that appeals to one's taste
7 : METHAMPHETAMINE; *also* **:** a related stimulant drug and especially an amphetamine
synonym see HASTE
— speed·ster \'spēd-stər\ *noun*
— at speed *chiefly British* **:** FAST, RAPIDLY
— up to speed : operating at full effectiveness or potential

²speed *verb* **sped** \'sped\ *or* **speed·ed; speed·ing** (before 12th century)
intransitive verb
1 a *archaic* **:** to prosper in an undertaking **b** *archaic* **:** GET ALONG, FARE
2 a : to make haste ⟨*sped* to her bedside⟩ **b :** to go or drive at excessive or illegal speed
3 : to move, work, or take place faster **:** ACCELERATE ⟨the heart *speeds* up⟩
transitive verb
1 a *archaic* **:** to cause or help to prosper **:** AID **b :** to further the success of
2 a : to cause to move quickly **:** HASTEN **b :** to wish Godspeed to **c :** to increase the speed of **:** ACCELERATE
3 : to send out ⟨*speed* an arrow⟩
— speed·er *noun*

¹speed·ball \'spēd-ˌbȯl\ *noun* (1905)
1 *slang* **:** a dose of cocaine mixed with heroin or morphine or an amphetamine and usually taken by injection
2 : a game which resembles soccer but in which a ball caught in the air may be passed with the hands and in which a score is made by kicking or heading the ball between the goalposts or by a successful forward pass over the goal line
3 : one that is outstandingly fast

²speedball *intransitive verb* (1970)
slang **:** to take a speedball especially by injection

speed·boat \-ˌbōt\ *noun* (1911)
: a fast launch or motorboat
— speed·boat·ing \-ˌbō-tiŋ\ *noun*

speed bump *noun* (1972)
: a low raised ridge across a roadway (as in a parking lot) to limit vehicle speed

speed freak *noun* (1967)
: one who habitually misuses amphetamines and especially methamphetamine

speed limit *noun* (1902)
: the maximum or minimum speed permitted by law in a given area under specified circumstances

speedo \'spē-(ˌ)dō\ *noun, plural* **speed·os** [by shortening] (1934)
chiefly British **:** SPEEDOMETER

speed of light (1823)
: a fundamental physical constant that is the speed at which electromagnetic radiation propagates in a vacuum and that has a value fixed by international convention of 299,792,458 meters per second — symbol *c*

speed·om·e·ter \spi-'dä-mə-tər\ *noun* (1903)
1 : an instrument for indicating speed **:** TACHOMETER
2 : an instrument for indicating distance traversed as well as speed of travel; *also* **:** ODOMETER

speed–read·ing \'spēd-ˌrē-diŋ\ *noun* (1962)
: a method of reading rapidly by skimming
— speed–read *transitive verb*

speed shop *noun* (1953)
: a shop that sells custom automotive equipment especially to hot-rodders

speed skating *noun* (1885)
: the sport of racing on skates
— speed skater *noun*

speed trap *noun* (1925)
: a stretch of road policed by often concealed officers or devices (as radar) so as to catch speeders

speed·up \'spēd-ˌəp\ *noun* (1921)
1 : ACCELERATION
2 : an employer's demand for accelerated output without increased pay

speed·way \'spēd-ˌwā\ *noun* (1894)
1 : a public road on which fast driving is allowed; *specifically* **:** EXPRESSWAY
2 : a racecourse for automobiles or motorcycles
3 : a sprint race for motorcycles

speed·well \'spēd-ˌwel\ *noun* (1578)
: a perennial European herb (*Veronica officinalis*) of the snapdragon family that is naturalized in North America and has small bluish flowers in axillary racemes; *also* **:** any plant of the same genus

speedy \'spē-dē\ *adjective* **speed·i·er; -est** (14th century)
: marked by swiftness of motion or action; *also* **:** PROMPT 2
synonym see FAST
— speed·i·ly \'spē-dᵊl-ē\ *adverb*
— speed·i·ness \'spē-dē-nəs\ *noun*

speel \'spē(ə)l\ *verb* [origin unknown] (1513)
chiefly Scottish **:** CLIMB

speer *or* **speir** \'spir\ *verb* [Middle English (Scots) *speren,* from Old English *spyrian* to seek after; akin to Old English *spor* spoor] (before 12th century)
chiefly Scottish **:** ASK, INQUIRE

spe·le·ol·o·gy \ˌspē-lē-'ä-lə-jē, ˌspe-\ *noun* [Latin *speleum* cave (from Greek *spēlaion*) + International Scientific Vocabulary *-o-* + *-logy* — more at SPELUNKER] (1895)
: the scientific study or exploration of caves
— spe·le·o·log·i·cal \ˌspē-lē-ə-'lä-ji-kəl, ˌspe-\ *adjective*
— spe·le·ol·o·gist \ˌspē-lē-'ä-lə-jist, ˌspe-\ *noun*

¹spell \'spel\ *verb* **spelled** \'speld, 'spelt\; **spell·ing** [Middle English, from Middle French *espeller,* of Germanic origin; akin to Old English *spell* talk] (14th century)
transitive verb
1 : to read slowly and with difficulty — often used with *out*
2 : to find out by study **:** come to understand — often used with *out* ⟨it requires some pains to *spell* out those decorations —F. J. Mather⟩
3 a (1) **:** to name the letters of in order; *also* **:** to write or print the letters of in order (2) **:** to write or print the letters of in a particular way ⟨you can *spell* it either way⟩ ⟨I *spelled* it wrong⟩ **b :** to make up (a word) ⟨what word do these letters *spell*⟩ **c :** WRITE 1b ⟨*catnip* is *spelled* as one word⟩
4 : to add up to **:** MEAN ⟨crop failure was likely to *spell* stark famine —Stringfellow Barr⟩
intransitive verb
: to form words with letters ⟨teach children to *spell*⟩; *also* **:** to spell words in a certain way ⟨*spells* the way he speaks⟩

²spell *noun* [Middle English, talk, tale, from Old English; akin to Old High German *spel* talk, tale] (1579)
1 a : a spoken word or form of words held to have magic power **:** INCANTATION **b :** a state of enchantment
2 : a strong compelling influence or attraction

³spell \'spel\ *transitive verb* **spelled** \'speld\; **spell·ing** (circa 1623)
: to put under a spell

⁴spell *noun* [probably alteration of Middle English *spale* substitute, from Old English *spala*] (1593)
1 a *archaic* **:** a shift of workers **b :** one's turn at work
2 a : a period spent in a job or occupation **b** *chiefly Australian* **:** a period of rest from work, activity, or use
3 a : an indeterminate period of time ⟨waited a *spell* before advancing⟩; *also* **:** a continuous period of time ⟨did a *spell* in prison⟩ **b :** a stretch of a specified type of weather
4 : a period of bodily or mental distress or disorder ⟨a *spell* of coughing⟩ ⟨fainting *spells*⟩

⁵spell *verb* **spelled** \'speld\; **spell·ing** [Middle English *spelen,* from Old English *spelian;* akin to Old English *spala* substitute] (1595)
transitive verb
1 : to take the place of for a time **:** RELIEVE ⟨he and the other assistant . . . *spelled* each other —Mary McCarthy⟩
2 : REST
intransitive verb
1 : to work in turns
2 *chiefly Australian* **:** to rest from work or activity for a time

spell·bind \'spel-ˌbīnd\ *transitive verb* **-bound** \-ˌbau̇nd\; **-bind·ing** [back-formation from *spellbound*] (1808)

: to bind or hold by or as if by a spell or charm **:** FASCINATE

spell·bind·er \-ˌbīn-dər\ *noun* (1888)
: a speaker of compelling eloquence; *also* **:** one that compels attention

spell·bind·ing \-ˌbīn-diŋ\ *adjective* (circa 1934)
: holding the attention as if by a spell
— spell·bind·ing·ly \-diŋ-lē\ *adverb*

spell·bound \-ˌbau̇nd\ *adjective* (1799)
: held by or as if by a spell

spell·er \ˈspe-lər\ *noun* (15th century)
1 : a person who spells words especially in a certain way ⟨a poor *speller*⟩
2 : a book with exercises for teaching spelling

spell·ing \ˈspe-liŋ\ *noun* (15th century)
1 : the forming of words from letters according to accepted usage **:** ORTHOGRAPHY
2 a : a sequence of letters composing a word **b :** the way in which a word is spelled

spelling bee *noun* (1872)
: a spelling contest in which contestants are eliminated as soon as they misspell a word

spelling pronunciation *noun* (1901)
: a pronunciation of a word based on its spelling alone

spell out *transitive verb* (1940)
1 : to make plain ⟨*spelled out* the orders in detail⟩
2 : to write or print in letters and in full ⟨numbers are to be *spelled out*⟩

[1]**spelt** \ˈspelt\ *noun* [Middle English, from Old English, from Late Latin *spelta,* of Germanic origin; perhaps akin to Middle High German *spelte* split piece of wood, Old High German *spaltan* to split — more at SPLIT] (before 12th century)
: a wheat (*Triticum aestivum spelta*) with lax spikes and spikelets containing two light red kernels

[2]**spelt** \ˈspelt\ *chiefly British past and past participle of* SPELL

spel·ter \ˈspel-tər\ *noun* [probably alteration of Middle Dutch *speauter*] (1661)
: ZINC; *especially* **:** zinc cast in slabs for commercial use

spe·lunk·er \spi-ˈləŋ kər, ˈspē-ˌ\ *noun* [Latin *spelunca* cave, from Greek *spēlynx;* akin to Greek *spēlaion* cave] (1944)
: one who makes a hobby of exploring and studying caves

spe·lunk·ing \-kiŋ\ *noun* (1944)
: the hobby or practice of exploring caves

spence \ˈspen(t)s\ *noun* [Middle English, from Middle French *despense,* from Medieval Latin *dispensa,* from Latin, feminine of *dispensus,* past participle of *dispendere* to weigh out — more at DISPENSE] (14th century)
chiefly dialect British **:** PANTRY

[1]**spen·cer** \ˈspen(t)-sər\ *noun* [George John, 2d earl *Spencer* (died 1834) English politician] (1795)
: a short waist-length jacket

[2]**spencer** *noun* [probably from the name *Spencer*] (1840)
: a trysail abaft the foremast or mainmast

Spen·ce·ri·an \spen-ˈsir-ē-ən\ *adjective* [Platt R. *Spencer* (died 1864) American calligrapher] (1863)
: of or relating to a form of slanting handwriting

Spen·ce·ri·an·ism \spen-ˈsir-ē-ə-ˌni-zəm\ *noun* (1881)
: the synthetic philosophy of Herbert Spencer that has as its central idea the mechanistic evolution of the cosmos from relative simplicity to relative complexity

spend \ˈspend\ *verb* **spent** \ˈspent\; **spend·ing** [Middle English, from Old English & Old French; Old English *spendan,* from Latin *expendere* to expend; Old French *despendre,* from Latin *dispendere* to weigh out — more at DISPENSE] (13th century)
transitive verb
1 : to use up or pay out **:** EXPEND
2 a : EXHAUST, WEAR OUT ⟨the hurricane gradually *spent* itself⟩ **b :** to consume wastefully **:** SQUANDER ⟨the waters are not ours to *spend* —J. R. Ellis⟩
3 : to cause or permit to elapse **:** PASS ⟨*spend* the night⟩
4 : GIVE UP, SACRIFICE
intransitive verb
1 : to expend or waste wealth or strength
2 : to become expended or consumed
3 : to have an orgasm
— spend·able \ˈspen-də-bəl\ *adjective*
— spend·er *noun*

spending money *noun* (15th century)
: POCKET MONEY

spend·thrift \ˈspen(d)-ˌthrift\ *noun* (1601)
: a person who spends improvidently or wastefully
— spendthrift *adjective*

Spen·gle·ri·an \ˌshpeŋ-ˈglir-ē-ən, ˌspeŋ-, -ˈlir-\ *adjective* (1922)
: of or relating to the theory of world history developed by Oswald Spengler which holds that all major cultures undergo similar cyclical developments from birth to maturity to decay
— Spenglerian *noun*

Spen·se·ri·an stanza \spen-ˈsir-ē-ən-\ *noun* [Edmund *Spenser*] (1817)
: a stanza consisting of eight verses of iambic pentameter and an alexandrine with a rhyme scheme *ababbcbcc*

spent \ˈspent\ *adjective* [Middle English, from past participle of *spenden* to spend] (15th century)
1 a : used up **:** CONSUMED **b :** exhausted of active or required components or qualities often for a particular purpose ⟨*spent* nuclear fuel⟩
2 : drained of energy or effectiveness **:** EXHAUSTED
3 : exhausted of spawn or sperm ⟨a *spent* salmon⟩

sperm \ˈspərm\ *noun, plural* **sperm** *or* **sperms** [Middle English, from Middle French *esperme,* from Late Latin *spermat-, sperma,* from Greek, literally, seed, from *speirein* to sow; probably akin to Armenian *p'aratem* I dispense] (14th century)
1 a : the male fecundating fluid **:** SEMEN **b :** a male gamete
2 : a product (as spermaceti or oil) of the sperm whale

sperm- *or* **spermo-** *or* **sperma-** *or* **spermi-** *combining form* [Greek *sperm-, spermo-,* from *sperma*]
: seed **:** germ **:** sperm ⟨*sperma*theca⟩ ⟨*spermi*cide⟩

sper·ma·ce·ti \ˌspər-mə-ˈsē-tē, -ˈse-\ *noun* [Middle English *sperma cete,* from Medieval Latin *sperma ceti* whale sperm] (15th century)
: a waxy solid obtained from the oil of cetaceans and especially sperm whales and used in ointments, cosmetics, and candles

sper·ma·go·ni·um \ˌspər-mə-ˈgō-nē-əm\ *noun, plural* **-nia** \-nē-ə\ [New Latin] (1861)
: a flask-shaped or depressed receptacle in which spermatia are produced in some fungi and lichens

sper·ma·ry \ˈspər-mə-rē, ˈspərm-rē\ *noun, plural* **-ries** [New Latin *spermarium* from Greek *sperma*] (circa 1859)
: an organ in which male gametes are developed

spermat- *or* **spermato-** *combining form* [Middle French, from Late Latin, from Greek, from *spermat-, sperma*]
: seed **:** spermatozoon ⟨*spermat*id⟩ ⟨*spermato*cyte⟩

sper·ma·the·ca \ˌspər-mə-ˈthē-kə\ *noun* [New Latin] (1826)
: a sac for sperm storage in the female reproductive tract of various lower animals and especially insects

sper·mat·ic \(ˌ)spər-ˈma-tik\ *adjective* (15th century)
1 : relating to sperm or a spermary
2 : resembling, carrying, or full of sperm

spermatic cord *noun* (1797)
: a cord that suspends the testis within the scrotum and contains the vas deferens and vessels and nerves of the testis

sper·ma·tid \ˈspər-mə-təd\ *noun* (1889)
: one of the haploid cells that are formed by division of the secondary spermatocytes and that differentiate into spermatozoa

sper·ma·ti·um \(ˌ)spər-ˈmā-sh(ē-)əm\ *noun, plural* **-tia** \-sh(ē-)ə\ [New Latin, from Greek *spermation,* diminutive of *spermat-, sperma*] (1856)
: a nonmotile cell functioning as a male gamete in some lower plants (as red algae)
— sper·ma·tial \-sh(ē-)əl\ *adjective*

sper·mato·cyte \(ˌ)spər-ˈma-tə-ˌsīt\ *noun* (1886)
: a cell giving rise to sperm cells; *especially* **:** a cell that is derived from a spermatogonium and ultimately gives rise to four haploid spermatids

sper·mato·gen·e·sis \(ˌ)spər-ˌma-tə-ˈje-nə-səs\ *noun* [New Latin] (1881)
: the process of male gamete formation including formation of a primary spermatocyte from a spermatogonium, meiotic division of the spermatocyte, and transformation of the four resulting spermatids into spermatozoa
— sper·mato·gen·ic \-ˈje-nik\ *adjective*

sper·mato·go·ni·um \-ˈgō-nē-əm\ *noun, plural* **-nia** \-nē-ə\ [New Latin] (1861)
: a primitive male germ cell
— sper·mato·go·ni·al \-nē-əl\ *adjective*

sper·mato·phore \(ˌ)spər-ˈma-tə-ˌfōr, -ˌfȯr\ *noun* [International Scientific Vocabulary] (circa 1849)
: a capsule, packet, or mass enclosing spermatozoa extruded by the male and conveyed to the female in the insemination of various lower animals (as insects)

sper·mato·phyte \-ˌfīt\ *noun* [ultimately from New Latin *spermat-* + Greek *phyton* plant — more at PHYT-] (1897)
: any of a group (Spermatophyta) of higher plants comprising those that produce seeds and including the gymnosperms and angiosperms
— sper·mato·phyt·ic \-ˌma-tə-ˈfi-tik\ *adjective*

sper·ma·to·zo·an \(ˌ)spər-ˌma-tə-ˈzō-ən, ˌspər-mə-tə-\ *noun* (circa 1900)
: SPERMATOZOON
— spermatozoan *adjective*

sper·ma·to·zo·id \-ˈzō-əd\ *noun* [International Scientific Vocabulary, from New Latin *spermatozoon*] (1857)
: a male gamete of a plant motile by anterior cilia and usually produced in an antheridium

sper·ma·to·zo·on \-ˈzō-ˌän, -ˈzō-ən\ *noun, plural* **-zoa** \-ˈzō-ə\ [New Latin] (circa 1839)
1 : a motile male gamete of an animal usually with rounded or elongate head and a long posterior flagellum
2 : SPERMATOZOID
— sper·ma·to·zo·al \-ˈzō-əl\ *adjective*

sperm cell *noun* (1851)
: a male gamete **:** a male germ cell

sper·mi·cide \ˈspər-mə-ˌsīd\ *noun* (1929)
: a preparation or substance (as nonoxynol-9) used to kill sperm
— sper·mi·cid·al \ˌspər-mə-ˈsī-dəl\ *adjective*

sper·mio·gen·e·sis \ˌspər-mē-ō-ˈje-nə-səs\ *noun* [New Latin, from *spermium* spermatozoon + *-o-* + Latin *genesis*] (1916)
: SPERMATOGENESIS; *specifically* **:** transformation of a spermatid into a spermatozoon

sperm nucleus *noun* (1887)

\ə\ **abut** \ᵊ\ **kitten** \ər\ **further** \a\ **ash** \ā\ **ace** \ä\ **mop, mar** \au̇\ **out** \ch\ **chin** \e\ **bet** \ē\ **easy** \g\ **go** \i\ **hit** \ī\ **ice** \j\ **job** \ŋ\ **sing** \ō\ **go** \ȯ\ **law** \ȯi\ **boy** \th\ **thin** \t͟h\ **the** \ü\ **loot** \u̇\ **foot** \y\ **yet** \zh\ **vision** *see also* Guide to Pronunciation

: either of two nuclei that derive from the generative nucleus of a pollen grain and function in the fertilization of a seed plant

sperm oil *noun* (1839)
: a pale yellow oil from the sperm whale

sper·mo·phile \ˈspər-mə-ˌfīl\ *noun* [ultimately from Greek *sperma* seed + *philos* loving] (1824)
: GROUND SQUIRREL

sperm whale \ˈspərm-\ *noun* [short for *spermaceti whale*] (1834)
: a large toothed whale (*Physeter macrocephalus* synonym *P. catodon*) with a large closed cavity in the head containing a fluid mixture of spermaceti and oil

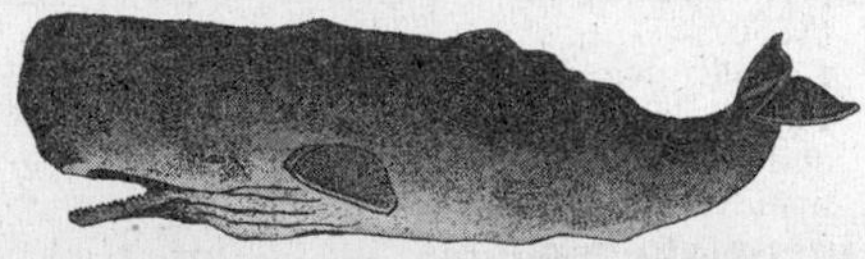

sperm whale

-spermy *noun combining form* [Greek *sperma* seed, sperm]
: state of exhibiting or resulting from (such) a fertilization ⟨agamo*spermy*⟩

sper·ry·lite \ˈsper-i-ˌlīt\ *noun* [Francis L. *Sperry,* 19th century Canadian chemist + English *-lite*] (1889)
: a mineral consisting of an arsenide of platinum that is found near Sudbury, Ontario, and occurs in grains and minute isometric crystals of a bluish white color

spes·sar·tite \ˈspe-sər-ˌtīt\ *or* **spes·sar·tine** \-ˌtēn\ *noun* [French, from *Spessart* mountain range, Germany] (1868)
: a manganese aluminum garnet usually containing other elements (as iron and magnesium) in minor amounts

[1]spew \ˈspyü\ *verb* [Middle English, from Old English *spīwan;* akin to Old High German *spīwan* to spit, Latin *spuere,* Greek *ptyein*] (before 12th century)
intransitive verb
1 : VOMIT
2 : to come forth in a flood or gush
3 : to ooze out as if under pressure **:** EXUDE
transitive verb
1 : VOMIT
2 : to send or cast forth with vigor or violence or in great quantity — often used with *out*
— **spew·er** *noun*

[2]spew *noun* (15th century)
1 : matter that is vomited **:** VOMIT
2 : material that exudes or is extruded

sphag·nous \ˈsfag-nəs\ *adjective* (circa 1828)
: of, relating to, or abounding in sphagnum

sphag·num \ˈsfag-nəm\ *noun* [New Latin, from Latin *sphagnos,* a moss, from Greek] (1741)
1 : any of an order (Sphagnales, containing a single genus *Sphagnum*) of atypical mosses that grow only in wet acid areas where their remains become compacted with other plant debris to form peat
2 : a mass of sphagnum plants

sphal·er·ite \ˈsfa-lə-ˌrīt\ *noun* [German *Sphalerit,* from Greek *sphaleros* deceitful, from *sphallein* to cause to fall; from its often being mistaken for galena — more at SPILL] (circa 1868)
: a widely distributed ore of zinc composed essentially of zinc sulfide

S phase *noun* [*s*ynthesis] (1945)
: the period in the cell cycle during which DNA replication takes place — compare G_1 PHASE, G_2 PHASE, M PHASE

sphene \ˈsfēn\ *noun* [French *sphène,* from Greek *sphēn* wedge] (1815)
: a mineral that is a silicate of calcium and titanium and often contains other elements

sphen·odon \ˈsfē-nə-ˌdän, ˈsfe-\ *noun* [New Latin, genus name, from Greek *sphēn* wedge + *odōn, odous* tooth — more at TOOTH] (1878)
: TUATARA
— **sphen·odont** \-ˌdänt\ *adjective*

[1]sphe·noid \ˈsfē-ˌnȯid\ *or* **sphe·noi·dal** \sfi-ˈnȯi-d^{ə}l\ *adjective* [New Latin *sphenoides,* from Greek *sphēnoeidēs* wedge-shaped, from *sphēn* wedge] (1732)
1 : of, relating to, or being a winged compound bone of the base of the cranium
2 *usually sphenoidal* **:** having a wedged shape

[2]sphenoid *noun* (1828)
: a sphenoid bone

sphe·nop·sid \sfi-ˈnäp-səd\ *noun* [ultimately from Greek *sphēn* wedge + New Latin *-opsis*] (1957)
: any of a class (Sphenopsida or Sphenophyta) of the tracheophytes characterized by jointed ribbed stems, small leaves usually in whorls at distinct stem nodes, and sporangia in sporangiophores and made up of the equisetums and extinct related forms

spher- *or* **sphero-** *also* **sphaer-** *or* **sphaero-** *combining form* [Latin *sphaer-,* from Greek *sphair-, sphairo-,* from *sphaira* sphere]
: sphere ⟨*spher*ule⟩ ⟨*sphero*meter⟩

spher·al \ˈsfir-əl\ *adjective* (1545)
1 : SPHERICAL
2 : of or relating to the spheres of ancient astronomy

[1]sphere \ˈsfir\ *noun* [Middle English *spere* globe, celestial sphere, from Middle French *espere,* from Latin *sphaera,* from Greek *sphaira,* literally, ball; perhaps akin to Greek *spairein* to quiver — more at SPURN] (14th century)
1 a (1) **:** the apparent surface of the heavens of which half forms the dome of the visible sky (2) **:** any of the concentric and eccentric revolving spherical transparent shells in which according to ancient astronomy stars, sun, planets, and moon are set **b :** a globe depicting such a sphere; *broadly* **:** GLOBE a
2 a : a globular body **:** BALL **b :** PLANET, STAR **c** (1) **:** a solid that is bounded by a surface consisting of all points at a given distance from a point constituting its center — see VOLUME table (2) **:** the bounding surface of a sphere
3 : natural, normal, or proper place; *especially* **:** social order or rank
4 a *obsolete* **:** ORBIT **b :** an area or range over or within which someone or something acts, exists, or has influence or significance
— **spher·ic** \ˈsfir-ik, ˈsfer-\ *adjective, archaic*
— **sphe·ric·i·ty** \sfir-ˈi-sə-tē\ *noun*

[2]sphere *transitive verb* **sphered; spher·ing** (1607)
1 : to place in a sphere or among the spheres **:** ENSPHERE
2 : to form into a sphere

sphere of influence (1885)
: a territorial area within which the political influence or the interests of one nation are held to be more or less paramount

spher·i·cal \ˈsfir-i-kəl, ˈsfer-\ *adjective* (15th century)
1 : having the form of a sphere or of one of its segments
2 : relating to or dealing with a sphere or its properties
— **spher·i·cal·ly** \-k(ə-)lē\ *adverb*

spherical aberration *noun* (1868)
: aberration that is caused by the spherical form of a lens or mirror and that gives different foci for central and marginal rays

spherical angle *noun* (1678)
: the angle between two intersecting arcs of great circles of a sphere measured by the plane angle formed by the tangents to the arcs at the point of intersection

spherical coordinate *noun* (circa 1864)
: one of three coordinates that are used to locate a point in space and that comprise the radius of the sphere on which the point lies in a system of concentric spheres, the angle formed by the point, the center, and a given axis of the sphere, and the angle between the plane of the first angle and a reference plane through the given axis of the sphere

spherical geometry *noun* (1728)
: the geometry of figures on a sphere

spherical polygon *noun* (circa 1825)
: a figure analogous to a plane polygon that is formed on a sphere by arcs of great circles

spherical triangle *noun* (1585)
: a spherical polygon of three sides

spherical trigonometry *noun* (1728)
: trigonometry applied to spherical triangles and polygons

spher·oid \ˈsfir-ˌȯid, ˈsfer-\ *noun* (1570)
: a figure resembling a sphere; *also* **:** an object of approximately spherical shape
— **sphe·roi·dal** \sfi-ˈrȯi-d^{ə}l\ *also* **spher·oid** *adjective*
— **sphe·roi·dal·ly** \-d^{ə}l-ē\ *adverb*

sphe·rom·e·ter \sfir-ˈä-mə-tər\ *noun* [International Scientific Vocabulary] (circa 1828)
: an instrument for measuring the curvature of a surface

sphe·ro·plast \ˈsfir-ə-ˌplast, ˈsfer-\ *noun* (circa 1920)
: a bacterium or yeast cell that is modified by nutritional or environmental factors or by artificial means (as the use of a lysozyme) and that is characterized by partial loss of the cell wall and by increased osmotic sensitivity

spher·ule \ˈsfir-ü(ə)l, ˈsfer-, -ˌyü(ə)l\ *noun* (1665)
: a little sphere or spherical body

spher·u·lite \ˈsfir-yə-ˌlīt, ˈsfer-, -ə-ˌlīt\ *noun* (1823)
: a usually spherical crystalline body of radiating crystal fibers often found in vitreous volcanic rocks
— **spher·u·lit·ic** \ˌsfir-yə-ˈli-tik, ˌsfer-, -ə-ˌli-\ *adjective*

sphery \ˈsfir-ē\ *adjective* (1590)
1 : of, relating to, or suggestive of the celestial bodies
2 : ROUND, SPHERICAL

sphinc·ter \ˈsfiŋ(k)-tər\ *noun* [Late Latin, from Greek *sphinktēr,* literally, band, from *sphingein* to bind tight] (1578)
: an annular muscle surrounding and able to contract or close a bodily opening
— **sphinc·ter·ic** \sfiŋ(k)-ˈter-ik\ *adjective*

sphin·gid \ˈsfin-jəd\ *noun* [ultimately from Greek *sphing-, sphinx* sphinx] (circa 1909)
: HAWKMOTH

sphin·go·sine \ˈsfiŋ-gə-ˌsēn\ *noun* [Greek *sphingos* (genitive of *sphinx*) + English [2]*-ine;* from riddles it posed to its first investigators] (1884)
: a long-chain unsaturated amino alcohol $C_{18}H_{37}O_2N$ found especially in nervous tissue and cell membranes

sphinx \ˈsfiŋ(k)s\ *noun, plural* **sphinx·es** *or* **sphin·ges** \ˈsfin-ˌjēz\ [Latin, from Greek *Sphinx, Sphix*]
1 a *capitalized* **:** a winged female monster in Greek mythology having a woman's head and a lion's body and noted for killing anyone unable to answer its riddle **b :** an enigmatic or mysterious person
2 : an ancient Egyptian image in the form of a recumbent lion having a man's head, a ram's head, or a hawk's head
3 : HAWKMOTH
— **sphinx·like** \-ˌlīk\ *adjective*

sphyg·mo·graph \ˈsfig-mə-ˌgraf\ *noun* [Greek *sphygmos* pulse + International Scientific Vocabulary *-graph*] (circa 1859)
: an instrument that records graphically the movements or character of the pulse

sphyg·mo·ma·nom·e·ter \ˌsfig-mō-mə-ˈnä-mə-tər\ *noun* [Greek *sphygmos* pulse (from

sphyzein to throb) + International Scientific Vocabulary *manometer*] (circa 1891)
: an instrument for measuring blood pressure and especially arterial blood pressure
— sphyg·mo·ma·nom·e·try \-mə-'nä-mə-trē\ *noun*

spi·ca \'spī-kə\ *noun, plural* **spi·cae** \-,kē\ *or* **spicas** [Latin, ear of grain — more at SPINE] (circa 1731)
: a bandage that is applied in successive V-shaped crossings and is used to immobilize a limb especially at a joint

Spi·ca \'spī-kə\ *noun* [Latin, literally, ear of grain]
: a star of the first magnitude in the constellation Virgo

spi·cate \'spī-,kāt\ *adjective* [Latin *spicatus,* past participle of *spicare* to arrange in the shape of heads of grain, from *spica*] (1668)
: arranged in the form of a spike ⟨a *spicate* inflorescence⟩

1spic·ca·to \spi-'kä-(,)tō\ *adjective* [Italian, past participle of *spiccare* to detach, pick off] (circa 1724)
: performed with a slight lifting of the bow after each note — used as a direction in music

2spiccato *noun, plural* **-tos** (circa 1903)
: a spiccato technique, performance, or passage

1spice \'spīs\ *noun* [Middle English, from Old French *espice,* from Late Latin *species* spices, from Latin, species, from *specere* to look — more at SPY] (13th century)
1 : any of various aromatic vegetable products (as pepper or nutmeg) used to season or flavor foods
2 a *archaic* **:** a small portion, quantity, or admixture **:** DASH **b :** something that gives zest or relish ⟨variety's the very *spice* of life —William Cowper⟩
3 : a pungent or fragrant odor **:** PERFUME ◆
— spice·less \-ləs\ *adjective*

2spice *transitive verb* **spiced; spic·ing** (14th century)
1 : to season with spices
2 : to add zest or relish to ⟨cynicism *spiced* with humor —J. W. Dawson⟩ — often used with *up*

spice·bush \'spīs-,bu̇sh\ *noun* (1770)
: an aromatic shrub (*Lindera benzoin*) of the laurel family that bears dense clusters of small yellow flowers followed by scarlet or yellow berries

spic·ery \'spī-sə-rē, 'spīs-rē\ *noun, plural* **-er·ies** (13th century)
1 : SPICES
2 *archaic* **:** a repository of spices
3 : a spicy quality

spick *or* **spic** *or* **spik** \'spik\ *noun* [by shortening & alteration from *spiggoty,* of unknown origin] (1916)
: SPANISH AMERICAN — usually taken to be offensive

spick–and–span *or* **spic–and–span** \,spik-ən(d)-'span, ,spik-ᵊŋ-\ *adjective* [short for *spick-and-span-new,* from obsolete English *spick* spike + English *and* + *span-new* brand-new] (1665)
1 : FRESH, BRAND-NEW
2 : spotlessly clean

spic·ule \'spi-(,)kyü(ə)l\ *noun* [New Latin *spicula* & Latin *spiculum;* New Latin *spicula,* alteration of Latin *spiculum* head of a spear or arrow, diminutive of *spicum, spica* ear of grain] (1785)
1 : a slender pointed usually hard body; *especially* **:** one of the minute calcareous or siliceous bodies that support the tissue of various invertebrates (as a sponge)
2 : a spikelike short-lived prominence appearing close to the chromosphere of the solar atmosphere
— spic·u·lar \'spi-kyə-lər\ *adjective*
— spic·u·la·tion \,spi-kyə-'lā-shən\ *noun*

spicy \'spī-sē\ *adjective* **spic·i·er; -est** (1562)
1 : having the quality, flavor, or fragrance of spice
2 : producing or abounding in spices
3 : LIVELY, SPIRITED
4 : PIQUANT, RACY; *especially* **:** somewhat scandalous or salacious ⟨*spicy* gossip⟩
— spic·i·ly \-sə-lē\ *adverb*
— spic·i·ness \-sē-nəs\ *noun*

spi·der \'spī-dər\ *noun* [Middle English *spyder,* alteration of *spithre;* akin to Old English *spinnan* to spin] (15th century)
1 : any of an order (Araneae synonym Araneida) of arachnids having the abdomen usually unsegmented and constricted at the base, chelicerae modified into poison fangs, and two or more pairs of abdominal spinnerets for spinning threads of silk used in making cocoons for their eggs, nests for themselves, or webs to catch prey
2 : a cast-iron frying pan originally made with short feet to stand among coals on the hearth
3 : any of various devices consisting of a frame or skeleton with radiating arms or members
— spi·der·ish \-də-rish\ *adjective*
— spi·der·like \-dər-,līk\ *adjective*

spider crab *noun* (circa 1710)
: any of a family (Majidae) of crabs with extremely long legs and nearly triangular bodies which they often cover with kelp

spider mite *noun* (1870)
: any of several small web-spinning mites (family Tetranychidae) that attack forage and crop plants — called also *red spider*

spider monkey *noun* (1764)
: any of a genus (*Ateles*) of New World monkeys with long slender limbs, the thumb absent or rudimentary, and a very long prehensile tail

spider plant *noun* (1944)
: a widely grown houseplant (*Chlorophytum comosum* variety *variegatum*) of the lily family having long narrow green leaves usually striped with white or ivory and producing white flowers and tufts of plantlets on long hanging stems

spi·der·web \'spī-dər-,web\ *noun* (circa 1649)
1 : the network of silken thread spun by most spiders and used as a resting place and as a trap for small prey
2 : something that resembles or suggests a spiderweb

spi·der·wort \-,wərt, -,wȯrt\ *noun* (1629)
: any of a genus (*Tradescantia* of the family Commelinaceae, the spiderwort family) of American monocotyledonous plants with ephemeral often blue or violet flowers

spi·dery \'spī-də-rē\ *adjective* (1837)
1 a : resembling a spider in form or manner **b :** resembling a spiderweb; *especially* **:** composed of fine threads or lines in a weblike arrangement ⟨*spidery* lace⟩
2 : infested with spiders

spie·gel·ei·sen \'spē-gə-,lī-zᵊn\ *also* **spie·gel** \'spē-gəl\ *noun* [German *Spiegeleisen,* from *Spiegel* mirror + *Eisen* iron] (1868)
: a composition of iron that contains 15 to 30 percent manganese and 4.5 to 6.5 percent carbon

1spiel \'spē(ə)l\ *verb* [German *spielen* to play, from Old High German *spilōn;* akin to Old English *spilian* to revel] (1870)
intransitive verb
1 : to play music
2 : to talk volubly or extravagantly
transitive verb
: to utter, express, or describe volubly or extravagantly
— spiel·er \'spē-lər\ *noun*

2spiel *noun* (1896)
: a voluble line of often extravagant talk **:** PITCH

1spi·er \'spī(-ə)r\ *noun* (13th century)
: SPY

2spier \'spir\ *chiefly Scottish variant of* SPEER

spiff \'spif\ *verb* [English dialect spiff *dandified*] (1877)
: SPRUCE — usually used with *up* ⟨*spiff* up your wardrobe⟩
— spiffed–up \'spift-'əp\ *adjective*

spiffy \'spi-fē\ *adjective* **spiff·i·er; -est** (1853)
: fine looking **:** SMART ⟨a *spiffy* sports jacket⟩
— spiff·i·ly \-fə-lē\ *adverb*
— spiff·i·ness \-fē-nəs\ *noun*

spig·ot \'spi-gət, -kət\ *noun* [Middle English] (14th century)
1 a : SPILE 2 **b :** the plug of a faucet or cock **c :** FAUCET
2 : something resembling a spigot especially in regulating availability or flow (as of money)

1spike \'spīk\ *noun* [Middle English, probably from Middle Dutch; akin to Middle High German *spīcher* spike] (13th century)
1 : a very large nail
2 a : one of a row of pointed irons placed (as on the top of a wall) to prevent passage **b** (1) **:** one of several metal projections set in the sole and heel of a shoe to improve traction (2) *plural* **:** a pair of shoes having spikes attached to the soles or soles and heels **c :** SPINDLE 1e
3 : something resembling a spike: as **a :** a young mackerel not over six inches (15.2 centimeters) long **b :** an unbranched antler of a young deer
4 *plural* **:** shoes with spike heels
5 : the act or an instance of spiking (as in volleyball)
6 a : a pointed element in a graph or tracing **b :** an unusually high and sharply defined maximum (as of amplitude in a wave train)
7 *slang* **:** HYPODERMIC NEEDLE
8 : a momentary sharp increase and fall in the record of an electric potential; *also* **:** ACTION POTENTIAL
9 : an abrupt sharp increase in prices or rates
— spike·like \-,līk\ *adjective*

2spike *transitive verb* **spiked; spik·ing** (1624)
1 : to fasten or furnish with spikes

◇ WORD HISTORY

spice The Latin word *species* originally meant "appearance," but it developed a distinctive range of secondary meanings. In philosophers' parlance *species* referred to a division of a larger class, called a *genus:* a *species* was a particular "appearance" that a member of a *genus* might take, and hence one kind or sort of something. These meanings are the basis of English *species* and *genus* in modern biological classification. They did not arise within Latin, however, but rather are exact translations of the Greek philosophical use of *eidos* "kind, sort" (literally, "form, appearance") and *genos* "class" (literally, "race, kin"). Imperial legal Latin of the 2d century A.D. reflects another Greek philosophical distinction, between "form, appearance" (*eidos*) and "matter" (*hylē*): again translating *eidos* as *species,* jurists used the word to refer to agricultural products, the "form" taken by grain, oil, and wine after their conversion from "matter," that is, unprocessed plants. On this basis Late Latin *species* began to denote any taxed commodity, especially those imported from the East, such as drugs and aromatic seasonings. *Species* in the sense "seasoning" developed via the spoken Latin of Gaul into Old French *espice,* borrowed into Middle English as *spice* by the 13th century.

2 a : to disable (a muzzle-loading cannon) temporarily by driving a spike into the vent **b :** to suppress or block completely ⟨*spiked* the rumor⟩
3 a : to pierce or impale with or on a spike **b :** to reject (as a story) for publication or broadcast for editorial reasons
4 a : to add an alcoholic beverage to (a drink) **b :** to add something highly reactive (as a radioactive tracer) to **c :** to add vitality, zest, or spice to **:** LIVEN ⟨*spiked* the speech with humor⟩ ⟨*spike* the broth with peppers⟩
5 : to drive (as a volleyball) sharply downward with a hard blow; *also* **:** to throw down sharply ⟨*spiked* the ball in the end zone⟩
6 : to undergo a sudden sharp increase in (temperature or fever) ⟨the patient *spiked* a fever of 103°⟩
— spik·er *noun*

³spike *noun* [Middle English *spik,* from Latin *spica* — more at SPINE] (14th century)
1 : an ear of grain
2 : an elongated inflorescence similar to a raceme but having the flowers sessile on the main axis — see INFLORESCENCE illustration

spiked \ˈspīkt, ˈspī-kəd\ *adjective* (1601)
1 : having an inflorescence that is a spike
2 : having a sharp projecting point

spike heel *noun* (1926)
: a very high tapering heel used on women's shoes

spike lavender \ˈspīk-\ *noun* [alteration of English dialect *spick* lavender] (1607)
: a European mint (*Lavandula latifolia*) related to true lavender

spike·let \ˈspī-klət\ *noun* (1851)
: a small or secondary spike; *specifically* **:** one of the small few-flowered bracted spikes that make up the compound inflorescence of a grass or sedge

spike·nard \ˈspīk-ˌnärd\ *noun* [Middle English, from Middle French or Medieval Latin; Middle French *spicanarde,* from Medieval Latin *spica nardi,* literally, spike of nard] (14th century)
1 a : a fragrant ointment of the ancients **b :** a Himalayan aromatic plant (*Nardostachys jatamansi*) of the valerian family from which spikenard is believed to have been derived
2 : an American herb (*Aralia racemosa*) of the ginseng family with aromatic root and panicled umbels

spike–tooth harrow \ˈspīk-ˈtüth-\ *noun* (1926)
: a harrow with straight steel teeth set in horizontal bars

spiky *also* **spikey** \ˈspī-kē\ *adjective* **spik·i·er; -est** (1720)
1 : of, relating to, or characterized by spikes
2 : sharply irritating or acerbic (as in temper or manner)
3 [from the alleged harshness of such views] *British* **:** strongly favoring Anglo-Catholic teaching or practice
— spik·i·ly \-kə-lē\ *adverb*
— spik·i·ness \-kē-nəs\ *noun*

¹spile \ˈspī(ə)l\ *noun* [probably from Dutch *spijl* stake] (1513)
1 : ¹PILE 1
2 : a small plug used to stop the vent of a cask **:** BUNG
3 : a spout inserted in a tree to draw off sap

²spile *transitive verb* **spiled; spil·ing** (1691)
1 : to plug with a spile
2 : to supply with a spile

¹spill \ˈspil\ *verb* **spilled** \ˈspild, ˈspilt\ *also* **spilt** \ˈspilt\; **spill·ing** [Middle English, from Old English *spillan;* akin to Old English *spildan* to destroy and perhaps to Latin *spolium* animal skin, Greek *sphallein* to cause to fall] (before 12th century)
transitive verb
1 a *archaic* **:** KILL, DESTROY **b :** to cause (blood) to flow
2 : to cause or allow especially accidentally or unintentionally to fall, flow, or run out so as to be lost or wasted
3 a : to relieve (a sail) from the pressure of the wind so as to reef or furl it **b :** to relieve the pressure of (wind) on a sail by coming about or by adjusting the sail with lines
4 : to throw off or out ⟨a horse *spilled* him⟩
5 : to let out **:** DIVULGE ⟨*spill* a secret⟩
intransitive verb
1 a : to flow, run, or fall out, over, or off and become wasted, scattered, or lost **b :** to cause or allow something to spill
2 : to spread profusely or beyond bounds ⟨crowds *spilled* into the streets⟩
3 : to fall from one's place (as on a horse)
— spill·able \ˈspi-lə-bəl\ *adjective*
— spill·er *noun*
— spill the beans : to divulge secret or hidden information

²spill *noun* (circa 1845)
1 : the act or an instance of spilling; *especially* **:** a fall from a horse or vehicle or an erect position
2 : something spilled

³spill *noun* [Middle English *spille*] (14th century)
1 : a wooden splinter
2 : a slender piece: as **a :** a metallic rod or pin **b** (1) **:** a small roll or twist of paper or slip of wood for lighting a fire (2) **:** a roll or cone of paper serving as a container **c :** a peg or pin for plugging a hole **:** SPILE

spill·age \ˈspi-lij\ *noun* (1924)
1 : the act or process of spilling
2 : the quantity that spills **:** material lost or scattered by spilling

spil·li·kin \ˈspi-li-kən\ *noun* [probably alteration of obsolete Dutch *spelleken* small peg] (1734)
1 : JACKSTRAW 2
2 *plural* **:** JACKSTRAW 1

spill·over \ˈspil-ˌō-vər\ *noun, often attributive* (1920)
1 : the act or an instance of spilling over
2 : a quantity that spills over
3 : an extension of something especially when due to an excess

spill·way \-ˌwā\ *noun* (1889)
: a passage for surplus water to run over or around an obstruction (as a dam)

spilth \ˈspilth\ *noun* (1607)
1 : the act or an instance of spilling
2 a : something spilled **b :** REFUSE, RUBBISH

¹spin \ˈspin\ *verb* **spun** \ˈspən\; **spin·ning** [Middle English *spinnen,* from Old English *spinnan;* akin to Old High German *spinnan* to spin and perhaps to Lithuanian *spęsti* to set (a trap)] (before 12th century)
intransitive verb
1 : to draw out and twist fiber into yarn or thread
2 : to form a thread by extruding a viscous rapidly hardening fluid — used especially of a spider or insect
3 a : to revolve rapidly **:** GYRATE **b :** to feel as if in a whirl **:** REEL ⟨my head is *spinning*⟩
4 : to move swiftly especially on or as if on wheels or in a vehicle
5 : to fish with spinning bait **:** TROLL
6 a *of an airplane* **:** to fall in a spin **b :** to plunge helplessly and out of control
transitive verb
1 a : to draw out and twist into yarns or threads **b :** to produce by drawing out and twisting a fibrous material
2 : to form (as a web or cocoon) by spinning
3 a : to stretch out or extend (as a story) lengthily **:** PROTRACT — usually used with *out* **b :** to evolve, express, or fabricate by processes of mind or imagination ⟨*spin* a yarn⟩
4 : to cause to whirl **:** impart spin to ⟨*spin* a top⟩
5 : to shape into threadlike form in manufacture; *also* **:** to manufacture by a whirling process

²spin *noun* (1831)
1 a : the act of spinning or twirling something; *also* **:** an instance of spinning or of spinning something ⟨doing axels and *spins*⟩ ⟨an assortment of *spins* and lobs⟩ **b :** the whirling motion imparted (as to a ball or top) by spinning **c :** an excursion or ride in a vehicle especially on wheels ⟨go for a *spin*⟩
2 a : an aerial maneuver or flight condition consisting of a combination of roll and yaw with the longitudinal axis of the airplane inclined steeply downward **b :** a plunging descent or downward spiral **c :** a state of mental confusion ⟨in a *spin*⟩
3 a : a quantum characteristic of an elementary particle that is visualized as the rotation of the particle on its axis and that is responsible for measurable angular momentum and magnetic moment **b :** the angular momentum associated with such rotation whose magnitude is quantized and which may assume either of two possible directions; *also* **:** the angular momentum of a system of such particles derived from the spins and orbital motions of the particles
4 : a special point of view, emphasis, or interpretation ⟨put the most favorable *spin* on the findings⟩
— spin·less \ˈspin-ləs\ *adjective*

spi·na bi·fi·da \ˌspī-nə-ˈbi-fə-də\ *noun* [New Latin, literally, spine split in two] (1720)
: a congenital cleft of the vertebral column with hernial protrusion of the meninges and sometimes the spinal cord

spin·ach \ˈspi-nich\ *noun* [Middle English *spynache,* from Middle French *espinache, espinage,* from Old Spanish *espinaca,* from Arabic *isfānākh,* from Persian] (15th century)
1 : an Asian herb (*Spinacia oleracea*) of the goosefoot family cultivated for its edible leaves which form in a dense basal rosette
2 a : something unwanted, pretentious, or spurious **b :** an untidy overgrowth
— spin·ach·like \-nich-ˌlīk\ *adjective*
— spin·achy \-ni-chē\ *adjective*

¹spi·nal \ˈspī-nᵊl\ *adjective* (1578)
1 : of, relating to, or situated near the backbone
2 a : of, relating to, or affecting the spinal cord ⟨*spinal* reflexes⟩ **b :** having the spinal cord functionally isolated (as by surgical section) from the brain ⟨experiments on *spinal* animals⟩
3 : of, relating to, or resembling a spine

²spinal *noun* (1944)
: a spinal anesthetic

spinal canal *noun* (1845)
: VERTEBRAL CANAL

spinal column *noun* (1836)
: the axial skeleton of the trunk and tail of a vertebrate consisting of an articulated series of vertebrae and protecting the spinal cord — called also *backbone*

spinal cord *noun* (1836)
: the cord of nervous tissue that extends from the brain lengthwise along the back in the vertebral canal, gives off the pairs of spinal nerves, carries impulses to and from the brain, and serves as a center for initiating and coordinating many reflex acts — see BRAIN illustration

spinal ganglion *noun* (circa 1860)
: a ganglion on the dorsal root of each spinal nerve that is one of a series of ganglia lodging cell bodies of sensory neurons

spi·nal·ly \ˈspī-nᵊl-ē\ *adverb* (1885)
: with respect to or along the spine

spinal nerve *noun* (circa 1793)
: any of the paired nerves which leave the spinal cord of a craniate vertebrate, supply muscles of the trunk and limbs, and connect with the nerves of the sympathetic nervous system, which arise by a short motor ventral root and a short sensory dorsal root, and of which there are 31 pairs in humans classified according to the part of the spinal cord from which they

arise into 8 cervical pairs, 12 thoracic pairs, 5 lumbar pairs, 5 sacral pairs, and one coccygeal pair

[1]**spin·dle** \'spin-dᵊl\ *noun* [Middle English *spindel,* from Old English *spinel;* akin to Old English *spinnan* to spin] (12th century)
1 a : a round stick with tapered ends used to form and twist the yarn in hand spinning **b :** the long slender pin by which the thread is twisted in a spinning wheel **c :** any of various rods or pins holding a bobbin in a textile machine (as a spinning frame) **d :** the pin in a loom shuttle **e :** a device usually consisting of a long upright pin in a base on which papers can be stuck for filing — called also *spindle file*
2 : something shaped like a spindle: as **a :** a spindle-shaped network of chiefly microtubular fibers along which the chromosomes are distributed during mitosis and meiosis **b :** MUSCLE SPINDLE
3 a : the bar or shaft usually of square section that carries the knobs and actuates the latch or bolt of a lock **b** (1) **:** a turned often decorative piece (as in a baluster) (2) **:** NEWEL **c** (1) **:** a revolving piece especially when thinner than a shaft (2) **:** a horizontal or vertical axle revolving on pin or pivot ends **d :** the part of an axle on which a vehicle wheel turns

[2]**spindle** *verb* **spin·dled; spin·dling** \'spin(d)-liŋ, 'spin-dᵊl-iŋ\ (1577)
intransitive verb
1 : to shoot or grow into a long slender stalk
2 : to grow to stalk or stem rather than to flower or fruit
transitive verb
1 : to impale, thrust, or perforate on the spike of a spindle file
2 : to make or equip (as a piece of furniture) with spindles
— spin·dler \'spin(d)-lər, 'spin-dᵊl-ər\ *noun*

spindle cell *noun* (1878)
: a fusiform cell (as in some tumors)

spin·dle–legged \'spin-dᵊl-,(l)e-gəd *also* -,(l)a- *or* -,(l)egd\ *adjective* (1710)
: having long slender legs

spin·dle–shanked \-,shaŋ(k)t\ *adjective* (circa 1600)
: SPINDLE-LEGGED

spindle tree *noun* (1548)
: any of various often evergreen shrubs, small trees, or vines (genus *Euonymus*) of the staff-tree family

spin·dling \'spin(d)-liŋ, -lən; 'spin-dᵊl-iŋ, -ən\ *adjective* (1750)
: SPINDLY

spin·dly \'spin(d)-lē, 'spin-dᵊl-ē\ *adjective* **spin·dli·er; -est** (1651)
1 : of a disproportionately tall or long and thin appearance that often suggests physical weakness ⟨*spindly* legs⟩
2 : frail or flimsy in appearance or structure

spin doctor *noun* (1984)
: a person (as a political aide) responsible for ensuring that others interpret an event from a particular point of view

spin·drift \'spin-,drift\ *noun* [alteration of Scots *speendrift,* from *speen* to drive before a strong wind + English *drift*] (1823)
1 : sea spray; *especially* **:** spray blown from waves during a gale
2 : fine wind-borne snow or sand

spine \'spīn\ *noun* [Middle English, thorn, spinal column, from Latin *spina;* perhaps akin to Latin *spica* ear of grain] (15th century)
1 a : SPINAL COLUMN **b :** something resembling a spinal column or constituting a central axis or chief support **c :** the part of a book to which the pages are attached and on the cover of which usually appear the title and author's and publisher's names
2 : a stiff pointed plant process; *especially* **:** one that is a modified leaf or leaf part
3 : a sharp rigid process on an animal: as **a :** SPICULE **b :** a stiff unsegmented fin ray of a fish **c :** a pointed prominence on a bone
— spined \'spīnd\ *adjective*
— spine·like \-,līk\ *adjective*

spine–chill·ing \-,chi-liŋ\ *adjective* (1946)
: alarmingly or eerily frightening

spi·nel *or* **spi·nelle** \spə-'nel\ *noun* [Italian *spinella,* diminutive of *spina* thorn, from Latin] (1528)
1 : a hard crystalline mineral consisting of an oxide of magnesium and aluminum that varies from colorless to ruby-red to black and is used as a gem
2 : any of a group of minerals that are essentially oxides of magnesium, ferrous iron, zinc, or manganese

spine·less \'spīn-ləs\ *adjective* (1827)
1 : free from spines, thorns, or prickles
2 a : having no spinal column **:** INVERTEBRATE **b :** lacking strength of character
— spine·less·ly *adverb*
— spine·less·ness *noun*

spin·et \'spi-nət *also* spi-'net\ *noun* [Italian *spinetta,* perhaps from diminutive of *spina* thorn, from Latin; from the manner of plucking its strings] (1664)
1 : an early harpsichord having a single keyboard and only one string for each note
2 a : a compactly built small upright piano **b :** a small electronic organ

spin fishing *noun* (1950)
: SPINNING

spi·ni·fex \'spī-nə-,feks\ *noun* [New Latin, from Latin *spina* + *facere* to make — more at DO] (1846)
: any of several Australian grasses (genera *Spinifex* and *Triodia*) with spiny seeds or stiff sharp leaves

spin·na·ker \'spi-ni-kər\ *noun* [origin unknown] (1866)
: a large triangular sail set on a long light pole and used when running before the wind

spin·ner \'spi-nər\ *noun* (13th century)
1 : one that spins
2 : a fisherman's lure consisting of a spoon, blade, or set of wings that revolves when drawn through the water
3 : a conical sheet metal fairing that is attached to an airplane propeller boss and revolves with it
4 : a movable arrow that is spun on its dial to indicate the number or kind of moves a player may make in a board game

spin·ner·et \,spi-nə-'ret\ *noun* (1826)
1 : an organ (as of a spider or caterpillar) for producing threads of silk from the secretion of silk glands
2 *or* **spin·ner·ette :** a small metal plate, thimble, or cap with fine holes through which a chemical solution (as of cellulose) is forced in the spinning of man-made filaments (as of rayon or nylon)

spin·ney \'spi-nē\ *noun, plural* **spinneys** [Middle French *espinaye* thorny thicket, from *espine* thorn, from Latin *spina*] (1597)
chiefly British **:** a small wood with undergrowth

spin·ning \'spi-niŋ\ *noun* (1855)
: a method of fishing in which a lure is cast by use of a light flexible rod, a spinning reel, and a light line

spinning frame *noun* (1825)
: a machine that draws, twists, and winds yarn

spinning jen·ny \-,je-nē\ *noun* [*Jenny,* nickname for *Jane*] (1783)
: an early multiple-spindle machine for spinning wool or cotton

spinning reel *noun* (1950)
: a fishing reel with a nonmoving spool on which the line is wound by means of a revolving arm which can be disengaged to allow the line to spiral freely off the spool during casting

spinning rod *noun* (1870)
: a light flexible fishing rod used with a spinning reel

spinning wheel *noun* (15th century)
: a small domestic hand-driven or foot-driven machine for spinning yarn or thread

spin–off \'spin-,ȯf\ *noun* (1950)
1 : the distribution by a business to its stockholders of particular assets and especially of stock of another company; *also* **:** the new company created by such a distribution
2 : a collateral or derived product or effect **:** BY-PRODUCT; *also* **:** a number of such products ⟨the *spin-off* from the space program⟩
3 : something that is imitative or derivative of an earlier work; *especially* **:** a television show starring a character popular in a secondary role of an earlier show

spin off (1950)
transitive verb
: to establish or produce as a spin-off ⟨the company *spun off* its computer division⟩ ⟨*spin off* a new TV series⟩
intransitive verb
: to establish or become a spin-off

spin·or \'spi-nər, -,nȯr\ *noun* [International Scientific Vocabulary *spin* + *-or* (as in *vector*)] (1931)
: a vector whose components are complex numbers in a two-dimensional or four-dimensional space and which is used especially in the mathematics of the theory of relativity

spi·nose \'spī-,nōs\ *adjective* (1660)
: SPINY 1 ⟨a fly with black *spinose* legs⟩
— spi·nos·i·ty \spī-'nä-sə-tē\ *noun*

spi·nous \'spī-nəs\ *adjective* (15th century)
1 : SPINY 1, 3 ⟨*spinous* appendages⟩ ⟨a *spinous* larva⟩
2 : difficult or unpleasant to handle or meet **:** THORNY

spin·out \'spin-,au̇t\ *noun* (1955)
: a rotational skid by an automobile that usually causes it to leave the roadway

spin out *intransitive verb* (1951)
: to make a rotational skid in an automobile

Spi·no·zism \spi-'nō-,zi-zəm\ *noun* (1728)
: the philosophy of Baruch Spinoza who taught that reality is one substance with an infinite number of attributes of which only thought and extension are capable of being apprehended by the human mind
— Spi·no·zist \-zist\ *noun*
— Spi·no·zis·tic \,spi-nō-'zis-tik\ *adjective*

spin·ster \'spin(t)-stər\ *noun* (14th century)
1 : a woman whose occupation is to spin
2 a *archaic* **:** an unmarried woman of gentle family **b :** an unmarried woman and especially one past the common age for marrying
3 : a woman who seems unlikely to marry ◆
— spin·ster·hood \-,hu̇d\ *noun*
— spin·ster·ish \-st(ə-)rish\ *adjective*
— spin·ster·ly *adjective*

◇ WORD HISTORY

spinster Before the industrial revolution spinning was a humble but all-important task generally carried out in the home. When the word *spinster* appeared in 14th century English, it meant a person of either sex who spun, especially as a regular occupation. Since spinning was traditionally regarded as women's work, *spinster* gradually came to be applied exclusively to women spinners. By the 17th century the word had also become an established legal designation for an unmarried woman. Not until the 18th century did *spinster* refer to a single woman who, according to the mores of her time, is regarded as having passed the usual or desirable age for marrying.

spin·thar·i·scope \spin-ˈthar-ə-ˌskōp\ *noun* [Greek *spintharis* spark + English *-scope*] (1903)
: an instrument for visual detection of alpha particles that consists of a fluorescent screen and a magnifying lens system

spin the bottle *noun* (1955)
: a kissing game in which one has as a partner the person a bottle points to when it stops spinning

spin·to \ˈspēn-(ˌ)tō, ˈspin-\ *noun, plural* **spin·tos** [Italian, literally, pushed, from past participle of *spingere* to push, from (assumed) Vulgar Latin *expingere,* from Latin *ex-* + *pangere* to fasten — more at PACT] (1944)
: a singing voice having both lyric and dramatic qualities
— **spinto** *adjective*

spi·nule \ˈspī-(ˌ)nyü(ə)l\ *noun* [Latin *spinula,* diminutive of *spina* thorn — more at SPINE] (1752)
: a minute spine
— **spi·nu·lose** \ˈspī-nyə-ˌlōs\ *adjective*

spiny \ˈspī-nē\ *adjective* **spin·i·er; -est** (1586)
1 : covered or armed with spines; *broadly* **:** bearing spines, prickles, or thorns
2 : abounding with difficulties, obstacles, or annoyances **:** THORNY ⟨*spiny* problems⟩
3 : slender and pointed like a spine
— **spin·i·ness** *noun*

spiny anteater *noun* (1827)
: ECHIDNA

spiny–head·ed worm \ˈspī-nē-ˌhe-dəd-\ *noun* (1946)
: any of a small phylum (Acanthocephala) of unsegmented parasitic worms that have a proboscis bearing hooks by which attachment is made to the intestinal wall of the host

spiny lobster *noun* (1819)
: any of several edible crustaceans (family Palinuridae and especially genus *Panulirus*) distinguished from the true lobsters by the simple unenlarged first pair of legs and claws and the spiny carapace

spi·ra·cle \ˈspir-i-kəl, ˈspī-ri-\ *noun* [Middle English, from Latin *spiraculum,* from *spirare* to breathe] (15th century)
1 : a breathing hole **:** VENT
2 : a breathing orifice: as **a :** BLOWHOLE 2 **b :** an external tracheal aperture of a terrestrial arthropod that in an insect is usually one of a series of small apertures located along each side of the thorax and abdomen — see INSECT illustration
— **spi·rac·u·lar** \spə-ˈra-kyə-lər, spī-\ *adjective*

¹spi·ral \ˈspī-rəl\ *adjective* [Medieval Latin *spiralis,* from Latin *spira* coil — more at SPIRE] (1551)
1 a : winding around a center or pole and gradually receding from or approaching it ⟨the *spiral* curve of a watch spring⟩ **b :** HELICAL **c :** SPIRAL-BOUND ⟨a *spiral* notebook⟩
2 : of or relating to the advancement to higher levels through a series of cyclical movements
— **spi·ral·ly** \-rə-lē\ *adverb*

²spiral *noun* (1656)
1 a : the path of a point in a plane moving around a central point while continuously receding from or approaching it **b :** a three-dimensional curve (as a helix) with one or more turns about an axis
2 : a single turn or coil in a spiral object
3 a : something having a spiral form **b** (1) **:** a spiral flight (2) **:** a kick or pass in which a football rotates on its long axis while moving through the air
4 : a continuously spreading and accelerating increase or decrease ⟨wage *spirals*⟩

³spiral *verb* **-raled** *or* **-ralled; -ral·ing** *or* **-ral·ling** (1834)
intransitive verb
: to go and especially to rise or fall in a spiral course ⟨costs *spiraled* upward⟩
transitive verb
1 : to form into a spiral
2 : to cause to spiral

spiral binding *noun* (1944)
: a book or notebook binding in which a continuous spiral wire or plastic strip is passed through holes along one edge

spi·ral–bound \ˈspī-rəl-ˌbau̇nd\ *adjective* (1941)
: having a spiral binding

spiral cleavage *noun* (1892)
: holoblastic cleavage that is typical of protostomes and that is characterized by arrangement of the blastomeres of each upper tier over the cell junctions of the next lower tier so that the blastomeres spiral around the pole to pole axis of the embryo — compare RADIAL CLEAVAGE

spiral galaxy *noun* (1913)
: a galaxy exhibiting a central nucleus or barred structure from which extend concentrations of matter forming curved arms — called also *spiral nebula*

spiral of Ar·chi·me·des \-ˌär-kə-ˈmē-dēz\ [*Archimedes*] (circa 1856)
: a plane curve that is generated by a point moving away from or toward a fixed point at a constant rate while the radius vector from the fixed point rotates at a constant rate and that has the equation $\rho = a\ \theta$ in polar coordinates

spiral spring *noun* (1690)
: a spring consisting of a wire coiled usually in a flat spiral or in a helix

spi·rant \ˈspī-rənt\ *noun* [International Scientific Vocabulary, from Latin *spirant-, spirans,* present participle of *spirare* to breathe] (1862)
: a consonant (as \f\, \s\, \sh\) uttered with friction of the breath against some part of the oral passage **:** FRICATIVE
— **spirant** *adjective*

¹spire \ˈspīr\ *noun* [Middle English, from Old English *spīr;* akin to Middle Dutch *spier* blade of grass] (before 12th century)
1 : a slender tapering blade or stalk (as of grass)
2 : the upper tapering part of something (as a tree or antler) **:** PINNACLE
3 a : a tapering roof or analogous pyramidal construction surmounting a tower **b :** STEEPLE

²spire *intransitive verb* **spired; spir·ing** (14th century)
: to rise like a spire

³spire *noun* [Latin *spira* coil, from Greek *speira;* perhaps akin to Greek *sparton* rope, esparto] (1545)
1 a : SPIRAL **b :** COIL
2 : the inner or upper part of a spiral gastropod shell consisting of all the whorls except the whorl in contact with the body

⁴spire *intransitive verb* **spired; spir·ing** (1591)
: to rise in or as if in a spiral

spi·rea *or* **spi·raea** \spī-ˈrē-ə\ *noun* [New Latin *Spiraea,* from Latin, a plant, from Greek *speiraia*] (1669)
1 : any of a genus (*Spiraea*) of shrubs of the rose family with small perfect usually white or pink flowers in dense racemes, corymbs, cymes, or panicles
2 : any of several garden plants resembling spireas; *especially* **:** a shrub (*Astilbe japonica*) of the saxifrage family

spired \ˈspīrd\ *adjective* (1610)
1 : having a spire ⟨a *spired* church⟩
2 : tapering usually to a sharp point ⟨*spired* cedars⟩

spi·ril·lum \spī-ˈri-ləm\ *noun, plural* **-ril·la** \-ˈri-lə\ [New Latin, from diminutive of Latin *spira* coil] (circa 1875)
: any of a genus (*Spirillum*) of long curved bacteria having tufts of flagella; *broadly* **:** a spiral filamentous bacterium (as a spirochete)

¹spir·it \ˈspir-ət\ *noun* [Middle English, from Old French or Latin; Old French, from Latin *spiritus,* literally, breath, from *spirare* to blow, breathe] (13th century)
1 : an animating or vital principle held to give life to physical organisms
2 : a supernatural being or essence: as **a** *capitalized* **:** HOLY SPIRIT **b :** SOUL 2a **c :** an often malevolent being that is bodiless but can become visible; *specifically* **:** GHOST 2 **d :** a malevolent being that enters and possesses a human being
3 : temper or disposition of mind or outlook especially when vigorous or animated ⟨in high *spirits*⟩
4 : the immaterial intelligent or sentient part of a person
5 a : the activating or essential principle influencing a person ⟨acted in a *spirit* of helpfulness⟩ **b :** an inclination, impulse, or tendency of a specified kind **:** MOOD
6 a : a special attitude or frame of mind ⟨the money-making *spirit* was for a time driven back —J. A. Froude⟩ **b :** the feeling, quality, or disposition characterizing something ⟨undertaken in a *spirit* of fun⟩
7 : a lively or brisk quality in a person or a person's actions
8 : a person having a character or disposition of a specified nature
9 : a mental disposition characterized by firmness or assertiveness ⟨denied the charge with *spirit*⟩
10 a : DISTILLATE 1: as (1) **:** the liquid containing ethyl alcohol and water that is distilled from an alcoholic liquid or mash — often used in plural (2) **:** any of various volatile liquids obtained by distillation or cracking (as of petroleum, shale, or wood) — often used in plural **b :** a usually volatile organic solvent (as an alcohol, ester, or hydrocarbon)
11 a : prevailing tone or tendency ⟨*spirit* of the age⟩ **b :** general intent or real meaning ⟨*spirit* of the law⟩
12 : an alcoholic solution of a volatile substance ⟨*spirit* of camphor⟩
13 : enthusiastic loyalty ⟨school *spirit*⟩
14 *capitalized, Christian Science* **:** GOD 1b
synonym see COURAGE

²spirit *transitive verb* (1608)
1 : to infuse with spirit; *especially* **:** ANIMATE ⟨hope and apprehension of feasibleness *spirits* all industry —John Goodman⟩
2 : to carry off usually secretly or mysteriously ⟨was hustled into a . . . motorcar and *spirited* off to the country —W. L. Shirer⟩

spir·it·ed \ˈspir-ə-təd\ *adjective* (1592)
: full of energy, animation, or courage ⟨a *spirited* discussion⟩
— **spir·it·ed·ly** *adverb*
— **spir·it·ed·ness** *noun*

spirit gum *noun* (1886)
: a solution (as of gum arabic in ether) used especially for attaching false hair to the skin

spir·it·ism \ˈspir-ə-ˌti-zəm\ *noun* (1856)
: SPIRITUALISM 2a
— **spir·it·ist** \-tist\ *noun*
— **spir·it·is·tic** \ˌspir-ə-ˈtis-tik\ *adjective*

spir·it·less \ˈspir-ət-ləs\ *adjective* (1597)
: lacking animation, cheerfulness, or courage
— **spir·it·less·ly** *adverb*
— **spir·it·less·ness** *noun*

spirit level *noun* (1768)
: LEVEL 1

spirit of hartshorn (circa 1684)
: AMMONIA WATER

spirit of wine (1753)
: ALCOHOL 1c

spir·i·to·so \ˌspir-ə-ˈtō-(ˌ)sō, -(ˌ)zō\ *adjective* [Italian, from *spirito* spirit, from Latin *spiritus*] (circa 1724)
: ANIMATED — used as a direction in music

spir·it·ous \ˈspir-ə-təs\ *adjective* (1605)
1 *archaic* **:** PURE, REFINED
2 : SPIRITUOUS

spirit rapping *noun* (1852)
: communication by raps held to be from the spirits of the dead

spirits of hartshorn (circa 1909)
: AMMONIA WATER

spirits of turpentine (circa 1792)
: TURPENTINE 2a

spirits of wine (1710)
: rectified spirit : ALCOHOL 1c

[1]**spir·i·tu·al** \'spir-i-chə-wəl, -i-chəl, -ich-wəl\ *adjective* [Middle English, from Middle French & Late Latin; Middle French *spirituel,* from Late Latin *spiritualis,* from Latin, of breathing, of wind, from *spiritus*] (14th century)
1 : of, relating to, consisting of, or affecting the spirit **:** INCORPOREAL ⟨man's *spiritual* needs⟩
2 a : of or relating to sacred matters ⟨*spiritual* songs⟩ **b :** ecclesiastical rather than lay or temporal ⟨*spiritual* authority⟩ ⟨lords *spiritual*⟩
3 : concerned with religious values
4 : related or joined in spirit ⟨our *spiritual* home⟩ ⟨his *spiritual* heir⟩
5 a : of or relating to supernatural beings or phenomena **b :** of, relating to, or involving spiritualism **:** SPIRITUALISTIC
— spir·i·tu·al·ly *adverb*
— spir·i·tu·al·ness *noun*

[2]**spiritual** *noun* (1582)
1 *plural* **:** things of a spiritual, ecclesiastical, or religious nature
2 : a religious song usually of a deeply emotional character that was developed especially among blacks in the southern U.S.
3 *capitalized* **:** any of a party of 13th and 14th century Franciscans advocating strict observance of a rule of poverty for their order

spiritual bouquet *noun* (1926)
: a card notifying the recipient of a number of devotional acts performed by a Roman Catholic on behalf of a person on special occasions (as name days or anniversaries) or for the soul of someone recently deceased especially as an expression of sympathy

spir·i·tu·al·ism \'spir-i-chə-wə-ˌli-zəm, -i-chə-ˌli-, -ich-wə-ˌli-\ *noun* (1796)
1 : the view that spirit is a prime element of reality
2 a : a belief that spirits of the dead communicate with the living usually through a medium **b** *capitalized* **:** a movement comprising religious organizations emphasizing spiritualism
— spir·i·tu·al·ist \-list\ *noun, often capitalized*
— spir·i·tu·al·is·tic \ˌspir-i-chə-wə-'lis-tik, -i-chə-'lis-, -ich-wə-'lis-\ *adjective*

spir·i·tu·al·i·ty \ˌspir-i-chə-'wa-lə-tē\ *noun, plural* **-ties** (15th century)
1 : something that in ecclesiastical law belongs to the church or to a cleric as such
2 : CLERGY
3 : sensitivity or attachment to religious values
4 : the quality or state of being spiritual

spir·i·tu·al·ize \'spir-i-chə-wə-ˌlīz, -i-chə-ˌlīz, -ich-wə-ˌlīz\ *transitive verb* **-ized; -iz·ing** (1631)
1 : to make spiritual; *especially* **:** to purify from the corrupting influences of the world
2 : to give a spiritual meaning to or understand in a spiritual sense
— spir·i·tu·al·i·za·tion \ˌspir-i-chə-wə-lə-'zā-shən, -i-chə-lə-, -ich-wə-lə-\ *noun*

spir·i·tu·al·ty \'spir-i-chə-wəl-tē, -i-chəl-tē, -ich-wəl-tē\ *noun* [Middle English *spiritualte,* from Middle French *spiritualté,* from Medieval Latin *spiritualitat-, spiritualitas,* from Late Latin *spiritualis* spiritual] (14th century)
1 : SPIRITUALITY 1
2 : CLERGY

spi·ri·tu·el *or* **spi·ri·tu·elle** \ˌspir-i-chə-'wel, spē-rē-tw(^{y})el\ *adjective* [*spirituel* from French, literally, spiritual; *spirituelle* from French, feminine of *spirituel*] (1673)
: having or marked by a refined and especially sprightly or witty nature

spir·i·tu·ous \'spir-i-chə-wəs, -ich-wəs\ *adjective* [probably from French *spiritueux,* from Latin *spiritus* spirit] (1681)
: containing or impregnated with alcohol obtained by distillation ⟨*spirituous* liquors⟩

spirit varnish *noun* (1850)
: a varnish in which a volatile liquid (as alcohol) is the solvent

spirit writing *noun* (1864)
: automatic writing held to be produced under the influence of spirits

spi·ro·chet·al \ˌspī-rə-'kē-t^{ə}l\ *adjective* (1915)
: caused by spirochetes

spi·ro·chete *also* **spi·ro·chaete** \'spī-rə-ˌkēt\ *noun* [New Latin *Spirochaeta,* genus of bacteria, from Latin *spira* coil + Greek *chaitē* long hair — more at SPIRE] (circa 1877)
: any of an order (Spirochaetales) of slender spirally undulating bacteria including those causing syphilis and relapsing fever

spi·ro·chet·osis \ˌspī-rə-ˌkē-'tō-səs\ *noun, plural* **-oses** \-ˌsēz\ [New Latin] (1906)
: infection with or a disease caused by spirochetes

spi·ro·gy·ra \ˌspī-rə-'jī-rə\ *noun* [New Latin, from Latin *spira* + Greek *gyros* ring, circle] (1875)
: any of a genus (*Spirogyra*) of freshwater green algae with spiral chloroplasts

spi·rom·e·ter \spī-'rä-mə-tər\ *noun* [International Scientific Vocabulary *spiro-* (from Latin *spirare* to breathe) + *-meter*] (1846)
: an instrument for measuring the air entering and leaving the lungs
— spi·ro·met·ric \ˌspī-rə-'me-trik\ *adjective*
— spi·rom·e·try \spī-'rä-mə-trē\ *noun*

spirt *variant of* SPURT

spiry \'spīr-ē\ *adjective* (1602)
: resembling a spire; *especially* **:** being tall, slender, and tapering ⟨*spiry* trees⟩

[1]**spit** \'spit\ *noun* [Middle English, from Old English *spitu;* akin to Old High German *spiz* spit, *spizzi* pointed] (before 12th century)
1 : a slender pointed rod for holding meat over a fire
2 : a small point of land especially of sand or gravel running into a body of water

[2]**spit** *transitive verb* **spit·ted; spit·ting** (13th century)
: to fix on or as if on a spit **:** IMPALE

[3]**spit** *verb* **spit** *or* **spat** \'spat\; **spit·ting** [Middle English *spitten,* from Old English *spittan;* akin to Middle High German *spiutzen* to spit] (before 12th century)
transitive verb
1 a : to eject (as saliva) from the mouth **:** EXPECTORATE **b** (1) **:** to express (unpleasant or malicious feelings) by or as if by spitting (2) **:** to utter with a spitting sound or scornful expression ⟨*spat* out his words⟩ **c :** to emit as if by spitting; *especially* **:** to emit (precipitation) in driving particles or in flurries ⟨*spit* rain⟩
2 : to set to burning ⟨*spit* a fuse⟩
intransitive verb
1 a (1) **:** to eject saliva as an expression of aversion or contempt (2) **:** to exhibit contempt **b :** to eject matter (as saliva) from the mouth **:** EXPECTORATE
2 : to rain or snow slightly or in flurries
3 : to make a noise suggesting expectoration **:** SPUTTER
— spit it out : to say what is in the mind without further delay

[4]**spit** *noun* (14th century)
1 a (1) **:** SPITTLE, SALIVA (2) **:** the act or an instance of spitting **b** (1) **:** a frothy secretion exuded by spittlebugs (2) **:** SPITTLEBUG
2 : perfect likeness
3 : a sprinkle of rain or flurry of snow

spit·al \'spi-t^{ə}l\ *noun* [Middle English *spitel,* modification of Medieval Latin *hospitale* — more at HOSPITAL] (14th century)
archaic **:** LAZARETTO, HOSPITAL

spit and polish *noun* [from the practice of polishing objects such as shoes by spitting on them and then rubbing them with a cloth] (1895)
: extreme attention to cleanliness, orderliness, smartness of appearance, and ceremony often at the expense of operational efficiency
— spit–and–polish *adjective*

spit·ball \'spit-ˌbȯl\ *noun* (1846)
1 : paper chewed and rolled into a ball to be thrown or shot as a missile
2 : a baseball pitch delivered after the ball has been moistened with saliva or sweat

spit curl *noun* [probably from its being sometimes plastered down with saliva] (1831)
: a spiral curl that is usually plastered on the forehead, temple, or cheek

[1]**spite** \'spīt\ *noun* [Middle English, short for *despite*] (14th century)
1 : petty ill will or hatred with the disposition to irritate, annoy, or thwart
2 : an instance of spite
synonym see MALICE
— in spite of : in defiance or contempt of **:** without being prevented by

[2]**spite** *transitive verb* **spit·ed; spit·ing** (circa 1555)
1 a : ANNOY, OFFEND **b :** to fill with spite
2 : to treat maliciously (as by shaming or thwarting)

spite·ful \'spīt-fəl\ *adjective* (15th century)
: filled with or showing spite **:** MALICIOUS
— spite·ful·ly \-fə-lē\ *adverb*
— spite·ful·ness *noun*

spit·fire \'spit-ˌfīr\ *noun* (1680)
: a quick-tempered or highly emotional person

[1]**spit·ter** \'spi-tər\ *noun* (14th century)
: one that spits

[2]**spitter** *noun* (1908)
: SPITBALL 2

spitting cobra *noun* (1910)
: either of two African cobras (*Naja nigricollis* and *Hemachatus haemachatus*) that in defense typically eject their venom toward the victim without striking

spitting image *noun* [alteration of *spit and image*] (1901)
: IMAGE 3b

spit·tle \'spi-t^{ə}l\ *noun* [Middle English *spetil,* from Old English *spǣtl;* akin to Old English *spittan* to spit] (before 12th century)
1 : SALIVA
2 : [4]SPIT 1b(1)

spit·tle·bug \-ˌbəg\ *noun* (1882)
: any of a family (Cercopidae) of leaping homopterous insects whose larvae secrete froth — called also *froghopper*

spittle insect *noun* (1891)
: SPITTLEBUG

spit·toon \spi-'tün, spə-\ *noun* [[4]*spit* + *-oon* (as in *balloon*)] (1823)
: a receptacle for spit — called also *cuspidor*

spit up *verb* (1779)
: REGURGITATE, VOMIT

spitz \'spits\ *noun* [German, from *spitz* pointed, from Old High German *spizzi;* from the shape of its ears and muzzle — more at SPIT] (1842)
: a member of any of several breeds of stocky heavy-coated dogs of northern origin with erect ears and usually a heavily furred tail carried over the back

spitz

spiv \'spiv\ *noun* [alteration of English dialect *spiff* flashy dresser, from *spiff* dandified] (circa 1934)
1 *British* **:** one who lives by his wits without regular employment

2 *British* **:** SLACKER

splanch·nic \ˈsplaŋk-nik\ *adjective* [New Latin *splanchnicus,* from Greek *splanchnikos,* from *splanchna,* plural, viscera; akin to Greek *splēn* spleen] (1681)
: of or relating to the viscera **:** VISCERAL

[1]**splash** \ˈsplash\ *verb* [alteration of *plash*] (circa 1727)
transitive verb
1 a (1) **:** to dash a liquid or thinly viscous substance upon or against (2) **:** to soil or stain with splashed liquid **b :** to mark or overlay with patches of contrasting color or texture **c :** to display prominently
2 a : to cause (a liquid or thinly viscous substance) to spatter about especially with force **b :** to scatter in the manner of a splashed liquid
intransitive verb
1 a : to strike and dash about a liquid or semiliquid substance **b :** to move in or into a liquid or semiliquid substance and cause it to spatter
2 a (1) **:** to become spattered about (2) **:** to spread or scatter in the manner of splashed liquid **b :** to fall, strike, or move with a splashing sound ⟨a brook *splashing* over rocks⟩
— splash·er *noun*

[2]**splash** *noun* (1736)
1 a (1) **:** splashed liquid or semiliquid substance; *also* **:** impounded water released suddenly (2) **:** a spot or daub from or as if from splashed liquid ⟨a mud *splash* on the fender⟩ **b :** a colored patch
2 a : the action of splashing **b :** a short plunge
3 : a sound produced by or as if by a liquid falling, moving, being hurled, or oscillating
4 a : a vivid impression created especially by ostentatious activity or appearance **b :** ostentatious display
5 : a small amount **:** SPRINKLING

splash·board \ˈsplash-ˌbōrd, -ˌbȯrd\ *noun* (1826)
1 : DASHBOARD 1
2 : a panel to protect against splashes

splash·down \ˈsplash-ˌdau̇n\ *noun* (1959)
: the landing of a manned spacecraft in the ocean
— splash down *intransitive verb*

splash guard *noun* (1926)
: a flap suspended behind a rear wheel to prevent tire splash from muddying windshields of following vehicles

splashy \ˈspla-shē\ *adjective* **splash·i·er; -est** (1856)
1 : that can be easily splashed about
2 : moving or being moved with a splash or splashing sounds
3 : tending to or exhibiting ostentatious display
4 : consisting of, being, or covered with colored splashes
— splash·i·ly \ˈspla-shə-lē\ *adverb*
— splash·i·ness \ˈspla-shē-nəs\ *noun*

[1]**splat** \ˈsplat\ *noun* [obsolete *splat* to spread flat] (1833)
: a single flat thin often ornamental member of a back of a chair

[2]**splat** *noun* [imitative] (1897)
: a splattering or slapping sound

[1]**splat·ter** \ˈspla-tər\ *verb* [probably blend of *splash* and *spatter*] (1785)
transitive verb
: SPATTER
intransitive verb
: to scatter or fall in or as if in drops

[2]**splatter** *noun* (1819)
: SPATTER, SPLASH

[1]**splay** \ˈsplā\ *verb* [Middle English, short for *displayen* — more at DISPLAY] (15th century)
transitive verb
1 : to cause to spread outward
2 : to make oblique **:** BEVEL
intransitive verb
1 : to extend apart or outward especially in an awkward manner
2 : SLOPE, SLANT

[2]**splay** *noun* (circa 1508)
1 : a slope or bevel especially of the sides of a door or window
2 : SPREAD, EXPANSION

[3]**splay** *adjective* (1767)
1 : turned outward ⟨*splay* knees⟩
2 : AWKWARD, UNGAINLY

splay·foot \ˈsplā-ˌfu̇t\ *noun* (1548)
: a foot abnormally flattened and spread out; *specifically* **:** FLATFOOT
— splay·foot·ed \-ˌfu̇-təd\ *adjective*

spleen \ˈsplēn\ *noun* [Middle English *splen,* from Middle French or Latin; Middle French *esplen,* from Latin *splen,* from Greek *splēn;* akin to Latin *lien* spleen, Sanskrit *plīhan*] (14th century)
1 : a highly vascular ductless organ that is located in the left abdominal region near the stomach or intestine of most vertebrates and is concerned with final destruction of red blood cells, filtration and storage of blood, and production of lymphocytes
2 *obsolete* **:** the seat of emotions or passions
3 *archaic* **:** MELANCHOLY
4 : feelings of anger or ill will often suppressed
5 *obsolete* **:** a sudden impulse or whim **:** CAPRICE
synonym see MALICE

spleen·ful \-fəl\ *adjective* (1588)
: full of or affected with spleen **:** SPLENETIC

spleen·wort \-ˌwərt, -ˌwȯrt\ *noun* [from the belief in its power to cure disorders of the spleen] (1578)
: any of a large genus (*Asplenium*) of chiefly evergreen ferns having linear or oblong sori

spleeny \ˈsplē-nē\ *adjective* (1604)
1 : full of or displaying spleen
2 *New England* **:** peevish and irritable with hypochondriac inclinations

splen- *or* **spleno-** *combining form* [Latin, from Greek *splēn-, splēno-,* from *splēn*]
: spleen ⟨*splen*ectomy⟩ ⟨*spleno*megaly⟩

splen·dent \ˈsplen-dənt\ *adjective* [Middle English, from Late Latin *splendent-, splendens,* from Latin, present participle of *splendēre*] (15th century)
1 : SHINING, GLOSSY ⟨*splendent* luster⟩
2 : ILLUSTRIOUS, BRILLIANT ⟨*splendent* genius⟩

splen·did \ˈsplen-dəd\ *adjective* [Latin *splendidus,* from *splendēre* to shine; perhaps akin to Middle Irish *lainn* bright] (1634)
1 : possessing or displaying splendor: as **a :** SHINING, BRILLIANT **b :** marked by showy magnificence
2 : ILLUSTRIOUS, GRAND
3 a : EXCELLENT ⟨a *splendid* opportunity⟩ **b :** being out of the ordinary **:** SINGULAR ☆
— splen·did·ly *adverb*
— splen·did·ness *noun*

splen·dif·er·ous \splen-ˈdi-f(ə-)rəs\ *adjective* [*splendor* + *-i-* + *-ferous*] (1843)
: extraordinarily or showily impressive
— splen·dif·er·ous·ly *adverb*
— splen·dif·er·ous·ness *noun*

splen·dor \ˈsplen-dər\ *noun* [Middle English *splendure,* from Anglo-French *splendur,* from Latin *splendor,* from *splendēre*] (15th century)
1 a : great brightness or luster **:** BRILLIANCY **b :** MAGNIFICENCE, POMP
2 : something splendid
— splen·dor·ous *also* **splen·drous** \-d(ə-)rəs\ *adjective*

splen·dour \-dər\ *chiefly British variant of* SPLENDOR

sple·nec·to·my \spli-ˈnek-tə-mē\ *noun, plural* **-mies** [International Scientific Vocabulary] (circa 1859)
: surgical removal of the spleen
— sple·nec·to·mize \-ˌmīz\ *transitive verb*

sple·net·ic \spli-ˈne-tik, *archaic* ˈsple-nə-(ˌ)tik\ *adjective* [Late Latin *spleneticus,* from Latin *splen* spleen] (1697)
1 *archaic* **:** given to melancholy
2 : marked by bad temper, malevolence, or spite
— splenetic *noun*
— sple·net·i·cal·ly \spli-ˈne-ti-k(ə-)lē\ *adverb*

splen·ic \ˈsple-nik\ *adjective* [Latin *splenicus,* from Greek *splēnikos,* from *splēn* spleen] (1619)
: of, relating to, or located in the spleen ⟨*splenic* blood flow⟩

sple·ni·us \ˈsplē-nē-əs\ *noun, plural* **-nii** \-nē-ˌī\ [New Latin, from Latin *splenium* plaster, compress, from Greek *splēnion,* from *splēn*] (1732)
: either of two flat oblique muscles on each side of the back of the neck and upper thoracic region

spleno·meg·a·ly \ˌsple-nō-ˈme-gə-lē\ *noun, plural* **-lies** [International Scientific Vocabulary] (circa 1900)
: enlargement of the spleen

spleu·chan \ˈsplü-k̲ən, ˈsplyü-\ *noun* [Scottish Gaelic *spliùcan* & Irish *spliúchán*] (1785)
Scottish & Irish **:** a pouch especially for tobacco or money

[1]**splice** \ˈsplīs\ *transitive verb* **spliced; splic·ing** [obsolete Dutch *splissen;* akin to Middle Dutch *splitten* to split] (circa 1525)
1 a : to unite (as two ropes) by interweaving the strands **b :** to unite (as lengths of magnetic tape) by lapping two ends together or by applying a piece that laps upon two ends and making fast
2 : to unite, link, or insert as if by splicing
3 : to combine (genetic material) from either the same organism or different organisms ⟨use enzymes to cut and *splice* genes⟩
— splic·er *noun*

[2]**splice** *noun* (1627)
1 : a joining or joint made by splicing something
2 : MARRIAGE, WEDDING

splice 1

spliff \ˈsplif\ *noun* [origin unknown] (1936)
: JOINT 4

spline \ˈsplīn\ *noun* [origin unknown] (1756)
1 : a thin wood or metal strip used in building construction
2 : a key that is fixed to one of two connected mechanical parts and fits into a keyway in the other; *also* **:** a keyway for such a key
3 : a function that is defined on an interval, is used to approximate a given function, and is composed of pieces of simple functions defined on subintervals and joined at their endpoints with a suitable degree of smoothness

[1]**splint** \ˈsplint\ *also* **splent** \ˈsplent\ *noun* [Middle English, from Middle Low German *splinte, splente;* probably akin to Middle Dutch *splitten* to split] (14th century)
1 : a small plate or strip of metal used in making armor

☆ SYNONYMS
Splendid, resplendent, gorgeous, glorious, sublime, superb mean extraordinarily or transcendently impressive. SPLENDID implies outshining the usual or customary ⟨the wedding was a *splendid* occasion⟩. RESPLENDENT suggests a glowing or blazing splendor ⟨*resplendent* in her jewelry⟩. GORGEOUS implies a rich splendor especially in display of color ⟨a *gorgeous* red dress⟩. GLORIOUS suggests radiance that heightens beauty or distinction ⟨a *glorious* sunset⟩. SUBLIME implies an exaltation or elevation almost beyond human comprehension ⟨a vision of *sublime* beauty⟩. SUPERB suggests a magnificence or excellence reaching the highest conceivable degree ⟨her singing was *superb*⟩.

2 a : a thin strip of wood suitable for interweaving (as into baskets) **b :** SPLINTER **c :** material or a device used to protect and immobilize a body part (as a broken arm)
3 : a bony enlargement on the upper part of the cannon bone of a horse usually on the inside of the leg
[2]splint *transitive verb* (15th century)
1 : to support and immobilize (as a broken bone) with a splint
2 : to brace with or as if with splints
splint bone *noun* (1704)
: one of the slender rudimentary metacarpal or metatarsal bones on either side of the cannon bone in the limbs of the horse and related animals
[1]splin·ter \'splin-tər\ *noun* [Middle English, from Middle Dutch; akin to Middle Low German *splinte* splint] (14th century)
1 a : a thin piece split or broken off lengthwise **:** SLIVER **b :** a small needlelike particle
2 : a group or faction broken away from a parent body
— splinter *adjective*
— splin·tery \'splin-tə-rē, -trē\ *adjective*
[2]splinter *verb* **splin·tered; splin·ter·ing** \'splin-tə-riŋ, -triŋ\ (1582)
transitive verb
1 : to split or rend into long thin pieces **:** SHIVER
2 : to split into fragments, parts, or factions
intransitive verb
: to become splintered
[1]split \'split\ *verb* **split; split·ting** [Dutch *splitten,* from Middle Dutch; akin to Middle High German *splīzen* to split and probably to Old High German *spaltan* to split] (1593)
transitive verb
1 a : to divide lengthwise usually along a grain or seam or by layers **b :** to affect as if by cleaving or forcing apart ⟨the river *splits* the town in two⟩
2 a (1) **:** to tear or rend apart **:** BURST (2) **:** to subject (an atom or atomic nucleus) to artificial disintegration by fission **b :** to affect as if by breaking up or tearing apart **:** SHATTER ⟨a roar that *split* the air⟩
3 : to divide into parts or portions: as **a :** to divide between persons **:** SHARE **b :** to divide into factions, parties, or groups **c :** to mark (a ballot) or cast or register (a vote) so as to vote for candidates of different parties **d** (1) **:** to divide or break down (a chemical compound) into constituents ⟨*split* a fat into glycerol and fatty acids⟩ (2) **:** to remove by such separation ⟨*split* off carbon dioxide⟩ **e :** to divide (stock) by issuing a larger number of shares to existing shareholders usually without increase in total par value
4 : to separate (the parts of a whole) by interposing something ⟨*split* an infinitive⟩
5 : LEAVE ⟨*split* the party⟩ ⟨*split* town⟩
intransitive verb
1 a : to become split lengthwise or into layers **b :** to break apart **:** BURST
2 a : to become divided up or separated off ⟨*split* into factions⟩ **b :** to sever relations or connections **c :** LEAVE; *especially* **:** to leave without delay
3 *British* **:** to betray confidence **:** act as an informer — usually used with *on*
4 : to apportion shares
synonym see TEAR
— split hairs : to make oversubtle or trivial distinctions
— split one's sides : to laugh heartily
[2]split *noun* (1597)
1 a : a narrow break made by or as if by splitting **b :** an arrangement of bowling pins left standing with space for pins between them
2 : a piece split off or made thin by splitting
3 a : a division into or between divergent or antagonistic elements or forces **b :** a faction formed in this way
4 a : the act or process of splitting (as the stock of a corporation) **b :** the act of lowering oneself to the floor or leaping into the air with legs extended at right angles to the trunk
5 : a product of division by or as if by splitting
6 : a wine bottle holding one quarter the usual amount or about .1875 liters (6 to 6.5 ounces); *also* **:** the quantity held by a split
7 : an ice cream sundae served over slices of fruit (as banana)
8 : the recorded time at or for a specific part of a race
[3]split *adjective* (1648)
1 : DIVIDED, FRACTURED
2 : prepared for use by splitting ⟨*split* bamboo⟩ ⟨*split* hides⟩
3 : HETEROZYGOUS — used especially by breeders of cage birds sometimes with *for*
4 : widely spaced
split–brain \'split-'brān\ *adjective* (1958)
: having the optic chiasma and corpus callosum severed ⟨behavior in *split-brain* animals⟩
split decision *noun* (1952)
: a decision in a boxing match reflecting a division of opinion among the referee and judges
split end *noun* (1955)
: an offensive football end who lines up usually several yards to the side of the formation
split–fingered fastball \'split-'fiŋ-gərd-\ *noun* (1979)
: a fast baseball pitch thrown with the ball gripped as for a forkball
split infinitive *noun* (1897)
: an infinitive with *to* having a modifier between the *to* and the verbal (as in "to really start") ▢
split–lev·el \'split-'le-vəl\ *adjective* (1946)
: divided vertically so that the floor level of rooms in one part is approximately midway between the levels of two successive stories in an adjoining part ⟨a *split-level* house⟩
— split–lev·el \-,le-vəl\ *noun*
split pea *noun* (1736)
: a dried hulled pea in which the cotyledons usually split apart
split personality *noun* (1927)
: SCHIZOPHRENIA; *also* **:** MULTIPLE PERSONALITY
split rail *noun* (1826)
: a fence rail split from a log
split screen *noun* (1944)
: a film or video technique in which the frame is divided into discrete nonoverlapping images; *also* **:** the visual composition based on this technique
split second *noun* (1912)
: a fractional part of a second **:** FLASH
split shift *noun* (1943)
: a shift of working hours divided into two or more working periods at times (as morning and evening) separated by more than normal periods of time off (as for lunch or rest)
split·ter \'spli-tər\ *noun* (1648)
1 : one that splits
2 : one who classifies organisms into numerous named groups based on relatively minor variations or characters — compare LUMPER
3 : SPLIT-FINGERED FASTBALL
split ticket *noun* (1836)
: a ballot cast by a voter who votes for candidates of more than one party
split·ting \'spli-tiŋ\ *adjective* (1593)
: that splits or causes to split: as **a :** causing a piercing sensation ⟨a *splitting* headache⟩ **b :** very fast or quick **c :** SIDESPLITTING ⟨a *splitting* laugh⟩
splore \'splōr, 'splȯr\ *noun* [origin unknown] (1785)
1 *Scottish* **:** FROLIC, CAROUSAL
2 *Scottish* **:** COMMOTION
[1]splotch \'spläch\ *noun* [perhaps blend of *spot* and *blotch*] (1601)
: SPOT, BLOTCH
— splotchy \'splä-chē\ *adjective*
[2]splotch *transitive verb* (1654)
: to mark with a splotch **:** cover with splotches
[1]splurge \'splərj\ *noun* [perhaps blend of *splash* and *surge*] (1830)
: an ostentatious effort, display, or expenditure
[2]splurge *verb* **splurged; splurg·ing** (1843)
intransitive verb
1 : to make a splurge
2 : to indulge oneself extravagantly — often used with *on* ⟨*splurge* on a new dress⟩
transitive verb
: to spend extravagantly or ostentatiously
[1]splut·ter \'splə-tər\ *noun* [probably alteration of *sputter*] (1677)
1 : a confused noise (as of hasty speaking)
2 : a splashing or sputtering sound
[2]splutter (1818)
intransitive verb
1 : to make a noise as if spitting
2 : to speak hastily and confusedly
transitive verb
: to utter hastily or confusedly **:** STAMMER
— splut·ter·er \'splə-tər-ər\ *noun*
splut·tery \'splə-tə-rē\ *adjective* (1866)
: marked by spluttering
Spode \'spōd\ *noun* (1869)
: ceramic ware (as bone china, stone china, or Parian ware) made at the works established by Josiah Spode in 1770 at Stoke in Staffordshire, England
spod·u·mene \'spä-jə-,mēn\ *noun* [probably from French *spodumène,* from German *Spodumen,* from Greek *spodoumenos,* present participle of *spodousthai* to be burnt to ashes, from *spodos* ashes] (1893)
: a white to yellowish, purplish, or emerald-green monoclinic mineral that is a silicate of lithium and aluminum and occurs in prismatic crystals often of great size
[1]spoil \'spȯi(ə)l\ *noun* [Middle English *spoile,* from Middle French *espoille,* from Latin *spolia,* plural of *spolium* animal skin — more at SPILL] (14th century)
1 a : plunder taken from an enemy in war or a victim in robbery **:** LOOT **b :** public offices made the property of a successful party — usually used in plural **c :** something valuable or desirable gained through special effort or opportunism or in return for a favor — usually used in plural
2 a : SPOLIATION, PLUNDERING **b :** the act of damaging **:** HARM, IMPAIRMENT
3 : an object of plundering **:** PREY
4 : earth and rock excavated or dredged

▢ USAGE

split infinitive The split infinitive was discovered and named in the 19th century. 19th century writers seem to have made greater use of this construction than earlier writers; the frequency of occurrence attracted the disapproving attention of grammarians, many of whom thought it to be a modern corruption. The construction had in fact been in occasional use since the 14th century; only its frequency had changed. Even though there has never been a rational basis for objecting to the split infinitive, the subject has become a fixture of folk belief about grammar. You can hardly publish a sentence containing one without hearing about it from somebody. Modern commentators know the split infinitive is not a vice, but they are loath to drop such a popular subject. They usually say it's all right to split an infinitive in the interest of clarity. Since clarity is the usual reason for splitting, this advice means merely that you can split them whenever you need to.

5 : an object damaged or flawed in the making ☆

²spoil *verb* **spoiled** \'spȯi(ə)ld, 'spȯi(ə)lt\ *also* **spoilt** \'spȯi(ə)lt\; **spoil·ing** [Middle English, from Middle French *espoillier,* from Latin *spoliare,* from *spolium*] (14th century)
transitive verb
1 a *archaic* **:** DESPOIL, STRIP **b :** PILLAGE, ROB
2 *archaic* **:** to seize by force
3 a : to damage seriously **:** RUIN **b :** to impair the quality or effect of ⟨a quarrel *spoiled* the celebration⟩
4 a : to impair the disposition or character of by overindulgence or excessive praise **b :** to pamper excessively **:** CODDLE
intransitive verb
1 : to practice plunder and robbery
2 : to lose valuable or useful qualities usually as a result of decay
3 : to have an eager desire ⟨*spoiling* for a fight⟩
synonym see DECAY, INDULGE
— spoil·able \'spȯi-lə-bəl\ *adjective*

spoil·age \'spȯi-lij\ *noun* (1597)
1 : the act or process of spoiling; *especially* **:** the process of decay in foodstuffs
2 : something spoiled or wasted
3 : loss by spoilage

spoil·er \'spȯi-lər\ *noun* (15th century)
1 a : one that spoils **b :** one (as a political candidate) having little or no chance of winning but capable of depriving a rival of success
2 a : a long narrow plate along the upper surface of an airplane wing that may be raised for reducing lift and increasing drag — see AIRPLANE illustration **b :** an air deflector on an automobile to reduce the tendency to lift off the road at high speeds

spoils·man \'spȯi(ə)lz-mən\ *noun* (1846)
: one who serves a party for a share of the spoils; *also* **:** one who sanctions such practice

spoil·sport \'spȯi(ə)l-ˌspōrt, -ˌspȯrt\ *noun* (1821)
: one who spoils the sport or pleasure of others

spoils system *noun* (1838)
: a practice of regarding public offices and their emoluments as plunder to be distributed to members of the victorious party

¹spoke \'spōk\ *past & archaic past participle of* SPEAK

²spoke *noun* [Middle English, from Old English *spāca;* akin to Middle Dutch *spike* spike — more at SPIKE] (before 12th century)
1 a : any of the small radiating bars inserted in the hub of a wheel to support the rim **b :** something resembling the spoke of a wheel
2 : any of the projecting handles of a steering wheel of a boat

³spoke *transitive verb* **spoked; spok·ing** (before 12th century)
: to furnish with or as if with spokes

spo·ken \'spō-kən\ *adjective* [past participle of *speak*] (1595)
1 : delivered by word of mouth **:** ORAL
2 : characterized by speaking in (such) a manner — used in combination ⟨soft-*spoken*⟩ ⟨plain*spoken*⟩

spoke·shave \'spōk-ˌshāv\ *noun* [²*spoke*] (1510)
: a drawknife or small transverse plane with end handles for planing convex or concave surfaces

spokes·man \'spōks-mən\ *noun* [probably irregular from *spoke,* obsolete past participle of *speak*] (1537)
: a person who speaks as the representative of another or others often in a professional capacity
— spokes·man·ship \-ˌship\ *noun*

spokes·peo·ple \-ˌpē-pəl\ *noun plural* (1972)
: people serving as spokesmen or spokeswomen

spokes·per·son \-ˌpər-sᵊn\ *noun* (1972)
: SPOKESMAN

spokes·wom·an \-ˌwu̇-mən\ *noun* (1654)
: a woman who speaks as the representative of another or others often in a professional capacity

spo·li·ate \'spō-lē-ˌāt\ *transitive verb* **-at·ed; -at·ing** [Latin *spoliatus,* past participle of *spoliare*] (circa 1727)
: DESPOIL
— spo·li·a·tor \-ˌā-tər\ *noun*

spo·li·a·tion \ˌspō-lē-'ā-shən\ *noun* [Middle English, from Latin *spoliation-, spoliatio,* from *spoliare* to plunder — more at SPOIL] (15th century)
1 a : the act of plundering **b :** the state of having been plundered especially in war
2 : the act of injuring especially beyond reclaim

spon·dee \'spän-ˌdē\ *noun* [Middle English *sponde,* from Middle French or Latin; Middle French *spondee,* from Latin *spondeum,* from Greek *spondeios,* from *spondeios* of a libation, from *spondē* libation, from *spendein* to make a libation; from its use in music accompanying libations — more at SPOUSE] (14th century)
: a metrical foot consisting of two long or stressed syllables
— spon·da·ic \spän-'dā-ik\ *adjective or noun*

spon·dy·li·tis \ˌspän-də-'lī-təs\ *noun* [New Latin, from Greek *sphondylos, spondylos* vertebra] (circa 1849)
: inflammation of the vertebrae

¹sponge \'spənj\ *noun* [Middle English, from Old English, from Latin *spongia,* from Greek] (before 12th century)
1 a (1) **:** an elastic porous mass of interlacing horny fibers that forms the internal skeleton of various marine animals (phylum Porifera) and is able when wetted to absorb water (2) **:** a piece of sponge (as for scrubbing and cleaning) (3) **:** a porous rubber or cellulose product used similarly to a sponge **b :** any of a phylum (Porifera) of aquatic chiefly marine lower invertebrate animals that are essentially double-walled cell colonies that are permanently attached as adults
2 : a pad (as of folded gauze) used in surgery and medicine (as to remove discharge or apply medication)
3 : one who lives on others **:** SPONGER
4 a : raised dough (as for yeast bread) **b :** a whipped dessert usually containing whites of eggs or gelatin **c :** a metal (as platinum) obtained in porous form usually by reduction without fusion ⟨titanium *sponge*⟩ **d :** the egg mass of a crab
5 : an absorbent contraceptive device that is impregnated with spermicide and inserted into the vagina before sexual intercourse to cover the cervix
synonym see PARASITE

²sponge *verb* **sponged; spong·ing** (14th century)
transitive verb
1 : to cleanse, wipe, or moisten with or as if with a sponge
2 : to erase or destroy with or as if with a sponge — often used with *out*
3 : to get by sponging on another
4 : to absorb with or as if with or in the manner of a sponge
intransitive verb
1 : to absorb, soak up, or imbibe like a sponge
2 : to get something from or live on another by imposing on hospitality or good nature
3 : to dive or dredge for sponges
— spong·er *noun*

sponge cake *noun* (1805)
: a light cake made without shortening

sponge cloth *noun* (1862)
: any of various soft porous fabrics especially in a loose honeycomb weave

sponge rubber *noun* (1886)
: cellular rubber resembling a natural sponge in structure used especially for cushions, vibration dampeners, weather stripping, and gaskets

sponge·ware \'spənj-ˌwar, -ˌwer\ *noun* (1943)
: a typically 19th century earthenware with background color spattered or dabbed (as with a sponge) and usually a freehand central design

spon·gin \'spən-jən\ *noun* [German, from Latin *spongia* sponge] (circa 1868)
: a scleroprotein that is the chief constituent of flexible fibers in sponge skeletons

spongy \'spən-jē\ *adjective* **spong·i·er; -est** (14th century)
1 : resembling a sponge: **a :** soft and full of cavities ⟨*spongy* ice⟩ **b :** elastic, porous, and absorbent
2 a : not firm or solid **b :** being in the form of a metallic sponge ⟨*spongy* iron⟩
3 : moist and soft like a sponge full of water ⟨a *spongy* moor⟩
— spong·i·ness *noun*

spongy parenchyma *noun* (1884)
: a spongy layer of irregular chlorophyll-bearing cells interspersed with air spaces that fills the interior part of a leaf below the palisade layer — called also *spongy layer, spongy tissue*

spon·son \'spän(t)-sən\ *noun* [perhaps by shortening & alteration from *expansion*] (1835)
1 a : a projection (as a gun platform) from the side of a ship or a tank **b :** an air chamber along a canoe to increase stability and buoyancy
2 : a light air-filled structure or a winglike part protruding from the hull of a seaplane to steady it on water

¹spon·sor \'spän(t)-sər\ *noun* [Late Latin, from Latin, guarantor, surety, from *spondēre* to promise — more at SPOUSE] (1651)
1 : one who presents a candidate for baptism or confirmation and undertakes responsibility for the person's religious education or spiritual welfare
2 : one who assumes responsibility for some other person or thing
3 : a person or an organization that pays for or plans and carries out a project or activity; *especially* **:** one that pays the cost of a radio or television program usually in return for advertising time during its course
— spon·so·ri·al \spän-'sōr-ē-əl, -'sȯr-\ *adjective*
— spon·sor·ship \'spän(t)-sər-ˌship\ *noun*

²sponsor *transitive verb* **spon·sored; spon·sor·ing** \'spän(t)s-(ə-)riŋ\ (1884)
: to be or stand sponsor for

spon·ta·ne·i·ty \ˌspän-tə-'nē-ə-tē, -'nā-; -tᵊn-'ē-, -'ā-\ *noun* (1651)
1 : the quality or state of being spontaneous
2 : voluntary or undetermined action or movement; *also* **:** its source

spon·ta·ne·ous \spän-'tā-nē-əs\ *adjective* [Late Latin *spontaneus,* from Latin *sponte* of one's free will, voluntarily] (1656)
1 : proceeding from natural feeling or native tendency without external constraint

☆ SYNONYMS
Spoil, plunder, booty, prize, loot mean something taken from another by force or craft. SPOIL, more commonly SPOILS, applies to what belongs by right or custom to the victor in war or political contest ⟨the *spoils* of political victory⟩. PLUNDER applies to what is taken not only in war but in robbery, banditry, grafting, or swindling ⟨a bootlegger's *plunder*⟩. BOOTY implies plunder to be shared among confederates ⟨thieves dividing up their *booty*⟩. PRIZE applies to spoils captured on the high seas or territorial waters of the enemy ⟨the wartime right of seizing *prizes* at sea⟩. LOOT applies especially to what is taken from victims of a catastrophe ⟨picked through the ruins for *loot*⟩.

2 : arising from a momentary impulse
3 : controlled and directed internally **:** SELF-ACTING ⟨*spontaneous* movement characteristic of living things⟩
4 : produced without being planted or without human labor **:** INDIGENOUS
5 : developing without apparent external influence, force, cause, or treatment
6 : not apparently contrived or manipulated **:** NATURAL ☆
— **spon·ta·ne·ous·ly** *adverb*
— **spon·ta·ne·ous·ness** *noun*

spontaneous combustion *noun* (1795)
: self-ignition of combustible material through chemical action (as oxidation) of its constituents — called also *spontaneous ignition*

spontaneous generation *noun* (1665)
: ABIOGENESIS

spontaneous recovery *noun* (1943)
: reappearance of an extinguished conditioned response without positive reinforcement

spon·toon \spän-ˈtün\ *noun* [French *sponton,* from Italian *spuntone,* from *punta* sharp point, from (assumed) Vulgar Latin *puncta* — more at POINT] (1598)
: a short pike formerly borne by subordinate officers of infantry

[1]**spoof** \ˈspüf\ *transitive verb* [*Spoof,* a hoaxing game invented by Arthur Roberts (died 1933) English comedian] (1889)
1 : DECEIVE, HOAX
2 : to make good-natured fun of

[2]**spoof** *noun* (1889)
1 : HOAX, DECEPTION
2 : a light humorous parody
— **spoof·ery** \ˈspü-f(ə-)rē\ *noun*
— **spoofy** \ˈspü-fē\ *adjective*

[1]**spook** \ˈspük\ *noun* [Dutch; akin to Middle Low German *spōk* ghost] (1801)
1 : GHOST, SPECTER
2 : an undercover agent **:** SPY
— **spook·ish** \ˈspü-kish\ *adjective*

[2]**spook** (1883)
transitive verb
1 : HAUNT 3
2 : to make frightened or frantic **:** SCARE; *especially* **:** to startle into violent activity (as stampeding)
intransitive verb
: to become spooked ⟨cattle *spooking* at shadows⟩

spook·ery \ˈspü-k(ə-)rē\ *noun, plural* **-er·ies** (1893)
: the quality of being spooky; *also* **:** something (as a story) that involves spooks

spooky \ˈspü-kē\ *adjective* **spook·i·er; -est** (1854)
1 : relating to, resembling, or suggesting spooks
2 : NERVOUS, SKITTISH ⟨a *spooky* horse⟩
— **spook·i·ly** \-kə-lē\ *adverb*
— **spook·i·ness** \-kē-nəs\ *noun*

[1]**spool** \ˈspül\ *noun* [Middle English *spole,* from Middle French or Middle Dutch; Middle French *espole,* from Middle Dutch *spoele;* akin to Old High German *spuola* spool] (14th century)
1 : a cylindrical device which has a rim or ridge at each end and an axial hole for a pin or spindle and on which material (as thread, wire, or tape) is wound
2 : material or the amount of material wound on a spool

[2]**spool** (1603)
transitive verb
1 : to wind on a spool
2 : WIND ⟨*spool* the thread off the bobbin⟩
intransitive verb
1 : to wind itself on a spool
2 : WIND

[1]**spoon** \ˈspün\ *noun* [Middle English, from Old English *spōn* splinter, chip; akin to Old High German *spān* splinter, chip] (14th century)
1 : an eating or cooking implement consisting of a small shallow bowl with a relatively long handle
2 : something (as a tool or fishing lure) that resembles a spoon in shape

[2]**spoon** (1715)
transitive verb
: to take up and usually transfer in a spoon
intransitive verb
[perhaps from the Welsh custom of an engaged man's presenting his fiancée with an elaborately carved wooden spoon] **:** to make love by caressing, kissing, and talking amorously **:** NECK

spoon·bill \ˈspün-ˌbil\ *noun* (circa 1678)
1 : any of several wading birds (family Threskiornithidae) related to the ibises that have an expanded bill that is flattened and rounded at the tip
2 : any of several broad-billed ducks (as the shoveler)

spoonbill cat *noun* (circa 1882)
: a paddlefish (*Polyodon spathula*)

spoon–billed \ˈspün-ˌbild\ *adjective* (1668)
: having the bill or snout expanded and spatulate at the end

spoon bread *noun* (1916)
: soft bread made of cornmeal mixed with milk, eggs, and shortening and served with a spoon

spoo·ner·ism \ˈspü-nə-ˌri-zəm\ *noun* [William A. *Spooner* (died 1930) English clergyman & educator] (1900)
: a transposition of usually initial sounds of two or more words (as in *tons of soil* for *sons of toil*) ◆

spoon–feed \ˈspün-ˌfēd\ *transitive verb* **-fed** \-ˌfed\; **-feed·ing** (1615)
1 : to feed by means of a spoon
2 a : to present (information) so completely as to preclude independent thought ⟨*spoon-feed* material to students⟩ **b :** to present information to in this manner

spoon·ful \ˈspün-ˌfu̇l\ *noun, plural* **spoon·fuls** \-ˌfu̇lz\ *also* **spoons·ful** \ˈspünz-ˌfu̇l\ (14th century)
: as much as a spoon will hold; *specifically* **:** TEASPOONFUL
usage see [2]-FUL

spoony *or* **spoon·ey** \ˈspü-nē\ *adjective* **spoon·i·er; -est** [English slang *spoon* simpleton] (circa 1812)
1 : SILLY, FOOLISH; *especially* **:** unduly sentimental
2 : being sentimentally in love

[1]**spoor** \ˈspu̇r, ˈspōr, ˈspȯr\ *noun, plural* **spoor** *or* **spoors** [Afrikaans, from Dutch; akin to Old English *spor* footprint, spoor, *spurnan* to kick — more at SPURN] (1823)
1 : a track, a trail, a scent, or droppings especially of a wild animal
2 : a trace by which the progress of someone or something may be followed

[2]**spoor** (1850)
transitive verb
: to track by a spoor
intransitive verb
: to track something by its spoor

spor- *or* **spori-** *or* **sporo-** *combining form* [New Latin *spora*]
: seed **:** spore ⟨*sporocyst*⟩ ⟨*sporangium*⟩ ⟨*sporicidal*⟩

spo·rad·ic \spə-ˈra-dik\ *adjective* [Medieval Latin *sporadicus,* from Greek *sporadikos,* from *sporadēn* here and there, from *sporad-, sporas* scattered; akin to Greek *speirein* to sow — more at SPERM] (circa 1689)
: occurring occasionally, singly, or in scattered instances
synonym see INFREQUENT
— **spo·rad·i·cal·ly** \-di-k(ə-)lē\ *adverb*

sporadic E layer *noun* (1949)
: a layer of ionization occurring irregularly within the E region of the ionosphere

spo·ran·gio·phore \spə-ˈran-jē-ə-ˌfōr, -ˌfȯr\ *noun* (1875)
: a stalk or similar structure bearing sporangia

spo·ran·gi·um \spə-ˈran-jē-əm\ *noun, plural* **-gia** \-jē-ə\ [New Latin, from *spor-* + Greek *angeion* vessel — more at ANGI-] (1821)
: a structure within which spores are produced
— **spo·ran·gial** \-j(ē-)əl\ *adjective*

[1]**spore** \ˈspōr, ˈspȯr\ *noun* [New Latin *spora* seed, spore, from Greek, act of sowing, seed, from *speirein* to sow — more at SPERM] (1836)
: a primitive usually unicellular often environmentally resistant dormant or reproductive body produced by plants and some microorganisms and capable of development into a new individual either directly or after fusion with another spore
— **spored** \ˈspōrd, ˈspȯrd\ *adjective*

[2]**spore** *intransitive verb* **spored; spor·ing** (1903)
: to produce or reproduce by spores

spore case *noun* (1836)
: a case containing spores **:** SPORANGIUM

spo·ri·cid·al \ˌspōr-ə-ˈsī-dᵊl, ˌspȯr-\ *adjective* (1939)
: tending to kill spores
— **spo·ri·cide** \ˈspōr-ə-ˌsīd, ˈspȯr-\ *noun*

spo·ro·carp \ˈspōr-ə-ˌkärp, ˈspȯr-\ *noun* [International Scientific Vocabulary] (1849)
: a structure (as in red algae, fungi, or mosses) in or on which spores are produced

spo·ro·cyst \-ˌsist\ *noun* [International Scientific Vocabulary] (1861)
1 : a case or cyst secreted by some sporozoans preliminary to sporogony; *also* **:** a sporozoan encysted in such a case

☆ SYNONYMS

Spontaneous, impulsive, instinctive, automatic, mechanical mean acting or activated without deliberation. SPONTANEOUS implies lack of prompting and connotes naturalness ⟨a *spontaneous* burst of applause⟩. IMPULSIVE implies acting under stress of emotion or spirit of the moment ⟨*impulsive* acts of violence⟩. INSTINCTIVE stresses spontaneous action involving neither judgment nor will ⟨blinking is an *instinctive* reaction⟩. AUTOMATIC implies action engaging neither the mind nor the emotions and connotes a predictable response ⟨his denial was *automatic*⟩. MECHANICAL stresses the lifeless, often perfunctory character of the response ⟨a *mechanical* teaching method⟩.

◇ WORD HISTORY

spoonerism William Archibald Spooner (1844–1930) was an English clergyman and scholar who served as warden of New College at Oxford. Known for his wit, humor and devotion to Oxford University, Spooner had a tendency while delivering sermons to transpose the initial letters or syllables of adjoining words. He once announced that "Kinquering Kongs their titles take" would be the theme for that day's sermon. On another occasion he assured his flock that "Yoo, in deed; the Lord is a shoving leopard." Even more puzzling to his parishioners was his observation that "We all know what it is to have a half-warmed fish within us." Spooner's actual, unintentional lapses so amused his listeners that they were inspired to devise their own transpositions and to attribute them to the minister. Toward the end of the 19th century *spoonerism* became a general term for such verbal slips.

2 : a saccular body that is the first asexual reproductive form of a digenetic trematode, develops from a miracidium, and buds off cells from its inner surface which develop into rediae

spo·ro·gen·e·sis \ˌspōr-ə-ˈje-nə-səs, ˌspȯr-\ *noun* [New Latin] (circa 1890)
1 : reproduction by spores
2 : spore formation
— spo·rog·e·nous \spə-ˈrä-jə-nəs, spȯ-\ *also* **spo·ro·gen·ic** \ˌspōr-ə-ˈje-nik, ˌspȯr-\ *adjective*

spo·ro·go·ni·um \ˌspōr-ə-ˈgō-nē-əm, ˌspȯr-\ *noun, plural* **-nia** \-nē-ə\ [New Latin, from *spor-* + *-gonium* (as in *archegonium*)] (1875)
: the sporophyte of a moss or liverwort consisting typically of a stalk bearing a spore-producing capsule which remains permanently attached to the gametophyte

spo·rog·o·ny \spə-ˈrä-gə-nē, spȯ-\ *noun* [International Scientific Vocabulary] (1888)
: reproduction by spores; *specifically* **:** formation of spores typically containing sporozoites that is characteristic of some sporozoans and that results from the encystment and subsequent division of a zygote
— spo·ro·gon·ic \ˌspōr-ə-ˈgä-nik, ˌspȯr-\ *adjective*

spo·ro·phore \ˈspōr-ə-ˌfōr, ˈspȯr-ə-ˌfȯr\ *noun* [International Scientific Vocabulary] (1849)
: the spore-producing organ of a fungus or slime mold

spo·ro·phyll \-ˌfil\ *noun* [International Scientific Vocabulary] (1888)
: a spore-bearing and usually greatly modified leaf

spo·ro·phyte \-ˌfīt\ *noun* [International Scientific Vocabulary] (1886)
: the individual or generation of a plant exhibiting alternation of generations that bears asexual spores — compare GAMETOPHYTE
— spo·ro·phyt·ic \ˌspōr-ə-ˈfi-tik, ˌspȯr-\ *adjective*

spo·ro·pol·len·in \ˌspōr-ə-ˈpä-lə-nən, ˌspȯr-\ *noun* [International Scientific Vocabulary *spor-* + *pollen* + [1]*-in*] (1931)
: a relatively chemically inert polymer that makes up the outer layer of pollen grains and some spores

spo·ro·tri·cho·sis \spə-ˌrä-tri-ˈkō-səs; ˌspōr-ə-tri-, ˌspȯr-\ *noun* [New Latin, from *Sporotrich-, Sporothrix,* genus name, from *spor-* + Greek *trich-, thrix* hair] (1908)
: infection with or disease caused by a fungus (*Sporothrix schenckii*) that is characterized by nodules and abscesses in the superficial lymph nodes, skin, and subcutaneous tissues and that is usually transmitted by entry of the fungus through a skin abrasion or wound

-sporous *adjective combining form* [New Latin *spora* spore]
: having (such or so many) spores ⟨homo*sporous*⟩

spo·ro·zo·an \ˌspōr-ə-ˈzō-ən, ˌspȯr-\ *noun* [New Latin *Sporozoa,* from *spor-* + *-zoa*] (1888)
: any of a large class (Sporozoa) of strictly parasitic protozoans that have a complicated life cycle usually involving both asexual and sexual generations often in different hosts and include important pathogens (as malaria parasites and babesias)
— sporozoan *adjective*

spo·ro·zo·ite \-ˈzō-ˌīt\ *noun* [New Latin *Sporozoa* + International Scientific Vocabulary *-ite*] (1888)
: a usually motile infective form of some sporozoans that is a product of sporogony and initiates an asexual cycle in the new host

spor·ran \ˈspȯr-ən, ˈspär-\ *noun* [Scottish Gaelic *sporan* purse] (1752)
: a pouch usually of skin with the hair or fur on that is worn in front of the kilt with Scots Highland dress

[1]**sport** \ˈspōrt, ˈspȯrt\ *verb* [Middle English, to divert, disport, short for *disporten*] (15th century)
intransitive verb
1 a : to amuse oneself **:** FROLIC ⟨lambs *sporting* in the meadow⟩ **b :** to engage in a sport
2 a : to mock or ridicule something **b :** to speak or act in jest **:** TRIFLE
3 [[2]*sport*] **:** to deviate or vary abruptly from type (as by bud variation) **:** MUTATE
transitive verb
1 : to display or wear usually ostentatiously **:** BOAST
2 [[2]*sport*] **:** to put forth as a sport or bud variation

1 sporran

[2]**sport** *noun* (15th century)
1 a : a source of diversion **:** RECREATION **b :** sexual play **c** (1) **:** physical activity engaged in for pleasure (2) **:** a particular activity (as an athletic game) so engaged in
2 a : PLEASANTRY, JEST **b :** often mean-spirited jesting **:** MOCKERY, DERISION
3 a : something tossed or driven about in or as if in play **b :** LAUGHINGSTOCK
4 a : SPORTSMAN **b :** a person considered with respect to living up to the ideals of sportsmanship ⟨a good *sport*⟩ ⟨a poor *sport*⟩ **c :** a companionable person
5 : an individual exhibiting a sudden deviation from type beyond the normal limits of individual variation usually as a result of mutation especially of somatic tissue
synonym see FUN

[3]**sport** *or* **sports** *adjective* (1582)
: of, relating to, or suitable for sports; *especially* **:** styled in a manner suitable for casual or informal wear ⟨*sport* coats⟩

sport fish *noun* (1944)
: a fish important for the sport it affords anglers

sport·fish·er·man \ˈspōrt-ˌfi-shər-mən, ˈspȯrt-\ *noun* (1954)
: a motorboat equipped for sportfishing

sport·fish·ing \-ˌfi-shiŋ\ *noun* (1910)
: fishing done with a rod and reel for sport or recreation

sport·ful \-fəl\ *adjective* (15th century)
1 a : productive of sport or amusement **:** ENTERTAINING, DIVERTING **b :** PLAYFUL, FROLICSOME
2 : done in sport
— sport·ful·ly \-fə-lē\ *adverb*
— sport·ful·ness *noun*

spor·tif \ˈspōr-tif, ˈspȯr-\ *adjective* [French, sporting, of sports, from *sport* sport, from English] (1920)
: SPORTY

sport·ing \ˈspōr-tiŋ, ˈspȯr-\ *adjective* (1799)
1 a : of, relating to, used, or suitable for sport **b :** marked by or calling for sportsmanship **c :** involving such risk as a sports contender may expect to take or encounter ⟨a *sporting* chance⟩
2 : of or relating to dissipation and especially gambling
3 : tending to mutate freely
— sport·ing·ly \-tiŋ-lē\ *adverb*

sporting house *noun* (1894)
: BORDELLO

sport·ive \ˈspōr-tiv, ˈspȯr-\ *adjective* (1590)
1 a : FROLICSOME, PLAYFUL **b :** ARDENT, WANTON
2 : of or relating to sports and especially field sports
— sport·ive·ly *adverb*
— sport·ive·ness *noun*

sports car *noun* (1928)
: a low small usually 2-passenger automobile designed for quick response, easy maneuverability, and high-speed driving

sports·cast \ˈspōrts-ˌkast, ˈspȯrts-\ *noun* [*sport* + broad*cast*] (1941)
: a radio or television broadcast of a sports event or of information about sports
— sports·cast·er \-ˌkas-tər\ *noun*

sports·man \ˈspōrts-mən, ˈspȯrts-\ *noun* (circa 1707)
1 : a person who engages in sports (as in hunting or fishing)
2 : a person who shows sportsmanship
— sports·man·like \-ˌlīk\ *adjective*
— sports·man·ly \-lē\ *adjective*

sports·man·ship \-ˌship\ *noun* (1745)
: conduct (as fairness, respect for one's opponent, and graciousness in winning or losing) becoming to one participating in a sport

sports·wear \ˈspōrts-ˌwar, ˈspȯrts-, -ˌwer\ *noun* (1912)
: clothing suitable for recreation; *broadly* **:** clothing designed for casual or informal wear

sports·wom·an \-ˌwu̇-mən\ *noun* (1754)
: a woman who engages in sports

sports·writ·er \ˈspōrts-ˌrī-tər, ˈspȯrts-\ *noun* (1927)
: one who writes about sports especially for a newspaper
— sports·writ·ing \-tiŋ\ *noun*

sporty \ˈspōr-tē, ˈspȯr-\ *adjective* **sport·i·er; -est** (1889)
1 : of, relating to, or typical of sports, sportsmen, sportswomen, or sportswear ⟨a *sporty* life⟩ ⟨clothes of a *sporty* cut⟩
2 *of an automobile* **:** resembling a sports car in styling or performance
— sport·i·ly \ˈspōr-tᵊl-ē, ˈspȯr-\ *adverb*
— sport·i·ness \ˈspōr-tē-nəs, ˈspȯr-\ *noun*

spor·u·late \ˈspōr-yə-ˌlāt, ˈspȯr-, -ə-ˌlāt\ *intransitive verb* **-lat·ed; -lat·ing** [back-formation from *sporulation*] (circa 1891)
: to undergo sporulation

spor·u·la·tion \ˌspōr-yə-ˈlā-shən, ˌspȯr-, -ə-ˈlā-\ *noun* [International Scientific Vocabulary, from New Latin *sporula,* diminutive of *spora* spore] (1876)
: the formation of spores; *especially* **:** division into many small spores (as after encystment)
— spor·u·la·tive \ˈspōr-yə-ˌlā-tiv, ˈspȯr-, -ə-ˌlā-\ *adjective*

-spory *noun combining form* [-*sporous* + [2]-*y*]
: quality or state of having (such) spores ⟨homo*spory*⟩

[1]**spot** \ˈspät\ *noun* [Middle English; akin to Middle Dutch *spotte* stain, speck, Old Norse *spotti* small piece] (13th century)
1 : a taint on character or reputation **:** FAULT ⟨the only *spot* on the family name⟩
2 a : a small area visibly different (as in color, finish, or material) from the surrounding area **b** (1) **:** an area marred or marked (as by dirt) (2) **:** a circumscribed surface lesion of disease (as measles) or decay ⟨*spots* of rot⟩ ⟨rust *spots* on a leaf⟩ **c :** a conventionalized design used on playing cards to distinguish the suits and indicate values
3 : an object having a specified number of spots or a specified numeral on its surface
4 : a small quantity or amount **:** BIT
5 a : a particular place, area, or part **b :** a small extent of space
6 *plural usually* **spot :** a small croaker (*Leiostomus xanthurus*) of the Atlantic coast with a black spot behind the opercula
7 a : a particular position (as in an organization or a hierarchy) **b :** a place or appearance on an entertainment program
8 : SPOTLIGHT
9 : a position usually of difficulty or embarrassment
10 : a brief announcement or advertisement broadcast between scheduled radio or television programs
11 : a brief segment or report on a broadcast especially of news

— **on the spot** **1 :** at once **:** IMMEDIATELY **2 :** at the place of action **3 a :** in a responsible or accountable position **b :** in a difficult or trying situation

[2]spot *verb* **spot·ted; spot·ting** (14th century)
transitive verb
1 : to stain the character or reputation of **:** DISGRACE
2 : to mark in or with a spot **:** STAIN
3 : to locate or identify by a spot
4 a : to single out **:** IDENTIFY; *especially* **:** to note as a known criminal or a suspicious person **b :** DETECT, NOTICE ⟨*spot* a mistake⟩ **c** (1) **:** to locate accurately ⟨*spot* an enemy position⟩ (2) **:** to cause to strike accurately ⟨*spot* the battery's fire⟩
5 a : to lie at intervals in or over **:** STUD **b :** to place at intervals or in a desired spot ⟨*spot* field telephones⟩ **c :** to fix in or as if in the beam of a spotlight **d :** to schedule in a particular spot or at a particular time
6 : to remove a spot from
7 : to allow as a handicap
intransitive verb
1 : to become stained or discolored in spots
2 : to cause a spot
3 : to act as a spotter; *especially* **:** to locate targets
4 : to experience abnormal and sporadic bleeding in small amounts from the uterus
— **spot·ta·ble** \'spä-tə-bəl\ *adjective*

[3]spot *adjective* (1881)
1 a : being, originating, or done on the spot or in or for a particular spot ⟨*spot* coverage of the news⟩ **b :** available for immediate delivery after sale ⟨*spot* commodities⟩ **c** (1) **:** paid out upon delivery ⟨*spot* cash⟩ (2) **:** involving immediate cash payment ⟨*spot* transaction⟩ **d** (1) **:** broadcast between scheduled programs ⟨*spot* announcements⟩ (2) **:** originating in a local station for a national advertiser **e :** performing occasionally when needed ⟨chance of making the . . . varsity as a *spot* starter and relief pitcher — *N. Y. Times*⟩
2 : made at random or restricted to a few places or instances ⟨a *spot* check⟩; *also* **:** selected at random or as a sample

spot–check \'spät-,chek\ (1943)
transitive verb
: to sample or investigate quickly or at random
intransitive verb
: to make a spot check

spot·less \'spät-ləs\ *adjective* (14th century)
: having no spot: **a :** free from impurity **:** IMMACULATE ⟨*spotless* kitchens⟩ **b :** PURE, UNBLEMISHED ⟨*spotless* reputation⟩
— **spot·less·ly** *adverb*
— **spot·less·ness** *noun*

[1]spot·light \'spät-,līt\ *noun* (1904)
1 a : a projected spot of light used to illuminate brilliantly a person, object, or group on a stage **b :** conspicuous public notice ⟨held the political *spotlight*⟩
2 a : a light designed to direct a narrow intense beam of light on a small area **b :** something that illuminates brilliantly

[2]spotlight *transitive verb* **-light·ed** *or* **-lit; -light·ing** (1922)
: to illuminate with or as if with a spotlight

spot pass *noun* (1948)
: a pass (as in football or basketball) made to a predetermined spot on the field or court rather than directly to a player

spot·ted \'spä-təd\ *adjective* (13th century)
1 : marked with spots
2 : being sullied **:** TARNISHED
3 : characterized by the appearance of spots

spotted alfalfa aphid *noun* (1958)
: a highly destructive Old World aphid (*Therioaphis maculata*) that is established in the U.S. in warmer areas and causes yellowing and stunting of affected plants

spotted cucumber beetle *noun* (1923)
: a rather slender greenish yellow beetle (*Diabrotica undecimpunctata howardi*) that feeds as an adult on various ornamental and crop plants and is a vector of wilt disease especially of cucumbers and melons

spotted fever *noun* (1650)
: any of various eruptive fevers: as **a :** TYPHUS **b :** ROCKY MOUNTAIN SPOTTED FEVER

spotted owl *noun* (1910)
: a rare large dark brown dark-eyed owl (*Strix occidentalis*) that has barred and spotted underparts and is found in humid old growth forests and thickly wooded canyons from British Columbia to southern California and central Mexico

spotted salamander *noun* (1922)
: a common salamander (*Ambystoma maculatum*) of eastern North America with glossy black skin spotted with yellow or orange on the back

spotted sea trout *noun* (1902)
: a weakfish (*Cynoscion nebulosus*) that is a valuable food and sport fish of the southern Atlantic and Gulf coasts of the U.S. — called also *sea trout, speckled trout, spotted weakfish*

spotted turtle *noun* (circa 1909)
: a freshwater turtle (*Clemmys guttata*) of the eastern U.S. that has a blackish carapace with round yellow spots

spot·ter \'spä-tər\ *noun* (1611)
1 : one that makes or applies a spot (as for identification)
2 : one that looks or keeps watch: as **a :** one that locates enemy targets **b :** a civilian who watches for approaching airplanes
3 : one that removes spots
4 : one that places something on or in a desired spot

spot test *noun* (1922)
1 : a test limited to a few key or sample points or a relatively small percentage of random spots
2 : a test conducted on the spot to yield immediate results

spot·ty \'spä-tē\ *adjective* **spot·ti·er; -est** (14th century)
1 : marked with spots **:** SPOTTED
2 : lacking uniformity especially in quality ⟨the performance was *spotty*⟩; *also* **:** irregularly or sparsely distributed ⟨*spotty* attendance⟩ ⟨*spotty* data⟩
— **spot·ti·ly** \'spä-tᵊl-ē\ *adverb*
— **spot·ti·ness** \'spä-tē-nəs\ *noun*

spou·sal \'spau̇-zəl, -səl\ *noun* [Middle English *spousaille,* from Middle French *espousailles* espousal] (14th century)
: NUPTIALS — usually used in plural

[1]spouse \'spau̇s *also* 'spau̇z\ *noun* [Middle English, from Old French *espous* (masculine) & *espouse* (feminine), from Latin *sponsus* betrothed man, groom & *sponsa* betrothed woman, bride, both from *sponsus,* past participle of *spondēre* to promise, betroth; akin to Greek *spendein* to pour a libation, Hittite *šipant-*] (13th century)
: married person **:** HUSBAND, WIFE
— **spou·sal** \'spau̇-zəl, -səl\ *adjective*

[2]spouse \'spau̇z, 'spau̇s\ *transitive verb* **spoused; spous·ing** (13th century)
archaic **:** WED

[1]spout \'spau̇t\ *verb* [Middle English; akin to Middle Dutch *spoiten* to spout, Old English *spīwan* to spew] (14th century)
transitive verb
1 : to eject (as liquid) in a stream ⟨wells *spouting* oil⟩
2 a : to speak or utter readily, volubly, and at length **b :** to speak or utter in a pompous or oratorical manner **:** DECLAIM
intransitive verb
1 : to issue with force or in a jet **:** SPURT
2 : to eject material (as liquid) in a jet
3 : DECLAIM
— **spout·er** *noun*

[2]spout *noun* (14th century)
1 : a pipe or conductor through which a liquid is discharged or conveyed in a stream: as **a :** a pipe for carrying rainwater from a roof **b :** a projecting tube or lip from which a liquid (as water) issues
2 : a discharge or jet of liquid from or as if from a pipe; *especially* **:** WATERSPOUT
3 *archaic* **:** PAWNSHOP
— **spout·ed** \'spau̇-təd\ *adjective*

sprach·ge·fühl \'shpräk-gə-,fu̅e̅l\ *noun* [German, from *Sprache* language + *Gefühl* feeling] (1894)
1 : the character of a language
2 : an intuitive sense of what is linguistically appropriate

sprad·dle \'spra-dᵊl\ *verb* **sprad·dled; sprad·dling** \'sprad-liŋ, 'spra-dᵊl-iŋ\ [perhaps blend of *straddle* and *sprawl*] (1632)
intransitive verb
1 : SPRAWL
2 : to go or walk with a straddling gait **:** STRADDLE
transitive verb
1 : SPRAWL
2 : to spread (the legs) in walking **:** STRADDLE

sprag \'sprag\ *noun* [origin unknown] (1878)
: a pointed stake or steel bar let down from a halted vehicle (as a wagon) to prevent it from rolling

[1]sprain \'sprān\ *noun* [origin unknown] (1601)
1 : a sudden or violent twist or wrench of a joint with stretching or tearing of ligaments
2 : a sprained condition

[2]sprain *transitive verb* (1622)
: to injure by a sudden or severe twist

[1]sprang \'spraŋ\ *past of* SPRING

[2]sprang *noun* [probably from Norwegian, a kind of embroidery] (1951)
: a weaving technique in which threads or cords are intertwined and twisted over one another to form an openwork mesh

sprat \'sprat\ *noun* [alteration of Middle English *sprot,* from Old English *sprott*] (1537)
1 a : a small European herring (*Sprattus sprattus*) related to the common herring — called also *brisling* **b :** a small or young herring or similar fish (as an anchovy)
2 : a young, small, or insignificant person

sprawl \'sprȯl\ *verb* [Middle English, from Old English *sprēawlian*] (before 12th century)
intransitive verb
1 a *archaic* **:** to lie thrashing or tossing about **b :** to creep or clamber awkwardly
2 : to lie or sit with arms and legs spread out
3 : to spread or develop irregularly
transitive verb
: to cause to spread out carelessly or awkwardly
— **sprawl** *noun*

[1]spray \'sprā\ *noun* [Middle English] (13th century)
1 : a usually flowering branch or shoot
2 : a decorative flat arrangement of flowers and foliage (as on a coffin)
3 : something (as a jeweled pin) resembling a spray

[2]spray *noun* [obsolete English *spray* to sprinkle, from Middle Dutch *sprayen*] (1621)
1 : water flying in small drops or particles blown from waves or thrown up by a waterfall
2 a : a jet of vapor or finely divided liquid **b :** a device (as an atomizer or sprayer) by which a spray is dispersed or applied **c** (1) **:** an application of a spray or by spraying (2) **:** a substance (as paint) so applied

[3]spray (1829)
transitive verb
1 : to disperse or apply as a spray
2 : to project spray on or into
intransitive verb

1 : to break up into spray
2 : to disperse or apply a spray
3 : to emit a stream or spray of urine ⟨a cat may *spray* to mark its territory⟩
— spray·er *noun*

spray can *noun* (1958)
: a pressurized container from which aerosols are dispensed

spray gun *noun* (1920)
: an apparatus resembling a gun for applying a substance (as paint or insecticide) in the form of a spray

[1]**spread** \'spred\ *verb* **spread; spread·ing** [Middle English *spreden,* from Old English *-sprǣdan;* akin to Old High German *spreiten* to spread] (13th century)
transitive verb
1 a : to open or expand over a larger area ⟨*spread* out the map⟩ **b :** to stretch out **:** EXTEND ⟨*spread* its wings for flight⟩
2 a : to distribute over an area ⟨*spread* fertilizer⟩ **b :** to distribute over a period or among a group ⟨*spread* the work over a few weeks⟩ **c :** to apply on a surface ⟨*spread* butter on bread⟩ **d** (1) **:** to cover or overlay something with ⟨*spread* the cloth on the table⟩ (2) *archaic* **:** to cover completely **e** (1) **:** to prepare or furnish for dining **:** SET ⟨*spread* the table⟩ (2) **:** SERVE ⟨*spread* the afternoon tea⟩
3 a : to make widely known ⟨*spread* the news⟩ **b :** to extend the range or incidence of ⟨*spread* a disease⟩ **c :** DIFFUSE, EMIT ⟨flowers *spreading* their fragrance⟩
4 : to push apart by weight or force
intransitive verb
1 a : to become dispersed, distributed, or scattered **b :** to become known or disseminated ⟨panic *spread* rapidly⟩
2 : to grow in length or breadth **:** EXPAND
3 : to move apart (as from pressure or weight) **:** SEPARATE
— spread·abil·i·ty \ˌspre-də-'bi-lə-tē\ *noun*
— spread·able \'spre-də-bəl\ *adjective*

[2]**spread** *noun* (1626)
1 a : the act or process of spreading **b :** extent of spreading
2 : something spread out: as **a :** a surface area **:** EXPANSE **b** *West* (1) **:** RANCH (2) **:** a herd of animals **c** (1) **:** a prominent display in a periodical (2) **:** two facing pages (as of a newspaper) usually with matter running across the fold; *also* **:** the matter occupying these pages
3 : something spread on or over a surface: as **a :** a food to be spread (as on bread or crackers) ⟨a cheese *spread*⟩ **b :** a sumptuous meal **:** FEAST **c :** a cloth cover for a table or bed
4 : distance between two points **:** GAP
5 : a commodities market transaction in which a participant hedges with simultaneous long and short options in different commodities or different delivery dates in the same commodity

[1]**spread–ea·gle** \'spred-ˌē-gəl\ *verb* **-ea·gled; -ea·gling** \-ˌē-g(ə-)liŋ\ (1826)
intransitive verb
1 : to execute a spread eagle (as in skating)
2 : to stand or move with arms and legs stretched out **:** SPRAWL
transitive verb
1 : to stretch out into the position of a spread eagle
2 : to spread over

[2]**spread–eagle** *adjective* [from the spread eagle on the Great Seal of the U.S.] (1858)
: marked by bombast and boastful exaggeration especially of the greatness of the U.S. ⟨*spread-eagle* oratory⟩

spread eagle *noun* (1570)
1 : a representation of an eagle with wings raised and legs extended
2 : something resembling or suggestive of a spread eagle; *specifically* **:** a skating figure executed with the skates heel to heel in a straight line

spread·er \'spre-dər\ *noun* (15th century)
: one that spreads: as **a :** an implement for scattering material **b :** a small knife for spreading butter **c :** a device (as a bar) holding two linear elements (as lines, guys, rails) apart and usually taut

spread formation *noun* (circa 1949)
: an offensive football formation in which the pass receivers are spread out across the field

spreading factor *noun* (1932)
: HYALURONIDASE

spread·sheet \'spred-ˌshēt\ *noun* (1982)
: an accounting program for a computer; *also* **:** the ledger layout modeled by such a program

spree \'sprē\ *noun* [origin unknown] (1804)
: an unrestrained indulgence in or outburst of an activity ⟨went on a buying *spree*⟩; *especially* **:** BINGE, CAROUSAL

sprent \'sprent\ *adjective* [Middle English *spreynt,* from past participle of *sprengen* to sprinkle] (14th century)
archaic **:** sprinkled over

sprier *comparative of* SPRY

spriest *superlative of* SPRY

[1]**sprig** \'sprig\ *noun* [Middle English *sprigge*] (14th century)
1 a : a small shoot **:** TWIG ⟨a *sprig* of parsley⟩ **b :** a small division of grass used for propagation
2 a : HEIR **b :** YOUTH **c :** a small specimen
3 : an ornament resembling a sprig, stemmed flower, or leaf
4 : a small headless nail **:** BRAD

[2]**sprig** *transitive verb* **sprigged; sprig·ging** (1713)
1 : to drive sprigs or brads into
2 : to mark or adorn with the representation of plant sprigs
3 : to propagate (a grass) by means of stolons or small divisions

spright·ful \'sprīt-fəl\ *adjective* [obsolete *spright*] (1595)
archaic **:** full of life or spirit **:** SPRIGHTLY
— spright·ful·ly \-fə-lē\ *adverb*
— spright·ful·ness *noun*

spright·ly \-lē\ *adjective* **spright·li·er; -est** [obsolete *spright* (sprite), alteration of *sprite*] (1596)
: marked by a gay lightness and vivacity **:** SPIRITED
synonym see LIVELY
— spright·li·ness *noun*
— sprightly *adverb*

[1]**spring** \'spriŋ\ *verb* **sprang** \'spraŋ\ *or* **sprung** \'sprəŋ\; **sprung; spring·ing** \'spriŋ-iŋ\ [Middle English, from Old English *springan;* akin to Old High German *springan* to jump and perhaps to Greek *sperchesthai* to hasten] (before 12th century)
intransitive verb
1 a (1) **:** DART, SHOOT (2) **:** to be resilient or elastic; *also* **:** to move by elastic force ⟨the lid *sprang* shut⟩ **b :** to become warped
2 : to issue with speed and force or as a stream
3 a : to grow as a plant **b :** to issue by birth or descent **c :** to come into being **:** ARISE ⟨towns *sprang* up across the plains⟩ **d** *archaic* **:** DAWN **e :** to begin to blow — used with *up* ⟨a breeze quickly *sprang* up⟩
4 a : to make a leap or series of leaps **b :** to leap or jump up suddenly
5 : to stretch out in height **:** RISE
6 : PAY — used with *for* ⟨I'll *spring* for the drinks⟩
transitive verb
1 : to cause to spring
2 a : to undergo or bring about the splitting or cracking of ⟨wind *sprang* the mast⟩ **b :** to undergo the opening of (a leak)
3 a : to cause to operate suddenly ⟨*spring* a trap⟩ **b :** to apply or insert by bending **c :** to bend by force
4 : to leap over
5 : to produce or disclose suddenly or unexpectedly
6 : to make lame
7 : to release or cause to be released from confinement or custody ☆

[2]**spring** *noun, often attributive* (before 12th century)
1 a : a source of supply; *especially* **:** a source of water issuing from the ground **b :** an ultimate source especially of action or motion
2 : SPRING TIDE
3 : a time or season of growth or development; *specifically* **:** the season between winter and summer comprising in the northern hemisphere usually the months of March, April, and May or as reckoned astronomically extending from the March equinox to the June solstice
4 : an elastic body or device that recovers its original shape when released after being distorted
5 a : the act or an instance of leaping up or forward **:** BOUND **b** (1) **:** capacity for springing **:** RESILIENCE (2) **:** ENERGY, BOUNCE
6 : the point or plane at which an arch or vault curve springs from its impost
— spring·like \-ˌlīk\ *adjective*

[3]**spring** *transitive verb* **sprung** \'sprəŋ\; **spring·ing** \'spriŋ-iŋ\ (1884)
: to fit with springs

spring·ald \'spriŋ-əld\ *or* **spring·al** \-əl\ *noun* [probably from Middle English, a kind of catapult, from Middle French *espringale*] (1501)
: a young man **:** STRIPLING

spring beauty *noun* (1821)
: any of a genus (*Claytonia*) of herbs of the purslane family; *especially* **:** one (*C. virginica*) that sends up in early spring a 2-leaved stem bearing delicate pink flowers

spring·board \'spriŋ-ˌbōrd, -ˌbȯrd\ *noun* (1799)
1 : a flexible board usually secured at one end and used for gymnastic stunts or diving
2 : a point of departure **:** JUMPING-OFF PLACE

spring·bok \'spriŋ-ˌbäk\ *noun, plural* **springbok** *or* **springboks** [Afrikaans, from *spring* to jump + *bok* male goat] (1775)
: a swift and graceful southern African gazelle (*Antidorcas marsupialis*) noted for its habit of springing lightly and suddenly into the air

spring bolt *noun* (1634)
: a bolt retracted by pressure and shot by a spring when the pressure is released

spring chicken *noun* (1879)
: a young person ⟨is no *spring chicken*⟩

☆ SYNONYMS
Spring, arise, rise, originate, derive, flow, issue, emanate, proceed, stem mean to come up or out of something into existence. SPRING implies rapid or sudden emerging ⟨an idea that *springs* to mind⟩. ARISE and RISE may both convey the fact of coming into existence or notice but RISE often stresses gradual growth or ascent ⟨new questions have *arisen*⟩ ⟨slowly *rose* to prominence⟩. ORIGINATE implies a definite source or starting point ⟨the fire *originated* in the basement⟩. DERIVE implies a prior existence in another form ⟨the holiday *derives* from an ancient Roman feast⟩. FLOW adds to SPRING a suggestion of abundance or ease of inception ⟨words *flowed* easily from her pen⟩. ISSUE suggests emerging from confinement through an outlet ⟨blood *issued* from the cut⟩. EMANATE applies to the coming of something immaterial (as a thought) from a source ⟨reports *emanating* from the capital⟩. PROCEED stresses place of origin, derivation, parentage, or logical cause ⟨advice that *proceeds* from the best of intentions⟩. STEM implies originating by dividing or branching off from something as an outgrowth or subordinate development ⟨industries *stemming* from space research⟩.

spring–clean·ing \'spriŋ-'klē-niŋ\ *noun* [[2]*spring*] (1857)
: the act or process of doing a thorough cleaning of a place

springe \'sprinj\ *noun* [Middle English *sprenge, springe;* akin to Old English *springan* to spring] (13th century)
1 : a noose fastened to an elastic body to catch small game
2 : SNARE, TRAP

springbok

spring·er \'spriŋ-ər\ *noun* (1611)
1 : a stone or other solid laid at the impost of an arch — see ARCH illustration
2 : one that springs
3 : SPRINGER SPANIEL
4 : a cow nearly ready to calve

springer spaniel *noun* (1885)
: a medium-sized sporting dog of either of two breeds that is often used for finding and flushing small game: **a :** ENGLISH SPRINGER SPANIEL **b :** WELSH SPRINGER SPANIEL

spring fever *noun* (1843)
: a lazy or restless feeling often associated with the onset of spring

Spring·field rifle \'spriŋ-ˌfēld-\ *noun* [*Springfield,* Mass.] (1888)
: a .30 caliber bolt-action rifle used by U.S. troops especially in World War I

spring·form pan \'spriŋ-ˌfȯrm-\ *noun* (1927)
: a pan or mold with an upright detachable rim fastened to the bottom of the pan with a clamp or spring

spring·head \'spriŋ-ˌhed\ *noun* (1561)
: FOUNTAINHEAD

spring·house \-ˌhau̇s\ *noun* (1755)
: a small building situated over a spring and used for cool storage (as of dairy products or meat)

spring·ing \'spriŋ-iŋ\ *noun* (1590)
1 : SPRING 5
2 : a point where an arch rises from its support

spring–load \'spriŋ-'lōd\ *transitive verb* (1944)
: to load or secure by means of spring tension or compression

spring peeper *noun* (1906)
: a small brown tree frog (*Hyla crucifer*) of the eastern U.S. and Canada that has a shrill piping call and breeds in ponds and streams in the spring

spring peeper

spring roll *noun* (1943)
: EGG ROLL; *also* **:** any of various similar appetizers in oriental cuisine

spring·tail \'spriŋ-ˌtāl\ *noun* (circa 1797)
: any of an order (Collembola) of small primitive wingless insects usually with a forked structure on the fourth or fifth abdominal segment that is used for jumping — called also *collembolan*

spring·tide \-ˌtīd\ *noun* (1530)
: SPRINGTIME

spring tide *noun* (1548)
: a tide of greater-than-average range around the times of new and full moon

spring·time \'spriŋ-ˌtīm\ *noun* (15th century)
1 : the season of spring
2 : YOUTH 1a
3 : an early or flourishing stage of development

spring wagon *noun* (1794)
: a light farm wagon equipped with springs

spring·wa·ter \'spriŋ-ˌwȯ-tər, -ˌwä-\ *noun* (15th century)
: water from a spring

spring·wood \-ˌwu̇d\ *noun* (1884)
: the softer more porous portion of an annual ring of wood that develops early in the growing season — compare SUMMERWOOD

springy \'spriŋ-ē\ *adjective* **spring·i·er; -est** (1641)
1 : having an elastic quality **:** RESILIENT
2 : having or showing a lively and energetic movement ⟨walks with a *springy* step⟩
synonym see ELASTIC
— spring·i·ly \'spriŋ-ə-lē\ *adverb*
— spring·i·ness \'spriŋ-ē-nəs\ *noun*

[1]sprin·kle \'spriŋ-kəl\ *verb* **sprin·kled; sprin·kling** \-k(ə-)liŋ\ [Middle English *sprenklen, sprinclen;* akin to Middle High German *spreckel, sprenkel* spot] (14th century)
transitive verb
1 : to scatter in drops or particles
2 a : to scatter over **b :** to scatter at intervals in or among **:** DOT ⟨*sprinkled* the speech with quips⟩ **c :** to wet lightly
intransitive verb
1 : to scatter a liquid in fine drops
2 : to rain lightly in scattered drops
— sprin·kler \-k(ə-)lər\ *noun*

[2]sprinkle *noun* (1641)
1 : the act or an instance of sprinkling; *especially* **:** a light rain
2 : SPRINKLING
3 *plural* **:** small particles of candy used as a topping (as on ice cream) **:** JIMMIES

sprin·klered \'spriŋ-klərd\ *adjective* (1927)
: having a sprinkler system

sprinkler system *noun* (circa 1909)
: a system for protecting a building against fire by means of overhead pipes which convey an extinguishing fluid (as water) to heat-activated outlets

sprin·kling \'spriŋ-kliŋ\ *noun* (1594)
1 : a limited quantity or amount **:** MODICUM
2 : a small quantity falling in scattered drops or particles
3 : a small number distributed at random **:** SCATTERING

[1]sprint \'sprint\ *intransitive verb* [of Scandinavian origin; akin to Swedish dialect *sprinta* to jump, hop; akin to Old High German *sprinzan* to jump up] (circa 1864)
: to run or go at top speed especially for a short distance
— sprint·er *noun*

[2]sprint *noun* (circa 1865)
1 : the act or an instance of sprinting
2 a : DASH 6b **b :** a burst of speed

sprint car *noun* (circa 1965)
: a rugged racing automobile that is midway in size between midget racers and ordinary racers, has about the same horsepower as the larger racers, and is usually raced on a dirt track

sprit \'sprit\ *noun* [Middle English *spret, sprit,* from Old English *sprēot* pole, spear; akin to Old English *-sprūtan* to sprout] (14th century)
: a spar that crosses a fore-and-aft sail diagonally

sprite \'sprīt\ *noun* [Middle English *sprit,* from Middle French *esprit,* from Latin *spiritus* spirit — more at SPIRIT] (14th century)
1 a *archaic* **:** SOUL **b :** a disembodied spirit **:** GHOST
2 a : ELF, FAIRY **b :** an elfish person

sprit·sail \'sprit-ˌsāl, -səl\ *noun* (15th century)
1 : a sail extended by a sprit
2 : a sail formerly set on a yard beneath the bowsprit

spritz \'sprits, 'shprits\ *verb* [German *spritzen* to squirt, spray] (1902)
transitive verb
: SPRAY
intransitive verb
: to disperse or apply a spray
— spritz *noun*

spritz·er \'sprit-sər, 'shprit-\ *noun* [German, from *spritzen*] (1945)
: a beverage of usually white wine and soda water

sprock·et \'sprä-kət\ *noun* [origin unknown] (1750)
1 : a toothed wheel whose teeth engage the links of a chain
2 : a cylinder with teeth around the circumference at either end that project through perforations in something (as motion-picture film) to move it through a mechanism (as a projector)

[1]sprout \'sprau̇t\ *verb* [Middle English *spruten,* from Old English *-sprūtan;* akin to Old High German *sproizan* to sprout, Lithuanian *sprausti* to squeeze, thrust] (13th century)
intransitive verb
1 : to grow, spring up, or come forth as or as if a sprout
2 : to send out new growth
transitive verb
: to send forth or up **:** cause to develop **:** GROW

[2]sprout *noun* (13th century)
1 a : SHOOT 1a; *especially* **:** a young shoot (as from a seed or root) **b** *plural* (1) *chiefly British* **:** BRUSSELS SPROUT 2 (2) **:** edible sprouts especially from recently germinated seeds (as of alfalfa or mung beans)
2 : something resembling a sprout: as **a :** a young person **b :** SCION

sprouting broccoli *noun* (1852)
: BROCCOLI 2a(2)

[1]spruce \'sprüs\ *verb* **spruced; spruc·ing** (1594)
transitive verb
: to make spruce — often used with *up*
intransitive verb
: to make oneself spruce ⟨*spruce* up a bit⟩

[2]spruce *adjective* **spruc·er; spruc·est** [perhaps from obsolete English *Spruce leather* leather imported from Prussia] (1599)
: neat or smart in appearance **:** TRIM
— spruce·ly *adverb*
— spruce·ness *noun*

[3]spruce *noun* [obsolete *Spruce* Prussia, from Middle English, alteration of *Pruce,* from Old French] (1670)
1 a : any of a genus (*Picea*) of evergreen trees of the pine family with a conical head of dense foliage and soft light wood **b :** any of several coniferous trees (as Douglas fir) of similar habit
2 : the wood of a spruce ◇

spruce beer *noun* (1500)

◇ WORD HISTORY

spruce Between the 14th and 17th centuries Prussia was called *Pruce* or *Spruce* in English, and hence *Spruce* was used attributively to characterize goods imported from Prussia, as *spruce canvas, spruce iron,* and *spruce leather.* Perhaps the most important of these Prussian products was a tall, straight conifer, called the *spruce tree* or *spruce,* that was especially desirable for use as the mast of a ship. By the middle of the 17th century *Spruce* as a name for the country had largely been supplanted by *Prussia.* By that time, however, *spruce* had become well established as the name for the tree. The particular species (*Picea abies*) that was originally called *spruce* is not limited to Prussia. Another of its native countries gave it its present common name, *Norway spruce,* which is something of an etymological paradox. *Spruce* is now the common name applied to any member of the genus *Picea.*

: a beverage flavored with spruce; *especially* : one made from spruce twigs and leaves boiled with molasses or sugar and fermented with yeast

spruce budworm *noun* (1884)
: a tortricid moth (*Choristoneura fumiferana*) whose larva feeds on evergreen trees (as spruce and balsam fir) in the northern U.S. and Canada

spruce pine *noun* (1684)
: an American tree (as some pines and spruces or the common eastern hemlock) of the pine family with light, soft, or weak wood

sprucy \'sprü-sē\ *adjective* **spruc·i·er; -est** (1774)
: SPRUCE

[1]sprue \'sprü\ *noun* [origin unknown] (1880)
1 : the hole through which metal or plastic is poured into the gate and thence into a mold
2 : the waste piece cast in a sprue

[2]sprue *noun* [Dutch *spruw;* akin to Middle Low German *sprūwe,* a kind of tumor] (1888)
1 : CELIAC DISEASE
2 : a disease of tropical regions that is of unknown cause and is characterized by fatty diarrhea and malabsorption of nutrients — called also *tropical sprue*

sprung *past and past participle of* SPRING

sprung rhythm *noun* (1877)
: a poetic rhythm designed to approximate the natural rhythm of speech and characterized by the frequent juxtaposition of single accented syllables and the occurrence of mixed types of feet

spry \'sprī\ *adjective* **spri·er** *or* **spry·er** \'sprī(-ə)r\; **spri·est** *or* **spry·est** \'sprī-əst\ [perhaps of Scandinavian origin; akin to Swedish dialect *sprygg* spry] (1746)
: NIMBLE 1 ⟨a *spry* 75-year-old⟩
— **spry·ly** *adverb*
— **spry·ness** *noun*

[1]spud \'spəd\ *verb* **spud·ded; spud·ding** (1652)
transitive verb
1 : to dig with a spud
2 : to begin to drill (an oil well)
intransitive verb
: to use a spud

[2]spud *noun* [Middle English *spudde* dagger] (1667)
1 : a tool or device (as for digging, lifting, or cutting) having the characteristics of a spade and a chisel
2 : POTATO

[1]spume \'spyüm\ *noun* [Middle English, from Middle French, from Latin *spuma* — more at FOAM] (14th century)
: frothy matter on liquids : FOAM, SCUM
— **spu·mous** \'spyü-məs\ *adjective*
— **spumy** \-mē\ *adjective*

[2]spume *intransitive verb* **spumed; spum·ing** (14th century)
: FROTH, FOAM

spu·mo·ni *or* **spu·mo·ne** \spu̇-'mō-nē\ *noun* [Italian *spumone,* augmentative of *spuma* foam, from Latin] (1924)
: ice cream in layers of different colors, flavors, and textures often with candied fruits and nuts

spun *past and past participle of* SPIN

spun·bond·ed \'spən-ˌbän-dəd\ *adjective* (1961)
: of or relating to a nonwoven polymeric material that resembles cloth or fabric

spun glass *noun* (1779)
1 : blown glass that has slender threads of glass incorporated in it
2 : FIBERGLASS

[1]spunk \'spəŋk\ *noun* [Scottish Gaelic *spong* sponge, tinder, from Latin *spongia* sponge] (1582)
1 a : a woody tinder : PUNK **b :** any of various fungi used to make tinder
2 : METTLE, PLUCK
3 : SPIRIT, LIVELINESS

[2]spunk *intransitive verb* (1840)
dialect **:** to show spirit — usually used with *up*

spunk·ie \'spəŋ-kē\ *noun* (1727)
Scottish : IGNIS FATUUS 1

spunky \'spəŋ-kē\ *adjective* **spunk·i·er; -est** (1786)
: full of spunk : SPIRITED
— **spunk·i·ly** \-kə-lē\ *adverb*
— **spunk·i·ness** \-kē-nəs\ *noun*

spun sugar *noun* (1846)
: sugar boiled to long threads and gathered up and shaped or heaped on a stick as a candy

spun yarn *noun* (14th century)
1 : a textile yarn spun from staple-length fiber
2 : a small rope or stuff formed of two or more rope yarns loosely twisted and used for seizings especially on board ship

[1]spur \'spər\ *noun* [Middle English *spure,* from Old English *spura;* akin to Old English *spurnan* to kick — more at SPURN] (before 12th century)
1 a : a pointed device secured to a rider's heel and used to urge on the horse **b** *plural* [from the acquisition of spurs by a person achieving knighthood] **:** recognition and reward for achievement ⟨won his academic *spurs* as the holder of a chair in a university —James Mountford⟩
2 : a goad to action : STIMULUS
3 : something projecting like or suggesting a spur: as **a :** a projecting root or branch of a tree, shrub, or vine **b** (1) **:** a stiff sharp spine (as on the wings or legs of a bird or insect); *especially* **:** one on a cock's leg (2) **:** a gaff for a gamecock **c :** a hollow projecting appendage of a corolla or calyx (as in larkspur or columbine) **d :** a bony outgrowth (as on the heel of the foot) **e :** CLIMBING IRON
4 a : an angular projection, offshoot, or branch extending out beyond or away from a main body or formation; *especially* **:** a ridge or lesser elevation that extends laterally from a mountain or mountain range **b :** a railroad track that branches off from a main line
5 : a reinforcing buttress of masonry in a fortification
synonym see MOTIVE
— **on the spur of the moment :** on impulse : SUDDENLY

[2]spur *verb* **spurred; spur·ring** (13th century)
transitive verb
1 : to urge (a horse) on with spurs
2 : to incite to action or accelerated growth or development : STIMULATE
3 : to put spurs on
intransitive verb
: to spur one's horse on

spurge \'spərj\ *noun* [Middle English, from Middle French, purge, spurge, from *espurgier* to purge, from Latin *expurgare* — more at EXPURGATE] (14th century)
: any of a genus (*Euphorbia* of the family Euphorbiaceae, the spurge family) of chiefly herbs and shrubs with a bitter milky juice

spur gear *noun* (1823)
: a gear wheel with radial teeth parallel to its axis — called also *spur wheel*

spur gear

spurge laurel *noun* (1597)
: a low Eurasian shrub (*Daphne laureola*) with oblong evergreen leaves and axillary racemes of yellowish flowers

spu·ri·ous \'spyu̇r-ē-əs\ *adjective* [Late Latin & Latin; Late Latin *spurius* false, from Latin, of illegitimate birth, from *spurius,* noun, bastard] (1598)
1 : of illegitimate birth : BASTARD
2 : outwardly similar or corresponding to something without having its genuine qualities : FALSE
3 a : of falsified or erroneously attributed origin : FORGED **b :** of a deceitful nature or quality
— **spu·ri·ous·ly** *adverb*
— **spu·ri·ous·ness** *noun*

[1]spurn \'spərn\ *verb* [Middle English, from Old English *spurnan;* akin to Old High German *spurnan* to kick, Latin *spernere* to spurn, Greek *spairein* to quiver] (before 12th century)
intransitive verb
1 *obsolete* **a :** STUMBLE **b :** KICK
2 *archaic* **:** to reject something disdainfully
transitive verb
1 : to tread sharply or heavily upon : TRAMPLE
2 : to reject with disdain or contempt : SCORN
synonym see DECLINE
— **spurn·er** *noun*

[2]spurn *noun* (14th century)
1 a : KICK **b** *obsolete* **:** STUMBLE
2 a : disdainful rejection **b :** contemptuous treatment

spur-of-the-moment *adjective* (1948)
: occurring or developing without premeditation : hastily extemporized ⟨a *spur-of-the-moment* decision⟩

spurred \'spərd\ *adjective* (15th century)
1 : wearing spurs
2 : having one or more spurs ⟨a *spurred* violet⟩

spur·rey *or* **spur·ry** \'spər-ē, 'spə-rē\ *noun, plural* **spurreys** *or* **spurries** [Dutch *spurrie,* from Medieval Latin *spergula*] (1577)
: a small white-flowered European weed (*Spergula arvensis*) of the pink family with whorled filiform leaves; *also* **:** any of several related and similar herbs

[1]spurt \'spərt\ *verb* [perhaps akin to Middle High German *spürzen* to spit, Old English *-sprūtan* to sprout — more at SPROUT] (1570)
intransitive verb
: to gush forth : SPOUT
transitive verb
: to expel in a stream or jet : SQUIRT

[2]spurt *noun* (circa 1775)
: a sudden gush : JET

[3]spurt *noun* [origin unknown] (circa 1591)
1 : a short period of time : MOMENT
2 a : a sudden brief burst of effort or activity **b :** a sharp or sudden increase in business activity

[4]spurt *intransitive verb* (1664)
: to make a spurt

spur·tle \'spər-tᵊl\ *noun* [origin unknown] (1756)
chiefly Scottish **:** a wooden stick for stirring porridge

sput·nik \'spu̇t-nik, 'spət-, 'spüt-\ *noun* [Russian, literally, traveling companion, from *s, so* with + *put'* path] (1957)
: SATELLITE 2b

[1]sput·ter \'spə-tər\ *verb* [akin to Dutch *sputteren* to sputter] (1598)
transitive verb
1 : to spit or squirt from the mouth with explosive sounds
2 : to utter hastily or explosively in confusion or excitement
3 : to dislodge (atoms) from the surface of a material by collision with high energy particles; *also* **:** to deposit (a metallic film) by such a process
intransitive verb
1 : to spit or squirt particles of food or saliva noisily from the mouth
2 : to speak explosively or confusedly in anger or excitement
3 : to make explosive popping sounds
— **sput·ter·er** *noun*

[2]sputter *noun* (1673)
1 : confused and excited speech or discussion
2 : the act or sound of sputtering

spu·tum \'spyü-təm, 'spü-\ *noun, plural* **spu·ta** \-tə\ [Latin, from neuter of *sputus,* past participle of *spuere* to spit — more at SPEW] (circa 1693)
: expectorated matter made up of saliva and often discharges from the respiratory passages

[1]**spy** \ˈspī\ *verb* **spied; spy·ing** [Middle English *spien,* from Old French *espier,* of Germanic origin; akin to Old High German *spehōn* to spy; akin to Latin *specere* to look, look at, Greek *skeptesthai & skopein* to watch, look at, consider] (13th century)
transitive verb
1 : to watch secretly usually for hostile purposes
2 : to catch sight of **:** SEE
3 : to search or look for intensively
intransitive verb
1 : to observe or search for something **:** LOOK
2 : to watch secretly as a spy

[2]**spy** *noun, plural* **spies** (13th century)
1 : one that spies: **a :** one who keeps secret watch on a person or thing to obtain information **b :** a person employed by one nation to secretly convey classified information of strategic importance to another nation; *also* **:** a person who conveys the trade secrets of one company to another
2 : an act of spying

spy·glass \ˈspī-ˌglas\ *noun* (1706)
: a small telescope

spy·mas·ter \ˈspī-ˌmas-tər\ *noun* (1938)
: the head of a ring of spies **:** a director of intelligence

squab \ˈskwäb\ *noun, plural* **squabs** [probably of Scandinavian origin; akin to Swedish dialect *skvabb* anything soft and thick] (1664)
1 a : COUCH **b :** a cushion for a chair or couch
2 *or plural* **squab :** a fledgling bird; *specifically* **:** a fledgling pigeon about four weeks old
3 : a short fat person
— squab *adjective*

[1]**squab·ble** \ˈskwä-bəl\ *noun* [probably of Scandinavian origin; akin to Swedish dialect *skvabbel* dispute] (1602)
: a noisy altercation or quarrel usually over trifles

[2]**squabble** *intransitive verb* **squab·bled; squab·bling** \-b(ə-)liŋ\ (1604)
: to quarrel noisily and usually over trifles
— squab·bler \-b(ə-)lər\ *noun*

[1]**squad** \ˈskwäd\ *noun* [Middle French *esquade,* from Old Spanish & Old Italian; Old Spanish *escuadra* & Old Italian *squadra,* ultimately from (assumed) Vulgar Latin *exquadrare* to make square — more at SQUARE] (1649)
1 : a small organized group of military personnel; *especially* **:** a tactical unit that can be easily directed in the field
2 : a small group engaged in a common effort or occupation

[2]**squad** *transitive verb* **squad·ded; squad·ding** (circa 1802)
: to arrange in squads

squad car *noun* (1938)
: a police automobile connected by a two-way radio with headquarters — called also *cruiser, prowl car*

squad·ron \ˈskwä-drən\ *noun* [Italian *squadrone,* augmentative of *squadra* squad, from Old Italian] (1562)
: a unit of military organization: **a :** a cavalry unit higher than a troop and lower than a regiment **b :** a naval unit consisting of two or more divisions and sometimes additional vessels **c (1) :** a unit of the U.S. Air Force higher than a flight and lower than a group **(2) :** a military flight formation

squadron leader *noun* (1919)
: a commissioned officer in the British air force who ranks with a major in the army

squad room *noun* (1943)
1 : a room in a barracks used to billet soldiers
2 : a room in a police station where members of the force assemble

squa·lene \ˈskwā-ˌlēn\ *noun* [International Scientific Vocabulary, from Latin *squalus,* a sea fish — more at WHALE] (1916)
: an acyclic hydrocarbon $C_{30}H_{50}$ that is widely distributed in nature (as a major component of sebum and in shark-liver oils) and is a precursor of sterols (as cholesterol)

squal·id \ˈskwä-ləd\ *adjective* [Latin *squalidus* rough, dirty, from *squalēre* to be covered with scales or dirt, from *squalus* dirty; perhaps akin to *squama* scale] (1596)
1 : marked by filthiness and degradation from neglect or poverty
2 : SORDID
synonym see DIRTY
— squal·id·ly *adverb*
— squal·id·ness *noun*

[1]**squall** \ˈskwȯl\ *verb* [probably of Scandinavian origin; akin to Old Norse *skval* useless chatter] (circa 1631)
intransitive verb
: to cry out raucously **:** SCREAM
transitive verb
: to utter in a strident voice
— squall·er *noun*

[2]**squall** *noun* (1709)
: a raucous cry

[3]**squall** *noun* [probably of Scandinavian origin; akin to Swedish *skval* rushing water] (1699)
1 : a sudden violent wind often with rain or snow
2 : a short-lived commotion

[4]**squall** *intransitive verb* (circa 1890)
: to blow a squall

squally \ˈskwȯ-lē\ *adjective* **squall·i·er; -est** (1719)
1 : marked by squalls
2 : GUSTY

squa·lor \ˈskwä-lər *also* ˈskwā- *or* ˈskwȯ-\ *noun* [Latin, from *squalēre*] (1621)
: the quality or state of being squalid

squa·ma \ˈskwā-mə, ˈskwä-\ *noun, plural* **squa·mae** \ˈskwā-ˌmē, ˈskwä-ˌmī\ [Latin] (circa 1706)
: SCALE; *also* **:** a structure resembling a scale

squa·mate \-ˌmāt\ *adjective* (1826)
: SCALY ⟨*squamate* reptiles⟩

squa·ma·tion \skwə-ˈmā-shən\ *noun* (1881)
1 : the state of being scaly
2 : the arrangement of scales on an animal

[1]**squa·mo·sal** \skwə-ˈmō-səl, -zəl\ *noun* (1848)
: a squamosal bone

[2]**squamosal** *adjective* (circa 1852)
1 : SQUAMOUS
2 : of, relating to, or being a bone of the skull of many vertebrates corresponding to the squamous portion of the temporal bone of most mammals including humans

squa·mous \ˈskwā-məs *also* ˈskwä-\ *adjective* [Middle English, from Latin *squamosus,* from *squama* scale] (15th century)
1 a : covered with or consisting of scales **:** SCALY **b :** of, relating to, or being a stratified epithelium that consists at least in its outer layers of small scalelike cells
2 : of, relating to, or being the anterior upper portion of the temporal bone of most mammals including humans

squamous cell *noun* (circa 1947)
: a cell of or derived from squamous epithelium

squa·mu·lose \ˈskwā-myə-ˌlōs, ˈskwä-\ *adjective* [Latin *squamula,* diminutive of *squama*] (1846)
: being or having a thallus made up of small leafy lobes ⟨a *squamulose* lichen⟩

[1]**squan·der** \ˈskwän-dər\ *verb* **squan·dered; squan·der·ing** \-d(ə-)riŋ\ [origin unknown] (1536)
transitive verb
1 : to spend extravagantly or foolishly **:** DISSIPATE
2 : to cause to disperse **:** SCATTER
3 : to lose (as an advantage or opportunity) through negligence or inaction
intransitive verb
: DISPERSE, SCATTER
— squan·der·er \-dər-ər\ *noun*

[2]**squander** *noun* (1709)
: an act of squandering

[1]**square** \ˈskwar, ˈskwer\ *noun* [Middle English, from Middle French *esquarre,* from (assumed) Vulgar Latin *exquadra,* from *exquadrare* to square, from Latin *ex-* + *quadrare* to square — more at QUADRATE] (13th century)
1 : an instrument having at least one right angle and two straight edges used especially to lay out or test right angles
2 : a rectangle with all four sides equal
3 : any of the quadrilateral spaces marked out on a board for playing games
4 : the product of a number multiplied by itself
5 a : an open place or area formed at the meeting of two or more streets **b :** BLOCK 6a
6 : a solid object or piece approximating a cube or having a square as its largest face
7 : an unopened cotton flower with its enclosing bracts
8 : a person who is conventional or conservative in taste or way of life
— on the square 1 : at right angles **2 :** in a fair open manner **:** HONESTLY
— out of square : not at an exact right angle

[2]**square** *adjective* **squar·er; squar·est** (14th century)
1 a : having four equal sides and four right angles **b :** forming a right angle ⟨*square* corner⟩ **c :** having a square base ⟨a *square* pyramid⟩
2 : raised to the second power
3 a : being approximately a cube ⟨*square* cabinet⟩ **b :** having a shape that is broad for the height and rectangular rather than curving in outline ⟨*square* shoulders⟩ ⟨a *square,* thick, hard-working man —Maria Edgeworth⟩ **c :** rectangular and equilateral in section ⟨*square* tower⟩
4 a : being or converted to a unit of area equal in measure to a square each side of which measures one unit of a specified unit of length ⟨a *square* foot⟩ — see METRIC SYSTEM table, WEIGHT table **b :** being of a specified length in each of two equal dimensions ⟨10 feet *square*⟩
5 a : exactly adjusted **:** precisely constructed or aligned **b :** JUST, FAIR ⟨a *square* deal⟩ ⟨*square* in all his dealings⟩ **c :** leaving no balance **:** SETTLED **d :** EVEN, TIED **e :** SUBSTANTIAL, SATISFYING ⟨*square* meal⟩ **f :** being unsophisticated, conservative, or conventional
6 : set at right angles with the mast and keel — used of the yards of a square-rigged ship
— square·ness *noun*

[3]**square** *verb* **squared; squar·ing** (14th century)
transitive verb
1 a : to make square or rectangular ⟨*square* a building stone⟩ **b :** to test for deviation from a right angle, straight line, or plane surface
2 : to bring approximately to a right angle ⟨*squared* his shoulders⟩
3 a : to multiply (a number) by itself **:** raise to the second power **b :** to find a square equal in area to ⟨*square* a circle⟩
4 : to regulate or adjust by or to some standard or principle ⟨*square* our actions by the opinions of others —John Milton⟩
5 a : BALANCE, SETTLE ⟨*square* an account⟩ **b :** to even the score of
6 : to mark off into squares
7 a : to set right **:** bring into agreement **b :** BRIBE, FIX
intransitive verb
1 : to agree precisely **:** CORRESPOND ⟨your actions should *square* with your words⟩
2 : to settle matters; *especially* **:** to pay the bill
— squar·er *noun*

[4]**square** *adverb* (circa 1582)

1 : in a straightforward or honest manner
2 a : so as to face or be face to face **b :** at right angles
3 : with nothing intervening **:** DIRECTLY ⟨ran *square* into it⟩
4 : in a firm manner ⟨looked her *square* in the eye⟩
5 : in a square shape

square away (1849)
intransitive verb
1 : to square the yards so as to sail before the wind
2 : to put everything in order or in readiness
3 : to take up a fighting stance
transitive verb
: to put in order or in readiness

square bracket *noun* (circa 1888)
: BRACKET 3a

square dance *noun* (1870)
: a dance for four couples who form the sides of a square
— **square–dance** *intransitive verb*
— **square dancer** *noun*
— **square dancing** *noun*

square knot *noun* (circa 1867)
: a knot made of two reverse half-knots and typically used to join the ends of two cords — see KNOT illustration

square·ly \'skwar-lē, 'skwer-\ *adverb* (1564)
1 : in a straightforward or honest manner ⟨we must *squarely* face the issue⟩
2 a : EXACTLY, PRECISELY ⟨*squarely* in the middle⟩ **b :** so as to make solid contact ⟨hit the ball *squarely*⟩ ⟨feet *squarely* planted⟩
3 : in a square form or manner **:** so as to be square ⟨a *squarely* cut dress⟩
4 : in a plain or unequivocal manner ⟨the responsibility lies *squarely* with us⟩ ⟨align ourselves *squarely* with our allies⟩

square matrix *noun* (1858)
: a mathematical matrix with the same number of rows and columns

square measure *noun* (1728)
: a unit or system of units for measuring area — see METRIC SYSTEM table, WEIGHT table

square off *intransitive verb* (1837)
: to take a fighting stance **:** prepare to fight; *also* **:** FIGHT

square of opposition (1864)
: a square figure on which may be demonstrated the logical relationships of contraries, contradictories, subcontraries, and subalterns and superalterns

square one *noun* [from the use of numbered squares in some board games] (1960)
: the initial stage or starting point ⟨the failure set us back to *square one*⟩

square rig *noun* (circa 1875)
: a sailing-ship rig in which the principal sails are extended on yards fastened to the masts horizontally and at their center

square–rigged \'skwar-'rigd, 'skwer-\ *adjective* (1769)
: having or equipped with a square rig

square–rig·ger \-,ri-gər\ *noun* (1855)
: a square-rigged craft

square-rigger

square root *noun* (1557)
: a factor of a number that when squared gives the number ⟨the *square root* of 9 is ±3⟩

square sail \'skwar-,sāl, 'skwer-, -səl\ *noun* (1600)
: a 4-sided sail extended on a yard suspended at the middle from a mast

square shooter *noun* (circa 1914)
: a just or honest person

square–shoul·dered \'skwar-'shōl-dərd, 'skwer-\ *adjective* (1825)
: having shoulders of a rectangular outline that are straight across the back

square–toed \-'tōd\ *adjective* (1785)
1 : having a toe that is square
2 : OLD-FASHIONED, CONSERVATIVE

square wave *noun* (1932)
: the rectangular waveform of a quantity that varies periodically and abruptly from one to the other of two uniform values

squar·ish \'skwar-ish, 'skwer-\ *adjective* (1742)
: somewhat square in form or appearance
— **squar·ish·ly** *adverb*
— **squar·ish·ness** *noun*

[1]**squash** \'skwäsh, 'skwȯsh\ *verb* [Middle French *esquasser,* from (assumed) Vulgar Latin *exquassare,* from Latin *ex-* + *quassare* to shake — more at QUASH] (1565)
transitive verb
1 : to press or beat into a pulp or a flat mass **:** CRUSH
2 : PUT DOWN, SUPPRESS ⟨*squash* a revolt⟩
intransitive verb
1 : to flatten out under pressure or impact
2 : to proceed with a splashing or squelching sound
3 : SQUEEZE, PRESS
— **squash·er** *noun*

[2]**squash** *noun* (1590)
1 *obsolete* **:** something soft and easily crushed; *specifically* **:** an unripe pod of peas
2 : the sudden fall of a heavy soft body or the sound of such a fall
3 : a squelching sound made by walking on oozy ground or in water-soaked boots
4 : a crushed mass
5 *British* **:** sweetened citrus fruit juice often served with added soda water
6 : a singles or doubles game played in a 4-wall court with a long-handled racket and a rubber ball that can be hit off any number of walls

[3]**squash** *adverb* (1766)
: with a squash or a squashing sound

[4]**squash** *noun, plural* **squash·es** *or* **squash** [by shortening & alteration from earlier *isquoutersquash,* from Narraganset *askútasquash*] (1634)
: any of various fruits of plants (genus *Cucurbita*) of the gourd family widely cultivated as vegetables; *also* **:** a plant and especially a vine that bears squashes — compare SUMMER SQUASH, WINTER SQUASH

squash bug *noun* (circa 1846)
: a large black American bug (*Anasa tristis* of the family Coreidae) injurious to plants of the gourd family

squash racquets *noun plural but singular in construction* (1886)
: SQUASH 6

squash tennis *noun* (1901)
: a singles racket game resembling squash played with an inflated ball the size of a tennis ball

squashy \'skwä-shē, 'skwȯ-\ *adjective* **squash·i·er; -est** (1698)
1 : easily squashed **:** very soft ⟨*squashy* cushions⟩
2 : softly wet **:** BOGGY
3 : soft because overripe ⟨*squashy* melons⟩
— **squash·i·ly** \-shə-lē\ *adverb*
— **squash·i·ness** \-shē-nəs\ *noun*

[1]**squat** \'skwät\ *verb* **squat·ted; squat·ting** [Middle English *squatten,* from Middle French *esquatir,* from *es-* ex- (from Latin *ex-*) + *quatir* to press, from (assumed) Vulgar Latin *coactire* to press together, from Latin *coactus,* past participle of *cogere* to drive together — more at COGENT] (15th century)
transitive verb
1 : to cause (oneself) to crouch or sit on the ground
2 : to occupy as a squatter
intransitive verb
1 : to crouch close to the ground as if to escape observation ⟨a hare *squatting* in the grass⟩
2 : to assume or maintain a position in which the body is supported on the feet and the knees are bent so that the buttocks rest on or near the heels
3 : to be or become a squatter

[2]**squat** *adjective* **squat·ter; squat·test** (15th century)
1 : sitting with the haunches close above the heels
2 a : low to the ground **b :** marked by disproportionate shortness or thickness
— **squat·ly** *adverb*
— **squat·ness** *noun*

[3]**squat** *noun* (1580)
1 a : the act of squatting **b :** the posture of one that squats
2 a : a place where one squats **b :** the lair of a small animal ⟨*squat* of a hare⟩
3 : a lift in weight lifting in which the lifter performs a knee bend while holding a barbell on the shoulders; *also* **:** a competitive event involving this lift
4 *chiefly British* **:** an empty house or building that is occupied by squatters
5 *slang* **:** DIDDLY-SQUAT

[1]**squat·ter** \'skwä-tər\ *intransitive verb* [imitative] (1785)
: to go along through or as if through water

[2]**squatter** *noun* (1788)
: one that squats: as **a :** one that settles on property without right or title or payment of rent **b :** one that settles on public land under government regulation with the purpose of acquiring title

squatter sovereignty *noun* (1854)
: POPULAR SOVEREIGNTY 2

squat·ty \'skwä-tē\ *adjective* **squat·ti·er; -est** (1881)
1 : low to the ground
2 : DUMPY, THICKSET

squaw \'skwȯ\ *noun* [Massachuset *squa, ussqua* woman] (1634)
1 : an American Indian woman
2 : WOMAN, WIFE — usually used disparagingly

squaw·fish \-,fish\ *noun* (1881)
: any of several large cyprinid fishes (genus *Ptychocheilus*) of western North America

[1]**squawk** \'skwȯk\ *intransitive verb* [probably blend of *squall* and *squeak*] (1821)
1 : to utter a harsh abrupt scream
2 : to complain or protest loudly or vehemently
— **squawk·er** *noun*

[2]**squawk** *noun* (1850)
1 : a harsh abrupt scream
2 : a noisy complaint

squawk box *noun* (1945)
: an intercom speaker

squaw man *noun* (1866)
: a white man married to an Indian woman and usually living as one of her tribe

squaw·root \'skwȯ-,rüt, -,rut\ *noun* (circa 1848)
: a North American scaly herb (*Conopholis americana*) of the broomrape family parasitic on oak and hemlock roots

[1]**squeak** \'skwēk\ *verb* [Middle English *squeken*] (14th century)
intransitive verb
1 : to utter or make a short shrill cry or noise
2 : SQUEAL 2a
3 : to pass, succeed, or win by a narrow margin ⟨just *squeaked* by in the election⟩
transitive verb
: to utter in a shrill piping tone

[2]**squeak** *noun* (1700)
1 : a sharp shrill cry or sound
2 : ESCAPE ⟨a close *squeak*⟩
— **squeaky** \'skwē-kē\ *adjective*

squeak·er \'skwē-kər\ *noun* (1671)
1 : one that squeaks
2 : a contest (as a game or an election) won by a small margin

squeaky–clean *adjective* (1968)
1 : completely clean ⟨*squeaky-clean* hair⟩

2 : completely free from moral taint of any kind ⟨a *squeaky-clean* reputation⟩

[1]**squeal** \ˈskwē(ə)l\ *verb* [Middle English *squelen*] (14th century)
intransitive verb
1 : to make a shrill cry or noise
2 a : to turn informer **b :** COMPLAIN, PROTEST
transitive verb
1 : to utter or express with or as if with a squeal
2 : to cause to make a loud shrill noise ⟨*squealing* the tires⟩
— **squeal·er** *noun*

[2]**squeal** *noun* (1747)
: a shrill sharp cry or noise

squea·mish \ˈskwē-mish\ *adjective* [Middle English *squaymisch,* modification of Anglo-French *escoymous*] (15th century)
1 a : easily nauseated **:** QUEASY **b :** affected with nausea
2 a : excessively fastidious or scrupulous in conduct or belief **b :** easily offended or disgusted
— **squea·mish·ly** *adverb*
— **squea·mish·ness** *noun*

[1]**squee·gee** \ˈskwē-ˌjē\ *noun* [origin unknown] (1844)
: a blade of leather or rubber set on a handle and used for spreading, pushing, or wiping liquid material on, across, or off a surface (as a window); *also* **:** a smaller similar device or a small rubber roller with handle used by a photographer or lithographer

[2]**squeegee** *transitive verb* **squee·geed; squee·gee·ing** (1883)
: to smooth, wipe, or treat with a squeegee

[1]**squeeze** \ˈskwēz\ *verb* **squeezed; squeez·ing** [alteration of obsolete English *quease,* from Middle English *queysen,* from Old English *cwȳsan;* akin to Icelandic *kveisa* stomach cramps] (circa 1601)
transitive verb
1 a : to exert pressure especially on opposite sides of **:** COMPRESS **b :** to extract or emit under pressure **c :** to force or thrust by compression
2 a (1) **:** to get by extortion (2) **:** to deprive by extortion **b :** to cause economic hardship to **c :** to reduce the amount of ⟨*squeezes* profits⟩
3 : to crowd into a limited area
4 : to gain or win by a narrow margin
5 : to force (another player) to discard in bridge so as to unguard a suit
6 : to score by means of a squeeze play
intransitive verb
1 : to give way before pressure
2 : to exert pressure; *also* **:** to practice extortion or oppression
3 : to force one's way ⟨*squeeze* through a door⟩
4 : to pass, win, or get by narrowly
— **squeez·abil·i·ty** \ˌskwē-zə-ˈbi-lə-tē\ *noun*
— **squeez·able** \ˈskwē-zə-bəl\ *adjective*
— **squeez·er** *noun*

[2]**squeeze** *noun* (1611)
1 a : an act or instance of squeezing **:** COMPRESSION **b :** HANDCLASP; *also* **:** EMBRACE
2 a : a quantity squeezed out from something ⟨a *squeeze* of lemon⟩ **b :** a group crowded together **:** CROWD
3 : a profit taken by a middleman on goods or transactions
4 : a financial pressure caused by narrowing margins or by shortages
5 : a forced discard in bridge
6 : SQUEEZE PLAY
7 *slang* **:** MAIN SQUEEZE

squeeze bottle *noun* (1950)
: a bottle of flexible plastic that dispenses its contents when it is squeezed

squeeze off (circa 1949)
transitive verb
: to fire (a round) by squeezing the trigger
intransitive verb
: to fire a weapon by squeezing the trigger

squeeze play *noun* (1905)
1 : a baseball play in which a runner on third base starts for home plate as the ball is being pitched and the batter attempts to bunt to give the runner a chance to score
2 : the exertion of pressure in order to extort a concession or gain a goal

[1]**squelch** \ˈskwelch\ *verb* [origin unknown] (1624)
transitive verb
1 a : to fall or stamp on so as to crush **b** (1) **:** to completely suppress **:** QUELL (2) **:** SILENCE
2 : to emit or move with a sucking sound
intransitive verb
1 : to emit a sucking sound
2 : to splash through water, slush, or mire
— **squelch·er** *noun*

[2]**squelch** *noun* (1895)
1 : a sound of or as if of semiliquid matter under suction ⟨the *squelch* of mud⟩
2 : the act of suppressing; *especially* **:** a retort that silences an opponent
— **squelchy** *adjective*

sque·teague \skwi-ˈtēg\ *noun, plural* **squeteague** [of Algonquian origin; akin to Mohegan *cheegut* weakfish] (1803)
: any of several weakfishes (especially *Cynoscion regalis*)

[1]**squib** \ˈskwib\ *noun* [origin unknown] (circa 1525)
1 a : a short humorous or satiric writing or speech **b :** a short news item; *especially* **:** FILLER
2 a : a small firecracker **b :** a broken firecracker in which the powder burns with a fizz
3 : a small electric or pyrotechnic device used to ignite a charge

[2]**squib** *verb* **squibbed; squib·bing** (circa 1580)
intransitive verb
1 : to speak, write, or publish squibs
2 : to fire a squib
transitive verb
1 a : to utter in an offhand manner **b :** to make squibs against **:** LAMPOON
2 : to shoot off **:** FIRE
3 : to kick (a football) just far enough on a kickoff to be legally recoverable by the kicking team

squib kick *noun* (circa 1956)
: ONSIDE KICK

[1]**squid** \ˈskwid\ *noun, plural* **squid** *or* **squids** [origin unknown] (1613)
: any of an order (Teuthoidea) of cephalopods having eight short arms and two usually longer tentacles, a long tapered body, a caudal fin on each side, and usually a slender internal chitinous support

[2]**squid** *intransitive verb* **squid·ded; squid·ding** (circa 1859)
: to fish with or for squid

SQUID \ˈskwid\ *noun* [*s*uperconducting *qu*antum *i*nterference *d*evice] (1967)
: an instrument for detecting and measuring very weak magnetic fields

squiffed \ˈskwift\ *or* **squif·fy** \ˈskwi-fē\ *adjective* [origin unknown] (circa 1855)
: INTOXICATED, DRUNK

[1]**squig·gle** \ˈskwi-gəl\ *verb* **squig·gled; squig·gling** \-g(ə-)liŋ\ [blend of *squirm* and *wriggle*] (circa 1816)
intransitive verb
1 : SQUIRM, WRIGGLE
2 : to write or paint hastily **:** SCRIBBLE
transitive verb
1 : SCRIBBLE
2 : to form or cause to form in squiggles

[2]**squiggle** *noun* (1900)
: a short wavy twist or line **:** CURLICUE; *especially* **:** an illegible scrawl
— **squig·gly** \-g(ə-)lē\ *adjective*

squil·gee \ˈskwē-ˌjē, ˈskwil-ˌjē\ *variant of* SQUEEGEE

squill \ˈskwil\ *noun* [Middle English, from Latin *squilla, scilla,* from Greek *skilla*] (14th century)
1 a : a Mediterranean bulbous herb (*Urginea maritima*) of the lily family — called also *sea onion*; compare RED SQUILL 1 **b** (1) **:** the dried sliced bulb scales of a squill used as an expectorant, cardiac stimulant, and diuretic (2) **:** RED SQUILL 2
2 : SCILLA

squil·la \ˈskwi-lə\ *noun, plural* **squillas** *or* **squil·lae** \ˈskwi-ˌlē, -ˌlī\ [New Latin, from Latin, shrimp, crayfish] (1658)
: any of various stomatopod crustaceans (especially genus *Squilla*) that burrow in mud or beneath stones in shallow water along the seashore

[1]**squinch** \ˈskwinch\ *verb* [probably blend of *squint* and *pinch*] (1835)
transitive verb
1 : to screw up (the eyes or face) **:** SQUINT
2 a : to make more compact **b :** to cause to crouch down or draw together
intransitive verb
1 : FLINCH
2 : to crouch down or draw together
3 : SQUINT

[2]**squinch** *noun* [alteration of earlier *scunch* back part of the side of an opening] (circa 1840)
: a support (as an arch, lintel, or corbeling) carried across the corner of a room under a superimposed mass

[1]**squin·ny** \ˈskwi-nē\ *verb* **squin·nied; squin·ny·ing** [probably from obsolete English *squin* asquint, from Middle English *skuin*] (1605)
: SQUINT

[2]**squinny** *noun* (circa 1881)
: SQUINT
— **squinny** *adjective*

[1]**squint** \ˈskwint\ *adjective* [Middle English *asquint*] (1579)
1 *of an eye* **:** looking or tending to look obliquely or askance (as with envy or disdain)
2 *of the eyes* **:** not having the visual axes parallel **:** CROSSED

[2]**squint** (1599)
intransitive verb
1 a : to have an indirect bearing, reference, or aim **b :** to deviate from a true line
2 a : to look in a squint-eyed manner **b :** to be cross-eyed **c :** to look or peer with eyes partly closed
transitive verb
: to cause (an eye) to squint
— **squint·er** *noun*
— **squint·ing·ly** \ˈskwin-tiŋ-lē\ *adverb*

[3]**squint** *noun* (circa 1652)
1 : STRABISMUS
2 : an instance of squinting
3 : HAGIOSCOPE
— **squinty** \ˈskwin-tē\ *adjective*

squint–eyed \ˈskwint-ˌīd\ *adjective* (1589)
1 : having eyes that squint; *specifically* **:** affected with cross-eye
2 : looking askance (as in envy)

squinting modifier *noun* (1924)
: a modifier (as *often* in "getting dressed often is a nuisance") so placed in a sentence that it can be interpreted as modifying either what precedes or what follows

[1]**squire** \ˈskwīr\ *noun* [Middle English *squier,* from Old French *esquier* — more at ESQUIRE] (13th century)
1 : a shield bearer or armor bearer of a knight
2 a : a male attendant especially on a great personage **b :** a man who devotedly attends a lady **:** GALLANT
3 a : a member of the British gentry ranking below a knight and above a gentleman **b :** an owner of a country estate; *especially* **:** the

principal landowner in a village or district **c** (1) **:** JUSTICE OF THE PEACE (2) **:** LAWYER (3) **:** JUDGE
— **squir·ish** \'skwīr-ish\ *adjective*

²**squire** *transitive verb* **squired; squir·ing** (14th century)
: to attend as a squire **:** ESCORT

squire·ar·chy *also* **squir·ar·chy** \'skwīr-ˌär-kē\ *noun, plural* **-chies** (1796)
: the class of landed gentry or landed proprietors

squirm \'skwərm\ *intransitive verb* [origin unknown] (circa 1691)
: to twist about like a worm **:** FIDGET
— **squirm** *noun*
— **squirmy** \'skwər-mē\ *adjective*

¹**squir·rel** \'skwər(-ə)l, 'skwə-rəl, *chiefly British* 'skwir-əl\ *noun, plural* **squirrels** *also* **squirrel** [Middle English *squirel,* from Middle French *esquireul,* from (assumed) Vulgar Latin *scuriolus,* diminutive of *scurius,* alteration of Latin *sciurus,* from Greek *skiouros,* probably from *skia* shadow + *oura* tail — more at SHINE, ASS] (14th century)
1 : any of various small or medium-sized rodents (family Sciuridae, the squirrel family): as **a :** any of numerous New or Old World arboreal forms having a long bushy tail and strong hind legs **b :** GROUND SQUIRREL
2 : the fur of a squirrel ◆

²**squirrel** *transitive verb* **-reled** *or* **-relled; -rel·ing** *or* **-rel·ling** [from the squirrel's habit of storing up gathered nuts and seeds for winter use] (1925)
: to store up for future use — often used with *away*

squirrel cage *noun* (1821)
1 : a cage for a small animal (as a squirrel) that contains a rotatable cylinder for exercising
2 : something resembling the working of a squirrel cage in repetitiveness or endlessness

squirrel corn *noun* (1843)
: a North American herb (*Dicentra canadensis*) of the fumitory family with much-divided leaves and a scapose raceme of cream-colored flowers

squir·rel·ly \'skwər(-ə)-lē, 'skwə-rə-\ *adjective* (1928)
: NUTTY 3

squirrel monkey *noun* (1773)
: a small soft-haired South American monkey (*Saimiri sciureus*) that has a long tail not used for grasping and is colored chiefly yellowish gray with a white face and black muzzle

squirrel rifle *noun* [from its being suitable only for small game] (1834)
: a small-caliber rifle — called also *squirrel gun*

¹**squirt** \'skwərt\ *verb* [Middle English; akin to Low German *swirtjen* to squirt] (15th century)
intransitive verb
: to come forth in a sudden rapid stream from a narrow opening **:** SPURT
transitive verb
: to cause to squirt
— **squirt·er** *noun*

²**squirt** *noun* (15th century)
1 a : an instrument (as a syringe) for squirting a liquid **b :** a small quick stream **:** JET **c :** the action or an instance of squirting
2 a : an impudent youngster **b :** KID

squirt gun *noun* (1803)
: WATER PISTOL

squirting cucumber *noun* (1802)
: a Mediterranean plant (*Ecballium elaterium*) of the gourd family with oblong fruit that bursts from the peduncle when ripe and forcibly ejects the seeds

squish \'skwish\ *verb* [alteration of *squash*] (circa 1647)
transitive verb
1 : SQUASH
2 : SQUELCH, SUCK
intransitive verb
: SQUELCH, SUCK ⟨their wet tennis shoes *squished* —Frank Noel⟩
— **squish** *noun*

squishy \'skwi-shē\ *adjective* **squish·i·er; -est** (circa 1847)
: being soft, yielding, and damp
— **squish·i·ness** *noun*

squoosh \'skwu̇sh, 'skwüsh\ *verb* [by alteration] (1942)
: SQUASH

squush \'skwəsh\ *verb* [by alteration] (1837)
: SQUASH

Sra·nan \'srä-nən\ *noun* [Sranan, short for *Sranan Tongo,* literally, Suriname tongue] (1953)
: an English-based creole widely spoken in Suriname

sri \'srē, 'shrē\ *noun* [Sanskrit *śrī,* literally, beauty, majesty; akin to Greek *kreiōn* ruler, master] (1799)
— used as a conventional title of respect when addressing or speaking of a distinguished Indian

sRNA \'es-ˌär-ˌen-'ā\ *noun* [*s*oluble *RNA*] (1957)
: TRANSFER RNA

SRO \ˌes-(ˌ)är-'ō\ *noun* [*s*ingle-*r*oom *o*ccupancy] (1941)
: a house, apartment building, or residential hotel in which low-income or welfare tenants live in single rooms

SS \ˌes-'es\ *noun* [German, abbreviation for *Schutzstaffel,* literally, protection echelon] (1932)
: a unit of Nazis created to serve as bodyguard to Hitler and later expanded to take charge of intelligence, central security, policing action, and the mass extermination of those they considered inferior or undesirable

SST \ˌes-ˌes-'tē\ *noun* [*s*uper*s*onic *t*ransport] (1961)
: SUPERSONIC TRANSPORT

¹**-st** — see -EST

²**-st** *symbol*
— used after the figure 1 to indicate the ordinal number *first* ⟨1*st*⟩ ⟨91*st*⟩

¹**stab** \'stab\ *noun* [Middle English *stabbe*] (15th century)
1 : a wound produced by a pointed weapon
2 a : a thrust of a pointed weapon **b :** a jerky thrust
3 : EFFORT, TRY

²**stab** *verb* **stabbed; stab·bing** (1530)
transitive verb
1 : to wound or pierce by the thrust of a pointed weapon
2 : THRUST, DRIVE
intransitive verb
: to thrust or give a wound with or as if with a pointed weapon
— **stab·ber** *noun*

¹**sta·bile** \'stā-ˌbīl, -ˌbil\ *adjective* [Latin *stabilis* — more at STABLE] (1896)
: STATIONARY, STABLE

²**sta·bile** \-ˌbēl\ *noun* [probably from French, from Latin *stabilis,* adjective] (1937)
: an abstract sculpture or construction similar in appearance to a mobile but made to be stationary

sta·bil·i·ty \stə-'bi-lə-tē\ *noun, plural* **-ties** (14th century)
1 : the quality, state, or degree of being stable: as **a :** the strength to stand or endure **:** FIRMNESS **b :** the property of a body that causes it when disturbed from a condition of equilibrium or steady motion to develop forces or moments that restore the original condition **c :** resistance to chemical change or to physical disintegration
2 : residence for life in one monastery

sta·bi·lize \'stā-bə-ˌlīz\ *verb* **-lized; -liz·ing** (1861)
transitive verb
1 : to make stable, steadfast, or firm
2 : to hold steady: as **a :** to maintain the stability of (as an airplane) by means of a stabilizer **b :** to limit fluctuations of (as prices) **c :** to establish a minimum price for
intransitive verb
: to become stable, firm, or steadfast
— **sta·bi·li·za·tion** \ˌstā-bə-lə-'zā-shən\ *noun*

sta·bi·liz·er \'stā-bə-ˌlī-zər\ *noun* (circa 1909)
: one that stabilizes something: as **a :** a substance added to another substance (as an explosive or plastic) or to a system (as an emulsion) to prevent or retard an unwanted alteration of physical state **b :** a gyroscope device to keep ships steady in a heavy sea **c :** an airfoil providing stability for an airplane; *specifically* **:** the fixed horizontal member of the tail assembly — see AIRPLANE illustration

¹**sta·ble** \'stā-bəl\ *noun* [Middle English, from Old French *estable,* from Latin *stabulum,* from *stare* to stand — more at STAND] (13th century)
1 : a building in which domestic animals are sheltered and fed; *especially* **:** such a building having stalls or compartments ⟨horse *stable*⟩
2 a : the racehorses of one owner **b :** a group of athletes (as boxers) or performers under one management **c :** the racing cars of one owner **d :** GROUP, COLLECTION
— **sta·ble·man** \-mən, -ˌman\ *noun*

²**stable** *verb* **sta·bled; sta·bling** \-b(ə-)liŋ\ (14th century)
transitive verb
: to put or keep in a stable
intransitive verb
: to dwell in or as if in a stable

³**stable** *adjective* **sta·bler** \-b(ə-)lər\; **sta·blest** \-b(ə-)ləst\ [Middle English, from Middle French *estable,* from Latin *stabilis,* from *stare* to stand] (13th century)
1 a : firmly established **:** FIXED, STEADFAST **b :** not changing or fluctuating **:** UNVARYING **c :** PERMANENT, ENDURING
2 a : steady in purpose **:** firm in resolution **b :** not subject to insecurity or emotional illness **:** SANE, RATIONAL ⟨a *stable* personality⟩
3 a (1) **:** placed so as to resist forces tending to cause motion or change of motion (2) **:** designed so as to develop forces that restore the original condition when disturbed from a condition of equilibrium or steady motion **b** (1) **:** not readily altering in chemical makeup or physical state ⟨*stable* emulsions⟩ (2) **:** not spontaneously radioactive
synonym see LASTING
— **sta·ble·ness** \-bəl-nəs\ *noun*
— **sta·bly** \-b(ə-)lē\ *adverb*

stable fly *noun* (1862)
: a biting dipteran fly (*Stomoxys calcitrans*) that resembles the common housefly and is abundant about stables

◇ WORD HISTORY

squirrel The ultimate source of English *squirrel* is Greek *skiouros,* a word first attested in the early centuries A.D. whose literal meaning is probably "shadow-tailed." This name would appear to come from the Greeks' observation that when the animal sits upright with its bushy tail behind its back, it looks as if it could be shading itself from the sun. The Greek word entered Latin as *sciurus,* which underwent a variety of permutations in the spoken Latin of the Roman provinces. One outcome is presumed to be a diminutive form *scuriolus,* which developed into Old French *escuiruel.* This was eventually borrowed into Middle English as *squirel* or *squerel.* The Old English word for "squirrel" was *ācwern,* with a Germanic and Indo-European pedigree; *ācwern* passed as *ocquerne* or *aquerne* into early Middle English, but no trace of the word is found after the 13th century.

sta·ble·mate \'stā-bəl-ˌmāt\ *noun* (1926)
1 : an animal stabled with another
2 : a member of a stable

sta·bler \-b(ə-)lər\ *noun* (15th century)
: one who keeps a stable

sta·bling \-b(ə-)liŋ\ *noun* (15th century)
: accommodation for animals in a building; *also* **:** the building for this

stab·lish \'sta-blish\ *verb* [Middle English, short for *establissen*] (14th century)
archaic **:** ESTABLISH
— stab·lish·ment \-mənt\ *noun, archaic*

stac·ca·to \stə-'kä-(ˌ)tō\ *adjective* [Italian, from past participle of *staccare* to detach, from *s-* ex- (from Latin *ex-*) + *attaccare* to attack, attach, perhaps from Old French *estachier* — more at ATTACH] (circa 1724)
1 a : cut short or apart in performing **:** DISCONNECTED ⟨*staccato* notes⟩ **b :** marked by short clear-cut playing or singing of tones or chords ⟨a *staccato* style⟩
2 : ABRUPT, DISJOINTED
— staccato *adverb*
— staccato *noun*

staccato mark *noun* (circa 1903)
: a pointed vertical stroke or a dot placed over or under a musical note to be produced staccato

¹stack \'stak\ *noun* [Middle English *stak,* from Old Norse *stakkr;* akin to Russian *stog* stack and probably to Old English *staca* stake] (14th century)
1 : a large usually conical pile (as of hay, straw, or grain in the sheaf) left standing in the field for storage
2 a : an orderly pile or heap **b :** a large quantity or number
3 : an English unit of measure especially for firewood that is equal to 108 cubic feet
4 a : a number of flues embodied in one structure rising above a roof **b :** a vertical pipe (as to carry off smoke) **c :** the exhaust pipe of an internal combustion engine
5 : a structure of bookshelves for compact storage of books — usually used in plural
6 : a pile of chips sold to or won by a poker player
7 a : a memory or a section of memory in a computer for temporary storage in which the last item stored is the first retrieved; *also* **:** a data structure that simulates a stack ⟨a push-down *stack*⟩ **b :** a computer memory consisting of arrays of memory elements stacked one on top of another

²stack (14th century)
transitive verb
1 a : to arrange in a stack **:** PILE **b :** to pile in or on ⟨*stacked* the table with books⟩ ⟨*stack* the dishwasher⟩
2 a : to arrange secretly for cheating ⟨*stack* a deck of cards⟩ **b :** to arrange or fix so as to make a particular result likely ⟨the odds are *stacked* against us⟩ ⟨will *stack* juries to suit themselves —Patrice Horn⟩
3 a : to assign (an airplane) by radio to a particular altitude and position within a group circling before landing **b :** to put into a waiting line ⟨another dozen rigs are *stacked* up and waiting —P. H. Hutchins, Jr.⟩
4 : COMPARE — used with *against* ⟨such a crime is nothing when *stacked* against a murder —Pete Censky⟩
intransitive verb
: to form a stack
— stack·er *noun*

stack·able \'sta-kə-bəl\ *adjective* (1958)
: easily arranged in a stack

stacked \'stakt\ *adjective* (1942)
of a woman **:** being shapely and having large breasts

stack up *intransitive verb* (1896)
1 : to add up **:** TOTAL
2 : MEASURE UP, COMPARE — usually used with *against*

stac·te \'stak-tē\ *noun* [Middle English *stacten,* from Latin *stacte,* from Greek *staktē,* from feminine of *staktos* oozing out in drops, from *stazein* to drip] (1535)
: a sweet spice used by the ancient Jews in preparing incense

stad·dle \'sta-dᵊl\ *noun* [Middle English *stathel* base, support, bottom of a stack, from Old English *statho* base; akin to Old High German *stān* to stand — more at STAND] (15th century)
: a base (as of piling) for a stack of hay or straw

stade \'stād\ *noun* [Middle French *estade,* from Latin *stadium*] (1537)
: STADIUM 1a

sta·dia \'stā-dē-ə\ *noun* [Italian, probably from Latin, plural of *stadium*] (1865)
: a surveying method for determination of distances and differences of elevation by means of a telescopic instrument having two horizontal lines through which the marks on a graduated rod are observed; *also* **:** the instrument or rod

sta·di·um \'stā-dē-əm\ *noun, plural* **-dia** \-dē-ə\ *or* **-di·ums** [Middle English, from Latin, from Greek *stadion*] (14th century)
1 a : any of various ancient Greek units of length ranging in value from 607 to 738 feet (about 185 to 225 meters) **b :** an ancient Roman unit of length equal to 607 feet (185 meters)
2 a : a course for footraces in ancient Greece originally one stadium in length **b :** a tiered structure with seats for spectators surrounding an ancient Greek running track **c :** a large usually unroofed building with tiers of seats for spectators at sports events
3 [New Latin, from Latin] **:** a stage in a life history; *especially* **:** one between successive molts ◆

stadt·hold·er \'stat-ˌhōl-dər\ *noun* [part translation of Dutch *stadhouder,* from *stad* place + *houder* holder] (1668)
1 : a viceroy in a province of the Netherlands
2 : a chief executive officer of the provinces that formed a union leading to establishment of the Netherlands
— stadt·hold·er·ate \-də rət\ *noun*
— stadt·hold·er·ship \-dər-ˌship\ *noun*

¹staff \'staf\ *noun, plural* **staffs** \'stafs, 'stavz\ *or* **staves** \'stavz, 'stāvz\ [Middle English *staf,* from Old English *stæf;* akin to Old High German *stab* staff, Sanskrit *stabhnāti* he supports] (before 12th century)
1 a : a long stick carried in the hand for support in walking **b :** a supporting rod: as (1) *archaic* **:** SHAFT 1a(1) (2) **:** a crosspiece in a ladder or chair **:** RUNG (3) **:** FLAGSTAFF (4) **:** a pivoted arbor **c :** CLUB, CUDGEL
2 a : CROSIER **b :** a rod carried as a symbol of office or authority
3 : the horizontal lines with their spaces on which music is written — called also *stave*
4 : any of various graduated sticks or rules used for measuring **:** ROD
5 *plural staffs* **a :** the officers chiefly responsible for the internal operations of an institution or business **b :** a group of officers appointed to assist a civil executive or commanding officer **c :** military or naval officers not eligible for operational command **d :** the personnel who assist a director in carrying out an assigned task **e** *plural* **staff :** a member of a staff
— staff *adjective*

²staff *transitive verb* (1859)
1 : to supply with a staff or with workers
2 : to serve as a staff member of

³staff *noun* [probably from German *staffieren* to trim] (1892)
: a building material having a plaster of Paris base and used in exterior wall coverings of temporary buildings

staff·er \'sta-fər\ *noun* (1941)
: a member of a staff (as of a newspaper)

staff officer *noun* (1777)
: a commissioned officer assigned to a military commander's staff — compare LINE OFFICER

staff of life (1638)
: a staple of diet; *especially* **:** BREAD

Staf·ford·shire bull terrier \'sta-fərd-ˌshir-, -shər-\ *noun* [*Staffordshire,* England] (1901)
: any of a breed of compact muscular terriers that have a short stiff glossy coat

Staffordshire bull terrier

staff sergeant *noun* (1851)
: a noncommissioned officer ranking in the army above a sergeant and below a platoon sergeant or sergeant first class, in the air force above a sergeant and below a technical sergeant, and in the marine corps above a sergeant and below a gunnery sergeant

staff sergeant major *noun* (1967)
: a noncommissioned officer in the army ranking above a master sergeant

staff tree *noun* (circa 1633)
: any of a genus (*Celastrus* of the family Celastraceae, the staff-tree family) of mostly twining shrubby plants including the common bittersweet

¹stag \'stag\ *noun, plural* **stags** [Middle English *stagge,* from Old English *stagga;* akin to Old Norse andar*steggi* drake] (12th century)
1 *or plural* **stag :** an adult male red deer; *also* **:** the male of various other deer (especially genus *Cervus*)
2 *chiefly Scottish* **:** a young horse; *especially* **:** a young unbroken stallion
3 : a male animal castrated after maturity — compare STEER 1
4 : a young adult male domestic fowl
5 a : a social gathering of men only **b :** one who attends a dance or party without a companion

²stag *verb* **stagged; stag·ging** [*stag* (informer)] (circa 1823)
transitive verb
British **:** to spy on
intransitive verb
: to attend a dance or party without a companion

³stag *adjective* (1843)
1 a : restricted to men ⟨a *stag* party⟩ **b :** intended or suitable for a gathering of men only; *especially* **:** PORNOGRAPHIC ⟨*stag* movies⟩
2 : unaccompanied by someone of the opposite sex ⟨*stag* women⟩
— stag *adverb*

◇ WORD HISTORY
stadium In ancient Greece a *stadion* was a unit of measurement equal to about 185 meters. One of the most important events in the ancient Olympic Games was a footrace exactly one *stadion* long. The course on which the race was run, including artificial embankments around the sides to accommodate spectators, also became known as a *stadion.* The Romans borrowed the word as *stadium.* When *stadium* first appeared in English-language contexts, in the 14th century, it was with reference to the Greco-Roman unit of measurement or to the racecourse at Olympia or one modeled after it. Over time *stadium* has come to be used of variously sized arenas surrounded by tiered seating and used for a variety of athletic contests.

stag beetle *noun* (1681)
: any of a family (Lucanidae) of mostly large lamellicorn beetles having males with long and often branched mandibles suggesting the antlers of a stag

[1]stage \'stāj\ *noun* [Middle English, from Middle French *estage,* from (assumed) Vulgar Latin *staticum,* from Latin *stare* to stand — more at STAND] (14th century)
1 a : one of a series of positions or stations one above the other **:** STEP **b :** the height of the surface of a river above an arbitrary zero point ⟨flood *stage*⟩
2 a (1) **:** a raised platform (2) **:** the part of a theater on which the acting takes place and which often includes the wings (3) **:** the acting profession **:** the theater as an occupation or activity (4) **:** SOUNDSTAGE **b :** a center of attention or scene of action
3 a : a scaffold for workmen **b :** the small platform of a microscope on which an object is placed for examination
4 a : a place of rest formerly provided for those traveling by stagecoach **:** STATION **b :** the distance between two stopping places on a road **c :** STAGECOACH
5 a : a period or step in a progress, activity, or development; *especially* **:** one of the distinguishable periods of growth and development of a plant or animal ⟨the larval *stage* of an insect⟩ **b :** one passing through a (specified) stage
6 : an element or part of an electronic device (as an amplifier)
7 : one of two or more sections of a rocket that have their own fuel and engine
— stage·ful \-ˌfu̇l\ *noun*
— stage·like \-ˌlīk\ *adjective*
— on the stage : in or into the acting profession

[2]stage *transitive verb* **staged; stag·ing** (1879)
1 : to produce (as a play) on a stage
2 : to produce or cause to happen for public view or public effect ⟨*stage* a track meet⟩ ⟨*stage* a hunger strike⟩
— stage·able \'stā-jə-bəl\ *adjective*

[3]stage *adjective* (1824)
: intended to represent a type or stereotype ⟨a *stage* Irishman⟩ ⟨a *stage* French accent⟩

stage business *noun* (1825)
: BUSINESS 6

stage·coach \'stāj-ˌkōch\ *noun* (1658)
: a horse-drawn passenger and mail coach running on a regular schedule between established stops

stage·craft \-ˌkraft\ *noun* (1882)
: the effective management of theatrical devices or techniques

stage direction *noun* (1790)
: a description (as of a character or setting) or direction (as to indicate stage business) provided in the text of a play

stage director *noun* (1782)
1 : DIRECTOR C
2 : STAGE MANAGER

stage fright *noun* (1878)
: nervousness felt at appearing before an audience

stage·hand \'stāj-ˌhand\ *noun* (circa 1902)
: a stage worker who handles scenery, properties, or lights

stage left *noun* (1931)
: the left part of a stage from the viewpoint of one who faces the audience

stage–man·age \-ˌma-nij\ *transitive verb* [back-formation from *stage manager*] (1879)
1 a : to arrange or exhibit so as to achieve a desired effect **b :** to arrange or direct from behind the scenes
2 : to act as stage manager for
— stage management *noun*

stage manager *noun* (1805)
: one who supervises the physical aspects of a stage production, assists the director during rehearsals, and is in charge of the stage during a performance

stag·er \'stā-jər\ *noun* (1570)
: an experienced person **:** VETERAN

stage right *noun* (1931)
: the right part of a stage from the viewpoint of one who faces the audience

stage set *noun* (1861)
: scenery and properties designed and arranged for a particular scene in a play

stage·struck \'stāj-ˌstrək\ *adjective* (1813)
: fascinated by the stage; *especially* **:** having an ardent desire to become an actor

stage whisper *noun* (circa 1865)
1 : a loud whisper by an actor that is audible to the spectators but is supposed for dramatic effect not to be heard by one or more of the actors
2 : an audible whisper
— stage–whisper *verb*

stag·fla·tion \ˌstag-'flā-shən\ *noun* [blend of *stagnation* and *inflation*] (1965)
: persistent inflation combined with stagnant consumer demand and relatively high unemployment
— stag·fla·tion·ary \-shə-ˌner-ē\ *adjective*

[1]stag·ger \'sta-gər\ *verb* **stag·gered; stag·ger·ing** \-g(ə-)riŋ\ [alteration of earlier *stacker,* from Middle English *stakeren,* from Old Norse *stakra,* frequentative of *staka* to push; perhaps akin to Old English *staca* stake — more at STAKE] (15th century)
intransitive verb
1 a : to reel from side to side **:** TOTTER **b :** to move on unsteadily
2 : to waver in purpose or action **:** HESITATE
3 : to rock violently ⟨the ship *staggered*⟩
transitive verb
1 : to cause to doubt or hesitate **:** PERPLEX
2 : to cause to reel or totter
3 : to arrange in any of various zigzags, alternations, or overlappings of position or time ⟨*stagger* work shifts⟩ ⟨*stagger* teeth on a cutter⟩
— stag·ger·er \-gər-ər\ *noun*

[2]stagger *noun* (1577)
1 *plural but singular or plural in construction* **:** an abnormal condition of domestic mammals and birds associated with damage to the central nervous system and marked by incoordination and a reeling unsteady gait
2 : a reeling or unsteady gait or stance
3 : an arrangement in which the leading edge of the upper wing of a biplane is advanced over that of the lower

[3]stagger *adjective* (1875)
: marked by an alternating or overlapping pattern

stag·ger·bush \'sta-gər-ˌbu̇sh\ *noun* (1847)
: a shrubby heath (*Lyonia mariana*) of the eastern U.S. that is poisonous to livestock

staggering *adjective* (1530)
: so great as to cause one to stagger **:** ASTONISHING, OVERWHELMING ⟨a *staggering* feat⟩ ⟨*staggering* medical bills⟩
— stag·ger·ing·ly \'sta-g(ə-)riŋ-lē\ *adverb*

stag·gery \'sta-g(ə-)rē\ *adjective* (1778)
: UNSTEADY

stag·gy \'sta-gē\ *adjective* (1918)
: having the appearance of a mature male — used of female or castrated male domestic animals

stag·horn coral \'stag-ˌhȯrn-\ *noun* (1884)
: any of several large branching corals (genus *Acropora* and especially *A. cervicornis*) that somewhat resemble antlers

staghorn sumac *noun* (circa 1868)
: a sumac (*Rhus typhina*) of eastern North America that is a shrub or small tree with velvety-pubescent branches and flower stalks, leaves turning brilliant red in fall, and dense panicles of greenish yellow flowers followed by hairy crimson fruits

stag·hound \'stag-ˌhau̇nd\ *noun* (1707)
: a hound formerly used in hunting the stag and other large animals; *specifically* **:** a large heavy hound resembling the English foxhound

staghorn sumac

stag·ing \'stā-jiŋ\ *noun* (14th century)
1 : SCAFFOLDING
2 a : the business of running stagecoaches **b :** the act of journeying in stagecoaches
3 : the putting of a play on the stage
4 a : the moving of troops or matériel forward in several stages **b :** the assembling of troops or matériel in transit in a particular place

staging area *noun* (1943)
: an area in which participants in a new operation or mission are assembled and readied

Stag·i·rite \'sta-jə-ˌrīt\ *noun* [Greek *Stagiritēs,* from *Stagira,* city in ancient Macedonia] (circa 1620)
: a native or resident of Stagira ⟨Aristotle the *Stagirite*⟩

stag·nant \'stag-nənt\ *adjective* (1666)
1 a : not flowing in a current or stream ⟨*stagnant* water⟩ **b :** STALE ⟨long disuse had made the air *stagnant* and foul —Bram Stoker⟩
2 : not advancing or developing
— stag·nan·cy \-nən(t)-sē\ *noun*
— stag·nant·ly *adverb*

stag·nate \'stag-ˌnāt\ *intransitive verb* **stag·nat·ed; stag·nat·ing** [Latin *stagnatus,* past participle of *stagnare,* from *stagnum* body of standing water] (1669)
: to become or remain stagnant
— stag·na·tion \stag-'nā-shən\ *noun*

stagy \'stā-jē\ *or* **stag·ey** *adjective* **stag·i·er; -est** (1860)
: of or characteristic of the stage; *especially* **:** marked by pretense or artificiality **:** THEATRICAL
— stag·i·ly \-jə-lē\ *adverb*
— stag·i·ness \-jē-nəs\ *noun*

[1]staid \'stād\ *adjective* [from past participle of [3]*stay*] (1557)
: marked by settled sedateness and often prim self-restraint **:** SOBER, GRAVE
synonym see SERIOUS
— staid·ly *adverb*
— staid·ness *noun*

[2]staid *past and past participle of* STAY

[1]stain \'stān\ *verb* [Middle English *steynen,* partly from Middle French *desteindre* to discolor & partly of Scandinavian origin; akin to Old Norse *steina* to paint — more at DISTAIN] (14th century)
transitive verb
1 : to suffuse with color
2 : DISCOLOR, SOIL
3 a : to taint with guilt, vice, or corruption **b :** to bring reproach on
4 : to color (as wood, glass, or cloth) by processes affecting chemically or otherwise the material itself
intransitive verb
: to receive a stain
— stain·able \'stā-nə-bəl\ *adjective*
— stain·er \'stā-nər\ *noun*

[2]stain *noun* (1583)
1 a : a soiled or discolored spot **b :** a natural spot of color contrasting with the ground
2 : a taint of guilt **:** STIGMA
3 a : a preparation (as of dye or pigment) used in staining; *especially* **:** one capable of penetrating the pores of wood **b :** a dye or mixture of dyes used in microscopy to make visible minute and transparent structures, to differentiate tissue elements, or to produce specific chemical reactions
— stain·proof \-ˌprüf\ *adjective*

stain·abil·i·ty \ˌstā-nə-ˈbi-lə-tē\ *noun* (1890)
: the capacity of cells and cell parts to stain specifically and consistently with particular dyes and stains

stained glass *noun* (1791)
: glass colored or stained (as by fusing metallic oxides into it) for decorative applications (as in windows)

[1]stain·less \ˈstān-ləs\ *adjective* (1586)
1 a : free from stain or stigma **b :** highly resistant to stain or corrosion
2 : made from materials resistant to stain
— **stain·less·ly** *adverb*

[2]stainless *noun* (1945)
: tableware made of stainless steel

stainless steel *noun* (1920)
: an alloy of steel with chromium and sometimes another element (as nickel or molybdenum) that is practically immune to rusting and ordinary corrosion

stair \ˈstar, ˈster\ *noun* [Middle English *steir,* from Old English *stǣger;* akin to Old English & Old High German *stīgan* to rise, Greek *steichein* to walk] (before 12th century)
1 : a series of steps or flights of steps for passing from one level to another — often used in plural but singular or plural in construction ⟨a narrow private *stairs* —Lewis Mumford⟩
2 : a single step of a stairway

stair·case \-ˌkās\ *noun* (1624)
1 : the structure containing a stairway
2 : a flight of stairs with the supporting framework, casing, and balusters

stair·way \-ˌwā\ *noun* (1767)
: one or more flights of stairs usually with landings to pass from one level to another

stair·well \-ˌwel\ *noun* (1920)
: a vertical shaft in which stairs are located

[1]stake \ˈstāk\ *noun* [Middle English, from Old English *staca;* akin to Middle Low German *stake* stake and perhaps to Latin *tignum* beam] (before 12th century)
1 : a pointed piece of wood or other material driven or to be driven into the ground as a marker or support
2 a : a post to which a person is bound for execution by burning **b :** execution by burning at a stake
3 a : something that is staked for gain or loss **b :** the prize in a contest **c :** an interest or share in an undertaking (as a commercial venture)
4 : a Mormon territorial jurisdiction comprising a group of wards
5 : an upright stick at the side or end of a vehicle to retain the load
6 : GRUBSTAKE
— **at stake :** at issue **:** in jeopardy

[2]stake *transitive verb* **staked; stak·ing** (14th century)
1 : to mark the limits of by or as if by stakes
2 : to tether to a stake
3 : BET, WAGER
4 : to fasten up or support (as plants) with stakes
5 : to back financially
6 : GRUBSTAKE
— **stake a claim :** to assert a title or right to something by or as if by placing stakes to satisfy a legal requirement

stake body *noun* (1907)
: an open motortruck body consisting of a platform with stakes inserted along the outside edges to retain a load

stake·hold·er \ˈstāk-ˌhōl-dər\ *noun* (1708)
: a person entrusted with the stakes of bettors

stake·out \ˈstā-ˌkau̇t\ *noun* (circa 1942)
: a surveillance maintained by the police of an area or a person suspected of criminal activity

stake out *transitive verb* (1951)
1 : to assign (as a police officer) to an area usually to conduct a surveillance
2 : to maintain a stakeout of

stake race *noun* (1896)
: a horse race in which the prize offered is made up at least in part of money (as entry fees) put up by the owners of the horses entered

stake truck *noun* (1907)
: a truck having a stake body

Sta·kha·nov·ite \stə-ˈkä-nə-ˌvīt\ *noun* [translation of Russian *stakhanovets,* from Alexei G. *Stakhanov* (died 1977) Russian miner] (1935)
: a Soviet industrial worker awarded recognition and special privileges for output beyond production norms
— **Sta·kha·nov·ism** \-ˌvi-zəm\ *noun*

sta·lac·tite \stə-ˈlak-ˌtīt *also* ˈsta-lək-\ *noun* [New Latin *stalactites,* from Greek *stalaktos* dripping, from *stalassein* to let drip] (1677)
: a deposit of calcium carbonate (as calcite) resembling an icicle hanging from the roof or sides of a cavern
— **sta·lac·tit·ic** \ˌsta-(ˌ)lak-ˈti-tik, stə-ˌ\ *adjective*

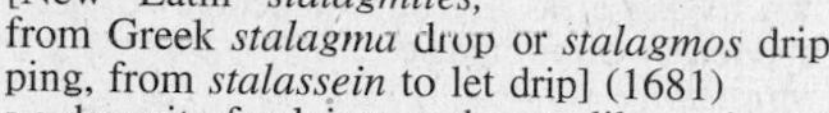
stalactite and stalagmite

sta·lag \ˈstä-ˌläg\ *noun* [German, short for *Stammlager* base camp, from *Stamm* base + *Lager* camp] (1940)
: a German prison camp for noncommissioned officers or enlisted men; *broadly* **:** PRISON CAMP 2

sta·lag·mite \stə-ˈlag-ˌmīt *also* ˈsta-ləg-\ *noun* [New Latin *stalagmites,* from Greek *stalagma* drop or *stalagmos* dripping, from *stalassein* to let drip] (1681)
: a deposit of calcium carbonate like an inverted stalactite formed on the floor of a cave by the drip of calcareous water
— **sta·lag·mit·ic** \ˌsta-(ˌ)lag-ˈmi-tik, stə-ˌ\ *adjective*

[1]stale \ˈstā(ə)l\ *adjective* **stal·er; stal·est** [Middle English, aged (of ale), not fresh; akin to Middle Dutch *stel* stale] (15th century)
1 : tasteless or unpalatable from age
2 : tedious from familiarity
3 : impaired in legal force or effect by reason of being allowed to rest without timely use, action, or demand ⟨a *stale* affidavit⟩ ⟨a *stale* debt⟩
4 : impaired in vigor or effectiveness
— **stale·ly** \ˈstā(ə)l-lē\ *adverb*
— **stale·ness** *noun*

[2]stale *verb* **staled; stal·ing** (1599)
transitive verb
1 : to make stale
2 *archaic* **:** to make common **:** CHEAPEN
intransitive verb
: to become stale

[3]stale *intransitive verb* **staled; stal·ing** [Middle English; akin to Middle Low German *stallen* to urinate, *stal* urine of horses] (15th century)
: URINATE — used chiefly of camels and horses

[4]stale *noun* (1548)
: urine of a domestic animal (as a horse)

[1]stale·mate \ˈstā(ə)l-ˌmāt\ *noun* [obsolete English *stale* stalemate + English [1]*mate*] (1765)
1 : a drawing position in chess in which only the king can move and although not in check can move only into check
2 : a drawn contest **:** DEADLOCK; *also* **:** the state of being stalemated

[2]stalemate *transitive verb* (1765)
: to bring into a stalemate

Sta·lin·ism \ˈstä-lə-ˌni-zəm, ˈsta-\ *noun* (1927)
: the political, economic, and social principles and policies associated with Stalin; *especially* **:** the theory and practice of communism developed by Stalin from Marxism-Leninism and marked especially by rigid authoritarianism, widespread use of terror, and often emphasis on Russian nationalism
— **Sta·lin·ist** \-nist\ *noun or adjective*
— **Sta·lin·ize** \-ˌnīz\ *transitive verb*
— **Sta·lin·oid** \-ˌnȯid\ *noun or adjective*

[1]stalk \ˈstȯk\ *noun* [Middle English *stalke;* akin to Old English *stela* stalk, support] (14th century)
1 : a slender upright object or supporting or connecting part; *especially* **:** PEDUNCLE ⟨the *stalk* of a crinoid⟩
2 a : the main stem of an herbaceous plant often with its dependent parts **b :** a part of a plant (as a petiole, stipe, or peduncle) that supports another
— **stalked** \ˈstȯkt\ *adjective*
— **stalk·less** \ˈstȯ-kləs\ *adjective*
— **stalky** \ˈstȯ-kē\ *adjective*

[2]stalk *verb* [Middle English, from Old English *bestealcian;* akin to Old English *stelan* to steal — more at STEAL] (14th century)
intransitive verb
1 : to pursue quarry or prey stealthily
2 : to walk stiffly or haughtily
transitive verb
1 : to pursue by stalking
2 : to go through (an area) in search of prey or quarry ⟨*stalk* the woods for deer⟩
— **stalk·er** *noun*

[3]stalk *noun* (14th century)
1 : the act of stalking
2 : a stalking gait

stalk·ing–horse \ˈstȯ-kiŋ-ˌhȯrs\ *noun* (1519)
1 : a horse or a figure like a horse behind which a hunter stalks game
2 : something used to mask a purpose
3 : a candidate put forward to divide the opposition or to conceal someone's real candidacy

[1]stall \ˈstȯl\ *noun* [Middle English, from Old English *steall;* akin to Old High German *stal* place, stall and perhaps to Latin *locus* (Old Latin *stlocus*) place] (before 12th century)
1 a : a compartment for a domestic animal in a stable or barn **b :** a space marked off for parking a motor vehicle
2 a : a seat in the chancel of a church with back and sides wholly or partly enclosed **b :** a church pew **c** *chiefly British* **:** a front orchestra seat in a theater — usually used in plural
3 : a booth, stand, or counter at which articles are displayed for sale
4 : a protective sheath for a finger or toe
5 : a small compartment ⟨a shower *stall*⟩

[2]stall (14th century)
transitive verb
1 : to put into or keep in a stall
2 *obsolete* **:** to install in office
3 a : to bring to a standstill **:** BLOCK; *especially* **:** MIRE **b :** to cause (an engine) to stop usually inadvertently **c :** to cause (an aircraft or airfoil) to go into a stall
intransitive verb
1 : to come to a standstill (as from mired wheels or engine failure)
2 : to experience a stall in flying

[3]stall *noun* (1916)
: the condition of an airfoil or aircraft in which excessive angle of attack causes disruption of airflow with attendant loss of lift

[4]stall *noun* [alteration of *stale* lure] (1903)
: a ruse to deceive or delay

[5]stall (1903)
intransitive verb
: to play for time **:** DELAY
transitive verb
: to hold off, divert, or delay by evasion or deception

stall–feed \ˈstȯl-ˌfēd\ *transitive verb* **-fed** \-ˌfed\; **-feed·ing** (1554)

: to feed in a stall especially so as to fatten ⟨*stall-feed* an ox⟩

stall·hold·er \-ˌhōl-dər\ *noun* (1881)
chiefly British **:** one who manages a stall at which articles are sold

stal·lion \ˈstal-yən\ *noun* [Middle English *stalion,* from Middle French *estalon,* of Germanic origin; akin to Old High German *stal* stall] (14th century)
: an uncastrated male horse **:** a male horse kept for breeding; *also* **:** a male animal (as a dog or a sheep) kept primarily as a stud

¹stal·wart \ˈstȯl-wərt\ *adjective* [Middle English, alteration of *stalworth,* from Old English *stǣlwierthe* serviceable] (15th century)
: marked by outstanding strength and vigor of body, mind, or spirit ⟨*stalwart* common sense⟩
synonym see STRONG
— stal·wart·ly *adverb*
— stal·wart·ness *noun*

²stalwart *noun* (15th century)
1 : a stalwart person
2 : an unwavering partisan

stal·worth \ˈstȯl-(ˌ)wərth\ *archaic variant of* STALWART

sta·men \ˈstā-mən\ *noun, plural* **stamens** *also* **sta·mi·na** \ˈstā-mə-nə, ˈsta-\ [Latin, warp, thread, from *stare* to stand — more at STAND] (1668)
: a microsporophyll of a seed plant; *specifically* **:** the pollen-producing male organ of a flower that consists of an anther and a filament — see FLOWER illustration

stamin- *combining form* [Latin *stamin-, stamen*]
: stamen ⟨*stamin*odium⟩

stam·i·na \ˈsta-mə-nə\ *noun* [Latin, plural of *stamen* warp, thread of life spun by the Fates] (1726)
: STAYING POWER, ENDURANCE ◆

sta·mi·nate \ˈstā-mə-nət, ˈsta-, -ˌnāt\ *adjective* (circa 1850)
1 : having or producing stamens
2 *of a diclinous flower* **:** having stamens but no pistils

sta·mi·no·di·um \ˌstā-mə-ˈnō-dē-əm, ˌsta-\ *noun, plural* **-dia** \-dē-ə\ [New Latin, from *stamin-* + *-odium* thing resembling, from Greek *-ōdēs* like] (circa 1821)
: an abortive or sterile stamen

stam·mel \ˈsta-məl\ *noun* [probably from *stamin* a woolen fabric] (1530)
1 *obsolete* **:** a coarse woolen clothing fabric usually dyed red and used sometimes for undershirts of penitents
2 *archaic* **:** the bright red color of stammel

stam·mer \ˈsta-mər\ *verb* **stam·mered; stam·mer·ing** \ˈsta-mə-riŋ, ˈstam-riŋ\ [Middle English *stameren,* from Old English *stamerian;* akin to Old High German *stamalōn* to stammer, Old Norse *stemma* to hinder, damn up — more at STEM] (before 12th century)
intransitive verb
: to make involuntary stops and repetitions in speaking **:** HALT — compare STUTTER
transitive verb
: to utter with involuntary stops or repetitions
— stammer *noun*
— stam·mer·er \-mər-ər\ *noun*

¹stamp \ˈstamp; *verb transitive & verb intransitive are also* ˈstämp *or* ˈstȯmp\ *verb* [Middle English; akin to Old High German *stampfōn* to stamp and perhaps to Greek *stembein* to shake up] (13th century)
transitive verb
1 : to pound or crush with a pestle or a heavy instrument
2 a (1) **:** to strike or beat forcibly with the bottom of the foot (2) **:** to bring down (the foot) forcibly **b :** to extinguish or destroy by or as if by stamping with the foot — usually used with *out* ⟨*stamp* out cancer⟩
3 a : IMPRESS, IMPRINT ⟨*stamp* "paid" on the bill⟩ **b :** to attach a stamp to
4 : to cut out, bend, or form with a stamp or die
5 a : to provide with a distinctive character ⟨*stamped* with a dreary, institutionalized look —Bernard Taper⟩ **b :** CHARACTERIZE
intransitive verb
1 : POUND 1
2 : to strike or thrust the foot forcibly or noisily downward

²stamp *noun* (15th century)
1 : a device or instrument for stamping
2 : the impression or mark made by stamping or imprinting
3 a : a distinctive character, indication, or mark **b :** a lasting imprint
4 : the act of stamping
5 : a stamped or printed paper affixed in evidence that a tax has been paid; *also* **:** POSTAGE STAMP
— stamp·less *adjective*

¹stam·pede \(ˌ)stam-ˈpēd\ *noun* [American Spanish *estampida,* from Spanish, crash, from *estampar* to stamp, of Germanic origin; akin to Old High German *stampfōn* to stamp] (1828)
1 : a wild headlong rush or flight of frightened animals
2 : a mass movement of people at a common impulse
3 : an extended festival combining a rodeo with exhibitions, contests, and social events

²stampede *verb* **stam·ped·ed; stam·ped·ing** (1838)
transitive verb
1 : to cause to run away in headlong panic
2 : to cause (as a group of people) to act on sudden or rash impulse
intransitive verb
1 : to flee headlong in panic
2 : to act on mass impulse
— stam·ped·er *noun*

stamp·er \ˈstam-pər, ˈstäm-, ˈstȯm-; *compare* ¹STAMP\ *noun* (14th century)
: one that stamps: as **a :** a worker who performs an industrial stamping operation **b :** an implement for pounding or stamping **c :** any of various stamping machines

stamping ground \ˈstam-piŋ, ˈstäm-, ˈstȯm-\ *noun* (1786)
: a favorite or habitual resort; *also* **:** familiar territory

stamp·ing mill \ˈstam-piŋ-\ *noun* (1552)
: STAMP MILL

stamp mill \ˈstamp-\ *noun* (1749)
: a mill in which ore is crushed with stamps; *also* **:** a machine for stamping ore

stamp tax *noun* (1797)
: a tax collected by means of a stamp purchased and affixed (as to a deck of playing cards); *specifically* **:** such a tax on a document (as a deed or promissory note) — called also *stamp duty*

stance \ˈstan(t)s\ *noun* [Middle French *estance* position, posture, stay, from (assumed) Vulgar Latin *stantia,* from Latin *stant-, stans,* present participle of *stare* to stand] (14th century)
1 *chiefly Scottish* **a :** STATION **b :** SITE
2 a : a way of standing or being placed **:** POSTURE **b :** intellectual or emotional attitude ⟨took an antiwar *stance*⟩
3 a : the position of the feet of a golfer or batter preparatory to making a swing **b :** the position of both body and feet from which an athlete starts or operates

¹stanch \ˈstȯnch, ˈstänch, ˈstanch\ *transitive verb* [Middle English *staunchen,* from Middle French *estancher,* perhaps from (assumed) Vulgar Latin *stanticare,* from Latin *stant-, stans,* present participle] (14th century)
1 : to check or stop the flowing of ⟨*stanched* her tears⟩; *also* **:** to stop the flow of blood from (a wound)
2 *archaic* **:** ALLAY, EXTINGUISH
3 a : to stop or check in its course ⟨trying to *stanch* the crime wave⟩ **b :** to make watertight **:** stop up
— stanch·er *noun*

²stanch *variant of* STAUNCH

stan·chion \ˈstan-chən\ *noun* [Middle English *stanchon,* from Middle French *estanchon,* from Old French, diminutive of *estance* stay, prop] (15th century)
1 : an upright bar, post, or support (as for a roof)
2 : a device that fits loosely around the neck of an animal (as a cow) and limits forward and backward motion (as in a stall)
— stan·chioned \-chənd\ *adjective*

¹stand \ˈstand\ *verb* **stood** \ˈstu̇d\; **stand·ing** [Middle English, from Old English *standan;* akin to Old High German *stantan, stān* to stand, Latin *stare,* Greek *histanai* to cause to stand, set, *histasthai* to stand, be standing] (before 12th century)
intransitive verb
1 a : to support oneself on the feet in an erect position **b :** to be a specified height when fully erect ⟨*stands* six feet two⟩ **c :** to rise to an erect position
2 a : to take up or maintain a specified position or posture ⟨*stand* aside⟩ ⟨can you *stand* on your head⟩ **b :** to maintain one's position ⟨*stand* firm⟩
3 : to be in a particular state or situation ⟨*stands* accused⟩
4 : to hold a course at sea
5 *obsolete* **:** HESITATE
6 a : to have or maintain a relative position in or as if in a graded scale ⟨*stands* first in the class⟩ **b :** to be in a position to gain or lose because of an action taken or a commitment made ⟨*stands* to make quite a profit⟩
7 *chiefly British* **:** to be a candidate **:** RUN
8 a : to rest or remain upright on a base or lower end ⟨a clock *stood* on the mantle⟩ **b :** to occupy a place or location ⟨the house *stands* on a knoll⟩
9 a : to remain stationary or inactive ⟨the car *stood* in the garage for a week⟩ **b :** to gather slowly and remain ⟨tears *standing* in her eyes⟩
10 : AGREE, ACCORD — used chiefly in the expression *it stands to reason*
11 a : to exist in a definite written or printed form ⟨copy a passage exactly as it *stands*⟩ **b :** to remain valid or efficacious ⟨the order given last week still *stands*⟩
12 *of a male animal* **:** to be available as a sire — used especially of horses
transitive verb

◇ WORD HISTORY

stamina The warp in a loom is the lengthwise series of threads through which the crosswise woof or weft is woven. In Latin the word for "warp" was *stamen.* In addition, *stamen* developed the transferred sense "thread," a meaning that had a divine as well as a human aspect. In Greco-Roman myth the three Fates were often pictured as spinning, measuring, and cutting the *stamina* or threads that symbolically represented mortal lives. When *stamina* was borrowed from Latin in the 17th century, both the literal meaning "warp" and the idea of the threads of fate went into the making of the English word. Like its Latin source, *stamina* was originally a plural. *Stamina* were the essential elements or qualities of something considered as being like the warp, the foundation on which a textile is woven. The English word was also used for the innate capacities that were once believed to determine, like the threads of the Fates, the duration of life. From this use developed the modern sense "endurance, staying power." In this sense, too, *stamina* was originally treated as plural, but it has now been universally reinterpreted as singular.

1 a : to endure or undergo successfully ⟨this book will *stand* the test of time⟩ **b :** to tolerate without flinching **:** bear courageously ⟨*stands* pain well⟩ **c :** to endure the presence or personality of ⟨can't *stand* the boss⟩ **d :** to derive benefit or enjoyment from ⟨you look like you could *stand* a drink⟩
2 : to remain firm in the face of ⟨*stand* a siege⟩
3 : to submit to ⟨*stand* trial⟩
4 a : to perform the duty of ⟨*stand* guard⟩ **b :** to participate in (a military formation)
5 : to pay the cost of (a treat) **:** pay for ⟨I'll *stand* you a dinner⟩ ⟨*stand* drinks⟩
6 : to cause to stand **:** set upright
7 : to make available for breeding ⟨*stand* a stallion⟩
synonym see BEAR
— **stand·er** *noun*
— **stand a chance :** to have a chance
— **stand for 1 :** to be a symbol for **:** REPRESENT **2 :** to put up with **:** PERMIT
— **stand on 1 :** to depend on **2 :** to insist on ⟨never *stands on* ceremony⟩
— **stand one's ground :** to maintain one's position
— **stand on one's own feet :** to think or act independently
— **stand treat :** to pay the cost of food, drink, or entertainment for others in a group

²stand *noun* (1592)
1 : an act of stopping or staying in one place
2 a : a halt for defense or resistance **b :** an often defensive effort of some duration or degree of success ⟨a goal-line *stand*⟩ **c** (1) **:** a stop made to give a performance ⟨a 6-game *stand* at home⟩ (2) **:** a town where such a stop is made
3 a : a place or post where one stands **b :** a strongly or aggressively held position especially on a debatable issue
4 a : the place taken by a witness for testifying in court **b** *plural* (1) **:** a section of the tiered seats for spectators of a sport or spectacle (2) **:** the occupants of such seats **c :** a raised platform (as for a speaker) serving as a point of vantage
5 a : a small often open-air structure for a small retail business ⟨a vegetable *stand*⟩ ⟨a hot dog *stand*⟩ **b :** a site fit for business opportunity
6 : a place where a passenger vehicle stops or parks ⟨a taxi *stand*⟩
7 : HIVE 2
8 : a frame on or in which something may be placed for support
9 : a group of plants growing in a continuous area
10 : a standing posture

stand–alone \ˈstan-də-ˈlōn\ *adjective* (1966)
: SELF-CONTAINED; *especially* **:** operating or capable of operating independently of a computer system ⟨a *stand-alone* word processor⟩

¹stan·dard \ˈstan-dərd\ *noun* [Middle English, from Old French *estandard* rallying point, standard, of Germanic origin; akin to Old English *standan* to stand and to Old English *ord* point — more at ODD] (12th century)
1 : a conspicuous object (as a banner) formerly carried at the top of a pole and used to mark a rallying point especially in battle or to serve as an emblem
2 a : a long narrow tapering flag that is personal to an individual or corporation and bears heraldic devices **b :** the personal flag of the head of a state or of a member of a royal family **c :** an organization flag carried by a mounted or motorized military unit **d :** BANNER
3 : something established by authority, custom, or general consent as a model or example **:** CRITERION
4 : something set up and established by authority as a rule for the measure of quantity, weight, extent, value, or quality
5 a : the fineness and legally fixed weight of the metal used in coins **b :** the basis of value in a monetary system
6 : a structure built for or serving as a base or support
7 a : a shrub or herb grown with an erect main stem so that it forms or resembles a tree **b :** a fruit tree grafted on a stock that does not induce dwarfing
8 a : the large odd upper petal of a papilionaceous flower (as the pea) **b :** one of the three inner usually erect and incurved petals of an iris
9 : a musical composition (as a song) that has become a part of the standard repertoire ☆
— **stan·dard·less** *adjective*

²standard *adjective* (1622)
1 a : constituting or conforming to a standard especially as established by law or custom ⟨*standard* weight⟩ **b :** sound and usable but not of top quality ⟨*standard* beef⟩
2 a : regularly and widely used, available, or supplied ⟨*standard* automobile equipment⟩ **b :** well-established and very familiar ⟨the *standard* opera⟩
3 : having recognized and permanent value ⟨a *standard* reference work⟩
4 : substantially uniform and well established by usage in the speech and writing of the educated and widely recognized as acceptable ⟨*standard* pronunciation is subject to regional variations⟩
— **stan·dard·ly** *adverb*

stan·dard–bear·er \ˈstan-dərd-ˌbar-ər, -ˌber-\ *noun* (15th century)
1 : one that bears a standard or banner
2 : the leader of an organization, movement, or party

stan·dard·bred \-ˌbred\ *noun, often capitalized* (1921)
: any of a breed of trotting and pacing horses developed in the U.S., noted for speed and stamina, and used especially in harness racing

standardbred

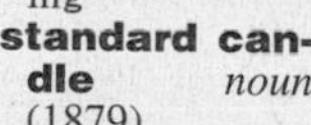
standard candle *noun* (1879)
: CANDELA

standard deviation *noun* (1894)
1 : a measure of the dispersion of a frequency distribution that is the square root of the arithmetic mean of the squares of the deviation of each of the class frequencies from the arithmetic mean of the frequency distribution; *also* **:** a similar quantity found by dividing by one less than the number of squares in the sum of squares instead of taking the arithmetic mean
2 : a parameter that indicates the way in which a probability function or a probability density function is centered around its mean and that is equal to the square root of the moment in which the deviation from the mean is squared

Standard English *noun* (1836)
: the English that with respect to spelling, grammar, pronunciation, and vocabulary is substantially uniform though not devoid of regional differences, that is well established by usage in the formal and informal speech and writing of the educated, and that is widely recognized as acceptable wherever English is spoken and understood

standard error *noun* (1897)
: the standard deviation of the probability function or probability density function of a random variable and especially of a statistic; *specifically* **:** the standard error of the mean of a sample from a population with a normal distribution that is equal to the standard deviation of the normal distribution divided by the square root of the sample size

standard gauge *noun* (1871)
: a railroad gauge of 4 feet 8½ inches

stan·dard·ise *British variant of* STANDARDIZE

stan·dard·ize \ˈstan-dər-ˌdīz\ *transitive verb* **-ized; -iz·ing** (1873)
1 : to compare with a standard
2 : to bring into conformity with a standard
— **stan·dard·i·za·tion** \ˌstan-dər-də-ˈzā-shən\ *noun*

standard of living (1902)
1 : the necessities, comforts, and luxuries enjoyed or aspired to by an individual or group
2 : a minimum of necessities, comforts, or luxuries held essential to maintaining a person or group in customary or proper status or circumstances

standard operating procedure *noun* (1952)
: established or prescribed methods to be followed routinely for the performance of designated operations or in designated situations — called also *standing operating procedure*

standard position *noun* (1950)
: the position of an angle with its vertex at the origin of a rectangular-coordinate system and its initial side coinciding with the positive x-axis

standard schnauzer *noun* (circa 1934)
: any of a breed of medium-sized schnauzers that attain a height at the highest point of the shoulder blades of 18 to 20 inches (46 to 51 centimeters) and have a pepper-and-salt or black coat

standard score *noun* (1928)
: an individual test score expressed as the deviation from the mean score of the group in units of standard deviation

standard time *noun* (1879)
: the time of a region or country that is established by law or general usage as civil time; *specifically* **:** the mean solar time of a meridian that is a multiple of 15 arbitrarily applied to a local area or to one of the 24 time zones and designated as a number of hours earlier or later than Greenwich time

stand·away \ˈstan-də-ˌwā\ *adjective* (1948)
: standing out from the body ⟨a *standaway* skirt⟩

¹stand·by \ˈstan(d)-ˌbī\ *noun, plural* **stand·bys** \-ˌbīz\ (1796)
1 a : one to be relied on especially in emergencies **b :** a favorite or reliable choice or resource
2 : one that is held in reserve ready for use **:** SUBSTITUTE

☆ SYNONYMS
Standard, criterion, gauge, yardstick, touchstone mean a means of determining what a thing should be. STANDARD applies to any definite rule, principle, or measure established by authority ⟨*standards* of behavior⟩. CRITERION may apply to anything used as a test of quality whether formulated as a rule or principle or not ⟨questioned the critic's *criteria* for excellence⟩. GAUGE applies to a means of testing a particular dimension (as thickness, depth, diameter) or figuratively a particular quality or aspect ⟨polls as a *gauge* of voter dissatisfaction⟩. YARDSTICK is an informal substitute for CRITERION that suggests quantity more often than quality ⟨housing construction as a *yardstick* of economic growth⟩. TOUCHSTONE suggests a simple test of the authenticity or value of something intangible ⟨fine service is one *touchstone* of a first-class restaurant⟩.

\ə\ **abut** \ᵊ\ **kitten** \ər\ **further** \a\ **ash** \ā\ **ace** \ä\ **mop, mar** \au̇\ **out** \ch\ **chin** \e\ **bet** \ē\ **easy** \g\ **go** \i\ **hit** \ī\ **ice** \j\ **job** \ŋ\ **sing** \ō\ **go** \ȯ\ **law** \ȯi\ **boy** \th\ **thin** \t͟h\ **the** \ü\ **loot** \u̇\ **foot** \y\ **yet** \zh\ **vision** *see also* Guide to Pronunciation

— **on standby :** ready or available for immediate action or use
²standby *adjective* (1882)
1 : held near at hand and ready for use ⟨a *standby* power plant⟩ ⟨*standby* equipment⟩
2 : relating to the act or condition of standing by ⟨*standby* duty⟩ ⟨a *standby* period⟩
3 : of, relating to, or traveling by an airline service in which the passenger must wait for an available unreserved seat ⟨*standby* passengers⟩ ⟨a *standby* ticket⟩
³standby *adverb* (1971)
: on a standby basis ⟨fly *standby*⟩
stand by (13th century)
intransitive verb
1 : to be present; *also* **:** to remain apart or aloof
2 : to be or to get ready to act
transitive verb
: to remain loyal or faithful to **:** DEFEND
stand–down \ˈstan(d)-ˌdau̇n\ *noun* (circa 1919)
: a relaxation of status of a military unit or force from an alert or operational posture
stand down *intransitive verb* (1681)
1 : to leave the witness stand
2 *chiefly British* **a :** to go off duty **b :** to withdraw from a contest or from a position of leadership
stand•ee \stan-ˈdē\ *noun* (1856)
: one who occupies standing room
stand–in \ˈstan-ˌdin\ *noun* (circa 1928)
1 : someone employed to occupy an actor's place while lights and camera are readied
2 : SUBSTITUTE
stand in *intransitive verb* (1904)
: to act as a stand-in
— **stand in with :** to be in a specially favored position with
¹stand•ing \ˈstan-diŋ\ *adjective* (14th century)
1 a : not yet cut or harvested ⟨*standing* timber⟩ ⟨*standing* grain⟩ **b :** upright on the feet or base **:** ERECT ⟨the *standing* audience⟩
2 a : not being used or operated ⟨a *standing* factory⟩ **b :** not flowing **:** STAGNANT ⟨*standing* water⟩
3 a : remaining at the same level, degree, or amount for an indeterminate period ⟨a *standing* offer⟩ **b :** continuing in existence or use indefinitely
4 : established by law or custom
5 : not movable
6 : done from a standing position ⟨a *standing* jump⟩ ⟨a *standing* ovation⟩
²standing *noun* (15th century)
1 a : a place to stand in **:** LOCATION **b :** a position from which one may assert or enforce legal rights and duties
2 a : length of service or experience especially as determining rank, pay, or privilege **b :** position or condition in society or in a profession; *especially* **:** good reputation **c :** position relative to a standard of achievement or to achievements of competitors
3 : maintenance of position or condition **:** DURATION ⟨a custom of long *standing*⟩
standing army *noun* (1603)
: a permanent army of paid soldiers
standing committee *noun* (circa 1636)
: a permanent committee especially of a legislative body
standing crop *noun* (1861)
: the total amount or number of living things or of one kind of living thing (as an uncut farm crop, the fish in a pond, or organisms in an ecosystem) in a particular situation at any given time
standing O \-ˈō\ *noun* [ovation] (1975)
: a standing ovation
standing order *noun* (1737)
: an instruction or prescribed procedure in force permanently or until changed or canceled; *especially* **:** any of the rules for the guidance and government of parliamentary procedure which endure through successive sessions until vacated or repealed
standing room *noun* (1603)
: space for standing; *especially* **:** accommodation available for spectators or passengers after all seats are filled
standing wave *noun* (1896)
: a single-frequency mode of vibration of a body or physical system in which the amplitude varies from place to place, is constantly zero at fixed points, and has maxima at other points
stan•dish \ˈstan-dish\ *noun* [Middle English *standyshe*] (14th century)
: a stand for writing materials **:** INKSTAND
¹stand•off \ˈstan-ˌdȯf\ *adjective* (1837)
1 : STANDOFFISH
2 : used for holding something at a distance from a surface ⟨a *standoff* insulator⟩
²standoff *noun* (1843)
1 a : TIE, DEADLOCK ⟨the two teams played to a *standoff*⟩ **b :** a counterbalancing effect
2 : the act of standing off
stand off (1601)
intransitive verb
1 : to stay at a distance from something
2 : to sail away from the shore
transitive verb
1 : to keep from advancing **:** REPEL
2 : PUT OFF, STALL
stand•off•ish \stan-ˈdȯ-fish\ *adjective* (1860)
: somewhat cold and reserved
— **stand•off•ish•ly** *adverb*
— **stand•off•ish•ness** *noun*
stand oil *noun* (1908)
: a thickened drying oil; *especially* **:** linseed oil heated to about 600° F (315° C)
stand•out \ˈstan-ˌdau̇t\ *noun, often attributive* (1928)
: one that is prominent or conspicuous especially because of excellence
stand out *intransitive verb* (15th century)
1 a : to appear as if in relief **:** PROJECT **b :** to be prominent or conspicuous
2 : to steer away from shore
3 : to be stubborn in resolution or resistance
stand•pat \ˈstan(d)-ˈpat\ *adjective* (1904)
: stubbornly conservative **:** resisting or opposing change
stand pat *intransitive verb* [⁴*pat*] (1882)
1 : to play one's hand as dealt in draw poker without drawing
2 : to oppose or resist change
— **stand•pat•ter** \ˈstan(d)-ˌpa-tər, -ˈpa-tər\ *noun*
— **stand•pat•tism** \-ˌpa-ˌti-zəm\ *noun*
stand•pipe \ˈstan(d)-ˌpīp\ *noun* (circa 1850)
: a high vertical pipe or reservoir that is used to secure a uniform pressure in a water-supply system
stand•point \-ˌpȯint\ *noun* (1829)
: a position from which objects or principles are viewed and according to which they are compared and judged
stand•still \-ˌstil\ *noun* (1702)
: a state characterized by absence of motion or of progress **:** STOP
¹stand–up \ˈstan-ˌdəp\ *adjective* (1812)
1 a : ERECT, UPRIGHT **b :** stiffened to stay upright without folding over ⟨a *stand-up* collar⟩
2 : performed in, performing in, or requiring a standing position ⟨a *stand-up* bar⟩; *especially* **:** of, relating to, performing, or being a monologue of jokes, gags, or satirical comments delivered usually while standing alone on a stage or in front of a camera ⟨*stand-up* comedy⟩ ⟨a *stand-up* comedian⟩
²stand–up *noun* (1971)
: stand-up comedy
stand up (before 12th century)
intransitive verb
1 : to rise to a standing position
2 : to remain sound and intact under stress, attack, or close scrutiny
transitive verb
: to fail to keep an appointment with
— **stand up for :** to defend against attack or criticism
— **stand up to 1 :** to meet fairly and fully
2 : to face boldly
— **stand up with :** to be best man or maid of honor for at a wedding ceremony
stand–up•per \ˈstan-ˈdə-pər\ *noun* (1973)
: a television news report or interview by an on-camera reporter standing usually at the scene of an occurrence
stane \ˈstān\ *Scottish variant of* STONE
Stan•ford–Bi•net test \ˌstan-fərd-bi-ˈnā-\ *noun* [*Stanford* University + Alfred *Binet* (died 1911) French psychologist] (1918)
: an intelligence test prepared at Stanford University as a revision of the Binet-Simon scale and commonly used with children — called also *Stanford-Binet*
¹stang \ˈstaŋ\ *transitive verb* [Middle English, from Old Norse *stanga* to prick; akin to Old Norse *stinga* to sting] (14th century)
chiefly Scottish **:** STING
²stang *noun* (1513)
chiefly Scottish **:** PANG
³stang *variant of* SATANG
stan•hope \ˈsta-nəp\ *noun* [Fitzroy *Stanhope* (died 1864) British clergyman] (1825)
: a gig, buggy, or phaeton typically having a high seat and closed back
sta•nine \ˈstā-ˌnīn\ *noun* [*sta*ndard (score) + *nine*] (1944)
: any of the nine classes into which a set of normalized standard scores arranged according to rank in educational testing are divided, which include the bottom 4% and the top 4% of the scores in the first and ninth classes and the middle 20% in the fifth, and which have a standard deviation of 2 and a mean of 5
Sta•ni•slav•ski method \ˌsta-nə-ˈslaf-ski-, -ˈslav-\ *noun* [Konstantin *Stanislavsky*] (1941)
: a technique in acting by which an actor strives to empathize with the character being portrayed so as to effect a realistic interpretation
¹stank \ˈstaŋk\ *past of* STINK
²stank *noun* [Middle English, from Middle French *estanc,* from *estancher* to dam up, stanch — more at STANCH] (14th century)
1 *dialect British* **a :** POND, POOL **b :** a ditch containing water
2 *British* **:** a small dam **:** WEIR
stan•na•ry \ˈsta-nə-rē\ *noun, plural* **-ries** [Middle English *stannarie,* from Medieval Latin *stannaria* tin mine, from Late Latin *stannum* tin] (15th century)
: any of the regions in England containing establishments for the working of tin — usually used in plural
stan•nic \ˈsta-nik\ *adjective* [probably from French *stannique,* from Late Latin *stannum* tin, from Latin *stagnum,* an alloy of silver and lead] (1790)
: of, relating to, or containing tin especially with a valence of four
stan•nite \ˈsta-ˌnīt\ *noun* [Late Latin *stannum*] (1868)
: a mineral that is a steel-gray or iron-black sulfide of copper, iron, and tin with a metallic luster and occurs in granular masses
stan•nous \ˈsta-nəs\ *adjective* [International Scientific Vocabulary, from Late Latin *stannum*] (1849)
: of, relating to, or containing tin especially with a valence of two
stan•za \ˈstan-zə\ *noun* [Italian, stay, abode, room, stanza, from (assumed) Vulgar Latin *stantia* stay — more at STANCE] (1588)
: a division of a poem consisting of a series of lines arranged together in a usually recurring pattern of meter and rhyme **:** STROPHE
— **stan•za•ic** \stan-ˈzā-ik\ *adjective*
sta•pe•dec•to•my \ˌstā-pi-ˈdek-tə-mē\ *noun, plural* **-mies** [International Scientific Vocabulary, from New Latin *staped-, stapes*] (1894)
: surgical removal and prosthetic replacement of part or all of the stapes to relieve deafness

sta·pe·di·al \stā-ˈpē-dē-əl, stə-\ *adjective* (circa 1859)
: of, relating to, or located near the stapes

sta·pe·lia \stə-ˈpēl-yə\ *noun* [New Latin, from J. B. van *Stapel* (died 1636) Dutch botanist] (circa 1785)
: any of a genus (*Stapelia*) of African perennial herbs of the milkweed family with succulent typically leafless toothed stems like cactus joints and showy but usually putrid-smelling flowers

sta·pes \ˈstā-(ˌ)pēz\ *noun, plural* **stapes** *or* **sta·pe·des** \ˈstā-pə-ˌdēz\ [New Latin *staped-, stapes,* from Medieval Latin, stirrup, alteration of Late Latin *stapia*] (1670)
: the innermost ossicle of the ear of mammals — called also *stirrup;* see EAR illustration

staph \ˈstaf\ *noun* (circa 1933)
: STAPHYLOCOCCUS

staph·y·li·nid \ˌsta-fə-ˈlī-nəd\ *noun* [New Latin *Staphylinidae,* ultimately from Greek *staphylē* bunch of grapes] (circa 1891)
: ROVE BEETLE
— staphylinid *adjective*

staph·y·lo·coc·cal \ˌsta-f(ə-)lō-ˈkä-kəl\ *also* **staph·y·lo·coc·cic** \-ˈkä-kik, -ˈkäk-sik\ *adjective* (1900)
: of, relating to, caused by, or being a staphylococcus

staph·y·lo·coc·cus \-ˈkä-kəs\ *noun, plural* **-coc·ci** \-ˈkä-ˌkī, -(ˌ)kē; -ˈkäk-ˌsī, -(ˌ)sē\ [New Latin, from Greek *staphylē* bunch of grapes + New Latin *-coccus*] (1887)
: any of various nonmotile gram-positive spherical bacteria (especially genus *Staphylococcus*) that occur singly, in pairs or tetrads, or in irregular clusters and include pathogens which infect skin and mucous membranes

¹sta·ple \ˈstā-pəl\ *noun* [Middle English *stapel* post, staple, from Old English *stapol* post; akin to Middle Dutch *stapel* step, heap, emporium, Old English *steppan* to step] (13th century)
1 : a U-shaped metal loop both ends of which are driven into a surface to hold the hook, hasp, or bolt of a lock, secure a rope, or fix a wire in place
2 : a small U-shaped wire both ends of which are driven through layers of thin and easily penetrable material (as paper) and usually clinched to hold the layers together

²staple *transitive verb* **sta·pled; sta·pling** \-p(ə-)liŋ\ (14th century)
: to provide with or secure by staples

³staple *noun* [Middle English, from Middle French *estaple,* from Middle Dutch *stapel* emporium] (14th century)
1 : a town used as a center for the sale or exportation of commodities in bulk
2 : a place of supply **:** SOURCE
3 : a chief commodity or production of a place
4 a : a commodity for which the demand is constant **b :** something having widespread and constant use or appeal **c :** the sustaining or principal element **:** SUBSTANCE
5 : RAW MATERIAL
6 a : textile fiber (as wool and rayon) of relatively short length that when spun and twisted forms a yarn rather than a filament **b :** the length of a piece of such textile fiber

⁴staple *adjective* (1615)
1 : used, needed, or enjoyed constantly usually by many individuals
2 : produced regularly or in large quantities ⟨*staple* crops such as wheat and rice⟩
3 : PRINCIPAL, CHIEF

¹sta·pler \ˈstā-p(ə-)lər\ *noun* (circa 1513)
: one that deals in staple goods or in staple fiber

²stapler *noun* (circa 1909)
: one that inserts staples; *especially* **:** a small usually hand-operated device for inserting wire staples

¹star \ˈstär\ *noun, often attributive* [Middle English *sterre,* from Old English *steorra;* akin to Old High German *sterno* star, Latin *stella,* Greek *astēr, astron*] (before 12th century)
1 a : a natural luminous body visible in the sky especially at night **b :** a self-luminous gaseous celestial body of great mass which produces energy by means of nuclear fusion reactions, whose shape is usually spheroidal, and whose size may be as small as the earth or larger than the earth's orbit
2 a (1) **:** a planet or a configuration of the planets that is held in astrology to influence one's destiny or fortune — usually used in plural (2) **:** a waxing or waning fortune or fame ⟨her *star* was rising⟩ **b** *obsolete* **:** DESTINY
3 a : a conventional figure with five or more points that represents a star; *especially* **:** ASTERISK **b :** an often star-shaped ornament or medal worn as a badge of honor, authority, or rank or as the insignia of an order **c :** one of a group of conventional stars used to place something in a scale of value
4 : something resembling a star ⟨was hit on the head and saw *stars*⟩
5 a : the principal member of a theatrical or operatic company who usually plays the chief roles **b :** a highly publicized theatrical or motion-picture performer **c :** an outstandingly talented performer ⟨a track *star*⟩ **d :** a person who is preeminent in a particular field
— star·less \-ləs\ *adjective*
— star·like \-ˌlīk\ *adjective*

²star *verb* **starred; star·ring** (1718)
transitive verb
1 : to sprinkle or adorn with stars
2 a : to mark with a star as being preeminent **b :** to mark with an asterisk
3 : to advertise or display prominently **:** FEATURE ⟨the movie *stars* a famous stage personality⟩
intransitive verb
1 : to play the most prominent or important role
2 : to perform outstandingly

³star *adjective* (1832)
1 : of, relating to, or being a star ⟨received *star* billing⟩
2 : of outstanding excellence **:** PREEMINENT ⟨a *star* athlete⟩

star apple *noun* (1683)
: a tropical American tree (*Chrysophyllum cainito*) of the sapodilla family grown in warm regions for ornament or fruit; *also* **:** its usually green to purple apple-shaped edible fruit

¹star·board \ˈstär-bərd\ *noun* [Middle English *sterbord,* from Old English *stēorbord,* from *stēor-* steering oar + *bord* ship's side — more at STEER, BOARD] (before 12th century)
: the right side of a ship or aircraft looking forward — compare PORT ◆

²starboard *adjective* (15th century)
: of, relating to, or situated to starboard

³starboard *transitive verb* (1598)
: to turn or put (a helm or rudder) to the right

¹starch \ˈstärch\ *transitive verb* [Middle English *sterchen,* probably from (assumed) Old English *stercan* to stiffen; akin to Old English *stearc* stiff — more at STARK] (15th century)
: to stiffen with or as if with starch

²starch *noun* (15th century)
1 : a white odorless tasteless granular or powdery complex carbohydrate $(C_6H_{10}O_5)_x$ that is the chief storage form of carbohydrate in plants, is an important foodstuff, and is used also in adhesives and sizes, in laundering, and in pharmacy and medicine
2 : a stiff formal manner **:** FORMALITY
3 : resolute vigor

star–cham·ber \ˈstär-ˈchām-bər\ *adjective* [*Star Chamber,* a court existing in England from the 15th century until 1641] (1800)
: characterized by secrecy and often being irresponsibly arbitrary and oppressive

starchy \ˈstär-chē\ *adjective* **starch·i·er; -est** (1802)
1 : containing, consisting of, or resembling starch
2 : consisting of or marked by formality or stiffness
— starch·i·ly \-chə-lē\ *adverb*
— starch·i·ness \-chē-nəs\ *noun*

star–crossed \ˈstär-ˌkrȯst\ *adjective* (1592)
: not favored by the stars **:** ILL-FATED ⟨a pair of *star-crossed* lovers take their life —Shakespeare⟩

star·dom \ˈstär-dəm\ *noun* (1865)
: the status or position of a star

star·dust \ˈstär-ˌdəst\ *noun* (1927)
: a feeling or impression of romance, magic, or ethereality

¹stare \ˈstar, ˈster\ *verb* **stared; star·ing** [Middle English, from Old English *starian;* akin to Old High German *starēn* to stare, Greek *stereos* solid, Lithuanian *starinti* to stiffen] (before 12th century)
intransitive verb
1 : to look fixedly often with wide-open eyes
2 : to show oneself conspicuously ⟨the error *stared* from the page⟩
3 *of hair* **:** to stand on end **:** BRISTLE; *also* **:** to appear rough and lusterless
transitive verb
1 : to have an effect on by staring
2 : to look at with a searching or earnest gaze
— star·er *noun*
— stare one in the face : to be undeniably and forcefully evident or apparent

²stare *noun* (15th century)
: the act or an instance of staring ⟨a blank *stare*⟩

sta·re de·ci·sis \ˌster-ē-di-ˈsī-səs, ˌstar-, ˌstär-\ *noun* [Latin, to stand by decided matters] (1782)
: a doctrine or policy of following rules or principles laid down in previous judicial decisions unless they contravene the ordinary principles of justice

stare down *transitive verb* (1925)
: to cause to waver or submit by or as if by staring

sta·rets \ˈstär-əts, -yəts\ *noun, plural* **star·tsy** \ˈstärt-sē\ [Russian, from *staryĭ* old — more at STOUR] (1917)
: a spiritual director or religious teacher in the Eastern Orthodox Church; *specifically* **:** a spiritual adviser who is not necessarily a priest

◇ WORD HISTORY

starboard Appearances to the contrary notwithstanding, the *star-* in *starboard* has nothing to do with the stars. The word *starboard* descends from Old English *stēorbord,* a compound of *stēor* "steering oar, rudder" and *bord* "side of a ship." Early Germanic craft such as those used by the Angles and Saxons who invaded Britain were steered by means of a long oar attached to the right of the ship; the right side was hence the "steering side." *Stēor* is related to the Modern English verb *steer. Bord* is the ancestor of Modern English *board,* though the sense "side of a ship" is now rare outside of compounds such as *overboard. Starboard* used to be contrasted with *larboard,* in Middle English *laddebord,* the first element of which has never been identified with certainty. English-speaking navies replaced *larboard* in the 19th century with *port,* first attested in the verb phrase *to port the helm* (turn the ship to the left) in the 16th century. *Port,* too, is a nautical mystery, since there is no sure connection between the left side of a ship and *port* meaning "harbor" or "gate."

who is recognized for his piety, and who is turned to by monks or laymen for spiritual guidance

star facet *noun* (1751)
: one of the eight small triangular facets which abut on the table in the bezel of a brilliant

star·fish \'stär-ˌfish\ *noun* (1538)
: any of a class (Asteroidea) of echinoderms having a body of usually five radially disposed arms about a central disk and feeding largely on mollusks (as oysters) — called also *sea star*

star·flow·er \-ˌflau̇(-ə)r\ *noun* (1629)
: any of several plants having star-shaped pentamerous flowers; *especially* **:** any of a genus (*Trientalis,* especially *T. borealis*) of perennial herbs of the primrose family

star fruit *noun* (1974)
: CARAMBOLA 1

star·gaze \-ˌgāz\ *intransitive verb* [back-formation from *stargazer*] (1626)
1 : to gaze at stars
2 : to gaze raptly or contemplatively

star·gaz·er \-ˌgā-zər\ *noun* (1560)
1 : one who gazes at the stars: as **a :** ASTROLOGER **b :** ASTRONOMER
2 : any of various marine bony fishes (families Uranoscopidae and Dactyloscopidae) with the eyes on top of a blocky or conical head

star·gaz·ing \-ˌgā-ziŋ\ *noun* (1576)
1 : the act or practice of a stargazer
2 a : absorption in chimerical or impractical ideas **:** WOOLGATHERING **b :** the quality or state of being absentminded

star grass *noun* (1687)
: any of various grassy plants with stellate flowers or arrangement of leaves: as **a :** any of a genus (*Hypoxis*) of herbs of the amaryllis family **b :** either of two colicroots (*Aletris farinosa* and *A. aurea*)

¹stark \'stärk\ *adjective* [Middle English, stiff, strong, from Old English *stearc;* akin to Old High German *starc* strong, Lithuanian *starinti* to stiffen — more at STARE] (before 12th century)
1 a : rigid in or as if in death **b :** rigidly conforming (as to a pattern or doctrine) **:** ABSOLUTE ⟨*stark* discipline⟩
2 *archaic* **:** STRONG, ROBUST
3 : UTTER, SHEER ⟨*stark* nonsense⟩
4 a : BARREN, DESOLATE **b** (1) **:** having few or no ornaments **:** BARE ⟨a *stark* white room⟩ (2) **:** HARSH, BLUNT ⟨the *stark* realities of death⟩
5 : sharply delineated ⟨a *stark* contrast⟩
— stark·ly *adverb*
— stark·ness *noun*

²stark *adverb* (13th century)
1 : in a stark manner
2 : to an absolute or complete degree **:** WHOLLY ⟨*stark* naked⟩ ⟨*stark* mad⟩

stark·ers \'stär-kərz\ *adjective* [alteration of ¹*stark*] (circa 1923)
chiefly British **:** completely unclothed **:** NAKED
word history see NUTS

star·let \'stär-lət\ *noun* (1920)
: a young movie actress being coached and publicized for starring roles

star·light \-ˌlīt\ *noun* (14th century)
: the light given by the stars

star·ling \'stär-liŋ\ *noun* [Middle English, from Old English *stærlinc,* from *stær* starling + *-ling, -linc* -ling; akin to Old High German *stara* starling, Latin *sturnus*] (before 12th century)
: any of a family (Sturnidae, especially genus *Sturnus*) of usually dark gregarious oscine birds; *especially* **:** a dark brown or in summer glossy greenish black European bird (*S. vulgaris*) naturalized nearly worldwide and often considered a pest

star·lit \'stär-ˌlit\ *adjective* (circa 1835)
: lighted by the stars

star–nosed mole \'stär-ˌnōz(d)-\ *noun* (1026)
: a common black long-tailed semiaquatic mole (*Condylura cristata*) of the northeastern U.S. and adjacent Canada that has a series of pink fleshy projections surrounding the nostrils

star-nosed mole

star–of–Beth·le·hem \-'beth-li-ˌhem, -lē-(h)əm\ *noun* (1573)
: any of various Old World bulbous herbs (genus *Ornithogalum*) of the lily family with basal leaves resembling grass; *especially* **:** one (*O. umbellatum*) with white flowers that is naturalized in the eastern U.S.

star of Bethlehem
: a star which according to Christian tradition guided the Magi to the infant Jesus in Bethlehem

Star of Da·vid \-'dā-vəd\ (circa 1936)
: MAGEN DAVID

star route *noun* [so called from the asterisk used to designate such routes in postal publications] (1880)
: a mail-delivery route in a rural or thinly populated area served by a private carrier under contract who takes mail from one post office to another or from a railroad station to a post office and usually also delivers mail to private mailboxes along the route

star·ry \'stär-ē\ *adjective* **star·ri·er; -est** (14th century)
1 a : adorned with stars; *especially* **:** STAR-STUDDED **b :** of, relating to, or consisting of stars **:** STELLAR **c :** shining like stars **:** SPARKLING **d :** having parts arranged like the rays of a star **:** STELLATE
2 : as high as or seemingly as high as the stars ⟨*starry* speculations⟩
3 : STARRY-EYED

star·ry–eyed \'stär-ē-ˌīd\ *adjective* (1904)
: regarding an object or a prospect in an overly favorable light; *specifically* **:** characterized by dreamy, impracticable, or utopian thinking **:** VISIONARY

Stars and Bars *noun plural but singular in construction* (1861)
: the first flag of the Confederate States of America having three bars of red, white, and red respectively and a blue union with white stars in a circle representing the seceded states

Stars and Stripes *noun plural but singular in construction* (1777)
: the flag of the United States having 13 alternately red and white horizontal stripes and a blue union with white stars representing the states

star sapphire *noun* (circa 1817)
: a sapphire that when cut with a convex surface and polished exhibits asterism

star shell *noun* (circa 1876)
1 : a shell that on bursting releases a shower of brilliant stars and is used for signaling
2 : a shell with an illuminating projectile

star·ship \'stär-ˌship\ *noun* (1934)
: a spacecraft designed for interstellar travel

star–span·gled \'stär-ˌspaŋ-gəld\ *adjective* (1591)
: STAR-STUDDED

star·struck \-ˌstrək\ *adjective* (1968)
: particularly taken with celebrities (as movie stars)

star–stud·ded \'stär-ˌstə-dəd\ *adjective* (1955)
: abounding in or covered with stars ⟨a *star-studded* cast⟩ ⟨a *star-studded* uniform⟩

star system *noun* (1902)
: the practice of casting famous performers in principal roles (as in motion pictures or the theater) especially in order to capitalize on their popular appeal

¹start \'stärt\ *verb* [Middle English *sterten;* akin to Middle High German *sterzen* to stand up stiffly, move quickly] (14th century)
intransitive verb
1 a : to move suddenly and violently **:** SPRING ⟨*started* angrily to his feet⟩ **b :** to react with a sudden brief involuntary movement ⟨*started* when a shot rang out⟩
2 a : to issue with sudden force ⟨blood *starting* from the wound⟩ **b :** to come into being, activity, or operation ⟨when does the movie *start*⟩ ⟨the rain *started* up again⟩
3 : to protrude or seem to protrude ⟨eyes *starting* from their sockets⟩
4 : to become loosened or forced out of place ⟨one of the planks has *started*⟩
5 a : to begin a course or journey ⟨*started* toward the door⟩ ⟨just *starting* out⟩ **b :** to range from a specified initial point ⟨the rates *start* at $10⟩
6 : to begin an activity or undertaking; *especially* **:** to begin work
7 : to be a participant in a game or contest; *especially* **:** to be in the starting lineup
transitive verb
1 : to cause to leave a place of concealment **:** FLUSH ⟨*start* a rabbit⟩
2 *archaic* **:** STARTLE, ALARM
3 : to bring up for consideration or discussion
4 : to bring into being ⟨*start* a rumor⟩
5 : to cause to become loosened or displaced
6 : to begin the use of ⟨*start* a fresh loaf of bread⟩
7 a : to cause to move, act, or operate ⟨*start* the motor⟩ **b :** to cause to enter a game or contest; *especially* **:** to put in the starting lineup **c :** to care for or train during the early stages of growth and development ⟨*started* plants⟩ ⟨a well-*started* coonhound⟩
8 : to do or experience the first stages or actions of ⟨*started* studying music at the age of five⟩
synonym see BEGIN
— start something *also* **start anything :** to make trouble ⟨always trying to *start something*⟩ ⟨don't *start anything*⟩
— to start with 1 : at the beginning **:** INITIALLY **2 :** in any event

²start *noun* (14th century)
1 a : a sudden involuntary bodily movement or reaction ⟨woke with a *start*⟩ **b :** a brief and sudden action or movement **c :** a sudden capricious impulse or outburst
2 : a beginning of movement, activity, or development ⟨a false *start*⟩ ⟨housing *starts*⟩
3 : HEAD START
4 : a place of beginning
5 : the act or an instance of being a competitor in a race or a member of a starting lineup in a game ⟨undefeated in six *starts* —*Current Biography*⟩

start·er \'stär-tər\ *noun* (1622)
1 : a person who initiates or sets going: as **a :** an official who gives the signal to begin a race **b :** one who dispatches vehicles
2 a : one that engages in a competition; *especially* **:** a member of a starting lineup **b :** one that begins to engage in an activity or process
3 : one that causes something to begin operating: as **a :** a device for starting an engine; *especially* **:** an electric motor used to start an internal combustion engine **b :** material containing microorganisms used to induce a desired fermentation
4 : something that is the beginning of a process, activity, or series; *especially* **:** APPETIZER
— for starters : to begin with

star thistle *noun* (1578)
1 : a widely naturalized spiny European knapweed (*Centaurea calcitrapa*) with purple flowers — called also *caltrops*
2 : any of various knapweeds related to the star thistle

starting block *noun* (1937)
: a device that consists of two blocks mounted on either side of an adjustable frame which is usually anchored to the ground and that provides a runner with a rigid surface against which to brace the feet at the start of a race

starting block

starting gate *noun* (1898)
1 : a mechanically operated barrier used as a starting device for a race
2 : a barrier that when knocked aside by a competitor (as a skier) starts an electronic timing device

¹star·tle \ˈstär-tᵊl\ *verb* **star·tled; star·tling** \ˈstärt-liŋ, ˈstär-tᵊl-iŋ\ [Middle English *stertlen,* frequentative of *sterten* to start] (1530)
intransitive verb
: to move or jump suddenly (as in surprise or alarm) ⟨the baby *startles* easily⟩
transitive verb
: to frighten or surprise suddenly and usually not seriously
— star·tle·ment \-mənt\ *noun*

²startle *noun* (1714)
: a sudden mild shock (as of surprise or alarm)

star·tling *adjective* (1714)
: causing momentary fright, surprise, or astonishment
— star·tling·ly \ˈstärt-liŋ-lē, ˈstär-tᵊl-iŋ-\ *adverb*

start–up \ˈstärt ˌəp\ *noun, often attributive* (1845)
1 : the act or an instance of setting in operation or motion
2 : a fledgling business enterprise

star turn *noun* (1898)
chiefly British **:** the featured skit or number in a theatrical production; *broadly* **:** the most widely publicized person or item in a group

star·va·tion \stär-ˈvā-shən\ *noun* (1778)
1 : the act or an instance of starving
2 : the state of being starved

starvation wages *noun* (1898)
: wages insufficient to provide the ordinary necessities of life

starve \ˈstärv\ *verb* **starved; starv·ing** [Middle English *sterven* to die, starve, from Old English *steorfan* to die; akin to Old High German *sterban* to die, Lithuanian *starinti* to stiffen — more at STARE] (15th century)
intransitive verb
1 a : to perish from lack of food **b :** to suffer extreme hunger
2 a *archaic* **:** to die of cold **b** *British* **:** to suffer greatly from cold
3 : to suffer or perish from deprivation ⟨*starved* for affection⟩
transitive verb
1 a : to kill with hunger **b :** to deprive of nourishment **c :** to cause to capitulate by or as if by depriving of nourishment
2 : to destroy by or cause to suffer from deprivation
3 *archaic* **:** to kill with cold

¹starve·ling \ˈstärv-liŋ\ *noun* (1546)
: one that is thin from or as if from lack of food

²starveling *adjective* (1597)
: being a starveling; *also* **:** marked by poverty or inadequacy ⟨*starveling* wages⟩

¹stash \ˈstash\ *transitive verb* [origin unknown] (1785)
: to store in a usually secret place for future use — often used with *away*

²stash *noun* (circa 1914)
1 : hiding place **:** CACHE
2 : something stored or hidden away ⟨a *stash* of narcotics⟩

sta·sis \ˈstā-səs, ˈsta-\ *noun, plural* **sta·ses** \ˈstā-ˌsēz, ˈsta-\ [New Latin, from Greek, act or condition of standing, stopping, from *histasthai* to stand — more at STAND] (1745)
1 : a slowing or stoppage of the normal flow of a bodily fluid or semifluid: as **a :** slowing of the current of circulating blood **b :** reduced motility of the intestines with retention of feces
2 : a state of static balance or equilibrium **:** STAGNATION

-stasis *noun combining form, plural* **-stases** [New Latin, from Greek *stasis*]
1 : stoppage **:** slowing ⟨hemo*stasis*⟩ ⟨bacterio*stasis*⟩
2 : stable state ⟨homeo*stasis*⟩

stat \ˈstat\ *noun* (circa 1961)
: STATISTIC

-stat *noun combining form* [New Latin *-stata,* from Greek *-statēs* one that stops or steadies, from *histanai* to cause to stand — more at STAND]
1 : stabilizing agent or device ⟨thermo*stat*⟩
2 : instrument for reflecting (something specified) constantly in one direction ⟨helio*stat*⟩
3 : agent causing inhibition of growth without destruction ⟨bacterio*stat*⟩

sta·tant \ˈstā-tᵊnt\ *adjective* [Latin *status,* past participle + English *-ant*] (circa 1500)
: standing in profile with all feet on the ground — used of a heraldic animal

¹state \ˈstāt\ *noun, often attributive* [Middle English *stat,* from Old French & Latin; Old French *estat,* from Latin *status,* from *stare* to stand — more at STAND] (13th century)
1 a : mode or condition of being ⟨a *state* of readiness⟩ **b** (1) **:** condition of mind or temperament ⟨in a highly nervous *state*⟩ (2) **:** a condition of abnormal tension or excitement
2 a : a condition or stage in the physical being of something ⟨insects in the larval *state*⟩ ⟨the gaseous *state* of water⟩ **b :** any of various conditions characterized by definite quantities (as of energy, angular momentum, or magnetic moment) in which an atomic system may exist
3 a : social position; *especially* **:** high rank **b** (1) **:** elaborate or luxurious style of living (2) **:** formal dignity **:** POMP — usually used with *in*
4 a : a body of persons constituting a special class in a society **:** ESTATE 3 **b** *plural* **:** the members or representatives of the governing classes assembled in a legislative body **c** *obsolete* **:** a person of high rank (as a noble)
5 a : a politically organized body of people usually occupying a definite territory; *especially* **:** one that is sovereign **b :** the political organization of such a body of people **c :** a government or politically organized society having a particular character ⟨a police *state*⟩ ⟨the welfare *state*⟩
6 : the operations or concerns of the government of a country
7 a : one of the constituent units of a nation having a federal government ⟨the fifty *states*⟩ **b** *plural, capitalized* **:** The United States of America
8 : the territory of a state

²state *transitive verb* **stat·ed; stat·ing** (1647)
1 : to set by regulation or authority
2 : to express the particulars of especially in words **:** REPORT; *broadly* **:** to express in words
— stat·able *or* **state·able** \ˈstā-tə-bəl\ *adjective*

state aid *noun* (1856)
: public monies appropriated by a state government for the partial support or improvement of a public local institution

state bank *noun* (1815)
1 : CENTRAL BANK
2 : a bank chartered by and operating under the laws of a state of the U.S.

state bird *noun* (1910)
: a bird selected (as by the legislature) as an emblem of a state of the U.S.

state capitalism *noun* (1903)
: an economic system in which private capitalism is modified by a varying degree of government ownership and control

state church *noun, often S&C capitalized* (1726)
: ESTABLISHED CHURCH

state college *noun* (1831)
: a college that is financially supported by a state government, often specializes in a branch of technical or professional education, and often forms part of the state university

state·craft \ˈstāt-ˌkraft\ *noun* (1642)
: the art of conducting state affairs

stat·ed \ˈstā-təd\ *adjective* (circa 1641)
1 : FIXED, REGULAR ⟨the president shall, at *stated* times, receive . . . a compensation —*U.S. Constitution*⟩
2 : set down explicitly **:** DECLARED ⟨our *stated* intention⟩
— stat·ed·ly *adverb*

stated clerk *noun* (circa 1909)
: an executive officer of a Presbyterian general assembly, synod, or presbytery ranking below the moderator

state flower *noun* (1898)
: a flowering plant selected (as by the legislature) as an emblem of a state of the U.S.

state·hood \ˈstāt-ˌhu̇d\ *noun* (1868)
: the condition of being a state; *especially* **:** the status of being one of the states of the U.S.

state·house \-ˌhau̇s\ *noun* (1638)
: the building in which a state legislature sits

state·less \-ləs\ *adjective* (1609)
1 : having no state
2 : lacking the status of a national ⟨a *stateless* refugee⟩
— state·less·ness *noun*

state·ly \-lē\ *adjective* **state·li·er; -est** (15th century)
1 a : marked by lofty or imposing dignity **b :** HAUGHTY, UNAPPROACHABLE
2 : impressive in size or proportions
synonym see GRAND
— state·li·ness *noun*
— stately *adverb*

state·ment \ˈstāt-mənt\ *noun* (circa 1775)
1 : something stated: as **a :** a single declaration or remark **:** ASSERTION **b :** a report of facts or opinions
2 : the act or process of stating or presenting orally or on paper
3 : PROPOSITION 2a
4 : the presentation of a theme in a musical composition
5 : a summary of a financial account showing the balance due
6 : an opinion, comment, or message conveyed indirectly usually by nonverbal means ⟨monuments are *statements* in form and space —O. B. Hardison, Jr.⟩
7 : an instruction in a computer program

state of the art (1910)
: the level of development (as of a device, procedure, process, technique, or science) reached at any particular time usually as a result of modern methods
— state–of–the–art *adjective*

state of war (1880)
1 a : a state of actual armed hostilities regardless of a formal declaration of war **b :** a legal state created and ended by official declaration regardless of actual armed hostilities and usually characterized by operation of the rules of war
2 : the period of time during which a state of war is in effect

\ə\ **abut** \ᵊ\ kitten \ər\ **further** \a\ **ash** \ā\ **ace** \ä\ mop, mar \au̇\ **out** \ch\ **chin** \e\ **bet** \ē\ **easy** \g\ **go** \i\ **hit** \ī\ **ice** \j\ **job** \ŋ\ **sing** \ō\ **go** \ȯ\ **law** \ȯi\ **boy** \th\ **thin** \t͟h\ **the** \ü\ **loot** \u̇\ **foot** \y\ **yet** \zh\ **vision** *see also* Guide to Pronunciation

sta·ter \'stā-tər, stä-'ter\ *noun* [Middle English, from Late Latin, from Greek *statēr*, literally, a unit of weight, from *histanai* to cause to stand, weigh — more at STAND] (14th century)
: an ancient gold or silver coin of the Greek city-states

state·room \'stāt-ˌrüm, -ˌru̇m\ *noun* (1660)
1 : CABIN 1a(1)
2 : a private room on a railroad car with one or more berths and a toilet

state's attorney *noun* (1809)
: a legal officer (as a district attorney) appointed or elected to represent a state in court proceedings within a district — called also *state attorney*

state's evidence *noun, often S capitalized* (1827)
: a participant in a crime or an accomplice who gives evidence for the prosecution especially in return for a reduced sentence; *also* **:** the evidence given — used chiefly in the phrase *turn state's evidence*

States General *noun plural* (1585)
1 : the assembly of the three orders of clergy, nobility, and third estate in France before the Revolution
2 : the legislature of the Netherlands from the 15th century to 1796

¹state·side \'stāt-ˌsīd\ *adjective, often capitalized* [*(United) States* + *side*] (1944)
: being in, going to, coming from, or characteristic of the 48 conterminous states of the U.S. ⟨transferred from Europe to *stateside* duty⟩

²stateside *adverb, often capitalized* (1945)
: in or to the continental U.S.

states·man \'stāts-mən\ *noun* (1592)
1 : one versed in the principles or art of government; *especially* **:** one actively engaged in conducting the business of a government or in shaping its policies
2 : one who exercises political leadership wisely and without narrow partisanship
— states·man·like \-ˌlīk\ *adjective*
— states·man·ly \-lē\ *adjective*
— states·man·ship \-ˌship\ *noun*

state socialism *noun* (1879)
: an economic system with limited socialist characteristics introduced by usually gradual political action

states' right·er \'stāts-'rī-tər\ *noun* (1945)
: one who advocates strict interpretation of the U.S. constitutional guarantee of states' rights

states' rights *noun plural* (1858)
: all rights not vested by the Constitution of the U.S. in the federal government nor forbidden by it to the separate states

state tree *noun* (1917)
: a tree selected (as by the legislature) as an emblem of a state of the U.S.

state university *noun* (1831)
: a university maintained and administered by one of the states of the U.S. as part of the state public educational system

¹state·wide \'stāt-'wīd\ *adjective* (1911)
: affecting or extending throughout all parts of a state

²statewide *adverb* (circa 1934)
: throughout the state

¹stat·ic \'sta-tik\ *adjective* [New Latin *staticus*, from Greek *statikos* causing to stand, skilled in weighing, from *histanai* to cause to stand, weigh — more at STAND] (1638)
1 : exerting force by reason of weight alone without motion
2 : of or relating to bodies at rest or forces in equilibrium
3 : showing little change ⟨a *static* population⟩
4 a : characterized by a lack of movement, animation, or progression **b :** producing an effect of repose or quiescence ⟨a *static* design⟩
5 a : standing or fixed in one place **:** STATIONARY **b** *of water* **:** stored in a tank but not under pressure
6 : of, relating to, or producing stationary charges of electricity **:** ELECTROSTATIC
7 : of, relating to, or caused by radio static
— stat·i·cal \-ti-kəl\ *adjective*
— stat·i·cal·ly \-ti-k(ə-)lē\ *adverb*

²static *noun* [*static electricity*] (1913)
1 : noise produced in a radio or television receiver by atmospheric or various natural or man-made electrical disturbances; *also* **:** the electrical disturbances producing this noise
2 : heated opposition or criticism
— stat·icky \'sta-ti-kē\ *adjective*

-static *adjective combining form* [New Latin *-staticus*, from *staticus*]
1 : of or relating to a position or state ⟨ortho*static*⟩
2 : inhibiting the growth of ⟨fungi*static*⟩

stat·i·ce \'sta-tə-(ˌ)sē\ *noun* [New Latin, genus of herbs, from Latin, an astringent plant, from Greek *statikē*, from feminine of *statikos* causing to stand, astringent] (1745)
: SEA LAVENDER

static electricity *noun* (1876)
: electricity that consists of isolated motionless charges (as those produced by friction)

static line *noun* (1930)
: a cord attached to a parachute pack and to an airplane to open the parachute after a jumper clears the plane

stat·ics \'sta-tiks\ *noun plural but singular or plural in construction* (1837)
: mechanics dealing with the relations of forces that produce equilibrium among material bodies

static tube *noun* (1923)
: a tube used for indicating static as distinct from impact pressure in a stream of fluid

¹sta·tion \'stā-shən\ *noun* [Middle English *stacioun*, from Middle French *station*, from Latin *station-, statio*, from *stare* to stand — more at STAND] (14th century)
1 a : the place or position in which something or someone stands or is assigned to stand or remain **b :** any of the places in a manufacturing operation at which one part of the work is done **c :** equipment used usually by one person for performing a particular job
2 : the act or manner of standing **:** POSTURE
3 : a stopping place: as **a** (1) **:** a regular stopping place in a transportation route (2) **:** the building connected with such a stopping place **:** DEPOT 3 **b :** one of the stations of the cross
4 a : a post or sphere of duty or occupation **b :** a stock farm or ranch especially of Australia or New Zealand
5 : STANDING, RANK ⟨a woman of high *station*⟩
6 : a place for specialized observation and study of scientific phenomena ⟨a seismological *station*⟩ ⟨a marine biological *station*⟩
7 : a place established to provide a public service: as **a** (1) **:** FIRE STATION (2) **:** POLICE STATION **b :** a branch post office
8 : SERVICE STATION 1
9 a : a complete assemblage of radio or television equipment for transmitting or receiving **b :** the place in which such a station is located

²station *transitive verb* **sta·tioned; sta·tion·ing** \'stā-sh(ə-)niŋ\ (1748)
: to assign to or set in a station or position **:** POST

sta·tion·al \'stā-shnəl, -shə-nᵊl\ *adjective* (1902)
: of, relating to, or being a mass formerly celebrated by the pope at designated churches in Rome on appointed holy days

sta·tion·ary \'stā-shə-ˌner-ē\ *adjective* (1626)
1 : fixed in a station, course, or mode **:** IMMOBILE
2 : unchanging in condition

stationary bicycle *noun* (1962)
: an exercise apparatus that can be pedaled like a bicycle — called also *stationary bike*

stationary front *noun* (circa 1940)
: the boundary between two air masses neither of which is replacing the other

stationary wave *noun* (1856)
: STANDING WAVE

station break *noun* (1937)
: a pause in a radio or television broadcast for announcement of the identity of the network or station; *also* **:** an announcement or advertisement during this pause

sta·tio·ner \'stā-sh(ə-)nər\ *noun* [Middle English *staciouner*, from Medieval Latin *stationarius*, from *station-, statio* shop, from Latin, station] (14th century)
1 *archaic* **a :** BOOKSELLER **b :** PUBLISHER
2 : one that sells stationery

sta·tio·nery \'stā-shə-ˌner-ē\ *noun* [*stationer*] (circa 1688)
1 : materials (as paper, pens, and ink) for writing or typing
2 : letter paper usually accompanied with matching envelopes

station house *noun* (1833)
: a house at a post or station; *especially* **:** POLICE STATION

sta·tion·mas·ter \'stā-shən-ˌmas-tər\ *noun* (1856)
: an official in charge of the operation of a railroad station

stations of the cross *often S&C capitalized* (circa 1890)
1 : a series of usually 14 images or pictures especially in a church that represent the stages of Christ's passion and death
2 : a devotion involving commemorative meditation before the stations of the cross

station wagon *noun* (1904)
: an automobile that has a passenger compartment which extends to the back of the vehicle, that has no trunk, that has one or more rear seats which can be folded down to make space for light cargo, and that has a tailgate or liftgate

stat·ism \'stā-ˌti-zəm\ *noun* (1880)
: concentration of economic controls and planning in the hands of a highly centralized government

stat·ist \'stā-tist\ *noun* (1946)
: an advocate of statism
— statist *adjective*

sta·tis·tic \stə-'tis-tik\ *noun* [singular of *statistics*] (1880)
1 : a single term or datum in a collection of statistics
2 a : a quantity (as the mean of a sample) that is computed from a sample; *specifically* **:** ESTIMATE 3b **b :** a random variable that takes on the possible values of a statistic

sta·tis·ti·cal \-ti-kəl\ *adjective* (1787)
: of, relating to, based on, or employing the principles of statistics
— sta·tis·ti·cal·ly \-k(ə-)lē\ *adverb*

statistical mechanics *noun plural but usually singular in construction* (1885)
: a branch of mechanics dealing with the application of the principles of statistics to the mechanics of a system consisting of a large number of parts having motions that differ by small steps over a large range

stat·is·ti·cian \ˌsta-tə-'sti-shən\ *noun* (1825)
: one versed in or engaged in compiling statistics

sta·tis·tics \stə-'tis-tiks\ *noun plural but singular or plural in construction* [German *Statistik* study of political facts and figures, from New Latin *statisticus* of politics, from Latin *status* state] (1770)
1 : a branch of mathematics dealing with the collection, analysis, interpretation, and presentation of masses of numerical data
2 : a collection of quantitative data

sta·tive \'stā-tiv\ *adjective* (1874)
: expressing a state, condition, or relation — compare ACTIVE 3b

stato- *combining form* [International Scientific Vocabulary, from Greek *statos* stationary, from *histasthai* to stand — more at STAND]
1 : resting ⟨*stato*blast⟩
2 : equilibrium ⟨*stato*cyst⟩

stato·blast \'sta-tə-ˌblast\ *noun* [International Scientific Vocabulary] (1855)
1 : a bud in a freshwater bryozoan that overwinters in a chitinous envelope and develops into a new individual in spring
2 : GEMMULE b

stato·cyst \-ˌsist\ *noun* [International Scientific Vocabulary] (1902)
: an organ of equilibrium occurring especially among invertebrate animals and consisting usually of a fluid-filled vesicle in which are suspended statoliths

stato·lith \'sta-tᵊl-ˌith\ *noun* [International Scientific Vocabulary] (1900)
1 : any of the usually calcareous bodies suspended in a statocyst
2 : any of various starch grains or other solid bodies in the plant cytoplasm that are held to be responsible by changes in their position for changes in orientation of a part or organ

sta·tor \'stā-tər\ *noun* [New Latin, from Latin, one that stands, from *stare* to stand — more at STAND] (1902)
: a stationary part in a machine in or about which a rotor revolves

stato·scope \'sta-tə-ˌskōp\ *noun* [International Scientific Vocabulary] (circa 1900)
: a sensitive aneroid barometer for recording small changes in atmospheric pressure; *especially* **:** one used for indicating small changes in the altitude of an aircraft

[1]**stat·u·ary** \'sta-chə-ˌwer-ē\ *noun, plural* **-ar·ies** (1542)
1 : SCULPTOR
2 a : the art of making statues **b :** a collection of statues **:** STATUES

[2]**statuary** *adjective* (1627)
: of, relating to, or suitable for statues

stat·ue \'sta-(ˌ)chü\ *noun* [Middle English, from Middle French, from Latin *statua,* from *statuere* to set up — more at STATUTE] (14th century)
: a three-dimensional representation usually of a person, animal, or mythical being that is produced by sculpturing, modeling, or casting

Statue of Liberty (circa 1900)
1 : a large copper statue of a woman holding a torch aloft in her right hand located on Liberty Island in New York harbor
2 : a trick played in football in which the ballcarrier takes the ball from the raised hand of a teammate who is faking a pass

stat·u·esque \ˌsta-chə-'wesk\ *adjective* (1834)
: resembling a statue especially in dignity, shapeliness, or stillness; *especially* **:** tall and shapely
— stat·u·esque·ly *adverb*

stat·u·ette \ˌsta-chə-'wet\ *noun* (1843)
: a small statue

stat·ure \'sta-chər\ *noun* [Middle English, from Middle French, from Latin *statura,* from *status,* past participle of *stare* to stand — more at STAND] (14th century)
1 : natural height (as of a person) in an upright position
2 : quality or status gained by growth, development, or achievement

sta·tus \'stā-təs, 'sta-\ *noun, plural* **sta·tus·es** *often attributive* [Latin — more at STATE] (1791)
1 : the condition of a person or thing in the eyes of the law
2 a : position or rank in relation to others ⟨the *status* of a father⟩ **b :** relative rank in a hierarchy of prestige; *especially* **:** high prestige
3 : state of affairs

status offender *noun* (circa 1976)
: a young offender (as a runaway or a truant) who is under the jurisdiction of a court for repeated offenses that are not crimes

status quo \-'kwō\ *noun* [Latin, state in which] (1833)
: the existing state of affairs ⟨seeks to preserve the *status quo*⟩

status quo an·te \-'an-tē\ *noun* [Latin, state in which previously] (1877)
: the state of affairs that existed previously

sta·tusy \'stā-tə-sē, 'sta-\ *adjective* (1962)
: having, showing, or conferring prestige

stat·ut·able \'sta-chə-tə-bəl, 'sta-ˌchü-\ *adjective* (1636)
: made, regulated, or imposed by or in conformity to statute **:** STATUTORY ⟨*statutable* tonnage⟩

stat·ute \'sta-(ˌ)chüt, -chət\ *noun* [Middle English, from Old French *statut,* from Late Latin *statutum* law, regulation, from Latin, neuter of *statutus,* past participle of *statuere* to set up, station, from *status* position, state] (14th century)
1 : a law enacted by the legislative branch of a government
2 : an act of a corporation or of its founder intended as a permanent rule
3 : an international instrument setting up an agency and regulating its scope or authority
synonym see LAW

statute book *noun* (1593)
: the whole body of legislation of a given jurisdiction whether or not published as a whole — usually used in plural

statute mile *noun* (1862)
: MILE 1a

statute of limitations (1768)
: a statute assigning a certain time after which rights cannot be enforced by legal action or offenses cannot be punished

stat·u·to·ry \'sta-chə-ˌtōr-ē, -ˌtȯr-\ *adjective* (1766)
1 : of or relating to statutes
2 : enacted, created, or regulated by statute ⟨a *statutory* age limit⟩
— stat·u·to·ri·ly \ˌsta-chə-'tōr-ə-lē, -'tȯr-\ *adverb*

statutory rape *noun* (1898)
: sexual intercourse with a person who is below the statutory age of consent

[1]**staunch** \'stȯnch, 'stänch\ *variant of* STANCH

[2]**staunch** *adjective* [Middle English, from Middle French *estanche,* feminine of *estanc,* from Old French, from *estancher* to stanch — more at STANCH] (15th century)
1 a : WATERTIGHT, SOUND **b :** strongly built **:** SUBSTANTIAL
2 : steadfast in loyalty or principle
synonym see FAITHFUL
— staunch·ly *adverb*
— staunch·ness *noun*

stau·ro·lite \'stȯr-ə-ˌlīt\ *noun* [French, from Greek *stauros* cross + French *-lite* — more at STEER] (circa 1815)
: a mineral consisting of a basic silicate of iron and aluminum in prismatic orthorhombic crystals often twinned so as to resemble a cross
— stau·ro·lit·ic \ˌstȯr-ə-'li-tik\ *adjective*

[1]**stave** \'stāv\ *noun* [Middle English, back-formation from *staves,* plural of *staf* staff] (13th century)
1 : [1]STAFF 1, 2
2 : any of the narrow strips of wood or narrow iron plates placed edge to edge to form the sides, covering, or lining of a vessel (as a barrel) or structure
3 : STANZA
4 : [1]STAFF 3

[2]**stave** *verb* **staved** *or* **stove** \'stōv\; **stav·ing** (circa 1595)
transitive verb
1 : to break in the staves of (a cask)
2 : to smash a hole in ⟨*stove* in the boat⟩; *also* **:** to crush or break inward ⟨*staved* in several ribs⟩
3 : to drive or thrust away
intransitive verb
1 *archaic* **:** to become stove in — used of a boat or ship
2 : to walk or move rapidly

stave off *transitive verb* (1624)
1 : to fend off ⟨*staving off* creditors⟩
2 : to ward off (as something adverse) **:** FORESTALL ⟨trying to *stave off* disaster⟩

staves *plural of* STAFF

staves·acre \'stāv-ˌzā-kər\ *noun* [by folk etymology from Middle English *staphisagre,* from Medieval Latin *staphis agria,* from Greek, literally, wild raisin] (15th century)
: a Eurasian delphinium (*Delphinium staphisagria*); *also* **:** its violently emetic and cathartic seeds

[1]**stay** \'stā\ *noun* [Middle English, from Old English *stæg;* akin to Old Norse *stag* stay, Old English *stēle* steel] (before 12th century)
1 : a large strong rope usually of wire used to support a mast
2 : [1]GUY

[2]**stay** (1627)
transitive verb
1 : to secure upright with or as if with stays
2 : to incline (a mast) forward, aft, or to one side by the stays
intransitive verb
: to go about **:** TACK

[3]**stay** *verb* **stayed** \'stād\ *also* **staid** \'stād\; **stay·ing** [Middle English, from Middle French *ester* to stand, stay, from Latin *stare* — more at STAND] (15th century)
intransitive verb
1 : to stop going forward **:** PAUSE
2 : to stop doing something **:** CEASE
3 : to continue in a place or condition **:** REMAIN ⟨*stayed* up all night⟩ ⟨went for a short vacation but *stayed* on for weeks⟩ ⟨*stay* put till I come back⟩
4 : to stand firm
5 : to take up residence **:** LODGE
6 : to keep even in a contest or rivalry ⟨*stay* with the leaders⟩
7 : to call a poker bet without raising
8 *obsolete* **:** to be in waiting or attendance
transitive verb
1 : to wait for **:** AWAIT
2 : to stick or remain with (as a race or trial of endurance) to the end — usually used in the phrase *stay the course*
3 : to remain during ⟨*stayed* the whole time⟩
4 a : to stop or delay the proceeding or advance of by or as if by interposing an obstacle **:** HALT ⟨*stay* an execution⟩ **b :** to check the course of (as a disease) **c :** ALLAY, PACIFY ⟨*stayed* tempers⟩ **d :** to quiet the hunger of temporarily
synonym see DEFER

[4]**stay** *noun* (1536)
1 a : the action of halting **:** the state of being stopped **b :** a stopping or suspension of procedure or execution by judicial or executive order
2 *obsolete* **:** SELF-CONTROL, MODERATION
3 : a residence or sojourn in a place
4 : capacity for endurance

[5]**stay** *noun* [Middle French *estaie,* of Germanic origin; akin to Old High German *stān* to stand — more at STAND] (1515)
1 : one that serves as a prop **:** SUPPORT
2 : a corset stiffened with bones — usually used in plural

[6]**stay** *transitive verb* (1548)
1 : to provide physical or moral support for **:** SUSTAIN
2 : to fix on something as a foundation

stay–at–home \'stā-ət-'hōm\ *adjective* (1806)
: remaining habitually in one's residence, locality, or country
— stay–at–home *noun*

stay·er \'stā-ər\ *noun* (circa 1580)
: one that stays; *especially* **:** one that upholds or supports

staying power *noun* (1859)
: capacity for endurance **:** STAMINA

stay·sail \'stā-ˌsāl, -səl\ *noun* (1669)
: a fore-and-aft sail hoisted on a stay — see SAIL illustration

STD \ˌes-(ˌ)tē-'dē\ *noun* [*s*exually *t*ransmitted *d*isease] (1976)
: any of various diseases transmitted by direct sexual contact that include the classic venereal diseases (as syphilis, gonorrhea, and chancroid) and other diseases (as hepatitis A, hepatitis B, giardiasis, and AIDS) that are often or sometimes contracted by other than sexual means

¹stead \'sted\ *noun* [Middle English *stede,* from Old English; akin to Old High German *stat* place, Old English *standan* to stand — more at STAND] (before 12th century)
1 *obsolete* **:** LOCALITY, PLACE
2 : ADVANTAGE — used chiefly in the phrase *to stand one in good stead*
3 : the office, place, or function ordinarily occupied or carried out by someone or something else ⟨acted in his brother's *stead*⟩

²stead *transitive verb* (13th century)
: to be of avail to **:** HELP

stead·fast \'sted-ˌfast *also* -fəst\ *adjective* [Middle English *stedefast,* from Old English *stedefæst,* from *stede* + *fæst* fixed, fast] (before 12th century)
1 a : firmly fixed in place **:** IMMOVABLE **b :** not subject to change ⟨the *steadfast* doctrine of original sin —Ellen Glasgow⟩
2 : firm in belief, determination, or adherence **:** LOYAL
synonym see FAITHFUL
— stead·fast·ly *adverb*
— stead·fast·ness \-ˌfas(t)-nəs, -fəs(t)-\ *noun*

stead·ing \'ste-dᵊn, 'stē-, -diŋ\ *noun* [Middle English *steding,* from *stede* place, farm] (15th century)
1 : a small farm
2 *chiefly Scottish* **:** the service buildings or area of a farm

¹steady \'ste-dē\ *adjective* **steadi·er; -est** [Middle English *stedy,* from *stede*] (14th century)
1 a : direct or sure in movement **:** UNFALTERING **b :** firm in position **:** FIXED **c :** keeping nearly upright in a seaway ⟨a *steady* ship⟩
2 : showing little variation or fluctuation **:** STABLE, UNIFORM ⟨a *steady* breeze⟩ ⟨*steady* prices⟩
3 a : not easily disturbed or upset ⟨*steady* nerves⟩ **b** (1) **:** constant in feeling, principle, purpose, or attachment (2) **:** DEPENDABLE **c :** not given to dissipation **:** SOBER ☆
— steadi·ly \'ste-dᵊl-ē\ *adverb*
— steadi·ness \'ste-dē-nəs\ *noun*

²steady *verb* **stead·ied; steady·ing** (1530)
transitive verb
: to make or keep steady
intransitive verb
: to become steady
— steadi·er *noun*

³steady *adverb* (circa 1605)
1 : in a steady manner **:** STEADILY
2 : on the course set — used as a direction to the helmsman of a ship

⁴steady *noun, plural* **stead·ies** (1792)
: one that is steady; *specifically* **:** a boyfriend or girlfriend with whom one goes steady

steady state *noun* (1885)
: a state or condition of a system or process (as one of the energy states of an atom) that does not change in time; *broadly* **:** a condition that changes only negligibly over a specified time

steady state theory *noun* (1948)
: a theory in astronomy: the universe has always existed and has always been expanding with hydrogen being created continuously — compare BIG BANG THEORY

steak \'stāk\ *noun* [Middle English *steke,* from Old Norse *steik;* akin to Old Norse *steikja* to roast on a stake, *stik* stick, stake — more at STICK] (15th century)
1 a : a slice of meat cut from a fleshy part of a beef carcass **b :** a similar slice of a specified meat other than beef ⟨ham *steak*⟩ **c :** a cross-section slice of a large fish ⟨swordfish *steak*⟩
2 : ground beef prepared for cooking or for serving in the manner of a steak ⟨hamburger *steak*⟩

steak house *noun* (1762)
: a restaurant whose specialty is beefsteak

steak knife *noun* (1895)
: a table knife with a sharp often serrated blade

steak tar·tare \-tär-'tär\ *noun* [French *tartare* Tartar] (1911)
: highly seasoned ground beef eaten raw

¹steal \'stē(ə)l\ *verb* **stole** \'stōl\; **sto·len** \'stō-lən\; **steal·ing** [Middle English *stelen,* from Old English *stelan;* akin to Old High German *stelan* to steal] (before 12th century)
intransitive verb
1 : to take the property of another wrongfully and especially as an habitual or regular practice
2 : to come or go secretly, unobtrusively, gradually, or unexpectedly
3 : to steal or attempt to steal a base
transitive verb
1 a : to take or appropriate without right or leave and with intent to keep or make use of wrongfully ⟨*stole* a car⟩ **b :** to take away by force or unjust means ⟨they've *stolen* our liberty⟩ **c :** to take surreptitiously or without permission ⟨*steal* a kiss⟩ **d :** to appropriate to oneself or beyond one's proper share **:** make oneself the focus of ⟨*steal* the show⟩
2 a : to move, convey, or introduce secretly **:** SMUGGLE **b :** to accomplish in a concealed or unobserved manner ⟨*steal* a visit⟩
3 a : to seize, gain, or win by trickery, skill, or daring ⟨a basketball player adept at *stealing* the ball⟩ ⟨*stole* the election⟩ **b** *of a base runner* **:** to reach (a base) safely solely by running and usually catching the opposing team off guard ☆
— steal·able \'stē-lə-bəl\ *adjective*
— steal·er *noun*
— steal a march on : to gain an advantage on unobserved
— steal one's thunder : to grab attention from another especially by anticipating an idea, plan, or presentation; *also* **:** to claim credit for another's idea

²steal *noun* (circa 1825)
1 : the act or an instance of stealing
2 : a fraudulent or questionable political deal
3 : BARGAIN 2 ⟨it's a *steal* at that price⟩

stealth \'stelth\ *noun* [Middle English *stelthe;* akin to Old English *stelan* to steal] (13th century)
1 a *archaic* **:** THEFT **b** *obsolete* **:** something stolen
2 : the act or action of proceeding furtively, secretly, or imperceptibly ⟨the state moves by *stealth* to gather information —Nat Hentoff⟩
3 : the state of being furtive or unobtrusive
4 : an aircraft-design characteristic consisting of oblique angular construction and avoidance of vertical surfaces that is intended to produce a very weak radar return

stealthy \'stel-thē\ *adjective* **stealth·i·er; -est** (1605)
1 : slow, deliberate, and secret in action or character
2 : intended to escape observation **:** FURTIVE
synonym see SECRET
— stealth·i·ly \-thə-lē\ *adverb*
— stealth·i·ness \-thē-nəs\ *noun*

¹steam \'stēm\ *noun* [Middle English *stem,* from Old English *stēam;* akin to Dutch *stoom* steam] (before 12th century)
1 : a vapor arising from a heated substance
2 a : the invisible vapor into which water is converted when heated to the boiling point **b :** the mist formed by the condensation on cooling of water vapor
3 a : water vapor kept under pressure so as to supply energy for heating, cooking, or mechanical work; *also* **:** the power so generated **b :** active force **:** POWER, MOMENTUM ⟨got there under his own *steam*⟩ ⟨sales began to pick up *steam*⟩; *also* **:** normal force ⟨at full *steam*⟩ **c :** pent-up emotional tension ⟨needed to let off a little *steam*⟩
4 a : STEAMER 2a **b :** travel by or a trip in a steamer

²steam (15th century)
transitive verb
1 : to give out as fumes **:** EXHALE
2 : to apply steam to; *especially* **:** to expose to the action of steam (as for softening or cooking)
intransitive verb
1 : to rise or pass off as vapor
2 : to give off steam or vapor
3 a : to move or travel by the agency of steam **b :** to move or proceed with energy or force
4 : to be angry **:** BOIL ⟨*steaming* over the insult⟩

steam·boat \'stēm-ˌbōt\ *noun* (1785)
: a boat driven by steam power; *specifically* **:** a shallow-draft vessel used on inland waterways

steamboat Gothic *noun* [from its use in homes of retired steamboat captains in imitation of the style of river steamboats] (1941)
: an elaborately ornamented architectural style used in homes built in the middle 19th century in the Ohio and Mississippi river valleys

steam boiler *noun* (1805)
: a boiler for producing steam

steam chest *noun* (1797)
: the chamber from which steam is distributed to a cylinder of a steam engine

steamed \'stēmd\ *adjective* (1802)
1 : cooked by steam
2 : ANGRY 1 — often used with *up*

steam engine *noun* (1751)
: an engine driven or worked by steam; *specifically* **:** a reciprocating engine having a piston driven in a closed cylinder by steam

steam·er \'stē-mər\ *noun* (1814)
1 : a vessel in which articles are subjected to steam
2 a : a ship propelled by steam **b :** an engine, machine, or vehicle operated or propelled by steam
3 : one that steams
4 : SOFT-SHELL CLAM

steamer rug *noun* (1890)

☆ SYNONYMS

Steady, even, equable mean not varying throughout a course or extent. STEADY implies lack of fluctuation or interruption of movement ⟨*steady* progress⟩. EVEN suggests a lack of variation in quality or character ⟨an *even* distribution⟩. EQUABLE implies lack of extremes or of sudden sharp changes ⟨maintain an *equable* temper⟩.

Steal, pilfer, filch, purloin mean to take from another without right or without detection. STEAL may apply to any surreptitious taking of something and differs from the other terms by commonly applying to intangibles as well as material things ⟨*steal* jewels⟩ ⟨*stole* a look at the gifts⟩. PILFER implies stealing repeatedly in small amounts ⟨*pilfered* from his employer⟩. FILCH adds a suggestion of snatching quickly and surreptitiously ⟨*filched* an apple from the tray⟩. PURLOIN stresses removing or carrying off for one's own use or purposes ⟨printed a *purloined* document⟩.

: a warm covering for the lap and feet especially of a person sitting on a ship's deck

steamer trunk *noun* (1886)
: a trunk suitable for use in a stateroom of a steamer; *especially* **:** a shallow trunk that may be stowed beneath a berth

steam·fit·ter \ˈstēm-ˌfi-tər\ *noun* (circa 1890)
: one that installs or repairs equipment (as steam pipes) for heating, ventilating, or refrigerating systems
— **steam fitting** *noun*

steam iron *noun* (circa 1943)
: a pressing iron with a compartment holding water that is converted to steam by the iron's heat and emitted through the soleplate onto the fabric being pressed

[1]**steam·roll·er** \ˈstēm-ˌrō-lər\ *noun* (1866)
1 : a steam-driven road roller; *broadly* **:** ROAD ROLLER
2 : a crushing force especially when ruthlessly applied to overcome opposition

[2]**steamroller** *or* **steam·roll** \-ˌrōl\ (1879)
transitive verb
1 : to overwhelm usually by greatly superior force ⟨*steamroller* the opposition⟩
2 : to bring or advance by overwhelming force or pressure ⟨*steamrollered* the bill through the legislature⟩
intransitive verb
: to move or proceed with irresistible force

steam·ship \ˈstēm-ˌship\ *noun* (1790)
: STEAMER 2a

steamship round *noun* (1964)
: a large beef roast consisting of the whole round with rump and heel

steam shovel *noun* (1879)
: a power shovel operated by steam; *broadly* **:** POWER SHOVEL

steam table *noun* (1861)
: a table having openings to hold containers of cooked food over steam or hot water circulating beneath them

steam turbine *noun* (1900)
: a turbine that is driven by the pressure of steam discharged at high velocity against the turbine vanes

steam up *transitive verb* (1860)
: to make angry or excited **:** AROUSE

steamy \ˈstē-mē\ *adjective* **steam·i·er; -est** (1644)
1 : consisting of, characterized by, or full of steam
2 : intensely or uncomfortably hot: as **a :** hot and humid ⟨a *steamy* afternoon⟩ **b :** sensually hot **:** EROTIC ⟨a *steamy* love scene⟩
— **steam·i·ly** \-mə-lē\ *adverb*
— **steam·i·ness** \-mē-nəs\ *noun*

ste·ap·sin \stē-ˈap-sən\ *noun* [Greek *stear* hard fat + English *-psin* (as in *pepsin*)] (1896)
: the lipase in pancreatic juice

stea·rate \ˈstē-ə-ˌrāt, ˈsti(-ə)r-ˌāt\ *noun* (1841)
: a salt or ester of stearic acid

stea·ric acid \stē-ˈar-ik-, ˈsti(ə)r-ik-\ *noun* (1831)
: a white crystalline fatty acid $C_{18}H_{36}O_2$ obtained by saponifying tallow or other hard fats containing stearin; *also* **:** a commercial mixture of stearic and palmitic acids

stea·rin \ˈstē-ə-rən, ˈstir-ən\ *noun* [French *stéarine,* from Greek *stear*] (1817)
: an ester of glycerol and stearic acid

ste·a·tite \ˈstē-ə-ˌtīt\ *noun* [Latin *steatitis,* a precious stone, from Greek, from *steat-, stear*] (1758)
1 : a massive talc having a grayish green or brown color **:** SOAPSTONE
2 : an electrically insulating porcelain composed largely of steatite
— **ste·a·tit·ic** \ˌstē-ə-ˈti-tik\ *adjective*

steato- *combining form* [Greek, from *steat-, stear;* perhaps akin to Sanskrit *styāyate* it hardens]
: fat ⟨*steato*rrhea⟩

ste·a·to·py·gia \ˌstē-a-tə-ˈpi-j(ē-)ə, -ˈpī-\ *noun* [New Latin, from *steat-, stear* + Greek *pygē* buttocks] (1879)
: an excessive development of fat on the buttocks that occurs especially among women of the Hottentots and some black peoples
— **ste·a·to·py·gous** \-gəs\ *or* **ste·a·to·py·gic** \ˌstē-ə-ˈtä-pə-jik, -tə-ˈpī-jik\ *adjective*

ste·at·or·rhea \(ˌ)stē-ˌa-tə-ˈrē-ə\ *noun* [New Latin] (circa 1859)
: an excess of fat in the stools

stedfast *variant of* STEADFAST

steed \ˈstēd\ *noun* [Middle English *stede,* from Old English *stēda* stallion; akin to Old English *stōd* stud — more at STUD] (before 12th century)
: HORSE; *especially* **:** a spirited horse (as for war)

steek \ˈstēk\ *verb* [Middle English *steken* to pierce, fix, enclose; akin to Old English *stician* to pierce — more at STICK] (13th century)
chiefly Scottish **:** SHUT, CLOSE

[1]**steel** \ˈstē(ə)l\ *noun* [Middle English *stele,* from Old English *stȳle, stēle;* akin to Old High German *stahal* steel and perhaps to Sanskrit *stakati* he resists] (before 12th century)
1 : commercial iron that contains carbon in any amount up to about 1.7 percent as an essential alloying constituent, is malleable when under suitable conditions, and is distinguished from cast iron by its malleability and lower carbon content
2 : an instrument or implement of or characteristically of steel: as **a :** a thrusting or cutting weapon **b :** an instrument (as a fluted round rod with a handle) for sharpening knives **c :** a piece of steel for striking sparks from flint
3 : a quality (as hardness of mind or spirit) that suggests steel ⟨nerves of *steel*⟩
4 a : the steel manufacturing industry **b** *plural* **:** shares of stock in steel companies

[2]**steel** *transitive verb* (13th century)
1 : to overlay, point, or edge with steel
2 a : to cause to resemble steel (as in looks or hardness) **b :** to fill with resolution or determination

[3]**steel** *adjective* (13th century)
1 : made of steel
2 : of or relating to the production of steel
3 : resembling steel

steel band *noun* (1949)
: a band of steel drums

steel blue *noun* (1817)
1 : a grayish blue
2 : any of the blue colors assumed by steel at various temperatures in tempering

steel drum *noun* (1952)
: a musical instrument originally developed in Trinidad that is played by hammering raised and tuned portions of the bottom of an oil drum

steel engraving *noun* (1824)
1 : the art or process of engraving on steel
2 : an impression taken from an engraved steel plate

steel guitar *noun* (1925)
1 : HAWAIIAN GUITAR
2 : PEDAL STEEL
— **steel guitarist** *noun*

steel·head \ˈste(ə)l-ˌhed\ *noun, plural* **steelhead** *also* **steelheads** (circa 1882)
: an anadromous rainbow trout

steel·ie *also* **steely** \ˈstē-lē\ *noun, plural* **steel·ies** (1922)
: a steel playing marble

steel·mak·er \ˈstē(ə)l-ˌmā-kər\ *noun* (1839)
: a manufacturer of steel
— **steel·mak·ing** \-kiŋ\ *noun*

steel–trap \ˈstē(ə)l-ˈtrap\ *adjective* (1945)
: QUICK, INCISIVE ⟨a *steel-trap* mind⟩

steel wool *noun* (1896)
: an abrasive material composed of long fine steel shavings and used especially for scouring and burnishing

steel·work \ˈstē(ə)l-ˌwərk\ *noun* (1681)
1 : work in steel
2 *plural but singular or plural in construction* **:** an establishment where steel is made

steel·work·er \-ˌwər-kər\ *noun* (1884)
: one who works in steel and especially in the manufacturing of it

steely \ˈstē-lē\ *adjective* **steel·i·er; -est** (1509)
1 : resembling or suggesting steel (as in hardness, color, strength, or coldness) ⟨*steely* determination⟩ ⟨*steely* blue⟩
2 : made of steel
— **steel·i·ness** *noun*

steel·yard \ˈstē(ə)l-ˌyärd, ˈstil-yərd\ *noun* [probably from [3]*steel* + [4]*yard* (rod)] (1639)
: a balance in which an object to be weighed is suspended from the shorter arm of a lever and the weight determined by moving a counterpoise along a graduated scale on the longer arm until equilibrium is attained

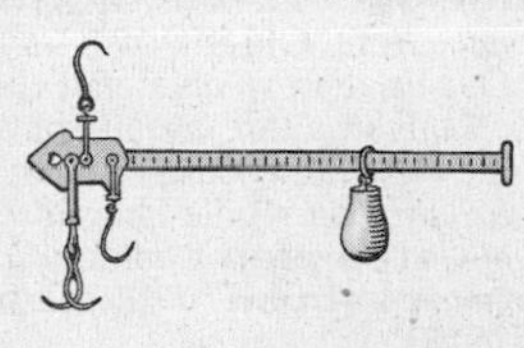
steelyard

steen·bok \ˈstēn-ˌbäk, ˈstān-\ *or* **stein·bok** \ˈstīn-, ˈstān-\ *noun* [Afrikaans *steenbok;* akin to Old English *stānbucca* ibex, *stān* stone, *bucca* buck] (1775)
: a small slender antelope (*Raphicerus campestris*) with long legs that inhabits chiefly grasslands of eastern and southern Africa

[1]**steep** \ˈstēp\ *adjective* [Middle English *stepe,* from Old English *stēap* high, steep, deep; akin to Middle High German *stief* steep] (before 12th century)
1 : LOFTY, HIGH — used chiefly of a sea
2 : making a large angle with the plane of the horizon
3 a : mounting or falling precipitously ⟨the stairs were very *steep*⟩ **b :** being or characterized by a rapid and intensive decline or increase
4 : extremely or excessively high ⟨*steep* prices⟩ ☆
— **steep·ish** \ˈstē-pish\ *adjective*
— **steep·ly** *adverb*
— **steep·ness** *noun*

[2]**steep** *noun* (1555)
: a precipitous place

[3]**steep** *verb* [Middle English *stepen;* akin to Swedish *stöpa* to steep] (14th century)
transitive verb
1 : to soak in a liquid at a temperature under the boiling point (as for softening, bleaching, or extracting an essence)
2 : to cover with or plunge into a liquid (as in bathing, rinsing, or soaking)
3 : to saturate with or subject thoroughly to (some strong or pervading influence) ⟨practices *steeped* in tradition⟩

☆ SYNONYMS
Steep, abrupt, precipitous, sheer mean having an incline approaching the perpendicular. STEEP implies such sharpness of pitch that ascent or descent is very difficult ⟨a *steep* hill⟩ ⟨a *steep* dive⟩. ABRUPT implies a sharper pitch and a sudden break in the level ⟨a beach with an *abrupt* drop-off⟩. PRECIPITOUS applies to an incline approaching the vertical ⟨the river winds through a *precipitous* gorge⟩. SHEER suggests an unbroken perpendicular expanse ⟨*sheer* cliffs that daunted the climbers⟩.

intransitive verb
: to undergo the process of soaking in a liquid
synonym see SOAK
— **steep·er** *noun*

[4]**steep** *noun* (15th century)
1 : the state or process of being steeped
2 : a bath or solution in which something is steeped

steep·en \'stē-pən, -pᵊm\ *verb* **steep·ened; steep·en·ing** \'stē-pə-niŋ, 'stēp-niŋ\ (1847)
intransitive verb
: to become steeper
transitive verb
: to make steeper

stee·ple \'stē-pəl\ *noun* [Middle English *stepel,* from Old English *stēpel* tower; akin to Old English *stēap* steep] (before 12th century)
: a tall structure usually having a small spire at the top and surmounting a church tower; *broadly* **:** a whole church tower
— **stee·pled** \-pəld\ *adjective*

stee·ple·bush \'stē-pəl-,bu̇sh\ *noun* (circa 1818)
: HARDHACK

stee·ple·chase \-,chās\ *noun* [from the use of church steeples as landmarks to guide the riders] (1793)
1 a : a horse race across country **b :** a horse race over a closed course with obstacles (as hedges and walls)
2 : a footrace of usually 3000 meters over hurdles and a water jump
— **stee·ple·chas·er** \-,chā-sər\ *noun*
— **stee·ple·chas·ing** \-siŋ\ *noun*

stee·ple·jack \-,jak\ *noun* (1881)
: one whose work is building smokestacks, towers, or steeples or climbing up the outside of such structures to paint and make repairs

[1]**steer** \'stir\ *noun* [Middle English, from Old English *stēor* young ox; akin to Old High German *stior* young ox] (before 12th century)
1 : a male bovine animal castrated before sexual maturity — compare STAG 3
2 : an ox less than four years old

[2]**steer** *verb* [Middle English *steren,* from Old English *stīeran;* akin to Old English *stēorsteering* oar, Greek *stauros* stake, cross, *stylos* pillar, Sanskrit *sthavira, sthūra* stout, thick] (12th century)
transitive verb
1 : to control the course of **:** DIRECT; *especially* **:** to guide by mechanical means (as a rudder)
2 : to set and hold to (a course)
intransitive verb
1 : to direct the course (as of a ship or automobile)
2 : to pursue a course of action
3 : to be subject to guidance or direction ⟨an automobile that *steers* well⟩
synonym see GUIDE
— **steer·able** \'stir-ə-bəl\ *adjective*
— **steer·er** *noun*
— **steer clear :** to keep entirely away — often used with *of* ⟨trying to *steer clear* of controversy⟩

[3]**steer** *noun* (1894)
: a hint as to procedure **:** TIP

[4]**steer** *dialect British variant of* STIR

steer·age \'stir-ij\ *noun* (15th century)
1 : the act or practice of steering; *broadly* **:** DIRECTION
2 [from its originally being located near the rudder] **:** a section of inferior accommodations in a passenger ship for passengers paying the lowest fares

steer·age·way \-,wā\ *noun* (1769)
: a rate of motion sufficient to make a ship or boat respond to movements of the rudder

steering column *noun* (1903)
: the column that encloses the connections to the steering gear of a vehicle (as an automobile)

steering committee *noun* (1887)
: a managing or directing committee; *specifically* **:** a committee that determines the order in which business will be taken up in a U.S. legislative body

steering gear *noun* (1869)
: a mechanism by which something is steered

steering wheel *noun* (1750)
: a handwheel by means of which one steers

steers·man \'stirz-mən\ *noun* (before 12th century)
: one who steers **:** HELMSMAN

[1]**steeve** \'stēv\ *transitive verb* **steeved; steev·ing** [Middle English *steven,* probably from Spanish *estibar* or Portuguese *estivar* to pack tightly, from Latin *stipare* to press together — more at STIFF] (15th century)
archaic **:** to stow especially in a ship's hold

[2]**steeve** *verb* **steeved; steev·ing** [origin unknown] (circa 1644)
intransitive verb
of a bowsprit **:** to incline upward at an angle with the horizon or the line of the keel
transitive verb
: to set (a bowsprit) at an upward inclination

stego·saur \'ste-gə-,sȯr\ *noun* [New Latin *Stegosauria,* from *Stegosaurus*] (1901)
: any of a suborder (Stegosauria) of quadrupedal ornithischian dinosaurs with strongly developed dorsal plates and spikes

stego·sau·rus \,ste-gə-'sȯr-əs\ *noun* [New Latin, from Greek *stegos* roof + *sauros* lizard — more at THATCH] (1892)
: any of a genus (*Stegosaurus*) of stegosaurs known from the Upper Jurassic rocks especially of Colorado and Wyoming

stein \'stīn\ *noun* [probably from German *Steingut* stoneware, from *Stein* stone + *Gut* goods] (1855)
: a large mug (as of earthenware) used especially for beer; *also* **:** the quantity of beer that a stein holds

ste·la \'stē-lə\ *or* **ste·le** \'stē-lē\ *noun, plural* **ste·lae** \-(,)lē\ [Latin & Greek; Latin *stela,* from Greek *stēlē;* akin to Old High German *stollo* pillar, Greek *stellein* to set up] (1776)
: a usually carved or inscribed stone slab or pillar used for commemorative purposes

ste·lar \'stē-lər, -,lär\ *adjective* (1901)
: of, relating to, or constituting a stele

stele \'stē(ə)l, 'stē-lē\ *noun* [New Latin, from Greek *stēlē* stela, pillar] (1895)
: the usually cylindrical central vascular portion of the axis of a vascular plant

stel·la \'ste-lə\ *noun* [Latin, star; from the star on the reverse] (1879)
: an experimental international coin based on the metric system that was issued by the U.S. in 1879 and 1880 and was worth about four dollars

stel·lar \'ste-lər\ *adjective* [Late Latin *stellaris,* from Latin *stella* star — more at STAR] (circa 1656)
1 a : of or relating to the stars **:** ASTRAL **b :** composed of stars
2 : of or relating to a theatrical or film star ⟨*stellar* names⟩
3 a : PRINCIPAL, LEADING ⟨a *stellar* role⟩ **b :** OUTSTANDING ⟨a *stellar* student⟩ ⟨a *stellar* performance⟩

stellar wind *noun* (1965)
: plasma continuously ejected from a star's surface into surrounding space

stel·late \'ste-,lāt\ *adjective* [Latin *stella*] (1661)
: resembling a star (as in shape) ⟨a *stellate* leaf⟩

Stel·ler's jay \'ste-lərz, 'shte-\ *noun* [Georg W. *Steller* (died 1746) German naturalist] (1828)
: a jay (*Cyanocitta stelleri*) of western North America with a high crest and black and dark blue plumage

Steller's sea cow *noun* (1814)
: an extinct very large aquatic sirenian (*Hydrodamalis gigas*) formerly common near the Asian coast of the Bering Sea

[1]**stem** \'stem\ *noun* [Middle English, from Old English *stefn, stemn* stem of a plant or ship; akin to Old High German *stam* plant stem and probably to Greek *stamnos* wine jar, *histanai* to set — more at STAND] (before 12th century)
1 a : the main trunk of a plant; *specifically* **:** a primary plant axis that develops buds and shoots instead of roots **b :** a plant part (as a branch, petiole, or stipe) that supports another (as a leaf or fruit) **c :** the complete fruiting stalk of a banana plant with its bananas
2 a : the main upright member at the bow of a ship **b :** the bow or prow of a ship — compare STERN
3 : a line of ancestry **:** STOCK; *especially* **:** a fundamental line from which others have arisen
4 : the part of an inflected word that remains unchanged except by phonetic changes or variations throughout an inflection
5 : something held to resemble a plant stem: as **a :** a main or heavy stroke of a letter **b :** the short perpendicular line extending from the head of a musical note **c :** the part of a tobacco pipe from the bowl outward **d :** the cylindrical support of a piece of stemware (as a goblet) **e :** a shaft of a watch used for winding
— **from stem to stern :** THROUGHOUT, THOROUGHLY

[2]**stem** *transitive verb* **stemmed; stem·ming** [Middle English (Scots) *stemmen* to keep a course, from [1]*stem* (of a ship)] (1593)
1 : to make headway against (as an adverse tide, current, or wind)
2 : to check or go counter to (something adverse)
— **stem·mer** *noun*

[3]**stem** *verb* **stemmed; stem·ming** [[1]*stem* (of a plant)] (1724)
transitive verb
1 : to remove the stem from
2 : to make stems for (as artificial flowers)
intransitive verb
: to occur or develop as a consequence **:** have or trace an origin ⟨her success *stems* from hard work⟩
synonym see SPRING
— **stem·mer** *noun*

[4]**stem** *verb* **stemmed; stem·ming** [Middle English *stemmen* to dam up, from Old Norse *stemma;* akin to Middle High German *stemmen* to dam up and probably to Lithuanian *stumti* to shove] (14th century)
transitive verb
1 a : to stop or dam up (as a river) **b :** to stop or check by or as if by damming; *especially* **:** STANCH ⟨*stem* a flow of blood⟩
2 : to turn (a ski) in stemming
intransitive verb
1 : to restrain or check oneself; *also* **:** to become checked or stanched
2 : to slide the heel of one ski or of both skis outward usually in making or preparing to make a turn

[5]**stem** *noun* (circa 1700)
1 : CHECK, DAM
2 : an act or instance of stemming on skis

stem cell *noun* (1885)
: an unspecialized cell that gives rise to differentiated cells ⟨hematopoietic *stem cells* in bone marrow⟩

stem christie *noun, often C capitalized* (1936)
: a turn in skiing begun by stemming a ski and completed by bringing the skis parallel into a christie

stem·less \'stem-ləs\ *adjective* (1796)
: having no stem **:** ACAULESCENT

stem·ma \'ste-mə\ *noun, plural* **stem·ma·ta** \-mə-tə\ [Latin, wreath, pedigree (from the wreaths placed on ancestral images), from Greek, wreath, from *stephein* to crown, enwreathe] (1826)
1 : a simple eye present in some insects
2 : a scroll (as among the ancient Romans) containing a genealogical list

3 : a tree showing the relationships of the manuscripts of a literary work
— **stem·mat·ic** \ste-'ma-tik, stə-\ *adjective*

stemmed \'stemd\ *adjective* (1576)
: having a stem — usually used in combination ⟨long-*stemmed* roses⟩

stem·my \'ste-mē\ *adjective* **stem·mi·er; -est** (1863)
: abounding in stems

stem rust *noun* (1899)
1 : a rust attacking the stem of a plant; *especially* **:** a destructive disease especially of wheat caused by a rust fungus (*Puccinia graminis*) which produces reddish brown lesions in the uredospore stage and black lesions in the teliospore stage and has any of several plants of the barberry family as an intermediate host
2 : the fungus causing stem rust

stem turn *noun* (1922)
: a skiing turn executed by stemming an outside ski

stem·ware \'stem-,war, -,wer\ *noun* (1926)
: glass hollowware mounted on a stem

stem–wind·er \-,wīn-dər\ *noun* (1875)
1 : a stem-winding watch
2 [from the superiority of the stem-winding watch over the older key-wound watch] **:** one that is first-rate of its kind; *especially* **:** a stirring speech

stem–wind·ing \-diŋ\ *adjective* (1867)
: wound by an inside mechanism turned by the knurled knob at the outside end of the stem ⟨a *stem-winding* watch⟩

Sten \'sten\ *noun* [R. V. *S*heppard, 20th century English army officer + H. J. *T*urpin, 20th century English civil servant + *En*gland] (1942)
: a light simple 9-millimeter British submachine gun

sten- *or* **steno-** *combining form* [Greek, from *stenos*]
: close **:** narrow **:** little ⟨*steno*bathic⟩

stench \'stench\ *noun* [Middle English, from Old English *stenc;* akin to Old English *stincan* to emit a smell — more at STINK] (before 12th century)
: STINK
— **stench·ful** \-fəl\ *adjective*
— **stenchy** \'sten-chē\ *adjective*

[1]**sten·cil** \'sten(t)-səl\ *noun* [Middle English *stanselen* to ornament with sparkling colors, from Middle French *estanceler,* from *estancele* spark, from (assumed) Vulgar Latin *stincilla,* alteration of Latin *scintilla*] (1707)
1 : an impervious material (as a sheet of paper, thin wax, or woven fabric) perforated with lettering or a design through which a substance (as ink, paint, or metallic powder) is forced onto a surface to be printed
2 : something (as a pattern, design, or print) that is produced by means of a stencil
3 : a printing process that uses a stencil

[2]**stencil** *transitive verb* **sten·ciled** *or* **sten·cilled; sten·cil·ing** *or* **sten·cil·ling** \-s(ə-)liŋ\ (circa 1828)
1 : to mark or paint with a stencil
2 : to produce by stencil
— **sten·cil·er** *or* **sten·cil·ler** \-s(ə-)lər\ *noun*

steno \'ste-(,)nō\ *noun, plural* **sten·os** (1913)
1 : STENOGRAPHER
2 : STENOGRAPHY

steno·bath·ic \,ste-nə-'ba-thik\ *adjective* [*sten-* + Greek *bathos* depth] (1902)
of a pelagic organism **:** living within narrow limits of depth

ste·nog·ra·pher \stə-'nä-grə-fər\ *noun* (1809)
1 : a writer of shorthand
2 : a person employed chiefly to take and transcribe dictation

ste·nog·ra·phy \-fē\ *noun* (1602)
1 : the art or process of writing in shorthand
2 : shorthand especially written from dictation or oral discourse
3 : the making of shorthand notes and subsequent transcription of them
— **steno·graph·ic** \,ste-nə-'gra-fik\ *adjective*
— **steno·graph·i·cal·ly** \-fi-k(ə-)lē\ *adverb*

steno·ha·line \,ste-nō-'hā-,līn, -'ha-,līn\ *adjective* [International Scientific Vocabulary *sten-* + Greek *halinos* of salt, from *hals* salt — more at SALT] (circa 1920)
of an aquatic organism **:** unable to withstand wide variation in salinity of the surrounding water

ste·nosed \stə-'nōzd, -'nōst\ *adjective* [from past participle of *stenose* to affect with stenosis] (1897)
: affected with stenosis

ste·no·sis \stə-'nō-səs\ *noun, plural* **-no·ses** \-,sēz\ [New Latin, from Greek *stenōsis* act of narrowing, from *stenoun* to narrow, from *stenos* narrow] (circa 1860)
: a narrowing or constriction of the diameter of a bodily passage or orifice
— **ste·not·ic** \-'nä-tik\ *adjective*

steno·ther·mal \,ste-nə-'thər-məl\ *adjective* (1881)
: capable of surviving over only a narrow range of temperatures ⟨*stenothermal* fish⟩
— **steno·therm** \'ste-nə-,thərm\ *noun*

steno·top·ic \,ste-nə-'tä-pik\ *adjective* [probably from German *stenotop* stenotopic, from *sten-* + Greek *topos* place] (1945)
: having a narrow range of adaptability to changes in environmental conditions

steno·type \'ste-nə-,tīp\ *noun* [*steno-* (as in *stenography*) + *type*] (1922)
: a small machine somewhat like a typewriter used to record speech by means of phonograms
— **stenotype** *transitive verb*
— **steno·typ·ist** \-,tī-pist\ *noun*
— **ste·no·ty·py** \-,tī-pē\ *noun*

sten·tor \'sten-,tȯr, -tər\ *noun* [Latin, from Greek *Stentōr* Stentor, a Greek herald in the Trojan War noted for his loud voice] (1609)
1 : a person having a loud voice
2 : any of a widely distributed genus (*Stentor*) of ciliate protozoans having a trumpet-shaped body with the mouth at the broad end and with the narrow end often attached to the substrate

sten·to·ri·an \sten-'tōr-ē-ən, -'tȯr-\ *adjective* (1605)
: extremely loud
synonym see LOUD

[1]**step** \'step\ *noun* [Middle English, from Old English *stæpe;* akin to Old High German *stapfo* step, *stampfōn* to stamp] (before 12th century)
1 : a rest for the foot in ascending or descending: as **a :** one of a series of structures consisting of a riser and a tread **b :** a ladder rung
2 a (1) **:** an advance or movement made by raising the foot and bringing it down elsewhere (2) **:** a combination of foot or foot and body movements constituting a unit or a repeated pattern ⟨a dance *step*⟩ (3) **:** manner of walking **:** STRIDE **b :** FOOTPRINT 1 **c :** the sound of a footstep ⟨heard *steps* in the hall⟩
3 a : the space passed over in one step **b :** a short distance ⟨just a *step* away from the bank⟩ **c :** the height of one stair
4 *plural* **:** COURSE, WAY ⟨directed his *steps* toward the river⟩
5 a : a degree, grade, or rank in a scale **b :** a stage in a process ⟨was guided through every *step* of my career⟩
6 : a frame on a ship designed to receive an upright shaft; *especially* **:** a block supporting the heel of a mast
7 : an action, proceeding, or measure often occurring as one in a series ⟨taking *steps* to improve the situation⟩
8 : a steplike offset or part usually occurring in a series
9 : an interval in a musical scale
— **step·like** \-,līk\ *adjective*
— **stepped** \'stept\ *adjective*
— **in step 1 :** with each foot moving to the same time as the corresponding foot of others or in time to music **2 :** in harmony or agreement
— **out of step :** not in step ⟨*out of step* with the times⟩

[2]**step** *verb* **stepped; step·ping** (before 12th century)
intransitive verb
1 a : to move by raising the foot and bringing it down elsewhere or by moving each foot in succession **b :** DANCE
2 a : to go on foot **:** WALK **b** *obsolete* **:** ADVANCE, PROCEED **c :** to be on one's way **:** LEAVE — often used with *along* **d :** to move briskly ⟨kept us *stepping*⟩
3 : to press down with the foot ⟨*step* on the brake⟩
4 : to come as if at a single step ⟨*stepped* into a good job⟩
transitive verb
1 : to take by moving the feet in succession ⟨*step* three paces⟩
2 a : to move (the foot) in any direction **:** SET ⟨the first man to *step* foot on the moon⟩ **b :** to traverse on foot
3 : to go through the steps of **:** PERFORM ⟨*step* a minuet⟩
4 : to make erect by fixing the lower end in a step ⟨*step* the mast⟩
5 : to measure by steps ⟨*step* off 50 yards⟩
6 a : to provide with steps **b :** to make steps in ⟨*step* a key⟩
7 : to construct or arrange in or as if in steps ⟨craggy peaks with terraces *stepped* up the sides —*Time*⟩
— **step on it :** to increase one's speed **:** hurry up

step- *combining form* [Middle English, from Old English *stēop-;* akin to Old High German *stiof-* step-, Old English *astēpan* to deprive, bereave]
: related by virtue of a remarriage (as of a parent) and not by blood ⟨*step*parent⟩ ⟨*step*sister⟩

step·broth·er \'step-,brə-t͟hər\ *noun* (15th century)
: a son of one's stepparent by a former marriage

step–by–step \,step-bī-'step\ *adjective or adverb* (1701)
: marked by successive degrees usually of limited extent **:** GRADUAL

step·child \'step-,chīld\ *noun* (before 12th century)
1 : a child of one's wife or husband by a former marriage
2 : one that fails to receive proper care or attention ⟨is no longer a *stepchild* in the family of nations —F. R. Smith⟩

step dance *noun* (1887)
: a dance in which steps are emphasized rather than gesture or posture

step·daugh·ter \'step-,dȯ-tər\ *noun* (before 12th century)
: a daughter of one's wife or husband by a former marriage

step–down \'step-,dau̇n\ *noun* (1922)
: a decrease or reduction in size or amount ⟨a *step down* in dosage⟩

step down (1890)
intransitive verb
: RETIRE, RESIGN
transitive verb
1 : to lower (a voltage) by means of a transformer
2 : to decrease or reduce especially by one or more steps
— **step–down** \'step-,dau̇n\ *adjective*

step·fam·i·ly \'step-,fam-lē, -,fa-mə-\ *noun* (1966)

: a family in which there is a stepparent

step·fa·ther \ˈstep-ˌfä-t͟hər\ *noun* (before 12th century)
: the husband of one's mother by a subsequent marriage

step function *noun* (circa 1929)
: a mathematical function of a single real variable that remains constant within each of a series of adjacent intervals but changes in value from one interval to the next

steph·a·no·tis \ˌste-fə-ˈnō-təs\ *noun* [New Latin, from Greek *stephanōtis* fit for a crown, from *stephanos* crown, from *stephein* to crown] (1843)
: any of a genus (*Stephanotis* and especially *S. floribunda*) of Old World tropical woody vines of the milkweed family with fragrant white flowers the corolla of which has a cylindrical dilated tube and spreading limb

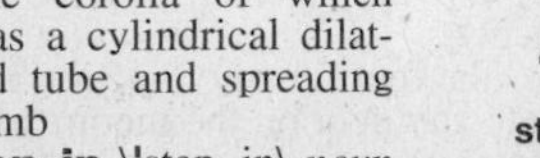
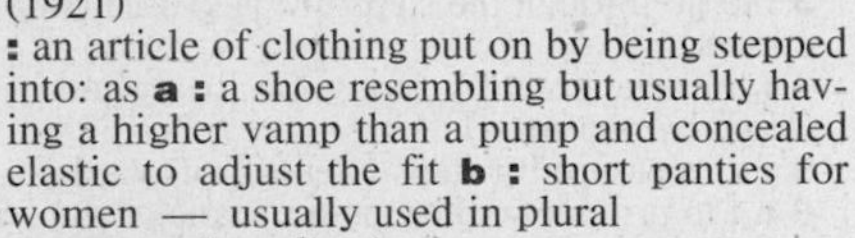
stephanotis

step–in \ˈstep-ˌin\ *noun* (1921)
: an article of clothing put on by being stepped into: as **a :** a shoe resembling but usually having a higher vamp than a pump and concealed elastic to adjust the fit **b :** short panties for women — usually used in plural
— **step–in** *adjective*

step in *intransitive verb* (15th century)
1 : to intervene in an affair or dispute
2 : to make a brief informal visit

step·lad·der \ˈstep-ˌla-dər\ *noun* (1751)
: a ladder that has broad flat steps and two pairs of legs connected by a hinge at the top and that opens at the bottom to become freestanding

step·moth·er \-ˌmə-t͟hər\ *noun* (before 12th century)
: the wife of one's father by a subsequent marriage

step out *intransitive verb* (circa 1533)
1 : to go away from a place usually for a short distance and for a short time
2 : to go or march at a vigorous or increased pace
3 : DIE
4 : to lead an active social life
5 : to be unfaithful — usually used with *on*

step·par·ent \ˈstep-ˌpar-ənt, -ˌper-\ *noun* (circa 1890)
: the spouse of one's mother or father by a subsequent marriage

step·par·ent·ing *noun* (1979)
: the raising of a child by a stepparent

steppe \ˈstep\ *noun* [Russian *step'*] (1671)
1 : one of the vast usually level and treeless tracts in southeastern Europe or Asia
2 : arid land with xerophilous vegetation found usually in regions of extreme temperature range and loess soil

stepped–up \ˈstept-ˈəp\ *adjective* (1902)
: increased in intensity **:** ACCELERATED, INTENSIFIED ⟨*stepped-up* security⟩

step·per \ˈste-pər\ *noun* (1835)
: one (as a fast horse or a dancer) that steps

stepper motor *noun* (1961)
: a motor whose driveshaft rotates in small steps rather than continuously — called also *stepping motor*

step·ping–stone \ˈste-piŋ-ˌstōn\ *noun* (14th century)
1 : a stone on which to step (as in crossing a stream)
2 : a means of progress or advancement

step·sis·ter \ˈstep-ˌsis-tər\ *noun* (15th century)
: a daughter of one's stepparent by a former marriage

step·son \-ˌsən\ *noun* (before 12th century)
: a son of one's husband or wife by a former marriage

step stool *noun* (1946)
: a stool with one or two steps that often fold away beneath the seat

step turn *noun* (1941)
: a skiing turn executed in a downhill traverse by lifting the upper ski from the ground, placing it in the desired direction, weighting it, and bringing the other ski parallel

step–up \ˈstep-ˌəp\ *noun* (1922)
: an increase or advance in size or amount

step up (1902)
transitive verb
1 : to increase (a voltage) by means of a transformer
2 : to increase, augment, or advance especially by one or more steps ⟨*step up* production⟩
intransitive verb
1 : to come forward
2 : to undergo an increase ⟨business is *stepping up*⟩
3 : to receive a promotion
— **step–up** \ˈstep-ˌəp\ *adjective*

step·wise \-ˌwīz\ *adjective* (1902)
1 : marked by or proceeding in steps
2 : moving by step to adjacent musical tones

-ster *noun combining form* [Middle English, from Old English *-estre* female agent; akin to Middle Dutch *-ster*]
1 : one that does or handles or operates ⟨spin*ster*⟩ ⟨tap*ster*⟩ ⟨team*ster*⟩
2 : one that makes or uses ⟨song*ster*⟩ ⟨pun*ster*⟩
3 : one that is associated with or participates in ⟨game*ster*⟩ ⟨gang*ster*⟩
4 : one that is ⟨young*ster*⟩

ster·co·ra·ceous \ˌstər-kə-ˈrā-shəs\ *adjective* [Latin *stercor-, stercus* excrement] (1731)
: relating to, being, or containing feces

ster·cu·lia gum \ˌstər-ˈk(y)ü-lē-ə-\ *noun* [New Latin *Sterculia,* genus of trees] (circa 1943)
: KARAYA GUM

stere- *or* **stereo-** *combining form* [New Latin, from Greek, from *stereos* solid — more at STARE]
1 : solid **:** solid body ⟨*stereo*gram⟩
2 a : stereoscopic ⟨*stere*opsis⟩ **b :** having or dealing with three dimensions of space ⟨*stereo*chemistry⟩

[1]**ste·reo** \ˈster-ē-ˌō, ˈstir-\ *noun, plural* **ste·re·os** (circa 1823)
1 : STEREOTYPE
2 [by shortening] **a :** stereophonic reproduction **b :** a stereophonic sound system

[2]**stereo** *adjective* (1876)
1 a : STEREOSCOPIC **b :** produced by or as if by means of a stereotype
2 : STEREOPHONIC

ste·reo·chem·is·try \ˌster-ē-ō-ˈke-mə-strē, ˌstir-\ *noun* [International Scientific Vocabulary] (1890)
1 : a branch of chemistry that deals with the spatial arrangement of atoms and groups in molecules
2 : the spatial arrangement of atoms and groups in a compound and its relation to the properties of the compound
— **ste·reo·chem·i·cal** \-ˈke-mi-kəl\ *adjective*

ste·reo·gram \ˈster-ē-ə-ˌgram, ˈstir-\ *noun* [International Scientific Vocabulary] (1868)
1 : a diagram or picture representing objects with an impression of solidity or relief
2 : STEREOGRAPH

ste·reo·graph \-ˌgraf\ *noun* [International Scientific Vocabulary] (1859)
: a pair of stereoscopic pictures or a picture composed of two superposed stereoscopic images that gives a three-dimensional effect when viewed with a stereoscope or special spectacles
— **stereograph** *transitive verb*

ste·reo·graph·ic \ˌster-ē-ə-ˈgra-fik\ *adjective* (1704)
: of, relating to, or being a delineation of the form of a solid body (as the earth) on a plane ⟨*stereographic* projection⟩
— **ste·re·og·ra·phy** \ˌster-ē-ˈä-grə-fē\ *noun*

ste·reo·iso·mer \ˌster-ē-ō-ˈī-sə-mər, ˌstir-\ *noun* [International Scientific Vocabulary] (1894)
: any of a group of isomers in which atoms are linked in the same order but differ in their spatial arrangement
— **ste·reo·iso·mer·ic** \-ˌī-sə-ˈmer-ik\ *adjective*
— **ste·reo·isom·er·ism** \-ī-ˈsä-mə-ˌri-zəm\ *noun*

ste·re·ol·o·gy \ˌster-ē-ˈä-lə-jē, ˌstir-\ *noun* [International Scientific Vocabulary] (1963)
: a branch of science concerned with inferring the three-dimensional properties of objects or matter ordinarily observed two-dimensionally
— **ste·reo·log·i·cal** \-ē-ə-ˈlä-ji-kəl\ *adjective*
— **ste·reo·log·i·cal·ly** \-ji-k(ə-)lē\ *adverb*

ste·reo·mi·cro·scope \-ˈmī-krə-ˌskōp\ *noun* (1948)
: a microscope having a set of optics for each eye to make an object appear in three dimensions
— **ste·reo·mi·cro·scop·ic** \-ˌmī-krə-ˈskä-pik\ *adjective*
— **ste·reo·mi·cro·scop·i·cal·ly** \-pi-k(ə-)lē\ *adverb*

ste·reo·phon·ic \ˌster-ē-ə-ˈfä-nik, ˌstir-\ *adjective* [International Scientific Vocabulary] (1927)
: of, relating to, or constituting sound reproduction involving the use of separated microphones and two transmission channels to achieve the sound separation of a live hearing
— **ste·reo·phon·i·cal·ly** \-ni-k(ə-)lē\ *adverb*
— **ste·reo·pho·ny** \ˌster-ē-ˈä-fə-nē, ˌstir-; ˈster-ē-ə-ˌfō-nē, ˈstir-\ *noun*

ste·reo·pho·tog·ra·phy \ˌster-ē-ō-fə-ˈtä-grə-fē, ˌstir-\ *noun* [International Scientific Vocabulary] (1903)
: stereoscopic photography
— **ste·reo·pho·to·graph·ic** \-ˌfō-tə-ˈgra-fik\ *adjective*

ste·re·op·sis \ˌster-ē-ˈäp-səs, ˌstir-\ *noun* [New Latin, from *stere-* + Greek *opsis* vision, appearance — more at OPTIC] (circa 1911)
: stereoscopic vision

ste·re·op·ti·con \-ˈäp-ti-kən\ *noun* [New Latin, from *stere-* + Greek *optikon,* neuter of *optikos* optic] (1863)
1 : a projector for transparent slides often made double so as to produce dissolving views
2 : STEREOSCOPE

ste·reo·reg·u·lar \ˌster-ē-ō-ˈre-gyə-lər, ˌstir-\ *adjective* (1958)
: of, relating to, or involving stereochemical regularity in the repeating units of a polymeric structure
— **ste·reo·reg·u·lar·i·ty** \-ˌre-gyə-ˈlar-ə-tē\ *noun*

ste·reo·scope \ˈster-ē-ə-ˌskōp, ˈstir-\ *noun* (1838)
: an optical instrument with two eyeglasses for helping the observer to combine the images of two pictures taken from points of view a little way apart and thus to get the effect of solidity or depth

ste·reo·scop·ic \ˌster-ē-ə-ˈskä-pik, ˌstir-\ *adjective* (1855)
1 : of or relating to stereoscopy or the stereoscope
2 : characterized by stereoscopy ⟨*stereoscopic* vision⟩
— **ste·reo·scop·i·cal·ly** \-pi-k(ə-)lē\ *adverb*

ste·re·os·co·py \ˌster-ē-ˈäs-kə-pē, ˌstir-; ˈster-ē-ə-ˌskō-pē, ˈstir-\ *noun* [International Scientific Vocabulary] (circa 1859)
1 : a science that deals with stereoscopic effects and methods
2 : the seeing of objects in three dimensions

ste·reo·spe·cif·ic \ˌster-ē-ō-spi-ˈsi-fik, ˌstir-\ *adjective* (1949)
: being, produced by, or involved in a stereochemically specific process ⟨many enzymes act as *stereospecific* catalysts⟩ ⟨*stereospecific* plastics⟩
— **ste·reo·spe·cif·i·cal·ly** \-fi-k(ə-)lē\ *adverb*
— **ste·reo·spec·i·fic·i·ty** \-ˌspe-sə-ˈfi-sə-tē\ *noun*

ste·reo·tac·tic \ˌster-ē-ə-ˈtak-tik, ˌstir-\ *adjective* (1950)
: STEREOTAXIC

ste·reo·tax·ic \ˌster-ē-ə-ˈtak-sik, ˌstir-\ *adjective* [New Latin *stereotaxis* stereotaxic technique, from *stere-* + *-taxis*] (1908)
: of, relating to, or being a technique or apparatus used in neurological research or surgery for directing the tip of a delicate instrument (as a needle or an electrode) in three planes in attempting to reach a specific locus in the brain
— **ste·reo·tax·i·cal·ly** \-si-k(ə-)lē\ *adverb*

[1]**ste·reo·type** \ˈster-ē-ə-ˌtīp, ˈstir-\ *transitive verb* (1804)
1 : to make a stereotype from
2 a : to repeat without variation **:** make hackneyed **b :** to develop a mental stereotype about
— **ste·reo·typ·er** *noun*

[2]**stereotype** *noun* [French *stéréotype,* from *stéré-* stere- + *type*] (1817)
1 : a plate cast from a printing surface
2 : something conforming to a fixed or general pattern; *especially* **:** a standardized mental picture that is held in common by members of a group and that represents an oversimplified opinion, prejudiced attitude, or uncritical judgment
word history see CLICHÉ
— **ste·reo·typ·i·cal** \ˌster-ē-ə-ˈti-pi-kəl\ *also* **ste·reo·typ·ic** \-pik\ *adjective*
— **ste·reo·typ·i·cal·ly** \-pi-k(ə-)lē\ *adverb*

ste·reo·typed *adjective* (1849)
: lacking originality or individuality
synonym see TRITE

ste·reo·ty·py \ˈster-ē-ə-ˌtī-pē, ˈstir-\ *noun, plural* **-pies** (circa 1889)
: frequent almost mechanical repetition of the same posture, movement, or form of speech (as in schizophrenia)

ste·ric \ˈster-ik, ˈstir-\ *adjective* [International Scientific Vocabulary *stere-* + [1]*-ic*] (1898)
: relating to or involving the arrangement of atoms in space **:** SPATIAL
— **ste·ri·cal·ly** \ˈster-i-k(ə-)lē, ˈstir-\ *adverb*

ste·rig·ma \stə-ˈrig-mə\ *noun, plural* **-ma·ta** \-mə-tə\ *also* **-mas** [New Latin, from Greek *stērigma* support, from *stērizein* to prop; perhaps akin to Greek *stereos* solid — more at STARE] (circa 1866)
: one of the slender stalks at the top of the basidium of some fungi from the tips of which the basidiospores are formed; *broadly* **:** a stalk or filament that bears conidia or spermatia

ster·il·ant \ˈster-ə-lənt\ *noun* (1941)
: a sterilizing agent

ster·ile \ˈster-əl, *chiefly British* -ˌīl\ *adjective* [Middle English *steryle,* from Latin *sterilis;* akin to Gothic *stairo* barren animal, Sanskrit *starī* sterile cow] (15th century)
1 a : failing to bear or incapable of producing fruit or spores **b :** failing to produce or incapable of producing offspring ⟨a *sterile* hybrid⟩ **c :** incapable of germinating ⟨*sterile* spores⟩ **d** *of a flower* **:** neither perfect nor pistillate
2 a : unproductive of vegetation ⟨a *sterile* arid region⟩ **b :** free from living organisms and especially microorganisms **c :** lacking in stimulating emotional or intellectual quality **:** LIFELESS ⟨a *sterile* work of art⟩
— **ster·ile·ly** \-ə(l)-lē\ *adverb*
— **ste·ril·i·ty** \stə-ˈri-lə-tē\ *noun*

ster·il·ize \ˈster-ə-ˌlīz\ *transitive verb* **-ized; -iz·ing** (1695)
: to make sterile: as **a :** to cause (land) to become unfruitful **b** (1) **:** to deprive of the power of reproducing (2) **:** to make incapable of germination **c :** to make powerless or useless usually by restraining from a normal function, relation, or participation **d :** to free from living microorganisms
— **ster·il·i·za·tion** \ˌster-ə-lə-ˈzā-shən\ *noun*
— **ster·il·iz·er** \ˈster-ə-ˌlī-zər\ *noun*

[1]**ster·ling** \ˈstər-liŋ\ *noun* [Middle English, silver penny, probably from (assumed) Old English *steorling,* from Old English *steorra* star + [1]*-ling* — more at STAR] (14th century)
1 : British money
2 : sterling silver or articles of it

[2]**sterling** *adjective* (15th century)
1 a : of, relating to, or calculated in terms of British sterling **b :** payable in sterling
2 a *of silver* **:** having a fixed standard of purity usually defined legally as represented by an alloy of 925 parts of silver with 75 parts of copper **b :** made of sterling silver
3 : conforming to the highest standard ⟨*sterling* character⟩
— **ster·ling·ly** \-liŋ-lē\ *adverb*
— **ster·ling·ness** *noun*

sterling area *noun* (1932)
: a former group of countries with currencies tied to the British pound sterling

[1]**stern** \ˈstərn\ *adjective* [Middle English *sterne,* from Old English *styrne;* akin to Old English *starian* to stare — more at STARE] (before 12th century)
1 a : having a definite hardness or severity of nature or manner **:** AUSTERE **b :** expressive of severe displeasure **:** HARSH
2 : forbidding or gloomy in appearance
3 : INEXORABLE ⟨*stern* necessity⟩
4 : STURDY, STOUT ⟨a *stern* resolve⟩
synonym see SEVERE
— **stern·ly** *adverb*
— **stern·ness** \ˈstərn-nəs\ *noun*

[2]**stern** *noun* [Middle English, rudder, probably of Scandinavian origin; akin to Old Norse *stjōrn* act of steering; akin to Old English *stieran* to steer — more at STEER] (14th century)
1 : the rear end of a boat
2 : a hinder or rear part **:** the last or latter part

ster·nal \ˈstər-nᵊl\ *adjective* (1756)
: of or relating to the sternum

stern chase *noun* [[2]*stern*] (1627)
: a chase in which a pursuing ship follows in the path of another

stern chaser *noun* (1815)
: a gun so placed as to be able to fire astern at a pursuing ship

stern·fore·most \ˈstərn-ˈfōr-ˌmōst, -ˈfȯr-\ *adverb* (1840)
: with the stern in advance **:** BACKWARD

ster·nite \ˈstər-ˌnīt\ *noun* [International Scientific Vocabulary, from Greek *sternon* chest] (1868)
: the ventral part or shield of a somite of an arthropod; *especially* **:** the chitinous plate that forms the ventral surface of an abdominal or occasionally a thoracic segment of an insect

stern·most \ˈstərn-ˌmōst\ *adjective* (1622)
: farthest astern

ster·no·cos·tal \ˌstər-nō-ˈkäs-tᵊl\ *adjective* [New Latin *stern*um + English *-o-* + *costal*] (1785)
: of, relating to, or situated between the sternum and ribs

stern·post \ˈstərn-ˌpōst\ *noun* (15th century)
: the principal member at the stern of a ship extending from keel to deck

stern sheets *noun plural* (15th century)
: the space in the stern of an open boat not occupied by the thwarts

ster·num \ˈstər-nəm\ *noun, plural* **sternums** *or* **ster·na** \-nə\ [New Latin, from Greek *sternon* chest, breastbone; akin to Old High German *stirna* forehead, Latin *sternere* to spread out — more at STREW] (1667)
: a compound ventral bone or cartilage of most vertebrates other than fishes that connects the ribs or the shoulder girdle or both and in humans consists of the manubrium, gladiolus, and xiphoid process — called also *breastbone*

ster·nu·ta·tion \ˌstər-nyə-ˈtā-shən\ *noun* [Middle English *sternutacion,* from Latin *sternutation-, sternutatio,* from *sternutare* to sneeze, frequentative of *sternuere* to sneeze; akin to Greek *ptarnysthai* to sneeze] (15th century)
: the act, fact, or noise of sneezing

ster·nu·ta·tor \ˈstər-nyə-ˌtā-tər\ *noun* (1922)
: an agent that induces sneezing and often lacrimation and vomiting

stern·ward \ˈstərn-wərd\ *or* **stern·wards** \-wərdz\ *adverb* (1832)
: AFT

stern·way \ˈstərn-ˌwā\ *noun* (1769)
: movement of a ship backward or with stern foremost

stern–wheel·er \-ˈhwē-lər, -ˈwē-\ *noun* (1855)
: a steamboat driven by a single paddle wheel at the stern

ste·roid \ˈstir-ˌȯid *also* ˈster-\ *noun* [International Scientific Vocabulary *ster*ol + *-oid*] (1926)
: any of numerous compounds containing a 17-carbon 4-ring system and including the sterols and various hormones and glycosides
— **steroid** *or* **ste·roi·dal** \stə-ˈrȯi-dᵊl\ *adjective*

ste·roido·gen·e·sis \stə-ˌrȯi-də-ˈje-nə-səs; ˌstir-ˌȯid- *also* ˌster-\ *noun* [New Latin] (1951)
: synthesis of steroids
— **ste·roido·gen·ic** \-ˈje-nik\ *adjective*

ste·rol \ˈstir-ˌȯl, ˈster-, -ˌōl\ *noun* [International Scientific Vocabulary, from *-sterol* (as in *cholesterol*)] (1913)
: any of various solid steroid alcohols (as cholesterol) widely distributed in animal and plant lipids

-sterone *combining form* [*ster*ol + *-one*]
: steroid hormone ⟨andro*sterone*⟩

ster·tor \ˈstər-tər, -ˌtȯr\ *noun* [New Latin, from Latin *stertere* to snore; akin to *sternuere* to sneeze] (1804)
: the act or fact of producing a snoring sound **:** SNORING

ster·to·rous \ˈstər-tə-rəs\ *adjective* (1802)
: characterized by a harsh snoring or gasping sound
— **ster·to·rous·ly** *adverb*

stet \ˈstet\ *transitive verb* **stet·ted; stet·ting** [Latin, let it stand, from *stare* to stand — more at STAND] (1755)
: to direct retention of (a word or passage previously ordered to be deleted or omitted from a manuscript or printer's proof) by annotating usually with the word *stet*

stetho·scope \ˈste-thə-ˌskōp *also* -t͟hə-\ *noun* [French *stéthoscope,* from Greek *stēthos* chest + French *-scope*] (1820)
: an instrument used to detect and study sounds produced in the body
— **stetho·scop·ic** \ˌste-thə-ˈskä-pik *also* -t͟hə-\ *adjective*

Stet·son \ˈstet-sən\ *trademark*
— used for a broad-brimmed high-crowned felt hat

[1]**ste·ve·dore** \ˈstē-və-ˌdōr, -ˌdȯr *also* ˈstēv-ˌ\ *noun* [Spanish *estibador,* from *estibar* to pack, from Latin *stipare* to press together — more at STIFF] (1788)
: one who works at or is responsible for loading and unloading ships in port

[2]**stevedore** *verb* **-dored; -dor·ing** (1862)
transitive verb

\ə\ **abut** \ᵊ\ **kitten** \ər\ **further** \a\ **ash** \ā\ **ace** \ä\ **mop, mar** \au̇\ **out** \ch\ **chin** \e\ **bet** \ē\ **easy** \g\ **go** \i\ **hit** \ī\ **ice** \j\ **job** \ŋ\ **sing** \ō\ **go** \ȯ\ **law** \ȯi\ **boy** \th\ **thin** \t͟h\ **the** \ü\ **loot** \u̇\ **foot** \y\ **yet** \zh\ **vision** *see also* Guide to Pronunciation

: to handle (cargo) as a stevedore; *also* **:** to load or unload the cargo of (a ship) in port
intransitive verb
: to work as a stevedore

stevedore knot *noun* (circa 1863)
: a stopper knot similar to a figure eight knot but with one or more extra turns — called also *stevedore's knot;* see KNOT illustration

Ste•ven•graph \'stē-vən-ˌgraf\ *or* **Ste•vens•graph** \-vənz-\ *noun* [Thomas *Stevens,* 19th century American weaver] (1879)
: a woven silk picture

¹stew \'stü, 'styü\ *noun* [Middle English *stewe* heated room for a steam bath, from Middle French *estuve,* from (assumed) Vulgar Latin *extufa* — more at STOVE] (13th century)
1 *obsolete* **:** a utensil used for boiling
2 : a hot bath
3 a : WHOREHOUSE **b :** a district of bordellos — usually used in plural
4 a : fish or meat usually with vegetables prepared by stewing **b** (1) **:** a heterogeneous mixture (2) **:** a state of heat and congestion
5 : a state of excitement, worry, or confusion

²stew (14th century)
transitive verb
: to boil slowly or with simmering heat
intransitive verb
1 : to become cooked by stewing
2 : to swelter especially from confinement in a hot or stuffy atmosphere
3 : to be in a state of suppressed agitation, worry, or resentment

³stew *noun* [short for *stewardess*] (1970)
: FLIGHT ATTENDANT

¹stew•ard \'stü-ərd, 'styü-; 'st(y)ů(-ə)rd\ *noun* [Middle English, from Old English *stīweard,* from *stī, stig* hall, sty + *weard* ward — more at STY, WARD] (before 12th century)
1 : one employed in a large household or estate to manage domestic concerns (as the supervision of servants, collection of rents, and keeping of accounts)
2 : SHOP STEWARD
3 : a fiscal agent
4 a : an employee on a ship, airplane, bus, or train who manages the provisioning of food and attends passengers **b :** one appointed to supervise the provision and distribution of food and drink in an institution
5 : one who actively directs affairs **:** MANAGER
◇

²steward (1621)
transitive verb
: to act as a steward for **:** MANAGE
intransitive verb
: to perform the duties of a steward

stew•ard•ess \'stü-ər-dəs, 'styü-; 'st(y)ů(-ə)r-dəs\ *noun* (1631)
: a woman who performs the duties of a steward; *especially* **:** one who attends passengers (as on an airplane)
usage see -ESS

stew•ard•ship \'stü-ərd-ˌship, 'styü-; 'st(y)ů(-ə)rd-\ *noun* (15th century)
1 : the office, duties, and obligations of a steward
2 : the conducting, supervising, or managing of something; *especially* **:** the careful and responsible management of something entrusted to one's care ⟨*stewardship* of our natural resources⟩

stewed *adjective* (circa 1737)
: DRUNK 1a

stew•pan \'stü-ˌpan, 'styü-\ *noun* (1651)
: a pan used for stewing

stib•nite \'stib-ˌnīt\ *noun* [alteration of obsolete English *stibine* stibnite, from French, from Latin *stibium* antimony, from Greek *stibi,* from Egyptian *stm*] (circa 1854)
: a mineral that consists of the trisulfide of antimony and occurs in orthorhombic lead-gray crystals of metallic luster or in massive form

sticho•myth•ia \ˌsti-kə-'mi-thē-ə\ *also* **sti•chom•y•thy** \sti-'kä-mə-thē\ *noun* [Greek *stichomythia,* from *stichomythein* to speak dialogue in alternate lines, from *stichos* row, verse + *mythos* speech, myth; akin to Greek *steichein* to walk, go — more at STAIR] (1861)
: dialogue especially of altercation or dispute delivered by two actors in alternating lines (as in classical Greek drama)
— **sticho•myth•ic** \ˌsti-kə-'mi-thik\ *adjective*

¹stick \'stik\ *noun* [Middle English *stik,* from Old English *sticca;* akin to Old Norse *stik* stick, Old English *stician* to stick] (before 12th century)
1 : a woody piece or part of a tree or shrub: as **a :** a usually dry or dead severed shoot, twig, or slender branch **b :** a cut or broken branch or piece of wood gathered for fuel or construction material
2 a : a long slender piece of wood or metal: as (1) **:** a club or staff used as a weapon (2) **:** WALKING STICK **b :** an implement used for striking or propelling an object in a game **c :** something used to force compliance **d :** a baton symbolizing an office or dignity; *also* **:** a person entitled to bear such a baton
3 : a piece of the materials composing something (as a building)
4 a : any of various implements resembling a stick in shape, origin, or use: as (1) **:** COMPOSING STICK (2) **:** an airplane lever operating the elevators and ailerons (3) **:** the gearshift lever of an automobile **b :** STICKFUL
5 : something prepared (as by cutting, molding, or rolling) in a relatively long and slender often cylindrical form ⟨a *stick* of candy⟩ ⟨a *stick* of butter⟩
6 a : PERSON, CHAP **b :** a dull, inert, stiff, or spiritless person
7 *plural* **:** remote usually rural districts regarded especially as backward, dull, or unsophisticated **:** BOONDOCKS
8 : an herbaceous stalk resembling a woody stick ⟨celery *sticks*⟩
9 : ¹MAST 1; *also* **:** ¹YARD 4
10 : a piece of furniture
11 a : a number of bombs arranged for release from a bombing plane in a series across a target **b :** a number of parachutists dropping together
12 *slang* **:** a marijuana cigarette
13 a : punishment or the threat of punishment used to force compliance or cooperation ⟨choosing between the carrot and the *stick*⟩ **b** *British* **:** CRITICISM, ABUSE
— **stick•like** \-ˌlīk\ *adjective*

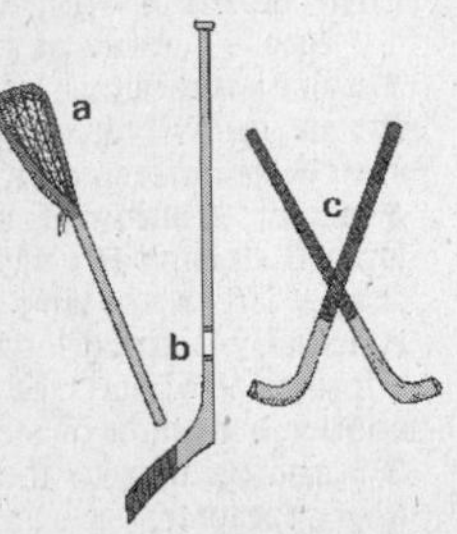

stick 2b: *a* lacrosse, *b* ice hockey, *c* field hockey

²stick *transitive verb* (1573)
1 : to arrange (lumber) in stacks
2 : to provide a stick as a support for

³stick *verb* **stuck** \'stək\; **stick•ing** [Middle English *stikken,* from Old English *stician;* akin to Old High German *sticken* to prick, Latin *instigare* to urge on, goad, Greek *stizein* to tattoo] (before 12th century)
transitive verb
1 a : to pierce with something pointed **:** STAB **b :** to kill by piercing
2 : to push or thrust so as or as if to pierce
3 a : to fasten by thrusting in **b :** IMPALE **c :** PUSH, THRUST
4 : to put or set in a specified place or position
5 : to furnish with things fastened on by or as if by piercing
6 : to attach by or as if by causing to adhere to a surface
7 a : to compel to pay especially by trickery ⟨got *stuck* with the bar bill⟩ **b :** OVERCHARGE
8 a : to halt the movement or action of **b :** BAFFLE, STUMP
9 a : CHEAT, DEFRAUD **b :** to saddle with something disadvantageous or disagreeable ⟨is still *stuck* with that lousy car⟩
intransitive verb
1 : to hold to something firmly by or as if by adhesion: **a :** to become fixed in place by means of a pointed end **b :** to become fast by or as if by miring or by gluing or plastering ⟨*stuck* in the mud⟩
2 a : to remain in a place, situation, or environment **b :** to hold fast or adhere resolutely **:** CLING **c :** to remain effective **d :** to keep close in a chase or competition
3 : to become blocked, wedged, or jammed
4 a : BALK, SCRUPLE **b :** to find oneself baffled **c :** to be unable to proceed
5 : PROJECT, PROTRUDE ☆
— **stick in one's craw :** to irritate, nag at, or obsess one
— **stick it to :** to treat harshly or unfairly
— **stick one's neck out :** to make oneself vulnerable by taking a risk
— **stick to one's guns :** to maintain one's position especially in face of opposition
— **stuck on :** infatuated with

⁴stick *noun* (1633)
1 : a thrust with a pointed instrument **:** STAB
2 a : DELAY, STOP **b :** IMPEDIMENT
3 : adhesive quality or substance

stick around *intransitive verb* (circa 1912)
: to stay or wait about **:** LINGER

stick•ball \'stik-ˌbȯl\ *noun* (1934)
: baseball adapted for play in streets or small areas and using a broomstick and a lightweight ball

stick•er \'sti-kər\ *noun* (15th century)
1 : one that pierces with a point
2 a : one that adheres or causes adhesion **b :** a slip of paper with adhesive back that can be fastened to a surface

sticker price *noun* (1969)
: a manufacturer's suggested retail price that is printed on a sticker and affixed to a new automobile

☆ SYNONYMS
Stick, adhere, cohere, cling, cleave mean to become closely attached. STICK implies attachment by affixing or by being glued together ⟨couldn't get the label to *stick*⟩. ADHERE is often interchangeable with *stick* but sometimes implies a growing together ⟨antibodies *adhering* to a virus⟩. COHERE suggests a sticking together of parts so that they form a unified mass ⟨eggs will make the mixture *cohere*⟩. CLING implies attachment by hanging on with arms or tendrils ⟨*clinging* to a capsized boat⟩. CLEAVE stresses strength of attachment ⟨the wet shirt *cleaved* to his back⟩.

◇ WORD HISTORY
steward The Old English word *stig* or *stī* meant "sty or pen" (for hogs) and "hall" (for people). We have no evidence, however, that *stigweard* or *stīweard,* a compound of *stig* or *stī* with *weard* "keeper, guard," ever referred to a keeper of pigs or pigsties. It meant rather the keeper of the hall, the supervisor of the household's domestic concerns, and thus an official of importance. One important concern of a medieval English hall or household was the dispensing, as well as the husbanding, of wealth, both of which were the responsibility of the steward. Even today the English sovereign has a Lord Steward of the Household, who is a peer and a member of the Privy Council, although his real power is now, unlike formerly, restricted to the royal household.

stick figure *noun* (1949)
1 : a drawing showing the head of a human being or animal as a circle and all other parts as straight lines
2 : a fictional character lacking depth and believability

stick·ful \'stik-ˌfu̇l\ *noun* (1683)
: as much set type as fills a composing stick

stick·han·dle \'stik-ˌhan-dᵊl\ *intransitive verb* (1929)
: to maneuver a puck (as in hockey) or a ball (as in lacrosse) with a stick
— stick·han·dler \-ˌhan(d)-lər, -ˌhan-dᵊl-ər\ *noun*

sticking plaster *noun* (1655)
chiefly British **:** an adhesive plaster especially for closing superficial wounds

sticking point *noun* (1946)
: an item (as in negotiations) resulting or likely to result in an impasse

stick insect *noun* (1854)
: any of various usually wingless insects (especially family Phasmatidae) with a long cylindrical body resembling a stick

stick–in–the–mud \'stik-ən-t͟hə-ˌməd\ *noun* (1733)
: one who is slow, old-fashioned, or unprogressive; *especially* **:** an old fogy

stick·it \'sti-kət\ *adjective* [Scots, from past participle of English ³*stick*] (1787)
1 *Scottish* **:** UNFINISHED
2 *chiefly Scottish* **:** having failed especially in an intended profession

stick·le \'sti-kəl\ *intransitive verb* **stick·led; stick·ling** \-k(ə-)liŋ\ [alteration of Middle English *stightlen,* frequentative of *stighten* to arrange, from Old English *stihtan;* akin to Old Norse *stētta* to found, support] (1642)
1 : to contend especially stubbornly and usually on insufficient grounds
2 : to feel scruples **:** SCRUPLE

stick·le·back \'sti-kəl-ˌbak\ *noun* [Middle English *stykylbak,* from Old English *sticel* goad + Middle English *bak* back; akin to Old English *stician* to stick] (15th century)
: any of a family (Gasterosteidae) of small scaleless bony fishes having two or more free spines in front of the dorsal fin and including marine, anadromous, and freshwater forms

stick·ler \'sti-k(ə-)lər\ *noun* (1644)
1 : one who insists on exactness or completeness in the observance of something ⟨a *stickler* for the rules⟩
2 : something that baffles or puzzles **:** POSER, STICKER

stick·man \'stik-ˌman, -mən\ *noun* (circa 1931)
: one who handles a stick: as **a :** one who supervises the play at a dice table, calls the decisions, and retrieves the dice **b :** a player in any of various games (as hockey or lacrosse) played with a stick

stick out (1567)
intransitive verb
1 a : to jut out **:** PROJECT **b :** to be prominent or conspicuous
2 : to be persistent (as in a demand or an opinion)
transitive verb
: ENDURE, LAST — often used with *it* ⟨*stuck* it out to the end⟩

stick·pin \'stik-ˌpin\ *noun* (1895)
: an ornamental pin; *especially* **:** one worn in a necktie

stick·seed \-ˌsēd\ *noun* (1843)
: any of various weedy herbs (genera *Lappula* and *Hackelia*) of the borage family with bristly adhesive fruit

stick shift *noun* (1959)
: a manually operated gearshift for a motor vehicle usually mounted on the floor

stick·tight \'stik-ˌtīt\ *noun* (circa 1884)
: BUR MARIGOLD

stick–to–it·ive·ness \stik-'tü-ə-tiv-nəs\ *noun* [from the phrase *stick to it*] (1867)
: dogged perseverance **:** TENACITY

stick·um \'sti-kəm\ *noun* [³*stick* + *-um* (probably alteration of *'em* them)] (circa 1909)
: a substance that adheres or causes adhesion

stick·up \'stik-ˌəp\ *noun* (1904)
: a robbery at gunpoint **:** HOLDUP

stick up (15th century)
intransitive verb
: to stand upright or on end **:** PROTRUDE
transitive verb
: to rob at gunpoint
— stick up for : to speak or act in defense of **:** SUPPORT

stick·weed \'stik-ˌwēd\ *noun* (1743)
: any of several plants (as a beggar's-lice) with adhesive seeds

stick·work \-ˌwərk\ *noun* (1903)
: the use of one's stick in offensive and defensive techniques (as in hockey)

sticky \'sti-kē\ *adjective* **stick·i·er; -est** (circa 1735)
1 a : ADHESIVE **b** (1) **:** VISCOUS, GLUEY (2) **:** coated with a sticky substance
2 : HUMID, MUGGY; *also* **:** CLAMMY
3 : tending to stick
4 a : DISAGREEABLE, UNPLEASANT **b :** AWKWARD, STIFF **c :** DIFFICULT, PROBLEMATIC
5 : excessively sentimental **:** CLOYING
— stick·i·ly \'sti-kə-lē\ *adverb*
— stick·i·ness \'sti-kē-nəs\ *noun*

sticky wicket *noun* (1926)
: a difficult or delicate problem or situation

stic·tion \'stik-shən\ *noun* [*st*atic + fr*iction*] (1946)
: the force required to cause one body in contact with another to begin to move

¹stiff \'stif\ *adjective* [Middle English *stif,* from Old English *stīf;* akin to Middle Dutch *stijf* stiff, Latin *stipare* to press together, Greek *steibein* to tread on] (before 12th century)
1 a : not easily bent **:** RIGID **b :** lacking in suppleness or responsiveness ⟨*stiff* muscles⟩ **c :** impeded in movement — used of a mechanism **d :** DRUNK 1a
2 a : FIRM, RESOLUTE **b :** STUBBORN, UNYIELDING **c :** PROUD **d** (1) **:** marked by reserve or decorum (2) **:** lacking in ease or grace **:** STILTED
3 : hard fought **:** PUGNACIOUS, SHARP
4 a (1) **:** exerting great force ⟨a *stiff* wind⟩ (2) **:** FORCEFUL, VIGOROUS **b :** POTENT ⟨a *stiff* dose⟩
5 : of a dense or glutinous consistency **:** THICK
6 a : HARSH, SEVERE ⟨a *stiff* penalty⟩ **b :** ARDUOUS, RUGGED ⟨*stiff* terrain⟩
7 : not easily heeled over by an external force (as the wind) ⟨a *stiff* ship⟩
8 : EXPENSIVE, STEEP ⟨paid a *stiff* price⟩ ☆
— stiff·ish \'sti-fish\ *adjective*
— stiff·ly *adverb*
— stiff·ness *noun*

²stiff *adverb* (13th century)
1 : in a stiff manner **:** STIFFLY
2 : to an extreme degree **:** SEVERELY ⟨scared *stiff*⟩ ⟨bored *stiff*⟩

³stiff *noun* (circa 1859)
1 : CORPSE
2 a : TRAMP, BUM **b :** HAND, LABORER **c :** PERSON ⟨a lucky *stiff*⟩
3 : FLOP, FAILURE

⁴stiff *transitive verb* (1950)
1 a : to refuse to pay or tip ⟨*stiffed* the doctor for the fee⟩ ⟨*stiffed* the waiter⟩ **b :** CHEAT ⟨*stiffed* him in a business deal⟩ **c :** STICK 7a ⟨*stiffed* us with the bar bill⟩
2 : SNUB 3 ⟨*stiffed* sportswriters after the game⟩

stiff–arm \'stif-ˌärm\ *verb or noun* (1909)
: STRAIGHT-ARM

stiff·en \'sti-fən\ *verb* **stiff·ened; stiff·en·ing** \'sti-fə-niŋ, 'stif-niŋ\ (15th century)
transitive verb
: to make stiff or stiffer
intransitive verb
: to become stiff or stiffer
— stiff·en·er \'sti-fə-nər, 'stif-nər\ *noun*

stiff–necked \'stif-'nekt\ *adjective* (1526)
1 : HAUGHTY, STUBBORN
2 : STILTED

stiff upper lip *noun* [from the phrase *keep a stiff upper lip*] (1815)
: a steady and determined attitude or manner in the face of trouble
— stiff–upper–lip *adjective*

¹sti·fle \'stī-fəl\ *noun* [Middle English] (14th century)
: the joint next above the hock in the hind leg of a quadruped (as a horse or dog) corresponding to the human knee — see HORSE illustration

²stifle *verb* **sti·fled; sti·fling** \-f(ə-)liŋ\ [alteration of Middle English *stuflen*] (1513)
transitive verb
1 a : to kill by depriving of oxygen **:** SUFFOCATE **b** (1) **:** SMOTHER (2) **:** MUFFLE
2 a : to cut off (as the voice or breath) **b :** to withhold from circulation or expression **:** REPRESS ⟨*stifled* our anger⟩ **c :** DETER, DISCOURAGE
intransitive verb
: to become suffocated by or as if by lack of oxygen **:** SMOTHER
— sti·fler \-f(ə-)lər\ *noun*
— sti·fling·ly \-f(ə-)liŋ-lē\ *adverb*

stig·ma \'stig-mə\ *noun, plural* **stig·ma·ta** \stig-'mä-tə, 'stig-mə-tə\ *or* **stig·mas** [Latin *stigmat-, stigma* mark, brand, from Greek, from *stizein* to tattoo — more at STICK] (circa 1593)
1 a *archaic* **:** a scar left by a hot iron **:** BRAND **b :** a mark of shame or discredit **:** STAIN **c :** an identifying mark or characteristic; *specifically* **:** a specific diagnostic sign of a disease
2 a *stigmata plural* **:** bodily marks or pains resembling the wounds of the crucified Christ and sometimes accompanying religious ecstasy **b :** PETECHIA
3 a : a small spot, scar, or opening on a plant or animal **b :** the usually apical part of the pistil of a flower which receives the pollen grains and on which they germinate — see FLOWER illustration ◆
— stig·mal \'stig-məl\ *adjective*

> ☆ SYNONYMS
> **Stiff, rigid, inflexible** mean difficult to bend. STIFF may apply to any degree of this condition ⟨stretching will prevent your muscles from becoming *stiff*⟩. RIGID applies to something so stiff that it cannot be bent without breaking ⟨a *rigid* surfboard⟩. INFLEXIBLE stresses lack of suppleness or pliability ⟨ski boots with *inflexible* soles⟩.

> ◇ WORD HISTORY
> **stigma** In ancient Greece and Rome runaway slaves and criminals were branded with a hot iron or needle as a sign of disgrace. The brand was called in Greek *stigma,* a derivative of the verb *stizein* "to tattoo." *Stigma* was taken into English in its original sense in the late 16th century. Soon, however, the word was being used figuratively to mean "a mark of shame or discredit." The English plural form for this sense is generally *stigmas.* The Greek and Latin plural form *stigmata* has been used since the Middle Ages specifically for the wounds that some people apparently bear on their hands and feet, and sometimes on their side and brow. Believers take these to be visible signs of the bearer's participation in Christ's sufferings on the cross. This religious sense gave rise in the 17th century to a medical sense, the disease-induced reddish spots known as petechiae.

stig·mas·ter·ol \stig-ˈmas-tə-ˌrȯl, -ˌrōl\ *noun* [New Latin Physo*stigma* (genus including the Calabar bean) + International Scientific Vocabulary *sterol*] (1907)
: a crystalline sterol $C_{29}H_{48}O$ obtained especially from the oils of Calabar beans and soybeans

¹stig·mat·ic \stig-ˈma-tik\ *noun* (1594)
: one marked with stigmata

²stigmatic *adjective* (1607)
1 : having or conveying a social stigma
2 : of or relating to supernatural stigmata
3 : ANASTIGMATIC — used especially of a bundle of light rays intersecting at a single point
— stig·mat·i·cal·ly \-ti-k(ə-)lē\ *adverb*

stig·ma·tist \ˈstig-mə-tist, stig-ˈmä-\ *noun* (1607)
: STIGMATIC

stig·ma·tize \ˈstig-mə-ˌtīz\ *transitive verb* **-tized; -tiz·ing** (1585)
1 a *archaic* **:** BRAND **b :** to describe or identify in opprobrious terms
2 : to mark with stigmata
— stig·ma·ti·za·tion \ˌstig-mə-tə-ˈzā-shən\ *noun*

stil·bene \ˈstil-ˌbēn\ *noun* [International Scientific Vocabulary, from Greek *stilbein* to glitter] (circa 1868)
: an aromatic hydrocarbon $C_{14}H_{12}$ used as a phosphor and in making dyes; *also* **:** a compound derived from stilbene

stil·bes·trol \stil-ˈbes-ˌtrȯl, -ˌtrōl\ *noun* [*stil*bene + *estr*us + ¹*-ol*] (1938)
1 : a crystalline synthetic derivative $C_{14}H_{12}O_2$ of stilbene that differs from the related diethylstilbestrol in lack of the ethyl groups and in possession of but slight estrogenic activity
2 : DIETHYLSTILBESTROL

stil·bite \ˈstil-ˌbīt\ *noun* [French, from Greek *stilbein*] (1815)
: a mineral consisting of a hydrous silicate of aluminum, calcium, and sodium and often occurring in sheaflike aggregations of crystals

¹stile \ˈstī(ə)l\ *noun* [Middle English, from Old English *stigel;* akin to Old English *stǣger* stair — more at STAIR] (before 12th century)
: a step or set of steps for passing over a fence or wall; *also* **:** TURNSTILE

²stile *noun* [probably from Dutch *stijl* post] (1678)
: one of the vertical members in a frame or panel into which the secondary members are fitted

sti·let·to \stə-ˈle-(ˌ)tō\ *noun, plural* **-tos** *or* **-toes** [Italian, diminutive of *stilo* stylus, dagger, from Latin *stilus* stylus — more at STYLE] (1611)
1 : a slender dagger with a blade thick in proportion to its breadth
2 : a pointed instrument for piercing holes for eyelets or embroidery
3 : STILETTO HEEL

stiletto heel *noun* (1953)
: a high thin heel on women's shoes that is narrower than a spike heel

¹still \ˈstil\ *adjective* [Middle English *stille,* from Old English; akin to Old High German *stilli* still and perhaps to Old English *steall* stall — more at STALL] (before 12th century)
1 a : devoid of or abstaining from motion **b** *archaic* **:** SEDENTARY **c :** not effervescent ⟨*still* wine⟩ **d** (1) **:** of, relating to, or being a static photograph as contrasted with a motion picture (2) **:** designed for taking still photographs ⟨a *still* camera⟩ (3) **:** engaged in taking still photographs ⟨a *still* photographer⟩
2 a : uttering no sound **:** QUIET **b :** SUBDUED, MUTED
3 a : CALM, TRANQUIL **b :** free from noise or turbulence
— still·ness *noun*

²still (before 12th century)
intransitive verb
: to become motionless or silent **:** QUIET
transitive verb
1 a : ALLAY, CALM **b :** to put an end to **:** SETTLE
2 : to arrest the motion of
3 : SILENCE

³still *adverb* (before 12th century)
1 : without motion ⟨sit *still*⟩
2 *archaic* **a :** ALWAYS, CONTINUALLY **b :** in a progressive manner **:** INCREASINGLY
3 — used as a function word to indicate the continuance of an action or condition ⟨*still* lives there⟩ ⟨drink it while it's *still* hot⟩
4 : in spite of that **:** NEVERTHELESS ⟨those who take the greatest care *still* make mistakes⟩
5 a : EVEN 2c ⟨a *still* more difficult problem⟩ **b :** YET 1a

⁴still *noun* (13th century)
1 : QUIET, SILENCE
2 : a static photograph; *specifically* **:** a photograph of actors or scenes of a motion picture for publicity or documentary purposes

⁵still *verb* [Middle English *stillen,* short for *distillen* to distill] (13th century)
: DISTILL

⁶still *noun* (1533)
1 : DISTILLERY
2 : apparatus used in distillation comprising either the chamber in which the vaporization is carried out or the entire equipment

still alarm *noun* (1875)
: a fire alarm transmitted (as by telephone call) without sounding the signal apparatus

still and all *adverb* (1829)
: NEVERTHELESS, STILL

still·birth \ˈstil-ˌbərth, -ˈbərth\ *noun* (1785)
: the birth of a dead fetus

still·born \-ˈbȯrn\ *adjective* (1593)
1 : dead at birth
2 : failing from the start **:** ABORTIVE ⟨a *stillborn* venture⟩
— still·born \-ˌbȯrn\ *noun*

still–hunt \-ˌhənt\ (1858)
intransitive verb
: to ambush or stalk a quarry; *especially* **:** to pursue game noiselessly usually without a dog
transitive verb
: to lie in wait for **:** approach by stealth

still hunt *noun* (1828)
: a quiet pursuing or ambushing of game

still life *noun, plural* **still lifes** (1695)
1 : a picture consisting predominantly of inanimate objects
2 : the category of graphic arts concerned with inanimate subject matter

still·man \ˈstil-mən\ *noun* (circa 1864)
: one who owns or operates a still

still·room \ˈstil-ˌrüm, -ˌru̇m\ *noun* [⁶*still*] (circa 1710)
British **:** a room connected with the kitchen where liqueurs, preserves, and cakes are kept and beverages (as tea) are prepared

still water *noun* (1832)
: a part of a stream where no current is visible

¹stil·ly \ˈstil-lē\ *adverb* (before 12th century)
: in a calm manner **:** QUIETLY

²stilly \ˈsti-lē\ *adjective* [⁴*still* + ¹*-y*] (1776)
: STILL, QUIET

¹stilt \ˈstilt\ *noun* [Middle English *stilte;* akin to Old High German *stelza* stilt] (15th century)
1 a : one of two poles each with a rest or strap for the foot used to elevate the wearer above the ground in walking **b :** a pile or post serving as one of the supports of a structure above ground or water level
2 *plural also* **stilt :** any of various notably long-legged 3-toed shorebirds (genera *Himantopus* and *Cladorhynchus*) that are related to the avocets, frequent inland ponds and marshes, and nest in small colonies

²stilt *transitive verb* (1649)
: to raise on or as if on stilts

stilt·ed \ˈstil-təd\ *adjective* (1820)
1 a : POMPOUS, LOFTY **b :** FORMAL, STIFF
2 : having the curve beginning at some distance above the impost ⟨a *stilted* arch⟩
— stilt·ed·ly *adverb*
— stilt·ed·ness *noun*

Stil·ton \ˈstil-tᵊn\ *noun* [*Stilton,* Huntingdonshire, England] (1826)
: a blue-veined cheese with wrinkled rind made of whole cows' milk enriched with cream

stime \ˈstīm\ *noun* [Middle English (northern dialect)] (14th century)
chiefly Scottish & Irish **:** GLIMMER; *also* **:** GLIMPSE

stim·u·lant \ˈstim-yə-lənt\ *noun* (circa 1728)
1 : an agent (as a drug) that produces a temporary increase of the functional activity or efficiency of an organism or any of its parts
2 : STIMULUS
3 : an alcoholic beverage — not used technically
— stimulant *adjective*

stim·u·late \-ˌlāt\ *verb* **-lat·ed; -lat·ing** [Latin *stimulatus,* past participle of *stimulare,* from *stimulus* goad; perhaps akin to Latin *stilus* stem, stylus — more at STYLE] (1619)
transitive verb
1 : to excite to activity or growth or to greater activity **:** ANIMATE, AROUSE
2 a : to function as a physiological stimulus to **b :** to arouse or affect by a stimulant (as a drug)
intransitive verb
: to act as a stimulant or stimulus
synonym see PROVOKE
— stim·u·la·tion \ˌstim-yə-ˈlā-shən\ *noun*
— stim·u·la·tive \ˈstim-yə-ˌlā-tiv\ *adjective*
— stim·u·la·tor \-ˌlā-tər\ *noun*
— stim·u·la·to·ry \-lə-ˌtōr-ē, -ˌtȯr-\ *adjective*

stim·u·lus \ˈstim-yə-ləs\ *noun, plural* **-li** \-ˌlī, -ˌlē\ [Latin] (1684)
: something that rouses or incites to activity: as **a :** INCENTIVE **b :** STIMULANT 1 **c :** an agent (as an environmental change) that directly influences the activity of a living organism or one of its parts (as by exciting a sensory organ or evoking muscular contraction or glandular secretion)

¹sting \ˈstiŋ\ *verb* **stung** \ˈstəŋ\; **sting·ing** \ˈstiŋ-iŋ\ [Middle English, from Old English *stingan;* akin to Old Norse *stinga* to sting and probably to Greek *stachys* spike of grain, *stochos* target, aim] (before 12th century)
transitive verb
1 : to prick painfully: as **a :** to pierce or wound with a poisonous or irritating process **b :** to affect with sharp quick pain or smart ⟨hail *stung* their faces⟩
2 : to cause to suffer acutely ⟨*stung* with remorse⟩
3 : OVERCHARGE, CHEAT
intransitive verb
1 : to wound one with or as if with a sting
2 : to feel a keen burning pain or smart; *also* **:** to cause such pain
— sting·ing·ly \-iŋ-lē\ *adverb*

²sting *noun* (before 12th century)
1 a : the act of stinging; *specifically* **:** the thrust of a stinger into the flesh **b :** a wound or pain caused by or as if by stinging
2 : STINGER 2
3 : a sharp or stinging element, force, or quality
4 : an elaborate confidence game; *specifically* **:** such a game worked by undercover police in order to trap criminals

sting·a·ree \ˈstiŋ-ə-rē *also* ˈstiŋ-rē\ *noun* [by alteration] (1836)
: STINGRAY

sting·er \ˈstiŋ-ər\ *noun* (circa 1552)
1 : one that stings; *specifically* **:** a sharp blow or remark
2 : a sharp organ (as of a bee, scorpion, or stingray) that is usually connected with a poison gland or otherwise adapted to wound by piercing and injecting a poison
3 : a cocktail usually consisting of brandy and white crème de menthe

stinging nettle *noun* (1525)

: NETTLE 1; *especially* **:** one (*Urtica dioica*) established in North America that has broad coarsely toothed leaves

sting·less \'stiŋ-ləs\ *adjective* (1554)
: having no sting or stinger

sting·ray \-ˌrā\ *noun* (1624)
: any of numerous rays (as of the family Dasyatidae) with one or more large sharp barbed dorsal spines near the base of the whiplike tail capable of inflicting severe wounds

stingray

stin·gy \'stin-jē\ *adjective* **stin·gi·er; -est** [perhaps from (assumed) English dialect *stinge,* noun, sting; akin to Old English *stingan* to sting] (1659)
1 : not generous or liberal **:** sparing or scant in giving or spending
2 : meanly scanty or small ☆
— stin·gi·ly \-jə-lē\ *adverb*
— stin·gi·ness \-jē-nəs\ *noun*

[1]**stink** \'stiŋk\ *intransitive verb* **stank** \'staŋk\ *or* **stunk** \'stəŋk\; **stunk; stink·ing** [Middle English, from Old English *stincan;* akin to Old High German *stinkan* to emit a smell] (before 12th century)
1 : to emit a strong offensive odor
2 : to be offensive; *also* **:** to be in bad repute
3 : to possess something to an offensive degree ⟨*stinking* with wealth⟩
4 : to be extremely bad in quality
— stinky \'stiŋ-kē\ *adjective*

[2]**stink** *noun* (13th century)
1 : a strong offensive odor **:** STENCH
2 : a public outcry against something

stink·ard \'stiŋ-kərd\ *noun* (circa 1600)
: a mean or contemptible person

stink bomb *noun* (1915)
: a small bomb charged usually with chemicals that gives off a foul odor on bursting

stink·bug \'stiŋk-ˌbəg\ *noun* (1877)
: any of various hemipterous bugs (especially family Pentatomidae) that emit a disagreeable odor

stink·er \'stiŋ-kər\ *noun* (1607)
1 a : one that stinks **b :** an offensive or contemptible person **c :** something of very poor quality
2 : any of several large petrels that have an offensive odor
3 *slang* **:** something extremely difficult ⟨the examination was a real *stinker*⟩

stink·horn \'stiŋk-ˌhórn\ *noun* (1724)
: any of various fetid basidiomycetous fungi (order Phallales, especially *Phallus impudicus*) having spores dispersed by insects

[1]**stink·ing** *adjective* (before 12th century)
1 : strong and offensive to the sense of smell
2 *slang* **:** offensively drunk
synonym see MALODOROUS
— stink·ing·ly \'stiŋ-kiŋ-lē\ *adverb*

[2]**stinking** *adverb* (1887)
: to an extreme degree ⟨got *stinking* drunk⟩

stinking smut *noun* (circa 1891)
: [1]BUNT

stink·pot \'stiŋk-ˌpät\ *noun* (1669)
1 : an earthen jar filled with fetid material and formerly sometimes thrown as a stink bomb on an enemy's deck
2 : MUSK TURTLE
3 : STINKER 1
4 *slang* **:** MOTORBOAT

stink up *transitive verb* (1941)
: to cause to stink or be filled with a stench

stink·weed \'stiŋk-ˌwēd\ *noun* (1753)
: any of various strong-scented or fetid plants; *especially* **:** PENNYCRESS

stink·wood \-ˌwu̇d\ *noun* (1731)
1 : any of several trees with a wood of unpleasant odor; *especially* **:** a southern African tree (*Ocotea bullata*) of the laurel family yielding a valued cabinet wood
2 : the wood of a stinkwood

[1]**stint** \'stint\ *verb* [Middle English, from Old English *styntan* to blunt, dull; akin to Old Norse *stuttr* scant] (13th century)
intransitive verb
1 *archaic* **:** STOP, DESIST
2 : to be sparing or frugal
transitive verb
1 *archaic* **:** to put an end to **:** STOP
2 a : to restrain within certain limits **:** CONFINE **b :** to restrict with respect to a share or allowance
3 : to assign a task to (a person)
— stint·er *noun*

[2]**stint** *noun* (circa 1530)
1 a : a definite quantity of work assigned **b :** a period of time spent at a particular activity ⟨served a brief *stint* as a waiter⟩
2 : RESTRAINT, LIMITATION
synonym see TASK

[3]**stint** *noun, plural* **stints** *also* **stint** [Middle English *stynte*] (15th century)
: any of several small sandpipers

stipe \'stīp\ *noun* [New Latin *stipes,* from Latin, tree trunk; akin to Latin *stipare* to press together — more at STIFF] (1785)
: a usually short plant stalk: as **a :** the stem supporting the cap of a fungus **b :** a part that is similar to a stipe and connects the holdfast and blade of a frondose alga **c :** the petiole of a fern frond **d :** a prolongation of the receptacle beneath the ovary of a seed plant
— stiped \'stīpt\ *adjective*

sti·pend \'stī-ˌpend, -pənd\ *noun* [Middle English, alteration of *stipendy,* from Latin *stipendium,* from *stip-, stips* gift + *pendere* to weigh, pay] (15th century)
: a fixed sum of money paid periodically for services or to defray expenses

[1]**sti·pen·di·ary** \stī-'pen-dē-ˌer-ē\ *noun, plural* **-ar·ies** (15th century)
: one who receives a stipend

[2]**stipendiary** *adjective* (circa 1545)
1 : receiving or compensated by wages or salary ⟨a *stipendiary* curate⟩
2 : of or relating to a stipend

sti·pes \'stī-ˌpēz\ *noun, plural* **stip·i·tes** \'sti-pə-ˌtēz\ [New Latin *stipit-, stipes,* from Latin, tree trunk] (1760)
: PEDUNCLE; *especially* **:** the second basal segment of a maxilla of an insect or crustacean

[1]**stip·ple** \'sti-pəl\ *transitive verb* **stip·pled; stip·pling** \-p(ə-)liŋ\ [Dutch *stippelen* to spot, dot] (circa 1762)
1 : to engrave by means of dots and flicks
2 a : to make by small short touches (as of paint or ink) that together produce an even or softly graded shadow **b :** to apply (as paint) by repeated small touches
3 : SPECKLE, FLECK
— stip·pler \-p(ə-)lər\ *noun*

[2]**stipple** *noun* (1837)
: production of gradation of light and shade in graphic art by stippling small points, larger dots, or longer strokes; *also* **:** an effect produced in this way

stip·u·lar \'sti-pyə-lər\ *adjective* (1793)
: of, resembling, or provided with stipules ⟨*stipular* glands⟩

[1]**stip·u·late** \'sti-pyə-ˌlāt\ *verb* **-lat·ed; -lat·ing** [Latin *stipulatus,* past participle of *stipulari* to demand some term in an agreement] (circa 1624)
intransitive verb
1 : to make an agreement or covenant to do or forbear something **:** CONTRACT
2 : to demand an express term in an agreement — used with *for*
transitive verb
1 : to specify as a condition or requirement of an agreement or offer
2 : to give a guarantee of
— stip·u·la·tor \-ˌlā-tər\ *noun*

[2]**stip·u·late** \'sti-pyə-lət\ *adjective* [New Latin *stipula*] (1776)
: having stipules

stip·u·la·tion \ˌsti-pyə-'lā-shən\ *noun* (circa 1552)
1 : an act of stipulating
2 : something stipulated; *especially* **:** a condition, requirement, or item specified in a legal instrument
— stip·u·la·to·ry \'sti-pyə-lə-ˌtōr-ē, -ˌtór-\ *adjective*

stip·ule \'sti-(ˌ)pyü(ə)l\ *noun* [New Latin *stipula,* from Latin, stalk; akin to Latin *stipes* tree trunk] (1793)
: either of a pair of appendages borne at the base of the leaf in many plants

[1]**stir** \'stər\ *verb* **stirred; stir·ring** [Middle English, from Old English *styrian;* akin to Old High German *stōren* to scatter] (before 12th century)
transitive verb
1 a : to cause an especially slight movement or change of position of **b :** to disturb the quiet of **:** AGITATE
2 a : to disturb the relative position of the particles or parts of especially by a continued circular movement **b :** to mix by or as if by stirring
3 : BESTIR, EXERT
4 : to bring into notice or debate **:** RAISE
5 a : to rouse to activity **:** evoke strong feelings in **b :** to call forth (as a memory) **:** EVOKE **c :** PROVOKE
intransitive verb
1 a : to make a slight movement **b :** to begin to move (as in rousing)
2 : to begin to be active
3 : to be active or busy
4 : to pass an implement through a substance with a circular movement
5 : to be able to be stirred
— stir·rer *noun*

[2]**stir** *noun* (14th century)
1 a : a state of disturbance, agitation, or brisk activity **b :** widespread notice and discussion **:** IMPRESSION ⟨the book caused quite a *stir*⟩
2 : a slight movement
3 : a stirring movement

[3]**stir** *noun* [origin unknown] (1851)
slang **:** PRISON

stir·about \'stər-ə-ˌbau̇t\ *noun* (1682)
: a porridge of Irish origin consisting of oatmeal or cornmeal boiled in water or milk and stirred

stir–cra·zy \'stər-'krā-zē\ *adjective* [[3]*stir*] (circa 1908)

☆ SYNONYMS

Stingy, close, niggardly, parsimonious, penurious, miserly mean being unwilling or showing unwillingness to share with others. STINGY implies a marked lack of generosity ⟨a *stingy* child, not given to sharing⟩. CLOSE suggests keeping a tight grip on one's money and possessions ⟨folks who are very *close* when charity calls⟩. NIGGARDLY implies giving or spending the very smallest amount possible ⟨the *niggardly* amount budgeted for the town library⟩. PARSIMONIOUS suggests a frugality so extreme as to lead to stinginess ⟨a *parsimonious* life-style notably lacking in luxuries⟩. PENURIOUS implies niggardliness that gives an appearance of actual poverty ⟨the *penurious* eccentric bequeathed a fortune⟩. MISERLY suggests a sordid avariciousness and a morbid pleasure in hoarding ⟨a *miserly* couple devoid of social conscience⟩.

\ə\ **abut** \ᵊ\ kitten \ər\ **further** \a\ ash \ā\ ace \ä\ mop, mar \au̇\ **out** \ch\ **chin** \e\ bet \ē\ **easy** \g\ go \i\ hit \ī\ ice \j\ job \ŋ\ sing \ō\ go \ó\ law \ói\ boy \th\ **thin** \t͟h\ **the** \ü\ loot \u̇\ foot \y\ yet \zh\ vision *see also* Guide to Pronunciation *slang* **:** distraught because of prolonged confinement

[1]**stir–fry** \-ˈfrī\ *transitive verb* (1958)
: to fry quickly over high heat in a lightly oiled pan (as a wok) while stirring continuously

[2]**stir–fry** \-ˌfrī\ *noun* (1959)
: a dish of something stir-fried

stirk \ˈstərk\ *noun* [Middle English, from Old English *stirc;* akin to Middle Low German *sterke* young cow and perhaps to Gothic *stairo* sterile animal — more at STERILE] (before 12th century)
British **:** a young bull or cow especially between one and two years old

Stir·ling engine \ˈstər-liŋ-\ *noun* [Robert *Stirling* (died 1878) Scottish engineer] (1896)
: an external combustion engine having an enclosed working fluid (as helium) that is alternately compressed and expanded to operate a piston

Stir·ling's formula \ˈstər-liŋz-\ *noun* [James *Stirling* (died 1770) Scottish mathematician] (1926)
: a formula

$$\sqrt{2\pi n}\, n^n e^{-n}$$

that approximates the value of the factorial of a very large number *n*

stirp \ˈstərp\ *noun* [Latin *stirp-, stirps*] (1502)
: a line descending from a common ancestor **:** STOCK, LINEAGE

stirps \ˈstirps, ˈstərps\ *noun, plural* **stir·pes** \ˈstir-ˌpās, ˈstər-(ˌ)pēz\ [Latin, literally, stem, stock] (1681)
1 : a branch of a family; *also* **:** the person from whom it is descended
2 a : a group of animals equivalent to a superfamily **b :** a race or fixed variety of plants

stir·ring \ˈstər-iŋ\ *adjective* (before 12th century)
1 : ACTIVE, BUSTLING
2 : ROUSING, INSPIRING ⟨a *stirring* speech⟩

stir·rup \ˈstər-əp *also* ˈstir-əp *or* ˈstə-rəp\ *noun* [Middle English *stirop,* from Old English *stigrāp,* from *stig-* (akin to Old High German *stīgan* to go up) + *rāp* rope — more at STAIR, ROPE] (before 12th century)
1 : either of a pair of small light frames or rings for receiving the foot of a rider that are attached by a strap to a saddle and used to aid in mounting and as a support while riding
2 : a piece resembling a stirrup (as a support or clamp in carpentry and machinery)
3 : a rope secured to a yard and attached to a thimble in its lower end for supporting a footrope
4 : STAPES

stirrup cup *noun* (1681)
1 : a small serving of drink (as wine) taken by a rider about to depart; *also* **:** the vessel in which it is served
2 : a farewell cup

stirrup leather *noun* (14th century)
: the looped strap suspending a stirrup

stirrup pump *noun* (1939)
: a portable hand pump held in position by a foot bracket and used for throwing a jet or spray of liquid

[1]**stitch** \ˈstich\ *noun* [Middle English *stiche,* from Old English *stice;* akin to Old English *stician* to stick] (before 12th century)
1 : a local sharp and sudden pain especially in the side
2 a : one in-and-out movement of a threaded needle in sewing, embroidering, or suturing **b :** a portion of thread left in the material or suture left in the tissue after one stitch
3 : a least part especially of clothing
4 : a single loop of thread or yarn around an implement (as a knitting needle or crochet hook)
5 : a stitch or series of stitches formed in a particular way ⟨a basting *stitch*⟩
— in stitches : in a state of uncontrollable laughter

[2]**stitch** (13th century)
transitive verb
1 a : to fasten, join, or close with or as if with stitches **b :** to make, mend, or decorate with or as if with stitches
2 : to unite by means of staples
intransitive verb
: SEW
— stitch·er *noun*

stitch·ery \ˈstich-rē, ˈsti-chə-\ *noun* (1607)
: NEEDLEWORK

stitch·wort \ˈstich-ˌwərt, -ˌwȯrt\ *noun* (13th century)
: any of several chickweeds (genus *Stellaria*)

stithy \ˈsti-t͟hē, -thē\ *noun, plural* **stith·ies** [Middle English, from Old Norse *stethi;* akin to Old English *stede* stead] (13th century)
1 *archaic* **:** ANVIL
2 *archaic* **:** SMITHY 1

sti·ver \ˈstī-vər\ *noun* [Dutch *stuiver*] (1502)
1 : a unit of value and coin of the Netherlands equal to 1/20 gulden
2 : something of little value

stoa \ˈstō-ə\ *noun* [Greek; akin to Greek *stylos* pillar — more at STEER] (1603)
: an ancient Greek portico usually walled at the back with a front colonnade designed to afford a sheltered promenade

stoat \ˈstōt\ *noun, plural* **stoats** *also* **stoat** [Middle English *stote*] (15th century)
: the common Holarctic ermine (*Mustela erminea*) especially in its brown summer coat

stob \ˈstäb\ *noun* [Middle English, stump; akin to Middle English *stubb* stub] (15th century)
chiefly dialect **:** STAKE, POST

stoc·ca·do \stə-ˈkä-(ˌ)dō\ *noun, plural* **-dos** [Italian *stoccata*] (1582)
archaic **:** a thrust with a rapier

sto·chas·tic \stə-ˈkas-tik, stō-\ *adjective* [Greek *stochastikos* skillful in aiming, from *stochazesthai* to aim at, guess at, from *stochos* target, aim, guess — more at STING] (1923)
1 : RANDOM; *specifically* **:** involving a random variable ⟨a *stochastic* process⟩
2 : involving chance or probability **:** PROBABILISTIC ⟨a *stochastic* model of radiation-induced mutation⟩
— sto·chas·ti·cal·ly \-ti-k(ə-)lē\ *adverb*

[1]**stock** \ˈstäk\ *noun* [Middle English *stok,* from Old English *stocc;* akin to Old High German *stoc* stick] (before 12th century)
1 a *archaic* **:** STUMP **b** *archaic* **:** a log or block of wood **c** (1) *archaic* **:** something without life or consciousness (2) **:** a dull, stupid, or lifeless person
2 : a supporting framework or structure: as **a** *plural* **:** the frame or timbers holding a ship during construction **b** *plural* **:** a device for publicly punishing offenders consisting of a wooden frame with holes in which the feet or feet and hands can be locked **c** (1) **:** the wooden part by which a shoulder arm is held during firing (2) **:** the butt of an implement (as a whip or fishing rod) (3) **:** BITSTOCK, BRACE **d :** a long beam on a field gun forming the third support point in firing
3 a : the main stem of a plant **:** TRUNK **b** (1) **:** a plant or plant part united with a scion in grafting and supplying mostly underground parts to a graft (2) **:** a plant from which slips or cuttings are taken
4 : the crosspiece of an anchor — see ANCHOR illustration
5 a : the original (as a person, race, or language) from which others derive **:** SOURCE **b** (1) **:** the descendants of one individual **:** FAMILY, LINEAGE (2) **:** a compound organism — compare CLONE **c :** an infraspecific group usually having unity of descent **d** (1) **:** a related group of languages (2) **:** a language family
6 a (1) **:** the equipment, materials, or supplies of an establishment (2) **:** LIVESTOCK **b :** a store or supply accumulated; *especially* **:** the inventory of goods of a merchant or manufacturer
7 a *archaic* **:** a supply of capital **:** FUNDS; *especially* **:** money or capital invested or available for investment or trading **b** (1) **:** the part of a tally formerly given to the creditor in a transaction (2) **:** a debt or fund due (as from a government) for money loaned at interest; *also, British* **:** capital or a debt or fund bearing interest in perpetuity and not ordinarily redeemable as to principal **c** (1) **:** the proprietorship element in a corporation usually divided into shares and represented by transferable certificates (2) **:** a portion of such stock of one or more companies (3) **:** STOCK CERTIFICATE
8 : any of a genus (Matthiola) of Old World herbs or subshrubs of the mustard family with racemes of usually sweet-scented flowers
9 : a wide band or scarf worn about the neck especially by some clergymen
10 a : liquid in which meat, fish, or vegetables have been simmered that is used as a basis for soup, gravy, or sauce **b** (1) **:** raw material from which something is manufactured (2) **:** paper used for printing **c :** the portion of a pack of cards not distributed to the players at the beginning of a game
11 a (1) **:** an estimate or evaluation of something ⟨take *stock* of the situation⟩ (2) **:** the estimation in which someone or something is held ⟨his *stock* with the electorate remains high —*Newsweek*⟩ **b :** confidence or faith placed in someone or something ⟨put little *stock* in his testimony⟩
12 : the production and presentation of plays by a stock company
13 : STOCK CAR 1
— in stock : on hand **:** in the store and ready for delivery
— out of stock : having no more on hand **:** completely sold out

[2]**stock** (15th century)
transitive verb
1 : to make (a domestic animal) pregnant
2 : to fit to or with a stock
3 : to provide with stock or a stock **:** SUPPLY ⟨*stock* a stream with trout⟩
4 : to procure or keep a stock of
5 : to graze (livestock) on land
intransitive verb
1 : to send out new shoots
2 : to put in stock or supplies ⟨*stock* up on canned goods⟩

[3]**stock** *adjective* (1625)
1 a : kept regularly in stock ⟨comes in *stock* sizes⟩ ⟨a *stock* model⟩ **b :** commonly used or brought forward **:** STANDARD ⟨the *stock* answer⟩
2 a : kept for breeding purposes **:** BROOD ⟨a *stock* mare⟩ **b :** devoted to the breeding and rearing of livestock ⟨a *stock* farm⟩ **c :** used or intended for livestock ⟨a *stock* train⟩ **d :** used in herding livestock ⟨a *stock* horse⟩ ⟨a *stock* dog⟩
3 : of or relating to a stock company
4 : employed in handling, checking, or taking care of the stock of merchandise on hand ⟨a *stock* clerk⟩

[1]**stock·ade** \stä-ˈkād\ *noun* [Spanish *estacada,* from *estaca* stake, pale, of Germanic origin; akin to Old English *staca* stake] (1614)
1 : a line of stout posts set firmly to form a defense
2 a : an enclosure or pen made with posts and stakes **b :** an enclosure in which prisoners are kept

[2]**stockade** *transitive verb* **stock·ad·ed; stock·ad·ing** (1677)
: to fortify or surround with a stockade

stockade fence (1985)
: a solid fence of half-round boards pointed at the top

stock·breed·er \ˈstäk-ˌbrē-dər\ *noun* (1815)
: a person engaged in the breeding and care of livestock for the market, for show purposes, or for racing

stock·bro·ker \-ˌbrō-kər\ *noun* (circa 1706)

: a broker who executes orders to buy and sell securities and often also acts as a security dealer
— **stock·bro·ker·age** \-k(ə-)rij\ *noun*
— **stock·brok·ing** \-ˌbrō-kiŋ\ *noun*

stock car *noun* (1858)
1 : a latticed railroad boxcar for carrying livestock
2 : a racing car having the basic chassis of a commercially produced assembly-line model

stock certificate *noun* (1863)
: an instrument evidencing ownership of one or more shares of the capital stock of a corporation

stock company *noun* (1827)
1 : a corporation or joint-stock company of which the capital is represented by stock
2 : a theatrical company attached to a repertory theater; *especially* **:** one without outstanding stars

stock dividend *noun* (circa 1902)
1 : the payment by a corporation of a dividend in the form of shares usually of its own stock without change in par value — compare STOCK SPLIT
2 : the stock distributed in a stock dividend

stock·er \ˈstä-kər\ *noun* (1881)
1 : a young animal (as a steer or heifer) suitable for being fed and fattened for market
2 : an animal (as a heifer) suitable for use in a breeding establishment
3 : STOCK CAR 2

stock exchange *noun* (1773)
1 : a place where security trading is conducted on an organized system
2 : an association of people organized to provide an auction market among themselves for the purchase and sale of securities

stock·fish \-ˌfish\ *noun* [Middle English *stokfish,* from Middle Dutch *stocvisch,* from *stoc* stick + *visch* fish] (13th century)
: fish (as cod, haddock, or hake) dried hard in the open air without salt

stock·hold·er \ˈstäk-ˌhōl-dər\ *noun* (circa 1776)
: an owner of corporate stock

stock·i·nette *or* **stock·i·net** \ˌstä-kə-ˈnet\ *noun* [alteration of earlier *stocking net*] (1784)
: a soft elastic usually cotton fabric used especially for bandages and infants' wear

stock·ing \ˈstä-kiŋ\ *noun* [obsolete *stock* to cover with a stocking] (1583)
1 a : a usually knit close-fitting covering for the foot and leg **b :** SOCK
2 : something resembling a stocking; *especially* **:** a ring of distinctive color on the lower part of the leg of an animal
— **stock·inged** \-kiŋd\ *adjective*
— **in one's stocking feet :** having on stockings but no shoes

stocking cap *noun* (circa 1897)
: a long knitted cone-shaped cap with a tassel or pom-pom worn especially for winter sports or play

stocking stuffer *noun* (1948)
: a small gift suitable for placing in a Christmas stocking

stock–in–trade \ˌstäk-ən-ˈtrād, ˈstäk-ən-ˌ\ *noun* (circa 1771)
1 : the equipment, merchandise, or materials necessary to or used in a trade or business
2 : something that resembles the standard equipment of a tradesman or business ⟨humor was her *stock-in-trade* as a writer⟩

stock·ish \ˈstä-kish\ *adjective* (1596)
: like a stock **:** STUPID

stock·ist \ˈstä-kist\ *noun* (1910)
British **:** one (as a retailer) that stocks goods

stock·job·ber \ˈstäk-ˌjä-bər\ *noun* (circa 1626)
: one who deals in stocks: as **a :** a member of the London Stock Exchange who deals speculatively with brokers or other jobbers and usually specializes in one class of securities — called also *jobber* **b :** STOCKBROKER — usually used disparagingly

stock·job·bing \-ˌjä-biŋ\ *noun* (1692)
: speculative exchange dealings

stock·keep·er \ˈstäk-ˌkē-pər\ *noun* (circa 1786)
1 : one (as a herdsman or shepherd) having the charge or care of livestock
2 : one that keeps and records stock (as in a warehouse) **:** one that keeps an inventory of goods on hand, shipped, or received

stock·man \-mən, -ˌman\ *noun* (1806)
: one occupied as an owner or worker in the raising of livestock (as cattle or sheep)

stock market *noun* (1809)
1 : STOCK EXCHANGE 1
2 a : a market for particular stocks **b :** the market for stocks throughout a country

stock option *noun* (1945)
1 : an option contract involving stock
2 : a right granted by a corporation to officers or employees as a form of compensation that allows purchase of corporate stock at a fixed price at a specified time with reimbursement derived from the difference between purchase and market prices

1**stock·pile** \ˈstäk-ˌpīl\ *noun* (1872)
: a storage pile: as **a :** a reserve supply of something essential accumulated within a country for use during a shortage **b :** a gradually accumulated reserve of something ⟨avert *stockpiles* of unsold cars —Bert Pierce⟩

2**stockpile** *transitive verb* (1921)
1 : to place or store in or on a stockpile
2 : to accumulate a stockpile of ⟨*stockpile* war materials in Europe —A. O. Wolfers⟩
— **stock·pil·er** *noun*

stock·pot \ˈstäk-ˌpät\ *noun* (1853)
1 : a pot in which soup stock is prepared
2 : an abundant supply **:** REPOSITORY

stock·room \-ˌrüm, -ˌrům\ *noun* (1825)
: a storage place for supplies or goods used in a business

stock saddle *noun* (1886)
: WESTERN SADDLE

stock split *noun* (1950)
: a division of corporate stock by the issuance to existing shareholders of a specified number of new shares with a corresponding lowering of par value for each outstanding share — compare STOCK DIVIDEND

stock–still \ˈstäk-ˈstil\ *adjective* (15th century)
: very still **:** MOTIONLESS ⟨stood *stock-still*⟩

stock·tak·ing \ˈstäk-ˌtā-kiŋ\ *noun* (circa 1858)
1 : INVENTORY 3
2 : the action of estimating a situation at a given moment

stocky \ˈstä-kē\ *adjective* **stock·i·er; -est** (1622)
: compact, sturdy, and relatively thick in build
— **stock·i·ly** \ˈstä-kə-lē\ *adverb*
— **stock·i·ness** \ˈstä-kē-nəs\ *noun*

stock·yard \ˈstäk-ˌyärd\ *noun* (1802)
: a yard for stock; *specifically* **:** one in which transient cattle, sheep, swine, or horses are kept temporarily for slaughter, market, or shipping

1**stodge** \ˈstäj\ *transitive verb* **stodged; stodg·ing** [origin unknown] (1674)
British **:** to stuff full especially with food

2**stodge** *noun* (1823)
British **:** something or someone stodgy

stodgy \ˈstä-jē\ *adjective* **stodg·i·er; -est** (1858)
1 : having a rich filling quality **:** HEAVY ⟨*stodgy* bread⟩
2 : moving in a slow plodding way especially as a result of physical bulkiness
3 : BORING, DULL ⟨out on a peaceful rather *stodgy* Sunday boat trip —Edna Ferber⟩
4 : extremely old-fashioned **:** HIDEBOUND ⟨received a pompously Victorian letter from his *stodgy* father —E. E. S. Montagu⟩
5 a : DRAB **b :** DOWDY
— **stodg·i·ly** \ˈstä-jə-lē\ *adverb*
— **stodg·i·ness** \ˈstä-jē-nəs\ *noun*

sto·gie *or* **sto·gy** \ˈstō-gē\ *noun, plural* **sto·gies** [*Conestoga,* Pa.] (1853)
1 : a stout coarse shoe **:** BROGAN
2 : an inexpensive slender cylindrical cigar; *broadly* **:** CIGAR

1**sto·ic** \ˈstō-ik\ *noun* [Middle English, from Latin *stoicus,* from Greek *stōïkos,* literally, of the portico, from *Stoa* (*Poikilē*) the Painted Portico, portico at Athens where Zeno taught] (14th century)
1 *capitalized* **:** a member of a school of philosophy founded by Zeno of Citium about 300 B.C. holding that the wise man should be free from passion, unmoved by joy or grief, and submissive to natural law
2 : one apparently or professedly indifferent to pleasure or pain ◆

2**stoic** *or* **sto·i·cal** \-i-kəl\ *adjective* (15th century)
1 *capitalized* **:** of, relating to, or resembling the Stoics or their doctrines ⟨*Stoic* logic⟩
2 : not affected by or showing passion or feeling; *especially* **:** firmly restraining response to pain or distress ⟨a *stoic* indifference to cold⟩
synonym see IMPASSIVE
— **sto·ical·ly** \-i-k(ə-)lē\ *adverb*

stoi·chio·met·ric \ˌstȯi-kē-ō-ˈme-trik\ *adjective* (1892)
: of, relating to, used in, or marked by stoichiometry
— **stoi·chio·met·ri·cal·ly** \-tri-k(ə-)lē\ *adverb*

stoi·chi·om·e·try \ˌstȯi-kē-ˈä-mə-trē\ *noun* [Greek *stoicheion* element (from *stoichos* row) + English *-metry;* akin to Greek *steichein* to walk, go — more at STAIR] (1807)
1 : a branch of chemistry that deals with the application of the laws of definite proportions and of the conservation of mass and energy to chemical activity
2 a : the quantitative relationship between constituents in a chemical substance **b :** the quantitative relationship between two or more substances especially in processes involving physical or chemical change

sto·i·cism \ˈstō-ə-ˌsi-zəm\ *noun* (1626)
1 *capitalized* **:** the philosophy of the Stoics
2 : indifference to pleasure or pain **:** IMPASSIVENESS

◇ WORD HISTORY

stoic In Athens about the year 300 B.C. the philosopher Zeno began to teach a radically new doctrine that was to spread its influence throughout the Greco-Roman world for at least the next five hundred years. He gave lectures at a public hall adjacent to the agora (marketplace) called the *Stoa Poikilē* ("Painted Colonnade"). In later years Zeno's philosophical school itself became known as the *Stoa,* and the derivative adjective *stōïkos,* literally, "of the colonnade," was used to refer to the school. Basically what he taught was that happiness and well-being do not depend on material things or on one's situation in life, but on one's reasoning faculty. Through reason, one can emulate the calm and order of the universe by learning to accept events with a stern and tranquil mind. Zeno's Greek followers elaborated and systematized his teachings, which were later adopted and popularized by the Roman Stoics Seneca, Epictetus, and Marcus Aurelius. Greek *stōïkos* became *stoicus* in Latin, and by the 14th century had found a place in English. By the 16th century the word was also being used as a common noun for "one apparently or professedly indifferent to pleasure or pain."

stoke \ˈstōk\ *verb* **stoked; stok·ing** [Dutch *stoken;* akin to Middle Dutch *stuken* to push] (1683)
transitive verb
1 : to poke or stir up (as a fire) **:** supply with fuel
2 : to feed abundantly
intransitive verb
: to stir up or tend a fire (as in a furnace) **:** supply a furnace with fuel

stoked \ˈstōkt\ *adjective* (1965)
slang **:** being in an enthusiastic or exhilarated state

stoke·hold \ˈstōk-ˌhōld\ *noun* (1887)
: the boiler room of a ship

stok·er \ˈstō-kər\ *noun* (1660)
1 : one employed to tend a furnace and supply it with fuel; *specifically* **:** one that tends a marine steam boiler
2 : a machine for feeding a fire

Stokes' aster \ˈstōks-\ *noun* [Jonathan *Stokes* (died 1831) English botanist] (circa 1890)
: a perennial composite herb (*Stokesia laevis*) of the southern U.S. often grown for its large showy heads of usually blue flowers — called also *sto·ke·sia* \stō-ˈkē-zh(ē-)ə, -zē-ə\

[1]stole \ˈstōl\ *past of* STEAL

[2]stole *noun* [Middle English, from Old English, from Latin *stola,* from Greek *stolē* equipment, robe, from *stellein* to set up, make ready] (before 12th century)
1 : a long loose garment **:** ROBE
2 : an ecclesiastical vestment consisting of a long usually silk band worn traditionally around the neck by bishops and priests and over the left shoulder by deacons
3 : a long wide scarf or similar covering worn by women usually across the shoulders

stolen *past participle of* STEAL

stol·id \ˈstä-ləd\ *adjective* [Latin *stolidus* dull, stupid] (circa 1600)
: having or expressing little or no sensibility **:** UNEMOTIONAL
synonym see IMPASSIVE
— sto·lid·i·ty \stä-ˈli-də-tē, stə-\ *noun*
— stol·id·ly \ˈstä-ləd-lē\ *adverb*

stol·len \ˈshtō-lən, ˈshtȯ-, ˈstə-, *or with* s *for* sh\ *noun, plural* **stollen** *or* **stollens** [German, literally, post, support, from Old High German *stollo* — more at STELA] (1906)
: a sweet yeast bread of German origin containing fruit and nuts

sto·lon \ˈstō-lən, -ˌlän\ *noun* [New Latin *stolon-, stolo,* from Latin, branch, sucker; akin to Old English *stela* stalk, Armenian *steɫn* branch] (1601)
1 a : a horizontal branch from the base of a plant that produces new plants from buds at its tip or nodes (as in the strawberry) — called also *runner* **b :** a hypha (as of rhizopus) produced on the surface and connecting a group of conidiophores

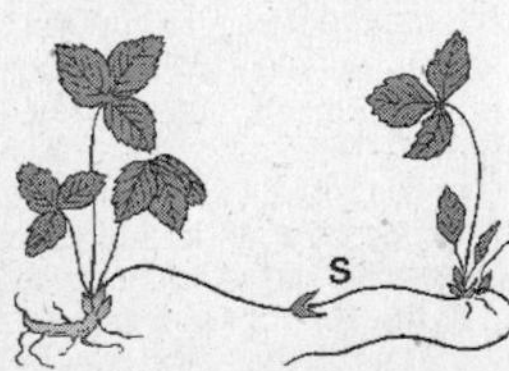

s stolon 1a

2 : an extension of the body wall (as of a hydrozoan or bryozoan) that develops buds giving rise to new zooids which usually remain united by the stolon

sto·lon·if·er·ous \ˌstō-lə-ˈni-f(ə-)rəs\ *adjective* (circa 1777)
: bearing or developing stolons

sto·ma \ˈstō-mə\ *noun, plural* **sto·ma·ta** \-mə-tə\ *also* **stomas** [New Latin, from Greek *stomat-, stoma* mouth] (circa 1684)
1 : any of various small simple bodily openings especially in a lower animal
2 : one of the minute openings in the epidermis of a plant organ (as a leaf) through which gaseous interchange takes place; *also* **:** the opening with its associated cellular structures
3 : an artificial permanent opening especially in the abdominal wall made in surgical procedures

[1]stom·ach \ˈstə-mək, -mik\ *noun* [Middle English *stomak,* from Middle French *estomac,* from Latin *stomachus* gullet, esophagus, stomach, from Greek *stomachos,* from *stoma* mouth; akin to Middle Breton *staffn* mouth, Avestan *staman-*] (14th century)
1 a (1) **:** a dilatation of the alimentary canal of a vertebrate communicating anteriorly with the esophagus and posteriorly with the duodenum (2) **:** one of the compartments of a ruminant stomach ⟨the abomasum is the fourth *stomach* of a ruminant⟩ **b :** a cavity in an invertebrate animal that is analogous to a stomach **c :** the part of the body that contains the stomach **:** BELLY, ABDOMEN
2 a : desire for food caused by hunger **:** APPETITE **b :** INCLINATION, DESIRE ⟨had no *stomach* for an argument⟩; *also* **:** COURAGE, GUTS
3 *obsolete* **a :** SPIRIT, VALOR **b :** PRIDE **c :** SPLEEN, RESENTMENT

[2]stomach *transitive verb* (1523)
1 *archaic* **:** to take offense at
2 : to bear without overt reaction or resentment **:** put up with ⟨couldn't *stomach* office politics⟩

stom·ach·ache \-ˌāk\ *noun* (1763)
: pain in or in the region of the stomach

stom·ach·er \ˈstə-mi-kər, -chər\ *noun* (15th century)
: the center front section of a waist or underwaist or a usually heavily embroidered or jeweled separate piece for the center front of a bodice worn by men and women in the 15th and 16th centuries and later by women only

[1]sto·mach·ic \stə-ˈma-kik\ *adjective* (circa 1656)
: of or relating to the stomach ⟨*stomachic* vessels⟩

[2]stomachic *noun* (1735)
: a stimulant or tonic for the stomach

stom·achy \ˈstə-mə-kē, -mi-\ *adjective* (circa 1825)
1 *dialect British* **:** IRASCIBLE, IRRITABLE
2 : having a large stomach

sto·mal \ˈstō-məl\ *adjective* (circa 1941)
: of, relating to, or situated near a surgical stoma ⟨a *stomal* ulcer⟩

stomat- *or* **stomato-** *combining form* [New Latin, from Greek, from *stomat-, stoma*]
: mouth **:** stoma ⟨*stomat*itis⟩

sto·ma·tal \ˈstō-mə-tᵊl\ *adjective* (1861)
: of, relating to, or constituting plant stomata ⟨*stomatal* openings⟩ ⟨*stomatal* transpiration⟩

sto·mate \ˈstō-ˌmāt\ *noun* [irregular from New Latin *stomat-, stoma*] (1835)
: STOMA 2

sto·ma·ti·tis \ˌstō-mə-ˈtī-təs\ *noun, plural* **-tit·i·des** \-ˈti-tə-ˌdēz\ *or* **-ti·tis·es** \-ˈtī-tə-səz\ [New Latin] (1859)
: any of numerous inflammatory diseases of the mouth

sto·mato·pod \stō-ˈma-tə-ˌpäd\ *noun* [New Latin *Stomatopoda,* from *stomat-* + Greek *pod-, pous* foot — more at FOOT] (1877)
: any of an order (Stomatopoda) of marine crustaceans (as a squilla) that have gills on the abdominal appendages
— stomatopod *adjective*

sto·mo·de·um *or* **sto·mo·dae·um** \ˌstō-mə-ˈdē-əm\ *noun, plural* **-dea** \-ˈde-ə\ *or* **-daea** \-ˈdē-ə\ *also* **-deums** *or* **-daeums** [New Latin, from Greek *stoma* mouth + *hodaion,* neuter of *hodaios* being on the way, from *hodos* way] (1876)
: the embryonic anterior ectodermal part of the alimentary canal or tract
— sto·mo·de·al *or* **sto·mo·dae·al** \-ˈdē-əl\ *adjective*

[1]stomp \ˈstämp, ˈstȯmp\ *verb* [by alteration] (1803)
: STAMP 1, 2

[2]stomp *noun* (circa 1899)
1 : STAMP 4
2 : a jazz dance marked by heavy stamping

stomping ground *noun* (1854)
: STAMPING GROUND

-stomy *noun combining form* [International Scientific Vocabulary, from Greek *stoma* mouth, opening]
: surgical operation establishing a usually permanent opening into (such) a part ⟨entero*stomy*⟩

[1]stone \ˈstōn\ *noun* [Middle English, from Old English *stān;* akin to Old High German *stein* stone, Old Church Slavonic *stěna* wall, and perhaps to Sanskrit *styāyate* it hardens — more at STEATO-] (before 12th century)
1 : a concretion of earthy or mineral matter: **a** (1) **:** such a concretion of indeterminate size or shape (2) **:** ROCK **b :** a piece of rock for a specified function: as (1) **:** a building block (2) **:** a paving block (3) **:** a precious stone **:** GEM (4) **:** GRAVESTONE (5) **:** GRINDSTONE (6) **:** WHETSTONE (7) **:** a surface upon which a drawing, text, or design to be lithographed is drawn or transferred
2 : something resembling a small stone: as **a :** CALCULUS 3a **b :** the hard central portion of a drupaceous fruit (as a peach) **c :** a hard stony seed (as of a date)
3 *plural usually* **stone :** any of various units of weight; *especially* **:** an official British unit equal to 14 pounds (6.3 kilograms)
4 a : CURLING STONE **b :** a round playing piece used in various games (as backgammon or go)
5 : a stand or table with a smooth flat top on which to impose or set type

[2]stone *transitive verb* **stoned; ston·ing** (13th century)
1 : to hurl stones at; *especially* **:** to kill by pelting with stones
2 *archaic* **:** to make hard or insensitive to feeling
3 : to face, pave, or fortify with stones
4 : to remove the stones or seeds of (a fruit)
5 a : to rub, scour, or polish with a stone **b :** to sharpen with a whetstone
— ston·er *noun*

[3]stone *adverb* (13th century)
: ENTIRELY, UTTERLY — used as an intensive; often used in combination ⟨*stone*-broke⟩ ⟨*stone*-cold soup⟩ ⟨*stone*-dead⟩

[4]stone *adjective* (14th century)
1 : of, relating to, or made of stone
2 : ABSOLUTE, UTTER ⟨pure *stone* craziness —Edwin Shrake⟩

Stone Age *noun* (1864)
: the first known period of prehistoric human culture characterized by the use of stone tools

stone–blind \ˈstōn-ˈblīnd\ *adjective* (14th century)
: totally blind

stone·boat \-ˌbōt\ *noun* (1859)
: a flat sledge or drag for transporting heavy articles (as stones)

stone canal *noun* (1887)
: a tube in many echinoderms that contains calcareous deposits and leads from the madreporite to the ring of the water-vascular system surrounding the mouth

stone cell *noun* (1875)
: SCLEREID

stone·chat \-ˌchat\ *noun* [[2]*chat*] (circa 1783)
: an Old World oscine songbird (*Saxicola torquata*) related to the thrushes; *also* **:** any of various related birds (genus *Saxicola*)

stone china *noun* (1823)
: a hard dense opaque feldspathic pottery developed in England; *broadly* **:** IRONSTONE CHINA

stone–cold \ˈstōn-ˈkōld\ *adverb* (1592)
: ABSOLUTELY ⟨*stone-cold* sober⟩

stone crab *noun* (1709)
: a large brownish edible crab (*Menippe mercenaria*) found on the southern coast of the U.S. and in the Caribbean area

stone·crop \ˈstōn-ˌkräp\ *noun* (before 12th century)

1 : SEDUM; *especially* : an Old World creeping evergreen sedum (*Sedum acre*) with pungent fleshy leaves and yellow flowers
2 : any of various plants of the orpine family related to the sedums

stone·cut·ter \-ˌkə-tər\ *noun* (1540)
1 : one that cuts, carves, or dresses stone
2 : a machine for dressing stone
— **stone·cut·ting** \-ˌkə-tiŋ\ *noun*

stoned \ˈstōnd\ *adjective* (1952)
1 : DRUNK 1a
2 : being under the influence of a drug (as marijuana) taken especially for pleasure : HIGH

stone–deaf \-ˈdef\ *adjective* (1837)
: totally deaf

stone–faced \-ˌfāst\ *adjective* (1932)
: showing no emotion : EXPRESSIONLESS

stone·fish \ˈstōn-ˌfish\ *noun* (1896)
: any of several small spiny venomous scorpion fishes (especially genus *Synanceja*) common about coral reefs of the tropical Indo-Pacific

stonefish

stone fly *noun* (15th century)
: any of an order (Plecoptera) of insects with an aquatic carnivorous nymph having gills and an adult having long antennae, two pairs of membranous wings, and usually long cerci

stone fruit *noun* (circa 1534)
: a fruit with a stony endocarp : DRUPE

stone–ground \ˈstōn-ˈgrau̇nd\ *adjective* (1905)
: ground in a buhrstone mill ⟨*stone-ground* flour⟩

stone·ma·son \ˈstōn-ˌmā-sᵊn\ *noun* (1758)
: a mason who builds with stone
— **stone·ma·son·ry** \-rē\ *noun*

stone roller *noun* (1878)
1 : HOG SUCKER
2 : a common cyprinid fish (*Campostoma anomalum*) found especially in clear streams of the central U.S.

stone's throw *noun* (1581)
: a short distance ⟨lives within a *stone's throw* of town⟩

stone·wall \ˈston -ˌwȯl\ (1880)
intransitive verb
1 *chiefly British* : to engage in obstructive parliamentary debate or delaying tactics
2 : to be uncooperative, obstructive, or evasive
transitive verb
: to refuse to comply or cooperate with
— **stone·wall·er** *noun*

stone wall *noun* (before 12th century)
1 : a fence made of stones; *especially* : one built of rough stones without mortar to enclose a field
2 : an immovable block or obstruction (as in public affairs)

stone·ware \-ˌwar, -ˌwer\ *noun* (1683)
: a strong opaque ceramic ware that is high-fired, well vitrified, and nonporous

stone·washed \-ˌwȯsht, -ˌwäsht *also* -ˌwȯrsht *or* -ˌwärsht\ *adjective* (1982)
: subjected to a washing process during manufacture that includes the use of abrasive stones especially to create a softer fabric ⟨*stone-washed* denim jeans⟩

stone·work \-ˌwərk\ *noun* (before 12th century)
1 : a structure or part built of stone : MASONRY
2 : the shaping, preparation, or setting of stone

stone·wort \-ˌwərt, -ˌwȯrt\ *noun* (1816)
: any of various freshwater green algae (order Charales) that have a thallus differentiated into rhizoids and stems with whorls of branchlets and that are often encrusted with calcareous deposits

stony *also* **ston·ey** \ˈstō-nē\ *adjective* **ston·i·er; -est** (before 12th century)
1 : abounding in or having the nature of stone : ROCKY
2 a : insensitive to pity or human feeling : OBDURATE **b** : manifesting no movement or reaction : DUMB, EXPRESSIONLESS **c** : fearfully gripping : PETRIFYING
3 *archaic* : consisting of or made of stones
4 *British* : stone-broke
— **ston·i·ly** \ˈstō-n°l-ē\ *adverb*
— **ston·i·ness** \ˈstō-nē-nəs\ *noun*

stony·heart·ed \ˈstō-nē-ˌhär-təd\ *adjective* (1569)
: UNFEELING, CRUEL

stood *past and past participle of* STAND

[1]stooge \ˈstüj\ *noun* [origin unknown] (1913)
1 a : one who plays a subordinate or compliant role to a principal **b** : PUPPET 3
2 : STRAIGHT MAN
3 : STOOL PIGEON

[2]stooge *intransitive verb* **stooged; stoog·ing** (1939)
: to act as a stooge ⟨congressmen who *stooge* for the oil and mineral interests —*New Republic*⟩

stook \ˈstük, ˈstu̇k\ *noun* [Middle English *stouk;* akin to Old English *stocc* stock — more at STOCK] (15th century)
chiefly British : [1]SHOCK
— **stook** *transitive verb, chiefly British*

[1]stool \ˈstül\ *noun* [Middle English, from Old English *stōl;* akin to Old High German *stuol* chair, Old Church Slavonic *stolŭ* seat, throne] (before 12th century)
1 a : a seat usually without back or arms supported by three or four legs or by a central pedestal **b** : a low bench or portable support for the feet or knees : FOOTSTOOL
2 : a seat used as a symbol of office or authority; *also* : the rank, dignity, office, or rule of a chieftain
3 a : a seat used while defecating or urinating **b** : a discharge of fecal matter
4 a : a stump or group of stumps of a tree especially when producing suckers **b** : a plant crown from which shoots grow out **c** : a shoot or growth from a stool
5 : STOOL PIGEON

[2]stool *intransitive verb* (1770)
: to throw out shoots in the manner of a stool

stool·ie \ˈstü-lē\ *noun* (1924)
: STOOL PIGEON 2

stool pigeon *noun* [probably from the early practice of fastening the decoy bird to a stool] (1836)
1 : a pigeon used as a decoy to draw others within a net
2 : a person acting as a decoy or informer; *especially* : a spy sent into a group to report (as to the police) on its activities

[1]stoop \ˈstüp\ *verb* [Middle English *stoupen,* from Old English *stūpian;* akin to Old English *stēap* steep, deep — more at STEEP] (before 12th century)
intransitive verb
1 a : to bend the body or a part of the body forward and downward sometimes simultaneously bending the knees **b** : to stand or walk with a forward inclination of the head, body, or shoulders
2 : YIELD, SUBMIT
3 a : to descend from a superior rank, dignity, or status **b** : to lower oneself morally
4 a *archaic* : to move down from a height : ALIGHT **b** : to fly or dive down swiftly usually to attack prey
transitive verb
1 : DEBASE, DEGRADE
2 : to bend (a part of the body) forward and downward

[2]stoop *noun* (1571)
1 a : an act of bending the body forward **b** : a temporary or habitual forward bend of the back and shoulders
2 : the descent of a bird especially on its prey
3 : a lowering of oneself

[3]stoop *noun* [Dutch *stoep;* akin to Old English *stæpe* step — more at STEP] (1755)
: a porch, platform, entrance stairway, or small veranda at a house door

stoop·ball \ˈstüp-ˌbȯl\ *noun* (1941)
: a variation of baseball in which a player throws a ball against a stoop or building and runs to base while other players attempt to retrieve the rebound and put the runner out

stoop labor *noun* (1949)
: the hard labor done or required to plant, cultivate, and harvest a crop and especially a crop of vegetables

[1]stop \ˈstäp\ *verb* **stopped; stop·ping** [Middle English *stoppen,* from Old English *-stoppian,* from (assumed) Vulgar Latin *stuppare* to stop with tow, from Latin *stuppa* tow, from Greek *styppē*] (13th century)
transitive verb
1 a : to close by filling or obstructing **b** : to hinder or prevent the passage of **c** : to get in the way of : be wounded or killed by ⟨easy to *stop* a bullet along a lonely . . . road —Harvey Fergusson⟩
2 a : to close up or block off (an opening) : PLUG **b** : to make impassable : CHOKE, OBSTRUCT **c** : to cover over or fill in (a hole or crevice)
3 a : to cause to give up or change a course of action **b** : to keep from carrying out a proposed action : RESTRAIN, PREVENT
4 a : to cause to cease : CHECK, SUPPRESS **b** : DISCONTINUE
5 a : to deduct or withhold (a sum due) **b** : to instruct one's bank to refuse (payment) or refuse payment of (as a check)
6 a : to arrest the progress or motion of : cause to halt ⟨*stopped* the car⟩ **b** : PARRY **c** : to check by means of a weapon : BRING DOWN, KILL **d** : to beat in a boxing match by a knockout; *broadly* : DEFEAT **e** : BAFFLE, NONPLUS
7 : to change the pitch of (as a violin string) by pressing with the finger or (as a wind instrument) by closing one or more finger holes or by thrusting the hand or a mute into the bell
8 : to hold an honor card and enough protecting cards to be able to block (a bridge suit) before an opponent can run many tricks
intransitive verb
1 a : to cease activity or operation **b** : to come to an end especially suddenly : CLOSE, FINISH
2 a : to cease to move on : HALT **b** : PAUSE, HESITATE
3 a : to break one's journey : STAY **b** *chiefly British* : REMAIN **c** : to make a brief call : drop in
4 : to become choked : CLOG ☆

☆ SYNONYMS
Stop, cease, quit, discontinue, desist mean to suspend or cause to suspend activity. STOP applies to action or progress or to what is operating or progressing and may imply suddenness or definiteness ⟨*stopped* at the red light⟩. CEASE applies to states, conditions, or existence and may add a suggestion of gradualness and a degree of finality ⟨by nightfall the fighting had *ceased*⟩. QUIT may stress either finality or abruptness in stopping or ceasing ⟨the engine faltered, sputtered, then *quit* altogether⟩. DISCONTINUE applies to the stopping of an accustomed activity or practice ⟨we have *discontinued* the manufacture of that item⟩. DESIST implies forbearance or restraint as a motive for stopping or ceasing ⟨*desisted* from further efforts to persuade them⟩.

— **stop·pa·ble** \'stä-pə-bəl\ *adjective*

²stop *noun* (15th century)
1 a : CESSATION, END **b :** a pause or breaking off in speech
2 a (1) **:** a graduated set of organ pipes of similar design and tone quality (2) **:** a corresponding set of vibrators or reeds of a reed organ (3) **:** STOP KNOB — often used figuratively in phrases like *pull out all the stops* to suggest holding nothing back **b :** a means of regulating the pitch of a musical instrument
3 a : something that impedes, obstructs, or brings to a halt **:** IMPEDIMENT, OBSTACLE **b :** the aperture of a camera lens; *also* **:** a marking of a series (as of f-numbers) on a camera for indicating settings of the diaphragm **c :** a drain plug **:** STOPPER
4 : a device for arresting or limiting motion
5 : the act of stopping **:** the state of being stopped **:** CHECK
6 a : a halt in a journey **:** STAY ⟨made a brief *stop* to refuel⟩ **b :** a stopping place ⟨a bus *stop*⟩
7 a *chiefly British* **:** any of several punctuation marks **b** — used in telegrams and cables to indicate a period **c :** a pause or break in a verse that marks the end of a grammatical unit
8 a : an order stopping payment (as of a check or note) by a bank **b :** STOP ORDER
9 : a consonant characterized by complete closure of the breath passage in the course of articulation — compare CONTINUANT
10 : a depression in the face of an animal at the junction of forehead and muzzle

³stop *adjective* (1594)
: serving to stop **:** designed to stop ⟨*stop* line⟩ ⟨*stop* signal⟩

stop–and–go \,stäp-ən-'gō, -ᵊm-, *attributively* -,gō\ *adjective* (1925)
: of, relating to, or involving frequent stops; *especially* **:** controlled or regulated by traffic lights ⟨*stop-and-go* driving⟩

stop bath *noun* (1898)
: an acid bath used to check photographic development of a negative or print

stop·cock \'stäp-,käk\ *noun* (1584)
: a cock for stopping or regulating flow (as through a pipe)

stop down *transitive verb* (circa 1891)
: to reduce the effective aperture of (a lens) by means of a diaphragm

stope \'stōp\ *noun* [probably from Low German *stope,* literally, step; akin to Old English *stæpe* step — more at STEP] (1747)
: a usually steplike excavation underground for the removal of ore that is formed as the ore is mined in successive layers

stop·gap \'stäp-,gap\ *noun, often attributive* (1684)
: something that serves as a temporary expedient **:** MAKESHIFT ⟨*stopgap* measures⟩
synonym see RESOURCE

stop knob *noun* (1887)
: one of the handles by which an organist draws or shuts off a particular stop

stop·light \'stäp-,līt\ *noun* (1926)
1 : a light on the rear of a motor vehicle that is illuminated when the driver presses the brake pedal
2 : TRAFFIC SIGNAL

stop order *noun* (circa 1891)
: an order to a broker to buy or sell respectively at the market when the price of a security advances or declines to a designated level

stop out *intransitive verb* [after *drop out*] (1973)
: to withdraw temporarily from enrollment at a college or university
— **stop–out** \'stäp-,au̇t\ *noun*

stop·over \'stäp-,ō-vər\ *noun* (1885)
1 : a stop at an intermediate point in one's journey
2 : a stopping place on a journey

stop·page \'stä-pij\ *noun* (15th century)
: the act of stopping **:** the state of being stopped **:** HALT, OBSTRUCTION

stop payment *noun* (circa 1919)
: a depositor's order to a bank to refuse to honor a specified check drawn by him or her

¹stop·per \'stä-pər\ *noun* (15th century)
1 : one that brings to a halt or causes to stop operating or functioning **:** CHECK: as **a :** a playing card that will stop the running of a suit **b :** a baseball pitcher depended on to win important games or to stop a losing streak; *also* **:** an effective relief pitcher
2 : one that closes, shuts, or fills up; *specifically* **:** something (as a bung or cork) used to plug an opening

²stopper *transitive verb* **stop·pered; stop·per·ing** \-p(ə-)riŋ\ (circa 1769)
: to close or secure with or as if with a stopper

stopper knot *noun* (1860)
: a knot used to prevent a rope from passing through a hole or opening

¹stop·ple \'stä-pəl\ *noun* [Middle English *stoppell,* from *stoppen* to stop] (14th century)
: something that closes an aperture **:** STOPPER, PLUG

²stopple *transitive verb* **stop·pled; stop·pling** \-p(ə-)liŋ\ (1795)
: STOPPER

stop·watch \'stäp-,wäch\ *noun* (1737)
: a watch with a hand or a digital display that can be started and stopped at will for exact timing (as of a race)

stor·age \'stōr-ij, 'stȯr-\ *noun* (circa 1613)
1 a : space or a place for storing **b :** an amount stored **c :** MEMORY 4
2 a : the act of storing **:** the state of being stored; *especially* **:** the safekeeping of goods in a depository (as a warehouse) **b :** the price charged for keeping goods in a storehouse
3 : the production by means of electric energy of chemical reactions that when allowed to reverse themselves generate electricity again without serious loss

storage battery *noun* (1881)
: a cell or connected group of cells that converts chemical energy into electrical energy by reversible chemical reactions and that may be recharged by passing a current through it in the direction opposite to that of its discharge — called also *storage cell*

sto·rax \'stōr-,aks, 'stȯr-\ *noun* [Middle English, from Late Latin, alteration of Latin *styrax,* from Greek] (14th century)
1 a : a fragrant balsam obtained from the bark of an Asian tree (*Liquidambar orientalis*) of the witch-hazel family that is used as an expectorant and sometimes in perfumery — called also *Levant storax* **b :** a balsam from the sweet gum that is similar to storax
2 : any of a genus (*Styrax* of the family Styracaceae, the storax family) of trees or shrubs with usually hairy leaves and white flowers in drooping racemes — compare BENZOIN

¹store \'stōr, 'stȯr\ *transitive verb* **stored; stor·ing** [Middle English, from Old French *estorer* to construct, restore, store, from Latin *instaurare* to renew, restore] (13th century)
1 : LAY AWAY, ACCUMULATE ⟨*store* vegetables for winter use⟩ ⟨an organism that absorbs and *stores* DDT⟩
2 : FURNISH, SUPPLY; *especially* **:** to stock against a future time ⟨*store* a ship with provisions⟩
3 : to place or leave in a location (as a warehouse, library, or computer memory) for preservation or later use or disposal
4 : to provide storage room for **:** HOLD ⟨elevators for *storing* surplus wheat⟩
— **stor·able** \'stōr-ə-bəl, 'stȯr-\ *adjective*

²store *noun* (13th century)
1 a : something that is stored or kept for future use **b** *plural* **:** articles (as of food) accumulated for some specific object and drawn upon as needed **:** STOCK, SUPPLIES **c :** something that is accumulated **d :** a source from which things may be drawn as needed **:** a reserve fund
2 : STORAGE — usually used with *in* ⟨when placing eggs in *store* —*Dublin Sunday Independent*⟩
3 : VALUE, IMPORTANCE ⟨set great *store* by a partner's opinion⟩
4 : a large quantity, supply, or number **:** ABUNDANCE
5 a : STOREHOUSE, WAREHOUSE **b** *chiefly British* **:** MEMORY 4
6 : a business establishment where usually diversified goods are kept for retail sale ⟨grocery *store*⟩ — compare SHOP
— **in store :** in readiness **:** in preparation ⟨there's a surprise *in store* for you⟩

³store *adjective* (1602)
1 *or* **stores :** of, relating to, kept in, or used for a store
2 : purchased from a store as opposed to being natural or homemade **:** MANUFACTURED, READY-MADE ⟨*store* clothes⟩ ⟨*store* bread⟩

store–bought \'stōr-,bȯt, 'stȯr-\ *adjective* (1905)
: STORE 2

store cheese *noun* [from its being a staple article stocked in grocery stores] (1863)
: CHEDDAR

¹store·front \'stōr-,frənt, 'stȯr-\ *noun* (1880)
1 : the front side of a store or store building facing a street
2 : a building, room, or suite of rooms having a storefront

²storefront *adjective* (1937)
1 : of, relating to, or characteristic of a storefront church ⟨a *storefront* evangelist⟩
2 : occupying a room or suite of rooms in a store building at street level and immediately behind a storefront ⟨a *storefront* school⟩
3 : of, relating to, or being outreach professional services ⟨*storefront* lawyers⟩ ⟨a *storefront* clinic⟩

storefront church *noun* (1937)
: a city church that utilizes storefront quarters as a meeting place

store·house \'stōr-,hau̇s, 'stȯr-\ *noun* (14th century)
1 : a building for storing goods (as provisions) **:** MAGAZINE, WAREHOUSE
2 : an abundant supply or source **:** REPOSITORY

store·keep·er \-,kē-pər\ *noun* (1618)
1 : one that has charge of supplies (as military stores)
2 : one that operates a retail store

store·room \-,rüm, -,ru̇m\ *noun* (1746)
1 : a room or space for the storing of goods or supplies
2 : STOREHOUSE 2

store·ship \-,ship\ *noun* (1693)
: a ship used to carry supplies

store·wide \-'wīd\ *adjective* (circa 1937)
: including all or most merchandise in a store ⟨a *storewide* sale⟩

¹sto·ried \'stōr-ēd, 'stȯr-\ *adjective* (14th century)
1 : decorated with designs representing scenes from story or history ⟨a *storied* tapestry⟩
2 : having an interesting history **:** celebrated in story or history ⟨a *storied* institution⟩

²storied *or* **sto·reyed** \'stōr-ēd, 'stȯr-\ *adjective* (1624)
: having stories — used in combination ⟨a two-*storied* house⟩

stork \'stȯrk\ *noun* [Middle English, from Old English *storc;* akin to Old High German *storah* stork and probably to Old English *stearc* stiff — more at STARK] (before 12th century)
: any of various large mostly Old World wading birds (family Ciconiidae) that have long stout bills and are related to the ibises and herons

storks·bill \'stȯrks-,bil\ *noun* (1562)
: any of several plants of the geranium family with elongate beaked fruits: **a :** PELARGONIUM **b :** ALFILARIA; *also* **:** any of several related plants (genus *Erodium*)

¹storm \'stȯrm\ *noun, often attributive* [Middle English, from Old English; akin to Old

stork

High German *sturm* storm, Old English *styrian* to stir] (before 12th century)
1 a : a disturbance of the atmosphere marked by wind and usually by rain, snow, hail, sleet, or thunder and lightning **b :** a heavy fall of rain, snow, or hail **c** (1) **:** wind having a speed of 64 to 72 miles (103 to 116 kilometers) per hour (2) **:** WHOLE GALE — see BEAUFORT SCALE table **d :** a serious disturbance of any element of nature
2 : a disturbed or agitated state **:** a sudden or violent commotion
3 : a heavy discharge of objects (as missiles)
4 : a tumultuous outburst
5 a : PAROXYSM, CRISIS **b :** a sudden heavy influx or onset
6 : a violent assault on a defended position
7 *plural* **:** STORM WINDOW
— by storm : by or as if by employing a bold swift frontal movement especially with the intent of defeating or winning over quickly ⟨took the literary world *by storm*⟩
— up a storm : in a remarkable or energetic fashion — used as an intensifier ⟨dancing *up a storm*⟩ ⟨can write *up a storm*⟩

[2]storm (15th century)
intransitive verb
1 a : to blow with violence **b :** to rain, hail, snow, or sleet vigorously
2 : to attack by storm ⟨*stormed* ashore at zero hour⟩
3 : to be in or to exhibit a violent passion **:** RAGE ⟨*storming* at the unusual delay⟩
4 : to rush about or move impetuously, violently, or angrily ⟨the mob *stormed* through the streets⟩
transitive verb
: to attack, take, or win over by storm ⟨*storm* a fort⟩
synonym see ATTACK

storm and stress *noun, often both Ss capitalized* (1855)
: STURM UND DRANG

storm·bound \'stȯrm-ˌbau̇nd\ *adjective* (1830)
: cut off from outside communication by a storm or its effects **:** stopped or delayed by storms

storm cellar *noun* (circa 1902)
: a cellar or covered excavation designed for protection from dangerous windstorms (as tornadoes)

storm door *noun* (1878)
: an additional door placed outside an ordinary outside door for protection against severe weather

storm petrel *noun* (circa 1833)
: any of various small petrels (family Hydrobatidae), *especially* **:** a small sooty black white-marked petrel (*Hydrobates pelagicus*) frequenting the north Atlantic and Mediterranean

storm petrel

storm trooper *noun* (1933)
1 : a member of a private Nazi army notorious for aggressiveness, violence, and brutality
2 : one that resembles a Nazi storm trooper

storm window *noun* (circa 1888)
: a sash placed outside an ordinary window as a protection against severe weather — called also *storm sash*

stormy \'stȯr-mē\ *adjective* **storm·i·er; -est** (12th century)
1 : relating to, characterized by, or indicative of a storm ⟨a *stormy* day⟩ ⟨a *stormy* autumn⟩
2 : marked by turmoil or fury ⟨a *stormy* life⟩ ⟨a *stormy* conference⟩
— storm·i·ly \'stȯr-mə-lē\ *adverb*
— storm·i·ness \-mē-nəs\ *noun*

stormy petrel *noun* (circa 1776)
1 : STORM PETREL
2 a : one fond of strife **b :** a harbinger of trouble

[1]sto·ry \'stōr-ē, 'stȯr-\ *noun, plural* **stories** [Middle English *storie,* from Old French *estorie,* from Latin *historia* — more at HISTORY] (13th century)
1 *archaic* **:** HISTORY 1, 3
2 a : an account of incidents or events **b :** a statement regarding the facts pertinent to a situation in question **c :** ANECDOTE; *especially* **:** an amusing one
3 a : a fictional narrative shorter than a novel; *specifically* **:** SHORT STORY **b :** the intrigue or plot of a narrative or dramatic work
4 : a widely circulated rumor
5 : LIE, FALSEHOOD
6 : LEGEND, ROMANCE
7 : a news article or broadcast
8 : MATTER, SITUATION

[2]story *transitive verb* **sto·ried; sto·ry·ing** (15th century)
1 *archaic* **:** to narrate or describe in story
2 : to adorn with a story or a scene from history

[3]story *also* **sto·rey** \'stōr-ē, 'stȯr-ē\ *noun, plural* **stories** *also* **storeys** [Middle English *storie,* from Medieval Latin *historia* picture, story of a building, from Latin, history, tale; probably from pictures adorning the windows of medieval buildings] (14th century)
1 a : the space in a building between two adjacent floor levels or between a floor and the roof **b :** a set of rooms in such a space **c :** a unit of measure equal to the height of the story of a building ⟨one *story* high⟩
2 : a horizontal division of a building's exterior not necessarily corresponding exactly with the stories within

sto·ry·board \-ˌbōrd, -ˌbȯrd\ *noun* (1942)
: a panel or series of panels on which a set of sketches is arranged depicting consecutively the important changes of scene and action in a series of shots (as for a film, television show, or commercial)
— storyboard *transitive verb*

[1]sto·ry·book \-ˌbu̇k\ *noun* (1711)
: a book of stories ⟨*storybooks* for children⟩

[2]storybook *adjective* (1844)
: FAIRY-TALE

story line *noun* (1941)
: the plot of a story or drama

sto·ry·tell·er \'stōr-ē-ˌte-lər, 'stȯr-\ *noun* (1709)
: a teller of stories: as **a :** a relater of anecdotes **b :** a reciter of tales (as in a children's library) **c :** LIAR, FIBBER **d :** a writer of stories
— sto·ry·tell·ing *noun*

stoss \'stäs, 'stȯs, 'stōs, 'shtōs\ *adjective* [German *stoss-,* from *stossen* to push, from Old High German *stōzen;* akin to Gothic *stautan* to strike — more at CONTUSION] (1878)
: facing toward the direction from which an overriding glacier impinges ⟨the *stoss* slope of a hill⟩

sto·tin·ka \stō-'tiŋ-kə, stə-\ *noun, plural* **-tin·ki** \-kē\ [Bulgarian] (circa 1892)
— see *lev* at MONEY table

stound \'stau̇nd, 'stünd\ *noun* [Middle English, from Old English *stund;* akin to Old High German *stunta* time, hour] (before 12th century)
archaic **:** TIME, WHILE

stoup \'stüp\ *noun* [Middle English *stowp,* probably of Scandinavian origin; akin to Old Norse *staup* cup] (14th century)
1 a : a beverage container (as a glass or tankard) **b :** FLAGON
2 : a basin for holy water at the entrance of a church

[1]stour \'stu̇r\ *adjective* [Middle English *stor,* from Old English *stōr;* akin to Old High German *stuori* large, Russian *staryĭ* old, Old English *standan* to stand] (before 12th century)
1 *chiefly Scottish* **:** STRONG, HARDY
2 *chiefly Scottish* **:** STERN, HARSH

[2]stour *noun* [Middle English, from Middle French *estour,* of Germanic origin; akin to Old High German *sturm* storm, battle — more at STORM] (14th century)
1 a *archaic* **:** BATTLE, CONFLICT **b** *dialect British* **:** TUMULT, UPROAR
2 *chiefly Scottish* **:** DUST, POWDER

[1]stout \'stau̇t\ *adjective* [Middle English, from Middle French *estout,* of Germanic origin; akin to Old High German *stolz* proud; perhaps akin to Old High German *stelza* stilt — more at STILT] (14th century)
1 : strong of character: as **a :** BRAVE, BOLD **b :** FIRM, DETERMINED; *also* **:** OBSTINATE, UNCOMPROMISING
2 : physically or materially strong: **a :** STURDY, VIGOROUS **b :** STAUNCH, ENDURING **c :** sturdily constructed **:** SUBSTANTIAL
3 : FORCEFUL ⟨a *stout* attack⟩; *also* **:** VIOLENT ⟨a *stout* wind⟩
4 : bulky in body **:** FAT
synonym see STRONG
— stout·ish \'stau̇-tish\ *adjective*
— stout·ly *adverb*
— stout·ness *noun*

[2]stout *noun* (1677)
1 : a very dark full-bodied ale with a distinctive malty flavor
2 a : a fat person **b :** a clothing size designed for the large figure

stout·en \'stau̇-tᵊn\ *verb* **stout·ened; stout·en·ing** \'stau̇t-niŋ, 'stau̇-tᵊn-iŋ\ (1834)
transitive verb
: to make stout ⟨*stouten* a resolve⟩
intransitive verb
: to become stout

stout·heart·ed \'stau̇t-'här-təd\ *adjective* (1552)
: having a stout heart or spirit: **a :** COURAGEOUS **b :** STUBBORN
— stout·heart·ed·ly *adverb*
— stout·heart·ed·ness *noun*

[1]stove \'stōv\ *noun* [Middle English, heated room, steam room, from Middle Dutch or Middle Low German, from (assumed) Vulgar Latin *extufa,* ultimately from Latin *ex-* + Greek *typhein* to smoke — more at DEAF] (1591)
1 a : a portable or fixed apparatus that burns fuel or uses electricity to provide heat (as for cooking or heating) **b :** a device that generates heat for special purposes (as for heating tools or heating air for a hot blast) **c :** KILN
2 *chiefly British* **:** a hothouse especially for the cultivation of tropical exotics; *broadly* **:** GREENHOUSE

[2]stove *past and past participle of* STAVE

stove·pipe \'stōv-ˌpīp\ *noun* (1699)
1 : pipe of large diameter usually of sheet steel used as a stove chimney or to connect a stove with a flue
2 : SILK HAT

sto·ver \'stō-vər\ *noun* [Middle English, modification of Anglo-French *estovers* necessary supplies, from Old French *estoveir* to be necessary, ultimately from Latin *est opus* there is need] (14th century)

1 *chiefly dialect English* **:** FODDER
2 **:** mature cured stalks of grain with the ears removed that are used as feed for livestock
stow \'stō\ *transitive verb* [Middle English, to place, from *stowe* place, from Old English *stōw;* akin to Old Frisian *stō* place, Greek *stylos* pillar — more at STEER] (14th century)
1 : HOUSE, LODGE
2 a : to put away for future use **:** STORE **b** *obsolete* **:** to lock up for safekeeping **:** CONFINE
3 a : to dispose in an orderly fashion **:** ARRANGE, PACK **b :** LOAD
4 *slang* **:** to put aside **:** STOP
5 a *archaic* **:** CROWD **b :** to eat or drink up **:** CONSUME — usually used with *away* ⟨*stowed* away a huge dinner⟩
stow·age \-ij\ *noun* (14th century)
1 a : an act or process of stowing **b :** goods in storage or to be stowed
2 a : storage capacity **b :** a place or receptacle for storage
3 : the state of being stored
stow·away \'stō-ə-ˌwā\ *noun* (1850)
: one that stows away
stow away *intransitive verb* (1879)
: to secrete oneself aboard a vehicle as a means of obtaining transportation
STP \ˌes-ˌtē-'pē\ *noun* [probably from *STP,* a trademark for a motor fuel additive] (1967)
: a psychedelic drug chemically related to mescaline and amphetamine
stra·bis·mus \strə-'biz-məs\ *noun* [New Latin, from Greek *strabismos* condition of squinting, from *strabizein* to squint, from *strabos* squint-eyed; akin to Greek *strephein* to twist] (circa 1684)
: inability of one eye to attain binocular vision with the other because of imbalance of the muscles of the eyeball — called also *squint*
— stra·bis·mic \-mik\ *adjective*
[1]**strad·dle** \'stra-dᵊl\ *verb* **strad·dled; strad·dling** \'strad-liŋ, 'stra-dᵊl-iŋ\ [irregular from *stride*] (1565)
intransitive verb
1 : to stand, sit, or walk with the legs wide apart; *especially* **:** to sit astride
2 : to spread out irregularly **:** SPRAWL
3 : to favor or seem to favor two apparently opposite sides
4 : to execute a commodities market spread
transitive verb
1 : to stand, sit, or be astride of ⟨*straddle* a horse⟩
2 : to be noncommittal in regard to ⟨*straddle* an issue⟩
— strad·dler \'strad-lər, 'stra-dᵊl-ər\ *noun*
— straddle the fence : to be in a position of neutrality or indecision
[2]**straddle** *noun* (1611)
1 : the act or position of one who straddles
2 : a noncommittal or equivocal position
3 : SPREAD 5
Stra·di·va·ri \ˌstra-də-'vär-ē, -'var-, -'ver-\ *noun* (circa 1903)
: STRADIVARIUS
Strad·i·var·i·us \ˌstra-də-'var-ē-əs, -'ver-\ *noun, plural* **-var·ii** \-ē-ˌī\ [Latinized form of *Stradivari*] (1833)
: a stringed instrument (as a violin) made by Antonio Stradivari of Cremona
strafe \'strāf, *especially British* 'sträf\ *transitive verb* **strafed; straf·ing** [German *Gott strafe England* God punish England, German propaganda slogan during World War I] (1915)
: to rake (as ground troops) with fire at close range and especially with machine-gun fire from low-flying aircraft
— strafe *noun*
— straf·er *noun*
[1]**strag·gle** \'stra-gəl\ *intransitive verb* **strag·gled; strag·gling** \-g(ə-)liŋ\ [Middle English *straglen*] (15th century)
1 : to wander from the direct course or way **:** ROVE, STRAY
2 : to trail off from others of its kind ⟨little cabins *straggling* off into the woods⟩
— strag·gler \-g(ə-)lər\ *noun*
[2]**straggle** *noun* (1865)
: a straggling group (as of persons or objects)
strag·gly \'stra-g(ə-)lē\ *adjective* **strag·gli·er; -est** (1862)
: spread out or scattered irregularly ⟨a *straggly* beard⟩
[1]**straight** \'strāt\ *adjective* [Middle English *streght, straight,* from past participle of *strecchen* to stretch — more at STRETCH] (14th century)
1 a : free from curves, bends, angles, or irregularities ⟨*straight* hair⟩ ⟨*straight* timber⟩ **b :** generated by a point moving continuously in the same direction and expressed by a linear equation ⟨a *straight* line⟩ ⟨the *straight* segment of a curve⟩
2 a : lying along or holding to a direct or proper course or method ⟨a *straight* thinker⟩ **b :** CANDID, FRANK ⟨a *straight* answer⟩ **c :** coming directly from a trustworthy source ⟨a *straight* tip on the horses⟩ **d** (1) **:** having the elements in an order ⟨the *straight* sequence of events⟩ (2) **:** CONSECUTIVE ⟨12 *straight* days⟩ **e :** having the cylinders arranged in a single straight line ⟨a *straight* 8-cylinder engine⟩ **f :** PLUMB, VERTICAL ⟨the picture isn't quite *straight*⟩
3 a : exhibiting honesty and fairness ⟨*straight* dealing⟩ **b :** properly ordered or arranged ⟨set the kitchen *straight*⟩ ⟨set us *straight* on that issue⟩; *also* **:** CORRECT ⟨get the facts *straight*⟩ **c :** free from extraneous matter **:** UNMIXED ⟨*straight* whiskey⟩ **d :** marked by no exceptions or deviations in support of a principle or party ⟨votes a *straight* Democratic ticket⟩ **e :** having a fixed price for each regardless of the number sold **f :** not deviating from an indicated pattern ⟨writes *straight* humor⟩ ⟨a *straight*-A student⟩ **g** (1) **:** exhibiting no deviation from what is established or accepted as usual, normal, or proper **:** CONVENTIONAL; *also* **:** SQUARE 5f (2) **:** not using or under the influence of drugs or alcohol **h :** HETEROSEXUAL
4 : being the only form of remuneration ⟨on *straight* commission⟩
— straight·ish \'strā-tish\ *adjective*
— straight·ly *adverb*
— straight·ness *noun*
[2]**straight** *adverb* (14th century)
: in a straight manner
[3]**straight** *transitive verb* (15th century)
chiefly Scottish **:** STRAIGHTEN
[4]**straight** *noun* (1645)
1 : something that is straight: as **a :** a straight line or arrangement **b :** STRAIGHTAWAY; *especially* **:** HOMESTRETCH **c :** a true or honest report or course
2 a : a sequence (as of shots, strokes, or moves) resulting in a perfect score in a game or contest **b :** first place at the finish of a horse race **:** WIN
3 : a poker hand containing five cards in sequence but not of the same suit — see POKER illustration
4 : a person who adheres to conventional attitudes and mores
straight–ahead \'strāt-ə-'hed\ *adjective* (1836)
: relating to or being music performed in an unembellished manner typical of a given idiom or performer; *broadly* **:** STRAIGHTFORWARD
straight and narrow *noun* [probably alteration of *strait and narrow;* from the admonition of Matthew 7:14 (Authorized Version), "strait is the gate and narrow is the way which leadeth unto life"] (1930)
: the way of propriety and rectitude — used with *the*
straight angle *noun* (1601)
: an angle whose sides lie in opposite directions from the vertex in the same straight line and which equals two right angles
straight–arm \'strāt-ˌärm\ *noun* (1903)
: an act or instance of warding off a football tackler with the arm fully extended from the shoulder, elbow locked, and the palm of the hand placed firmly against any part of his body — called also *stiff-arm*
— straight–arm *verb*
straight–arrow \-'ar-(ˌ)ō\ *adjective* [from the expression *straight as an arrow*] (circa 1969)
: rigidly proper and conventional
[1]**straight·away** \ˌstrāt-ə-'wā\ *adverb* (1662)
: without hesitation or delay
[2]**straight·away** \'strāt-ə-ˌwā\ *adjective* (1874)
1 : proceeding in a straight line **:** continuous in direction
2 : IMMEDIATE
[3]**straight·away** \'strā-tə-ˌwā\ *noun* (1878)
: a straight course: as **a :** the straight part of a closed racecourse **:** STRETCH **b :** a straight and unimpeded stretch of road or way
straight·bred \'strāt-'bred\ *adjective* (1898)
: produced by breeding a single breed, strain, or type ⟨*straightbred* cattle⟩
— straight·bred \-ˌbred\ *noun*
straight chain *noun* (1890)
: an open chain of atoms having no side chains — usually hyphenated when used attributively
straight·edge \'strāt-ˌej\ *noun* (1812)
: a bar or piece of material (as of wood, metal, or plastic) with a straight edge for testing straight lines and surfaces or drawing straight lines
straight·en \'strā-tᵊn\ *verb* **straight·ened; straight·en·ing** \'strāt-niŋ, 'strā-tᵊn-iŋ\ (1542)
transitive verb
: to make straight — usually used with *up* or *out*
intransitive verb
: to become straight — usually used with *up* or *out*
word history see STRAITEN
— straight·en·er \'strāt-nər, 'strā-tᵊn-ər\ *noun*
straight face *noun* (circa 1890)
: a face giving no evidence of emotion and especially of merriment
— straight–faced \'strāt-'fāst\ *adjective*
— straight–faced·ly \-'fā-səd-lē, -'fāst-lē\ *adverb*
straight flush *noun* (1864)
: a poker hand containing five cards of the same suit in sequence — see POKER illustration
[1]**straight·for·ward** \ˌstrāt-'fȯr-wərd, 'strāt-ˌ\ *adjective* (1806)
1 a : free from evasiveness or obscurity **:** EXACT, CANDID ⟨a *straightforward* account⟩ **b :** CLEAR-CUT, PRECISE
2 : proceeding in a straight course or manner **:** DIRECT, UNDEVIATING
— straight·for·ward·ly *adverb*
— straight·for·ward·ness *noun*
[2]**straightforward** *also* **straight·for·wards** \-wərdz\ *adverb* (1809)
: in a straightforward manner
straight–line \'strāt-'līn\ *adjective* (1843)
1 : being a mechanical linkage or equivalent device designed to produce or copy motion in a straight line
2 : having the principal parts arranged in a straight line
3 : marked by a uniform spread and especially in equal segments over a given term ⟨*straight-line* amortization⟩ ⟨*straight-line* depreciation⟩
4 : occurring, measured, or made in or along a straight line ⟨*straight-line* motion⟩ ⟨*straight-line* extrapolation⟩
straight man *noun* (1923)
: a member of a comedy team who feeds lines to his partner who in turn replies with usually humorous quips
straight off *adverb* (1873)
: at once **:** IMMEDIATELY
straight–out \'strāt-'au̇t\ *adjective* (1848)

1 : FORTHRIGHT, BLUNT ⟨gave him a *straight-out* answer⟩
2 : OUTRIGHT, THOROUGHGOING
— straight–out *adverb*

straight poker *noun* (1864)
: poker in which the players bet on the five cards dealt to them and have a showdown without drawing — compare DRAW POKER, STUD POKER

straight razor *noun* (1938)
: a razor with a rigid steel cutting blade hinged to a case that forms a handle when the razor is open for use

straight·way \'strāt-'wā, -,wā\ *adverb* (15th century)
1 : in a direct course **:** DIRECTLY ⟨fell *straightway* down the stairs⟩
2 : RIGHT AWAY, IMMEDIATELY, STRAIGHTAWAY ⟨*straightway* the clouds began to part⟩

[1]strain \'strān\ *noun* [Middle English *streen* progeny, lineage, from Old English *strēon* gain, acquisition; akin to Old High German gi*striuni* gain, Latin *struere* to heap up — more at STREW] (13th century)
1 a : LINEAGE, ANCESTRY **b :** a group of presumed common ancestry with clear-cut physiological but usually not morphological distinctions ⟨a high-yielding *strain* of winter wheat⟩; *broadly* **:** a specified infraspecific group (as a stock, line, or ecotype) **c :** KIND, SORT ⟨discussions of a lofty *strain*⟩
2 a : inherited or inherent character, quality, or disposition ⟨a *strain* of madness in the family⟩ **b :** TRACE, STREAK ⟨a *strain* of fanaticism⟩
3 a : TUNE, AIR **b :** a passage of verbal or musical expression **c :** a stream or outburst of forceful or impassioned speech
4 a : the tenor, pervading note, burden, or tone of an utterance or of a course of action or conduct **b :** MOOD, TEMPER

[2]strain *verb* [Middle English, from Middle French *estraindre,* from Latin *stringere* to bind or draw tight, press together; akin to Greek *strang-, stranx* drop squeezed out, *strangalē* halter] (14th century)
transitive verb
1 a : to draw tight **:** cause to fit firmly ⟨*strain* the bandage over the wound⟩ **b :** to stretch to maximum extension and tautness ⟨*strain* a canvas over a frame⟩
2 a : to exert (as oneself) to the utmost **b :** to injure by overuse, misuse, or excessive pressure ⟨*strained* his back⟩ **c :** to cause a change of form or size in (a body) by application of external force
3 : to squeeze or clasp tightly: as **a :** HUG **b :** to compress painfully **:** CONSTRICT
4 a : to cause to pass through a strainer **:** FILTER **b :** to remove by straining ⟨*strain* lumps out of the gravy⟩
5 : to stretch beyond a proper limit ⟨that story *strains* my credulity⟩
6 *obsolete* **:** to squeeze out **:** EXTORT
intransitive verb
1 a : to make violent efforts **:** STRIVE ⟨has to *strain* to reach the high notes⟩ **b :** to pull against resistance ⟨a dog *straining* at its leash⟩ **c :** to contract the muscles forcefully in attempting to defecate — often used in the phrase *strain at stool*
2 : to pass through or as if through a strainer ⟨the liquid *strains* readily⟩
3 : to make great difficulty or resistance **:** BALK
word history see STRAITEN
— strain a point : to go beyond a usual, accepted, or proper limit or rule

[3]strain *noun* (1558)
1 : an act of straining or the condition of being strained: as **a :** bodily injury from excessive tension, effort, or use ⟨heart *strain*⟩; *especially* **:** one resulting from a wrench or twist and involving undue stretching of muscles or ligaments ⟨back *strain*⟩ **b :** excessive or difficult exertion or labor **c :** excessive physical or mental tension; *also* **:** a force, influence, or factor causing such tension ⟨her responsibilities were a constant *strain*⟩ **d :** deformation of a material body under the action of applied forces
2 : an unusual reach, degree, or intensity **:** PITCH
3 *archaic* **:** a strained interpretation of something said or written

strained \'strānd\ *adjective* (circa 1542)
1 : done or produced with excessive effort
2 : pushed by antagonism near to open conflict ⟨*strained* relations⟩

strain·er \'strā-nər\ *noun* (14th century)
: one that strains: as **a :** a device (as a sieve) to retain solid pieces while a liquid passes through **b :** any of various devices for stretching or tightening something

strain gauge *noun* (1910)
: EXTENSOMETER

[1]strait \'strāt\ *adjective* [Middle English, from Old French *estreit,* from Latin *strictus* strait, strict, from past participle of *stringere*] (13th century)
1 *archaic* **:** STRICT, RIGOROUS
2 *archaic* **a :** NARROW **b :** limited in space or time **c :** closely fitting **:** CONSTRICTED, TIGHT
3 a : causing distress **:** DIFFICULT **b :** limited as to means or resources
— strait·ly *adverb*
— strait·ness *noun*

[2]strait *adverb* (13th century)
obsolete **:** in a close or tight manner

[3]strait *noun* (14th century)
1 a *archaic* **:** a narrow space or passage **b :** a comparatively narrow passageway connecting two large bodies of water — often used in plural but singular in construction **c :** ISTHMUS
2 : a situation of perplexity or distress — often used in plural ⟨in dire *straits*⟩
synonym see JUNCTURE

strait·en \'strā-t^{ə}n\ *transitive verb* **strait·ened; strait·en·ing** \'strāt-niŋ, 'strā-t^{ə}n-iŋ\ (circa 1552)
1 a : to make strait or narrow **b :** to hem in **:** CONFINE
2 *archaic* **:** to restrict in freedom or scope **:** HAMPER
3 : to subject to distress, privation, or deficiency ⟨in *straitened* circumstances⟩ ◆

[1]strait·jack·et *also* **straight·jack·et** \'strāt-,ja-kət\ *noun* (1814)
1 : a cover or overgarment of strong material (as canvas) used to bind the body and especially the arms closely in restraining a violent prisoner or patient
2 : something that restricts or confines like a straitjacket

[2]straitjacket *also* **straightjacket** *transitive verb* (1863)
: to confine in or as if in a straitjacket

strait·laced *or* **straight·laced** \'strāt-'lāst\ *adjective* (1554)
1 : excessively strict in manners, morals, or opinion
2 : wearing or having a bodice or stays tightly laced
— strait·laced·ly \-'lā-səd-lē, -'lāst-lē\ *adverb*
— strait·laced·ness \-'lās(t)-nəs, -'lā-səd-nəs\ *noun*

Straits dollar \'strāts-\ *noun* [*Straits* Settlements, former British crown colony] (1908)
: a dollar formerly issued by British Malaya and used in much of southern and eastern Asia and the East Indies

strake \'strāk\ *noun* [Middle English; akin to Old English *streccan* to stretch — more at STRETCH] (14th century)
1 : a continuous band of hull planking or plates on a ship; *also* **:** the width of such a band
2 : STREAK, STRIPE

stra·mash \strə-'mash\ *noun* [origin unknown] (1803)
1 *chiefly Scottish* **:** DISTURBANCE, RACKET
2 *chiefly Scottish* **:** CRASH, SMASHUP

stra·mo·ni·um \strə-'mō-nē-əm\ *noun* [New Latin] (1663)
1 : the dried leaves of the jimsonweed or of a related plant (genus *Datura*) that contain toxic alkaloids (as atropine) and are used in medicine similarly to belladonna
2 : JIMSONWEED

[1]strand \'strand\ *noun* [Middle English, from Old English; akin to Old Norse *strǫnd* shore] (before 12th century)
: the land bordering a body of water **:** SHORE, BEACH

[2]strand (1621)
transitive verb
1 : to run, drive, or cause to drift onto a strand **:** run aground
2 : to leave in a strange or an unfavorable place especially without funds or means to depart
3 : to leave (a base runner) on base at the end of an inning in baseball
intransitive verb
: to become stranded

[3]strand *noun* [Middle English *stronde, strande*] (13th century)
1 *Scottish & dialect English* **:** STREAM
2 *Scottish & dialect English* **:** SEA

[4]strand *noun* [Middle English *strond*] (15th century)
1 a : fibers or filaments twisted, plaited, or laid parallel to form a unit for further twisting or plaiting into yarn, thread, rope, or cordage **b :** one of the wires twisted together or laid parallel to form a wire rope or cable **c :** something (as a molecular chain) resembling a strand
2 : an element (as a yarn or thread) of a woven or plaited material
3 : an elongated or twisted and plaited body resembling a rope ⟨a *strand* of pearls⟩
4 : one of the elements interwoven in a complex whole

[5]strand *transitive verb* (1841)
1 : to break a strand of (a rope) accidentally
2 a : to form (as a rope) from strands **b :** to play out, twist, or arrange in a strand

strand·ed \'stran-dəd\ *adjective* (1875)
: having a strand or strands especially of a specified kind or number — usually used in combination ⟨the double-*stranded* molecule of DNA⟩
— strand·ed·ness *noun*

strand·line \'stran(d)-,līn\ *noun* (1903)
: SHORELINE; *especially* **:** a shoreline above the present water level

strange \'strānj\ *adjective* **strang·er; strang·est** [Middle English, from Old

◇ WORD HISTORY

straiten In straitened circumstances money is tight, and one is apt to feel anxiety, worry, and strain. Appropriately so. The word *straiten,* a 16th century addition to our language, is based on the early French adjective *estreit,* which descends from the Latin word *strictus,* the past participle of the verb *stringere* "to bind tight." *Straiten,* then, is etymologically unrelated to *straighten* (which instead comes ultimately from an old past participle of *stretch*). But *straiten* is related to *strain.* The word *strain* came into Middle English as *streinen, strainen* from the medieval French verb *estreindre,* the direct descendant of Latin *stringere* "to bind tight," the past participle of which underlies *straiten.*

\ə\ **a**but \$^{ə}$\ kitten \ər\ **f**ur**th**er \a\ **a**sh \ā\ **a**ce \ä\ m**o**p, m**a**r \au̇\ **ou**t \ch\ **ch**in \e\ b**e**t \ē\ **ea**sy \g\ **g**o \i\ h**i**t \ī\ **i**ce \j\ **j**ob \ŋ\ si**ng** \ō\ g**o** \ȯ\ l**aw** \ȯi\ b**oy** \th\ **th**in \t͟h\ **th**e \ü\ l**oo**t \u̇\ f**oo**t \y\ **y**et \zh\ vi**si**on *see also* Guide to Pronunciation

French *estrange,* from Latin *extraneus,* literally, external, from *extra* outside — more at EXTRA-] (13th century)
1 a *archaic* **:** of, relating to, or characteristic of another country **:** FOREIGN **b :** not native to or naturally belonging in a place **:** of external origin, kind, or character
2 a : not before known, heard, or seen **:** UNFAMILIAR **b :** exciting wonder or awe **:** EXTRAORDINARY
3 a : discouraging familiarities **:** RESERVED, DISTANT **b :** ILL AT EASE
4 : UNACCUSTOMED 2 ⟨she was *strange* to his ways⟩
5 : having the quantum characteristic of strangeness ⟨*strange* quark⟩ ⟨*strange* particle⟩
☆
— strange·ly *adverb*

strange·ness \'strānj-nəs\ *noun* (14th century)
1 : the quality or state of being strange
2 : a quantum characteristic of subatomic particles that accounts for the relatively long lifetime of certain particles, is conserved in interactions involving electromagnetism or the strong force, and has a value of zero for most known particles

¹strang·er \'strān-jər\ *noun* [Middle English, from Middle French *estrangier* foreign, foreigner, from *estrange*] (14th century)
1 : one who is strange: as **a** (1) **:** FOREIGNER (2) **:** a resident alien **b :** one in the house of another as a guest, visitor, or intruder **c :** a person or thing that is unknown or with whom one is unacquainted **d :** one who does not belong to or is kept from the activities of a group **e :** one not privy or party to an act, contract, or title **:** one that interferes without right
2 : one ignorant of or unacquainted with someone or something

²stranger *adjective* (15th century)
: of, relating to, or being a stranger **:** FOREIGN

³stranger *transitive verb* (1605)
obsolete **:** ESTRANGE, ALIENATE

strange woman *noun* [from the expression frequently used in Proverbs (Authorized Version)] (1535)
: PROSTITUTE

stran·gle \'straŋ-gəl\ *verb* **stran·gled; stran·gling** \-g(ə-)liŋ\ [Middle English, from Middle French *estrangler,* from Latin *strangulare,* from Greek *strangalan,* from *strangalē* halter — more at STRAIN] (14th century)
transitive verb
1 a : to choke to death by compressing the throat with something (as a hand or rope) **:** THROTTLE **b :** to obstruct seriously or fatally the normal breathing of **c :** STIFLE
2 : to suppress or hinder the rise, expression, or growth of
intransitive verb
1 : to become strangled
2 : to die from or as if from interference with breathing
— stran·gler \-g(ə-)lər\ *noun*

stran·gle·hold \'straŋ-gəl-,hōld\ *noun* (1893)
1 : an illegal wrestling hold by which one's opponent is choked
2 : a force or influence that chokes or suppresses freedom of movement or expression

strangler fig *noun* (1933)
: any of various vines and trees especially of the mulberry family (as *Ficus aurea* of the southeastern U.S.) and Saint-John's-wort family (as *Clusia rosea* of tropical America) that start as epiphytes but send down roots to the ground around the host tree

stran·gles \'straŋ-gəlz\ *noun plural but singular or plural in construction* [plural of obsolete *strangle* act of strangling] (circa 1706)
: an infectious febrile disease of horses caused by a bacterium (*Streptococcus equi*) and marked by inflammation and congestion of mucous membranes

stran·gu·late \'straŋ-gyə-,lāt\ *verb* **-lat·ed; -lat·ing** [Latin *strangulatus,* past participle of *strangulare*] (1665)
transitive verb
: STRANGLE, CONSTRICT
intransitive verb
: to become constricted so as to stop circulation ⟨the hernia will *strangulate*⟩

stran·gu·la·tion \,straŋ-gyə-'lā-shən\ *noun* (1542)
1 : the action or process of strangling or strangulating
2 : the state of being strangled or strangulated; *especially* **:** excessive or pathological constriction or compression of a bodily tube (as a blood vessel or a loop of intestine) that interrupts its ability to act as a passage

stran·gu·ry \'straŋ-gyə-rē, -,gyůr-ē\ *noun, plural* **-ries** [Middle English, from Latin *stranguria,* from Greek *strangouria,* from *strang-, stranx* drop squeezed out + *ourein* to urinate, from *ouron* urine — more at STRAIN, URINE] (14th century)
: a slow and painful spasmodic discharge of urine drop by drop

¹strap \'strap\ *noun* [alteration of *strop,* from Middle English, band or loop of leather or rope, from Old English, thong for securing an oar, from Latin *struppus* band, strap, from Greek *strophos* twisted band, from *strephein* to twist] (1601)
1 a : a narrow usually flat strip or thong of a flexible material and especially leather used for securing, holding together, or wrapping **b :** something made of a strap forming a loop ⟨a boot *strap*⟩ **c :** a strip of leather used for flogging **d :** STROP
2 : a band, plate, or loop of metal for binding objects together or for clamping an object in position
3 : a shoe fastened with a usually buckled strap
4 *Irish* **:** TROLLOP 2

²strap *transitive verb* **strapped; strap·ping** (1711)
1 a (1) **:** to secure with or attach by means of a strap (2) **:** to support (as a sprained joint) with overlapping strips of adhesive plaster **b :** BIND, CONSTRICT
2 : to beat or punish with a strap
3 : STROP
4 : to cause to suffer from an extreme scarcity ⟨is often *strapped* for cash⟩

strap·hang·er \'strap-,haŋ-ər\ *noun* (1905)
: a standing passenger in a subway, streetcar, bus, or train who clings for support to one of the short straps or similar devices placed along the aisle
— strap·hang \-,haŋ\ *intransitive verb*

strap·less \-ləs\ *adjective* (1846)
: having no strap; *specifically* **:** made or worn without shoulder straps ⟨a *strapless* evening gown⟩
— strapless *noun*

strap·pa·do \strə-'pā-(,)dō, -'pä-\ *noun* [modification of Italian *strappata,* literally, sharp pull] (1560)
: a punishment or torture in which the subject is hoisted by rope and allowed to fall its full length; *also* **:** a machine used to inflict this torture

strap·per \'stra-pər\ *noun* (1675)
: one that is unusually large or robust

¹strap·ping \'stra-piŋ\ *adjective* (1657)
: having a vigorously sturdy constitution

²strapping *noun* (1818)
1 : material for a strap
2 : STRAPS

strass \'stras\ *noun* [French *stras, strass*] (1820)
: PASTE 3

strat·a·gem \'stra-tə-jəm, -,jem\ *noun* [Italian *stratagemma,* from Latin *strategema,* from Greek *stratēgēma,* from *stratēgein* to be a general, maneuver, from *stratēgos* general, from *stratos* camp, army (akin to Latin *stratus,* past participle, spread out) + *agein* to lead — more at STRATUM, AGENT] (15th century)
1 a : an artifice or trick in war for deceiving and outwitting the enemy **b :** a cleverly contrived trick or scheme for gaining an end
2 : skill in ruses or trickery
synonym see TRICK

stra·te·gic \strə-'tē-jik\ *adjective* (1825)
1 : of, relating to, or marked by strategy ⟨a *strategic* retreat⟩
2 a : necessary to or important in the initiation, conduct, or completion of a strategic plan **b :** required for the conduct of war and not available in adequate quantities domestically ⟨*strategic* materials⟩ **c :** of great importance within an integrated whole or to a planned effect ⟨emphasized *strategic* points⟩
3 : designed or trained to strike an enemy at the sources of his military, economic, or political power ⟨a *strategic* bomber⟩
— stra·te·gi·cal \-ji-kəl\ *adjective*
— stra·te·gi·cal·ly \-ji-k(ə-)lē\ *adverb*

strat·e·gist \'stra-tə-jist\ *noun* (1838)
: one skilled in strategy

strat·e·gize \-,jīz\ *intransitive verb* **-gized; -giz·ing** (1921)
: to devise a strategy or course of action

strat·e·gy \-jē\ *noun, plural* **-gies** [Greek *stratēgia* generalship, from *stratēgos*] (1810)
1 a (1) **:** the science and art of employing the political, economic, psychological, and military forces of a nation or group of nations to afford the maximum support to adopted policies in peace or war (2) **:** the science and art of military command exercised to meet the enemy in combat under advantageous conditions **b :** a variety of or instance of the use of strategy
2 a : a careful plan or method **:** a clever stratagem **b :** the art of devising or employing plans or stratagems toward a goal
3 : an adaptation or complex of adaptations (as of behavior, metabolism, or structure) that serves or appears to serve an important function in achieving evolutionary success ⟨foraging *strategies* of insects⟩

strath \'strath\ *noun* [Scottish Gaelic *srath*] (1540)
: a flat wide river valley or the low-lying grassland along it

strath·spey \,strath-'spā\ *noun, plural* **strathspeys** [*Strath Spey,* district of Scotland] (circa 1653)

☆ SYNONYMS
Strange, singular, unique, peculiar, eccentric, erratic, odd, queer, quaint, outlandish mean departing from what is ordinary, usual, or to be expected. STRANGE stresses unfamiliarity and may apply to the foreign, the unnatural, the unaccountable ⟨a journey filled with *strange* sights⟩. SINGULAR suggests individuality or puzzling strangeness ⟨a *singular* feeling of impending disaster⟩. UNIQUE implies singularity and the fact of being without a known parallel ⟨a career *unique* in the annals of science⟩. PECULIAR implies a marked distinctiveness ⟨the *peculiar* status of America's first lady⟩. ECCENTRIC suggests a wide divergence from the usual or normal especially in behavior ⟨the *eccentric* eating habits of preschoolers⟩. ERRATIC stresses a capricious and unpredictable wandering or deviating ⟨a friend's suddenly *erratic* behavior⟩. ODD applies to a departure from the regular or expected ⟨an *odd* sense of humor⟩. QUEER suggests a dubious sometimes sinister oddness ⟨*queer* happenings offering no ready explanation⟩. QUAINT suggests an old-fashioned but pleasant oddness ⟨a *quaint* fishing village⟩. OUTLANDISH applies to what is uncouth, bizarre, or barbaric ⟨the *outlandish* getups of heavy metal bands⟩.

: a Scottish dance that is similar to but slower than the reel; *also* **:** the music for this dance

strati- *combining form* [New Latin *stratum*]
: stratum ⟨*strati*form⟩

strat·i·fi·ca·tion \ˌstra-tə-fə-ˈkā-shən\ *noun* (circa 1617)
1 a : the act or process of stratifying **b :** the state of being stratified
2 : a stratified formation

strat·i·fi·ca·tion·al grammar \ˌstra-tə-fə-ˈkā-shnəl-, -shə-n^{ə}l-\ *noun* (1962)
: a grammar based on the theory that language consists of a series of hierarchically related strata linked together by representational rules

stratified charge engine *noun* (1962)
: an internal-combustion engine in whose cylinders the combustion of fuel in a layer of rich fuel-air mixture promotes ignition in a greater volume of lean mixture

strat·i·form \ˈstra-tə-ˌfȯrm\ *adjective* (1805)
: having a stratified formation

strat·i·fy \ˈstra-tə-ˌfī\ *verb* **-fied; -fy·ing** [New Latin *stratificare,* from *stratum* + Latin *-ificare* -ify] (1661)
transitive verb
1 : to form, deposit, or arrange in strata
2 a : to divide or arrange into classes, castes, or social strata **b :** to divide into a series of graded statuses
intransitive verb
: to become arranged in strata

strati·graph·ic \ˌstra-tə-ˈgra-fik\ *adjective* (1877)
: of, relating to, or determined by stratigraphy

stra·tig·ra·phy \strə-ˈti-grə-fē\ *noun* [International Scientific Vocabulary] (1865)
1 : geology that deals with the origin, composition, distribution, and succession of strata
2 : the arrangement of strata

strato- *combining form* [New Latin *stratus*]
: stratus and ⟨*strato*cumulus⟩

stra·toc·ra·cy \strə-ˈtä-krə-sē\ *noun, plural* **-cies** [Greek *stratos* army — more at STRATAGEM] (1652)
: a military government

stra·to·cu·mu·lus \ˌstrā-tō-ˈkyü-myə-ləs, ˌstra-\ *noun* [New Latin] (circa 1891)
: stratified cumulus consisting of large balls or rolls of dark cloud which often cover the whole sky especially in winter — see CLOUD illustration

strato·sphere \ˈstra-tə-ˌsfir\ *noun* [French *stratosphère,* from New Latin *stratum* + *-o-* + French *sphère* sphere, from Latin *sphaera*] (1909)
1 : the part of the earth's atmosphere which extends from about 7 miles (11 kilometers) above the surface to 31 miles (50 kilometers) and in which temperature increases gradually to about 32° F (0° C) and clouds rarely form
2 : a very high or the highest region on or as if on a graded scale ⟨construction costs in the *stratosphere*⟩ ⟨the celebrity *stratosphere*⟩
— strato·spher·ic \ˌstra-tə-ˈsfir-ik, -ˈsfer-\ *adjective*

stra·to·vol·ca·no \ˌstra-tō-väl-ˈkā-(ˌ)nō, ˌstrā-, -vȯl-\ *noun* [New Latin *stratum* + English *-o-* + *volcano*] (1937)
: a volcano composed of explosively erupted cinders and ash with occasional lava flows

stra·tum \ˈstrā-təm, ˈstra-\ *noun, plural* **stra·ta** \ˈstrā-tə, ˈstra-\ [New Latin, from Latin, spread, bed, from neuter of *stratus,* past participle of *sternere* to spread out — more at STREW] (1599)
1 : a bed or layer artificially made
2 a : a sheetlike mass of sedimentary rock or earth of one kind lying between beds of other kinds **b :** a region of the sea or atmosphere that is analogous to a stratum of the earth **c :** a layer of tissue ⟨deep *stratum* of the skin⟩ **d :** a layer in which archaeological material (as artifacts, skeletons, and dwelling remains) is found on excavation
3 a : a part of a historical or sociological series representing a period or a stage of development **b :** a socioeconomic level of society comprising persons of the same or similar status especially with regard to education or culture
4 : one of a series of layers, levels, or gradations in an ordered system ⟨*strata* of thought⟩
5 : a statistical subpopulation ▢

stra·tus \ˈstrā-təs, ˈstra-\ *noun, plural* **stra·ti** \ˈstrā-ˌtī, ˈstra-\ [New Latin, from Latin, past participle of *sternere*] (circa 1803)
: a cloud form extending over a large area at altitudes of usually 2000 to 7000 feet (600 to 2100 meters) — see CLOUD illustration

stra·vage *or* **stra·vaig** \strə-ˈvāg\ *intransitive verb* [probably by shortening & alteration from *extravagate*] (1773)
chiefly Scottish **:** ROAM

[1]straw \ˈstrȯ\ *noun* [Middle English, from Old English *strēaw;* akin to Old High German *strō* straw, Old English *strewian* to strew] (before 12th century)
1 a : stalks of grain after threshing; *broadly* **:** dry stalky plant residue used like grain straw (as for bedding or packing) **b :** a natural or artificial heavy fiber used for weaving, plaiting, or braiding
2 : a dry coarse stem especially of a cereal grass
3 a (1) **:** something of small worth or significance (2) **:** something too insubstantial to provide support or help in a desperate situation ⟨clutching at *straws*⟩ **b :** CHAFF 2
4 a : something (as a hat) made of straw **b :** a tube (as of paper, plastic, or glass) for sucking up a beverage
— strawy \ˈstrȯ-ē\ *adjective*
— straw in the wind : a slight fact that is an indication of a coming event

[2]straw *adjective* (15th century)
1 : made of straw ⟨a *straw* rug⟩
2 : of, relating to, or used for straw ⟨a *straw* barn⟩
3 : of the color of straw ⟨*straw* hair⟩
4 : of little or no value **:** WORTHLESS
5 : of, relating to, resembling, or being a straw man
6 : of, relating to, or concerned with the discovery of preferences by means of a straw vote

straw·ber·ry \ˈstrȯ-ˌber-ē, -b(ə-)rē\ *noun, often attributive* [Middle English *strawberie,* from Old English *strēawberige,* from *strēaw* straw + *berige* berry; perhaps from the appearance of the achenes on the surface] (before 12th century)
: the juicy edible usually red fruit of any of several low-growing temperate herbs (genus *Fragaria*) of the rose family that is technically an enlarged pulpy receptacle bearing numerous achenes; *also* **:** a plant whose fruits are strawberries

strawberry

strawberry bush *noun* (circa 1856)
1 : a shrubby North American spindle tree (*Euonymus americanus*) with crimson pods and seeds with a scarlet aril
2 : [2]WAHOO

strawberry mark *noun* (1847)
: a tumor of the skin filled with small blood vessels and appearing usually as a red and elevated birthmark

strawberry roan *noun* (1955)
: a roan horse with a light red ground color

strawberry shrub *noun* (circa 1890)
: any of a genus (*Calycanthus* of the family Calycanthaceae, the strawberry-shrub family) of shrubs with fragrant brownish red flowers

strawberry tomato *noun* (circa 1847)
: GROUND-CHERRY; *especially* **:** a stout hairy annual herb (*Physalis pruinosa*) of eastern North America with sweet globular yellow fruits

strawberry tree *noun* (15th century)
: a small European evergreen tree (*Arbutus unedo*) of the heath family with racemose white flowers and fruits like strawberries

straw boss *noun* (1894)
1 : an assistant to a foreman in charge of supervising and expediting the work of a small gang of workers
2 : a member of a group of workers who supervises the work of the others in addition to doing his or her own job

straw·flow·er \ˈstrȯ-ˌflau̇(-ə)r\ *noun* (circa 1922)
: any of several everlasting flowers; *especially* **:** an Australian perennial composite herb (*Helichrysum bracteatum*) that is widely cultivated for its heads of chaffy brightly colored long-keeping flowers

straw·hat \ˌstrȯ-ˌhat\ *adjective* [from the former fashion of wearing straw hats in summer] (1935)
: of, relating to, or being summer theater

straw man *noun* (1896)
1 : a weak or imaginary opposition (as an argument or adversary) set up only to be easily confuted
2 : a person set up to serve as a cover for a usually questionable transaction

straw vote *noun* (1866)
: an unofficial vote taken (as at a chance gathering) to indicate the relative strength of opposing candidates or issues — called also *straw poll*

straw wine *noun* (1824)
: a sweet wine produced by partially drying the grapes on beds of straw prior to vinification

straw yellow *noun* (circa 1796)
: a pale yellow

[1]stray \ˈstrā\ *noun* [Middle English, from Old French *estraié,* past participle of *estraier*] (13th century)
1 a : a domestic animal that is wandering at large or is lost **b :** a person or thing that strays
2 [Middle English, from *straien* to stray] *archaic* **:** the act of going astray

[2]stray *intransitive verb* [Middle English *straien,* from Middle French *estraier,* from (assumed) Vulgar Latin *extravagare,* from Latin *extra-* outside + *vagari* to wander — more at EXTRA-] (14th century)
: WANDER: as **a :** to wander from company, restraint, or proper limits **b :** to roam about without fixed direction or purpose **c :** to move in a winding course **:** MEANDER **d :** to move without conscious or intentional effort ⟨eyes *straying* absently around the room⟩ **e :** to become distracted from an argument or chain of

▢ USAGE
stratum The plural *strata* has occasionally been used as a singular since the 18th century and is sometimes given the plural *stratas* ⟨there was a *strata* of Paris which mere criticism of books fails to get hold of —Ezra Pound⟩ ⟨a Roman burial ground suggests *stratas* of corruption and decay —Connie Fletcher, *Booklist*⟩. Current evidence shows senses 2, 3b, and 4 so used, with 3b the most common. Singular *strata* is persistent but not frequent. It may someday establish itself as a singular like *agenda,* but its current status is something less than established.

\ə\ **abut** \$^{ə}$\ kitten \ər\ **further** \a\ **ash** \ā\ **ace** \ä\ **mop, mar** \au̇\ **out** \ch\ **chin** \e\ **bet** \ē\ **easy** \g\ **go** \i\ **hit** \ī\ **ice** \j\ **job** \ŋ\ **sing** \ō\ **go** \ȯ\ **law** \ȯi\ **boy** \th\ **thin** \t͟h\ **the** \ü\ **loot** \u̇\ **foot** \y\ **yet** \zh\ **vision** *see also* Guide to Pronunciation

thought ⟨*strayed* from the point⟩ **f :** to wander accidentally from a fixed or chosen route **g :** ERR, SIN
— **stray·er** *noun*
³stray *adjective* (1607)
1 : having strayed or escaped from a proper or intended place ⟨a *stray* cow⟩ ⟨hit by a *stray* bullet⟩ ⟨fixed a few *stray* hairs⟩
2 : occurring at random or sporadically ⟨a few *stray* thoughts⟩
3 : not serving any useful purpose **:** UNWANTED ⟨*stray* light⟩
¹streak \'strēk\ *noun* [Middle English *streke,* from Old English *strica;* akin to Old High German *strich* line, Latin *striga* row — more at STRIKE] (before 12th century)
1 : a line or mark of a different color or texture from the ground **:** STRIPE
2 a : the color of the fine powder of a mineral obtained by scratching or rubbing against a hard white surface and constituting an important distinguishing character **b :** inoculum implanted in a line on a solid medium **c :** any of several virus diseases of plants (as the potato, tomato, or raspberry) resembling mosaic but usually producing at least some linear markings
3 a : a narrow band of light **b :** a lightning bolt
4 a : a slight admixture **:** TRACE ⟨had a mean *streak* in him⟩ **b :** a brief run (as of luck) **c :** a consecutive series ⟨was on a winning *streak*⟩
5 : a narrow layer (as of ore)
6 : an act or instance of streaking
²streak (1595)
transitive verb
: to make streaks on or in ⟨tears *streaking* her face⟩
intransitive verb
1 : to move swiftly **:** RUSH ⟨a jet *streaking* across the sky⟩
2 : to have a streak (as of winning or outstanding performances)
3 : to run naked through a public place
— **streak·er** *noun*
streak camera *noun* (1959)
: a camera for recording very fast or short-lived phenomena (as fluorescence or shock waves)
streaked \'strēkt, 'strē-kəd\ *adjective* (1596)
1 : marked with stripes or linear discolorations
2 : physically or mentally disturbed **:** UPSET
streak·ing \'strē-kiŋ\ *noun* (circa 1964)
: the lightening (as by chemicals) of a few long strands of hair to produce a streaked effect
streaky \'strē-kē\ *adjective* **streak·i·er; -est** (1745)
1 : marked with streaks ⟨*streaky* bacon⟩
2 : APPREHENSIVE ⟨nervous and *streaky*⟩
3 : apt to vary (as in effectiveness) **:** UNRELIABLE
— **streak·i·ness** *noun*
¹stream \'strēm\ *noun* [Middle English *streme,* from Old English *strēam;* akin to Old High German *stroum* stream, Greek *rhein* to flow] (before 12th century)
1 : a body of running water (as a river or brook) flowing on the earth; *also* **:** any body of flowing fluid (as water or gas)
2 a : a steady succession (as of words or events) ⟨kept up an endless *stream* of chatter⟩ **b :** a constantly renewed supply **c :** a continuous moving procession ⟨a *stream* of traffic⟩
3 : an unbroken flow (as of gas or particles of matter)
4 : a ray of light
5 a : a prevailing attitude or group ⟨has always run against the *stream* of current fashion⟩ **b :** a dominant influence or line of development
6 *British* **:** TRACK 3c
²stream (13th century)
intransitive verb
1 a : to flow in or as if in a stream **b :** to leave a bright trail ⟨a meteor *streamed* through the sky⟩
2 a : to exude a bodily fluid profusely ⟨her eyes were *streaming*⟩ **b :** to become wet with a discharge of bodily fluid ⟨*streaming* with perspiration⟩
3 : to trail out at full length ⟨her hair *streaming* back as she ran⟩
4 : to pour in large numbers ⟨complaints came *streaming* in⟩
transitive verb
1 : to emit freely or in a stream ⟨his eyes *streamed* tears⟩
2 : to display (as a flag) by waving
stream·bed \'strēm-ˌbed\ *noun* (1857)
: the channel occupied or formerly occupied by a stream
stream·er \'strē-mər\ *noun* (13th century)
1 a : a flag that streams in the wind; *especially* **:** PENNANT **b :** any long narrow wavy strip resembling or suggesting a banner floating in the wind **c :** BANNER 2
2 a : a long extension of the solar corona visible only during a total solar eclipse **b** *plural* **:** AURORA BOREALIS
stream·ing \'strē-miŋ\ *noun* (14th century)
1 : an act or instance of flowing; *specifically* **:** CYCLOSIS
2 *British* **:** TRACKING
stream·let \'strēm-lət\ *noun* (circa 1552)
: a small stream
¹stream·line \'strēm-ˌlīn\ *noun* (1868)
1 : the path of a particle in a fluid relative to a solid body past which the fluid is moving in smooth flow without turbulence
2 a : a contour designed to minimize resistance to motion through a fluid (as air) **b :** a smooth or flowing line designed as if for decreasing air resistance
²streamline *transitive verb* (1913)
1 : to design or construct with a streamline
2 : to bring up to date **:** MODERNIZE
3 a : to put in order **:** ORGANIZE **b :** to make simpler or more efficient
stream·lined \-ˌlīnd\ *adjective* (1913)
1 a : contoured to reduce resistance to motion through a fluid (as air) **b :** stripped of nonessentials **:** COMPACT **c :** effectively integrated **:** ORGANIZED
2 : having flowing lines
3 : brought up to date **:** MODERNIZED
4 : of or relating to streamline flow
streamline flow *noun* (circa 1918)
: an uninterrupted flow (as of air) past a solid body in which the direction at every point remains unchanged with the passage of time
stream·lin·er \'strēm-ˌlī-nər\ *noun* (1934)
: one that is streamlined; *especially* **:** a streamlined train
stream of consciousness (1855)
1 : the continuous unedited chronological flow of conscious experience through the mind
2 : INTERIOR MONOLOGUE
stream·side \'strēm-ˌsīd\ *noun* (1844)
: the land bordering on a stream
streek \'strēk\ *transitive verb* [Middle English (northern dialect) *streken;* akin to Old English *streccan* to stretch] (13th century)
1 *chiefly Scottish* **:** STRETCH, EXTEND
2 *chiefly Scottish* **:** to lay out (a dead body)
¹streel \'strē(ə)l\ *intransitive verb* [Irish *straoill-, sraoill-* to tear apart, trail, trudge, from Old Irish *sroiglid* he scourges, from *sroigell* scourge, from Latin *flagellum* — more at FLAGELLATE] (1805)
1 *chiefly Irish* **:** to saunter idly and aimlessly
2 *chiefly Irish* **:** to trail or float in the manner of a streamer
²streel *noun* [Irish *straoill, sraoill,* from *straoill-,* verb] (1842)
chiefly Irish **:** an untidy slovenly person
¹street \'strēt\ *noun* [Middle English *strete,* from Old English *strǣt,* from Late Latin *strata* paved road, from Latin, feminine of *stratus,* past participle — more at STRATUM] (before 12th century)
1 a : a thoroughfare especially in a city, town, or village that is wider than an alley or lane and that usually includes sidewalks **b :** the part of a street reserved for vehicles **c :** a thoroughfare with abutting property ⟨lives on a fashionable *street*⟩
2 : the people occupying property on a street ⟨the whole *street* knew about the accident⟩
3 : a promising line of development or a channeling of effort
4 *capitalized* **:** a district (as Wall Street or Fleet Street) identified with a particular profession
5 : an environment (as in a depressed neighborhood or section of a city) of prostitution, poverty, dereliction, or crime
— **on the street** *or* **in the street 1 :** idle, homeless, or out of a job **2 :** out of prison **:** at liberty
— **up one's street** *or* **down one's street :** suited to one's abilities or taste
²street *adjective* (15th century)
: of or relating to the streets: as **a :** adjoining or giving access to a street ⟨the *street* door⟩ **b :** carried on or taking place in the street ⟨*street* fighting⟩ **c :** living or working on the streets ⟨a *street* peddler⟩ ⟨*street* people⟩ **d :** located in, used for, or serving as a guide to the streets ⟨a *street* map⟩ **e :** performing in or heard on the street ⟨a *street* band⟩ **f** (1) **:** suitable for wear or use on the street ⟨*street* clothes⟩ (2) **:** not touching the ground — used of a woman's dress in lengths reaching the knee, calf, or ankle **g :** of or relating to the street environment ⟨*street* drugs⟩
street arab \-'ar-əb, -'ā-ˌrab\ *noun, often A capitalized* (1859)
: a homeless vagabond and especially an outcast boy or girl in the streets of a city **:** GAMIN
street·car \'strēt-ˌkär\ *noun* (1862)
: a vehicle on rails used primarily for transporting passengers and typically operating on city streets
street·light \-ˌlīt\ *noun* (1906)
: a light usually mounted on a pole and constituting one of a series spaced at intervals along a public street or highway — called also *streetlamp*
street railway *noun* (1861)
: a line operating streetcars or buses
streets \'strēts\ *adverb* (1898)
chiefly British **:** by a considerable margin ⟨a nice woman, *streets* above these other callers —Katherine Mansfield⟩
street·scape \'strēt-ˌskāp\ *noun* (1924)
1 : the appearance or view of a street
2 : a work of art depicting a view of a street
word history see LANDSCAPE
street–smart \-ˌsmärt\ *adjective* (1974)
: STREETWISE
street smarts *noun plural* (1972)
: the quality of being streetwise
street theater *noun* (1967)
: drama dealing with controversial social and political issues that is usually performed outdoors
street virus *noun* (circa 1911)
: a naturally occurring rabies virus as distinguished from virus attenuated in the laboratory
street·walk·er \'strēt-ˌwȯ-kər\ *noun* (1592)
: PROSTITUTE; *especially* **:** one who solicits in the streets — compare CALL GIRL
— **street·walk·ing** \-kiŋ\ *noun*
street·wise \-ˌwīz\ *adjective* (1965)
: possessing the skills and attitudes necessary to survive in an often violent urban environment
strength \'streŋ(k)th, 'stren(t)th\ *noun, plural* **strengths** \'streŋ(k)ths, 'stren(t)ths, 'streŋks\ [Middle English *strengthe,* from Old English *strengthu;* akin to Old High German *strengi* strong — more at STRONG] (before 12th century)

1 : the quality or state of being strong **:** capacity for exertion or endurance
2 : power to resist force **:** SOLIDITY, TOUGHNESS
3 : power of resisting attack **:** IMPREGNABILITY
4 a : legal, logical, or moral force **b :** a strong attribute or inherent asset ⟨the *strengths* and the weaknesses of the book are evident⟩
5 a : degree of potency of effect or of concentration **b :** intensity of light, color, sound, or odor **c :** vigor of expression
6 : force as measured in numbers **:** effective numbers of any body or organization ⟨an army at full *strength*⟩
7 : one regarded as embodying or affording force or firmness **:** SUPPORT
8 : maintenance of or a rising tendency in a price level **:** firmness of prices
9 : BASIS — used in the phrase *on the strength of*
synonym see POWER
— from strength to strength : vigorously forward **:** from one high point to the next

strength·en \ˈstreŋ(k)-thən, ˈstren(t)-\ *verb* **strength·ened; strength·en·ing** \ˈstreŋ(k)th-niŋ, ˈstren(t)th-; ˈstreŋ(k)-thə-, ˈstren(t)-\ (15th century)
transitive verb
: to make stronger
intransitive verb
: to become stronger
— strength·en·er \ˈstreŋ(k)th-nər, ˈstren(t)th-; ˈstreŋ(k)-thə-, ˈstren(t)-\ *noun*

stren·u·ous \ˈstren-yə-wəs\ *adjective* [Latin *strenuus*] (1599)
1 a : vigorously active **:** ENERGETIC **b :** FERVENT, ZEALOUS
2 : marked by or calling for energy or stamina **:** ARDUOUS
synonym see VIGOROUS
— stren·u·os·i·ty \ˌstren-yə-ˈwä-sə-tē\ *noun*
— stren·u·ous·ly \ˈstren-yə-wəs-lē\ *adverb*
— stren·u·ous·ness *noun*

strep \ˈstrep\ *noun, often attributive* (1927)
: STREPTOCOCCUS

strep throat *noun* (circa 1927)
: an inflammatory sore throat caused by hemolytic streptococci and marked by fever, prostration, and toxemia — called also *septic sore throat*

strepto- *combining form* [New Latin, from Greek, from *streptos* twisted, from *strephein* to twist]
1 : twisted **:** twisted chain ⟨*strepto*coccus⟩
2 : streptococcus ⟨*strepto*kinase⟩

strep·to·ba·cil·lus \ˌstrep-tō-bə-ˈsi-ləs\ *noun* [New Latin] (1897)
: any of a genus (*Streptobacillus*) of nonmotile gram-negative rod-shaped bacteria in which the individual cells are often joined in a chain; *especially* **:** one (*S. moniliformis*) that is the causative agent of one form of rat-bite fever

strep·to·coc·cal \ˌstrep-tə-ˈkä-kəl\ *also* **strep·to·coc·cic** \-ˈkä-kik, -ˈkäk-sik\ *adjective* (1877)
: of, relating to, caused by, or being streptococci ⟨a *streptococcal* sore throat⟩ ⟨*streptococcal* organisms⟩

strep·to·coc·cus \-ˈkä-kəs\ *noun, plural* **-coc·ci** \-ˈkä-ˌkī, -(ˌ)kē; -ˈkäk-ˌsī, -(ˌ)sē\ [New Latin] (1877)
: any of a genus (*Streptococcus*) of spherical or ovoid chiefly nonmotile and parasitic gram-positive bacteria that divide only in one plane, occur in pairs or chains, and include important pathogens of humans and domestic animals; *broadly* **:** a coccus occurring in chains

strep·to·ki·nase \ˌstrep-tō-ˈkī-ˌnās, -ˌnāz\ *noun* (1944)
: a proteolytic enzyme from hemolytic streptococci active in promoting dissolution of blood clots

strep·to·ly·sin \ˌstrep-tə-ˈlī-sᵊn\ *noun* (1904)
: an antigenic hemolysin produced by streptococci

strep·to·my·ces \-ˈmī-ˌsēz\ *noun, plural* **streptomyces** [New Latin, from *strepto-* + Greek *mykēs* fungus — more at MYC-] (1951)
: any of a genus (*Streptomyces*) of mostly soil streptomycetes including some that form antibiotics as by-products of their metabolism

strep·to·my·cete \-ˈmī-ˌsēt, -ˌmī-ˈsēt\ *noun* [New Latin *Streptomycet-, Streptomyces,* genus name] (1948)
: any of a family (Streptomycetaceae) of actinomycetes (as a streptomyces) that form vegetative mycelia which rarely break up into bacillary forms, have conidia borne on sporophores, and are typically aerobic soil saprophytes but include a few parasites of plants and animals

strep·to·my·cin \-ˈmī-sᵊn\ *noun* (1944)
: an antibiotic organic base $C_{21}H_{39}N_7O_{12}$ produced by a soil actinomycete (*Streptomyces griseus*), active against many bacteria, and used especially in the treatment of infections (as tuberculosis) by gram-negative bacteria

strep·to·thri·cin \-ˈthrī-sᵊn, -ˈthri-\ *noun* [New Latin *Streptothric-, Streptothrix,* genus of bacteria, from *strepto-* + Greek *trich-, thrix* hair] (1926)
: any of a group of related basic antibiotics produced by a soil actinomycete (*Streptomyces lavendulae*) and active against bacteria and to some degree against fungi

¹stress \ˈstres\ *noun* [Middle English *stresse* stress, distress, short for *destresse* — more at DISTRESS] (14th century)
1 : constraining force or influence: as **a :** a force exerted when one body or body part presses on, pulls on, pushes against, or tends to compress or twist another body or body part; *especially* **:** the intensity of this mutual force commonly expressed in pounds per square inch **b :** the deformation caused in a body by such a force **c :** a physical, chemical, or emotional factor that causes bodily or mental tension and may be a factor in disease causation **d :** a state resulting from a stress; *especially* **:** one of bodily or mental tension resulting from factors that tend to alter an existent equilibrium **e :** STRAIN, PRESSURE ⟨the environment is under *stress* to the point of collapse —Joseph Shoben⟩
2 : EMPHASIS, WEIGHT ⟨lay *stress* on a point⟩
3 *archaic* **:** intense effort or exertion
4 : intensity of utterance given to a speech sound, syllable, or word producing relative loudness
5 a : relative force or prominence of sound in verse **b :** a syllable having relative force or prominence
6 : ACCENT 6a

²stress *transitive verb* (1545)
1 : to subject to physical or psychological stress
2 : to subject to phonetic stress **:** ACCENT
3 : to lay stress on **:** EMPHASIZE

stressed–out \ˈstrest-ˈau̇t\ *adjective* (1983)
: suffering from high levels of physical or especially psychological stress

stress fracture *noun* (1952)
: a usually hairline fracture of a bone that has been subjected to repeated stress

stress·ful \ˈstres-fəl\ *adjective* (1853)
: full of or tending to induce stress
— stress·ful·ly \-fə-lē\ *adverb*

stress·less \-ləs\ *adjective* (1885)
: having no stress; *specifically* **:** having no accent ⟨a *stressless* syllable⟩
— stress·less·ness *noun*

stress mark *noun* (1888)
: a mark used with (as before, after, or over) a written syllable in the respelling of a word to show that this syllable is to be stressed when spoken **:** ACCENT MARK

stress·or \ˈstre-sər, -ˌsȯr\ *noun* (1950)
: a stimulus that causes stress

stress test *noun* (1975)
: an electrocardiographic test of heart function before, during, and after a controlled period of increasingly strenuous exercise (as on a treadmill)

¹stretch \ˈstrech\ *verb* [Middle English *strecchen,* from Old English *streccan;* akin to Old High German *strecchan* to stretch, Old English *stræc* firm, severe] (before 12th century)
transitive verb
1 : to extend (as one's limbs or body) in a reclining position
2 : to reach out **:** EXTEND ⟨*stretched* out her arms⟩
3 : to extend in length ⟨*stretched* his neck to see what was going on⟩
4 : to fell with or as if with a blow
5 : to cause the limbs of (a person) to be pulled especially in torture
6 : to draw up (one's body) from a cramped, stooping, or relaxed position
7 : to pull taut ⟨canvas *stretched* on a frame⟩
8 a : to enlarge or distend especially by force **b :** to extend or expand as if by physical force ⟨*stretch* one's mind with a good book⟩ **c :** STRAIN ⟨*stretched* his already thin patience⟩
9 : to cause to reach or continue (as from one point to another or across a space) ⟨*stretch* a wire between two posts⟩
10 a : to amplify or enlarge beyond natural or proper limits ⟨the rules can be *stretched* this once⟩ **b :** to expand (as by improvisation) to fulfill a larger function ⟨*stretching* a dollar⟩
11 : to extend (a hit) to an extra base usually by fast or daring running ⟨*stretch* a single into a double⟩
intransitive verb
1 a : to become extended in length or breadth or both **:** SPREAD ⟨broad plains *stretching* to the sea⟩ **b :** to extend over a continuous period
2 : to become extended without breaking
3 a : to extend one's body or limbs **b :** to lie down at full length
— stretch·abil·i·ty \ˌstre-chə-ˈbi-lə-tē\ *noun*
— stretch·able \ˈstre-chə-bəl\ *adjective*
— stretchy \-chē\ *adjective*
— stretch a point : to go beyond what is strictly warranted in making a claim or concession
— stretch one's legs 1 : to extend the legs **2 :** to take a walk in order to relieve stiffness caused by prolonged sitting

²stretch *noun* (1541)
1 a : an exercise of something (as the understanding or the imagination) beyond ordinary or normal limits **b :** an extension of the scope or application of something ⟨a *stretch* of language⟩
2 : the extent to which something may be stretched
3 : the act of stretching **:** the state of being stretched
4 a : an extent in length or area ⟨a *stretch* of woods⟩ **b :** a continuous period of time ⟨can write for eight hours at a *stretch*⟩
5 : a walk to relieve fatigue
6 : a term of imprisonment
7 a : either of the straight sides of a race course; *especially* **:** HOMESTRETCH **b :** a final stage
8 : the capacity for being stretched **:** ELASTICITY
9 : STRETCH LIMO

³stretch *adjective* (1954)
: easily stretched **:** ELASTIC ⟨a *stretch* wig⟩

stretch·er \ˈstre-chər\ *noun* (15th century)
1 a : one that stretches; *especially* **:** a device or machine for stretching or expanding something **b :** an exaggerated story **:** a tall tale

2 a : a brick or stone laid with its length parallel to the face of the wall **b :** a timber or rod used especially when horizontal as a tie in framed work
3 : a device for carrying a sick, injured, or dead person
4 : a rod or bar extending between two legs of a chair or table

stretch limo *noun* (1971)
: a long limousine that is luxuriously furnished (as with a TV or bar) — called also *stretch limousine*

stretch marks *noun plural* (1956)
: striae on the skin (as of the hips, abdomen, and breasts) from excessive stretching and rupture of elastic fibers especially due to pregnancy or obesity

stretch–out \'strech-ˌau̇t\ *noun* (1930)
1 : a system of industrial operation in which workers are required to do extra work with slight or with no additional pay
2 : the act of stretching out **:** the state of being stretched out
3 : an economizing measure that spreads a limited quantity over a larger field than originally intended: as **a :** a slackening of production schedules so that a quantity of goods will be produced over a longer period than initially planned **b :** a restructuring of a loan repayment schedule over an extended period of time

stretch receptor *noun* (1936)
: MUSCLE SPINDLE

stretch runner *noun* (1922)
: a racehorse that makes a strong bid in the homestretch

stret·to \'stre-(ˌ)tō\ *also* **stret·ta** \-tə\ *noun, plural* **stret·ti** \-(ˌ)tē\ *or* **strettos** [*stretto* from Italian, from *stretto* narrow, close, from Latin *strictus,* past participle; *stretta* from Italian, from feminine of *stretto* — more at STRICT] (circa 1740)
1 a : the overlapping of answer with subject in a musical fugue **b :** the part of a fugue characterized by this overlapping
2 : a concluding passage performed in a quicker tempo

streu·sel \'strü-səl, -zəl, 'strȯi-, 'shtrȯi-\ *noun* [German, literally, something strewn, from Middle High German *ströusel,* from *ströuwen* to strew, from Old High German *strewen*] (1909)
: a crumbly mixture of fat, sugar, and flour and sometimes nuts and spices that is used as topping or filling for cake

strew \'strü\ *transitive verb* **strewed; strewed** *or* **strewn** \'strün\; **strew·ing** [Middle English *strewen, strowen,* from Old English *strewian, strēowian;* akin to Old High German *strewen* to strew, Latin *struere* to heap up, *sternere* to spread out, Greek *stornynai*] (before 12th century)
1 : to spread by scattering
2 : to cover by or as if by scattering something ⟨*strewing* the highways with litter⟩
3 : to become dispersed over as if scattered
4 : to spread abroad **:** DISSEMINATE

strew·ment \'strü-mənt\ *noun* (1602)
archaic **:** something (as flowers) strewed or designed for strewing

stria \'strī-ə\ *noun, plural* **stri·ae** \'strī-ˌē\ [Latin, furrow, channel — more at STRIKE] (1563)
1 : STRIATION 2
2 : a stripe or line (as in the skin) distinguished from the surrounding area by color, texture, or elevation — compare STRETCH MARKS

¹stri·ate \'strī-ət, -ˌāt\ *adjective* (1670)
: STRIATED

²stri·ate \-ˌāt\ *transitive verb* **stri·at·ed; stri·at·ing** (1709)
: to mark with striations or striae

stri·at·ed \'strī-ˌā-təd\ *adjective* (1646)
1 : marked with striations or striae
2 : of, relating to, or being striated muscle

striated muscle *noun* (1866)
: muscle tissue that is marked by transverse dark and light bands, that is made up of elongated multinuclear fibers, and that includes skeletal muscle, cardiac muscle, and most muscle of arthropods — compare SMOOTH MUSCLE, VOLUNTARY MUSCLE

stri·a·tion \strī-'ā-shən\ *noun* (circa 1847)
1 a : the fact or state of being striated **b :** arrangement of striations or striae
2 : a minute groove, scratch, or channel especially when one of a parallel series
3 : any of the alternate dark and light cross bands of a myofibril of striated muscle

strick \'strik\ *noun* [Middle English *stric, strik,* probably of Low German or Dutch origin; akin to Middle Low German *strik* rope, Middle Dutch *stric*] (14th century)
: a bunch of hackled flax, jute, or hemp

strick·en \'stri-kən\ *adjective* [Middle English *striken,* from past participle of *striken* to strike] (14th century)
1 a : afflicted or overwhelmed by or as if by disease, misfortune, or sorrow **b :** made incapable or unfit
2 : hit or wounded by or as if by a missile

strick·le \'stri-kəl\ *noun* [Middle English *strikell* a piece of wood for leveling a measure of grain; akin to Old English *strīcan* to stroke — more at STRIKE] (1688)
: a foundry tool for smoothing the surface of a core or mold
— strickle *transitive verb*

strict \'strikt\ *adjective* [Middle English *stricte,* from Latin *strictus,* from past participle of *stringere* to bind tight — more at STRAIN] (15th century)
1 *archaic* **a :** TIGHT, CLOSE; *also* **:** INTIMATE **b :** NARROW
2 a : stringent in requirement or control ⟨under *strict* orders⟩ **b :** severe in discipline ⟨a *strict* teacher⟩
3 a : inflexibly maintained or adhered to ⟨*strict* secrecy⟩ **b :** rigorously conforming to principle or a norm or condition
4 : EXACT, PRECISE ⟨in the *strict* sense of the word⟩
5 : of narrow erect habit of growth ⟨a *strict* inflorescence⟩
synonym see RIGID
— strict·ly \'strik(t)-lē\ *adverb*
— strict·ness \-nəs\ *noun*

strict liability *noun* (1896)
: liability imposed without regard to fault

stric·ture \'strik-chər\ *noun* [Middle English, from Late Latin *strictura,* from Latin *strictus,* past participle] (14th century)
1 a : an abnormal narrowing of a bodily passage; *also* **:** the narrowed part **b :** a constriction of the breath passage in the production of a speech sound
2 : something that closely restrains or limits **:** RESTRICTION ⟨moral *strictures*⟩
3 : an adverse criticism **:** CENSURE

¹stride \'strīd\ *verb* **strode** \'strōd\; **strid·den** \'stri-dᵊn\; **strid·ing** \'strī-diŋ\ [Middle English, from Old English *strīdan;* akin to Middle Low German *striden* to straddle, Old High German *strītan* to quarrel] (before 12th century)
intransitive verb
1 : to stand astride
2 : to move with or as if with long steps
3 : to take a very long step
transitive verb
1 : BESTRIDE, STRADDLE
2 : to step over
3 : to move over or along with or as if with long measured steps
— strid·er \'strī-dər\ *noun*

²stride *noun* (before 12th century)
1 a : a cycle of locomotor movements (as of a horse) completed when the feet regain the initial relative positions; *also* **:** the distance traversed in a stride **b :** the most effective natural pace **:** maximum competence or capability — often used in the phrase *hit one's stride*
2 : a long step
3 : an act of striding
4 : a stage of progress **:** ADVANCE
5 : a manner of striding
6 : STRIDE PIANO
— in stride 1 : without interference with regular activities **2 :** without emotional reaction ⟨took the news *in stride*⟩

stri·dence \'strī-dᵊn(t)s\ *noun* (1890)
: STRIDENCY

stri·den·cy \'strī-dᵊn(t)-sē\ *noun* (1865)
: the quality or state of being strident

stri·dent \'strī-dᵊnt\ *adjective* [Latin *strident-, stridens,* present participle of *stridere, stridēre* to make a harsh noise] (circa 1656)
: characterized by harsh, insistent, and discordant sound ⟨a *strident* voice⟩; *also* **:** commanding attention by a loud or obtrusive quality ⟨*strident* slogans⟩
synonym see LOUD, VOCIFEROUS
— stri·dent·ly *adverb*

stride piano *noun* [from *stride bass* left hand part consisting of large skips] (1952)
: a style of jazz piano playing in which the right hand plays the melody while the left hand alternates between a single note and a chord played an octave or more higher

stri·dor \'strī-dər, -ˌdȯr\ *noun* [Latin, from *stridere, stridēre*] (1632)
1 : a harsh, shrill, or creaking noise
2 : a harsh vibrating sound heard during respiration in cases of obstruction of the air passages

strid·u·late \'stri-jə-ˌlāt\ *intransitive verb* **-lat·ed; -lat·ing** [back-formation from *stridulation,* from French, high-pitched sound, from Latin *stridulus* shrill] (1838)
: to make a shrill creaking noise by rubbing together special bodily structures — used especially of male insects (as crickets or grasshoppers)
— strid·u·la·tion \ˌstri-jə-'lā-shən\ *noun*
— strid·u·la·to·ry \'stri-jə-lə-ˌtōr-ē, -ˌtȯr-\ *adjective*

strid·u·lous \'stri-jə-ləs\ *adjective* [Latin *stridulus,* from *stridere, stridēre*] (1611)
: making a shrill creaking sound
— strid·u·lous·ly *adverb*

strife \'strīf\ *noun* [Middle English *strif,* from Old French *estrif,* probably from *estriver* to struggle — more at STRIVE] (13th century)
1 a : bitter sometimes violent conflict or dissension ⟨political *strife*⟩ **b :** an act of contention **:** FIGHT, STRUGGLE
2 : exertion or contention for superiority
3 *archaic* **:** earnest endeavor
synonym see DISCORD
— strife·less \'strī-fləs\ *adjective*

strig·il \'stri-jəl\ *noun* [Latin *strigilis;* akin to Latin *stringere* to touch lightly] (1581)
: an instrument used by ancient Greeks and Romans for scraping moisture off the skin after bathing or exercising

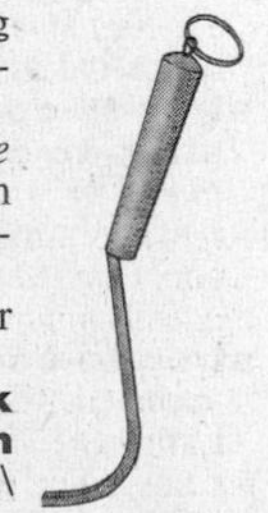
strigil

stri·gose \'strī-ˌgōs\ *adjective* [New Latin *strigosus,* from *striga* row of bristles, from Latin, furrow] (1793)
: having appressed bristles or scales ⟨a *strigose* leaf⟩

¹strike \'strīk\ *verb* **struck** \'strək\; **struck** *also* **strick·en** \'stri-kən\; **strik·ing** \'strī-kiŋ\ [Middle English, from Old English *strīcan* to stroke, go; akin to Old High German *strīhhan* to stroke, Latin *stringere* to touch lightly, *striga, stria* furrow] (before 12th century)
intransitive verb
1 : to take a course **:** GO ⟨*struck* off through the brush⟩
2 a : to aim and usually deliver a blow, stroke, or thrust (as with the hand, a weapon, or a

tool) **b :** to arrive with detrimental effect ⟨disaster *struck*⟩ **c :** to attempt to undermine or harm something as if by a blow ⟨*struck* at . . . cherished notions —R. P. Warren⟩
3 : to come into contact forcefully ⟨two ships *struck* in mid channel⟩
4 : to delete something
5 : to lower a flag usually in surrender
6 a : to become indicated by a clock, bell, or chime ⟨the hour had just *struck*⟩ **b :** to make known the time by sounding ⟨the clock *struck* as they entered⟩
7 : PIERCE, PENETRATE ⟨the wind seemed to *strike* through our clothes⟩
8 a : to engage in battle **b :** to make a military attack
9 : to become ignited
10 : to discover something ⟨*struck* on a new plan of attack⟩
11 a : to pull on a fishing rod in order to set the hook **b** *of a fish* **:** to seize the bait
12 : DART, SHOOT
13 a *of a plant cutting* **:** to take root **b** *of a seed* **:** GERMINATE
14 : to make an impression
15 : to stop work in order to force an employer to comply with demands
16 : to make a beginning ⟨the need to *strike* vigorously for success⟩
17 : to thrust oneself forward ⟨he *struck* into the midst of the argument⟩
18 : to work diligently **:** STRIVE
transitive verb
1 a : to strike at **:** HIT **b :** to drive or remove by or as if by a blow **c :** to attack or seize with a sharp blow (as of fangs or claws) ⟨*struck* by a snake⟩ **d :** INFLICT ⟨*strike* a blow⟩ **e :** to produce by or as if by a blow or stroke ⟨Moses *struck* water from the rock⟩ **f :** to separate by a sharp blow ⟨*strike* off flints⟩
2 a : to haul down **:** LOWER ⟨*strike* the sails⟩ **b :** to dismantle and take away **c :** to strike the tents of (a camp)
3 : to afflict suddenly ⟨*stricken* by a heart attack⟩
4 a : to engage in (a battle) **:** FIGHT **b :** to make a military attack on
5 : DELETE, CANCEL ⟨*strike* the last paragraph⟩
6 a : to penetrate painfully **:** PIERCE **b :** to cause to penetrate **c :** to send down or out ⟨trees *struck* roots deep into the soil⟩
7 a : to level (as a measure of grain) by scraping off what is above the rim **b :** to smooth or form with a strickle
8 : to indicate by sounding
9 a (1) **:** to bring into forceful contact (2) **:** to shake (hands) in confirming an agreement (3) **:** to thrust suddenly **b :** to come into contact or collision with **c** *of light* **:** to fall on **d** *of a sound* **:** to become audible to
10 a : to affect with a mental or emotional state or a strong emotion ⟨*struck* with horror at the sight⟩ **b :** to affect a person with (a strong emotion) ⟨words that *struck* fear in the listeners⟩ **c :** to cause to become by or as if by a sudden blow ⟨*struck* him dead⟩
11 a : to produce by stamping **b** (1) **:** to produce (as fire) by or as if by striking (2) **:** to cause to ignite by friction
12 : to make and ratify the terms of ⟨*strike* a bargain⟩
13 a : to play or produce by stroking keys or strings ⟨*struck* a series of chords on the piano⟩ **b :** to produce as if by playing an instrument ⟨his voice *struck* a note of concern⟩
14 a : to hook (a fish) by a sharp pull on the line **b** *of a fish* **:** to snatch at (a bait)
15 a : to occur to ⟨the answer *struck* me suddenly⟩ **b :** to appear to especially as a revelation or as remarkable **:** IMPRESS
16 : BEWITCH
17 : to arrive at by or as if by computation ⟨*strike* a balance⟩
18 a : to come to **:** ATTAIN **b :** to come upon **:** DISCOVER ⟨*strike* gold⟩
19 : to engage in a strike against (an employer)
20 : TAKE ON, ASSUME ⟨*strike* a pose⟩
21 a : to place (a plant cutting) in a medium for growth and rooting **b :** to so propagate (a plant)
22 : to make one's way along
23 : to cause (an arc) to form (as between electrodes of an arc lamp)
24 *of an insect* **:** to oviposit on or in
synonym see AFFECT
— strike it rich : to become rich usually suddenly

[2]**strike** *noun* (15th century)
1 : STRICKLE
2 : an act or instance of striking
3 a : a work stoppage by a body of workers to enforce compliance with demands made on an employer **b :** a temporary stoppage of activities in protest against an act or condition
4 : the direction of the line of intersection of a horizontal plane with an uptilted geological stratum
5 a : a pull on a fishing rod to strike a fish **b :** a pull on a line by a fish in striking
6 : a stroke of good luck; *especially* **:** a discovery of a valuable mineral deposit
7 a : a pitched ball that is in the strike zone or is swung at and is not hit fair **b :** a perfectly thrown ball or pass
8 : DISADVANTAGE, HANDICAP
9 : an act or instance of knocking down all the bowling pins with the first bowl
10 : establishment of roots and plant growth
11 : cutaneous myiasis (as of sheep)
12 a : a military attack; *especially* **:** an air attack on a single objective **b :** a group of airplanes taking part in such an attack

strike·bound \ˈstrīk-ˌbau̇nd\ *adjective* (1943)
: subjected to a strike

strike·break·er \-ˌbrā-kər\ *noun* (1904)
: a person hired to replace a striking worker

strike·break·ing \-kiŋ\ *noun* (1905)
: action designed to break up a strike

strike force *noun* (1968)
: a team of federal agents assigned to investigate organized crime in a specific area

strike off *transitive verb* (1821)
1 : to produce in an effortless manner
2 : to depict clearly and exactly

strike·out \ˈstrīk-ˌau̇t\ *noun* (1887)
: an out in baseball resulting from a batter's being charged with three strikes

strike out (1712)
intransitive verb
1 : to enter upon a course of action
2 : to set out vigorously
3 : to make an out in baseball by a strikeout
4 : to finish bowling a string with consecutive strikes; *specifically* **:** to bowl three strikes in the last frame
5 : FAIL 2c
transitive verb
of a baseball pitcher **:** to retire (as a batter) by a strikeout

strike·over \ˈstrīk-ˌō-vər\ *noun* (1938)
: an act or instance of striking a typewriter character on a spot occupied by another character

strik·er \ˈstrī-kər\ *noun* (1581)
1 : one that strikes: as **a :** a player in any of several games who is striking or attempting to strike a ball **b :** the hammer of the striking mechanism of a clock or watch **c :** a blacksmith's helper who swings the sledgehammer **d :** a worker on strike
2 : a junior enlisted man in the U.S. Navy who has declared an occupational specialty

strike up (circa 1562)
intransitive verb
: to begin to sing or play or to be sung or played
transitive verb
1 : to cause to begin singing or playing ⟨*strike up* the band⟩
2 : to cause to begin ⟨*strike up* a conversation⟩

strike zone *noun* (1948)
: the area over home plate through which a pitched baseball must pass to be called a strike

strik·ing \ˈstrī-kiŋ\ *adjective* (1752)
: attracting attention or notice through unusual or conspicuous qualities ⟨a place of *striking* beauty⟩
synonym see NOTICEABLE
— strik·ing·ly \-kiŋ-lē\ *adverb*

striking distance *noun* (1751)
: a distance from which something can be easily reached or attained ⟨almost within *striking distance* of their goal⟩

striking price *noun* (1961)
: an agreed-upon price at which an option contract can be exercised — called also *strike price*

[1]**string** \ˈstriŋ\ *noun* [Middle English, from Old English *streng;* akin to Old High German *strang* rope, Latin *stringere* to bind tight — more at STRAIN] (before 12th century)
1 : a small cord used to bind, fasten, or tie
2 a *archaic* **:** a cord (as a tendon or ligament) of an animal body **b :** a plant fiber (as a leaf vein)
3 a : the gut, wire, or nylon cord of a musical instrument **b** *plural* (1) **:** the stringed instruments of an orchestra (2) **:** the players of such instruments
4 a : a group of objects threaded on a string ⟨a *string* of fish⟩ ⟨a *string* of pearls⟩ **b** (1) **:** a series of things arranged in or as if in a line ⟨a *string* of cars⟩ ⟨a *string* of names⟩ (2) **:** a sequence of like items (as bits, characters, or words) **c :** a group of business properties scattered geographically ⟨a *string* of newspapers⟩ **d :** the animals and especially horses belonging to or used by one individual
5 a : a means of recourse **:** EXPEDIENT **b :** a group of players ranked according to skill or proficiency
6 : SUCCESSION 3a ⟨a *string* of successes⟩
7 a : one of the inclined sides of a stair supporting the treads and risers **b :** STRINGCOURSE
8 a : BALKLINE 1 **b :** the action of lagging for break in billiards
9 : LINE 13
10 *plural* **a :** contingent conditions or obligations **b :** CONTROL, DOMINATION
11 : a scanty bikini — called also *string bikini*
12 : a hypothetical one-dimensional object that is infinitely thin but has a length of 10^{-33} centimeters, that vibrates as it moves through space, and whose mode of vibration manifests itself as a subatomic particle
— string·less \ˈstriŋ-ləs\ *adjective*
— on the string : subject to one's influences

[2]**string** *verb* **strung** \ˈstrəŋ\; **string·ing** \ˈstriŋ-iŋ\ (15th century)
transitive verb
1 a : to equip with strings **b :** to tune the strings of
2 : to make tense **:** key up
3 a : to thread on or as if on a string **b :** to thread with objects **c :** to tie, hang, or fasten with string **d :** to put together (as words or ideas) like objects threaded on a string
4 : to hang by the neck — used with *up*
5 : to remove the strings of ⟨*string* beans⟩
6 a : to extend or stretch like a string ⟨*string* wires from tree to tree⟩ **b :** to set out in a line or series — often used with *out*
7 : FOOL, HOAX ⟨cowboys *stringing* tenderfeet with tall tales —Carl Van Doren⟩ — often used with *along*
intransitive verb
1 : to move, progress, or lie in a string
2 : to form into strings

3 : LAG 3
[3]string *adjective* (1712)
: of or relating to stringed musical instruments ⟨the *string* section⟩
string along (1914)
transitive verb
: to keep waiting
intransitive verb
: GO ALONG, AGREE
string bass *noun* (circa 1927)
: DOUBLE BASS
string bean *noun* (1759)
1 : a bean of one of the older varieties of kidney bean that have stringy fibers on the lines of separation of the pods; *broadly* **:** SNAP BEAN
2 : a very tall thin person

string bean 1

string·course \-ˌkōrs, -ˌkȯrs\ *noun* (1825)
: a horizontal band (as of bricks) in a building forming a part of the design
stringed \'striŋd\ *adjective* (before 12th century)
1 : having strings ⟨*stringed* instruments⟩
2 : produced by strings
strin·gen·cy \'strin-jən(t)-sē\ *noun* (1844)
: the quality or state of being stringent
strin·gen·do \strin-'jen-(ˌ)dō\ *adverb* [Italian, verbal of *stringere* to press, from Latin, to bind tight — more at STRAIN] (1853)
: with quickening of tempo (as to a climax) — used as a direction in music
strin·gent \'strin-jənt\ *adjective* [Latin *stringent-, stringens,* present participle of *stringere*] (1736)
1 : TIGHT, CONSTRICTED
2 : marked by rigor, strictness, or severity especially with regard to rule or standard
3 : marked by money scarcity and credit strictness
synonym see RIGID
— strin·gent·ly *adverb*
string·er \'striŋ-ər\ *noun* (14th century)
1 : one that strings
2 : a string, wire, or chain often with snaps on which fish are strung by a fisherman
3 : a narrow vein or irregular filament of mineral traversing a rock mass of different material
4 a : a long horizontal timber to connect uprights in a frame or to support a floor **b :** STRING 7a **c :** a tie in a truss
5 a : a longitudinal member extending from bent to bent of a railroad bridge and carrying the track **b :** a longitudinal member (as in an airplane fuselage or wing) to reinforce the skin
6 a : a news correspondent who is paid space rates **b :** a reporter who works for a publication or news agency on a part-time basis; *broadly* **:** CORRESPONDENT
7 : one estimated to be of specified excellence or efficiency — used in combination ⟨first-*stringer*⟩ ⟨second-*stringer*⟩
string·halt \'striŋ-ˌhȯlt\ *noun* (circa 1534)
: a condition of lameness in a horse's hind legs caused by muscular spasms
— string·halt·ed \-ˌhȯl-təd\ *adjective*
string·ing \'striŋ-iŋ\ *noun* (1812)
1 : lines of inlay in furniture decoration
2 : the material with which a racket is strung
string line *noun* (1867)
: BALKLINE 1
string·piece \'striŋ-ˌpēs\ *noun* (1769)
: the heavy squared timber lying along the top of the piles forming a dock front or timber pier
string quartet *noun* (1875)
1 : a composition for string quartet
2 : a quartet of performers on stringed instruments usually including a first and second violin, a viola, and a cello
string tie *noun* (1895)
: a narrow necktie
stringy \'striŋ-ē\ *adjective* **string·i·er; -est** (1669)
1 a : containing, consisting of, or resembling fibrous matter or string ⟨*stringy* hair⟩ **b :** lean and sinewy in build **:** WIRY
2 : capable of being drawn out to form a string **:** ROPY ⟨a *stringy* precipitate⟩
— string·i·ness *noun*
stringy·bark \'striŋ-ē-ˌbärk\ *noun* (1799)
1 : any of several Australian eucalypti with fibrous inner bark
2 : the bark of a stringybark
[1]strip \'strip\ *verb* **stripped** \'stript\ *also* **stript; strip·ping** [Middle English *strippen,* from Old English *-strīepan;* akin to Old High German *stroufen* to strip] (13th century)
transitive verb
1 a : to remove clothing, covering, or surface matter from **b :** to deprive of possessions **c :** to divest of honors, privileges, or functions
2 a : to remove extraneous or superficial matter from ⟨a prose style *stripped* to the bones⟩ **b :** to remove furniture, equipment, or accessories from ⟨*strip* a ship for action⟩
3 : to make bare or clear (as by cutting or grazing)
4 : to finish a milking of by pressing the last available milk from the teats ⟨*strip* a cow⟩
5 a : to remove cured leaves from the stalks of (tobacco) **b :** to remove the midrib from (tobacco leaves)
6 : to tear or damage the thread of (a separable part or fitting)
7 : to separate (components) from a mixture or solution
8 : to press eggs or milt out of (a fish)
intransitive verb
1 a : to take off clothes **b :** to perform a striptease
2 : PEEL 1
— strip·pa·ble \'stri-pə-bəl\ *adjective*
[2]strip *noun* [Middle English, perhaps from Middle Low German *strippe* strap] (15th century)
1 a : a long narrow piece of a material **b :** a long narrow area of land or water
2 : AIRSTRIP
3 : a commercially developed area especially along a highway
4 : COMIC STRIP
5 : STRIPTEASE
strip–chart recorder \'strip-ˌchärt-\ *noun* (1950)
: a device used for the continuous graphic recording of time-dependent data
— strip–chart recording *noun*
strip–crop·ping \'strip-ˌkrä-piŋ\ *noun* (1936)
: the growing of a cultivated crop (as corn) in strips alternating with strips of a sod-forming crop (as hay) arranged to follow an approximate contour of the land and minimize erosion
— strip–crop \'strip-ˌkräp\ *verb*
[1]stripe \'strīp\ *noun* [Middle English; akin to Middle Dutch *stripe* strip, stripe] (15th century)
: a stroke or blow with a rod or lash
[2]stripe *transitive verb* **striped** \'strīpt\; **strip·ing** [probably from Middle Dutch; akin to Middle High German *strīfe* stripe] (15th century)
: to make stripes on or variegate with stripes
[3]stripe *noun* (1626)
1 a : a line or long narrow section differing in color or texture from parts adjoining **b** (1) **:** a textile design consisting of lines or bands against a plain background (2) **:** a fabric with a striped design
2 : a narrow strip of braid or embroidery usually in the shape of a bar, arc, or chevron that is worn (as on the sleeve of a military uniform) to indicate rank or length of service
3 : a distinct variety or sort **:** TYPE ⟨persons of the same political *stripe*⟩
— stripe·less \'strī-pləs\ *adjective*
striped \'strīpt, 'strī pəd\ *adjective* (1616)
: having stripes or streaks
striped bass *noun* (1818)
: a large anadromous food and sport fish (*Morone saxatilis* of the family Percichthyidae) that occurs along the Atlantic coast of the U.S. and has been introduced along the Pacific coast — called also *rockfish*
striped skunk *noun* (1882)
: a common North American skunk (*Mephitis mephitis*) usually with white on the top of the head that extends posteriorly in two narrowly separated stripes
strip·er \'strī-pər\ *noun* (1937)
: STRIPED BASS
strip·ing \'strī-piŋ\ *noun* (1677)
1 a : the stripes marked or painted on something **b :** a design of stripes
2 : the act or process of marking with stripes
strip·ling \'stri-pliŋ\ *noun* [Middle English] (14th century)
: YOUTH 2a
strip mine *noun* (1926)
: a mine that is worked from the earth's surface by the stripping of overburden; *especially* **:** a coal mine situated along the outcrop of a flat dipping bed
— strip–mine *verb*
— strip miner *noun*
stripped–down \'strip(t)-'dau̇n\ *adjective* (1928)
: lacking any extra features **:** SIMPLE 2b
strip·per \'stri-pər\ *noun* (1581)
1 : one that strips
2 : STRIPTEASER
3 : a machine that separates a desired part of an agricultural crop
4 : an oil well that produces 10 barrels or less per day
strip poker *noun* (1919)
: a poker game in which players pay their losses by removing articles of clothing
strip search *noun* (1947)
: a search for something concealed on a person made after removal of the person's clothing
— strip–search *verb*
strip·tease \'strip-ˌtēz\ *noun* (1936)
: a burlesque act in which a performer removes clothing piece by piece
strip·teas·er \-ˌtē-zər\ *noun* (1930)
: one who performs a striptease
stripy \'strī-pē\ *adjective* **strip·i·er; -est** (1513)
: marked by stripes or streaks
strive \'strīv\ *intransitive verb* **strove** \'strōv\ *also* **strived** \'strīvd\; **striv·en** \'stri-vən\ *or* **strived; striv·ing** \'strī-viŋ\ [Middle English, from Old French *estriver,* of Germanic origin; akin to Middle High German *streben* to endeavor] (13th century)
1 : to devote serious effort or energy **:** ENDEAVOR
2 : to struggle in opposition **:** CONTEND
synonym see ATTEMPT
— striv·er \'strī-vər\ *noun*
strobe \'strōb\ *noun* [by shortening & alteration] (1942)
1 : STROBOSCOPE
2 : a device that utilizes a flashtube for high-speed illumination (as in photography)
3 : STROBOTRON
strobe light *noun* (1947)
: STROBE
stro·bi·la \strō-'bī-lə, 'strō-bə-\ *noun, plural* **-lae** \-(ˌ)lē\ [New Latin, from Greek *strobilē* plug of lint shaped like a pinecone, from *strobilos* pinecone] (1864)
: a linear series of similar animal structures (as the proglottids of a tapeworm) produced by budding
stro·bi·la·tion \ˌstrō-bə-'lā-shən\ *noun* [New Latin *strobila*] (1878)

: asexual reproduction (as in various coelenterates and tapeworms) by transverse division of the body into segments which develop into separate individuals, zooids, or proglottids

stro·bi·lus \strō-ˈbī-ləs, ˈstrō-bə-\ *noun, plural* **-li** \-ˌlī\ [New Latin, from Late Latin, pinecone, from Greek *strobilos* twisted object, top, pinecone, from *strobos* action of whirling; akin to *strephein* to twist] (circa 1753)
1 : an aggregation of sporophylls resembling a cone (as in the club mosses and horsetails)
2 : the cone of a gymnosperm

stro·bo·scope \ˈstrō-bə-ˌskōp\ *noun* [Greek *strobos* whirling + International Scientific Vocabulary *-scope*] (1896)
: an instrument for determining the speed of cyclic motion (as rotation or vibration) that causes the motion to appear slowed or stopped: as **a :** a revolving disk with holes around the edge through which an object is viewed **b :** a device that uses a flashtube to intermittently illuminate a moving object **c :** a cardboard disk with marks to be viewed under intermittent light

stro·bo·scop·ic \ˌstrō-bə-ˈskä-pik\ *adjective* (circa 1846)
: of, utilizing, or relating to a stroboscope or a strobe
— stro·bo·scop·i·cal·ly \-pi-k(ə-)lē\ *adverb*

stro·bo·tron \ˈstrō-bə-ˌträn\ *noun* [*strobo*scope + *-tron*] (1937)
: a gas-filled electron tube used especially as a source of bright flashes of light for a stroboscope

strode *past of* STRIDE

[1]stroke \ˈstrōk\ *transitive verb* **stroked; strok·ing** [Middle English, from Old English *strācian;* akin to Old High German *strīhhan* to stroke — more at STRIKE] (before 12th century)
1 : to rub gently in one direction; *also* **:** CARESS
2 : to flatter or pay attention to in a manner designed to reassure or persuade
— strok·er *noun*

[2]stroke *noun* [Middle English; akin to Old English *strīcan* to stroke — more at STRIKE] (13th century)
1 : the act of striking; *especially* **:** a blow with a weapon or implement
2 : a single unbroken movement; *especially* **:** one of a series of repeated or to-and-fro movements
3 a : a controlled swing intended to hit a ball or shuttlecock; *also* **:** a striking of the ball **b :** such a stroke charged to a player as a unit of scoring in golf
4 a : a sudden action or process producing an impact ⟨*stroke* of lightning⟩ **b :** an unexpected result ⟨*stroke* of luck⟩
5 : sudden diminution or loss of consciousness, sensation, and voluntary motion caused by rupture or obstruction (as by a clot) of an artery of the brain — called also *apoplexy*
6 a : one of a series of propelling beats or movements against a resisting medium ⟨a *stroke* of the oar⟩ **b :** a rower who sets the pace for a crew
7 a : a vigorous or energetic effort ⟨a *stroke* of genius⟩ **b :** a delicate or clever touch in a narrative, description, or construction
8 : HEARTBEAT
9 : the movement or the distance of the movement in either direction of a mechanical part (as a piston) having a reciprocating motion
10 : the sound of a bell being struck ⟨at the *stroke* of twelve⟩
11 [[1]*stroke*] **:** an act of stroking or caressing
12 a : a mark or dash made by a single movement of an implement **b :** one of the lines of a letter of the alphabet

[3]stroke *verb* **stroked; strok·ing** (1597)
transitive verb
1 a : to mark with a short line ⟨*stroke* the *t*'s⟩ **b :** to cancel by drawing a line through ⟨*stroked* out his name⟩
2 : to set the stroke for (a rowing crew); *also* **:** to set the stroke for the crew of (a rowing boat)
3 : HIT; *especially* **:** to propel (a ball) with a controlled swinging blow
intransitive verb
1 : to execute a stroke
2 : to row at a certain number of strokes a minute

stroke play *noun* (1905)
: golf competition scored by total number of strokes

stroll \ˈstrōl\ *verb* [probably from German dialect *strollen*] (1680)
intransitive verb
1 : to walk in a leisurely or idle manner **:** RAMBLE
2 : to go from place to place in search of work or profit ⟨*strolling* players⟩ ⟨*strolling* musicians⟩
transitive verb
: to walk at leisure along or about
— stroll *noun*

stroll·er \ˈstrō-lər\ *noun* (1608)
1 a : an itinerant actor **b :** VAGRANT, TRAMP
2 : one that strolls
3 : a collapsible carriage designed as a chair in which a small child may be pushed

stro·ma \ˈstrō-mə\ *noun, plural* **stro·ma·ta** \-mə-tə\ [New Latin *stromat-, stroma,* from Latin, bed covering, from Greek *strōmat-, strōma,* from *stornynai* to spread out — more at STREW] (1832)
1 a : a compact mass of fungal hyphae producing perithecia or pycnidia **b :** the colorless proteinaceous matrix of a chloroplast in which the chlorophyll-containing lamellae are embedded
2 a : the supporting framework of an animal organ typically consisting of connective tissue **b :** the spongy protoplasmic framework of some cells (as a red blood cell)
— stro·mal \-məl\ *adjective*

stro·mat·o·lite \strō-ˈma-tᵊl-ˌīt\ *noun* [Latin *stromat-, stroma* bed covering + English *-o-* + *-lite*] (1930)
: a laminated sedimentary fossil formed from layers of blue-green algae
— stro·mat·o·lit·ic \-ˌma-tᵊl-ˈi-tik\ *adjective*

strong \ˈstrȯŋ\ *adjective* **stron·ger** \ˈstrȯŋ-gər *also* -ər\; **stron·gest** \ˈstrȯŋ-gəst *also* -əst\ [Middle English, from Old English *strang;* akin to Old High German *strengi* strong, Latin *stringere* to bind tight — more at STRAIN] (before 12th century)
1 : having or marked by great physical power
2 : having moral or intellectual power
3 : having great resources (as of wealth or talent)
4 : of a specified number ⟨an army ten thousand *strong*⟩
5 a : striking or superior of its kind ⟨a *strong* resemblance⟩ **b :** effective or efficient especially in a specified direction ⟨*strong* on watching other people work —A. Alvarez⟩
6 : FORCEFUL, COGENT ⟨*strong* evidence⟩ ⟨*strong* talk⟩
7 : not mild or weak **:** EXTREME, INTENSE: as **a :** rich in some active agent ⟨*strong* beer⟩ **b** *of a color* **:** high in chroma **c :** ionizing freely in solution ⟨*strong* acids and bases⟩ **d :** magnifying by refracting greatly ⟨*strong* lens⟩
8 *obsolete* **:** FLAGRANT
9 : moving with rapidity or force ⟨*strong* wind⟩
10 : ARDENT, ZEALOUS ⟨a *strong* supporter⟩
11 a : not easily injured or disturbed **:** SOLID **b :** not easily subdued or taken ⟨a *strong* fort⟩
12 : well established **:** FIRM ⟨*strong* beliefs⟩
13 : not easily upset or nauseated ⟨a *strong* stomach⟩
14 : having an offensive or intense odor or flavor **:** RANK
15 : tending to steady or higher prices ⟨a *strong* market⟩
16 : of, relating to, or being a verb that is inflected by a change in the root vowel (as *strive, strove, striven*) rather than by regular affixation ☆
— strong *adverb*
— strong·ish \ˈstrȯŋ-ish\ *adjective*
— strong·ly \ˈstrȯŋ-lē\ *adverb*

strong anthropic principle *noun* (1985)
: ANTHROPIC PRINCIPLE b

[1]strong–arm \ˈstrȯŋ-ˈärm\ *adjective* (1897)
: having or using undue force

[2]strong–arm *transitive verb* (1903)
1 a : to use force on **:** ASSAULT **b :** BULLY, INTIMIDATE
2 : to rob by force

strong·box \ˈstrȯŋ-ˌbäks\ *noun* (1684)
: a strongly made chest or case for money or valuables

strong breeze *noun* (1867)
: wind having a speed of 25 to 31 miles (40 to 50 kilometers) per hour — see BEAUFORT SCALE table

strong drink *noun* (14th century)
: intoxicating liquor

strong force *noun* (1969)
: a fundamental physical force that acts on hadrons and is responsible for the binding together of protons and neutrons in the atomic nucleus and for processes of particle creation in high-energy collisions and that is the strongest known fundamental physical force but acts only over distances comparable to those between nucleons in an atomic nucleus — called also *strong interaction, strong nuclear force;* compare ELECTROMAGNETISM 2a, GRAVITY 3a(2), WEAK FORCE

strong gale *noun* (circa 1867)
: wind having a speed of 47 to 54 miles (76 to 87 kilometers) per hour — see BEAUFORT SCALE table

strong·hold \ˈstrȯŋ-ˌhōld\ *noun* (15th century)
1 : a fortified place
2 a : a place of security or survival ⟨one of the last *strongholds* of the ancient Gaelic language —George Holmes⟩ **b :** a place dominated by a particular group or marked by a particular characteristic ⟨a Republican *stronghold*⟩ ⟨*strongholds* of snobbery —Lionel Trilling⟩

strong·man \ˈstrȯŋ-ˌman\ *noun* (1859)
: one who leads or controls by force of will and character or by military methods

strong–mind·ed \ˈstrȯŋ-ˈmīn-dəd\ *adjective* (1791)

☆ SYNONYMS
Strong, stout, sturdy, stalwart, tough, tenacious mean showing power to resist or to endure. STRONG may imply power derived from muscular vigor, large size, structural soundness, intellectual or spiritual resources ⟨*strong* arms⟩ ⟨the defense has a *strong* case⟩. STOUT suggests an ability to endure stress, pain, or hard use without giving way ⟨*stout* hiking boots⟩. STURDY implies strength derived from vigorous growth, determination of spirit, solidity of construction ⟨a *sturdy* table⟩ ⟨people of *sturdy* independence⟩. STALWART suggests an unshakable dependability ⟨*stalwart* environmentalists⟩. TOUGH implies great firmness and resiliency ⟨a *tough* political opponent⟩. TENACIOUS suggests strength in seizing, retaining, clinging to, or holding together ⟨*tenacious* farmers clinging to an age-old way of life⟩.

: having a vigorous mind; *especially* **:** marked by independence of thought and judgment
— **strong–mind·ed·ly** *adverb*
— **strong–mind·ed·ness** *noun*

strong room *noun* (1761)
: a room for money or valuables specially constructed to be fireproof and burglarproof

strong safety *noun* (1970)
: a safety in football who plays opposite the strong side of an offensive formation

strong side *noun* (circa 1951)
: the side of a football formation having the greater number of players; *specifically* : the side on which the tight end plays

strong suit *noun* (1857)
1 : a long suit containing high cards
2 : something in which one excels **:** FORTE ⟨details of legislation have never been my *strong suit* —Tip O'Neill⟩

stron·gyle \'strän-ˌjīl, -jəl\ *noun* [New Latin *Strongylus,* genus of worms, from Greek *strongylos* round, compact; akin to Greek *stranx* drop squeezed out — more at STRAIN] (1847)
: any of various nematode worms (family Strongylidae) related to the hookworms and parasitic especially in the alimentary tract and tissues of horses

stron·gy·loi·di·a·sis \ˌsträn-jə-ˌlȯi-'dī-ə-səs\ *also* **stron·gy·loi·do·sis** \-'dō-səs\ *noun* [New Latin, from *Strongyloides,* genus name, from *Strongylus*] (1905)
: infestation with or disease caused by any of a genus (*Strongyloides*) of strongyles that sometimes parasitize the intestines of vertebrates including humans

stron·tian·ite \'strän(t)-shə-ˌnīt\ *noun* (1794)
: a mineral consisting of a carbonate of strontium and occurring in various forms and colors

stron·tium \'strän(t)-sh(ē-)əm, 'strän-tē-əm\ *noun* [New Latin, from *strontia* strontium oxide, from obsolete English *strontian,* from *Strontian,* village in Scotland] (1808)
: a soft malleable ductile metallic element of the alkaline-earth group occurring only in combination and used especially in color TV tubes, in crimson fireworks, and in the production of some ferrites — see ELEMENT table

strontium 90 *noun* (1952)
: a heavy radioactive isotope of strontium of mass number 90 that has a half-life of 29 years and that is present in the fallout from nuclear explosions and is hazardous because like calcium it can be assimilated in biological processes and deposited in the bones of human beings and animals — called also *radiostrontium*

¹strop \'sträp\ *noun* [Middle English — more at STRAP] (before 12th century)
: STRAP: **a :** a short rope with its ends spliced to form a circle **b :** a usually leather band for sharpening a razor

²strop *transitive verb* **stropped; stropping** (1841)
: to sharpen (a razor) on a strop

stro·phan·thin \strō-'fan(t)-thən\ *noun* [International Scientific Vocabulary, from New Latin *Strophanthus,* from Greek *strophos* twisted band (from *strephein* to twist) + *anthos* flower — more at ANTHOLOGY] (1873)
: any of several glycosides (as ouabain) or mixtures of glycosides from African plants (genera *Strophanthus* and *Acokanthera*) of the dogbane family; *especially* **:** a bitter toxic glycoside $C_{36}H_{54}O_{14}$ from a woody vine (*Strophanthus kombé*) used similarly to digitalis

stro·phe \'strō-(ˌ)fē\ *noun* [Greek *strophē,* literally, act of turning, from *strephein* to turn, twist] (1603)
1 a : a rhythmic system composed of two or more lines repeated as a unit; *especially* **:** such a unit recurring in a series of strophic units **b :** STANZA
2 a : the movement of the classical Greek chorus while turning from one side to the other of the orchestra **b :** the part of a Greek choral ode sung during the strophe of the dance

stro·phic \'strō-fik, 'strä-\ *adjective* (1848)
1 : relating to, containing, or consisting of strophes
2 *of a song* **:** using the same music for successive stanzas — compare THROUGH-COMPOSED

strop·py \'strä-pē\ *adjective* [perhaps by shortening & alteration from *obstreperous*] (1951)
British **:** TOUCHY, BELLIGERENT

stroud \'strau̇d\ *noun* [probably from *Stroud,* town in England] (1683)
1 *also* **stroud·ing** \'strau̇-diŋ\ **:** a coarse woolen cloth formerly used in trade with North American Indians
2 : a blanket or garment of stroud

strove *past & chiefly dialect past participle of* STRIVE

strow \'strō\ *transitive verb* **strowed; strown** \'strōn\ *or* **strowed; strow·ing** [Middle English — more at STREW] (14th century)
archaic **:** SCATTER

stroy *verb* [Middle English *stroyen,* short for *destroyen*] (13th century)
obsolete **:** DESTROY

¹struck \'strək\ *past and past participle of* STRIKE

²struck *adjective* (1894)
: closed by or subjected to a labor strike ⟨a *struck* factory⟩ ⟨a *struck* employer⟩

struc·tur·al \'strək-chə-rəl, 'strək-shrəl\ *adjective* (1835)
1 : of or relating to the physical makeup of a plant or animal body
2 a : of, relating to, or affecting structure ⟨*structural* stability⟩ **b :** used in building structures ⟨*structural* clay⟩ **c :** involved in or caused by structure especially of the economy ⟨*structural* unemployment⟩
3 : of, relating to, or resulting from the effects of folding or faulting of the earth's crust **:** TECTONIC
4 : concerned with or relating to structure rather than history or comparison ⟨*structural* linguistics⟩
— **struc·tur·al·ly** *adverb*

structural formula *noun* (1872)
: an expanded molecular formula showing the arrangement within the molecule of atoms and of bonds

structural gene *noun* (1959)
: a gene that codes for the amino acid sequence of a protein (as an enzyme) or for a ribosomal RNA or transfer RNA

struc·tur·al·ism \'strək-chə-rə-ˌli-zəm, 'strək-shrə-\ *noun* (1907)
1 : psychology concerned especially with resolution of the mind into structural elements
2 : structural linguistics
3 : an anthropological movement associated especially with Claude Lévi-Strauss that seeks to analyze social relationships in terms of highly abstract relational structures often expressed in a logical symbolism
4 : a method of analysis (as of a literary text or a political system) that is related to cultural anthropology and that focuses on recurring patterns of thought and behavior
— **struc·tur·al·ist** \-list\ *noun or adjective*

structural isomer *noun* (1926)
: one of two or more compounds that contain the same number and kinds of atoms but that differ significantly in their geometric arrangement

struc·tur·al·ize \'strək-chə-rə-ˌlīz, 'strək-shrə-ˌlīz\ *transitive verb* **-ized; -iz·ing** (circa 1931)
: to organize or incorporate into a structure
— **struc·tur·al·i·za·tion** \ˌstrək-chə-rə-lə-'zā-shən, ˌstrək-shrə-\ *noun*

structural steel *noun* (1895)
1 : rolled steel in structural shapes
2 : steel suitable for structural shapes

struc·tur·a·tion \ˌstrək-chə-'rā-shən, -shə-'rā-\ *noun* (1925)
: the interrelation of parts in an organized whole

¹struc·ture \'strək-chər\ *noun* [Middle English, from Latin *structura,* from *structus,* past participle of *struere* to heap up, build — more at STREW] (15th century)
1 : the action of building **:** CONSTRUCTION
2 a : something (as a building) that is constructed **b :** something arranged in a definite pattern of organization ⟨a rigid totalitarian *structure* —J. L. Hess⟩ ⟨leaves and other plant *structures*⟩
3 : manner of construction **:** MAKEUP ⟨Gothic in *structure*⟩
4 a : the arrangement of particles or parts in a substance or body ⟨soil *structure*⟩ ⟨molecular *structure*⟩ **b :** organization of parts as dominated by the general character of the whole ⟨economic *structure*⟩ ⟨personality *structure*⟩
5 : the aggregate of elements of an entity in their relationships to each other
— **struc·ture·less** \-ləs\ *adjective*
— **struc·ture·less·ness** \-nəs\ *noun*

²structure *transitive verb* **struc·tured; struc·tur·ing** \'strək-chə-riŋ, 'strək-shriŋ\ (circa 1693)
1 : to form into or according to a structure
2 : CONSTRUCT

struc·tured *adjective* (1966)
: of, relating to, or being a method of computer programming in which each step of the solution to a problem is contained in a separate subprogram

stru·del \'strü-dᵊl, 'shtrü-\ *noun* [German, literally, whirlpool] (circa 1893)
: a pastry made from a thin sheet of dough rolled up with filling and baked ⟨apple *strudel*⟩

¹strug·gle \'strə-gəl\ *intransitive verb* **struggled; strug·gling** \-g(ə-)liŋ\ [Middle English *struglen*] (14th century)
1 : to make strenuous or violent efforts against opposition **:** CONTEND
2 : to proceed with difficulty or with great effort ⟨*struggled* through the high grass⟩ ⟨*struggling* to make a living⟩
— **strug·gler** \-g(ə-)lər\ *noun*

²struggle *noun* (1692)
1 : CONTEST, STRIFE ⟨armed *struggle*⟩ ⟨a power *struggle*⟩
2 : a violent effort or exertion **:** an act of strongly motivated striving

struggle for existence (1832)
: the automatic competition (as for food, space, or light) of members of a natural population that tends to eliminate less efficient individuals and thereby increase the chance of the more efficient to pass on inherited adaptive traits

¹strum \'strəm\ *verb* **strummed; strumming** [imitative] (1777)
transitive verb
1 a : to brush the fingers over the strings of (a musical instrument) in playing ⟨*strum* a guitar⟩; *also* **:** ³THRUM 1 **b :** to play (music) on a stringed instrument ⟨*strum* a tune⟩
2 : to cause to sound vibrantly ⟨winds *strummed* the rigging —H. A. Chippendale⟩
intransitive verb
1 : to strum a stringed instrument
2 : to sound vibrantly
— **strum·mer** *noun*

²strum *noun* (circa 1793)
: an act, instance, or sound of strumming

stru·ma \'strü-mə\ *noun, plural* **stru·mae** \-(ˌ)mē, -ˌmī\ *or* **strumas** [Latin, swelling of the lymph glands] (1565)
: GOITER

strum·pet \'strəm-pət\ *noun* [Middle English] (14th century)
: PROSTITUTE 1a

strung \'strəŋ\ *past and past participle of* STRING

strung out *adjective* (circa 1959)
1 : physically debilitated (as from long-term drug addiction)

2 : addicted to a drug
3 : intoxicated or stupefied from drug use

strunt \ˈstrənt\ *intransitive verb* [by alteration] (1786)
Scottish **:** STRUT

¹strut \ˈstrət\ *verb* **strut·ted; strut·ting** [Middle English *strouten,* from Old English *strūtian* to exert oneself; akin to Middle High German *strozzen* to be swollen] (13th century)
intransitive verb
1 : to become turgid **:** SWELL
2 a : to walk with a proud gait **b :** to walk with a pompous and affected air
transitive verb
: to parade (as clothes) with a show of pride
— strut·ter *noun*
— strut one's stuff : to display one's best work **:** SHOW OFF

²strut *noun* (1587)
1 : a structural piece designed to resist pressure in the direction of its length
2 : a pompous step or walk
3 : arrogant behavior **:** SWAGGER

³strut *transitive verb* **strut·ted; strut·ting** (circa 1828)
: to provide, stiffen, support, or hold apart with or as if with a strut

stru·thi·ous \ˈstrü-thē-əs, -t͟hē-\ *adjective* [Late Latin *struthio* ostrich, irregular from Greek *strouthos*] (1773)
: of or relating to the ostriches and related birds

strych·nine \ˈstrik-ˌnīn, -nən, -ˌnēn\ *noun* [French, from New Latin *Strychnos,* from Latin, nightshade, from Greek] (1819)
: a bitter poisonous alkaloid $C_{21}H_{22}N_2O_2$ that is obtained from nux vomica and related plants (genus *Strychnos*) and is used as a poison (as for rodents) and medicinally as a stimulant to the central nervous system

Stu·art \ˈstü-ərt, ˈstyü-; ˈst(y)u̇(-ə)rt\ *adjective* [Robert *Stewart* (Robert II of Scotland) (died 1390)] (1873)
: of or relating to the Scottish royal house to which belonged the rulers of Scotland from 1371 to 1603 and of Great Britain from 1603 to 1649 and from 1660 to 1714
— Stuart *noun*

¹stub \ˈstəb\ *noun* [Middle English *stubb,* from Old English *stybb;* akin to Old Norse *stūfr* stump, Greek *stypos* stem, *typtein* to beat — more at TYPE] (before 12th century)
1 a : STUMP 2 **b :** a short piece remaining on a stem or trunk where a branch has been lost
2 : something made or worn to a short or blunt shape; *especially* **:** a pen with a short blunt nib
3 : a short blunt part left after a larger part has been broken off or used up ⟨pencil *stub*⟩
4 : something cut short or stunted
5 a : a small part of a leaf (as of a checkbook) attached to the backbone for memoranda of the contents of the part torn away **b :** the part of a ticket returned to the user

²stub *transitive verb* **stubbed; stub·bing** (15th century)
1 a : to grub up by the roots **b :** to clear (land) by grubbing out rooted growth **c :** to hew or cut down (a tree) close to the ground
2 : to extinguish (as a cigarette) by crushing
3 : to strike (one's foot or toe) against an object

stub·ble \ˈstə-bəl\ *noun, often attributive* [Middle English *stuble,* from Old French *estuble,* from Latin *stupula* stalk, straw, alteration of *stipula* — more at STIPULE] (14th century)
1 : the basal part of herbaceous plants and especially cereal grasses remaining attached to the soil after harvest
2 : a rough surface or growth resembling stubble; *especially* **:** a short growth of beard
— stub·bled \-bəld\ *adjective*
— stub·bly \-b(ə-)lē\ *adjective*

stubble mulch *noun* (1942)
: a lightly tilled mulch of plant residue used to prevent erosion, conserve moisture, and add organic matter to the soil

stub·born \ˈstə-bərn\ *adjective* [Middle English *stuborn*] (14th century)
1 a (1) **:** unreasonably or perversely unyielding **:** MULISH (2) **:** justifiably unyielding **:** RESOLUTE **b :** suggestive or typical of a strong stubborn nature ⟨a *stubborn* jaw⟩
2 : performed or carried on in an unyielding, obstinate, or persistent manner ⟨*stubborn* effort⟩
3 : difficult to handle, manage, or treat ⟨a *stubborn* cold⟩
4 : LASTING ⟨*stubborn* facts⟩
synonym see OBSTINATE
— stub·born·ly *adverb*
— stub·born·ness \-bər(n)-nəs\ *noun*

stub·by \ˈstə-bē\ *adjective* **stub·bi·er; -est** (15th century)
1 : abounding with stubs
2 a : resembling a stub **:** being short and thick ⟨*stubby* fingers⟩ **b :** being short and thickset **:** SQUAT **c :** being short, broad, or blunt (as from use or wear)

stuc·co \ˈstə-(ˌ)kō\ *noun, plural* **stuccos** *or* **stuccoes** [Italian, of Germanic origin; akin to Old High German *stucki* piece, crust, Old English *stocc* stock — more at STOCK] (1598)
1 a : a fine plaster used in decoration and ornamentation (as of interior walls) **b :** a material usually made of portland cement, sand, and a small percentage of lime and applied in a plastic state to form a hard covering for exterior walls
2 : STUCCOWORK
— stuc·coed \-(ˌ)kōd\ *adjective*

stuc·co·work \ˈstə-kō-ˌwərk\ *noun* (1686)
: work done in stucco

stuck *past and past participle of* STICK

stuck–up \ˈstək-ˈəp\ *adjective* (1829)
: superciliously self-important **:** CONCEITED

¹stud \ˈstəd\ *noun, often attributive* [Middle English *stod,* from Old English *stōd;* akin to Old Church Slavonic *stado* flock and probably to Old High German *stān* to stand — more at STAND] (before 12th century)
1 a : a group of animals and especially horses kept primarily for breeding **b :** a place (as a farm) where a stud is kept
2 : STUDHORSE; *broadly* **:** a male animal kept for breeding
3 a : a young man **:** GUY; *especially* **:** one who is virile and promiscuous **b :** a tough person **c :** HUNK 2
— at stud : for breeding as a stud ⟨retired racehorses *at stud*⟩

²stud *noun* [Middle English *stode,* from Old English *studu;* akin to Middle High German *stud* prop, Old English *stōw* place — more at STOW] (before 12th century)
1 a : one of the smaller uprights in the framing of the walls of a building to which sheathing, paneling, or laths are fastened **:** SCANTLING **b :** height from floor to ceiling
2 a : a boss, rivet, or nail with a large head used (as on a shield or belt) for ornament or protection **b :** a solid button with a shank or eye on the back inserted (as through an eyelet in a garment) as a fastener or ornament **c :** a small often round earring made for a pierced ear
3 a : any of various infixed pieces (as a rod or pin) projecting from a machine and serving chiefly as a support or axis **b :** one of the metal cleats inserted in a snow tire to increase traction

³stud *transitive verb* **stud·ded; stud·ding** (circa 1506)
1 : to furnish (as a building or wall) with studs
2 : to adorn, cover, or protect with studs
3 : to set, mark, or decorate conspicuously often at intervals ⟨a sky *studded* with stars⟩ ⟨a career *studded* with honors⟩

stud·book \ˈstəd-ˌbu̇k\ *noun* (1803)
: an official record (as in a book) of the pedigree of purebred animals (as horses or dogs)

stud·ding \ˈstə-diŋ\ *noun* (1588)
: the studs of a building or wall

stud·ding sail \ˈstə-diŋ-ˌsāl, ˈstən(t)-səl\ *noun* [origin unknown] (1549)
: a light sail set at the side of a principal square sail of a ship in free winds

stu·dent \ˈstü-dᵊnt, ˈstyü-, *chiefly Southern* -dənt\ *noun, often attributive* [Middle English, from Latin *student-, studens,* from present participle of *studēre* to study — more at STUDY] (14th century)
1 : SCHOLAR, LEARNER; *especially* **:** one who attends a school
2 : one who studies **:** an attentive and systematic observer

student body *noun* (1906)
: the students at an educational institution

student government *noun* (1948)
: the organization and management of student life by various student organizations

student lamp *noun* (1873)
: a desk reading lamp with a tubular shaft, one or two arms for a shaded light, and originally an oil reservoir

stu·dent·ship \ˈstü-dᵊnt-ˌship, ˈstyü-, -dənt-\ *noun* (circa 1782)
1 *British* **:** a grant for university study
2 : the state of being a student

stu·dent's t distribution \ˈstü-dᵊn(t)s-, -dən(t)s-\ *noun, often S capitalized* [*Student,* pen name of W. S. Gossett (died 1937) British statistician] (1929)
: T DISTRIBUTION

Student's t–test *noun* (1935)
: T-TEST

student teacher *noun* (1909)
: a student who is engaged in practice teaching

student teaching *noun* (1929)
: PRACTICE TEACHING

student union *noun* (1949)
: a building on a college campus that is devoted to student activities and that usually contains lounges, auditoriums, offices, and game rooms

stud·horse \ˈstəd-ˌhȯrs\ *noun* (before 12th century)
: a stallion kept especially for breeding

stud·ied \ˈstə-dēd\ *adjective* (15th century)
1 : carefully considered or prepared **:** THOUGHTFUL
2 : KNOWLEDGEABLE, LEARNED
3 : produced or marked by conscious design or premeditation **:** CALCULATED ⟨*studied* indifference⟩
— stud·ied·ly *adverb*
— stud·ied·ness *noun*

stu·dio \ˈstü-dē-(ˌ)ō, ˈstyü-\ *noun, plural* **-dios** [Italian, literally, study, from Latin *studium*] (1819)
1 a : the working place of a painter, sculptor, or photographer **b :** a place for the study of an art (as dancing, singing, or acting)
2 : a place where motion pictures are made
3 : a place maintained and equipped for the transmission of radio or television programs
4 : a place where audio recordings are made
5 : STUDIO APARTMENT

studio apartment *noun* (1903)
: a small apartment consisting typically of a main room, kitchenette, and bathroom

studio couch *noun* (1931)
: an upholstered usually backless couch that can be made to serve as a double bed by sliding from underneath it the frame of a single cot

stu·di·ous \'stü-dē-əs, 'styü-\ *adjective* (14th century)
1 : assiduous in the pursuit of learning
2 a : of, relating to, or concerned with study **b :** favorable to study ⟨a *studious* environment⟩
3 a : diligent or earnest in intent ⟨made a *studious* effort⟩ **b :** marked by or suggesting purposefulness or diligence ⟨a *studious* expression on his face⟩ **c :** deliberately or consciously planned ⟨spoke with a *studious* accent⟩
— **stu·di·ous·ly** *adverb*
— **stu·di·ous·ness** *noun*

stud·ly \'stəd-lē\ *adjective* [[1]*stud*] (1972)
slang **:** HUNKY

stud poker *noun* [[1]*stud*] (1864)
: poker in which each player is dealt the first card facedown and the other four cards faceup with a round of betting taking place after each of the last four rounds of dealing

[1]**study** \'stə-dē\ *noun, plural* **stud·ies** [Middle English *studie,* from Old French *estudie,* from Latin *studium,* from *studēre* to devote oneself, study; probably akin to Latin *tundere* to beat — more at CONTUSION] (14th century)
1 : a state of contemplation **:** REVERIE
2 a : application of the mental faculties to the acquisition of knowledge ⟨years of *study*⟩ **b :** such application in a particular field or to a specific subject ⟨the *study* of Latin⟩ **c :** careful or extended consideration ⟨the proposal is under *study*⟩ **d** (1) **:** a careful examination or analysis of a phenomenon, development, or question (2) **:** the published report of such a study
3 : a building or room devoted to study or literary pursuits
4 : PURPOSE, INTENT
5 a : a branch or department of learning **:** SUBJECT **b :** the activity or work of a student ⟨returning to her *studies* after vacation⟩ **c :** an object of study or deliberation ⟨every gesture a careful *study* —Marcia Davenport⟩ **d :** something attracting close attention or examination
6 : a person who learns or memorizes something (as a part in a play) — usually used with a qualifying adjective ⟨he's a fast *study*⟩
7 : a literary or artistic production intended as a preliminary outline, an experimental interpretation, or an exploratory analysis of specific features or characteristics
8 : a musical composition for the practice of a point of technique

[2]**study** *verb* **stud·ied; study·ing** (14th century)
intransitive verb
1 a : to engage in study **b :** to undertake formal study of a subject
2 *dialect* **:** MEDITATE, REFLECT
3 : ENDEAVOR, TRY
transitive verb
1 : to read in detail especially with the intention of learning
2 : to engage in the study of ⟨*study* biology⟩
3 : PLOT, DESIGN
4 : to consider attentively or in detail
synonym see CONSIDER
— **studi·er** \'stə-dē-ər\ *noun*

study hall *noun* (1846)
1 : a room in a school set aside for study
2 : a period in a student's day set aside for study and homework

[1]**stuff** \'stəf\ *noun* [Middle English, from Middle French *estoffe,* from Old French, from *estoffer* to equip, stock, of Germanic origin; akin to Old High German *stopfōn* to stop up, from (assumed) Vulgar Latin *stuppare* — more at STOP] (14th century)
1 : materials, supplies, or equipment used in various activities: as **a** *obsolete* **:** military baggage **b :** PERSONAL PROPERTY
2 : material to be manufactured, wrought, or used in construction ⟨clear half-inch pine *stuff* —Emily Holt⟩
3 : a finished textile suitable for clothing; *especially* **:** wool or worsted material
4 a : literary or artistic production **b :** writing, discourse, talk, or ideas of little value **:** TRASH
5 a : an unspecified material substance or aggregate of matter ⟨volcanic rock is curious *stuff*⟩ **b :** something (as a drug or food) consumed or introduced into the body by humans **c :** a matter to be considered ⟨the truth was heady *stuff*⟩ ⟨long-term policy *stuff*⟩ **d :** a group or scattering of miscellaneous objects or articles ⟨pick that *stuff* up off the floor⟩; *also* **:** nonphysical miscellaneous material
6 a : fundamental material **:** SUBSTANCE ⟨*stuff* of greatness⟩ **b :** subject matter ⟨a teacher who knows her *stuff*⟩
7 : special knowledge or capability ⟨showing their *stuff*⟩
8 a : spin imparted to a thrown or hit ball to make it curve or change course **b :** the movement of a baseball pitch out of its apparent line of flight **:** the liveliness of a pitch ⟨greatest pitcher of my time . . . had tremendous *stuff* —Ted Williams⟩
9 : DUNK SHOT
— **stuff·less** *adjective*

[2]**stuff** *transitive verb* (15th century)
1 a : to fill by packing things in **:** CRAM ⟨the boy *stuffed* his pockets with candy⟩ **b :** to fill to satiety **:** SURFEIT ⟨*stuffed* themselves with turkey⟩ **c :** to prepare (meat or vegetables) by filling or lining with a stuffing **d :** to fill (as a cushion) with a soft material **e :** to fill out the skin of (an animal) for mounting
2 a : to fill by intellectual effort ⟨*stuffing* their heads with facts⟩ **b :** to pack full of something immaterial ⟨a book *stuffed* with information⟩
3 : to fill or block up (as nasal passages)
4 a : to cause to enter or fill **:** THRUST ⟨*stuffed* a lot of clothing into a laundry bag⟩ **b :** to put (as a ball or puck) into a goal forcefully from close range
5 — used in the imperative to express contempt ⟨if they didn't like it, *stuff* 'em —Eric Clapton⟩; often used in the phrases *stuff it* and *get stuffed*
6 : to stop abruptly in a football game ⟨*stuff* the run⟩

stuffed shirt *noun* (1913)
: a smug, conceited, and usually pompous person often with an inflexibly conservative or reactionary attitude

stuff·er \'stə-fər\ *noun* (1611)
1 : one that stuffs
2 : an enclosure (as a leaflet) inserted in an envelope in addition to a bill, statement, or notice
3 : a series of extra threads or yarn running lengthwise in a fabric to add weight and bulk and to form a backing especially for carpets

stuff·ing \'stə-fiŋ\ *noun* (15th century)
: material used to stuff; *especially* **:** a seasoned mixture used to stuff food (as meat, vegetables, or eggs)

stuffing box *noun* (1798)
: a device that prevents leakage along a moving part (as a connecting rod) passing through a hole in a vessel (as a cylinder) containing steam, water, or oil and that consists of a box or chamber made by enlarging the hole and a gland to compress the contained packing

stuffy \'stə-fē\ *adjective* **stuff·i·er; -est** (1813)
1 : lacking in vitality or interest **:** STODGY, DULL
2 : ILL-NATURED, ILL-HUMORED
3 a : oppressive to the breathing **:** CLOSE **b :** stuffed up ⟨a *stuffy* nose⟩
4 : narrowly inflexible in standards of conduct **:** SELF-RIGHTEOUS
— **stuff·i·ly** \'stə-fə-lē\ *adverb*
— **stuff·i·ness** \'stə-fē-nəs\ *noun*

stul·ti·fy \'stəl-tə-ˌfī\ *transitive verb* **-fied; -fy·ing** [Late Latin *stultificare* to make foolish, from Latin *stultus* foolish; akin to Latin *stolidus* stolid] (1766)
1 : to allege or prove to be of unsound mind and hence not responsible
2 : to cause to appear or be stupid, foolish, or absurdly illogical
3 a : to impair, invalidate, or make ineffective **:** NEGATE **b :** to have a dulling or inhibiting effect on
— **stul·ti·fi·ca·tion** \ˌstəl-tə-fə-'kā-shən\ *noun*

[1]**stum·ble** \'stəm-bəl\ *verb* **stum·bled; stum·bling** \-b(ə-)liŋ\ [Middle English, probably of Scandinavian origin; akin to Norwegian dialect *stumle* to stumble; akin to Old Norse *stemma* to hinder — more at STEM] (14th century)
intransitive verb
1 a : to fall into sin or waywardness **b :** to make an error **:** BLUNDER **c :** to come to an obstacle to belief
2 : to trip in walking or running
3 a : to walk unsteadily or clumsily **b :** to speak or act in a hesitant or faltering manner
4 a : to come unexpectedly or by chance ⟨*stumble* onto the truth⟩ **b :** to fall or move carelessly
transitive verb
1 : to cause to stumble **:** TRIP
2 : BEWILDER, CONFOUND
— **stum·bler** \-b(ə-)lər\ *noun*
— **stum·bling·ly** \-b(ə-)liŋ-lē\ *adverb*

[2]**stumble** *noun* (1547)
: an act or instance of stumbling

stum·ble·bum \'stəm-bəl-ˌbəm\ *noun* (1932)
: a clumsy or inept person; *especially* **:** an inept boxer

stum·bling block \'stəm-bliŋ-\ *noun* (1588)
1 : an obstacle to progress
2 : an impediment to belief or understanding **:** PERPLEXITY

[1]**stump** \'stəmp\ *noun* [Middle English *stumpe;* akin to Old High German *stumpf* stump and perhaps to Middle English *stampen* to stamp] (14th century)
1 a : the basal portion of a bodily part remaining after the rest is removed **b :** a rudimentary or vestigial bodily part
2 : the part of a plant and especially a tree remaining attached to the root after the trunk is cut
3 : a remaining part **:** STUB
4 : one of the pointed rods stuck in the ground to form a cricket wicket
5 : a place or occasion for public speaking (as for a cause or candidate); *also* **:** the circuit followed by a maker of such speeches — used especially in the phrase *on the stump*

[2]**stump** (1596)
transitive verb
1 : to reduce to a stump **:** TRIM
2 a : DARE, CHALLENGE **b :** to frustrate the progress or efforts of **:** BAFFLE
3 : to clear (land) of stumps
4 : to travel over (a region) making political speeches or supporting a cause
5 a : to walk over heavily or clumsily **b :** STUB 3
intransitive verb
1 : to walk heavily or clumsily
2 : to go about making political speeches or supporting a cause
— **stump·er** *noun*

[3]**stump** *noun* [French or Flemish; French *estompe,* from Flemish *stomp,* literally, stub, from Middle Dutch; akin to Old High German *stumpf* stump] (1778)
: a short thick roll of leather, felt, or paper usually pointed at both ends and used for shading or blending a drawing in crayon, pencil, charcoal, pastel, or chalk

[4]**stump** *transitive verb* (1807)
: to tone or treat (a drawing) with a stump

stump·age \'stəm-pij\ *noun* (1835)
1 : the value of standing timber
2 : uncut marketable timber; *also* **:** the right to cut it

stump–tailed macaque \'stəmp-'tāl(d)-\ *noun* (1938)
: a dark reddish brown naked-faced short-tailed macaque (*Macaca arctoides* synonym *M. speciosa*) that is found in eastern Asia — called also *stump-tailed monkey*

stump work *noun* (1904)
: embroidery with intricate padded designs or scenes in high relief popular especially in the 17th century

stumpy \'stəm-pē\ *adjective* (1600)
1 : being short and thick **:** STUBBY
2 : full of stumps

[1]stun \'stən\ *transitive verb* **stunned; stun·ning** [Middle English, modification of Middle French *estoner* — more at ASTONISH] (14th century)
1 : to make senseless, groggy, or dizzy by or as if by a blow **:** DAZE
2 : to shock with noise
3 : to overcome especially with paralyzing astonishment or disbelief

[2]stun *noun* (1727)
: the effect of something that stuns **:** SHOCK

stung *past and past participle of* STING

stun gun *noun* (1967)
: a weapon designed to stun or immobilize (as by electric shock) rather than kill or injure the one affected

stunk *past and past participle of* STINK

stun·ner \'stə-nər\ *noun* (1829)
: one that stuns or is stunning

stun·ning \'stə-niŋ\ *adjective* (1667)
1 : causing astonishment or disbelief
2 : strikingly impressive especially in beauty or excellence
— stun·ning·ly \-niŋ-lē\ *adverb*

[1]stunt \'stənt\ *transitive verb* [English dialect *stunt* stubborn, stunted, abrupt, probably of Scandinavian origin; akin to Old Norse *stuttr* scant — more at STINT] (1659)
: to hinder the normal growth, development, or progress of
— stunt·ed·ness *noun*

[2]stunt *noun* (1725)
1 : one (as an animal) that is stunted
2 : a check in growth
3 : a plant disease in which dwarfing occurs

[3]stunt *noun* [origin unknown] (1878)
1 : an unusual or difficult feat requiring great skill or daring; *especially* **:** one performed or undertaken chiefly to gain attention or publicity
2 : a shifting or switching of the positions by defensive players at the line of scrimmage in football to disrupt the opponent's blocking efforts

[4]stunt *intransitive verb* (1917)
: to perform or engage in a stunt

stunt·man \'stənt-,man\ *noun* (1927)
: a man who performs stunts; *especially* **:** one who doubles for an actor during the filming of stunts and dangerous scenes

stunt·wom·an \-,wu̇-mən\ *noun* (1948)
: a woman who doubles for an actress during the filming of stunts and dangerous scenes

stu·pa \'stü-pə\ *noun* [Sanskrit *stūpa*] (1876)
: a usually dome-shaped structure (as a mound) serving as a Buddhist shrine

stupa

[1]stupe \'stüp, 'styüp\ *noun* [Middle English, from Latin *stuppa* coarse part of flax, tow, from Greek *styppē*] (14th century)
: a hot wet often medicated cloth applied externally (as to stimulate circulation)

[2]stupe *noun* [short for *stupid*] (1762)
: a stupid person **:** DOLT

stu·pe·fac·tion \,stü-pə-'fak-shən, ,styü-\ *noun* [Middle English *stupefaccioun,* from Medieval Latin *stupefaction-, stupefactio,* from Latin *stupefacere*] (15th century)
: the act of stupefying **:** the state of being stupefied

stu·pe·fy \'stü-pə-,fī, 'styü-\ *transitive verb* **-fied; -fy·ing** [Middle English *stupifien,* from Middle French *stupefier,* modification of Latin *stupefacere,* from *stupēre* to be astonished + *facere* to make, do — more at DO] (15th century)
1 : to make stupid, groggy, or insensible
2 : ASTONISH, ASTOUND
— stu·pe·fy·ing·ly \-iŋ-lē\ *adverb*

stu·pen·dous \stu̇-'pen-dəs, styu̇-\ *adjective* [Latin *stupendus,* gerundive of *stupēre*] (1666)
1 : causing astonishment or wonder **:** AWESOME, MARVELOUS
2 : of amazing size or greatness **:** TREMENDOUS
synonym see MONSTROUS
— stu·pen·dous·ly *adverb*
— stu·pen·dous·ness *noun*

[1]stu·pid \'stü-pəd, 'styü-\ *adjective* [Middle French *stupide,* from Latin *stupidus,* from *stupēre* to be numb, be astonished — more at TYPE] (1541)
1 a : slow of mind **:** OBTUSE **b :** given to unintelligent decisions or acts **:** acting in an unintelligent or careless manner **c :** lacking intelligence or reason **:** BRUTISH
2 : dulled in feeling or sensation **:** TORPID ⟨still *stupid* from the sedative⟩
3 : marked by or resulting from unreasoned thinking or acting **:** SENSELESS
4 a : lacking interest or point **b :** VEXATIOUS, EXASPERATING ⟨this *stupid* flashlight won't work⟩ ☆
— stu·pid·ly *adverb*
— stu·pid·ness *noun*

[2]stupid *noun* (1712)
: a stupid person

stu·pid·i·ty \stu̇-'pi-də-tē, styu̇-\ *noun, plural* **-ties** (1541)
1 : the quality or state of being stupid
2 : a stupid idea or act

stu·por \'stü-pər, 'styü-\ *noun* [Middle English, from Latin, from *stupēre*] (14th century)
1 : a condition of greatly dulled or completely suspended sense or sensibility ⟨drunken *stupor*⟩
2 : a state of extreme apathy or torpor resulting often from stress or shock **:** DAZE
synonym see LETHARGY

stu·por·ous \'stü-p(ə-)rəs, 'styü-\ *adjective* (1892)
: marked or affected by or as if by stupor

stur·dy \'stər-dē\ *adjective* **stur·di·er; -est** [Middle English, brave, stubborn, from Middle French *estourdi* stunned, from past participle of *estourdir* to stun, from (assumed) Vulgar Latin *exturdire,* from Latin *ex-* + (assumed) Vulgar Latin *turdus* simpleton, from Latin, thrush — more at THRUSH] (14th century)
1 a : firmly built or constituted **:** STOUT **b :** HARDY **c :** sound in design or execution **:** SUBSTANTIAL
2 a : marked by or reflecting physical strength or vigor **b :** FIRM, RESOLUTE **c :** RUGGED, STABLE ◇
synonym see STRONG
— stur·di·ly \'stər-dᵊl-ē\ *adverb*
— stur·di·ness \'stər-dē-nəs\ *noun*

stur·geon \'stər-jən\ *noun* [Middle English, from Middle French *estourjon,* of Germanic origin; akin to Old English *styria* sturgeon] (13th century)
: any of a family (Acipenseridae) of usually large elongate anadromous or freshwater bony fishes which are widely distributed in the north temperate zone and whose roe is made into caviar

Sturm und Drang \,shtu̇rm-u̇nt-'dräŋ, ,stu̇r-, -ənt-\ *noun* [German, literally, storm and stress, from *Sturm und Drang* (1776), drama by Friedrich von Klinger (died 1831) German novelist and dramatist] (1845)
1 : a late 18th century German literary movement characterized by works containing rousing action and high emotionalism that often deal with the individual's revolt against society
2 : TURMOIL

sturt \'stərt\ *noun* [Middle English, contention, alteration of *strut;* akin to Old English *strūtian* to exert oneself — more at STRUT] (14th century)
chiefly Scottish **:** CONTENTION

[1]stut·ter \'stə-tər\ *verb* [frequentative of English dialect *stut* to stutter, from Middle English *stutten;* akin to Dutch *stotteren* to stutter, Gothic *stautan* to strike — more at CONTUSION] (circa 1570)
intransitive verb
1 : to speak with involuntary disruption or blocking of speech (as by spasmodic repetition or prolongation of vocal sounds)
2 : to move or act in a halting or spasmodic

☆ SYNONYMS

Stupid, dull, dense, crass, dumb mean lacking in power to absorb ideas or impressions. STUPID implies a slow-witted or dazed state of mind that may be either congenital or temporary ⟨*stupid* students just keeping the seats warm⟩ ⟨*stupid* with drink⟩. DULL suggests a slow or sluggish mind such as results from disease, depression, or shock ⟨monotonous work that leaves the mind *dull*⟩. DENSE implies a thickheaded imperviousness to ideas ⟨too *dense* to take a hint⟩. CRASS suggests a grossness of mind precluding discrimination or delicacy ⟨a *crass,* materialistic people⟩. DUMB applies to an exasperating obtuseness or lack of comprehension ⟨too *dumb* to figure out what's going on⟩.

◇ WORD HISTORY

sturdy The usual modern meanings of *sturdy,* such as "firmly constituted" or "marked by physical strength or vigor," lie at the end of a long historical trail. Though these senses were already present in the English of Chaucer, Middle English also shows a wide range of other meanings now obsolete, such as "bold, valiant," "violent, furious," "severe, harsh," and "refractory, obstinate." (The latter sense long survived in the collocation *sturdy beggar,* referring to someone who lived by begging but who was able-bodied and hard to manage.) The source of *sturdy* was Old French *estourdi,* the past participle of *estourdir* "to stun, daze, befuddle." Though *estourdi* usually meant "dazed, stunned," it also is attested in the senses "giddy" or "foolish," and the association of giddiness with recklessness and recklessness with valor presumably led to the Middle English meaning "bold, valiant." Old French *estourdir,* in turn, most likely developed from an unattested spoken Latin verb *exturdire,* a derivative of Latin *turdus* "thrush." The thrush in folk belief was considered giddy or foolish, perhaps from the notion that it gorged itself on overripe and hence slightly alcoholic grapes until it became dizzy.

manner ⟨the old jalopy bucks and *stutters* uphill —William Cleary⟩
transitive verb
: to say, speak, or sound with or as if with a stutter
— stut·ter·er \-tər-ər\ *noun*

[2]stutter *noun* (circa 1847)
1 : an act or instance of stuttering
2 : a speech disorder involving stuttering accompanied by fear and anxiety

[1]sty \'stī\ *noun, plural* **sties** *also* **styes** [Middle English, from Old English *stig;* akin to Old Norse *-stī* sty] (before 12th century)
1 : a pen or enclosed housing for swine
2 : an unkempt filthy place ⟨this house is a *sty*⟩

[2]sty *verb* **stied** *or* **styed; sty·ing** (before 12th century)
transitive verb
: to lodge or keep in a sty
intransitive verb
: to live in a sty

[3]sty *or* **stye** \'stī\ *noun, plural* **sties** *or* **styes** [short for obsolete English *styan,* from (assumed) Middle English, alteration of Old English *stīgend,* from *stīgan* to go up, rise — more at STAIR] (1617)
: an inflamed swelling of a sebaceous gland at the margin of an eyelid

sty·gian \'sti-j(ē-)ən\ *adjective, often capitalized* [Latin *stygius,* from Greek *stygios,* from *Styg-, Styx* Styx] (1566)
1 : of or relating to the river Styx
2 : extremely dark, gloomy, or forbidding

styl- *or* **styli-** *or* **stylo-** *combining form* [Latin *stilus* spike, stem — more at STYLE]
: style **:** styloid process ⟨*stylo*podium⟩ ⟨*stylo*graphy⟩

sty·lar \'stī-lər, -ˌlär\ *adjective* [[1]*style*] (circa 1928)
: of or relating to the style of a plant ovary

[1]style \'stī(ə)l\ *noun* [Middle English *stile, style,* from Latin *stilus* spike, stem, stylus, style of writing; perhaps akin to Latin in*stigare* to goad — more at STICK] (14th century)
1 : DESIGNATION, TITLE
2 a : a distinctive manner of expression (as in writing or speech) ⟨writes with more attention to *style* than to content⟩ ⟨the flowery *style* of 18th century prose⟩ **b :** a distinctive manner or custom of behaving or conducting oneself ⟨the formal *style* of the court⟩ ⟨his *style* is abrasive⟩; *also* **:** a particular mode of living ⟨in high *style*⟩ **c :** a particular manner or technique by which something is done, created, or performed ⟨a unique *style* of horseback riding⟩ ⟨the classical *style* of dance⟩
3 a : STYLUS **b :** GNOMON 1b **c :** a filiform prolongation of a plant ovary bearing a stigma at its apex — see FLOWER illustration **d :** a slender elongated process (as a bristle) on an animal
4 : a distinctive quality, form, or type of something ⟨a new dress *style*⟩ ⟨the Greek *style* of architecture⟩
5 a : the state of being popular **:** FASHION ⟨clothes that are always in *style*⟩ **b :** fashionable elegance **c :** beauty, grace, or ease of manner or technique ⟨an awkward moment she handled with *style*⟩
6 : a convention with respect to spelling, punctuation, capitalization, and typographic arrangement and display followed in writing or printing
synonym see FASHION
— style·less \'stī(ə)l-ləs\ *adjective*
— style·less·ness *noun*

[2]style *transitive verb* **styled; styl·ing** (circa 1580)
1 : to call or designate by an identifying term **:** NAME
2 a : to give a particular style to **b :** to design, make, or arrange in accord with the prevailing mode
— styl·er *noun*

-style *adjective or adverb combining form*
: being in the style of ⟨a Beaujolais-*style* wine⟩

style·book *noun* (1708)
: a book explaining, describing, or illustrating a prevailing, accepted, or authorized style

sty·let \stī-'let, 'stī-lət\ *noun* [French, from Middle French *stilet* stiletto, from Old Italian *stiletto* — more at STILETTO] (1697)
1 a : a slender surgical probe **b :** a thin wire inserted into a catheter to maintain rigidity or into a hollow needle to maintain patency **c :** a pointed instrument (as for graving)
2 : a relatively rigid elongated organ or appendage (as a piercing mouthpart) of an animal
3 : STILETTO

sty·li·form \'stī-lə-ˌfȯrm\ *adjective* [New Latin *stiliformis,* from Latin *stilus* + *-formis* -form] (1578)
: resembling a style **:** bristle-shaped ⟨a *styliform* copulatory organ⟩

styl·ing \'stī-liŋ\ *noun* (1928)
: the way in which something is styled

styl·ise *British variant of* STYLIZE

styl·ish \'stī-lish\ *adjective* (1785)
: having style; *specifically* **:** conforming to current fashion
— styl·ish·ly *adverb*
— styl·ish·ness *noun*

styl·ist \'stī-list\ *noun* (1795)
1 a : a master or model of style; *especially* **:** a writer or speaker who is eminent in matters of style **b :** one (as a writer or singer) noted for a distinctive style
2 a : one who develops, designs, or advises on styles **b :** HAIRSTYLIST

sty·lis·tic \stī-'lis-tik\ *adjective* (1860)
: of or relating especially to literary or artistic style
— sty·lis·ti·cal·ly \-ti-k(ə-)lē\ *adverb*

sty·lis·tics \stī-'lis-tiks\ *noun plural but singular or plural in construction* (circa 1883)
1 : an aspect of literary study that emphasizes the analysis of various elements of style (as metaphor and diction)
2 : the study of the devices in a language that produce expressive value

sty·lite \'stī-ˌlīt\ *noun* [Late Greek *stylitēs,* from Greek *stylos* pillar — more at STEER] (circa 1638)
: a Christian ascetic living atop a pillar
— sty·lit·ic \stī-'li-tik\ *adjective*

styl·ize \'stī(-ə)-ˌlīz\ *transitive verb* **styl·ized; styl·iz·ing** (1898)
: to conform to a conventional style; *specifically* **:** to represent or design according to a style or stylistic pattern rather than according to nature or tradition
— styl·i·za·tion \ˌstī-lə-'zā-shən\ *noun*

sty·lo·bate \'stī-lə-ˌbāt\ *noun* [Latin *stylobates,* from Greek *stylobatēs,* from *stylos* pillar + *bainein* to walk, go — more at COME] (1694)
: a continuous flat coping or pavement supporting a row of architectural columns

sty·log·ra·phy \stī-'lä-grə-fē\ *noun* (circa 1840)
: a mode of writing or tracing lines by means of a style or similar instrument

sty·loid \'stī(-ə)-ˌlȯid\ *adjective* (1709)
: resembling a style **:** STYLIFORM — used especially of slender pointed skeletal processes (as on the ulna)

sty·lo·po·di·um \ˌstī-lə-'pō-dē-əm\ *noun, plural* **-dia** \-dē-ə\ [New Latin, from *styl-* + Greek *podion* small foot, base — more at PEW] (circa 1832)
: a disk-shaped or conical expansion at the base of the style in plants of the carrot family

sty·lus \'stī-ləs\ *noun, plural* **sty·li** \'stī(-ə)-ˌlī\ *also* **sty·lus·es** \'stī-lə-səz\ [Latin *stylus, stilus* spike, stylus — more at STYLE] (1807)
: an instrument for writing, marking, or incising: as **a :** an instrument used by the ancients in writing on clay or waxed tablets **b :** a hard-pointed pen-shaped instrument for marking on stencils used in a reproducing machine **c** (1) **:** NEEDLE 3c (2) **:** a cutting tool used to produce an original record groove during disc recording **d :** a pen-shaped pointing device used for entering positional information (as from a graphics tablet) into a computer

sty·mie \'stī-mē\ *transitive verb* **sty·mied; sty·mie·ing** [Scots *stimie, stymie* to obstruct a golf shot by interposition of the opponent's ball] (1902)
: to present an obstacle to **:** stand in the way of

styp·tic \'stip-tik\ *adjective* [Middle English *stiptik,* from Latin *stypticus,* from Greek *styptikos,* from *styphein* to contract] (14th century)
: tending to contract or bind **:** ASTRINGENT; *especially* **:** tending to check bleeding
— styptic *noun*

styptic pencil *noun* (1908)
: a stick of a medicated styptic substance for use especially in shaving to stop the bleeding from small cuts

sty·rax \'stī-ˌraks\ *noun* [Latin — more at STORAX] (1558)
: STORAX

sty·rene \'stī-ˌrēn\ *noun* [International Scientific Vocabulary, from Latin *styrax*] (1885)
: a fragrant liquid unsaturated hydrocarbon C_8H_8 used chiefly in making synthetic rubber, resins, and plastics and in improving drying oils; *also* **:** any of various synthetic plastics made from styrene by polymerization or copolymerization

Sty·ro·foam \'stī-rə-ˌfōm\ *trademark*
— used for an expanded rigid polystyrene plastic

Styx \'stiks\ *noun* [Latin *Styg-, Styx,* from Greek]
: the principal river of the underworld in Greek mythology

su·able \'sü-ə-bəl\ *adjective* (circa 1623)
: liable to be sued in court
— su·abil·i·ty \ˌsü-ə-'bi-lə-tē\ *noun*
— su·ably \-blē\ *adverb*

sua·sion \'swā-zhən\ *noun* [Middle English, from Latin *suasion-, suasio,* from *suadēre* to urge, persuade — more at SWEET] (14th century)
: the act of influencing or persuading
— sua·sive \'swā-siv, -ziv\ *adjective*
— sua·sive·ly *adverb*
— sua·sive·ness *noun*

suave \'swäv\ *adjective* **suav·er; -est** [Middle French, pleasant, sweet, from Latin *suavis* — more at SWEET] (1847)
1 : smoothly though often superficially gracious and sophisticated
2 : smooth in texture, performance, or style ☆
— suave·ly *adverb*
— suave·ness *noun*

☆ SYNONYMS
Suave, urbane, diplomatic, bland, smooth, politic mean pleasantly tactful and well-mannered. SUAVE suggests a specific ability to deal with others easily and without friction ⟨a *suave* public relations coordinator⟩. URBANE implies high cultivation and poise coming from wide social experience ⟨an *urbane* traveler⟩. DIPLOMATIC stresses an ability to deal with ticklish situations tactfully ⟨a *diplomatic* negotiator⟩. BLAND emphasizes mildness of manner and absence of irritating qualities ⟨a *bland* master of ceremonies⟩. SMOOTH suggests often a deliberately assumed suavity ⟨a *smooth* salesman⟩. POLITIC implies shrewd as well as tactful and suave handling of people ⟨a cunningly *politic* manager⟩.

— **sua·vi·ty** \'swä-və-tē\ *noun*

[1]sub \'səb\ *noun* (1830)
: SUBSTITUTE

[2]sub *verb* **subbed; sub·bing** (1853)
intransitive verb
: to act as a substitute
transitive verb
1 *British* **:** to read and edit as a copy editor **:** SUBEDIT
2 : SUBCONTRACT 1

[3]sub *noun* (1916)
: SUBMARINE

sub- *prefix* [Middle English, from Latin, under, below, secretly, from below, up, near, from *sub* under, close to — more at UP]
1 : under **:** beneath **:** below ⟨*sub*soil⟩ ⟨*sub*aqueous⟩
2 a : subordinate **:** secondary **:** next lower than or inferior to ⟨*sub*station⟩ ⟨*sub*editor⟩ **b :** subordinate portion of **:** subdivision of ⟨*sub*committee⟩ ⟨*sub*species⟩ **c :** with repetition (as of a process) so as to form, stress, or deal with subordinate parts or relations ⟨*sub*let⟩ ⟨*sub*contract⟩
3 : less than completely, perfectly, or normally **:** somewhat ⟨*sub*acute⟩ ⟨*sub*clinical⟩
4 a : almost **:** nearly ⟨*sub*erect⟩ **b :** falling nearly in the category of and often adjoining **:** bordering on ⟨*sub*arctic⟩

sub·ad·o·les·cent
sub·agen·cy
sub·agent
sub·al·lo·ca·tion
sub·ar·ea
sub·au·di·ble
sub·av·er·age
sub·base·ment
sub·ba·sin
sub·block
sub·branch
sub·caste
sub·cat·e·go·ri·za·tion
sub·cat·e·go·rize
sub·cat·e·go·ry
sub·ceil·ing
sub·cel·lar
sub·chap·ter
sub·chief
sub·clan
sub·clus·ter
sub·code
sub·col·lec·tion
sub·col·lege
sub·col·le·giate
sub·col·o·ny
sub·com·mis·sion
sub·com·po·nent
sub·cor·date
sub·co·ri·a·ceous
sub·coun·ty
sub·cult
sub·cu·ra·tive
sub·dean
sub·de·ci·sion
sub·de·part·ment
sub·de·vel·op·ment
sub·di·a·lect
sub·di·rec·tor
sub·dis·ci·pline
sub·dis·trict
sub·econ·o·my
sub·erect
sub·file
sub·frame
sub·gen·er·a·tion
sub·genre
sub·goal
sub·gov·ern·ment
sub·hu·mid
sub·in·dus·try
sub·in·hib·i·to·ry
sub·lan·guage
sub·lev·el
sub·li·brar·i·an
sub·li·cense
sub·lit·er·a·cy
sub·lit·er·ate
sub·lot
sub·man·ag·er
sub·mar·ket
sub·max·i·mal
sub·menu
sub·min·i·mal
sub·min·is·ter
sub·na·tion·al
sub·net·work
sub·niche
sub·op·ti·mal
sub·op·ti·mi·za·tion
sub·op·ti·mize
sub·op·ti·mum
sub·or·bic·u·lar
sub·or·ga·ni·za·tion
sub·pan·el
sub·par
sub·para·graph
sub·par·al·lel
sub·part
sub·pe·ri·od
sub·phase
sub·pri·mate
sub·pro·cess
sub·prod·uct
sub·proj·ect
sub·pro·le·tar·i·at
sub·ra·tio·nal
sub·sat·u·rat·ed
sub·sat·u·ra·tion
sub·scale
sub·science
sub·sea
sub·sec·re·tary
sub·sec·tor
sub·seg·ment
sub·sei·zure
sub·sense
sub·sen·tence
sub·se·ries
sub·site
sub·skill
sub·so·ci·ety
sub·spe·cial·ist
sub·spe·cial·ize
sub·spe·cial·ty
sub·state
sub·sys·tem
sub·task
sub·tax·on
sub·test
sub·theme
sub·ther·a·peu·tic
sub·top·ic
sub·trea·sury
sub·trend
sub·tribe
sub·type
sub·unit
sub·va·ri·ety
sub·vas·sal
sub·vis·i·ble
sub·vi·su·al
sub·world
sub·writ·er
sub·ze·ro
sub·zone

sub·ac·id \ˌsəb-'a-səd\ *adjective* [Latin *subacidus,* from *sub-* + *acidus* acid] (1765)
: somewhat acrimonious **:** CUTTING ⟨*subacid* comments⟩
— **sub·ac·id·ly** *adverb*
— **sub·ac·id·ness** *noun*

sub·acute \-ə-'kyüt\ *adjective* (1822)
1 : having a tapered but not sharply pointed form ⟨*subacute* leaves⟩
2 a : falling between acute and chronic in character especially when closer to acute ⟨*subacute* endocarditis⟩ **b :** less marked in severity or duration than a corresponding acute state ⟨*subacute* pain⟩
— **sub·acute·ly** *adverb*

subacute sclerosing pan·en·ceph·a·li·tis \-ˌpan-in-ˌse-fə-'lī-təs\ *noun* [*panencephalitis* from New Latin, from *pan-* + *encephalitis*] (1950)
: a central nervous system disease of children and young adults caused by infection of the brain by measles virus or a closely related virus and marked by intellectual deterioration, convulsions, and paralysis

sub·adult \ˌsəb-ə-'dəlt; ˌsəb-'a-ˌdəlt\ *noun* (1923)
: an individual that has passed through the juvenile period but not yet attained typical adult characteristics
— **subadult** *adjective*

sub·aer·i·al \-'ar-ē-əl, -'er-; -ā-'ir-ē-əl\ *adjective* (1833)
: situated, formed, or occurring on or immediately adjacent to the surface of the earth ⟨*subaerial* erosion⟩ ⟨*subaerial* roots⟩
— **sub·aer·i·al·ly** \-ē-ə-lē\ *adverb*

sub·bah·dar *or* **su·ba·dar** \ˌsü-bə-'där\ *noun* [Persian *ṣūbadār*] (1698)
1 : a governor of a province
2 : the chief Indian officer of a company of Indian troops in the British army of India

sub·al·pine \ˌsəb-'al-ˌpīn\ *adjective* (circa 1656)
1 : of or relating to the region about the foot and lower slopes of the Alps
2 : of, relating to, or inhabiting high upland slopes and especially the zone just below the timberline

[1]sub·al·tern \sə-'bȯl-tərn, *especially British* 'sə-bəl-tərn\ *adjective* [Late Latin *subalternus,* from Latin *sub-* + *alternus* alternate, from *alter* other (of two) — more at ALTER] (1570)
1 : particular with reference to a related universal proposition ⟨"some S is P" is a *subaltern* proposition to "all S is P"⟩
2 : SUBORDINATE

[2]subaltern *noun* (1605)
1 : a person holding a subordinate position; *specifically* **:** a junior officer (as in the British army)
2 : a particular proposition that follows immediately from a universal

sub·ant·arc·tic \ˌsəb-ant-'ärk-tik, -'är-tik\ *adjective* (1875)
: of, relating to, characteristic of, or being a region just outside the antarctic circle

sub·api·cal \-'ā-pi-kəl *also* -'a-\ *adjective* (1846)
: situated below or near an apex

sub·aquat·ic \-ə-'kwä-tik, -'kwa-\ *adjective* [International Scientific Vocabulary] (1844)
: somewhat aquatic ⟨a marginal *subaquatic* flora⟩

sub·aque·ous \-'ā-kwē-əs, -'a-\ *adjective* (1677)
: existing, formed, or taking place in or under water

sub·arach·noid \-ə-'rak-ˌnȯid\ *also* **sub·arach·noid·al** \-rak-'nȯi-dᵊl\ *adjective* (1843)
: of, relating to, occurring, or situated under the arachnoid membrane ⟨*subarachnoid* hemorrhage⟩

sub·arc·tic \-'ärk-tik, -'är-tik\ *adjective* [International Scientific Vocabulary] (1854)
: of, relating to, characteristic of, or being regions immediately outside of the arctic circle or regions similar to these in climate or conditions of life
— **subarctic** *noun*

sub·as·sem·bly \-ə-'sem-blē\ *noun* (1919)
: an assembled unit designed to be incorporated with other units in a finished product

sub·at·mo·spher·ic \-ˌat-mə-'sfir-ik, -'sfer-\ *adjective* (1941)
: less or lower than that of the atmosphere ⟨*subatmospheric* pressure⟩

sub·atom·ic \-ə-'tä-mik\ *adjective* (1903)
1 : of or relating to the inside of the atom
2 : of, relating to, or being particles smaller than atoms

sub·au·di·tion \-ȯ-'di-shən\ *noun* [Late Latin *subaudition-, subauditio,* from *subaudire* to understand, from Latin *sub-* + *audire* to hear — more at AUDIBLE] (1798)
: the act of understanding or supplying something not expressed **:** a reading between the lines

sub·base \'səb-ˌbās\ *noun* (1826)
: underlying support placed below what is normally construed as a base: as **a :** the lowest member horizontally of an architectural base or of a baseboard or pedestal **b :** pervious fill (as crushed stone) placed under a roadbed

sub·bi·tu·mi·nous \ˌsəb-bə-'tü-mə-nəs, -bī-, -'tyü-\ *adjective* (1908)
: of, relating to, or being coal of lower rank than bituminous coal but higher than lignite

sub·cab·i·net \-'kab-nit, -'ka-bə-\ *adjective* (1954)
: of, relating to, or being a high administrative position in the U.S. government that ranks below the cabinet level

sub·cap·su·lar \-'kap-sə-lər\ *adjective* (1889)
: situated or occurring beneath or within a capsule ⟨*subcapsular* cataracts⟩

sub·cel·lu·lar \-'sel-yə-lər\ *adjective* (1948)
: of less than cellular scope or level of organization ⟨*subcellular* particles⟩ ⟨*subcellular* studies⟩

sub·cen·ter \-'sen-tər\ *noun* (circa 1925)
: a secondary center; *especially* **:** a center (as for shopping) located outside the main business area of a city

sub·cen·tral \-trəl\ *adjective* (1822)
1 : nearly but not quite central
2 : located under a center
— **sub·cen·tral·ly** \-trə-lē\ *adverb*

sub·chas·er \'səb-ˌchā-sər\ *noun* (circa 1918)
: a small maneuverable patrol or escort vessel used for antisubmarine warfare

sub·class \'səb-ˌklas\ *noun* (1819)
: a primary division of a class: as **a :** a category in biological classification ranking below a class and above an order **b :** SUBSET

sub·clas·si·fi·ca·tion \ˌsəb-ˌkla-sə-fə-'kā-shən\ *noun* (1873)
1 : a primary division of a classification

\ə\ **abut** \ᵊ\ **kitten** \ər\ **further** \a\ **ash** \ā\ **ace** \ä\ **mop, mar** \aü\ **out** \ch\ **chin** \e\ **bet** \ē\ **easy** \g\ **go** \i\ **hit** \ī\ **ice** \j\ **job** \ŋ\ **sing** \ō\ **go** \ȯ\ **law** \ȯi\ **boy** \th\ **thin** \t͟h\ **the** \ü\ **loot** \u̇\ **foot** \y\ **yet** \zh\ **vision** *see also* Guide to Pronunciation

2 **:** arrangement into or assignment to subclassifications
— **sub·clas·si·fy** \-ˈkla-sə-ˌfī\ *transitive verb*

[1]**sub·cla·vi·an** \-ˈklā-vē-ən\ *adjective* [New Latin *subclavius,* from *sub-* + *clavicula* clavicle] (1646)
: of, relating to, being, or performed on a part (as an artery, vein, or nerve) located under the clavicle ⟨*subclavian* angioplasty⟩

[2]**subclavian** *noun* (1719)
: a subclavian part (as an artery, vein, or nerve)

subclavian artery *noun* (1688)
: the proximal part of the main artery of the arm or forelimb

subclavian vein *noun* (1770)
: the proximal part of the main vein of the arm or forelimb

sub·cli·max \ˌsəb-ˈklī-ˌmaks\ *noun* (1916)
: a stage or community in an ecological succession immediately preceding a climax; *especially* **:** one held in relative stability throughout edaphic or biotic influences or by fire

sub·clin·i·cal \-ˈkli-ni-kəl\ *adjective* (circa 1935)
: not detectable or producing effects that are not detectable by the usual clinical tests ⟨a *subclinical* infection⟩ ⟨*subclinical* cancer⟩
— **sub·clin·i·cal·ly** \-k(ə-)lē\ *adverb*

sub·com·mit·tee \ˈsəb-kə-ˌmi-tē, ˌsəb-kə-ˈ\ *noun* (circa 1607)
: a subdivision of a committee usually organized for a specific purpose

sub·com·mu·ni·ty \ˌsəb-kə-ˈmyü-nə-tē\ *noun* (1966)
: a distinct grouping within a community

sub·com·pact \ˈsəb-ˈkäm-ˌpakt\ *noun* (1967)
: an automobile smaller than a compact

[1]**sub·con·scious** \ˌsəb-ˈkän(t)-shəs, ˈsəb-\ *adjective* (circa 1834)
: existing in the mind but not immediately available to consciousness ⟨a *subconscious* motive⟩
— **sub·con·scious·ly** *adverb*
— **sub·con·scious·ness** *noun*

[2]**subconscious** *noun* (1886)
: the mental activities just below the threshold of consciousness

sub·con·ti·nent \ˌsəb-ˈkän-tᵊn-ənt, -ˈkänt-nənt\ *noun* (1863)
: a large landmass smaller than a continent; *especially* **:** a major subdivision of a continent ⟨the Indian *subcontinent*⟩
— **sub·con·ti·nen·tal** \ˌsəb-ˌkän-tᵊn-ˈen-tᵊl\ *adjective*

[1]**sub·con·tract** \ˌsəb-ˈkän-ˌtrakt\ *noun* (1817)
: a contract between a party to an original contract and a third party; *especially* **:** one to provide all or a specified part of the work or materials required in the original contract

[2]**sub·con·tract** \ˌsəb-ˈkän-ˌtrakt, ˌsəb-kən-ˈ\ (1842)
intransitive verb
: to let out or undertake work under a subcontract
transitive verb
1 **:** to engage a third party to perform under a subcontract all or part of (work included in an original contract) — sometimes used with *out*
2 **:** to undertake (work) under a subcontract

sub·con·trac·tor \ˌsəb-ˈkän-ˌtrak-tər, ˌsəb-kən-ˈ\ *noun* (1842)
: an individual or business firm contracting to perform part or all of another's contract

sub·con·tra·oc·tave \ˌsəb-ˌkän-trə-ˈäk-tiv, -təv, -ˌtāv\ *noun* (circa 1901)
: the musical octave that begins on the fourth C below middle C — see PITCH illustration

sub·con·trary \ˌsəb-ˈkän-ˌtrer-ē\ *noun* (1685)
: a proposition so related to another that though both may be true they cannot both be false
— **subcontrary** *adjective*

sub·cool \-ˈkül\ *transitive verb* (1916)
: SUPERCOOL

sub·cor·ti·cal \-ˈkȯr-ti-kəl\ *adjective* (1887)
: of, relating to, involving, or being nerve centers below the cerebral cortex ⟨*subcortical* lesions⟩

sub·crit·i·cal \-ˈkri-ti-kəl\ *adjective* (1930)
1 **:** less or lower than critical in respect to a specified factor
2 **a :** of insufficient size to sustain a chain reaction ⟨a *subcritical* mass of fissionable material⟩ **b :** designed for use with fissionable material of subcritical mass ⟨a *subcritical* reactor⟩

sub·crust·al \-ˈkrəs-tᵊl\ *adjective* (1897)
: situated or occurring below a crust and especially the crust of the earth

sub·cul·ture \ˈsəb-ˌkəl-chər\ *noun* (1886)
1 **a :** a culture (as of bacteria) derived from another culture **b :** an act or instance of producing a subculture
2 **:** an ethnic, regional, economic, or social group exhibiting characteristic patterns of behavior sufficient to distinguish it from others within an embracing culture or society ⟨a criminal *subculture*⟩
— **sub·cul·tur·al** \-ˈkəlch-rəl, -ˈkəl-chə-\ *adjective*
— **sub·cul·tur·al·ly** *adverb*
— **subculture** *transitive verb*

sub·cu·ta·ne·ous \ˌsəb-kyu̇-ˈtā-nē-əs\ *adjective* [Late Latin *subcutaneus,* from Latin *sub-* + *cutis* skin — more at HIDE] (1651)
: being, living, used, or made under the skin ⟨*subcutaneous* parasites⟩
— **sub·cu·ta·ne·ous·ly** *adverb*

sub·cu·tis \ˌsəb-ˈkyü-təs\ *noun* [New Latin, from Late Latin, beneath the skin, from Latin *sub-* + *cutis*] (1900)
: the deeper part of the dermis

sub·dea·con \-ˈdē-kən\ *noun* [Middle English *subdecon,* from Late Latin *subdiaconus,* from Latin *sub-* + Late Latin *diaconus* deacon — more at DEACON] (14th century)
: a cleric ranking below a deacon: as **a :** a cleric in the lowest of the former major orders of the Roman Catholic Church **b :** an Eastern Orthodox or Armenian cleric in minor orders **c :** a clergyman performing the liturgical duties of a subdeacon

sub·deb \ˈsəb-ˌdeb\ *noun* (1917)
: SUBDEBUTANTE

sub·deb·u·tante \ˌsəb-ˈde-byu̇-ˌtänt\ *noun* (1919)
: a young girl who is about to become a debutante; *broadly* **:** a girl in her middle teens

sub·der·mal \-ˈdər-məl\ *adjective* (1887)
: SUBCUTANEOUS
— **sub·der·mal·ly** \-mə-lē\ *adverb*

sub·di·vide \ˌsəb-də-ˈvīd, ˈsəb-də-ˌ\ *verb* [Middle English, from Late Latin *subdividere,* from Latin *sub-* + *dividere* to divide] (15th century)
transitive verb
1 **:** to divide the parts of into more parts
2 **:** to divide into several parts; *especially* **:** to divide (a tract of land) into building lots
intransitive verb
: to separate or become separated into subdivisions
— **sub·di·vid·able** \-ˈvī-də-bəl, -ˌvī-\ *adjective*
— **sub·di·vid·er** *noun*

sub·di·vi·sion \ˈsəb-də-ˌvi-zhən\ *noun* (15th century)
1 **:** an act or instance of subdividing
2 **:** something produced by subdividing; *especially* **:** a tract of land surveyed and divided into lots for purposes of sale
3 **:** a category in botanical classification ranking below a division and above a class

sub·dom·i·nant \ˌsəb-ˈdäm-nənt, -ˈdä-mə-\ *noun* (1793)
1 **:** the fourth tone of a diatonic scale
2 **:** something partly but incompletely dominant; *especially* **:** an ecologically important life form subordinate in influence to the dominants of a community
— **subdominant** *adjective*

sub·duc·tion \(ˌ)səb-ˈdək-shən\ *noun* [Late Latin *subduction-, subductio* withdrawal, from Latin *subducere* to withdraw, from *sub-* + *ducere* to draw — more at TOW] (1970)
: the action or process of the edge of one crustal plate descending below the edge of another
— **sub·duct** \(ˌ)səb-ˈdəkt\ *verb*

sub·due \səb-ˈdü, -ˈdyü\ *transitive verb* **sub·dued; sub·du·ing** [Middle English *sodewen, subduen* (influenced in form and meaning by Latin *subdere* to subject), from Middle French *soduire* to seduce (influenced in meaning by Latin *seducere* to seduce), from Latin *subducere*] (14th century)
1 **:** to conquer and bring into subjection **:** VANQUISH
2 **:** to bring under control especially by an exertion of the will **:** CURB ⟨*subdued* my foolish fears⟩
3 **:** to bring under cultivation
4 **:** to reduce the intensity or degree of **:** tone down
synonym see CONQUER
— **sub·du·er** *noun*

sub·dued \-ˈdüd, -ˈdyüd\ *adjective* (1591)
: lacking in vitality, intensity, or strength ⟨*subdued* colors⟩
— **sub·dued·ly** \-ˈdü(-ə)d-lē, -ˈdyü(-ə)d-\ *adverb*

sub·dur·al \ˌsəb-ˈdu̇r-əl, -ˈdyu̇r-; ˈsəb-ˌ\ *adjective* [*sub-* + *dura (mater)*] (1875)
: situated or occurring beneath the dura mater or between the dura mater and the arachnoid membrane ⟨*subdural* space⟩ ⟨*subdural* hematomas⟩

sub·ed·i·tor \-ˈe-də-tər\ *noun* (1835)
chiefly British **:** COPY EDITOR
— **sub·ed·it** \-ˈe-dət\ *transitive verb, chiefly British*
— **sub·ed·i·to·ri·al** \ˌsəb-ˌe-də-ˈtōr-ē-əl, -ˈtȯr-\ *adjective, chiefly British*

sub·em·ployed \ˌsəb-im-ˈplȯid\ *adjective* (1967)
: UNDEREMPLOYED

sub·em·ploy·ment \-ˈplȯi-mənt\ *noun* (1967)
: a condition of inadequate employment in a labor force including unemployment and underemployment

sub·en·try \ˈsəb-ˌen-trē\ *noun* (circa 1891)
: an entry (as in a catalog or an account) made under a more general entry

sub·epi·der·mal \ˌsəb-ˌe-pə-ˈdər-məl\ *adjective* (1853)
: lying beneath or constituting the innermost part of the epidermis

su·ber·in \ˈsü-bə-rən\ *noun* [French *subérine,* from Latin *suber* cork tree, cork] (1830)
: a complex fatty substance found especially in the cell walls of cork

su·ber·i·za·tion \ˌsü-bə-rə-ˈzā-shən\ *noun* (1882)
: conversion of the cell walls into corky tissue by infiltration with suberin
— **su·ber·ized** \ˈsü-bə-ˌrīzd\ *adjective*

sub·fam·i·ly \ˈsəb-ˌfam-lē, -ˌfa-mə-\ *noun* [International Scientific Vocabulary] (1833)
1 **:** a category in biological classification ranking below a family and above a genus
2 **:** a subgroup of languages within a language family

sub·field \-ˌfēld\ *noun* (circa 1949)
1 **:** a subset of a mathematical field that is itself a field
2 **:** a subdivision of a field (as of study)

sub·floor \-ˌflōr, -ˌflȯr\ *noun* (1893)

: a rough floor laid as a base for a finished floor

sub·fos·sil \\-ˌfä-səl\\ *adjective* [International Scientific Vocabulary] (1832)
: of less than typical fossil age but partially fossilized
— subfossil *noun*

sub·freez·ing \\-ˈfrē-ziŋ\\ *adjective* (1949)
: being or marked by temperature below the freezing point (as of water) ⟨*subfreezing* weather⟩

sub·fusc \\(ˌ)səb-ˈfəsk, ˈsəb-ˌ\\ *adjective* [Latin *subfuscus* brownish, dusky, from *sub-* + *fuscus* dark brown — more at DUSK] (1710)
chiefly British **:** DRAB, DUSKY

sub·ge·nus \\ˈsəb-ˌjē-nəs\\ *noun* [New Latin] (1813)
: a category in biological classification ranking below a genus and above a species

sub·gla·cial \\ˌsəb-ˈglā-shəl\\ *adjective* (1820)
: of or relating to the bottom of a glacier or the area immediately underlying a glacier
— sub·gla·cial·ly \\-shə-lē\\ *adverb*

sub·grade \\ˈsəb-ˌgrād\\ *noun* (1898)
: a surface of earth or rock leveled off to receive a foundation (as of a road)

sub·graph \\ˈsəb-ˌgraf\\ *noun* (1963)
: a graph all of whose points and lines are contained in a larger graph

sub·group \\-ˌgrüp\\ *noun* (1845)
1 : a subordinate group whose members usually share some common differential quality
2 : a subset of a mathematical group that is itself a group

sub·gum \\ˈsəb-ˈgəm\\ *noun* [Chinese (Guangdong) *sahp-gám,* literally, assorted, mixed] (1938)
: a dish of Chinese origin prepared with a mixture of vegetables (as peppers, water chestnuts, and mushrooms)

sub·head \\-ˌhed\\ *noun* (1673)
1 : a heading of a subdivision (as in an outline)
2 : a subordinate caption, title, or headline

sub·head·ing \\-ˌhe-diŋ\\ *noun* (1889)
: SUBHEAD

[1]**sub·hu·man** \\ˌsəb-ˈhyü-mən, -ˈyü-\\ *adjective* (1793)
: less than human: as **a :** failing to attain the level (as of morality or intelligence) associated with normal human beings **b :** unsuitable to or unfit for human beings ⟨*subhuman* living conditions⟩ **c :** of or relating to an infrahuman taxonomic group ⟨the *subhuman* primates⟩

[2]**subhuman** *noun* (1937)
: a subhuman being

sub·in·dex \\ˌsəb-ˈin-ˌdeks\\ *noun* (1923)
: an index to a division of a main classification

sub·in·feu·da·tion \\ˌsəb-ˌin-fyü-ˈdā-shən\\ *noun* [*sub-* + *infeudation* enfeoffment] (circa 1730)
: the subdivision of a feudal estate by a vassal who in turn becomes feudal lord over his tenants
— sub·in·feu·date *transitive verb*

sub·in·ter·val \\ˌsəb-ˈin-tər-vəl\\ *noun* (1927)
: an interval that is a subdivision or a subset of an interval

sub·ir·ri·ga·tion \\-ˌir-ə-ˈgā-shən\\ *noun* (1880)
: irrigation below the surface (as by a periodic rise of the water table or by a system of underground porous pipes)
— sub·ir·ri·gate \\-ˈir-ə-ˌgāt\\ *transitive verb*

su·bi·to \\ˈsü-bi-ˌtō\\ *adverb* [Italian, from Latin, suddenly, from *subitus* sudden — more at SUDDEN] (circa 1724)
: IMMEDIATELY, SUDDENLY — used as a direction in music

sub·ja·cen·cy \\ˌsəb-ˈjā-sᵊn(t)-sē\\ *noun* (circa 1891)
: the quality or state of being subjacent

sub·ja·cent \\-sᵊnt\\ *adjective* [Latin *subjacent-, subjacens,* present participle of *subjacēre* to lie under, from *sub-* + *jacēre* to lie — more at ADJACENT] (1597)
: lying under or below; *also* **:** lower than though not directly below ⟨hills and *subjacent* valleys⟩
— sub·ja·cent·ly *adverb*

[1]**sub·ject** \\ˈsəb-jikt, -(ˌ)jekt\\ *noun* [Middle English, from Middle French, from Latin *subjectus* one under authority & *subjectum* subject of a proposition, from masculine & neuter respectively of *subjectus,* past participle of *subicere* to subject, literally, to throw under, from *sub-* + *jacere* to throw — more at JET] (14th century)
1 : one that is placed under authority or control: as **a :** VASSAL **b** (1) **:** one subject to a monarch and governed by the monarch's law (2) **:** one who lives in the territory of, enjoys the protection of, and owes allegiance to a sovereign power or state
2 a : that of which a quality, attribute, or relation may be affirmed or in which it may inhere **b :** SUBSTRATUM; *especially* **:** material or essential substance **c :** the mind, ego, or agent of whatever sort that sustains or assumes the form of thought or consciousness
3 a : a department of knowledge or learning **b :** MOTIVE, CAUSE **c** (1) **:** one that is acted on ⟨the helpless *subject* of their cruelty⟩ (2) **:** an individual whose reactions or responses are studied (3) **:** a dead body for anatomical study and dissection **d** (1) **:** something concerning which something is said or done ⟨the *subject* of the essay⟩ (2) **:** something represented or indicated in a work of art **e** (1) **:** the term of a logical proposition that denotes the entity of which something is affirmed or denied; *also* **:** the entity denoted (2) **:** a word or word group denoting that of which something is predicated **f :** the principal melodic phrase on which a musical composition or movement is based
synonym see CITIZEN
— sub·ject·less \\-ləs\\ *adjective*

[2]**subject** *adjective* (14th century)
1 : owing obedience or allegiance to the power or dominion of another
2 a : suffering a particular liability or exposure ⟨*subject* to temptation⟩ **b :** having a tendency or inclination **:** PRONE ⟨*subject* to colds⟩
3 : contingent on or under the influence of some later action ⟨the plan is *subject* to discussion⟩
synonym see LIABLE

[3]**sub·ject** \\səb-ˈjekt, ˈsəb-ˌjekt\\ *transitive verb* (14th century)
1 a : to bring under control or dominion **:** SUBJUGATE **b :** to make (as oneself) amenable to the discipline and control of a superior
2 : to make liable **:** PREDISPOSE
3 : to cause or force to undergo or endure (something unpleasant, inconvenient, or trying) ⟨was *subjected* to constant verbal abuse⟩
— sub·jec·tion \\səb-ˈjek-shən\\ *noun*

[1]**sub·jec·tive** \\(ˌ)səb-ˈjek-tiv\\ *adjective* (15th century)
1 : of, relating to, or constituting a subject: as **a** *obsolete* **:** of, relating to, or characteristic of one that is a subject especially in lack of freedom of action or in submissiveness **b :** being or relating to a grammatical subject; *especially* **:** NOMINATIVE
2 : of or relating to the essential being of that which has substance, qualities, attributes, or relations
3 a : characteristic of or belonging to reality as perceived rather than as independent of mind **:** PHENOMENAL — compare OBJECTIVE 1b **b :** relating to or being experience or knowledge as conditioned by personal mental characteristics or states
4 a (1) **:** peculiar to a particular individual **:** PERSONAL ⟨*subjective* judgments⟩ (2) **:** modified or affected by personal views, experience, or background ⟨a *subjective* account of the incident⟩ **b :** arising from conditions within the brain or sense organs and not directly caused by external stimuli ⟨*subjective* sensations⟩ **c :** arising out of or identified by means of one's perception of one's own states and processes ⟨a *subjective* symptom of disease⟩ — compare OBJECTIVE 1c
5 : lacking in reality or substance **:** ILLUSORY
— sub·jec·tive·ly *adverb*
— sub·jec·tive·ness *noun*
— sub·jec·tiv·i·ty \\-ˌjek-ˈti-və-tē\\ *noun*

[2]**subjective** *noun* (1817)
: something that is subjective; *also* **:** NOMINATIVE

subjective complement *noun* (1923)
: a grammatical complement relating to the subject of an intransitive verb (as *sick* in "he had fallen sick")

sub·jec·tiv·ise *British variant of* SUBJECTIVIZE

sub·jec·tiv·ism \\(ˌ)səb-ˈjek-ti-ˌvi-zəm\\ *noun* (circa 1857)
1 a : a theory that limits knowledge to subjective experience **b :** a theory that stresses the subjective elements in experience
2 a : a doctrine that the supreme good is the realization of a subjective experience or feeling (as pleasure) **b :** a doctrine that individual feeling or apprehension is the ultimate criterion of the good and the right
— sub·jec·tiv·ist \\-vist\\ *noun*
— sub·jec·tiv·is·tic \\-ˌjek-ti-ˈvis-tik\\ *adjective*

sub·jec·tiv·ize \\-ti-ˌvīz\\ *transitive verb* **-ized; -iz·ing** (1868)
: to make subjective
— sub·jec·tiv·i·za·tion \\-jek-ti-və-ˈzā-shən\\ *noun*

subject matter *noun* (1598)
: matter presented for consideration in discussion, thought, or study

sub·join \\(ˌ)səb ˈjȯin\\ *transitive verb* [Middle French *subjoindre,* from Latin *subjungere* to join beneath, add, from *sub-* + *jungere* to join — more at YOKE] (1573)
: ANNEX, APPEND ⟨*subjoined* a statement of expenses to her report⟩

sub ju·di·ce \\ˌsu̇b-ˈyü-di-ˌkā, ˈsəb-ˈjü-də-(ˌ)sē\\ *adverb* [Latin] (1613)
: before a judge or court **:** not yet judicially decided

sub·ju·gate \\ˈsəb-ji-ˌgāt\\ *transitive verb* **-gat·ed; -gat·ing** [Middle English, from Latin *subjugatus,* past participle of *subjugare,* from *sub-* + *jugum* yoke — more at YOKE] (15th century)
1 : to bring under control and governance as a subject **:** CONQUER
2 : to make submissive **:** SUBDUE
— sub·ju·ga·tion \\ˌsəb-ji-ˈgā-shən\\ *noun*
— sub·ju·ga·tor \\ˈsəb-ji-ˌgā-tər\\ *noun*

sub·junc·tion \\(ˌ)səb-ˈjəŋ(k)-shən\\ *noun* (1633)
1 : an act of subjoining or the state of being subjoined
2 : something subjoined

[1]**sub·junc·tive** \\səb-ˈjəŋ(k)-tiv\\ *adjective* [Late Latin *subjunctivus,* from Latin *subjunctus,* past participle of *subjungere* to join beneath, subordinate] (1530)
: of, relating to, or constituting a verb form or set of verb forms that represents a denoted act or state not as fact but as contingent or possible or viewed emotionally (as with doubt or desire) ⟨the *subjunctive* mood⟩

[2]**subjunctive** *noun* (1622)
1 : the subjunctive mood of a language

2 : a form of verb or verbal in the subjective mood ■

sub·king·dom \'səb-ˌkiŋ-dəm\ *noun* (1825) **:** a category in biological classification ranking below a kingdom and above a phylum

sub·late \ˌsə-'blāt\ *transitive verb* **sub·lat·ed; sub·lat·ing** [Latin *sublatus* (past participle of *tollere* to take away, lift up), from *sub-* up + *latus,* past participle of *ferre* to carry — more at SUB-, TOLERATE, BEAR] (1838)
1 : NEGATE, DENY
2 : to negate or eliminate (as an element in a dialectic process) but preserve as a partial element in a synthesis
— sub·la·tion \-'blā-shən\ *noun*

¹sub·lease \'səb-'lēs, -ˌlēs\ *noun* (1826) **:** a lease by a tenant or lessee of part or all of leased premises to another person but with the original tenant retaining some right or interest under the original lease

²sublease *transitive verb* (circa 1843) **:** to make or obtain a sublease of

¹sub·let \'səb-'let\ *verb* **-let; -let·ting** (1766)
transitive verb
1 : SUBLEASE
2 : SUBCONTRACT 1
intransitive verb
: to lease or rent all or part of a leased or rented property

²sub·let \-ˌlet\ *noun* (1906) **:** property and especially housing obtained by or available through a sublease

sub·le·thal \ˌsəb-'lē-thəl\ *adjective* (1895) **:** less than but usually only slightly less than lethal ⟨a *sublethal* dose⟩
— sub·le·thal·ly \-thə-lē\ *adverb*

sub·lieu·ten·ant \ˌsəb-lü-'te-nənt, *British* -le(f)-'ten-\ *noun* (1804) **:** a commissioned officer in the British navy ranking immediately below lieutenant

¹sub·li·mate \'sə-blə-ˌmāt\ *transitive verb* **-mat·ed; -mat·ing** [Middle English, from Medieval Latin *sublimatus,* past participle of *sublimare*] (15th century)
1 a : SUBLIME 1 **b** *archaic* **:** to improve or refine as if by subliming
2 : to divert the expression of (an instinctual desire or impulse) from its primitive form to one that is considered more socially or culturally acceptable
— sub·li·ma·tion \ˌsə-blə-'mā-shən\ *noun*

²sub·li·mate \'sə-blə-ˌmāt, -mət\ *noun* (circa 1626) **:** a chemical product obtained by sublimation

¹sub·lime \sə-'blīm\ *verb* **sub·limed; sub·lim·ing** [Middle English, from Middle French *sublimer,* from Medieval Latin *sublimare* to refine, sublime, from Latin, to elevate, from *sublimis*] (14th century)
transitive verb
1 : to cause to pass directly from the solid to the vapor state and condense back to solid form
2 [French *sublimer,* from Latin *sublimare*] **a** (1) **:** to elevate or exalt especially in dignity or honor (2) **:** to render finer (as in purity or excellence) **b :** to convert (something inferior) into something of higher worth
intransitive verb
: to pass directly from the solid to the vapor state
— sub·lim·able \-'blī-mə-bəl\ *adjective*
— sub·lim·er *noun*

²sublime *adjective* **sub·lim·er; -est** [Latin *sublimis,* literally, high, elevated] (1586)
1 a : lofty, grand, or exalted in thought, expression, or manner **b :** of outstanding spiritual, intellectual, or moral worth **c :** tending to inspire awe usually because of elevated quality (as of beauty, nobility, or grandeur) or transcendent excellence
2 a *archaic* **:** high in place **b** *obsolete* **:** lofty of mien **:** HAUGHTY **c** *capitalized* **:** SUPREME — used in a style of address **d :** COMPLETE, UTTER ⟨*sublime* ignorance⟩
synonym see SPLENDID
— sub·lime·ly *adverb*
— sub·lime·ness *noun*

sub·lim·i·nal \(ˌ)sə-'bli-mə-nᵊl\ *adjective* [*sub-* + Latin *limin-, limen* threshold] (1886)
1 : inadequate to produce a sensation or a perception
2 : existing or functioning below the threshold of consciousness ⟨the *subliminal* mind⟩ ⟨*subliminal* advertising⟩
— sub·lim·i·nal·ly *adverb*

sub·lim·i·ty \sə-'bli-mə-tē\ *noun, plural* **-ties** (15th century)
1 : the quality or state of being sublime
2 : something sublime or exalted

sub·line \'səb-ˌlīn\ *noun* (1942) **:** an inbred or selectively cultured line (as of cells) within a strain

sub·lin·gual \ˌsəb-'liŋ-gwəl, -gyə-wəl\ *adjective* [New Latin *sublingualis,* from Latin *sub-* + *lingua* tongue — more at TONGUE] (1661) **:** situated or administered under the tongue ⟨*sublingual* tablets⟩ ⟨*sublingual* glands⟩

sub·lit·er·ary \-'li-tə-ˌrer-ē\ *adjective* (1936) **:** relating to or being subliterature

sub·lit·er·a·ture \-'li-tə-rə-ˌchu̇r, -'li-trə-ˌchu̇r, -'li-tə(r)-ˌchu̇r, -chər, -ˌtyu̇r, -ˌtu̇r\ *noun* (1952) **:** popular writing (as mystery or adventure stories) considered inferior to standard literature

¹sub·lit·to·ral \-'li-tə-rəl; ˌsəb-ˌli-tə-'ral, -'räl\ *adjective* (1846)
1 : situated, occurring, or formed on the aquatic side of a shoreline or littoral zone
2 : constituting the sublittoral

²sublittoral *noun* (circa 1935) **:** the deeper part of the littoral portion of a body of water: **a :** the region in a lake between the deepest-growing rooted vegetation and the part of the lake below the thermocline **b :** the region in an ocean between the lowest point exposed by a low tide and the margin of the continental shelf

sub·lu·na·ry \ˌsəb-'lü-nə-rē, 'səb-lü-ˌner-ē\ *also* **sub·lu·nar** \ˌsəb-'lü-nər *also* -ˌnär\ *adjective* [modification of Late Latin *sublunaris,* from Latin *sub-* + *luna* moon — more at LUNAR] (1592) **:** of, relating to, or characteristic of the terrestrial world ⟨dull *sublunary* lovers —John Donne⟩

sub·lux·a·tion \ˌsə-ˌblək-'sā-shən\ *noun* (circa 1688) **:** partial dislocation (as of one of the bones in a joint)

sub·ma·chine gun \ˌsəb-mə-'shēn-ˌgən\ *noun* (1920) **:** a portable automatic firearm that uses pistol-type ammunition and is fired from the shoulder or hip

¹sub·man·dib·u·lar \ˌsəb-man-'di-byə-lər\ *adjective* (1875)
1 : of, relating to, situated in, or performed in the region below the lower jaw
2 : of, relating to, or associated with the salivary glands inside of and near the lower edge of the mandible on each side

²submandibular *noun* (1974) **:** a submandibular part (as an artery or bone)

sub·mar·gin·al \ˌsəb-'märj-nəl, -'mär-jə-nᵊl\ *adjective* (1829)
1 : adjacent to a margin or a marginal part or structure ⟨*submarginal* spots on an insect wing⟩
2 : falling below a necessary minimum ⟨*submarginal* economic conditions⟩

¹sub·ma·rine \'səb-mə-ˌrēn, ˌsəb-mə-'\ *adjective* (1648) **:** UNDERWATER; *especially* **:** UNDERSEA ⟨*submarine* plants⟩ ⟨*submarine* minerals⟩

²submarine *noun* (1703)
1 : something that functions or operates underwater; *specifically* **:** a naval vessel designed to operate underwater
2 : a large sandwich on a long split roll with any of a variety of fillings (as meatballs or cold cuts, cheese, lettuce, and tomato) — called also *grinder, hero, hoagie, Italian sandwich, poor boy, sub, torpedo*

³submarine *verb* **-rined; -rin·ing** (1914)
transitive verb
: to attack by or as if by a submarine **:** attack from beneath
intransitive verb
: to dive or slide under something

sub·ma·ri·ner \ˌsəb-mə-'rē-nər, -'mar-ə-nər\ *noun* (1914) **:** a member of a submarine crew

sub·max·il·lary \ˌsəb-'mak-sə-ˌler-ē, 'səb-, *chiefly British* ˌsəb-mak-'si-lə-rē\ *adjective or noun* (1787) **:** SUBMANDIBULAR

sub·me·di·ant \ˌsəb-'mē-dē-ənt\ *noun* (1806) **:** the sixth tone of a diatonic scale

sub·merge \səb-'mərj\ *verb* **sub·merged; sub·merg·ing** [Latin *submergere,* from *sub-* + *mergere* to plunge — more at MERGE] (1611)
transitive verb
1 : to put under water

□ USAGE

subjunctive Various observers of the language have been predicting the disappearance of the subjunctive for more than a century now. While the range of the subjunctive is much reduced from what it was four or five hundred years ago, it is probably little different from what it was a century ago. One reason for its assumed extinction is that it is nearly invisible. Its chief characteristic for most verbs is lack of inflection, which can be detected only when it occurs in a context calling for an inflection (as the present third person singular). The present subjunctive of *think* is *think;* in "lest she think . . ." we can identify the subjunctive by the missing *-s,* but in "lest you think . . ." it has no identifying mark. So one can often use the subjunctive without even knowing it. The subjunctive turns up chiefly in three contexts. First, it occurs in a handful of fixed phrases like *far be it from me, if need be,* or *as it were.* Second, it occurs regularly in American English in the clause following what one grammarian calls "suasive" verbs, such as *ask, demand, suggest, move* ⟨I move the meeting *be* adjourned⟩ ⟨we suggested that he *retire* early⟩. Both these uses are regular in American English and draw no comment. The third context—the one everyone talks about—is after verbs of wishing and in hypothetical or contrary-to-fact conditional clauses. Here the contrast lies chiefly between the indicative *was* and the subjunctive *were* ⟨I wish I *were* twenty-two again —F. Scott Fitzgerald⟩ ⟨if I *were* elected King of Scotland tomorrow —F. Scott Fitzgerald⟩ ⟨my birthday is two-column front page news as if I *were* 80 instead of 40 —F. Scott Fitzgerald⟩. Here *wish, if,* and *as if* have triggered the subjunctive. But the same words are often followed by the indicative ⟨I wish I *was* in print —F. Scott Fitzgerald⟩ ⟨if I *was* Vassar, I wouldn't take you —F. Scott Fitzgerald⟩ ⟨keep the matter as secret as if there *was* something disgraceful in the connexion —Washington Irving⟩. Clearly there is a choice to be made here, and if Fitzgerald could use either form, so can others. We do not know why in the late 16th century *was* began to be used in contexts where before only *were* had been used. It just happened. A curious demonstration of the power of the conditional conjunctions, such as *as if, as though, if,* is their ability to produce a subjunctive where none is needed ⟨it sounds just a little as though Mr. Vance thought I *were* trying to slip something over on him —Ernest Hemingway⟩ ⟨a clause in her contract letting her leave CBS Inc. if she *weren't* given an anchor position —*Wall Street Journal*⟩.

2 **:** to cover or overflow with water
3 **:** to make obscure or subordinate **:** SUPPRESS ⟨personal lives *submerged* by professional responsibilities⟩
intransitive verb
: to go under water
— **sub·mer·gence** \-ˈmər-jən(t)s\ *noun*
— **sub·merg·ible** \-ˈmər-jə-bəl\ *adjective*

submerged *adjective* (1799)
1 **:** covered with water
2 **:** SUBMERSED b
3 **:** sunk in poverty and misery ⟨the *submerged* masses⟩ ⟨the *submerged* tenth of the population⟩
4 **:** HIDDEN, SUPPRESSED ⟨*submerged* emotions⟩

sub·merse \səb-ˈmərs\ *transitive verb* **sub·mersed; sub·mers·ing** [Latin *submersus,* past participle of *submergere*] (1837)
: SUBMERGE
— **sub·mer·sion** \-ˈmər-zhən, -shən\ *noun*

submersed *adjective* (circa 1727)
: SUBMERGED: as **a :** covered with water **b :** growing or adapted to grow underwater ⟨*submersed* weeds⟩

[1]**sub·mers·ible** \səb-ˈmər-sə-bəl\ *adjective* (1866)
: capable of being submerged

[2]**submersible** *noun* (1900)
: something that is submersible; *especially* **:** a usually small underwater craft used especially for deep-sea research

sub·meta·cen·tric \ˌsəb-ˌme-tə-ˈsen-trik\ *adjective* (1962)
: having the centromere situated so that one chromosome is somewhat shorter than the other
— **submetacentric** *noun*

sub·mi·cro·gram \ˌsəb-ˈmī-krə-ˌgram\ *adjective* (1946)
: relating to or having a mass of less than one microgram ⟨*submicrogram* quantities of a chemical⟩

sub·mi·cron \-ˌkrän\ *adjective* (1948)
1 **:** being less than a micron in a (specified) measurement and especially in diameter ⟨a *submicron* particle⟩
2 **:** having or consisting of submicron particles ⟨a *submicron* metal powder⟩

sub·mi·cro·scop·ic \ˌsəb-ˌmī-krə-ˈskä-pik\ *adjective* [International Scientific Vocabulary] (1912)
1 **:** too small to be seen in an ordinary light microscope
2 **:** of, relating to, or dealing with the very minute ⟨the *submicroscopic* world⟩
— **sub·mi·cro·scop·i·cal·ly** \-pi-k(ə-)lē\ *adverb*

sub·mil·li·me·ter \ˌsəb-ˈmi-lə-ˌmē-tər\ *adjective* (1955)
: being less than a millimeter in diameter or wavelength ⟨a *submillimeter* particle⟩ ⟨a *submillimeter* radio wave⟩

sub·min·i·a·ture \ˌsəb-ˈmi-nē-ə-ˌchu̇r, -ˈmi-ni-ˌchu̇r, -ˈmin-yə-, -chər, -ˌtyu̇r, -ˌtu̇r\ *adjective* [International Scientific Vocabulary] (1947)
: very small — used especially of a very compact assembly of electronic equipment

sub·miss \səb-ˈmis\ *adjective* [Latin *submissus,* from past participle of *submittere*] (1570)
archaic **:** SUBMISSIVE, HUMBLE

sub·mis·sion \səb-ˈmi-shən\ *noun* [Middle English, from Middle French, from Latin *submission-, submissio* act of lowering, from *submittere*] (14th century)
1 **a :** a legal agreement to submit to the decision of arbitrators **b :** an act of submitting something (as for consideration or inspection); *also* **:** something submitted (as a manuscript)
2 **:** the condition of being submissive, humble, or compliant
3 **:** an act of submitting to the authority or control of another

sub·mis·sive \-ˈmi-siv\ *adjective* (circa 1586)
: submitting to others
— **sub·mis·sive·ly** *adverb*
— **sub·mis·sive·ness** *noun*

sub·mit \sə-ˈmit\ *verb* **sub·mit·ted; sub·mit·ting** [Middle English *submitten,* from Latin *submittere* to lower, submit, from *sub-* + *mittere* to send] (14th century)
transitive verb
1 **a :** to yield to governance or authority **b :** to subject to a condition, treatment, or operation ⟨the metal was *submitted* to analysis⟩
2 **:** to present or propose to another for review, consideration, or decision ⟨*submit* a question to the court⟩ ⟨*submit* a bid on a contract⟩ ⟨*submit* a report⟩; *also* **:** to deliver formally ⟨*submitted* my resignation⟩
3 **:** to put forward as an opinion or contention ⟨we *submit* that the charge is not proved⟩
intransitive verb
1 **a :** to yield oneself to the authority or will of another **:** SURRENDER **b :** to permit oneself to be subjected to something ⟨had to *submit* to surgery⟩
2 **:** to defer to or consent to abide by the opinion or authority of another
synonym see YIELD
— **sub·mit·tal** \-ˈmi-tᵊl\ *noun*

sub·mi·to·chon·dri·al \ˌsəb-ˌmī-tə-ˈkän-drē-əl\ *adjective* (1963)
: relating to, composed of, or being parts and especially fragments of mitochondria ⟨*submitochondrial* membranes⟩ ⟨*submitochondrial* particles⟩

sub·mu·co·sa \ˌsəb-myü-ˈkō-zə\ *noun* [New Latin] (1885)
: a supporting layer of loose connective tissue directly under a mucous membrane
— **sub·mu·co·sal** \-zəl\ *adjective*

sub·mul·ti·ple \-ˈməl-tə-pəl\ *noun* (1758)
: an exact divisor of a number ⟨8 is a *submultiple* of 72⟩

sub·mu·ni·tion \(ˌ)səb-myü-ˈni-shən\ *noun* (1975)
: any of a group of smaller weapons carried as a warhead by a missile or projectile and expelled as the carrier approaches its target

sub·nor·mal \-ˈnȯr-məl\ *adjective* [International Scientific Vocabulary] (circa 1890)
1 **:** lower or smaller than normal
2 **:** having less of something and especially of intelligence than is normal
— **sub·nor·mal·i·ty** \ˌsəb-nȯr-ˈma-lə-tē\ *noun*
— **sub·nor·mal·ly** \ˌsəb-ˈnȯr-mə-lē\ *adverb*

sub·nu·cle·ar \ˌsəb-ˈnü-klē-ər, -ˈnyü-, ÷-kyə-lər\ *adjective* (1937)
: of, relating to, or being a particle smaller than the atomic nucleus

sub·oce·an·ic \ˌsəb-ˌō-shē-ˈa-nik\ *adjective* (1858)
: situated, taking place, or formed beneath the ocean or its bottom ⟨*suboceanic* oil resources⟩

sub·or·bit·al \ˌsəb-ˈȯr-bə-tᵊl\ *adjective* (circa 1827)
1 **:** situated beneath the eye or the orbit of the eye
2 **:** being or involving less than one orbit (as of the earth or moon) ⟨a spacecraft's *suborbital* flight⟩; *also* **:** intended for suborbital flight ⟨a *suborbital* rocket⟩

sub·or·der \ˈsəb-ˌȯr-dər\ *noun* (1826)
: a subdivision of an order ⟨a soil *suborder*⟩; *especially* **:** a taxonomic category ranking between an order and a family

[1]**sub·or·di·nate** \sə-ˈbȯr-dᵊn-ət, -ˈbȯrd-nət\ *adjective* [Middle English *subordinat,* from Medieval Latin *subordinatus,* past participle of *subordinare* to subordinate, from Latin *sub-* + *ordinare* to order — more at ORDAIN] (15th century)
1 **:** placed in or occupying a lower class, rank, or position **:** INFERIOR
2 **:** submissive to or controlled by authority
3 **a :** of, relating to, or constituting a clause that functions as a noun, adjective, or adverb **b :** SUBORDINATING
— **sub·or·di·nate·ly** *adverb*
— **sub·or·di·nate·ness** *noun*

[2]**subordinate** *noun* (1640)
: one that is subordinate

[3]**sub·or·di·nate** \sə-ˈbȯr-dᵊn-ˌāt\ *transitive verb* **-nat·ed; -nat·ing** [Medieval Latin *subordinatus*] (1597)
1 **:** to make subject or subservient
2 **:** to treat as of less value or importance ⟨stylist . . . whose crystalline prose *subordinates* content to form —Susan Heath⟩
— **sub·or·di·na·tion** \-ˌbȯr-dᵊn-ˈā-shən\ *noun*
— **sub·or·di·na·tive** \-ˈbȯr-dᵊn-ˌā-tiv\ *adjective*

subordinating *adjective* (1857)
: introducing and linking a subordinate clause to a main clause ⟨*subordinating* conjunction⟩

sub·or·di·na·tor \-ˌā-tər\ *noun* (1959)
: one that subordinates; *especially* **:** a subordinating conjunction

sub·orn \sə-ˈbȯrn\ *transitive verb* [Middle French *suborner,* from Latin *subornare,* from *sub-* secretly + *ornare* to furnish, equip — more at ORNATE] (1534)
1 **:** to induce secretly to do an unlawful thing
2 **:** to induce to commit perjury; *also* **:** to obtain (perjured testimony) from a witness
— **sub·or·na·tion** \ˌsə-ˌbȯr-ˈnā-shən\ *noun*
— **sub·orn·er** *noun*

sub·phy·lum \ˈsəb-ˌfī-ləm\ *noun* [New Latin] (circa 1934)
: a category in biological classification ranking below a phylum and above a class

sub·plot \ˈsəb-ˌplät\ *noun* (1916)
1 **:** a subordinate plot in fiction or drama
2 **:** a subdivision of an experimental plot of land

[1]**sub·poe·na** \sə-ˈpē-nə, ÷-nē\ *noun* [Middle English *suppena,* from Latin *sub poena* under penalty] (15th century)
: a writ commanding a person designated in it to appear in court under a penalty for failure

[2]**subpoena** *transitive verb* **-naed; -na·ing** (1640)
: to serve or summon with a writ of subpoena

subpoena ad tes·ti·fi·can·dum \ˌad-ˌtes-tə-fi-ˈkan-dəm\ *noun* [New Latin, under penalty to give testimony] (circa 1769)
: a writ commanding a person to appear in court to testify as a witness

subpoena du·ces te·cum \-ˌdü-səs-ˈtē-kəm\ *noun* [New Latin, under penalty you shall bring with you] (1768)
: a writ commanding a person to produce in court certain designated documents or evidence

sub·po·lar \ˌsəb-ˈpō-lər\ *adjective* (1826)
: SUBANTARCTIC, SUBARCTIC

sub·pop·u·la·tion \ˈsəb-ˌpä-pyə-ˈlā-shən\ *noun* (1944)
: an identifiable fraction or subdivision of a population

sub·po·tent \ˌsəb-ˈpō-tᵊnt\ *adjective* (circa 1909)
: less potent than normal ⟨*subpotent* drugs⟩
— **sub·po·ten·cy** \-ˈpō-tᵊn(t)-sē\ *noun*

sub·prin·ci·pal \-ˈprin(t)-s(ə-)pəl, -sə-bəl\ *noun* (1597)
1 **:** an assistant principal (as of a school)
2 **:** a secondary or bracing rafter

sub·prob·lem \ˈsəb-ˌprä-bləm\ *noun* (1906)
: a problem that is contingent on or forms a part of another more inclusive problem

sub·pro·fes·sion·al \ˌsəb-prə-ˈfesh-nəl, -ˈfe-shə-nᵊl\ *adjective* (1941)
: functioning or qualified to function below the professional level but distinctly above the clerical or labor level and usually under the supervision of a professionally trained person
— **subprofessional** *noun*

sub·pro·gram \ˈsəb-ˌprō-ˌgram, -grəm\ *noun* (1947)
: a semi-independent portion of a program (as for a computer)

sub·re·gion \ˈsəb-ˌrē-jən\ *noun* [International Scientific Vocabulary] (1864)
1 : a subdivision of a region
2 : one of the primary divisions of a biogeographic region
— **sub·re·gion·al** \-ˌrēj-nəl, -ˌrē-jə-nᵊl\ *adjective*

sub·rep·tion \(ˌ)səb-ˈrep-shən\ *noun* [Late Latin *subreption-, subreptio,* from Latin, act of stealing, from *subripere, surripere* to take away secretly — more at SURREPTITIOUS] (1600)
: a deliberate misrepresentation; *also* **:** an inference drawn from it
— **sub·rep·ti·tious** \ˌsəb-ˌrep-ˈti-shəs\ *adjective*
— **sub·rep·ti·tious·ly** *adverb*

sub·ring \ˈsəb-ˌriŋ\ *noun* (1937)
: a subset of a mathematical ring which is itself a ring

sub·ro·gate \ˈsə-brō-ˌgāt\ *transitive verb* **-gat·ed; -gat·ing** [Middle English, from Latin *subrogatus,* past participle of *subrogare, surrogare* — more at SURROGATE] (15th century)
: to put in the place of another; *especially* **:** to substitute (as a second creditor) for another with regard to a legal right or claim

sub·ro·ga·tion \ˌsə-brō-ˈgā-shən\ *noun* (15th century)
: the act of subrogating; *specifically* **:** the assumption by a third party (as a second creditor or an insurance company) of another's legal right to collect a debt or damages

sub–rosa *adjective* (1923)
: SECRETIVE, PRIVATE

sub ro·sa \ˌsəb-ˈrō-zə\ *adverb* [New Latin, literally, under the rose; from the ancient association of the rose with secrecy] (1654)
: in confidence **:** SECRETLY

sub·rou·tine \ˈsəb-(ˌ)rü-ˌtēn\ *noun* [International Scientific Vocabulary] (circa 1946)
: a subordinate routine; *specifically* **:** a sequence of computer instructions for performing a specified task that can be used repeatedly

sub–Sa·ha·ran \ˌsəb-sə-ˈhar-ən, -ˈher-, -ˈhär-\ *adjective* (1955)
: of, relating to, or being the part of Africa south of the Sahara

¹sub·sam·ple \ˈsəb-ˌsam-pəl, ˌsəb-ˈsam-\ *transitive verb* (circa 1899)
: to draw samples from (a previously selected group or population) **:** sample a sample of

²subsample *noun* (circa 1899)
: a sample or specimen obtained by subsampling

sub·sat·el·lite \ˌsəb-ˈsa-tᵊl-ˌīt\ *noun* (1956)
: an object carried into orbit in and subsequently released from a satellite or spacecraft

sub·scribe \səb-ˈskrīb\ *verb* **sub·scribed; sub·scrib·ing** [Middle English, from Latin *subscribere,* literally, to write beneath, from *sub-* + *scribere* to write — more at SCRIBE] (15th century)
transitive verb
1 : to write (one's name) underneath **:** SIGN
2 a : to sign (as a document) with one's own hand in token of consent or obligation **b :** to attest by signing **c :** to pledge (a gift or contribution) by writing one's name with the amount
3 : to assent to **:** SUPPORT
intransitive verb
1 : to sign one's name to a document
2 a : to give consent or approval to something written by signing ⟨unwilling to *subscribe* to the agreement⟩ **b :** to set one's name to a paper in token of promise to give something (as a sum of money); *also* **:** to give something in accordance with such a promise **c :** to enter one's name for a publication or service; *also* **:** to receive a periodical or service regularly on order **d :** to agree to purchase and pay for securities especially of a new offering ⟨*subscribed* for 1000 shares⟩
3 : to feel favorably disposed ⟨I *subscribe* to your sentiments⟩
synonym see ASSENT
— **sub·scrib·er** *noun*

sub·script \ˈsəb-ˌskript\ *noun* [Latin *subscriptus,* past participle of *subscribere*] (1895)
: a distinguishing symbol (as a letter or numeral) written immediately below or below and to the right or left of another character
— **subscript** *adjective*

sub·scrip·tion \səb-ˈskrip-shən\ *noun* [Middle English *subscripcioun* signature, from Latin *subscription-, subscriptio,* from *subscribere*] (15th century)
1 a : the act of signing one's name (as in attesting or witnessing a document) **b :** the acceptance (as of ecclesiastical articles of faith) attested by the signing of one's name
2 : something that is subscribed: as **a :** an autograph signature; *also* **:** a paper to which a signature is attached **b :** a sum subscribed or pledged
3 : an arrangement for providing, receiving, or making use of something of a continuing or periodic nature on a prepayment plan: as **a :** a purchase by prepayment for a certain number of issues (as of a periodical) **b :** application to purchase securities of a new issue **c :** a method of offering or presenting a series of public performances **d** *British* **:** membership dues

subscription TV *noun* (1953)
: pay-TV that broadcasts programs directly over the air to customers provided with a special receiver — called also *subscription television*; compare PAY-CABLE, PAY-TV

sub·sec·tion \ˈsəb-ˌsek-shən\ *noun* (1621)
1 : a subdivision or a subordinate division of a section
2 : a subordinate part or branch

¹sub·se·quence \ˈsəb-sə-ˌkwen(t)s, -si-kwən(t)s\ *noun* (circa 1500)
: the quality or state of being subsequent; *also* **:** a subsequent event

²sub·se·quence \ˈsəb-ˌsē-kwən(t)s, -ˌkwen(t)s\ *noun* (1908)
: a mathematical sequence that is part of another sequence

sub·se·quent \ˈsəb-si-kwənt, -sə-ˌkwent\ *adjective* [Middle English, from Latin *subsequent-, subsequens,* present participle of *subsequi* to follow close, from *sub-* near + *sequi* to follow — more at SUB-, SUE] (15th century)
: following in time, order, or place
— **subsequent** *noun*
— **sub·se·quent·ly** \-ˌkwent-lē, -kwənt-\ *adverb*

subsequent to *preposition* (1647)
: at a time later or more recent than **:** SINCE

sub·serve \(ˌ)səb-ˈsərv\ *transitive verb* [Latin *subservire* to serve, be subservient, from *sub-* + *servire* to serve] (1661)
1 : to promote the welfare or purposes of
2 : to serve as an instrument or means in carrying out

sub·ser·vi·ence \səb-ˈsər-vē-ən(t)s\ *noun* (circa 1676)
1 : a subservient or subordinate place or function
2 : obsequious servility

sub·ser·vi·en·cy \-ən(t)-sē\ *noun* (1651)
: SUBSERVIENCE

sub·ser·vi·ent \-ənt\ *adjective* [Latin *subservient-, subserviens,* present participle of *subservire*] (1632)
1 : serving to promote some end
2 : useful in an inferior capacity **:** SUBORDINATE
3 : obsequiously submissive **:** TRUCKLING ☆
— **sub·ser·vi·ent·ly** *adverb*

sub·set \ˈsəb-ˌset\ *noun* (1902)
: a set each of whose elements is an element of an inclusive set

sub·shrub \-ˌshrəb, *especially Southern* -ˌsrəb\ *noun* (1851)
: a perennial plant having woody stems except for the terminal part of the new growth which is killed back annually; *also* **:** a low shrub

sub·side \səb-ˈsīd\ *intransitive verb* **sub·sid·ed; sub·sid·ing** [Latin *subsidere,* from *sub-* + *sidere* to sit down, sink; akin to Latin *sedēre* to sit — more at SIT] (1607)
1 : to sink or fall to the bottom **:** SETTLE
2 : to tend downward **:** DESCEND; *especially* **:** to flatten out so as to form a depression
3 : to let oneself settle down **:** SINK ⟨*subsided* into a chair⟩
4 : to become quiet or less ⟨as the fever *subsides*⟩ ⟨my anger *subsided*⟩
synonym see ABATE
— **sub·si·dence** \səb-ˈsī-dᵊn(t)s, ˈsəb-sə-dən(t)s\ *noun*

sub·sid·i·ar·i·ty \ˌsəb-si-dē-ˈer-ə-tē, səb-ˌsi-\ *noun* (1936)
1 : the quality or state of being subsidiary
2 : a principle in social organization: functions which subordinate or local organizations perform effectively belong more properly to them than to a dominant central organization

¹sub·sid·i·ary \səb-ˈsi-dē-ˌer-ē, -ˈsi-də-rē\ *adjective* [Latin *subsidiarius,* from *subsidium* reserve troops] (1543)
1 a : furnishing aid or support **:** AUXILIARY ⟨*subsidiary* details⟩ **b :** of secondary importance ⟨a *subsidiary* stream⟩
2 : of, relating to, or constituting a subsidy ⟨a *subsidiary* payment to an ally⟩
— **sub·sid·i·ari·ly** \-ˌsi-dē-ˈer-ə-lē\ *adverb*

²subsidiary *noun, plural* **-ar·ies** (1603)
: one that is subsidiary; *especially* **:** a company wholly controlled by another

sub·si·dise *British variant of* SUBSIDIZE

sub·si·dize \ˈsəb-sə-ˌdīz, -zə-\ *transitive verb* **-dized; -diz·ing** (1795)
: to furnish with a subsidy: as **a :** to purchase the assistance of by payment of a subsidy **b :** to aid or promote (as a private enterprise) with public money ⟨*subsidize* soybean farmers⟩ ⟨*subsidize* public transportation⟩
— **sub·si·di·za·tion** \ˌsəb-sə-də-ˈzā-shən, -zə-\ *noun*
— **sub·si·diz·er** *noun*

sub·si·dy \ˈsəb-sə-dē, -zə-\ *noun, plural* **-dies** [Middle English, from Latin *subsidium* reserve troops, support, assistance, from *sub-* near + *sedēre* to sit — more at SUB-, SIT] (14th century)
: a grant or gift of money: as **a :** a sum of money formerly granted by the British Parliament to the crown and raised by special taxation **b :** money granted by one state to another **c :** a grant by a government to a private person or company to assist an enterprise deemed advantageous to the public

sub·sist \səb-ˈsist\ *verb* [Late Latin *subsistere* to exist, from Latin, to come to a halt, remain, from *sub-* + *sistere* to come to a stand; akin to Latin *stare* to stand — more at STAND] (1549)

☆ SYNONYMS

Subservient, servile, slavish, obsequious mean showing or characterized by extreme compliance or abject obedience. SUBSERVIENT implies the cringing manner of one very conscious of a subordinate position ⟨domestic help was expected to be properly *subservient*⟩. SERVILE suggests the mean or fawning behavior of a slave ⟨a political boss and his entourage of *servile* hangers-on⟩. SLAVISH suggests abject or debased servility ⟨the *slavish* status of migrant farm workers⟩. OBSEQUIOUS implies fawning or sycophantic compliance and exaggerated deference of manner ⟨waiters who are *obsequious* in the presence of celebrities⟩.

intransitive verb
1 a : to have existence **:** BE **b :** PERSIST, CONTINUE
2 : to have or acquire the necessities of life (as food and clothing); *especially* **:** to nourish oneself ⟨*subsisting* on roots, berries and grubs⟩
3 a : to hold true **b :** to be logically conceivable as the subject of true statements
transitive verb
: to support with provisions

sub·sis·tence \səb-ˈsis-tən(t)s\ *noun* [Middle English, from Late Latin *subsistentia,* from *subsistent-, subsistens,* present participle of *subsistere*] (15th century)
1 a (1) **:** real being **:** EXISTENCE (2) **:** the condition of remaining in existence **:** CONTINUATION, PERSISTENCE **b :** an essential characteristic quality of something that exists **c :** the character possessed by whatever is logically conceivable
2 : means of subsisting: as **a :** the minimum (as of food and shelter) necessary to support life **b :** a source or means of obtaining the necessities of life
— sub·sis·tent \-tənt\ *adjective*

subsistence farming *noun* (1939)
1 : farming or a system of farming that provides all or almost all the goods required by the farm family usually without any significant surplus for sale
2 : farming or a system of farming that produces a minimum and often inadequate return to the farmer — called also *subsistence agriculture*
— subsistence farmer *noun*

sub·so·cial \ˌsəb-ˈsō-shəl\ *adjective* (circa 1909)
: incompletely social; *especially* **:** tending to associate gregariously but lacking fixed or complex social organization ⟨*subsocial* insects⟩

1sub·soil \ˈsəb-ˌsoil\ *noun* (1799)
: the stratum of weathered material that underlies the surface soil

2subsoil *transitive verb* (1840)
: to turn, break, or stir the subsoil of
— sub·soil·er *noun*

sub·so·lar point \ˌsəb-ˈsō-lər-\ *noun* (circa 1908)
: the point on the surface of the earth or a planet at which the sun is at the zenith

sub·son·ic \ˌsəb-ˈsä-nik\ *adjective* [International Scientific Vocabulary] (1937)
1 : of, relating to, or being a speed less than that of sound in air
2 : moving, capable of moving, or utilizing air currents moving at a subsonic speed
3 : INFRASONIC 1
— sub·son·i·cal·ly \-ni-k(ə-)lē\ *adverb*

sub·space \ˈsəb-ˌspās\ *noun* (1927)
: a subset of a space; *especially* **:** one that has the essential properties (as those of a vector space or topological space) of the including space

sub spe·cie ae·ter·ni·ta·tis \su̇b-ˈspe-kē-ˌā-ˌī-ˌter-nə-ˈtä-təs\ *adverb* [New Latin, literally, under the aspect of eternity] (1895)
: in its essential or universal form or nature

sub·spe·cies \ˈsəb-ˌspē-shēz, -sēz\ *noun* [New Latin] (1699)
: a subdivision of a species: as **a :** a category in biological classification that ranks immediately below a species and designates a population of a particular geographical region genetically distinguishable from other such populations of the same species and capable of interbreeding successfully with them where its range overlaps theirs **b :** a named subdivision (as a race or variety) of a taxonomic species **c :** SUBGROUP 1 ⟨*subspecies* of economy fares —Michael DiPaola⟩
— sub·spe·cif·ic \ˌsəb-spi-ˈsi-fik\ *adjective*

sub·stage \ˈsəb-ˌstāj\ *noun* (1888)
: an attachment to a microscope by means of which accessories (as mirrors, diaphragms, or condensers) are held in place beneath the stage of the instrument

sub·stance \ˈsəb-stən(t)s\ *noun* [Middle English, from Middle French, from Latin *substantia,* from *substant-, substans,* present participle of *substare* to stand under, from *sub-* + *stare* to stand — more at STAND] (14th century)
1 a : essential nature **:** ESSENCE **b :** a fundamental or characteristic part or quality **c** *Christian Science* **:** GOD 1b
2 a : ultimate reality that underlies all outward manifestations and change **b :** practical importance **:** MEANING, USEFULNESS ⟨the . . . bill—which will be without *substance* in the sense that it will authorize nothing more than a set of ideas —Richard Reeves⟩
3 a : physical material from which something is made or which has discrete existence **b :** matter of particular or definite chemical constitution **c :** something (as drugs or alcoholic beverages) deemed harmful and usually subject to legal restriction ⟨possession of a controlled *substance*⟩ ⟨has a *substance* problem⟩
4 : material possessions **:** PROPERTY ⟨a family of *substance*⟩
— sub·stance·less \-ləs\ *adjective*
— in substance : in respect to essentials **:** FUNDAMENTALLY

substance abuse *noun* (1982)
: excessive use of a drug (as alcohol, narcotics, or cocaine) **:** use of a drug without medical justification
— substance abuser *noun*

substance P *noun* (1934)
: a neuropeptide that consists of 11 amino-acid residues, that is widely distributed in the brain, spinal cord, and peripheral nervous system, and that acts across nerve synapses to produce prolonged postsynaptic excitation

sub·stan·dard \ˌsəb-ˈstan-dərd\ *adjective* (1897)
: deviating from or falling short of a standard or norm: as **a :** of a quality lower than that prescribed by law **b :** conforming to a pattern of linguistic usage existing within a speech community but not that of the prestige group in that community **c :** constituting a greater than normal risk to an insurer

sub·stan·tial \səb-ˈstan(t)-shəl\ *adjective* (14th century)
1 a : consisting of or relating to substance **b :** not imaginary or illusory **:** REAL, TRUE **c :** IMPORTANT, ESSENTIAL
2 : ample to satisfy and nourish **:** FULL ⟨a *substantial* meal⟩
3 a : possessed of means **:** WELL-TO-DO **b :** considerable in quantity **:** significantly great ⟨earned a *substantial* wage⟩
4 : firmly constructed **:** STURDY
5 : being largely but not wholly that which is specified ⟨a *substantial* lie⟩
— substantial *noun*
— sub·stan·ti·al·i·ty \-ˌstan(t)-shē-ˈa-lə-tē\ *noun*
— sub·stan·tial·ly \-ˈstan(t)-sh(ə-)lē\ *adverb*
— sub·stan·tial·ness \-ˈstan(t)-shəl-nəs\ *noun*

sub·stan·tia ni·gra \səb-ˌstan(t)-shē-ə-ˈnī-grə, -ˈni-\ *noun, plural* **sub·stan·ti·ae ni·grae** \-chē-ˌē-ˈnī-(ˌ)grē, -ˈni-\ [New Latin, literally, black substance] (1882)
: a layer of deeply pigmented gray matter situated in the midbrain and containing the cell bodies of a tract of dopamine-producing nerve cells whose secretion tends to be deficient in Parkinson's disease

sub·stan·ti·ate \səb-ˈstan(t)-shē-ˌāt\ *transitive verb* **-at·ed; -at·ing** (1657)
1 : to give substance or form to **:** EMBODY
2 : to establish by proof or competent evidence **:** VERIFY ⟨*substantiate* a charge⟩
synonym see CONFIRM
— sub·stan·ti·a·tion \-ˌstan(t)-shē-ˈā-shən\ *noun*
— sub·stan·ti·a·tive \-ˈstan(t)-shē-ˌā-tiv\ *adjective*

sub·stan·ti·val \ˌsəb-stən-ˈtī-vəl\ *adjective* (circa 1832)
: of, relating to, or serving as a substantive
— sub·stan·ti·val·ly \-və-lē\ *adverb*

1sub·stan·tive \ˈsəb-stən-tiv\ *noun* [Middle English *substantif,* from Middle French, from *substantif,* adjective, having or expressing substance, from Late Latin *substantivus*] (14th century)
: NOUN; *broadly* **:** a word or word group functioning syntactically as a noun
— sub·stan·tiv·ize \-ti-ˌvīz\ *transitive verb*

2sub·stan·tive \ˈsəb-stən-tiv; *2c & 3 also* səb-ˈstan-tiv\ *adjective* [Middle English, from Late Latin *substantivus* having substance, from Latin *substantia*] (14th century)
1 : being a totally independent entity
2 a : real rather than apparent **:** FIRM; *also* **:** PERMANENT, ENDURING **b :** belonging to the substance of a thing **:** ESSENTIAL **c :** expressing existence ⟨the *substantive* verb is the verb *to be*⟩ **d :** requiring or involving no mordant ⟨a *substantive* dyeing process⟩
3 a : having the nature or function of a grammatical substantive ⟨a ˈ*substantive* phrase⟩ **b :** relating to or having the character of a noun or pronominal term in logic
4 : considerable in amount or numbers **:** SUBSTANTIAL
5 : creating and defining rights and duties ⟨*substantive* law⟩ — compare PROCEDURAL
6 : having substance **:** involving matters of major or practical importance to all concerned ⟨*substantive* discussions among world leaders⟩
— sub·stan·tive·ly *adverb*
— sub·stan·tive·ness *noun*

substantive due process *noun* (1954)
: DUE PROCESS 2

substantive right *noun* (1939)
: a right (as of life, liberty, property, or reputation) held to exist for its own sake and to constitute part of the normal legal order of society

sub·sta·tion \ˈsəb-ˌstā-shən\ *noun* (1881)
: a subordinate or subsidiary station: as **a :** a branch post office **b :** a subsidiary station in which electric current is transformed **c :** a police station serving a particular area

sub·stit·u·ent \səb-ˈsti-chə-wənt, -ˈstich-wənt\ *noun* [Latin *substituent-, substituens,* present participle of *substituere*] (circa 1896)
: an atom or group that replaces another atom or group in a molecule
— substituent *adjective*

sub·sti·tut·able \ˈsəb-stə-ˌtü-tə-bəl, -ˈtyü-\ *adjective* (1805)
: capable of being substituted
— sub·sti·tut·abil·i·ty \ˌsəb-stə-ˌtü-tə-ˈbi-lə-tē, -ˌtyü-\ *noun*

1sub·sti·tute \ˈsəb-stə-ˌtüt, -ˌtyüt\ *noun* [Middle English, from Latin *substitutus,* past participle of *substituere* to put in place of, from *sub-* + *statuere* to set up, place — more at STATUTE] (15th century)
: a person or thing that takes the place or function of another
— substitute *adjective*

2substitute *verb* **-tut·ed; -tut·ing** (1588)
transitive verb
1 a : to put or use in the place of another **b :** to introduce (an atom or group) as a substituent; *also* **:** to alter (as a compound) by introduction of a substituent ⟨a *substituted* benzene ring⟩
2 : to take the place of **:** REPLACE
intransitive verb
: to serve as a substitute

sub·sti·tu·tion \ˌsəb-stə-ˈtü-shən, -ˈtyü-\ *noun*

\ə\ **abut** \ᵊ\ **kitten** \ər\ **further** \a\ **ash** \ā\ **ace** \ä\ **mop, mar** \au̇\ **out** \ch\ **chin** \e\ **bet** \ē\ **easy** \g\ **go** \i\ **hit** \ī\ **ice** \j\ **job** \ŋ\ **sing** \ō\ **go** \ȯ\ **law** \ȯi\ **boy** \th\ **thin** \t͟h\ **the** \ü\ **loot** \u̇\ **foot** \y\ **yet** \zh\ **vision** *see also* Guide to Pronunciation

[Middle English *substitucion,* from Middle French *substitution,* from Late Latin *substitution-, substitutio,* from *substituere*] (14th century)
1 a : the act, process, or result of substituting one thing for another **b :** replacement of one mathematical entity by another of equal value
2 : one that is substituted for another
— **sub·sti·tu·tion·al** \-shnəl, -shə-nᵊl\ *adjective*
— **sub·sti·tu·tion·al·ly** *adverb*
— **sub·sti·tu·tion·ary** \-shə-ˌner-ē\ *adjective*

substitution cipher *noun* (1936)
: a cipher in which the letters of the plaintext are systematically replaced by substitute letters — compare TRANSPOSITION CIPHER

sub·sti·tu·tive \ˈsəb-stə-ˌtü-tiv, -ˌtyü-\ *adjective* (1668)
: serving or suitable as a substitute
— **sub·sti·tu·tive·ly** *adverb*

sub·strate \ˈsəb-ˌstrāt\ *noun* [Medieval Latin *substratum*] (1807)
1 : SUBSTRATUM
2 : the base on which an organism lives ⟨the soil is the *substrate* of most seed plants⟩
3 : a substance acted upon (as by an enzyme)

sub·stra·tum \ˈsəb-ˌstrā-təm, -ˌstra-, ˌsəb-ˈ\ *noun, plural* **-stra·ta** \-tə\ [Medieval Latin, from Latin, neuter of *substratus,* past participle of *substernere* to spread under, from *sub-* + *sternere* to spread — more at STREW] (1631)
: an underlying support **:** FOUNDATION: as **a :** substance that is a permanent subject of qualities or phenomena **b :** the material of which something is made and from which it derives its special qualities **c :** a layer beneath the surface soil; *specifically* **:** SUBSOIL **d :** SUBSTRATE 2

sub·struc·ture \ˈsəb-ˌstrək-chər\ *noun* (1726)
: an underlying or supporting part of a structure
— **sub·struc·tur·al** \-chə-rəl, -shrəl\ *adjective*

sub·sume \səb-ˈsüm\ *transitive verb* **sub·sumed; sub·sum·ing** [New Latin *subsumere,* from Latin *sub-* + *sumere* to take up — more at CONSUME] (1825)
: to include or place within something larger or more comprehensive **:** encompass as a subordinate or component element ⟨red, green, and yellow are *subsumed* under the term "color"⟩
— **sub·sum·able** \-ˈsü-mə-bəl\ *adjective*

sub·sump·tion \səb-ˈsəm(p)-shən\ *noun* [New Latin *subsumption-, subsumptio,* from *subsumere*] (1651)
: the act or process of subsuming

¹sub·sur·face \ˈsəb-ˌsər-fəs\ *noun* (1778)
: earth material (as rock) near but not exposed at the surface of the ground

²sub·sur·face \ˌsəb-ˈsər-fəs\ *adjective* (1875)
: of, relating to, or being something located beneath a surface and especially underground

sub·teen \ˈsəb-ˈtēn\ *noun* (1951)
: a preadolescent child

sub·tem·per·ate \ˌsəb-ˈtem-p(ə-)rət\ *adjective* (1852)
: of or occurring in the colder parts of the temperate zones

sub·ten·an·cy \-ˈte-nən(t)-sē\ *noun* (circa 1861)
: the state of being a subtenant

sub·ten·ant \-ˈte-nənt\ *noun* (15th century)
: one who rents from a tenant

sub·tend \səb-ˈtend\ *transitive verb* [Latin *subtendere* to stretch beneath, from *sub-* + *tendere* to stretch — more at THIN] (1570)
1 a : to be opposite to and extend from one side to the other of ⟨a hypotenuse *subtends* a right angle⟩ **b :** to fix the angular extent of with respect to a fixed point or object taken as the vertex ⟨the angle *subtended* at the eye by an object of given width and a fixed distance away⟩ ⟨a central angle *subtended* by an arc⟩ **c :** to determine the measure of by marking off the endpoints of ⟨a chord *subtends* an arc⟩
2 a : to underlie so as to include **b :** to occupy an adjacent and usually lower position to and often so as to embrace or enclose ⟨a bract that *subtends* a flower⟩

sub·ter·fuge \ˈsəb-tər-ˌfyüj\ *noun* [Late Latin *subterfugium,* from Latin *subterfugere* to escape, evade, from *subter-* secretly (from *subter* underneath; akin to Latin *sub* under) + *fugere* to flee — more at UP, FUGITIVE] (1573)
1 : deception by artifice or stratagem in order to conceal, escape, or evade
2 : a deceptive device or stratagem
synonym see DECEPTION

sub·ter·mi·nal \ˌsəb-ˈtərm-nəl, -ˈtər-mə-nᵊl\ *adjective* (1828)
: situated or occurring near but not precisely at an end ⟨a *subterminal* band of color on the tail feathers⟩

sub·ter·ra·nean \ˌsəb-tə-ˈrā-nē-ən, -nyən\ *also* **sub·ter·ra·neous** \-nē-əs, -nyəs\ *adjective* [Latin *subterraneus,* from *sub-* + *terra* earth — more at THIRST] (1603)
1 : being, lying, or operating under the surface of the earth
2 : existing or working in secret **:** HIDDEN
— **sub·ter·ra·nean·ly** *also* **sub·ter·ra·neous·ly** *adverb*

sub·text \ˈsəb-ˌtekst\ *noun* (1950)
: the implicit or metaphorical meaning (as of a literary text)
— **sub·tex·tu·al** \ˌsəb-ˈteks-chə-wəl, -chəl\ *adjective*

sub·thresh·old \ˌsəb-ˈthre-ˌshōld, -ˈthresh-ˌhōld\ *adjective* (1942)
: inadequate to produce a response ⟨*subthreshold* dosage⟩ ⟨a *subthreshold* stimulus⟩

sub·tile \ˈsə-tᵊl, ˈsəb-tᵊl\ *adjective* **sub·til·er** \ˈsət-lər, ˈsə-tᵊl-ər; ˈsəb-tə-lər\; **sub·til·est** \ˈsət-ləst, ˈsə-tᵊl-əst; ˈsəb-tə-ləst\ [Middle English, from Latin *subtilis*] (14th century)
1 : SUBTLE, ELUSIVE ⟨a *subtile* aroma⟩
2 a : CUNNING, CRAFTY **b :** SAGACIOUS, DISCERNING
— **sub·tile·ly** \ˈsət-lē, ˈsə-tᵊl-(l)ē; ˈsəb-tə-lē\ *adverb*
— **sub·tile·ness** \ˈsə-tᵊl-nəs, ˈsəb-tᵊl-\ *noun*

sub·til·i·sin \ˌsəb-ˈti-lə-sən\ *noun* [New Latin *subtilis,* specific epithet of *Bacillus subtilis,* species to which *Bacillus amyloliquefaciens* was once thought to belong] (1953)
: an extracellular protease produced by a soil bacillus (*Bacillus amyloliquefaciens*)

sub·til·ize \ˈsə-tᵊl-ˌīz, ˈsəb-tə-ˌlīz\ *verb* **-ized; -iz·ing** (1592)
intransitive verb
: to act or think subtly
transitive verb
: to make subtile
— **sub·til·i·za·tion** \ˌsə-tᵊl-ə-ˈzā-shən, ˌsəb-tə-lə-\ *noun*

sub·til·ty \ˈsə-tᵊl-tē, ˈsəb-tᵊl-\ *noun, plural* **-ties** (14th century)
: SUBTLETY

¹sub·ti·tle \ˈsəb-ˌtī-tᵊl\ *noun* (1825)
1 : a secondary or explanatory title
2 : a printed statement or fragment of dialogue appearing on the screen between the scenes of a silent motion picture or appearing as a translation at the bottom of the screen during the scenes of a motion picture or television show in a foreign language

²subtitle *transitive verb* (1891)
: to give a subtitle to

sub·tle \ˈsə-tᵊl\ *adjective* **sub·tler** \ˈsət-lər, ˈsə-tᵊl-ər\; **sub·tlest** \ˈsət-ləst, ˈsə-tᵊl-əst\ [Middle English *sutil, sotil,* from Middle French *soutil,* from Latin *subtilis,* literally, finely textured, from *sub-* + *tela* cloth on a loom; akin to Latin *texere* to weave — more at TECHNICAL] (14th century)
1 a : DELICATE, ELUSIVE ⟨a *subtle* fragrance⟩ **b :** difficult to understand or perceive **:** OBSCURE ⟨*subtle* differences in sound⟩
2 a : PERCEPTIVE, REFINED ⟨a writer's sharp and *subtle* moral sense⟩ **b :** having or marked by keen insight and ability to penetrate deeply and thoroughly ⟨a *subtle* scholar⟩
3 a : highly skillful **:** EXPERT ⟨a *subtle* craftsman⟩ **b :** cunningly made or contrived **:** INGENIOUS
4 : ARTFUL, CRAFTY ⟨a *subtle* rogue⟩
5 : operating insidiously ⟨*subtle* poisons⟩
— **sub·tle·ness** \ˈsə-tᵊl-nəs\ *noun*
— **sub·tly** \ˈsət-lē, ˈsə-tᵊl-(l)ē\ *adverb*

sub·tle·ty \ˈsə-tᵊl-tē\ *noun, plural* **-ties** [Middle English *sutilte,* from Middle French *sutilté,* from Latin *subtilitat-, subtilitas,* from *subtilis*] (14th century)
1 : the quality or state of being subtle
2 : something subtle

sub·ton·ic \ˌsəb-ˈtä-nik\ *noun* [from its being a half tone below the upper tonic] (circa 1854)
: LEADING TONE

¹sub·to·tal \ˈsəb-ˌtō-tᵊl\ *noun* (1906)
: the sum of part of a series of figures

²sub·to·tal \ˌsəb-ˈtō-tᵊl\ *adjective* (1908)
: somewhat less than complete **:** nearly total ⟨*subtotal* thyroidectomy⟩
— **sub·to·tal·ly** *adverb*

sub·tract \səb-ˈtrakt\ *verb* [Latin *subtractus,* past participle of *subtrahere* to draw from beneath, withdraw, from *sub-* + *trahere* to draw] (1557)
transitive verb
: to take away by or as if by deducting ⟨*subtract* 5 from 9⟩ ⟨*subtract* funds from the project⟩
intransitive verb
: to perform a subtraction
— **sub·tract·er** *noun*

sub·trac·tion \səb-ˈtrak-shən\ *noun* [Middle English *subtraccion,* from Late Latin *subtraction-, subtractio,* from Latin *subtrahere*] (15th century)
: an act, operation, or instance of subtracting: as **a :** the withdrawing or withholding of a right to which an individual is entitled **b :** the operation of deducting one number from another

sub·trac·tive \-ˈtrak-tiv\ *adjective* (1690)
1 : tending to subtract
2 : constituting or involving subtraction

sub·tra·hend \ˈsəb-trə-ˌhend\ *noun* [Latin *subtrahendus,* gerundive of *subtrahere*] (1674)
: a number that is to be subtracted from a minuend

sub·trop·i·cal \ˌsəb-ˈträ-pi-kəl\ *also* **sub·trop·ic** \-pik\ *adjective* [International Scientific Vocabulary] (1842)
: of, relating to, or being the regions bordering on the tropical zone ⟨*subtropical* environment⟩ ⟨*subtropical* grasses⟩

sub·trop·ics \-piks\ *noun plural* (1886)
: subtropical regions

su·bu·late \ˈsü-byə-lət, ˈsə-, -ˌlāt\ *adjective* [New Latin *subulatus,* from Latin *subula* awl, from *suere* to sew — more at SEW] (circa 1760)
: linear and tapering to a fine point ⟨a *subulate* leaf⟩

sub·um·brel·la \ˌsəb-(ˌ)əm-ˈbre-lə\ *noun* (1878)
: the concave undersurface of a jellyfish

sub·urb \ˈsə-ˌbərb\ *noun* [Middle English, from Latin *suburbium,* from *sub-* near + *urbs* city — more at SUB-] (14th century)
1 a : an outlying part of a city or town **b :** a smaller community adjacent to or within commuting distance of a city **c** *plural* **:** the residential area on the outskirts of a city or large town
2 *plural* **:** the near vicinity **:** ENVIRONS
— **sub·ur·ban** \sə-ˈbər-bən\ *adjective or noun*
— **sub·ur·ban·ite** \-bə-ˌnīt\ *noun*

sub·ur·ban·ise *British variant of* SUBURBANIZE

sub·ur·ban·ize \sə-ˈbər-bə-ˌnīz\ *transitive verb* **-ized; -iz·ing** (1893)

: to make suburban **:** give a suburban character to

— **sub·ur·ban·i·za·tion** \-ˌbər-bə-nə-'zā-shən\ *noun*

sub·ur·bia \sə-'bər-bē-ə\ *noun* [New Latin, from English *suburb*] (1895)

1 : the suburbs of a city

2 : people who live in the suburbs

3 : suburban life

sub·ven·tion \səb-'ven(t)-shən\ *noun* [Middle English *subvencion,* from Old French & Late Latin; Old French *subvenciòn,* from Late Latin *subvention-, subventio* assistance, from Latin *subvenire* to come up, come to the rescue, from *sub-* up + *venire* to come — more at SUB-, COME] (15th century)

: the provision of assistance or financial support: as **a :** ENDOWMENT **b :** a subsidy from a government or foundation

— **sub·ven·tion·ary** \-shə-ˌner-ē\ *adjective*

sub·ver·sion \səb-'vər-zhən, -shən\ *noun* [Middle English, from Middle French, from Late Latin *subversion-, subversio,* from Latin *subvertere*] (14th century)

1 : the act of subverting **:** the state of being subverted; *especially* **:** a systematic attempt to overthrow or undermine a government or political system by persons working secretly from within

2 *obsolete* **:** a cause of overthrow or destruction

— **sub·ver·sion·ary** \-zhə-ˌner-ē, -shə-\ *adjective*

— **sub·ver·sive** \-'vər-siv, -ziv\ *adjective or noun*

— **sub·ver·sive·ly** *adverb*

— **sub·ver·sive·ness** *noun*

sub·vert \səb-'vərt\ *transitive verb* [Middle English, from Middle French *subvertir,* from Latin *subvertere,* literally, to turn from beneath, from *sub-* + *vertere* to turn — more at WORTH] (14th century)

1 : to overturn or overthrow from the foundation **:** RUIN

2 : to pervert or corrupt by an undermining of morals, allegiance, or faith

— **sub·vert·er** *noun*

sub·vi·ral \ˌsəb-'vī-rəl\ *adjective* (1963)

: relating to, being, or caused by a piece or a structural part (as a protein) of a virus ⟨*subviral* infection⟩

sub·vo·cal \-'vō-kəl\ *adjective* (1924)

: characterized by the occurrence in the mind of words in speech order with or without inaudible articulation of the speech organs

— **sub·vo·cal·ly** \-kə-lē\ *adverb*

sub·vo·cal·i·za·tion \ˌsəb-ˌvō-kə-lə-'zā-shən\ *noun* (1947)

: the act or process of inaudibly articulating speech with the speech organs

— **sub·vo·cal·ize** \ˌsəb-'vō-kə-ˌlīz\ *verb*

sub·way \'səb-ˌwā\ *noun* (1825)

: an underground way: as **a :** a passage under a street (as for pedestrians, power cables, or water or gas mains) **b :** a usually electric underground railway **c :** UNDERPASS

— **subway** *intransitive verb*

suc·ce·da·ne·um \ˌsək-sə-'dā-nē-əm\ *noun, plural* **-ne·ums** *or* **-nea** \-nē-ə\ [New Latin, from Latin, neuter of *succedaneus* substituted, from *succedere* to follow after] (1641)

: SUBSTITUTE

— **suc·ce·da·ne·ous** \-nē-əs\ *adjective*

suc·ce·dent \sək-'sē-dᵊnt\ *adjective* [Middle English, from Latin *succedent-, succedens,* present participle of *succedere*] (15th century)

: coming next **:** SUCCEEDING, SUBSEQUENT

suc·ceed \sək-'sēd\ *verb* [Middle English *succeden,* from Latin *succedere* to go up, follow after, succeed, from *sub-* near + *cedere* to go — more at SUB-] (14th century)

intransitive verb

1 a : to come next after another in office or position or in possession of an estate; *especially* **:** to inherit sovereignty, rank, or title **b :** to follow after another in order

2 a : to turn out well **b :** to attain a desired object or end

3 *obsolete* **:** to pass to a person by inheritance

transitive verb

1 : to follow in sequence and especially immediately

2 : to come after as heir or successor

synonym see FOLLOW

— **suc·ceed·er** *noun*

suc·cès de scan·dale \sək-ˌsā-də-skä^{n}-'däl, (ˌ)sük-\ *noun* [French, literally, success of scandal] (1896)

: something (as a work of art) that wins popularity or notoriety because of its scandalous nature; *also* **:** the reception accorded such a piece

suc·cès d'es·time \-ˌdes-'tēm\ *noun* [French, literally, success of esteem] (1859)

: something (as a work of art) that wins critical respect but not popular success; *also* **:** the reception accorded such a piece

suc·cès fou \-'fü\ *noun* [French, literally, mad success] (1878)

: an extraordinary success

suc·cess \sək-'ses\ *noun* [Latin *successus,* from *succedere*] (1537)

1 *obsolete* **:** OUTCOME, RESULT

2 a : degree or measure of succeeding **b :** favorable or desired outcome; *also* **:** the attainment of wealth, favor, or eminence

3 : one that succeeds

suc·cess·ful \-fəl\ *adjective* (1588)

1 : resulting or terminating in success

2 : gaining or having gained success

— **suc·cess·ful·ly** \-fə-lē\ *adverb*

— **suc·cess·ful·ness** *noun*

suc·ces·sion \sək-'se-shən\ *noun* [Middle English, from Middle French or Latin; Middle French, from Latin *succession-, successio,* from *succedere*] (14th century)

1 a : the order in which or the conditions under which one person after another succeeds to a property, dignity, title, or throne **b :** the right of a person or line to succeed **c :** the line having such a right

2 a : the act or process of following in order **:** SEQUENCE **b** (1) **:** the act or process of one person's taking the place of another in the enjoyment of or liability for rights or duties or both (2) **:** the act or process of a person's becoming beneficially entitled to a property or property interest of a deceased person **c :** the continuance of corporate personality **d :** unidirectional change in the composition of an ecosystem as the available competing organisms and especially the plants respond to and modify the environment

3 a : a number of persons or things that follow each other in sequence **b :** a group, type, or series that succeeds or displaces another

— **suc·ces·sion·al** \-'sesh-nəl, -'se-shə-nᵊl\ *adjective*

— **suc·ces·sion·al·ly** *adverb*

succession duty *noun* (1853)

chiefly British **:** INHERITANCE TAX

suc·ces·sive \sək-'se-siv\ *adjective* (15th century)

1 : following in order **:** following each other without interruption

2 : characterized by or produced in succession

— **suc·ces·sive·ly** *adverb*

— **suc·ces·sive·ness** *noun*

suc·ces·sor \sək-'se-sər\ *noun* [Middle English *successour,* from Old French, from Latin *successor,* from *succedere*] (14th century)

: one that follows; *especially* **:** one who succeeds to a throne, title, estate, or office

suc·ci·nate \'sək-sə-ˌnāt\ *noun* (1790)

: a salt or ester of succinic acid

succinate dehydrogenase *noun* (1962)

: an iron-containing flavoprotein enzyme that catalyzes often reversibly the dehydrogenation of succinic acid to fumaric acid in the Krebs cycle and that is widely distributed especially in animal tissues, bacteria, and yeast — called also *succinic dehydrogenase*

suc·cinct \(ˌ)sək-'siŋ(k)t, sə-'siŋ(k)t\ *adjective* [Middle English, from Latin *succinctus,* past participle of *succingere* to gird from below, tuck up, from *sub-* + *cingere* to gird — more at CINCTURE] (15th century)

1 *archaic* **a :** being girded **b :** close-fitting

2 : marked by compact precise expression without wasted words

synonym see CONCISE

— **suc·cinct·ly** \-'siŋ(k)t-lē, -'siŋ-klē\ *adverb*

— **suc·cinct·ness** \-'siŋt-nəs, -'siŋk-nəs\ *noun*

suc·cin·ic acid \(ˌ)sək-'si-nik-\ *noun* [French *succinique,* from Latin *succinum* amber] (circa 1790)

: a crystalline dicarboxylic acid $C_4H_6O_4$ found widely in nature and active in energy-yielding metabolic reactions

suc·ci·nyl \'sək-sə-nᵊl, -ˌnil\ *noun* [International Scientific Vocabulary] (circa 1868)

: either of two groups of succinic acid: **a :** a bivalent group $OCCH_2CH_2CO$ **b :** a univalent group $HOOCCH_2CH_2CO$

suc·ci·nyl·cho·line \ˌsək-sə-nᵊl-'kō-ˌlēn, -ˌnil-\ *noun* (1950)

: a basic compound that is used intravenously chiefly in the form of a hydrated chloride $C_{14}H_{30}Cl_2N_2O_4 \cdot 2H_2O$ as a muscle relaxant in surgery

[1]**suc·cor** \'sə-kər\ *noun* [Middle English *succur,* from earlier *sucurs,* taken as plural, from Old French *sucors,* from Medieval Latin *succursus,* from Latin *succurrere* to run up, run to help, from *sub-* up + *currere* to run — more at CAR] (13th century)

1 : RELIEF; *also* **:** AID, HELP

2 : something that furnishes relief

[2]**succor** *transitive verb* **suc·cored; suc·cor·ing** \'sə-k(ə-)riŋ\ (13th century)

: to go to the aid of **:** RELIEVE

— **suc·cor·er** \'sə-kər-ər\ *noun*

suc·co·ry \'sə-k(ə-)rē\ *noun* [alteration of Middle English *cicoree*] (1533)

: CHICORY

suc·co·tash \'sə-kə-ˌtash\ *noun* [Narraganset *msíckquatash* boiled corn kernels] (1751)

: lima or shell beans and green corn cooked together

suc·cour \'sə-kər\ *chiefly British variant of* SUCCOR

suc·cu·ba \'sə-kyə-bə\ *noun, plural* **-bae** \-ˌbē; -ˌbī\ [Latin, paramour] (1559)

: SUCCUBUS

suc·cu·bus \-bəs\ *noun, plural* **-bi** \-ˌbī, -ˌbē\ [Middle English, from Medieval Latin, alteration of Latin *succuba* paramour, from *succubare* to lie under, from *sub-* + *cubare* to lie, recline] (14th century)

: a demon assuming female form to have sexual intercourse with men in their sleep — compare INCUBUS

suc·cu·lence \'sə-kyə-lən(t)s\ *noun* (1787)

1 : the state of being succulent

2 : succulent feed ⟨wild game subsisting on *succulence*⟩

[1]**suc·cu·lent** \-lənt\ *adjective* [Latin *suculentus,* from *sucus* juice, sap; perhaps akin to Latin *sugere* to suck — more at SUCK] (1601)

1 a : full of juice **:** JUICY **b :** moist and tasty **:** TOOTHSOME **c** *of a plant* **:** having fleshy tissues that conserve moisture

2 : rich in interest

— **suc·cu·lent·ly** *adverb*

[2]**succulent** *noun* (1825)

: a succulent plant (as a cactus)

suc·cumb \sə-'kəm\ *intransitive verb* [French & Latin; French *succomber,* from Latin *succumbere,* from *sub-* + *-cumbere* to lie down; akin to Latin *cubare* to lie] (1604)

1 : to yield to superior strength or force or overpowering appeal or desire
2 : to be brought to an end (as death) by the effect of destructive or disruptive forces
synonym see YIELD

[1]**such** \'səch, 'sich\ *adjective* [Middle English, from Old English *swilc;* akin to Old High German *sulīh* such, Old English *swā* so, *gelīk* like — more at SO, LIKE] (before 12th century)
1 a : of a kind or character to be indicated or suggested ⟨a bag *such* as a doctor carries⟩ **b :** having a quality to a degree to be indicated ⟨his excitement was *such* that he shouted⟩
2 : of the character, quality, or extent previously indicated or implied ⟨in the past few years many *such* women have shifted to full-time jobs⟩
3 : of so extreme a degree or quality ⟨never heard *such* a hubbub⟩
4 : of the same class, type, or sort ⟨other *such* clinics throughout the state⟩
5 : not specified

[2]**such** *pronoun* (before 12th century)
1 : such a person or thing
2 : someone or something stated, implied, or exemplified ⟨*such* was the result⟩
3 : someone or something similar **:** similar persons or things ⟨tin and glass and *such*⟩ ■
— as such : intrinsically considered **:** in itself ⟨*as such* the gift was worth little⟩

[3]**such** *adverb* (before 12th century)
1 a : to such a degree **:** so ⟨*such* tall buildings⟩ ⟨*such* a fine person⟩ **b :** VERY, ESPECIALLY ⟨hasn't been in *such* good spirits lately⟩
2 : in such a way ⟨related *such* that each excludes the other⟩

[1]**such and such** *adjective* (13th century)
: not named or specified

[2]**such and such** *pronoun* (15th century)
: something not specified

[1]**such·like** \'səch-ˌlīk\ *adjective* (15th century)
: of like kind **:** SIMILAR

[2]**suchlike** *pronoun* (15th century)
: SUCH 3

[1]**suck** \'sək\ *verb* [Middle English *suken,* from Old English *sūcan;* akin to Old High German *sūgan* to suck, Latin *sugere*] (before 12th century)
transitive verb
1 a : to draw (as liquid) into the mouth through a suction force produced by movements of the lips and tongue ⟨*sucked* milk from his mother's breast⟩ **b :** to draw something from or consume by such movements ⟨*suck* an orange⟩ ⟨*suck* a lollipop⟩ **c :** to apply the mouth to in order to or as if to suck out a liquid ⟨*sucked* his burned finger⟩
2 a : to draw by or as if by suction ⟨when a receding wave *sucks* the sand from under your feet —Kenneth Brower⟩ ⟨inadvertently *sucked* into the . . . intrigue —Martin Levin⟩ **b :** to take in and consume by or as if by suction ⟨a vacuum cleaner *sucking* up dirt⟩ ⟨*suck* up a few beers⟩ ⟨opponents say that malls *suck* the life out of downtown areas —Michael Knight⟩
intransitive verb
1 : to draw something in by or as if by exerting a suction force; *especially* **:** to draw milk from a breast or udder with the mouth
2 : to make a sound or motion associated with or caused by suction ⟨his pipe *sucked* wetly⟩ ⟨flanks *sucked* in and out, the long nose resting on his paws —Virginia Woolf⟩
3 : to act in an obsequious manner ⟨when they want votes . . . the candidates come *sucking* around —W. G. Hardy⟩ ⟨*sucked* up to the boss⟩
4 *slang* **:** to be objectionable or inadequate ⟨our lifestyle *sucks* —*Playboy*⟩ ⟨people who went said it *sucked* —H. S. Thompson⟩

[2]**suck** *noun* (13th century)
1 : a sucking movement or force
2 : the act of sucking

[1]**suck·er** \'sə-kər\ *noun* (14th century)
1 a : one that sucks especially a breast or udder **:** SUCKLING **b :** a device for creating or regulating suction (as a piston or valve in a pump) **c :** a pipe or tube through which something is drawn by suction **d** (1) **:** an organ in various animals for adhering or holding (2) **:** a mouth (as of a leech) adapted for sucking or adhering
2 : a shoot from the roots or lower part of the stem of a plant
3 : any of numerous chiefly North American freshwater bony fishes (family Catostomidae) closely related to the carps but distinguished from them especially by the structure of the mouth which usually has thick soft lips
4 : LOLLIPOP 1
5 a : a person easily cheated or deceived **b :** a person irresistibly attracted by something specified ⟨a *sucker* for ghost stories⟩ **c —** used as generalized term of reference ⟨see if you can get that *sucker* working again⟩

[2]**sucker** *verb* **suck·ered; suck·er·ing** \'sə-k(ə-)riŋ\ (circa 1661)
transitive verb
1 : to remove suckers from ⟨*sucker* tobacco⟩
2 : HOODWINK
intransitive verb
: to send out suckers

sucker punch *transitive verb* (1964)
: to punch (a person) suddenly without warning and often without apparent provocation
— sucker punch *noun*

suck in *transitive verb* (1840)
1 : DUPE, HOODWINK
2 : to contract, flatten, and tighten (the abdomen) especially by inhaling deeply

suck·ing *adjective* (before 12th century)
: not yet weaned; *broadly* **:** very young

sucking louse *noun* (circa 1907)
: any of an order (Anoplura) of wingless insects comprising the true lice with mouthparts adapted to sucking body fluids

suck·le \'sə-kəl\ *verb* **suck·led; suck·ling** \-k(ə-)liŋ\ [Middle English *suklen,* probably back-formation from *suklyng*] (14th century)
transitive verb
1 a : to give milk to from the breast or udder ⟨a mother *suckling* her child⟩ **b :** to nurture as if by giving milk from the breast ⟨was *suckled* on pulp magazines⟩
2 : to draw milk from the breast or udder of ⟨lambs *suckling* the ewes⟩
intransitive verb
: to draw milk from the breast or udder

suck·ling \'sə-kliŋ\ *noun* [Middle English *suklyng,* from *suken* to suck] (13th century)
: a young unweaned animal

su·crase \'sü-ˌkrās, -ˌkrāz\ *noun* [International Scientific Vocabulary, from French *sucre* sugar, from Middle French — more at SUGAR] (circa 1900)
: INVERTASE

su·cre \'sü-(ˌ)krā\ *noun* [Spanish, from Antonio José de *Sucre*] (1886)
— see MONEY table

su·crose \'sü-ˌkrōs, -ˌkrōz\ *noun* [International Scientific Vocabulary, from French *sucre* sugar] (1857)
: a sweet crystalline dextrorotatory disaccharide sugar $C_{12}H_{22}O_{11}$ that occurs naturally in most plants and is obtained commercially especially from sugarcane or sugar beets

[1]**suc·tion** \'sək-shən\ *noun* [Late Latin *suction-, suctio,* from Latin *sugere* to suck — more at SUCK] (1626)
1 : the act or process of sucking
2 a : the act or process of exerting a force upon a solid, liquid, or gaseous body by reason of reduced air pressure over part of its surface **b :** force so exerted
3 : a device (as a pipe or fitting) used in a machine that operates by suction
— suc·tion·al \-shə-nᵊl, -shnəl\ *adjective*

[2]**suction** *transitive verb* (1954)
: to remove (as from a body cavity or passage) by suction

suction cup *noun* (1942)
: a cup-shaped device in which a partial vacuum can be produced when applied to a surface

suction pump *noun* (1825)
: a common pump in which the liquid to be raised is pushed by atmospheric pressure into the partial vacuum under a retreating valved piston on the upstroke and reflux is prevented by a check valve in the pipe

suction stop *noun* (1887)
: a voice stop in the formation of which air behind the articulation is rarefied with consequent inrush of air when articulation is broken

suc·to·ri·al \ˌsək-'tōr-ē-əl, -'tȯr-\ *adjective* [New Latin *suctorius,* from Latin *sugere*] (1833)
: adapted for sucking; *especially* **:** serving to draw up fluid or to adhere by suction ⟨*suctorial* mouths⟩

suc·to·ri·an \-ē-ən\ *noun* [New Latin *Suctoria,* from neuter plural of *suctorius* suctorial] (circa 1842)
: any of a class or subclass (Suctoria) of complex protozoans which are ciliated only early in development and in which the mature form is fixed to the substrate, lacks locomotor organelles or a mouth, and obtains food through specialized suctorial tentacles

Su·dan grass \sü-'dan-, -'dän-\ *noun* [the *Sudan,* region in Africa] (1911)
: a vigorous tall-growing annual sorghum grass (*Sorghum bicolor sudanese*) widely grown for hay and fodder

Su·dan·ic \sü-'da-nik\ *noun* [the *Sudan* region] (1912)
: a group of languages neither Bantu nor Afro-Asiatic spoken in central and western Africa that were formerly considered a residual category and are now divided among the Niger-Congo and other families
— Sudanic *adjective*

su·da·to·ri·um \ˌsü-də-'tōr-ē-əm, -'tȯr-\ *noun* [Latin, from *sudare* to sweat — more at SWEAT] (circa 1757)
: a sweat room in a bath

su·da·to·ry \'sü-də-ˌtōr-ē, -ˌtȯr-\ *noun, plural* **-ries** (1615)
: SUDATORIUM

sudd \'səd\ *noun* [Arabic, literally, obstruction] (1874)
: floating vegetable matter that forms obstructive masses especially in the upper White Nile

[1]**sud·den** \'sə-dᵊn\ *adjective* [Middle English *sodain,* from Middle French, from Latin *subitaneus,* from *subitus* sudden, from past participle of *subire* to come up, from *sub-* up + *ire* to go — more at SUB-, ISSUE] (14th century)
1 a : happening or coming unexpectedly ⟨a *sudden* shower⟩ **b :** changing angle or character all at once
2 : marked by or manifesting abruptness or haste
3 : made or brought about in a short time **:** PROMPT
synonym see PRECIPITATE
— sud·den·ly *adverb*
— sud·den·ness \'sə-dᵊn-(n)əs\ *noun*

[2]**sudden** *noun* (1559)
obsolete **:** an unexpected occurrence **:** EMERGENCY
— all of a sudden *or* **on a sudden :** sooner than was expected **:** at once

sudden death *noun* (14th century)
1 : unexpected death that is instantaneous or occurs within minutes from any cause other than violence ⟨*sudden death* following coronary occlusion⟩

□ USAGE
such For reasons that are hard to understand, commentators on usage disapprove of *such* used as a pronoun. Dictionaries, however, recognize it as standard; all of the citations upon which our definitions of this word are based are clearly standard.

2 : extra play to break a tie in a sports contest in which the first to go ahead wins

sudden infant death syndrome *noun* (1970)
: death of an apparently healthy infant usually before one year of age that is of unknown cause and occurs especially during sleep — called also *crib death;* abbreviation *SIDS*

su·do·rif·er·ous \ˌsü-də-ˈri-f(ə-)rəs\ *adjective* [Late Latin *sudorifer,* from Latin *sudor* sweat (from *sudare* to sweat) + *-ifer* -iferous — more at SWEAT] (1597)
: producing or conveying sweat ⟨*sudoriferous* glands⟩ ⟨a *sudoriferous* duct⟩

su·do·rif·ic \-ˈri-fik\ *adjective* [New Latin *sudorificus,* from Latin *sudor*] (1626)
: causing or inducing sweat **:** DIAPHORETIC ⟨*sudorific* herbs⟩
— sudorific *noun*

Su·dra \ˈsü-drə, ˈshü-\ *noun* [Sanskrit *śūdra*] (1630)
: a Hindu of a lower caste traditionally assigned to menial occupations
— Sudra *adjective*

¹suds \ˈsədz\ *noun plural but singular or plural in construction* [probably from Middle Dutch *sudse* marsh; akin to Old English *sēothan* to seethe — more at SEETHE] (1581)
1 : water impregnated with soap or a synthetic detergent compound and worked up into froth; *also* **:** the lather or froth on such water
2 a : FOAM, FROTH **b :** BEER
— suds·less \-ləs\ *adjective*

²suds (1834)
transitive verb
: to wash in suds
intransitive verb
: to form suds

suds·er \ˈsəd-zər\ *noun* (1967)
: SOAP OPERA

sudsy \ˈsəd-zē\ *adjective* **suds·i·er; -est** (1866)
1 : full of suds **:** FROTHY, FOAMY
2 : SOAPY 4

sue \ˈsü\ *verb* **sued; su·ing** [Middle English, from Middle French *suivre,* from (assumed) Vulgar Latin *sequere,* from Latin *sequi* to follow, come or go after; akin to Greek *hepesthai* to follow, Sanskrit *sacate* he accompanies] (14th century)
transitive verb
1 *obsolete* **:** to make petition to or for
2 *archaic* **:** to pay court or suit to **:** WOO
3 a : to seek justice or right from (a person) by legal process; *specifically* **:** to bring an action against **b :** to proceed with and follow up (a legal action) to proper termination
intransitive verb
1 : to make a request or application **:** PLEAD — usually used with *for* or *to*
2 : to pay court **:** WOO
3 : to take legal proceedings in court
— su·er *noun*

¹suede *or* **suède** \ˈswād\ *noun* [French *gants de Suède* Swedish gloves] (1884)
1 : leather with a napped surface
2 : a fabric finished with a nap to simulate suede

²suede *verb* **sued·ed; sued·ing** (1921)
transitive verb
: to give a suede finish or nap to (a fabric or leather)
intransitive verb
: to give cloth or leather a suede finish

su·et \ˈsü-ət\ *noun* [Middle English *sewet,* from (assumed) Anglo-French, diminutive of Anglo-French *sue,* from Latin *sebum* tallow] (14th century)
: the hard fat about the kidneys and loins in beef and mutton that yields tallow

suf·fer \ˈsə-fər\ *verb* **suf·fered; suf·fer·ing** \-f(ə-)riŋ\ [Middle English *suffren,* from Old French *souffrir,* from (assumed) Vulgar Latin *sufferire,* from Latin *sufferre,* from *sub-* up + *ferre* to bear — more at SUB-, BEAR] (13th century)
transitive verb
1 a : to submit to or be forced to endure ⟨*suffer* martyrdom⟩ **b :** to feel keenly **:** labor under ⟨*suffer* thirst⟩
2 : UNDERGO, EXPERIENCE
3 : to put up with especially as inevitable or unavoidable
4 : to allow especially by reason of indifference ⟨the eagle *suffers* little birds to sing —Shakespeare⟩
intransitive verb
1 : to endure death, pain, or distress
2 : to sustain loss or damage
3 : to be subject to disability or handicap
synonym see BEAR
— suf·fer·able \ˈsə-f(ə-)rə-bəl\ *adjective*
— suf·fer·able·ness *noun*
— suf·fer·ably \-blē\ *adverb*
— suf·fer·er \ˈsə-fər-ər\ *noun*

suf·fer·ance \ˈsə-f(ə-)rən(t)s\ *noun* (14th century)
1 : patient endurance **:** LONG-SUFFERING
2 : PAIN, MISERY
3 : consent or sanction implied by a lack of interference or failure to enforce a prohibition
4 : power or ability to withstand **:** ENDURANCE

suffering *noun* (14th century)
1 : the state or experience of one that suffers
2 : PAIN
synonym see DISTRESS

suf·fice \sə-ˈfīs *also* -ˈfīz\ *verb* **suf·ficed; suf·fic·ing** [Middle English, from Middle French *suffis-,* stem of *suffire,* from Latin *sufficere* to provide, be adequate, from *sub-* + *facere* to make, do — more at DO] (14th century)
intransitive verb
1 : to meet or satisfy a need **:** be sufficient ⟨a brief note will *suffice*⟩ — often used with an impersonal *it* ⟨*suffice* it to say that they are dedicated, serious personalities —Cheryl Aldridge⟩
2 : to be competent or capable
transitive verb
: to be enough for
— suf·fic·er *noun*

suf·fi·cien·cy \sə-ˈfi-shən(t)-sē\ *noun* (15th century)
1 : sufficient means to meet one's needs **:** COMPETENCY; *also* **:** a modest but adequate scale of living
2 : the quality or state of being sufficient **:** ADEQUACY

suf·fi·cient \sə-ˈfi-shənt\ *adjective* [Middle English, from Latin *sufficient-, sufficiens,* from present participle of *sufficere*] (14th century)
1 a : enough to meet the needs of a situation or a proposed end ⟨*sufficient* provisions for a month⟩ **b :** being a sufficient condition
2 *archaic* **:** QUALIFIED, COMPETENT ☆
— suf·fi·cient·ly *adverb*

sufficient condition *noun* (1914)
1 : a proposition whose truth assures the truth of another proposition
2 : a state of affairs whose existence assures the existence of another state of affairs

¹suf·fix \ˈsə-fiks\ *noun* [New Latin *suffixum,* from Latin, neuter of *suffixus,* past participle of *suffigere* to fasten underneath, from *sub-* + *figere* to fasten — more at FIX] (1778)
: an affix occurring at the end of a word, base, or phrase — compare PREFIX
— suf·fix·al \ˈsə-fik-səl, (ˌ)sə-ˈfik-səl\ *adjective*

²suf·fix \ˈsə-fiks, (ˌ)sə-ˈfiks\ *transitive verb* (1778)
: to attach as a suffix
— suf·fix·a·tion \ˌsə-fik-ˈsā-shən\ *noun*

suf·fo·cate \ˈsə-fə-ˌkāt\ *verb* **-cat·ed; -cat·ing** [Middle English, from Latin *suffocatus,* past participle of *suffocare* to choke, stifle, from *sub-* + *fauces* throat] (15th century)
transitive verb
1 a : to stop the respiration of (as by strangling or asphyxiation) **b :** to deprive of oxygen **c :** to make uncomfortable by want of cool fresh air
2 : to impede or stop the development of
intransitive verb
1 : to become suffocated: **a :** to die from being unable to breathe **b :** to be uncomfortable through lack of air
2 : to become checked in development
— suf·fo·cat·ing·ly \-ˌkā-tiŋ-lē\ *adverb*
— suf·fo·ca·tion \ˌsə-fə-ˈkā-shən\ *noun*
— suf·fo·ca·tive \ˈsə-fə-ˌkā-tiv\ *adjective*

Suf·folk \ˈsə-fək, -ˌfōk\ *noun* [*Suffolk,* England] (1831)
1 : any of a breed of chestnut-colored draft horses of English origin — called also *Suffolk punch*
2 : any of a breed of large hornless black-faced sheep of English origin raised chiefly for mutton

¹suf·fra·gan \ˈsə-fri-gən, -jən\ *noun* [Middle English, from Middle French, from Medieval Latin *suffraganeus,* from *suffragium* support, prayer] (14th century)
1 : a diocesan bishop (as in the Roman Catholic Church and the Church of England) subordinate to a metropolitan
2 : an Anglican or Episcopal bishop assisting a diocesan bishop and not having the right of succession

²suffragan *adjective* (15th century)
1 : of or being a suffragan
2 : subordinate to a metropolitan or archiepiscopal see

suf·frage \ˈsə-frij, *sometimes* -fə-rij\ *noun* [in sense 1, from Middle English, from Middle French, from Medieval Latin *suffragium,* from Latin, vote, political support, from *suffragari* to support with one's vote; in other senses, from Latin *suffragium*] (14th century)
1 : a short intercessory prayer usually in a series
2 : a vote given in deciding a controverted question or in the choice of a person for an office or trust
3 : the right of voting **:** FRANCHISE; *also* **:** the exercise of such right

suf·frag·ette \ˌsə-fri-ˈjet\ *noun* (1906)
: a woman who advocates suffrage for women

suf·frag·ist \ˈsəf-ri-jist\ *noun* (1822)
: one who advocates extension of suffrage especially to women

suf·fuse \sə-ˈfyüz\ *transitive verb* **suf·fused; suf·fus·ing** [Latin *suffusus,* past participle of *suffundere,* literally, to pour beneath, from *sub-* + *fundere* to pour — more at FOUND] (1590)
: to spread over or through in the manner of fluid or light **:** FLUSH, FILL
synonym see INFUSE
— suf·fu·sion \-ˈfyü-zhən\ *noun*
— suf·fu·sive \-ˈfyü-siv, -ziv\ *adjective*

Su·fi \ˈsü-(ˌ)fē\ *noun* [Arabic *ṣūfīy,* perhaps from *ṣūf* wool] (1653)

☆ SYNONYMS
Sufficient, enough, adequate, competent mean being what is necessary or desirable. SUFFICIENT suggests a close meeting of a need ⟨*sufficient* savings⟩. ENOUGH is less exact in suggestion than SUFFICIENT ⟨do you have *enough* food?⟩. ADEQUATE may imply barely meeting a requirement ⟨the service was *adequate*⟩. COMPETENT suggests measuring up to all requirements without question or being adequately adapted to an end ⟨had no *competent* notion of what was going on⟩.

: a Muslim mystic
— **Sufi** *adjective*
— **Su·fic** \-fik\ *adjective*
— **Su·fism** \-ˌfi-zəm\ *noun*

¹sug·ar \ˈshu̇-gər\ *noun* [Middle English *sugre, sucre,* from Middle French *sucre,* from Medieval Latin *zuccarum,* from Old Italian *zucchero,* from Arabic *sukkar,* from Persian *shakar,* from Sanskrit *śarkarā;* akin to Sanskrit *śarkara* pebble — more at CROCODILE] (14th century)
1 a : a sweet crystallizable material that consists wholly or essentially of sucrose, is colorless or white when pure tending to brown when less refined, is obtained commercially from sugarcane or sugar beet and less extensively from sorghum, maples, and palms, and is important as a source of dietary carbohydrate and as a sweetener and preservative of other foods **b :** any of various water-soluble compounds that vary widely in sweetness and include the oligosaccharides (as sucrose)
2 : a unit (as a spoonful, cube, or lump) of sugar
3 : a sugar bowl
— **sug·ar·less** \-ləs\ *adjective*

²sugar *verb* **sug·ared; sug·ar·ing** \ˈshu̇-g(ə-)riŋ\ (15th century)
transitive verb
1 : to make palatable or attractive **:** SWEETEN
2 : to sprinkle or mix with sugar
intransitive verb
1 : to form or be converted into sugar
2 : to become granular **:** GRANULATE
3 : to make maple syrup or maple sugar

sugar apple *noun* (1738)
: a tropical American tree (*Annona squamosa*) of the custard apple family; *also* **:** its edible sweet pulpy fruit with thick green scaly rind and shining black seeds

sugar beet *noun* (1817)
: a white-rooted beet grown for the sugar in its roots

sug·ar·ber·ry \ˈshu̇-gər-ˌber-ē\ *noun* (circa 1818)
: any of several hackberries (especially *Celtis laevigata* and *C. occidentalis*) with sweet edible fruits

sugar bush *noun* (1823)
: a woods in which sugar maples predominate

sug·ar·cane \ˈshu̇-gər-ˌkān\ *noun* (15th century)
: a stout tall perennial grass (*Saccharum officinarum*) that has a large terminal panicle and is widely grown in warm regions as a source of sugar

sugarcane

sug·ar·coat \ˈshu̇-gər-ˌkōt\ *transitive verb* [back-formation from *sugarcoated*] (1870)
1 : to coat with sugar
2 : to make superficially attractive or palatable ⟨tried to *sugarcoat* an unpleasant truth⟩

sugar daddy *noun* (1926)
1 : a well-to-do usually older man who supports or spends lavishly on a mistress or girlfriend
2 : a generous benefactor of a cause

sug·ar·house \ˈshu̇-gər-ˌhau̇s\ *noun* (1600)
: a building where sugar is made or refined, *specifically* **:** one where maple sap is boiled and maple syrup and maple sugar are made

sugaring off *noun* (1836)
1 : the act or process of converting maple syrup into sugar
2 : a party held at the time of sugaring off

sug·ar·loaf \ˈshu̇-gər-ˌlōf\ *noun* (15th century)
1 : refined sugar molded into a cone
2 : a hill or mountain shaped like a sugarloaf
— **sugar–loaf** *adjective*

sugar maple *noun* (1731)
1 : a maple (*Acer saccharum*) of eastern North America with 3- to 5-lobed leaves, hard close-grained wood much used for cabinetwork, and sap that is the chief source of maple syrup and maple sugar — called also *rock maple, hard maple*
2 : any of several maples (especially *Acer nigrum* and *A. grandidentatum*) sometimes considered subspecies of the sugar maple

sugar off *intransitive verb* (1836)
1 : to complete the process of boiling down the syrup in making maple sugar until it is thick enough to crystallize
2 : to approach or reach the state of granulation

sugar orchard *noun* (1833)
chiefly New England **:** SUGAR BUSH

sugar pea *noun* (1707)
: SNOW PEA

sugar pine *noun* (1846)
: a very tall pine (*Pinus lambertiana*) found from Oregon to Baja California and having needles in clusters of five, cones up to 18 inches (46 cm) long, and soft reddish brown wood; *also* **:** its wood

sug·ar·plum \ˈshu̇-gər-ˌpləm\ *noun* (circa 1668)
: a small candy in the shape of a ball or disk **:** SWEETMEAT

sugar snap pea *noun* (1979)
: SNAP PEA

sug·ary \ˈshu̇-g(ə-)rē\ *adjective* (1591)
1 a : exaggeratedly sweet **:** HONEYED ⟨his *sugary* deprecating voice —D. H. Lawrence⟩ **b :** cloyingly sweet **:** SENTIMENTAL
2 : containing, resembling, or tasting of sugar

sug·gest \səg-ˈjest, sə-ˈjest\ *transitive verb* [Latin *suggestus,* past participle of *suggerere* to pile up, furnish, suggest, from *sub-* + *gerere* to carry] (1526)
1 a *obsolete* **:** to seek to influence **:** SEDUCE **b :** to call forth **:** EVOKE **c :** to mention or imply as a possibility ⟨*suggested* that he might bring his family⟩ **d :** to propose as desirable or fitting ⟨*suggest* a stroll⟩ **e :** to offer for consideration or as a hypothesis ⟨*suggest* a solution to a problem⟩
2 a : to call to mind by thought or association ⟨the explosion . . . *suggested* sabotage —F. L. Paxson⟩ **b :** to serve as a motive or inspiration for ⟨a play *suggested* by a historic incident⟩ ☆
— **sug·gest·er** *noun*

sug·gest·ible \səg-ˈjes-tə-bəl, sə-ˈjes-\ *adjective* (1890)
: easily influenced by suggestion
— **sug·gest·ibil·i·ty** \-ˌjes-tə-ˈbi-lə-tē\ *noun*

sug·ges·tion \səg-ˈjes-chən, sə-ˈjes-, -ˈjesh-\ *noun* (14th century)
1 a : the act or process of suggesting **b :** something suggested
2 a : the process by which a physical or mental state is influenced by a thought or idea ⟨the power of *suggestion*⟩ **b :** the process by which one thought leads to another especially through association of ideas
3 : a slight indication **:** TRACE ⟨a *suggestion* of a smile⟩

sug·ges·tive \səg-ˈjes-tiv, sə-ˈjes-\ *adjective* (1631)
1 a : giving a suggestion **:** INDICATIVE ⟨*suggestive* of a past era⟩ **b :** full of suggestions **:** stimulating thought ⟨provided a *suggestive* . . . commentary on the era —Lloyd Morris⟩ **c :** stirring mental associations **:** EVOCATIVE
2 : suggesting or tending to suggest something improper or indecent **:** RISQUÉ
— **sug·ges·tive·ly** *adverb*
— **sug·ges·tive·ness** *noun*

sui·cid·al \ˌsü-ə-ˈsī-dᵊl\ *adjective* (1777)
1 a : dangerous especially to life **b :** destructive to one's own interests
2 : relating to or of the nature of suicide
3 : marked by an impulse to commit suicide
— **sui·cid·al·ly** *adverb*

¹sui·cide \ˈsü-ə-ˌsīd\ *noun* [Latin *sui* (genitive) of oneself + English *-cide;* akin to Old English & Old High German *sīn* his, Latin *suus* one's own, *sed, se* without, Sanskrit *sva* oneself, one's own] (1643)
1 a : the act or an instance of taking one's own life voluntarily and intentionally especially by a person of years of discretion and of sound mind **b :** ruin of one's own interests ⟨political *suicide*⟩
2 : one that commits or attempts suicide

²suicide *verb* **sui·cid·ed; sui·cid·ing** (1841)
intransitive verb
: to commit suicide
transitive verb
: to put (oneself) to death

suicide squad *noun* [from the fact that kickoffs are more dangerous than other plays] (circa 1966)
: a special squad used on kickoffs in football

suicide squeeze *noun* (1955)
: a squeeze play in which the runner runs all out at the pitch without knowing whether the batter will contact the ball

sui ge·ner·is \ˌsü-ˌī-ˈje-nə-rəs; ˌsü-ē-ˈje-, -ˈge-\ *adjective* [Latin, of its own kind] (1787)
: constituting a class alone **:** UNIQUE, PECULIAR

sui ju·ris \ˌsü-ˌī-ˈju̇r-əs, ˌsü-ē-ˈyu̇r-\ *adjective* [Latin, of one's own right] (1675)
: having full legal rights or capacity

su·int \ˈsü-ənt, ˈswint\ *noun* [French, from Middle French, from *suer* to sweat, from Latin *sudare* — more at SWEAT] (1791)
: dried perspiration of sheep deposited in the wool and rich in potassium salts

¹suit \ˈsüt\ *noun* [Middle English *siute* act of following, retinue, sequence, set, from Old French, act of following, retinue, from (assumed) Vulgar Latin *sequita,* from feminine of *sequitus,* past participle of *sequere* to follow — more at SUE] (14th century)
1 *archaic* **:** SUITE 1
2 a : recourse or appeal to a feudal superior for justice or redress **b :** an action or process in a court for the recovery of a right or claim
3 : an act or instance of suing or seeking by entreaty **:** APPEAL; *specifically* **:** COURTSHIP
4 : SUITE 2 — used chiefly of armor, sails, and counters in games
5 : a set of garments: as **a :** an outer costume of two or more pieces **b :** a costume to be worn for a special purpose or under particular conditions ⟨gym *suit*⟩
6 a : all the playing cards in a pack bearing the same symbol **b :** all the dominoes bearing the same number **c :** all the cards or counters in a particular suit held by one player ⟨a 5-card *suit*⟩ **d :** the suit led ⟨follow *suit*⟩
7 *slang* **:** a business executive — usually used in plural

☆ SYNONYMS
Suggest, imply, hint, intimate, insinuate mean to convey an idea indirectly. SUGGEST may stress putting into the mind by association of ideas, awakening of a desire, or initiating a train of thought ⟨a film title that *suggests* its subject matter⟩. IMPLY is close to SUGGEST but may indicate a more definite or logical relation of the unexpressed idea to the expressed ⟨measures *implying* that bankruptcy was imminent⟩. HINT implies the use of slight or remote suggestion with a minimum of overt statement ⟨*hinted* that she might have a job lined up⟩. INTIMATE stresses delicacy of suggestion without connoting any lack of candor ⟨*intimates* that there is more to the situation than meets the eye⟩. INSINUATE applies to the conveying of a usually unpleasant idea in a sly underhanded manner ⟨*insinuated* that there were shady dealings⟩.

²suit (14th century)
transitive verb
1 a : to be becoming to **b :** to be proper for **:** BEFIT
2 : to outfit with clothes **:** DRESS
3 : ACCOMMODATE, ADAPT ⟨*suit* the action to the word⟩
4 : to meet the needs or desires of **:** PLEASE ⟨*suits* me fine⟩
intransitive verb
1 : to be in accordance **:** AGREE ⟨the position *suits* with your abilities⟩
2 : to be appropriate or satisfactory ⟨these prices don't *suit*⟩
3 : to put on specially required clothing (as a uniform or protective garb) — usually used with *up*

suit·able \'sü-tə-bəl\ *adjective* (1582)
1 *obsolete* **:** SIMILAR, MATCHING
2 a : adapted to a use or purpose **b :** satisfying propriety **:** PROPER **c :** ABLE, QUALIFIED
synonym see FIT
— suit·abil·i·ty \ˌsü-tə-'bi-lə-tē\ *noun*
— suit·able·ness *noun*
— suit·ably \-blē\ *adverb*

suit·case \'süt-ˌkās\ *noun* (1897)
: a bag or case carried by hand and designed to hold a traveler's clothing and personal articles

suite \'swēt, *2d is also* 'süt\ *noun* [French, alteration of Old French *siute* — more at SUIT] (1673)
1 : RETINUE; *especially* **:** the personal staff accompanying a ruler, diplomat, or dignitary on official business
2 : a group of things forming a unit or constituting a collection **:** SET: as **a :** a group of rooms occupied as a unit **b** (1) **:** a 17th and 18th century instrumental musical form consisting of a series of dances in the same or related keys (2) **:** a modern instrumental composition in several movements of different character (3) **:** a long orchestral concert arrangement in suite form of material drawn from a longer work (as a ballet) **c :** a collection of minerals or rocks having some characteristic in common (as type or origin) **d :** a set of matched furniture

suit·er \'sü-tər\ *noun* (1952)
: a suitcase for holding a specified number of suits — usually used in combination ⟨a two-*suiter*⟩

suit·ing \'sü-tiŋ\ *noun* (1883)
1 : fabric for suits
2 : a suit of clothes

suit·or \'sü-tər\ *noun* [Middle English, follower, pleader, from Anglo-French, from Latin *secutor* follower, from *sequi* to follow — more at SUE] (15th century)
1 : one that petitions or entreats
2 : a party to a suit at law
3 : one who courts a woman or seeks to marry her
4 : one who seeks to take over a business

su·ki·ya·ki \skē-'yä-kē; ˌsü-kē-', ˌsü-\ *noun* [Japanese, from *suki-* slice + *yaki* broil] (1920)
: a dish consisting of thin slices of meat, bean curd, and vegetables cooked in soy sauce and sugar

suk·kah \'sü-kə\ *noun* [Hebrew *sukkah*] (1875)
: a booth or shelter with a roof of branches and leaves that is used especially for meals during the Sukkoth

Suk·koth *or* **Suk·kot** \'sü-kəs, -ˌkōt, -ˌkōth, -ˌkōs\ *noun* [Hebrew *sukkōth,* plural of *sukkāh*] (1882)
: a Jewish harvest festival beginning on the 15th of Tishri and commemorating the temporary shelters used by the Jews during their wandering in the wilderness

sul·cate \'səl-ˌkāt\ *adjective* [Latin *sulcatus,* past participle of *sulcare* to furrow, from *sulcus*] (1760)
: scored with usually longitudinal furrows ⟨a *sulcate* seedpod⟩

sul·cus \'səl-kəs\ *noun, plural* **sul·ci** \-ˌkī, -ˌkē, -ˌsī\ [Latin; akin to Old English *sulh* plow, Greek *holkos* furrow, *helkein* to pull] (1662)
: FURROW, GROOVE; *especially* **:** a shallow furrow on the surface of the brain separating adjacent convolutions

sulf- *or* **sulfo-** *combining form* [French *sulf-, sulfo-,* from Latin *sulfur*]
: sulfur **:** containing sulfur ⟨*sulf*ide⟩

sul·fa \'səl-fə\ *adjective* [short for *sulfanilamide*] (1940)
1 : related chemically to sulfanilamide
2 : of, relating to, or containing sulfa drugs

sul·fa·di·a·zine \ˌsəl-fə-'dī-ə-ˌzēn\ *noun* [*sulfa* + *diazine* ($C_4H_4N_2$)] (1940)
: a sulfa drug $C_{10}H_{10}N_4O_2S$ that is used especially in the treatment of meningitis, pneumonia, and intestinal infections

sulfa drug *noun* (1940)
: any of various synthetic organic bacteria-inhibiting drugs that are sulfonamides closely related chemically to sulfanilamide

sul·fa·nil·amide \ˌsəl-fə-'ni-lə-ˌmīd, -məd\ *noun* [*sulfanil*ic + *amide*] (1937)
: a crystalline sulfonamide $C_6H_8N_2O_2S$ that is the amide of sulfanilic acid and the parent compound of most of the sulfa drugs

sul·fa·nil·ic acid \ˌsəl-fə-'ni-lik-\ *noun* [International Scientific Vocabulary *sulf-* + *anil*ine + *-ic*] (1856)
: a crystalline acid $C_6H_7NO_3S$ obtained from aniline

sul·fa·tase \'səl-fə-ˌtās, -ˌtāz\ *noun* [¹*sulfate*] (1924)
: any of various esterases that accelerate the hydrolysis of sulfuric esters and that are found in animal tissues and in microorganisms

¹sul·fate \'səl-ˌfāt\ *noun* [French, from Latin *sulfur*] (1790)
1 : a salt or ester of sulfuric acid
2 : a bivalent group or anion SO_4 characteristic of sulfuric acid and the sulfates

²sulfate *transitive verb* **sul·fat·ed; sul·fat·ing** (1802)
: to treat or combine with sulfuric acid or a sulfate

sulf·hy·dryl \ˌsəlf-'(h)ī-drəl\ *noun* [International Scientific Vocabulary] (circa 1901)
: THIOL 2 — used chiefly in molecular biology

sul·fide \'səl-ˌfīd\ *noun* (1836)
1 : any of various organic compounds characterized by a sulfur atom attached to two carbon atoms
2 : a binary compound (as CuS) of sulfur usually with a more electropositive element or group **:** a salt of hydrogen sulfide

sul·fin·py·ra·zone \ˌsəl-fən-'pī-rə-ˌzōn\ *noun* [*sulfin*ic acid (RSO_2H) + *pyr-* + *az*ole + *-one*] (1958)
: a uricosuric drug $C_{23}H_{20}N_2O_3S$ used in long-term treatment of chronic gout

sul·fite \'səl-ˌfīt\ *noun* [French *sulfite,* alteration of *sulfate*] (1790)
: a salt or ester of sulfurous acid
— sul·fit·ic \ˌsəl-'fi-tik\ *adjective*

sulfon- *combining form* [International Scientific Vocabulary *sulfonic*]
: sulfonic ⟨*sulfon*amide⟩

sul·fon·amide \ˌsəl-'fä-nə-ˌmīd, -məd; -'fō-nə-ˌmīd\ *noun* (1881)
: any of various amides (as sulfanilamide) of a sulfonic acid; *also* **:** SULFA DRUG

¹sul·fo·nate \'səl-fə-ˌnāt\ *noun* (1876)
: a salt or ester of a sulfonic acid

²sulfonate *transitive verb* **-nat·ed; -nat·ing** (1882)
: to introduce the SO_3H group into; *broadly* **:** to treat (an organic substance) with sulfuric acid
— sul·fo·na·tion \ˌsəl-fə-'nā-shən\ *noun*

sul·fone \'səl-ˌfōn\ *noun* (1872)
: any of various compounds containing the sulfonyl group with its sulfur atom having two bonds with carbon

sul·fon·ic \ˌsəl-'fä-nik, -'fō-\ *adjective* (circa 1891)
: of, relating to, being, or derived from the univalent acid group SO_3H

sulfonic acid *noun* (1873)
: any of numerous acids that contain the SO_3H group and may be derived from sulfuric acid by replacement of a hydroxyl group by either an inorganic anion or a univalent organic group

sul·fo·ni·um \ˌsəl-'fō-nē-əm\ *noun* [New Latin, from *sulf-* + *-onium*] (1885)
: a univalent group or cation SH_3 or derivative SR_3

sul·fo·nyl \'səl-fə-ˌnil\ *noun* (1920)
: the bivalent group SO_2

sul·fo·nyl·urea \ˌsəl-fə-ˌnil-'yu̇r-ē-ə\ *noun* [New Latin, from International Scientific Vocabulary *sulfonyl* + New Latin *urea*] (1956)
: any of several hypoglycemic compounds related to the sulfonamides and used in the oral treatment of diabetes

sulf·ox·ide \ˌsəl-'fäk-ˌsīd\ *noun* [International Scientific Vocabulary] (circa 1894)
: any of a class of organic compounds characterized by an SO group with its sulfur atom having two bonds with carbon

sul·fur *also* **sul·phur** \'səl-fər\ *noun* [Middle English *sulphur* brimstone, from Latin *sulpur, sulphur, sulfur*] (14th century)
1 : a nonmetallic element that occurs either free or combined especially in sulfides and sulfates, is a constituent of proteins, exists in several allotropic forms including yellow orthorhombic crystals, resembles oxygen chemically but is less active and more acidic, and is used especially in the chemical and paper industries, in rubber vulcanization, and in medicine for treating skin diseases — see ELEMENT table
2 : something (as scathing language) that suggests sulfur ■
— sul·fury *or* **sul·phury** \-ē\ *adjective*

sulfur bacterium *noun* (1891)
: any of various bacteria (especially genus *Thiobacillus*) capable of metabolizing sulfur compounds

sulfur dioxide *noun* (1869)
: a heavy pungent toxic gas SO_2 that is easily condensed to a colorless liquid, is used especially in making sulfuric acid, in bleaching, as a preservative, and as a refrigerant, and is a major air pollutant especially in industrial areas

sul·fu·ric \ˌsəl-'fyu̇r-ik\ *adjective* (1790)
: of, relating to, or containing sulfur especially with a higher valence than sulfurous compounds ⟨*sulfuric* esters⟩

sulfuric acid *or* **sul·phu·ric acid** \ˌsəl-'fyu̇r-ik-\ *noun* (1790)
: a heavy corrosive oily dibasic strong acid H_2SO_4 that is colorless when pure and is a vigorous oxidizing and dehydrating agent
usage see SULFUR

sul·fu·rize \'səl-fə-ˌrīz, -fyə-\ *transitive verb* **-rized; -riz·ing** (1794)
: to treat with sulfur or a sulfur compound

sul·fu·rous *also* **sul·phu·rous** \'səl-fə-rəs,

□ USAGE
sulfur The spelling *sulfur* predominates in U.S. technical usage, while both *sulfur* and *sulphur* are common in general usage. British usage tends to favor *sulphur* for all applications. The same pattern is seen in most of the words derived from *sulfur.*

\ə\ abut \ᵊ\ kitten \ər\ further \a\ ash \ā\ ace \ä\ mop, mar \au̇\ out \ch\ chin \e\ bet \ē\ easy \g\ go \i\ hit \ī\ ice \j\ job \ŋ\ sing \ō\ go \ȯ\ law \ȯi\ boy \th\ thin \t͟h\ the \ü\ loot \u̇\ foot \y\ yet \zh\ vision *see also* Guide to Pronunciation

-fyə- *also especially for 1b* ˌsəl-ˈfyu̇r-əs\ *adjective* (15th century)
1 a : of, relating to, or containing sulfur especially with a lower valence than sulfuric compounds ⟨*sulfurous* esters⟩ **b :** resembling or emanating from sulfur and especially burning sulfur
2 a : of, relating to, or dealing with the fire of hell **:** INFERNAL **b :** SCATHING, VIRULENT ⟨*sulfurous* denunciations⟩ **c :** PROFANE, BLASPHEMOUS ⟨*sulfurous* language⟩
usage see SULFUR
— sul·fu·rous·ly *adverb*
— sul·fu·rous·ness *noun*

sulfurous acid *noun* (1790)
: a weak unstable dibasic acid H_2SO_3 known in solution and through its salts and used as a reducing and bleaching agent

sul·fu·ryl \ˈsəl-fə-ˌril, -fyə-\ *noun* [International Scientific Vocabulary] (1867)
: SULFONYL — used especially in names of inorganic compounds

¹sulk \ˈsəlk\ *intransitive verb* [back-formation from *sulky*] (1781)
: to be moodily silent

²sulk *noun* (1804)
1 : the state of one sulking — often used in plural ⟨had a case of the *sulks*⟩
2 : a sulky mood or spell ⟨in a *sulk*⟩

¹sulky \ˈsəl-kē\ *adjective* [probably alteration of obsolete *sulke* sluggish] (1744)
1 a : sulking or given to spells of sulking **b :** relating to or indicating a sulk ⟨a *sulky* expression⟩
2 [²*sulky*] **:** having wheels and usually a seat for the driver ⟨a *sulky* plow⟩
synonym see SULLEN
— sulk·i·ly \-kə-lē\ *adverb*
— sulk·i·ness \-kē-nəs\ *noun*

²sulky *noun, plural* **sulkies** [probably from ¹*sulky;* from its having room for only one person] (1756)
: a light 2-wheeled vehicle (as for harness racing) having a seat for the driver only and usually no body

sul·lage \ˈsə-lij\ *noun* [probably from Middle French *soiller, souiller* to soil — more at SOIL] (1553)
: REFUSE, SEWAGE

sul·len \ˈsə-lən\ *adjective* [Middle English *solain* solitary, probably from (assumed) Anglo-French *solein,* alteration of Old French *soltain,* from Late Latin *solitaneus* private, ultimately from Latin *solus* alone] (14th century)
1 a : gloomily or resentfully silent or repressed **b :** suggesting a sullen state **:** LOWERING
2 : dull or somber in sound or color
3 : DISMAL, GLOOMY
4 : moving sluggishly ☆
— sul·len·ly *adverb*
— sul·len·ness \ˈsə-lə(n)-nəs\ *noun*

¹sul·ly \ˈsə-lē\ *transitive verb* **sul·lied; sul·ly·ing** [Middle English *sollyen,* probably from Middle French *soiller* to soil] (15th century)
: to make soiled or tarnished **:** DEFILE

²sully *noun, plural* **sullies** (1602)
archaic **:** SOIL, STAIN

sulph- *or* **sulpho-** *chiefly British variant of* SULF-

sul·phate, sul·phide *chiefly British variant of* SULFATE, SULFIDE

sulphur butterfly *noun* (1879)
: any of numerous butterflies (especially *Colias* and related genera of the family Pieridae) having the wings usually yellow or orange with a black border — called also *sulphur*

sul·phu·re·ous *adjective* (circa 1552)
: SULFUROUS

sul·phu·rise *British variant of* SULFURIZE

sulphur yellow *noun* (1816)
: a brilliant greenish yellow

Sul·pi·cian \ˌsəl-ˈpi-shən\ *noun* [French *sulpicien,* from Compagnie de Saint-*Sulpice* Society of Saint Sulpice] (1786)
: a member of the Society of Priests of Saint Sulpice founded by Jean Jacques Olier in Paris, France, in 1642 and dedicated to the teaching of seminarians

sul·tan \ˈsəl-tᵊn\ *noun* [Middle French, from Arabic *sulṭān*] (1555)
: a king or sovereign especially of a Muslim state
— sul·tan·ic \ˌsəl-ˈta-nik\ *adjective*

sul·ta·na \(ˌ)səl-ˈta-nə\ *noun* [Italian, feminine of *sultano* sultan, from Arabic *sulṭān*] (1585)
1 : a female member of a sultan's family; *especially* **:** a sultan's wife
2 a : a pale yellow seedless grape grown for raisins and wine **b :** the raisin of a sultana

sul·tan·ate \ˈsəl-tᵊn-ˌāt\ *noun* (1822)
1 : a state or country governed by a sultan
2 : the office, dignity, or power of a sultan

sul·tan·ess \ˈsəl-tᵊn-əs\ *noun* (1611)
archaic **:** SULTANA 1

sul·try \ˈsəl-trē\ *adjective* **sul·tri·er; -est** [obsolete English *sulter* to swelter, alteration of English *swelter*] (1594)
1 a : very hot and humid **:** SWELTERING ⟨a *sultry* day⟩ **b :** burning hot **:** TORRID
2 a : hot with passion or anger **b :** exciting or capable of exciting strong sexual desire ⟨*sultry* glances⟩
— sul·tri·ly \-trə-lē\ *adverb*
— sul·tri·ness \-trē-nəs\ *noun*

¹sum \ˈsəm\ *noun* [Middle English *summe,* from Old French, from Latin *summa,* from feminine of *summus* highest; akin to Latin *super* over — more at OVER] (14th century)
1 : an indefinite or specified amount of money
2 : the whole amount **:** AGGREGATE
3 : the utmost degree **:** SUMMIT ⟨reached the *sum* of human happiness⟩
4 a : a summary of the chief points or thoughts **:** SUMMATION ⟨the *sum* of this criticism follows —C. W. Hendel⟩ **b :** GIST ⟨the *sum* and substance of an argument⟩
5 a (1) **:** the result of adding numbers ⟨*sum* of 5 and 7 is 12⟩ (2) **:** the limit of the sum of the first *n* terms of an infinite series as *n* increases indefinitely **b :** numbers to be added; *broadly* **:** a problem in arithmetic **c** (1) **:** DISJUNCTION 2 (2) **:** UNION 2d
— sum·ma·bil·i·ty \ˌsə-mə-ˈbi-lə-tē\ *noun*
— sum·ma·ble \ˈsə-mə-bəl\ *adjective*
— in sum : in short **:** BRIEFLY

²sum *verb* **summed; sum·ming** (14th century)
transitive verb
1 : to calculate the sum of **:** TOTAL
2 : SUMMARIZE
intransitive verb
: to reach a sum **:** AMOUNT

su·mac *also* **su·mach** \ˈshü-ˌmak, ˈsü-\ *noun* [Middle English *sumac,* from Middle French, from Arabic *summāq*] (14th century)
1 : a material used in tanning and dyeing that consists of dried powdered leaves and flowers of various sumacs
2 : any of a genus (*Rhus*) of trees, shrubs, and woody vines of the cashew family that have pinnately compound leaves turning to brilliant colors in the autumn, dioecious flowers, spikes or loose clusters of red or whitish berries, and in some cases foliage poisonous to the touch — compare POISON IVY, POISON OAK

Su·me·ri·an \sü-ˈmer-ē-ən, -ˈmir-\ *noun* (1878)
1 : a native of Sumer
2 : the language of the Sumerians that has no known linguistic affinities
— Sumerian *adjective*

Su·me·rol·o·gy \ˌsü-mə-ˈrä-lə-jē\ *noun* (1897)
: the study of Sumerian culture, language, and history
— Su·me·rol·o·gist \-jist\ *noun*

sum·ma \ˈsu̇-mə, ˈsü-, ˈsə-\ *noun, plural* **sum·mae** \ˈsu̇-ˌmī, ˈsü-, -ˌmā; ˈsə-ˌmē, -ˌmī\ [Medieval Latin, from Latin, sum] (1725)
: a comprehensive treatise; *especially* **:** one by a scholastic philosopher

sum·ma cum lau·de \ˌsu̇-mə-(ˌ)ku̇m-ˈlau̇-də, ˌsü-, -ˈlau̇-dē; ˌsə-mə-ˌkəm-ˈlȯ-dē\ *adverb or adjective* [Latin, with highest praise] (1900)
: with highest distinction ⟨graduated *summa cum laude*⟩ — compare CUM LAUDE, MAGNA CUM LAUDE

sum·mand \ˈsə-ˌmand, ˌsə-ˈmand\ *noun* [Medieval Latin *summandus,* gerund of *summare* to sum, from *summa*] (1893)
: a term in a summation **:** ADDEND

sum·ma·rise *British variant of* SUMMARIZE

sum·ma·ri·za·tion \ˌsə-mə-rə-ˈzā-shən, ˌsəm-rə-\ *noun* (1865)
1 : the act of summarizing
2 : SUMMARY

sum·ma·rize \ˈsə-mə-ˌrīz\ *verb* **-rized; -riz·ing** (1871)
transitive verb
: to tell in or reduce to a summary
intransitive verb
: to make a summary
— sum·ma·riz·able \ˌsə-mə-ˈrī-zə-bəl\ *adjective*
— sum·ma·riz·er *noun*

¹sum·ma·ry \ˈsə-mə-rē *also* ˈsəm-rē *or* -ˌmer-ē\ *adjective* [Middle English, from Medieval Latin *summarius,* from Latin *summa* sum] (15th century)
1 : COMPREHENSIVE; *especially* **:** covering the main points succinctly
2 a : done without delay or formality **:** quickly executed ⟨a *summary* dismissal⟩ **b :** of, relating to, or using a summary proceeding ⟨a *summary* trial⟩
synonym see CONCISE
— sum·mar·i·ly \(ˌ)sə-ˈmer-ə-lē\ *adverb*

²sum·ma·ry \ˈsə-mə-rē *also* ˈsəm-rē\ *noun, plural* **-ries** (1509)
: an abstract, abridgment, or compendium especially of a preceding discourse

sum·mate \ˈsə-ˌmāt\ *verb* **sum·mat·ed; sum·mat·ing** [back-formation from *summation*] (1900)
transitive verb
: to add together **:** SUM UP
intransitive verb
: to form a sum or cumulative effect

sum·ma·tion \(ˌ)sə-ˈmā-shən\ *noun* (1760)
1 : the act or process of forming a sum **:** ADDITION
2 : SUM, TOTAL
3 : cumulative action or effect; *especially* **:** the process by which a sequence of stimuli that are individually inadequate to produce a response are cumulatively able to induce a nerve impulse

☆ SYNONYMS
Sullen, glum, morose, surly, sulky, crabbed, saturnine, gloomy mean showing a forbidding or disagreeable mood. SULLEN implies a silent ill humor and a refusal to be sociable ⟨remained *sullen* amid the festivities⟩. GLUM suggests a silent dispiritedness ⟨a *glum* candidate left to ponder a stunning defeat⟩. MOROSE adds to GLUM an element of bitterness or misanthropy ⟨*morose* job seekers who are inured to rejection⟩. SURLY implies gruffness and sullenness of speech or manner ⟨a typical *surly* teenager⟩. SULKY suggests childish resentment expressed in peevish sullenness ⟨grew *sulky* after every spat⟩. CRABBED applies to a forbidding morose harshness of manner ⟨the school's notoriously *crabbed* headmaster⟩. SATURNINE describes a heavy forbidding aspect or suggests a bitter disposition ⟨a *saturnine* cynic always finding fault⟩. GLOOMY implies a depression in mood making for seeming sullenness or glumness ⟨a *gloomy* mood ushered in by bad news⟩.

4 : a final part of an argument reviewing points made and expressing conclusions
— **sum·ma·tion·al** \-shnəl, -shə-nᵊl\ *adjective*

sum·ma·tive \'sə-mə-tiv, -ˌmā-\ *adjective* (1881)
: ADDITIVE, CUMULATIVE

[1]**sum·mer** \'sə-mər\ *noun* [Middle English *sumer,* from Old English *sumor;* akin to Old High German & Old Norse *sumer* summer, Sanskrit *samā* year, season] (before 12th century)
1 : the season between spring and autumn comprising in the northern hemisphere usually the months of June, July, and August or as reckoned astronomically extending from the June solstice to the September equinox
2 : the warmer half of the year
3 : YEAR ⟨a girl of seventeen *summers*⟩
4 : a period of maturing powers
— **sum·mer·like** \-ˌlīk\ *adjective*

[2]**summer** *adjective* (14th century)
1 : of, relating to, or suitable for summer ⟨*summer* vacation⟩ ⟨a *summer* home⟩
2 : sown in the spring and harvested in the same year as sown ⟨*summer* wheat⟩ — compare WINTER

[3]**summer** *verb* **sum·mered; sum·mer·ing** \'sə-mə-riŋ, 'səm-riŋ\ (15th century)
intransitive verb
: to pass the summer
transitive verb
: to keep or carry through the summer; *especially* **:** to provide (as cattle or sheep) with pasture during the summer

summer cypress *noun* (1767)
: a densely branched Eurasian herb (*Kochia scoparia*) of the goosefoot family grown for its foliage which turns red in autumn

sum·mer·house \'sə-mər-ˌhau̇s\ *noun* (before 12th century)
1 : a country house for summer residence
2 : a covered structure in a garden or park designed to provide a shady resting place in summer

summer kitchen *noun* (1874)
: a small building or shed that is usually adjacent to a house and is used as a kitchen in warm weather

sum·mer·long \-ˌlȯŋ, -'lȯŋ\ *adjective* (1960)
: lasting through the summer

sum·mers \'sə-mərz\ *adverb* (1907)
: during the summers ⟨worked *summers* as a waiter⟩

sum·mer·sault *variant of* SOMERSAULT

summer savory *noun* (circa 1573)
: an annual European mint (*Satureja hortensis*) used in cookery — compare WINTER SAVORY

summer school *noun* (1860)
: a school or school session conducted in summer enabling students to accelerate progress toward a degree, to make up credits lost through absence or failure, or to round out professional education

summer squash *noun* (1815)
: any of various garden squashes derived from a variety (*Cucurbita pepo* variety *melopepo*) and used as a vegetable while immature and before hardening of the seeds and rind

summer stock *noun* (1927)
: theatrical productions of stock companies presented during the summer

summer theater *noun* (1801)
: a theater that presents several different plays or musicals during the summer

sum·mer·time \'sə-mər-ˌtīm\ *noun* (14th century)
: the summer season or a period like summer

summer time *noun* (1916)
chiefly British **:** DAYLIGHT SAVING TIME

sum·mer·wood \'sə-mər-ˌwu̇d\ *noun* (1902)
: the harder less porous portion of an annual ring of wood that develops late in the growing season — compare SPRINGWOOD

sum·mery \'sə-mə-rē, 'səm-rē\ *adjective* (1824)
: of, resembling, or fit for summer

sum·ming–up \ˌsə-miŋ-'əp\ *noun, plural* **sum·mings–up** \-miŋz-\ (1790)
: the act or statement of one who sums up

[1]**sum·mit** \'sə-mət\ *noun* [Middle English *somete,* from Middle French, from Old French, diminutive of *sum* top, from Latin *summum,* neuter of *summus* highest — more at SUM] (15th century)
1 : TOP, APEX; *especially* **:** the highest point **:** PEAK
2 : the topmost level attainable ⟨the *summit* of human fame⟩
3 a : the highest level of officials; *especially* **:** the diplomatic level of heads of government **b :** a conference of highest-level officials (as heads of government) ⟨an economic *summit*⟩
☆

[2]**summit** *intransitive verb* (1972)
: to participate in a summit conference

sum·mit·eer \ˌsə-mə-'tir\ *noun* (1957)
: one who takes part in a summit

sum·mit·ry \'sə-mə-trē\ *noun* (1958)
: the use of a summit conference for international negotiation

sum·mon \'sə-mən\ *transitive verb* **sum·moned; sum·mon·ing** \'sə-mə-niŋ, 'səm-niŋ\ [Middle English *somonen,* from Old French *somondre,* from (assumed) Vulgar Latin *summonere,* alteration of Latin *summonēre* to remind secretly, from *sub-* secretly + *monēre* to warn — more at SUB-, MIND] (13th century)
1 : to issue a call to convene **:** CONVOKE
2 : to command by service of a summons to appear in court
3 : to call upon for specified action
4 : to bid to come **:** send for ⟨*summon* a physician⟩
5 : to call forth **:** EVOKE — often used with *up*
☆
— **sum·mon·able** \'sə-mə-nə-bəl\ *adjective*
— **sum·mon·er** \'sə-mə-nər, 'səm-nər\ *noun*

[1]**sum·mons** \'sə-mənz\ *noun, plural* **sum·mons·es** [Middle English *somouns,* from Old French *somonse,* from past participle of *somondre*] (13th century)
1 : the act of summoning; *especially* **:** a call by authority to appear at a place named or to attend to a duty
2 : a warning or citation to appear in court: as **a :** a written notification to be served on a person as a warning to appear in court at a day specified to answer to the plaintiff **b :** a subpoena to appear as a witness
3 : something (as a call) that summons

[2]**summons** *transitive verb* (1683)
: SUMMON 2

sum·mum bo·num \ˌsu̇-məm-'bō-nəm, ˌsü-, ˌsə-\ *noun* [Latin] (1563)
: the supreme good from which all others are derived

su·mo \'sü-(ˌ)mō\ *noun* [Japanese *sumō*] (1880)
: a Japanese form of wrestling in which a contestant loses if he is forced out of the ring or if any part of his body except the soles of his foot touches the ground

sump \'səmp\ *noun* [Middle English *sompe* swamp — more at SWAMP] (1653)
1 : a pit or reservoir serving as a drain or receptacle for liquids: as **a :** CESSPOOL **b :** a pit at the lowest point in a circulating or drainage system (as the oil-circulating system of an internal combustion engine) **c** *chiefly British* **:** OIL PAN
2 *British* **:** CRANKCASE
3 [German *Sumpf,* literally, marsh, from Middle High German — more at SWAMP] **a :** the lowest part of a mine shaft into which water drains **b :** an excavation ahead of regular work in driving a mine tunnel or sinking a mine shaft
4 : something resembling a sump **:** SINK 2

sump pump *noun* (circa 1899)
: a pump (as in a basement) to remove accumulations of liquid from a sump pit

sump·ter \'səm(p)-tər\ *noun* [Middle English, short for *sumpter horse,* from *sumpter* driver of a packhorse, from Middle French *sometier,* from (assumed) Vulgar Latin *sagmatarius,* from Late Latin *sagmat-, sagma* packsaddle, from Greek; akin to Greek *sattein* to pack, stuff] (15th century)
: a pack animal

sump·tu·ary \'səm(p)-chə-ˌwer-ē\ *adjective* [Latin *sumptuarius,* from *sumptus* expense, from *sumere* to take, spend — more at CONSUME] (1600)
1 : relating to personal expenditures and especially to prevent extravagance and luxury ⟨conservative *sumptuary* tastes —John Cheever⟩
2 : designed to regulate extravagant expenditures or habits especially on moral or religious grounds ⟨*sumptuary* laws⟩ ⟨*sumptuary* tax⟩

sump·tu·ous \'səm(p)(t)-shə-wəs, -shəs, -shwəs\ *adjective* [Middle English, from Middle French *sumptueux,* from Latin *sumptuosus,* from *sumptus*] (15th century)
: extremely costly, rich, luxurious, or magnificent ⟨*sumptuous* banquets⟩ ⟨a *sumptuous* residence⟩; *also* **:** MAGNIFICENT 4
— **sump·tu·ous·ly** *adverb*
— **sump·tu·ous·ness** *noun*

sum total *noun* (14th century)
1 : a total arrived at through the counting of sums
2 : total result **:** TOTALITY

sum–up \'səm-ˌəp\ *noun* (1894)
: SUMMARY

☆ SYNONYMS

Summit, peak, pinnacle, climax, apex, acme, culmination mean the highest point attained or attainable. SUMMIT implies the topmost level attainable ⟨at the *summit* of the Victorian social scene⟩. PEAK suggests the highest among other high points ⟨an artist working at the *peak* of her powers⟩. PINNACLE suggests a dizzying and often insecure height ⟨the *pinnacle* of worldly success⟩. CLIMAX implies the highest point in an ascending series ⟨the war was the *climax* to a series of hostile actions⟩. APEX implies the point where all ascending lines converge ⟨the *apex* of Dutch culture⟩. ACME implies a level of quality representing the perfection of a thing ⟨a statue that was once deemed the *acme* of beauty⟩. CULMINATION suggests the outcome of a growth or development representing an attained objective ⟨the *culmination* of years of effort⟩.

Summon, call, cite, convoke, convene, muster mean to demand the presence of. SUMMON implies the exercise of authority ⟨was *summoned* to answer charges⟩. CALL may be used less formally for SUMMON ⟨*called* the legislature into special session⟩. CITE implies a summoning to court usually to answer a charge ⟨*cited* for drunken driving⟩. CONVOKE implies a summons to assemble for deliberative or legislative purposes ⟨*convoked* a Vatican council⟩. CONVENE is somewhat less formal than CONVOKE ⟨*convened* the students⟩. MUSTER suggests a calling up of a number of things that form a group in order that they may be exhibited, displayed, or utilized as a whole ⟨*mustered* the troops⟩.

sum up (15th century)
transitive verb
1 : to be the sum of **:** bring to a total ⟨10 victories *summed up* his record⟩
2 a : to present or show succinctly **:** SUMMARIZE ⟨*sum up* the evidence presented⟩ **b :** to assess and then describe briefly **:** size up
intransitive verb
: to present a summary or recapitulation

¹sun \'sən\ *noun* [Middle English *sunne,* from Old English; akin to Old High German *sunna* sun, Latin *sol* — more at SOLAR] (before 12th century)
1 a *often capitalized* **:** the luminous celestial body around which the earth and other planets revolve, from which they receive heat and light, and which has a mean distance from earth of 93,000,000 miles (150,000,000 kilometers), a linear diameter of 864,000 miles (1,390,000 kilometers), a mass 332,000 times greater than earth, and a mean density about one fourth that of earth **b :** a celestial body like the sun
2 : the heat or light radiated from the sun
3 : one resembling the sun (as in warmth or brilliance)
4 : the rising or setting of the sun ⟨from *sun* to *sun*⟩
5 : GLORY, SPLENDOR
— in the sun : in the public eye
— under the sun : in the world **:** on earth

²sun *verb* **sunned; sun·ning** (15th century)
transitive verb
: to expose to or as if to the rays of the sun
intransitive verb
: to sun oneself

sun·baked \'sən-ˌbākt\ *adjective* (1628)
1 : heated, parched, or compacted especially by excessive sunlight
2 : baked by exposure to sunshine

sun·bath \'sən-ˌbath, -ˌbȧth\ *noun* (1866)
: an exposure to sunlight or a sunlamp

sun·bathe \-ˌbāth\ *intransitive verb* [back-formation from *sunbather*] (1600)
: to take a sunbath
— sun·bath·er \-ˌbā-thər\ *noun*

sun·beam \-ˌbēm\ *noun* (before 12th century)
: a ray of sunlight

sun bear *noun* (1842)
: a small forest-dwelling bear (*Ursus malayanus* synonym *Helarctos malayanus*) that is found from Myanmar to Borneo and has short glossy black fur with a lighter muzzle and often an orange or white breast mark

sun·bird \-ˌbərd\ *noun* (1826)
: any of numerous small brilliantly colored oscine birds (family Nectariniidae) of the tropical Old World somewhat resembling hummingbirds

sun·block \-ˌbläk\ *noun* (1972)
: a preparation designed to block out more of the sun's rays than a sunscreen; *also* **:** its active ingredient (as para-aminobenzoic acid)

sun·bon·net \-ˌbä-nət\ *noun* (1824)
: a woman's bonnet with a wide brim framing the face and usually having a ruffle at the back to protect the neck from the sun

sunbonnet

sun·bow \-ˌbō\ *noun* (1816)
: an arch resembling a rainbow made by the sun shining through vapor or mist

¹sun·burn \-ˌbərn\ *verb* **-burned** \-ˌbərnd\ *or* **-burnt** \-ˌbərnt\; **-burn·ing** [back-formation from *sunburned,* from *sun* + *burned*] (1530)
transitive verb
: to burn or discolor by the sun
intransitive verb
: to become sunburned

²sunburn *noun* (1652)
: inflammation of the skin caused by overexposure to sunlight

sun·burst \'sən-ˌbərst\ *noun, often attributive* (1816)
1 : a flash of sunlight especially through a break in clouds
2 a : a jeweled brooch representing a sun surrounded by rays **b :** a design in the form of rays diverging from a central point

sun·choke \-ˌchōk\ *noun* (1980)
: JERUSALEM ARTICHOKE

sun·dae \'sən-dē, -(ˌ)dā\ *noun* [probably alteration of *Sunday*] (1897)
: ice cream served with topping (as crushed fruit, syrup, nuts, or whipped cream)

sun dance *noun, often S&D capitalized* (1849)
: a solo or group solstice rite of American Indians

¹Sun·day \'sən-dē, -(ˌ)dā\ *noun* [Middle English, from Old English *sunnandæg* (akin to Old High German *sunnūntag*), from *sunne* sun + *dæg* day] (before 12th century)
: the first day of the week **:** the Christian analogue of the Jewish Sabbath
— Sun·days \-dēz, -(ˌ)dāz\ *adverb*

²Sunday *adjective* (14th century)
1 : of, relating to, or associated with Sunday
2 [from the practice of wearing one's best clothes on Sunday to attend church] **:** BEST ⟨*Sunday* suit⟩
3 : AMATEUR ⟨*Sunday* painters⟩

Sun·day–go–to–meet·ing \'sən-dē-ˌgō-tə-'mē-tiŋ\ *adjective* (1831)
: appropriate for Sunday churchgoing

Sunday punch *noun* (1929)
1 : a powerful or devastating blow; *especially* **:** a knockout punch
2 : something capable of delivering a powerful or devastating blow to the opposition ⟨saving his *Sunday punch* for the end of the campaign —*Newsweek*⟩

Sunday school *noun* (1783)
: a school held on Sunday for religious education; *also* **:** the teachers and pupils of such a school

sun·deck \'sən-ˌdek\ *noun* (1897)
1 : the usually upper deck of a ship that is exposed to the most sun
2 : a roof, deck, or terrace for sunning

sun·der \'sən-dər\ *verb* **sun·dered; sun·der·ing** \-d(ə-)riŋ\ [Middle English, from Old English *gesundrian, syndrian;* akin to Old High German *suntarōn* to sunder, Old English *sundor* apart, Latin *sine* without, Sanskrit *sanutar* away] (before 12th century)
transitive verb
: to break apart or in two **:** separate by or as if by violence or by intervening time or space
intransitive verb
: to become parted, disunited, or severed
synonym see SEPARATE

sun·dew \'sən-(ˌ)dü, -(ˌ)dyü\ *noun* (1578)
: any of a genus (*Drosera* of the family Droseraceae, the sundew family) of bog-inhabiting insectivorous herbs having viscid glands on the leaves

sun·di·al \-ˌdī(-ə)l\ *noun* (1599)
: an instrument to show the time of day by the shadow of a gnomon on a usually horizontal plate or on a cylindrical surface

sun disk *noun* (1877)
: an ancient Near Eastern symbol consisting of a disk with conventionalized wings emblematic of the sun-god (as Ra in Egypt)

sun disk

sun dog *noun* (1635)
1 : PARHELION
2 : a small nearly round halo on the parhelic circle most frequently just outside the halo of 22 degrees

sun·down \'sən-ˌdau̇n\ *noun* (1620)
: SUNSET 2

sun·down·er \-ˌdau̇-nər\ *noun* (1868)
1 [from his habit of arriving at a place where he hopes to obtain food and lodging too late to do any work] *Australian* **:** HOBO, TRAMP
2 *chiefly British* **:** a drink taken at sundown

sun·dress \-ˌdres\ *noun* (1942)
: a dress with an abbreviated bodice usually exposing the shoulders, arms, and back

sun·dries \'sən-drēz\ *noun plural* [[1]sundry] (1755)
: miscellaneous small articles, details, or items

sun·drops \'sən-ˌdräps\ *noun plural but singular or plural in construction* (1784)
: any of several day-flowering herbs (genus *Oenothera*) of the evening-primrose family

¹sun·dry \'sən-drē\ *adjective* [Middle English, different for each, from Old English *syndrig,* from *sundor* apart — more at SUNDER] (13th century)
: MISCELLANEOUS, VARIOUS ⟨*sundry* articles⟩

²sundry *pronoun, plural in construction* (15th century)
: an indeterminate number ⟨recommended for reading by all and *sundry* —Edward Huberman⟩

sun·fish \-ˌfish\ *noun* (1629)
1 : OCEAN SUNFISH
2 : any of numerous American freshwater bony fishes (family Centrarchidae, especially genus *Lepomis*) usually with a deep compressed body and metallic luster

Sunfish *trademark*
— used for a light sailboat that has one sail and is designed for use by no more than two people

sun·flow·er \-ˌflau̇(-ə)r\ *noun* (1597)
: any of a genus (*Helianthus*) of New World composite plants with large yellow-rayed flower heads bearing edible seeds that yield an edible oil

sung \'səŋ\ *past and past participle of* SING

Sung \'su̇ŋ\ *noun* [Chinese (Beijing) *Sòng*] (1673)
: a Chinese dynasty dated A.D. 960–1280 and marked by cultural refinement and achievements in philosophy, literature, and art
— Sung *adjective*

sun·glass \'sən-ˌglas\ *noun* (1804)
1 : a convex lens for converging the sun's rays
2 *plural* **:** glasses to protect the eyes from the sun

sung mass *noun* (1931)
: HIGH MASS

sun god *noun, often capitalized* (1592)
: a god that represents or personifies the sun in various religions

sun goddess *noun, often capitalized* (1861)
: a goddess that represents or personifies the sun in various religions

sun–grebe \'sən-ˌgrēb\ *noun* (circa 1889)
: any of a small family (Heliornithidae) of African, Asian, and American tropical birds related to the rails

¹sunk \'səŋk\ *past and past participle of* SINK

²sunk *adjective* (1719)
1 : depressed in spirits
2 : DONE FOR, RUINED

sunk·en \'səŋ-kən\ *adjective* [Middle English *sonkyn,* past participle of *sinken* to sink] (14th century)
1 : SUBMERGED; *especially* **:** lying at the bottom of a body of water
2 a : HOLLOW, RECESSED ⟨*sunken* cheeks⟩ **b :** lying in a depression ⟨a *sunken* garden⟩ **c :** settled below the normal level **d :** constructed below the normal floor level ⟨a *sunken* living room⟩

sunk fence *noun* (circa 1771)
: a ditch with a retaining wall used to divide lands without defacing a landscape — called also *ha-ha*

sun·lamp \'sən-ˌlamp\ *noun* (1885)
: an electric lamp designed to emit radiation of wavelengths from ultraviolet to infrared

sun·less \-ləs\ *adjective* (1589)
: lacking sunshine : DARK, CHEERLESS

sun·light \-ˌlīt\ *noun* (13th century)
: the light of the sun : SUNSHINE

sun·lit \-ˌlit\ *adjective* (1822)
: lighted by or as if by the sun

sunn \ˈsən\ *noun* [Hindi *san,* from Sanskrit *śaṇa*] (1774)
: an East Indian leguminous plant (*Crotalaria juncea*) with slender branches, simple leaves, and yellow flowers; *also* : its valuable fiber resembling hemp that is lighter and stronger than jute

sun·na *also* **sun·nah** \ˈsu̇(n)-(ˌ)nə, ˈsə(n)-\ *noun, often capitalized* [Arabic *sunnah*] (1728)
: the body of Islamic custom and practice based on Muhammad's words and deeds

sunn hemp *noun* (1849)
: SUNN

Sun·ni \ˈsu̇(n)-(ˌ)nē\ *noun* [Arabic *sunnīy,* from *sunnah*] (1595)
1 : the Muslims of the branch of Islam that adheres to the orthodox tradition and acknowledges the first four caliphs as rightful successors of Muhammad — compare SHIA
2 : a Sunni Muslim
— **Sunni** *adjective*

Sun·nism \ˈsu̇(n)-ˌni-zəm\ *noun* (1892)
: the religious system or distinctive tenets of the Sunni

Sun·nite \-ˌnīt\ *noun* (1718)
: SUNNI 2
— **Sun·nite** \ˈsu̇(n)-ˌnīt\ *adjective*

sun·ny \ˈsə-nē\ *adjective* **sun·ni·er; -est** (14th century)
1 : marked by brilliant sunlight : full of sunshine
2 : CHEERFUL, OPTIMISTIC ⟨a *sunny* disposition⟩
3 : exposed to, brightened, or warmed by the sun ⟨a *sunny* room⟩
— **sun·ni·ly** \ˈsə-nᵊl-ē\ *adverb*
— **sun·ni·ness** \ˈsə-nē-nəs\ *noun*

sun·ny–side up \ˌsə-nē-ˌsīd-ˈəp\ *adjective* (circa 1901)
of an egg : fried on one side only

sun·porch \-ˌpōrch, -ˌpȯrch\ *noun* (1918)
: a screened-in or glassed-in porch with a sunny exposure

sun·rise \ˈsən-ˌrīz\ *noun* (15th century)
1 : the apparent rising of the sun above the horizon; *also* : the accompanying atmospheric effects
2 : the time when the upper limb of the sun appears above the horizon as a result of the diurnal rotation of the earth

sun·roof \-ˌrüf, -ˌru̇f\ *noun* (1952)
: a panel in an automobile roof that can be opened

sun·room \-ˌrüm, -ˌru̇m\ *noun* (1917)
: a glass-enclosed porch or living room with a sunny exposure — called also *sun parlor*

sun·scald \-ˌskȯld\ *noun* (1855)
: an injury of woody plants (as fruit or forest trees) characterized by localized death of the tissues and sometimes by cankers and caused when it occurs in the summer by the combined action of both the heat and light of the sun and in the winter by the combined action of sun and low temperature to produce freezing of bark and underlying tissues

sun·screen \-ˌskrēn\ *noun* (1738)
: a screen to protect against sun; *especially* : a substance (as para-aminobenzoic acid) used in suntan preparations to protect the skin from excessive ultraviolet radiation
— **sun·screen·ing** *adjective*

sun·seek·er \-ˌsē-kər\ *noun* (1954)
: a person who travels to an area of warmth and sun especially in winter

¹sun·set \-ˌset\ *noun* (14th century)
1 : the apparent descent of the sun below the horizon; *also* : the accompanying atmospheric effects
2 : the time when the upper limb of the sun disappears below the horizon as a result of the diurnal rotation of the earth
3 : a period of decline; *especially* : old age

²sunset *adjective* (1976)
: stipulating the periodic review of government agencies and programs in order to continue their existence ⟨*sunset* law⟩

sun·shade \ˈsən-ˌshād\ *noun* (1842)
: something used as a protection from the sun's rays: as **a** : PARASOL **b** : AWNING

¹sun·shine \-ˌshīn\ *noun* (13th century)
1 a : the sun's light or direct rays **b** : the warmth and light given by the sun's rays **c** : a spot or surface on which the sun's light shines
2 : something (as a person, condition, or influence) that radiates warmth, cheer, or happiness
— **sun·shiny** \-ˌshī-nē\ *adjective*

²sunshine *adjective* (1972)
: forbidding or restricting closed meetings of legislative or executive bodies and sometimes providing for public access to records ⟨*sunshine* law⟩

sun·spot \-ˌspät\ *noun* (1868)
: any of the dark spots that appear from time to time on the sun's surface and are usually visible only through a telescope

sun·stroke \-ˌstrōk\ *noun* (1851)
: heatstroke caused by direct exposure to the sun

sun·struck \-ˌstrək\ *adjective* (1794)
: affected or touched by the sun

sun·suit \-ˌsüt\ *noun* (1929)
: an outfit worn usually for sunbathing and play

sun·tan \-ˌtan\ *noun* (1904)
1 : a browning of the skin from exposure to the rays of the sun
2 *plural* : a tan-colored summer uniform
— **sun·tanned** \-ˌtand\ *adjective*

sun·up \-ˌəp\ *noun* (1712)
: SUNRISE

¹sun·ward \ˈsən-wərd\ *or* **sun·wards** \-wərdz\ *adverb* (1611)
: toward the sun

²sunward *adjective* (1769)
: facing the sun

sun·wise \ˈsən-ˌwīz\ *adverb* (circa 1864)
: CLOCKWISE

¹sup \ˈsəp\ *verb* **supped; sup·ping** [Middle English *suppen,* from Old English *sūpan, suppan;* akin to Old High German *sūfan* to drink, sip, Old English *sopp* sop] (before 12th century)
transitive verb
: to take or drink in swallows or gulps
intransitive verb
chiefly dialect : to take food and especially liquid food into the mouth a little at a time

²sup *noun* (circa 1570)
: a mouthful especially of liquor or broth : SIP; *also* : a small quantity of liquid ⟨pour me just a *sup* of tea⟩

³sup *intransitive verb* **supped; sup·ping** [Middle English *soupen, suppen,* from Old French *souper,* from *soupe* sop, soup — more at SOUP] (14th century)
1 : to eat the evening meal
2 : to make one's supper — used with *on* or *off* ⟨*sup* on roast beef⟩

¹su·per \ˈsü-pər\ *adjective* [*super-*] (1837)
1 a — used as a generalized term of approval ⟨a *super* cook⟩ **b** : of high grade or quality
2 : very large or powerful ⟨a *super* atomic bomb⟩
3 : exhibiting the characteristics of its type to an extreme or excessive degree ⟨*super* secrecy⟩

²super *noun* (1838)
1 [by shortening] **a** : SUPERNUMERARY; *especially* : a supernumerary actor **b** : SUPERINTENDENT, SUPERVISOR; *especially* : the superintendent of an apartment building
2 [short for obsolete *superhive*] : a removable upper story of a beehive
3 [¹*super*] : a superfine grade or extra large size
4 [origin unknown] : a thin loosely woven open-meshed starched cotton fabric used especially for reinforcing books

³super *adverb* [*super-*] (1946)
1 : VERY, EXTREMELY ⟨a *super* fast car⟩
2 : to an excessive degree

super- *prefix* [Latin, over, above, in addition, from *super* over, above, on top of — more at OVER]
1 a (1) : over and above : higher in quantity, quality, or degree than : more than ⟨*super*human⟩ (2) : in addition : extra ⟨*super*tax⟩ **b** (1) : exceeding or so as to exceed a norm ⟨*super*heat⟩ (2) : in or to an extreme or excessive degree or intensity ⟨*super*subtle⟩ **c** : surpassing all or most others of its kind ⟨*super*highway⟩
2 a : situated or placed above, on, or at the top of ⟨*super*lunary⟩; *specifically* : situated on the dorsal side of **b** : next above or higher ⟨*super*tonic⟩
3 : having the (specified) ingredient present in a large or unusually large proportion ⟨*super*phosphate⟩
4 : constituting a more inclusive category than that specified ⟨*super*family⟩
5 : superior in status, title, or position ⟨*super*power⟩

su·per·ab·sor·bent
su·per·achiev·er
su·per·ac·tiv·i·ty
su·per·ad·di·tion
su·per·ad·min·is·tra·tor
su·per·agent
su·per·am·bi·tious
su·per·ath·lete
su·per·bad
su·per·bank
su·per·bil·lion·aire
su·per·bitch
su·per·board
su·per·bomb
su·per·bomb·er
su·per·bright
su·per·bu·reau·crat
su·per·cab·i·net
su·per·car
su·per·car·ri·er
su·per·cau·tious
su·per·cen·ter
su·per·chic
su·per·church
su·per·civ·i·li·za·tion
su·per·civ·i·lized
su·per·clean
su·per·club
su·per·co·los·sal
su·per·com·fort·able
su·per·com·pet·i·tive
su·per·con·fi·dent
su·per·con·glom·er·ate
su·per·con·ser·va·tive
su·per·con·ve·nient
su·per·cop
su·per·cor·po·ra·tion
su·per·crim·i·nal
su·per·cute
su·per·de·luxe
su·per·dip·lo·mat
su·per·ef·fec·tive
su·per·ef·fi·cien·cy
su·per·ef·fi·cient
su·per·ego·ist
su·per·elite
su·per·em·i·nence
su·per·em·i·nent
su·per·em·i·nent·ly
su·per·ex·pen·sive
su·per·ex·press
su·per·fan
su·per·farm
su·per·fast
su·per·firm
su·per·flack
su·per·fund
su·per·good
su·per·gov·ern·ment
su·per·growth
su·per·hard·en
su·per·heavy
su·per·her·o·ine
su·per·hit
su·per·hot
su·per·hype
su·per·in·su·lat·ed
su·per·in·tel·lec·tu·al
su·per·in·tel·li·gence
su·per·in·tel·li·gent
su·per·in·ten·si·ty
su·per·jock
su·per·jum·bo
su·per·large
su·per·law·yer
su·per·light
su·per·lob·by·ist
su·per·loy·al·ist
su·per·lux·u·ri·ous
su·per·lux·u·ry
su·per·ma·cho
su·per·ma·jor·i·ty
su·per·male
su·per·mas·cu·line
su·per·mas·sive
su·per·mil·i·tant
su·per·mil·lion·aire
su·per·mind
su·per·min·is·ter
su·per·mod·el
su·per·mod·ern
su·per·mom
su·per·na·tion
su·per·na·tion·al
su·per·nu·tri·tion
su·per·or·gan·ic
su·per·or·gasm

\ə\ **abut** \ᵊ\ **kitten** \ər\ **further** \a\ **ash** \ā\ **ace** \ä\ **mop, mar** \au̇\ **out** \ch\ **chin** \e\ **bet** \ē\ **easy** \g\ **go** \i\ **hit** \ī\ **ice** \j\ **job** \ŋ\ **sing** \ō\ **go** \ȯ\ **law** \ȯi\ **boy** \th\ **thin** \t͟h\ **the** \ü\ **loot** \u̇\ **foot** \y\ **yet** \zh\ **vision** *see also* Guide to Pronunciation

su•per•pa•tri•ot
su•per•pa•tri•ot•ic
su•per•pa•tri•o•tism
su•per•per•son
su•per•per•son•al
su•per•phe•nom•e•non
su•per•pimp
su•per•plane
su•per•play•er
su•per•po•lite
su•per•port
su•per•pow•er•ful
su•per•pre•mi•um
su•per•pro
su•per•prof•it
su•per•qual•i•ty
su•per•race
su•per•real
su•per•re•al•ism
su•per•re•gion•al
su•per•rich
su•per•road
su•per•ro•man•tic
su•per•ro•man•ti•cism
su•per•safe
su•per•sale
su•per•sales•man
su•per•scale
su•per•school
su•per•scout
su•per•se•cre•cy
su•per•se•cret
su•per•sell
su•per•sell•er
su•per•sex•u•al•i•ty
su•per•sharp
su•per•show
su•per•sing•er
su•per•size
su•per•sized
su•per•sleuth
su•per•slick
su•per•smart
su•per•smooth
su•per•soft
su•per•so•phis•ti•cat•ed
su•per•spe•cial
su•per•spe•cial•ist
su•per•spe•cial•i•za•tion
su•per•spe•cial•ized
su•per•spec•ta•cle
su•per•spec•tac•u•lar
su•per•spec•u•la•tion
su•per•spy
su•per•state
su•per•sta•tion
su•per•stim•u•late
su•per•stock
su•per•stra•tum
su•per•strength
su•per•strike
su•per•strong
su•per•stud
su•per•sub•tle
su•per•sub•tle•ty
su•per•sur•geon
su•per•sweet
su•per•tank•er
su•per•ter•rif•ic
su•per•thick
su•per•thin
su•per•thril•ler
su•per•tight
su•per•vir•ile
su•per•vir•tu•o•so
su•per•wave
su•per•weap•on
su•per•wide
su•per•wife

su•per•a•ble \'sü-p(ə-)rə-bəl\ *adjective* [Latin *superabilis,* from *superare* to surmount — more at INSUPERABLE] (1629)
: capable of being overcome or conquered
— **su•per•a•ble•ness** *noun*
— **su•per•a•bly** \-blē\ *adverb*

su•per•abound \ˌsü-pər-ə-'bau̇nd\ *intransitive verb* [Middle English, from Late Latin *superabundare,* from Latin *super-* + *abundare* to abound] (14th century)
: to abound or prevail in greater measure or to excess

su•per•abun•dant \-'bən-dənt\ *adjective* [Middle English, from Late Latin *superabundant-, superabundans,* from present participle of *superabundare*] (15th century)
: more than ample **:** EXCESSIVE
— **su•per•abun•dance** \-dən(t)s\ *noun*
— **su•per•abun•dant•ly** *adverb*

su•per•add \ˌsü-pər-'ad\ *transitive verb* [Middle English, from Latin *superaddere,* from *super-* + *addere* to add] (15th century)
: to add especially in a way that compounds an effect ⟨the loss of his job was *superadded* to the loss of his house⟩

su•per•agen•cy \'sü-pər-ˌā-jən(t)-sē\ *noun* (1943)
: a large complex governmental agency especially when set up to supervise other agencies

su•per•al•loy \ˌsü-pər-'a-ˌlȯi, -ə-'lȯi\ *noun* (1948)
: any of various high-strength often complex alloys resistant to high temperature

su•per•al•tern \ˌsü-pər-'ȯl-tərn\ *noun* [*super-* + sub*altern*] (1921)
: a universal proposition in traditional logic that is a ground for the immediate inference of a corresponding subaltern

su•per•an•nu•ate \ˌsü-pər-'an-yə-ˌwāt\ *verb* **-at•ed; -at•ing** [back-formation from *superannuated*] (1649)
transitive verb
1 : to make, declare, or prove obsolete or out-of-date
2 : to retire and pension because of age or infirmity
intransitive verb
1 : to become retired
2 : to become antiquated
— **su•per•an•nu•a•tion** \-ˌan-yə-'wā-shən\ *noun*

superannuated *adjective* [Medieval Latin *superannuatus,* past participle of *superannuari* to be too old, from Latin *super-* + *annus* year — more at ANNUAL] (1740)
: incapacitated or disqualified for active duty by advanced age

su•perb \su̇-'pərb\ *adjective* [Latin *superbus* excellent, proud, from *super* above + *-bus* (akin to Old English *bēon* to be) — more at OVER, BE] (1549)
: marked to the highest degree by grandeur, excellence, brilliance, or competence
synonym see SPLENDID
— **su•perb•ly** *adverb*
— **su•perb•ness** *noun*

Su•per•ball \'sü-pər-ˌbȯl\ *trademark*
— used for a toy rubber ball with a high bounce

su•per•block \-ˌbläk\ *noun* (1928)
: a very large commercial or residential block barred to through traffic, crossed by pedestrian walks and sometimes access roads, and often spotted with grassed malls

Super Bowl *noun* [from the *Super Bowl,* annual championship game of the National Football League] (1969)
: a contest or event that is the most important or prestigious of its kind ⟨the 2,500-mile, 24-day *Super Bowl* of bicycle racing —John Krakauer⟩

¹su•per•cal•en•der \'sü-pər-ˌka-lən-dər\ *transitive verb* (1888)
: to process (paper) in a supercalender

²supercalender *noun* (1894)
: a stack of highly polished calender rolls used to give an extra finish to paper

su•per•car•go \ˌsü-pər-'kär-(ˌ)gō, 'sü-pər-ˌ\ *noun* [Spanish *sobrecargo,* from *sobre-* over (from Latin *super-*) + *cargo* cargo] (1697)
: an officer on a merchant ship in charge of the commercial concerns of the voyage

supercede *variant of* SUPERSEDE

su•per•charge \'sü-pər-ˌchärj\ *transitive verb* (1876)
1 : to charge greatly or excessively (as with vigor or tension)
2 : to supply a charge to the intake of (as an engine) at a pressure higher than that of the surrounding atmosphere
3 : PRESSURIZE 1
— **supercharge** *noun*

su•per•charg•er \-ˌchär-jər\ *noun* (1921)
: a device (as a blower or compressor) for pressurizing the cabin of an airplane or for increasing the volume air charge of an internal combustion engine over that which would normally be drawn in through the pumping action of the pistons

su•per•cil•i•ary \ˌsü-pər-'si-lē-ˌer-ē\ *adjective* [New Latin *superciliaris,* from Latin *supercilium*] (1732)
: of, relating to, or adjoining the eyebrow **:** SUPRAORBITAL

su•per•cil•ious \-'si-lē-əs, -'sil-yəs\ *adjective* [Latin *superciliosus,* from *supercilium* eyebrow, haughtiness, from *super-* + *-cilium* eyelid (akin to *celare* to hide) — more at HELL] (1598)
: coolly and patronizingly haughty ◆
synonym see PROUD
— **su•per•cil•ious•ly** *adverb*
— **su•per•cil•ious•ness** *noun*

su•per•city \'sü-pər-ˌsi-tē\ *noun* (1925)
: MEGALOPOLIS

su•per•class \-ˌklas\ *noun* (circa 1891)
: a category in biological classification ranking below a phylum or division and above a class

su•per•clu•ster \'sü-pər-ˌkləs-tər\ *noun* (1926)
: a group of gravitationally associated clusters of galaxies

su•per•coil \-ˌkȯi(ə)l\ *noun* (1965)
: SUPERHELIX
— **supercoil** *verb*

su•per•col•lid•er \-kə-ˌlī-dər\ *noun* (1984)
: a very large collider capable of accelerating particles to very high energies

su•per•com•put•er \-kəm-ˌpyü-tər\ *noun* (1968)
: a large very fast mainframe used especially for scientific computations

su•per•con•duct \ˌsü-pər-kən-'dəkt\ *intransitive verb* (1952)
: to exhibit superconductivity
— **su•per•con•duc•tive** \-'dək-tiv\ *adjective*

su•per•con•duc•tiv•i•ty \-ˌkän-ˌdək-'ti-və-tē, -kən-\ *noun* (1913)
: a complete disappearance of electrical resistance in a substance especially at very low temperatures
— **su•per•con•duc•tor** \-kən-'dək-tər\ *noun*

su•per•con•ti•nent \'sü-pər-ˌkän-tᵊn-ənt, -ˌkänt-nənt\ *noun* (1960)
: a former large continent from which other continents are held to have broken off and drifted away

su•per•cool \ˌsü-pər-'kül\ (1906)
transitive verb
: to cool below the freezing point without solidification or crystallization
intransitive verb
: to become supercooled

su•per•crit•i•cal \-'kri-ti-kəl\ *adjective* (1934)
: being or having a temperature above a critical temperature ⟨*supercritical* fluid⟩

su•per•cur•rent \'sü-pər-ˌkər-ənt, -ˌkə-rənt\ *noun* (1940)
: a current of electricity flowing in a superconductor

su•per–du•per \'sü-pər-'dü-pər\ *adjective* [reduplication of ¹*super*] (1940)
: of the greatest excellence, size, effectiveness, or impressiveness

su•per•ego \ˌsü-pər-'ē-(ˌ)gō *also* -'e-(ˌ)gō\ *noun* [New Latin, translation of German *Über-ich,* from *über* over + *ich* I] (1919)
: the one of the three divisions of the psyche in psychoanalytic theory that is only partly conscious, represents internalization of parental conscience and the rules of society, and functions to reward and punish through a system of moral attitudes, conscience, and a sense of guilt — compare EGO, ID

su•per•el•e•vate \-'e-lə-ˌvāt\ *transitive verb* (circa 1945)
: BANK 1c

su•per•el•e•va•tion \-ˌe-lə-'vā-shən\ *noun* (1889)

◇ WORD HISTORY
supercilious Anyone who has ever been on the receiving end of disapprovingly raised eyebrows will have no trouble understanding the etymology of *supercilious.* The Latin word for "eyebrow" was *supercilium,* formed from *super-* "above, over" and *cilium* "eyelid." The raising of the eyebrows to express an attitude of haughty disdain was evidently as well-known in ancient times as it is now, inasmuch as Latin *supercilium* was also used figuratively to mean "haughtiness, disdainful superiority." The derivative adjective in Latin was *superciliosus,* which was adopted in English as *supercilious* in the 16th century.

1 : the vertical distance between the heights of inner and outer edges of highway pavement or railroad rails
2 : additional elevation

su·per·er·o·ga·tion \ˌsü-pə-ˌrer-ə-ˈgā-shən\ *noun* [Medieval Latin *supererogation-, supererogatio,* from *supererogare* to perform beyond the call of duty, from Late Latin, to expend in addition, from Latin *super-* + *erogare* to expend public funds after asking the consent of the people, from *e-* + *rogare* to ask — more at RIGHT] (1526)
: the act of performing more than is required by duty, obligation, or need

su·per·erog·a·to·ry \ˌsü-pə-ri-ˈrä-gə-ˌtōr-ē, -ˌtȯr-\ *adjective* (1593)
1 : observed or performed to an extent not enjoined or required
2 : exceeding what is needed **:** SUPERFLUOUS

su·per·fam·i·ly \ˈsü-pər-ˌfam-lē, -ˈfa-mə-\ *noun* (circa 1890)
: a category of biological classification ranking below an order and above a family

su·per·fat·ted \-ˌfa-təd\ *adjective* (1891)
: containing extra oil or fat ⟨*superfatted* soap⟩

su·per·fe·cun·da·tion \ˌsü-pər-ˌfe-kən-ˈdā-shən, -ˌfē-\ *noun* (circa 1855)
1 : successive fertilization of two or more ova from the same ovulation especially by different sires
2 : fertilization at one time of a number of ova excessive for the species

su·per·fe·ta·tion \-fē-ˈtā-shən\ *noun* [Medieval Latin *superfetation-, superfetatio,* from Latin *superfetare* to conceive while already pregnant, from *super-* + *fetus* act of bearing young, offspring — more at FETUS] (1603)
: a progressive accumulation or accretion reaching an extreme or excessive degree

su·per·fi·cial \ˌsü-pər-ˈfi-shəl\ *adjective* [Middle English, from Late Latin *superficialis,* from Latin *superficies*] (15th century)
1 a (1) **:** of or relating to a surface (2) **:** lying on, not penetrating below, or affecting only the surface ⟨*superficial* wounds⟩ **b** *British, of a unit of measure* **:** SQUARE ⟨*superficial* foot⟩
2 a : concerned only with the obvious or apparent **:** SHALLOW **b :** lying on the surface **:** EXTERNAL **c :** presenting only an appearance without substance or significance ☆
— su·per·fi·cial·ly \-ˈfi-sh(ə-)lē\ *adverb*

superficial fascia *noun* (1876)
: the thin layer of loose fatty connective tissue underlying the skin and binding it to the parts beneath — called also *hypodermis*

su·per·fi·ci·al·i·ty \ˌsü-pər-ˌfi-shē-ˈa-lə-tē\ *noun, plural* **-ties** (1530)
1 : the quality or state of being superficial
2 : something superficial

su·per·fi·cies \-ˈfi-(ˌ)shēz, -shē-ˌēz\ *noun, plural* **superficies** [Latin, surface, from *super-* + *facies* face, aspect — more at FACE] (1530)
1 : a surface of a body or a region of space
2 : the external aspects or appearance of a thing

su·per·fine \ˌsü-pər-ˈfīn\ *adjective* (1575)
1 : overly refined or nice
2 : of extremely fine size or texture ⟨*superfine* toothbrush bristles⟩ ⟨*superfine* sugar⟩
3 : of high quality or grade — used especially of merchandise

su·per·fix \ˈsü-pər-ˌfiks\ *noun* [*super-* + *-fix* (as in *prefix*)] (circa 1948)
: a morpheme consisting of a pattern of stress, intonation, or juncture features that are associated with the syllables of a word or phrase (as the distinctive stress patterns of the noun *subject* and the verb *subject*)

su·per·flu·id \ˌsü-pər-ˈflü-əd\ *noun* (1938)
: an unusual state of matter noted only in liquid helium cooled to near absolute zero and characterized by apparently frictionless flow (as through fine holes)
— superfluid *adjective*
— su·per·flu·id·i·ty \-flü-ˈi-də-tē\ *noun*

su·per·flu·i·ty \ˌsü-pər-ˈflü-ə-tē\ *noun, plural* **-ties** [Middle English *superfluitee,* from Middle French *superfluité,* from Late Latin *superfluitat-, superfluitas,* from Latin *superfluus*] (14th century)
1 a : EXCESS, OVERSUPPLY **b :** something unnecessary or superfluous
2 : immoderate and especially luxurious living, habits, or desires

su·per·flu·ous \sü-ˈpər-flü-əs\ *adjective* [Middle English, from Latin *superfluus,* literally, running over, from *superfluere* to overflow, from *super-* + *fluere* to flow — more at FLUID] (15th century)
1 a : exceeding what is sufficient or necessary **:** EXTRA **b :** not needed **:** UNNECESSARY
2 *obsolete* **:** marked by wastefulness **:** EXTRAVAGANT
— su·per·flu·ous·ly *adverb*
— su·per·flu·ous·ness *noun*

su·per·gene \ˈsü-pər-ˌjēn\ *noun* (circa 1949)
: a group of linked genes acting as an allelic unit especially when due to the suppression of crossing-over

su·per·gi·ant \-ˌjī-ənt\ *noun* (1926)
: something that is extremely large; *especially* **:** a star of very great intrinsic luminosity and enormous size
— supergiant *adjective*

su·per·glue *noun* (1946)
: a very strong glue; *specifically* **:** a glue whose chief ingredient is a cyanoacrylate that becomes adhesive through polymerization rather than evaporation of a solvent

su·per·graph·ics \-ˌgra-fiks\ *noun plural but singular or plural in construction* (1969)
: billboard-sized graphic shapes usually of bright color and simple design

su·per·grav·i·ty \ˌsü-pər-ˈgra-və-tē\ *noun* (1976)
: any of various theories in physics that are based on supersymmetry and attempt to unify general relativity and quantum theory and that state that the principle transmitter of gravity is the graviton

su·per·group \ˈsü-pər-ˌgrüp\ *noun* (1968)
: a rock group made up of prominent former members of other rock groups; *also* **:** an extremely successful rock group

[1]**su·per·heat** \ˌsü-pər-ˈhēt\ *transitive verb* (1859)
1 : to heat (a vapor not in contact with its own liquid) so as to cause to remain free from suspended liquid droplets ⟨*superheated* steam⟩
2 : to heat (a liquid) above the boiling point without converting into vapor
— su·per·heat·er *noun*

[2]**su·per·heat** \ˈsü-pər-ˌhēt, ˌsü-pər-ˈ\ *noun* (1884)
: the extra heat imparted to a vapor in superheating it from a dry and saturated condition; *also* **:** the corresponding rise of temperature

su·per·heat·ed \-ˌhē-təd, -ˈhē-\ *adjective* (1857)
: very hot; *also* **:** exceedingly emotional or intense

su·per·hea·vy·weight \ˌsü-pər-ˈhe-vē-ˌwāt\ *noun* (1971)
: an athlete (as an Olympic weightlifter, boxer, or wrestler) who competes in the heaviest class or division

su·per·he·lix \ˈsü-pər-ˌhē-liks\ *noun* (1964)
: a helix (as of DNA) which has its axis arranged in a helical coil
— su·per·he·li·cal \ˌsü-pər-ˈhe-li-kəl, -ˈhē-\ *adjective*

su·per·hero \-ˌhir-(ˌ)ō, -ˌhē-(ˌ)rō\ *noun* (1917)
: a fictional hero having extraordinary or superhuman powers; *also* **:** an exceptionally skillful or successful person

su·per·het·ero·dyne \ˌsü-pər-ˈhe-tə-rə-ˌdīn, -ˈhe-trə-\ *adjective* [*super*sonic + *heterodyne*] (1922)
: used in or being a radio receiver in which an incoming signal is mixed with a locally generated frequency to produce an ultrasonic signal at a fixed frequency that is then rectified, amplified, and rectified again to reproduce the sound
— superheterodyne *noun*

su·per·high frequency \ˈsü-pər-ˌhī-\ *noun* (circa 1945)
: a radio frequency in the next to the highest range of the radio spectrum — see RADIO FREQUENCY table

su·per·high·way \ˌsü-pər-ˈhī-ˌwā, ˈsü-pər-ˌ\ *noun* (1925)
: a multilane highway (as an expressway or turnpike) designed for high-speed traffic

su·per·hu·man \ˌsü-pər-ˈhyü-mən, -ˈyü-\ *adjective* (1633)
1 : being above the human **:** DIVINE ⟨*superhuman* beings⟩
2 : exceeding normal human power, size, or capability **:** HERCULEAN ⟨a *superhuman* effort⟩; *also* **:** having such power, size, or capability
— su·per·hu·man·i·ty \-hyü-ˈma-nə-tē, -yü-\ *noun*
— su·per·hu·man·ly \-ˈhyü-mən-lē, -ˈyü-\ *adverb*
— su·per·hu·man·ness \-mən-nəs\ *noun*

su·per·im·pose \ˌsü-pər-im-ˈpōz\ *transitive verb* (1794)
: to place or lay over or above something ⟨*superimposed* images⟩ ⟨*superimposed* a formula on the stories⟩
— su·per·im·pos·able \-ˈpō-zə-bəl\ *adjective*
— su·per·im·po·si·tion \-ˌim-pə-ˈzi-shən\ *noun*

su·per·in·cum·bent \-in-ˈkəm-bənt\ *adjective* [Latin *superincumbent-, superincumbens,* present participle of *superincumbere* to lie on top of, from *super-* + *incumbere* to lie down on — more at INCUMBENT] (1664)
: lying or resting and usually exerting pressure on something else
— su·per·in·cum·bent·ly *adverb*

su·per·in·di·vid·u·al \ˌsü-pər-ˌin-də-ˈvij-wəl, -ˈvi-jə wəl, -ˈvi-jəl\ *adjective* (1916)
: of, relating to, or being an organism, entity, or complex of more than individual complexity or nature

su·per·in·duce \-in-ˈdüs, -ˈdyüs\ *transitive verb* [Latin *superinducere,* from *super-* + *inducere* to lead in — more at INDUCE] (circa 1555)
1 : to introduce as an addition over or above something already existing
2 : BRING ON, INDUCE
— su·per·in·duc·tion \-ˈdək-shən\ *noun*

su·per·in·fec·tion \-in-ˈfek-shən\ *noun* (circa 1922)
: reinfection or a second infection with a microbial agent (as a bacterium, fungus, or virus)
— su·per·in·fect \-ˈfekt\ *transitive verb*

su·per·in·tend \ˌsü-p(ə-)rin-ˈtend, ˌsü-pərn-\ *transitive verb* [Late Latin *superintendere,*

☆ SYNONYMS
Superficial, shallow, cursory mean lacking in depth or solidity. SUPERFICIAL implies a concern only with surface aspects or obvious features ⟨a *superficial* analysis of the problem⟩. SHALLOW is more generally derogatory in implying lack of depth in knowledge, reasoning, emotions, or character ⟨a light, *shallow,* and frivolous review⟩. CURSORY suggests a lack of thoroughness or a neglect of details ⟨gave the letter only a *cursory* reading⟩.

from Latin *super-* + *intendere* to stretch out, direct — more at INTEND] (circa 1615)
: to have or exercise the charge and oversight of **:** DIRECT

su·per·in·ten·dence \-'ten-dən(t)s\ *noun* (1603)
: the act or function of superintending or directing **:** SUPERVISION

su·per·in·ten·den·cy \-dən(t)-sē\ *noun, plural* **-cies** (1598)
: the office, post, or jurisdiction of a superintendent; *also* **:** SUPERINTENDENCE

su·per·in·ten·dent \-dənt\ *noun* [Medieval Latin *superintendent-, superintendens,* from Late Latin, present participle of *superintendere*] (1554)
: one who has executive oversight and charge
— **superintendent** *adjective*

[1]**su·pe·ri·or** \sü-'pir-ē-ər\ *adjective* [Middle English, from Middle French *superieur,* from Latin *superior,* comparative of *superus* upper, from *super* over, above — more at OVER] (14th century)
1 : situated higher up **:** UPPER
2 : of higher rank, quality, or importance
3 : courageously or serenely indifferent (as to something painful or disheartening)
4 a : greater in quantity or numbers ⟨escaped by *superior* speed⟩ **b :** excellent of its kind **:** BETTER ⟨her *superior* memory⟩
5 : being a superscript
6 a *of an animal structure* **:** situated above or anterior or dorsal to another and especially a corresponding part ⟨a *superior* artery⟩ **b** *of a plant structure* **:** situated above or near the top of another part: as (1) *of a calyx* **:** attached to and apparently arising from the ovary (2) *of an ovary* **:** free from the calyx or other floral envelope
7 : more comprehensive ⟨a genus is *superior* to a species⟩
8 : affecting or assuming an air of superiority **:** SUPERCILIOUS
— **su·pe·ri·or·ly** *adverb*

[2]**superior** *noun* (15th century)
1 : one who is above another in rank, station, or office; *especially* **:** the head of a religious house or order
2 : one that surpasses another in quality or merit
3 : SUPERSCRIPT

superior conjunction *noun* (1833)
: a conjunction in which the primary of an orbiting celestial body is aligned between the celestial body and an observer

superior court *noun* (1686)
1 : a court of general jurisdiction intermediate between the inferior courts (as a justice of the peace court) and the higher appellate courts
2 : a court with juries having original jurisdiction

superior general *noun, plural* **superiors general** (1775)
: the superior of a religious order or congregation

su·pe·ri·or·i·ty \sü-,pir-ē-'ör-ə-tē, ,sü-, -'är-\ *noun, plural* **-ties** (15th century)
: the quality or state of being superior; *also* **:** a superior characteristic

superiority complex *noun* (circa 1924)
: an exaggerated opinion of oneself

superior planet *noun* (1583)
: a planet whose orbit lies outside that of the earth

superior vena cava *noun* (1901)
: the branch of the vena cava of a vertebrate that brings blood back from the head and anterior part of the body to the heart

su·per·ja·cent \,sü-pər-'jā-sᵊnt\ *adjective* [Latin *superjacent-, superjacens,* present participle of *superjacēre* to lie over or upon, from *super-* + *jacēre* to lie; akin to Latin *jacere* to throw — more at JET] (1610)
: lying above or upon **:** OVERLYING ⟨*superjacent* rocks⟩

su·per·jet \'sü-pər-,jet\ *noun* (1958)
: a very large jet airplane

[1]**su·per·la·tive** \sü-'pər-lə-tiv\ *adjective* [Middle English *superlatif,* from Middle French, from Late Latin *superlativus,* from Latin *superlatus* (past participle of *superferre* to carry over, raise high), from *super-* + *latus,* past participle of *ferre* to carry — more at TOLERATE, BEAR] (14th century)
1 : of, relating to, or constituting the degree of grammatical comparison that denotes an extreme or unsurpassed level or extent
2 a : surpassing all others **:** SUPREME **b :** of very high quality **:** EXCELLENT
3 : EXCESSIVE, EXAGGERATED
— **su·per·la·tive·ly** *adverb*
— **su·per·la·tive·ness** *noun*

[2]**superlative** *noun* (15th century)
1 a : the superlative degree of comparison in a language **b :** a superlative form of an adjective or adverb
2 : the superlative or utmost degree of something **:** ACME
3 : a superlative person or thing
4 : an admiring sometimes exaggerated expression especially of praise ■

su·per·lin·er \'sü-pər-,lī-nər\ *noun* (1919)
: a fast luxurious passenger liner of great size

su·per·lu·na·ry \,sü-pər-'lü-nə-rē\ *or* **su·per·lu·nar** \-nər, -,när\ *adjective* [Latin *super-* + *luna* moon — more at LUNAR] (1614)
archaic **:** being above the moon **:** CELESTIAL

su·per·man \'sü-pər-,man\ *noun* [translation of German *Übermensch,* from *über* over, super- + *Mensch* man] (1903)
1 : a superior man that according to Nietzsche has learned to forgo fleeting pleasures and attain happiness and dominance through the exercise of creative power
2 : a person of extraordinary or superhuman power or achievements

su·per·mar·ket \-,mär-kət\ *noun, often attributive* (1933)
1 : a self-service retail market selling especially foods and household merchandise
2 : something resembling a supermarket especially in the variety or volume of its goods or services

su·per·mi·cro \-,mī-(,)krō\ *noun* (1982)
: a very fast and powerful microcomputer

su·per·mini·com·put·er \-'mi-nē-kəm-,pyü-tər\ *noun* (1980)
: a very fast and powerful minicomputer — called also *supermini*

su·per·nal \sü-'pər-nᵊl\ *adjective* [Middle English, from Middle French, from Latin *supernus,* from *super* over, above — more at OVER] (15th century)
1 a : being or coming from on high **b :** HEAVENLY, ETHEREAL ⟨*supernal* melodies⟩ **c :** superlatively good ⟨*supernal* trumpet playing⟩
2 : located in or belonging to the sky
— **su·per·nal·ly** \-nᵊl-ē\ *adverb*

su·per·na·tant \,sü-pər-'nā-tᵊnt\ *noun* [Latin *supernatant-, supernatans,* present participle of *supernatare* to float, from *super-* + *natare* to swim — more at NATANT] (1922)
: the usually clear liquid overlying material deposited by settling, precipitation, or centrifugation
— **supernatant** *adjective*

su·per·nat·u·ral \,sü-pər-'na-chə-rəl, -'nach-rəl\ *adjective* [Medieval Latin *supernaturalis,* from Latin *super-* + *natura* nature] (15th century)
1 : of or relating to an order of existence beyond the visible observable universe; *especially* **:** of or relating to God or a god, demigod, spirit, or devil
2 a : departing from what is usual or normal especially so as to appear to transcend the laws of nature **b :** attributed to an invisible agent (as a ghost or spirit)
— **supernatural** *noun*
— **su·per·nat·u·ral·ly** \-'na-chər-ə-lē, -'nach-rə-, -'na-chər-lē\ *adverb*
— **su·per·nat·u·ral·ness** *noun*

su·per·nat·u·ral·ism \,sü-pər-'na-chə-rə-,li-zəm, -'nach-rə-\ *noun* (1799)
1 : the quality or state of being supernatural
2 : belief in a supernatural power and order of existence
— **su·per·nat·u·ral·ist** \-list\ *noun or adjective*
— **su·per·nat·u·ral·is·tic** \-,na-chə-rə-'lis-tik, -,nach-rə-\ *adjective*

su·per·na·ture \'sü-pər-,nā-chər\ *noun* [back-formation from *supernatural*] (1844)
: the realm of the supernatural

su·per·nor·mal \,sü-pər-'nör-məl\ *adjective* (1868)
1 : exceeding the normal or average
2 : being beyond normal human powers **:** PARANORMAL
— **su·per·nor·mal·i·ty** \-nör-'ma-lə-tē\ *noun*
— **su·per·nor·mal·ly** \-'nör-mə-lē\ *adverb*

su·per·no·va \,sü-pər-'nō-və\ *noun* [New Latin] (1926)
1 : the explosion of a very large star in which the star may reach a maximum intrinsic luminosity one billion times that of the sun
2 : one that explodes into prominence or popularity; *also* **:** SUPERSTAR

[1]**su·per·nu·mer·ary** \,sü-pər-'nü-mə-,rer-ē, -'nyü-, -mə-rē; -'n(y)üm-rē\ *adjective* [Late Latin *supernumerarius,* from Latin *super-* + *numerus* number] (1605)
1 a : exceeding the usual, stated, or prescribed number ⟨a *supernumerary* tooth⟩ **b :** not enumerated among the regular components of a group and especially of a military organization
2 : exceeding what is necessary, required, or desired
3 : more numerous

[2]**supernumerary** *noun, plural* **-ar·ies** (1639)
1 : a supernumerary person or thing
2 : an actor employed to play a walk-on

su·per·or·der \'sü-pər-,ör-dər\ *noun* (circa 1890)
: a category of biological classification ranking below a class and above an order

su·per·or·di·nate \,sü-pər-'örd-nət, -'ör-dᵊn-ət, -'ör-dᵊn-,āt\ *adjective* [*super-* + *subordinate*] (1620)
: superior in rank, class, or status

□ USAGE
superlative "Miss Fletcher and I were very thick, but I am the thinnest of the two," wrote Jane Austen in 1796, unmindful of having committed one of the newest improprieties of the grammarians, using the superlative of two. Just how this subject became a staple of handbooks and schoolbooks for two centuries is not entirely clear, for two of the earliest and most widely copied commentators found it a trivial matter. However, one man's preference became another man's rule soon enough, and we can still find warnings against using the superlative of two in present-day handbooks for student writers. Jane Austen's little joke is perfectly understandable even though she used the superlative, and it does not become any more understandable (or any funnier) if the comparative is substituted for the superlative. Thus the rule serves no useful purpose. Nor has it reflected real usage accurately; examples of the superlative used of two abound in literature from Shakespeare to the present. When a rule serves no purpose other than to divide those who observe it from those who don't, it has little authority.

su·per·or·gan·ism \-ˈór-gə-ˌni-zəm\ *noun* (circa 1899)
: an organized society (as of a social insect) that functions as an organic whole

su·per·ovu·la·tion \-ˌäv-yə-ˈlā-shən, -ˌōv-\ *noun* (1927)
: production of exceptional numbers of ova at one time
— **su·per·ovu·late** \-ˈäv-yə-ˌlāt, -ˈōv-\ *verb*

su·per·ox·ide \-ˈäk-ˌsīd\ *noun* (circa 1847)
: the univalent anion O_2^- or a compound containing it ⟨potassium *superoxide* KO_2⟩

superoxide dis·mut·ase \-dis-ˈmyü-ˌtās, -ˌtāz\ *noun* [*dismut*ation (simultaneous oxidation and reduction) + *-ase*] (1969)
: a metal-containing enzyme that reduces potentially harmful free radicals of oxygen formed during normal metabolic cell processes to oxygen and hydrogen peroxide

su·per·par·a·sit·ism \ˌsü-pər-ˈpar-ə-ˌsī-ˌti-zəm, -sə-\ *noun* (circa 1899)
: parasitization of a host by more than one parasitic individual usually of one kind — used especially of parasitic insects

su·per·phos·phate \ˌsü-pər-ˈfäs-ˌfāt\ *noun* (1797)
1 : an acid phosphate
2 : a soluble mixture of phosphates used as fertilizer and made from insoluble mineral phosphates by treatment with sulfuric acid

su·per·phys·i·cal \-ˈfi-zi-kəl\ *adjective* (circa 1603)
: being above or beyond the physical world or explanation on physical principles

su·per·plas·tic \-ˈplas-tik\ *adjective* (1947)
1 : capable of plastic deformation under low stress at an elevated temperature — used of metals and alloys
2 : of or relating to superplastic materials ⟨*superplastic* forming⟩
— **su·per·plas·tic·i·ty** \-pla-ˈsti-sə-tē\ *noun*

su·per·pose \ˌsü-pər-ˈpōz\ *transitive verb* **-posed; -pos·ing** [probably from French *superposer,* back-formation from *superposition,* from Late Latin *superposition-, superpositio,* from Latin *superponere* to superpose, from *super-* + *ponere* to place — more at POSITION] (1823)
1 : to place or lay over or above whether in or not in contact **:** SUPERIMPOSE
2 : to lay (as a geometric figure) upon another so as to make all like parts coincide
— **su·per·pos·able** \-ˈpō-zə-bəl\ *adjective*
— **su·per·po·si·tion** \-pə-ˈzi-shən\ *noun*

su·per·posed \-ˈpōzd\ *adjective* (1823)
: situated vertically over another layer or part

su·per·pow·er \ˈsü-pər-ˌpau̇(-ə)r\ *noun* (1922)
1 : excessive or superior power
2 a : an extremely powerful nation; *specifically* **:** one of a very few dominant states in an era when the world is divided politically into these states and their satellites **b :** an international governing body able to enforce its will upon the most powerful states
— **su·per·pow·ered** \-ˌpau̇(-ə)rd\ *adjective*

su·per·sat·u·rate \ˌsü-pər-ˈsa-chə-ˌrāt\ *transitive verb* (1788)
: to add to (a solution) beyond saturation

su·per·sat·u·rat·ed *adjective* (1854)
: containing an amount of a substance greater than that required for saturation as a result of having been cooled from a higher temperature to a temperature below that at which saturation occurs ⟨a *supersaturated* solution⟩ ⟨air *supersaturated* with water vapor⟩

su·per·sat·u·ra·tion \-ˌsa-chə-ˈrā-shən\ *noun* (1836)
: the state of being supersaturated

su·per·scribe \ˈsü-pər-ˌskrīb, ˌsü-pər-ˈ\ *transitive verb* **-scribed; -scrib·ing** [Middle English, from Latin *superscribere,* from *super-* + *scribere* to write — more at SCRIBE] (15th century)
1 : to write (as a name or address) on the outside or cover of **:** ADDRESS
2 : to write or engrave on the top or outside

su·per·script \ˈsü-pər-ˌskript\ *noun* [Latin *superscriptus,* past participle of *superscribere*] (1901)
: a distinguishing symbol (as a numeral or letter) written immediately above or above and to the right or left of another character
— **superscript** *adjective*

su·per·scrip·tion \ˌsü-pər-ˈskrip-shən\ *noun* [Middle English, from Middle French, from Late Latin *superscription-, superscriptio,* from Latin *superscribere*] (14th century)
1 : something written or engraved on the surface of, outside, or above something else **:** INSCRIPTION; *also* **:** ADDRESS
2 : the act of superscribing

su·per·sede \ˌsü-pər-ˈsēd\ *transitive verb* **-sed·ed; -sed·ing** [Middle English *superceden,* from Middle French *superseder* to refrain from, from Latin *supersedēre* to be superior to, refrain from, from *super-* + *sedēre* to sit — more at SIT] (15th century)
1 a : to cause to be set aside **b :** to force out of use as inferior
2 : to take the place, room, or position of
3 : to displace in favor of another **:** SUPPLANT
synonym see REPLACE
— **su·per·sed·er** *noun*

su·per·se·de·as \-ˈsē-dē-əs\ *noun, plural* **supersedeas** [Middle English, from Latin, you shall refrain, from *supersedēre*] (14th century)
1 : a common-law writ commanding a stay of legal proceedings that is issued under various conditions and especially to stay an officer from proceeding under another writ
2 : an order staying proceedings of an inferior court

su·per·se·dure \-ˈsē-jər\ *noun* (1788)
: the act or process of superseding; *especially* **:** the replacement of an old or inferior queen bee by a young or superior queen

su·per·sen·si·ble \ˌsü-pər-ˈsen(t)-sə-bəl\ *adjective* (1798)
: being above or beyond that which is apparent to the senses **:** SPIRITUAL

su·per·sen·si·tive \-ˈsen(t)-sə-tiv, -ˈsen(t)-stiv\ *adjective* (1839)
: HYPERSENSITIVE ⟨a *supersensitive* palate⟩
— **su·per·sen·si·tive·ly** *adverb*
— **su·per·sen·si·tiv·i·ty** \-ˌsen(t)-sə-ˈti-və-tē\ *noun*

su·per·sen·so·ry \-ˈsen(t)-sə-rē, -ˈsen(t)s-rē\ *adjective* (1883)
: SUPERSENSIBLE

su·per·ser·vice·able \-ˈsər-və-sə-bəl\ *adjective* (1605)
: offering unwanted services **:** OFFICIOUS

su·per·ses·sion \ˌsü-pər-ˈse-shən\ *noun* [Medieval Latin *supersession-, supersessio,* from Latin *supersedēre*] (1790)
: the act of superseding **:** the state of being superseded

[1]**su·per·son·ic** \-ˈsä-nik\ *adjective* [Latin *super-* + *sonus* sound — more at SOUND] (1919)
1 : ULTRASONIC
2 : of, being, or relating to speeds from one to five times the speed of sound in air — compare SONIC
3 : moving, capable of moving, or utilizing air currents moving at supersonic speed
4 : relating to supersonic airplanes or missiles ⟨the *supersonic* age⟩
— **su·per·son·i·cal·ly** \-ni-k(ə-)lē\ *adverb*

[2]**supersonic** *noun* (circa 1924)
1 : a supersonic wave or frequency
2 : a supersonic airplane

su·per·son·ics \ˌsü-pər-ˈsä-niks\ *noun plural but singular in construction* (1925)
: the science of supersonic phenomena

supersonic transport *noun* (1961)
: a supersonic transport airplane

su·per·star \ˈsü-pər-ˌstär\ *noun* (1924)
1 : a star (as in sports or the movies) who is considered extremely talented, has great public appeal, and can usually command a high salary
2 : one that is very prominent or is a prime attraction
— **su·per·star·dom** \-dəm\ *noun*

su·per·sti·tion \ˌsü-pər-ˈsti-shən\ *noun* [Middle English *supersticion,* from Middle French, from Latin *superstition-, superstitio,* from *superstit-, superstes* standing over (as witness or survivor), from *super-* + *stare* to stand — more at STAND] (13th century)
1 a : a belief or practice resulting from ignorance, fear of the unknown, trust in magic or chance, or a false conception of causation **b :** an irrational abject attitude of mind toward the supernatural, nature, or God resulting from superstition
2 : a notion maintained despite evidence to the contrary

su·per·sti·tious \-ˈsti-shəs\ *adjective* [Middle English *supersticious,* from Middle French *supersticieux,* from Latin *superstitiosus,* from *superstitio*] (14th century)
: of, relating to, or swayed by superstition
— **su·per·sti·tious·ly** *adverb*

su·per·store \ˈsü-pər-ˌstōr, -ˌstȯr\ *noun* (1943)
: a very large store offering a wide variety of merchandise for sale

su·per·string \-ˌstriŋ\ *noun* (1983)
: a hypothetical string obeying the rules of supersymmetry whose vibrations manifest themselves as particles existing in ten dimensions of which only four are evident

su·per·struc·ture \-ˌstrək-chər\ *noun* (1641)
1 a : an entity, concept, or complex based on a more fundamental one **b :** social institutions (as the law or politics) that are in Marxist theory erected upon the economic base
2 : a structure built as a vertical extension of something else: as **a :** all of a building above the basement **b :** the structural part of a ship above the main deck
— **su·per·struc·tur·al** \ˌsü-pər-ˈstrək-chə-rəl, -ˈstrək-shrəl\ *adjective*

su·per·sub·stan·tial \ˌsü-pər-səb-ˈstan(t)-shəl\ *adjective* [Late Latin *supersubstantialis,* from Latin *super-* + *substantia* substance] (1534)
: being above material substance **:** of a transcending substance

su·per·symmetry \-ˈsi-mə-trē\ *noun* (1974)
: the correspondence between fermions and bosons of identical mass that is postulated to have existed during the opening moments of the big bang and that relates gravity to the other forces of nature
— **su·per·sym·met·ric** \-sə-ˈme-trik\ *adjective*

su·per·sys·tem \ˈsü-pər-ˌsis-təm\ *noun* (circa 1928)
: a system that is made up of systems

su·per·tax \-ˌtaks\ *noun* (1906)
: SURTAX

su·per·ton·ic \ˌsü-pər-ˈtä-nik\ *noun* (1806)
: the second tone of a diatonic scale

su·per·vene \ˌsü-pər-ˈvēn\ *intransitive verb* **-vened; -ven·ing** [Latin *supervenire,* from *super-* + *venire* to come — more at COME] (circa 1648)

\ə\ **abut** \ᵊ\ **kitten** \ər\ **further** \a\ **ash** \ā\ **ace** \ä\ **mop, mar** \au̇\ **out** \ch\ **chin** \e\ **bet** \ē\ **easy** \g\ **go** \i\ **hit** \ī\ **ice** \j\ **job** \ŋ\ **sing** \ō\ **go** \ȯ\ **law** \ȯi\ **boy** \th\ **thin** \t͟h\ **the** \ü\ **loot** \u̇\ **foot** \y\ **yet** \zh\ **vision** *see also* Guide to Pronunciation

: to follow or result as an additional, adventitious, or unlooked-for development
synonym see FOLLOW
— **su·per·ven·tion** \-'ven(t)-shən\ *noun*
su·per·ve·nient \-'vē-nyənt\ *adjective* [Latin *supervenient-, superveniens,* present participle of *supervenire*] (1594)
: coming or occurring as something additional, extraneous, or unexpected
su·per·vise \'sü-pər-ˌvīz\ *transitive verb* **-vised; -vis·ing** [Medieval Latin *supervisus,* past participle of *supervidēre,* from Latin *super-* + *vidēre* to see — more at WIT] (1588)
: SUPERINTEND, OVERSEE
su·per·vi·sion \ˌsü-pər-'vi-zhən\ *noun* (1640)
: the action, process, or occupation of supervising; *especially* : a critical watching and directing (as of activities or a course of action)
su·per·vi·sor \'sü-pər-ˌvī-zər\ *noun* (15th century)
: one that supervises; *especially* : an administrative officer in charge of a business, government, or school unit or operation
— **su·per·vi·so·ry** \ˌsü-pər-'vī-zə-rē, -'vīz-rē\ *adjective*
su·per·wom·an \'sü-pər-ˌwu̇-mən\ *noun* (1906)
: an exceptional woman; *especially* : a woman who succeeds in having a career and raising a family
su·pi·nate \'sü-pə-ˌnāt\ *verb* **-nat·ed; -nat·ing** [Latin *supinatus,* past participle of *supinare* to lay backward or on the back, from *supinus*] (1831)
transitive verb
: to cause to undergo supination
intransitive verb
: to undergo supination
su·pi·na·tion \ˌsü-pə-'nā-shən\ *noun* (1666)
1 : rotation of the forearm and hand so that the palm faces forward or upward and the radius lies parallel to the ulna; *also* : a corresponding movement of the foot and leg
2 : the position resulting from supination
su·pi·na·tor \'sü-pə-ˌnā-tər\ *noun* [New Latin, from Latin *supinare*] (1615)
: a muscle that produces the motion of supination
¹su·pine \'sü-ˌpīn\ *noun* [Middle English *supyn,* from Late Latin *supinum,* from Latin, neuter of *supinus,* adjective] (15th century)
1 : a Latin verbal noun having an accusative of purpose in *-um* and an ablative of specification in *-u*
2 : an English infinitive with *to*
²su·pine \su̇-'pīn, *attributive also* 'sü-ˌpīn\ *adjective* [Middle English *suppyne,* from Latin *supinus;* akin to Latin *sub* under, up to — more at UP] (15th century)
1 a : lying on the back or with the face upward **b :** marked by supination
2 : exhibiting indolent or apathetic inertia or passivity; *especially* : mentally or morally slack
3 *archaic* **:** leaning or sloping backward
synonym see PRONE, INACTIVE
usage see PRONE
— **su·pine·ly** \su̇-'pīn-lē\ *adverb*
— **su·pine·ness** \-'pīn-nəs\ *noun*
sup·per \'sə-pər\ *noun* [Middle English, from Old French *souper,* from *souper* to sup — more at SUP] (13th century)
1 a : the evening meal especially when dinner is taken at midday **b :** a social affair featuring a supper; *especially* : an evening social especially for raising funds ⟨a church *supper*⟩
2 : the food served as a supper ⟨eat your *supper*⟩
3 : a light meal served late in the evening
supper club *noun* (1925)
: NIGHTCLUB
sup·plant \sə-'plant\ *transitive verb* [Middle English, from Middle French *supplanter,* from Latin *supplantare* to overthrow by tripping up, from *sub-* + *planta* sole of the foot — more at PLACE] (14th century)
1 : to supersede (another) especially by force or treachery
2 a (1) *obsolete* **:** UPROOT (2) **:** to eradicate and supply a substitute for ⟨efforts to *supplant* the vernacular⟩ **b :** to take the place of and serve as a substitute for especially by reason of superior excellence or power
synonym see REPLACE
— **sup·plan·ta·tion** \(ˌ)sə-ˌplan-'tā-shən\ *noun*
— **sup·plant·er** \sə-'plan-tər\ *noun*
¹sup·ple \'sə-pəl *also* 'sü-\ *adjective* **sup·pler** \-p(ə-)lər\; **sup·plest** \-p(ə-)ləst\ [Middle English *souple,* from Middle French, from Latin *supplic-, supplex* entreating for mercy, suppliant, perhaps from *sub-* + *-plic-* (akin to *plicare* to fold) — more at PLY] (14th century)
1 a : compliant often to the point of obsequiousness **b :** readily adaptable or responsive to new situations
2 a : capable of being bent or folded without creases, cracks, or breaks **:** PLIANT ⟨*supple* leather⟩ **b :** able to perform bending or twisting movements with ease and grace **:** LIMBER ⟨*supple* legs of a dancer⟩ **c :** easy and fluent without stiffness or awkwardness ⟨sang with a lively, *supple* voice —Douglas Watt⟩
synonym see ELASTIC
— **sup·ple·ly** \-pə(l)-lē\ *or* **sup·ply** \-p(ə-)lē\ *adverb*
— **sup·ple·ness** \-pəl-nəs\ *noun*
²supple *verb* **sup·pled; sup·pling** \-p(ə-)liŋ\ (14th century)
transitive verb
1 : to make pacific or complaisant ⟨*supple* the tempers of your race —Laurence Sterne⟩
2 : to alleviate with a salve
3 : to make flexible or pliant
intransitive verb
: to become soft and pliant
sup·ple·jack \'sə-pəl-ˌjak *also* 'sü-\ *noun* (circa 1725)
: any of various woody climbers having tough pliant stems; *especially* : a southern U.S. vine (*Berchemia scandens*) of the buckthorn family
¹sup·ple·ment \'sə-plə-mənt\ *noun* [Middle English, from Latin *supplementum,* from *supplēre* to fill up, complete — more at SUPPLY] (14th century)
1 : something that completes or makes an addition ⟨dietary *supplements*⟩
2 : a part added to or issued as a continuation of a book or periodical to correct errors or make additions
3 : an angle or arc that when added to a given angle or arc equals 180°
²sup·ple·ment \'sə-plə-ˌment\ *transitive verb* (1829)
: to add or serve as a supplement to ⟨does odd jobs to *supplement* his income⟩
— **sup·ple·men·ta·tion** \ˌsə-plə-ˌmen-'tā-shən, -mən-\ *noun*
— **sup·ple·ment·er** \'sə-plə-ˌmen-tər\ *noun*
sup·ple·men·tal \ˌsə-plə-'men-tᵊl\ *adjective* (1605)
1 : serving to supplement
2 : NONSCHEDULED ⟨a *supplemental* airline⟩
— **supplemental** *noun*
sup·ple·men·ta·ry \ˌsə-plə-'men-tə-rē, -'men-trē\ *adjective* (1667)
1 : added or serving as a supplement **:** ADDITIONAL ⟨*supplementary* reading⟩
2 : being or relating to a supplement or a supplementary angle
supplementary angle *noun* (circa 1924)
: one of two angles or arcs whose sum is 180° — usually used in plural
sup·ple·tion \sə-'plē-shən\ *noun* [Medieval Latin *suppletion-, suppletio* act of supplementing, from Latin *supplēre*] (1914)
: the occurrence of phonemically unrelated allomorphs of the same morpheme (as *went* as the past tense of *go* or *better* as the comparative form of *good*)
— **sup·ple·tive** \sə-'plē-tiv, 'sə-plə-\ *adjective*
sup·ple·to·ry \sə-'plē-tə-rē; 'sə-plə-ˌtōr-ē, -ˌtȯr-\ *adjective* [Latin *supplēre*] (1628)
: supplying deficiencies **:** SUPPLEMENTARY
sup·pli·ance \'sə-plē-ən(t)s\ *noun* (circa 1611)
: ENTREATY, SUPPLICATION
¹sup·pli·ant \-ənt\ *noun* [Middle English, from Middle French, from present participle of *supplier* to supplicate, from Latin *supplicare*] (15th century)
: one who supplicates
²suppliant *adjective* [Middle French, present participle] (circa 1586)
1 : humbly imploring **:** ENTREATING ⟨a *suppliant* sinner seeking forgiveness —O. J. Baab⟩
2 : expressing supplication ⟨upraised to the heavens . . . *suppliant* arms —William Styron⟩
— **sup·pli·ant·ly** *adverb*
¹sup·pli·cant \'sə-pli-kənt\ *adjective* (1597)
: SUPPLIANT
²supplicant *noun* (1597)
: SUPPLIANT
sup·pli·cate \'sə-plə-ˌkāt\ *verb* **-cat·ed; -cat·ing** [Middle English, from Latin *supplicatus,* past participle of *supplicare,* from *supplic-, supplex* suppliant — more at SUPPLE] (15th century)
intransitive verb
: to make a humble entreaty; *especially* : to pray to God
transitive verb
1 : to ask humbly and earnestly of
2 : to ask for earnestly and humbly
synonym see BEG
— **sup·pli·ca·tion** \ˌsə-plə-'kā-shən\ *noun*
sup·pli·ca·to·ry \'sə-pli-kə-ˌtōr-ē, -ˌtȯr-\ *adjective* (15th century)
: expressing supplication **:** SUPPLIANT ⟨a *supplicatory* prayer⟩
¹sup·ply \sə-'plī\ *verb* **sup·plied; sup·ply·ing** [Middle English *supplien,* from Middle French *soupleier,* from Latin *supplēre* to fill up, supplement, supply, from *sub-* up + *plēre* to fill — more at SUB-, FULL] (14th century)
transitive verb
1 : to add as a supplement
2 a : to provide for **:** SATISFY ⟨laws by which the material wants of men are *supplied* —*Bulletin of Bates College*⟩ **b :** to make available for use **:** PROVIDE ⟨*supplied* the necessary funds⟩ **c :** to satisfy the needs or wishes of **d :** to furnish (organs, tissues, or cells) with a vital element (as blood or nerve fibers)
3 : to substitute for another in; *specifically* : to serve as a supply in (a church or pulpit)
intransitive verb
: to serve as a supply or substitute
— **sup·pli·er** \-'plī(-ə)r\ *noun*
²supply *noun, plural* **supplies** (15th century)
1 *obsolete* **:** ASSISTANCE, SUCCOR
2 a *obsolete* **:** REINFORCEMENTS — often used in plural **b :** a clergyman filling a vacant pulpit temporarily **c :** the quantity or amount (as of a commodity) needed or available ⟨beer was in short *supply* in that hot weather —Nevil Shute⟩ **d :** PROVISIONS, STORES — usually used in plural
3 : the act or process of filling a want or need ⟨engaged in the *supply* of raw materials to industry⟩
4 : the quantities of goods or services offered for sale at a particular time or at one price
5 : something that maintains or constitutes a supply
sup·ply–side \sə-'plī-'sīd\ *adjective* (1976)
: of, relating to, or being an economic theory that reduction of tax rates encourages more

earnings, savings, and investment and thereby expands economic activity and the total taxable national income

— **sup·ply–sid·er** \-ˈsī-dər\ *noun*

[1]**sup·port** \sə-ˈpōrt, -ˈpȯrt\ *transitive verb* [Middle English, from Middle French *supporter,* from Late Latin *supportare,* from Latin, to carry, from *sub-* + *portare* to carry — more at FARE] (14th century)

1 : to endure bravely or quietly **:** BEAR

2 a (1) **:** to promote the interests or cause of (2) **:** to uphold or defend as valid or right **:** ADVOCATE (3) **:** to argue or vote for **b** (1) **:** ASSIST, HELP (2) **:** to act with (a star actor) (3) **:** to bid in bridge so as to show support for **c :** to provide with substantiation **:** CORROBORATE ⟨*support* an alibi⟩

3 a : to pay the costs of **:** MAINTAIN **b :** to provide a basis for the existence or subsistence of ⟨the island could probably *support* three —A. B. C. Whipple⟩

4 a : to hold up or serve as a foundation or prop for **b :** to maintain (a price) at a desired level by purchases or loans; *also* **:** to maintain the price of by purchases or loans

5 : to keep from fainting, yielding, or losing courage **:** COMFORT

6 : to keep (something) going ☆

— **sup·port·abil·i·ty** \sə-ˌpōr-tə-ˈbi-lə-tē, -ˌpȯr-\ *noun*

— **sup·port·able** \-ˈpōr-tə-bəl, -ˈpȯr-\ *adjective*

— **sup·port·ive** \-ˈpōr-tiv, -ˈpȯr-\ *adjective*

— **sup·port·ive·ness** \-nəs\ *noun*

[2]**support** *noun* (14th century)

1 : the act or process of supporting **:** the condition of being supported

2 : one that supports

3 : sufficient strength in a suit bid by one's partner in bridge to justify raising the suit

sup·port·er *noun* (15th century)

: one that supports or acts as a support: as **a :** ADHERENT, PARTISAN **b :** one of two figures (as of men or animals) placed one on each side of an escutcheon and exterior to it **c :** GARTER 1 **d :** ATHLETIC SUPPORTER

support group *noun* (1969)

: a group of people with common experiences and concerns who provide emotional and moral support for one another

support hose *noun* (1963)

: elastic stockings

support level *noun* (1953)

: a price level on a declining market at which a security resists further decline due to increased attractiveness to traders and investors — called also *support area*

support system *noun* (1980)

: a network of people who provide an individual with practical or emotional support

sup·pos·able \sə-ˈpō-zə-bəl\ *adjective* (1643)

: capable of being supposed **:** CONCEIVABLE

— **sup·pos·ably** \-blē\ *adverb*

sup·pos·al \-ˈpō-zəl\ *noun* (14th century)

1 : the act or process of supposing

2 : something supposed **:** HYPOTHESIS, SUPPOSITION

sup·pose \sə-ˈpōz, *oftenest after "I"* ˈspōz\ *verb* **sup·posed; sup·pos·ing** [Middle English, from Middle French *supposer,* from Medieval Latin *supponere* (perfect indicative *supposui*), from Latin, to put under, substitute, from *sub-* + *ponere* to put — more at POSITION] (14th century)

transitive verb

1 a : to lay down tentatively as a hypothesis, assumption, or proposal ⟨*suppose* a fire broke out⟩ ⟨*suppose* you bring the salad⟩ **b** (1) **:** to hold as an opinion **:** BELIEVE ⟨they *supposed* they were early⟩ (2) **:** to think probable or in keeping with the facts ⟨seems reasonable to *suppose* that he would profit⟩

2 a : CONCEIVE, IMAGINE **b :** to have a suspicion of

3 : PRESUPPOSE

intransitive verb

: CONJECTURE, OPINE

sup·posed \sə-ˈpōzd; *1b & 2a usually* -ˈpō-zəd, *3 & 4 often* -ˈpōst\ *adjective* (1566)

1 a : PRETENDED ⟨twelve hours are *supposed* to elapse between Acts I and II —A. S. Sullivan⟩ **b :** ALLEGED ⟨trusted my *supposed* friends⟩

2 a : held as an opinion **:** BELIEVED; *also* **:** mistakenly believed **:** IMAGINED ⟨the sight which makes *supposed* terror true —Shakespeare⟩ **b :** considered probable or certain **:** EXPECTED ⟨it was not *supposed* that everybody could master the technical aspects —J. C. Murray⟩ **c :** UNDERSTOOD ⟨you will be *supposed* to refer to my grandaunt —G. B. Shaw⟩

3 : made or fashioned by intent or design ⟨what's that button *supposed* to do⟩

4 a : required by or as if by authority ⟨soldiers are *supposed* to obey their commanding officers⟩ **b :** given permission **:** PERMITTED ⟨was not *supposed* to have visitors⟩

— **sup·pos·ed·ly** \-ˈpō-zəd-lē *also* -ˈpōzd-lē\ *adverb*

sup·pos·ing \sə-ˈpō-ziŋ\ *conjunction* (circa 1843)

: if by way of hypothesis **:** on the assumption that

sup·po·si·tion \ˌsə-pə-ˈzi-shən\ *noun* [Middle English, from Late Latin *supposition-, suppositio,* from Latin, act of placing beneath, from *supponere*] (15th century)

1 : something that is supposed **:** HYPOTHESIS

2 : the act of supposing

— **sup·po·si·tion·al** \-ˈzish-nəl, -ˈzi-shə-nᵊl\ *adjective*

sup·po·si·tious \-ˈzi-shəs\ *adjective* [by contraction] (1624)

: SUPPOSITITIOUS

sup·pos·i·ti·tious \sə-ˌpä-zə-ˈti-shəs\ *adjective* [Latin *suppositicius,* from *suppositus,* past participle of *supponere* to substitute] (1611)

1 a : fraudulently substituted **:** SPURIOUS **b** *of a child* (1) **:** falsely presented as a genuine heir (2) **:** ILLEGITIMATE

2 [influenced in meaning by *supposition*] **a :** IMAGINARY **b :** of the nature of or based on a supposition **:** HYPOTHETICAL

— **sup·pos·i·ti·tious·ly** *adverb*

sup·pos·i·to·ry \sə-ˈpä-zə-ˌtōr-ē, -ˌtȯr-\ *noun, plural* **-ries** [Middle English, from Medieval Latin *suppositorium,* from Late Latin, neuter of *suppositorius* placed beneath, from Latin *supponere* to put under] (14th century)

: a solid but readily meltable cone or cylinder of usually medicated material for insertion into a bodily passage or cavity (as the rectum)

sup·press \sə-ˈpres\ *transitive verb* [Middle English, from Latin *suppressus,* past participle of *supprimere,* from *sub-* + *premere* to press — more at PRESS] (14th century)

1 : to put down by authority or force **:** SUBDUE

2 : to keep from public knowledge: as **a :** to keep secret **b :** to stop or prohibit the publication or revelation of

3 a : to exclude from consciousness **b :** to keep from giving vent to **:** CHECK

4 *obsolete* **:** to press down

5 a : to restrain from a usual course or action **:** ARREST ⟨*suppress* a cough⟩ **b :** to inhibit the growth or development of **:** STUNT

6 : to inhibit the genetic expression of ⟨*suppress* a mutation⟩

— **sup·press·ibil·i·ty** \-ˌpre-sə-ˈbi-lə-tē\ *noun*

— **sup·press·ible** \-ˈpre-sə-bəl\ *adjective*

— **sup·pres·sive** \-ˈpre-siv\ *adjective*

— **sup·pres·sive·ness** \-nəs\ *noun*

sup·pres·sant \sə-ˈpre-sᵊnt\ *noun* (1942)

: an agent (as a drug) that tends to suppress or reduce in intensity rather than eliminate something (as appetite)

sup·pres·sion \sə-ˈpre-shən\ *noun* (15th century)

1 : an act or instance of suppressing **:** the state of being suppressed

2 : the conscious intentional exclusion from consciousness of a thought or feeling

sup·pres·sor \-ˈpre-sər\ *noun* (1560)

: one that suppresses; *especially* **:** a mutant gene that suppresses the expression of another nonallelic mutant gene when both are present

suppressor T cell *noun* (1972)

: a T cell that suppresses the immune response of B cells and other T cells to an antigen — called also *suppressor cell*

sup·pu·rate \ˈsə-pyə-ˌrāt\ *intransitive verb* **-rat·ed; -rat·ing** [Latin *suppuratus,* past participle of *suppurare,* from *sub-* + *pur-, pus* pus — more at FOUL] (1656)

: to form or discharge pus

— **sup·pu·ra·tion** \ˌsə-pyə-ˈrā-shən\ *noun*

— **sup·pu·ra·tive** \ˈsə-pyə-rə-tiv, -ˌrā-; ˈsə-prə-tiv\ *adjective*

su·pra \ˈsü-prə, -ˌprä\ *adverb* [Latin] (1526)

: earlier in this writing **:** ABOVE

supra- *prefix* [Latin, from *supra* above, beyond, earlier; akin to Latin *super* over — more at OVER]

1 : SUPER- 2a ⟨*supra*orbital⟩

2 : transcending ⟨*supra*molecular⟩

su·pra·lim·i·nal \ˌsü-prə-ˈli-mə-nᵊl, -ˌprä-\ *adjective* [*supra-* + Latin *limin-, limen* threshold] (1892)

1 : existing above the threshold of consciousness

2 : adequate to evoke a response or induce a sensation ⟨*supraliminal* stimulus⟩

su·pra·mo·lec·u·lar \-mə-ˈle-kyə-lər\ *adjective* (circa 1909)

: more complex than a molecule; *also* **:** composed of many molecules

su·pra·na·tion·al \-ˈnash-nəl, -ˈna-shə-nᵊl\ *adjective* (1908)

: transcending national boundaries, authority, or interests ⟨a *supranational* authority, regulating ocean usage —N. H. Jacoby⟩

— **su·pra·na·tion·al·ism** \-ˈnash-nə-ˌli-zəm, -ˈna-shə-nᵊl-ˌi-\ *noun*

— **su·pra·na·tion·al·ist** \-list, -ist\ *noun*

— **su·pra·na·tion·al·i·ty** \-ˌna-shə-ˈna-lə-tē\ *noun*

su·pra·op·tic \-ˈäp-tik\ *adjective* (1921)

: situated above the optic chiasma; *also* **:** being a small nucleus of closely packed neurons overlying the optic chiasma and intimately connected with the neurohypophysis

su·pra·or·bit·al \-ˈȯr-bə-tᵊl\ *adjective* [New Latin *supraorbitalis,* from Latin *supra-* + Medieval Latin *orbita* orbit] (1828)

: situated or occurring above the orbit of the eye

☆ SYNONYMS

Support, uphold, advocate, back, champion mean to favor actively one that meets opposition. SUPPORT is least explicit about the nature of the assistance given ⟨*supports* waterfront development⟩. UPHOLD implies extended support given to something attacked ⟨*upheld* the legitimacy of the military action⟩. ADVOCATE stresses urging or pleading ⟨*advocated* prison reform⟩. BACK suggests supporting by lending assistance to one failing or falling ⟨refusing to *back* the call for sanctions⟩. CHAMPION suggests publicly defending one unjustly attacked or too weak to advocate his or her own cause ⟨*championed* the rights of children⟩.

su·pra·ra·tio·nal \-ˈrash-nəl, -ˈra-shə-nᵊl\ *adjective* (1894)
: transcending the rational **:** based on or involving factors not to be comprehended by reason alone ⟨the stars inspire *suprarational* dreams —R. J. Dubos⟩

[1]su·pra·re·nal \-ˈrē-nᵊl\ *adjective* [New Latin *suprarenalis,* from Latin *supra-* + *renes* kidneys] (1828)
: situated above or anterior to the kidneys

[2]suprarenal *noun* (1841)
: a suprarenal part; *especially* **:** ADRENAL GLAND

suprarenal gland *noun* (1876)
: ADRENAL GLAND

su·pra·seg·men·tal \ˌsü-prə-seg-ˈmen-tᵊl, -ˌprä-\ *adjective* (1941)
: of or relating to significant features (as stress, pitch, or juncture) that occur simultaneously with vowels and consonants in an utterance

su·pra·ven·tric·u·lar \-ven-ˈtri-kyə-lər, -vən-\ *adjective* (1951)
: relating to or being a rhythmic abnormality of the heart caused by impulses originating above the ventricles ⟨*supraventricular* tachycardia⟩

su·pra·vi·tal \-ˈvī-tᵊl\ *adjective* [International Scientific Vocabulary] (1919)
: constituting or relating to the staining of living tissues or cells surviving after removal from a living body by dyes that penetrate living substance but induce more or less rapid degenerative changes — compare INTRAVITAL 2
— su·pra·vi·tal·ly \-tᵊl-ē\ *adverb*

su·prem·a·cist \sə-ˈpre-mə-sist, sü-\ *noun* (1949)
1 : an advocate or adherent of group supremacy
2 : WHITE SUPREMACIST

su·prem·a·cy \sə-ˈpre-mə-sē, sü- *also* -ˈprē-\ *noun, plural* **-cies** [*supreme* + *-acy* (as in *primacy*)] (1537)
: the quality or state of being supreme; *also* **:** supreme authority or power

su·prem·a·tism \-mə-ˌti-zəm\ *noun, often capitalized* [Russian *suprematizm,* from French *suprématie* supremacy + Russian *-izm* -ism] (1933)
: an early 20th-century art movement in Russia producing abstract works featuring flat geometric forms
— su·prem·a·tist \-tist\ *adjective or noun, often capitalized*

su·preme \sə-ˈprēm, sü-\ *adjective* [Latin *supremus,* superlative of *superus* upper — more at SUPERIOR] (circa 1533)
1 : highest in rank or authority ⟨the *supreme* commander⟩
2 : highest in degree or quality ⟨*supreme* endurance in war and in labour —R. W. Emerson⟩
3 : ULTIMATE, FINAL ⟨the *supreme* sacrifice⟩
— su·preme·ly *adverb*
— su·preme·ness *noun*

Supreme Being *noun* (1699)
: GOD 1

supreme court *noun, often S&C capitalized* (1709)
1 : the highest judicial tribunal in a political unit (as a nation or state)
2 : a court of original jurisdiction in New York state subordinate to a final court of appeals

Supreme Soviet *noun* (1936)
: the highest legislative body of a nation (as Russia or the former Soviet Union)

su·pre·mo \sə-ˈprē-(ˌ)mō, sü-\ *noun, plural* **-mos** [Spanish & Italian, from *supremo,* adjective, supreme, from Latin *supremus*] (1937)
British **:** one who is highest in rank or authority

suq \ˈsük\ *variant of* SOUK

sur- *prefix* [Middle English, from Old French, from Latin *super-*]
1 : over **:** SUPER- ⟨*sur*print⟩ ⟨*sur*tax⟩
2 : above **:** up ⟨*sur*base⟩

su·ra \ˈsu̇r-ə\ *noun* [Arabic *sūrah,* literally, row] (1661)
: a chapter of the Koran

su·rah \ˈsu̇r-ə\ *noun* [probably alteration of *surat,* a cotton produced in Surat, India] (1873)
: a soft twilled fabric of silk or rayon

sur·base \ˈsər-ˌbās\ *noun* (1678)
: a molding just above the base of a wall, pedestal, or podium

[1]sur·cease \(ˌ)sər-ˈsēs, ˈsər-ˌ\ *verb* **sur·ceased; sur·ceas·ing** [Middle English *sursesen, surcesen,* from Middle French *sursis,* past participle of *surseoir,* from Latin *supersedēre* — more at SUPERSEDE] (15th century)
intransitive verb
: to desist from action; *also* **:** to come to an end **:** CEASE
transitive verb
: to put an end to **:** DISCONTINUE

[2]sur·cease \ˈsər-ˌsēs, (ˌ)sər-ˈ\ *noun* (1586)
: CESSATION; *especially* **:** a temporary respite or end ⟨to borrow from my books *surcease* of sorrow —E. A. Poe⟩

[1]sur·charge \ˈsər-ˌchärj\ *transitive verb* [Middle English, from Middle French *surchargier,* from *sur-* + *chargier* to charge, from Old French — more at CHARGE] (15th century)
1 a : OVERCHARGE **b :** to charge an extra fee **c :** to show an omission in (an account) for which credit ought to have been given
2 *British* **:** OVERSTOCK
3 : to fill or load to excess ⟨the atmosphere . . . was *surcharged* with war hysteria —H. A. Chippendale⟩
4 a : to mark a surcharge on (a stamp) **b :** OVERPRINT ⟨*surcharge* a banknote⟩

[2]surcharge *noun* (15th century)
1 a : an additional tax, cost, or impost **b :** an extra fare ⟨a sleeping car *surcharge*⟩ **c :** an instance of surcharging an account
2 : an excessive load or burden
3 : the action of surcharging **:** the state of being surcharged
4 a (1) **:** an overprint on a stamp; *specifically* **:** one that alters the denomination (2) **:** a stamp bearing such an overprint **b :** an overprint on a currency note

sur·cin·gle \ˈsər-ˌsiŋ-gəl\ *noun* [Middle English *sursengle,* from Middle French *surcengle,* from *sur-* + *cengle* girdle, from Latin *cingulum* — more at CINGULUM] (14th century)
1 : a belt, band, or girth passing around the body of a horse to bind a saddle or pack fast to the horse's back
2 *archaic* **:** the cincture of a cassock

sur·coat \ˈsər-ˌkōt\ *noun* [Middle English *surcote,* from Middle French, from *sur-* + *cote* coat] (13th century)
: an outer coat or cloak; *specifically* **:** a tunic worn over armor

[1]surd \ˈsərd\ *adjective* [Latin *surdus* deaf, silent, stupid] (1551)
1 : lacking sense **:** IRRATIONAL ⟨the *surd* mystery and the strange forces of existence —D. C. Williams⟩
2 : VOICELESS — used of speech sounds

[2]surd *noun* (1557)
1 a : an irrational root (as $\sqrt{3}$) **b :** IRRATIONAL NUMBER
2 : a surd speech sound

[1]sure \ˈshu̇r, *especially Southern* ˈshōr\ *adjective* **sur·er; sur·est** [Middle English, from Middle French *sur,* from Latin *securus* secure] (13th century)
1 *obsolete* **:** safe from danger or harm
2 : firmly established **:** STEADFAST ⟨a *sure* hold⟩
3 : RELIABLE, TRUSTWORTHY
4 : marked by or given to feelings of confident certainty ⟨I'm *sure* I'm right⟩
5 : admitting of no doubt **:** INDISPUTABLE ⟨spoke from *sure* knowledge⟩
6 a : bound to happen **:** INEVITABLE ⟨*sure* disaster⟩ **b :** BOUND, DESTINED ⟨is *sure* to win⟩
7 : careful to remember, attend to, or find out something ⟨be *sure* to lock the door⟩ ☆
— sure·ness *noun*
— for sure : without doubt or question **:** CERTAINLY
— to be sure : it must be acknowledged **:** ADMITTEDLY

[2]sure *adverb* (14th century)
: SURELY ▩

sure–enough \ˈshu̇r-ə-ˈnəf\ *adjective* (circa 1846)
: ACTUAL, GENUINE, REAL

sure enough *adverb* (circa 1545)
: as one might expect **:** CERTAINLY

sure·fire \ˈshu̇r-ˈfīr\ *adjective* (circa 1909)
: certain to get successful or expected results ⟨a *surefire* recipe⟩

sure·foot·ed \-ˈfu̇-təd\ *adjective* (1633)
: not liable to stumble, fall, or err
— sure·foot·ed·ly *adverb*
— sure·foot·ed·ness *noun*

sure–hand·ed \-ˈhan-dəd\ *adjective* (1930)
: proficient and confident in performance especially using the hands
— sure–hand·ed·ness *noun*

sure·ly \ˈshu̇r-lē, *especially Southern* ˈshōr-\ *adverb* (14th century)

☆ SYNONYMS

Sure, certain, positive, cocksure mean having no doubt or uncertainty. SURE usually stresses the subjective or intuitive feeling of assurance ⟨felt *sure* that I had forgotten something⟩. CERTAIN may apply to a basing of a conclusion or conviction on definite grounds or indubitable evidence ⟨police are *certain* about the cause of the fire⟩. POSITIVE intensifies sureness or certainty and may imply opinionated conviction or forceful expression of it ⟨I'm *positive* that's the person I saw⟩. COCKSURE implies presumptuous or careless positiveness ⟨you're always so *cocksure* about everything⟩.

□ USAGE

sure Most commentators consider the adverb *sure* to be something less than completely standard; *surely* is usually recommended as a substitute. Our current evidence shows, however, that *sure* and *surely* have become differentiated in use. *Sure* is used in much more informal contexts than *surely.* It is used as a simple intensive ⟨I can never know how much I bored her, but, be certain, she *sure* amused me —Norman Mailer⟩ and, because it connotes strong affirmation, it is used when the speaker or writer expects to be agreed with ⟨it's a moot point whether politicians are less venal than in Twain's day. But they're *sure* as the devil more intrusive —Alan Abelson⟩ ⟨he *sure* gets them to play —D. S. Looney⟩. *Surely,* like *sure,* is used as a simple intensive ⟨I *surely* don't want to leave the impression that I had an unhappy childhood —E. C. Welsh⟩ but it occurs in more formal contexts than *sure.* Unlike *sure* it may be used neutrally—the reader or hearer may or may not agree ⟨it would *surely* be possible, within a few years, to program a computer to construct a grammar —Noam Chomsky⟩ and it is often used when the writer is trying to persuade ⟨*surely* a book on the avant-garde cannot be so conventional —Karl Shapiro⟩.

1 : in a sure manner: **a** *archaic* **:** without danger or risk of injury or loss **:** SAFELY **b** (1) **:** with assurance **:** CONFIDENTLY ⟨answered quickly and *surely*⟩ (2) **:** without doubt **:** CERTAINLY ⟨they will *surely* be heard from in the future —R. J. Lifton⟩
2 : INDEED, REALLY — often used as an intensive ⟨you *surely* don't believe that⟩
usage see [2]SURE

sure thing *noun* (1836)
: one that is certain to succeed **:** a sure bet

sure·ty \ˈshu̇r(-ə)-tē\ *noun, plural* **-ties** [Middle English *surte,* from Middle French *surté,* from Latin *securitat-, securitas* security, from *securus*] (14th century)
1 : the state of being sure: as **a :** sure knowledge **:** CERTAINTY **b :** confidence in manner or behavior **:** ASSURANCE
2 a : a formal engagement (as a pledge) given for the fulfillment of an undertaking **:** GUARANTEE **b :** ground of confidence or security
3 : one who has become legally liable for the debt, default, or failure in duty of another
— sure·ty·ship \-ˌship\ *noun*

surety bond *noun* (1911)
: a bond guaranteeing performance of a contract or obligation

[1]**surf** \ˈsərf\ *noun* [origin unknown] (1685)
1 : the swell of the sea that breaks upon the shore
2 : the foam, splash, and sound of breaking waves

[2]**surf** *intransitive verb* (1926)
: to ride the surf (as on a surfboard)
— surf·er *noun*

[1]**sur·face** \ˈsər-fəs\ *noun* [French, from *sur-* + *face* face, from Old French — more at FACE] (circa 1604)
1 : the exterior or upper boundary of an object or body
2 : a plane or curved two-dimensional locus of points (as the boundary of a three-dimensional region) ⟨plane *surface*⟩ ⟨*surface* of a sphere⟩
3 a : the external or superficial aspect of something **b :** an external part or layer
— on the surface : to all outward appearances

[2]**surface** *adjective* (1664)
1 a : of, located on, or designed for use at the surface of something **b :** situated, transported, or employed on the surface of the earth ⟨*surface* mail⟩ ⟨*surface* vehicles⟩
2 : appearing to be such on the surface only **:** SUPERFICIAL ⟨*surface* friendships⟩

[3]**surface** *verb* **sur·faced; sur·fac·ing** (1778)
transitive verb
1 : to give a surface to: as **a :** to plane or make smooth **b :** to apply the surface layer to ⟨*surface* a highway⟩
2 : to bring to the surface
intransitive verb
1 : to work on or at the surface
2 : to come to the surface
3 : to come into public view **:** SHOW UP
— sur·fac·er *noun*

surface–active *adjective* (1920)
: altering the properties and especially lowering the tension at the surface of contact between phases ⟨soaps and wetting agents are typical *surface-active* substances⟩

surface of revolution (1840)
: a surface formed by the revolution of a plane curve about a line in its plane

sur·face–rip·ened \ˈsər-fəs-ˌrī-pənd, -ˌrī-pᵊmd\ *adjective* (1945)
of cheese **:** ripened by the action of microorganisms (as molds) on the surface

surface structure *noun* (1964)
: a formal representation of the phonetic form of a sentence; *also* **:** the structure which such a representation describes

surface tension *noun* (1876)
: the attractive force exerted upon the surface molecules of a liquid by the molecules beneath that tends to draw the surface molecules into the bulk of the liquid and makes the liquid assume the shape having the least surface area

surface–to–air *adjective* (1949)
: launched from the ground against a target in the air

surfacing *noun* (1882)
: material forming or used to form a surface

sur·fac·tant \(ˌ)sər-ˈfak-tənt, ˈsər-ˌ\ *noun* [*surf*ace-*act*ive + *-ant*] (1950)
: a surface-active substance (as a detergent)
— surfactant *adjective*

surf and turf *noun* (1973)
: seafood and steak served as a single course

surf·bird \ˈsərf-ˌbərd\ *noun* (1839)
: a shorebird (*Aphriza virgata*) of the Pacific coasts of America that has a black-tipped white tail

surf·board \-ˌbōrd, -ˌbȯrd\ *noun* (circa 1826)
: a long narrow buoyant board (as of lightweight wood or fiberglass-covered foam) used in the sport of surfing
— surfboard *intransitive verb*
— surf·board·er *noun*

surf·boat \-ˌbōt\ *noun* (1847)
: a boat for use in heavy surf

surf casting *noun* (1928)
: a method of fishing in which artificial or natural bait is cast into the open ocean or in a bay where waves break on a beach
— surf caster *noun*

surf clam *noun* (1884)
: any of various typically rather large surf-dwelling edible clams (family Mactridae); *especially* **:** a common clam (*Spisula solidissima*) of the Atlantic coast chiefly from Nova Scotia to South Carolina

[1]**sur·feit** \ˈsər-fət\ *noun* [Middle English *surfait,* from Middle French, from *surfaire* to overdo, from *sur-* + *faire* to do, from Latin *facere* — more at DO] (14th century)
1 : an overabundant supply **:** EXCESS
2 : an intemperate or immoderate indulgence in something (as food or drink)
3 : disgust caused by excess

[2]**surfeit** (14th century)
transitive verb
: to feed, supply, or give to surfeit
intransitive verb
archaic **:** to indulge to satiety in a gratification (as indulgence of the appetite or senses)
synonym see SATIATE
— sur·feit·er *noun*

surf fish *noun* (1882)
: SURFPERCH

sur·fi·cial \ˌsər-ˈfi-shəl\ *adjective* [*sur*face + *-icial* (as in *superficial*)] (1892)
: of or relating to a surface ⟨*surficial* geologic processes⟩

surf·ing \ˈsər-fiŋ\ *noun* (1926)
: the sport of riding the surf especially on a surfboard

surf·perch \ˈsərf-ˌpərch\ *noun* (1885)
: any of a family (Embiotocidae) of small or medium-sized viviparous bony fishes chiefly of shallow water along the Pacific coast of North America that resemble the perches

[1]**surge** \ˈsərj\ *verb* **surged; surg·ing** [Middle French *sourge-,* stem of *sourdre* to rise, surge, from Latin *surgere* to go straight up, rise, from *sub-* up + *regere* to lead straight — more at SUB-, RIGHT] (1511)
intransitive verb
1 : to rise and fall actively **:** TOSS ⟨a ship *surging* in heavy seas⟩
2 : to rise and move in waves or billows **:** SWELL
3 : to slip around a windlass, capstan, or bitts — used especially of a rope
4 : to rise suddenly to an excessive or abnormal value ⟨the stock market *surgeed* to a record high⟩
5 : to move with a surge or in surges ⟨felt the blood *surging* into his face —Harry Hervey⟩
transitive verb
: to let go or slacken gradually (as a rope)

[2]**surge** *noun* (1520)
1 : a swelling, rolling, or sweeping forward like that of a wave or series of waves ⟨a *surge* of interest⟩
2 a : a large wave or billow **:** SWELL **b** (1) **:** a series of such swells or billows (2) **:** the resulting elevation of water level
3 : the tapered part of a windlass barrel or a capstan
4 a : a movement (as a slipping or slackening) of a rope or cable **b :** a sudden jerk or strain caused by such a movement
5 : a transient sudden rise of current or voltage in an electrical circuit

sur·geon \ˈsər-jən\ *noun* [Middle English *surgien,* from Anglo-French, from Old French *cirurgien,* from *cirurgie* surgery] (14th century)
: a medical specialist who practices surgery

sur·geon·fish \-ˌfish\ *noun* (1871)
: any of a family (Acanthuridae) of tropical bony fishes that have toxic flesh and typically a movable spine on each side of the body near the base of the tail capable of inflicting a painful wound

surgeon general *noun, plural* **surgeons general** (1706)
: the chief medical officer of a branch of the armed services or of a public health service

surgeon's knot *noun* (1733)
: any of several knots used in tying ligatures or surgical stitches; *especially* **:** a reef knot in which the first knot has two turns — see KNOT illustration

sur·gery \ˈsərj-rē, ˈsər-jə-\ *noun, plural* **-ger·ies** [Middle English *surgerie,* from Middle French *cirurgie, surgerie,* from Latin *chirurgia,* from Greek *cheirourgia,* from *cheirourgos* surgeon, from *cheirourgos* doing by hand, from *cheir* hand + *ergon* work — more at CHIR-, WORK] (14th century)
1 : a branch of medicine concerned with diseases and conditions requiring or amenable to operative or manual procedures
2 : alterations made as if by surgery ⟨literary *surgery*⟩
3 a *British* **:** a physician's or dentist's office **b :** a room or area where surgery is performed
4 a : the work done by a surgeon **b :** OPERATION

sur·gi·cal \ˈsər-ji-kəl\ *adjective* [*surg*eon + *-ical*] (1770)
1 a : of or relating to surgeons or surgery ⟨*surgical* skills⟩ **b :** used in or in connection with surgery **c :** characteristic of or resembling surgery or a surgeon especially in control or incisiveness ⟨*surgical* precision⟩
2 : following or resulting from surgery ⟨*surgical* fevers⟩
— sur·gi·cal·ly \-k(ə-)lē\ *adverb*

su·ri·mi \su̇-ˈrē-mē\ *noun* [Japanese, chopped meat or fish] (1976)
: a fish product made from inexpensive whitefish and often processed to resemble more expensive seafood (as crabmeat)

sur·jec·tion \(ˌ)sər-ˈjek-shən\ *noun* [probably from *sur-* + *-jection* (as in *projection*)] (1964)
: a mathematical function that is an onto mapping — compare BIJECTION, INJECTION 3

sur·jec·tive \-ˈjek-tiv\ *adjective* (1964)
: ONTO ⟨a set of *surjective* functions⟩

sur·ly \ˈsər-lē\ *adjective* **sur·li·er; -est** [alteration of Middle English *sirly* lordly, imperious, from *sir*] (circa 1572)
1 *obsolete* **:** ARROGANT, IMPERIOUS
2 : irritably sullen and churlish in mood or manner **:** CRABBED

\ə\ **abut** \ᵊ\ **kitten** \ər\ **further** \a\ **ash** \ā\ **ace** \ä\ **mop, mar** \au̇\ **out** \ch\ **chin** \e\ **bet** \ē\ **easy** \g\ **go** \i\ **hit** \ī\ **ice** \j\ **job** \ŋ\ **sing** \ō\ **go** \ȯ\ **law** \ȯi\ **boy** \th\ **thin** \t͟h\ **the** \ü\ **loot** \u̇\ **foot** \y\ **yet** \zh\ **vision** *see also* Guide to Pronunciation

3 : menacing or threatening in appearance ⟨*surly* weather⟩ ◇
synonym see SULLEN
— **sur·li·ly** \-lə-lē\ *adverb*
— **sur·li·ness** \-lē-nəs\ *noun*
— **surly** *adverb*

¹sur·mise \sər-ˈmīz, ˈsər-ˌ\ *noun* (1569)
: a thought or idea based on scanty evidence **:** CONJECTURE

²sur·mise \sər-ˈmīz\ *transitive verb* **sur·mised; sur·mis·ing** [Middle English, to accuse, from Middle French *surmis,* past participle of *surmetre,* from Latin *supermittere* to throw on, from *super-* + *mittere* to send] (1700)
: to imagine or infer on slight grounds

sur·mount \sər-ˈmau̇nt\ *transitive verb* [Middle English, from Middle French *surmonter,* from *sur-* + *monter* to mount] (14th century)
1 *obsolete* **:** to surpass in quality or attainment **:** EXCEL
2 : to prevail over **:** OVERCOME ⟨*surmount* an obstacle⟩
3 : to get to the top of **:** CLIMB
4 : to stand or lie at the top of
— **sur·mount·able** \-ˈmau̇n-tə-bəl\ *adjective*

¹sur·name \ˈsər-ˌnām\ *noun* (14th century)
1 : an added name derived from occupation or other circumstance **:** NICKNAME 1
2 : the name borne in common by members of a family

²surname *transitive verb* (15th century)
: to give a surname to

sur·pass \sər-ˈpas\ *transitive verb* [Middle French *surpasser,* from *sur-* + *passer* to pass] (1555)
1 : to become better, greater, or stronger than **:** EXCEED
2 : to go beyond **:** OVERSTEP
3 : to transcend the reach, capacity, or powers of
synonym see EXCEED
— **sur·pass·able** \-ˈpa-sə-bəl\ *adjective*

sur·pass·ing *adjective* (circa 1580)
: greatly exceeding others **:** of a very high degree
— **sur·pass·ing·ly** \-ˈpa-siŋ-lē\ *adverb*

¹sur·plice \ˈsər-pləs\ *noun* [Middle English *surplis,* from Old French *surpliz,* from Medieval Latin *superpellicium,* from *super-* + *pellicium* coat of skins, from Latin, neuter of *pellicius* made of skins, from *pellis* skin — more at FELL] (13th century)
: a loose white outer ecclesiastical vestment usually of knee length with large open sleeves

²surplice *adjective* (1845)
: having a diagonally overlapping neckline or closing ⟨a *surplice* collar⟩ ⟨*surplice* sweaters⟩

sur·plus \ˈsər-(ˌ)pləs\ *noun* [Middle English, from Middle French, from Medieval Latin *superplus,* from Latin *super-* + *plus* more — more at PLUS] (14th century)
1 a : the amount that remains when use or need is satisfied **b :** an excess of receipts over disbursements
2 : the excess of a corporation's net worth over the par or stated value of its capital stock
— **surplus** *adjective*

sur·plus·age \-(ˌ)plə-sij\ *noun* (15th century)
1 : SURPLUS 1a
2 a : excessive or nonessential matter **b :** matter introduced in legal pleading which is not necessary or relevant to the case

surplus value *noun* (1887)
: the difference in Marxist theory between the value of work done or of commodities produced by labor and the usually subsistence wages paid by the employer

sur·print \ˈsər-ˌprint\ *transitive verb or noun* (1917)
: OVERPRINT

sur·pris·al \sə(r)-ˈprī-zəl\ *noun* (1591)
: the action of surprising **:** the state of being surprised

¹sur·prise *also* **sur·prize** \sə(r)-ˈprīz\ *noun* [Middle English, from Middle French, from feminine of *surpris,* past participle of *surprendre* to take over, surprise, from *sur-* + *prendre* to take — more at PRIZE] (15th century)
1 a : an attack made without warning **b :** a taking unawares
2 : something that surprises
3 : the state of being surprised **:** ASTONISHMENT

²surprise *also* **surprize** *verb* **sur·prised; sur·pris·ing** (15th century)
transitive verb
1 : to attack unexpectedly; *also* **:** to capture by an unexpected attack
2 a : to take unawares **b :** to detect or elicit by a taking unawares
3 : to strike with wonder or amazement especially because unexpected
intransitive verb
: to cause astonishment or surprise ⟨her success didn't *surprise*⟩ ☆
— **sur·pris·er** *noun*

sur·pris·ing *adjective* (1645)
: of a nature that excites surprise

sur·pris·ing·ly \sə(r)-ˈprī-ziŋ-lē\ *adverb* (1661)
1 : in a surprising manner **:** to a surprising degree ⟨a *surprisingly* fast runner⟩
2 : it is surprising that ⟨*surprisingly,* voter turnout was high⟩

sur·ra \ˈsu̇r-ə\ *noun* [Marathi *sūra* wheezing sound] (1883)
: a severe Old World febrile and hemorrhagic disease of domestic animals that is caused by a flagellate protozoan (*Trypanosoma evansi*) and is transmitted by biting insects

sur·re·al \sə-ˈrē(-ə)l, -ˈri-əl *also* -ˈrā-əl\ *adjective* [back-formation from *surrealism*] (1937)
1 : having the intense irrational reality of a dream
2 : SURREALISTIC
— **sur·re·al·ly** *adverb*

sur·re·al·ism \sə-ˈrē-ə-ˌli-zəm, -ˈri- *also* -ˈrā-\ *noun* [French *surréalisme,* from *sur-* + *réalisme* realism] (1925)
: the principles, ideals, or practice of producing fantastic or incongruous imagery or effects in art, literature, film, or theater by means of unnatural juxtapositions and combinations
— **sur·re·al·ist** \-list\ *noun or adjective*

sur·re·al·is·tic \-ˌrē-ə-ˈlis-tik, -ˌri- *also* -ˌrā-\ *adjective* (1925)
1 : of or relating to surrealism
2 : having a strange dreamlike atmosphere or quality like that of a surrealist painting
— **sur·re·al·is·ti·cal·ly** \-ti-k(ə-)lē\ *adverb*

sur·re·but·ter \ˌsər-(r)i-ˈbə-tər\ *noun* (circa 1601)
: the reply in common law pleading of a plaintiff to a defendant's rebutter

sur·re·join·der \-(r)i-ˈjȯin-dər\ *noun* (circa 1543)
: the reply in common law pleading of a plaintiff to a defendant's rejoinder

¹sur·ren·der \sə-ˈren-dər\ *verb* **-dered; -der·ing** \-d(ə-)riŋ\ [Middle English, from Middle French *surrendre,* from *sur-* + *rendre* to give back, yield — more at RENDER] (15th century)
transitive verb
1 a : to yield to the power, control, or possession of another upon compulsion or demand ⟨*surrendered* the fort⟩ **b :** to give up completely or agree to forgo especially in favor of another
2 a : to give (oneself) up into the power of another especially as a prisoner **b :** to give (oneself) over to something (as an influence)
intransitive verb
: to give oneself up into the power of another **:** YIELD
synonym see RELINQUISH

²surrender *noun* (15th century)
1 a : the action of yielding one's person or giving up the possession of something especially into the power of another **b :** the relinquishment by a patentee of rights or claims under a patent **c :** the delivery of a principal into lawful custody by bail — called also *surrender by bail* **d :** the voluntary cancellation of the legal liability of an insurance company by the insured and beneficiary for a consideration **e :** the delivery of a fugitive from justice by one government to another
2 : an instance of surrendering

sur·rep·ti·tious \ˌsər-əp-ˈti-shəs, ˌsə-rəp-, sə-ˌrep-\ *adjective* [Middle English, from Latin *surrepticius,* from *surreptus,* past participle of *surripere* to snatch secretly, from *sub-* + *rapere* to seize — more at RAPID] (15th century)
1 : done, made, or acquired by stealth **:** CLANDESTINE
2 : acting or doing something clandestinely **:** STEALTHY
synonym see SECRET
— **sur·rep·ti·tious·ly** *adverb*

sur·rey \ˈsər-ē, ˈsə-rē\ *noun, plural* **surreys** [*Surrey,* England] (circa 1891)
: a four-wheel two-seated horse-drawn pleasure carriage

sur·ro·ga·cy \ˈsər-ə-gə-sē\ *noun* (1984)
: the practice of serving as a surrogate mother

¹sur·ro·gate \ˈsər-ə-ˌgāt, ˈsə-rə-\ *transitive verb* **-gat·ed; -gat·ing** [Latin *surrogatus,* past participle of *surrogare* to choose in place of another, substitute, from *sub-* + *rogare* to ask — more at RIGHT] (1533)
: to put in the place of another: **a :** to appoint as successor, deputy, or substitute for oneself **b :** SUBSTITUTE

²sur·ro·gate \-ˌgāt, -gət\ *noun, often attributive* (1603)
1 a : one appointed to act in place of another **:** DEPUTY **b :** a local judicial officer in some states (as New York) who has jurisdiction over the probate of wills, the settlement of estates, and the appointment and supervision of guardians
2 : one that serves as a substitute
3 : SURROGATE MOTHER

surrogate mother *noun* (1978)
: a woman who becomes pregnant usually by artificial insemination or surgical implantation of a fertilized egg for the purpose of carrying the fetus to term for another woman

☆ SYNONYMS
Surprise, astonish, astound, amaze, flabbergast mean to impress forcibly through unexpectedness. SURPRISE stresses causing an effect through being unexpected at a particular time or place rather than by being essentially unusual or novel ⟨*surprised* to find them at home⟩. ASTONISH implies surprising so greatly as to seem incredible ⟨a discovery that *astonished* the world⟩. ASTOUND stresses the shock of astonishment ⟨too *astounded* to respond⟩. AMAZE suggests an effect of bewilderment ⟨*amazed* by the immense size of the place⟩. FLABBERGAST may suggest thorough astonishment and bewilderment or dismay ⟨*flabbergasted* by his angry refusal⟩.

◇ WORD HISTORY
surly To the highborn an association between good manners and aristocratic birth might seem obvious and inescapable, but the fact that the two do not inevitably go together is evidenced by the word *surly.* In Middle English the word was spelled *sirly,* thus making clearer its connection to *sir,* the traditional title of respect. *Sirly* had much the same meaning as *lordly* has today, that is, "proud, haughty." Although its meaning has evolved into "sullen," *surly* still refers to a disposition that is less than noble.

— **surrogate motherhood** *noun*

¹sur·round \sə-ˈrau̇nd\ *transitive verb* [Middle English, to overflow, from Middle French *suronder,* from Late Latin *superundare,* from Latin *super-* + *unda* wave; influenced in meaning by ⁵*round* — more at WATER] (circa 1616)
1 a (1) **:** to enclose on all sides **:** ENVELOP ⟨the crowd *surrounded* her⟩ (2) **:** to enclose so as to cut off communication or retreat **:** INVEST **b :** to form or be a member of the entourage of ⟨flatterers who *surround* the king⟩ **c :** to constitute part of the environment of ⟨*surrounded* by poverty⟩ **d :** to extend around the margin or edge of **:** ENCIRCLE ⟨a wall *surrounds* the old city⟩
2 : to cause to be surrounded by something ⟨*surrounded* himself with friends⟩

²surround *noun* (1825)
: something (as a border or ambient environment) that surrounds ⟨from urban centre to rural *surround* —Emrys Jones⟩

sur·round·ings \sə-ˈrau̇n-diŋz\ *noun plural* (1861)
: the circumstances, conditions, or objects by which one is surrounded **:** ENVIRONMENT

sur·roy·al \ˈsər-ˌrȯi(-ə)l\ *noun* [Middle English *surryal,* from *sur-* + *ryall* royal antler] (15th century)
: one of the terminal tines above the royal antler of a large deer (as a stag) usually grown by four years of age

sur·sum cor·da \ˌsu̇r-səm-ˈkȯr-də, -ˌdä\ *noun* [Late Latin, (lift) up (your) hearts; from the opening words] (1537)
1 *often S&C capitalized* **:** a versicle that in traditional eucharistic liturgies exhorts the faithful to enthusiastic worship
2 : something inspiriting

sur·tax \ˈsər-ˌtaks\ *noun* (1881)
1 : an extra tax or charge
2 : a graduated income tax in addition to the normal income tax imposed on the amount by which one's net income exceeds a specified sum

sur·tout \(ˌ)sər-ˈtü, ˈsər-ˌ\ *noun* [French, from *sur* over (from Latin *super*) + *tout* all, from Latin *totus* whole — more at OVER] (1686)
: a man's long close-fitting overcoat

sur·veil \sər-ˈvā(ə)l\ *transitive verb* **sur·veilled; sur·veil·ling** [back-formation from *surveillance*] (1949)
: to subject to surveillance

sur·veil·lance \sər-ˈvā-lən(t)s *also* -ˈvāl-yən(t)s *or* -ˈvā-ən(t)s\ *noun* [French, from *surveiller* to watch over, from *sur-* + *veiller* to watch, from Latin *vigilare,* from *vigil* watchful — more at VIGIL] (1802)
: close watch kept over someone or something (as by a detective); *also* **:** SUPERVISION

sur·veil·lant \-ˈvā-lənt *also* -ˈvāl-yənt *or* -ˈvā-ənt\ *noun* (1819)
: one that exercises surveillance

¹sur·vey \sər-ˈvā, ˈsər-ˌ\ *verb* **sur·veyed; sur·vey·ing** [Middle English, from Middle French *surveeir* to look over, from *sur-* + *veeir* to see — more at VIEW] (15th century)
transitive verb
1 a : to examine as to condition, situation, or value **:** APPRAISE **b :** to query (someone) in order to collect data for the analysis of some aspect of a group or area
2 : to determine and delineate the form, extent, and position of (as a tract of land) by taking linear and angular measurements and by applying the principles of geometry and trigonometry
3 : to view or consider comprehensively
4 : INSPECT, SCRUTINIZE ⟨he *surveyed* us in a lordly way —Alan Harrington⟩
intransitive verb
: to make a survey

²sur·vey \ˈsər-ˌvā, sər-ˈ\ *noun, plural* **sur·veys** (1548)
1 : the act or an instance of surveying: as **a :** a broad treatment of a subject **b :** POLL 5a
2 : something that is surveyed

survey course \ˈsər-ˌvā-\ *noun* (1916)
: a course treating briefly the chief topics of a broad field of knowledge

sur·vey·ing \sər-ˈvā-iŋ\ *noun* (1682)
: a branch of applied mathematics that teaches the art of determining the area of any portion of the earth's surface, the lengths and directions of the bounding lines, and the contour of the surface and of accurately delineating the whole on paper

sur·vey·or \sər-ˈvā-ər\ *noun* (15th century)
: one that surveys; *especially* **:** one whose occupation is surveying land

sur·viv·able \sər-ˈvī-və-bəl\ *adjective* (1955)
: resulting in or permitting survival
— **sur·viv·abil·i·ty** \-ˌvī-və-ˈbi-lə-tē\ *noun*

sur·viv·al \sər-ˈvī-vəl\ *noun, often attributive* (1598)
1 a : a living or continuing longer than another person or thing **b :** the continuation of life or existence ⟨problems of *survival* in arctic conditions⟩
2 : one that survives

sur·viv·al·ist \-və-list\ *noun* (1970)
: one who views survival as a primary objective; *especially* **:** one who has prepared to survive in the anarchy of an anticipated breakdown of society
— **survivalist** *adjective*

survival of the fittest (1864)
: NATURAL SELECTION

sur·viv·ance \sər-ˈvī-vən(t)s\ *noun* (circa 1623)
: SURVIVAL

sur·vive \sər-ˈvīv\ *verb* **sur·vived; sur·viv·ing** [Middle English, from Middle French *survivre* to outlive, from Latin *supervivere,* from *super-* + *vivere* to live — more at QUICK] (15th century)
intransitive verb
1 : to remain alive or in existence **:** live on
2 : to continue to function or prosper
transitive verb
1 : to remain alive after the death of ⟨he is *survived* by his wife⟩
2 : to continue to exist or live after ⟨*survived* the earthquake⟩
3 : to continue to function or prosper despite **:** WITHSTAND
— **sur·vi·vor** \-ˈvī-vər\ *noun*

sur·viv·er \-ˈvī-vər\ *noun* (1602)
archaic **:** one that survives **:** SURVIVOR

sur·vi·vor·ship \-ˈvī-vər-ˌship\ *noun* (circa 1625)
1 : the legal right of the survivor of persons having joint interests in property to take the interest of the person who has died
2 : the state of being a survivor
3 : the probability of surviving to a particular age; *also* **:** the number or proportion of survivors (as of an age group)

Su·san B. An·tho·ny Day \ˈsü-zᵊn-ˌbē-ˈan(t)-thə-nē-\ *noun* (circa 1951)
: February 15 observed to commemorate the birth of Susan B. Anthony

sus·cep·ti·bil·i·ty \sə-ˌsep-tə-ˈbi-lə-tē\ *noun, plural* **-ties** (1644)
1 : the quality or state of being susceptible; *especially* **:** lack of ability to resist some extraneous agent (as a pathogen or drug) **:** SENSITIVITY
2 a : a susceptible temperament or constitution **b** *plural* **:** FEELINGS, SENSIBILITIES
3 a : the ratio of the magnetization in a substance to the corresponding magnetizing force **b :** the ratio of the electric polarization to the electric intensity in a polarized dielectric

sus·cep·ti·ble \sə-ˈsep-tə-bəl\ *adjective* [Late Latin *susceptibilis,* from Latin *susceptus,* past participle of *suscipere* to take up, admit, from *sub-, sus-* up + *capere* to take — more at SUB-, HEAVE] (1605)
1 : capable of submitting to an action, process, or operation ⟨a theory *susceptible* to proof⟩
2 : open, subject, or unresistant to some stimulus, influence, or agency
3 : IMPRESSIONABLE, RESPONSIVE
synonym see LIABLE
— **sus·cep·ti·ble·ness** *noun*
— **sus·cep·ti·bly** \-blē\ *adverb*

sus·cep·tive \-tiv\ *adjective* (15th century)
1 : RECEPTIVE
2 : SUSCEPTIBLE
— **sus·cep·tive·ness** *noun*
— **sus·cep·tiv·i·ty** \sə-ˌsep-ˈti-və-tē\ *noun*

su·shi \ˈsü-shē *also* ˈsu̇-\ *noun* [Japanese] (1893)
: cold rice dressed with vinegar, formed into any of various shapes, and garnished especially with bits of raw fish or shellfish

su·slik \ˈsü-slik\ *noun* [Russian] (1774)
1 : any of several rather large short-tailed ground squirrels (genus *Citellus*) of eastern Europe or northern Asia
2 : the mottled grayish black fur of a suslik

¹sus·pect \ˈsəs-ˌpekt, sə-ˈspekt\ *adjective* [Middle English, from Middle French, from Latin *suspectus,* from past participle of *suspicere*] (14th century)
1 : regarded or deserving to be regarded with suspicion **:** SUSPECTED ⟨investigates *suspect* employees⟩
2 : DOUBTFUL, QUESTIONABLE ⟨whose skills are *suspect* —Peter Vecsey⟩

²sus·pect \ˈsəs-ˌpekt\ *noun* (1591)
: one who is suspected; *especially* **:** one suspected of a crime

³sus·pect \sə-ˈspekt\ *verb* [Middle English, from Latin *suspectare,* frequentative of *suspicere* to look up at, regard with awe, suspect, from *sub-, sus-* up, secretly + *specere* to look at — more at SUB-, SPY] (15th century)
transitive verb
1 : to imagine (one) to be guilty or culpable on slight evidence or without proof ⟨*suspect* him of giving false information⟩
2 : to have doubts of **:** DISTRUST
3 : to imagine to exist or be true, likely, or probable
intransitive verb
: to imagine something to be true or likely

sus·pend \sə-ˈspend\ *verb* [Middle English, from Old French *suspendre* to hang up, interrupt, from Latin *suspendere,* from *sub-, sus-* up + *pendere* to cause to hang, weigh] (14th century)
transitive verb
1 : to debar temporarily from a privilege, office, or function ⟨*suspend* a student from school⟩
2 a : to cause to stop temporarily ⟨*suspend* bus service⟩ **b :** to set aside or make temporarily inoperative ⟨*suspend* the rules⟩
3 : to defer to a later time on specified conditions ⟨*suspend* sentence⟩
4 : to hold in an undetermined or undecided state awaiting further information ⟨*suspend* judgment⟩ ⟨*suspend* disbelief⟩
5 a : HANG; *especially* **:** to hang so as to be free on all sides except at the point of support ⟨*suspend* a ball by a thread⟩ **b :** to keep from falling or sinking by some invisible support (as buoyancy) ⟨dust *suspended* in the air⟩
6 a : to keep fixed or lost (as in wonder or contemplation) **b :** to keep waiting in suspense or indecision
7 : to hold (a musical note) over into the following chord
intransitive verb
1 : to cease operation temporarily
2 : to stop payment or fail to meet obligations
3 : HANG
synonym see DEFER

suspended animation *noun* (1795)
: temporary suspension of the vital functions (as in persons nearly drowned)

sus·pend·er \sə-ˈspen-dər\ *noun* (1524)
1 : one that suspends
2 : a device by which something may be suspended: as **a :** one of two supporting bands worn across the shoulders to support trousers, skirt, or belt — usually used in plural and often with *pair* **b** *British* **:** a fastener attached to a garment or garter to hold up a stocking or sock; *also* **:** a device consisting of garter and fastener
— sus·pend·ered \-dərd\ *adjective*

sus·pense \sə-ˈspen(t)s\ *noun* [Middle English, from Middle French, from *suspendre*] (15th century)
1 : the state of being suspended **:** SUSPENSION
2 a : mental uncertainty **:** ANXIETY **b :** pleasant excitement as to a decision or outcome ⟨a novel of *suspense*⟩
3 : the state or character of being undecided or doubtful **:** INDECISIVENESS
— sus·pense·ful \-fəl\ *adjective*
— sus·pense·ful·ly \-fə-lē\ *adverb*
— sus·pense·ful·ness \-fəl-nəs\ *noun*
— sus·pense·less \-ləs\ *adjective*

suspense account *noun* (1869)
: an account for the temporary entry of charges or credits or especially of doubtful accounts receivable pending determination of their ultimate disposition

sus·pens·er \sə-ˈspen(t)-sər\ *noun* (circa 1960)
: a suspenseful film

sus·pen·sion \sə-ˈspen(t)-shən\ *noun* [Middle English *suspensyon,* from Middle French *suspension,* from Late Latin *suspension-, suspensio,* from Latin *suspendere*] (15th century)
1 : the act of suspending **:** the state or period of being suspended: as **a :** temporary removal from office or privileges **b :** temporary withholding (as of belief or decision) **c :** temporary abrogation of a law or rule **d** (1) **:** the holding over of one or more musical tones of a chord into the following chord producing a momentary discord and suspending the concord which the ear expects; *specifically* **:** such a dissonance which resolves downward — compare ANTICIPATION, RETARDATION (2) **:** the tone thus held over **e :** stoppage of payment of business obligations **:** FAILURE — used especially of a business or a bank **f :** a rhetorical device whereby the principal idea is deferred to the end of a sentence or longer unit
2 a : the act of hanging **:** the state of being hung **b** (1) **:** the state of a substance when its particles are mixed with but undissolved in a fluid or solid (2) **:** a substance in this state (3) **:** a system consisting of a solid dispersed in a solid, liquid, or gas usually in particles of larger than colloidal size — compare EMULSION
3 : something suspended
4 a : a device by which something (as a magnetic needle) is suspended **b :** the system of devices (as springs) supporting the upper part of a vehicle on the axles **c :** the act, process, or manner in which the pendulum of a timepiece is suspended

suspension bridge *noun* (1821)
: a bridge that has its roadway suspended from two or more cables usually passing over towers and securely anchored at the ends — see BRIDGE illustration

suspension points *noun plural* (1919)
: usually three spaced periods used to show the omission of a word or word group from a written context

sus·pen·sive \sə-ˈspen(t)-siv\ *adjective* (15th century)
1 : stopping temporarily **:** SUSPENDING ⟨a *suspensive* veto⟩
2 : characterized by suspense, suspended judgment, or indecisiveness
3 : characterized by suspension
— sus·pen·sive·ly *adverb*

sus·pen·sor \sə-ˈspen(t)-sər\ *noun* [New Latin, from Latin *suspendere*] (1832)
: a suspending part or structure: as **a :** a group or chain of cells that is produced from the zygote of a heterosporous plant and serves to push the embryo which arises at its extremity deeper into the embryo sac and into contact with the food supply of the megaspore **b :** either of a pair of gametangia-bearing hyphal outgrowths in fungi (order Mucorales) that extend from two sexually compatible hyphae and support the resulting zygospore

¹sus·pen·so·ry \sə-ˈspen(t)-sə-rē, -ˈspen(t)s-rē\ *adjective* (15th century)
1 : held in suspension; *also* **:** fitted or serving to suspend
2 : temporarily leaving undetermined **:** SUSPENSIVE 1

²suspensory *noun, plural* **-ries** (15th century)
: something that suspends or holds up; *especially* **:** a fabric supporter for the scrotum

suspensory ligament *noun* (1831)
: a ligament or fibrous membrane suspending an organ or part; *especially* **:** a ringlike fibrous membrane connecting the ciliary body and the lens of the eye and holding the lens in place — see EYE illustration

¹sus·pi·cion \sə-ˈspi-shən\ *noun* [Middle English, from Latin *suspicion-, suspicio,* from *suspicere* to suspect — more at SUSPECT] (14th century)
1 a : the act or an instance of suspecting something wrong without proof or on slight evidence **:** MISTRUST **b :** a state of mental uneasiness and uncertainty **:** DOUBT
2 : a barely detectable amount **:** TRACE ⟨just a *suspicion* of garlic⟩
synonym see UNCERTAINTY

²suspicion *transitive verb* **sus·pi·cioned; sus·pi·cion·ing** \-ˈspi-sh(ə-)niŋ\ (circa 1637)
chiefly dialect **:** SUSPECT

sus·pi·cious \sə-ˈspi-shəs\ *adjective* (14th century)
1 : tending to arouse suspicion **:** QUESTIONABLE
2 : disposed to suspect **:** DISTRUSTFUL ⟨*suspicious* of strangers⟩
3 : expressing or indicative of suspicion ⟨a *suspicious* glance⟩
— sus·pi·cious·ly *adverb*
— sus·pi·cious·ness *noun*

sus·pi·ra·tion \ˌsəs-pə-ˈrā-shən\ *noun* (15th century)
: a long deep breath **:** SIGH

sus·pire \sə-ˈspīr\ *intransitive verb* **sus·pired; sus·pir·ing** [Middle English, from Latin *suspirare,* from *sub-* + *spirare* to breathe] (15th century)
: to draw a long deep breath **:** SIGH

suss \ˈsəs\ *transitive verb* [by shortening & alteration from *suspect*] (1966)
1 *chiefly British* **:** to inspect or investigate so as to gain more knowledge — usually used with *out*
2 *chiefly British* **:** FIGURE OUT — usually used with *out*

Sus·sex spaniel \ˈsə-siks-, -ˌseks-\ *noun* [*Sussex,* England] (1856)
: any of a breed of short-legged short-necked long-bodied spaniels of English origin with a flat or slightly wavy golden liver-colored coat

sus·tain \sə-ˈstān\ *transitive verb* [Middle English *sustenen,* from Old French *sustenir,* from Latin *sustinēre* to hold up, sustain, from *sub-, sus-* up + *tenēre* to hold — more at SUB-, THIN] (13th century)
1 : to give support or relief to
2 : to supply with sustenance **:** NOURISH
3 : KEEP UP, PROLONG
4 : to support the weight of **:** PROP; *also* **:** to carry or withstand (a weight or pressure)
5 : to buoy up ⟨*sustained* by hope⟩
6 a : to bear up under **b :** SUFFER, UNDERGO ⟨*sustained* heavy losses⟩
7 a : to support as true, legal, or just **b :** to allow or admit as valid ⟨the court *sustained* the motion⟩
8 : to support by adequate proof **:** CONFIRM ⟨testimony that *sustains* our contention⟩ ■
— sus·tained·ly \-ˈstā-nəd-lē, -ˈstānd-lē\ *adverb*
— sus·tain·er *noun*

sus·tain·able \səs-ˈstā-nə-bəl\ *adjective* (circa 1727)
1 : capable of being sustained
2 a : of, relating to, or being a method of harvesting or using a resource so that the resource is not depleted or permanently damaged ⟨*sustainable* techniques⟩ ⟨*sustainable* agriculture⟩ **b :** of or relating to a lifestyle involving the use of sustainable methods ⟨*sustainable* society⟩
— sus·tain·abil·i·ty \-ˌstā-nə-ˈbi-lə-tē\ *noun*

sustained yield *noun* (circa 1905)
: production of a biological resource (as timber or fish) under management procedures which insure replacement of the part harvested by regrowth or reproduction before another harvest occurs
— sustained–yield *adjective*

sus·tain·ing *adjective* (1605)
1 : serving to sustain
2 : aiding in the support of an organization through a special fee ⟨a *sustaining* member⟩

sus·te·nance \ˈsəs-tə-nən(t)s\ *noun* [Middle English, from Old French, from *sustenir*] (14th century)
1 a : means of support, maintenance, or subsistence **:** LIVING **b :** FOOD, PROVISIONS; *also* **:** NOURISHMENT
2 a : the act of sustaining **:** the state of being sustained **b :** a supplying or being supplied with the necessaries of life
3 : something that gives support, endurance, or strength

sus·ten·tac·u·lar cell \ˌsəs-tən-ˈta-kyə-lər-, -ˌten-\ *noun* [New Latin *sustentaculum* supporting part, from Latin, prop, from *sustentare*] (1901)
: a supporting epithelial cell (as a Sertoli cell or a cell of the olfactory epithelium) that lacks a specialized function (as nerve-impulse conduction)

sus·ten·ta·tion \-ˈtā-shən\ *noun* [Middle English, from Middle French, from Latin *sustentation-, sustentatio* act of holding up, from *sustentare* to hold up, frequentative of *sustinēre* to sustain] (14th century)
1 : the act of sustaining **:** the state of being sustained: as **a :** MAINTENANCE, UPKEEP **b :** PRESERVATION, CONSERVATION **c :** maintenance of life, growth, or morale **d :** provision with sustenance

□ USAGE
sustain Sense 6b of *sustain* has been around since the 1400s and has been given without stigma in dictionaries at least since Johnson's of 1755, yet a few commentators have found fault with it. The earliest of these was evidently offended by the frequency with which the word was used in English newspapers in the 1860s. More recent commentators content themselves with finding something vaguely pretentious about the use; a few call it formal, which is odd, since it is commonly found in journalistic surroundings ⟨the company *sustained* operating losses totalling more than $30 million —Richard A. Lester *(New Republic)*⟩ ⟨from injuries *sustained* before and during arrest —E. J. Kahn, Jr. *(New Yorker)*⟩ and especially in sportswriting ⟨in a 1983 game against the Eagles, he *sustained* a concussion —Jill Lieber *(Sports Illustrated)*⟩. It has also been considered a bit of a cliché ⟨is man ever hurt in a motor smash? No. He *sustains* an injury —Flann O'Brien⟩ but is nonetheless standard.

2 : something that sustains **:** SUPPORT
— **sus·ten·ta·tive** \ˈsəs-tən-ˌtā-tiv, sə-ˈsten-tə-tiv\ *adjective*

Su·su \ˈsü-(ˌ)sü\ *noun, plural* **Susu** *or* **Susus** (1786)
1 : a member of a West African people of Mali, Guinea, and the area along the northern border of Sierra Leone
2 : the Mande language of the Susu people

su·sur·ra·tion \ˌsü-sə-ˈrā-shən\ *noun* (14th century)
: a whispering sound **:** MURMUR

su·sur·rous \su̇-ˈsər-əs, -ˈsə-rəs\ *adjective* (1859)
: full of whispering sounds

su·sur·rus \su̇-ˈsər-əs, -ˈsə-rəs\ *noun* [Latin, hum, whisper — more at SWARM] (1826)
: a whispering or rustling sound
— **su·sur·rant** \-ˈsər-ənt, -ˈsə-rənt\ *adjective*

sut·ler \ˈsət-lər\ *noun* [obsolete Dutch *soeteler,* from Low German *suteler* sloppy worker, camp cook] (1599)
: a civilian provisioner to an army post often with a shop on the post

su·tra \ˈsü-trə\ *noun* [Sanskrit *sūtra* precept, literally, thread; akin to Latin *suere* to sew — more at SEW] (1801)
1 : a precept summarizing Vedic teaching; *also* **:** a collection of these precepts
2 : a discourse of the Buddha

sut·tee \(ˌ)sə-ˈtē, ˈsə-ˌtē\ *noun* [Sanskrit *satī* wife who performs suttee, literally, devoted woman, from feminine of *sat* true, good; akin to Old English *sōth* true — more at SOOTH] (1786)
: the act or custom of a Hindu widow willingly being cremated on the funeral pyre of her husband as an indication of her devotion to him; *also* **:** a woman cremated in this way

¹su·ture \ˈsü-chər\ *noun* [Middle French & Latin; Middle French, from Latin *sutura* seam, suture, from *sutus,* past participle of *suere* to sew — more at SEW] (1541)
1 a : a strand or fiber used to sew parts of the living body; *also* **:** a stitch made with a suture **b :** the act or process of sewing with sutures
2 a : a uniting of parts **b :** the seam or seamlike line along which two things or parts are sewed or united
3 a : the line of union in an immovable articulation (as between the bones of the skull); *also* **:** such an articulation **b :** a furrow at the junction of adjacent bodily parts; *especially* **:** a line of dehiscence (as on a fruit)
— **su·tur·al** \ˈsü-chə-rəl, ˈsüch-rəl\ *adjective*
— **su·tur·al·ly** \-rə-lē\ *adverb*

²suture *transitive verb* **su·tured; su·tur·ing** \ˈsü-chə-riŋ, ˈsüch-riŋ\ (1777)
: to unite, close, or secure with sutures ⟨*suture* a wound⟩

su·zer·ain \ˈsü-zə-rən, -ˌrān; ˈsüz-rən\ *noun* [French, from (assumed) Middle French *suserain,* from Middle French *sus* up (from Latin *sursum,* from *sub-* up + *versum* -ward, from neuter of *versus,* past participle of *vertere* to turn) + *-erain* (as in *soverain* sovereign) — more at SUB-, WORTH] (1807)
1 : a superior feudal lord to whom fealty is due **:** OVERLORD
2 : a dominant state controlling the foreign relations of a vassal state but allowing it sovereign authority in its internal affairs

su·zer·ain·ty \-tē\ *noun* [French *suzeraineté,* from Middle French *suserenete,* from (assumed) Middle French *suserain*] (1823)
: the dominion of a suzerain **:** OVERLORDSHIP

sved·berg \ˈsved-ˌbərg, ˈsfed-, -ˌber-ē\ *noun* [The *Svedberg*] (1939)
: a unit of time amounting to 10^{-13} second that is used to measure the sedimentation velocity of a colloidal solution (as of a protein) in an ultracentrifuge and to determine molecular weight by substitution in an equation — called also *svedberg unit*

svelte \ˈsvelt, ˈsfelt\ *adjective* [French, from Italian *svelto,* from past participle of *svellere* to pluck out, modification of Latin *evellere,* from *e-* + *vellere* to pluck — more at VULNERABLE] (circa 1817)
1 a : SLENDER, LITHE **b :** having clean lines **:** SLEEK
2 : URBANE, SUAVE
— **svelte·ly** *adverb*
— **svelte·ness** *noun*

Sven·ga·li \sven-ˈgä-lē, sfen-\ *noun* [*Svengali,* maleficent hypnotist in the novel *Trilby* (1894) by George du Maurier] (1919)
: one who attempts usually with evil intentions to persuade or force another to do his bidding

¹swab \ˈswäb\ *noun* [probably from obsolete Dutch *swabbe;* akin to Low German *swabber* mop] (1653)
1 a : MOP; *especially* **:** a yarn mop **b** (1) **:** a wad of absorbent material usually wound around one end of a small stick and used for applying medication or for removing material from an area (2) **:** a specimen taken with a swab **c :** a sponge or cloth patch attached to a long handle and used to clean the bore of a firearm
2 a : a useless or contemptible person **b :** SAILOR, GOB

²swab *transitive verb* **swabbed; swab·bing** [back-formation from *swabber*] (1719)
1 : to clean with or as if with a swab
2 : to apply medication to with a swab ⟨*swabbed* the wound with iodine⟩

swab·ber \ˈswä-bər\ *noun* [akin to Low German *swabber* mop, Middle English *swabben* to sway] (1592)
1 : one that swabs
2 : SWAB 2a

swab·bie *also* **swab·by** \ˈswä-bē\ *noun, plural* **swabbies** (1944)
slang **:** SWAB 2b

swad·dle \ˈswä-dᵊl\ *transitive verb* **swad·dled; swad·dling** \ˈswäd liŋ, ˈswä-dᵊl-iŋ\ [Middle English *swadelen, swathelen,* probably alteration of *swedelen, swethelen,* from *swethel* swaddling band, from Old English; akin to Old English *swathian* to swathe] (14th century)
1 a : to wrap (an infant) with swaddling clothes **b :** ENVELOP, SWATHE ⟨*swaddled* ourselves in sleeping bags⟩
2 : RESTRAIN, RESTRICT ⟨marriage . . . *swaddled* him in a domesticity he came to loathe —Nina Auerbach⟩

swaddling clothes *noun plural* (1535)
1 : narrow strips of cloth wrapped around an infant to restrict movement
2 : limitations or restrictions imposed on the immature or inexperienced

¹swag \ˈswag\ *verb* **swagged; swag·ging** [probably of Scandinavian origin; akin to Old Norse *sveggja* to cause to sway; akin to Old High German *swingan* to swing] (1530)
intransitive verb
1 : SWAY, LURCH
2 : SAG
transitive verb
1 : to adorn with swags
2 : to arrange (as drapery) in swags

²swag *noun* (1660)
1 : SWAY
2 a : something (as a decoration) hanging in a curve between two points **:** FESTOON **b :** a suspended cluster (as of evergreen branches)
3 a : goods acquired by unlawful means **:** LOOT **b :** SPOILS, PROFITS
4 : a depression in the earth
5 *chiefly Australian* **:** a pack of personal belongings

¹swage \ˈswāj, ˈswej\ *noun* [Middle English, ornamental border, from Middle French *souage*] (circa 1812)
: a tool used by metalworkers for shaping their work by holding it on the work or the work on it and striking with a hammer or sledge

²swage *transitive verb* **swaged; swag·ing** (1831)
: to shape by or as if by means of a swage

swage block *noun* (1843)
: a perforated cast-iron or steel block with grooved sides that is used in heading bolts and swaging bars by hand

¹swag·ger \ˈswa-gər\ *verb* **swag·gered; swag·ger·ing** \-g(ə-)riŋ\ [probably from ¹*swag* + *-er* (as in *chatter*)] (1590)
intransitive verb
1 : to conduct oneself in an arrogant or superciliously pompous manner; *especially* **:** to walk with an air of overbearing self-confidence
2 : BOAST, BRAG
transitive verb
: to force by argument or threat **:** BULLY
— **swag·ger·er** \-gər-ər\ *noun*
— **swag·ger·ing·ly** \-g(ə-)riŋ-lē\ *adverb*

²swagger *noun* (1725)
1 a : an act or instance of swaggering **b :** arrogant or conceitedly self-assured behavior **c :** ostentatious display or bravado
2 : a self-confident outlook **:** COCKINESS

³swagger *adjective* (1879)
: marked by elegance or showiness **:** POSH

swagger stick *noun* (1887)
: a short light stick usually covered with leather and tipped with metal at each end and intended for carrying in the hand (as by military officers)

swag·gie \ˈswa-gē\ *noun* [by shortening & alteration] (1891)
chiefly Australian **:** SWAGMAN

swag·man \ˈswag-mən\ *noun* (1851)
chiefly Australian **:** DRIFTER; *especially* **:** one who carries a swag when traveling

Swa·hi·li \swä-ˈhē-lē\ *noun, plural* **Swahili** *or* **Swahilis** [Arabic *sawāḥil,* plural of *sāḥil* coast] (1814)
1 : a member of a Bantu-speaking people of Zanzibar and the adjacent coast
2 : a Bantu language that is a trade and governmental language over much of East Africa and in the Congo region

swain \ˈswān\ *noun* [Middle English *swein* boy, servant, from Old Norse *sveinn;* akin to Old English *swan* swain, Latin *suus* one's own — more at SUICIDE] (14th century)
1 : RUSTIC, PEASANT; *specifically* **:** SHEPHERD
2 : a male admirer or suitor
— **swain·ish** \ˈswā-nish\ *adjective*
— **swain·ish·ness** *noun*

Swain·son's hawk \ˈswān(t)-sənz-\ *noun* [William *Swainson* (died 1855) English naturalist] (1895)
: a buteo (*Buteo swainsonii*) chiefly of western North America and South America having pointed wings and usually a dark breast

swale \ˈswā(ə)l\ *noun* [origin unknown] (1584)
: a low-lying or depressed and often wet stretch of land; *also* **:** a shallow depression on a golf fairway or green

¹swal·low \ˈswä-(ˌ)lō\ *noun* [Middle English *swalowe,* from Old English *swealwe;* akin to Old High German *swalawa* swallow] (before 12th century)
1 : any of numerous small widely distributed oscine birds (family Hirundinidae) that have a short bill, long pointed wings, and often a deeply forked tail and that feed on insects caught on the wing
2 : any of several swifts that superficially resemble swallows

swallow 1

²swallow *verb* [Middle English *swalowen,* from Old English *swelgan;* akin to Old High German *swelgan* to swallow] (before 12th century)
transitive verb
1 : to take through the mouth and esophagus into the stomach
2 : to envelop or take in as if by swallowing **:** ABSORB
3 : to accept without question, protest, or resentment ⟨*swallow* an insult⟩ ⟨a hard story to *swallow*⟩
4 : TAKE BACK, RETRACT ⟨had to *swallow* my words⟩
5 : to keep from expressing or showing **:** REPRESS ⟨*swallowed* my anger⟩
6 : to utter (as words) indistinctly
intransitive verb
1 : to receive something into the body through the mouth and esophagus
2 : to perform the action characteristic of swallowing something especially under emotional stress
— swal·low·able \'swä-lō-ə-bəl\ *adjective*
— swal·low·er \'swä-lə-wər\ *noun*

³swallow *noun* (14th century)
1 : the passage connecting the mouth to the stomach
2 : a capacity for swallowing
3 a : an act of swallowing **b :** an amount that can be swallowed at one time
4 : an aperture in a block on a ship between the sheave and frame through which the rope reeves

swal·low·tail \'swä-lō-,tāl, -lə-\ *noun* (1703)
1 : a deeply forked and tapering tail (as of a swallow)
2 : TAILCOAT
3 : any of various large butterflies (family Papilionidae and especially genus *Papilio*) with each hind wing usually having an elongated process
— swal·low–tailed \-,tāld\ *adjective*

swallowtail 3

swam *past of* SWIM

swa·mi \'swä-mē\ *noun* [Hindi *svāmī,* from Sanskrit *svāmin* owner, lord, from *sva* one's own — more at SUICIDE] (1895)
1 : a Hindu ascetic or religious teacher; *specifically* **:** a senior member of a religious order — used as a title
2 : one that resembles or emulates a swami **:** PUNDIT, SEER

¹swamp \'swämp, 'swȯmp\ *noun* [perhaps alteration of Middle English *sompe,* from Middle Dutch *somp* morass; akin to Middle High German *sumpf* marsh, Greek *somphos* spongy] (1624)
1 : a wetland often partially or intermittently covered with water; *especially* **:** one dominated by woody vegetation
2 : a tract of swamp
3 : a difficult or troublesome situation or subject
— swamp *adjective*

²swamp (1784)
transitive verb
1 a : to fill with or as if with water **:** INUNDATE, SUBMERGE **b :** to overwhelm numerically or by an excess of something **:** FLOOD ⟨*swamped* with work⟩
2 : to open by removing underbrush and debris
intransitive verb
: to become submerged

swamp buggy *noun* (1941)
: a vehicle designed to travel over swampy terrain; *especially* **:** a four-wheel motor vehicle with oversize tires

swamp·er \'swäm-pər, 'swȯm-\ *noun* (1775)
1 a : an inhabitant of swamps or lowlands **b :** one familiar with swampy terrain
2 : a general assistant **:** HANDYMAN, HELPER

swamp·land \'swämp-,land, 'swȯmp-\ *noun* (1662)
: SWAMP 1

swampy \'swäm-pē, 'swȯm-\ *adjective* **swamp·i·er; -est** (1649)
: consisting of, suggestive of, or resembling swamp **:** MARSHY
— swamp·i·ness *noun*

¹swan \'swän\ *noun, plural* **swans** [Middle English, from Old English; akin to Middle High German *swan* and perhaps to Latin *sonus* sound — more at SOUND] (before 12th century)
1 *plural also* **swan :** any of various large heavy-bodied long-necked mostly pure white aquatic birds (family Anatidae and especially genus *Cygnus*) that are related to but larger than the geese
2 : one that resembles or is likened to a swan
3 *capitalized* **:** the constellation Cygnus

²swan *intransitive verb* **swanned; swan·ning** (1942)
: to wander aimlessly or idly **:** DALLY

³swan *intransitive verb* **swanned; swan·ning** [perhaps euphemism for *swear*] (1784)
dialect **:** DECLARE, SWEAR

swan boat *noun* (1953)
: a small boat usually for children or sightseers pedaled by an operator who sits aft in a large model of a swan

swan dive *noun* (1898)
: a front dive executed with the head back, back arched, and arms spread sideways and then brought together above the head to form a straight line with the body as the diver enters the water

¹swank \'swaŋk\ *adjective* [Middle Low German or Middle Dutch *swanc* supple; akin to Old High German *swingan* to swing] (1773)
Scottish **:** full of life or energy **:** ACTIVE

²swank *intransitive verb* [perhaps akin to Middle High German *swanken* to sway; akin to Middle Dutch *swanc* supple] (circa 1809)
: SHOW OFF, SWAGGER; *also* **:** BOAST 1

³swank *or* **swanky** \'swaŋ-kē\ *adjective* **swank·er** *or* **swank·i·er; -est** (circa 1842)
1 : characterized by showy display **:** OSTENTATIOUS ⟨a *swank* limousine⟩
2 : fashionably elegant **:** SMART ⟨a *swank* restaurant⟩
— swank·i·ly \-kə-lē\ *adverb*
— swank·i·ness \-kē-nəs\ *noun*

⁴swank *noun* (circa 1854)
1 : arrogance or ostentation of dress or manner **:** PRETENTIOUSNESS, SWAGGER
2 : ELEGANCE, FASHIONABLENESS

swan·nery \'swä-nə-rē, 'swän-rē\ *noun, plural* **-ner·ies** (1754)
: a place where swans are bred or kept

swans·down \'swänz-,dau̇n\ *noun* (1606)
1 : the soft downy feathers of the swan often used as trimming on articles of dress
2 : a heavy cotton flannel that has a thick nap on the face and is made with sateen weave

swan·skin \'swän-,skin\ *noun* (1610)
1 : the skin of a swan with the down or feathers on it
2 : fabric resembling flannel and having a soft nap or surface

swan song *noun* (1831)
1 : a song of great sweetness said to be sung by a dying swan
2 : a farewell appearance or final act or pronouncement

¹swap \'swäp\ *verb* **swapped; swap·ping** [Middle English *swappen* to strike; from the practice of striking hands in closing a business deal] (14th century)
transitive verb
1 a : to give in trade **:** BARTER **b :** EXCHANGE 2
2 : to take turns in telling ⟨*swap* stories⟩
intransitive verb
: to make an exchange
— swap·per *noun*

²swap *noun* (1625)
: an act, instance, or process of exchanging one thing for another

swap meet *noun* (1965)
: a gathering for the sale or barter of usually secondhand objects

swa·raj \swə-'räj\ *noun* [Hindi *svarāj,* from Sanskrit *sva* own + Hindi *rāj* rule — more at SUICIDE, RAJ] (1908)
: national or local self-government in India
— swa·raj·ist \-'rä-jist\ *noun*

sward \'swȯrd\ *noun* [Middle English, from Old English *sweard, swearth* skin, rind; akin to Middle High German *swart* skin, hide] (15th century)
1 : a portion of ground covered with grass
2 : the grassy surface of land
— sward·ed \'swȯr-dəd\ *adjective*

swarf \'swȯrf\ *noun* [of Scandinavian origin; akin to Old Norse *svarf* file dust; akin to Old English *sweorfan* to file away — more at SWERVE] (1587)
: material (as metallic particles and abrasive fragments) removed by a cutting or grinding tool

¹swarm \'swȯrm\ *noun* [Middle English, from Old English *swearm;* akin to Old High German *swaram* swarm and probably to Latin *susurrus* hum] (before 12th century)
1 a (1) **:** a great number of honeybees emigrating together from a hive in company with a queen to start a new colony elsewhere (2) **:** a colony of honeybees settled in a hive **b :** an aggregation of free-floating or free-swimming unicellular organisms — usually used of zoospores
2 a : a large number of animate or inanimate things massed together and usually in motion **:** THRONG ⟨*swarms* of sightseers⟩ ⟨a *swarm* of locusts⟩ ⟨a *swarm* of meteors⟩ **b :** a number of similar geological features or phenomena close together in space or time ⟨a *swarm* of dikes⟩ ⟨an earthquake *swarm*⟩

²swarm (14th century)
intransitive verb
1 : to form and depart from a hive in a swarm
2 a : to move or assemble in a crowd **:** THRONG **b :** to hover about in the manner of a bee in a swarm
3 : to contain a swarm **:** TEEM
transitive verb
1 : to fill with a swarm
2 : to beset or surround in a swarm ⟨players *swarming* the quarterback⟩
— swarm·er *noun*

³swarm *verb* [origin unknown] (14th century)
intransitive verb
: to climb with the hands and feet; *specifically* **:** SHIN ⟨*swarm* up a pole⟩
transitive verb
: to climb up **:** MOUNT

swarm spore *noun* (1859)
: any of various minute motile sexual or asexual spores; *especially* **:** ZOOSPORE

swart \'swȯrt\ *adjective* [Middle English, from Old English *sweart;* akin to Old High German *swarz* black, Latin *sordes* dirt] (before 12th century)
1 a : SWARTHY **b** *archaic* **:** producing a swarthy complexion
2 : BANEFUL, MALIGNANT
— swart·ness *noun*

swar·thy \'swȯr-t͟hē, -thē\ *adjective* **swar·thi·er; -est** [alteration of obsolete *swarty,* from *swart*] (1587)
: of a dark color, complexion, or cast
— swar·thi·ness *noun*

¹swash \'swäsh, 'swȯsh\ *verb* [probably imitative] (1556)
intransitive verb
1 : BLUSTER, SWAGGER
2 : to make violent noisy movements

3 : to move with a splashing sound
transitive verb
: to cause to splash
²swash *noun* (1593)
1 : SWAGGER
2 : a narrow channel of water lying within a sandbank or between a sandbank and the shore
3 : a dashing of water against or on something; *especially* **:** the rush of water up a beach from a breaking wave
³swash *noun* [obsolete English *swash* slanting] (1683)
: an extended flourish on a printed character
⁴swash *adjective* (1683)
: having one or more swashes ⟨*swash* capitals⟩
swash·buck·le \'swäsh-ˌbə-kəl, 'swȯsh-\ *intransitive verb* **-led; -ling** \-ˌbə-k(ə-)liŋ\ [back-formation from *swashbuckler*] (1897)
: to act the part of a swashbuckler
swash·buck·ler \-ˌbə-klər\ *noun* [¹*swash* + *buckler*] (1560)
1 : a swaggering or daring soldier or adventurer
2 : a novel or drama dealing with a swashbuckler
swash·buck·ling \-ˌbə-k(ə-)liŋ\ *adjective* [*swashbuckler*] (circa 1693)
1 : acting in the manner of a swashbuckler
2 : characteristic of, marked by, or done by swashbucklers
swash·er \'swä-shər, 'swȯ-\ *noun* (1589)
: SWASHBUCKLER
swas·ti·ka \'swäs-ti-kə *also* swä-'stē-\ *noun* [Sanskrit *svastika,* from *svasti* well-being, from *su-* well + *as-* to be; akin to Sanskrit *asti* he is, Old English *is;* from its being regarded as a good luck symbol] (1871)
1 : a symbol or ornament in the form of a Greek cross with the ends of the arms extended at right angles all in the same rotary direction
2 : a swastika used as a symbol of anti-Semitism or of Nazism
¹swat \'swät\ *transitive verb* **swat·ted; swat·ting** [English dialect, to squat, alteration of English *squat*] (circa 1796)
: to hit with a sharp slapping blow usually with an instrument (as a bat or swatter)
²swat *noun* (circa 1800)
1 : a powerful or crushing blow
2 : a long hit in baseball; *especially* **:** HOME RUN
swatch \'swäch\ *noun* [origin unknown] (1647)
1 a : a sample piece (as of fabric) or a collection of samples **b :** a characteristic specimen
2 : PATCH
3 : a small collection
4 : SWATH 2
swath \'swäth, 'swȯth\ *or* **swathe** \'swäth, 'swȯth, 'swāth\ *noun* [Middle English, from Old English *swæth* footstep, trace; akin to Middle High German *swade* swath] (14th century)
1 a : a row of cut grain or grass left by a scythe or mowing machine **b :** the sweep of a scythe or a machine in mowing or the path cut in one course
2 : a long broad strip or belt
3 : a stroke of or as if of a scythe
4 : a space devastated as if by a scythe
¹swathe \'swäth, 'swȯth, 'swāth\ *or* **swath** \'swäth, 'swäth, 'swȯth, 'swȯth\ *noun* [Middle English, from (assumed) Old English *swæth;* akin to Old English *swathian* to swathe] (before 12th century)
1 : a band used in swathing
2 : an enveloping medium
²swathe \'swäth, 'swȯth, 'swāth\ *transitive verb* **swathed; swath·ing** [Middle English, from Old English *swathian*] (12th century)
1 : to bind, wrap, or swaddle with or as if with a bandage
2 : ENVELOP

swath·er \'swä-thər, -thər\ *noun* (circa 1875)
: a harvesting machine that cuts and windrows grain and seed crops; *also* **:** a mower attachment that windrows the swath
swathing clothes *noun plural* [Middle English] (14th century)
obsolete **:** SWADDLING CLOTHES
swats \'swäts\ *noun plural* [probably from Old English *swātan,* plural, beer] (1508)
Scottish **:** DRINK; *especially* **:** new ale
swat·ter \'swä-tər\ *noun* (1912)
: one that swats; *especially* **:** FLYSWATTER
S wave *noun* [*s*econdary] (1937)
: a wave (as from an earthquake) in which the propagated disturbance is a shear in an elastic medium (as the earth) — compare PRESSURE WAVE
¹sway \'swā\ *noun* (14th century)
1 : the action or an instance of swaying or of being swayed **:** an oscillating, fluctuating, or sweeping motion
2 : an inclination or deflection caused by or as if by swaying
3 a : a controlling influence **b :** sovereign power **:** DOMINION **c :** the ability to exercise influence or authority **:** DOMINANCE
synonym see POWER
²sway *verb* [alteration of earlier *swey* to fall, swoon, from Middle English *sweyen,* probably of Scandinavian origin; akin to Old Norse *sveigja* to sway; akin to Lithuanian *svaigti* to become dizzy] (circa 1500)
intransitive verb
1 a : to swing slowly and rhythmically back and forth from a base or pivot **b :** to move gently from an upright to a leaning position
2 : to hold sway **:** act as ruler or governor
3 : to fluctuate or veer between one point, position, or opinion and another
transitive verb
1 a : to cause to sway **:** set to swinging, rocking, or oscillating **b :** to cause to bend downward to one side **c :** to cause to turn aside **:** DEFLECT, DIVERT
2 *archaic* **a :** WIELD **b :** GOVERN, RULE
3 a : to cause to vacillate **b :** to exert a guiding or controlling influence on
4 : to hoist in place ⟨*sway* up a mast⟩
synonym see SWING, AFFECT
— sway·er *noun*
sway·backed \'swā-ˌbakt\ *also* **sway·back** \-ˌbak\ *adjective* (1680)
: having an abnormally hollow or sagging back ⟨a *swaybacked* mare⟩
— sway·back *noun*
sway bar *noun* (1949)
: a bar that torsionally couples the right and left front-wheel suspensions of an automobile to reduce roll and sway
Swa·zi \'swä-zē\ *noun, plural* **Swazi** *or* **Swazis** (1878)
1 : a member of a Bantu people of southeastern Africa
2 : a Bantu language of the Swazi people
¹swear \'swar, 'swer\ *verb* **swore** \'swōr, 'swȯr\; **sworn** \'swōrn, 'swȯrn\; **swear·ing** [Middle English *sweren,* from Old English *swerian;* akin to Old High German *swerien* to swear and perhaps to Old Church Slavonic *svarŭ* quarrel] (before 12th century)
transitive verb
1 : to utter or take solemnly (an oath)
2 a : to assert as true or promise under oath ⟨a *sworn* affidavit⟩ **b :** to assert or promise emphatically or earnestly ⟨*swore* to uphold the Constitution⟩
3 a : to put to an oath **:** administer an oath to **b :** to bind by an oath ⟨*swore* them to secrecy⟩
4 *obsolete* **:** to invoke the name of (a sacred being) in an oath
5 : to bring into a specified state by swearing ⟨*swore* his life away⟩
intransitive verb
1 : to take an oath

2 : to use profane or obscene language **:** CURSE
— swear·er *noun*
— swear by : to place great confidence in
— swear for : to give assurance for **:** GUARANTEE
— swear off : to vow to abstain from **:** RENOUNCE ⟨*swear off* smoking⟩
²swear *noun* (14th century)
: OATH, SWEARWORD
swear in *transitive verb* (1536)
: to induct into office by administration of an oath
swear out *transitive verb* (1895)
: to procure (a warrant for arrest) by making a sworn accusation
swear·word \'swar-ˌwərd, 'swer-\ *noun* (1883)
: a profane or obscene oath or word
¹sweat \'swet\ *verb* **sweat** *or* **sweat·ed; sweat·ing** [Middle English *sweten,* from Old English *swǣtan,* from *swāt* sweat; akin to Old High German *sweiz* sweat, Latin *sudare* to sweat, Greek *hidrōs* sweat] (before 12th century)
intransitive verb
1 a : to excrete moisture in visible quantities through the openings of the sweat glands **:** PERSPIRE **b :** to labor or exert oneself so as to cause perspiration
2 a : to emit or exude moisture ⟨cheese *sweats* in ripening⟩ **b :** to gather surface moisture in beads as a result of condensation ⟨stones *sweat* at night⟩ **c** (1) **:** FERMENT (2) **:** PUTREFY
3 : to undergo anxiety or mental or emotional distress
4 : to become exuded through pores or a porous surface **:** OOZE
transitive verb
1 : to emit or seem to emit from pores **:** EXUDE
2 : to manipulate or produce by hard work or drudgery
3 : to get rid of or lose (weight) by or as if by sweating or being sweated
4 : to make wet with perspiration
5 a : to cause to excrete moisture from the skin **b :** to drive hard **:** OVERWORK **c :** to exact work from at low wages and under unfair or unhealthful conditions **d** *slang* **:** to give the third degree to
6 : to cause to exude or lose moisture; *especially* **:** to subject (as tobacco leaves) to fermentation
7 a : to extract something valuable from by unfair or dishonest means **:** FLEECE **b :** to remove particles of metal from (a coin) by abrasion
8 a : to heat (as solder) so as to melt and cause to run especially between surfaces to unite them; *also* **:** to unite by such means ⟨*sweat* a pipe joint⟩ **b :** to heat so as to extract an easily fusible constituent ⟨*sweat* bismuth ore⟩ **c :** to sauté in a covered vessel until natural juices are exuded
9 *slang* **:** to worry about ⟨doesn't *sweat* the small stuff —Barry McDermott⟩
— sweat blood : to work or worry intensely ⟨in preparing speeches each *sweats blood* in his own way —Stewart Cockburn⟩
²sweat *noun* (13th century)
1 : hard work **:** DRUDGERY
2 : the fluid excreted from the sweat glands of the skin **:** PERSPIRATION
3 : moisture issuing from or gathering in drops on a surface
4 a : the condition of one sweating or sweated **b :** a spell of sweating
5 : a state of anxiety or impatience
6 *plural* **a :** SWEAT SUIT **b :** SWEATPANTS
— no sweat *slang* **:** with little or no difficulty **:** EASILY; *also* **:** EASY — often used interjectionally

sweat·band \'swet-,band\ *noun* (1891)
1 : a usually leather band lining the inner edge of a hat or cap to prevent sweat damage
2 : a band of material worn around the head or wrist to absorb sweat

sweat bee *noun* (1894)
: any of various small black or brownish bees (family Halictidae) that are attracted to perspiration

sweat·box \-,bäks\ *noun* (1864)
1 : a place in which one is made to sweat; *especially* **:** a narrow box or cell in which a prisoner is placed for punishment
2 : a device for sweating something (as hides in tanning or dried figs)

sweat·ed \'swe-təd\ *adjective* (1882)
: of, subjected to, or produced under sweatshop conditions ⟨*sweated* labor⟩ ⟨*sweated* goods⟩

sweat equity *noun* (1966)
: equity in a property resulting from labor invested in improvements that increase its value; *also* **:** the labor so invested

sweat·er \'swe-tər\ *noun* (15th century)
1 : one that sweats or causes sweating
2 : a knitted or crocheted jacket or pullover

sweat·er·dress \-,dres\ *noun* (1952)
: a knitted or crocheted dress

sweater girl *noun* (1940)
: a woman with a shapely bust

sweat·er–vest \'swe-tər-,vest\ *noun* (1952)
: a sleeveless pullover or buttoned sweater

sweat gland *noun* (1845)
: a simple tubular gland of the skin that secretes perspiration, is widely distributed in nearly all parts of the human skin, and consists typically of an epithelial tube extending spirally from a minute pore on the surface of the skin into the dermis or subcutaneous tissues where it ends in a convoluted tuft

sweat lodge *noun* (1850)
: a hut, lodge, or cavern heated by steam from water poured on hot stones and used especially by American Indians for ritual or therapeutic sweating

sweat out *transitive verb* (1589)
1 : to work one's way painfully through or to
2 : to endure or wait through the course of

sweat·pants \'swet-,pan(t)s\ *noun plural* (1925)
: pants having a drawstring or elastic waist and elastic cuffs at the ankle that are worn especially by athletes in warming up

sweat·shirt \-,shərt\ *noun* (1925)
: a loose collarless pullover usually of heavy cotton jersey

sweat·shop \-,shäp\ *noun* (1892)
: a shop or factory in which workers are employed for long hours at low wages and under unhealthy conditions

sweat suit *noun* (1930)
: a suit worn usually for exercise that consists of a sweatshirt and sweatpants

sweat test *noun* (1978)
: a test for cystic fibrosis that involves measuring the subject's sweat for abnormally high sodium chloride content

sweaty \'swe-tē\ *adjective* **sweat·i·er; -est** (14th century)
1 : causing sweat ⟨a *sweaty* day⟩ ⟨*sweaty* work⟩
2 : wet or stained with or smelling of sweat
— sweat·i·ly \'swe-tᵊl-ē\ *adverb*
— sweat·i·ness \'swe-tē-nəs\ *noun*

swede \'swēd\ *noun* [Low German or obsolete Dutch] (1589)
1 *capitalized* **a :** a native or inhabitant of Sweden **b :** a person of Swedish descent
2 *chiefly British* **:** RUTABAGA

Swe·den·bor·gian \,swē-dᵊn-'bȯr-j(ē-)ən, -'bȯr-gē-ən\ *adjective* (1807)
: of or relating to the teachings of Emanuel Swedenborg or the Church of the New Jerusalem based on his teachings
— Swedenborgian *noun*
— Swe·den·bor·gian·ism \-j(ē-)ə-,ni-zəm, -gē-ə-\ *noun*

Swed·ish \'swē-dish\ *noun* (1605)
1 : the North Germanic language spoken in Sweden and a part of Finland
2 *plural in construction* **:** the people of Sweden
— Swedish *adjective*

Swedish massage *noun* (1911)
: massage involving a system of active and passive exercise of muscles and joints

¹sweep \'swēp\ *verb* **swept** \'swept\; **sweep·ing** [Middle English *swepen;* akin to Old English *swāpan* to sweep, Old High German *sweifen* to wander] (14th century)
transitive verb
1 a : to remove from a surface with or as if with a broom or brush ⟨*swept* the crumbs from the table⟩ **b :** to destroy completely **:** WIPE OUT — usually used with *away* ⟨everything she cherished, might be *swept* away overnight —Louis Bromfield⟩ **c :** to remove or take with a single continuous forceful action ⟨*swept* the books off the desk⟩ **d :** to remove from sight or consideration ⟨the problem can't be *swept* under the rug⟩ **e :** to drive or carry along with irresistible force ⟨a wave of protest that *swept* the opposition into office⟩
2 a : to clean with or as if with a broom or brush **b :** to clear by repeated and forcible action **c :** to move across or along swiftly, violently, or overwhelmingly ⟨fire *swept* the business district —*American Guide Series: Maryland*⟩ **d :** to win an overwhelming victory in or on ⟨*sweep* the elections⟩ **e :** to win all the games or contests of ⟨*sweep* a double-header⟩ ⟨*sweep* a series⟩
3 : to touch in passing with a swift continuous movement
4 : to trace or describe the locus or extent of (as a line, circle, or angle)
5 : to cover the entire range of ⟨his eyes *swept* the horizon⟩
intransitive verb
1 a : to clean a surface with or as if with a broom **b :** to move swiftly, forcefully, or devastatingly ⟨the wind *swept* through the treetops⟩
2 : to go with stately or sweeping movements ⟨proudly *swept* into the room⟩
3 : to move or extend in a wide curve or range
— sweep one off one's feet : to gain immediate and unquestioning support, approval, or acceptance by a person
— sweep the board *or* **sweep the table 1 :** to win all the bets on the table **2 :** to win everything **:** beat all competitors

²sweep *noun* (1548)
1 : something that sweeps or works with a sweeping motion: as **a :** a long pole or timber pivoted on a tall post and used to raise and lower a bucket in a well **b :** a triangular cultivator blade that cuts off weeds under the soil surface **c :** a windmill sail

sweep 1a

2 a : an instance of sweeping; *especially* **:** a clearing out or away with or as if with a broom **b :** the removal from the table in one play in casino of all the cards by pairing or combining **c :** an overwhelming victory **d :** a winning of all the contests or prizes in a competition **e :** a wide-ranging search of an area (as by police)
3 a : a movement of great range and force **b :** a curving or circular course or line **c :** the compass of a sweeping movement **:** SCOPE **d :** a broad unbroken area or extent **e :** an end run in football in which one or more linemen pull back and run interference for the ballcarrier
4 : CHIMNEY SWEEP
5 : SWEEPSTAKES
6 : obliquity with respect to a reference line ⟨*sweep* of an airplane wing⟩; *especially* **:** SWEEPBACK
7 *plural* **:** a television ratings period during which surveys are taken to determine advertising rates
synonym see RANGE

sweep·back \'swēp-,bak\ *noun* (1914)
: the backward slant of an airplane wing in which the outer portion of the wing is downstream from the inner portion

sweep·er \'swē-pər\ *noun* (15th century)
1 : one that sweeps
2 : a lone back in soccer who plays between the line of the defenders and the goal

sweep hand *noun* (1943)
: SWEEP-SECOND HAND

¹sweep·ing *noun* (14th century)
1 : the act or action of one that sweeps ⟨gave the room a good *sweeping*⟩
2 *plural* **:** things collected by sweeping **:** REFUSE

²sweeping *adjective* (1610)
1 a : moving or extending in a wide curve or over a wide area **b :** having a curving line or form
2 a : EXTENSIVE ⟨*sweeping* reforms⟩ **b :** marked by wholesale and indiscriminate inclusion ⟨*sweeping* generalities⟩
— sweep·ing·ly \'swē-piŋ-lē\ *adverb*
— sweep·ing·ness *noun*

sweep–sec·ond hand \'swēp-,se-kənd-, -kənt-\ *noun* (circa 1940)
: a hand marking seconds on a timepiece mounted concentrically with the other hands and read from the same dial as the minute hand

sweep·stakes \-,stāks\ *noun plural but singular or plural in construction, also* **sweep–stake** \-,stāk\ [Middle English *swepestake* one who wins all the stakes in a game, from *swepen* to sweep + *stake*] (1785)
1 a : a race or contest in which the entire prize may be awarded to the winner; *specifically* **:** STAKE RACE **b :** CONTEST, COMPETITION
2 : any of various lotteries

sweepy \'swē-pē\ *adjective* **sweep·i·er; -est** (1697)
: sweeping in motion, line, or force

¹sweet \'swēt\ *adjective* [Middle English *swete,* from Old English *swēte;* akin to Old High German *suozi* sweet, Latin *suadēre* to urge, *suavis* sweet, Greek *hēdys*] (before 12th century)
1 a (1) **:** pleasing to the taste (2) **:** being or inducing the one of the four basic taste sensations that is typically induced by disaccharides and is mediated especially by receptors in taste buds at the front of the tongue — compare BITTER, SALT, SOUR **b** (1) *of a beverage* **:** containing a sweetening ingredient **:** not dry (2) *of wine* **:** retaining a portion of natural sugar
2 a : pleasing to the mind or feelings **:** AGREEABLE, GRATIFYING — often used as a generalized term of approval ⟨how *sweet* it is⟩ **b :** marked by gentle good humor or kindliness **c :** FRAGRANT **d** (1) **:** delicately pleasing to the ear or eye (2) **:** played in a straightforward melodic style ⟨*sweet* jazz⟩ **e :** SACCHARINE, CLOYING
3 : much loved **:** DEAR
4 a : not sour, rancid, decaying, or stale **:** WHOLESOME ⟨*sweet* milk⟩ **b :** not salt or salted **:** FRESH ⟨*sweet* butter⟩ **c** *of land* **:** free from excessive acidity **d :** free from noxious gases and odors **e :** free from excess of acid, sulfur, or corrosive salts ⟨*sweet* crude oil⟩

5 : SKILLFUL, PROFICIENT ⟨a *sweet* golf swing⟩
6 — used as an intensive ⟨take your own *sweet* time⟩
— **sweet·ly** *adverb*
— **sweet·ness** *noun*
— **sweet on :** having a crush on

²sweet *adverb* (13th century)
: in a sweet manner

³sweet *noun* (14th century)
1 : something that is sweet to the taste: as **a :** a food (as a candy or preserve) having a high sugar content ⟨fill up on *sweets*⟩ **b** *British* **:** DESSERT **c** *British* **:** HARD CANDY
2 : a sweet taste sensation
3 : a pleasant or gratifying experience, possession, or state
4 : DARLING, SWEETHEART
5 a *archaic* **:** FRAGRANCE **b** *plural, archaic* **:** things having a sweet smell

sweet alyssum *noun* (1822)
: a widely cultivated perennial European herb (*Lobularia maritima*) of the mustard family having narrow leaves and clusters of small fragrant usually white or pink flowers

sweet–and–sour \ˌswē-tᵊn-ˈsau̇(ə)r\ *adjective* (1928)
: seasoned with a sauce containing sugar and vinegar or lemon juice ⟨*sweet-and-sour* shrimp⟩

sweet basil *noun* (circa 1647)
: a basil (*Ocimum basilicum*) with whitish or purple flowers that includes several cultivars (as bush basil)

sweet bay *noun* (1716)
1 : LAUREL 1
2 : a magnolia (*Magnolia virginiana*) of the eastern U.S. that has fragrant white flowers and leaves with glaucous undersides

sweet birch *noun* (1785)
: a common birch (*Betula lenta*) of the eastern U.S. that has spicy brown bark when young, hard dark-colored wood, and a volatile oil in its bark resembling wintergreen — called also *black birch*

sweet·bread \ˈswēt-ˌbred\ *noun* (1565)
: the thymus or pancreas of a young animal (as a calf) used for food

sweet·bri·er *also* **sweet·bri·ar** \-ˌbrī(-ə)r\ *noun* (1538)
: an Old World rose (especially *Rosa eglanteria*) with stout recurved prickles and white to deep rosy pink single flowers — called also *eglantine*

sweet cherry *noun* (circa 1901)
: a white-flowered Eurasian cherry (*Prunus avium*) widely grown for its large sweet-flavored fruits; *also* **:** its fruit

sweet chocolate *noun* (1897)
: chocolate that contains added sugar

sweet cic·e·ly \-ˈsi-s(ə-)lē\ *noun* [*cicely* from Latin *seselis,* from Greek] (1668)
: any of a genus (*Osmorhiza*) of American and eastern Asian herbs of the carrot family that typically have thick fleshy roots and grow in moist woodlands

sweet clover *noun* (1868)
: any of a genus (*Melilotus*) of Old World legumes that have trifoliolate leaves and are widely grown for soil improvement or hay

sweet corn *noun* (1646)
: an Indian corn (especially *Zea mays rugosa*) with kernels containing a high percentage of sugar and adapted for table use when the kernels are unripe

sweet·en \ˈswē-tᵊn\ *verb* **sweet·ened; sweet·en·ing** \ˈswēt-niŋ, ˈswē-tᵊn-iŋ\ (circa 1552)
transitive verb
1 : to make sweet
2 : to soften the mood or attitude of
3 : to make less painful or trying
4 : to free from a harmful or undesirable quality or substance; *especially* **:** to remove sulfur compounds from ⟨*sweeten* natural gas⟩
5 : to make more valuable or attractive: as **a :** to increase (a pot not won on the previous deal) by anteing prior to another deal **b :** to place additional securities as collateral for (a loan)
intransitive verb
: to become sweet
— **sweet·en·er** \ˈswēt-nər, ˈswē-tᵊn-ər\ *noun*

sweetening *noun* (1819)
: something that sweetens

sweet fern *noun* (1654)
: a small North American shrub (*Comptonia peregrina*) of the wax-myrtle family with aromatic leaves

sweet flag *noun* (1784)
: a perennial marsh herb (*Acorus calamus*) of the arum family with long narrow leaves and an aromatic rootstock — called also *calamus*

sweet gum *noun* (1700)
1 : a North American tree (*Liquidambar styraciflua*) of the witch-hazel family with palmately lobed leaves, corky branches, and hard wood
2 : heartwood of the sweet gum or reddish brown lumber sawed from it

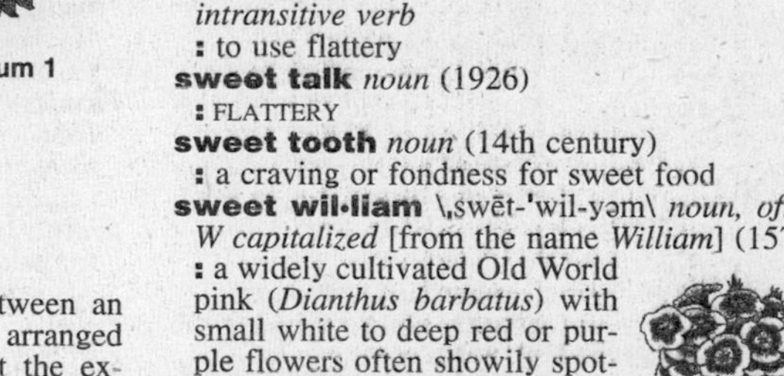
sweet gum 1

¹sweet·heart \ˈswēt-ˌhärt\ *noun* (14th century)
1 : DARLING
2 : one who is loved
3 : a generally likable person
4 : a remarkable one of its kind

²sweetheart *adjective* (1942)
: of or relating to an agreement between an employer and a labor union official arranged privately for their benefit usually at the expense of the workers ⟨a *sweetheart* contract⟩; *broadly* **:** arranged in private for the benefit of a few at the expense of many ⟨a *sweetheart* deal⟩

sweetheart neckline *noun* (1941)
: a neckline for women's clothing that is high in back and low in front where it is scalloped to resemble the top of a heart

sweet·ie \ˈswē-tē\ *noun* (1705)
1 *plural, British* **:** SWEET 1a
2 : SWEETHEART

sweetie pie *noun* (1928)
: SWEETHEART

sweet·ing \ˈswē-tiŋ\ *noun* (13th century)
1 *archaic* **:** SWEETHEART
2 : a sweet apple

sweet·ish \-tish\ *adjective* (1580)
1 : somewhat sweet
2 : unpleasantly sweet
— **sweet·ish·ly** *adverb*

sweet marjoram *noun* (1565)
: a perennial marjoram (*Origanum majorana* synonym *Majorana hortensis*) with dense spikelike flower clusters

sweet·meat \ˈswēt-ˌmēt\ *noun* (14th century)
: a food rich in sugar: as **a :** a candied or crystallized fruit **b :** CANDY, CONFECTION

sweetness and light *noun* (1704)
1 : a harmonious combination of beauty and enlightenment viewed as a hallmark of culture
2 a : amiable reasonableness of disposition ⟨they were all *sweetness and light*⟩ **b :** an untroubled or harmonious state or condition ⟨but all was not *sweetness and light*⟩

sweet orange *noun* (1538)
1 a : an orange (*Citrus sinensis*) that is probably native to southeastern Asia, has a fruit with a pithy central axis, and is the source of the widely cultivated oranges of commerce **b :** a cultivated orange derived from the sweet orange and usually having fruit with a relatively thin skin and sweet juicy edible pulp
2 : the fruit of a sweet orange

sweet pea *noun* (1732)
1 : a widely cultivated Italian legume (*Lathyrus odoratus*) having slender usually climbing stems, ovate leaves, and large fragrant flowers
2 : the flower of a sweet pea

sweet pepper *noun* (1814)
: a large mild thick-walled capsicum fruit; *also* **:** a pepper plant bearing this fruit

sweet potato *noun* (1750)
1 : a tropical vine (*Ipomoea batatas*) related to the morning glory with variously shaped leaves and purplish flowers; *also* **:** its large thick sweet and nutritious tuberous root that is cooked and eaten as a vegetable — compare YAM 2
2 : OCARINA

sweet·shop \ˈswēt-ˌshäp\ *noun* (1879)
chiefly British **:** a candy store

sweet·sop \-ˌsäp\ *noun* (1696)
: SUGAR APPLE

sweet sorghum *noun* (1867)
: SORGO

sweet spot *noun* (circa 1949)
: the area around the center of mass of a bat, racket, or head of a club that is the most effective part with which to hit a ball

sweet–talk \ˈswēt-ˌtȯk\ (1928)
transitive verb
: CAJOLE, COAX
intransitive verb
: to use flattery

sweet talk *noun* (1926)
: FLATTERY

sweet tooth *noun* (14th century)
: a craving or fondness for sweet food

sweet wil·liam \ˌswēt-ˈwil-yəm\ *noun, often W capitalized* [from the name *William*] (1573)
: a widely cultivated Old World pink (*Dianthus barbatus*) with small white to deep red or purple flowers often showily spotted, banded, or mottled and borne in flat bracteate heads on erect stalks

sweet william

¹swell \ˈswel\ *verb* **swelled; swelled** *or* **swol·len** \ˈswō-lən\; **swell·ing** [Middle English, from Old English *swellan;* akin to Old High German *swellan* to swell] (before 12th century)
intransitive verb
1 a : to expand (as in size, volume, or numbers) gradually beyond a normal or original limit ⟨the population *swelled*⟩ **b :** to become distended or puffed up ⟨her ankle is badly *swollen*⟩ **c :** to form a bulge or rounded elevation
2 a : to become filled with pride and arrogance **b :** to behave or speak in a pompous, blustering, or self-important manner **c :** to play the swell
3 : to become distended with emotion
transitive verb
1 : to affect with a powerful or expansive emotion
2 : to increase the size, number, or intensity of
synonym see EXPAND

²swell *noun* (1606)
1 : a long often massive and crestless wave or succession of waves often continuing beyond or after its cause (as a gale)
2 a : the condition of being protuberant **b :** a rounded elevation
3 a : the act or process of swelling **b** (1) **:** a gradual increase and decrease of the loudness of a musical sound; *also* **:** a sign indicating a swell (2) **:** a device used in an organ for governing loudness

4 a *archaic* **:** an impressive, pompous, or fashionable air or display **b :** a person dressed in the height of fashion **c :** a person of high social position or outstanding competence

³swell *adjective* (1785)
1 a : STYLISH **b :** socially prominent
2 : EXCELLENT — used as a generalized term of enthusiasm

swell box *noun* (circa 1801)
: a chamber in an organ containing a set of pipes and having shutters that open or shut to regulate the volume of tone

swelled head *noun* (1891)
: an exaggerated opinion of oneself **:** SELF-CONCEIT
— swelled–head·ed \'sweld-ˌhe-dəd\ *adjective*
— swelled–head·ed·ness *noun*

swell·fish \'swel-ˌfish\ *noun* (1807)
: PUFFER 2a

swell–front \'swel-ˌfrənt\ *adjective* (1914)
: BOWFRONT 1

swell·head \-ˌhed\ *noun* (1845)
: one who has a swelled head
— swell·head·ed \-ˌhe-dəd\ *adjective*
— swell·head·ed·ness *noun*

swell·ing \'swe-liŋ\ *noun* (before 12th century)
1 : something that is swollen; *specifically* **:** an abnormal bodily protuberance or localized enlargement
2 : the condition of being swollen

¹swel·ter \'swel-tər\ *verb* **swel·tered; swel·ter·ing** \-t(ə-)riŋ\ [Middle English *sweltren,* frequentative of *swelten* to die, be overcome by heat, from Old English *sweltan* to die; akin to Gothic *swiltan* to die] (14th century)
intransitive verb
1 : to suffer, sweat, or be faint from heat
2 : to become exceedingly hot ⟨in summer, the place *swelters*⟩
transitive verb
1 : to oppress with heat
2 *archaic* **:** EXUDE ⟨*sweltered* venom —Shakespeare⟩

²swelter *noun* (1851)
1 : a state of oppressive heat
2 : WELTER
3 : an excited or overwrought state of mind **:** SWEAT ⟨in a *swelter*⟩

swel·ter·ing *adjective* (1586)
: oppressively hot
— swel·ter·ing·ly \-t(ə-)riŋ-lē\ *adverb*

swept \'swept\ *adjective* [*swept,* past participle of *sweep*] (1903)
: slanted backward

swept–back \'swep(t)-'bak\ *adjective* (1914)
: possessing sweepback

swerve \'swərv\ *verb* **swerved; swerv·ing** [Middle English, from Old English *sweorfan* to wipe, file away; akin to Old High German *swerban* to wipe off, Welsh *chwerfu* to whirl] (14th century)
intransitive verb
: to turn aside abruptly from a straight line or course **:** DEVIATE
transitive verb
: to cause to turn aside or deviate ☆
— swerve *noun*

swev·en \'swe-vən\ *noun* [Middle English, from Old English *swefn* sleep, dream, vision — more at SOMNOLENT] (before 12th century)
archaic **:** DREAM, VISION

swid·den \'swi-dᵊn\ *noun, often attributive* [English dialect, burned clearing, probably from Old Norse *svithinn,* past participle of *svitha* to burn, singe] (circa 1868)
: a temporary agricultural plot produced by cutting back and burning off vegetative cover

¹swift \'swift\ *adjective* [Middle English, from Old English; akin to Old English *swīfan* to revolve — more at SWIVEL] (before 12th century)
1 : moving or capable of moving with great speed
2 : occurring suddenly or within a very short time
3 : quick to respond **:** READY
synonym see FAST

²swift *adverb* (14th century)
: SWIFTLY ⟨*swift*-flowing⟩

³swift *noun* (15th century)
1 : any of several lizards (especially of the genus *Sceloporus*) that run swiftly
2 : a reel for winding yarn or thread
3 : any of numerous small plainly colored birds (family Apodidae) that are related to the hummingbirds but superficially much resemble swallows

swift fox *noun* (1869)
: a small fox (*Vulpes velox*) with large ears that occurs on the plains of western North America

swift·let \'swif(t)-lət\ *noun* (circa 1890)
: any of various cave-dwelling swifts (genus *Collocalia*) of Asia including one (*C. unicolor*) that produces the nest used in bird's nest soup

swift·ly *adverb* (before 12th century)
: in a swift manner **:** with speed **:** QUICKLY

swift·ness \'swif(t)-nəs\ *noun* (before 12th century)
1 : the quality or state of being swift **:** CELERITY
2 : the fact of being swift

¹swig \'swig\ *noun* [origin unknown] (circa 1623)
: a quantity drunk at one time

²swig *verb* **swigged; swig·ging** (circa 1650)
transitive verb
: to drink in long drafts ⟨*swig* cider⟩
intransitive verb
: to take a swig **:** DRINK
— swig·ger *noun*

¹swill \'swil\ *verb* [Middle English *swilen,* from Old English *swillan*] (before 12th century)
transitive verb
1 : WASH, DRENCH
2 : to drink great drafts of **:** GUZZLE
3 : to feed (as a pig) with swill
intransitive verb
1 : to drink or eat freely, greedily, or to excess
2 : SWASH
— swill·er *noun*

²swill *noun* (1553)
1 : something suggestive of slop or garbage **:** REFUSE
2 a : a semiliquid food for animals (as swine) composed of edible refuse mixed with water or skimmed or sour milk **b :** GARBAGE

¹swim \'swim\ *verb* **swam** \'swam\; **swum** \'swəm\; **swim·ming** [Middle English *swimmen,* from Old English *swimman;* akin to Old High German *swimman* to swim] (before 12th century)
intransitive verb
1 a : to propel oneself in water by natural means (as movements of the limbs, fins, or tail) **b :** to play in the water (as at a beach or swimming pool)
2 : to move with a motion like that of swimming **:** GLIDE ⟨a cloud *swam* slowly across the moon⟩
3 a : to float on a liquid **:** not sink **b :** to surmount difficulties **:** not go under ⟨sink or *swim,* live or die, survive or perish —Daniel Webster⟩
4 : to become immersed in or flooded with or as if with a liquid ⟨potatoes *swimming* in gravy⟩
5 : to have a floating or reeling appearance or sensation
transitive verb
1 a : to cross by propelling oneself through water ⟨*swim* a stream⟩ **b :** to execute in swimming
2 : to cause to swim or float
— swim·mer *noun*

²swim *noun* (1599)
1 : a smooth gliding motion
2 : an act or period of swimming
3 : a temporary dizziness or unconsciousness
4 a : an area frequented by fish **b :** the main current of activity ⟨in the *swim*⟩

³swim *adjective* (1924)
: of, relating to, or used in or for swimming ⟨a *swim* meet⟩

swim bladder *noun* (1837)
: the air bladder of a fish

swim fin *noun* (1947)
: FLIPPER 1b

swim·ma·ble \'swi-mə-bəl\ *adjective* (1852)
: that can be swum

swim·mer·et \ˌswi-mə-'ret, 'swi-mə-ˌ\ *noun* (1840)
: one of a series of small unspecialized appendages under the abdomen of many crustaceans that are best developed in some decapods (as a lobster) and usually function in locomotion or reproduction

swimmer's itch *noun* (1928)
: a severe urticarial reaction to the presence in the skin of larval schistosomes that are not normally human parasites

¹swimming *adjective* (before 12th century)
1 [present participle of *swim*] **:** that swims ⟨a *swimming* bird⟩
2 [gerund of *swim*] **:** adapted to or used in or for swimming

²swimming *noun* (14th century)
: the act, art, or sport of one that swims and dives

swim·ming·ly \'swi-miŋ-lē\ *adverb* (1622)
: very well **:** SPLENDIDLY

swimming pool *noun* (1899)
: a pool suitable for swimming; *especially* **:** a tank (as of concrete or plastic) made for swimming

swim·my \'swi-mē\ *adjective* **swim·mi·er; -est** (1836)
1 : verging on, causing, or affected by dizziness or giddiness
2 *of vision* **:** UNSTEADY, BLURRED
— swim·mi·ly \'swi-mə-lē\ *adverb*

swim·suit \'swim-ˌsüt\ *noun* (1926)
: a suit for swimming or bathing

swim·wear \-ˌwar, -ˌwer\ *noun* (1935)
: clothing suitable for wear while swimming or bathing

¹swin·dle \'swin-dᵊl\ *verb* **swin·dled; swin·dling** \'swin(d)-liŋ, 'swin-dᵊl-iŋ\ [back-formation from *swindler,* from German *Schwindler* giddy person, from *schwindeln* to be dizzy, from Old High German *swintilōn,* frequentative of *swintan* to diminish, vanish; akin to Old English *swindan* to vanish] (circa 1782)
intransitive verb
: to obtain money or property by fraud or deceit
transitive verb

☆ SYNONYMS
Swerve, veer, deviate, depart, digress, diverge mean to turn aside from a straight course. SWERVE may suggest a physical, mental, or moral turning away from a given course, often with abruptness ⟨*swerved* to avoid hitting the dog⟩. VEER implies a major change in direction ⟨at that point the path *veers* to the right⟩. DEVIATE implies a turning from a customary or prescribed course ⟨never *deviated* from her daily routine⟩. DEPART suggests a deviation from a traditional or conventional course or type ⟨occasionally *departs* from his own guidelines⟩. DIGRESS applies to a departing from the subject of one's discourse ⟨a professor prone to *digress*⟩. DIVERGE may equal DEPART but usually suggests a branching of a main path into two or more leading in different directions ⟨after school their paths *diverged*⟩.

: to take money or property from by fraud or deceit ◆
synonym see CHEAT
— **swin·dler** \ˈswin(d)-lər, ˈswin-dᵊl-ər\ *noun*

[2]**swindle** *noun* (1833)
: an act or instance of swindling **:** FRAUD

swine \ˈswīn\ *noun, plural* **swine** [Middle English, from Old English *swīn;* akin to Old High German *swīn* swine, Latin *sus* — more at SOW] (before 12th century)
1 : any of various stout-bodied short-legged omnivorous mammals (family Suidae) with a thick bristly skin and a long mobile snout; *especially* **:** a domesticated member of the species (*Sus scrofa*) that includes the European wild boar
2 : a contemptible person

swine·herd \-ˌhərd\ *noun* (before 12th century)
: one who tends swine

[1]**swing** \ˈswiŋ\ *verb* **swung** \ˈswəŋ\; **swing·ing** \ˈswiŋ-iŋ\ [Middle English, to beat, fling, hurl, rush, from Old English *swingan* to beat, fling oneself, rush; akin to Old High German *swingan* to fling, rush] (13th century)
transitive verb
1 a : to cause to move vigorously through a wide arc or circle ⟨*swing* an ax⟩ **b :** to cause to sway to and fro **c** (1) **:** to cause to turn on an axis (2) **:** to cause to face or move in another direction ⟨*swing* the car into a side road⟩
2 : to suspend so as to permit swaying or turning
3 : to convey by suspension ⟨cranes *swinging* cargo into the ship's hold⟩
4 a (1) **:** to influence decisively ⟨*swing* a lot of votes⟩ (2) **:** to bring around by influence **b :** to handle successfully **:** MANAGE ⟨wasn't able to *swing* a new car on his income⟩ ⟨*swing* a deal⟩
5 : to play or sing (as a melody) in the style of swing music
intransitive verb
1 : to move freely to and fro especially in suspension from an overhead support
2 a : to die by hanging **b :** to hang freely from a support
3 : to move in or describe a circle or arc: **a :** to turn on a hinge or pivot **b :** to turn in place **c :** to convey oneself by grasping a fixed support ⟨*swing* aboard the train⟩
4 a : to have a steady pulsing rhythm **b :** to play or sing with a lively compelling rhythm; *specifically* **:** to play swing music
5 : to shift or fluctuate from one condition, form, position, or object of attention or favor to another ⟨*swing* constantly from optimism to pessimism and back —Sinclair Lewis⟩
6 a : to move along rhythmically **b :** to start up in a smooth vigorous manner ⟨ready to *swing* into action⟩
7 : to hit or aim at something with a sweeping arm movement
8 a : to be lively and up-to-date **b :** to engage freely in sex ☆ ☆

[2]**swing** *noun* (14th century)
1 : an act or instance of swinging **:** swinging movement: as **a** (1) **:** a stroke or blow delivered with a sweeping arm movement ⟨a batter with a powerful *swing*⟩ (2) **:** a sweeping or rhythmic movement of the body or a bodily part (3) **:** a dance figure in which two dancers revolve with joined arms or hands (4) **:** jazz dancing in moderate tempo with a lilting syncopation **b** (1) **:** the regular movement of a freely suspended object (as a pendulum) along an arc and back (2) **:** back and forth sweep ⟨the *swing* of the tides⟩ **c** (1) **:** steady pulsing rhythm (as in poetry or music) (2) **:** a steady vigorous movement characterizing an activity or creative work **d** (1) **:** a trend toward a high or low point in a fluctuating cycle (as of business activity) (2) **:** an often periodic shift from one condition, form, position, or object of attention or favor to another
2 a : liberty of action **b** (1) **:** the driving power of something swung or hurled (2) **:** steady vigorous advance **:** driving speed ⟨a train approaching at full *swing*⟩
3 : the progression of an activity, process, or phase of existence ⟨the work is in full *swing*⟩
4 : the arc or range through which something swings
5 : something that swings freely from or on a support; *especially* **:** a seat suspended by a rope or chains for swinging to and fro on for pleasure
6 a : a curving course or outline **b :** a course from and back to a point **:** a circular tour
7 : jazz played usually by a large dance band and characterized by a steady lively rhythm, simple harmony, and a basic melody often submerged in improvisation
8 : a short pass in football thrown to a back running to the outside

[3]**swing** *adjective* (1934)
1 : of or relating to musical swing ⟨a *swing* band⟩ ⟨*swing* music⟩
2 : that may swing often decisively either way on an issue or in an election ⟨*swing* voters⟩ ⟨a *swing* state⟩

[1]**swinge** \ˈswinj\ *transitive verb* **swinged; swinge·ing** [Middle English *swengen* to shake, from Old English *swengan;* akin to Old English *swingan*] (12th century)
chiefly dialect **:** BEAT, SCOURGE

[2]**swinge** *transitive verb* **swinged; swinge·ing** [alteration of *singe*] (1590)
dialect **:** SINGE, SCORCH

[1]**swinge·ing** *also* **swing·ing** \ˈswin-jiŋ\ *adjective* [from present participle of [1]*swinge*] (circa 1590)
chiefly British **:** very large, high, or severe ⟨*swingeing* fines⟩ ⟨*swingeing* taxes⟩

[2]**swingeing** *or* **swinging** *adverb* (1690)
chiefly British **:** VERY, SUPERLATIVELY

[1]**swing·er** \ˈswiŋ-ər\ *noun* (1543)
: one that swings: as **a :** a lively up-to-date person who indulges in what is considered fashionable **b :** one who engages freely in sex

[2]**swing·er** \ˈswin-jər\ *noun* [[1]*swinge*] (1599)
: WHOPPER 1

swing·ing \ˈswiŋ-iŋ\ *adjective* [present participle of [1]*swing*] (1956)
: being lively and up-to-date; *also* **:** abounding in swingers and swinging entertainment ⟨a *swinging* coffeehouse⟩

[1]**swing·ing·ly** \ˈswin-jiŋ-lē\ *adverb* (1672)
chiefly British **:** VERY, EXTREMELY

[2]**swing·ing·ly** \ˈswiŋ-iŋ-lē\ *adverb* (1882)
: in a swinging manner **:** with a swinging movement

swin·gle·tree \ˈswiŋ-gəl-(ˌ)trē\ *noun* [Middle English *swyngyll tre,* from *swyngyll* rod for beating flax (from Middle Dutch *swengel*) + *tre* tree] (15th century)
: WHIFFLETREE

swing·man \ˈswiŋ-ˌman, -mən\ *noun* (1965)
: a player capable of playing effectively in two different positions and especially of playing both guard and forward on a basketball team

swing shift *noun* (1940)
1 : the work shift between the day and night shifts (as from 4 P.M. to midnight)
2 : a group of workers in a factory operating seven days a week that work as needed to permit the regular shift workers to have one or more free days per week

swingy \ˈswiŋ-ē\ *adjective* **swing·i·er; -est** (1915)
: marked by swing

swin·ish \ˈswī-nish\ *adjective* (13th century)
: of, suggesting, or characteristic of swine **:** BEASTLY
— **swin·ish·ly** *adverb*
— **swin·ish·ness** *noun*

[1]**swink** \ˈswiŋk\ *intransitive verb* [Middle English, from Old English *swincan;* akin to Old High German *swingan* to rush — more at SWING] (before 12th century)
archaic **:** TOIL, SLAVE

[2]**swink** *noun* (12th century)
archaic **:** LABOR, DRUDGERY

[1]**swipe** \ˈswīp\ *noun* [probably alteration of *sweep*] (1739)
1 : a strong sweeping blow
2 : a sharp often critical remark ⟨took a parting *swipe* at management⟩

[2]**swipe** *verb* **swiped; swip·ing** (circa 1825)
intransitive verb
: to strike or move with a sweeping motion
transitive verb
1 : to strike or wipe with a sweeping motion
2 : STEAL, PILFER

swipes \ˈswīps\ *noun plural* [origin unknown] (circa 1796)
British **:** poor, thin, or spoiled beer; *also* **:** BEER

[1]**swirl** \ˈswər(-ə)l\ *verb* [Middle English] (14th century)

☆ SYNONYMS

Swing, wave, flourish, brandish, thrash mean to wield or cause to move to and fro or up and down. SWING implies regular or uniform movement ⟨*swing* the rope back and forth⟩. WAVE usually implies smooth or continuous motion ⟨*waving* the flag⟩. FLOURISH suggests vigorous, ostentatious, graceful movement ⟨*flourished* the winning lottery ticket⟩. BRANDISH implies threatening or menacing motion ⟨*brandishing* a knife⟩. THRASH suggests vigorous, abrupt, violent movement ⟨an infant *thrashing* his arms about⟩.

Swing, sway, oscillate, vibrate, fluctuate, waver, undulate mean to move from one direction to its opposite. SWING implies a movement of something attached at one end or one side ⟨the door suddenly *swung* open⟩. SWAY implies a slow swinging or teetering movement ⟨trees *swaying* in the breeze⟩. OSCILLATE stresses a usually regular alternation of direction ⟨an *oscillating* fan⟩. VIBRATE suggests the rapid oscillation of an elastic body under stress or impact ⟨the *vibrating* strings of a piano⟩. FLUCTUATE suggests constant irregular changes of level, intensity, or value ⟨*fluctuating* interest rates⟩. WAVER stresses irregular motion suggestive of reeling or tottering ⟨the exhausted runner *wavered* before collapsing⟩. UNDULATE suggests a gentle wavelike motion ⟨an *undulating* sea of grass⟩.

◇ WORD HISTORY

swindle One would hardly think that someone whose head is whirling could be convincing enough to perpetrate a swindle. However, the original meaning of the German noun *Schwindler* was "giddy person." In the same way that *giddy* has been extended in English to describe someone who is frivolous or foolish, *Schwindler* was extended to individuals given to habitual flights of fancy. The Germans applied the word as well to a fantastic schemer, then to a participant in shaky business deals, and finally to a cheat. The word *Schwindler* is a derivative of the verb *schwindeln,* first used to mean "to be dizzy" and then "to cheat." *Swindler* appeared first in English in the second half of the 18th century, and *swindle* was created from it by removal of the suffix. At first they were slang terms associated with the criminal underworld, but they soon became standard.

intransitive verb
1 a : to move with an eddying or whirling motion **b :** to pass in whirling confusion
2 : to have a twist or convolution
transitive verb
: to cause to swirl
— **swirl·ing·ly** \'swər-liŋ-lē\ *adverb*

²swirl *noun* (15th century)
1 a : a whirling mass or motion **:** EDDY **b :** whirling confusion ⟨a *swirl* of events⟩
2 : a twisting shape, mark, or pattern
3 : an act or instance of swirling

swirly \'swər-lē\ *adjective* **swirl·i·er; -est** (1785)
1 *Scottish* **:** KNOTTED, TWISTED
2 : that swirls **:** SWIRLING ⟨the *swirly* water of the rapids⟩

¹swish \'swish\ *verb* [imitative] (1756)
intransitive verb
: to move, pass, swing, or whirl with the sound of a swish
transitive verb
1 : to move, cut, or strike with a swish ⟨the horse *swished* its tail⟩
2 : to make (a basketball shot) so that the ball falls through the rim without touching it ⟨*swished* a three-point jumper⟩
— **swish·er** *noun*
— **swish·ing·ly** \'swi-shiŋ-lē\ *adverb*

²swish *noun* (1820)
1 a : a prolonged hissing sound (as of a whip cutting the air) **b :** a light sweeping or brushing sound (as of a full silk skirt in motion)
2 : a swishing movement
3 : an effeminate homosexual — usually used disparagingly

³swish *adjective* [origin unknown] (1879)
: SMART, FASHIONABLE

swishy \'swi-shē\ *adjective* **swish·i·er; -est** (1828)
1 : producing a swishing sound
2 : characterized by effeminate behavior

¹Swiss \'swis\ *noun* [Middle French *Suisse,* from Middle High German *Swīzer,* from *Swīz* Switzerland] (1515)
1 *plural* **Swiss a :** a native or inhabitant of Switzerland **b :** one that is of Swiss descent
2 *often not capitalized* **:** any of various fine sheer fabrics of cotton originally made in Switzerland; *especially* **:** DOTTED SWISS
3 : SWISS CHEESE

²Swiss *adjective* (1530)
: of, relating to, or characteristic of Switzerland or the Swiss

Swiss chard *noun* (1832)
: a beet (*Beta vulgaris cicla*) having large leaves and succulent stalks often cooked as a vegetable — called also *chard*

Swiss cheese *noun* (1822)
: a hard cheese characterized by elastic texture, mild nutlike flavor, and large holes that form during ripening

Swiss steak *noun* (1924)
: a slice of steak pounded with flour and braised usually with vegetables and seasonings

¹switch \'swich\ *noun* [perhaps from Middle Dutch *swijch* twig] (1592)
1 : a slender flexible whip, rod, or twig ⟨a riding *switch*⟩
2 : an act of switching: as **a :** a blow with a switch **b :** a shift from one to another **c :** a change from the usual
3 : a tuft of long hairs at the end of the tail of an animal (as a cow) — see COW illustration
4 a : a device made usually of two movable rails and necessary connections and designed to turn a locomotive or train from one track to another **b :** a railroad siding
5 : a device for making, breaking, or changing the connections in an electrical circuit
6 : a heavy strand of hair used in addition to a person's own hair for some coiffures

²switch (circa 1611)
transitive verb
1 : to strike or beat with or as if with a switch
2 : WHISK, LASH ⟨a cat *switching* its tail⟩
3 a (1) **:** to turn from one railroad track to another **:** SHUNT (2) **:** to move (cars) to different positions on the same track within terminal areas **b :** to make a shift in or exchange of ⟨*switch* seats⟩
4 a : to shift to another electrical circuit by means of a switch **b :** to operate an electrical switch so as to turn (as a light) off or on
intransitive verb
1 : to lash from side to side
2 : to make a shift or exchange
— **switch·able** \'swi-chə-bəl\ *adjective*
— **switch·er** *noun*

¹switch·back \'swich-ˌbak\ *noun* (1863)
: a zigzag road, trail, or section of railroad tracks for climbing a steep hill

²switchback *adjective* (1887)
: resembling a switchback (as in taking a zigzag course) ⟨a *switchback* career⟩

³switchback *intransitive verb* (1903)
: to follow a zigzag course especially in ascending or descending

switch·blade \-ˌblād\ *noun* (1932)
: a pocketknife having the blade spring-operated so that pressure on a release catch causes it to fly open — called also *switchblade knife*

switch·board \-ˌbōrd, -ˌbȯrd\ *noun* (1873)
: an apparatus (as in a telephone exchange) consisting of a panel on which are mounted electric switches so arranged that a number of circuits may be connected, combined, and controlled

switch engine *noun* (1867)
: a railroad engine used in switching cars

switch·er·oo \ˌswi-chə-'rü\ *noun, plural* **-oos** [alteration of *switch*] (1933)
slang **:** a surprising variation **:** REVERSAL

switch·grass \'swich-ˌgras\ *noun* (1840)
: a panic grass (*Panicum virgatum*) of the western U.S. that is used for hay

switch–hit \-'hit\ *intransitive verb* **-hit; -hit·ting** [back-formation from *switch-hitter*] (1938)
: to bat right-handed against a left-hander and left-handed against a right-hander in baseball

switch–hit·ter \-'hi-tər\ *noun* (1948)
1 : a baseball player who switch-hits
2 *slang* **:** BISEXUAL

switch knife *noun* (1950)
: SWITCHBLADE

switch·man \'swich-mən\ *noun* (1843)
: one who attends a switch (as in a railroad yard)

switch·yard \-ˌyärd\ *noun* (1943)
: a usually enclosed area for the switching facilities of a power station

swith \'swith\ *adverb* [Middle English, strongly, quickly, from Old English *swīthe* strongly, from *swīth* strong; akin to Gothic *swinths* strong, Old English *gesund* sound — more at SOUND] (13th century)
chiefly dialect **:** INSTANTLY, QUICKLY

swith·er \'swi-t͟hər\ *intransitive verb* [origin unknown] (1501)
dialect chiefly British **:** DOUBT, WAVER
— **swither** *noun, dialect chiefly British*

Swit·zer \'swit-sər\ *noun* [Middle High German *Swīzer*] (1549)
: SWISS

¹swiv·el \'swi-vəl\ *noun, often attributive* [Middle English; akin to Old English *swīfan* to revolve, Old High German *swebōn* to roll, heave] (14th century)
: a device joining two parts so that one or both can pivot freely (as on a bolt or pin)

²swivel *verb* **-eled** *or* **-elled; -el·ing** *or* **-el·ling** \'swi-və-liŋ, 'swiv-liŋ\ (1794)
transitive verb
: to turn on or as if on a swivel ⟨*swiveled* his eyes in various directions⟩
intransitive verb
: to swing or turn on or as if on a swivel

swivel chair *noun* (1860)
: a chair that swivels on its base

swiv·el–hipped \'swi-vəl-ˌhipt\ *adjective* (1947)
: moving with or characterized by movement with a twisting motion of the hips

swiv·et \'swi-vət\ *noun* [origin unknown] (circa 1892)
: a state of extreme agitation

¹swiz·zle \'swi-zəl\ *noun* [origin unknown] (1813)
: an iced whiskey sour churned with a swizzle stick until the glass or pitcher becomes frosted

²swizzle *verb* **swiz·zled; swiz·zling** \'swi-zə-liŋ, 'swiz-liŋ\ (circa 1847)
intransitive verb
: to drink especially to excess **:** GUZZLE
transitive verb
: to mix or stir with or as if with a swizzle stick
— **swiz·zler** \'swi-zə-lər, 'swiz-lər\ *noun*

swizzle stick *noun* (1879)
: a stick used to stir mixed drinks

swob *archaic variant of* SWAB

swollen *past participle of* SWELL

¹swoon \'swün\ *intransitive verb* [Middle English *swounen*] (13th century)
1 a : FAINT **b :** to become enraptured ⟨*swooning* with joy⟩
2 : DROOP, FADE
— **swoon·er** *noun*
— **swoon·ing·ly** \'swü-niŋ-lē\ *adverb*

²swoon *noun* (13th century)
1 a : a partial or total loss of consciousness **b :** a state of bewilderment or ecstasy **:** DAZE, RAPTURE
2 : a state of suspended animation **:** TORPOR

¹swoop \'swüp\ *verb* [alteration of Middle English *swopen* to sweep, from Old English *swāpan* — more at SWEEP] (1566)
intransitive verb
: to move with a sweep ⟨the eagle *swooped* down on its prey⟩
transitive verb
: to gain or carry off in or as if in a swoop — usually used with *up*
— **swoop·er** *noun*

²swoop *noun* (1605)
: an act or instance of swooping

swoop·stake \'swüp-ˌstāk\ *adverb* [from alteration of *sweepstake*] (1602)
obsolete **:** in an indiscriminate manner

¹swoosh \'swüsh, 'swu̇sh\ *verb* [imitative] (1867)
intransitive verb
1 : to make or move with a rushing sound ⟨a car *swooshed* by⟩
2 : GUSH, SWIRL
transitive verb
: to discharge or transport with a rushing sound

²swoosh *noun* (1885)
: an act or instance of swooshing

swop *chiefly British variant of* SWAP

sword \'sōrd, 'sȯrd\ *noun, often attributive* [Middle English, from Old English *sweord;* akin to Old High German *swert* sword] (before 12th century)
1 : a weapon (as a cutlass or rapier) with a long blade for cutting or thrusting that is often used as a symbol of honor or authority
2 a : an agency or instrument of destruction or combat **b :** the use of force ⟨the pen is mightier than the *sword* —E. G. Bulwer-Lytton⟩
3 : coercive power
4 : something that resembles a sword
— **sword·like** \-ˌlīk\ *adjective*
— **at swords' points :** mutually antagonistic **:** ready to fight

sword cane *noun* (1837)
: a cane in which a sword blade is concealed

sword dance *noun* (1604)
1 : a dance performed by men in a circle holding a sword in the right hand and grasping the tip of a neighbor's sword in the left hand
2 : a dance performed over or around swords
— **sword dancer** *noun*

sword fern *noun* (circa 1829)
: any of several ferns with long narrow more or less sword-shaped fronds: as **a :** a tropical fern (*Nephrolepis exaltata*) from which the Boston fern has been developed **b :** a fern (*Polystichum munitum*) of western North America with a large fleshy rhizome

sword·fish \'sōrd-ˌfish, 'sȯrd-\ *noun* (15th century)
: a very large oceanic bony fish (*Xiphias gladius* of the family Xiphiidae) that has a long swordlike beak formed by the bones of the upper jaw and is an important food and game fish

swordfish

sword grass *noun* (1598)
: any of various grasses or sedges having leaves with a sharp or toothed edge

sword knot *noun* (1694)
: an ornamental cord or tassel tied to the hilt of a sword

sword of Dam·o·cles \-'da-mə-ˌklēz\ *often S capitalized* (1820)
: an impending disaster

sword·play \'sōrd-ˌplā, 'sȯrd-\ *noun* (1627)
1 : the art or skill of wielding a sword especially in fencing
2 : an exhibition of swordplay
— sword·play·er *noun*

swords·man \'sōrdz-mən, 'sȯrdz-\ *noun* (circa 1680)
1 : one skilled in swordplay; *especially* **:** a saber fencer
2 *archaic* **:** a soldier armed with a sword

swords·man·ship \-ˌship\ *noun* (circa 1852)
: SWORDPLAY

sword·tail \'sōrd-ˌtāl, 'sȯrd-\ *noun* (circa 1928)
: a small brightly marked Central American live-bearer (*Xiphophorus helleri* of the family Poeciliidae) often kept in tropical aquariums and bred in many colors

swore *past of* SWEAR

sworn *past participle of* SWEAR

[1]swot \'swät\ *noun* [English dialect, sweat, from Middle English *swot,* from Old English *swāt* — more at SWEAT] (1850)
British **:** GRIND 2b

[2]swot *intransitive verb* **swot·ted; swot·ting** (circa 1860)
British **:** GRIND 4

[1]swound \'swau̇nd, 'swünd\ *noun* [Middle English, alteration of *swoun* swoon, from *swounen* to swoon] (15th century)
archaic **:** SWOON 1a

[2]swound *intransitive verb* (1530)
archaic **:** SWOON

swum *past participle of* SWIM

swung *past and past participle of* SWING

swung dash *noun* (1951)
: a character ~ used in printing to conserve space by representing part or all of a previously spelled-out word

syb·a·rite \'si-bə-ˌrīt\ *noun* (circa 1555)
1 [from the notorious luxury of the Sybarites] **:** VOLUPTUARY, SENSUALIST
2 *capitalized* **:** a native or resident of the ancient city of Sybaris
— syb·a·rit·ic \ˌsi-bə-'ri-tik\ *adjective*
— syb·a·rit·i·cal·ly \-ti-k(ə-)lē\ *adverb*
— syb·a·rit·ism \'si-bə-ˌrī-ˌtī-zəm\ *noun*

syc·a·mine \'si-kə-ˌmīn, -mən\ *noun* [Latin *sycaminus,* from Greek *sykaminos,* of Semitic origin; akin to Hebrew *shiqmāh* mulberry tree, sycamore] (1526)
: a tree of the Bible that is usually considered a mulberry (*Morus nigra*)

syc·a·more \'si-kə-ˌmōr, -ˌmȯr\ *noun* [Middle English *sicamour,* from Middle French *sicamor,* from Latin *sycomorus,* from Greek *sykomoros,* probably modification of a Semitic word akin to Hebrew *shiqmāh* sycamore] (14th century)
1 *also* **syc·o·more** \'si-kə-ˌ\ **:** a fig tree (*Ficus sycomorus*) of Africa and the Middle East that is the sycamore of Scripture and has sweet and edible fruit similar but inferior to the common fig
2 : a Eurasian maple (*Acer pseudoplatanus*) with long racemes of showy yellowish green flowers that is widely planted as a shade tree
3 : [2]PLANE; *especially* **:** a very large spreading tree (*Platanus occidentalis*) of eastern and central North America with 3- to 5-lobed broadly ovate leaves

syce \'sīs\ *noun* [Hindi *sāīs,* from Arabic *sā'is*] (1653)
: an attendant (as a groom) especially in India

sy·cee \'sī-ˌsē\ *noun* [Chinese (Guangdong) *sai-sì,* literally, fine silk] (1711)
: silver money made in the form of ingots and formerly used in China

sy·co·ni·um \sī-'kō-nē-əm\ *noun, plural* **-nia** \-nē-ə\ [New Latin, from Greek *sykon* fig + New Latin *-ium*] (circa 1856)
: the multiple fleshy fruit of a fig in which the ovaries are borne within an enlarged succulent concave or hollow receptacle

sy·co·phan·cy \'si-kə-fən(t)-sē *also* 'sī- & -ˌfan(t)-sē\ *noun* (1672)
: obsequious flattery; *also* **:** the character or behavior of a sycophant

sy·co·phant \-fənt *also* -ˌfant\ *noun* [Latin *sycophanta* slanderer, swindler, from Greek *sykophantēs* slanderer, from *sykon* fig + *phainein* to show — more at FANCY] (1575)
: a servile self-seeking flatterer
synonym see PARASITE
— sycophant *adjective*

sy·co·phan·tic \ˌsi-kə-'fan-tik *also* ˌsī-\ *adjective* (1676)
: of, relating to, or characteristic of a sycophant **:** FAWNING, OBSEQUIOUS
— sy·co·phan·ti·cal·ly \-'fan-ti-k(ə-)lē\ *adverb*

sy·co·phant·ish \ˌsi-kə-'fan-tish *also* ˌsī-\ *adjective* (1794)
: SYCOPHANTIC
— sy·co·phant·ish·ly *adverb*

sy·co·phant·ism \'si-kə-fən-ˌti-zəm *also* 'sī- & -ˌfan-\ *noun* (1821)
: SYCOPHANCY

sy·co·phant·ly \-lē\ *adverb* (1672)
: in a sycophantic manner

sy·co·sis \sī-'kō-səs\ *noun* [New Latin, from Greek *sykōsis,* from *sykon* fig] (circa 1827)
: a chronic inflammatory disorder of the hair follicles marked by papules, pustules, and tubercles with crusting

sy·e·nite \'sī-ə-ˌnīt\ *noun* [Latin *Syenites* (*lapis*) stone of Syene, from *Syene,* ancient city in Egypt] (circa 1796)
: an igneous rock composed chiefly of feldspar
— sy·e·nit·ic \ˌsī-ə-'ni-tik\ *adjective*

sy·li \'sē-lē\ *noun, plural* **sylis** [Susu *sílí,* literally, elephant] (1974)
: the monetary unit of Guinea from 1972 to 1986

syl·la·bary \'si-lə-ˌber-ē\ *noun, plural* **-bar·ies** [New Latin *syllabarium,* from Latin *syllaba* syllable] (1586)
: a table or listing of syllables; *specifically* **:** a series or set of written characters each one of which is used to represent a syllable

[1]syl·lab·ic \sə-'la-bik\ *adjective* [Late Latin *syllabicus,* from Greek *syllabikos,* from *syllabē* syllable] (1728)
1 : constituting a syllable or the nucleus of a syllable: **a :** not accompanied in the same syllable by a vowel ⟨a *syllabic* consonant⟩ **b :** having vowel quality more prominent than that of another vowel in the syllable ⟨the first vowel of a falling diphthong, as \ȯ\ in \ȯi\, is *syllabic*⟩
2 : of, relating to, or denoting syllables ⟨*syllabic* accent⟩
3 : characterized by distinct enunciation or separation of syllables
4 : of, relating to, or constituting a type of verse distinguished primarily by count of syllables rather than by rhythmical arrangement of accents or quantities
— syl·lab·i·cal·ly \-bi-k(ə-)lē\ *adverb*

[2]syllabic *noun* (1880)
: a syllabic character or sound

syl·lab·i·cate \sə-'la-bə-ˌkāt\ *transitive verb* **-cat·ed; -cat·ing** (circa 1654)
: SYLLABIFY

syl·lab·i·ca·tion \sə-ˌla-bə-'kā-shən\ *noun* (15th century)
: the act, process, or method of forming or dividing words into syllables

syl·la·bic·i·ty \ˌsi-lə-'bi-sə-tē\ *noun* (1933)
: the state of being or the power of forming a syllable

syl·lab·i·fi·ca·tion \sə-ˌla-bə-fə-'kā-shən\ *noun* (1838)
: SYLLABICATION

syl·lab·i·fy \sə-'la-bə-ˌfī\ *transitive verb* **-fied; -fy·ing** [Latin *syllaba* syllable] (circa 1859)
: to form or divide into syllables

[1]syl·la·ble \'si-lə-bəl\ *noun* [Middle English, from Middle French *sillabe,* from Latin *syllaba,* from Greek *syllabē,* from *syllambanein* to gather together, from *syn-* + *lambanein* to take — more at LATCH] (14th century)
1 : a unit of spoken language that is next bigger than a speech sound and consists of one or more vowel sounds alone or of a syllabic consonant alone or of either with one or more consonant sounds preceding or following
2 : one or more letters (as *syl, la,* and *ble*) in a word (as *syl·la·ble*) usually set off from the rest of the word by a centered dot or a hyphen and roughly corresponding to the syllables of spoken language and treated as helps to pronunciation or as guides to placing hyphens at the end of a line
3 : the smallest conceivable expression or unit of something **:** JOT
4 : SOL-FA SYLLABLES

[2]syllable *transitive verb* **syl·la·bled; syl·la·bling** \-b(ə-)liŋ\ (15th century)
1 : to give a number or arrangement of syllables to (a word or verse)
2 : to express or utter in or as if in syllables

syl·la·bub \'si-lə-ˌbəb\ *noun* [origin unknown] (circa 1537)
1 : a drink made by curdling milk or cream with an acid beverage (as wine or cider)
2 : a sweetened drink or topping made of milk or cream beaten with wine or liquor and sometimes further thickened with gelatin and served as a dessert

syl·la·bus \-bəs\ *noun, plural* **-bi** \-ˌbī, -ˌbē\ *or* **-bus·es** [Late Latin, alteration of Latin *sillybus* label for a book, from Greek *sillybos*] (circa 1656)
1 : a summary outline of a discourse, treatise, or course of study or of examination requirements
2 : HEADNOTE 2

syl·lep·sis \sə-'lep-səs\ *noun, plural* **-lep·ses** \-ˌsēz\ [Latin, from Greek *syllēpsis,* from *syllambanein*] (circa 1550)
1 : the use of a word to modify or govern syntactically two or more words with only one of which it formally agrees in gender, number, or case

2 : the use of a word in the same grammatical relation to two adjacent words in the context with one literal and the other metaphorical in sense
— **syl·lep·tic** \-'lep-tik\ *adjective*

syl·lo·gism \'si-lə-ˌji-zəm\ *noun* [Middle English *silogisme,* from Middle French, from Latin *syllogismus,* from Greek *syllogismos,* from *syllogizesthai* to syllogize, from *syn-* + *logizesthai* to calculate, from *logos* reckoning, word — more at LEGEND] (14th century)
1 : a deductive scheme of a formal argument consisting of a major and a minor premise and a conclusion (as in "every virtue is laudable; kindness is a virtue; therefore kindness is laudable")
2 : a subtle, specious, or crafty argument
3 : deductive reasoning
— **syl·lo·gis·tic** \ˌsi-lə-'jis-tik\ *adjective*
— **syl·lo·gis·ti·cal·ly** \-ti-k(ə-)lē\ *adverb*

syl·lo·gist \'si-lə-jist\ *noun* (1799)
: one who applies or is skilled in syllogistic reasoning

syl·lo·gize \'si-lə-ˌjīz\ *verb* **-gized; -giz·ing** [Middle English *sylogysen,* from Late Latin *syllogizare,* from Greek *syllogizesthai*] (15th century)
intransitive verb
: to reason by means of syllogisms
transitive verb
: to deduce by syllogism ⟨*syllogizes* moral laws⟩

sylph \'silf\ *noun* [New Latin *sylphus*] (1657)
1 : an elemental being in the theory of Paracelsus that inhabits air
2 : a slender graceful woman or girl
— **sylph·like** \'sil-ˌflīk\ *adjective*

sylph·id \'sil-fəd\ *noun* (1680)
: a young or diminutive sylph

sylva, sylviculture *variant of* SILVA, SILVICULTURE

[1]syl·van \'sil-vən\ *noun* (1565)
: one that frequents groves or woods

[2]sylvan *adjective* [Medieval Latin *silvanus, sylvanus,* from Latin *silva, sylva* wood] (circa 1583)
1 a : living or located in the woods or forest **b :** of, relating to, or characteristic of the woods or forest
2 a : made, shaped, or formed of woods or trees **b :** abounding in woods, groves, or trees **:** WOODED

syl·va·nite \'sil-və-ˌnīt\ *noun* [French *sylvanite,* from New Latin *sylvanium* tellurium, from *Transylvania,* region in Romania] (1796)
: a mineral that is a gold silver telluride and often occurs in crystals resembling written characters

syl·vat·ic \sil-'va-tik\ *adjective* [Latin *silvaticus* of the woods, wild — more at SAVAGE] (1661)
1 : SYLVAN ⟨*sylvatic* rodents⟩
2 : occurring in or affecting wild animals ⟨*sylvatic* diseases⟩

syl·vite \'sil-ˌvīt\ *also* **syl·vine** \-ˌvēn\ *noun* [alteration of *sylvine,* from French, from New Latin *sal digestivus Sylvii* digestive salt of Sylvius, from *Sylvius* latinized name of Jacques Dubois (died 1555) French physician] (1868)
: a mineral that is a natural potassium chloride and occurs in colorless cubes or crystalline masses

sym- — see SYN-

sym·bi·ont \'sim-bē-ˌänt\ *noun* [probably from German, modification of Greek *symbiount-, symbiōn,* present participle of *symbioun*] (1887)
: an organism living in symbiosis; *especially* **:** the smaller member of a symbiotic pair

sym·bi·o·sis \ˌsim-bē-'ō-səs, -ˌbī-\ *noun, plural* **-bi·o·ses** \-ˌsēz\ [New Latin, from German *Symbiose,* from Greek *symbiōsis* state of living together, from *symbioun* to live together, from *symbios* living together, from *sym-* + *bios* life — more at QUICK] (1622)
1 : the living together in more or less intimate association or close union of two dissimilar organisms
2 : the intimate living together of two dissimilar organisms in a mutually beneficial relationship; *especially* **:** MUTUALISM
3 : a cooperative relationship (as between two persons or groups) ⟨the *symbiosis* . . . between the resident population and the immigrants —John Geipel⟩
— **sym·bi·ot·ic** \-'ä-tik\ *adjective*
— **sym·bi·ot·i·cal·ly** \-ti-k(ə-)lē\ *adverb*

sym·bi·ote \'sim-bē-ˌōt, -ˌbī-\ *noun* [French, from Greek *symbiōtēs* companion, from *symbioun* to live together] (circa 1909)
: SYMBIONT

[1]sym·bol \'sim-bəl\ *noun* [in sense 1, from Late Latin *symbolum,* from Late Greek *symbolon,* from Greek, token, sign; in other senses from Latin *symbolum* token, sign, symbol, from Greek *symbolon,* literally, token of identity verified by comparing its other half, from *symballein* to throw together, compare, from *syn-* + *ballein* to throw — more at DEVIL] (15th century)
1 : an authoritative summary of faith or doctrine **:** CREED
2 : something that stands for or suggests something else by reason of relationship, association, convention, or accidental resemblance; *especially* **:** a visible sign of something invisible ⟨the lion is a *symbol* of courage⟩
3 : an arbitrary or conventional sign used in writing or printing relating to a particular field to represent operations, quantities, elements, relations, or qualities
4 : an object or act representing something in the unconscious mind that has been repressed ⟨phallic *symbols*⟩
5 : an act, sound, or object having cultural significance and the capacity to excite or objectify a response

[2]symbol *verb* **-boled** *or* **-bolled; -bol·ing** *or* **-bol·ling** (1832)
: SYMBOLIZE

sym·bol·ic \sim-'bä-lik\ *also* **sym·bol·i·cal** \-li-kəl\ *adjective* (1610)
1 a : using, employing, or exhibiting a symbol **b :** consisting of or proceeding by means of symbols
2 : of, relating to, or constituting a symbol
3 : characterized by or terminating in symbols ⟨*symbolic* thinking⟩
4 : characterized by symbolism ⟨a *symbolic* dance⟩
— **sym·bol·i·cal·ly** \-li-k(ə-)lē\ *adverb*

symbolic logic *noun* (1856)
: a science of developing and representing logical principles by means of a formalized system consisting of primitive symbols, combinations of these symbols, axioms, and rules of inference

sym·bol·ise *British variant of* SYMBOLIZE

sym·bol·ism \'sim-bə-ˌli-zəm\ *noun* (1654)
1 : the art or practice of using symbols especially by investing things with a symbolic meaning or by expressing the invisible or intangible by means of visible or sensuous representations: as **a :** artistic imitation or invention that is a method of revealing or suggesting immaterial, ideal, or otherwise intangible truth or states **b :** the use of conventional or traditional signs in the representation of divine beings and spirits
2 : a system of symbols or representations

sym·bol·ist \'sim-bə-list\ *noun* (1812)
1 : one who employs symbols or symbolism
2 : one skilled in the interpretation or explication of symbols
3 : one of a group of writers and artists in France after 1880 reacting against realism, concerning themselves with general truths instead of actualities, exalting the metaphysical and the mysterious, and aiming to unify and blend the arts and the functions of the senses
— **symbolist** *adjective*

sym·bol·is·tic \ˌsim-bə-'lis-tik\ *adjective* (circa 1864)
: SYMBOLIC

sym·bol·i·za·tion \ˌsim-bə-lə-'zā-shən\ *noun* (1603)
1 : an act or instance of symbolizing
2 : the human capacity to develop a system of meaningful symbols

sym·bol·ize \'sim-bə-ˌlīz\ *verb* **-ized; -iz·ing** (1603)
transitive verb
1 : to serve as a symbol of
2 : to represent, express, or identify by a symbol
intransitive verb
: to use symbols or symbolism
— **sym·bol·iz·er** *noun*

sym·bol·o·gy \sim-'bä-lə-jē\ *noun, plural* **-gies** [*symbol* + *-logy*] (1840)
1 : the art of expression by symbols
2 : the study or interpretation of symbols
3 : a system of symbols

sym·met·al·lism \(ˌ)si(m)-'me-tᵊl-ˌi-zəm\ *noun* [*syn-* + *-metallism* (as in *bimetallism*)] (circa 1895)
: a system of coinage in which the unit of currency consists of a particular weight of an alloy of two or more metals

sym·met·ri·cal \sə-'me-tri-kəl\ *or* **sym·met·ric** \-trik\ *adjective* (1751)
1 : having, involving, or exhibiting symmetry
2 : having corresponding points whose connecting lines are bisected by a given point or perpendicularly bisected by a given line or plane ⟨*symmetrical* curves⟩
3 *symmetric* **:** being such that the terms or variables may be interchanged without altering the value, character, or truth ⟨*symmetric* equations⟩ ⟨R is a *symmetric* relation if *a*R*b* implies *b*R*a*⟩
4 a : capable of division by a longitudinal plane into similar halves ⟨*symmetrical* plant parts⟩ **b :** having the same number of members in each whorl of floral leaves ⟨*symmetrical* flowers⟩
5 : affecting corresponding parts simultaneously and similarly ⟨*symmetrical* rash⟩
6 : exhibiting symmetry in a structural formula; *especially* **:** being a derivative with groups substituted symmetrically in the molecule
— **sym·met·ri·cal·ly** \-tri-k(ə-)lē\ *adverb*
— **sym·met·ri·cal·ness** \-kəl-nəs\ *noun*

symmetric group *noun* (1897)
: a permutation group that is composed of all of the permutations of *n* things

symmetric matrix *noun* (circa 1949)
: a matrix that is its own transpose

sym·me·trize \'si-mə-ˌtrīz\ *transitive verb* **-trized; -triz·ing** (1796)
: to make symmetrical
— **sym·me·tri·za·tion** \ˌsi-mə-trə-'zā-shən\ *noun*

sym·me·try \'si-mə-trē\ *noun, plural* **-tries** [Latin *symmetria,* from Greek, from *symmetros* symmetrical, from *syn-* + *metron* measure — more at MEASURE] (1541)
1 : balanced proportions; *also* **:** beauty of form arising from balanced proportions
2 : the property of being symmetrical; *especially* **:** correspondence in size, shape, and relative position of parts on opposite sides of a dividing line or median plane or about a center or axis — compare BILATERAL SYMMETRY, RADIAL SYMMETRY
3 : a rigid motion of a geometric figure that determines a one-to-one mapping onto itself
4 : the property of remaining invariant under certain changes (as of orientation in space, of the sign of the electric charge, of parity, or of the direction of time flow) — used of physical phenomena and of equations describing them

sympath- *or* **sympatho-** *combining form* [International Scientific Vocabulary, from *sympathetic*]
: sympathetic nerve ⟨*sympatho*lytic⟩

sym·pa·thec·to·my \ˌsim-pə-'thek-tə-mē\ *noun, plural* **-mies** [International Scientific Vocabulary] (1900)
: surgical interruption of sympathetic nerve pathways
— sym·pa·thec·to·mized \-ˌmīzd\ *adjective*

1sym·pa·thet·ic \ˌsim-pə-'the-tik\ *adjective* [New Latin *sympatheticus,* from Latin *sympathia* sympathy] (1644)
1 : existing or operating through an affinity, interdependence, or mutual association
2 a : not discordant or antagonistic **b :** appropriate to one's mood, inclinations, or disposition **c :** marked by kindly or pleased appreciation
3 : given to, marked by, or arising from sympathy, compassion, friendliness, and sensitivity to others' emotions ⟨a *sympathetic* gesture⟩
4 : favorably inclined **:** APPROVING ⟨not *sympathetic* to the idea⟩
5 a : showing empathy **b :** arousing sympathy or compassion ⟨a *sympathetic* role in the play⟩
6 a : of or relating to the sympathetic nervous system **b :** mediated by or acting on the sympathetic nerves
7 : relating to musical tones produced by sympathetic vibration or to strings so tuned as to sound by sympathetic vibration
— sym·pa·thet·i·cal·ly \-ti-k(ə-)lē\ *adverb*

2sympathetic *noun* (1808)
: a sympathetic structure; *especially* **:** SYMPATHETIC NERVOUS SYSTEM

sympathetic magic *noun* (1905)
: magic based on the assumption that a person or thing can be supernaturally affected through its name or an object representing it

sympathetic nervous system *noun* (circa 1891)
: the part of the autonomic nervous system that contains chiefly adrenergic fibers and tends to depress secretion, decrease the tone and contractility of smooth muscle, and increase heart rate — compare PARASYMPATHETIC NERVOUS SYSTEM

sympathetic strike *noun* (1895)
: SYMPATHY STRIKE

sympathetic vibration *noun* (1898)
: a vibration produced in one body by the vibrations of exactly the same period in a neighboring body

sym·pa·thin \'sim-pə-thən\ *noun* [International Scientific Vocabulary] (1931)
: a substance (as norepinephrine) that is secreted by sympathetic nerve endings and acts as a chemical mediator

sym·pa·thise *chiefly British variant of* SYMPATHIZE

sym·pa·thize \'sim-pə-ˌthīz\ *intransitive verb* **-thized; -thiz·ing** (1591)
1 : to be in keeping, accord, or harmony
2 : to react or respond in sympathy
3 : to share in suffering or grief **:** COMMISERATE ⟨*sympathize* with a friend in trouble⟩; *also* **:** to express such sympathy
4 : to be in sympathy intellectually ⟨*sympathize* with a proposal⟩
— sym·pa·thiz·er *noun*

sym·pa·tho·lyt·ic \ˌsim-pə-thō-'li-tik\ *adjective* [International Scientific Vocabulary] (1943)
: tending to oppose the physiological results of sympathetic nervous activity or of sympathomimetic drugs
— sympatholytic *noun*

sym·pa·tho·mi·met·ic \-mə-'me-tik, -(ˌ)mī-\ *adjective* [International Scientific Vocabulary] (1910)
: simulating sympathetic nervous action in physiological effect
— sympathomimetic *noun*

sym·pa·thy \'sim-pə-thē\ *noun, plural* **-thies** [Latin *sympathia,* from Greek *sympatheia,* from *sympathēs* having common feelings, sympathetic, from *syn-* + *pathos* feelings, emotion, experience — more at PATHOS] (1579)
1 a : an affinity, association, or relationship between persons or things wherein whatever affects one similarly affects the other **b :** mutual or parallel susceptibility or a condition brought about by it **c :** unity or harmony in action or effect
2 a : inclination to think or feel alike **:** emotional or intellectual accord **b :** feeling of loyalty **:** tendency to favor or support ⟨republican *sympathies*⟩
3 a : the act or capacity of entering into or sharing the feelings or interests of another **b :** the feeling or mental state brought about by such sensitivity ⟨have *sympathy* for the poor⟩
4 : the correlation existing between bodies capable of communicating their vibrational energy to one another through some medium
synonym see ATTRACTION, PITY

sympathy strike *noun* (1912)
: a strike in which the strikers have no direct grievance against their own employer but attempt to support or aid usually another group of workers on strike

sym·pat·ric \sim-'pa-trik\ *adjective* [*syn-* + Greek *patra* fatherland, from *patēr* father — more at FATHER] (circa 1904)
: occurring in the same area; *specifically* **:** occupying the same range without loss of identity from interbreeding ⟨*sympatric* species⟩ — compare ALLOPATRIC
— sym·pat·ri·cal·ly \-tri-k(ə-)lē\ *adverb*
— sym·pat·ry \'sim-ˌpa-trē\ *noun*

sym·pet·al·ous \(ˌ)sim-'pe-tᵊl-əs\ *adjective* (circa 1877)
: GAMOPETALOUS
— sym·pet·aly \-tᵊl-ē, 'sim-ˌ\ *noun*

sym·phon·ic \sim-'fä-nik\ *adjective* (1856)
1 : HARMONIOUS, SYMPHONIOUS
2 : relating to or having the form or character of a symphony ⟨*symphonic* music⟩
3 : suggestive of a symphony especially in form, interweaving of themes, or harmonious arrangement ⟨a *symphonic* drama⟩
— sym·phon·i·cal·ly \-ni-k(ə-)lē\ *adverb*

symphonic poem *noun* (1873)
: an extended programmatic composition for symphony orchestra usually freer in form than a symphony

sym·pho·ni·ous \sim-'fō-nē-əs\ *adjective* (1652)
: agreeing especially in sound **:** HARMONIOUS
— sym·pho·ni·ous·ly *adverb*

sym·pho·nist \'sim(p)-fə-nist\ *noun* (1767)
1 : a member of a symphony orchestra
2 : a composer of symphonies

sym·pho·ny \-nē\ *noun, plural* **-nies** [Middle English *symphonie,* from Middle French, from Latin *symphonia,* from Greek *symphōnia,* from *symphōnos* concordant in sound, from *syn-* + *phōnē* voice, sound — more at BAN] (15th century)
1 : consonance of sounds
2 a : RITORNELLO 1 **b :** SINFONIA 1 **c (1) :** a usually long and complex sonata for symphony orchestra **(2) :** a musical composition (as for organ) resembling such a symphony in complexity or variety
3 : consonance or harmony of color (as in a painting)
4 a : SYMPHONY ORCHESTRA **b :** a symphony orchestra concert
5 : something that in its harmonious complexity or variety suggests a symphonic composition

symphony orchestra *noun* (circa 1881)
: a large orchestra of winds, strings, and percussion that plays symphonic works

sym·phy·se·al \ˌsim(p)-fə-'sē-əl\ *also* **sym·phys·i·al** \sim-'fi-zē-əl\ *adjective* [Greek *symphyse-, symphysis* symphysis] (circa 1836)
: of, relating to, or constituting a symphysis

sym·phy·sis \'sim(p)-fə-səs\ *noun, plural* **-phy·ses** \-ˌsēz\ [New Latin, from Greek, state of growing together, from *symphyesthai* to grow together, from *syn-* + *phyein* to make grow, bring forth — more at BE] (circa 1578)
1 : an immovable or more or less movable articulation of various bones in the median plane of the body
2 : an articulation in which the bony surfaces are connected by pads of fibrous cartilage without a synovial membrane

sym·po·di·al \sim-'pō-dē-əl\ *adjective* [New Latin *sympodium* apparent main axis formed from secondary axes, from Greek *syn-* + *podion* base — more at -PODIUM] (1875)
: having or involving the formation of an apparent main axis from successive secondary axes ⟨*sympodial* branching of a cyme⟩

sym·po·si·arch \sim-'pō-zē-ˌärk\ *noun* [Greek *symposiarchos,* from *symposion* symposium + *-archos* -arch] (1603)
: one who presides over a symposium

sym·po·si·ast \-zē-ˌast, -əst\ *noun* [Greek *symposiazein* to take part in a symposium, from *symposion*] (circa 1656)
: a contributor to a symposium

sym·po·sium \sim-'pō-zē-əm *also* -zh(ē-)əm\ *noun, plural* **-sia** \-zē-ə, -zh(ē-)ə\ *or* **-siums** [Latin, from Greek *symposion,* from *sympinein* to drink together, from *syn-* + *pinein* to drink — more at POTABLE] (1711)
1 a : a convivial party (as after a banquet in ancient Greece) with music and conversation **b :** a social gathering at which there is free interchange of ideas
2 a : a formal meeting at which several specialists deliver short addresses on a topic or on related topics — compare COLLOQUIUM **b :** a collection of opinions on a subject; *especially* **:** one published by a periodical **c :** DISCUSSION
◆

symp·tom \'sim(p)-təm\ *noun* [Late Latin *symptomat-, symptoma,* from Greek *symptōmat-, symptōma* happening, attribute, symptom, from *sympiptein* to happen, from *syn-* + *piptein* to fall — more at FEATHER] (1541)
1 a : subjective evidence of disease or physical disturbance; *broadly* **:** something that indicates the presence of bodily disorder **b :** an evident reaction by a plant to a pathogen

◇ WORD HISTORY
symposium In ancient Greece the evening meal was often followed by a drinking party, attended only by men, that usually featured songs, games, and performances by hired entertainers. This party was called in Greek *symposion,* a derivative of *sympotēs* "drinking companion," itself formed from the prefix *syn-* "together" and *po-,* a variant stem of the verb *pinein* "to drink." For Greeks with somewhat loftier aspirations, the drinking became subordinate to the entertainment, which at its most exalted might take the form of philosophical discussion. The Platonic dialogue known as the *Symposium* is one example of a Greek literary genre that used conversation at a drinking party as a frame for extended treatment of a serious topic—in the case of Plato's work, love and beauty. The Greek word has been borrowed into English via Latin, but the *symposion* as philosophical discussion has influenced the modern use of English *symposium* more than actual Greek practice, and the stuffy academic connotations the word now has could hardly be further from the riotous excess more typical of the ancient *symposion.*

2 a : something that indicates the existence of something else ⟨*symptoms* of the time⟩ **b :** a slight indication **:** TRACE
synonym see SIGN
— symp·tom·less \-ləs\ *adjective*
symp·tom·at·ic \ˌsim(p)-tə-ˈma-tik\ *adjective* (1698)
1 a : being a symptom of a disease **b :** having the characteristics of a particular disease but arising from another cause
2 : concerned with or affecting symptoms
3 : CHARACTERISTIC, INDICATIVE ⟨his behavior was *symptomatic* of his character⟩
— symp·tom·at·i·cal·ly \-ti-k(ə-)lē\ *adverb*
symp·tom·atol·o·gy \ˌsim(p)-tə-mə-ˈtä-lə-jē\ *noun* (1798)
1 : the symptom complex of a disease
2 : a branch of medical science concerned with symptoms of diseases
— symp·tom·at·o·log·i·cal \-ˌma-tᵊl-ˈä-ji-kəl\ *or* **symp·tom·at·o·log·ic** \-ˈä-jik\ *adjective*
— symp·tom·at·o·log·i·cal·ly \-ji-k(ə-)lē\ *adverb*
syn- *or* **sym-** *prefix* [Middle English, from Old French, from Latin, from Greek, from *syn* with, together with]
1 : with **:** along with **:** together ⟨*syn*clinal⟩ ⟨*sym*petalous⟩
2 : at the same time ⟨*syn*esthesia⟩
syn·ae·re·sis *variant of* SYNERESIS
syn·aes·the·sia *variant of* SYNESTHESIA
syn·aes·the·sis \ˌsi-nəs-ˈthē-səs\ *noun* [Greek *synaisthēsis* joint perception, from *synaisthanesthai* to perceive simultaneously, from *syn-* + *aisthanesthai* to perceive — more at AUDIBLE] (1922)
: harmony of different or opposing impulses produced by a work of art
syn·a·gogue *or* **syn·a·gog** \ˈsi-nə-ˌgäg\ *noun* [Middle English *synagoge,* from Old French, from Late Latin *synagoga,* from Greek *synagōgē* assembly, synagogue, from *synagein* to bring together, from *syn-* + *agein* to lead — more at AGENT] (13th century)
1 : a Jewish congregation
2 : the house of worship and communal center of a Jewish congregation
— syn·a·gog·al \ˌsi-nə-ˈgä-gəl\ *adjective*
syn·a·loe·pha *or* **syn·a·le·pha** \ˌsi-nə-ˈlē-fə\ *noun* [New Latin, from Greek *synaloiphē,* from *synaleiphein* to clog up, coalesce, unite two syllables into one, from *syn-* + *aleiphein* to anoint — more at ALIPHATIC] (1540)
: the reduction to one syllable of two vowels of adjacent syllables (as in *th' army* for *the army*)
¹syn·apse \ˈsi-ˌnaps, sə-ˈnaps\ *noun* [New Latin *synapsis,* from Greek, juncture, from *synaptein* to fasten together, from *syn-* + *haptein* to fasten] (1899)
: the point at which a nervous impulse passes from one neuron to another
²synapse *intransitive verb* **syn·apsed; syn·aps·ing** (1910)
1 : to form a synapse
2 : to come together in synapsis
syn·ap·sid \sə-ˈnap-səd\ *noun* [New Latin *Synapsida,* from Greek *syn-* + *apsid-, apsis* arch, vault — more at APSIS] (1956)
: any of a subclass (Synapsida) of extinct reptiles existing during the Pennsylvanian, Permian, and Jurassic, having a single pair of lateral temporal openings in the skull, and usually held to be ancestral to mammals
— synapsid *adjective*
syn·ap·sis \sə-ˈnap-səs\ *noun, plural* **-ap·ses** \-ˌsēz\ [New Latin] (circa 1892)
: the association of homologous chromosomes that is characteristic of the first meiotic prophase
syn·ap·tic \sə-ˈnap-tik\ *adjective* [New Latin *synapsis*] (1895)
1 : of or relating to a synapsis
2 : of or relating to a synapse
— syn·ap·ti·cal·ly \-ti-k(ə-)lē\ *adverb*
syn·ap·to·ne·mal complex \sə-ˌnap-tə-ˈnē-məl-\ *noun* [*synap*tic + *-o-* or *-i-* + Greek *nēma* thread — more at NEMAT-] (1958)
: a complex tripartite protein structure that spans the region between synapsed chromosomes in meiotic prophase — called also *syn·ap·ti·ne·mal complex* \sə-ˌnap-tə-ˈnē-məl-\
syn·ap·to·some \sə-ˈnap-tə-ˌsōm\ *noun* [*synaptic* + *-o-* + ³*-some*] (1964)
: a nerve ending that is isolated from homogenized nerve tissue
— syn·ap·to·som·al \-ˌnap-tə-ˈsō-məl\ *adjective*
syn·ar·thro·di·al \ˌsin-är-ˈthrō-dē-əl\ *adjective* [New Latin *synarthrodia* synarthrosis] (1830)
: of, relating to, or being a synarthrosis
syn·ar·thro·sis \-ˈthrō-səs\ *noun, plural* **-thro·ses** \-ˌsēz\ [Greek *synarthrōsis,* from *syn-* + *arthrōsis* arthrosis] (1578)
: an immovable articulation in which the bones are united by intervening fibrous connective tissues
¹sync *also* **synch** \ˈsiŋk\ *transitive verb* **synced** *also* **synched** \ˈsiŋ(k)t\; **sync·ing** *also* **synch·ing** \ˈsiŋ-kiŋ\ (1929)
: SYNCHRONIZE
²sync *also* **synch** *noun* (1937)
: SYNCHRONIZATION, SYNCHRONISM
— sync *adjective*
syn·car·pous \(ˌ)sin-ˈkär-pəs\ *adjective* (circa 1830)
: having the carpels of the gynoecium united in a compound ovary
— syn·car·py \ˈsin-ˌkär-pē\ *noun*
syn·cat·e·gor·e·mat·ic \ˌsin-ˌka-tə-ˌgȯr-ə-ˈma-tik, -ˌgȯr-ē-\ *adjective* [Late Latin *syncategoremat-, syncategorema* syncategorematic term, from Greek *synkatēgorēma,* from *synkatēgorein* to predicate jointly, from *syn-* + *katēgorein* to predicate — more at CATEGORY] (1827)
: forming a meaningful expression only in conjunction with a denotative expression (as a content word) ⟨logical operators and function words are *syncategorematic*⟩
— syn·cat·e·gor·e·mat·i·cal·ly \-ti-k(ə-)lē\ *adverb*
¹syn·chro \ˈsiŋ-(ˌ)krō, ˈsin-\ *noun, plural* **synchros** [*synchronous*] (1943)
: SELSYN
²synchro *adjective* [*synchro-*] (1947)
: adapted to synchronization
synchro- *combining form* [*synchronized* & *synchronous*]
: synchronized **:** synchronous ⟨*synchro*mesh⟩
syn·chro·cy·clo·tron \ˌsiŋ-(ˌ)krō-ˈsī-klə-ˌträn, ˌsin-\ *noun* (1947)
: a modified cyclotron that achieves greater energies for the charged particles by compensating for the variation in mass that the particles experience with increasing velocity
syn·chro·mesh \ˈsiŋ-krō-ˌmesh, ˈsin-\ *adjective* (1928)
: designed for effecting synchronized shifting of gears
— synchromesh *noun*
syn·chro·nal \ˈsiŋ-krə-nᵊl, ˈsin-\ *adjective* (1660)
: SYNCHRONOUS
syn·chro·ne·ity \ˌsiŋ-krə-ˈnē-ə-tē, ˌsin-, -ˈnā-\ *noun* [*synchronous* + *-eity* (as in *spontaneity*)] (circa 1909)
: the state of being synchronous
syn·chron·ic \sin-ˈkrä-nik, siŋ-\ *adjective* (1833)
1 : SYNCHRONOUS
2 a : DESCRIPTIVE 4 ⟨*synchronic* linguistics⟩ **b :** concerned with events existing in a limited time period and ignoring historical antecedents
— syn·chron·i·cal \-ni-kəl\ *adjective*
— syn·chron·i·cal·ly \-ni-k(ə-)lē\ *adverb*
syn·chro·nic·i·ty \siŋ-krə-ˈni-sə-tē, sin-\ *noun* (circa 1889)
1 : the quality or fact of being synchronous
2 : the coincidental occurrence of events and especially psychic events (as similar thoughts in widely separated persons or a mental image of an unexpected event before it happens) that seem related but are not explained by conventional mechanisms of causality — used especially in the psychology of C. G. Jung
syn·chro·ni·sa·tion, syn·chro·nise *British variant of* SYNCHRONIZATION, SYNCHRONIZE
syn·chro·nism \ˈsiŋ-krə-ˌni-zəm, ˈsin-\ *noun* (1588)
1 : the quality or state of being synchronous **:** SIMULTANEOUSNESS
2 : chronological arrangement of historical events and personages so as to indicate coincidence or coexistence; *also* **:** a table showing such concurrences
— syn·chro·nis·tic \ˌsiŋ-krə-ˈnis-tik, ˌsin-\ *adjective*
syn·chro·ni·za·tion \ˌsiŋ-krə-nə-ˈzā-shən, ˌsin-\ *noun* (1828)
1 : the act or result of synchronizing
2 : the state of being synchronous
syn·chro·nize \ˈsiŋ-krə-ˌnīz, ˈsin-\ *verb* **-nized; -niz·ing** (circa 1624)
intransitive verb
: to happen at the same time
transitive verb
1 : to represent or arrange (events) to indicate coincidence or coexistence
2 : to make synchronous in operation
3 : to make (motion picture sound) exactly simultaneous with the action
— syn·chro·niz·er *noun*
synchronized swimming *noun* (1950)
: swimming in which the movements of one or more swimmers are synchronized with a musical accompaniment so as to form changing patterns
syn·chro·nous \ˈsiŋ-krə-nəs, ˈsin-\ *adjective* [Late Latin *synchronos,* from Greek, from *syn-* + *chronos* time] (1669)
1 : happening, existing, or arising at precisely the same time
2 : recurring or operating at exactly the same periods
3 : involving or indicating synchronism
4 a : having the same period; *also* **:** having the same period and phase **b :** GEOSTATIONARY
5 : of, used in, or being digital communication (as between computers) in which a common timing signal is established that dictates when individual bits can be transmitted, in which characters are not individually delimited, and which allows for very high rates of data transfer
synonym see CONTEMPORARY
— syn·chro·nous·ly *adverb*
— syn·chro·nous·ness *noun*
synchronous motor *noun* (1897)
: an electric motor having a speed strictly proportional to the frequency of the operating current
syn·chro·ny \ˈsiŋ-krə-nē, ˈsin-\ *noun, plural* **-nies** (1848)
: synchronistic occurrence, arrangement, or treatment
syn·chro·scope \-ˌskōp\ *noun* (1907)
: any of several devices for showing whether two associated machines or moving parts are operating in synchronism with each other
syn·chro·tron \ˈsiŋ-krə-ˌträn, ˈsin-\ *noun* (1945)
1 : an apparatus for imparting very high speeds to charged particles by means of a combination of a high-frequency electric field and a low-frequency magnetic field
2 : SYNCHROTRON RADIATION
synchrotron radiation *noun* [from its having been first observed in a synchrotron] (1956)
: radiation emitted by high-energy charged relativistic particles (as electrons) when they are accelerated by a magnetic field (as in a nebula)

syn·cli·nal \(ˌ)sin-'klī-nᵊl\ *adjective* [Greek *syn-* + *klinein* to lean — more at LEAN] (1833)
1 : inclined down from opposite directions so as to meet
2 : having or relating to a folded rock structure in which the sides dip toward a common line or plane

syn·cline \'sin-ˌklīn\ *noun* [back-formation from *synclinal*] (1873)
: a trough of stratified rock in which the beds dip toward each other from either side — compare ANTICLINE

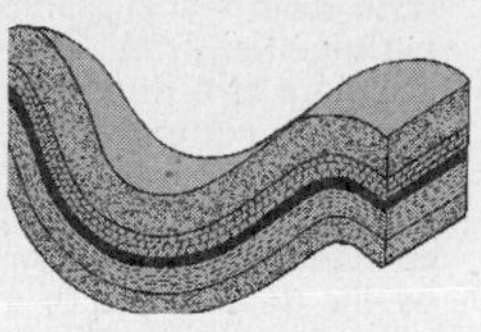
cross section of strata showing syncline

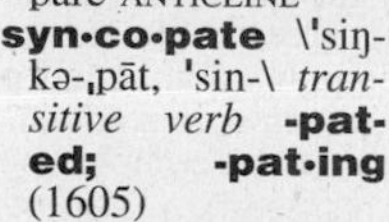

syn·co·pate \'siŋ-kə-ˌpāt, 'sin-\ *transitive verb* **-pat·ed; -pat·ing** (1605)
1 a : to shorten or produce by syncope ⟨*syncopate suppose* to *s'pose*⟩ **b :** to cut short **:** CLIP, ABBREVIATE
2 : to modify or affect (musical rhythm) by syncopation
— syn·co·pa·tor \-ˌpā-tər\ *noun*

syncopated *adjective* (1665)
1 : cut short **:** ABBREVIATED
2 : marked by or exhibiting syncopation ⟨*syncopated* rhythm⟩

syn·co·pa·tion \ˌsiŋ-kə-'pā-shən, ˌsin-\ *noun* (1597)
1 : a temporary displacement of the regular metrical accent in music caused typically by stressing the weak beat
2 : a syncopated rhythm, passage, or dance step
— syn·co·pa·tive \'siŋ-kə-ˌpā-tiv, 'sin-\ *adjective*

syn·co·pe \'siŋ-kə-(ˌ)pē, 'sin-\ *noun* [Late Latin, from Greek *synkopē*, literally, cutting short, from *synkoptein* to cut short, from *syn-* + *koptein* to cut — more at CAPON] (circa 1550)
1 : loss of consciousness resulting from insufficient blood flow to the brain **:** FAINT
2 : the loss of one or more sounds or letters in the interior of a word (as in *fo'c'sle* for *forecastle*)
— syn·co·pal \-kə-pəl\ *adjective*

syn·cret·ic \sin-'kre-tik, siŋ-\ *adjective* (1840)
: characterized or brought about by syncretism **:** SYNCRETISTIC

syn·cre·tise *British variant of* SYNCRETIZE

syn·cre·tism \'siŋ-krə-ˌti-zəm, 'sin-\ *noun* [New Latin *syncretismus*, from Greek *synkrētismos* federation of Cretan cities, from *syn-* + *Krēt-*, *Krēs* Cretan] (1618)
1 : the combination of different forms of belief or practice
2 : the fusion of two or more originally different inflectional forms
— syn·cre·tist \-tist\ *noun or adjective*
— syn·cre·tis·tic \ˌsiŋ-krə-'tis-tik, ˌsin-\ *adjective*

syn·cre·tize \'siŋ-krə-ˌtīz, 'sin-\ *transitive verb* **-tized; -tiz·ing** (circa 1891)
: to attempt to unite and harmonize especially without critical examination or logical unity

syn·cy·tium \sin-'si-sh(ē-)əm\ *noun, plural* **-tia** \-sh(ē-)ə\ [New Latin, from *syn-* + *cyt-*] (1877)
1 : a multinucleate mass of protoplasm (as in the plasmodium of a slime mold) resulting from fusion of cells
2 : COENOCYTE 1
— syn·cy·tial \-'si-sh(ē-)əl\ *adjective*

syn·dac·ty·lism \(ˌ)sin-'dak-tə-ˌli-zəm\ *noun* (1889)
: SYNDACTYLY

syn·dac·ty·ly \-lē\ *noun* [New Latin *syndactylia*, from *syn-* + Greek *daktylos* finger] (1864)
: a union of two or more digits that is normal in some animals (as various marsupials) and occurs as a human hereditary disorder marked by webbing of two or more fingers or toes

syn·des·mo·sis \ˌsin-ˌdez-'mō-səs, -ˌdes-\ *noun, plural* **-mo·ses** \-ˌsēz\ [New Latin, from Greek *syndesmos* fastening, ligament, from *syndein*] (1726)
: an articulation in which the contiguous surfaces of the bones are rough and are bound together by a ligament

syn·det·ic \sin-'de-tik\ *adjective* [Greek *syndetikos*, from *syndein* to bind together — more at ASYNDETON] (1621)
: CONNECTIVE, CONNECTING ⟨*syndetic* pronoun⟩; *also* **:** marked by a conjunctive ⟨*syndetic* relative clause⟩
— syn·det·i·cal·ly \-ti-k(ə-)lē\ *adverb*

syn·dic \'sin-dik\ *noun* [French, from Late Latin *syndicus* representative of a corporation, from Greek *syndikos* assistant at law, advocate, representative of a state, from *syn-* + *dikē* judgment, case at law — more at DICTION] (1601)
1 : a municipal magistrate in some countries
2 : an agent of a university or corporation

syn·di·cal \-di-kəl\ *adjective* (1864)
1 : of or relating to a syndic or to a committee that assumes the powers of a syndic
2 : of or relating to syndicalism

syn·di·cal·ism \'sin-di-kə-ˌli-zəm\ *noun* [French *syndicalisme*, from *chambre syndicale* trade union] (1907)
1 : a revolutionary doctrine by which workers seize control of the economy and the government by direct means (as a general strike)
2 : a system of economic organization in which industries are owned and managed by the workers
3 : a theory of government based on functional rather than territorial representation
— syn·di·cal·ist \-list\ *adjective or noun*

¹syn·di·cate \'sin-di-kət\ *noun* [French *syndicat*, from *syndic*] (1624)
1 a : a council or body of syndics **b :** the office or jurisdiction of a syndic
2 : an association of persons officially authorized to undertake a duty or negotiate business
3 a : a group of persons or concerns who combine to carry out a particular transaction **b :** CARTEL 2 **c :** a loose association of racketeers in control of organized crime
4 : a business concern that sells materials for publication in a number of newspapers or periodicals simultaneously
5 : a group of newspapers under one management

²syn·di·cate \'sin-də-ˌkāt\ *verb* **-cat·ed; -cat·ing** (1882)
transitive verb
1 : to subject to or manage as a syndicate
2 a : to sell (as a cartoon) to a syndicate or for publication in many newspapers or periodicals at once **b :** to sell (as a series of television programs) directly to local stations
intransitive verb
: to unite to form a syndicate
— syn·di·ca·tion \ˌsin-də-'kā-shən\ *noun*
— syn·di·ca·tor \'sin-də-ˌkā-tər\ *noun*

syn·drome \'sin-ˌdrōm *also* -drəm\ *noun* [New Latin, from Greek *syndromē* combination, syndrome, from *syn-* + *dramein* to run — more at DROMEDARY] (1541)
1 : a group of signs and symptoms that occur together and characterize a particular abnormality
2 : a set of concurrent things (as emotions or actions) that usually form an identifiable pattern

¹syne \'sīn\ *adverb* [Middle English (northern), probably from Old Norse *sīthan;* akin to Old English *siththan* since — more at SINCE] (14th century)
chiefly Scottish **:** since then **:** AGO

²syne *conjunction or preposition* (14th century)
Scottish **:** SINCE

syn·ec·do·che \sə-'nek-də-(ˌ)kē\ *noun* [Latin, from Greek *synekdochē*, from *syn-* + *ekdochē* sense, interpretation, from *ekdechesthai* to receive, understand, from *ex* from + *dechesthai* to receive; akin to Greek *dokein* to seem good — more at EX-, DECENT] (15th century)
: a figure of speech by which a part is put for the whole (as *fifty sail* for *fifty ships*), the whole for a part (as *society* for *high society*), the species for the genus (as *cutthroat* for *assassin*), the genus for the species (as *a creature* for *a man*), or the name of the material for the thing made (as *boards* for *stage*)
— syn·ec·doch·ic \ˌsi-ˌnek-'dä-kik\ *adjective*
— syn·ec·doch·i·cal \-'dä-ki-kəl\ *adjective*
— syn·ec·doch·i·cal·ly \-ki-k(ə-)lē\ *adverb*

syn·ecol·o·gy \ˌsi-ni-'kä-lə-jē, ˌsi-ne-\ *noun* [German *Synökologie*, from *syn-* syn- + *Ökologie* ecology] (1910)
: a branch of ecology that deals with the structure, development, and distribution of ecological communities
— syn·eco·log·i·cal \ˌsi-ˌnē-kə-'lä-ji-kəl, ˌsi-ˌne-\ *adjective*

syn·er·e·sis \sə-'ner-ə-səs, -'nir-, *especially for 2* ˌsi-nə-'rē-\ *noun* [Late Latin *synaeresis*, from Greek *synairesis*, from *synairein* to contract, from *syn-* + *hairein* to take] (circa 1577)
1 : SYNIZESIS
2 : the separation of liquid from a gel caused by contraction

syn·er·get·ic \ˌsi-nər-'je-tik\ *adjective* [Greek *synergētikos*, from *synergein* to work with, cooperate, from *synergos* working together, from *syn-* + *ergon* work — more at WORK] (circa 1836)
: SYNERGIC

syn·er·gic \sə-'nər-jik\ *adjective* (circa 1859)
: working together **:** COOPERATING
— syn·er·gi·cal·ly \-ji-k(ə-)lē\ *adverb*

syn·er·gid \sə-'nər-jəd, 'si-nər-\ *noun* [New Latin *synergida*, from Greek *synergos* working together] (1898)
: one of two small cells lying near the micropyle of the embryo sac of an angiosperm

syn·er·gism \'si-nər-ˌji-zəm\ *noun* [New Latin *synergismus*, from Greek *synergos*] (1910)
: interaction of discrete agencies (as industrial firms), agents (as drugs), or conditions such that the total effect is greater than the sum of the individual effects

syn·er·gist \-jist\ *noun* (1876)
: something (as a chemical or a muscle) that enhances the effectiveness of an active agent; *broadly* **:** either member of a synergistic pair

syn·er·gis·tic \ˌsi-nər-'jis-tik\ *adjective* (circa 1847)
1 : having the capacity to act in synergism ⟨*synergistic* drugs⟩
2 : of, relating to, or resembling synergism ⟨a *synergistic* reaction⟩
— syn·er·gis·ti·cal·ly \-ti-k(ə-)lē\ *adverb*

syn·er·gy \'si-nər-jē\ *noun* [New Latin *synergia*, from Greek *synergos* working together] (1660)
: SYNERGISM; *broadly* **:** combined action or operation

syn·e·sis \'si-nə-səs\ *noun* [New Latin, from Greek, understanding, sense, from *synienai* to bring together, understand, from *syn-* + *hienai* to send — more at JET] (circa 1891)
: a grammatical construction in which agreement or reference is according to sense rather than strict syntax (as *anyone* and *them* in "if anyone calls, tell them I am out")

syn·es·the·sia \ˌsi-nəs-ˈthē-zh(ē-)ə\ *noun* [New Latin, from *syn-* + *-esthesia* (as in *anesthesia*)] (circa 1891)
: a concomitant sensation; *especially* **:** a subjective sensation or image of a sense (as of color) other than the one (as of sound) being stimulated
— syn·es·thet·ic \-ˈthe-tik\ *adjective*

syn·fu·el \ˈsin-ˌfyü(-ə)l\ *noun* [*syn*thetic + *fuel*] (1976)
: a liquid or gaseous fuel derived especially from a fossil fuel that is a solid (as coal) or part of a solid (as tar sand or oil shale)

syn·ga·my \ˈsiŋ-gə-mē\ *noun* [International Scientific Vocabulary] (1904)
: sexual reproduction by union of gametes

syn·gas \ˈsin-ˌgas\ *noun* (1975)
: SYNTHESIS GAS

syn·ge·ne·ic \ˌsin-jə-ˈnē-ik\ *adjective* [*syn-* + *-geneic* (as in *isogeneic*)] (1961)
: genetically identical or similar especially with respect to antigens or immunological reactions ⟨*syngeneic* tumor cells⟩ ⟨grafts between *syngeneic* mice⟩ — compare ALLOGENEIC, XENOGENEIC

syn·i·ze·sis \ˌsi-nə-ˈzē-səs\ *noun* [Late Latin, from Greek *synizēsis,* from *synizein* to sit down together, collapse, blend, from *syn-* + *hizein* to sit down; akin to Latin *sidere* to sit down — more at SUBSIDE] (1846)
: contraction of two syllables into one by uniting in pronunciation two adjacent vowels

syn·kary·on \sin-ˈkar-ē-ˌän, -ē-ən\ *noun* [New Latin, from Greek *syn-* + *karyon* nut — more at CAREEN] (1905)
: a cell nucleus formed by the fusion of two preexisting nuclei

syn·od \ˈsi-nəd *also* -ˌäd\ *noun* [Middle English *sinod,* from Late Latin *synodus,* from Late Greek *synodos,* from Greek, meeting, assembly, from *syn-* + *hodos* way, journey] (14th century)
1 : an ecclesiastical governing or advisory council: as **a :** the governing assembly of an Episcopal province **b :** a Presbyterian governing body ranking between the presbytery and the general assembly **c :** a regional or national organization of Lutheran congregations
2 : the ecclesiastical district governed by a synod
— syn·od·al \ˈsi-nə-dᵊl, -ˌnä-; sə-ˈnä-dᵊl\ *adjective*

syn·od·ic \sə-ˈnä-dik\ *or* **syn·od·i·cal** \-di-kəl\ *adjective* (1561)
1 : of or relating to a synod **:** SYNODAL
2 *usually* **synodic** [Greek *synodikos,* from *synodos* meeting, conjunction] **:** relating to conjunction; *especially* **:** relating to the period between two successive conjunctions of the same celestial bodies (as the moon and the sun)

synodic month *noun* (1654)
: a lunar month

syn·o·nym \ˈsi-nə-ˌnim\ *noun* [Middle English *sinonyme,* from Latin *synonymum,* from Greek *synōnymon,* from neuter of *synōnymos* synonymous, from *syn-* + *onyma* name — more at NAME] (15th century)
1 : one of two or more words or expressions of the same language that have the same or nearly the same meaning in some or all senses
2 : a symbolic or figurative name **:** METONYM
3 : a taxonomic name rejected as being incorrectly applied or incorrect in form — compare HOMONYM
— syn·o·nym·ic \ˌsi-nə-ˈni-mik\ *also* **syn·o·nym·i·cal** \-mi-kəl\ *adjective*
— syn·o·nym·i·ty \-ˈni-mə-tē\ *noun*

syn·on·y·mist \sə-ˈnä-nə-mist\ *noun* (circa 1753)
: one who lists, studies, or discriminates synonyms

syn·on·y·mize \-ˌmīz\ *transitive verb* **-mized; -miz·ing** (circa 1595)
1 a : to give or analyze the synonyms of (a word) **b :** to provide (as a dictionary) with synonymies
2 : to demonstrate (a taxonomic name) to be a synonym

syn·on·y·mous \-məs\ *adjective* (1610)
1 : having the character of a synonym; *also* **:** alike in meaning or significance
2 : having the same connotations, implications, or reference ⟨to runners, Boston is *synonymous* with marathon —*Runners World*⟩
— syn·on·y·mous·ly *adverb*

syn·on·y·my \-mē\ *noun, plural* **-mies** (1683)
1 a : a list or collection of synonyms often defined and discriminated from each other **b :** the study or discrimination of synonyms
2 : the scientific names that have been used in different publications to designate a taxonomic group (as a species); *also* **:** a list of these
3 : the quality or state of being synonymous

syn·op·sis \sə-ˈnäp-səs\ *noun, plural* **-op·ses** \-ˌsēz\ [Late Latin, from Greek, literally, comprehensive view, from *synopsesthai* to be going to see together, from *syn-* + *opsesthai* to be going to see — more at OPTIC] (1611)
1 : a condensed statement or outline (as of a narrative or treatise) **:** ABSTRACT
2 : the abbreviated conjugation of a verb in one person only

syn·op·size \-ˌsīz\ *transitive verb* **-sized; -siz·ing** (1882)
1 : EPITOMIZE
2 : to make a synopsis of (as a novel)

syn·op·tic \sə-ˈnäp-tik\ *also* **syn·op·ti·cal** \-ti-kəl\ *adjective* [Greek *synoptikos,* from *synopsesthai*] (1763)
1 : affording a general view of a whole
2 : manifesting or characterized by comprehensiveness or breadth of view
3 : presenting or taking the same or common view; *specifically, often capitalized* **:** of or relating to the first three Gospels of the New Testament
4 : relating to or displaying conditions (as of the atmosphere or weather) as they exist simultaneously over a broad area
— syn·op·ti·cal·ly \-ti-k(ə-)lē\ *adverb*

syn·os·to·sis \ˌsi-ˌnäs-ˈtō-səs\ *noun, plural* **-to·ses** \-ˌsēz\ [New Latin] (circa 1848)
: union of two or more separate bones to form a single bone

sy·no·via \sə-ˈnō-vē-ə, sī-\ *noun* [New Latin] (1726)
: a transparent viscid lubricating fluid secreted by a membrane of an articulation, bursa, or tendon sheath

sy·no·vi·al \-vē-əl\ *adjective* (1756)
: of, relating to, secreting, or being synovia ⟨*synovial* membranes⟩ ⟨*synovial* fluid⟩

sy·no·vi·tis \ˌsī-nə-ˈvī-təs\ *noun* (circa 1836)
: inflammation of a synovial membrane

syn·tac·tic \sin-ˈtak-tik\ *or* **syn·tac·ti·cal** \-ti-kəl\ *adjective* [New Latin *syntacticus,* from Greek *syntaktikos* arranging together, from *syntassein*] (1577)
: of, relating to, or according to the rules of syntax or syntactics
— syn·tac·ti·cal·ly \-ti-k(ə-)lē\ *adverb*

syn·tac·tics \-tiks\ *noun plural but singular or plural in construction* (1937)
: a branch of semiotic that deals with the formal relations between signs or expressions in abstraction from their signification and their interpreters

syn·tag·ma \sin-ˈtag-mə\ *noun, plural* **-mas** *or* **-ma·ta** \-mə-tə\ [Greek, from *syntassein*] (1937)
: a syntactic element
— syn·tag·mat·ic \ˌsin-ˌtag-ˈma-tik\ *adjective*

syn·tax \ˈsin-ˌtaks\ *noun* [French or Late Latin; French *syntaxe,* from Late Latin *syntaxis,* from Greek, from *syntassein* to arrange together, from *syn-* + *tassein* to arrange] (1574)
1 a : the way in which linguistic elements (as words) are put together to form constituents (as phrases or clauses) **b :** the part of grammar dealing with this
2 : a connected or orderly system **:** harmonious arrangement of parts or elements
3 : syntactics especially as dealing with the formal properties of languages or calculi

synth \ˈsin(t)th\ *noun, often attributive* (1976)
: SYNTHESIZER 2

syn·the·sis \ˈsin(t)-thə-səs\ *noun, plural* **-the·ses** \-ˌsēz\ [Greek, from *syntithenai* to put together, from *syn-* + *tithenai* to put, place — more at DO] (1589)
1 a : the composition or combination of parts or elements so as to form a whole **b :** the production of a substance by the union of chemical elements, groups, or simpler compounds or by the degradation of a complex compound **c :** the combining of often diverse conceptions into a coherent whole; *also* **:** the complex so formed
2 a : deductive reasoning **b :** the dialectic combination of thesis and antithesis into a higher stage of truth
3 : the frequent and systematic use of inflected forms as a characteristic device of a language
— syn·the·sist \-sist\ *noun*

synthesis gas *noun* (circa 1941)
: a mixture of carbon monoxide and hydrogen used especially in chemical synthesis

syn·the·size \-ˌsīz\ *verb* **-sized; -siz·ing** (1830)
transitive verb
1 : to combine or produce by synthesis
2 : to make a synthesis of
intransitive verb
: to make a synthesis

syn·the·siz·er \-ˌsī-zər\ *noun* (1869)
1 : one that synthesizes ⟨an expert *synthesizer* of diverse views⟩
2 : a usually computerized electronic apparatus for the production and control of sound (as for producing music)

syn·the·tase \ˈsin-thə-ˌtās, -ˌtāz\ *noun* [*synthetic* + *-ase*] (1947)
: an enzyme that catalyzes the linking together of two molecules usually using the energy derived from the concurrent splitting off of a pyrophosphate group from a triphosphate (as ATP) — called also *ligase*

¹syn·thet·ic \sin-ˈthe-tik\ *adjective* [Greek *synthetikos* of composition, component, from *syntithenai* to put together] (1697)
1 : relating to or involving synthesis **:** not analytic
2 : attributing to a subject something determined by observation rather than analysis of the nature of the subject and not resulting in self-contradiction if negated — compare ANALYTIC
3 : characterized by frequent and systematic use of inflected forms to express grammatical relationships
4 a (1) **:** of, relating to, or produced by chemical or biochemical synthesis; *especially* **:** produced artificially ⟨*synthetic* drugs⟩ ⟨*synthetic* silk⟩ (2) **:** of or relating to a synfuel **b :** devised, arranged, or fabricated for special situations to imitate or replace usual realities **c :** FACTITIOUS, BOGUS
— syn·thet·i·cal·ly \-ti-k(ə-)lē\ *adverb*

²synthetic *noun* (1946)
: something resulting from synthesis rather than occurring naturally; *especially* **:** a product (as a drug or plastic) of chemical synthesis

synthetic division *noun* (1904)
: a simplified method for dividing a polynomial by another polynomial of the first degree by writing down only the coefficients of the several powers of the variable and changing the sign of the constant term in the divisor so as to replace the usual subtractions by additions

synthetic geometry *noun* (1889)
: elementary euclidean geometry or projective geometry as distinguished from analytic geometry

synthetic resin *noun* (1907)
: RESIN 2

syph \ˈsif\ *noun* (circa 1914)
slang : SYPHILIS

syph·i·lis \ˈsi-f(ə-)ləs\ *noun* [New Latin, from *Syphilus,* hero of the poem *Syphilis sive Morbus Gallicus* (*Syphilis or the French disease*) (1530) by Girolamo Fracastoro (died 1553) Italian poet, physician, and astronomer] (1718)
: a chronic contagious usually venereal and often congenital disease caused by a spirochete (*Treponema pallidum*) and if left untreated producing chancres, rashes, and systemic lesions in a clinical course with three stages continued over many years — compare PRIMARY SYPHILIS, SECONDARY SYPHILIS, TERTIARY SYPHILIS ◆
— **syph·i·lit·ic** \ˌsi-fə-ˈli-tik\ *adjective or noun*

sy·phon *variant of* SIPHON

Sy·rette \sə-ˈret\ *trademark*
— used for a small collapsible tube fitted with a hypodermic needle for injecting a single dose of a medicinal agent

Syr·i·ac \ˈsir-ē-ˌak\ *noun* [Latin *syriacus* Syrian, from Greek *syriakos,* from *Syria,* ancient country in Asia] (1605)
1 : a literary language based on an eastern Aramaic dialect and used as the literary and liturgical language by several eastern Christian churches
2 : Aramaic spoken by Christian communities
— **Syriac** *adjective*

Syr·i·an hamster \ˈsir-ē-ən-\ *noun* [*Syria,* Asia] (circa 1949)
: GOLDEN HAMSTER

sy·rin·ga \sə-ˈriŋ-gə\ *noun* [New Latin, genus name, from Greek *syring-, syrinx* panpipe] (1664)
: MOCK ORANGE 1

[1]**sy·ringe** \sə-ˈrinj *also* ˈsir-inj\ *noun* [Middle English *syring,* from Medieval Latin *syringa,* from Late Latin, injection, from Greek *syring-, syrinx* panpipe, tube] (14th century)
: a device used to inject fluids into or withdraw them from something (as the body or its cavities): as **a :** a device that consists of a nozzle of varying length and a compressible rubber bulb and is used for injection or irrigation **b :** an instrument (as for the injection of medicine or the withdrawal of bodily fluids) that consists of a hollow barrel fitted with a plunger and a hollow needle **c :** a gravity device consisting of a reservoir fitted with a long rubber tube ending with an exchangeable nozzle that is used for irrigation of the vagina or bowel

[2]**syringe** *transitive verb* **sy·ringed; sy·ring·ing** (1610)
: to irrigate or spray with or as if with a syringe

sy·rin·go·my·e·lia \sə-ˌriŋ-gō-mī-ˈē-lē-ə\ *noun* [New Latin, from Greek *syring-, syrinx* tube, fistula + New Latin *myel-* + *-ia*] (1897)
: a chronic progressive disease of the spinal cord associated with sensory disturbances, muscle atrophy, and spasticity
— **sy·rin·go·my·el·ic** \-ˈe-lik\ *adjective*

syr·inx \ˈsir-iŋ(k)s\ *noun, plural* **sy·rin·ges** \sə-ˈriŋ-ˌgēz, -ˈrin-ˌjēz\ *or* **syr·inx·es** (1606)
1 [Late Latin, from Greek] **:** PANPIPE
2 [New Latin, from Greek] **:** the vocal organ of birds that is a special modification of the lower part of the trachea or of the bronchi or of both

syr·phid fly \ˈsər-fəd-, ˈsir-\ *noun* [New Latin *Syrphidae,* from *Syrphus,* genus of flies, from Greek *syrphos* gnat] (circa 1891)
: any of a family (Syrphidae) of dipteran flies which frequent flowers and some of whose larvae prey on plant lice — called also *syrphid*
— **syrphid** *adjective*

syr·up \ˈsər-əp, ˈsir-əp, ˈsə-rəp\ *noun* [Middle English *sirup,* from Middle French *sirop,* from Medieval Latin *syrupus,* from Arabic *sharāb*] (14th century)
1 a : a thick sticky solution of sugar and water often flavored or medicated **b :** the concentrated juice of a fruit or plant
2 : cloying sweetness or sentimentality
— **syr·upy** *adjective*

sys·op \ˈsis-(ˌ)äp\ *noun* [*sys*tem *op*erator] (1983)
: the administrator of a computer bulletin board

sys·tal·tic \sis-ˈtȯl-tik, -ˈtal-\ *adjective* [Late Latin *systalticus,* from Greek *systaltikos,* from *systellein* to contract — more at SYSTOLE] (1676)
: marked by regular contraction and dilatation : PULSING

sys·tem \ˈsis-təm\ *noun* [Late Latin *systemat-, systema,* from Greek *systēmat-, systēma,* from *synistanai* to combine, from *syn-* + *histanai* to cause to stand — more at STAND] (1603)
1 : a regularly interacting or interdependent group of items forming a unified whole ⟨a number *system*⟩: as **a** (1) **:** a group of interacting bodies under the influence of related forces ⟨a gravitational *system*⟩ (2) **:** an assemblage of substances that is in or tends to equilibrium ⟨a thermodynamic *system*⟩ **b** (1) **:** a group of body organs that together perform one or more vital functions ⟨the digestive *system*⟩ (2) **:** the body considered as a functional unit **c :** a group of related natural objects or forces ⟨a river *system*⟩ **d :** a group of devices or artificial objects or an organization forming a network especially for distributing something or serving a common purpose ⟨a telephone *system*⟩ ⟨a heating *system*⟩ ⟨a highway *system*⟩ ⟨a data processing *system*⟩ **e :** a major division of rocks usually larger than a series and including all formed during a period or era **f :** a form of social, economic, or political organization or practice ⟨the capitalist *system*⟩
2 : an organized set of doctrines, ideas, or principles usually intended to explain the arrangement or working of a systematic whole ⟨the Newtonian *system* of mechanics⟩
3 a : an organized or established procedure ⟨the touch *system* of typing⟩ **b :** a manner of classifying, symbolizing, or schematizing ⟨a taxonomic *system*⟩ ⟨the decimal *system*⟩
4 : harmonious arrangement or pattern **:** ORDER ⟨bring *system* out of confusion —Ellen Glasgow⟩
5 : an organized society or social situation regarded as stultifying **:** ESTABLISHMENT 2 — usually used with *the*
synonym see METHOD
— **sys·tem·less** \-ləs\ *adjective*

sys·tem·at·ic \ˌsis-tə-ˈma-tik\ *adjective* [Late Latin *systematicus,* from Greek *systēmatikos,* from *systēmat-, systēma*] (circa 1680)
1 : relating to or consisting of a system
2 : presented or formulated as a coherent body of ideas or principles ⟨*systematic* thought⟩
3 a : methodical in procedure or plan ⟨a *systematic* approach⟩ ⟨a *systematic* scholar⟩ **b :** marked by thoroughness and regularity ⟨*systematic* efforts⟩
4 : of, relating to, or concerned with classification; *specifically* **:** TAXONOMIC
— **sys·tem·at·i·cal·ly** \-ti-k(ə-)lē\ *adverb*
— **sys·tem·at·ic·ness** \-tik-nəs\ *noun*

systematic error *noun* (1891)
: an error that is not determined by chance but is introduced by an inaccuracy (as of observation or measurement) inherent in the system

sys·tem·at·ics \ˌsis-tə-ˈma-tiks\ *noun plural but singular in construction* (1888)
1 : the science of classification
2 a : a system of classification **b :** the classification and study of organisms with regard to their natural relationships **:** TAXONOMY

systematic theology *noun* (1836)
: a branch of theology concerned with summarizing the doctrinal traditions of a religion (as Christianity) especially with a view to relating the traditions convincingly to the religion's present-day setting

sys·tem·a·tise *British variant of* SYSTEMATIZE

sys·tem·a·tism \ˈsis-tə-mə-ˌti-zəm, sis-ˈte-mə-\ *noun* (1846)
: the practice of forming intellectual systems

sys·tem·a·tist \ˈsis-tə-mə-tist, sis-ˈte-mə-\ *noun* (1700)
1 : a maker or follower of a system
2 : a specialist in taxonomy **:** TAXONOMIST

sys·tem·a·tize \ˈsis-tə-mə-ˌtīz\ *transitive verb* **-tized; -tiz·ing** (circa 1767)
: to arrange in accord with a definite plan or scheme **:** order systematically ⟨the need to *systematize* their work⟩
synonym see ORDER
— **sys·tem·a·ti·za·tion** \ˌsis-tə-mə-tə-ˈzā-shən, sis-ˌte-mə-\ *noun*
— **sys·tem·a·tiz·er** *noun*

[1]**sys·tem·ic** \sis-ˈte-mik\ *adjective* (1803)
: of, relating to, or common to a system: as **a :** affecting the body generally **b :** supplying those parts of the body that receive blood through the aorta rather than through the pulmonary artery **c :** of, relating to, or being a pesticide that as used is harmless to the plant or higher animal but when absorbed into its sap or bloodstream makes the entire organism toxic to pests (as an insect or fungus)
— **sys·tem·i·cal·ly** \-mi-k(ə-)lē\ *adverb*

[2]**systemic** *noun* (1951)
: a systemic pesticide

systemic lupus er·y·the·ma·to·sus \-ˌer-ə-ˌthē-mə-ˈtō-səs\ *noun* (1951)
: an inflammatory connective tissue disease of unknown cause that occurs chiefly in women, is characterized especially by fever, skin rash, and arthritis, often by acute hemolytic anemia, by small hemorrhages in the skin and mucous membranes, by inflammation of the pericardium, and in serious cases by involvement of the kidneys and central nervous system

sys·tem·ize \ˈsis-tə-ˌmīz\ *transitive verb* **-ized; -iz·ing** (1778)
: SYSTEMATIZE

◇ WORD HISTORY

syphilis In 1530 the Italian physician and poet Girolamo Fracastoro (circa 1478–1553) published a medical poem entitled *Syphilis, sive morbus Gallicus* ("Syphilis or the French Disease"). The "hero" of the poem is an unfortunate swineherd named Syphilus, whose name in Greek literally means "swine lover." The poem is actually a mythological tale that purports to tell how the disease first came into being. In the course of the poem Syphilus offends the sun god; for his punishment he becomes the first person to be afflicted with the fearsome contagion. In a way the poem also serves as a kind of digest of all that was then known about the disease. It discusses the venereal nature of syphilis and its symptoms and stages. Fracastoro's title gave us our name for the disease, but his work is also notable for contributing to the time-honored tradition of blaming the disease on some other country than one's own.

— **sys·tem·i·za·tion** \ˌsis-tə-mə-ˈzā-shən\ *noun*

systems analysis *noun* (1950)
: the act, process, or profession of studying an activity (as a procedure, a business, or a physiological function) typically by mathematical means in order to define its goals or purposes and to discover operations and procedures for accomplishing them most efficiently
— **systems analyst** *noun*

sys·to·le \ˈsis-tə-(ˌ)lē\ *noun* [Greek *systolē,* from *systellein* to contract, from *syn-* + *stellein* to send] (1578)
: a rhythmically recurrent contraction; *especially* : the contraction of the heart by which the blood is forced onward and the circulation kept up
— **sys·tol·ic** \sis-ˈtä-lik\ *adjective*

syz·y·gy \ˈsi-zə-jē\ *noun, plural* **-gies** [Late Latin *syzygia* conjunction, from Greek, from *syzygos* yoked together, from *syn-* + *zygon* yoke — more at YOKE] (circa 1847)
: the nearly straight-line configuration of three celestial bodies (as the sun, moon, and earth during a solar or lunar eclipse) in a gravitational system

Szech·uan *or* **Szech·wan** \ˈsech-ˌwän, ˈsesh-\ *adjective* [*Szechwan* or *Szechuan* (Sichuan), province in China] (1956)
: of, relating to, or being a style of Chinese cooking that is spicy, oily, and especially peppery

T *is the twentieth letter of the English alphabet. It came through Etruscan and Latin from the Greek* tau, *which itself was taken from Phoenician* tāw. *In modern English,* t *usually represents an alveolar stop, that is, a consonant produced by touching the tongue against the upper gum ridge (or alveolus) so as to stop the air flow through the mouth. In many other languages it is dental, or pronounced by pressing the tongue against the teeth. Unlike the otherwise identical sound of* d, *it is voiceless, that is, pronounced without vibration of the vocal cords. With* i *it often represents the sound of* sh, *as in* nation *and* palatial; *before* u *it often represents the sound of* ch, *as in* nature *and* mutual. *With* h *it forms the digraph* th, *which has two sounds, one voiceless, as in* thin, *the other voiced, as in* then. T *is frequently silent before syllabic* n, *as in* chasten *and* soften. *Small* t *came into being when medieval scribes began forming the miniaturized variant of the letter with two pen strokes, thus allowing the vertical stroke to emerge above the letter's crossbar.*

t \'tē\ *noun, plural* **t's** *or* **ts** \'tēz\ *often capitalized, often attributive* (before 12th century)
1 a : the 20th letter of the English alphabet **b :** a graphic representation of this letter **c :** a speech counterpart of orthographic *t*
2 : a graphic device for reproducing the letter *t*
3 : one designated *t* especially as the 20th in order or class
4 : something shaped like the letter T
5 : T FORMATION
6 : TECHNICAL FOUL
— to a T [short for *to a tittle*] **:** to perfection

't \t\ *pronoun*
: IT ⟨my country, *'tis* of thee —S. F. Smith⟩

ta \'tä\ *noun* [baby talk] (1772)
British **:** THANKS

Taal \'täl\ *noun* [Afrikaans, from Dutch, language; akin to Old English *talu* talk — more at TALE] (1896)
: AFRIKAANS — usually used with *the*

[1]tab \'tab\ *noun, often attributive* [origin unknown] (1607)
1 a : a short projecting device: as (1) **:** a small flap or loop by which something may be grasped or pulled (2) **:** a projection from a card used as an aid in filing **b :** a small insert, addition, or remnant **c :** APPENDAGE, EXTENSION; *especially* **:** one of a series of small pendants forming a decorative border or edge of a garment **d :** a small auxiliary airfoil hinged to a control surface (as a trailing edge) to help stabilize an airplane in flight — see AIRPLANE illustration
2 [partly short for [1]*table;* partly from sense 1] **a :** close surveillance **:** WATCH ⟨keep *tabs* on trends⟩ **b :** a creditor's statement **:** BILL, CHECK **c :** COST ⟨the *tab* for the new program⟩
3 [by shortening] **a :** TABLOID **b :** TABLET
4 [short for *tabulator*] **:** a device (as on a typewriter) for arranging data in columns

[2]tab *transitive verb* **tabbed; tab·bing** (1872)
1 : to furnish or ornament with tabs
2 : to single out **:** DESIGNATE
3 : TABULATE

ta·ba·nid \tə-'bā-nəd, -'ba-\ *noun* [ultimately from Latin *tabanus* horsefly] (circa 1891)
: HORSEFLY

tab·ard \'ta-bərd *also* -ˌbärd\ *noun* [Middle English, from Middle French *tabart*] (14th century)
: a short loose-fitting sleeveless or short-sleeved coat or cape: as **a :** a tunic worn by a knight over his armor and emblazoned with his arms **b :** a herald's official cape or coat emblazoned with his lord's arms **c :** a woman's sleeveless outer garment often with side slits

Ta·bas·co \tə-'bas-(ˌ)kō\ *trademark*
— used for a pungent condiment sauce made from hot peppers

tab·bou·leh \tə-'bü-lə, -lē\ *noun* [Arabic *tabbūla*] (1955)
: a salad of Lebanese origin consisting chiefly of cracked wheat, tomatoes, parsley, mint, onions, lemon juice, and olive oil

[1]tab·by \'ta-bē\ *noun, plural* **tabbies** [French *tabis,* from Medieval Latin *attabi,* from Arabic *'attābī,* from Al-*'Attābīya,* quarter in Baghdad] (1638)
1 a *archaic* **:** a plain silk taffeta especially with moiré finish **b :** a plain-woven fabric
2 [[2]*tabby*] **a :** a domestic cat with a striped and mottled coat **b :** a domestic cat; *especially* **:** a female cat ◆

tabby 2a

[2]tabby *adjective* (1661)
1 : of, relating to, or made of tabby
2 : striped and mottled with darker color **:** BRINDLED ⟨a *tabby* cat⟩

[3]tabby *noun* [Gullah *tabi,* ultimately from Spanish *tapia* adobe wall] (1775)
: a cement made of lime, sand or gravel, and oyster shells and used chiefly along the coast of Georgia and South Carolina in the 17th and 18th centuries

[1]tab·er·na·cle \'ta-bər-ˌna-kəl\ *noun* [Middle English, from Old French, from Late Latin *tabernaculum,* from Latin, tent, from *taberna* hut] (13th century)
1 a *often capitalized* **:** a tent sanctuary used by the Israelites during the Exodus **b** *archaic* **:** a dwelling place **c** *archaic* **:** a temporary shelter **:** TENT
2 : a receptacle for the consecrated elements of the Eucharist; *especially* **:** an ornamental locked box used for reserving the Communion hosts
3 : a house of worship; *specifically* **:** a large building or tent used for evangelistic services
— tab·er·nac·u·lar \ˌta-bər-'na-kyə-lər\ *adjective*

[2]tabernacle *intransitive verb* **tab·er·na·cled; tab·er·na·cling** \-ˌna-k(ə-)liŋ\ (1653)
: to take up temporary residence; *especially* **:** to inhabit a physical body

ta·bes \'tā-(ˌ)bēz\ *noun, plural* **tabes** [Latin, wasting disease, decay, from *tabēre* to decay — more at THAW] (1651)
: wasting accompanying a chronic disease
— ta·bet·ic \tə-'be-tik\ *adjective or noun*

tabes dor·sa·lis \-dȯr-'sa-ləs, -'sā-, -'sä-\ *noun* [New Latin, dorsal tabes] (circa 1681)
: a syphilitic disorder of the nervous system marked by wasting, pain, lack of coordination of voluntary movements and reflexes, and disorders of sensation, nutrition, and vision — called also *locomotor ataxia*

ta·bla \'tä-blə\ *noun* [Hindi *tablā,* from Arabic *ṭabla*] (1865)
: a pair of small different-sized hand drums used especially in music of India

tabla

tab·la·ture \'ta-blə-ˌchu̇r, -chər, -ˌtyu̇r, -ˌtu̇r\ *noun* [Middle French, from Medieval Latin *tabulatus* tablet, from Latin *tabula*] (1574)
: an instrumental notation indicating the string, fret, key, or finger to be used instead of the tone to be sounded

[1]ta·ble \'tā-bəl\ *noun, often attributive* [Middle English, from Old English *tabule* & Old French *table;* both from Latin *tabula* board, tablet, list] (before 12th century)
1 : TABLET 1a
2 a *plural* **:** BACKGAMMON **b :** one of the two leaves of a backgammon board or either half of a leaf
3 a : a piece of furniture consisting of a smooth flat slab fixed on legs **b** (1) **:** a supply or source of food (2) **:** an act or instance of assembling to eat **:** MEAL ⟨sit down to *table*⟩ **c** (1) **:** a group of people assembled at or as if at a table (2) **:** a legislative or negotiating session ⟨the bargaining *table*⟩
4 : STRINGCOURSE
5 a : a systematic arrangement of data usually in rows and columns for ready reference **b :** a condensed enumeration **:** LIST ⟨a *table* of contents⟩
6 : something that resembles a table especially in having a plane surface: as **a :** the upper flat surface of a cut precious stone **b** (1) **:** TABLELAND (2) **:** a horizontal stratum ⟨water *table*⟩
— under the table 1 : into a stupor ⟨can drink you *under the table*⟩ **2 :** in a covert manner ⟨took money *under the table*⟩

[2]table *transitive verb* **ta·bled; ta·bling** \-b(ə-)liŋ\ (15th century)
1 : to enter in a table
2 a *British* **:** to place on the agenda **b :** to remove (as a parliamentary motion) from consideration indefinitely **c :** to put on a table

[3]table *adjective* (1547)

◇ WORD HISTORY

tabby A silk cloth with a moiré or wavy pattern was once made in a quarter of the city of Baghdad, now the capital of the modern nation of Iraq. In Arabic the cloth was known as *'attābī;* from *Al-'Attābīya,* the name of its place of origin. The name for the cloth passed into Medieval Latin as *attabi* and then into French as *tabis.* When the word was borrowed into English as *tabby* in the 17th century, it was solely with reference to a plain-woven silk cloth with a moiré pattern. Before long, however, people noted a resemblance between the cloth's pattern and the striped or mottled markings on the coats of their domestic cats. *Tabby* has been the standard name for this type of cat ever since.

\ə\ **abut** \ᵊ\ **kitten** \ər\ **further** \a\ **ash** \ā\ **ace** \ä\ **mop, mar** \au̇\ **out** \ch\ **chin** \e\ **bet** \ē\ **easy** \g\ **go** \i\ **hit** \ī\ **ice** \j\ **job** \ŋ\ **sing** \ō\ **go** \ȯ\ **law** \ȯi\ **boy** \th\ **thin** \t͟h\ **the** \ü\ **loot** \u̇\ **foot** \y\ **yet** \zh\ **vision** *see also* Guide to Pronunciation

1 : suitable for a table or for use at a table ⟨a *table* lamp⟩
2 : suitable for serving at a table ⟨*table* grapes⟩
3 : proper for conduct at a table ⟨*table* manners⟩

tab·leau \'ta-ˌblō, ta-'blō\ *noun, plural* **tab·leaux** \-ˌblōz, -'blōz\ *also* **tableaus** [French, from Middle French *tablel,* diminutive of *table,* from Old French] (1699)
1 : a graphic description or representation **:** PICTURE ⟨winsome *tableaux* of old-fashioned literary days —J. D. Hart⟩
2 : a striking or artistic grouping
3 [short for *tableau vivant* (from French, literally, living picture)] **:** a depiction of a scene usually presented on a stage by silent and motionless costumed participants

tableau curtain *noun* (1881)
: a stage curtain that opens in the center and has its sections drawn upward as well as to the side

ta·ble·cloth \'tā-bəl-ˌklȯth\ *noun* (15th century)
: a covering spread over a dining table before the tableware is set

ta·ble d'hôte \ˌtä-bəl-'dōt, ˌta-\ *noun* [French, literally, host's table] (1617)
1 : a meal served to all guests at a stated hour and fixed price
2 : a complete meal of several courses offered at a fixed price

ta·ble·ful \'tā-bəl-ˌfu̇l\ *noun* (1535)
: as much or as many as a table can hold or accommodate

ta·ble–hop \'tā-bəl-ˌhäp\ *intransitive verb* (1942)
: to move from table to table (as in a restaurant) in order to chat with friends
— ta·ble–hop·per *noun*

ta·ble·land \-bə(l)-ˌland\ *noun* (1697)
: a broad level elevated area **:** PLATEAU

table linen *noun* (1629)
: linen (as tablecloths and napkins) for the table

ta·ble·mate \'tā-bəl-ˌmāt\ *noun* (1624)
: a dining companion

table of organization (circa 1918)
: a table listing the number and duties of personnel and the major items of equipment authorized for a military unit

table salt *noun* (1878)
: salt suitable for use at the table and in cooking

ta·ble·spoon \'tā-bəl-ˌspün\ *noun* (1763)
1 : a large spoon used for serving
2 : a unit of measure used especially in cookery equal to ½ fluid ounce (15 milliliters)

ta·ble·spoon·ful \ˌtā-bəl-'spün-ˌfu̇l, 'tā-bəl-ˌ\ *noun, plural* **tablespoonfuls** \-ˌfu̇lz\ *also* **ta·ble·spoons·ful** \-'spünz-ˌfu̇l, -ˌspünz-\ (1772)
1 : enough to fill a tablespoon
2 : TABLESPOON 2

table sugar *noun* (1958)
: SUGAR 1a; *especially* **:** granulated white sugar

tab·let \'ta-blət\ *noun* [Middle English *tablett,* from Middle French *tablete,* diminutive of *table* table] (14th century)
1 a : a flat slab or plaque suited for or bearing an inscription **b :** a thin slab or one of a set of portable sheets used for writing **c :** PAD 4
2 a : a compressed or molded block of a solid material **b :** a small mass of medicated material ⟨an aspirin *tablet*⟩
3 : GRAPHICS TABLET

table talk *noun* (1569)
: informal conversation at or as if at a dining table; *especially* **:** the social talk of a celebrity recorded for publication

table tennis *noun* (1901)
: a game resembling tennis that is played on a tabletop with wooden paddles and a small hollow plastic ball

ta·ble·top \'tā-bəl-ˌtäp\ *noun* (1807)
1 : the top of a table
2 : a photograph of small objects or a miniature scene arranged on a table
— table·top *adjective*

ta·ble·ware \-ˌwar, -ˌwer\ *noun* (1832)
: utensils (as of china, glass, or silver) for table use

table wine *noun* (circa 1827)
: an unfortified wine averaging 12 percent alcohol by volume and usually suitable for serving with food

[1]tab·loid \'ta-ˌblȯid\ *adjective* [from *Tabloid,* a trademark] (1901)
1 : compressed or condensed into small scope ⟨*tabloid* criticism⟩
2 : of, relating to, or characteristic of tabloids; *especially* **:** featuring stories of violence, crime, or scandal presented in a sensational manner

[2]tabloid *noun* (1906)
1 : DIGEST, SUMMARY
2 : a newspaper that is about half the page size of an ordinary newspaper and that contains news in condensed form and much photographic matter

[1]ta·boo *also* **ta·bu** \tə-'bü, ta-\ *adjective* [Tongan *tabu*] (1777)
1 : forbidden to profane use or contact because of what are held to be dangerous supernatural powers
2 a : banned on grounds of morality or taste ⟨the subject is *taboo*⟩ **b :** banned as constituting a risk ⟨the area beyond is *taboo,* still alive with explosives —Robert Leckie⟩

[2]taboo *also* **tabu** *noun, plural* **taboos** *also* **tabus** (1777)
1 : a prohibition against touching, saying, or doing something for fear of immediate harm from a supernatural force
2 : a prohibition imposed by social custom or as a protective measure
3 : belief in taboos

[3]taboo *also* **tabu** *transitive verb* (1777)
1 : to set apart as taboo especially by marking with a ritualistic symbol
2 : to avoid or ban as taboo

ta·bor *also* **ta·bour** \'tā-bər\ *noun* [Middle English, from Old French] (14th century)
: a small drum with one head of soft calfskin used to accompany a pipe or fife played by the same person

tabor

ta·bor·er *also* **ta·bour·er** \-bər-ər\ *noun* (15th century)
: one that plays on the tabor

tab·o·ret *or* **tab·ou·ret** \ˌta-bə-'ret, -'rā\ *noun* [French *tabouret,* literally, small drum, from Middle French, diminutive of *tabor, tabour* drum] (1630)
1 : a cylindrical seat or stool without arms or back
2 : a small portable stand or cabinet

ta·bou·li *variant of* TABBOULEH

Ta·briz \tə-'brēz\ *noun, plural* **Tabriz** [*Tabriz,* Iran] (1904)
: a Persian rug usually having a cotton warp, firm wool pile, and a medallion design

tab·u·lar \'ta-byə-lər\ *adjective* [Latin *tabularis* of boards, from *tabula* board, tablet] (circa 1656)
1 : having a flat surface **:** LAMINAR ⟨a *tabular* crystal⟩
2 a : of, relating to, or arranged in a table; *specifically* **:** set up in rows and columns **b :** computed by means of a table

ta·bu·la ra·sa \ˌta-byə-lə-'rä-zə, -sə\ *noun, plural* **ta·bu·lae ra·sae** \-ˌlī-'rä-ˌzī, -ˌsī\ [Latin, smoothed or erased tablet] (1607)
1 : the mind in its hypothetical primary blank or empty state before receiving outside impressions
2 : something existing in its original pristine state

tab·u·late \'ta-byə-ˌlāt\ *transitive verb* **-lat·ed; -lat·ing** [Latin *tabula* tablet] (1734)
1 : to put into tabular form
2 : to count, record, or list systematically
— tab·u·la·tion \ˌta-byə-'lā-shən\ *noun*
— tab·u·la·tor \'ta-byə-ˌlā-tər\ *noun*

ta·bun \'tä-ˌbu̇n\ *noun* [German] (1948)
: a liquid organophosphate $C_5H_{11}N_2O_2P$ that acts as a nerve gas

tac·a·ma·hac \'ta-kə-mə-ˌhak\ *noun* [Spanish *tacamahaca,* from Nahuatl *tecamac*] (1739)
: BALSAM POPLAR

TACAN \'ta-ˌkan\ *noun, often attributive* [*tac*tical *a*ir *n*avigation] (1955)
: a system of navigation that uses ultrahigh frequency signals to determine the distance and bearing of an aircraft from a transmitting station

ta·cet \'tä-ˌket; 'tā-sət, 'ta-sət\ [Latin, literally, (it) is silent, from *tacēre* to be silent — more at TACIT] (circa 1724)
— used as a direction in music to indicate that an instrument is not to play during a movement or long section

tach \'tak\ *noun* (circa 1930)
: TACHOMETER

tach·i·nid \'ta-kə-nəd, -ˌnid\ *noun* [New Latin *Tachinidae,* from *Tachina,* genus of flies, from Greek *tachinos* fleet, from *tachys* swift] (1888)
: any of a family (Tachinidae) of bristly usually grayish or black dipteran flies whose parasitic larvae are often used in the biological control of insect pests
— tachinid *adjective*

tach·ism \'ta-ˌshi-zəm\ *noun, often capitalized* [French *tachisme,* from *tache* stain, spot, blob, from Middle French *teche, tache,* of Germanic origin; akin to Old Saxon *tēkan* sign, Old High German *zeihhan* — more at TOKEN] (1955)
: ACTION PAINTING
— tach·ist \'ta-shist\ *also* **ta·chiste** \ta-'shēst\ *adjective or noun, often capitalized*

ta·chis·to·scope \tə-'kis-tə-ˌskōp, ta-\ *noun* [Greek *tachistos* (superlative of *tachys* swift) + International Scientific Vocabulary *-scope*] (circa 1890)
: an apparatus for the brief exposure of visual stimuli that is used in the study of learning, attention, and perception
— ta·chis·to·scop·ic \-ˌkis-tə-'skä-pik\ *adjective*
— ta·chis·to·scop·i·cal·ly \-pi-k(ə-)lē\ *adverb*

ta·chom·e·ter \ta-'kä-mə-tər, tə-\ *noun* [Greek *tachos* speed + English *-meter*] (1810)
: a device for indicating speed of rotation

tachy- *combining form* [Greek, from *tachys*]
: rapid **:** accelerated ⟨*tachy*cardia⟩

tachy·ar·rhyth·mia \ˌta-kē-ā-'ri<u>th</u>-mē-ə\ *noun* [New Latin] (1926)
: arrhythmia characterized by a rapid irregular heartbeat

tachy·car·dia \ˌta-ki-'kär-dē-ə\ *noun* [New Latin] (1889)
: relatively rapid heart action whether physiological (as after exercise) or pathological — compare BRADYCARDIA

tachy·on \'ta-kē-ˌän\ *noun* [*tachy-* + [2]*-on*] (1967)
: a hypothetical particle held to travel only faster than light

tac·it \'ta-sət\ *adjective* [French or Latin; French *tacite,* from Latin *tacitus* silent, from past participle of *tacēre* to be silent; akin to Old High German *dagēn* to be silent] (circa 1604)
1 : expressed or carried on without words or speech
2 a : implied or indicated but not actually expressed ⟨*tacit* consent⟩ **b** (1) **:** arising without express contract or agreement (2) **:** arising by operation of law ⟨*tacit* mortgage⟩
— tac·it·ly *adverb*
— tac·it·ness *noun*

tac·i·turn \ˈta-sə-ˌtərn\ *adjective* [French or Latin; French *taciturne,* from Latin *taciturnus,* from *tacitus*] (1771)
: temperamentally disinclined to talk
synonym see SILENT
— tac·i·tur·ni·ty \ˌta-sə-ˈtər-nə-tē\ *noun*

[1]tack \ˈtak\ *verb* [Middle English *takken,* from *tak*] (14th century)
transitive verb
1 : ATTACH; *especially* **:** to fasten or affix with tacks
2 : to join in a slight or hasty manner
3 a : to add as a supplement **b :** to add (a rider) to a parliamentary bill
4 : to change the direction of (a sailing ship) when sailing close-hauled by turning the bow to the wind and shifting the sails so as to fall off on the other side at about the same angle as before
intransitive verb
1 a : to tack a sailing ship **b** *of a ship* **:** to change to an opposite tack by turning the bow to the wind **c :** to follow a course against the wind by a series of tacks
2 a : to follow a zigzag course **b :** to modify one's policy or attitude abruptly
— tack·er *noun*

[2]tack *noun* [Middle English *tak* something that attaches; akin to Middle Dutch *tac* sharp point] (1574)
1 : a small short sharp-pointed nail usually having a broad flat head
2 a : the direction of a ship with respect to the trim of her sails ⟨starboard *tack*⟩ **b :** the run of a sailing ship on one tack **c :** a change when close-hauled from the starboard to the port tack or vice versa **d :** a zigzag movement on land **e :** a course or method of action; *especially* **:** one sharply divergent from that previously followed
3 : any of various usually temporary stitches
4 : the lower forward corner of a fore-and-aft sail
5 : a sticky or adhesive quality or condition

[3]tack *noun* [origin unknown] (1841)
: HARDTACK 1

[4]tack *noun* [perhaps short for *tackle*] (1924)
: stable gear; *especially* **:** articles of harness (as saddle and bridle) for use on a saddle horse

tack·board \ˈtak-ˌbōrd, -ˌbȯrd\ *noun* (circa 1927)
: a board (as of cork) for tacking up notices and display materials

tack claw *noun* (circa 1876)
: a small hand tool for removing tacks

tack·i·fy \ˈta-kə-ˌfī\ *transitive verb* **-fied; -fy·ing** (1942)
: to make (as a resin adhesive) tacky or more tacky
— tack·i·fi·er \-ˌfī(-ə)r\ *noun*

tack·i·ly \ˈta-kə-lē\ *adverb* (1903)
: in a tacky manner **:** so as to be tacky

tack·i·ness \ˈta-kē-nəs\ *noun* (1883)
: the quality or state of being tacky

[1]tack·le \ˈta-kəl, *nautical often* ˈtā-\ *noun* [Middle English *takel;* akin to Middle Dutch *takel* ship's rigging] (13th century)
1 : a set of the equipment used in a particular activity **:** GEAR ⟨fishing *tackle*⟩
2 a : a ship's rigging **b :** an assemblage of ropes and pulleys arranged to gain mechanical advantage for hoisting and pulling
3 a : the act or an instance of tackling **b** (1) **:** either of two offensive football players positioned on each side of the center and between guard and end (2) **:** either of two football players positioned on the inside of a defensive line

tackle 2b

[2]tackle *verb* **tack·led; tack·ling** \-k(ə-)liŋ\ (1714)
transitive verb
1 : to attach or secure with or as if with tackle
2 a : to seize, take hold of, or grapple with especially with the intention of stopping or subduing **b :** to seize and throw down or stop (an opposing player with the ball) in football
3 : to set about dealing with ⟨*tackle* the problem⟩
intransitive verb
: to tackle an opposing player in football
— tack·ler \-k(ə-)lər\ *noun*

tack·ling \ˈta-kliŋ, *nautical often* ˈtā-\ *noun* (15th century)
: TACKLE, GEAR

[1]tacky \ˈta-kē\ *adjective* **tack·i·er; -est** [[2]*tack*] (1788)
: somewhat sticky to the touch ⟨*tacky* varnish⟩; *also* **:** characterized by tack **:** ADHESIVE

[2]tacky *adjective* **tacki·er; -est** [*tacky* a low-class person] (1862)
1 a : characterized by lack of good breeding **:** COMMON ⟨couldn't run around downtown . . . in a bikini, which was *tacky* —Cyra McFadden⟩ **b :** SHABBY, SEEDY ⟨a *tacky* town whose citrus groves were blighted by smoke —Bryce Nelson⟩
2 : not having or exhibiting good taste ⟨hopelessly *tacky* bric-a-brac —T. C. Boyle⟩: as **a :** marked by lack of style **:** DOWDY **b :** marked by cheap showiness **:** GAUDY

ta·co \ˈtä-(ˌ)kō\ *noun, plural* **tacos** \-(ˌ)kōz\ [Mexican Spanish] (1934)
: a usually fried tortilla that is folded or rolled and stuffed with a mixture (as of seasoned meat, cheese, and lettuce)

tac·o·nite \ˈta-kə-ˌnīt\ *noun* [*Taconic* Range, mountains in U.S.] (1892)
: a flintlike rock high enough in iron content to constitute a low-grade iron ore

tact \ˈtakt\ *noun* [French, sense of touch, from Latin *tactus,* from *tangere* to touch — more at TANGENT] (1797)
1 : sensitive mental or aesthetic perception ⟨converted the novel into a play with remarkable skill and *tact*⟩
2 : a keen sense of what to do or say in order to maintain good relations with others or avoid offense ☆

tact·ful \ˈtakt-fəl\ *adjective* (1864)
: having or showing tact
— tact·ful·ly \-fə-lē\ *adverb*
— tact·ful·ness *noun*

[1]tac·tic \ˈtak-tik\ *noun* [New Latin *tactica,* from Greek *taktikē,* from feminine of *taktikos*] (1640)
1 : a device for accomplishing an end
2 : a method of employing forces in combat

[2]tactic *adjective* [New Latin *tacticus,* from Greek *taktikos*] (1871)
: of or relating to arrangement or order

-tactic *adjective combining form* [Greek *taktikos*]
1 : of, relating to, or having (such) an arrangement or pattern ⟨para*tactic*⟩
2 : showing orientation or movement directed by a (specified) force or agent ⟨geo*tactic*⟩

tac·ti·cal \ˈtak-ti-kəl\ *adjective* (1570)
1 : of or relating to combat tactics: as **a** (1) **:** of or occurring at the battlefront ⟨*tactical* defense⟩ ⟨*tactical* first strike⟩ (2) **:** using or being weapons or forces employed at the battlefront ⟨*tactical* missiles⟩ **b** *of an air force* **:** of, relating to, or designed for air attack in close support of friendly ground forces
2 a : of or relating to tactics: as (1) **:** of or relating to small-scale actions serving a larger purpose (2) **:** made or carried out with only a limited or immediate end in view **b :** adroit in planning or maneuvering to accomplish a purpose
— tac·ti·cal·ly \-k(ə-)lē\ *adverb*

tac·ti·cian \tak-ˈti-shən\ *noun* (1798)
: one versed in tactics

tac·tics \ˈtak-tiks\ *noun plural but singular or plural in construction* [New Latin *tactica,* plural, from Greek *taktika,* from neuter plural of *taktikos* of order, of tactics, fit for arranging, from *tassein* to arrange, place in battle formation] (1626)
1 a : the science and art of disposing and maneuvering forces in combat **b :** the art or skill of employing available means to accomplish an end
2 : a system or mode of procedure
3 : the study of the grammatical relations within a language including morphology and syntax

tac·tile \ˈtak-tᵊl, -ˌtīl\ *adjective* [French or Latin; French, from Latin *tactilis,* from *tangere* to touch — more at TANGENT] (1615)
1 : perceptible by touch **:** TANGIBLE
2 : of or relating to the sense of touch
— tac·tile·ly \-tə-lē, -ˌtīl-lē\ *adverb*

tactile corpuscle *noun* (1873)
: an end organ of touch

tac·til·i·ty \tak-ˈti-lə-tē\ *noun* (1659)
1 : the capability of being felt or touched
2 : responsiveness to stimulation of the sense of touch

tac·tion \ˈtak-shən\ *noun* [Latin *taction-, tactio,* from *tangere*] (circa 1623)
: TOUCH

tact·less \ˈtakt-ləs\ *adjective* (circa 1847)
: marked by lack of tact
— tact·less·ly *adverb*
— tact·less·ness *noun*

tac·tu·al \ˈtak-chə-wəl, -chəl\ *adjective* [Latin *tactus* sense of touch — more at TACT] (1642)
: TACTILE 2
— tac·tu·al·ly *adverb*

tad \ˈtad\ *noun* [probably from English dialect, toad, from Middle English *tode* — more at TOAD] (circa 1877)
1 : a small child; *especially* **:** BOY
2 : a small or insignificant amount or degree **:** BIT ⟨might give him some water and a *tad* to eat —C. T. Walker⟩
— a tad : SOMEWHAT, RATHER ⟨looked *a tad* bigger than me —Larry Hodgson⟩

tad·pole \ˈtad-ˌpōl\ *noun* [Middle English *taddepol,* from *tode* toad + *polle* head] (15th century)
: a larval amphibian; *specifically* **:** a frog or toad larva that has a rounded body with a long tail bordered by fins and external gills soon replaced by

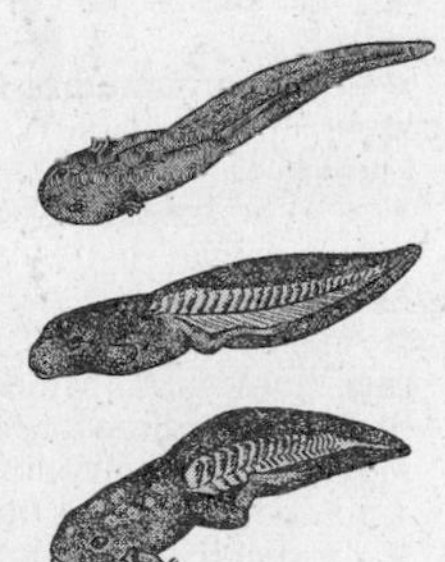

tadpole

☆ SYNONYMS
Tact, address, poise, savoir faire mean skill and grace in dealing with others. TACT implies delicate and considerate perception of what is appropriate ⟨questions showing a lack of *tact*⟩. ADDRESS stresses dexterity and grace in dealing with new and trying situations and may imply success in attaining one's ends ⟨brought it off with remarkable *address*⟩. POISE may imply both tact and address but stresses self-possession and ease in meeting difficult situations ⟨answered the accusations with unruffled *poise*⟩. SAVOIR FAIRE is likely to stress worldly experience and a sure awareness of what is proper or expedient ⟨the *savoir faire* of a seasoned traveler⟩.

internal gills and that undergoes a metamorphosis to the adult

Ta·dzhik *variant of* TAJIK

tae·di·um vi·tae \ˌtē-dē-əm-ˈvī-ˌtē, ˌtī-dē-əm-ˈwē-ˌtī\ *noun* [Latin] (1759)
: weariness or loathing of life

tae kwon do \ˈtī-ˈkwän-ˈdō\ *noun, often T&K&D capitalized* [Korean *t'aekwŏndo,* from *t'ae-* to trample + *kwŏn* fist + *to* way] (1967)
: a Korean martial art resembling karate

tael \ˈtā(ə)l\ *noun* [Portuguese, from Malay *tahil*] (1588)
1 : any of various Chinese units of value based on the value of a tael weight of silver
2 : any of various units of weight of eastern Asia

tae·nia \ˈtē-nē-ə\ *noun, plural* **-ni·ae** \-nē-ˌī, -ˌē\ *or* **-nias** [Latin, ribbon, fillet, from Greek *tainia;* akin to Greek *teinein* to stretch — more at THIN] (1563)
1 : a band on a Doric order separating the frieze from the architrave
2 : TAPEWORM
3 : an ancient Greek fillet
4 [New Latin, from Latin, fillet, band] **:** a band of nervous tissue or muscle

tae·ni·a·sis \tē-ˈnī-ə-səs\ *noun* [New Latin, from Latin *taenia* tapeworm] (circa 1890)
: infestation with or disease caused by tapeworms

taf·fe·ta \ˈta-fə-tə\ *noun* [Middle English, from Middle French *taffetas,* from Old Italian *taffettà,* from Turkish *tafta,* from Persian *tāftah* woven] (14th century)
: a crisp plain-woven lustrous fabric of various fibers used especially for women's clothing

taf·fe·tized \ˈta-fə-ˌtīzd\ *adjective* (1949)
of cloth **:** having a crisp finish

taff·rail \ˈtaf-ˌrāl, -rəl\ *noun* [modification of Dutch *tafereel,* from Middle Dutch, panel, from Old French *tablel* — more at TABLEAU] (circa 1704)
1 : the upper part of the stern of a wooden ship
2 : a rail around the stern of a ship

taf·fy \ˈta-fē\ *noun, plural* **taffies** [origin unknown] (circa 1817)
1 : a boiled candy usually of molasses or brown sugar that is pulled until porous and light-colored
2 : insincere flattery

¹tag \ˈtag\ *noun* [Middle English *tagge,* probably of Scandinavian origin; akin to Swedish *tagg* barb] (15th century)
1 : a loose hanging piece of cloth **:** TATTER
2 : a metal or plastic binding on an end of a shoelace
3 : a piece of hanging or attached material; *specifically* **:** a loop, knot, or tassel on a garment
4 a : a brief quotation used for rhetorical emphasis or sententious effect **b :** a recurrent or characteristic verbal expression **c :** TAG LINE 1
5 a : a cardboard, plastic, or metal marker used for identification or classification ⟨license *tags*⟩ **b :** a descriptive or identifying epithet **c :** something used for identification or location **:** FLAG **d :** LABEL 3d **e :** PRICE TAG
6 : a detached fragmentary piece **:** BIT

²tag *verb* **tagged; tag·ging** (15th century)
transitive verb
1 : to provide or mark with or as if with a tag: as **a :** to supply with an identifying marker or price ⟨*tagged* every item in the store⟩ ⟨was *tagged* at $4.95⟩ **b :** to provide with a name or epithet **:** LABEL, BRAND ⟨*tagged* him a has-been⟩ **c :** to put a ticket on (a motor vehicle) for a traffic violation
2 : to attach as an addition **:** APPEND
3 : to follow closely and persistently
4 : to hold to account; *especially* **:** to charge with violating the law ⟨was *tagged* for . . . assault —Burt Woolis⟩
5 : LABEL 2
intransitive verb
: to keep close ⟨*tagging* at their heels —Corey Ford⟩

³tag *noun* [origin unknown] (1738)
1 : a game in which the player who is it chases others and tries to touch one of them who then becomes it
2 : an act or instance of tagging a runner in baseball

⁴tag *transitive verb* **tagged; tag·ging** (1878)
1 a : to touch in or as if in a game of tag **b :** to put out (a runner) in baseball by a touch with the ball or the gloved hand containing the ball
2 : to hit solidly
3 : to choose usually for a special purpose **:** SELECT
4 : to make a hit or run off (a pitcher) in baseball

Ta·ga·log \tə-ˈgä-ˌlȯg\ *noun, plural* **Tagalog** *or* **Tagalogs** [Tagalog] (circa 1808)
1 : a member of a people of central Luzon
2 : an Austronesian language of the Tagalog people — compare PILIPINO

tag·along \ˈta-gə-ˌlȯŋ\ *noun* (1935)
: one that persistently and often annoyingly follows the lead of another

tag along *intransitive verb* (1900)
: to follow another's lead especially in going from one place to another

tag·board \ˈtag-ˌbōrd, -ˌbȯrd\ *noun* (1904)
: strong cardboard used especially for making shipping tags

tag end *noun* (1807)
1 : the last part
2 : a miscellaneous or random bit

ta·glia·tel·le \ˌtäl-yä-ˈte-(ˌ)lā\ *noun* [Italian, from *tagliare* to cut, from Late Latin *taliare* — more at TAILOR] (1899)
: FETTUCCINE

tag line *noun* (1926)
1 : a final line (as in a play or joke); *especially* **:** one that serves to clarify a point or create a dramatic effect
2 : a reiterated phrase identified with an individual, group, or product **:** SLOGAN

tag question *noun* (1933)
: a question (as *isn't it* in "it's fine, isn't it?") added to a statement or command (as to gain the assent of or challenge the person addressed); *also* **:** a sentence ending in a tag question

tag, rag, and bobtail *or* **tagrag and bobtail** \ˌtag-ˌrag-ən-ˈbäb-ˌtāl, -ˌrag-ᵊŋ-\ *noun* (1645)
: RABBLE

tag sale *noun* [from the price tag on each item] (1955)
: GARAGE SALE

tag team *noun* [⁴*tag*] (1952)
: a team of two or more professional wrestlers who spell each other during a match

tag up *intransitive verb* (1942)
: to touch a base before running in baseball after a fly ball is caught

ta·hi·ni \tə-ˈhē-nē, tä-\ *noun* [Arabic dialect *ṭaḥīna,* from *ṭaḥana* to grind] (1950)
: a smooth paste of sesame seeds

Ta·hi·tian \tə-ˈhē-shən\ *noun* (1825)
1 : a native or inhabitant of Tahiti
2 : the Polynesian language of the Tahitians
— Tahitian *adjective*

tahr \ˈtär\ *noun* [Nepali *thār*] (1835)
: any of a genus (*Hemitragus*) of wild Asian goats; *especially* **:** one (*H. jemlahicus*) of the Himalayas having a reddish brown to dark brown coat and a long shaggy mane

tah·sil \tä-ˈsē(ə)l\ *noun* [Urdu *taḥsīl,* from Arabic, collection of revenue] (1846)
: a district administration or revenue subdivision in India

Tai \ˈtī\ *noun, plural* **Tai** (1693)
: a widespread group of peoples in southeast Asia associated ethnically with valley paddy-rice culture

tai chi chuan *or* **t'ai chi ch'uan** \ˈtī-ˈjē-chü-ˈän, ˈtī-ˈchē-\ *noun, often T and both Cs capitalized* [Chinese (Beijing) *tàijíquán,* from *tàijí* the Absolute in Chinese cosmology + *quán* fist, boxing] (1954)
: an ancient Chinese discipline of meditative movements practiced as a system of exercises — called also *tai chi, t'ai chi*

tai·ga \ˈtī-gə\ *noun* [Russian *taĭga*] (1888)
: a moist subarctic forest dominated by conifers (as spruce and fir) that begins where the tundra ends

¹tail \ˈtā(ə)l\ *noun, often attributive* [Middle English, from Old English *tægel;* akin to Old High German *zagal* tail, Middle Irish *dúal* lock of hair] (before 12th century)
1 : the rear end or a process or prolongation of the rear end of the body of an animal
2 : something resembling an animal's tail in shape or position: as **a :** the luminous train of a comet **b :** the rear part of an airplane consisting usually of horizontal and vertical stabilizing surfaces with attached control surfaces
3 : RETINUE
4 *plural* **a :** TAILCOAT **b :** full evening dress for men
5 a : BUTTOCKS, BUTT **b :** a female sexual partner — usually considered vulgar
6 : the back, last, lower, or inferior part of something
7 : TAILING 1 — usually used in plural
8 : the reverse of a coin — usually used in plural ⟨*tails,* I win⟩
9 : one (as a detective) who follows or keeps watch on someone
10 : the blank space at the bottom of a page
11 : a location immediately or not far behind ⟨had a posse on his *tail*⟩
— tailed \ˈtā(ə)ld\ *adjective*
— tail·less \ˈtā(ə)l-ləs\ *adjective*
— tail·like \-ˌlīk\ *adjective*

²tail (1523)
transitive verb
1 : to connect end to end
2 a : to remove the tail of (an animal) **:** DOCK **b :** to remove the stem or bottom part of ⟨topping and *tailing* gooseberries⟩
3 a : to make or furnish with a tail **b :** to follow or be drawn behind like a tail
4 : to follow for purposes of surveillance
intransitive verb
1 : to form or move in a straggling line
2 : to grow progressively smaller, fainter, or more scattered **:** ABATE — usually used with *off* ⟨productivity is *tailing* off —Tom Nicholson⟩
3 : to swing or lie with the stern in a named direction — used of a ship at anchor
4 : ²TAG
— tail·er *noun*

³tail *noun* [Middle English, from Middle French, from Old French, from *taillier*] (14th century)
: ENTAIL 1a

⁴tail *adjective* [Middle English *taille,* from Anglo-French *taylé,* from Old French *taillié,* past participle of *taillier* to cut, limit — more at TAILOR] (15th century)
: limited as to tenure **:** ENTAILED

tail·back \ˈtā(ə)l-ˌbak\ *noun* (1930)
: the offensive football back farthest from the line of scrimmage

tail·board \-ˌbōrd, -ˌbȯrd\ *noun* (1805)
chiefly British **:** TAILGATE 1

tail·bone \-ˌbōn\ *noun* (circa 1577)
1 : a caudal vertebra
2 : COCCYX

tail·coat \-ˌkōt\ *noun* (1847)
: a coat with tails; *especially* **:** a man's full-dress coat with two long tapering skirts at the back
— tail·coat·ed \-ˌkō-təd\ *adjective*

tail covert *noun* (1815)
: one of the coverts of the tail quills

tail end *noun* (14th century)

1 : BUTTOCKS, RUMP
2 : the hindmost end
3 : the concluding period ⟨the *tail end* of the session⟩

tail·end·er \ˈtā(ə)l-ˌen-dər\ *noun* (1885)
: one positioned at the end or in last place ⟨the *tailenders* in a race⟩

tail fin *noun* (1681)
1 : the terminal fin of a fish or cetacean
2 : FIN 2b

¹tail·gate \ˈtā(ə)l-ˌgāt\ *noun* (1868)
1 : a board or gate at the rear of a vehicle that can be removed or let down (as for loading)
2 [from the custom of seating trombonists at the rear of trucks carrying jazz bands in parades] **:** a jazz trombone style marked by much use of slides to and from long sustained tones

²tailgate *verb* **tail·gat·ed; tail·gat·ing** (1949)
intransitive verb
1 : to drive dangerously close behind another vehicle
2 : to hold a tailgate picnic
transitive verb
: to drive dangerously close behind
— tail·gat·er *noun*

³tailgate *adjective* (1965)
: relating to or being a picnic set up on the tailgate especially of a station wagon

tail·ing \ˈtā-liŋ\ *noun* (1764)
1 : residue separated in the preparation of various products (as grain or ores) — usually used in plural
2 : the part of a projecting stone or brick inserted in a wall

taille \ˈtä-yə, ˈtī-, ˈtā(ə)l\ *noun* [Middle French, from Old French, from *taillier* to cut, tax] (circa 1533)
: a tax formerly levied by a French king or seigneur on his subjects or on lands held of him

tail·light \ˈtā(ə)l-ˌlīt\ *noun* (1844)
: a usually red warning light mounted at the rear of a vehicle — called also *taillamp*

¹tai·lor \ˈtā-lər\ *noun* [Middle English *taillour,* from Old French *tailleur,* from *taillier* to cut, from Late Latin *taliare,* from Latin *talea* twig, cutting] (14th century)
: a person whose occupation is making or altering outer garments

²tailor (1719)
intransitive verb
: to do the work of a tailor
transitive verb
1 a : to make or fashion as the work of a tailor **b :** to make or adapt to suit a special need or purpose
2 : to fit with clothes
3 : to style with trim straight lines and finished handwork

tai·lor·bird \ˈtā-lər-ˌbərd\ *noun* (1769)
: any of a genus (*Orthotomus* of the family Sylviidae) of chiefly Asian warblers that stitch leaves together to support and hide their nests

tai·lored \ˈtā-lərd\ *adjective* (1862)
1 : fashioned or fitted to resemble a tailor's work
2 : CUSTOM-MADE
3 : having the look of one fitted by a custom tailor

tai·lor·ing \ˈtā-lə-riŋ\ *noun* (1662)
1 a : the business or occupation of a tailor **b :** the work or workmanship of a tailor
2 : the making or adapting of something to suit a particular purpose

¹tai·lor–made \ˌtā-lər-ˈmād\ *adjective* (1832)
1 : made by a tailor or with a tailor's care and style
2 : made or fitted especially to a particular use or purpose
3 : factory made rather than hand-rolled ⟨*tailor-made* cigarettes⟩

²tailor–made *noun* (1892)
: one that is tailor-made; *specifically* **:** a woman's garment styled for a trim fit and with stiff straight lines

tail·piece \ˈtā(ə)l-ˌpēs\ *noun* (1601)
1 : a piece added at the end
2 : a device from which the strings of a stringed instrument are stretched to the pegs — see VIOLIN illustration
3 : an ornament placed below the text matter of a page

tail·pipe \-ˌpīp\ *noun* (1922)
: an outlet by which the exhaust gases are removed from an engine (as of an automobile or jet aircraft)

tail·plane \-ˌplān\ *noun* (1909)
: the horizontal tail surfaces of an airplane including the stabilizer and the elevator

tail·race \ˈtā(ə)l-ˌrās\ *noun* (1776)
: a race for conveying water away from a point of industrial application (as a waterwheel or turbine) after use

tail·slide \-ˌslīd\ *noun* (1916)
: an aerobatic maneuver in which an aircraft that has been pulled into a steep climb stalls and then loses altitude by dropping backward

tail·spin \-ˌspin\ *noun* (circa 1917)
1 : SPIN 2a
2 : a mental or emotional letdown or collapse
3 : a sustained and usually severe decline or downturn ⟨stock prices in a *tailspin*⟩

tail·wa·ter \-ˌwȯ-tər, -ˌwä-\ *noun* (1759)
1 : water below a dam or waterpower development
2 : excess surface water draining especially from a field under cultivation

tail·wind \-ˌwind\ *noun* (1897)
: a wind having the same general direction as a course of movement (as of an aircraft)

Tai·no \ˈtī-(ˌ)nō\ *noun, plural* **Taino** *or* **Tainos** [Spanish] (circa 1895)
1 : a member of an aboriginal Arawakan people of the Greater Antilles and the Bahamas
2 : the extinct language of the Taino people

¹taint \ˈtānt\ *verb* [Middle English *tainten* to color & *taynten* to attaint; Middle English *tainten,* from Anglo-French *teinter,* from Middle French *teint,* past participle of *teindre,* from Latin *tingere;* Middle English *taynten,* from Middle French *ataint,* past participle of *ataindre* — more at TINGE, ATTAIN] (1573)
transitive verb
1 : to contaminate morally **:** CORRUPT ⟨scholarship *tainted* by envy⟩
2 : to affect with putrefaction **:** SPOIL
3 : to touch or affect slightly with something bad ⟨persons *tainted* with prejudice⟩
intransitive verb
1 *obsolete* **:** to become weak
2 : to become affected with putrefaction **:** SPOIL
synonym see CONTAMINATE

²taint *noun* (1601)
: a contaminating mark or influence
— taint·less \-ləs\ *adjective*

'tain't \ˈtānt\ (1818)
: it ain't

¹tai·pan \ˈtī-ˌpan, ˈtī-ˈpän\ *noun* [Chinese (Guangdong) *daaih-bāan,* from *daaih* big + *bāan* class] (1834)
: a powerful businessman and especially formerly a foreigner living and operating in Hong Kong or China

²tai·pan \ˈtī-ˌpan\ *noun* [Wik Munkan (Australian aboriginal language of northern Queensland) *dhayban*] (1933)
: an exceedingly venomous elapid snake (*Oxyuranus scutellatus*) of northern Australia and New Guinea

Ta·jik \tä-ˈjik, tə-, -ˈjēk\ *noun* (1815)
1 : a member of a Persian-speaking ethnic group living in Tajikistan, Afghanistan, and adjacent areas of central Asia
2 : the form of Persian spoken by the Tajiks

ta·ka \ˈtä-kə, -(ˌ)kä\ *noun* [Bengali *ṭākā* rupee, taka, from Sanskrit *ṭaṅka* stamped coin] (1972)
— see MONEY table

ta·ka·he \tä-ˈkä-(ˌ)hā\ *noun* [Maori] (1851)
: a flightless bird (*Notornis mantelli* synonym *Porphyrio mantelli*) of New Zealand that is related to the gallinules

¹take \ˈtāk\ *verb* **took** \ˈtu̇k\; **tak·en** \ˈtā-kən\; **tak·ing** [Middle English, from Old English *tacan,* from Old Norse *taka;* akin to Middle Dutch *taken* to take] (before 12th century)
transitive verb
1 : to get into one's hands or into one's possession, power, or control: as **a :** to seize or capture physically ⟨*took* them as prisoners⟩ **b :** to get possession of (as fish or game) by killing or capturing **c** (1) **:** to move against (as an opponent's piece in chess) and remove from play (2) **:** to win in a card game ⟨able to *take* 12 tricks⟩ **d :** to acquire by eminent domain
2 : GRASP, GRIP ⟨*take* the ax by the handle⟩
3 a : to catch or attack through the effect of a sudden force or influence ⟨*taken* with a fit of laughing⟩ ⟨*taken* ill⟩ **b :** to catch or come upon in a particular situation or action ⟨was *taken* unawares⟩ **c :** to gain the approval or liking of **:** CAPTIVATE, DELIGHT ⟨was quite *taken* with her at their first meeting⟩
4 a : to receive into one's body (as by swallowing, drinking, or inhaling) ⟨*take* a pill⟩ **b :** to expose oneself to (as sun or air) for pleasure or physical benefit **c :** to partake of **:** EAT ⟨*takes* dinner about seven⟩
5 a : to bring or receive into a relation or connection ⟨*takes* just four students a year⟩ ⟨it's time he *took* a wife⟩ **b :** to copulate with
6 : to transfer into one's own keeping: **a :** APPROPRIATE **b :** to obtain or secure for use (as by lease, subscription, or purchase) ⟨*take* a cottage for the summer⟩ ⟨I'll *take* the red one⟩ ⟨*took* an ad in the paper⟩
7 a : ASSUME ⟨gods often *took* the likeness of a human being⟩ ⟨when the college *took* its present form⟩ **b** (1) **:** to enter into or undertake the duties of ⟨*take* a job⟩ ⟨*take* office⟩ ⟨*took* command of the fleet⟩ (2) **:** to move onto or into **:** move into position on ⟨the home team *took* the field⟩ **c** (1) **:** to bind oneself by ⟨*take* the oath of office⟩ (2) **:** to make (a decision) especially with finality or authority **d :** to impose upon oneself ⟨*take* the trouble to do good work⟩ **e** (1) **:** to adopt as one's own ⟨*take* a stand on the issue⟩ ⟨*take* an interest⟩ (2) **:** to align or ally oneself with ⟨mother *took* his side⟩ **f :** to assume as if rightfully one's own or as if granted ⟨*take* the credit⟩ **g :** to have or assume as a proper part of or accompaniment to itself ⟨transitive verbs *take* an object⟩
8 a : to secure by winning in competition ⟨*took* first place⟩ **b :** DEFEAT
9 : to pick out **:** CHOOSE, SELECT
10 : to adopt, choose, or avail oneself of for use: as **a :** to have recourse to as an instrument for doing something ⟨*take* a scythe to the weeds⟩ **b :** to use as a means of transportation or progression ⟨*take* the bus⟩ **c :** to have recourse to for safety or refuge ⟨*take* shelter⟩ **d :** to go along, into, or through ⟨*took* a different route⟩ **e** (1) **:** to proceed to occupy ⟨*take* a seat in the rear⟩ (2) **:** to use up (as space or time) ⟨*takes* a long time to dry⟩ (3) **:** NEED, REQUIRE ⟨*takes* a size nine shoe⟩ ⟨it *takes* two to start a fight⟩
11 a : to obtain by deriving from a source **:** DRAW ⟨*takes* its title from the name of the hero⟩ **b** (1) **:** to obtain as the result of a special procedure **:** ASCERTAIN ⟨*take* the temperature⟩ ⟨*take* a census⟩ (2) **:** to get in or as if in writing ⟨*take* notes⟩ ⟨*take* an inventory⟩ (3) **:** to get by drawing or painting or by photography ⟨*take* a

\ə\ **abut** \ᵊ\ **kitten** \ər\ **further** \a\ **ash** \ā\ **ace** \ä\ **mop, mar** \au̇\ **out** \ch\ **chin** \e\ **bet** \ē\ **easy** \g\ **go** \i\ **hit** \ī\ **ice** \j\ **job** \ŋ\ **sing** \ō\ **go** \ȯ\ **law** \ȯi\ **boy** \th\ **thin** \t͟h\ **the** \ü\ **loot** \u̇\ **foot** \y\ **yet** \zh\ **vision** *see also* Guide to Pronunciation

snapshot) (4) **:** to get by transference from one surface to another ⟨*take* a proof⟩ ⟨*take* fingerprints⟩
12 : to receive or accept whether willingly or reluctantly ⟨*take* a bribe⟩ ⟨will you *take* this call⟩ ⟨*take* a bet⟩: as **a** (1) **:** to submit to **:** ENDURE ⟨*take* a cut in pay⟩ (2) **:** WITHSTAND ⟨it will *take* a lot of punishment⟩ (3) **:** SUFFER ⟨*took* a direct hit⟩ **b** (1) **:** to accept as true **:** BELIEVE ⟨I'll *take* your word for it⟩ (2) **:** FOLLOW ⟨*take* my advice⟩ (3) **:** to accept with the mind in a specified way ⟨*took* the news hard⟩ **c :** to indulge in and enjoy ⟨was *taking* his ease on the porch⟩ **d :** to receive or accept as a return (as in payment, compensation, or reparation) **e :** to accept in a usually professional relationship — often used with *on* ⟨agreed to *take* him on as a client⟩ **f :** to refrain from hitting at (a pitched ball)
13 a (1) **:** to let in **:** ADMIT ⟨the boat was *taking* water fast⟩ (2) **:** ACCOMMODATE ⟨the suitcase wouldn't *take* another thing⟩ **b :** to be affected injuriously by (as a disease) **:** CONTRACT ⟨*take* cold⟩; *also* **:** to be seized by ⟨*take* a fit⟩ ⟨*take* fright⟩ **c :** to absorb or become impregnated with (as dye); *also* **:** to be effectively treated by ⟨a surface that *takes* a fine polish⟩
14 a : APPREHEND, UNDERSTAND ⟨how should I *take* your remark⟩ **b :** CONSIDER, SUPPOSE ⟨I *take* it you're not going⟩ **c :** RECKON, ACCEPT ⟨*taking* a stride at 30 inches⟩ **d :** FEEL, EXPERIENCE ⟨*take* pleasure⟩ ⟨*take* an instant dislike to someone⟩ ⟨*take* offense⟩
15 : to lead, carry, or cause to go along to another place ⟨this bus will *take* you into town⟩ ⟨*took* an umbrella with her⟩
16 a : REMOVE ⟨*take* eggs from a nest⟩ **b** (1) **:** to put an end to (life) (2) **:** to remove by death ⟨was *taken* in his prime⟩ **c :** SUBTRACT ⟨*take* two from four⟩ **d :** EXACT ⟨the weather *took* its toll⟩
17 a : to undertake and make, do, or perform ⟨*take* a walk⟩ ⟨*take* aim⟩ ⟨*take* legal action⟩ ⟨*take* a test⟩ ⟨*take* a look⟩ **b :** to participate in ⟨*take* a meeting⟩
18 a : to deal with ⟨*take* first things first⟩ **b :** to consider or view in a particular relation ⟨*taken* together, the details were significant⟩; *especially* **:** to consider as an example ⟨*take* style, for instance⟩ **c** (1) **:** to apply oneself to the study of ⟨*take* music lessons⟩ ⟨*take* French⟩ (2) **:** to study for especially successfully ⟨*taking* a degree in engineering⟩ ⟨*took* holy orders⟩
19 : to obtain money from especially fraudulently ⟨*took* me for all I had⟩
20 : to pass or attempt to pass through, along, or over ⟨*took* the curve too fast⟩
intransitive verb
1 : to obtain possession: as **a :** CAPTURE **b :** to receive property under law as one's own
2 : to lay hold **:** CATCH, HOLD
3 : to establish a take especially by uniting or growing ⟨90 percent of the grafts *take*⟩
4 a : to betake oneself **:** set out **:** GO ⟨*take* after a purse snatcher⟩ **b** *chiefly dialect* — used as an intensifier or redundantly with a following verb ⟨*took* and swung at the ball⟩
5 a : to take effect **:** ACT, OPERATE ⟨hoped the lesson he taught would *take*⟩ **b :** to show the natural or intended effect ⟨dry fuel *takes* readily⟩
6 : CHARM, CAPTIVATE ⟨a *taking* smile⟩
7 : DETRACT
8 : to be seized or attacked in a specified way **:** BECOME ⟨*took* sick⟩
usage see BRING ☆
— **tak•er** *noun*
— **take a back seat :** to have or assume a secondary position or status
— **take a bath :** to suffer a heavy financial loss
— **take account of :** to take into account
— **take advantage of 1 :** to use to advantage **:** profit by **2 :** to impose on **:** EXPLOIT
— **take after :** to resemble in features, build, character, or disposition
— **take apart 1 :** to disconnect the pieces of **:** DISASSEMBLE **2 :** to treat roughly or harshly **:** tear into
— **take a powder :** to leave hurriedly
— **take care :** to be careful or watchful **:** exercise caution or prudence
— **take care of :** to attend to or provide for the needs, operation, or treatment of
— **take charge :** to assume care, custody, command, or control
— **take effect 1 :** to become operative **2 :** to be effective
— **take exception :** OBJECT ⟨*took exception* to the remark⟩
— **take five** *or* **take ten :** to take a break especially from work
— **take for :** to suppose to be; *especially* **:** to suppose mistakenly to be
— **take for granted 1 :** to assume as true, real, or expected **2 :** to value too lightly
— **take heart :** to gain courage or confidence
— **take hold 1 :** GRASP, GRIP, SEIZE **2 :** to become attached or established **:** take effect
— **take into account :** to make allowance for
— **take in vain :** to use (a name) profanely or without proper respect
— **take issue :** DISAGREE
— **take it on the chin :** to suffer from the results of a situation
— **take kindly to :** to show an inclination to accept or approve
— **take notice of :** to observe or treat with special attention
— **take one's time :** to be leisurely about doing something
— **take part :** JOIN, PARTICIPATE, SHARE
— **take place :** HAPPEN, OCCUR
— **take root 1 :** to become rooted **2 :** to become fixed or established
— **take shape :** to assume a definite or distinctive form
— **take stock :** to make an assessment
— **take the cake :** to carry off the prize **:** rank first
— **take the count 1** *of a boxer* **:** to be counted out **2 :** to go down in defeat
— **take the floor :** to rise (as in a meeting or a legislative assembly) to make a formal address
— **take to 1 :** to go to or into ⟨*take to* the woods⟩ **2 :** to apply or devote oneself to (as a practice, habit, or occupation) ⟨*take to* begging⟩ **3 :** to adapt oneself to **:** respond to ⟨*takes to* water like a duck⟩ **4 :** to conceive a liking for
— **take to task :** to call to account for a shortcoming **:** CRITICIZE
— **take turns :** ALTERNATE

²take *noun* (1654)
1 : something that is taken: **a :** the amount of money received **:** PROCEEDS, RECEIPTS, INCOME **b :** SHARE, CUT ⟨wanted a bigger *take*⟩ **c :** the number or quantity (as of animals, fish, or pelts) taken at one time **:** CATCH, HAUL **d :** a section or installment done as a unit or at one time **e** (1) **:** a scene filmed or televised at one time without stopping the camera (2) **:** a sound recording made during a single recording period; *especially* **:** a trial recording
2 : an act or the action of taking: as **a :** the action of killing, capturing, or catching (as game or fish) **b** (1) **:** the uninterrupted photographing or televising of a scene (2) **:** the making of a sound recording
3 a : a local or systemic reaction indicative of successful vaccination (as against smallpox) **b :** a successful union (as of a graft)
4 : a visible response or reaction (as to something unexpected) ⟨a delayed *take*⟩
5 : a distinct or personal point of view, outlook, or assessment ⟨was asked for her *take* on recent developments⟩; *also* **:** a distinct treatment or variation ⟨a new *take* on an old style⟩
— **on the take :** illegally paid for favors

take–away \ˈtāk-ə-ˌwā\ *adjective* (1964)
British **:** TAKE-OUT
— **take–away** *noun*

take back *transitive verb* (1775)
: to make a retraction of **:** WITHDRAW

take–charge \ˈtāk-ˈchärj\ *adjective* (1954)
: having the qualities of a forceful leader ⟨a *take-charge* executive⟩

¹take•down \ˈtāk-ˌdau̇n\ *noun* (1893)
1 : the action or an act of taking down
2 : something (as a rifle) having takedown construction

²take•down \ˈtāk-ˈdau̇n\ *adjective* (1907)
: constructed so as to be readily taken apart ⟨a *takedown* rifle⟩

take down (15th century)
transitive verb
1 : to lower without removing ⟨*took down* his pants⟩
2 a : to pull to pieces ⟨*take down* a building⟩ **b :** DISASSEMBLE ⟨*take* a rifle *down*⟩
3 : to lower the spirit or vanity of
4 a : to write down **b :** to record by mechanical means
intransitive verb
: to become seized or attacked especially by illness

take–home pay \ˈtāk-ˌhōm-\ *noun* (1943)
: income remaining from salary or wages after deductions (as for income-tax withholding)

take–in \ˈtā-ˌkin\ *noun* (1778)
: an act of taking in especially by deceiving

take in *transitive verb* (circa 1515)
1 : to draw into a smaller compass ⟨*take in* the slack of a line⟩: **a :** FURL **b :** to make (a garment) smaller by enlarging seams or tucks
2 a : to receive as a guest or lodger **b :** to give shelter to **c :** to take to a police station as a prisoner
3 : to receive as payment or proceeds
4 : to receive (work) into one's house to be done for pay ⟨*take in* washing⟩
5 : to encompass within its limits
6 a : to include in an itinerary **b :** ATTEND ⟨*take in* a movie⟩
7 : to receive into the mind **:** PERCEIVE
8 : DECEIVE, DUPE

taken *past participle of* TAKE

take•off \ˈtā-ˌkȯf\ *noun* (1846)
1 : an imitation especially in the way of caricature
2 a : a spot at which one takes off **b :** a starting point **:** point of departure
3 a : a rise or leap from a surface in making a jump or flight or an ascent in an aircraft or in the launching of a rocket **b :** an action of starting out **c :** a rapid rise in activity, growth, or popularity ⟨an economic *takeoff*⟩
4 : an action of removing something
5 : a mechanism for transmission of the power of an engine or vehicle to operate some other mechanism

☆ SYNONYMS
Take, seize, grasp, clutch, snatch, grab mean to get hold of by or as if by catching up with the hand. TAKE is a general term applicable to any manner of getting something into one's possession or control ⟨*take* some salad from the bowl⟩. SEIZE implies a sudden and forcible movement in getting hold of something tangible or an apprehending of something fleeting or elusive when intangible ⟨*seized* the suspect⟩. GRASP stresses a laying hold so as to have firmly in possession ⟨*grasp* the handle and pull⟩. CLUTCH suggests avidity or anxiety in seizing or grasping and may imply less success in holding ⟨*clutching* her purse⟩. SNATCH suggests more suddenness or quickness but less force than SEIZE ⟨*snatched* a doughnut and ran⟩. GRAB implies more roughness or rudeness than SNATCH ⟨*grabbed* roughly by the arm⟩.

take off (14th century)
transitive verb
1 : REMOVE ⟨*take* your shoes *off*⟩
2 a : RELEASE ⟨*take* the brake *off*⟩ **b :** DISCONTINUE, WITHDRAW ⟨*took off* the morning train⟩ **c :** to take or allow as a discount **:** DEDUCT ⟨*took* 10 percent *off*⟩ **d :** to spend (a period of time) away from a usual occupation or activity ⟨*took* two weeks *off*⟩
3 *slang* **:** ROB
intransitive verb
1 : to take away **:** DETRACT
2 a : to start off or away **:** SET OUT, DEPART **b** (1) **:** to branch off (as from a main stream or stem) (2) **:** to take a point of origin **c :** to begin a leap or spring **d :** to leave the surface **:** begin flight **e :** to embark on rapid activity, development, or growth **f :** to spring into wide use or popularity

take on (15th century)
transitive verb
1 a : to begin to perform or deal with **:** UNDERTAKE ⟨*took on* new responsibilities⟩ **b :** to contend with as an opponent ⟨*took on* the neighborhood bully⟩
2 : ENGAGE, HIRE
3 a : to assume or acquire as or as if one's own ⟨the city's plaza *takes on* a carnival air —W. T. LeViness⟩ **b :** to have as a mathematical domain or range ⟨what values does the function *take on*⟩
intransitive verb
: to show one's feelings especially of grief or anger in a demonstrative way ⟨they cried and *took on* something terrible —Bob Hope⟩

take·out \ˈtā-ˌkau̇t\ *noun* (circa 1917)
1 : the action or an act of taking out
2 a : something taken out or prepared to be taken out **b** (1) **:** an article (as in a newspaper) printed on consecutive pages so as to be conveniently removed (2) **:** an intensive study or report
3 : CARRYOUT 1

take–out \ˈtā-ˌkau̇t\ *adjective* (1965)
: of, relating to, or being food not to be consumed on the premises ⟨*take-out* counter⟩ ⟨a *take-out* supper⟩

take out (13th century)
transitive verb
1 a (1) **:** DEDUCT, SEPARATE (2) **:** EXCLUDE, OMIT (3) **:** WITHDRAW, WITHHOLD **b :** to find release for **:** VENT ⟨*take out* their resentments on one another —J. W. Aldridge⟩ **c** (1) **:** ELIMINATE (2) **:** KILL, DESTROY (3) **:** KNOCK OUT
2 : to take as an equivalent in another form ⟨*took* the debt *out* in trade⟩
3 a : to obtain from the proper authority ⟨*take out* a charter⟩ **b :** to arrange for (insurance)
4 : to overcall (a bridge partner) in a different suit
intransitive verb
: to start on a course **:** SET OUT
— take it out on : to expend anger, vexation, or frustration in harassment of

take·out double \ˈtā-ˌkau̇t-\ *noun* (circa 1944)
: a double made in bridge to convey information to and request a bid from one's partner

take·over \ˈtā-ˌkō-vər\ *noun* (circa 1917)
: the action or an act of taking over

take over (1884)
transitive verb
: to assume control or possession of or responsibility for ⟨military leaders *took over* the government⟩
intransitive verb
1 : to assume control or possession
2 : to become dominant

take–up \ˈtā-ˌkəp\ *noun* (1838)
: the action of taking up

take up (14th century)
transitive verb
1 : PICK UP, LIFT
2 a : to begin to occupy (land) **b :** to gather from a number of sources ⟨*took up* a collection⟩
3 a : to accept or adopt for the purpose of assisting **b :** to accept or adopt as one's own ⟨*took up* the life of a farmer⟩ ⟨*took up* Irish citizenship⟩ **c :** to absorb or incorporate into itself ⟨plants *taking up* nutrients⟩
4 a : to enter upon (as a business, hobby, or subject of study) ⟨*take up* skiing⟩ ⟨*took up* the trumpet⟩ ⟨had *taken up* Marxism⟩ **b :** to proceed to consider or deal with ⟨*take up* one problem at a time⟩
5 : to establish oneself in ⟨*took up* residence in town⟩
6 : to occupy entirely or exclusively **:** fill up ⟨the meeting was *taken up* with old business⟩
7 : to make tighter or shorter ⟨*take up* the slack⟩
8 : to respond favorably to (as a person offering a bet, challenge, or proposal) ⟨*took* me *up* on it⟩
9 : to begin again or take over from another ⟨we must *take* the good work *up* again⟩
intransitive verb
1 : to make a beginning where another has left off
2 : to become shortened **:** draw together **:** SHRINK
— take up the cudgels : to engage vigorously in a defense or dispute
— take up with 1 : to become interested or absorbed in **2 :** to begin to associate or consort with

ta·kin \ˈtä-ˌkēn\ *noun* [Mishmi (Tibeto-Burman language of northeast India)] (1850)
: a large heavily built ruminant (*Budorcas taxicolor*) of Tibet and adjacent areas of Asia that is related to the goats but in some respects resembles the antelopes

takin

tak·ings \ˈtā-kiŋz\ *noun plural* (1632)
chiefly British **:** receipts especially of money

¹ta·la \ˈtä-lə\ *noun* [Sanskrit *tāla,* literally, hand-clapping] (1891)
: one of the ancient traditional rhythmic patterns of Indian music — compare RAGA

²ta·la \ˈtä-lə, -(ˌ)lä\ *noun, plural* **tala** [Samoan, from English *dollar*] (1967)
— see MONEY table

Tal·bot \ˈtȯl-bət, ˈtal-\ *noun* [probably from *Talbot,* name of a Norman family in England] (1562)
: a large heavy mostly white hound with pendulous ears and drooping flews held to be ancestral to the bloodhound

talc \ˈtalk\ *noun* [Middle French *talc* mica, from Medieval Latin *talk,* from Arabic *ṭalq*] (1610)
1 : a very soft mineral that is a basic silicate of magnesium, has a soapy feel, and is used especially in making talcum powder
2 : TALCUM POWDER
— talc·ose \ˈtal-ˌkōs\ *adjective*

tal·cum powder \ˈtal-kəm-\ *noun* [Medieval Latin *talcum* mica, alteration of earlier *talk*] (circa 1890)
1 : powdered talc
2 : a toilet powder composed of perfumed talc or talc and a mild antiseptic

tale \ˈtā(ə)l\ *noun* [Middle English, from Old English *talu;* akin to Old Norse *tala* talk] (before 12th century)
1 *obsolete* **:** DISCOURSE, TALK
2 a : a series of events or facts told or presented **:** ACCOUNT **b** (1) **:** a report of a private or confidential matter ⟨dead men tell no *tales*⟩ (2) **:** a libelous report or piece of gossip
3 a : a usually imaginative narrative of an event **:** STORY **b :** an intentionally untrue report **:** FALSEHOOD ⟨always preferred the *tale* to the truth —Sir Winston Churchill⟩
4 a : COUNT, TALLY **b :** TOTAL

tale·bear·er \-ˌbar-ər, -ˌber-\ *noun* (15th century)
: one that spreads gossip or rumors; *also* **:** TATTLETALE
— tale·bear·ing \-iŋ\ *adjective or noun*

tal·ent \ˈta-lənt\ *noun* [Middle English, from Old English *talente,* from Latin *talenta,* plural of *talentum* unit of weight or money, from Greek *talanton* pan of a scale, weight; akin to Greek *tlēnai* to bear; in senses 2–5, from the parable of the talents in Matthew 25:14–30 — more at TOLERATE] (before 12th century)
1 a : any of several ancient units of weight **b :** a unit of value equal to the value of a talent of gold or silver
2 *archaic* **:** a characteristic feature, aptitude, or disposition of a person or animal
3 : the natural endowments of a person
4 a : a special often creative or artistic aptitude **b :** general intelligence or mental power **:** ABILITY
5 : a person of talent or a group of persons of talent in a field or activity ◆
synonym see GIFT
— tal·ent·ed \-lən-təd\ *adjective*
— tal·ent·less \-lənt-ləs\ *adjective*

talent scout *noun* (1936)
: a person engaged in discovering and recruiting people of talent for a specialized field or activity

talent show *noun* (1953)
: a show consisting of a series of individual performances (as singing) by amateurs who may be selected for special recognition as performing talent

ta·ler \ˈtä-lər\ *noun* [German — more at DOLLAR] (circa 1905)
: any of numerous silver coins issued by various German states from the 15th to the 19th centuries

tales·man \ˈtā(ə)lz-mən, ˈtā-lēz-\ *noun* [Middle English *tales* talesmen, from Medieval Latin *tales de circumstantibus* such (persons) of the bystanders; from the wording of the writ summoning them] (1679)
1 : a person added to a jury usually from among bystanders to make up a deficiency in the available number of jurors
2 : a member of a large pool of persons called for jury duty from which jurors are selected

◇ WORD HISTORY
talent The ancient Greek word *talanton,* originally denoting the pan of a scale, was also applied to a unit of weight and then a monetary unit. Borrowed into English through Latin as *talent,* the word has developed several figurative meanings based on its use in the New Testament parable of the talents (Matthew 25:14–30). In the parable a master gives money to his servants: five talents to one, two talents to another, and finally a single talent to a third. The first two doubled their capital, and the master was pleased. The third servant, however, buried his talent in the earth for fear of losing it. This made the master very angry since he expected his servant at least to have invested his talent with bankers so that it would gain interest. The master then took back the one talent and gave it to the servant who had ten. Traditionally the talent has been taken as a metaphor for a God-given endowment that if not used would be lost. The English word *talent* has had this metaphorical sense since the 15th century, and several centuries later it gave rise to the extended meaning "mental ability or aptitude."

tale–tell·er \ˈtā(ə)l-ˌte-lər\ *noun* (14th century)
1 : one who tells tales or stories
2 : TALEBEARER
— **tale–tell·ing** \-ˌte-liŋ\ *adjective or noun*

tali *plural of* TALUS

tali·pes \ˈta-lə-ˌpēz\ *noun* [New Latin, from Latin *talus* ankle + *pes* foot — more at FOOT] (circa 1841)
: CLUBFOOT

tal·is·man \ˈta-ləs-mən, -ləz-\ *noun, plural* **-mans** [French *talisman* or Spanish *talismán* or Italian *talismano;* all from Arabic *ṭilsam,* from Middle Greek *telesma,* from Greek, consecration, from *telein* to initiate into the mysteries, complete, from *telos* end — more at TELOS] (1638)
1 : an object held to act as a charm to avert evil and bring good fortune
2 : something producing apparently magical or miraculous effects
— **tal·is·man·ic** \ˌta-ləs-ˈma-nik, -ləz-\ *adjective*
— **tal·is·man·i·cal·ly** \-ni-k(ə-)lē\ *adverb*

[1]**talk** \ˈtȯk\ *verb* [Middle English; akin to Old English *talu* tale] (13th century)
transitive verb
1 : to deliver or express in speech **:** UTTER
2 : to make the subject of conversation or discourse **:** DISCUSS ⟨*talk* business⟩
3 : to influence, affect, or cause by talking ⟨*talked* them into going⟩
4 : to use (a language) for conversing or communicating **:** SPEAK
intransitive verb
1 a : to express or exchange ideas by means of spoken words **b :** to convey information or communicate in any way (as with signs or sounds) ⟨can make a trumpet *talk*⟩
2 : to use speech **:** SPEAK
3 a : to speak idly **:** PRATE **b :** GOSSIP **c :** to reveal secret or confidential information
4 : to give a talk **:** LECTURE
— **talk·er** *noun*
— **talk back :** to answer impertinently
— **talk sense :** to voice rational, logical, or sensible thoughts
— **talk through one's hat :** to voice irrational, illogical, or erroneous ideas
— **talk turkey :** to speak frankly or bluntly

[2]**talk** *noun* (15th century)
1 : the act or an instance of talking **:** SPEECH
2 : a way of speaking **:** LANGUAGE
3 : pointless or fruitless discussion **:** VERBIAGE
4 : a formal discussion, negotiation, or exchange of views — often used in plural
5 a : MENTION, REPORT **b :** RUMOR, GOSSIP
6 : the topic of interested comment, conversation, or gossip ⟨it's the *talk* of the town⟩
7 a : ADDRESS, LECTURE **b :** written analysis or discussion presented in an informal or conversational manner
8 : communicative sounds or signs resembling or functioning as talk ⟨bird *talk*⟩

talk·a·thon \ˈtȯ-kə-ˌthän\ *noun* (1934)
: a long session of discussion or speechmaking

talk·a·tive \ˈtȯ-kə-tiv\ *adjective* (15th century)
: given to talking; *also* **:** full of talk ☆
— **talk·a·tive·ly** *adverb*
— **talk·a·tive·ness** *noun*

talk down (1901)
intransitive verb
: to speak in a condescending or oversimplified fashion
transitive verb
: to disparage or belittle by talking

talk·ie \ˈtȯ-kē\ *noun* (1913)
: a motion picture with a synchronized sound track

talking book *noun* (1932)
: a phonograph or tape recording of a reading of a book or magazine designed chiefly for the use of the blind

talking head *noun* (1968)
: the televised head and shoulders shot of a person talking; *also* **:** a television personality who appears in such shots

talking machine *noun* (1890)
: an early phonograph

talking point *noun* (circa 1914)
: something that lends support to an argument; *also* **:** a subject of discussion

talk·ing–to \ˈtȯ-kiŋ-ˌtü\ *noun* (1884)
: REPRIMAND, LECTURE

talk out *transitive verb* (1954)
: to clarify or settle by oral discussion

talk over *transitive verb* (1734)
: to review or consider in conversation **:** DISCUSS

talk radio *noun* (1972)
: radio programming consisting of call-in shows

talk show *noun* (1965)
: a radio or television program in which usually well-known persons engage in discussions or are interviewed

talk up (1722)
transitive verb
: to discuss favorably **:** ADVOCATE, PROMOTE
intransitive verb
: to speak up plainly or directly

talky \ˈtȯ-kē\ *adjective* **talk·i·er; -est** (1815)
1 : TALKATIVE
2 : containing too much talk
— **talk·i·ness** \-nəs\ *noun*

tall \ˈtȯl\ *adjective* [Middle English, probably from Old English *getæl* quick, ready; akin to Old High German *gizal* quick] (15th century)
1 *obsolete* **:** BRAVE, COURAGEOUS
2 a : high in stature **b :** of a specified height ⟨five feet *tall*⟩
3 a : of considerable height ⟨*tall* trees⟩ **b :** long from bottom to top ⟨a *tall* book⟩ **c :** of a higher growing variety or species of plant
4 a : large or formidable in amount, extent, or degree ⟨a *tall* order to fill⟩ **b :** POMPOUS, HIGHFLOWN ⟨*tall* talk about the vast mysteries of life —W. A. White⟩ **c :** highly exaggerated **:** INCREDIBLE, IMPROBABLE ⟨a *tall* story⟩
synonym see HIGH
— **tall** *adverb*
— **tall·ish** \ˈtȯ-lish\ *adjective*
— **tall·ness** \ˈtȯl-nəs\ *noun*

tal·lage \ˈta-lij\ *noun* [Middle English *taillage, tallage,* from Old French *taillage,* from *taillier* to cut, limit, tax — more at TAILOR] (14th century)
: an impost or due levied by a lord upon his tenants

tall·boy \ˈtȯl-ˌbȯi\ *noun* (1769)
1 a : HIGHBOY **b :** a double chest of drawers usually with the upper section slightly smaller than the lower
2 *British* **:** CLOTHESPRESS

tall fescue *noun* (circa 1762)
: a European fescue (*Festuca elatior*) with erect smooth stems 3 to 4 feet (about 1 meter) high that has been introduced into North America — called also *tall fescue grass;* compare FESCUE FOOT

tall·grass prairie \ˈtȯl-ˌgras-\ *noun* (1920)
: PRAIRIE 2a

tal·lith \ˈtä-ləs, ˈta-, -lət, -ləth\ *or* **tal·lis** \-ləs\ *noun* [Hebrew *ṭallīth* cover, cloak] (1613)
: a shawl with fringed corners worn over the head or shoulders by Jewish men especially during morning prayers

tallith

tall oil \ˈtäl-, ˈtȯl-\ *noun* [part translation of German *Tallöl,* part translation of Swedish *tallolja,* from *tall* pine + *olja* oil] (circa 1926)
: a resinous by-product from the manufacture of chemical wood pulp used especially in making soaps, coatings, and oils

[1]**tal·low** \ˈta-(ˌ)lō\ *noun* [Middle English *talgh, talow;* akin to Middle Dutch *talch* tallow] (14th century)
: the white nearly tasteless solid rendered fat of cattle and sheep used chiefly in soap, candles, and lubricants
— **tal·lowy** \ˈta-lə-wē\ *adjective*

[2]**tallow** *transitive verb* (15th century)
: to grease or smear with tallow

tall ship *noun* (circa 1548)
: a sailing vessel with at least two masts; *especially* **:** SQUARE-RIGGER

[1]**tal·ly** \ˈta-lē\ *noun, plural* **tallies** [Middle English *talye,* from Medieval Latin *talea, tallia,* from Latin *talea* twig, cutting] (15th century)
1 : a device (as a notched rod or mechanical counter) for visibly recording or accounting especially business transactions
2 a : a recorded reckoning or account (as of items or charges) ⟨keep a daily *tally* of accidents⟩ **b :** a score or point made (as in a game)
3 a : a part that corresponds to an opposite or companion member **:** COMPLEMENT **b :** a state of correspondence or agreement

[2]**tally** *verb* **tal·lied; tal·ly·ing** (15th century)
transitive verb
1 a : to record on or as if on a tally **:** TABULATE **b :** to list or check off (as a cargo) by items **c :** to register (as a score) in a contest
2 : to make a count of **:** RECKON
3 : to cause to correspond
intransitive verb
1 a : to make a tally by or as if by tabulating **b :** to register a point in a contest **:** SCORE
2 : CORRESPOND, MATCH

tal·ly·ho \ˌta-lē-ˈhō\ *noun, plural* **-hos** [probably from French *taïaut,* a cry used to excite hounds in deer hunting] (1772)
1 : a call of a huntsman at sight of the fox
2 [*Tally-ho,* name of a coach formerly plying between London and Birmingham] **:** a four-in-hand coach

tal·ly·man \ˈta-lē-mən, -ˌman\ *noun* (1654)
1 *British* **:** one who sells goods on the installment plan
2 : one who tallies, checks, or keeps an account or record (as of receipt of goods)

Tal·mud \ˈtäl-ˌmu̇d, ˈtal-məd\ *noun* [Late Hebrew *talmūdh,* literally, instruction] (1532)
: the authoritative body of Jewish tradition comprising the Mishnah and Gemara
— **Tal·mu·dic** \tal-ˈmü-dik, -ˈmyü-, -ˈmə-; täl-ˈmu̇-\ *adjective*
— **tal·mud·ism** \ˈtäl-ˌmu̇-ˌdi-zəm, ˈtal-, -mə-\ *noun, often capitalized*

Tal·mud·ist \ˈtäl-ˌmu̇-dist, ˈtal-, -mə-\ *noun* (1569)
: a specialist in Talmudic studies

tal·on \ˈta-lən\ *noun* [Middle English, from Middle French, heel, spur, from (assumed) Vulgar Latin *talon-, talo,* from Latin *talus* ankle, anklebone] (15th century)
1 a : the claw of an animal and especially of a bird of prey **b :** a finger or hand of a human being
2 : a part or object shaped like or suggestive of a heel or claw: as **a :** an ogee molding **b :** the shoulder of the bolt of a lock on which the key acts to shoot the bolt

☆ SYNONYMS
Talkative, loquacious, garrulous, voluble mean given to talk or talking. TALKATIVE may imply a readiness to engage in talk or a disposition to enjoy conversation ⟨a *talkative* neighbor⟩. LOQUACIOUS suggests the power of expressing oneself articulately, fluently, or glibly ⟨a *loquacious* spokesperson⟩. GARRULOUS implies prosy, rambling, or tedious loquacity ⟨*garrulous* traveling companions⟩. VOLUBLE suggests a free, easy, and unending loquacity ⟨a *voluble* raconteur⟩.

3 a : cards laid aside in a pile in solitaire **b :** STOCK 10c
— **tal·oned** \-lənd\ *adjective*

¹ta·lus \'tā-ləs, 'ta-\ *noun* [French, from Latin *talutium* slope indicating presence of gold under the soil] (1645)
1 : a slope formed especially by an accumulation of rock debris
2 : rock debris at the base of a cliff

²ta·lus \'tā-ləs\ *noun, plural* **ta·li** \'tā-,lī\ [New Latin, from Latin] (circa 1693)
1 : the human tarsal bone that bears the weight of the body and that together with the tibia and fibula forms the ankle joint
2 : the entire ankle

tam \'tam\ *noun* (1895)
: TAM-O'-SHANTER

ta·ma·le \tə-'mä-lē\ *noun* [Mexican Spanish *tamales,* plural of *tamal* tamale, from Nahuatl *tamalli* steamed cornmeal dough] (1854)
: ground meat seasoned usually with chili, rolled in cornmeal dough, wrapped in corn husks, and steamed

ta·man·dua \tə-'man-də-wə, -,man-də-'wä\ *noun* [Portuguese *tamanduá,* from Tupi] (1834)
: either of two arboreal anteaters (*Tamandua mexicana* and *T. tetradactyla*) of Central and South America having a nearly hairless tail

tam·a·rack \'ta-mə-,rak, 'tam-rak\ *noun* [origin unknown] (1805)
1 : any of several American larches; *especially* **:** a larch (*Larix laricina*) of northern North America that inhabits usually moist or wet areas
2 : the wood of a tamarack

tam·a·rau \,ta-mə-'raù\ *noun* [Tagalog *tamaráw*] (1898)
: a small dark sturdily built buffalo (*Bubalus mindorensis*) native to Mindoro

ta·ma·ri \tə-'mär-ē\ *noun* [Japanese] (1965)
: an aged soy sauce prepared with little or no added wheat

tam·a·ril·lo \,ta-mə-'ri-(,)lō\ *noun* [alteration of *tomatillo*] (1966)
: the reddish edible fruit of an arborescent shrub (*Cyphomandra betacea*) of the nightshade family that is native to South America but is grown commercially elsewhere; *also* **:** the shrub itself

tam·a·rin \'ta mə-rən, -,ran\ *noun* [French, from Carib] (1780)
: any of numerous small chiefly South American monkeys (genera *Saguinus* and *Leontopithecus*) that are related to the marmosets and have silky fur, a long tail, and lower canine teeth that are longer than the incisors

tam·a·rind \'ta-mə-rənd, -,rind\ *noun* [Spanish & Portuguese *tamarindo,* from Arabic *tamr hindī,* literally, Indian date] (1533)
: a tropical Old World leguminous tree (*Tamarindus indica*) with hard yellowish wood, pinnate leaves, and red-striped yellow flowers; *also* **:** its fruit which has an acid pulp often used for preserves or in a cooling laxative drink

tam·a·risk \'ta-mə-,risk\ *noun* [Middle English *tamarisc,* from Late Latin *tamariscus,* from Latin *tamaric-, tamarix*] (14th century)
: any of a genus (*Tamarix* of the family Tamaricaceae, the tamarisk family) of chiefly Old World desert shrubs and trees having tiny narrow leaves and masses of minute flowers with five stamens and a one-celled ovary

tam·ba·la \täm-'bä-lə\ *noun, plural* **-la** *or* **-las** [Nyanja (Bantu language of Malawi), literally, cockerel] (1970)
— see *kwacha* at MONEY table

¹tam·bour \'tam-,bùr, tam-'\ *noun* [Middle French, drum, from Arabic *ṭanbūr,* modification of Persian *tabīr*] (15th century)
1 : ¹DRUM 1
2 a : an embroidery frame; *especially* **:** a set of two interlocking hoops between which cloth is stretched before stitching **b :** embroidery made on a tambour frame
3 : a shallow metallic cup or drum with a thin elastic membrane supporting a writing lever used to transmit and register slight motions (as arterial pulsations)
4 : a rolling top or front (as of a rolltop desk) of narrow strips of wood glued on canvas

²tambour (1774)
transitive verb
: to embroider (cloth) with tambour
intransitive verb
: to work at a tambour frame
— **tam·bour·er** *noun*

tam·bou·ra *or* **tam·bu·ra** \tam-'bùr-ə\ *noun* [Persian *ṭambūra*] (1585)
: an Asian musical instrument resembling a lute in construction but without frets and used to produce a drone accompaniment to singing

tam·bou·rine \,tam-bə-'rēn\ *noun* [Middle French *tambourin,* diminutive of *tambour*] (1579)
: a small drum; *especially* **:** a shallow one-headed drum with loose metallic disks at the sides played especially by shaking or striking with the hand

tambourine

¹tame \'tām\ *adjective* **tam·er; tam·est** [Middle English, from Old English *tam;* akin to Old High German *zam* tame, Latin *domare* to tame, Greek *damnanai*] (before 12th century)
1 : reduced from a state of native wildness especially so as to be tractable and useful to humans **:** DOMESTICATED ⟨*tame* animals⟩
2 : made docile and submissive **:** SUBDUED
3 : lacking spirit, zest, interest, or the capacity to excite **:** INSIPID ⟨a *tame* campaign⟩
— **tame·ly** *adverb*
— **tame·ness** *noun*

²tame *verb* **tamed; tam·ing** (14th century)
transitive verb
1 a : to reduce from a wild to a domestic state **b :** to subject to cultivation **c :** to bring under control **:** HARNESS
2 : to deprive of spirit **:** HUMBLE, SUBDUE ⟨the once revolutionary . . . party, long since *tamed* —*Times Literary Supplement*⟩
3 : to tone down **:** SOFTEN ⟨*tamed* the language in the play⟩
intransitive verb
: to become tame
— **tam·able** *or* **tame·able** \'tā-mə-bəl\ *adjective*
— **tam·er** *noun*

tame·less \'tām-ləs\ *adjective* (circa 1598)
: not tamed or not capable of being tamed

Tam·il \'ta-məl, 'tä-\ *noun* (1734)
1 : a Dravidian language of Tamil Nadu state, India, and of northern and eastern Sri Lanka
2 : a Tamil-speaking person or a descendant of Tamil-speaking ancestors

Tam·ma·ny \'ta-mə-nē\ *adjective* [*Tammany Hall,* headquarters of the Tammany Society, political organization in New York City] (1872)
: of, relating to, or constituting a group or organization exercising or seeking municipal political control by methods often associated with corruption and bossism
— **Tam·ma·ny·ism** \-,i-zəm\ *noun*

Tam·muz \'tä-,mùz\ *noun* [Hebrew *Tammūz*] (1614)
: the 10th month of the civil year or the 4th month of the ecclesiastical year in the Jewish calendar — see MONTH table

Tam o' Shan·ter *noun*
1 \,ta-mə-'shan-tər\ **:** the hero of Burns's poem *Tam o' Shanter*
2 *usually* **tam–o'–shanter** \'ta-mə-,\ **:** a woolen cap of Scottish origin with a tight headband, wide flat circular crown, and usually a pompon in the center

ta·mox·i·fen \tə-'mäk-sə-,fen\ *noun* [perhaps from *trans-* + *amine* + *oxi-* (alteration of *oxy*) + *-fen* (alteration of *phenyl*)] (1972)
: an estrogen antagonist $C_{26}H_{29}NO$ used especially to treat postmenopausal breast cancer

¹tamp \'tamp\ *transitive verb* [probably back-formation from obsolete *tampion, tampin* plug, from Middle English, from Middle French *tapon, tampon,* from (assumed) Old French *taper* to plug, of Germanic origin; akin to Old English *tæppa* tap] (1834)
1 : to drive in or down by a succession of light or medium blows ⟨*tamp* wet concrete⟩
2 : to put a check on **:** REDUCE, LESSEN
— **tamp·er** *noun*

²tamp *noun* (1920)
: a tool for tamping

tam·per \'tam-pər\ *intransitive verb* **tam·pered; tam·per·ing** \-p(ə-)riŋ\ [probably from Middle French *temprer* to temper, mix, meddle — more at TEMPER] (1567)
1 : to carry on underhand or improper negotiations (as by bribery)
2 a : to interfere so as to weaken or change for the worse — used with *with* **b :** to try foolish or dangerous experiments — used with *with*
— **tam·per·er** \-pər-ər\ *noun*
— **tam·per·proof** \'tam-pər-,prüf\ *adjective*

tam·pi·on \'tam-pē-ən, 'täm-\ *noun* [obsolete *tampion, tampin* plug — more at TAMP] (circa 1625)
: a wooden plug or a metal or canvas cover for the muzzle of a gun

¹tam·pon \'tam-,pän\ *noun* [French, literally, plug, from Middle French — more at TAMP] (1848)
: a plug (as of cotton) introduced into a body cavity usually to absorb secretions (as from menstruation) or to arrest hemorrhaging

²tampon *transitive verb* (1860)
: to plug with a tampon

tam–tam \'tam-,tam, 'täm-,täm\ *noun* [Hindi *ṭamṭam*] (1782)
1 : TOM-TOM
2 : GONG; *especially* **:** one of a tuned set in a gamelan orchestra

Tam·worth \'tam-(,)wərth\ *noun* [*Tamworth,* borough in Staffordshire, England] (1860)
: any of a breed of large long-bodied red swine developed in England especially for the production of bacon

¹tan \'tan\ *verb* **tanned; tan·ning** [Middle English *tannen,* from Middle French *tanner,* from Medieval Latin *tannare,* from *tanum, tannum* tanbark] (14th century)
transitive verb
1 a : to convert (hide) into leather by treatment with an infusion of tannin-rich bark or other agent of similar effect **b :** to convert (protein) to leather or a similar substance
2 : to make (skin) tan especially by exposure to the sun
3 : THRASH, WHIP
intransitive verb
: to get or become tanned

²tan *adjective* **tan·ner; tan·nest** (1586)
1 : of, relating to, or used for tan or tanning
2 : of the color tan

³tan *noun* [French, tanbark, from Old French, from Medieval Latin *tanum*] (1674)
1 : a tanning material or its active agent (as tannin)
2 : a brown color imparted to the skin by exposure to the sun or wind
3 : a light yellowish brown
4 *plural* **:** tan-colored articles of clothing

tan·a·ger \'ta-ni-jər\ *noun* [New Latin *tanagra,* from Portuguese *tangará,* from Tupi] (1688)
: any of numerous chiefly tropical American oscine birds (family Thraupidae) that are often

brightly colored and usually unmusical and inhabit mostly woodlands

tan·bark \'tan-ˌbärk\ *noun* (1799)
1 : a bark rich in tannin bruised or cut into small pieces and used in tanning
2 : a surface (as a circus ring) covered with spent tanbark

¹tan·dem \'tan-dəm\ *noun* [Latin, at last, at length (taken to mean "lengthwise"), from *tam* so; akin to Old English *thæt* that] (circa 1785)
1 a (1) **:** a 2-seated carriage drawn by horses harnessed one before the other (2) **:** a team so harnessed **b :** TANDEM BICYCLE **c :** a vehicle (as a motortruck) having close-coupled pairs of axles
2 : a group of two or more arranged one behind the other or used or acting in conjunction ◆
— in tandem 1 : in a tandem arrangement
2 : in partnership or conjunction

²tandem *adverb* (circa 1795)
: one after or behind another ⟨ride *tandem*⟩

³tandem *adjective* (1801)
1 : consisting of things or having parts arranged one behind the other
2 : working or occurring in conjunction with each other

tandem bicycle *noun* (circa 1890)
: a bicycle for usually two persons sitting tandem

tan·door \tän-'dủr\ *noun* [Hindi *tandūr, tannūr,* from Persian *tanūr,* from Arabic *tannūr*] (1840)
: a cylindrical clay oven in which food is cooked over charcoal

tan·doori \tän-'dủr-ē\ *adjective* [Hindi *tandurī,* from *tandūr*] (circa 1968)
: cooked in a tandoor ⟨*tandoori* chicken⟩
— tandoori *noun*

¹tang \'taŋ\ *noun* [Middle English, of Scandinavian origin; akin to Old Norse *tangi* point of land, tang] (15th century)
1 : a projecting shank, prong, fang, or tongue (as on a knife, file, or sword) to connect with the handle
2 a : a sharp distinctive often lingering flavor **b :** a pungent odor **c :** something having the effect of a tang (as in stimulation of the senses)
3 a : a faint suggestion **:** TRACE **b :** a distinguishing characteristic that sets apart or gives a special individuality
— tanged \'taŋd\ *adjective*

²tang *transitive verb* (1566)
1 : to furnish with a tang
2 : to affect with a tang

³tang *verb* [imitative] (1556)
: CLANG, RING

⁴tang *noun* (circa 1625)
: a sharp twanging sound

Tang *or* **T'ang** \'täŋ\ *noun* [Chinese (Beijing) *Táng*] (1669)
: a Chinese dynasty dated A.D. 618–907 and marked by wide contacts with other cultures and by the development of printing and the flourishing of poetry and art

tan·ge·lo \'tan-jə-ˌlō\ *noun, plural* **-los** [*tan*gerine + pom*elo*] (1904)
: the fruit of a hybrid (*Citrus tangelo*) between a tangerine or mandarin orange and a grapefruit; *also* **:** the tree

tan·gen·cy \'tan-jən(t)-sē\ *noun* (1819)
: the quality or state of being tangent

¹tan·gent \-jənt\ *adjective* [Latin *tangent-, tangens,* present participle of *tangere* to touch; perhaps akin to Old English *thaccian* to touch gently, stroke] (1594)
1 a : meeting a curve or surface in a single point if a sufficiently small interval is considered ⟨straight line *tangent* to a curve⟩ **b** (1) **:** having a common tangent line at a point ⟨*tangent* curves⟩ (2) **:** having a common tangent plane at a point ⟨*tangent* surfaces⟩
2 : diverging from an original purpose of course **:** IRRELEVANT ⟨*tangent* remarks⟩

²tangent *noun* [New Latin *tangent-, tangens,* from *linea tangens* tangent line] (1594)
1 a : the trigonometric function that for an acute angle is the ratio between the leg opposite to the angle when it is considered part of a right triangle and the leg adjacent **b :** a trigonometric function that is equal to the sine divided by the cosine for all real numbers θ for which the cosine is not equal to zero and is exactly equal to the tangent of an angle of measure θ in radians
2 : a line that is tangent; *specifically* **:** a straight line that is the limiting position of a secant of a curve through a fixed point and a variable point on the curve as the variable point approaches the fixed point — see CIRCLE illustration
3 : an abrupt change of course **:** DIGRESSION ⟨the speaker went off on a *tangent*⟩
4 : a small upright flat-ended metal pin at the inner end of a clavichord key that strikes the string to produce the tone

tan·gen·tial \tan-'jen(t)-shəl\ *adjective* (1630)
1 : of, relating to, or of the nature of a tangent
2 : acting along or lying in a tangent ⟨*tangential* forces⟩
3 a : DIVERGENT, DIGRESSIVE **b :** touching lightly **:** INCIDENTAL, PERIPHERAL ⟨*tangential* involvement⟩
— tan·gen·tial·ly \-'jen(t)-sh(ə-)lē\ *adverb*

tangent plane *noun* (1856)
: the plane through a point of a surface that contains the tangent lines to all the curves on the surface through the same point

tan·ger·ine \'tan-jə-ˌrēn, ˌtan-jə-'\ *noun* [French *Tanger* Tangier, Morocco] (1842)
1 a : any of various mandarins that have usually deep orange skin and pulp; *broadly* **:** MANDARIN 3b **b :** a tree producing tangerines
2 : a moderate to strong reddish orange

¹tan·gi·ble \'tan-jə-bəl\ *adjective* [Late Latin *tangibilis,* from Latin *tangere* to touch] (1589)
1 a : capable of being perceived especially by the sense of touch **:** PALPABLE **b :** substantially real **:** MATERIAL
2 : capable of being precisely identified or realized by the mind ⟨her grief was *tangible*⟩
3 : capable of being appraised at an actual or approximate value ⟨*tangible* assets⟩
synonym see PERCEPTIBLE
— tan·gi·bil·i·ty \ˌtan-jə-'bi-lə-tē\ *noun*
— tan·gi·ble·ness \'tan-jə-bəl-nəs\ *noun*
— tan·gi·bly \-blē\ *adverb*

²tangible *noun* (1890)
: something tangible; *especially* **:** a tangible asset

¹tan·gle \'taŋ-gəl\ *verb* **tan·gled; tan·gling** \-g(ə-)liŋ\ [Middle English *tangilen,* probably of Scandinavian origin; akin to Swedish dialect *taggla* to tangle] (14th century)
transitive verb
1 : to involve so as to hamper, obstruct, or embarrass
2 : to seize and hold in or as if in a snare **:** ENTRAP
3 : to unite or knit together in intricate confusion
intransitive verb
1 : to interact in a contentious or conflicting way
2 : to become entangled

²tangle *noun* (1615)
1 : a tangled twisted mass **:** SNARL
2 a : a complicated or confused state or condition **b :** a state of perplexity or complete bewilderment
3 : a serious altercation **:** DISPUTE

³tangle *noun* [of Scandinavian origin; akin to Old Norse *thǫngull* tangle, *thang* kelp] (1536)
: a large seaweed

tan·gled \'taŋ-gəld\ *adjective* (1590)
1 : existing in or giving the appearance of a state of utter disorder
2 : very involved **:** exceedingly complex

tan·gle·ment \-gəl-mənt\ *noun* (1831)
: ENTANGLEMENT

tan·gly \'taŋ-g(ə-)lē\ *adjective* (1813)
: full of tangles or knots **:** INTRICATE

¹tan·go \'taŋ-(ˌ)gō\ *noun, plural* **tangos** [American Spanish] (1913)
: a ballroom dance of Latin-American origin in 2/4 time with a basic pattern of step-step-step-step-close and characterized by long pauses and stylized body positions; *also* **:** the music for this dance

²tango *intransitive verb* (1913)
: to dance the tango

Tan·go (1952)
— a communications code word for the letter *t*

tan·gram \'taŋ-grəm, 'tan-\ *noun* [perhaps from Chinese (Beijing) *táng* Chinese + English *-gram*] (circa 1864)
: a Chinese puzzle made by cutting a square of thin material into five triangles, a square, and a rhomboid which are capable of being recombined in many different figures

tangram

tangy \'taŋ-ē\ *adjective* **tang·i·er; -est** (1875)
: having or suggestive of a tang

¹tank \'taŋk\ *noun* [Portuguese *tanque,* alteration of *estanque,* from *estancar* to stanch, perhaps from (assumed) Vulgar Latin *stanticare* — more at STANCH] (circa 1616)
1 *dialect* **:** POND, POOL; *especially* **:** one built as a water supply
2 : a usually large receptacle for holding, transporting, or storing liquids (as water or fuel)
3 : an enclosed heavily armed and armored combat vehicle that moves on tracks
4 : a prison cell or enclosure used especially for receiving prisoners
5 : TANK TOP
— tank·ful \-ˌfủl\ *noun*
— tank·like \-ˌlīk\ *adjective*

²tank (1863)
transitive verb
1 : to place, store, or treat in a tank
2 : to make no effort to win **:** lose intentionally ⟨*tanked* the match⟩
intransitive verb
: to lose intentionally **:** give up in competition

tan·ka \'täŋ-kə\ *noun* [Japanese] (circa 1877)
: an unrhymed Japanese verse form of five lines containing 5, 7, 5, 7, and 7 syllables respectively; *also* **:** a poem in this form — compare HAIKU

tank·age \'taŋ-kij\ *noun* (1866)
1 a : the aggregate of tanks required for a purpose **b :** the capacity or contents of a tank

◇ WORD HISTORY

tandem When a pair of horses pulls a carriage, the two are usually harnessed side by side. But there is a type of two-wheeled, two-seated carriage that is drawn by two horses harnessed one after the other. This carriage, a *tandem,* apparently owes its name to a rather contorted Latin-English pun. The Latin word *tandem* means "at length, at last, finally." We do not know the identity of the punster who first suggested that horses arranged lengthwise should be thought of as "at length" and the carriage called a *tandem,* but the individual need not have been a scholar. The Latin word is not a rare one and would have been known to any student of the language. First attested in a slang dictionary published in 1785, *tandem* in the course of the 19th century came to be used in a general way of any arrangement of things or of people one behind the other.

2 : dried animal residues usually freed from the fat and gelatin and used as fertilizer and feedstuff
3 : the act or process of putting or storing in tanks

tan·kard \ˈtaŋ-kərd\ *noun* [Middle English] (15th century)
: a tall one-handled drinking vessel; *especially* **:** a silver or pewter mug with a lid

tankard

tank destroyer *noun* (1941)
: a highly mobile lightly armored vehicle usually on a half-track or a tank chassis and mounting a cannon

tanked \ˈtaŋ(k)t\ *adjective* (1893)
slang **:** DRUNK 1a — often used with *up*

tank·er \ˈtaŋ-kər\ *noun* (1900)
1 a : a cargo ship fitted with tanks for carrying liquid in bulk **b :** a vehicle on which a tank is mounted to carry fluids; *also* **:** a cargo airplane for transporting fuel
2 : a member of a military tank crew

tank suit *noun* (1940)
: a one-piece bathing suit with usually wide shoulder straps

tank top *noun* (1950)
: a sleeveless collarless shirt with usually wide shoulder straps and no front opening

tank town *noun* [from the fact that formerly trains stopped at such towns only to take on water] (1906)
: a small town

tan·nage \ˈta-nij\ *noun* (1662)
: the act, process, or result of tanning

tan·nate \ˈta-ˌnāt\ *noun* [French, from *tannin*] (1802)
: a compound of a tannin

¹tan·ner \ˈta-nər\ *noun* (before 12th century)
1 : one that tans hides
2 : one who acquires or seeks to acquire a suntan

²tanner *noun* [origin unknown] (circa 1811)
British **:** SIXPENCE

tanner crab *noun* [probably from New Latin *tanneri,* specific epithet of *Chionoecetes tanneri,* from Zera L. *Tanner* (died 1906) American naval officer] (1947)
: any of several spider crabs (genus *Chionoecetes*); *especially* **:** SNOW CRAB

tan·nery \ˈta-nə-rē, ˈtan-rē\ *noun, plural* **-ner·ies** (1736)
: a place where tanning is carried on

tan·nic \ˈta-nik\ *adjective* [French *tannique,* from *tannin*] (circa 1859)
1 : of, resembling, or derived from tan or a tannin
2 *of wine* **:** containing an abundance of tannins **:** markedly astringent

tannic acid *noun* (1836)
: TANNIN 1

tan·nin \ˈta-nən\ *noun* [French, from *tanner* to tan] (1802)
1 : any of various soluble astringent complex phenolic substances of plant origin used especially in tanning, dyeing, the making of ink, and in medicine
2 : a substance that has a tanning effect

tan·ning \ˈta-niŋ\ *noun* (15th century)
1 : the art or process by which a skin is tanned
2 : a browning of the skin especially by exposure to sun
3 : a sound spanking
4 : a natural darkening and hardening of the cuticle of an insect immediately after molting

tan·nish \ˈta-nish\ *adjective* (1935)
: somewhat tan

tan oak *noun* (circa 1925)
: a U.S. Pacific coast evergreen tree (*Lithocarpus densiflora*) of the beech family that has erect staminate catkins and is rich in tannins

Ta·no·an \tə-ˈnō-ən, ˈtä-nə-wən\ *noun* [*Tano,* a group of former pueblos in New Mexico] (1891)
: a family of American Indian languages spoken in New Mexico and Arizona
— **Tanoan** *adjective*

tan·sy \ˈtan-zē\ *noun, plural* **tansies** [Middle English *tanesey,* from Middle French *tanesie,* from Medieval Latin *athanasia,* from Greek, immortality, from *athanatos* immortal, from *a-* + *thanatos* death — more at THANATOS] (15th century)
: a common aromatic Old World composite herb (*Tanacetum vulgare*) that is naturalized in North America and has bitter-tasting finely divided leaves; *broadly* **:** a plant of the same genus

tansy ragwort *noun* (circa 1900)
: an Old World yellow-flowered senecio (*Senecio jacobaea*) that is naturalized in North America and is toxic to livestock

tan·ta·late \ˈtan-tᵊl-ˌāt\ *noun* (1849)
: a salt containing tantalum in combination with oxygen

tan·ta·lise *British variant of* TANTALIZE

tan·ta·lite \ˈtan-tᵊl-ˌīt\ *noun* (1805)
: a mineral consisting of a heavy dark lustrous oxide of tantalum and usually other metals (as iron or manganese)

tan·ta·lize \ˈtan-tᵊl-ˌīz\ *verb* **-lized; -liz·ing** [*Tantalus*] (1597)
transitive verb
: to tease or torment by or as if by presenting something desirable to the view but continually keeping it out of reach
intransitive verb
: to cause one to be tantalized
— **tan·ta·liz·er** *noun*

tantalizing *adjective* (circa 1683)
: possessing a quality that arouses or stimulates desire or interest; *also* **:** mockingly or teasingly out of reach
— **tan·ta·liz·ing·ly** \-ˌī-ziŋ-lē\ *adverb*

tan·ta·lum \ˈtan-tᵊl-əm\ *noun* [New Latin, from Latin *Tantalus;* from its inability to absorb acid] (1802)
: a hard ductile gray white acid-resisting metallic element of the vanadium family found combined in rare minerals (as tantalite and columbite) — see ELEMENT table

Tan·ta·lus \ˈtan-tᵊl-əs\ *noun* [Latin, from Greek *Tantalos*]
1 : a legendary king of Lydia condemned to stand up to the chin in a pool of water in Hades and beneath fruit-laden boughs only to have the water or fruit recede at each attempt to drink or eat
2 *not capitalized* **:** a locked cellarette with contents visible but not obtainable without a key

tan·ta·mount \ˈtan-tə-ˌmau̇nt\ *adjective* [obsolete *tantamount,* noun, equivalent, from Anglo-French *tant amunter* to amount to as much] (1641)
: equivalent in value, significance, or effect

tan·ta·ra \tan-ˈtar-ə, -ˈtär-\ *noun* [Latin *taratantara,* of imitative origin] (1584)
: the blare of a trumpet or horn

¹tan·tivy \tan-ˈti-vē\ *adverb* [origin unknown] (1641)
: at a gallop

²tantivy *noun, plural* **-tiv·ies** (circa 1658)
1 : a rapid gallop or ride
2 : TANTARA

tan·tra \ˈtən-trə, ˈtän-, ˈtan-\ *noun, often capitalized* [Sanskrit, literally, warp, from *tanoti* he stretches, weaves; akin to Greek *teinein* to stretch — more at THIN] (1799)
: one of the later Hindu or Buddhist scriptures dealing especially with techniques and rituals including meditative and sexual practices; *also* **:** the rituals or practices outlined in the tantra
— **tan·tric** \-trik\ *adjective, often capitalized*
— **Tan·trism** \-ˌtri-zəm\ *noun*
— **Tan·trist** \-trist\ *noun*

tan·trum \ˈtan-trəm\ *noun* [origin unknown] (1714)
: a fit of bad temper

ta·nu·ki \tä-ˈnü-kē\ *noun* [Japanese, raccoon dog] (circa 1929)
: the fur of a raccoon dog; *also* **:** RACCOON DOG

tan·yard \ˈtan-ˌyärd\ *noun* (1666)
: the section or part of a tannery housing tanning vats

tan·za·nite \ˈtan-zə-ˌnīt\ *noun* [*Tanzania,* Africa] (1968)
: a mineral that is a deep blue variety of zoisite and is used as a gemstone

Tao \ˈdau̇, ˈtau̇\ *noun* [Chinese (Beijing) *dào,* literally, way] (1736)
1 a : the unconditional and unknowable source and guiding principle of all reality as conceived by Taoists **b :** the process of nature by which all things change and which is to be followed for a life of harmony
2 *often not capitalized* **:** the path of virtuous conduct as conceived by Confucians
3 *often not capitalized* **:** the art or skill of doing something in harmony with the essential nature of the thing ⟨the *tao* of archery⟩

Tao·ism \-ˌi-zəm\ *noun* [*Tao*] (1838)
1 : a Chinese mystical philosophy traditionally founded by Lao-tzu in the 6th century B.C. that teaches conformity to the Tao by unassertive action and simplicity
2 : a religion developed from Taoist philosophy and folk and Buddhist religion and concerned with obtaining long life and good fortune often by magical means
— **Tao·ist** \-ist\ *adjective or noun*
— **Tao·is·tic** \dau̇-ˈis-tik, tau̇-\ *adjective*

¹tap \ˈtap\ *noun* [Middle English *tappe,* from Old English *tæppa;* akin to Old High German *zapho* tap] (before 12th century)
1 a : a plug for a hole (as in a cask) **:** SPIGOT **b :** a device consisting of a spout and valve attached to the end of a pipe to control the flow of a fluid **:** FAUCET
2 a : a liquor drawn through a tap **b :** the procedure of removing fluid (as from a body cavity)
3 : a tool for forming an internal screw thread
4 : an intermediate point in an electric circuit where a connection may be made
5 : WIRETAP
— **on tap 1 :** ready to be drawn from a large container (as a cask or keg) ⟨ale *on tap*⟩ **2 :** broached or furnished with a tap **3 :** on hand **:** AVAILABLE ⟨services instantly *on tap* —Hugh Dwan⟩ **4 :** coming up ⟨other matches *on tap* —H. W. Wind⟩

²tap *transitive verb* **tapped; tap·ping** (15th century)
1 : to let out or cause to flow by piercing or by drawing a plug from the containing vessel ⟨*tap* wine from a cask⟩
2 a : to pierce so as to let out or draw off a fluid ⟨*tap* maple trees⟩ **b :** to draw out, from, or upon ⟨*tap* new sources of revenue⟩
3 a : to cut in on (a telephone or telegraph wire) to get information **b :** to cut in (an electrical circuit) on another circuit
4 : to form an internal screw thread in by means of a tap
5 : to get money from as a loan or gift
6 : to connect (a street gas or water main) with a local supply
— **tap·per** *noun*
— **tap into :** to make a usually advantageous connection with ⟨clothing that . . . *taps into* the currents of popular culture —John Duka⟩

³tap *verb* **tapped; tap·ping** [Middle English *tappen,* from Middle French *taper* to strike with the flat of the hand, of Germanic origin; akin to Middle High German *tāpe* paw, blow dealt with the paw] (13th century)
transitive verb

\ə\ **abut** \ᵊ\ **kitten** \ər\ **further** \a\ **ash** \ā\ **ace** \ä\ **mop, mar** \au̇\ **out** \ch\ **chin** \e\ **bet** \ē\ **easy** \g\ **go** \i\ **hit** \ī\ **ice** \j\ **job** \ŋ\ **sing** \ō\ **go** \ȯ\ **law** \ȯi\ **boy** \th\ **thin** \t͟h\ **the** \ü\ **loot** \u̇\ **foot** \y\ **yet** \zh\ **vision** *see also* Guide to Pronunciation

1 : to strike lightly especially with a slight sound
2 : to give a light blow with ⟨*tap* a pencil on the table⟩
3 : to bring about by repeated light blows ⟨*tap* out a story on the typewriter⟩
4 : to repair by putting a tap on
5 : SELECT, DESIGNATE ⟨was *tapped* for police commissioner⟩; *specifically* **:** to elect to membership (as in a fraternity)
intransitive verb
1 : to strike a light audible blow **:** RAP
2 : to walk with light audible steps
3 : TAP-DANCE
— tap·per *noun*

[4]tap *noun* (14th century)
1 a : a light usually audible blow; *also* **:** its sound **b :** one of several usually rapid drumbeats on a snare drum
2 : HALF SOLE
3 : a small metal plate for the sole or heel of a shoe
4 : TAP DANCE 1
5 : FLAP 7

[1]ta·pa \ˈtä-pə, ˈta-\ *noun* [Marquesan & Tahitian] (1823)
: a coarse cloth made in the Pacific islands from the pounded bark especially of the paper mulberry and usually decorated with geometric patterns

[2]tapa *noun* [Spanish, literally, cover, lid, probably of Germanic origin; akin to Old English *tæppa* tap] (1953)
: an hors d'oeuvre served with drinks in Spanish bars — usually used in plural

tap dance *noun* (1928)
1 : a step dance tapped out audibly by means of shoes with hard soles or soles and heels to which taps have been added
2 : something suggesting a tap dance; *especially* **:** an action or discourse intended to rationalize or distract ⟨does a clever *tap dance* to explain —Campbell Geeslin⟩
— tap–dance *intransitive verb*
— tap dancer *noun*
— tap dancing *noun*

[1]tape \ˈtāp\ *noun* [Middle English, from Old English *tæppe*] (before 12th century)
1 : a narrow woven fabric
2 : a string or ribbon stretched breast-high above the finish line of a race
3 : a narrow flexible strip or band: as **a :** ADHESIVE TAPE **b :** MAGNETIC TAPE
4 : TAPE RECORDING

[2]tape *verb* **taped; tap·ing** (1609)
transitive verb
1 : to fasten, tie, bind, cover, or support with tape
2 : to record on tape and especially magnetic tape ⟨*tape* an interview⟩
intransitive verb
: to record something on tape and especially magnetic tape

[3]tape *adjective* (1947)
1 : recorded on tape ⟨*tape* music⟩
2 : intended for use with recording (as magnetic) tape ⟨a *tape* cartridge⟩

tape deck *noun* (1949)
: a device used to play back and often to record on magnetic tape that usually has to be connected to an audio system

tape grass *noun* (circa 1818)
: any of several submerged aquatic monocotyledonous plants (genus *Vallisneria* of the family Hydrocharitaceae) with long ribbonlike leaves — called also *eelgrass, wild celery*

tape measure *noun* (1845)
: a narrow strip (as of a limp cloth or steel tape) marked off in units (as inches or centimeters) for measuring

[1]ta·per \ˈtā-pər\ *noun* [Middle English, from Old English *tapor* candle, wick, perhaps modification of Latin *papyrus* papyrus] (before 12th century)
1 a : a slender candle **b :** a long waxed wick used especially for lighting candles, lamps, pipes, or fires **c :** a feeble light
2 a : a tapering form or figure **b :** gradual diminution of thickness, diameter, or width in an elongated object **c :** a gradual decrease

[2]taper *adjective* (15th century)
1 : progressively narrowed toward one end
2 : furnished with or adjusted to a scale **:** GRADUATED ⟨*taper* freight rates⟩

[3]taper *verb* **ta·pered; ta·per·ing** \ˈtā-p(ə-)riŋ\ (1610)
intransitive verb
1 : to become progressively smaller toward one end
2 : to diminish gradually
transitive verb
: to cause to taper

[4]tap·er \ˈtā-pər\ *noun* [[1]*tape*] (circa 1920)
: one that applies or dispenses tape

tape–re·cord \ˌtā-pri-ˈkȯrd, ˈtā-pri-ˌ\ *transitive verb* [back-formation from *tape recording*] (1950)
: to make a recording of on magnetic tape

tape recorder *noun* (1932)
: a device for recording on and playing back magnetic tape

tape recording *noun* (1940)
: magnetic recording on magnetic tape; *also* **:** a recording made by this process

ta·per·er \ˈtā-pər-ər\ *noun* (15th century)
: one who bears a taper in a religious procession

taper off *verb* (1848)
: TAPER ⟨business growth had *tapered off* seriously while the unemployment rate had climbed —*Current Biography*⟩

ta·per·stick \ˈtā-pər-ˌstik\ *noun* (1546)
: a candlestick that holds tapers

tap·es·tried \ˈta-pə-strēd\ *adjective* (1769)
1 : covered or decorated with or as if with tapestry
2 : woven or depicted in tapestry

tap·es·try \ˈta-pə-strē\ *noun, plural* **-tries** [Middle English *tapistry,* modification of Middle French *tapisserie,* from *tapisser* to carpet, cover with tapestry, from Old French *tapis* carpet, from Greek *tapēt-, tapēs* rug, carpet] (15th century)
1 a : a heavy handwoven reversible textile used for hangings, curtains, and upholstery and characterized by complicated pictorial designs **b :** a nonreversible imitation of tapestry used chiefly for upholstery **c :** embroidery on canvas resembling woven tapestry ⟨needlepoint *tapestry*⟩
2 : something resembling tapestry (as in complexity or richness of design)

tapestry carpet *noun* (1852)
: a carpet in which the designs are printed in colors on the threads before the fabric is woven

ta·pe·tum \tə-ˈpē-təm\ *noun, plural* **ta·pe·ta** \-ˈpē-tə\ [New Latin, from Latin *tapete* carpet, tapestry, from Greek *tapēt-, tapēs* rug, carpet] (1713)
1 : any of various membranous layers or areas especially of the choroid coat and retina of the eye
2 : a layer of nutritive cells that invests the sporogenous tissue in the sporangium of higher plants

tape·worm \ˈtāp-ˌwərm\ *noun* [from its shape] (1752)
: any of a class (Cestoda) of platyhelminthic worms parasitic especially in the intestines of vertebrates — called also *cestode*

tap·hole \ˈtap-ˌhōl\ *noun* (1594)
: a hole for a tap; *specifically* **:** a hole at or near the bottom of a furnace or ladle through which molten metal, matte, or slag can be tapped

ta·phon·o·my \tə-ˈfä-nə-mē, ta-\ *noun* [Greek *taphē* burial + English *-nomy*] (1940)
: the study of the processes (as burial, decay, and preservation) that affect animal and plant remains as they become fossilized
— taph·o·nom·ic \ˌta-fə-ˈnä-mik\ *adjective*
— ta·phon·o·mist \tə-ˈfä-nə-mist, ta-\ *noun*

tap–in \ˈtap-ˌin\ *noun* (1948)
1 : TIP-IN
2 : a short putt in golf

tap·i·o·ca \ˌta-pē-ˈō-kə\ *noun* [Spanish & Portuguese, from Tupi *tipióka*] (1707)
1 : a usually granular preparation of cassava starch used especially in puddings and as a thickening in liquid food; *also* **:** a dish (as pudding) containing tapioca
2 : CASSAVA

ta·pir \ˈtā-pər *also* ˈtā-ˌpir *or* tə-ˈpir\ *noun, plural* **tapir** *or* **tapirs** [Portuguese *tapira,* from Tupi *tapiíra*] (1774)
: any of a genus (*Tapirus*) of chiefly nocturnal perissodactyl ungulates of tropical America and Myanmar to Sumatra that have the snout and upper lip prolonged into a short flexible proboscis

tapir

tap·is \ˈta-ˌpē, ta-ˈpē\ *noun* [Middle English, from Middle French — more at TAPESTRY] (15th century)
archaic **:** a small tapestry used for hangings and floor and table coverings
— on the tapis : under consideration

tap–off \ˈtap-ˌȯf\ *noun* (circa 1932)
: [2]TIP-OFF

tap out *intransitive verb* (1939)
: to run out of money by betting

tap pants *noun plural* (1977)
: a loose-fitting woman's undergarment of a style similar to shorts formerly worn for tap dancing

tapped out *adjective* (1950)
: out of money **:** BROKE

tap·pet \ˈta-pət\ *noun* [irregular from [3]*tap*] (1745)
: a lever or projection moved by some other piece (as a cam) or intended to tap or touch something else to cause a particular motion

tapping *noun* (1597)
: the act, process, or means by which something is tapped

tap·pit hen \ˈta-pət-\ *noun* [Scots *tappit,* alteration of English *topped*] (1721)
Scottish **:** a drinking vessel with a knob on the lid

tap·room \ˈtap-ˌrüm, -ˌru̇m\ *noun* (1807)
: BARROOM

tap·root \-ˌrüt, -ˌru̇t\ *noun* [[1]*tap*] (1601)
1 : a primary root that grows vertically downward and gives off small lateral roots
2 : the central element or position in a line of growth or development

taps \ˈtaps\ *noun plural but singular or plural in construction* [probably alteration of earlier *taptoo* tattoo — more at TATTOO] (1824)
: the last bugle call at night blown as a signal that lights are to be put out; *also* **:** a similar call blown at military funerals and memorial services

tap·sal–tee·rie \ˌtap-səl-ˈtē-rē\ *adverb* [by alteration] (1784)
Scottish **:** TOPSY-TURVY

tap·ster \ˈtap-stər\ *noun* (before 12th century)
: BARTENDER

tap water *noun* (1881)
: water as it comes from a tap (as in a home)

[1]tar \ˈtär\ *noun* [Middle English *terr, tarr,* from Old English *teoru;* akin to Old English *trēow* tree — more at TREE] (before 12th century)
1 a : a dark brown or black bituminous usually odorous viscous liquid obtained by destructive distillation of organic material (as wood, coal, or peat) **b :** a substance in some respects resembling tar; *especially* **:** a condensable residue present in smoke from burning tobacco

that contains combustion by-products (as resins, acids, phenols, and essential oils)
2 [short for *tarpaulin*] **:** SAILOR

²tar *transitive verb* **tarred; tar·ring** (13th century)
1 : to cover with tar
2 : to defile as if with tar ⟨least *tarred* by the scandal —*Newsweek*⟩
— tar and feather : to smear (a person) with tar and cover with feathers as a punishment or indignity
— tar with the same brush : to mark or stain with the same fault or characteristic

³tar *or* **tarre** \ˈtär\ *transitive verb* **tarred; tar·ring; tars** *or* **tarres** [Middle English *terren, tarren,* from Old English *tyrwan*] (before 12th century)
: to urge to action — usually used with *on*

tar·a·did·dle *or* **tar·ra·did·dle** \ˌtar-ə-ˈdi-dᵊl, ˈtar-ə-ˌ\ *noun* [origin unknown] (circa 1796)
1 : FIB
2 : pretentious nonsense

Tar·a·hu·ma·ra \ˌtar-ə-hü-ˈmär-ə\ *noun, plural* **Tarahumara** *or* **Tarahumaras** [Spanish] (1874)
1 : a member of an American Indian people living in the state of Chihuahua, Mexico
2 : the Uto-Aztecan language of the Tarahumara people

tar·an·tel·la \ˌtar-ən-ˈte-lə\ *noun* [Italian, from *Taranto,* Italy] (1782)
: a lively folk dance of southern Italy in ⁶⁄₈ time

tar·an·tism \ˈtar-ən-ˌti-zəm\ *noun* [New Latin *tarantismus,* from *Taranto,* Italy] (circa 1656)
: a dancing mania or malady of late medieval Europe

ta·ran·tu·la \tə-ˈran-chə-lə, -tə-lə; -ˈranch-lə, -ˈrant-\ *noun, plural* **ta·ran·tu·las** *also* **ta·ran·tu·lae** \-ˌlē\ [Medieval Latin, from Old Italian *tarantola,* from *Taranto*] (1561)
1 : a European wolf spider (*Lycosa tarentula*) popularly held to be the cause of tarantism
2 : any of a family (Theraphosidae) of large hairy American spiders that are typically rather sluggish and capable of biting sharply though most forms are not significantly poisonous to humans

Ta·ras·can \tə-ˈras-kən, -ˈräs-\ *noun* [Spanish *tarasco*] (1922)
1 : a member of an American Indian people of the state of Michoacán, Mexico
2 : the language of the Tarascan people

tar baby *noun* [from the tar baby that trapped Brer Rabbit in an Uncle Remus story by Joel Chandler Harris] (1924)
: something from which it is nearly impossible to extricate oneself

tar·boosh *also* **tar·bush** \tär-ˈbüsh, ˈtär-ˌ\ *noun* [Arabic *ṭarbūsh*] (1702)
: a red hat similar to the fez worn especially by Muslim men

tar·di·grade \ˈtär-də-ˌgrād\ *noun* [ultimately from Latin *tardigradus* slow-moving, from *tardus* slow + *gradi* to step, go — more at GRADE] (1860)
: any of a phylum (Tardigrada) of microscopic arthropods with four pairs of stout legs that live usually in water or damp moss — called also *water bear*

tar·di·ly \ˈtär-dᵊl-ē\ *adverb* (1597)
1 : at a slow pace
2 : LATE

tar·dive dyskinesia \ˈtär-div-\ *noun* [*tardive* tending toward late development (from French, feminine of *tardif,* from Middle French) + *dyskinesia*] (1964)
: a central nervous system disorder characterized by twitching of the face and tongue and involuntary motor movements of the trunk and limbs and occurring especially as a side effect of prolonged use of antipsychotic drugs (as phenothiazine)

tar·do \ˈtär-(ˌ)dō\ *adjective* [Italian, from Latin *tardus*] (circa 1843)
: SLOW — used as a direction in music

¹tar·dy \ˈtär-dē\ *adjective* **tar·di·er; -est** [alteration of earlier *tardif,* from Middle French, from (assumed) Vulgar Latin *tardivus,* from Latin *tardus*] (15th century)
1 : moving slowly **:** SLUGGISH
2 : delayed beyond the expected or proper time **:** LATE
— tar·di·ness \ˈtär-dē-nəs\ *noun*

²tardy *noun, plural* **tardies** (1960)
: an instance of being tardy (as to a class)

¹tare \ˈtar, ˈter\ *noun* [Middle English; probably akin to Middle Dutch *tarwe* wheat] (14th century)
1 a : the seed of a vetch **b :** any of several vetches (especially *Vicia sativa* and *V. hirsuta*)
2 : a weed of grainfields usually held to be the darnel
3 *plural* **:** an undesirable element

²tare *noun* [Middle English, from Middle French, from Old Italian *tara,* from Arabic *ṭarḥa,* literally, that which is removed] (15th century)
1 : a deduction from the gross weight of a substance and its container made in allowance for the weight of the container; *also* **:** the weight of the container
2 : COUNTERWEIGHT

³tare *transitive verb* **tared; tar·ing** (1812)
: to ascertain or mark the tare of; *especially* **:** to weigh so as to determine the tare

targe \ˈtärj\ *noun* [Middle English, from Old French] (14th century)
: a light shield used especially by the Scots

¹tar·get \ˈtär-gət\ *noun, often attributive* [Middle English, from Middle French *targette,* diminutive of *targe* light shield, of Germanic origin; akin to Old Norse *targa* shield] (15th century)
1 : a small round shield
2 a : a mark to shoot at **b :** a target marked by shots fired at it **c :** something or someone fired at or marked for attack **d :** a goal to be achieved
3 a : an object of ridicule or criticism **b :** something or someone to be affected by an action or development
4 a : a railroad day signal that is attached to a switch stand and indicates whether the switch is open or closed **b :** a sliding sight on a surveyor's leveling rod
5 a : the metallic surface (as of platinum or tungsten) upon which the stream of electrons within an X-ray tube is focused and from which the X rays are emitted **b :** a body, surface, or material bombarded with nuclear particles or electrons; *especially* **:** fluorescent material on which desired visual effects are produced in electronic devices (as in radar) ◇
— off target : not valid **:** INACCURATE
— on target : precisely correct or valid especially in interpreting or addressing a problem or vital issue

²target *transitive verb* (1837)
1 : to make a target of; *especially* **:** to set as a goal
2 : to direct or use toward a target

tar·get·able \ˈtär-gə-tə-bəl\ *adjective* (1964)
: capable of being aimed at a target ⟨missiles with *targetable* warheads⟩

target date *noun* (1945)
: the date set for an event or for the completion of a project, goal, or quota

target language *noun* (1953)
1 : a language into which another language is to be translated — compare SOURCE LANGUAGE
2 : a language other than one's native language that is being learned

Tar·gum \ˈtär-ˌgu̇m, -ˌgüm\ *noun* [Late Hebrew *targūm,* from Aramaic, translation] (1587)
: an Aramaic translation or paraphrase of a portion of the Old Testament

Tar·heel \ˈtär-ˌhēl\ *noun* (1864)
: a native or resident of North Carolina — used as a nickname

¹tar·iff \ˈtar-əf\ *noun* [Italian *tariffa,* from Arabic *ta'rīf* notification] (1592)
1 a : a schedule of duties imposed by a government on imported or in some countries exported goods **b :** a duty or rate of duty imposed in such a schedule
2 : a schedule of rates or charges of a business or a public utility
3 : PRICE, CHARGE

²tariff *transitive verb* (circa 1828)
: to subject to a tariff

tar·la·tan \ˈtär-lə-tᵊn\ *noun* [French *tarlatane*] (circa 1741)
: a sheer cotton fabric in open plain weave usually heavily sized for stiffness

tar·mac \ˈtär-ˌmak\ *noun* [from *Tarmac,* a trademark] (1919)
: a tarmacadam road, apron, or runway

Tarmac *trademark*
— used for a bituminous binder for roads

tar·mac·ad·am \ˌtär-mə-ˈka-dəm\ *noun* (1882)
1 : a pavement constructed by spraying or pouring a tar binder over layers of crushed stone and then rolling
2 : a material of tar and aggregates mixed in a plant and shaped on the roadway

tarn \ˈtärn\ *noun* [Middle English *tarne,* of Scandinavian origin; akin to Old Norse *tjǫrn* small lake] (14th century)
: a small steep-banked mountain lake or pool

tar·na·tion \ˌtär-ˈnā-shən\ *noun* [alteration of *darnation,* euphemism for *damnation*] (1790)
: DAMNATION — often used as an interjection or intensive; often used with *in* ⟨*tarnation* strike me —James Joyce⟩ ⟨where in *tarnation* you from? —Jessamyn West⟩

¹tar·nish \ˈtär-nish\ *verb* [Middle French *terniss-,* stem of *ternir,* probably of Germanic origin; akin to Old High German *ternen* to hide] (1598)
transitive verb
1 : to dull or destroy the luster of by or as if by air, dust, or dirt **:** SOIL, STAIN
2 a : to detract from the good quality of **:** VITIATE ⟨his fine dreams now slightly *tarnished*⟩ **b :** to bring disgrace on **:** SULLY
intransitive verb
: to become tarnished
— tar·nish·able \-ni-shə-bəl\ *adjective*

²tarnish *noun* (1713)
: something that tarnishes; *especially* **:** a film of chemically altered material on the surface of a metal (as silver)

tarnished plant bug *noun* (circa 1890)

◇ WORD HISTORY

target Old French *targe,* a name for a small leather-covered shield carried by a foot soldier, is probably of Frankish origin and hence akin to Old Norse *targa* "round shield." *Targe* and its diminutive form *targette* were both borrowed into Middle English. Our pronunciation of the modern word *target* with \g\ rather than \j\, as well as occasional Middle English variants such as *targatte,* shows the influence of forms with a "hard *g*" such as Middle French *targuete* or late Old Provençal *targueta.* Not until the later 18th century was *target* used with reference to something to be aimed at, originally a shieldlike object used in archery practice. During the 19th century the word acquired the general sense "a mark to shoot at," either literal or figurative.

: a common hemipterous bug (*Lygus lineolaris*) of eastern North America that causes injury to plants by sucking sap from buds, leaves, and fruits and that carries plant diseases

ta·ro \'tär-(,)ō, 'tar-, 'ter-\ *noun, plural* **taros** [Tahitian & Maori] (1769)
: a plant (*Colocasia esculenta*) of the arum family grown throughout the tropics for its edible starchy tuberous rootstocks and in temperate regions for ornament; *also* : its rootstock

tar·ok \'tar-,äk\ *noun* [Italian *tarocchi* tarots] (1739)
: an old card game popular in central Europe and played with a pack containing 40, 52, or 56 cards equivalent to modern playing cards plus the 22 tarots

tar·ot \'tar-(,)ō\ *noun* [Middle French, from Italian *tarocchi* (plural)] (circa 1623)
: any of a set of usually 78 playing cards including 22 pictorial cards used for fortune-telling; *also* : the 22 pictorial cards serving as trumps in tarok

tarp \'tärp\ *noun* (1906)
: TARPAULIN

tar paper *noun* (1891)
: a heavy paper coated or impregnated with tar for use especially in building

tar·pau·lin \tär-'pȯ-lən, 'tär-pə-; ÷tär-'pōl-yən\ *noun* [probably from [1]*tar* + *-palling, -pauling* (from *pall*)] (1605)
1 : a piece of material (as durable plastic) used for protecting exposed objects or areas
2 : SAILOR

tar pit *noun* (1839)
: an area in which natural bitumens collect and are exposed at the earth's surface and which tends to trap animals and preserve their hard parts (as bones or teeth)

tar·pon \'tär-pən\ *noun, plural* **tarpon** *or* **tarpons** [origin unknown] (1685)
: a large silvery elongate anadromous bony fish (*Megalops atlanticus* of the family Elopidae) that occurs especially in the Gulf of Mexico, Caribbean, and warm coastal waters of the Atlantic, reaches a length of about six feet (two meters), and is often caught for sport

tar·ra·gon \'tar-ə-gən\ *noun* [Middle French *targon,* from Medieval Latin *tarchon,* from Arabic *ṭarkhūn*] (1538)
: a small widely cultivated perennial artemisia (*Artemisia dracunculus*) having pungent narrow usually entire leaves; *also* : its leaves used as a flavoring

tarre *variant of* TAR

tar·ri·ance \'tar-ē-ən(t)s\ *noun* (15th century)
: the act or an instance of tarrying

[1]**tar·ry** \'tar-ē\ *intransitive verb* **tar·ried; tar·ry·ing** [Middle English *tarien*] (14th century)
1 a : to delay or be tardy in acting or doing **b** : to linger in expectation : WAIT
2 : to abide or stay in or at a place

[2]**tarry** *noun, plural* **tarries** (14th century)
: STAY, SOJOURN

[3]**tar·ry** \'tär-ē\ *adjective* (1552)
: of, resembling, or covered with tar

[1]**tar·sal** \'tär-səl\ *adjective* (1817)
1 : of or relating to the tarsus
2 : being or relating to plates of dense connective tissue that serve to stiffen the eyelids

[2]**tarsal** *noun* (1881)
: a tarsal part (as a bone or cartilage)

tar sand *noun* (1899)
: a natural impregnation of sand or sandstone with petroleum from which the lighter portions have escaped

tar·si·er \'tär-sē-ər, -sē-,ā\ *noun* [French, from *tarse* tarsus, from New Latin *tarsus*] (1774)
: any of a family (Tarsiidae) of small chiefly nocturnal and arboreal carnivorous primates of the Malay Archipelago that have large round eyes, long legs, and a long nearly hairless tail

tar·so·meta·tar·sus \'tär-(,)sō-'me-tə-,tär-səs\ *noun* [New Latin, from *tarsus* + *-o-* + *metatarsus*] (1854)
: the large compound bone of the tarsus of a bird; *also* : the segment of the limb it supports

tar·sus \'tär-səs\ *noun, plural* **tar·si** \-,sī, -,sē\ [New Latin, from Greek *tarsos* wickerwork mat, flat of the foot, ankle, edge of the eyelid; akin to Greek *tersesthai* to become dry — more at THIRST] (1676)
1 : the part of the foot of a vertebrate between the metatarsus and the leg; *also* : the small bones that support this part of the limb
2 : the tarsal plate of the eyelid
3 : the distal part of the limb of an arthropod
4 : TARSOMETATARSUS

[1]**tart** \'tärt\ *adjective* [Middle English, from Old English *teart* sharp, severe; akin to Middle High German *traz* spite] (14th century)
1 : agreeably sharp or acid to the taste
2 : marked by a biting, acrimonious, or cutting quality
— **tart·ish** \'tär-tish\ *adjective*
— **tart·ly** *adverb*
— **tart·ness** *noun*

[2]**tart** *noun* [Middle English *tarte,* from Middle French] (15th century)
1 : a dish baked in a pastry shell : PIE: as **a** : a small pie or pastry shell without a top containing jelly, custard, or fruit **b** : a small pie made of pastry folded over a filling
2 : PROSTITUTE

tar·tan \'tär-tᵊn\ *noun* [probably from Middle French *tiretaine* linsey-woolsey] (circa 1500)
1 : a plaid textile design of Scottish origin consisting of stripes of varying width and color usually patterned to designate a distinctive clan
2 a : a twilled woolen fabric with tartan design **b** : a fabric with tartan design
3 : a garment of tartan design

[1]**tar·tar** \'tär-tər\ *noun* [Middle English, from Medieval Latin *tartarum*] (14th century)
1 : a substance consisting essentially of cream of tartar that is derived from the juice of grapes and deposited in wine casks together with yeast and other suspended matters as a pale or dark reddish crust or sediment; *especially* : a recrystallized product yielding cream of tartar on further purification
2 : an incrustation on the teeth consisting of salivary secretion, food residue, and various salts (as calcium carbonate)

[2]**tartar** *noun* [Middle English *Tartre,* from Middle French *Tartare,* probably from Medieval Latin *Tartarus,* modification of Persian *Tātār* — more at TATAR] (14th century)
1 *capitalized* : a native or inhabitant of Tatary
2 *capitalized* : TATAR 2
3 *often capitalized* : a person of irritable or violent temper
4 : one that proves to be unexpectedly formidable
— **Tartar** *adjective*
— **Tar·tar·i·an** \tär-'tar-ē-ən, -'ter-\ *adjective*

Tar·tar·e·an \tär-'tar-ē-ən, -'ter-\ *adjective* [Latin *tartareus,* from Greek *tartareios,* from *Tartaros* Tartarus] (circa 1623)
: of, relating to, or resembling Tartarus : INFERNAL

tartar emetic *noun* (1704)
: a poisonous efflorescent crystalline salt $KSbOC_4H_4O_6 \cdot \frac{1}{2}H_2O$ of sweetish metallic taste that is used in dyeing as a mordant and in medicine especially in the treatment of amebiasis

tar·tar·ic acid \(,)tär-'tar-ik-\ *noun* (1810)
: a strong dicarboxylic acid $C_4H_6O_6$ of plant origin that occurs in three optically isomeric forms, is usually obtained from tartar, and is used especially in food and medicines, in photography, and in making salts and esters

tar·tar sauce *or* **tar·tare sauce** \'tär-tər-\ *noun* [French *sauce tartare*] (1855)
: a sauce made principally of mayonnaise and chopped pickles

Tar·ta·rus \'tär-tə-rəs\ *noun* [Latin, from Greek *Tartaros*]
: a section of Hades reserved for punishment of the wicked

tart·let \'tärt-lət\ *noun* (15th century)
: a small tart

tar·trate \'tär-,trāt\ *noun* [International Scientific Vocabulary, from French *tartre* tartar, from Medieval Latin *tartarum*] (1794)
: a salt or ester of tartaric acid

Tar·tuffe \,tär-'tu̇f, -'tüf\ *noun* [French *Tartufe*]
: a religious hypocrite and protagonist in Molière's play *Tartuffe*

tart up *transitive verb* (1938)
: DRESS UP, FANCY UP ⟨*tarted up* pubs and restaurants for the spenders —Arnold Ehrlich⟩

tarty \'tär-(,)tē\ *adjective* [[2]*tart*] (1918)
: resembling or suggestive of a prostitute (as in clothing or manner)

[1]**task** \'task\ *noun* [Middle English *taske,* from Old North French *tasque,* from Medieval Latin *tasca* tax or service imposed by a feudal superior, from *taxare* to tax] (14th century)
1 a : a usually assigned piece of work often to be finished within a certain time **b** : something hard or unpleasant that has to be done **c** : DUTY, FUNCTION
2 : subjection to adverse criticism : REPRIMAND — used in the expressions *to take, call,* or *bring to task* ☆

[2]**task** *transitive verb* (15th century)
1 *obsolete* : to impose a tax on
2 : to assign a task to
3 : to oppress with great labor ⟨*tasks* his mind with petty details⟩

task force *noun* (1941)
: a temporary grouping under one leader for the purpose of accomplishing a definite objective

task·mas·ter \'task-,mas-tər\ *noun* (1530)
: one that imposes a task or burdens another with labor

task·mis·tress \-,mis-trəs\ *noun* (1603)
: a woman who is a taskmaster

Tas·ma·nian devil \(,)taz-'mā-nē-ən-, -nyən-\ *noun* (1867)
: a powerful heavily built carnivorous terrestrial Tasmanian marsupial (*Sarcophilus harrisii*) that is about the size of a badger and has a chiefly black coat marked with white on the chest

Tasmanian wolf *noun* (circa 1890)
: a somewhat doglike carnivorous marsupial (*Thylacinus cynocephalus*) that formerly inhabited Tasmania but is now considered extinct — called also *Tasmanian tiger*

tasse \'tas\ *noun* [perhaps from Middle French *tasse* purse, pouch] (circa 1548)
: one of a series of overlapping metal plates in a suit of armor that form a short skirt over the body below the waist — see ARMOR illustration

☆ SYNONYMS
Task, duty, job, chore, stint, assignment mean a piece of work to be done. TASK implies work imposed by a person in authority or an employer or by circumstance ⟨charged with a variety of *tasks*⟩. DUTY implies an obligation to perform or responsibility for performance ⟨the *duties* of a lifeguard⟩. JOB applies to a piece of work voluntarily performed; it may sometimes suggest difficulty or importance ⟨the *job* of turning the company around⟩. CHORE implies a minor routine activity necessary for maintaining a household or farm ⟨every child was assigned *chores*⟩. STINT implies a carefully allotted or measured quantity of assigned work or service ⟨a two-month *stint* as a reporter⟩. ASSIGNMENT implies a definite limited task assigned by one in authority ⟨a reporter's *assignment*⟩.

[1]tas·sel \'ta-səl, *oftenest of corn* 'tä-, 'tȯ-\ *noun* [Middle English, clasp, tassel, from Middle French, from (assumed) Vulgar Latin *tassellus,* alteration of Latin *taxillus* small die; akin to Latin *talus* anklebone, die] (14th century)
1 : a dangling ornament made by laying parallel a bunch of cords or threads of even length and fastening them at one end
2 : something resembling a tassel; *especially* **:** the terminal male inflorescence of some plants and especially Indian corn

tassel 1

[2]tassel *verb* **-seled** *or* **-selled; -sel·ing** *or* **-sel·ling** \-s(ə-)liŋ\ (14th century)
transitive verb
: to adorn with tassels
intransitive verb
: to put forth tassel inflorescences

[1]taste \'tāst\ *verb* **tast·ed; tast·ing** [Middle English, to touch, test, taste, from Middle French *taster,* from (assumed) Vulgar Latin *taxitare,* frequentative of Latin *taxare* to touch, feel — more at TAX] (14th century)
transitive verb
1 : to become acquainted with by experience ⟨has *tasted* the frustration of defeat⟩
2 : to ascertain the flavor of by taking a little into the mouth
3 : to eat or drink especially in small quantities
4 : to perceive or recognize as if by the sense of taste
5 *archaic* **:** APPRECIATE, ENJOY
intransitive verb
1 : to eat or drink a little
2 : to test the flavor of something by taking a small part into the mouth
3 : to have perception, experience, or enjoyment **:** PARTAKE — often used with *of*
4 : to have a specific flavor ⟨the apple *tastes* sour⟩

[2]taste *noun* (14th century)
1 *obsolete* **:** TEST
2 a *obsolete* **:** the act of tasting **b :** a small amount tasted **c :** a small amount **:** BIT; *especially* **:** a sample of experience ⟨first *taste* of success⟩
3 : the one of the special senses that perceives and distinguishes the sweet, sour, bitter, or salty quality of a dissolved substance and is mediated by taste buds on the tongue
4 : the objective sweet, sour, bitter, or salty quality of a dissolved substance as perceived by the sense of taste
5 a : a sensation obtained from a substance in the mouth that is typically produced by the stimulation of the sense of taste combined with those of touch and smell **:** FLAVOR **b :** the distinctive quality of an experience ⟨that gruesome scene left a bad *taste* in my mouth⟩
6 : individual preference **:** INCLINATION
7 a : critical judgment, discernment, or appreciation **b :** manner or aesthetic quality indicative of such discernment or appreciation

taste bud *noun* (1879)
: an end organ mediating the sensation of taste and lying chiefly in the epithelium of the tongue

taste·ful \'tāst-fəl\ *adjective* (1611)
1 : TASTY 1a
2 : having, exhibiting, or conforming to good taste
— taste·ful·ly \-fə-lē\ *adverb*
— taste·ful·ness *noun*

taste·less \'tāst-ləs\ *adjective* (1603)
1 a : having no taste **:** INSIPID ⟨*tasteless* vegetables⟩ **b :** arousing no interest **:** DULL ⟨the tale comes to a flat and *tasteless* end —R. C. Carpenter⟩
2 : not having or exhibiting good taste
— taste·less·ly *adverb*
— taste·less·ness *noun*

taste·mak·er \-ˌmā-kər\ *noun* (1954)
: one who sets the standards of what is currently popular or fashionable

tast·er \'tā-stər\ *noun* (15th century)
1 : one that tastes: as **a :** one that tests (as tea) for quality by tasting **b :** a person who is able to taste the chemical phenylthiocarbamide
2 : a device for tasting or sampling

tasty \'tā-stē\ *adjective* **tast·i·er; -est** (1617)
1 a : having a marked and appetizing flavor **b :** strikingly attractive or interesting ⟨a *tasty* bit of gossip⟩
2 : TASTEFUL
synonym see PALATABLE
— tast·i·ly \-stə-lē\ *adverb*
— tast·i·ness \-stē-nəs\ *noun*

tat \'tat\ *verb* **tat·ted; tat·ting** [back-formation from *tatting*] (1882)
intransitive verb
: to work at tatting
transitive verb
: to make by tatting

ta·ta·mi \tä-'tä-mē, ta-\ *noun, plural* **-mi** *or* **-mis** [Japanese] (1614)
: straw matting used as a floor covering in a Japanese home

Ta·tar \'tä-tər\ *noun* [Persian *Tātār,* of Turkic origin; akin to Turkish *Tatar* Tatar] (1811)
1 : a member of any of a group of Turkic peoples found mainly in the Tatar Republic of Russia and parts of Siberia and central Asia
2 : any of the Turkic languages spoken by the Tatar peoples

ta·ter \'tā-tər\ *noun* [by shortening & alteration] (1759)
dialect **:** POTATO

[1]tat·ter \'ta-tər\ (14th century)
transitive verb
: to make ragged
intransitive verb
: to become ragged

[2]tatter *noun* [Middle English, of Scandinavian origin; akin to Old Norse *tǫturr* tatter; akin to Old English *tætteca* rag, Old High German *zotta* matted hair, tuft] (15th century)
1 : a part torn and left hanging **:** SHRED
2 *plural* **:** tattered clothing **:** RAGS

[1]tat·ter·de·ma·lion \ˌta-tər-di-'mal-yən, -'mal-, -'ma-lē-ən\ *noun* [origin unknown] (1611)
: a person dressed in ragged clothing **:** RAGAMUFFIN

[2]tatterdemalion *adjective* (1614)
1 a : ragged or disreputable in appearance **b :** being in a decayed state or condition **:** DILAPIDATED
2 : BEGGARLY, DISREPUTABLE

tat·tered \'ta-tərd\ *adjective* (14th century)
1 : wearing ragged clothes ⟨a *tattered* barefoot boy⟩
2 : torn into shreds **:** RAGGED ⟨a *tattered* flag⟩
3 a : broken down **:** DILAPIDATED ⟨decaying houses along *tattered* paved streets —P. B. Martin⟩ **b :** being in a shattered condition **:** DISRUPTED ⟨led their *tattered* party to victory⟩

tat·ter·sall \'ta-tər-ˌsȯl, -səl\ *noun* [*Tattersall's* horse market, London, England] (1891)
1 : a pattern of colored lines forming squares of solid background
2 : a fabric woven or printed in a tattersall pattern

tat·tie \'ta-tē\ *noun* [by shortening & alteration] (1788)
Scottish **:** POTATO

tat·ting \'ta-tiŋ\ *noun* [origin unknown] (1842)
1 : a delicate handmade lace formed usually by looping and knotting with a single cotton thread and a small shuttle
2 : the act or process of making tatting

tatting 1

[1]tat·tle \'ta-tᵊl\ *noun* (circa 1529)
1 : idle talk **:** CHATTER
2 : GOSSIP

[2]tattle *verb* **tat·tled; tat·tling** \'tat-liŋ, 'ta-tᵊl-iŋ\ [Middle Dutch *tatelen;* akin to Middle English *tateren* to tattle] (1547)
intransitive verb
1 : CHATTER, PRATE
2 : to tell secrets **:** BLAB
transitive verb
: to utter or disclose in gossip or chatter

tat·tler \'tat-lər, 'ta-tᵊl-ər\ *noun* (1550)
1 : TATTLETALE
2 : any of various slender long-legged shorebirds (as the willet, yellowlegs, and redshank) with a loud and frequent call

tat·tle·tale \'ta-tᵊl-ˌtāl\ *noun* (1888)
: one that tattles **:** INFORMER

[1]tat·too \ta-'tü\ *noun, plural* **tattoos** [alteration of earlier *taptoo,* from Dutch *taptoe,* from the phrase *tap toe!* taps shut!] (circa 1627)
1 : a rapid rhythmic rapping
2 a : a call sounded shortly before taps as notice to go to quarters **b :** outdoor military exercise given by troops as evening entertainment

[2]tattoo (1780)
transitive verb
: to beat or rap rhythmically on **:** drum on
intransitive verb
: to give a series of rhythmic taps

[3]tattoo *transitive verb* [Tahitian *tatau,* noun, tattoo] (1769)
1 : to mark or color (the skin) with tattoos
2 : to mark the skin with (a tattoo) ⟨*tattooed* a flag on his chest⟩
— tat·too·er *noun*
— tat·too·ist \-'tü-ist\ *noun*

[4]tattoo *noun, plural* **tattoos** (1777)
1 : the act of tattooing **:** the fact of being tattooed
2 : an indelible mark or figure fixed upon the body by insertion of pigment under the skin or by production of scars

tat·ty \'ta-tē\ *adjective* **tat·ti·er; -est** [perhaps akin to Old English *tætteca* rag — more at TATTER] (1513)
: rather worn, frayed, or dilapidated **:** SHABBY
— tat·ti·ness \-nəs\ *noun*

tau \'tau̇, 'tȯ\ *noun* [Middle English *taw,* from Latin *tau,* from Greek, of Semitic origin; akin to Hebrew *tāw* taw] (14th century)
1 : the 19th letter of the Greek alphabet — see ALPHABET table
2 : a short-lived elementary particle of the lepton family that exists in positive and negative charge states and has a mass about 3500 times heavier than an electron — called also *tau particle*

tau cross *noun* (15th century)
: a T-shaped cross sometimes having expanded ends and foot — see CROSS illustration

taught *past and past participle of* TEACH

[1]taunt \'tȯnt, 'tänt\ *noun* (circa 1529)
: a sarcastic challenge or insult

[2]taunt *transitive verb* [perhaps from Middle French *tenter* to try, tempt — more at TEMPT] (1539)
: to reproach or challenge in a mocking or insulting manner **:** jeer at
synonym see RIDICULE
— taunt·er *noun*
— taunt·ing·ly \'tȯn-tiŋ-lē, 'tän-\ *adverb*

taupe \'tōp\ *noun* [French, literally, mole, from Latin *talpa*] (circa 1909)
: a brownish gray

Tau·re·an \'tȯr-ē-ən\ *noun* (1911)
: TAURUS 2b

[1]tau·rine \'tȯ-ˌrīn\ *adjective* [Latin *taurinus,* from *taurus* bull; akin to Greek *tauros* bull, Middle Irish *tarb*] (1613)
: of or relating to a bull **:** BOVINE

\ə\ abut \ᵊ\ kitten \ər\ further \a\ ash \ā\ ace \ä\ mop, mar \au̇\ out \ch\ chin \e\ bet \ē\ easy \g\ go \i\ hit \ī\ ice \j\ job \ŋ\ sing \ō\ go \ȯ\ law \ȯi\ boy \th\ thin \th\ the \ü\ loot \u̇\ foot \y\ yet \zh\ vision *see also* Guide to Pronunciation

[2]tau·rine \'tȯ-ˌrēn\ *noun* [International Scientific Vocabulary, from Latin *taurus;* from its having been discovered in ox bile] (1845)
: a colorless crystalline cysteine derivative $C_2H_7NO_3S$ of neutral reaction found in the juices of muscle especially in invertebrates, nerve tissue, and bile

tau·ro·cho·lic acid \ˌtȯr-ə-'kō-lik-, -'kä-\ *noun* [Latin *taurus* + International Scientific Vocabulary *-o-* + *cholic (acid)*] (1857)
: a deliquescent acid $C_{26}H_{45}NO_7S$ occurring as the sodium salt in the bile of the ox, humans, and various carnivores

Tau·rus \'tȯr-əs\ *noun* [Middle English, from Latin (genitive *Tauri*), literally, bull]
1 : a zodiacal constellation that contains the Pleiades and Hyades and is represented pictorially by a bull's forequarters
2 a : the 2d sign of the zodiac in astrology **b :** one born under the sign of Taurus — see ZODIAC table

[1]taut \'tȯt\ *adjective* [Middle English *toght*] (14th century)
1 a : having no give or slack **:** tightly drawn **b :** HIGH-STRUNG, TENSE ⟨*taut* nerves⟩
2 a : kept in proper order or condition ⟨a *taut* ship⟩ **b** (1) **:** not loose or flabby (2) **:** marked by economy of structure and detail ⟨a *taut* story⟩
— **taut·ly** *adverb*
— **taut·ness** *noun*

[2]taut *transitive verb* [origin unknown] (1721)
Scottish **:** MAT, TANGLE

taut- *or* **tauto-** *combining form* [Late Latin, from Greek, from *tauto* the same, contraction of *to auto*]
: same ⟨*tauto*merism⟩ ⟨*tauto*nym⟩

taut·en \'tȯ-tᵊn\ *verb* **taut·ened; taut·en·ing** \'tȯt-niŋ, 'tȯ-tᵊn-iŋ\ (circa 1814)
transitive verb
: to make taut
intransitive verb
: to become taut

tau·tog \'tȯ-ˌtȯg, -ˌtäg, tȯ-'\ *noun* [Narraganset *tautaŭog,* plural] (1643)
: an edible fish (*Tautoga onitis*) of the wrasse family found along the Atlantic coast of the U.S. and adjacent Canada — called also *blackfish*

tau·to·log·i·cal \ˌtȯ-tᵊl-'ä-ji-kəl\ *adjective* (1620)
: TAUTOLOGOUS
— **tau·to·log·i·cal·ly** \-k(ə-)lē\ *adverb*

tau·tol·o·gous \tȯ-'tä-lə-gəs\ *adjective* [Greek *tautologos,* from *taut-* + *legein* to say — more at LEGEND] (1714)
1 : involving or containing rhetorical tautology **:** REDUNDANT
2 : true by virtue of its logical form alone
— **tau·tol·o·gous·ly** *adverb*

tau·tol·o·gy \tȯ-'tä-lə-jē\ *noun, plural* **-gies** [Late Latin *tautologia,* from Greek, from *tautologos*] (1574)
1 a : needless repetition of an idea, statement, or word **b :** an instance of tautology
2 : a tautologous statement

tau·to·mer \'tȯ-tə-mər\ *noun* [International Scientific Vocabulary, from *tautomeric*] (1903)
: any of the forms of a tautomeric compound

tau·to·mer·ic \ˌtȯ-tə-'mer-ik\ *adjective* [International Scientific Vocabulary] (circa 1890)
: of, relating to, or marked by tautomerism

tau·tom·er·ism \tȯ-'tä-mə-ˌri-zəm\ *noun* (circa 1890)
: isomerism in which the isomers change into one another with great ease so that they ordinarily exist together in equilibrium

taut·o·nym \'tȯ-tə-ˌnim\ *noun* (1899)
: a taxonomic binomial in which the generic name and specific epithet are alike and which is common in zoology especially to designate a typical form but is forbidden to botany under the International Code of Botanical Nomenclature
— **tau·ton·y·my** \tȯ-'tä-nə-mē\ *noun*

tav·ern \'ta-vərn\ *noun* [Middle English *taverne,* from Old French, from Latin *taberna* hut, shop] (14th century)
1 : an establishment where alcoholic beverages are sold to be drunk on the premises
2 : INN

ta·ver·na \tä-'ver-nə\ *noun* [New Greek *taberna,* probably from Late Greek, drinking establishment, from Latin *taberna*] (1914)
: a café in Greece

tav·ern·er \'tav-ə(r)-nər\ *noun* (14th century)
: one who keeps a tavern

[1]taw \'tȯ\ *transitive verb* [Middle English, to prepare for use, from Old English *tawian;* akin to Old High German *zawjan* to hasten, Gothic *taujan* to do, make] (before 12th century)
: to dress (skins) usually by a dry process (as with alum or salt)

[2]taw \'täf, 'tȯf, 'täv, 'tȯv\ *noun* [Hebrew *tāw,* literally, mark, cross] (1701)
: the 23d letter of the Hebrew alphabet — see ALPHABET table

[3]taw *noun* [origin unknown] (1709)
1 a : a marble used as a shooter **b :** RINGTAW
2 : the line from which players shoot at marbles
3 : a square-dance partner

[4]taw *intransitive verb* (1863)
: to shoot a marble

[1]taw·dry \'tȯ-drē, 'tä-\ *adjective* **taw·dri·er; -est** [*tawdry lace* a tie of lace for the neck, from *Saint Audrey* (Saint Etheldreda) (died 679) queen of Northumbria] (1676)
: cheap and gaudy in appearance or quality; *also* **:** IGNOBLE
synonym see GAUDY
— **taw·dri·ly** \-drə-lē\ *adverb*
— **taw·dri·ness** \-drē-nəs\ *noun*

[2]tawdry *noun* (circa 1680)
: cheap showy finery

[1]taw·ny \'tȯ-nē, 'tä-nē\ *adjective* **taw·ni·er; -est** [Middle English, from Middle French *tanné,* past participle of *tanner* to tan] (14th century)
1 : of the color tawny
2 : of a warm sandy color like that of well-tanned skin ⟨the lion's *tawny* coat⟩
— **taw·ni·ness** *noun*

[2]tawny *noun, plural* **tawnies** (15th century)
: a brownish orange to light brown color

taw·pie \'tȯ-pē\ *noun* [of Scandinavian origin; akin to Norwegian *tåpe* simpleton] (1728)
chiefly Scottish **:** a foolish or awkward young person

tawse *also* **taws** \'tȯz\ *noun plural but singular or plural in construction* [probably from plural of obsolete *taw* tawed leather] (circa 1585)
British **:** a leather strap slit into strips at the end and used especially for disciplining children

[1]tax \'taks\ *transitive verb* [Middle English, to estimate, assess, tax, from Old French *taxer,* from Medieval Latin *taxare,* from Latin, to feel, estimate, censure, frequentative of *tangere* to touch — more at TANGENT] (14th century)
1 : to assess or determine judicially the amount of (costs in a court action)
2 : to levy a tax on
3 *obsolete* **:** to enter (a name) in a list ⟨there went out a decree . . . that all the world should be *taxed* —Luke 2:1 (Authorized Version)⟩
4 : CHARGE, ACCUSE ⟨*taxed* him with neglect of duty⟩; *also* **:** CENSURE
5 : to make onerous and rigorous demands on ⟨the job *taxed* her strength⟩
— **tax·able** \'tak-sə-bəl\ *adjective*
— **tax·er** *noun*

[2]tax *noun, often attributive* (14th century)
1 a : a charge usually of money imposed by authority on persons or property for public purposes **b :** a sum levied on members of an organization to defray expenses
2 : a heavy demand

tax- *or* **taxo-** *also* **taxi-** *combining form* [Greek *taxi-,* from *taxis*]
: arrangement ⟨*tax*eme⟩ ⟨*taxi*dermy⟩

taxa *plural of* TAXON

tax·a·tion \tak-'sā-shən\ *noun* (14th century)
1 : the action of taxing; *especially* **:** the imposition of taxes
2 : revenue obtained from taxes
3 : the amount assessed as a tax

tax base *noun* (circa 1943)
: the wealth (as real estate or income) within a jurisdiction that is liable to taxation

tax·eme \'tak-ˌsēm\ *noun* [*tax-*] (1933)
: a minimum grammatical feature of selection, order, stress, pitch, or phonetic modification
— **tax·e·mic** \tak-'sē-mik\ *adjective*

tax–ex·empt \ˌtaks-ig-'zem(p)t\ *adjective* (1923)
1 : exempted from a tax
2 : bearing interest that is free from federal or state income tax

[1]taxi \'tak-sē\ *noun, plural* **tax·is** \-sēz\ *also* **tax·ies** (circa 1907)
: TAXICAB; *also* **:** a similarly operated boat or aircraft ◇

[2]taxi *verb* **tax·ied; taxi·ing** *or* **taxy·ing; tax·is** *or* **tax·ies** (1916)
intransitive verb
1 a *of an aircraft* **:** to go at low speed along the surface of the ground or water **b :** to operate an aircraft on the ground under its own power
2 : to ride in a taxicab
transitive verb
1 : to transport by or as if by taxi
2 : to cause (an aircraft) to taxi

taxi·cab \'tak-sē-ˌkab\ *noun* [*taxi*meter *cab*] (circa 1907)
: an automobile that carries passengers for a fare usually determined by the distance traveled

taxi dancer *noun* (circa 1927)
: a girl employed by a dance hall, café, or cabaret to dance with patrons who pay a certain amount for each dance

taxi·der·my \'tak-sə-ˌdər-mē\ *noun* [*tax-* + *derm-* + [2]*-y*] (1820)

◇ WORD HISTORY

taxi *Taxi* in the sense "an automobile that carries passengers for a fare usually determined by distance" is shortened from *taxicab,* itself shortened from *taximeter cab.* The taximeter is the mechanical device in a taxi that is activated when the ride starts and automatically shows the fare due at the end of the ride. Meters of this kind were first used in 19th century Germany under the name *Taxameter,* attached to horsedrawn vehicles for hire. *Taxameter* was made up irregularly in terms of the rules of Greco-Latin word formation from *Taxe* "fee, tax," Latinized to *taxa* (which happens to be the original Medieval Latin form from which *Taxe* was ultimately borrowed), and *-meter,* which as in English is used in names of measuring instruments. (The regular outcome would have been *Taxometer.*) The word soon showed up in French as *taxamètre* and in English as *taxameter,* but both languages then altered the internal *-a-* to *-i-,* yielding French *taximètre* and English *taximeter.* Though these forms violate no rules of the classical languages, they assume a nonexistent connection with Greek *taxis* "arrangement." Etymological confusion notwithstanding, *taxi* has become an international word, finding its way into languages as diverse as Arabic (*taksī*) and Japanese (*takushī*).

: the art of preparing, stuffing, and mounting the skins of animals and especially vertebrates
— **taxi·der·mic** \ˌtak-sə-ˈdər-mik\ *adjective*
— **taxi·der·mist** \ˈtak-sə-ˌdər-mist\ *noun*

taxi·man \ˈtak-sē-mən\ *noun* (1909)
chiefly British **:** the operator of a taxi

taxi·me·ter \ˈtak-sē-ˌmē-tər\ *noun* [French *taximètre,* modification of German *Taxameter,* from Medieval Latin *taxa* tax, charge (from *taxare* to tax) + German *-meter*] (1894)
: an instrument for use in a hired vehicle (as a taxicab) for automatically showing the fare due
word history see TAXI

tax·ing \ˈtak-siŋ\ *adjective* (1841)
: ONEROUS, WEARING ⟨a *taxing* operatic role⟩
— **tax·ing·ly** \-siŋ-lē\ *adverb*

tax·is \ˈtak-səs\ *noun, plural* **tax·es** \-ˌsēz\ [Greek, literally, arrangement, order, from *tassein* to arrange] (1758)
1 : reflex translational or orientational movement by a freely motile and usually simple organism in relation to a source of stimulation (as a light or a temperature or chemical gradient)
2 : a reflex reaction involving a taxis

-taxis *noun combining form, plural* **-taxes** [New Latin, from Greek, from *taxis*]
1 : arrangement **:** ordering ⟨thermo*taxis*⟩
2 : physiological taxis ⟨chemo*taxis*⟩

taxi stand *noun* (1922)
: a place where taxis may park while awaiting hire

taxi·way \ˈtak-sē-ˌwā\ *noun* (circa 1933)
: a usually paved strip for taxiing (as from the terminal to a runway) at an airport

tax·on \ˈtak-ˌsän\ *noun, plural* **taxa** \-sə\ *also* **tax·ons** [New Latin, from International Scientific Vocabulary *taxonomy*] (1929)
1 : a taxonomic group or entity
2 : the name applied to a taxonomic group in a formal system of nomenclature

tax·on·o·my \tak-ˈsa-nə-me\ *noun* [French *taxonomie,* from *tax-* + *-nomie* -nomy] (circa 1828)
1 : the study of the general principles of scientific classification **:** SYSTEMATICS
2 : CLASSIFICATION; *especially* **:** orderly classification of plants and animals according to their presumed natural relationships
— **tax·o·nom·ic** \ˌtak-sə-ˈnä-mik\ *adjective*
— **tax·o·nom·i·cal·ly** \-mi-k(ə-)lē\ *adverb*
— **tax·on·o·mist** \tak-ˈsä-nə-mist\ *noun*

tax·pay·er \ˈtaks-ˌpā-ər\ *noun* (1816)
: one that pays or is liable for a tax

tax·pay·ing \-ˌpā-iŋ\ *adjective* (1832)
: of, relating to, or subject to the paying of a tax

tax selling *noun* (1963)
: concerted selling of securities late in the year to establish gains and losses for income-tax purposes

tax shelter *noun* (1952)
: a strategy, investment, or tax code provision that reduces tax liability
— **tax–shel·tered** \ˈtaks-ˌshel-tərd\ *adjective*

tax stamp *noun* (circa 1929)
: a stamp marked on or affixed to a taxable item as evidence that the tax has been paid

tax·us \ˈtak-səs\ *noun, plural* **tax·us** \-səs\ [New Latin, genus comprising the yews, from Latin, yew] (circa 1945)
: YEW 1a

-taxy *noun combining form* [Greek *-taxia,* from *taktos,* verbal of *tassein* to arrange]
: -TAXIS ⟨epi*taxy*⟩

Tay·lor's series \ˈtā-lərz-\ *noun* [Brook *Taylor* (died 1731) English mathematician] (1842)
: a power series that gives the expansion of a function $f(x)$ in the neighborhood of a point a provided that in the neighborhood the function is continuous, all its derivatives exist, and the series converges to the function in which case it has the form

$$f(x) = f(a) + \frac{f^{[1]}(a)}{1!}(x-a) + \frac{f^{[2]}(a)}{2!}(x-a)^2 + \ldots + \frac{f^{[n]}(a)}{n!}(x-a)^n + \ldots$$

where $f^{[n]}(a)$ is the derivative of nth order of $f(x)$ evaluated at a — called also *Taylor series*

Tay–Sachs disease \ˈtā-ˈsaks-\ *noun* [Warren *Tay* (died 1927) British physician & Bernard P. *Sachs* (died 1944) American neurologist] (1907)
: a hereditary disorder of lipid metabolism typically affecting individuals of eastern European Jewish ancestry that is characterized by the accumulation of lipids especially in nervous tissue due to a deficiency of hexosaminidase and causes death in early childhood — called also *Tay-Sachs*

taz·za \ˈtät-sə\ *noun* [Italian, cup, from Arabic *ṭassah*] (1824)
: a shallow cup or vase on a pedestal

TB \ˌtē-ˈbē\ *noun* [*TB* (abbreviation for *tubercle bacillus*)] (1912)
: TUBERCULOSIS

T–ball \ˈtē-ˌbȯl\ *noun* [[2]tee] (1976)
: baseball modified for youngsters in which the ball is batted from a tee rather than being pitched

T–bar \ˈtē-ˌbär\ *noun* (1948)
: a ski lift having a series of T-shaped bars each of which pulls two skiers — called also *T-bar lift*

T–bill \ˈtē-ˌbil\ *noun* [*T*reasury] (1973)
: a U.S. treasury note

T–bone \ˈtē-ˌbōn\ *noun* (circa 1916)
: a small steak from the thin end of the short loin containing a T-shaped bone and a small piece of tenderloin; *also* **:** this bone — see BEEF illustration

TCDD \ˌtē-(ˌ)sē-(ˌ)dē-ˈdē\ *noun* [*t*etra- + *c*hlor- + *d*ibenzo- (containing two benzene rings) + *d*ioxin] (1971)
: a carcinogenic dioxin $C_{12}H_4O_2Cl_4$ found especially as a contaminant in 2,4,5-T

T cell *noun* [*t*hymus-derived *cell*] (1970)
: any of several lymphocytes (as a helper T cell) that differentiate in the thymus, possess highly specific cell-surface antigen receptors, and include some that control the initiation or suppression of cell-mediated and humoral immunity (as by the regulation of T and B cell maturation and proliferation) and others that lyse antigen-bearing cells — called also *T lymphocyte;* compare B CELL

tchotch·ke \ˈchäch-kə, ˈtsäts-\ *noun* [Yiddish *tshatshke* trinket, from obsolete Polish *czaczko*] (1971)
: KNICKKNACK, TRINKET

t distribution *noun* (circa 1957)
: a probability density function that is used especially in testing hypotheses concerning means of normal distributions whose standard deviations are unknown and that is the distribution of a random variable

$$t = \frac{u\sqrt{n}}{v}$$

where u and v are themselves independent random variables and u has a normal distribution with mean 0 and a standard deviation of 1 and v^2 has a chi-square distribution with n degrees of freedom — called also *student's t distribution*

tea \ˈtē\ *noun* [Chinese (Xiamen) *t'e*] (circa 1655)
1 a : a shrub (*Camellia sinensis* of the family Theaceae, the tea family) cultivated especially in China, Japan, and the East Indies **b :** the leaves, leaf buds, and internodes of the tea plant prepared and cured for the market, classed according to method of manufacture into one set of types (as green tea, black tea, or oolong), and graded according to leaf size into another (as orange pekoe, pekoe, or souchong)
2 : an aromatic beverage prepared from tea leaves by infusion with boiling water
3 : any of various plants somewhat resembling tea in properties; *also* **:** an infusion of their leaves used medicinally or as a beverage
4 a : refreshments usually including tea with sandwiches, crackers, or cookies served in late afternoon **b :** a reception at which tea is served
5 *slang* **:** MARIJUANA
— **tea·like** \-ˌlīk\ *adjective*

tea bag *noun* (1935)
: a bag usually of filter paper holding enough tea for an individual serving

tea ball *noun* (1895)
: a perforated metal ball that holds tea leaves and is used in brewing tea in a pot or cup

tea·ber·ry \ˈtē-ˌber-ē\ *noun* [from the use of its leaves as a substitute for tea] (1818)
: CHECKERBERRY

tea caddy *noun* (1790)
: CADDY

tea cake *noun* (1824)
1 : a small flat cake usually made with raisins
2 : COOKIE

tea cart *noun* (1817)
: TEA WAGON

teach \ˈtēch\ *verb* **taught** \ˈtȯt\; **teach·ing** [Middle English *techen* to show, instruct, from Old English *tǣcan;* akin to Old English *tācn* sign — more at TOKEN] (before 12th century)
transitive verb
1 a : to cause to know something ⟨*taught* them a trade⟩ **b :** to cause to know how ⟨is *teaching* me to drive⟩ **c :** to accustom to some action or attitude ⟨*teach* students to think for themselves⟩ **d :** to cause to know the disagreeable consequences of some action ⟨I'll *teach* you to come home late⟩
2 : to guide the studies of
3 : to impart the knowledge of ⟨*teach* algebra⟩
4 a : to instruct by precept, example, or experience **b :** to make known and accepted ⟨experience *teaches* us our limitations⟩
5 : to conduct instruction regularly in ⟨*teach* school⟩
intransitive verb
: to provide instruction **:** act as a teacher ☆
usage see LEARN

teach·able \ˈtē-chə-bəl\ *adjective* (15th century)
1 a : capable of being taught **b :** apt and willing to learn
2 : favorable to teaching
— **teach·able·ness** *noun*
— **teach·ably** \-blē\ *adverb*

☆ SYNONYMS
Teach, instruct, educate, train, discipline, school mean to cause to acquire knowledge or skill. TEACH applies to any manner of imparting information or skill so that others may learn ⟨*taught* us a lot about our planet⟩. INSTRUCT suggests methodical or formal teaching ⟨*instructs* raw recruits in military drill⟩. EDUCATE implies development of the mind ⟨more things than formal schooling serve to *educate* a person⟩. TRAIN stresses instruction and drill with a specific end in view ⟨*trained* foreign pilots to operate the new aircraft⟩. DISCIPLINE implies training in habits of order and precision ⟨a *disciplined* mind⟩. SCHOOL implies training or disciplining especially in what is hard to master ⟨*schooled* the horse in five gaits⟩.

teach•er \'tē-chər\ *noun* (14th century)
1 : one that teaches; *especially* **:** one whose occupation is to instruct
2 : a Mormon ranking above a deacon in the Aaronic priesthood

teach•er•ly \-lē\ *adjective* (circa 1683)
: resembling, characteristic of, or befitting a teacher

teachers college *noun* (circa 1911)
: a college for the training of teachers usually offering a full 4-year course and granting a bachelor's degree

teacher's pet *noun* (1914)
1 : a pupil who has won his teacher's special favor
2 : one who has ingratiated himself with an authority

teach–in \'tēch-ˌin\ *noun* (1965)
: an extended meeting usually held on a college campus for lectures, debates, and discussions to raise awareness of or express a position on a social or political issue

¹teaching *noun* (13th century)
1 : the act, practice, or profession of a teacher
2 : something taught; *especially* **:** DOCTRINE ⟨the *teachings* of Confucius⟩

²teaching *adjective* (1642)
: of, relating to, used for, or engaged in teaching ⟨a *teaching* aid⟩ ⟨the *teaching* profession⟩ ⟨a *teaching* assistant⟩

teaching hospital *noun* (circa 1951)
: a hospital that is affiliated with a medical school and provides means for medical education

tea•cup \'tē-ˌkəp\ *noun* (1700)
: a small cup usually with a handle used with a saucer for hot beverages
— teacupful *noun*

tea dance *noun* (1885)
: a dance held in the late afternoon

tea garden *noun* (1802)
1 : a public garden where tea and light refreshments are served
2 : a tea plantation

tea gown *noun* (1878)
: a semiformal gown of fine materials in graceful flowing lines worn especially for afternoon entertaining at home

tea•house \'tē-ˌhau̇s\ *noun* (1689)
: a public house or restaurant where tea and light refreshments are sold

teak \'tēk\ *noun* [Portuguese *teca,* from Malayalam *tēkka*] (1698)
1 : a tall East Indian timber tree (*Tectona grandis*) of the vervain family
2 : the hard yellowish brown wood of teak used especially for furniture and shipbuilding

tea•ket•tle \'tē-ˌke-tᵊl\ *noun* (1705)
: a covered kettle with a handle and spout for boiling water

teak•wood \'tēk-ˌwu̇d\ *noun* (1783)
: TEAK 2

teal \'tē(ə)l\ *noun, plural* **teal** *or* **teals** [Middle English *tele;* akin to Middle Dutch *teling* teal] (14th century)
1 : any of various widely distributed small short-necked dabblers (genus *Anas*) — compare GREEN-WINGED TEAL
2 : TEAL BLUE

teal 1

teal blue *noun* (1938)
: a dark greenish blue

¹team \'tēm\ *noun* [Middle English *teme,* from Old English *tēam* offspring, lineage, group of draft animals; akin to Old High German *zoum* rein, Old English *tēon* to draw, pull — more at TOW] (before 12th century)
1 a : two or more draft animals harnessed to the same vehicle or implement; *also* **:** these with their harness and attached vehicle **b :** a draft animal often with harness and vehicle
2 *obsolete* **:** LINEAGE, RACE
3 : a group of animals: as **a :** a brood especially of young pigs or ducks **b :** a matched group of animals for exhibition
4 : a number of persons associated together in work or activity: as **a :** a group on one side (as in football or a debate) **b :** CREW, GANG

²team (1552)
transitive verb
1 : to yoke or join in a team; *also* **:** to put together in a coordinated ensemble
2 : to convey or haul with a team
intransitive verb
1 : to drive a team or motortruck
2 : to form a team or association **:** COLLABORATE ⟨*teamed* up to write a book⟩

³team *adjective* (1886)
: of or performed by a team ⟨a *team* effort⟩; *also* **:** marked by devotion to teamwork rather than individual achievement ⟨a *team* player⟩

team foul *noun* (1966)
: one of a designated number of personal fouls the players on a basketball team may commit during a given period of play before the opposing team begins receiving bonus free throws

team handball *noun* (1970)
: a game developed from soccer which is played between two teams of seven players each and in which the ball is thrown, caught, and dribbled with the hands

team•mate \'tēm-ˌmāt\ *noun* (1915)
: a fellow member of a team

team•ster \'tēm(p)-stər\ *noun* (1777)
: one who drives a team or motortruck especially as an occupation

team•work \'tēm-ˌwərk\ *noun* (circa 1828)
: work done by several associates with each doing a part but all subordinating personal prominence to the efficiency of the whole

tea party *noun* (1778)
1 : an afternoon social gathering at which tea is served
2 [from the Boston Tea Party, name applied to the occasion in 1773 when a shipment of tea was thrown into Boston harbor in protest against the tax on imports] **:** an exciting disturbance or proceeding

tea•pot \'tē-ˌpät\ *noun* (1705)
: a vessel with a spout and a handle in which tea is brewed and from which it is served

tea•poy \'tē-ˌpȯi\ *noun* [Hindi *tipāī*] (1828)
1 : a 3-legged ornamental stand
2 [influenced by *tea*] **:** a stand or table containing a tea chest or caddy and used for supporting a tea set; *also* **:** TEA CADDY

¹tear \'tir\ *noun* [Middle English, from Old English *tæhher, tēar;* akin to Old High German *zahar* tear, Greek *dakry*] (before 12th century)
1 a : a drop of clear saline fluid secreted by the lacrimal gland and diffused between the eye and eyelids to moisten the parts and facilitate their motion **b** *plural* **:** a secretion of profuse tears that overflow the eyelids and dampen the face
2 : a transparent drop of fluid or hardened fluid matter (as resin)
3 *plural* **:** an act of weeping or grieving ⟨broke into *tears*⟩
— tear•less *adjective*

²tear *intransitive verb* (before 12th century)
: to fill with tears **:** shed tears ⟨eyes *tearing* in the November wind —Saul Bellow⟩

³tear \'tar, 'ter\ *verb* **tore** \'tōr, 'tȯr\; **torn** \'tōrn, 'tȯrn\; **tear•ing** [Middle English *teren,* from Old English *teran;* akin to Old High German *zeran* to destroy, Greek *derein* to skin, Sanskrit *dṛṇāti* he bursts, tears] (before 12th century)
transitive verb
1 a : to separate parts of or pull apart by force **:** REND **b :** to wound by or as if by tearing **:** LACERATE ⟨*tear* the skin⟩
2 : to divide or disrupt by the pull of contrary forces ⟨a mind *torn* with doubts⟩
3 a : to remove by force **:** WRENCH — often used with *off* ⟨*tear* a cover off a box⟩ **b :** to remove as if by wrenching ⟨*tear* your thoughts away from the scene⟩
4 : to make or effect by or as if by tearing ⟨*tear* a hole in the wall⟩
intransitive verb
1 : to separate on being pulled **:** REND ⟨this cloth *tears* easily⟩
2 a : to move or act with violence, haste, or force ⟨went *tearing* down the street⟩ **b :** to smash or penetrate something with violent force ⟨the bullet *tore* through his leg⟩ ☆
— tear•able \'tar-ə-bəl, 'ter-\ *adjective*
— tear•er *noun*
— tear at : to cause anguish to **:** DISTRESS ⟨her grief *tore at* his heart⟩
— tear into : to attack without restraint or caution
— tear it : to cause frustration, defeat, or an end to plans or hopes ⟨that *tears it*⟩
— tear one's hair : to pull one's hair as an expression of grief, rage, frustration, desperation, or anxiety

⁴tear \'tar, 'ter\ *noun* (1611)
1 a : damage from being torn; *especially* **:** a hole or flaw made by tearing **b :** the act of tearing
2 a : a tearing pace **:** HURRY **b :** SPREE ⟨go on a *tear*⟩ **c :** a run of unusual success ⟨the team was on a *tear*⟩

tear•away \'tar-ə-ˌwā, 'ter-\ *noun* (1950)
British **:** a rebellious and unruly or reckless young person

tear away *transitive verb* (circa 1699)
: to remove (as oneself) reluctantly

tear•down \'tar-ˌdau̇n, 'ter-\ *noun* (1926)
: the act or process of disassembling

tear down *transitive verb* (1614)
1 a : to cause to decompose or disintegrate **b :** VILIFY, DENIGRATE
2 : to take apart **:** DISASSEMBLE

tear•drop \'tir-ˌdräp\ *noun* (1789)
1 : ¹TEAR 1a
2 : something shaped like a dropping tear; *specifically* **:** a pendent gem (as on an earring)

tear•ful \'tir-fəl\ *adjective* (circa 1586)
1 : flowing with or accompanied by tears ⟨*tearful* entreaties⟩
2 : causing tears **:** TEARY
— tear•ful•ly \-fə-lē\ *adverb*
— tear•ful•ness *noun*

tear•gas \-ˌgas\ *transitive verb* (1946)
: to use tear gas on

tear gas *noun* (1917)
: a solid, liquid, or gaseous substance that on dispersion in the atmosphere blinds the eyes with tears and is used chiefly in dispelling mobs

tear•ing \'tar-iŋ, 'ter-\ *adjective* (1606)
1 : causing continued or repeated pain or distress
2 : HASTY, VIOLENT
3 *chiefly British* **:** SPLENDID

tear•jerk•er \'tir-ˌjər-kər\ *noun* (1912)

☆ SYNONYMS
Tear, rip, rend, split, cleave, rive mean to separate forcibly. TEAR implies pulling apart by force and leaving jagged edges ⟨*tear* up the letter⟩. RIP implies a pulling apart in one rapid uninterrupted motion often along a line or joint ⟨*ripped* the shirt on a nail⟩. REND implies very violent or ruthless severing or sundering ⟨an angry mob *rent* the prisoner's clothes⟩. SPLIT implies a cutting or breaking apart in a continuous, straight, and usually lengthwise direction or in the direction of grain or layers ⟨*split* logs for firewood⟩. CLEAVE implies very forceful splitting or cutting with a blow ⟨a bolt of lightning *cleaved* the giant oak⟩. RIVE occurs most often in figurative use ⟨a political party *riven* by conflict⟩.

: an extravagantly pathetic story, song, play, film, or broadcast
— **tear–jerk·ing** \-kiŋ\ *adjective*

tea·room \ˈtē-ˌrüm, -ˌrům\ *noun* (1778)
: a small restaurant or café with service and decor designed primarily for a female clientele

tea rose *noun* (1850)
: a garden bush rose (*Rosa odorata*) of Chinese origin that includes several cultivars and is valued especially for its abundant large usually tea-scented blossoms — compare HYBRID TEA ROSE

tear sheet *noun* (circa 1924)
: a sheet torn from a publication

tear·stain \ˈtir-ˌstān\ *noun* (1922)
: a spot or streak left by tears
— **tear·stained** \-ˌstānd\ *adjective*

tear up *transitive verb* (1699)
: to damage, remove, or effect an opening in ⟨*tore up* the street to lay a new water main⟩

teary \ˈtir-ē\ *adjective* **tear·i·er; -est** (14th century)
1 a : wet or stained with tears **:** TEARFUL **b :** consisting of tears or drops resembling tears
2 : causing tears **:** PATHETIC ⟨a *teary* story⟩

¹tease \ˈtēz\ *transitive verb* **teased; teas·ing** [Middle English *tesen,* from Old English *tǣsan;* akin to Old High German *zeisan* to tease] (before 12th century)
1 a : to disentangle and lay parallel by combing or carding ⟨*tease* wool⟩ **b :** TEASEL
2 : to tear in pieces; *especially* **:** to shred (a tissue or specimen) for microscopic examination
3 a : to disturb or annoy by persistent irritating or provoking especially in a petty or mischievous way **b :** to annoy with petty persistent requests **:** PESTER; *also* **:** to obtain by repeated coaxing **c :** to persuade to acquiesce especially by persistent small efforts **:** COAX **d :** to manipulate or influence as if by teasing
4 : to comb (hair) by taking hold of a strand and pushing the short hairs toward the scalp with the comb
5 : to tantalize especially by arousing desire or curiosity without intending to satisfy it
synonym see WORRY
— **teas·ing·ly** \ˈtē-ziŋ-lē\ *adverb*

²tease *noun* (1693)
1 : the act of teasing **:** the state of being teased
2 : one that teases

¹tea·sel \ˈtē-zəl\ *noun* [Middle English *tesel,* from Old English *tǣsel;* akin to Old English *tǣsan* to tease] (before 12th century)
1 a : an Old World prickly herb (*Dipsacus fullonum* of the family Dipsacaceae, the teasel family) with flower heads that are covered with stiff hooked bracts and are used in the woolen industry — called also *fuller's teasel* **b :** a plant of the same genus as the teasel
2 a : a flower head of the fuller's teasel used when dried to raise a nap on woolen cloth **b :** a wire substitute for the teasel

teasel 1a

²teasel *transitive verb* **tea·seled** *or* **tea·selled; tea·sel·ing** *or* **tea·sel·ling** \ˈtēz-liŋ, ˈtē-zə-\ (1543)
: to nap (cloth) with teasels

tease out *transitive verb* (1828)
: to obtain by or as if by disentangling or freeing with a pointed instrument

teas·er \ˈtē-zər\ *noun* (1659)
1 : one that teases
2 : an advertising or promotional device intended to arouse interest or curiosity especially in something to follow

tea service *noun* (1809)
: TEA SET

tea set *noun* (1849)
: a matching set of metalware or china (as a teapot, sugar bowl, creamer, and often plates, cups, and saucers) for serving tea and sometimes coffee at table

tea shop *noun* (1856)
chiefly British **:** a small restaurant or café **:** TEAROOM

tea·spoon \ˈtē-ˌspün, -ˈspün\ *noun* (1686)
1 : a small spoon that is used especially for eating soft foods and stirring beverages and that holds one third of a tablespoon
2 : a unit of measure especially in cookery equal to ⅙ fluid ounces or ⅓ tablespoon (5 milliliters)

tea·spoon·ful \-ˌfůl\ *noun, plural* **tea·spoonfuls** \-ˌfůlz\ *also* **tea·spoons·ful** \-ˌspünz-ˌfůl, -ˈspünz-\ (1731)
1 : as much as a teaspoon can hold
2 : TEASPOON 2
usage see ²-FUL

teat \ˈtit, ˈtēt\ *noun* [Middle English *tete,* from Old French, of Germanic origin; akin to Old English *tit* teat, Middle High German *zitze*] (13th century)
1 : the protuberance through which milk is drawn from an udder or breast **:** NIPPLE
2 : a small projection or a nib (as on a mechanical part)
— **teat·ed** \ˈti-təd, ˈtē-\ *adjective*

tea table *noun* (1688)
: a table used or spread for tea; *specifically* **:** a small table for serving afternoon tea

tea·time \ˈtē-ˌtīm\ *noun* (1741)
: the customary time for tea **:** late afternoon or early evening

tea towel *noun* (1871)
: a cloth for drying dishes

tea tray *noun* (1773)
: a tray that accommodates a tea set

tea wagon *noun* (1921)
: a small table on wheels used in serving tea

Te·bet \tā-ˈvāt, -ˈvāth, ˈtā-ˌves\ *noun* [Hebrew *Ṭēbhēth*] (14th century)
: the 4th month of the civil year or the 10th month of the ecclesiastical year in the Jewish calendar — see MONTH table

teched \ˈtecht\ *adjective* [alteration of *touched*] (1921)
: mentally unbalanced

tech·ie \ˈte-kē\ *noun* [by shortening & alteration] (1982)
: TECHNICIAN 1

tech·ne·tium \tek-ˈnē-sh(ē-)əm\ *noun* [New Latin, from Greek *technētos* artificial, from *technasthai* to devise by art, from *technē*] (circa 1946)
: a metallic element obtained by bombarding molybdenum with deuterons or neutrons and in the fission of uranium — see ELEMENT table

tech·ne·tron·ic \ˌtek-nə-ˈträ-nik\ *adjective* [*techn*ological + *electronic*] (1967)
: shaped or influenced by the changes wrought by advances in technology and communications ⟨our modern *technetronic* society⟩

tech·nic \ˈtek-nik, *for 1 also* tek-ˈnēk\ *noun* (1855)
1 : TECHNIQUE 1
2 *plural but singular or plural in construction* **:** TECHNOLOGY 1a

tech·ni·cal \ˈtek-ni-kəl\ *adjective* [Greek *technikos* of art, skillful, from *technē* art, craft, skill; akin to Greek *tektōn* builder, carpenter, Latin *texere* to weave, Sanskrit *takṣati* he fashions] (1617)
1 a : having special and usually practical knowledge especially of a mechanical or scientific subject ⟨a *technical* consultant⟩ **b :** marked by or characteristic of specialization ⟨*technical* language⟩
2 a : of or relating to a particular subject **b :** of or relating to a practical subject organized on scientific principles ⟨a *technical* school⟩ **c :** TECHNOLOGICAL 1
3 a : based on or marked by a strict or legal interpretation **b :** LEGAL 6
4 : of or relating to technique
5 : of, relating to, or produced by ordinary commercial processes without being subjected to special purification ⟨*technical* sulfuric acid⟩
6 : relating to or caused by the functioning of the market as a discrete mechanism not influenced by macroeconomic factors ⟨*technical* rally⟩ ⟨*technical* analysis⟩
— **tech·ni·cal·ly** \-k(ə-)lē\ *adverb*

technical foul *noun* (circa 1929)
: a foul (as in basketball) that involves no physical contact with an opponent and that usually is incurred by unsportsmanlike conduct — called also *technical*; compare PERSONAL FOUL

tech·ni·cal·i·ty \ˌtek-nə-ˈka-lə-tē\ *noun, plural* **-ties** (1814)
1 : something technical; *especially* **:** a detail meaningful only to a specialist ⟨a legal *technicality*⟩
2 : the quality or state of being technical

tech·ni·cal·ize \ˈtek-ni-kə-ˌlīz\ *transitive verb* **-ized; -iz·ing** (1852)
: to give a technical slant to
— **tech·ni·cal·i·za·tion** \ˌtek-ni-kə-lə-ˈzā-shən\ *noun*

technical knockout *noun* (1921)
: the termination of a boxing match when a boxer is unable or is declared by the referee to be unable (as because of injuries) to continue the fight

technical sergeant *noun* (circa 1956)
: a noncommissioned officer in the air force ranking above a staff sergeant and below a master sergeant

tech·ni·cian \tek-ˈni-shən\ *noun* (1833)
1 : a specialist in the technical details of a subject or occupation ⟨a computer *technician*⟩
2 : one who has acquired the technique of an art or other area of specialization ⟨a superb *technician* and a musician of integrity —Irving Kolodin⟩

Tech·ni·col·or \ˈtek-ni-ˌkə-lər\ *trademark*
— used for a process of color cinematography

tech·nique \tek-ˈnēk\ *noun* [French, from *technique* technical, from Greek *technikos*] (1817)
1 : the manner in which technical details are treated (as by a writer) or basic physical movements are used (as by a dancer); *also* **:** ability to treat such details or use such movements ⟨good piano *technique*⟩
2 a : a body of technical methods (as in a craft or in scientific research) **b :** a method of accomplishing a desired aim

techno- *combining form* [*technology*]
: technical **:** technological ⟨*techno*cracy⟩

tech·no·bab·ble \ˈtek-nə-ˌba-bəl\ *noun* (1981)
: technical jargon

tech·noc·ra·cy \tek-ˈnä-krə-sē\ *noun* (circa 1919)
: government by technicians; *specifically* **:** management of society by technical experts

tech·no·crat \ˈtek-nə-ˌkrat\ *noun* (1932)
1 : an adherent of technocracy
2 : a technical expert; *especially* **:** one exercising managerial authority

tech·no·crat·ic \ˌtek-nə-ˈkra-tik\ *adjective* (1932)
: of, relating to, or suggestive of a technocrat or a technocracy

tech·no·log·i·cal \ˌtek-nə-ˈlä-ji-kəl\ *also* **tech·no·log·ic** \-ˈlä-jik\ *adjective* (1800)
1 : of, relating to, or characterized by technology ⟨*technological* advances⟩
2 : resulting from improvements in technical processes that increase productivity of machines and eliminates manual operations or operations done by older machines ⟨*technological* unemployment⟩
— **tech·no·log·i·cal·ly** \-ji-k(ə-)lē\ *adverb*

tech·nol·o·gize \tek-ˈnä-lə-ˌjīz\ *transitive verb* **-gized; -giz·ing** (1954)

: to affect or alter by technology
tech·nol·o·gy \-jē\ *noun, plural* **-gies** [Greek *technologia* systematic treatment of an art, from *technē* art, skill + *-o-* + *-logia* -logy] (1859)
1 a : the practical application of knowledge especially in a particular area **:** ENGINEERING 2 ⟨medical *technology*⟩ **b :** a capability given by the practical application of knowledge ⟨a car's fuel-saving *technology*⟩
2 : a manner of accomplishing a task especially using technical processes, methods, or knowledge ⟨new *technologies* for information storage⟩
3 : the specialized aspects of a particular field of endeavor ⟨educational *technology*⟩
— tech·nol·o·gist \-jist\ *noun*
tech·no·phile \ˈtek-nə-ˌfīl\ *noun* (1968)
: an enthusiast of technology
tech·no·pho·bia \ˌtek-nə-ˈfō-bē-ə\ *noun* (1965)
: fear or dislike of advanced technology or complex devices and especially computers
— tech·no·phobe \ˈtek-nə-ˌfōb\ *noun*
— tech·no·pho·bic \ˌtek-nə-ˈfō-bik\ *adjective*
tech·no–pop \ˈtek-(ˌ)nō-ˌpäp\ *noun* (1980)
: pop music featuring extensive use of synthesizers
tech·no·struc·ture \ˈtek-nō-ˌstrək-chər\ *noun* (1967)
: the network of professionally skilled managers (as scientists, engineers, and administrators) that tends to control the economy both within and beyond individual corporate groups
tech·no–thrill·er \ˈtek-nō-ˌthri-lər, -nə-\ *noun* (1989)
: a thriller whose plot relies on modern technology
tec·ton·ic \tek-ˈtä-nik\ *adjective* [Late Latin *tectonicus,* from Greek *tektonikos* of a builder, from *tektōn* builder — more at TECHNICAL] (1894)
: of or relating to tectonics
— tec·ton·i·cal·ly \-ni-k(ə-)lē\ *adverb*
tec·ton·ics \-niks\ *noun plural but singular or plural in construction* (1899)
1 : geological structural features as a whole
2 a : a branch of geology concerned with the structure of the crust of a planet (as earth) or moon and especially with the formation of folds and faults in it **b :** DIASTROPHISM
tec·to·nism \ˈtek-tə-ˌni-zəm\ *noun* [International Scientific Vocabulary] (1948)
: DIASTROPHISM
tec·tum \ˈtek-təm\ *noun, plural* **tec·ta** \-tə\ [New Latin, from Latin, roof, dwelling, from neuter of *tectus,* past participle of *tegere* to cover — more at THATCH] (circa 1905)
: a bodily structure resembling or serving as a roof; *especially* **:** the dorsal part of the midbrain
— tec·tal \ˈtek-tᵊl\ *adjective*
ted \ˈted\ *transitive verb* **ted·ded; ted·ding** [(assumed) Middle English *tedden,* probably from Old Norse *tethja* to manure; akin to Old Norse *tath* spread dung, Old High German *zetten* to spread] (15th century)
: to spread or turn from the swath and scatter (as new-mown grass) for drying
ted·der \ˈte-dər\ *noun* (15th century)
: one that teds; *specifically* **:** a machine for stirring and spreading hay to hasten drying and curing
ted·dy \ˈte-dē\ *noun, plural* **teddies** [origin unknown] (1924)
: CHEMISE 1
ted·dy bear \ˈte-dē-ˌ\ *noun* [*Teddy,* nickname of Theodore Roosevelt; from a cartoon depicting the president sparing the life of a bear cub while hunting] (1906)
: a stuffed toy bear
teddy boy *noun* [*Teddy,* nickname for Edward] (1954)
: a young British hoodlum affecting Edwardian dress
Te De·um \ˌtā-ˈdā-əm, ˌtē-ˈdē-\ *noun, plural* **Te Deums** [Middle English, from Late Latin *te deum laudamus* thee, God, we praise; from the opening words of the hymn] (before 12th century)
: a liturgical Christian hymn of praise to God
te·dious \ˈtē-dē-əs, ˈtē-jəs\ *adjective* [Middle English, from Late Latin *taediosus,* from Latin *taedium*] (15th century)
: tiresome because of length or dullness **:** BORING ⟨a *tedious* public ceremony⟩
— te·dious·ly *adverb*
— te·dious·ness *noun*
te·di·um \ˈtē-dē-əm\ *noun* [Latin *taedium* disgust, irksomeness, from *taedēre* to disgust, weary] (1662)
1 : the quality or state of being tedious **:** TEDIOUSNESS; *also* **:** BOREDOM
2 : a tedious period of time ⟨long *tediums* of strained anxiety —H. G. Wells⟩
¹tee \ˈtē\ *noun* [Middle English] (15th century)
1 : the letter *t*
2 : something shaped like a capital T
3 : a mark aimed at in various games (as curling)
— to a tee : EXACTLY, PRECISELY
²tee *noun* [origin unknown] (1673)
1 a : a small mound or a peg on which a golf ball is placed before being struck at the beginning of play on a hole **b :** a device for holding a football in position for kicking
2 : the area from which a golf ball is struck at the beginning of play on a hole
³tee *transitive verb* **teed; tee·ing** (1673)
: to place (a ball) on a tee — often used with *up*
teed off *adjective* [probably from *tee off* (on)] (1951)
: ANGRY, ANNOYED
¹teem \ˈtēm\ *verb* [Middle English *temen,* from Old English *tīman, tǣman;* akin to Old English *tēam* offspring — more at TEAM] (before 12th century)
transitive verb
archaic **:** BRING FORTH **:** give birth to **:** PRODUCE
intransitive verb
1 *obsolete* **:** to become pregnant **:** CONCEIVE
2 a : to become filled to overflowing **:** ABOUND **b :** to be present in large quantity
— teem·ing·ly \ˈtē-miŋ-lē\ *adverb*
— teem·ing·ness *noun*
²teem *transitive verb* [Middle English *temen,* from Old Norse *tœma;* akin to Old English *tōm* empty] (14th century)
: EMPTY, POUR ⟨*teem* molten metal into a mold⟩
¹teen \ˈtēn\ *noun* [Middle English *tene,* from Old English *tēona* injury, grief; akin to Old Norse *tjōn* loss, damage] (14th century)
archaic **:** MISERY, AFFLICTION
²teen *noun* (1818)
: TEENAGER
— teen *adjective*
teen·age \ˈtē-ˌnāj\ *or* **teen·aged** \-ˌnājd\ *adjective* (1921)
: of, being, or relating to people in their teens
— teen·ag·er \-ˌnā-jər\ *noun*
teen·er \ˈtē-nər\ *noun* (1894)
: a teenage person
teens \ˈtēnz\ *noun plural* [*-teen* (as in *thirteen*)] (1604)
: the numbers 13 to 19 inclusive; *specifically* **:** the years 13 to 19 in a lifetime or century
teen·sy \ˈtēn(t)-sē\ *adjective* **teen·si·er; -est** [baby-talk alteration of *teeny*] (1899)
: TINY
teen·sy–ween·sy \ˌtēn(t)-sē-ˈwēn(t)-sē\ *adjective* [baby-talk alteration of *teeny-weeny*] (circa 1906)
: TINY
tee·ny \ˈtē-nē\ *adjective* **tee·ni·er; -est** [by alteration] (1825)
: TINY
teeny·bop \ˈtē-nē-ˌbäp\ *adjective* [back-formation from *teenybopper*] (1967)
: of, relating to, or being a teenybopper
teeny·bop·per \-ˌbä-pər\ *noun* [*teeny* teenager + *-bopper,* perhaps from ⁴*bop*] (1966)
1 : a teenage girl
2 : a young teenager who is enthusiastically devoted to pop music and to current fads
tee·ny–wee·ny \ˌtē-nē-ˈwē-nē\ *adjective* [*teeny + weeny*] (circa 1879)
: TINY
tee off *intransitive verb* (1895)
1 : to drive from a tee
2 : BEGIN, START
3 : to hit hard
4 : to make an angry denunciation — often used with *on*
tee·pee *variant of* TEPEE
tee shirt *variant of* T-SHIRT
¹tee·ter \ˈtē-tər\ *intransitive verb* [Middle English *titeren* to totter, reel; akin to Old High German *zittarōn* to shiver] (1844)
1 a : to move unsteadily **:** WOBBLE **b :** WAVER, VACILLATE ⟨*teetered* on the brink of bankruptcy⟩
2 : SEESAW
²teeter *noun* (1863)
: SEESAW 2b
tee·ter·board \-ˌbōrd, -ˌbȯrd\ *noun* (1855)
1 : SEESAW 2b
2 : a board placed on a raised support so that a person standing on one end of the board is thrown into the air if another jumps on the opposite end
tee·ter–tot·ter \ˈtē-tər-ˌtä-tər\ *noun* (circa 1905)
: SEESAW 2b
teeth *plural of* TOOTH
teethe \ˈtēth\ *intransitive verb* **teethed; teeth·ing** [back-formation from *teething*] (15th century)
: to cut one's teeth **:** grow teeth
teeth·er \ˈtē-thər\ *noun* (1946)
: an object (as a teething ring) designed for a baby to bite on during teething
teeth·ing \ˈtē-thiŋ\ *noun* [*teeth*] (1732)
1 : the first growth of teeth
2 : the phenomena accompanying growth of teeth through the gums
teething ring *noun* (1872)
: a usually rubber or plastic ring for a teething infant to bite on
teeth·ridge \ˈtēth-ˌrij\ *noun* (1928)
: the inner surface of the gums of the upper front teeth
tee·to·tal \ˈtē-ˈtō-tᵊl, -ˌtō-\ *adjective* [*t*otal + *total* (abstinence)] (1834)
1 : of, relating to, or practicing teetotalism
2 : TOTAL, COMPLETE
— tee·to·tal·ly \-tᵊl-ē\ *adverb*
tee·to·tal·er *or* **tee·to·tal·ler** \-ˈtō-tᵊl-ər\ *noun* (1834)
: one who practices or advocates teetotalism
tee·to·tal·ism \-tᵊl-ˌi-zəm\ *noun* (1834)
: the principle or practice of complete abstinence from alcoholic drinks
— tee·to·tal·ist \-tᵊl-ist\ *noun*
tee·to·tum \ˈtē-ˈtō-təm\ *noun* [¹*tee* + Latin *totum* all, from neuter of *totus* whole; from the letter *T* inscribed on one side as an abbreviation of *totum* (take) all] (1720)
: a small top usually inscribed with letters and used in put-and-take
teff \ˈtef\ *noun* [Amharic *ṭef*] (1790)
: an economically important African cereal grass (*Eragrostis tef* synonym *E. abyssinica*) that is grown for its grain which yields a white flour and as a forage and hay crop
te·fil·lin \tē-ˈfi-lən *also* -ləm\ *noun plural but sometimes singular in construction* [Late Hebrew *tĕphillīn,* from Aramaic, attachments] (1613)
: the phylacteries worn by Jews
Tef·lon \ˈte-ˌflän\ *trademark*
— used for synthetic fluorine-containing resins used especially for molding articles and for coatings to prevent sticking (as of food in cookware)

teg·men \'teg-mən\ *noun, plural* **teg·mi·na** \-mə-nə\ [New Latin *tegmin-, tegmen,* from Latin, covering, from *tegere* to cover — more at THATCH] (1807)
: a superficial layer or cover usually of a plant or animal part

teg·men·tal \teg-'men-tᵊl\ *adjective* (circa 1890)
: of, relating to, or associated with an integument or a tegmentum

teg·men·tum \teg-'men-təm\ *noun, plural* **-men·ta** \-'men-tə\ [New Latin, from Latin *tegumentum, tegmentum,* covering, from *tegere*] (1832)
: an anatomical covering **:** TEGMEN; *especially* **:** the part of the ventral midbrain above the substantia nigra formed of longitudinal white fibers with arched transverse fibers and gray matter

teg·u·ment \'te-gyə-mənt\ *noun* [Middle English, from Latin *tegumentum*] (15th century)
: INTEGUMENT

te·iid \'tē-əd, 'tī-\ *noun* [New Latin *Teiidae,* from *Teius,* genus of lizards, from Portuguese *teiu,* a lizard, from Tupi *tejú*] (1956)
: any of a family (Teiidae) of mostly tropical American lizards (as the race runner) with a flat elongate scaly forked tongue
— **teiid** *adjective*

tek·tite \'tek-,tīt\ *noun* [International Scientific Vocabulary, from Greek *tēktos* molten, from *tēkein* to melt — more at THAW] (1909)
: a glassy body of probably meteoritic origin and of rounded but indefinite shape
— **tek·tit·ic** \tek-'ti-tik\ *adjective*

tel- *or* **telo-** *combining form* [International Scientific Vocabulary, from Greek *telos* — more at TELOS]
: end ⟨*tel*angiectasia⟩

tel·a·mon \'te-lə-,män\ *noun, plural* **tel·a·mo·nes** \,te-lə-'mō-(,)nēz\ [Latin, from Greek *telamōn* bearer, supporter; akin to Greek *tlēnai* to bear — more at TOLERATE] (circa 1706)
: ATLAS 5

tel·an·gi·ec·ta·sia \,te-,lan-jē-,ek-'tā-zh(ē-)ə, ,tē-, tə-\ *or* **tel·an·gi·ec·ta·sis** \-'ek-tə-səs\ *noun, plural* **-ta·sias** *or* **-ta·ses** \-tə-,sēz\ [New Latin, from *tel-* + *angi-* + *ectasia, ectasis* (as in *atelectasis*)] (1831)
: an abnormal dilatation of capillary vessels and arterioles that often forms an angioma
— **tel·an·gi·ec·tat·ic** \-,ek-'ta-tik\ *adjective*

tele \'te-lē\ *noun* (1936)
British **:** TELEVISION

tele- *or* **tel-** *combining form* [New Latin, from Greek *tēle-, tēl-,* from *tēle* far off — more at PALE-]
1 : distant **:** at a distance **:** over a distance ⟨*tele*gram⟩
2 a : telegraph ⟨*tele*typewriter⟩ **b :** television ⟨*tele*cast⟩ **c :** telecommunication ⟨*tele*marketing⟩

tele·cast \'te-li-,kast\ *verb* **-cast** *also* **-cast·ed; -cast·ing** [*tele-* + broad*cast*] (1937)
transitive verb
: to broadcast by television
intransitive verb
: to broadcast a television program
— **telecast** *noun*
— **tele·cast·er** *noun*

tele·com·mu·ni·ca·tion \,te-li-kə-,myü-nə-'kā-shən\ *noun* [International Scientific Vocabulary] (1932)
1 : communication at a distance (as by telephone)
2 : a science that deals with telecommunication — usually used in plural

tele·com·mute \'te-li-kə-,myüt\ *intransitive verb* (1974)
: to work at home by the use of an electronic linkup with a central office
— **tele·com·mut·er** *noun*

tele·con·fer·enc·ing \'te-li-,kän-f(ə-)rən(t)-siŋ, -fərn(t)-\ *noun* (1974)
: the holding of a conference among people remote from one another by means of telecommunication devices (as telephones or computer terminals)
— **tele·con·fer·ence** \-f(ə-)rən(t)s, -fərn(t)s\ *noun*

Tele·copi·er \'te-lə-,kä-pē-ər\ *trademark*
— used for transmitting and receiving equipment for producing facsimile copies of documents

tele·course \'te-li-,kōrs, -,kȯrs\ *noun* (1950)
: a course of study conducted over television; *especially* **:** such a course taken at home for academic credit

tele·fac·sim·i·le \,te-li-fak-'si-mə-(,)lē\ *noun* (1952)
: FACSIMILE 2

tele·film \'te-li-,film\ *noun* (1939)
: a motion picture made to be telecast

tele·ge·nic \,te-lə-'je-nik, -'jē-\ *adjective* (1939)
: well-suited to the medium of television; *especially* **:** having an appearance and manner that are markedly attractive to television viewers

¹tele·gram \'te-lə-,gram, *Southern also* -grəm\ *noun* (circa 1852)
: a telegraphic dispatch

²telegram \-,gram\ *transitive verb* **-grammed; -gram·ming** (1864)
: TELEGRAPH

¹tele·graph \-,graf\ *noun* [French *télégraphe,* from *télé-* tele- (from Greek *tēle-*) + *-graphe* -graph] (1794)
1 : an apparatus for communication at a distance by coded signals; *especially* **:** an apparatus, system, or process for communication at a distance by electric transmission over wire
2 : TELEGRAM

²telegraph *transitive verb* (1805)
1 a : to send or communicate by or as if by telegraph **b :** to send a telegram to **c :** to send by means of a telegraphic order ⟨*telegraph* flowers to a sick friend⟩
2 : to make known by signs especially unknowingly and in advance ⟨*telegraph* a punch⟩
— **te·leg·ra·pher** \tə-'le-grə-fər\ *noun*
— **te·leg·ra·phist** \-fist\ *noun*

tele·graph·ese \,te-lə-gra-'fēz, -'fēs\ *noun* (1885)
: language characterized by the terseness and ellipses that are common in telegrams

tele·graph·ic \,te-lə-'gra-fik\ *adjective* (1794)
1 : of or relating to the telegraph
2 : CONCISE, TERSE
— **tele·graph·i·cal·ly** \-fi-k(ə-)lē\ *adverb*

te·leg·ra·phy \tə-'le-grə-fē\ *noun* (1795)
: the use or operation of a telegraph apparatus or system for communication

tele·ki·ne·sis \,te-li-kə-'nē-səs, -kī-\ *noun* [New Latin] (1890)
: the production of motion in objects (as by a spiritualistic medium) without contact or other physical means
— **tele·ki·net·ic** \-'ne-tik\ *adjective*
— **tele·ki·net·i·cal·ly** \-ti-k(ə-)lē\ *adverb*

Te·lem·a·chus \tə-'le-mə-kəs\ *noun* [Latin, from Greek *Tēlemachos*]
: the son of Odysseus and Penelope who contrives with his father to slay his mother's suitors

tel·e·mark \'te-lə-,märk\ *noun, often capitalized* [Norwegian, from *Telemark,* region in Norway] (1904)
: a turn in skiing in which the outside ski is advanced considerably ahead of the other ski and then turned inward at a steadily widening angle until the turn is completed

tele·mar·ket·ing \,te-lə-'mär-kə-tiŋ\ *noun* (1980)
: the marketing of goods or services by telephone
— **tele·mar·ket·er** \-tər\ *noun*

¹tele·me·ter \'te-lə-,mē-tər\ *noun* [International Scientific Vocabulary] (1860)
1 : an instrument for measuring the distance of an object from an observer
2 : an electrical apparatus for measuring a quantity (as pressure, speed, or temperature), transmitting the result especially by radio to a distant station, and there indicating or recording the quantity measured

²telemeter (1925)
transitive verb
: to transmit (as the measurement of a quantity) by telemeter
intransitive verb
: to telemeter the measurement of a quantity

te·lem·e·try \tə-'le-mə-trē\ *noun* (1885)
1 : the science or process of telemetering data
2 : data transmitted by telemetry
3 : BIOTELEMETRY
— **tele·met·ric** \,te-lə-'me-trik\ *adjective*
— **tele·met·ri·cal·ly** \-tri-k(ə-)lē\ *adverb*

tel·en·ceph·a·lon \,te-len-'se-fə-,län, -lən\ *noun* [New Latin] (1897)
: the anterior subdivision of the embryonic forebrain or the corresponding part of the adult forebrain that includes the cerebral hemispheres and associated structures
— **tel·en·ce·phal·ic** \-,len(t)-sə-'fa-lik\ *adjective*

tel·e·o·log·i·cal \,te-lē-ə-'lä-ji-kəl, ,tē-\ *also* **tel·e·o·log·ic** \-'lä-jik\ *adjective* (1798)
: exhibiting or relating to design or purpose especially in nature
— **tel·e·o·log·i·cal·ly** \-ji-k(ə-)lē\ *adverb*

tel·e·ol·o·gy \,te-lē-'ä-lə-jē, ,tē-\ *noun* [New Latin *teleologia,* from Greek *tele-, telos* end, purpose + *-logia* -logy — more at WHEEL] (1740)
1 a : the study of evidences of design in nature **b :** a doctrine (as in vitalism) that ends are immanent in nature **c :** a doctrine explaining phenomena by final causes
2 : the fact or character attributed to nature or natural processes of being directed toward an end or shaped by a purpose
3 : the use of design or purpose as an explanation of natural phenomena
— **tel·e·ol·o·gist** \-jist\ *noun*

tel·e·on·o·my \,te-lē-'ä-nə-mē, ,tē-\ *noun* [*teleo-* (as in *teleology*) + *-nomy*] (1958)
: the quality of apparent purposefulness in living organisms that derives from their evolutionary adaptation
— **tel·e·o·nom·ic** \,te-lē-ə-'nä-mik, ,tē-\ *adjective*

tel·e·ost \'te-lē-,äst, 'tē-\ *noun* [ultimately from Greek *teleios* complete, perfect (from *telos* end) + *osteon* bone — more at OSSEOUS] (1862)
: BONY FISH
— **teleost** *adjective*
— **tel·e·os·te·an** \,te-lē-'äs-tē-ən, ,tē-\ *adjective*

tele·path \'te-lə-,path\ *noun* (1904)
: one who is able to communicate by telepathy

te·lep·a·thy \tə-'le-pə-thē\ *noun* (1882)
: communication from one mind to another by extrasensory means
— **tele·path·ic** \,te-lə-'pa-thik\ *adjective*
— **tele·path·i·cal·ly** \-thi-k(ə-)lē\ *adverb*

¹tele·phone \'te-lə-,fōn\ *noun, often attributive* (1849)
: an instrument for reproducing sounds at a distance; *specifically* **:** one in which sound is converted into electrical impulses for transmission by wire

²telephone *verb* **-phoned; -phon·ing** (1877)
transitive verb
1 : to speak to or attempt to reach by telephone
2 : to send by telephone

intransitive verb
: to communicate by telephone
— tele·phon·er *noun*
telephone book *noun* (1915)
: a book listing the names, addresses, and telephone numbers of telephone customers
telephone booth *noun* (circa 1895)
: an enclosure within which one may stand or sit while making a telephone call
telephone box *noun* (1904)
British **:** a public telephone booth
telephone directory *noun* (1907)
: TELEPHONE BOOK
telephone number *noun* (1885)
: a number assigned to a telephone line for a specific location that is used to call that location
telephone tag *noun* (1980)
: telephoning back and forth by parties trying to reach each other without success
tele·phon·ic \ˌte-lə-ˈfä-nik\ *adjective* (1877)
: of, relating to, or conveyed by a telephone
— tele·phon·i·cal·ly \-ni-k(ə-)lē\ *adverb*
te·le·pho·nist \tə-ˈle-fə-nist, ˈte-lə-ˌfō-nist\ *noun* (1880)
British **:** a telephone switchboard operator
te·le·pho·ny \tə-ˈle-fə-nē *also* ˈte-lə-ˌfō-\ *noun* (1835)
: the use or operation of an apparatus for transmission of sounds between widely removed points with or without connecting wires
[1]**tele·pho·to** \ˌte-lə-ˈfō-(ˌ)tō\ *adjective* (circa 1895)
: being a camera lens system designed to give a large image of a distant object; *also* **:** relating to or being photography in which a telephoto lens is used
[2]**telephoto** *noun, plural* **-tos** (1904)
1 : a telephoto lens
2 : a photograph taken with a camera having a telephoto lens
Telephoto *trademark*
— used for an apparatus for transmitting photographs electrically or for a photograph so transmitted
tele·pho·tog·ra·phy \-fə-ˈtä-grə-fē\ *noun* [International Scientific Vocabulary] (1892)
: the photography of distant objects (as by a camera provided with a telephoto lens)
tele·play \ˈte-li-ˌplā\ *noun* (1952)
: a play written for television
tele·port \ˈte-lə-ˌpōrt, -ˌpȯrt\ *transitive verb* [back-formation from *teleportation*] (1947)
: to transfer by teleportation
tele·por·ta·tion \ˌte-lə-ˌpōr-ˈtā-shən, -ˌpȯr-, -pər-\ *noun* [*tele-* + trans*portation*] (1931)
: the act or process of moving an object or person by psychokinesis
tele·print·er \ˈte-lə-ˌprin-tər\ *noun* (1929)
: a device capable of producing hard copy from signals received over a communications circuit; *especially* **:** TELETYPEWRITER
tele·pro·cess·ing \-ˈprä-ˌse-siŋ, -ˈprō-, -sə-siŋ\ *noun* (1962)
: computer processing via remote terminals
Tele·Promp·Ter \ˈte-lə-ˌpräm(p)-tər\ *trademark*
— used for a device for unrolling a magnified script in front of a speaker on television
[1]**tele·scope** \ˈte-lə-ˌskōp\ *noun, often attributive* [New Latin *telescopium*, from Greek *tēleskopos* farseeing, from *tēle-* tele- + *skopos* watcher; akin to Greek *skopein* to look — more at SPY] (1648)
1 : a usually tubular optical instrument for viewing distant objects by means of the refraction of light rays through a lens or the reflection of light rays by a concave mirror — compare REFLECTOR, REFRACTOR
2 : any of various tubular magnifying optical instruments
3 : RADIO TELESCOPE
[2]**telescope** *verb* **-scoped; -scop·ing** (1867)
intransitive verb
1 : to become forced together lengthwise with one part entering another as the result of collision
2 : to slide or pass one within another like the cylindrical sections of a collapsible hand telescope
3 : to become compressed or condensed
transitive verb
1 : to cause to telescope
2 : COMPRESS, CONDENSE ⟨the book arbitrarily *telescopes* time and space, and as arbitrarily extends them —Phoebe Adams⟩
tele·scop·ic \ˌte-lə-ˈskä-pik\ *adjective* (1705)
1 a : of, relating to, or performed with a telescope **b :** suitable for seeing or magnifying distant objects
2 : seen or discoverable only by a telescope ⟨*telescopic* stars⟩
3 : able to discern objects at a distance
4 : having parts that telescope
— tele·scop·i·cal·ly \-pi-k(ə-)lē\ *adverb*
tel·e·sis \ˈte-lə-səs\ *noun, plural* **-e·ses** \-ˌsēz\ [New Latin, from Greek, fulfillment, from *telein* to complete, from *telos* end — more at TELOS] (1896)
: progress that is intelligently planned and directed **:** the attainment of desired ends by the application of intelligent human effort to the means
tele·text \ˈte-lə-ˌtekst\ *noun* (1974)
: a system for broadcasting text over an unused portion of a television signal and displaying it on a decoder-equipped television set — compare VIDEOTEX
tele·thon \ˈte-lə-ˌthän\ *noun* [*tele-* + *-thon*] (1949)
: a long television program usually to solicit funds especially for a charity
Tele·type \ˈte-lə-ˌtīp\ *trademark*
— used for a teletypewriter
Tele·type·set·ter \ˌte-lə-ˈtīp-ˌse-tər\ *trademark*
— used for a telegraphic apparatus for the automatic operation of a keyboard typesetting machine
tele·type·writ·er \-ˌrī-tər\ *noun* (1903)
: a printing device resembling a typewriter that is used to send and receive telephonic signals
te·leu·to·spore \tə-ˈlü-tə-ˌspōr, -ˌspȯr\ *noun* [Greek *teleutē* end (akin to Greek *telos* end) + International Scientific Vocabulary *spore* — more at TELOS] (1874)
: TELIOSPORE
tel·evan·ge·list \ˌte-li-ˈvan-jə-list\ *noun* (1973)
: an evangelist who conducts regularly televised religious programs
— tel·evan·ge·lism \-ˌli-zəm\ *noun*
tele·view \ˈte-li-ˌvyü\ *intransitive verb* (1935)
: to observe or watch by means of a television receiver
— tele·view·er *noun*
tele·vise \ˈte-lə-ˌvīz\ *verb* **-vised; -vis·ing** [back-formation from *television*] (1927)
transitive verb
: to broadcast (as a baseball game) by television
intransitive verb
: to broadcast by television
tele·vi·sion \ˈte-lə-ˌvi-zhən *especially British* ˌte-lə-ˈ\ *noun, often attributive* [French *télévision*, from *télé-* tele- + *vision* vision] (1907)
1 : an electronic system of transmitting transient images of fixed or moving objects together with sound over a wire or through space by apparatus that converts light and sound into electrical waves and reconverts them into visible light rays and audible sound
2 : a television receiving set
3 a : the television broadcasting industry **b :** television as a medium of communication
television tube *noun* (1937)
: PICTURE TUBE
tele·vi·su·al \ˌte-lə-ˈvi-zhə-wəl, -zhəl; -ˈvizh-wəl\ *adjective* (1926)
chiefly British **:** of, relating to, or suitable for broadcast by television
[1]**tel·ex** \ˈte-ˌleks\ *noun* [*tel*eprinter + *ex*change] (1932)
1 : a communication service involving teletypewriters connected by wire through automatic exchanges; *also* **:** a teletypewriter used in telex
2 : a message sent by telex
[2]**telex** *transitive verb* (1960)
1 : to send (as a message) by telex
2 : to communicate with by telex
te·lic \ˈte-lik, ˈtē-\ *adjective* [Greek *telikos*, from *telos* end — more at TELOS] (1889)
: tending toward an end
— te·li·cal·ly \-li-k(ə-)lē\ *adverb*
te·lio·spore \ˈtē-lē-ə-ˌspōr, -ˌspȯr\ *noun* [Greek *teleios* complete (from *telos* end) + English *spore*] (1905)
: a chlamydospore that is the final stage in the life cycle of a rust fungus and that after nuclear fusion gives rise to the basidium
te·li·um \ˈtē-lē-əm\ *noun, plural* **te·lia** \-lē-ə\ [New Latin, from Greek *teleios* complete] (circa 1905)
: a teliospore-producing sorus or pustule on the host plant of a rust fungus
— te·li·al \ˈtē-lē-əl\ *adjective*
[1]**tell** \ˈtel\ *verb* **told** \ˈtōld\; **tell·ing** [Middle English, from Old English *tellan;* akin to Old High German *zellen* to count, tell, Old English *talu* tale] (before 12th century)
transitive verb
1 : COUNT, ENUMERATE
2 a : to relate in detail **:** NARRATE **b :** to give utterance to **:** SAY ⟨who dares think one thing, and another *tell* —Alexander Pope⟩
3 a : to make known **:** DIVULGE, REVEAL **b :** to express in words ⟨she never *told* her love —Shakespeare⟩
4 a : to report to **:** INFORM **b :** to assure emphatically ⟨they did not do it, I *tell* you⟩
5 : ORDER, DIRECT ⟨*told* me to wait⟩
6 : to find out by observing **:** RECOGNIZE ⟨you can *tell* it's a masterpiece⟩
intransitive verb
1 : to give an account
2 : to act as an informer — often used with *on* ⟨I'll get even with you if you ever *tell* on me —*Inside Detective*⟩
3 : to have a marked effect
4 : to serve as evidence or indication
synonym see REVEAL
[2]**tell** *noun* [Arabic *tall*] (1864)
: HILL, MOUND; *specifically* **:** an ancient mound in the Middle East composed of remains of successive settlements
tell–all \ˈtel-ˈȯl\ *noun* (1954)
: a written account (as a biography) that contains revealing and often scandalous information
— tell–all *adjective*
tell·er \ˈte-lər\ *noun* (14th century)
1 : one that relates or communicates ⟨a *teller* of stories⟩
2 : one that reckons or counts: as **a :** one appointed to count votes **b :** a member of a bank's staff concerned with the direct handling of money received or paid out
tell·ing \ˈte-liŋ\ *adjective* (1851)
: carrying great weight and producing a marked effect **:** EFFECTIVE, EXPRESSIVE ⟨the most *telling* evidence⟩
synonym see VALID
— tell·ing·ly \-liŋ-lē\ *adverb*
tell off *transitive verb* (1804)
1 : to number and set apart; *especially* **:** to assign to a special duty ⟨*told off* a detail and put them to opening a trench —J. F. Dobie⟩
2 : REPRIMAND, EXCORIATE ⟨*told* him *off* for his arrogance⟩
tell·tale \ˈtel-ˌtāl\ *noun* (circa 1548)
1 a : TALEBEARER, INFORMER **b :** an outward sign **:** INDICATION

2 : a device for indicating or recording something: as **a :** a wind-direction indicator often in the form of a ribbon **b :** a strip of metal on the front wall of a racquets or squash court above which the ball must be hit
— **telltale** *adjective*

tellur- *or* **telluro-** *combining form* [Latin *tellur-, tellus* — more at THILL]
1 : earth ⟨*telluric*⟩
2 [New Latin *tellurium*] **:** tellurium ⟨*telluride*⟩

tel·lu·ric \tə-ˈlu̇r-ik, te-\ *adjective* (1836)
1 : of or relating to the earth **:** TERRESTRIAL
2 : being or relating to a usually natural electric current flowing near the earth's surface

tel·lu·ride \ˈtel-yə-ˌrīd\ *noun* [International Scientific Vocabulary] (1849)
: a binary compound of tellurium with a more electropositive element or group

tel·lu·ri·um \tə-ˈlu̇r-ē-əm, te-\ *noun* [New Latin, from Latin *tellur-, tellus* earth] (1800)
: a semimetallic element related to selenium and sulfur that occurs in a silvery white brittle crystalline form of metallic luster, in a dark amorphous form, or combined with metals and that is used especially in alloys — see ELEMENT table

tel·lu·rom·e·ter \ˌtel-yə-ˈrä-mə-tər\ *noun* (1957)
: a device that measures distance by means of microwaves

tel·ly \ˈte-lē\ *noun, plural* **tellys** *also* **tellies** [by shortening & alteration] (1939)
chiefly British **:** TELEVISION

telo- — see TEL-

telo·cen·tric \ˌte-lə-ˈsen-trik, ˌtē-\ *adjective* [International Scientific Vocabulary *tel-* + *centromere* + *-ic*] (1939)
: having the centromere terminally situated so that there is only one chromosomal arm ⟨a *telocentric* chromosome⟩
— **telocentric** *noun*

te·lome \ˈtē-ˌlōm\ *noun* [International Scientific Vocabulary *tel-* + *-ome*] (1935)
: a hypothetical plant structure in a theory of the evolution of leaves and sporophylls in vascular plants that consists of one of the vegetative or reproductive terminal branchlets of a dichotomously branched axis

telo·mere \ˈte-lə-ˌmir, ˈtē-\ *noun* [International Scientific Vocabulary] (1940)
: the natural end of a eukaryotic chromosome

telo·phase \ˈte-lə-ˌfāz, ˈtē-\ *noun* [International Scientific Vocabulary] (1895)
1 : the final stage of mitosis and of the second division of meiosis in which the spindle disappears and the nuclear envelope reforms around each set of chromosomes
2 : the final stage in the first division of meiosis that may be missing in some organisms and is characterized by the gathering at opposite poles of the cell of half the original number of chromosomes including one from each homologous pair

te·los \ˈte-ˌläs, ˈtē-\ *noun* [Greek; probably akin to Greek *tellein* to accomplish, *tlēnai* to bear — more at TOLERATE] (1904)
: an ultimate end

te·lo·tax·is \ˌte-lə-ˈtak-səs, ˌtē-\ *noun* [New Latin] (1934)
: a taxis in which an organism orients itself in respect to a stimulus (as a light source) as though that were the only stimulus acting on it

tel·son \ˈtel-sən\ *noun* [New Latin, from Greek, end of a plowed field; perhaps akin to Greek *telos* end] (1855)
: the terminal segment of the body of an arthropod or segmented worm; *especially* **:** that of a crustacean forming the middle lobe of the tail

Tel·u·gu \ˈte-lə-ˌgü\ *noun, plural* **Telugu** *or* **Telugus** (1789)
1 : a member of the largest group of people in Andhra Pradesh, India
2 : the Dravidian language of the Telugu people

tem·blor \ˈtem-blər; ˈtem-ˌblȯr, -ˌblōr, tem-ˈ\ *noun* [Spanish, literally, trembling, from *temblar* to tremble, from Medieval Latin *tremulare* — more at TREMBLE] (1876)
: EARTHQUAKE

tem·er·ar·i·ous \ˌte-mə-ˈrer-ē-əs, -ˈrar-\ *adjective* [Latin *temerarius,* from *temere*] (1532)
: marked by temerity **:** rashly or presumptuously daring
— **tem·er·ar·i·ous·ly** *adverb*
— **tem·er·ar·i·ous·ness** *noun*

te·mer·i·ty \tə-ˈmer-ə-tē\ *noun, plural* **-ties** [Middle English *temeryte,* from Latin *temeritas,* from *temere* blindly, recklessly; akin to Old High German *demar* darkness, Latin *tenebrae,* Sanskrit *tamas*] (15th century)
1 : unreasonable or foolhardy contempt of danger or opposition **:** RASHNESS, RECKLESSNESS
2 : an act or instance of temerity ☆

¹temp \ˈtemp\ *noun* (1886)
1 : TEMPERATURE 2a, c
2 : a temporary worker

²temp *intransitive verb* (1973)
: to work as a temp

tem·peh \ˈtem-ˌpā\ *noun* [Javanese *témpé*] (1961)
: an Asian food prepared by fermenting soybeans with a rhizopus

¹tem·per \ˈtem-pər\ *transitive verb* **tem·pered; tem·per·ing** \-p(ə-)riŋ\ [Middle English, from Old English & Old French; Old English *temprian* & Old French *temprer,* from Latin *temperare* to moderate, mix, temper; probably akin to Latin *tempor-, tempus* time] (before 12th century)
1 : to dilute, qualify, or soften by the addition or influence of something else **:** MODERATE ⟨*temper* justice with mercy⟩
2 *archaic* **a :** to exercise control over **:** GOVERN, RESTRAIN **b :** to cause to be well disposed **:** MOLLIFY ⟨*tempered* and reconciled them both —Richard Steele⟩
3 : to bring to a suitable state by mixing in or adding a usually liquid ingredient: as **a :** to mix (clay) with water or a modifier (as grog) and knead to a uniform texture **b :** to mix oil with (colors) in making paint ready for use
4 a (1) **:** to soften (as hardened steel or cast iron) by reheating at a lower temperature (2) **:** to harden (as steel) by reheating and cooling in oil **b :** to anneal or toughen (glass) by a process of gradually heating and cooling
5 : to make stronger and more resilient through hardship **:** TOUGHEN ⟨troops *tempered* in battle⟩
6 a : to put in tune with something **:** ATTUNE **b :** to adjust the pitch of (a note, chord, or instrument) to a temperament
— **tem·per·able** \-p(ə-)rə-bəl\ *adjective*
— **tem·per·er** \-pər-ər\ *noun*

²temper *noun* (14th century)
1 a *archaic* **:** a suitable proportion or balance of qualities **:** a middle state between extremes **:** MEAN, MEDIUM ⟨virtue is . . . a just *temper* between propensities —T. B. Macaulay⟩ **b** *archaic* **:** CHARACTER, QUALITY ⟨the *temper* of the land you design to sow —John Mortimer⟩ **c :** characteristic tone **:** TREND, TENDENCY ⟨the *temper* of the times⟩ **d :** high quality of mind or spirit **:** COURAGE, METTLE
2 a : the state of a substance with respect to certain desired qualities (as hardness, elasticity, or workability); *especially* **:** the degree of hardness or resiliency given steel by tempering **b :** the feel and relative solidity of leather
3 a : a characteristic cast of mind or state of feeling **:** DISPOSITION **b :** calmness of mind **:** COMPOSURE, EQUANIMITY **c :** state of feeling or frame of mind at a particular time usually dominated by a single strong emotion **d :** heat of mind or emotion **:** proneness to anger **:** PASSION
4 : a substance (as a metal) added to or mixed with something else (as another metal) to modify the properties of the latter
synonym see DISPOSITION

tem·pera \ˈtem-pə-rə\ *noun* [Italian *tempera,* literally, temper, from *temperare* to temper, from Latin] (1832)
1 : a process of painting in which an albuminous or colloidal medium (as egg yolk) is employed as a vehicle instead of oil; *also* **:** a painting done in tempera
2 : POSTER COLOR

tem·per·a·ment \ˈtem-p(ə-)rə-mənt, -pər-mənt\ *noun* [Middle English, from Latin *temperamentum,* from *temperare* to mix, temper] (15th century)
1 *obsolete* **a :** constitution of a substance, body, or organism with respect to the mixture or balance of its elements, qualities, or parts **:** MAKEUP **b :** COMPLEXION 1
2 *obsolete* **a :** CLIMATE **b :** TEMPERATURE 2
3 a : the peculiar or distinguishing mental or physical character determined by the relative proportions of the humors according to medieval physiology **b :** characteristic or habitual inclination or mode of emotional response ⟨a nervous *temperament*⟩ **c :** extremely high sensibility; *especially* **:** excessive sensitiveness or irritability
4 a : the act or process of tempering or modifying **:** ADJUSTMENT, COMPROMISE **b :** middle course **:** MEAN
5 : the process of slightly modifying the musical intervals of the pure scale to produce a set of 12 equally spaced tones to the octave which enables a keyboard instrument to play in all keys
synonym see DISPOSITION

tem·per·a·men·tal \ˌtem-p(ə-)rə-ˈmen-tᵊl, ˌtem-pər-ˈ\ *adjective* (1646)
1 : of, relating to, or arising from temperament **:** CONSTITUTIONAL ⟨*temperamental* peculiarities⟩
2 a : marked by excessive sensitivity and impulsive changes of mood ⟨a *temperamental* child⟩ **b :** unpredictable in behavior or performance
— **tem·per·a·men·tal·ly** \-tᵊl-ē\ *adverb*

tem·per·ance \ˈtem-p(ə-)rən(t)s, -pərn(t)s\ *noun* [Middle English, from Latin *temperantia,* from *temperant-, temperans,* present participle of *temperare* to moderate, be moderate] (14th century)
1 : moderation in action, thought, or feeling **:** RESTRAINT
2 a : habitual moderation in the indulgence of the appetites or passions **b :** moderation in or abstinence from the use of intoxicating drink

☆ SYNONYMS
Temerity, audacity, hardihood, effrontery, nerve, cheek, gall, chutzpah mean conspicuous or flagrant boldness. TEMERITY suggests boldness arising from rashness and contempt of danger ⟨had the *temerity* to refuse⟩. AUDACITY implies a disregard of restraints commonly imposed by convention or prudence ⟨an entrepreneur with *audacity* and vision⟩. HARDIHOOD suggests firmness in daring and defiance ⟨admired for her *hardihood*⟩. EFFRONTERY implies shameless, insolent disregard of propriety or courtesy ⟨outraged at his *effrontery*⟩. NERVE, CHEEK, GALL, and CHUTZPAH are informal equivalents for EFFRONTERY ⟨the *nerve* of that guy⟩ ⟨has the *cheek* to call herself a singer⟩ ⟨had the *gall* to demand proof⟩ ⟨the *chutzpah* needed for a career in show business⟩.

tem·per·ate \'tem-p(ə-)rət\ *adjective* [Middle English *temperat,* from Latin *temperatus,* from past participle of *temperare*] (14th century)
1 : marked by moderation: as **a :** keeping or held within limits **:** not extreme or excessive **:** MILD **b :** moderate in indulgence of appetite or desire **c :** moderate in the use of intoxicating liquors **d :** marked by an absence or avoidance of extravagance, violence, or extreme partisanship **:** RESTRAINED
2 a : having a moderate climate **b :** found in or associated with a moderate climate ⟨*temperate* insects⟩
3 : existing as a prophage in infected cells and rarely causing lysis ⟨*temperate* bacteriophages⟩
— tem·per·ate·ly *adverb*
— tem·per·ate·ness *noun*

temperate rain forest *noun* (circa 1930)
: woodland of a usually rather mild climatic area within the temperate zone that receives heavy rainfall, usually includes numerous kinds of trees, and is distinguished from a tropical rain forest by the presence of a dominant tree

temperate zone *noun, often T&Z capitalized* (1551)
: the area or region between the tropic of Cancer and the arctic circle or between the tropic of Capricorn and the antarctic circle

tem·per·a·ture \'tem-pə(r)-ˌchu̇r, -p(ə-)rə-, -chər, -ˌtyu̇r, -ˌtu̇r\ *noun* [Latin *temperatura* mixture, moderation, from *temperatus,* past participle of *temperare*] (1533)
1 *archaic* **a :** COMPLEXION 1 **b :** TEMPERAMENT 3b
2 a : degree of hotness or coldness measured on a definite scale — compare THERMOMETER **b :** the degree of heat that is natural to the body of a living being **c :** abnormally high body heat **d :** relative state of emotional warmth ⟨aware of a change in the *temperature* of our friendship —Christopher Isherwood⟩

temperature inversion *noun* (1921)
: INVERSION 6

tem·pered \'tem-pərd\ *adjective* (14th century)
1 a : having the elements mixed in satisfying proportions **:** TEMPERATE **b :** qualified, lessened, or diluted by the mixture or influence of an additional ingredient **:** MODERATED ⟨a pale gleam of *tempered* sunlight fell through the leaves —W. H. Hudson (died 1922)⟩
2 : treated by tempering; *especially, of glass* **:** treated so as to impart increased strength and the property of shattering into pellets when broken
3 : having a specified temper — used in combination ⟨short-*tempered*⟩
4 : conforming to adjustment by temperament — used of a musical interval, intonation, semitone, or scale

¹tem·pest \'tem-pəst\ *noun* [Middle English, from Old French *tempeste,* from (assumed) Vulgar Latin *tempesta,* alteration of Latin *tempestas* season, weather, storm, from *tempus* time] (13th century)
1 : a violent storm
2 : TUMULT, UPROAR

²tempest *transitive verb* (14th century)
: to raise a tempest in or around

tem·pes·tu·ous \tem-'pes-chə-wəs, -'pesh-\ *adjective* [Late Latin *tempestuosus,* from Old Latin *tempestus* season, weather, storm, from *tempus*] (1509)
: of, relating to, or resembling a tempest **:** TURBULENT, STORMY ⟨*tempestuous* weather⟩ ⟨a *tempestuous* relationship⟩
— tem·pes·tu·ous·ly *adverb*
— tem·pes·tu·ous·ness *noun*

Tem·plar \'tem-plər\ *noun* [Middle English *templer,* from Old French *templier,* from Medieval Latin *templarius,* from Latin *templum* temple] (13th century)
1 : a knight of a religious military order established in the early 12th century in Jerusalem for the protection of pilgrims and the Holy Sepulcher
2 *not capitalized* **:** a barrister or student of law in London
3 : KNIGHT TEMPLAR 2

tem·plate *also* **tem·plet** \'tem-plət\ *noun* [probably from French *templet,* diminutive of *temple,* part of a loom, probably from Latin *templum*] (1677)
1 : a short piece or block placed horizontally in a wall under a beam to distribute its weight or pressure (as over a door)
2 a (1) **:** a gauge, pattern, or mold (as a thin plate or board) used as a guide to the form of a piece being made (2) **:** a molecule (as of DNA) that serves as a pattern for the generation of another macromolecule (as messenger RNA) **b :** OVERLAY c
3 : something that establishes or serves as a pattern

¹tem·ple \'tem-pəl\ *noun* [Middle English, from Old English & Old French; Old English *tempel* & Old French *temple,* both from Latin *templum* space marked out for observation of auguries, temple, small timber; probably akin to Greek *temenos* sacred precinct, *temnein* to cut — more at TOME] (before 12th century)
1 : an edifice for religious exercises: as **a** *often capitalized* **:** one of three successive national sanctuaries in ancient Jerusalem **b :** a building for Mormon sacred ordinances **c :** a Reform or Conservative synagogue
2 : a local lodge of any of various fraternal orders; *also* **:** the building housing it
3 : a place devoted to a special purpose
— tem·pled \-pəld\ *adjective*

²temple *noun* [Middle English, from Middle French, from (assumed) Vulgar Latin *tempula,* alteration of Latin *tempora* (plural) temples] (14th century)
1 : the flattened space on each side of the forehead of some mammals including humans
2 : one of the side supports of a pair of glasses jointed to the bows and passing on each side of the head

tem·po \'tem-(ˌ)pō\ *noun, plural* **tem·pi** \-(ˌ)pē\ *or* **tempos** [Italian, literally, time, from Latin *tempus*] (circa 1724)
1 : the rate of speed of a musical piece or passage indicated by one of a series of directions (as largo, presto, or allegro) and often by an exact metronome marking
2 : rate of motion or activity **:** PACE

¹tem·po·ral \'tem-p(ə-)rəl\ *adjective* [Middle English, from Latin *temporalis,* from *tempor-, tempus* time] (14th century)
1 a : of or relating to time as opposed to eternity **b :** of or relating to earthly life **c :** lay or secular rather than clerical or sacred **:** CIVIL ⟨lords *temporal*⟩
2 : of or relating to grammatical tense or a distinction of time
3 a : of or relating to time as distinguished from space **b :** of or relating to the sequence of time or to a particular time **:** CHRONOLOGICAL
— tem·po·ral·ly *adverb*

²temporal *noun* [Middle French, from *temporal,* adjective] (1541)
: a temporal part (as a bone or muscle)

³temporal *adjective* [Middle French, from Late Latin *temporalis,* from Latin *tempora* temples] (1597)
: of or relating to the temples or the sides of the skull behind the orbits

temporal bone *noun* (1771)
: a compound bone of the side of the skull of some mammals including humans

tem·po·ral·i·ty \ˌtem-pə-'ra-lə-tē\ *noun, plural* **-ties** (14th century)
1 a : civil or political as distinguished from spiritual or ecclesiastical power or authority **b :** an ecclesiastical property or revenue — often used in plural
2 : the quality or state of being temporal

tem·po·ral·ize \'tem-p(ə-)rə-ˌlīz\ *transitive verb* **-ized; -iz·ing** (1828)
1 : SECULARIZE
2 : to place or define in time relations

temporal lobe *noun* (circa 1891)
: a large lobe of each cerebral hemisphere that is situated in front of the occipital lobe and contains a sensory area associated with the organ of hearing

temporal summation *noun* (1950)
: sensory summation that involves the addition of single stimuli over a short period of time

tem·po·rar·i·ly \ˌtem-pə-'rer-ə-lē\ *adverb* (1534)
: during a limited time

¹tem·po·rary \'tem-pə-ˌrer-ē\ *adjective* [Latin *temporarius,* from *tempor-, tempus* time] (circa 1564)
: lasting for a limited time
— tem·po·rar·i·ness *noun*

²temporary *noun, plural* **-rar·ies** (1848)
: one serving for a limited time ⟨adding several *temporaries* as typists during the summer⟩

temporary duty *noun* (1945)
: temporary military service away from one's permanent duty station

tem·po·rise *British variant of* TEMPORIZE

tem·po·rize \'tem-pə-ˌrīz\ *intransitive verb* **-rized; -riz·ing** [Middle French *temporiser,* from Medieval Latin *temporizare* to pass the time, from Latin *tempor-, tempus*] (1579)
1 : to act to suit the time or occasion **:** yield to current or dominant opinion **:** COMPROMISE
2 : to draw out discussions or negotiations so as to gain time ⟨you'd have to *temporize* until you found out how she wanted to be advised —Mary Austin⟩
— tem·po·ri·za·tion \ˌtem-pə-rə-'zā-shən\ *noun*
— tem·po·riz·er *noun*

tem·po·ro·man·dib·u·lar \'tem-pə-rō-man-'di-byə-lər\ *adjective* [³*temporal* + *-o-* + *mandibular*] (1889)
: of, relating to, being, or affecting the joint between the temporal bone and the mandible ⟨*temporomandibular* dysfunction⟩

tempt \'tem(p)t\ *transitive verb* [Middle English, from Old French *tempter, tenter,* from Latin *temptare, tentare* to feel, try] (13th century)
1 : to entice to do wrong by promise of pleasure or gain
2 a *obsolete* **:** to make trial of **:** TEST **b :** to try presumptuously **:** PROVOKE **c :** to risk the dangers of
3 a : to induce to do something **b :** to cause to be strongly inclined ⟨was *tempted* to call it quits⟩
synonym see LURE
— tempt·able \'tem(p)-tə-bəl\ *adjective*

temp·ta·tion \tem(p)-'tā-shən\ *noun* (13th century)
1 : the act of tempting or the state of being tempted especially to evil **:** ENTICEMENT
2 : something tempting **:** a cause or occasion of enticement

tempt·er \'tem(p)-tər\ *noun* (14th century)
: one that tempts or entices

tempt·ing *adjective* (1596)
: having an appeal **:** ENTICING ⟨a *tempting* offer⟩
— tempt·ing·ly \-tiŋ-lē\ *adverb*

tempt·ress \'tem(p)-trəs\ *noun* (1594)
: a woman who tempts or entices

tem·pu·ra \'tem-pə-rə, -ˌrä; tem-'pu̇r-ə\ *noun* [Japanese *tenpura*] (1920)
: seafood or vegetables dipped in batter and fried in deep fat

ten \'ten\ *noun* [Middle English, from Old English *tīene,* from *tīen,* adjective, ten; akin to Old High German *zehan* ten, Latin *decem,* Greek *deka*] (before 12th century)
1 — see NUMBER table
2 : the 10th in a set or series ⟨wears a *ten*⟩
3 : something having 10 units or members

4 : a 10-dollar bill
— **ten** *adjective*
— **ten** *pronoun, plural in construction*

ten·a·ble \'te-nə-bəl\ *adjective* [Middle French, from Old French, from *tenir* to hold, from Latin *tenēre* — more at THIN] (1579)
: capable of being held, maintained, or defended **:** DEFENSIBLE, REASONABLE
— **ten·a·bil·i·ty** \ˌte-nə-'bi-lə-tē\ *noun*
— **ten·a·ble·ness** *noun*
— **ten·a·bly** \'te-nə-blē\ *adverb*

ten·ace \'te-ˌnās, te-'nās, 'te-nəs\ *noun* [modification of Spanish *tenaza,* literally, forceps, probably from Latin *tenacia,* neuter plural of *tenax*] (1655)
: a combination of two high or relatively high cards (as ace and queen) of the same suit in one hand with one ranking two degrees below the other

te·na·cious \tə-'nā-shəs\ *adjective* [Latin *tenac-, tenax* tending to hold fast, from *tenēre* to hold] (1607)
1 a : not easily pulled apart **:** COHESIVE, TOUGH ⟨a *tenacious* metal⟩ **b :** tending to adhere or cling especially to another substance **:** STICKY ⟨*tenacious* burs⟩ ⟨*tenacious* clay⟩
2 a : persistent in maintaining or adhering to something valued or habitual ⟨a man very *tenacious* of his rights⟩ **b :** RETENTIVE ⟨a *tenacious* memory⟩
synonym see STRONG
— **te·na·ci·ous·ly** *adverb*
— **te·na·cious·ness** *noun*

te·nac·i·ty \tə-'na-sə-tē\ *noun* (1526)
: the quality or state of being tenacious
synonym see COURAGE

te·nac·u·lum \tə-'na-kyə-ləm\ *noun, plural* **-la** \-lə\ *or* **-lums** [New Latin, from Late Latin, instrument for holding, from Latin *tenēre*] (circa 1693)
1 : a slender sharp-pointed hook attached to a handle and used mainly in surgery for seizing and holding parts (as arteries)
2 : an adhesive animal structure

ten·an·cy \'te-nən(t)-sē\ *noun, plural* **-cies** (1590)
: a holding of an estate or a mode of holding an estate **:** the temporary possession or occupancy of something (as a house) that belongs to another; *also* **:** the period of a tenant's occupancy or possession

[1]**ten·ant** \'te-nənt\ *noun* [Middle English, from Middle French, from present participle of *tenir* to hold] (14th century)
1 a : one who holds or possesses real estate or sometimes personal property (as an annuity) by any kind of right **b :** one who has the occupation or temporary possession of lands or tenements of another; *specifically* **:** one who rents or leases (as a house) from a landlord
2 : OCCUPANT, DWELLER
— **ten·ant·less** \-ləs\ *adjective*

[2]**tenant** *transitive verb* (1634)
: to hold or occupy as a tenant **:** INHABIT
— **ten·ant·able** \-nən-tə-bəl\ *adjective*

tenant farmer *noun* (1748)
: a farmer who works land owned by another and pays rent either in cash or in shares of produce

ten·ant·ry \'te-nən-trē\ *noun, plural* **-ries** (14th century)
1 : TENANCY
2 : a body of tenants

ten–cent store \'ten-'sent-\ *noun* (1901)
: FIVE-AND-TEN

tench \'tench\ *noun, plural* **tench** *or* **tench·es** [Middle English, from Middle French *tenche,* from Late Latin *tinca*] (14th century)
: a cyprinid fish (*Tinca tinca*) native to Eurasia but introduced in the U.S. and noted for its ability to survive outside water

Ten Commandments *noun plural* (13th century)
: the ethical commandments of God given according to biblical accounts to Moses by voice and by writing on stone tablets on Mount Sinai

[1]**tend** \'tend\ *verb* [Middle English, short for *attenden* to attend] (14th century)
intransitive verb
1 *archaic* **:** LISTEN
2 : to pay attention **:** apply oneself ⟨*tend* to your own affairs⟩
3 : to act as an attendant **:** SERVE
4 *obsolete* **:** AWAIT
transitive verb
1 *archaic* **:** to attend as a servant
2 a : to apply oneself to the care of **:** watch over **b :** to have or take charge of as a caretaker or overseer **c :** CULTIVATE, FOSTER **d :** to manage the operations of **:** MIND ⟨*tend* the store⟩
3 : to stand by (as a rope) in readiness to prevent mischance (as fouling)

[2]**tend** *intransitive verb* [Middle English, from Middle French *tendre* to stretch, from Latin *tendere* — more at THIN] (14th century)
1 : to move, direct, or develop one's course in a particular direction ⟨cannot tell where society is *tending*⟩
2 : to exhibit an inclination or tendency **:** CONDUCE ⟨*tends* to be optimistic⟩

ten·dance \'ten-dən(t)s\ *noun* [short for *attendance*] (1573)
1 : watchful care
2 *archaic* **:** persons in attendance **:** RETINUE

ten·den·cious *chiefly British variant of* TENDENTIOUS

ten·den·cy \'ten-dən(t)-sē\ *noun, plural* **-cies** [Medieval Latin *tendentia,* from Latin *tendent-, tendens,* present participle of *tendere*] (1628)
1 a : direction or approach toward a place, object, effect, or limit **b :** a proneness to a particular kind of thought or action
2 a : the purposeful trend of something written or said **:** AIM **b :** deliberate but indirect advocacy ☆

ten·den·tious \ten-'den(t)-shəs\ *adjective* (1900)
: marked by a tendency in favor of a particular point of view **:** BIASED
— **ten·den·tious·ly** *adverb*
— **ten·den·tious·ness** *noun*

[1]**ten·der** \'ten-dər\ *adjective* [Middle English, from Old French *tendre,* from Latin *tener;* perhaps akin to Latin *tenuis* thin, slight — more at THIN] (13th century)
1 a : having a soft or yielding texture **:** easily broken, cut, or damaged **:** DELICATE, FRAGILE ⟨*tender* feet⟩ **b :** easily chewed **:** SUCCULENT
2 a : physically weak **:** not able to endure hardship **b :** IMMATURE, YOUNG ⟨children of *tender* years⟩ **c :** incapable of resisting cold **:** not hardy
3 : marked by, responding to, or expressing the softer emotions **:** FOND, LOVING ⟨a *tender* lover⟩
4 a : showing care **:** CONSIDERATE, SOLICITOUS ⟨*tender* regard⟩ **b :** highly susceptible to impressions or emotions **:** IMPRESSIONABLE ⟨a *tender* conscience⟩
5 a : appropriate or conducive to a delicate or sensitive constitution or character **:** GENTLE, MILD ⟨*tender* breeding⟩ ⟨*tender* irony⟩ **b :** delicate or soft in quality or tone ⟨never before heard the piano sound so *tender* —Elva S. Daniels⟩
6 *obsolete* **:** DEAR, PRECIOUS
7 a : sensitive to touch or palpation ⟨the bruise was still *tender*⟩ **b :** sensitive to injury or insult **:** TOUCHY ⟨*tender* pride⟩ **c :** demanding careful and sensitive handling **:** TICKLISH ⟨a *tender* situation⟩ **d** *of a ship* **:** [5]CRANK
— **ten·der·ly** *adverb*
— **ten·der·ness** *noun*

[2]**tender** *verb* **ten·dered; ten·der·ing** \-d(ə-)riŋ\ (14th century)
transitive verb
1 : to make tender **:** SOFTEN, WEAKEN
2 *archaic* **:** to regard or treat with tenderness
intransitive verb
: to become tender

[3]**tender** *verb* **ten·dered; ten·der·ing** \-d(ə-)riŋ\ [Middle French *tendre* to stretch, stretch out, offer — more at TEND] (1535)
transitive verb
1 : to make a tender of
2 : to present for acceptance **:** OFFER ⟨*tendered* my resignation⟩
intransitive verb
: to make a bid or tender ⟨*tender* for a building contract⟩ ⟨*tendered* for 6% of the stock⟩

[4]**tender** *noun, often attributive* (circa 1543)
1 : an unconditional offer of money or service in satisfaction of a debt or obligation made to save a penalty or forfeiture for nonpayment or nonperformance
2 : an offer or proposal made for acceptance: as **a :** an offer of a bid for a contract **b :** a public expression of willingness to buy not less than a specified number of shares of a stock at a fixed price from stockholders usually in an attempt to gain control of the issuing company
3 : something that may be offered in payment; *specifically* **:** MONEY

[5]**tender** *noun* [[1]*tender*] (1596)
obsolete **:** CONSIDERATION, REGARD

[6]**tend·er** \'ten-dər\ *noun* (1675)
: one that tends: as **a** (1) **:** a ship employed to attend other ships (as to supply provisions) (2) **:** a boat for communication between shore and a larger ship (3) **:** a warship that provides logistic support **b :** a car attached to a steam locomotive for carrying a supply of fuel and water

ten·der·foot \'ten-dər-ˌfu̇t\ *noun, plural* **ten·der·feet** \-ˌfēt\ *also* **ten·der·foots** \-ˌfu̇ts\ (1849)
1 : a newcomer in a comparatively rough or newly settled region; *especially* **:** one not hardened to frontier or outdoor life
2 : an inexperienced beginner **:** NOVICE ⟨a political *tenderfoot*⟩

ten·der·heart·ed \'ten-dər-ˌhär-təd\ *adjective* (15th century)
: easily moved to love, pity, or sorrow **:** COMPASSIONATE, IMPRESSIONABLE
— **ten·der·heart·ed·ly** *adverb*
— **ten·der·heart·ed·ness** *noun*

ten·der·ize \'ten-də-ˌrīz\ *transitive verb* **-ized; -iz·ing** (1930)
: to make (meat or meat products) tender by applying a process or substance that breaks down connective tissue
— **ten·der·i·za·tion** \ˌten-d(ə-)rə-'zā-shən\ *noun*

☆ SYNONYMS

Tendency, trend, drift, tenor, current mean movement in a particular direction. TENDENCY implies an inclination sometimes amounting to an impelling force ⟨a general *tendency* toward inflation⟩. TREND applies to the general direction maintained by a winding or irregular course ⟨the long-term *trend* of the stock market is upward⟩. DRIFT may apply to a tendency determined by external forces ⟨the *drift* of the population away from large cities⟩ or it may apply to an underlying or obscure trend of meaning or discourse ⟨got the *drift* of her argument⟩. TENOR stresses a clearly perceptible direction and a continuous, undeviating course ⟨the *tenor* of the times⟩. CURRENT implies a clearly defined but not necessarily unalterable course ⟨an encounter that changed the *current* of my life⟩.

— **ten·der·iz·er** \'ten-də-ˌrī-zər\ *noun*

ten·der·loin \'ten-dər-ˌlȯin\ *noun* (circa 1828)
1 : a strip of tender meat consisting of a large internal muscle of the loin on each side of the vertebral column
2 [from its making possible a luxurious diet for a corrupt police officer] **:** a district of a city largely devoted to vice

ten·der–mind·ed \'ten-dər-ˌmīn-dəd\ *adjective* (1605)
: marked by idealism, optimism, and dogmatism

ten·der·om·e·ter \ˌten-də-'rä-mə-tər\ *noun* (1938)
: a device for determining the maturity and tenderness of samples of fruits and vegetables

ten·di·ni·tis *or* **ten·don·itis** \ˌten-də-'nī-təs\ *noun* [*tendinitis* from New Latin, from *tendin, tendo* + *-itis; tendonitis* from *tendon* + *-itis*] (circa 1900)
: inflammation of a tendon

ten·di·nous \'ten-də-nəs\ *adjective* [New Latin *tendinosus,* from *tendin-, tendo* tendon, alteration of Medieval Latin *tendon-, tendo*] (1658)
1 : consisting of tendons **:** SINEWY ⟨*tendinous* tissue⟩
2 : of, relating to, or resembling a tendon

ten·don \'ten-dən\ *noun* [Medieval Latin *tendon-, tendo,* from Latin *tendere* to stretch — more at THIN] (1541)
: a tough cord or band of dense white fibrous connective tissue that unites a muscle with some other part (as a bone) and transmits the force which the muscle exerts

tendon of Achil·les \-ə-'ki-lēz\ (circa 1885)
: ACHILLES TENDON

ten·dresse \tä^{n}-dres\ *noun* [Middle English, from Middle French, from *tendre* tender] (14th century)
: FONDNESS

ten·dril \'ten-drəl\ *noun* [probably modification of Middle French *tendron, tendrum* bud, cartilage, from (assumed) Vulgar Latin *tenerumen,* from Latin *tener* tender — more at TENDER] (1538)
1 : a leaf, stipule, or stem modified into a slender spirally coiling sensitive organ serving to attach a climbing plant to its support
2 : something suggestive of a tendril ⟨hair hanging in *tendrils*⟩ ⟨creeping *tendrils* of fog⟩
— **ten·driled** *or* **ten·drilled** \-drəld\ *adjective*
— **ten·dril·ous** \-drə-ləs\ *adjective*

tendril 1

[1]-tene *adjective combining form* [French *-tène,* from Latin *taenia* ribbon, band — more at TAENIA]
: having (such or so many) chromosomal filaments ⟨poly*tene*⟩ ⟨pachy*tene*⟩

[2]-tene *noun combining form*
: stage of meiotic prophase characterized by (such) chromosomal filaments ⟨diplo*tene*⟩ ⟨pachy*tene*⟩

Ten·e·brae \'te-nə-ˌbrā, -ˌbrī, -ˌbrē\ *noun plural but singular or plural in construction* [Medieval Latin, from Latin, darkness — more at TEMERITY] (1651)
: a church service observed during the final part of Holy Week commemorating the sufferings and death of Christ

ten·e·brif·ic \ˌte-nə-'bri-fik\ *adjective* [Latin *tenebrae* darkness] (1785)
1 : GLOOMY
2 : causing gloom or darkness

te·ne·bri·o·nid \tə-'ne-brē-ə-nəd, ˌte-nə-'brī-ə-nəd\ *noun* [New Latin *Tenebrionidae,* from *Tenebrion-, Tenebrio,* type genus, from Latin, one that shuns the light, from *tenebrae* darkness — more at TEMERITY] (1902)
: any of a family (Tenebrionidae) of firm-bodied mostly dark-colored vegetable-feeding beetles which often have vestigial and functionless wings and whose larvae are usually hard cylindrical worms (as a mealworm) — called also *darkling beetle*
— **tenebrionid** *adjective*

te·neb·ri·ous \tə-'ne-brē-əs\ *adjective* [by alteration] (1594)
: TENEBROUS

ten·e·brism \'te-nə-ˌbri-zəm\ *noun, often capitalized* [Latin *tenebrae* darkness] (1954)
: a style of painting especially associated with the Italian painter Caravaggio and his followers in which most of the figures are engulfed in shadow but some are dramatically illuminated by a concentrated beam of light usually from an identifiable source
— **ten·e·brist** \-brist\ *noun or adjective, often capitalized*

ten·e·brous \'te-nə-brəs\ *adjective* [Middle English, from Middle French *tenebreus,* from Latin *tenebrosus,* from *tenebrae*] (15th century)
1 : shut off from the light **:** DARK, MURKY
2 : hard to understand **:** OBSCURE
3 : causing gloom

1080 *also* **ten–eighty** \(ˌ)ten-'ā-tē\ *noun* [from its laboratory serial number] (1945)
: a poisonous preparation of sodium fluoroacetate used as a rodenticide and pesticide

ten·e·ment \'te-nə-mənt\ *noun* [Middle English, from Middle French, from Medieval Latin *tenementum,* from Latin *tenēre* to hold — more at THIN] (14th century)
1 : any of various forms of corporeal property (as land) or incorporeal property that is held by one person from another
2 : DWELLING
3 a : a house used as a dwelling **:** RESIDENCE **b :** APARTMENT, FLAT **c :** TENEMENT HOUSE

tenement house *noun* (1858)
: APARTMENT HOUSE; *especially* **:** one meeting minimum standards of sanitation, safety, and comfort and usually located in a city

te·nes·mus \tə-'nez-məs\ *noun* [Latin, from Greek *teinesmos,* from *teinein* to stretch, strain — more at THIN] (1527)
: a distressing but ineffectual urge to evacuate the rectum or bladder

te·net \'te-nət *also* 'tē-nət\ *noun* [Latin, he holds, from *tenēre* to hold] (circa 1600)
: a principle, belief, or doctrine generally held to be true; *especially* **:** one held in common by members of an organization, movement, or profession

ten·fold \'ten-ˌfōld, -'fōld\ *adjective* (before 12th century)
1 : being 10 times as great or as many
2 : having 10 units or members
— **ten·fold** \-'fōld\ *adverb*

ten–gallon hat *noun* (1927)
: COWBOY HAT

te·nia, te·ni·a·sis *variant of* TAENIA, TAENIASIS

ten·ner \'te-nər\ *noun* (1845)
1 : a 10-pound note
2 : a 10-dollar bill

Ten·nes·see walking horse \'te-nə-ˌsē-\ *noun* [*Tennessee,* state of U.S.] (1938)
: any of an American breed of large easy-gaited saddle horses largely of standardbred and Morgan ancestry — called also *Tennessee walker*

Tennessee walking horse

ten·nies \'te-nēz\ *noun plural* [by shortening & alteration] (1951)
: TENNIS SHOES, SNEAKERS

ten·nis \'te-nəs\ *noun, often attributive* [Middle English *tenetz, tenys,* probably from Anglo-French *tenetz,* 2d person plural imperative of *tenir* to hold — more at TENABLE] (15th century)
1 : COURT TENNIS
2 : an indoor or outdoor game that is played with rackets and a light elastic ball by two players or pairs of players on a level court (as of clay or grass) divided by a low net

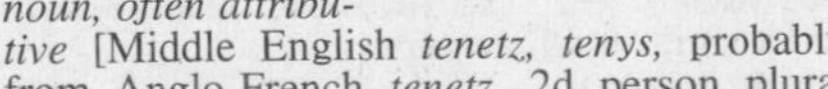

tennis elbow *noun* (1883)
: inflammation and pain over the outer side of the elbow usually resulting from excessive strain on and twisting of the forearm

tennis shoe *noun* (1886)
: a lightweight usually low-cut sneaker

ten·nist \'te-nist\ *noun* [blend of *tennis* and *-ist*] (1932)
: a tennis player

[1]ten·on \'te-nən\ *noun* [Middle English, from Middle French, from *tenir* to hold — more at TENABLE] (15th century)
: a projecting member in a piece of wood or other material for insertion into a mortise to make a joint — see DOVETAIL illustration

[2]tenon *transitive verb* (1596)
1 : to unite by a tenon
2 : to cut or fit for insertion in a mortise

[1]ten·or \'te-nər\ *noun* [Middle English, from Middle French, from Latin *tenor* uninterrupted course, from *tenēre* to hold — more at THIN] (14th century)
1 a : the drift of something spoken or written **:** PURPORT **b :** an exact copy of a writing **:** TRANSCRIPT **c :** the concept, object, or person meant in a metaphor
2 a : the melodic line usually forming the cantus firmus in medieval music **b :** the voice part next to the lowest in a 4-part chorus **c :** the highest natural adult male singing voice; *also* **:** a person having this voice **d :** a member of a family of instruments having a range next lower than that of the alto
3 : a continuance in a course, movement, or activity
4 : habitual condition **:** CHARACTER
synonym see TENDENCY

[2]tenor *adjective* (1522)
: relating to or having the range or part of a tenor

ten·or·ist \'te-nə-rist\ *noun* (1865)
: one who sings tenor or plays a tenor instrument

te·no·syn·o·vi·tis \'te-nō-ˌsi-nə-'vī-təs, ˌtē-\ *noun* [New Latin, from Greek *tenōn* tendon (akin to Greek *teinein* to stretch) + New Latin *synovitis* — more at THIN] (circa 1860)
: inflammation of a tendon sheath

ten·our \'te-nər\ *chiefly British variant of* TENOR

ten·pen·ny \'ten-'pe-nē, *British* -pə-nē\ *adjective* (1592)
: amounting to, worth, or costing 10 pennies

tenpenny nail *noun* [from its original price per hundred] (15th century)
: a nail three inches (7.6 centimeters) long

ten·pin \'ten-ˌpin\ *noun* (1807)
1 : a bottle-shaped bowling pin 15 inches high
2 *plural but singular in construction* **:** a bowling game using 10 tenpins and a large ball 27 inches in circumference and allowing each player to bowl 2 balls in each of 10 frames

ten·pound·er \'ten-'pau̇n-dər\ *noun* (1699)
: LADYFISH 2

ten·rec \'ten-ˌrek\ *noun* [French, from Malagasy *tàndraka*] (1785)
: any of numerous small often spiny mammalian insectivores (family Tenrecidae) chiefly of Madagascar

[1]tense \'ten(t)s\ *noun* [Middle English *tens* time, tense, from Middle French, from Latin *tempus*] (14th century)
1 : a distinction of form in a verb to express distinctions of time or duration of the action or state it denotes
2 a : a set of inflectional forms of a verb that express distinctions of time **b :** an inflectional form of a verb expressing a specific time distinction

[2]tense *adjective* **tens·er; tens·est** [Latin *tensus,* from past participle of *tendere* to stretch — more at THIN] (1670)
1 : stretched tight **:** made taut **:** RIGID
2 a : feeling or showing nervous tension **b :** marked by strain or suspense
3 : produced with the muscles involved in a relatively tense state ⟨the vowels \ē\ and \ü\ in contrast with the vowels \i\ and \u̇\ are *tense*⟩
— tense·ly *adverb*
— tense·ness *noun*

[3]tense *verb* **tensed; tens·ing** (1676)
transitive verb
: to make tense
intransitive verb
: to become tense

ten·sile \'ten(t)-səl *also* 'ten-ˌsīl\ *adjective* [New Latin *tensilis,* from Latin *tensus,* past participle] (1626)
1 : capable of tension **:** DUCTILE
2 : of, relating to, or involving tension ⟨*tensile* stress⟩
— ten·sil·i·ty \ten-'si-lə-tē\ *noun*

tensile strength *noun* (circa 1864)
: the greatest longitudinal stress a substance can bear without tearing apart

ten·si·om·e·ter \ˌten(t)-sē-'ä-mə-tər\ *noun* [*tension*] (1912)
1 : a device for measuring tension (as of structural material)
2 : an instrument for determining the moisture content of soil
3 : an instrument for measuring the surface tension of liquids
— ten·sio·met·ric \-sē-ō-'me-trik\ *adjective*
— ten·si·om·e·try \-sē-'ä-mə-trē\ *noun*

[1]ten·sion \'ten(t)-shən\ *noun* [Middle French or Latin; Middle French, from Latin *tension-, tensio,* from *tendere*] (1533)
1 a : the act or action of stretching or the condition or degree of being stretched to stiffness **:** TAUTNESS **b :** STRESS 1b
2 a : either of two balancing forces causing or tending to cause extension **b :** the stress resulting from the elongation of an elastic body
3 a : inner striving, unrest, or imbalance often with physiological indication of emotion **b :** a state of latent hostility or opposition between individuals or groups **c :** a balance maintained in an artistic work between opposing forces or elements
4 : a device to produce a desired tension (as in a loom)
— ten·sion·al \'ten(t)-sh(ə-)nəl\ *adjective*
— ten·sion·less \'ten(t)-shən-ləs\ *adjective*

[2]tension *transitive verb* **ten·sioned; ten·sion·ing** \'ten(t)-sh(ə-)niŋ\ (1891)
: to subject to tension; *especially* **:** to tighten to a desired or appropriate degree
— ten·sion·er \-sh(ə-)nər\ *noun*

ten·si·ty \'ten(t)-sə-tē\ *noun, plural* **-ties** (circa 1658)
: the quality or state of being tense **:** TENSENESS

ten·sive \'ten(t)-siv\ *adjective* (1702)
: of, relating to, or causing tension

ten·sor \'ten(t)-sər, 'ten-ˌsȯr\ *noun* [New Latin, from Latin *tendere*] (circa 1704)
1 : a muscle that stretches a part
2 : a generalized vector with more than three components each of which is a function of the coordinates of an arbitrary point in space of an appropriate number of dimensions

ten–speed \'ten-ˌspēd\ *noun* (1971)
: a bicycle with 10 gear combinations

tens place \'tenz-\ *noun* (1937)
: the place two to the left of the decimal point in a number expressed in the Arabic system of writing numbers

ten–strike \'ten-ˌstrīk\ *noun* (1840)
1 : a strike in tenpins
2 : a highly successful stroke or achievement

[1]tent \'tent\ *noun* [Middle English *tente,* from Old French, from Latin *tenta,* feminine of *tentus,* past participle of *tendere* to stretch — more at THIN] (14th century)
1 : a collapsible shelter of fabric (as nylon or canvas) stretched and sustained by poles and used for camping outdoors or as a temporary building
2 : DWELLING
3 a : something that resembles a tent or that serves as a shelter; *especially* **:** a canopy or enclosure placed over the head and shoulders to retain vapors or oxygen during medical administration **b :** the web of a tent caterpillar
— tent·less \'tent-ləs\ *adjective*
— tent·like \-ˌlīk\ *adjective*

[2]tent (1607)
intransitive verb
1 : to reside for the time being **:** LODGE
2 : to live in a tent
transitive verb
1 : to cover with or as if with a tent
2 : to lodge in tents

[3]tent *transitive verb* [Middle English, from *tent* attention, short for *attent,* from Old French *attente,* from *atendre* to attend] (14th century)
chiefly Scottish **:** to attend to

ten·ta·cle \'ten-ti-kəl\ *noun* [New Latin *tentaculum,* from Latin *tentare* to feel, touch — more at TEMPT] (1762)
1 : any of various elongate flexible usually tactile or prehensile processes borne by animals and especially invertebrates chiefly on the head or about the mouth
2 a : something that resembles a tentacle especially in or as if in grasping or feeling out **b :** a sensitive hair or emergence on a plant (as the sundew)
— ten·ta·cled \-kəld\ *adjective*

ten·tac·u·lar \ten-'ta-kyə-lər\ *adjective* [New Latin *tentaculum*] (1828)
1 : of, relating to, or resembling tentacles
2 : equipped with tentacles

tent·age \'ten-tij\ *noun* (1603)
: a collection of tents **:** tent equipment

ten·ta·tive \'ten-tə-tiv\ *adjective* [Medieval Latin *tentativus,* from Latin *tentatus,* past participle of *tentare, temptare* to feel, try] (1626)
1 : not fully worked out or developed ⟨*tentative* plans⟩
2 : HESITANT, UNCERTAIN ⟨a *tentative* smile⟩
— tentative *noun*
— ten·ta·tive·ly *adverb*
— ten·ta·tive·ness *noun*

tent caterpillar *noun* (1854)
: any of several destructive gregarious caterpillars (genus *Malacosoma* and especially *M. americanum* of the family Lasiocampidae) that construct large silken webs on trees

tent·ed \'ten-təd\ *adjective* (1604)
1 : covered with a tent or tents
2 : shaped like a tent

[1]ten·ter \'ten-tər\ *noun* [Middle English *teyntur, tentowre*] (14th century)
1 : a frame or endless track with hooks or clips along two sides that is used for drying and stretching cloth
2 *archaic* **:** TENTERHOOK

[2]tenter *noun* (1846)
: one who lives in or occupies a tent

ten·ter·hook \'ten-tər-ˌhu̇k\ *noun* (15th century)
: a sharp hooked nail used especially for fastening cloth on a tenter
— on tenterhooks : in a state of uneasiness, strain, or suspense

tenth \'ten(t)th\ *noun, plural* **tenths** \'ten(t)s, 'ten(t)ths\ (15th century)
1 — see NUMBER table
2 a : a musical interval embracing an octave and a third **b :** the tone at this interval
— tenth *adjective or adverb*

tenth–rate \'tenth-'rāt\ *adjective* (1834)
: of the lowest character or quality

tent stitch *noun* (1639)
: a short stitch slanting to the right that is used in embroidery to form even lines of solid background

tenty *also* **tent·ie** \'ten-tē\ *adjective* [[3]*tent*] (circa 1555)
Scottish **:** ATTENTIVE, WATCHFUL

ten·u·is \'ten-yə-wəs\ *noun, plural* **-u·es** \-yə-ˌwēz, -ˌwās\ [Medieval Latin, from Latin, thin, slight] (1650)
: an unaspirated voiceless stop

te·nu·ity \te-'nü-ə-tē, tə-, -'nyü-\ *noun* [Latin *tenuitas,* from *tenuis* thin, tenuous] (circa 1536)
1 : lack of substance or strength
2 : lack of thickness **:** SLENDERNESS, THINNESS
3 : lack of density **:** rarefied quality or state

ten·u·ous \'ten-yə-wəs\ *adjective* [Latin *tenuis* thin, slight, tenuous — more at THIN] (1597)
1 : not dense **:** RARE ⟨a *tenuous* fluid⟩
2 : not thick **:** SLENDER ⟨a *tenuous* rope⟩
3 a : having little substance or strength **:** FLIMSY, WEAK ⟨*tenuous* influences⟩ **b :** SHAKY 2a ⟨*tenuous* reasons⟩ ⟨on grounds that were *tenuous*⟩
synonym see THIN
— ten·u·ous·ly *adverb*
— ten·u·ous·ness *noun*

ten·ure \'ten-yər *also* -ˌyu̇r\ *noun* [Middle English, from Middle French *teneüre, tenure,* from Medieval Latin *tenitura,* from (assumed) Vulgar Latin *tenitus,* past participle of Latin *tenēre* to hold — more at THIN] (15th century)
1 : the act, right, manner, or term of holding something (as a landed property, a position, or an office); *especially* **:** a status granted after a trial period to a teacher that gives protection from summary dismissal
2 : GRASP, HOLD
— ten·ur·able \-ə-bəl\ *adjective*
— te·nur·ial \te-'nyu̇r-ē-əl\ *adjective*
— te·nur·ial·ly \-ə-lē\ *adverb*

ten·ured \'ten-yərd\ *adjective* (1965)
: having tenure ⟨*tenured* faculty members⟩

ten·ure–track \'ten-yər-ˌtrak *also* -ˌyu̇r-\ *adjective* (1976)
: relating to or being a teaching position that may lead to a grant of tenure

te·nu·to \tā-'nü-(ˌ)tō\ *adverb or adjective* [Italian, from past participle of *tenere* to hold, from Latin *tenēre*] (1762)
: in a manner so as to hold a tone or chord to its full value — used as a direction in music

te·o·cal·li \ˌtē-ə-'ka-lē, ˌtā-ə-'kä-\ *noun* [Nahuatl *teōcalli,* from *teōtl* god + *calli* house] (1613)
: an ancient temple of Mexico or Central America usually built upon the summit of a truncated pyramidal mound; *also* **:** the mound itself

te·o·sin·te \ˌtā-ō-'sin-tē\ *noun* [Mexican Spanish, from Nahuatl *teōcintli,* from *teōtl* god + *cintli* dried ears of maize] (circa 1877)
: a tall annual grass (*Zea mexicana* synonym *Euchlaena mexicana*) of Mexico and Central America that is closely related to and sometimes considered ancestral to maize

te·pa \'te-pə\ *noun* [*tri-* + *e*thylene + *p*hosphor- + *a*mide] (1953)
: a soluble crystalline compound $C_6H_{12}N_3OP$ that is used especially as a chemical sterilizing agent of insects, a palliative in some kinds of cancer, and in finishing and flame-proofing textiles

te·pa·ry bean \'te-pə-rē-\ *noun* [origin unknown] (1912)
: an annual twining bean (*Phaseolus acutifolius* variety *latifolius*) that is native to the

southwestern U.S. and Mexico and is cultivated for its roundish white, yellow, brown, or bluish black edible seeds; *also* **:** the seed

te·pee \'tē-(,)pē\ *noun* [Dakota *tʰípi,* from *tʰi-* to dwell] (1743)
: a conical tent usually consisting of skins and used especially by American Indians of the Plains

tepee

teph·ra \'te-frə\ *noun* [New Latin, from Greek, ashes; akin to Sanskrit *dahati* it burns — more at FOMENT] (1944)
: solid material ejected into the air during a volcanic eruption; *especially* **:** ²ASH 2b

tep·id \'te-pəd\ *adjective* [Middle English *teped,* from Latin *tepidus,* from *tepēre* to be moderately warm; akin to Sanskrit *tapati* it heats, Old Irish *tess* heat] (14th century)
1 : moderately warm **:** LUKEWARM ⟨a *tepid* bath⟩
2 a : lacking in passion, force, or zest ⟨a *tepid* joke⟩ **b :** marked by an absence of enthusiasm or conviction ⟨a *tepid* interest⟩
— te·pid·i·ty \tə-'pi-də-tē, te-\ *noun*
— tep·id·ly \'te-pəd-lē\ *adverb*
— tep·id·ness *noun*

TEPP \,tē-(,)ē-(,)pē-'pē\ *noun* [*te*tra- + *e*thyl + *p*yro*p*hosphate] (1948)
: a mobile hygroscopic corrosive liquid organophosphate $C_8H_{20}O_7P_2$ that is a powerful anticholinesterase and is used especially as an insecticide

te·qui·la \tə-'kē-lə, tā-\ *noun* [Spanish, from *Tequila,* town in Jalisco state, Mexico] (1849)
: a Mexican liquor distilled from pulque

tequila sunrise *noun* (1965)
: a cocktail consisting of tequila, orange juice, and grenadine

ter- *combining form* [Latin, from *ter;* akin to Greek & Sanskrit *tris* three times, Latin *tres* three — more at THREE]
: three times **:** threefold **:** three ⟨*ter*centenary⟩

tera- *combining form* [International Scientific Vocabulary, from Greek *terat-, teras* monster]
: trillion ⟨*tera*watt⟩

te·rai \tə-'rī\ *noun* [*Tarai,* lowland belt of India] (1888)
: a wide-brimmed double felt sun hat worn especially in subtropical regions

ter·aph \'ter-əf\ *noun, plural* **ter·a·phim** \'ter-ə-,fim\ [Hebrew *tĕrāphīm* (plural in form but singular in meaning)] (14th century)
: an image of a Semitic household god

terat- *or* **terato-** *combining form* [Greek, from *terat-, teras* marvel, portent, monster]
: developmental malformation ⟨*terato*genic⟩

te·rato·car·ci·no·ma \,ter-ə-tō-,kär-sᵊn-'ō-mə\ *noun* (1946)
: a malignant teratoma; *especially* **:** one involving germinal cells of the testis

te·rato·gen \tə-'ra-tə-jən\ *noun* (1959)
: a teratogenic agent

ter·a·to·gen·e·sis \,ter-ə-tə-'je-nə-səs\ *noun* [New Latin] (1901)
: production of developmental malformations

ter·a·to·gen·ic \-'je-nik\ *adjective* (1879)
: of, relating to, or causing developmental malformations ⟨*teratogenic* substances⟩ ⟨*teratogenic* effects⟩
— ter·a·to·ge·nic·i·ty \-jə-'ni-sə-tē\ *noun*

ter·a·to·log·i·cal \,ter-ə-tᵊl-'ä-ji-kəl\ *or* **ter·a·to·log·ic** \-jik\ *adjective* (1857)
1 : abnormal in growth or structure
2 : of or relating to teratology

ter·a·tol·o·gy \,ter-ə-'tä-lə-jē\ *noun* (circa 1842)
: the study of malformations or serious deviations from the normal type in organisms
— ter·a·tol·o·gist \-jist\ *noun*

ter·a·to·ma \,ter-ə-'tō-mə\ *noun* [New Latin] (1879)
: a tumor made up of a heterogeneous mixture of tissues

tera·watt \'ter-ə-,wät\ *noun* (1970)
: a unit of power equal to one trillion watts

ter·bi·um \'tər-bē-əm\ *noun* [New Latin, from *Ytterby,* Sweden] (1843)
: a usually trivalent metallic element of the rare-earth group — see ELEMENT table

terce \'tərs\ *noun, often capitalized* [Middle English, third, terce — more at TIERCE] (14th century)
: the third of the canonical hours

ter·cel \'tər-səl\ *variant of* TIERCEL

ter·cen·te·na·ry \,tər-(,)sen-'te-nə-rē, (,)tər-'sen-tᵊn-,er-ē\ *noun, plural* **-ries** (1855)
: a 300th anniversary or its celebration
— tercentenary *adjective*

ter·cen·ten·ni·al \,tər-(,)sen-'te-nē-əl\ *adjective or noun* (1872)
: TERCENTENARY

ter·cet \'tər-sət\ *noun* [Italian *terzetto,* from diminutive of *terzo* third, from Latin *tertius* — more at THIRD] (1598)
: a unit or group of three lines of verse: **a :** one of the 3-line stanzas in terza rima **b :** one of the two groups of three lines forming the sestet in an Italian sonnet

ter·e·binth \'ter-ə-,bin(t)th\ *noun* [Middle English *terebynt,* from Middle French *terebinthe,* from Latin *terebinthus* — more at TURPENTINE] (14th century)
: a small European tree (*Pistacia terebinthus*) of the cashew family yielding turpentine

te·re·do \tə-'rē-(,)dō, -'rā-\ *noun, plural* **tere·dos** *or* **te·red·i·nes** \-'re-dᵊn-,ēz\ [Middle English, from Latin *teredin-, teredo,* from Greek *terēdōn;* akin to Greek *tetrainein* to bore — more at THROW] (14th century)
: SHIPWORM

tere·phthal·ate \,ter-ə(f)-'tha-,lāt\ *noun* (1868)
: a salt or ester of terephthalic acid; *especially* **:** a dimethyl-ester that is a major starting material for polyester fibers and coatings

tere·phthal·ic acid \,ter-ə(f)-'tha-lik-\ *noun* [International Scientific Vocabulary *tere*bene, mixture of terpenes from distilled turpentine + *phthalic acid*] (1857)
: a *p*-dicarboxylic acid $C_8H_6O_4$ that is obtained especially by oxidation of xylene and is used chiefly in the synthesis of polyesters

te·rete \tə-'rēt, te-\ *adjective* [Latin *teret-, teres* well turned, rounded; akin to Latin *terere* to rub — more at THROW] (circa 1619)
: approximately cylindrical but usually tapering at both ends ⟨a *terete* seedpod⟩

Te·reus \'tir-,yüs, 'tē-,rüs\ *noun* [Latin, from Greek *Tēreus*]
: the husband of Procne who rapes his sister-in-law Philomela

ter·gite \'tər-,gīt\ *noun* [New Latin *tergum*] (1885)
: the dorsal plate or dorsal portion of the covering of a metameric segment of an arthropod; *especially* **:** one on the abdomen

ter·gi·ver·sate \'tər-jə-vər-,sāt; ,tər-'ji-vər-,sāt, -'gi-; ,tər-jə-'vər-\ *intransitive verb* **-sat·ed; -sat·ing** [Latin *tergiversatus,* past participle of *tergiversari* to show reluctance, from *tergum* back + *versare* to turn, frequentative of *vertere* to turn — more at WORTH] (1654)
: to engage in tergiversation
— ter·gi·ver·sa·tor \-,sā-tər\ *noun*

ter·gi·ver·sa·tion \,tər-,ji-vər-'sā-shən, -,gi-; ,tər-ji-(,)vər-\ *noun* (1570)
1 : evasion of straightforward action or clear-cut statement **:** EQUIVOCATION
2 : desertion of a cause, position, party, or faith

ter·gum \'tər-gəm\ *noun, plural* **ter·ga** \-gə\ [New Latin, from Latin, back] (circa 1826)
: the dorsal part or plate of a segment of an arthropod **:** TERGITE, NOTUM
— ter·gal \-gəl\ *adjective*

ter·i·ya·ki \,ter-ē-'yä-kē\ *noun* [Japanese, from *teri* glaze + *yaki* broil] (1962)
: a Japanese dish of meat or fish that is grilled or broiled after being soaked in a seasoned soy sauce marinade

¹term \'tərm\ *noun* [Middle English *terme* boundary, end, from Old French, from Latin *terminus;* akin to Greek *termōn* boundary, end, Sanskrit *tarman* top of a post] (13th century)
1 a : END, TERMINATION; *also* **:** a point in time assigned to something (as a payment) **b :** the time at which a pregnancy of normal length terminates ⟨had her baby at full *term*⟩
2 a : a limited or definite extent of time; *especially* **:** the time for which something lasts **:** DURATION, TENURE ⟨*term* of office⟩ ⟨lost money in the short *term*⟩ **b :** the whole period for which an estate is granted; *also* **:** the estate or interest held by one for a term **c :** the time during which a court is in session
3 *plural* **:** provisions that determine the nature and scope of an agreement **:** CONDITIONS ⟨*terms* of sale⟩ ⟨liberal credit *terms*⟩
4 a : a word or expression that has a precise meaning in some uses or is peculiar to a science, art, profession, or subject ⟨legal *terms*⟩ **b** *plural* **:** expression of a specified kind ⟨described in glowing *terms*⟩
5 a : a unitary or compound expression connected with another by a plus or minus sign **b :** an element of a fraction or proportion or of a series or sequence
6 *plural* **a :** mutual relationship **:** FOOTING ⟨on good *terms*⟩ **b :** AGREEMENT, CONCORD ⟨come to *terms*⟩
7 : any of the three substantive elements of a syllogism
8 : a quadrangular pillar often tapering downward and adorned on the top with the figure of a head or the upper part of the body
9 : division in a school year during which instruction is regularly given to students
— in terms of : with respect to or in relation to ⟨thinks of everything *in terms of* money⟩
— on one's own terms : in accordance with one's wishes **:** in one's own way ⟨prefers to live *on his own terms*⟩

²term *transitive verb* (circa 1557)
: to apply a term to **:** CALL, NAME

¹ter·ma·gant \'tər-mə-gənt\ *noun* [Middle English]
1 *capitalized* **:** a deity erroneously ascribed to Islam by medieval European Christians and represented in early English drama as a violent character
2 : an overbearing or nagging woman **:** SHREW

²termagant *adjective* (1596)
: OVERBEARING, SHREWISH

term·er \'tər-mər\ *noun* (1634)
: a person serving for a specified term (as in a political office or in prison) ⟨a first *termer*⟩

ter·mi·na·ble \'tər-mə-nə-bəl, 'tərm-nə-\ *adjective* [Middle English, from *terminen* to terminate, from Middle French *terminer,* from Latin *terminare*] (15th century)
: capable of being terminated
— ter·mi·na·ble·ness *noun*
— ter·mi·na·bly \-blē\ *adverb*

¹ter·mi·nal \'tərm-nəl, 'tər-mə-nᵊl\ *adjective* [Latin *terminalis,* from *terminus*] (1744)
1 a : of or relating to an end, extremity, boundary, or terminus ⟨a *terminal* pillar⟩ **b :** growing at the end of a branch or stem ⟨a *terminal* bud⟩
2 a : of, relating to, or occurring in a term or each term ⟨*terminal* payments⟩ **b :** leading ultimately to death **:** FATAL ⟨*terminal* cancer⟩ **c :** approaching or close to death **:** being in the final stages of a fatal disease ⟨a *terminal* patient⟩ **d :** extremely or hopelessly severe ⟨*terminal* boredom⟩
3 a : occurring at or constituting the end of a period or series **:** CONCLUDING ⟨the *terminal*

moments of life⟩ **b :** not intended as preparation for further academic work ⟨a *terminal* curriculum⟩
synonym see LAST
— ter·mi·nal·ly *adverb*

²terminal *noun* (1831)
1 : a part that forms the end **:** EXTREMITY, TERMINATION
2 : a terminating usually ornamental detail **:** FINIAL
3 : a device attached to the end of a wire or cable or to an electrical apparatus for convenience in making connections
4 a : either end of a carrier line having facilities for the handling of freight and passengers **b :** a freight or passenger station that is central to a considerable area or serves as a junction at any point with other lines **c :** a town or city at the end of a carrier line **:** TERMINUS
5 : a combination of a keyboard and output device (as a video display unit) by which data can be entered into or output from a computer or electronic communications system

terminal leave *noun* (1944)
: a final leave consisting of accumulated unused leave granted to a member of the armed forces just prior to separation or discharge from service

terminal side *noun* (1927)
: a straight line that has been rotated around a point on another line to form an angle measured in a clockwise or counterclockwise direction — compare INITIAL SIDE

¹ter·mi·nate \ˈtər-mə-nət\ *adjective* [Middle English, from Latin *terminatus,* past participle of *terminare,* from *terminus*] (15th century)
: coming to an end or capable of ending

²ter·mi·nate \-ˌnāt\ *verb* **-nat·ed; -nat·ing** (1610)
intransitive verb
1 : to extend only to a limit (as a point or line); *especially* **:** to reach a terminus
2 : to form an ending
3 : to come to an end in time
transitive verb
1 a : to bring to an end **:** CLOSE ⟨*terminate* a marriage by divorce⟩ ⟨*terminate* a transmission line⟩ **b :** to form the conclusion of ⟨review questions *terminate* each chapter⟩ **c :** to discontinue the employment of ⟨workers *terminated* because of slow business⟩
2 : to serve as an ending, limit, or boundary of
synonym see CLOSE

terminating decimal *noun* (circa 1909)
: a decimal which can be expressed in a finite number of figures or for which all figures to the right of some place are zero — compare REPEATING DECIMAL

ter·mi·na·tion \ˌtər-mə-ˈnā-shən\ *noun* (circa 1500)
1 : end in time or existence **:** CONCLUSION ⟨the *termination* of life⟩
2 : the last part of a word; *especially* **:** an inflectional ending
3 : the act of terminating
4 : a limit in space or extent **:** BOUND
5 : OUTCOME, RESULT
— ter·mi·na·tion·al \-shnəl, -shə-nᵊl\ *adjective*

ter·mi·na·tive \ˈtər-mə-ˌnā-tiv\ *adjective* (15th century)
: tending or serving to terminate **:** ENDING
— ter·mi·na·tive·ly *adverb*

ter·mi·na·tor \-ˌnā-tər\ *noun* (1770)
1 : the dividing line between the illuminated and the unilluminated part of the moon's or a planet's disk
2 : one that terminates

ter·mi·nol·o·gy \ˌtər-mə-ˈnä-lə-jē\ *noun, plural* **-gies** [Medieval Latin *terminus* term, expression (from Latin, boundary, limit) + English *-o-* + *-logy*] (1801)
1 : the technical or special terms used in a business, art, science, or special subject
2 : nomenclature as a field of study
— ter·mi·no·log·i·cal \-mə-nᵊl-ˈä-ji-kəl\ *adjective*
— ter·mi·no·log·i·cal·ly \-ji-k(ə-)lē\ *adverb*

term insurance *noun* (1897)
: insurance for a specified period that provides for no payment to the insured except on losses during the period and that becomes void upon its expiration

ter·mi·nus \ˈtər-mə-nəs\ *noun, plural* **-ni** \-ˌnī, -ˌnē\ *or* **-nus·es** [Latin, boundary, end — more at TERM] (circa 1617)
1 : a final goal **:** a finishing point
2 : a post or stone marking a boundary
3 : either end of a transportation line or travel route; *also* **:** the station, town, or city at such a place **:** TERMINAL
4 : an extreme point or element **:** TIP ⟨the *terminus* of a glacier⟩

terminus ad quem \-ˌäd-ˈkwem\ *noun* [New Latin, literally, limit to which] (circa 1555)
1 : a goal, object, or course of action **:** DESTINATION, PURPOSE
2 : a final limiting point in time

terminus a quo \-ˌä-ˈkwō\ *noun* [New Latin, literally, limit from which] (circa 1555)
1 : a point of origin
2 : a first limiting point in time

ter·mi·tar·i·um \ˌtər-mə-ˈter-ē-əm, -ˌmī-\ *noun, plural* **-ia** \-ē-ə\ [New Latin] (1863)
: a termites' nest

ter·mi·tary \ˈtər-mə-ˌter-ē, -ˌmī-ˌter-ē\ *noun, plural* **-tar·ies** (1826)
: TERMITARIUM

ter·mite \ˈtər-ˌmīt\ *noun* [New Latin *Termit-, Termes,* genus of termites, from Late Latin, a worm that eats wood, alteration of Latin *tarmit-, tarmes;* akin to Greek *tetrainein* to bore — more at THROW] (1781)
: any of numerous pale-colored soft-bodied social insects (order Isoptera) that live in colonies consisting usually of winged sexual forms, wingless sterile workers, and soldiers, feed on wood, and include some which are very destructive to wooden structures and trees — called also *white ant*

term·less \ˈtərm-ləs\ *adjective* (circa 1541)
1 : having no term or end **:** BOUNDLESS, UNENDING
2 : UNCONDITIONED, UNCONDITIONAL

term paper *noun* (1926)
: a major written assignment in a school or college course representative of a student's achievement during a term

tern \ˈtərn\ *noun* [of Scandinavian origin; akin to Danish *terne* tern] (1678)
: any of various chiefly marine birds (subfamily Sterninae of the family Laridae and especially genus *Sterna*) that differ from the related gulls in usually smaller size, a more slender build, a sharply pointed bill, narrower wings, and an often forked tail

ter·na·ry \ˈtər-nə-rē\ *adjective* [Middle English, from Latin *ternarius,* from *terni* three each; akin to Latin *tres* three — more at THREE] (15th century)
1 a : of, relating to, or proceeding by threes **b :** having three elements, parts, or divisions **:** THREEFOLD **c :** arranged in threes ⟨*ternary* petals⟩
2 : using three as the base ⟨a *ternary* logarithm⟩
3 a : being or consisting of an alloy of three elements **b :** of, relating to, or containing three different elements, atoms, radicals, or groups ⟨sulfuric acid is a *ternary* acid⟩
4 : third in order or rank

ter·nate \ˈtər-ˌnāt, -nət\ *adjective* [New Latin *ternatus,* from Medieval Latin, past participle of *ternare* to treble, from Latin *terni*] (1760)
: arranged in threes or in subdivisions so arranged ⟨a *ternate* leaf⟩
— ter·nate·ly *adverb*

terne \ˈtərn\ *noun* [*terneplate*] (1891)
1 : an alloy of lead and tin typically in a ratio of four to one that is used as a coating in producing terneplate
2 : TERNEPLATE

terne·plate \-ˌplāt\ *noun* [probably from French *terne* dull (from Middle French, from *ternir* to tarnish) + English *plate*] (circa 1858)
: sheet iron or steel coated with an alloy of about four parts lead to one part tin

ter·pene \ˈtər-ˌpēn\ *noun* [International Scientific Vocabulary *terp-* (from German *Terpentin* turpentine, from Medieval Latin *terbentina*) + *-ene* — more at TURPENTINE] (1873)
: any of various isomeric hydrocarbons $C_{10}H_{16}$ found present in essential oils (as from conifers) and used especially as solvents and in organic synthesis; *broadly* **:** any of numerous hydrocarbons $(C_5H_8)_n$ found especially in essential oils, resins, and balsams
— ter·pene·less \-ləs\ *adjective*
— ter·pe·noid \ˈtər-pə-ˌnȯid, ˌtər-ˈpē-\ *adjective or noun*

ter·pin·e·ol \ˌtər-ˈpi-nē-ˌȯl, -ˌōl\ *noun* [International Scientific Vocabulary, from *terpine* $(C_{10}H_{18}(OH)_2)$] (1894)
: any of three fragrant isomeric alcohols $C_{10}H_{17}OH$ found in essential oils or made artificially and used especially in perfume or as solvents

ter·poly·mer \ˌtər-ˈpä-lə-mər\ *noun* (1947)
: a polymer (as a complex resin) that results from copolymerization of three discrete monomers

Terp·sich·o·re \ˌtərp-ˈsi-kə-(ˌ)rē\ *noun* [Latin, from Greek *Terpsichorē*]
: the Greek Muse of dancing and choral song

terp·si·cho·re·an \ˌtərp-(ˌ)si-kə-ˈrē-ən; -sə-ˈkōr-ē-, -ˈkȯr-\ *adjective* (1825)
: of or relating to dancing

ter·ra \ˈter-ə\ *noun, plural* **ter·rae** \-(ˌ)ē, -ˌī\ [New Latin, from Latin, land] (1946)
: any of the relatively light-colored highland areas on the surface of the moon or a planet

¹ter·race \ˈter-əs\ *noun* [Middle French, pile of earth, platform, terrace, from Old Provençal *terrassa,* from *terra* earth, from Latin, earth, land; akin to Latin *torrēre* to parch — more at THIRST] (1515)
1 a : a colonnaded porch or promenade **b :** a flat roof or open platform **c :** a relatively level paved or planted area adjoining a building
2 a : a raised embankment with the top leveled **b :** one of usually a series of horizontal ridges made in a hillside to increase cultivatable land, conserve moisture, or minimize erosion
3 : a level ordinarily narrow plain usually with steep front bordering a river, lake, or sea; *also* **:** a similar undersea feature
4 a : a row of houses or apartments on raised ground or a sloping site **b :** a group of row houses **c :** a strip of park in the middle of a street often planted with trees or shrubs **d :** STREET
5 : a section of a British soccer stadium set aside for standing spectators

²terrace *transitive verb* **ter·raced; ter·rac·ing** (1615)
1 : to provide (as a building or hillside) with a terrace
2 : to make into a terrace

ter·ra–cot·ta \ˌter-ə-ˈkä-tə\ *noun, often attributive* [Italian *terra cotta,* literally, baked earth] (1722)
1 : a glazed or unglazed fired clay used especially for statuettes and vases and architectural purposes (as roofing, facing, and relief ornamentation)
2 : a brownish orange

ter·ra fir·ma \-ˈfər-mə *also* -ˈfir-\ *noun* [New Latin, literally, solid land] (1693)
: dry land : solid ground

ter·rain \tə-ˈrān *also* te-\ *noun* [French, land, ground, from (assumed) Vulgar Latin *terranum,* alteration of Latin *terrenum,* from neuter of *terrenus* of earth — more at TERRENE] (1766)
1 a (1) : a geographical area (2) : a piece of land : GROUND **b** : the physical features of a tract of land
2 : TERRANE 1
3 a : a field of knowledge or interest : TERRITORY **b** : ENVIRONMENT, MILIEU

ter·ra in·cog·ni·ta \ˈter-ə-ˌin-ˌkäg-ˈnē-tə, -in-ˈkäg-nə-tə\ *noun, plural* **ter·rae in·cog·ni·tae** \ˈter-ˌī-ˌin-ˌkäg-ˈnē-ˌtī, -in-ˈkäg-nə-ˌtī\ [Latin] (1616)
: unknown territory : an unexplored country or field of knowledge

Ter·ra·my·cin \ˌter-ə-ˈmī-sᵊn\ *trademark*
— used for oxytetracycline

ter·rane \tə-ˈrān, te-\ *noun* [alteration of *terrain*] (1864)
1 : the area or surface over which a particular rock or group of rocks is prevalent
2 : TERRAIN 1a

ter·ra·pin \ˈter-ə-pən, ˈtar-\ *noun* [of Algonquian origin; akin to Delaware (dialect of New York) *tó·lpe·w,* a kind of turtle] (1613)
: any of various aquatic turtles (family Emydidae); *especially* : DIAMONDBACK TERRAPIN

terrapin

terr·aque·ous \te-ˈrā-kwē-əs, tə-, -ˈra-\ *adjective* [Latin *terra* land + English *aqueous*] (circa 1658)
: consisting of land and water

ter·rar·i·um \tə-ˈrar-ē-əm, -ˈrer-\ *noun, plural* **-ia** \-ē-ə\ *or* **-i·ums** [New Latin, from Latin *terra* + *-arium* (as in *vivarium*)] (1890)
: a usually transparent enclosure for keeping or raising plants or usually small animals (as turtles) indoors

ter·raz·zo \tə-ˈra-(ˌ)zō, -ˈrät-(ˌ)sō\ *noun* [Italian, literally, terrace, perhaps from Old Provençal *terrassa*] (1897)
: a mosaic flooring consisting of small pieces of marble or granite set in mortar and given a high polish

¹ter·rene \te-ˈrēn, tə-; ˈter-ˌēn\ *adjective* [Middle English, from Latin *terrenus* of earth, from *terra* earth] (14th century)
: MUNDANE, EARTHLY

²terrene *noun* (1667)
: a land area : EARTH, TERRAIN

ter·re·plein \ˈter-ə-ˌplān\ *noun* [Middle French, from Old Italian *terrapieno,* from Medieval Latin *terraplenum,* from *terra plenus* filled with earth] (1591)
: the level space behind a parapet of a rampart where guns are mounted

ter·res·tri·al \tə-ˈres-t(r)ē-əl; -ˈres-chəl, -ˈresh-\ *adjective* [Middle English, from Latin *terrestris,* from *terra* earth — more at TERRACE] (15th century)
1 a : of or relating to the earth or its inhabitants ⟨*terrestrial* magnetism⟩ **b** : mundane in scope or character : PROSAIC
2 a : of or relating to land as distinct from air or water ⟨*terrestrial* transportation⟩ **b** (1) : living on or in or growing from land ⟨*terrestrial* plants⟩ (2) : of or relating to terrestrial organisms ⟨*terrestrial* habits⟩
3 : belonging to the class of planets that are like the earth (as in density and silicate composition) ⟨the *terrestrial* planets Mercury, Venus, and Mars⟩
— **terrestrial** *noun*
— **ter·res·tri·al·ly** *adverb*

ter·ret \ˈter-ət\ *noun* [Middle English *turette,* alteration of *toret,* from Middle French, from Old French, diminutive of *tour* circuit, ring — more at TURN] (15th century)
: one of the rings on the top of a harness pad through which the reins pass

ter·ri·ble \ˈter-ə-bəl\ *adjective* [Middle English, from Middle French, from Latin *terribilis,* from *terrēre* to frighten — more at TERROR] (15th century)
1 a : exciting extreme alarm or intense fear : TERRIFYING **b** : formidable in nature : AWESOME ⟨a *terrible* responsibility⟩ **c** : DIFFICULT ⟨in a *terrible* bind⟩
2 : EXTREME, GREAT ⟨a *terrible* disappointment⟩
3 : extremely bad: as **a** : strongly repulsive : OBNOXIOUS ⟨a *terrible* smell⟩ **b** : notably unattractive or objectionable **c** : of very poor quality ⟨a *terrible* movie⟩
— **ter·ri·ble·ness** *noun*
— **ter·ri·bly** \-blē\ *adverb*

ter·ric·o·lous \te-ˈri-kə-ləs, tə-\ *adjective* [Latin *terricola* earth dweller, from *terra* earth + *colere* to inhabit — more at WHEEL] (circa 1836)
: living on or in the ground

ter·ri·er \ˈter-ē-ər\ *noun* [Middle English, from Middle French (*chien*) *terrier,* literally, earth dog, from *terrier* of earth, from Medieval Latin *terrarius,* from Latin *terra*] (15th century)
: any of various usually small dogs originally used by hunters to dig for small furred game and engage the quarry underground or drive it out

ter·rif·ic \tə-ˈri-fik\ *adjective* [Latin *terrificus,* from *terrēre* to frighten] (1667)
1 a : very bad : FRIGHTFUL **b** : exciting or fit to excite fear or awe
2 : EXTRAORDINARY ⟨*terrific* speed⟩
3 : unusually fine : MAGNIFICENT ⟨*terrific* weather⟩
— **ter·rif·i·cal·ly** \-fi-k(ə-)lē\ *adverb*

ter·ri·fy \ˈter-ə-ˌfī\ *transitive verb* **-fied; -fy·ing** [Latin *terrificare,* from *terrificus*] (1575)
1 a : to drive or impel by menacing : SCARE **b** : DETER, INTIMIDATE
2 : to fill with terror

ter·ri·fy·ing \-ˌfī-iŋ\ *adjective* (circa 1586)
1 : causing terror or apprehension
2 : of a formidable nature
— **ter·ri·fy·ing·ly** \-iŋ-lē\ *adverb*

ter·rig·e·nous \te-ˈri-jə-nəs, tə-\ *adjective* [Latin *terrigena* earthborn, from *terra* earth + *gignere* to beget — more at KIN] (1882)
: being or relating to oceanic sediment derived directly from the destruction of rocks on the earth's surface

ter·rine \tə-ˈrēn, ter-ˈēn\ *noun* [French — more at TUREEN] (circa 1706)
1 a : TUREEN 1 **b** : a usually earthenware dish in which foods are cooked and served
2 : a mixture of chopped meat, fish, or vegetables cooked and served in a terrine

¹ter·ri·to·ri·al \ˌter-ə-ˈtōr-ē-əl, -ˈtȯr-\ *adjective* (1625)
1 a : NEARBY, LOCAL **b** : serving outlying areas : REGIONAL
2 a : of or relating to a territory ⟨*territorial* government⟩ **b** : of or relating to or organized chiefly for home defense ⟨a *territorial* army⟩ **c** : of or relating to private property
3 a : of or relating to an assigned or preempted area ⟨*territorial* commanders⟩ **b** : exhibiting territoriality ⟨*territorial* birds⟩
— **ter·ri·to·ri·al·ly** \-ē-ə-lē\ *adverb*

²territorial *noun* (1907)
: a member of a territorial military unit

territorial court *noun* (1857)
: a court in a U.S. territory that has jurisdiction over local and federal cases

ter·ri·to·ri·al·ism \ˌter-ə-ˈtōr-ē-ə-ˌli-zəm, -ˈtȯr-\ *noun* (1881)
1 : LANDLORDISM
2 : the principle established in 1555 requiring the inhabitants of a territory of the Holy Roman Empire to conform to the religion of their ruler or to emigrate
— **ter·ri·to·ri·al·ist** \-list\ *noun*

ter·ri·to·ri·al·i·ty \-ˌtōr-ē-ˈa-lə-tē, -ˌtȯr-\ *noun* (1894)
1 : territorial status
2 a : persistent attachment to a specific territory **b** : the pattern of behavior associated with the defense of a territory

ter·ri·to·ri·al·ize \-ˈtōr-ē-ə-ˌlīz, -ˈtȯr-\ *transitive verb* **-ized; -iz·ing** (1818)
: to organize on a territorial basis
— **ter·ri·to·ri·al·i·za·tion** \-ˌtōr-ē-ə-lə-ˈzā-shən, -ˌtȯr-\ *noun*

territorial waters *noun plural* (1841)
: the waters under the sovereign jurisdiction of a nation or state including both marginal sea and inland waters

ter·ri·to·ry \ˈter-ə-ˌtōr-ē, -ˌtȯr-\ *noun, plural* **-ries** [Middle English, from Latin *territorium,* literally, land around a town, from *terra* land — more at TERRACE] (15th century)
1 a : a geographical area belonging to or under the jurisdiction of a governmental authority **b** : an administrative subdivision of a country **c** : a part of the U.S. not included within any state but organized with a separate legislature **d** : a geographical area (as a colonial possession) dependent on an external government but having some degree of autonomy
2 a : an indeterminate geographical area **b** : a field of knowledge or interest
3 a : an assigned area; *especially* : one in which a sales representative or distributor operates **b** : an area often including a nesting or denning site and a variable foraging range that is occupied and defended by an animal or group of animals
— **go with the territory** *or* **come with the territory** : to be a natural or unavoidable aspect or accompaniment of a particular situation, position, or field ⟨criticism *goes with the territory* in this job⟩

ter·ror \ˈter-ər\ *noun* [Middle English, from Middle French *terreur,* from Latin *terror,* from *terrēre* to frighten; akin to Greek *trein* to be afraid, flee, *tremein* to tremble — more at TREMBLE] (14th century)
1 : a state of intense fear
2 a : one that inspires fear : SCOURGE **b** : a frightening aspect ⟨the *terrors* of invasion⟩ **c** : a cause of anxiety : WORRY **d** : an appalling person or thing; *especially* : BRAT
3 : REIGN OF TERROR
4 : violence (as bombing) committed by groups in order to intimidate a population or government into granting their demands ⟨insurrection and revolutionary *terror*⟩
synonym see FEAR
— **ter·ror·less** \-ləs\ *adjective*

ter·ror·ise *chiefly British variant of* TERRORIZE

ter·ror·ism \ˈter-ər-ˌi-zəm\ *noun* (1795)
: the systematic use of terror especially as a means of coercion
— **ter·ror·ist** \-ər-ist\ *adjective or noun*
— **ter·ror·is·tic** \ˌter-ər-ˈis-tik\ *adjective*

ter·ror·ize \ˈter-ər-ˌīz\ *transitive verb* **-ized; -iz·ing** (1823)
1 : to fill with terror or anxiety : SCARE
2 : to coerce by threat or violence
— **ter·ror·i·za·tion** \ˌter-ər-ə-ˈzā-shən\ *noun*

ter·ry \ˈter-ē\ *noun, plural* **terries** [perhaps modification of French *tiré,* past participle of *tirer* to draw] (1784)
1 : the loop forming the pile in uncut pile fabrics
2 : an absorbent fabric with such loops — called also *terry cloth*

terse \ˈtərs\ *adjective* **ters·er; ters·est** [Latin *tersus* clean, neat, from past participle of *tergēre* to wipe off] (1601)
1 : smoothly elegant : POLISHED
2 : devoid of superfluity ⟨a *terse* summary⟩; *also* : SHORT, BRUSQUE ⟨dismissed me with a *terse* "no"⟩
synonym see CONCISE
— **terse·ly** *adverb*
— **terse·ness** *noun*

¹ter·tian \'tər-shən\ *adjective* [Middle English *tercian,* from Latin *tertianus,* literally, of the third, from *tertius* third — more at THIRD] (14th century)
: recurring at approximately 48-hour intervals — used of malaria

²tertian *noun* (14th century)
: a tertian fever (as vivax malaria)

¹ter·tia·ry \'tər-shē-,er-ē, -shə-rē\ *noun, plural* **-ries** (circa 1550)
1 [Medieval Latin *tertiarius,* from Latin, of a third] **:** a member of a monastic third order especially of lay people
2 *capitalized* **:** the Tertiary period or system of rocks

²tertiary *adjective* [Latin *tertiarius* of or containing a third, from *tertius* third] (circa 1656)
1 a : of third rank, importance, or value **b** *chiefly British* **:** of, relating to, or being higher education **c :** of, relating to, or constituting the third strongest of the three or four degrees of stress recognized by most linguists (as the stress of the third syllable of *basketball team*)
2 *capitalized* **:** of, relating to, or being the first period of the Cenozoic era or the corresponding system of rocks marked by the formation of high mountains (as the Alps, Caucasus, and Himalayas) and the dominance of mammals on land — see GEOLOGIC TIME table
3 a : involving or resulting from the substitution of three atoms or groups ⟨a *tertiary* salt⟩ ⟨*tertiary* amine⟩ **b :** being or containing a carbon atom having bonds to three other carbon atoms ⟨an acid containing a *tertiary* carbon⟩ ⟨*tertiary* alcohols⟩
4 : occurring in or being a third stage: as **a :** being or relating to the recovery of oil and gas from old wells by means of the underground application of heat and chemicals **b :** being or relating to the purification of wastewater by removal of fine particles, nitrates, and phosphates

tertiary color *noun* (circa 1864)
1 : a color produced by mixing two secondary colors
2 : a color produced by an equal mixture of a primary color with a secondary color adjacent to it on the color wheel

tertiary syphilis *noun* (1875)
: the third stage of syphilis that develops after the disappearance of the secondary symptoms and is marked by ulcers in and gummas under the skin and commonly by involvement of the skeletel, cardiovascular, and nervous systems

ter·ti·um quid \,tər-shē-əm-'kwid, ,tər-tē-\ *noun* [Late Latin, literally, third something; from its failing to fit into a dichotomy] (circa 1724)
1 : a middle course or an intermediate component ⟨where there are two systems of law and two orders of courts, there must . . . be some *tertium quid* to deal with conflicts of law and jurisdiction —Ernest Baker⟩
2 : a third party of ambiguous status ⟨there was a man and his wife and a *tertium quid* —Rudyard Kipling⟩

ter·va·lent \,tər-'vā-lənt, 'tər-\ *adjective* (circa 1903)
: TRIVALENT

ter·za ri·ma \,tert-sə-'rē-mə\ *noun* [Italian, literally, third rhyme] (1819)
: a verse form consisting of tercets usually in iambic pentameter in English poetry with an interlaced rhyme scheme (as *aba, bcb, cdc*)

tes·la \'tes-lə\ *noun* [Nikola *Tesla*] (1958)
: a unit of magnetic flux density in the meter-kilogram-second system equivalent to one weber per square meter

tes·sel·late \'te-sə-,lāt\ *transitive verb* **-lat·ed; -lat·ing** [Late Latin *tessellatus,* past participle of *tessellare* to pave with tesserae, from Latin *tessella,* diminutive of *tessera*] (1789)
: to form into or adorn with mosaic

tes·sel·lat·ed \-,lā-təd\ *adjective* (1695)
: having a checkered appearance

tes·sel·la·tion \,te-sə-'lā-shən\ *noun* (1660)
1 : MOSAIC; *especially* **:** a covering of an infinite geometric plane without gaps or overlaps by congruent plane figures of one type or a few types
2 : an act of tessellating **:** the state of being tessellated

tes·sera \'te-sə-rə\ *noun, plural* **-ser·ae** \-,rē, -,rī\ [Latin, probably ultimately from Greek *tessares* four; from its having four corners — more at FOUR] (circa 1656)
1 : a small tablet (as of wood, bone, or ivory) used by the ancient Romans as a ticket, tally, voucher, or means of identification
2 : a small piece (as of marble, glass, or tile) used in mosaic work

tes·ser·act \'te-sə-,rakt\ *noun* [Greek *tessares* four + *aktis* ray — more at ACTIN-] (1888)
: the four-dimensional analogue of a cube

tes·si·tu·ra \,te-sə-'tür-ə\ *noun* [Italian, literally, texture, from Latin *textura*] (circa 1891)
: the general range of a melody or voice part; *specifically* **:** the part of the register in which most of the tones of a melody or voice part lie

¹test \'test\ *noun* [Middle English, vessel in which metals were assayed, cupel, from Middle French, from Latin *testum* earthen vessel; akin to Latin *testa* earthen pot, shell] (14th century)
1 a *chiefly British* **:** CUPEL **b** (1) **:** a critical examination, observation, or evaluation **:** TRIAL; *specifically* **:** the procedure of submitting a statement to such conditions or operations as will lead to its proof or disproof or to its acceptance or rejection ⟨a *test* of a statistical hypothesis⟩ (2) **:** a basis for evaluation **:** CRITERION **c :** an ordeal or oath required as proof of conformity with a set of beliefs
2 a : a means of testing: as (1) **:** a procedure, reaction, or reagent used to identify or characterize a substance or constituent (2) **:** something (as a series of questions or exercises) for measuring the skill, knowledge, intelligence, capacities, or aptitudes of an individual or group **b :** a positive result in such a test
3 : a result or value determined by testing
4 : TEST MATCH ◆

²test (1689)
transitive verb
1 : to put to test or proof **:** TRY
2 : to require a doctrinal oath of
intransitive verb
1 a : to undergo a test **b :** to be assigned a standing or evaluation on the basis of tests ⟨*tested* positive for cocaine⟩ ⟨the cake *tested* done⟩
2 : to apply a test as a means of analysis or diagnosis — used with *for* ⟨*test* for mechanical aptitude⟩
— test·abil·i·ty \,tes-tə-'bi-lə-tē\ *noun*
— test·able \'tes-tə-bəl\ *adjective*
— test the waters *also* **test the water**
: to make a preliminary test or survey (as of reaction or interest) before embarking on a course of action

³test *noun* [Latin *testa* shell] (circa 1842)
: an external hard or firm covering (as a shell) of many invertebrates (as a foraminifer or a mollusk)

tes·ta \'tes-tə\ *noun, plural* **tes·tae** \-,tē, -,tī\ [New Latin, from Latin, shell] (1796)
: the hard external coating or integument of a seed

tes·ta·ceous \tes-'tā-shəs\ *adjective* [Latin *testaceus,* from *testa* shell, earthen pot, brick] (1646)
1 : having a shell ⟨a *testaceous* protozoan⟩
2 : of any of the several light colors of bricks

tes·ta·cy \'tes-tə-sē\ *noun, plural* **-cies** (circa 1864)
: the state of being testate

tes·ta·ment \'tes-tə-mənt\ *noun* [Middle English, from Late Latin & Latin; Late Latin *testamentum* covenant with God, holy scripture, from Latin, last will, from *testari* to be a witness, call to witness, make a will, from *testis* witness; akin to Latin *tres* three & to Latin *stare* to stand; from the witness's standing by as a third party in a litigation — more at THREE, STAND] (14th century)
1 a *archaic* **:** a covenant between God and the human race **b** *capitalized* **:** either of two main divisions of the Bible
2 a : a tangible proof or tribute **b :** an expression of conviction **:** CREDO
3 a : an act by which a person determines the disposition of his or her property after death **b :** WILL
— tes·ta·men·ta·ry \,tes-tə-'men-tə-rē, -'men-trē\ *adjective*

tes·tate \'tes-,tāt, -tət\ *adjective* [Middle English, from Latin *testatus,* past participle of *testari* to make a will] (15th century)
: having left a valid will ⟨she died *testate*⟩

tes·ta·tor \'tes-,tā-tər, tes-'\ *noun* [Middle English *testatour,* from Anglo-French, from Late Latin *testator,* from Latin *testari*] (15th century)
: a person who dies leaving a will or testament in force

tes·ta·trix \'tes-,tā-triks, tes-'\ *noun* [Late Latin, feminine of *testator*] (1591)
: a female testator

test ban *noun* (1958)
: a self-imposed partial or complete ban on the testing of nuclear weapons that is mutually agreed to by countries possessing such weapons

test bed *noun* (1914)
: a vehicle (as an airplane) used for testing new equipment (as engines or weapons systems)

test case *noun* (1894)
1 : a representative case whose outcome is likely to serve as a precedent
2 : a proceeding brought by agreement or on an understanding of the parties to obtain a decision as to the constitutionality of a statute

test·cross \'tes(t)-,krȯs\ *noun* (1934)
: a genetic cross between a homozygous recessive individual and a corresponding suspected heterozygote to determine the genotype of the latter
— testcross *transitive verb*

test-drive \'tes(t)-,drīv\ *transitive verb* **-drove** \-,drōv\; **-driv·en** \-,dri-vən\; **-driv·ing** \-,drī-viŋ\ (1950)

◇ WORD HISTORY

test Latin *testu* or *testum* was a generally used word for an earthenware pot. In the Middle Ages its descendant in Old French, *test,* denoted a specific type of vessel that was used in the assaying of precious metals, namely, a cupel. The cupel or test is a shallow, porous cup. When impure silver or gold is heated in it, the impurities are absorbed in the porous material, leaving a relatively pure button of silver or gold. As the name for a cupel, *test* was borrowed into later Middle English as *test* or *teste.* By about 1600 it was being used figuratively: *to put to the test* or *to bring to the test* was to make trial of something, to determine its quality or genuineness, as a precious metal might be tried in a cupel. The sense "trial, critical examination" soon became the most common meaning of *test.* French *têt,* the modern outcome of Old French *test,* has retained only its limited reference to an assayer's cupel; but Latin *testa* "earthenware jar, shard, shell," a feminine noun closely related to *testu,* went off in a quite different figurative direction, developing in French into *tête,* the usual word for "head."

: to drive (a motor vehicle) in order to evaluate performance
— **test-drive** *noun*

test·ed \'tes-təd\ *adjective* (1748)
: subjected to or qualified through testing — often used in combination ⟨time-*tested* principles⟩

test·ee \tes-'tē\ *noun* (1930)
: one who takes an examination

¹tes·ter \'tē-stər, 'tes-\ *noun* [Middle English, from Middle French *testiere* headpiece, head covering, from *teste* head, from Late Latin *testa* skull, from Latin, shell] (14th century)
: the canopy over a bed, pulpit, or altar

²tes·ter \'tes-tər\ *noun* [modification of Middle French *testart,* from *teston*] (1546)
: TESTON a

³test·er \'tes-tər\ *noun* (1661)
: one that tests or is used for testing

test-fly \-,flī\ *transitive verb* **-flew** \-,flü\; **-flown** \-,flōn\; **-fly·ing** (1936)
: to subject to a flight test ⟨*test-fly* an experimental plane⟩

tes·ti·cle \'tes-ti-kəl\ *noun* [Middle English *testicule,* from Latin *testiculus,* diminutive of *testis*] (15th century)
: TESTIS; *especially* : one of a higher mammal usually with its enclosing structures
— **tes·tic·u·lar** \tes-'ti-kyə-lər\ *adjective*

tes·ti·fy \'tes-tə-,fī\ *verb* **-fied; -fy·ing** [Middle English *testifien,* from Latin *testificari,* from *testis* witness] (14th century)
intransitive verb
1 a : to make a statement based on personal knowledge or belief : bear witness **b :** to serve as evidence or proof
2 : to express a personal conviction
3 : to make a solemn declaration under oath for the purpose of establishing a fact (as in a court)
transitive verb
1 a : to bear witness to : ATTEST **b :** to serve as evidence of : PROVE
2 *archaic* **a :** to make known (a personal conviction) **b :** to give evidence of : SHOW
3 : to declare under oath before a tribunal or officially constituted public body
— **tes·ti·fi·er** \-,fī(-ə)r\ *noun*

¹tes·ti·mo·ni·al \,tes-tə-'mō-nē-əl, -nyəl\ *adjective* (15th century)
1 : of, relating to, or constituting testimony
2 : expressive of appreciation or esteem ⟨a *testimonial* dinner⟩

²testimonial *noun* (15th century)
1 : EVIDENCE, TESTIMONY
2 a : a statement testifying to benefits received **b :** a character reference : letter of recommendation
3 : an expression of appreciation : TRIBUTE

tes·ti·mo·ny \'tes-tə-,mō-nē\ *noun, plural* **-nies** [Middle English, from Late Latin & Latin; Late Latin *testimonium* Decalogue, from Latin, evidence, witness, from *testis* witness — more at TESTAMENT] (14th century)
1 a (1) **:** the tablets inscribed with the Mosaic law (2) **:** the ark containing the tablets **b :** a divine decree attested in the Scriptures
2 a : firsthand authentication of a fact : EVIDENCE **b :** an outward sign **c :** a solemn declaration usually made orally by a witness under oath in response to interrogation by a lawyer or authorized public official
3 a : an open acknowledgment **b :** a public profession of religious experience

test·ing *adjective* (1878)
: requiring maximum effort or ability ⟨a most difficult and *testing* problem —Ernest Bevin⟩

tes·tis \'tes-təs\ *noun, plural* **tes·tes** \'tes-,tēz\ [Latin, witness, testis] (circa 1704)
: a typically paired male reproductive gland that produces sperm and that in most mammals is contained within the scrotum at sexual maturity

test-mar·ket \'tes(t)-,mär-kət\ *transitive verb* (1953)
: to subject (a product) to trial in a limited market

test match *noun* (1862)
1 : any of a series of championship cricket matches played between teams representing Australia and England
2 : a championship game or series (as of cricket) played between teams representing different countries

tes·ton \'tes-,tän\ *or* **tes·toon** \tes-'tün\ *noun* [Middle French, from Old Italian *testone,* augmentative of *testa* head, from Late Latin, skull — more at TESTER] (1536)
: any of several old European coins: as **a :** a shilling of Henry VIII of England decreasing in value to ninepence and then to sixpence in Shakespeare's time **b :** a French silver coin of the 16th century worth between 10 and 14½ sous

tes·tos·ter·one \te-'stäs-tə-,rōn\ *noun* [*testis* + *-o-* + *-sterone*] (1935)
: a hormone that is a hydroxy steroid ketone $C_{19}H_{28}O_2$ produced especially by the testes or made synthetically and that is responsible for inducing and maintaining male secondary sex characters

test pattern *noun* (circa 1946)
: a fixed picture broadcast by a television station to assist viewers in adjusting their receivers

test pilot *noun* (1917)
: a pilot who specializes in putting new or experimental airplanes through maneuvers designed to test them (as for strength) by producing strains in excess of normal

test-tube *adjective* (1935)
: produced by fertilization in laboratory apparatus and implantation in the uterus, by fertilization and growth in laboratory apparatus, or sometimes by artificial insemination ⟨*test-tube* babies⟩

test tube *noun* (1846)
: a plain or lipped tube usually of thin glass closed at one end and used especially in chemistry and biology

tes·tu·do \tes-'tü-(,)dō, -'tyü-\ *noun, plural* **-dos** [Latin *testudin-, testudo,* literally, tortoise, tortoise shell; akin to Latin *testa* shell] (1609)
: a cover of overlapping shields or a shed wheeled up to a wall used by the ancient Romans to protect an attacking force

tes·ty \'tes-tē\ *adjective* **tes·ti·er; -est** [Middle English *testif,* from Anglo-French, headstrong, from Old French *teste* head — more at TESTER] (1526)
1 : easily annoyed : IRRITABLE
2 : marked by impatience or ill humor ⟨*testy* remarks⟩
— **tes·ti·ly** \-tə-lē\ *adverb*
— **tes·ti·ness** \-tē-nəs\ *noun*

Tet \'tet\ *noun* [Vietnamese *tết*] (1885)
: the Vietnamese New Year observed for three days beginning at the first new moon after January 20

tet·a·nal \'te-tᵊn-əl\ *adjective* (1939)
: of, relating to, or derived from tetanus ⟨*tetanal* toxin⟩

te·tan·ic \te-'ta-nik\ *adjective* (circa 1727)
: of, relating to, being, or tending to produce tetany or tetanus
— **te·tan·i·cal·ly** \-ni-k(ə-)lē\ *adverb*

tet·a·nize \'te-tᵊn-,īz\ *transitive verb* **-nized; -niz·ing** (1849)
: to induce tetanus in ⟨*tetanize* a muscle⟩
— **tet·a·ni·za·tion** \,te-tᵊn-ə-'zā-shən, ,tet-nə-\ *noun*

tet·a·nus \'te-tᵊn-əs, 'tet-nəs\ *noun* [Middle English, from Latin, from Greek *tetanos,* from *tetanos* stretched, rigid; akin to Greek *teinein* to stretch — more at THIN] (14th century)
1 a : an acute infectious disease characterized by tonic spasm of voluntary muscles especially of the jaw and caused by the specific toxin of a bacterium (*Clostridium tetani*) which is usually introduced through a wound — compare LOCKJAW **b :** the bacterium that causes tetanus
2 : prolonged contraction of a muscle resulting from rapidly repeated motor impulses

tet·a·ny \'te-tᵊn-ē, 'tet-nē\ *noun* [International Scientific Vocabulary, from Latin *tetanus*] (circa 1885)
: a condition of physiologic calcium imbalance marked by tonic spasm of muscles and often associated with deficient parathyroid secretion

te·tar·to·he·dral \te-,tär-tə-'hē-drəl\ *adjective* [Greek *tetartos* fourth; akin to Greek *tettares* four — more at FOUR] (circa 1858)
of a crystal **:** having one fourth the number of planes required by complete symmetry — compare HEMIHEDRAL, HOLOHEDRAL

tetched *variant of* TECHED

tetchy \'te-chē\ *adjective* **tetch·i·er; -est** [perhaps from obsolete *tetch* habit] (1592)
: irritably or peevishly sensitive : TOUCHY ⟨the *tetchy* manner of two women living in the same house —Elizabeth Taylor (died 1975)⟩
— **tetch·i·ly** \-chə-lē\ *adverb*
— **tetch·i·ness** \-chē-nəs\ *noun*

¹tête-à-tête \,tet-ə-'tet, ,tāt-ə-'tāt, *2 is also* 'tet-ə-,tet\ *noun* [French, literally, head to head] (1697)
1 : a private conversation between two persons
2 : a short piece of furniture (as a sofa) intended to seat two persons especially facing each other

²tête-à-tête \,tet-ə-'tet, ,tāt-ə-'tāt\ *adverb* (1700)
: in private

³tête-à-tête \,tet-ə-,tet, ,tāt-ə-,tāt\ *adjective* (1728)
: FACE-TO-FACE, PRIVATE

tête-bêche \'tet-'besh\ *adjective* [French, noun, pair of inverted stamps, from *tête* head + *-bêche,* alteration of Middle French *bechevet* head against foot] (circa 1913)
: of or relating to a pair of stamps inverted in relation to one another either through a printing error or intentionally

teth \'tāt, 'tāth, 'tās\ *noun* [Hebrew *ṭēth*] (circa 1823)
: the 9th letter of the Hebrew alphabet — see ALPHABET table

¹teth·er \'te-t͟hər\ *noun* [Middle English *tethir,* probably of Scandinavian origin; akin to Old Norse *tjōthr* tether; akin to Old High German *zeotar* pole of a wagon] (14th century)
1 : something (as a rope or chain) by which an animal is fastened so that it can range only within a set radius
2 : the limit of one's strength or resources ⟨at the end of my *tether*⟩

²tether *transitive verb* **teth·ered; teth·er·ing** \-t͟h(ə-)riŋ\ (15th century)
: to fasten or restrain by or as if by a tether

teth·er·ball \'te-t͟hər-,bȯl\ *noun* (circa 1900)
: a game played with a ball suspended by a string from an upright pole in which the object for each contestant is to wrap the string around the pole by striking the ball in a direction opposite to that of the other contestant; *also* **:** the ball used in this game

Te·thys \'tē-thəs\ *noun* [Latin, from Greek *Tēthys*]
: a Titaness and wife of Oceanus

tet·ra \'te-trə\ *noun* [by shortening from New Latin *Tetragonopterus,* former genus name, from Late Latin *tetragonum* quadrangle + Greek *pteron* wing — more at TETRAGONAL, FEATHER] (1931)
: any of numerous small often brightly colored South American characin fishes often bred in tropical aquariums

tetra- *or* **tetr-** *combining form* [Middle English, from Latin, from Greek; akin to Greek *tettares* four — more at FOUR]
1 : four : having four : having four parts ⟨*tetra*valent⟩

2 : containing four atoms or groups (of a specified kind) ⟨*tetra*chloride⟩

tet·ra·caine \'te-trə-ˌkān\ *noun* [*tetra-* + *-caine*] (circa 1935)
: a crystalline basic ester $C_{15}H_{24}N_2O_2$ that is closely related chemically to procaine and is used chiefly in the form of its hydrochloride as a local anesthetic

tet·ra·chlo·ride \ˌte-trə-'klōr-ˌīd, -'klȯr-\ *noun* (1866)
: a chloride containing four atoms of chlorine

tet·ra·chord \'te-trə-ˌkȯrd\ *noun* [Greek *tetrachordon,* from neuter of *tetrachordos* of four strings, from *tetra-* + *chordē* string — more at YARN] (1603)
: a diatonic series of four tones with an interval of a perfect fourth between the first and last

tet·ra·cy·cline \ˌte-trə-'sī-ˌklēn\ *noun* [International Scientific Vocabulary *tetracyclic* having four fused hydrocarbon rings + ²*-ine*] (1952)
: a yellow crystalline broad-spectrum antibiotic $C_{22}H_{24}N_2O_8$ produced by streptomyces or synthetically

tet·rad \'te-ˌtrad\ *noun* [Greek *tetrad-, tetras,* from *tetra-*] (1653)
: a group or arrangement of four: as **a :** a group of four cells produced by the successive divisions of a mother cell ⟨a *tetrad* of spores⟩ **b :** a group of four synapsed chromatids that become visibly evident in the pachytene stage of meiotic prophase
— **te·trad·ic** \te-'tra-dik\ *adjective*

tet·ra·drachm \'te-trə-ˌdram\ *noun* [Greek *tetradrachmon,* from *tetra-* + *drachmē* drachma] (circa 1580)
: an ancient Greek silver coin worth four drachmas

tet·ra·dy·na·mous \ˌte-trə-'dī-nə-məs\ *adjective* [International Scientific Vocabulary *tetra-* + Greek *dynamis* power — more at DYNAMIC] (1830)
: having six stamens four of which are longer than the others ⟨*tetradynamous* plants of the mustard family⟩

tet·ra·eth·yl lead \ˌte-trə-ˌe-thəl-'led\ *noun* (1923)
: a heavy oily poisonous liquid $Pb(C_2H_5)_4$ used as an antiknock agent

tet·ra·flu·o·ride \ˌte-trə-'flȯr-ˌīd, -'flu̇r-\ *noun* (circa 1909)
: a fluoride containing four atoms of fluorine

te·trag·o·nal \te-'tra-gə-nᵊl\ *adjective* [Late Latin *tetragonalis* having four angles and four sides, from *tetragonum* quadrangle, from Greek *tetragōnon,* from neuter of *tetragōnos* tetragonal, from *tetra-* + *gōnia* angle — more at -GON] (1868)
: of, relating to, or characteristic of the tetragonal system
— **te·trag·o·nal·ly** \-nᵊl-ē\ *adverb*

tetragonal system *noun* (1879)
: a crystal system characterized by three axes at right angles of which only the two lateral axes are equal

tet·ra·gram·ma·ton \ˌte-trə-'gra-mə-ˌtän\ *noun* [Middle English, from Greek, from neuter of *tetragrammatos* having four letters, from *tetra-* + *grammat-, gramma* letter — more at GRAM] (15th century)
: the four Hebrew letters usually transliterated YHWH or JHVH that form a biblical proper name of God — compare YAHWEH

tet·ra·he·dral \ˌte-trə-'hē-drəl\ *adjective* (1794)
1 : being a polyhedral angle with four faces
2 : relating to, forming, or having the form of a tetrahedron
— **tet·ra·he·dral·ly** \-drə-lē\ *adverb*

tet·ra·he·drite \-ˌdrīt\ *noun* [German *Tetraëdrit,* from Late Greek *tetraedros* having four faces] (1868)
: a fine-grained gray mineral that consists of a sulfide of copper, iron, and antimony, often contains other elements (as silver), and occurs in tetrahedral crystals or in massive form

tet·ra·he·dron \ˌte-trə-'hē-drən\ *noun, plural* **-drons** *or* **-dra** \-drə\ [New Latin, from Late Greek *tetraedron,* neuter of *tetraedros* having four faces, from Greek *tetra-* + *hedra* seat, face — more at SIT] (1570)
: a polyhedron that has four faces

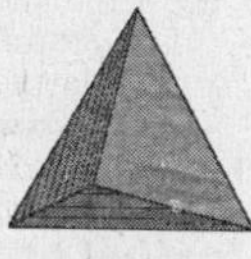
tetrahedron

tet·ra·hy·dro·can·nab·i·nol \-ˌhī-drə-kə-'na-bə-ˌnȯl, -ˌnōl\ *noun* [*tetrahydro-* (combined with four atoms of hydrogen) + *cannabin* + ¹*-ol*] (1940)
: THC

tet·ra·hy·dro·fu·ran \-'fyu̇r-ˌan, -fyu̇-'ran\ *noun* [*tetrahydro-* + *furan*] (circa 1943)
: a flammable liquid heterocyclic ether C_4H_8O that is derived from furan and used as a solvent and as an intermediate in organic synthesis

tet·ra·hy·me·na \ˌte-trə-'hī-mə-nə\ *noun* [New Latin, from *tetra-* + Greek *hymēn* membrane] (1962)
: any of a genus (*Tetrahymena*) of ciliate protozoans

te·tral·o·gy \te-'trä-lə-jē, -'tra-\ *noun, plural* **-gies** [Greek *tetralogia,* from *tetra-* + *-logia* -logy] (1656)
1 : a group of four dramatic pieces presented consecutively on the Attic stage at the Dionysiac festival
2 : a series of four connected works (as operas or novels)

tet·ra·mer \'te-trə-mər\ *noun* (1929)
: a molecule (as an enzyme or a polymer) that consists of four structural subunits (as peptide chains or condensed monomers)
— **tet·ra·mer·ic** \ˌte-trə-'mer-ik\ *adjective*

te·tram·er·ous \te-'tra-mə-rəs\ *adjective* [New Latin *tetramerus,* from Greek *tetramerēs,* from *tetra-* + *meros* part — more at MERIT] (1826)
: having or characterized by the presence of four parts or of parts arranged in sets or multiples of four ⟨*tetramerous* flowers⟩

te·tram·e·ter \te-'tra-mə-tər\ *noun* [Greek *tetrametron,* from neuter of *tetrametros* having four measures, from *tetra-* + *metron* measure — more at MEASURE] (1612)
: a line of verse consisting either of four dipodies (as in classical iambic, trochaic, and anapestic verse) or four metrical feet (as in modern English verse)

tet·ra·meth·yl·lead \ˌte-trə-ˌme-thəl-'led\ *noun* (1964)
: a volatile poisonous liquid $Pb(CH_3)_4$ used as an antiknock agent

¹**tet·ra·ploid** \'te-trə-ˌplȯid\ *adjective* [International Scientific Vocabulary] (1912)
: having or being a chromosome number four times the monoploid number ⟨a *tetraploid* cell⟩
— **tet·ra·ploi·dy** \-ˌplȯi-dē\ *noun*

²**tetraploid** *noun* (1921)
: a tetraploid individual

tet·ra·pod \'te-trə-ˌpäd\ *noun* [New Latin *tetrapodus,* from Greek *tetrapod-, tetrapous* four-footed, from *tetra-* + *pod-, pous* foot — more at FOOT] (circa 1891)
: a vertebrate (as a frog, bird, or cat) with two pairs of limbs

tet·ra·pyr·role \ˌte-trə-'pir-ˌōl\ *noun* (circa 1928)
: a chemical group consisting of four pyrrole rings joined either in a straight chain or in a ring (as in chlorophyll)

te·trarch \'te-ˌträrk, 'tē-\ *noun* [Middle English, from Latin *tetrarcha,* from Greek *tetrarchēs,* from *tetra-* + *-archēs* -arch] (14th century)
1 : a governor of the fourth part of a province
2 : a subordinate prince
— **te·trar·chic** \te-'trär-kik, tē-\ *adjective*

te·trar·chy \'te-ˌträr-kē, 'tē-\ *noun, plural* **-chies** (circa 1630)
: government by four persons ruling jointly

tet·ra·spore \'te-trə-ˌspōr, -ˌspȯr\ *noun* [International Scientific Vocabulary] (1857)
: one of the haploid asexual spores developed meiotically in the red algae usually in groups of four
— **tet·ra·spor·ic** \ˌte-trə-'spōr-ik, -'spȯr-\ *adjective*

tet·ra·va·lent \ˌte-trə-'vā-lənt\ *adjective* [International Scientific Vocabulary] (1868)
: having a valence of four

te·traz·zi·ni \ˌte-trə-'zē-nē\ *adjective, often capitalized* [Luisa *Tetrazzini* (died 1940) Italian opera singer] (1951)
: prepared with pasta and a white sauce seasoned with sherry and served au gratin ⟨chicken *tetrazzini*⟩

tet·ra·zo·li·um \ˌte-trə-'zō-lē-əm\ *noun* [New Latin, from International Scientific Vocabulary *tetrazole* (CH_2N_4)] (1895)
: a univalent cation or group CH_3N_4 that is analogous to ammonium; *also* **:** any of several of its derivatives used especially as electron acceptors to test for metabolic activity in living cells

tet·rode \'te-ˌtrōd\ *noun* (1902)
: a vacuum tube with a cathode, an anode, a control grid, and an additional grid or other electrode

te·tro·do·tox·in \te-ˌtrō-də-'täk-sən\ *noun* [International Scientific Vocabulary *tetrodo-* (from New Latin *Tetrodon,* genus of tropical marine fishes) + *toxin*] (1911)
: a neurotoxin $C_{11}H_{17}N_3O_8$ found especially in puffer fish that blocks nerve conduction by suppressing permeability of the nerve fiber to sodium ions

te·trox·ide \te-'träk-ˌsīd\ *noun* [International Scientific Vocabulary] (1866)
: a compound of an element or group with four atoms of oxygen

tet·ryl \'te-trəl\ *noun* [International Scientific Vocabulary] (1909)
: a pale yellow crystalline explosive $C_7H_5N_5O_8$ used especially as a detonator

tet·ter \'te-tər\ *noun* [Middle English *teter,* from Old English; akin to Old High German *zittaroh* tetter, Sanskrit *dadru* leprosy, *dṛṇāti* he tears — more at TEAR] (before 12th century)
: any of various vesicular skin diseases (as ringworm, eczema, and herpes)

Teu·ton \'tü-tᵊn, 'tyü-\ *noun* [Latin *Teutoni,* plural] (circa 1741)
1 : a member of an ancient probably Germanic or Celtic people
2 : a member of a people speaking a language of the Germanic branch of the Indo-European language family; *especially* **:** GERMAN

¹**Teu·ton·ic** \tü-'tä-nik, tyü-\ *noun* (1612)
: GERMANIC

²**Teutonic** *adjective* (1618)
: of, relating to, or characteristic of the Teutons
— **Teu·ton·i·cal·ly** \-ni-k(ə-)lē\ *adverb*

Teu·ton·ism \'tü-tᵊn-ˌi-zəm, 'tyü-\ *noun* (1854)
: GERMANISM

Teu·ton·ist \-ist\ *noun* (1882)
: GERMANIST

teu·ton·ize \-ˌīz\ *transitive verb* **-ized; -iz·ing** *often capitalized* (1845)
: GERMANIZE

tex·as \'tek-səs, -siz\ *noun* [*Texas,* state of U.S.; from the naming of cabins on Mississippi steamboats after states, the officers' cabins being the largest] (1857)

: a structure on the awning deck of a steamer that contains the officers' cabins and has the pilothouse in front or on top

Texas fever *noun* (1866)
: an infectious disease of cattle transmitted by the cattle tick and caused by a protozoan (*Babesia bigemina*) that multiplies in the blood and destroys the red blood cells

Texas Independence Day *noun* (circa 1928)
: March 2 observed as the anniversary of the declaration of independence of Texas from Mexico in 1836 and also as the birthday of Sam Houston

Texas leaguer *noun* [*Texas League,* a baseball minor league] (1905)
: a fly in baseball that falls too far out to be caught by an infielder and too close in to be caught by an outfielder

Texas longhorn *noun* (1908)
1 : LONGHORN 1a
2 : any of a breed of relatively small cattle developed in the U.S. from descendants of the original longhorns and typically having a 40 inch (one meter) horn spread and a highly variable color pattern

Texas Ranger *noun* (1846)
: a member of a formerly mounted police force in Texas

Tex–Mex \'teks-'meks\ *adjective* [*Texas* + *Mexico*] (1949)
: of, relating to, or being the Mexican-American culture or cuisine existing or originating in especially southern Texas ⟨*Tex-Mex* cooking⟩ ⟨*Tex-Mex* music⟩

text \'tekst\ *noun* [Middle English, from Middle French *texte,* from Medieval Latin *textus,* from Latin, texture, context, from *texere* to weave — more at TECHNICAL] (14th century)
1 a (1) **:** the original words and form of a written or printed work (2) **:** an edited or emended copy of an original work **b :** a work containing such text
2 a : the main body of printed or written matter on a page **b :** the principal part of a book exclusive of front and back matter **c :** the printed score of a musical composition
3 a (1) **:** a verse or passage of Scripture chosen especially for the subject of a sermon or for authoritative support (as for a doctrine) (2) **:** a passage from an authoritative source providing an introduction or basis (as for a speech) **b :** a source of information or authority
4 : THEME, TOPIC
5 a : the words of something (as a poem) set to music **b :** matter chiefly in the form of words that is treated as data for processing by computerized equipment ⟨a *text*-editing typewriter⟩
6 : a type suitable for printing running text
7 : TEXTBOOK
8 a : something written or spoken considered as an object to be examined, explicated, or deconstructed **b :** something likened to a text ⟨the surfaces of daily life are *texts* to be explicated —Michiko Kakutani⟩ ⟨he ceased to be a teacher as he became a *text* —D. J. Boorstin⟩

[1]text·book \'teks(t)-,bu̇k\ *noun* (1779)
: a book used in the study of a subject: as **a :** one containing a presentation of the principles of a subject **b :** a literary work relevant to the study of a subject

[2]textbook *adjective* (1905)
: of, suggesting, or suitable to a textbook; *especially* **:** CLASSIC ⟨a *textbook* example of bureaucratic waste⟩

text·book·ish \-,bu̇-kish\ *adjective* (1914)
: of, relating to, or having the characteristics of a textbook ⟨the style is heavy and *textbookish —Nation*⟩

text edition *noun* (1895)
: an edition of a book prepared for use especially in schools and colleges — compare TRADE EDITION

tex·tile \'tek-,stīl, 'teks-tᵊl\ *noun* [Latin, from neuter of *textilis* woven, from *texere*] (1626)
1 : CLOTH 1a; *especially* **:** a woven or knit cloth
2 : a fiber, filament, or yarn used in making cloth

tex·tu·al \'teks-chə-wəl, -chəl\ *adjective* [Middle English, from Medieval Latin *textus* text] (15th century)
: of, relating to, or based on a text
— tex·tu·al·ly *adverb*

textual critic *noun* (1938)
: a practitioner of textual criticism

textual criticism *noun* (1859)
1 : the study of a literary work that aims to establish the original text
2 : a critical study of literature emphasizing a close reading and analysis of the text

[1]tex·tu·ary \'teks-chə-,wer-ē\ *noun, plural* **-ar·ies** [Medieval Latin *textus*] (1608)
: one who is well informed in the Bible or in biblical scholarship

[2]textuary *adjective* (1646)
: TEXTUAL

[1]tex·ture \'teks-chər\ *noun* [Latin *textura,* from *textus,* past participle of *texere* to weave — more at TECHNICAL] (1578)
1 a : something composed of closely interwoven elements; *specifically* **:** a woven cloth **b :** the structure formed by the threads of a fabric
2 a : essential part **:** SUBSTANCE **b :** identifying quality **:** CHARACTER
3 a : the disposition or manner of union of the particles of a body or substance **b :** the visual or tactile surface characteristics and appearance of something ⟨the *texture* of an oil painting⟩
4 a : a composite of the elements of prose or poetry ⟨all these words . . . meet violently to form a *texture* impressive and exciting —John Berryman⟩ **b :** a pattern of musical sound created by tones or lines played or sung together
5 a : basic scheme or structure **b :** overall structure
— tex·tur·al \-chə-rəl\ *adjective*
— tex·tur·al·ly \-rə-lē\ *adverb*
— tex·tured \-chərd\ *adjective*
— tex·ture·less \-chər-ləs\ *adjective*

[2]texture *transitive verb* **tex·tured; tex·tur·ing** (1694)
: to give a particular texture to

tex·tur·ize \'teks-chə-,rīz\ *transitive verb* **-ized; -iz·ing** (circa 1950)
: TEXTURE ⟨*texturize* a polyester yarn⟩

tex·tus re·cep·tus \,tek-stəs-ri-'sep-təs\ *noun* [New Latin, literally, received text] (1856)
: the generally accepted text of a literary work (as the Greek New Testament)

T formation *noun* (1930)
: an offensive football formation in which the fullback lines up behind the center and quarterback with one halfback stationed on each side of the fullback

T4 cell \,tē-'fōr-, -'fȯr-\ *noun* [*T cell* + *CD4*] (1983)
: any of the T cells (as a helper T cell) that bear the CD4 molecular marker and become severely depleted in AIDS — called also *T4 lymphocyte*

T–group \'tē-,grüp\ *noun* [*t*raining *group*] (1950)
: a group of people under the leadership of a trainer who seek to develop self-awareness and sensitivity to others by verbalizing feelings uninhibitedly at group sessions — compare ENCOUNTER GROUP

[1]-th — see -ETH

[2]-th *or* **-eth** *adjective suffix* [Middle English *-the, -te,* from Old English *-tha, -ta;* akin to Old High German *-do* -th, Latin *-tus,* Greek *-tos,* Sanskrit *-tha*]
— used in forming ordinal numbers ⟨hundred*th*⟩ ⟨fortie*th*⟩

[3]-th *noun suffix* [Middle English, from Old English; akin to Old High German *-ida,* suffix forming abstract nouns, Latin *-ta,* Greek *-tē,* Sanskrit *-tā*]
1 : act or process ⟨spil*th*⟩
2 : state or condition ⟨dear*th*⟩

[4]-th *symbol* [[2]*-th*]
— used with the figures 4, 5, 6, 7, 8, 9, and 0 and related Roman numerals to indicate an ordinal number ⟨25*th*⟩ ⟨50*th* wedding anniversary⟩ ⟨XXV*th* Olympiad⟩

[1]Thai \'tī\ *noun, plural* **Thai** *or* **Thais** (1808)
1 : a family of languages including Thai and Shan spoken in southeast Asia and China
2 a : a native or inhabitant of Thailand **b :** one who is descended from a Thai
3 : the official language of Thailand

[2]Thai *adjective* (1808)
: of or relating to Thailand, its people, or their language or culture

thal·a·mus \'tha-lə-məs\ *noun, plural* **-mi** \-,mī, -,mē\ [New Latin, from Greek *thalamos* chamber] (1756)
: the largest subdivision of the diencephalon that consists chiefly of an ovoid mass of nuclei in each lateral wall of the third ventricle and functions in the integration of sensory information — see BRAIN illustration
— tha·lam·ic \thə-'la-mik\ *adjective*

thal·as·sae·mia *chiefly British variant of* THALASSEMIA

thal·as·se·mia \,tha-lə-'sē-mē-ə\ *noun* [New Latin, from Greek *thalassa* sea + New Latin *-emia*] (1932)
: any of a group of inherited disorders of hemoglobin synthesis affecting the globin chain that are characterized usually by mild to severe hemolytic anemia, are caused by a series of allelic genes, and tend to occur especially in individuals of Mediterranean, black, or southeast Asian ancestry; *especially* **:** COOLEY'S ANEMIA
— thal·as·se·mic \-mik\ *adjective or noun*

thalassemia major *noun* [New Latin, greater thalassemia] (1944)
: COOLEY'S ANEMIA

thalassemia minor *noun* [New Latin, lesser thalassemia] (1944)
: a mild form of thalassemia associated with the heterozygous condition for the gene involved

tha·las·sic \thə-'la-sik\ *adjective* [French *thalassique,* from Greek *thalassa* sea] (1883)
: of, relating to, or situated or developed about inland seas ⟨*thalassic* civilizations of the Aegean⟩

thal·as·soc·ra·cy \,tha-lə-'sä-krə-sē\ *noun* [Greek *thalassokratia,* from *thalassa* + *-kratia* -cracy] (1846)
: maritime supremacy

tha·las·so·crat \thə-'la-sə-,krat\ *noun* (1846)
: one who has maritime supremacy

tha·ler \'tä-lər\ *variant of* TALER

Tha·lia \thə-'lī-ə\ *noun* [Latin, from Greek *Thaleia*]
1 : the Greek Muse of comedy
2 : one of the three Graces

tha·lid·o·mide \thə-'li-də-,mīd, -məd\ *noun* [ph*thal*ic acid + *-id-* (from *imide*) + *-o-* + *imide*] (1958)
: a sedative and hypnotic drug $C_{13}H_{10}N_2O_4$ that has been the cause of malformation of infants born to mothers using it during pregnancy

thal·li·um \'tha-lē-əm\ *noun* [New Latin, from Greek *thallos* green shoot; from the bright green line in its spectrum] (1861)
: a sparsely but widely distributed poisonous metallic element that resembles lead in physical properties and is used chiefly in the form of compounds in photoelectric cells or as a pesticide — see ELEMENT table

thal·loid \'tha-,lȯid\ *adjective* (1857)
: of, relating to, resembling, or consisting of a thallus ⟨*thalloid* liverworts⟩

thal·lo·phyte \'tha-lə-ˌfīt\ *noun* [ultimately from Greek *thallos* + *phyton* plant — more at PHYT-] (1854)
: any of a primary division (Thallophyta) of the plant kingdom comprising plants with single-celled sex organs or with many-celled sex organs of which all cells give rise to gametes, including the algae, fungi, and lichens, and usually held to be a heterogeneous assemblage
— **thal·lo·phyt·ic** \ˌtha-lə-'fi-tik\ *adjective*

thal·lus \'tha-ləs\ *noun, plural* **thal·li** \'tha-ˌlī, -ˌlē\ *or* **thal·lus·es** [New Latin, from Greek *thallos,* from *thallein* to sprout; akin to Armenian *dalar* green, fresh, Albanian *dal* I come forth] (1829)
: a plant body that is characteristic of thallophytes, lacks differentiation into distinct members (as stem, leaves, and roots), and does not grow from an apical point

¹than \t͟hən, 't͟han\ *conjunction* [Middle English *than, then* then, than — more at THEN] (before 12th century)
1 a — used as a function word to indicate the second member or the member taken as the point of departure in a comparison expressive of inequality; used with comparative adjectives and comparative adverbs ⟨older *than* I am⟩ ⟨easier said *than* done⟩ **b** — used as a function word to indicate difference of kind, manner, or identity; used especially with some adjectives and adverbs that express diversity ⟨anywhere else *than* at home⟩
2 : rather than — usually used only after *prefer, preferable,* and *preferably*
3 : other than
4 : WHEN — used especially after *scarcely* and *hardly*

²than *preposition* (1560)
: in comparison with ⟨you are older *than* me⟩
■

than·a·tol·o·gy \ˌtha-nə-'tä-lə-jē\ *noun* [Greek *thanatos* + English *-logy*] (circa 1842)
: the description or study of the phenomena of death and of psychological mechanisms for coping with them
— **than·a·to·log·i·cal** \ˌtha-nə-tᵊl-'ä-ji kəl\ *adjective*
— **than·a·tol·o·gist** \-ə-'tä-lə-jist\ *noun*

Than·a·tos \'tha-nə-ˌtäs\ *noun* [Greek, death; akin to Sanskrit *adhvanīt* it vanished] (1935)
: DEATH INSTINCT

thane \'thān\ *noun* [Middle English *theyn,* from Old English *thegn;* akin to Old High German *thegan* thane and perhaps to Greek *tiktein* to bear, beget] (before 12th century)
1 : a free retainer of an Anglo-Saxon lord; *especially* **:** one resembling a feudal baron by holding lands of and performing military service for the king
2 : a Scottish feudal lord
— **thane·ship** \-ˌship\ *noun*

thank \'thaŋk\ *transitive verb* [Middle English, from Old English *thancian;* akin to Old English *thanc* gratitude — more at THANKS] (before 12th century)
1 : to express gratitude to ⟨*thanked* her for the present⟩ — used in the phrase *thank you* usually without a subject to politely express gratitude ⟨*thank* you for your consideration⟩, used in such phrases as *thank God, thank goodness* usually without a subject to express gratitude or more often only the speaker's or writer's pleasure or satisfaction in something
2 : to hold responsible ⟨had only himself to *thank* for his loss⟩
— **thank·er** *noun*

thank·ful \'thaŋk-fəl\ *adjective* (before 12th century)
1 : conscious of benefit received ⟨for what we are about to receive make us truly *thankful*⟩
2 : expressive of thanks ⟨*thankful* service⟩
3 : well pleased **:** GLAD ⟨was *thankful* that it didn't rain⟩
— **thank·ful·ness** *noun*

thank·ful·ly \-f(ə-)lē\ *adverb* (before 12th century)
1 : in a thankful manner ⟨spoke *thankfully*⟩
2 : as makes one thankful ⟨graceless stadiums . . . *thankfully* going out of fashion —R. G. Echevarriá⟩ ⟨*thankfully,* those opinions are advanced with graceful prose —Ken Auletta⟩

thank·less \'thaŋ-kləs\ *adjective* (1536)
1 : not expressing or feeling gratitude **:** UNGRATEFUL ⟨how sharper than a serpent's tooth it is to have a *thankless* child —Shakespeare⟩
2 : not likely to obtain thanks **:** UNAPPRECIATED ⟨a *thankless* task⟩
— **thank·less·ly** *adverb*
— **thank·less·ness** *noun*

thanks \'thaŋ(k)s\ *noun plural* [plural of Middle English *thank,* from Old English *thanc* thought, gratitude; akin to Old High German *dank* gratitude, Latin *tongēre* to know] (before 12th century)
1 : kindly or grateful thoughts **:** GRATITUDE
2 : an expression of gratitude ⟨return *thanks* before the meal⟩ — often used in an utterance containing no verb and serving as a courteous and somewhat informal expression of gratitude ⟨many *thanks*⟩
— **no thanks to :** not as a result of any benefit conferred by ⟨he feels better now, *no thanks to* you⟩
— **thanks to :** with the help of **:** owing to ⟨arrived early, *thanks to* good weather⟩

thanks·giv·ing \thaŋ(k)s-'gi-viŋ *also* 'thaŋ(k)s-ˌ\ *noun* (1533)
1 : the act of giving thanks
2 : a prayer expressing gratitude
3 a : a public acknowledgment or celebration of divine goodness **b** *capitalized* **:** THANKSGIVING DAY

Thanksgiving Day *noun* (1674)
: a day appointed for giving thanks for divine goodness: as **a :** the fourth Thursday in November observed as a legal holiday in the U.S. **b :** the second Monday in October observed as a legal holiday in Canada

thank·wor·thy \'thaŋ-ˌkwər-t͟hē\ *adjective* (14th century)
: worthy of thanks or gratitude **:** MERITORIOUS

thank–you \'thaŋ-ˌkyü\ *noun* [from the phrase *thank you* used in expressing gratitude] (1792)
: a polite expression of one's gratitude

thank–you–ma'am \'thaŋk-yu̇-ˌmam, -(y)ē-\ *noun* [probably from its causing a nodding of the head] (1849)
: a bump or depression in a road; *especially* **:** a ridge or hollow made across a road on a hillside to cause water to run off

¹that \'t͟hat, t͟hət\ *pronoun, plural* **those** \'t͟hōz\ [Middle English, from Old English *thæt,* neuter demonstrative pronoun & definite article; akin to Old High German *daz,* neuter demonstrative pronoun & definite article, Greek *to,* Latin is*tud,* neuter demonstrative pronoun] (before 12th century)
1 a : the person, thing, or idea indicated, mentioned, or understood from the situation ⟨*that* is my father⟩ **b :** the time, action, or event specified ⟨after *that* I went to bed⟩ **c :** the kind or thing specified as follows ⟨the purest water is *that* produced by distillation⟩ **d :** one or a group of the indicated kind ⟨*that's* a cat — quick and agile⟩
2 a : the one farther away or less immediately under observation or discussion ⟨*those* are maples and these are elms⟩ **b :** the former one
3 a — used as a function word after *and* to indicate emphatic repetition of the idea expressed by a previous word or phrase ⟨he was helpful, and *that* to an unusual degree⟩ **b** — used as a function word immediately before or after a word group consisting of a verbal auxiliary or a form of the verb *be* preceded by *there* or a personal pronoun subject to indicate emphatic repetition of the idea expressed by a previous verb or predicate noun or predicate adjective ⟨is she capable? She *is that*⟩
4 a : the one **:** the thing **:** the kind **:** SOMETHING, ANYTHING ⟨the truth of *that* which is true⟩ ⟨the senses are *that* whereby we experience the world⟩ ⟨what's *that* you say⟩ **b** *plural* **:** some persons ⟨*those* who think the time has come⟩
— **all that :** everything of the kind indicated ⟨tact, discretion, and *all that*⟩
— **at that 1 :** in spite of what has been said or implied **2 :** in addition **:** ²BESIDES

²that \t͟hət, 't͟hat\ *conjunction* (before 12th century)
1 a (1) — used as a function word to introduce a noun clause that is usually the subject or object of a verb or a predicate nominative ⟨said *that* he was afraid⟩ (2) — used as a function word to introduce a subordinate clause that is anticipated by the expletive *it* occurring as subject of the verb ⟨it is unlikely *that* he'll be in⟩ (3) — used as a function word to introduce a subordinate clause that is joined as complement to a noun or adjective ⟨we are certain *that* this is true⟩ ⟨the fact *that* you are here⟩ (4) — used as a function word to introduce a subordinate clause modifying an adverb or adverbial expression ⟨will go anywhere *that* he is invited⟩ **b** — used as a function word to introduce an exclamatory clause expressing a strong emotion especially of surprise, sorrow, or indignation ⟨*that* it should come to this!⟩
2 a (1) — used as a function word to introduce a subordinate clause expressing purpose or desired result ⟨cutting down expenses *that* her son might inherit an unencumbered estate —W. B. Yeats⟩ (2) — used as a function word to introduce a subordinate clause expressing a reason or cause ⟨rejoice *that* you are lightened of a load —Robert Browning⟩ (3) — used as a function word to introduce a subordinate clause expressing consequence, result, or effect ⟨are of sufficient importance *that* they cannot be neglected —Hannah Wormington⟩ **b** — used as a function word to introduce an exclamatory clause expressing a wish ⟨oh, *that* he would come⟩
3 — used as a function word after a subordinating conjunction without modifying its

□ USAGE

than After about 200 years of innocent if occasional use, the preposition *than* was called into question by 18th century grammarians. Some 200 years of elaborate and sometimes tortuous reasoning have led to these present-day inconsistent conclusions: *than whom* is standard but clumsy ⟨Beelzebub . . . *than* whom, Satan except, none higher sat —John Milton⟩ ⟨T. S. Eliot, *than* whom nobody could have been more insularly English —Anthony Burgess⟩; *than me* may be acceptable in speech ⟨a man no mightier *than* thyself or me —Shakespeare⟩ ⟨why should a man be better *than* me because he's richer *than* me —William Faulkner, in a talk to students⟩; *than* followed by a third-person objective pronoun (*her, him, them*) is usually frowned upon. Surveyed opinion tends to agree with these conclusions. Our evidence shows that the conjunction is more common than the preposition, that *than whom* is chiefly limited to writing, and that *me* is more common after the preposition than the third-person objective pronouns. You have the same choice Shakespeare had: you can use *than* either as a conjunction or as a preposition.

meaning ⟨if *that* thy bent of love be honorable —Shakespeare⟩

³that *adjective, plural* **those** (13th century)
1 a : being the person, thing, or idea specified, mentioned, or understood **b :** being the one specified — usually used for emphasis ⟨*that* rarity among leaders⟩ ⟨*that* brother of yours⟩ **c :** so great a **:** SUCH
2 : the farther away or less immediately under observation or discussion ⟨this chair or *that* one⟩

⁴that \thət, ˈthat\ *pronoun* [Middle English, from Old English *thæt,* neuter relative pronoun, from *thæt,* neuter demonstrative pronoun] (before 12th century)
1 — used as a function word to introduce a restrictive relative clause and to serve as a substitute within that clause for the substantive modified by the clause ⟨the house *that* Jack built⟩ ⟨I'll make a ghost of him *that* lets me —Shakespeare⟩
2 a : at which **:** in which **:** on which **:** by which **:** with which **:** to which ⟨each year *that* the lectures are given⟩ **b :** according to what **:** to the extent of what — used after a negative ⟨has never been here *that* I know of⟩
3 a *archaic* **:** that which **b** *obsolete* **:** the person who ◻ ◻

⁵that \ˈthat\ *adverb* (15th century)
1 : to such an extent ⟨a nail about *that* long⟩
2 : VERY, EXTREMELY — usually used with the negative ⟨did not take the festival *that* seriously —Eric Goldman⟩

that·away \ˈtha-də-ˌwā\ *adverb* [alteration of *that way*] (1839)
: in that direction

¹thatch \ˈthach\ *transitive verb* [Middle English *thecchen,* from Old English *theccan* to cover; akin to Old High German *decchen* to cover, Latin *tegere,* Greek *stegein* to cover, *stegos* roof, Sanskrit *sthagati* he covers] (14th century)
: to cover with or as if with thatch
— thatch·er *noun*

²thatch *noun* (14th century)
1 a : a plant material (as straw) used as a sheltering cover especially of a house **b :** a sheltering cover (as a house roof) made of such material **c :** a mat of undecomposed plant material (as grass clippings) accumulated next to the soil in a grassy area (as a lawn)
2 : something likened to the thatch of a house; *especially* **:** the hair of one's head

thau·ma·turge \ˈthȯ-mə-ˌtərj\ *noun* [French, from New Latin *thaumaturgus,* from Greek *thaumatourgos* working miracles, from *thaumat-, thauma* miracle + *ergon* work — more at THEATER, WORK] (1715)
: THAUMATURGIST

thau·ma·tur·gic \ˌthȯ-mə-ˈtər-jik\ *adjective* (1680)
1 : performing miracles
2 : of, relating to, or dependent on thaumaturgy

thau·ma·tur·gist \ˈthȯ-mə-ˌtər-jist\ *noun* (1829)
: a performer of miracles; *especially* **:** MAGICIAN

thau·ma·tur·gy \-jē\ *noun* (circa 1727)
: the performance of miracles; *specifically* **:** MAGIC

¹thaw \ˈthȯ\ *verb* [Middle English, from Old English *thawian;* akin to Old High German *douwen* to thaw, Greek *tēkein* to melt, Latin *tabēre* to waste away] (before 12th century)
transitive verb
: to cause to thaw
intransitive verb
1 a : to go from a frozen to a liquid state **:** MELT **b :** to become free of the effect (as stiffness, numbness, or hardness) of cold as a result of exposure to warmth
2 : to be warm enough to melt ice and snow — used with *it* in reference to the weather
3 : to abandon aloofness, reserve, or hostility **:** UNBEND
4 : to become mobile, active, or susceptible to change

²thaw *noun* (15th century)
1 : the action, fact, or process of thawing
2 : a period of weather warm enough to thaw ice ⟨the January *thaw*⟩
3 : the action or process of becoming less aloof, less hostile, or more genial ⟨a *thaw* in international relations⟩

THC \ˌtē-(ˌ)āch-ˈsē\ *noun* [*t*etra*h*ydro*c*annabinol] (1967)
: a physiologically active chemical $C_{21}H_{30}O_2$ from hemp plant resin that is the chief intoxicant in marijuana — called also *tetrahydrocannabinol*

¹the *before consonants usually* thə, *before vowels usually* thē, *especially Southern before vowels also* thə; *for emphasis before titles and names or to suggest uniqueness often* ˈthē\ *definite article* [Middle English, from Old English *thē,* masculine demonstrative pronoun & definite article, alteration (influenced by oblique cases — as *thæs,* genitive — & neuter, *thæt*) of *sē;* akin to Greek *ho,* masculine demonstrative pronoun & definite article — more at THAT] (before 12th century)
1 a — used as a function word to indicate that a following noun or noun equivalent is definite or has been previously specified by context or by circumstance ⟨put *the* cat out⟩ **b** — used as a function word to indicate that a following noun or noun equivalent is a unique or a particular member of its class ⟨*the* President⟩ ⟨*the* Lord⟩ **c** — used as a function word before nouns that designate natural phenomena or points of the compass ⟨*the* night is cold⟩ **d** — used as a function word before a noun denoting time to indicate reference to what is present or immediate or is under consideration ⟨in *the* future⟩ **e** — used as a function word before names of some parts of the body or of the clothing as an equivalent of a possessive adjective ⟨how's *the* arm today⟩ **f** — used as a function word before the name of a branch of human endeavor or proficiency ⟨*the* law⟩ **g** — used as a function word in prepositional phrases to indicate that the noun in the phrase serves as a basis for computation ⟨sold by *the* dozen⟩ **h** — used as a function word before a proper name (as of a ship or a well-known building) ⟨*the* Mayflower⟩ **i** — used as a function word before the plural form of a numeral that is a multiple of ten to denote a particular decade of a century or of a person's life ⟨life in *the* twenties⟩ **j** — used as a function word before the name of a commodity or any familiar appurtenance of daily life to indicate reference to the individual thing, part, or supply thought of as at hand ⟨talked on *the* telephone⟩ **k** — used as a function word to designate one of a class as the best, most typical, best known, or most worth singling out ⟨this is *the* life⟩ ⟨*the* Pill⟩; sometimes used before a personal name to denote the most prominent bearer of that name
2 a (1) — used as a function word with a noun modified by an adjective or by an attributive noun to limit the application of the modified noun to that specified by the adjective or by the attributive noun ⟨*the* right answer⟩ ⟨Peter *the* Great⟩ (2) — used as a function word before an absolute adjective or an ordinal number ⟨nothing but *the* best⟩ ⟨due on *the* first⟩ **b** (1) — used as a function word before a noun to limit its application to that specified by a succeeding element in the sentence ⟨*the* poet Wordsworth⟩ ⟨*the* days of our youth⟩ ⟨didn't have *the* time to write⟩ (2) — used as a function word after a person's name to indicate a characteristic trait or notorious activity specified by the succeeding noun ⟨Jack *the* Ripper⟩
3 a — used as a function word before a singular noun to indicate that the noun is to be understood generically ⟨*the* dog is a domestic animal⟩ **b** — used as a function word before a singular substantivized adjective to indicate an abstract idea ⟨an essay on *the* sublime⟩
4 — used as a function word before a noun or a substantivized adjective to indicate reference to a group as a whole ⟨*the* elite⟩

²the *adverb* [Middle English, from Old English *thȳ* by that, instrumental of *thæt* that] (before 12th century)
1 : than before **:** than otherwise — used before a comparative ⟨none *the* wiser for attending⟩
2 a : to what extent ⟨*the* sooner the better⟩ **b :** to that extent ⟨the sooner *the* better⟩
3 : beyond all others ⟨likes this *the* best⟩

³the *preposition* [¹*the*] (15th century)
: PER 2 ⟨a dollar *the* dozen⟩

the- *or* **theo-** *combining form* [Middle English *theo-,* from Latin, from Greek *the-, theo-,* from *theos*]
: god **:** God ⟨*the*ism⟩ ⟨*theo*centric⟩

¹the·ater *or* **the·atre** \ˈthē-ə-tər, ˈthi-ə-, *oftenest in Southern* ˈthē-ˌā- *also* thē-ˈā-\ *noun* [Middle English *theatre,* from Middle French, from Latin *theatrum,* from Greek *theatron,* from *theasthai* to view, from *thea* act of seeing; akin to Greek *thauma* miracle] (14th century)
1 a : an outdoor structure for dramatic performances or spectacles in ancient Greece and Rome **b :** a building for dramatic performances **c :** a building or area for showing motion pictures
2 : a place or sphere of enactment of usually significant events or action ⟨the *theater* of public life⟩
3 a : a place rising by steps or gradations ⟨a woody *theater* of stateliest view —John Milton⟩ **b :** a room often with rising tiers of seats for assemblies (as for lectures or surgical demonstrations)
4 a : dramatic literature **:** PLAYS **b :** dramatic representation as an art or profession **:** DRAMA
5 a : dramatic or theatrical quality or effectiveness **b :** SPECTACLE 1a

²theater *adjective* (1977)
: of, relating to, or appropriate for use in a theater of operations ⟨*theater* nuclear weapons⟩

the·ater·go·er \ˈthē-ə-tər-ˌgō(-ə)r\ *noun* (1870)
: a person who frequently goes to the theater
— the·ater·go·ing \-ˌgō-iŋ, -gȯ(-)iŋ\ *noun or adjective*

theater–in–the–round *noun, plural* **theaters–in–the–round** (1948)
: a theater in which the stage is located in the center of the auditorium — called also *arena theater*

theater of operations (circa 1879)

☐ USAGE

that *That, which, who:* In current usage *that* refers to persons or things, *which* chiefly to things and rarely to subhuman entities, *who* chiefly to persons and sometimes to animals. The notion that *that* should not be used to refer to persons is without foundation; such use is entirely standard. Because *that* has no genitive form or construction, *of which* or *whose* must be substituted for it in contexts that call for the genitive. See in addition WHOSE.

that *That, which:* Although some handbooks say otherwise, *that* and *which* are both regularly used to introduce restrictive clauses in edited prose. *Which* is also used to introduce nonrestrictive clauses. *That* was formerly used to introduce nonrestrictive clauses; such use is virtually nonexistent in present-day edited prose, though it may occasionally be found in poetry.

: the part of a theater of war in which active combat operations are conducted

theater of the absurd (1961)
: theater that seeks to represent the absurdity of human existence in a meaningless universe by bizarre or fantastic means

theater of war (circa 1890)
: the entire land, sea, and air area that is or may become involved directly in war operations

The·a·tine \'thē-ə-ˌtīn, -ˌtēn\ *noun* [New Latin *Theatinus,* from Latin *Teatinus* inhabitant of Chieti, from *Teate* Chieti, Italy] (circa 1598)
: a priest of the Order of Clerks Regular founded in 1524 in Italy by Saint Cajetan and Gian Pietro Caraffa to reform Catholic morality and combat Lutheranism
— Theatine *adjective*

1the·at·ri·cal \thē-'a-tri-kəl\ *also* **the·at·ric** \-trik\ *adjective* (1558)
1 : of or relating to the theater or the presentation of plays ⟨a *theatrical* costume⟩
2 : marked by pretense or artificiality of emotion
3 a : HISTRIONIC ⟨a *theatrical* gesture⟩ **b :** marked by extravagant display or exhibitionism
synonym see DRAMATIC
— the·at·ri·cal·ism \-kə-ˌli-zəm\ *noun*
— the·at·ri·cal·i·ty \-ˌa-trə-'ka-lə-tē\ *noun*
— the·at·ri·cal·ly \-'a-tri-k(ə-)lē\ *adverb*

2theatrical *noun* (circa 1683)
1 *plural* **a :** the performance of plays **b :** DRAMATICS
2 *British* **:** a professional actor
3 *plural* **:** showy or extravagant gestures

the·at·ri·cal·ize \thē-'a-tri-kə-ˌlīz\ *transitive verb* **-ized; -iz·ing** (1778)
1 : to adapt to the theater **:** DRAMATIZE
2 : to display in showy fashion
— the·at·ri·cal·i·za·tion \-ˌa tri kə-lə-'zā-shən\ *noun*

the·at·rics \thē-'a-triks\ *noun plural* (1807)
1 : THEATRICAL 1
2 : staged or contrived effects

the·be \'thā-(ˌ)bā\ *noun, plural* **thebe** [Tswana, literally, shield] (1967)
— see *pula* at MONEY table

the·ca \'thē-kə\ *noun, plural* **the·cae** \'thē-ˌsē, -ˌkē\ [New Latin, from Greek *thēkē* case — more at TICK] (circa 1666)
: an enveloping sheath or case of an animal or animal part
— the·cal \'thē-kəl\ *adjective*

-thecium *noun combining form, plural* **-thecia** [New Latin, from Greek *thēkion,* diminutive of *thēkē* case]
: small containing structure ⟨endo*thecium*⟩

1the·co·dont \'thē-kə-ˌdänt\ *adjective* [International Scientific Vocabulary *thec-* (from New Latin *theca*) + *-odont*] (1840)
: having the teeth inserted in sockets

2thecodont *noun* (1840)
: any of an order (Thecodontia) of Triassic diapsid thecodont reptiles that were presumably on the common ancestral line of the dinosaurs, birds, and crocodiles

thé dan·sant \tā-dä^{n}-sä^{n}\ *noun, plural* **thés dansants** *same*\ [French] (circa 1845)
: TEA DANCE

thee \'thē\ *pronoun, archaic objective case of* THOU
1 a — used especially in ecclesiastical or literary language and by Friends especially among themselves in contexts where the objective case form would be expected **b** — used by Friends especially among themselves in contexts where the subjective case form would be expected
2 : THYSELF
word history see YOU

thee·lin \'thē-(ə-)lən\ *noun* [irregular from Greek *thēlys* female; akin to Greek *thēlē* nipple — more at FEMININE] (1930)
: ESTRONE

theft \'theft\ *noun* [Middle English *thiefthe,* from Old English *thīefth;* akin to Old English *thēof* thief] (before 12th century)
1 a : the act of stealing; *specifically* **:** the felonious taking and removing of personal property with intent to deprive the rightful owner of it **b :** an unlawful taking (as by embezzlement or burglary) of property
2 *obsolete* **:** something stolen
3 : a stolen base in baseball

thegn \'thān\ *noun* [Old English — more at THANE] (1848)
: THANE 1

thegn·ly \-lē\ *adjective* (1876)
: of, relating to, or befitting a thegn

their \thər, 'ther, 'thar\ *adjective* [Middle English, from *their,* pronoun, from Old Norse *theirra,* genitive plural demonstrative & personal pronoun; akin to Old English *thæt* that] (13th century)
1 : of or relating to them or themselves especially as possessors, agents, or objects of an action ⟨*their* furniture⟩ ⟨*their* verses⟩ ⟨*their* being seen⟩
2 : his or her **:** HIS, HER, ITS — used with an indefinite third person singular antecedent ⟨anyone in *their* senses —W. H. Auden⟩
usage see ANYBODY, EVERYBODY, NOBODY, SOMEBODY, THEY

theirs \'therz, 'tharz\ *pronoun, singular or plural in construction* (14th century)
1 : that which belongs to them — used without a following noun as a pronoun equivalent in meaning to the adjective *their*
2 : his or hers **:** HIS, HERS — used with an indefinite third person singular antecedent ⟨I will do my part if everybody else will do *theirs*⟩

their·selves \thər-'selvz, (ˌ)ther-, (ˌ)thar-, *Southern also* -'sevz\ *pronoun plural* (14th century)
chiefly dialect **:** THEMSELVES

the·ism \'thē-ˌi-zəm\ *noun* (1678)
: belief in the existence of a god or gods; *specifically* **:** belief in the existence of one God viewed as the creative source of man and the world who transcends yet is immanent in the world
— the·ist \-ist\ *noun or adjective*
— the·is·tic \thē-'is-tik\ *also* **the·is·ti·cal** \-ti-kəl\ *adjective*
— the·is·ti·cal·ly \-ti-k(ə-)lē\ *adverb*

T–help·er cell \'tē-'hel-pər-\ *noun* (1980)
: HELPER T CELL

1them \(th)əm, 'them, *after* p, b, v, f, *also* ^{ə}m\ *pronoun, objective case of* THEY
usage see ANYBODY, EVERYBODY, NOBODY, SOMEBODY

2them \'them\ *adjective* (1594)
nonstandard **:** THOSE — used chiefly in the speech of less educated people and in the familiar speech of educated people especially when they are being humorous

the·mat·ic \thi-'ma-tik\ *adjective* [Greek *thematikos,* from *themat-, thema* theme] (1861)
1 a : of or relating to the stem of a word **b** *of a vowel* **:** being the last part of a word stem before an inflectional ending
2 : of, relating to, or constituting a theme
— the·mat·i·cal·ly \-ti-k(ə-)lē\ *adverb*

thematic apperception test *noun* (1941)
: a projective technique that is widely used in clinical psychology to make personality, psychodynamic, and diagnostic assessments based on the subject's verbal responses to a series of black and white pictures

theme \'thēm\ *noun* [Middle English *teme, theme,* from Middle French & Latin; Middle French *teme,* from Latin *thema,* from Greek, literally, something laid down, from *tithenai* to place — more at DO] (14th century)
1 a : a subject or topic of discourse or of artistic representation **b :** a specific and distinctive quality, characteristic, or concern ⟨the campaign has lacked a *theme*⟩
2 : STEM 4
3 : a written exercise **:** COMPOSITION ⟨a research *theme*⟩
4 : a melodic subject of a musical composition or movement
— themed \'thēmd\ *adjective*

theme park *noun* (1960)
: an amusement park in which the structures and settings are based on a central theme

theme song *noun* (1929)
1 : a melody recurring so often in a musical play that it characterizes the production or one of its characters
2 : a song used as a signature

them·selves \thəm-'selvz, them-\ *pronoun plural* (14th century)
1 a : those identical ones that are they — compare THEY 1a; used reflexively, for emphasis, or in absolute constructions ⟨nations that govern *themselves*⟩ ⟨they *themselves* were present⟩ ⟨*themselves* busy, they disliked idleness in others⟩ **b :** himself or herself **:** HIMSELF, HERSELF — used with an indefinite third person singular antecedent ⟨nobody can call *themselves* oppressed —Leonard Wibberley⟩
2 : their normal, healthy, or sane condition ⟨were *themselves* again after a night's rest⟩
usage see THEY

1then \'then\ *adverb* [Middle English *than, then* then, than, from Old English *thonne, thænne;* akin to Old High German *denne* then, than, Old English *thæt* that] (before 12th century)
1 : at that time
2 a : soon after that **:** next in order of time ⟨walked to the door, *then* turned⟩ **b :** following next after in order of position, narration, or enumeration **:** being next in a series ⟨first came the clowns, *then* came the elephants⟩ **c :** in addition **:** BESIDES ⟨*then* there is the interest to be paid⟩
3 a (1) **:** in that case ⟨take it, *then,* if you want it so much⟩ (2) — used after *but* to qualify or offset a preceding statement ⟨she lost the race, but *then* she never really expected to win⟩ **b :** according to that **:** as may be inferred ⟨your mind is made up, *then*⟩ **c :** as it appears **:** by way of summing up ⟨the cause of the accident, *then,* is established⟩ **d :** as a necessary consequence ⟨if the angles are equal, *then* the complements are equal⟩
— and then some : with much more in addition ⟨would require all his strength *and then some*⟩

2then *noun* (14th century)
: that time ⟨since *then,* he's been more cautious⟩

3then *adjective* (1584)
: existing or acting at or belonging to the time mentioned ⟨the *then* secretary of state⟩

then and there *adverb* (15th century)
: on the spot **:** IMMEDIATELY ⟨wanted the money right *then and there*⟩

the·nar \'thē-ˌnär, -nər\ *adjective* [New Latin, palm of the hand, from Greek; akin to Old High German *tenar* palm of the hand] (circa 1857)
: of, relating to, involving, or constituting the ball of the thumb or the intrinsic musculature of the thumb ⟨*thenar* muscles⟩

thence \'then(t)s *also* 'then(t)s\ *adverb* [Middle English *thannes,* from *thanne* from that place, from Old English *thanon;* akin to Old High German *thanan* from that place, Old English *thænne* then — more at THEN] (14th century)
1 : from that place
2 *archaic* **:** from that time **:** THENCEFORTH
3 : from that fact or circumstance **:** THEREFROM
— from thence : from that place

thence·forth \-ˌfōrth, -ˌfȯrth\ *adverb* (14th century)
: from that time forward

thence·for·ward \then(t)s-ˈfȯr-wərd *also* then(t)s-\ *also* **thence·for·wards** \-wərdz\ *adverb* (15th century)
: onward from that place or time

theo- — see THE-

theo·bro·mine \ˌthē-ə-ˈbrō-ˌmēn, -mən\ *noun* [New Latin *Theobroma,* genus that includes the cacao, from *the-* + Greek *brōma* food, from *bibrōskein* to devour — more at VORACIOUS] (1842)
: a bitter alkaloid $C_7H_8N_4O_2$ closely related to caffeine that occurs especially in cacao beans and has stimulant and diuretic properties

theo·cen·tric \-ˈsen-trik\ *adjective* (1886)
: having God as the central interest and ultimate concern ⟨a *theocentric* culture⟩
— **theo·cen·tric·i·ty** \-ˌsen-ˈtri-sə-tē\ *noun*
— **theo·cen·trism** \-ˈsen-ˌtri-zəm\ *noun*

the·oc·ra·cy \thē-ˈä-krə-sē\ *noun, plural* **-cies** [Greek *theokratia,* from *the-* + *-kratia* -cracy] (1622)
1 : government of a state by immediate divine guidance or by officials who are regarded as divinely guided
2 : a state governed by a theocracy

theo·crat \ˈthē-ə-ˌkrat\ *noun* (1827)
1 : one who rules in or lives under a theocratic form of government
2 : one who favors a theocratic form of government

theo·crat·ic \ˌthē-ə-ˈkra-tik\ *also* **theo·crat·i·cal** \-ti-kəl\ *adjective* (1690)
: of, relating to, or being a theocracy
— **theo·crat·i·cal·ly** \-ti-k(ə-)lē\ *adverb*

the·od·i·cy \thē-ˈä-də-sē\ *noun, plural* **-cies** [modification of French *théodicée,* from *théo-* the- (from Latin *theo-*) + Greek *dikē* judgment, right — more at DICTION] (1797)
: defense of God's goodness and omnipotence in view of the existence of evil

the·od·o·lite \thē-ˈä-dᵊl-ˌīt\ *noun* [New Latin *theodelitus*] (1571)
: a surveyor's instrument for measuring horizontal and usually also vertical angles

the·og·o·ny \thē-ˈä-gə-nē\ *noun, plural* **-nies** [Greek *theogonia,* from *the-* + *-gonia* -gony] (1612)
: an account of the origin and descent of the gods
— **theo·gon·ic** \ˌthē-ə-ˈgä-nik\ *adjective*

theo·lo·gian \ˌthē-ə-ˈlō-jən\ *noun* (15th century)
: a specialist in theology

theo·log·i·cal \-ˈlä-ji-kəl\ *also* **theo·log·ic** \-jik\ *adjective* (15th century)
1 : of or relating to theology
2 : preparing for a religious vocation ⟨a *theological* student⟩
— **theo·log·i·cal·ly** \-ji-k(ə-)lē\ *adverb*

theological virtue *noun* (1526)
: one of the three spiritual graces faith, hope, and charity drawing the soul to God according to scholastic theology

the·ol·o·gise *British variant of* THEOLOGIZE

the·ol·o·gize \thē-ˈä-lə-ˌjīz\ *verb* **-gized; -giz·ing** (1649)
transitive verb
: to make theological **:** give a religious significance to
intransitive verb
: to theorize theologically
— **the·ol·o·giz·er** *noun*

theo·logue *also* **theo·log** \ˈthē-ə-ˌlȯg, -ˌläg\ *noun* [Latin *theologus* theologian, from Greek *theologos,* from *the-* + *legein* to speak — more at LEGEND] (1663)
: a theological student or specialist

the·ol·o·gy \thē-ˈä-lə-jē\ *noun, plural* **-gies** [Middle English *theologie,* from Latin *theologia,* from Greek, from *the-* + *-logia* -logy] (14th century)
1 : the study of religious faith, practice, and experience; *especially* **:** the study of God and of God's relation to the world
2 a : a theological theory or system ⟨Thomist *theology*⟩ ⟨a *theology* of atonement⟩ **b :** a distinctive body of theological opinion ⟨Catholic *theology*⟩
3 : a usually 4-year course of specialized religious training in a Roman Catholic major seminary

the·on·o·mous \thē-ˈä-nə-məs\ *adjective* [*the-* + *-nomous* (as in *autonomous*)] (1947)
: governed by God **:** subject to God's authority

the·on·o·my \-mē\ *noun* [German *theonomie,* from *theo-* the- (from Latin) + *-nomie* -nomy] (1890)
: the state of being theonomous **:** government by God

the·oph·a·ny \thē-ˈä-fə-nē\ *noun, plural* **-nies** [Medieval Latin *theophania,* from Late Greek *theophaneia,* from Greek *the-* + *-phaneia* (as in *epiphaneia* appearance) — more at EPIPHANY] (circa 1633)
: a visible manifestation of a deity
— **theo·phan·ic** \ˌthē-ə-ˈfa-nik\ *adjective*

the·oph·yl·line \thē-ˈä-fə-lən\ *noun* [International Scientific Vocabulary *theo-* (from New Latin *thea* tea) + *phyll-* + ²*-ine*] (circa 1894)
: a feebly basic bitter crystalline compound $C_7H_8N_4O_2$ from tea leaves that is isomeric with theobromine and is used in medicine especially as a bronchodilator

the·or·bo \thē-ˈȯr-(ˌ)bō\ *noun, plural* **-bos** [modification of Italian *tiorba, teorba*] (1605)
: a stringed instrument of the 17th century resembling a large lute but having an extra set of long bass strings

theorbo

the·o·rem \ˈthē-ə-rəm, ˈthi(-ə)r-əm\ *noun* [Late Latin *theorema,* from Greek *theōrēma,* from *theōrein* to look at, from *theōros* spectator, from *thea* act of seeing — more at THEATER] (1551)
1 : a formula, proposition, or statement in mathematics or logic deduced or to be deduced from other formulas or propositions
2 : an idea accepted or proposed as a demonstrable truth often as a part of a general theory **:** PROPOSITION ⟨the *theorem* that the best defense is offense⟩
3 : STENCIL
4 : a painting produced especially on velvet by the use of stencils for each color
— **the·o·rem·at·ic** \ˌthē-ə-rə-ˈma-tik, ˌthi(-ə)r-ə-\ *adjective*

the·o·ret·i·cal \ˌthē-ə-ˈre-ti-kəl, ˌthi(-ə)r-ˈe-\ *also* **the·o·ret·ic** \-tik\ *adjective* [Late Latin *theoreticus,* from Greek *theōrētikos,* from *theōrein* to look at] (1601)
1 a : relating to or having the character of theory **:** ABSTRACT **b :** confined to theory or speculation often in contrast to practical applications **:** SPECULATIVE ⟨*theoretical* physics⟩
2 : given to or skilled in theorizing ⟨a brilliant *theoretical* physicist⟩
3 : existing only in theory **:** HYPOTHETICAL ⟨gave as an example a *theoretical* situation⟩

the·o·ret·i·cal·ly \-ti-k(ə-)lē\ *adverb* (1701)
1 : in a theoretical way
2 : according to an ideal or assumed set of facts or principles **:** in theory

the·o·re·ti·cian \ˌthē-ə-rə-ˈti-shən, -re-; ˌthi(-ə)r-ə-\ *noun* (1886)
: THEORIST

the·o·rise *British variant of* THEORIZE

the·o·rist \ˈthē-ə-rist, ˈthi(-ə)r-ist\ *noun* (1646)
: a person who theorizes

the·o·rize \ˈthē-ə-ˌrīz\ *verb* **-rized; -riz·ing** (1638)
intransitive verb
: to form a theory **:** SPECULATE
transitive verb
1 : to form a theory about
2 : to propose as a theory
— **the·o·ri·za·tion** \ˌthē-ə-rə-ˈzā-shən, ˌthi(-ə)r-ə-\ *noun*
— **the·o·riz·er** *noun*

the·o·ry \ˈthē-ə-rē, ˈthi(-ə)r-ē\ *noun, plural* **-ries** [Late Latin *theoria,* from Greek *theōria,* from *theōrein*] (1592)
1 : the analysis of a set of facts in their relation to one another
2 : abstract thought **:** SPECULATION
3 : the general or abstract principles of a body of fact, a science, or an art ⟨music *theory*⟩
4 a : a belief, policy, or procedure proposed or followed as the basis of action ⟨her method is based on the *theory* that all children want to learn⟩ **b :** an ideal or hypothetical set of facts, principles, or circumstances — often used in the phrase *in theory* ⟨in *theory,* we have always advocated freedom for all⟩
5 : a plausible or scientifically acceptable general principle or body of principles offered to explain phenomena ⟨wave *theory* of light⟩
6 a : a hypothesis assumed for the sake of argument or investigation **b :** an unproved assumption **:** CONJECTURE **c :** a body of theorems presenting a concise systematic view of a subject ⟨*theory* of equations⟩
synonym see HYPOTHESIS

theory of games (1951)
: GAME THEORY

theory of numbers (1811)
: NUMBER THEORY

the·os·o·phist \thē-ˈä-sə-fist\ *noun* (1656)
1 : an adherent of theosophy
2 *capitalized* **:** a member of a theosophical society

the·os·o·phy \-fē\ *noun* [Medieval Latin *theosophia,* from Late Greek, from Greek *the-* + *sophia* wisdom — more at -SOPHY] (1650)
1 : teaching about God and the world based on mystical insight
2 *often capitalized* **:** the teachings of a modern movement originating in the U.S. in 1875 and following chiefly Buddhist and Brahmanic theories especially of pantheistic evolution and reincarnation
— **theo·soph·i·cal** \ˌthē-ə-ˈsä-fi-kəl\ *adjective*
— **theo·soph·i·cal·ly** \-k(ə-)lē\ *adverb*

Theo·to·kos \ˌthē-ə-ˈtō-(ˌ)kōs, ˌthā-, -kəs\ *noun* [Late Greek, from Greek *the-* + *tokos* childbirth; akin to Greek *tiktein* to beget — more at THANE]
: VIRGIN MARY

ther·a·peu·sis \ˌther-ə-ˈpyü-səs\ *noun* [New Latin, from Greek, treatment, from *therapeuein*] (circa 1857)
: THERAPEUTICS

ther·a·peu·tic \-ˈpyü-tik\ *adjective* [Greek *therapeutikos,* from *therapeuein* to attend, treat, from *theraps* attendant] (1646)
1 : of or relating to the treatment of disease or disorders by remedial agents or methods ⟨a *therapeutic* rather than a diagnostic specialty⟩
2 : providing or assisting in a cure **:** CURATIVE, MEDICINAL ⟨*therapeutic* diets⟩ ⟨a *therapeutic* investigation of government waste⟩
— **ther·a·peu·ti·cal·ly** \-ti-k(ə-)lē\ *adverb*

therapeutic index *noun* (1926)
: a measure of the relative desirability of a drug for the attaining of a particular medical end that is usually expressed as the ratio of the largest dose producing no toxic symptoms to the smallest dose routinely producing cures

ther·a·peu·tics \ˌther-ə-ˈpyü-tiks\ *noun plural but singular or plural in construction* (1671)
: a branch of medical science dealing with the application of remedies to diseases

ther·a·pist \ˈther-ə-pist\ *noun* (1886)
: one specializing in therapy; *especially* **:** a person trained in methods of treatment and rehabilitation other than the use of drugs or surgery ⟨a speech *therapist*⟩

the·rap·sid \thə-ˈrap-səd\ *noun* [New Latin *Therapsida,* from *ther-* mammal (from Greek

thēr wild animal) + *apsid-, apsis* arch, vault — more at FIERCE, APSIS] (1912)
: any of an order (Therapsida) of Permian and Triassic reptiles that are considered ancestors of the mammals
— therapsid *adjective*

ther·a·py \'ther-ə-pē\ *noun, plural* **-pies** [New Latin *therapia,* from Greek *therapeia,* from *therapeuein*] (circa 1846)
: therapeutic treatment especially of bodily, mental, or behavioral disorder

Ther·a·va·da \,ther-ə-'vä-də\ *noun* [Pali *theravāda,* literally, doctrine of the elders] (1882)
: a conservative branch of Buddhism comprising sects chiefly in Sri Lanka, Myanmar, Thailand, Laos, and Cambodia and adhering to the original Pali scriptures alone and to the nontheistic ideal of nirvana for a limited select number — compare MAHAYANA

[1]**there** \'thar, 'ther\ *adverb* [Middle English, from Old English *thǣr;* akin to Old High German *dār* there, Old English *thæt* that] (before 12th century)
1 : in or at that place ⟨stand over *there*⟩ — often used interjectionally
2 : to or into that place **:** THITHER ⟨went *there* after church⟩
3 : at that point or stage ⟨stop right *there* before you say something you'll regret⟩
4 : in that matter, respect, or relation ⟨*there* is where I disagree with you⟩
5 — used interjectionally to express satisfaction, approval, encouragement or sympathy, or defiance ⟨*there,* it's finished⟩

[2]**there** \'thar, 'ther, *1 is also* thər\ *pronoun* (before 12th century)
1 — used as a function word to introduce a sentence or clause ⟨*there* shall come a time⟩
2 — used as an indefinite substitute for a name ⟨hi *there*⟩

[3]**there** *same as* [1]\ *noun* (1588)
1 : that place or position ⟨there is no here and no *there* . . . in pure space —James Ward⟩
2 : that point ⟨you take it from *there*⟩

[4]**there** *same as* [1]\ *adjective* (1590)
1 — used for emphasis especially after a demonstrative pronoun or a noun modified by a demonstrative adjective ⟨those men *there* can tell you⟩
2 *nonstandard* — used for emphasis after a demonstrative adjective but before the noun modified ⟨I bet I cussed that *there* blamed mule five hundred times —Elizabeth M. Roberts⟩
3 : capable of being relied on for support or aid ⟨she is always *there* for him⟩
4 : fully conscious, rational, or aware ⟨not all *there*⟩

there·abouts *also* **there·about** \,thar-ə-'baù̇t(s), 'thar-ə-,, ,ther-ə-'baù̇t(s), 'ther-ə-,\ *adverb* (before 12th century)
1 : near that place or time
2 : near that number, degree, or quantity ⟨a boy of 18 or *thereabouts*⟩

there·af·ter \tha-'raf-tər, the-\ *adverb* (before 12th century)
1 : after that
2 *archaic* **:** according to that **:** ACCORDINGLY

there·at \-'rat\ *adverb* (before 12th century)
1 : at that place
2 : at that occurrence

there·by \thar-'bī, ther-, 'thar-,, 'ther-,\ *adverb* (before 12th century)
1 : by that **:** by that means ⟨*thereby* lost her chance to win⟩
2 : connected with or with reference to that ⟨*thereby* hangs a tale —Shakespeare⟩

there·for \thar-'fȯr, ther-\ *adverb* (12th century)
: for or in return for that ⟨ordered a change and gave his reasons *therefor*⟩

there·fore \'thar-,fōr, 'ther-, -,fȯr\ *adverb* (14th century)
1 a : for that reason **:** CONSEQUENTLY **b :** because of that **c :** on that ground
2 : to that end

there·from \thar-'frəm, ther-, -'främ\ *adverb* (13th century)
: from that or it

there·in \tha-'rin, the-\ *adverb* (before 12th century)
1 : in or into that place, time, or thing
2 : in that particular or respect ⟨*therein* lies the problem⟩

there·in·af·ter \,thar-in-'af-tər, ,ther-\ *adverb* (1818)
: in the following part of that matter (as writing, document, or speech)

there·in·to \tha-'rin-(,)tü, the-\ *adverb* (14th century)
archaic **:** into that or it

the·re·min \'ther-ə-mən\ *noun* [Lev *Theremin* (born 1896) Russian engineer & inventor] (1927)
: a purely melodic electronic musical instrument typically played by moving a hand between two projecting electrodes

there·of \tha-'rəv, -'räv, the-\ *adverb* (before 12th century)
1 : of that or it
2 : from that cause or particular **:** THEREFROM

there·on \-'ròn, -'rän\ *adverb* (before 12th century)
1 : on that ⟨a text with a commentary *thereon*⟩
2 *archaic* **:** THEREUPON

there·to \thar-'tü, ther-\ *adverb* (before 12th century)
: to that ⟨a text and the notes *thereto*⟩

there·to·fore \'thar-tə-,fōr, 'ther-, -,fȯr; ,thar-tə-', ,ther-\ *adverb* (14th century)
: up to that time ⟨a *theretofore* unknown author⟩

there·un·der \tha-'rən-dər, the-\ *adverb* (before 12th century)
: under that

there·un·to \-'rən-(,)tü; ,thar-ən-'tü, ,ther-\ *adverb* (14th century)
archaic **:** THERETO

there·up·on \'thar-ə-,pòn, 'ther-, -,pän; ,thar-ə-', ,ther-\ *adverb* (13th century)
1 : on that matter
2 : THEREFORE
3 : immediately after that

there·with \thar-'with, ther-, -'with\ *adverb* (before 12th century)
1 : with that
2 *archaic* **:** THEREUPON, FORTHWITH

there·with·al \'thar-wi-,thòl, 'ther-, -thòl\ *adverb* (14th century)
1 *archaic* **:** BESIDES
2 : THEREWITH

the·ri·ac \'thir-ē-,ak\ *noun* [New Latin *theriaca*] (1568)
1 : THERIACA
2 : CURE-ALL

the·ri·a·ca \thi-'rī-ə-kə\ *noun* [New Latin, from Latin, antidote against poison — more at TREACLE] (1562)
: a mixture of many drugs and honey formerly held to be an antidote to poison
— the·ri·a·cal \-kəl\ *adjective*

the·rio·mor·phic \,thir-ē-ō-'mòr-fik\ *adjective* [Greek *thēriomorphos,* from *thērion* beast + *morphē* form — more at TREACLE] (1882)
: having an animal form ⟨*theriomorphic* gods⟩

therm \'thərm\ *noun* [Greek *thermē* heat, from *thermos* hot; akin to Latin *formus* warm, Sanskrit *gharma* heat] (1888)
: a unit used to measure quantity of heat that equals 100,000 British thermal units

therm- *or* **thermo-** *combining form* [Greek, from *thermē*]
1 : heat ⟨*thermo*stat⟩
2 : thermoelectric ⟨*thermo*pile⟩

-therm *noun combining form* [Greek *thermē* heat]
: animal having a (specified) body temperature ⟨ecto*therm*⟩

[1]**ther·mal** \'thər-məl\ *adjective* [Greek *thermē*] (1756)
1 [Latin *thermae* public baths, from Greek *thermai,* plural of *thermē*] **:** of, relating to, or marked by the presence of hot springs ⟨*thermal* waters⟩
2 a : of, relating to, or caused by heat ⟨*thermal* stress⟩ ⟨*thermal* insulation⟩ **b :** being or involving a state of matter dependent upon temperature ⟨*thermal* conductivity⟩ ⟨*thermal* agitation of molecular structure⟩
3 : designed (as with insulating air spaces) to prevent the dissipation of body heat ⟨*thermal* underwear⟩
4 : having energies of the order of those due to thermal agitation ⟨*thermal* neutrons⟩
— ther·mal·ly \-mə-lē\ *adverb*

[2]**thermal** *noun* (1933)
: a rising body of warm air

ther·mal·ize \'thər-mə-,līz\ *transitive verb* **-ized; -iz·ing** (1948)
: to change the effective speed of (a particle) to a thermal value ⟨*thermalize* a neutron⟩
— ther·mal·i·za·tion \,thər-mə-lə-'zā-shən\ *noun*

thermal pollution *noun* (1966)
: the discharge of heated liquid (as wastewater from a factory) into natural waters at a temperature harmful to the environment

thermal printer *noun* (1966)
: a dot matrix printer (as for a computer) in which heat is applied to the pins of the matrix to form dots on usually heat-sensitive paper

ther·mic \'thər-mik\ *adjective* (1842)
: [1]THERMAL 2 ⟨*thermic* energy⟩
— ther·mi·cal·ly \-mi-k(ə-)lē\ *adverb*

therm·ion·ic \,thər-(,)mī-'ä-nik\ *adjective* [*thermion* charged particle from an incandescent source, from *therm-* + *ion*] (1909)
: relating to, using, or being the emission of charged particles (as electrons) by an incandescent material

therm·ion·ics \,thər-(,)mī-'ä-niks\ *noun plural but singular in construction* (1909)
: physics dealing with thermionic phenomena

therm·is·tor \'thər-,mis-tər\ *noun* [*therm*al re*sistor*] (1940)
: an electrical resistor making use of a semiconductor whose resistance varies sharply in a known manner with the temperature

Ther·mit \'thər-mət, -,mīt\ *trademark*
— used for thermite

ther·mite \'thər-,mīt\ *noun* [*therm-* + [1]*-ite*] (1900)
: a mixutre of aluminum powder and a metal oxide (as iron oxide) that when ignited evolves a great deal of heat and is used in welding and in incendiary bombs

ther·mo·chem·is·try \,thər-mō-'ke-mə-strē\ *noun* (1844)
: a branch of chemistry that deals with the interrelation of heat with chemical reaction or physical change of state
— ther·mo·chem·i·cal \-'ke-mi-kəl\ *adjective*
— ther·mo·chem·ist \-'ke-mist\ *noun*

ther·mo·cline \'thər-mə-,klīn\ *noun* (1898)
: the region in a thermally stratified body of water which separates warmer oxygen-rich surface water from cold oxygen-poor deep water and in which temperature decreases rapidly with depth

ther·mo·cou·ple \'thər-mə-,kə-pəl\ *noun* (1890)
: a device for measuring temperature in which a pair of wires of dissimilar metals (as copper and iron) are joined and the free ends of the wires are connected to an instrument (as a voltmeter) that measures the difference in potential created at the junction of the two metals

ther·mo·du·ric \ˌthər-mō-ˈdu̇r-ik, -ˈdyu̇r-\ *adjective* [*therm-* + Latin *durare* to last — more at DURING] (1927)
: able to survive high temperatures; *specifically* **:** able to survive pasteurization — used of microorganisms

ther·mo·dy·nam·ic \ˌthər-mō-dī-ˈna-mik, -də-\ *also* **ther·mo·dy·nam·i·cal** \-mi-kəl\ *adjective* (1849)
1 : of or relating to thermodynamics
2 : being or relating to a system of atoms, molecules, colloidal particles, or larger bodies considered as an isolated group in the study of thermodynamic processes
— **ther·mo·dy·nam·i·cal·ly** \-mi-k(ə-)lē\ *adverb*

ther·mo·dy·nam·ics \-miks\ *noun plural but singular or plural in construction* (1854)
1 : physics that deals with the mechanical action or relations of heat
2 : thermodynamic processes and phenomena
— **ther·mo·dy·nam·i·cist** \-ˈna-mə-sist\ *noun*

ther·mo·elec·tric \ˌthər-mō-i-ˈlek-trik\ *adjective* (1823)
: of, relating to, or dependent on phenomena that involve relations between the temperature and the electrical condition in a metal or in contacting metals

ther·mo·elec·tric·i·ty \ˌthər-mō-i-ˌlek-ˈtri-sə-tē, -ˈtris-tē\ *noun* (1823)
: electricity produced by the direct action of heat (as by the unequal heating of a circuit composed of two dissimilar metals)

ther·mo·el·e·ment \-ˈe-lə-mənt\ *noun* (circa 1888)
: a device for measuring small currents consisting of a wire heating element and a thermocouple in electrical contact with it

ther·mo·form \ˈthər-mə-ˌfȯrm\ *transitive verb* (1956)
: to give a final shape to (as a plastic) with the aid of heat and usually pressure
— **ther·mo·form·able** \-ˌfȯr-mə-bəl\ *adjective*

ther·mo·gram \-ˌgram\ *noun* (1883)
1 : the record made by a thermograph
2 : a photographic record made by thermography

ther·mo·graph \-ˌgraf\ *noun* [International Scientific Vocabulary] (1843)
1 : THERMOGRAM
2 : a self-recording thermometer
3 : the apparatus used in thermography

ther·mog·ra·phy \(ˌ)thər-ˈmä-grə-fē\ *noun* (1840)
1 : a process of writing or printing involving the use of heat; *especially* **:** a raised-printing process in which matter printed by letterpress is dusted with powder and heated to make the lettering rise
2 : a technique for detecting and measuring variations in the heat emitted by various regions of the body and transforming them into visible signals that can be recorded photographically (as for diagnosing abnormal or diseased underlying conditions); *also* **:** a similar technique used elsewhere (as on buildings)
— **ther·mo·graph·ic** \ˌthər-mə-ˈgra-fik\ *adjective*
— **ther·mo·graph·i·cal·ly** \-fi-k(ə-)lē\ *adverb*

ther·mo·ha·line \ˌthər-mō-ˈhā-ˌlīn, -ˈha-\ *adjective* [*therm-* + Greek *hal-, hals* salt — more at SALT] (1942)
: involving or dependent upon the conjoint effect of temperature and salinity ⟨*thermohaline* circulation in the eastern Pacific⟩

ther·mo·junc·tion \-ˈjəŋ(k)-shən\ *noun* (1889)
: a junction of two dissimilar conductors used to produce a thermoelectric current

ther·mo·la·bile \-ˈlā-ˌbīl, -bəl\ *adjective* [International Scientific Vocabulary] (1904)
: unstable when heated; *specifically* **:** subject to loss of characteristic properties on being heated to or above 55°C ⟨many immune bodies, enzymes, and vitamins are *thermolabile*⟩
— **ther·mo·la·bil·i·ty** \-lā-ˈbi-lə-tē\ *noun*

ther·mo·lu·mi·nes·cence \-ˌlü-mə-ˈne-sᵊn(t)s\ *noun* [International Scientific Vocabulary] (1897)
1 : phosphorescence developed in a previously excited substance upon gentle heating
2 : the determination of the age of old material (as pottery) by the amount of thermoluminescence it produces — called also *thermoluminescence dating*
— **ther·mo·lu·mi·nes·cent** \-sᵊnt\ *adjective*

ther·mo·mag·net·ic \-mag-ˈne-tik\ *adjective* (1823)
: of or relating to the effects of heat upon the magnetic properties of substances or to the effects of a magnetic field upon thermal conduction

ther·mom·e·ter \thə(r)-ˈmä-mə-tər\ *noun* [French *thermomètre,* from Greek *thermē* heat + French *-o-* + *-mètre* -meter — more at THERM] (1633)
: an instrument for determining temperature consisting typically of a glass bulb attached to a fine tube of glass with a numbered scale and containing a liquid (as mercury or colored alcohol) that is sealed in and rises and falls with changes of temperature
— **ther·mo·met·ric** \ˌthər-mə-ˈme-trik\ *adjective*
— **ther·mo·met·ri·cal·ly** \-tri-k(ə-)lē\ *adverb*

ther·mom·e·try \thə(r)-ˈmä-mə-trē\ *noun* [International Scientific Vocabulary] (1858)
: the measurement of temperature

ther·mo·nu·cle·ar \ˌthər-mō-ˈnü-klē-ər, -ˈnyü-, ÷-ˈn(y)ü-kyə-lər\ *adjective* [International Scientific Vocabulary] (1938)
1 : of or relating to the transformations in the nucleus of atoms of low atomic weight (as hydrogen) that require a very high temperature for their inception (as in the hydrogen bomb or in the sun) ⟨*thermonuclear* reaction⟩ ⟨*thermonuclear* weapon⟩
2 : of, utilizing, or relating to a thermonuclear bomb ⟨*thermonuclear* war⟩ ⟨*thermonuclear* attack⟩

ther·mo·pe·ri·od·ic·i·ty \-ˌpir-ē-ə-ˈdi-sə-tē\ *noun* (1944)
: THERMOPERIODISM

ther·mo·pe·ri·od·ism \ˌthər-mō-ˈpir-ē-ə-ˌdi-zəm\ *noun* (1937)
: the sum of the responses of an organism and especially a plant to appropriately fluctuating temperatures

ther·mo·phil·ic \ˌthər-mə-ˈfi-lik\ *also* **ther·moph·i·lous** \(ˌ)thər-ˈmä-fə-ləs\ *or* **ther·mo·phile** \ˈthər-mə-ˌfīl\ *adjective* (1894)
: of, relating to, or being an organism growing at a high temperature ⟨*thermophilic* fermentation⟩ ⟨*thermophilic* bacteria⟩
— **thermophile** *noun*

ther·mo·pile \ˈthər-mə-ˌpīl\ *noun* [[4]*pile*] (1849)
: an apparatus that consists of a number of thermocouples combined so as to multiply the effect and is used for generating electric currents or for determining intensities of radiation

ther·mo·plas·tic \ˌthər-mə-ˈplas-tik\ *adjective* (1883)
: capable of softening or fusing when heated and of hardening again when cooled ⟨*thermoplastic* synthetic resins⟩ — compare THERMOSETTING
— **thermoplastic** *noun*
— **ther·mo·plas·tic·i·ty** \-ˌpla-ˈsti-sə-tē\ *noun*

ther·mo·re·cep·tor \ˌthər-mō-ri-ˈsep-tər\ *noun* (1937)
: a sensory end organ that is stimulated by heat or cold

ther·mo·reg·u·la·tion \-ˌre-gyə-ˈlā-shən\ *noun* [International Scientific Vocabulary] (1927)
: the maintenance or regulation of temperature; *specifically* **:** the maintenance of a particular temperature of the living body
— **ther·mo·reg·u·late** \-ˈre-gyə-ˌlāt\ *verb*

ther·mo·reg·u·la·tor \-ˈre-gyə-ˌlā-tər\ *noun* [International Scientific Vocabulary] (1875)
: a device (as a thermostat) for the regulation of temperature

ther·mo·reg·u·la·to·ry \-ˈre-gyə-lə-ˌtōr-ē, -ˌtȯr-\ *adjective* (1941)
: tending to maintain a body at a particular temperature whatever its environmental temperature ⟨*thermoregulatory* adjustments⟩

ther·mo·rem·a·nent \-ˈre-mə-nənt\ *adjective* (1951)
: being or relating to magnetic remanence (as in a rock cooled from a molten state or in a baked clay object containing magnetic minerals) that indicates the strength and direction of the earth's magnetic field at a former time
— **ther·mo·rem·a·nence** \-nən(t)s\ *noun*

ther·mos \ˈthər-məs\ *noun* [from *Thermos,* a trademark] (1907)
: a container (as a bottle or jar) with a vacuum between an inner and outer wall used to keep material and especially liquids either hot or cold for considerable periods

ther·mo·scope \ˈthər-mə-ˌskōp\ *noun* [New Latin *thermoscopium,* from *therm-* + *-scopium* -scope] (1804)
: an instrument for indicating changes of temperature by accompanying changes in volume (as of a gas)

ther·mo·set \ˈthər-mō-ˌset\ *noun* (1947)
: a thermosetting resin or plastic

ther·mo·set·ting \-ˌse-tiŋ\ *adjective* (circa 1931)
: capable of becoming permanently rigid when heated or cured ⟨a *thermosetting* resin⟩ — compare THERMOPLASTIC

ther·mo·sphere \ˈthər-mə-ˌsfir\ *noun* [International Scientific Vocabulary] (1950)
: the part of the earth's atmosphere that begins at about 50 miles (80 kilometers) above the earth's surface, extends to outer space, and is characterized by steadily increasing temperature with height
— **ther·mo·spher·ic** \ˌthər-mə-ˈsfir-ik, -ˈsfer-\ *adjective*

ther·mo·sta·ble \ˌthər-mō-ˈstā-bəl\ *adjective* (1904)
: stable when heated; *specifically* **:** retaining characteristic properties on being moderately heated ⟨a *thermostable* bacterial protease⟩
— **ther·mo·sta·bil·i·ty** \-stə-ˈbi-lə-tē\ *noun*

[1]**ther·mo·stat** \ˈthər-mə-ˌstat\ *noun* (1831)
: an automatic device for regulating temperature (as by controlling the supply of gas or electricity to a heating apparatus); *also* **:** a similar device for actuating fire alarms or for controlling automatic sprinklers
— **ther·mo·stat·ic** \ˌthər-mə-ˈsta-tik\ *adjective*
— **ther·mo·stat·i·cal·ly** \-ti-k(ə-)lē\ *adverb*

[2]**thermostat** *transitive verb* **-stat·ed** \-ˌsta-təd\ *also* **-stat·ted; -stat·ing** *also* **-stat·ting** (1924)
: to provide with or control the temperature of by a thermostat

ther·mo·tac·tic \ˌthər-mə-ˈtak-tik\ *adjective* (1839)
: of, relating to, or exhibiting thermotaxis

ther·mo·tax·is \-ˈtak-səs\ *noun* [New Latin] (circa 1891)
1 : the regulation of body temperature
2 : a taxis in which a temperature gradient constitutes the directive factor

ther·mo·trop·ic \-ˈträ-pik\ *adjective* [International Scientific Vocabulary] (1885)
: of, relating to, or exhibiting thermotropism

ther·mot·ro·pism \(ˌ)thər-ˈmä-trə-ˌpi-zəm\ *noun* [International Scientific Vocabulary] (circa 1890)

: a tropism in which a temperature gradient determines the orientation

-thermy *noun combining form* [New Latin *-thermia,* from Greek *thermē* heat — more at THERM]
1 : state of heat ⟨endo*thermy*⟩
2 : generation of heat ⟨dia*thermy*⟩

the·ro·pod \ˈthir-ə-ˌpäd\ *noun* [New Latin *Theropoda,* from Greek *thēr* wild animal + *pod-, pous* foot — more at FIERCE, FOOT] (circa 1891)
: any of a suborder (Theropoda) of carnivorous bipedal saurischian dinosaurs (as a tyrannosaur or allosaurus) usually having small forelimbs

Ther·si·tes \(ˌ)thər-ˈsī-(ˌ)tēz\ *noun* [Latin, from Greek *Thersitēs*]
: a Greek warrior at Troy known as a carping critic and slain by Achilles for mocking him

the·sau·rus \thi-ˈsȯr-əs\ *noun, plural* **-sau·ri** \-ˈsȯr-ˌī, -ˌē\ *or* **-sau·rus·es** \-ˈsȯr-ə-səz\ [New Latin, from Latin, treasure, collection, from Greek *thēsauros*] (circa 1823)
1 : TREASURY, STOREHOUSE
2 a : a book of words or of information about a particular field or set of concepts; *especially* **:** a book of words and their synonyms **b :** a list of subject headings or descriptors usually with a cross-reference system for use in the organization of a collection of documents for reference and retrieval
— the·sau·ral \-ˈsȯr-əl\ *adjective*

these *plural of* THIS

The·seus \ˈthē-ˌsüs, -sē-əs\ *noun* [Latin, from Greek *Theseus*]
: a king of Athens in Greek mythology who kills Procrustes and the Minotaur before defeating the Amazons and marrying their queen

the·sis \ˈthē-səs, *British especially for 1* ˈthe-sis\ *noun, plural* **the·ses** \ˈthē-ˌsēz\ [in sense 1, Middle English, from Late Latin & Greek; Late Latin, lowering of the voice, from Greek, downbeat, more important part of a foot, literally, act of laying down; in other senses, Latin, from Greek, literally, act of laying down, from *tithenai* to put, lay down — more at DO] (14th century)
1 a (1) **:** the unstressed part of a poetic foot especially in accentual verse (2) **:** the longer part of a poetic foot especially in quantitative verse **b :** the accented part of a musical measure **:** DOWNBEAT — compare ARSIS
2 a : a position or proposition that a person (as a candidate for scholastic honors) advances and offers to maintain by argument **b :** a proposition to be proved or one advanced without proof **:** HYPOTHESIS
3 : the first and least adequate stage of dialectic — compare SYNTHESIS
4 : a dissertation embodying results of original research and especially substantiating a specific view; *especially* **:** one written by a candidate for an academic degree

¹thes·pi·an \ˈthes-pē-ən\ *adjective* (1675)
1 *capitalized* **:** of or relating to Thespis
2 *often capitalized* [from the tradition that Thespis was the originator of the actor's role] **:** relating to the drama **:** DRAMATIC

²thespian *noun* (1827)
: ACTOR

Thes·sa·lo·nians \ˌthe-sə-ˈlō-nyənz, -nē-ənz\ *noun plural but singular in construction* [*Thessalonian* inhabitant of ancient Thessalonica, irregular from *Thessalonica*]
: either of two letters written by Saint Paul to the Christians of Thessalonica and included as books in the New Testament — see BIBLE table

the·ta \ˈthā-tə, *chiefly British* ˈthē-tə\ *noun* [Greek *thēta,* of Semitic origin; akin to Hebrew *ṭēth* teth] (1603)
1 : the 8th letter of the Greek alphabet — see ALPHABET table
2 : THETA RHYTHM

theta rhythm *noun* (1944)
: a relatively high amplitude brain wave pattern between approximately 4 and 9 hertz that is characteristic especially of the hippocampus but occurs in many regions of the brain including the cortex — called also *theta wave*

thet·ic \ˈthe-tik, ˈthē-\ *adjective* [Greek *thetikos* of a proposition, from *tithenai* to lay down — more at DO] (1815)
: constituting or beginning with a poetic thesis ⟨a *thetic* syllable⟩
— thet·i·cal·ly \-ti-k(ə-)lē\ *adverb*

The·tis \ˈthē-təs\ *noun* [Latin, from Greek]
: a sea goddess who marries Peleus and becomes the mother of Achilles

the·ur·gist \ˈthē-(ˌ)ər-jist\ *noun* (1652)
: WONDER-WORKER, MAGICIAN

the·ur·gy \ˈthē-(ˌ)ər-jē\ *noun* [Late Latin *theurgia,* from Late Greek *theourgia,* from *theourgos* miracle worker, from Greek *the-* + *ergon* work — more at WORK] (1569)
: the art or technique of compelling or persuading a god or beneficent or supernatural power to do or refrain from doing something
— the·ur·gic \thē-ˈər-jik\ *or* **the·ur·gi·cal** \-ji-kəl\ *adjective*

thew \ˈthü, ˈthyü\ *noun* [Middle English, personal quality, virtue, from Old English *thēaw;* akin to Old High German *thau* custom] (15th century)
1 a : muscular power or development **b :** STRENGTH, VITALITY
2 : MUSCLE, SINEW — usually used in plural

they \ˈt͟hā\ *pronoun, plural in construction* [Middle English, from Old Norse *their,* masculine plural demonstrative & personal pronoun; akin to Old English *thæt* that] (13th century)
1 a : those ones — used as third person pronoun serving as the plural of *he, she,* or *it* or referring to a group of two or more individuals not all of the same sex ⟨*they* dance well⟩ **b :** ¹HE 2 — often used with an indefinite third person singular antecedent ⟨everyone knew where *they* stood —E. L. Doctorow⟩ ⟨nobody has to go to school if *they* don't want to —*N. Y. Times*⟩
2 : PEOPLE 2 — used in a generic sense ⟨as lazy as *they* come⟩ ▫ ▫

they'd \ˈt͟hād\ (1676)
: they had **:** they would

they'll \ˈt͟hā(ə)l, t͟hel\ (1607)
: they will **:** they shall

they're \t͟hər, ˈt͟her\ (circa 1594)
: they are

they've \ˈt͟hāv\ (1611)
: they have

thi- *or* **thio-** *combining form* [International Scientific Vocabulary, from Greek *thei-, theio-* sulfur, from *theion*]
: containing sulfur ⟨*thi*amine⟩ ⟨*thio*cyanate⟩

thia·ben·da·zole \ˌthī-ə-ˈben-də-ˌzōl\ *noun* [*thia*zole + *benz-* + imide + *azole*] (1961)
: a drug $C_{10}H_7N_3S$ used in the control of parasitic nematode worms, in the treatment of fungus infections, and as an agricultural fungicide

thi·ami·nase \thī-ˈa-mə-ˌnās, ˈthī-ə-mə-, -ˌnāz\ *noun* [International Scientific Vocabulary] (1938)
: an enzyme that catalyzes the breakdown of thiamine

thi·a·mine \ˈthī-ə-mən, -ˌmēn\ *also* **thi·a·min** \-mən\ *noun* [*thiamine* alteration of *thiamin,* from *thi-* + *-amin* (as in *vitamin*)] (1937)
: a vitamin $(C_{12}H_{17}N_4OS)Cl$ of the B complex that is essential to normal metabolism and nerve function and is widespread in plants and animals — called also *vitamin B_1*

thi·a·zide \ˈthī-ə-ˌzīd, -zəd\ *noun* [*thi-* + di*az*ine + diox*ide*] (1959)
: any of a group of drugs used as oral diuretics especially in the control of high blood pressure

thi·a·zine \ˈthī-ə-ˌzēn\ *noun* [International Scientific Vocabulary] (1900)
: any of various compounds that are characterized by a ring composed of four carbon atoms, one sulfur atom, and one nitrogen atom and include some important as dyes and others as tranquilizers — compare PHENOTHIAZINE

thi·a·zole \ˈthī-ə-ˌzōl\ *noun* [International Scientific Vocabulary] (1888)
1 : a colorless basic liquid C_3H_3NS consisting of a 5-membered ring and having an odor like pyridine
2 : any of various thiazole derivatives including some used in medicine and others important as chemical accelerators

¹thick \ˈthik\ *adjective* [Middle English *thikke,* from Old English *thicce;* akin to Old High German *dicki* thick, Old Irish *tiug*] (before 12th century)
1 a : having or being of relatively great depth or extent from one surface to its opposite ⟨a *thick* plank⟩ **b :** heavily built **:** THICKSET
2 a : close-packed with units or individuals ⟨the air was *thick* with snow⟩ **b :** occurring in large numbers **:** NUMEROUS **c :** viscous in consistency ⟨*thick* syrup⟩ **d :** SULTRY, STUFFY **e :** marked by haze, fog, or mist ⟨*thick* weather⟩ **f :** impenetrable to the eye **:** PROFOUND ⟨*thick* darkness⟩ **g :** extremely intense ⟨*thick* silence⟩
3 : measuring in thickness ⟨12 inches *thick*⟩
4 a : imperfectly articulated **:** INDISTINCT ⟨*thick* speech⟩ **b :** plainly apparent **:** DECIDED ⟨a *thick* French accent⟩ **c :** producing inarticulate speech ⟨a *thick* tongue⟩
5 : OBTUSE, STUPID
6 : associated on close terms **:** INTIMATE ⟨was quite *thick* with his pastor⟩
7 : exceeding bounds of propriety or fitness **:** EXCESSIVE ⟨called it a bit *thick* to be fired without warning⟩
— thick·ish \ˈthi-kish\ *adjective*
— thick·ly *adverb*

▫ USAGE

they *They* used as an indefinite subject (sense 2) is sometimes objected to on the grounds that it does not have an antecedent. Not every pronoun requires an antecedent, however. The indefinite *they* is used in all varieties of contexts and is standard.

they *They, their, them, themselves:* English lacks a common-gender third person singular pronoun that can be used to refer to indefinite pronouns (as *everyone, anyone, someone*). Writers and speakers have supplied this lack by using the plural pronouns ⟨and every one to rest *themselves* betake —Shakespeare⟩ ⟨I would have everybody marry if *they* can do it properly —Jane Austen⟩ ⟨it is too hideous for anyone in *their* senses to buy —W. H. Auden⟩. The plural pronouns have also been put to use as pronouns of indefinite number to refer to singular nouns that stand for many persons ⟨'tis meet that some more audience than a mother, since nature makes *them* partial, should o'erhear the speech —Shakespeare⟩ ⟨a person can't help *their* birth —W. M. Thackeray⟩ ⟨no man goes to battle to be killed. — But *they* do get killed —G. B. Shaw⟩. The use of *they, their, them,* and *themselves* as pronouns of indefinite gender and indefinite number is well established in speech and writing, even in literary and formal contexts. This gives you the option of using the plural pronouns where you think they sound best, and of using the singular pronouns (as *he, she, he or she,* and their inflected forms) where you think they sound best. See in addition ANYBODY, EVERYBODY, HE, NOBODY, SOMEBODY.

²thick *adverb* (before 12th century)
: in a thick manner : THICKLY
³thick *noun* (13th century)
1 : the most crowded or active part ⟨in the *thick* of the battle⟩
2 : the part of greatest thickness ⟨the *thick* of the thumb⟩
thick and thin *noun* (14th century)
: every difficulty and obstacle — used especially in the phrase *through thick and thin*
thick·en \'thi-kən\ *verb* **thick·ened; thick·en·ing** \'thik-niŋ, 'thi-kə-\ (15th century)
transitive verb
1 a : to make thick, dense, or viscous in consistency ⟨*thicken* gravy with flour⟩ **b** : to make close or compact
2 : to increase the depth or diameter of
3 : to make inarticulate : BLUR ⟨alcohol *thickened* his speech⟩
intransitive verb
1 a : to become dense ⟨the mist *thickened*⟩ **b** : to become concentrated in numbers, mass, or frequency
2 : to grow blurred or obscure
3 : to grow broader or bulkier
4 : to grow complicated or keen ⟨the plot *thickens*⟩
— **thick·en·er** \'thik-nər, 'thi-kə-\ *noun*
thickening *noun* (circa 1580)
1 : the act of making or becoming thick
2 : a thickened part or place
3 : something used to thicken
thick·et \'thi-kət\ *noun* [(assumed) Middle English *thikket,* from Old English *thiccet,* from *thicce* thick] (before 12th century)
1 : a dense growth of shrubbery or small trees : COPPICE
2 : something resembling a thicket in density or impenetrability : TANGLE
— **thick·ety** \-kə-tē\ *adjective*
thick·et·ed \'thi-kə-təd\ *adjective* (circa 1624)
: dotted or covered with thickets
thick·head \'thik-ˌhed\ *noun* (1824)
: a stupid person : BLOCKHEAD
thick·head·ed \-ˌhe-dəd\ *adjective* (1707)
1 : having a thick head
2 : sluggish and obtuse of mind
thick·ness \-nəs\ *noun* (before 12th century)
1 : the smallest of three dimensions ⟨length, width, and *thickness*⟩
2 : the quality or state of being thick
3 a : viscous consistency ⟨boiled to the *thickness* of honey⟩ **b** : the condition of being smoky, foul, or foggy
4 : the thick part of something
5 : CONCENTRATION, DENSITY
6 : STUPIDITY, DULLNESS
7 : LAYER, PLY, SHEET ⟨a single *thickness* of canvas⟩
thick·set \-ˌset\ *adjective* (14th century)
1 : closely placed; *also* : growing thickly ⟨a *thickset* wood⟩
2 : having a thick body : BURLY
thick–skinned \-ˌskind\ *adjective* (1545)
1 : having a thick skin : PACHYDERMATOUS
2 a : CALLOUS, INSENSITIVE **b** : impervious to criticism
thick–wit·ted \-ˌwi-təd\ *adjective* (1634)
: dull or slow of mind : STUPID
thief \'thēf\ *noun, plural* **thieves** \'thēvz\ [Middle English *theef,* from Old English *thēof;* akin to Old High German *diob* thief] (before 12th century)
: one that steals especially stealthily or secretly; *also* : one who commits theft or larceny
thieve \'thēv\ *verb* **thieved; thiev·ing** (before 12th century)
: STEAL, ROB
thiev·ery \'thēv-rē, 'thē-və-\ *noun, plural* **-er·ies** (1568)
: the act or practice or an instance of stealing : THEFT
thiev·ish \'thē-vish\ *adjective* (1538)
1 : given to stealing
2 : of, relating to, or characteristic of a thief
— **thiev·ish·ly** *adverb*
— **thiev·ish·ness** *noun*
thigh \'thī\ *noun* [Middle English, from Old English *thēoh;* akin to Old High German *dioh* thigh, Lithuanian *taukai,* plural, fat] (before 12th century)
1 a : the proximal segment of the vertebrate hind limb extending from the hip to the knee and supported by a single large bone **b** : the segment of the leg immediately distal to the thigh in a bird or in a quadruped in which the true thigh is obscured **c** : the femur of an insect
2 : something resembling or covering a thigh
— **thighed** \'thīd\ *adjective*
thigh·bone \'thī-ˌbōn\ *noun* (15th century)
: FEMUR 1
thig·mo·tax·is \ˌthig-mə-'tak-səs\ *noun* [New Latin, from Greek *thigma* touch (from *thinganein* to touch) + New Latin *-taxis;* akin to Latin *fingere* to shape — more at DOUGH] (1903)
: a taxis in which contact especially with a solid body is the directive factor
thig·mot·ro·pism \thig-'mä-trə-ˌpi-zəm\ *noun* [Greek *thigma* + International Scientific Vocabulary *-o-* + *-tropism*] (circa 1900)
: a tropism in which contact especially with a solid or a rigid surface is the orienting factor
thill \'thil\ *noun* [Middle English *thille,* perhaps from Old English, plank; akin to Old English *thel* board, Old High German *dili,* and probably to Latin *tellus* earth] (14th century)
: a shaft of a vehicle
thim·ble \'thim-bəl\ *noun* [Middle English *thymbyl,* probably alteration of Old English *thȳmel* covering for the thumb, from *thūma* thumb] (15th century)
1 : a pitted cap or cover worn on the finger to push the needle in sewing
2 a : a grooved ring of thin metal used to fit in a spliced loop in a rope as protection from chafing **b** : a lining (as of metal) for an opening (as in a roof or wall) through which a stovepipe or chimney passes
thim·ble·ber·ry \-ˌber-ē\ *noun* (1788)
: any of several American raspberries or blackberries (especially *Rubus occidentalis, R. parviflorus,* and *R. odoratus*) having thimble-shaped fruit
thim·ble·ful \-ˌfu̇l\ *noun* (1607)
1 : as much as a thimble will hold
2 : a very small quantity
¹thim·ble·rig \-ˌrig\ *noun* (1826)
1 : a swindling trick in which a small ball or pea is quickly shifted from under one to another of three small cups to fool the spectator guessing its location
2 : one who manipulates the cup in thimblerig : THIMBLERIGGER
²thimblerig *transitive verb* (1839)
1 : to cheat by trickery
2 : to swindle by thimblerig
— **thim·ble·rig·ger** *noun*
thim·ble·weed \'thim-bəl-ˌwēd\ *noun* (1833)
: any of various anemones (as *Anemone virginiana* and *A. cylindrica*)
thi·mer·o·sal \thī-'mer-ə-ˌsal\ *noun* [probably from *thi-* + *mercury* + *-o-* + *sal*icylate] (1949)
: a crystalline organic mercurial antiseptic $C_9H_9HgNaO_2S$ used especially for its antifungal and bacteriostatic properties
¹thin \'thin\ *adjective* **thin·ner; thin·nest** [Middle English *thinne,* from Old English *thynne;* akin to Old High German *dunni* thin, Latin *tenuis* thin, *tenēre* to hold, *tendere* to stretch, Greek *teinein*] (before 12th century)
1 a : having little extent from one surface to its opposite ⟨*thin* paper⟩ **b** : measuring little in cross section or diameter ⟨*thin* rope⟩
2 : not dense in arrangement or distribution ⟨*thin* hair⟩
3 : not well fleshed : LEAN
4 a : more fluid or rarefied than normal ⟨*thin* air⟩ **b** : having less than the usual number : SCANTY ⟨*thin* attendance⟩ **c** : few in number : SCARCE **d** : scantily supplied **e** : characterized by a paucity of bids or offerings ⟨a *thin* market⟩
5 a : lacking substance or strength ⟨*thin* broth⟩ ⟨a *thin* plot⟩ **b** *of a soil* : INFERTILE, POOR
6 a : FLIMSY, UNCONVINCING ⟨a *thin* disguise⟩ **b** : disappointingly poor or hard ⟨had a *thin* time of it⟩
7 : somewhat feeble, shrill, and lacking in resonance ⟨a *thin* voice⟩
8 : lacking in intensity or brilliance ⟨*thin* light⟩
9 : lacking sufficient photographic density or contrast ☆
— **thin·ly** *adverb*
— **thin·ness** \'thin-nəs\ *noun*
— **thin·nish** \'thi-nish\ *adjective*
²thin *verb* **thinned; thin·ning** (before 12th century)
transitive verb
: to make thin or thinner: **a** : to reduce in thickness or depth : ATTENUATE **b** : to make less dense or viscous **c** : DILUTE, WEAKEN **d** : to cause to lose flesh ⟨*thinned* by weeks of privation⟩ **e** : to reduce in number or bulk
intransitive verb
1 : to become thin or thinner
2 : to become weak
³thin *adverb* **thin·ner; thin·nest** (13th century)
: in a thin manner : THINLY — used especially in combinations ⟨*thin*-clad⟩ ⟨*thin*-flowing⟩
¹thine \'thīn\ *adjective* [Middle English *thin,* from Old English *thīn*] (before 12th century)
archaic : THY — used especially before a word beginning with a vowel or *h*
²thine *pronoun, singular or plural in construction* [Middle English *thin,* from Old English *thīn,* from *thīn* thy — more at THY] (before 12th century)
archaic : that which belongs to thee — used without a following noun as a pronoun equivalent in meaning to the adjective *thy;* used especially in ecclesiastical or literary language and still surviving in the speech of Friends especially among themselves
thin film *noun* (1944)
: a very thin layer of a substance on a supporting material; *especially* : a coating (as of a semiconductor) that is deposited in a layer one atom or one molecule thick
thing \'thiŋ\ *noun* [Middle English, from Old English, thing, assembly; akin to Old High German *ding* thing, assembly, Gothic *theihs* time] (before 12th century)
1 a : a matter of concern : AFFAIR ⟨many *things* to do⟩ **b** *plural* : state of affairs in general or within a specified or implied sphere ⟨*things* are improving⟩ **c** : a particular state of affairs : SITUATION ⟨look at this *thing* another way⟩ **d** : EVENT, CIRCUMSTANCE ⟨that shooting was a terrible *thing*⟩

☆ SYNONYMS
Thin, slender, slim, slight, tenuous mean not thick, broad, abundant, or dense. THIN implies comparatively little extension between surfaces or in diameter, or it may imply lack of substance, richness, or abundance ⟨*thin* wire⟩ ⟨a *thin* soup⟩. SLENDER implies leanness or spareness often with grace and good proportion ⟨the *slender* legs of a Sheraton chair⟩. SLIM applies to slenderness that suggests fragility or scantiness ⟨a *slim* volume of poetry⟩ ⟨a *slim* chance⟩. SLIGHT implies smallness as well as thinness ⟨a *slight* build⟩. TENUOUS implies extreme thinness, sheerness, or lack of substance and firmness ⟨a *tenuous* thread⟩.

2 a : DEED, ACT, ACCOMPLISHMENT ⟨do great *things*⟩ **b :** a product of work or activity ⟨likes to build *things*⟩ **c :** the aim of effort or activity ⟨the *thing* is to get well⟩
3 a : a separate and distinct individual quality, fact, idea, or usually entity **b :** the concrete entity as distinguished from its appearances **c :** a spatial entity **d :** an inanimate object distinguished from a living being
4 a *plural* **:** POSSESSIONS, EFFECTS ⟨pack your *things*⟩ **b :** whatever may be possessed or owned or be the object of a right **c :** an article of clothing ⟨not a *thing* to wear⟩ **d** *plural* **:** equipment or utensils especially for a particular purpose ⟨bring the tea *things*⟩
5 : an object or entity not precisely designated or capable of being designated ⟨use this *thing*⟩
6 a : DETAIL, POINT ⟨checks every little *thing*⟩ **b :** a material or substance of a specified kind ⟨avoid fatty *things*⟩
7 a : a spoken or written observation or point **b :** IDEA, NOTION ⟨says the first *thing* he thinks of⟩ **c :** a piece of news or information ⟨couldn't get a *thing* out of him⟩
8 : INDIVIDUAL ⟨not a living *thing* in sight⟩
9 : the proper or fashionable way of behaving, talking, or dressing — used with *the*
10 a : a mild obsession or phobia ⟨has a *thing* about driving⟩; *also* **:** the object of such an obsession or phobia **b :** something (as an activity) that makes a strong appeal to the individual **:** FORTE, SPECIALTY ⟨letting students do their own *thing* —*Newsweek*⟩ ⟨I think travelling is very much a novelist's *thing* —Philip Larkin⟩
◇

thing·am·a·bob \ˈthiŋ-ə-mə-ˌbäb\ *noun* (1832)
: THINGAMAJIG

thing·am·a·jig *or* **thing·um·a·jig** \ˈthiŋ-ə-mə-ˌjig\ *noun* [alteration of earlier *thingum,* from *thing*] (1828)
: something that is hard to classify or whose name is unknown or forgotten

thing–in–itself *noun, plural* **things–in–themselves** [translation of German *Ding an sich*] (1798)
: NOUMENON

thing·ness \ˈthiŋ nəs\ *noun* (1896)
: the quality or state of objective existence or reality

thing·um·my \ˈthiŋ-ə-mē\ *noun, plural* **-mies** [alteration of earlier *thingum*] (1796)
: THINGAMAJIG

¹think \ˈthiŋk\ *verb* **thought** \ˈthȯt\; **think·ing** [Middle English *thenken,* from Old English *thencan;* akin to Old High German *denken* to think, Latin *tongēre* to know — more at THANKS] (before 12th century)
transitive verb
1 : to form or have in the mind
2 : to have as an intention ⟨*thought* to return early⟩
3 a : to have as an opinion ⟨*think* it's so⟩ **b :** to regard as **:** CONSIDER ⟨*think* the rule unfair⟩
4 a : to reflect on **:** PONDER ⟨*think* the matter over⟩ **b :** to determine by reflecting ⟨*think* what to do next⟩
5 : to call to mind **:** REMEMBER ⟨he never *thinks* to ask how we do⟩
6 : to devise by thinking — usually used with *up* ⟨*thought* up a plan to escape⟩
7 : to have as an expectation **:** ANTICIPATE ⟨we didn't *think* we'd have any trouble⟩
8 a : to center one's thoughts on ⟨talks and *thinks* business⟩ **b :** to form a mental picture of
9 : to subject to the processes of logical thought ⟨*think* things out⟩
intransitive verb
1 a : to exercise the powers of judgment, conception, or inference **:** REASON **b :** to have in the mind or call to mind a thought
2 a : to have the mind engaged in reflection **:** MEDITATE **b :** to consider the suitability ⟨*thought* of her for president⟩
3 : to have a view or opinion ⟨*thinks* of himself as a poet⟩
4 : to have concern — usually used with *of* ⟨a man must *think* first of his family⟩
5 : to consider something likely **:** SUSPECT ⟨may happen sooner than you *think*⟩ ☆ ☆
— think·er *noun*
— think better of : to reconsider and make a wiser decision
— think much of : to view with satisfaction **:** APPROVE — usually used in negative constructions ⟨I didn't *think much of* the new car⟩

²think *noun* (1834)
: an act of thinking ⟨has another *think* coming⟩

³think *adjective* (1906)
: relating to, requiring, or stimulating thinking

think·able \ˈthiŋ-kə-bəl\ *adjective* (1854)
1 : capable of being comprehended or reasoned about ⟨the ultimate nature of Deity is scarcely *thinkable*⟩
2 : conceivably possible
— think·able·ness *noun*
— think·ably \-blē\ *adverb*

¹thinking *noun* (14th century)
1 : the action of using one's mind to produce thoughts
2 a : OPINION, JUDGMENT ⟨I'd like to know your *thinking* on this⟩ **b :** thought that is characteristic (as of a period, group, or person) ⟨the current student *thinking* on fraternities⟩

²thinking *adjective* (1681)
: marked by use of the intellect **:** RATIONAL ⟨*thinking* citizens⟩
— think·ing·ly \ˈthiŋ-kiŋ-lē\ *adverb*
— think·ing·ness *noun*

thinking cap *noun* (1874)
: a state or mood in which one thinks — usually used in the phrase *put one's thinking cap on*

think piece *noun* (1941)
: a piece of writing meant to be thought-provoking and speculative that consists chiefly of background material and personal opinion and analysis

think tank *noun* (1959)
: an institute, corporation, or group organized for interdisciplinary research (as in technological and social problems) — called also *think factory*

thin–layer chromatography *noun* (1957)
: chromatography in which the stationary phase is an absorbent medium (as alumina or silica gel) arranged as a thin layer on a rigid support (as a glass plate)

thin·ner \ˈthi-nər\ *noun* (1832)
: one that thins; *specifically* **:** a volatile liquid (as turpentine) used especially to thin paint

thin–skinned \ˈthin-ˌskind\ *adjective* (1598)
1 : having a thin skin or rind
2 : unduly susceptible to criticism or insult **:** TOUCHY

thio- — see THI-

thio acid \ˈthī-ō-\ *noun* [International Scientific Vocabulary, from *thi-*] (circa 1891)
: an acid in which oxygen is partly or wholly replaced by sulfur

thio·cy·a·nate \ˌthī-ō-ˈsī-ə-ˌnāt, -nət\ *noun* [International Scientific Vocabulary] (1877)
: a compound that consists of the chemical group SCN bonded by the sulfur atom to a group or an atom other than a hydrogen atom

thi·ol \ˈthī-ˌȯl, -ˌōl\ *noun* [International Scientific Vocabulary *thi-* + ¹*-ol*] (circa 1890)
1 : any of various compounds having the general formula RSH which are analogous to alcohols but in which sulfur replaces the oxygen of the hydroxyl group and which have disagreeable odors
2 : the functional group — SH characteristic of thiols
— thi·o·lic \thī-ˈō-lik\ *adjective*

thion- *combining form* [International Scientific Vocabulary, from Greek *theion*]
: sulfur ⟨me*thion*ine⟩

☆ SYNONYMS

Think, conceive, imagine, fancy, realize, envisage, envision mean to form an idea of. THINK implies the entrance of an idea into one's mind with or without deliberate consideration or reflection ⟨I just *thought* of a good joke⟩. CONCEIVE suggests the forming and bringing forth and usually developing of an idea, plan, or design ⟨*conceived* of a new marketing approach⟩. IMAGINE stresses a visualization ⟨*imagine* you're at the beach⟩. FANCY suggests an imagining often unrestrained by reality but spurred by desires ⟨*fancied* himself a super athlete⟩. REALIZE stresses a grasping of the significance of what is conceived or imagined ⟨*realized* the enormity of the task ahead⟩. ENVISAGE and ENVISION imply a conceiving or imagining that is especially clear or detailed ⟨*envisaged* a totally computerized operation⟩ ⟨*envisioned* a cure for the disease⟩.

Think, cogitate, reflect, reason, speculate, deliberate mean to use one's powers of conception, judgment, or inference. THINK is general and may apply to any mental activity, but used alone often suggests attainment of clear ideas or conclusions ⟨teaches students how to *think*⟩. COGITATE implies deep or intent thinking ⟨*cogitated* on the mysteries of nature⟩. REFLECT suggests unhurried consideration of something recalled to the mind ⟨*reflecting* on fifty years of married life⟩. REASON stresses consecutive logical thinking ⟨able to *reason* brilliantly in debate⟩. SPECULATE implies reasoning about things theoretical or problematic ⟨*speculated* on the fate of the lost explorers⟩. DELIBERATE suggests slow or careful reasoning before forming an opinion or reaching a conclusion or decision ⟨the jury *deliberated* for five hours before reaching a verdict⟩.

◇ WORD HISTORY

thing Old English *thing* and its cognates in other early Germanic languages meant "assembly, court." This meaning has long fallen into disuse in English, German, and Dutch, but the Scandinavian languages preserve it, notably in the names of national legislatures, such as the *Althing* ("general assembly") in Iceland and the *Storting* ("big assembly") in Norway. The passage of Old English *thing* from "assembly" to a meaning as nebulous as "thing" seems peculiar, but is actually fairly straightforward given some intermediary steps. The sequence runs from "assembly, court" to "matter before an assembly, case before a court" to the more general "matter, affair, concern" and finally to "thing" in its many and varied modern applications. Parallel shifts of meaning can be found in the histories of English *sake* and its German cognate *Sache* "thing, affair," both of whose antecedents meant "action at law," and French *chose* "thing," from Latin *causa* "legal case." Before it meant "assembly" the common Germanic word represented by Old English *thing* most likely meant simply "something fixed, fixed time" (hence, "meeting at a fixed time"). A hint of this earlier meaning is given by the cognate word *theihs* "time, occasion" in Gothic, the oldest fully attested Germanic language.

thio·pen·tal \ˌthī-ō-ˈpen-ˌtal, -ˌtȯl\ *noun* [*thi-* + *pento*barbit*al*] (1947)
: a barbiturate $C_{11}H_{18}N_2O_2S$ used in the form of its sodium salt especially as an intravenous anesthetic — compare PENTOTHAL

thio·phene \ˈthī-ə-ˌfēn\ *noun* [International Scientific Vocabulary *thi-* + *phene* benzene] (1883)
: a heterocyclic liquid C_4H_4S from coal tar that resembles benzene

thi·o·rid·a·zine \ˌthī-ə-ˈri-də-ˌzēn, -zən\ *noun* [*thi-* + pipe*rid*ine + phenothi*azine*] (1959)
: a phenothiazine tranquilizer used as the hydrochloride $C_{21}H_{26}N_2S_2{\cdot}HCl$ for relief of anxiety states and in the treatment of schizophrenia

thio·sul·fate \-ˈsəl-ˌfāt\ *noun* [International Scientific Vocabulary] (1873)
: a salt containing the anion $S_2O_3^{2-}$

thio·te·pa \ˌthī-ə-ˈtē-pə\ *noun* (1953)
: a sulfur analogue of tepa $C_6H_{12}N_3PS$ that is used especially as an antineoplastic agent

thio·ura·cil \ˌthī-ō-ˈyu̇r-ə-ˌsil\ *noun* [International Scientific Vocabulary] (1905)
: a bitter crystalline compound $C_4H_4N_2OS$ that depresses the function of the thyroid gland

thio·urea \-yu̇-ˈrē-ə\ *noun* [New Latin] (1894)
: a colorless crystalline bitter compound $CS(NH_2)_2$ analogous to and resembling urea that is used especially as a photographic and organic chemical reagent; *also* **:** a substituted derivative of this compound

thir \thər, ˈthi(ə)r, ˈthu̇r\ *pronoun or adjective* [Middle English (northern), perhaps irregular from Middle English *this*] (14th century)
dialect British **:** THESE

thi·ram \ˈthī-ˌram\ *noun* [probably by alteration from *thiuram* the chemical group NH_2CS] (1949)
: a compound $C_6H_{12}N_2S_4$ used as a fungicide and seed disinfectant

[1]**third** \ˈthərd\ *adjective* [Middle English *thridde, thirde,* from Old English *thridda, thirdda;* akin to Latin *tertius* third, Greek *tritos, treis* three — more at THREE] (before 12th century)
1 a : being next after the second in place or time ⟨the *third* taxi in line⟩ **b :** ranking next after the second of a grade or degree in authority or precedence ⟨*third* mate⟩ **c :** being the forward speed or gear next higher than second especially in a motor vehicle
2 a : being one of three equal parts into which something is divisible **b :** being the last in each group of three in a series ⟨take out every *third* card⟩
— third *or* **third·ly** *adverb*

[2]**third** *noun* (14th century)
1 : one of three equal parts of something
2 a — see NUMBER table **b :** one that is next after second in rank, position, authority, or precedence ⟨the *third* in line⟩
3 a : the musical interval embracing three diatonic degrees **b :** a tone at this interval; *specifically* **:** MEDIANT **c :** the harmonic combination of two tones a third apart
4 *plural* **:** merchandise whose quality falls below the manufacturer's standard for seconds
5 : THIRD BASE
6 : the third forward gear or speed especially of a motor vehicle

third base *noun* (1845)
1 : the base that must be touched third by a base runner in baseball
2 : the player position for defending the area around third base
— third baseman *noun*

third–class *adjective* (1839)
: of or relating to a class, rank, or grade next below the second
— third–class *adverb*

third class *noun* (1844)
1 : the third and usually next below second class in a classification
2 : the least expensive class of accommodations (as on a passenger ship)
3 a : a class of U.S. mail comprising printed matter exclusive of regularly issued periodicals and merchandise less than 16 ounces (454 grams) in weight that is not sealed against inspection **b :** a similar class of Canadian mail with different weight limits

third degree *noun* (1900)
: the subjection of a prisoner to mental or physical torture to extract a confession

third–degree burn *noun* (1930)
: a severe burn characterized by destruction of the skin through its deeper layers and possibly into underlying tissues, loss of fluid, and sometimes shock

third dimension *noun* (1858)
1 : THICKNESS, DEPTH; *also* **:** a dimension that adds the effect of solidity to a two-dimensional system
2 : a quality that confers reality or lifelikeness ⟨night sounds that stick in the mind and give a *third dimension* to the memory —Adie Suehsdorf⟩
— third–dimensional *adjective*

third estate *noun, often T&E capitalized* (1604)
: the third of the traditional political orders; *specifically* **:** the commons

third force *noun* [translation of French *troisième force*] (1936)
: a grouping (as of political parties or international powers) intermediate between two opposing political forces

third·hand \ˈthərd-ˈhand\ *adjective* (1599)
1 : received from or through two intermediaries ⟨*thirdhand* information⟩
2 a : acquired after being used by two previous owners **b :** dealing in thirdhand merchandise

third house *noun* (1849)
: a legislative lobby

third market *noun* (1964)
: the over-the-counter market in listed securities

third order *noun, often T&O capitalized* (1629)
1 : an organization composed of lay people living in secular society under a religious rule and directed by a religious order
2 : a congregation especially of teaching or nursing sisters affiliated with a religious order

third party *noun* (1801)
1 a : a major political party operating over a limited period of time in addition to two other major parties in a nation or state normally characterized by a two-party system **b :** MINOR PARTY
2 : a person other than the principals ⟨a *third party* to a divorce proceeding⟩

third person *noun* (circa 1586)
1 a : a set of linguistic forms (as verb forms, pronouns, and inflectional affixes) referring to one that is neither the speaker or writer of the utterance in which they occur nor the one to whom that utterance is addressed **b :** a linguistic form belonging to such a set
2 : reference of a linguistic form to one that is neither the speaker or writer of the utterance in which it occurs nor the one to whom that utterance is addressed

third rail *noun* (1890)
: a metal rail through which electric current is led to the motors of an electric vehicle (as a subway car)

third–rate \ˈthərd-ˈrāt\ *adjective* (1814)
: of third quality or value; *specifically* **:** worse than second-rate
— third–rat·er \-ˈrā-tər\ *noun*

third reading *noun* (circa 1571)
: the final stage of the consideration of a legislative bill before a vote on its final disposition

third–stream *adjective* (1962)
: of, relating to, or being music that incorporates elements of classical music and jazz

third ventricle *noun* (circa 1860)
: the median unpaired ventricle of the brain bounded by parts of the telencephalon and diencephalon

third world *noun, often T&W capitalized* [translation of French *tiers monde*] (1963)
1 : a group of nations especially in Africa and Asia not aligned with either the Communist or the non-Communist blocs
2 : an aggregate of minority groups within a larger predominant culture
3 : the aggregate of the underdeveloped nations of the world
— third world·er \-ˈwər(-ə)l-dər\ *noun, often T&W capitalized*

[1]**thirl** \ˈthər(-ə)l\ *noun* [Middle English, from Old English *thyrel,* from *thurh* through — more at THROUGH] (before 12th century)
dialect **:** HOLE, PERFORATION, OPENING

[2]**thirl** *transitive verb* (before 12th century)
dialect British **:** PIERCE, PERFORATE

[1]**thirst** \ˈthərst\ *noun* [Middle English, from Old English *thurst;* akin to Old High German *durst* thirst, Latin *torrēre* to dry, parch, Greek *tersesthai* to become dry] (before 12th century)
1 a : a sensation of dryness in the mouth and throat associated with a desire for liquids; *also* **:** the bodily condition (as of dehydration) that induces this sensation **b :** a desire or need to drink
2 : an ardent desire **:** CRAVING, LONGING

[2]**thirst** *intransitive verb* (before 12th century)
1 : to feel thirsty **:** suffer thirst
2 : to crave vehemently and urgently
synonym see LONG
— thirst·er *noun*

thirst·i·ly \ˈthər-stə-lē\ *adverb* (1549)
: with or on account of thirst

thirsty \ˈthər-stē\ *adjective* **thirst·i·er; -est** (before 12th century)
1 a : feeling thirst **b :** deficient in moisture **:** PARCHED ⟨*thirsty* land⟩ **c :** highly absorbent ⟨*thirsty* towels⟩
2 : having a strong desire **:** AVID ⟨*thirsty* for knowledge⟩
— thirst·i·ness \-stē-nəs\ *noun*

thir·teen \ˌthər(t)-ˈtēn, ˈthər(t)-\ *noun* [Middle English *thrittene,* from *thrittene,* adjective, from Old English *thrēotīne;* akin to Old English *tīen* ten — more at TEN] (14th century)
— see NUMBER table
— thirteen *adjective*
— thirteen *pronoun, plural in construction*
— thir·teenth \-ˈtēn(t)th, -ˌtēn(t)th\ *adjective or noun*

thir·ty \ˈthər-tē\ *noun, plural* **thirties** [Middle English *thritty,* from *thritty,* adjective, from Old English *thrītig,* from *thrītig* group of 30, from *thrīe* three + *-tig* group of ten; akin to Old English *tīen* ten] (before 12th century)
1 — see NUMBER table
2 *plural* **:** the numbers 30 to 39; *specifically* **:** the years 30 to 39 in a lifetime or century
3 : a sign of completion **:** END — usually written 30 ⟨wrote *thirty* on the last page of the story⟩
4 : the second point scored by a side in a game of tennis
— thir·ti·eth \-tē-əth\ *adjective or noun*
— thirty *adjective*
— thirty *pronoun, plural in construction*
— thir·ty·ish \-ish\ *adjective*

thir·ty–eight \ˌthər-tē-ˈāt\ *noun* (circa 1934)
1 — see NUMBER table
2 : a handgun nominally of .38 caliber — usually written .38
— thirty–eight *adjective*
— thirty–eight *pronoun, plural in construction*

thir·ty–sec·ond note \-ˈse-kən(d)-ˌnōt, -kən(t)-\ *noun* (circa 1890)
: a musical note with the time value of 1⁄32 of a whole note — see NOTE illustration

thirty–second rest *noun* (circa 1903)
: a musical rest corresponding in time value to a thirty-second note

thir·ty–thir·ty \ˌthər-tē-ˈthər-tē\ *noun* (1929)
: a rifle that fires a .30 caliber cartridge having a 30 grain powder charge — usually written .30–30

thir·ty–three \ˌthər-tē-ˈthrē\ *noun* (1902)
1 — see NUMBER table
2 : a microgroove phonograph record designed to be played at 33⅓ revolutions per minute — usually written 33
— thirty–three *adjective*
— thirty–three *pronoun, plural in construction*

thir·ty–two \-ˈtü\ *noun* (1904)
1 — see NUMBER table
2 : a .32 caliber handgun — usually written .32
— thirty–two *adjective*
— thirty–two *pronoun, plural in construction*

thir·ty–two·mo \-(ˌ)mō\ *noun, plural* **-mos** (circa 1841)
: the size of a piece of paper cut 32 from a sheet; *also* **:** a book, a page, or paper of this size

¹this \ˈthis, thəs\ *pronoun, plural* **these** \ˈthēz\ [Middle English, pronoun & adjective, from Old English *thes* (masculine), *this* (neuter); akin to Old High German *dese* this, Old English *thæt* that] (before 12th century)
1 a (1) **:** the person, thing, or idea that is present or near in place, time, or thought or that has just been mentioned ⟨*these* are my hands⟩ (2) **:** what is stated in the following phrase, clause, or discourse ⟨I can only say *this:* it wasn't here yesterday⟩ **b :** this time or place ⟨expected to return before *this*⟩
2 a : the one nearer or more immediately under observation or discussion ⟨*this* is iron and that is tin⟩ **b :** the one more recently referred to

²this *adjective, plural* **these** (before 12th century)
1 a : being the person, thing, or idea that is present or near in place, time, or thought or that has just been mentioned ⟨*this* book is mine⟩ ⟨early *this* morning⟩ **b :** constituting the immediately following part of the present discourse **c :** constituting the immediate past or future ⟨friends all *these* years⟩ **d :** being one not previously mentioned — used especially in narrative to give a sense of immediacy or vividness ⟨then *this* guy runs in⟩
2 : being the nearer at hand or more immediately under observation or discussion ⟨*this* car or that one⟩

³this \ˈthis\ *adverb* (15th century)
: to the degree or extent indicated by something in the immediate context or situation ⟨didn't expect to wait *this* long⟩

This·be \ˈthiz-bē\ *noun* [Latin, from Greek *Thisbē*]
: a legendary young woman of Babylon who dies for love of Pyramus

this·tle \ˈthi-səl\ *noun* [Middle English *thistel,* from Old English; akin to Old High German *distill* thistle] (before 12th century)
: any of various prickly composite plants (especially genera *Carduus, Cirsium,* and *Onopordum*) with often showy heads of mostly tubular flowers; *also* **:** any of various other prickly plants
— this·tly \ˈthi-s(ə-)lē\ *adjective*

thistle

this·tle·down \ˈthi-səl-ˌdau̇n\ *noun* (1561)
: the typically plumose pappus from the ripe flower head of a thistle

thistle tube *noun* (circa 1891)
: a funnel tube usually of glass with a bulging top and flaring mouth

this–world·li·ness \ˈthis-ˈwərld-lē-nəs\ *noun* (1887)
: interest in, concern with, or devotion to things of this world especially as opposed to a future stage of existence (as after death)

this–world·ly \-lē\ *adjective* (1883)
: characterized by or manifesting this-worldliness

¹thith·er \ˈthi-thər *also* ˈthi-\ *adverb* [Middle English, from Old English *thider;* akin to Old Norse *thathra* there, Old English *thæt* that] (before 12th century)
: to that place **:** THERE

²thither *adjective* (1830)
: being on the other and farther side **:** more remote

thith·er·to \-ˌtü; ˌthi-thər-ˈ, ˌthi-\ *adverb* (15th century)
: until that time

thith·er·ward \ˈthi-thər-wərd, ˈthi-\ *also* **thith·er·wards** \-wərdz\ *adverb* (before 12th century)
: toward that place **:** THITHER

thix·ot·ro·py \thik-ˈsä-trə-pē\ *noun* [International Scientific Vocabulary *thixo-* (from Greek *thixis* act of touching, from *thinganein* to touch) + *-tropy* — more at THIGMOTAXIS] (1927)
: the property of various gels of becoming fluid when disturbed (as by shaking)
— thixo·tro·pic \ˌthik-sə-ˈtrō-pik, -ˈträ-\ *adjective*

tho *variant of* THOUGH

¹thole \ˈthōl\ *verb* **tholed; thol·ing** [Middle English, from Old English *tholian* — more at TOLERATE] (before 12th century)
chiefly dialect **:** ENDURE

²thole *noun* [Middle English *tholle,* from Old English *thol;* akin to Old Norse *thollr* fir tree, peg, Greek *tylos* knob, callus] (before 12th century)
1 : either of a pair of pins set in the gunwale of a boat to hold an oar in place
2 : PEG, PIN

tho·lei·ite \ˈtō-lə-ˌīt, ˈthō-\ *noun* [German *Tholeiit,* from *Tholey,* village in Saarland, Germany + German *it* -itc] (1866)
: a basaltic rock that is rich in aluminum and low in potassium, is found typically in the ocean floor, and is probably derived from the earth's mantle
— tho·lei·it·ic \ˌtō-lə-ˈi-tik, ˌthō-\ *adjective*

thole·pin \ˈthōl-ˌpin\ *noun* (1598)
: THOLE 1

Thom·as \ˈtä-məs\ *noun* [Greek *Thōmas,* from Hebrew *tʾōm* twin]
: an apostle who demanded proof of Christ's resurrection

Thom·as Jef·fer·son's Birthday \ˈtä-məs-ˈje-fər-sənz-\ *noun* (circa 1928)
: April 13 observed as a legal holiday in Alabama

Tho·mism \ˈtō-ˌmi-zəm\ *noun* [New Latin *Thomista* Thomist, from Saint *Thomas* Aquinas] (circa 1731)
: the scholastic philosophical and theological system of Saint Thomas Aquinas
— Tho·mist \-mist\ *noun or adjective*
— Tho·mis·tic \tō-ˈmis-tik\ *adjective*

Thomp·son submachine gun \ˈtäm(p)-sən-\ *noun* [John T. *Thompson* (died 1940) American army officer] (1920)
: a .45 caliber submachine gun with a drum or stick magazine, a pistol grip, and a detachable buttstock

Thom·son's gazelle \ˈtäm(p)-sənz-\ *noun* [Joseph *Thomson* (died 1895) Scottish explorer] (1897)
: a small gazelle (*Gazella thomsoni*) of eastern Africa that is tan above and white below with a broad black stripe on each side

thong \ˈthȯŋ\ *noun* [Middle English, from Old English *thwong;* akin Old Norse *thvengr* thong] (before 12th century)
1 : a strip especially of leather or hide
2 : a sandal held on the foot by a thong fitting between the toes and connected to a strap across the top or around the sides of the foot
— thonged \ˈthȯŋd\ *adjective*

Thor \ˈthȯr\ *noun* [Old Norse *Thōrr*]
: the Norse god of thunder, weather, and crops

tho·rac·ic \thə-ˈra-sik\ *adjective* (circa 1658)
: of, relating to, located within, or involving the thorax
— tho·rac·i·cal·ly \-si-k(ə-)lē\ *adverb*

thoracic duct *noun* (circa 1741)
: the main trunk of the system of lymphatic vessels that lies along the front of the spinal column and opens into the left subclavian vein

tho·ra·cot·o·my \ˌthōr-ə-ˈkä-tə-mē, ˌthȯr-\ *noun, plural* **-mies** [Latin *thorac-, thorax* + International Scientific Vocabulary *-tomy*] (circa 1857)
: surgical incision of the chest wall

tho·rax \ˈthōr-ˌaks, ˈthȯr-\ *noun, plural* **tho·rax·es** *or* **tho·ra·ces** \ˈthōr-ə-ˌsēz, ˈthȯr-\ [Middle English, from Latin *thorac-, thorax* breastplate, thorax, from Greek *thōrak-, thōrax*] (15th century)
1 : the part of the mammalian body between the neck and the abdomen; *also* **:** its cavity in which the heart and lungs lie
2 : the middle of the three chief divisions of the body of an insect; *also* **:** the corresponding part of a crustacean or an arachnid

Tho·ra·zine \ˈthōr-ə-ˌzēn, ˈthȯr-\ *trademark*
— used for chlorpromazine

tho·ria \ˈthōr-ē-ə, ˈthȯr-\ *noun* [New Latin, from *thor*ium + *-a*] (circa 1841)
: a powdery white oxide of thorium ThO_2 used especially as a catalyst and in crucibles and refractories and optical glass

tho·ri·a·nite \-ē-ə-ˌnīt\ *noun* [irregular from *thoria*] (1904)
: a strongly radioactive mineral that is an oxide of thorium and often contains rare-earth metals

tho·rite \ˈthōr-ˌīt, ˈthȯr-\ *noun* [Swedish *thorit,* from New Latin *thorium*] (1832)
: a rare mineral that is a brown to black or sometimes orange-yellow silicate of thorium resembling zircon

tho·ri·um \ˈthōr-ē-əm, ˈthȯr-\ *noun* [New Latin, from Old Norse *Thōrr* Thor] (1832)
: a radioactive metallic element that occurs combined in minerals and is usually associated with rare earths — see ELEMENT table

thorn \ˈthȯrn\ *noun, often attributive* [Middle English, from Old English; akin to Old High German *dorn* thorn, Sanskrit *tṛṇa* grass, blade of grass] (before 12th century)
1 : a woody plant bearing sharp impeding processes (as briers, prickles, or spines); *especially* **:** HAWTHORN
2 a : a sharp rigid process on a plant; *specifically* **:** a short, indurated, sharp-pointed, and leafless modified branch **b :** any of various sharp spinose structures on an animal
3 : the runic letter þ used in Old English, Middle English, and Icelandic to represent either of the fricatives \th\ or \th\
4 : something that causes distress or irritation — often used in the phrase *thorn in one's side*
— thorned \ˈthȯrnd\ *adjective*
— thorn·less \ˈthȯrn-ləs\ *adjective*
— thorn·like \-ˌlīk\ *adjective*

thorn apple *noun* (1578)
1 : JIMSONWEED; *also* **:** any plant of the same genus
2 : the fruit of a hawthorn; *also* **:** HAWTHORN

thorn·back \ˈthȯrn-ˌbak\ *noun* (14th century)
: any of various rays having spines on the back

thorn·bush \-ˌbu̇sh\ *noun* (1535)
1 : any of various spiny or thorny shrubs or small trees
2 : a low growth of thorny shrubs especially of dry tropical regions

thorny \'thȯr-nē\ *adjective* **thorn·i·er; -est** (before 12th century)
1 : full of thorns
2 : full of difficulties or controversial points **:** TICKLISH ⟨a *thorny* problem⟩
— **thorn·i·ness** *noun*

thoro *nonstandard variant of* THOROUGH

[1]**thor·ough** \'thər-(ˌ)ō, *sometimes* 'thȯr-; 'thə-(ˌ)rō\ *preposition* [Middle English *thorow,* from Old English *thurh, thuruh,* preposition & adverb] (before 12th century)
archaic **:** THROUGH

[2]**thorough** *adverb* (before 12th century)
archaic **:** THROUGH

[3]**thorough** *adjective* (15th century)
1 : carried through to completion **:** EXHAUSTIVE ⟨a *thorough* search⟩
2 a : marked by full detail ⟨a *thorough* description⟩ **b :** careful about detail **:** PAINSTAKING ⟨a *thorough* scholar⟩ **c :** complete in all respects ⟨*thorough* pleasure⟩ **d :** having full mastery (as of an art) ⟨a *thorough* musician⟩
3 : passing through
— **thor·ough·ly** *adverb*
— **thor·ough·ness** *noun*

thor·ough·bass \'thər-ə-ˌbās, 'thə-rə-\ *noun* (1662)
: CONTINUO

thor·ough·brace \-ˌbrās\ *noun* (1837)
: any of several leather straps supporting the body of a carriage and serving as springs

[1]**thor·ough·bred** \-ˌbred\ *adjective* (1701)
1 : thoroughly trained or skilled
2 : bred from the best blood through a long line **:** PUREBRED ⟨*thoroughbred* dogs⟩
3 a *capitalized* **:** of, relating to, or being a member of the Thoroughbred breed of horses **b :** having characteristics resembling those of a Thoroughbred

[2]**thoroughbred** *noun* (1842)
1 *capitalized* **:** any of an English breed of light speedy horses kept chiefly for racing that originated from crosses between English mares of uncertain ancestry and Arabian stallions
2 : a purebred or pedigreed animal
3 : one that has characteristics resembling those of a Thoroughbred

Thoroughbred

thor·ough·fare \-ˌfar, -ˌfer\ *noun* (14th century)
1 : a way or place for passage: as **a :** a street open at both ends **b :** a main road
2 a : PASSAGE, TRANSIT **b :** the conditions necessary for passing through

thor·ough·go·ing \ˌthər-ə-'gō-iŋ, ˌthə-rə-, -'gȯ(-)iŋ\ *adjective* (1800)
: marked by thoroughness or zeal **:** THOROUGH, COMPLETE

thor·ough–paced \-'pāst\ *adjective* (1646)
1 : THOROUGH, COMPLETE
2 : thoroughly trained **:** ACCOMPLISHED

thor·ough·pin \'thər-ə-ˌpin, 'thə-rə-\ *noun* (1789)
: a synovial swelling just above the hock of a horse on both sides of the leg and slightly anterior to the hamstring tendon that is often associated with lameness

thor·ough·wort \-ˌwərt, -ˌwȯrt\ *noun* (1814)
: BONESET

thorp \'thȯrp\ *noun* [Middle English, from Old English, perhaps from Old Norse; akin to Old High German *dorf* village, Latin *trabs* beam, roof] (before 12th century)
archaic **:** VILLAGE, HAMLET

those [Middle English, from *those* these, from Old English *thās,* plural of *thes* this — more at THIS] *plural of* THAT

[1]**thou** \'t͟haů\ *pronoun* [Middle English, from Old English *thū;* akin to Old High German *dū* thou, Latin *tu,* Greek *sy*] (before 12th century)
archaic **:** the one addressed ⟨*thou* shalt have no other gods before me —Exodus 20:3 (Authorized Version)⟩ — used especially in ecclesiastical or literary language and by Friends as the universal form of address to one person; compare THEE, THINE, THY, YE, YOU
word history see YOU

[2]**thou** *transitive verb* (15th century)
: to address as *thou*

[3]**thou** \'thaů\ *noun, plural* **thou** [short for *thousand*] (1867)
: a thousand of something (as dollars)

[1]**though** \'t͟hō\ *conjunction* [Middle English, adverb & conjunction, of Scandinavian origin; akin to Old Norse *thō* nevertheless; akin to Old English *thēah* nevertheless, Old High German *doh*] (before 12th century)
1 : in spite of the fact that **:** WHILE ⟨*though* they know the war is lost, they continue to fight —Bruce Bliven (died 1977)⟩
2 : in spite of the possibility that **:** even if ⟨*though* I may fail, I will try⟩

[2]**though** *adverb* (13th century)
: HOWEVER, NEVERTHELESS ⟨It's hard work. I enjoy it *though*⟩

[1]**thought** \'thȯt\ *past and past participle of* THINK

[2]**thought** *noun* [Middle English, from Old English *thōht;* akin to Old English *thencan* to think — more at THINK] (before 12th century)
1 a : the action or process of thinking **:** COGITATION **b :** serious consideration **:** REGARD **c** *archaic* **:** RECOLLECTION, REMEMBRANCE
2 a : reasoning power **b :** the power to imagine **:** CONCEPTION
3 : something that is thought: as **a :** an individual act or product of thinking **b :** a developed intention or plan ⟨had no *thought* of leaving home⟩ **c :** something (as an opinion or belief) in the mind ⟨he spoke his *thoughts* freely⟩ **d :** the intellectual product or the organized views and principles of a period, place, group, or individual ⟨contemporary Western *thought*⟩
synonym see IDEA
— **a thought :** a little **:** SOMEWHAT ⟨a *thought* too much vinegar in the dressing⟩

thought experiment *noun* (1945)
: GEDANKENEXPERIMENT

thought·ful \'thȯt-fəl\ *adjective* (13th century)
1 a : absorbed in thought **:** MEDITATIVE **b :** characterized by careful reasoned thinking
2 a : having thoughts **:** HEEDFUL ⟨became *thoughtful* about religion⟩ **b :** given to or chosen or made with heedful anticipation of the needs and wants of others
— **thought·ful·ly** \-fə-lē\ *adverb*
— **thought·ful·ness** *noun*

thought·less \-ləs\ *adjective* (1592)
1 a : insufficiently alert **:** CARELESS **b :** RECKLESS, RASH
2 : devoid of thought **:** INSENSATE
3 : lacking concern for others **:** INCONSIDERATE
— **thought·less·ly** *adverb*
— **thought·less·ness** *noun*

thought–out \-'aůt\ *adjective* (1870)
: produced or arrived at through mental effort and especially through careful and thorough consideration

thought·way \-ˌwā\ *noun* (circa 1944)
: a way of thinking that is characteristic of a particular group, time, or culture

thou·sand \'thaů-zᵊn(d)\ *noun, plural* **thousands** *or* **thousand** [Middle English, from Old English *thūsend;* akin to Old High German *dūsunt* thousand, Lithuanian *tūkstantis,* and probably to Sanskrit *tavas* strong, Latin *tumēre* to swell — more at THUMB] (before 12th century)
1 — see NUMBER table
2 : a very large number ⟨*thousands* of ants⟩
— **thousand** *adjective*
— **thou·sand·fold** \-zᵊn(d)-ˌfōld\ *adjective or adverb*
— **thou·sandth** \-zᵊn(t)th\ *adjective or noun*

Thousand Island dressing *noun* [*Thousand Islands,* islands in the Saint Lawrence River] (1924)
: mayonnaise with chili sauce and seasonings (as chopped pimientos, green peppers, and onion)

thousand–leg·ger \ˌthaů-zᵊn(d)-'le-gər, -'lā-\ *noun* (1914)
: MILLIPEDE

thousands place *noun* (1937)
: the place four to the left of the decimal point in a number expressed in the Arabic system of writing numbers

Thra·cian \'thrā-shən\ *noun* (1569)
1 : a native or inhabitant of Thrace
2 : the Indo-European language of the ancient Thracians — see INDO-EUROPEAN LANGUAGES table
— **Thracian** *adjective*

[1]**thrall** \'thrȯl\ *noun* [Middle English *thral,* from Old English *thræl,* from Old Norse *thræll*] (before 12th century)
1 a : a servant slave **:** BONDMAN; *also* **:** SERF **b :** a person in moral or mental servitude
2 a : a state of servitude or submission ⟨in *thrall* to his emotions⟩ **b :** a state of complete absorption ⟨mountains could hold me in *thrall* with a subtle attraction of their own —Elyne Mitchell⟩
— **thrall** *adjective*
— **thrall·dom** *or* **thral·dom** \'thrȯl-dəm\ *noun*

[2]**thrall** *transitive verb* (13th century)
archaic **:** ENTHRALL, ENSLAVE

[1]**thrash** \'thrash\ *verb* [alteration of *thresh*] (1588)
transitive verb
1 : to separate the seeds of from the husks and straw by beating **:** THRESH 1
2 a : to beat soundly with or as if with a stick or whip **:** FLOG **b :** to defeat decisively or severely ⟨*thrashed* the visiting team⟩
3 : to swing, beat, or strike in the manner of a rapidly moving flail ⟨*thrashing* his arms⟩
4 a : to go over again and again ⟨*thrash* the matter over inconclusively⟩ **b :** to hammer out **:** FORGE ⟨*thrash* out a plan⟩
intransitive verb
1 : THRESH 1
2 : to deal blows or strokes like one using a flail or whip
3 : to move or stir about violently **:** toss about ⟨*thrash* in bed with a fever⟩
synonym see SWING

[2]**thrash** *noun* (1840)
: an act of thrashing

[1]**thrash·er** \'thra-shər\ *noun* (1632)
: one that thrashes or threshes

[2]**thrasher** *noun* [perhaps alteration of dialect *thresher* thrush] (circa 1814)
: any of various American oscine birds (family Mimidae and especially genus *Toxostoma*) related to the mockingbird that resemble thrushes but have a curved bill and long tail

thra·son·i·cal \thrā-'sä-ni-kəl, thrə-\ *adjective* [Latin *Thrason-, Thraso* Thraso, braggart soldier in the comedy *Eunuchus* by Terence] (1564)
: of, relating to, resembling, or characteristic of Thraso **:** BRAGGING, BOASTFUL
— **thra·son·i·cal·ly** \-k(ə-)lē\ *adverb*

[1]**thraw** \'thrȧ\ *verb* [Middle English — more at THROW] (before 12th century)
transitive verb
1 *chiefly Scottish* **:** to cause to twist or turn
2 *chiefly Scottish* **:** CROSS, THWART
intransitive verb
1 *chiefly Scottish* **:** TWIST, TURN
2 *chiefly Scottish* **:** to be in disagreement

[2]**thraw** *noun* (circa 1585)
1 *chiefly Scottish* **:** TWIST, TURN

2 *chiefly Scottish* **:** ill humor

thra·wart \'thrȧ-wərt\ *adjective* [Middle English (Scots), alteration of Middle English *fraward, froward* froward] (15th century)
1 *chiefly Scottish* **:** STUBBORN
2 *Scottish* **:** CROOKED

thrawn \'thrȧn\ *adjective* [Middle English (Scots) *thrawin,* from past participle of Middle English *thrawen* to twist] (15th century)
chiefly Scottish **:** lacking in pleasing or attractive qualities: as **a :** PERVERSE, RECALCITRANT **b :** CROOKED, MISSHAPEN
— **thrawn·ly** *adverb, chiefly Scottish*

[1]**thread** \'thred\ *noun* [Middle English *thred,* from Old English *thrǣd;* akin to Old High German *drāt* wire, Old English *thrāwan* to cause to twist or turn — more at THROW] (before 12th century)
1 a : a filament, a group of filaments twisted together, or a filamentous length formed by spinning and twisting short textile fibers into a continuous strand **b :** a piece of thread
2 a : any of various natural filaments ⟨the *threads* of a spiderweb⟩ **b :** a slender stream (as of water) **c :** a streak of light or color **d :** a projecting helical rib (as in a fitting or on a pipe) by which parts can be screwed together **:** SCREW THREAD
3 : something continuous or drawn out: as **a :** a train of thought **b :** a continuing element ⟨a *thread* of melancholy marked all his writing⟩
4 : a tenuous or feeble support
5 *plural* **:** CLOTHING
— **thread·less** \-ləs\ *adjective*
— **thread·like** \-ˌlīk\ *adjective*

[2]**thread** (14th century)
transitive verb
1 a : to pass a thread through the eye of (a needle) **b :** to arrange a thread, yarn, or lead-in piece in working position for use in (a machine)
2 a (1) **:** to pass something through in the manner of a thread ⟨*thread* a pipe with wire⟩ (2) **:** to pass (as a tape, line, or film) into or through something ⟨*threaded* a fresh film into the camera⟩ **b :** to make one's way through or between ⟨*threading* narrow alleys⟩; *also* **:** to make (one's way) usually cautiously through a hazardous situation
3 : to put together on or as if on a thread **:** STRING ⟨*thread* beads⟩
4 : to interweave with or as if with threads **:** INTERSPERSE ⟨dark hair *threaded* with silver⟩
5 : to form a screw thread on or in
intransitive verb
1 : to make one's way
2 : to form a thread
— **thread·er** *noun*

thread·bare \'thred-ˌbar, -ˌber\ *adjective* (14th century)
1 : having the nap worn off so that the thread shows **:** SHABBY
2 : exhausted of interest or freshness
synonym see TRITE
— **thread·bare·ness** *noun*

thread·fin \-ˌfin\ *noun* (circa 1890)
: any of various bony fishes (family Polynemidae and especially genus *Polydactylus*) having filamentous rays on the lower part of the pectoral fin

thread·worm \-ˌwərm\ *noun* (1802)
: a long slender nematode worm

thready \'thre-dē\ *adjective* (1597)
1 : consisting of or bearing fibers or filaments ⟨a *thready* bark⟩
2 a : resembling a thread **:** FILAMENTOUS **b :** tending to form or draw out into strands **:** ROPY
3 : lacking in fullness, body, or vigor **:** THIN ⟨a *thready* voice⟩
— **thread·i·ness** *noun*

threap \'thrēp\ *transitive verb* [Middle English *threpen,* from Old English *thrēapian*] (before 12th century)
1 *chiefly Scottish* **:** SCOLD, CHIDE
2 *chiefly Scottish* **:** to maintain persistently

[1]**threat** \'thret\ *noun* [Middle English *thret* coercion, threat, from Old English *thrēat* coercion; akin to Middle High German *drōz* annoyance, Latin *trudere* to push, thrust] (before 12th century)
1 : an expression of intention to inflict evil, injury, or damage
2 : one that threatens
3 : an indication of something impending ⟨the sky held a *threat* of rain⟩

[2]**threat** *verb* (before 12th century)
archaic **:** THREATEN

threat·en \'thre-tᵊn\ *verb* **threat·ened; threat·en·ing** \'thret-niŋ, 'thre-tᵊn-iŋ\ (13th century)
transitive verb
1 : to utter threats against
2 a : to give signs or warning of **:** PORTEND ⟨the clouds *threatened* rain⟩ **b :** to hang over dangerously **:** MENACE
3 : to announce as intended or possible ⟨the workers *threatened* a strike⟩
intransitive verb
1 : to utter threats
2 : to portend evil
— **threat·en·er** \'thret-nər, 'thre-tᵊn-ər\ *noun*
— **threat·en·ing·ly** \'thret-niŋ-lē, 'thre-tᵊn-iŋ-\ *adverb*

threatened *adjective* (1960)
: having an uncertain chance of continued survival ⟨a *threatened* species of owls⟩; *specifically* **:** likely to become an endangered species

three \'thrē\ *noun* [Middle English, from *three,* adjective, from Old English *thrīe* (masculine), *thrēo* (feminine & neuter); akin to Old High German *drī* three, Latin *tres,* Greek *treis*] (before 12th century)
1 — see NUMBER table
2 : the third in a set or series ⟨the *three* of hearts⟩
3 : something having three units or members
— **three** *adjective*
— **three** *pronoun, plural in construction*

three–bag·ger \-'ba-gər\ *noun* (1881)
: TRIPLE

three–ball \-ˌbȯl\ *adjective* (circa 1890)
: relating to or being a golf match in which three players compete against one another with each playing a single ball

three–card monte \'thrē-'kärd-\ *noun* (1854)
: a gambling game in which the dealer shows three cards, shuffles them, places them face down, and invites spectators to bet they can identify the location of a particular card

three–col·or \'thrē-'kə-lər\ *adjective* (1893)
: being or relating to a printing or photographic process wherein three primary colors are used to reproduce all the colors of the subject

3–D \'thrē-'dē\ *noun* [*D,* abbreviation of *dimensional*] (1951)
: the three-dimensional form; *also* **:** an image or a picture produced in it
— **3–D** *adjective*

three–deck·er \'thrē-'de-kər\ *noun* (1795)
1 : a wooden warship carrying guns on three decks
2 : TRIPLE-DECKER

three–dimensional *adjective* (circa 1891)
1 : of or relating to three dimensions
2 : giving the illusion of depth or varying distances — used of an image or a pictorial representation especially when this illusion is enhanced by stereoscopic means
3 : describing or being described in well-rounded completeness ⟨a *three-dimensional* analysis of multiple historical processes —L. L. Snyder⟩
4 : true to life **:** LIFELIKE

three·fold \'thrē-ˌfōld, -'fōld\ *adjective* (before 12th century)
1 : having three units or members **:** TRIPLE
2 : being three times as great or as many
— **three·fold** \-'fōld\ *adverb*

three–gait·ed \-'gā-təd\ *adjective* (1948)
of a horse **:** trained to use the walk, trot, and canter

three–hand·ed \-'han-dəd\ *adjective* (1719)
: played by three players ⟨*three-handed* bridge⟩

Three Hours *noun* (circa 1891)
: a service of devotion between noon and three o'clock on Good Friday

three–legged \'thrē-'legd, -'lāgd; -'le-gəd, -'lā-\ *adjective* (1596)
: having three legs ⟨a *three-legged* stool⟩

three–legged race *noun* (1903)
: a race between pairs of competitors with each pair having their adjacent legs bound together

three–line octave *noun* (circa 1931)
: the musical octave that begins on the second C above middle C — see PITCH illustration

three–mile limit *noun* (circa 1891)
: the limit of the marginal sea of three miles included in the territorial waters of a state

three of a kind (circa 1897)
: three cards of the same rank in one hand — see POKER illustration

three·pence \'thre-pən(t)s, 'thri-, 'thrə-, *US also* 'thrē-pen(t)s\ *noun* (1589)
1 *plural* **threepence** *or* **three·penc·es :** a coin worth threepence
2 : the sum of three British pennies

three·pen·ny \'thre-p(ə-)nē, 'thri-, 'thrə-, *US also* 'thrē-ˌpe-nē\ *adjective* (15th century)
1 : costing or worth threepence
2 : POOR

three–phase *adjective* (circa 1900)
: of, relating to, or operating by means of a combination of three circuits energized by alternating electromotive forces that differ in phase by one third of a cycle

three–piece *adjective* (circa 1909)
: consisting of or made in three pieces ⟨a *three-piece* suit⟩

three–point landing *noun* (1918)
: an airplane landing in which the two main wheels of the landing gear and the tail wheel or skid or nose wheel touch the ground simultaneously

three–quarter *adjective* (1677)
: extending to three-quarters of the normal full length ⟨a *three-quarter* sleeve⟩

three–quarter–bound *adjective* (circa 1951)
of a book **:** bound like a half-bound book but having the material on the spine extended to cover about one third of the boards
— **three–quarter binding** *noun*

three–ring circus *noun* (1904)
1 : a circus with simultaneous performances in three rings
2 : something wild, confusing, engrossing, or entertaining

three R's *noun plural* [from the facetiously used phrase *reading, 'riting, and 'rithmetic*] (1828)
1 : the fundamentals taught in elementary school; *especially* **:** reading, writing, and arithmetic
2 : the fundamental skills in a field of endeavor

three·score \'thrē-'skōr, -'skȯr\ *adjective* (14th century)
: being three times twenty **:** SIXTY

three·some \'thrē-səm\ *noun* (14th century)
1 : a group of three persons or things **:** TRIO
2 : a golf match in which one person plays his ball against the ball of two others playing each stroke alternately

three–spined stickleback \'thrē-'spīn(d)-\ *noun* (1769)
: a stickleback (*Gasterosteus aculeatus*) chiefly of fresh and brackish waters that typically has three dorsal spines

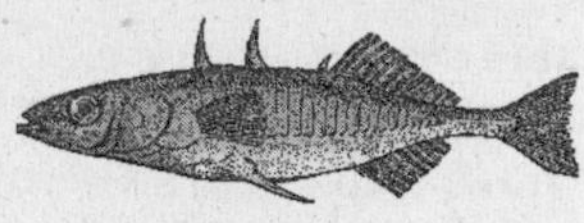
three-spined stickleback

three–toed sloth \'thrē-'tōd-\ *noun* (1879)
: any of a genus (*Bradypus*) of sloths having three clawed digits on each foot and nine vertebrae in the neck — compare TWO-TOED SLOTH

three–wheel·er \-,hwē-lər, -,wē-\ *noun* (1886)
: any of various vehicles having three wheels

thre·node \'thrē-,nōd, 'thre-\ *noun* (1858)
: THRENODY
— thre·nod·ic \thri-'nä-dik\ *adjective*
— thren·o·dist \'thre-nə-dist\ *noun*

thren·o·dy \'thre-nə-dē\ *noun, plural* **-dies** [Greek *thrēnōidia,* from *thrēnos* dirge + *aeidein* to sing — more at DRONE, ODE] (1634)
: a song of lamentation for the dead **:** ELEGY

thre·o·nine \'thrē-ə-,nēn\ *noun* [probably from *threonic acid* ($C_4H_8O_5$)] (1936)
: a colorless crystalline essential amino acid $C_4H_9NO_3$

thresh \'thrash, 'thresh\ *verb* [Middle English *thresshen,* from Old English *threscan;* akin to Old High German *dreskan* to thresh] (before 12th century)
transitive verb
1 : to separate seed from (a harvested plant) mechanically; *also* **:** to separate (seed) in this way
2 : THRASH 4
3 : to strike repeatedly
intransitive verb
1 : to thresh grain
2 : THRASH 2, 3

thresh·er *noun* (14th century)
1 : one that threshes; *especially* **:** THRESHING MACHINE
2 : THRESHER SHARK

thresher shark *noun* (1888)
: a large nearly cosmopolitan shark (*Alopias vulpinus*) having a greatly elongated curved upper lobe of its tail with which it is said to thresh the water to round up the fish on which it feeds — see SHARK illustration

threshing machine *noun* (1775)
: a machine for separating grain crops into grain or seeds and straw

thresh·old \'thresh-,hōld, 'thre-,shōld\ *noun* [Middle English *thresshold,* from Old English *threscwald;* akin to Old Norse *threskjǫldr* threshold, Old English *threscan* to thresh] (before 12th century)
1 : the plank, stone, or piece of timber that lies under a door **:** SILL
2 a : GATE, DOOR **b** (1) **:** END, BOUNDARY; *specifically* **:** the end of a runway (2) **:** the place or point of entering or beginning **:** OUTSET ⟨on the *threshold* of a new age⟩
3 a : the point at which a physiological or psychological effect begins to be produced **b :** a level, point, or value above which something is true or will take place and below which it is not or will not

threw *past of* THROW

thrice \'thrīs\ *adverb* [Middle English *thrie, thries,* from Old English *thriga;* akin to Old Frisian *thria* three times, Old English *thrīe* three] (13th century)
1 : three times
2 a : in a threefold manner or degree **b :** to a high degree

thrift \'thrift\ *noun* [Middle English, from Old Norse, prosperity, from *thrīfask* to thrive] (13th century)
1 : healthy and vigorous growth
2 : careful management especially of money
3 *chiefly Scottish* **:** gainful occupation
4 : any of a genus (*Armeria*) of the plumbago family of perennial evergreen acaulescent herbs; *especially* **:** a scapose herb (*A. maritima*) with pink or white flower heads
5 : a savings bank or savings and loan association — called also *thrift institution*

thrift·less \'thrift-ləs\ *adjective* (1568)
1 : lacking usefulness or worth
2 : careless, wasteful, or incompetent in handling money or resources **:** IMPROVIDENT
— thrift·less·ly *adverb*
— thrift·less·ness *noun*

thrift shop *noun* (1944)
: a shop that sells secondhand articles and especially clothes and is often run for charitable purposes

thrifty \'thrif-tē\ *adjective* **thrift·i·er; -est** (15th century)
1 : thriving by industry and frugality **:** PROSPEROUS
2 : growing vigorously
3 : given to or marked by economy and good management
synonym see SPARING
— thrift·i·ly \-tə-lē\ *adverb*
— thrift·i·ness \-tē-nəs\ *noun*

thrill \'thril\ *verb* [Middle English *thirlen, thrillen* to pierce, from Old English *thyrlian,* from *thyrel* hole, from *thurh* through — more at THROUGH] (1592)
transitive verb
1 a : to cause to experience a sudden sharp feeling of excitement **b :** to cause to have a shivering or tingling sensation
2 : to cause to vibrate or tremble perceptibly
intransitive verb
1 : to move or pass so as to cause a sudden wave of emotion
2 : to become thrilled: **a :** to experience a sudden sharp excitement **b :** TINGLE, THROB
3 : TREMBLE, VIBRATE ◆
— thrill *noun*
— thrill·ing·ly \'thri-liŋ-lē\ *adverb*

thril·ler \'thri-lər\ *noun* (1889)
: one that thrills; *especially* **:** a work of fiction or drama designed to hold the interest by the use of a high degree of intrigue, adventure, or suspense

thrips \'thrips\ *noun, plural* **thrips** [Latin, woodworm, from Greek] (1795)
: any of an order (Thysanoptera) of small to minute sucking insects most of which feed often destructively on plant juices

thrive \'thrīv\ *intransitive verb* **throve** \'thrōv\ *or* **thrived; thriv·en** \'thri-vən\ *also* **thrived; thriv·ing** \'thrī-viŋ\ [Middle English, from Old Norse *thrīfask,* probably reflexive of *thrīfa* to grasp] (13th century)
1 : to grow vigorously **:** FLOURISH
2 : to gain in wealth or possessions **:** PROSPER
3 : to progress toward or realize a goal
— thriv·er \'thrī-vər\ *noun*

thriving *adjective* (1607)
: characterized by success or prosperity
— thriv·ing·ly \'thrī-viŋ-lē\ *adverb*

thro \'thrü\ *preposition* (15th century)
archaic **:** THROUGH

¹throat \'thrōt\ *noun* [Middle English *throte,* from Old English; akin to Old High German *drozza* throat] (before 12th century)
1 a (1) **:** the part of the neck in front of the spinal column (2) **:** the passage through the neck to the stomach and lungs **b** (1) **:** VOICE (2) **:** the seat of the voice
2 : something resembling the throat especially in being an entrance, a passageway, a constriction, or a narrowed part: as **a :** the orifice of a tubular organ especially of a plant **b :** the opening in the vamp of a shoe at the instep **c :** the part of a tennis racket that connects the head with the shaft
3 : the curved part of an anchor's arm where it joins the shank — see ANCHOR illustration
— at each other's throats : in open and aggressive conflict

²throat *transitive verb* (circa 1611)
1 : to utter in the throat **:** MUTTER
2 : to sing or enunciate in a throaty voice

throat·ed \'thrō-təd\ *adjective* (circa 1530)
: having a throat especially of a specified kind — usually used in combination ⟨white-*throated*⟩

throat·latch \'thrōt-,lach\ *noun* (1794)
1 : a strap of a bridle or halter passing under a horse's throat
2 : the part of a horse's throat around which the throatlatch passes — see HORSE illustration

throaty \'thrō-tē\ *adjective* **throat·i·er; -est** (circa 1645)
1 : uttered or produced from low in the throat ⟨a *throaty* voice⟩
2 : heavy, thick, and deep as if from the throat ⟨*throaty* notes of a horn⟩
— throat·i·ly \'thrō-tᵊl-ē\ *adverb*
— throat·i·ness \'thrō-tē-nəs\ *noun*

¹throb \'thräb\ *intransitive verb* **throbbed; throb·bing** [Middle English *throbben*] (14th century)
1 : to pulsate or pound with abnormal force or rapidity
2 : to beat or vibrate rhythmically
— throb·ber *noun*

²throb *noun* (1579)
: BEAT, PULSE

throe \'thrō\ *noun* [Middle English *thrawe, throwe,* from Old English *thrawu, thrēa* threat, pang; akin to Old High German *drawa* threat] (13th century)
1 : PANG, SPASM ⟨death *throes*⟩ ⟨*throes* of childbirth⟩
2 *plural* **:** a hard or painful struggle ⟨the *throes* of revolutionary social change —M. D. Geismar⟩

thromb- *or* **thrombo-** *combining form* [Greek *thrombos* clot]
: blood clot **:** clotting of blood ⟨*thromb*in⟩ ⟨*thrombo*plastic⟩

throm·bin \'thräm-bən\ *noun* [International Scientific Vocabulary] (1898)
: a proteolytic enzyme that is formed from prothrombin and facilitates the clotting of blood by catalyzing conversion of fibrinogen to fibrin

throm·bo·cyte \-bə-,sīt\ *noun* [International Scientific Vocabulary] (1893)
: BLOOD PLATELET; *also* **:** an invertebrate cell with similar function
— throm·bo·cyt·ic \,thräm-bə-'si-tik\ *adjective*

throm·bo·cy·to·pe·nia \,thräm-bə-,sī-tə-'pē-nē-ə, -nyə\ *noun* [New Latin, from International Scientific Vocabulary *thrombocyte* + New Latin *-o-* + *-penia*] (1923)
: persistent decrease in the number of blood platelets that is often associated with hemorrhagic conditions
— throm·bo·cy·to·pe·nic \-nik\ *adjective*

◇ WORD HISTORY

thrill In the 14th century a person who was *thrilled* might not live long enough to tell anyone about it, because the Middle English verb *thrillen,* an altered form of *thirlen* with the first vowel and *r* transposed, meant "to pierce, stab, penetrate," as with a sword or spear. The older form *thirlen* continued Old English *thyrlian,* a verbal derivative of *thyrel* "hole." (The word *thyrel* also enters into Old English *nosthyrl,* literally, "nose hole," the ancestor of Modern English *nostril.*) In early Modern English a connection was made between the physical sensation of being pierced and the experience of being strongly moved by some emotion, so that *thrill* came to mean "to cause (someone) to feel excited." The old literal sense of the word has become obsolete, and "to be thrilled" now usually denotes a quite positive feeling.

throm·bo·em·bo·lism \ˌthräm-bō-ˈem-bə-ˌli-zəm\ *noun* (1907)
: the blocking of a blood vessel by a particle that has broken away from a blood clot at its site of formation
— **throm·bo·em·bol·ic** \-em-ˈbä-lik\ *adjective*

throm·bo·ki·nase \ˌthräm-bō-ˈkī-ˌnās, -ˌnāz\ *noun* [International Scientific Vocabulary] (1908)
: THROMBOPLASTIN

throm·bo·lyt·ic \ˌthräm-bə-ˈli-tik\ *adjective* (1929)
: destroying or breaking up a thrombus ⟨a *thrombolytic* agent⟩ ⟨*thrombolytic* therapy⟩

throm·bo·phle·bi·tis \ˌthräm-bō-fli-ˈbī-təs\ *noun* [New Latin] (circa 1890)
: inflammation of a vein with formation of a thrombus

throm·bo·plas·tic \-ˈplas-tik\ *adjective* [International Scientific Vocabulary] (1911)
: initiating or accelerating the clotting of blood

throm·bo·plas·tin \-ˈplas-tən\ *noun* [International Scientific Vocabulary, from *thromboplastic*] (1911)
: a complex enzyme found especially in blood platelets that functions in the conversion of prothrombin into thrombin in the clotting of blood

throm·bo·sis \thräm-ˈbō-səs, thrəm-\ *noun, plural* **-bo·ses** \-ˌsēz\ [New Latin, from Greek *thrombōsis* clotting, from *thrombousthai* to become clotted, from *thrombos* clot] (1866)
: the formation or presence of a blood clot within a blood vessel
— **throm·bot·ic** \-ˈbä-tik\ *adjective*

throm·box·ane \thräm-ˈbäk-ˌsān\ *noun* [*thromb-* + *ox-* + *-ane*] (1975)
: any of several substances that are formed from endoperoxides, cause constriction of vascular and bronchial smooth muscle, and promote blood coagulation

throm·bus \ˈthräm-bəs\ *noun, plural* **throm·bi** \-ˌbī, -ˌbē\ [New Latin, from Greek *thrombos* clot] (circa 1693)
: a clot of blood formed within a blood vessel and remaining attached to its place of origin — compare EMBOLUS

¹throne \ˈthrōn\ *noun* [Middle English *trone, throne,* from Old French *trone,* from Latin *thronus,* from Greek *thronos* — more at FIRM] (13th century)
1 a : the chair of state of a sovereign or high dignitary (as a bishop) **b :** the seat of a deity
2 : royal power and dignity **:** SOVEREIGNTY
3 *plural* **:** an order of angels — see CELESTIAL HIERARCHY

²throne *verb* **throned; thron·ing** (14th century)
transitive verb
1 : to seat on a throne
2 : to invest with kingly rank or power
intransitive verb
1 : to sit on a throne
2 : to hold kingly power

throne room *noun* (1864)
: a formal audience room containing the throne of a sovereign

¹throng \ˈthrȯŋ\ *noun* [Middle English *thrang, throng,* from Old English *thrang, gethrang;* akin to Old English *thringan* to press, crowd, Old High German *dringan,* Lithuanian *trenkti* to jolt] (before 12th century)
1 a : a multitude of assembled persons **b :** a large number **:** HOST
2 a : a crowding together of many persons **b :** PRESSURE ⟨this *throng* of business —S. R. Crockett⟩
synonym see CROWD

²throng *verb* **thronged; throng·ing** \ˈthrȯŋ-iŋ\ (1534)
transitive verb
1 : to crowd upon **:** PRESS
2 : to crowd into **:** PACK ⟨shoppers *thronging* the streets⟩
intransitive verb
: to crowd together in great numbers

thros·tle \ˈthrä-səl\ *noun* [Middle English, from Old English — more at THRUSH] (before 12th century)
: ¹THRUSH 1; *specifically* **:** SONG THRUSH

¹throt·tle \ˈthrä-tᵊl\ *verb* **throt·tled; throt·tling** \ˈthrät-liŋ, ˈthrä-tᵊl-iŋ\ [Middle English *throtlen,* from *throte* throat] (15th century)
transitive verb
1 a (1) **:** to compress the throat of **:** CHOKE (2) **:** to kill by such action **b :** to prevent or check expression or activity of **:** SUPPRESS
2 a : to decrease the flow of (as steam or fuel to an engine) by a valve **b :** to regulate and especially to reduce the speed of (as an engine) by such means **c :** to vary the thrust of (a rocket engine) during flight
intransitive verb
: to throttle something (as an engine) — usually used with *back* or *down* ⟨the pilot *throttled* back⟩
— **throt·tler** \ˈthrät-lər, ˈthrä-tᵊl-ər\ *noun*

²throttle *noun* [perhaps from (assumed) Middle English, diminutive of Middle English *throte* throat] (circa 1547)
1 a : THROAT 1a **b :** TRACHEA 1
2 a : a valve for regulating the supply of a fluid (as steam) to an engine; *especially* **:** the valve controlling the volume of vaporized fuel charge delivered to the cylinders of an internal combustion engine **b :** the lever controlling this valve **c :** the condition of being throttled
— **at full throttle :** at full speed

throt·tle·able \ˈthrä-tᵊl-ə-bəl\ *adjective* (1960)
: capable of having the thrust varied — used of a rocket engine

throt·tle·hold \ˈthrä-tᵊl-ˌhōld\ *noun* (1935)
: a vicious, strangling, or stultifying control

¹through \ˈthrü\ *preposition* [Middle English *thurh, thruh, through,* from Old English *thurh;* akin to Old High German *durh* through, Latin *trans* across, beyond, Sanskrit *tarati* he crosses over] (before 12th century)
1 a (1) — used as a function word to indicate movement into at one side or point and out at another and especially the opposite side of ⟨drove a nail *through* the board⟩ (2) **:** by way of ⟨left *through* the door⟩ (3) — used as a function word to indicate passage from one end or boundary to another ⟨a highway *through* the forest⟩ ⟨a road *through* the desert⟩ (4) **:** without stopping for **:** PAST ⟨drove *through* a red light⟩ **b** — used as a function word to indicate passage into and out of a treatment, handling, or process ⟨the matter has already passed *through* her hands⟩
2 — used as a function word to indicate means, agency, or intermediacy: as **a :** by means of **:** by the agency of **b :** because of ⟨failed *through* ignorance⟩ **c :** by common descent from or relationship with ⟨related *through* their grandfather⟩
3 a : over the whole surface or extent of **:** THROUGHOUT ⟨homes scattered *through* the valley⟩ **b** — used as a function word to indicate movement within a large expanse ⟨flew *through* the air⟩ **c** — used as a function word to indicate exposure to a specified set of conditions ⟨put him *through* hell⟩
4 — used as a function word to indicate a period of time: as **a :** during the entire period of ⟨all *through* her life⟩ **b :** from the beginning to the end of ⟨the tower stood *through* the earthquake⟩ **c :** to and including ⟨Monday *through* Friday⟩
5 a — used as a function word to indicate completion or exhaustion ⟨got *through* the book⟩ ⟨went *through* the money in a year⟩ **b** — used as a function word to indicate acceptance or approval especially by an official body ⟨got the bill *through* the legislature⟩

²through \ˈthrü\ *adverb* (before 12th century)
1 : from one end or side to the other
2 a : from beginning to end **b :** to completion, conclusion, or accomplishment ⟨see it *through*⟩
3 : to the core **:** COMPLETELY ⟨soaked *through*⟩
4 : into the open **:** OUT ⟨break *through*⟩

³through \ˈthrü\ *adjective* (1523)
1 a : extending from one surface to another ⟨a *through* mortise⟩ **b :** admitting free or continuous passage **:** DIRECT ⟨a *through* road⟩
2 a (1) **:** going from point of origin to destination without change or reshipment ⟨a *through* train⟩ (2) **:** of or relating to such movement ⟨a *through* ticket⟩ **b :** initiated at and destined for points outside a local zone ⟨*through* traffic⟩
3 a : arrived at completion or accomplishment ⟨is *through* with the job⟩ **b :** WASHED-UP, FINISHED

through and through *adverb* (15th century)
: in every way **:** THOROUGHLY

through–com·posed \ˌthrü-kəm-ˈpōzd\ *adjective* [translation of German *durchkomponiert*] (1884)
of a song **:** having new music provided for each stanza — compare STROPHIC

through·ith·er *or* **through·oth·er** \ˈthrü-(ə-)t͟hər\ *adverb* [¹*through* + *other*] (1596)
chiefly Scottish **:** in confusion **:** PROMISCUOUSLY

through·ly \ˈthrü-lē\ *adverb* (15th century)
archaic **:** in a thorough manner

¹through·out \thrü-ˈau̇t\ *adverb* (13th century)
1 : in or to every part **:** EVERYWHERE ⟨of one color *throughout*⟩
2 : during the whole time or action **:** from beginning to end ⟨remained loyal *throughout*⟩

²throughout *preposition* (13th century)
1 : all the way from one end to the other of **:** in or to every part of ⟨cities *throughout* the United States⟩
2 : during the whole course or period of ⟨troubled her *throughout* her life⟩

through·put \ˈthrü-ˌpu̇t\ *noun* (circa 1915)
: OUTPUT, PRODUCTION ⟨the *throughput* of a computer⟩

through street *noun* (1930)
: a street on which the through movement of traffic is given preference

throve *past of* THRIVE

¹throw \ˈthrō\ *verb* **threw** \ˈthrü\; **thrown** \ˈthrōn\; **throw·ing** [Middle English *thrawen, throwen* to cause to twist, throw, from Old English *thrāwan* to cause to twist or turn; akin to Old High German *drāen* to turn, Latin *terere* to rub, Greek *tribein* to rub, *tetrainein* to bore, pierce] (14th century)
transitive verb
1 a : to propel through the air by a forward motion of the hand and arm ⟨*throw* a baseball⟩ **b :** to propel through the air in any manner ⟨a rifle that can *throw* a bullet a mile⟩
2 a : to cause to fall ⟨*threw* his opponent⟩ **b :** to cause to fall off **:** UNSEAT ⟨the horse *threw* its rider⟩ **c :** to get the better of **:** OVERCOME ⟨the problem didn't *throw* her⟩
3 a : to fling (oneself) precipitately ⟨*threw* herself down on the sofa⟩ **b :** to drive or impel violently **:** DASH ⟨the ship was *thrown* on a reef⟩
4 a (1) **:** to put in a particular position or condition ⟨*threw* her arms around him⟩ (2) **:** to put on or off hastily or carelessly ⟨*threw* on a coat⟩ **b :** to bring to bear **:** EXERT ⟨*threw* all his efforts into the boy's defense⟩ **c :** BUILD, CONSTRUCT ⟨*threw* a pontoon bridge over the river⟩
5 : to form or shape on a potter's wheel
6 : to deliver (a blow) in or as if in boxing
7 : to twist two or more filaments of into a thread or yarn

8 a : to make a cast of (dice or a specified number on dice) **b :** ROLL 1a ⟨*throw* a bowling ball⟩
9 : to give up **:** ABANDON
10 : to send forth **:** PROJECT ⟨the setting sun *threw* long shadows⟩
11 : to make (oneself) dependent **:** commit (oneself) for help, support, or protection ⟨*threw* himself on the mercy of the court⟩
12 : to indulge in **:** give way to ⟨*threw* a temper tantrum⟩
13 : to bring forth **:** give birth to **:** SIRE, PRODUCE ⟨*throws* a good crop⟩ ⟨*threw* large litters⟩
14 : to lose intentionally ⟨*throw* a game⟩
15 : to move (a lever) so as to connect or disconnect parts of a clutch or switch; *also* **:** to make or break (a connection) with a lever
16 : to give by way of entertainment ⟨*throw* a party⟩
intransitive verb
: CAST, HURL ☆
— throw·er \'thrō(-ə)r\ *noun*
— throw one's weight around *or* **throw one's weight about :** to exercise influence or authority especially to an excessive degree or in an objectionable manner
— throw together 1 : to put together in a hurried and usually careless manner ⟨a bookshelf hastily *thrown together*⟩ **2 :** to bring into casual association ⟨different kinds of people are *thrown together* —Richard Sennett⟩

²throw *noun* (1530)
1 a : an act of throwing, hurling, or flinging **b** (1) **:** an act of throwing dice (2) **:** the number thrown with a cast of dice **c :** a method of throwing an opponent in wrestling or judo
2 : the distance a missile may be thrown or light rays may be projected
3 : an undertaking involving chance or danger **:** RISK, VENTURE
4 : the amount of vertical displacement produced by a geological fault
5 a : the extreme movement given to a pivoted or reciprocating piece by a cam, crank, or eccentric **:** STROKE **b :** the length of the radius of a crank or the virtual crank radius of an eccentric or cam
6 a : a light coverlet (as for a bed) **b :** a woman's scarf or light wrap
— a throw : for each one **:** APIECE ⟨copies are to be sold at $5 *a throw* —Harvey Breit⟩

¹throw·away \'thrō-ə-ˌwā\ *noun* (1903)
1 : one that is or is designed to be thrown away: as **a :** a free handbill or circular **b :** a line of dialogue (as in a play) de-emphasized by casual delivery; *especially* **:** a joke or witticism delivered casually
2 : something made or done without care or interest

²throwaway *adjective* (1928)
1 : designed to be thrown away **:** DISPOSABLE ⟨*throwaway* containers⟩
2 : written or spoken (as in a play) in a low-key or unemphatic manner ⟨*throwaway* lines⟩
3 : NONCHALANT, CASUAL

throw away *transitive verb* (1530)
1 a : to get rid of as worthless or unnecessary **b :** DISCARD 2b
2 a : to use in a foolish or wasteful manner **:** SQUANDER **b :** to fail to take advantage of **:** WASTE
3 : to make (as a line in a play) unemphatic by casual delivery

throw·back \'thrō-ˌbak\ *noun* (1888)
1 a : reversion to an earlier type or phase **:** ATAVISM **b :** an instance or product of atavistic reversion
2 : one that is suggestive of or suited to an earlier time or style ⟨his manners were a *throwback* to a more polite era⟩

throw back (1840)
transitive verb
1 : to delay the progress or advance of **:** CHECK
2 : to cause to rely **:** make dependent ⟨they are *thrown back* upon . . . native intelligence —Michael Novak⟩
3 : REFLECT
intransitive verb
: to revert to an earlier type or phase

throw down *transitive verb* (14th century)
1 : to cause to fall **:** OVERTHROW
2 : PRECIPITATE
3 : to cast off **:** DISCARD

throw–in \'thrō-ˌin\ *noun* (1881)
: an act or instance of throwing a ball in: as **a :** a throw made from the touchline in soccer to put the ball back in play after it has gone into touch **b :** a throw from an outfielder to the infield in baseball **c :** an inbounds pass in basketball

throw in (1678)
transitive verb
1 : to add as a gratuity or supplement
2 : to introduce or interject in the course of something **:** CONTRIBUTE ⟨they *throw in* some . . . sound effects on several songs —Tom Phillips⟩
3 : DISTRIBUTE 3b
4 : ENGAGE ⟨*throw in* the clutch⟩
intransitive verb
: to enter into association or partnership **:** JOIN ⟨agrees to *throw in* with a crooked ex-cop —*Newsweek*⟩
— throw in the towel *also* **throw in the sponge :** to abandon a struggle or contest **:** acknowledge defeat **:** GIVE UP

throw off (1618)
transitive verb
1 a : to free oneself from **:** get rid of ⟨*threw off* his inhibitions⟩ **b :** to cast off often in a hurried or vigorous manner **:** ABANDON ⟨*threw off* all restraint⟩ **c :** DISTRACT, DIVERT ⟨dogs *thrown off* by a false scent⟩
2 : EMIT, GIVE OFF ⟨stacks *throwing off* plumes of smoke⟩
3 : to produce in an offhand manner **:** execute with speed or facility ⟨some little . . . tune that the composer had *thrown off* —James Hilton⟩
4 a : to cause to depart from an expected or desired course ⟨mistakes *threw* his calculations *off* a bit⟩ **b :** to cause to make a mistake **:** MISLEAD
intransitive verb
1 : to begin hunting
2 : to make derogatory comments

throw out *transitive verb* (1526)
1 a : to remove from a place, office, or employment usually in a sudden or unexpected manner **b :** to get rid of as worthless or unnecessary
2 : to give expression to **:** UTTER ⟨*threw out* a remark . . . that utterly confounded him —Jean Stafford⟩
3 : to dismiss from acceptance or consideration **:** REJECT ⟨the testimony was *thrown out*⟩
4 : to make visible or manifest **:** DISPLAY ⟨the signal was *thrown out* for the . . . fleet to prepare for action —Archibald Duncan⟩
5 : to leave behind **:** OUTDISTANCE
6 : to give forth from within **:** EMIT
7 a : to send out **b :** to cause to project **:** EXTEND
8 : CONFUSE, DISCONCERT ⟨automobiles in line blocking the road . . . *threw* the whole schedule *out* —F. D. Roosevelt⟩
9 : to cause to stand out **:** make prominent
10 : to make a throw that enables a teammate to put out (a base runner)

throw over *transitive verb* (1835)
1 : to forsake despite bonds of attachment or duty
2 : to refuse to accept **:** REJECT

throw pillow *noun* (1956)
: a small pillow used especially as a decorative accessory

throw rug *noun* (1928)
: SCATTER RUG

throw·ster \'thrō-stər\ *noun* (15th century)
: one who throws textile filaments

throw up (15th century)
transitive verb
1 : to raise quickly
2 : GIVE UP, QUIT ⟨the urge . . . to *throw up* all intellectual work —Norman Mailer⟩
3 : to build hurriedly ⟨new houses *thrown up* almost overnight⟩
4 : VOMIT
5 : to bring forth **:** PRODUCE
6 : to make distinct especially by contrast **:** cause to stand out
7 : to mention repeatedly by way of reproach
intransitive verb
: VOMIT
— throw up one's hands : to admit defeat ⟨in the end *throws up his hands* in despair —Frank Conroy⟩

throw weight *noun* (1969)
: the maximum payload of an ICBM

thru *variant of* THROUGH

¹thrum \'thrəm\ *noun* [Middle English, from Old English *-thrum* (in *tungethrum* ligament of the tongue); akin to Old High German *drum* fragment] (14th century)
1 a (1) **:** a fringe of warp threads left on the loom after the cloth has been removed (2) **:** one of these warp threads **b :** a tuft or short piece of rope yarn used in thrumming canvas — usually used in plural **c :** BIT, PARTICLE
2 : a hair, fiber, or threadlike leaf on a plant; *also* **:** a tuft or fringe of such structures
— thrum *adjective*

²thrum *transitive verb* **thrummed; thrum·ming** (15th century)
1 : to furnish with thrums **:** FRINGE
2 : to insert short pieces of rope yarn or spun yarn in (a piece of canvas) to make a rough surface or a mat which can be wrapped about rigging to prevent chafing

³thrum *verb* **thrummed; thrum·ming** [imitative] (1592)
intransitive verb
1 : to play or pluck a stringed instrument idly **:** STRUM
2 : to sound with a monotonous hum
transitive verb
1 : to play (as a stringed instrument) in an idle or relaxed manner
2 : to recite tiresomely or monotonously

⁴thrum *noun* (1798)
: the monotonous sound of thrumming

¹thrush \'thrəsh\ *noun* [Middle English *thrusche,* from Old English *thrysce;* akin to Old English *throstle* thrush, Old High German *droscala,* Latin *turdus*] (before 12th century)
1 : any of numerous small or medium-sized oscine birds (families Turdidae and Muscicapidae) which are mostly of a plain color often with spotted underparts and many of which are excellent singers
2 : a bird held to resemble a thrush

²thrush *noun* [probably of Scandinavian origin; akin to Danish & Norwegian *trøske* thrush] (1665)

☆ SYNONYMS
Throw, cast, toss, fling, hurl, pitch, sling mean to cause to move swiftly through space by a propulsive movement or a propelling force. THROW is general and interchangeable with the other terms but may specifically imply a distinctive motion with bent arm ⟨can *throw* a fastball and a curve⟩. CAST usually implies lightness in the thing thrown and sometimes a scattering ⟨*cast* it to the winds⟩. TOSS suggests a light or careless or aimless throwing and may imply an upward motion ⟨*tossed* the coat on the bed⟩. FLING stresses a violent throwing ⟨*flung* the ring back in his face⟩. HURL implies power as in throwing a massive weight ⟨*hurled* himself at the intruder⟩. PITCH suggests throwing carefully at a target ⟨*pitch* horseshoes⟩. SLING stresses either the use of whirling momentum in throwing or directness of aim ⟨*slung* the bag over his shoulder⟩.

1 : a disease that is caused by a fungus (*Candida albicans*), occurs especially in infants and children, and is marked by white patches in the oral cavity; *broadly* **:** CANDIDIASIS ⟨vaginal *thrush*⟩
2 : a suppurative disorder of the feet in various animals (as a horse)

1thrust \ˈthrəst\ *verb* **thrust; thrust·ing** [Middle English *thrusten, thristen,* from Old Norse *thrȳsta;* probably akin to Old Norse *thrjōta* to tire, Old English *thrēat* coercion — more at THREAT] (13th century)
transitive verb
1 : to push or drive with force **:** SHOVE
2 : to cause to enter or pierce something by or as if by pushing ⟨*thrust* a dagger into his heart⟩
3 : EXTEND, SPREAD
4 : STAB, PIERCE
5 a : to put (as an unwilling person) forcibly into a course of action or position ⟨was *thrust* into the job⟩ **b :** to introduce often improperly into a position **:** INTERPOLATE
6 : to press, force, or impose the acceptance of upon someone ⟨*thrust* new responsibilities upon her⟩
intransitive verb
1 a : to force an entrance or passage **b :** to push forward **:** press onward **c :** to push upward **:** PROJECT
2 : to make a thrust, stab, or lunge with or as if with a pointed weapon ⟨*thrust* at them with a knife⟩

2thrust *noun* (circa 1586)
1 a : a push or lunge with a pointed weapon **b** (1) **:** a verbal attack (2) **:** a military assault
2 a : a strong continued pressure **b :** the sideways force or pressure of one part of a structure against another part (as of an arch against an abutment) **c :** the force produced by a propeller or by a jet or rocket engine that drives a vehicle (as an aircraft) forward **d :** a nearly horizontal geological fault
3 a : a forward or upward push **b :** a movement (as by a group of people) in a specified direction
4 a : salient or essential element or meaning **b :** principal concern or objective

thrust·er *also* **thrust·or** \ˈthrəs-tər\ *noun* (1597)
: one that thrusts; *especially* **:** an engine (as a jet engine) that develops thrust by expelling a jet of fluid or a stream of particles

thrust·ful \ˈthrəst-fəl\ *adjective* (1909)
British **:** characterized by thrust **:** AGGRESSIVE ⟨*thrustful* young man on the make —*Current Literature*⟩

thrust stage *noun* [*thrust,* past participle of *1thrust*] (1965)
: a stage that projects beyond the proscenium so that the audience sits around the projection; *also* **:** a forestage that is extended into the auditorium to increase the stage area

thru·way \ˈthrü-ˌwā\ *noun* (1930)
: EXPRESSWAY

1thud \ˈthəd\ *noun* [imitative] (1787)
1 : BLOW
2 : a dull sound **:** THUMP

2thud *intransitive verb* **thud·ded; thud·ding** (1796)
: to move or strike so as to make a thud

thug \ˈthəg\ *noun* [Hindi *ṭhag,* literally, thief] (1810)
: a brutal ruffian or assassin **:** GANGSTER, KILLER ◆
— thug·gery \ˈthə-g(ə-)rē\ *noun*
— thug·gish \ˈthə-gish\ *adjective*

thu·ja \ˈthü-jə, ˈthyü-\ *noun* [New Latin *Thuja,* from Medieval Latin *thuia,* a cedar, from Greek *thyia,* from *thyein* to sacrifice — more at THYME] (circa 1760)
: any of a genus (*Thuja*) of evergreen shrubs and trees (as an arborvitae) of the cypress family having scalelike closely imbricated or compressed leaves

1Thu·le \ˈthü-lē, ˈthyü-\ *noun* [Middle English *Tyle,* from Old English, from Latin *Thule, Thyle,* from Greek *Thoulē, Thylē*] (before 12th century)
: the northernmost part of the habitable ancient world

2Thu·le \ˈtü-lē\ *adjective* [*Thule,* Greenland] (1925)
: of, relating to, or being the culture existing in the arctic lands from Alaska to Greenland from about 500 A.D. to 1400 A.D.

thu·li·um \ˈthü-lē-əm, ˈthyü-\ *noun* [New Latin, from Latin *Thule*] (1879)
: a trivalent metallic element of the rare-earth group — see ELEMENT table

1thumb \ˈthəm\ *noun* [Middle English *thoume, thoumbe,* from Old English *thūma;* akin to Old High German *thūmo* thumb, Latin *tumēre* to swell] (before 12th century)
1 : the short thick digit of the human hand that is analogous in position to the big toe and differs from the other fingers in having only two phalanges, allowing greater freedom of movement, and being opposable to each of them; *also* **:** a corresponding digit in lower animals
2 : the part of a glove or mitten that covers the thumb
3 : a convex molding **:** OVOLO
— all thumbs : extremely awkward or clumsy
— under one's thumb *or* **under the thumb :** under control **:** in a state of subservience ⟨her father did not have her that much *under his thumb* —Hamilton Basso⟩

2thumb (circa 1647)
transitive verb
1 a : to leaf through (pages) with the thumb **:** TURN **b :** to soil or wear by or as if by repeated thumbing ⟨a badly *thumbed* book⟩
2 : to request or obtain (a ride) in a passing automobile by signaling with the thumb
intransitive verb
1 : to turn over pages ⟨*thumb* through a book⟩
2 : to travel by thumbing rides **:** HITCHHIKE ⟨*thumbed* across the country⟩
— thumb one's nose 1 : to place the thumb at one's nose and extend the fingers as a gesture of scorn or defiance **2 :** to express disdain or defiance ⟨*thumb their nose* at opulence —*Sales Management*⟩

thumb·hole \ˈthəm-ˌhōl\ *noun* (1859)
1 : an opening in which to insert the thumb
2 : a hole in a wind musical instrument opened or closed by the thumb

thumb index *noun* (1903)
: a series of notches cut in the fore edge of a book to facilitate reference

1thumb·nail \ˈthəm-ˌnāl, -ˈnā(ə)l\ *noun* (1604)
: the nail of the thumb

2thumbnail *adjective* (1852)
: CONCISE, BRIEF ⟨a *thumbnail* sketch⟩

thumb piano *noun* (1949)
: MBIRA

thumb·print \ˈthəm-ˌprint\ *noun* (1900)
: an impression made by the thumb; *especially* **:** a print made by the inside of the first joint

thumb·screw \-ˌskrü\ *noun* (1794)
1 : a screw having a flat-sided or knurled head so that it may be turned by the thumb and forefinger
2 : an instrument of torture for compressing the thumb by a screw

thumbscrew 1

thumbs–down \ˈthəmz-ˈdau̇n\ *noun* (1889)
: an instance or gesture of rejection, disapproval, or condemnation

thumbs–up \-ˈəp\ *noun* (circa 1917)
: an instance or gesture of approval or encouragement

1thumb·tack \ˈthəm-ˌtak\ *noun* (1884)
: a tack with a broad flat head for pressing into a surface with the thumb

2thumbtack *transitive verb* (1914)
: to fasten with a thumbtack

thumb·wheel \ˈthəm-ˌhwēl, -ˌwēl\ *noun* (1903)
: a control for various devices consisting of a partially exposed wheel that can be turned by moving the exposed edge with a finger

1thump \ˈthəmp\ *verb* [imitative] (1548)
transitive verb
1 : to strike or beat with or as if with something thick or heavy so as to cause a dull sound
2 : POUND, KNOCK
3 : WHIP, THRASH
4 : to produce (music) mechanically or in a mechanical manner — usually used with *out* ⟨*thumped* out a tune on the piano⟩
intransitive verb
1 a : to inflict a thump **b :** to make or move with a thumping sound
2 : to make a vigorous endorsement ⟨got a couple of . . . senators to *thump* for him —*N.Y. Herald Tribune*⟩
— thump·er *noun*

2thump *noun* (1552)
: a blow or knock with or as if with something blunt or heavy; *also* **:** the sound made by such a blow

thump·ing *adjective* [*thumping,* present participle of *1thump*] (1576)
: impressively large, great, or excellent ⟨a *thumping* majority⟩

1thun·der \ˈthən-dər\ *noun* [Middle English *thoner, thunder,* from Old English *thunor;* akin to Old High German *thonar* thunder, Latin *tonare* to thunder] (before 12th century)
1 : the sound that follows a flash of lightning and is caused by sudden expansion of the air in the path of the electrical discharge
2 : a loud utterance or threat
3 : BANG, RUMBLE ⟨the *thunder* of big guns⟩

2thunder *verb* **thun·dered; thun·der·ing** \-d(ə-)riŋ\ (before 12th century)
intransitive verb

◇ WORD HISTORY

thug In the 1830s British officials in India launched a vigorous campaign against bands of highway robbers known as *thugs* (from Hindi *ṭhag* "thief"). The *thugs* were supposedly members of a centuries-old hereditary criminal caste who befriended travelers and after winning their trust strangled them and plundered their goods. With the aid of *thugs* who confessed in return for reduced sentences, over a thousand men were convicted under an extraordinary legal measure that made simple membership in a *thug* gang punishable by life imprisonment. Numerous popular accounts of *thuggee,* as the phenomenon was called, that appeared in England, especially the novel *Confessions of a Thug* (1839) by an Indian civil servant named Meadows Taylor, made *thug* a well-known word that was soon generalized to refer to any brutal ruffian. It now appears that *thuggee* was largely a product of British imagination, created by combining the embellished accounts of *thug* informers with Britons' own preconceptions of mysterious evil lurking below the surface of exotic Asian societies. An actual increase in the 1830s of vagrant groups of armed men preying on travelers probably resulted from the disbanding of militias in the service of local rulers, as the British subdued one native state after another.

1 a : to produce thunder — usually used impersonally ⟨it *thundered*⟩ **b :** to give forth a sound that resembles thunder ⟨horses *thundered* down the road⟩
2 : ROAR, SHOUT
transitive verb
1 : to utter loudly **:** ROAR
2 : to strike with a sound likened to thunder
— thun·der·er \-dər-ər\ *noun*

thun·der·bird \'thən-dər-ˌbərd\ *noun* (1871)
: a bird that causes lightning and thunder in American Indian myth

thun·der·bolt \-ˌbōlt\ *noun* (15th century)
1 a : a single discharge of lightning with the accompanying thunder **b :** an imaginary elongated mass cast as a missile to earth in the lightning flash
2 a : a person or thing that resembles lightning in suddenness, effectiveness, or destructive power **b :** a vehement threat or censure

thun·der·clap \-ˌklap\ *noun* (14th century)
1 : a clap of thunder
2 : something sharp, loud, or sudden like a clap of thunder

thun·der·cloud \-ˌklau̇d\ *noun* (1697)
: a cloud charged with electricity and producing lightning and thunder

thunder egg *noun* (1941)
: chalcedony in rounded concretionary nodules

thun·der·head \-ˌhed\ *noun* (1853)
: a rounded mass of cumulus cloud often appearing before a thunderstorm

thun·der·ing *adjective* [*thundering,* present participle of [2]*thunder*] (1543)
: awesomely great, intense, or unusual
— thun·der·ing·ly \-d(ə-)riŋ-lē\ *adverb*

thunder lizard *noun* [translation of New Latin *Brontosaurus*] (1960)
: BRONTOSAURUS

thun·der·ous \'thən-d(ə-)rəs\ *adjective* (1582)
1 a : producing thunder **b :** making or accompanied by a noise like thunder ⟨*thunderous* applause⟩
2 : THUNDERING
— thun·der·ous·ly *adverb*

thun·der·show·er \-ˌshau̇(-ə)r\ *noun* (1699)
: a shower accompanied by lightning and thunder

thun·der·stone \-ˌstōn\ *noun* (1598)
1 *archaic* **:** THUNDERBOLT 1b
2 : any of various stones (as a meteorite or an ancient artifact) that are the probable source of the imaginary thunderbolt

thun·der·storm \-ˌstȯrm\ *noun* (1652)
: a storm accompanied by lightning and thunder

thun·der·strike \-ˌstrīk\ *transitive verb* **-struck** \-ˌstrək\; **-struck** *also* **-strick·en** \-ˌstri-kən\; **-strik·ing** \-ˌstrī-kiŋ\ (circa 1586)
1 : to strike dumb **:** ASTONISH ⟨was *thunderstruck* at the news⟩
2 *archaic* **:** to strike by or as if by lightning

thun·der·stroke \-ˌstrōk\ *noun* (1587)
: a stroke of or as if of lightning with the attendant thunder

[1]**thunk** \'thəŋk\ *dialect past and past participle of* THINK

[2]**thunk** *noun* [imitative] (1947)
: a flat hollow sound

thu·ri·ble \'thu̇r-ə-bəl, 'thyu̇r-, 'thər-\ *noun* [Middle English *turrible,* from Middle French *thurible,* from Latin *thuribulum,* from *thur-, thus* incense, from Greek *thyos* incense, sacrifice, from *thyein* to sacrifice — more at THYME] (15th century)
: CENSER

thu·ri·fer \-ə-fər\ *noun* [New Latin, from Latin *thurifer,* adjective, incense-bearing, from *thur-, thus* + *-ifer* -iferous] (1853)
: one who carries a censer in a liturgical service

Thu·rin·ger \'thu̇r-ən-jər, 'thyu̇r-\ *noun* [German *Thüringerwurst,* from *Thüringer* Thuringian + *Wurst* sausage] (circa 1923)
: a mildly seasoned fresh or smoked sausage

Thu·rin·gian \thu̇-'rin-j(ē-)ən, thyu̇r-\ *noun* (1618)
1 : a member of an ancient Germanic people whose kingdom was overthrown by the Franks in the 6th century
2 : a native or inhabitant of Thuringia
— Thuringian *adjective*

thurl \'thər(-ə)l\ *noun* [perhaps from English dialect, gaunt]
: the hip joint in cattle — see COW illustration

Thurs·day \'thərz-dē, -(ˌ)dā\ *noun* [Middle English, from Old English *thursdæg,* from Old Norse *thōrsdagr;* akin to Old English *thunresdæg* Thursday, Old Norse *Thōrr* Thor, Old English *thunor* thunder — more at THUNDER] (before 12th century)
: the fifth day of the week
— Thurs·days \-dēz, -(ˌ)dāz\ *adverb*

thus \'t͟həs\ *adverb* [Middle English, from Old English; akin to Old Saxon *thus* thus] (before 12th century)
1 : in this or that manner or way
2 : to this degree or extent **:** SO
3 : because of this or that **:** HENCE, CONSEQUENTLY
4 : as an example

thus·ly \-lē\ *adverb* (1865)
: in this manner **:** THUS ▪

[1]**thwack** \'thwak\ *transitive verb* [imitative] (circa 1530)
: to strike with or as if with something flat or heavy **:** WHACK

[2]**thwack** *noun* (1587)
: a heavy blow **:** WHACK; *also* **:** the sound of or as if of such a blow

[1]**thwart** \'thwȯrt\ *transitive verb* [Middle English *thwerten,* from *thwert,* adverb] (13th century)
1 a : to run counter to so as to effectively oppose or baffle **:** CONTRAVENE **b :** to oppose successfully **:** defeat the hopes or aspirations of
2 : to pass through or across
synonym see FRUSTRATE
— thwart·er *noun*

[2]**thwart** \'thwȯrt, *nautical often* 'thȯrt\ *adverb* [Middle English *thwert,* from Old Norse *thvert,* from neuter of *thverr* transverse, oblique; akin to Old High German *dwerah* transverse, oblique] (14th century)
: ATHWART

[3]**thwart** *adjective* (15th century)
: situated or placed across something else
: TRANSVERSE
— thwart·ly *adverb*

[4]**thwart** *noun* [alteration of obsolete *thought, thoft,* from Middle English *thoft,* from Old English *thofte;* akin to Old High German *dofta* rower's seat] (circa 1736)
: a rower's seat extending athwart a boat

thwart·wise \-ˌwīz\ *adverb or adjective* (1589)
: CROSSWISE

thy \'t͟hī\ *adjective* [Middle English *thin, thy,* from Old English *thīn,* genitive of *thū* thou — more at THOU] (12th century)
archaic **:** of or relating to thee or thyself especially as possessor or agent or as object of an action — used especially in ecclesiastical or literary language and sometimes by Friends especially among themselves

Thy·es·te·an \thī-'es-tē-ən\ *adjective* [*Thyestes,* brother of Atreus who unwittingly ate the flesh of his children] (1667)
: of or relating to the eating of human flesh
: CANNIBAL

thy·la·cine \'thī-lə-ˌsīn\ *noun* [New Latin *Thylacinus,* genus of marsupials, from Greek *thylakos* sack, pouch] (1838)
: TASMANIAN WOLF

thy·la·koid \'thī-lə-ˌkȯid\ *noun* [International Scientific Vocabulary *thylak-* (from Greek *thylakos* sack) + *-oid*] (1966)
: any of the membranous disks of lamellae within plant chloroplasts that are composed of protein and lipid and are the sites of the photochemical reactions of photosynthesis

thym- *or* **thymo-** *combining form* [New Latin *thymus*]
: thymus ⟨*thymic*⟩ ⟨*thymocyte*⟩

thyme \'tīm *also* 'thīm\ *noun* [Middle English, from Middle French *thym,* from Latin *thymum,* from Greek *thymon,* probably from *thyein* to make a burnt offering, sacrifice; akin to Latin *fumus* smoke — more at FUME] (14th century)
1 : any of a genus (*Thymus*) of Eurasian mints with small pungent aromatic leaves; *especially* **:** a garden herb (*T. vulgaris*)
2 : thyme leaves used as a seasoning

thy·mec·to·my \thī-'mek-tə-mē\ *noun, plural* **-mies** (circa 1905)
: excision of the thymus
— thy·mec·to·mize \-ˌmīz\ *transitive verb*

-thymia *noun combining form* [New Latin, from Greek, from *thymos* mind]
: condition of mind and will ⟨cyclo*thymia*⟩

thy·mic \'thī-mik\ *adjective* (circa 1656)
: of or relating to the thymus

thy·mi·dine \'thī-mə-ˌdēn\ *noun* [*thymine* + *-idine*] (1912)
: a nucleoside $C_{10}H_{14}N_2O_5$ that is composed of thymine and deoxyribose and occurs as a structural part of DNA

thy·mine \'thī-ˌmēn\ *noun* [International Scientific Vocabulary, from New Latin *thymus*] (1894)
: a pyrimidine base $C_5H_6N_2O_2$ that is one of the four bases coding genetic information in the polynucleotide chain of DNA — compare ADENINE, CYTOSINE, GUANINE, URACIL

thy·mo·cyte \'thī-mə-ˌsīt\ *noun* [International Scientific Vocabulary] (circa 1923)
: a cell of the thymus; *especially* **:** a thymic lymphocyte

thy·mol \'thī-ˌmȯl, -ˌmōl\ *noun* [International Scientific Vocabulary, from Latin *thymum* thyme] (1857)
: a crystalline phenol $C_{10}H_{14}O$ of aromatic odor and antiseptic properties found especially in thyme oil or made synthetically and used chiefly as a fungicide and preservative

thy·mo·sin \'thī-mə-sən\ *noun* [Greek *thymos* thymus + English [1]*-in*] (1966)
: a mixture of polypeptides isolated from the thymus; *also* **:** any of these

thy·mus \'thī-məs\ *noun* [New Latin, from Greek *thymos* warty excrescence, thymus] (circa 1693)

☐ USAGE
thusly *Thusly* seems to have been invented in the mid-1860s by one of two American humorists, either Artemus Ward or Josh Billings ⟨they are cheerful, and why should it not be *thusly* with us? —Artemus Ward⟩. It was not a word meant to be taken seriously, but in the intervening 130 years or so it has gradually crept into standard use, especially but not exclusively in journalism. In standard writing it almost always follows a verb ⟨in the realm (or category, for booksellers segregate it *thusly*) of women's literature —Joyce Carol Oates⟩. By far its most common use is, like *as follows,* to stand before a colon and introduce a quotation ⟨who is quoted *thusly:* "Political appointees lack job security, but it sure is fun while it lasts" —Herb Caen⟩ ⟨exhorting the poet . . . *thusly:* "Some Homer of the cotton fields should sing the saga of the mule . . ." —Maxine Kumin⟩. The critics disapprove of it and characterize it in a bewildering variety of ways: superfluous, incorrect, unacceptable, semiliterate, pretentious, mistaken, nonstandard, an abomination, an illiteracy. Those about to commit *thusly* to writing should be aware of this attitude. One can ignore it or can substitute *thus* or the often more natural-sounding "in this way" or "as follows."

: a glandular structure of largely lymphoid tissue that functions especially in the development of the body's immune system, is present in the young of most vertebrates typically in the upper anterior chest or at the base of the neck, and tends to atrophy in the adult

thymy *or* **thym·ey** \'tī-mē *also* 'thī-\ *adjective* (1727)
: abounding in or fragrant with thyme

thyr- *or* **thyro-** *combining form* [*thyroid*]
: thyroid ⟨*thyro*toxicosis⟩ ⟨*thyr*oxine⟩

thy·ra·tron \'thī-rə-ˌträn\ *noun* [from *Thyratron,* a trademark] (1929)
: a gas-filled hot-cathode electron tube in which the grid controls only the start of a continuous current thus giving the tube a trigger effect

thy·ris·tor \thī-'ris-tər\ *noun* [*thyr*atron + tran*sistor*] (1958)
: any of several semiconductor devices that act as switches, rectifiers, or voltage regulators

thy·ro·cal·ci·to·nin \ˌthī-rō-ˌkal-sə-'tō-nən\ *noun* (1963)
: CALCITONIN

thy·ro·glob·u·lin \-'glä-byə-lən\ *noun* [International Scientific Vocabulary] (circa 1905)
: an iodine-containing protein of the thyroid gland that is the precursor of thyroxine and triiodothyronine

[1]thy·roid \'thī-ˌrȯid\ *also* **thy·roi·dal** \thī-'rȯi-dᵊl\ *adjective* [New Latin *thyroides,* from Greek *thyreoeidēs* shield-shaped, thyroid, from *thyreos* shield shaped like a door, from *thyra* door — more at DOOR] (circa 1741)
1 a : of, relating to, or being a large bilobed endocrine gland of craniate vertebrates lying at the base of the neck and producing especially the hormones thyroxine and triiodothyronine **b :** suggestive of a disordered thyroid ⟨a *thyroid* personality⟩
2 : of, relating to, or being the chief cartilage of the larynx

[2]thyroid *noun* (1840)
1 : a thyroid gland or cartilage; *also* **:** a part (as an artery or nerve) associated with either of these
2 : a preparation of mammalian thyroid gland used in treating thyroid disorders

thy·roid·ec·to·my \ˌthī-ˌrȯi-'dek-tə-mē, ˌthī-rə-\ *noun, plural* **-mies** (1889)
: surgical removal of thyroid gland tissue
— thy·roid·ec·to·mized \-ˌmīzd\ *adjective*

thy·roid·itis \ˌthī-ˌrȯi-'dī-təs, ˌthī-rə-\ *noun* [New Latin] (circa 1885)
: inflammation of the thyroid gland

thyroid–stimulating hormone *noun* (1941)
: THYROTROPIN

thy·ro·tox·i·co·sis \ˌthī-rō-ˌtäk-sə-'kō-səs\ *noun* [New Latin] (circa 1911)
: HYPERTHYROIDISM

thy·ro·tro·pic \ˌthī-rə-'trō-pik, -'trä-\ *also* **thy·ro·tro·phic** \-'trō-fik\ *adjective* (circa 1923)
: exerting or characterized by a direct influence on the secretory activity of the thyroid gland ⟨*thyrotropic* functions⟩

thy·ro·tro·pin \ˌthī-rə-'trō-pən\ *also* **thy·ro·tro·phin** \-fən\ *noun* [*thyrotropic, thyrotrophic*] (1939)
: a hormone secreted by the anterior pituitary that stimulates the thyroid gland — called also *thyroid-stimulating hormone, thyrotropic hormone*

thyrotropin–releasing hormone *noun* (1968)
: a tripeptide hormone synthesized in the hypothalamus that stimulates secretion of thyrotropin by the anterior lobe of the pituitary gland — called also *thyrotropin-releasing factor*

thy·rox·ine *or* **thy·rox·in** \thī-'räk-ˌsēn, -sən\ *noun* [International Scientific Vocabulary] (1918)
: an iodine-containing hormone $C_{15}H_{11}I_4NO_4$ that is an amino acid produced by the thyroid gland as a product of the cleavage of thyroglobulin, increases metabolic rate, and is used to treat thyroid disorders

thyrse \'thərs\ *noun* [New Latin *thyrsus,* from Latin, thyrsus] (1744)
: an inflorescence (as in the lilac and horse chestnut) in which the main axis is racemose and the secondary and later axes are cymose

thyr·sus \'thər-səs\ *noun, plural* **thyr·si** \-ˌsī, -ˌsē\ [Latin, from Greek *thyrsos*] (1591)
: a staff surmounted by a pinecone or by a bunch of vine or ivy leaves with grapes or berries that is carried by Bacchus and by satyrs and others engaging in bacchic rites

thy·sa·nu·ran \ˌthī-sə-'nu̇r-ən, -'nyu̇r-\ *noun* [ultimately from Greek *thysanos* tassel + *oura* tail — more at ASS] (1835)
: BRISTLETAIL
— thysanuran *adjective*

thy·self \t͟hī-'self\ *pronoun* (before 12th century)
archaic **:** YOURSELF — used especially in ecclesiastical or literary language and sometimes by Friends especially among themselves

[1]ti \'tē\ *noun* [Tahitian, Marquesan, Samoan, & Maori] (1832)
: any of several Asian and Pacific trees or shrubs (genus *Cordyline*) of the agave family with leaves in terminal tufts

[2]ti *noun* [alteration of *si*] (1839)
: the seventh tone of the diatonic scale in solmization

ti·ara \tē-'ar-ə, -'er-, -'är-\ *noun* [Latin, royal Persian headdress, from Greek] (1616)
1 : a 3-tiered crown worn by the pope
2 : a decorative jeweled or flowered headband or semicircle for formal wear by women

tiara 2

Ti·bet·an \tə-'be-tᵊn\ *noun* (1747)
1 a : a member of the predominant people of Tibet and adjacent areas of Asia **b :** a native or inhabitant of Tibet
2 : the Tibeto-Burman language of the Tibetan people
— Tibetan *adjective*

Tibetan terrier *noun* (1905)
: any of a breed of terriers resembling Old English sheepdogs but having a curled well-feathered tail

Tibetan terrier

Ti·beto–Bur·man \tə-ˌbe-tō-'bər-mən\ *noun* (1901)
1 : a language family that includes Tibetan, Burmese, and related languages of southern and eastern Asia
2 : a member of a people speaking a Tibeto-Burman language

tib·ia \'ti-bē-ə\ *noun, plural* **-i·ae** \-bē-ˌē, -bē-ˌī\ *also* **-i·as** [Latin] (circa 1706)
1 : the inner and usually larger of the two bones of the vertebrate hind limb between the knee and ankle
2 : the fourth joint of the leg of an insect between the femur and tarsus
— tib·i·al \-bē-əl\ *adjective*

tib·io·fib·u·la \ˌti-bē-ō-'fi-byə-lə\ *noun* [New Latin] (circa 1909)
: a bone especially in frogs and toads that is formed by fusion of the tibia and fibula

tic \'tik\ *noun* [French] (circa 1834)
1 : local and habitual spasmodic motion of particular muscles especially of the face **:** TWITCHING
2 : a frequent usually unconscious quirk of behavior or speech ⟨"you know" is a verbal *tic*⟩

ti·cal \ti-'käl, 'ti-kəl\ *noun, plural* **ticals** *or* **tical** [Thai, from Portuguese, from Malay *tikal,* a monetary unit] (1662)
: BAHT

tic dou·lou·reux \'tik-ˌdü-lə-'rü, -'rə(r)\ *noun* [French, painful twitch] (1800)
: TRIGEMINAL NEURALGIA

[1]tick \'tik\ *noun* [Middle English *tyke, teke;* akin to Middle High German *zeche* tick, Armenian *tiz*] (14th century)
1 : any of a superfamily (Ixodoidea of the order Acarina) of bloodsucking arachnids that are larger than the related mites, attach themselves to warm-blooded vertebrates to feed, and include important vectors of infectious diseases
2 : any of various usually wingless parasitic dipteran flies — compare SHEEP KED

[1]tick 1

[2]tick *noun* [Middle English *tike,* probably from Middle Dutch (akin to Old High German *ziahha* tick), from Latin *theca* cover, from Greek *thēkē* case; akin to Greek *tithenai* to place — more at DO] (15th century)
1 : the fabric case of a mattress, pillow, or bolster; *also* **:** a mattress consisting of a tick and its filling
2 : TICKING

[3]tick *noun* [Middle English *tek* pat, light stroke; akin to Middle High German *zic* light push] (1680)
1 a : a light rhythmic audible tap or beat; *also* **:** a series of such ticks **b** *chiefly British* **:** the time taken by the tick of a clock **:** MOMENT
2 : a small spot or mark; *especially* **:** one used to direct attention to something, to check an item on a list, or to represent a point on a scale

[4]tick (1721)
intransitive verb
1 : to make the sound of a tick or a series of ticks
2 : to operate as a functioning mechanism **:** RUN ⟨tried to understand what made him *tick*⟩ ⟨the motor was *ticking* over quietly⟩
transitive verb
1 : to mark with a written tick **:** CHECK — usually used with *off* ⟨*ticked* off each item in the list⟩
2 : to mark, count, or announce by or as if by ticking beats ⟨a meter *ticking* off the cab fare⟩
3 : to touch with a momentary glancing blow ⟨*ticked* the ball⟩

[5]tick *noun* [short for [1]*ticket*] (1642)
chiefly British **:** CREDIT, TRUST; *also* **:** a credit account

tick–borne \'tik-ˌbōrn, -ˌbȯrn\ *adjective* (1921)
: capable of being transmitted by the bites of ticks ⟨*tick-borne* encephalitis⟩

[1]ticked \'tikt\ *adjective* (circa 1688)
1 : marked with ticks **:** FLECKED
2 : having or made of hair banded with two or more colors ⟨a *ticked* cat⟩ ⟨a *ticked* coat⟩

[2]ticked *adjective* [*tick off*] (circa 1959)
: ANGRY, UPSET

tick·er \'ti-kər\ *noun* (1828)
: something that ticks or produces a ticking sound: as **a :** WATCH **b :** a telegraphic receiving instrument that automatically prints off information (as stock quotations or news) on a paper ribbon **c** *slang* **:** HEART

ticker tape *noun* (1902)
: the paper ribbon on which a telegraphic ticker prints off its information

\ə\ abut \ᵊ\ kitten \ər\ further \a\ ash \ā\ ace \ä\ mop, mar \au̇\ out \ch\ chin \e\ bet \ē\ easy \g\ go \i\ hit \ī\ ice \j\ job \ŋ\ sing \ō\ go \ȯ\ law \ȯi\ boy \th\ thin \t͟h\ the \ü\ loot \u̇\ foot \y\ yet \zh\ vision *see also* Guide to Pronunciation

[1]tick·et \'ti-kət\ *noun* [Middle French *etiquet, estiquette* notice attached to something, from *estiquier* to attach, from Middle Dutch *steken* to stick; akin to Old High German *sticken* to prick — more at STICK] (1529)
1 a : a document that serves as a certificate, license, or permit; *especially* **:** a mariner's or airman's certificate **b :** TAG, LABEL
2 a : a certificate or token showing that a fare or admission fee has been paid **b :** a means of access or passage ⟨education is the *ticket* to a good job⟩
3 : a list of candidates for nomination or election **:** SLATE
4 : the correct or desirable thing ⟨cooperation, that's the *ticket* —K. E. Trombley⟩
5 : a slip or card recording a transaction or undertaking or giving instructions ⟨a savings deposit *ticket*⟩
6 : a summons or warning issued to a traffic-law violator
— tick·et·less \-ləs\ *adjective*
[2]ticket *transitive verb* (1611)
1 : to attach a ticket to **:** LABEL; *also* **:** DESIGNATE
2 : to furnish or serve with a ticket ⟨*ticketed* for illegal parking⟩
ticket agency *noun* (1923)
: an agency selling transportation or theater and entertainment tickets
ticket agent *noun* (1861)
: one who sells transportation or theater and entertainment tickets
ticket office *noun* (circa 1667)
: an office of a transportation company, theatrical or entertainment enterprise, or ticket agency where tickets are sold and reservations made
tick·et–of–leave \,ti-kət-ə(v)-'lēv\ *noun, plural* **tickets–of–leave** (1732)
: a license or permit formerly given in the United Kingdom and the Commonwealth to a convict under imprisonment to go at large and to get work subject to certain specific conditions
tick fever *noun* (circa 1897)
1 : TEXAS FEVER
2 : a febrile disease (as Rocky Mountain spotted fever) transmitted by the bites of ticks
[1]tick·ing \'ti-kiŋ\ *noun* [[2]*tick*] (1649)
: a strong linen or cotton fabric used in upholstering and as a covering for a mattress or pillow
[2]ticking *noun* [[3]*tick*] (1885)
: ticked marking on a bird or mammal or on individual hairs
[1]tick·le \'ti-kəl\ *verb* **tick·led; tick·ling** \-k(ə-)liŋ\ [Middle English *tikelen;* akin to Old English *tinclian* to tickle] (14th century)
transitive verb
1 a : to excite or stir up agreeably **:** PLEASE ⟨music . . . does more than *tickle* our sense of rhythm —Edward Sapir⟩ **b :** to provoke to laughter or merriment **:** AMUSE ⟨were *tickled* by the clown's antics⟩
2 : to touch (as a body part) lightly so as to excite the surface nerves and cause uneasiness, laughter, or spasmodic movements
intransitive verb
1 : to have a tingling or prickling sensation ⟨my back *tickles*⟩
2 : to excite the surface nerves to prickle
[2]tickle *noun* (1801)
1 : the act of tickling
2 : a tickling sensation
3 : something that tickles
tick·ler \'ti-k(ə-)lər\ *noun* (1680)
1 : a person or device that tickles
2 : a device for jogging the memory; *specifically* **:** a file that serves as a reminder and is arranged to bring matters to timely attention
tick·lish \'ti-k(ə-)lish\ *adjective* (1581)
1 a : TOUCHY, OVERSENSITIVE ⟨*ticklish* about his baldness⟩ **b :** easily overturned ⟨a canoe is a *ticklish* craft⟩
2 : requiring delicate handling ⟨a *ticklish* subject⟩
3 : sensitive to tickling
— tick·lish·ly *adverb*
— tick·lish·ness *noun*
tick off *transitive verb* [[4]*tick*] (circa 1919)
1 : REPRIMAND, REBUKE ⟨his father *ticked* him *off* for his impudence⟩
2 : to make angry or indignant ⟨the cancellation really *ticked* me *off*⟩
tick·seed \'tik-,sēd\ *noun* [[1]*tick*] (circa 1562)
: COREOPSIS
tick·tack *or* **tic·tac** \'tik-,tak\ *noun* [reduplication of *tick*] (1549)
1 : a ticking or tapping beat like that of a clock or watch
2 : a contrivance used by children to tap on a window from a distance
tick·tack·toe *also* **tic–tac–toe** \,tik-,tak-'tō\ *noun* [*tic-tac-toe,* former game in which players with eyes shut brought a pencil down on a slate marked with numbers and scored the number hit] (circa 1866)
: a game in which two players alternately put Xs and Os in compartments of a figure formed by two vertical lines crossing two horizontal lines and each tries to get a row of three Xs or three Os before the opponent does
tick·tock \'tik-'täk, -,täk\ *noun* [imitative] (1848)
: the ticking sound of a clock
tick trefoil *noun* [[1]*tick*] (1857)
: any of various leguminous plants (genus *Desmodium*) with trifoliolate leaves and rough sticky loments
[1]ticky–tacky \,ti-kē-'ta-kē\ *also* **ticky–tack** \-'tak\ *noun, plural* **ticky–tackies** *also* **ticky–tacks** [reduplication of *tacky*] (1962)
: sleazy or shoddy material used especially in the construction of look-alike tract houses; *also* **:** something built of ticky-tacky
[2]ticky–tacky *also* **ticky–tack** *adjective* (1964)
1 : of an uninspired or monotonous sameness
2 : TACKY
3 : built of ticky-tacky
tid·al \'tī-dᵊl\ *adjective* (1807)
1 a : of, relating to, caused by, or having tides ⟨*tidal* cycles⟩ ⟨*tidal* erosion⟩ **b :** periodically rising and falling or flowing and ebbing ⟨*tidal* waters⟩
2 : dependent (as to the time of arrival or departure) upon the state of the tide ⟨a *tidal* steamer⟩
— tid·al·ly \-dᵊl-ē\ *adverb*
tidal wave *noun* (1830)
1 a : an unusually high sea wave that sometimes follows an earthquake **b :** an unusual rise of water alongshore due to strong winds
2 : something overwhelming especially in quantity or volume ⟨a *tidal wave* of tourists⟩
tid·bit \'tid-,bit\ *noun* [perhaps from *tit-* (as in *titmouse*) + *bit*] (circa 1640)
1 : a choice morsel of food
2 : a choice or pleasing bit (as of information)
tid·dle·dy·winks \'ti-dᵊl-dē-,wiŋ(k)s\ *or* **tid·dly·winks** \'ti-dᵊl-ē-, 'tid-lē-\ *noun plural but singular in construction* [probably from English dialect *tiddly* little, alteration of *little*] (1898)
: a game whose object is to snap small disks from a flat surface into a small container
tid·dler \'ti-dᵊl-ər, 'tid-lər\ *noun* [probably from English dialect *tiddly* little] (1885)
British **:** a small fish (as a stickleback or minnow)
tid·dly \'ti-dᵊl-ē, 'tid-lē\ *adjective* [*tiddly* an alcoholic drink, probably from English dialect *tiddly*] (1905)
chiefly British **:** slightly drunk
[1]tide \'tīd\ *noun* [Middle English, time, from Old English *tīd;* akin to Old High German *zīt* time and perhaps to Greek *daiesthai* to divide] (before 12th century)
1 a *obsolete* **:** a space of time **:** PERIOD **b :** a fit or opportune time **:** OPPORTUNITY **c :** an ecclesiastical anniversary or festival; *also* **:** its season — usually used in combination ⟨Easter*tide*⟩
2 a (1) **:** the alternate rising and falling of the surface of the ocean and of water bodies (as gulfs and bays) connected with the ocean that occurs usually twice a day and is caused by the gravitational attraction of the sun and moon occurring unequally on different parts of the earth (2) **:** a less marked rising and falling of an inland body of water (3) **:** a periodic movement in the earth's crust caused by the same forces that produce ocean tides (4) **:** a tidal distortion on one celestial body caused by the gravitational attraction of another (5) **:** one of the tidal movements of the atmosphere resembling those of the ocean and produced by gravitation or diurnal temperature changes **b :** FLOOD TIDE 1
3 a : something that fluctuates like the tides of the sea ⟨the *tide* of public opinion⟩ **b :** a surging movement of a group ⟨a *tide* of opportunists⟩
4 a : a flowing stream **:** CURRENT **b :** the waters of the ocean **c :** the overflow of a flooding stream
— tide·less \-ləs\ *adjective*
[2]tide *verb* **tid·ed; tid·ing** (1593)
intransitive verb
: to flow as or in a tide **:** SURGE
transitive verb
: to cause to float with or as if with the tide
[3]tide *intransitive verb* **tid·ed; tid·ing** [Middle English, from Old English *tīdan;* akin to Middle Dutch *tiden* to go, come, Old English *tīd* time] (before 12th century)
archaic **:** BETIDE, BEFALL
tide·land \'tīd-,land, -lənd\ *noun* (1802)
1 : land overflowed during flood tide
2 : land underlying the ocean and lying beyond the low-water limit of the tide but being within the territorial waters of a nation — often used in plural
tide·mark \-,märk\ *noun* (1799)
1 a : a high-water or sometimes low-water mark left by tidal water or a flood **b :** a mark placed to indicate this point
2 : the point to which something has attained or below which it has receded ⟨the *tidemark* of tolerance has risen —*New Republic*⟩
tide over *transitive verb* [[2]*tide*] (1821)
: to enable to surmount or endure a difficulty ⟨money to *tide* us *over* the emergency⟩
tide table *noun* (1594)
: a table that indicates the height of the tide at one place at different times of day throughout one year
tide·wa·ter \-,wò-tər, -,wä-\ *noun* (1772)
1 : water overflowing land at flood tide; *also* **:** water affected by the ebb and flow of the tide
2 : low-lying coastal land
tide·way \-,wā\ *noun* (1798)
: a channel in which the tide runs
tid·ing \'tī-diŋ\ *noun* [Middle English, from Old English *tīdung,* from *tīdan* to betide] (12th century)
: a piece of news — usually used in plural ⟨good *tidings*⟩
[1]ti·dy \'tī-dē\ *adjective* **ti·di·er; -est** [Middle English, timely, in good condition, from *tide* time] (13th century)
1 : properly filled out **:** PLUMP
2 : adequately satisfactory **:** ACCEPTABLE, FAIR ⟨a *tidy* solution to their problem⟩
3 a : neat and orderly in appearance or habits **:** well ordered and cared for **b :** METHODICAL, PRECISE ⟨a *tidy* mind⟩
4 : LARGE, SUBSTANTIAL ⟨a *tidy* profit⟩
— ti·di·ly \'tī-dᵊl-ē\ *adverb*
— ti·di·ness \'tī-dē-nəs\ *noun*
[2]tidy *verb* **ti·died; ti·dy·ing** (1821)
transitive verb
: to put in order ⟨*tidy* up a room⟩

intransitive verb
: to make things tidy ⟨*tidying* up after supper⟩
— ti·di·er *noun*

^{3}**tidy** *noun, plural* **tidies** (circa 1828)
1 : a usually compartmentalized receptacle for various small objects
2 : a piece of fancywork used to protect the back, arms, or headrest of a chair or sofa from wear or soil

ti·dy·tips \'tī-dē-ˌtips\ *noun plural but singular or plural in construction* (1888)
: an annual California composite herb (*Layia platyglossa*) having yellow-rayed flower heads often tipped with white

^{1}**tie** \'tī\ *noun* [Middle English *teg, tye,* from Old English *tēag;* akin to Old Norse *taug* rope, Old English *tēon* to pull — more at TOW] (before 12th century)
1 a : a line, ribbon, or cord used for fastening, uniting, or drawing something closed; *especially* **:** SHOELACE **b** (1) **:** a structural element (as a rod or angle iron) holding two pieces together **:** a tension member in a construction (2) **:** any of the transverse supports to which railroad rails are fastened to keep them in line
2 : something that serves as a connecting link: as **a :** a moral or legal obligation to someone or something typically constituting a restraining power, influence, or duty **b :** a bond of kinship or affection
3 : a curved line that joins two musical notes of the same pitch to denote a single tone sustained through the time value of the two
4 a : an equality in number (as of votes or scores) **b :** equality in a contest; *also* **:** a contest that ends in a draw
5 : a method or style of tying or knotting
6 : something that is knotted or is to be knotted when worn: as **a :** NECKTIE **b :** a low laced shoe **:** OXFORD
— tie·less \-ləs\ *adjective*

^{2}**tie** *verb* **tied; ty·ing** \'tī-iŋ\ *or* **tie·ing** (before 12th century)
transitive verb
1 a : to fasten, attach, or close by means of a tie **b :** to form a knot or bow in ⟨*tie* your scarf⟩ **c :** to make by tying constituent elements ⟨*tied* a wreath⟩ ⟨*tie* a fishing fly⟩
2 a : to place or establish in relationship **:** CONNECT **b :** to unite in marriage **c :** to unite (musical notes) by a tie **d :** to join (power systems) electrically
3 : to restrain from independence or freedom of action or choice **:** constrain by or as if by authority, influence, agreement, or obligation
4 a (1) **:** to make or have an equal score with in a contest (2) **:** to equalize (the score) in a game or contest (3) **:** to equalize the score of (a game) **b :** to provide or offer something equal to **:** EQUAL
intransitive verb
: to make a tie: as **a :** to make a bond or connection **b :** to make an equal score **c :** to become attached **d :** to close by means of a tie
— tie into : to attack with vigor
— tie one on *slang* **:** to get drunk
— tie the knot : to perform a marriage ceremony; *also* **:** to get married

tie–and–dye \'tī-ən-ˌdī\ *noun* (1928)
: TIE-DYEING

tie·back \'tī-ˌbak\ *noun* (1926)
1 : a decorative strip or device of cloth, cord, or metal for draping a curtain to the side of a window
2 : a curtain with a tieback — usually used in plural

tie·break·er \'tī-ˌbrā-kər\ *noun* (circa 1932)
: a contest used to select a winner from among contestants with tied scores at the end of a previous contest

tied cottage *noun* (1899)
British **:** a cottage or house owned by an employer (as a farmer) and reserved for occupancy by an employee

tie–down \-ˌdau̇n\ *noun* (circa 1942)
: a fitting or a system of lines and fittings used to secure something (as an aircraft or cargo)

tie–dye \'tī-ˌdī\ *noun* (circa 1939)
1 : TIE-DYEING
2 : a tie-dyed garment or fabric

tie–dyed *adjective* (1904)
: having patterns produced by tie-dyeing ⟨*tie-dyed* shirts⟩

tie–dye·ing *noun* (1904)
: a hand method of producing patterns in textiles by tying portions of the fabric or yarn so that they will not absorb the dye

tie–in \'tī-ˌin\ *noun* (1925)
1 : something that ties in, relates, or connects especially in a promotional campaign
2 : a book that inspired or was inspired by a motion picture or television program

tie in (1793)
transitive verb
: to bring into connection with something relevant: as **a :** to make the final connection of ⟨*tied in* the new branch pipeline⟩ **b :** to coordinate in such a manner as to produce balance and unity ⟨the illustrations were *tied in* with the text⟩ **c :** to use as a tie-in especially in advertising
intransitive verb
: to become tied in

tie–line \'tī-ˌlīn\ *noun* (1923)
: a telephone line that directly connects two or more private branch exchanges

tie·mann·ite \'tē-mə-ˌnīt\ *noun* [German *Tiemannit,* from W. *Tiemann* (died 1899) German scientist] (1868)
: a mineral that consists of mercuric selenide and occurs in dark gray or nearly black masses of metallic luster

tie·pin \'tī-ˌpin\ *noun* (1780)
: an ornamental straight pin that has usually a sheath for the point and is used to hold the ends of a necktie in place

^{1}**tier** \'tir\ *noun* [Middle French *tire* rank, from Old French — more at ATTIRE] (1569)
1 a : a row, rank, or layer of articles; *especially* **:** one of two or more rows, levels, or ranks arranged one above another **b :** a group of political or geographical divisions that form a row across the map ⟨the southern *tier* of states⟩
2 : CLASS, CATEGORY

^{2}**tier** (circa 1889)
transitive verb
: to place or arrange in tiers
intransitive verb
: to rise in tiers

^{3}**ti·er** \'tī(-ə)r\ *noun* (1633)
: one that ties

^{1}**tierce** \'tirs\ *variant of* TERCE

^{2}**tierce** *noun* [Middle English *terce, tierce,* from Middle French, from feminine of *terz,* adjective, third, from Latin *tertius* — more at THIRD] (15th century)
1 *obsolete* **:** THIRD 2
2 : a sequence of three playing cards of the same suit

tier·cel \'tir-səl\ *noun* [Middle English *tercel,* from Middle French, from (assumed) Vulgar Latin *tertiolus,* from diminutive of Latin *tertius* third] (14th century)
: a male hawk

tiered \'tird\ *adjective* (1807)
: having or arranged in tiers, rows, or layers — often used in combination ⟨triple-*tiered*⟩

tie–rod \'tī-ˌräd\ *noun* (1839)
: a rod (as of steel) used as a connecting member or brace

tie silk *noun* (circa 1915)
: a silk fabric of firm resilient pliable texture used for neckties and for blouses and accessories

tie tack *or* **tie tac** \-ˌtak\ *noun* (1954)
: an ornamented pin with a receiving button or clasp that is used to attach the two parts of a necktie together or to attach a necktie to a shirt

tie–up \'tī-ˌəp\ *noun* (1851)
1 a : a cow stable; *also* **:** a space for a single cow in a stable **b :** a mooring place for a boat
2 : a slowdown or stoppage of traffic, business, or operation (as by a mechanical breakdown)
3 : CONNECTION, ASSOCIATION ⟨helpful financial *tie-ups*⟩

tie up (1530)
transitive verb
1 : to attach, fasten, or bind securely; *also* **:** to wrap up and fasten
2 a : to connect closely **:** JOIN ⟨*tie up* the loose ends⟩ **b :** to cause to be linked so as to depend on or relate to something
3 a : to place or invest in such a manner as to make unavailable for other purposes ⟨their money was *tied up* in stocks⟩ **b :** to restrain from normal movement, operation, or progress ⟨traffic was *tied up* for miles⟩
4 a : to keep busy ⟨was *tied up* in conference all day⟩ **b :** to preempt the use of ⟨*tied up* the phone for an hour⟩
intransitive verb
1 : DOCK ⟨the ferry *ties up* at the south slip⟩
2 : to assume a definite relationship ⟨this *ties up* with what I told you before⟩

^{1}**tiff** \'tif\ *intransitive verb* [origin unknown] (1700)
: to have a petty quarrel

^{2}**tiff** *noun* (1754)
: a petty quarrel

tif·fa·ny \'ti-fə-nē\ *noun, plural* **-nies** [perhaps from obsolete French *tiphanie* Epiphany, from Late Latin *theophania,* from Late Greek, ultimately from Greek *theos* god + *phainein* to show] (1601)
1 : a sheer silk gauze formerly used for clothing and trimmings
2 : a plain-woven open-mesh cotton fabric (as cheesecloth)

Tif·fa·ny \'ti-fə-nē\ *adjective* (1936)
: being glass or an article of glass made by or in the manner of Louis C. Tiffany; *especially* **:** made of pieces of stained glass ⟨a *Tiffany*-style lamp⟩ ⟨a *Tiffany* window⟩

tif·fin \'ti-fən\ *noun* [probably alteration of *tiffing,* gerund of obsolete English *tiff* to eat between meals] (1800)
chiefly British **:** a light midday meal **:** LUNCHEON

ti·ger \'tī-gər\ *noun, plural* **tigers** [Middle English *tigre,* from Old English *tiger* & Old French *tigre,* both from Latin *tigris,* from Greek, probably of Iranian origin; akin to Avestan *tighra-* pointed; akin to Greek *stizein* to tattoo — more at STICK] (before 12th century)
1 *plural also* **tiger a :** a large Asian carnivorous mammal (*Panthera tigris*) of the cat family having a usually tawny coat transversely striped with black **b :** any of several large wildcats (as the jaguar or cougar) **c :** a domestic cat with striped pattern **d** *Australian* **:** TASMANIAN WOLF

tiger 1a

2 a : a fierce, daring, or aggressive person or quality ⟨aroused the *tiger* in him⟩ ⟨a *tiger* for work⟩ **b :** one (as a situation) that is formidable or impossible to control ⟨how the *tiger* of inflation can be tamed —J. A. Davenport⟩ — often used in the phrases *ride a tiger* and *have a tiger by the tail*
3 *British* **:** a groom in livery
— ti·ger·ish \-g(ə-)rish\ *adjective*
— ti·ger·ish·ly *adverb*

— **ti·ger·ish·ness** *noun*
— **ti·ger·like** \-gər-ˌlīk\ *adjective*

tiger beetle *noun* (1826)
: any of numerous active carnivorous beetles (family Cicindelidae) having larvae that tunnel in the soil

tiger cat *noun* (1699)
1 : any of various wildcats (as the serval, ocelot, or margay) of moderate size and variegated coloration
2 : a striped or sometimes blotched tabby cat

ti·ger·eye \ˈtī-gər-ˌī\ *or* **ti·ger's–eye** \-gərz-\ *noun* (circa 1891)
: a usually yellowish to grayish brown chatoyant stone that is much used for ornament and is a silicified crocidolite

tiger lily *noun* (1824)
: a common Asian garden lily (*Lilium lancifolium* synonym *L. tigrinum*) having nodding orange-colored flowers densely spotted with black; *also* **:** any of various lilies with similar flowers

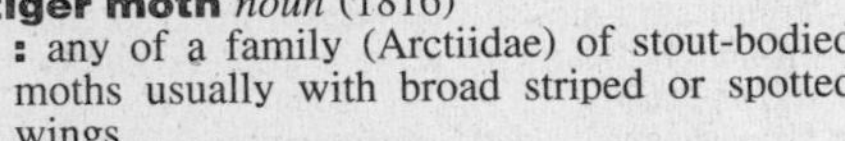
tiger lily

tiger maple *noun* (1952)
: maple lumber having a distinct irregularly striped pattern and much used for furniture

tiger moth *noun* (1816)
: any of a family (Arctiidae) of stout-bodied moths usually with broad striped or spotted wings

tiger salamander *noun* (circa 1909)
: a large widely distributed North American salamander (*Ambystoma tigrinum*) that is variably colored with contrasting blotches, spots, or bars

tiger shark *noun* (circa 1785)
: a large gray or brown stocky-bodied requiem shark (*Galeocerdo cuvieri*) that is nearly cosmopolitan especially in warm seas and can be dangerous to humans — see SHARK illustration

tiger shrimp *noun* (1979)
: a large shrimp (*Penaeus monodon*) of the Indian and Pacific oceans that is often farmed and widely sold as food

tiger swallowtail *noun* (circa 1890)
: a large widely distributed swallowtail (*Papilio glaucus*) of eastern North America that in the male is largely yellow with black margins and black stripes on the wings and in the female is usually similarly marked in the north but is very often all or mostly black in the south; *also* **:** any of several closely related black and yellow swallowtails (as *P. rutulus* of western North America)

[1]**tight** \ˈtīt\ *adjective* [alteration of earlier *thight* close set, dense, of Scandinavian origin; akin to Old Norse *thēttr* tight; akin to Middle High German *dīhte* thick, Sanskrit *tanakti* it causes to coagulate] (1507)
1 a : so close in structure as to prevent passage or escape (as of liquid, gas, or light) ⟨a *tight* ship⟩ ⟨a *tight* seal⟩ — compare LIGHTPROOF, WATERTIGHT **b :** fitting very close to the body ⟨*tight* jeans⟩; *also* **:** too snug ⟨*tight* shoes⟩ **c** (1) **:** closely packed **:** very full ⟨a *tight* bale of hay⟩ (2) **:** barely allowing time for completion ⟨a *tight* schedule⟩ ⟨*tight* deadlines⟩ **d :** having elements close together ⟨a *tight* formation⟩ ⟨a *tight* line of type⟩ **e :** allowing little or no room for free motion or movement ⟨a *tight* connection⟩ ⟨a *tight* crawl space⟩; *also* **:** having a small radius ⟨a *tight* turn⟩
2 a : strongly fixed or held **:** SECURE ⟨a *tight* jar lid⟩ ⟨a *tight* grip on the ladder⟩ **b** (1) **:** not slack or loose **:** TAUT ⟨kept the reins *tight*⟩ ⟨a *tight* knot⟩ ⟨a *tight* drumhead⟩; *also* **:** marked by firmness and muscle tone ⟨a *tight* stomach⟩ (2) **:** marked by unusual tension (as in the face or body) ⟨lips *tight* with anger⟩ ⟨a family *tight* with fear⟩
3 *chiefly dialect* **a :** CAPABLE, COMPETENT **b :** having a graceful or shapely form **:** COMELY
4 a : difficult to cope with ⟨in a *tight* spot financially⟩ **b :** relatively difficult to obtain ⟨money is *tight* just now⟩; *also* **:** characterized by such difficulty ⟨a *tight* job market⟩ **c :** not liberal in giving **:** STINGY ⟨*tight* with a penny⟩
5 : characterized by little difference in the relative positions of contestants with respect to final outcome **:** CLOSE ⟨a *tight* race for mayor⟩
6 : somewhat drunk
7 a : characterized by firmness or strictness in control or application or in attention to details ⟨*tight* zoning codes⟩ ⟨*tight* security⟩ ⟨ran a *tight* newsroom⟩ ⟨keeps a *tight* hand on her investments⟩ **b :** marked by control or discipline in expression or style **:** having little or no extraneous matter ⟨*tight* writing⟩ **c :** characterized by a polished style and precise arrangements in music performance
8 : having a close personal or working relationship **:** INTIMATE ⟨in *tight* with the boss⟩
9 : being such that the subject fills the frame ⟨filming a *tight* close-up⟩
— **tight·ly** *adverb*
— **tight·ness** *noun*

[2]**tight** *adverb* (1680)
1 : FAST, TIGHTLY, FIRMLY ⟨the door was shut *tight*⟩
2 : in a sound manner **:** SOUNDLY ⟨sleep *tight*⟩

tight·en \ˈtī-tᵊn\ *verb* **tight·ened; tight·en·ing** \ˈtīt-niŋ, ˈtī-tᵊn-iŋ\ (circa 1727)
transitive verb
: to make tight or tighter
intransitive verb
: to become tight or tighter
— **tight·en·er** \ˈtīt-nər, ˈtī-tᵊn-ər\ *noun*
— **tighten one's belt :** to practice strict economy

tight end *noun* (1962)
: an offensive football end who lines up within two yards of the tackle

tight·fist·ed \ˈtīt-ˈfis-təd\ *adjective* (1844)
: reluctant to part with money
— **tight·fist·ed·ness** \-nəs\ *noun*

tight–knit \-ˈnit\ *adjective* (1946)
: closely integrated and bound in love or friendship ⟨a *tight-knit* family⟩

tight–lipped \-ˈlipt\ *adjective* (1876)
1 : having the lips closed tight (as in determination)
2 : reluctant to speak **:** TACITURN

tight–mouthed \-ˈmau̇thd, -ˈmau̇tht\ *adjective* (1926)
: CLOSEMOUTHED

tight·rope \ˈtīt-ˌrōp\ *noun* (1801)
1 : a rope or wire stretched taut for acrobats to perform on
2 : a dangerously precarious situation — usually used in the phrase *walk a tightrope*

tights \ˈtīts\ *noun plural* (circa 1837)
: a skintight garment covering the body from the neck down or from the waist down; *also, British* **:** PANTY HOSE

tight·wad \ˈtīt-ˌwäd\ *noun* (1906)
: a close or miserly person

tight·wire \-ˌwīr\ *noun* (1928)
: a tightrope made of wire

ti·glon \ˈtī-glən\ *noun* [*tiger* + *lion*] (1942)
: a hybrid between a male tiger and a female lion

ti·gon \ˈtī-gən\ *noun* [*tiger* + *lion*] (1926)
: TIGLON

Ti·gre \ˈti-(ˌ)grā, ˈtē-; tē-ˈ\ *noun* (1878)
: a Semitic language of northern Ethiopia

ti·gress \ˈtī-grəs\ *noun* (1611)
: a female tiger; *also* **:** a tigerish woman

Ti·gri·nya \tə-ˈgrē-nyə\ *noun* (1878)
: a Semitic language of northern Ethiopia

tike *variant of* TYKE

ti·ki \ˈtē-kē\ *noun* [Maori & Marquesan, from *Tiki,* first man or creator of first man] (1777)
: a wood or stone image of a Polynesian supernatural power

til \ˈtil\ *noun* [Hindi, from Sanskrit *tila*] (1840)
: SESAME

'til *or* **til** \ˈtil\ *variant of* [1]TILL 2, [2]TILL

ti·la·pia \tə-ˈlä-pē-ə, -ˈlā-\ *noun* [New Latin, genus name] (1849)
: any of a genus (*Tilapia*) of African freshwater cichlid fishes often raised for food

til·bury \ˈtil-ˌber-ē, -b(ə-)rē\ *noun, plural* **-bur·ies** [*Tilbury,* 19th century English coach builder] (1814)
: a light 2-wheeled carriage

til·de \ˈtil-də\ *noun* [Spanish, from Medieval Latin *titulus* tittle] (circa 1864)
1 : a mark ˜ placed especially over the letter *n* (as in Spanish *señor* sir) to denote the sound \n^y\ or over vowels (as in Portuguese *irmã* sister) to indicate nasality
2 : the mark ˜ used to indicate negation in logic and the geometric relation "is similar to" in mathematics

[1]**tile** \ˈtī(ə)l\ *noun, often attributive* [Middle English, from Old English *tigele,* from Latin *tegula* tile; akin to Latin *tegere* to cover — more at THATCH] (before 12th century)
1 *plural* **tiles** *or* **tile a :** a flat or curved piece of fired clay, stone, or concrete used especially for roofs, floors, or walls and often for ornamental work **b :** a hollow or a semicircular and open earthenware or concrete piece used in constructing a drain **c :** a hollow building unit made of fired clay or of shale or gypsum
2 : TILING
3 : HAT; *especially* **:** a high silk hat
4 : a thin piece of resilient material (as cork, linoleum, or rubber) used especially for covering floors or walls
5 : a thin piece resembling a ceramic tile that usually bears a mark or letter and is used as a playing piece in a board game (as mah-jongg)
— **on the tiles** *British* **:** engaged in late-night carousing

[2]**tile** *transitive verb* **tiled; til·ing** (14th century)
1 : to cover with tiles
2 : to install drainage tile in
— **til·er** *noun*

tile·fish \ˈtī(ə)l-ˌfish\ *noun* [*tile-* modification of New Latin *Lopholatilus*] (1881)
: any of various marine bony fishes (family Malacanthidae) used as food; *especially* **:** a large fish (*Lopholatilus chamaeleonticeps*) of deep waters of the Atlantic and Gulf of Mexico with a fleshy appendage on the head and yellow spots on the upper body and some of its fins

til·ing \ˈtī-liŋ\ *noun* (15th century)
1 : the action or work of one who tiles
2 a : TILES **b :** a surface of tiles

[1]**till** \tᵊl, təl, ˈtil\ *preposition* [Middle English, from Old English *til;* akin to Old Norse *til* to, till, Old English *til* good] (before 12th century)
1 *chiefly Scottish* **:** TO
2 : UNTIL

[2]**till** *conjunction* (12th century)
: UNTIL

[3]**till** \ˈtil\ *transitive verb* [Middle English *tilien, tillen,* from Old English *tilian;* akin to Old English *til* good, suitable, Old High German *zil* goal] (13th century)
: to work by plowing, sowing, and raising crops **:** CULTIVATE
— **till·able** \ˈti-lə-bəl\ *adjective*

[4]**till** \ˈtil\ *noun* [Anglo-French *tylle*] (15th century)
1 a : a box, drawer, or tray in a receptacle (as a cabinet or chest) used especially for valuables **b :** a money drawer in a store or bank; *also* **:** CASH REGISTER

2 a : the money contained in a till **b :** a supply of especially ready money

[5]**till** \'til\ *noun* [origin unknown] (1842)
: unstratified glacial drift consisting of clay, sand, gravel, and boulders intermingled

till·age \'ti-lij\ *noun* (15th century)
1 : the operation of tilling land
2 : cultivated land

til·land·sia \tə-'lan(d)-zē-ə\ *noun* [New Latin, from Elias *Tillands* (died 1693) Finnish botanist] (1759)
: any of a large genus (*Tillandsia*) of chiefly epiphytic plants of the pineapple family native to tropical and subtropical America

[1]**til·ler** \'ti-lər\ *noun* [from (assumed) Middle English, from Old English *telgor, telgra* twig, shoot; akin to Old High German *zelga* twig, Old Irish *dlongaid* he splits] (before 12th century)
: STALK, SPROUT; *especially* **:** one from the base of a plant or from the axils of its lower leaves

[2]**til·ler** *intransitive verb* **til·lered; til·ler·ing** \'ti-lə-riŋ, 'til-riŋ\ (1677)
of a plant **:** to put forth tillers

[3]**till·er** \'ti-lər\ *noun* (13th century)
: one that tills **:** CULTIVATOR

[4]**til·ler** \'ti-lər\ *noun* [Middle English *tiler* stock of a crossbow, from Middle French *telier,* literally, beam of a loom, from Medieval Latin *telarium,* from Latin *tela* web — more at TOIL] (circa 1625)
: a lever used to turn the rudder of a boat from side to side; *broadly* **:** a device or system that plays a part in steering something

til·ler·man \'ti-lər-mən\ *noun* (circa 1934)
: one in charge of a tiller **:** STEERSMAN

Til·sit \'til-sət\ *also* **Til·sit·er** \-sə-tər\ *noun* [German *Tilsiter,* from *Tilsit* (now Sovetsk, Russia)] (circa 1932)
: a semisoft porous light yellow cheese with a flavor that ranges from mild to sharp

[1]**tilt** \'tilt\ *noun* [Middle English *teld, telte* tent, canopy, from Old English *teld;* akin to Old High German *zelt* tent] (15th century)
: a canopy for a wagon, boat, or stall

[2]**tilt** *transitive verb* (15th century)
: to cover or provide with a tilt

[3]**tilt** *noun* [[4]*tilt*] (1511)
1 a : a contest on horseback in which two combatants charging with lances or similar weapons try to unhorse each other **:** JOUST **b :** a tournament of tilts
2 a : DISPUTE, CONTENTION **b :** SPEED — used in the phrase *full tilt*
3 a : the act of tilting **:** the state or position of being tilted **b :** a sloping surface **c :** SLANT, BIAS ⟨a *tilt* toward military involvement⟩
4 : any of various contests resembling or suggesting tilting with lances
— tilt *adjective*

[4]**tilt** *verb* [Middle English *tulten, tilten* to cause to fall; akin to Swedish *tulta* to waddle] (1594)
transitive verb
1 : to cause to have an inclination
2 a : to point or thrust in or as if in a tilt ⟨*tilt* a lance⟩ **b :** to charge against ⟨*tilt* an adversary⟩
intransitive verb
1 a : to move or shift so as to lean or incline **:** SLANT **b :** to incline, tend, or become drawn toward an opinion, course of action, or one side of a controversy
2 a : to engage in a combat with lances **:** JOUST **b :** to make an impetuous attack ⟨*tilt* at social evils⟩
— tilt·able \'til-tə-bəl\ *adjective*
— tilt·er *noun*

tilth \'tilth\ *noun* [Middle English, from Old English, from *tilian* to till] (before 12th century)
1 : cultivated land **:** TILLAGE
2 : the state of aggregation of a soil especially in relation to its suitability for crop growth

tilt·me·ter \'tilt-,mē-tər\ *noun* (1932)
: an instrument to measure the tilting of the earth's surface

tilt–ro·tor \-,rō-tər\ *noun* (1980)
: an aircraft that has rotors at the end of each wing which can be oriented vertically for vertical takeoffs and landings, horizontally for forward flight, or to any position in between

tilt·yard \'tilt-,yärd\ *noun* (1528)
: a yard or place for tilting contests

tim·bal \'tim-bəl\ *noun* [French *timbale,* from Middle French, alteration of *tamballe,* modification of Old Spanish *atabal,* from Arabic *aṭ-ṭablative* the drum] (1680)
: KETTLEDRUM

tim·bale \'tim-bəl; tim-'bäl, tam-\ *noun* [French, literally, kettledrum] (1824)
1 : a creamy mixture (as of meat or vegetables) baked in a mold; *also* **:** the mold in which it is baked
2 : a small pastry shell filled with a cooked timbale mixture

[1]**tim·ber** \'tim-bər\ *noun* [Middle English, from Old English, building, wood; akin to Old High German *zimbar* wood, room, Greek *demein* to build, *domos* course of stones or bricks] (before 12th century)
1 a : growing trees or their wood **b** — used interjectionally to warn of a falling tree
2 : wood suitable for building or for carpentry
3 : MATERIAL, STUFF; *especially* **:** personal qualification for a particular position or status
4 a : a large squared or dressed piece of wood ready for use or forming part of a structure **b** *British* **:** LUMBER 2a **c :** a curving frame branching outward from the keel of a ship and bending upward in a vertical direction that is usually composed of several pieces united **:** RIB
— timber *adjective*

[2]**timber** *transitive verb* **tim·bered; tim·ber·ing** \-b(ə-)riŋ\ (before 12th century)
: to frame, cover, or support with timbers

tim·ber·doo·dle \,tim-bər-'dü-dᵊl\ *noun* [[1]*timber* + *doodle* cock] (1856)
: the American woodcock

tim·bered \'tim-bərd\ *adjective* (15th century)
1 : having walls framed by exposed timbers
2 : having a specified structure or constitution
3 : covered with growing timber **:** WOODED

tim·ber·head \'tim-bər-,hed\ *noun* (1794)
1 : the top end of a ship's timber used above the gunwale (as for belaying ropes)
2 : a bollard bolted to the deck where the end of a timber would come

timber hitch *noun* (circa 1815)
: a knot used to secure a line to a log or spar — see KNOT illustration

tim·ber·ing \'tim-b(ə-)riŋ\ *noun* (15th century)
: a set or arrangement of timbers

tim·ber·land \-bər-,land\ *noun* (1654)
: wooded land especially with marketable timber

tim·ber·line \-,līn\ *noun* (1867)
: the upper limit of arboreal growth in mountains or high latitudes — called also *tree line*

tim·ber·man \-mən\ *noun* (1889)
: LUMBERMAN

timber rattlesnake *noun* (1895)
: a widely distributed rattlesnake (*Crotalus horridus*) of the eastern U.S.

timber wolf *noun* (1860)
: GRAY WOLF

tim·ber·work \'tim-bər-,wərk\ *noun* (14th century)
: timber construction

tim·bre *also* **tim·ber** \'tam-bər, 'tim-; 'tam(brᵊ)\ *noun* [French, from Middle French, bell struck by a hammer, from Old French, drum, from Middle Greek *tymbanon* kettledrum, from Greek *tympanon* — more at TYMPANUM] (1849)
: the quality given to a sound by its overtones: as **a :** the resonance by which the ear recognizes and identifies a voiced speech sound **b :** the quality of tone distinctive of a particular singing voice or musical instrument
— tim·bral \'tam-brəl, 'tim-\ *adjective*

tim·brel \'tim-brəl\ *noun* [diminutive of obsolete English *timbre* tambourine, from Middle English, from Old French, drum] (circa 1520)
: a small hand drum or tambourine
— tim·brelled \-brəld\ *adjective*

[1]**time** \'tīm\ *noun* [Middle English, from Old English *tīma;* akin to Old Norse *tīmiles* time, Old English *tīd* — more at TIDE] (before 12th century)
1 a : the measured or measurable period during which an action, process, or condition exists or continues **:** DURATION **b :** a nonspatial continuum that is measured in terms of events which succeed one another from past through present to future **c :** LEISURE ⟨*time* for reading⟩
2 : the point or period when something occurs **:** OCCASION
3 a : an appointed, fixed, or customary moment or hour for something to happen, begin, or end ⟨arrived ahead of *time*⟩ **b :** an opportune or suitable moment ⟨decided it was *time* to retire⟩ — often used in the phrase *about time* ⟨about *time* for a change⟩
4 a : an historical period **:** AGE **b :** a division of geologic chronology **c :** conditions at present or at some specified period — usually used in plural ⟨*times* are hard⟩ ⟨move with the *times*⟩ **d :** the present time ⟨issues of the *time*⟩
5 a : LIFETIME **b :** a period of apprenticeship **c :** a term of military service **d :** a prison sentence
6 : SEASON ⟨very hot for this *time* of year⟩
7 a : rate of speed **:** TEMPO **b :** the grouping of the beats of music **:** RHYTHM
8 a : a moment, hour, day, or year as indicated by a clock or calendar ⟨what *time* is it⟩ **b :** any of various systems (as sidereal or solar) of reckoning time
9 a : one of a series of recurring instances or repeated actions ⟨you've been told many *times*⟩ **b** *plural* (1) **:** added or accumulated quantities or instances ⟨five *times* greater⟩ (2) **:** equal fractional parts of which an indicated number equal a comparatively greater quantity ⟨seven *times* smaller⟩ ⟨three *times* closer⟩ **c :** TURN ⟨three *times* at bat⟩
10 : finite as contrasted with infinite duration
11 : a person's experience during a specified period or on a particular occasion ⟨a good *time*⟩ ⟨a hard *time*⟩
12 a : the hours or days required to be occupied by one's work ⟨make up *time*⟩ ⟨on company *time*⟩ **b :** an hourly pay rate ⟨straight *time*⟩ **c :** wages paid at discharge or resignation ⟨pick up your *time* and get out⟩
13 a : the playing time of a game **b :** TIME-OUT
14 : a period during which something is used or available for use ⟨computer *time*⟩
— at the same time : NEVERTHELESS, YET ⟨slick and *at the same time* strangely unprofessional —Gerald Weaks⟩
— at times : at intervals **:** OCCASIONALLY
— for the time being : for the present
— from time to time : once in a while **:** OCCASIONALLY
— in no time : very quickly or soon
— in time 1 : sufficiently early **2 :** EVENTUALLY **3 :** in correct tempo ⟨learn to play *in time*⟩
— on time 1 a : at the appointed time **b :** on schedule **2 :** on the installment plan
— time and again : FREQUENTLY, REPEATEDLY

²time *verb* **timed; tim·ing** (14th century)
transitive verb
1 a : to arrange or set the time of **:** SCHEDULE **b :** to regulate (a watch) to keep correct time
2 : to set the tempo, speed, or duration of ⟨*timed* his leap perfectly —Neil Amdur⟩
3 : to cause to keep time with something
4 : to determine or record the time, duration, or rate of ⟨*time* a horse⟩
5 : to dispose (as a mechanical part) so that an action occurs at a desired instant or in a desired way
intransitive verb
: to keep or beat time

³time *adjective* (circa 1711)
1 a : of or relating to time **b :** recording time
2 : timed to ignite or explode at a specific moment ⟨a *time* bomb⟩
3 a : payable on a specified future day or a certain length of time after presentation for acceptance ⟨a *time* draft⟩ ⟨*time* deposits⟩ **b :** based on installment payments ⟨a *time* sale⟩

time and a half *noun* (1888)
: payment for work (as overtime or holiday work) at one and a half times the worker's regular wage rate

time bomb *noun* (1893)
1 : a bomb so made as to explode at a predetermined time
2 : something with a potentially dangerous or detrimental delayed reaction

time capsule *noun* (1938)
1 : a container holding historical records or objects representative of current culture that is deposited (as in a cornerstone) for preservation until discovery by some future age
2 : something resembling a time capsule ⟨sunken vessels are archaeological *time capsules* —Philip Trupp⟩

time card *noun* (circa 1891)
: a card used with a time clock to record an employee's starting and quitting times each day or on each job

time chart *noun* (circa 1830)
1 : a chart showing the standard times in various parts of the world with reference to a specified time at a specified place
2 : TIME LINE 1

time clock *noun* (1887)
: a clock that stamps starting and quitting times on an employee's time card

time–con·sum·ing \'tīm-kən-'sü-miŋ\ *adjective* (1890)
1 : using or taking up a great deal of time ⟨*time-consuming* chores⟩
2 : wasteful of time ⟨*time-consuming* tactics⟩

timed \'tīmd\ *adjective* (circa 1760)
1 : made to occur at or in a set time ⟨a *timed* explosion⟩
2 : done or taking place at a time of a specified sort ⟨an ill-*timed* arrival⟩

time dilation *noun* (1934)
: a slowing of time in accordance with the theory of relativity that occurs in a system in motion relative to an outside observer and that becomes apparent especially as the speed of the system approaches that of light — called also *time dilatation*

timed–re·lease \'tīmd-ri-'lēs\ *or* **time–re·lease** \'tīm-\ *adjective* (1966)
: consisting of or containing a drug that is released in small amounts over time (as by dissolution of a coating) usually in the gastrointestinal tract ⟨*timed-release* capsules⟩

time exposure *noun* (1893)
: exposure of a photographic film for a definite time usually of more than one half second; *also* **:** a photograph taken by such exposure

time frame *noun* (1964)
: a period of time especially with respect to some action or project

time–hon·ored \'tīm-,ä-nərd\ *adjective* (1593)
: honored because of age or long usage ⟨*time-honored* traditions⟩

time immemorial *noun* (1602)
1 : a time antedating a period legally fixed as the basis for a custom or right
2 : time so long past as to be indefinite in history or tradition — called also *time out of mind*

time·keep·er \'tīm-,kē-pər\ *noun* (1686)
1 : TIMEPIECE
2 : a clerk who keeps records of the time worked by employees
3 : a person appointed to mark and announce the time in an athletic game or contest
— time·keep·ing \-piŋ\ *noun*

time killer *noun* (1751)
1 : a person with free time
2 : something that passes the time **:** DIVERSION

time lag *noun* (1892)
: an interval of time between two related phenomena (as a cause and its effect)

time–lapse \'tīm-,laps\ *adjective* (1927)
: of, relating to, or constituting a motion picture made so that when projected a slow action (as the opening of a flower bud) appears to be speeded up

time·less \'tīm-ləs\ *adjective* (circa 1560)
1 *archaic* **:** PREMATURE, UNTIMELY
2 a : having no beginning or end **:** ETERNAL **b :** not restricted to a particular time or date ⟨the *timeless* themes of love, solitude, joy, and nature —*Writer*⟩
3 : not affected by time **:** AGELESS
— time·less·ly *adverb*
— time·less·ness *noun*

time line *noun* (1951)
1 : a table listing important events for successive years within a particular historical period
2 *usually* **time·line** \'tīm-,līn\ **:** a schedule of events and procedures **:** TIMETABLE 2

time lock *noun* (circa 1871)
: a lock controlled by clockwork to prevent its being opened before a set time

¹time·ly \'tīm-lē\ *adverb* (before 12th century)
1 *archaic* **:** EARLY, SOON
2 : in time **:** OPPORTUNELY ⟨the question was not . . . *timely* raised in the state court —W. O. Douglas⟩

²timely *adjective* **time·li·er; -est** (13th century)
1 : coming early or at the right time **:** OPPORTUNE
2 : appropriate or adapted to the times or the occasion ⟨a *timely* book⟩
— time·li·ness *noun*

time machine *noun* (1895)
: a hypothetical device that permits travel into the past and future

time–of–flight \'tīm-ə(v)-'flīt\ *adjective* (1945)
: of, relating to, being, or done with an instrument (as a mass spectrometer) that separates particles (as ions) according to the time required for them to traverse a tube of a certain length

time·ous \'tī-məs\ *adjective* (circa 1520)
chiefly British **:** TIMELY
— time·ous·ly *adverb*

time–out \'tīm-'au̇t\ *noun* (1926)
: a brief suspension of activity **:** BREAK; *especially* **:** a suspension of play in an athletic game

time out of mind (15th century)
: TIME IMMEMORIAL 2

time·piece \'tīm-,pēs\ *noun* (1765)
: a device (as a clock or watch) to measure or show progress of time; *especially* **:** one that does not chime

time·pleas·er \-,plē-zər\ *noun* (1601)
obsolete **:** TIMESERVER

tim·er \'tī-mər\ *noun* (1841)
: one that times: as **a :** TIMEPIECE; *especially* **:** a stopwatch for timing races **b :** TIMEKEEPER **c :** a device (as a clock) that indicates by a sound the end of an interval of time or that starts or stops a device at predetermined times

time reversal *noun* (1955)
: a formal operation in mathematical physics that reverses the order in which a sequence of events occurs

times \'tīmz\ *preposition* (14th century)
: multiplied by ⟨two *times* two is four⟩

time·sav·ing \'tīm-,sā-viŋ\ *adjective* (1865)
: intended or serving to expedite something ⟨*timesaving* kitchen appliances⟩
— time–sav·er \-,sā-vər\ *noun*

time·scale \-,skā(ə)l\ *noun* (1890)
: an arrangement of events used as a measure of the relative or absolute duration or antiquity of a period of history or geologic or cosmic time

time series *noun* (1919)
: a set of data collected sequentially usually at fixed intervals of time

time·serv·er \-,sər-vər\ *noun* (1584)
: a person whose behavior is adjusted to the pattern of the times or to please superiors **:** TEMPORIZER

¹time·serv·ing \-viŋ\ *noun* (1621)
: the behavior or practice of a timeserver

²timeserving *adjective* (1630)
: marked by or revealing a lack of independence or integrity ⟨a mean, *timeserving* little man, grovelling odiously before the wealthy people —Peter Forster⟩

time–shar·ing \'tīm-,sher-iŋ, -,shar-\ *noun* (1953)
1 : simultaneous use of a central computer by many users at remote locations
2 *or* **time–share** \-,sher, -,shar\ **:** joint ownership or rental of a vacation lodging (as a condominium) by several persons with each occupying the premises in turn for short periods
— time–share *transitive verb*

time sheet *noun* (circa 1909)
1 : a sheet for recording the time of arrival and departure of workers and for recording the amount of time spent on each job
2 : a sheet for summarizing hours worked by each worker during a pay period

time signature *noun* (1875)
: a sign used in music to indicate meter and usually written as a fraction with the bottom number indicating the kind of note used as a unit of time and the top number indicating the number of units in each measure

times sign *noun* (1948)
: the symbol × used to indicate multiplication

time stamp *noun* (1892)
: a device for recording the date and time of day that letters or papers are received or sent out
— time–stamp *transitive verb*

time·ta·ble \'tīm-,tā-bəl\ *noun* (1838)
1 : a table of departure and arrival times of trains, buses, or airplanes
2 a : a schedule showing a planned order or sequence **b :** PROGRAM 3
— time–table *transitive verb*

time–test·ed \-,tes-təd\ *adjective* (1930)
: having effectiveness that has been proved over a long period of time ⟨*time-tested* methods⟩

time trial *noun* (circa 1949)
: a competitive event (as in auto racing) in which individuals are successively timed over a set course or distance

time warp *noun* (1954)
: an anomaly, discontinuity, or suspension held to occur in the progress of time

time·work \'tīm-,wərk\ *noun* (1829)
: work paid for at a standard rate for the hour or the day
— time·work·er \-,wər-kər\ *noun*

time·worn \-,wōrn, -,wȯrn\ *adjective* (1729)
1 : worn or impaired by time ⟨*timeworn* mansions⟩
2 a : AGE-OLD, ANCIENT ⟨*timeworn* procedures⟩ **b :** HACKNEYED, STALE ⟨a *timeworn* joke⟩

time zone *noun* (1892)
: a geographical region within which the same standard time is used

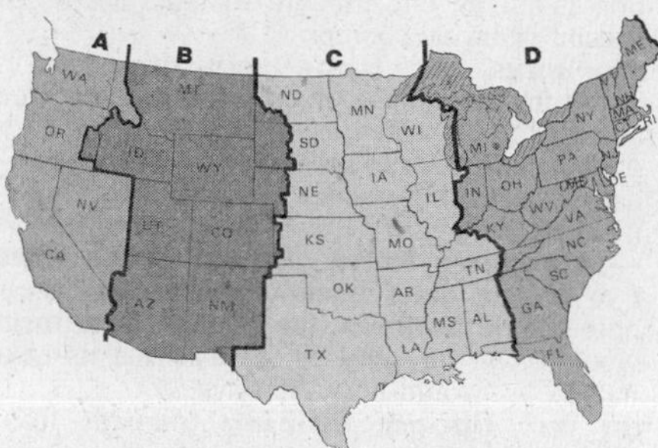

time zones in the conterminous United States: ***A*** **Pacific time,** ***B*** **mountain time,** ***C*** **central time,** ***D*** **eastern time**

tim·id \'ti-məd\ *adjective* [Latin *timidus,* from *timēre* to fear] (1549)
1 : lacking in courage or self-confidence ⟨a *timid* person⟩
2 : lacking in boldness or determination ⟨a *timid* policy⟩
— ti·mid·i·ty \tə-'mi-də-tē\ *noun*
— tim·id·ly \'ti-məd-lē\ *adverb*
— tim·id·ness *noun*

tim·ing \'tī-miŋ\ *noun* (1597)
1 : selection or the ability to select for maximum effect of the precise moment for beginning or doing something ⟨haven't got my *timing* down⟩
2 : observation and recording (as by a stopwatch) of the elapsed time of an act, action, or process

ti·moc·ra·cy \tī-'mä-krə-sē\ *noun* [Middle French *tymocracie,* from Medieval Latin *timocratia,* from Greek *timokratia,* from *timē* price, value, honor + *-kratia* -cracy; akin to Greek *tiein* to honor, Sanskrit *cāyati* he respects] (1586)
1 : government in which a certain amount of property is necessary for office
2 : government in which love of honor is the ruling principle
— ti·mo·crat·ic \,tī-mə-'kra-tik\ *or* **ti·mo·crat·i·cal** \-ti-kəl\ *adjective*

ti·mo·lol \'tī-mə-,lōl, -,lȯl\ *noun* [*tim-* (of unknown origin) + *-olol* (as in *propranolol*)] (1973)
: a beta-blocker $C_{13}H_{24}N_4O_3S$ used in the form of its maleate salt to treat glaucoma and to reduce the risk of second heart attacks

tim·o·rous \'ti-mə-rəs, 'tim-rəs\ *adjective* [Middle English, from Middle French *timoureus,* from Medieval Latin *timorosus,* from Latin *timor* fear, from *timēre* to fear] (15th century)
1 : of a timid disposition **:** FEARFUL
2 : expressing or suggesting timidity ⟨proceed with doubtful and *timorous* steps —Edward Gibbon⟩
— tim·o·rous·ly *adverb*
— tim·o·rous·ness *noun*

tim·o·thy \'ti-mə-thē\ *noun* [probably after *Timothy* Hanson, 18th century American farmer said to have introduced it from New England to the southern states] (1747)
: a European grass (*Phleum pratense*) that has long cylindrical spikes and is widely grown for hay

Tim·o·thy \'ti-mə-thē\ *noun* [Latin *Timotheus,* from Greek *Timotheos*]
1 : a disciple of the apostle Paul
2 : either of two letters written with regard to pastoral care in the early Church and included as books in the New Testament — see BIBLE table

tim·pa·ni \'tim-pə-nē\ *noun plural but singular or plural in construction* [Italian, plural of *timpano* kettledrum, from Latin *tympanum* drum — more at TYMPANUM] (circa 1854)
: a set of two or more kettledrums played by one performer in an orchestra or band

tim·pa·nist \-nist\ *noun* (1906)
: one who plays the timpani

[1]**tin** \'tin\ *noun* [Middle English, from Old English; akin to Old High German *zin* tin] (before 12th century)
1 : a soft faintly bluish white lustrous low-melting crystalline metallic element that is malleable and ductile at ordinary temperatures and that is used as a protective coating, in tinfoil, and in soft solders and alloys — see ELEMENT table
2 a : a box, can, pan, vessel, or a sheet made of tinplate; *broadly* **:** such a container of any metal (as aluminum) **b :** a metal container and its contents ⟨a *tin* of tomatoes⟩
— tin *adjective*
— tin·ful \-,fu̇l\ *noun*

[2]**tin** *transitive verb* **tinned; tin·ning** (14th century)
1 : to cover or plate with tin or a tin alloy
2 : to put up or pack in tins **:** CAN

tin·a·mou \'ti-nə-,mü\ *noun* [French, from Carib *tinamu*] (circa 1783)
: any of a family (Tinamidae) of South and Central American game birds that have a deeply keeled sternum and a rudimentary tail and that produce eggs with a surface resembling enamel

tin can *noun* (1770)
1 : a can made of tinplate; *broadly* **:** CAN 1c
2 *slang* **:** DESTROYER 2

[1]**tinct** \'tiŋ(k)t\ *adjective* [Latin *tinctus,* past participle] (1579)
: COLORED, TINGED

[2]**tinct** *noun* (1602)
: TINCTURE, TINGE

tinc·to·ri·al \tiŋ(k)-'tōr-ē-əl, -'tȯr-\ *adjective* [Latin *tinctorius,* from *tingere* to tinge] (1655)
: of or relating to colors or to dyeing or staining; *also* **:** imparting color
— tinc·to·ri·al·ly \-ē-ə-lē\ *adverb*

[1]**tinc·ture** \'tiŋ(k)-chər\ *noun* [Middle English, from Latin *tinctura* act of dyeing, from *tinctus,* past participle of *tingere* to tinge] (14th century)
1 a *archaic* **:** a substance that colors, dyes, or stains **b :** COLOR, TINT
2 a : a characteristic quality **:** CAST **b :** a slight admixture **:** TRACE
3 *obsolete* **:** an active principle or extract
4 : a heraldic metal, color, or fur
5 : a solution of a medicinal substance in an alcoholic menstruum

[2]**tincture** *transitive verb* **tinc·tured; tinc·tur·ing** \'tiŋ(k)-chə-riŋ, -shriŋ\ (1616)
1 : to tint or stain with a color **:** TINGE
2 a : to infuse or instill with a property or entity **:** IMPREGNATE **b :** to imbue with a quality **:** AFFECT ⟨writing *tinctured* with wit and wisdom⟩

tin·der \'tin-dər\ *noun* [Middle English, from Old English *tynder;* akin to Old High German *zuntra* tinder, Old English *tendan* to kindle] (before 12th century)
1 : a very flammable substance adaptable for use as kindling
2 : something that serves to incite or inflame

tin·der·box \'tin-dər-,bäks\ *noun* (1530)
1 a : a metal box for holding tinder and usually a flint and steel for striking a spark **b :** a highly inflammable object or place
2 : a potentially explosive place or situation

[1]**tine** \'tīn\ *noun* [Middle English *tind,* from Old English; akin to Old High German *zint* point, tine] (before 12th century)
1 : a slender pointed projecting part **:** PRONG
2 : a pointed branch of an antler
— tined \'tīnd\ *adjective*

[2]**tine** *verb* **tined** \'tīnd\ *or* **tint** \'tint\; **tin·ing** \'tī-niŋ\ [Middle English, of Scandinavian origin; akin to Old Norse *tȳna* to lose, destroy, *tjōn* injury, loss — more at TEEN] (13th century)
transitive verb
dialect British **:** LOSE
intransitive verb
dialect British **:** to become lost

tin·ea \'ti-nē-ə\ *noun* [Middle English, from Medieval Latin, from Latin, worm, moth] (14th century)
: any of several fungal diseases of the skin; *especially* **:** RINGWORM
— tin·e·al \-nē-əl\ *adjective*

tinea cru·ris \-'krůr-əs\ *noun* [New Latin, literally, tinea of the leg] (circa 1923)
: a fungal infection involving especially the groin and perineum

tin ear *noun* (1935)
: a deafened or insensitive ear

tin·foil \'tin-,fȯil\ *noun* (15th century)
: a paper-thin metal sheeting usually of aluminum or tin-lead alloy

ting *noun* [*ting,* verb, from Middle English *tingen,* of imitative origin] (1602)
: a high-pitched sound like that made by a light stroke on a crystal goblet
— ting *intransitive verb*

[1]**tinge** \'tinj\ *transitive verb* **tinged; tinge·ing** *or* **ting·ing** \'tin-jiŋ\ [Middle English, from Latin *tingere* to dip, moisten, tinge; akin to Greek *tengein* to moisten and probably to Old High German *dunkōn* to dip] (15th century)
1 a : to color with a slight shade or stain **:** TINT **b :** to affect or modify with a slight odor or taste
2 : to affect or modify in character

[2]**tinge** *noun* (1752)
1 : a slight staining or suffusing shade or color
2 : an affective or modifying property or influence **:** TOUCH

tin·gle \'tiŋ-gəl\ *intransitive verb* **tin·gled; tin·gling** \-g(ə-)liŋ\ [Middle English, alteration of *tinklen* to tinkle, tingle] (14th century)
1 a : to feel a ringing, stinging, prickling, or thrilling sensation **b :** to cause such a sensation
2 : TINKLE
— tingle *noun*
— tin·gling·ly \-g(ə-)liŋ-lē\ *adverb*
— tin·gly \-g(ə-)lē\ *adjective*

tin hat *noun* (1903)
: a metal helmet

tin·horn \'tin-,hȯrn\ *noun* (1885)
: one (as a gambler) who pretends to have money, ability, or influence

[1]**tin·ker** \'tiŋ-kər\ *noun* [Middle English *tinkere*] (14th century)
1 a : a usually itinerant mender of household utensils **b :** an unskillful mender **:** BUNGLER
2 *chiefly Irish* **:** GYPSY

[2]**tinker** *verb* **tin·kered; tin·ker·ing** \-k(ə-)riŋ\ (1592)
intransitive verb
: to work in the manner of a tinker; *especially* **:** to repair, adjust, or work with something in an unskilled or experimental manner **:** FIDDLE
transitive verb
: to repair, adjust, or experiment with
— tin·ker·er \-kər-ər\ *noun*

tinker's damn *also* **tinker's dam** \-'dam\ *noun* [probably from the tinkers' reputation for blasphemy] (1839)
: a minimum amount or degree (as of care) ⟨didn't give a *tinker's damn* about poetry —James Blish⟩

Tin·ker·toy \'tiŋ-kər-,tȯi\ *trademark*
— used for a construction toy of fitting parts

[1]**tin·kle** \'tiŋ-kəl\ *verb* **tin·kled; tin·kling** \-k(ə-)liŋ\ [Middle English, frequentative of *tinken* to tinkle, of imitative origin] (15th century)
intransitive verb

\ə\ **abut** \ᵊ\ **kitten** \ər\ **further** \a\ **ash** \ā\ **ace** \ä\ **mop, mar** \au̇\ **out** \ch\ **chin** \e\ **bet** \ē\ **easy** \g\ **go** \i\ **hit** \ī\ **ice** \j\ **job** \ŋ\ **sing** \ō\ **go** \ȯ\ **law** \ȯi\ **boy** \th\ **thin** \t͟h\ **the** \ü\ **loot** \u̇\ **foot** \y\ **yet** \zh\ **vision** *see also* Guide to Pronunciation

1 : to make or emit a tinkle or a sound suggestive of a tinkle
2 : URINATE
transitive verb
1 : to sound or make known (the time) by a tinkle
2 a : to cause to make a tinkle **b :** to produce by tinkling ⟨*tinkle* a tune⟩

[2]tinkle *noun* (1725)
1 : a jingling effect in verse or prose
2 : a series of short high ringing or clinking sounds

tin·kly \'tiŋ-k(ə-)lē\ *adjective* (1892)
: that tinkles **:** TINKLING

tin liz·zie \-'li-zē\ *noun* [from *Tin Lizzie,* nickname for the Model T Ford automobile] (1915)
: a small inexpensive early automobile

tin·ner \'ti-nər\ *noun* (1512)
1 : a tin miner
2 : TINSMITH

tin·ni·tus \'ti-nə-təs, tə-'nī-təs\ *noun* [Latin, ringing, tinnitus, from *tinnire* to ring, of imitative origin] (1843)
: a sensation of noise (as a ringing or roaring) that is caused by a bodily condition (as a disturbance of the auditory nerve or wax in the ear) and can usually be heard only by the one affected

tin·ny \'ti-nē\ *adjective* **tin·ni·er; -est** (1552)
1 : of, abounding in, or yielding tin
2 a : resembling tin **b :** LIGHT, CHEAP
3 : thin in tone ⟨a *tinny* voice⟩
— tin·ni·ly \'ti-nᵊl-ē\ *adverb*
— tin·ni·ness \'ti-nē-nəs\ *noun*

Tin Pan Alley *noun* (1908)
: a district that is a center for composers and publishers of popular music; *also* **:** the body of such composers and publishers

tin·plate \'tin-'plāt\ *noun* (1677)
: thin sheet iron or steel coated with tin

tin–plate *transitive verb* (1890)
: to plate or coat (as a metal sheet) with tin

tin–pot \'tin-'pät\ *adjective* (1838)
: SMALL-TIME

[1]tin·sel \'tin(t)-səl *also* 'tin-zəl\ *noun* [Middle French *estincelle, estancele, etincelle* spark, glitter, spangle — more at STENCIL] (1538)
1 : a thread, strip, or sheet of metal, paper, or plastic used to produce a glittering and sparkling appearance in fabrics, yarns, or decorations
2 : something superficially attractive or glamorous but of little real worth ⟨disfigured by no gaudy *tinsel* of rhetoric or declamation —Thomas Jefferson⟩

[2]tinsel *adjective* (1575)
1 : made of or covered with tinsel
2 a : cheaply gaudy **:** TAWDRY **b :** SPECIOUS ⟨*tinsel* promises⟩

[3]tinsel *transitive verb* **tin·seled** *or* **tin·selled; tin·sel·ing** *or* **tin·sel·ling** \'tin(t)-s(ə-)liŋ *also* 'tin-zə-liŋ\ (1594)
1 : to interweave, overlay, or adorn with or as if with tinsel
2 : to impart a specious brightness to

tin·sel·ly \'tin(t)-s(ə-)lē *also* 'tin-zə-lē\ *adjective* (1811)
: TINSEL

tin·smith \'tin-,smith\ *noun* (1812)
: a worker who makes or repairs things of sheet metal (as tinplate)
— tin·smith·ing \-,smi-thiŋ\ *noun*

tin·stone \'tin-,stōn\ *noun* (1602)
: CASSITERITE

[1]tint \'tint\ *noun* [alteration of earlier *tinct,* from Latin *tinctus* act of dyeing, from *tingere* to tinge] (1717)
1 a : a usually slight or pale coloration **:** HUE **b :** any of various lighter or darker shades of a color **:** TINGE
2 : a variation of a color produced by adding white to it and characterized by a low saturation with relatively high lightness
3 : a usually slight modifying quality or characteristic **:** TOUCH
4 : a shaded effect in engraving produced by fine parallel lines close together
5 : a panel of light color serving as background
6 : dye for the hair
— tint·er *noun*

[2]tint *transitive verb* (1791)
: to impart or apply a tint to **:** COLOR

tint·ing *noun* (circa 1841)
1 : the act or process of one that tints
2 : the engraved or colored tint produced by tinting

tin·tin·nab·u·lary \,tin-tə-'na-byə-,ler-ē\ *adjective* [Latin *tintinnabulum* bell] (1787)
: of, relating to, or characterized by bells or their sounds

tin·tin·nab·u·la·tion \,tin-tə-,na-byə-'lā-shən\ *noun* [Latin *tintinnabulum* bell, from *tintinnare* to ring, jingle, from *tinnire*] (1831)
1 : the ringing or sounding of bells
2 : a jingling or tinkling sound as if of bells

tint·less \'tint-ləs\ *adjective* (1789)
archaic **:** having no tints **:** lacking color

tin·type \-,tīp\ *noun* (1864)
: FERROTYPE 1

tin·ware \'tin-,war, -,wer\ *noun* (1758)
: articles and especially utensils made of tinplate

tin·work \-,wərk\ *noun* (circa 1934)
: work in tin

ti·ny \'tī-nē\ *adjective* **ti·ni·er; -est** [alteration of Middle English *tine*] (1598)
: very small or diminutive **:** MINUTE
synonym see SMALL
— ti·ni·ly \'tī-nᵊl-ē\ *adverb*
— ti·ni·ness \'tī-nē-nəs\ *noun*

[1]tip \'tip\ *verb* **tipped; tip·ping** [Middle English] (14th century)
transitive verb
1 : OVERTURN, UPSET — usually used with *over*
2 a : CANT, TILT **b :** to raise and tilt forward in salute ⟨*tipped* his hat⟩
intransitive verb
1 : to become tipped **:** TOPPLE
2 : LEAN, SLANT
— tip the scales 1 : to register weight ⟨*tips the scales* at 285 pounds⟩ **2 :** to shift the balance of power or influence ⟨*tipped the scales* in favor of a declaration of war —S. F. Bemis⟩

[2]tip *noun* (1673)
1 : the act or an instance of tipping **:** TILT
2 *chiefly British* **:** a place for depositing something (as rubbish) by dumping

[3]tip *transitive verb* **tipped; tip·ping** [Middle English *tipped* having a tip, from [4]*tip*] (14th century)
1 a : to furnish with a tip **b** (1) **:** to cover or adorn the tip of (2) **:** to blend (furs) for improved appearance by brushing the tips of the hair with dye
2 : to affix (an insert) in a book — often used with *in*
3 : to remove the ends of ⟨*tip* raspberries⟩

[4]tip *noun* [Middle English; akin to Middle High German *zipf* tip, Old English *tæppa* tap — more at TAP] (15th century)
1 : the usually pointed end of something
2 : a small piece or part serving as an end, cap, or point
— tipped \'tipt\ *adjective*
— on the tip of one's tongue 1 : about to be uttered ⟨it was *on the tip of my tongue* to tell him exactly what I thought⟩ **2 :** just eluding recall

[5]tip *noun* [Middle English *tippe;* akin to Low German *tippen* to tap] (15th century)
: a light touch or blow

[6]tip *verb* **tipped; tip·ping** (1567)
transitive verb
1 : to strike lightly **:** TAP
2 : to give (a baseball) a glancing blow
intransitive verb
: TIPTOE

[7]tip *noun* [origin unknown] (1567)
1 : a piece of advice or expert or authoritative information
2 : a piece of advance or confidential information given by one thought to have access to special or inside sources

[8]tip *transitive verb* **tipped; tip·ping** (1883)
1 : to impart a piece of information or advice about or to — often used with *off*
2 *chiefly British* **:** to mention as a likely candidate, prospective winner, or profitable investment **:** TOUT 4
— tip one's hand *also* **tip one's mitt :** to declare one's intentions or reveal one's opinions or resources ⟨the Justice Department wouldn't *tip its hand* by saying what its next move . . . would be —*Newsweek*⟩

[9]tip *verb* **tipped; tip·ping** [perhaps from [6]*tip*] (circa 1610)
transitive verb
1 : GIVE, PRESENT
2 : to give a gratuity to
intransitive verb
: to bestow a gratuity

[10]tip *noun* (1755)
: a gift or a sum of money tendered for a service performed or anticipated **:** GRATUITY

tip·cart \'tip-,kärt\ *noun* (circa 1877)
: a cart whose body can be tipped on the frame to empty its contents

tip·cat \-,kat\ *noun* [[6]*tip*] (1676)
: a game in which one player using a bat strikes lightly a tapered wooden peg and as it flies up strikes it again to drive it as far as possible while fielders try to recover it; *also* **:** the peg used in this game

ti·pi \'tē-(,)pē\ *variant of* TEPEE

tip–in \'tip-,in\ *noun* [[6]*tip*] (1948)
: a goal in basketball made by deflecting a rebound into the basket with the fingertips

[1]tip–off \'tip-,ȯf\ *noun* [[8]*tip*] (1901)
1 : WARNING, TIP
2 : a telltale sign

[2]tip–off *noun* [[6]*tip*] (1922)
: the act or an instance of putting the ball in play in basketball by a jump ball

tip of the iceberg [from an iceberg being mostly submerged] (1969)
: the earliest, most obvious, or most superficial manifestation of some phenomenon

tip·per \'ti-pər\ *noun* (1819)
: one that tips

tip·pet \'ti-pət\ *noun* [Middle English *tipet*] (14th century)
1 : a long hanging end of cloth attached to a sleeve, cap, or hood
2 : a shoulder cape of fur or cloth often with hanging ends
3 : a long black scarf worn over the robe by Anglican clergymen during morning and evening prayer

[1]tip·ple \'ti-pəl\ *verb* **tip·pled; tip·pling** \-p(ə-)liŋ\ [back-formation from obsolete *tippler* barkeeper] (1560)
intransitive verb
: to drink liquor especially by habit or to excess
transitive verb
: to drink (liquor) especially continuously in small amounts
— tip·pler \-p(ə-)lər\ *noun*

[2]tipple *noun* (1581)
: DRINK 1, 2

[3]tipple *noun* [English dialect *tipple* to tip over, frequentative of English [1]*tip*] (1880)
1 : a place where or an apparatus by which cars (as for coal) are loaded or emptied
2 : a coal-screening plant

tip·py \'ti-pē\ *adjective* **tip·pi·er; -est** (1886)
: liable to tip ⟨a *tippy* boat⟩

tip·staff \'tip-,staf\ *noun, plural* **tip·staves** \-,stavz, -,stāvz\ [obsolete *tipstaff* staff tipped with metal] (1535)
: an officer (as a constable or bailiff) who bears a staff

tip·ster \'tip-stər\ *noun* (1862)

: one who gives or sells tips especially for gambling or speculation

tip·sy \'tip-sē\ *adjective* **tip·si·er; -est** [[1]*tip* + *-sy* (as in *tricksy*)] (1577)
1 : unsteady, staggering, or foolish from the effects of liquor **:** FUDDLED
2 : UNSTEADY, ASKEW ⟨a *tipsy* angle⟩
— tip·si·ly \-sə-lē\ *adverb*
— tip·si·ness \-sē-nəs\ *noun*

[1]**tip·toe** \'tip-ˌtō, -'tō\ *noun* (14th century)
: the position of being balanced on the balls of the feet and toes with the heels raised — usually used with *on; also* **:** the ends of the toes
— on tiptoe : ALERT, AROUSED ⟨the contest of skill that puts one *on tiptoe* to win —*Deerfield (Wisconsin) Independent*⟩

[2]**tiptoe** *adverb* (1592)
: on or as if on tiptoe

[3]**tiptoe** *adjective* (1593)
1 : standing or walking on or as if on tiptoe
2 : CAUTIOUS, STEALTHY

[4]**tiptoe** *intransitive verb* **tip·toed; tip·toe·ing** (circa 1661)
1 : to stand or raise oneself on tiptoe
2 : to walk or proceed quietly or cautiously on or as if on tiptoe

[1]**tip–top** \'tip-'täp, -ˌtäp\ *noun* [[4]*tip* + *top*] (1702)
: the highest point

[2]**tip–top** *adjective* (1722)
: EXCELLENT, FIRST-RATE ⟨*tip-top* working conditions⟩

[3]**tip–top** *adverb* (1882)
: very well

ti·rade \'tī-ˌrād *also* ti-'\ *noun* [French, shot, tirade, from Middle French, from Old Italian *tirata,* from *tirare* to draw, shoot] (1802)
: a protracted speech usually marked by intemperate, vituperative, or harshly censorious language

tir·a·mi·su \ˌtir-ə-'mē-(ˌ)sü, -'mi-; -mē-'sü\ *noun* [Italian *tiramisù,* from *tirami su!*, literally, pull me up!] (1982)
: a dessert made with ladyfingers, mascarpone, chocolate, and espresso

[1]**tire** \'tīr\ *verb* **tired; tir·ing** [Middle English *tyren,* from Old English *tēorian, tȳrian*] (before 12th century)
intransitive verb
: to become weary
transitive verb
1 : to exhaust or greatly decrease the physical strength of **:** FATIGUE
2 : to wear out the patience of **:** BORE ☆

[2]**tire** *noun* [Middle English, short for *attire*] (14th century)
1 *obsolete* **:** ATTIRE
2 *archaic* **:** a woman's headband or hair ornament

[3]**tire** *transitive verb* **tired; tir·ing** (14th century)
1 *obsolete* **:** ATTIRE
2 *archaic* **:** to adorn (the hair) with an ornament

[4]**tire** *noun, often attributive* [Middle English, probably from [2]*tire*] (15th century)
1 : a metal hoop forming the tread of a wheel
2 : a rubber cushion that fits around a wheel (as of an automobile) and usually contains compressed air

tired \'tīrd\ *adjective* (15th century)
1 : drained of strength and energy **:** fatigued often to the point of exhaustion
2 : obviously worn by hard use **:** RUN-DOWN
3 : TRITE, HACKNEYED
— tired·ly *adverb*
— tired·ness *noun*

tire·less \'tīr-ləs\ *adjective* (1591)
: seemingly incapable of tiring **:** INDEFATIGABLE
— tire·less·ly *adverb*
— tire·less·ness *noun*

Ti·re·si·as \tī-'rē-sē-əs, -zē-\ *noun* [Latin, from Greek *Teiresias*]
: a blind seer of Thebes who in one Greek myth is changed into a woman for several years and then changed back to a man

tire·some \'tīr-səm\ *adjective* (circa 1520)
: WEARISOME, TEDIOUS
— tire·some·ly *adverb*
— tire·some·ness *noun*

tir·ing–house \'tī-riŋ-ˌhau̇s\ *noun* [[3]*tire*] (1590)
: a section of a theater reserved for the actors and used especially for dressing for stage entrances

tir·ing–room \-ˌrüm, -ˌru̇m\ *noun* [[3]*tire*] (1623)
: a dressing room especially in a theater

tirl \'tərl\ *verb* [alteration of [1]*trill*] (circa 1550)
intransitive verb
chiefly Scottish **:** to make a rattling sound (as with a door latch)
transitive verb
chiefly Scottish **:** TWIRL

tiro *chiefly British variant of* TYRO

'tis \'tiz, (ˌ)tiz, təz\ [contraction] (15th century)
: it is

ti·sane \ti-'zan, -'zän\ *noun* [Middle English, from Middle French, from Latin *ptisana,* from Greek *ptisanē,* literally, crushed barley, from *ptissein* to crush — more at PESTLE] (14th century)
: an infusion (as of dried herbs) used as a beverage or for medicinal effects

Tish·ah–b'Ab \'ti-shə-ˌbäv, -ˌbȯv\ *noun* [Hebrew *tish'āh bĕ Ābh* ninth in Ab] (circa 1902)
: a Jewish holiday observed with fasting on the 9th of Ab in commemoration of the destruction of the temples at Jerusalem

Tish·ri \'tish-rē\ *noun* [Hebrew *tishrī*] (circa 1771)
: the 1st month of the civil year or the 7th month of the ecclesiastical year in the Jewish calendar — see MONTH table

tis·sue \'ti-(ˌ)shü, *chiefly British* 'tis-(ˌ)yü\ *noun* [Middle English *tissu,* a rich fabric, from Old French, from past participle of *tistre* to weave, from Latin *texere* — more at TECHNICAL] (1563)
1 a : a fine lightweight often sheer fabric **b :** MESH, NETWORK, WEB ⟨a *tissue* of lies⟩
2 : a piece of soft absorbent tissue paper used especially as a handkerchief or for removing cosmetics
3 : an aggregate of cells usually of a particular kind together with their intercellular substance that form one of the structural materials of a plant or an animal
— tis·su·ey \'ti-shə-wē\ *adjective*

tissue culture *noun* (1912)
: the process or technique of making body tissue grow in a culture medium outside the organism; *also* **:** a culture of tissue (as epithelium)

tissue fluid *noun* (1900)
: a fluid that permeates the spaces between individual cells, that is in osmotic contact with the blood and lymph, and that serves in interstitial transport of nutrients and waste

tissue paper *noun* (1777)
: a thin gauzy paper used especially for protecting something (as by covering or wrapping)

tissue plasminogen activator *noun* (1981)
: a clot-dissolving enzyme with an affinity for fibrin that is produced naturally in blood vessel linings and is used in a genetically engineered form to prevent damage to heart muscle following a heart attack — abbreviation *TPA*

tis·su·lar \'ti-shə-lər\ *adjective* [*tissue* + *-lar* (as in *cellular*)] (circa 1935)
: of, relating to, or affecting organismic tissue ⟨*tissular* grafts⟩ ⟨*tissular* lesions⟩

[1]**tit** \'tit\ *noun* [Middle English, from Old English — more at TEAT] (before 12th century)
1 : TEAT
2 : BREAST — usually used in plural; usually considered vulgar

[2]**tit** *noun* (circa 1706)
: TITMOUSE; *broadly* **:** any of various small plump often long-tailed birds

ti·tan \'tī-tᵊn\ *noun* [Greek] (circa 1741)
1 *capitalized* **:** any of a family of giants born of Uranus and Gaea and ruling the earth until overthrown by the Olympian gods
2 : one that is gigantic in size or power **:** one that stands out for greatness of achievement

titan- *or* **titani-** *combining form* [New Latin *titanium*]
: titanium ⟨*titan*ate⟩

ti·ta·nate \'tī-tᵊn-ˌāt\ *noun* (1839)
1 : any of various multiple oxides of titanium dioxide with other metallic oxides
2 : a titanium ester of the general formula $Ti(OR)_4$

ti·tan·ess \'tī-tᵊn-əs\ *noun, often capitalized* (1596)
: a female titan

ti·ta·nia \tī-'tā-nē-ə, tə-, -'tān-yə *also* -'ta-nē-ə *or* -'tan-yə\ *noun* [New Latin] (1922)
: TITANIUM DIOXIDE

Ti·ta·nia \tə-'tān-yə, -'tän-; tī-'tān-\ *noun*
: the wife of Oberon and queen of the fairies in Shakespeare's *A Midsummer Night's Dream*

ti·tan·ic \tī-'ta-nik *also* tə-\ *adjective* [Greek *titanikos* of the Titans] (1709)
: having great magnitude, force, or power **:** COLOSSAL
— ti·tan·i·cal·ly \-ni-k(ə-)lē\ *adverb*

ti·ta·nif·er·ous \ˌtī-tᵊn-'i-f(ə-)rəs\ *adjective* (circa 1828)
: containing or yielding titanium ⟨*titaniferous* minerals⟩

ti·tan·ism \'tī-tᵊn-ˌi-zəm\ *noun, often capitalized* [from the Titans' rebellion against their father Uranus] (1867)
: defiance of and revolt against social or artistic conventions

ti·ta·ni·um \tī-'tā-nē-əm, tə- *also* -'ta-nē-əm, -'tan-yəm\ *noun* [New Latin, from Greek *Titan*] (circa 1796)
: a silvery gray light strong metallic element found combined in ilmenite and rutile and used especially in alloys (as steel) and combined in refractory materials and in coatings — see ELEMENT table

titanium dioxide *noun* (1877)
: an oxide TiO_2 of titanium that occurs in rutile, anatase, or ilmenite and is used especially as a pigment

titanium white *noun* (1920)
: TITANIUM DIOXIDE; *also* **:** a brilliant white lead-free pigment consisting of titanium dioxide often together with barium sulfate and zinc oxide

tit·bit \'tit-ˌbit\ *variant of* TIDBIT

☆ SYNONYMS
Tire, weary, fatigue, exhaust, jade, fag mean to make or become unable or unwilling to continue. TIRE implies a draining of one's strength or patience ⟨the long ride *tired* us out⟩. WEARY stresses tiring until one is unable to endure more of the same thing ⟨*wearied* of the constant arguing⟩. FATIGUE suggests causing great lassitude through excessive strain or undue effort ⟨*fatigued* by the day's chores⟩. EXHAUST implies complete draining of strength by hard exertion ⟨shoveling snow *exhausted* him⟩. JADE suggests the loss of all freshness and eagerness ⟨*jaded* by overindulgence⟩. FAG implies a drooping with fatigue ⟨shoppers all *fagged* out by the Christmas rush⟩.

ti·ter \'tī-tər\ *noun* [French *titre* title, proportion of gold or silver in a coin, from Old French *title* inscription, title] (1888)
: the strength of a solution or the concentration of a substance (as an antibody) in solution as determined by titration

tit·fer \'tit-fər\ *noun* [by shortening & alteration from *tit for tat,* rhyming slang for *hat*] (circa 1930)
British **:** HAT

tit for tat \,tit-fər-'tat\ [alteration of earlier *tip for tap,* from *tip* (blow) + *for* + *tap*] (1556)
: an equivalent given in return (as for an injury) **:** retaliation in kind
— tit–for–tat *adjective*

tith·able \'tī-t͟hə-bəl\ *adjective* (15th century)
: subject or liable to payment of tithes

[1]tithe \'tīt͟h\ *verb* **tithed; tith·ing** [Middle English, from Old English *teogothian,* from *teogotha* tenth] (before 12th century)
transitive verb
1 : to pay or give a tenth part of especially for the support of the church
2 : to levy a tithe on
intransitive verb
: to give a tenth of one's income as a tithe

[2]tithe *noun* [Middle English, from Old English *teogotha* tenth; akin to Middle Low German *tegede* tenth, Old English *tīen* ten — more at TEN] (before 12th century)
1 : a tenth part of something paid as a voluntary contribution or as a tax especially for the support of a religious establishment
2 : the obligation represented by individual tithes
3 : TENTH; *broadly* **:** a small part
4 : a small tax or levy

tith·er \'tī-t͟hər\ *noun* (14th century)
1 : one that pays tithes
2 : one that collects or advocates the payment of tithes

tith·ing \'tī-t͟hiŋ\ *noun* [Middle English, from Old English *tēothung,* from *teogothian, tēothian* to tithe, take one tenth] (before 12th century)
: a small administrative division preserved in parts of England apparently originally made up of ten men with their families

ti·tho·nia \tə-'thō-nyə, tī-, -nē-ə\ *noun* [New Latin, probably from Latin *Tithonia,* poetical name of Aurora] (1940)
: any of a genus (*Tithonia*) of tall composite herbs or shrubs of Mexico and Central America that have flower heads resembling sunflowers and that are sometimes grown as ornamentals

[1]ti·ti \'tī-,tī\ *noun* [origin unknown] (1827)
1 : a tree (*Cliftonia monophylla* of the family Cyrillaceae) of the southeastern U.S. with leathery leaves and racemes of fragrant white flowers
2 : LEATHERWOOD 2

[2]ti·ti \ti-'tē\ *noun* [American Spanish *tití*] (1832)
: any of a genus (*Callicebus*) of small South American monkeys having long thick variably colored fur and a tail that is not prehensile

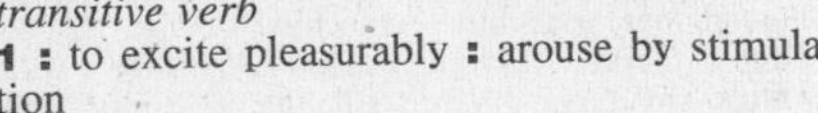
titi

ti·tian \'ti-shən\ *adjective, often capitalized* [*Titian* (Tiziano Vecellio)] (1896)
: of a brownish orange color

tit·il·late \'ti-tᵊl-,āt\ *verb* **-lat·ed; -lat·ing** [Latin *titillatus,* past participle of *titillare*] (1620)
transitive verb
1 : to excite pleasurably **:** arouse by stimulation
2 : TICKLE
intransitive verb
: to act as a stimulant to pleasurable excitement
— tit·il·la·tion \,ti-tᵊl-'ā-shən\ *noun*
— tit·il·la·tive \'ti-tᵊl-,ā-tiv\ *adjective*

tit·il·lat·ing \'ti-tᵊl-,ā-tiŋ\ *adjective* (circa 1714)
: pleasantly stimulating or exciting ⟨*titillating* reading⟩; *also* **:** EROTIC
— tit·il·lat·ing·ly \-tiŋ-lē\ *adverb*

tit·i·vate *or* **tit·ti·vate** \'ti-tə-,vāt\ *verb* **-vat·ed; -vat·ing** [perhaps from [1]*tidy* + reno*vate*] (1824)
transitive verb
: to make smart or spruce
intransitive verb
: SMARTEN, SPRUCE
— tit·i·va·tion \,ti-tə-'vā-shən\ *noun*

tit·lark \'tit-,lärk\ *noun* [*tit-* (as in *titmouse*) + *lark*] (1688)
: PIPIT

[1]ti·tle \'tī-tᵊl\ *noun* [Middle English, from Middle French, from Latin *titulus* inscription, title] (14th century)
1 a *obsolete* **:** INSCRIPTION **b :** written material introduced into a motion picture or television program to give credits, explain an action, or represent dialogue — usually used in plural
2 a : all the elements constituting legal ownership **b :** a legally just cause of exclusive possession **c :** the instrument (as a deed) that is evidence of a right
3 a : something that justifies or substantiates a claim **b :** an alleged or recognized right
4 a : a descriptive or general heading (as of a chapter in a book) **b :** the heading which names an act or statute **c :** the heading of a legal action or proceeding
5 a : the distinguishing name of a written, printed, or filmed production **b :** a similar distinguishing name of a musical composition or a work of art
6 : a descriptive name **:** APPELLATION
7 : a division of an instrument, book, or bill; *especially* **:** one larger than a section or article
8 a : an appellation of dignity, honor, distinction, or preeminence attached to a person or family by virtue of rank, office, precedent, privilege, attainment, or lands **b :** a person holding a title especially of nobility
9 : a written work as distinguished from a particular copy ⟨published 25 *titles* last year⟩
10 : CHAMPIONSHIP 1 ⟨won the batting *title*⟩

[2]title *transitive verb* **ti·tled; ti·tling** \'tīt-liŋ, 'tī-tᵊl-iŋ\ (14th century)
1 : to provide a title for
2 : to designate or call by a title **:** TERM, STYLE

[3]title *adjective* (1886)
: of or relating to a title: as **a :** having the same name as the title of a production ⟨did the *title* role in *Hamlet*⟩ **b :** having the same title as or providing the title for the collection or production of which it forms a part ⟨the *title* song⟩ **c :** of, relating to, or involving a championship ⟨a *title* match⟩ **d :** of, relating to, or used with the titles that introduce a motion picture or television program ⟨*title* music⟩

ti·tled \'tī-tᵊld\ *adjective* (1593)
: having a title especially of nobility

title deed *noun* (circa 1768)
: the deed constituting the evidence of a person's legal ownership

ti·tle·hold·er \'tī-tᵊl-,hōl-dər\ *noun* (1904)
: one that holds a title; *specifically* **:** CHAMPION

title page *noun* (1594)
: a page of a book bearing the title and usually the names of the author and publisher and the place and sometimes date of publication

ti·tlist \'tī-tᵊl-ist, 'tīt-list\ *noun* (1924)
: TITLEHOLDER

tit·mouse \'tit-,mau̇s\ *noun, plural* **tit·mice** \-,mīs\ [by folk etymology from Middle English *titmose,* from (assumed) Middle English *tit* any small object or creature + Middle English *mose* titmouse, from Old English *māse;* akin to Old High German *meisa* titmouse] (14th century)
: any of numerous small widely distributed insectivorous oscine birds (family Paridae and especially genus *Parus*) that are related to the chickadees and have small bills and relatively long tails

Ti·to·ism \'tē-(,)tō-,i-zəm\ *noun* (1949)
: the political, economic, and social policies associated with Tito; *specifically* **:** nationalistic policies and practices followed by a communist state or group independently of and often in opposition to the U.S.S.R.
— Ti·to·ist \-,ō-ist\ *noun or adjective*

ti·trant \'tī-trənt\ *noun* (1939)
: a substance (as a reagent solution of precisely known concentration) that is added in titration

ti·trate \'tī-,trāt\ *verb* **ti·trat·ed; ti·trat·ing** [*titer*] (circa 1859)
transitive verb
: to subject to titration
intransitive verb
: to perform titration
— ti·trat·able \-,trā-tə-bəl\ *adjective*
— ti·tra·tor \-,trā-tər\ *noun*

ti·tra·tion \tī-'trā-shən\ *noun* (circa 1859)
: a method or the process of determining the concentration of a dissolved substance in terms of the smallest amount of a reagent of known concentration required to bring about a given effect in reaction with a known volume of the test solution

ti·tre \'tī-tər\ *variant of* TITER

ti·tri·met·ric \,tī-trə-'me-trik\ *adjective* [*titration* + *-i-* + *-metric*] (1902)
: employing or determined by titration

tit–tat–toe \,ti-,ta(t)-'tō\ *variant of* TICKTACKTOE

tit·ter \'ti-tər\ *intransitive verb* [imitative] (circa 1619)
: to laugh in a nervous, affected, or partly suppressed manner **:** GIGGLE, SNICKER
— titter *noun*

tit·tie \'ti-tē\ *noun* [probably baby talk alteration of *sister*] (circa 1700)
chiefly Scottish **:** SISTER

tit·tle \'ti-tᵊl\ *noun* [Middle English *titel,* from Medieval Latin *titulus,* from Latin, title] (14th century)
1 : a point or small sign used as a diacritical mark in writing or printing
2 : a very small part

tit·tle–tat·tle \'ti-tᵊl-,ta-tᵊl\ *noun* [reduplication of [1]*tattle*] (circa 1529)
: GOSSIP, PRATTLE
— tittle–tattle *intransitive verb*

[1]tit·tup \'ti-təp\ *noun* [imitative of the sound of a horse's hooves] (1703)
: lively, gay, or restless behavior

[2]tittup *intransitive verb* **-tupped** *or* **-tuped; -tup·ping** *or* **-tup·ing** (1785)
: to move in a lively manner often with an exaggerated or affected action

[1]tit·u·lar \'ti-chə-lər, 'tich-lər\ *adjective* [Latin *titulus* title] (1611)
1 a : existing in title only; *especially* **:** bearing a title derived from a defunct ecclesiastical jurisdiction (as an episcopal see) ⟨a *titular* bishop⟩ **b :** having the title and usually the honors belonging to an office or dignity without the duties, functions, or responsibilities ⟨the *titular* head of a political party⟩
2 : bearing a title **:** TITLED
3 : of, relating to, or constituting a title ⟨the *titular* hero of the play⟩
— tit·u·lar·ly *adverb*

[2]titular *noun* (1613)
: a person holding a title

Ti·tus \'tī-təs\ *noun* [Late Latin, from Greek *Titos*]
1 : an early Christian convert who assisted Paul in his missionary work

2 : a letter written on the subject of pastoral care in the early Church and included as a book in the New Testament — see BIBLE table

Tiu \'tē-(,)ü\ *noun* [Old English *Tīw* — more at DEITY]
: an ancient Germanic god of war identified with Tyr

tiz·zy \'ti-zē\ *noun, plural* **tizzies** [origin unknown] (1935)
: a highly excited and distracted state of mind

TKO \,tē-,kā-'ō\ *noun* [*t*echnical *k*nock*o*ut] (1941)
: TECHNICAL KNOCKOUT

Tlin·git \'tliŋ-kət, -gət *also* 'kliŋ-\ *noun, plural* **Tlingit** *or* **Tlingits** (1865)
1 : a member of a group of American Indian peoples of the islands and coast of southern Alaska
2 : the language of the Tlingit peoples
— Tlingit *adjective*

T lymphocyte *noun* [*t*hymus-derived] (1972)
: T CELL

T–man \'tē-,man\ *noun* [*T*reasury *man*] (1937)
: a special agent of the U.S. Treasury Department

tme·sis \(tə-)'mē-səs\ *noun* [Late Latin, from Greek *tmēsis* act of cutting, from *temnein* to cut — more at TOME] (1550)
: separation of parts of a compound word by the intervention of one or more words (as *what place soever* for *whatsoever place*)

TNT \,tē-(,)en-'tē\ *noun* [*t*ri*n*itro*t*oluene] (1915)
: a flammable toxic compound $C_7H_5N_3O_6$ used as a high explosive and in chemical synthesis

¹to \tə, tü, 'tü\ *preposition* [Middle English, from Old English *tō;* akin to Old High German *zuo* to, Latin *do*nec as long as, until] (before 12th century)
1 a — used as a function word to indicate movement or an action or condition suggestive of movement toward a place, person, or thing reached ⟨drove *to* the city⟩ ⟨went back *to* the original idea⟩ **b** — used as a function word to indicate direction ⟨a mile *to* the south⟩ ⟨turned his back *to* the door⟩ ⟨a tendency *to* silliness⟩ **c** — used as a function word to indicate contact or proximity ⟨applied polish *to* the table⟩ ⟨put her hand *to* her heart⟩ **d** (1) — used as a function word to indicate the place or point that is the far limit ⟨100 miles *to* the nearest town⟩ (2) — used as a function word to indicate the limit of extent ⟨stripped *to* the waist⟩ **e** — used as a function word to indicate relative position ⟨perpendicular *to* the floor⟩
2 a — used as a function word to indicate purpose, intention, tendency, result, or end ⟨came *to* our aid⟩ ⟨drink *to* his health⟩ **b** — used as a function word to indicate the result of an action or a process ⟨broken all *to* pieces⟩ ⟨go *to* seed⟩ ⟨*to* their surprise, the train left on time⟩
3 — used as a function word to indicate position or relation in time: as **a :** BEFORE ⟨five minutes *to* five⟩ **b :** TILL ⟨from eight *to* five⟩ ⟨up *to* now⟩
4 — used as a function word to indicate addition, attachment, connection, belonging, possession, accompaniment, or response ⟨the key *to* the door⟩ ⟨danced *to* live music⟩ ⟨comes *to* her call⟩
5 — used as a function word (1) to indicate the extent or degree (as of completeness or accuracy) ⟨loyal *to* a man⟩ ⟨generous *to* a fault⟩ or the extent and result (as of an action or a condition) ⟨beaten *to* death⟩ (2) to indicate the last or an intermediate point of a series ⟨moderate *to* cool temperatures⟩
6 a — used as a function word (1) to indicate a relation to one that serves as a standard ⟨inferior *to* her earlier works⟩ (2) to indicate similarity, correspondence, dissimilarity, or proportion ⟨compared him *to* a god⟩ **b** — used as a function word to indicate agreement or conformity ⟨add salt *to* taste⟩ ⟨*to* my knowledge⟩ **c** — used as a function word to indicate a proportion in terms of numbers or quantities ⟨400 *to* the box⟩ ⟨odds of ten *to* one⟩
7 a — used as a function word (1) to indicate the application of an adjective or a noun ⟨agreeable *to* everyone⟩ ⟨attitude *to* friends⟩ ⟨title *to* the property⟩ (2) to indicate the relation of a verb to its complement or to a complementary element ⟨refers *to* the traditions⟩ ⟨refers us *to* the traditions⟩ (3) to indicate the receiver of an action or the one for which something is done or exists ⟨spoke *to* his mother⟩ ⟨gives a dollar *to* the man⟩ and often used with a reflexive pronoun to indicate exclusiveness (as of possession) or separateness ⟨had the house *to* themselves⟩ ⟨thought *to* herself⟩ **b** — used as a function word to indicate agency ⟨falls *to* his opponent's blows⟩
8 — used as a function word to indicate that the following verb is an infinitive ⟨wants *to* go⟩ and often used by itself at the end of a clause in place of an infinitive suggested by the preceding context ⟨knows more than she seems *to*⟩

²to \'tü\ *adverb* (before 12th century)
1 a — used as a function word to indicate direction toward ⟨feathers wrong end *to*⟩ ⟨run *to* and fro⟩ **b :** close to the wind ⟨the gale having gone over, we came *to* —R. H. Dana⟩
2 a : into contact especially with the frame — used of a door or a window ⟨the door snapped *to*⟩ **b** — used as a function word to indicate physical application or attachment ⟨he . . . hath set *to* his seal —John 3:33 (Authorized Version)⟩
3 — used as a function word to indicate application or attention ⟨will stand *to* —Shakespeare⟩
4 : to a state of consciousness or awareness ⟨brings her *to* with smelling salts⟩
5 : at hand **:** BY ⟨get to see 'em close *to* —Richard Llewellyn⟩

toad \'tōd\ *noun* [Middle English *tode,* from Old English *tāde, tādige*] (before 12th century)
1 : any of numerous anuran amphibians (especially family Bufonidae) that as compared with the related frogs are generally more terrestrial in habit though returning to water to lay their eggs, squatter and shorter in build and with weaker hind limbs, and rough, dry, and warty rather than smooth and moist of skin

toad 1

2 : a contemptible person or thing

toad·eat·er \-,ē-tər\ *noun* (circa 1572)
archaic **:** TOADY

toad·fish \-,fish\ *noun* (1704)
: any of a family (Batrachoididae) of chiefly marine bony fishes with jugular pelvic fins, a large thick head, a wide mouth, and usually scaleless slimy skin

toad·flax \-,flaks\ *noun* (1578)
: BUTTER-AND-EGGS; *also* **:** any of several related plants (especially genus *Linaria*) of the snapdragon family

toad·stone \-,stōn\ *noun* (1558)
: a stone or similar object held to have formed in the head or body of a toad and formerly often worn as a charm or antidote to poison

toad·stool \-,stül\ *noun* (14th century)
: a fungus having an umbrella-shaped pileus **:** MUSHROOM; *especially* **:** a poisonous or inedible one as distinguished from an edible mushroom

¹toady \'tō-dē\ *noun, plural* **toad·ies** [by shortening & alteration from *toadeater*] (1826)
: one who flatters in the hope of gaining favors **:** SYCOPHANT
synonym see PARASITE

²toady *intransitive verb* **toad·ied; toady·ing** (circa 1859)
: to behave as a toady **:** engage in sycophancy
synonym see FAWN
— toady·ism \-ē-,i-zəm\ *noun*

¹to–and–fro \,tü-ən-'frō\ *adjective* (1749)
: forward and backward ⟨*to-and-fro* motion⟩

²to–and–fro *noun* (1847)
: activity involving alternating movement in opposite directions ⟨the busy *to-and-fro* of the holiday shoppers⟩

to and fro *adverb* (14th century)
: from one place to another

¹toast \'tōst\ *verb* [Middle English *tosten,* from Middle French *toster,* from Late Latin *tostare* to roast, from Latin *tostus,* past participle of *torrēre* to dry, parch — more at THIRST] (14th century)
transitive verb
1 : to warm thoroughly
2 : to make (as bread) crisp, hot, and brown by heat
intransitive verb
: to become toasted; *especially* **:** to warm thoroughly

²toast *noun* (15th century)
1 a : sliced bread browned on both sides by heat **b :** food prepared with toasted bread
2 [from the use of pieces of spiced toast to flavor drinks] **a :** someone or something in honor of which persons usually drink **b :** one that is highly admired ⟨she's the *toast* of society⟩
3 [³*toast*] **:** an act of proposing or of drinking in honor of a toast
4 : a rhyming narrative poem existing in oral tradition among black Americans

³toast *transitive verb* (1640)
: to propose or drink to as a toast

toast·er \'tō-stər\ *noun* (1582)
: one that toasts; *especially* **:** an electrical appliance for toasting

toaster oven *noun* (1961)
: a portable electrical appliance that can function as an oven or a toaster

toast·mas·ter \'tōs(t)-,mas-tər\ *noun* (1749)
: one who presides at a banquet and introduces the after-dinner speakers

toast·mis·tress \-,mis-trəs\ *noun* (1921)
: a woman who acts as toastmaster

toasty \'tō-stē\ *adjective* **toast·i·er; -est** (1953)
: pleasantly or comfortably warm ⟨felt snug and *toasty* by the fire⟩

to·bac·co \tə-'ba-(,)kō\ *noun, plural* **-cos** [Spanish *tabaco,* probably from Taino, roll of tobacco leaves] (circa 1565)
1 : any of a genus (*Nicotiana*) of chiefly American plants of the nightshade family with viscid foliage and tubular flowers; *especially* **:** a tall erect annual tropical American herb (*N. tabacum*) cultivated for its leaves
2 : the leaves of cultivated tobacco prepared for use in smoking or chewing or as snuff
3 : manufactured products of tobacco (as cigars or cigarettes); *also* **:** smoking as a practice ⟨has sworn off *tobacco*⟩

tobacco 1

tobacco budworm *noun* (1918)
: a noctuid moth (*Heliothis virescens* synonym *Helicoverpa virescens*) whose small rusty often green-striped caterpillar feeds on buds and young leaves especially of tobacco and cotton

tobacco hornworm *noun* (circa 1909)
: a hawkmoth (*Manduca sexta*) whose large usually green larva is a hornworm that feeds on plants of the nightshade family

\ə\ **abut** \ᵊ\ **kitten** \ər\ **further** \a\ **ash** \ā\ **ace** \ä\ **mop, mar** \au̇\ **out** \ch\ **chin** \e\ **bet** \ē\ **easy** \g\ **go** \i\ **hit** \ī\ **ice** \j\ **job** \ŋ\ **sing** \ō\ **go** \ȯ\ **law** \ȯi\ **boy** \th\ **thin** \t͟h\ **the** \ü\ **loot** \u̇\ **foot** \y\ **yet** \zh\ **vision** *see also* Guide to Pronunciation

tobacco juice *noun* (1833)
: saliva colored brown by tobacco or snuff

tobacco mosaic *noun* (1914)
: any of a complex of virus diseases of plants of the nightshade family and especially of tobacco

to·bac·co·nist \tə-ˈba-kə-nist\ *noun* [irregular from *tobacco* + *-ist*] (1657)
: a dealer in tobacco especially at retail

tobacco road *noun, often T&R capitalized* [from *Tobacco Road,* novel (1932) by Erskine Caldwell and play (1933) by Jack Kirkland (died 1969) American playwright] (1937)
: a squalid poverty-stricken rural area or community

to–be \tə-ˈbē\ *adjective* (circa 1600)
: that is to be **:** FUTURE — usually used postpositively and often in combination ⟨a bride-*to-be*⟩

To·bi·as \tə-ˈbī-əs\ *noun* [Greek *Tobias*]
1 : a Jewish hero who with divine aid marries his kinswoman Sarah in spite of a jealous evil spirit and restores his father Tobit's sight
2 : TOBIT 2

To·bit \ˈtō-bət\ *noun* [Greek *Tōbit*]
1 : the elderly father of Tobias
2 : a book of Scripture included in the Roman Catholic canon of the Old Testament and in the Protestant Apocrypha — see BIBLE table

¹to·bog·gan \tə-ˈbä-gən\ *noun* [Canadian French *tobogan,* of Algonquian origin; akin to Micmac *tobâgun* drag made of skin] (circa 1820)
1 : a long flat-bottomed light sled made usually of thin boards curved up at one end with usually low handrails at the sides

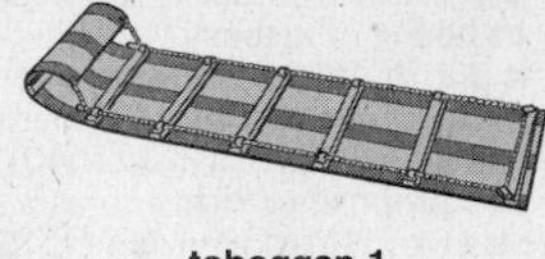
toboggan 1

2 : a downward course or a sharp decline

²toboggan *intransitive verb* (1846)
1 : to coast on or as if on a toboggan
2 : to decline suddenly and sharply
— to·bog·gan·er *noun*
— to·bog·gan·ist \-gə-nist\ *noun*

to·bog·gan·ing *noun* (1849)
: the act, art, or sport of riding a toboggan

to·by jug \ˈtō-bē-\ *noun, often capitalized* [*Toby,* nickname from the name *Tobias*] (1840)
: a small jug, pitcher, or mug shaped somewhat like a stout man with a cocked hat for the brim — called also *toby*

toc·ca·ta \tə-ˈkä-tə\ *noun* [Italian, from *toccare* to touch, from (assumed) Vulgar Latin — more at TOUCH] (circa 1724)
: a musical composition usually for organ or harpsichord in a free style and characterized by full chords, rapid runs, and high harmonies

To·char·i·an \tō-ˈkar-ē-ən, -ˈker-, -ˈkär-\ *noun* [Greek *Tocharoi*] (circa 1926)
1 a : a language of central Asia known from documents from the sixth to eighth centuries A.D. **b :** a branch of the Indo-European language family containing Tocharian — see INDO-EUROPEAN LANGUAGES table
2 : a member of a people of presumably European origin dwelling in central Asia during the first millennium of the Christian era

Tocharian A *noun* (1926)
: the eastern dialect of Tocharian — see INDO-EUROPEAN LANGUAGES table

Tocharian B *noun* (1926)
: the western dialect of Tocharian — see INDO-EUROPEAN LANGUAGES table

toch·er \ˈtä-kər\ *noun* [Middle English (Scots) *tochir,* from Scottish Gaelic *tochar*] (15th century)
chiefly Scottish **:** DOWRY 2, 3

to·coph·er·ol \tō-ˈkä-fə-ˌrȯl, -ˌrōl\ *noun* [International Scientific Vocabulary, ultimately from Greek *tokos* childbirth, offspring (akin to Greek *tiktein* to beget) + *pherein* to carry, bear — more at THANE, BEAR] (1936)
: any of several fat-soluble oily phenolic compounds with varying degrees of antioxidant vitamin E activity; *especially* **:** one $C_{29}H_{50}O_2$ of high vitamin E potency obtained from germ oils or by synthesis

toc·sin \ˈtäk-sən\ *noun* [Middle French *toquassen,* from Old Provençal *tocasenh,* from *tocar* to touch, ring a bell (from assumed Vulgar Latin *toccare*) + *senh* sign, bell, from Medieval Latin & Latin *signum;* Medieval Latin, bell, from Late Latin, ringing of a bell, from Latin, mark, sign — more at TOUCH, SIGN] (1586)
1 : an alarm bell or the ringing of it
2 : a warning signal

¹tod \ˈtäd\ *noun* [Middle English] (12th century)
chiefly Scottish **:** FOX

²tod *noun* [Middle English *todd, todde;* probably akin to Old High German *zotta* tuft of hair] (15th century)
1 *archaic* **:** any of various units of weight for wool; *especially* **:** one equal to 28 pounds (13 kilograms)
2 *British* **:** a bushy clump (as of ivy)

¹to·day \tə-ˈdā\ *adverb* (before 12th century)
1 : on or for this day
2 : at the present time

²today *noun* (1535)
: the present day, time, or age ⟨*today's* youth⟩

³today *adjective* (1966)
: of or characteristic of today **:** NOW

tod·dle \ˈtä-dᵊl\ *intransitive verb* **tod·dled; tod·dling** \ˈtäd-liŋ, ˈtä-dᵊl-iŋ\ [origin unknown] (circa 1600)
1 : to walk with short tottering steps in the manner of a young child
2 : to take a stroll **:** SAUNTER
— toddle *noun*

tod·dler \ˈtäd-lər, ˈtä-dᵊl-ər\ *noun* (1793)
: one that toddles; *especially* **:** a young child
— tod·dler·hood \-ˌhu̇d\ *noun*

tod·dy \ˈtä-dē\ *noun, plural* **toddies** [Hindi *tāṛī* juice of the palmyra palm, from *tār* palmyra palm, from Sanskrit *tāla*] (circa 1610)
1 : the fresh or fermented sap of various chiefly East Indian palms
2 : a usually hot drink consisting of liquor (as rum), water, sugar, and spices

to–do \tə-ˈdü\ *noun, plural* **to–dos** \-ˈdüz\ (circa 1576)
: BUSTLE, STIR, FUSS

¹toe \ˈtō\ *noun* [Middle English *to,* from Old English *tā;* akin to Old High German *zēha* toe] (before 12th century)
1 a (1) **:** one of the terminal members of a vertebrate's foot (2) **:** the fore end of a foot or hoof **b :** a terminal segment of a limb of an invertebrate **c :** the forepart of something worn on the foot ⟨the *toe* of a boot⟩
2 : a part that by its position or form is felt to resemble a toe ⟨the *toe* of Italy⟩: as **a :** a lateral projection at one end or between the ends of a piece (as a rod or bolt) **b :** the lowest part (as of an embankment, dam, or cliff)
3 : TOE DANCE
— toe·less \-ləs\ *adjective*
— on one's toes : ALERT 1
— toe to toe : facing one another

²toe *verb* **toed; toe·ing** (1607)
transitive verb
1 : to furnish with a toe ⟨*toe* a sock⟩
2 : to touch, reach, or drive with the toe ⟨*toe* a football⟩
3 : to drive (as a nail) obliquely; *also* **:** to clinch or fasten by or with nails or rods so driven
intransitive verb
1 : TIPTOE
2 : to stand, walk, or be placed so that the toes assume an indicated position or direction ⟨*toe* in⟩
— toe the line *or* **toe the mark :** to conform rigorously to a rule or standard

toea \ˈtȯi-ə\ *noun, plural* **toea** [Hiri Motu (pidgin of Papua New Guinea based on Motu, an Austronesian language), a kind of shell] (1975)
— see *kina* at MONEY table

toe cap *noun* (1797)
: a piece of leather covering the toe of a shoe and reinforcing or decorating it

toed \ˈtōd\ *adjective* [¹*toe*] (1611)
1 : having a toe or toes especially of a specified kind or number — usually used in combination ⟨five-*toed*⟩ ⟨round-*toed* shoes⟩
2 [from past participle of ²*toe*] **:** driven obliquely ⟨a *toed* nail⟩; *also* **:** secured by diagonal or oblique nailing

toe dance *noun* (1898)
: a dance executed on the tips of the toes by means of a ballet slipper with a reinforced toe
— toe–dance *intransitive verb*
— toe dancer *noun*
— toe dancing *noun*

toe·hold \ˈtō-ˌhōld\ *noun* (1880)
1 a : a hold or place of support for the toes (as in climbing) **b** (1) **:** a means of progressing (as in surmounting barriers) (2) **:** a slight footing ⟨used his money to get a *toehold,* then a foothold, then a near stranglehold on the political economy —R. W. Armstrong⟩
2 : a wrestling hold in which the aggressor bends or twists his opponent's foot

toe–in \ˈtō-ˌin\ *noun* (1928)
1 : CAMBER 3
2 : adjustment of the front wheels of an automotive vehicle so that they are closer together at the front than at the back

¹toe·nail \ˈtō-ˌnāl, -ˈnā(ə)l\ *noun* (1841)
: a nail of a toe

²toenail *transitive verb* (1900)
: to fasten by toed nails **:** TOE

toe·piece \ˈtō-ˌpēs\ *noun* (1879)
: a piece designed to form a toe (as of a shoe) or cover the toes of the foot

toe·plate \-ˌplāt\ *noun* (1894)
: a tab attached to the toe of a shoe (as to prevent wear due to heavy use)

toe–to–toe *adjective or adverb* (circa 1942)
: slugging it out at or as if at close range ⟨a *toe-to-toe* confrontation over the new policy⟩

toff \ˈtäf\ *noun* [probably alteration of *tuft* titled college student] (1851)
chiefly British **:** DANDY, SWELL

tof·fee *or* **tof·fy** \ˈtȯ-fē, ˈtä-\ *noun, plural* **toffees** *or* **toffies** [alteration of *taffy*] (circa 1825)
: candy of brittle but tender texture made by boiling sugar and butter together

tof·fee–nosed \-ˈnōzd\ *adjective* (circa 1925)
chiefly British **:** STUCK-UP

toft \ˈtȯft, ˈtäft\ *noun* [Middle English, from Old English, from Old Norse *topt;* probably akin to Greek *dapedon* floor, *demein* to build, *pedon* ground — more at TIMBER, PED] (before 12th century)
British **:** a site for a dwelling and its outbuildings; *also* **:** an entire holding comprising a homestead and additional land

to·fu \ˈtō-(ˌ)fü\ *noun* [Japanese *tōfu*] (1771)
: BEAN CURD

tog \ˈtäg, ˈtȯg\ *transitive verb* **togged; tog·ging** [*togs*] (circa 1785)
: to dress especially in fine clothing — usually used with *up* or *out*

to·ga \ˈtō-gə\ *noun* [Latin; akin to Latin *tegere* to cover — more at THATCH] (1600)
: the loose outer garment worn in public by citizens of ancient Rome; *also* **:** a similar loose wrap or a professional, official, or academic gown
— to·gaed \-gəd\ *adjective*

toga

to·ga vi·ri·lis \ˌtō-gə-və-ˈrē-ləs, -ˈri-\ *noun, plural* **to·gae vi·ri·les** \ˈtō-ˌgī-və-ˈrē-ˌlās, -ˈri-\ [Latin, men's toga] (1600)
: the white toga of manhood assumed by boys of ancient Rome at age 15

[1]to·geth·er \tə-ˈge-t͟hər\ *adverb* [Middle English *togedere,* from Old English *togædere,* from *tō* to + *gædere* together; akin to Middle High German *gater* together, Old English *gaderian* to gather] (before 12th century)
1 a : in or into one place, mass, collection, or group ⟨the men get *together* every Thursday for poker⟩ **b :** in a body **:** as a group ⟨students and faculty *together* presented the petition⟩
2 a : in or into contact (as connection, collision, or union) ⟨mix these ingredients *together*⟩ **b :** in or into association or relationship ⟨colors that go well *together*⟩ ⟨went to school *together*⟩
3 a : at one time **:** SIMULTANEOUSLY ⟨events that happened *together*⟩ **b :** in succession **:** without intermission ⟨was depressed for days *together*⟩
4 a : by combined action **:** JOINTLY ⟨*together* we forced the door⟩ **b :** in or into agreement or harmony ⟨the soloist and the orchestra weren't quite *together*⟩ **c :** in or into a unified or coherent structure or an integrated whole ⟨can't even put a simple sentence *together*⟩ ⟨pull yourself *together*⟩
5 a : with each other — used as an intensive after certain verbs ⟨join *together*⟩ ⟨add *together*⟩ **b :** as a unit **:** in the aggregate ⟨these arguments taken *together* make a convincing case⟩ **c :** considered as a whole **:** counted or summed up ⟨all *together*, there were 21 entries⟩
— to·geth·er·ness *noun*
— together with : in addition to **:** in association with

[2]together *adjective* (1963)
1 : appropriately prepared, organized, or balanced
2 : composed in mind or manner **:** SELF-POSSESSED ⟨a warm, sensitive, reasonably *together* girl —*East Village Other*⟩

Tog·gen·burg \ˈtä-gən-ˌbərg\ *noun* [*Toggenburg,* district in northeastern Switzerland] (1886)
: any of a breed of brown hornless dairy goats of Swiss origin with white stripes on the face

tog·gery \ˈtä-g(ə-)rē, ˈtó-\ *noun* [*togs*] (circa 1811)
: CLOTHING

[1]tog·gle \ˈtä-gəl\ *noun* [origin unknown] (circa 1775)
1 : a piece or device for holding or securing: as **a :** a pin inserted in a nautical knot to make it more secure or easier to slip **b :** a crosspiece attached to the end of or to a loop in something (as a chain, rope, line, strap, or belt) usually to prevent slipping, to serve in twisting or tightening, or to hold something attached
2 : a device consisting of two bars jointed together end to end but not in line so that when a force is applied to the joint tending to straighten it pressure will be exerted on the parts adjacent or fixed to the outer ends of the bars; *also* **:** a device with a joint using a toggle

[2]toggle *transitive verb* **tog·gled; tog·gling** \-g(ə-)liŋ\ (1836)
1 : to fasten with or as if with a toggle
2 : to furnish with a toggle

toggle bolt *noun* (1906)
: a bolt that has a nut with wings which close for passage through a small hole and spring open after passing through the hole to keep the bolt from slipping back through

toggle switch *noun* (circa 1924)
: an electric switch operated by pushing a projecting lever through a small arc

togs \ˈtägz, ˈtógz\ *noun plural* [plural of English slang *tog* coat, short for obsolete English cant *togeman, togman*] (1779)
: CLOTHING; *especially* **:** a set of clothes and accessories for a specified use ⟨riding *togs*⟩

togue \ˈtōg\ *noun* [Canadian French] (1839)
: LAKE TROUT

[1]toil \ˈtói(ə)l\ *noun* [Middle English *toile,* from Anglo-French *toyl,* from Old French *toeil* battle, confusion, from *toeillier*] (14th century)
1 *archaic* **a :** STRUGGLE, BATTLE **b :** laborious effort
2 : long strenuous fatiguing labor
synonym see WORK
— toil·ful \-fəl\ *adjective*
— toil·ful·ly \-fə-lē\ *adverb*

[2]toil *verb* [Middle English, to argue, struggle, from Anglo-French *toiller,* from Old French *toeillier* to stir, disturb, dispute, from Latin *tudiculare* to crush, grind, from *tudicula* machine for crushing olives, diminutive of *tudes* hammer; akin to Latin *tundere* to beat — more at CONTUSION] (14th century)
intransitive verb
1 : to work hard and long **:** LABOR
2 : to proceed with laborious effort **:** PLOD ⟨*toiling* wearily up the hill⟩
transitive verb
1 *archaic* **:** OVERWORK
2 *archaic* **:** to get or accomplish with great effort
— toil·er \ˈtói-lər\ *noun*

[3]toil *noun* [Middle French *toile* cloth, net, from Latin *tela* cloth on a loom — more at SUBTLE] (circa 1529)
1 : a net to trap game
2 : something by which one is held fast or inextricably involved **:** SNARE, TRAP — usually used in plural ⟨caught in the *toils* of the law⟩

toile \ˈtwäl\ *noun* [French, cloth, linen, from Middle French] (1794)
1 : any of many plain or simple twill weave fabrics; *especially* **:** LINEN
2 : a mock-up model of a garment

toile de Jouy \ˌtwäl-də-ˈzhwē\ *noun* [French, literally, cloth of Jouy, from *Jouy*-en-Josas, France] (circa 1920)
: an 18th century French scenic pattern usually printed on cotton, linen, or silk in one color on a light ground; *broadly* **:** a similar printed fabric

[1]toi·let \ˈtói-lət\ *noun* [Middle French *toilette* cloth put over the shoulders while dressing the hair or shaving, diminutive of *toile* cloth] (circa 1695)
1 *archaic* **:** DRESSING TABLE
2 : the act or process of dressing and grooming oneself
3 a (1) **:** BATHROOM, LAVATORY 2 (2) **:** PRIVY **b :** a fixture that consists usually of a water-flushed bowl and seat and is used for defecation and urination
4 : cleansing in preparation for or in association with a medical or surgical procedure ⟨a pharyngeal *toilet*⟩ ◆

[2]toilet (1840)
intransitive verb
1 : to dress and groom oneself
2 : to use the toilet — usually used of a child
transitive verb
1 : DRESS, GARB
2 : to help (as a child or sick person) use the toilet

toilet paper *noun* (1884)
: a thin sanitary absorbent paper for bathroom use chiefly after defecation and urination

toilet powder *noun* (1840)
: a fine powder usually with soothing or antiseptic ingredients for sprinkling or rubbing (as after bathing) over the skin

toi·let·ry \ˈtói-lə-trē\ *noun, plural* **-ries** (1892)
: an article or preparation (as toothpaste, shaving cream, or cologne) used in making one's toilet — usually used in plural

toilet soap *noun* (1839)
: a mild soap that is often perfumed and colored and stabilized with preservatives

toi·lette \twä-ˈlet\ *noun* [French, from Middle French] (1681)
1 : TOILET 2
2 a : formal or fashionable attire or style of dressing **b :** a particular costume or outfit
word history see TOILET

toilet training *noun* (1940)
: the process of training a child to control bladder and bowel movements and to use the toilet
— toilet train *transitive verb*

toilet water *noun* (1855)
: a perfumed liquid containing a high percentage of alcohol for use in or after a bath or as a skin freshener

toil·some \-səm\ *adjective* (1581)
: marked by or full of toil or fatigue **:** LABORIOUS
— toil·some·ly *adverb*
— toil·some·ness *noun*

toil·worn \-ˌwōrn, -ˌwórn\ *adjective* (1751)
: showing the effects of or worn out with toil ⟨*toilworn* hands⟩

to–ing and fro–ing \ˈtü-iŋ-ən(d)-ˈfrō-iŋ\ *noun, plural* **to–ings and fro–ings** [*to and fro*] (1847)
: a passing back and forth

to·ka·mak *also* **to·ko·mak** \ˈtō-kə-ˌmak, ˈtä-\ *noun* [Russian, from *to*roidal'naya *kam*era s *ak*sial'nym magnitnym polem (toroidal chamber with an axial magnetic field)] (1965)
: a toroidal device for producing controlled nuclear fusion that involves the confining and heating of a gaseous plasma by means of an electric current and magnetic field

To·kay \tō-ˈkā\ *noun* (1710)
1 : a naturally sweet wine from the area around Tokaj, Hungary
2 : a blend of Angelica, port, and sherry made in California

toke \ˈtōk\ *noun* [American Spanish *toque,* from Spanish, touch, test, from *tocar* to touch, from Vulgar Latin *toccare* — more at TOUCH] (1968)
slang **:** a puff on a marijuana cigarette or pipe
— toke *intransitive verb, slang*

[1]to·ken \ˈtō-kən\ *noun* [Middle English, from Old English *tācen, tācn* sign, token; akin to Old High German *zeihhan* sign, Greek *deiknynai* to show — more at DICTION] (before 12th century)
1 : an outward sign or expression ⟨his tears were *tokens* of his grief⟩

◇ WORD HISTORY

toilet Latin *tela* "cloth on a loom," ultimately a derivative of *texere* "to weave," developed in Old French into *toile.* From *toile* was formed the diminutive *toilette* in Middle French, which could mean simply "small piece of cloth" though it also came to be used in more specific senses, such as a wrapper for clothing or a cover for a dressing table on which items used in dressing oneself could be placed. By the 17th century, *toilette* could refer to the dressing table itself or the act of grooming that takes place at it. English borrowed the word in the late 17th century and developed the spelling *toilet* from it. Both forms have flourished with some overlap and some differentiation of meaning. *Toilet,* however, has had its other meanings overshadowed by one particular sense development, from "dressing room" through "dressing room with bath facilities" through "lavatory" and finally to "water closet." Amusingly, while English uses a French word to designate euphemistically this homely fixture, French has repaid the compliment by calling it *le water-closet* or *le w.c.*

2 a : SYMBOL, EMBLEM ⟨a white flag is a *token* of surrender⟩ **b :** an instance of a linguistic expression
3 : a distinguishing feature **:** CHARACTERISTIC
4 a : SOUVENIR, KEEPSAKE **b :** a small part representing the whole **:** INDICATION ⟨this is only a *token* of what we hope to accomplish⟩ **c :** something given or shown as a guarantee (as of authority, right, or identity)
5 a : a piece resembling a coin issued as money by some person or body other than a de jure government **b :** a piece resembling a coin issued for use (as for fare on a bus) by a particular group on specified terms
6 : a member of a group (as a minority) that is included within a larger group through tokenism; *especially* **:** a token employee
synonym see SIGN
— by the same token : for the same reason

²token *adjective* (1547)
1 : done or given as a token especially in partial fulfillment of an obligation or engagement ⟨a *token* payment⟩
2 a : representing no more than a symbolic effort **:** MINIMAL, PERFUNCTORY ⟨*token* resistance⟩ ⟨*token* integration⟩ **b :** serving or intended to show absence of discrimination ⟨a *token* female employee⟩

to·ken·ism \'tō-kə-ˌni-zəm\ *noun* (1961)
: the policy or practice of making only a symbolic effort (as to desegregate)

token money *noun* (1889)
: money of regular government issue (as paper currency or coins) having a greater face value than intrinsic value

To·khar·i·an *variant of* TOCHARIAN

to·ko·no·ma \ˌtō-kə-'nō-mə\ *noun* [Japanese] (1871)
: a niche or alcove in the wall of a Japanese house for the display of a decorative object

Tok Pis·in \'tȯk-'pi-zən, -sən\ *noun* [Tok Pisin, literally, pidgin talk] (1974)
: an English-based creole that is a national language of Papua New Guinea

tol- *or* **tolu-** *combining form* [International Scientific Vocabulary, from *toluene*]
: toluene ⟨*tol*yl⟩

to·la \'tō-lə, tō-'lä\ *noun* [Hindi *tolā*, from Sanskrit *tulā* weight; akin to Latin *tollere* to lift up] (1614)
: a unit of weight of India equal to 180 grains troy or 0.4114 ounce (11.7 grams)

tol·booth \'tō(l)-ˌbüth, 'täl-, 'tōl-\ *noun* [Middle English *tolbothe, tollbothe* tollbooth, town hall, jail] (15th century)
1 *Scottish* **:** a town or market hall
2 *Scottish* **:** JAIL, PRISON

tol·bu·ta·mide \täl-'byü-tə-ˌmīd\ *noun* [*tol-* + *but*yric + *amide*] (1956)
: a sulfonylurea $C_{12}H_{18}N_2O_3S$ used in the treatment of diabetes

told *past and past participle of* TELL

tole \'tōl\ *noun, often attributive* [French *tôle*, from Middle French dialect *taule*, from Latin *tabula* board, tablet] (1927)
: sheet metal and especially tinplate for use in domestic and ornamental wares in which it is usually japanned or painted and often elaborately decorated; *also* **:** objects made of tole

To·le·do \tə-'lē-(ˌ)dō\ *noun, plural* **-dos** (1596)
: a finely tempered sword of a kind made in Toledo, Spain

tol·er·a·ble \'tä-lə-rə-bəl, 'täl-rə-; 'tä-lər-bəl\ *adjective* (15th century)
1 : capable of being borne or endured ⟨*tolerable* pain⟩
2 : moderately good or agreeable **:** PASSABLE ⟨a *tolerable* singing voice⟩
— tol·er·a·bil·i·ty \ˌtä-lə-rə-'bi-lə-tē, ˌtäl-rə-\ *noun*
— tol·er·a·bly \'tä-lə-rə-blē, 'täl-rə-; 'tä-lər-blē\ *adverb*

tol·er·ance \'tä-lə-rən(t)s, 'täl-rən(t)s\ *noun* (15th century)
1 : capacity to endure pain or hardship **:** ENDURANCE, FORTITUDE, STAMINA
2 a : sympathy or indulgence for beliefs or practices differing from or conflicting with one's own **b :** the act of allowing something **:** TOLERATION
3 : the allowable deviation from a standard; *especially* **:** the range of variation permitted in maintaining a specified dimension in machining a piece
4 a (1) **:** the capacity of the body to endure or become less responsive to a substance (as a drug) or a physiological insult with repeated use or exposure ⟨immunological *tolerance* to a virus⟩ ⟨an addict's increasing *tolerance* for a drug⟩ (2) **:** relative capacity of an organism to grow or thrive when subjected to an unfavorable environmental factor **b :** the maximum amount of a pesticide residue that may lawfully remain on or in food

tol·er·ant \-rənt\ *adjective* (1784)
1 : inclined to tolerate; *especially* **:** marked by forbearance or endurance
2 : exhibiting tolerance (as for a drug or an environmental factor)
— tol·er·ant·ly *adverb*

tol·er·ate \'tä-lə-ˌrāt\ *transitive verb* **-at·ed; -at·ing** [Latin *toleratus*, past participle of *tolerare* to endure, put up with; akin to Old English *tholian* to bear, Latin *tollere* to lift up, *latus* carried (suppletive past participle of *ferre*), Greek *tlēnai* to bear] (1531)
1 : to exhibit physiological tolerance for (as a drug)
2 a : to suffer to be or to be done without prohibition, hindrance, or contradiction **b :** to put up with
synonym see BEAR
— tol·er·a·tive \-ˌrā-tiv\ *adjective*
— tol·er·a·tor \-ˌrā-tər\ *noun*

tol·er·a·tion \ˌtä-lə-'rā-shən\ *noun* (1531)
1 a : the act or practice of tolerating something **b :** a government policy of permitting forms of religious belief and worship not officially established
2 : TOLERANCE 4a(1)

tol·i·dine \'tä-lə-ˌdēn\ *noun* [International Scientific Vocabulary *tol-* + *-idine*] (1879)
: any of several isomeric aromatic diamines $C_{14}H_{16}N_2$ that are homologues of benzidine and used especially as dye intermediates

¹toll \'tōl\ *noun* [Middle English, from Old English, from (assumed) Vulgar Latin *tolonium*, alteration of Late Latin *telonium* customhouse, from Greek *tolōnion*, from *telōnēs* collector of tolls, from *telos* tax, toll; perhaps akin to Greek *tlēnai* to bear] (before 12th century)
1 : a tax or fee paid for some liberty or privilege (as of passing over a highway or bridge)
2 : compensation for services rendered: as **a :** a charge for transportation **b :** a charge for a long-distance telephone call
3 : a grievous or ruinous price ⟨inflation has taken its *toll*⟩; *especially* **:** cost in life or health ⟨the death *toll* from the hurricane⟩

²toll (14th century)
intransitive verb
: to take or levy toll
transitive verb
1 a : to exact part of as a toll **b :** to take as toll
2 : to exact a toll from (someone)

³toll *or* **tole** \'tōl\ *transitive verb* **tolled** *or* **toled; toll·ing** *or* **tol·ing** [Middle English *tollen, tolen;* akin to Old English for*tyllan* to seduce] (13th century)
1 : ALLURE, ENTICE
2 a : to entice (game) to approach **b :** to attract (fish) with scattered bait **c :** to lead or attract (domestic animals) to a desired point

⁴toll *verb* [Middle English, perhaps from *tollen* to entice] (15th century)
transitive verb
1 : to sound (a bell) by pulling the rope
2 a : to give signal or announcement of ⟨the clock *tolled* each hour⟩ **b :** to announce by tolling ⟨church bells *tolled* the death of the bishop⟩ **c :** to call to or from a place or occasion ⟨bells *tolled* the congregation to church⟩
intransitive verb
: to sound with slow measured strokes ⟨the bell *tolls* solemnly⟩

⁵toll *noun* (15th century)
: the sound of a tolling bell

toll·booth \'tōl-ˌbüth\ *noun* [Middle English *tolbothe, tollbothe* tollbooth, town hall, jail, from *tol, toll* toll + *bothe* booth] (14th century)
: a booth (as on a highway or bridge) where tolls are paid

toll call *noun* (1928)
: a long-distance telephone call at charges above a local rate

toll·gate \'tōl-ˌgāt\ *noun* (1773)
: a point where the driver of a vehicle must pay a toll

toll·house \-ˌhau̇s\ *noun* (15th century)
: a house or booth where tolls are taken

Toll House *trademark*
— used for cookies containing chocolate morsels

toll·way \-ˌwā\ *noun* (1949)
: TURNPIKE 2a(1)

Tol·tec \'tōl-ˌtek, 'täl-\ *noun* [Spanish *tolteca*, from Nahuatl *tōltēcah*, plural of *tōltēcatl*, literally, person from *Tōllān* (now Tula de Allende, Mexico)] (1787)
: a member of a people that dominated central and southern Mexico prior to the Aztecs
— Tol·tec·an \-ən\ *adjective*

tol·u·ene \'täl-yə-ˌwēn\ *noun* [French *toluène*, from *tolu* balsam from the tropical American tree *Myroxylon balsamum*, from Spanish *tolú*, from Santiago de *Tolú*, Colombia] (1871)
: a liquid aromatic hydrocarbon C_7H_8 that resembles benzene but is less volatile, flammable, and toxic and is used as a solvent, in organic synthesis, and as an antiknock agent for gasoline

to·lu·i·dine \tə-'lü-ə-ˌdēn\ *noun* [International Scientific Vocabulary] (1850)
: any of three isomeric amino derivatives of toluene C_7H_9N that are analogous to aniline and are used as dye intermediates

toluidine blue *noun* (1898)
: a basic thiazine dye that is related to methylene blue and is used as a biological stain

tol·u·ol \'täl-yə-ˌwȯl, -ˌwōl\ *noun* (circa 1848)
: toluene especially of commercial grade

tol·yl \'tä-ləl\ *noun* [International Scientific Vocabulary] (circa 1868)
: any of three univalent groups $CH_3C_6H_4$ derived from toluene

tom \'täm\ *noun* [*Tom*, nickname for *Thomas*] (1762)
1 : the male of various animals: as **a :** TOMCAT **b :** a male turkey
2 *capitalized* **:** UNCLE TOM 2

¹tom·a·hawk \'tä-mi-ˌhȯk\ *noun* [Virginia Algonquian *tomahack*] (circa 1612)
: a light ax used as a missile and as a hand weapon especially by North American Indians

²tomahawk *transitive verb* (circa 1650)
: to cut, strike, or kill with a tomahawk

to·mal·ley \tə-'ma-lē; 'tä-ˌma-lē, -mə-lē\ *noun, plural* **-leys** [Carib *tumali* sauce of lobster livers] (circa 1666)
: the liver of the lobster

Tom and Jer·ry \ˌtäm-ən(d)-'jer-ē\ *noun* [Corinthian *Tom & Jerry* Hawthorne, characters in *Life in London* (1821) by Pierce Egan (died 1849) English sportswriter] (1845)
: a hot drink that is a combination of a toddy and an eggnog

to·ma·til·lo \ˌtō-mə-'tē-(ˌ)yō, -'tēl-(ˌ)yō\ *noun, plural* **-los** [Spanish, diminutive of *tomate*] (circa 1913)
: the small round pale green or yellow or purplish edible viscid fruit of a Mexican groundcherry (*Physalis ixocarpa*); *also* **:** the plant

to·ma·to \tə-'mā-(ˌ)tō; *chiefly British, eastern New England, northeastern Virginia, and sometimes elsewhere in cultivated speech*

-'mä- *or* -'mä-; *chiefly Northern* -'ma-\ *noun, plural* **-toes** [alteration of earlier *tomate,* from Spanish, from Nahuatl *tomatl*] (1604)
1 : the usually large rounded typically red or yellow pulpy berry of a tomato
2 : any of a genus (*Lycopersicon*) of South American herbs of the nightshade family; *especially* **:** one (*L. lycopersicum* synonym *L. esculentum*) that is more or less perennial in its native habitat but is widely cultivated as an annual for its edible fruit ▫

to·ma·to·ey \-tə-wē\ *adjective* (1972)
1 : of, relating to, or characteristic of a tomato
2 : richly flavored with tomatoes

tomato fruitworm *noun* (circa 1891)
: CORN EARWORM

tomato hornworm *noun* (1921)
: a hawkmoth (*Manduca quinquemaculata*) whose larva is a hornworm feeding on plants of the nightshade family and especially tobacco and tomato

¹tomb \'tüm\ *noun* [Middle English *tombe,* from Anglo-French *tumbe,* from Late Latin *tumba* sepulchral mound, from Greek *tymbos;* perhaps akin to Latin *tumēre* to be swollen — more at THUMB] (13th century)
1 a : an excavation in which a corpse is buried **:** GRAVE **b :** a place of interment
2 : a house, chamber, or vault for the dead
3 : a building or structure resembling a tomb (as in appearance)
— tomb·less \-ləs\ *adjective*

²tomb *transitive verb* (14th century)
: BURY, ENTOMB

tom·bac \'täm-,bak\ *noun* [French, from Dutch *tombak,* from Malay *tĕmbaga* copper] (1602)
: an alloy essentially of copper and zinc and sometimes tin or arsenic that is used especially for cheap jewelry and gilding

tom·bo·lo \'tōm-bə-,lō, 'täm-\ *noun, plural* **-los** [Italian, from Latin *tumulus* mound, tumulus] (1899)
: a sand or gravel bar connecting an island with the mainland or another island

tom·boy \'täm-,bȯi\ *noun* (1592)
: a girl who behaves in a manner usually considered boyish **:** HOYDEN
— tom·boy·ish \-ish\ *adjective*
— tom·boy·ish·ness *noun*

tomb·stone \'tüm-,stōn\ *noun* (1565)
: GRAVESTONE

tom·cat \'täm-,kat\ *noun* (1789)
: a male domestic cat

tom·cod \-,käd\ *noun* (1722)
: either of two small fishes (*Microgadus tomcod* of the Atlantic and *M. proximus* of the Pacific) of the cod family

Tom Col·lins \'täm-'kä-lənz\ *noun* [from the name *Tom Collins*] (circa 1909)
: a collins with a base of gin

Tom, Dick, and Har·ry \,täm-,dik-ən(d)-'har-ē\ *noun, plural* **Toms, Dicks, and Harrys** (1815)
: the common man **:** ANYONE — often used with *every* ⟨helps every *Tom, Dick, and Harry* in need⟩

tome \'tōm\ *noun* [Middle French or Latin; Middle French, from Latin *tomus,* from Greek *tomos* section, roll of papyrus, tome, from *temnein* to cut; akin to Middle Irish *tamnaid* he lops, Polish *ciąć* to cut, and perhaps to Latin *tondēre* to shear] (1519)
1 : a volume forming part of a larger work
2 : BOOK; *especially* **:** a large or scholarly book

-tome *noun combining form* [Greek *tomos*]
1 : part **:** segment ⟨myo*tome*⟩
2 : cutting instrument ⟨micro*tome*⟩

to·men·tose \tō-'men-,tōs, 'tō-mən-\ *adjective* [New Latin *tomentosus,* from *tomentum*] (1698)
: covered with densely matted hairs ⟨a *tomentose* leaf⟩

to·men·tum \tō-'men-təm\ *noun, plural* **-ta** \-tə\ [New Latin, from Latin, cushion stuffing] (1699)
: pubescence composed of densely matted woolly hairs

¹tom·fool \'täm-'fül\ *noun* (circa 1709)
: a great fool **:** BLOCKHEAD

²tomfool *adjective* (1819)
: extremely foolish, stupid, or doltish

tom·fool·ery \,täm-'fül-rē, -'fü-lə-\ *noun* (1812)
: playful or foolish behavior

Tom·my \'tä-mē\ *noun, plural* **Tommies** [*Thomas* Atkins, name used as model in official army forms] (1892)
: a British soldier

Tommy At·kins \-'at-kənz\ *noun* (1883)
: TOMMY

tom·my–gun \'tä-mē-,gən\ *transitive verb* (1942)
: to shoot with a tommy gun

tommy gun *noun* [by shortening & alteration] (1929)
: THOMPSON SUBMACHINE GUN; *broadly* **:** SUBMACHINE GUN

tom·my·rot \'tä-mē-,rät\ *noun* [English dialect *tommy* fool + English *rot*] (1884)
: utter foolishness or nonsense

to·mo·gram \'tō-mə-,gram\ *noun* (1936)
: a roentgenogram made by tomography

to·mog·ra·phy \tō-'mä-grə-fē\ *noun* [Greek *tomos* section + International Scientific Vocabulary *-graphy* — more at TOME] (1935)
: a method of producing a three-dimensional image of the internal structures of a solid object (as the human body or the earth) by the observation and recording of the differences in the effects on the passage of waves of energy impinging on those structures — compare COMPUTED TOMOGRAPHY
— to·mo·graph·ic \,tō-mə-'gra-fik\ *adjective*

¹to·mor·row \tə-'mär-(,)ō, -'mȯr-\ *adverb* [Middle English *to morgen,* from Old English *tō morgen,* from *tō* to + *morgen* morrow, morning — more at MORN] (13th century)
: on or for the day after today ⟨will do it *tomorrow*⟩

²tomorrow *noun* (14th century)
1 : the day after the present ⟨the court will recess until *tomorrow*⟩
2 : FUTURE 1a ⟨the world of *tomorrow*⟩

tom·pi·on \'täm-pē-ən\ *variant of* TAMPION

Tom Thumb \'täm-'thəm\ *noun*
1 : a legendary English dwarf
2 : a dwarf type, race, or individual

tom·tit \'täm-,tit, täm-'\ *noun* [probably short for *tomtitmouse,* from the name *Tom* + *titmouse*] (1700)
: any of various small active birds

tom–tom \'täm-,täm, 'təm-,təm\ *noun* [Hindi *ṭamṭam*] (1693)
1 : a usually long and narrow small-headed drum commonly beaten with the hands
2 : a monotonous beating, rhythm, or rhythmical sound

-tomy *noun combining form* [New Latin *-tomia,* from Greek, from *-tomos* that cuts, from *temnein* to cut — more at TOME]
: incision **:** section ⟨laparo*tomy*⟩

¹ton \'tən\ *noun, plural* **tons** *also* **ton** [Middle English *tunne* unit of weight or capacity — more at TUN] (14th century)
1 a : a unit of internal capacity for ships equal to 100 cubic feet — called also *register ton* **b :** a unit approximately equal to the volume of a long ton weight of seawater used in reckoning the displacement of ships and equal to 35 cubic feet **c :** a unit of volume for cargo freight usually reckoned at 40 cubic feet — called also *measurement ton*
2 : any of various units of weight: **a** — see WEIGHT table **b :** METRIC TON
3 : a great quantity **:** LOT — often used in plural ⟨ate *tons* of hamburgers⟩ ⟨has *tons* of money⟩

²ton \'tō^{n}\ *noun* [French, literally, tone, from Latin *tonus*] (1765)
1 : the prevailing fashion **:** VOGUE
2 : the quality or state of being smart or fashionable

ton·al \'tō-n^{ə}l\ *adjective* (1776)
1 : of or relating to tone, tonality, or tonicity
2 : having tonality
— ton·al·ly \-n^{ə}l-ē\ *adverb*

to·nal·i·ty \tō-'na-lə-tē\ *noun, plural* **-ties** (1838)
1 : tonal quality
2 a : KEY 7 **b :** the organization of all the tones and harmonies of a piece of music in relation to a tonic
3 : the arrangement or interrelation of the tones of a work of visual art

ton·do \'tän-(,)dō\ *noun, plural* **ton·di** \-(,)dē\ [Italian, from *tondo* round, short for *rotondo,* from Latin *rotundus* — more at ROTUND] (1890)
1 : a circular painting
2 : a sculptured medallion

¹tone \'tōn\ *noun* [Middle English, from Latin *tonus* tension, tone, from Greek *tonos,* literally, act of stretching; akin to Greek *teinein* to stretch — more at THIN] (14th century)
1 : vocal or musical sound of a specific quality ⟨spoke in low *tones*⟩ ⟨masculine *tones*⟩; *especially* **:** musical sound with respect to timbre and manner of expression
2 a : a sound of definite pitch and vibration **b :** WHOLE STEP
3 : accent or inflection expressive of a mood or emotion
4 : the pitch of a word often used to express differences of meaning
5 : a particular pitch or change of pitch constituting an element in the intonation of a phrase or sentence ⟨high *tone*⟩ ⟨low *tone*⟩ ⟨mid *tone*⟩ ⟨low-rising *tone*⟩ ⟨falling *tone*⟩
6 : style or manner of expression in speaking or writing ⟨seemed wise to adopt a conciliatory *tone*⟩
7 a (1) **:** color quality or value (2) **:** a tint or shade of color **b :** the color that appreciably modifies a hue or white or black ⟨gray walls of greenish *tone*⟩
8 : the effect in painting of light and shade together with color
9 a : the state of a living body or of any of its organs or parts in which the functions are

▫ USAGE

tomato It is ironic that the most renowned modern shibboleth, like the ancient one, should be the name of a humble plant with edible fruit (*shibboleth* in Hebrew means "head of grain" or "stream"). The original pronunciation of this Spanish loanword was with stressed \ä\, as was also the case for *potato.* The older \ä\ pronunciation of *potato* is found in Robert Burns' poem " Holy Willie's Prayer" and persists in some Scottish dialects. Perhaps because *potato* has been part of English longer than *tomato,* its pronunciation has been anglicized in most dialects, with the stressed \ä\ vowel becoming \ā\, like the *a* in *Plato* (from Greek) and *dado* (from Italian). The pronunciation of *tomato* began the same journey of anglicization but was halted halfway between, so that British and some Eastern American dialects have the older \ä\ while the others have the newer \ā\. *Tomato,* being the name of a New World plant, might have entered the English language first in the Americas, where it would have had more time to undergo anglicization in American mouths. The various pronunciations of *tomato* are all acceptable in standard English.

healthy and performed with due vigor **b :** normal tension or responsiveness to stimuli; *specifically* **:** muscular tonus
10 a : healthy elasticity **:** RESILIENCY **b :** general character, quality, or trend ⟨a city's upbeat *tone*⟩ **c :** frame of mind **:** MOOD

²tone *verb* **toned; ton·ing** (1660)
transitive verb
1 : INTONE
2 : to give a particular intonation or inflection to
3 a : to impart tone to **:** STRENGTHEN ⟨medicine to *tone* up the system⟩ **b :** to soften or reduce in intensity, color, appearance, or sound **:** MELLOW — often used with *down* **c :** to change the normal silver image of (as a photographic print) into a colored image
intransitive verb
1 : to assume a pleasing color quality or tint
2 : to blend or harmonize in color

tone·arm \'tōn-ˌärm\ *noun* (1913)
: the movable part of a phonograph or record player that carries the pickup and permits the needle to follow the record groove

tone color *noun* (1881)
: TIMBRE

toned \'tōnd\ *adjective* (15th century)
1 : having tone or a specified tone **:** characterized or distinguished by a tone
2 *of paper* **:** having a slight tint

tone–deaf \'tōn-ˌdef\ *adjective* (1894)
: relatively insensitive to differences in musical pitch
— tone deafness *noun*

tone language *noun* (circa 1909)
: a language (as Chinese or Zulu) in which variations in tone distinguish words or phrases of different meaning that otherwise would sound alike

tone·less \'tōn-ləs\ *adjective* (1773)
: lacking in tone, modulation, or expression
— tone·less·ly *adverb*
— tone·less·ness *noun*

to·neme \'tō-ˌnēm\ *noun* (1923)
: an intonation phoneme in a tone language
— to·ne·mic \tō-'nē-mik\ *adjective*

tone poem *noun* (1902)
: SYMPHONIC POEM
— tone poet *noun*

ton·er \'tō-nər\ *noun* (1888)
: one that tones or is a source of tones: as **a :** a solution used to impart color to a silver photographic image **b :** a substance used to develop a latent xerographic image **c :** a liquid cosmetic for cleansing the skin and contracting the pores

tone row *noun* (1936)
: the chosen sequence of tones that serves as the basis for a work of serial music; *specifically* **:** TWELVE-TONE ROW

to·net·ic \tō-'ne-tik\ *adjective* (1922)
1 : relating to linguistic tones or to tone languages
2 : of or relating to intonation ⟨*tonetic* notation⟩
— to·net·i·cal·ly \-ti-k(ə-)lē\ *adverb*

to·net·ics \-tiks\ *noun plural but singular in construction* (1921)
: the use or study of linguistic tones

to·nette \tō-'net\ *noun* [¹*tone* + *-ette*] (1939)
: a simple fipple flute with a range somewhat larger than an octave that is often used in elementary music education

ton·ey *variant of* TONY

¹tong \'täŋ, 'tȯŋ\ *noun* [Chinese (Guangdong) *tòhng,* literally, hall] (1883)
: a secret society or fraternal organization especially of Chinese in the U.S. formerly notorious for gang warfare

²tong *verb* [*tongs*] (1901)
transitive verb
: to take, gather, hold, or handle with tongs ⟨*tong* oysters⟩
intransitive verb
: to use tongs especially in taking or gathering something
— tong·er \'täŋ-ər, 'tȯŋ-\ *noun*

ton·ga \'täŋ-gə\ *noun* [Hindi *tāṅgā*] (1874)
: a light 2-wheeled vehicle for two or four persons drawn by one horse and common in India

Ton·gan \'täŋ-gən *also* -ən\ *noun* (1853)
1 : a member of a Polynesian people of Tonga
2 : the Polynesian language of the Tongans
— Tongan *adjective*

tongs \'täŋz, 'tȯŋz\ *noun plural but singular or plural in construction* [Middle English *tonges,* plural of *tonge,* from Old English *tang;* akin to Old High German *zanga* tongs and perhaps to Greek *daknein* to bite] (before 12th century)
: any of numerous grasping devices consisting commonly of two pieces joined at one end by a pivot or hinged like scissors

¹tongue \'təŋ\ *noun* [Middle English *tunge,* from Old English; akin to Old High German *zunga* tongue, Latin *lingua*] (before 12th century)
1 a : a fleshy movable process of the floor of the mouths of most vertebrates that bears sensory end organs and small glands and functions especially in taking and swallowing food and in humans as a speech organ **b :** a part of various invertebrate animals that is analogous to the tongue
2 : the flesh of a tongue (as of the ox or sheep) used as food
3 : the power of communication through speech
4 a : LANGUAGE; *especially* **:** a spoken language **b :** manner or quality of utterance with respect to tone or sound, the sense of what is expressed, or the intention of the speaker ⟨she has a clever *tongue*⟩ ⟨a sharp *tongue*⟩ **c :** ecstatic usually unintelligible utterance accompanying religious excitation — usually used in plural **d :** the cry of or as if of a hound pursuing or in sight of game — used especially in the phrase *to give tongue*
5 : a tapering flame ⟨*tongues* of fire⟩
6 : a long narrow strip of land projecting into a body of water
7 : something resembling an animal's tongue in being elongated and fastened at one end only: as **a :** a movable pin in a buckle **b :** a metal ball suspended inside a bell so as to strike against the sides as the bell is swung **c :** the pole of a vehicle **d :** the flap under the lacing or buckles of a shoe at the throat of the vamp
8 a : the rib on one edge of a board that fits into a corresponding groove in an edge of another board to make a flush joint **b :** FEATHER 4
— tongue·like \-ˌlīk\ *adjective*

²tongue *verb* **tongued; tongu·ing** \'təŋ-iŋ\ (14th century)
transitive verb
1 *archaic* **:** SCOLD
2 : to touch or lick with or as if with the tongue
3 a : to cut a tongue on ⟨*tongue* a board⟩ **b :** to join (as boards) by means of a tongue and groove ⟨*tongue* flooring together⟩
4 : to articulate (notes) by tonguing
intransitive verb
1 : to project in a tongue
2 : to articulate notes on a wind instrument by successively interrupting the stream of wind with the action of the tongue

tongue and groove *noun* (circa 1876)
: a joint made by a tongue on one edge of a board fitting into a corresponding groove on the edge of another board
— tongue–and–groove *adjective*

tongued \'təŋd\ *adjective* (14th century)
: having a tongue especially of a specified kind — often used in combination ⟨sharp-*tongued*⟩

tongue–in–cheek *adjective* (1933)
: characterized by insincerity, irony, or whimsical exaggeration

tongue in cheek *adverb* (circa 1934)
: with insincerity, irony, or whimsical exaggeration

tongue–lash \'təŋ-ˌlash\ *verb* [back-formation from *tongue-lashing*] (1881)
: CHIDE, SCOLD
— tongue–lash·ing *noun*

tongue·less \'təŋ-ləs\ *adjective* (14th century)
1 : having no tongue
2 : lacking power of speech **:** MUTE

¹tongue–tie \'təŋ-ˌtī\ *transitive verb* [back-formation from *tongue-tied*] (1555)
: to deprive of speech or the power of distinct articulation

²tongue–tie *noun* (circa 1852)
: limited mobility of the tongue due to shortness of its frenulum

tongue–tied \'təŋ-ˌtīd\ *adjective* (1529)
1 : unable or disinclined to speak freely (as from shyness)
2 : affected with tongue-tie

tongue twister *noun* (1904)
: a word, phrase, or sentence difficult to articulate because of a succession of similar consonantal sounds (as in "twin-screw steel cruiser")

-tonia *noun combining form* [New Latin, from *tonus*]
: condition or degree of tonus ⟨myo*tonia*⟩

¹ton·ic \'tä-nik\ *adjective* [Greek *tonikos,* from *tonos* tension, tone] (1649)
1 a : characterized by tonus ⟨*tonic* contraction of muscle⟩; *also* **:** marked by prolonged muscular contraction ⟨*tonic* convulsions⟩ **b :** producing or adapted to produce healthy muscular condition and reaction of organs (as muscles)
2 a : increasing or restoring physical or mental tone **:** REFRESHING **b :** yielding a tonic substance
3 : relating to or based on the first tone of a scale ⟨*tonic* harmony⟩
4 *of a syllable* **:** bearing a principal stress or accent
5 : of or relating to speech tones or to languages using them to distinguish words otherwise identical
— ton·i·cal·ly \'tä-ni-k(ə-)lē\ *adverb*

²tonic *noun* (1799)
1 a : an agent (as a drug) that increases body tone **b :** one that invigorates, restores, refreshes, or stimulates ⟨a day in the country was a *tonic* for him⟩ **c :** a liquid preparation for the scalp or hair **d** *chiefly New England* **:** a carbonated flavored beverage **e :** TONIC WATER
2 : the first tone of a diatonic scale **:** KEYNOTE
3 : a voiced sound

tonic accent *noun* (1867)
1 : relative phonetic prominence (as from greater stress or higher pitch) of a spoken syllable or word
2 : accent depending on pitch rather than stress

to·nic·i·ty \tō-'ni-sə-tē\ *noun* (1824)
1 : the property of possessing tone; *especially* **:** healthy vigor of body or mind
2 : muscular tonus

tonic sol–fa *noun* (1852)
: a system of solmization based on key relationships that replaces the normal notation with sol-fa syllables or their initials

tonic water *noun* (1926)
: a carbonated beverage flavored with a small amount of quinine, lemon, and lime

¹to·night \tə-'nīt\ *adverb* (before 12th century)
: on this present night or the night following this present day ⟨will do it *tonight*⟩

²tonight *noun* (14th century)
: the present night or the night following this present day

ton·ka bean \'täŋ-kə-\ *noun* [probably from Tupi *tonka*] (1796)
: the coumarin-containing seed of any of several tropical American leguminous trees (genus *Dipteryx*) that is used in perfumes and as an artificial vanilla flavoring; *also* **:** a tree bearing tonka beans

ton·nage \'tə-nij\ *noun* [in sense 1, from Middle English, from Middle French, from *tonne*

tun; in other senses, from [1]*ton* — more at TUNNEL] (15th century)
1 : a duty formerly levied on every tun of wine imported into England
2 a : a duty or impost on vessels based on cargo capacity **b :** a duty on goods per ton transported
3 : ships in terms of the total number of tons registered or carried or of their carrying capacity
4 a : the cubical content of a merchant ship in units of 100 cubic feet **b :** the displacement of a warship
5 a : total weight in tons shipped, carried, or produced **b :** impressively large amount or weight

tonne \'tən\ *noun* [French, from *tonne* tun, from Old French — more at TUNNEL] (1877)
: METRIC TON

ton·neau \'tä-,nō, tə-'nō\ *noun, plural* **ton·neaus** [French, literally, tun, from Old French *tonel* — more at TUNNEL] (1901)
1 : the rear seating compartment of an automobile; *also* **:** the entire seating compartment
2 : a shape of watch case or dial resembling a barrel in profile

ton·ner \'tə-nər\ *noun* (1851)
: an object (as a ship) having a specified tonnage — used in combination ⟨a thousand-*tonner*⟩

to·nom·e·ter \tō-'nä-mə-tər\ *noun* [Greek *tonos* tone + English *-meter*] (1725)
1 : an instrument or device for determining the exact pitch or the vibration rate of tones
2 : an instrument for measuring tension or pressure and especially intraocular pressure
— to·nom·e·try \tō-'nä-mə-trē\ *noun*

to·no·plast \'tō-nə-,plast\ *noun* [International Scientific Vocabulary *tono-* (from Greek *tonos* tension) + *-plast* — more at TONE] (circa 1888)
: a semipermeable membrane surrounding a plant-cell vacuole

ton·sil \'tän(t)-səl\ *noun* [Latin *tonsillae,* plural, tonsils] (1601)
1 : either of a pair of prominent masses of lymphoid tissue that lie one on each side of the throat between the anterior and posterior pillars of the fauces
2 : any of various masses of lymphoid tissue that are similar to tonsils
— ton·sil·lar \'tän(t)-s(ə-)lər\ *adjective*

tonsill- *combining form* [Latin *tonsillae*]
: tonsil ⟨*tonsill*ectomy⟩

ton·sil·lec·to·my \,tän(t)-sə-'lek-tə-mē\ *noun, plural* **-mies** (1899)
: the surgical removal of the tonsils

ton·sil·li·tis \-'lī-təs\ *noun* [New Latin] (1801)
: inflammation of the tonsils

ton·so·ri·al \tän-'sōr-ē-əl, -'sȯr-\ *adjective* [Latin *tonsorius,* from *tondēre*] (1813)
: of or relating to a barber or the work of a barber

[1]**ton·sure** \'tän(t)-shər\ *noun* [Middle English, from Medieval Latin *tonsura,* from Latin, act of shearing, from *tonsus,* past participle of *tondēre* to shear — more at TOME] (14th century)
1 : the Roman Catholic or Eastern rite of admission to the clerical state by the clipping or shaving of a portion of the head
2 : the shaven crown or patch worn by monks and other clerics
3 : a bald spot resembling a tonsure

[2]**tonsure** *transitive verb* **ton·sured; ton·sur·ing** \'tän(t)-sh(ə-)riŋ\ (1706)
: to shave the head of; *especially* **:** to confer the tonsure upon

ton·tine \'tän-,tēn, tän-'\ *noun* [French, from Lorenzo *Tonti* (died 1695) Italian banker] (1765)
: a joint financial arrangement whereby the participants usually contribute equally to a prize that is awarded entirely to the participant who survives all the others

to·nus \'tō-nəs\ *noun* [New Latin, from Latin, tension, tone] (1876)
: TONE 9a; *especially* **:** a state of partial contraction characteristic of normal muscle

tony \'tō-nē\ *adjective* **ton·i·er; -est** (1877)
: marked by an aristocratic or high-toned manner or style ⟨*tony* private schools⟩

To·ny \'tō-nē\ *noun, plural* **Tonys** [*Tony,* nickname of Antoinette Perry (died 1946) American actress & producer] (1947)
: a medallion awarded annually by a professional organization for notable achievement in the theater

too \'tü\ *adverb* [Middle English, from Old English *tō* to, too — more at TO] (before 12th century)
1 : BESIDES, ALSO ⟨sell the house and furniture *too*⟩
2 a : to an excessive degree **:** EXCESSIVELY ⟨*too* large a house for us⟩ **b :** to such a degree as to be regrettable ⟨this time he has gone *too* far⟩ **c :** VERY ⟨didn't seem *too* interested⟩
3 : SO 2d ⟨"I didn't do it." "You did *too.*"⟩

took *past of* TAKE

[1]**tool** \'tül\ *noun* [Middle English, from Old English *tōl;* akin to Old English *tawian* to prepare for use — more at TAW] (before 12th century)
1 a : a handheld device that aids in accomplishing a task **b** (1) **:** the cutting or shaping part in a machine or machine tool (2) **:** a machine for shaping metal **:** MACHINE TOOL
2 a : something (as an instrument or apparatus) used in performing an operation or necessary in the practice of a vocation or profession ⟨a scholar's books are his *tools*⟩ **b :** a means to an end ⟨a book's cover can be a marketing *tool*⟩ **c :** PENIS — often considered vulgar
3 : one that is used or manipulated by another
4 *plural* **:** natural ability ⟨has all the *tools*⟩
synonym see IMPLEMENT

[2]**tool** (1812)
transitive verb
1 a : to cause (a vehicle) to go **:** DRIVE **b :** to convey in a vehicle
2 : to shape, form, or finish with a tool; *especially* **:** to letter or ornament (as leather or gold) by means of hand tools
3 : to equip (as a plant or industry) with tools, machines, and instruments for production
intransitive verb
1 : DRIVE, RIDE
2 : to equip a plant or industry with the means (as machines, machine tools, and instruments) of production — often used with *up*

[3]**tool** *noun* (1881)
: a design (as on the binding of a book) made by tooling

tool·box \'tül-,bäks\ *noun* (1832)
: a chest for tools

tool·hold·er \-,hōl-dər\ *noun* (circa 1876)
: a short steel bar having a shank at one end by which it is clamped to a machine and a clamp at the other end to hold small interchangeable cutting bits

tool·house \-,hau̇s\ *noun* (1818)
: TOOLSHED

tool·mak·er \'tül-,mā-kər\ *noun* (1844)
: one that makes tools; *especially* **:** a machinist who specializes in the construction, repair, maintenance, and calibration of the tools, jigs, fixtures, and instruments of a machine shop

tool·mak·ing \-kiŋ\ *noun* (1893)
: the action, process, or art of making tools; *also* **:** the trade of a toolmaker

tool·room \'tül-,rüm, -,ru̇m\ *noun* (1878)
: a room where tools are kept; *especially* **:** a room in a machine shop in which tools are made, stored, and issued for use by workers

tool·shed \-,shed\ *noun* (1840)
: an outbuilding for storing tools

tools of ignorance [from the notion that a smart athlete would not play such a grueling position] (1939)
: a baseball catcher's equipment

tool subject *noun* (1925)
: a subject studied to gain competence in a skill used in other subjects

toom \'tüm\ *adjective* [Middle English, from Old English *tōm* — more at TEEM] (before 12th century)
chiefly Scottish **:** EMPTY

toon \'tün\ *noun* [Hindi *tūn,* from Sanskrit *tunna*] (1810)
: a southeast Asian and Australian tree (*Cedrela toona*) of the mahogany family with fragrant dark red wood and flowers that yield a dye; *also* **:** its wood

[1]**toot** \'tüt\ *verb* [probably imitative] (circa 1510)
intransitive verb
1 a : to sound a short blast ⟨the horn *tooted*⟩ **b :** to sound a note or call suggesting the short blast of a wind instrument
2 : to blow or sound an instrument (as a horn) especially so as to produce short blasts
transitive verb
: to cause to sound ⟨*toot* a whistle⟩
— toot·er *noun*

[2]**toot** *noun* (1641)
: a short blast (as on a horn); *also* **:** a sound resembling such a blast

[3]**toot** *noun* [Scots *toot* to drink heavily] (circa 1790)
: a drinking bout **:** SPREE

tooth \'tüth\ *noun, plural* **teeth** \'tēth\ [Middle English, from Old English *tōth;* akin to Old High German *zand* tooth, Latin *dent-, dens,* Greek *odont-, odous*] (before 12th century)
1 a : one of the hard bony appendages that are borne on the jaws or in many of the lower vertebrates on other bones in the walls of the mouth or pharynx and serve especially for the prehension and mastication of food and as weapons of offense and defense **b :** any of various usually hard and sharp processes especially about the mouth of an invertebrate
2 : TASTE, LIKING
3 : a projection resembling or suggesting the tooth of an animal in shape, arrangement, or action ⟨a saw *tooth*⟩: as **a :** any of the regular projections on the circumference or sometimes the face of a wheel that engage with corresponding projections on another wheel especially to transmit force **:** COG **b :** a small sharp-pointed marginal lobe or process on a plant
4 a : something that injures, tortures, devours, or destroys **b** *plural* **:** effective means of enforcement
5 : a roughness of surface produced by mechanical or artificial means
— tooth·like \'tüth-,līk\ *adjective*
— in the teeth of 1 : in or into direct contact or collision with ⟨found themselves sailing *in the teeth of* a hurricane —*Current Biography*⟩ **2 :** in direct opposition to ⟨rule had . . . been imposed by conquest *in the teeth of* obstinate resistance —A. J. Toynbee⟩

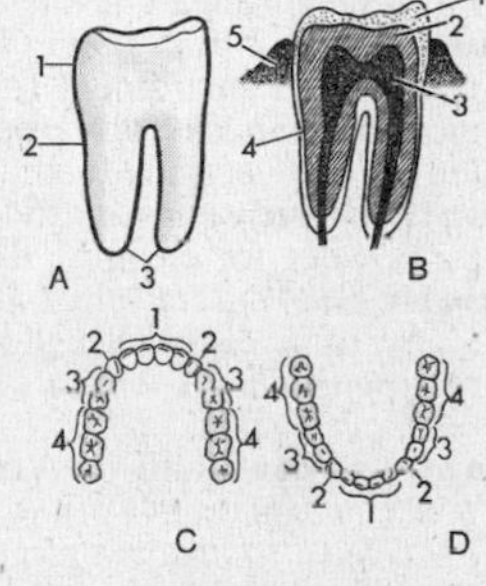

tooth 1a: ***A*** **outside of a molar:** ***1*** **crown,** ***2*** **neck,** ***3*** **roots;** ***B*** **cross section of a molar:** ***1*** **enamel,** ***2*** **dentin,** ***3*** **pulp,** ***4*** **cementum,** ***5*** **gum;** ***C*** **dentition of adult human, upper;** ***D*** **dentition of adult human, lower:** ***1*** **incisors,** ***2*** **canines,** ***3*** **bicuspids,** ***4*** **molars**

— to the teeth : FULLY, COMPLETELY ⟨armed *to the teeth*⟩
tooth·ache \'tüth-,āk\ *noun* (14th century)
: pain in or about a tooth
tooth and nail *adverb* (1550)
: with every available means : ALL OUT ⟨fight *tooth and nail*⟩
tooth·brush \'tüth-,brəsh\ *noun* (1690)
: a brush for cleaning the teeth
tooth·brush·ing *noun* (1920)
: the action of using a toothbrush to clean teeth
toothed \'tütht, *uncompounded also* 'tü-thəd\ *adjective* (14th century)
: having teeth especially of a specified kind or number — often used in combination ⟨buck-*toothed*⟩
toothed whale \'tüht-, 'tü-thəd-\ *noun* (1843)
: any of a suborder (Odontoceti) of cetaceans (as a dolphin, porpoise, or killer whale) bearing usually numerous simple conical teeth — compare BALEEN WHALE
tooth fairy *noun* (1962)
: a fairy believed by children to leave money while they sleep in exchange for a tooth that has come out
tooth·less \'tüth-ləs\ *adjective* (14th century)
1 : having no teeth
2 a : lacking in sharpness or bite ⟨spoke in *toothless* generalities —Arthur Hepner⟩ **b :** lacking in means of enforcement or coercion : INEFFECTUAL
tooth·paste \-,pāst\ *noun* (1832)
: a paste for cleaning the teeth
tooth·pick \-,pik\ *noun* (15th century)
: a pointed instrument (as a slender tapering piece of wood) for removing food particles lodged between the teeth
tooth powder *noun* (1542)
: a powder for cleaning the teeth
tooth shell *noun* (circa 1711)
: any of a class (Scaphopoda) of burrowing marine mollusks with a tapering tubular shell; *also* : this shell
tooth·some \'tüth-səm\ *adjective* (1551)
1 a : AGREEABLE, ATTRACTIVE **b :** sexually attractive ⟨a *toothsome* blonde⟩
2 : of palatable flavor and pleasing texture : DELICIOUS ⟨crisp *toothsome* fried chicken⟩
synonym see PALATABLE
— tooth·some·ly *adverb*
— tooth·some·ness *noun*
tooth·wort \-,wərt, -,wȯrt\ *noun* (1597)
1 : a European parasitic plant (*Lathraea squamaria*) of the broomrape family having a rootstock covered with tooth-shaped scales
2 : any of various cresses (genus *Dentaria*) including several cultivated for their showy flowers
toothy \'tü-thē\ *adjective* **tooth·i·er; -est** (1530)
1 : having or showing prominent teeth ⟨*toothy* grin⟩
2 : TOOTHSOME 2
— tooth·i·ly \-thə-lē\ *adverb*
too·tle \'tü-tᵊl\ *verb* **too·tled; too·tling** \'tüt-liŋ, 'tü-tᵊl-iŋ\ [frequentative of [1]*toot*] (1820)
intransitive verb
1 : to toot gently, repeatedly, or continuously
2 : to drive or move along in a leisurely manner
transitive verb
: to toot continuously on
— tootle *noun*
— too·tler \'tüt-lər, 'tü-tᵊl-ər\ *noun*
too–too \'tü-'tü\ *adjective* (1881)
1 : going beyond the bounds of convention, good taste, or common sense : EXTREME
2 : LA-DI-DA
toot·sie \'tu̇t-sē\ *noun* [origin unknown] (1905)
1 : DEAR, SWEETHEART
2 : PROSTITUTE
toot·sy *also* **toot·sie** \'tu̇t-sē\ *noun, plural* **tootsies** [baby-talk alteration of *foot*] (1854)
: FOOT
[1]**top** \'täp\ *noun* [Middle English, from Old English; akin to Old High German *zopf* tip, tuft of hair] (before 12th century)
1 a (1) **:** the highest point, level, or part of something : SUMMIT, CROWN (2) **:** the head or top of the head — used especially in the phrase *top to toe* (3) **:** the head of a plant and especially one with edible roots ⟨beet *tops*⟩ (4) **:** a garment worn on the upper body **b** (1) **:** the highest or uppermost region or part (2) **:** the upper end, edge, or surface
2 : a fitted, integral, or attached part or unit serving as an upper piece, lid, or covering
3 a : a platform surrounding the head of a lower mast that serves to spread the topmast rigging, strengthen the mast, and furnish a standing place for men aloft **b :** a comparable part of the superstructure; *especially* : such a part on a warship used as a fire-control station or antiaircraft gun platform
4 a : the highest degree or pitch conceivable or attained : ACME, PINNACLE **b :** the loudest or highest range of a sound
5 a : the part that is nearest in space or time to the source or beginning **b :** the first half of an inning in baseball
6 a (1) **:** the highest position (as in rank or achievement) (2) **:** a person or thing at the top **b** *plural* **:** aces and kings in a hand or the three highest honors in a suit
7 : the choicest part : CREAM, PICK
8 : a forward spin given to a ball (as in golf or billiards) by striking it on or near the top or above the center; *also* : the stroke so given
— topped \'täpt\ *adjective*
— off the top of one's head : in an impromptu manner ⟨sat down and wrote the . . . story *off the top of his head* —Jerome Beatty, Jr.⟩
— on top of 1 a : in control of ⟨acted like a man *on top of* his job —*Newsweek*⟩ **b :** informed about ⟨tried to keep *on top of* new developments⟩ **2 :** in sudden and unexpected proximity to ⟨the deadline was *on top of* them⟩ **3 :** in addition to
— on top of the world : in a position of eminent success, happiness, or fame
— over the top : beyond the bounds of what is expected, usual, normal, or appropriate
[2]**top** *verb* **topped; top·ping** (1509)
transitive verb
1 : to remove or cut the top of: as **a :** to shorten or remove the top of (a plant) : PINCH 1b **b :** to remove the most volatile parts from (as crude petroleum)
2 a : to cover with a top or on the top : provide, form, or serve as a top for **b :** to supply with a decorative or protective finish or final touch **c :** to resupply or refill to capacity — usually used with *off* ⟨*topped* off the tank⟩ **d :** to complete the basic structure of (as a high-rise building) by putting on a cap or uppermost section — usually used with *out* or *off* **e :** to bring to an end or climax — usually used with *off* ⟨the event was *topped* off with a relay race —Paula Rodenas⟩
3 a : to be or become higher than : OVERTOP ⟨*tops* the previous record⟩ **b :** to be superior to : EXCEL, SURPASS **c :** to gain ascendancy over : DOMINATE
4 a : to rise to, reach, or be at the top of **b :** to go over the top of : CLEAR, SURMOUNT
5 : to strike (a ball) above the center thereby imparting topspin
intransitive verb
1 : to make an end, finish, or conclusion
2 : to reach a summit or crest — usually used with *off* or *out*
[3]**top** *adjective* (1593)
1 : of, relating to, or being at the top : UPPERMOST
2 : CHIEF, LEADING ⟨one of the world's *top* journalists⟩
3 : of the highest quality, amount, or degree ⟨*top* value⟩ ⟨*top* form⟩
4 : having a quantum characteristic whose existence was postulated on the basis of the discovery of the bottom quark ⟨*top* quark⟩
[4]**top** *noun* [Middle English, from Old English, plaything] (14th century)
: a commonly cylindrical or conoidal device that has a tapering usually steel-shod point on which it is made to spin and that is used especially as a toy
top- *or* **topo-** *combining form* [Middle English, from Late Latin, from Greek, from *topos*]
: place : locality ⟨*topo*logy⟩ ⟨*top*onymy⟩
to·paz \'tō-,paz\ *noun* [Middle English *topace*, from Old French, from Latin *topazus*, from Greek *topazos*] (13th century)
1 a : a mineral that is essentially a silicate of aluminum and usually occurs in orthorhombic translucent or transparent crystals or in white translucent masses **b :** a usually yellow to brownish yellow transparent mineral topaz used as a gem **c :** a yellow sapphire **d :** a yellow quartz
2 : either of two large brilliantly colored South American hummingbirds (*Topaza pella* and *T. pyra*)
top banana *noun* [from a burlesque routine involving three comedians in which the one that gets the punch line also gets a banana] (1952)
: the leading comedian in a burlesque show; *broadly* : KINGPIN 2
top billing *noun* (1945)
1 : prominent emphasis, featuring, or advertising
2 : the position at the top of a theatrical bill usually featuring the star's name
top boot *noun* (1768)
: a high boot often with light-colored leather bands around the upper part
top·coat \'täp-,kōt\ *noun* (1804)
1 : a lightweight overcoat
2 : OVERCOAT 2
top·cross \-,krȯs\ *noun* (1890)
: a cross between a superior or purebred male and inferior female stock to improve the average quality of the progeny; *also* : the product of such a cross
top dog *noun* (1900)
: a person or group in a position of authority especially through victory in a hard-fought competition
top dollar *noun* (1970)
: the highest amount being paid for a commodity or service ⟨willing to pay *top dollar* to get them —Dean Failey⟩
top–down \'täp-'dau̇n\ *adjective* [from the phrase *from the top down*] (1941)
1 : controlled, directed, or instituted from the top level ⟨a *top-down* corporate structure⟩
2 : proceeding by breaking large general aspects (as of a problem) into smaller more detailed constituents : working from the general to the specific ⟨*top-down* programming⟩ ⟨*top-down* design⟩
top drawer *noun* (1905)
: the highest level of society, authority, or excellence
— top–draw·er \-'drȯ(-ə)r\ *adjective*
top–dress \'täp-,dres\ *transitive verb* (1733)
: to apply material to (as land or a road) without working it in; *especially* : to scatter fertilizer over (land)
top·dress·ing \-,dre-siŋ\ *noun* (1744)
: a material used to top-dress soil
[1]**tope** \'tōp\ *intransitive verb* **toped; top·ing** [obsolete English *tope*, interjection used to wish good health before drinking] (1667)
: to drink liquor to excess
[2]**tope** *noun* [origin unknown] (1686)
: a small cosmopolitan shark (*Galeorhinus galeus*) with a liver very rich in vitamin A
[3]**tope** *noun* [perhaps from Panjabi *top* hat] (1815)

: STUPA

to·pee *or* **to·pi** \tō-'pē, 'tō-(,)pē\ *noun* [Hindi *ṭopī*] (1835)
: a lightweight helmet-shaped hat made of pith or cork

top·er \'tō-pər\ *noun* (1673)
: one that topes; *especially* : DRUNKARD

top·flight \'täp-'flīt\ *adjective* (1931)
: of, relating to, or being the highest level of achievement, excellence, or eminence
— **top flight** *noun*

[1]**Top 40** *noun plural* (1966)
: the forty best-selling phonograph records for a given period

[2]**Top 40** *adjective* (1966)
: constituting, playing, listing, or relating to the Top 40 ⟨*Top 40* hits⟩ ⟨*Top 40* stations⟩ ⟨*Top 40* charts⟩

top·ful *or* **top·full** \'täp-'fu̇l\ *adjective* (1553)
: BRIMFUL

[1]**top·gal·lant** \(,)täp-'ga-lənt, tə-'ga-\ *adjective* [[1]*top* + [2]*gallant*] (1514)
: of, relating to, or being a part next above the topmast and below the royal mast ⟨*topgallant* sails⟩ ⟨the *topgallant* mast⟩

[2]**topgallant** *noun* (1581)
1 *archaic* : the topmost point : SUMMIT ⟨the high *topgallant* of my joy —Shakespeare⟩
2 : a topgallant mast or sail

top gun *noun* (1976)
: one who is at the top (as in ability, rank, or prestige)

top–ham·per \'täp-'ham-pər\ *noun* (1791)
1 : matter or weight (as spars or rigging) in the upper part of a ship
2 : unnecessary cumbersome matter

top hat *noun* (1881)
: a tall-crowned hat usually of beaver or silk

top–heavy \'täp-,he-vē\ *adjective* (circa 1533)
1 : having the top part too heavy for the lower part
2 : having too high a proportion of administrators ⟨a *top-heavy* bureaucracy⟩
3 : oversupplied with one element at the expense of others : lacking balance ⟨a novel *top-heavy* with description⟩

To·phet \'tō-fət\ *noun* [Middle English, shrine south of ancient Jerusalem where human sacrifices were performed to Moloch in Jeremiah 7:31, Gehenna, from Late Latin *Topheth*, from Hebrew *topheth*] (14th century)
: HELL, GEHENNA

top–hole \'täp-'hōl\ *adjective* (1908)
chiefly British : EXCELLENT, FIRST-CLASS

to·phus \'tō-fəs\ *noun, plural* **to·phi** \'tō-,fī, -,fē\ [Latin, tufa] (1607)
: a deposit of urates in tissues (as cartilage) characteristic of gout

to·pi \'tō-pē\ *noun, plural* **topi** *or* **topis** [perhaps from Swahili] (1894)
: a sub-Saharan antelope (*Damaliscis lunatus* synonym *D. korrigum*) having a glossy reddish brown coat with purplish black and yellowish markings

[1]**to·pi·ary** \'tō-pē-,er-ē\ *adjective* [Latin *topiarius*, from *topia* ornamental gardening, irregular from Greek *topos* place] (1592)
: of, relating to, or being the practice or art of training, cutting, and trimming trees or shrubs into odd or ornamental shapes; *also* : characterized by such work

[2]**topiary** *noun, plural* **-ar·ies** (1908)
1 : topiary art or gardening; *also* : a topiary garden
2 a : a plant shaped by topiary art **b** : topiary plants

top·ic \'tä-pik\ *noun* [Latin *Topica* Topics (work by Aristotle), from Greek *Topika*, from *topika*, neuter plural of *topikos* of a place, of a topos, from *topos* place, topos] (1603)
1 a : one of the general forms of argument employed in probable reasoning **b** : ARGUMENT, REASON
2 a : a heading in an outlined argument or exposition **b** : the subject of a discourse or of a section of a discourse

top·i·cal \-pi-kəl\ *adjective* (1588)
1 : designed for or involving local application and action (as on the body) ⟨a *topical* anesthetic⟩ ⟨a *topical* remedy⟩
2 a : of, relating to, or arranged by topics ⟨set down in *topical* form⟩ **b** : referring to the topics of the day or place : of local or temporary interest ⟨a *topical* novel⟩ ⟨*topical* references⟩
— **top·i·cal·ly** \-k(ə-)lē\ *adverb*

top·i·cal·i·ty \,tä-pə-'ka-lə-tē\ *noun, plural* **-ties** (1904)
1 : the quality or state of being topical
2 : an item of topical interest

topic sentence *noun* (1885)
: a sentence that states the main thought of a paragraph or of a larger unit of discourse and is usually placed at or near the beginning

top·kick \'täp-'kik\ *noun* (1918)
: FIRST SERGEANT 1

top·knot \-,nät\ *noun* (circa 1688)
1 : an ornament (as a knot of ribbons or a pompom) forming a headdress or worn as part of a coiffure
2 : a crest of feathers or tuft of hair on the top of the head

top·less \-ləs\ *adjective* (1589)
1 *archaic* : so high as to reach up beyond sight ⟨and burnt the *topless* towers of Ilium —Christopher Marlowe⟩
2 : being without a top
3 a : wearing no clothing on the upper body **b** : featuring topless waitresses or entertainers ⟨*topless* bars⟩ **c** : being a place where topless women are permitted ⟨a *topless* beach⟩
— **top·less·ness** \-nəs\ *noun*

top·line \-,līn\ *noun* (circa 1909)
: the outline of the top of the body of an animal (as a dog or horse)

top·lofty \'täp-,lȯf-tē\ *also* **top·loft·i·cal** \täp-'lȯf-ti-kəl\ *adjective* [probably from the phrase *top loft*] (1823)
: very superior in air or attitude
— **top·loft·i·ly** \täp-'lȯf-tə-lē\ *adverb*
— **top·loft·i·ness** \'täp-,lȯf-tē-nəs\ *noun*

top·mast \'täp-,mast, -məst\ *noun* (15th century)
: the mast that is next above the lower mast and is topmost in a fore-and-aft rig

top milk *noun* (1891)
: the upper layer of milk in a container enriched by whatever cream has risen

top·min·now \'täp-,mi-(,)nō\ *noun* (1884)
: any of several live-bearers (family Poeciliidae) or killifish (family Cyprinodontidae)

top·most \'täp-,mōst\ *adjective* (1697)
: highest of all : UPPERMOST

top–notch \-'näch\ *adjective* (1900)
: of the highest quality : FIRST-RATE
— **top·notch·er** \-'nä-chər\ *noun*

to·po·cen·tric \,tä-pə-'sen-trik, ,tō-\ *adjective* (circa 1942)
: relating to, measured from, or as if observed from a particular point on the earth's surface : having or relating to such a point as origin ⟨*topocentric* coordinates⟩ — compare GEOCENTRIC

to·pog·ra·pher \tə-'pä-grə-fər\ *noun* (1603)
: one skilled in topography

to·po·graph·ic \,tä-pə-'gra-fik, ,tō-\ *adjective* (1632)
: of, relating to, or concerned with topography ⟨*topographic* maps⟩

to·po·graph·i·cal \-fi-kəl\ *adjective* (circa 1576)
1 : TOPOGRAPHIC
2 : of, relating to, or concerned with the artistic representation of a particular locality ⟨a *topographical* poem⟩ ⟨*topographical* painting⟩
— **to·po·graph·i·cal·ly** \-k(ə-)lē\ *adverb*

to·pog·ra·phy \tə-'pä-grə-fē\ *noun* [Middle English *topographie*, from Late Latin *topographia*, from Greek, from *topographein* to describe a place, from *topos* place + *graphein* to write — more at CARVE] (15th century)
1 a : the art or practice of graphic delineation in detail usually on maps or charts of natural and man-made features of a place or region especially in a way to show their relative positions and elevations **b** : topographical surveying
2 a : the configuration of a surface including its relief and the position of its natural and man-made features **b** : the physical or natural features of an object or entity and their structural relationships

to·po·log·i·cal \,tä-pə-'lä-ji-kəl\ *adjective* (1715)
1 : of or relating to topology
2 : being or involving properties unaltered under a homeomorphism ⟨continuity and connectedness are *topological* properties⟩
— **to·po·log·i·cal·ly** \-k(ə-)lē\ *adverb*

topological group *noun* (1946)
: a mathematical group which is also a topological space, whose multiplicative operation is continuous such that given any neighborhood of a product there exist neighborhoods of the elements composing the product with the property that any pair of elements representing each of these neighborhoods form a product belonging to the given neighborhood, and whose operation of taking inverses is continuous such that for any neighborhood of the inverse of an element there exists a neighborhood of the element itself in which every element has its inverse in the other neighborhood

topologically equivalent *adjective* (1915)
: related by a homeomorphism

topological space *noun* (1926)
: a set with a collection of subsets satisfying the conditions that both the empty set and the set itself belong to the collection, the union of any number of the subsets is also an element of the collection, and the intersection of any finite number of the subsets is an element of the collection

topological transformation *noun* (1946)
: HOMEOMORPHISM

to·pol·o·gy \tə-'pä-lə-jē, tä-\ *noun, plural* **-gies** [International Scientific Vocabulary] (1850)
1 : topographic study of a particular place; *specifically* : the history of a region as indicated by its topography
2 a (1) : a branch of mathematics concerned with those properties of geometric configurations (as point sets) which are unaltered by elastic deformations (as a stretching or a twisting) that are homeomorphisms (2) : the set of all open subsets of a topological space **b** : CONFIGURATION ⟨*topology* of a molecule⟩ ⟨*topology* of a magnetic field⟩
— **to·pol·o·gist** \-jist\ *noun*

top·o·nym \'tä-pə-,nim, 'tō-\ *noun* [International Scientific Vocabulary, back-formation from *toponymy*] (1899)
: PLACE-NAME

top·o·nym·ic \,tä-pə-'ni-mik, ,tō-\ *adjective* (1896)
: of or relating to toponyms or toponymy
— **top·o·nym·i·cal** \-mi-kəl\ *adjective*

to·pon·y·my \tə-'pä-nə-mē, tō-\ *noun* [International Scientific Vocabulary, from *top-* + Greek *onyma, onoma* name — more at NAME] (1876)
: the place-names of a region or language or especially the etymological study of them
— **to·pon·y·mist** \-mist\ *noun*

to·pos \'tō-,päs, 'tä-\ *noun, plural* **to·poi** \-,pȯi\ [Greek, short for *koinos topos*, literally, common place] (1936)
: a traditional or conventional literary or rhetorical theme or topic

top·per \'tä-pər\ *noun* (1688)
1 : one that puts on or takes off tops

2 : one that is at or on the top
3 a : SILK HAT **b :** OPERA HAT
4 : something (as a joke) that caps everything preceding
5 : a woman's usually short and loose-fitting lightweight outer coat
¹top·ping \'tä-piŋ\ *noun* (14th century)
1 : something that forms a top; *especially* **:** a garnish (as a sauce, bread crumbs, or whipped cream) placed on top of a food for flavor or decoration
2 : the action of one that tops
3 : something removed by topping
²topping *adjective* (circa 1685)
1 : highest in rank or eminence
2 *New England* **:** PROUD, ARROGANT
3 *chiefly British* **:** EXCELLENT
top·ple \'tä-pəl\ *verb* **top·pled; top·pling** \-p(ə-)liŋ\ [frequentative of ²*top*] (1590)
intransitive verb
: to fall from or as if from being top-heavy
transitive verb
1 : to cause to topple
2 a : OVERTHROW 2 **b :** DEFEAT 3
top round *noun* (1903)
: meat from the inner part of a round of beef
¹tops \'täps\ *adjective* [plural of ¹*top*] (1935)
: topmost in quality, ability, popularity, or importance — used predicatively ⟨is *tops* in his field⟩
²tops *adverb* (1956)
: at the very most ⟨will cost $50, *tops*⟩
top·sail \'täp-ˌsāl, -səl\ *also* **top·s'l** \-səl\ *noun* (14th century)
1 : the sail next above the lowermost sail on a mast in a square-rigged ship
2 : the sail set above and sometimes on the gaff in a fore-and-aft rigged ship
top secret *adjective* (1944)
1 : protected by a high degree of secrecy ⟨a *top secret* weapon⟩ ⟨a *top secret* meeting⟩
2 a : containing or being information whose unauthorized disclosure could result in exceptionally grave danger to the nation ⟨*top secret* messages⟩ — compare CONFIDENTIAL, SECRET **b :** of or relating to top secret documents ⟨a *top secret* clearance⟩
top sergeant *noun* (1898)
: FIRST SERGEANT 1
top–shelf \'täp-'shelf\ *adjective* (circa 1892)
: of the best quality
¹top·side \'täp-'sīd\ *noun* (1815)
1 *plural* **:** the top portion of the outer surface of a ship on each side above the waterline
2 : the highest level of authority
3 : the upper portion of the ionosphere
²topside *adverb or adjective* (1873)
1 : to or on the top or surface
2 : in a position of authority
3 : on deck
Top–Sid·er \'täp-ˌsī-dər\ *trademark*
— used for a low casual shoe having a rubber sole
top·soil \'täp-ˌsȯil\ *noun* (1836)
: surface soil usually including the organic layer in which plants have most of their roots and which the farmer turns over in plowing
top·spin \-ˌspin\ *noun* [¹*top*] (1902)
: a rotary motion imparted to a ball that causes it to rotate forward in the direction it is traveling
top·stitch \'täp-ˌstich\ *transitive verb* (1934)
: to make a line of stitching on the outside of (a garment) close to a seam
top·sy–tur·vi·ness \ˌtäp-sē-'tər-vē-nəs\ *noun* (1842)
: the quality or state of being topsy-turvy
¹top·sy–tur·vy \ˌtäp-sē-'tər-vē\ *adverb* [probably ultimately from *tops* (plural of ¹*top*) + obsolete English *terve* to turn upside down] (1528)
1 : in utter confusion or disorder
2 : with the top or head downward **:** UPSIDE DOWN
²topsy–turvy *adjective* (1612)
: turned topsy-turvy **:** totally disordered
— **top·sy–tur·vi·ly** \-'tər-və-lē\ *adverb*
— **top·sy–tur·vy·dom** \-vē-dəm\ *noun*
³topsy–turvy *noun* (1655)
: TOPSY-TURVINESS
top up (1937)
transitive verb
British **:** to make up to the full quantity, capacity, or amount
intransitive verb
British **:** to replenish a supply
— **top–up** \'täp-ˌəp\ *noun, British*
top·work \'täp-ˌwərk\ *transitive verb* (1882)
: to graft scions of another variety on the main branches of (as fruit trees) usually to obtain more desirable fruit
toque \'tōk\ *noun* [Middle French, soft hat with a narrow brim worn especially in the 16th century, from Old Spanish *toca* headdress] (1505)
1 : a woman's small hat without a brim made in any of various soft close-fitting shapes
2 : TUQUE
3 : a tall brimless hat worn by a chef — called also *toque blanche*
tor \'tȯr\ *noun* [Middle English, from Old English *torr*] (before 12th century)
: a high craggy hill
To·rah \'tȯr-ə, 'tȯr-; 'tȯi-rə\ *noun* [Hebrew *tōrāh*] (1577)
1 : the five books of Moses constituting the Pentateuch
2 : the body of wisdom and law contained in Jewish Scripture and other sacred literature and oral tradition
3 : a leather or parchment scroll of the Pentateuch used in a synagogue for liturgical purposes
¹torch \'tȯrch\ *noun, often attributive* [Middle English *torche,* from Old French, bundle of twisted straw or tow, torch, from (assumed) Vulgar Latin *torca;* akin to Latin *torquēre* to twist — more at TORTURE] (14th century)
1 : a burning stick of resinous wood or twist of tow used to give light and usually carried in the hand **:** FLAMBEAU
2 : something (as tradition, wisdom, or knowledge) likened to a torch as giving light or guidance
3 : any of various portable devices for emitting an unusually hot flame — compare BLOWTORCH
4 *chiefly British* **:** FLASHLIGHT 2
5 : INCENDIARY 1a
²torch *transitive verb* (1901)
: to set fire to with or as if with a torch
torch·bear·er \-ˌbar-ər, -ˌber-\ *noun* (1538)
1 : one that carries a torch
2 : someone in the forefront of a campaign, crusade, or movement
tor·chère \ˌtȯr-'sher\ *noun* [French, from *torche* torch] (1904)
1 : a tall ornamental stand for a candlestick or candelabra
2 : an electric floor lamp giving indirect light
torch·light \-ˌlīt\ *noun* (15th century)
1 : light given by torches
2 : TORCH
tor·chon \'tȯr-ˌshän\ *noun* [French, duster, from Old French, bundle of twisted straw, from *torche*] (1865)
: a coarse bobbin or machine-made lace made with fan-shaped designs forming a scalloped edge
torch singer *noun* (circa 1932)
: a singer of torch songs
torch song *noun* [from the phrase *to carry a torch for* (to be in love)] (1930)
: a popular sentimental song of unrequited love
torch·wood \'tȯrch-ˌwu̇d\ *noun* (1833)
1 : any of a genus (*Amyris*) of tropical American trees and shrubs of the rue family with hard heavy fragrant resinous streaky yellowish brown wood
2 : the wood of a torchwood
torchy \'tȯr-chē\ *adjective* **torch·i·er; -est** (1941)
: of, relating to, or characteristic of a torch song or torch singer
tore *past of* TEAR
to·re·a·dor \'tȯr-ē-ə-ˌdȯr, 'tōr-, 'tär-\ *noun* [Spanish, from *torear* to fight bulls, from *toro* bull, from Latin *taurus* — more at TAURINE] (1618)
: TORERO, BULLFIGHTER
to·re·ro \tə-'rer-(ˌ)ō\ *noun, plural* **-ros** [Spanish, from Latin *taurarius* bullfighter, from Latin *taurus* bull] (1728)
: a matador or a member of the attending cuadrilla
to·reu·tics \tə-'rü-tiks\ *noun plural but singular in construction* [*toreutic,* adjective, from Greek *toreutikos,* from *toreuein* to bore through, chase, from *toreus* boring tool; akin to Greek *tetrainein* to bore — more at THROW] (1847)
: the art or process of working in metal especially by embossing or chasing
— **to·reu·tic** \-'rü-tik\ *adjective*
tori *plural of* TORUS
to·ric \'tȯr-ik, 'tȯr-\ *adjective* (circa 1898)
: of, relating to, or shaped like a torus or segment of a torus ⟨a *toric* lens⟩
to·rii \'tȯr-ē-ˌē, 'tȯr-\ *noun, plural* **torii** [Japanese] (1727)
: a Japanese gateway of light construction commonly built at the approach to a Shinto shrine
¹tor·ment \'tȯr-ˌment\ *noun* [Middle English, from Old French, from Latin *tormentum* torture; akin to *torquēre* to twist — more at TORTURE] (14th century)
1 : the infliction of torture (as by rack or wheel)
2 : extreme pain or anguish of body or mind **:** AGONY
3 : a source of vexation or pain
²tor·ment \tȯr-'ment, 'tȯr-ˌ\ *transitive verb* (14th century)
1 : to cause severe usually persistent or recurrent distress of body or mind to ⟨cattle *tormented* by flies⟩
2 : DISTORT, TWIST
synonym see AFFLICT
tor·men·til \'tȯr-mən-ˌtil\ *noun* [Middle English *turmentill,* from Medieval Latin *tormentilla,* from Latin *tormentum;* from its use in allaying pain] (15th century)
: a yellow-flowered Eurasian cinquefoil (*Potentilla erecta* synonym *P. tormentilla*) with a root sometimes used in tanning and dyeing
tor·men·tor *also* **tor·ment·er** \tȯr-'men-tər, 'tȯr-ˌ\ *noun* (14th century)
1 : one that torments
2 : a fixed curtain or flat on each side of a theater stage that prevents the audience from seeing into the wings
torn *past participle of* TEAR
tor·na·dic \tȯr-'nā-dik, -'na-\ *adjective* (1884)
: relating to, characteristic of, or constituting a tornado
tor·na·do \tȯr-'nā-(ˌ)dō\ *noun, plural* **-does** *or* **-dos** [modification of Spanish *tronada* thunderstorm, from *tronar* to thunder, from Latin *tonare* — more at THUNDER] (1556)
1 *archaic* **:** a tropical thunderstorm
2 a : a squall accompanying a thunderstorm in Africa **b :** a violent destructive whirling wind accompanied by a funnel-shaped cloud that progresses in a narrow path over the land
3 : a violent windstorm **:** WHIRLWIND
tor·nil·lo \tȯr-'nē-(ˌ)yō, -'ni-(ˌ)lō\ *noun, plural* **-los** [Spanish, literally, small lathe, screw, diminutive of *torno* lathe, from Latin *tornus* — more at TURN] (circa 1844)
: SCREWBEAN 1
to·roid \'tōr-ˌȯid, 'tȯr-\ *noun* [New Latin *torus*] (circa 1900)
1 : a surface generated by a plane closed curve rotated about a line that lies in the same plane as the curve but does not intersect it

2 : a body whose surface has the form of a toroid

to·roi·dal \tō-ˈrȯi-dᵊl\ *adjective* (circa 1889)
: of, relating to, or shaped like a torus or toroid **:** doughnut-shaped ⟨a *toroidal* resistance coil⟩
— **to·roi·dal·ly** \-dᵊl-ē\ *adverb*

[1]**tor·pe·do** \tȯr-ˈpē-(ˌ)dō\ *noun, plural* **-does** [Latin, literally, stiffness, numbness, from *torpēre* to be sluggish or numb — more at TORPID] (circa 1520)
1 : ELECTRIC RAY
2 : a weapon for destroying ships by rupturing their hulls below the waterline: as **a :** a submarine mine **b :** a thin cylindrical self-propelled underwater projectile
3 : a small firework that explodes when thrown against a hard object
4 : a professional gunman or assassin
5 : SUBMARINE 2 ◆

[2]**torpedo** *transitive verb* **tor·pe·doed; tor·pe·do·ing** \-ˈpē-də-wiŋ\ (circa 1879)
1 : to hit or sink (a ship) with a naval torpedo **:** strike or destroy by torpedo
2 : to destroy or nullify altogether **:** WRECK ⟨*torpedo* a plan⟩

torpedo boat *noun* (1810)
: a boat designed for launching torpedoes; *specifically* **:** a small very fast boat with one or more torpedo tubes

torpedo bomber *noun* (1930)
: a military airplane designed to carry torpedoes — called also *torpedo plane*

torpedo tube *noun* (circa 1891)
: a tube from which torpedoes are fired

tor·pid \ˈtȯr-pəd\ *adjective* [Latin *torpidus,* from *torpēre* to be sluggish or numb; akin to Lithuanian *tirpti* to become numb] (1613)
1 a : having lost motion or the power of exertion or feeling **:** DORMANT, NUMB **b :** sluggish in functioning or acting ⟨a *torpid* frog⟩ ⟨a *torpid* mind⟩
2 : lacking in energy or vigor **:** APATHETIC, DULL
— **tor·pid·i·ty** \tȯr-ˈpi-də-tē\ *noun*

tor·por \ˈtȯr-pər\ *noun* [Latin, from *torpēre*] (1607)
1 : APATHY, DULLNESS
2 : a state of mental and motor inactivity with partial or total insensibility **:** extreme sluggishness or stagnation of function
synonym see LETHARGY

[1]**torque** \ˈtȯrk\ *noun* [French, from Latin *torques,* from *torquēre* to twist — more at TORTURE] (1695)
: a usually metal collar or neck chain worn by the ancient Gauls, Germans, and Britons

[2]**torque** *noun* [Latin *torquēre* to twist] (circa 1884)
1 : a force that produces or tends to produce rotation or torsion ⟨an automobile engine delivers *torque* to the drive shaft⟩; *also* **:** a measure of the effectiveness of such a force that consists of the product of the force and the perpendicular distance from the line of action of the force to the axis of rotation
2 : a turning or twisting force

[3]**torque** *transitive verb* **torqued; torqu·ing** (1959)
: to impart torque to **:** cause to twist (as about an axis)
— **torqu·er** *noun*

torque converter *noun* (1927)
: a device for transmitting and amplifying torque especially by hydraulic means

torr \ˈtȯr\ *noun, plural* **torr** [Evangelista *Torricelli*] (1949)
: a unit of pressure equal to 1/760 of an atmosphere (about 133.3 pascals)

[1]**tor·rent** \ˈtȯr-ənt, ˈtär-\ *noun* [Middle French, from Latin *torrent-, torrens,* from *torrent-, torrens,* adjective, burning, seething, rushing, from present participle of *torrēre* to parch, burn — more at THIRST] (1582)
1 : a tumultuous outpouring **:** RUSH
2 : a violent stream of a liquid (as water or lava)
3 : a channel of a mountain stream

[2]**torrent** *adjective* (1667)
: TORRENTIAL

tor·ren·tial \tȯ-ˈren(t)-shəl, tə-\ *adjective* (1849)
1 a : relating to or having the character of a torrent ⟨*torrential* rains⟩ **b :** caused by or resulting from action of rapid streams ⟨*torrential* gravel⟩
2 : resembling a torrent in violence or rapidity of flow
— **tor·ren·tial·ly** \-ˈren(t)-sh(ə-)lē\ *adverb*

tor·rid \ˈtȯr-əd, ˈtär-\ *adjective* [Latin *torridus,* from *torrēre*] (1545)
1 a : parched with heat especially of the sun **:** HOT ⟨*torrid* sands⟩ **b :** giving off intense heat **:** SCORCHING
2 : ARDENT, PASSIONATE ⟨*torrid* love letters⟩
— **tor·rid·i·ty** \tō-ˈri-də-tē\ *noun*
— **tor·rid·ly** \ˈtȯr-əd-lē, ˈtär-\ *adverb*
— **tor·rid·ness** *noun*

torrid zone *noun, often T&Z capitalized* (1586)
: the region of the earth between the tropic of Cancer and the tropic of Capricorn

tor·sade \tȯr-ˈsäd, -ˈsad\ *noun* [French, from obsolete French *tors* twisted, from (assumed) Vulgar Latin *torsus,* alteration of Latin *tortus,* past participle of *torquēre* to twist] (1882)
: a twisted cord or ribbon used especially as a hat ornament

tor·sion \ˈtȯr-shən\ *noun* [Late Latin *torsion-, torsio* torment, alteration of Latin *tortio,* from *torquēre* to twist] (1543)
1 : the twisting or wrenching of a body by the exertion of forces tending to turn one end or part about a longitudinal axis while the other is held fast or turned in the opposite direction; *also* **:** the state of being twisted
2 : the twisting of a bodily organ or part on its own axis
3 : the reactive torque that an elastic solid exerts by reason of being under torsion
— **tor·sion·al** \ˈtȯr-shnəl, -shə-nᵊl\ *adjective*
— **tor·sion·al·ly** *adverb*

torsion bar *noun* (1948)
: a long metal element in an automobile suspension that has one end held rigidly to the frame end and the other twisted and connected to the axle and that acts as a spring

tor·so \ˈtȯr-(ˌ)sō\ *noun, plural* **torsos** *or* **tor·si** \ˈtȯr-ˌsē\ [Italian, literally, stalk, from Latin *thyrsus* stalk, thyrsus] (1722)
1 : a sculptured representation of the trunk of a human body
2 : something (as a piece of writing) that is mutilated or left unfinished
3 : the human trunk

tort \ˈtȯrt\ *noun* [Middle English, injury, from Middle French, from Medieval Latin *tortum,* from Latin, neuter of *tortus* twisted, from past participle of *torquēre*] (1586)
: a wrongful act other than a breach of contract for which relief may be obtained in the form of damages or an injunction

torte \ˈtȯrt-ə, ˈtȯrt\ *noun, plural* **tor·ten** \ˈtȯr-tᵊn\ *or* **tortes** [German; probably from Italian *torta* cake, from Late Latin, round loaf of bread] (1555)
: a cake made with many eggs and often grated nuts or dry bread crumbs and usually covered with a rich frosting

tor·tel·li·ni \ˌtȯr-tᵊl-ˈē-nē\ *noun* [Italian, plural of *tortellino* pasta round, diminutive of *tortello,* from *torta* cake] (circa 1911)
: pasta cut in rounds, folded around a filling (as of meat or cheese), formed into rings, and boiled

tor·ti·col·lis \ˌtȯr-tə-ˈkä-ləs\ *noun* [New Latin, from Latin *tortus* twisted + *-i-* + *collum* neck — more at COLLAR] (circa 1811)
: a twisting of the neck to one side that results in abnormal carriage of the head and is usually caused by muscle spasms — called also *wryneck*

tor·ti·lla \tȯr-ˈtē-yə\ *noun* [American Spanish, from Spanish, diminutive of *torta* cake, from Late Latin, round loaf of bread] (circa 1699)
: a round thin cake of unleavened cornmeal or wheat flour bread usually eaten hot with a topping or filling (as of ground meat or cheese)

tor·tious \ˈtȯr-shəs\ *adjective* (1544)
: implying or involving tort
— **tor·tious·ly** *adverb*

tor·toise \ˈtȯr-təs\ *noun* [Middle English *tortu, tortuce,* from Middle French *tortue* — more at TURTLE] (14th century)
1 : any of a family (Testudinidae) of terrestrial turtles; *broadly* **:** TURTLE
2 : someone or something regarded as slow or laggard

tortoise beetle *noun* (circa 1711)
: any of various chrysomelid beetles (subfamily Cassidinae) with leaf-eating larvae

[1]**tor·toise·shell** \ˈtȯr-tə-ˌshel, -əsh-ˌshel\ *noun* (1632)
1 : the mottled horny substance of the shell of some turtles (as the hawksbill turtle) used in inlaying and in making various ornamental articles
2 : any of several showy nymphalid butterflies (genus *Nymphalis*)

[2]**tortoiseshell** *adjective* (1651)
1 : made of or resembling tortoiseshell especially in mottled brown and yellow coloring
2 : of, relating to, or being a color pattern of the domestic cat consisting of patches of black, orange, and cream

tor·to·ni \tȯr-ˈtō-nē\ *noun* [probably from *Tortoni* 19th century Italian restaurateur in Paris] (1922)
: ice cream made of heavy cream often with minced almonds and chopped maraschino cherries and often flavored with rum

tor·tri·cid \ˈtȯr-trə-səd\ *noun* [New Latin *Tortricidae,* from *Tortric-, Tortrix*] (circa 1891)

◇ WORD HISTORY

torpedo In 1776, David Bushnell, a Connecticut native and supporter of the American cause in the Revolutionary War, devised and built a wooden submarine with a detachable powder magazine that was to be hooked to an enemy ship and set to explode after the submarine had steered clear. According to later accounts, Bushnell dubbed his magazine, with a mixture of learning and whimsy, with the Latin word *torpedo* "electric ray," a fish that delivers an electric shock. The following year Bushnell experimented with floating mines—powder kegs set off by a spring-released flintlock—but neither his submarine nor his mines succeeded in damaging a British ship. Both of Bushnell's innovations were taken up in the early 1800s by the American inventor Robert Fulton, who also employed *torpedo* as the name for a "submarine bomb." Though Fulton's schemes came to naught as well, *torpedo* became fixed as the term for a naval mine. This situation changed in the late 1860s, when the British engineer Robert Whitehead perfected the cylindrical, self-propelled underwater projectile to which the word *torpedo* soon referred exclusively. Bushnell's crude experiments bore awful fruit in World War I, when submarine-launched torpedoes sank hundreds of ships and changed the character of warfare.

: any of a family (Tortricidae) of small stout-bodied moths many of whose larvae feed in fruits
— **tortricid** *adjective*

tor·trix \'tȯr-triks\ *noun* [New Latin *Tortric-, Tortrix,* genus of moths, from Latin *tortus,* past participle of *torquēre* to twist; from the habit of twisting or rolling leaves to make a nest] (circa 1797)
: a tortricid moth

tor·tu·os·i·ty \ˌtȯr-chə-'wä-sə-tē\ *noun, plural* **-ties** (1603)
1 : the quality or state of being tortuous
2 : something winding or twisted **:** BEND

tor·tu·ous \'tȯrch-wəs, 'tȯr-chə-\ *adjective* [Middle English, from Middle French *tortueux,* from Latin *tortuosus,* from *tortus* twist, from *torquēre* to twist] (15th century)
1 : marked by repeated twists, bends, or turns **:** WINDING ⟨a *tortuous* path⟩
2 a : marked by devious or indirect tactics **:** CROOKED, TRICKY ⟨a *tortuous* conspiracy⟩ **b :** CIRCUITOUS, INVOLVED ⟨the *tortuous* jargon of legal forms⟩
— **tor·tu·ous·ly** *adverb*
— **tor·tu·ous·ness** *noun*

¹tor·ture \'tȯr-chər\ *noun* [French, from Late Latin *tortura,* from Latin *tortus,* past participle of *torquēre* to twist; probably akin to Old High German *drāhsil* turner, Greek *atraktos* spindle] (1540)
1 a : anguish of body or mind **:** AGONY **b :** something that causes agony or pain
2 : the infliction of intense pain (as from burning, crushing, or wounding) to punish, coerce, or afford sadistic pleasure
3 : distortion or overrefinement of a meaning or an argument **:** STRAINING

²torture *transitive verb* **tor·tured; tor·tur·ing** \'tȯrch-riŋ, 'tȯr-chə-\ (1588)
1 : to cause intense suffering to **:** TORMENT
2 : to punish or coerce by inflicting excruciating pain
3 : to twist or wrench out of shape **:** DISTORT, WARP
synonym see AFFLICT
— **tor·tur·er** \'tȯr-chər-ər\ *noun*

tor·tur·ous \'tȯrch-rəs, 'tȯr-chə-\ *adjective* (15th century)
1 a : causing torture ⟨*torturous* inquisitions⟩ **b :** very unpleasant or painful ⟨a *torturous* day⟩ ⟨*torturous* self-doubts⟩
2 : painfully difficult or slow ⟨the *torturous* course of the negotiations⟩
— **tor·tur·ous·ly** *adverb*

tor·u·la \'tȯr-yə-lə, 'tär-, -ə-lə\ *noun, plural* **-lae** \-ˌlē, -ˌlī\ *also* **-las** [New Latin, from Latin *torus* protuberance] (1861)
: any of various fungi and especially yeasts that lack sexual spores, do not produce alcoholic fermentations, and are typically acid formers — called also *torula yeast*

to·rus \'tōr-əs, 'tȯr-\ *noun, plural* **to·ri** \'tōr-ˌī, 'tȯr-, -ˌē\ [New Latin, from Latin, protuberance, bulge, torus molding] (1563)
1 : a large molding of convex profile commonly occurring as the lowest molding in the base of a column — see BASE illustration
2 : the thickening of a membrane closing a wood-cell pit (as of gymnosperm tracheids) having the secondary cell wall arched over the pit cavity
3 : a doughnut-shaped surface generated by a circle rotated about an axis in its plane that does not intersect the circle; *broadly* **:** TOROID
4 : a smooth rounded anatomical protuberance (as a bony ridge on the skull)

To·ry \'tōr-ē, 'tȯr-\ *noun, plural* **Tories** [Irish *tóraidhe* outlaw, robber, from Middle Irish *tóir* pursuit] (1646)
1 : a dispossessed Irishman subsisting as an outlaw chiefly in the 17th century
2 *obsolete* **:** BANDIT, OUTLAW
3 a : a member or supporter of a major British political group of the 18th and early 19th centuries favoring at first the Stuarts and later royal authority and the established church and seeking to preserve the traditional political structure and defeat parliamentary reform — compare WHIG **b :** CONSERVATIVE 1b
4 : an American upholding the cause of the British Crown against the supporters of colonial independence during the American Revolution **:** LOYALIST
5 *often not capitalized* **:** an extreme conservative especially in political and economic principles
— **Tory** *adjective*

Tory Democracy *noun* (1879)
: a political philosophy advocating preservation of established institutions and traditional principles combined with political democracy and a social and economic program designed to benefit the common man

To·ry·ism \'tōr-ē-ˌi-zəm, 'tȯr-\ *noun* (1682)
1 : the principles and practices of or associated with Tories
2 : the British Tory party or its members

tosh \'täsh\ *noun* [origin unknown] (1528)
: sheer nonsense **:** BOSH, TWADDLE

¹toss \'tȯs, 'täs\ *verb* [probably of Scandinavian origin; akin to Swedish dialect *tossa* to spread, scatter] (1506)
transitive verb
1 a : to fling or heave continuously about, to and fro, or up and down ⟨a ship *tossed* by waves⟩ **b :** BANDY 2b, c, d **c :** to mix lightly until well coated with a dressing or until the elements are thoroughly combined ⟨*toss* a salad⟩
2 : to make uneasy **:** stir up **:** DISTURB
3 a : to throw with a quick, light, or careless motion or with a sudden jerk ⟨*toss* a ball around⟩ **b :** to throw up in the air ⟨*tossed* by a bull⟩ **c :** MATCH 5a **d :** to send as if by throwing ⟨*tossed* in jail⟩ ⟨*tossed* out of the game⟩ **e :** to get rid of **:** THROW AWAY
4 a : to fling or lift with a sudden motion ⟨*tosses* her head angrily⟩ **b :** to tilt suddenly so as to empty by drinking ⟨*tossed* his glass⟩; *also* **:** to consume by drinking ⟨*toss* down a drink⟩
5 : to accomplish, provide, or produce readily or easily ⟨*toss* off a few verses⟩
6 : THROW 16 ⟨*toss* a party⟩
7 : VOMIT 1 — often used in the phrase *toss one's cookies*
intransitive verb
1 a : to move restlessly or turbulently; *especially* **:** to twist and turn repeatedly ⟨*tossed* sleeplessly all night⟩ **b :** to move with a quick or spirited gesture
2 : to decide an issue by flipping a coin
synonym see THROW
— **toss·er** *noun*

²toss *noun* (1634)
1 : the state or fact of being tossed
2 : an act or instance of tossing: as **a :** an abrupt tilting or upward fling **b :** a deciding by chance and especially by flipping a coin **c :** THROW, PITCH

toss·pot \-ˌpät\ *noun* (1568)
: DRUNKARD, SOT

toss–up \-ˌəp\ *noun* (1812)
1 : TOSS 2b
2 : an even chance
3 : something that offers no clear basis for choice

tos·ta·da \tō-'stä-də\ *also* **tos·ta·do** \-(ˌ)dō\ *noun* [Mexican Spanish *tostada,* from Spanish, feminine of *tostado,* past participle of *tostar* to toast, roast, from Late Latin *tostare* — more at TOAST] (1939)
: a tortilla fried in deep fat

¹tot \'tät\ *noun* [origin unknown] (1725)
1 : a small child **:** TODDLER
2 : a small drink or allowance of liquor **:** SHOT

²tot *verb* **tot·ted; tot·ting** [*tot.,* abbreviation of *total*] (circa 1772)
transitive verb
: to add together **:** TOTAL — usually used with *up* ⟨*tots* up the score⟩
intransitive verb
: ADD

¹to·tal \'tō-tᵊl\ *adjective* [Middle English, from Middle French, from Medieval Latin *totalis,* from Latin *totus* whole, entire] (15th century)
1 : comprising or constituting a whole **:** ENTIRE ⟨the *total* amount⟩
2 : ABSOLUTE, UTTER ⟨a *total* failure⟩
3 : involving a complete and unified effort especially to achieve a desired effect ⟨*total* war⟩ ⟨*total* theater⟩
synonym see WHOLE

²total *noun* (1557)
1 : a product of addition **:** SUM
2 : an entire quantity **:** AMOUNT

³total *adverb* (1601)
: TOTALLY

⁴total *transitive verb* **to·taled** *or* **to·talled; to·tal·ing** *or* **to·tal·ling** (1716)
1 : to add up **:** COMPUTE
2 : to amount to **:** NUMBER
3 : to make a total wreck of **:** DEMOLISH ⟨*totaled* the car⟩

total depravity *noun* (1794)
: a state of corruption due to original sin held in Calvinism to infect every part of man's nature and to make the natural man unable to know or obey God

total eclipse *noun* (1671)
: an eclipse in which one celestial body is completely obscured by the shadow or body of another

to·tal·ism \'tō-tᵊl-ˌi-zəm\ *noun* (1941)
: TOTALITARIANISM
— **to·tal·is·tic** \ˌtō-tᵊl-'is-tik\ *adjective*

¹to·tal·i·tar·i·an \(ˌ)tō-ˌta-lə-'ter-ē-ən\ *adjective* [Italian *totalitario,* from *totalità* totality] (1926)
1 a : of or relating to centralized control by an autocratic leader or hierarchy **:** AUTHORITARIAN, DICTATORIAL; *especially* **:** DESPOTIC **b :** of or relating to a political regime based on subordination of the individual to the state and strict control of all aspects of the life and productive capacity of the nation especially by coercive measures (as censorship and terrorism)
2 a : advocating or characteristic of totalitarianism **b :** completely regulated by the state especially as an aid to national mobilization in an emergency **c :** exercising autocratic powers **:** tending toward monopoly

²totalitarian *noun* (circa 1934)
: an advocate or practitioner of totalitarianism

to·tal·i·tar·i·an·ism \(ˌ)tō-ˌta-lə-'ter-ē-ə-ˌni-zəm\ *noun* (1926)
1 : centralized control by an autocratic authority
2 : the political concept that the citizen should be totally subject to an absolute state authority

to·tal·i·tar·i·an·ize \-ˌnīz\ *transitive verb* **-ized; -iz·ing** (1935)
: to make totalitarian ⟨a society *totalitarianized* by the military-industrial complex —W. F. Buckley (born 1925)⟩

to·tal·i·ty \tō-'ta-lə-tē\ *noun, plural* **-ties** (1598)
1 : an aggregate amount **:** SUM, WHOLE
2 a : the quality or state of being total **:** WHOLENESS **b :** the phase of an eclipse during which it is total **:** state of total eclipse

to·tal·iza·tor *or* **to·tal·isa·tor** \'tō-tᵊl-ə-ˌzā-tər\ *noun* (1879)
: PARI-MUTUEL 2

to·tal·ize \'tō-tᵊl-ˌīz\ *transitive verb* **-ized; -iz·ing** (1818)
1 : to add up **:** TOTAL
2 : to express as a whole

to·tal·iz·er \-ˌī-zər\ *noun* (1887)
: one that totalizes: as **a :** PARI-MUTUEL 2 **b :** a device (as a meter) that records a remaining total (as of fuel)

to·tal·ly \'tō-tᵊl-ē\ *adverb* (1509)
: in a total manner **:** to a total or complete degree **:** WHOLLY, ENTIRELY

total recall *noun* (1926)

: the faculty of remembering with complete clarity and in complete detail

[1]tote \'tōt\ *transitive verb* **tot·ed; tot·ing** [perhaps from an English-based creole; akin to Gullah & Krio *tot* to carry] (1677)
1 : to carry by hand **:** bear on the person **:** LUG, PACK
2 : HAUL, CONVEY ◆
— tot·er \'tō-tər\ *noun*

[2]tote *noun* (circa 1772)
1 : BURDEN, LOAD
2 : TOTE BAG

[3]tote *transitive verb* **tot·ed; tot·ing** [English dialect *tote,* noun, total] (1888)
: ADD, TOTAL — usually used with *up* ⟨*toted* up his accomplishments —G. P. Morrill⟩

[4]tote *noun* [short for *totalizator*] (1891)
: PARI-MUTUEL 2

tote bag *noun* (1900)
: a large 2-handled open-topped bag (as of canvas)

tote board *noun* [[4]*tote*] (circa 1949)
: an electrically operated board (as at a racetrack) on which pertinent information (as betting odds and race results) is posted

to·tem \'tō-təm\ *noun* [Ojibwa *oto·te·man* his totem] (circa 1776)
1 a : an object (as an animal or plant) serving as the emblem of a family or clan and often as a reminder of its ancestry; *also* **:** a usually carved or painted representation of such an object **b :** a family or clan identified by a common totemic object
2 : something that serves as an emblem or revered symbol

to·tem·ic \tō-'te-mik\ *adjective* (1846)
1 : of, relating to, suggestive of, or characteristic of a totem or totemism ⟨a *totemic* animal⟩
2 : based on or practicing totemism ⟨*totemic* clan structure⟩

to·tem·ism \'tō-tə-ˌmi-zəm\ *noun* (1791)
1 : belief in kinship with or a mystical relationship between a group or an individual and a totem
2 : a system of social organization based on totemic affiliations

to·tem·is·tic \ˌtō-tə-'mis-tik\ *adjective* (1873)
: TOTEMIC

totem pole *noun* (1897)
1 : a pole or pillar carved and painted with a series of totemic symbols representing family lineage and often mythical or historical incidents and erected before the houses of Indian tribes of the northwest coast of North America
2 : an order of rank **:** HIERARCHY

totem pole 1

toth·er *or* **t'oth·er** \'tə-t͟hər\ *pronoun or adjective* [Middle English *tother,* alteration (resulting from misdivision of *thet other* the other, from *thet* the— from Old English *thæt* — + *other*) of *other* — more at THAT] (13th century)
chiefly dialect **:** the other

to·ti·po·ten·cy \tō-'ti-pə-tən(t)-sē, ˌtō-tə-'pō-tᵊn-\ *noun* (1909)
: ability to generate or regenerate a whole organism from a part

to·ti·po·tent \-tənt, -tᵊnt\ *adjective* [Latin *totus* whole, entire + English *-i-* + *potent*] (circa 1899)
: capable of developing into a complete organism or differentiating into any of its cells or tissues ⟨*totipotent* blastomeres⟩

[1]tot·ter \'tä-tər\ *intransitive verb* [Middle English *toteren*] (15th century)
1 a : to tremble or rock as if about to fall **:** SWAY **b :** to become unstable **:** threaten to collapse
2 : to move unsteadily **:** STAGGER, WOBBLE

[2]totter *noun* (1747)
: an unsteady gait **:** WOBBLE

tot·ter·ing *adjective* (1534)
1 a : being in an unstable condition ⟨a *tottering* building⟩ **b :** walking unsteadily
2 : lacking firmness or stability **:** INSECURE ⟨a *tottering* regime⟩
— tot·ter·ing·ly \-tə-riŋ-lē\ *adverb*

tot·tery \'tä-tə-rē\ *adjective* (circa 1755)
: of an infirm or precarious nature

Toua·reg *variant of* TUAREG

tou·can \'tü-ˌkan, -ˌkän, tü-'\ *noun* [French, from Portuguese *tucano,* from Tupi *tukána*] (1568)
: any of a family (Ramphastidae) of fruit-eating birds of tropical America with brilliant coloring and a very large but light and thin-walled beak

[1]touch \'təch\ *verb* [Middle English, from Old French *tuchier,* from (assumed) Vulgar Latin *toccare* to knock, strike a bell, touch, probably of imitative origin] (14th century)
transitive verb
1 : to bring a bodily part into contact with especially so as to perceive through the tactile sense **:** handle or feel gently usually with the intent to understand or appreciate ⟨loved to *touch* the soft silk⟩
2 : to strike or push lightly especially with the hand or foot or an implement
3 : to lay hands upon (one afflicted with scrofula) with intent to heal
4 *archaic* **a :** to play on (a stringed instrument) **b :** to perform (a melody) by playing or singing
5 a : to take into the hands or mouth ⟨never *touches* alcohol⟩ **b :** to put hands upon in any way or degree ⟨don't *touch* anything before the police come⟩; *especially* **:** to commit violence upon ⟨swears he never *touched* the child⟩
6 : to deal with **:** become involved with ⟨a sticky situation and I wouldn't *touch* it with a 10-foot pole⟩
7 : to induce to give or lend ⟨*touched* him for ten dollars⟩
8 : to cause to be briefly in contact or conjunction with something ⟨*touched* her spurs to the horse⟩ ⟨*touched* his hand to his hat⟩
9 a (1) **:** to meet without overlapping or penetrating **:** ADJOIN (2) **:** to get to **:** REACH ⟨the speedometer needle *touched* 80⟩ **b :** to be tangent to **c :** to rival in quality or value ⟨nothing can *touch* that cloth for durability⟩
10 : to speak or tell of especially in passing ⟨barely *touched* the incident in the speech⟩
11 a : to relate to **:** CONCERN **b :** to have an influence on **:** AFFECT
12 a : to leave a mark or impression on ⟨few reagents will *touch* gold⟩; *also* **:** TINGE **b :** to harm slightly by or as if by contact **:** TAINT, BLEMISH ⟨fruit *touched* by frost⟩ **c :** to give a delicate tint, line, or expression to ⟨a smile *touched* her lips⟩
13 : to draw or delineate with light strokes
14 a : to hurt the feelings of **:** WOUND **b :** to move to sympathetic feeling
intransitive verb
1 a : to feel something with a body part (as the hand or foot) **b :** to lay hand or finger on a person to cure disease (as scrofula)
2 : to be in contact
3 : to come close **:** VERGE ⟨your actions *touch* on treason⟩
4 : to have a bearing **:** RELATE — used with *on* or *upon*
5 a : to make a brief or incidental stop on shore during a trip by water ⟨*touched* at several ports⟩ **b :** to treat a topic in a brief or casual manner — used with *on* or *upon* ⟨*touched* upon many points⟩
synonym see AFFECT
— touch·able \'tə-chə-bəl\ *adjective*
— touch·er *noun*
— touch base : to come in contact or communication ⟨coming in from the cold to *touch base* with civilization —Carla Hunt⟩

[2]touch *noun* (14th century)
1 : a light stroke, tap, or push
2 : the act or fact of touching
3 : the special sense by which pressure or traction exerted on the skin or mucous membrane is perceived
4 : mental or moral sensitiveness, responsiveness, or tact ⟨has a wonderful *touch* with children⟩
5 : a specified sensation that arises in response to stimulation of the tactile receptors **:** FEEL ⟨the velvety *touch* of velour⟩
6 a : the act of rubbing gold or silver on a touchstone to test its quality **b :** TEST, TRIAL — used chiefly in the phrase *put to the touch*
7 a : a visible effect **:** MARK ⟨a *touch* of the tropical sun⟩ **b :** WEAKNESS, DEFECT
8 : something slight of its kind: as **a :** a light attack ⟨a *touch* of fever⟩ **b :** a small quantity or indication **:** HINT ⟨a *touch* of spring in the air⟩ **c :** a transient emotion ⟨a momentary *touch* of compunction⟩ **d :** a near approach **:** CLOSE CALL ⟨beaten in the championships by a mere *touch*⟩
9 a *archaic* **:** the playing of an instrument (as a lute or piano) with the fingers; *also* **:** musical notes or strains so produced **b :** particular action of a keyboard with reference to the resistance of its keys to pressure ⟨piano with a stiff *touch*⟩
10 : control of the hands: as **a :** a manner or method of touching or striking especially the keys of a keyboard instrument **b :** ability to control the distance something (as a ball) is propelled ⟨developed his putting *touch*⟩
11 : a set of changes in change ringing that is less than a peal
12 a : an effective and subtle detail ⟨applies the finishing *touches* to the story⟩ **b :** distinctive and often effective manner or method ⟨the *touch* of a master⟩ **c :** a characteristic or distinguishing trait or quality
13 *slang* **:** an act of soliciting or getting a gift or loan
14 : the state or fact of being in contact or communication or of having awareness ⟨lost *touch* with her cousin⟩ ⟨let's keep in *touch*⟩ ⟨out of *touch* with modern times⟩

◇ WORD HISTORY

tote *Tote,* meaning "to carry by hand," first appears in the spelling *toat* in a document of 1677 from Gloucester County, Virginia. It is next documented a century later from Boston and subsequently turns up throughout the eastern U.S. Today *tote* is fairly general in American English. The origin of this word has prompted much debate. Since early in the century it has been known that *tote* has close phonetic counterparts in western Bantu languages of Africa, such as Kikongo *tota* "pick up" and Kimbundu *tuta* "carry." *Tote* has a long record of use in African-American vernacular and occurs in Sea Islands Creole (Gullah) as well as in English-based creoles of Sierra Leone and Cameroon. Though the case for an English creole, if not Bantu, origin for *tote* looks strong, it has been objected that the early appearance of the word in New England and its general use in places such as 19th century Maine (where it usually meant "haul" rather than "carry by hand") argue against an African origin. *Tote* may have diffused into both colonial American English and Atlantic English creoles from some earlier speech form, perhaps 17th century nautical English, though unless further evidence surfaces we can only speculate on the prehistory of the word.

15 : the area outside of the touchlines in soccer or outside of and including the touchlines in rugby
— a touch : SOMEWHAT, RATHER ⟨aimed *a touch* too low and missed⟩
touch–and–go \ˌtəch-ən(d)-ˈgō\ *noun* (1953)
: an airplane landing followed immediately by application of power and a takeoff and usually executed as one of a series for practice at landings
touch and go *adjective* (1815)
: unpredictable as to outcome **:** UNCERTAIN ⟨it was *touch and go* there for a while⟩
touch·back \ˈtəch-ˌbak\ *noun* (circa 1890)
: a situation in football in which the ball is down behind the goal line after a kick or intercepted forward pass after which it is put in play by the team defending the goal on its own 20-yard line — compare SAFETY
touch·down \ˈtəch-ˌdau̇n\ *noun* (1876)
1 : the act of touching a football to the ground behind an opponent's goal; *specifically* **:** the act of scoring six points in American football by being lawfully in possession of the ball on, above, or behind an opponent's goal line when the ball is declared dead
2 : the act or moment of touching down (as with an airplane or spacecraft)
touch down (1864)
transitive verb
: to place (the ball in rugby) by hand on the ground on or over an opponent's goal line in scoring a try or behind one's own goal line as a defensive measure
intransitive verb
: to reach the ground **:** LAND
tou·ché \tü-ˈshā\ *interjection* [French, from past participle of *toucher* to touch, from Old French *tuchier*] (1904)
— used to acknowledge a hit in fencing or the success or appropriateness of an argument, an accusation, or a witty point
touched \ˈtəcht\ *adjective* (14th century)
1 : emotionally stirred (as with gratitude)
2 : slightly unbalanced mentally
touch football *noun* (1933)
: football played informally and chiefly characterized by the substitution of touching for tackling
touch·hole \ˈtəch-ˌhōl\ *noun* (1501)
: the vent in muzzle-loading guns through which the charge is ignited
¹touch·ing *preposition* (14th century)
: in reference to **:** CONCERNING
²touching *adjective* (1601)
: capable of arousing emotions of tenderness or compassion
synonym see MOVING
— touch·ing·ly \ˈtə-chiŋ-lē\ *adverb*
touch·line \ˈtəch-ˌlīn\ *noun* (1868)
: either of the lines that bound the long sides of the field of play in rugby and soccer
touch·mark \-ˌmärk\ *noun* (1904)
: an identifying maker's mark impressed on pewter
touch–me–not \ˈtəch-mē-ˌnät\ *noun* [from the bursting of the ripe pods and scattering of their seeds when touched] (1659)
: either of two North American impatiens growing in moist areas: as **a :** one (*Impatiens capensis*) typically having crimson-spotted orange flowers **b :** one (*I. pallida*) having yellow to white flowers sometimes spotted with brownish red
touch off *transitive verb* (circa 1765)
1 : to describe or characterize with precision
2 a : to cause to explode by or as if by touching with fire **b :** to provoke or initiate with sudden intensity ⟨the verdict *touched off* local riots⟩
touch pad *noun* (1979)
: a keypad for an electronic device (as a microwave oven) that consists of a flat surface divided into several differently marked areas which are touched to choose options
touch screen *noun* (1974)
: a display screen on which the user selects options (as from a menu) by touching the screen
touch·stone \ˈtəch-ˌstōn\ *noun* (1530)
1 : a black siliceous stone related to flint and formerly used to test the purity of gold and silver by the streak left on the stone when rubbed by the metal
2 : a test or criterion for determining the quality or genuineness of a thing
3 : a fundamental or quintessential part or feature
synonym see STANDARD
touch system *noun* (1918)
: a method of typing that assigns a particular finger to each key and makes it possible to type without looking at the keyboard
Touch–Tone \ˈtəch-ˈtōn, -ˌtōn\ *trademark*
— used for a telephone having push buttons that produce tones corresponding to numbers
touch–type \ˈtəch-ˌtīp\ *intransitive verb* (1943)
: to type by the touch system
— touch typist *noun*
touch–up \ˈtəch-ˌəp\ *noun* (1885)
: an act or instance of touching up
touch up *transitive verb* (1715)
1 : to improve or perfect by small additional strokes or alterations **:** fix the minor and usually visible defects or damages of
2 : to stimulate by or as if by a flick of a whip
touch·wood \ˈtəch-ˌwu̇d\ *noun* (1579)
: ³PUNK
touchy \ˈtə-chē\ *adjective* **touch·i·er; -est** (1605)
1 : marked by readiness to take offense on slight provocation
2 a *of a body part* **:** acutely sensitive or irritable **b** *of a chemical* **:** highly explosive or inflammable
3 : calling for tact, care, or caution in treatment ⟨a *touchy* subject⟩
— touch·i·ly \ˈtə-chə-lē\ *adverb*
— touch·i·ness \ˈtə-chē-nəs\ *noun*
touchy–feely \ˌtə-chē-ˈfē-lē\ *adjective* (1968)
: characterized by or encouraging interpersonal touching especially in the free expression of emotions ⟨*touchy-feely* therapy⟩; *also* **:** openly or excessively emotional and personal ⟨*touchy-feely* management⟩
¹tough \ˈtəf\ *adjective* [Middle English, from Old English *tōh;* akin to Old High German *zāhi* tough] (before 12th century)
1 a : strong or firm in texture but flexible and not brittle **b :** not easily chewed
2 : GLUTINOUS, STICKY
3 : characterized by severity or uncompromising determination
4 : capable of enduring strain, hardship, or severe labor
5 : very hard to influence **:** STUBBORN
6 : difficult to accomplish, resolve, endure, or deal with ⟨a *tough* question⟩ ⟨*tough* luck⟩
7 : stubbornly fought ⟨a *tough* contest⟩
8 : UNRULY, ROWDYISH
9 : marked by absence of softness or sentimentality
synonym see STRONG
— tough·ly *adverb*
— tough·ness *noun*
²tough *adverb* (14th century)
: in a tough manner ⟨talks *tough* and insensitively —A. E. Stevenson (died 1965)⟩
³tough *transitive verb* (1830)
: to bear unflinchingly **:** ENDURE — usually used with *out* especially in the phrase *tough it out*
⁴tough *noun* (1866)
: a tough person **:** ROWDY
tough·en \ˈtə-fən\ *verb* **tough·ened; tough·en·ing** \ˈtə-fə-niŋ, ˈtəf-niŋ\ (1582)
transitive verb
: to make tough
intransitive verb
: to become tough
tough·ie *also* **toughy** \ˈtə-fē\ *noun, plural* **tough·ies** (1921)
: one that is tough: as **a :** a loud rough rowdy person **b :** a difficult problem or question
tough–mind·ed \ˈtəf-ˌmīn-dəd\ *adjective* (1909)
: realistic or unsentimental in temper or outlook
— tough–mind·ed·ness *noun*
tou·pee \tü-ˈpā\ *noun* [French *toupet* forelock, from Old French, diminutive of *top, toup,* of Germanic origin; akin to Old High German *zopf* tuft of hair — more at TOP] (1728)
1 : a curl or lock of hair made into a topknot on a periwig or natural coiffure; *also* **:** a periwig with such a topknot
2 : a wig or section of hair worn to cover a bald spot
¹tour \ˈtu̇r, *1 is also* ˈtau̇r\ *noun* [Middle English, from Middle French, from Old French *tourn, tour* lathe, circuit, turn — more at TURN] (14th century)
1 a : one's turn in an orderly schedule **:** SHIFT **b :** a period during which an individual or unit is on a specific duty or at one place ⟨a *tour* of duty⟩
2 a : a journey for business, pleasure, or education often involving a series of stops and ending at the starting point; *also* **:** something resembling such a tour ⟨a *tour* of the history of philosophy⟩ **b :** a brief turn **:** ROUND **c :** a series of professional tournaments (as in golf or tennis)
²tour (1789)
intransitive verb
: to make a tour
transitive verb
1 : to make a tour of
2 : to present (as a theatrical production) on a tour
tou·ra·co \ˈtu̇r-ə-ˌkō\ *noun, plural* **-cos** [origin unknown] (circa 1743)
: any of a family (Musophagidae) of African birds that are related to the cuckoos and have a long tail, a short stout often colored bill, and red wing feathers
tour·bil·lion \tu̇r-ˈbil-yən\ *or* **tour·bil·lon** \tür-bē-yōⁿ\ *noun* [Middle English, from Middle French *tourbillon,* from Latin *turbin-, turbo* — more at TURBINE] (15th century)
1 : WHIRLWIND
2 : a vortex especially of a whirlwind or whirlpool
tour de force \ˌtu̇r-də-ˈfōrs, -ˈfȯrs\ *noun, plural* **tours de force** *same*\ [French] (1802)
: a feat of strength, skill, or ingenuity
tour·er \ˈtu̇r-ər\ *noun* (1926)
1 : TOURING CAR
2 : one that tours
Tou·rette's syndrome \tu̇-ˈrets-\ *noun* [Georges Gille de la *Tourette* (died 1904) French physician] (1970)
: a rare disease characterized by involuntary tics and by uncontrollable verbalization involving especially echolalia and the use of obscene language — called also *Tou·rette syndrome* \-ˈret-\
tour·ing \ˈtu̇r-iŋ\ *noun* (1794)
1 : participation in a tour
2 : cross-country skiing for pleasure
touring car *noun* (1903)
: an automobile suitable for distance driving: as **a :** a vintage automobile with two cross seats, usually four doors, and a folding top **:** PHAETON 2 **b :** a modern usually 2-door sedan as distinguished from a sports car

touring car a

tour·ism \ˈtu̇r-ˌi-zəm\ *noun* (1811)
1 : the practice of traveling for recreation
2 : the guidance or management of tourists

3 a : the promotion or encouragement of touring **b :** the accommodation of tourists

tour·ist \'tu̇r-ist\ *noun* (1780)
1 : one that makes a tour for pleasure or culture
2 : TOURIST CLASS
— tourist *adjective or adverb*

tourist card *noun* (1948)
: a citizenship identity card issued to a tourist usually for a stated period of time in lieu of a passport or a visa

tourist class *noun* (1935)
: economy accommodations (as on a ship)

tourist court *noun* (1937)
: MOTEL

tour·is·tic \tu̇r-'is-tik\ *adjective* (1848)
: of or relating to a tour, tourism, or tourists
— tour·is·ti·cal·ly \-ti-k(ə-)lē\ *adverb*

tourist trap *noun* (1939)
: a place that attracts and exploits tourists

tour·isty \'tu̇r-əs-tē\ *adjective* (1906)
: of or relating to tourists: as **a :** patronized by tourists **b :** of a type appealing to tourists

tour·ma·line \'tu̇r-mə-lən, -ˌlēn\ *noun* [Sinhalese *toramalli* carnelian] (1759)
: a mineral of variable color that consists of a complex borosilicate and makes a striking gem when transparent and cut

tour·na·ment \'tu̇r-nə-mənt *also* 'tər- *or* 'tȯr-\ *noun* [Middle English *tornement,* from Old French *torneiement,* from *torneier*] (14th century)
1 a : a knightly sport of the middle ages between mounted combatants armed with blunted lances or swords and divided into two parties contesting for a prize or favor bestowed by the lady of the tournament **b :** the whole series of knightly sports, jousts, and tilts occurring at one time and place
2 : a series of games or contests that make up a single unit of competition (as on the professional golf tour), the championship play-offs of a league or conference, or an invitational event

tour·ne·dos \ˌtu̇r-nə-'dō\ *noun, plural* **tour·ne·dos** \-'dō(z)\ [French, from *tourner* to turn + *dos* back] (1877)
: a small fillet of beef usually cut from the tip of the tenderloin

¹tour·ney \'tu̇r-nē, 'tər- *also* 'tȯr-\ *intransitive verb* **tour·neyed; tour·ney·ing** [Middle English, from Middle French *torneier,* from Old French, from *torn, tourn* lathe, circuit] (14th century)
: to perform in a tournament

²tourney *noun, plural* **tourneys** (14th century)
: TOURNAMENT

tour·ni·quet \'tu̇r-ni-kət, 'tər-\ *noun* [French, turnstile, tourniquet, from *tourner* to turn, from Old French — more at TURN] (1695)
: a device (as a bandage twisted tight with a stick) to check bleeding or blood flow

¹touse \'tau̇z\ *transitive verb* **toused; tous·ing** [Middle English *-tousen;* akin to Old High German *zirzūsōn* to pull to pieces] (1598)
: RUMPLE, TOUSLE

²touse *noun* (1795)
: a noisy disturbance

¹tou·sle \'tau̇-zəl, -səl\ *transitive verb* **tou·sled; tou·sling** \'tau̇z-liŋ, 'tau̇s-; 'tau̇-zə-, -sə-\ [Middle English *touselen,* frequentative of *-tousen*] (15th century)
: DISHEVEL, RUMPLE

²tou·sle \'tau̇-zəl, *1 is also* 'tü-\ *noun* (1788)
1 *Scottish* **:** rough dalliance **:** TUSSLE
2 : a tangled mass (as of hair)

¹tout \'tau̇t, *in sense 4 also* 'tüt\ *verb* [Middle English *tuten* to peer; probably akin to Old English *tōtian* to stick out, Norwegian *tyte*] (circa 1700)
transitive verb
1 : to spy on **:** WATCH
2 a *British* **:** to spy out information about (as a racing stable or horse) **b :** to give a tip or solicit bets on (a racehorse)
3 : to solicit, peddle, or persuade importunately ⟨not meant to *tout* you off the movie —Russell Baker⟩
4 : to praise or publicize loudly or extravagantly ⟨*touted* as the . . . most elaborate suburban shopping development —*Wall Street Journal*⟩
intransitive verb
1 : to solicit patronage
2 a *chiefly British* **:** to spy on racehorses in training to gain information for betting **b :** to give a tip or solicit bets on a racehorse

²tout *noun* (1853)
: one who touts: as **a :** one who solicits patronage **b** *chiefly British* **:** one who spies out racing information for betting purposes **c :** one who gives tips or solicits bets on a racehorse

tout·er \'tau̇-tər\ *noun* (circa 1754)
: one that touts

to·va·rich *or* **to·va·rish** \tə-'vär-ish, -ich\ *noun* [Russian *tovarishch*] (circa 1917)
: COMRADE

¹tow \'tō\ *verb* [Middle English, from Old English *togian;* akin to Old English *tēon* to draw, pull, Old High German *ziohan* to draw, pull, Latin *ducere* to draw, lead] (before 12th century)
transitive verb
: to draw or pull along behind **:** HAUL
intransitive verb
: to move in tow ⟨trailers that *tow* behind the family auto —Bob Munger⟩

²tow *noun* (1600)
1 : a rope or chain for towing
2 a : the act or an instance of towing **b :** the fact or state of being towed
3 a : something towed (as a boat or car) **b :** a group of barges lashed together and usually pushed
4 a : something (as a tugboat) that tows **b :** SKI TOW
— in tow 1 : under guidance or protection ⟨taken *in tow* by a friendly native⟩ **2 :** in the position of a dependent or devoted follower or admirer

³tow *noun* [Middle English, from Old English *tow-* spinning; akin to Old Norse *tō* tuft of wool for spinning, Old English *tawian* to prepare for use — more at TAW] (14th century)
1 : short or broken fiber (as of flax, hemp, or synthetic material) that is used especially for yarn, twine, or stuffing
2 a : yarn or cloth made of tow **b :** a loose essentially untwisted strand of synthetic fibers

⁴tow *noun* [Middle English (Scots), probably from Old English *toh-* (in *tohlīne* towline); akin to Old English *togian* to tow] (15th century)
chiefly Scottish & dialect English **:** ROPE

tow·age \'tō-ij\ *noun* (14th century)
1 : the act of towing
2 : a charge for towing

¹to·ward \'tō(-ə)rd, 'tȯ(-ə)rd\ *adjective* [Middle English *toward,* from Old English *tōweard* facing, imminent, from *tō,* preposition, to + *-weard* -ward] (before 12th century)
1 *also* **to·wards** \'tō(-ə)rdz, 'tȯ(-ə)rdz\ [Middle English *towardes,* from Old English *tōweardes,* preposition, toward, from *tōweard,* adjective] **a :** coming soon **:** IMMINENT **b :** happening at the moment **:** AFOOT
2 a *obsolete* **:** quick to learn **:** APT **b :** PROPITIOUS, FAVORING ⟨a *toward* breeze⟩

²to·ward *or* **to·wards** \'tō(-ə)rd(z), 'tȯ(-ə)rd(z), tə-'wȯrd(z), 'twȯrd(z), 'twōrd(z)\ *preposition* (before 12th century)
1 : in the direction of ⟨driving *toward* town⟩
2 a : along a course leading to ⟨a long stride *toward* disarmament⟩ **b :** in relation to ⟨an attitude *toward* life⟩
3 a : at a point in the direction of **:** NEAR ⟨a cottage somewhere up *toward* the lake⟩ **b :** in such a position as to be in the direction of ⟨your back was *toward* me⟩
4 : not long before ⟨*toward* the end of the afternoon⟩
5 a : in the way of help or assistance in ⟨did all he could *toward* raising campaign funds⟩ **b :** for the partial payment of ⟨proceeds go *toward* the establishment of a scholarship⟩

to·ward·li·ness \'tōrd-lē-nəs, 'tȯrd-\ *noun* (1566)
archaic **:** the quality or state of being toward or towardly

to·ward·ly \'tō(-ə)rd-lē, 'tȯ(-ə)rd-\ *adjective* (1513)
archaic
1 : PLEASANT, AFFABLE
2 : FAVORABLE, PROPITIOUS
3 : developing favorably **:** PROMISING
— towardly *adverb*

tow·boat \'tō-ˌbōt\ *noun* (1815)
1 : TUGBOAT
2 : a compact shallow-draft boat with a squared bow designed and fitted for pushing tows of barges on inland waterways

¹tow·el \'tau̇(-ə)l\ *noun* [Middle English *towaille,* from Old French *toaille,* of Germanic origin; akin to Old High German *dwahila* towel; akin to Old High German *dwahan* to wash] (13th century)
: an absorbent cloth or paper for wiping or drying

²towel *verb* **-eled** *or* **-elled; -el·ing** *or* **-el·ling** (circa 1839)
transitive verb
: to rub or dry (as the body) with a towel
intransitive verb
: to use a towel to dry oneself

tow·el·ette \ˌtau̇-(ə-)'let\ *noun* (1902)
: a small usually premoistened piece of material used for personal cleansing (as of the hands)

tow·el·ing *or* **tow·el·ling** \'tau̇-(ə-)liŋ\ *noun* (1583)
: a cotton or linen fabric often used for making towels

¹tow·er \'tau̇(-ə)r\ *noun* [Middle English *tour, tor,* from Old English *torr* & Old French *tor, tur,* both from Latin *turris,* from Greek *tyrris, tyrsis*] (before 12th century)
1 : a building or structure typically higher than its diameter and high relative to its surroundings that may stand apart (as a campanile), or be attached (as a church belfry) to a larger structure, and that may be fully walled in or of skeleton framework (as an observation or transmission tower)
2 : a towering citadel **:** FORTRESS
3 : one that provides support or protection **:** BULWARK ⟨a *tower* of strength⟩
— tow·ered \'tau̇(-ə)rd\ *adjective*
— tow·er·like \'tau̇(-ə)r-ˌlīk\ *adjective*

²tower *intransitive verb* (1582)
1 : to reach or rise to a great height
2 : to exhibit superior qualities **:** SURPASS ⟨her intellect *towered* over the others'⟩

tower block *noun* (1966)
chiefly British **:** a tall building (as a high-rise apartment building)

tower house *noun* (1687)
: a medieval fortified castle (as in Scotland)

tow·er·ing *adjective* (1592)
1 : impressively high or great **:** IMPOSING ⟨*towering* pines⟩
2 : reaching a high point of intensity **:** OVERWHELMING ⟨a *towering* rage⟩
3 : going beyond proper bounds **:** EXCESSIVE ⟨*towering* ambitions⟩
— tow·er·ing·ly \'tau̇(-ə)r-iŋ-lē\ *adverb*

Tower of Babel (circa 1887)
: BABEL 2
tow·head \'tō-ˌhed\ *noun* (1829)
1 : a low alluvial island or shoal in a river : SANDBAR
2 : a head of hair resembling tow especially in being flaxen or tousled; *also* **:** a person having such a head of hair
— tow·head·ed \-ˌhe-dəd\ *adjective*
to·whee \'tō-ˌhē, 'tō-(ˌ)ē, tō-'hē\ *noun* [imitative] (circa 1729)
1 : a common finch (*Pipilo erythrophthalmus*) of eastern North America having the male black, white, and rufous — called also *chewink*
2 : any of the North American finches belonging to the same genus (*Pipilo*) as the towhee
to wit \tə-'wit\ *adverb* [Middle English *to witen,* literally, to know — more at WIT] (14th century)
: that is to say **:** NAMELY
tow·line \'tō-ˌlīn\ *noun* (1719)
: TOWROPE
tow·mond \'tō-ˌmänd\ *noun* [Middle English *towlmonyth,* from Old English *twelf mōnath,* from *twelf* twelve + *mōnath* month] (15th century)
Scottish **:** YEAR, TWELVEMONTH
town \'tau̇n\ *noun* [Middle English, from Old English *tūn* enclosure, village, town; akin to Old High German *zūn* enclosure, Old Irish *dún* fortress] (before 12th century)
1 *dialect English* **:** a cluster or aggregation of houses recognized as a distinct place with a place-name **:** HAMLET
2 a : a compactly settled area as distinguished from surrounding rural territory **b :** a compactly settled area usually larger than a village but smaller than a city **c :** a large densely populated urban area **:** CITY **d :** an English village having a periodic fair or market
3 : a neighboring city, capital city, or metropolis
4 : the city or urban life as contrasted with the country
5 : the inhabitants of a city or town
6 : a New England territorial and political unit usually containing both rural and unincorporated urban areas under a single town government; *also* **:** a New England community governed by a town meeting
7 : a group of prairie dog burrows
— town *adjective*
— on the town : in usually carefree pursuit of entertainment or amusement (as city nightlife) especially as a relief from routine
town car *noun* (1907)
: a 4-door automobile with a usually open driver's compartment and a separate enclosed passenger compartment
town clerk *noun* (14th century)
: a public officer charged with recording the official proceedings and vital statistics of a town
town crier *noun* (1602)
: a town officer who makes public proclamations
town·ee \tau̇-'nē\ *noun* (1897)
chiefly British **:** TOWNIE
town hall *noun* (15th century)
: a public building used for town-government offices and meetings
town house *noun* (1803)
1 : a house in town; *specifically* **:** the city residence of one having a countryseat or having a chief residence elsewhere ⟨stayed at their *town house* during the social season⟩
2 : a usually single-family house of two or sometimes three stories that is usually connected to a similar house by a common sidewall; *also* **:** ROW HOUSE
town·ie *or* **towny** \'tau̇-nē\ *noun, plural* **townies** (1852)
: TOWNSMAN; *especially* **:** a permanent inhabitant of a town as distinguished from a member of another group (as the academic community)
town·let \'tau̇n-lət\ *noun* (circa 1552)
: a very small town
town manager *noun* (1922)
: an official appointed to direct the administration of a town government
town meeting *noun* (1636)
: a meeting of inhabitants or taxpayers constituting the legislative authority of a town
town·scape \-ˌskāp\ *noun* (1880)
1 : a representation of an urban scene
2 : a town or city viewed as a scene
towns·folk \'tau̇nz-ˌfōk\ *noun plural* (1737)
: TOWNSPEOPLE
town·ship \'tau̇n-ˌship\ *noun* (15th century)
1 : an ancient unit of administration in England identical in area with or a division of a parish
2 a : TOWN 6 **b :** a unit of local government in some northeastern and north central states usually having a chief administrative officer or board **c :** an unorganized subdivision of the county in Maine, New Hampshire, and Vermont **d :** an electoral and administrative district of the county in the southern U.S.
3 : a division of territory in surveys of U.S. public land containing 36 sections or 36 square miles
4 : an area in the Republic of South Africa segregated for occupation by persons of non-European descent
towns·man \'tau̇nz-mən\ *noun* (before 12th century)
1 a : a native or resident of a town or city **b :** an urban or urbane person
2 : a fellow citizen of a town
towns·peo·ple \-ˌpē-pəl\ *noun plural* (1648)
1 : the inhabitants of a town or city **:** TOWNSMEN
2 : town-dwelling or town-bred persons
towns·wom·an \-ˌwu̇-mən\ *noun* (1684)
1 : a woman who is a native or resident of a town or city
2 : a woman born or residing in the same town or city as another
tow·path \'tō-ˌpath, -ˌpȧth\ *noun* (1788)
: a path (as along a canal) traveled especially by draft animals towing boats — called also *towing path*
tow·rope \-ˌrōp\ *noun* (1743)
: a line used in towing something (as a boat)
tow sack \'tō-'sak\ *noun* [³*tow*] (1926)
Midland & Southern **:** GUNNYSACK
tow truck *noun* (1944)
: a truck with winches and hoist mechanisms for freeing stuck vehicles and towing wrecked or disabled vehicles
tox- *or* **toxi-** *or* **toxo-** *combining form* [Late Latin, from Latin *toxicum* poison]
: poisonous **:** poison ⟨*tox*emia⟩
tox·a·phene \'täk-sə-ˌfēn\ *noun* [from *Toxaphene,* a trademark] (1947)
: an insecticide with the approximate empirical formula $C_{10}H_{10}Cl_8$ that is a complex mixture of chlorinated compounds
tox·e·mia \täk-'sē-mē-ə\ *noun* [New Latin] (circa 1860)
: an abnormal condition associated with the presence of toxic substances in the blood
— tox·e·mic \-mik\ *adjective*
[1]**tox·ic** \'täk-sik\ *adjective* [Late Latin *toxicus,* from Latin *toxicum* poison, from Greek *toxikon* arrow poison, from neuter of *toxikos* of a bow, from *toxon* bow, arrow] (1664)
1 : of, relating to, or caused by a poison or toxin
2 : affected by a poison or toxin ⟨*toxic* pregnant women⟩
3 : POISONOUS
— tox·ic·i·ty \täk-'si-sə-tē\ *noun*
[2]**toxic** *noun* (1890)
: a toxic substance
toxic- *or* **toxico-** *combining form* [New Latin, from Latin *toxicum*]
: poison ⟨*toxico*logy⟩
tox·i·cant \'täk-si-kənt\ *noun* [Medieval Latin *toxicant-, toxicans,* present participle of *toxicare* to poison, from Latin *toxicum*] (circa 1882)
: a toxic agent; *especially* **:** PESTICIDE
tox·i·co·log·i·cal \ˌtäk-si-kə-'lä-ji-kəl\ *or* **tox·i·co·log·ic** \-jik\ *adjective* (1839)
: of or relating to toxicology or toxins
— tox·i·co·log·i·cal·ly \-ji-k(ə-)lē\ *adverb*
tox·i·col·o·gy \-'kä-lə-jē\ *noun* (circa 1799)
: a science that deals with poisons and their effect and with the problems involved (as clinical, industrial, or legal)
— tox·i·col·o·gist \-jist\ *noun*
tox·i·co·sis \ˌtäk-sə-'kō-səs\ *noun, plural* **-co·ses** \-ˌsēz\ [New Latin] (circa 1857)
: a pathological condition caused by the action of a poison or toxin
toxic shock syndrome *noun* (1978)
: an acute disease that is characterized by fever, diarrhea, nausea, diffuse erythema, and shock, that is associated especially with the presence of a bacterium (*Staphylococcus aureus*), and that occurs especially in menstruating females using tampons — called also *toxic shock*
toxi·gen·ic \ˌtäk-sə-'je-nik\ *adjective* (circa 1923)
: producing toxin ⟨*toxigenic* bacteria and fungi⟩
— toxi·ge·nic·i·ty \ˌtäk-si-jə-'ni-sə-tē\ *noun*
tox·in \'täk-sən\ *noun* [International Scientific Vocabulary] (1886)
: a poisonous substance that is a specific product of the metabolic activities of a living organism and is usually very unstable, notably toxic when introduced into the tissues, and typically capable of inducing antibody formation
tox·in–an·ti·tox·in \'täk-sən-'an-ti-ˌtäk-sən\ *noun* (1904)
: a mixture of toxin and antitoxin used especially formerly in immunizing against a disease (as diphtheria) for which they are specific
tox·oid \'täk-ˌsȯid\ *noun* [International Scientific Vocabulary] (circa 1894)
: a toxin of a pathogenic organism treated so as to destroy its toxicity but leave it capable of inducing the formation of antibodies on injection
tox·oph·i·lite \täk-'sä-fə-ˌlīt\ *noun* [Greek *toxon* bow, arrow + *philos* dear, loving] (1794)
: a person fond of or expert at archery
— toxophilite *adjective*
— tox·oph·i·ly \-lē\ *noun*
toxo·plas·ma \ˌtäk-sə-'plaz-mə\ *noun* [New Latin] (1926)
: any of a genus (*Toxoplasma*) of sporozoans that are typically serious pathogens of vertebrates
— toxo·plas·mic \-mik\ *adjective*
toxo·plas·mo·sis \-ˌplaz-'mō-səs\ *noun, plural* **-mo·ses** \-ˌsēz\ [New Latin] (1926)
: infection of humans, other mammals, or birds with disease caused by a toxoplasma (*Toxoplasma gondii*) that invades the tissues and may seriously damage the central nervous system especially of infants
[1]**toy** \'tȯi\ *noun* [Middle English *toye*] (15th century)
1 *obsolete* **a :** flirtatious or seductive behavior **b :** PASTIME; *also* **:** a sportive or amusing act **:** ANTIC
2 a : something (as a preoccupation) that is paltry or trifling **b :** a literary or musical trifle or diversion **c :** TRINKET, BAUBLE
3 : something for a child to play with
4 : something diminutive; *especially* **:** a diminutive animal (as of a small breed or variety)
5 : something that can be toyed with
6 *Scottish* **:** a headdress of linen or woolen hanging down over the shoulders and formerly worn by old women of the lower classes

— **toy·like** \-ˌlīk\ *adjective*

²toy *intransitive verb* (circa 1529)
1 : to act or deal with something lightly or without vigor or purpose
2 : to engage in flirtation
3 : to amuse oneself as if with a toy **:** PLAY
synonym see TRIFLE
— **toy·er** \'tȯi-ər\ *noun*

³toy *adjective* (1801)
1 : resembling a toy especially in diminutive size
2 : designed or made for use as a toy ⟨a *toy* stove⟩

toy Man·ches·ter terrier \-'man-ˌches-tər-, -chə-stər-\ *noun* (1935)
: any of a breed of toy dogs developed from the Manchester terrier that have erect ears of moderate size and weigh not more than 12 pounds (5.4 kilograms) — called also *toy Manchester*

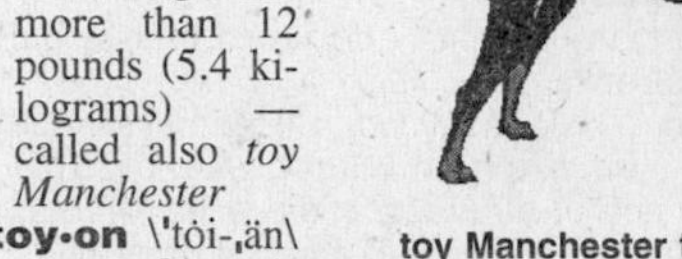
toy Manchester terrier

toy·on \'tȯi-ˌän\ *noun* [American Spanish *tollon*] (1848)
: a chiefly Californian ornamental evergreen shrub (*Heteromeles arbutifolia*) of the rose family having white flowers succeeded by persistent usually bright red berries

toy poodle *noun* (1935)
: any of a breed of toy dogs developed from the standard poodle that are not more than 10 inches (25 centimeters) high at the withers

TPN \ˌtē-ˌpē-'en\ *noun* [*tri*phosphopyridine *n*ucleotide] (1938)
: NADP

tra·be·at·ed \'trā-bē-ˌā-təd\ *also* **tra·be·ate** \-ˌāt\ *adjective* [Latin *trabs, trabes* beam — more at THORP] (1843)
: designed or constructed with horizontal beams or lintels
— **tra·be·a·tion** \ˌtrā-bē-'ā-shən\ *noun*

tra·bec·u·la \trə-'be-kyə-lə\ *noun, plural* **-lae** \-ˌlē, -ˌlī\ *also* **-las** [New Latin, from Latin, little beam, diminutive of *trabs, trabes* beam] (circa 1866)
1 : a small bar, rod, bundle of fibers, or septal membrane in the framework of a body organ or part
2 : a fold, ridge, or bar projecting into or extending from a plant part; *especially* **:** a row of cells bridging an intercellular space
— **tra·bec·u·lar** \-lər\ *adjective*
— **tra·bec·u·late** \-lət\ *adjective*

¹trace \'trās\ *noun* [Middle English, from Middle French, from *tracier* to trace] (14th century)
1 *archaic* **:** a course or path that one follows
2 a : a mark or line left by something that has passed; *also* **:** FOOTPRINT **b :** a path, trail, or road made by the passage of animals, people, or vehicles
3 a : a sign or evidence of some past thing **:** VESTIGE **b :** ENGRAM
4 : something (as a line) traced or drawn: as **a :** the marking made by a recording instrument (as a seismograph or kymograph) **b :** the ground plan of a military installation or position either on a map or on the ground
5 a : the intersection of a line or plane with a plane **b :** the usually bright line or spot that moves across the screen of a cathode-ray tube; *also* **:** the path taken by such a line or spot
6 a : a minute and often barely detectable amount or indication ⟨a *trace* of a smile⟩ **b :** an amount of a chemical constituent not always quantitatively determinable because of minuteness ☆
— **trace·less** \-ləs\ *adjective*

²trace *verb* **traced; trac·ing** [Middle English, from Middle French *tracier,* from (assumed) Vulgar Latin *tractiare* to drag, from Latin *tractus,* past participle of *trahere* to pull] (14th century)
transitive verb
1 a : DELINEATE, SKETCH **b :** to form (as letters or figures) carefully or painstakingly **c :** to copy (as a drawing) by following the lines or letters as seen through a transparent superimposed sheet **d :** to impress or imprint (as a design or pattern) with a tracer **e :** to record a tracing of in the form of a curved, wavy, or broken line ⟨*trace* the heart action⟩ **f :** to adorn with linear ornamentation (as tracery or chasing)
2 *archaic* **:** to travel over **:** TRAVERSE
3 a : to follow the footprints, track, or trail of **b :** to follow or study out in detail or step by step ⟨*trace* the history of the labor movement⟩ **c :** to discover by going backward over the evidence step by step ⟨*trace* your ancestry⟩ **d :** to discover signs, evidence, or remains of
4 : to lay out the trace of (a military installation)
intransitive verb
1 : to make one's way; *especially* **:** to follow a track or trail
2 : to be traceable historically
— **trace·abil·i·ty** \ˌtrā-sə-'bi-lə-tē\ *noun*
— **trace·able** \'trā-sə-bəl\ *adjective*

³trace *noun* [Middle English *trais,* plural, traces, from Middle French, plural of *trait* pull, draft, trace — more at TRAIT] (14th century)
1 : either of two straps, chains, or lines of a harness for attaching a draft animal to something (as a vehicle) to be drawn
2 : LEADER 1e(2)
3 : one or more vascular bundles supplying a leaf or twig
4 : a connecting bar or rod pivoted at each end to another piece and used for transmitting motion

trace element *noun* (1932)
: a chemical element present in minute quantities; *especially* **:** one used by organisms and held essential to their physiology

trac·er \'trā-sər\ *noun* (circa 1552)
1 : one that traces, tracks down, or searches out: as **a :** a person who traces missing persons or property and especially goods lost in transit **b :** an inquiry sent out in tracing a shipment lost in transit
2 : one who traces designs, patterns, or markings
3 : a device (as a stylus) used in tracing
4 a : ammunition containing a chemical composition to mark the flight of projectiles by a trail of smoke or fire **b :** a substance used to trace the course of a chemical or biological process; *especially* **:** LABEL 3d

trac·ery \'trā-sə-rē, 'trās-rē\ *noun, plural* **-er·ies** (1669)
1 : architectural ornamental work with branching lines; *especially* **:** decorative openwork in the head of a Gothic window
2 : a decorative interlacing of lines suggestive of Gothic tracery
— **trac·er·ied** \-rēd\ *adjective*

tracery 1

trache- *or* **tracheo-** *combining form* [New Latin, from Medieval Latin *trachea*]
1 : trachea ⟨*trache*itis⟩ ⟨*tracheo*tomy⟩
2 : tracheal and ⟨*tracheo*bronchial⟩

tra·chea \'trā-kē-ə\ *noun, plural* **-che·ae** \-kē-ˌē, -kē-ˌī\ *also* **-che·as** [Middle English, from Medieval Latin, from Late Latin *trachia,* from Greek *tracheia* (*artēria*) rough (artery), from feminine of *trachys* rough] (14th century)
1 : the main trunk of the system of tubes by which air passes to and from the lungs in vertebrates
2 [New Latin, from Medieval Latin] **:** VESSEL 3b; *also* **:** one of its constituent cellular elements
3 [New Latin] **:** one of the air-conveying tubules forming the respiratory system of most insects and many other arthropods
— **tra·che·al** \-kē-əl\ *adjective*

tra·che·ary \'trā-kē-ˌer-ē\ *adjective* (1885)
: of, relating to, or being plant tracheae ⟨*tracheary* elements⟩

tra·che·ate \-kē-ˌāt, -ət\ *or* **tra·che·at·ed** \-ˌā-təd\ *adjective* (1877)
: having tracheae as breathing organs

tra·cheid \'trā-kē-əd, -ˌkēd\ *noun* [International Scientific Vocabulary] (1875)
: a long tubular pitted cell that is peculiar to xylem, functions in conduction and support, and has tapering closed ends and thickened lignified walls

tra·che·i·tis \ˌtrā-kē-'ī-təs\ *noun* [New Latin] (1859)
: inflammation of the trachea

tra·cheo·bron·chi·al \ˌtrā-kē-ō-'bräŋ-kē-əl\ *adjective* (1896)
: of or relating to both trachea and bronchi ⟨*tracheobronchial* lesions⟩

tra·che·ole \'trā-kē-ˌōl\ *noun* [New Latin *tracheola,* diminutive of *trachea*] (1901)
: one of the minute delicate endings of a branched trachea of an insect
— **tra·che·o·lar** \trā-'kē-ə-lər\ *adjective*

tra·cheo·phyte \'trā-kē-ə-ˌfīt\ *noun* [New Latin *Tracheophyta,* from *trache-* + Greek *phyton* plant; akin to Greek *phyein* to bring forth — more at BE] (1937)
: any of a division (Tracheophyta) comprising green plants (as ferns and seed plants) with a vascular system that contains tracheids or tracheary elements

tra·che·os·to·my \ˌtrā-kē-'äs-tə-mē\ *noun, plural* **-mies** (circa 1923)
: the surgical formation of an opening into the trachea through the neck especially to allow the passage of air

tra·che·ot·o·my \ˌtrā-kē-'ä-tə-mē\ *noun, plural* **-mies** (circa 1726)
: the surgical operation of cutting into the trachea especially through the skin

tra·cho·ma \trə-'kō-mə\ *noun* [New Latin, from Greek *trachōma,* from *trachys* rough] (circa 1693)
: a chronic contagious bacterial conjunctivitis marked by inflammatory granulations on the conjunctival surfaces, caused by a chlamydia (*Chlamydia trachomatis*), and commonly resulting in blindness if left untreated

tra·chyte \'tra-ˌkīt, 'trā-\ *noun* [French, from Greek *trachys* rough] (1821)
: a usually light-colored volcanic rock consisting chiefly of potash feldspar

> ☆ SYNONYMS
> **Trace, vestige, track** mean a perceptible sign made by something that has passed. TRACE may suggest any line, mark, or discernible effect ⟨a snowfield pockmarked with the *traces* of caribou⟩. VESTIGE applies to a tangible reminder such as a fragment or remnant of what is past and gone ⟨boulders that are *vestiges* of the last ice age⟩. TRACK implies a continuous line that can be followed ⟨the fossilized *tracks* of dinosaurs⟩.

tra·chyt·ic \trə-ˈki-tik\ *adjective* (1827)
: of or relating to a texture of igneous rocks in which lath-shaped feldspar crystals are in almost parallel lines

trac·ing \ˈtrā-siŋ\ *noun* (15th century)
1 : the act of one that traces
2 : something that is traced: as **a :** a copy made on a superimposed transparent sheet **b :** a graphic record made by an instrument (as a seismograph) that registers some movement

tracing paper *noun* (1824)
: a semitransparent paper for tracing drawings; *also* **:** a thin paper containing a clothing pattern to be transferred to fabric (as through carbon paper) by tracing

tracing wheel *noun* (circa 1891)
: a usually toothed wheel with a handle that is used on tracing paper to trace a pattern

¹track \ˈtrak\ *noun* [Middle English *trak,* from Middle French *trac,* perhaps of Germanic origin; akin to Middle Dutch *tracken, trecken* to pull, haul — more at TREK] (15th century)
1 a : detectable evidence (as the wake of a ship, a line of footprints, or a wheel rut) that something has passed **b :** a path made by repeated footfalls **:** TRAIL **c :** a course laid out especially for racing **d :** the parallel rails of a railroad **e** (1) **:** one of a series of parallel or concentric paths along which material (as music or information) is recorded (as on a phonograph record or magnetic tape) (2) **:** a group of grooves on a phonograph record containing recorded sound (3) **:** material recorded especially on a track **f :** a usually metal way (as a groove) serving as a guide (as for a movable lighting fixture)
2 : a footprint whether recent or fossil ⟨the huge *track* of a dinosaur⟩
3 a : the course along which something moves **b :** a way of life, conduct, or action **c :** one of several curricula of study to which students are assigned according to their needs or levels of ability **d :** the projection on the earth's surface of the path along which something (as a missile or an airplane) has flown
4 a : a sequence of events **:** a train of ideas **:** SUCCESSION **b :** an awareness of a fact, progression, or condition ⟨keep *track* of the costs⟩ ⟨lose *track* of the time⟩
5 a : the width of a wheeled vehicle from wheel to wheel and usually from the outside of the rims **b :** the tread of an automobile tire **c :** either of two endless metal belts on which a tracklaying vehicle travels
6 : track-and-field sports; *especially* **:** those performed on a running track
synonym see TRACE
— track·less \ˈtrak-ləs\ *adjective*
— in one's tracks : where one stands or is at the moment **:** on the spot ⟨was stopped *in his tracks*⟩

²track (1565)
transitive verb
1 a : to follow the tracks or traces of **:** TRAIL **b :** to search for by following evidence until found ⟨*track* down the source⟩
2 a : to follow by vestiges **:** TRACE **b :** to observe or plot the moving path of (as a spacecraft or missile) instrumentally
3 : to travel over **:** TRAVERSE ⟨*track* a desert⟩
4 a : to make tracks upon **b :** to carry (as mud) on the feet and deposit
5 : to keep track of (as a trend) **:** FOLLOW
intransitive verb
1 : TRAVEL ⟨comet *tracks* eastward⟩
2 a *of a phonograph needle* **:** to follow the groove undulations of a recording **b** *of a pair of wheels* (1) **:** to maintain a constant distance apart on the straightaway (2) **:** to fit a track or rails **c** *of a rear wheel of a vehicle* **:** to follow accurately the corresponding fore wheel on a straightaway
3 : to leave tracks (as on a floor)
— track·er *noun*

track·age \ˈtra-kij\ *noun* (1880)
1 : lines of railway track
2 a : a right to use the tracks of another railroad line **b :** the charge for such right

track–and–field \ˌtrak-ən(d)-ˈfē(ə)ld\ *adjective* (1905)
: of, relating to, or being any of various competitive athletic events (as running, jumping, and weight throwing) performed on a running track and on the adjacent field

track·ball \ˈtrak-ˌbȯl\ *noun* (1967)
: a ball that is mounted usually in a computer console so as to be only partially exposed and is rotated to control the movement of a cursor on a display

tracked \ˈtrakt\ *adjective* (1926)
1 : traveling on endless metal belts instead of wheels
2 : moving along a rail ⟨a *tracked* air-cushion vehicle⟩

track·ing \ˈtra-kiŋ\ *noun* (circa 1929)
: the assigning of students to a curricular track

tracking shot *noun* (circa 1940)
: a scene photographed from a moving dolly

track·lay·er \ˈtrak-ˌlā-ər, -ˌle(-ə)r\ *noun* (circa 1861)
1 : a worker engaged in tracklaying
2 : a tracklaying vehicle

¹track·lay·ing \-ˌlā-iŋ\ *noun* (1857)
: the laying of tracks on a railway line

²tracklaying *adjective* (1884)
: of, relating to, or being a vehicle that travels on two or more endless usually metal belts

trackless trolley *noun* (1921)
: TROLLEYBUS

track lighting *noun* (1972)
: adjustable lamps mounted along an electrified metal track

track·man \ˈtrak-mən, -ˌman\ *noun* (1922)
: a runner on a track team

track record *noun* [¹*track* (track-and-field sports)] (1952)
: a record of past performance often taken as an indicator of likely future performance

track·side \ˈtrak-ˌsīd\ *adjective* (1886)
: of, relating to, or situated in the area immediately adjacent to a track
— trackside *noun*

track·suit \ˈtrak-ˌsüt\ *noun* (1922)
: a suit of clothing consisting usually of a jacket and pants that is often worn by athletes when working out

track·walk·er \-ˌwȯ-kər\ *noun* (1872)
: a worker employed to walk over and inspect a section of railroad tracks

track·way \-ˌwā\ *noun* (1818)
1 : a beaten or trodden path
2 : a series of fossil footprints of a dinosaur

¹tract \ˈtrakt\ *noun, often capitalized* [Middle English *tracte,* from Medieval Latin *tractus,* from Latin, action of drawing, extension; perhaps from its being sung without a break by one voice] (14th century)
: verses of Scripture (as from the Psalms) used between the gradual and the Gospel at some masses (as during penitential seasons)

²tract *noun* [Middle English *tracte,* from Latin *tractus* action of drawing, extension, from *trahere* to pull, draw] (15th century)
1 : extent or lapse of time
2 : an area either large or small: as **a :** an indefinite stretch of land **b :** a defined area of land
3 a : a system of body parts or organs that act together to perform some function ⟨the digestive *tract*⟩ **b :** a bundle of nerve fibers having a common origin, termination, and function

³tract *noun* [Middle English, treatise, modification of Latin *tractatus* tractate] (1760)
: a pamphlet or leaflet of political or religious propaganda; *also* **:** a piece of writing that is suggestive of such a tract

trac·ta·ble \ˈtrak-tə-bəl\ *adjective* [Latin *tractabilis,* from *tractare* to handle, treat] (1502)
1 : capable of being easily led, taught, or controlled **:** DOCILE ⟨a *tractable* horse⟩
2 : easily handled, managed, or wrought **:** MALLEABLE
synonym see OBEDIENT
— trac·ta·bil·i·ty \ˌtrak-tə-ˈbi-lə-tē\ *noun*
— trac·ta·ble·ness \ˈtrak-tə-bəl-nəs\ *noun*
— trac·ta·bly \-blē\ *adverb*

Trac·tar·i·an \trak-ˈter-ē-ən\ *noun* [from *Tracts for the Times,* series of pamphlets expounding the Oxford movement] (circa 1839)
: a promoter or supporter of the Oxford movement

Trac·tar·i·an·ism \-ē-ə-ˌni-zəm\ *noun* (1840)
: a system of High Church principles set forth in a series of tracts at Oxford (1833–41)

trac·tate \ˈtrak-ˌtāt\ *noun* [Latin *tractatus,* from *tractare* to draw out, handle, treat — more at TREAT] (15th century)
: TREATISE, DISSERTATION

tract house *noun* (1956)
: any of many similarly designed houses built on a tract of land

trac·tion \ˈtrak-shən\ *noun* [Medieval Latin *traction-, tractio,* from Latin *trahere*] (1615)
1 : the act of drawing **:** the state of being drawn; *also* **:** the force exerted in drawing
2 : the drawing of a vehicle by motive power; *also* **:** the motive power employed
3 a : the adhesive friction of a body on a surface on which it moves ⟨the *traction* of a wheel on a rail⟩ **b :** a pulling force exerted on a skeletal structure (as in a fracture) by means of a special device ⟨a *traction* splint⟩; *also* **:** a state of tension created by such a pulling force ⟨a leg in *traction*⟩
— trac·tion·al \-shnəl, -shə-nᵊl\ *adjective*

trac·tive \ˈtrak-tiv\ *adjective* [Latin *tractus,* past participle] (1615)
1 : serving to draw
2 : of or relating to traction **:** TRACTIONAL

trac·tor \ˈtrak-tər\ *noun* [New Latin, from Latin *trahere*] (1900)
1 a : a 4-wheeled or tracklaying automotive vehicle used especially for drawing farm equipment **b :** a smaller 2-wheeled apparatus controlled through handlebars by a walking operator **c :** ³TRUCK 3f
2 : an airplane having the propeller forward of the main supporting surfaces

trad \ˈtrad\ *adjective* (1958)
chiefly British **:** TRADITIONAL

¹trade \ˈtrād\ *noun* [Middle English, from Middle Low German; akin to Old High German *trata* track, course, Old English *tredan* to tread] (14th century)
1 a *obsolete* **:** a path traversed **:** WAY **b** *archaic* **:** a track or trail left by a person or animal **:** TREAD 1
2 : a customary course of action **:** PRACTICE ⟨thy sin's not accidental, but a *trade* —Shakespeare⟩
3 a : the business or work in which one engages regularly **:** OCCUPATION **b :** an occupation requiring manual or mechanical skill **:** CRAFT **c :** the persons engaged in an occupation, business, or industry
4 a *obsolete* **:** dealings between persons or groups **b** (1) **:** the business of buying and selling or bartering commodities **:** COMMERCE (2) **:** BUSINESS, MARKET ⟨novelties for the tourist *trade*⟩ ⟨did a good *trade* in small appliances⟩
5 a : an act or instance of trading **:** TRANSACTION; *also* **:** an exchange of property usually without use of money **b :** a firm's customers **:** CLIENTELE **c :** the group of firms engaged in a business or industry
6 : TRADE WIND — usually used in plural
7 : a publication intended for persons in the entertainment business — usually used in plural
synonym see BUSINESS

²trade *verb* **trad·ed; trad·ing** (1553)
intransitive verb
1 *obsolete* **:** to have dealings **:** NEGOTIATE
2 a : to engage in the exchange, purchase, or sale of goods **b :** to make one's purchases **:** SHOP ⟨*trades* at his store⟩
3 : to give one thing in exchange for another
4 : SELL 3
transitive verb
1 *archaic* **:** to do business with
2 a : to give in exchange for another commodity **:** BARTER; *also* **:** to make an exchange of ⟨*traded* places⟩ **b :** to engage in frequent buying and selling of (as stocks or commodities) usually in search of quick profits
— trad·able *also* **trade·able** \'trā-də-bəl\ *adjective*
— trade on : to take often unscrupulous advantage of **:** EXPLOIT ⟨*traded on* their influence . . . in securing special favors —T. C. Pease⟩

³trade *adjective* (1633)
1 : of, relating to, or used in trade
2 a : intended for or limited to persons in a business or industry ⟨a *trade* publication⟩ ⟨*trade* sales⟩ **b :** serving others in the same business rather than the ultimate user or consumer ⟨a *trade* printing house⟩
3 *also* **trades :** of, composed of, or representing the trades or trade unions ⟨a *trade* committee⟩
4 : of a larger softcover format than mass-market paperbacks and usually sold only in bookstores ⟨*trade* paperbacks⟩; *also* **:** of or relating to the publishing of such books

trade acceptance *noun* (1916)
: a time draft or bill of exchange for the amount of a specific purchase drawn by the seller on the buyer, bearing the buyer's acceptance, and often noting the place of payment (as a bank)

trade agreement *noun* (circa 1921)
1 : an international agreement on conditions of trade in goods and services
2 : an agreement resulting from collective bargaining

trade book *noun* (circa 1945)
1 : a book intended for general readership
2 : TRADE EDITION

trade·craft \'trād-ˌkraft\ *noun* (1961)
: the techniques and procedures of espionage

trade discount *noun* (1901)
: a deduction from the list price of goods allowed by a manufacturer or wholesaler to a retailer

trade dollar *noun* (1873)
: a U.S. silver dollar weighing 420 grains .900 fine issued 1873–85 for use in oriental trade

trade down *intransitive verb* (1942)
1 : to trade something in (as an automobile) for something less expensive or valuable of its kind
2 : to stock or purchase lower-priced items **:** ECONOMIZE

trade edition *noun* (1930)
: an edition of a book intended for general distribution — compare TEXT EDITION

trade–in \'trād-ˌin\ *noun* (1917)
: an item of merchandise (as an automobile or refrigerator) taken as payment or part payment for a purchase

trade in *transitive verb* (1923)
1 : to turn in as payment or part payment for a purchase or bill ⟨*trade* the old car *in* on a new one⟩
2 : EXCHANGE 2

trade language *noun* (1662)
: a restructured language (as a lingua franca or pidgin) used especially in commercial communication

trade–last \'trād-ˌlast\ *noun* (1891)
: a complimentary remark by a third person that a hearer offers to repeat to the person complimented if he or she will first report a compliment made about the hearer

¹trade·mark \-ˌmärk\ *noun* (1838)
1 : a device (as a word) pointing distinctly to the origin or ownership of merchandise to which it is applied and legally reserved to the exclusive use of the owner as maker or seller
2 : a distinguishing characteristic or feature firmly associated with a person or thing ⟨derringers . . . became almost a *trademark* of gamblers —Elmer Keith⟩ ⟨wearing his *trademark* bow tie and derby hat⟩

²trademark *transitive verb* (1906)
: to secure trademark rights for **:** register the trademark of

¹trade name *noun* (1861)
1 a : the name used for an article among traders **b :** an arbitrarily adopted name that is given by a manufacturer or merchant to an article or service to distinguish it as produced or sold by him and that may be used and protected as a trademark
2 : the name or style under which a concern does business

²trade name *transitive verb* (1945)
: to designate with a trade name

trade–off \'trād-ˌȯf\ *noun* (1961)
1 : a balancing of factors all of which are not attainable at the same time ⟨the education versus experience *trade-off* which governs personnel practices —H. S. White⟩
2 : a giving up of one thing in return for another **:** EXCHANGE
— trade off *transitive verb*

trad·er \'trā-dər\ *noun* (1585)
1 : a person whose business is buying and selling or barter: as **a :** MERCHANT **b :** a person who buys and sells (as stocks or commodities futures) in search of short-term profits
2 : a ship engaged in the coastal or foreign trade

trade route *noun* (1876)
1 : a route followed by traders (as in caravans)
2 : one of the sea-lanes ordinarily used by merchant ships

trad·es·can·tia \ˌtra-də-'skan(t)-sh(ē-)ə\ *noun* [New Latin, genus name, from John *Tradescant* (died 1638) English traveler & gardener] (circa 1909)
: SPIDERWORT

trade school *noun* (1889)
: a secondary school teaching the skilled trades

trade secret *noun* (1895)
: a formula, process, or device used in a business that is not published or divulged and that thereby gives an advantage over competitors

trades·man \'trādz-mən\ *noun* (1597)
1 : a worker in a skilled trade **:** CRAFTSMAN
2 : one who runs a retail store **:** SHOPKEEPER

trades·peo·ple \-ˌpē-pəl\ *noun plural* (1728)
: people engaged in trade

trade union *also* **trades union** *noun* (1835)
: LABOR UNION
— trade unionism *noun*
— trade unionist *noun*

trade up *intransitive verb* (1926)
1 : to trade something in (as an automobile) for something more expensive or valuable of its kind
2 : to stock or purchase higher-priced items

trade wind *noun* (1650)
: a wind blowing almost constantly in one direction; *especially* **:** a wind blowing almost continually toward the equator from the northeast in the belt between the northern horse latitudes and the doldrums and from the southeast in the belt between the southern horse latitudes and the doldrums — usually used in plural

trading post *noun* (1796)
1 : a station of a trader or trading company established in a sparsely settled region where trade in products of local origin (as furs) is carried on
2 : ⁶POST 3b

trading stamp *noun* (1897)
: a printed stamp of value given as a premium to a retail customer to be redeemed in merchandise when accumulated in numbers

tra·di·tion \trə-'di-shən\ *noun* [Middle English *tradicioun,* from Middle French & Latin; Middle French *tradition,* from Latin *tradition-, traditio* action of handing over, tradition — more at TREASON] (14th century)
1 : an inherited, established, or customary pattern of thought, action, or behavior (as a religious practice or a social custom)
2 : the handing down of information, beliefs, and customs by word of mouth or by example from one generation to another without written instruction
3 : cultural continuity in social attitudes, customs, and institutions
4 : characteristic manner, method, or style
word history see TREASON
— tra·di·tion·al \-'dish-nəl, -'di-shə-nᵊl\ *adjective*
— tra·di·tion·al·ly *adverb*
— tra·di·tion·less \-'di-shən-ləs\ *adjective*

tra·di·tion·al·ism \trə-'dish-nə-ˌli-zəm, -'di-shə-nᵊl-ˌi-\ *noun* (circa 1859)
1 : adherence to the doctrines or practices of a tradition
2 : the beliefs of those opposed to modernism, liberalism, or radicalism
— tra·di·tion·al·ist \-list, -ist\ *noun or adjective*
— tra·di·tion·al·is·tic \-ˌdish-nə-'lis-tik, -ˌdi-shə-nᵊl-'is-\ *adjective*

tra·di·tion·al·ize \trə-'dish-nə-ˌlīz, -'di-shə-nᵊl-ˌīz\ *transitive verb* **-ized; -iz·ing** (1882)
: to make traditional **:** imbue with traditions or traditionalism

tra·di·tion·ary \trə-'di-shə-ˌner-ē\ *adjective* (1661)
: TRADITIONAL

tra·duce \trə-'düs, -'dyüs\ *transitive verb* **tra·duced; tra·duc·ing** [Latin *traducere* to lead across, transfer, degrade, from *tra-, trans-* trans- + *ducere* to lead — more at TOW] (1592)
1 : to expose to shame or blame by means of falsehood and misrepresentation
2 : VIOLATE, BETRAY ⟨*traduce* a principle of law⟩
synonym see MALIGN
— tra·duce·ment \-mənt\ *noun*
— tra·duc·er *noun*

¹traf·fic \'tra-fik\ *noun, often attributive* [Middle French *trafique,* from Old Italian *traffico,* from *trafficare* to traffic] (1506)
1 a : import and export trade **b :** the business of bartering or buying and selling **c :** illegal or disreputable usually commercial activity ⟨the drug *traffic*⟩
2 a : communication or dealings especially between individuals or groups **b :** EXCHANGE ⟨a lively *traffic* in ideas —F. L. Allen⟩
3 *archaic* **:** WARES, GOODS
4 a : the movement (as of vehicles or pedestrians) through an area or along a route **b :** the vehicles, pedestrians, ships, or planes moving along a route **c :** the information or signals transmitted over a communications system **:** MESSAGES
5 a : the passengers or cargo carried by a transportation system **b :** the business of transporting passengers or freight
6 : the volume of customers visiting a business establishment
synonym see BUSINESS
— the traffic will bear : existing conditions will allow or permit ⟨charge what *the traffic will bear*⟩

²traffic *verb* **traf·ficked; traf·fick·ing** (1540)
intransitive verb
: to carry on traffic
transitive verb
1 : to travel over ⟨heavily *trafficked* highways⟩
2 : TRADE, BARTER
— traf·fick·er *noun*

traf·fic·abil·i·ty \ˌtra-fi-kə-ˈbi-lə-tē\ *noun* (1899)
: the quality of a terrain that permits passage (as of vehicles and troops)
— traf·fic·able \ˈtra-fi-kə-bəl\ *adjective*

traffic circle *noun* (1942)
: ROTARY 2

traffic cone *noun* (1953)
: a conical marker used on a road or highway (as for indicating an area under repair)

traffic court *noun* (1919)
: a minor court for disposition of petty prosecutions for violations of statutes, ordinances, and local regulations governing the use of highways and motor vehicles

traffic engineering *noun* (1931)
: engineering dealing with the design of streets and control of traffic
— traffic engineer *noun*

traffic island *noun* (1931)
: ISLAND 2a

traffic light *noun* (1912)
: an electrically operated visual signal (as a system of colored lights) for controlling traffic

traffic manager *noun* (1862)
1 : a supervisor of the traffic functions of a commercial or industrial organization
2 : the director of a large telegraph office

traffic signal *noun* (1917)
: a signal (as a traffic light) for controlling traffic

trag·a·canth \ˈtra-jə-ˌkan(t)th, ˈtra-gə-, -kən(t)th; *also* ˈtra-gə-ˌsan(t)th\ *noun* [Middle French *tragacanthe,* from Latin *tragacantha,* from Greek *tragakantha,* from *tragos* goat + *akantha* thorn] (1573)
: a gum obtained from various Asian or East European leguminous plants (genus *Astragalus* and especially *A. gummifer*) that swells in water and is used chiefly as an emulsifying, suspending, or thickening agent

tra·ge·di·an \trə-ˈjē-dē-ən\ *noun* [Middle English *tragedien,* from Middle French, from *tragedie*] (14th century)
1 : a writer of tragedies
2 : an actor specializing in tragic roles

tra·ge·di·enne \trə-ˌjē-dē-ˈen\ *noun* [French *tragédienne,* from Middle French, from *tragedie*] (1851)
: an actress who plays tragic roles

trag·e·dy \ˈtra-jə-dē\ *noun, plural* **-dies** [Middle English *tragedie,* from Middle French, from Latin *tragoedia,* from Greek *tragōidia,* from *tragos* goat (akin to Greek *trōgein* to gnaw) + *aeidein* to sing — more at TROGLODYTE, ODE] (14th century)
1 a : a medieval narrative poem or tale typically describing the downfall of a great man **b :** a serious drama typically describing a conflict between the protagonist and a superior force (as destiny) and having a sorrowful or disastrous conclusion that excites pity or terror **c :** the literary genre of tragic dramas
2 a : a disastrous event **:** CALAMITY **b :** MISFORTUNE
3 : tragic quality or element

trag·ic \ˈtra-jik\ *also* **trag·i·cal** \-ji-kəl\ *adjective* [Middle English, from Latin *tragicus,* from Greek *tragikos,* irregular from *tragōidia* tragedy] (15th century)
1 : of, marked by, or expressive of tragedy ⟨the *tragic* significance of the atomic bomb —H. S. Truman⟩
2 a : dealing with or treated in tragedy ⟨the *tragic* hero⟩ **b :** appropriate to or typical of tragedy
3 a : regrettably serious or unpleasant **:** DEPLORABLE, LAMENTABLE ⟨a *tragic* mistake⟩ **b :** marked by a sense of tragedy
— trag·i·cal·ly \-ji-k(ə-)lē\ *adverb*

tragic flaw *noun* (1913)
: a flaw in character that brings about the downfall of the hero of a tragedy

tragic irony *noun* (1833)
: IRONY 3b

tragi·com·e·dy \ˌtra-ji-ˈkä-mə-dē\ *noun* [Middle French *tragicomedie,* from Old Italian *tragicomedia,* from Old Spanish, from Latin *tragicomoedia,* from *tragicus* + *comoedia* comedy] (circa 1580)
: a drama or a situation blending tragic and comic elements
— tragi·com·ic \-ˈkä-mik\ *also* **tragi·com·i·cal** \-mi-kəl\ *adjective*

tra·gus \ˈtrā-gəs\ *noun, plural* **tra·gi** \-ˌgī, -ˌjī\ [New Latin, from Greek *tragos,* a part of the ear, literally, goat] (circa 1693)
: the prominence in front of the external opening of the ear

¹trail \ˈtrā(ə)l\ *verb* [Middle English, from Middle French *trailler* to tow, from (assumed) Vulgar Latin *tragulare,* from Latin *tragula* sledge, dragnet; akin to Latin *trahere* to pull] (14th century)
intransitive verb
1 a : to hang down so as to drag along or sweep the ground **b :** to extend over a surface in a loose or straggling manner ⟨a vine that *trails* over the ground⟩ **c :** to grow to such length as to droop over toward the ground ⟨*trailing* branches of a weeping birch⟩
2 a : to walk or proceed draggingly, heavily, or wearily **:** PLOD, TRUDGE **b :** to lag behind **:** do poorly in relation to others
3 : to move, flow, or extend slowly in thin streams ⟨smoke *trailing* from chimneys⟩
4 a : to extend in an erratic or uneven course or line **:** STRAGGLE **b :** DWINDLE ⟨voice *trailing* off⟩
5 : to follow a trail **:** track game
transitive verb
1 a : to draw or drag loosely along a surface **:** allow to sweep the ground **b :** HAUL, TOW
2 a : to drag (as a limb or the body) heavily or wearily **b :** to carry or bring along as an addition, burden, or encumbrance **c :** to draw along in one's wake
3 a : to follow upon the scent or trace of **:** TRACK **b :** to follow in the footsteps of **:** PURSUE **c :** to follow along behind **d :** to lag behind (as a competitor)
synonym see CHASE

²trail *noun* (14th century)
1 : something that trails or is trailed: as **a :** a trailing plant **b :** the train of a gown **c :** a trailing arrangement (as of flowers) **:** SPRAY **d :** the part of a gun carriage that rests on the ground when the piece is unlimbered
2 a : something that follows or moves along as if being drawn along **:** TRAIN ⟨a *trail* of admirers⟩ **b** (1) **:** the streak produced by a meteor (2) **:** a continuous line produced photographically by permitting the image of a celestial body (as a star) to move over the plate **c :** a chain of consequences **:** AFTERMATH ⟨the . . . movement left a *trail* of bitterness and prejudice behind it —Paul Blanshard⟩
3 a : a trace or mark left by something that has passed or been drawn along **:** SCENT, TRACK ⟨a *trail* of blood⟩ **b** (1) **:** a track made by passage especially through a wilderness (2) **:** a marked or established path or route especially through a forest or mountainous region **c :** a course followed or to be followed ⟨hit the campaign *trail*⟩
— trail·less \ˈtrā(ə)l-ləs\ *adjective*

trail bike *noun* (1966)
: a small motorcycle designed for off-road use

trail·blaz·er \ˈtrā(ə)l-ˌblā-zər\ *noun* (1908)
1 : one that blazes a trail to guide others **:** PATHFINDER
2 : PIONEER 2 ⟨a *trailblazer* in astrophysics⟩

trail·blaz·ing \-ziŋ\ *adjective* (1951)
: making or pointing a new way ⟨*trailblazing* legislation⟩

trail·break·er \-ˌbrā-kər\ *noun* (1925)
: TRAILBLAZER

¹trail·er \ˈtrā-lər\ *noun* (1590)
1 : one that trails
2 : a trailing plant
3 : a nonautomotive vehicle designed to be hauled by road: as **a :** a vehicle for transporting something ⟨a boat *trailer*⟩; *especially* **:** SEMITRAILER 1 **b :** a vehicle designed to serve wherever parked as a temporary dwelling or place of business **c :** MOBILE HOME
4 a : PREVIEW 2 **b :** a short blank strip of film attached to the end of a reel

²trailer (1938)
intransitive verb
1 : to live or travel in or with a trailer
2 : to be transportable by trailer ⟨a light boat that *trailers* easily⟩
transitive verb
: to transport (as a boat) by means of a trailer
— trail·er·able \-lə-rə-bəl\ *adjective*
— trail·er·ing *noun*

trail·er·ist \-lə-rist\ *noun* (1950)
1 : a person traveling or vacationing with a trailer
2 : TRAILERITE 1

trail·er·ite \-ˌrīt\ *noun* (1936)
1 : a person living in a mobile home
2 : TRAILERIST 1

trailer park *noun* (1942)
: an area equipped to accommodate mobile homes — called also *trailer camp, trailer court*

trail·head \ˈtrāl-ˌhed\ *noun* (1948)
: the point at which a trail begins

trailing arbutus *noun* (1785)
: ARBUTUS 2

trailing edge *noun* (1909)
: the rearmost edge of an object that moves and especially of an airfoil

trail mix *noun* (1979)
: a mixture of seeds, nuts, and dried fruits eaten as a snack especially by hikers

trail·side \ˈtrāl-ˌsīd\ *adjective* (1923)
: of, relating to, or situated in the area immediately adjacent to a trail

¹train \ˈtrān\ *noun* [Middle English *traine,* from Middle French, from Old French, from *traïr* to betray, from Latin *tradere* — more at TRAITOR] (14th century)
obsolete **:** SCHEME, TRICK

²train *noun* [Middle English, from Middle French, from Old French, from *trainer* to draw, drag] (15th century)
1 : a part of a gown that trails behind the wearer
2 a : RETINUE, SUITE **b :** a moving file of persons, vehicles, or animals
3 : the vehicles, personnel, and sometimes animals that furnish supply, maintenance, and evacuation services to a combat unit
4 a : order of occurrence leading to some result — often used in the phrase *in train* ⟨this humiliating process had been in *train* for decades —Paul Fussell⟩ **b :** an orderly succession ⟨a *train* of thought⟩ **c :** accompanying or resultant circumstances **:** AFTERMATH ⟨consequences the discovery will bring in its *train*⟩
5 : a line of combustible material laid to lead fire to a charge
6 : a series of moving mechanical parts (as gears) that transmit and modify motion
7 a : a connected line of railroad cars with or without a locomotive **b :** an automotive tractor with one or more trailer units
8 : a series of parts or elements that together constitute a system for producing a result and

especially for carrying on a process (as of manufacture) automatically ◇
— **train·ful** \'trān-,fu̇l\ *noun*

[3]train *verb* [Middle English, from Middle French *trainer,* from Old French, from (assumed) Vulgar Latin *traginare;* akin to Latin *trahere* to draw] (15th century)
transitive verb
1 : TRAIL, DRAG
2 : to direct the growth of (a plant) usually by bending, pruning, and tying
3 a : to form by instruction, discipline, or drill **b :** to teach so as to make fit, qualified, or proficient
4 : to make prepared (as by exercise) for a test of skill
5 : to aim at an object or objective **:** DIRECT ⟨*trained* his camera on the deer⟩ ⟨*training* every effort toward success⟩
intransitive verb
1 : to undergo instruction, discipline, or drill
2 : to go by train
synonym see TEACH
— **train·abil·i·ty** \,trā-nə-'bi-lə-tē\ *noun*
— **train·able** \'trā-nə-bəl\ *adjective*

train·band \'trān-,band\ *noun* [alteration of *trained band*] (1630)
: a 17th or 18th century militia company in England or America

train·bear·er \'trān-,bar-ər, -,ber-\ *noun* (1722)
: an attendant who holds up (as on a ceremonial occasion) the train of a robe or gown

train case *noun* (1948)
: a small boxlike piece of luggage used especially for toilet articles

train dispatcher *noun* (1857)
: a railroad employee who directs the movement of trains within a division and coordinates their movement from one division to another

train·ee \trā-'nē\ *noun* (1841)
: one that is being trained especially for a job
— **train·ee·ship** \-'nē-,ship\ *noun*

train·er \'trā-nər\ *noun* (1598)
1 : one that trains
2 : one (as a machine or vehicle) used in training
3 : a person who treats the ailments and minor injuries of the members of an athletic team

train·ing *noun* (1548)
1 a : the act, process, or method of one that trains **b :** the skill, knowledge, or experience acquired by one that trains
2 : the state of being trained

training college *noun* (circa 1829)
British **:** TEACHERS COLLEGE

training school *noun* (1829)
1 : a school preparing students for a particular occupation
2 : a correctional institution for the custody and reeducation of juvenile delinquents

training table *noun* (1893)
: a table where athletes under a training regimen eat meals planned to help in their conditioning

training wheels *noun plural* (1964)
: a pair of small wheels connected to the rear axle of a bicycle to help a beginning bicyclist maintain balance

train·load \'trān-,lōd\ *noun* (1876)
: the full freight or passenger capacity of a railroad train; *also* **:** a load that fills a train

train·man \'trān-mən, -,man\ *noun* (1877)
: a member of a train crew supervised by a conductor

train oil \'trān-\ *noun* [obsolete *train* train oil, from Middle English *trane,* from Middle Dutch *trane* or Middle Low German *trān;* akin to Old High German *trahan* tear] (circa 1553)
: oil from a marine animal (as a whale)

traipse \'trāps\ *verb* **traipsed; traips·ing** [origin unknown] (1647)
intransitive verb
: to go on foot **:** WALK ⟨*traipsed* over to the restaurant⟩ ⟨children *traipsing* at her heels⟩; *also* **:** to walk or travel about without apparent plan but with or without a purpose ⟨a week *traipsing* through the Ozarks⟩ ⟨*traipsing* from office to office⟩
transitive verb
: TRAMP, WALK
synonym see WANDER
— **traipse** *noun*

trait \'trāt, *British usually* 'trā\ *noun* [Middle French, literally, act of drawing, from Latin *tractus* — more at TRACT] (1589)
1 a : a stroke of or as if of a pencil **b :** TOUCH, TRACE
2 a : a distinguishing quality (as of personal character) **:** PECULIARITY **b :** an inherited characteristic

trai·tor \'trā-tər\ *noun* [Middle English *traitre,* from Old French, from Latin *traditor,* from *tradere* to hand over, deliver, betray, from *trans-, tra-* trans- + *dare* to give — more at DATE] (13th century)
1 : one who betrays another's trust or is false to an obligation or duty
2 : one who commits treason
word history see TREASON

trai·tor·ous \'trā-tə-rəs, 'trā-trəs\ *adjective* (14th century)
1 : guilty or capable of treason
2 : constituting treason ⟨*traitorous* activities⟩
synonym see FAITHLESS
— **trai·tor·ous·ly** *adverb*

trai·tress \'trā-trəs\ *or* **trai·tor·ess** \'trā-tə-rəs, 'trā-trəs\ *noun* (14th century)
: a female traitor

tra·ject \trə-'jekt\ *transitive verb* [Latin *trajectus,* past participle of *traicere*] (1657)
: TRANSMIT
— **tra·jec·tion** \-'jek-shən\ *noun*

tra·jec·to·ry \trə-'jek-t(ə-)rē\ *noun, plural* **-ries** [New Latin *trajectoria,* from feminine of *trajectorius* of passing, from Latin *traicere* to cause to cross, cross, from *trans-, tra-* trans- + *jacere* to throw — more at JET] (1696)
1 : the curve that a body (as a planet or comet in its orbit or a rocket) describes in space
2 : a path, progression, or line of development resembling a physical trajectory

Tra·keh·ner \tra-'kā-nər\ *noun* [German, from *Trakehnen,* site of the Prussian royal stud in East Prussia] (1926)
: any of a breed of large powerful saddle horses that originated in East Prussia and excel in dressage and jumping

[1]tram \'tram\ *noun* [English dialect, shaft of a wheelbarrow, probably from Low German *traam,* literally, beam] (circa 1517)
1 : any of various vehicles: as **a :** a boxlike wagon running on rails (as in a mine) **b** *chiefly British* **:** STREETCAR **c :** a carrier that travels on an overhead cable or rails
2 *plural, chiefly British* **:** a streetcar line

[2]tram *transitive verb* **trammed; tram·ming** (1874)
: to haul in a tram or over a tramway

tram·car \'tram-,kär\ *noun* (1873)
1 *chiefly British* **:** STREETCAR
2 : TRAM 1a

tram·line \-,līn\ *noun* (1886)
British **:** a streetcar line

[1]tram·mel \'tra-məl\ *noun* [Middle English *tramayle,* a kind of net, from Middle French *tremail,* from Late Latin *tremaculum,* from Latin *tres* three + *macula* mesh, spot — more at THREE] (15th century)
1 : a net for catching birds or fish; *especially* **:** one having three layers with the middle one finer-meshed and slack so that fish passing through carry some of the center net through the coarser opposite net and are trapped
2 : an adjustable pothook for a fireplace crane
3 : a shackle used for making a horse amble
4 : something impeding activity, progress, or freedom **:** RESTRAINT — usually used in plural
5 a : an instrument for drawing ellipses **b :** a compass for drawing large circles that consists of a beam with two sliding parts — usually used in plural **c :** any of various gauges used for aligning or adjusting machine parts

[2]trammel *transitive verb* **-meled** *or* **-melled; -mel·ing** *or* **-mel·ling** \'tra-mə-liŋ, 'tram-liŋ\ (1605)
1 : to catch or hold in or as if in a net **:** ENMESH
2 : to prevent or impede the free play of **:** CONFINE
synonym see HAMPER

[1]tra·mon·tane \trə-'män-,tān, ,tra-mən-'\ *noun* (1593)
: one dwelling in a tramontane region; *broadly* **:** FOREIGNER

[2]tramontane *adjective* [Italian *tramontano,* from Latin *transmontanus,* from *trans-* + *mont-, mons* mountain — more at MOUNT] (1596)
1 : TRANSALPINE
2 : lying on or coming from the other side of a mountain range

[1]tramp \'tramp, *verb intransitive 1 & verb transitive 1 are also* 'trämp, 'trȯmp\ *verb* [Middle English; akin to Middle Low German *trampen* to stamp] (14th century)
intransitive verb
1 : to walk, tread, or step especially heavily
2 a : to travel about on foot **:** HIKE **b :** to journey as a tramp
transitive verb
1 : to tread on forcibly and repeatedly
2 : to travel or wander through or over on foot ⟨have *tramped* all the woods on their property⟩
— **tramp·er** *noun*

[2]tramp \'tramp, *3, 4 are also* 'trämp, 'trȯmp\ *noun* (1664)
1 a : a foot traveler **b :** a begging or thieving vagrant **c :** a woman of loose morals; *specifically* **:** PROSTITUTE
2 : a walking trip **:** HIKE
3 : the succession of sounds made by the beating of feet on a surface (as a road, pavement, or floor)
4 : an iron plate to protect the sole of a shoe

◇ WORD HISTORY

train The word *train* in the sense "a connected series of railroad cars" has given rise to words with the same meaning in a number of European languages, though sometimes in a rather roundabout way. Railways and trains with their associated vocabulary were British inventions, spreading from Britain to the continent. In English, *train* in its railway sense was shortened from phrases such as *train of carriages* or *train of wagons,* with *train* used in its long-established sense "a set of objects linked in succession." French had given English the word centuries before and now borrowed this particular meaning back. *Train* is first noted in French with the sense "series of railroad cars" in 1829, about five years after the meaning first appeared in English. Spanish and Italian borrowed the French or English word directly as, respectively, *tren* and *treno.* In central Europe, however, a different kind of borrowing took place. Because French *train* is a transparent derivative of the verb *trainer* "to pull, drag," German translated *train* with the noun *Zug,* which functions as a derivative of the verb *ziehen* "to pull, draw"—appropriate enough since a locomotive pulls a series of cars.

5 : a ship not making regular trips but taking cargo when and where it offers and to any port — called also *tramp steamer*

[3]tramp \'tramp\ *adjective* (1873)
: having no fixed abode, connection, or destination ⟨a *tramp* dog⟩

tramp art *noun* (1974)
: a style of wood carving flourishing in the U.S. from about 1875 to 1930 that is characterized by ornate layered whittling often of cigar boxes or fruit crates; *also* **:** an object carved in this style

tram•ple \'tram-pəl\ *verb* **tram•pled; tram•pling** \-p(ə-)liŋ\ [Middle English, frequentative of *trampen* to tramp] (14th century)
intransitive verb
1 : TRAMP; *especially* **:** to tread heavily so as to bruise, crush, or injure
2 : to inflict injury or destruction especially contemptuously or ruthlessly — usually used with *on, over,* or *upon* ⟨*trampling* on the rights of others⟩
transitive verb
: to crush, injure, or destroy by or as if by treading
— trample *noun*
— tram•pler \-p(ə-)lər\ *noun*

tram•po•line \ˌtram-pə-'lēn, 'tram-pə-ˌ\ *noun* [Italian *trampolino* springboard, from *trampoli* stilts, of Germanic origin; akin to Middle Low German *trampen* to stamp] (1928)
: a resilient sheet or web (as of nylon) supported by springs in a metal frame and used as a springboard and landing area in tumbling
— tram•po•lin•er \-'lē-nər, -ˌlē-\ *noun*
— tram•po•lin•ist \-nist\ *noun*

tram•po•lin•ing *noun* (1949)
: the sport of jumping and tumbling on a trampoline

tram•way \'tram-ˌwā\ *noun* (1825)
1 a : a railway for trams **b** *British* **:** a streetcar line
2 : an overhead cable for trams

[1]trance \'tran(t)s\ *noun* [Middle English, from Middle French *transe,* from *transir* to pass away, swoon, from Latin *transire* to pass, pass away — more at TRANSIENT] (14th century)
1 : a state of partly suspended animation or inability to function
2 : a somnolent state (as of deep hypnosis)
3 : a state of profound abstraction or absorption
— trance•like \-ˌlīk\ *adjective*

[2]trance *transitive verb* **tranced; tranc•ing** (circa 1598)
: ENTRANCE, ENRAPTURE

tranche \'trä^{n}sh\ *noun* [French, literally, slice, from Old French, from *trenchier, trancher* to cut — more at TRENCH] (1930)
: a bond series issued for sale in a foreign country

tran•gam \'traŋ-gəm\ *noun* [origin unknown] (circa 1658)
archaic **:** TRINKET, GIMCRACK

tran•quil \'traŋ-kwəl, 'tran-\ *adjective* [Latin *tranquillus*] (1604)
1 a : free from agitation of mind or spirit ⟨a *tranquil* self-assurance⟩ **b :** free from disturbance or turmoil ⟨a *tranquil* scene⟩
2 : unvarying in aspect **:** STEADY, STABLE
synonym see CALM
— tran•quil•ly \-kwə-lē\ *adverb*
— tran•quil•ness *noun*

tran•quil•ize *also* **tran•quil•lize** \'traŋ-kwə-ˌlīz, 'tran-\ *verb* **-ized** *also* **-lized; -iz•ing** *also* **-liz•ing** (1623)
transitive verb
: to make tranquil or calm **:** PACIFY; *especially* **:** to relieve of mental tension and anxiety by means of drugs
intransitive verb
1 : to become tranquil **:** RELAX 1
2 : to make one tranquil

tran•quil•iz•er *also* **tran•quil•liz•er** \-ˌlī-zər\ *noun* (1800)
1 : one that tranquilizes
2 : a drug used to reduce mental disturbance (as anxiety and tension)

tran•quil•li•ty *or* **tran•quil•i•ty** \tran-'kwi-lə-tē, traŋ-\ *noun* (14th century)
: the quality or state of being tranquil

trans \'tran(t)s, 'tranz\ *adjective* (1892)
: characterized by having certain groups of atoms on opposite sides of the longitudinal axis of a double bond or of the plane of a ring in a molecule

trans- *prefix* [Latin *trans-, tra-* across, beyond, through, so as to change, from *trans* across, beyond — more at THROUGH]
1 : on or to the other side of **:** across **:** beyond ⟨*trans*atlantic⟩
2 a : beyond (a specified chemical element) in the periodic table ⟨*trans*uranium⟩ **b** *usually italic* **:** trans ⟨*trans*-dichloro-ethylene⟩ — compare CIS- 2
3 : through ⟨*trans*cutaneous⟩
4 : so or such as to change or transfer ⟨*trans*literate⟩ ⟨*trans*location⟩ ⟨*trans*amination⟩ ⟨*trans*ship⟩

trans•act \tran-'zakt, tran(t)-'sakt\ *verb* [Latin *transactus,* past participle of *transigere* to drive through, complete, transact, from *trans-* + *agere* to drive, do — more at AGENT] (circa 1585)
intransitive verb
: to carry on business
transitive verb
1 : to carry to completion ⟨*transact* a sale⟩
2 : to carry on the operation or management of **:** DO ⟨*transact* business⟩
— trans•ac•tor \-'zak-tər, -'sak-tər\ *noun*

trans•ac•ti•nide \tran(t)-'sak-tə-ˌnīd, tran-'zak-\ *adjective* (1969)
: of, relating to, or being actual or hypothetical elements with atomic weights higher than those of the actinide series ⟨*transactinide* chemistry⟩

trans•ac•tion \tran-'zak-shən, tran(t)-'sak-\ *noun* (1647)
1 a : something transacted; *especially* **:** an exchange or transfer of goods, services, or funds ⟨electronic *transactions*⟩ **b** *plural* **:** the often published record of the meeting of a society or association
2 a : an act, process, or instance of transacting **b :** a communicative action or activity involving two parties or things that reciprocally affect or influence each other
— trans•ac•tion•al \-shnəl, -shə-n^{ə}l\ *adjective*

transactional analysis *noun* (1961)
: a system of psychotherapy involving analysis of individual episodes of social interaction for insight that will aid communication

trans•al•pine \tran(t)s-'al-ˌpīn, tranz-\ *adjective* [Latin *transalpinus,* from *trans-* + *Alpes* the Alps] (1590)
: situated on the north side of the Alps ⟨*Transalpine* Gaul⟩ — compare CISALPINE

trans•am•i•nase \tran(t)s-'a-mə-ˌnās, tranz-, -ˌnāz\ *noun* (1940)
: an enzyme promoting transamination — called also *aminotransferase*

trans•am•i•na•tion \ˌtran(t)s-ˌa-mə-'nā-shən, ˌtranz-\ *noun* (1939)
: a reversible oxidation-reduction reaction in which an amino group is transferred typically from an alpha-amino acid to the carbonyl carbon atom of an alpha-keto acid

trans•at•lan•tic \ˌtran(t)s-ət-'lan-tik, ˌtranz-\ *adjective* (1779)
1 a : crossing or extending across the Atlantic Ocean ⟨a *transatlantic* cable⟩ **b :** relating to or involving crossing the Atlantic Ocean ⟨*transatlantic* air fares⟩
2 a : situated or originating from beyond the Atlantic Ocean **b :** of, relating to, or involving countries on both sides of the Atlantic Ocean and especially the U.S. and Great Britain ⟨*transatlantic* cooperation⟩

trans•ax•le \tran(t)s-'ak-səl, tranz-\ *noun* [*trans*mission + *axle*] (1958)
: a unit that consists of a combination of transmission and front axle used in front-wheel-drive automobiles

trans•ceiv•er \tran(t)-'sē-vər\ *noun* [*trans*mitter + *receiver*] (1934)
: a radio transmitter-receiver that uses many of the same components for both transmission and reception

tran•scend \tran(t)-'send\ *verb* [Middle English, from Latin *transcendere* to climb across, transcend, from *trans-* + *scandere* to climb — more at SCAN] (14th century)
transitive verb
1 a : to rise above or go beyond the limits of **b :** to triumph over the negative or restrictive aspects of **:** OVERCOME **c :** to be prior to, beyond, and above (the universe or material existence)
2 : to outstrip or outdo in some attribute, quality, or power
intransitive verb
: to rise above or extend notably beyond ordinary limits
synonym see EXCEED

tran•scen•dence \-'sen-dən(t)s\ *noun* (1601)
: the quality or state of being transcendent

tran•scen•den•cy \-dən(t)-sē\ *noun* (1615)
: TRANSCENDENCE

tran•scen•dent \-dənt\ *adjective* [Latin *transcendent-, transcendens,* present participle of *transcendere*] (1598)
1 a : exceeding usual limits **:** SURPASSING **b :** extending or lying beyond the limits of ordinary experience **c** *in Kantian philosophy* **:** being beyond the limits of all possible experience and knowledge
2 : being beyond comprehension
3 : transcending the universe or material existence
— tran•scen•dent•ly *adverb*

tran•scen•den•tal \ˌtran(t)-ˌsen-'den-t^{ə}l, -sən-\ *adjective* (1624)
1 a : TRANSCENDENT 1b **b :** SUPERNATURAL **c :** ABSTRUSE, ABSTRACT **d :** of or relating to transcendentalism
2 a : incapable of being the root of an algebraic equation with rational coefficients ⟨π is a *transcendental* number⟩ **b :** being, involving, or representing a function (as sin x, log x, e^x) that cannot be expressed by a finite number of algebraic operations ⟨*transcendental* curves⟩
3 *in Kantian philosophy* **a :** of or relating to experience as determined by the mind's make-up **b :** transcending experience but not human knowledge
4 : TRANSCENDENT 1a
— tran•scen•den•tal•ly \-t^{ə}l-ē\ *adverb*

tran•scen•den•tal•ism \-t^{ə}l-ˌi-zəm\ *noun* (1803)
1 : a philosophy that emphasizes the a priori conditions of knowledge and experience or the unknowable character of ultimate reality or that emphasizes the transcendent as the fundamental reality
2 : a philosophy that asserts the primacy of the spiritual and transcendental over the material and empirical
3 : the quality or state of being transcendental; *especially* **:** visionary idealism
— tran•scen•den•tal•ist \-t^{ə}l-ist\ *adjective or noun*

transcendental meditation *noun* (1966)
: a technique of meditation in which a mantra is chanted in order to foster calm, creativity, and spiritual well-being

trans•con•ti•nen•tal \ˌtran(t)s-ˌkän-t^{ə}n-'en-t^{ə}l, ˌtranz-\ *adjective* (1853)
: extending or going across a continent ⟨a *transcontinental* railroad⟩

tran•scribe \tran(t)-'skrīb\ *transitive verb* **tran•scribed; tran•scrib•ing** [Latin *transcribere,* from *trans-* + *scribere* to write — more at SCRIBE] (1552)

1 a : to make a written copy of **b :** to make a copy of (dictated or recorded matter) in longhand or on a machine (as a typewriter) **c :** to paraphrase or summarize in writing **d :** WRITE DOWN, RECORD
2 a : to represent (speech sounds) by means of phonetic symbols **b :** TRANSLATE 2a **c :** to transfer (data) from one recording form to another **d :** to record (as on magnetic tape) for later broadcast
3 : to make a musical transcription of
4 : to cause (as DNA) to undergo genetic transcription
— **tran·scrib·er** *noun*

tran·script \'tran(t)-,skript\ *noun* [Middle English, from Medieval Latin *transcriptum,* from Latin, neuter of *transcriptus,* past participle of *transcribere*] (14th century)
1 a : a written, printed, or typed copy; *especially* **:** a usually typewritten copy of dictated or recorded material **b :** an official or legal and often published copy ⟨a court reporter's *transcript*⟩; *especially* **:** an official copy of a student's educational record
2 : a representation (as of experience) in an art form
3 : a sequence of RNA produced by transcription from a DNA template

tran·scrip·tase \tran-'skrip-,tās, -,tāz\ *noun* [*transcrip*tion + *-ase*] (1963)
: RNA POLYMERASE; *also* **:** REVERSE TRANSCRIPTASE

tran·scrip·tion \tran(t)-'skrip-shən\ *noun* (1598)
1 : an act, process, or instance of transcribing
2 : COPY, TRANSCRIPT: as **a :** an arrangement of a musical composition for some instrument or voice other than the original **b :** a recording (as on magnetic tape) made especially for use in radiobroadcasting
3 : the process of constructing a messenger RNA molecule using a DNA molecule as a template with resulting transfer of genetic information to the messenger RNA — compare TRANSLATION 2, REVERSE TRANSCRIPTION
— **tran·scrip·tion·al** \-shnəl, -shə-nᵊl\ *adjective*
— **tran·scrip·tion·al·ly** *adverb*

tran·scrip·tion·ist \-shə-nist\ *noun* (1963)
: one that transcribes; *especially* **:** a typist who transcribes dictated medical reports

trans·cul·tur·al \tran(t)s-'kəl-chə-rəl, tranz-, -'kəlch-rəl\ *adjective* (1951)
: involving, encompassing, or extending across two or more cultures ⟨*transcultural* problems⟩

trans·cu·ta·ne·ous \,tran(t)s-kyu̇-'tā-nē-əs\ *adjective* (circa 1941)
: passing, entering, or made by penetration through the skin ⟨*transcutaneous* infection⟩ ⟨*transcutaneous* inoculation⟩

trans·der·mal \tran(t)s-'dər-məl, tranz-\ *adjective* (1944)
: relating to, being, or supplying a medication in a form for absorption through the skin into the bloodstream ⟨*transdermal* drug delivery⟩ ⟨*transdermal* nitroglycerin⟩ ⟨*transdermal* nicotine patch⟩

trans·dis·ci·plin·ary \-'di-sə-plə-,ner-ē\ *adjective* (1948)
: INTERDISCIPLINARY

trans·duce \tran(t)s-'düs, tranz-, -'dyüs\ *transitive verb* **trans·duced; trans·duc·ing** [Latin *transducere* to lead across, transfer, from *trans-* + *ducere* to lead — more at TOW] (1947)
1 : to convert (as energy or a message) into another form ⟨essentially sense organs *transduce* physical energy into a nervous signal⟩
2 : to bring about the transfer of (as a gene) from one microorganism to another by means of a viral agent

trans·duc·er \-'dü-sər, -'dyü-\ *noun* (1924)
: a device that is actuated by power from one system and supplies power usually in another form to a second system ⟨a loudspeaker is a *transducer* that transforms electrical signals into sound energy⟩

trans·duc·tion \-'dək-shən\ *noun* [Latin *transducere*] (1947)
: the action or process of transducing; *especially* **:** the transfer of genetic determinants from one microorganism to another by a viral agent (as a bacteriophage)
— **trans·duc·tant** \-tənt\ *noun*
— **trans·duc·tion·al** \-shnəl, -shə-nᵊl\ *adjective*

¹tran·sect \tran(t)-'sekt\ *transitive verb* [*trans-* + inter*sect*] (1634)
: to cut transversely
— **tran·sec·tion** \-'sek-shən\ *noun*

²tran·sect \'tran(t)-,sekt\ *noun* (1905)
: a sample area (as of vegetation) usually in the form of a long continuous strip

tran·sept \'tran(t)-,sept\ *noun* [New Latin *transeptum,* from Latin *trans-* + *septum, saeptum* enclosure, wall] (circa 1542)
: the part of a cruciform church that crosses at right angles to the greatest length between the nave and the apse or choir; *also* **:** either of the projecting ends of a transept
— **tran·sep·tal** \tran(t)-'sep-tᵊl\ *adjective*

trans·fec·tion \tran(t)s-'fek-shən, tranz-\ *noun* [*trans-* + in*fection*] (1964)
: infection of a cell with isolated viral nucleic acid followed by production of the complete virus in the cell; *also* **:** the incorporation of exogenous DNA into a cell
— **trans·fect** \-'fekt\ *transitive verb*

¹trans·fer \tran(t)s-'fər, 'tran(t)s-,\ *verb* **trans·ferred; trans·fer·ring** [Middle English *transferren,* from Latin *transferre,* from *trans-* + *ferre* to carry — more at BEAR] (14th century)
transitive verb
1 a : to convey from one person, place, or situation to another **:** TRANSPORT **b :** to cause to pass from one to another **:** TRANSMIT **c :** TRANSFORM, CHANGE
2 : to make over the possession or control of **:** CONVEY
3 : to print or otherwise copy from one surface to another by contact
intransitive verb
1 : to move to a different place, region, or situation; *especially* **:** to withdraw from one educational institution to enroll at another
2 : to change from one vehicle or transportation line to another
— **trans·fer·abil·i·ty** \(,)tran(t)s-,fər-ə-'bi-lə-tē\ *noun*
— **trans·fer·able** *also* **trans·fer·ra·ble** \tran(t)s-'fər-ə-bəl\ *adjective*
— **trans·fer·al** \-əl\ *noun*
— **trans·fer·rer** \-ər\ *noun*

²trans·fer \'tran(t)s-,fər\ *noun* (1674)
1 a : conveyance of right, title, or interest in real or personal property from one person to another **b :** removal or acquisition of property by mere delivery with intent to transfer title
2 a : an act, process, or instance of transferring **:** TRANSFERENCE 2 **b :** the carryover or generalization of learned responses from one type of situation to another
3 : one that transfers or is transferred; *especially* **:** a graphic image transferred by contact from one surface to another
4 : a place where a transfer is made (as of trains to ferries or as where one form of power is changed to another)
5 : a ticket entitling a passenger on a public conveyance to continue the trip on another route

trans·fer·ase \'tran(t)s-(,)fər-,ās, -,āz\ *noun* (1948)
: an enzyme that promotes transfer of a group from one molecule to another

trans·fer·ee \,tran(t)s-(,)fər-'ē\ *noun* (circa 1736)
1 : a person to whom a conveyance is made
2 : one who is transferred

trans·fer·ence \tran(t)s-'fər-ən(t)s, 'tran(t)s-(,)\ *noun* (1681)
1 : an act, process, or instance of transferring **:** CONVEYANCE, TRANSFER
2 : the redirection of feelings and desires and especially of those unconsciously retained from childhood toward a new object (as a psychoanalyst conducting therapy)
— **trans·fer·en·tial** \,tran(t)s-fə-'ren(t)-shəl\ *adjective*

transfer factor *noun* (1956)
: a substance that is produced and secreted by a lymphocyte functioning in cell-mediated immunity and that upon incorporation into a lymphocyte which has not been sensitized confers on it the same immunological specificity as the sensitized cell

trans·fer·or \,tran(t)s-(,)fər-'ȯr\ *noun* (1875)
: one that conveys a title, right, or property

transfer payment *noun* (circa 1945)
1 : a public expenditure made for a purpose (as unemployment compensation) other than procuring goods or services — usually used in plural
2 *plural* **:** money (as welfare payments) that is received by individuals and that is neither compensation for goods or services currently supplied nor income from investments

trans·fer·rin \tran(t)s-'fer-ən\ *noun* [*trans-* + Latin *ferrum* iron] (1947)
: a beta globulin in blood plasma capable of combining with ferric ions and transporting iron in the body

transfer RNA \'tran(t)s-,fər-\ *noun* (1961)
: a relatively small RNA that transfers a particular amino acid to a growing polypeptide chain at the ribosomal site of protein synthesis during translation — compare MESSENGER RNA

trans·fig·u·ra·tion \(,)tran(t)s-,fi-gyə-'rā-shən, -gə-\ *noun* (14th century)
1 a : a change in form or appearance **:** METAMORPHOSIS **b :** an exalting, glorifying, or spiritual change
2 *capitalized* **:** August 6 observed as a Christian feast in commemoration of the transfiguration of Christ on a mountaintop in the presence of three disciples

trans·fig·ure \tran(t)s-'fi-gyər, *especially British* -'fi-gər\ *transitive verb* **-ured; -ur·ing** [Middle English, from Latin *transfigurare,* from *trans-* + *figurare* to shape, fashion, from *figura* figure] (14th century)
: to give a new and typically exalted or spiritual appearance to **:** transform outwardly and usually for the better
synonym see TRANSFORM

trans·fi·nite \(,)tran(t)s-'fī-,nīt\ *adjective* [German *transfinit,* from *trans-* (from Latin) + *finit* finite, from Latin *finitus*] (1902)
1 : going beyond or surpassing any finite number, group, or magnitude
2 : being or relating to cardinal and ordinal numbers of sets with an infinite number of elements

trans·fix \tran(t)s-'fiks\ *transitive verb* [Latin *transfixus,* past participle of *transfigere,* from *trans-* + *figere* to fasten, pierce — more at FIX] (1590)
1 : to pierce through with or as if with a pointed weapon **:** IMPALE
2 : to hold motionless by or as if by piercing
— **trans·fix·ion** \-'fik-shən\ *noun*

¹trans·form \tran(t)s-'fȯrm\ *verb* [Middle English, from Latin *transformare,* from *trans-* + *formare* to form, from *forma* form] (14th century)
transitive verb
1 a : to change in composition or structure **b :** to change the outward form or appearance of

c : to change in character or condition **:** CONVERT
2 : to subject to mathematical transformation
3 : to cause (a cell) to undergo genetic transformation
intransitive verb
: to become transformed **:** CHANGE ☆
— trans·form·able \-'fȯr-mə-bəl\ *adjective*
— trans·for·ma·tive \-'fȯr-mə-tiv\ *adjective*

²trans·form \'tran(t)s-,fȯrm\ *noun* (1853)
1 : a mathematical element obtained from another by transformation
2 : TRANSFORMATION 3a(1), (2)
3 : a linguistic structure (as a sentence) produced by means of a transformation ⟨"the duckling is killed by the farmer" is a *transform* of "the farmer kills the duckling"⟩

trans·for·ma·tion \,tran(t)s-fər-'mā-shən, -fȯr-\ *noun* (15th century)
1 : an act, process, or instance of transforming or being transformed
2 : false hair worn especially by a woman to replace or supplement natural hair
3 a (1) **:** the operation of changing (as by rotation or mapping) one configuration or expression into another in accordance with a mathematical rule; *especially* **:** a change of variables or coordinates in which a function of new variables or coordinates is substituted for each original variable or coordinate (2) **:** the formula that effects a transformation **b :** FUNCTION 5a **c :** an operation that converts (as by insertion, deletion, or permutation) one grammatical string (as a sentence) into another; *also* **:** a formal statement of such an operation
4 a : genetic modification of a bacterium by incorporation of free DNA from another ruptured bacterial cell — compare TRANSDUCTION **b :** genetic modification of a cell by the uptake and incorporation of exogenous DNA

trans·for·ma·tion·al \-shnəl, -shə-nᵊl\ *adjective* (1894)
: of, relating to, characterized by, or concerned with transformation and especially linguistic transformation
— trans·for·ma·tion·al·ly *adverb*

transformational grammar *noun* (1961)
: a grammar that generates the deep structures of a language and converts these to the surface structures by means of transformations

trans·for·ma·tion·al·ist \,tran(t)s-fər-'mā-shnə-list, -shə-nᵊl-ist\ *noun* (1964)
: an exponent of transformational grammar

trans·form·er \tran(t)s-'fȯr-mər\ *noun* (1596)
: one that transforms; *specifically* **:** a device employing the principle of mutual induction to convert variations of current in a primary circuit into variations of voltage and current in a secondary circuit

trans·fuse \tran(t)s-'fyüz\ *transitive verb* **trans·fused; trans·fus·ing** [Middle English, from Latin *transfusus,* past participle of *transfundere,* from *trans-* + *fundere* to pour — more at FOUND] (15th century)
1 a : to cause to pass from one to another **:** TRANSMIT **b :** to diffuse into or through **:** PERMEATE ⟨sunlight *transfuses* the bay⟩
2 a : to transfer (as blood) into a vein of a person or animal **b :** to subject (a patient) to transfusion
— trans·fus·ible *or* **trans·fus·able** \-'fyü-zə-bəl\ *adjective*

trans·fu·sion \tran(t)s-'fyü-zhən\ *noun* (1578)
1 : an act, process, or instance of transfusing; *especially* **:** the process of transfusing fluid into a vein or artery
2 : something transfused
— trans·fu·sion·al \-'fyüzh-nəl, -'fyü-zhə-nᵊl\ *adjective*

trans·gen·ic \tran(t)s-'je-nik\ *adjective* (1982)
: having chromosomes into which one or more heterologous genes have been incorporated either artificially or naturally ⟨*transgenic* mice⟩

trans·gress \tran(t)s-'gres, tranz-\ *verb* [Middle French *transgresser,* from Latin *transgressus,* past participle of *transgredi* to step beyond or across, from *trans-* + *gradi* to step — more at GRADE] (1526)
transitive verb
1 : to go beyond limits set or prescribed by **:** VIOLATE ⟨*transgress* divine law⟩
2 : to pass beyond or go over (a limit or boundary)
intransitive verb
1 : to violate a command or law **:** SIN
2 : to go beyond a boundary or limit
— trans·gres·sive \-'gre-siv\ *adjective*
— trans·gres·sor \-'gre-sər\ *noun*

trans·gres·sion \-'gre-shən\ *noun* (15th century)
: an act, process, or instance of transgressing: as **a :** infringement or violation of a law, command, or duty **b :** the spread of the sea over land areas and the consequent unconformable deposit of sediments on older rocks

tran·ship *variant of* TRANSSHIP

trans·his·tor·i·cal \,tran(t)s-(h)is-'tȯr-i-kəl, ,tranz-, -'tär-\ *adjective* (1909)
: transcending historical bounds

trans·hu·mance \tran(t)s-'hyü-mən(t)s, tranz-, -'yü-\ *noun* [French, from *transhumer* to practice transhumance, from Spanish *trashumar,* from *tras-* trans- (from Latin *trans-*) + Latin *humus* earth — more at HUMBLE] (circa 1901)
: seasonal movement of livestock and especially sheep between mountain and lowland pastures either under the care of herders or in company with the owners
— trans·hu·mant \-mənt\ *adjective or noun*

tran·sience \'tran(t)-sh(ē-)ən(t)s; 'tran-zē-ən(t)s, 'tran(t)-sē-; 'tran-zhən(t)s, -jən(t)s\ *noun* (1745)
: the quality or state of being transient

tran·sien·cy \-sh(ē-)ən(t)-sē, -zē-ən(t)-, -sē-ən(t)-; -zhən(t)-sē, -jən(t)-\ *noun* (1652)
: TRANSIENCE

¹tran·sient \-sh(ē-)ənt, -zē-ənt, -sē-; -zhənt, -jənt\ *adjective* [Latin *transeunt-, transiens,* present participle of *transire* to go across, pass, from *trans-* + *ire* to go — more at ISSUE] (1599)
1 a : passing especially quickly into and out of existence **:** TRANSITORY **b :** passing through or by a place with only a brief stay or sojourn
2 : affecting something or producing results beyond itself ☆
— tran·sient·ly *adverb*

²transient *noun* (1652)
1 : one that is transient: as **a :** a transient guest **b :** a person traveling about usually in search of work
2 a : a temporary oscillation that occurs in a circuit because of a sudden change of voltage or of load **b :** a transient current or voltage

transient ischemic attack *noun* (1966)
: a brief episode of cerebral ischemia that is usually characterized by temporary blurring of vision, slurring of speech, numbness, paralysis, or syncope and that is often predictive of more serious cerebral accidents — abbreviation *TIA*

trans·il·lu·mi·nate \,tran(t)s-ə-'lü-mə-,nāt, ,tranz-\ *transitive verb* (1900)
: to cause light to pass through; *especially* **:** to pass light through (a body part) for medical examination
— trans·il·lu·mi·na·tion \-,lü-mə-'nā-shən\ *noun*
— trans·il·lu·mi·na·tor \-'lü-mə-,nā-tər\ *noun*

tran·sis·tor \tran-'zis-tər, tran(t)-'sis-\ *noun* [¹*trans*fer + re*sistor;* from its transferring an electrical signal across a resistor] (1948)
1 : a solid-state electronic device that is used to control the flow of electricity in electronic equipment and consists of a small block of a semiconductor (as germanium) with at least three electrodes
2 : a transistorized radio

tran·sis·tor·ise *British variant of* TRANSISTORIZE

tran·sis·tor·ize \-tə-,rīz\ *transitive verb* **-ized; -iz·ing** (circa 1952)
: to equip (a device) with transistors
— tran·sis·tor·i·za·tion \-,zis-tə-rə-'zā-shən\ *noun*

¹tran·sit \'tran(t)-sət, 'tran-zət\ *noun* [Middle English *transite,* from Latin *transitus,* from *transire* to go across, pass] (15th century)
1 a : an act, process, or instance of passing through or over **:** PASSAGE **b :** CHANGE, TRANSITION **c** (1) **:** conveyance of persons or things from one place to another (2) **:** usually local transportation especially of people by public conveyance; *also* **:** vehicles or a system engaged in such transportation
2 a : passage of a celestial body over the meridian of a place or through the field of a telescope **b :** passage of a smaller body (as Venus) across the disk of a larger (as the sun)
3 : a theodolite with the telescope mounted so that it can be transited

²transit (15th century)
intransitive verb
: to make a transit
transitive verb
1 a : to pass over or through **:** TRAVERSE **b :** to cause to pass over or through
2 : to pass across (a meridian, a celestial body, or the field of view of a telescope)
3 : to turn (a telescope) over about the horizontal transverse axis in surveying

☆ SYNONYMS

Transform, metamorphose, transmute, convert, transmogrify, transfigure mean to change a thing into a different thing. TRANSFORM implies a major change in form, nature, or function ⟨*transformed* a small company into a corporate giant⟩. METAMORPHOSE suggests an abrupt or startling change induced by or as if by magic or a supernatural power ⟨awkward girls *metamorphosed* into graceful ballerinas⟩. TRANSMUTE implies transforming into a higher element or thing ⟨attempted to *transmute* lead into gold⟩. CONVERT implies a change fitting something for a new or different use or function ⟨*converted* the study into a nursery⟩. TRANSMOGRIFY suggests a strange or preposterous metamorphosis ⟨a story in which a frog is *transmogrified* into a prince⟩. TRANSFIGURE implies a change that exalts or glorifies ⟨joy *transfigured* her face⟩.

Transient, transitory, ephemeral, momentary, fugitive, fleeting, evanescent mean lasting or staying only a short time. TRANSIENT applies to what is actually short in its duration or stay ⟨a hotel catering primarily to *transient* guests⟩. TRANSITORY applies to what is by its nature or essence bound to change, pass, or come to an end ⟨fame in the movies is *transitory*⟩. EPHEMERAL implies striking brevity of life or duration ⟨many slang words are *ephemeral*⟩. MOMENTARY suggests coming and going quickly and therefore being merely a brief interruption of a more enduring state ⟨my feelings of guilt were only *momentary*⟩. FUGITIVE and FLEETING imply passing so quickly as to make apprehending difficult ⟨let a *fugitive* smile flit across his face⟩ ⟨*fleeting* moments of joy⟩. EVANESCENT suggests a quick vanishing and an airy or fragile quality ⟨the story has an *evanescent* touch of whimsy that is lost in translation⟩.

tran·si·tion \tran(t)-ˈsi-shən, tran-ˈzi-, *chiefly British* tran(t)-ˈsi-zhən\ *noun* [Latin *transition-, transitio,* from *transire*] (1551)
1 a : passage from one state, stage, subject, or place to another **:** CHANGE **b :** a movement, development, or evolution from one form, stage, or style to another
2 a : a musical modulation **b :** a musical passage leading from one section of a piece to another
3 : an abrupt change in energy state or level (as of an atomic nucleus or a molecule) usually accompanied by loss or gain of a single quantum of energy
— tran·si·tion·al \-ˈsish-nəl, -ˈsizh-, -ˈzish-; -ˈsi-shə-n^{ə}l, -ˈzi-, -zhə-\ *adjective*
— tran·si·tion·al·ly *adverb*

transition metal *noun* [from their being transitional between the more highly electropositive and the less highly electropositive elements] (1940)
: any of various metallic elements (as chromium, iron, and nickel) that have valence electrons in two shells instead of only one — called also *transition element*

tran·si·tive \ˈtran(t)-sə-tiv, ˈtran-zə-; ˈtran(t)s-tiv\ *adjective* [Late Latin *transitivus,* from Latin *transitus,* past participle of *transire*] (1590)
1 : characterized by having or containing a direct object ⟨a *transitive* verb⟩ ⟨a *transitive* construction⟩
2 : being or relating to a relation with the property that if the relation holds between a first element and a second and between the second element and a third, it holds between the first and third elements ⟨equality is a *transitive* relation⟩
3 : of, relating to, or characterized by transition
— tran·si·tive·ly *adverb*
— tran·si·tive·ness *noun*
— tran·si·tiv·i·ty \ˌtran(t)-sə-ˈti-və-tē, ˌtran-zə-\ *noun*

tran·si·to·ry \ˈtran(t)-sə-ˌtōr-ē, ˈtran-zə-, -ˌtȯr-\ *adjective* [Middle English *transitorie,* from Middle French *transitoire,* from Late Latin *transitorius,* from Latin, of or allowing passage, from *transire*] (14th century)
1 : tending to pass away **:** not persistent
2 : of brief duration **:** TEMPORARY ⟨the *transitory* nature of earthly joy⟩
synonym see TRANSIENT
— tran·si·to·ri·ly \ˌtran(t)-sə-ˈtōr-ə-lē, ˌtran-zə-, -ˈtȯr-\ *adverb*
— tran·si·to·ri·ness \ˈtran(t)-sə-ˌtōr-ē-nəs, ˈtran-zə-, -ˌtȯr-\ *noun*

trans·late \tran(t)s-ˈlāt, tranz-; ˈtran(t)s-ˌlāt, ˈtranz-\ *verb* **trans·lat·ed; trans·lat·ing** [Middle English, from Latin *translatus* (past participle of *transferre* to transfer, translate), from *trans-* + *latus,* past participle of *ferre* to carry — more at TOLERATE, BEAR] (14th century)
transitive verb
1 a : to bear, remove, or change from one place, state, form, or appearance to another **:** TRANSFER, TRANSFORM ⟨a country boy *translated* to the city⟩ ⟨*translate* ideas into action⟩ **b :** to convey to heaven or to a nontemporal condition without death **c :** to transfer (a bishop) from one see to another
2 a : to turn into one's own or another language **b :** to transfer or turn from one set of symbols into another **:** TRANSCRIBE **c** (1) **:** to express in different terms and especially different words **:** PARAPHRASE (2) **:** to express in more comprehensible terms **:** EXPLAIN, INTERPRET
3 : ENRAPTURE
4 : to subject to mathematical translation
5 : to subject (as genetic information) to translation in protein synthesis
intransitive verb
1 : to practice translation or make a translation; *also* **:** to admit of or be adaptable to translation ⟨a word that doesn't *translate* easily⟩
2 : to undergo a translation
3 : LEAD, RESULT — usually used with *into* ⟨tax cuts *translate* into bigger savings⟩
— trans·lat·abil·i·ty \(ˌ)tran(t)s-ˌlā-tə-ˈbi-lə-tē, (ˌ)tranz-\ *noun*
— trans·lat·able \tran(t)s-ˈlā-tə-bəl, tranz-\ *adjective*
— trans·la·tor \-ˈlā-tər\ *noun*

trans·la·tion \tran(t)s-ˈlā-shən, tranz-\ *noun* (14th century)
1 : an act, process, or instance of translating: as **a :** a rendering from one language into another; *also* **:** the product of such a rendering **b :** a change to a different substance, form, or appearance **:** CONVERSION **c** (1) **:** a transformation of coordinates in which the new axes are parallel to the old ones (2) **:** uniform motion of a body in a straight line
2 : the process of forming a protein molecule at a ribosomal site of protein synthesis from information contained in messenger RNA — compare TRANSCRIPTION 3
— trans·la·tion·al \-shnəl, -shə-n^{ə}l\ *adjective*

trans·la·tive \-ˈlā-tiv\ *adjective* (circa 1682)
1 : of, relating to, or involving removal or transference from one person or place to another
2 : of, relating to, or serving to translate from one language or system into another

trans·la·to·ry \ˈtran(t)s-lə-ˌtōr-ē, ˈtranz-, -ˌtȯr-; tran(t)s-ˈlā-tə-rē, tranz-\ *adjective* (1849)
: of, relating to, or involving uniform motion in one direction

trans·lit·er·ate \tran(t)s-ˈli-tə-ˌrāt, tranz-\ *transitive verb* **-at·ed; -at·ing** [*trans-* + Latin *littera* letter] (1861)
: to represent or spell in the characters of another alphabet
— trans·lit·er·a·tion \(ˌ)tran(t)s-ˌli-tə-ˈrā-shən, (ˌ)tranz-\ *noun*

trans·lo·ca·tion \ˌtran(t)s-lō-ˈkā-shən, ˌtranz-\ *noun* (1624)
: a change of location **:** DISPLACEMENT: as **a :** the conduction of soluble material (as metabolic products) from one part of a plant to another **b :** transfer of part of a chromosome to a different position especially on a nonhomologous chromosome; *especially* **:** the exchange of parts between nonhomologous chromosomes
— trans·lo·cate \ˈtrans(t)s-lō-ˌkāt, ˈtranz-, (ˌ)tran(t)s-ˈ, (ˌ)tranz-ˈ\ *verb*

trans·lu·cence \tran(t)s-ˈlü-s^{ə}n(t)s, tranz-\ *noun* (1755)
: the quality or state of being translucent

trans·lu·cen·cy \-s^{ə}n(t)-sē\ *noun, plural* **-cies** (circa 1610)
1 : TRANSLUCENCE
2 : something that is translucent

trans·lu·cent \-s^{ə}nt\ *adjective* [Latin *translucent-, translucens,* present participle of *translucēre* to shine through, from *trans-* + *lucēre* to shine — more at LIGHT] (1607)
1 : permitting the passage of light: **a :** CLEAR, TRANSPARENT ⟨*translucent* water⟩ **b :** transmitting and diffusing light so that objects beyond cannot be seen clearly
2 : free from disguise or falseness ⟨his *translucent* patriotism —*Newsweek*⟩
synonym see CLEAR
— trans·lu·cent·ly *adverb*

trans·ma·rine \ˌtran(t)s-mə-ˈrēn, ˌtranz-\ *adjective* [Latin *transmarinus,* from *trans-* + *mare* sea — more at MARINE] (1583)
1 : being or coming from beyond or across the sea ⟨a *transmarine* people⟩
2 : passing over or extending across the sea

trans·mem·brane \(ˌ)tran(t)s-ˈmem-ˌbrān, (ˌ)tranz-\ *adjective* (1944)
: taking place or existing across a membrane ⟨a *transmembrane* potential⟩

trans·mi·grate \(ˌ)tran(t)s-ˈmī-ˌgrāt, (ˌ)tranz-, ˈtran(t)s-ˌ, ˈtranz-ˌ\ *verb* [Latin *transmigratus,* past participle of *transmigrare* to migrate to another place, from *trans-* + *migrare* to migrate] (circa 1559)
transitive verb
: to cause to go from one state of existence or place to another
intransitive verb
1 *of the soul* **:** to pass at death from one body or being to another
2 : MIGRATE
— trans·mi·gra·tion \ˌtran(t)s-mī-ˈgrā-shən, ˌtranz-\ *noun*
— trans·mi·gra·tor \(ˌ)tran(t)s-ˈmī-ˌgrā-tər, (ˌ)tranz-, ˈtran(t)s-ˌ, ˈtranz-\ *noun*
— trans·mi·gra·to·ry \tran(t)s-ˈmī-grə-ˌtōr-ē, tranz-, -ˌtȯr-\ *adjective*

trans·mis·si·ble \tran(t)s-ˈmi-sə-bəl, tranz-\ *adjective* (1644)
: capable of being transmitted ⟨*transmissible* diseases⟩
— trans·mis·si·bil·i·ty \(ˌ)tran(t)s-ˌmi-sə-ˈbi-lə-tē, (ˌ)tranz-\ *noun*

trans·mis·sion \tran(t)s-ˈmi-shən, tranz-\ *noun* [Latin *transmission-, transmissio,* from *transmittere* to transmit] (1611)
1 : an act, process, or instance of transmitting ⟨*transmission* of a nerve impulse across a synapse⟩
2 : the passage of radio waves in the space between transmitting and receiving stations; *also* **:** the act or process of transmitting by radio or television
3 : an assembly of parts including the speed-changing gears and the propeller shaft by which the power is transmitted from an automobile engine to a live axle; *also* **:** the speed-changing gears in such an assembly
4 : something that is transmitted **:** MESSAGE
— trans·mis·sive \-ˈmi-siv\ *adjective*
— trans·mis·siv·i·ty \ˌtran(t)s-(ˌ)mi-ˈsi-və-tē, ˌtranz-\ *noun*

trans·mis·som·e·ter \ˌtran(t)s-(ˌ)mi-ˈsä-mə-tər, ˌtranz-\ *noun* (circa 1931)
: an instrument for measuring the transmission of light through a fluid (as the atmosphere)

trans·mit \tran(t)s-ˈmit, tranz-\ *verb* **trans·mit·ted; trans·mit·ting** [Middle English *transmitten,* from Latin *transmittere,* from *trans-* + *mittere* to send] (15th century)
transitive verb
1 a : to send or convey from one person or place to another **:** FORWARD **b :** to cause or allow to spread: as (1) **:** to convey by or as if by inheritance or heredity **:** HAND DOWN (2) **:** to convey (infection) abroad or to another
2 a (1) **:** to cause (as light or force) to pass or be conveyed through space or a medium (2) **:** to admit the passage of **:** CONDUCT ⟨glass *transmits* light⟩ **b :** to send out (a signal) either by radio waves or over a wire
intransitive verb
: to send out a signal either by radio waves or over a wire
— trans·mit·ta·ble \-ˈmi-tə-bəl\ *adjective*
— trans·mit·tal \-ˈmi-t^{ə}l\ *noun*

trans·mit·tance \-ˈmi-t^{ə}n(t)s\ *noun* (circa 1855)
1 : TRANSMISSION
2 : the fraction of radiant energy that having entered a layer of absorbing matter reaches its farther boundary

trans·mit·ter \tran(t)s-ˈmi-tər, tranz-; ˈtran(t)s-ˌ, ˈtranz-\ *noun* (1727)
: one that transmits: as **a :** an apparatus for transmitting radio or television signals **b :** NEUROTRANSMITTER

trans·mog·ri·fy \tran(t)s-ˈmä-grə-ˌfī, tranz-\ *verb* **-fied; -fy·ing** [origin unknown] (1656)

transitive verb
: to change or alter greatly and often with grotesque or humorous effect
intransitive verb
: to become transmogrified
synonym see TRANSFORM
— **trans·mog·ri·fi·ca·tion** \(ˌ)tran(t)s-ˌmä-grə-fə-ˈkā-shən, (ˌ)tranz-\ *noun*

trans·mon·tane \(ˌ)tran(t)s-ˈmän-ˌtān, (ˌ)tranz-; ˌtran(t)s-(ˌ)män-ˈ, ˌtranz-\ *adjective* [Latin *transmontanus*] (1727)
: TRAMONTANE

trans·moun·tain \ˌtran(t)s-ˈmau̇n-tᵊn, ˌtranz-\ *adjective* (1929)
: crossing or extending over or through a mountain ⟨a *transmountain* road⟩ ⟨a *transmountain* tunnel⟩

trans·mu·ta·tion \ˌtran(t)s-myu̇-ˈtā-shən, ˌtranz-\ *noun* [Middle English *transmutacioun,* from Middle French or Latin; Middle French *transmutation,* from Latin *transmutation-, transmutatio,* from *transmutare*] (14th century)
: an act or instance of transmuting or being transmuted: as **a :** the conversion of base metals into gold or silver **b :** the conversion of one element or nuclide into another either naturally or artificially
— **trans·mut·a·tive** \tran(t)s-ˈmyü-tə-tiv, tranz-\ *adjective*

trans·mute \tran(t)s-ˈmyüt, tranz-\ *verb* **trans·mut·ed; trans·mut·ing** [Middle English, from Latin *transmutare,* from *trans-* + *mutare* to change — more at MUTABLE] (15th century)
transitive verb
1 : to change or alter in form, appearance, or nature and especially to a higher form
2 : to subject (as an element) to transmutation
intransitive verb
: to undergo transmutation
synonym see TRANSFORM
— **trans·mut·able** \-ˈmyü-tə-bəl\ *adjective*

trans·na·tion·al \(ˌ)tran(t)s-ˈnash-nəl, (ˌ)tranz-, -ˈna-shə-nᵊl\ *adjective* (1921)
: extending or going beyond national boundaries ⟨*transnational* corporations⟩
— **trans·na·tion·al·ism** \-ˈnash-nə-ˌli-zəm, -ˈna-shə-nᵊl-ˌi-zəm\ *noun*

trans·nat·u·ral \-ˈna-chə-rəl, -ˈnach-rəl\ *adjective* (1569)
: being above or beyond nature

trans·oce·an·ic \ˌtran(t)s-ˌō-shē-ˈa-nik, ˌtranz-\ *adjective* (1827)
1 : lying or dwelling beyond the ocean
2 : crossing or extending across the ocean ⟨a *transoceanic* telephone cable⟩

tran·som \ˈtran(t)-səm\ *noun* [Middle English *traunsom, transyn,* probably alteration of *traversayn,* from Middle French *traversin,* from *traverse* traverse] (15th century)
1 : a transverse piece in a structure **:** CROSSPIECE: as **a :** LINTEL **b :** a horizontal crossbar in a window, over a door, or between a door and a window or fanlight above it **c :** the horizontal bar or member of a cross or gallows **d :** any of several transverse timbers or beams secured to the sternpost of a boat; *also* **:** the planking forming the stern of a square-ended boat

T: **transom 1b**

2 : a window above a door or other window built on and commonly hinged to a transom
— **over the transom :** without solicitation or prior arrangement ⟨the manuscript arrived *over the transom*⟩

tran·son·ic *also* **trans·son·ic** \tran(t)s-ˈsä-nik, tran-ˈsä-\ *adjective* [*trans-* + super*sonic*] (1945)
1 : being or relating to speeds near that of sound in air or about 741 miles (1185 kilometers) per hour at sea level and especially to speeds slightly below the speed of sound at which the speed of airflow varies from subsonic to supersonic at different points along the surface of a body in motion relative to the surrounding air
2 : moving, capable of moving, or utilizing air currents moving at a transonic speed

trans·pa·cif·ic \ˌtran(t)s-pə-ˈsi-fik\ *adjective* (1891)
1 a : crossing or extending across the Pacific Ocean ⟨*transpacific* airlines⟩ **b :** relating to or involving crossing the Pacific Ocean ⟨*transpacific* airfares⟩
2 a : situated or occurring beyond the Pacific Ocean **b :** of, relating to, or involving countries on both sides of the Pacific Ocean ⟨the *transpacific* economy⟩

trans·par·ence \tran(t)s-ˈpar-ən(t)s, -ˈper-\ *noun* (1594)
: TRANSPARENCY 1

trans·par·en·cy \-ən(t)-sē\ *noun, plural* **-cies** (1615)
1 : the quality or state of being transparent
2 : something transparent; *especially* **:** a picture (as on film) viewed by light shining through it or by projection

trans·par·ent \-ənt\ *adjective* [Middle English, from Medieval Latin *transparent-, transparens,* present participle of *transparēre* to show through, from Latin *trans-* + *parēre* to show oneself] (15th century)
1 a (1) **:** having the property of transmitting light without appreciable scattering so that bodies lying beyond are seen clearly **:** PELLUCID (2) **:** allowing the passage of a specified form of radiation (as X rays or ultraviolet light) **b :** fine or sheer enough to be seen through **:** DIAPHANOUS
2 a : free from pretense or deceit **:** FRANK **b :** easily detected or seen through **:** OBVIOUS **c :** readily understood
synonym see CLEAR
— **trans·par·ent·ly** *adverb*
— **trans·par·ent·ness** *noun*

trans·par·ent·ize \-ən-ˌtīz\ *transitive verb* **-ized; -iz·ing** (1925)
: to make transparent or more nearly transparent ⟨*transparentize* tracing paper⟩

trans·per·son·al \(ˌ)tran(t)s-ˈpərs-nəl, -ˈpər-sə-nᵊl\ *adjective* (circa 1906)
1 : extending or going beyond the personal or individual
2 : of, relating to, or being psychology concerned especially with esoteric mental experience (as mysticism and altered states of consciousness) beyond the usual limits of ego and personality

tran·spic·u·ous \tran(t)s-ˈpi-kyə-wəs\ *adjective* [New Latin *transpicuus,* from Latin *transpicere* to look through, from *trans-* + *specere* to look, see — more at SPY] (1638)
: clearly seen through or understood

trans·pierce \tran(t)s-ˈpirs\ *transitive verb* [Middle French *transpercer,* from Old French, from *trans-* (from Latin) + *percer* to pierce] (1592)
: to pierce through **:** PENETRATE

tran·spi·ra·tion \ˌtran(t)s-pə-ˈrā-shən\ *noun* (1551)
: the act or process or an instance of transpiring; *especially* **:** the passage of watery vapor from a living body through a membrane or pores
— **tran·spi·ra·tion·al** \-shnəl, -shə-nᵊl\ *adjective*

tran·spire \tran(t)-ˈspīr\ *verb* **tran·spired; tran·spir·ing** [Middle French *transpirer,* from Latin *trans-* + *spirare* to breathe] (1597)
transitive verb
: to pass off or give passage to (a fluid) through pores or interstices; *especially* **:** to excrete (as water) in the form of a vapor through a living membrane (as the skin)
intransitive verb
1 : to give off vaporous material; *specifically* **:** to give off or exude watery vapor especially from the surfaces of leaves
2 : to pass in the form of a vapor from a living body
3 a : to be revealed **:** come to light **b :** to become known or apparent **:** DEVELOP
4 : to take place **:** GO ON, OCCUR ■

trans·pla·cen·tal \ˌtran(t)s-plə-ˈsen-tᵊl\ *adjective* [International Scientific Vocabulary] (circa 1929)
: passing through or occurring by way of the placenta ⟨*transplacental* immunization⟩
— **trans·pla·cen·tal·ly** \-tᵊl-ē\ *adverb*

1trans·plant \ˌtran(t)s-ˈplant\ *verb* [Middle English *transplaunten,* from Late Latin *transplantare,* from Latin *trans-* + *plantare* to plant] (15th century)
transitive verb
1 : to lift and reset (a plant) in another soil or situation
2 : to remove from one place or context and settle or introduce elsewhere **:** RELOCATE
3 : to transfer (an organ or tissue) from one part or individual to another
intransitive verb
: to admit of being transplanted
— **trans·plant·abil·i·ty** \ˌtran(t)s-ˌplan-tə-ˈbi-lə-tē\ *noun*
— **trans·plant·able** \tran(t)s-ˈplan-tə-bəl\ *adjective*
— **trans·plan·ta·tion** \ˌtran(t)s-ˌplan-ˈtā-shən\ *noun*
— **trans·plant·er** \tran(t)s-ˈplan-tər\ *noun*

2trans·plant \ˈtran(t)s-ˌplant\ *noun* (1756)
1 : a person or thing that is transplanted
2 : the act or process of transplanting

trans·po·lar \(ˌ)tran(t)s-ˈpō-lər\ *adjective* (1850)
: crossing or extending across either of the polar regions

tran·spon·der \tran(t)-ˈspän-dər\ *noun* [*trans*mitter + res*ponder*] (circa 1944)
: a radio or radar set that upon receiving a designated signal emits a radio signal of its own and that is used especially for the detection, identification, and location of objects

trans·pon·tine \tran(t)s-ˈpän-ˌtīn\ *adjective* [*trans-* + Latin *pont-, pons* bridge — more at FIND] (1844)
1 : situated on the farther side of a bridge
2 *British* **:** situated on the south side of the Thames

1trans·port \tran(t)s-ˈpōrt, -ˈpȯrt, ˈtran(t)s-ˌ\ *transitive verb* [Middle English, from Middle French or Latin; Middle French *transporter,* from Latin *transportare,* from *trans-* + *portare* to carry — more at FARE] (14th century)

□ USAGE

transpire Sense 4 of *transpire* is the frequent whipping boy of those who suppose sense 3 to be the only meaning of the word. Sense 4 appears to have developed in the late 18th century; it was well enough known to have been used by Abigail Adams in a letter to her husband in 1775 ⟨there is nothing new *transpired* since I wrote you last —Abigail Adams⟩. Noah Webster recognized the new sense in his dictionary of 1828. *Transpire* was evidently a popular word with 19th century journalists; sense 4 turns up in such pretentiously worded statements as "The police drill will transpire under shelter to-day in consequence of the moist atmosphere prevailing." Around 1870 the sense began to be attacked as a misuse on the grounds of etymology, and modern critics echo the damnation of 1870. Sense 4 has been in existence for about two centuries; it is firmly established as standard; it occurs now primarily in serious prose, not the ostentatiously flamboyant prose typical of 19th century journalism.

1 : to transfer or convey from one place to another ⟨*transporting* ions across a living membrane⟩
2 : to carry away with strong and often intensely pleasant emotion
3 : to send to a penal colony overseas
synonym see BANISH
— **trans·port·abil·i·ty** \(ˌ)tran(t)s-ˌpōr-tə-ˈbi-lə-tē, -ˌpȯr-\ *noun*
— **trans·port·able** \tran(t)s-ˈpōr-tə-bəl, -ˈpȯr-\ *adjective*

[2]trans·port \ˈtran(t)s-ˌpōrt, -ˌpȯrt\ *noun* (1611)
1 : an act or process of transporting **:** TRANSPORTATION
2 : strong or intensely pleasurable emotion ⟨*transports* of joy⟩
3 a : a ship for carrying soldiers or military equipment **b :** a vehicle (as a truck or airplane) used to transport persons or goods **c :** TRANSPORTATION 3
4 : a transported convict
5 : a mechanism for moving magnetic tape past a recording head
synonym see ECSTASY

trans·por·ta·tion \ˌtran(t)s-pər-ˈtā-shən\ *noun* (1540)
1 : an act, process, or instance of transporting or being transported
2 : banishment to a penal colony
3 a : means of conveyance or travel from one place to another **b :** public conveyance of passengers or goods especially as a commercial enterprise
— **trans·por·ta·tion·al** \-shnəl, -shə-nᵊl\ *adjective*

transport café \-ka-ˈfā, -ˈkaf\ *noun* (1938)
British **:** a roadside restaurant frequented chiefly by truck drivers

trans·port·er \tran(t)s-ˈpōr-tər, -ˈpȯr-, ˈtran(t)s-ˌ\ *noun* (1535)
: one that transports; *especially* **:** a vehicle for transporting large or heavy loads

transposable element *noun* (1979)
: a segment of genetic material that is capable of changing its location in the genome or in some bacteria of undergoing transfer between an extrachromosomal plasmid and a chromosome

[1]trans·pose \tran(t)s-ˈpōz\ *transitive verb* **trans·posed; trans·pos·ing** [Middle English, from Middle French *transposer,* from Latin *transponere* (perfect indicative *transposui*) to change the position of, from *trans-* + *ponere* to put, place — more at POSITION] (14th century)
1 : to change in form or nature **:** TRANSFORM
2 : to render into another language, style, or manner of expression **:** TRANSLATE
3 : to transfer from one place or period to another **:** SHIFT
4 : to change the relative place or normal order of **:** alter the sequence of ⟨*transpose* letters to change the spelling⟩
5 : to write or perform (a musical composition) in a different key
6 : to bring (a term) from one side of an algebraic equation to the other with change of sign
synonym see REVERSE
— **trans·pos·able** \-ˈpō-zə-bəl\ *adjective*

[2]trans·pose \ˈtran(t)s-ˌpōz\ *noun* (1937)
: a matrix formed from another matrix by interchanging the rows and columns

trans·po·si·tion \ˌtran(t)s-pə-ˈzi-shən\ *noun* [Medieval Latin *transposition-, transpositio,* from Latin *transponere* to transpose] (1538)
1 : an act, process, or instance of transposing or being transposed
2 a : the transfer of any term of an equation from one side over to the other side with a corresponding change of the sign **b :** a mathematical permutation or interchange of two letters or symbols
— **trans·po·si·tion·al** \-ˈzish-nəl, -ˈzi-shə-nᵊl\ *adjective*

transposition cipher *noun* (1939)
: a cipher in which the letters of the plaintext are systematically rearranged into another sequence — compare SUBSTITUTION CIPHER

trans·po·son \tran(t)s-ˈpō-zän\ *noun* [*transpose* + [2]*-on*] (1974)
: a transposable element especially when it contains genetic material controlling functions other than those related to its relocation

trans·sex·u·al \(ˌ)tran(t)s-ˈsek-sh(ə-)wəl, -shəl\ *noun* (1957)
: a person with a psychological urge to belong to the opposite sex that may be carried to the point of undergoing surgery to modify the sex organs to mimic the opposite sex
— **trans·sex·u·al·ism** \-sh(ə-)wə-ˌli-zəm, -shə-ˌli-\ *noun*
— **trans·sex·u·al·i·ty** \-ˌsek-shə-ˈwa-lə-tē\ *noun*

trans·shape \tran(sh)-ˈshāp, tran(t)s-\ *transitive verb* (1575)
archaic **:** to change into another shape **:** TRANSFORM

trans·ship \tran(sh)-ˈship, tran(t)s-\ (1792)
transitive verb
: to transfer for further transportation from one ship or conveyance to another
intransitive verb
: to change from one ship or conveyance to another
— **trans·ship·ment** \-mənt\ *noun*

trans·tho·rac·ic \ˌtran(t)s-thə-ˈra-sik\ *adjective* (1905)
: done or made by way of the thoracic cavity
— **trans·tho·rac·i·cal·ly** \-si-k(ə-)lē\ *adverb*

tran·sub·stan·tial \ˌtran(t)-səb-ˈstan(t)-shəl\ *adjective* (1567)
: changed or capable of being changed from one substance to another

tran·sub·stan·ti·ate \ˌtran(t)-səb-ˈstan(t)-shē-ˌāt\ *verb* **-at·ed; -at·ing** [Medieval Latin *transubstantiatus,* past participle of *transubstantiare,* from Latin *trans-* + *substantia* substance] (1533)
transitive verb
1 : to change into another substance **:** TRANSMUTE
2 : to effect transubstantiation in (sacramental bread and wine)
intransitive verb
: to undergo transubstantiation

tran·sub·stan·ti·a·tion \-ˌstan(t)-shē-ˈā-shən\ *noun* (14th century)
1 : an act or instance of transubstantiating or being transubstantiated
2 : the miraculous change by which according to Roman Catholic and Eastern Orthodox dogma the eucharistic elements at their consecration become the body and blood of Christ while keeping only the appearances of bread and wine

tran·su·date \ˈtran(t)-sü-ˌdāt, -syü-; ˈtran-zü-, -zyü-\ *noun* (1876)
: a transuded substance

tran·su·da·tion \ˌtran(t)-sü-ˈdā-shən, -syü-; ˌtran-zü-, -zyü-\ *noun* (1617)
1 : the act or process of transuding or being transuded
2 : TRANSUDATE

tran·sude \tran(t)-ˈsüd, -ˈsyüd; tran-ˈzüd, -ˈzyüd\ *verb* **tran·sud·ed; tran·sud·ing** [New Latin *transudare,* from Latin *trans-* + *sudare* to sweat — more at SWEAT] (1664)
intransitive verb
: to pass through a membrane or permeable substance **:** EXUDE
transitive verb
: to permit passage of

[1]trans·ura·nic \ˌtran(t)-syu̇-ˈra-nik, -shə-, -ˈrā-; ˌtran-zyu̇-, -zhə-\ *or* **trans·ura·ni·um** \-ˈrā-nē-əm\ *adjective* (1935)
: of, relating to, or being an element with an atomic number greater than that of uranium

[2]transuranic *noun* (1950)
: a transuranic element

trans·val·u·ate \(ˌ)tran(t)s-ˈval-yə-ˌwāt, (ˌ)tranz-\ *transitive verb* **-at·ed; -at·ing** [back-formation from *transvaluation*] (1912)
: TRANSVALUE

trans·val·u·a·tion \ˌtran(t)s-ˌval-yə-ˈwā-shən, ˌtranz-\ *noun* (1898)
: the act or process of transvaluing

trans·val·ue \(ˌ)tran(t)s-ˈval-(ˌ)yü, (ˌ)tranz-\ *transitive verb* **-val·ued; -valu·ing** (1911)
: to reevaluate especially on a basis that repudiates accepted standards

trans·ver·sal \tran(t)s-ˈvər-səl, tranz-\ *noun* [*transversal,* adjective, transverse, from Middle English, from Medieval Latin *transversalis,* from Latin *transversus*] (circa 1847)
: a line that intersects a system of lines

[1]trans·verse \tran(t)s-ˈvərs, tranz-, ˈtran(t)s-ˌ, ˈtranz-ˌ\ *adjective* [Latin *transversus,* from *trans-* + *-versus* (as in *adversus* adverse)] (1621)
1 : acting, lying, or being across **:** set crosswise
2 : made at right angles to the anterior-posterior axis of the body ⟨a *transverse* section⟩
— **trans·verse·ly** *adverb*

[2]trans·verse \ˈtran(t)s-ˌvərs, ˈtranz-\ *noun* (circa 1633)
: something (as a piece, section, or part) that is transverse

transverse colon *noun* (circa 1860)
: the middle portion of the colon that extends across the abdominal cavity

transverse process *noun* (1696)
: a lateral process of a vertebra

transverse wave *noun* (1922)
: a wave in which the vibrating element moves in a direction perpendicular to the direction of advance of the wave

trans·ves·tite \tran(t)s-ˈves-ˌtīt, tranz-\ *noun* [German *Transvestit,* from Latin *trans-* + *vestire* to clothe — more at VEST] (circa 1922)
: a person and especially a male who adopts the dress and often the behavior typical of the opposite sex especially for purposes of emotional or sexual gratification
— **trans·ves·tism** \-ˈves-ˌti-zəm\ *noun*
— **transvestite** *adjective*

[1]trap \ˈtrap\ *noun* [Middle English, from Old English *treppe* & Old French *trape* (of Germanic origin); akin to Middle Dutch *trappe* trap, stair, Old English *treppan* to tread] (before 12th century)
1 : a device for taking game or other animals; *especially* **:** one that holds by springing shut suddenly
2 a : something by which one is caught or stopped unawares; *also* **:** a position or situation from which it is difficult or impossible to escape **b :** a football play in which a defensive player is allowed to cross the line of scrimmage and then is blocked from the side while the ballcarrier advances through the spot vacated by the defensive player
3 a : a device for hurling clay pigeons into the air **b :** SAND TRAP **c :** a piece of leather or section of interwoven leather straps between the thumb and forefinger of a baseball glove that forms an extension of the pocket
4 *slang* **:** MOUTH
5 : a light usually one-horse carriage with springs
6 : any of various devices for preventing passage of something often while allowing other matter to proceed; *especially* **:** a device for

\ə\ **abut** \ᵊ\ **kitten** \ər\ **further** \a\ **ash** \ā\ **ace** \ä\ **mop, mar** \au̇\ **out** \ch\ **chin** \e\ **bet** \ē\ **easy** \g\ **go** \i\ **hit** \ī\ **ice** \j\ **job** \ŋ\ **sing** \ō\ **go** \ȯ\ **law** \ȯi\ **boy** \th\ **thin** \t͟h\ **the** \ü\ **loot** \u̇\ **foot** \y\ **yet** \zh\ **vision** *see also* Guide to Pronunciation

drains or sewers consisting of a bend or partitioned chamber in which the liquid forms a seal to prevent the passage of sewer gas
7 *plural* **:** a group of percussion instruments (as a bass drum, snare drums, and cymbals) used especially in a dance or jazz band
8 : an arrangement of rock strata that favors the accumulation of oil and gas
9 *plural* [*speed trap*] **:** a measured stretch of a course over which electronic timing devices measure the speed of a vehicle (as a racing car or dragster)

[2]trap *verb* **trapped; trap·ping** (14th century)
transitive verb
1 a : to catch or take in or as if in a trap **:** ENTRAP **b :** to place in a restricted position **:** CONFINE ⟨*trapped* in the burning wreck⟩
2 : to provide or set (a place) with traps
3 a : STOP, HOLD ⟨these mountains *trap* rains and fogs generated over the ocean —*American Guide Series: California*⟩ **b :** to separate out (as water from steam)
4 a : to catch (as a baseball) immediately after a bounce **b :** to block out (a defensive football player) by means of a trap
intransitive verb
: to engage in trapping animals (as for furs)
synonym see CATCH
— trap·per *noun*

[3]trap *transitive verb* **trapped; trap·ping** [Middle English *trappen,* from *trappe* cloth, perhaps modification of Middle French *drap* — more at DRAB] (14th century)
: to adorn with or as if with trappings

[4]trap *noun* [Swedish *trapp,* from *trappa* stair, from Middle Low German *trappe;* akin to Middle Dutch *trappe* stair] (1794)
: TRAPROCK

trap·door \'trap-'dōr, -'dȯr\ *noun* (14th century)
: a lifting or sliding door covering an opening (as in a roof, ceiling, or floor)

trap–door spider *noun* (1826)
: any of various often large burrowing spiders (especially family Ctenizidae) that construct a tubular subterranean silk-lined nest topped with a hinged lid

trap-door spider

tra·peze \tra-'pēz *also* trə-\ *noun* [French *trapèze,* from New Latin *trapezium*] (1861)
: a gymnastic or acrobatic apparatus consisting of a short horizontal bar suspended by two parallel ropes

tra·pez·ist \-'pē-zist\ *noun* (1875)
: a performer on the trapeze — called also *trapeze artist*

tra·pe·zi·um \trə-'pē-zē-əm, tra-\ *noun, plural* **-zi·ums** *or* **-zia** \-zē-ə\ [New Latin, from Greek *trapezion,* literally, small table, diminutive of *trapeza* table, from *tra-* four (akin to *tettares* four) + *peza* foot; akin to Greek *pod-, pous* foot — more at FOUR, FOOT] (1570)
1 a : a quadrilateral having no two sides parallel **b** *British* **:** TRAPEZOID 1b
2 : a bone in the wrist at the base of the thumb

tra·pe·zi·us \-zē-əs\ *noun* [New Latin, from *trapezium;* from the pair on the back forming together the figure of a trapezium] (circa 1704)
: a large flat triangular superficial muscle of each side of the upper back

tra·pe·zo·he·dron \trə-ˌpē-zō-'hē-drən, ˌtra-pə-\ *noun, plural* **-drons** *or* **-dra** \-drə\ [New Latin, from *trapezium* + *-o-* + *-hedron*] (1822)
: a crystalline form whose faces are trapeziums

trap·e·zoid \'tra-pə-ˌzȯid\ *noun* [New Latin *trapezoïdes,* from Greek *trapezoeidēs* trapezium-shaped, from *trapeza* table] (circa 1706)
1 a *British* **:** TRAPEZIUM 1a **b :** a quadrilateral having only two sides parallel
2 : a bone in the wrist at the base of the forefinger
— trap·e·zoi·dal \ˌtra-pə-'zȯi-dᵊl\ *adjective*

trapezoid 1b

trap·line \'trap-ˌlīn\ *noun* (1920)
: a line or series of traps; *also* **:** the route along which such a line of traps is set

trap·ping \'tra-piŋ\ *noun* [Middle English, from gerund of *trappen* to adorn] (14th century)
1 : CAPARISON 1 — usually used in plural
2 *plural* **:** outward decoration or dress **:** ornamental equipment; *also* **:** outward signs ⟨conventional men with all the *trappings* . . . of banality —Robert Plank⟩

Trap·pist \'tra-pist\ *noun* [French *trappiste,* from La *Trappe,* France] (1814)
: a member of a reformed branch of the Roman Catholic Cistercian Order established by the Abbot de Rancé in 1664 at the monastery of La Trappe in Normandy
— Trappist *adjective*

trap·rock \'trap-'räk\ *noun* [[4]*trap*] (1813)
: any of various dark-colored fine-grained igneous rocks (as basalt) used especially in road making

traps \'traps\ *noun plural* [Middle English *trappe* cloth — more at TRAP] (1813)
: personal belongings **:** LUGGAGE

trap·shoot·er \'trap-ˌshü-tər\ *noun* (1875)
: a person who engages in trapshooting

trap·shoot·ing \-ˌshü-tiŋ\ *noun* (1875)
: shooting at clay pigeons sprung from a trap into the air away from the shooter

tra·pun·to \trə-'pün-(ˌ)tō, -'pu̇n-\ *noun, plural* **-tos** [Italian, from past participle of *trapungere* to embroider, from *tra-* across (from Latin *trans-*) + *pungere* to prick, from Latin — more at TRANS-, PUNGENT] (circa 1924)
: a decorative quilted design in high relief worked through at least two layers of cloth by outlining the design in running stitch and padding it from the underside

[1]trash \'trash\ *noun* [of Scandinavian origin; akin to Norwegian *trask* trash; akin to Old Norse *tros* fallen leaves and twigs, Old English *trus*] (circa 1518)
1 : something worth little or nothing: as **a :** JUNK, RUBBISH **b** (1) **:** empty talk **:** NONSENSE (2) **:** inferior or worthless writing or artistic matter; *also* **:** such material intended purely for entertainment
2 : something in a crumbled or broken condition or mass; *especially* **:** debris from pruning or processing plant material
3 : a worthless person; *also* **:** such persons as a group **:** RIFFRAFF

[2]trash (circa 1859)
transitive verb
1 : VANDALIZE, DESTROY
2 : ATTACK, ASSAULT
3 : SPOIL, RUIN ⟨*trashing* the environment⟩
4 : to subject to criticism or invective; *especially* **:** to disparage strongly ⟨a film *trashed* by the critics⟩
5 : THROW AWAY 1 ⟨standards of reality and truth were *trashed* —Edwin Diamond⟩
intransitive verb
: to trash something or someone

trash fish *noun* (1944)
1 : ROUGH FISH
2 : any of various sea fishes that have no market value as human food but are sometimes processed for use in animal feed, fertilizers, and paints

trash·man \'trash-ˌman, -mən\ *noun* (1951)
: a worker who collects and hauls away trash

trashy \'tra-shē\ *adjective* **trash·i·er; -est** (circa 1620)
: resembling or containing trash **:** of inferior quality
— trash·i·ness *noun*

trat·to·ria \ˌträ-tə-'rē-ə\ *noun, plural* **-ri·as** *or* **-rie** \-'rē-ˌā\ [Italian, from *trattore* restaurateur, from French *traiteur,* from *traiter* to treat, from Old French *traitier* — more at TREAT] (1832)
: RESTAURANT; *specifically* **:** a usually small modest Italian restaurant
word history see RESTAURANT

trau·ma \'trau̇-mə, 'trȯ-\ *noun, plural* **traumas** *also* **trau·ma·ta** \-mə-tə\ [Greek *traumat-, trauma* wound, alteration of *trōma;* akin to Greek *titrōskein* to wound, *tetrainein* to pierce — more at THROW] (circa 1693)
1 a : an injury (as a wound) to living tissue caused by an extrinsic agent ⟨surgical *trauma*⟩ **b :** a disordered psychic or behavioral state resulting from mental or emotional stress or physical injury
2 : an agent, force, or mechanism that causes trauma
— trau·mat·ic \trə-'ma-tik, trȯ-, trau̇-\ *adjective*
— trau·mat·i·cal·ly \-ti-k(ə-)lē\ *adverb*

trau·ma·tise *British variant of* TRAUMATIZE

trau·ma·tism \'trau̇-mə-ˌti-zəm, 'trȯ-\ *noun* (1857)
: the development or occurrence of trauma; *also* **:** TRAUMA

trau·ma·tize \-ˌtīz\ *transitive verb* **-tized; -tiz·ing** (1903)
: to inflict a trauma upon
— trau·ma·ti·za·tion \ˌtrau̇-mə-tə-'zā-shən, ˌtrȯ-\ *noun*

[1]tra·vail \trə-'vā(ə)l, 'tra-ˌvāl\ *noun* [Middle English, from Old French, from *travaillier* to torture, labor, from (assumed) Vulgar Latin *trepaliare* to torture, from Late Latin *trepalium* instrument of torture, from Latin *tripalis* having three stakes, from *tri-* + *palus* stake — more at POLE] (13th century)
1 a : work especially of a painful or laborious nature **:** TOIL **b :** a physical or mental exertion or piece of work **:** TASK, EFFORT **c :** AGONY, TORMENT
2 : LABOR, PARTURITION
synonym see WORK
word history see TRAVEL

[2]travail *same as* [1]*; in prayer-book communion service usually* 'tra-ˌvāl\ *intransitive verb* [Middle English, from Old French *travaillier*] (13th century)
1 : to labor hard **:** TOIL
2 : LABOR 3

[1]trav·el \'tra-vəl\ *verb* **-eled** *or* **-elled; -el·ing** *or* **-el·ling** \'tra-və-liŋ, 'trav-liŋ\ [Middle English *travailen* to labor, journey, from Old French *travaillier* to labor] (14th century)
intransitive verb
1 a : to go on or as if on a trip or tour **:** JOURNEY **b** (1) **:** to go as if by traveling **:** PASS ⟨the news *traveled* fast⟩ (2) **:** ASSOCIATE ⟨*travels* with a sophisticated crowd⟩ **c :** to go from place to place as a sales representative or business agent
2 a : to move or undergo transmission from one place to another ⟨goods *traveling* by plane⟩ **b :** to move in a given direction or path or through a given distance ⟨the stylus *travels* in a groove⟩ **c :** to move rapidly ⟨a car that can really *travel*⟩
3 : to walk or run with a basketball in violation of the rules
transitive verb
1 a : to journey through or over **b :** to follow (a course or path) as if by traveling
2 : to traverse (a specified distance)

3 : to cover (a place or region) as a commercial traveler ◇
— **travel light :** to travel with a minimum of equipment or baggage

²travel *noun* (14th century)
1 a : the act of traveling **:** PASSAGE **b :** a journey especially to a distant or unfamiliar place **:** TOUR, TRIP — often used in plural
2 *plural* **:** an account of one's travels
3 : the number traveling **:** TRAFFIC
4 a : MOVEMENT, PROGRESSION ⟨the *travel* of satellites around the earth⟩ **b :** the motion of a piece of machinery; *especially* **:** reciprocating motion

travel agency *noun* (1927)
: an agency engaged in selling and arranging personal transportation and accommodations for travelers — called also *travel bureau*

travel agent *noun* (1925)
: a person engaged in selling and arranging transportation, tours, or trips for travelers

trav·eled *or* **trav·elled** \ˈtra-vəld\ *adjective* (15th century)
1 : experienced in travel ⟨a widely *traveled* journalist⟩
2 : used by travelers ⟨a well-*traveled* highway⟩

trav·el·er *or* **trav·el·ler** \ˈtra-və-lər, ˈtrav-lər\ *noun* (14th century)
1 : one that travels: as **a :** one that goes on a trip or journey **b :** TRAVELING SALESMAN
2 a : an iron ring sliding along a rope, bar, or rod of a ship **b :** a rod on the deck on which such a ring slides
3 : any of various devices for handling something that is being transported laterally

traveler's check *noun* (1891)
: a draft purchased from a bank or express company and signed by the purchaser at the time of purchase and again at the time of cashing as a precaution against forgery

trav·el·ing *or* **trav·el·ling** \ˈtra-və-liŋ, ˈtrav-liŋ\ *adjective* (14th century)
1 : that travels ⟨a *traveling* opera company⟩ ⟨a *traveling* executive⟩
2 : carried, used by, or accompanying a traveler ⟨a *traveling* alarm clock⟩ ⟨a *traveling* companion⟩

traveling bag *noun* (1838)
: SUITCASE

traveling case *noun* (1744)
: a usually rigid and box-shaped suitcase

traveling fellowship *noun* (1789)
: a fellowship whose terms permit or direct the holder to travel or go abroad for study or research

traveling salesman *noun* (1885)
: a traveling representative of a business concern who solicits orders usually in an assigned territory

trav·el·ogue *or* **trav·el·og** \ˈtra-və-ˌlȯg, -ˌläg\ *noun* [*travel* + *-logue*] (1903)
1 : a talk or lecture on travel usually accompanied by a film or slides
2 : a narrated motion picture about travel
3 : a piece of writing about travel

travel trailer *noun* (1961)
: a trailer drawn especially by a passenger automobile and equipped for use (as while traveling) as a dwelling

tra·vers·al \trə-ˈvər-səl *also* tra-ˈ *or* ˈtra-ˌ\ *noun* (1897)
: the act or an instance of traversing

¹tra·verse \ˈtra-vərs *also* -ˌvərs, *especially for 6 & 8 also* trə-ˈ *or* tra-ˈ\ *noun* [Middle English *travers,* from Middle French *traverse,* from *traverser* to cross, from Late Latin *transversare,* from Latin *transversus* transverse — more at TRANSVERSE] (14th century)
1 : something that crosses or lies across
2 : OBSTACLE, ADVERSITY
3 : a formal denial of a matter of fact alleged by the opposite party in a legal pleading
4 a : a compartment or recess formed by a partition, curtain, or screen **b :** a gallery or loft of communication from side to side in a large building
5 : a route or way across or over: as **a :** a zigzag course of a sailing ship with contrary winds **b :** a curving or zigzag way up a steep grade **c :** the course followed in traversing
6 : the act or an instance of traversing **:** CROSSING
7 : a protective projecting wall or bank of earth in a trench
8 a : a lateral movement (as of the saddle of a lathe carriage); *also* **:** a device for imparting such movement **b :** the lateral movement of a gun about a pivot or on a carriage to change direction of fire
9 : a line surveyed across a plot of ground

²tra·verse \trə-ˈvərs *also* tra-ˈ *or* ˈtra-(ˌ)\ *verb* **tra·versed; tra·vers·ing** (14th century)
transitive verb
1 a : to go against or act in opposition to **:** OPPOSE, THWART **b :** to deny (as an allegation of fact or an indictment) formally at law
2 a : to go or travel across or over **b :** to move or pass along or through ⟨light rays *traversing* a crystal⟩
3 : to make a study of **:** EXAMINE
4 : to lie or extend across **:** CROSS ⟨the bridge *traverses* a brook⟩
5 a : to move to and fro over or along **b :** to ascend, descend, or cross (a slope or gap) at an angle **c :** to move (a gun) to right or left on a pivot
6 : to make or carry out a survey of by using traverses
intransitive verb
1 : to move back and forth or from side to side
2 : to move or turn laterally **:** SWIVEL
3 a : to climb at an angle or in a zigzag course **b :** to ski across rather than straight down a hill
4 : to make a survey by using traverses
— **tra·vers·able** \-ˈvər-sə-bəl, -(ˌ)vər-\ *adjective*
— **tra·vers·er** *noun*

³tra·verse \ˈtra-(ˌ)vərs, trə-ˈ, tra-ˈ\ *adjective* (15th century)
: lying across **:** TRANSVERSE

trav·erse jury \ˈtrav-ərs-\ *noun* (1823)
: PETIT JURY

traverse rod *noun* (1948)
: a metal rod or track with a pulley mechanism for drawing curtains

trav·er·tine \ˈtra-vər-ˌtēn, -tən\ *noun* [French *travertin,* from Italian *travertino, trevertino,* from Latin *tiburtinus,* adjective, of travertine, literally, of Tibur (Tivoli)] (1797)
: a mineral consisting of a massive usually layered calcium carbonate (as aragonite or calcite) formed by deposition from spring waters or especially from hot springs

¹trav·es·ty \ˈtra-və-stē\ *transitive verb* **-tied; -ty·ing** (1673)
: to make a travesty of **:** PARODY

²travesty *noun, plural* **-ties** [obsolete English *travesty,* disguised, parodied, from French *travesti,* past participle of *travestir* to disguise, from Italian *travestire,* from *tra-* across (from Latin *trans-*) + *vestire* to dress, from Latin — more at VEST] (1674)
1 : a burlesque translation or literary or artistic imitation usually grotesquely incongruous in style, treatment, or subject matter
2 : a debased, distorted, or grossly inferior imitation ⟨a *travesty* of justice⟩
synonym see CARICATURE

tra·vois \trə-ˈvȯi, ˈtra-ˌvȯi\ *noun, plural* **travois** *also* **tra·voises** \-ˈvȯiz, -ˌvȯiz\ [American French *travail,* from Canadian French, shaft of a cart, from Middle French *traveil* catafalque, prop, from Late Latin *trepalium* instrument of torture — more at TRAVAIL] (1847)
: a simple vehicle used by Plains Indians consisting of two trailing poles serving as shafts and bearing a platform or net for the load

travois

¹trawl \ˈtrȯl\ *verb* [probably from obsolete Dutch *tragelen*] (1561)
intransitive verb
1 : to fish with a trawl
2 : TROLL 2
transitive verb
: to catch (fish) with a trawl

²trawl *noun* (1759)
1 : a large conical net dragged along the sea bottom in gathering fish or other marine life
2 : SETLINE

trawl·er \ˈtrȯ-lər\ *noun* (1630)
1 : a person who fishes by trawling
2 : a boat used in trawling

trawl·er·man \-mən\ *noun* (1633)
: a fisherman who trawls or one who mans a trawler

tray \ˈtrā\ *noun* [Middle English, from Old English *trīg, trēg;* akin to Old Swedish *trø* wooden grain measure and probably to Old English *trēow* tree — more at TREE] (before 12th century)
: an open receptacle with a flat bottom and a low rim for holding, carrying, or exhibiting articles
— **tray·ful** \-ˌfu̇l\ *noun*

treach·er·ous \ˈtre-chə-rəs, ˈtrech-rəs\ *adjective* (14th century)
1 : characterized by or manifesting treachery **:** PERFIDIOUS
2 a : likely to betray trust **:** UNRELIABLE ⟨a *treacherous* memory⟩ **b :** providing insecure footing or support ⟨*treacherous* quicksand⟩ **c :** marked by hidden dangers, hazards, or perils
synonym see FAITHLESS
— **treach·er·ous·ly** *adverb*
— **treach·er·ous·ness** *noun*

treach·ery \-rē\ *noun, plural* **-er·ies** [Middle English *trecherie,* from Old French, from *trechier, trichier* to deceive, from (assumed) Vulgar Latin *triccare* — more at TRICK] (13th century)

◇ WORD HISTORY

travel For many of us in the U.S. travel is usually for pleasure and work is not very taxing physically, so that we are unlikely to associate travel with hard labor, and labor with torture. However, travel, labor, and torture are all bound up in the history of the word *travel.* The ultimate source of both *travel* and *travail* is an unattested spoken Latin verb *trepaliare* "to torture," a derivative of Late Latin *trepalium,* a name for an instrument of torture. *Trepaliare* developed into Old French *travaillier,* which meant "to torture, torment" and intransitively "to suffer, labor." Middle English borrowed the Old French verb as well as its noun derivative *travail,* and Modern English *travail* preserves the original sense "torment." But the difficulties of getting from place to place in the Middle Ages, when any journey was a wearisome effort and a pilgrimage to a distant religious shrine a real act of devotion, led medieval Englishmen to apply *travailen* "to labor" to the act of making a trip. Shift in stress and a weakening of the vowel in the second syllable eventually led to a distinction between *travail* and *travel.*

1 : violation of allegiance or of faith and confidence **:** TREASON
2 : an act of perfidy or treason

trea·cle \'trē-kəl\ *noun* [Middle English *triacle,* from Middle French, from Latin *theriaca,* from Greek *thēriakē* antidote against a poisonous bite, from feminine of *thēriakos* of a wild animal, from *thērion* wild animal, diminutive of *thēr* wild animal — more at FIERCE] (14th century)
1 : a medicinal compound formerly in wide use as a remedy against poison
2 *chiefly British* **a :** MOLASSES **b :** a blend of molasses, invert sugar, and corn syrup used as syrup at the table — called also *golden syrup*
3 : something (as a tone of voice) heavily sweet and cloying

trea·cly \-k(ə-)lē\ *adjective* (1733)
: resembling treacle (as in quality or appearance) ⟨*treacly* sentimentality⟩

¹tread \'tred\ *verb* **trod** \'träd\ *also* **treaded; trod·den** \'trä-dᵊn\ *or* **trod; tread·ing** [Middle English *treden,* from Old English *tredan;* akin to Old High German *tretan* to tread] (before 12th century)
transitive verb
1 a : to step or walk on or over **b :** to walk along **:** FOLLOW
2 a : to beat or press with the feet **:** TRAMPLE **b :** to subdue or repress as if by trampling **:** CRUSH
3 : to copulate with — used of a male bird
4 a : to form by treading **:** BEAT ⟨*tread* a path⟩ **b :** to execute by stepping or dancing ⟨*tread* a measure⟩
intransitive verb
1 : to move on foot **:** WALK
2 a : to set foot **b :** to put one's foot **:** STEP
3 : COPULATE
— tread·er *noun*
— tread on one's toes : to give offense (as by encroaching on one's rights or feelings)
— tread water : to keep the body nearly upright in the water and the head above water by a treading motion of the feet usually aided by the hands

²tread *noun* (13th century)
1 : a mark (as a footprint or the imprint of a tire) made by or as if by treading
2 a (1) **:** the action of treading (2) **:** an act or instance of treading **:** STEP **b :** manner of stepping **c :** the sound of treading
3 a : the part of a shoe or boot sole that touches the ground; *also* **:** the pattern on the bottom of a sole **b** (1) **:** the part of a wheel or tire that makes contact with a road or rail (2) **:** the pattern of ridges or grooves made or cut in the face of a tire
4 : the distance between the points of contact with the ground of the two front wheels or the two rear wheels of a vehicle
5 a : the upper horizontal part of a step **b :** the width of such a tread
— tread·less \-ləs\ *adjective*

¹trea·dle \'tre-dᵊl\ *noun* [Middle English *tredel* step of a stair, from Old English, from *tredan*] (15th century)
: a swiveling or lever device pressed by the foot to drive a machine

²treadle *verb* **trea·dled; trea·dling** \'tred-liŋ, 'tre-dᵊl-iŋ\ (1891)
intransitive verb
: to operate a treadle
transitive verb
: to operate (as a machine) by a treadle

tread·mill \'tred-ˌmil\ *noun* (1822)
1 a : a mill worked by persons treading on steps on the periphery of a wide wheel having a horizontal axis and used formerly in prison punishment **b :** a mill worked by an animal treading an endless belt **c :** a device having an endless belt on which an individual walks or runs in place for exercise or physiological testing
2 : a wearisome or monotonous routine resembling continued activity on a treadmill

trea·son \'trē-zᵊn\ *noun* [Middle English *tresoun,* from Old French *traison,* from Latin *tradition-, traditio* act of handing over, from *tradere* to hand over, betray — more at TRAITOR] (13th century)
1 : the betrayal of a trust **:** TREACHERY
2 : the offense of attempting by overt acts to overthrow the government of the state to which the offender owes allegiance or to kill or personally injure the sovereign or the sovereign's family ◆

trea·son·able \'trēz-nə-bəl, 'trē-zᵊn-ə-bəl\ *adjective* (14th century)
: relating to, consisting of, or involving treason
— trea·son·ably \-blē\ *adverb*

trea·son·ous \'trēz-nəs, 'trē-zᵊn-əs\ *adjective* (1593)
: TREASONABLE

trea·sur·able \'tre-zhə-rə-bəl, 'trā-; 'trezh-rə-, 'trāzh-rə-\ *adjective* (1607)
: worthy of being treasured **:** PRECIOUS

¹trea·sure \'tre-zhər, 'trā-\ *noun* [Middle English *tresor,* from Old French, from Latin *thesaurus* — more at THESAURUS] (12th century)
1 a (1) **:** wealth (as money, jewels, or precious metals) stored up or hoarded ⟨buried *treasure*⟩ (2) **:** wealth of any kind or in any form **:** RICHES **b :** a store of money in reserve
2 : something of great worth or value; *also* **:** a person esteemed as rare or precious
3 : a collection of precious things

²treasure *transitive verb* **trea·sured; trea·sur·ing** \-zh(ə-)riŋ\ (14th century)
1 : to collect and store up (something of value) for future use **:** HOARD
2 : to hold or keep as precious **:** CHERISH, PRIZE ⟨she *treasured* those memories⟩
synonym see APPRECIATE

trea·sure–house \'tre-zhər-ˌhau̇s, 'trā-\ *noun* (15th century)
1 : a building where treasure is kept **:** TREASURY
2 : a place or source (as a collection) where many things of value can be found

trea·sur·er \'tre-zhə-rər, 'trā-; 'trezh-rər, 'trāzh-\ *noun* (14th century)
1 : a guardian of a collection of treasures **:** CURATOR
2 : an officer entrusted with the receipt, care, and disbursement of funds: as **a :** a governmental officer charged with receiving, keeping, and disbursing public revenues **b :** the executive financial officer of a club, society, or business corporation
— trea·sur·er·ship \-ˌship\ *noun*

treasure trove \-ˌtrōv\ *noun* [Anglo-French *tresor trové,* literally, found treasure] (1523)
1 : treasure that anyone finds; *specifically* **:** gold or silver in the form of money, plate, or bullion which is found hidden and whose ownership is not known
2 : a valuable discovery, resource, or collection

trea·sury \'tre-zh(ə-)rē, 'trā-\ *noun, plural* **-sur·ies** [Middle English *tresorie,* from Old French, from *tresor* treasure] (14th century)
1 a : a place in which stores of wealth are kept **b :** the place of deposit and disbursement of collected funds; *especially* **:** one where public revenues are deposited, kept, and disbursed **c :** funds kept in such a depository
2 *obsolete* **:** TREASURE
3 *capitalized* **a :** a governmental department in charge of finances and especially the collection, management, and expenditure of public revenues **b :** the building in which the business of such a governmental department is transacted
4 *capitalized* **:** a government security (as a note or bill) issued by the Treasury
5 : a repository for treasures ⟨a *treasury* of poems⟩

treasury note *noun* (1890)
1 : a currency note issued by the U.S. Treasury in payment for silver bullion purchased under the Sherman Silver Purchase Act of 1890
2 : a U.S. government bond usually with a maturity of not less than one year or more than seven years

treasury of merits (1884)
: the superabundant satisfaction of Christ for human sins and the excess of merit of the saints which according to Roman Catholic theology is effective for salvation of others and is available for dispensation through indulgences

treasury stock *noun* (1903)
: issued stock reacquired by a corporation and held as an asset

¹treat \'trēt\ *verb* [Middle English *treten,* from Old French *traitier,* from Latin *tractare* to drag about, handle, deal with, frequentative of *trahere* to drag, pull] (14th century)
intransitive verb
1 : to discuss terms of accommodation or settlement **:** NEGOTIATE
2 : to deal with a matter especially in writing **:** DISCOURSE — usually used with *of* ⟨a book *treating* of conservation⟩
3 : to pay another's expenses (as for a meal or drink) especially as a compliment or as an expression of regard or friendship
transitive verb
1 a : to deal with in speech or writing **:** EXPOUND **b :** to present or represent artistically **c :** to deal with **:** HANDLE ⟨food is plentiful and *treated* with imagination —Cecil Beaton⟩
2 a : to bear oneself toward **:** USE ⟨*treat* a horse cruelly⟩ **b :** to regard and deal with in a specified manner — usually used with *as*
3 a : to provide with free food, drink, or entertainment **b :** to provide with enjoyment or gratification
4 : to care for or deal with medically or surgically ⟨*treat* a disease⟩
5 : to act upon with some agent especially to improve or alter ⟨*treat* a metal with acid⟩
— treat·er *noun*

²treat *noun* (1651)
1 : an entertainment given without expense to those invited
2 : an especially unexpected source of joy, delight, or amusement

treat·able \'trē-tə-bəl\ *adjective* (14th century)
: capable of being treated **:** yielding or responsive to treatment ⟨a *treatable* disease⟩
— treat·abil·i·ty \ˌtrē-tə-'bi-lə-tē\ *noun*

trea·tise \'trē-təs *also* -təz\ *noun* [Middle English *tretis,* from Anglo-French *tretiz,* from Old French *traitier* to treat] (14th century)

◇ WORD HISTORY

treason In some circumstances *treason* may seem to be subverting the forces of *tradition*, yet both of these words come from the same Latin source. Latin *traditio* means "transmission of knowledge" or "teaching," but these senses developed from its more literal meaning "the act of handing over a person or thing." Thus *tradition* is maintained by passing information from one generation to another, whereas *treason* is committed when someone who has been entrusted with information passes it on to someone who is not supposed to have it. The difference in form between the two words is accounted for by the fact that *treason* came to us via Old French, after many generations of speakers had altered Latin *traditio* into Old French *traison*, while *tradition* was borrowed into English directly from Latin. Latin *traditio* is itself a derivative of the verb *tradere*, meaning "to hand over, deliver." Through another derivative, *traditor*, this verb is also the ultimate source of English *traitor*.

1 : a systematic exposition or argument in writing including a methodical discussion of the facts and principles involved and conclusions reached ⟨a *treatise* on higher education⟩
2 *obsolete* **:** ACCOUNT, TALE

treat·ment \'trēt-mənt\ *noun* (circa 1560)
1 a : the act or manner or an instance of treating someone or something **:** HANDLING, USAGE **b :** the techniques or actions customarily applied in a specified situation
2 a : a substance or technique used in treating **b :** an experimental condition

trea·ty \'trē-tē\ *noun, plural* **treaties** [Middle English *tretee,* from Middle French *traité,* from Medieval Latin *tractatus,* from Latin, handling, treatment, from *tractare* to treat, handle] (14th century)
1 : the action of treating and especially of negotiating
2 a : an agreement or arrangement made by negotiation: (1) **:** PRIVATE TREATY (2) **:** a contract in writing between two or more political authorities (as states or sovereigns) formally signed by representatives duly authorized and usually ratified by the lawmaking authority of the state **b :** a document in which such a contract is set down

treaty port *noun* (1863)
: any of numerous ports and inland cities in China, Japan, and Korea formerly open by treaty to foreign commerce

[1]tre·ble \'tre-bəl\ *noun* [Middle English, perhaps from Middle French, trio, from *treble,* adjective] (14th century)
1 a : the highest voice part in harmonic music **:** SOPRANO **b :** one that performs a treble part; *also* **:** a member of a family of instruments having the highest range **c :** a high-pitched or shrill voice, tone, or sound **d :** the upper half of the whole vocal or instrumental tonal range — compare BASS **e :** the higher portion of the audio frequency range in sound recording and broadcasting
2 : something treble in construction, uses, amount, number, or value

[2]treble *adjective* [Middle English, from Middle French, from Latin *triplus* — more at TRIPLE] (14th century)
1 a : having three parts or uses **:** THREEFOLD **b :** triple in number or amount
2 a : relating to or having the range or part of a treble **b :** HIGH-PITCHED, SHRILL **c :** of, relating to, or having the range of treble in sound recording and broadcasting ⟨*treble* frequencies⟩
— tre·bly \'tre-b(ə-)lē\ *adverb*

[3]treble *verb* **tre·bled; tre·bling** \'tre-b(ə-)liŋ\ (14th century)
transitive verb
: to increase threefold
intransitive verb
1 : to sing treble
2 : to grow to three times the size, amount, or number

treble clef *noun* [[1]*treble;* from its use for the notation of treble parts] (circa 1854)
1 : a clef that places G above middle C on the second line of the staff
2 : TREBLE STAFF

treble staff *noun* (circa 1854)
: the musical staff carrying the treble clef

treb·u·chet \,tre-byə-'shet, -bə-, -'chet\ *or* **treb·uc·ket** \,tre-bə-'ket\ *noun* [Middle English *trebochet,* from Middle French *trebuchet*] (15th century)
: a medieval military engine for hurling heavy missiles (as rocks)

tre·cen·to \trā-'chen-(,)tō\ *noun* [Italian, literally, three hundred, from Latin *tres* three + *centum* hundred — more at THREE, HUNDRED] (1841)
: the 14th century; *specifically* **:** the 14th century in Italian literature and art

tre·de·cil·lion \,trē-di-'sil-yən\ *noun, often attributive* [Latin *tredecim* thirteen (from *tres* three + *decem* ten) + English *-illion* (as in *million*) — more at THREE, TEN] (circa 1934)
— see NUMBER table

[1]tree \'trē\ *noun* [Middle English, from Old English *trēow;* akin to Old Norse *trē* tree, Greek *drys,* Sanskrit *dāru* wood] (before 12th century)
1 a : a woody perennial plant having a single usually elongate main stem generally with few or no branches on its lower part **b :** a shrub or herb of arborescent form ⟨rose *trees*⟩ ⟨a banana *tree*⟩
2 a (1) **:** a piece of wood (as a post or pole) usually adapted to a particular use or forming part of a structure or implement (2) *archaic* **:** the cross on which Jesus was crucified **b** *archaic* **:** GALLOWS
3 : something in the form of or resembling a tree: as **a :** a diagram or graph that branches usually from a simple stem without forming loops or polygons ⟨genealogical *tree*⟩ **b :** a much-branched system of channels especially in an animal body ⟨the vascular *tree*⟩
4 : SADDLETREE
— tree·less \-ləs\ *adjective*
— tree·like \-,līk\ *adjective*

[2]tree *transitive verb* **treed; tree·ing** (1575)
1 a : to drive to or up a tree ⟨*treed* by a bull⟩ ⟨dogs *treeing* game⟩ **b :** to put into a position of extreme disadvantage **:** CORNER; *especially* **:** to bring to bay
2 : to furnish or fit (as a shoe) with a tree

treed \'trēd\ *adjective* (1860)
: planted or grown with trees **:** WOODED

tree ear *noun* (1967)
: any of several brown ear-shaped basidiomycetous fungi (genus *Auricularia*); *especially* **:** one (*A. polytricha*) used especially in Chinese cooking

tree farm *noun* (1941)
: an area of forest land managed to ensure continuous commercial production

tree fern *noun* (1832)
: any of various ferns (especially families Cyatheaceae and Marattiaceae) of arborescent habit with a woody stem

tree frog *noun* (1738)
: any of numerous small anuran amphibians (especially family Hylidae) of usually arboreal habits that typically have adhesive suckers on the toes

tree·hop·per \'trē-,hä-pər\ *noun* (circa 1839)
: any of a family (Membracidae) of small leaping homopterous insects that feed on the sap especially of shrubs and trees

tree house *noun* (circa 1899)
: a structure (as a playhouse) built among the branches of a tree

tree line *noun* (1893)
: TIMBERLINE

treen \'trēn\ *noun, singular or plural in construction* [*treen* wooden, from Middle English, from Old English *trēowen,* from *treow* tree, wood] (1927)
: small woodenware — called also *treenware*

tree·nail *also* **tre·nail** \'trē-nᵊl\ *noun* (13th century)
: a wooden peg made usually of dry compressed timber so as to swell in its hole when moistened

tree of heaven (1845)
: a Chinese ailanthus (*Ailanthus altissima* synonym *A. glandulosa*) that has foliage similar to that of the sumacs, has ill-scented staminate flowers, and is grown as a shade and ornamental tree

tree of life (1880)
: a conventionalized and often ornate representation of a tree used as a decorative motif

tree peony *noun* (circa 1891)
: a shrubby Chinese peony (*Paeonia suffruticosa*) that has large showy flowers and is the source of many horticultural varieties

tree ring *noun* (1919)
: ANNUAL RING

tree shrew *noun* (circa 1893)
: any of an order (Scandentia) of arboreal insectivorous mammals of southeastern Asia often classified as true insectivores or sometimes as primitive primates

tree shrew

tree sparrow *noun* (circa 1770)
1 : a European sparrow (*Passer montanus*) that has a black spot on the ear coverts
2 : a North American sparrow (*Spizella arborea*) that has a single dark spot on the breast and breeds in northern North America and winters in the U.S.

tree surgeon *noun* (1908)
: a specialist in tree surgery — compare ARBORIST

tree surgery *noun* (1902)
: operative treatment of diseased trees especially for control of decay; *broadly* **:** practices forming part of the professional care of specimen or shade trees

tree swallow *noun* (1873)
: an American swallow (*Tachycineta bicolor* synonym *Iridoprocne bicolor*) with iridescent greenish blue upperparts and white underparts

tree toad *noun* (1778)
: TREE FROG

tree tomato *noun* (circa 1881)
: TAMARILLO

tree·top \'trē-,täp\ *noun* (1530)
1 : the topmost part of a tree
2 *plural* **:** the height or line marked by the tops of a group of trees

tre·foil \'trē-,fȯil, 'tre-\ *noun* [Middle English, from Middle French *trefeuil,* from Latin *trifolium,* from *tri-* + *folium* leaf — more at BLADE] (15th century)
1 a : CLOVER; *broadly* **:** any of several trifoliolate leguminous herbs **b :** a trifoliolate leaf
2 : an ornament or symbol in the form of a stylized trifoliolate leaf

trefoil 2

tre·ha·lose \tri-'hä-,lōs, -,lōz\ *noun* [International Scientific Vocabulary *trehala* (a sweet substance constituting the pupal covering of a beetle) + [2]*-ose*] (1862)
: a crystalline disaccharide $C_{12}H_{22}O_{11}$ stored instead of starch by many fungi and found in the blood of many insects

treil·lage \tre-'yäzh\ *noun* [French, from Middle French, from *treille* vine arbor, from Latin *trichila*] (1698)
: latticework for vines **:** TRELLIS 1

[1]trek \'trek\ *intransitive verb* **trekked; trek·king** [Afrikaans, from Dutch *trecken* to pull, haul, migrate; akin to Old High German *trechan* to pull] (1821)
1 *chiefly South African* **a :** to travel by ox wagon **b :** to migrate by ox wagon or in a train of such
2 : to make one's way arduously; *broadly* **:** to go on a journey
— trek·ker *noun*

[2]trek *noun* [Afrikaans, from Dutch *treck* pull, haul, from *trecken*] (1835)
1 *chiefly South African* **:** a journey by ox wagon; *especially* **:** an organized migration by a group of settlers

2 : a trip or movement especially when involving difficulties or complex organization

¹trel·lis \'tre-ləs\ *noun* [Middle English *trelis,* from Middle French *treliz* fabric of coarse weave, trellis, from (assumed) Vulgar Latin *trilicius* woven with triple thread, from Latin *tri-* + *licium* thread] (15th century)
1 : a frame of latticework used as a screen or as a support for climbing plants
2 : a construction (as a summerhouse) chiefly of latticework
3 : an arrangement that forms or gives the effect of a lattice ⟨a *trellis* of interlacing streams⟩
— trel·lised \'tre-ləst\ *adjective*

²trellis *transitive verb* (15th century)
1 : to provide with a trellis; *especially* **:** to train (as a vine) on a trellis
2 : to cross or interlace on or through **:** INTERWEAVE

trel·lis·work \'tre-ləs-ˌwərk\ *noun* (1712)
: LATTICEWORK

trem·a·tode \'tre-mə-ˌtōd\ *noun* [ultimately from Greek *trēmatōdēs* pierced with holes, from *trēmat-, trēma* hole, from *tetrainein* to bore — more at THROW] (circa 1859)
: any of a class (Trematoda) of parasitic platyhelminthic flatworms including the flukes
— trematode *adjective*

¹trem·ble \'trem-bəl\ *intransitive verb* **trem·bled; trem·bling** \-b(ə-)liŋ\ [Middle English, from Middle French *trembler,* from Medieval Latin *tremulare,* from Latin *tremulus* tremulous, from *tremere* to tremble; akin to Greek *tremein* to tremble] (14th century)
1 : to shake involuntarily (as with fear or cold) **:** SHIVER
2 : to move, sound, pass, or come to pass as if shaken or tremulous ⟨the building *trembled* from the blast⟩
3 : to be affected with fear or doubt ⟨*tremble* for the safety of another⟩
— trem·bler \-b(ə-)lər\ *noun*

²tremble *noun* (1609)
1 : an act or instance of trembling; *especially* **:** a fit or spell of involuntary shaking or quivering
2 *plural but singular in construction* **:** severe poisoning of livestock and especially cattle by a toxic alcohol present in a snakeroot (*Eupatorium rugosum*) and rayless goldenrod that is characterized by muscular tremors, weakness, and constipation

trem·bly \'trem-b(ə-)lē\ *adjective* (1848)
: marked by trembling **:** TREMULOUS

tre·men·dous \tri-'men-dəs\ *adjective* [Latin *tremendus,* from gerundive of *tremere*] (1632)
1 : being such as may excite trembling or arouse dread, awe, or terror
2 a : notable by reason of extreme size, power, greatness, or excellence — often used as a generalized term of approval **b :** unusually large **:** HUGE
synonym see MONSTROUS
— tre·men·dous·ly *adverb*
— tre·men·dous·ness *noun*

trem·o·lite \'tre-mə-ˌlīt\ *noun* [French *trémolite,* from *Tremola,* valley in Switzerland] (1799)
: a white or gray mineral of the amphibole group that is a silicate of calcium and magnesium
— trem·o·lit·ic \ˌtre-mə-'li-tik\ *adjective*

trem·o·lo \'tre-mə-ˌlō\ *noun, plural* **-los** [Italian, from *tremolo* tremulous, from Latin *tremulus*] (circa 1801)
1 a : the rapid reiteration of a musical tone or of alternating tones to produce a tremulous effect **b :** vocal vibrato especially when prominent or excessive
2 : a mechanical device in an organ for causing a tremulous effect

trem·or \'tre-mər\ *noun* [Middle English *tremour* terror, from Middle French, from Latin *tremor* trembling, from *tremere*] (1615)
1 : a trembling or shaking usually from physical weakness, emotional stress, or disease
2 : a quivering or vibratory motion; *especially* **:** a discrete small movement following or preceding a major seismic event
3 a : a feeling of uncertainty or insecurity **b :** a cause of such a feeling

trem·u·lant \'trem-yə-lənt\ *adjective* [Medieval Latin *tremulant-, tremulans,* present participle of *tremulare* — more at TREMBLE] (1837)
: TREMULOUS, TREMBLING

trem·u·lous \-ləs\ *adjective* [Latin *tremulus* — more at TREMBLE] (1611)
1 : characterized by or affected with trembling or tremors
2 : affected with timidity **:** TIMOROUS
3 : such as is or might be caused by nervousness or shakiness ⟨a *tremulous* smile⟩ ⟨*tremulous* handwriting⟩
4 : exceedingly sensitive **:** easily shaken or disordered
— trem·u·lous·ly *adverb*
— trem·u·lous·ness *noun*

¹trench \'trench\ *noun* [Middle English *trenche* track cut through a wood, from Middle French, act of cutting, from *trenchier* to cut, probably from (assumed) Vulgar Latin *trinicare* to cut in three, from Latin *trini* three each — more at TRINE] (15th century)
1 a : a long cut in the ground **:** DITCH; *especially* **:** one used for military defense often with the excavated dirt thrown up in front **b** *plural* **:** a place, position, or level at which an activity is carried on in a manner likened to trench warfare — often used in the phrase *in the trenches*
2 : a long, narrow, and usually steep-sided depression in the ocean floor — compare TROUGH
3 : TRENCH COAT

²trench (15th century)
transitive verb
1 : to make a cut in **:** CARVE
2 a : to protect with or as if with a trench **b :** to cut a trench in **:** DITCH
intransitive verb
1 a : ENTRENCH, ENCROACH ⟨*trenching* on other domains which were more vital —Sir Winston Churchill⟩ **b :** to come close **:** VERGE
2 : to dig a trench

tren·chan·cy \'tren-chən(t)-sē\ *noun* (1866)
: the quality or state of being trenchant

tren·chant \-chənt\ *adjective* [Middle English, from Middle French, present participle of *trenchier*] (14th century)
1 : KEEN, SHARP
2 : vigorously effective and articulate ⟨a *trenchant* analysis⟩; *also* **:** CAUSTIC ⟨*trenchant* remarks⟩
3 a : sharply perceptive **:** PENETRATING **b :** CLEAR-CUT, DISTINCT ⟨the *trenchant* divisions between right and wrong —Edith Wharton⟩
— tren·chant·ly *adverb*

trench coat *noun* (1916)
1 : a waterproof overcoat with a removable lining designed for wear in trenches
2 : a usually double-breasted raincoat with deep pockets, wide belt, and often straps on the shoulders

trenched \'trencht\ *adjective* (1541)
1 : furrowed or drained by trenches
2 : provided with protective trenches

¹tren·cher \'tren-chər\ *noun* [Middle English, from Middle French *trencheoir,* from *trenchier* to cut] (14th century)
: a wooden platter for serving food

²trencher *adjective* (15th century)
1 : of or relating to a trencher or to meals
2 *archaic* **:** having the nature of a parasite **:** SYCOPHANTIC

³trench·er \'tren-chər\ *noun* [²*trench*] (circa 1864)
: one that digs trenches; *specifically* **:** a usually self-propelled excavating machine typically employing a bucket conveyor and used to dig trenches especially for pipelines and cables

tren·cher·man \'tren-chər-mən\ *noun* (1590)
1 : a hearty eater
2 *archaic* **:** HANGER-ON, SPONGER

trench fever *noun* (1915)
: a disease that is marked by fever and pain in muscles, bones, and joints and that is caused by a bacterium (*Rochalimaea quintana*) transmitted by the human body louse (*Pediculus humanus*)

trench foot *noun* (1915)
: a painful foot disorder resembling frostbite and resulting from exposure to cold and wet

trench mouth *noun* (1918)
1 : VINCENT'S ANGINA
2 : VINCENT'S INFECTION

trench warfare *noun* (1917)
: warfare in which the opposing forces attack and counterattack from a relatively permanent system of trenches protected by barbed-wire entanglements

¹trend \'trend\ *intransitive verb* [Middle English, to turn, revolve, from Old English *trendan;* akin to Middle High German *trendel* disk, spinning top] (1598)
1 a : to extend in a general direction **:** follow a general course ⟨mountain ranges *trending* north and south⟩ **b :** to veer in a new direction **:** BEND ⟨coastline that *trends* westward⟩
2 a : to show a tendency **:** INCLINE ⟨prices *trending* upward⟩ **b :** to become deflected **:** SHIFT ⟨opinions *trending* toward conservatism⟩

²trend *noun* (1777)
1 : a line of general direction or movement ⟨the *trend* of the coast turned toward the west⟩
2 a : a prevailing tendency or inclination **:** DRIFT **b :** a general movement **:** SWING ⟨the *trend* toward suburban living⟩ **c :** a current style or preference **:** VOGUE ⟨new fashion *trends*⟩ **d :** a line of development **:** APPROACH
3 : the general movement in the course of time of a statistically detectable change; *also* **:** a statistical curve reflecting such a change
synonym see TENDENCY

trend·set·ter \'tren(d)-ˌse-tər\ *noun* (1960)
: one that sets a trend

trend·set·ting \-ˌse-tiŋ\ *adjective* (1960)
: that sets a trend ⟨a *trendsetting* fashion look⟩

trendy \'tren-dē\ *adjective* **trend·i·er; -est** (1962)
1 : very fashionable **:** UP-TO-DATE ⟨he's a *trendy* dresser —*Sunday Mirror*⟩
2 : marked by ephemeral, superficial, or faddish appeal or taste ⟨a newspaper of *trendy* triviality —J.H. Plumb⟩
— trend·i·ly \-də-lē\ *adverb*
— trend·i·ness \-dē-nəs\ *noun*
— trendy \-dē\ *noun*

¹tre·pan \tri-'pan\ *transitive verb* **tre·panned; tre·pan·ning** [Middle English, from *trepane* trephine] (15th century)
1 : to use a trephine on (the skull)
2 : to remove a disk or cylindrical core (as from metal for testing)
— trep·a·na·tion \ˌtre-pə-'nā-shən\ *noun*

²tre·pan \'trē-ˌpan, tri-'pan\ *noun* [Middle English *trepane* trephine, from Medieval Latin *trepanum,* from Greek *trypanon* auger, from *trypan* to bore] (circa 1877)
: a heavy tool used in boring mine shafts

³tre·pan \tri-'pan\ *noun* [origin unknown] (1641)
1 *archaic* **:** TRICKSTER
2 *archaic* **:** a deceptive device **:** SNARE

⁴tre·pan \tri-'pan\ *transitive verb* **tre·panned; tre·pan·ning** (circa 1656)
archaic **:** ENTRAP, LURE

tre·pang \tri-'paŋ, 'trē-ˌ\ *noun* [Malay *tĕripang*] (1783)
: any of several large sea cucumbers (especially genera *Actinopyga* and *Holothuria*) that are taken mostly in the southwestern Pacific and

are boiled, dried, and used especially by the Chinese for making soup — called also *bêche-de-mer*

treph·i·na·tion \ˌtre-fə-ˈnā-shən\ *noun* (1874)
: an act or instance of perforating the skull with a surgical instrument

tre·phine \ˈtrē-ˌfīn\ *noun* [French *tréphine,* from obsolete English *trefine, trafine,* from Latin *tres fines* three ends, from *tres* three + *fines,* plural of *finis* end — more at THREE] (1628)
: a surgical instrument for cutting out circular sections (as of bone or corneal tissue)
— trephine *transitive verb*

trep·id \ˈtre-pəd\ *adjective* [Latin *trepidus*] (1650)
: TIMOROUS, FEARFUL

trep·i·dant \ˈtre-pə-dənt\ *adjective* [Latin *trepidant-, trepidans,* present participle of *trepidare*] (1892)
: TIMID, TREMBLING

trep·i·da·tion \ˌtre-pə-ˈdā-shən\ *noun* [Latin *trepidation-, trepidatio,* from *trepidare* to tremble, from *trepidus* agitated; probably akin to Old English *thrafian* to urge, push, Greek *trapein* to press grapes] (1605)
1 *archaic* **:** a tremulous motion **:** TREMOR
2 : timorous uncertain agitation **:** APPREHENSION
synonym see FEAR

trep·o·ne·ma \ˌtre-pə-ˈnē-mə\ *noun, plural* **-ma·ta** \-mə-tə\ *or* **-mas** [New Latin *Treponemat-, Treponema,* from Greek *trepein* to turn + *nēma* thread, from *nēn* to spin — more at NEEDLE] (circa 1908)
: any of a genus (*Treponema*) of spirochetes that are pathogenic in humans and other warm-blooded animals and include organisms causing syphilis and yaws
— trep·o·ne·mal \-məl\ *adjective*

trep·o·ne·ma·to·sis \-ˌnē-mə-ˈtō-səs, -ˌne-\ *noun, plural* **-to·ses** \-ˌsēz\ [New Latin] (1927)
: infection with or disease caused by treponemata

trep·o·neme \ˈtre-pə-ˌnēm\ *noun* (1919)
: TREPONEMA

¹tres·pass \ˈtres-pəs, -ˌpas\ *noun* [Middle English *trespas,* from Old French, crossing, trespass, from *trespasser* to go across] (14th century)
1 a : a violation of moral or social ethics **:** TRANSGRESSION; *especially* **:** SIN **b :** an unwarranted infringement
2 a : an unlawful act committed on the person, property, or rights of another; *especially* **:** a wrongful entry on real property **b :** the legal action for injuries resulting from trespass

²trespass \-ˌpas *also* -pəs\ *verb* [Middle English, from Middle French *trespasser,* from Old French, literally, to go across, from *tres* across (from Latin *trans*) + *passer* to pass — more at THROUGH, PASS] (14th century)
intransitive verb
1 a : ERR, SIN **b :** to make an unwarranted or uninvited incursion
2 : to commit a trespass; *especially* **:** to enter unlawfully upon the land of another
transitive verb
: VIOLATE ⟨*trespass* the bounds of good taste⟩
☆
— tres·pass·er *noun*

tress \ˈtres\ *noun* [Middle English *tresse,* from Middle French *trece*] (14th century)
1 : a long lock of hair; *especially* **:** the long unbound hair of a woman — usually used in plural
2 *archaic* **:** a plait of hair **:** BRAID

tressed \ˈtrest\ *adjective* (14th century)
1 *obsolete* **:** being braided **:** PLAITED
2 : having tresses — usually used in combination ⟨golden-*tressed*⟩

tres·tle *also* **tres·sel** \ˈtre-səl *also* ˈtrā-\ *noun* [Middle English *trestel,* from Middle French, from (assumed) Vulgar Latin *transtellum,* from Latin *transtillum,* diminutive of *transtrum* traverse beam, from *translation* across — more at THROUGH] (14th century)
1 : HORSE 2b
2 : a braced frame serving as a support
3 : a braced framework of timbers, piles, or steelwork for carrying a road or railroad over a depression

trestle table *noun* (circa 1891)
: a table supported on trestles

tres·tle·work \-ˌwərk\ *noun* (1848)
: a system of connected trestles supporting a structure (as a railroad bridge)

tre·tin·o·in \tre-ˈti-nə-win\ *noun* [perhaps from *trans-* + *retinoic* acid] (1980)
: RETINOIC ACID

trews \ˈtrüz\ *noun plural* [Scottish Gaelic *triubhas*] (circa 1568)
1 *chiefly British* **:** ³PANT 1; *especially* **:** tight-fitting trousers usually of tartan
2 : close-cut tartan shorts worn under the kilt in Highland dress

trey \ˈtrā\ *noun, plural* **treys** [Middle English *treye, treis,* from Middle French *treie, treis,* from Latin *tres* three] (14th century)
1 : the side of a die or domino that has three spots
2 : a card numbered three or having three main pips

tri- *combining form* [Middle English, from Latin (from *tri-, tres*) & Greek, from *tri-, treis* — more at THREE]
1 : three **:** having three elements or parts ⟨*tri*graph⟩
2 : into three ⟨*tri*sect⟩
3 a : thrice ⟨*tri*weekly⟩ **b :** every third ⟨*tri*monthly⟩

tri·able \ˈtrī-ə-bəl\ *adjective* (15th century)
: liable or subject to judicial or quasi-judicial examination or trial

tri·ac·e·tate \(ˌ)trī-ˈa-sə-ˌtāt\ *noun* [International Scientific Vocabulary] (1885)
1 : an acetate containing three CH_3COO- groups
2 : a textile fiber or fabric consisting of cellulose that is completely or almost completely acetylated

tri·ad \ˈtrī-ˌad *also* -əd\ *noun* [Latin *triad-, trias,* from Greek, from *treis* three] (1546)
1 : a union or group of three **:** TRINITY
2 : a chord of three tones consisting of a root with its third and fifth and constituting the harmonic basis of tonal music
— tri·ad·ic \trī-ˈa-dik\ *adjective*
— tri·ad·i·cal·ly \-di-k(ə-)lē\ *adverb*

tri·age \trē-ˈäzh, ˈtrē-ˌ\ *noun* [French, sorting, sifting, from *trier* to sort, from Old French — more at TRY] (1918)
: the sorting of and allocation of treatment to patients and especially battle and disaster victims according to a system of priorities designed to maximize the number of survivors; *broadly* **:** the assigning of priority order to projects on the basis of where funds and resources can be best used or are most needed
— triage *transitive verb*

¹tri·al \ˈtrī(-ə)l\ *noun* [Anglo-French, from *trier* to try] (15th century)
1 a : the action or process of trying or putting to the proof **:** TEST **b :** a preliminary contest (as in a sport)
2 : the formal examination before a competent tribunal of the matter in issue in a civil or criminal cause in order to determine such issue
3 : a test of faith, patience, or stamina through subjection to suffering or temptation; *broadly* **:** a source of vexation or annoyance
4 a : a tryout or experiment to test quality, value, or usefulness **b :** one of a number of repetitions of an experiment
5 : ATTEMPT

²trial *adjective* (1555)
1 : of, relating to, or used in a trial
2 : made or done as a test or experiment
3 : used or tried out in a test or experiment

trial and error *noun* (1806)
: a finding out of the best way to reach a desired result or a correct solution by trying out one or more ways or means and by noting and eliminating errors or causes of failure; *also* **:** the trying of one thing or another until something succeeds

trial balance *noun* (1838)
: a list of the debit and credit balances of accounts in a double-entry ledger at a given date prepared primarily to test their equality

trial balloon *noun* (1935)
: a project or scheme tentatively announced in order to test public opinion

trial court *noun* (1890)
: the court before which issues of fact and law are first determined as distinguished from an appellate court

trial examiner *noun* (1949)
: a person appointed to hold hearings and to investigate and report facts sometimes with recommendations to an administrative or quasi-judicial agency or tribunal

trial horse *noun* (1901)
: one set up as an opponent for a champion in trial competitions or workouts

trial jury *noun* (1884)
: a jury impaneled to try a cause **:** PETIT JURY

trial lawyer *noun* (circa 1914)
: a lawyer who engages chiefly in the trial of cases before courts of original jurisdiction

tri·a·logue \ˈtrī-ə-ˌlȯg, -ˌläg\ *noun* [*tri-* + *-alogue* (as in *dialogue*)] (1532)
: a scene, discourse, or colloquy in which three persons share

trial run *noun* (1903)
: a testing exercise **:** EXPERIMENT

tri·am·cin·o·lone \ˌtrī-ˌam-ˈsi-nᵊl-ˌōn\ *noun* [*tri-* + *-amcin-* (alteration of *Am*erican *Cyan*imid Company) + ¹*-ol* + *-one*] (1957)
: a glucocorticoid drug $C_{21}H_{27}FO_6$ used especially in treating psoriasis and allergic skin and respiratory disorders

tri·an·gle \ˈtrī-ˌaŋ-gəl\ *noun* [Middle English, from Latin *triangulum,* from neuter of *triangulus* triangular, from *tri-* + *angulus* angle] (14th century)
1 : a polygon having three sides — compare SPHERICAL TRIANGLE
2 a : a percussion instrument consisting of a rod of steel bent into the form of a triangle open at one angle and sounded by striking with a small metal rod **b :** a drafting instrument consisting of a thin flat right-angled

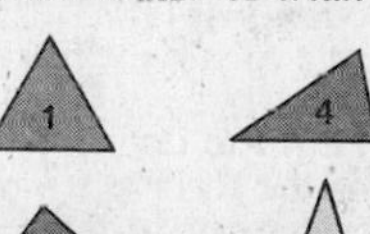

triangle 1: ***1*** **equilateral,** ***2*** **acute,** ***3*** **obtuse,** ***4*** **scalene,** ***5*** **isosceles,** ***6*** **right triangle**

☆ SYNONYMS
Trespass, encroach, infringe, invade mean to make inroads upon the property, territory, or rights of another. TRESPASS implies an unwarranted, unlawful, or offensive intrusion ⟨hunters *trespassing* on farmland⟩. ENCROACH suggests gradual or stealthy entrance upon another's territory or usurpation of another's rights or possessions ⟨the *encroaching* settlers gradually displaced the native peoples⟩. INFRINGE implies an encroachment clearly violating a right or prerogative ⟨a product that *infringes* an existing patent⟩. INVADE implies a hostile and injurious entry into the territory or sphere of another ⟨accused of *invading* their privacy⟩.

triangle of wood or plastic with acute angles of 45 degrees or of 30 degrees and 60 degrees **3 :** a situation in which one member of a couple is involved in a love affair with a third person

triangle inequality *noun* [from its application to the distances between three points in a coordinate system] (1941)
: an inequality stating that the absolute value of a sum is less than or equal to the sum of the absolute value of the terms

tri·an·gu·lar \trī-ˈaŋ-gyə-lər\ *adjective* [Late Latin *triangularis,* from Latin *triangulum*] (1541)
1 a : of, relating to, or having the form of a triangle ⟨a *triangular* plot of land⟩ **b :** having a triangular base or principal surface ⟨a *triangular* table⟩ ⟨a *triangular* pyramid⟩
2 a (1) **:** of, relating to, or involving three elements ⟨a *triangular* agreement⟩ (2) *of a military group* **:** based primarily on three units **b :** of or relating to a love triangle
— tri·an·gu·lar·i·ty \(ˌ)trī-ˌaŋ-gyə-ˈlar-ə-tē\ *noun*
— tri·an·gu·lar·ly \trī-ˈaŋ-gyə-lər-lē\ *adverb*

triangular number *noun* (1796)
: any of the numbers (as 1, 3, 6, 10, 15) that represent the number of dots in the figures formed starting with one dot and adding rows one after another to form triangles each row of which has one more dot than the one before and are of the general form

$$\frac{n(n+1)}{2} \text{ where } n = 1, 2, 3, \ldots$$

1tri·an·gu·late \trī-ˈaŋ-gyə-lət\ *adjective* [Medieval Latin *triangulatus,* past participle of *triangulare* to make triangles, from Latin *triangulum*] (1766)
: consisting of or marked with triangles

2tri·an·gu·late \-ˌlāt\ *transitive verb* **-lat·ed; -lat·ing** (1833)
1 : to survey, map, or determine by triangulation
2 a : to divide into triangles **b :** to give triangular form to

tri·an·gu·la·tion \(ˌ)trī-ˌaŋ-gyə-ˈlā-shən\ *noun* (1818)
: the measurement of the elements necessary to determine the network of triangles into which any part of the earth's surface is divided in surveying; *broadly* **:** any similar trigonometric operation for finding a position or location by means of bearings from two fixed points a known distance apart

tri·ar·chy \ˈtrī-ˌär-kē\ *noun, plural* **-chies** [Greek *triarchia,* from *tri-* + *-archia* -archy] (circa 1656)
1 : government by three persons **:** TRIUMVIRATE
2 : a country under three rulers

Tri·as·sic \trī-ˈa-sik\ *adjective* [International Scientific Vocabulary, from Latin *trias* triad; from the three subdivisions of the European Triassic — more at TRIAD] (1841)
: of, relating to, or being the earliest period of the Mesozoic era or the corresponding system of rocks marked by the first appearance of the dinosaurs — see GEOLOGIC TIME table
— Triassic *noun*

tri·ath·lete \trī-ˈath-ˌlēt, ÷-ˈa-thə-\ *noun* (1982)
: an athlete who competes in a triathlon

tri·ath·lon \trī-ˈath-lən, -ˌlän, ÷-ˈa-thə-\ *noun* [*tri-* + *-athlon* (as in *decathlon*)] (1978)
: an athletic contest that is a long-distance race consisting of three phases (as swimming, bicycling, and running)

tri·atom·ic \ˌtrī-ə-ˈtä-mik\ *adjective* [International Scientific Vocabulary] (1862)
: having three atoms in the molecule ⟨ozone is *triatomic* oxygen⟩

tri·ax·i·al \(ˌ)trī-ˈak-sē-əl\ *adjective* [International Scientific Vocabulary] (1886)
: having or involving three axes
— tri·ax·i·al·i·ty \-ˌak-sē-ˈa-lə-tē\ *noun*

tri·azine \ˈtrī-ə-ˌzēn, trī-ˈa-ˌzēn\ *noun* [International Scientific Vocabulary] (1894)
: any of three compounds $C_3H_3N_3$ containing a ring composed of three carbon and three nitrogen atoms; *also* **:** any of various derivatives of these including several used as herbicides

trib·al \ˈtrī-bəl\ *adjective* (1632)
: of, relating to, or characteristic of a tribe ⟨*tribal* customs⟩
— trib·al·ly \-bə-lē\ *adverb*

trib·al·ism \-bə-ˌli-zəm\ *noun* (1886)
1 : tribal consciousness and loyalty; *especially* **:** exaltation of the tribe above other groups
2 : strong in-group loyalty

tri·ba·sic \(ˌ)trī-ˈbā-sik\ *adjective* (1837)
: having three replaceable hydrogen atoms — used of acids

tribe \ˈtrīb\ *noun* [Middle English, from Latin *tribus,* a division of the Roman people, tribe] (13th century)
1 a : a social group comprising numerous families, clans, or generations together with slaves, dependents, or adopted strangers **b :** a political division of the Roman people originally representing one of the three original tribes of ancient Rome **c :** PHYLE
2 : a group of persons having a common character, occupation, or interest
3 : a category of taxonomic classification ranking below a subfamily; *also* **:** a natural group irrespective of taxonomic rank ⟨the cat *tribe*⟩ ⟨rose *tribe*⟩

tribes·man \ˈtrībz-mən\ *noun* (1798)
: a member of a tribe

tribes·peo·ple \-ˌpē-pəl\ *noun plural* (1888)
: members of a tribe

tribo- *combining form* [French, from Greek *tribein* to rub; probably akin to Latin *terere* to rub — more at THROW]
: friction ⟨*tribo*luminescence⟩

tri·bo·elec·tric·i·ty \ˌtrī-bō-i-ˌlek-ˈtri-sə-tē, ˌtri-, -ˈtris-tē\ *noun* (circa 1917)
: a charge of electricity generated by friction (as by rubbing glass with silk)
— tri·bo·elec·tric \-ˈlek-trik\ *adjective*

tri·bol·o·gy \trī-ˈbä-lə-jē, tri-\ *noun* (1966)
: a study that deals with the design, friction, wear, and lubrication of interacting surfaces in relative motion (as in bearings or gears)
— tri·bo·log·i·cal \ˌtrī-bə-ˈlä-ji-kəl, ˌtri-\ *adjective*
— tri·bol·o·gist \trī-ˈbä-lə-jist, tri-\ *noun*

tri·bo·lu·mi·nes·cence \ˌtrī-bō-ˌlü-mə-ˈne-sᵊn(t)s, ˈtri-\ *noun* [International Scientific Vocabulary] (1889)
: luminescence due to friction
— tri·bo·lu·mi·nes·cent \-sᵊnt\ *adjective*

tri·brach \ˈtrī-ˌbrak\ *noun* [Latin *tribrachys,* from Greek, having three short syllables, from *tri-* + *brachys* short — more at BRIEF] (1589)
: a metrical foot of three short syllables of which two belong to the thesis and one to the arsis
— tri·brach·ic \trī-ˈbra-kik\ *adjective*

trib·u·late \ˈtri-byə-ˌlāt\ *transitive verb* **-lat·ed; -lat·ing** [Late Latin *tribulatus,* past participle of *tribulare*] (circa 1637)
: to cause to endure tribulation

trib·u·la·tion \ˌtri-byə-ˈlā-shən\ *noun* [Middle English *tribulacion,* from Old French, from Latin *tribulation-, tribulatio,* from *tribulare* to press, oppress, from *tribulum* drag used in threshing, from *terere* to rub — more at THROW] (13th century)
: distress or suffering resulting from oppression or persecution; *also* **:** a trying experience

tri·bu·nal \trī-ˈbyü-nᵊl, tri-\ *noun* [Latin, platform for magistrates, from *tribunus* tribune] (1526)
1 : 2TRIBUNE
2 : a court or forum of justice
3 : something that decides or determines ⟨the *tribunal* of public opinion⟩

tri·bu·nate \ˈtri-byə-ˌnāt, tri-ˈbyü-nət\ *noun* (1546)
: the office, function, or term of office of a tribune

1tri·bune \ˈtri-ˌbyün, tri-ˈ\ *noun* [Middle English, from Latin *tribunus,* from *tribus* tribe] (14th century)
1 : a Roman official under the monarchy and the republic with the function of protecting the plebeian citizen from arbitrary action by the patrician magistrates
2 : an unofficial defender of the rights of the individual
— tri·bune·ship \-ˌship\ *noun*

2tribune *noun* [French, from Italian *tribuna,* from Latin *tribunal*] (circa 1771)
: a dais or platform from which an assembly is addressed

1trib·u·tary \ˈtri-byə-ˌter-ē\ *adjective* (14th century)
1 : paying tribute to another to acknowledge submission, to obtain protection, or to purchase peace **:** SUBJECT
2 : paid or owed as tribute
3 : channeling material or supplies into something more inclusive **:** CONTRIBUTORY

2tributary *noun, plural* **-tar·ies** (15th century)
1 : a ruler or state that pays tribute to a conqueror
2 : a stream feeding a larger stream or a lake

trib·ute \ˈtri-(ˌ)byüt, -byət\ *noun* [Middle English *tribut,* from Latin *tributum,* from neuter of *tributus,* past participle of *tribuere* to allot, bestow, grant, pay, from *tribus* tribe] (14th century)
1 a : a payment by one ruler or nation to another in acknowledgment of submission or as the price of protection; *also* **:** the tax levied for such a payment **b** (1) **:** an excessive tax, rental, or tariff imposed by a government, sovereign, lord, or landlord (2) **:** an exorbitant charge levied by a person or group having the power of coercion **c :** the liability to pay tribute
2 a : something given or contributed voluntarily as due or deserved; *especially* **:** a gift or service showing respect, gratitude, or affection ⟨floral *tribute*⟩ **b :** something (as material evidence or a formal attestation) that indicates the worth, virtue, or effectiveness of the one in question ⟨the product is a *tribute* to their ingenuity⟩
synonym see ENCOMIUM

tri·car·box·yl·ic \ˌtrī-ˌkär-ˌbäk-ˈsi-lik\ *adjective* (1894)
: containing three carboxyl groups in the molecule ⟨*tricarboxylic* acid⟩

tricarboxylic acid cycle *noun* (1945)
: KREBS CYCLE

1trice \ˈtrīs\ *transitive verb* **triced; tric·ing** [Middle English *trisen, tricen* to pull, trice, from Middle Dutch *trisen* to hoist, from *trise* windlass] (15th century)
: to haul up or in and lash or secure (as a sail) with a small rope

2trice *noun* [Middle English *trise,* literally, pull, from *trisen*] (15th century)
: a brief space of time **:** INSTANT — used chiefly in the phrase *in a trice*

tri·ceps \ˈtrī-ˌseps\ *noun, plural* **triceps** [New Latin *tricipit-, triceps,* from Latin, three-headed, from *tri-* + *capit-, caput* head — more at HEAD] (circa 1704)
: a muscle that arises from three heads; *especially* **:** the great extensor muscle along the back of the upper arm

tri·cer·a·tops \(ˌ)trī-ˈser-ə-ˌtäps\ *noun, plural* **-tops** *also* **-tops·es** \-ˌtäp-səz\ [New Latin, from *tri-* + Greek *kerat-, keras* horn + *ōps* face — more at HORN, EYE] (1892)
: any of a genus (*Triceratops*) of large herbivorous Cretaceous dinosaurs with three horns, a bony hood or crest on the neck, and hoofed toes

-trices *plural of* -TRIX

trich- *or* **tricho-** *combining form* [New Latin, from Greek, from *trich-, thrix* hair]
: hair **:** filament ⟨*tricho*gyne⟩

tri·chi·a·sis \tri-ˈkī-ə-səs\ *noun* [Late Latin, from Greek, from *trich-* + *-iasis*] (1661)
: a turning inward of the eyelashes often causing irritation of the eyeball

tri·chi·na \tri-ˈkī-nə\ *noun, plural* **-nae** \-(ˌ)nē\ *also* **-nas** [New Latin, from Greek *trichinos* made of hair, from *trich-, thrix* hair] (1857)
: a small slender nematode worm (*Trichinella spiralis*) that in the larval state is parasitic in the voluntary muscles of flesh-eating mammals (as humans and swine)
— **tri·chi·nal** \-nᵊl\ *adjective*

trich·i·nize \ˈtri-kə-ˌnīz\ *transitive verb* **-nized; -niz·ing** (1864)
: to infest with trichinae ⟨*trichinized* pork⟩

trich·i·no·sis \ˌtri-kə-ˈnō-səs\ *noun* [New Latin] (1866)
: infestation with or disease caused by trichinae and marked especially by muscular pain, dyspnea, fever, and edema

tri·chi·nous \ˈtri-kə-nəs, tri-ˈkī-\ *adjective* [International Scientific Vocabulary] (1857)
1 : infested with trichinae ⟨*trichinous* meat⟩
2 : of, relating to, or involving trichinae or trichinosis ⟨*trichinous* infection⟩

tri·chlor·fon *also* **tri·chlor·phon** \(ˌ)trī-ˈklōr-ˌfän, -ˈklȯr-\ *noun* [*tri-* + *chlor-* + *-fon* (irregular from *phosphonate* — a salt derived from phosphine)] (1960)
: a crystalline compound $C_4H_8Cl_3O_4P$ used especially as an insecticide

tri·chlo·ro·ace·tic acid \ˌtrī-ˌklōr-ō-ə-ˈsē-tik-, -ˌklȯr-\ *noun* [International Scientific Vocabulary] (1885)
: a strong vesicant pungent acid $C_2Cl_3HO_2$ used in weed control and in medicine as a caustic and astringent

tri·chlo·ro·eth·y·lene \-ˈe-thə-ˌlēn\ *noun* (circa 1919)
: a nonflammable liquid C_2HCl_3 used especially as a solvent and in dry cleaning and for the removal of grease from metal

tricho·cyst \ˈtri-kə-ˌsist\ *noun* (1859)
: any of the minute lassoing or stinging organelles of protozoans and especially of many ciliates

tricho·gyne \-ˌjīn, -ˌgīn\ *noun* [International Scientific Vocabulary] (circa 1875)
: a slender terminal prolongation of the ascogonium of a fungus or lichen that may serve as a fertilization tube; *also* **:** a similar reproductive structure in a red alga

tri·chol·o·gist \tri-ˈkä-lə-jist\ *noun* (1887)
: a person who cares for and dresses hair **:** HAIRDRESSER
— **tri·chol·o·gy** \-jē\ *noun*

tri·chome \ˈtri-ˌkōm, ˈtrī-\ *noun* [German *Trichom,* from Greek *trichōma* growth of hair, from *trichoun* to cover with hair, from *trich-, thrix* hair] (1875)
: a filamentous outgrowth; *especially* **:** an epidermal hair structure on a plant

tricho·mo·na·cide \ˌtri-kə-ˈmō-nə-ˌsīd\ *noun* [*trichomona*d + *-cide*] (1949)
: an agent used to destroy trichomonads
— **tricho·mo·na·cid·al** \-ˌmō-nə-ˈsī-dᵊl\ *adjective*

tricho·mo·nad \ˌtri-kə-ˈmō-ˌnad, -nəd\ *noun* [New Latin *Trichomonad-, Trichomonas,* from *trich-* + Late Latin *monad-, monas* monad] (1861)
: any of a genus (*Trichomonas*) of flagellated protozoans parasitic in many animals
— **trichomonad** *or* **tricho·mo·nal** \-ˈmō-nᵊl\ *adjective*

tricho·mo·ni·a·sis \ˌtri-kə-mə-ˈnī-ə-səs\ *noun, plural* **-a·ses** \-ˌsēz\ [New Latin, from *Trichomonas* + *-iasis*] (1915)
: infection with or disease caused by trichomonads: as **a :** a human sexually transmitted disease occurring especially as vaginitis with a persistent discharge and caused by a trichomonad (*Trichomonas vaginalis*) that may also invade the male urethra and bladder **b :** a venereal disease of domestic cattle marked by abortion and sterility **c :** one or more diseases of various birds resembling blackhead — called also *roup*

tri·chop·ter·an \tri-ˈkäp-tə-rən\ *noun* [ultimately from Greek *trich-, thrix* hair + *pteron* wing — more at FEATHER] (circa 1842)
: CADDIS FLY
— **trichopteran** *adjective*

tricho·the·cene \ˌtri-kə-ˈthē-ˌsēn\ *noun* [New Latin *Trichothecium* (from *trich-* + *-thecium*) + English *-ene*] (1971)
: any of several mycotoxins that are produced by imperfect fungi (genera *Fusarium* and *Trichothecium*) and that include some contaminants of livestock feed and some held to be found in yellow rain

tri·chot·o·mous \trī-ˈkä-tə-məs\ *adjective* [Late Greek *trichotomein* to trisect, from Greek *tricha* in three (akin to *treis* three) + *-tomein* (akin to *temnein* to cut) — more at THREE, TOME] (1800)
: divided or dividing into three parts or into threes ⟨*trichotomous* branching⟩
— **tri·chot·o·mous·ly** *adverb*

tri·chot·o·my \-mē\ *noun, plural* **-mies** (1610)
: division into three parts, elements, or classes

tri·chro·mat \ˈtrī-krō-ˌmat, (ˌ)trī-ˈ\ *noun* [back-formation from *trichromatic*] (1929)
: a person with trichromatism

tri·chro·mat·ic \ˌtrī-krō ˈma-tik\ *adjective* (circa 1890)
1 : of, relating to, or consisting of three colors ⟨*trichromatic* light⟩
2 a : relating to or being the theory that human color vision involves three types of retinal sensory receptors **b :** characterized by trichromatism ⟨*trichromatic* vision⟩

tri·chro·ma·tism \(ˌ)trī-ˈkrō-mə-ˌti-zəm\ *noun* (circa 1895)
: color vision based on the perception of three primary colors and especially red, green, and blue

[1]**trick** \ˈtrik\ *noun* [Middle English *trik,* from Old North French *trique,* from *trikier* to deceive, cheat, from (assumed) Vulgar Latin *triccare,* alteration of Latin *tricari* to behave evasively, shuffle, from *tricae* complications, trifles] (15th century)
1 a : a crafty procedure or practice meant to deceive or defraud **b :** a mischievous act **:** PRANK **c :** an indiscreet or childish action **d :** a deceptive, dexterous, or ingenious feat; *especially* **:** one designed to puzzle or amuse ⟨a juggler's *tricks*⟩
2 a : a habitual peculiarity of behavior or manner ⟨a horse with the *trick* of shying⟩ **b :** a characteristic and identifying feature ⟨a *trick* of speech⟩ **c :** a delusive appearance especially when caused by art or legerdemain **:** an optical illusion ⟨a mere *trick* of the light⟩
3 a : a quick or artful way of getting a result **:** KNACK **b :** a technical device (as of an art or craft) ⟨the *tricks* of stage technique⟩
4 : the cards played in one round of a card game often used as a scoring unit
5 a : a turn of duty at the helm usually lasting for two hours **b :** SHIFT 4b(1) **c :** a trip taken as part of one's employment **d :** a sexual act performed by a prostitute; *also* **:** JOHN 2
6 : an attractive child or woman ⟨a cute little *trick*⟩ ☆

[2]**trick** *transitive verb* (circa 1500)
1 : to dress or adorn fancifully or ornately **:** ORNAMENT ⟨*tricked* out in a gaudy uniform⟩
2 : to deceive by cunning or artifice **:** CHEAT
synonym see DUPE

[3]**trick** *adjective* (circa 1530)
1 : TRIG
2 a : of or relating to or involving tricks or trickery ⟨*trick* photography⟩ ⟨*trick* dice⟩ **b :** skilled in or used for tricks ⟨a *trick* horse⟩
3 a : somewhat defective and unreliable ⟨a *trick* lock⟩ **b :** inclined to give way unexpectedly ⟨a *trick* knee⟩

trick·er \ˈtri-kər\ *noun* (circa 1553)
: one that tricks **:** TRICKSTER

trick·ery \ˈtri-k(ə-)rē\ *noun* (1800)
: the practice of crafty underhanded ingenuity to deceive or cheat
synonym see DECEPTION

trick·ish \ˈtri-kish\ *adjective* (1705)
: given to or characterized by tricks or trickery **:** TRICKY
— **trick·ish·ly** *adverb*
— **trick·ish·ness** *noun*

[1]**trick·le** \ˈtri-kəl\ *intransitive verb* **trick·led; trick·ling** \-k(ə-)liŋ\ [Middle English, of imitative origin] (14th century)
1 a : to issue or fall in drops **b :** to flow in a thin gentle stream
2 a : to move or go one by one or little by little **b :** to dissipate slowly ⟨his enthusiasm *trickled* away⟩

[2]**trickle** *noun* (1580)
: a thin, slow, or intermittent stream or movement

trickle–down *adjective* (1944)
1 : relating to or working on the principle of trickle-down theory
2 : relating to or being an effect caused gradually by remote or indirect influences

trickle–down theory *noun* (1954)
: a theory that financial benefits given to big business will in turn pass down to smaller businesses and consumers

trick or treat *noun* (circa 1941)
: a children's Halloween practice of asking for treats from door to door under threat of playing tricks on householders who refuse
— **trick–or–treat** *intransitive verb*
— **trick–or–treater** *noun*

trick·ster \ˈtrik-stər\ *noun* (1711)
: one who tricks: as **a :** a dishonest person who defrauds others by trickery **b :** a person (as a stage magician) skilled in the use of tricks and illusion **c :** a deceptive character appearing in various forms in the folklore of many cultures

tricksy \ˈtrik-sē\ *adjective* **tricks·i·er; -est** [*tricks,* plural of *trick*] (1552)
1 *archaic* **:** smartly attired **:** SPRUCE
2 : full of tricks **:** PRANKISH
3 a *archaic* **:** having the craftiness of a trickster **b :** difficult to cope with or handle **:** TRYING ⟨a *tricksy* job⟩
— **tricks·i·ness** *noun*

☆ SYNONYMS
Trick, ruse, stratagem, maneuver, artifice, wile, feint mean an indirect means to gain an end. TRICK may imply deception, roguishness, illusion, and either an evil or harmless end ⟨the *tricks* of the trade⟩. RUSE stresses an attempt to mislead by a false impression ⟨the *ruses* of smugglers⟩. STRATAGEM implies a ruse used to entrap, outwit, circumvent, or surprise an opponent or enemy ⟨the *stratagem*-filled game⟩. MANEUVER suggests adroit and skillful avoidance of difficulty ⟨last-minute *maneuvers* to avert bankruptcy⟩. ARTIFICE implies ingenious contrivance or invention ⟨the clever *artifices* of the stage⟩. WILE suggests an attempt to entrap or deceive with false allurements ⟨used all of his *wiles* to ingratiate himself⟩. FEINT implies a diversion or distraction of attention away from one's real intent ⟨a *feint* toward the enemy's left flank⟩.

\ə\ **abut** \ᵊ\ **kitten** \ər\ **further** \a\ **ash** \ā\ **ace** \ä\ **mop, mar** \au̇\ **out** \ch\ **chin** \e\ **bet** \ē\ **easy** \g\ **go** \i\ **hit** \ī\ **ice** \j\ **job** \ŋ\ **sing** \ō\ **go** \ȯ\ **law** \ȯi\ **boy** \th\ **thin** \t͟h\ **the** \ü\ **loot** \u̇\ **foot** \y\ **yet** \zh\ **vision** *see also* Guide to Pronunciation

tricky \'tri-kē\ *adjective* **trick·i·er; -est** (1786)
1 : inclined to or marked by trickery
2 a : giving a deceptive impression of easiness, simplicity, or order **:** TICKLISH ⟨a *tricky* path through the swamp⟩ **b :** TRICK 3
3 : requiring skill, knack, or caution (as in doing or handling) **:** DIFFICULT; *also* **:** INGENIOUS ⟨a *tricky* rhythm⟩
synonym see SLY
— trick·i·ly \'tri-kə-lē\ *adverb*
— trick·i·ness \'tri-kē-nəs\ *noun*

tri·clad \'trī-ˌklad\ *noun* [New Latin *Tricladida,* from *tri-* + Greek *klados* branch — more at CLAD-] (1888)
: any of an order (Tricladida) of turbellarian flatworms (as a planarian) that have the intestine composed of a median anterior division and two lateral posterior divisions with side branches
— triclad *adjective*

tri·clin·ic \(ˌ)trī-'kli-nik\ *adjective* [International Scientific Vocabulary] (1854)
: having three unequal axes intersecting at oblique angles — used especially of a crystal

tri·clin·i·um \trī-'kli-nē-əm\ *noun, plural* **-ia** \-nē-ə\ [Latin, from Greek *triklinion,* from *tri-* + *klinein* to lean, recline — more at LEAN] (1646)
1 : a couch extending round three sides of a table used by the ancient Romans for reclining at meals
2 : a dining room furnished with a triclinium

tric·o·lette \ˌtri-kə-'let\ *noun* [*trico*t + *-lette* (as in *flannelette*)] (1919)
: a usually silk or rayon knitted fabric used especially for women's clothing

¹tri·col·or \'trī-ˌkə-lər *also* 'trē-, *especially British* 'tri-kə-lər\ *adjective* [French *tricolore* three-colored, from Late Latin *tricolor,* from Latin *tri-* + *color* color] (1795)
1 a *or* **tri·col·ored** \'trī-'kəl-ərd\ **:** having, using, or marked with three colors **b** *of a dog* **:** having a coat of black, tan, and white
2 : of, relating to, or characteristic of a tricolor or a nation whose flag is a tricolor; *especially* **:** FRENCH

²tricolor *noun* (1797)
1 : a flag of three colors arranged in equal horizontal or vertical bands ⟨the French *tricolor*⟩
2 : a tricolor animal; *especially* **:** a tricolor dog

tri·corn \'trī-ˌkȯrn\ *adjective* [Latin *tricornis*] (circa 1844)
: having three horns or corners

tri·corne *or* **tri·corn** \'trī-ˌkȯrn\ *noun* [French *tricorne,* from *tricorne* three-cornered, from Latin *tricornis,* from *tri-* + *cornu* horn — more at HORN] (1876)
: COCKED HAT 1

tri·cor·nered \'trī-'kȯ(r)-nərd\ *adjective* (1819)
: having three corners

tri·cot \'trē-(ˌ)kō, 'trī-kət\ *noun* [French, from *tricoter* to knit, from Middle French, to agitate, hop, ultimately from Old French *estriquier* to stroke, of Germanic origin; akin to Old English *strīcan* to stroke — more at STRIKE] (1872)
1 : a plain warp-knitted fabric (as of nylon, wool, rayon, silk, or cotton) with a close inelastic knit and used especially in clothing (as underwear)
2 : a twilled clothing fabric of wool with fine warp ribs or of wool and cotton with fine weft ribs

tri·co·tine \ˌtri-kə-'tēn, ˌtrē-\ *noun* [French, from *tricot*] (circa 1899)
: a sturdy suiting woven of tightly twisted yarns in a double twill

tric·trac \'trik-ˌtrak\ *noun* [French, of imitative origin] (1687)
: an old form of backgammon played with pegs

¹tri·cus·pid \(ˌ)trī-'kəs-pəd\ *adjective* [Latin *tricuspid-, tricuspis,* from *tri-* + *cuspid-, cuspis* point] (1834)
: having three cusps ⟨a *tricuspid* molar⟩

²tricuspid *noun* (circa 1860)
: a tricuspid anatomical structure; *especially* **:** a tooth having three cusps

tricuspid valve *noun* (1670)
: a valve of three flaps that prevents reflux of blood from the right ventricle to the right atrium

tri·cy·cle \'trī-sə-kəl, -ˌsi-kəl\ *noun* [French, from *tri-* + Greek *kyklos* wheel — more at WHEEL] (1868)
: a 3-wheeled vehicle propelled by pedals or a motor

tri·cy·clic \(ˌ)trī-'sī-klik, -'si-klik\ *adjective* [*tri-* + *cyclic*] (1891)
: being a chemical with three usually fused rings in the molecular structure and especially a tricyclic antidepressant

tricyclic antidepressant *noun* (1966)
: any of a group of antidepressant drugs (as imipramine, amitriptyline, desipramine, and nortriptyline) that potentiate the action of catecholamines and do not inhibit the action of monoamine oxidase — called also *tricyclic*

¹tri·dent \'trī-dᵊnt\ *adjective* [Latin *trident-, tridens*] (1589)
: having three teeth, processes, or points

²trident *noun* [Latin *trident-, tridens,* from *trident-, tridens* having three teeth, from *tri-* + *dent-, dens* tooth — more at TOOTH] (1599)
1 : a 3-pronged spear serving in classical mythology as the attribute of a sea god (as Neptune)
2 : a 3-pronged spear (as for fishing)

Tri·den·tine \trī-'den-ˌtīn, -ˌtēn; 'trī-dᵊn-, 'tri-\ *adjective* [New Latin *Tridentinus,* from Latin *Tridentum* Trent, Italy] (1561)
: of or relating to the Roman Catholic Church council held at Trent from 1545 to 1563

tri·di·men·sion·al \ˌtrī-də-'mench-nəl, -ˌdī-, -'men(t)-shə-nᵊl\ *adjective* [International Scientific Vocabulary] (1858)
: of, relating to, or concerned with three dimensions ⟨*tridimensional* space⟩
— tri·di·men·sion·al·i·ty \-ˌmen(t)-shə-'na-lə-tē\ *noun*

trid·u·um \'tri-jə-wəm, 'tri-dyə-\ *noun* [Latin, space of three days, from *tri-* + *-duum* (akin to *dies* day) — more at DEITY] (1873)
: a period of three days of prayer usually preceding a Roman Catholic feast

tried \'trīd\ *adjective* [Middle English, from past participle of *trien* to try, test] (15th century)
1 : found good, faithful, or trustworthy through experience or testing ⟨a *tried* recipe⟩
2 : subjected to trials or distress ⟨a kind but much-*tried* father⟩

tried-and-true *adjective* (1935)
: proved good, desirable, or feasible **:** shown or known to be worthy ⟨a *tried-and-true* sales technique⟩

tri·ene \'trī-ˌēn\ *noun* (1917)
: a chemical compound containing three double bonds

tri·en·ni·al \(ˌ)trī-'e-nē-əl\ *adjective* (1562)
1 : occurring or being done every three years ⟨the *triennial* convention⟩
2 : consisting of or lasting for three years ⟨a *triennial* contract⟩
— triennial *noun*
— tri·en·ni·al·ly \-nē-ə-lē\ *adverb*

tri·en·ni·um \trī-'e-nē-əm\ *noun, plural* **-ni·ums** *or* **-nia** \-nē-ə\ [Latin, from *tri-* + *annus* year — more at ANNUAL] (1847)
: a period of three years

tri·er \'trī(-ə)r\ *noun* (14th century)
1 : someone or something that tries
2 : an implement (as a tapered hollow tube) used in obtaining samples of bulk material for examination and testing

tri·er·arch \'trī(-ə)-ˌrärk\ *noun* [Latin *trierarchus,* from Greek *triērarchos,* from *triērēs* trireme (from *tri-* + *-ērēs* — akin to Latin *remus* oar) + *-archos* -arch — more at ROW] (circa 1656)
1 : the commander of a trireme
2 : an Athenian citizen who had to fit out a trireme for the public service

tri·er·ar·chy \-ˌrär-kē\ *noun* (1837)
: the ancient Athenian plan whereby individual citizens furnished and maintained triremes as a civic duty

tri·eth·yl \(ˌ)trī-'e-thəl\ *adjective* [International Scientific Vocabulary] (1858)
: containing three ethyl groups in the molecule

tri·fec·ta \trī-'fek-tə, 'trī-ˌ\ *noun* [*tri-* + *perfecta*] (1974)
: a variation of the perfecta in which a bettor wins by selecting the first three finishers of a race in the correct order of finish

tri·fid \'trī-ˌfid, -fəd\ *adjective* [Latin *trifidus* split into three, from *tri-* + *findere* to split — more at BITE] (1753)
: being deeply and narrowly cleft into three teeth, processes, or points ⟨a spoon with a *trifid* top⟩

¹tri·fle \'trī-fəl\ *noun* [Middle English *trufle, trifle,* from Old French *trufe, trufle* mockery] (13th century)
1 : something of little value, substance, or importance
2 : a dessert of many varieties typically including plain or sponge cake, sherry, rum, or brandy, jam or jelly, fruit, custard, and whipped cream
— a trifle : to some small degree **:** SLIGHTLY ⟨*a trifle* annoyed⟩

²trifle *verb* **tri·fled; tri·fling** \-f(ə-)liŋ\ [Middle English *truflen, triflen,* from Old French *trufer, trufler* to mock, trick] (14th century)
intransitive verb
1 a : to talk in a jesting or mocking manner or with intent to delude or mislead **b :** to treat someone or something as unimportant
2 : to handle something idly
transitive verb
: to spend or waste in trifling or on trifles ☆
— tri·fler \-f(ə-)lər\ *noun*

tri·fling \'trī-fliŋ\ *adjective* (1535)
: lacking in significance or solid worth: as **a :** FRIVOLOUS ⟨*trifling* talk⟩ **b :** TRIVIAL ⟨a *trifling* gift⟩ **c** *chiefly dialect* **:** LAZY, SHIFTLESS ⟨a *trifling* fellow⟩

tri·fluo·per·a·zine \ˌtrī-ˌflü-ō-'per-ə-ˌzēn, -zən\ *noun* [*tri-* + *fluor-* + *piperazine*] (circa 1957)
: a phenothiazine tranquilizer $C_{21}H_{24}F_3N_3S$ used to treat psychotic conditions and especially schizophrenia

tri·flu·ra·lin \trī-'flu̇r-ə-lən\ *noun* [*tri-* + *fluor-* + *aniline*] (circa 1961)
: an herbicide $C_{13}H_{16}F_3N_3O_4$ used in the control of weeds

¹tri·fo·cal \(ˌ)trī-'fō-kəl\ *adjective* (1826)
: having three focal lengths

²trifocal *noun* (1899)
1 *plural* **:** eyeglasses with trifocal lenses
2 : a trifocal glass or lens

☆ SYNONYMS
Trifle, toy, dally, flirt, coquet mean to deal with or act toward without serious purpose. TRIFLE may imply playfulness, unconcern, indulgent contempt ⟨to *trifle* with a lover's feelings⟩. TOY implies acting without full attention or serious exertion of one's powers ⟨a political novice *toying* with great issues⟩. DALLY suggests indulging in thoughts or plans merely as an amusement ⟨*dallying* with the idea of building a boat someday⟩. FLIRT implies an interest or attention that soon passes to another object ⟨*flirted* with one fashionable ism after another⟩. COQUET implies attracting interest or admiration without serious intention ⟨companies that *coquet* with environmentalism solely for public relations⟩.

tri·fo·li·ate \(ˌ)trī-ˈfō-lē-ət\ *adjective* [*tri-* + Latin *folium* leaf — more at BLADE] (1753)
1 : having three leaves ⟨a *trifoliate* plant⟩
2 : TRIFOLIOLATE

trifoliate orange *noun* (circa 1900)
: a Chinese citrus (*Poncirus trifoliata*) with trifoliolate leaves that is widely grown for ornament and especially as a stock for budding oranges

tri·fo·li·o·late \(ˌ)trī-ˈfō-lē-ə-ˌlāt\ *adjective* [International Scientific Vocabulary *tri-* + Late Latin *foliolum* leaflet, diminutive of Latin *folium* leaf] (circa 1828)
: having three leaflets ⟨a *trifoliolate* leaf⟩ — see LEAF illustration

tri·fo·li·um \trī-ˈfō-lē-əm\ *noun* [New Latin, from Latin, trefoil — more at TREFOIL] (1541)
: CLOVER 1

tri·fo·ri·um \trī-ˈfōr-ē-əm, -ˈfȯr-\ *noun, plural* **-ria** \-ē-ə\ [Medieval Latin] (1703)
: a gallery forming an upper story to the aisle of a church and typically an arcaded story between the nave arches and clerestory

tri·form \ˈtrī-ˌfȯrm\ *adjective* [Middle English *triforme,* from Latin *triformis,* from *tri-* + *forma* form] (15th century)
: having a triple form or nature

tri·fur·cate \(ˌ)trī-ˈfər-kət, -ˌkāt; ˈtrī-(ˌ)fər-ˌkāt\ *adjective* [Latin *trifurcus,* from *tri-* + *furca* fork] (circa 1831)
: having three branches or forks **:** TRICHOTOMOUS
— **tri·fur·cate** \ˈtrī-(ˌ)fər-ˌkāt, trī-ˈfər-\ *intransitive verb*
— **tri·fur·ca·tion** \ˌtrī-(ˌ)fər-ˈkā-shən\ *noun*

[1]trig \ˈtrig\ *adjective* [Middle English, trusty, nimble, of Scandinavian origin; akin to Old Norse *tryggr* faithful; akin to Old English *trēowe* faithful — more at TRUE] (1513)
1 : stylishly or jauntily trim
2 : extremely precise **:** PRIM
3 *dialect chiefly British* **:** FIRM, VIGOROUS

[2]trig *noun* [by shortening] (circa 1895)
: TRIGONOMETRY

tri·gem·i·nal \trī-ˈje-mə-nᵊl\ *adjective* [New Latin *trigeminus* trigeminal nerve, from Latin, threefold, from *tri-* + *geminus* twin] (1872)
: of or relating to the trigeminal nerve

trigeminal nerve *noun* (1830)
: either of a pair of large mixed nerves that are the 5th cranial nerves and supply motor and sensory fibers mostly to the face — called also *trigeminal*

trigeminal neuralgia *noun* (1874)
: an intense paroxysmal neuralgia involving one or more branches of the trigeminal nerve

[1]trig·ger \ˈtri-gər\ *noun* [alteration of earlier *tricker,* from Dutch *trekker,* from Middle Dutch *trecker* one that pulls, from *trecken* to pull — more at TREK] (1621)
1 a : a piece (as a lever) connected with a catch or detent as a means of releasing it; *especially* **:** the part of the action moved by the finger to fire a gun **b :** a similar movable part by which a mechanism is actuated ⟨*trigger* of a spray gun⟩
2 : something that acts like a mechanical trigger in initiating a process or reaction
— **trigger** *adjective*
— **trig·gered** \-gərd\ *adjective*

[2]trigger *verb* **trig·gered; trig·ger·ing** \ˈtri-g(ə-)riŋ\ (1916)
transitive verb
1 a : to release or activate by means of a trigger; *especially* **:** to fire by pulling a mechanical trigger ⟨*trigger* a rifle⟩ **b :** to cause the explosion of ⟨*trigger* a missile with a proximity fuze⟩
2 : to initiate, actuate, or set off by a trigger ⟨an indiscreet remark that *triggered* a fight⟩ ⟨a stimulus that *triggered* a reflex⟩
intransitive verb
: to release a mechanical trigger

trig·ger·fish \ˈtri-gər-ˌfish\ *noun* (1849)
: any of various deep-bodied bony fishes (family Balistidae, especially genus *Balistes*) of warm seas having an anterior dorsal fin with two or three stout erectile spines

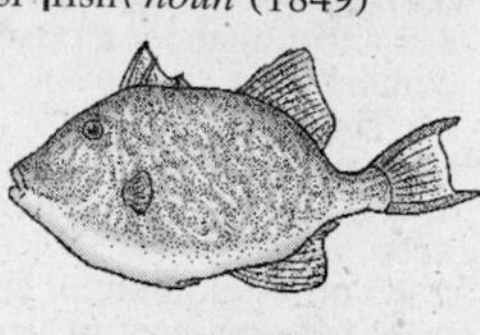
triggerfish

trig·ger–hap·py \-ˌha-pē\ *adjective* (1943)
1 : irresponsible in the use of firearms; *especially* **:** inclined to shoot before clearly identifying the target
2 a : inclined to be irresponsible in matters that might precipitate war **b :** aggressively belligerent in attitude

trig·ger·man \-mən, -ˌman\ *noun* (circa 1930)
: a gunman who shoots the victim (as in a gangland murder)

tri·glyc·er·ide \(ˌ)trī-ˈgli-sə-ˌrīd\ *noun* [International Scientific Vocabulary] (circa 1860)
: an ester formed from glycerol by reacting all three of its alcohol hydroxy groups with fatty acids

tri·glyph \ˈtrī-ˌglif\ *noun* [Latin *triglyphus,* from Greek *triglyphos,* from *tri-* + *glyphein* to carve — more at CLEAVE] (1563)
: a slightly projecting rectangular tablet in a Doric frieze with two vertical channels of V section and two corresponding chamfers or half channels on the vertical sides
— **tri·glyph·ic** \trī-ˈgli-fik\ *or* **tri·glyph·i·cal** \-fi-kəl\ *adjective*

tri·gon \ˈtrī-ˌgän\ *noun* [Latin *trigonum,* from Greek *trigōnon,* from neuter of *trigōnos* triangular, from *tri-* + *gōnia* angle — more at -GON] (1563)
: TRIPLICITY 1

tri·go·nal \trī-ˈgō-nᵊl\ *adjective* (1878)
: of, relating to, or being the division of the hexagonal crystal system or the forms belonging to it characterized by a vertical axis of threefold symmetry
— **tri·go·nal·ly** \-nᵊl-ē\ *adverb*

trig·o·no·met·ric \ˌtri-gə-nə-ˈme-trik\ *also* **trig·o·no·met·ri·cal** \-tri-kəl\ *adjective* (1690)
: of, relating to, or being in accordance with trigonometry
— **trig·o·no·met·ri·cal·ly** \-tri-k(ə-)lē\ *adverb*

trigonometric function *noun* (1909)
1 : a function (as the sine, cosine, tangent, cotangent, secant, or cosecant) of an arc or angle most simply expressed in terms of the ratios of pairs of sides of a right-angled triangle — called also *circular function*
2 : the inverse (as the arcsine, arccosine, or arctangent) of a trigonometric function

trig·o·nom·e·try \ˌtri-gə-ˈnä-mə-trē\ *noun* [New Latin *trigonometria,* from Greek *trigōnon* + *-metria* -metry] (1614)
: the study of the properties of triangles and trigonometric functions and of their applications

tri·gram \ˈtrī-ˌgram\ *noun* (1606)
1 : TRIGRAPH 2
2 : any of the eight possible combinations of three whole or broken lines used especially in Chinese divination

tri·graph \ˈtrī-ˌgraf\ *noun* (circa 1836)
1 : three letters spelling a single consonant, vowel, or diphthong ⟨*eau* of *beau* is a *trigraph*⟩
2 : a cluster of three successive letters ⟨*the, ion,* and *ing* are high frequency *trigraphs*⟩
— **tri·graph·ic** \(ˌ)trī-ˈgra-fik\ *adjective*

tri·halo·meth·ane \(ˌ)trī-ˌha-lə-ˈme-ˌthān, *British usually* -ˈmē-\ *noun* (1969)
: any of various derivatives CHX_3 of methane (as chloroform) that have three halogen atoms per molecule and are formed especially during the chlorination of drinking water

tri·he·dral \-ˈhē-drəl\ *adjective* (1789)
1 : having three faces ⟨*trihedral* angle⟩
2 : of or relating to a trihedral angle
— **trihedral** *noun*

tri·hy·brid \-ˈhī-brəd\ *noun* (1903)
: an individual or strain that is heterozygous for three pairs of genes
— **trihybrid** *adjective*

tri·hy·droxy \ˌtrī-hī-ˈdräk-sē, -hə-\ *adjective* [International Scientific Vocabulary] (1903)
: containing three hydroxyl groups in the molecule

tri·io·do·thy·ro·nine \ˌtrī-ˌī-ə-dō-ˈthī-rə-ˌnēn\ *noun* [*tri-* + *iod-* + *thyronine* (an amino acid of which thyroxine is a derivative)] (1952)
: an iodine-containing hormone $C_{15}H_{12}I_3NO_4$ that is an amino acid derived from thyroxine

tri·jet \ˈtrī-ˌjet\ *noun* (1967)
: an aircraft powered with three jet engines

trike \ˈtrīk\ *noun* [by shortening & alteration] (1884)
: TRICYCLE

tri·lat·er·al \(ˌ)trī-ˈla-tə-rəl, -ˈla-trəl\ *adjective* [Latin *trilaterus,* from *tri-* + *later-, latus* side] (1660)
: having three sides or parties ⟨*trilateral* business ventures⟩ ⟨*trilateral* discussions⟩

tril·by \ˈtril-bē\ *noun, plural* **trilbies** [from the fact that such a hat was worn in the London stage version of the novel *Trilby* (1894) by George du Maurier] (1897)
chiefly British **:** a soft felt hat with indented crown

tri·lin·e·ar \(ˌ)trī-ˈli-nē-ər\ *adjective* (1715)
: of, relating to, or involving three lines ⟨*trilinear* coordinates⟩

tri·lin·gual \-ˈliŋ-gwəl *also* -ˈliŋ-gyə-wəl\ *adjective* (1834)
: consisting of, having, or expressed in three languages; *also* **:** familiar with or able to use three languages
— **tri·lin·gual·ly** *adverb*

[1]tri·lit·er·al \-ˈli-t(ə-)rəl\ *adjective* [*tri-* + Latin *littera* letter] (1751)
: consisting of three letters and especially of three consonants ⟨*triliteral* roots in Semitic languages⟩
— **tri·lit·er·al·ism** \-t(ə-)rə-ˌli-zəm\ *noun*

[2]triliteral *noun* (circa 1828)
: a root or word that is triliteral

[1]trill \ˈtril\ *verb* [Middle English, probably of Scandinavian origin; akin to Swedish *trilla* to roll; akin to Middle Dutch *trillen* to vibrate] (14th century)
intransitive verb
1 : to flow in a small stream or in drops **:** TRICKLE
2 : TWIRL, REVOLVE
transitive verb
: to cause to flow in a small stream

[2]trill *noun* [Italian *trillo* probably of imitative origin] (1649)
1 a : the alternation of two musical tones a diatonic second apart — called also *shake* **b :** VIBRATO **c :** a rapid reiteration of the same tone especially on a percussion instrument
2 : a sound resembling a musical trill **:** WARBLE
3 a : the rapid vibration of one speech organ against another (as of the tip of the tongue against the teethridge) **b :** a speech sound made by a trill

[3]trill (circa 1667)
intransitive verb
: to play or sing with a trill **:** QUAVER
transitive verb
: to utter as or with or as if with a trill ⟨*trill* the *r*⟩
— **trill·er** *noun*

tril·lion \ˈtri(l)-yən\ *noun* [French, from *tri-* + *-illion* (as in *million*)] (circa 1690)

1 — see NUMBER table
2 **:** a very large number
— **trillion** *adjective*
— **tril·lionth** \-yən(t)th\ *adjective or noun*

tril·li·um \'tri-lē-əm\ *noun* [New Latin, from Swedish *trilling* triplet; from its three leaves] (circa 1760)
: any of a genus (*Trillium*) of herbs of the lily family with an erect stem bearing a whorl of three leaves and a large solitary typically spring-blooming flower

trillium

tri·lo·bate \(,)trī-'lō-,bāt\ *adjective* (1785)
: TRILOBED

tri·lobed \'trī-'lōbd\ *adjective* (1826)
: having three lobes ⟨a *trilobed* leaf⟩

tri·lo·bite \'trī-lə-,bīt\ *noun* [ultimately from Greek *trilobos* three-lobed, from *tri-* + *lobos* lobe] (1832)
: any of numerous extinct Paleozoic marine arthropods (group Trilobita) having the segments of the body divided by furrows on the dorsal surface into three lobes

trilobite

tril·o·gy \'tri-lə-jē\ *noun, plural* **-gies** [Greek *trilogia,* from *tri-* + *-logia* -logy] (circa 1661)
: a series of three dramas or literary works or sometimes three musical compositions that are closely related and develop a single theme

[1]**trim** \'trim\ *verb* **trimmed; trim·ming** [probably from (assumed) Middle English *trimmen* to prepare, put in order, from Old English *trymian, tryman* to strengthen, arrange, from *trum* strong, firm; probably akin to Old English *trēo* tree, wood — more at TREE] (circa 1521)
transitive verb
1 : to embellish with or as if with ribbons, lace, or ornaments ⟨*trim* the Christmas tree⟩ ⟨the coat was *trimmed* with fur⟩
2 a : to administer a beating to **:** THRASH **b :** DEFEAT ⟨*trimmed* me at chess⟩
3 a : to make trim and neat especially by cutting or clipping ⟨*trim* the hedges⟩ **b :** to free of excess or extraneous matter by or as if by cutting ⟨*trim* a budget⟩ ⟨*trim* down the inventory⟩ **c :** to remove by or as if by cutting ⟨*trimmed* thousands from federal payrolls —*Grit*⟩
4 a (1) **:** to cause (as a ship) to assume a desirable position in the water by arrangement of ballast, cargo, or passengers (2) **:** to adjust (as an airplane or submarine) for horizontal movement or for motion upward or downward **b :** to adjust (as cargo or a sail) to a desired position
5 : to adjust (as one's opinions) for reasons of expediency — often used in the phrase *trim one's sails*
intransitive verb
1 a : to maintain neutrality between opposing parties or to favor each equally **b :** to change one's views for reasons of expediency
2 : to assume or cause a boat to assume a desired position in the water ⟨a boat that *trims* badly⟩
— **trim one's sails :** to adjust oneself or one's actions to prevailing conditions

[2]**trim** *adjective* **trim·mer; trim·mest** (circa 1521)
1 *obsolete* **:** EXCELLENT, FINE
2 : ready for service or use; *also* **:** in good physical condition ⟨keeps *trim* by jogging⟩
3 : exhibiting neatness, good order, or compactness of line or structure ⟨*trim* houses⟩
— **trim·ly** *adverb*
— **trim·ness** *noun*

[3]**trim** *adverb* (1529)
: in a trim manner **:** TRIMLY — used chiefly in combination ⟨the *trim*-cut forest vistas —W. M. Thackeray⟩

[4]**trim** *noun* (1590)
1 : suitable or excellent condition ⟨tries to keep in *trim*⟩
2 a : one's clothing or appearance **b :** material used for ornament or trimming **c :** the lighter woodwork in the finish of a building especially around openings **d :** the interior furnishings of an automobile
3 a : the position of a ship or boat especially with reference to the horizontal; *also* **:** the difference between the draft of a ship forward and that aft **b :** the relation between the plane of a sail and the direction of the ship **c :** the buoyancy status of a submarine **d :** the attitude of a lighter-than-air craft relative to a fore-and-aft horizontal plane **e :** the attitude with respect to wind axes at which an airplane will continue in level flight with free controls
4 : something that is trimmed off or cut out
5 : a haircut that neatens a previous haircut

tri·ma·ran \'trī-mə-,ran, ,trī-mə-'\ *noun* [*tri-* + *catamaran*] (1949)
: a fast pleasure sailboat with three hulls side by side

tri·mer \'trī-mər\ *noun* [International Scientific Vocabulary] (circa 1930)
: a polymer formed from three molecules of a monomer
— **tri·mer·ic** \trī-'mer-ik\ *adjective*

trim·er·ous \'tri-mə-rəs\ *adjective* [New Latin *trimerus,* from Greek *tri-* + *meros* part — more at MERIT] (1826)
: having the parts in threes — used of a flower and often written *3-merous*

tri·mes·ter \(,)trī-'mes-tər, 'trī-,\ *noun* [French *trimestre,* from Latin *trimestris* of three months, from *tri-* + *mensis* month — more at MOON] (1821)
1 : a period of three or about three months; *especially* **:** any of three periods of approximately three months each into which a human pregnancy is divided
2 : one of three terms into which the academic year is sometimes divided

trim·e·ter \'tri-mə-tər\ *noun* [Latin *trimetrus,* from Greek *trimetros* having three measures, from *tri-* + *metron* measure — more at MEASURE] (1567)
: a line of verse consisting of three dipodies or three metrical feet

tri·meth·o·prim \trī-'me-thə-,prim\ *noun* [*tri-* + *meth-* + *-prim* (by shortening & alteration from *pyrimidine*)] (1964)
: a synthetic antibacterial and antimalarial drug $C_{14}H_{18}N_4O_3$ often used with a sulfa drug

tri·met·ro·gon \trī-'me-trə-,gän\ *noun* [*tri-* + Greek *metron* measure + English *-gon*] (1943)
: a system of aerial mapping involving the use of sets of one vertical and two oblique aerial photographs taken simultaneously over the area being mapped

trim·mer \'tri-mər\ *noun* (1555)
1 a (1) **:** one that trims articles (2) **:** one that stows coal or freight on a ship so as to distribute the weight properly **b :** an instrument or machine with which trimming is done **c :** a circuit element (as a capacitor) used to tune a circuit to a desired frequency
2 : a beam that receives the end of a header in floor framing
3 : a person who modifies a policy, position, or opinion especially out of expediency

trim·ming *noun* (circa 1518)
1 : DEFEAT, BEATING
2 : the act of one who trims
3 a : a decorative accessory or additional item ⟨*trimmings* for a hat⟩ **b :** an additional garnishing ⟨turkey and all the *trimmings*⟩

tri·month·ly \(,)trī-'mən(t)th-lē\ *adjective* (1856)
: occurring every three months

tri·mor·phic \(,)trī-'mȯr-fik\ *adjective* [Greek *trimorphos* having three forms, from *tri-* + *-morphos* -morphous] (1866)
: occurring in or having three distinct forms

tri·mo·tor \'trī-,mō-tər, -'mō-\ *noun* (1923)
: an airplane powered by three engines

trim size *noun* (circa 1929)
: the actual size (as of a book page) after excess material required in production has been cut off

Tri·mur·ti \tri-'mu̇r-tē\ *noun* [Sanskrit *-trimūrti,* from *trimūrti* having three forms, from *tri-* tri- + *mūrti* body, form] (1810)
: the great triad of Hindu gods comprising Brahma, Vishnu, and Siva

tri·nal \'trī-nᵊl\ *adjective* [Late Latin *trinalis,* from Latin *trini* three each] (1590)
: THREEFOLD

[1]**trine** \'trīn\ *adjective* [Middle English, from Middle French *trin,* from Latin *trinus,* from *trini* three each; akin to Latin *tres* three — more at THREE] (14th century)
1 : THREEFOLD, TRIPLE
2 : of, relating to, or being the favorable astrological aspect of two celestial bodies 120 degrees apart

[2]**trine** *noun* (1552)
1 : a group of three **:** TRIAD
2 : the trine astrological aspect of two celestial bodies

trine immersion *noun* (1637)
: the practice of immersing a candidate for baptism three times in the names of the members of the Trinity

trin·i·tar·i·an \,tri-nə-'ter-ē-ən\ *adjective* (1628)
1 *capitalized* **:** of or relating to the Trinity, the doctrine of the Trinity, or adherents to that doctrine
2 : having three parts or aspects **:** THREEFOLD

Trinitarian *noun* (1628)
1 : a member of a religious teaching and nursing order for men founded in France in 1198 by John of Matha and Philip of Valois
2 : one who subscribes to the doctrine of the Trinity
— **Trin·i·tar·i·an·ism** \-ē-ə-,ni-zəm\ *noun*

tri·ni·tro·tol·u·ene \,trī-,nī-trō-'täl-yə-,wēn\ *noun* [International Scientific Vocabulary] (circa 1900)
: TNT

Trin·i·ty \'tri-nə-tē\ *noun* [Middle English *trinite,* from Old French *trinité,* from Late Latin *trinitat-, trinitas* state of being threefold, from Latin *trinus* threefold] (13th century)
1 : the unity of Father, Son, and Holy Spirit as three persons in one Godhead according to Christian dogma
2 *not capitalized* **:** a group of three closely related persons or things
3 : the Sunday after Whitsunday observed as a feast in honor of the Trinity

Trin·i·ty·tide \-,tīd\ *noun* (1511)
: the season of the church year between Trinity Sunday and Advent

[1]**trin·ket** \'triŋ-kət\ *noun* [origin unknown] (circa 1533)
1 : a small ornament (as a jewel or ring)
2 : a small article of equipment
3 : a thing of little value **:** TRIFLE

[2]**trinket** *intransitive verb* [perhaps from [1]*trinket*] (1646)
: to deal clandestinely **:** INTRIGUE
— **trin·ket·er** *noun*

trin·ket·ry \-kə-trē\ *noun* (1810)
: small items of personal ornament

trin·oc·u·lar \(,)trī-'nä-kyə-lər\ *adjective* [*tri-* + *binocular*] (1960)
: relating to or being a binocular microscope equipped with a lens for photographic recording during direct visual observation

[1]**tri·no·mi·al** \trī-'nō-mē-əl\ *noun* [*tri-* + *-nomial* (as in *binomial*)] (1674)
1 : a polynomial of three terms
2 : a biological taxonomic name of three terms of which the first designates the genus, the

second the species, and the third the subspecies or variety

²trinomial *adjective* (circa 1704)

1 : consisting of three mathematical terms

2 : of, relating to, or being a biological trinomial

tri·nu·cle·o·tide \(,)trī-ˈnü-klē-ə-ˌtīd, -ˈnyü-\ *noun* (1918)

: a nucleotide consisting of three mononucleotides in combination **:** CODON

trio \ˈtrē-(ˌ)ō\ *noun, plural* **tri·os** [French, from Italian, from *tri-* (from Latin)] (circa 1724)

1 a : a musical composition for three voice parts or three instruments **b :** the secondary or episodic division of a minuet or scherzo, a march, or of various dance forms

2 : the performers of a musical or dance trio

3 : a group or set of three

tri·ode \ˈtrī-ˌōd\ *noun* (1922)

: an electron tube with an anode, a cathode, and a controlling grid

tri·ol \ˈtrī-ˌȯl, -ˌōl\ *noun* (1936)

: a chemical compound containing three hydroxyl groups

tri·o·let \ˈtrē-ə-lət, ˈtrī-\ *noun* [French, from Middle French, literally, clover leaf, clover, ultimately from Greek *triphyllon,* from *tri-* tri- + *phyllon* leaf — more at BLADE] (1651)

: a poem or stanza of eight lines in which the first line is repeated as the fourth and seventh and the second line as the eighth with a rhyme scheme of *ABaAabAB*

tri·ose \ˈtrī-ˌōs, -ˌōz\ *noun* [International Scientific Vocabulary] (1894)

: either of two simple sugars $C_3H_6O_3$ containing three carbon atoms

tri·ox·ide \(ˌ)trī-ˈäk-ˌsīd\ *noun* [International Scientific Vocabulary] (circa 1868)

: an oxide containing three atoms of oxygen

¹trip \ˈtrip\ *verb* **tripped; trip·ping** [Middle English *trippen,* from Middle French *triper,* of Germanic origin; akin to Old English *treppan* to tread — more at TRAP] (14th century)

intransitive verb

1 a : to dance, skip, or caper with light quick steps **b :** to walk with light quick steps

2 : to catch the foot against something so as to stumble

3 : to make a mistake or false step (as in morality or accuracy)

4 : to stumble in articulation when speaking

5 : to make a journey

6 : to run past the pallet of an escapement without previously locking — used of a tooth of the escapement wheel of a watch

7 a : to actuate a mechanism **b :** to become operative

8 : to get high on a psychedelic drug (as LSD) **:** TURN ON — often used with *out*

transitive verb

1 a : to cause to stumble — often used with *up* **b :** to cause to fail **:** OBSTRUCT — often used with *up*

2 : to detect in a misstep, fault, or blunder; *also* **:** EXPOSE — usually used with *up*

3 *archaic* **:** to perform (as a dance) lightly or nimbly

4 : to raise (an anchor) from the bottom so as to hang free

5 a : to pull (a yard) into a perpendicular position for lowering **b :** to hoist (a topmast) far enough to enable the fid to be withdrawn preparatory to housing or lowering

6 : to release or operate (a mechanism) especially by releasing a catch or detent

— trip the light fantastic : DANCE

²trip *noun* (15th century)

1 : a stroke or catch by which a wrestler is made to lose footing

2 a : VOYAGE, JOURNEY **b :** a single round or tour on a business errand

3 : ERROR, MISSTEP

4 : a quick light step

5 : a faltering step caused by stumbling

6 a : the action of tripping mechanically **b** (1) **:** a device for tripping a mechanism (as a catch or detent) (2) **:** TUP 2

7 : an intense visionary experience undergone by a person who has taken a psychedelic drug (as LSD)

8 : absorption in or obsession with an interest, attitude, or state of mind ⟨a guilt *trip*⟩ ⟨on a nostalgia *trip*⟩

9 : SCENE, LIFESTYLE

tri·pack \ˈtrī-ˌpak\ *noun* (1911)

: a combination of three superposed films or emulsions each sensitive to a different primary color for simultaneous exposure in one camera

tri·par·tite \(ˌ)trī-ˈpär-ˌtīt\ *adjective* [Middle English, from Latin *tripartitus,* from *tri-* + *partitus* divided — more at PARTITE] (15th century)

1 : divided into or composed of three parts

2 : having three corresponding parts or copies

3 : made between or involving three parties ⟨a *tripartite* treaty⟩

tripe \ˈtrīp\ *noun* [Middle English, from Middle French] (14th century)

1 : stomach tissue of a ruminant and especially of the ox used as food

2 : something poor, worthless, or offensive

¹trip–ham·mer \ˈtrip-ˌha-mər\ *noun* (1781)

: a massive power hammer having a head that is tripped and allowed to fall by cam or lever action

²trip–hammer *adjective* (circa 1864)

: suggesting a trip-hammer in loud pounding or persistent action

tri·phe·nyl·meth·ane \ˌtrī-ˌfe-nᵊl-ˈme-ˌthān, -ˌfē-\ *noun* [International Scientific Vocabulary] (circa 1885)

: a crystalline hydrocarbon $CH(C_6H_5)_3$ that is the parent compound of many dyes

tri·phos·phate \(ˌ)trī-ˈfäs-ˌfāt\ *noun* (circa 1826)

: a salt or acid that contains three phosphate groups — compare ATP, GTP

tri·phos·pho·pyr·i·dine nucleotide \ˌtrī-ˌfäs-fō-ˈpir-ə-ˌdēn-\ *noun* (1937)

: NADP

triph·thong \ˈtrif-ˌthȯŋ, ˈtrip-\ *noun* [*tri-* + *-phthong* (as in *diphthong*)] (circa 1599)

1 : a phonological unit consisting of three successive vocalic sounds in one syllable

2 : TRIGRAPH

— triph·thon·gal \trif-ˈthȯŋ-gəl, trip-, -əl\ *adjective*

tri·pin·nate \(ˌ)trī-ˈpi-ˌnāt\ *adjective* (circa 1760)

: bipinnate with each division pinnate

— tri·pin·nate·ly *adverb*

tri·plane \ˈtrī-ˌplān\ *noun* (1909)

: an airplane with three main supporting surfaces superposed

¹tri·ple \ˈtri-pəl\ *verb* **tri·pled; tri·pling** \-p(ə-)liŋ\ [Middle English, from Late Latin *triplare,* from Latin *triplus,* adjective] (14th century)

transitive verb

1 : to make three times as great or as many

2 a : to score (a base runner) by a triple **b :** to bring about the scoring of (a run) by a triple

intransitive verb

1 : to become three times as great or as numerous

2 : to make a triple in baseball

²triple *noun* [Middle English, from Latin *triplus,* adjective] (15th century)

1 a : a triple sum, quantity, or number **b :** a combination, group, or series of three

2 : a base hit that allows the batter to reach third base safely

3 : TRIFECTA

³triple *adjective* [Middle French or Latin; Middle French, from Latin *triplus,* from *tri-* + *-plus* multiplied by — more at -FOLD] (1550)

1 : being three times as great or as many

2 : having three units or members

3 : having a threefold relation or character ⟨worked as a double or even *triple* agent —*Time*⟩

4 : three times repeated **:** TREBLE

5 : marked by three beats per musical measure ⟨*triple* meter⟩

6 a : having units of three components ⟨*triple* feet⟩ **b** *of rhyme* **:** involving correspondence of three syllables (as in *unfortunate-importunate*)

triple bond *noun* (1889)

: a chemical bond in which three pairs of electrons are shared by two atoms in a molecule — compare DOUBLE BOND, SINGLE BOND

triple counterpoint *noun* (circa 1869)

: three-part musical counterpoint so written that any part may be transposed above or below any other

Triple Crown *noun* (circa 1897)

1 : an unofficial title in horse racing representing the championship achieved by a horse that wins the three classic races for a designated category

2 : the unofficial title signifying the achievement of a baseball player who at the end of a season leads the league in batting average, home runs, and runs batted in

tri·ple–deck·er \ˌtri-pəl-ˈde-kər\ *noun, often attributive* (1938)

: something having three basic components or levels: as **a :** TRILOGY **b :** a sandwich consisting of three pieces of bread and two layers of filling **c :** a 3-story dwelling with an apartment on each floor

tri·ple–head·er \-ˈhe-dər\ *noun* (circa 1949)

: a program consisting of three consecutive games, contests, or events

triple jump *noun* (1964)

: a jump for distance in track-and-field athletics usually from a running start and combining a hop, a stride, and a jump in succession

— triple jumper *noun*

triple play *noun* (1869)

: a play in baseball by which three players are put out

triple point *noun* (1872)

: the condition of temperature and pressure under which the gaseous, liquid, and solid phases of a substance can exist in equilibrium

triple sec \-ˈsek\ *noun* [from *Triple Sec,* a trademark] (1943)

: a colorless orange-flavored liqueur

tri·ple–space \ˌtri-pəl-ˈspās\ (circa 1939)

transitive verb

: to type (text) leaving two blank lines between lines of copy

intransitive verb

: to type on every third line

trip·let \ˈtri-plət\ *noun* [²*triple*] (1656)

1 : a unit of three lines of verse

2 a : a combination, set, or group of three **b :** a group of three elementary particles (as positive, negative, and neutral pions) with different charge states but otherwise similar properties **c :** an atom or molecule with an even number of electrons that have a net magnetic moment **d :** CODON

3 : one of three children or offspring born at one birth

4 : a group of three musical notes or tones performed in the time of two of the same value

tri·ple·tail \ˈtri-pəl-ˌtāl\ *noun* (circa 1803)

: a large marine bony fish (*Lobotes surinamensis* of the family Lobotidae) of warm and tropical waters that has long dorsal and anal fins which extend backward and together with the caudal fin appear like a 3-lobed tail

triple threat *noun* (1924)

1 : a football player adept at running, kicking, and passing

2 : a person adept in three different fields of activity
— **triple–threat** *adjective*

tri·ple–tongue \ˌtri-pəl-ˈtəŋ\ *intransitive verb* (1879)
: to articulate the notes of triplets in fast tempo on a wind instrument by using the tongue positions especially for *t, k, t* for the notes of each successive triplet

¹tri·plex \ˈtri-ˌpleks, ˈtrī-\ *noun* (1601)
: something (as an apartment) that is triplex

²triplex *adjective* [Latin, from *tri-* + *-plex* -fold — more at -FOLD] (1655)
1 : THREEFOLD, TRIPLE ⟨*triplex* windows⟩
2 : having three apartments, floors, or sections ⟨*triplex* buildings⟩ ⟨*triplex* apartments⟩ ⟨a *triplex* theater⟩

¹trip·li·cate \ˈtri-pli-kət\ *adjective* [Middle English, from Latin *triplicatus,* past participle of *triplicare* to triple, from *triplic-, triplex* threefold] (15th century)
: consisting of or existing in three corresponding or identical parts or examples ⟨*triplicate* invoices⟩

²trip·li·cate \-plə-ˌkāt\ *transitive verb* **-cat·ed; -cat·ing** (circa 1623)
1 : to make triple or threefold
2 : to prepare in triplicate
— **trip·li·ca·tion** \ˌtri-plə-ˈkā-shən\ *noun*

³trip·li·cate \-pli-kət\ *noun* (1810)
: three copies all alike — used with *in* ⟨typed in *triplicate*⟩

tri·plic·i·ty \tri-ˈpli-sə-tē, trī-\ *noun, plural* **-ties** [Middle English *triplicite,* from Late Latin *triplicitas* condition of being threefold, from Latin *triplic-, triplex*] (14th century)
1 : one of the groups of three signs each distant 120 degrees from the other two into which the signs of the zodiac are divided — called also *trigon*
2 : the quality or state of being triple or threefold

trip·lo·blas·tic \ˌtri-plō-ˈblas-tik\ *adjective* [Latin *triplus* + English *-o-* + *-blastic*] (circa 1888)
: having three primary germ layers

trip·loid \ˈtri-ˌplȯid\ *adjective* [International Scientific Vocabulary, from Latin *triplus* triple] (1911)
: having or being a chromosome number three times the monoploid number
— **trip·loid** *noun*
— **trip·loi·dy** \-ˌplȯi-dē\ *noun*

tri·ply \ˈtri-p(ə-)lē\ *adverb* (1660)
: in a triple degree, amount, or manner

tri·pod \ˈtrī-ˌpäd\ *noun* [Latin *tripod-, tripus,* from Greek *tripod-, tripous,* from *tripod-, tripous,* adjective, three-footed, from *tri-* + *pod-, pous* foot — more at FOOT] (1603)
1 : a vessel (as a cauldron) resting on three legs
2 : a stool, table, or altar with three legs
3 : a three-legged stand (as for a camera)
— **tripod** *or* **tri·po·dal** \ˈtri-pə-dᵊl, ˈtrī-ˌpä-\ *adjective*

trip·o·li \ˈtri-pə-lē\ *noun* [French, from *Tripoli,* region of Africa] (circa 1601)
1 : an earth consisting of very friable soft schistose deposits of silica and including diatomite and kieselguhr
2 : an earth consisting of friable dustlike silica not of diatomaceous origin

tri·pos \ˈtrī-ˌpäs\ *noun* [modification of Latin *tripus*] (1589)
1 *archaic* **:** TRIPOD
2 [from the three-legged stool occupied by a participant in a disputation at the degree ceremonies] **:** a final honors examination at Cambridge university originally in mathematics

trip·per \ˈtri-pər\ *noun* (1813)
1 *chiefly British* **:** one that takes a trip **:** TOURIST
2 : a tripping device (as for operating a railroad signal)

trip·ping·ly \ˈtri-piŋ-lē\ *adverb* (1590)
: in a nimble or lively manner ⟨the new name . . . may not roll *trippingly* off the tongue —Paul B. Carroll⟩

trip·py \ˈtri-pē\ *adjective* (1968)
: of, relating to, or suggesting a trip on psychedelic drugs or the culture associated with such drugs

trip·tych \ˈtrip-(ˌ)tik\ *noun* [Greek *triptychos* having three folds, from *tri-* + *ptychē* fold] (1731)
1 : an ancient Roman writing tablet with three waxed leaves hinged together
2 a : a picture (as an altarpiece) or carving in three panels side by side **b :** something composed or presented in three parts or sections; *especially* **:** TRILOGY

trip wire *noun* (1916)
1 : a low-placed concealed wire used especially in warfare to trip an enemy or trespasser and usually to trigger an alarm or explosive device when moved
2 : something (as a small military force) intended to function like a trip wire (as to set a larger military force in motion)

tri·que·trous \trī-ˈkwē-trəs, -ˈkwe-\ *adjective* [Latin *triquetrus* three-cornered, from *tri-* + *-quetrus;* probably akin to Old High German *waz* sharp — more at WHET] (circa 1879)
: having three acute angles ⟨*triquetrous* stems⟩

tri·ra·di·ate \(ˌ)trī-ˈrā-dē-ət, -dē-ˌāt\ *adjective* (1846)
: having three rays or radiating branches ⟨a *triradiate* sponge spicule⟩

tri·reme \ˈtrī-ˌrēm\ *noun* [Latin *triremis,* from *tri-* + *remus* oar — more at ROW] (1601)
: an ancient galley having three banks of oars

tri·sac·cha·ride \(ˌ)trī-ˈsa-kə-ˌrīd\ *noun* [International Scientific Vocabulary] (circa 1899)
: a sugar that yields on complete hydrolysis three monosaccharide molecules

tri·sect \ˈtrī-ˌsekt, trī-ˈ\ *transitive verb* [*tri-* + inter*sect*] (1695)
: to divide into three usually equal parts
— **tri·sec·tion** \ˈtrī-ˌsek-shən, trī-ˈ\ *noun*
— **tri·sec·tor** \ˈtrī-ˌsek-tər, trī-ˈ\ *noun*

tri·shaw \ˈtrī-ˌshȯ\ *noun* [*tri-* + rick*shaw*] (1946)
: PEDICAB

tris·kai·deka·pho·bia \ˌtris-ˌkī-ˌde-kə-ˈfō-bē-ə, ˌtris-kə-\ *noun* [New Latin, from Greek *treiskaideka* thirteen (from *treis* three + *kai* and + *deka* ten) + New Latin *phobia* — more at THREE, TEN] (circa 1911)
: fear of the number 13

tri·skel·i·on \trī-ˈske-lē-ən, tri-\ *or* **tri·skele** \ˈtrī-ˌskēl, ˈtri-\ *noun* [*triskelion* from New Latin, from Greek *triskelēs* three-legged, from *tri-* + *skelos* leg; *triskele* from Greek *triskelēs* — more at ISOSCELES] (1857)
: a figure composed of three usually curved or bent branches radiating from a center

triskelion

tris·mus \ˈtriz-məs\ *noun* [New Latin, from Greek *trismos* gnashing (of teeth), from *trizein* to squeak, gnash; akin to Latin *stridēre* to creak — more at STRIDENT] (circa 1693)
: spasm of the muscles of mastication **:** LOCKJAW

tris·oc·ta·he·dron \ˌtri-ˌsäk-tə-ˈhē-drən\ *noun* [Greek *tris* thrice + English *octahedron* — more at TER-] (circa 1847)
: a solid (as a crystal) having 24 congruent faces meeting on the edges of a regular octahedron

tri·so·di·um phosphate \ˌtrī-ˈsō-dē-əm-\ *noun* (1923)
: a crystalline compound Na_3PO_4 that is used especially in cleaning compositions

tri·so·my \ˈtrī-ˌsō-mē\ *noun, plural* **-mies** [*tri-* + ³*-some* + ²*-y*] (1930)
: the condition (as in Down's syndrome) of having one or a few chromosomes triploid in an otherwise diploid set
— **tri·so·mic** \(ˌ)trī-ˈsō-mik\ *adjective or noun*

Tris·tan \ˈtris-tən, -ˌtän, -ˌtan\ *noun*
: TRISTRAM

tri·state \ˈtrī-ˈstāt, -ˌstāt\ *adjective* (1900)
: of, relating to, or consisting of three adjoining states ⟨the *tristate* area⟩

triste \ˈtrēst\ *adjective* [French, from Latin *tristis*] (1756)
: SAD, MOURNFUL; *also* **:** WISTFUL

tri·stea·rin \(ˌ)trī-ˈstē-ə-rən, -ˈsti(-ə)r-ən\ *noun* [International Scientific Vocabulary] (circa 1856)
: the crystallizable triglyceride $C_{57}H_{110}O_6$ of stearic acid that is found especially in hard fats

tris·te·za \tri-ˈstā-zə\ *noun* [Portuguese, literally, sadness, from Latin *tristitia,* from *tristis* sad] (circa 1902)
: a highly infectious virus disease of citrus trees grafted on sour orange rootstocks that is characterized by rotting of the rootlets and eventually causes the death of the trees

trist·ful \ˈtrist-fəl\ *adjective* [Middle English *trist* sad, from Middle French *triste*] (15th century)
: SAD, MELANCHOLY
— **trist·ful·ly** \-fə-lē\ *adverb*
— **trist·ful·ness** *noun*

tri·stim·u·lus \(ˌ)trī-ˈstim-yə-ləs\ *adjective* (1933)
: of or relating to values giving the amounts of the three colored lights red, green, and blue that when combined additively produce a match for the color being considered

Tris·tram \ˈtris-trəm\ *noun* [Middle English *Tristrem,* from Anglo-French *Tristan,* from Old Welsh *Trystan*]
: the lover of Isolde of Ireland and husband of Isolde of Brittany in medieval legend

tri·sub·sti·tut·ed \ˈtrī-ˈsəb-stə-ˌtü-təd, -ˌtyü-\ *adjective* (circa 1899)
: having three substituent atoms or groups in the molecule

tri·sul·fide \-ˈsəl-ˌfīd\ *noun* (1866)
: a compound of an element or radical with three atoms of sulfur

tri·syl·lab·ic \ˌtrī-sə-ˈla-bik\ *adjective* [Latin *trisyllabus,* from Greek *trisyllabos,* from *tri-* + *syllabē* syllable] (1637)
: having three syllables ⟨a *trisyllabic* word⟩

tri·syl·la·ble \ˈtrī-ˌsi-lə-bəl, (ˌ)trī-ˈ\ *noun* (1589)
: a word of three syllables

trite \ˈtrīt\ *adjective* **trit·er; trit·est** [Latin *tritus,* from past participle of *terere* to rub, wear away — more at THROW] (1548)
: hackneyed or boring from much use **:** not fresh or original ☆
— **trite·ly** *adverb*
— **trite·ness** *noun*

tri·the·ism \ˈtrī-thē-ˌi-zəm\ *noun* (1678)

☆ SYNONYMS
Trite, hackneyed, stereotyped, threadbare mean lacking the freshness that evokes attention or interest. TRITE applies to a once effective phrase or idea spoiled from long familiarity ⟨"you win some, you lose some" is a *trite* expression⟩. HACKNEYED stresses being worn out by overuse so as to become dull and meaningless ⟨all of the metaphors and images in the poem are *hackneyed*⟩. STEREOTYPED implies falling invariably into the same pattern or form ⟨views of minorities that are *stereotyped* and out-of-date⟩. THREADBARE applies to what has been used until its possibilities of interest have been totally exhausted ⟨a mystery novel with a *threadbare* plot⟩.

: the doctrine that the Father, Son, and Holy Spirit are three distinct Gods
— **tri·the·ist** \-(ˌ)thē-ist\ *noun or adjective*
— **tri·the·is·tic** \ˌtrī-thē-'is-tik\ *or* **tri·the·is·ti·cal** \-'is-ti-kəl\ *adjective*

tri·thing \'trī-thiŋ\ *noun* [Middle English, alteration of (assumed) Old English *thrithing, thriding*] (13th century)
archaic **:** [3]RIDING 1

tri·ti·at·ed \'tri-tē-ˌā-təd, 'tri-shē-\ *adjective* (1953)
: containing and especially labeled with tritium

trit·i·ca·le \ˌtri-tə-'kā-lē\ *noun* [New Latin, blend of *Triticum,* genus of wheat, and *Secale,* genus of rye] (1952)
: an amphidiploid hybrid between wheat and rye that has a high yield and rich protein content

tri·ti·um \'tri-tē-əm, 'tri-shē-\ *noun* [New Latin, from Greek *tritos* third — more at THIRD] (1933)
: a radioactive isotope of hydrogen with atoms of three times the mass of ordinary light hydrogen atoms

trit·o·ma \'tri-tə-mə\ *noun* [New Latin, genus name, from Greek *tritomos* thrice cut, from *tri-* + *temnein* to cut; from their trimerous flowers — more at TOME] (1804)
: any of a genus (*Khiphofia*) of African herbs of the lily family that are often grown for their spikes of showy red or yellow flowers

[1]**tri·ton** \'trī-tᵊn\ *noun* [Latin, from Greek *Tritōn*]
1 *capitalized* **:** a son of Poseidon described as a demigod of the sea with the lower part of his body like that of a fish
2 [New Latin, genus name, from Latin *Triton*] **:** any of various large marine gastropod mollusks (especially family Ranellidae) with a heavy elongated conical shell; *also* **:** the shell

[2]**tri·ton** \'trī-ˌtän\ *noun* [*tri*tium + [2]*-on*] (1934)
: the nucleus of tritium

tri·tone \'trī-ˌtōn\ *noun* [Greek *tritonon,* from *tri-* + *tonos* tone] (1609)
: a musical interval of three whole steps

[1]**trit·u·rate** \'tri-chə-ˌrāt\ *transitive verb* **-rat·ed; -rat·ing** [Late Latin *trituratus,* past participle of *triturare* to thresh, from Latin *tritura* act of rubbing, threshing, from *tritus,* past participle — more at TRITE] (circa 1755)
1 : CRUSH, GRIND
2 : to pulverize and comminute thoroughly by rubbing or grinding
— **trit·u·ra·ble** \'tri-chə-rə-bəl\ *adjective*
— **trit·u·ra·tor** \-ˌrā-tər\ *noun*

[2]**trit·u·rate** \-rət\ *noun* (circa 1891)
: a triturated substance **:** TRITURATION 2

trit·u·ra·tion \ˌtri-chə-'rā-shən\ *noun* (1646)
1 : the act or process of triturating **:** the state of being triturated **:** COMMINUTION
2 : a triturated medicinal powder made by triturating a substance with a diluent

[1]**tri·umph** \'trī-əm(p)f\ *noun, plural* **tri·umphs** \-əm(p)fs, -əm(p)s\ [Middle English *triumphe,* from Middle French, from Latin *triumphus*] (14th century)
1 : a ceremony attending the entering of Rome by a general who had won a decisive victory over a foreign enemy — compare OVATION 1
2 : the joy or exultation of victory or success
3 a : a victory or conquest by or as if by military force **b :** a notable success
— **tri·um·phal** \trī-'əm(p)-fəl\ *adjective*

[2]**triumph** *intransitive verb* (1508)
1 : to obtain victory **:** PREVAIL
2 a : to receive the honor of a triumph **b :** to celebrate victory or success boastfully or exultingly

tri·um·phal·ism \trī-'əm(p)-fə-ˌli-zəm\ *noun* (1964)
: the doctrine, attitude, or belief that one religious creed is superior to all others
— **tri·um·phal·ist** \-fə-list\ *noun or adjective*

tri·um·phant \trī-'əm(p)-fənt\ *adjective* (15th century)
1 : VICTORIOUS, CONQUERING
2 *archaic* **:** of or relating to a triumph
3 : rejoicing for or celebrating victory
4 : notably successful
— **tri·um·phant·ly** *adverb*

tri·um·vir \trī-'əm-vər\ *noun, plural* **-virs** *also* **-vi·ri** \-və-ˌrī, -ˌrē\ [Latin, back-formation from *triumviri,* plural, commission of three men, from *trium virum* of three men] (circa 1580)
: one of a commission or ruling body of three

tri·um·vi·rate \-və-rət\ *noun* (1584)
1 : a body of triumvirs
2 : the office or government of triumvirs
3 : a group or association of three

[1]**tri·une** \'trī-ˌyün\ *noun, often capitalized* [Latin *tri-* + *unus* one — more at ONE] (1605)
: TRINITY 1

[2]**triune** *adjective* (1635)
: three in one: **a** *often capitalized* **:** of or relating to the Trinity ⟨the *triune* God⟩ **b :** consisting of three parts, members, or aspects

tri·va·lent \(ˌ)trī-'vā-lənt\ *adjective* [International Scientific Vocabulary] (1868)
1 : having a chemical valence of three
2 : reacting immunologically with three different combining sites (as of antigens or antibodies) ⟨*trivalent* oral polio vaccine⟩

triv·et \'tri-vət\ *noun* [Middle English *trevet,* from Old English *trefet,* probably modification of Late Latin *triped-, tripes,* from Latin, three-footed, from *tri-* + *ped-, pes* foot — more at FOOT] (before 12th century)
1 : a three-legged stand **:** TRIPOD
2 : a usually metal stand with short feet for use under a hot dish at table

triv·ia \'tri-vē-ə\ *noun plural but singular or plural in construction* [(assumed) New Latin, back-formation from Latin *trivialis*] (1920)
: unimportant matters **:** trivial facts or details; *also, singular in construction* **:** a quizzing game involving obscure facts

triv·i·al \'tri-vē-əl\ *adjective* [Latin *trivialis* found everywhere, commonplace, from *trivium* crossroads, from *tri-* + *via* way — more at WAY] (1589)
1 : COMMONPLACE, ORDINARY
2 a : of little worth or importance **b :** relating to or being the mathematically simplest case; *specifically* **:** characterized by having all variables equal to zero ⟨a *trivial* solution to a linear equation⟩
3 : SPECIFIC 4 ◆
— **triv·i·al·ist** \-ə-list\ *noun*
— **triv·i·al·ly** \-ə-lē\ *adverb*

triv·i·al·ise *British variant of* TRIVIALIZE

triv·i·al·i·ty \ˌtri-vē-'a-lə-tē\ *noun, plural* **-ties** (1598)
1 : the quality or state of being trivial
2 : something trivial **:** TRIFLE

triv·i·al·ize \'tri-vē-ə-ˌlīz\ *transitive verb* **-ized; -iz·ing** (1846)
: to make trivial **:** reduce to triviality
— **triv·i·al·i·za·tion** \ˌtri-vē-ə-lə-'zā-shən\ *noun*

trivial name *noun* (1759)
1 : SPECIFIC EPITHET
2 : a common or vernacular name of an organism or chemical

triv·i·um \'tri-vē-əm\ *noun, plural* **triv·ia** \-vē-ə\ [Medieval Latin, from Latin, meeting of three ways, crossroads] (1804)
: a group of studies consisting of grammar, rhetoric, and logic and forming the lower division of the seven liberal arts in medieval universities — compare QUADRIVIUM

[1]**tri·week·ly** \(ˌ)trī-'wē-klē\ *adjective* (1832)
1 : occurring or appearing three times a week
2 : occurring or appearing every three weeks
— **triweekly** *adverb*

[2]**triweekly** *noun, plural* **-lies** (1851)
: a triweekly publication

-trix *noun suffix, plural* **-trices** *or* **-trixes** [Middle English, from Latin, feminine of *-tor,* suffix denoting an agent]
1 : female that does or is associated with a (specified) thing ⟨avia*trix*⟩
2 : geometric line, point, or surface ⟨genera*trix*⟩

tRNA \ˌtē-ˌär-ˌen-'ā, 'tē-ˌär-ˌen-ˌā\ *noun* (1962)
: TRANSFER RNA

tro·car *also* **tro·char** \'trō-ˌkär\ *noun* [French *trocart,* from *trois* three (from Latin *tres*) + *carre* side of a sword blade, from *carrer* to make square, from Latin *quadrare* — more at THREE, QUADRATE] (circa 1706)
: a sharp-pointed surgical instrument fitted with a cannula and used especially to insert the cannula into a body cavity as a drainage outlet

tro·cha·ic \trō-'kā-ik\ *adjective* [Middle French *trochaïque,* from Latin *trochaicus,* from Greek *trochaikos,* from *trochaios* trochee] (1589)
: of, relating to, or consisting of trochees
— **trochaic** *noun*

tro·chan·ter \trō-'kan-tər\ *noun* [Greek *trochantēr;* akin to Greek *trechein* to run] (1615)
1 : a rough prominence at the upper part of the femur of many vertebrates
2 : the second segment counting from the base of the leg of an insect
— **tro·chan·ter·al** \-tə-rəl\ *adjective*
— **tro·chan·ter·ic** \ˌtrō-kən-'ter-ik, -ˌkan-\ *adjective*

tro·che \'trō-kē, *British usually* 'trōsh\ *noun* [alteration of earlier *trochisk,* from Late Latin *trochiscus,* from Greek *trochiskos,* from diminutive of *trochos* wheel] (1597)
: a usually circular medicinal tablet or lozenge for slow dissolution in the mouth; *especially* **:** one used as a demulcent

tro·chee \'trō-(ˌ)kē\ *noun* [probably from Middle French *trochée,* from Latin *trochaeus,* from Greek *trochaios,* from *trochaios* running, from *trochē* run, course, from *trechein* to run; akin to Greek *trochos* wheel, Old Irish *droch*] (1589)
: a metrical foot consisting of one long syllable followed by one short syllable or of one stressed syllable followed by one unstressed syllable (as in *apple*)

◇ WORD HISTORY

trivial The ancient Romans used the adjective *trivialis* to refer to what is commonplace or ordinary. Literally meaning "pertaining to a crossroads," *trivialis* is a derivative of *trivium,* which was compounded from *tri-* "three" and *via* "way, road." Thus *trivium* might be rendered literally as "a place where three roads meet." The extended sense of *trivialis* rested on the belief that things found at such a public place as a crossroads, where all the world may pass by, are generally common things. The notion that people often stop where roads meet to pass the time of day with inconsequential talk may also have influenced the development of this sense. This is the source for the modern senses of *trivial* such as "commonplace" or "of little importance." It was not until the 1920s that the word *trivia* began being used for "unimportant matters." Its form is just like that of the plural of Latin *trivium,* but its meaning is dependent on Modern English senses of *trivial.*

troch·lea \'trä-klē-ə\ *noun* [New Latin, from Latin, block of pulleys, from Greek *trochileia,* from *trochilos* sheave, from *trochos* wheel] (circa 1693)
: an anatomical structure that is held to resemble a pulley; *especially* **:** the articular surface on the medial condyle of the humerus that articulates with the ulna

troch·le·ar \-ər\ *adjective* (circa 1681)
1 : of, relating to, or being a trochlea
2 : of, relating to, or being a trochlear nerve

trochlear nerve *noun* (circa 1890)
: either of the 4th pair of cranial nerves that supply some of the eye muscles with motor fibers — called also *trochlear*

tro·choid \'trō-ˌkȯid, 'trä-ˌkȯid\ *noun* [Greek *trochoeidēs* like a wheel, from *trochos* wheel] (circa 1704)
: the curve generated by a point on the radius of a circle or the radius extended as the circle rolls on a fixed straight line
— tro·choi·dal \trō-'kȯi-dᵊl, trä-\ *adjective*

tro·cho·phore \'trä-kə-ˌfōr, -ˌfȯr\ *noun* [ultimately from Greek *trochos* wheel + *pherein* to carry — more at BEAR] (1892)
: a free-swimming ciliate larva occurring in several invertebrate groups (as the polychaete worms and mollusks)

trod *past and past participle of* TREAD

trodden *past participle of* TREAD

trof·fer \'trä-fər, 'trȯ-\ *noun* [blend of *trough* and *coffer*] (1942)
: an inverted trough serving as a support and reflector usually for a fluorescent lighting unit

trog·lo·dyte \'trä-glə-ˌdīt\ *noun* [Latin *troglodytae,* plural, from Greek *trōglodytai,* from *trōglē* hole, cave (akin to Greek *trōgein* to gnaw, Armenian *aracem* I lead to pasture, graze) + *dyein* to enter] (1558)
1 : a member of a primitive people dwelling in caves
2 : a person resembling a troglodyte (as in reclusive habits or outmoded or reactionary attitudes)
— trog·lo·dyt·ic \ˌträ-glə-'di-tik\ *adjective*

tro·gon \'trō-ˌgän\ *noun* [New Latin, genus name, from Greek *trōgōn,* present participle of *trōgein* to gnaw] (1792)
: any of numerous nonpasserine tropical birds (family Trogonidae) with brilliant often iridescent plumage

troi·ka \'trȯi-kə\ *noun* [Russian *troĭka,* from *troe* three; akin to Old English *thrīe* three] (1842)
1 : a Russian vehicle drawn by three horses abreast; *also* **:** a team for such a vehicle
2 : a group of three; *especially* **:** an administrative or ruling body of three

troi·lite \'trō-ə-ˌlīt, 'trȯi-ˌlīt\ *noun* [German *Troilit,* from Domenico *Troili,* 18th century Italian scientist + German *-it* -ite] (circa 1868)
: a mineral that is a variety of pyrrhotite and that is widely but sparsely distributed (as on earth, in meteorites, and in lunar soil samples)

Troi·lus \'trȯi-ləs, 'trō-ə-ləs\ *noun* [Middle English, from Latin, from Greek *Trōïlos*]
: a son of Priam who in medieval legend loved Cressida and lost her to Diomedes

[1]Tro·jan \'trō-jən\ *noun* [Middle English, from Latin *trojanus* of Troy, from *Troia, Troja* Troy, from Greek *Trōïa*] (14th century)
1 : a native or inhabitant of Troy
2 : one who shows qualities (as pluck, endurance, or determined energy) attributed to the defenders of ancient Troy
3 : a gay, irresponsible, or disreputable companion

[2]Trojan *adjective* (14th century)
1 : of, relating to, or resembling ancient Troy or its inhabitants
2 : of, relating to, or constituting a Trojan horse

Trojan horse *noun* [from the large hollow wooden horse filled with Greek soldiers and introduced within the walls of Troy by a stratagem] (1837)
1 : someone or something intended to defeat or subvert from within
2 : a seemingly useful computer program that contains concealed instructions which when activated perform an illicit or malicious action (as destroying data files); *also* **:** the concealed instructions of such a program — compare VIRUS

Trojan War *noun*
: a 10-year war between the Greeks and Trojans brought on by the abduction of Helen by Paris and ended with the destruction of Troy

[1]troll \'trōl\ *verb* [Middle English] (15th century)
transitive verb
1 : to cause to move round and round **:** ROLL
2 a : to sing the parts of (as a round or catch) in succession **b :** to sing loudly **c :** to celebrate in song
3 a : to fish for by trolling **b :** to fish by trolling in ⟨*troll* lakes⟩ **c :** to pull through the water in trolling ⟨*troll* a lure⟩
intransitive verb
1 : to move around **:** RAMBLE
2 : to fish by trailing a lure or baited hook from a moving boat
3 : to sing or play in a jovial manner
4 : to speak rapidly
— troll·er *noun*

[2]troll *noun* (1869)
: a lure or a line with its lure and hook used in trolling

[3]troll *noun* [Norwegian *troll* & Danish *trold,* from Old Norse *troll* giant, demon; probably akin to Middle High German *trolle* lout] (1616)
: a dwarf or giant in Scandinavian folklore inhabiting caves or hills

[1]trol·ley *also* **trol·ly** \'trä-lē\ *noun, plural* **trolleys** *also* **trollies** [probably from [1]*troll*] (1823)
1 *dialect English* **:** a cart of any of various kinds
2 a : a device that carries electric current from an overhead wire to an electrically driven vehicle **b :** a streetcar powered electrically through a trolley — called also *trolley car*
3 : a wheeled carriage running on an overhead rail or track
4 *chiefly British* **:** a cart or wheeled stand used for conveying something (as food or books)

[2]trolley *also* **trolly** *verb* **trol·leyed** *also* **trol·lied; trol·ley·ing** *also* **trol·ly·ing** (1882)
transitive verb
: to convey by a trolley
intransitive verb
: to ride on a trolley

trol·ley·bus \'trä-lē-ˌbəs\ *noun* (1921)
: a bus powered electrically from two overhead wires

trol·lop \'trä-ləp\ *noun* [perhaps irregular from *trull*] (1621)
: a vulgar or disreputable woman; *especially* **:** one who engages in sex promiscuously or for money

Trombe wall \'trȯmb-, 'trämb-, 'trō^{n}b-\ *noun* [Félix *Trombe* 20th century French designer] (1978)
: a masonry wall that is usually separated from the outdoors by a glass wall and is designed to absorb solar heat and release it into the interior of a building

trom·bone \träm-'bōn, (ˌ)trəm-', 'träm-ˌ\ *noun* [Italian, augmentative of *tromba* trumpet, of Germanic origin; akin to Old High German *trumba, trumpa* trumpet] (circa 1724)
: a brass instrument consisting of a long cylindrical metal tube with two turns and having a movable slide or valves for varying the tone and a usual range one octave lower than that of the trumpet
— trom·bon·ist \-'bō-nist, -ˌbō-\ *noun*

trombone

trom·mel \'trä-məl\ *noun* [German, drum, from Middle High German *trummel,* diminutive of *trumme* drum — more at DRUM] (circa 1877)
: a usually cylindrical or conical revolving screen used especially for screening or sizing rock, ore, or coal

tromp \'trämp, 'trȯmp\ *verb* [by alteration] (1883)
intransitive verb
1 : TRAMP 1 ⟨a lot of knocking on doors, *tromping* from room to room —Sara Davidson⟩
2 : to step hard **:** STAMP ⟨*tromped* on the brake⟩
transitive verb
1 : TRAMP
2 : STAMP ⟨*tromps* the accelerator to the floor —Jim Becker⟩
3 a : to give a physical beating to **b :** to defeat decisively

trompe l'oeil \(ˌ)trȯmp-'ləi, trō^{n}p-lœi\ *noun, often attributive* [French *trompe-l'œil,* literally, deceive the eye] (1889)
1 : a style of painting in which objects are depicted with photographically realistic detail; *also* **:** the use of similar technique in interior decorating
2 : a trompe l'oeil painting or effect

-tron *noun suffix* [Greek, suffix denoting an instrument; akin to Old English *-thor,* suffix denoting an instrument, Latin *-trum*]
1 : vacuum tube ⟨magne*tron*⟩
2 : device for the manipulation of subatomic particles ⟨cyclo*tron*⟩

tro·na \'trō-nə\ *noun* [Swedish, probably from Arabic *natrūn* natron — more at NATRON] (1799)
: a gray-white or yellowish white monoclinic mineral consisting of a hydrous acid sodium carbonate

[1]troop \'trüp\ *noun* [Middle French *trope, troupe* company, herd, of Germanic origin; akin to Old English *thorp, throp* village — more at THORP] (1545)
1 a : a group of soldiers **b :** a cavalry unit corresponding to an infantry company **c** *plural* **:** ARMED FORCES, SOLDIERS
2 : a collection of people or things **:** CREW 2
3 : a flock of mammals or birds
4 : the basic organizational unit of Boy Scouts or Girl Scouts under an adult leader

[2]troop *intransitive verb* (1565)
1 : to move or gather in crowds
2 : to go one's way **:** WALK
3 : to spend time together **:** ASSOCIATE
4 : to move in large numbers

troop·er \'trü-pər\ *noun* (1640)
1 a (1) **:** an enlisted cavalryman (2) **:** the horse of a cavalryman **b :** PARATROOPER **c :** SOLDIER
2 a : a mounted police officer **b :** a state police officer
3 : TROUPER 2

troop·ship \'trüp-ˌship\ *noun* (1862)
: a ship for carrying troops **:** TRANSPORT

trop- *or* **tropo-** *combining form* [International Scientific Vocabulary, from Greek *tropos*]
1 : turn **:** turning **:** change ⟨*tropo*sphere⟩
2 : tropism ⟨*trop*ic⟩

trope \'trōp\ *noun* [Latin *tropus,* from Greek *tropos* turn, way, manner, style, trope, from *trepein* to turn] (1533)
1 : the use of a word or expression in a figurative sense **:** FIGURE OF SPEECH
2 : a phrase or verse added as an embellishment or interpolation to the sung parts of the Mass in the medieval period

troph- *or* **tropho-** *combining form* [French, from Greek, from *trophē* nourishment]
: nutritive ⟨*tropho*blast⟩

troph·al·lax·is \ˌtrō-fə-ˈlak-səs\ *noun* [New Latin, from *troph-* + Greek *allaxis* exchange, from *allassein* to change, exchange, from *allos* other — more at ELSE] (1918)
: exchange of food (as from special glands) between social insects (as ants or termites)

tro·phic \ˈtrō-fik\ *adjective* [French *trophique,* from Greek *trophikos,* from *trophē* nourishment, from *trephein* to nourish] (1873)
1 : of or relating to nutrition **:** NUTRITIONAL ⟨*trophic* disorders⟩
2 : [3]TROPIC
— tro·phi·cal·ly \-fi-k(ə-)lē\ *adverb*

-trophic *adjective combining form* [New Latin *-trophia,* -trophy]
1 a : of, relating to, or characterized by (such) nutrition ⟨ecto*trophic*⟩ **b :** requiring or utilizing (such) a kind of nutrition ⟨hetero*trophic*⟩
2 : -TROPIC 2 ⟨gonado*trophic*⟩

trophic level *noun* (1942)
: one of the hierarchical strata of a food web characterized by organisms which are the same number of steps removed from the primary producers

tro·pho·blast \ˈtrō-fə-ˌblast\ *noun* [International Scientific Vocabulary] (1889)
: a thin layer of ectoderm that forms the wall of many mammalian blastulas and functions in the nutrition and implantation of the embryo
— tro·pho·blas·tic \ˌtrō-fə-ˈblas-tik\ *adjective*

tro·pho·zo·ite \ˌtrō-fə-ˈzō-ˌīt\ *noun* [*troph-* + *zo-* + [1]*-ite*] (circa 1909)
: a protozoan of a vegetative form as distinguished from one of a reproductive or resting form

tro·phy \ˈtrō-fē\ *noun, plural* **trophies** [Middle French *trophee,* from Latin *tropaeum, trophaeum,* from Greek *tropaion,* from neuter of *tropaios* of a turning, of a rout, from *tropē* turn, rout, from *trepein* to turn] (15th century)
1 : something gained or given in victory or conquest especially when preserved or mounted as a memorial
2 a : a memorial of an ancient Greek or Roman victory raised on the field of battle or in case of a naval victory on the nearest land **b :** a representation of such a memorial (as on a medal); *also* **:** an architectural ornament representing a group of military weapons
3 : a game animal or fish suitable for mounting as a trophy — usually used attributively
◆
— trophy *transitive verb*

-trophy *noun combining form* [New Latin *-trophia,* from Greek, from *-trophos* nourishing, from *trephein*]
: nutrition **:** nurture **:** growth ⟨dys*trophy*⟩

[1]**trop·ic** \ˈträ-pik\ *noun* [Middle English *tropik,* from Latin *tropicus* of the solstice, from Greek *tropikos,* from *tropē* turn] (1527)
1 : either of the two parallels of terrestrial latitude at a distance of about 23½ degrees north or south of the equator where the sun is directly overhead when it reaches its most northerly or southerly point in the sky — compare TROPIC OF CANCER, TROPIC OF CAPRICORN
2 *plural, often capitalized* **:** the region lying between the tropics

[2]**tropic** *adjective* (1551)
: of, relating to, or occurring in the tropics **:** TROPICAL

[3]**tro·pic** \ˈtrō-pik\ *adjective* [*trop-*] (1903)
1 : of, relating to, or characteristic of tropism or of a tropism
2 *of a hormone* **:** influencing the activity of a specified gland

-tropic *adjective combining form* [French *-tropique,* from Greek *-tropos* -tropous]
1 : turning, changing, or tending to turn or change in a (specified) manner or in response to a (specified) stimulus ⟨geo*tropic*⟩
2 : attracted to or acting upon (something specified) ⟨neuro*tropic*⟩

trop·i·cal *for 1* ˈträ-pi-kəl, *for 2* ˈtrō- *also* ˈträ-\ *adjective* (1527)
1 a : of, relating to, occurring in, or suitable for use in the tropics **b :** of, being, or characteristic of a region or climate that is frost-free with temperatures high enough to support year-round plant growth given sufficient moisture ⟨*tropical* Florida⟩
2 [Latin *tropicus,* from Greek *tropikos,* from *tropos* trope] **:** FIGURATIVE 2
— trop·i·cal·ly \-pi-k(ə-)lē\ *adverb*

tropical aquarium *noun* (circa 1948)
: an aquarium kept at a uniform warmth and used especially for tropical fish

tropical cyclone *noun* (1920)
: a cyclone in the tropics characterized by winds rotating at the rate of 74 miles (119 kilometers) an hour or more

tropical fish *noun* (1931)
: any of various small usually showy fishes of tropical origin often kept in the tropical aquarium

trop·i·cal·ize \ˈträ-pi-kə-ˌlīz\ *transitive verb* **-ized; -iz·ing** (1885)
1 : to make tropical (as in character, conditions, or appearance)
2 : to fit or adapt for use in a tropical climate especially by measures designed to combat the effects of fungi and moisture

tropical oil *noun* (1988)
: any of several oils (as coconut oil and palm oil) that are high in saturated fatty acids

tropical rain forest *noun* (1926)
: RAIN FOREST 1

tropical sprue *noun* (circa 1955)
: SPRUE 2

tropical storm *noun* (circa 1945)
: a tropical cyclone with strong winds of less than hurricane intensity

tropic bird *noun* (circa 1681)
: any of a genus (*Phaethon* of the family Phaethontidae) of web-footed birds that are related to the pelicans, are found chiefly in tropical seas often far from land, and have mostly white satiny plumage marked with a little black, a greatly elongated central pair of tail feathers, and a bright-colored bill

tropic bird

tropic of Cancer [from the sign of the zodiac which its celestial projection intersects] (1555)
: the parallel of latitude that is approximately 23½ degrees north of the equator and that is the northernmost latitude reached by the overhead sun

tropic of Capricorn [from the sign of the zodiac which its celestial projection intersects] (1545)
: the parallel of latitude that is approximately 23½ degrees south of the equator and that is the southernmost latitude reached by the overhead sun

-tropin *or* **-trophin** *noun combining form* [*-tropin,* alteration of *-trophin,* from *-trophic* + [1]*-in*]
: hormone ⟨gonado*tropin*⟩ ⟨somato*tropin*⟩

tro·pism \ˈtrō-ˌpi-zəm\ *noun* [International Scientific Vocabulary *-tropism*] (1899)
1 a : involuntary orientation by an organism or one of its parts that involves turning or curving by movement or by differential growth and is a positive or negative response to a source of stimulation **b :** a reflex reaction involving a tropism
2 : an innate tendency to react in a definite manner to stimuli; *broadly* **:** a natural inclination **:** PROPENSITY
— tro·pis·tic \trō-ˈpis-tik\ *adjective*

-tropism *noun combining form* [International Scientific Vocabulary, from *trop-*]
: tropism ⟨helio*tropism*⟩

tropo- — see TROP-

tro·po·col·la·gen \ˌträ-pə-ˈkä-lə-jən, ˌtrō-\ *noun* (1954)
: a subunit of collagen fibrils consisting of three polypeptide strands arranged in a helix

tro·po·log·i·cal \ˌtrō-pə-ˈlä-ji-kəl, ˌträ-\ *also* **tro·po·log·ic** \-jik\ *adjective* (14th century)
1 : of, relating to, or involving biblical interpretation stressing moral metaphor; *also* **:** MORAL
2 : characterized or varied by tropes **:** FIGURATIVE
— tro·po·log·i·cal·ly \-ji-k(ə-)lē\ *adverb*

tro·po·my·o·sin \ˌträ-pə-ˈmī-ə-sən, ˌtrō-\ *noun* (1946)
: a protein of muscle that forms a complex with troponin regulating the interaction of actin and myosin in muscular contraction

tro·po·nin \ˈtrō-pə-nən, ˈträ-, -ˌnin\ *noun* [by shortening & alteration from *tropomyosin*] (1966)
: a protein of muscle that together with tropomyosin forms a regulatory protein complex controlling the interaction of actin and myosin and that when combined with calcium ions permits muscular contraction

tro·po·pause \ˈtrō-pə-ˌpȯz, ˈträ-\ *noun* [International Scientific Vocabulary *tropo*sphere + *pause*] (1918)
: the region at the top of the troposphere; *also* **:** a comparable layer of a celestial body

tro·po·sphere \ˈtrō-pə-ˌsfir, ˈträ-\ *noun* [International Scientific Vocabulary] (1909)
: the lowest densest part of the earth's atmosphere in which most weather changes occur and temperature generally decreases rapidly with altitude and which extends from the surface to the bottom of the stratosphere
— tro·po·spher·ic \ˌtrō-pə-ˈsfir-ik, ˌträ-, -ˈsfer-\ *adjective*

tro·po·tax·is \ˌtrō-pə-ˈtak-səs, ˌträ-\ *noun* [New Latin] (1934)
: a taxis in which an organism orients itself by the simultaneous comparison of stimuli of different intensity acting on separate end organs

-tropous *adjective combining form* [Greek *-tropos,* from *trepein* to turn]
: turning or curving in (such) a way **:** exhibiting (such) a tropism ⟨ana*tropous*⟩

-tropy *noun combining form* [French *-tropie,* from Greek *-tropia,* from *-tropos*]
1 : condition of exhibiting (such) a behavior ⟨allo*tropy*⟩

◇ WORD HISTORY

trophy After a military victory, the ancient Greeks had a custom of hanging captured enemy arms and armor from a stake or tree somewhere on the battlefield, ideally at the spot where the enemy had first turned in retreat. The Greek name for such a memorial was *tropaion,* from the neuter form of the adjective *tropaios* "of or relating to a turn or rout," a derivative of *tropē* "turn, rout." Eventually *tropaion* was applied to much more elaborate and permanent victory memorials of stone or bronze, ones not necessarily erected on the actual battlefield. From Hellenistic Greece the ancient Romans took the custom as well as the word, which was Latinized as *tropaeum,* later *trophaeum.* English borrowed the word via French as *trophe* in the 16th century. From its original classical reference, the meaning of *trophy* has been generalized to apply to anything that could be displayed as a reminder of success, whether in war, sports, or life.

2 : change in a (specified) way or in response to a (specified) stimulus ⟨thixo*tropy*⟩

¹trot \'trät\ *noun* [Middle English, from Middle French, from *troter* to trot, of Germanic origin; akin to Old High German *trottōn* to tread, Old English *tredan*] (14th century)
1 a (1) **:** a moderately fast gait of a quadruped (as a horse) in which the legs move in diagonal pairs (2) **:** a jogging gait of a human that falls between a walk and a run **b :** a ride on horseback
2 : an old woman
3 : a literal translation of a foreign text
4 *plural* **:** DIARRHEA

²trot *verb* **trot·ted; trot·ting** (14th century)
intransitive verb
1 : to ride, drive, or proceed at a trot ⟨the fox *trotted* over the knoll⟩
2 : to proceed briskly **:** HURRY
transitive verb
1 : to cause to go at a trot
2 : to traverse at a trot

³trot *noun* (1883)
: TROTLINE; *also* **:** one of the short lines with hooks that are attached to it at intervals

¹troth \'träth, 'trȯth, 'trōth, *or with* th\ *noun* [Middle English *trouth,* from Old English *trēowth* — more at TRUTH] (13th century)
1 : loyal or pledged faithfulness **:** FIDELITY
2 : one's pledged word; *also* **:** BETROTHAL

²troth *transitive verb* (15th century)
: PLEDGE, BETROTH

¹troth·plight \'träth-ˌplīt, 'trȯth-, 'trōth-\ *transitive verb* (15th century)
: BETROTH

²trothplight *noun* (1513)
: BETROTHAL

trot·line \'trät-ˌlīn\ *noun* [probably from ²*trot*] (1826)
: SETLINE; *especially* **:** a comparatively short setline used near shore or along streams

trot out *transitive verb* (1838)
1 : to lead out and show the paces of (as a horse)
2 : to bring forward for display ⟨always *trots out* some new excuse⟩

Trots·ky·ism \'trät-skē-ˌi-zəm, 'trȯt-\ *noun* (1925)
: the political, economic, and social principles advocated by Trotsky; *especially* **:** the theory and practice of communism developed by or associated with Trotsky and usually including adherence to the concept of worldwide revolution as opposed to socialism in one country
— Trots·ky·ist \-skē-ist\ *noun or adjective*
— Trots·ky·ite \-skē-ˌīt\ *noun or adjective*

trot·ter \'trä-tər\ *noun* (14th century)
1 : one that trots; *specifically* **:** a standardbred horse trained for harness racing
2 : a pig's foot used as food

trou·ba·dour \'trü-bə-ˌdōr, -ˌdȯr, -ˌdu̇r\ *noun* [French, from Old Provençal *trobador,* from *trobar* to compose, probably from (assumed) Vulgar Latin *tropare,* from Latin *tropus* trope] (circa 1741)
: one of a class of lyric poets and poet-musicians often of knightly rank who flourished from the 11th to the end of the 13th century chiefly in the south of France and the north of Italy and whose major theme was courtly love — compare TROUVÈRE

¹trou·ble \'trə-bəl\ *verb* **trou·bled; trou·bling** \'trə-b(ə-)liŋ\ [Middle English, from Old French *tourbler, troubler,* from (assumed) Vulgar Latin *turbulare,* from *turbulus* agitated, alteration of Latin *turbulentus* — more at TURBULENT] (13th century)
transitive verb
1 a : to agitate mentally or spiritually **:** WORRY, DISTURB **b** (1) *archaic* **:** MISTREAT, OPPRESS (2) **:** to produce physical disorder in **:** AFFLICT ⟨*troubled* by a cold⟩ **c :** to put to exertion or inconvenience
2 : to put into confused motion ⟨the wind *troubled* the sea⟩
intransitive verb
1 : to become mentally agitated **:** WORRY ⟨refused to *trouble* over trifles⟩
2 : to make an effort **:** be at pains ⟨did not *trouble* to come⟩
— trou·bler \-b(ə-)lər\ *noun*

²trouble *noun* (13th century)
1 : the quality or state of being troubled especially mentally
2 : public unrest or disturbance ⟨there's *trouble* brewing downtown⟩
3 : an instance of trouble ⟨used to disguise her frustrations and despair by making light of her *troubles —Current Biography*⟩
4 : a state or condition of distress, annoyance, or difficulty ⟨in *trouble* with the law⟩ ⟨heading for *trouble*⟩ ⟨got into financial *trouble*⟩: as **a :** a condition of physical distress or ill health **:** AILMENT ⟨back *trouble*⟩ ⟨heart *trouble*⟩ **b :** a condition of mechanical malfunction ⟨engine *trouble*⟩ **c :** a condition of doing something badly or only with great difficulty ⟨has *trouble* reading⟩ ⟨has *trouble* breathing⟩ **d :** pregnancy out of wedlock ⟨got a girl in *trouble*⟩
5 : an effort made **:** PAINS ⟨took the *trouble* to do it right⟩
6 a : a cause of distress, annoyance, or inconvenience ⟨don't mean to be any *trouble*⟩ ⟨what's the *trouble?*⟩ **b :** a negative feature **:** DRAWBACK ⟨the *trouble* with you is you're too honest⟩ ⟨the main *trouble* with electronic systems is the overreliance on them —John Perham⟩ **c :** the unhappy or sad fact ⟨the *trouble* is, I need the money⟩

trou·bled \'trə-bəld\ *adjective* (14th century)
1 a : CONCERNED, WORRIED ⟨*troubled* feelings about the decision⟩ **b :** emotionally or mentally disturbed ⟨a home for *troubled* children⟩
2 : characterized by or indicative of trouble ⟨our *troubled* cities⟩ ⟨a gray and *troubled* sky⟩

trou·ble·mak·er \'trə-bəl-ˌmā-kər\ *noun* (circa 1914)
: a person who consciously or unconsciously causes trouble
— trou·ble·mak·ing \-kiŋ\ *adjective or noun*

trou·ble·shoot \-ˌshüt\ *verb* **-shot** \-ˌshät\; **-shoot·ing** [back-formation from *troubleshooter*] (1918)
intransitive verb
: to operate or serve as a troubleshooter ⟨is *troubleshooting* for an electronics firm⟩
transitive verb
: to investigate or deal with in the role of troubleshooter ⟨*troubleshoots* TV receivers⟩

trou·ble·shoot·er \-ˌshü-tər\ *noun* (1905)
1 : a skilled worker employed to locate trouble and make repairs in machinery and technical equipment
2 : an expert in resolving diplomatic or political disputes **:** a mediator of disputes that are at an impasse
3 : a person skilled at solving or anticipating problems or difficulties

trou·ble·some \-səm\ *adjective* (1542)
1 : DIFFICULT, BURDENSOME
2 : giving trouble or anxiety **:** VEXATIOUS
— trou·ble·some·ly *adverb*
— trou·ble·some·ness *noun*

trou·blous \'trə-b(ə-)ləs\ *adjective* (15th century)
1 : full of trouble **:** STORMY ⟨these *troublous* times⟩
2 : causing trouble **:** TROUBLESOME ⟨inflation is a *troublous* matter⟩
— trou·blous·ly *adverb*
— trou·blous·ness *noun*

trough \'trȯf, 'trȯth, *by bakers often* 'trō\ *noun, plural* **troughs** \'trȯfs, 'trȯvz; 'trȯths, 'trȯ(th)z; 'trōz\ [Middle English, from Old English *trog;* akin to Old High German *trog* trough, Old English *trēow* tree, wood — more at TREE] (before 12th century)
1 a : a long shallow often V-shaped receptacle for the drinking water or feed of domestic animals **b :** any of various domestic or industrial containers
2 a : a conduit, drain, or channel for water; *especially* **:** a gutter along the eaves of a building **b :** a long and narrow or shallow channel or depression (as between waves or hills); *especially* **:** a long but shallow depression in the bed of the sea — compare TRENCH
3 : the minimum point of a complete cycle of a periodic function: as **a :** an elongated area of low barometric pressure **b :** the low point in a business cycle

trounce \'trau̇n(t)s\ *transitive verb* **trounced; trounc·ing** [origin unknown] (1868)
: to thrash or punish severely; *especially* **:** to defeat decisively

¹troupe \'trüp\ *noun* [French, from Middle French — more at TROOP] (1825)
: COMPANY, TROOP; *especially* **:** a group of theatrical performers

²troupe *intransitive verb* **trouped; troup·ing** (1900)
: to travel in a troupe; *also* **:** to perform as a member of a theatrical troupe

troup·er \'trü-pər\ *noun* (1890)
1 : a member of a troupe; *especially* **:** ACTOR
2 : a person who deals with and persists through difficulty or hardship without complaint ⟨you're a real *trouper* to wait so long⟩

trou·pi·al \'trü-pē-əl\ *noun* [French *troupiale,* from *troupe;* from its living in flocks] (1825)
: a large showy oriole (*Icterus icterus*) of Central and South America; *also* **:** any of various related birds (family Icteridae)

¹trou·ser \'trau̇-zər\ *noun* [alteration of earlier *trouse,* from Scottish Gaelic *triubhas*] (1613)
: ³PANT 1 — usually used in plural

²trouser *adjective* (circa 1771)
1 : of, relating to, or designed for trousers ⟨*trouser* pockets⟩
2 : of or relating to a male dramatic role played by a woman

trouser suit *noun* (1939)
chiefly British **:** PANTSUIT

trous·seau \'trü-(ˌ)sō, trü-'\ *noun, plural* **trous·seaux** \-(ˌ)sōz, -'sōz\ *or* **trous·seaus** [French, from Old French, diminutive of *trousse* bundle, from *trousser* to truss] (1833)
: the personal possessions of a bride usually including clothes, accessories, and household linens and wares

trout \'trau̇t\ *noun, plural* **trout** *also* **trouts** [Middle English, from Old English *trūht,* from Late Latin *trocta, tructa,* a fish with sharp teeth, from Greek *trōktēs,* literally, gnawer, from *trōgein* to gnaw — more at TROGLODYTE] (before 12th century)
1 : any of various salmonid food and sport fishes that are mostly smaller than the typical salmons and are anadromous or restricted to cool clear fresh waters: **a :** any of various Old or New World fishes (genera *Salmo* and *Oncorhynchus*) — compare BROWN TROUT, RAINBOW TROUT **b :** ¹CHAR
2 : any of various fishes (as the largemouth bass) held to resemble the true trouts

trout lily *noun* [probably from its speckled leaves] (circa 1898)
: DOGTOOTH VIOLET

trout–perch \'trau̇t-ˌpərch\ *noun* (1883)
: a small freshwater fish (*Percopsis omiscomaycus* of the family Percopsidae) chiefly of northern North America having a scaleless head and large eyes

trouty \'trau̇-tē\ *adjective* **trout·i·er; -est** (1676)
: containing or likely to contain abundant trout

trou·vère \trü-'ver\ *noun* [French, from Old French *troveor, troverre,* from *trover* to compose, find, from (assumed) Vulgar Latin *tropare* — more at TROUBADOUR] (1795)
: one of a school of poets who flourished from the 11th to the 14th centuries and who composed mostly narrative works (as chansons de geste and fabliaux) — compare TROUBADOUR

trove \'trōv\ *noun* [short for *treasure trove*] (1888)
1 : DISCOVERY, FIND
2 : a valuable collection **:** TREASURE; *also* **:** HAUL, COLLECTION

tro·ver \'trō-vər\ *noun* [Middle French *trover* to find, from Old French] (1594)
: a common law action to recover the value of goods wrongfully converted to another's own use

trow \'trō\ *verb* [Middle English, from Old English *trēowan;* akin to Old English *trēowe* faithful, true — more at TRUE] (before 12th century)
1 *obsolete* **:** BELIEVE
2 *archaic* **:** THINK

[1]**trow·el** \'trau̇(-ə)l\ *noun* [Middle English *truel,* from Middle French *truelle,* from Late Latin *truella,* from Latin *trulla* ladle] (14th century)
: any of various hand tools used to apply, spread, shape, or smooth loose or plastic material; *also* **:** a scoop-shaped or flat-bladed garden tool for taking up and setting small plants

trowel: *1* gardener's, *2* plasterer's, *3* bricklayer's

[2]**trowel** *transitive verb* **-eled** *or* **-elled; -el·ing** *or* **-el·ling** (circa 1670)
: to smooth, mix, or apply with or as if with a trowel
— trow·el·er *noun*

troy \'trȯi\ *adjective* [Middle English *troye,* from *Troyes,* France] (15th century)
: expressed in troy weight

troy weight *noun* (15th century)
: a series of units of weight based on a pound of 12 ounces and an ounce of 480 grains — see WEIGHT table

tru·an·cy \'trü-ən(t)-sē\ *noun, plural* **-cies** (1784)
: an act or instance of playing truant **:** the state of being truant

[1]**tru·ant** \'trü-ənt\ *noun* [Middle English, vagabond, idler, from Old French, vagrant, of Celtic origin; akin to Old Irish *trógán* wretch, *trúag* wretched] (15th century)
: one who shirks duty; *especially* **:** one who stays out of school without permission

[2]**truant** *adjective* (1561)
1 : shirking responsibility
2 : being, resembling, or characteristic of a truant

[3]**truant** *intransitive verb* (1580)
: to idle away time especially while playing truant

truant officer *noun* (1872)
: ATTENDANCE OFFICER

tru·ant·ry \'trü-ən-trē\ *noun, plural* **-ries** (15th century)
: TRUANCY

[1]**truce** \'trüs\ *noun* [Middle English *trewes,* plural of *trewe* agreement, from Old English *trēow* fidelity; akin to Old English *trēowe* faithful — more at TRUE] (13th century)
1 : a suspension of fighting especially of considerable duration by agreement of opposing forces **:** ARMISTICE, CEASE-FIRE
2 : a respite especially from a disagreeable or painful state or action

[2]**truce** *verb* **truced; truc·ing** (1569)
intransitive verb
: to make a truce
transitive verb
: to end with a truce

[1]**truck** \'trək\ *verb* [Middle English *trukken,* from Old French *troquer*] (13th century)
transitive verb
1 : to give in exchange **:** SWAP
2 : to barter or dispose of by barter
intransitive verb
1 : to exchange commodities **:** BARTER
2 : to negotiate or traffic especially in an underhanded way **:** have dealings

[2]**truck** *noun* (1553)
1 : BARTER
2 : commodities appropriate for barter or for small trade
3 : close association or connection ⟨will have no *truck* with crooks⟩
4 : payment of wages in goods instead of cash
5 : vegetables grown for market
6 : heterogeneous small articles often of little value; *also* **:** RUBBISH

[3]**truck** *noun* [probably back-formation from *truckle* small wheel — more at TRUCKLE BED] (1611)
1 : a small wheel; *specifically* **:** a small strong wheel for a gun carriage
2 : a small wooden cap at the top of a flagstaff or masthead usually having holes for reeving flag or signal halyards
3 : a wheeled vehicle for moving heavy articles: as **a :** a strong horse-drawn or automotive vehicle for hauling **b :** a small barrow consisting of a rectangular frame having at one end a pair of handles and at the other end a pair of small heavy wheels and a projecting edge to slide under a load — called also *hand truck* **c :** a small heavy rectangular frame supported on four wheels for moving heavy objects **d :** a small flat-topped car pushed or pulled by hand **e :** a shelved stand mounted on casters **f :** an automotive vehicle with a short chassis equipped with a swivel for attaching a trailer and used especially for the highway hauling of freight; *also* **:** a truck with attached trailer
4 a *British* **:** an open railroad freight car **b :** a swiveling carriage consisting of a frame with one or more pairs of wheels and springs to carry and guide one end (as of a railroad car) in turning sharp curves
— truck·ful \-ˌfu̇l\ *noun*

[4]**truck** (1748)
transitive verb
: to load or transport on a truck
intransitive verb
1 : to transport goods by truck
2 : to be employed in driving a truck
3 : to roll along especially in an easy untroubled way

truck·age \'trə-kij\ *noun* (1830)
1 : conveyance by truck
2 : money paid for conveyance on a truck

[1]**truck·er** \'trə-kər\ *noun* [[1]*truck*] (1598)
1 : one that barters
2 *Scottish* **:** PEDDLER

[2]**trucker** *noun* [[4]*truck*] (1878)
1 : one whose business is transporting goods by truck
2 : a truck driver

truck farm *noun* [[2]*truck*] (1866)
: a farm devoted to the production of vegetables for the market
— truck farmer *noun*

truck·ing *noun* (1809)
: the process or business of transporting goods on trucks

truck·le \'trə-kəl\ *intransitive verb* **truck·led; truck·ling** \-k(ə-)liŋ\ [from the lower position of the truckle bed] (1667)
: to act in a subservient manner **:** SUBMIT
synonym see FAWN
— truck·ler \-k(ə-)lər\ *noun*

truckle bed *noun* [*truckle* small wheel, pulley, from Middle English *trocle,* from Latin *trochlea* block of pulleys — more at TROCHLEA] (15th century)
: TRUNDLE BED

truck·line \'trək-ˌlīn\ *noun* (1924)
: a transportation line using trucks

truck·load \-'lōd, -ˌlōd\ *noun* (1862)
1 : a load or amount that fills or could fill a truck
2 : the minimum weight required for shipping at truckload rates

truck·man \-mən\ *noun* (1787)
1 : [2]TRUCKER
2 : a member of a fire department unit that operates a hook and ladder truck

truck·mas·ter \-ˌmas-tər\ *noun* (1637)
archaic **:** an officer in charge of trade with Indians especially among the early settlers

truck stop *noun* (circa 1951)
: a facility especially for truckers that is usually by a highway and that includes a diner, fuel pumps, and a garage

truck system *noun* (1830)
: the system of paying wages in goods instead of cash

tru·cu·lence \'trə-kyə-lən(t)s *also* 'trü-\ *noun* (circa 1727)
: the quality or state of being truculent

tru·cu·len·cy \-lən(t)-sē\ *noun* (1569)
: TRUCULENCE

tru·cu·lent \-lənt\ *adjective* [Latin *truculentus,* from *truc-, trux* savage; perhaps akin to Middle Irish *trú* doomed person] (circa 1540)
1 : feeling or displaying ferocity **:** CRUEL, SAVAGE
2 : DEADLY, DESTRUCTIVE
3 : scathingly harsh **:** VITRIOLIC
4 : aggressively self-assertive **:** BELLIGERENT
— tru·cu·lent·ly *adverb*

[1]**trudge** \'trəj\ *verb* **trudged; trudg·ing** [origin unknown] (1547)
intransitive verb
: to walk or march steadily and usually laboriously ⟨*trudged* through deep snow⟩
transitive verb
: to trudge along or over
— trudg·er *noun*

[2]**trudge** *noun* (1835)
: a long tiring walk **:** TRAMP

trud·gen stroke \'trə-jən-\ *noun* [John *Trudgen* (died 1902) English swimmer] (1893)
: a swimming stroke consisting of alternating overarm strokes and a scissors kick

[1]**true** \'trü\ *adjective* **tru·er; tru·est** [Middle English *trewe,* from Old English *trēowe* faithful; akin to Old High German *gitriuwi* faithful, Old Irish *derb* sure, and probably to Sanskrit *dāruṇa* hard, *dāru* wood — more at TREE] (before 12th century)
1 a : STEADFAST, LOYAL **b :** HONEST, JUST **c** *archaic* **:** TRUTHFUL
2 a (1) **:** being in accordance with the actual state of affairs ⟨*true* description⟩ (2) **:** conformable to an essential reality (3) **:** fully realized or fulfilled ⟨dreams come *true*⟩ **b :** IDEAL, ESSENTIAL **c :** being that which is the case rather than what is manifest or assumed ⟨the *true* dimension of the problem⟩ **d :** CONSISTENT ⟨*true* to character⟩
3 a : properly so called ⟨*true* love⟩ ⟨the *true* faith⟩ ⟨the *true* stomach of ruminant mammals⟩ **b** (1) **:** possessing the basic characters of and belonging to the same natural group as ⟨a whale is a *true* but not a typical mammal⟩ (2) **:** TYPICAL ⟨the *true* cats⟩
4 : LEGITIMATE, RIGHTFUL ⟨our *true* and lawful king⟩
5 a : that is fitted or formed or that functions accurately **b :** conformable to a standard or pattern **:** ACCURATE
6 : determined with reference to the earth's axis rather than the magnetic poles ⟨*true* north⟩
7 : logically necessary
8 : NARROW, STRICT ⟨in the *truest* sense⟩
9 : corrected for error
— true·ness *noun*

[2]**true** *adverb* (14th century)
1 : in accordance with fact or reality
2 a : without deviation ⟨the bullet flew straight and *true*⟩ **b :** without variation from type ⟨breed *true*⟩

[3]**true** *noun* (1812)
1 : TRUTH, REALITY — usually used with *the*

\ə\ **abut** \ᵊ\ **kitten** \ər\ **further** \a\ **ash** \ā\ **ace** \ä\ **mop, mar** \au̇\ **out** \ch\ **chin** \e\ **bet** \ē\ **easy** \g\ **go** \i\ **hit** \ī\ **ice** \j\ **job** \ŋ\ **sing** \ō\ **go** \ȯ\ **law** \ȯi\ **boy** \th\ **thin** \t͟h\ **the** \ü\ **loot** \u̇\ **foot** \y\ **yet** \zh\ **vision** *see also* Guide to Pronunciation

2 : the quality or state of being accurate (as in alignment or adjustment) — used in the phrases *in true* and *out of true*

⁴true *transitive verb* **trued; true·ing** *also* **tru·ing** (1841)
: to make level, square, balanced, or concentric **:** bring or restore to a desired mechanical accuracy or form ⟨*true* up a board⟩ ⟨*true* up an engine cylinder⟩

true believer *noun* (circa 1820)
1 : one who professes absolute belief in something
2 : a zealous supporter of a particular cause

true bill *noun* (1769)
: a bill of indictment endorsed by a grand jury as warranting prosecution of the accused

true–blue *adjective* (1674)
1 : marked by unswerving loyalty (as to a party)
2 : GENUINE ⟨a *true-blue* romantic⟩

true blue *noun* [from the association of blue with constancy] (1762)
: one who is true-blue

true·born \'trü-'bȯrn\ *adjective* (1591)
: genuinely such by birth ⟨a *trueborn* Englishman —Shakespeare⟩

true bug *noun* (1895)
: BUG 1c

true–false test \'trü-'fȯls-\ *noun* (1924)
: a test consisting of a series of statements to be marked as true or false

true·heart·ed \-'här-təd\ *adjective* (15th century)
: FAITHFUL, LOYAL
— true·heart·ed·ness *noun*

true–life \'trü-'līf\ *adjective* (1926)
: true to life ⟨a *true-life* story⟩

true·love \'trü-ˌləv\ *noun* (14th century)
: one truly beloved or loving **:** SWEETHEART

true lover's knot *noun* (1615)
: a complicated ornamental knot not readily untied and symbolic of mutual love — called also *truelove knot;* see KNOT illustration

true·pen·ny \'trü-ˌpe-nē\ *noun* (1595)
: an honest or trusty person

true rib *noun* (1741)
: any of the ribs having costal cartilages connected directly with the sternum and in humans constituting the first seven pairs

true seal *noun* (1923)
: HAIR SEAL

truf·fle \'trə-fəl, 'trü-\ *noun* [modification of Middle French *truffe,* from Old Provençal *trufa,* from (assumed) Vulgar Latin *tufera;* akin to Latin *tuber* swelling, truffle — more at TUBER] (1591)
1 : the usually dark and rugose edible subterranean fruiting body of several European ascomycetous fungi (especially genus *Tuber*); *also* **:** one of these fungi
2 : a candy made of chocolate, butter, sugar, and sometimes liqueur shaped into balls and often coated with cocoa

truf·fled \-fəld\ *adjective* (1837)
: cooked, stuffed, or garnished with truffles

trug \'trəg\ *noun* [origin unknown] (circa 1864)
chiefly British **:** a shallow rectangular garden-ing basket

tru·ism \'trü-ˌi-zəm\ *noun* (1708)
: an undoubted or self-evident truth; *especially* **:** one too obvious for mention
— tru·is·tic \trü-'is-tik\ *adjective*

trull \'trəl\ *noun* [German *Trulle,* from Middle High German; akin to Middle High German *trolle* lout — more at TROLL] (1519)
: PROSTITUTE, STRUMPET

tru·ly \'trü-lē\ *adverb* (13th century)
1 a : INDEED — often used as an intensive ⟨*truly,* she is fair⟩ or interjectionally to express astonishment or doubt **b :** without feigning, falsity, or inaccuracy in truth or fact
2 : in all sincerity **:** SINCERELY — often used with *yours* as a complimentary close
3 : in agreement with fact **:** TRUTHFULLY
4 : with exactness of construction or operation
5 : in a proper or suitable manner

¹trump \'trəmp\ *noun* [Middle English *trompe,* from Old French, of Germanic origin; akin to Old High German *trumba, trumpa* trumpet] (14th century)
1 a : TRUMPET **b** *chiefly Scottish* **:** JEW'S HARP
2 : a sound of or as if of trumpeting ⟨the *trump* of doom⟩

²trump *noun* [alteration of ¹*triumph*] (1529)
1 a : a card of a suit any of whose cards will win over a card that is not of this suit **b :** the suit whose cards are trumps for a particular hand — often used in plural
2 : a decisive overriding factor or final resource
3 : a dependable and exemplary person

³trump (1586)
transitive verb
1 : to get the better of **:** OUTDO
2 : to play a trump on (a card or trick) when another suit was led
intransitive verb
: to play a trump when another suit was led

trump card *noun* (1822)
: ²TRUMP 1a, 2

trumped–up \'trəm(p)t-'əp\ *adjective* (1728)
: fraudulently concocted **:** SPURIOUS ⟨*trumped-up* charges⟩

trum·pery \'trəm-p(ə-)rē\ *noun* [Middle English *tromperie* deceit, from Middle French, from *tromper* to deceive] (15th century)
1 a : worthless nonsense **b :** trivial or useless articles **:** JUNK ⟨a wagon loaded with household *trumpery* —Washington Irving⟩
2 *archaic* **:** tawdry finery
— trumpery *adjective*

¹trum·pet \'trəm-pət\ *noun* [Middle English *trompette,* from Middle French, from Old French *trompe* trump] (14th century)
1 a : a wind instrument consisting of a conical or cylindrical usually metal tube, a cup-shaped mouthpiece, and a flared bell; *specifically* **:** a valved brass instrument having a cylindrical tube with two turns and a usual range from F sharp below middle C upward for 2½ octaves **b :** a musical instrument (as a cornet) resembling a trumpet

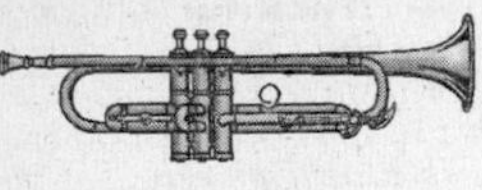
trumpet 1a

2 : a trumpet player
3 : something that resembles a trumpet or its tonal quality: as **a :** a funnel-shaped instrument (as a megaphone) for collecting, directing, or intensifying sound **b** (1) **:** a stentorian voice (2) **:** a penetrating cry (as of an elephant)
— trum·pet·like *adjective*

²trumpet (1530)
intransitive verb
1 : to blow a trumpet
2 : to make a sound suggestive of that of a trumpet
transitive verb
: to sound or proclaim on or as if on a trumpet

trumpet creeper *noun* (1818)
: a North American woody vine (*Campsis radicans* of the family Bignoniaceae, the trumpet-creeper family) having pinnate leaves and large typically red trumpet-shaped flowers

trumpet creeper

trum·pet·er \'trəm-pə-tər\ *noun* (15th century)
1 a : a trumpet player; *specifically* **:** one that gives signals with a trumpet **b :** one that praises or advocates **:** EULOGIST, SPOKESMAN
2 a : any of a genus (*Psophia* of the family Psophiidae) of several large gregarious long-legged long-necked South American birds related to the cranes **b :** any of an Asian breed of pigeons with a rounded crest and heavily feathered feet

trumpeter swan *noun* (1834)
: a rare pure white North American wild swan (*Cygnus buccinator*) that is noted for its sonorous voice and is sometimes considered a subspecies (*C. cygnus buccinator*) of the whooper swan

trumpet flower *noun* (circa 1731)
1 : any of various plants (as a trumpet creeper or a datura) with trumpet-shaped flowers
2 : the flower of a trumpet flower

trumpet honeysuckle *noun* (1731)
: a North American honeysuckle (*Lonicera sempervirens*) with coral-red or orange flowers having a trumpet-shaped corolla

trumpet vine *noun* (1709)
: TRUMPET CREEPER

trump up *transitive verb* (1695)
1 : to concoct especially with intent to deceive **:** FABRICATE, INVENT
2 *archaic* **:** to cite as support for an action or claim

¹trun·cate \'trəŋ-ˌkāt, 'trən-\ *adjective* [Latin *truncatus,* past participle of *truncare* to shorten, from *truncus* trunk] (1716)
: having the end square or even ⟨the *truncate* leaves of the tulip tree⟩

²truncate *transitive verb* **trun·cat·ed; trun·cat·ing** (circa 1727)
1 : to shorten by or as if by cutting off
2 : to replace (an edge or corner of a crystal) by a plane
— trun·ca·tion \trəŋ-'kā-shən, trən-\ *noun*

trun·cat·ed \-ˌkā-təd\ *adjective* (circa 1704)
1 : having the apex replaced by a plane section and especially by one parallel to the base ⟨*truncated* cone⟩
2 a : cut short **:** CURTAILED **b :** lacking an expected or normal element (as a syllable) at the beginning or end **:** CATALECTIC

¹trun·cheon \'trən-chən\ *noun* [Middle English *tronchoun,* from Middle French *tronchon,* from (assumed) Vulgar Latin *truncion-, truncio,* from Latin *truncus* trunk] (14th century)
1 : a shattered spear or lance
2 a *obsolete* **:** CLUB, BLUDGEON **b :** BATON 2 **c :** a police officer's billy club

²truncheon *transitive verb* (1597)
archaic **:** to beat with a truncheon

¹trun·dle \'trən-dᵊl\ *verb* **trun·dled; trun·dling** \'trən(d)-liŋ, 'trən-dᵊl-iŋ\ [²*trundle*] (1598)
transitive verb
1 a : to propel by causing to rotate **:** ROLL **b** *archaic* **:** to cause to revolve **:** SPIN
2 : to transport in or as if in a wheeled vehicle **:** HAUL, WHEEL
intransitive verb
1 : to progress by revolving
2 : to move on or as if on wheels **:** ROLL
— trun·dler \'trən(d)-lər, 'trən-dᵊl-ər\ *noun*

²trundle *noun* [from *trundle* small wheel, alteration of earlier *trendle,* from Middle English, circle, ring, wheel, from Old English *trendel;* akin to Old English *trendan* to revolve — more at TREND] (1851)
1 : TRUNDLE BED
2 : the motion or sound of something rolling
3 : a round or oval wooden tub

trundle bed *noun* (1542)
: a low bed usually on casters that can be rolled or slid under a higher bed when not in use — called also *truckle bed*

trun·dle–tail \'trən-dᵊl-ˌtāl\ *noun* (15th century)
archaic **:** a curly-tailed dog

trunk \'trəŋk\ *noun* [Middle English *tronke* box, trunk, from Middle French *tronc,* from Latin *truncus* trunk, torso] (15th century)
1 a : the main stem of a tree apart from limbs and roots — called also *bole* **b** (1) **:** the human or animal body apart from the head and appendages **:** TORSO (2) **:** the thorax of an insect **c :** the central part of anything; *specifically* **:** the shaft of a column or pilaster

2 a (1) **:** a large rigid piece of luggage used usually for transporting clothing and personal effects (2) **:** the luggage compartment of an automobile **b** (1) **:** a superstructure over a ship's hatches usually level with the poop deck (2) **:** the part of the cabin of a boat projecting above the deck (3) **:** the housing for a centerboard or rudder
3 : PROBOSCIS; *especially* **:** the long muscular proboscis of the elephant
4 *plural* **:** men's shorts worn chiefly for sports
5 a : a usually major channel or passage (as a chute or shaft) **b :** a circuit between two telephone exchanges for making connections between subscribers; *broadly* **:** a usually electronic path over which information is transmitted (as between computer memories)
6 a : the principal channel of a tributary system ⟨an arterial *trunk*⟩ ⟨*trunk* of a river⟩ **b :** TRUNK LINE
— **trunk·ful** \ˈtrəŋk-ˌfu̇l\ *noun*

trunked \ˈtrəŋ(k)t\ *adjective* (1640)
: having a trunk especially of a specified kind — usually used in combination ⟨a gray-*trunked* tree⟩

trunk·fish \ˈtrəŋk-ˌfish\ *noun* (circa 1804)
: any of numerous small often bright-colored bony fishes (family Ostraciidae) of tropical seas with the body and head enclosed in a bony carapace

trunk hose \ˈtrəŋk-\ *noun plural* [probably from obsolete English *trunk* to truncate] (1637)
: short full breeches reaching about halfway down the thigh that were worn chiefly in the late 16th and early 17th centuries

trunk line *noun* (1843)
1 : a transportation system (as an airline, railroad, or highway) handling long-distance through traffic
2 a : a main supply channel (as for gas or oil) **b :** TRUNK 5b

trun·nel \ˈtrə-nᵊl\ *variant of* TREENAIL

trun·nion \ˈtrən-yən\ *noun* [French *trognon* core, stump] (circa 1625)
: a pin or pivot on which something can be rotated or tilted; *especially* **:** either of two opposite gudgeons on which a cannon is swiveled

¹truss \ˈtrəs\ *transitive verb* [Middle English, from Old French *trousser, tourser* to bundle, pack, from (assumed) Vulgar Latin *torsare,* from *torsus* twisted — more at TORSADE] (13th century)
1 a : to secure tightly **:** BIND **b :** to arrange for cooking by binding close the wings or legs of (a fowl)
2 : to support, strengthen, or stiffen by or as if by a truss
— **truss·er** *noun*

²truss *noun* (13th century)
1 : an iron band around a lower mast with an attachment by which a yard is secured to the mast
2 a : BRACKET 1 **b :** an assemblage of members (as beams) forming a rigid framework
3 : a device worn to reduce a hernia by pressure
4 : a compact flower or fruit cluster

truss bridge *noun* (1840)
: a bridge supported mainly by trusses — see BRIDGE illustration

truss·ing \ˈtrə-siŋ\ *noun* (1840)
1 : the members forming a truss
2 : the trusses and framework of a structure

¹trust \ˈtrəst\ *noun* [Middle English, probably of Scandinavian origin; akin to Old Norse *traust* trust; akin to Old English *trēowe* faithful — more at TRUE] (13th century)
1 a : assured reliance on the character, ability, strength, or truth of someone or something **b :** one in which confidence is placed
2 a : dependence on something future or contingent **:** HOPE **b :** reliance on future payment for property (as merchandise) delivered **:** CREDIT
3 a : a property interest held by one person for the benefit of another **b :** a combination of firms or corporations formed by a legal agreement; *especially* **:** one that reduces or threatens to reduce competition
4 *archaic* **:** TRUSTWORTHINESS
5 a (1) **:** a charge or duty imposed in faith or confidence or as a condition of some relationship (2) **:** something committed or entrusted to one to be used or cared for in the interest of another **b :** responsible charge or office **c :** CARE, CUSTODY ⟨the child committed to her *trust*⟩
— **in trust :** in the care or possession of a trustee

²trust (13th century)
intransitive verb
1 a : to place confidence **:** DEPEND ⟨*trust* in God⟩ ⟨*trust* to luck⟩ **b :** to be confident **:** HOPE
2 : to sell or deliver on credit
transitive verb
1 a : to commit or place in one's care or keeping **:** ENTRUST **b :** to permit to stay or go or to do something without fear or misgiving
2 a : to rely on the truthfulness or accuracy of **:** BELIEVE **b :** to place confidence in **:** rely on **c :** to hope or expect confidently
3 : to extend credit to
— **trust·abil·i·ty** \ˌtrəs-tə-ˈbi-lə-tē\ *noun*
— **trust·able** \ˈtrəs-tə-bəl\ *adjective*
— **trust·er** *noun*
— **trust·ing·ly** \ˈtrəs-tiŋ-lē\ *adverb*
— **trust·ing·ness** *noun*

trust·bust·er \ˈtrəs(t)-ˌbəs-tər\ *noun* (1903)
: one who seeks to break up business trusts; *specifically* **:** a federal official who prosecutes trusts under the antitrust laws
— **trust–bust·ing** \-tiŋ\ *noun*

trust company *noun* (1834)
: an incorporated trustee; *broadly* **:** a corporation that functions as a corporate and personal trustee and usually also engages in the normal activities of a commercial bank

¹trust·ee \ˌtrəs-ˈtē\ *noun* (1647)
1 a : one to whom something is entrusted **b :** a country charged with the supervision of a trust territory
2 a : a natural or legal person to whom property is legally committed to be administered for the benefit of a beneficiary (as a person or a charitable organization) **b :** one (as a corporate director) occupying a position of trust and performing functions comparable to those of a trustee

²trustee *verb* **trust·eed; trust·ee·ing** (1818)
transitive verb
: to commit to the care of a trustee
intransitive verb
: to serve as trustee

trust·ee·ship \ˌtrəs-ˈtē-ˌship\ *noun* (circa 1736)
1 : the office or function of a trustee
2 : supervisory control by one or more countries over a trust territory

trust·ful \ˈtrəst-fəl\ *adjective* (1834)
: full of trust **:** CONFIDING
— **trust·ful·ly** \-fə-lē\ *adverb*
— **trust·ful·ness** *noun*

trust fund *noun* (1780)
: property (as money or securities) settled or held in trust

trust·less \ˈtrəst-ləs\ *adjective* (circa 1530)
1 : not deserving of trust **:** FAITHLESS
2 : DISTRUSTFUL

trust territory *noun* (1945)
: a non-self-governing territory placed under an administrative authority by the Trusteeship Council of the United Nations

trust·wor·thy \ˈtrəst-ˌwər-t͟hē\ *adjective* (1829)
: worthy of confidence **:** DEPENDABLE
— **trust·wor·thi·ly** \-t͟hə-lē\ *adverb*
— **trust·wor·thi·ness** *noun*

¹trusty \ˈtrəs-tē\ *adjective* **trust·i·er; -est** (14th century)
: TRUSTWORTHY, DEPENDABLE
— **trust·i·ness** *noun*

²trusty \ˈtrəs-tē *also* ˌtrəs-ˈtē\ *noun, plural* **trust·ies** (1573)
: a trusty or trusted person; *specifically* **:** a convict considered trustworthy and allowed special privileges

truth \ˈtrüth\ *noun, plural* **truths** \ˈtrüt͟hz, ˈtrüths\ [Middle English *trewthe,* from Old English *trēowth* fidelity; akin to Old English *trēowe* faithful — more at TRUE] (before 12th century)
1 a *archaic* **:** FIDELITY, CONSTANCY **b :** sincerity in action, character, and utterance
2 a (1) **:** the state of being the case **:** FACT (2) **:** the body of real things, events, and facts **:** ACTUALITY (3) *often capitalized* **:** a transcendent fundamental or spiritual reality **b :** a judgment, proposition, or idea that is true or accepted as true ⟨*truths* of thermodynamics⟩ **c :** the body of true statements and propositions
3 a : the property (as of a statement) of being in accord with fact or reality **b** *chiefly British* **:** TRUE 2 **c :** fidelity to an original or to a standard
4 *capitalized, Christian Science* **:** GOD
— **in truth :** in accordance with fact **:** ACTUALLY

truth·ful \ˈtrüth-fəl\ *adjective* (1596)
: telling or disposed to tell the truth
— **truth·ful·ly** \-fə-lē\ *adverb*
— **truth·ful·ness** *noun*

truth serum *noun* (1925)
: a hypnotic or anesthetic held to induce a subject under questioning to talk freely

truth set *noun* (1940)
: a mathematical or logical set containing all the elements that make a given statement of relationships true when substituted in it ⟨the equation $x + 7 = 10$ has as its *truth set* the single number 3⟩

truth table *noun* (1921)
: a table that shows the truth-value of a compound statement for every truth-value of its component statements; *also* **:** a similar table (as for a computer logic circuit) showing the value of the output for each value of each input

TRUTH TABLE

a statement p	a statement q	not p denial $\sim p$	p and q conjunction $p \cdot q$	p or q (inclusive) inclusive disjunction $p \vee q$	p or q (exclusive) exclusive disjunction	if p then q conditional $p \supset q$	p if and only if q biconditional $p \equiv q$
T	T	F	T	T	F	T	T
T	F		F	T	T	F	F
F	T	T	F	T	T	T	F
F	F		F	F	F	T	T

T = true   F = false

truth–value *noun* (1903)
: the truth or falsity of a proposition or statement

¹try \ˈtrī\ *verb* **tried; try·ing** [Middle English *trien,* from Anglo-French *trier,* from Old French, to pick out, sift, probably from Late

Latin *tritare* to grind, frequentative of Latin *terere* to rub — more at THROW] (14th century)
transitive verb
1 a : to examine or investigate judicially **b** (1) **:** to conduct the trial of (2) **:** to participate as counsel in the judicial examination of
2 a : to put to test or trial ⟨*try* one's luck⟩ — often used with *out* **b :** to subject to something (as undue strain or excessive hardship or provocation) that tests the powers of endurance **c :** DEMONSTRATE, PROVE
3 a *obsolete* **:** PURIFY, REFINE **b :** to melt down and procure in a pure state **:** RENDER ⟨*try* out whale oil from blubber⟩
4 : to fit or finish with accuracy
5 : to make an attempt at — often used with an infinitive
intransitive verb
: to make an attempt ■
synonym see AFFLICT, ATTEMPT
— try one's hand : to attempt something for the first time

²try *noun, plural* **tries** (1607)
1 : an experimental trial **:** ATTEMPT
2 : a play in rugby that is similar to a touchdown in football, scores usually four points, and entitles the scoring side to attempt a placekick at the goal for additional points; *also* **:** the score made on a try

try for point (1929)
: an attempt made after scoring a touchdown in football to score one or two additional points by kicking the ball over the crossbar or again carrying it into the opponents' end zone

try·ing \ˈtrī-iŋ\ *adjective* (1718)
: severely straining the powers of endurance
— try·ing·ly \-iŋ-lē\ *adverb*

try on *transitive verb* (1693)
1 : to put on (a garment) in order to test the fit
2 : to use or test experimentally
— try–on \ˈtrī-ˌȯn, -ˌän\ *noun*

try·out \ˈtrī-ˌau̇t\ *noun* (1903)
: an experimental performance or demonstration: as **a :** a test of the ability (as of an athlete or actor) to fill a part or meet standards **b :** a performance of a play prior to its official opening to determine response and discover weaknesses

try out *intransitive verb* (1909)
: to compete for a position especially on an athletic team or for a part in a play

try·pano·some \tri-ˈpa-nə-ˌsōm\ *noun* [New Latin *Trypanosoma,* from Greek *trypanon* auger + New Latin *-soma* -some — more at TREPAN] (1903)
: any of a genus (*Trypanosoma*) of parasitic flagellate protozoans that infest the blood of various vertebrates including humans, are usually transmitted by the bite of an insect, and include some that cause serious disease (as sleeping sickness)

try·pano·so·mi·a·sis \tri-ˌpa-nə-sə-ˈmī-ə-səs\ *noun, plural* **-a·ses** \-ˌsēz\ [New Latin] (1902)
: infection with or disease caused by trypanosomes

try–pot \ˈtrī-ˌpät\ *noun* (1795)
: a metallic pot used on a whaler or on shore to render whale oil from blubber

tryp·sin \ˈtrip-sən\ *noun* [perhaps from Greek *tryein* to wear down + English pe*psin;* akin to Latin *terere* to rub — more at THROW] (1876)
: a proteolytic enzyme from pancreatic juice active in an alkaline medium — compare CHYMOTRYPSIN

tryp·sin·o·gen \trip-ˈsi-nə-jən\ *noun* [International Scientific Vocabulary] (circa 1890)
: the inactive substance released by the pancreas into the duodenum to form trypsin

trypt·amine \ˈtrip-tə-ˌmēn\ *noun* [*trypto*phan + *amine*] (1929)
: a crystalline amine $C_{10}H_{12}N_2$ derived from tryptophan

tryp·tic \ˈtrip-tik\ *adjective* [International Scientific Vocabulary, from *trypsin*] (1888)
: of, relating to, or produced by trypsin or its action

tryp·to·phan \ˈtrip-tə-ˌfan\ *also* **tryp·to·phane** \-ˌfān\ *noun* [International Scientific Vocabulary *tryptic* + *-o-* + *-phane*] (1890)
: a crystalline essential amino acid $C_{11}H_{12}N_2O_2$ that is widely distributed in proteins

try·sail \ˈtrī-ˌsāl, -səl\ *noun* [obsolete *at try* lying to] (1769)
: a fore-and-aft sail bent to a gaff and hoisted on a lower mast or a small mast close abaft

try square *noun* (circa 1877)
: an instrument consisting of two straightedges fixed at right angles to each other and used for laying off right angles and testing whether work is square

¹tryst \ˈtrist, *especially British* ˈtrīst\ *noun* [Middle English, from Middle French *triste* watch post, probably of Scandinavian origin; akin to Old Norse *traust* trust] (14th century)
1 : an agreement (as between lovers) to meet
2 : an appointed meeting or meeting place

²tryst *intransitive verb* (14th century)
: to make or keep a tryst
— tryst·er \ˈtris-tər, ˈtrīs-\ *noun*

try·works \ˈtrī-ˌwərks\ *noun plural* (1792)
: a brick furnace in which try-pots are placed; *also* **:** the furnace with the pots

tsa·de \ˈ(t)sä-də, -dē\ *variant of* SADHE

tsar \ˈzär, ˈ(t)sär\ *variant of* CZAR

tset·se fly \ˈ(t)set-sē-, ˈtet-, ˈ(t)sēt-, ˈtēt-\ *noun* [Afrikaans, from Tswana *tsètsè* fly] (1865)
: any of several dipteran flies (genus *Glossina*) that occur in Africa south of the Sahara and include vectors of human and animal trypanosomes — called also *tsetse;* compare SLEEPING SICKNESS

Tshi \ˈchwē, chə-ˈwē, ˈtwē, ˈchē\ *variant of* TWI

Tshi·lu·ba \chi-ˈlü-bə\ *noun* (circa 1961)
: a Bantu language used as a lingua franca in southeastern Zaire

T–shirt \ˈtē-ˌshərt\ *noun* (1920)
: a collarless short-sleeved or sleeveless usually cotton undershirt; *also* **:** an outer shirt of similar design
— T–shirt·ed *adjective*

T square *noun* (1785)
: a ruler with a crosspiece or head at one end used in making parallel lines

tsu·na·mi \(t)su̇-ˈnä-mē\ *noun, plural* **tsunamis** *also* **tsunami** [Japanese, from *tsu* harbor + *nami* wave] (1897)
: a great sea wave produced by submarine earth movement or volcanic eruption **:** TIDAL WAVE
— tsu·na·mic \-mik\ *adjective*

tsu·tsu·ga·mu·shi disease \ˌ(t)süt-sə-gə-ˈmü-shē-, ˌtüt-\ *noun* [Japanese *tsutsugamushi* scrub typhus mite, from *tsutsuga* sickness + *mushi* insect] (1906)
: an acute febrile bacterial disease that is caused by a rickettsia (*Rickettsia tsutsugamushi*) transmitted by mite larvae, resembles louse-borne typhus, and is widespread in the western Pacific area — called also *scrub typhus, tsutsugamushi*

Tswa·na \ˈ(t)swä-nə\ *noun, plural* **Tswana** *or* **Tswanas** (1937)
1 : the language of the Tswana people
2 : a member of a Bantu-speaking people of Botswana and the Republic of South Africa

t–test \ˈtē-ˈtest\ *noun* (1932)
: a statistical test involving confidence limits for the random variable *t* of a t distribution and used especially in testing hypotheses about means of normal distributions when the standard deviations are unknown

T_3 *also* **T–3** \ˈtē-ˈthrē\ *noun* [*T,* abbreviation for *thyronine* + *3,* number of iodine atoms attached to the thyronine nucleus] (1956)
: TRIIODOTHYRONINE

Tua·reg \ˈtwä-ˌreg\ *noun, plural* **Tuareg** *or* **Tuaregs** [Arabic *Tawāriq*] (1821)
: a member of a nomadic people of the central and western Sahara and along the middle Niger from Tombouctou to Nigeria

tu·a·ta·ra \ˌtü-ə-ˈtär-ə\ *noun* [Maori *tuatàra*] (1890)
: a large spiny quadrupedal reptile (*Sphenodon punctatum*) of islands off the coast of New Zealand that has a vestigial third eye in the middle of the forehead representing the pineal gland and that is the only surviving rhynchocephalian

tuatara

¹tub \ˈtəb\ *noun* [Middle English *tubbe,* from Middle Dutch; akin to Middle Low German *tubbe* tub] (14th century)
1 a : a wide low vessel originally formed with wooden staves, round bottom, and hoops **b :** a small round container in which a product is sold ⟨a *tub* of oleo⟩
2 : an old or slow boat
3 : BATHTUB; *also* **:** BATH
4 : the amount that a tub will hold
— tub·ful \-ˌfu̇l\ *noun*
— tub·like \-ˌlīk\ *adjective*

²tub *verb* **tubbed; tub·bing** (1610)
transitive verb
1 : to wash or bathe in a tub
2 : to put or store in a tub
intransitive verb
1 : BATHE
2 : to undergo washing
— tub·ba·ble \ˈtə-bə-bəl\ *adjective*
— tub·ber *noun*

tu·ba \ˈtü-bə, ˈtyü-\ *noun* [Italian, from Latin, trumpet] (1852)
: a large low-pitched brass instrument usually oval in shape and having a conical tube, a cup-shaped mouthpiece, and a usual range an octave lower than that of the euphonium
— tu·ba·ist \-bə(-i)st\ *or* **tub·ist** \-bist\ *noun*

tuba

tub·al \ˈtü-bəl, ˈtyü-\ *adjective* (circa 1736)

□ USAGE
try Transitive sense 5 of *try* is commonly followed by an infinitive phrase. *And* has been used where *to* might be expected since the late 17th century; the combination *try and* is of just about the same age as this sense of *try. Try and* is often found in informal circumstances ⟨*try and* write straight English —Ernest Hemingway (letter)⟩ ⟨I'm going to *try and* see him today —E. B. White (letter)⟩. It is also common in general prose, the sort that does not aspire to great elegance ⟨to *try and* soften his father's anger —George Eliot⟩ ⟨I am glad of the opportunity to *try and* get this point cleared up —William Empson⟩ ⟨he always dressed rapidly, so as to *try and* conserve his night warmth till the sun rose —Doris Lessing⟩. *Try and* is, however, not as adaptable as *try to.* If *try* needs to be inflected, *try and* cannot be used ⟨like *trying to* ward off a flood with a teacup —Gloria Naylor⟩. A negative such as *don't* can be prefixed to *try and,* but if *not* follows *try, to* must be used ⟨*try* not *to* take her out shopping —*McCall's*⟩. Almost all disparaging criticism of *try and* comes from American critics; British commentators have generally been tolerant. There appears to be no rational basis for hostility to the expression and no need to avoid it in appropriate surroundings.

: of, relating to, or involving a tube and especially a fallopian tube ⟨*tubal* infection⟩

tubal ligation *noun* (1948)
: ligation of the fallopian tubes to prevent passage of ova from the ovaries to the uterus used as a method of female sterilization

tub•by \'tə-bē\ *adjective* **tub•bi•er; -est** (circa 1807)
1 : sounding dull and without proper resonance or freedom of sound ⟨a *tubby* violin⟩
2 : PUDGY, FAT

tube \'tüb, 'tyüb\ *noun* [French, from Latin *tubus;* akin to Latin *tuba* trumpet] (1651)
1 : any of various usually cylindrical structures or devices: as **a :** a hollow elongated cylinder; *especially* **:** one to convey fluids **b :** a soft tubular container whose contents (as toothpaste) can be removed by squeezing **c** (1) **:** TUNNEL (2) *British* **:** SUBWAY b **d :** the basically cylindrical section between the mouthpiece and bell that is the fundamental part of a wind instrument
2 a : a slender channel within a plant or animal body **:** DUCT — compare FALLOPIAN TUBE **b :** the narrow basal portion of a gamopetalous corolla or a calyx with united sepals
3 : INNER TUBE
4 a : ELECTRON TUBE; *especially* **:** VACUUM TUBE **b :** CATHODE-RAY TUBE; *especially* **:** a television picture tube **c :** TELEVISION
5 : an article of clothing shaped like a tube ⟨*tube* top⟩ ⟨*tube* socks⟩
— tubed \'tübd, 'tyübd\ *adjective*
— tube•like \'tüb-ˌlīk, 'tyüb-\ *adjective*
— down the tube *or* **down the tubes**
: into a state of collapse, deterioration, or ruin

tube foot *noun* (1888)
: one of the small flexible tubular processes of most echinoderms that are extensions of the water-vascular system and are used especially in locomotion and grasping

tube•less \'tü-bləs, 'tyü-\ *adjective* (1855)
: lacking a tube; *specifically* **:** being a pneumatic tire that does not depend on an inner tube for airtightness

tube nucleus *noun* (1939)
: the one of the two nuclei formed by mitotic division of a microspore during the formation of a pollen grain that is held to control subsequent growth of the pollen tube and that does not divide again — compare GENERATIVE NUCLEUS

tube pan *noun* (1926)
: a round cake pan having a hollow tube in the center

tu•ber \'tü-bər, 'tyü-\ *noun* [Latin, swelling, truffle; perhaps akin to Latin *tumēre* to swell — more at THUMB] (1668)
1 a : a short fleshy usually underground stem bearing minute scale leaves each of which bears a bud in its axil and is potentially able to produce a new plant — compare BULB, CORM **b :** a fleshy root or rhizome resembling a tuber
2 : an anatomical prominence **:** TUBEROSITY

tu•ber•cle \'tü-bər-kəl, 'tyü-\ *noun* [Latin *tuberculum,* diminutive of *tuber*] (1578)
1 : a small knobby prominence or excrescence especially on a plant or animal **:** NODULE: as **a :** a protuberance near the head of a rib that articulates with the transverse process of a vertebra **b :** any of several prominences in the central nervous system **c :** NODULE b
2 : a small abnormal discrete lump in the substance of an organ or in the skin; *especially* **:** the specific lesion of tuberculosis

tubercle bacillus *noun* (circa 1890)
: a bacterium (*Mycobacterium tuberculosis*) that is a major cause of tuberculosis

tubercul- *combining form* [New Latin, from Latin *tuberculum*]
1 : tubercle ⟨*tubercul*ar⟩
2 : tubercle bacillus ⟨*tubercul*in⟩
3 : tuberculosis ⟨*tubercul*oid⟩

¹tu•ber•cu•lar \tu̇-'bər-kyə-lər, tyu̇-\ *adjective* (1799)
1 a : of, relating to, or affected with tuberculosis **:** TUBERCULOUS **b :** caused by the tubercle bacillus ⟨*tubercular* meningitis⟩
2 : characterized by lesions that are or resemble tubercles ⟨*tubercular* leprosy⟩
3 : relating to, resembling, or constituting a tubercle **:** TUBERCULATED

²tubercular *noun* (1925)
: a person with tuberculosis

tu•ber•cu•lat•ed \tu̇-'bər-kyə-ˌlā-təd, tyu̇-\ *also* **tu•ber•cu•late** \-lət\ *adjective* (1771)
: having tubercles **:** characterized by or beset with tubercles

tu•ber•cu•lin \tu̇-'bər-kyə-lən, tyu̇-\ *noun* [International Scientific Vocabulary] (circa 1890)
: a sterile liquid containing the growth products of or specific substances extracted from the tubercle bacillus and used in the diagnosis of tuberculosis especially in children and cattle

tuberculin test *noun* (circa 1900)
: a test for hypersensitivity to tuberculin as an indication of past or present tubercular infection

tu•ber•cu•loid \tu̇-'bər-kyə-ˌlȯid, tyu̇-\ *adjective* [International Scientific Vocabulary] (circa 1923)
: resembling tuberculosis especially in the presence of tubercles ⟨*tuberculoid* leprosy⟩

tu•ber•cu•lo•sis \tu̇-ˌbər-kyə-'lō-səs, tyu̇-\ *noun, plural* **-lo•ses** \-ˌsēz\ [New Latin] (1860)
: a highly variable communicable disease of humans and some other vertebrates caused by the tubercle bacillus and rarely in the U.S. by a related mycobacterium (*Mycobacterium bovis*) and characterized by toxic symptoms or allergic manifestations which in humans primarily affect the lungs

tu•ber•cu•lous \tu̇-'bər-kyə-ləs, tyu̇-\ *adjective* (1891)
1 : constituting or affected with tuberculosis ⟨a *tuberculous* process⟩
2 : caused by or resulting from the presence or products of the tubercle bacillus ⟨*tuberculous* peritonitis⟩

tube•rose \'tü-ˌbrōz, 'tyü-, *also* -bə-ˌrōz, -bə-ˌrōs\ *noun* [New Latin *tuberosa,* specific epithet, from Latin, feminine of *tuberosus* tuberous, from *tuber* tuber] (1664)
: a Mexican bulbous herb (*Polianthes tuberosa*) of the agave family cultivated for its spike of fragrant white single or double flowers

tu•ber•os•i•ty \ˌtü-bə-'rä-sə-tē, ˌtyü-\ *noun, plural* **-ties** (1611)
: a rounded prominence; *especially* **:** a large prominence on a bone usually serving for the attachment of muscles or ligaments

tu•ber•ous \'tü-b(ə-)rəs, 'tyü-\ *adjective* (1650)
1 : consisting of, bearing, or resembling a tuber
2 : of, relating to, or being a plant tuber or tuberous root of a plant

tuberous root *noun* (circa 1668)
: a thick fleshy storage root like a tuber but lacking buds or scale leaves
— tuberous–rooted *adjective*

tube worm *noun* (circa 1819)
: a worm that lives in a tube: as **a :** any of various polychaetes or oligochaetes **b :** a pogonophoran and especially an extremely large one that is found near deep-ocean hydrothermal vents

tu•bi•fex \'tü-bə-ˌfeks, 'tyü-\ *noun* [New Latin *Tubific-, Tubifex,* from Latin *tubus* tube + *facere* to make — more at DO] (1948)
: any of a genus (*Tubifex*) of slender reddish oligochaete worms that live in tubes in fresh or brackish water and are widely used as food for aquarium fish

tu•bi•fi•cid \tü-'bi-fə-səd, tyü-; ˌt(y)ü-bə-'fi-səd\ *noun* [New Latin *Tubificidae,* from *Tubific-, Tubifex*] (1950)
: any of a family (Tubificidae) of aquatic oligochaetes including the tubifex worms
— tubificid *adjective*

tub•ing \'tü-biŋ, 'tyü-\ *noun* (1845)
1 : material in the form of a tube; *also* **:** a length or piece of tube
2 : a series or system of tubes

tu•bo•cu•ra•rine \ˌtü-bō-kyu̇-'rär-ən, ˌtyü-, -ˌēn\ *noun* [International Scientific Vocabulary *tubo-* (from Latin *tubus* tube) + *curare* + *-ine;* from its being shipped in sections of hollow bamboo] (1898)
: a toxic alkaloid or its crystalline hydrated hydrochloride salt $C_{37}H_{42}Cl_2N_2O_6{\cdot}5H_2O$ that is obtained chiefly from the bark and stems of a South American vine (*Chondrodendron tomentosum* of the family Menispermaceae) and in its dextrorotatory form constitutes the chief active constituent of curare and is used especially as a skeletal muscle relaxant

tub–thump•er \'təb-ˌthəm-pər\ *noun* (1662)
: a vociferous supporter (as of a cause)
— tub–thump \-ˌthəmp\ *verb*

tu•bu•lar \'tü-byə-lər, 'tyü-\ *adjective* (1673)
1 a : having the form of or consisting of a tube ⟨a *tubular* calyx⟩ **b :** made or provided with tubes
2 : of, relating to, or sounding as if produced through tubes

tu•bule \'tü-(ˌ)byü(ə)l, 'tyü-\ *noun* [Latin *tubulus,* diminutive of *tubus*] (1677)
: a small tube; *especially* **:** a slender elongated anatomical channel

tu•bu•lin \'tü-byə-lən, 'tyü-\ *noun* [*tubule* + ¹*-in*] (1968)
: a globular protein that polymerizes to form microtubules

tu•chun \'dü-'jün, -'jœn\ *noun* [Chinese (Beijing) *dūjūn*] (1917)
1 : a Chinese military governor (as of a province)
2 : a Chinese warlord

¹tuck \'tək\ *verb* [Middle English *tuken* to pull up sharply, scold, from Old English *tūcian* to ill-treat; akin to Old High German *zuhhen* to jerk, Old English *togian* to pull — more at TOW] (15th century)
transitive verb
1 a : to pull up into a fold **b :** to make a tuck in
2 : to put into a snug often concealing or isolating place ⟨cottage *tucked* away in the hill⟩
3 a : to push in the loose end of so as to hold tightly ⟨*tuck* in your shirt⟩ **b :** to cover by tucking in bedclothes — usually used with *in*
4 : EAT — usually used with *away* or *in* ⟨*tucked* away a big lunch⟩
5 : to put into a tuck position
intransitive verb
1 : to draw together into tucks or folds
2 : to eat or drink heartily — usually used with *into* ⟨*tucked* into their beer and pretzels⟩
3 : to fit snugly

²tuck *noun* (1532)
1 : a fold stitched into cloth to shorten, decorate, or control fullness
2 : the part of a vessel where the ends of the lower planks meet under the stern
3 a : an act or instance of tucking **b :** something tucked or to be tucked in
4 a : a body position (as in diving) in which the knees are bent, the thighs drawn tightly to the chest, and the hands clasped around the shins **b :** a skiing position in which the skier squats forward and holds the ski poles under the arms and parallel to the ground
5 : a cosmetic surgical operation for the removal of excess skin or fat from a body part ⟨a tummy *tuck*⟩

3tuck *noun* [Middle English (Scots) *tuicke* beat, stroke] (15th century)
: a sound of or as if of a drumbeat

4tuck *noun* [Middle French *estoc,* from Old French, tree trunk, sword point, of Germanic origin; akin to Old English *stocc* stump of a tree — more at STOCK] (1508)
archaic **:** RAPIER

5tuck *noun* [probably from 2*tuck*] (1878)
: VIGOR, ENERGY ⟨seemed to kind of take the *tuck* all out of me —Mark Twain⟩

tuck·a·hoe \ˈtə-kə-ˌhō\ *noun* [Virginia Algonquian *tockawhoughe*] (1612)
1 : either of two arums (*Peltandra virginica* and *Orontium aquaticum*) of the U.S. with rootstocks used as food by American Indians
2 : the large edible sclerotium of a subterranean fungus (*Poria cocos*)

1tuck·er \ˈtə-kər\ *noun* (1688)
1 : a piece of lace or cloth in the neckline of a dress
2 : one that tucks
3 *chiefly Australian* **:** FOOD

2tucker *transitive verb* **tuck·ered; tuck·er·ing** \ˈtə-k(ə-)riŋ\ [obsolete English *tuck* to reproach + *-er* (as in 1*batter*)] (1833)
: EXHAUST — often used with *out*

tuck·er–bag \ˈtə-kər-ˌbag\ *noun* (1885)
chiefly Australian **:** a bag used especially by travelers in the bush to hold food

tuck·et \ˈtə-kət\ *noun* [probably from obsolete English *tuk* to beat the drum, sound the trumpet] (1593)
: a fanfare on a trumpet

tuck–point \ˈtək-ˌpȯint\ *transitive verb* (1881)
: to finish (the mortar joints between bricks or stones) with a narrow ridge of putty or fine lime mortar

tuckshop \-ˌshäp\ *noun* [British *tuck* food, confectionery] (1857)
British **:** a confectioner's shop **:** CONFECTIONERY

Tu·dor \ˈtü-dər, ˈtyü-\ *adjective* [Henry *Tudor* (Henry VII of England)] (1779)
1 : of or relating to the English royal house that ruled from 1485 to 1603
2 : of, relating to, or characteristic of the Tudor period
— Tudor *noun*

Tudor arch *noun* (1815)
: a low elliptical 3-, 4-, or 5-centered arch; *especially* **:** a 4-centered pointed arch — see ARCH illustration

Tues·day \ˈtüz-dē, ˈtyüz-, -(ˌ)dā\ *noun* [Middle English *tiwesday,* from Old English *tīwesdæg* (akin to Old High German *zīostag* Tuesday), from Old English *Tīw* Tiu + *dæg* day — more at DEITY] (before 12th century)
: the third day of the week
— Tues·days \-dēz, -(ˌ)dāz\ *adverb*

tu·fa \ˈtü-fə, ˈtyü-\ *noun* [Italian *tufo,* from Latin *tofus*] (1770)
1 : TUFF
2 : a porous rock formed as a deposit from springs or streams
— tu·fa·ceous \tü-ˈfā-shəs, tyü-\ *adjective*

tuff \ˈtəf\ *noun* [Middle French *tuf,* from Old Italian *tufo*] (1815)
: a rock composed of the finer kinds of volcanic detritus usually fused together by heat
— tuff·a·ceous \ˌtə-ˈfā-shəs\ *adjective*

tuf·fet \ˈtə-fət\ *noun* [alteration of 1*tuft*] (1553)
1 : TUFT 1a
2 : a low seat

1tuft \ˈtəft\ *noun* [Middle English, modification of Middle French *tufe*] (14th century)
1 a : a small cluster of elongated flexible outgrowths attached or close together at the base and free at the opposite ends; *especially* **:** a growing bunch of grasses or close-set plants **b :** a bunch of soft fluffy threads cut off short and used as ornament
2 : CLUMP, CLUSTER
3 : MOUND
4 : any of the projections of yarns drawn through a fabric or making up a fabric so as to produce a surface of raised loops or cut pile
— tuft·ed \ˈtəf-təd\ *adjective*
— tufty \ˈtəf-tē\ *adjective*

2tuft (1535)
transitive verb
1 a : to provide or adorn with a tuft **b :** to make (a fabric) of or with tufts
2 : to make (as a mattress) firm by stitching at intervals and sewing on tufts
intransitive verb
: to form into or grow in tufts
— tuft·er *noun*

1tug \ˈtəg\ *verb* **tugged; tug·ging** [Middle English *tuggen;* akin to Old English *togian* to pull — more at TOW] (14th century)
intransitive verb
1 : to pull hard
2 : to struggle in opposition **:** CONTEND
3 : to exert oneself laboriously **:** LABOR
transitive verb
1 : to pull or strain hard at
2 a : to move by pulling hard **:** HAUL **b :** to carry with difficulty **:** LUG
3 : to tow with a tugboat
— tug·ger *noun*

2tug *noun* (15th century)
1 a : 3TRACE 1 **b :** a short leather strap or loop **c :** a rope or chain used for pulling
2 a : an act or instance of tugging **:** PULL **b :** a strong pulling force
3 a : a straining effort **b :** a struggle between two people or opposite forces
4 : TUGBOAT

tug·boat \ˈtəg-ˌbōt\ *noun* (1830)
: a strongly built powerful boat used for towing and pushing — called also *towboat*

tug–of–war \ˌtəg-ə(v)-ˈwȯr\ *noun, plural* **tugs–of–war** (1677)
1 : a struggle for supremacy or control usually involving two antagonists
2 : a contest in which two teams pull against each other at opposite ends of a rope with the object of pulling the middle of the rope over a mark on the ground

tu·grik *or* **tu·ghrik** \ˈtü-grik\ *noun* [Mongolian *tögrig,* literally, circle, wheel] (1927)
— see MONEY table

tuille \ˈtwē(ə)l\ *noun* [Middle English *toile,* from Middle French *tuille* tile, from Latin *tegula* — more at TILE] (15th century)
: one of the hinged plates before the thigh in plate armor — see ARMOR illustration

tu·ition \tə-ˈwi-shən, tyü-\ *noun* [Middle English *tuicioun* protection, from Old French *tuicion,* from Latin *tuition-, tuitio,* from *tueri* to look at, look after] (15th century)
1 *archaic* **:** CUSTODY, GUARDIANSHIP
2 : the act or profession of teaching **:** INSTRUCTION ⟨pursued his studies under private *tuition*⟩
3 : the price of or payment for instruction
— tu·ition·al \-ˈwish-nəl, -ˈwi-shə-nᵊl\ *adjective*

tu·la·re·mia \ˌtü-lə-ˈrē-mē-ə, ˌtyü-\ *noun* [New Latin, from *Tulare* County, Calif.] (1921)
: an infectious disease especially of wild rabbits, rodents, humans, and some domestic animals that is caused by a bacterium (*Francisella tularensis*), is transmitted especially by the bites of insects, and in humans is marked by symptoms (as fever) of toxemia
— tu·la·re·mic \-mik\ *adjective*

tu·le \ˈtü-lē\ *noun* [Spanish, from Nahuatl *tōllin*] (1837)
: either of two large New World bulrushes (*Scirpus californicus* and *S. acutus*)

tu·lip \ˈtü-ləp, ˈtyü-\ *noun* [New Latin *tulipa,* from Turkish *tülbent* turban — more at TURBAN] (1578)
: any of a genus (*Tulipa*) of Eurasian bulbous herbs of the lily family that have linear or broadly lanceolate leaves and are widely grown for their showy flowers; *also* **:** the flower or bulb of a tulip ◆

tulip tree *noun* (1705)
1 : a tall North American timber tree (*Liriodendron tulipifera*) of the magnolia family having large greenish yellow tulip-shaped flowers and soft white wood used especially for cabinetwork and woodenware — called also *tulip poplar, yellow poplar*
2 : any of various trees other than the tulip tree with tulip-shaped flowers

tu·lip·wood \ˈtü-ləp-ˌwu̇d, ˈtyü-\ *noun* (1843)
1 : wood of the North American tulip tree
2 a : any of several showily striped or variegated woods; *especially* **:** the rose-colored wood of a Brazilian tree (*Physocalymma scaberrimum* of the family Lythraceae) that is much used by cabinetmakers for inlaying **b :** a tree that yields tulipwood

tulle \ˈtül\ *noun* [French, from *Tulle,* France] (circa 1818)
: a sheer often stiffened silk, rayon, or nylon net used chiefly for veils or ballet costumes

1tum·ble \ˈtəm-bəl\ *verb* **tum·bled; tum·bling** \-b(ə-)liŋ\ [Middle English, frequentative of *tumben* to dance, from Old English *tumbian;* akin to Old High German *tūmōn* to reel] (14th century)
intransitive verb
1 a : to fall suddenly and helplessly **b :** to suffer a sudden downfall, overthrow, or defeat **c :** to decline suddenly and sharply (as in price) **:** DROP ⟨the stock market *tumbled*⟩ **d :** to fall into ruin **:** COLLAPSE
2 a : to perform gymnastic feats in tumbling **b :** to turn end over end in falling or flight
3 : to roll over and over, to and fro, or end over end **:** TOSS
4 : to issue forth hurriedly and confusedly
5 : to come by chance **:** STUMBLE
6 : to come to understand **:** CATCH ON ⟨didn't *tumble* to the seriousness of the problem⟩
transitive verb
1 : to cause to tumble (as by pushing or toppling)
2 a : to throw together in a confused mass **b :** RUMPLE, DISORDER
3 : to whirl in a tumbling barrel

2tumble *noun* (1634)
1 a : a disordered mass of objects or material **b :** a disorderly state
2 : an act or instance of tumbling

tum·ble·bug \ˈtəm-bəl-ˌbəg\ *noun* (1805)

◇ WORD HISTORY

tulip Although the tulip is often associated with Holland, where it has been cultivated diligently, the Dutch and other Western European peoples originally introduced the plant from Turkey in the later 16th century. The first known reference to the flower is in the Latin letters of Augier Ghislain de Busbecq, the Holy Roman Empire's ambassador to the Turkish court, who, while traveling to Constantinople in 1554 alludes to the "huge abundance of flowers . . . narcissi, hyacinths, and those which the Turks call *tulipan.*" Busbecq appears to have heard the 16th century word corresponding to Modern Turkish *tülbent* "muslin, gauzy fabric." That word was once also applied to turbans wound from such fabric, which Busbecq's interpreter may have compared to the tulip. (*Tülbent* in the sense "turban" had earlier been borrowed into Romance languages and is the ultimate source of English *turban.*) It is not known for certain how Busbecq's word lost its *n* and became *tulipa*—the form on which modern English *tulip* is based—though *tulpe* and *tulipa* appear in an early Dutch dictionary, and the Dutch word may have influenced other languages. Curiously, the Turks both in Busbecq's time and today employ an entirely different word for the tulip, *lâle.*

: any of various scarabaeid beetles (especially genera *Scarabaeus, Canthon, Copris,* or *Phanaeus*) that roll dung into small balls, bury them in the ground, and lay eggs in them

tum·ble·down \'təm-bəl-'daủn\ *adjective* (1818)
: DILAPIDATED, RAMSHACKLE ⟨a *tumbledown* house at the edge of town —Sherwood Anderson⟩

tumble dry *transitive verb* (1962)
: to dry (as clothes) by tumbling in a dryer
— tumble dryer *noun*
— tumble drying *noun*

tum·bler \'təm-blər\ *noun* (14th century)
1 : one that tumbles: as **a :** one that performs tumbling feats **:** ACROBAT **b :** any of various domestic pigeons that tumble or somersault backward in flight or on the ground
2 : a drinking glass without foot or stem and originally with pointed or convex base
3 a : a movable obstruction in a lock (as a lever, latch, wheel, slide, or pin) that must be adjusted to a particular position (as by a key) before the bolt can be thrown **b :** a piece on which the mainspring acts in a gun's lock
4 : a device or mechanism for tumbling (as a revolving cage in which clothes are dried)
5 : a worker that operates a tumbler
— tum·bler·ful \-ˌfủl\ *noun*

tum·ble·weed \'təm-bəl-ˌwēd\ *noun* (1887)
: a plant (as Russian thistle or any of several amaranths) that breaks away from its roots in the autumn and is driven about by the wind as a light rolling mass

¹tum·bling \'təm-b(ə-)liŋ\ *noun* (1604)
: the skill, practice, or sport of executing gymnastic feats (as somersaults and handsprings) without the use of apparatus

²tumbling *adjective* (circa 1916)
: tipped or slanted out of the vertical — used especially of a cattle brand

tumbling barrel *noun* (circa 1890)
: a revolving cask in which objects or materials undergo a process (as drying or polishing) by being whirled about

tumbling verse *noun* (1585)
: an early modern English type of verse having four stresses but no prevailing type of foot and no regular number of syllables

tum·brel *or* **tum·bril** \'təm-brəl\ *noun* [Middle English *tombrel,* from Old French *tumberel* tipcart, from *tomber* to tumble, of Germanic origin; akin to Old High German *tūmōn* to reel — more at TUMBLE] (15th century)
1 : a farm tipcart
2 : a vehicle carrying condemned persons (as political prisoners during the French Revolution) to a place of execution

tu·me·fac·tion \ˌtü-mə-'fak-shən, ˌtyü-\ *noun* [Middle French, from Latin *tumefacere* to cause to swell, from *tumēre* to swell + *facere* to make, do — more at THUMB, DO] (1597)
1 : an action or process of swelling or becoming tumorous
2 : SWELLING

tu·mes·cence \tü-'me-sᵊn(t)s, tyü-\ *noun* (1859)
: the quality or state of being tumescent; *especially* **:** readiness for sexual activity marked especially by vascular congestion of the sex organs

tu·mes·cent \-sᵊnt\ *adjective* [Latin *tumescent-, tumescens,* present participle of *tumescere* to swell up, inchoative of *tumēre* to swell] (1882)
: somewhat swollen ⟨*tumescent* tissue⟩

tu·mid \'tü-məd, 'tyü-\ *adjective* [Latin *tumidus,* from *tumēre*] (1541)
1 : marked by swelling **:** SWOLLEN, ENLARGED ⟨a badly infected *tumid* leg⟩
2 : PROTUBERANT, BULGING ⟨sails *tumid* in the breeze⟩
3 : BOMBASTIC, TURGID

tum·my \'tə-mē\ *noun, plural* **tummies** [baby-talk alteration of *stomach*] (1867)
: STOMACH 1c

tu·mor \'tü-mər, 'tyü-\ *noun* [Latin *tumor,* from *tumēre*] (1597)
1 : a swollen or distended part
2 : an abnormal benign or malignant mass of tissue that is not inflammatory, arises without obvious cause from cells of preexistent tissue, and possesses no physiologic function
— tu·mor·al \-mə-rəl\ *adjective*
— tu·mor·like \-mər-ˌlīk\ *adjective*

tu·mor·i·gen·ic \ˌtü-mə-rə-'je-nik, ˌtyü-\ *adjective* (1941)
: producing or tending to produce tumors; *also* **:** CARCINOGENIC
— tu·mor·i·gen·e·sis \-'je-nə-səs\ *noun*
— tu·mor·i·ge·nic·i·ty \-jə-'ni-sə-tē\ *noun*

tumor necrosis factor *noun* (1975)
: a protein that is produced by monocytes and macrophages in response especially to endotoxins and that activates leukocytes and has antitumor activity

tu·mor·ous \'tü-mə-rəs, 'tyü-; 'tüm-rəs, 'tyüm-\ *adjective* (1547)
: of, relating to, or resembling a tumor

tu·mour \'tyü-mər\ *chiefly British variant of* TUMOR

¹tump \'təmp\ *noun* [origin unknown] (1589)
1 *chiefly dialect English* **:** MOUND, HUMMOCK
2 : a clump of vegetation

²tump *verb* [perhaps akin to British dialect *tumpoke* to fall head over heels] (1967)
intransitive verb
chiefly Southern **:** to tip or turn over especially accidentally — usually used with *over* ⟨sooner or later everybody *tumps* over. Nothing to worry about if you don't get caught under the canoe —Don Kennard⟩
transitive verb
chiefly Southern **:** to cause to tip over **:** OVERTURN, UPSET — usually used with *over*

tump·line \'təmp-ˌlīn\ *noun* [*tump,* of Algonquian origin; akin to Eastern Abenaki *mádûmbí* pack strap] (1796)
: a sling formed by a strap slung over the forehead or chest and used for carrying or helping to support a pack on the back or in hauling loads

tu·mult \'tü-ˌməlt, 'tyü- *also* 'tə-\ *noun* [Middle English *tumulte,* from Middle French, from Latin *tumultus;* perhaps akin to Sanskrit *tumula* noisy] (15th century)
1 a : disorderly agitation or milling about of a crowd usually with uproar and confusion of voices **:** COMMOTION **b :** a turbulent uprising **:** RIOT
2 : HUBBUB, DIN
3 a : violent agitation of mind or feelings **b :** a violent outburst

tu·mul·tu·ary \tủ-'məl-chə-ˌwer-ē, tyủ-, tə-\ *adjective* (1590)
: attended or marked by tumult, riot, lawlessness, confusion, or impetuosity

tu·mul·tu·ous \tủ-'məl-chə-wəs, tyủ-, tə-, -chəs; -'məlch-wəs\ *adjective* (circa 1548)
1 : marked by tumult
2 : tending or disposed to cause or incite a tumult
3 : marked by violent or overwhelming turbulence or upheaval
— tu·mul·tu·ous·ly *adverb*
— tu·mul·tu·ous·ness *noun*

tu·mu·lus \'tü-myə-ləs, 'tyü-, 'tə-\ *noun, plural* **-li** \-ˌlī, -ˌlē\ [Latin; akin to Latin *tumēre* to swell — more at THUMB] (1686)
: an artificial hillock or mound (as over a grave); *especially* **:** an ancient grave **:** BARROW

tun \'tən\ *noun* [Middle English *tunne,* from Old English, from Medieval Latin *tunna*] (before 12th century)
1 : a large cask especially for wine
2 : any of various units of liquid capacity; *especially* **:** one equal to 252 gallons

¹tu·na \'tü-nə\ *noun* [Spanish, from Taino] (circa 1555)
1 : any of various flat-jointed prickly pears (genus *Opuntia*); *especially* **:** one (*O. tuna*) of tropical America
2 : the edible fruit of a tuna

²tu·na \'tü-nə, 'tyü-\ *noun, plural* **tuna** *or* **tunas** [American Spanish, alteration of Spanish *atún,* modification of Arabic *tūn,* from Latin *thunnus,* from Greek *thynnos*] (1881)
1 : any of numerous large vigorous scombroid food and sport fishes (as an albacore or a bluefin tuna)
2 : the flesh of a tuna especially when canned for use as food — called also *tuna fish*

tun·able \'tü-nə-bəl, 'tyü-\ *adjective* (circa 1500)
1 *archaic* **a :** TUNEFUL **b :** sounding in tune **:** CONCORDANT
2 : capable of being tuned
— tun·abil·i·ty \ˌtü-nə-'bi-lə-tē, ˌtyü-\ *noun*
— tun·able·ness \'tü-nə-bəl-nəs, 'tyü-\ *noun*
— tun·ably \-blē\ *adverb*

tun·dish \'tən-ˌdish\ *noun* [Middle English, funnel for filling a tun] (14th century)
1 : FUNNEL 1a
2 : a reservoir in the top part of a mold into which molten metal is poured

tun·dra \'tən-drə *also* 'tün-\ *noun* [Russian, of Lappish origin; akin to Kola Lappish *tundar* hill] (circa 1841)
: a level or rolling treeless plain that is characteristic of arctic and subarctic regions, consists of black mucky soil with a permanently frozen subsoil, and has a dominant vegetation of mosses, lichens, herbs, and dwarf shrubs; *also* **:** a similar region confined to mountainous areas above timberline

¹tune \'tün, 'tyün\ *noun* [Middle English, alteration of *tone*] (14th century)
1 a *archaic* **:** quality of sound **:** TONE **b :** manner of utterance **:** INTONATION; *specifically* **:** phonetic modulation
2 a : a succession of pleasing musical tones **:** MELODY **b :** a dominant theme
3 : correct musical pitch or consonance — used chiefly in the phrases *in tune* and *out of tune*
4 a *archaic* **:** a frame of mind **:** MOOD **b :** AGREEMENT, HARMONY ⟨in *tune* with the times⟩ **c :** general attitude **:** APPROACH ⟨changed his *tune* when the going got rough⟩
5 : AMOUNT, EXTENT ⟨custom-made to the *tune* of $40 to $50 apiece —*American Fabrics*⟩

²tune *verb* **tuned; tun·ing** (15th century)
transitive verb
1 : to adjust in musical pitch or cause to be in tune ⟨*tuned* her guitar⟩
2 a : to bring into harmony **:** ATTUNE **b :** to adjust for precise functioning — often used with *up* ⟨*tune* up an engine⟩ **c :** to make more precise, intense, or effective
3 : to adjust with respect to resonance at a particular frequency: as **a :** to adjust (a radio or television receiver) to respond to waves of a particular frequency — often used with *in* **b :** to establish radio contact with ⟨*tune* in a directional beacon⟩
4 : to adjust the frequency of the output of (a device) to a chosen frequency or range of frequencies; *also* **:** to alter the frequency of (radiation)
intransitive verb
1 : to become attuned
2 : to adjust a radio or television receiver to respond to waves of a particular frequency

tuned–in \'tünd-'in, 'tyünd-\ *adjective* (1963)
: TURNED-ON

tune·ful \'tün-fəl, 'tyün-\ *adjective* (1591)
: MELODIOUS, MUSICAL
— tune·ful·ly \-fə-lē\ *adverb*
— tune·ful·ness *noun*

tune in (1913)
transitive verb
: to listen to or view a broadcast of ⟨*tuned in* the weather report⟩
intransitive verb
: to listen to or view a broadcast ⟨*tune in* next week for the conclusion⟩

tune·less \'tün-ləs, 'tyün-\ *adjective* (1591)
1 : not tuneful
2 : not producing music
— tune·less·ly *adverb*

tune out (1910)
transitive verb
: to become unresponsive to : IGNORE
intransitive verb
: to dissociate oneself from what is happening or one's surroundings

tun·er \'tü-nər, 'tyü-\ *noun* (circa 1801)
1 : one that tunes ⟨a piano *tuner*⟩
2 : something used for tuning; *specifically* **:** the part of a receiving set that converts radio signals into audio or video signals

tune·smith \'tün-ˌsmith, 'tyün-\ *noun* (1926)
: a composer especially of popular songs

tune–up \-ˌəp\ *noun* (1933)
1 : a general adjustment to insure operation at peak efficiency
2 : a preliminary trial : WARM-UP

tung \'təŋ\ *noun* (1914)
: TUNG TREE

tung oil *noun* [part translation of Chinese (Beijing) *tóngyóu*] (1881)
: a pale yellow pungent drying oil obtained from the seeds of tung trees and used chiefly in quick-drying varnishes and paints and as a waterproofing agent

tung·state \'təŋ-ˌstāt\ *noun* (1800)
: a salt or ester of a tungstic acid and especially of H_2WO_4

tung·sten \'təŋ-stən\ *noun* [Swedish, from *tung* heavy + *sten* stone] (1796)
: a gray-white heavy high-melting ductile hard polyvalent metallic element that resembles chromium and molybdenum in many of its properties and is used especially for electrical purposes and in hardening alloys (as steel) — called also *wolfram*; see ELEMENT table

tung·stic acid \'təŋ-stik-\ *noun* [*tungsten*] (1796)
: a yellow crystalline powder WO_3 that is the trioxide of tungsten; *also* **:** an acid (as H_2WO_4) derived from this

tung tree *noun* [Chinese (Beijing) *tóng*] (1895)
: any of several trees (genus *Aleurites*) of the spurge family whose seeds yield tung oil; *especially* **:** an Asian tree (*A. fordii*) widely grown in warm regions

Tun·gus \tůŋ-'güz, tən-\ *noun, plural* **Tungus** *or* **Tun·gus·es** [Russian] (1674)
1 : a member of an indigenous people of central and southeastern Siberia
2 : the Tungusic language of the Tungus people

Tun·gu·sic \-'gü-zik\ *noun* (circa 1867)
: a family of Altaic languages spoken in Manchuria and northward
— Tungusic *adjective*

tu·nic \'tü-nik, 'tyü-\ *noun* [Latin *tunica,* of Semitic origin; akin to Hebrew *kuttōneth* coat] (circa 1609)
1 a : a simple slip-on garment made with or without sleeves and usually knee-length or longer, belted at the waist, and worn as an under or outer garment by men and women of ancient Greece and Rome **b :** SURCOAT
2 : an enclosing or covering membrane or tissue ⟨the *tunic* of a seed⟩
3 : a long usually plain close-fitting jacket with high collar worn especially as part of a uniform
4 : TUNICLE
5 a : a short overskirt **b :** a hip-length or longer blouse or jacket

tu·ni·ca \'tü-ni-kə, 'tyü-\ *noun, plural* **tu·ni·cae** \-nə-ˌkē, -ˌkī, -ˌsē\ [Latin, tunic, membrane] (circa 1828)
: an enveloping membrane or layer of body tissue

[1]tu·ni·cate \'tü-ni-kət, 'tyü-, -nə-ˌkāt\ *also* **tu·ni·cat·ed** \-nə-ˌkā-təd\ *adjective* [Latin *tunicatus,* from *tunica*] (circa 1623)
1 a : having or covered with a tunic or tunica **b :** having, arranged in, or made up of concentric layers ⟨a *tunicate* bulb⟩
2 : of or relating to the tunicates

[2]tu·ni·cate \-ni-kət, -nə-ˌkāt\ *noun* [New Latin *Tunicata,* from neuter plural of Latin *tunicatus*] (1889)
: any of a subphylum (Urochordata synonym Tunicata) of marine chordate animals that have a thick secreted covering layer, a greatly reduced nervous system, a heart able to reverse the direction of blood flow, and only in the larval stage a notochord

tu·ni·cle \'tü-ni-kəl, 'tyü-\ *noun* [Middle English, from Latin *tunicula,* diminutive of *tunica*] (15th century)
: a short vestment worn by a subdeacon over the alb during mass and by a bishop under the dalmatic at pontifical ceremonies

tuning fork *noun* (1799)
: a 2-pronged metal implement that gives a fixed tone when struck and is useful for tuning musical instruments and ascertaining standard pitch

tuning pipe *noun* (1897)
: PITCH PIPE; *specifically* **:** one of a set of pitch pipes used especially for tuning stringed musical instruments

[1]tun·nel \'tə-nᵊl\ *noun* [Middle English *tonel* tube-shaped net, from Middle French, tun, from Old French, from *tonne* tun, from Medieval Latin *tunna*] (1548)
1 : a hollow conduit or recess : TUBE, WELL
2 a : a covered passageway; *specifically* **:** a horizontal passageway through or under an obstruction **b :** a subterranean gallery (as in a mine) **c :** BURROW
— tun·nel·like \-nᵊl-ˌ(l)īk\ *adjective*

[2]tunnel *verb* **-neled** *or* **-nelled; -nel·ing** *or* **-nel·ling** \'tən-liŋ, 'tə-nᵊl-iŋ\ (1795)
intransitive verb
1 : to make or use a tunnel
2 *physics* **:** to pass through a potential barrier ⟨electrons *tunneling* through an insulator between semiconductors⟩
transitive verb
: to make a tunnel or similar opening through or under; *also* **:** to make (one's way) by or as if by making a tunnel
— tun·nel·er \'tən-lər, 'tə-nᵊl-ər\ *noun*

tunnel vision *noun* (circa 1942)
1 : constriction of the visual field resulting in loss of peripheral vision
2 : extreme narrowness of viewpoint : NARROWMINDEDNESS; *also* **:** single-minded concentration on one objective
— tun·nel–vi·sioned \-'vi-zhənd\ *adjective*

tun·ny \'tə-nē\ *noun, plural* **tunnies** *also* **tunny** [modification of Middle French *thon* or Old Italian *tonno;* both from Old Provençal *ton,* from Latin *thunnus* — more at TUNA] (circa 1530)
: TUNA

[1]tup \'təp\ *noun* [Middle English *tupe*] (14th century)
1 *chiefly British* **:** RAM 1a
2 : a heavy metal body (as the weight of a pendulum)

[2]tup *transitive verb* **tupped; tup·ping** (1604)
chiefly British **:** to copulate with (a ewe)

tu·pe·lo \'tü-pə-ˌlō, 'tyü-\ *noun, plural* **-los** [perhaps from Creek *ito opilwa* swamp tree] (circa 1730)
1 : any of a genus (*Nyssa* of the family Nyssaceae) of North American and Asian deciduous trees that have simple alternate leaves, small greenish dioecious stalked flowers, and a rounded drupe; *especially* **:** BLACK GUM
2 : the pale soft wood of a tupelo

Tu·pi \tü-'pē, 'tü-(ˌ)\ *noun, plural* **Tupi** *or* **Tu·pis** (1842)
1 : the language of the Tupi people
2 : a member of a group of Tupi-Guaranian peoples of Brazil living especially in the Amazon valley

Tu·pi·an \tü-'pē-ən, 'tü-(ˌ)\ *adjective* (1902)
: of, relating to, or constituting the Tupi or other Tupi-Guaranian peoples or their languages

Tu·pi–Gua·ra·ni \tü-ˌpē-ˌgwär-ə-'nē, 'tü-(ˌ)pē-\ *noun* (1850)
1 : a member of a group of South American Indian peoples spread over an area from eastern Brazil to the Peruvian Andes and from the Guianas to Uruguay
2 : TUPI-GUARANIAN

Tupi–Gua·ra·ni·an \-'nē-ən\ *noun* (circa 1902)
: a language family widely distributed in tropical South America

-tu·ple \ˌtə-pəl, ˌtü-\ *noun combining form* [quin*tuple,* sex*tuple*]
: set of (so many) elements — usually used of sets with ordered elements ⟨the ordered 2-*tuple (a, b)*⟩

tup·pence *variant of* TWOPENCE

tuque \'tük, 'tyük\ *noun* [Canadian French, from French *toque* — more at TOQUE] (1871)
: a warm knitted usually pointed stocking cap

tu quo·que \'tü-'kwō-kwē, 'tyü-, -'kō-\ *noun* [Latin, you too] (1614)
: a retort charging an adversary with being or doing what he criticizes in others

tu·ra·co \'tůr-ə-(ˌ)kō\ *noun, plural* **-cos** [origin unknown] (1743)
: any of a family (Musophagidae) of African birds that are related to the cuckoos and have a long tail, a short stout often colored bill, and red wing feathers

Tu·ra·ni·an \tů-'rā-nē-ən, tyů-, -'rä-\ *noun* [Persian *Tūrān* Turkestan, the region north of the Amu Darya] (circa 1777)
1 : a member of any of various peoples speaking Ural-Altaic languages
2 : URAL-ALTAIC
— Turanian *adjective*

tur·ban \'tər-bən\ *noun* [Middle French *turbant,* from Italian *turbante,* from Turkish *tülbent,* from Persian *dulband*] (1588)
1 : a headdress worn chiefly in countries of the eastern Mediterranean and southern Asia especially by Muslims and made of a cap around which is wound a long cloth
2 : a headdress resembling a turban; *specifically* **:** a woman's close-fitting hat without a brim
3 : a rolled stuffed fillet of fish ⟨*turban* of sole⟩
word history see TULIP
— tur·baned *or* **tur·banned** \-bənd\ *adjective*

turban 1

tur·bel·lar·i·an \ˌtər-bə-'ler-ē-ən, -'lar-\ *noun* [ultimately from Latin *turbellae* (plural) bustle, stir, diminutive of *turba* confusion, crowd; from the tiny eddies created in water by the cilia] (1883)
: any of a class (Turbellaria) of mostly aquatic and free-living flatworms (as a planarian)
— turbellarian *adjective*

tur·bid \'tər-bəd\ *adjective* [Latin *turbidus* confused, turbid, from *turba* confusion, crowd, probably from Greek *tyrbē* confusion] (1626)
1 a : thick or opaque with or as if with roiled sediment ⟨a *turbid* stream⟩ **b :** heavy with smoke or mist
2 a : deficient in clarity or purity : FOUL, MUDDY ⟨*turbid* depths of degradation and misery —C. I. Glicksberg⟩ **b :** characterized by or producing obscurity (as of mind or emotions) ⟨an emotionally *turbid* response⟩

— **tur·bid·i·ty** \ˌtər-ˈbi-də-tē\ *noun*
— **tur·bid·ly** \ˈtər-bəd-lē\ *adverb*
— **tur·bid·ness** *noun*

tur·bi·dim·e·ter \ˌtər-bə-ˈdi-mə-tər\ *noun* [International Scientific Vocabulary *turbid*ity + *-meter*] (1905)
1 : an instrument for measuring and comparing the turbidity of liquids by viewing light through them and determining how much light is cut off
2 : NEPHELOMETER
— **tur·bi·di·met·ric** \ˌtər-bə-də-ˈme-trik, ˌtər-ˌbi-də-\ *adjective*
— **tur·bi·di·met·ri·cal·ly** \-tri-k(ə-)lē\ *adverb*
— **tur·bi·dim·e·try** \ˌtər-bə-ˈdi-mə-trē\ *noun*

tur·bi·dite \ˈtər-bə-ˌdīt\ *noun* [*turbid*ity current (a current flowing down a slope and spreading out on the ocean floor) + ¹*-ite*] (1957)
: a sedimentary deposit consisting of material that has moved down the steep slope at the edge of a continental shelf; *also* **:** a rock formed from this deposit

¹**tur·bi·nate** \ˈtər-bə-nət, -ˌnāt\ *also* **tur·bi·nat·ed** \-ˌnā-təd\ *adjective* [Latin *turbinatus,* from *turbin-, turbo*] (1661)
1 : shaped like a top or an inverted cone ⟨*turbinate* seed capsule⟩
2 : relating to or being a turbinate

²**turbinate** *noun* (circa 1803)
: one of usually several thin plicated membrane-covered bony or cartilaginous plates on the walls of the nasal chambers

tur·bine \ˈtər-bən, -ˌbīn\ *noun* [French, from Latin *turbin-, turbo* top, whirlwind, whirl, from *turba* confusion — more at TURBID] (1842)
: a rotary engine actuated by the reaction or impulse or both of a current of fluid (as water, steam, or air) subject to pressure and usually made with a series of curved vanes on a central rotating spindle

tur·bo \ˈtər-(ˌ)bō\ *noun, plural* **turbos** [*turbo*] (1904)
1 : TURBINE
2 [by shortening] **:** TURBOCHARGER

turbo- *combining form* [*turbine*]
1 : coupled directly to a driving turbine ⟨*turbo*fan⟩
2 : consisting of or incorporating a turbine ⟨*turbo*jet engine⟩

tur·bo·car \ˈtər-bō-ˌkär\ *noun* (1950)
: an automotive vehicle propelled by a gas turbine

tur·bo·charged \-ˌchärjd\ *adjective* (1945)
: equipped with a turbocharger

tur·bo·charg·er \-ˌchär-jər\ *noun* (1934)
: a centrifugal blower driven by exhaust gas turbines and used to supercharge an engine

tur·bo·elec·tric \ˌtər-bō-i-ˈlek-trik\ *adjective* (1904)
: using or being a turbine generator that produces electricity usually for motive power

tur·bo·fan \ˈtər-bō-ˌfan\ *noun* (1911)
1 : a fan that is directly connected to and driven by a turbine and is used to supply air for cooling, ventilation, or combustion
2 : a jet engine having a turbofan

tur·bo·gen·er·a·tor \ˌtər-bō-ˈje-nə-ˌrā-tər\ *noun* (1898)
: an electric generator driven by a turbine

tur·bo·jet \-ˌjet\ *noun* (1945)
1 : an airplane powered by turbojet engines
2 : TURBOJET ENGINE

turbojet engine *noun* (1944)
: a jet engine in which a turbine drives a compressor that supplies air to a burner and hot gases from the burner drive the turbine before being discharged rearward

tur·bo·ma·chin·ery \ˌtər-bō-mə-ˈshēn-rē, -ˈshē-nə-\ *noun* (1948)
: machinery consisting of, incorporating, or constituting a turbine

tur·bo·prop \ˈtər-bō-ˌpräp\ *noun* (1945)
1 : TURBOPROP ENGINE
2 : an airplane powered by turboprop engines

turboprop engine *noun* (1947)
: a jet engine designed to produce thrust principally by means of a propeller driven by a turbine with additional thrust usually obtained by the rearward discharge of hot exhaust gases

tur·bo·shaft \ˈtər-bō-ˌshaft\ *noun* (1958)
: a gas turbine engine that is similar in operation to a turboprop engine but instead of being used to power a propeller is used through a transmission system for powering other devices (as helicopter rotors and pumps)

tur·bot \ˈtər-bət\ *noun, plural* **turbot** *also* **turbots** [Middle English, from Old French *tourbot*] (14th century)
1 : a large European flatfish (*Psetta maxima*) that is a popular food fish and has a brownish upper surface marked with scattered tubercles and a white undersurface
2 : any of various flatfishes resembling the turbot

tur·bu·lence \ˈtər-byə-lən(t)s\ *noun* (1598)
: the quality or state of being turbulent: as **a :** wild commotion **b :** irregular atmospheric motion especially when characterized by up-and-down currents **c :** departure in a fluid from a smooth flow

tur·bu·len·cy \-lən(t)-sē\ *noun, plural* **-cies** (1607)
archaic **:** TURBULENCE

tur·bu·lent \-lənt\ *adjective* [Latin *turbulentus,* from *turba* confusion, crowd — more at TURBID] (1538)
1 : causing unrest, violence, or disturbance
2 a : characterized by agitation or tumult **:** TEMPESTUOUS **b :** exhibiting physical turbulence
— **tur·bu·lent·ly** *adverb*

turbulent flow *noun* (1895)
: a fluid flow in which the velocity at a given point varies erratically in magnitude and direction — compare LAMINAR FLOW

Tur·co- *or* **Tur·ko-** \ˈtər-(ˌ)kō\ *combining form* [*Turco-* from Medieval Latin *Turcus* Turk; *Turko-* from *Turk*]
1 : Turkic **:** Turkish **:** Turk ⟨*Turco*phil⟩
2 : Turkish and ⟨*Turco*-Greek⟩

turd \ˈtərd\ *noun* [Middle English *tord, turd,* from Old English *tord;* akin to Middle Dutch *tort* dung and probably to Old English *teran* to tear — more at TEAR] (before 12th century)
1 : a piece of excrement — sometimes considered vulgar
2 : a contemptible person — usually considered vulgar

tu·reen \tə-ˈrēn, tyü-\ *noun* [French *terrine,* from Middle French, from feminine of *terrin* of earth, from (assumed) Vulgar Latin *terrinus,* from Latin *terra* earth — more at TERRACE] (circa 1706)
1 : a deep and usually covered bowl from which foods (as soup) are served
2 : CASSEROLE 1

¹**turf** \ˈtərf\ *noun, plural* **turfs** \ˈtərfs\ *also* **turves** \ˈtərvz\ [Middle English, from Old English; akin to Old High German *zurba* turf, Sanskrit *darbha* tuft of grass] (before 12th century)
1 a : the upper stratum of soil bound by grass and plant roots into a thick mat; *also* **:** a piece of this **b :** an artificial substitute for this (as on a playing field)
2 a : PEAT 2 **b :** a piece of peat dried for fuel
3 a : a track or course for horse racing **b :** the sport or business of horse racing
4 a : territory considered by a teenage gang to be under its control **b :** TERRITORY 2 ⟨have to play two of the last three games on hostile *turf* —Joe Klein⟩ ⟨points toward the biographer's *turf,* not the critic's —Hugh Kenner⟩; *also* **:** a sphere of activity or influence ⟨people who could hurt him on his own foreign-policy *turf* —*Wall Street Journal*⟩
— **turfy** \ˈtər-fē\ *adjective*

²**turf** *transitive verb* (15th century)
1 : to cover with turf
2 *chiefly British* **:** to eject forcibly **:** KICK — usually used with *out*

turf accountant *noun* (1915)
British **:** BOOKMAKER 2

turf·man \ˈtərf-mən\ *noun* (1818)
: a devotee of horse racing; *especially* **:** a person who owns and races horses

turf·ski \-ˌskē\ *noun* (1967)
: a short ski with rollers on the bottom that can be used to ski down a grassy slope
— **turf·ski·ing** *noun*

tur·ges·cent \ˌtər-ˈje-sᵊnt\ *adjective* [Latin *turgescent-, turgescens,* present participle of *turgescere* to swell, inchoative of *turgēre* to be swollen] (circa 1727)
: becoming turgid, distended, or inflated **:** SWELLING
— **tur·ges·cence** \-sᵊn(t)s\ *noun*

tur·gid \ˈtər-jəd\ *adjective* [Latin *turgidus,* from *turgēre* to be swollen] (1620)
1 : being in a state of distension **:** SWOLLEN, TUMID ⟨*turgid* limbs⟩; *especially* **:** exhibiting turgor
2 : excessively embellished in style or language **:** BOMBASTIC, POMPOUS
— **tur·gid·i·ty** \ˌtər-ˈji-də-tē\ *noun*
— **tur·gid·ly** \ˈtər-jəd-lē\ *adverb*
— **tur·gid·ness** *noun*

tur·gor \ˈtər-gər, -ˌgȯr\ *noun* [Late Latin, turgidity, swelling, from Latin *turgēre*] (1876)
: the normal state of turgidity and tension in living cells; *especially* **:** the distension of the protoplasmic layer and wall of a plant cell by the fluid contents

Tu·ring machine \ˈtu̇r-iŋ-, ˈtyu̇r-\ *noun* [A. M. *Turing* (died 1954) English mathematician] (1937)
: a hypothetical computing machine that has an unlimited amount of information storage

tu·ris·ta \tü-ˈrē-stə\ *noun* [Spanish, tourist] (1962)
: intestinal sickness and diarrhea commonly affecting a tourist in a foreign country; *especially* **:** MONTEZUMA'S REVENGE

Turk \ˈtərk\ *noun* [Middle English, from Middle French or Turkish; Middle French *Turc,* from Medieval Latin or Turkish; Medieval Latin *Turcus,* from Turkish *Türk*] (14th century)
1 : a member of any of numerous Asian peoples speaking Turkic languages who live in a region extending from the Balkans to eastern Siberia and western China
2 : a native or inhabitant of Turkey
3 *archaic* **:** one who is cruel or tyrannical
4 : MUSLIM; *specifically* **:** a Muslim subject of the Turkish sultan
5 : a Turkish horse; *specifically* **:** a Turkish strain of Arab and crossbred horses
6 *often not capitalized* **:** a usually young dynamic person eager for change; *especially* **:** YOUNG TURK

tur·key \ˈtər-kē\ *noun, plural* **turkeys** [*Turkey,* country in western Asia and southeastern Europe; from confusion with the guinea fowl, supposed to be imported from Turkish territory] (1555)
1 *plural also* **turkey :** a large North American gallinaceous bird (*Meleagris gallopavo*) that is domesticated in most parts of the world
2 : FAILURE, FLOP; *especially* **:** a theatrical production that has failed

turkey 1

\ə\ abut \ᵊ\ kitten \ər\ further \a\ ash \ā\ ace \ä\ mop, mar \au̇\ out \ch\ chin \e\ bet \ē\ easy \g\ go \i\ hit \ī\ ice \j\ job \ŋ\ sing \ō\ go \ȯ\ law \ȯi\ boy \th\ thin \th̲\ the \ü\ loot \u̇\ foot \y\ yet \zh\ vision *see also* Guide to Pronunciation

3 : three successive strikes in bowling
4 : a stupid, foolish, or inept person

tur·key–cock \'tər-kē-ˌkäk\ *noun* (1578)
1 : GOBBLER
2 : a strutting pompous person

turkey oak *noun* (1709)
: a small oak (*Quercus laevis* synonym *Q. catesbaei*) of the southeastern U.S.; *also* **:** a Eurasian oak (*Q. cerris*)

Tur·key red \'tər-kē-\ *noun* [*Turkey*] (1789)
: a brilliant durable red produced on cotton by means of alizarin in connection with an aluminum mordant and fatty matter

tur·key shoot \'tər-kē-\ *noun* (1845)
: a marksmanship contest using a moving target with a turkey offered as a prize

turkey trot *noun* (1908)
: a ragtime dance danced with the feet well apart and with a characteristic rise on the ball of the foot followed by a drop upon the heel

turkey vulture *noun* (1823)
: an American vulture (*Cathartes aura*) with a red head and whitish bill — called also *turkey buzzard*

Turk·ic \'tər-kik\ *adjective* (1859)
1 a : of, relating to, or constituting a family of Altaic languages including Turkish **b :** of or relating to the peoples speaking Turkic
2 : TURKISH 1
— Turkic *noun*

¹Turk·ish \'tər-kish\ *adjective* (1545)
1 : of, relating to, or characteristic of Turkey, the Turks, or Turkish
2 : TURKIC 1a

²Turkish *noun* (1677)
: the Turkic language of the Republic of Turkey

Turkish bath *noun* (1644)
: a bath in which the bather passes through a series of steam rooms of increasing temperature and then receives a rubdown, massage, and cold shower

Turkish coffee *noun* (1854)
: a sweetened decoction of pulverized coffee

Turkish delight *noun* (1877)
: a jellylike or gummy confection usually cut in cubes and dusted with sugar — called also *Turkish paste*

Turkish towel *noun* (1862)
: a towel made of cotton terry cloth

Turk·ism \'tər-ˌki-zəm\ *noun* (1595)
: the customs, beliefs, institutions, and principles of the Turks

Tur·ko·man *or* **Tur·co·man** \'tər-kə-mən\ *noun, plural* **Turkomans** *or* **Turcomans** [Medieval Latin *Turcomannus,* from Persian *Turkmān,* from *turkmān* resembling a Turk, from *Turk*] (circa 1520)
: a member of a Turkic-speaking traditionally nomadic people living chiefly in Turkmenistan, Afghanistan, and Iran
— Turkoman *or* **Turcoman** *adjective*

Turk's head *noun* (1833)
: a turban-shaped knot worked on a rope with a piece of small line — see KNOT illustration

tur·mer·ic \'tər-mə-rik *also* 'tü-mə- *or* 'tyü-\ *noun* [modification of Middle French *terre merite* saffron, from Medieval Latin *terra merita,* literally, deserving or deserved earth] (1545)
1 : an East Indian perennial herb (*Curcuma domestica* synonym *C. longa*) of the ginger family with a large aromatic yellow rhizome
2 : the cleaned, boiled, dried, and usually pulverized rhizome of the turmeric plant used as a coloring agent, a flavoring, or a stimulant
3 : a yellow to reddish brown dyestuff obtained from turmeric

tur·moil \'tər-ˌmȯil\ *noun* [origin unknown] (1526)
: a state or condition of extreme confusion, agitation, or commotion

¹turn \'tərn\ *verb* [Middle English; partly from Old English *tyrnan* & *turnian* to turn, from Medieval Latin *tornare,* from Latin, to turn on a lathe, from *tornus* lathe, from Greek *tornos;* partly from Old French *torner, tourner* to turn, from Medieval Latin *tornare;* akin to Latin *terere* to rub — more at THROW] (before 12th century)
transitive verb
1 a : to cause to move around an axis or a center **:** make rotate or revolve ⟨*turn* a wheel⟩ ⟨*turn* a crank⟩ **b** (1) **:** to cause to move around so as to effect a desired end (as of locking, opening, or shutting) ⟨*turned* the knob till the door opened⟩ (2) **:** to affect or alter the functioning of (as a mechanical device) by such movement ⟨*turned* the oven to a higher temperature⟩ **c :** to execute or perform by rotating or revolving ⟨*turn* handsprings⟩ **d :** to twist out of line or shape **:** WRENCH ⟨had *turned* his ankle⟩
2 a (1) **:** to cause to change position by moving through an arc of a circle ⟨*turned* her chair to the fire⟩ (2) **:** to cause to move around a center so as to show another side of ⟨*turn* the page⟩ (3) **:** to cause (as a scale) to move so as to register weight **b :** to revolve mentally **:** think over **:** PONDER
3 a : to reverse the sides or surfaces of **:** INVERT ⟨*turn* pancakes⟩ ⟨*turn* the shirt inside out⟩: as (1) **:** to dig or plow so as to bring the lower soil to the surface (2) **:** to make (as a garment) over by reversing the material and resewing ⟨*turn* a collar⟩ (3) **:** to invert feet up and face down (as a character, rule, or slug) in setting type **b :** to reverse or upset the order or disposition of ⟨everything was *turned* topsy-turvy⟩ **c :** to disturb or upset the mental balance of **:** DERANGE, UNSETTLE ⟨a mind *turned* by grief⟩ **d :** to set in another especially contrary direction
4 a : to bend or change the course of **:** DIVERT **b :** to cause to retreat ⟨used fire hoses to *turn* the mob⟩ **c :** to alter the drift, tendency, or expected result of **d :** to bend a course around or about **:** ROUND ⟨*turned* the corner at full speed⟩
5 a (1) **:** to direct or point (as the face) in a specified way or direction (2) **:** to present by a change in direction or position ⟨*turning* his back to his guests⟩ **b :** to bring to bear (as by aiming, pointing, or focusing) **:** TRAIN ⟨*turned* the light into the dark doorway⟩ ⟨*turned* a questioning eye toward her⟩ **c :** to direct (as the attention or mind) toward or away from something **d :** to direct the employment of **:** APPLY, DEVOTE ⟨*turned* his skills to the service of mankind⟩ **e** (1) **:** to cause to rebound or recoil ⟨*turns* their argument against them⟩ (2) **:** to make antagonistic **:** PREJUDICE ⟨*turn* a child against its mother⟩ **f** (1) **:** to cause to go in a particular direction ⟨*turned* our steps homeward⟩ (2) **:** DRIVE, SEND ⟨*turn* cows to pasture⟩ ⟨*turning* hunters off his land⟩ (3) **:** to convey or direct into or out of a receptacle by inverting
6 a (1) **:** to make acid or sour (2) **:** to change the color of (as foliage) **b** (1) **:** CONVERT, TRANSFORM ⟨*turn* defeat into victory⟩ (2) **:** TRANSLATE, PARAPHRASE **c :** to cause to become of a specified nature or appearance ⟨*turned* him into a frog⟩ ⟨illness *turned* her hair white⟩ **d :** to exchange for something else ⟨*turn* coins into paper money⟩ **e :** to cause to defect to another side
7 a : to shape especially in a rounded form by applying a cutting tool while revolving in a lathe **b :** to give a rounded form to by any means ⟨*turn* the heel of a sock⟩ **c :** to shape or mold artistically, gracefully, or neatly ⟨a well *turned* phrase⟩
8 : to make a fold, bend, or curve in: **a :** to form by bending ⟨*turn* a lead pipe⟩ **b :** to cause (the edge of a blade) to bend back or over **:** BLUNT, DULL
9 a : to keep (as money or goods) moving; *specifically* **:** to dispose of (a stock) to make room for another **b :** to gain in the course of business ⟨*turning* a quick profit⟩ **c :** to make use of ⟨*turned* her education to advantage⟩ **d :** to carry to completion ⟨*turned* a double play⟩
10 : to engage in (an act of prostitution) ⟨*turn* tricks⟩
intransitive verb
1 a : to move around on an axis or through an arc of a circle **:** ROTATE **b :** to become giddy or dizzy **:** SPIN ⟨heights always made his head *turn*⟩ **c** (1) **:** to have as a decisive factor **:** HINGE ⟨the argument *turns* on a point of logic⟩ ⟨the outcome of the game *turned* on an interception⟩ (2) **:** to have a center (as of interest) in something specified
2 a : to direct one's course **b** (1) **:** to reverse a course or direction (2) **:** to have a reactive usually adverse effect **:** RECOIL **c :** to take a different course or direction ⟨*turned* toward home⟩ ⟨the main road *turns* sharply to the right⟩
3 a : to change position so as to face another way **b :** to face toward or away from someone or something **c :** to change one's attitude or reverse one's course of action to one of opposition or hostility ⟨felt the world had *turned* against him⟩ **d :** to make a sudden violent assault especially without evident cause ⟨dogs *turning* on their owners⟩
4 a : to direct one's attention or thoughts to or away from someone or something **b** (1) **:** to change one's religion (2) **:** to go over to another side or party **:** DEFECT **c :** to have recourse **:** REFER, RESORT ⟨*turned* to a friend for help⟩ ⟨*turned* to his notes for the exact figures⟩ **d :** to direct one's efforts or interests **:** devote or apply oneself ⟨*turned* to the study of the law⟩
5 a : to become changed, altered, or transformed: as (1) *archaic* **:** to become different (2) **:** to change color ⟨the leaves have *turned*⟩ (3) **:** to become sour, rancid, or tainted ⟨the milk had *turned*⟩ (4) **:** to be variable or inconstant (5) **:** to become mentally unbalanced **:** become deranged **b** (1) **:** to pass from one state to another **:** CHANGE ⟨water had *turned* to ice⟩ (2) **:** BECOME, GROW ⟨his hair had *turned* gray⟩ ⟨the weather *turned* bad⟩ ⟨just *turned* twenty⟩ (3) **:** to become someone or something specified by change from another state **:** change into ⟨*turn* pro⟩ ⟨doctors *turned* authors⟩
6 : to become curved or bent (as from pressure); *especially* **:** to become blunted by bending ⟨edge of the knife had *turned*⟩
7 : to operate a lathe
8 *of merchandise* **:** to be stocked and disposed of **:** change hands
— turn·able \'tər-nə-bəl\ *adjective*
— turn a blind eye : to refuse to see **:** be oblivious ⟨might *turn a blind eye* to the use of violence —Arthur Krock⟩
— turn a deaf ear : to refuse to listen
— turn a hair : to give a sign of distress or disturbance ⟨did not *turn a hair* when told of the savage murder —*Times Literary Supplement*⟩
— turn color 1 : to become of a different color **2 a :** BLUSH, FLUSH **b :** to grow pale
— turn loose 1 a : to set free ⟨*turned loose* the captured animal⟩ **b :** to free from all restraints ⟨*turned* them *loose* with a pile of theme paper to write whatever they liked —Elizabeth P. Schafer⟩ **2 :** to fire off **:** DISCHARGE **3 :** to open fire
— turn one's back on 1 : REJECT, DENY ⟨would be *turning one's back on* history —Pius Walsh⟩ **2 :** FORSAKE ⟨*turned his back on* his obligations⟩
— turn one's hand *or* **turn a hand :** to set to work **:** apply oneself
— turn one's head : to cause to become infatuated or conceited ⟨success had not *turned his head*⟩
— turn one's stomach : to disgust completely **:** SICKEN, NAUSEATE ⟨the foul smell *turned his stomach*⟩
— turn tail : to turn away so as to flee ⟨*turned tail* and ran⟩
— turn the other cheek : to respond to injury or unkindness with patience **:** forgo retaliation

— **turn the tables :** to bring about a reversal of the relative conditions or fortunes of two contending parties
— **turn the trick :** to bring about the desired result or effect
— **turn turtle :** CAPSIZE, OVERTURN

²turn *noun* [Middle English; partly from Old French *tourn, tour* lathe, circuit, turn (partly from Latin *tornus* lathe; partly from Old French *torner, tourner* to turn); partly from Middle English *turnen* to turn] (13th century)
1 a : the action or an act of turning about a center or axis **:** REVOLUTION, ROTATION **b :** any of various rotating or pivoting movements in dancing or gymnastics
2 a : the action or an act of giving or taking a different direction **:** change of course or posture ⟨illegal left *turn*⟩: as (1) **:** a drill maneuver in which troops in mass formation change direction without preserving alignment (2) **:** any of various shifts of direction in skiing (3) **:** an interruption of a curve in figure skating **b :** DEFLECTION, DEVIATION **c :** the action or an act of turning so as to face in the opposite direction **:** reversal of posture or course ⟨an about *turn*⟩ ⟨*turn* of the tide⟩ **d :** a change effected by turning over to another side ⟨*turn* of the cards⟩ **e :** a place at which something turns, turns off, or turns back **:** BEND, CURVE
3 : a short trip out and back or round about ⟨took a *turn* through the park⟩
4 : an act or deed affecting another especially when incidental or unexpected ⟨one good *turn* deserves another⟩
5 a : a period of action or activity **:** GO, SPELL **b :** a place, time, or opportunity accorded an individual or unit of a series in simple succession or in a scheduled order ⟨waiting her *turn* in line⟩ **c :** a period or tour of duty **:** SHIFT **d :** a short act or piece (as for a variety show); *also* **:** PERFORMANCE 3 **e** (1) **:** an event in any gambling game after which bets are settled (2) **:** the order of the last three cards in faro — used in the phrase *call the turn*
6 : something that revolves around a center: as **a** (1) **:** LATHE (2) **:** a catch or latch for a cupboard or cabinet door operated by turning a handle **b :** a musical ornament consisting of a group of four or more notes that wind about the principal note by including the notes next above and next below
7 : a special purpose or requirement — used chiefly in the phrase *serve one's turn*
8 a : an act of changing **:** ALTERATION, MODIFICATION ⟨a nasty *turn* in the weather⟩ **b :** a change in tendency, trend, or drift ⟨hoped for a *turn* in his luck⟩ ⟨a *turn* for the better⟩ **c :** the beginning of a new period of time ⟨the *turn* of the century⟩
9 a : distinctive quality or character **b** (1) **:** a skillful fashioning of language or arrangement of words (2) **:** a particular form of expression or peculiarity of phrasing **c :** the shape or mold in which something is fashioned **:** CAST
10 a : the state or manner of being coiled or twisted **b :** a single round (as of rope passed about an object or of wire wound on a core)
11 : natural or special ability or aptitude **:** BENT, INCLINATION ⟨a *turn* for logic⟩ ⟨an optimistic *turn* of mind⟩
12 : a special twist, construction, or interpretation ⟨gave the old yarn a new *turn*⟩
13 a : a disordering spell or attack (as of illness, faintness, or dizziness) **b :** a nervous start or shock
14 a : a complete transaction involving a purchase and sale of securities; *also* **:** a profit from such a transaction **b :** TURNOVER 7b
15 : something turned or to be turned: as **a :** a character or slug inverted in setting type **b :** a piece of type placed bottom up
— **at every turn :** on every occasion **:** CONTINUALLY ⟨they opposed her *at every turn*⟩
— **by turns 1 :** one after another in regular succession **2 :** VARIOUSLY, ALTERNATELY ⟨a book that is *by turns* pedantic, delightful, and infuriating⟩
— **in turn :** in due order of succession
— **on the turn :** at the point of turning ⟨tide is *on the turn*⟩
— **out of turn 1 :** not in due order of succession ⟨play *out of turn*⟩ **2 :** at a wrong time or place and usually imprudently ⟨talking *out of turn*⟩
— **to a turn :** to perfection

turn·about \ˈtər-nə-ˌbau̇t\ *noun* (1789)
1 : MERRY-GO-ROUND
2 a : a change or reversal of direction, trend, policy, role, or character **b :** a changing from one allegiance to another **c :** TURNCOAT, RENEGADE **d :** an act or instance of retaliating ⟨*turnabout* is fair play⟩

turn·around \-ˌrau̇nd\ *noun* (1926)
1 a : the process of readying a vehicle for departure after its arrival especially without any intervening delays; *also* **:** the time spent in this process **b :** the action of receiving, processing, and returning something
2 : a space permitting the turning around of a vehicle
3 : TURNABOUT 2a, b

turn around (1949)
intransitive verb
1 : to act in an abrupt, different, or surprising manner — used with *and* ⟨after three years he just *turned around* and left school⟩
2 : to become changed for the better
transitive verb
: to change for the better ⟨*turned* her life *around*⟩

turn away (13th century)
transitive verb
1 : DEFLECT, AVERT
2 a : to send away **:** REJECT, DISMISS **b :** REPEL **c :** to refuse admittance or acceptance to
intransitive verb
: to start to go away **:** DEPART

turn back (1535)
intransitive verb
1 a : to go in the reverse direction **b :** to stop going forward
2 : to refer to an earlier time or place
transitive verb
1 : to drive back or away
2 : to stop the advance of
3 : to fold back
4 : GIVE BACK, RETURN
— **turn back the clock :** to revert to or remind of a condition existing in the past

turn·buck·le \ˈtərn-ˌbə-kəl\ *noun* (circa 1877)
: a device that usually consists of a link with screw threads at both ends, that is turned to bring the ends closer together, and that is used for tightening a rod or stay

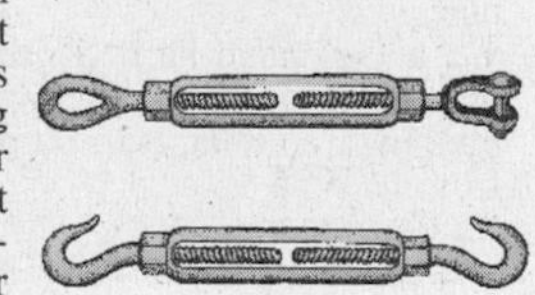
turnbuckles

turn·coat \-ˌkōt\ *noun* (1557)
: one who switches to an opposing side or party; *specifically* **:** TRAITOR

¹turn·down \ˈtərn-ˈdau̇n\ *adjective* (1840)
: capable of being turned down; *especially* **:** worn turned down ⟨*turndown* collar⟩

²turn·down \ˈtərn-ˌdau̇n\ *noun* (1849)
1 : something turned down; *also* **:** an instance of turning something (as a bed sheet) down ⟨hotel *turndown* service⟩
2 : REJECTION
3 : DOWNTURN

turn down (1601)
transitive verb
1 : to fold or double down
2 : to turn (a card) face downward
3 : to reduce the height or intensity of by turning a control ⟨*turn down* the radio⟩
4 : to decline to accept **:** REJECT ⟨*turned down* the offer⟩
intransitive verb
: to be capable of being folded or doubled down ⟨the collar *turns down*⟩

turned–on \ˈtərnd-ˈȯn, -ˈän\ *adjective* (1966)
: keenly aware of and responsive to what is new and fashionable **:** HIP

¹turn·er \ˈtər-nər\ *noun* (15th century)
: one that turns or is used for turning ⟨a pancake *turner*⟩; *especially* **:** a person who forms articles with a lathe

²tur·ner \ˈtər-nər, ˈtu̇r-\ *noun* [German, from *turnen* to perform gymnastic exercises, from Old High German *turnēn* to turn, from Medieval Latin *tornare* — more at TURN] (1854)
: a member of a turnverein **:** GYMNAST

Tur·ner's syndrome \ˈtər-nərz-\ *noun* [Henry Hubert *Turner* (died 1970) American physician] (1942)
: a genetically determined condition that is associated with the presence of only one complete X chromosome and no Y chromosome and that is characterized by a female phenotype with underdeveloped and infertile ovaries

turn·ery \ˈtər-nə-rē\ *noun, plural* **-er·ies** (1644)
: the work, products, or shop of a turner

turn–in \ˈtər-ˌnin\ *noun* (1902)
: something that turns in or is turned in

turn in (1535)
intransitive verb
1 : to make an entrance by turning from a road or path
2 : to go to bed ⟨*turned in* early⟩
transitive verb
1 : to deliver up **:** HAND OVER ⟨*turned in* his badge and quit⟩
2 a : to inform on **:** BETRAY **b :** to deliver to an authority ⟨urged the wanted man to *turn* himself *in*⟩
3 : to acquit oneself of **:** PUT ON, PRODUCE ⟨*turned in* a good performance⟩

turn·ing *noun* (14th century)
1 : the act or course of one that turns
2 : a place of a change in direction
3 a : a forming by use of a lathe; *broadly* **:** TURNERY **b** *plural* **:** waste produced in turning

turning chisel *noun* (circa 1877)
: a chisel used for shaping or finishing work in a lathe

turning point *noun* (1851)
: a point at which a significant change occurs

tur·nip \ˈtər-nəp\ *noun* [probably from ¹*turn* + *neep;* from the well-rounded root] (1533)
1 a : either of two biennial herbs of the mustard family with thick edible roots: (1) **:** one (*Brassica rapa rapifera*) with usually flattened roots and leaves that are cooked as a vegetable (2) **:** RUTABAGA **b :** the root of a turnip
2 : a large pocket watch

¹turn·key \ˈtərn-ˌkē\ *noun, plural* **turnkeys** (1654)
: one who has charge of a prison's keys

²turnkey *adjective* (1927)
: built, supplied, or installed complete and ready to operate ⟨a *turnkey* nuclear plant⟩ ⟨a *turnkey* computer system⟩; *also* **:** of or relating to a turnkey building or installation ⟨a *turnkey* contract⟩ ⟨*turnkey* vendors⟩

turn·off \ˈtər-ˌnȯf\ *noun* (circa 1852)
1 : a turning off
2 : a place where one turns off; *especially* **:** EXIT 4
3 : one that causes loss of interest or enthusiasm ⟨the music was a *turnoff*⟩

turn off (1564)
transitive verb

\ə\ **abut** \ᵊ\ **kitten** \ər\ **further** \a\ **ash** \ā\ **ace** \ä\ **mop, mar** \au̇\ **out** \ch\ **chin** \e\ **bet** \ē\ **easy** \g\ **go** \i\ **hit** \ī\ **ice** \j\ **job** \ŋ\ **sing** \ō\ **go** \ȯ\ **law** \ȯi\ **boy** \th\ **thin** \t͟h\ **the** \ü\ **loot** \u̇\ **foot** \y\ **yet** \zh\ **vision** *see also* Guide to Pronunciation

1 a : DISMISS, DISCHARGE **b :** to dispose of **:** SELL
2 : DEFLECT, EVADE
3 : PRODUCE, ACCOMPLISH
4 : to stop the flow of or shut off by or as if by turning a control ⟨*turn* the water *off*⟩
5 : HANG 1b
6 a : to remove (material) by the process of turning **b :** to shape or produce by turning
7 : to cause to lose interest **:** BORE ⟨economics *turns* me *off*⟩; *also* **:** to evoke a negative feeling in
intransitive verb
1 : to deviate from a straight course or from a main road ⟨*turn off* into a side road⟩
2 a *British* **:** to turn bad **:** SPOIL **b :** to change to a specified state **:** BECOME
3 : to lose interest **:** WITHDRAW

turn on (1833)
transitive verb
1 : to activate or cause to flow, operate, or function by or as if by turning a control ⟨*turn* the water *on* full⟩ ⟨*turn on* the power⟩
2 a : to cause to undergo an intense often visionary experience by taking a drug; *broadly* **:** to cause to get high **b :** to move pleasurably ⟨rock music *turns* her *on*⟩; *also* **:** to excite sexually **c :** to cause to gain knowledge or appreciation of something specified ⟨*turned* her *on* to ballet⟩
intransitive verb
: to become turned on
— turn–on \ˈtər-ˌȯn, -ˌnän\ *noun*

turn·out \ˈtər-ˌnau̇t\ *noun* (1688)
1 : an act of turning out
2 *chiefly British* **a :** STRIKE 3a **b :** STRIKER 1d
3 : the number of people who participate in or attend an event ⟨a heavy voter *turnout*⟩
4 a : a place where something (as a road) turns out or branches off **b :** a space adjacent to a highway in which vehicles may park or pull into to enable others to pass **c :** a railroad siding
5 : a clearing out and cleaning
6 a : a coach or carriage together with the horses, harness, and attendants **b :** EQUIPMENT, RIG **c :** manner of dress **:** GETUP
7 : net quantity of produce yielded

turn out (1546)
transitive verb
1 a : EXPEL, EVICT **b :** to put (as a horse) to pasture
2 a : to turn inside out ⟨*turning out* his pockets⟩ **b :** to empty the contents of especially for cleaning or rearranging; *also* **:** CLEAN
3 : to produce often rapidly or regularly by or as if by machine ⟨a writer *turning out* stories⟩
4 : to equip, dress, or finish in a careful or elaborate way
5 : to put out by turning a switch ⟨*turn out* the lights⟩
6 : to call (as the guard or a company) out from rest or shelter and into formation
intransitive verb
1 a : to come or go out from home in or as if in answer to a summons ⟨voters *turned out* in droves⟩ **b :** to get out of bed
2 a : to prove to be in the result or end ⟨the play *turned out* to be a flop⟩ ⟨it *turned out* that we were both wrong⟩ **b :** to become in maturity ⟨nobody thought he'd *turn out* like this⟩ **c :** END ⟨stories that *turn out* happily⟩

¹turn·over \ˈtər-ˌnō-vər\ *noun* (14th century)
1 : an act or result of turning over **:** UPSET
2 : a turning from one side, place, or direction to its opposite **:** SHIFT, REVERSAL
3 : a reorganization with a view to a shift in personnel **:** SHAKE-UP
4 : something that is turned over
5 : a filled pastry made by folding half of the crust over the other half
6 : the amount of business done; *especially* **:** the volume of shares traded on a stock exchange
7 a : movement (as of goods or people) into, through, and out of a place **b :** a cycle of purchase, sale, and replacement of a stock of goods; *also* **:** the ratio of sales for a stated period to average inventory **c :** the number of persons hired within a period to replace those leaving or dropped from a workforce; *also* **:** the ratio of this number to the number in the average force maintained
8 : the act or an instance of a team's losing possession of a ball through error or a minor violation of the rules (as in basketball or football)

²turnover *adjective* (circa 1849)
: capable of being turned over

turn over (14th century)
transitive verb
1 a : to turn from an upright position **:** OVERTURN **b :** ROTATE ⟨*turn over* a stiff valve with a wrench⟩; *also* **:** to cause (an internal combustion engine) to begin firing
2 : to search (as clothes or papers) by lifting or moving one by one
3 : to read or examine (as a book) slowly or idly
4 : DELIVER, SURRENDER ⟨I'm *turning* the job *over* to you⟩; *also* **:** to lose possession of ⟨*turned* the ball *over* three times⟩
5 a : to receive and dispose of (a stock of merchandise) **b :** to do business to the amount of ⟨*turning over* $1000 a week⟩
intransitive verb
1 : UPSET, CAPSIZE
2 a : ROTATE **b** *of an engine* **:** to have crankshaft rotation especially by external means (as by a starter) ⟨the engine *turned over* but didn't start⟩
3 a *of one's stomach* **:** to heave with nausea **b** *of one's heart* **:** to seem to leap or lurch convulsively with sudden fright
— turn over a new leaf : to make a change for the better especially in one's way of living

turn·pike \ˈtərn-ˌpīk\ *noun* [Middle English *turnepike* revolving frame bearing spikes and serving as a barrier, from *turnen* to turn + *pike*] (1533)
1 : TOLLGATE
2 a (1) **:** a road (as an expressway) for the use of which tolls are collected (2) **:** a road formerly maintained as a turnpike **b :** a main road; *especially* **:** a paved highway with crowned surface

turn·sole \ˈtərn-ˌsōl\ *noun* [Middle English *turnesole,* from Middle French *tournesol,* from Old Italian *tornasole,* from *tornare* to turn (from Medieval Latin) + *sole* sun, from Latin *sol* — more at TURN, SOLAR] (14th century)
1 : a European herb (*Chrozophora tinctoria*) of the spurge family with juice that is turned blue by ammonia; *also* **:** a purple dye obtained from it
2 : HELIOTROPE 1

turn·spit \-ˌspit\ *noun* (1570)
1 a : one that turns a spit; *specifically* **:** a small dog formerly used in a treadmill to turn a spit **b :** a roasting jack
2 : a rotatable spit

turn·stile \-ˌstīl\ *noun* (1643)
: a post with arms pivoted on the top set in a passageway so that persons can pass through only on foot one by one

turn·stone \-ˌstōn\ *noun* [from a habit of turning over stones to find food] (circa 1674)
: either of two shorebirds (genus *Arenaria*) that resemble the related plovers and sandpipers: **a :** a bird (*A. interpres*) of worldwide distribution that has black and chestnut upperparts and a black breast **b :** a North American bird (*A. melanocephala*) with black upperparts and breast

turn·ta·ble \-ˌtā-bəl\ *noun* (1835)
: a revolvable platform: as **a :** a platform with a track for turning wheeled vehicles **b :** LAZY SUSAN **c :** a rotating platform that carries a phonograph record

turnstone

turn to *intransitive verb* (1813)
: to apply oneself to work **:** act vigorously

¹turn·up \ˈtər-ˈnəp\ *adjective* (1685)
1 : turned up ⟨a *turnup* nose⟩
2 : made or fitted to be turned up ⟨a *turnup* collar⟩

²turn·up \ˈtər-ˌnəp\ *noun* (1688)
: something that is turned up

turn up (1563)
transitive verb
1 : FIND, DISCOVER
2 : to raise or increase by or as if by turning a control ⟨*turn up* the volume on the radio⟩
3 *British* **a :** to look up (as a word or fact) in a book **b :** to refer to or consult (a book)
4 : to turn (a card) face upward
intransitive verb
1 : to appear or come to light unexpectedly or after being lost ⟨new evidence has *turned up*⟩
2 a (1) **:** to turn out to be ⟨he *turned up* missing at roll call⟩ (2) **:** APPEAR 4 ⟨her name is always *turning up* in the newspapers⟩ **b :** to arrive or show up at an appointed or expected time or place ⟨*turned up* half an hour late⟩
3 : to happen or occur unexpectedly ⟨something always *turned up* to prevent their meeting⟩
4 *of a ship* **:** TACK 1b
— turn up one's nose : to show scorn or disdain

turn·ver·ein \ˈtərn-və-ˌrīn, ˈtu̇rn-\ *noun* [German, from *turnen* to perform gymnastic exercises + *Verein* club] (1852)
: an athletic club

tu·ro·phile \ˈtu̇r-ə-ˌfīl, ˈtyu̇r-\ *noun* [irregular from Greek *tyros* cheese + English *-phile* — more at BUTTER] (1938)
: a connoisseur of cheese **:** a cheese fancier

¹tur·pen·tine \ˈtər-pən-ˌtīn, ˈtər-pᵊm-\ *noun* [Middle English *terbentyne, turpentyne,* from Middle French & Medieval Latin; Middle French *terbentine, tourbentine,* from Medieval Latin *terbentina,* from Latin *terebinthina,* feminine of *terebinthinus* of terebinth, from *terebinthus* terebinth, from Greek *terebinthos*] (14th century)
1 a : a yellow to brown semifluid oleoresin obtained as an exudate from the terebinth **b :** an oleoresin obtained from various conifers (as some pines and firs)
2 a : an essential oil obtained from turpentines by distillation and used especially as a solvent and thinner — called also *gum turpentine* **b :** a similar oil obtained by distillation or carbonization of pinewood — called also *wood turpentine*

²turpentine *transitive verb* **-tined; -tin·ing** (1759)
1 : to apply turpentine to
2 : to extract turpentine from; *especially* **:** to tap (pine trees) in order to obtain turpentine

tur·pi·tude \ˈtər-pə-ˌtüd, -ˌtyüd\ *noun* [Middle French, from Latin *turpitudo,* from *turpis* vile, base] (15th century)
: inherent baseness **:** DEPRAVITY ⟨moral *turpitude*⟩; *also* **:** a base act

turps \ˈtərps\ *noun plural but singular in construction* [by shortening & alteration] (circa 1823)
: TURPENTINE

tur·quoise *also* **tur·quois** \ˈtər-ˌkȯiz, -ˌkwȯiz\ *noun* [Middle English *turkeis, turcas,* from Middle French *turquoyse,* from feminine of *turquoys* Turkish, from Old French, from *Turc* Turk] (14th century)

1 : a mineral that is a blue, bluish green, or greenish gray hydrous basic phosphate of copper and aluminum, takes a high polish, and is valued as a gem when skyblue
2 : a light greenish blue

turquoise blue *noun* (1799)
: a light greenish blue that is paler and slightly bluer than average turquoise

turquoise green *noun* (1886)
: a light bluish green

tur·ret \ˈtər-ət, ˈtə-rət, ˈtu̇r-ət\ *noun* [Middle English *touret,* from Middle French *torete, tourete,* from Old French, diminutive of *tor, tur* tower — more at TOWER] (14th century)
1 : a little tower; *specifically* **:** an ornamental structure at an angle of a larger structure
2 a : a pivoted and revolvable holder in a machine tool **b :** a device (as on a microscope or a television camera) holding several lenses
3 a : a tall building usually moved on wheels and formerly used for carrying soldiers and equipment for breaching or scaling a wall **b** (1) **:** a gunner's fixed or movable enclosure in an airplane (2) **:** a revolving armored structure on a warship that protects one or more guns mounted within it (3) **:** a similar upper structure usually for one gun on a tank

tur·ret·ed \ˈtər-ə-təd, ˈtə-rə-, ˈtu̇r-ə-\ *adjective* (1550)
: furnished with or as if with turrets

¹tur·tle \ˈtər-tᵊl\ *noun* [Middle English, from Old English *turtla,* from Latin *turtur*] (before 12th century)
archaic **:** TURTLEDOVE

²turtle *noun, plural* **turtles** *also* **turtle** *often attributive* [modification of French *tortue,* from Late Latin *tartarucha,* feminine of *tartaruchus* of Tartarus, from Greek *tartarouchos,* from *Tartaros* Tartarus] (1657)
: any of an order (Testudines) of terrestrial, freshwater, and marine reptiles that have a toothless horny beak and a shell of bony dermal plates usually covered with horny shields enclosing the trunk and into which the head, limbs, and tail usually may be withdrawn ◆

³turtle *noun* (1952)
: TURTLENECK

tur·tle·back \ˈtər-tᵊl-ˌbak\ *noun* (1881)
: a raised convex surface
— turtleback *or* **tur·tle–backed** \-ˌbakt\ *adjective*

tur·tle·dove \-ˌdəv\ *noun* (14th century)
: any of several small wild pigeons (genus *Streptopelia* and especially *S. turtur*) noted for plaintive cooing

tur·tle·head \-ˌhed\ *noun* (1857)
: any of a genus (*Chelone*) of perennial herbs of the snapdragon family with spikes of showy white or purple flowers

tur·tle·neck \-ˌnek\ *noun* (1897)
1 : a high close-fitting turnover collar used especially for sweaters
2 : a sweater with a turtleneck
— tur·tle·necked \-ˌnekt\ *adjective*

tur·tling \ˈtərt-liŋ, ˈtər-tᵊl-iŋ\ *noun* (1726)
: the action or process of catching turtles

turves *plural of* TURF

¹Tus·can \ˈtəs-kən\ *adjective* [Latin *tuscanus* Etruscan, from *Tusci* Etruscans] (1563)
1 : of or relating to one of the five classical orders of architecture that is of Roman origin and plain in style
2 : of, relating to, or characteristic of Tuscany, the Tuscans, or Tuscan

²Tuscan *noun* (1568)
1 a : the Italian language as spoken in Tuscany **b :** the standard literary dialect of Italian
2 : a native or inhabitant of Tuscany

Tus·ca·ro·ra \ˌtəs-kə-ˈrōr-ə, -ˈrȯr-\ *noun, plural* **Tuscarora** *or* **Tuscaroras** [Tuscarora *skarò·rə̨ʔ*] (1650)
1 : a member of an American Indian people originally of North Carolina and later of New York and Ontario
2 : the Iroquoian language of the Tuscarora people

tu·sche \ˈtüsh, ˈtü-shə\ *noun* [German, from *tuschen* to lay on color, from French *toucher,* literally, to touch, from Old French *tuchier* — more at TOUCH] (1885)
: a black liquid used in lithography for drawing and painting and in etching and the silkscreen process as a resist

¹tush \ˈtəsh\ *noun* [Middle English *tusch,* from Old English *tūsc;* akin to Old Frisian *tusk* tooth, Old English *tōth* tooth] (before 12th century)
: a long pointed tooth; *especially* **:** a horse's canine

²tush *interjection* [Middle English *tussch*] (15th century)
— used to express disdain or reproach

³tush \ˈtu̇sh\ *noun* [perhaps modification of Yiddish *tokhes,* from Hebrew *taḥath* under, beneath] (1962)
slang **:** BUTTOCKS

¹tusk \ˈtəsk\ *noun* [Middle English, alteration of *tux,* from Old English *tūx;* akin to Old English *tūsc* tush] (before 12th century)
1 : an elongated greatly enlarged tooth (as of an elephant or walrus) that projects when the mouth is closed and serves for digging food or as a weapon; *broadly* **:** a long protruding tooth
2 : one of the small projections on a tusk tenon
— tusked \ˈtəskt\ *adjective*
— tusk·like \ˈtəsk-ˌlīk\ *adjective*

²tusk *transitive verb* (1629)
: to dig up with a tusk; *also* **:** to gash with a tusk

tusk·er \ˈtəs-kər\ *noun* (1859)
: an animal with tusks; *especially* **:** a male elephant with two normally developed tusks

tusk tenon *noun* (circa 1825)
: a tenon strengthened by one or more smaller tenons underneath forming a steplike outline

tus·sah \ˈtə-sə, -ˌsȯ\ *or* **tus·sore** \-ˌsōr, -ˌsȯr\ *noun* [Hindi *tasar*] (1590)
: silk or silk fabric from the brownish fiber produced by larvae of some saturniid moths (as *Antheraea paphia*)

tus·sive \ˈtə-siv\ *adjective* [Latin *tussis* cough] (circa 1857)
: of, relating to, or involved in coughing

¹tus·sle \ˈtə-səl\ *noun* (1629)
1 : a physical contest or struggle **:** SCUFFLE
2 : an intense argument, controversy, or struggle

²tussle *intransitive verb* **tus·sled; tus·sling** \-s(ə-)liŋ\ [Middle English *tussillen,* frequentative of Middle English *-tusen, -tousen* to tousle — more at TOUSE] (1638)
: to struggle roughly **:** SCUFFLE

tus·sock \ˈtə-sək\ *noun* [origin unknown] (1580)
: a compact tuft especially of grass or sedge; *also* **:** an area of raised solid ground in a marsh or bog that is bound together by roots of low vegetation
— tus·socky \-sə-kē\ *adjective*

tussock grass *noun* (1842)
: a grass or sedge that typically grows in tussocks

tussock moth *noun* (1826)
: any of numerous dull-colored moths (family Lymantriidae) that usually have wingless females and larvae with long tufts of hair

tut *a t-sound made by suction rather than explosion; often read as* ˈtət\ *interjection* (circa 1529)
— used to express disapproval or disbelief

tu·tee \tü-ˈtē, tyü-\ *noun* [*tutor* + *-ee*] (circa 1927)
: one who is being tutored

tu·te·lage \ˈtü-tᵊl-ij, ˈtyü-\ *noun* [Latin *tutela* protection, guardian (from *tutari* to protect, frequentative of *tueri* to look at, guard) + English *-age*] (1605)
1 a : an act or process of serving as guardian or protector **:** GUARDIANSHIP **b :** hegemony over a foreign territory **:** TRUSTEESHIP 2
2 : the state of being under a guardian or tutor
3 a : instruction especially of an individual **b :** a guiding influence

tu·te·lar \ˈtü-tᵊl-ər, ˈtyü-, -ˌär\ *adjective or noun* (1600)
: TUTELARY

¹tu·te·lary \ˈtü-tᵊl-ˌer-ē, ˈtyü-\ *adjective* (1611)
1 : having the guardianship of a person or a thing ⟨a *tutelary* goddess⟩
2 : of or relating to a guardian

²tutelary *noun, plural* **-lar·ies** (1652)
: a tutelary power (as a deity)

¹tu·tor \ˈtü-tər, ˈtyü-\ *noun* [Middle English, from Middle French & Latin; Middle French *tuteur,* from Latin *tutor,* from *tueri*] (14th century)
: a person charged with the instruction and guidance of another: as **a :** a private teacher **b :** a teacher in a British university who gives individual instruction to undergraduates

²tutor (1592)
transitive verb
1 : to have the guardianship, tutelage, or care of
2 : to teach or guide usually individually in a special subject or for a particular purpose **:** COACH
intransitive verb
1 : to do the work of a tutor
2 : to receive instruction especially privately

tu·tor·age \ˈtü-tə-rij, ˈtyü-\ *noun* (1617)
: the function or work of a tutor

tu·tor·ess \ˈtü-tə-rəs, ˈtyü-\ *noun* (1614)
: a woman or girl who is a tutor

¹tu·to·ri·al \tü-ˈtōr-ē-əl, tyü-, -ˈtȯr-\ *adjective* (1822)
: of, relating to, or involving a tutor

²tutorial *noun* (1923)
1 : a class conducted by a tutor for one student or a small number of students
2 : a paper, book, film, or computer program that provides practical information about a specific subject

tu·tor·ship \ˈtü-tər-ˌship, ˈtyü-\ *noun* (1581)
1 : the office, function, or work of a tutor
2 : TUTELAGE 3

tu·toy·er \tue-twà-yā\ *transitive verb* [French, to address with the familiar pronoun *tu* thou,

◇ WORD HISTORY

turtle The word *turtle* appears to have developed in 17th century nautical English out of an earlier word *tortu,* or its French source *tortue,* from which English *tortoise* is also derived. French *tortue* as a name for this animal, like Spanish *tortuga* and Italian *tartaruga,* ultimately descends from Late Latin *tartarucha.* This word is short for *bestia tartarucha* (or a similar collocation), literally, "infernal beast" or "beast of the nether regions." Such a humble and generally inoffensive animal hardly seems to deserve this epithet, but in early Christian literature and art the turtle, thought because of its lowness to the ground to wallow in mud and filth, was taken as a symbol of heresy and evil. A mosaic of the late 3rd century A.D. in Aquileia, northern Italy, depicts a turtle partly withdrawn into its shell aggressively confronted by a rooster, symbolic of goodness. Both rooster and turtle as emblems of light and darkness were taken from the pre-Christian cult of Mithras popular in the Roman empire. This demonic characterization of the animal was pervasive enough to affect everyday speech, and the Romance descendants of *tartarucha* largely ousted the descendants of the classical Latin word for "turtle," *testudo.*

from Middle French, from *tu* thou (from Latin) + *toi* thee, from Latin *te* (accusative of *tu*) — more at THOU] (1697)
: to address familiarly
Tut·si \'tü-sē, 'tüt-\ *noun, plural* **Tutsi** *or* **Tutsis** (1950)
: a member of a Nilotic people of Rwanda and Burundi
[1]tut·ti \'tü-tē, 'tu̇-; 'tü-ˌtē, 'tu̇-\ *adjective or adverb* [Italian, masculine plural of *tutto* all, from (assumed) Vulgar Latin *tottus,* alteration of Latin *totus*] (circa 1724)
: with all voices or instruments performing together — used as a direction in music
[2]tutti *noun* (1816)
: a passage or section performed by all the performers
tut·ti–frut·ti \ˌtü-ti-'frü-tē, ˌtu̇-\ *noun* [Italian *tutti frutti* all fruits] (1834)
: a confection or ice cream containing chopped usually candied fruits
[1]tut–tut *two t-sounds made by suction rather than explosion; often read as* 'tət-'tət\ *interjection* (1591)
: TUT
[2]tut–tut \'tət-'tət\ *intransitive verb* **tut–tutted; tut–tut·ting** (1873)
: to express disapproval or disbelief by or as if by uttering tut ⟨the civilian press corps *tut-tutted* over the officers' equivocation, ignorance and bigotry —M. B. Duberman⟩
tu·tu \'tü-(ˌ)tü\ *noun* [French, from (baby talk) *tutu* backside] (1913)
: a short projecting skirt worn by a ballerina
tu–whit tu–whoo \tə-ˌ(h)wit-tə-'(h)wü\ *noun* [imitative] (1588)
: the cry of an owl
tux \'təks\ *noun* (1922)
: TUXEDO
tux·e·do \ˌtək-'sē-(ˌ)dō\ *noun, plural* **-dos** *or* **-does** [*Tuxedo* Park, N.Y.] (1889)
1 : a single-breasted or double-breasted usually black or blackish blue jacket
2 : semiformal evening clothes for men
— tux·e·doed \-(ˌ)dōd\ *adjective*

tutu

tu·yere \twē-'er\ *noun* [French *tuyère,* from Middle French, from *tuyau* pipe] (1781)
: a nozzle through which an air blast is delivered to a forge or blast furnace
TV \'tē-'vē\ *noun* [*t*ele*v*ision] (1947)
: TELEVISION
TV dinner *noun* [from its saving the television viewer from having to interrupt viewing to prepare and serve a meal] (1954)
: a quick-frozen packaged dinner (as of meat, potatoes, and a vegetable) that requires only heating before it is served
twa \'twȧ\ *or* **twae** \'twȧ, 'twē\ *Scottish variant of* TWO
[1]twad·dle \'twä-d^{ə}l\ *noun* [probably alteration of English dialect *twattle* idle talk] (1782)
1 : silly idle talk **:** DRIVEL
2 : one that twaddles **:** TWADDLER
[2]twaddle *verb* **twad·dled; twad·dling** \'twäd-liŋ, 'twä-d^{ə}l-iŋ\ (1826)
: PRATE, BABBLE
— twad·dler \'twäd-lər, 'twä-d^{ə}l-ər\ *noun*
[1]twain \'twān\ *adjective* [Middle English, from Old English *twēgen* — more at TWO] (before 12th century)
archaic **:** TWO
[2]twain *pronoun* (before 12th century)
: TWO; *especially* **:** two fathoms ⟨mark *twain*⟩
[3]twain *noun* (14th century)
1 : TWO
2 : COUPLE, PAIR
[1]twang \'twaŋ\ *noun* [imitative] (circa 1553)
1 : a harsh quick ringing sound like that of a plucked banjo string
2 a : nasal speech or resonance **b :** the characteristic speech of a region, locality, or group of people
3 a : an act of plucking **b :** PANG, TWINGE
— twangy \'twaŋ-ē\ *adjective*
[2]twang *verb* **twanged; twang·ing** \'twaŋ-iŋ\ (1567)
intransitive verb
1 : to sound with a twang ⟨the couch *twanged* when he sat down⟩
2 : to speak or sound with a nasal intonation
3 : to throb or twitch with pain or tension
transitive verb
1 : to cause to sound with a twang
2 : to utter or pronounce with a nasal twang
3 : to pluck the string of
— twang·er *noun*
[3]twang *noun* [alteration of *tang*] (1611)
1 : a persisting flavor, taste, or odor **:** TANG
2 : SUGGESTION, TRACE
'twas \'twəz, 'twäz\ [by contraction] (1588)
: it was
twat \'twät\ *noun* [origin unknown] (1656)
: VULVA — usually considered vulgar
tway·blade \'twā-ˌblād\ *noun* [English dialect *tway* two] (1578)
: any of various orchids (genera *Listera* and *Liparis*) often having two leaves
[1]tweak \'twēk\ *verb* [probably alteration of Middle English *twikken* to pull sharply, from Old English *twiccian* to pluck — more at TWITCH] (1601)
transitive verb
1 : to pinch and pull with a sudden jerk and twist **:** TWITCH ⟨*tweaked* a bud from the stem⟩
2 : to pinch (a person or a body part) lightly or playfully ⟨*tweaked* the baby's ear affectionately⟩
3 : to make small adjustments in or to ⟨*tweak* the controls⟩; *especially* **:** FINE-TUNE
intransitive verb
: PULL 1a, PLUCK
[2]tweak *noun* (1609)
: an act of tweaking **:** PINCH
twee \'twē\ *adjective* [baby-talk alteration of *sweet*] (1905)
chiefly British **:** affectedly or excessively dainty, delicate, cute, or quaint ⟨such a theme might sound *twee* or corny —*Times Literary Supplement*⟩
tweed \'twēd\ *noun* [probably short for Scots *tweedling, twidling* twilled cloth] (1841)
1 : a rough woolen fabric made usually in twill weaves and used especially for suits and coats
2 *plural* **:** tweed clothing; *specifically* **:** a tweed suit
Twee·dle·dum and Twee·dle·dee \ˌtwē-d^{ə}l-'dəm-ən(d)-ˌtwē-d^{ə}l-'dē\ *noun* [English *tweedle* to chirp + *dum* (imitative of a low musical note) *& dee* (imitative of a high musical note)] (1725)
: two individuals or groups that are practically indistinguishable
tweedy \'twē-dē\ *adjective* **tweed·i·er; -est** (1912)
1 : of or resembling tweed
2 a : given to wearing tweeds **b :** informal or suggestive of the outdoors in taste or habits
— tweed·i·ness *noun*
tween \'twēn\ *preposition* [Middle English *twene,* short for *betwene*] (14th century)
: BETWEEN
tweet \'twēt\ *noun* [imitative] (1845)
: a chirping note
— tweet *intransitive verb*
tweet·er \'twē-tər\ *noun* (1934)
: a small loudspeaker responsive only to the higher acoustic frequencies and reproducing sounds of high pitch — compare WOOFER
tweeze \'twēz\ *transitive verb* **tweezed; tweez·ing** [back-formation from *tweezers*] (1932)
: to pluck, remove, or handle with tweezers
twee·zer \'twē-zər\ *noun* (1904)
: TWEEZERS
twee·zers \-zərz\ *noun plural but singular or plural in construction* [obsolete English *tweeze,* noun, *etui,* short for obsolete English *etweese,* from plural of obsolete English *etwee,* from French *étui*] (1654)
: any of various small metal instruments that are usually held between the thumb and forefinger, are used for plucking, holding, or manipulating, and consist of two legs joined at one end
Twelfth Day *noun* [from its being the 12th day after Christmas] (before 12th century)
: EPIPHANY 1
Twelfth Night *noun* (before 12th century)
: the evening or sometimes the eve of Epiphany
twelve \'twelv\ *noun* [Middle English, from Old English *twelf;* akin to Old High German *zwelif* twelve, Old English *twā* two, *-leofan* (as in *endleofan* eleven) — more at TWO, ELEVEN] (before 12th century)
1 — see NUMBER table
2 *capitalized* **a :** the twelve original disciples of Jesus **b :** the books of the Minor Prophets in the Jewish Scriptures
3 : the 12th in a set or series
4 : something having 12 units or members
5 *plural* **:** TWELVEMO
— twelfth \'twelf(t)th\ *adjective or noun*
— twelve *adjective*
— twelve *pronoun, plural in construction*
twelve·mo \'twelv-(ˌ)mō\ *noun, plural* **-mos** (1819)
: the size of a piece of paper cut 12 from a sheet; *also* **:** a book, a page, or paper of this size
twelve·month \-ˌmən(t)th\ *noun* (13th century)
: YEAR
twelve–tone \-'tōn\ *adjective* (1926)
: of, relating to, or being serial music utilizing the 12 chromatic tones
twelve–tone row *noun* (1941)
: the 12 chromatic tones of the octave placed in a chosen fixed order and constituting with some permitted permutations and derivations the melodic and harmonic material of a serial musical piece
twen·ty \'twen-tē, 'twən-\ *noun, plural* **twenties** [Middle English, from *twenty,* adjective, from Old English *twēntig,* noun, group of 20, from *twēn-* (akin to Old English *twā* two) + *-tig* group of 10; akin to Old English *tīen* ten — more at TWO, TEN] (15th century)
1 — see NUMBER table
2 *plural* **:** the numbers 20 to 29; *specifically* **:** the years 20 to 29 in a lifetime or century
3 : a 20-dollar bill
— twen·ti·eth \-tē-əth\ *adjective or noun*
— twenty *adjective*
— twenty *pronoun, plural in construction*
twen·ty–four·mo \ˌtwen-tē-'fōr-(ˌ)mō, ˌtwən-, -'fȯr-\ *noun, plural* **-mos** (circa 1841)
: the size of a piece of paper cut 24 from a sheet; *also* **:** a book, a page, or paper of this size
twen·ty–one \ˌtwen-tē-'wən, ˌtwən-\ *noun* (1611)
1 — see NUMBER table
2 [translation of French *vingt-et-un*] **:** BLACKJACK
— twenty–one *adjective*
— twenty–one *pronoun, plural in construction*
twen·ty–twen·ty *or* **20/20** \-'twen-tē, -'twən-\ *adjective* [from the testing of vision by reading letters at a distance of 20 feet] (1875)
of the human eye **:** meeting a standard of normal visual acuity ⟨*twenty-twenty* vision⟩
twen·ty–two \-'tü\ *noun* (1526)
1 — see NUMBER table
2 : a .22-caliber firearm; *especially* **:** one firing rimfire cartridges — usually written .22
— twenty–two *adjective*
— twenty–two *pronoun, plural in construction*

'twere \'twər\ [by contraction] (1588)
: it were

twerp \'twərp\ *noun* [origin unknown] (circa 1923)
: a silly, insignificant, or contemptible person

Twi \'chwē, chə-'wē, 'twē, 'chē\ *noun* (1874)
1 : a dialect of Akan
2 : a literary language based on the Twi dialect and used by the Akan-speaking peoples (as the Ashanti)

twi- \'twī\ *prefix* [Middle English, from Old English; akin to Old High German *zwi-* twi-, Latin *bi-*, Greek *di-*, Old English *twā* two]
: two : double : doubly : twice ⟨*twi*-headed⟩

twice \'twīs\ *adverb* [Middle English *twiges, twies,* from Old English *twiga;* akin to Old English *twi-*] (12th century)
1 : on two occasions ⟨*twice* absent⟩
2 : two times : in doubled quantity or degree ⟨*twice* two is four⟩ ⟨*twice* as much⟩

twice–born \-'bȯrn\ *adjective* (15th century)
1 : born a second time
2 : having undergone a definite experience of fundamental moral and spiritual renewal : REGENERATE
3 : of or forming one of the three upper Hindu caste groups in which boys undergo an initiation symbolizing spiritual birth

twice–laid \-'lād\ *adjective* (circa 1593)
: made from the ends of rope and strands of used rope ⟨*twice-laid* rope⟩

twice–told \-'tōld\ *adjective* (1595)
: well known from repeated telling — used chiefly in the phrase *a twice-told tale*

[1]twid·dle \'twi-dᵊl\ *verb* **twid·dled; twid·dling** \'twid-liŋ, 'twi-dᵊl-iŋ\ [origin unknown] (circa 1540)
intransitive verb
1 : to play negligently with something : FIDDLE
2 : to turn or jounce lightly ⟨*twiddles* round and round in the water —J. B. S. Haldane⟩
transitive verb
: to rotate lightly or idly ⟨*twiddled* his cigar —James Lord⟩
— twiddle one's thumbs : to spend time idly : do nothing

[2]twiddle *noun* (1774)
: TURN, TWIST

[1]twig \'twig\ *noun* [Middle English *twigge,* from Old English; akin to Old High German *zwīg* twig, Old English *twā* two] (before 12th century)
1 : a small shoot or branch usually without its leaves
2 : a minute branch of a nerve or artery
— twigged \'twigd\ *adjective*
— twig·gy \'twi-gē\ *adjective*

[2]twig *verb* **twigged; twig·ging** [perhaps from Irish & Scottish Gaelic *tuig-* understand] (1764)
transitive verb
1 : NOTICE, OBSERVE
2 : to understand the meaning of : COMPREHEND
intransitive verb
: to gain a grasp : UNDERSTAND ⟨*twigged* instinctively about things —H. E. Bates⟩

[3]twig *noun* [origin unknown] (circa 1811)
British : FASHION, STYLE

twi·light \'twī-,līt\ *noun, often attributive* (15th century)
1 : the light from the sky between full night and sunrise or between sunset and full night produced by diffusion of sunlight through the atmosphere and its dust
2 a : an intermediate state that is not clearly defined ⟨lived in the *twilight* of neutrality —*Newsweek*⟩ **b :** a period of decline ⟨the *twilight* of a great career⟩

twilight glow *noun* (1819)
: airglow seen at twilight

Twilight of the Gods
: RAGNAROK

twilight zone *noun* (1909)
1 a : TWILIGHT 2a **b :** an area just beyond ordinary legal and ethical limits
2 : a world of fantasy, illusion, or unreality

twi·lit \'twī-,lit\ *adjective* [*twi*light + *lit*] (1869)
: lighted by or as if by twilight

twill \'twil\ *noun* [Middle English *twyll, twylle,* from Old English *twilic* having a double thread, part translation of Latin *bilic-, bilix,* from *bi-* + *licium* thread] (14th century)
1 : a fabric with a twill weave
2 : a textile weave in which the filling threads pass over one and under two or more warp threads to give an appearance of diagonal lines

twilled \'twild\ *adjective* (15th century)
: made with a twill weave

twill·ing \'twi-liŋ\ *noun* (circa 1859)
: twilled fabric; *also* : the process of making it

[1]twin \'twin\ *noun* [Middle English, from *twin* twofold] (14th century)
1 a : either of two offspring produced at a birth **b** *plural, capitalized* : GEMINI
2 : one of two persons or things closely related to or resembling each other
3 : a compound crystal composed of two adjoining crystals or parts of crystals of the same kind that share a common plane of atoms
— twin·ship \-,ship\ *noun*

[2]twin *verb* **twinned; twin·ning** (14th century)
transitive verb
1 : to bring together in close association : COUPLE
2 : DUPLICATE, MATCH
intransitive verb
1 : to bring forth twins
2 : to grow as a twin crystal

[3]twin *adjective* [Middle English, twofold, double, from Old English *twinn;* akin to Old Norse *tvinnr* two by two, Old English *twā* two] (1590)
1 : born with one other or as a pair at one birth ⟨*twin* brother⟩ ⟨*twin* girls⟩
2 a : made up of two similar, related, or connected members or parts : DOUBLE **b :** paired in a close or necessary relationship : MATCHING **c :** having or consisting of two identical units **d :** being one of a pair

twin bed *noun* (1919)
: one of a pair of matching single beds

twin·ber·ry \'twin-,ber-ē\ *noun* [from the occurrence of the berries in pairs] (1821)
1 : a shrubby North American honeysuckle (*Lonicera involucrata*) with yellowish involucrate flowers
2 : PARTRIDGEBERRY

twin bill *noun* (circa 1939)
: DOUBLEHEADER

twin·born \'twin-'bȯrn\ *adjective* (1598)
: born at the same birth

twin double *noun* (1960)
: a system of betting (as on horse races) in which the bettor must pick the winners of four stipulated races in order to win — compare DAILY DOUBLE

[1]twine \'twīn\ *noun* [Middle English *twin,* from Old English *twīn;* akin to Middle Dutch *twijn* twine, Old English *twā* two] (before 12th century)
1 : a strong string of two or more strands twisted together
2 : a twined or interlaced part or object
3 : an act of twining, interlacing, or embracing
— twiny \'twī-nē\ *adjective*

[2]twine *verb* **twined; twin·ing** (13th century)
transitive verb
1 a : to twist together **b :** to form by twisting : WEAVE
2 a : INTERLACE ⟨the girl *twined* her hands —John Buchan⟩ **b :** to cause to encircle or enfold something **c :** to cause to be encircled
intransitive verb
1 : to coil about a support
2 : to stretch or move in a sinuous manner : MEANDER ⟨the river *twines* through the valley⟩
— twin·er *noun*

[3]twine *verb* **twined; twin·ing** [alteration of Scots *twin,* from Middle English *twinnen,* from *twin* double] (1722)
transitive verb
chiefly Scottish : to cause (one) to lose possession : DEPRIVE ⟨*twined* him of his nose —J. C. Ransom⟩
intransitive verb
chiefly Scottish : PART ⟨you and me must *twine* —R. L. Stevenson⟩

twin·flow·er \'twin-,flau̇(-ə)r\ *noun* (circa 1818)
: a low prostrate subshrub (*Linnaea borealis*) of the honeysuckle family that is found in cool regions of the northern hemisphere and has fragrant usually pink flowers

[1]twinge \'twinj\ *verb* **twinged; twing·ing** \'twin-jiŋ\ *or* **twinge·ing** [Middle English *twengen,* from Old English *twengan;* akin to Old High German *zwengen* to pinch] (before 12th century)
transitive verb
1 *dialect* : PLUCK, TWEAK
2 : to affect with a sharp pain or pang
intransitive verb
: to feel a sudden sharp local pain

[2]twinge *noun* (1608)
1 : a sudden sharp stab of pain
2 : a moral or emotional pang ⟨a *twinge* of conscience⟩

twi–night \'twī-,nīt\ *adjective* [*twi*light + *night*] (1946)
: of, relating to, or being a baseball doubleheader in which the first game is played in the late afternoon and the second continues into the evening

[1]twin·kle \'twiŋ-kəl\ *verb* **twin·kled; twin·kling** \-k(ə-)liŋ\ [Middle English, from Old English *twinclian;* akin to Middle High German *zwinken* to blink] (before 12th century)
intransitive verb
1 : to shine with a flickering or sparkling light : SCINTILLATE
2 a : to flutter the eyelids **b :** to appear bright especially with merriment ⟨his eyes *twinkled*⟩
3 : to flutter or flit rapidly
transitive verb
1 : to cause to shine with fluctuating light
2 : to flicker or flirt rapidly ⟨*twinkled* the straight, red-lacquered toes —Glenway Wescott⟩
— twin·kler \-k(ə-)lər\ *noun*

[2]twinkle *noun* (1548)
1 : a wink of the eyelids
2 : the instant's duration of a wink : TWINKLING
3 : an intermittent radiance : FLICKER, SPARKLE
4 : a rapid flashing motion : FLIT
— twin·kly \-k(ə-)lē\ *adjective*

twin·kling \'twiŋ-kliŋ\ *noun* (14th century)
: the time required for a wink : INSTANT ⟨the kettle will boil in a *twinkling* —*Punch*⟩

twin·set \'twin-'set\ *noun* (1937)
: a combination of a matching pullover and cardigan worn together

twin–size \'twin-'sīz\ *adjective* [*twin bed*] (1926)
: having the dimensions 39 inches by 75 inches (about 99 centimeters by 191 centimeters) — used of a bed; compare FULL-SIZE, KING-SIZE, QUEEN-SIZE

[1]twirl \'twər-(-ə)l\ *verb* [perhaps of Scandinavian origin; akin to Norwegian dialect *tvirla* to twirl; akin to Old High German *dweran* to stir] (1598)
intransitive verb
1 : to revolve rapidly
2 : to pitch in a baseball game
transitive verb
1 : to cause to rotate rapidly
2 : PITCH 2a

— **twirl·er** \'twər-lər\ *noun*
²twirl *noun* (1598)
1 : an act of twirling
2 : COIL, WHORL
— **twirly** \'twər-lē\ *adjective*
twirp *variant of* TWERP
¹twist \'twist\ *verb* [Middle English, from Old English *-twist* rope; akin to Middle Dutch *twist* quarrel, twine, Old English *twā* two] (15th century)
transitive verb
1 a : to unite by winding ⟨*twisting* strands together⟩ **b :** to make by twisting strands together ⟨*twist* thread from yarn⟩ **c :** to mingle by interlacing
2 : TWINE, COIL
3 a : to wring or wrench so as to dislocate or distort; *especially* **:** SPRAIN ⟨*twisted* my ankle⟩ **b :** to alter the meaning of **:** DISTORT, PERVERT ⟨*twisted* the facts⟩ **c :** CONTORT ⟨*twisted* his face into a grin⟩ **d :** to pull off, turn, or break by torsion **e :** to cause to move with a turning motion **f :** to form into a spiral shape **g :** to cause to take on moral, mental, or emotional deformity **h :** to make (one's way) in a winding or devious manner to a destination or objective
intransitive verb
1 : to follow a winding course **:** SNAKE
2 a : to turn or change shape under torsion **b :** to assume a spiral shape **c :** SQUIRM, WRITHE **d :** to dance the twist
3 *of a ball* **:** to rotate while taking a curving path or direction
4 : TURN 3a ⟨*twisted* around to see behind him⟩
— **twist one's arm :** to bring strong pressure to bear on one
²twist *noun* (1555)
1 : something formed by twisting or winding: as **a :** a thread, yarn, or cord formed by twisting two or more strands together **b :** a strong tightly twisted sewing silk **c :** a baked piece of twisted dough **d :** tobacco leaves twisted into a thick roll **e :** a strip of citrus peel used to flavor a drink
2 a : an act of twisting **:** the state of being twisted **b :** a dance performed with strenuous gyrations especially of the hips **c :** the spin given the ball in any of various games **d :** a spiral turn or curve **e** (1) **:** torque or torsional stress applied to a body (as a rod or shaft) (2) **:** torsional strain (3) **:** the angle through which a thing is twisted
3 a : a turning off a straight course **b :** ECCENTRICITY, IDIOSYNCRASY **c :** a distortion of meaning or sense
4 a : an unexpected turn or development ⟨weird *twists* of fate —W. L. Shirer⟩ **b :** a clever device **:** TRICK ⟨questions demanding special *twists* of thinking —*New Yorker*⟩ **c :** a variant approach or method **:** GIMMICK ⟨a kind of *twist* on the old triangle theme —Dave Fedo⟩
5 : a front or back dive in which the diver twists sideways a half or full turn before entering the water
— **twisty** \'twis-tē\ *adjective*
twist drill *noun* (circa 1875)
: a drill having deep helical grooves extending from the point to the smooth portion of the shank
twist·ed \'twis-təd\ *adjective* (circa 1890)
: mentally or emotionally unsound or disturbed **:** SICK
twist·er \'twis-tər\ *noun* (1579)
1 : one that twists; *especially* **:** a ball with a forward and spinning motion
2 : a tornado, waterspout, or dust devil in which the rotatory ascending movement of a column of air is especially apparent
twist·ing \'twis-tiŋ\ *noun* (circa 1905)
: the use of misrepresentation or trickery to get someone to lapse a life insurance policy and buy another usually in another company
¹twit \'twit\ *noun* (1528)
1 : an act of twitting **:** TAUNT
2 : a silly annoying person **:** FOOL
²twit *transitive verb* **twit·ted; twit·ting** [Middle English *atwiten* to reproach, from Old English *ætwītan,* from *æt* at + *wītan* to reproach; akin to Old High German *wīzan* to punish, Old English *witan* to know] (1530)
1 : to subject to light ridicule or reproach **:** RALLY
2 : to make fun of as a fault
¹twitch \'twich\ *verb* [Middle English *twicchen;* akin to Old English *twiccian* to pluck, Old High German *gizwickan* to pinch] (14th century)
transitive verb
: to move or pull with a sudden motion **:** JERK
intransitive verb
1 : PULL 1a, PLUCK ⟨*twitched* at my sleeve⟩
2 : to move jerkily **:** QUIVER
— **twitch·er** *noun*
²twitch *noun* (1523)
1 : an act of twitching; *especially* **:** a short sudden pull or jerk
2 : a physical or mental pang
3 : a loop of rope or strap that is tightened over a horse's lip as a restraining device
4 a : a short spastic contraction of the muscle fibers **b :** a slight jerk of a body part
— **twitch·i·ly** \'twi-chə-lē\ *adverb*
— **twitchy** \'twi-chē\ *adjective*
³twitch *noun* [alteration of *quitch*] (1598)
: QUACK GRASS
¹twit·ter \'twi-tər\ *verb* [Middle English *twiteren;* akin to Old High German *zwizzirōn* to twitter] (14th century)
intransitive verb
1 : to utter successive chirping noises
2 a : to talk in a chattering fashion **b :** GIGGLE, TITTER
3 : to tremble with agitation **:** FLUTTER
transitive verb
1 : to utter in chirps or twitters ⟨the robin *twittered* its morning song⟩
2 : to shake rapidly back and forth **:** FLUTTER
²twitter *noun* (1678)
1 : a trembling agitation **:** QUIVER
2 : a small tremulous intermittent sound (as of birds)
3 a : a light chattering **b :** a light silly laugh **:** GIGGLE
— **twit·tery** \'twi-tə-rē\ *adjective*
twixt \'twikst\ *or* **'twixt** *preposition* [Middle English *twix,* short for *betwix, betwixt*] (14th century)
: BETWEEN
¹two \'tü\ *adjective* [Middle English *twa, two,* from Old English *twā* (feminine & neuter); akin to Old English *twēgen* two (masculine), *tū* (neuter), Old High German *zwēne,* Latin *duo,* Greek *dyo*] (before 12th century)
1 : being one more than one in number
2 : being the second — used postpositively ⟨section *two* of the instructions⟩
²two *pronoun, plural in construction* (before 12th century)
1 : two countable individuals not specified ⟨only *two* were found⟩
2 : a small approximate number of indicated things ⟨only a shot or *two* were fired⟩
³two *noun, plural* **twos** (circa 1585)
1 — see NUMBER table
2 : the second in a set or series ⟨the *two* of spades⟩
3 : a 2-dollar bill
4 : something having two units or members
two-bag·ger \-'ba-gər\ *noun* (1880)
: DOUBLE 1b
two-bit \'tü-'bit\ *adjective* (1802)
1 : of the value of two bits
2 : cheap or trivial of its kind **:** PETTY, SMALL-TIME
two bits *noun plural but singular or plural in construction* (1730)
1 : the value of a quarter of a dollar
2 : something of small worth or importance
¹two-by-four \,tü-bī-'fōr, -'fór\ *noun* (1884)
: a piece of lumber approximately 2 by 4 inches as sawed and usually 1⅝ by 3⅝ inches if dressed
²two-by-four *adjective* (1897)
1 : measuring two units (as inches) by four
2 : small or petty of its kind ⟨this house and its *two-by-four* garden —Philip Barry⟩
two cents *noun* (1947)
1 : a sum or object of very small value **:** practically nothing ⟨said angrily that for *two cents* he'd punch your nose⟩
2 *or* **two cents worth :** an opinion offered on a topic under discussion ⟨each speaker . . . is getting in his *two cents worth* —Dwight Macdonald⟩
two-cycle *adjective* (1902)
of an internal combustion engine **:** having a 2-stroke cycle
two-dimensional *adjective* (1883)
1 : having two dimensions
2 : lacking depth of characterization ⟨*two-dimensional* characters⟩
— **two-dimensionality** *noun*
two-edged sword \'tü-'ejd-, -'e-jəd-\ *noun* (1526)
: DOUBLE-EDGED SWORD
two-faced \'tü-'fāst\ *adjective* (1609)
1 : DOUBLE-DEALING, FALSE
2 : having two faces
— **two-faced·ness** \-'fāst-nəs, -'fā-səd-nəs\ *noun*
two·fer \'tü-fər\ *noun* [alteration of *two for (one)*] (1890)
1 : a cheap item of merchandise; *especially* **:** a cigar selling at two for a nickel
2 : a free coupon entitling the bearer to purchase two tickets to a specified theatrical production for the price of one
3 : two articles available for the price of one or about the price of one
two-fist·ed \-'fis-təd\ *adjective* (1774)
: marked by vigorous often virile energy **:** HARD-HITTING
two·fold \'tü-,fōld, -'fōld\ *adjective* (1559)
1 : having two parts or aspects
2 : being twice as great or as many
— **twofold** \-'fōld\ *adverb*
2,4-D \,tü-,fōr-'dē, -,fór-\ *noun* [*di-*] (circa 1945)
: a white crystalline irritant compound $C_8H_6Cl_2O_3$ used especially as a weed killer
2,4,5-T \-,fīv-'tē\ *noun* [*tri-*] (1946)
: an irritant compound $C_8H_5Cl_3O_3$ used especially as an herbicide and defoliant
two-hand·ed \'tü-'han-dəd\ *adjective* (15th century)
1 : used with both hands ⟨a *two-handed* sword⟩
2 : requiring two persons ⟨a *two-handed* saw⟩
3 *archaic* **:** STOUT, STRONG
4 a : having two hands **b :** efficient with either hand
two-line octave *noun* (circa 1931)
: the musical octave that begins on the first C above middle C — see PITCH illustration
two-party *adjective* (1925)
: characterized by two major political parties of comparable strength
two·pence \'tə-pən(t)s, *US also* 'tü-,pen(t)s\ *noun* (15th century)
1 : the sum of two British pennies
2 *plural* **twopence** *or* **two·pen·ces :** a coin worth twopence
two·pen·ny \'təp-nē, 'tə-pə-, *US also* 'tü-,pe-nē\ *adjective* (1532)
: costing or worth twopence
two-phase *adjective* (circa 1896)
: DIPHASIC
¹two-piece \'tü-'pēs\ *adjective* (1910)
: forming a clothing ensemble with matching top and bottom parts
²two-piece \'tü-,pēs\ *noun* (1930)
: a garment (as a bathing suit) that is two-piece
two-piec·er \'tü-'pē-sər\ *noun* (1943)
: TWO-PIECE
two-ply \-'plī\ *adjective* (circa 1847)

1 : consisting of two thicknesses
2 a : woven with two sets of warp thread and two of filling ⟨a *two-ply* carpet⟩ **b :** consisting of two strands ⟨*two-ply* yarn⟩

two–sid·ed \-'sī-dəd\ *adjective* (1884)
: having two sides **:** BILATERAL

two·some \'tü-səm\ *noun* (14th century)
1 : a group of two persons or things **:** COUPLE
2 : a golf singles match

two–spot·ted spider mite \'tü-'spä-təd-\ *noun* (1947)
: a widely distributed mite (*Tetranychus urticae*) that feeds on soft plant parts and is a serious pest in greenhouses and gardens

two–step \'tü-ˌstep\ *noun* (1895)
1 : a ballroom dance in 2/4 or 4/4 time having a basic pattern of step-close-step
2 : a piece of music for the two-step
— two–step *intransitive verb*

two–suit·er \-'sü-tər\ *noun* (1948)
: a man's suitcase designed to hold two suits and accessories

two–tailed \'tü-'tāl(d)\ *also* **two–tail** \-'tā(ə)l\ *adjective* (1945)
: being a statistical test for which the critical region consists of all values of the test statistic greater than a given value plus the values less than another given value — compare ONE-TAILED

two–time \'tü-ˌtīm\ *transitive verb* (1924)
1 : DOUBLE-CROSS
2 : to betray (a spouse or lover) by secret lovemaking with another
— two–tim·er *noun*

two–toed sloth \'tü-'tōd-\ *noun* (1781)
: any of a genus (*Choloepus*) of sloths having two clawed digits on each forefoot, three clawed digits on each hind foot, and usually six or seven vertebrae in the neck — compare THREE-TOED SLOTH

two–tone \'tü-'tōn\ *adjective* (1906)
: colored in two colors or in two shades of one color ⟨*two-tone* shoes⟩

two–toned \'tü-'tōnd\ *adjective* (1897)
: TWO-TONE

two–way *adjective* (1571)
1 : being a cock or valve that will connect a pipe or channel with either of two others
2 : moving or allowing movement in either direction ⟨a *two-way* bridge⟩
3 a : involving or allowing an exchange between two individuals or groups ⟨there must be good *two-way* communication —Jerrold Orne⟩; *especially* **:** designed for both sending and receiving messages ⟨*two-way* radio⟩ **b :** involving mutual responsibility or reciprocal relationships ⟨political alliance is a *two-way* thing —T. H. White (died 1986)⟩
4 : involving two participants ⟨a *two-way* race⟩
5 : usable in either of two manners ⟨a *two-way* lamp⟩

two–way street *noun* (1948)
: a situation or relationship requiring give-and-take ⟨marriage is a *two-way street*⟩

two–wheel·er \-'hwē-lər, -'wē-\ *noun* (1861)
: a 2-wheeled vehicle (as a bicycle)

two–winged fly \'tü-'wiŋ(d)-\ *noun* (1753)
: [4]FLY 2a

-ty *noun suffix* [Middle English *-te*, from Old French *-té*, from Latin *-tat-*, *-tas* — more at -ITY]
: quality **:** condition **:** degree ⟨apriori*ty*⟩

ty·coon \tī-'kün\ *noun* [Japanese *taikun*] (1857)
1 : SHOGUN
2 a : a top leader (as in politics) **b :** a businessman of exceptional wealth and power **:** MAGNATE ◆

ty·er \'tī(-ə)r\ *variant of* [3]TIER

tying *present participle of* TIE

tyke \'tīk\ *noun* [Middle English *tyke*, from Old Norse *tīk* bitch; akin to Middle Low German *tīke* bitch] (15th century)
1 : DOG; *especially* **:** an inferior or mongrel dog
2 a *chiefly British* **:** a clumsy, churlish, or eccentric person **b :** a small child

tym·bal \'tim-bəl\ *noun* [alteration of *timbal*] (1929)
: the vibrating membrane in the shrilling organ of a cicada

tym·pan \'tim-pən\ *noun* [in sense 1, from Middle English, from Old English *timpana*, from Latin *tympanum;* in other senses, from Medieval Latin & Latin *tympanum*] (before 12th century)
1 : DRUM
2 : a sheet (as of paper or cloth) placed between the impression surface of a press and the paper to be printed
3 : TYMPANUM 2

tym·pa·ni, tym·pa·nist *variant of* TIMPANI, TIMPANIST

tym·pan·ic \tim-'pa-nik\ *adjective* [Latin & New Latin *tympanum*] (1808)
: of, relating to, or being a tympanum

tympanic membrane *noun* (1860)
: a thin membrane that closes externally the cavity of the middle ear and functions in the mechanical reception of sound waves and in their transmission to the site of sensory reception — called also *eardrum*; see EAR illustration

tym·pa·ni·tes \ˌtim-pə-'nī-tēz\ *noun* [Middle English, from Late Latin, from Greek *tympanitēs*, from *tympanon*] (14th century)
: a distension of the abdomen caused by accumulation of gas in the intestinal tract or peritoneal cavity
— tym·pa·nit·ic \-'ni-tik\ *adjective*

tym·pa·num \'tim-pə-nəm\ *noun, plural* **-na** \-nə\ *also* **-nums** [Medieval Latin & Latin; Medieval Latin, eardrum, from Latin, drum, architectural panel, from Greek *tympanon* drum, kettledrum; perhaps akin to Greek *typtein* to beat] (1619)
1 a (1) **:** TYMPANIC MEMBRANE (2) **:** MIDDLE EAR **b :** a thin tense membrane covering an organ of hearing of an insect — see INSECT illustration **c :** a membranous resonator in a sound-producing organ
2 a : the recessed usually triangular face of a pediment within the frame made by the upper and lower cornices **b :** the space within an arch and above a lintel or a subordinate arch

1 tympanum 2a

tym·pa·ny \-nē\ *noun, plural* **-nies** [Medieval Latin *tympanias*, from Greek, from *tympanon*] (1528)
1 : TYMPANITES
2 : BOMBAST, TURGIDITY

Tyn·dar·e·us \tin-'dar-ē-əs\ *noun* [Latin, from Greek]
: a king of Sparta and husband of Leda in Greek mythology

tyne *variant of* TINE

typ·al \'tī-pəl\ *adjective* (1853)
1 : serving as a type **:** TYPICAL
2 : of or relating to a type

[1]type \'tīp\ *noun, often attributive* [Middle English, from Late Latin *typus*, from Latin & Greek; Latin *typus* image, from Greek *typos* blow, impression, model, from *typtein* to strike, beat; akin to Sanskrit *tupati* he injures and probably to Latin *stupēre* to be benumbed] (15th century)
1 a : a person or thing (as in the Old Testament) believed to foreshadow another (as in the New Testament) **b :** one having qualities of a higher category **:** MODEL **c :** a lower taxonomic category selected as a standard of reference for a higher category; *also* **:** a specimen or series of specimens on which a taxonomic species or subspecies is actually based
2 : a distinctive mark or sign
3 a (1) **:** a rectangular block usually of metal bearing a relief character from which an inked print can be made (2) **:** a collection of such blocks ⟨a font of *type*⟩ (3) **:** alphanumeric characters for printing ⟨the *type* for this book has been photoset⟩ **b :** TYPEFACE ⟨italic *type*⟩ **c :** printed letters **d :** matter set in type
4 a : qualities common to a number of individuals that distinguish them as an identifiable class: as (1) **:** the morphological, physiological, or ecological characters by which relationship between organisms may be recognized (2) **:** the form common to all instances of a linguistic element **b :** a typical and often superior specimen **c :** a member of an indicated class or variety of people ⟨the guests were mostly urban *types* —Lucy Cook⟩ **d :** a particular kind, class, or group ⟨oranges of the seedless *type*⟩ ⟨leaders of the new *type* . . . did England yeoman's service —G. M. Trevelyan⟩ **e :** something distinguishable as a variety **:** SORT ⟨what *type* of food do you like?⟩ ☆

☆ SYNONYMS
Type, kind, sort, nature, description, character mean a number of individuals thought of as a group because of a common quality or qualities. TYPE may suggest strong and clearly marked similarity throughout the items included so that each is typical of the group ⟨one of three basic body *types*⟩. KIND may suggest natural grouping ⟨a zoo seemingly having animals of every *kind*⟩. SORT often suggests some disparagement ⟨the *sort* of newspaper dealing in sensational stories⟩. NATURE may imply inherent, essential resemblance rather than obvious or superficial likenesses ⟨two problems of a similar *nature*⟩. DESCRIPTION implies a group marked by agreement in all details belonging to a type as described or defined ⟨not all acts of that *description* are actually illegal⟩. CHARACTER implies a group marked by distinctive likenesses peculiar to the type ⟨research on the subject so far has been of an elementary *character*⟩.

◇ WORD HISTORY
tycoon When the United States forced Japan to open full commercial and diplomatic relations with the West in 1854, the real ruler of the island nation was the shogun. Officially only a military deputy of the emperor, the shogun—a title shortened from *seii-taishōgun* "barbarian-subjugating generalissimo"—stood at the pinnacle of a feudal hierarchy based at Edo (later Tokyo) that effectively controlled the imperial court at Kyoto and ruled the country. Westerners in the initial period of diplomatic relations concluded that the shogun was a sort of secular emperor and the emperor something like the pope. Townsend Harris, the first American consul to Japan, got the idea that the shogun's correct title was *taikun*, a Japanese borrowing from Middle Chinese elements equivalent to Modern Beijing Chinese *dà* "great" and *jūn* "prince." This word, in the spelling *tycoon*, became quite popular in America immediately before and during the Civil War as a colloquialism meaning "top leader" or "potentate." After fading from use for several decades *tycoon* was revived in 1920s journalism with the narrower sense "a businessman of exceptional wealth and power," a usage that continues to be part of English.

²type *verb* **typed; typ·ing** (1596)
transitive verb
1 : to represent beforehand as a type **:** PREFIGURE
2 a : to produce a copy of **b :** to represent in terms of typical characteristics **:** TYPIFY
3 : to produce (as a character or document) on a typewriter; *also* **:** KEYBOARD
4 : to identify as belonging to a type: as **a :** to determine the natural type of (as a blood sample) **b :** TYPECAST
intransitive verb
: to write something on a typewriter or enter data into a computer by way of a keyboard
— type·able \'tī-pə-bəl\ *adjective*

type A *adjective* (1970)
: relating to, characteristic of, having, or being a personality that is marked by impatience, aggressiveness, and competitiveness and that is held to be associated with increased risk of cardiovascular disease ⟨*type A* behavior⟩

type·cast \'tīp-ˌkast\ *transitive verb* **-cast; -cast·ing** (1927)
1 : to cast (an actor or actress) in a part calling for the same characteristics as those possessed by the performer
2 : to cast (an actor or actress) repeatedly in the same type of role
3 : STEREOTYPE ⟨*typecast* as the grubby, unbeautiful place which is full of life —Miles Kington⟩

type·face \-ˌfās\ *noun* (1887)
1 : the face of printing type
2 : all type of a single design

type·found·er \-ˌfaun-dər\ *noun* (1797)
: one engaged in the design and production of metal printing type for hand composition
— type·found·ing \-diŋ\ *noun*

type genus *noun* (1840)
: the genus of a taxonomic family or subfamily from which the name of the family or subfamily is formed

type I error \'tīp-'wən-\ *noun* (1947)
: rejection of the null hypothesis in statistical testing when it is true

type·script \'tīp-ˌskript\ *noun* [*type* + manu*script*] (1893)
: a typewritten manuscript; *especially* **:** one intended for use as printer's copy

type·set \-ˌset\ *transitive verb* **-set; -set·ting** (1945)
: to set in type **:** COMPOSE

type·set·ter \-ˌse-tər\ *noun* (1883)
: one that sets type

type·set·ting \-ˌse-tiŋ\ *noun* (1846)
: the process of setting material in type or into a form to be used in printing; *also* **:** the process of producing graphic matter (as through a computer system)

type species *noun* (1840)
: the species of a genus with which the generic name is permanently associated

type specimen *noun* (circa 1891)
: a specimen or individual designated as type of a species or lesser group and serving as the final criterion of the characteristics of that group

type·style \'tīp-ˌstī(ə)l\ *noun* (1954)
: TYPEFACE

type II error \-'tü-\ *noun* (1947)
: acceptance of the null hypothesis in statistical testing when it is false

type·write \'tīp-ˌrīt\ *verb* **-wrote** \-ˌrōt\; **-writ·ten** \-ˌri-tᵊn\ [back-formation from *typewriter*] (1887)
transitive verb
: TYPE 3
intransitive verb
: TYPE

type·writ·er \-ˌrī-tər\ *noun* (1868)
1 : a machine for writing in characters similar to those produced by printer's type by means of keyboard-operated types striking a ribbon to transfer ink or carbon impressions onto the paper
2 : TYPIST

type·writ·ing \-ˌrī-tiŋ\ *noun* (1867)
1 : the act or study of or skill in using a typewriter
2 : writing produced with a typewriter

typ·ey *also* **typy** \'tī-pē\ *adjective* **typ·i·er; -est** [¹*type*] (1923)
: characterized by strict conformance to type; *also* **:** exhibiting superior bodily conformation ⟨a sound *typey* heifer⟩

typh·lo·sole \'ti-flə-ˌsōl\ *noun* [Greek *typhlos* blind + *sōlēn* pipe, channel — more at DEAF] (1859)
: a longitudinal fold of the intestinal wall that projects into the cavity especially in bivalve mollusks, annelids, and starfishes

Ty·pho·eus \tī-'fō-ˌyüs, -yəs\ *noun* [Latin, from Greek *Typhōeus*]
: TYPHON
— Ty·phoe·an \-'fē-ən\ *adjective*

¹ty·phoid \'tī-ˌfȯid, (ˌ)tī-'\ *adjective* [New Latin *typhus*] (1800)
1 : of, relating to, or suggestive of typhus
2 [²*typhoid*] **:** of, relating to, or constituting typhoid

²typhoid *noun* (1861)
1 : TYPHOID FEVER
2 : a disease of domestic animals resembling human typhus or typhoid

typhoid fever *noun* (1845)
: a communicable disease marked especially by fever, diarrhea, prostration, headache, and intestinal inflammation and caused by a bacterium (*Salmonella typhi*)

Typhoid Mary *noun, plural* **Typhoid Marys** [*Typhoid Mary,* nickname of Mary Mallon (died 1938) Irish cook in U.S. who was found to be a typhoid carrier] (1931)
: one that is by force of circumstances a center from which something undesirable spreads

Ty·phon \'tī-ˌfän\ *noun* [Latin, from Greek *Typhōn*]
: a monster with a tremendous voice who according to classical mythology was father of Cerberus, the Chimera, and the Sphinx

ty·phoon \tī-'fün\ *noun* [alteration (influenced by Chinese — Guangdong — *daaih-fùng,* from *daaih* big + *fùng* wind) of earlier *touffon,* from Arabic *ṭūfān* hurricane, from Greek *typhōn* violent storm] (1771)
1 : a tropical cyclone occurring in the region of the Philippines or the China sea
2 : WHIRLWIND 2a ⟨a veritable *typhoon* of interest and corporate investment —Norman Sklarewitz⟩

ty·phus \'tī-fəs\ *noun* [New Latin, from Greek *typhos* fever; akin to Greek *typhein* to smoke — more at DEAF] (1785)
: any of various bacterial diseases caused by rickettsias: as **a :** a severe human febrile disease that is caused by one (*Rickettsia prowazekii*) transmitted especially by body lice and is marked by high fever, stupor alternating with delirium, intense headache, and a dark red rash **b :** MURINE TYPHUS **c :** TSUTSUGAMUSHI DISEASE

typ·ic \'ti-pik\ *adjective* (1610)
: TYPICAL 1

typ·i·cal \'ti-pi-kəl\ *adjective* [Late Latin *typicalis,* from *typicus,* from Greek *typikos,* from *typos* model — more at TYPE] (1612)
1 : constituting or having the nature of a type **:** SYMBOLIC
2 a : combining or exhibiting the essential characteristics of a group ⟨*typical* suburban houses⟩ **b :** conforming to a type ⟨a specimen *typical* of the species⟩
synonym see REGULAR
— typ·i·cal·i·ty \ˌti-pə-'ka-lə-tē\ *noun*
— typ·i·cal·ness \'ti-pi-kəl-nəs\ *noun*

typ·i·cal·ly \'ti-pi-k(ə-)lē\ *adverb* (1605)
1 : in a typical manner ⟨*typically* American⟩
2 : on a typical occasion **:** in typical circumstances ⟨*typically,* members of our staff receive little . . . recognition —Brendan Gill⟩

typ·i·fy \'ti-pə-ˌfī\ *transitive verb* **-fied; -fy·ing** (1634)
1 : to represent in typical fashion **:** to constitute a typical mark or instance of ⟨realism . . . that *typified* his earlier work —*Current Biography*⟩
2 : to embody the essential or salient characteristics of **:** be the type of
— typ·i·fi·ca·tion \ˌti-pə-fə-'kā-shən\ *noun*

typ·ist \'tī-pist\ *noun* (1885)
: one who types or enters data into a computer especially as a job

ty·po \'tī-(ˌ)pō\ *noun, plural* **typos** [short for *typographical (error)*] (1892)
: an error (as of spelling) in typed or typeset material

ty·po·graph \'tī-pə-ˌgraf\ *transitive verb* (circa 1933)
: to produce (stamps) by letterpress

ty·pog·ra·pher \tī-'pä-grə-fər\ *noun* (1643)
: a person (as a compositor, printer, or designer) who specializes in the design, choice, and arrangement of type matter

ty·po·graph·ic \ˌtī-pə-'gra-fik\ *or* **ty·po·graph·i·cal** \-fi-kəl\ *adjective* (1593)
: of, relating to, or occurring or used in typography or typeset matter ⟨a *typographic* character⟩ ⟨a *typographical* error⟩
— ty·po·graph·i·cal·ly \-fi-k(ə-)lē\ *adverb*

ty·pog·ra·phy \tī-'pä-grə-fē\ *noun* [Medieval Latin *typographia,* from Greek *typos* impression, cast + *-graphia* -graphy — more at TYPE] (1610)
1 : letterpress printing
2 : the style, arrangement, or appearance of typeset matter

ty·po·log·i·cal \ˌtī-pə-'lä-ji-kəl\ *adjective* (1845)
: of or relating to typology or types
— ty·po·log·i·cal·ly \-ji-k(ə-)lē\ *adverb*

ty·pol·o·gy \tī-'pä-lə-jē\ *noun, plural* **-gies** (1845)
1 : a doctrine of theological types; *especially* **:** one holding that things in Christian belief are prefigured or symbolized by things in the Old Testament
2 : study of or analysis or classification based on types or categories
— ty·pol·o·gist \-jist\ *noun*

Tyr \'tir\ *noun* [Old Norse *Tȳr;* akin to Old English *Tīw* Tiu — more at DEITY]
: a god of war in Norse mythology

ty·ra·mine \'tī-rə-ˌmēn\ *noun* [International Scientific Vocabulary *tyr*osine + *amine*] (1910)
: a phenolic amine $C_8H_{11}NO$ found in various foods and beverages (as cheese and red wine) that has a sympathomimetic action and is derived from tyrosine

ty·ran·ni·cal \tə-'ra-ni-kəl, tī-\ *also* **ty·ran·nic** \-nik\ *adjective* [Latin *tyrannicus,* from Greek *tyrannikos,* from *tyrannos* tyrant] (15th century)
: being or characteristic of a tyrant or tyranny **:** DESPOTIC ⟨*tyrannical* rule⟩ ⟨a *tyrannical* ruler⟩
— ty·ran·ni·cal·ly \-ni-k(ə-)lē\ *adverb*
— ty·ran·ni·cal·ness \-kəl-nəs\ *noun*

ty·ran·ni·cide \tə-'ra-nə-ˌsīd, tī-\ *noun* [in sense 1, from French, from Latin *tyrannicidium,* from *tyrannus* + *-i-* + *-cidium* -cide (killing); in sense 2, from French, from Latin *tyrannicida,* from *tyrannus* + *-i-* + *-cida* -cide (killer)] (1650)
1 : the act of killing a tyrant
2 : the killer of a tyrant

tyr·an·nise *British variant of* TYRANNIZE

tyr·an·nize \'tir-ə-ˌnīz\ *verb* **-nized; -niz·ing** (15th century)
intransitive verb
: to exercise arbitrary oppressive power or severity ⟨some ways the living *tyrannize* over the dying —Thomas Powers⟩

transitive verb
: to treat tyrannically **:** OPPRESS
— tyr·an·niz·er *noun*

ty·ran·no·saur \tə-ˈra-nə-ˌsȯr, tī-\ *noun* [New Latin *Tyrannosaurus,* genus name, from Greek *tyrannos* tyrant + *sauros* lizard] (1924)
: a very large bipedal carnivorous dinosaur (*Tyrannosaurus rex*) with small forelegs that occurs in the Upper Cretaceous of North America

ty·ran·no·sau·rus \tə-ˌra-nə-ˈsȯr-əs, (ˌ)tī-\ *noun* [New Latin] (1905)
: TYRANNOSAUR

tyr·an·nous \ˈtir-ə-nəs\ *adjective* (15th century)
: marked by tyranny; *especially* **:** unjustly severe
— tyr·an·nous·ly *adverb*

tyr·an·ny \ˈtir-ə-nē\ *noun, plural* **-nies** [Middle English *tyrannie,* from Middle French, from Medieval Latin *tyrannia,* from Latin *tyrannus* tyrant] (14th century)
1 : oppressive power ⟨every form of *tyranny* over the mind of man —Thomas Jefferson⟩; *especially* **:** oppressive power exerted by government ⟨the *tyranny* of a police state⟩
2 a : a government in which absolute power is vested in a single ruler; *especially* **:** one characteristic of an ancient Greek city-state **b :** the office, authority, and administration of a tyrant
3 : a rigorous condition imposed by some outside agency or force ⟨living under the *tyranny* of the clock —Dixon Wecter⟩
4 : a tyrannical act

ty·rant \ˈtī-rənt\ *noun* [Middle English *tirant,* from Old French *tyran, tyrant,* from Latin *tyrannus,* from Greek *tyrannos*] (14th century)
1 a : an absolute ruler unrestrained by law or constitution **b :** a usurper of sovereignty
2 a : a ruler who exercises absolute power oppressively or brutally **b :** one resembling an oppressive ruler in the harsh use of authority or power

tyrant flycatcher *noun* (circa 1783)
: any of various large American flycatchers (family Tyrannidae) that are usually strictly insectivorous and have a flattened bill often hooked at the tip and usually bristly at the gape

tyre *chiefly British variant of* TIRE

Tyr·i·an purple \ˈtir-ē-ən-\ *noun* [*Tyre,* maritime city of ancient Phoenicia] (circa 1586)
: a crimson or purple dye that is related to indigo, obtained by the ancient Greeks and Romans from gastropod mollusks, and now made synthetically

ty·ro \ˈtī-(ˌ)rō\ *noun, plural* **tyros** *often attributive* [Medieval Latin, from Latin *tiro* young soldier, tyro] (1611)
: a beginner in learning **:** NOVICE
synonym see AMATEUR

ty·ro·ci·dine *also* **ty·ro·ci·din** \ˌtī-rə-ˈsī-dᵊn\ *noun* [*tyro*thricin + grami*cidin*] (1940)
: a basic polypeptide antibiotic produced by a soil bacillus (*Bacillus brevis*)

Ty·ro·le·an *also* **Ty·ro·li·an** \tə-ˈrō-lē-ən, tī-\ *adjective* (1805)
1 : of or relating to the Tirol
2 *of a hat* **:** of a style originating in the Tirol and marked by soft often green felt, a narrow brim and pointed crown, and an ornamental feather

ty·ros·i·nase \tə-ˈrä-sə-ˌnās, tī-, -ˌnāz\ *noun* (1896)
: a copper-containing enzyme that promotes the oxidation of phenols (as tyrosine) and is widespread in plants and animals

ty·ro·sine \ˈtī-rə-ˌsēn\ *noun* [International Scientific Vocabulary, irregular from Greek *tyros* cheese — more at BUTTER] (1857)
: a phenolic amino acid $C_9H_{11}NO_3$ that is a precursor of several important substances (as epinephrine and melanin)

ty·ro·thri·cin \ˌtī-rə-ˈthrī-sᵊn\ *noun* [New Latin *Tyrothoric-, Tyrothrix,* genus name formerly applied to various bacteria including *Bacillus brevis*] (1940)
: an antibiotic mixture that consists chiefly of tyrocidine and gramicidin, is usually extracted from a soil bacillus (*Bacillus brevis*) as a gray to brown powder, and is used for local applications especially for infection caused by gram-positive bacteria

tzaddik *noun, plural* **tzaddikim** *variant of* ZADDIK

tzar \ˈzär, ˈ(t)sär\ *variant of* CZAR

tzi·gane \(t)sē-ˈgän\ *noun* [French, from Hungarian *cigány*] (1763)
1 : GYPSY 1
2 : ROMANY 2

tzim·mes \ˈtsi-məs\ *noun* [Yiddish *tsimes,* from Middle High German *z, zuo* at, too + *imbīz* light meal] (1892)
: a sweetened combination of vegetables (as carrots and potatoes) or of meat and carrots often with dried fruits (as prunes) that is stewed or baked in a casserole

tzi·tzis, tzi·tzit *variant of* ZIZITH

U *is the twenty-first letter of the English alphabet. It is a cursive form of the letter* V, *with which it was once used interchangeably, both letters having the value of either the modern vowel* U *or the modern consonant* V. *By the 10th century,* V *came to be preferred as the capital form used at the beginning of a word, and* U *as the cursive form used in the middle of a word. Since the consonant sound usually occurred at the beginning and the vowel sound in the middle of words,* V *gradually became specialized to represent the former and* U *the latter, but small* u *and small* v *were used interchangeably as late as the 13th century. In dictionaries of English,* U *and* V *were not given separate alphabetical positions until about 1800. The vowel sounds of* u *in Old English were the ones which it still retains in most of the languages of Europe, that of the vowel in* tool, *and the vowel in* wood, *but Modern English* u *represents various other sounds as well.*

---

u \ˈyü\ *noun, plural* **u's** *or* **us** \ˈyüz\ *often capitalized, often attributive* (before 12th century)
1 a : the 21st letter of the English alphabet **b :** a graphic representation of this letter **c :** a speech counterpart of orthographic *u*
2 : a graphic device for reproducing the letter *u*
3 : one designated *u* especially as the 21st in order or class
4 [abbreviation for *unsatisfactory*] **a :** a grade rating a student's work as unsatisfactory **b :** one graded or rated with a U
5 : something shaped like the letter U

U \ˈyü\ *adjective* [*u*pper class] (1954)
: characteristic of the upper classes

Uban·gi \yü-ˈbaŋ-gē, ü-, -ˈbaŋ-ē\ *noun* [*Ubangi-Shari,* Africa] (1942)
: a woman of the district of Kyabé village in Chad with lips pierced and distended to unusual dimensions with wooden disks — not used technically

ubi·qui·none \yü-ˈbi-kwə-ˌnōn, ˌyü-bə-kwi-ˈnōn\ *noun* [blend of Latin *ubique* everywhere and English *quinone;* from its widespread occurrence in nature] (1958)
: any of a group of lipid-soluble quinones that contain a long isoprenoid side chain and that function in the part of cellular respiration comprising oxidative phosphorylation as electron-carrying coenzymes in the transport of electrons from organic substrates to oxygen especially along the chain of reactions leading from the Krebs cycle

ubiq·ui·tous \yü-ˈbi-kwə-təs\ *adjective* (1837)
: existing or being everywhere at the same time **:** constantly encountered **:** WIDESPREAD
— ubiq·ui·tous·ly *adverb*
— ubiq·ui·tous·ness *noun*

ubiq·ui·ty \-kwə-tē\ *noun* [Latin *ubique* everywhere, from *ubi* where + *-que,* enclitic generalizing particle; akin to Latin *quis* who and to Latin *-que* and — more at WHO, SESQUI-] (1597)
: presence everywhere or in many places especially simultaneously **:** OMNIPRESENCE

U–boat \ˈyü-ˌbōt\ *noun* [translation of German *U-boot,* short for *Unterseeboot,* literally, undersea boat] (1916)
: a German submarine

ud·der \ˈə-dər\ *noun* [Middle English, from Old English *ūder;* akin to Old High German *ūtar* udder, Latin *uber,* Greek *outhar,* Sanskrit *ūdhar*] (before 12th century)
1 : a large pendulous organ consisting of two or more mammary glands enclosed in a common envelope and each provided with a single nipple — see COW illustration
2 : MAMMARY GLAND

UFO \ˌyü-(ˌ)ef-ˈō\ *noun, plural* **UFO's** *or* **UFOs** \-ˈōz\ [*u*nidentified *f*lying *o*bject] (1953)
: an unidentified flying object; *especially* **:** FLYING SAUCER

ufol·o·gy \yü-ˈfä-lə-jē\ *noun, often UFO capitalized* [*UFO* + *-logy*] (1959)
: the study of unidentified flying objects
— ufo·log·i·cal \ˌyü-fə-ˈlä-ji-kəl\ *adjective, often UFO capitalized*
— ufol·o·gist \yü-ˈfä-lə-jist\ *noun, often UFO capitalized*

[1]Uga·rit·ic \ˌyü-gə-ˈri-tik, ˌü-gə-\ *noun* (1936)
: the Semitic language of ancient Ugarit closely related to Phoenician and Hebrew

[2]Ugaritic *adjective* (1938)
: of, relating to, or characteristic of the ancient city of Ugarit, its inhabitants, or Ugaritic

ugh *often read as* ˈəg *or* ˈək *or* ˈə\ *interjection* (1837)
— used to indicate the sound of a cough or grunt or to express disgust or horror

Ug·li \ˈə-glē\ *trademark*
— used for a tangelo

ug·li·fy \ˈə-gli-ˌfī\ *transitive verb* **-fied; -fy·ing** (1576)
: to make ugly
— ug·li·fi·ca·tion \ˌə-gli-fə-ˈkā-shən\ *noun*

ug·li·ness \ˈə-glē-nəs\ *noun* (14th century)
1 : the quality or state of being ugly
2 : something that is ugly

[1]ug·ly \ˈə-glē\ *adjective* **ug·li·er; -est** [Middle English, from Old Norse *uggligr,* from *uggr* fear; akin to Old Norse *ugga* to fear] (13th century)
1 : FRIGHTFUL, DIRE
2 a : offensive to the sight **:** HIDEOUS **b :** offensive or unpleasant to any sense
3 : morally offensive or objectionable ⟨corruption—the *ugliest* stain of all⟩
4 a : likely to cause inconvenience or discomfort ⟨the *ugly* truth⟩ **b :** SURLY, QUARRELSOME ⟨an *ugly* disposition⟩ ⟨the crowd got *ugly*⟩
— ug·li·ly \-glə-lē\ *adverb*

[2]ugly *adverb* (14th century)
: in an ugly manner ⟨was acting *ugly*⟩

Ugly American *noun* [*The Ugly American* (1958), collection of stories by Eugene Burdick (died 1965) and William J. Lederer (born 1912) American authors] (1965)
: an American in a foreign country whose behavior is offensive to the people of that country

ugly duckling *noun* [*The Ugly Duckling,* story by Hans Christian Andersen] (1883)
: one that appears very unpromising but often has great potential

Ugri·an \ˈyü-grē-ən, ˈü-\ *noun* [Old Russian *Ugre* Hungarians] (1841)
: a member of a division of the Finno-Ugric peoples that includes the Hungarians and two peoples of western Siberia
— Ugrian *adjective*

Ugric \-grik\ *adjective* (1854)
: of, relating to, or characteristic of the languages of the Ugrians

ug·some \ˈəg-səm\ *adjective* [Middle English, from *uggen* to fear, inspire fear, from Old Norse *ugga* to fear] (15th century)
archaic **:** FRIGHTFUL, LOATHSOME

uh–huh *two* m*'s or two* n*'s separated by the voiceless sound* h; ˈə^{n}-(ˌ)hə^{n}, (ˌ)ə^{n}-ˈ\ *interjection* (1899)
— used to indicate affirmation, agreement, or gratification

uh·lan \ˈü-ˌlän, ü-ˈ; ˈyü-lən, ˈü-\ *noun* [German, from Polish *ulan,* from Turkish *oğlan* boy, servant] (1753)
: any of a body of Prussian light cavalry originally modeled on Tatar lancers

uh–oh \ˈə-ˌō, *usually with strong glottal stops before the vowels*\ *interjection* (1971)
— used to indicate dismay or concern

uh–uh *two* m*'s or two* n*'s preceded by glottal stops;* ˈə^{n}-ˌə^{n}\ *interjection* (circa 1924)
— used to indicate negation

Ui·ghur *also* **Ui·gur** \ˈwē-ˌgu̇r\ *noun* [Uighur *Uighur*] (1747)
1 : a member of a Turkic people powerful in Mongolia and eastern Turkestan between the 8th and 12th centuries A.D. who constitute a majority of the population of Chinese Turkestan
2 : the Turkic language of the Uighurs
— Uighur *also* **Uigur** *adjective*

uil·leann pipes \ˈi-lən-\ *noun plural, often capitalized* [*uilleann* from Irish, genitive singular of *uillinn* elbow, from Old Irish *uilen;* akin to Old English *eln* ell — more at ELL] (1906)
: an Irish bagpipe with air supplied by a bellows held under and worked by the elbow

uin·ta·ite *also* **uin·tah·ite** \yu̇-ˈin-tə-ˌīt\ *noun* [*Uinta, Uintah,* mountains in Utah] (1888)
: a black lustrous asphalt occurring especially in Utah

Uit·land·er \ˈāt-ˌlan-dər, ˈau̇t-, -ˌlän-\ *noun* [Afrikaans, from Middle Dutch *utelander* foreigner, from *utelant* foreign territory, from *ute* out + *lant* land] (1892)
: FOREIGNER; *especially* **:** a British resident in the former republics of the Transvaal and Orange Free State

ukase \yü-ˈkās, -ˈkāz, ˈyü-ˌ; ü-ˈkäz\ *noun* [French & Russian; French, from Russian *ukaz,* from *ukazat'* to show, order; akin to Old Church Slavonic *u-* away, Latin *au-,* Sanskrit *ava-* and to Old Church Slavonic *kazati* to show] (1729)
1 : a proclamation by a Russian emperor or government having the force of law
2 : EDICT

uke \ˈyük\ *noun* (1921)
: UKULELE

uki·yo–e *also* **uki·yo–ye** \ü-ˌkē-ō-ˈyā, -ˈā\ *noun* [Japanese *ukiyo-e* genre picture, from *ukiyo* world, life + *e* picture] (1879)
: a Japanese art movement that flourished from the 17th to the 19th century and produced paintings and prints depicting the everyday life and interests of the common people; *also* **:** the paintings and prints themselves

Ukrai·ni·an \yü-ˈkrā-nē-ən *also* -ˈkrī-\ *noun* (1823)
1 : a native or inhabitant of Ukraine
2 : the Slavic language of the Ukrainian people
— Ukrainian *adjective*

uku·le·le *also* **uke·le·le** \ˌyü-kə-ˈlā-lē, ˌü-\ *noun* [Hawaiian *ʻukulele,* from *ʻuku* flea + *lele* jumping] (1896)

: a small guitar of Portuguese origin popularized in Hawaii in the 1880s and strung typically with four strings ◆

ukulele

-ular *adjective suffix* [Latin *-ularis,* from *-ulus, -ula, -ulum* -ule + *-aris* -ar]
: of, relating to, or resembling ⟨valv*ular*⟩

ul·cer \ˈəl-sər\ *noun* [Middle English, from Latin *ulcer-, ulcus;* akin to Greek *helkos* wound] (14th century)
1 : a break in skin or mucous membrane with loss of surface tissue, disintegration and necrosis of epithelial tissue, and often pus
2 : something that festers and corrupts like an open sore
— ulcer *verb*

ul·cer·ate \ˈəl-sə-ˌrāt\ *verb* **-at·ed; -at·ing** (15th century)
intransitive verb
: to become affected with or as if with an ulcer
transitive verb
: to affect with or as if with an ulcer

ul·cer·a·tion \ˌəl-sə-ˈrā-shən\ *noun* (14th century)
1 : the process of becoming ulcerated **:** the state of being ulcerated
2 : ULCER
— ul·cer·a·tive \ˈəl-sə-ˌrā-tiv, ˈəls-rə-, ˈəl-sə-rə-\ *adjective*

ulcerative colitis *noun* (circa 1928)
: a nonspecific inflammatory disease of the colon of unknown cause characterized by diarrhea with discharge of mucus and blood, cramping abdominal pain, and inflammation and edema of the mucous membrane with patches of ulceration

ul·cero·gen·ic \ˌəl-sə-rō-ˈje-nik\ *adjective* (1950)
: tending to produce or develop into ulcers or ulceration

ul·cer·ous \ˈəls-rəs, ˈəl-sə-\ *adjective* (1577)
1 : being or marked by an ulceration ⟨*ulcerous* lesions⟩
2 : affected with or as if with an ulcer **:** ULCERATED

-ule *noun suffix* [French & Latin; French, from Latin *-ulus,* masculine diminutive suffix, *-ula,* feminine diminutive suffix, *-ulum,* neuter diminutive suffix]
: little one ⟨duct*ule*⟩

ule·ma *or* **ula·ma** \ˌü-lə-ˈmä\ *noun* [Arabic, Turkish, & Persian; Turkish & Persian *ʿulemā,* from Arabic *ʿulamā*] (1688)
1 *plural in construction* **:** the body of mullahs
2 : MULLAH

-ulent *adjective suffix* [Latin *-ulentus*]
: that abounds in (a specified thing) ⟨floccu*lent*⟩

ulex·ite \ˈyü-lək-ˌsīt\ *noun* [George L. *Ulex* (died 1883) German chemist] (1867)
: a mineral consisting of a hydrous borate of sodium and calcium and usually occurring in loosely packed white fibers that transmit light lengthwise with nearly undiminished intensity

ul·lage \ˈə-lij\ *noun* [Middle English *ulage,* from Middle French *eullage* act of filling a cask, from *euillier* to fill a cask, from Old French *ouil* eye, bunghole, from Latin *oculus* eye — more at EYE] (15th century)
: the amount that a container (as a tank or cask) lacks of being full

ul·na \ˈəl-nə\ *noun* [New Latin, from Latin, elbow — more at ELL] (1541)
: the bone on the little-finger side of the human forearm; *also* **:** a corresponding part of the forelimb of vertebrates above fishes
— ul·nar \-nər, -ˌnär\ *adjective*

ul·ster \ˈəl-stər\ *noun* [*Ulster,* Ireland] (1876)
: a long loose overcoat of Irish origin made of heavy material (as frieze)

ul·te·ri·or \ˌəl-ˈtir-ē-ər\ *adjective* [Latin, farther, further, comparative of (assumed) Latin *ulter* situated beyond, from *uls* beyond; akin to Latin *ollus, ille,* that one, Old Irish ind*oll* beyond] (1646)
1 a : FURTHER, FUTURE **b :** more distant **:** REMOTER **c :** situated on the farther side **:** THITHER
2 : going beyond what is openly said or shown and especially what is proper ⟨*ulterior* motives⟩
— ul·te·ri·or·ly *adverb*

ul·ti·ma \ˈəl-tə-mə\ *noun* [Latin, feminine of *ultimus* last] (circa 1864)
: the last syllable of a word

ul·ti·ma·cy \ˈəl-tə-mə-sē\ *noun, plural* **-cies** (1842)
1 : the quality or state of being ultimate
2 : ULTIMATE 1

ul·ti·ma ra·tio \ˌu̇l-tə-mə-ˈrä-tē-ˌō\ *noun* [New Latin] (1780)
: the final argument; *also* **:** the last resort (as force)

[1]**ul·ti·mate** \ˈəl-tə-mət\ *adjective* [Medieval Latin *ultimatus* last, final, from Late Latin, past participle of *ultimare* to come to an end, be last, from Latin *ultimus* farthest, last, final, superlative of (assumed) Latin *ulter* situated beyond] (1654)
1 a : most remote in space or time **:** FARTHEST **b :** last in a progression or series ⟨their *ultimate* destination was Paris⟩ **c :** EVENTUAL ⟨they hoped for *ultimate* success⟩ **d :** the best or most extreme of its kind **:** UTMOST ⟨the *ultimate* sacrifice⟩
2 : arrived at as the last result ⟨the *ultimate* question⟩
3 a : BASIC, FUNDAMENTAL ⟨the *ultimate* nature of things —A. N. Whitehead⟩ **b :** ORIGINAL ⟨the *ultimate* source⟩ **c :** incapable of further analysis, division, or separation
4 : MAXIMUM
synonym see LAST
— ul·ti·mate·ness *noun*

[2]**ultimate** *noun* (1681)
1 : something ultimate; *especially* **:** FUNDAMENTAL
2 : ACME

[3]**ul·ti·mate** \-mət, -ˌmāt\ *verb* **-mat·ed; -mat·ing** (circa 1834)
: END

ul·ti·mate·ly \-mət-lē\ *adverb* (1652)
1 : in the end **:** FINALLY, FUNDAMENTALLY
2 : EVENTUALLY

ul·ti·ma Thu·le \ˌəl-tə-mə-ˈthü-lē, -ˈthyü-\ *noun* [Latin, farthest Thule] (1665)
: THULE

ul·ti·ma·tum \ˌəl-tə-ˈmā-təm, -ˈmä-\ *noun, plural* **-tums** *or* **-ta** \-tə\ [New Latin, from Medieval Latin, neuter of *ultimatus* final] (1731)
: a final proposition, condition, or demand; *especially* **:** one whose rejection will end negotiations and cause a resort to force or other direct action

ul·ti·mo \ˈəl-tə-ˌmō\ *adjective* [Latin *ultimo mense* in the last month] (1616)
: of or occurring in the month preceding the present

ul·ti·mo·gen·i·ture \ˌəl-tə-mō-ˈje-nə-ˌchu̇r, -ni-chər, -nə-ˌtyu̇r, -nə-ˌtu̇r\ *noun* [Latin *ultimus* last + English primo*geniture*] (1882)
: a system of inheritance by which the youngest child succeeds to the estate

[1]**ul·tra** \ˈəl-trə\ *adjective* [*ultra-*] (1818)
: going beyond others or beyond due limit **:** EXTREME

[2]**ultra** *noun* [*ultra-*] (1819)
: one that is ultra **:** EXTREMIST

ultra- *prefix* [Latin, from *ultra* beyond, adverb & preposition, from (assumed) Latin *ulter* situated beyond — more at ULTERIOR]
1 : beyond in space **:** on the other side **:** TRANS- ⟨*ultra*violet⟩
2 : beyond the range or limits of **:** transcending **:** SUPER- ⟨*ultra*microscopic⟩
3 : beyond what is ordinary, proper, or moderate **:** excessively **:** extremely ⟨*ultra*modern⟩

ul·tra·care·ful
ul·tra·ca·su·al
ul·tra·cau·tious
ul·tra·chic
ul·tra·civ·i·lized
ul·tra·clean
ul·tra·cold
ul·tra·com·mer·cial
ul·tra·com·pact
ul·tra·com·pe·tent
ul·tra·con·ser·va·tism
ul·tra·con·ser·va·tive
ul·tra·con·tem·po·rary
ul·tra·con·ve·nient
ul·tra·cool
ul·tra·crit·i·cal
ul·tra·dem·o·crat·ic
ul·tra·dense
ul·tra·dis·tance
ul·tra·dis·tant
ul·tra·dry
ul·tra·ef·fi·cient
ul·tra·en·er·get·ic
ul·tra·ex·clu·sive
ul·tra·fa·mil·iar
ul·tra·fast
ul·tra·fas·tid·i·ous
ul·tra·fem·i·nine
ul·tra·fine
ul·tra·glam·or·ous
ul·tra·haz·ard·ous
ul·tra·heat
ul·tra·heavy
ul·tra·high
ul·tra·hip
ul·tra·hot
ul·tra·hu·man
ul·tra·left
ul·tra·left·ism
ul·tra·left·ist
ul·tra·lib·er·al
ul·tra·lib·er·al·ism
ul·tra·light·weight
ul·tra·low
ul·tra·mas·cu·line
ul·tra·mil·i·tant
ul·tra·min·i·a·tur·ized
ul·tra·mod·ern
ul·tra·mod·ern·ist
ul·tra·na·tion·al·ism
ul·tra·na·tion·al·ist
ul·tra·na·tion·al·is·tic
ul·tra·or·tho·dox
ul·tra·par·a·dox·i·cal
ul·tra·pa·tri·ot·ic
ul·tra·phys·i·cal
ul·tra·pow·er·ful
ul·tra·prac·ti·cal
ul·tra·pre·cise
ul·tra·pre·ci·sion
ul·tra·pro·fes·sion·al
ul·tra·pro·gres·sive
ul·tra·pure
ul·tra·qui·et
ul·tra·rad·i·cal
ul·tra·rap·id
ul·tra·rare
ul·tra·rar·e·fied
ul·tra·ra·tio·nal
ul·tra·re·al·ism
ul·tra·re·al·ist
ul·tra·re·al·is·tic
ul·tra·re·fined
ul·tra·re·li·able
ul·tra·re·spect·able
ul·tra·rev·o·lu·tion·ary
ul·tra·rich
ul·tra·right
ul·tra·right·ist
ul·tra·ro·man·tic
ul·tra·roy·al·ist
ul·tra·safe
ul·tra·se·cret
ul·tra·seg·re·ga·tion·ist
ul·tra·sen·si·tive
ul·tra·se·ri·ous
ul·tra·sharp
ul·tra·sim·ple
ul·tra·slick
ul·tra·slow
ul·tra·small
ul·tra·smart
ul·tra·smooth
ul·tra·soft
ul·tra·so·phis·ti·cat·ed
ul·tra·thin
ul·tra·vac·u·um
ul·tra·vi·o·lence
ul·tra·vi·o·lent
ul·tra·vir·ile
ul·tra·vi·ril·i·ty
ul·tra·wide

◇ WORD HISTORY

ukulele In Hawaiian *ʻukulele* means literally "jumping flea," a creature with little resemblance to a small guitar. It is likely that this improbable name became attached to the instrument in a rather roundabout way. Edward Purvis, a former British army officer living in Hawaii as an official at the court of King Kalakaua, is said to have been given the nickname *ʻukulele* because he was a small and lively man. In 1879 Portuguese immigrants to the Hawaiian Islands from Madeira brought with them several musical instruments. Among these was the *machete,* a small four-stringed guitar. Purvis, being fond of music, was taken with the new instrument and soon learned to play it. When the *machete* became a Hawaiian favorite, it took the name of its popularizer, and its Portuguese name and origin were soon forgotten.

ul·tra·ba·sic \ˌəl-trə-ˈbā-sik\ *adjective* [International Scientific Vocabulary] (1881)
: extremely basic; *specifically* **:** very low in silica and rich in iron and magnesium minerals
— ultrabasic *noun*

ul·tra·cen·trif·u·gal \-ˌsen-ˈtri-fyə-gəl, -fi-gəl\ *adjective* (1930)
: of, relating to, or obtained by means of an ultracentrifuge
— ul·tra·cen·trif·u·gal·ly \-gə-lē\ *adverb*

¹ul·tra·cen·tri·fuge \-ˈsen-trə-ˌfyüj\ *noun* (1924)
: a high-speed centrifuge able to sediment colloidal and other small particles and used especially in determining sizes of such particles and molecular weights of large molecules

²ultracentrifuge *transitive verb* (1930)
: to subject to an ultracentrifuge
— ul·tra·cen·tri·fu·ga·tion \-ˌsen-trə-fyü-ˈgā-shən\ *noun*

ul·tra·fiche \ˈəl-trə-ˌfēsh\ *noun* (1969)
: a microfiche whose microimages are of printed matter reduced 90 or more times

ul·tra·fil·tra·tion \ˌəl-trə-fil-ˈtrā-shən\ *noun* (1908)
: filtration through a medium (as a semipermeable capillary wall) which allows small molecules (as of water) to pass but holds back larger ones (as of protein)
— ul·tra·fil·trate \-ˈfil-ˌtrāt\ *noun*

ultrahigh frequency *noun* (1932)
: a radio frequency between superhigh frequency and very high frequency — see RADIO FREQUENCY table

ul·tra·ism \ˈəl-trə-ˌi-zəm\ *noun* (1821)
1 : the principles of those who advocate extreme measures (as radicalism)
2 : an instance or example of radicalism
— ul·tra·ist \-trə-ist\ *adjective or noun*
— ul·tra·is·tic \ˌəl-trə-ˈis-tik\ *adjective*

¹ul·tra·light \ˈəl-trə-ˌlīt\ *adjective* (1974)
: extremely light in mass or weight ⟨an *ultralight* alloy⟩ ⟨an *ultralight* pullover⟩

²ultralight *noun* (1980)
: a very light recreational aircraft typically for one person that is powered by a small gasoline engine

ul·tra·maf·ic \ˌəl-trə-ˈma-fik\ *adjective* (1933)
: ULTRABASIC

ul·tra·mar·a·thon \-ˈmar-ə-ˌthän\ *noun* (1977)
: a footrace longer than a marathon
— ul·tra·mar·a·thon·er \-ˌthä-nər\ *noun*

¹ul·tra·ma·rine \-mə-ˈrēn\ *noun* [Medieval Latin *ultramarinus* coming from beyond the sea, from Latin *ultra-* + *mare* sea — more at MARINE] (1598)
1 a (1) **:** a blue pigment prepared by powdering lapis lazuli (2) **:** a similar pigment prepared from kaolin, soda ash, sulfur, and charcoal **b :** any of several related pigments
2 : a vivid blue

²ultramarine *adjective* (1652)
: situated beyond the sea

ul·tra·mi·cro \ˌəl-trə-ˈmī-(ˌ)krō\ *adjective* (1937)
: being or dealing with something smaller than micro

ul·tra·mi·cro·scope \ˌəl-trə-ˈmī-krə-ˌskōp\ *noun* [back-formation from *ultramicroscopic*] (1906)
: an apparatus for making visible by scattered light particles too small to be perceived by an ordinary microscope

ul·tra·mi·cro·scop·ic \-ˌmī-krə-ˈskä-pik\ *also* **ul·tra·mi·cro·scop·i·cal** \-pi-kəl\ *adjective* [International Scientific Vocabulary] (1870)
1 : too small to be seen with an ordinary microscope
2 : of or relating to an ultramicroscope
— ul·tra·mi·cro·scop·i·cal·ly \-pi-k(ə-)lē\ *adverb*

ul·tra·mi·cro·tome \-ˈmī-krə-ˌtōm\ *noun* (1946)
: a microtome for cutting extremely thin sections for electron microscopy
— ul·tra·mi·crot·o·my \-mī-ˈkrä-tə-mē\ *noun*

ul·tra·min·i·a·ture \-ˈmi-nē-ə-ˌchu̇r, -ˈmi-ni-ˌchu̇r, -ˈmin-yə-, -chər, -ˌtyu̇r, -ˌtu̇r\ *adjective* (1942)
: SUBMINIATURE

ul·tra·mon·tane \-ˈmän-ˌtān, -ˌmän-ˈ\ *adjective* [Medieval Latin *ultramontanus,* from Latin *ultra-* + *mont-, mons* mountain — more at MOUNT] (circa 1618)
1 : of or relating to countries or peoples beyond the mountains (as the Alps)
2 : favoring greater or absolute supremacy of papal over national or diocesan authority in the Roman Catholic Church
— ultramontane *noun, often capitalized*
— ul·tra·mon·tan·ism \-ˈmän-tᵊn-ˌi-zəm\ *noun*

ul·tra·short \-ˈshȯrt\ *adjective* (1926)
1 : having a wavelength below 10 meters ⟨*ultrashort* radiation⟩
2 : very short in duration ⟨an *ultrashort* pulse of light⟩

ul·tra·son·ic \-ˈsä-nik\ *adjective* (1923)
1 : having a frequency above the human ear's audibility limit of about 20,000 hertz — used of waves and vibrations
2 : utilizing, produced by, or relating to ultrasonic waves or vibrations ⟨*ultrasonic* testing of metal⟩
— ul·tra·son·i·cal·ly \-ni-k(ə-)lē\ *adverb*

ul·tra·son·ics \ˌəl-trə-ˈsä-niks\ *noun plural* (1924)
1 : ultrasonic vibrations or compressional waves
2 *singular in construction* **:** the study of ultrasonic vibrations and their associated phenomena
3 : ultrasonic devices

ul·tra·so·nog·ra·phy \-sə-ˈnä-grə-fē, -sō-\ *noun* [*ultrasonic* + *-o-* + *-graphy*] (1951)
: ULTRASOUND 2
— ul·tra·so·nog·ra·pher \-fər\ *noun*
— ul·tra·so·no·graph·ic \-ˌsō-nə-ˈgra-fik, -ˌsä-\ *adjective*

ul·tra·sound \ˈəl-trə-ˌsau̇nd\ *noun* (1923)
1 : vibrations of the same physical nature as sound but with frequencies above the range of human hearing
2 : the diagnostic or therapeutic use of ultrasound and especially a technique involving the formation of a two-dimensional image used for the examination and measurement of internal body structures and the detection of bodily abnormalities — called also *sonography*
3 : a diagnostic examination using ultrasound

ul·tra·struc·ture \ˈəl-trə-ˌstrək-chər\ *noun* (1939)
: biological structure and especially fine structure (as of a cell) not visible through an ordinary microscope
— ul·tra·struc·tur·al \ˌəl-trə-ˈstrək-chə-rəl, -ˈstrək-shrəl\ *adjective*
— ul·tra·struc·tur·al·ly *adverb*

ul·tra·vi·o·let \ˌəl-trə-ˈvī-(ə-)lət\ *adjective* (1840)
1 : situated beyond the visible spectrum at its violet end — used of radiation having a wavelength shorter than wavelengths of visible light and longer than those of X rays
2 : relating to, producing, or employing ultraviolet radiation
— ultraviolet *noun*

ul·tra vi·res \ˌəl-trə-ˈvī-(ˌ)rēz\ *adverb or adjective* [New Latin, literally, beyond power] (1793)
: beyond the scope or in excess of legal power or authority

ul·u·lant \ˈəl-yə-lənt\ *adjective* (1868)
: having a howling sound **:** WAILING ⟨dark wasteland . . . *ululant* with bitter wind —Rudi Blesh⟩

ul·u·late \-ˌlāt\ *intransitive verb* **-lat·ed; -lat·ing** [Latin *ululatus,* past participle of *ululare,* of imitative origin] (1623)
: HOWL, WAIL
— ul·u·la·tion \ˌəl-yə-ˈlā-shən\ *noun*

ul·va \ˈəl-və\ *noun* [New Latin, genus name, from Latin, sedge] (1706)
: SEA LETTUCE

Ulys·ses \yu̇-ˈli-(ˌ)sēz\ *noun* [Latin *Ulysses, Ulixes,* from Greek *Oulixes, Olysseus, Odysseus*]
: ODYSSEUS

um·bel \ˈəm-bəl\ *noun* [New Latin *umbella,* from Latin, umbrella — more at UMBRELLA] (1597)
: a racemose inflorescence typical of the carrot family in which the axis is very much contracted so that the pedicels appear to spring from the same point to form a flat or rounded flower cluster — see INFLORESCENCE illustration

um·bel·late \ˈəm-bə-ˌlāt, ˌəm-ˈbe-lət\ *adjective* (1760)
1 : bearing, consisting of, or arranged in umbels
2 : resembling an umbel in form

um·bel·li·fer \ˌəm-ˈbe-lə-fər\ *noun* [New Latin *Umbelliferae,* group name, feminine plural of *umbellifer* bearing umbels] (1718)
: a plant of the carrot family

um·bel·lif·er·ous \ˌəm-bə-ˈli-f(ə-)rəs\ *adjective* (1662)
: of or relating to the carrot family ⟨*umbelliferous* flower heads⟩

¹um·ber \ˈəm-bər\ *noun* [probably from obsolete English, shade, color, from Middle English *umbre* shade, shadow, from Middle French, from Latin *umbra* — more at UMBRAGE] (1568)
1 : a brown earth that is darker in color than ocher and sienna because of its content of manganese and iron oxides and is highly valued as a permanent pigment either in the raw or burnt state
2 a : a moderate to dark yellowish brown **b :** a moderate brown

²umber *transitive verb* **um·bered; um·ber·ing** \-b(ə-)riŋ\ (1610)
: to darken with or as if with umber

³umber *adjective* (1802)
: of, relating to, or having the characteristics of umber; *specifically* **:** of the color of umber

¹um·bil·i·cal \ˌəm-ˈbi-li-kəl, *British also* ˌəm-bə-ˈlī-kəl\ *adjective* (1541)
1 : of, relating to, or used at the navel
2 : of or relating to the central region of the abdomen

²umbilical *noun* (1774)
: UMBILICAL CORD 2

umbilical cord *noun* (1753)
1 : a cord arising from the navel that connects the fetus with the placenta; *also* **:** YOLK STALK
2 : a tethering or supply line (as for an astronaut outside a spacecraft or a diver underwater)

um·bil·i·cate \ˌəm-ˈbi-li-kət\ *or* **um·bil·i·cat·ed** \-lə-ˌkā-təd\ *adjective* (1698)
1 : depressed like a navel
2 : having an umbilicus
— um·bil·i·ca·tion \ˌəm-ˌbi-lə-ˈkā-shən\ *noun*

um·bi·li·cus \ˌəm-ˈbi-li-kəs, ˌəm-bə-ˈlī-\ *noun, plural* **um·bi·li·ci** \ˌəm-ˈbi-li-ˌkī, -ˌkē; ˌəm-bə-ˈlī-ˌkī, -ˌsī\ *or* **um·bi·li·cus·es** [Latin — more at NAVEL] (circa 1615)
1 a : NAVEL 1 **b :** any of several morphological depressions; *especially* **:** HILUM 1
2 : a central point **:** CORE, HEART

um·bles \ˈəm-bəlz\ *noun plural* [Middle English, alteration of *nombles,* from Middle French, plural of *nomble* fillet of beef, pork loin, modification of Latin *lumbulus,* diminutive of *lumbus* loin — more at LOIN] (15th century)
: the edible viscera of an animal and especially of a deer or hog

um·bo \ˈəm-(ˌ)bō\ *noun, plural* **um·bo·nes** \ˌəm-ˈbō-(ˌ)nēz\ *or* **umbos** [Latin; akin to Latin *umbilicus* — more at NAVEL] (1721)
1 : the boss of a shield
2 : a rounded elevation: as **a :** an inward projection of the tympanic membrane of the ear **b :** one of the lateral prominences just above the hinge of a bivalve shell
— um·bo·nal \ˈəm-bə-nᵊl, ˌəm-ˈbō-\ *adjective*
— um·bo·nate \ˈəm-bə-ˌnāt, ˌəm-ˈbō-nət\ *adjective*

um·bra \ˈəm-brə\ *noun, plural* **umbras** *or* **um·brae** \-(ˌ)brē, -ˌbrī\ [Latin] (1638)
1 : a shaded area
2 a : a conical shadow excluding all light from a given source; *specifically* **:** the conical part of the shadow of a celestial body excluding all light from the primary source **b :** the central dark part of a sunspot
— um·bral \-brəl\ *adjective*

um·brage \ˈəm-brij\ *noun* [Middle English, from Middle French, from Latin *umbraticum,* neuter of *umbraticus* of shade, from *umbratus,* past participle of *umbrare* to shade, from *umbra* shade, shadow; akin to Lithuanian *unksmė* shadow] (15th century)
1 : SHADE, SHADOW
2 : shady branches **:** FOLIAGE
3 a : an indistinct indication **:** vague suggestion **:** HINT **b :** a reason for doubt **:** SUSPICION
4 : a feeling of pique or resentment at some often fancied slight or insult ⟨took *umbrage* at the speaker's remarks⟩
synonym see OFFENSE

um·bra·geous \ˌəm-ˈbrā-jəs\ *adjective* (1587)
1 a : affording shade **b :** spotted with shadows
2 : inclined to take offense easily
— um·bra·geous·ly *adverb*
— um·bra·geous·ness *noun*

¹um·brel·la \ˌəm-ˈbre-lə, *especially Southern* ˈəm-ˌ\ *noun* [Italian *ombrella,* modification of Latin *umbella,* diminutive of *umbra*] (1611)
1 : a collapsible shade for protection against weather consisting of fabric stretched over hinged ribs radiating from a central pole; *especially* **:** a small one for carrying in the hand
2 : the bell-shaped or saucer-shaped largely gelatinous structure that forms the chief part of the body of most jellyfishes
3 : something which provides protection: as **a :** defensive air cover (as over a battlefront) **b :** a heavy barrage
4 : something which covers or embraces a broad range of elements or factors ⟨decided to expand . . . by building new colleges under a federation *umbrella* —Diane Ravitch⟩

²umbrella *transitive verb* **-laed; -la·ing** (circa 1800)
: to protect, cover, or provide with an umbrella

umbrella plant *noun* (1874)
: an African sedge (*Cyperus alternifolius*) that has large terminal whorls of slender leaves and is often grown as an ornamental

umbrella tree *noun* (1738)
1 : a magnolia (*Magnolia tripetala*) of the eastern U.S. having large leaves clustered at the ends of the branches
2 : any of various trees or shrubs resembling an umbrella especially in the arrangement of leaves or the shape of the crown

Um·bri·an \ˈəm-brē-ən\ *noun* (1601)
1 : a native or inhabitant of Umbria
2 : the Italic language of ancient Umbria — see INDO-EUROPEAN LANGUAGES table
— Umbrian *adjective*

Um·bun·du \ˌəm-ˈbün-(ˌ)dü\ *noun* (circa 1895)
: a Bantu language of central Angola

umi·ak \ˈü-mē-ˌak\ *noun* [Inuit *umiaq*] (1769)
: an open Eskimo boat made of a wooden frame covered with hide

¹um·laut \ˈüm-ˌlau̇t, ˈu̇m-\ *noun* [German, from *um-* around, transformation + *Laut* sound] (1852)
1 a : the change of a vowel that is caused by partial assimilation to a succeeding sound or that occurs as a reflex of the former presence of a succeeding sound which has been lost or altered (as to mark pluralization in *goose, geese* or *mouse, mice*) **b :** a vowel resulting from such partial assimilation
2 : a diacritical mark ¨ placed over a vowel to indicate a more central or front articulation — compare DIAERESIS

²umlaut *transitive verb* (1879)
1 : to produce by umlaut
2 : to write or print an umlaut over

¹ump \ˈəmp\ *noun* (1912)
: UMPIRE 2

²ump *intransitive verb* (1928)
: to act as umpire

¹um·pire \ˈəm-ˌpīr\ *noun* [Middle English *oumpere,* alteration (from misdivision of *a noumpere*) of *noumpere,* from Middle French *nomper* not equal, not paired, from *non-* + *per* equal, from Latin *par*] (15th century)
1 : one having authority to decide finally a controversy or question between parties: as **a :** one appointed to decide between arbitrators who have disagreed **b :** an impartial third party chosen to arbitrate disputes arising under the terms of a labor agreement
2 : an official in a sport who rules on plays
3 : a military officer who evaluates maneuvers
◆

²umpire *verb* **um·pired; um·pir·ing** (1609)
transitive verb
: to supervise or decide as umpire
intransitive verb
: to act as umpire

ump·teen \ˈəm(p)-ˌtēn, ˌəm(p)-ˈ\ *adjective* [blend of *umpty* (such and such) and *-teen* (as in *thirteen*)] (1918)
: very many **:** indefinitely numerous
— ump·teenth \-ˌtēn(t)th, -ˈtēn(t)th\ *adjective*

¹un- \ˌən, *often* ˈən *before* ˈ*-stressed syllable*\ *prefix* [Middle English, from Old English; akin to Old High German *un-* un-, Latin *in-,* Greek *a-, an-,* Old English *ne* not — more at NO]
1 : not **:** IN-, NON- — in adjectives formed from adjectives ⟨*un*ambitious⟩ ⟨*un*skilled⟩ or participles ⟨*un*dressed⟩, in nouns formed from nouns ⟨*un*availability⟩, and rarely in verbs formed from verbs ⟨*un*be⟩; sometimes in words that have a meaning that merely negates that of the base word and are thereby distinguished from words that prefix *in-* or a variant of it (as *im-*) to the same base word and have a meaning positively opposite to that of the base word ⟨*un*artistic⟩ ⟨*un*moral⟩
2 : opposite of **:** contrary to — in adjectives formed from adjectives ⟨*un*constitutional⟩ ⟨*un*graceful⟩ ⟨*un*mannered⟩ or participles ⟨*un*believing⟩ and in nouns formed from nouns ⟨*un*rest⟩

²un- *prefix* [Middle English, from Old English *un-, on-,* alteration of *and-* against — more at ANTE-]
1 : do the opposite of **:** reverse (a specified action) **:** DE- 1a, DIS- 1a — in verbs formed from verbs ⟨*un*bend⟩ ⟨*un*dress⟩ ⟨*un*fold⟩
2 a : deprive of **:** remove (a specified thing) from **:** remove — in verbs formed from nouns ⟨*un*frock⟩ ⟨*un*sex⟩ **b :** release from **:** free from — in verbs formed from nouns ⟨*un*hand⟩ **c :** remove from **:** extract from **:** bring out of — in verbs formed from nouns ⟨*un*bosom⟩ **d :** cause to cease to be — in verbs formed from nouns ⟨*un*man⟩
3 : completely ⟨*un*loose⟩

un·abrad·ed
un·ab·sorbed
un·ab·sor·bent
un·ac·a·dem·ic
un·ac·a·dem·i·cal·ly
un·ac·cent·ed
un·ac·cept·ed
un·ac·cli·mat·ed
un·ac·cli·ma·tized
un·ac·com·mo·dat·ed
un·ac·com·mo·dat·ing
un·ac·cred·it·ed
un·ac·cul·tur·at·ed
un·achieved
un·ac·knowl·edged
un·ac·quaint·ed
un·act·able
un·act·ed
un·ac·tor·ish
un·adapt·able
un·adapt·ed
un·ad·dressed
un·ad·ju·di·cat·ed
un·ad·just·ed
un·ad·mired
un·ad·mit·ted
un·adopt·able
un·adult
un·ad·ven·tur·ous
un·ad·ver·tised
un·aes·thet·ic
un·af·fect·ing
un·af·fec·tion·ate
un·af·fec·tion·ate·ly
un·af·fil·i·at·ed
un·af·flu·ent
un·af·ford·able
un·afraid
un·ag·gres·sive
un·aid·ed
un·air–con·di·tioned
un·akin
un·alien·at·ed
un·alike
un·al·le·vi·at·ed
un·al·lo·cat·ed
un·al·lur·ing
un·al·tered
un·am·bi·tious
un·ame·na·ble
un·amend·ed
un·ami·a·ble
un·am·or·tized
un·am·pli·fied
un·amus·ing
un·an·a·lyz·able
un·an·a·lyzed
un·an·no·tat·ed
un·an·nounced
un·apol·o·giz·ing
un·ap·par·ent
un·ap·peased
un·ap·pre·ci·at·ed
un·ap·pre·cia·tive
un·ap·pro·pri·at·ed
un·ap·proved
un·ar·gu·able
un·ar·gu·ably
un·ar·mored
un·ar·ro·gant
un·ar·tis·tic
un·as·pi·rat·ed
un·as·sailed
un·as·sem·bled
un·as·signed
un·as·sim·i·la·ble
un·as·sim·i·lat·ed
un·as·so·ci·at·ed
un·as·suaged
un·ath·let·ic
un·at·tain·able
un·at·tend·ed
un·at·ten·u·at·ed
un·at·test·ed
un·at·trib·ut·able
un·at·trib·ut·ed
un·at·tuned
un·au·dit·ed
un·au·then·tic
un·au·tho·rized
un·au·to·mat·ed
un·avail·abil·i·ty
un·avail·able
un·avowed
un·awak·ened
un·award·ed
un·awe·some
un·ban
un·bap·tized
un·barbed
un·bar·ri·cad·ed
un·be·hold·en
un·bel·lig·er·ent
un·be·loved
un·be·mused
un·billed
un·bit·ten
un·bit·ter
un·bleached
un·blem·ished
un·blend·ed
un·blood·ed
un·book·ish
un·bought

◇ WORD HISTORY
umpire The medieval French adjective *nonper* was a compound formed from *non* "not" and *per* "equal" and it meant "not paired, odd (of a number)" or "not equaled, incomparable." It was adopted into Middle English as *noumpere,* a noun meaning "arbitrator," in other words, someone not equal to the parties in a dispute and thus able to mediate between them. Early on, English speakers must have lost any sense of the word's French etymological components, because not long after *noumpere* appears in the late 14th century, a variant *oumpere* is also found in manuscripts. This new form resulted from misplacing the boundary between the noun and a preceding indefinite article: *a noumpere* was understood as *an oumpere.* By the 16th century no further trace of the word with an initial *n* is found, and spellings such as *umpier* and *umpire* reflect a change in the quality of the second vowel that must have begun shortly after the word was first borrowed. (The same change is evident in Modern English *briar, choir,* and *contrive* from Middle English *brere, quer,* and *contreven.*) In reference to an arbiter in a sporting event, *umpire* is first recorded in 1714.

un•bowd•ler•ized
un•brack•et•ed
un•brake
un•breach•able
un•break•able
un•bridge•able
un•bridged
un•briefed
un•bright
un•bril•liant
un•bruised
un•brushed
un•bud•get•ed
un•buf•fered
un•build•able
un•bulky
un•bu•reau•crat•ic
un•bur•ied
un•burn•able
un•burned
un•burnt
un•busi•ness•like
un•busy
un•but•tered
un•cal•ci•fied
un•cal•cined
un•cal•i•brat•ed
un•called
un•cal•loused
un•can•celed
un•can•did
un•can•did•ly
un•ca•non•i•cal
un•cap
un•cap•i•tal•ized
un•cap•tioned
un•cap•tur•able
un•cared–for
un•car•ing
un•car•pet•ed
un•case
un•cas•trat•ed
un•cat•a•loged
un•catch•able
un•catchy
un•cat•e•go•riz•able
un•caught
un•cen•sored
un•cen•so•ri•ous
un•cen•sured
un•cer•ti•fied
un•chal•lenge•able
un•chal•lenged
un•chal•leng•ing
un•changed
un•chan•neled
un•chap•er•oned
un•char•is•mat•ic
un•charm•ing
un•chart•ered
un•chau•vin•is•tic
un•check•able
un•checked
un•chew•able
un•chewed
un•chic
un•child•like
un•chlo•ri•nat•ed
un•cho•reo•graphed
un•chris•tened
un•chron•i•cled
un•chro•no•log•i•cal
un•church•ly
un•cil•i•at•ed
un•cin•e•mat•ic
un•clad
un•claimed
un•clar•i•fied
un•clas•si•fi•able
un•cleaned
un•clear
un•cli•chéd
un•clip
un•cloy•ing
un•co•alesce
un•coat•ed
un•coat•ing
un•cod•ed
un•cod•i•fied
un•co•erced
un•co•er•cive
un•co•er•cive•ly
un•col•lect•ed
un•col•lect•ible
un•co•lored
un•com•bat•ive
un•combed
un•com•bined
un•come•ly
un•com•ic
un•com•mer•cial•ized
un•com•pas•sion•ate
un•com•pel•ling
un•com•pen•sat•ed
un•com•pla•cent
un•com•plet•ed
un•com•pound•ed
un•com•pre•hend•ed
un•com•pu•ter•ized
un•con•cealed
un•con•fessed
un•con•fined
un•con•firmed
un•con•found•ed
un•con•fuse
un•con•ju•gat•ed
un•con•nect•ed
un•con•quered
un•con•se•crat•ed
un•con•strained
un•con•strict•ed
un•con•struc•tive
un•con•sumed
un•con•sum•mat•ed
un•con•tain•able
un•con•tam•i•nat•ed
un•con•tem•plat•ed
un•con•tem•po•rary
un•con•ten•tious
un•con•test•ed
un•con•tract•ed
un•con•tra•dict•ed
un•con•trived
un•con•trolled
un•con•tro•ver•sial
un•con•tro•ver•sial•ly
un•con•vert•ed
un•con•vinced
un•con•voyed
un•cooked
un•cooled
un•co•op•er•a•tive
un•co•or•di•nat•ed
un•copy•right•able
un•cor•rect•able
un•cor•rect•ed
un•cor•re•lat•ed
un•cor•rob•o•rat•ed
un•cor•rupt
un•count•able
un•cou•ra•geous
un•cov•e•nant•ed
un•coy
un•cracked
un•crate
un•cra•zy
un•cre•a•tive
un•cre•den•tialed
un•cred•it•ed
un•crip•pled
un•cropped
un•cross•able
un•crowd•ed
un•crush•able
un•cuff
un•cul•ti•va•ble
un•cul•ti•vat•ed
un•cul•tured
un•cured
un•cu•ri•ous
un•cur•rent
un•cur•tained
un•cus•tom•ar•i•ly
un•cus•tom•ary
un•cute
un•cyn•i•cal
un•cyn•i•cal•ly
un•dam•aged
un•damped
un•dance•able
un•dat•ed
un•dec•a•dent
un•de•cid•abil•i•ty
un•de•cid•able
un•de•cid•ed
un•de•ci•pher•able
un•de•ci•phered
un•de•clared
un•de•com•posed
un•dec•o•rat•ed
un•ded•i•cat•ed
un•de•feat•ed
un•de•fend•ed
un•de•filed
un•de•fin•able
un•de•fined
un•de•fo•li•at•ed
un•de•formed
un•del•e•gat•ed
un•de•liv•er•able
un•de•liv•ered
un•de•lud•ed
un•de•mand•ing
un•de•nom•i•na•tion•al
un•de•pend•able
un•de•scrib•able
un•de•served
un•de•serv•ing
un•des•ig•nat•ed
un•de•sired
un•de•tect•able
un•de•tect•ed
un•de•ter•min•able
un•de•ter•mined
un•de•terred
un•de•vel•oped
un•di•ag•nos•able
un•di•ag•nosed
un•di•a•lec•ti•cal
un•di•dac•tic
un•dif•fer•en•ti•at•ed
un•di•gest•ed
un•di•gest•ible
un•dig•ni•fied
un•di•lut•ed
un•di•min•ished
un•dimmed
un•dis•charged
un•dis•ci•plined
un•dis•closed
un•dis•cour•aged
un•dis•cov•er•able
un•dis•cov•ered
un•dis•crim•i•nat•ing
un•dis•cussed
un•dis•mayed
un•dis•pu•ta•ble
un•dis•put•ed
un•dis•solved
un•dis•tin•guished
un•dis•tort•ed
un•dis•tract•ed
un•dis•trib•ut•ed
un•dis•turbed
un•di•vid•ed
un•do•able
un•doc•ile
un•doc•tored
un•doc•tri•naire
un•doc•u•ment•ed
un•do•mes•tic
un•do•mes•ti•cat•ed
un•dot•ted
un•doubt•able
un•doubt•ing
un•drained
un•dra•ma•tized
un•drilled
un•dubbed
un•dulled
un•du•pli•cat•ed
un•dyed
un•dy•nam•ic
un•ea•ger
un•ear•marked
un•eat•able
un•eat•en
un•ec•cen•tric
un•eco•log•i•cal
un•ed•i•fy•ing
un•ed•u•ca•ble
un•ed•u•cat•ed
un•elab•o•rate
un•elect•able
un•elect•ed
un•elec•tri•fied
un•em•bar•rassed
un•em•bel•lished
un•em•bit•tered
un•em•phat•ic
un•em•phat•i•cal•ly
un•em•pir•i•cal
un•en•chant•ed
un•en•closed
un•en•cour•ag•ing
un•en•dear•ing
un•en•dur•able
un•en•dur•able•ness
un•en•dur•ably
un•en•force•able
un•en•forced
un•en•larged
un•en•light•ened
un•en•light•en•ing
un•en•riched
un•en•ter•pris•ing
un•en•thu•si•as•tic
un•en•thu•si•as•ti•cal•ly
un•en•vi•able
un•en•vi•ous
un•erot•ic
un•es•cap•able
un•es•tab•lished
un•eth•i•cal
un•eval•u•at•ed
un•ex•am•ined
un•ex•celled
un•ex•cit•able
un•ex•cit•ed
un•ex•cit•ing
un•ex•cused
un•ex•ot•ic
un•ex•pend•ed
un•ex•pired
un•ex•plain•able
un•ex•plained
un•ex•plod•ed
un•ex•plored
un•ex•posed
un•ex•pressed
un•ex•pur•gat•ed
un•ex•traor•di•nary
un•faked
un•fa•mous
un•fan•cy
un•fas•tid•i•ous
un•fea•si•ble
un•felt
un•fem•i•nine
un•fenced
un•fer•ment•ed
un•fer•tile
un•fer•til•ized
un•filled
un•fired
un•flam•boy•ant
un•flashy
un•fly•able
un•fond
un•forced
un•fore•see•able
un•fore•seen
un•for•est•ed
un•for•giv•able
un•forked
un•for•mu•lat•ed
un•forth•com•ing
un•for•ti•fied
un•fos•sil•if•er•ous
un•framed
un•free
un•free•dom
un•friv•o•lous
un•ful•fill•able
un•ful•filled
un•fun•ny
un•fur•nished
un•fused
un•gal•lant
un•gal•lant•ly
un•gar•nished
un•ge•nial
un•gen•teel
un•gen•tle
un•gen•tle•man•ly
un•gen•tri•fied
un•ger•mi•nat•ed
un•gift•ed
un•gim•micky
un•glam•or•ized
un•glam•or•ous
un•glazed
un•grace•ful
un•grace•ful•ly
un•grad•ed
un•grasp•able
un•ground
un•grouped
un•guess•able
un•guid•ed
un•hack•neyed
un•ham•pered
un•harmed
un•har•ness
un•har•vest•ed
un•hatched
un•healed
un•health•ful
un•heat•ed
un•hedged
un•heed•ed
un•heed•ing
un•help•ful
un•help•ful•ly
un•her•ald•ed
un•he•ro•ic
un•hin•dered
un•hip
un•his•tor•i•cal
un•ho•mog•e•nized
un•hon•ored
un•hope•ful
un•housed
un•hu•mor•ous
un•hurt
un•hy•dro•lyzed
un•hy•gien•ic
un•hy•phen•at•ed
un•hys•ter•i•cal
un•hys•ter•i•cal•ly
un•iden•ti•fi•able
un•iden•ti•fied
un•ideo•log•i•cal
un•id•i•om•at•ic
un•ig•nor•able
un•il•lu•mi•nat•ing
un•imag•i•na•tive
un•imag•i•na•tive•ly
un•im•mu•nized
un•im•paired
un•im•pas•sioned
un•im•ped•ed
un•im•por•tant
un•im•pos•ing
un•im•pressed
un•im•pres•sive
un•in•cor•po•rat•ed
un•in•dexed
un•in•dict•ed
un•in•dus•tri•al•ized
un•in•fect•ed
un•in•flat•ed
un•in•flect•ed
un•in•flu•enced
un•in•for•ma•tive
un•in•for•ma•tive•ly
un•in•formed
un•in•gra•ti•at•ing
un•in•hab•it•able
un•in•hab•it•ed
un•ini•ti•at•ed
un•in•jured
un•in•oc•u•lat•ed
un•in•spect•ed
un•in•spired
un•in•spir•ing
un•in•struct•ed
un•in•struc•tive
un•in•su•lat•ed
un•in•sur•able
un•in•sured
un•in•te•grat•ed
un•in•tel•lec•tu•al
un•in•tel•li•gent
un•in•tel•li•gent•ly
un•in•tel•li•gi•bil•i•ty
un•in•tel•li•gi•ble
un•in•tel•li•gi•ble•ness
un•in•tel•li•gi•bly
un•in•tend•ed
un•in•ten•tion•al
un•in•ten•tion•al•ly
un•in•ter•est•ing
un•in•ter•rupt•ed
un•in•ter•rupt•ed•ly
un•in•tim•i•dat•ed
un•in•ven•tive
un•in•vit•ed
un•in•vit•ing
un•in•volved
un•iron•i•cal•ly
un•ir•ra•di•at•ed
un•ir•ri•gat•ed
un•is•sued
un•jad•ed
un•joint•ed
un•jus•ti•fi•able
un•jus•ti•fi•ably
un•jus•ti•fied
un•kept
un•knot
un•knowl•edge•able
un•ko•sher
un•la•beled
un•la•dy•like
un•la•ment•ed
un•laun•dered
un•learn•able
un•leav•ened
un•lib•er•at•ed
un•li•censed
un•lik•able
un•lined
un•lis•ten•able
un•lit
un•lit•er•ary
un•liv•able
un•lo•cal•ized
un•lov•able
un•loved
un•lov•ing
un•lyr•i•cal
un•ma•cho
un•mag•ni•fied
un•ma•li•cious
un•ma•li•cious•ly
un•man•age•able
un•man•age•ably
un•man•aged
un•ma•nip•u•lat•ed
un•mapped
un•marked

un•mar•ket•able
un•marred
un•mas•cu•line
un•match•able
un•matched
un•meant
un•mea•sur•able
un•mea•sured
un•mech•a•nized
un•med•i•cat•ed
un•me•lo•di•ous
un•me•lo•di•ous•ness
un•mem•o•ra•ble
un•mem•o•ra•bly
un•mer•it•ed
un•mer•ry
un•met
un•me•tab•o•lized
un•mil•i•tary
un•milled
un•mined
un•mix
un•mix•able
un•mixed
un•mod•ern•ized
un•mod•i•fied
un•mod•ish
un•mo•lest•ed
un•mon•i•tored
un•mo•ti•vat•ed
un•mount•ed
un•mov•able
un•moved
un•mu•si•cal
un•name•able
un•named
un•need•ed
un•ne•go•tia•ble
un•neu•rot•ic
un•news•wor•thy
un•no•tice•able
un•no•ticed
un•nour•ish•ing
un•ob•jec•tion•able
un•ob•serv•able
un•ob•served
un•ob•struct•ed
un•ob•tain•able
un•of•fi•cial
un•of•fi•cial•ly
un•open•able
un•opened
un•op•posed
un•or•dered
un•orig•i•nal
un•or•na•ment•ed
un•os•ten•ta•tious
un•os•ten•ta•tious•ly
un•owned
un•ox•y•gen•at•ed
un•paint•ed
un•par•a•sit•ized
un•par•don•able
un•pass•able
un•pas•teur•ized
un•pas•to•ral
un•pat•ent•able
un•pa•tri•ot•ic
un•paved
un•pe•dan•tic
un•peeled
un•per•ceived
un•per•cep•tive
un•per•form•able
un•per•formed
un•per•suad•ed
un•per•sua•sive
un•per•turbed
un•pic•tur•esque
un•planned
un•plau•si•ble
un•play•able
un•pleased
un•pleas•ing
un•plowed
un•po•et•ic
un•po•liced
un•pol•ished
un•pol•lut•ed
un•posed
un•prac•ti•cal
un•pre•dict•abil•i•ty
un•pre•dict•able
un•pre•dict•ably
un•prej•u•diced
un•pre•med•i•tat•ed
un•pre•pared
un•pre•pared•ness
un•pre•pos•sess•ing
un•pressed
un•pres•sured
un•pres•sur•ized
un•pret•ty
un•priv•i•leged
un•prob•lem•at•ic
un•pro•cessed
un•pro•duced
un•pro•duc•tive
un•pro•fes•sion•al
un•pro•gram•ma•ble
un•pro•grammed
un•pro•gres•sive
un•prompt•ed
un•pro•nounce•able
un•pro•pi•tious
un•pros•per•ous
un•pro•tect•ed
un•prov•able
un•proved
un•prov•en
un•pro•voked
un•pruned
un•pub•li•cized
un•pub•lished
un•punc•tu•al
un•punc•tu•al•i•ty
un•punc•tu•at•ed
un•pun•ished
un•quan•ti•fi•able
un•quench•able
un•ques•tioned
un•raised
un•ranked
un•rat•ed
un•rav•ished
un•reach•able
un•reached
un•read•able
un•read•i•ness
un•ready
un•re•al•iz•able
un•re•al•ized
un•re•cep•tive
un•re•claim•able
un•re•claimed
un•rec•og•niz•able
un•rec•og•niz•ably
un•rec•og•nized
un•rec•on•cil•able
un•rec•on•ciled
un•re•cord•ed
un•re•cov•er•able
un•re•cov•ered
un•re•cy•cla•ble
un•re•deem•able
un•re•deemed
un•re•dressed
un•re•fined
un•re•flec•tive
un•re•formed
un•re•frig•er•at•ed
un•reg•is•tered
un•reg•u•lat•ed
un•re•hearsed
un•re•in•forced
un•re•lat•ed
un•re•laxed
un•re•li•abil•i•ty
un•re•li•able
un•re•lieved
un•re•liev•ed•ly
un•re•luc•tant
un•re•mark•able
un•re•mark•ably
un•re•mem•bered
un•rem•i•nis•cent
un•re•mov•able
un•re•peat•able
un•re•pen•tant
un•re•pen•tant•ly
un•re•port•ed
un•rep•re•sen•ta•tive
un•rep•re•sen•ta•tive•ness
un•rep•re•sent•ed
un•re•pressed
un•re•sis•tant
un•re•solv•able
un•re•solved
un•re•spect•able
un•re•spon•sive
un•re•spon•sive•ly
un•re•spon•sive•ness
un•rest•ful
un•re•stored
un•re•strict•ed
un•re•touched
un•re•turn•able
un•re•vealed
un•re•view•able
un•re•viewed
un•re•vised
un•rev•o•lu•tion•ary
un•re•ward•ed
un•re•ward•ing
un•rhe•tor•i•cal
un•rhymed
un•rhyth•mic
un•rid•able
un•ri•fled
un•rip•ened
un•ro•man•tic
un•ro•man•ti•cal•ly
un•ro•man•ti•cized
un•roofed
un•rushed
un•safe
un•sal•able
un•sal•a•ried
un•salt•ed
un•sal•vage•able
un•sanc•tioned
un•san•i•tary
un•sat•is•fied
un•scal•able
un•scarred
un•scent•ed
un•sched•uled
un•schol•ar•ly
un•screened
un•scrip•tur•al
un•sea•soned
un•sea•wor•thy
un•se•cured
un•seed•ed
un•seg•ment•ed
un•self–con•scious
un•self–con•scious•ly
un•self–con•scious•ness
un•sell•able
un•sen•sa•tion•al
un•sen•si•tized
un•sent
un•sen•ti•men•tal
un•sep•a•rat•ed
un•se•ri•ous
un•se•ri•ous•ness
un•served
un•ser•vice•able
un•sex•u•al
un•sexy
un•shad•ed
un•shak•able
un•shak•ably
un•shak•en
un•shape•ly
un•shared
un•sharp
un•shav•en
un•shelled
un•shock•able
un•shorn
un•showy
un•signed
un•sink•able
un•size
un•slaked
un•smart
un•smil•ing
un•smoothed
un•soiled
un•sol•dier•ly
un•so•lic•it•ed
un•solv•able
un•solved
un•sort•ed
un•sound•ed
un•sown
un•spe•cial•ized
un•spec•i•fi•able
un•spe•cif•ic
un•spec•i•fied
un•spec•tac•u•lar
un•spent
un•spir•i•tu•al
un•split
un•spoiled
un•spoilt
un•spo•ken
un•sprayed
un•stained
un•stan•dard•ized
un•star•tling
un•stat•ed
un•stayed
un•ster•ile
un•ster•il•ized
un•stint•ed
un•stitch
un•stop•per
un•strained
un•strat•i•fied
un•stuffy
un•styl•ish
un•sub•dued
un•sub•si•dized
un•sub•stan•ti•at•ed
un•sub•tle
un•sub•tly
un•suit•abil•ity
un•suit•able
un•suit•ably
un•suit•ed
un•sul•lied
un•su•per•vised
un•sup•port•able
un•sup•port•ed
un•sure
un•sur•pass•able
un•sur•passed
un•sur•prised
un•sus•cep•ti•ble
un•sus•pect•ed
un•sus•pect•ing
un•sus•pi•cious
un•sus•tain•able
un•sweet•ened
un•sym•pa•thet•ic
un•sym•pa•thet•i•cal•ly
un•syn•chro•nized
un•sys•tem•at•ic
un•sys•tem•at•i•cal•ly
un•sys•tem•a•tized
un•tact•ful
un•tagged
un•taint•ed
un•tal•ent•ed
un•tam•able
un•tamed
un•tanned
un•tar•nished
un•taxed
un•teach•able
un•tech•ni•cal
un•tem•pered
un•ten•ant•ed
un•tend•ed
un•ten•ured
un•test•able
un•test•ed
un•the•o•ret•i•cal
un•threat•en•ing
un•thrifty
un•till•able
un•tilled
un•to•geth•er
un•trace•able
un•tra•di•tion•al
un•tra•di•tion•al•ly
un•trained
un•tram•meled
un•trans•formed
un•trans•lat•abil•i•ty
un•trans•lat•able
un•trans•lat•ed
un•trav•eled
un•tra•versed
un•treat•ed
un•trendy
un•trimmed
un•trust•ing
un•trust•wor•thy
un•tuck
un•tuft•ed
un•typ•i•cal
un•typ•i•cal•ly
un•un•der•stand•able
un•us•able
un•uti•lized
un•vac•ci•nat•ed
un•var•ied
un•vary•ing
un•ven•ti•lat•ed
un•ver•bal•ized
un•ver•i•fi•able
un•versed
un•vi•a•ble
un•vis•it•ed
un•want•ed
un•war•like
un•war•rant•ed
un•wa•ver•ing
un•wa•ver•ing•ly
un•waxed
un•weaned
un•wear•able
un•weath•ered
un•wed
un•weight•ed
un•wel•come
un•white
un•willed
un•win•na•ble
un•wom•an•ly
un•won
un•work•abil•i•ty
un•work•able
un•worked
un•wor•ried
un•wound•ed
un•wo•ven
un•young

un•abashed \ˌən-ə-ˈbasht\ *adjective* (1571)
: not abashed **:** UNDISGUISED, UNAPOLOGETIC
— **un•abash•ed•ly** \-ˈbash-əd-lē\ *adverb*

un•abat•ed \ˌən ə-ˈbā-təd\ *adjective* (circa 1611)
: not abated **:** being at full strength or force
— **un•abat•ed•ly** *adverb*

un•able \ˌən-ˈa-bəl\ *adjective* (14th century)
: not able **:** INCAPABLE: as **a :** UNQUALIFIED, INCOMPETENT **b :** IMPOTENT, HELPLESS

un•abridged \ˌən-ə-ˈbrijd\ *adjective* (1599)
1 : not abridged **:** COMPLETE ⟨an *unabridged* reprint of a novel⟩
2 : being the most complete of its class **:** not based on one larger ⟨an *unabridged* dictionary⟩

un•ac•cept•able \ˌən-ik-ˈsep-tə-bəl, -ak-\ *adjective* (15th century)
: not acceptable **:** not pleasing or welcome
— **un•ac•cept•abil•i•ty** \-ˌsep-tə-ˈbi-lə-tē\ *noun*
— **un•ac•cept•ably** \-ˈsep-tə-blē\ *adverb*

un•ac•com•pa•nied \ˌən-ə-ˈkəmp-nēd, -ˈkämp-; -ˈkəm-pə-, -ˈkäm-\ *adjective* (1545)
: not accompanied; *especially* **:** being without instrumental accompaniment

un•ac•count•able \ˌən-ə-ˈkau̇n-tə-bəl\ *adjective* (1643)
1 : not to be accounted for **:** INEXPLICABLE, STRANGE
2 : not to be called to account **:** not responsible
— **un•ac•count•abil•i•ty** \-ˌkau̇n-tə-ˈbi-lə-tē\ *noun*

un•ac•count•ably \-ˈkau̇n-tə-blē\ *adverb* (1694)
1 : in an unaccountable manner ⟨looking *unaccountably* upset⟩ ⟨heat was *unaccountably* disappearing⟩
2 : for reasons that are hard to understand ⟨*unaccountably*, he stayed right there⟩

un•ac•count•ed \-ˈkau̇n-təd\ *adjective* (1689)

: not accounted **:** UNEXPLAINED — often used with *for*

un·ac·cus·tomed \ˌən-ə-ˈkəs-təmd\ *adjective* (1526)
1 : not customary **:** not usual or common
2 : not habituated — usually used with *to*
— un·ac·cus·tomed·ly \-təmd-lē\ *adverb*

una cor·da \ˌü-nə-ˈkȯr-də, -(ˌ)dä\ *adverb or adjective* [Italian, literally, one string] (circa 1849)
: with soft pedal depressed — used as a direction in piano music

un·adorned \ˌən-ə-ˈdȯrnd\ *adjective* (1634)
: not adorned **:** lacking embellishment or decoration **:** PLAIN, SIMPLE

un·adul·ter·at·ed \ˌən-ə-ˈdəl-tə-ˌrā-təd\ *adjective* (circa 1719)
1 : not adulterated **:** PURE ⟨*unadulterated* food⟩
2 : COMPLETE, UNQUALIFIED ⟨an *unadulterated* fool⟩
— un·adul·ter·at·ed·ly *adverb*

un·ad·vised \ˌən-əd-ˈvīzd\ *adjective* (14th century)
1 : done without due consideration **:** RASH ⟨*unadvised* and dangerous dealings with the terrorists⟩
2 : not prudent **:** ILL-ADVISED ⟨done with *unadvised* haste⟩
— un·ad·vis·ed·ly \-ˈvī-zəd-lē\ *adverb*

un·af·fect·ed \ˌən-ə-ˈfek-təd\ *adjective* (circa 1586)
1 : not influenced or changed mentally, physically, or chemically
2 : free from affectation **:** GENUINE
— un·af·fect·ed·ly *adverb*
— un·af·fect·ed·ness *noun*

un·ag·ing *or* **un·age·ing** \ˌən-ˈā-jiŋ\ *adjective* (1860)
: AGELESS

una·kite \ˈyü-nə-ˌkīt\ *noun* [*Unaka* Mountains, Tenn. & N.C. + [1]*-ite*] (1874)
: an altered igneous rock that is usually opaque with green, black, pink, and white flecks and is usually used as a gemstone

un·alien·able \ˌən-ˈāl-yə-nə-bəl, -ˈā-lē-ə-\ *adjective* (1611)
: INALIENABLE

un·aligned \ˌən-ə-ˈlīnd\ *adjective* (circa 1934)
: NONALIGNED

un·al·loyed \ˌən-ə-ˈlȯid\ *adjective* (1667)
: not alloyed **:** UNMIXED, UNQUALIFIED, PURE ⟨*unalloyed* metals⟩ ⟨*unalloyed* happiness⟩

un·al·ter·able \ˌən-ˈȯl-t(ə-)rə-bəl\ *adjective* (1611)
: not capable of being altered or changed ⟨an *unalterable* resolve⟩ ⟨*unalterable* hatred⟩
— un·al·ter·abil·i·ty \ˌən-ˌȯl-t(ə-)rə-ˈbi-lə-tē\ *noun*
— un·al·ter·able·ness \ˌən-ˈȯl-t(ə-)rə-bəl-nəs\ *noun*
— un·al·ter·ably \-blē\ *adverb*

un·am·big·u·ous \ˌən-am-ˈbi-gyə-wəs\ *adjective* (1751)
: not ambiguous **:** CLEAR, PRECISE
— un·am·big·u·ous·ly *adverb*

un·am·biv·a·lent \-ˈbi-və-lənt\ *adjective* (1945)
: not ambivalent **:** CLEAR-CUT, DEFINITE
— un·am·biv·a·lent·ly *adverb*

un–Amer·i·can \ˌən-ə-ˈmer-ə-kən\ *adjective* (1818)
: not American **:** not characteristic of or consistent with American customs, principles, or traditions

un·an·chored \ˌən-ˈaŋ-kərd\ *adjective* (1651)
1 : not anchored **:** not at anchor
2 : not having a firm basis or foundation

un·aneled \ˌən-ə-ˈnē(ə)ld\ *adjective* [*un-* + *aneled*, past participle of *anele* to anoint, from Middle English, from *an* on + *elen* to anoint, from *ele* oil, from Old English, from Latin *oleum* — more at OIL] (1602)
archaic **:** not having received extreme unction

un·anes·the·tized \ˌən-ə-ˈnes-thə-ˌtīzd\ *adjective* (1950)
: not having been subjected to an anesthetic

una·nim·i·ty \ˌyü-nə-ˈni-mə-tē\ *noun* (15th century)
: the quality or state of being unanimous

unan·i·mous \yu̇-ˈna-nə-məs\ *adjective* [Latin *unanimus*, from *unus* one + *animus* mind — more at ONE, ANIMATE] (1624)
1 : being of one mind **:** AGREEING
2 : formed with or indicating unanimity **:** having the agreement and consent of all
— unan·i·mous·ly *adverb*

un·an·swer·able \ˌən-ˈan(t)s-rə-bəl, -ˈan(t)-sə-\ *adjective* (1613)
: not capable of being answered; *also* **:** IRREFUTABLE
— un·an·swer·abil·i·ty \ˌən-ˌan(t)s-rə-ˈbi-lə-tē, -ˌan(t)-sə-\ *noun*
— un·an·swer·ably \ˌən-ˈan(t)s-rə-blē, -ˈan(t)-sə-\ *adverb*

un·an·swered \ˌən-ˈan(t)-sərd\ *adjective* (14th century)
1 : not answered ⟨*unanswered* letters⟩
2 : scored in succession during a period in which an opponent fails to score ⟨scored 20 *unanswered* points in the last quarter⟩

un·an·tic·i·pat·ed \ˌən-an-ˈti-sə-ˌpā-təd\ *adjective* (circa 1779)
: not anticipated **:** UNEXPECTED, UNFORESEEN
— un·an·tic·i·pat·ed·ly *adverb*

un·apol·o·get·ic \ˌən-ə-ˌpä-lə-ˈje-tik\ *adjective* (1834)
: not apologetic **:** offered, put forward, or being such without apology or qualification ⟨an *unapologetic* liberal⟩
— un·apol·o·get·i·cal·ly \-ti-k(ə-)lē\ *adverb*

un·ap·peal·able \ˌən-ə-ˈpē-lə-bəl\ *adjective* (1635)
: not appealable **:** not subject to appeal

un·ap·peal·ing \-ˈpē-liŋ\ *adjective* (circa 1846)
: not appealing **:** UNATTRACTIVE
— un·ap·peal·ing·ly \-liŋ-lē\ *adverb*

un·ap·peas·able \-ˈpē-zə-bəl\ *adjective* (1561)
: not to be appeased **:** IMPLACABLE
— un·ap·peas·ably \-blē\ *adverb*

un·ap·pe·tiz·ing \ˌən-ˈa-pə-ˌtī-ziŋ\ *adjective* (1884)
: not appetizing **:** INSIPID, UNATTRACTIVE
— un·ap·pe·tiz·ing·ly \-ziŋ-lē\ *adverb*

un·ap·pre·ci·a·tion \ˌən-ə-ˌprē-shē-ˈā-shən\ *noun* (1886)
: failure to appreciate something

un·ap·proach·able \ˌən-ə-ˈprō-chə-bəl\ *adjective* (1581)
1 : not approachable **:** physically inaccessible
2 : discouraging intimacies **:** RESERVED
— un·ap·proach·abil·i·ty \-ˌprō-chə-ˈbi-lə-tē\ *noun*
— un·ap·proach·ably \-ˈprō-chə-blē\ *adverb*

un·apt \ˌən-ˈapt\ *adjective* (circa 1513)
1 : INAPPROPRIATE, UNSUITABLE ⟨an *unapt* quote⟩
2 : not accustomed and not likely ⟨a teacher *unapt* to tolerate carelessness⟩
3 : DULL, BACKWARD ⟨*unapt* scholars⟩
— un·apt·ly \-ˈap(t)-lē\ *adverb*
— un·apt·ness \-nəs\ *noun*

un·arm \ˌən-ˈärm\ *transitive verb* (1560)
: DISARM

un·armed \-ˈärmd\ *adjective* (14th century)
1 : not armed or armored
2 : having no hard and sharp projections (as spines, spurs, or claws)

un·ar·tic·u·lat·ed \ˌən-är-ˈti-kyə-ˌlā-təd\ *adjective* (circa 1700)
: not articulated; *especially* **:** not carefully reasoned or analyzed

una·ry \ˈyü-nə-rē\ *adjective* [Latin *unus* one + English *-ary*] (1576)
: having, consisting of, or acting on a single element, item, or component **:** MONADIC

un·ashamed \ˌən-ə-ˈshāmd\ *adjective* (1600)
: not ashamed **:** being without guilt, self-consciousness, or doubt
— un·asham·ed·ly \-ˈshā-məd-lē\ *adverb*

un·asked \ˌən-ˈas(k)t\ *adjective* (13th century)
1 : not being asked **:** UNINVITED
2 : not asked ⟨*unasked* questions⟩
3 : not asked for ⟨*unasked* advice⟩

un·as·sail·able \ˌən-ə-ˈsā-lə-bəl\ *adjective* (1596)
: not assailable **:** not liable to doubt, attack, or question
— un·as·sail·abil·i·ty \-ˌsā-lə-ˈbi-lə-tē\ *noun*
— un·as·sail·able·ness \-ˈsā-lə-bəl-nəs\ *noun*
— un·as·sail·ably \-blē\ *adverb*

un·as·ser·tive \ˌən-ə-ˈsər-tiv\ *adjective* (1861)
: not assertive **:** MODEST, SHY
— un·as·ser·tive·ly *adverb*

un·as·sist·ed \ˌən-ə-ˈsis-təd\ *adjective* (1614)
1 : not assisted **:** lacking help
2 : made or performed without an assist ⟨an *unassisted* double play⟩

un·as·suage·able \ˌən-ə-ˈswā-jə-bəl\ *adjective* (circa 1611)
: not capable of being assuaged

un·as·sum·ing \ˌən-ə-ˈsü-miŋ\ *adjective* (1726)
: not assuming **:** not arrogant or presuming **:** MODEST, RETIRING
— un·as·sum·ing·ness *noun*

un·at·tached \ˌən-ə-ˈtacht\ *adjective* (1796)
1 a : not assigned or committed (as to a particular task, organization, or person); *especially* **:** not married or engaged **b :** not seized as security for a legal judgment
2 : not joined or united ⟨*unattached* buildings⟩

un·at·trac·tive \-ˈtrak-tiv\ *adjective* (circa 1775)
: not attractive **:** PLAIN, DULL
— un·at·trac·tive·ly *adverb*
— un·at·trac·tive·ness *noun*

un·avail·ing \-ˈvā-liŋ\ *adjective* (1670)
: not availing **:** FUTILE, USELESS
— un·avail·ing·ly \-liŋ-lē\ *adverb*
— un·avail·ing·ness *noun*

un·av·er·age \ˌən-ˈa-v(ə-)rij\ *adjective* (1962)
: not average **:** UNUSUAL, OUTSTANDING

un·avoid·able \ˌən-ə-ˈvȯi-də-bəl\ *adjective* (1577)
: not avoidable **:** INEVITABLE
— un·avoid·ably \-blē\ *adverb*

[1]**un·aware** \-ə-ˈwar, -ˈwer\ *adverb* (1592)
: UNAWARES

[2]**unaware** *adjective* (1704)
: not aware **:** IGNORANT
— un·aware·ly *adverb*
— un·aware·ness *noun*

un·awares \-ˈwarz, -ˈwerz\ *adverb* [*un-* + *aware* + *-s*, adverb suffix, from Middle English, from *-s*, genitive singular ending of nouns — more at -S] (1535)
1 : without design, attention, preparation, or premeditation
2 : without warning **:** SUDDENLY, UNEXPECTEDLY

un·backed \ˌən-ˈbakt\ *adjective* (1609)
: lacking support or aid

[1]**un·bal·ance** \-ˈba-lən(t)s\ *transitive verb* (1856)
: to put out of balance; *especially* **:** to derange mentally

[2]**unbalance** *noun* (1887)
: lack of balance **:** IMBALANCE

un·bal·anced \-lən(t)st\ *adjective* (1650)
: not balanced: as **a :** not in equilibrium **b :** mentally disordered or deranged **c :** not adjusted so as to make credits equal to debits ⟨an *unbalanced* account⟩

un·bal·last·ed \-ˈba-lə-stəd\ *adjective* (1657)
: not furnished with or steadied by ballast **:** UNSTEADY

un·ban·dage \-ˈban-dij\ *transitive verb* (1840)
: to remove a bandage from

un·bar \ˌən-ˈbär\ *transitive verb* (14th century)
: to remove a bar from **:** UNBOLT, OPEN

un·bar·bered \-'bär-bərd\ *adjective* (1845)
: having long and especially unkempt hair

un·barred \-'bärd\ *adjective* (1603)
1 : not secured by a bar **:** UNLOCKED
2 : not marked with bars

un·bat·ed \-'bā-təd\ *adjective* (1596)
1 : UNABATED
2 *archaic* **:** not blunted

un·be \-'bē\ *intransitive verb* (15th century)
archaic **:** to lack or cease to have being

un·bear·able \,ən-'bar-ə-bəl, -'ber-\ *adjective* (15th century)
: not bearable **:** UNENDURABLE
— **un·bear·ably** \-blē\ *adverb*

un·beat·able \-'bē-tə-bəl\ *adjective* (1897)
1 : not capable of being defeated
2 : possessing unsurpassable qualities
— **un·beat·ably** \-blē\ *adverb*

un·beat·en \-'bē-tᵊn\ *adjective* (13th century)
1 : not pounded or beaten **:** not whipped
2 : not traversed **:** UNTROD
3 : not defeated

un·beau·ti·ful \-'byü-ti-fəl\ *adjective* (15th century)
: not beautiful **:** UNATTRACTIVE
— **un·beau·ti·ful·ly** \-f(ə-)lē\ *adverb*

un·be·com·ing \,ən-bi-'kə-miŋ\ *adjective* (1598)
: not becoming ⟨an *unbecoming* dress⟩; *especially* **:** not according with the standards appropriate to one's position or condition of life ⟨*unbecoming* conduct⟩
synonym see INDECOROUS
— **un·be·com·ing·ly** \-miŋ-lē\ *adverb*
— **un·be·com·ing·ness** *noun*

un·be·knownst \,ən-bi-'nōn(t)st\ *also* **un·be·known** \-'nōn\ *adjective* [[1]*un-* + obsolete English *beknown* known; *unbeknownst,* irregular from *unbeknown*] (1636)
: happening without one's knowledge **:** UNKNOWN — usually used with *to*

un·be·lief \,ən-bə-'lēf\ *noun* (12th century)
: incredulity or skepticism especially in matters of religious faith

un·be·liev·able \-'lē-və-bəl\ *adjective* (1548)
: too improbable for belief ⟨the plot is unreal and *unbelievable*⟩; *also* **:** of such a superlative degree as to be hard to believe ⟨the destruction was *unbelievable*⟩ ⟨made an *unbelievable* catch in center field⟩
— **un·be·liev·ably** \-blē\ *adverb*

un·be·liev·er \-'lē-vər\ *noun* (1526)
1 : one that does not believe **:** an incredulous person **:** DOUBTER, SKEPTIC
2 : one that does not believe in a particular religious faith

un·be·liev·ing \-'lē-viŋ\ *adjective* (14th century)
: marked by unbelief **:** INCREDULOUS, SKEPTICAL
— **un·be·liev·ing·ly** \-viŋ-lē\ *adverb*

un·belt·ed \,ən-'bel-təd\ *adjective* (1814)
: not furnished with a belt

un·bend \-'bend\ *verb* **-bent** \-'bent\; **-bend·ing** (13th century)
transitive verb
1 : to free from flexure **:** make or allow to become straight ⟨*unbend* a bow⟩
2 : to cause (as the mind) to relax
3 a : to unfasten (as a sail) from a spar or stay **b :** to cast loose (as a rope) **:** UNTIE
intransitive verb
1 : to relax one's severity, stiffness, or austerity
2 : to cease to be bent **:** become straight

un·bend·able \-'ben-də-bəl\ *adjective* (circa 1775)
: SINGLE-MINDED, FIRM

un·bend·ing \-'ben-diŋ\ *adjective* [[1]*un-*] (circa 1688)
1 : not bending **:** UNYIELDING, INFLEXIBLE ⟨an *unbending* will⟩
2 : aloof or unsocial in manner **:** RESERVED

un·be·seem·ing \,ən-bi-'sē-miŋ\ *adjective* (1583)
: not befitting **:** UNBECOMING

un·bi·ased \,ən-'bī-əst\ *adjective* (1647)
1 : free from bias; *especially* **:** free from all prejudice and favoritism **:** eminently fair
2 : having an expected value equal to a population parameter being estimated ⟨an *unbiased* estimate of the population mean⟩
synonym see FAIR
— **un·bi·ased·ness** \-əs(t)-nəs\ *noun*

un·bib·li·cal \,ən-'bi-bli-kəl\ *adjective* (1828)
: contrary to or unsanctioned by the Bible

un·bid·den \-'bi-dᵊn\ *also* **un·bid** \-'bid\ *adjective* [Middle English *unbiden, unbeden,* from Old English *unbeden,* from *un-* + *beden,* past participle of *biddan* to entreat — more at BID] (before 12th century)
: not bidden **:** UNASKED, UNINVITED

un·bind \-'bīnd\ *transitive verb* **-bound** \-'bau̇nd\; **-bind·ing** (before 12th century)
1 : to remove a band from **:** free from fastenings **:** UNTIE, UNFASTEN
2 : to set free **:** RELEASE

un·bit·ted \-'bi-təd\ *adjective* [[2]*bit*] (circa 1586)
: UNBRIDLED, UNCONTROLLED

un·blenched \-'blencht\ *adjective* (1634)
: not disconcerted **:** UNDAUNTED

un·blessed *also* **un·blest** \,ən-'blest\ *adjective* (14th century)
1 : not blessed
2 : EVIL, ACCURSED

un·blind·ed \-'blīn-dəd\ *adjective* (circa 1611)
: not blinded; *especially* **:** free from illusion

un·blink·ing \-'bliŋ-kiŋ\ *adjective* (circa 1909)
1 : not blinking
2 : not showing signs of emotion, doubt, or confusion
— **un·blink·ing·ly** \-kiŋ-lē\ *adverb*

un·block \-'bläk\ *transitive verb* (1611)
: to free from being blocked

un·blush·ing \-'blə-shiŋ\ *adjective* (1595)
1 : not blushing
2 : SHAMELESS, UNABASHED
— **un·blush·ing·ly** \-shiŋ-lē\ *adverb*

un·bod·ied \-'bä-dēd\ *adjective* (1532)
1 : having no body **:** INCORPOREAL; *also* **:** freed from the body ⟨*unbodied* souls⟩
2 : FORMLESS

un·bolt \,ən-'bōlt\ *transitive verb* (circa 1598)
: to open or unfasten by withdrawing a bolt

[1]un·bolt·ed \-'bōl-təd\ *adjective* [[2]*bolt*] (circa 1580)
: not fastened by bolts

[2]unbolted *adjective* [[4]*bolt*] (1598)
: not sifted ⟨*unbolted* flour⟩

un·bon·net·ed \,ən-'bä-nə-təd\ *adjective* (1604)
: BAREHEADED

un·born \-'bȯrn\ *adjective* (before 12th century)
1 : not born **:** not brought into life
2 : still to appear **:** FUTURE
3 : existing without birth

un·bos·om \-'bu̇-zəm *also* -'bü-\ (1588)
transitive verb
1 : to give expression to **:** DISCLOSE, REVEAL
2 : to disclose the thoughts or feelings of (oneself)
intransitive verb
: to unbosom oneself

un·bound \-'bau̇nd\ *adjective* (before 12th century)
: not bound: as **a** (1) **:** not fastened (2) **:** not confined **b :** not having the leaves fastened together ⟨an *unbound* book⟩ **c :** not bound together with other issues ⟨*unbound* periodicals⟩ **d :** not held in chemical or physical combination

un·bound·ed \-'bau̇n-dəd\ *adjective* (1593)
1 : having no limit
2 : UNRESTRAINED, UNCONTROLLED
— **un·bound·ed·ness** *noun*

un·bowed \,ən-'bau̇d\ *adjective* (14th century)
1 : not bowed down
2 : not subdued ⟨bloodied but *unbowed*⟩

un·box \-'bäks\ *transitive verb* (1611)
: to remove from a box

un·brace \-'brās\ *transitive verb* (1593)
1 : to free or detach by or as if by untying or removing a brace or bond
2 : ENFEEBLE, WEAKEN

un·braid \-'brād\ *transitive verb* (circa 1828)
: to separate the strands of **:** UNRAVEL

un·branched \-'brancht\ *adjective* (1665)
1 : having no branches ⟨a straight *unbranched* trunk⟩
2 : not divided into branches ⟨a leaf with *unbranched* veins⟩

un·brand·ed \-'bran-dəd\ *adjective* (1886)
1 : not marked with the owner's name or mark ⟨*unbranded* cattle⟩
2 : not sold under a brand name

un·breath·able \-'brē-t͟hə-bəl\ *adjective* (1846)
: not fit for being breathed

un·bred \-'bred\ *adjective* (1662)
1 *obsolete* **:** ILL-BRED
2 : not taught **:** UNTRAINED
3 : not bred **:** never having been bred ⟨an *unbred* heifer⟩

un·bri·dle \,ən-'brī-dᵊl\ *transitive verb* (15th century)
: to free or loose from a bridle; *broadly* **:** to set loose **:** free from restraint

un·bri·dled \-'brī-dᵊld\ *adjective* (14th century)
1 : UNRESTRAINED ⟨*unbridled* enthusiasm⟩
2 : not confined by a bridle

un·broke \-'brōk\ *adjective* (14th century)
: UNBROKEN

un·bro·ken \-'brō-kən\ *adjective* (14th century)
: not broken: as **a :** not violated **b :** WHOLE, INTACT **c :** not subdued **:** UNTAMED; *especially* **:** not trained for service or use ⟨*unbroken* colts⟩ **d :** CONTINUOUS ⟨miles of *unbroken* forest⟩ **e :** not plowed **f :** not disorganized ⟨advanced in *unbroken* ranks⟩

un·buck·le \-'bə-kəl\ (14th century)
transitive verb
: to loose the buckle of **:** UNFASTEN
intransitive verb
1 : to loosen buckles
2 : RELAX

un·budge·able \-'bə-jə-bəl\ *adjective* (1929)
: not able to be budged or changed **:** INFLEXIBLE
— **un·budge·ably** \-blē\ *adverb*

un·budg·ing \-'bə-jiŋ\ *adjective* (circa 1934)
: not budging **:** resisting movement or change
— **un·budg·ing·ly** \-jiŋ-lē\ *adverb*

un·build \,ən-'bild\ *verb* **-built** \-'bilt\; **-build·ing** (1607)
transitive verb
: to pull down **:** DEMOLISH, RAZE
intransitive verb
: to destroy something

un·built \-'bilt\ *adjective* (15th century)
1 : not built **:** not yet constructed
2 : not built on ⟨an *unbuilt* plot⟩

un·bun·dle \-'bən-dᵊl\ (1969)
intransitive verb
: to give separate prices for equipment and supporting services
transitive verb
: to price separately

un·bur·den \-'bər-dᵊn\ *transitive verb* (1538)
1 : to free or relieve from a burden
2 : to relieve oneself of (as cares, fears, or worries) **:** cast off

un·bur·dened \-'bər-dᵊnd\ *adjective* (1548)
: not burdened **:** having no weight or load

un·but·ton \-'bə-tᵊn\ (14th century)
transitive verb
1 : to loose the buttons of
2 : to open by or as if by loosing buttons

intransitive verb
: to undo buttons

un·but·toned \-tᵊnd\ *adjective* (1583)
1 a : not buttoned **b :** not provided with buttons
2 : not under constraint **:** free and unrestricted in action and expression

un·cage \ˌən-ˈkāj\ *transitive verb* (1620)
: to release from or as if from a cage **:** free from restraint

un·cal·cu·lat·ed \-ˈkal-kyə-ˌlā-təd\ *adjective* (circa 1828)
: not planned or thought out beforehand **:** SPONTANEOUS

un·cal·cu·lat·ing \-ˌlā-tiŋ\ *adjective* (circa 1828)
: not based on or marked by calculation

un·called–for \ˌən-ˈkȯl(d)-ˌfȯr\ *adjective* (circa 1656)
1 : not called for or needed **:** UNNECESSARY
2 : being or offered without provocation or justification ⟨an *uncalled-for* display of temper⟩ ⟨*uncalled-for* insults⟩

un·can·ny \-ˈka-nē\ *adjective* (1843)
1 a : seeming to have a supernatural character or origin **:** EERIE, MYSTERIOUS **b :** being beyond what is normal or expected **:** suggesting superhuman or supernatural powers ⟨an *uncanny* sense of direction⟩
2 *chiefly Scottish* **:** SEVERE, PUNISHING
synonym see WEIRD
— un·can·ni·ly \-ˈka-nᵊl-ē\ *adverb*
— un·can·ni·ness \-ˈka-nē-nəs\ *noun*

un·caused \-ˈkȯzd\ *adjective* (circa 1628)
: having no antecedent cause

un·ceas·ing \-ˈsē-siŋ\ *adjective* (14th century)
: never ceasing **:** CONTINUOUS, INCESSANT
— un·ceas·ing·ly \-siŋ-lē\ *adverb*

un·cel·e·brat·ed \-ˈse-lə-ˌbrā-təd\ *adjective* (1660)
1 : not formally honored or commemorated
2 : not famous **:** OBSCURE

un·cer·e·mo·ni·ous \ˌən-ˌser-ə-ˈmō-nē-əs\ *adjective* (1598)
1 : not ceremonious **:** INFORMAL
2 : ABRUPT, RUDE ⟨an *unceremonious* dismissal⟩
— un·cer·e·mo·ni·ous·ly *adverb*
— un·cer·e·mo·ni·ous·ness *noun*

un·cer·tain \ˌən-ˈsər-tᵊn\ *adjective* (14th century)
1 : INDEFINITE, INDETERMINATE ⟨the time of departure is *uncertain*⟩
2 : not certain to occur **:** PROBLEMATICAL
3 : not reliable **:** UNTRUSTWORTHY
4 a : not known beyond doubt **:** DUBIOUS **b :** not having certain knowledge **:** DOUBTFUL **c :** not clearly identified or defined
5 : not constant **:** VARIABLE, FITFUL
— un·cer·tain·ly *adverb*
— un·cer·tain·ness \-tᵊn-(n)əs\ *noun*

un·cer·tain·ty \-tᵊn-tē\ *noun* (14th century)
1 : the quality or state of being uncertain **:** DOUBT
2 : something that is uncertain ☆

uncertainty principle *noun* (1929)
: a principle in quantum mechanics: it is impossible to discern simultaneously and with high accuracy both the position and the momentum of a particle (as an electron) — called also *Heisenberg uncertainty principle*

un·chain \ˌən-ˈchān\ *transitive verb* (1582)
: to free by or as if by removing a chain **:** set loose

un·chancy \-ˈchan(t)-sē\ *adjective* (1533)
1 *chiefly Scottish* **:** ILL-FATED
2 *chiefly Scottish* **:** DANGEROUS

un·change·able \-ˈchān-jə-bəl\ *adjective* (14th century)
: not changing or to be changed **:** IMMUTABLE
— un·change·abil·i·ty \ˌən-ˌchān-jə-ˈbi-lə-tē\ *noun*
— un·change·able·ness \ˌən-ˈchān-jə-bəl-nəs\ *noun*
— un·chan·ge·ably \-blē\ *adverb*

un·chang·ing \-ˈchān-jiŋ\ *adjective* (1593)
: CONSTANT, INVARIABLE
— un·chang·ing·ly \-jiŋ-lē\ *adverb*
— un·chang·ing·ness *noun*

un·char·ac·ter·is·tic \ˌən-ˌkar-ik-tə-ˈris-tik\ *adjective* (1753)
: not characteristic **:** not typical or distinctive
— un·char·ac·ter·is·ti·cal·ly \-ti-k(ə-)lē\ *adverb*

un·charge \ˌən-ˈchärj\ *transitive verb* (1602) *obsolete* **:** ACQUIT

un·charged \-ˈchärjd\ *adjective* (1815)
: not charged; *specifically* **:** having no electric charge

un·char·i·ta·ble \-ˈchar-ə-tə-bəl\ *adjective* (15th century)
: lacking in charity **:** severe in judging **:** HARSH
— un·char·i·ta·ble·ness *noun*
— un·char·i·ta·bly \-blē\ *adverb*

un·chart·ed \-ˈchär-təd\ *adjective* (circa 1847)
: not recorded or plotted on a map, chart, or plan; *broadly* **:** UNKNOWN ⟨*uncharted* territory⟩

un·chaste \-ˈchāst\ *adjective* (14th century)
: not chaste **:** lacking in chastity
— un·chaste·ly *adverb*
— un·chaste·ness \-ˈchās(t)-nəs\ *noun*

un·chas·ti·ty \-ˈchas-tə-tē\ *noun* (14th century)
: the quality or state of being unchaste

un·chiv·al·rous \-ˈshi-vəl-rəs\ *adjective* (circa 1846)
: not chivalrous **:** lacking in chivalry
— un·chiv·al·rous·ly *adverb*

un·choke \-ˈchōk\ *transitive verb* (1588)
: to clear of obstruction

un·chris·tian \-ˈkris-chən, -ˈkrish-\ *adjective* (1555)
1 : not of the Christian faith
2 a : contrary to the Christian spirit or character **b :** UNCIVILIZED, BARBAROUS

un·church \-ˈchərch\ *transitive verb* (circa 1620)
1 : to expel from a church **:** EXCOMMUNICATE
2 : to deprive of a church or of status as a church

un·churched \-ˈchərcht\ *adjective* (1681)
: not belonging to or connected with a church

unci *plural of* UNCUS

¹un·cial \ˈən(t)-shəl, -sē-əl\ *adjective* [Late Latin *unciales* (*litterae*) uncial (letters), from Latin, plural of *uncialis* weighing an ounce, from *uncia* twelfth part, ounce — more at OUNCE] (1712)
: written in the style or size of uncials
— un·cial·ly *adverb*

²uncial *noun* (circa 1775)
1 : a handwriting used especially in Greek and Latin manuscripts of the 4th to the 8th centuries A.D. and made with somewhat rounded separated majuscules but having cursive forms for some letters
2 : an uncial letter
3 : a manuscript written in uncial

ROMAN UNCIAL

uncials

un·ci·form \ˈən(t)-sə-ˌfȯrm\ *adjective* [New Latin *unciformis,* from Latin *uncus* hook + *-formis* -form — more at ANGLE] (circa 1734)
: hook-shaped **:** UNCINATE

un·ci·na·ri·a·sis \ˌən-ˌsi-nə-ˈrī-ə-səs\ *noun* [New Latin, from *Uncinaria,* genus that includes hookworms, from Latin *uncinus* hook] (1903)
: ANCYLOSTOMIASIS

un·ci·nate \ˈən(t)-sə-ˌnāt\ *adjective* [Latin *uncinatus,* from *uncinus* hook, from *uncus*] (circa 1760)
: bent at the tip like a hook **:** HOOKED ⟨an *uncinate* achene⟩

un·cir·cu·lat·ed \ˌən-ˈsər-kyə-ˌlā-təd\ *adjective* (1917)
: issued for use as money but kept out of circulation (as for preservation in a collection)

un·cir·cum·cised \ˌən-ˈsər-kəm-ˌsīzd\ *adjective* (14th century)
1 : not circumcised
2 : spiritually impure **:** HEATHEN
— un·cir·cum·ci·sion \ˌən-ˌsər-kəm-ˈsi-zhən\ *noun*

un·civ·il \ˌən-ˈsi-vəl\ *adjective* (1553)
1 : not civilized **:** BARBAROUS
2 : lacking in courtesy **:** ILL-MANNERED, IMPOLITE
3 : not conducive to civic harmony and welfare
— un·civ·il·ly \-və-lē\ *adverb*

un·civ·i·lized \-ˈsi-və-ˌlīzd\ *adjective* (1607)
1 : not civilized **:** BARBAROUS
2 : remote from settled areas **:** WILD

un·clamp \-ˈklamp\ *transitive verb* (1809)
: to loosen the clamp of **:** to free from a clamp

un·clar·i·ty \-ˈklar-ə-tē\ *noun, plural* **-ties** (1923)
: lack of clarity **:** AMBIGUITY, OBSCURITY

un·clasp \-ˈklasp\ (1530)
transitive verb
1 : to open the clasp of
2 : to open or cause to be opened (as a clenched hand)
intransitive verb
: to loosen a hold

un·clas·si·cal \-ˈkla-si-kəl\ *adjective* (1725)
: not classical; *especially* **:** unconcerned with the classics

un·clas·si·fied \-ˈkla-sə-ˌfīd\ *adjective* (1865)
1 : not placed or belonging in a class
2 : not subject to a security classification

un·cle \ˈəŋ-kəl\ *noun* [Middle English, from Old French, from Latin *avunculus* mother's brother; akin to Old English *ēam* uncle, Welsh *ewythr* uncle, Latin *avus* grandfather] (14th century)
1 a : the brother of one's father or mother **b :** the husband of one's aunt
2 : one who helps, advises, or encourages
3 — used as a cry of surrender ⟨was forced to cry *uncle*⟩
4 *capitalized* **:** UNCLE SAM

un·clean \ˌən-ˈklēn\ *adjective* (before 12th century)
1 : morally or spiritually impure
2 : infected with a harmful supernatural contagion; *also* **:** prohibited by ritual law for use or contact
3 : DIRTY, FILTHY
4 : lacking in clarity and precision of conception or execution
— un·clean·ness \-ˈklēn-nəs\ *noun*

¹un·clean·ly \-ˈklen-lē\ *adjective* (before 12th century)
: morally or physically unclean
— un·clean·li·ness *noun*

²un·clean·ly \-ˈklēn-lē\ *adverb* (before 12th century)
: in an unclean manner

un·clench \-ˈklench\ (1699)

☆ SYNONYMS

Uncertainty, doubt, dubiety, skepticism, suspicion, mistrust mean lack of sureness about someone or something. UNCERTAINTY may range from a falling short of certainty to an almost complete lack of conviction or knowledge especially about an outcome or result ⟨assumed the role of manager without hesitation or *uncertainty*⟩. DOUBT suggests both uncertainty and inability to make a decision ⟨plagued by *doubts* as to what to do⟩. DUBIETY stresses a wavering between conclusions ⟨felt some *dubiety* about its practicality⟩. SKEPTICISM implies unwillingness to believe without conclusive evidence ⟨an economic forecast greeted with *skepticism*⟩. SUSPICION stresses lack of faith in the truth, reality, fairness, or reliability of something or someone ⟨regarded the stranger with *suspicion*⟩. MISTRUST implies a genuine doubt based upon suspicion ⟨had a great *mistrust* of doctors⟩.

transitive verb
1 : to open from a clenched position
2 : to release from a grip
intransitive verb
: to become unclasped or relaxed

Un·cle Sam \ˌəŋ-kəl-ˈsam\ *noun* [expansion of *U.S.*, abbreviation of *United States*] (1813)
1 : the U.S. government
2 : the American nation or people

[1]Uncle Tom \-ˈtäm\ *noun* [*Uncle Tom*, pious and faithful black slave in *Uncle Tom's Cabin* (1851–52) by Harriet Beecher Stowe] (1922)
1 : a black who is overeager to win the approval of whites (as by obsequious behavior or uncritical acceptance of white values and goals)
2 : a member of a low-status group who is overly subservient to or cooperative with authority ⟨the worst floor managers and supervisors by far are women . . . Some of them are regular *Uncle Toms* Jane Fonda⟩
— **Uncle Tom·ism** \-ˈtä-ˌmi-zəm\ *noun*

[2]Uncle Tom *intransitive verb* **Uncle Tommed; Uncle Tom·ming** (1947)
: to behave like an Uncle Tom

un·climb·able \ˌən-ˈklī-mə-bəl\ *adjective* (1553)
: not able to be climbed
— **un·climb·able·ness** *noun*

un·clinch \ˌən-ˈklinch\ *transitive verb* (1598)
: UNCLENCH

un·cloak \-ˈklōk\ (1598)
transitive verb
1 : to remove a cloak or cover from
2 : REVEAL, UNMASK
intransitive verb
: to take off a cloak

un·clog \-ˈkläg\ *transitive verb* (1607)
: to free from a difficulty or obstruction

un·close \-ˈklōz\ (14th century)
transitive verb
1 : OPEN
2 : DISCLOSE, REVEAL
intransitive verb
: to become opened

un·closed \-ˈklōzd\ *adjective* (15th century)
: not closed or settled **:** not concluded

un·clothe \-ˈklōt͟h\ *transitive verb* (14th century)
1 : to strip of clothes
2 : DIVEST, UNCOVER

un·clothed \-ˈklōt͟hd\ *adjective* (15th century)
: not clothed

un·cloud·ed \-ˈklau̇-dəd\ *adjective* (1594)
: not covered by clouds **:** not darkened or obscured **:** CLEAR
— **un·cloud·ed·ly** *adverb*

un·club·ba·ble \ˌən-ˈklə-bə-bəl\ *adjective* (circa 1764)
: UNSOCIABLE 1

un·clut·ter \-ˈklə-tər\ *transitive verb* (1930)
: to remove clutter from **:** make neat and orderly

un·clut·tered \-tərd\ *adjective* (1925)
: not cluttered

[1]un·co \ˈəŋ-(ˌ)kō, -kə\ *adjective* [Middle English (Scots) *unkow*, alteration of Middle English *uncouth*] (15th century)
1 *chiefly Scottish* **a :** STRANGE, UNKNOWN **b :** UNCANNY, WEIRD
2 *chiefly Scottish* **:** EXTRAORDINARY

[2]unco *adverb* (1724)
: EXTREMELY, REMARKABLY, UNCOMMONLY

[3]unco *noun, plural* **uncos** (1785)
1 *plural, chiefly Scottish* **:** NEWS, TIDINGS
2 *chiefly Scottish* **:** STRANGER

un·cock \ˌən-ˈkäk\ *transitive verb* (circa 1775)
: to remove the hammer of (a firearm) from a cocked position

un·cof·fin \-ˈkȯ-fən\ *transitive verb* (1836)
: to remove from or as if from a coffin

un·cof·fined \-fənd\ *adjective* (1648)
: not placed in a coffin

un·coil \ˌən-ˈkȯi(ə)l\ (1713)
transitive verb
: to release from a coiled state **:** UNWIND
intransitive verb
: to become uncoiled

un·coiled \-ˈkȯi(ə)ld\ *adjective* (1713)
: not coiled

un·coined \-ˈkȯind\ *adjective* (15th century)
1 : not minted ⟨*uncoined* metal⟩
2 : not fabricated **:** NATURAL

un·com·fort·able \ˌən-ˈkəm(p)(f)-tə(r)-bəl, -ˈkəm(p)-fə(r)-tə-bəl, -ˈkəm-fə(r)-bəl\ *adjective* (1592)
1 : causing discomfort or annoyance ⟨an *uncomfortable* chair⟩ ⟨an *uncomfortable* performance⟩
2 : feeling discomfort **:** UNEASY ⟨was *uncomfortable* with them⟩
— **un·com·fort·ably** \-blē\ *adverb*

un·com·mer·cial \ˌən-kə-ˈmər-shəl\ *adjective* (1768)
1 : not engaged in or related to commerce
2 : not based on commercial principles
3 : not likely to result in financial success ⟨an *uncommercial* book⟩

un·com·mit·ted \ˌən-kə-ˈmi-təd\ *adjective* (1598)
: not committed; *specifically* **:** not pledged to a particular belief, allegiance, or program

un·com·mon \ˌən-ˈkä-mən\ *adjective* (1611)
1 : not ordinarily encountered **:** UNUSUAL
2 : REMARKABLE, EXCEPTIONAL
synonym see INFREQUENT
— **un·com·mon·ly** *adverb*
— **un·com·mon·ness** \-mən-nəs\ *noun*

un·com·mu·ni·ca·ble \ˌən-kə-ˈmyü-ni-kə-bəl\ *adjective* (14th century)
: INCOMMUNICABLE

un·com·mu·ni·ca·tive \-ˈmyü-nə-ˌkā-tiv, -ni-kə-tiv\ *adjective* (1691)
: not disposed to talk or impart information **:** RESERVED

un·com·pet·i·tive \-ˈpe-tə-tiv\ *adjective* (1885)
: not competitive **:** unable to compete
— **un·com·pet·i·tive·ness** *noun*

un·com·plain·ing \-ˈplā-niŋ\ *adjective* (1744)
: not complaining **:** PATIENT
— **un·com·plain·ing·ly** \-niŋ-lē\ *adverb*

un·com·pli·cat·ed \ˌən-ˈkäm-plə-ˌkā-təd\ *adjective* (circa 1775)
1 a : not complicated by something outside itself **b :** not involving medical complications ⟨an *uncomplicated* peptic ulcer⟩
2 : not complex **:** SIMPLE ⟨*uncomplicated* machinery⟩

un·com·pli·men·ta·ry \ˌən-ˌkäm-plə-ˈmen-tə-rē, -ˈmen-trē\ *adjective* (1842)
: not complimentary **:** DEROGATORY

un·com·pre·hend·ing \-pri-ˈhen-diŋ\ *adjective* (1838)
: not comprehending **:** lacking understanding
— **un·com·pre·hend·ing·ly** \-diŋ-lē\ *adverb*

un·com·pro·mis·able \ˌən-ˈkäm-prə-ˌmī-zə-bəl\ *adjective* (1958)
: not able to be compromised

un·com·pro·mis·ing \-ˌmī-ziŋ\ *adjective* (circa 1828)
: not making or accepting a compromise **:** making no concessions **:** INFLEXIBLE, UNYIELDING
— **un·com·pro·mis·ing·ly** \-ziŋ-lē\ *adverb*
— **un·com·pro·mis·ing·ness** \-nəs\ *noun*

un·con·ceiv·able \ˌən-kən-ˈsē-və-bəl\ *adjective* (circa 1607)
: INCONCEIVABLE

un·con·cern \ˌən-kən-ˈsərn\ *noun* (1711)
1 : lack of care or interest **:** INDIFFERENCE
2 : freedom from excessive concern or anxiety

un·con·cerned \-ˈsərnd\ *adjective* (circa 1635)
1 : not involved **:** not having any part or interest
2 : not anxious or upset **:** free of worry
synonym see INDIFFERENT
— **un·con·cerned·ly** \-ˈsər-nəd-lē, -ˈsərnd-lē\ *adverb*
— **un·con·cerned·ness** \-ˈsər-nəd-nəs, -ˈsərn(d)-nəs\ *noun*

un·con·di·tion·al \ˌən-kən-ˈdish-nəl, -ˈdi-shə-nᵊl\ *adjective* (1666)
1 : not conditional or limited **:** ABSOLUTE, UNQUALIFIED ⟨*unconditional* surrender⟩
2 : UNCONDITIONED 2
— **un·con·di·tion·al·ly** *adverb*

un·con·di·tioned \-ˈdi-shənd\ *adjective* (circa 1631)
1 : not subject to conditions or limitations
2 a : not dependent on or subjected to conditioning or learning **:** NATURAL ⟨*unconditioned* responses⟩ **b :** producing an unconditioned response ⟨*unconditioned* stimuli⟩

un·con·form·able \-ˈfȯr-mə-bəl\ *adjective* (1594)
1 : not conforming
2 : exhibiting geological unconformity
— **un·con·form·ably** \-blē\ *adverb*

un·con·for·mi·ty \-ˈfȯr-mə-tē\ *noun* (circa 1600)
1 *archaic* **:** lack of conformity
2 a : lack of continuity in deposition between rock strata in contact corresponding to a period of nondeposition, weathering, or erosion **b :** the surface of contact between unconformable strata

un·con·ge·nial \-ˈjē-nyəl, -nē-əl\ *adjective* (circa 1775)
1 : not sympathetic or compatible ⟨*uncongenial* roommates⟩
2 a : not fitted **:** UNSUITABLE ⟨a soil *uncongenial* to most crops⟩ **b :** not to one's taste **:** DISAGREEABLE ⟨an *uncongenial* task⟩
— **un·con·ge·nial·i·ty** \-ˌjē-nē-ˈa-lə-tē, -ˌjēn-ˈya-\ *noun*

un·con·quer·able \ˌən-ˈkäŋ-k(ə-)rə-bəl\ *adjective* (1598)
1 : incapable of being conquered **:** INDOMITABLE ⟨an *unconquerable* will⟩
2 : incapable of being surmounted ⟨*unconquerable* difficulties⟩
— **un·con·quer·ably** \-blē\ *adverb*

un·con·scio·na·ble \-ˈkän(t)-sh(ə-)nə-bəl\ *adjective* (1570)
1 : not guided or controlled by conscience **:** UNSCRUPULOUS ⟨an *unconscionable* villain⟩
2 a : EXCESSIVE, UNREASONABLE ⟨found an *unconscionable* number of defects in the car⟩ **b :** shockingly unfair or unjust ⟨*unconscionable* sales practices⟩
— **un·con·scio·na·bil·i·ty** \ˌən-ˌkän(t)-sh(ə-)nə-ˈbi-lə-tē\ *noun*
— **un·con·scio·na·ble·ness** \ˌən-ˈkän(t)-sh(ə-)nə-bəl-nəs\ *noun*
— **un·con·scio·na·bly** \-blē\ *adverb*

[1]un·con·scious \ˌən-ˈkän(t)-shəs\ *adjective* (1712)
1 a : not knowing or perceiving **:** not aware **b :** free from self-awareness
2 a : not possessing mind or consciousness ⟨*unconscious* matter⟩ **b** (1) **:** not marked by conscious thought, sensation, or feeling ⟨*unconscious* motivation⟩ (2) **:** of or relating to the unconscious **c :** having lost consciousness ⟨was *unconscious* for three days⟩
3 : not consciously held or deliberately planned or carried out ⟨*unconscious* bias⟩
— **un·con·scious·ly** *adverb*
— **un·con·scious·ness** *noun*

[2]unconscious *noun* (1912)
: the part of the psychic apparatus that does not ordinarily enter the individual's awareness and that is manifested especially by slips of the tongue or dissociated acts or in dreams

un·con·sid·ered \ˌən-kən-ˈsi-dərd\ *adjective* (1587)
1 : not considered or worth consideration
2 : not resulting from consideration

\ə\ **abut** \ᵊ\ **kitten** \ər\ **further** \a\ **ash** \ā\ **ace** \ä\ **mop, mar** \au̇\ **out** \ch\ **chin** \e\ **bet** \ē\ **easy** \g\ **go** \i\ **hit** \ī\ **ice** \j\ **job** \ŋ\ **sing** \ō\ **go** \ȯ\ **law** \ȯi\ **boy** \th\ **thin** \t͟h\ **the** \ü\ **loot** \u̇\ **foot** \y\ **yet** \zh\ **vision** *see also* Guide to Pronunciation

un·con·sol·i·dat·ed \-ˈsä-lə-ˌdā-təd\ *adjective* (circa 1775)
: loosely arranged ⟨*unconsolidated* subsidiaries⟩; *especially* : not stratified ⟨*unconsolidated* soil⟩

un·con·sti·tu·tion·al \ˌən-ˌkän(t)-stə-ˈtü-shnəl, -ˈtyü-, -shə-nᵊl\ *adjective* (1734)
: not according or consistent with the constitution of a body politic (as a nation)
— **un·con·sti·tu·tion·al·i·ty** \-ˌtü-shə-ˈna-lə-tē, -ˌtyü-\ *noun*
— **un·con·sti·tu·tion·al·ly** *adverb*

un·con·straint \ˌən-kən-ˈstrānt\ *noun* (1711)
: freedom from constraint : EASE

un·con·struct·ed \-ˈstrək-təd\ *adjective* (1970)
of clothing : manufactured without added material for padding, stiffening, or shape retention

un·con·trol·la·ble \-ˈtrō-lə-bəl\ *adjective* (1593)
1 *archaic* : free from control by a superior power : ABSOLUTE
2 : incapable of being controlled : UNGOVERNABLE
— **un·con·trol·la·bil·i·ty** \-ˌtrō-lə-ˈbi-lə-tē\ *noun*
— **un·con·trol·la·bly** \-ˈtrō-lə-blē\ *adverb*

un·con·ven·tion·al \-ˈvench-nəl, -ˈven(t)-shə-nᵊl\ *adjective* (1839)
: not conventional : not bound by or in accordance with convention : being out of the ordinary
— **un·con·ven·tion·al·i·ty** \-ˌven(t)-shə-ˈna-lə-tē\ *noun*
— **un·con·ven·tion·al·ly** *adverb*

un·con·vinc·ing \-ˈvin(t)-siŋ\ *adjective* (1653)
: not convincing : IMPLAUSIBLE
— **un·con·vinc·ing·ly** \-siŋ-lē\ *adverb*
— **un·con·vinc·ing·ness** *noun*

un·cool \ˌən-ˈkül\ *adjective* (1953)
1 : lacking in assurance
2 : failing to accord with the current styles (as of dress or behavior) of a particular group

un·cork \ˌən-ˈkȯrk\ *transitive verb* (1727)
1 : to draw a cork from
2 a : to release from a sealed or pent-up state ⟨*uncork* a surprise⟩ **b** : to let go : RELEASE ⟨*uncork* a wild pitch⟩

un·corked \-ˈkȯrkt\ *adjective* (1791)
: not provided with a cork

un·cor·set·ed \-ˈkȯr-sə-təd\ *adjective* (1856)
1 : not wearing a corset
2 : not controlled or inhibited ⟨*uncorseted* freedom⟩

un·count·ed \-ˈkau̇n-təd\ *adjective* (15th century)
1 : not counted
2 : INNUMERABLE

un·cou·ple \-ˈkə-pəl\ *transitive verb* (14th century)
1 : to release (dogs) from a couple
2 : DETACH, DISCONNECT ⟨*uncouple* railroad cars⟩
— **un·cou·pler** \-p(ə-)lər\ *noun*

un·couth \-ˈküth\ *adjective* [Middle English, from Old English *uncūth*, from *un-* + *cūth* familiar, known; akin to Old High German *kund* known, Old English *can* know — more at CAN] (before 12th century)
1 a *archaic* : not known or not familiar to one : seldom experienced : UNCOMMON, RARE **b** *obsolete* : MYSTERIOUS, UNCANNY
2 a : strange or clumsy in shape or appearance : OUTLANDISH **b** : lacking in polish and grace : RUGGED ⟨*uncouth* verse⟩ **c** : awkward and uncultivated in appearance, manner, or behavior
— **un·couth·ly** *adverb*
— **un·couth·ness** *noun*

un·cov·er \-ˈkə-vər\ (14th century)
transitive verb
1 : to make known : bring to light : DISCLOSE, REVEAL
2 : to expose to view by removing some covering
3 a : to take the cover from **b** : to remove the hat from
4 : to deprive of protection
intransitive verb
1 : to remove a cover or covering
2 : to take off the hat as a token of respect

un·cov·ered \-vərd\ *adjective* (15th century)
: not covered: as **a** : not supplied with a covering **b** : not covered by insurance or included in a social insurance or welfare program **c** : not covered by collateral ⟨an *uncovered* note⟩

un·cre·at·ed \ˌən-krē-ˈā-təd\ *adjective* (circa 1549)
1 : not existing by creation : ETERNAL, SELF-EXISTENT
2 : not yet created

un·crit·i·cal \ˌən-ˈkri-ti-kəl\ *adjective* (1659)
1 : not critical : lacking in discrimination
2 : showing lack or improper use of critical standards or procedures
— **un·crit·i·cal·ly** \-k(ə-)lē\ *adverb*

un·cross \-ˈkrȯs\ *transitive verb* (1599)
: to change from a crossed position

un·crown \-ˈkrau̇n\ *transitive verb* (14th century)
: to take the crown from : DEPOSE, DETHRONE

un·crum·ple \-ˈkrəm-pəl\ *transitive verb* (1611)
: to restore to an original smooth condition

un·crys·tal·lized \-ˈkris-tə-ˌlīzd\ *adjective* (1759)
: not crystallized; *also* : not finally or definitely formed

unc·tion \ˈəŋ(k)-shən\ *noun* [Middle English *unctioun*, from Latin *unction-*, *unctio*, from *unguere* to anoint — more at OINTMENT] (14th century)
1 : the act of anointing as a rite of consecration or healing
2 : something used for anointing : OINTMENT, UNGUENT
3 a : religious or spiritual fervor or the expression of such fervor **b** : exaggerated, assumed, or superficial earnestness of language or manner : UNCTUOUSNESS

unc·tu·ous \ˈəŋ(k)-chə-wəs, -chəs, -shwəs\ *adjective* [Middle English, from Middle French or Medieval Latin; Middle French *unctueux*, from Medieval Latin *unctuosus*, from Latin *unctus* act of anointing, from *unguere* to anoint] (14th century)
1 a : FATTY, OILY **b** : smooth and greasy in texture or appearance
2 : PLASTIC ⟨fine *unctuous* clay⟩
3 : full of unction; *especially* : revealing or marked by a smug, ingratiating, and false earnestness or spirituality
— **unc·tu·ous·ly** *adverb*
— **unc·tu·ous·ness** *noun*

un·curl \ˌən-ˈkər(-ə)l\ (1588)
intransitive verb
: to become straightened out from a curled or coiled position
transitive verb
: to straighten the curls of : UNROLL

un·cus \ˈəŋ-kəs\ *noun, plural* **un·ci** \ˈəŋ-ˌkī, -ˌkē; ˈən-ˌsī\ [New Latin, from Latin, hook — more at ANGLE] (1826)
: a hooked anatomical part or process

un·cut \ˌən-ˈkət\ *adjective* (1548)
1 : not cut down or cut into
2 : not shaped by cutting ⟨an *uncut* diamond⟩
3 *of a book* : not having the folds of the leaves slit
4 : not abridged, curtailed, or expurgated ⟨the *uncut* version of the film⟩
5 : not diluted or adulterated ⟨*uncut* vinegar⟩ ⟨*uncut* heroin⟩

un·daunt·able \ˌən-ˈdȯn-tə-bəl, -ˈdän-\ *adjective* (1587)
: incapable of being daunted : FEARLESS

un·daunt·ed \-təd\ *adjective* (1587)
: courageously resolute especially in the face of danger or difficulty : not discouraged
— **un·daunt·ed·ly** *adverb*

un·dead \ˌən-ˈded\ *noun, plural* **undead** (1897)
1 : VAMPIRE 1
2 : ZOMBIE 1b

un·de·bat·able \ˌən-di-ˈbā-tə-bəl\ *adjective* (1869)
: not subject to debate : INDISPUTABLE
— **un·de·bat·ably** \-blē\ *adverb*

un·de·ceive \ˌən-di-ˈsēv\ *transitive verb* (1598)
: to free from deception, illusion, or error

un·de·cil·lion \ˌən-di-ˈsil-yən\ *noun, often attributive* [Latin *undecim* eleven (from *unus* one + *decem* ten) + English *-illion* (as in *million*) — more at ONE, TEN] (1931)
— see NUMBER table

un·dec·y·le·nic acid \ˌən-ˌde-sə-ˈle-nik-, -ˈlē-\ *noun* [*undecylene* ($C_{11}H_{22}$)] (1879)
: an acid $C_{11}H_{20}O_2$ found in perspiration, obtained commercially from castor oil, and used in the treatment of fungal infections (as ringworm) of the skin

un·dem·o·crat·ic \ˌən-ˌde-mə-ˈkra-tik\ *adjective* (1839)
: not democratic : not agreeing with democratic practice or ideals
— **un·dem·o·crat·i·cal·ly** \-ti-k(ə-)lē\ *adverb*

un·de·mon·stra·tive \ˌən-di-ˈmän(t)-strə-tiv\ *adjective* (1846)
: restrained in expression of feeling : RESERVED
— **un·de·mon·stra·tive·ly** *adverb*
— **un·de·mon·stra·tive·ness** *noun*

un·de·ni·able \ˌən-di-ˈnī-ə-bəl\ *adjective* (1547)
1 : plainly true : INCONTESTABLE
2 : unquestionably excellent or genuine ⟨an applicant with *undeniable* references⟩
— **un·de·ni·able·ness** *noun*
— **un·de·ni·ably** \-blē\ *adverb*

[1]**un·der** \ˈən-dər\ *adverb* [Middle English, adverb & preposition, from Old English; akin to Old High German *untar* under, Latin *inferus* situated beneath, lower, *infra* below, Sanskrit *adha*] (before 12th century)
1 : in or into a position below or beneath something
2 : below or short of some quantity, level, or limit ⟨$10 or *under*⟩ — often used in combination ⟨*under*staffed⟩
3 : in or into a condition of subjection, subordination, or unconsciousness
4 : down to defeat, ruin, or death
5 : so as to be covered

[2]**under** *preposition* (before 12th century)
1 : below or beneath so as to be overhung, surmounted, covered, protected, or concealed by ⟨*under* sunny skies⟩ ⟨a soft heart *under* a stern exterior⟩ ⟨*under* cover of darkness⟩
2 a : subject to the authority, control, guidance, or instruction of ⟨served *under* the general⟩ ⟨*under* the terms of the contract⟩ **b** : receiving or undergoing the action or effect of ⟨*under* pressure⟩ ⟨courage *under* fire⟩ ⟨*under* the influence of alcohol⟩ ⟨the image of a point *under* a mapping⟩
3 : within the group or designation of ⟨*under* this heading⟩
4 : less or lower than (as in size, amount, or rank); *especially* : falling short of a standard or required degree ⟨*under* the legal age⟩ ⟨*under* par⟩

[3]**under** *adjective* (14th century)
1 a : lying or placed below, beneath, or on the ventral side — often used in combination ⟨*under*lip⟩ **b** : facing or protruding downward
2 : lower in rank or authority : SUBORDINATE
3 : lower than usual, proper, or desired in amount, quality, or degree ⟨an *under* dose of medicine⟩

un·der·achiev·er \ˌən-dər-ə-ˈchē-vər\ *noun* (1952)
: a person and especially a student who fails to attain a predicted level of achievement or does not do as well as expected

— **un·der·achieve** \-ə-ˈchēv\ *intransitive verb*
— **un·der·achieve·ment** \-mənt\ *noun*
un·der·act \ˌən-dər-ˈakt\ (circa 1623)
transitive verb
: to perform (a dramatic part) with restraint for effect **:** UNDERPLAY
intransitive verb
: to perform feebly or with restraint
un·der·ac·tive \-ˈak-tiv\ *adjective* (1959)
: characterized by an abnormally low level of activity ⟨an *underactive* thyroid gland⟩
— **un·der·ac·tiv·i·ty** \-ˈti-və-tē\ *noun*
un·der·age \ˌən-dər-ˈāj\ *adjective* (1594)
1 : of less than mature or legal age
2 : done by or involving underage persons ⟨*underage* drinking⟩
un·der·ap·pre·ci·at·ed \ˌən-dər-ə-ˈprē-shē-ˌā-təd\ *adjective* (1968)
: not duly appreciated
¹un·der·arm \ˈən-dər-ˈärm\ *adjective* (1816)
1 : UNDERHAND 4
2 : placed under or on the underside of the arm ⟨*underarm* seams⟩
²un·der·arm \ˌən-dər-ˈärm\ *adverb* (circa 1909)
: UNDERHAND
³un·der·arm \ˈən-dər-ˌärm\ *noun* (1923)
1 : ARMPIT 1
2 : the part of a garment that covers the underside of the arm
un·der·bel·ly \ˈən-dər-ˌbe-lē\ *noun* (1607)
1 : the underside of a body or mass
2 : a vulnerable area
un·der·bid \ˌən-dər-ˈbid\ *verb* **-bid; -bid·ding** (circa 1677)
transitive verb
1 : to bid less than (a competing bidder)
2 : to bid (a hand of cards) at less than the strength of the hand warrants
intransitive verb
: to bid too low
— **un·der·bid·der** *noun*
un·der·body \ˈən-dər-ˌbä-dē\ *noun* (1879)
: the lower part of something: as **a :** the lower part of an animal's body **:** UNDERPARTS **b :** the lower parts of the body of a vehicle
un·der·boss \ˈən-dər-ˌbȯs\ *noun* (1942)
: a boss ranking next below the head of a branch of a crime syndicate
un·der·bred \ˌən-dər-ˈbred\ *adjective* (1650)
: marked by lack of good breeding **:** ILL-BRED
un·der·brim \ˈən-dər-ˌbrim\ *noun* (1886)
: a facing on the underside of a hat brim
un·der·brush \ˈən-dər-ˌbrəsh\ *noun* (1775)
1 : shrubs, bushes, or small trees growing beneath large trees in a wood or forest **:** BRUSH
2 : a tangled, obstructing, or impeding mass
un·der·bud·get·ed \ˌən-dər-ˈbə-jə-təd\ *adjective* (1965)
: provided with an inadequate budget
un·der·cap·i·tal·ized \-ˈka-pə-tᵊl-ˌīzd, -ˈkap-tᵊl-\ *adjective* (1967)
: having too little capital for efficient operation
un·der·card \ˈən-dər-ˌkärd\ *noun* (1948)
: a program (as of boxing matches) supporting the featured match
un·der·car·riage \ˈən-dər-ˌkar-ij\ *noun* (circa 1796)
1 : a supporting framework (as of an automobile)
2 *chiefly British* **:** the landing gear of an airplane
un·der·charge \ˌən-dər-ˈchärj\ *transitive verb* (1633)
: to charge (as a person) too little
— **undercharge** \ˈən-dər-ˌ\ *noun*
un·der·class \ˈən-dər-ˌklas\ *noun* (1918)
: the lowest social stratum usually made up of disadvantaged minority groups
un·der·class·man \ˌən-dər-ˈklas-mən\ *noun* (1871)
: a member of the freshman or sophomore class in a school or college
un·der·clothes \ˈən-dər-ˌklō(th)z\ *noun plural* (circa 1859)
: UNDERWEAR
un·der·cloth·ing \-ˌklō-thiŋ\ *noun* (1835)
: UNDERWEAR
un·der·coat \-ˌkōt\ *noun* (1648)
1 : a coat or jacket worn under another
2 : a growth of short hair or fur partly concealed by the longer and usually coarser guard hairs of a mammal
3 a : a coat (as of paint) applied as a base for another coat **b :** UNDERCOATING
4 *dialect* **:** PETTICOAT
un·der·coat·ing \-ˌkō-tiŋ\ *noun* (1922)
: a usually asphalt-based waterproof coating applied to the underside of a vehicle
un·der·cool \ˌən-dər-ˈkül\ *transitive verb* (1902)
: SUPERCOOL
un·der·count \-ˈkau̇nt\ *transitive verb* (1951)
: to count fewer than the actual number of
— **undercount** *noun*
un·der·cov·er \-ˈkə-vər\ *adjective* (1920)
: acting or executed in secret; *specifically* **:** employed or engaged in spying or secret investigation ⟨an *undercover* agent⟩
un·der·croft \ˈən-dər-ˌkrȯft\ *noun* [Middle English, from *under* + *crofte* crypt, from Middle Dutch, from Medieval Latin *crupta,* from Latin *crypta* — more at CRYPT] (14th century)
: a subterranean room; *especially* **:** a vaulted chamber under a church
un·der·cur·rent \-ˌkər-ənt, -ˌkə-rənt\ *noun* (1683)
1 : a current below the upper currents or surface
2 : a hidden opinion, feeling, or tendency often contrary to the one publicly shown
— **undercurrent** *adjective*
¹un·der·cut \ˌən-dər-ˈkət\ *verb* **-cut; -cut·ting** (1598)
transitive verb
1 : to cut away the underpart of ⟨*undercut* a vein of ore⟩
2 : to cut away material from the underside of (an object) so as to leave an overhanging portion in relief
3 : to offer to sell at lower prices than or to work for lower wages than (a competitor)
4 : to cut obliquely into (a tree) below the main cut and on the side toward which the tree will fall
5 : to strike (a ball) with a downward glancing blow so as to give a backspin or elevation to the shot
6 : to undermine or destroy the force, value, or effectiveness of ⟨inflation *undercuts* consumer buying power⟩
intransitive verb
: to perform the action of cutting away beneath
²un·der·cut \ˈən-dər-ˌkət\ *noun* (1859)
1 *British* **:** TENDERLOIN 1
2 : the action or result of cutting away from the underside or lower part of something
3 : a notch cut in the base of a tree before felling to determine the direction of falling and to prevent splitting
un·der·de·vel·oped \ˌən-dər-di-ˈve-ləpt\ *adjective* (1892)
1 : not normally or adequately developed ⟨*underdeveloped* muscles⟩ ⟨an *underdeveloped* film⟩
2 : having a relatively low economic level of industrial production and standard of living (as from lack of capital) ⟨*underdeveloped* nations⟩
— **un·der·de·vel·op·ment** \-ləp-mənt\ *noun*
un·der·dog \ˈən-dər-ˌdȯg\ *noun* (1887)
1 : a loser or predicted loser in a struggle or contest
2 : a victim of injustice or persecution
un·der·done \ˌən-dər-ˈdən\ *adjective* (1683)
: not thoroughly cooked **:** RARE
un·der·draw·ers \ˈən-dər-ˌdrȯ(-ə)rz\ *noun plural* (1894)
: an article of underwear for the lower body
un·der·ed·u·cat·ed \ˌən-dər-ˈe-jə-ˌkā-təd\ *adjective* (1856)
: poorly educated
un·der·em·pha·sis \ˌən-dər-ˈem(p)-fə-səs\ *noun* (1964)
: less emphasis than is possible or desirable
un·der·em·pha·size \-ˌsīz\ *transitive verb* (1967)
: to fail to emphasize adequately
un·der·em·ployed \ˌən-dər-im-ˈplȯid\ *adjective* (1908)
: having less than full-time, regular, or adequate employment
un·der·em·ploy·ment \-ˈplȯi-mənt\ *noun* (1910)
1 : the condition in which people in a labor force are employed at less than full time or regular jobs or at jobs inadequate with respect to their training or economic needs
2 : the condition of being underemployed
un·der·es·ti·mate \ˌən-dər-ˈes-tə-ˌmāt\ *transitive verb* (1812)
1 : to estimate as being less than the actual size, quantity, or number
2 : to place too low a value on **:** UNDERRATE
— **un·der·es·ti·mate** \-mət\ *noun*
— **un·der·es·ti·ma·tion** \-ˌes-tə-ˈmā-shən\ *noun*
un·der·ex·pose \ˌən-dər-ik-ˈspōz\ *transitive verb* (1861)
: to expose insufficiently; *especially* **:** to expose (as film) to insufficient radiation (as light)
— **un·der·ex·po·sure** \-ˈspō-zhər\ *noun*
un·der·feed \ˌən-dər-ˈfēd\ *transitive verb* **-fed** \-ˈfed\; **-feed·ing** (1659)
: to feed with too little food
un·der·fi·nanced \-fə-ˈnan(t)st, -ˈfī-ˌ, -fī-ˈ\ *adjective* (1922)
: inadequately financed
un·der·foot \-ˈfu̇t\ *adverb* (13th century)
1 : under the foot especially against the ground ⟨trampled the flowers *underfoot*⟩
2 : below, at, or before one's feet ⟨warm sand *underfoot*⟩
3 : in the way ⟨children always getting *underfoot*⟩
un·der·fund \-ˈfənd\ *transitive verb* (1968)
: to provide insufficient funds for
un·der·fur \ˈən-dər-ˌfər\ *noun* (1877)
: an undercoat of fur especially when thick and soft
un·der·gar·ment \-ˌgär-mənt\ *noun* (1530)
: a garment to be worn under another
un·der·gird \ˌən-dər-ˈgərd\ *transitive verb* (1526)
1 *archaic* **:** to make secure underneath ⟨took measures to *undergird* the ship —Acts 27:17 (Revised Standard Version)⟩
2 : to form the basis or foundation of **:** STRENGTHEN, SUPPORT ⟨facts and statistics subtly *undergird* his commentary —Susan Q. Stranahan⟩
un·der·glaze \ˈən-dər-ˌglāz\ *adjective* (1879)
: applied or suitable for applying before the glaze is put on ⟨*underglaze* decorations⟩ ⟨*underglaze* colors⟩
— **underglaze** *noun*
un·der·go \ˌən-dər-ˈgō\ *transitive verb* **-went** \-ˈwent\; **-gone** \-ˈgȯn *also* -ˈgän\; **-go·ing** \-ˈgō-iŋ, -ˈgȯ(-)iŋ\ (14th century)
1 : to submit to **:** ENDURE
2 : to go through **:** EXPERIENCE
3 *obsolete* **:** UNDERTAKE
4 *obsolete* **:** to partake of
un·der·grad \ˈən-dər-ˌgrad\ *noun* (1827)
: UNDERGRADUATE

un·der·grad·u·ate \ˌən-dər-ˈgra-jə-wət, -ˌwāt; -ˈgraj-wət\ *noun* (1630)
: a student at a college or university who has not taken a first and especially a bachelor's degree
— undergraduate *adjective*

[1]**un·der·ground** \ˌən-dər-ˈgrau̇nd\ *adverb* (1571)
1 : beneath the surface of the earth
2 : in or into hiding or secret operation

[2]**underground** \ˈən-dər-ˌ\ *noun* (1594)
1 : a subterranean space or channel
2 : an underground city railway system
3 a : a movement or group organized in strict secrecy among citizens especially in an occupied country for maintaining communications, popular solidarity, and concerted resistive action pending liberation **b :** a clandestine conspiratorial organization set up for revolutionary or other disruptive purposes especially against a civil order **c :** an unofficial, unsanctioned, or illegal but informal movement or group; *especially* **:** a usually avant-garde group or movement that functions outside the establishment

[3]**un·der·ground** \ˈən-dər-ˌgrau̇nd\ *adjective* (1610)
1 : being, growing, operating, or situated below the surface of the ground
2 : conducted by secret means
3 a : existing outside the establishment ⟨an *underground* literary reputation⟩ **b :** existing outside the purview of tax collectors or statisticians ⟨the *underground* economy⟩
4 a : produced or published outside the establishment especially by the avant-garde ⟨*underground* movies⟩ ⟨*underground* newspapers⟩ **b :** of or relating to the avant-garde underground ⟨an *underground* moviemaker⟩ ⟨an *underground* theater⟩

un·der·ground·er \ˈən-dər-ˌgrau̇n-dər\ *noun* (1882)
: a member of the underground

Underground Railroad *noun* (1842)
: a system of cooperation among active antislavery people in the U.S. before 1863 by which fugitive slaves were secretly helped to reach the North or Canada — called also *Underground Railway*

un·der·growth \ˈən-dər-ˌgrōth\ *noun* (1600)
: low growth on the floor of a forest including seedlings and saplings, shrubs, and herbs

[1]**un·der·hand** \ˈən-dər-ˌhand\ *adverb* (1538)
1 a : in a clandestine manner **b** *archaic* **:** in a quiet or unobtrusive manner
2 : with the target seen below the hand holding the bow
3 : with an underhand motion ⟨bowl *underhand*⟩ ⟨pitch *underhand*⟩

[2]**underhand** *adjective* (1545)
1 : aimed so that the target is seen below the hand holding the bow ⟨*underhand* shooting at long range⟩
2 : UNDERHANDED
3 : done so as to evade notice
4 : made with the hand brought forward and up from below the shoulder level

[1]**un·der·hand·ed** \ˌən-dər-ˈhan-dəd\ *adverb* (circa 1822)
: UNDERHAND

[2]**underhanded** *adjective* (1853)
: marked by secrecy, chicanery, and deception **:** not honest and aboveboard **:** SLY
synonym see SECRET
— un·der·hand·ed·ly *adverb*
— un·der·hand·ed·ness *noun*

un·der·in·flat·ed \ˌən-dər-in-ˈflā-təd\ *adjective* (1928)
: not sufficiently inflated ⟨*underinflated* tires⟩
— un·der·in·fla·tion \-ˈflā-shən\ *noun*

un·der·in·sured \-ˈshu̇rd\ *adjective* (1893)
: not sufficiently insured

un·der·in·vest·ment \ˌən-dər-in-ˈves(t)-mənt\ *noun* (1940)
: an insufficient amount of investment

un·der·laid \ˌən-dər-ˈlād\ *adjective* (before 12th century)
1 : laid or placed underneath
2 : having something laid or lying underneath

[1]**un·der·lay** \-ˈlā\ *transitive verb* **-laid** \-ˈlād\; **-lay·ing** (before 12th century)
1 : to cover, line, or traverse the bottom of **:** give support to on the underside or below
2 : to raise or support by something laid under

[2]**un·der·lay** \ˈən-dər-ˌlā\ *noun* (1612)
: something that is or is designed to be laid under

un·der·lay·ment \ˌən-dər-ˈlā-mənt\ *noun* (1949)
: UNDERLAY

un·der·let \ˌən-dər-ˈlet\ *transitive verb* **-let; -let·ting** (1677)
1 : to let below the real value
2 : SUBLET

un·der·lie \-ˈlī\ *transitive verb* **-lay** \-ˈlā\; **-lain** \-ˈlān\; **-ly·ing** \-ˈlī-iŋ\ (before 12th century)
1 *archaic* **:** to be subject or amenable to
2 : to lie or be situated under
3 : to be at the basis of **:** form the foundation of **:** SUPPORT ⟨ideas *underlying* the revolution⟩
4 : to exist as a claim or security superior and prior to (another)

[1]**un·der·line** \ˈən-dər-ˌlīn, ˌən-dər-ˈ\ *transitive verb* (1721)
1 : to mark (as a word) with a line underneath
2 : to put emphasis on **:** STRESS
3 : to show clearly or emphatically

[2]**un·der·line** \ˈən-dər-ˌlīn\ *noun* (1886)
1 : the outline of a quadruped's underbody; *also* **:** the ventral surface of a quadruped's body
2 : a horizontal line placed underneath something

un·der·ling \ˈən-dər-liŋ\ *noun* (12th century)
: one who is under the orders of another **:** SUBORDINATE, INFERIOR

un·der·lip \ˌən-dər-ˈlip\ *noun* (1669)
: the lower lip

un·der·ly·ing \ˌən-dər-ˈlī-iŋ\ *adjective* (1611)
1 a : lying beneath or below ⟨the *underlying* rock is shale⟩ **b :** BASIC, FUNDAMENTAL ⟨an investigation of the *underlying* issues⟩
2 : evident only on close inspection **:** IMPLICIT
3 : anterior and prior in claim ⟨*underlying* mortgage⟩
4 : of or being present in deep structure ⟨*underlying* word order⟩

un·der·ly·ing·ly \-ˈlī-iŋ-lē\ *adverb* (1973)
: in deep structure

un·der·manned \ˌən-dər-ˈmand\ *adjective* (circa 1867)
: inadequately staffed

un·der·mine \-ˈmīn\ *transitive verb* (14th century)
1 : to excavate the earth beneath **:** form a mine under **:** SAP
2 : to wash away supporting material from under
3 : to subvert or weaken insidiously or secretly
4 : to weaken or ruin by degrees
synonym see WEAKEN

un·der·most \ˈən-dər-ˌmōst\ *adjective* (1555)
: lowest in relative position
— undermost *adverb*

[1]**un·der·neath** \ˌən-dər-ˈnēth\ *preposition* [Middle English *undernethe,* preposition & adverb, from Old English *underneothan,* from *under* + *neothan* below — more at BENEATH] (before 12th century)
1 a : directly beneath ⟨write the date *underneath* the address⟩ **b :** close under especially so as to be hidden ⟨treachery lying *underneath* a mask of friendliness⟩ ⟨wore a swimsuit *underneath* his slacks⟩
2 : under subjection to

[2]**underneath** *adverb* (before 12th century)
1 : under or below an object or a surface **:** BENEATH
2 : on the lower side
— underneath *adjective*

un·der·nour·ished \ˌən-dər-ˈnər-isht, -ˈnə-risht\ *adjective* (1910)
1 : supplied with less than the minimum amount of the foods essential for sound health and growth
2 : poorly supplied with vital elements or qualities ⟨*undernourished* independent libraries⟩
— un·der·nour·ish·ment \-ˈnər-ish-mənt, -ˈnə-rish-\ *noun*

un·der·nu·tri·tion \-nu̇-ˈtri-shən, -nyu̇-\ *noun* (circa 1889)
: deficient bodily nutrition due to inadequate food intake or faulty assimilation

un·der·paint·ing \ˈən-dər-ˌpān-tiŋ\ *noun* (1866)
: preliminary painting; *especially* **:** such painting done on a canvas or panel and covered completely or partially by the final layers of paint

un·der·pants \ˈən-dər-ˌpan(t)s\ *noun plural* (1925)
: a usually short undergarment for the lower trunk **:** DRAWERS
word history see PANTALOON

un·der·part \-ˌpärt\ *noun* (1662)
1 : a part lying on the lower side especially of a bird or mammal
2 : a subordinate or auxiliary part or role

un·der·pass \-ˌpas\ *noun* (1903)
: a crossing of a highway and another way (as a road or railroad) at different levels; *also* **:** the lower level of such a crossing

un·der·pay \ˌən-dər-ˈpā\ *transitive verb* **-paid** \-ˈpād\; **-pay·ing** (1817)
: to pay less than what is normal or required ⟨*underpay* taxes⟩
— un·der·pay·ment *noun*

un·der·pin \-ˈpin\ *transitive verb* (1522)
1 : SUPPORT, SUBSTANTIATE ⟨*underpin* a thesis with evidence⟩
2 : to form part of, strengthen, or replace the foundation of ⟨*underpin* a structure⟩ ⟨*underpin* a sagging building⟩

un·der·pin·ning \ˈən-dər-ˌpi-niŋ\ *noun* (1538)
1 : the material and construction (as a foundation) used for support of a structure
2 : something that serves as a foundation **:** BASIS, SUPPORT — often used in plural ⟨the philosophical *underpinnings* of educational methods⟩
3 : UNDERWEAR — usually used in plural
4 : a person's legs — usually used in plural

un·der·play \ˌən-dər-ˈplā\ (1896)
intransitive verb
: to play a role with subdued force
transitive verb
1 : to act or present (as a role or a scene) with restraint **:** PLAY DOWN
2 : to play a card lower than (a held high card)

un·der·plot \ˈən-dər-ˌplät\ *noun* (1668)
: SUBPLOT 1

un·der·pop·u·lat·ed \ˌən-dər-ˈpä-pyə-ˌlā-təd\ *adjective* (1884)
: having a lower density of population than is normal or desirable

un·der·pow·ered \-ˈpau̇(-ə)rd\ *adjective* (1905)
1 : driven by an engine of insufficient power
2 : having or supplied with insufficient power

un·der·pre·pared \-prə-ˈpard, -ˈperd\ *adjective* (1964)
: inadequately prepared

un·der·price \-ˈprīs\ *transitive verb* (1756)
1 : to price below what is normal or below the real value
2 : to undercut (a competitor) in prices

un·der·priv·i·leged \-ˈpriv-lijd, -ˈpri-və-\ *adjective* (1896)
1 : deprived through social or economic condition of some of the fundamental rights of all members of a civilized society
2 : of or relating to underprivileged people ⟨*underprivileged* areas of the city⟩

un·der·pro·duc·tion \-prə-ˈdək-shən\ *noun* (1887)
: the production of less than enough to satisfy the demand or of less than the usual amount

un·der·proof \ˌən-dər-ˈprüf\ *adjective* (1857)
: containing less alcohol than proof spirit

un·der·pub·li·cized \-ˈpə-blə-ˌsīzd\ *adjective* (1966)
: insufficiently publicized

un·der·rate \ˌən-də(r)-ˈrāt\ *transitive verb* (1650)
: to rate too low : UNDERVALUE

un·der·re·act \-rē-ˈakt\ *intransitive verb* (1965)
: to react with less than appropriate force or intensity

un·der·re·port \-ri-ˈpōrt, -ˈpȯrt\ *transitive verb* (1949)
: to report to be less than is actually the case : UNDERSTATE ⟨*underreports* his income⟩ ⟨the number of cases has been *underreported*⟩

un·der·rep·re·sent·ed \-ˌre-pri-ˈzen-təd\ *adjective* (1884)
: inadequately represented
— **un·der·rep·re·sen·ta·tion** \-ˌzen-ˈtā-shən, -zən-\ *noun*

[1]un·der·run \-ˈrən\ *transitive verb* **-ran** \-ˈran\; **-run; -run·ning** (1547)
1 : to pass along under in order to examine (a cable)
2 : to pass or extend under

[2]un·der·run \ˈən-də(r)-ˌrən\ *noun* (1926)
: the amount by which something produced (as a cut of lumber) falls below an estimate

un·der·sat·u·rat·ed \ˌən-dər-ˈsa-chə-ˌrā-təd\ *adjective* (circa 1828)
: less than normally or adequately saturated

[1]un·der·score \ˈən-dər-ˌskōr, -ˌskȯr\ *transitive verb* (1771)
1 : to draw a line under : UNDERLINE
2 : to make evident : EMPHASIZE, STRESS
3 : to provide (action on film) with accompanying music

[2]underscore *noun* (1901)
1 : a line drawn under a word or line especially for emphasis or to indicate intent to italicize
2 : music accompanying the action and dialogue of a film

[1]un·der·sea \ˌən-dər-ˈsē\ *adjective* (1613)
1 : being or carried on under the sea or under the surface of the sea ⟨*undersea* oil deposits⟩ ⟨*undersea* fighting⟩
2 : designed for use under the surface of the sea ⟨an *undersea* fleet⟩

[2]undersea *or* **un·der·seas** \-ˈsēz\ *adverb* (1684)
: under the sea : beneath the surface of the sea ⟨photographs taken *undersea*⟩

un·der·sec·re·tary \ˌən-dər-ˈse-krə-ˌter-ē, -ˈse-kə-\ *noun* (1687)
: a secretary immediately subordinate to a principal secretary ⟨*undersecretary* of health⟩

un·der·sell \ˌən-dər-ˈsel\ *transitive verb* **-sold** \-ˈsōld\; **-sell·ing** (1622)
1 : to sell articles cheaper than ⟨*undersell* a competitor⟩ ⟨we will not be *undersold*⟩
2 : to sell cheaper than ⟨imported cars that *undersell* domestic ones⟩

un·der·served \-ˈsərvd\ *adjective* (1710)
: provided with inadequate service

un·der·sexed \-ˈsekst\ *adjective* (1931)
: deficient in sexual desire

un·der·shirt \ˈən-dər-ˌshərt\ *noun* (1648)
: a collarless undergarment with or without sleeves
— **un·der·shirt·ed** *adjective*

un·der·shoot \ˌən-dər-ˈshüt\ *transitive verb* **-shot** \-ˈshät\; **-shoot·ing** (circa 1661)
1 : to shoot short of or below (a target)
2 : to fall short of (a runway) in landing an airplane

un·der·shorts \ˈən-dər-ˌshȯrts\ *noun plural* (1949)
: SHORT 4b

un·der·shot \ˈən-dər-ˌshät\ *adjective* (1610)
1 : moved by water passing beneath ⟨an *undershot* wheel⟩
2 : having the lower incisor teeth or lower jaw projecting beyond the upper when the mouth is closed

un·der·shrub \ˈən-dər-ˌshrəb, *especially Southern* -ˌsrəb\ *noun* (1598)
: SUBSHRUB

un·der·side \ˈən-dər-ˌsīd, ˌən-dər-ˈ\ *noun* (1680)
1 : the side or surface lying underneath
2 : a side usually hidden from sight; *specifically* **:** the more unpleasant or reprehensible side ⟨the *underside* of politics⟩

un·der·signed \ˈən-dər-ˌsīnd\ *noun, plural* **undersigned** (1643)
: one whose name is signed at the end of a document ⟨the *undersigned* all agree⟩

un·der·sized \ˌən-dər-ˈsīzd\ *also* **un·der·size** \-ˈsīz\ *adjective* (1706)
: of a size less than is common, proper, normal, or average ⟨*undersized* trout⟩

un·der·skirt \ˈən-dər-ˌskərt\ *noun* (1861)
: a skirt worn under an outer skirt; *especially* **:** PETTICOAT

un·der·slung \ˌən-dər-ˈsləŋ\ *adjective* (1903)
1 *of a vehicle frame* **:** suspended below the axles
2 : having a low center of gravity

un·der·spin \ˈən-dər-ˌspin\ *noun* (1901)
: BACKSPIN

un·der·staffed \ˌən-dər-ˈstaft\ *adjective* (1891)
: inadequately staffed
— **un·der·staff·ing** \-ˈsta-fiŋ\ *noun*

un·der·stand \ˌən-dər-ˈstand\ *verb* **-stood** \-ˈstu̇d\; **-stand·ing** [Middle English, from Old English *understandan,* from *under* + *standan* to stand] (before 12th century)
transitive verb
1 a : to grasp the meaning of ⟨*understand* Russian⟩ **b :** to grasp the reasonableness of ⟨his behavior is hard to *understand*⟩ **c :** to have thorough or technical acquaintance with or expertness in the practice of ⟨*understand* finance⟩ **d :** to be thoroughly familiar with the character and propensities of ⟨*understands* children⟩
2 : to accept as a fact or truth or regard as plausible without utter certainty ⟨we *understand* that he is returning from abroad⟩
3 : to interpret in one of a number of possible ways
4 : to supply in thought as though expressed ⟨"to be married" is commonly *understood* after the word *engaged*⟩
intransitive verb
1 : to have understanding : have the power of comprehension
2 : to achieve a grasp of the nature, significance, or explanation of something
3 : to believe or infer something to be the case
4 : to show a sympathetic or tolerant attitude toward something ☆
— **un·der·stand·abil·i·ty** \-ˌstan-də-ˈbi-lə-tē\ *noun*
— **un·der·stand·able** \-ˈstan-də-bəl\ *adjective*

un·der·stand·ably \ˌən-dər-ˈstan-də-blē\ *adverb* (circa 1921)
1 as can be easily understood **:** for understandable reasons ⟨is *understandably* nervous⟩

[1]un·der·stand·ing \ˌən-dər-ˈstan-diŋ\ *noun* (before 12th century)
1 : a mental grasp : COMPREHENSION
2 a : the power of comprehending; *especially* **:** the capacity to apprehend general relations of particulars **b :** the power to make experience intelligible by applying concepts and categories
3 a : friendly or harmonious relationship **b :** an agreement of opinion or feeling : adjustment of differences **c :** a mutual agreement not formally entered into but in some degree binding on each side
4 : EXPLANATION, INTERPRETATION
5 : SYMPATHY 3a

[2]understanding *adjective* (13th century)
1 *archaic* **:** KNOWING, INTELLIGENT
2 : endowed with understanding : TOLERANT, SYMPATHETIC
— **un·der·stand·ing·ly** \-ˈstan-diŋ-lē\ *adverb*

un·der·state \ˌən-dər-ˈstāt\ *transitive verb* (1824)
1 : to represent as less than is the case
2 : to state or present with restraint especially for effect
— **un·der·state·ment** \-mənt\ *noun*

un·der·stat·ed \-ˈstā-təd\ *adjective* (circa 1909)
: avoiding obvious emphasis or embellishment
— **un·der·stat·ed·ly** *adverb*

un·der·steer \ˈən-dər-ˌstir\ *noun* (1936)
: the tendency of an automobile to turn less sharply than the driver intends
— **un·der·steer** \ˌən-dər-ˈ\ *intransitive verb*

un·der·stood \ˌən-dər-ˈstu̇d\ *adjective* (1605)
1 : fully apprehended
2 : agreed upon
3 : IMPLICIT

un·der·sto·ry \ˈən-dər-ˌstōr-ē, -ˌstȯr-\ *noun* (1902)
1 : an underlying layer of vegetation; *specifically* **:** the vegetative layer and especially the trees and shrubs between the forest canopy and the ground cover
2 : the plants that form the understory

un·der·strap·per \-ˌstra-pər\ *noun* [[3]*under* + *strapper* one who harnesses horses] (circa 1704)
: a petty agent or subordinate : UNDERLING

un·der·strength \ˌən-dər-ˈstreŋ(k)th\ *adjective* (1925)
: deficient in strength; *especially* **:** lacking sufficient or prescribed personnel

[1]un·der·study \ˈən-dər-ˌstə-dē, ˌən-dər-ˈ\ (1874)
intransitive verb
: to study another actor's part in order to substitute in an emergency
transitive verb
: to prepare (as a part) as understudy; *also* **:** to prepare as understudy to (as an actor)

[2]un·der·study \ˈən-dər-ˌstə-dē\ *noun* (1882)
: one who is prepared to act another's part or take over another's duties

un·der·sup·ply \ˌən-dər-sə-ˈplī\ *noun* (1848)
: an inadequate supply or amount

[1]un·der·sur·face \ˈən-dər-ˌsər-fəs\ *noun* (1733)
: UNDERSIDE

[2]un·der·sur·face \ˌən-dər-ˈsər-fəs\ *adjective* (circa 1934)
: existing or moving below the surface

un·der·take \ˌən-dər-ˈtāk\ *verb* **-took** \-ˈtu̇k\; **-tak·en** \-ˈtā-kən\; **-tak·ing** (14th century)

☆ SYNONYMS
Understand, comprehend, appreciate mean to have a clear or complete idea of. UNDERSTAND and COMPREHEND are very often interchangeable. UNDERSTAND may, however, stress the fact of having attained a firm mental grasp of something ⟨orders that were fully *understood* and promptly obeyed⟩. COMPREHEND may stress the process of coming to grips with something intellectually ⟨I have trouble *comprehending* your reasons for doing this⟩. APPRECIATE implies a just evaluation or judgment of a thing's value or nature ⟨failed to *appreciate* the risks involved⟩.

transitive verb
1 : to take upon oneself **:** set about **:** ATTEMPT ⟨*undertake* a task⟩ ⟨*undertake* to learn to swim⟩
2 : to put oneself under obligation to perform; *also* **:** to accept as a charge or responsibility ⟨the lawyer who *undertook* the case⟩
3 : GUARANTEE, PROMISE
intransitive verb
archaic **:** to give surety or assume responsibility

un·der·tak·er \ˌən-dər-ˈtā-kər, *2 is* ˈən-dər-ˌ\ *noun* (1615)
1 : one that undertakes **:** one that takes the risk and management of business **:** ENTREPRENEUR
2 : one whose business is to prepare the dead for burial and to arrange and manage funerals
3 : an Englishman taking over forfeited lands in Ireland in the 16th and 17th centuries

un·der·tak·ing \ˈən-dər-ˌtā-kiŋ, ˌən-dər-ˈ; *1b is* ˈən-dər-ˌ *only*\ *noun* (15th century)
1 a : the act of one who undertakes or engages in a project or business **b :** the business of an undertaker
2 : something undertaken **:** ENTERPRISE
3 : PLEDGE, GUARANTEE

un·der·ten·ant \ˈən-dər-ˌte-nənt\ *noun* (1546)
: one who holds lands or tenements by a sublease

under–the–counter *adjective* [from the hiding of illicit wares under the counter of stores where they are sold] (1926)
: surreptitious and usually irregular or illicit ⟨*under-the-counter* liquor sales⟩

under–the–table *adjective* (1948)
: covert and usually unlawful ⟨*under-the-table* payoffs⟩

un·der·thrust \ˌən-dər-ˈthrəst\ *transitive verb* **-thrust; -thrust·ing** (1893)
: to insert (a faulted rock mass) into position under a passive rock mass

un·der·tone \ˈən-dər-ˌtōn\ *noun* (1806)
1 : a low or subdued utterance or accompanying sound
2 : a quality (as of emotion) underlying the surface of an utterance or action
3 : a subdued color; *specifically* **:** a color seen through and modifying another color

un·der·tow \-ˌtō\ *noun* (1817)
1 : the current beneath the surface that sets seaward or along the beach when waves are breaking upon the shore
2 : an underlying current, force, or tendency that is in opposition to what is apparent

un·der·trick \-ˌtrik\ *noun* (1903)
: any of the tricks by which a declarer in bridge falls short of making the contract

un·der·used \ˌən-dər-ˈyüzd\ *adjective* (1960)
: not fully used

un·der·uti·lize \ˌən-dər-ˈyü-tᵊl-ˌīz\ *transitive verb* (1951)
: to utilize less than fully or below the potential use
— un·der·uti·li·za·tion \ˈən-dər-ˌyü-tᵊl-ə-ˈzā-shən\ *noun*

un·der·val·u·a·tion \ˌən-dər-ˌval-yə-ˈwā-shən\ *noun* (1653)
1 : the act of undervaluing
2 : a value below the real worth

un·der·val·ue \-ˈval-(ˌ)yü\ *transitive verb* (1599)
1 : to value, rate, or estimate below the real worth ⟨*undervalue* stock⟩
2 : to treat as having little value ⟨was *undervalued* as a poet⟩

un·der·wa·ter \ˌən-dər-ˈwȯ-tər, -ˈwä-\ *adjective* (1627)
1 : lying, growing, worn, or operating below the surface of the water
2 : being below the waterline of a ship
— underwater *adverb*

un·der·way \ˌən-dər-ˈwā\ *adjective* (1743)
: occurring, performed, or used while traveling or in motion ⟨*underway* replenishment of fuel⟩

under way *adverb* [probably from Dutch *onderweg,* from Middle Dutch *onderwegen,* literally, under or among the ways] (1751)
1 : in motion **:** not at anchor or aground
2 : into motion from a standstill
3 : in progress **:** AFOOT ⟨preparations were *under way*⟩ ⟨the season got *under way* with a bang⟩

un·der·wear \ˈən-dər-ˌwar, -ˌwer\ *noun* (circa 1879)
: clothing or an article of clothing worn next to the skin and under other clothing

under weigh *adverb* [by folk etymology] (1777)
: UNDER WAY

¹un·der·weight \ˌən-dər-ˈwāt\ *noun* (1596)
: weight below normal, average, or requisite weight

²underweight *adjective* (1890)
: weighing less than the normal or requisite amount

un·der·whelm \-ˈhwelm, -ˈwelm\ *transitive verb* [*under* + *overwhelm*] (1949)
: to fail to impress or stimulate

¹un·der·wing \ˈən-dər-ˌwiŋ\ *noun* (1535)
1 : one of the posterior wings of an insect
2 : any of various noctuid moths (especially genus *Catocala*) that have the hind wings banded with contrasting colors (as red and black) — called also *underwing moth*
3 : the underside of a bird's wing

²underwing *adjective* (1896)
: placed or growing underneath the wing ⟨*underwing* rockets⟩

un·der·wood \ˈən-dər-ˌwu̇d\ *noun* (14th century)
: UNDERGROWTH, UNDERBRUSH

un·der·wool \-ˌwu̇l\ *noun* (1939)
: short woolly underfur

un·der·world \-ˌwərld\ *noun* (1608)
1 : the place of departed souls **:** HADES
2 *archaic* **:** EARTH
3 : the side of the earth opposite to one **:** ANTIPODES
4 : a social sphere below the level of ordinary life; *especially* **:** the world of organized crime

un·der·write \ˈən-də(r)-ˌrīt, ˌən-də(r)-ˈ\ *verb* **-wrote** \-ˌrōt, -ˈrōt\; **-writ·ten** \-ˌri-tᵊn, -ˈri-tᵊn\; **-writ·ing** \-ˌrī-tiŋ, -ˈrī-\ (15th century)
transitive verb
1 : to write under or at the end of something else
2 : to set one's name to (an insurance policy) for the purpose of thereby becoming answerable for a designated loss or damage on consideration of receiving a premium percent **:** insure on life or property; *also* **:** to assume liability for (a sum or risk) as an insurer
3 : to subscribe to **:** agree to
4 a : to agree to purchase (as security issue) usually on a fixed date at a fixed price with a view to public distribution **b :** to guarantee financial support of
intransitive verb
: to work as an underwriter

un·der·writ·er \ˈən-də(r)-ˌrī-tər\ *noun* (1622)
1 : one that underwrites **:** GUARANTOR
2 a : one that underwrites a policy of insurance **:** INSURER **b :** one who selects risks to be solicited or rates the acceptability of risks solicited
3 : one that underwrites a security issue

un·de·scend·ed \ˌən-di-ˈsen-dəd\ *adjective* (1701)
: retained within the inguinal region rather than descending into the scrotum ⟨an *undescended* testis⟩

un·de·sign·ing \ˌən-di-ˈzī-niŋ\ *adjective* (1697)
: having no ulterior or fraudulent purpose **:** SINCERE

¹un·de·sir·able \-ˈzī-rə-bəl\ *adjective* (1667)
: not desirable **:** UNWANTED
— un·de·sir·abil·i·ty \-ˌzī-rə-ˈbi-lə-tē\ *noun*
— un·de·sir·able·ness \-ˈzī-rə-bəl-nəs\ *noun*
— un·de·sir·ably \-blē\ *adverb*

²undesirable *noun* (1883)
: one that is undesirable

un·de·vi·at·ing \ˌən-ˈdē-vē-ˌā-tiŋ\ *adjective* (1732)
: keeping a true course **:** UNSWERVING ⟨served their country with *undeviating* loyalty and devotion⟩
— un·de·vi·at·ing·ly \-tiŋ-lē\ *adverb*

un·dies \ˈən-dēz\ *noun plural* [by shortening & alteration] (1900)
: UNDERWEAR; *especially* **:** women's underwear

un·dine \ˌən-ˈdēn, ˈən-ˌ\ *noun* [New Latin *undina,* from Latin *unda* wave — more at WATER] (1657)
: an elemental being in the theory of Paracelsus inhabiting water **:** WATER NYMPH

un·dip·lo·mat·ic \ˌən-ˌdi-plə-ˈma-tik\ *adjective* (circa 1828)
: not diplomatic; *especially* **:** TACTLESS
— un·dip·lo·mat·i·cal·ly \-ti-k(ə-)lē\ *adverb*

un·di·rect·ed \ˌən-də-ˈrek-təd, -dī-\ *adjective* (1596)
: not directed **:** not planned or guided ⟨*undirected* efforts⟩

un·dis·guised \ˌən-dis-ˈgīzd\ *adjective* (circa 1500)
: not disguised or concealed **:** FRANK, OPEN
— un·dis·guis·ed·ly \-ˈgī-zəd-lē\ *adverb*

un·dis·so·ci·at·ed \ˌən-di-ˈsō-shē-ˌā-təd, -sē-\ *adjective* (1899)
: not electrolytically dissociated

un·do \ˌən-ˈdü, ˈən-\ *verb* **-did** \-ˈdid\; **-done** \-ˈdən\; **-do·ing** \-ˈdü-iŋ\ (before 12th century)
transitive verb
1 : to open or loose by releasing a fastening
2 : to make of no effect or as if not done **:** make null **:** REVERSE
3 a : to ruin the worldly means, reputation, or hopes of **b :** to disturb the composure of **:** UPSET ⟨she's come *undone*⟩ **c :** SEDUCE 3
intransitive verb
: to come open or apart
— un·do·er \-ˈdü-ər\ *noun*

un·dock \-ˈdäk\ (1750)
intransitive verb
: to move away from a dock (as at sailing time)
transitive verb
: UNCOUPLE ⟨*undock* the shuttle from the space station⟩

un·dog·mat·ic \ˌən-dȯg-ˈma-tik, -däg-\ *adjective* (1857)
: not dogmatic **:** not committed to dogma
— un·dog·mat·i·cal·ly \-ti-k(ə-)lē\ *adverb*

un·do·ing \-ˈdü-iŋ\ *noun* (14th century)
1 : an act of loosening **:** UNFASTENING
2 : RUIN; *also* **:** a cause of ruin ⟨greed was to prove his *undoing*⟩
3 : ANNULMENT, REVERSAL

un·done \-ˈdən\ *adjective* (14th century)
: not done **:** not performed or finished

un·dou·ble \ˌən-ˈdə-bəl\ *verb* (circa 1611)
: UNFOLD, UNCLENCH

un·dou·bled \-ˈdə-bəld\ *adjective* (1598)
: not doubled

un·doubt·ed \-ˈdau̇-təd\ *adjective* (15th century)
: not doubted **:** GENUINE, UNDISPUTED
— un·doubt·ed·ly *adverb*

un·dra·mat·ic \ˌən-drə-ˈma-tik\ *adjective* (1754)
: lacking dramatic force or quality **:** UNSPECTACULAR
— un·dra·mat·i·cal·ly \-ti-k(ə-)lē\ *adverb*

un·drape \ˌən-ˈdrāp\ *transitive verb* (1869)
: to strip of drapery **:** UNVEIL

un·draw \-ˈdrȯ\ *transitive verb* **-drew** \-ˈdrü\; **-drawn** \-ˈdrȯn\; **-draw·ing** (1677)
: to draw aside (as a curtain) **:** OPEN

un·dreamed \-ˈdrem(p)t, -ˈdrēmd\ *also* **un·dreamt** \-ˈdrem(p)t\ *adjective* (1611)
: not dreamed **:** not thought of **:** UNIMAGINED ⟨technical advances *undreamed* of a few years ago⟩

^{1}**un·dress** \-ˈdres\ (1596)
transitive verb
1 : to remove the clothes or covering of **:** DIVEST, STRIP
2 : EXPOSE, REVEAL
intransitive verb
: to take off one's clothes **:** DISROBE

^{2}**undress** *noun* (1683)
1 : informal dress: as **a :** a loose robe or dressing gown **b :** ordinary dress — compare FULL DRESS
2 : the state of being undressed

un·dressed \ˌən-ˈdrest\ *adjective* (1535)
: not dressed: as **a :** partially, improperly, or informally clothed **b :** not fully processed or finished ⟨*undressed* hides⟩ **c :** not cared for or tended ⟨an *undressed* wound⟩ ⟨*undressed* fields⟩

un·drink·able \ˌən-ˈdriŋ-kə-bəl\ *adjective* (1611)
: unsuitable or unpleasant to drink

un·drunk \-ˈdrəŋk\ *adjective* (1637)
: not swallowed

un·due \-ˈdü, -ˈdyü\ *adjective* (14th century)
1 : not due **:** not yet payable
2 : exceeding or violating propriety or fitness **:** EXCESSIVE

un·du·lant \ˈən-jə-lənt, ˈən-dyə-, ˈən-də-\ *adjective* (circa 1830)
1 : rising and falling in waves
2 : having a wavy form, outline, or surface ⟨played her approach shot onto the *undulant* green⟩

undulant fever *noun* (1896)
: a persistent human brucellosis marked by remittent fever, pain and swelling in the joints, and great weakness and contracted by contact with infected domestic animals or consumption of their products

^{1}**un·du·late** \ˈən-jə-lət, ˈən-dyə-, ˈən-də-, -ˌlāt\ *or* **un·du·lat·ed** \-ˌlā-təd\ *adjective* [Latin *undulatus,* from (assumed) Latin *undula,* diminutive of Latin *unda* wave — more at WATER] (1658)
: having a wavy surface, edge, or markings ⟨the *undulate* margin of a leaf⟩

^{2}**un·du·late** \-ˌlāt\ *verb* **-lat·ed; -lat·ing** [Late Latin *undula* small wave, from (assumed) Latin] (1664)
intransitive verb
1 : to form or move in waves **:** FLUCTUATE
2 : to rise and fall in volume, pitch, or cadence
3 : to present a wavy appearance
transitive verb
: to cause to move in a wavy, sinuous, or flowing manner
synonym see SWING

un·du·la·tion \ˌən-jə-ˈlā-shən, ˌən-dyə-, ˌən-də-\ *noun* (1646)
1 a : a rising and falling in waves **b :** a wavelike motion to and fro in a fluid or elastic medium propagated continuously among its particles but with little or no permanent translation of the particles in the direction of the propagation **:** VIBRATION
2 : the pulsation caused by the vibrating together of two tones not quite in unison
3 : a wavy appearance, outline, or form **:** WAVINESS

un·du·la·to·ry \ˈən-jə-lə-ˌtōr-ē, ˈən-dyə-, ˈən-də-, -ˌtȯr-\ *adjective* (1728)
: of or relating to undulation **:** moving in or resembling waves **:** UNDULATING

undulatory theory *noun* (circa 1828)
: WAVE THEORY

un·du·ly \ˌən-ˈdü-lē, -ˈdyü-\ *adverb* (14th century)
: in an undue manner **:** EXCESSIVELY

un·du·ti·ful \-ˈdü-ti-fəl, -ˈdyü-\ *adjective* (1582)
: not dutiful
— **un·du·ti·ful·ly** \-fə-lē\ *adverb*
— **un·du·ti·ful·ness** *noun*

un·dy·ing \-ˈdī-iŋ\ *adjective* (14th century)
: not dying **:** IMMORTAL, PERPETUAL

un·earned \-ˈərnd\ *adjective* (13th century)
1 : not gained by labor, service, or skill ⟨*unearned* income⟩
2 : scored as a result of an error by the opposing team ⟨an *unearned* run⟩

unearned increment *noun* (1871)
: an increase in the value of property (as land) that is due to no labor or expenditure of the owner but to natural causes (as the increase of population) that create an increased demand for it

un·earth \ˌən-ˈərth\ *transitive verb* (15th century)
1 : to dig up out of or as if out of the earth **:** EXHUME ⟨*unearth* a hidden treasure⟩ ⟨*unearth* a forgotten photo album⟩
2 : to make known or public **:** bring to light ⟨*unearth* a plot⟩
synonym see DISCOVER

un·earth·ly \-lē\ *adjective* (1611)
: not earthly: as **a :** not terrestrial ⟨*unearthly* radio sources⟩ **b :** PRETERNATURAL, SUPERNATURAL ⟨an *unearthly* light⟩ **c :** WEIRD, EERIE ⟨*unearthly* howls⟩ **d :** not mundane **:** IDEAL ⟨*unearthly* love⟩ **e :** ABSURD 1, UNGODLY ⟨getting up at an *unearthly* hour⟩
— **un·earth·li·ness** *noun*

un·ease \ˌən-ˈēz\ *noun* (14th century)
: mental or spiritual discomfort: as **a :** vague dissatisfaction **:** MISGIVING **b :** ANXIETY, DISQUIET **c :** lack of ease (as in social relations) **:** EMBARRASSMENT

un·eas·i·ly \-ˈē-zə-lē\ *adverb* (14th century)
: in an uneasy manner

^{1}**un·easy** \-ˈē-zē\ *adjective* (13th century)
1 : causing physical or mental discomfort ⟨*uneasy* news of captures and killings —Marjory S. Douglas⟩
2 : not easy **:** DIFFICULT
3 : marked by lack of ease **:** AWKWARD, EMBARRASSED ⟨gave an *uneasy* laugh⟩
4 : APPREHENSIVE, WORRIED
5 : RESTLESS, UNQUIET
6 : PRECARIOUS, UNSTABLE ⟨an *uneasy* truce⟩
— **un·eas·i·ness** *noun*

^{2}**uneasy** *adverb* (1596)
: UNEASILY

un·eco·nom·ic \ˌən-ˌe-kə-ˈnä-mik, -ˌē-kə-\ *or* **un·eco·nom·i·cal** \-mi-kəl\ *adjective* (1840)
: not economically practicable ⟨*uneconomic* transportation routes⟩; *also* **:** COSTLY, WASTEFUL ⟨an *uneconomic* nuclear technology⟩

un·ed·it·ed \ˌən-ˈe-də-təd\ *adjective* (1829)
: not edited: as **a :** left unrevised **b :** not yet edited ⟨*unedited* books⟩ ⟨*unedited* films⟩

un·emo·tion·al \ˌən-i-ˈmō-shnəl, -shə-nᵊl\ *adjective* (1876)
: not emotional: as **a :** not easily aroused or excited **:** COLD **b :** involving a minimum of emotion **:** INTELLECTUAL
— **un·emo·tion·al·ly** *adverb*

un·em·ploy·able \ˌən-im-ˈplȯi-ə-bəl\ *adjective* (1887)
: not acceptable for employment
— **un·em·ploy·abil·i·ty** \-ˌplȯi-ə-ˈbi-lə-tē\ *noun*
— **unemployable** *noun*

un·em·ployed \-ˈplȯid\ *adjective* (1600)
: not employed: **a :** not being used **b :** not engaged in a gainful occupation **c :** not invested
— **unemployed** *noun*

un·em·ploy·ment \-ˈplȯi-mənt\ *noun* (1888)
1 : the state of being unemployed **:** involuntary idleness of workers; *also* **:** the rate of such unemployment
2 : UNEMPLOYMENT COMPENSATION

unemployment compensation *noun* (1944)
: compensation paid at regular intervals (as by a government agency) to an unemployed worker and especially one who has been laid off — called also *unemployment benefit*

unemployment insurance *noun* (1923)
: social insurance against involuntary unemployment that provides unemployment compensation for a limited period to unemployed workers; *also* **:** UNEMPLOYMENT COMPENSATION

un·en·cum·bered \ˌən-in-ˈkəm-bərd\ *adjective* (1722)
: free of encumbrance

un·end·ing \ˌən-ˈen-diŋ\ *adjective* (1661)
: never ending **:** ENDLESS
— **un·end·ing·ly** \-diŋ-lē\ *adverb*

un–En·glish \ˌən-ˈiŋ-glish *also* -ˈiŋ-lish\ *adjective* (1633)
1 : not characteristically English
2 : not agreeing with standard or generally accepted usage of the English language

^{1}**un·equal** \ˌən-ˈē-kwəl\ *adjective* (1565)
1 a : not of the same measurement, quantity, or number as another **b :** not like or not the same as another in degree, worth, or status
2 : not uniform **:** VARIABLE, UNEVEN
3 a : badly balanced or matched ⟨an *unequal* contest⟩ **b :** contracted between unequals ⟨*unequal* marriages⟩ **c** *archaic* **:** not equable
4 *archaic* **:** not equitable **:** UNJUST
5 : INADEQUATE, INSUFFICIENT ⟨*unequal* to the task⟩
— **un·equal·ly** \-kwə-lē\ *adverb*

^{2}**unequal** *noun* (1600)
: one that is not equal to another

^{3}**unequal** *adverb* (1602)
archaic **:** in an unequal manner ⟨*unequal* match'd —Shakespeare⟩

un·equaled *or* **un·equalled** \-kwəld\ *adjective* (1600)
: not equaled **:** UNPARALLELED

un·equiv·o·ca·bly \ˌən-i-ˈkwi-və-kə-blē\ *adverb* [by alteration] (1917)
nonstandard **:** UNEQUIVOCALLY

un·equiv·o·cal \ˌən-i-ˈkwi-və-kəl\ *adjective* (1784)
1 : leaving no doubt **:** CLEAR, UNAMBIGUOUS
2 : UNQUESTIONABLE ⟨production of *unequivocal* masterpieces —Carole Cook⟩

un·equiv·o·cal·ly \-kə-lē, -klē\ *adverb* (1794)
: in an unequivocal manner

un·err·ing \ˌən-ˈer-iŋ, ˌən-ˈər-\ *adjective* (1621)
: committing no error **:** FAULTLESS, UNFAILING
— **un·err·ing·ly** \-iŋ-lē\ *adverb*

un·es·sen·tial \ˌən-ə-ˈsen(t)-shəl\ *adjective* (circa 1656)
1 : not essential **:** DISPENSABLE, UNIMPORTANT
2 *archaic* **:** void of essence **:** INSUBSTANTIAL

un–Eu·ro·pe·an \ˌən-ˌyu̇r-ə-ˈpē-ən\ *adjective* (1846)
: not characteristically European

un·even \ˌən-ˈē-vən\ *adjective* (before 12th century)
1 a *archaic* **:** UNEQUAL 1a **b :** ODD 3a
2 a : not even **:** not level or smooth **:** RUGGED, RAGGED ⟨large *uneven* teeth⟩ ⟨*uneven* handwriting⟩ **b :** varying from the straight or parallel **c :** not uniform **:** IRREGULAR ⟨*uneven* combustion⟩ **d :** varying in quality ⟨an *uneven* performance⟩
3 : UNEQUAL 3a ⟨an *uneven* confrontation⟩
synonym see ROUGH
— **un·even·ly** *adverb*
— **un·even·ness** \-və(n)-nəs\ *noun*

un·event·ful \ˌən-i-ˈvent-fəl\ *adjective* (1800)
: marked by no noteworthy or untoward incidents **:** PLACID
— **un·event·ful·ly** \-fə-lē\ *adverb*
— **un·event·ful·ness** \-fəl-nəs\ *noun*

un·ex·am·pled \ˌən-ig-ˈzam-pəld\ *adjective* (1610)

: having no example or parallel **:** UNPRECEDENTED

un·ex·cep·tion·able \ˌən-ik-ˈsep-sh(ə-)nə-bəl\ *adjective* [*un-* + obsolete *exception* to take exception, object] (1664)
: not open to objection or criticism **:** beyond reproach **:** UNIMPEACHABLE
— **un·ex·cep·tion·able·ness** *noun*
— **un·ex·cep·tion·ably** \-blē\ *adverb*

un·ex·cep·tion·al \-shnəl, -shə-nᵊl\ *adjective* (1806)
: not out of the ordinary **:** COMMONPLACE

un·ex·er·cised \ˌən-ˈek-sər-ˌsīzd\ *adjective* (14th century)
1 : having terms that are not implemented ⟨*unexercised* options⟩
2 : not subjected to exercise ⟨*unexercised* muscles⟩

un·ex·pect·ed \ˌən-ik-ˈspek-təd\ *adjective* (circa 1586)
: not expected **:** UNFORESEEN
— **un·ex·pect·ed·ly** *adverb*
— **un·ex·pect·ed·ness** *noun*

un·ex·ploit·ed \ˌən-ik-ˈsplȯi-təd\ *adjective* (1888)
: not exploited or developed **:** not taken advantage of

un·ex·pres·sive \ˌən-ik-ˈspre-siv\ *adjective* (1600)
1 *obsolete* **:** INEFFABLE
2 : not expressive **:** failing to convey the feeling or meaning intended

un·fad·ing \ˌən-ˈfā-diŋ\ *adjective* (1652)
1 : not losing color or freshness
2 : not losing value or effectiveness
— **un·fad·ing·ly** \-diŋ-lē\ *adverb*

un·fail·ing \ˌən-ˈfā-liŋ\ *adjective* (14th century)
: not failing or liable to fail: **a :** CONSTANT, UNFLAGGING ⟨*unfailing* courtesy⟩ **b :** EVERLASTING, INEXHAUSTIBLE ⟨a subject of *unfailing* interest⟩ **c :** INFALLIBLE, SURE ⟨an *unfailing* test⟩
— **un·fail·ing·ly** \-liŋ-lē\ *adverb*

un·fair \ˌən-ˈfar, -ˈfer\ *adjective* (1700)
1 : marked by injustice, partiality, or deception **:** UNJUST
2 : not equitable in business dealings
— **un·fair·ness** *noun*

un·fair·ly *adverb* (1713)
: in an unfair manner

un·faith \ˌən-ˈfāth, ˈən-ˌ\ *noun* (15th century)
: absence of faith **:** DISBELIEF

un·faith·ful \ˌən-ˈfāth-fəl\ *adjective* (15th century)
: not faithful: **a :** not adhering to vows, allegiance, or duty **:** DISLOYAL **b :** not faithful to marriage vows **c :** INACCURATE, UNTRUSTWORTHY
— **un·faith·ful·ly** \-fə-lē\ *adverb*
— **un·faith·ful·ness** *noun*

un·fall·en \ˌən-ˈfȯ-lən\ *adjective* (1653)
: not morally fallen **:** INNOCENT 1a

un·fal·si·fi·able \ˌən-ˌfȯl-sə-ˈfī-ə-bəl\ *adjective* (circa 1934)
: not capable of being proved false ⟨*unfalsifiable* hypotheses⟩

un·fal·ter·ing \ˌən-ˈfȯl-t(ə-)riŋ\ *adjective* (1727)
: not wavering or weakening **:** FIRM
— **un·fal·ter·ing·ly** \-t(ə-)riŋ-lē\ *adverb*

un·fa·mil·iar \ˌən-fə-ˈmil-yər\ *adjective* (1594)
: not familiar: **a :** not well-known **:** STRANGE ⟨an *unfamiliar* place⟩ **b :** not well acquainted ⟨*unfamiliar* with the subject⟩
— **un·fa·mil·iar·i·ty** \-ˌmil-ˈyar-ə-tē, -ˌmi-lē-ˈyar-\ *noun*
— **un·fa·mil·iar·ly** \-ˈmil-yər-lē\ *adverb*

un·fash·ion·able \-ˈfa-sh(ə-)nə-bəl\ *adjective* (1648)
1 : not in keeping with the current fashion ⟨*unfashionable* clothes⟩
2 : not favored socially ⟨*unfashionable* neighborhoods⟩
— **un·fash·ion·able·ness** \-bəl-nəs\ *noun*
— **un·fash·ion·ably** \-blē\ *adverb*

un·fas·ten \-ˈfa-sᵊn\ *transitive verb* (15th century)
: to make loose: as **a :** UNPIN, UNBUCKLE **b :** UNDO ⟨*unfasten* a button⟩ **c :** DETACH ⟨*unfasten* a boat from its moorings⟩

un·fa·thered \-ˈfä-t͟hərd\ *adjective* (1597)
1 : having no father **:** ILLEGITIMATE, BASTARD
2 : having no known origin ⟨*unfathered* slanders⟩

un·fath·om·able \-ˈfa-t͟hə-mə-bəl\ *adjective* (1640)
: not capable of being fathomed: **a :** impossible to comprehend **b :** IMMEASURABLE

un·fa·vor·able \ˌən-ˈfā-v(ə-)rə-bəl, -ˈfā-vər-bəl\ *adjective* (1548)
1 a : OPPOSED, CONTRARY **b :** expressing disapproval **:** NEGATIVE ⟨*unfavorable* reviews⟩
2 : not propitious **:** DISADVANTAGEOUS
3 : not pleasing
— **un·fa·vor·able·ness** *noun*
— **un·fa·vor·ably** \-blē\ *adverb*

un·fa·vor·ite \-ˈfā-v(ə-)rət, -ˈfā-vərt\ *adjective* (circa 1934)
: not being a favorite; *especially* **:** being regarded with special disfavor or dislike

un·fazed \ˌən-ˈfāzd\ *adjective* (1945)
: not fazed **:** UNDAUNTED

un·feel·ing \-ˈfē-liŋ\ *adjective* (before 12th century)
1 : devoid of feeling **:** INSENSATE ⟨an *unfeeling* corpse⟩
2 : devoid of kindness or sympathy **:** HARDHEARTED, CRUEL ⟨an *unfeeling* wretch⟩
— **un·feel·ing·ly** \-liŋ-lē\ *adverb*
— **un·feel·ing·ness** *noun*

un·feigned \-ˈfānd\ *adjective* (14th century)
: not feigned or hypocritical **:** GENUINE
synonym see SINCERE
— **un·feigned·ly** \-ˈfā-nəd-lē, -ˈfānd-lē\ *adverb*

un·fet·ter \-ˈfe-tər\ *transitive verb* (14th century)
1 : to free from fetters ⟨*unfetter* a prisoner⟩
2 : EMANCIPATE, LIBERATE ⟨*unfetter* the mind from prejudice⟩

un·fet·tered \-tərd\ *adjective* (1601)
: FREE, UNRESTRAINED

un·fil·ial \ˌən-ˈfi-lē-əl, -ˈfil-yəl\ *adjective* (1611)
: not observing the obligations of a child to a parent **:** UNDUTIFUL
— **un·fil·ial·ly** *adverb*

un·fil·tered \ˌən-ˈfil-tərd\ *adjective* (circa 1775)
1 : not filtered ⟨*unfiltered* wine⟩; *also* **:** not modified, processed, or refined ⟨*unfiltered* commercial publicity material —Paul Grimes⟩
2 : lacking a filter ⟨an *unfiltered* cigarette⟩

un·find·able \ˌən-ˈfīn-də-bəl\ *adjective* (1791)
: not capable of being found

un·fin·ished \-ˈfi-nisht\ *adjective* (1539)
: not finished: **a :** not brought to an end or to the desired final state **b :** being in a rough state **:** UNPOLISHED **c :** subjected to no other processes (as bleaching or dyeing) after coming from the loom

¹un·fit \-ˈfit\ *adjective* (1545)
: not fit: **a :** not adapted to a purpose **:** UNSUITABLE **b :** not qualified **:** INCAPABLE, INCOMPETENT **c :** physically or mentally unsound
— **un·fit·ly** *adverb*
— **un·fit·ness** *noun*

²unfit *transitive verb* (1611)
: to make unfit **:** DISABLE, DISQUALIFY

un·fit·ted \ˌən-ˈfi-təd\ *adjective* (1592)
: not adapted **:** UNQUALIFIED

un·fit·ting \-ˈfi-tiŋ\ *adjective* (1534)
: not fitting **:** UNSUITABLE

un·fix \-ˈfiks\ *transitive verb* (1597)
1 : to loosen from a fastening **:** DETACH, DISENGAGE
2 : to make unstable **:** UNSETTLE

un·flag·ging \-ˈfla-giŋ\ *adjective* (1715)
1 : not flagging **:** TIRELESS
2 : UNRELENTING 2
— **un·flag·ging·ly** \-giŋ-lē\ *adverb*

un·flap·pa·ble \-ˈfla-pə-bəl\ *adjective* [¹*un-* + ¹*flap* (state of excitement) + *-able*] (1954)
: marked by assurance and self-control
— **un·flap·pa·bil·i·ty** \-ˌfla-pə-ˈbi-lə-tē\ *noun*
— **un·flap·pa·bly** \-ˈfla-pə-blē\ *adverb*

un·flat·ter·ing \-ˈfla-tə-riŋ\ *adjective* (1581)
: not flattering; *especially* **:** UNFAVORABLE
— **un·flat·ter·ing·ly** \-riŋ-lē\ *adverb*

un·fledged \ˌən-ˈflejd\ *adjective* (1611)
1 : not feathered **:** not ready for flight
2 : not fully developed **:** IMMATURE ⟨an *unfledged* writer⟩

un·flinch·ing \-ˈflin-chiŋ\ *adjective* (1728)
: not flinching or shrinking **:** STEADFAST, UNCOMPROMISING
— **un·flinch·ing·ly** \-chiŋ-lē\ *adverb*

un·fo·cused *also* **un·fo·cussed** \-ˈfō-kəst\ *adjective* (1886)
1 : not adjusted to a focus
2 : not concentrated on one point or objective ⟨*unfocused* rage⟩

un·fold \-ˈfōld\ (before 12th century)
transitive verb
1 a : to open the folds of **:** spread or straighten out **:** EXPAND ⟨*unfolded* the map⟩ **b :** to remove (as a package) from the folds **:** UNWRAP
2 : to open to the view **:** REVEAL; *especially* **:** to make clear by gradual disclosure and often by recital
intransitive verb
1 a : to open from a folded state **:** open out **:** EXPAND **b :** BLOSSOM
2 : DEVELOP, EVOLVE ⟨as the story *unfolds*⟩
3 : to open out gradually to the view or understanding **:** become known ⟨a panorama *unfolds* before their eyes⟩
— **un·fold·ment** \-ˈfōl(d)-mənt\ *noun*

un·fold·ed *adjective* (1683)
: not folded

un·for·get·ta·ble \ˌən-fər-ˈge-tə-bəl\ *adjective* (1806)
: incapable of being forgotten **:** MEMORABLE
— **un·for·get·ta·bly** \-ˈge-tə-blē\ *adverb*

un·for·giv·ing \ˌən-fər-ˈgi-viŋ\ *adjective* (1713)
1 : unwilling or unable to forgive
2 : having or making no allowance for error or weakness ⟨an *unforgiving* environment where false moves can prove fatal —Jaclyn Fierman⟩
— **un·for·giv·ing·ness** *noun*

un·formed \-ˈfȯrmd\ *adjective* (14th century)
: not arranged in regular shape, order, or relations; *especially* **:** IMMATURE, UNDEVELOPED

¹un·for·tu·nate \-ˈfȯrch-nət, -ˈfȯr-chə-\ *adjective* (15th century)
1 a : not favored by fortune **:** UNSUCCESSFUL, UNLUCKY ⟨an *unfortunate* young man⟩ **b :** marked or accompanied by or resulting in misfortune ⟨an *unfortunate* decision⟩
2 a : INFELICITOUS, UNSUITABLE ⟨an *unfortunate* choice of words⟩ **b :** DEPLORABLE, REGRETTABLE ⟨an *unfortunate* lack of taste⟩

²unfortunate *noun* (1683)
: an unfortunate person

un·for·tu·nate·ly \-lē\ *adverb* (circa 1548)
1 : in an unfortunate manner ⟨the marriage turned out *unfortunately*⟩
2 : it is unfortunate ⟨*unfortunately* for him your letter has let the cat out of the bag —G. B. Shaw⟩

un·found·ed \ˌən-ˈfau̇n-dəd\ *adjective* (1648)
: lacking a sound basis **:** GROUNDLESS, UNWARRANTED

un·freeze \-ˈfrēz\ *transitive verb* **-froze** \-ˈfrōz\; **-fro·zen** \-ˈfrō-zᵊn\; **-freez·ing** (1584)
1 : to cause to thaw
2 : to remove from a freeze

un·fre·quent·ed \ˌən-frē-ˈkwen-təd; ˌən-ˈfrē-kwən-\ *adjective* (1588)
: not often visited or traveled over

un·friend·ed \ˌən-ˈfren-dəd\ *adjective* (1513)
: having no friends **:** not befriended

un·friend·li·ness \-ˈfren(d)-lē-nəs\ *noun* (circa 1684)
: the quality or state of being unfriendly : HOSTILITY

un·friend·ly \-ˈfren(d)-lē\ *adjective* (15th century)
: not friendly: as **a** : HOSTILE, UNSYMPATHETIC **b** : INHOSPITABLE, UNFAVORABLE

un·frock \-ˈfräk\ *transitive verb* (1644)
: DEFROCK

un·froz·en \-ˈfrō-zᵊn\ *adjective* (1596)
: not frozen ⟨*unfrozen* ground⟩

un·fruit·ful \-ˈfrüt-fəl\ *adjective* (14th century)
: not fruitful: as **a** : not producing offspring : BARREN **b** : yielding no valuable result : UNPROFITABLE ⟨an *unfruitful* conference⟩
— **un·fruit·ful·ly** \-fə-lē\ *adverb*
— **un·fruit·ful·ness** *noun*

un·fund·ed \-ˈfən-dəd\ *adjective* (circa 1775)
1 : not funded : FLOATING ⟨an *unfunded* debt⟩
2 : not provided with funds ⟨*unfunded* schools⟩

un·furl \-ˈfər(-ə)l\ (1641)
transitive verb
: to release from a furled state
intransitive verb
: to open out from or as if from a furled state : UNFOLD

un·fussy \-ˈfə-sē\ *adjective* (1825)
: not fussy: as **a** : not particular : UNCONCERNED **b** : not cluttered with pretentious or nonessential matters : UNCOMPLICATED
— **un·fuss·i·ly** \-ˈfə-sə-lē\ *adverb*

un·gain·ly \-ˈgān-lē\ *adjective* (1611)
1 a : lacking in smoothness or dexterity : CLUMSY **b** : hard to handle : UNWIELDY
2 : having an awkward appearance : UGLY
— **un·gain·li·ness** *noun*

un·gen·er·os·i·ty \ˌən-ˌje-nə-ˈrä-sə tē, -ˈräs-tē\ *noun* (1757)
: lack of generosity

un·gen·er·ous \ˌən-ˈjen-rəs, -ˈje-nə-\ *adjective* (1641)
: not generous: **a** : PETTY, MEAN **b** : deficient in liberality : STINGY
— **un·gen·er·ous·ly** *adverb*

un·gird \-ˈgərd\ *transitive verb* (before 12th century)
: to divest of a restraining band or girdle : UNBIND

un·girt \-ˈgərt\ *adjective* (14th century)
1 : having the belt or girdle off or loose
2 : lacking in discipline or compactness : LOOSE, SLACK

un·glue \-ˈglü\ *transitive verb* (circa 1548)
: to separate by or as if by dissolving an adhesive

un·glued \-ˈglüd\ *adjective* (1922)
: UPSET, DISORDERED

un·god·li·ness \ˌən-ˈgäd-lē-nəs *also* -ˈgȯd-\ *noun* (1526)
: the quality or state of being ungodly

un·god·ly \-lē\ *adjective* (1526)
1 a : denying or disobeying God : IMPIOUS, IRRELIGIOUS **b** : contrary to moral law : SINFUL, WICKED
2 : OUTRAGEOUS ⟨gets up at an *ungodly* hour⟩

un·got·ten \-ˈgä-tᵊn\ *or* **un·got** \-ˈgät\ *adjective* (15th century)
1 *obsolete* : not begotten
2 : not obtained

un·gov·ern·able \-ˈgə-vər-nə-bəl\ *adjective* (1673)
: not capable of being governed, guided, or restrained
synonym see UNRULY

un·gra·cious \-ˈgrā-shəs\ *adjective* (13th century)
1 *archaic* : WICKED
2 : not courteous : RUDE
3 : not pleasing : DISAGREEABLE
— **un·gra·cious·ly** *adverb*
— **un·gra·cious·ness** *noun*

un·gram·mat·i·cal \ˌən-grə-ˈma-ti-kəl\ *adjective* (1654)
: not following rules of grammar
— **un·gram·mat·i·cal·i·ty** \-ˌma-tə-ˈka-lə-tē\ *noun*

un·grate·ful \ˌən-ˈgrāt-fəl\ *adjective* (1533)
1 : showing no gratitude : making a poor return
2 : DISAGREEABLE; *also* : THANKLESS
— **un·grate·ful·ly** \-fə-lē\ *adverb*
— **un·grate·ful·ness** *noun*

un·grudg·ing \-ˈgrə-jiŋ\ *adjective* (circa 1774)
: being without envy or reluctance

un·gual \ˈəŋ-gwəl, ˈən-\ *adjective* [Latin *unguis* nail, claw, hoof — more at NAIL] (1834)
: of, relating to, or resembling a nail, claw, or hoof

un·guard \ˌən-ˈgärd\ *transitive verb* [back-formation from *unguarded*] (1745)
: to leave unprotected

un·guard·ed \-ˈgär-dəd\ *adjective* (circa 1593)
1 : vulnerable to attack : UNPROTECTED
2 : free from guile or wariness : DIRECT, INCAUTIOUS
— **un·guard·ed·ly** *adverb*
— **un·guard·ed·ness** *noun*

un·guent \ˈəŋ-gwənt, ˈən-; ˈən-jənt\ *noun* [Middle English, from Latin *unguentum* — more at OINTMENT] (15th century)
: a soothing or healing salve : OINTMENT

un·guis \ˈəŋ-gwəs, ˈən-\ *noun, plural* **un·gues** \-ˌgwēz\ [Latin — more at NAIL] (circa 1790)
: a nail, claw, or hoof especially on a digit of a vertebrate

¹un·gu·late \ˈəŋ-gyə-lət, ˈən-, -ˌlāt\ *adjective* [Late Latin *ungulatus,* from Latin *ungula* hoof, from *unguis* nail, hoof] (1839)
1 : having hoofs
2 : of or relating to the ungulates

²ungulate *noun* [New Latin *Ungulata,* from Late Latin, neuter plural of *ungulatus*] (circa 1842)
: a hoofed typically herbivorous quadruped mammal (as a ruminant, swine, camel, hippopotamus, horse, tapir, rhinoceros, elephant, or hyrax) of a polyphyletic group formerly considered a major mammalian taxon (Ungulata)

un·hair \ˌən ˈhar, -ˈher\ *transitive verb* (14th century)
archaic : to deprive of hair

un·hal·low \-ˈha-(ˌ)lō\ *transitive verb* (1535)
archaic : to make profane

un·hal·lowed \-(ˌ)lōd\ *adjective* (before 12th century)
1 : not blessed : UNCONSECRATED, UNHOLY
2 a : unsanctioned by or showing lack of reverence for religion : IMPIOUS, PROFANE **b** : contrary to accepted standards : IMMORAL

un·hand \ˌən-ˈhand\ *transitive verb* (1602)
: to remove the hand from : let go

un·hand·some \-ˈhan(t)-səm\ *adjective* (1530)
: not handsome: as **a** : not beautiful : HOMELY **b** : UNBECOMING, UNSEEMLY **c** : lacking in courtesy or taste : RUDE
— **un·hand·some·ly** *adverb*

un·handy \-ˈhan-dē\ *adjective* (1664)
1 : hard to handle : INCONVENIENT
2 : lacking in skill or dexterity : AWKWARD
— **un·hand·i·ly** \-də-lē\ *adverb*
— **un·hand·i·ness** \-dē-nəs\ *noun*

un·hap·pi·ly \ˌən-ˈha-pə-lē\ *adverb* (14th century)
1 : UNFORTUNATELY 2 ⟨*unhappily,* medicine has not yet found a cure —Diana Trilling⟩
2 : in an unhappy manner : without pleasure ⟨practiced law *unhappily* for a few years⟩

un·hap·py \-ˈha-pē\ *adjective* (14th century)
1 : not fortunate : UNLUCKY
2 : not cheerful or glad : SAD, WRETCHED
3 a : causing or subject to misfortune : INAUSPICIOUS **b** : INFELICITOUS, INAPPROPRIATE
— **un·hap·pi·ness** *noun*

un·healthy \-ˈhel-thē\ *adjective* (1595)
1 : not conducive to health ⟨an *unhealthy* climate⟩
2 : not in good health : SICKLY, DISEASED
3 a : DANGEROUS, RISKY **b** : BAD, INJURIOUS **c** : morally contaminated : CORRUPT, UNWHOLESOME ⟨an *unhealthy* imagination⟩
— **un·health·i·ly** \-thə-lē\ *adverb*
— **un·health·i·ness** \-thē-nəs\ *noun*

un·heard \-ˈhərd\ *adjective* (14th century)
1 a : not perceived by the ear **b** : not given a hearing
2 *archaic* : UNHEARD-OF

un·heard–of \-ˌəv, -ˌäv\ *adjective* (1592)
: previously unknown; *especially* : UNPRECEDENTED

un·hes·i·tat·ing \-ˈhe-zə-ˌtā-tiŋ\ *adjective* (1753)
: not hesitating : not checked or qualified
— **un·hes·i·tat·ing·ly** \-tiŋ-lē\ *adverb*

un·hinge \-ˈhinj\ *transitive verb* (1616)
1 : to remove (as a door) from the hinges
2 : to make unstable : UNSETTLE, DISRUPT ⟨*unhinge* the balance of world peace⟩ ⟨pressure that would *unhinge* a less experienced person⟩

un·hitch \-ˈhich\ *transitive verb* (1706)
: to free from or as if from being hitched

un·ho·ly \ˌən-ˈhō-lē\ *adjective* (before 12th century)
1 : showing disregard for what is holy : WICKED
2 : SHOCKING, OUTRAGEOUS
— **un·ho·li·ness** *noun*

un·hood \-ˈhu̇d\ *transitive verb* (1575)
: to remove a hood or covering from

un·hook \-ˈhu̇k\ *transitive verb* (1611)
1 : to remove from a hook
2 : to unfasten by disengaging a hook
3 : to free from a habit or dependency

un·hoped \-ˈhōpt\ *adjective* (14th century)
archaic : not hoped for or expected

un·horse \-ˈhȯrs\ *transitive verb* (14th century)
: to dislodge from or as if from a horse

un·hou·seled \-ˈhau̇-zəld\ *adjective* (1532)
archaic : not having received the Eucharist especially shortly before death

un·hur·ried \-ˈhər-ēd, -ˈhə rēd\ *adjective* (circa 1774)
: not hurried : LEISURELY
— **un·hur·ried·ly** *adverb*

uni- *prefix* [Middle English, from Middle French, from Latin, from *unus* — more at ONE]
: one : single ⟨*unicellular*⟩

uni·al·gal \ˌyü-nē-ˈal-gəl\ *adjective* (1914)
: of, relating to, or derived from a single algal individual or cell ⟨a *unialgal* culture⟩

Uni·ate *or* **Uni·at** \ˈyü-nē-ˌat, ˈü-\ *noun* [Ukrainian *uniat, uniyat* one in favor of the union of the Greek and Roman Catholic churches, from *uniya* union, from Polish *unija,* from Late Latin *unio* — more at UNION] (1833)
: a Christian of a church adhering to an Eastern rite and discipline but submitting to papal authority
— **Uniate** *adjective*

uni·ax·i·al \ˌyü-nē-ˈak-sē-əl\ *adjective* (circa 1828)
1 : having only one axis
2 : of or relating to only one axis

uni·cam·er·al \ˌyü-ni-ˈkam-rəl, -ˈka-mə-\ *adjective* [*uni-* + Late Latin *camera* room, chamber — more at CHAMBER] (1853)
: having or consisting of a single legislative chamber
— **uni·cam·er·al·ly** *adverb*

uni·cel·lu·lar \ˌyü-ni-ˈsel-yə-lər\ *adjective* (1858)
: having or consisting of a single cell

uni·corn \ˈyü-nə-ˌkȯrn\ *noun* [Middle English *unicorne,* from Old French, from Late Latin *unicornis,* from Latin, having one horn, from *uni-* + *cornu* horn — more at HORN] (13th century)
: a mythical animal generally depicted with the body and head of a horse, the hind legs of a stag, the tail of a lion, and a single horn in the middle of the forehead ◆

unicorn

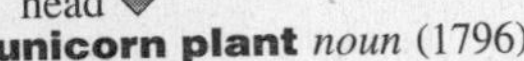
unicorn plant *noun* (1796)
: DEVIL'S CLAW

uni·cy·cle \ˈyü-ni-ˌsī-kəl\ *noun* [*uni* + *-cycle* (as in *tricycle*)] (1869)
: a vehicle that has a single wheel and is usually propelled by pedals
— **uni·cy·clist** \-ˌsī-k(ə-)list\ *noun*

unicycle

uni·di·men·sion·al \ˌyü-ni-də-ˈmench-nəl, -ˈmen(t)-sh(ə-)nᵊl *also* -ˌdī-\ *adjective* (1883)
: ONE-DIMENSIONAL
— **uni·di·men·sion·al·i·ty** \-ˌmen(t)-shə-ˈna-lə-tē\ *noun*

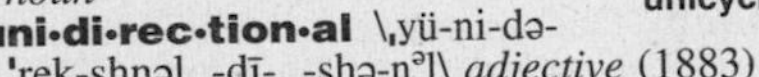
uni·di·rec·tion·al \ˌyü-ni-də-ˈrek-shnəl, -dī-, -shə-nᵊl\ *adjective* (1883)
1 : involving, functioning, moving, or responsive in a single direction
2 : not subject to change or reversal of direction
— **uni·di·rec·tion·al·ly** *adverb*

unidirectional current *noun* (1883)
: DIRECT CURRENT

uni·fi·ca·tion \ˌyü-nə-fə-ˈkā-shən\ *noun* (1851)
: the act, process, or result of unifying : the state of being unified

uni·fo·li·ate \-ˈfō-lē-ət\ *adjective* [*uni-* + Latin *folium* leaf — more at BLADE] (1849)
1 : having only one leaf
2 : UNIFOLIOLATE

uni·fo·li·o·late \-ˈfō-lē-ə-ˌlāt\ *adjective* [*uni-* + Late Latin *foliolum* leaflet, diminutive of Latin *folium* leaf] (circa 1859)
of a leaf **:** compound but having only a single leaflet and distinguishable from a simple leaf by the basal joint

¹uni·form \ˈyü-nə-ˌfȯrm\ *adjective* [Middle French *uniforme,* from Latin *uniformis,* from *uni-* + *-formis* -form] (1538)
1 : consistent in conduct or opinion ⟨*uniform* interpretation of laws⟩
2 : having always the same form, manner, or degree **:** not varying or variable
3 : of the same form with others **:** conforming to one rule or mode **:** CONSONANT
4 : presenting an unvaried appearance of surface, pattern, or color ⟨*uniform* red brick houses⟩
5 : relating to or being convergence of a series whose terms are functions in such manner that the absolute value of the difference between the sum of the first *n* terms of the series and the sum of all terms can be made arbitrarily small for all values of the domain of the functions by choosing the *n*th term sufficiently far along in the series
— **uni·form·ly** \ˈyü-nə-ˌfȯrm-lē, ˌyü-nə-ˈ\ *adverb*
— **uni·form·ness** \ˈyü-nə-ˌfȯrm-nəs\ *noun*

²uniform *transitive verb* (circa 1681)
1 : to bring into uniformity
2 : to clothe with a uniform

³uniform *noun* (1748)
: dress of a distinctive design or fashion worn by members of a particular group and serving as a means of identification; *broadly* **:** distinctive or characteristic clothing

Uniform (circa 1956)
— a communication code word for the letter *u*

uni·for·mi·tar·i·an \ˌyü-nə-ˌfȯr-mə-ˈter-ē-ən\ *noun* (1840)
1 : an adherent of the doctrine of uniformitarianism
2 : an advocate of uniformity
— **uniformitarian** *adjective*

uni·for·mi·tar·i·an·ism \-ē-ə-ˌni-zəm\ *noun* (1865)
: a geological doctrine that existing processes acting in the same manner as at present are sufficient to account for all geological changes — compare CATASTROPHISM

uni·for·mi·ty \ˌyü-nə-ˈfȯr-mə-tē\ *noun, plural* **-ties** (15th century)
1 : the quality or state of being uniform
2 : an instance of uniformity

uni·fy \ˈyü-nə-ˌfī\ *transitive verb* **-fied; -fy·ing** [Late Latin *unificare,* from Latin *uni-* + *-ficare* -fy] (1502)
: to make into a unit or a coherent whole : UNITE
— **uni·fi·able** \-ˌfī-ə-bəl\ *adjective*
— **uni·fi·er** \-ˌfī(-ə)r\ *noun*

uni·lat·er·al \ˌyü-ni-ˈla-tə-rəl, -ˈla-trəl\ *adjective* (1802)
1 a : done or undertaken by one person or party **b :** of, relating to, or affecting one side of a subject **:** ONE-SIDED **c :** constituting or relating to a contract or engagement by which an express obligation to do or forbear is imposed on only one party
2 a : having parts arranged on one side ⟨a *unilateral* raceme⟩ **b :** occurring on, performed on, or affecting one side of the body or one of its parts ⟨*unilateral* exophthalmos⟩
3 : UNILINEAL
4 : having only one side
— **uni·lat·er·al·ly** *adverb*

uni·lin·e·al \-ˈli-nē-əl\ *adjective* (1952)
: tracing descent through either the maternal or paternal line only

uni·lin·e·ar \ˌyü-ni-ˈli-nē-ər\ *adjective* (1851)
: developing in or involving a series of stages usually from the primitive to the more advanced ⟨a *unilinear* cultural sequence⟩

uni·lin·gual \ˌyü-ni-ˈliŋ-gwəl, -gyə-wəl\ *adjective* [*uni-* + Latin *lingua* tongue, language — more at TONGUE] (1866)
: composed in or using one language only

un·il·lu·sioned \ˌən-i-ˈlü-zhənd\ *adjective* (1926)
: free from illusion

uni·loc·u·lar \ˌyü-ni-ˈlä-kyə-lər\ *adjective* (1753)
: containing a single cavity

un·imag·in·able \ˌən-ə-ˈmaj-nə-bəl, -ˈma-jə-\ *adjective* (1611)
: not imaginable or comprehensible
— **un·imag·in·ably** \-blē\ *adverb*

un·im·peach·able \ˌən-im-ˈpē-chə-bəl\ *adjective* (1784)
: not impeachable **:** not to be called in question **:** not liable to accusation **:** IRREPROACHABLE, BLAMELESS
— **un·im·peach·ably** \-blē\ *adverb*

¹un·im·proved \-ˈprüvd\ *adjective* (1602)
obsolete **:** not reproved or admonished

²unimproved *adjective* (1665)
: not improved: as **a :** not tilled, built on, or otherwise improved for use ⟨*unimproved* land⟩ **b :** not used or employed advantageously ⟨wasted time and *unimproved* opportunities⟩ **c :** not selectively bred for better quality or productiveness

un·in·hib·it·ed \ˌən-in-ˈhi-bə-təd\ *adjective* (1880)
: free from inhibition; *also* **:** boisterously informal
— **un·in·hib·it·ed·ly** *adverb*
— **un·in·hib·it·ed·ness** *noun*

un·ini·tiate \ˌən-ə-ˈni-sh(ē-)ət\ *adjective* (1801)
: not initiated **:** INEXPERIENCED
— **uninitiate** *noun*

un·in·ter·est \ˌən-ˈin-trəst; -ˈin-tə-rəst, -tə-ˌrest, -tərst; -ˈin-ˌtrest\ *noun* (1890)
: lack of interest

uninterested *adjective* (1661)
: not interested **:** not having the mind or feelings engaged
usage see DISINTERESTED

uni·nu·cle·ate \ˌyü-ni-ˈnü-klē-ət, -ˈnyü-\ *adjective* (1885)
: having a single nucleus ⟨a *uninucleate* yeast cell⟩

¹union \ˈyün-yən\ *noun* [Middle English, from Middle French, from Late Latin *union-, unio* oneness, union, from Latin *unus* one — more at ONE] (15th century)
1 a : an act or instance of uniting or joining two or more things into one: as (1) **:** the formation of a single political unit from two or more separate and independent units (2) **:** a uniting in marriage; *also* **:** SEXUAL INTERCOURSE (3) **:** the growing together of severed parts **b :** a unified condition **:** COMBINATION, JUNCTION ⟨a gracious *union* of excellence and strength⟩
2 : something that is made one **:** something formed by a combining or coalition of parts or members: as **a :** a confederation of independent individuals (as nations or persons) for some common purpose **b** (1) **:** a political unit constituting an organic whole formed usually from previously independent units (as England and Scotland in 1707) which have surrendered their principal powers to the government of the whole or to a newly created government (as the U.S. in 1789) (2) *capitalized* **:** the federal union of states during the period of the U.S. Civil War **c** *capitalized* **:** an organization on a college or university campus providing recreational, social, cultural, and sometimes dining facilities; *also* **:** the building housing such an organization **d :** the set of all elements belonging to one or more of a given collection of two or more sets — called also *join, sum* **e :** LABOR UNION
3 a : a device emblematic of the union of two or more sovereignties borne on a national flag typically in the upper inner corner or constituting the whole design of the flag **b :** the upper inner corner of a flag
4 : any of various devices for connecting parts (as of a machine); *especially* **:** a coupling for pipes or pipes and fittings

◇ WORD HISTORY

unicorn The two-horned African rhinoceros was quite familiar in the ancient Mediterranean world, even to the urban populaces of Alexandria and Rome, where specimens are reported to have been exhibited. The Greeks dubbed the animal *rhinokerōs,* literally, "nose-horned one"—an appropriate name, since the rhinoceros's primary horn protrudes from above its snout, unlike the horns of other ungulates, which grow above the brow. The single-horned Indian rhinoceros, on the other hand, was never observed firsthand by ancient authors, and descriptions of it are rather garbled. By late antiquity legend had shaped out of the Indian rhinoceros a mythical animal, the *monokerōs* ("single-horned one"). This animal is described in the *Physiologus,* an anonymous 4th century Greek prototype of the medieval bestiary, as having most of the physical features we now associate with the unicorn. The *Physiologus* introduces as well the belief that the fierce unicorn can only be subdued by a virgin, and attaches to this trait a Christian interpretation, that the taming of the unicorn symbolizes Christ's incarnation in a virgin. Authors writing in Latin translated Greek *monokerōs* as *unicornis* or *unicornuus,* the ultimate source of English *unicorn.*

²union *adjective* (1707)
: of, relating to, dealing with, or constituting a union; *especially, capitalized* **:** of, relating to, or being the side favoring the Union in the U.S. Civil War ⟨*Union* troops⟩

union card *noun* (1874)
1 : a card certifying personal membership in good standing in a labor union
2 : something that resembles a union card especially in being necessary for employment or in providing evidence of in-group status

union church *noun* (1847)
: a local church uniting members of diverse denominational backgrounds in an interdenominational congregation

un·ion·i·sa·tion, union·ise *British variant of* UNIONIZATION, UNIONIZE

union·ism \ˈyün-yə-ˌni-zəm\ *noun* (1845)
: the principle or policy of forming or adhering to a union: as **a** *capitalized* **:** adherence to the policy of a firm federal union between the states of the United States especially during the Civil War period **b :** the principles, theory, advocacy, or system of trade unions

union·ist \-nist\ *noun* (1799)
: an advocate or supporter of union or unionism

union·i·za·tion \ˌyün-yə-nə-ˈzā-shən\ *noun* (1896)
1 : the quality or state of being unionized
2 : the action of unionizing

union·ize \ˈyün-yə-ˌnīz\ *verb* **-ized; -iz·ing** (1890)
transitive verb
: to organize into a labor union
intransitive verb
: to form or join a labor union

unionized *adjective* (1900)
: characterized by the presence of labor unions ⟨*unionized* states⟩

union jack *noun* (1674)
1 : a jack consisting of the union of a national ensign
2 *U&J capitalized* **:** the state flag of the United Kingdom consisting of the union of the British national ensign

union shop *noun* (1904)
: an establishment in which the employer by agreement is free to hire nonmembers as well as members of the union but retains nonmembers on the payroll only on condition of their becoming members of the union within a specified time

union suit *noun* (1892)
: an undergarment with shirt and drawers in one piece

union suit

uni·pa·ren·tal \ˌyü-ni-pə-ˈren-tᵊl\ *adjective* (1900)
: having or involving a single parent; *especially* **:** PARTHENOGENETIC
— uni·pa·ren·tal·ly \-tᵊl-ē\ *adverb*

unique \yu̇-ˈnēk\ *adjective* [French, from Latin *unicus,* from *unus* one — more at ONE] (1602)
1 : being the only one **:** SOLE ⟨his *unique* concern was his own comfort⟩ ⟨I can't walk away with a *unique* copy. Suppose I lost it? —Kingsley Amis⟩ ⟨the *unique* factorization of a number into prime factors⟩
2 a : being without a like or equal **:** UNEQUALED ⟨could stare at the flames, each one new, violent, *unique* —Robert Coover⟩ **b :** distinctively characteristic **:** PECULIAR 1 ⟨this is not a condition *unique* to California —Ronald Reagan⟩
3 : UNUSUAL ⟨a very *unique* ball-point pen⟩ ⟨we were fairly *unique*, the sixty of us, in that there wasn't one good mixer in the bunch —J. D. Salinger⟩ ■
synonym see STRANGE
— unique·ly *adverb*
— unique·ness *noun*

¹uni·sex \ˈyü-nə-ˌseks\ *noun* (1966)
: the state or condition of not being distinguishable (as by hair or clothing) as to sex

²unisex *adjective* (1968)
1 : not distinguishable as male or female ⟨a *unisex* face⟩
2 : suitable or designed for both males and females ⟨*unisex* clothes⟩

uni·sex·u·al \ˌyü-nə-ˈsek-sh(ə-)wəl, -shəl\ *adjective* (circa 1802)
1 : of, relating to, or restricted to one sex: **a :** male or female but not hermaphroditic **b :** DICLINOUS ⟨a *unisexual* flower⟩
2 : UNISEX
— uni·sex·u·al·i·ty \-ˌsek-shə-ˈwa-lə-tē\ *noun*

uni·son \ˈyü-nə-sən, -nə-zən\ *noun* [Middle French, from Medieval Latin *unisonus* having the same sound, from Latin *uni-* + *sonus* sound — more at SOUND] (1575)
1 a : identity in musical pitch; *specifically* **:** the interval of a perfect prime **b :** the state of being so tuned or sounded **c :** the writing, playing, or singing of parts in a musical passage at the same pitch or in octaves
2 : a harmonious agreement or union **:** CONCORD
— unison *adjective*
— in unison 1 : in perfect agreement **:** so as to harmonize exactly **2 :** at the same time **:** SIMULTANEOUSLY

¹unit \ˈyü-nət\ *noun* [back-formation from *unity*] (1570)
1 a : the first and least natural number **:** ONE **b :** a single quantity regarded as a whole in calculation
2 : a determinate quantity (as of length, time, heat, or value) adopted as a standard of measurement: as **a :** an amount of work used in education in calculating student credits **b :** an amount of a biologically active agent (as a drug or antigen) required to produce a specific result under strictly controlled conditions
3 a : a single thing, person, or group that is a constituent of a whole **b :** a part of a military establishment that has a prescribed organization (as of personnel and materiel) **c :** a piece or complex of apparatus serving to perform one particular function **d :** a part of a school course focusing on a central theme **e :** a local congregation of Jehovah's Witnesses

²unit *adjective* (1844)
: being, relating to, or measuring one unit

unit·age \ˈyü-nə-tij\ *noun* (1935)
1 : specifications of the amount constituting a unit
2 : amount in units

uni·tard \ˈyü-nə-ˌtärd\ *noun* [*uni-* + leo*tard*] (1961)
: a close-fitting one-piece garment for the torso, legs, feet, and often the arms

uni·tar·i·an \ˌyü-nə-ˈter-ē-ən\ *noun* [New Latin *unitarius,* from Latin *unitas* unity] (1687)
1 a *often capitalized* **:** one who believes that the deity exists only in one person **b** *capitalized* **:** a member of a denomination that stresses individual freedom of belief, the free use of reason in religion, a united world community, and liberal social action
2 : an advocate of unity or a unitary system
— unitarian *adjective, often capitalized*
— uni·tar·i·an·ism \-ē-ə-ˌni-zəm\ *noun, often capitalized*

uni·tary \ˈyü-nə-ˌter-ē\ *adjective* (1861)
1 a : of or relating to a unit **b :** based on or characterized by unity or units
2 : having the character of a unit **:** UNDIVIDED, WHOLE
— uni·tar·i·ly \ˌyü-nə-ˈter-ə-lē\ *adverb*

unit cell *noun* (1915)
: the simplest polyhedron that embodies all the structural characteristics of and by indefinite repetition makes up the lattice of a crystal

unit character *noun* (1902)
: a natural character inherited on an all-or-none basis; *especially* **:** one dependent on the presence or absence of a single gene

unit circle *noun* (1955)
: a circle having a radius of 1

¹unite \yu̇-ˈnīt\ *verb* **unit·ed; unit·ing** [Middle English, from Latin *unire,* from *unus* one — more at ONE] (15th century)
transitive verb
1 a : to put together to form a single unit **b :** to cause to adhere **c :** to link by a legal or moral bond
2 : to possess (as qualities) in combination
intransitive verb
1 a : to become one or as if one **b :** to become combined by or as if by adhesion or mixture
2 : to act in concert
synonym see JOIN
— unit·er *noun*

²unite \ˈyü-ˌnīt\ *noun* [obsolete *unite* united, from Middle English *unit,* from Late Latin *unitus,* past participle] (1604)
: an old British gold 20-shilling piece issued first by James I in 1604 for the newly united England and Scotland — called also *Jacobus*

unit·ed \yu̇-ˈnī-təd\ *adjective* (1552)
1 : made one **:** COMBINED
2 : relating to or produced by joint action
3 : being in agreement **:** HARMONIOUS
— unit·ed·ly *adverb*

United Nations Day *noun* (1947)
: October 24 observed in commemoration of the founding of the United Nations

Unit·ed States \yu̇-ˈnī-təd-, *especially Southern* ˈyü-ˌ\ *noun plural but singular or plural in construction* (1617)
: a federation of states especially when forming a nation in a usually specified territory ⟨advocating a *United States* of Europe⟩

uni·tive \ˈyü-nə-tiv, yu̇-ˈnī-\ *adjective* (1526)
: characterized by or tending to produce union

unit·ize \ˈyü-nə-ˌtīz\ *transitive verb* **-ized; -iz·ing** (1860)
1 : to form or convert into a unit
2 : to divide into units ⟨the added cost of *unitizing* bulk products⟩
— unit·i·za·tion \ˌyü-nə-tə-ˈza-shən\ *noun*

unit membrane *noun* [from its being the basic structural unit of the cell] (1959)

□ USAGE

unique Many commentators have objected to the comparison or modification (as by *somewhat* or *very*) of *unique;* the statement that a thing is either unique or it is not has often been repeated by them. Objections are based chiefly on the assumption that *unique* has but a single absolute sense, an assumption contradicted by information readily available in a dictionary. *Unique* dates back to the 17th century but was little used until the end of the 18th when, according to the Oxford English Dictionary, it was reacquired from French. H. J. Todd entered it as a foreign word in his edition (1818) of Johnson's Dictionary, characterizing it as "affected and useless." Around the middle of the 19th century it ceased to be considered foreign and came into considerable popular use. With popular use came a broadening of application beyond the original two meanings (here numbered 1 and 2a). In modern use both comparison and modification are widespread and standard but are confined to the extended senses 2b and 3. When sense 1 or sense 2a is intended, *unique* is used without qualifying modifiers.

: a 3-layered membrane that consists of an inner lipid layer surrounded by a protein layer on each side

unit rule *noun* (1884)
: a rule under which a delegation to a national political convention casts its entire vote as a unit as determined by a majority vote

uni·trust \'yü-ni-ˌtrəst\ *noun* (1970)
: a trust from which the beneficiary receives annually a fixed percentage of the fair market value of its assets

units place *noun* (1937)
: the place just to the left of the decimal point in a number expressed in the Arabic system of writing numbers

unit train *noun* (1964)
: a railway train that transports a single commodity directly from producer to consumer

unit trust *noun* (1936)
1 *British* **:** MUTUAL FUND
2 : an investment company whose portfolio consists of long-term bonds that are held to maturity

uni·ty \'yü-nə-tē\ *noun, plural* **-ties** [Middle English *unite,* from Middle French *unité,* from Latin *unitat-, unitas,* from *unus* one — more at ONE] (14th century)
1 a : the quality or state of not being multiple **:** ONENESS **b** (1) **:** a definite amount taken as one or for which 1 is made to stand in calculation ⟨in a table of natural sines the radius of the circle is regarded as *unity*⟩ (2) **:** IDENTITY ELEMENT
2 a : a condition of harmony **:** ACCORD **b :** continuity without deviation or change (as in purpose or action)
3 a : the quality or state of being made one **:** UNIFICATION **b :** a combination or ordering of parts in a literary or artistic production that constitutes a whole or promotes an undivided total effect; *also* **:** the resulting singleness of effect or symmetry and consistency of style and character
4 : a totality of related parts **:** an entity that is a complex or systematic whole
5 : any of three principles of dramatic structure derived by French classicists from Aristotle's *Poetics* and requiring a play to have a single action represented as occurring in one place and within one day
6 *capitalized* **:** a 20th century American religious movement that emphasizes spiritual sources of health and prosperity

¹uni·va·lent \ˌyü-ni-'vā-lənt\ *adjective* (1898)
1 : MONOVALENT 1
2 : being a chromosomal univalent

²univalent *noun* (1912)
: a chromosome that lacks a synaptic mate

uni·valve \'yü-ni-ˌvalv\ *noun* (1668)
1 : a mollusk with a shell consisting of one valve; *especially* **:** GASTROPOD
2 : the shell of a univalve
— **univalve** *adjective*

uni·var·i·ate \ˌyü-ni-'ver-ē-ət, -'var-\ *adjective* (1928)
: characterized by or depending on only one random variable ⟨a *univariate* linear model⟩

¹uni·ver·sal \ˌyü-nə-'vər-səl\ *adjective* [Middle English, from Middle French, from Latin *universalis,* from *universum* universe] (14th century)
1 : including or covering all or a whole collectively or distributively without limit or exception
2 a : present or occurring everywhere **b :** existent or operative everywhere or under all conditions ⟨*universal* cultural patterns⟩
3 a : embracing a major part or the greatest portion (as of mankind) ⟨a *universal* state⟩ ⟨*universal* practices⟩ **b :** comprehensively broad and versatile ⟨a *universal* genius⟩
4 a : affirming or denying something of all members of a class or of all values of a variable **b :** denoting every member of a class ⟨a *universal* term⟩
5 : adapted or adjustable to meet varied requirements (as of use, shape, or size) ⟨a *universal* gear cutter⟩
— **uni·ver·sal·ly** \-s(ə-)lē\ *adverb*
— **uni·ver·sal·ness** \-səl-nəs\ *noun*

²universal *noun* (1553)
1 : one that is universal: as **a :** a universal proposition in logic **b :** a predicable of traditional logic **c :** a general concept or term or something in reality to which it corresponds **:** ESSENCE
2 a : a behavior pattern or institution (as the family) existing in all cultures **b :** a culture trait characteristic of all normal adult members of a particular society

universal donor *noun* (1922)
: a person who belongs to ABO blood group O and can donate blood to any recipient

universal grammar *noun* (1751)
: the study of general principles believed to underlie the grammatical phenomena of all languages; *also* **:** such principles viewed as part of an innate human capacity for learning a language

uni·ver·sal·ism \ˌyü-nə-'vər-sə-ˌli-zəm\ *noun* (1805)
1 *often capitalized* **a :** a theological doctrine that all human beings will eventually be saved **b :** the principles and practices of a liberal Christian denomination founded in the 18th century originally to uphold belief in universal salvation and now united with Unitarianism
2 : something that is universal in scope
3 : the state of being universal **:** UNIVERSALITY
— **uni·ver·sal·ist** \-s(ə-)list\ *noun or adjective, often capitalized*

uni·ver·sal·is·tic \-ˌvər-sə-'lis-tik\ *adjective* (1872)
: of or relating to the whole **:** universal in scope or nature

uni·ver·sal·i·ty \-(ˌ)vər-'sa-lə-tē\ *noun* (14th century)
1 : the quality or state of being universal
2 : universal comprehensiveness in range

uni·ver·sal·ize \-'vər-sə-ˌlīz\ *transitive verb* **-ized; -iz·ing** (1642)
: to make universal **:** GENERALIZE
— **uni·ver·sal·i·za·tion** \-ˌvər-sə-lə-'zā-shən\ *noun*

universal joint *noun* (1676)
: a shaft coupling capable of transmitting rotation from one shaft to another not collinear with it — called also *universal coupling*

universal joint

universal motor *noun* (1925)
: an electric motor that can be used on either an alternating or a direct current supply

Universal Product Code *noun* (1974)
: a combination of a bar code and numbers by which a scanner can identify a product and usually assign a price

universal recipient *noun* (1922)
: a person who belongs to ABO blood group AB and can receive blood from any donor

Universal time *noun* (1882)
: GREENWICH MEAN TIME

uni·verse \'yü-nə-ˌvərs\ *noun* [Latin *universum,* from neuter of *universus* entire, whole, from *uni-* + *versus* turned toward, from past participle of *vertere* to turn — more at WORTH] (1589)
1 : the whole body of things and phenomena observed or postulated **:** COSMOS: as **a :** a systematic whole held to arise by and persist through the direct intervention of divine power **b :** the world of human experience **c** (1) **:** the entire celestial cosmos (2) **:** MILKY WAY GALAXY (3) **:** an aggregate of stars comparable to the Milky Way galaxy
2 : a distinct field or province of thought or reality that forms a closed system or self-inclusive and independent organization
3 : POPULATION 4
4 : a set that contains all elements relevant to a particular discussion or problem
5 : a great number or quantity ⟨a large enough *universe* of stocks . . . to choose from —G. B. Clairmont⟩

universe of discourse (1881)
: an inclusive class of entities that is tacitly implied or explicitly delineated as the subject of a statement, discourse, or theory

uni·ver·si·ty \ˌyü-nə-'vər-sə-tē, -'vər-stē\ *noun, plural* **-ties** [Middle English *universite,* from Middle French *université,* from Medieval Latin *universitat-, universitas,* from Latin *universus*] (14th century)
1 : an institution of higher learning providing facilities for teaching and research and authorized to grant academic degrees; *specifically* **:** one made up of an undergraduate division which confers bachelor's degrees and a graduate division which comprises a graduate school and professional schools each of which may confer master's degrees and doctorates
2 : the physical plant of a university

univ·o·cal \yü-'ni-və-kəl\ *adjective* [Late Latin *univocus,* from Latin *uni-* + *voc-, vox* voice — more at VOICE] (1599)
: having one meaning only
— **univ·o·cal·ly** \-k(ə-)lē\ *adverb*

un·joined \ˌən-'jȯind\ *adjective* (1538)
: not joined

un·just \ˌən-'jəst\ *adjective* (14th century)
1 : characterized by injustice **:** UNFAIR
2 *archaic* **:** DISHONEST, FAITHLESS
— **un·just·ly** *adverb*
— **un·just·ness** \-'jəs(t)-nəs\ *noun*

un·kempt \-'kem(p)t\ *adjective* [*un-* + *kempt* combed, neat] (1579)
1 : deficient in order or neatness ⟨*unkempt* individuals⟩ ⟨*unkempt* hotel rooms⟩; *also* **:** ROUGH, UNPOLISHED ⟨*unkempt* prose⟩
2 : not combed ⟨*unkempt* hair⟩

un·kenned \-'kend\ *adjective* (14th century)
chiefly dialect **:** UNKNOWN, STRANGE

un·ken·nel \-'ke-nᵊl\ *transitive verb* (1575)
1 a : to drive (as a fox) from a hiding place or den **b :** to free (dogs) from a kennel
2 : to bring out into the open **:** UNCOVER

un·kind \-'kīnd\ *adjective* (13th century)
1 : not pleasing or mild **:** INCLEMENT ⟨an *unkind* climate⟩
2 : lacking in kindness or sympathy **:** HARSH, CRUEL
— **un·kind·ness** \-'kīn(d)-nəs\ *noun*

¹un·kind·ly \-'kīn(d)-lē\ *adjective* (13th century)
: not kindly
— **un·kind·li·ness** *noun*

²unkindly *adverb* (14th century)
: in an unkind manner ⟨dwells *unkindly* long on his final decline —A. H. Johnston⟩

un·kink \ˌən-'kiŋk\ (1891)
transitive verb
: to free from kinks **:** STRAIGHTEN
intransitive verb
: to become lax or loose **:** RELAX

un·knit \-'nit\ *verb* **-knit** *or* **-knit·ted; -knit·ting** (before 12th century)
: UNDO, UNRAVEL

un·know·able \ˌən-'nō-ə-bəl\ *adjective* (14th century)
: not knowable; *especially* **:** lying beyond the limits of human experience or understanding
— **un·know·abil·i·ty** \-ˌnō-ə-'bi-lə-tē\ *noun*

¹un·know·ing \-'nō-iŋ\ *adjective* (14th century)
: not knowing
— **un·know·ing·ly** \-iŋ-lē\ *adverb*

²unknowing *noun* (14th century)
: IGNORANCE

¹un·known \-'nōn\ *adjective* (14th century)

: not known or not well-known; *also* **:** having an unknown value ⟨an *unknown* quantity⟩

[2]**unknown** *noun* (1597)

1 : one that is not known or not well-known; *especially* **:** a person who is little known (as to the public)

2 : something that requires discovery, identification, or clarification: as **a :** a symbol (as *x, y,* or *z*) in a mathematical equation representing an unknown quantity **b :** a specimen (as of bacteria or mixed chemicals) required to be identified as an exercise in appropriate laboratory techniques

Unknown Soldier *noun* (1923)

: an unidentified soldier whose body is selected to receive national honors as a representative of all of the same nation who died in a war and especially in one of the world wars

un·lace \ˌən-ˈlās\ *transitive verb* (14th century)

1 : to loose by undoing a lacing

2 *obsolete* **:** UNDO, DISGRACE

un·lade \-ˈlād\ *verb* **-lad·ed; -laded** *or* **-lad·en** \-ˈlā-dᵊn\; **-lad·ing** (14th century)

transitive verb

1 : to take the load or cargo from

2 : DISCHARGE, UNLOAD

intransitive verb

: to discharge cargo

un·lash \-ˈlash\ *transitive verb* (1748)

: to untie the lashing of

un·latch \-ˈlach\ (1642)

transitive verb

: to open or loose by lifting the latch

intransitive verb

: to become loosed or opened

un·law·ful \ˌən-ˈlȯ-fəl\ *adjective* (14th century)

1 : not lawful **:** ILLEGAL

2 : not morally right or conventional

— un·law·ful·ly \-f(ə-)lē\ *adverb*

— un·law·ful·ness \-fəl-nəs\ *noun*

un·lay \-ˈlā\ *verb* **-laid** \-ˈlād\; **-lay·ing** (1726)

transitive verb

: to untwist the strands of (as a rope)

intransitive verb

: UNTWIST

un·lead·ed \-ˈle-dəd\ *adjective* (circa 1891)

1 : not having leads between the lines in printing

2 : not treated or mixed with lead or lead compounds ⟨*unleaded* fuels⟩

un·learn \-ˈlərn\ *transitive verb* (15th century)

1 : to put out of one's knowledge or memory

2 : to undo the effect of **:** discard the habit of

un·learned \-ˈlər-nəd *for 1, 2,* -ˈlərnd *for 3*\ *adjective* (15th century)

1 : possessing inadequate learning or education; *especially* **:** deficient in scholarly attainments

2 : characterized by or revealing ignorance

3 : not gained by study or training

synonym see IGNORANT

un·leash \-ˈlēsh\ *transitive verb* (circa 1671)

: to free from or as if from a leash **:** let loose

[1]**un·less** \ən-ˈles, ˈən-ˌ\ *conjunction* [Middle English *unlesse,* alteration of *onlesse,* from *on* + *lesse* less] (1509)

1 : except on the condition that **:** under any other circumstance than

2 : without the accompanying circumstance or condition that **:** but that **:** BUT

[2]**unless** *preposition* (circa 1532)

: except possibly **:** EXCEPT

un·let·tered \ˌən-ˈle-tərd\ *adjective* (14th century)

1 a : lacking facility in reading and writing and ignorant of the knowledge to be gained from books **b :** ILLITERATE

2 : not marked with letters

synonym see IGNORANT

un·licked \-ˈlikt\ *adjective* (1593)

1 : lacking proper form or shape

2 *archaic* **:** not licked dry

[1]**un·like** \-ˈlīk\ *adjective* (13th century)

: not like: as **a :** marked by lack of resemblance **:** DIFFERENT ⟨the two books are quite *unlike*⟩ **b :** marked by inequality **:** UNEQUAL ⟨contributed *unlike* amounts⟩

— un·like·ness *noun*

[2]**unlike** *preposition* (1593)

: not like: as **a :** different from **b :** not characteristic of **c :** in a different manner from

un·like·li·hood \ˌən-ˈlī-klē-ˌhu̇d\ *noun* (1548)

1 : IMPROBABILITY

2 : something unlikely

un·like·li·ness \-nəs\ *noun* (1614)

: IMPROBABILITY

un·like·ly \-ˈlī-klē\ *adjective* (14th century)

1 : not likely **:** IMPROBABLE

2 : likely to fail **:** UNPROMISING

un·lim·ber \ˌən-ˈlim-bər\ (1802)

transitive verb

1 : to detach the limber from and so make ready ⟨*unlimber* a gun for action⟩

2 : to prepare for operation or performance ⟨*unlimbered* his banjo and began to play⟩

intransitive verb

: to perform the task of preparing something for action

un·lim·it·ed \-ˈli-mə-təd\ *adjective* (15th century)

1 : lacking any controls **:** UNRESTRICTED

2 : BOUNDLESS, INFINITE

3 : not bounded by exceptions **:** UNDEFINED

— un·lim·it·ed·ly *adverb*

un·link \-ˈliŋk\ (1600)

transitive verb

: to unfasten the links of **:** SEPARATE, DISCONNECT

intransitive verb

: to become detached

un·linked \-ˈliŋ(k)t\ *adjective* (1966)

: not belonging to the same genetic linkage group ⟨*unlinked* genes⟩

un·list·ed \-ˈlis-təd\ *adjective* (1644)

1 : not appearing on a list; *especially* **:** not appearing in a telephone book ⟨*unlisted* numbers⟩

2 : being or involving a security not listed formally on an organized exchange **:** OVER-THE-COUNTER

un·live \-ˈliv\ *transitive verb* (1614)

: ANNUL, REVERSE

un·load \ˌən-ˈlōd\ (1523)

transitive verb

1 a (1) **:** to take off **:** DELIVER (2) **:** to take the cargo from **b :** to give outlet to **:** pour forth ⟨*unloaded* her bitter feelings⟩

2 : to relieve of something burdensome, unwanted, or oppressive ⟨*unloaded* the pack animals⟩ ⟨*unloaded* himself to his friend⟩

3 : to draw the charge from ⟨*unloaded* the gun⟩

4 : to sell or dispose of especially in large quantities **:** DUMP

5 : to hit or propel with a great release of power ⟨*unloaded* his ninth homer⟩

intransitive verb

1 : to perform the act of unloading

2 : to release or deliver something especially with power ⟨*unloaded* on the ball⟩

— un·load·er *noun*

un·lock \-ˈläk\ (15th century)

transitive verb

1 : to unfasten the lock of

2 : OPEN, UNDO

3 : to free from restraints or restrictions ⟨the shock *unlocked* a flood of tears⟩

4 : to furnish a key to **:** DISCLOSE

intransitive verb

: to become unfastened or freed from restraints

un·looked–for \-ˈlu̇kt-ˌfȯr\ *adjective* (1535)

: not foreseen **:** UNEXPECTED

un·loose \ˌən-ˈlüs\ *transitive verb* (14th century)

1 : to relax the strain of ⟨*unloose* a grip⟩

2 : to release from or as if from restraints **:** set free

3 : to loosen the ties of ⟨*unloose* traditional social bonds⟩

un·loos·en \-ˈlü-sᵊn\ *transitive verb* (15th century)

: UNLOOSE

un·love·ly \-ˈləv-lē\ *adjective* (14th century)

: not likable **:** DISAGREEABLE, UNPLEASANT

— un·love·li·ness \-lē-nəs\ *noun*

un·luck·i·ly \ˌən-ˈlə-kə-lē\ *adverb* (1530)

: UNFORTUNATELY ⟨*unluckily,* it has a nasty way of turning to rain —Ambrose Bierce⟩ ⟨his ascent being *unluckily* a little out of the perpendicular —T. L. Peacock⟩

un·lucky \-ˈlə-kē\ *adjective* (1530)

1 : marked by adversity or failure ⟨an *unlucky* year⟩

2 : likely to bring misfortune **:** INAUSPICIOUS

3 : having or meeting with bad luck ⟨*unlucky* people⟩

4 : producing dissatisfaction **:** REGRETTABLE

— un·luck·i·ness \-ˈlə-kē-nəs\ *noun*

un·made \ˌən-ˈmād\ *adjective* (13th century)

: not made ⟨an *unmade* bed⟩

un·make \-ˈmāk\ *transitive verb* **-made** \-ˈmād\; **-mak·ing** (15th century)

1 : to cause to disappear **:** DESTROY

2 : to deprive of rank or office **:** DEPOSE

3 : to deprive of essential characteristics **:** change the nature of

un·man \ˌən-ˈman\ *transitive verb* (circa 1600)

1 : to deprive of manly vigor, fortitude, or spirit

2 : CASTRATE, EMASCULATE

synonym see UNNERVE

un·man·ly \-ˈman-lē\ *adjective* (circa 1547)

: not manly: as **a :** being of weak character **:** COWARDLY **b :** EFFEMINATE

— un·man·li·ness \-lē-nəs\ *noun*

un·manned \-ˈmand\ *adjective* (1544)

1 : not manned ⟨an *unmanned* spaceflight⟩

2 *of a hawk* **:** not trained

un·man·nered \-ˈma-nərd\ *adjective* (1594)

1 : marked by a lack of good manners **:** RUDE

2 : characterized by an absence of artificiality **:** UNAFFECTED

— un·man·nered·ly *adverb*

[1]**un·man·ner·ly** \-ˈma-nər-lē\ *adverb* (14th century)

: in an unmannerly fashion

[2]**unmannerly** *adjective* (14th century)

: not mannerly **:** DISCOURTEOUS

— un·man·ner·li·ness \-lē-nəs\ *noun*

un·mar·ried \-ˈmar-ēd\ *adjective* (14th century)

: not married: **a :** not now or previously married **b :** being divorced or widowed

— unmarried *noun*

un·mask \ˌən-ˈmask\ (1602)

transitive verb

1 : to remove a mask from

2 : to reveal the true nature of **:** EXPOSE

intransitive verb

: to remove one's mask

un·mean·ing \-ˈmē-niŋ\ *adjective* (1704)

1 : lacking intelligence **:** VAPID

2 : having no meaning **:** SENSELESS

un·me·di·at·ed \ˌən-ˈmē-dē-ˌā-təd\ *adjective* (1648)

: not mediated **:** not communicated or transformed by an intervening agency ⟨experience *unmediated* by artifice⟩

un·meet \-ˈmēt\ *adjective* (circa 1529)

: not meet **:** UNSUITABLE, IMPROPER

[1]**un·men·tion·able** \-ˈmen(t)-sh(ə-)nə-bəl\ *adjective* (1837)

: not fit or allowed to be mentioned or discussed **:** UNSPEAKABLE

[2]**unmentionable** *noun* (1928)

: one that is not to be mentioned or discussed: as **a** *plural* **:** [3]PANT 1 **b** *plural* **:** UNDERWEAR

un·mer·ci·ful \ˌən-ˈmər-si-fəl\ *adjective* (15th century)
1 : not merciful **:** MERCILESS
2 : EXCESSIVE, EXTREME ⟨chatted for an *unmerciful* length of time⟩
— un·mer·ci·ful·ly \-f(ə-)lē\ *adverb*

un·mind·ful \-ˈmīn(d)-fəl\ *adjective* (14th century)
: not conscientiously aware, attentive, or heedful **:** INATTENTIVE, CARELESS

un·mis·tak·able \ˌən-mə-ˈstā-kə-bəl\ *adjective* (1666)
: not capable of being mistaken or misunderstood **:** CLEAR
— un·mis·tak·ably \-blē\ *adverb*

un·mit·i·gat·ed \ˌən-ˈmi-tə-ˌgā-təd\ *adjective* (1599)
1 : not lessened **:** UNRELIEVED ⟨sufferings *unmitigated* by any hope of early relief⟩
2 : being so definitely what is stated as to offer little chance of change or relief ⟨an *unmitigated* evil⟩
— un·mit·i·gat·ed·ly *adverb*
— un·mit·i·gat·ed·ness *noun*

un·mold \ˌən-ˈmōld\ *transitive verb* (circa 1900)
: to remove from a mold

un·moor \-ˈmu̇r\ (15th century)
transitive verb
: to loosen from or as if from moorings
intransitive verb
: to cast off moorings

un·mor·al \-ˈmȯr-əl, -ˈmär-\ *adjective* (1841)
1 : having no moral perception or quality; *also* **:** not influenced or guided by moral considerations
2 : lying outside the bounds of morals or ethics **:** AMORAL
— un·mo·ral·i·ty \ˌən-mə-ˈra-lə-tē, -mȯ-\ *noun*

un·muf·fle \ˌən-ˈmə-fəl\ *transitive verb* (1611)
: to free from something that muffles

un·muz·zle \-ˈmə-zəl\ *transitive verb* (1601)
: to free from or as if from a muzzle

un·my·elin·at·ed \-ˈmī-ə-lə-ˌnā-təd\ *adjective* (1919)
: lacking a myelin sheath

un·nail \ˌən-ˈnā(ə)l\ *transitive verb* (15th century)
: to unfasten by removing nails

un·nat·u·ral \ˌən-ˈna-chə-rəl, -ˈnach-rəl\ *adjective* (15th century)
1 : not being in accordance with nature or consistent with a normal course of events
2 a : not being in accordance with normal human feelings or behavior **:** PERVERSE **b :** lacking ease and naturalness **:** CONTRIVED ⟨her manner was forced and *unnatural*⟩ **c :** inconsistent with what is reasonable or expected ⟨an *unnatural* alliance⟩
synonym see IRREGULAR
— un·nat·u·ral·ly \-ˈna-chə-rə-lē, -ˈnach-rə-, -ˈna-chər-\ *adverb*
— un·nat·u·ral·ness \-ˈna-chə-rəl-nəs, -ˈnach-rəl-\ *noun*

un·nec·es·sar·i·ly \ˌən-ˌne-sə-ˈser-ə-lē\ *adverb* (1594)
: not by necessity **:** to an unnecessary degree

un·nec·es·sary \ˌən-ˈne-sə-ˌser-ē\ *adjective* (1548)
: not necessary

un·nerve \-ˈnərv\ *transitive verb* (1601)
1 : to deprive of courage, strength, or steadiness
2 : to cause to become nervous **:** UPSET ☆
— un·nerv·ing·ly \-ˈnər-viŋ-lē\ *adverb*

un·nil·hex·i·um \ˌyü-nᵊl-ˈhek-sē-əm\ *noun* [New Latin, from *unnil-* (from Latin *unus* one + *nil* nothing, zero) + Greek *hex* six + New Latin *-ium* — more at ONE, NIL, SIX] (1981)
: the chemical element of atomic number 106 — see ELEMENT table

un·nil·pen·ti·um \-ˈpen-tē-əm\ *noun* [New Latin, from *unnil-* + Greek *pente* five + New Latin *-ium* — more at FIVE] (1981)
: the chemical element of atomic number 105 — see ELEMENT table

un·nil·qua·di·um \-ˈkwä-dē-əm\ *noun* [New Latin, from *unnil-* + *quadri-* + *-ium*] (1979)
: the chemical element of atomic number 104 — see ELEMENT table

un·num·bered \ˌən-ˈnəm-bərd\ *adjective* (14th century)
1 : INNUMERABLE
2 : not having an identifying number ⟨*unnumbered* pages⟩

un·ob·tru·sive \ˌən-əb-ˈtrü-siv, -ziv\ *adjective* (1743)
: not obtrusive **:** not blatant, arresting, or aggressive **:** INCONSPICUOUS
— un·ob·tru·sive·ly *adverb*
— un·ob·tru·sive·ness *noun*

un·oc·cu·pied \ˌən-ˈä-kyə-ˌpīd\ *adjective* (14th century)
: not occupied: as **a :** not busy **:** UNEMPLOYED **b :** not lived in **:** EMPTY

un·or·ga·nized \-ˈȯr-gə-ˌnīzd\ *adjective* (circa 1828)
1 : not organized: as **a :** not brought into a coherent or well-ordered whole **b :** not belonging to a labor union
2 : not having the characteristics of a living organism

un·or·tho·dox \-ˈȯr-thə-ˌdäks\ *adjective* (1657)
: not orthodox
— un·or·tho·dox·ly *adverb*

un·or·tho·doxy \-ˌdäk-sē\ *noun* (circa 1704)
1 : the quality or state of being unorthodox
2 : something (as an opinion or doctrine) that is unorthodox

un·pack \ˌən-ˈpak\ (15th century)
transitive verb
1 a : to remove the contents of ⟨*unpack* a trunk⟩ **b :** UNBURDEN, REVEAL ⟨must . . . *unpack* my heart with words —Shakespeare⟩
2 : to remove or undo from packing or a container ⟨*unpacked* his gear⟩
3 : to analyze the nature of by examining in detail **:** EXPLICATE ⟨*unpack* a concept⟩
intransitive verb
: to engage in unpacking a container
— un·pack·er *noun*

un·paged \-ˈpājd\ *adjective* (1874)
: having no page numbers

un·paid \-ˈpād\ *adjective* (14th century)
1 : not paid ⟨an *unpaid* volunteer⟩
2 : not paying a salary ⟨an *unpaid* position⟩

un·paired \-ˈpard, -ˈperd\ *adjective* (1648)
1 : not paired; *especially* **:** not matched or mated
2 : characterized by the absence of pairing ⟨electrons in the *unpaired* state⟩

un·pal·at·able \-ˈpa-lə-tə-bəl\ *adjective* (1682)
1 : not palatable **:** DISTASTEFUL
2 : UNPLEASANT, DISAGREEABLE
— un·pal·at·abil·i·ty \ˌən-ˌpa-lə-tə-ˈbi-lə-tē\ *noun*

un·par·al·leled \ˌən-ˈpar-ə-ˌleld, -ləld\ *adjective* (1594)
: having no parallel; *especially* **:** having no equal or match **:** unique in kind or quality

un·par·lia·men·ta·ry \ˌən-ˌpär-lə-ˈmen-tə-rē, -ˌpärl-yə-, -ˈmen-trē\ *adjective* (1626)
: contrary to the practice of parliamentary bodies

un·peg \ˌən-ˈpeg\ *transitive verb* (1602)
1 : to remove a peg from
2 : to unfasten by or as if by removing a peg

un·peo·ple \-ˈpē-pəl\ *transitive verb* (circa 1533)
: DEPOPULATE

un·peo·pled \ˌən-ˈpē-pəld\ *adjective* (circa 1586)
: not filled with or occupied by people

un·per·fect \-ˈpər-fikt\ *adjective* (14th century)
: IMPERFECT

un·per·son \ˈən-ˈpər-sᵊn, -ˌpər-\ *noun* (1949)
: an individual who usually for political or ideological reasons is removed completely from recognition or consideration

un·pick \ˌən-ˈpik\ *transitive verb* (circa 1775)
: to undo (as sewing) by taking out stitches

un·pile \-ˈpī(ə)l\ *transitive verb* (1611)
: to take or disentangle from a pile

un·pin \-ˈpin\ *transitive verb* (14th century)
1 : to remove a pin from
2 : to loosen, free, or unfasten by or as if by removing a pin

un·placed \ˌən-ˈplāst\ *adjective* (1512)
1 : not placed **:** not having a definite or assigned place ⟨taxonomically *unplaced* organisms⟩
2 : not finishing in one of the first three places in a horse race

un·pleas·ant \-ˈple-zᵊnt\ *adjective* (1538)
: not pleasant **:** not amiable or agreeable **:** DISPLEASING ⟨*unpleasant* odors⟩
— un·pleas·ant·ly *adverb*

un·pleas·ant·ness *noun* (1548)
1 : the quality or state of being unpleasant
2 : an unpleasant situation, experience, or event

un·plug \ˌən-ˈpləg\ *transitive verb* (circa 1775)
1 a : to take a plug out of **b :** to remove an obstruction from
2 a : to remove (as an electric plug) from a socket or receptacle **b :** to disconnect from an electric circuit by removing a plug ⟨*unplug* the refrigerator⟩

un·plumbed \-ˈpləmd\ *adjective* (1623)
1 : not tested with a plumb line
2 a : not measured with a plumb **b :** not thoroughly explored

un·po·lar·ized \-ˈpō-lə-ˌrīzd\ *adjective* (circa 1828)
: not polarized; *specifically* **:** having a random pattern of vibrations

un·po·lit·i·cal \ˌən-pə-ˈli-ti-kəl\ *adjective* (1643)
: APOLITICAL 1

un·pop·u·lar \ˌən-ˈpä-pyə-lər\ *adjective* (1647)
: not popular **:** viewed or received unfavorably by the public
— un·pop·u·lar·i·ty \ˌən-ˌpä-pyə-ˈlar-ə-tē\ *noun*

un·prec·e·dent·ed \ˌən-ˈpre-sə-ˌden-təd\ *adjective* (1623)
: having no precedent **:** NOVEL, UNEXAMPLED
— un·prec·e·dent·ed·ly *adverb*

un·preg·nant \ˌən-ˈpreg-nənt\ *adjective* (1602)
obsolete **:** INEPT 1

un·pre·tend·ing \ˌən-pri-ˈten-diŋ\ *adjective* (1697)
: UNPRETENTIOUS

un·pre·ten·tious \-ˈten(t)-shəs\ *adjective* (1859)
: free from ostentation, elegance, or affectation **:** MODEST ⟨*unpretentious* homes⟩
— un·pre·ten·tious·ly *adverb*
— un·pre·ten·tious·ness *noun*

☆ SYNONYMS
Unnerve, enervate, unman, emasculate mean to deprive of strength or vigor and the capacity for effective action. UNNERVE implies marked often temporary loss of courage, self-control, or power to act ⟨*unnerved* by the near collision⟩. ENERVATE suggests a gradual physical or moral weakening (as through luxury or indolence) until one is too feeble to make an effort ⟨a nation's youth *enervated* by affluence and leisure⟩. UNMAN implies a loss of manly vigor, fortitude, or spirit ⟨a soldier *unmanned* by the terrors of battle⟩. EMASCULATE stresses a depriving of characteristic force by removing something essential ⟨an amendment that *emasculates* existing safeguards⟩.

un·prin·ci·pled \ˌən-ˈprin(t)-s(ə-)pəld, -sə-bəld\ *adjective* (1644)
: lacking moral principles : UNSCRUPULOUS
— **un·prin·ci·pled·ness** *noun*

un·print·able \-ˈprin-tə-bəl\ *adjective* (1871)
: unfit to be printed

un·pro·fessed \ˌən-prə-ˈfest\ *adjective* (15th century)
: not professed ⟨an *unprofessed* aim⟩

un·prof·it·able \ˌən-ˈprä-fə-tə-bəl, -ˈpräf-tə-bəl\ *adjective* (14th century)
: not profitable : producing no gain, good, or result
— **un·prof·it·able·ness** *noun*
— **un·prof·it·ably** \-blē\ *adverb*

un·prom·is·ing \-ˈprä-mə-siŋ\ *adjective* (1663)
: appearing unlikely to prove worthwhile or result favorably
— **un·prom·is·ing·ly** \-siŋ-lē\ *adverb*

un·pro·nounced \ˌən-prə-ˈnau̇n(t)st\ *adjective* (1611)
: not pronounced; *especially* : MUTE

un·pub·lish·able \-ˈpə-bli-shə-bəl\ *adjective* (1815)
: UNPRINTABLE

un·qual·i·fied \ˌən-ˈkwä-lə-ˌfīd\ *adjective* (1556)
1 : not fit : not having requisite qualifications
2 : not modified or restricted by reservations : COMPLETE ⟨an *unqualified* denial⟩
— **un·qual·i·fied·ly** \-ˌfī(-ə)d-lē\ *adverb*

un·ques·tion·able \-ˈkwes-chə-nə-bəl, -ˈkwesh-\ *adjective* (1631)
: not questionable : INDISPUTABLE ⟨*unquestionable* evidence⟩
— **un·ques·tion·ably** \-blē\ *adverb*

un·ques·tion·ing \-ˈkwes-chə-niŋ, -ˈkwesh-\ *adjective* (circa 1828)
: not questioning : not expressing or marked by doubt or hesitation ⟨*unquestioning* obedience⟩
— **un·ques·tion·ing·ly** \-niŋ-lē\ *adverb*

un·qui·et \-ˈkwī-ət\ *adjective* (1523)
1 : not quiet : AGITATED, TURBULENT
2 : physically, emotionally, or mentally restless : UNEASY
— **un·qui·et·ly** *adverb*
— **un·qui·et·ness** *noun*

un·quote \ˈən-ˌkwōt *also* -ˌkōt\ *noun* (1915)
— used orally to indicate the end of a direct quotation

un·rav·el \ˌən-ˈra-vəl\ (1603)
transitive verb
1 a : to disengage or separate the threads of : DISENTANGLE **b** : to cause to come apart by or as if by separating the threads of
2 : to resolve the intricacy, complexity, or obscurity of : clear up ⟨*unravel* a mystery⟩
intransitive verb
: to become unraveled

un·read \-ˈred\ *adjective* (15th century)
1 : not read : left unexamined
2 : lacking the experience or the benefits of reading ⟨*unread* in political science⟩

un·re·al \-ˈrē(-ə)l, -ˈri(-ə)l\ *adjective* (1605)
: lacking in reality, substance, or genuineness : ARTIFICIAL, ILLUSORY; *also* : INCREDIBLE, FANTASTIC

un·re·al·is·tic \ˌən-ˌrē-ə-ˈlis-tik, -ˌri-ə-\ *adjective* (1865)
: not realistic : inappropriate to reality or fact
— **un·re·al·is·ti·cal·ly** \-ti-k(ə-)lē\ *adverb*

un·re·al·i·ty \ˌən-rē-ˈa-lə-tē\ *noun* (1751)
1 a : the quality or state of being unreal : lack of substance or validity **b** : something unreal, insubstantial, or visionary : FIGMENT
2 : ineptitude in dealing with reality

un·rea·son \ˌən-ˈrē-zᵊn, ˈən-ˌ\ *noun* (1827)
: the absence of reason or sanity : IRRATIONALITY, MADNESS

un·rea·son·able \-ˈrēz-nə-bəl, -ˈrē-zᵊn-ə-bəl\ *adjective* (14th century)
1 a : not governed by or acting according to reason ⟨*unreasonable* people⟩ **b** : not conformable to reason : ABSURD ⟨*unreasonable* beliefs⟩
2 : exceeding the bounds of reason or moderation ⟨working under *unreasonable* pressure⟩
— **un·rea·son·able·ness** \-nəs\ *noun*
— **un·rea·son·ably** \-blē\ *adverb*

un·rea·soned \-ˈrē-zᵊnd\ *adjective* (1790)
: not founded on reason or reasoning

un·rea·son·ing \-ˈrēz-niŋ, -ˈrē-zᵊn-iŋ\ *adjective* (1751)
: not reasoning; *especially* : not moderated or controlled by reason ⟨*unreasoning* fear⟩
— **un·rea·son·ing·ly** \-ˈrēz-niŋ-lē, -ˈrē-zᵊn-iŋ-\ *adverb*

un·re·con·struct·ed \ˌən-ˌrē-kən-ˈstrək-təd\ *adjective* (1867)
: not reconciled to some political, economic, or social change ⟨an *unreconstructed* rebel⟩; *also* : holding stubbornly to a particular belief, view, place, or style ⟨an *unreconstructed* New Yorker⟩

un·reel \ˌən-ˈrē(ə)l\ (1567)
transitive verb
1 : to unwind from a reel
2 : to perform successfully ⟨*unreeled* a 66-yard pass play⟩
3 : REEL OFF 2
intransitive verb
1 : to become unwound
2 : to be presented ⟨the dress rehearsal *unreeled* flawlessly⟩

un·reeve \-ˈrēv\ *transitive verb* **-rove** \-ˈrōv\ *or* **-reeved; -reev·ing** (1600)
: to withdraw (a rope) from an opening (as a ship's block or thimble)

un·re·gen·er·ate \ˌən-ri-ˈje-nə-rət, -ˈjen-rət\ *adjective* (1589)
1 : not regenerate ⟨the *unregenerate* condition of humanity⟩ ⟨*unregenerate* pagans⟩
2 a : not reformed : UNRECONSTRUCTED ⟨*unregenerate* liberals⟩ ⟨*unregenerate* Confederates⟩ **b** : OBSTINATE, STUBBORN ⟨struggling against *unregenerate* impulses⟩ ⟨his *unregenerate* competitiveness⟩
— **un·re·gen·er·ate·ly** *adverb*

un·re·lent·ing \-ˈlen-tiŋ\ *adjective* (1588)
1 : not softening or yielding in determination : HARD, STERN ⟨an *unrelenting* leader⟩
2 : not letting up or weakening in vigor or pace : CONSTANT ⟨the *unrelenting* struggle⟩
— **un·re·lent·ing·ly** \-tiŋ-lē\ *adverb*

un·re·marked \-ˈmärkt\ *adjective* (circa 1775)
: not remarked : UNNOTICED

un·re·mit·ting \-ˈmi-tiŋ\ *adjective* (1728)
: not remitting : CONSTANT, INCESSANT
— **un·re·mit·ting·ly** \-tiŋ-lē\ *adverb*

un·re·quit·ed \ˌən-ri-ˈkwī-təd\ *adjective* (circa 1542)
: not requited : not reciprocated or returned in kind ⟨*unrequited* love⟩

un·re·serve \-ˈzərv\ *noun* (1751)
: absence of reserve : FRANKNESS

un·re·served \-ˈzərvd\ *adjective* (1539)
1 : not limited or partial : ENTIRE, UNQUALIFIED ⟨*unreserved* enthusiasm⟩
2 : not cautious or reticent : FRANK, OPEN
3 : not set aside for special use
— **un·re·serv·ed·ly** \-ˈzər-vəd-lē\ *adverb*
— **un·re·served·ness** \-ˈzər-vəd-nəs, -ˈzərv(d)-nəs\ *noun*

un·rest \ˌən-ˈrest\ *noun* (14th century)
: a disturbed or uneasy state : TURMOIL

un·re·strained \ˌən-ri-ˈstrānd\ *adjective* (circa 1600)
1 : not restrained : IMMODERATE, UNCONTROLLED ⟨*unrestrained* proliferation of technology⟩
2 : free of constraint : SPONTANEOUS ⟨felt happy and *unrestrained*⟩
— **un·re·strain·ed·ly** \-ˈstrā-nəd-lē\ *adverb*
— **un·re·strained·ness** \-ˈstrā-nəd-nəs, -ˈstrān(d)-nəs\ *noun*

un·re·straint \-ˈstrānt\ *noun* (1804)
: freedom from or lack of restraint

un·rid·dle \ˌən-ˈri-dᵊl\ *transitive verb* (circa 1586)
: to find the explanation of : FIGURE OUT, SOLVE ⟨*unriddle* a puzzle⟩ ⟨no trouble *unriddling* three-part syllogisms —*New Yorker*⟩; *also* : to make understandable : EXPLAIN ⟨just trying to *unriddle* the man —Helen Dudar⟩

un·rig \-ˈrig\ *transitive verb* (circa 1580)
: to strip of rigging ⟨*unrig* a ship⟩

un·righ·teous \-ˈrī-chəs\ *adjective* (before 12th century)
1 : not righteous : SINFUL, WICKED
2 : UNJUST, UNMERITED ⟨intolerable and *unrighteous* interference in their lives —W. W. Wagar⟩
— **un·righ·teous·ly** *adverb*
— **un·righ·teous·ness** *noun*

un·ripe \-ˈrīp\ *adjective* (13th century)
1 : not ripe : IMMATURE
2 : not ready : UNPREPARED
— **un·ripe·ness** *noun*

un·ri·valed *or* **un·ri·valled** \ˌən-ˈrī-vəld\ *adjective* (1591)
: having no rival : INCOMPARABLE, SUPREME

un·roll \-ˈrōl\ (15th century)
transitive verb
1 : to unwind a roll of : open out : UNCOIL
2 : to spread out like a scroll for reading or inspection : UNFOLD, REVEAL
intransitive verb
: to be unrolled : UNWIND

un·roof \-ˈrüf, -ˈru̇f\ *transitive verb* (1598)
: to strip off the roof or covering of

¹un·round \ˌən-ˈrau̇nd\ *transitive verb* (1874)
1 : to pronounce (a sound) without lip rounding or with decreased lip rounding
2 : to spread (the lips) laterally ⟨necessary to *unround* the lips in pronouncing \ē\⟩

²unround *adjective* (1958)
: pronounced with the lips not rounded

un·ruf·fled \ˌən-ˈrə-fəld\ *adjective* (1659)
1 : poised and serene especially in the face of setbacks or confusion
2 : not ruffled : SMOOTH ⟨*unruffled* water⟩
synonym see COOL

un·ruly \-ˈrü-lē\ *adjective* [Middle English *unreuly,* from *un-* + *reuly* disciplined, from *reule* rule] (15th century)
: not readily ruled, disciplined, or managed ⟨an *unruly* crowd⟩ ⟨a mane of *unruly* hair⟩ ☆
— **un·rul·i·ness** *noun*

un·sad·dle \ˌən-ˈsa-dᵊl\ (14th century)
transitive verb
1 : to take the saddle from

☆ SYNONYMS
Unruly, ungovernable, intractable, refractory, recalcitrant, willful, headstrong mean not submissive to government or control. UNRULY implies lack of discipline or incapacity for discipline and often connotes waywardness or turbulence of behavior ⟨*unruly* children⟩. UNGOVERNABLE implies either an escape from control or guidance or a state of being unsubdued and incapable of controlling oneself or being controlled by others ⟨*ungovernable* rage⟩. INTRACTABLE suggests stubborn resistance to guidance or control ⟨*intractable* opponents of the hazardous-waste dump⟩. REFRACTORY stresses resistance to attempts to manage or to mold ⟨special schools for *refractory* children⟩. RECALCITRANT suggests determined resistance to or defiance of authority ⟨acts of sabotage by a *recalcitrant* populace⟩. WILLFUL implies an obstinate determination to have one's own way ⟨a *willful* disregard for the rights of others⟩. HEADSTRONG suggests self-will impatient of restraint, advice, or suggestion ⟨a *headstrong* young calvary officer⟩.

2 : to throw from the saddle
intransitive verb
: to remove the saddle from a horse

un·said \-ˈsed\ *adjective* (before 12th century)
: not said; *especially* **:** not spoken aloud

un·sat·is·fac·to·ry \-ˌsa-təs-ˈfak-t(ə-)rē\ *adjective* (circa 1650)
: not satisfactory
— un·sat·is·fac·to·ri·ly \-t(ə-)rə-lē\ *adverb*
— un·sat·is·fac·to·ri·ness \-t(ə-)rē-nəs\ *noun*

un·sat·u·rate \-ˈsa-chə-rət, -ˈsach-rət\ *noun* (1934)
: an unsaturated chemical compound

un·sat·u·rat·ed \-ˈsa-chə-ˌrā-təd\ *adjective* (1758)
: not saturated: as **a :** capable of absorbing or dissolving more of something ⟨an *unsaturated* solution⟩ **b :** able to form products by chemical addition; *especially* **:** containing double or triple bonds between carbon atoms

un·saved \ˌən-ˈsāvd\ *adjective* (1648)
: not saved; *especially* **:** not absolved from eternal punishment **:** not regenerate

un·sa·vory \-ˈsā-və-rē, -ˈsāv-rē\ *adjective* (13th century)
1 : INSIPID, TASTELESS
2 a : unpleasant to taste or smell **b :** DISAGREEABLE, DISTASTEFUL ⟨an *unsavory* assignment⟩; *especially* **:** morally offensive

un·say \-ˈsā, *Southern also* -ˈse\ *transitive verb* **-said** \-ˈsed\; **-say·ing** \-ˈsā-iŋ\ (15th century)
: to make as if not said **:** RECANT, RETRACT

un·say·able \-ˈsā-ə-bəl\ *adjective* (1870)
: not sayable **:** not easily expressed or related

un·scathed \-ˈskāth̲d\ *adjective* (14th century)
: wholly unharmed **:** not injured

un·schooled \-ˈsküld\ *adjective* (1589)
1 : not schooled **:** UNTAUGHT, UNTRAINED ⟨an *unschooled* woodsman⟩
2 : not artificial **:** NATURAL ⟨*unschooled* talent⟩

un·sci·en·tif·ic \ˌən-ˌsī-ən-ˈti-fik\ *adjective* (circa 1775)
: not scientific: as **a :** not used in scientific work **b :** not being in accord with the principles and methods of science ⟨*unscientific* management of woodlands⟩ ⟨an *unscientific* survey⟩ **c :** not showing scientific knowledge or familiarity with scientific methods
— un·sci·en·tif·i·cal·ly \-fi-k(ə-)lē\ *adverb*

un·scram·ble \ˌən-ˈskram-bəl\ *transitive verb* (circa 1920)
1 : to separate (as a conglomeration or tangle) into original components **:** RESOLVE, CLARIFY
2 : to restore (scrambled communication) to intelligible form
— un·scram·bler \-b(ə-)lər\ *noun*

un·screw \-ˈskrü\ (1605)
transitive verb
1 : to draw the screws from
2 : to loosen or withdraw by turning
intransitive verb
: to become or admit of being unscrewed

un·script·ed \-ˈskrip-təd\ *adjective* (circa 1950)
: not following a prepared script

un·scru·pu·lous \-ˈskrü-pyə-ləs\ *adjective* (1803)
: not scrupulous **:** UNPRINCIPLED
— un·scru·pu·lous·ly *adverb*
— un·scru·pu·lous·ness *noun*

un·seal \-ˈsē(ə)l\ *transitive verb* (15th century)
: to break or remove the seal of **:** OPEN

un·sealed \-ˈsē(ə)ld\ *adjective* (14th century)
: not sealed

un·seam \ˌən-ˈsēm\ *transitive verb* (1592)
: to open the seams of

un·search·able \-ˈsər-chə-bəl\ *adjective* (14th century)
: not capable of being searched or explored **:** INSCRUTABLE
— un·search·ably \-blē\ *adverb*

un·sea·son·able \-ˈsēz-nə-bəl, -ˈsē-zᵊn-ə-\ *adjective* (15th century)
1 : occurring at other than the proper time **:** UNTIMELY
2 : not being in season
3 a : not normal for the season of the year ⟨*unseasonable* weather⟩ **b :** marked by unseasonable weather ⟨an *unseasonable* summer⟩
— un·sea·son·able·ness *noun*
— un·sea·son·ably \-blē\ *adverb*

un·seat \-ˈsēt\ *transitive verb* (1596)
1 : to dislodge from one's seat especially on horseback
2 : to remove from a place or position; *especially* **:** to remove from political office

[1]un·seem·ly \-ˈsēm-lē\ *adjective* (14th century)
: not seemly: as **a :** not according with established standards of good form or taste ⟨*unseemly* bickering⟩ **b :** not suitable for time or place **:** INAPPROPRIATE, UNSEASONABLE
synonym see INDECOROUS
— un·seem·li·ness \-lē-nəs\ *noun*

[2]unseemly *adverb* (14th century)
: in an unseemly manner

un·seen \ˌən-ˈsēn\ *adjective* (13th century)
1 : not seen or perceived
2 : SIGHT 1 ⟨an *unseen* translation⟩

un·seg·re·gat·ed \-ˈse-gri-ˌgā-təd\ *adjective* (1905)
: not segregated; *especially* **:** free from racial segregation

un·se·lect·ed \ˌən-sə-ˈlek-təd\ *adjective* (circa 1891)
: not selected **:** chosen at random

un·se·lec·tive \-ˈlek-tiv\ *adjective* (circa 1925)
: not marked by selection **:** RANDOM, INDISCRIMINATE
— un·se·lec·tive·ly \-lē\ *adverb*

un·self·ish \ˌən-ˈsel-fish\ *adjective* (1698)
: not selfish **:** GENEROUS
— un·self·ish·ly *adverb*
— un·self·ish·ness *noun*

un·sell \-ˈsel\ *transitive verb* **-sold** \-ˈsōld\; **-sel·ling** (circa 1929)
1 : to dissuade from a belief in the truth, value, or desirability of something ⟨ads that *unsell* the public on energy consumption⟩
2 : to discourage a belief in the truth, value, or desirability of ⟨I *unsold* the coat he wanted and sold him another⟩

un·set \-ˈset\ *adjective* (14th century)
: not set: as **a :** not fixed in a setting **:** UNMOUNTED ⟨*unset* diamonds⟩ **b :** not firmed or solidified ⟨*unset* concrete⟩

un·set·tle \ˌən-ˈse-tᵊl\ (1598)
transitive verb
1 : to loosen or move from a settled state or condition **:** make unstable **:** DISORDER
2 : to perturb or agitate mentally or emotionally **:** DISCOMPOSE
intransitive verb
: to become unsettled

un·set·tled \-ˈse-tᵊld\ *adjective* (1591)
: not settled: as **a** (1) **:** not calm or tranquil **:** DISTURBED ⟨*unsettled* political conditions⟩ (2) **:** likely to vary widely especially in the near future **:** VARIABLE ⟨*unsettled* weather⟩ (3) **:** not settled down ⟨*unsettled* dust⟩ **b** (1) **:** not decided or determined **:** DOUBTFUL ⟨an *unsettled* state of mind⟩ (2) **:** not resolved or worked out **:** UNDECIDED ⟨an *unsettled* question⟩ **c :** characterized by irregularity ⟨an *unsettled* life⟩ **d :** not inhabited or populated ⟨*unsettled* land⟩ **e :** mentally unbalanced **f** (1) **:** not disposed of according to law ⟨an *unsettled* estate⟩ (2) **:** not paid or discharged ⟨*unsettled* debts⟩
— un·set·tled·ness \-tᵊl(d)-nəs\ *noun*

un·set·tling \ˌən-ˈset-liŋ, -ˈse-tᵊl-iŋ\ *adjective* (1665)
: having the effect of upsetting, disturbing, or discomposing ⟨*unsettling* images of the war⟩
— un·set·tling·ly \-lē\ *adverb*

un·set·tle·ment \-tᵊl-mənt\ *noun* (1648)
1 : an act, process, or instance of unsettling
2 : the quality or state of being unsettled

un·sew \ˌən-ˈsō\ *transitive verb* **-sewed; -sewn** \-ˈsōn\ *or* **-sewed; -sew·ing** (14th century)
: to undo the sewing of

un·sex \-ˈseks\ *transitive verb* (1605)
1 : to deprive of sex or sexual power
2 : to deprive of the qualities typical of one's sex

un·shack·le \-ˈsha-kəl\ *transitive verb* (1611)
: to free from shackles

un·shaped \-ˈshāpt\ *adjective* (1572)
: not shaped: as **a :** not dressed or finished to final form ⟨an *unshaped* timber⟩ **b :** imperfect in form or formulation ⟨*unshaped* ideas⟩

un·shap·en \-ˈshā-pən\ *adjective* [Middle English, from [1]*un-* + *shapen,* past participle of *shapen* to shape] (14th century)
: UNSHAPED

un·sheathe \ˌən-ˈshēth̲\ *transitive verb* (circa 1542)
: to draw from or as if from a sheath or scabbard

un·ship \-ˈship\ (15th century)
transitive verb
1 : to take out of a ship **:** DISCHARGE, UNLOAD
2 : to remove (as an oar or tiller) from position **:** DETACH
intransitive verb
: to become or admit of being detached or removed

un·shod \ˌən-ˈshäd\ *adjective* (before 12th century)
: not wearing or provided with shoes

[1]un·sight \-ˈsīt\ *transitive verb* (1615)
: to prevent from seeing

[2]unsight *adjective* (circa 1622)
: not sighted or examined

un·sight·ly \ˌən-ˈsīt-lē\ *adjective* (15th century)
: not pleasing to the sight **:** not comely
— un·sight·li·ness \-lē-nəs\ *noun*

un·skilled \-ˈskild\ *adjective* (1581)
1 : not skilled in a branch of work **:** lacking technical training ⟨an *unskilled* worker⟩
2 : not requiring skill ⟨*unskilled* jobs⟩
3 : marked by lack of skill ⟨produced *unskilled* poems⟩

un·skill·ful \-ˈskil-fəl\ *adjective* (1565)
: not skillful **:** lacking in skill or proficiency
— un·skill·ful·ly \-fə-lē\ *adverb*
— un·skill·ful·ness *noun*

un·slak·able \ˌən-ˈslā-kə-bəl\ *adjective* (1820)
: unable to be slaked **:** UNQUENCHABLE ⟨an *unslakable* thirst⟩ ⟨an *unslakable* desire for excellence⟩

un·sling \-ˈsliŋ\ *transitive verb* **-slung** \-ˈsləŋ\; **-sling·ing** \-ˈsliŋ-iŋ\ (1630)
1 : to take off the slings of especially aboard ship **:** release from slings
2 : to remove from being slung ⟨*unslung* the carbine⟩

un·snap \-ˈsnap\ *transitive verb* (1862)
: to loosen or free by or as if by undoing a snap

un·snarl \-ˈsnär(-ə)l\ *transitive verb* (1555)
: to disentangle a snarl in

un·so·cia·ble \ˌən-ˈsō-shə-bəl\ *adjective* (1600)
1 : having or showing a disinclination for social activity **:** SOLITARY, RESERVED
2 : not conducive to sociability
— un·so·cia·bil·i·ty \ˌən-ˌsō-shə-ˈbi-lə-tē\ *noun*
— un·so·cia·ble·ness \ˌən-ˈsō-shə-bəl-nəs\ *noun*
— un·so·cia·bly \-blē\ *adverb*

un·so·cial \-ˈsō-shəl\ *adjective* (1731)
: lacking a taste or desire for society or close association; *also* **:** marked by or arising from such a lack ⟨an *unsocial* disposition⟩
— un·so·cial·ly \-ˈsō-sh(ə-)lē\ *adverb*

un·sold \-ˈsōld\ *adjective* (14th century)
: not sold

un·so·phis·ti·cat·ed \ˌən-sə-ˈfis-tə-ˌkā-təd\ *adjective* (1630)
: not sophisticated: as **a :** not changed or corrupted **:** GENUINE **b** (1) **:** not worldly-wise **:** lacking social or economic sophistication (2) **:** lacking complexity of structure **:** SIMPLE, STRAIGHTFORWARD ⟨an *unsophisticated* analysis⟩ ⟨*unsophisticated* rhythms⟩
synonym see NATURAL

un·so·phis·ti·ca·tion \-ˌfis-tə-ˈkā-shən\ *noun* (1825)
: lack of or freedom from sophistication

un·sought \ˌən-ˈsȯt\ *adjective* (13th century)
: not searched for or sought out ⟨*unsought* compliments⟩

un·sound \-ˈsau̇nd\ *adjective* (14th century)
: not sound: as **a :** not healthy or whole **b :** not mentally normal **:** not wholly sane **c :** not firmly made, placed, or fixed **d :** not valid or true **:** INVALID, SPECIOUS
— un·sound·ly \-ˈsau̇n(d)-lē\ *adverb*

un·sound·ness \-ˈsau̇n(d)-nəs\ *noun* (1586)
1 : the quality or state of being unsound
2 : something (as a disease, injury, or defect) that causes one to be unsound

un·spar·ing \-ˈspar-iŋ, -ˈsper-\ *adjective* (circa 1586)
1 : not merciful or forbearing **:** HARD, RUTHLESS
2 : not frugal **:** LIBERAL, PROFUSE
— un·spar·ing·ly \-iŋ-lē\ *adverb*

un·speak \-ˈspēk\ *transitive verb* (1605)
obsolete **:** UNSAY

un·speak·able \-ˈspē-kə-bəl\ *adjective* (14th century)
1 a : incapable of being expressed in words **:** UNUTTERABLE **b :** inexpressibly bad **:** HORRENDOUS
2 : that may not or cannot be spoken ⟨the bawdy thoughts that come into one's head — the *unspeakable* words —L. P. Smith⟩ ⟨*unspeakable* collections of consonants —Rosemary Jellis⟩
— un·speak·ably \-blē\ *adverb*

un·sports·man·like \-ˈspōrts-mən-ˌlīk, -ˈspȯrts-\ *adjective* (1754)
: not characteristic of or exhibiting good sportsmanship **:** not sportsmanlike

un·spot·ted \-ˈspä-təd\ *adjective* (14th century)
: not spotted **:** free from spot or stain; *especially* **:** free from moral stain

un·sprung \-ˈsprəŋ\ *adjective* (1600)
: not sprung; *especially* **:** not equipped with springs

un·sta·ble \-ˈstā-bəl\ *adjective* (13th century)
: not stable **:** not firm or fixed **:** not constant: as **a :** not steady in action or movement **:** IRREGULAR ⟨an *unstable* pulse⟩ **b :** wavering in purpose or intent **:** VACILLATING **c :** lacking steadiness **:** apt to move, sway, or fall ⟨an *unstable* tower⟩ **d** (1) **:** liable to change or alteration ⟨an *unstable* economy⟩ ⟨*unstable* weather⟩ (2) **:** readily changing (as by decomposing) in chemical composition or biological activity **e :** characterized by lack of emotional control
synonym see INCONSTANT
— un·sta·ble·ness *noun*
— un·sta·bly \-b(ə-)lē\ *adverb*

[1]**un·steady** \ˌən-ˈste-dē\ *transitive verb* (1532)
: to make unsteady

[2]**unsteady** *adjective* (1598)
: not steady: as **a :** not firm or solid **:** not fixed in position **:** UNSTABLE **b :** marked by change or fluctuation **:** CHANGEABLE **c :** not uniform or even **:** IRREGULAR
— un·stead·i·ly \-ˈste-dᵊl-ē\ *adverb*
— un·stead·i·ness \-ˈste-dē-nəs\ *noun*

un·step \ˌən-ˈstep\ *transitive verb* (1853)
: to remove (a mast) from a step

un·stick \-ˈstik\ *transitive verb* **-stuck** \-ˈstək\; **-stick·ing** (1706)
: to release from a state of adhesion

un·stint·ing \-ˈstin-tiŋ\ *adjective* (1845)
: not restricting or holding back **:** giving or being given freely or generously ⟨an *unstinting* volunteer⟩ ⟨*unstinting* praise⟩
— un·stint·ing·ly *adverb*

un·stop \-ˈstäp\ *transitive verb* (14th century)
1 : to free from an obstruction **:** OPEN
2 : to remove a stopper from

un·stop·pa·ble \-ˈstä-pə-bəl\ *adjective* (1836)
: incapable of being stopped
— un·stop·pa·bly \-blē\ *adverb*

un·strap \-ˈstrap\ *transitive verb* (1828)
: to remove or loose a strap from

un·stressed \ˌən-ˈstrest\ *adjective* (1883)
1 : not bearing a stress or accent ⟨*unstressed* syllables⟩
2 : not subjected to stress ⟨*unstressed* wires⟩

un·string \-ˈstriŋ\ *transitive verb* **-strung** \-ˈstrəŋ\; **-string·ing** \-ˈstriŋ-iŋ\ (1611)
1 : to loosen or remove the strings of
2 : to remove from a string
3 : to make weak, disordered, or unstable ⟨was *unstrung* by the news⟩

un·struc·tured \-ˈstrək-chərd\ *adjective* (1936)
: lacking structure or organization: as **a :** not formally organized in a set or conventional pattern ⟨an *unstructured* question⟩ ⟨feel insecure in an *unstructured* situation⟩ **b :** not having a system or hierarchy typical of an organized society

un·stuck \ˌən-ˈstək\ *adjective* (1934)
: brought into a state of disarray, discomposure, or incoherence ⟨the deal came *unstuck*⟩

un·stud·ied \-ˈstə-dēd\ *adjective* (14th century)
: not studied: as **a :** not acquired by study **b :** not forced **:** not done or planned for effect

un·sub·stan·tial \ˌən-səb-ˈstan(t)-shəl\ *adjective* (15th century)
: not substantial **:** lacking substance, firmness, or strength
— un·sub·stan·ti·al·i·ty \-ˌstan(t)-shē-ˈa-lə-tē\ *noun*
— un·sub·stan·tial·ly \ˌən-səb-ˈstan(t)-sh(ə-)lē\ *adverb*

un·suc·cess \ˌən-sək-ˈses\ *noun* (circa 1586)
: lack of success **:** FAILURE

un·suc·cess·ful \-fəl\ *adjective* (1617)
: not successful **:** not meeting with or producing success
— un·suc·cess·ful·ly \-fə-lē\ *adverb*

un·sung \ˌən-ˈsəŋ\ *adjective* (15th century)
1 : not sung
2 : not celebrated or praised (as in song or verse) ⟨an *unsung* hero⟩

un·sur·pris·ing \ˌən-sə(r)-ˈprī-ziŋ\ *adjective* (1671)
: not surprising or unexpected

un·sur·pris·ing·ly \-ziŋ-lē\ *adverb* (1950)
1 : as is not surprising ⟨matters complicate, *unsurprisingly* —Stanley Kauffmann⟩
2 : in an unsurprising manner ⟨the story ended *unsurprisingly*⟩

un·swathe \-ˈswäth, -ˈswȯth, -ˈswāth\ *transitive verb* (14th century)
: to free from something that swathes

un·swear \-ˈswar, -ˈswer\ *verb* **-swore** \-ˈswōr, -ˈswȯr\; **-sworn** \-ˈswōrn, -ˈswȯrn\; **swear·ing** (1596)
intransitive verb
archaic **:** to unsay or retract something sworn
transitive verb
archaic **:** to recant or recall (as an oath) especially by a second oath

un·swerv·ing \-ˈswər-viŋ\ *adjective* (1694)
1 : not swerving or turning aside
2 : STEADY, UNFALTERING ⟨*unswerving* loyalty⟩

un·sym·met·ri·cal \ˌən-sə-ˈme-tri-kəl\ *adjective* (circa 1755)
: ASYMMETRIC
— un·sym·met·ri·cal·ly \-k(ə-)lē\ *adverb*

un·tan·gle \ˌən-ˈtaŋ-gəl\ *transitive verb* (1550)
: to loose from tangles or entanglement **:** straighten out
synonym see EXTRICATE

un·tapped \-ˈtapt\ *adjective* (circa 1775)
1 : not subjected to tapping ⟨an *untapped* keg⟩
2 : not drawn upon or utilized ⟨as yet *untapped* markets⟩

un·taught \-ˈtȯt\ *adjective* (14th century)
1 : not instructed or trained **:** IGNORANT
2 : NATURAL, SPONTANEOUS ⟨*untaught* kindness⟩

un·teach \-ˈtēch\ *transitive verb* **-taught** \-ˈtȯt\; **-teach·ing** (1532)
1 : to cause to unlearn something
2 : to teach the contrary of

un·ten·a·ble \-ˈte-nə-bəl\ *adjective* (1647)
1 : not able to be defended
2 : not able to be occupied
— un·ten·a·bil·i·ty \ˌən-ˌte-nə-ˈbi-lə-tē\ *noun*

un·tent·ed \-ˈten-təd\ *adjective* [[1]*un-* + obsolete English *tented,* past participle of *tent* to probe] (1605)
archaic **:** not probed or dressed ⟨the *untented* woundings of a father's curse —Shakespeare⟩

un·teth·er \-ˈte-thər\ *transitive verb* (circa 1775)
: to free from a tether

un·think \-ˈthiŋk\ *transitive verb* **-thought** \-ˈthȯt\; **-think·ing** (circa 1600)
: to put out of mind

un·think·able \-ˈthiŋ-kə-bəl\ *adjective* (15th century)
1 : not capable of being grasped by the mind
2 : being contrary to what is reasonable, desirable, or probable **:** being out of the question
— un·think·abil·i·ty \ˌən-ˌthiŋ-kə-ˈbi-lə-tē\ *noun*
— un·think·ably \ˌən-ˈthiŋ-kə-blē\ *adverb*

un·think·ing \ˌən-ˈthiŋ-kiŋ\ *adjective* (1676)
1 : not taking thought **:** HEEDLESS, UNMINDFUL
2 : not indicating thought or reflection
3 : not having the power of thought
— un·think·ing·ly \-kiŋ-lē\ *adverb*

un·thought \-ˈthȯt\ *adjective* (circa 1548)
: not anticipated **:** UNEXPECTED — often used with *of* ⟨an *unthought*-of development⟩

un·thread \ˌən-ˈthred\ *transitive verb* (1595)
1 : to draw or take out a thread from
2 : to loosen the threads or connections of
3 : to make one's way through ⟨*unthread* a maze⟩

un·throne \-ˈthrōn\ *transitive verb* (1611)
: to remove from or as if from a throne

un·ti·dy \-ˈtī-dē\ *adjective* (14th century)
1 a : not neat **:** SLOVENLY **b :** not neat or orderly in habits or procedure ⟨an *untidy* mind⟩
2 a : not neatly organized or carried out ⟨an *untidy* manuscript⟩ **b :** conducive to a lack of neatness ⟨*untidy* tasks like bathing the baby —*New Yorker*⟩
— un·ti·di·ly \-ˈtī-dᵊl-ē\ *adverb*
— un·ti·di·ness \-ˈtī-dē-nəs\ *noun*

un·tie \-ˈtī\ *verb* **-tied; -ty·ing** *or* **-tie·ing** (before 12th century)
transitive verb
1 : to free from something that ties, fastens, or restrains **:** UNBIND
2 a : to disengage the knotted parts of **b :** DISENTANGLE, RESOLVE ⟨*untie* a traffic jam⟩
intransitive verb
: to become loosened or unbound

[1]**un·til** \ən-ˈtil, -ˈtel; ˈən-ˌ, -tᵊl\ *preposition* [Middle English, from *un-* up to, until (akin to Old English *oth* to, until, Old High German *unt* up to, until, Old English *ende* end) + *til, till*] (13th century)
1 *chiefly Scottish* **:** TO
2 — used as a function word to indicate continuance (as of an action or condition) to a specified time ⟨stayed *until* morning⟩
3 : BEFORE 2 ⟨not available *until* tomorrow⟩ ⟨we don't open *until* ten⟩

[2]**until** *conjunction* (14th century)

: up to the time that : up to such time as ⟨play continued *until* it got dark⟩ ⟨never able to relax *until* he took up fishing⟩ ⟨ran *until* he was breathless⟩

[1]**un·time·ly** \ˌən-ˈtīm-lē\ *adverb* (13th century)
1 : at an inopportune time **:** UNSEASONABLY
2 : before the due, natural, or proper time **:** PREMATURELY ⟨went *untimely* to the grave⟩

[2]**untimely** *adjective* (1535)
1 : occurring or done before the due, natural, or proper time **:** too early **:** PREMATURE ⟨*untimely* death⟩
2 : INOPPORTUNE, UNSEASONABLE ⟨an *untimely* joke⟩ ⟨*untimely* frost⟩
— **un·time·li·ness** *noun*

un·time·ous \-ˈtī-məs\ *adjective* (15th century)
chiefly Scottish **:** UNTIMELY

un·tir·ing \-ˈtī-riŋ\ *adjective* (1822)
: not becoming tired **:** INDEFATIGABLE ⟨an *untiring* worker⟩
— **un·tir·ing·ly** *adverb*

un·ti·tled \-ˈtī-tᵊld\ *adjective* (1590)
1 *obsolete* **:** having no title or right to rule
2 : not named ⟨an *untitled* novel⟩
3 : not called by a title ⟨*untitled* nobility⟩

un·to \ˈən-(ˌ)tü\ *preposition* [Middle English, from *un-* up to, until + *to*] (14th century)
1 : TO
2 — used as a function word to indicate reference or concern ⟨they became a world *unto* themselves —Anne T. Fleming⟩

un·told \ˌən-ˈtōld\ *adjective* (15th century)
1 : too great or numerous to count **:** INCALCULABLE, VAST
2 a : not told or related **b :** kept secret

un·touch·abil·i·ty \ˌən-ˌtə-chə-ˈbi-lə-tē\ *noun* (1919)
: the quality or state of being untouchable; *especially* **:** the state of being an untouchable

[1]**un·touch·able** \ˌən-ˈtə-chə-bəl\ *adjective* (1567)
1 a : forbidden to the touch **:** not to be handled **b :** exempt from criticism or control
2 : lying beyond reach
3 : disagreeable or defiling to the touch

[2]**untouchable** *noun* (1909)
: one that is untouchable; *specifically* **:** a member of a large formerly segregated hereditary group in India having in traditional Hindu belief the quality of defiling by contact a member of a higher caste

un·touched \ˌən-ˈtəcht\ *adjective* (14th century)
1 : not subjected to touching **:** not handled
2 : not described or dealt with
3 a : not tasted **b :** being in the first or a primeval state or condition
4 : not influenced **:** UNAFFECTED

un·to·ward \ˌən-ˈtō(-ə)rd, -ˈtȯ(-ə)rd; ˌən-tə-ˈwȯrd\ *adjective* (1526)
1 : difficult to guide, manage, or work with **:** UNRULY, INTRACTABLE
2 a : marked by trouble or unhappiness **:** UNLUCKY **b :** not favorable **:** ADVERSE, UNPROPITIOUS ⟨*untoward* side effects⟩
3 : IMPROPER, INDECOROUS
— **un·to·ward·ly** *adverb*
— **un·to·ward·ness** *noun*

un·tread \ˌən-ˈtred\ *transitive verb* (1592)
archaic **:** to tread back **:** RETRACE

un·tried \-ˈtrīd\ *adjective* (1526)
1 : not tested or proved by experience or trial
2 : not tried in court

un·trod \-ˈträd\ *or* **un·trod·den** \-ˈträ-dᵊn\ *adjective* (1593)
: not trod **:** UNTRAVERSED

un·trou·bled \-ˈtrə-bəld\ *adjective* (15th century)
1 : not given trouble **:** not made uneasy
2 : CALM, TRANQUIL

un·true \-ˈtrü\ *adjective* (before 12th century)
1 : not faithful **:** DISLOYAL
2 : not according with a standard of correctness **:** not level or exact
3 : not according with the facts **:** FALSE
— **un·tru·ly** \-ˈtrü-lē\ *adverb*

un·truss \-ˈtrəs\ (1577)
transitive verb
1 *archaic* **:** UNTIE, UNFASTEN — used in the phrase *untruss one's points*
2 *archaic* **:** UNDRESS
intransitive verb
archaic **:** to unfasten or take off one's clothes and especially one's breeches

un·truth \ˌən-ˈtrüth, ˈən-ˌ\ *noun* (before 12th century)
1 *archaic* **:** DISLOYALTY
2 : lack of truthfulness **:** FALSITY
3 : something that is untrue **:** FALSEHOOD

un·truth·ful \-ˈtrüth-fəl\ *adjective* (circa 1843)
: not containing or telling the truth **:** FALSE, INACCURATE ⟨*untruthful* report⟩
synonym see DISHONEST
— **un·truth·ful·ly** \-fə-lē\ *adverb*
— **un·truth·ful·ness** *noun*

un·tune \-ˈtün, -ˈtyün\ *transitive verb* (1598)
1 : to put out of tune
2 : DISARRANGE, DISCOMPOSE

un·tu·tored \-ˈtü-tərd, -ˈtyü-\ *adjective* (1593)
1 a : having no formal learning or training **b :** NAIVE, UNSOPHISTICATED
2 : not produced or developed by instruction **:** NATIVE ⟨*untutored* shrewdness⟩
synonym see IGNORANT

un·twine \-ˈtwīn\ (15th century)
transitive verb
1 : to unwind the twisted or tangled parts of **:** DISENTANGLE
2 : to remove by unwinding
intransitive verb
: to become disentangled or unwound

un·twist \ˌən-ˈtwist\ (1538)
transitive verb
: to separate the twisted parts of **:** UNTWINE
intransitive verb
: to become untwined

un·twist·ed \-ˈtwis-təd\ *adjective* (1575)
: not twisted

un·used \-ˈyüzd, *in the phrase "unused to" usually* -ˈyüs(t)\ *adjective* (14th century)
1 : not habituated **:** UNACCUSTOMED ⟨*unused* to crowds⟩
2 : not used: as **a :** FRESH, NEW ⟨set an *unused* canvas on the easel⟩ **b :** not put to use **:** IDLE ⟨*unused* land⟩ **c :** not consumed **:** ACCRUED ⟨*unused* sick leave⟩

un·usu·al \-ˈyü-zhə-wəl, -zhəl; -ˈyüzh-wəl\ *adjective* (1582)
: not usual **:** UNCOMMON, RARE
— **un·usu·al·ly** *adverb*
— **un·usu·al·ness** *noun*

un·ut·ter·able \ˌən-ˈə-tə-rə-bəl\ *adjective* (circa 1586)
: being beyond the powers of description **:** INEXPRESSIBLE
— **un·ut·ter·ably** \-blē\ *adverb*

un·val·ued \-ˈval-(ˌ)yüd, -yəd\ *adjective* (1586)
1 *obsolete* **:** INVALUABLE
2 a : not important or prized **:** DISREGARDED **b :** not appraised

un·var·nished \-ˈvär-nisht\ *adjective* (1604)
1 a : not adorned or glossed **:** PLAIN, STRAIGHTFORWARD ⟨told the *unvarnished* truth⟩ **b :** ARTLESS, FRANK ⟨the *unvarnished* candor of old people and children —Janet Flanner⟩
2 : not coated with or as if with varnish **:** CRUDE, UNFINISHED

un·veil \ˌən-ˈvā(ə)l\ (1599)
transitive verb
1 : to remove a veil or covering from
2 : to make public **:** DIVULGE, REVEAL
intransitive verb
: to throw off a veil or protective cloak

un·veiled \-ˈvā(ə)ld\ *adjective* (1606)
: not veiled **:** OPEN, REVEALED

un·vo·cal \ˌən-ˈvō-kəl\ *adjective* (1773)
1 : not eloquent or outspoken **:** INARTICULATE
2 : not musical **:** DISCORDANT

un·voice \-ˈvȯis\ *transitive verb* (1637)
: DEVOICE

un·voiced \-ˈvȯist\ *adjective* (1859)
1 : not verbally expressed
2 : VOICELESS 2

un·war·rant·able \-ˈwȯr-ən-tə-bəl, -ˈwär-\ *adjective* (1612)
: not justifiable **:** INEXCUSABLE
— **un·war·rant·ably** \-blē\ *adverb*

un·wary \ˌən-ˈwar-ē, -ˈwer-\ *adjective* (1579)
: not alert **:** easily fooled or surprised **:** HEEDLESS, GULLIBLE
— **un·wari·ly** \-ˈwar-ə-lē, -ˈwer-\ *adverb*
— **un·wari·ness** \-ˈwar-ē-nəs, -ˈwer-\ *noun*

[1]**un·washed** \-ˈwȯsht, -ˈwäsht\ *adjective* (14th century)
1 : not cleaned with or as if with soap and water
2 : IGNORANT, PLEBEIAN
— **un·washed·ness** *noun*

[2]**unwashed** *noun* (1830)
: an ignorant or underprivileged group **:** RABBLE — usually used with *great* ⟨the great *unwashed*⟩

un·watch·able \ˌən-ˈwä-chə-bəl, -ˈwȯ-\ *adjective* (1886)
: not suitable or fit for watching **:** tending to discourage watching

un·wea·ried \-ˈwir-ēd\ *adjective* (13th century)
: not tired or jaded **:** FRESH
— **un·wea·ried·ly** *adverb*

un·weave \-ˈwēv\ *transitive verb* **-wove** \-ˈwōv\; **-wo·ven** \-ˈwō-vən\; **-weav·ing** (1542)
: DISENTANGLE, UNRAVEL

un·weet·ing \-ˈwē-tiŋ\ *adjective* (14th century)
archaic **:** UNWITTING
— **un·weet·ing·ly** \-tiŋ-lē\ *adverb, archaic*

un·weight \-ˈwāt\ (circa 1939)
transitive verb
: to reduce momentarily the force exerted by (as a ski) upon a surface by shifting the weight or position of one's body
intransitive verb
: to unweight something by shifting the weight or position of one's body

un·well \ˌən-ˈwel\ *adjective* (15th century)
1 : being in poor health **:** AILING, SICK
2 : undergoing menstruation

un·whole·some \-ˈhōl-səm\ *adjective* (13th century)
1 : detrimental to physical, mental, or moral well-being **:** UNHEALTHY ⟨*unwholesome* food⟩ ⟨*unwholesome* pastimes⟩
2 a : CORRUPT, UNSOUND **b :** offensive to the senses **:** LOATHSOME
— **un·whole·some·ly** *adverb*

un·wieldy \-ˈwē(ə)l-dē\ *adjective* (1530)
: not easily managed, handled, or used (as because of bulk, weight, complexity, or awkwardness) **:** CUMBERSOME
— **un·wield·i·ly** \-ˈwēl-də-lē\ *adverb*
— **un·wield·i·ness** \-dē-nəs\ *noun*

un·will·ing \-ˈwi-liŋ\ *adjective* (before 12th century)
: not willing: **a :** LOATH, RELUCTANT ⟨was *unwilling* to learn⟩ **b :** done or given reluctantly ⟨*unwilling* approval⟩ **c :** offering opposition **:** OBSTINATE ⟨an *unwilling* student⟩
— **un·will·ing·ly** \-liŋ-lē\ *adverb*
— **un·will·ing·ness** *noun*

un·wind \-ˈwīnd\ *verb* **-wound** \-ˈwau̇nd\; **-wind·ing** (14th century)
transitive verb
1 a : to cause to uncoil **:** wind off **:** UNROLL **b :** to free from or as if from a binding or wrapping **c :** to release from tension **:** RELAX
2 *archaic* **:** to trace to the end ⟨*unwinding* the labryinth and bringing the hero out —Laurence Sterne⟩
intransitive verb
1 : to become uncoiled or disentangled **:** UNFOLD
2 : to become released from tension

un·wis·dom \ˌən-ˈwiz-dəm\ *noun* (before 12th century)
: lack of wisdom **:** FOOLISHNESS, FOLLY

un·wise \-ˈwīz\ *adjective* (before 12th century)
: lacking wisdom or good sense **:** FOOLISH, IMPRUDENT
— **un·wise·ly** *adverb*

un·wish \-ˈwish\ *transitive verb* (1594)
obsolete **:** to wish away

un·wit·ting \-ˈwi-tiŋ\ *adjective* (before 12th century)
1 : not knowing **:** UNAWARE ⟨kept the truth from their *unwitting* friends⟩
2 : not intended **:** INADVERTENT ⟨an *unwitting* mistake⟩
— **un·wit·ting·ly** \-tiŋ-lē\ *adverb*

un·wont·ed \-ˈwȯn-təd, -ˈwōn- *also* -ˈwən- *or* -ˈwän-\ *adjective* (1553)
1 : being out of the ordinary **:** RARE, UNUSUAL
2 : not accustomed by experience
— **un·wont·ed·ly** *adverb*
— **un·wont·ed·ness** *noun*

un·world·ly \-ˈwər(-ə)l-dlē, -ˈwərl-lē\ *adjective* (1707)
1 : not of this world **:** UNEARTHLY; *specifically* **:** SPIRITUAL
2 a : not wise in the ways of the world **:** NAIVE **b :** not swayed by mundane considerations
— **un·world·li·ness** \-ˈwərl(d)-lē-nəs\ *noun*

un·worn \-ˈwōrn, -ˈwȯrn\ *adjective* (circa 1586)
1 a : not impaired by use **:** not worn away **b :** not worn **:** NEW
2 : not jaded **:** FRESH, ORIGINAL

un·wor·thy \ˌən-ˈwər-t͟hē\ *adjective* (13th century)
1 a : lacking in excellence or value **:** POOR, WORTHLESS **b :** BASE, DISHONORABLE
2 : not meritorious **:** UNDESERVING ⟨*unworthy* of attention⟩
3 : not deserved **:** UNMERITED ⟨*unworthy* treatment⟩
4 : inappropriate to one's condition or station ⟨actions *unworthy* of a gentleman⟩
— **un·wor·thi·ly** \-t͟hə-lē\ *adverb*
— **un·wor·thi·ness** \-t͟hē-nəs\ *noun*

un·wrap \-ˈrap\ *transitive verb* (14th century)
: to remove the wrapping from **:** DISCLOSE ⟨*unwrap* a package⟩ ⟨*unwrap* evidence in a criminal case⟩

un·wreathe \-ˈrēt͟h\ *transitive verb* (1591)
: UNCOIL, UNTWIST

un·writ·ten \-ˈri-tᵊn\ *adjective* (14th century)
1 : not expressed in writing **:** ORAL, TRADITIONAL
2 : containing no writing **:** BLANK

unwritten constitution *noun* (1890)
: a constitution not embodied in a single document but based chiefly on custom and precedent as expressed in statutes and judicial decisions

unwritten law *noun* (1641)
: law based chiefly on custom rather than legislative enactments

un·yield·ing \ˌən-ˈyē(ə)l-diŋ\ *adjective* (1658)
1 : characterized by lack of softness or flexibility
2 : characterized by firmness or obduracy
— **un·yield·ing·ly** \-diŋ-lē\ *adverb*

un·yoke \-ˈyōk\ (before 12th century)
transitive verb
1 : to free from a yoke or harness
2 : to take apart **:** DISJOIN
intransitive verb
1 *archaic* **:** to unharness a draft animal
2 *archaic* **:** to cease from work

un·zip \-ˈzip\ (1939)
transitive verb
: to zip open
intransitive verb
: to open by or as if by means of a zipper

¹up \ˈəp\ *adverb* [partly from Middle English *up* upward, from Old English *ūp;* partly from Middle English *uppe* on high, from Old English; both akin to Old High German *ūf* up and probably to Latin *sub* under, Greek *hypo* under, *hyper* over — more at OVER] (before 12th century)
1 a (1) **:** in or into a higher position or level; *especially* **:** away from the center of the earth (2) **:** from beneath the ground or water to the surface (3) **:** from below the horizon (4) **:** UPSTREAM (5) **:** in or into an upright position ⟨sit *up*⟩; *especially* **:** out of bed **b :** upward from the ground or surface ⟨pull *up* a daisy⟩ **c :** so as to expose a particular surface
2 : with greater intensity ⟨speak *up*⟩
3 a : in or into a better or more advanced state **b :** at an end ⟨your time is *up*⟩ **c :** in or into a state of greater intensity or excitement **d :** in a continual sequence **:** in continuance from a point or to a point ⟨from third grade *up*⟩ ⟨at prices of $10 and *up*⟩ ⟨*up* until now⟩
4 a (1) **:** into existence, evidence, prominence, or prevalence ⟨put *up* several new buildings⟩ (2) **:** into operation or practical form **b :** into consideration or attention ⟨bring *up* for discussion⟩
5 : into possession or custody
6 a : ENTIRELY, COMPLETELY ⟨button *up* your coat⟩ **b** — used as an intensifier ⟨clean *up* the house⟩
7 : in or into storage **:** BY ⟨lay *up* supplies⟩
8 a : so as to arrive or approach **b :** in a direction conventionally the opposite of down: (1) **:** to windward (2) **:** NORTHWARD (3) **:** to or at the top (4) **:** to or at the rear of a theatrical stage
9 : in or into parts
10 : to a stop — usually used with *draw, bring, fetch,* or *pull*
11 : for each side ⟨the score is 15 *up*⟩

²up *adjective* (before 12th century)
1 a : risen above the horizon ⟨the sun is *up*⟩ **b :** STANDING **c :** being out of bed **d :** relatively high ⟨the river is *up*⟩ ⟨was well *up* in her class⟩ **e :** being in a raised position **:** LIFTED ⟨windows are *up*⟩ **f :** being in a state of completion **:** CONSTRUCTED, BUILT **g :** having the face upward **h :** mounted on a horse ⟨a new jockey *up*⟩ **i :** grown above a surface ⟨the corn is *up*⟩ **j** (1) **:** moving, inclining, or directed upward ⟨the *up* escalator⟩ (2) **:** bound in a direction regarded as up
2 a (1) **:** marked by agitation, excitement, or activity (2) **:** positive or upbeat in mood or demeanor **b :** being above a former or normal level (as of quantity or intensity) ⟨attendance is *up*⟩ ⟨the wind is *up*⟩ **c :** exerting enough power (as for operation) ⟨sail when steam is *up*⟩ **d :** READY; *specifically* **:** highly prepared **e :** going on **:** taking place ⟨find out what is *up*⟩
3 a : risen from a lower position ⟨men *up* from the ranks⟩ **b :** being at the same level or point ⟨did not feel *up* to par⟩ **c** (1) **:** well informed **:** ABREAST ⟨*up* on the news⟩ (2) **:** being on schedule ⟨*up* on his homework⟩ **d :** being ahead of one's opponent
4 a : presented for or undergoing consideration ⟨contract *up* for negotiation⟩; *also* **:** charged before a court ⟨*up* for robbery⟩ **b :** being the one whose turn it is ⟨you're *up* next⟩
5 *of a quark* **:** having an electric charge of +⅔, zero charm, and zero strangeness — compare ⁴DOWN 6
— **up against :** confronted with **:** face-to-face with ⟨the problem we are *up against*⟩
— **up to 1 :** capable of performing or dealing with ⟨feels *up to* her role⟩ **2 :** engaged in ⟨what is he *up to*⟩ **3 :** being the responsibility of ⟨it's *up to* me⟩

³up *preposition* (1509)
1 a — used as a function word to indicate motion to or toward or situation at a higher point of ⟨went *up* the stairs⟩ **b :** up into or in the ⟨went *up* attic⟩
2 a : in a direction regarded as being toward or near the upper end or part of ⟨lives a few miles *up* the coast⟩ ⟨walked *up* the street⟩ **b :** toward or near a point closer to the source or beginning of ⟨sail *up* the river⟩
3 : in the direction opposite to ⟨sailed *up* the wind⟩

⁴up *noun* (1536)
1 : one in a high or advantageous position
2 : an upward slope
3 : a period or state of prosperity or success
4 : ³UPPER

⁵up *verb* **upped** \ˈəpt\ *or in intransitive verb 2* **up; upped; up·ping; ups** *or in intransitive verb 2* **up** (1643)
intransitive verb
1 a : to rise from a lying or sitting position **b :** to move upward **:** ASCEND
2 — used with *and* and another verb to indicate that the action of the following verb was either surprisingly or abruptly initiated ⟨he *up* and married a showgirl⟩
transitive verb
1 : RAISE, LIFT
2 a : to advance to a higher level: (1) **:** INCREASE (2) **:** PROMOTE 1a **b :** RAISE 8d, e

up–and–com·ing \ˌəp-ən(d)-ˈkə-miŋ, ˌəp-ᵊm-\ *adjective* (1926)
: gaining prominence and likely to advance or succeed
— **up–and–com·er** \-ˈkə-mər\ *noun*

up–and–down \-ˈdau̇n\ *adjective* (circa 1755)
1 : marked by alternate upward and downward movement, action, or surface
2 : PERPENDICULAR

up and down *adverb* (12th century)
1 : TO AND FRO ⟨paced *up and down*⟩
2 : alternately upward and downward ⟨jump *up and down*⟩
3 *archaic* **:** here and there especially throughout an area
4 : with regard to every particular **:** THOROUGHLY ⟨knew the territory *up and down*⟩
— **up and down** *preposition*

up–and–up \ˈəp-ən-ˈəp\ *noun* (1863)
: an honest or respectable course — used in the phrase *on the up and up*

Upa·ni·shad \ü-ˈpä-ni-ˌshäd, yü-ˈpa-nə-ˌshad\ *noun* [Sanskrit *upaniṣad*] (1805)
: one of a class of Vedic treatises dealing with broad philosophic problems
— **Upa·ni·shad·ic** \(ˌ)ü-ˌpä-ni-ˈshä-dik, (ˌ)yü-ˌpa-nə-ˈsha-dik\ *adjective*

upas \ˈyü-pəs\ *noun* [Indonesian Malay *pohon upas* poison tree] (1783)
1 : a tall tropical Asian tree (*Antiaris toxicaria*) of the mulberry family with a latex that contains poisonous glucosides used as an arrow poison; *also* **:** a poisonous concentrate of the juice or latex of a upas
2 : a poisonous or harmful influence or institution

¹up·beat \ˈəp-ˌbēt\ *noun* (1869)
1 : an unaccented beat in a musical measure; *specifically* **:** the last beat of the measure
2 : an increase in activity or prosperity ⟨business that is on the *upbeat*⟩

²upbeat *adjective* (1947)
: CHEERFUL, OPTIMISTIC

up–bow \ˈəp-ˌbō\ *noun* (circa 1890)
: a stroke in playing a bowed instrument in which the bow is moved across the strings from the tip to the heel

up·braid \ˌəp-ˈbrād\ *transitive verb* [Middle English *upbreyden,* from Old English *ūpbregdan,* probably from *ūp* up + *bregdan* to snatch, move suddenly — more at BRAID] (14th century)
1 : to criticize severely **:** find fault with
2 : to reproach severely **:** scold vehemently
synonym see SCOLD
— **up·braid·er** *noun*

up·bring·ing \ˈəp-ˌbriŋ-iŋ\ *noun* (1520)
: early training; *especially* **:** a particular way of bringing up a child ⟨had a strict *upbringing*⟩

up·build \ˌəp-ˈbild\ *transitive verb* **-built** \-ˈbilt\; **-build·ing** (1513)
: BUILD UP

up·cast \ˈəp-ˌkast\ *noun* (1890)
: something cast up

up·chuck \ˈəp-ˌchək\ *verb* (1936)
: VOMIT

up close *adverb or adjective* (1851)
: at close range

up·coast \ˈəp-ˈkōst\ *adverb* (1909)
: up the coast

up·com·ing \ˈəp-ˌkə-miŋ\ *adjective* (1944)
: FORTHCOMING, APPROACHING

up–coun·try \ˈəp-ˌkən-trē\ *adjective* (1835)
: of, relating to, or characteristic of an inland, upland, or outlying region
— **up–country** *noun*
— **up–country** \ˈəp-ˈ\ *adverb*

[1]up·date \ˌəp-ˈdāt\ *transitive verb* (1910)
: to bring up to date

[2]up·date \ˈəp-ˌdāt\ *noun* (1965)
1 : an act or instance of updating
2 : current information for updating something
3 : an up-to-date version, account, or report

up·do \ˈəp-(ˌ)dü\ *noun, plural* **updos** [*up*swept hair*do*] (1938)
: an upswept hairdo

up·draft \ˈəp-ˌdraft, -ˌdråft\ *noun* (circa 1887)
: an upward movement of gas (as air)

up·end \ˌəp-ˈend\ (1823)
transitive verb
1 : to set or stand on end; *also* **:** OVERTURN 1
2 a : to affect to the point of being upset or flurried ⟨a . . . literary shocker, designed to *upend* the credulous matrons —Wolcott Gibbs⟩ **b :** DEFEAT, BEAT
intransitive verb
: to rise on an end

up·field \ˈəp-ˈfē(ə)ld\ *adverb or adjective* (circa 1934)
: in or into the part of the field toward which the offensive team is headed

up–front \ˌəp-ˈfrənt, ˈəp-ˌ\ *adjective* (1945)
: being or coming in or at the front: as **a** (1) **:** being in a conspicuous or leading position (2) **:** FRANK, FORTHRIGHT **b :** playing in a front line (as in football) **c :** paid or payable in advance

up front *adverb* (1937)
1 : in or at the front
2 : in advance
3 : in an up-front manner **:** FRANKLY

[1]up·grade \ˈəp-ˌgrād\ *noun* (1873)
1 : an upward grade or slope
2 : INCREASE, RISE
3 : IMPROVEMENT 2b

[2]up·grade \ˈəp-ˌgrād, ˌəp-ˈ\ *transitive verb* (1901)
: to raise or improve the grade of: as **a :** to improve (livestock) by use of purebred sires **b :** to advance to a job requiring a higher level of skill especially as part of a training program **c :** to raise the quality of **d :** to raise the classification and usually the price of without improving the quality **e :** to extend the usefulness of (as a device)
— **up·grad·abil·i·ty** *or* **up·grade·abil·i·ty** \ˌəp-ˌgrā-də-ˈbi-lə-tē\ *noun*
— **up·grad·able** *or* **up·grade·able** \ˌəp-ˈgrā-də-bəl\ *adjective*

up·growth \ˈəp-ˌgrōth\ *noun* (1844)
: the process of growing upward **:** DEVELOPMENT; *also* **:** a product or result of this

up·heav·al \ˌəp-ˈhē-vəl, (ˌ)ə-ˈpē-\ *noun* (1838)
1 : the action or an instance of upheaving especially of part of the earth's crust
2 : extreme agitation or disorder **:** radical change; *also* **:** an instance of this

up·heave \ˌəp-ˈhēv, (ˌ)ə-ˈpēv\ (14th century)
transitive verb
: to heave up **:** LIFT
intransitive verb
: to move upward especially with power
— **up·heav·er** *noun*

[1]up·hill \ˈəp-ˌhil\ *noun* (1548)
: rising ground **:** ASCENT

[2]up·hill \-ˈhil\ *adverb* (1607)
1 : upward on a hill or incline
2 : against difficulties ⟨seemed to be talking *uphill* —Willa Cather⟩

[3]up·hill \-ˌhil\ *adjective* (1613)
1 : situated on elevated ground
2 a : going up **:** ASCENDING **b :** being the higher one or part especially of a set; *specifically* **:** being nearer the top of an incline
3 : DIFFICULT, LABORIOUS

up·hold \(ˌ)əp-ˈhōld\ *transitive verb* **-held** \-ˈheld\; **-hold·ing** (13th century)
1 a : to give support to **b :** to support against an opponent
2 a : to keep elevated **b :** to lift up
synonym see SUPPORT
— **up·hold·er** *noun*

up·hol·ster \(ˌ)əp-ˈhōl-stər, (ˌ)ə-ˈpōl-\ *transitive verb* **-stered; -ster·ing** \-st(ə-)riŋ\ [back-formation from *upholstery*] (1864)
: to furnish with or as if with upholstery
— **up·hol·ster·er** \-stər-ər\ *noun*

up·hol·stery \-st(ə-)rē\ *noun, plural* **-ster·ies** [Middle English *upholdester* upholsterer, from *upholden* to uphold, from *up* + *holden* to hold] (1649)
: materials (as fabric, padding, and springs) used to make a soft covering especially for a seat

up·keep \ˈəp-ˌkēp\ *noun* (1884)
1 : the act of maintaining in good condition **:** the state of being maintained in good condition
2 : the cost of maintaining in good condition

up·land \ˈəp-lənd, -ˌland\ *noun* (1566)
1 : high land especially at some distance from the sea **:** PLATEAU
2 : ground elevated above the lowlands along rivers or between hills
— **upland** *adjective*
— **up·land·er** \-lən-dər, -ˌlan-\ *noun*

upland cotton *noun* (1819)
: a widely cultivated American cotton plant (*Gossypium hirsutum*) having short- to medium-staple fibers

upland sandpiper *noun* (circa 1890)
: a large North American sandpiper (*Bartramia longicauda*) that frequents fields and prairies — called also *upland plover*

upland sandpiper

[1]up·lift \(ˌ)əp-ˈlift\ (14th century)
transitive verb
1 : to lift up **:** ELEVATE; *especially* **:** to cause (a portion of the earth's surface) to rise above adjacent areas
2 : to improve the spiritual, social, or intellectual condition of
intransitive verb
: RISE
— **up·lift·er** *noun*

[2]up·lift \ˈəp-ˌlift\ *noun* (circa 1845)
1 : an act, process, result, or cause of uplifting: as **a** (1) **:** the uplifting of a part of the earth's surface (2) **:** an uplifted mass of land **b :** a bettering of a condition especially spiritually, socially, or intellectually **c** (1) **:** influences intended to uplift (2) **:** a social movement to improve especially morally or culturally
2 : a brassiere designed to hold the breasts up

up·link \ˈəp-ˌliŋk\ *noun* (1968)
1 : a communications channel for transmissions to a spacecraft; *also* **:** the transmissions themselves
2 : a facility on earth for transmitting to a spacecraft

up·load \(ˌ)əp-ˈlōd, ˈəp-ˌ\ *transitive verb* (1983)
: to transfer (information) from a microcomputer to a remote computer usually with a modem

up·man·ship \ˈəp-mən-ˌship\ *noun* (1959)
: ONE-UPMANSHIP

up·mar·ket \ˈəp-ˈmär-kət\ *adjective* (1972)
: UPSCALE
— **upmarket** *adverb*

up·most \ˈəp-ˌmōst\ *adjective* (1560)
: UPPERMOST

[1]up·on \ə-ˈpȯn, -ˈpän\ *preposition* (13th century)
: ON

[2]up·on \ə-ˈpȯn, -ˈpän\ *adverb* (14th century)
1 *obsolete* **:** on the surface **:** on it
2 *obsolete* **:** THEREAFTER, THEREON

[1]up·per \ˈə-pər\ *adjective* [Middle English, comparative of [2]*up*] (14th century)
1 a : higher in physical position, rank, or order **b :** farther inland ⟨the *upper* Mississippi⟩
2 : constituting the branch of a bicameral legislature that is usually smaller and more restricted in membership and possesses greater traditional prestige than the lower house
3 a : constituting a stratum relatively near the earth's surface **b** *capitalized* **:** being a later epoch or series of the period or system named ⟨*Upper* Cretaceous⟩ ⟨*Upper* Paleolithic⟩
4 : NORTHERN ⟨*upper* Manhattan⟩

[2]upper *noun* (1789)
: one that is upper: as **a :** the parts of a shoe or boot above the sole **b :** an upper tooth or denture **c :** an upper berth
— **on one's uppers :** in straitened circumstances **:** DESTITUTE

[3]upper *noun* [*up* + [2]*-er*] (circa 1968)
1 : a stimulant drug; *especially* **:** AMPHETAMINE
2 : something that induces a state of good feeling or exhilaration

[1]up·per·case \ˌə-pər-ˈkās\ *adjective* [from the compositor's practice of keeping capital letters in the upper of a pair of type cases] (1738)
: CAPITAL 1

[2]uppercase *noun* (circa 1916)
: capital letters

[3]uppercase *transitive verb* **-cased; -cas·ing** (1949)
: to print or set in capital letters

upper case *noun* (1683)
: a type case containing capitals and usually small capitals, fractions, symbols, and accents

upper–class *adjective* (1837)
: of, relating to, or characteristic of the upper class

upper class *noun* (1839)
: a social class occupying a position above the middle class and having the highest status in a society

up·per·class·man \ˌə-pər-ˈklas-mən\ *noun* (1871)
: a member of the junior or senior class in a school or college

upper crust *noun* (1836)
: the highest social class or group; *especially* **:** the highest circle of the upper class
— **upper–crust** *adjective*

up·per·cut \ˈə-pər-ˌkət\ *noun* (1842)
: a swinging blow (as in boxing) directed upward with a bent arm
— **uppercut** *verb*

upper hand *noun* (15th century)
: MASTERY, ADVANTAGE, CONTROL ⟨was determined not to let the opposition get the *upper hand*⟩

up·per·most \ˈə-pər-ˌmōst\ *adverb* (15th century)
: in or into the highest or most prominent position
— **uppermost** *adjective*

up·per·part \-ˌpärt\ *noun* (1526)
: a part lying on the upper side (as of a bird)

upper respiratory *adjective* (1950)
: of, affecting, or being the part of the respiratory system that includes the nose, nasal passages, and nasopharynx ⟨*upper respiratory* tract⟩ ⟨*upper respiratory* infection⟩

up·pish \ˈə-pish\ *adjective* (circa 1734)
: UPPITY
— **up·pish·ly** *adverb*
— **up·pish·ness** *noun*

up·pi·ty \ˈə-pə-tē\ *adjective* [probably from *up* + *-ity* (as in *persnickity,* variant of *persnickety*)] (1880)
: putting on or marked by airs of superiority : ARROGANT, PRESUMPTUOUS ⟨*uppity* technicians⟩ ⟨a small *uppity* country⟩
— **up·pi·ti·ness** *also* **up·pi·ty·ness** *noun*

up·raise \(ˌ)əp-ˈrāz\ *transitive verb* (14th century)
: to raise or lift up : ELEVATE

up·rate \ˈəp-ˌrāt\ *transitive verb* (1965)
: UPGRADE; *specifically* : to improve the power output of (as an engine)

up·rear \-ˈrir\ (14th century)
transitive verb
1 : to lift up
2 : ERECT
intransitive verb
: RISE

[1]**up·right** \ˈəp-ˌrīt\ *adjective* (before 12th century)
1 a : PERPENDICULAR, VERTICAL **b** : erect in carriage or posture **c** : having the main axis or a main part perpendicular ⟨*upright* freezer⟩
2 : marked by strong moral rectitude ☆
— **up·right·ly** *adverb*
— **up·right·ness** *noun*

[2]**upright** *adverb* (1590)
: vertically upward : in an upright position

[3]**upright** *noun* (1683)
1 : the state of being upright : PERPENDICULAR ⟨a pillar out of *upright*⟩
2 : something that stands upright; *especially* : a football goalpost — usually used in plural
3 : UPRIGHT PIANO

upright piano *noun* (circa 1890)
: a piano with vertical frame and strings — compare GRAND PIANO

[1]**up·rise** \ˌəp-ˈrīz\ *intransitive verb* **up·rose** \-ˈrōz\; **up·ris·en** \-ˈri-zᵊn\; **up·ris·ing** \-ˈrī-ziŋ\ (14th century)
1 a : to rise to a higher position **b** (1) : STAND UP (2) : to get out of bed **c** : to come into view especially from below the horizon
2 : to rise up in sound
— **up·ris·er** \ˌəp-ˈrī-zər, ˈəp-ˌ\ *noun*

[2]**up·rise** \ˈəp-ˌrīz\ *noun* (15th century)
1 : an act or instance of uprising
2 : an upward slope

up·ris·ing \ˈəp-ˌrī-ziŋ\ *noun* (13th century)
: an act or instance of rising up; *especially* : a usually localized act of popular violence in defiance usually of an established government
synonym see REBELLION

up·riv·er \ˈəp-ˈri-vər\ *adverb or adjective* (1774)
: toward or at a point nearer the source of a river

up·roar \ˈəp-ˌrōr, -ˌrȯr\ *noun* [by folk etymology from Dutch *oproer,* from Middle Dutch, from *op* up (akin to Old English *ūp*) + *roer* motion; akin to Old English *hrēran* to stir] (1526)
: a state of commotion, excitement, or violent disturbance ◆

up·roar·i·ous \ˌəp-ˈrōr-ē-əs, -ˈrȯr-\ *adjective* (1819)
1 : marked by uproar
2 : very noisy and full
3 : extremely funny ⟨an *uproarious* comedy⟩
— **up·roar·i·ous·ly** *adverb*
— **up·roar·i·ous·ness** *noun*

up·root \(ˌ)əp-ˈrüt, -ˈru̇t\ *transitive verb* (circa 1620)
1 : to remove as if by pulling up
2 : to pull up by the roots
3 : to displace from a country or traditional habitat
synonym see EXTERMINATE
— **up·root·ed·ness** *noun*
— **up·root·er** *noun*

up·rush \ˈəp-ˌrəsh\ *noun* (1871)
1 : an upward rush (as of gas or liquid)
2 : a sudden increase

ups and downs *noun plural* (1659)
: alternating rise and fall especially in fortune

up·scale \ˈəp-ˈskāl\ *adjective* (1966)
: relating to, being, or appealing to affluent consumers; *also* : of a superior quality
— **upscale** *adverb or transitive verb*

[1]**up·set** \(ˌ)əp-ˈset\ *verb* **-set; -set·ting** (1677)
transitive verb
1 : to thicken and shorten (as a heated bar of iron) by hammering on the end : SWAGE
2 : to force out of the usual upright, level, or proper position : OVERTURN
3 a : to trouble mentally or emotionally : disturb the poise of **b** : to throw into disorder **c** : INVALIDATE **d** : to defeat unexpectedly
4 : to cause a physical disorder in; *specifically* : to make somewhat ill
intransitive verb
: to become overturned
synonym see DISCOMPOSE
— **up·set·ter** *noun*

[2]**up·set** \ˈəp-ˌset\ *noun* (1804)
1 : an act of overturning : OVERTURN
2 a (1) : an act of throwing into disorder : DERANGEMENT (2) : a state of disorder : CONFUSION **b** : an unexpected defeat
3 a : a minor physical disorder ⟨a stomach *upset*⟩ **b** : an emotional disturbance ⟨went through a big *upset* after his father's death⟩
4 a : a part of a rod (as the head on a bolt) that is upset **b** : the expansion of a bullet on striking

[3]**up·set** \(ˌ)əp-ˈset\ *adjective* (1805)
: emotionally disturbed or agitated ⟨was too *upset* to speak to him⟩

up·set price \ˈəp-ˌset-\ *noun* (1814)
: the minimum price set for property offered at auction or public sale

up·shift \ˈəp-ˌshift\ *intransitive verb* (1952)
: to shift an automotive vehicle into a higher gear
— **upshift** *noun*

up·shot \ˈəp-ˌshät\ *noun* (1604)
: the final result : OUTCOME

[1]**up·side** \ˈəp-ˌsīd\ *noun* [[2]*up* + [1]*side*] (1927)
1 : an upward trend (as of prices)
2 : a positive aspect

[2]**up·side** \ˈəp-ˈsīd\ *preposition* [perhaps from [1]*up* + *-side* (as in *alongside*)] (1929)
: up on or against the side of ⟨layin' in this death cell, writin' my time *upside* the wall —Lonnie Johnson⟩ ⟨if they wish to knock a thug *upside* the head, they do so —Robert MacKenzie⟩

up·side down \ˈəp-ˌsīd-ˈdau̇n\ *adverb* [alteration of Middle English *up so doun,* from *up* + *so* + *doun* down] (15th century)
1 : in such a way that the upper and the lower parts are reversed in position
2 : in or into great disorder
— **upside–down** *adjective*

upside–down cake *noun* (1930)
: a cake baked with its batter covering an arrangement of fruit (as pineapple) and served fruit side up

up·si·lon \ˈüp-sə-ˌlän, ˈyüp-, ˈəp-, -lən, *British usually* yüp-ˈsī-lən\ *noun* [Middle Greek *y psilon,* literally, simple *y;* from the desire to distinguish it from *oi,* which was pronounced the same in later Greek] (1621)
1 : the 20th letter of the Greek alphabet — see ALPHABET table
2 : any of a group of unstable electrically neutral elementary particles of the meson family that have a mass about 10 times that of a proton

up·spring \ˌəp-ˈspriŋ\ *intransitive verb* **-sprang** \-ˈspraŋ\ *or* **-sprung** \-ˈsprəŋ\; **-sprung; -spring·ing** \-ˈspriŋ-iŋ\ (before 12th century)
1 : to spring up
2 : to come into being

[1]**up·stage** \ˈəp-ˈstāj\ *adverb* (1870)
1 : toward or at the rear of a theatrical stage
2 : away from a motion-picture or television camera

[2]**upstage** *adjective* (1918)
1 [[3]*upstage*] : HAUGHTY
2 : of or relating to the rear of a stage

[3]**up·stage** \ˌəp-ˈstāj\ *transitive verb* (1921)
1 : to draw attention away from
2 : to force (an actor) to face away from the audience by staying upstage
3 : to treat snobbishly

[4]**up·stage** \ˈəp-ˌstāj\ *noun* (circa 1931)
: the part of a stage that is farthest from the audience or camera

[1]**up·stairs** \ˌəp-ˈstarz, -ˈsterz\ *adverb* (1596)
1 : up the stairs : on or to a higher floor
2 : to or at a high altitude or higher position ⟨kicked *upstairs* to company management⟩
3 : in the head : INTELLECTUALLY ⟨a little slow *upstairs* —Tom Clancy⟩

[2]**up·stairs** \ˈəp-ˈ\ *adjective* (1782)
: situated above the stairs especially on an upper floor ⟨an *upstairs* bedroom⟩

[3]**up·stairs** \ˈəp-ˈ, ˈəp-ˌ\ *noun plural but singular or plural in construction* (1842)
: the part of a building above the ground floor

up·stand·ing \ˌəp-ˈstan-diŋ, ˈəp-ˌ\ *adjective* (before 12th century)
1 : ERECT, UPRIGHT
2 : marked by integrity
— **up·stand·ing·ness** *noun*

[1]**up·start** \ˌəp-ˈstärt\ *intransitive verb* (14th century)
: to jump up (as to one's feet) suddenly

[2]**up·start** \ˈəp-ˌstärt\ *noun* (1555)

☆ SYNONYMS

Upright, honest, just, conscientious, scrupulous, honorable mean having or showing a strict regard for what is morally right. UPRIGHT implies a strict adherence to moral principles ⟨a stern and *upright* minister⟩. HONEST stresses adherence to such virtues as truthfulness, candor, fairness ⟨known for being *honest* in business dealings⟩. JUST stresses conscious choice and regular practice of what is right or equitable ⟨workers given *just* compensation⟩. CONSCIENTIOUS and SCRUPULOUS imply an active moral sense governing all one's actions and painstaking efforts to follow one's conscience ⟨*conscientious* in the completion of her assignments⟩ ⟨*scrupulous* in carrying out the terms of the will⟩. HONORABLE suggests a firm holding to codes of right behavior and the guidance of a high sense of honor and duty ⟨a difficult but *honorable* decision⟩.

◇ WORD HISTORY

uproar Despite appearances to the contrary, the *-roar* part of the word *uproar* has no etymological connection with the sound made by some animals. In Dutch *oproer* means "revolt, uprising," having been compounded from *op* "up," and *roer* "motion." When the word was borrowed into English, its Dutch meaning was initially retained, but through the process known as folk etymology its spelling and meaning were altered to fit already familiar English words. English speakers assumed that the *-roar* in *uproar* did indeed refer to loud cries or sounds, and so the word went from meaning "uprising" to meaning "a state of commotion."

1 : one that has risen suddenly (as from a low position to wealth or power) **:** PARVENU; *especially* **:** one that claims more personal importance than is warranted
2 : a start-up enterprise
— up·start \'əp-'\ *adjective*
up·state \'əp-ˌstāt\ *noun* (1901)
: the chiefly northerly sections of a state; *also* **:** the chiefly rural part of a state when the major metropolitan area is in the south
— up·state \-'stāt\ *adverb or adjective*
— up·stat·er \-'stā-tər\ *noun*
up·stream \'əp-'strēm\ *adverb or adjective* (1681)
1 : in the direction opposite to the flow of a stream
2 : toward a portion of the production stream closer to basic extractive or manufacturing processes ⟨make most of its money *upstream,* selling cheap crude . . . to refineries —John Quirt⟩
up·stroke \'əp-ˌstrōk\ *noun* (1828)
: a stroke (as of a pen) made in an upward direction
up·surge \'əp-ˌsərj\ *noun* (1917)
: a rapid or sudden rise
[1]up·sweep \'əp-ˌswēp\ *intransitive verb* **-swept** \-ˌswept\; **-sweep·ing** (1791)
: to sweep upward
[2]upsweep *noun* (circa 1891)
1 : an upward sweep
2 : an upswept hairdo
up·swept \'əp-ˌswept\ *adjective* (1938)
: swept upward; *especially* **:** brushed up to the top of the head ⟨*upswept* hair⟩
up·swing \'əp-ˌswiŋ\ *noun* (1924)
1 : an upward swing
2 : a marked increase or improvement — often used in the phrase *on the upswing*
up·take \'əp-ˌtāk\ *noun* [Scots *uptake* to understand] (1816)
1 : UNDERSTANDING, COMPREHENSION ⟨quick on the *uptake*⟩
2 : a flue leading upward
3 : an act or instance of absorbing and incorporating especially into a living organism
up–tem·po \'əp-ˌtem-(ˌ)pō\ *adjective* (1948)
: having a fast-moving tempo (as in jazz)
[1]up·throw \'əp-ˌthrō\ *transitive verb* **-threw** \-ˌthrü\; **-thrown** \-ˌthrōn\; **-throw·ing** (1600)
: to throw or thrust upward
[2]upthrow *noun* (1807)
: an upward displacement (as of a rock stratum) **:** UPHEAVAL, UPTHRUST
[1]up·thrust \'əp-ˌthrəst\ (1845)
transitive verb
: to thrust up; *especially* **:** to elevate (a part of the earth's surface) in an upthrust
intransitive verb
: to rise with an upward thrust
[2]upthrust *noun* (1846)
: an upward thrust; *specifically* **:** an uplift of part of the earth's crust
up·tick \'əp-ˌtik\ *noun* [[2]*up* + [3]*tick*] (1955)
: INCREASE, RISE
up·tight \'əp-'tīt, (ˌ)əp-'\ *adjective* (1934)
1 a : being tense, nervous, or uneasy **b :** ANGRY, INDIGNANT **c :** rigidly conventional
2 : being in financial difficulties
— up·tight·ness \(ˌ)əp-'tīt-nəs\ *noun*
up·tilt \ˌəp-'tilt\ *transitive verb* (1849)
: to tilt upward
up·time \'əp-ˌtīm\ *noun* (1958)
: time during which a piece of equipment (as a computer) is functioning or able to function
up to *preposition* (13th century)
1 — used as a function word to indicate extension as far as a specified place ⟨sank *up to* his knees in the mud⟩
2 — used as a function word to indicate a limit or boundary ⟨*up to* 50,000 copies a month⟩ ⟨worked *up to* the last minute⟩
up–to–date *adjective* (1888)
1 : extending up to the present time **:** including the latest information ⟨*up-to-date* maps⟩
2 : abreast of the times **:** MODERN ⟨*up-to-date* methods⟩
— up–to–date·ly *adverb*
— up–to–date·ness *noun*
up–to–the–minute *adjective* (1912)
1 : extending up to the immediate present **:** including the very latest information
2 : marked by complete up-to-dateness
[1]up·town \'əp-'taůn\ *adjective* (1838)
1 : of or relating to uptown
2 : UPSCALE, FASHIONABLE
— uptown *adverb*
[2]up·town \'əp-ˌtaůn\ *noun* (1844)
: the upper part of a town or city; *especially* **:** the residential district
up·trend \'əp-ˌtrend\ *noun* (1926)
: an upturn especially in business or economic activity
[1]up·turn \'əp-ˌtərn, ˌəp-'\ (1567)
transitive verb
1 : to turn up or over
2 : to direct upward
intransitive verb
: to turn upward
[2]up·turn \'əp-ˌtərn\ *noun* (1864)
: an upward turn especially toward better conditions or higher prices
[1]up·ward \'əp-wərd\ *or* **up·wards** \-wərdz\ *adverb* (before 12th century)
1 a : in a direction from lower to higher ⟨the kite rose *upward*⟩ **b** (1) **:** toward the source (as of a river) (2) **:** toward the interior (as of a region) **c :** in a higher position ⟨held out his hand, palm *upward*⟩ **d :** in the upper parts **:** toward the head **:** ABOVE ⟨from the waist *upward*⟩
2 : toward a higher or better condition or level ⟨young lawyers moving *upward*⟩
3 a : to an indefinitely greater amount, figure, or rank ⟨from $5 *upward*⟩ **b :** toward a greater amount or higher number, degree, or rate ⟨attendance figures have risen *upward*⟩
4 : toward or into later years ⟨from youth *upward*⟩
[2]upward *adjective* (1607)
1 : directed toward or situated in a higher place or level **:** ASCENDING
2 : rising to a higher pitch
— up·ward·ly *adverb*
— up·ward·ness *noun*
upward mobility *noun* (1949)
: the capacity or facility for rising to a higher social or economic position
— upwardly mobile *adjective*
upwards of *also* **upward of** *adverb* (1721)
: more than **:** in excess of ⟨they cost *upwards of* $25⟩
up·well \ˌəp-'wel\ *intransitive verb* (1885)
: to well up; *specifically* **:** to move or flow upward
up·well·ing \-'we-liŋ\ *noun* (1868)
: the process or an instance of rising or appearing to rise to the surface and flowing outward; *especially* **:** the process of upward movement to the surface of marine often nutrient-rich lower waters especially along some shores due to the offshore drift of surface water (as from the action of winds and the Coriolis force)
up·wind \'əp-'wind\ *adverb or adjective* (1838)
: in the direction from which the wind is blowing
[1]ur- *or* **uro-** *combining form* [New Latin, from Greek *our-, ouro-,* from *ouron* urine, from *ourein* to urinate — more at URINE]
1 : urine ⟨*ur*ic⟩
2 : urinary tract ⟨*uro*logy⟩
3 : urinary and ⟨*uro*genital⟩
4 : urea ⟨*ur*acil⟩
[2]ur- *or* **uro-** *combining form* [New Latin, from Greek *our-, ouro-,* from *oura* tail — more at ASS]
: tail ⟨*uro*pod⟩
[3]ur- \'ůr\ *prefix, often capitalized* [German, from Old High German *ir-, ur-* thoroughly (perfective prefix) — more at ABIDE]
1 : original **:** primitive ⟨*ur*-form⟩
2 : original version of ⟨*ur*text⟩
3 : prototypical **:** ARCH- ⟨*ur*-anticommunist⟩
ura·cil \'yůr-ə-ˌsil, -səl\ *noun* [International Scientific Vocabulary [1]*ur-* + *ac*etic + *-il* (substance relating to)] (1890)
: a pyrimidine base $C_4H_4N_2O_2$ that is one of the four bases coding genetic information in the polynucleotide chain of RNA — compare ADENINE, CYTOSINE, GUANINE, THYMINE
urae·us \yů-'rē-əs\ *noun, plural* **uraei** \-'rē-ˌī\ [New Latin, from Late Greek *ouraios,* a kind of snake] (1832)
: a representation of the sacred asp (*Naja haje*) appearing in ancient Egyptian art and especially on the headdress of rulers and serving as a symbol of sovereignty

uraeus

Ural–Al·ta·ic \ˌyůr-əl-al-'tā-ik\ *noun* (1853)
: a postulated language family comprising the Uralic and Altaic languages
— Ural–Altaic *adjective*
Ura·li·an \yů-'rā-lē-ən, -'ra-\ *adjective* (1801)
1 : of or relating to the Ural mountains
2 : URALIC
[1]Ural·ic \yů-'ra-lik\ *noun* (1861)
: a language family comprising the Finno-Ugric and Samoyed languages
[2]Uralic *adjective* (1880)
: of, relating to, or constituting the Finno-Ugric and Samoyed languages
Ura·nia \yů-'rā-nē-ə, -nyə\ *noun* [Latin, from Greek *Ourania*]
: the Greek Muse of astronomy
Ura·ni·an \yů-'rā-nē-ən, -nyən\ *adjective* (1844)
: of or relating to the planet Uranus
ura·ni·nite \yů-'rā-nə-ˌnīt\ *noun* [German *Uranin* uraninite (from New Latin *uranium*) + English *-ite*] (1879)
: a mineral that is basically a black octahedral or cubic oxide of uranium which contains thorium, lead, and rare earth elements and is the chief ore of uranium
ura·ni·um \yů-'rā-nē-əm\ *noun, often attributive* [New Latin, from *Uranus*] (circa 1797)
: a silvery heavy radioactive polyvalent metallic element that is found especially in pitchblende and uraninite and exists naturally as a mixture of three isotopes of mass number 234, 235, and 238 in the proportions of 0.006 percent, 0.71 percent, and 99.28 percent respectively — see ELEMENT table ◆
uranium hexa·flu·o·ride \-ˌhek-sə-'flòr-ˌīd, -'flůr-\ *noun* (1899)

◇ WORD HISTORY
uranium In Greek mythology the god of the heavens was known as *Ouranos,* literally "sky, heaven," and the muse of astronomy as *Ourania.* The Romans latinized these figures as *Uranus* and *Urania.* Uranus was the father of the god Saturn and the grandfather of Jupiter. In 1781 the English astronomer Sir William Herschel discovered by telescope the seventh planet of our solar system. Because the planets long familiar to astronomy were named after Roman gods, the German astronomer Johann Bode suggested the name *Uranus,* an especially apt choice since the fifth and sixth planets were Jupiter and Saturn respectively. In 1789 the German chemist Martin Klaproth discovered a new element and later gave it the name *uranium,* after the planet that had been recently discovered.

: a volatile compound UF_6 of uranium and fluorine that is used in one major process of enriching uranium in uranium 235

uranium 238 *noun* (1942)
: an isotope of uranium of mass number 238 that is the most abundant and stable uranium isotope, that is not fissionable but can be used to produce a fissionable isotope of plutonium, and that has a half-life of 4.51×10^9 years

uranium 235 *noun* (1940)
: a light isotope of uranium of mass number 235 that when bombarded with slow neutrons undergoes rapid fission into smaller atoms with the release of neutrons and energy and that is used in nuclear power plants and atomic bombs

ura·nog·ra·phy \ˌyu̇r-ə-ˈnä-grə-fē\ *noun* [Greek *ouranographia* description of the heavens, from *ouranos* sky + *-graphia* -graphy] (1675)
: the construction of celestial representations (as maps)

Ura·nus \ˈyu̇r-ə-nəs, yu̇-ˈrā-\ *noun* [Late Latin, from Greek *Ouranos*]
1 : the sky personified as a god and father of the Titans in Greek mythology
2 : the planet seventh in order from the sun — see PLANET table ■

ura·nyl \ˈyu̇r-ə-ˌnil, yu̇-ˈrā-nᵊl\ *noun* [International Scientific Vocabulary, from New Latin *uranium* + International Scientific Vocabulary *-yl*] (1850)
: a bivalent radical UO_2

urate \ˈyu̇r-ˌāt\ *noun* [French, from *urique* uric, from English *uric*] (1800)
: a salt of uric acid
— **urat·ic** \yu̇-ˈra-tik\ *adjective*

ur·ban \ˈər-bən\ *adjective* [Latin *urbanus,* from *urbs* city] (1619)
: of, relating to, characteristic of, or constituting a city

ur·bane \ˌər-ˈbān\ *adjective* [Latin *urbanus* urban, urbane] (circa 1623)
: notably polite or finished in manner : POLISHED
synonym see SUAVE
— **ur·bane·ly** *adverb*

ur·ban·i·sa·tion, ur·ban·ise *British variant of* URBANIZATION, URBANIZE

ur·ban·ism \ˈər-bə-ˌni-zəm\ *noun* (1889)
1 : the characteristic way of life of city dwellers
2 a : the study of the physical needs of urban societies **b :** CITY PLANNING
3 : URBANIZATION

ur·ban·ist \ˈər-bə-nist\ *noun* (1930)
: a specialist in city planning
— **ur·ban·is·tic** \ˌər-bə-ˈnis-tik\ *adjective*
— **ur·ban·is·ti·cal·ly** \-ti-k(ə-)lē\ *adverb*

ur·ban·ite \ˈər-bə-ˌnīt\ *noun* (1897)
: a person who lives in a city

ur·ban·i·ty \ˌər-ˈba-nə-tē\ *noun, plural* **-ties** (1535)
1 : the quality or state of being urbane
2 *plural* **:** urbane acts or conduct

ur·ban·i·za·tion \ˌər-bə-nə-ˈzā-shən\ *noun* (1888)
: the quality or state of being urbanized or the process of becoming urbanized

ur·ban·ize \ˈər-bə-ˌnīz\ *transitive verb* **-ized; -iz·ing** (1884)
1 : to cause to take on urban characteristics ⟨*urbanized* areas⟩
2 : to impart an urban way of life to ⟨*urbanize* migrants from rural areas⟩

ur·ban·ol·o·gy \ˌər-bə-ˈnä-lə-jē\ *noun* (1961)
: a study dealing with specialized problems of cities (as planning, education, sociology, and politics)
— **ur·ban·ol·o·gist** \-jist\ *noun*

urban renewal *noun* (1954)
: a construction program to replace or restore substandard buildings in an urban area

urban sprawl *noun* (1958)
: the spreading of urban developments (as houses and shopping centers) on undeveloped land near a city

ur·ce·o·late \ˌər-ˈsē-ə-lət, ˈər-sē-ə-ˌlāt\ *adjective* [New Latin *urceolatus,* from Latin *urceolus,* diminutive of *urceus* pitcher] (1760)
: shaped like an urn ⟨the *urceolate* corolla of a blueberry⟩

ur·chin \ˈər-chən\ *noun* [Middle English, from Middle French *herichon, heriçon,* from (assumed) Old French *eriz,* from Latin *ericius,* from *eris;* akin to Greek *chēr* hedgehog] (14th century)
1 *archaic* **:** HEDGEHOG 1a
2 : a mischievous youngster **:** SCAMP
3 : SEA URCHIN

urd \ˈu̇rd, ˈərd\ *noun* [Hindi] (circa 1934)
: an annual Asian legume (*Vigna mungo* synonym *Phaseolus mungo*) widely grown in warm regions for its edible blackish seed, for green manure, or for forage; *also* **:** the seed

Ur·du \ˈu̇r-(ˌ)dü, ˈər-\ *noun* [Hindi *urdū,* from Persian *zabān-e-urdū-e-muallā* language of the Exalted Camp (the imperial bazaar in Delhi)] (1796)
: an Indo-Aryan language that has the same colloquial basis as standard Hindi, is an official language of Pakistan, and is widely used by Muslims in urban areas of India

-ure *noun suffix* [Middle English, from Old French, from Latin *-ura*]
1 : act **:** process ⟨expos*ure*⟩
2 : office **:** function; *also* **:** body performing (such) a function ⟨legislat*ure*⟩

urea \yu̇-ˈrē-ə\ *noun* [New Latin, from French *urée,* from *urine*] (1806)
: a soluble weakly basic nitrogenous compound $CO(NH_2)_2$ that is the chief solid component of mammalian urine and an end product of protein decomposition, is synthesized from carbon dioxide and ammonia, and is used especially in synthesis (as of resins and plastics) and in fertilizers and animal rations

urea–formaldehyde *noun* (1928)
: a thermosetting synthetic resin made by condensing urea with formaldehyde

ure·ase \ˈyu̇r-ē-ˌās, -ˌāz\ *noun* (1892)
: an enzyme that catalyzes the hydrolysis of urea

ure·din·i·um \ˌyu̇r-ə-ˈdi-nē-əm\ *noun, plural* **-ia** \-nē-ə\ [New Latin, from Latin *uredin-, uredo* burning, blight, from *urere* to burn — more at EMBER] (1905)
: a usually reddish or black mass of hyphae and spores of a rust fungus forming pustules that rupture the host's cuticle
— **ure·din·i·al** \-nē-əl\ *adjective*

ure·do·spore \yu̇-ˈrē-də-ˌspōr, -ˌspȯr\ *also* **ure·dio·spore** \-ˈrē-dē-ə-\ *or* **ure·din·io·spore** \yu̇r-ə-ˈdi-nē-ə-\ *noun* [New Latin *ure*dium uredinium (from Latin *uredo*) + English *-o-* + *spore*] (1875)
: one of the thin-walled spores that are produced by the uredinial hyphae of rust fungi and spread the fungus vegetatively

ure·ide \ˈyu̇r-ē-ˌīd\ *noun* (1857)
: a cyclic or acyclic acyl derivative of urea

ure·mia \yu̇-ˈrē-mē-ə\ *noun* [New Latin] (circa 1857)
1 : accumulation in the blood of constituents normally eliminated in the urine that produces a severe toxic condition and usually occurs in severe kidney disease
2 : the toxic bodily condition associated with uremia ⟨the patient was in *uremia*⟩
— **ure·mic** \-mik\ *adjective*

ureo·tel·ic \yu̇-ˌrē-ə-ˈte-lik, ˌyu̇r-ē-ō-\ *adjective* [*urea* + *-o-* + *tel-* + *-ic;* from the fact that urea is the end product] (1924)
: excreting nitrogen mostly in the form of urea ⟨*ureotelic* mammals⟩
— **ureo·te·lism** \-ˈte-ˌli-zəm, ˌyu̇r-ē-ˈä-tᵊl-ˌi-zəm\ *noun*

ure·ter \ˈyu̇r-ə-tər\ *noun* [New Latin, from Greek *ourētēr,* from *ourein* to urinate — more at URINE] (1543)
: a duct that carries away the urine from a kidney to the bladder or cloaca
— **ure·ter·al** \yu̇-ˈrē-tə-rəl\ *or* **ure·ter·ic** \ˌyu̇r-ə-ˈter-ik\ *adjective*

ure·thane \ˈyu̇r-ə-ˌthān\ *or* **ure·than** \-ˌthan\ *noun* [French *uréthane,* from *ur-* [1]ur- + *éth-* eth- + *-ane*] (1838)
1 a : a crystalline compound $C_3H_7NO_2$ that is the ethyl ester of carbamic acid and is used especially as a solvent and medicinally as an antineoplastic agent **b :** an ester of carbamic acid other than the ethyl ester
2 : POLYURETHANE

ure·thra \yu̇-ˈrē-thrə\ *noun, plural* **-thras** *or* **-thrae** \-(ˌ)thrē\ [Late Latin, from Greek *ourēthra,* from *ourein* to urinate] (1634)
: the canal that in most mammals carries off the urine from the bladder and in the male serves also as a genital duct
— **ure·thral** \-thrəl\ *adjective*

ure·thri·tis \ˌyu̇r-i-ˈthrī-təs\ *noun* [New Latin] (circa 1823)
: inflammation of the urethra

ure·thro·scope \yu̇-ˈrē-thrə-ˌskōp\ *noun* [International Scientific Vocabulary] (1868)
: an instrument for viewing the interior of the urethra

[1]**urge** \ˈərj\ *verb* **urged; urg·ing** [Latin *urgēre* to press, push, entreat — more at WREAK] (1560)
transitive verb
1 : to present, advocate, or demand earnestly or pressingly ⟨his conviction was upheld on a theory never *urged* at his . . . trial —Leon Friedman⟩
2 : to undertake the accomplishment of with energy, swiftness, or enthusiasm ⟨*urge* the attack⟩
3 a : SOLICIT, ENTREAT **b :** to serve as a motive or reason for
4 : to force or impel in an indicated direction or into motion or greater speed ⟨the dog *urged* the sheep toward the gate⟩
5 : STIMULATE, PROVOKE
intransitive verb
: to declare, advance, or press earnestly a statement, argument, charge, or claim
— **urg·er** *noun*

[2]**urge** *noun* (circa 1618)
1 : the act or process of urging
2 : a force or impulse that urges; *especially* **:** a continuing impulse toward an activity or goal

ur·gen·cy \ˈər-jən(t)-sē\ *noun, plural* **-cies** (1540)
1 : the quality or state of being urgent **:** INSISTENCE
2 : a force or impulse that impels or constrains **:** URGE

□ USAGE
Uranus The name of a god and a celestial body can be both heavenly and earthy in the case of *Uranus.* Some critics perceive the variant \yu̇-ˈrā-nəs\ to be a vulgar pun, and yet evidence in the Merriam-Webster citation files suggests that this newer variant has been gaining popularity since the beginning of the 20th century. The Greek word for "heaven" from which Uranus descended has stress on the final syllable, a fact which supports neither side in the modern English dispute. For those who would avoid distractions from or misperceptions of their speech, the older variant with first-syllable stress is advised, although this pronunciation is itself just like the pronunciation of the rare word *urinous.*

ur·gent \ˈər-jənt\ *adjective* [Middle English, from Middle French, from Latin *urgent-, urgens,* present participle of *urgēre*] (15th century)
1 a : calling for immediate attention **:** PRESSING ⟨*urgent* appeals⟩ **b :** conveying a sense of urgency
2 : urging insistently **:** IMPORTUNATE
— **ur·gent·ly** *adverb*

-urgy *noun combining form* [New Latin *-urgia,* from Greek *-ourgia,* from *-ourgos* working, from *-o-* + *ergon* work — more at WORK]
: technique or art of dealing or working with (such) a product, matter, or tool ⟨metall*urgy*⟩

-uria *noun combining form* [New Latin, from Greek *-ouria,* from *ouron* urine, from *ourein* to urinate — more at URINE]
1 : presence of (a specified substance) in urine ⟨albumin*uria*⟩
2 : condition of having (such) urine ⟨poly*uria*⟩; *especially* **:** abnormal or diseased condition marked by the presence of (a specified substance) ⟨py*uria*⟩

uri·al \ˈu̇r-ē-əl, -ˌäl\ *noun* [Panjabi *hureāl*] (1860)
: an upland wild sheep (*Ovis vignei*) of southern and central Asia which is reddish brown and the males of which have a beard from the neck to the chest

urial

uric \ˈyu̇r-ik\ *adjective* (1797)
: of, relating to, or found in urine

uric acid *noun* (1800)
: a white odorless and tasteless nearly insoluble acid $C_5H_4N_4O_3$ that is the chief nitrogenous waste present in the urine especially of lower vertebrates (as birds and reptiles), is present in small quantity in human urine, and occurs pathologically in renal calculi and the tophi of gout

uri·co·su·ric \ˌyu̇r-i-kə-ˈshu̇r-ik, -ˈsu̇r-\ *adjective* [irregular from *uric*] (circa 1947)
: relating to or promoting the excretion of uric acid in the urine

uri·co·tel·ic \ˌyu̇r-i-kō-ˈte-lik\ *adjective* [*uric* + *-o-* + *tel-* + *-ic;* from the fact that uric acid is the end product] (1924)
: excreting nitrogen mostly in the form of uric acid ⟨birds are typical *uricotelic* animals⟩
— **uri·co·tel·ism** \-ˈte-ˌli-zəm, -ˈkä-tᵊl-ˌi-zəm\ *noun*

uri·dine \ˈyu̇r-ə-ˌdēn\ *noun* [International Scientific Vocabulary [1]*ur-* + *-idine*] (1911)
: a pyrimidine nucleoside $C_9H_{12}N_2O_6$ that is composed of uracil attached to ribose, is derived by hydrolysis from nucleic acids, and in the form of phosphate derivatives plays an important role in carbohydrate metabolism

Uri·el \ˈyu̇r-ē-əl\ *noun* [Hebrew *Ūrī'ēl*]
: one of the four archangels named in Hebrew tradition

Urim and Thum·mim \ˌyu̇r-ə-mən(d)-ˈthə-məm, ˌu̇r-; ˌu̇r-ēm-ən(d)-ˈtü-mēm\ *noun plural* [part translation of Hebrew *ūrīm wĕthummīm*] (1537)
: sacred lots used in early times by the Hebrews

urin- *or* **urino-** *combining form* [Middle English, from Old French, from Latin, from *urina* urine]
: [1]UR- ⟨*urino*genital⟩ ⟨*urinary*⟩

uri·nal \ˈyu̇r-ə-nᵊl, *British also* yu̇-ˈrī-nᵊl\ *noun* [Middle English, from Middle French, from Late Latin, from Latin *urina*] (13th century)
1 : a vessel for receiving urine
2 a : a building or enclosure with facilities for urinating **b :** a fixture used for urinating

uri·nal·y·sis \ˌyu̇r-ə-ˈna-lə-səs\ *noun, plural* **-y·ses** [New Latin, irregular from *urin-* + *analysis*] (1889)
: chemical analysis of urine

uri·nary \ˈyu̇r-ə-ˌner-ē\ *adjective* (1578)
1 : relating to, occurring in, affecting, or constituting the organs concerned with the formation and discharge of urine ⟨*urinary* system⟩ ⟨*urinary* calculi⟩
2 : of, relating to, or for urine
3 : excreted as or in urine

urinary bladder *noun* (1728)
: a membranous sac in many vertebrates that serves for the temporary retention of urine and discharges by the urethra

uri·nate \ˈyu̇r-ə-ˌnāt\ *intransitive verb* **-nat·ed; -nat·ing** (1599)
: to discharge urine **:** MICTURATE
— **uri·na·tion** \ˌyu̇r-ə-ˈnā-shən\ *noun*

urine \ˈyu̇r-ən\ *noun* [Middle English, from Middle French, from Latin *urina,* from *urinari* to dive; akin to Sanskrit *vār* water and perhaps to Sanskrit *varṣati* it rains, Greek *ourein* to urinate] (14th century)
: waste material that is secreted by the kidney in vertebrates, is rich in end products of protein metabolism together with salts and pigments, and forms a clear amber and usually slightly acid fluid in mammals but is semisolid in birds and reptiles
— **urin·ous** \ˈyu̇r-ə-nəs\ *adjective*

uri·no·gen·i·tal \ˌyu̇r-ə-nō-ˈje-nə-tᵊl\ *adjective* (1836)
: UROGENITAL

uri·nom·e·ter \ˌyu̇r-ə-ˈnä-mə-tər\ *noun* [International Scientific Vocabulary] (1843)
: a small hydrometer for determining the specific gravity of urine

urn \ˈərn\ *noun* [Middle English *urne,* from Latin *urna*] (14th century)
1 : a vessel that is typically an ornamental vase on a pedestal and that is used for various purposes (as preserving the ashes of the dead after cremation)
2 : a closed vessel usually with a spigot for serving a hot beverage ⟨a coffee *urn*⟩

uro- — see UR-

uro·ca·nic acid \ˌyu̇r-ə-ˈkā-nik-, -ˈka-\ *noun* [[1]*ur-* + *canine* + *-ic;* from its being first obtained from the urine of a dog] (circa 1903)
: a crystalline acid $C_6H_6N_2O_2$ that is normally present in human skin

uro·chor·date \ˌyu̇r-ə-ˈkȯr-dət, -ˌdāt\ *noun* [New Latin *Urochordata,* former group name, from [2]*ur-* + *chordatus* having a notochord, from *chorda* notochord, from Latin, string, cord — more at CORD] (1948)
: TUNICATE
— **urochordate** *adjective*

uro·chrome \ˈyu̇r-ə-ˌkrōm\ *noun* (1864)
: a yellow pigment to which the color of normal urine is principally due

uro·dele \ˈyu̇r-ə-ˌdēl\ *noun* [French *urodèle,* ultimately from Greek *oura* tail + *dēlos* evident, showing — more at ASS] (1842)
: any of an order (Caudata synonym Urodela) of amphibians (as newts) that have a tail throughout life
— **urodele** *adjective*

uro·gen·i·tal \ˌyu̇r-ō-ˈje-nə-tᵊl\ *adjective* [International Scientific Vocabulary] (1848)
: of, relating to, or being the organs or functions of excretion and reproduction

uro·ki·nase \ˌyu̇r-ō-ˈkī-ˌnās, -ˌnāz\ *noun* (1952)
: an enzyme that is produced by the kidney and found in urine, that activates plasminogen, and that is used therapeutically to dissolve blood clots (as in the heart)

uro·lith \ˈyu̇r-ə-ˌlith\ *noun* [International Scientific Vocabulary] (circa 1900)
: a calculus in the urinary tract

uro·lith·ia·sis \ˌyu̇r-ə-li-ˈthī-ə-səs\ *noun* [New Latin] (circa 1860)
: a condition that is characterized by the formation or presence of calculi in the urinary tract

uro·log·ic \ˌyu̇r-ə-ˈlä-jik\ *also* **uro·log·i·cal** \-ji-kəl\ *adjective* (1855)
: of or relating to the urinary tract or to urology

urol·o·gist \yu̇-ˈrä-lə-jist\ *noun* (1889)
: a physician who specializes in the urinary or urogenital tract
— **urol·o·gy** \-jē\ *noun*

-uronic *adjective suffix* [Greek *ouron* urine]
: connected with urine — in names of certain aldehyde-acids derived from sugars or compounds of such acids ⟨hyal*uronic*⟩

uron·ic acid \yu̇-ˈrä-nik-\ *noun* (1925)
: any of a class of acidic compounds of the general formula $HOOC(CHOH)_nCHO$ that contain both carboxylic and aldehydic groups, are oxidation products of sugars, and occur combined in many polysaccharides and in urine

uro·pod \ˈyu̇r-ə-ˌpäd\ *noun* [International Scientific Vocabulary [2]*ur-* + Greek *pod-, pous* foot — more at FOOT] (circa 1890)
: either of the flattened lateral appendages of the last abdominal segment of a crustacean; *broadly* **:** an abdominal appendage of a crustacean

uro·py·gi·al gland \ˌyu̇r-ə-ˈpī-jē-əl-\ *noun* (1870)
: a large gland that occurs in most birds, opens dorsally at the base of the tail feathers, and usually secretes an oily fluid which the bird uses in preening its feathers — called also *oil gland*

uro·py·gi·um \ˌyu̇r-ə-ˈpī-jē-əm\ *noun* [New Latin, from Greek *ouropygion,* from *ouro-* [2]ur- + *pygē* rump] (1771)
: the fleshy and bony prominence at the posterior extremity of a bird's body that supports the tail feathers

uro·style \ˈyu̇r-ə-ˌstīl\ *noun* [International Scientific Vocabulary [2]*ur-* + Greek *stylos* pillar — more at STEER] (1875)
: a long unsegmented bone that represents a number of fused vertebrae and forms the posterior part of the vertebral column of frogs and toads

Ur·sa Ma·jor \ˌər-sə-ˈmā-jər\ *noun* [Latin (genitive *Ursae Majoris*), literally, greater bear]
: a constellation that is the most conspicuous of the northern constellations, is situated near the north pole of the heavens, and contains the stars forming the Big Dipper two of which are in a line indicating the direction of the North Star — called also *Great Bear*

Ursa Mi·nor \-ˈmī-nər\ *noun* [Latin (genitive *Ursae Minoris*), literally, lesser bear]
: a constellation that includes the north pole of the heavens and the stars which form the Little Dipper with the North Star at the tip of the handle — called also *Little Bear*

ur·sine \ˈər-ˌsīn\ *adjective* [Latin *ursinus,* from *ursus* bear — more at ARCTIC] (circa 1550)
1 : of or relating to a bear or the bear family (Ursidae)
2 : suggesting or characteristic of a bear ⟨a lumbering *ursine* gait⟩

Ur·su·line \ˈər-sə-lən, -ˌlīn, -ˌlēn\ *noun* [New Latin *Ursulina,* from *Ursula* Saint Ursula, legendary Christian martyr] (1693)
: a member of any of several Roman Catholic teaching orders of nuns; *especially* **:** a member of a teaching order founded by Saint Angela Merici in Brescia, Italy, in 1535

ur·text \ˈu̇r-ˌtekst\ *noun* [German, from [3]*ur-* ur- + *Text* text] (circa 1932)
: the original text (as of a musical score)

ur·ti·car·ia \ˌər-tə-ˈkar-ē-ə, -ˈker-\ *noun* [New Latin, from Latin *urtica* nettle] (circa 1771)
: an allergic disorder marked by raised edematous patches of skin or mucous membrane and usually intense itching and caused by contact

with a specific precipitating factor (as a food, drug, or inhalant) either externally or internally

— **ur·ti·car·i·al** \-ē-əl\ *adjective*

ur·ti·cate \'ər-tə-ˌkāt\ *intransitive verb* **-cat·ed; -cat·ing** [Medieval Latin *urticatus,* past participle of *urticare* to sting, from Latin *urtica*] (1843)
: to produce wheals or itching; *especially* **:** to induce urticaria
— **ur·ti·ca·tion** \ˌər-tə-'kā-shən\ *noun*

urus \'yu̇r-əs\ *noun* [Latin, of Germanic origin; akin to Old High German *ūro* aurochs — more at AUROCHS] (1601)
: AUROCHS 1

uru·shi·ol \yu̇-'rü-shē-ˌȯl, u̇-', -ˌōl\ *noun* [International Scientific Vocabulary, from Japanese *urushi* lacquer + International Scientific Vocabulary [1]*-ol*] (1908)
: a mixture of pyrocatechol derivatives with saturated or unsaturated side chains of 15 or 17 carbon atoms that is an oily toxic irritant principle present in poison ivy and some related plants (genus *Rhus*) and in oriental lacquers derived from such plants

us \'əs\ *pronoun* [Middle English, from Old English *ūs;* akin to Old High German *uns* us, Latin *nos*] *objective case of* WE

us·able *also* **use·able** \'yü-zə-bəl\ *adjective* (14th century)
1 : capable of being used
2 : convenient and practicable for use
— **us·abil·i·ty** \ˌyü-zə-'bi-lə-tē\ *noun*
— **us·able·ness** \'yü-zə-bəl-nəs\ *noun*
— **us·ably** \-blē\ *adverb*

us·age \'yü-sij, -zij\ *noun* [Middle English, from Middle French, from *us* use] (14th century)
1 a : firmly established and generally accepted practice or procedure **b :** a uniform certain reasonable lawful practice existing in a particular locality or occupation and binding persons entering into transactions chiefly on the basis of presumed familiarity **c :** the way in which words and phrases are actually used (as in a particular form or sense) in a language community
2 a : the action, amount, or mode of using ⟨a decrease in the *usage* of electricity⟩ **b :** manner of treating ⟨suffered ill *usage* at the hands of his captors⟩
synonym see HABIT

us·ance \'yü-z[ə]n(t)s\ *noun* (14th century)
1 : USAGE 1a
2 : USE, EMPLOYMENT
3 a *obsolete* **:** USURY **b :** INTEREST
4 : the time allowed by custom for payment of a bill of exchange in foreign commerce

[1]use \'yüs\ *noun* [Middle English *us,* from Old French, from Latin *usus,* from *uti* to use] (13th century)
1 a : the act or practice of employing something **:** EMPLOYMENT, APPLICATION ⟨he made good *use* of his spare time⟩ **b :** the fact or state of being used ⟨a dish in daily *use*⟩ **c :** a method or manner of employing or applying something ⟨gained practice in the *use* of the camera⟩
2 a (1) **:** habitual or customary usage (2) **:** an individual habit or group custom **b :** a liturgical form or observance; *especially* **:** a liturgy having modifications peculiar to a local church or religious order
3 a : the privilege or benefit of using something ⟨gave him the *use* of her car⟩ **b :** the ability or power to use something (as a limb or faculty) **c :** the legal enjoyment of property that consists in its employment, occupation, exercise, or practice ⟨she had the *use* of the estate for life⟩
4 a : a particular service or end ⟨put learning to practical *use*⟩ **b :** the quality of being suitable for employment ⟨saving things that might be of *use*⟩ **c :** the occasion or need to employ ⟨took only what they had *use* for⟩
5 a : the benefit in law of one or more persons; *specifically* **:** the benefit or profit of property established in one other than the legal possessor **b :** a legal arrangement by which such benefits and profits are so established
6 : a favorable attitude **:** LIKING ⟨had no *use* for modern art⟩

[2]use \'yüz\ *verb* **used** \'yüzd, *in the phrase "used to" usually* 'yüs(t)\; **us·ing** \'yü-ziŋ\ (14th century)
transitive verb
1 *archaic* **:** ACCUSTOM, HABITUATE
2 : to put into action or service **:** avail oneself of **:** EMPLOY
3 : to consume or take (as liquor or drugs) regularly
4 : to carry out a purpose or action by means of **:** UTILIZE; *also* **:** MANIPULATE 2b ⟨*used* him only as a means up the corporate ladder⟩
5 : to expend or consume by putting to use — often used with *up*
6 : to behave toward **:** act with regard to **:** TREAT ⟨*used* the prisoners cruelly⟩
7 : STAND 1d ⟨the house could *use* a coat of paint⟩
intransitive verb
— used in the past with *to* to indicate a former fact or state ⟨claims winters *used* to be harder⟩ ⟨didn't *use* to smoke so much⟩ ☆

used \'yüzd, *in the phrase "used to" usually* 'yüs(t)\ *adjective* (14th century)
1 : employed in accomplishing something
2 : that has endured use; *specifically* **:** SECONDHAND ⟨a *used* car⟩
3 : ACCUSTOMED, HABITUATED

use·ful \'yüs-fəl\ *adjective* (1595)
1 : capable of being put to use; *especially* **:** serviceable for an end or purpose
2 : of a valuable or productive kind ⟨do something *useful* with your life⟩
— **use·ful·ly** \-fə-lē\ *adverb*

use·ful·ness *noun* (1617)
: the quality of having utility and especially practical worth or applicability

use·less \'yü-sləs\ *adjective* (1593)
: having or being of no use: **a :** INEFFECTUAL **b :** not able to give service or aid **:** INEPT
— **use·less·ly** *adverb*
— **use·less·ness** *noun*

us·er \'yü-zər\ *noun* (15th century)
: one that uses

user fee *noun* (1967)
: an excise tax often in the form of a license or supplemental charge levied to fund a public service — called also *user's fee*

us·er–friend·ly \ˌyü-zər-'fren(d)-lē\ *adjective* (1977)
: easy to learn, use, understand, or deal with
— **user–friendliness** *noun*

use up *transitive verb* (1833)
: to exhaust of strength or useful properties ⟨land that has been *used up*⟩

Ushak \ü-'shäk\ *variant of* OUSHAK

[1]ush·er \'ə-shər\ *noun* [Middle English *ussher,* from Middle French *ussier,* from (assumed) Vulgar Latin *ustiarius* doorkeeper, from Latin *ostium, ustium* door, mouth of a river — more at OSTIUM] (14th century)
1 a : an officer or servant who has the care of the door of a court, hall, or chamber **b :** an officer who walks before a person of rank **c :** one who escorts persons to their seats (as in a theater)
2 *archaic* **:** an assistant teacher

[2]usher *verb* **ush·ered; ush·er·ing** \'ə-sh(ə-)riŋ\ (1596)
transitive verb
1 : to conduct to a place
2 : to precede as an usher, forerunner, or harbinger
3 : to cause to enter **:** INTRODUCE ⟨a new theory *ushered* into the world⟩
intransitive verb
: to serve as an usher ⟨*usher* at a wedding⟩

ush·er·ette \ˌə-shə-'ret\ *noun* (1925)
: a girl or woman who is an usher (as in a theater)

usher in *transitive verb* (circa 1600)
1 : to serve to bring into being ⟨a discovery that *ushered in* a period of prosperity⟩
2 : to mark or observe the beginning of ⟨*ushered in* the new year with much merrymaking⟩
synonym see BEGIN

us·nea \'əs-nē-ə, 'əz-\ *noun* [New Latin, from Arabic *ushnah* moss] (circa 1597)
: any of a genus (*Usnea*) of widely distributed lichens (as old-man's beard) that have a grayish or yellow pendulous freely branched thallus

us·que·baugh \'əs-kwi-ˌbȯ, -ˌbä\ *noun* [Irish *uisce beathadh*] (1581)
Irish & Scottish **:** WHISKEY

[1]usu·al \'yü-zhə-wəl, -zhəl; 'yüzh-wəl\ *adjective* [Middle English, from Late Latin *usualis,* from Latin *usus* use] (14th century)
1 : accordant with usage, custom, or habit **:** NORMAL
2 : commonly or ordinarily used ⟨followed his *usual* route⟩
3 : found in ordinary practice or in the ordinary course of events **:** ORDINARY ☆
— **usu·al·ly** \'yü-zhə-wə-lē, -zhə-lē; 'yüzh-wə-lē, 'yüzh-lē\ *adverb*
— **usu·al·ness** \'yü-zhə-wəl-nəs, -zhəl-; 'yüzh-wəl-\ *noun*
— **as usual :** in the accustomed or habitual way ⟨*as usual* they were late⟩

[2]usual *noun* (1589)
: something usual

usu·fruct \'yü-zə-ˌfrəkt, -sə-\ *noun* [Latin *ususfructus,* from *usus et fructus* use and enjoyment] (circa 1630)
1 : the legal right of using and enjoying the fruits or profits of something belonging to another
2 : the right to use or enjoy something

[1]usu·fruc·tu·ary \ˌyü-zə-'frək-chə-ˌwer-ē, -sə-\ *noun* (circa 1618)

☆ SYNONYMS

Use, employ, utilize mean to put into service especially to attain an end. USE implies availing oneself of something as a means or instrument to an end ⟨willing to *use* any means to achieve her ends⟩. EMPLOY suggests the use of a person or thing that is available but idle, inactive, or disengaged ⟨looking for better ways to *employ* their skills⟩. UTILIZE may suggest the discovery of a new, profitable, or practical use for something ⟨an old wooden bucket *utilized* as a planter⟩.

Usual, customary, habitual, wonted, accustomed mean familiar through frequent or regular repetition. USUAL stresses the absence of strangeness or unexpectedness ⟨my *usual* order for lunch⟩. CUSTOMARY applies to what accords with the practices, conventions, or usages of an individual or community ⟨the *customary* waiting period before the application is approved⟩. HABITUAL suggests a practice settled or established by much repetition ⟨an *habitual* morning routine⟩. WONTED stresses habituation but usually applies to what is favored, sought, or purposefully cultivated ⟨his *wonted* determination⟩. ACCUSTOMED is less emphatic than WONTED or HABITUAL in suggesting fixed habit or invariable custom ⟨accepted the compliment with her *accustomed* modesty⟩.

1 : one having the usufruct of property
2 : one having the use or enjoyment of something

[2]usufructuary *adjective* (1710)
: of, relating to, or having the character of a usufruct

usu·rer \ˈyü-zhər-ər, ˈyüzh-rər\ *noun* (14th century)
: one that lends money especially at an exorbitant rate

usu·ri·ous \yu̇-ˈzhu̇r-ē-əs, -ˈzu̇r-\ *adjective* (1610)
1 : practicing usury
2 : involving usury **:** of the character of usury
— **usu·ri·ous·ly** *adverb*
— **usu·ri·ous·ness** *noun*

usurp \yu̇-ˈsərp *also* -ˈzərp\ *verb* [Middle English, from Middle French *usurper,* from Latin *usurpare* to take possession of without legal claim, from *usu* (ablative of *usus* use) + *rapere* to seize — more at RAPID] (14th century)
transitive verb
1 a : to seize and hold (as office, place, or powers) in possession by force or without right ⟨*usurp* a throne⟩ **b :** to take or make use of without right ⟨*usurped* the rights to her life story⟩
2 : to take the place of by or as if by force **:** SUPPLANT ⟨must not let stock responses based on inherited prejudice *usurp* careful judgment⟩
intransitive verb
: to seize or exercise authority or possession wrongfully
— **usur·pa·tion** \ˌyü-sər-ˈpā-shən *also* ˌyü-zər-\ *noun*
— **usurp·er** \yu̇-ˈsər-pər *also* -ˈzər-\ *noun*

usu·ry \ˈyü-zhə-rē, ˈyüzh-rē\ *noun, plural* **-ries** [Middle English, from Medieval Latin *usuria,* alteration of Latin *usura,* from *usus,* past participle of *uti* to use] (14th century)
1 *archaic* **:** INTEREST
2 : the lending of money with an interest charge for its use; *especially* **:** the lending of money at exorbitant interest rates
3 : an unconscionable or exorbitant rate or amount of interest; *specifically* **:** interest in excess of a legal rate charged to a borrower for the use of money

ut \ˈət, ˈüt, ˈu̇t\ *noun* [Middle English, first note in the diatonic scale, from Medieval Latin, from the syllable sung to this note in a medieval hymn to Saint John the Baptist] (14th century)
: a syllable used for the first note in the diatonic scale in an early solmization system and later replaced by *do*

Ute \ˈyüt\ *noun, plural* **Ute** *or* **Utes** [short for earlier *Utah, Utaw,* from American Spanish *Yuta*] (1776)
1 : a member of an American Indian people originally ranging through Utah, Colorado, Arizona, and New Mexico
2 : the Uto-Aztecan language of the Ute people

uten·sil \yu̇-ˈten(t)-səl, ˈyü-ˌ\ *noun* [Middle English, vessels for domestic use, from Middle French *utensile,* from Latin *utensilia,* from neuter plural of *utensilis* useful, from *uti* to use] (14th century)
1 : an implement, instrument, or vessel used in a household and especially a kitchen
2 : a useful tool or implement
synonym see IMPLEMENT

uter·ine \ˈyü-tə-ˌrīn, -rən\ *adjective* [Middle English, from Late Latin *uterinus,* from Latin *uterus*] (15th century)
1 : born of the same mother but by a different father
2 : of, relating to, or affecting the uterus ⟨*uterine* cancer⟩

uter·us \ˈyü-tə-rəs, ˈyü-trəs\ *noun, plural* **uteri** \ˈyü-tə-ˌrī\ *also* **uter·us·es** [Latin, belly, womb; probably akin to Greek *hoderos* belly, Sanskrit *udara*] (1615)
1 : an organ of the female mammal for containing and usually for nourishing the young during development previous to birth — called also *womb*
2 : a structure in some lower animals analogous to the uterus in which eggs or young develop

Uther \ˈü-thər, ˈyü-\ *noun*
: a legendary British king and father of Arthur

utile \ˈyü-tᵊl, ˈyü-ˌtīl\ *adjective* [Middle French, from Latin *utilis*] (15th century)
: USEFUL

uti·lise *British variant of* UTILIZE

[1]util·i·tar·i·an \(ˌ)yü-ˌti-lə-ˈter-ē-ən\ *noun* (circa 1780)
: an advocate or adherent of utilitarianism

[2]utilitarian *adjective* (1802)
1 : of or relating to or advocating utilitarianism
2 : marked by utilitarian views or practices
3 a : of, relating to, or aiming at utility **b :** exhibiting or preferring mere utility ⟨spare *utilitarian* furnishings⟩

util·i·tar·i·an·ism \-ē-ə-ˌni-zəm\ *noun* (1827)
1 : a doctrine that the useful is the good and that the determining consideration of right conduct should be the usefulness of its consequences; *specifically* **:** a theory that the aim of action should be the largest possible balance of pleasure over pain or the greatest happiness of the greatest number
2 : utilitarian character, spirit, or quality

[1]util·i·ty \yü-ˈti-lə-tē\ *noun, plural* **-ties** [Middle English *utilite,* from Middle French *utilité,* from Latin *utilitat-, utilitas,* from *utilis* useful, from *uti* to use] (14th century)
1 : fitness for some purpose or worth to some end
2 : something useful or designed for use
3 a : PUBLIC UTILITY **b** (1) **:** a service (as light, power, or water) provided by a public utility (2) **:** equipment or a piece of equipment to provide such service or a comparable service
4 : a program or routine designed to perform or facilitate especially routine operations (as copying files or editing text) on a computer

[2]utility *adjective* (1851)
1 : capable of serving as a substitute in various roles or positions ⟨a *utility* infielder⟩
2 a : kept to provide a useful product or service rather than for show or as a pet ⟨*utility* livestock⟩ ⟨a *utility* dog⟩ **b :** being of a usable but inferior grade ⟨*utility* beef⟩
3 : serving primarily for utility rather than beauty **:** UTILITARIAN
4 : designed or adapted for general use ⟨a *utility* knife⟩
5 : of or relating to a utility ⟨a *utility* company⟩

uti·lize \ˈyü-tᵊl-ˌīz\ *transitive verb* **-lized; -liz·ing** [French *utiliser,* from *utile*] (1807)
: to make use of **:** turn to practical use or account ⟨I'm a great person for *utilizing* waste power —Robert Frost⟩
synonym see USE
— **uti·liz·able** \-ˌī-zə-bəl\ *adjective*
— **uti·li·za·tion** \ˌyü-tᵊl-ə-ˈzā-shən\ *noun*
— **uti·liz·er** \ˈyü-tᵊl-ˌī-zər\ *noun*

[1]ut·most \ˈət-ˌmōst, *especially Southern* -məst\ *adjective* [Middle English, alteration of *utmest,* from Old English *ūtmest,* superlative adjective, from *ūt* out, adverb — more at OUT] (before 12th century)
1 : situated at the farthest or most distant point **:** EXTREME ⟨the *utmost* point of the earth —John Hunt⟩
2 : of the greatest or highest degree, quantity, number, or amount ⟨a matter of *utmost* concern⟩

[2]utmost *noun* (before 12th century)
1 : the most possible **:** the extreme limit **:** the highest attainable point or degree ⟨the *utmost* in reliability⟩
2 : the highest, greatest, or best of one's abilities, powers, and resources ⟨will do our *utmost* to help⟩

Uto–Az·tec·an \ˌyü-tō-ˈaz-ˌte-kən\ *noun* [*Ute* + *-o-* + *Aztec*] (1891)
: a family of American Indian languages spoken by peoples from the U.S. Great Basin south to Central America
— **Uto–Aztecan** *adjective*

uto·pia \yu̇-ˈtō-pē-ə\ *noun* [*Utopia,* imaginary and ideal country in *Utopia* (1516) by Sir Thomas More, from Greek *ou* not, no + *topos* place] (1610)
1 : an imaginary and indefinitely remote place
2 *often capitalized* **:** a place of ideal perfection especially in laws, government, and social conditions
3 : an impractical scheme for social improvement ◆

[1]uto·pi·an \-pē-ən\ *adjective, often capitalized* (1551)
1 : of, relating to, or having the characteristics of a utopia; *especially* **:** having impossibly ideal conditions especially of social organization
2 : proposing or advocating impractically ideal social and political schemes ⟨*utopian* idealists⟩
3 : impossibly ideal **:** VISIONARY ⟨recognised the *utopian* nature of his hopes —C. S. Kilby⟩
4 : believing in, advocating, or having the characteristics of utopian socialism ⟨*utopian* doctrines⟩ ⟨*utopian* novels⟩

[2]utopian *noun* (circa 1873)
1 : one that believes in the perfectibility of human society
2 : one that proposes or advocates utopian schemes

uto·pi·an·ism \-pē-ə-ˌni-zəm\ *noun* (circa 1661)
1 : a utopian idea or theory
2 *often capitalized* **:** the body of ideas, views, or aims of a utopian

utopian socialism *noun* (circa 1923)
: socialism based on a belief that social ownership of the means of production can be achieved by voluntary and peaceful surrender of their holdings by propertied groups
— **utopian socialist** *noun*

uto·pism \ˈyü-tə-ˌpi-zəm, yu̇-ˈtō-\ *noun* (1888)
: UTOPIANISM 2
— **utopist** \yu̇-ˈtō-pist\ *noun*
— **uto·pis·tic** \ˌyü-tə-ˈpis-tik, yu̇-ˌtō-\ *adjective*

utri·cle \ˈyü-tri-kəl\ *noun* [Latin *utriculus,* diminutive of *uter* leather bag] (1731)
: any of various small pouches or saccate parts of an animal or plant body: as **a :** the part of the membranous labyrinth of the inner ear into which the semicircular canals open **b :** a small usually indehiscent one-seeded fruit with thin membranous pericarp
— **utric·u·lar** \yu̇-ˈtri-kyə-lər\ *adjective*

◇ WORD HISTORY

utopia In 1516 Sir Thomas More, the English humanist and statesman, published his book *Utopia,* in which the social and economic conditions of Europe, outlined in Book I, are compared in Book II with those of an ideal society established on an imaginary island off the shore of the New World. That such an ideal state is unattainable in reality is implied by the name More gave to this island, *Utopia,* coined from Greek *ou* "no, not" and *topos* "place." In a poem prefaced to the book, More also puns on the name, suggesting *eutopos* "good place." In Modern English *utopia* has become, through the influence of More's classic, a generic term for any place of ideal perfection, especially in laws, government, and social conditions. Less optimistically *utopia* and its adjective derivative *utopian* have also come to refer to impractical schemes for social improvement.

utric·u·lus \yu̇-ˈtri-kyə-ləs\ *noun* [Latin, small bag] (1847)
: UTRICLE a

¹ut·ter \ˈə-tər\ *adjective* [Middle English, remote, from Old English *ūtera* outer, comparative adjective from *ūt* out, adverb — more at OUT] (15th century)
: carried to the utmost point or highest degree **:** ABSOLUTE, TOTAL ⟨*utter* darkness⟩ ⟨*utter* strangers⟩
— ut·ter·ly *adverb*

²utter *verb* [Middle English *uttren,* from *utter* outside, adverb, from Old English *ūtor,* comparative of *ūt* out] (15th century)
transitive verb
1 *obsolete* **:** to offer for sale
2 a : to send forth as a sound **b :** to give utterance to **:** PRONOUNCE, SPEAK **c :** to give public expression to **:** express in words
3 : to put (as currency) into circulation; *specifically* **:** to circulate (as a counterfeit note) as if legal or genuine ⟨*utter* false tokens⟩
4 : to put forth or out **:** DISCHARGE
intransitive verb
: to make a statement or sound
synonym see EXPRESS
— ut·ter·able \ˈə-tə-rə-bəl\ *adjective*
— ut·ter·er \ˈə-tər-ər\ *noun*

¹ut·ter·ance \ˈə-tə-rən(t)s, ˈə-trən(t)s\ *noun* [Middle English *uttraunce,* modification of Middle French *outrance* — more at OUTRANCE] (15th century)
archaic **:** the last extremity **:** BITTER END

²ut·ter·ance \ˈə-tə-rən(t)s *also* ˈə-trən(t)s\ *noun* (15th century)
1 : something uttered; *especially* **:** an oral or written statement **:** a stated or published expression
2 : vocal expression **:** SPEECH
3 : power, style, or manner of speaking

¹ut·ter·most \ˈə-tər-ˌmōst\ *adjective* [Middle English, alteration of *uttermest,* from ¹*utter* + *-mest* (as in *utmest* utmost)] (14th century)
1 : OUTERMOST
2 : EXTREME, UTMOST

²uttermost *noun* (14th century)
: UTMOST ⟨to the *uttermost* of our capacity —H. S. Truman⟩

U–turn \ˈyü-ˌtərn\ *noun* (1930)
1 : a turn resembling the letter U; *especially* **:** a 180-degree turn made by a vehicle in a road
2 : something (as a reversal of policy) resembling a U-turn

U–val·ue \ˈyü-ˌval-(ˌ)yü\ *noun* [unit] (1949)
: a measure of the heat transmission through a building part (as a wall or window) or a given thickness of a material (as insulation) with lower numbers indicating better insulating properties — compare R-VALUE

uva·rov·ite \yü-ˈvär-ə-ˌvīt, ü-\ *noun* [German *Uwarowit,* from Count Sergei S. *Uvarov* (died 1855) Russian statesman] (1837)
: an emerald green calcium-chromium garnet

uvea \ˈyü-vē-ə\ *noun* [Medieval Latin, from Latin *uva* grape] (1525)
: the posterior pigmented layer of the iris; *also* **:** the iris and ciliary body together with the choroid coat
— uve·al \-vē-əl\ *adjective*

uve·itis \ˌyü-vē-ˈī-təs\ *noun* [New Latin] (circa 1848)
: inflammation of the uvea of the eye

uvu·la \ˈyü-vyə-lə\ *noun, plural* **-las** *or* **-lae** \-ˌlē, -ˌlī\ [Middle English, from Medieval Latin, diminutive of Latin *uva* cluster of grapes, uvula; probably akin to Greek *oa* service tree, Old English *īw* yew — more at YEW] (14th century)
: the pendent fleshy lobe in the middle of the posterior border of the soft palate

uvu·lar \-lər\ *adjective* (1843)
1 : of or relating to the uvula ⟨*uvular* glands⟩
2 : produced with the aid of the uvula ⟨a *uvular* sound⟩

ux·o·ri·al \ˌək-ˈsōr-ē-əl, -ˈsȯr-; ˌəg-ˈzōr-, -ˈzȯr-\ *adjective* [Latin *uxorius*] (1800)
: of, relating to, or characteristic of a wife

ux·or·i·cide \ˌək-ˈsȯr-ə-ˌsīd, -ˈsär-; ˌəg-ˈzȯr-, -ˈzär-\ *noun* (1860)
1 [Medieval Latin *uxoricidium,* from Latin *uxor* wife + *-i-* + *-cidium* -cide] **:** murder of a wife by her husband
2 [Latin *uxor* + English *-i-* + *-cide*] **:** a wife murderer

ux·o·ri·ous \ˌək-ˈsōr-ē-əs, -ˈsȯr-; ˌəg-ˈzōr-, -ˈzȯr-\ *adjective* [Latin *uxorius* uxorious, uxorial, from *uxor* wife] (1598)
: excessively fond of or submissive to a wife
— ux·o·ri·ous·ly *adverb*
— ux·o·ri·ous·ness *noun*

Uz·bek \ˈu̇z-ˌbek, ˈəz-, u̇z-ˈ\ *or* **Uz·beg** \-ˌbeg, -ˈbeg\ *noun* (1616)
1 : a member of a Turkic people of Uzbekistan and adjacent regions of central Asia
2 : the Turkic language of the Uzbek people

V is the twenty-second letter of the English alphabet. V and U are varieties of the same character, U being the cursive form, and the two letters were once used indiscriminately (see U). The letter W, a doubled V called "double u," is a survival of this interchangeable use. V does not occur in the oldest English texts; its sound was represented in the middle of words by F. The letter was probably introduced into English during the later Middle Ages by French scribes. V is from the Latin alphabet, where it was used both as a vowel and as a consonant (first with the value of English w, and later with that of v). Latin derived the letter from a western alphabetic form of the Greek upsilon (see Y). In Greek it was used as a vowel, but it originated in a Phoenician character, wāw, which had the value of a consonant (see F). In English V represents the sound of the second consonant in love. Small v originated as a miniaturized form of the capital letter.

v \'vē\ *noun, plural* **v's** *or* **vs** \'vēz\ *often capitalized, often attributive* (15th century)
1 a : the 22d letter of the English alphabet **b :** a graphic representation of this letter **c :** a speech counterpart of orthographic *v*
2 : FIVE — see NUMBER table
3 : a graphic device for reproducing the letter *v*
4 : one designated *v* especially as the 22d in order or class
5 : something shaped like the letter V

va·can·cy \'vā-kən(t)-sē\ *noun, plural* **-cies** (1599)
1 *archaic* **:** an interval of leisure
2 : physical or mental inactivity or relaxation **:** IDLENESS
3 a : a vacating of an office, post, or piece of property **b :** the time such office or property is vacant
4 : a vacant office, post, or tenancy
5 : empty space **:** VOID; *specifically* **:** an unoccupied site for an atom or ion in a crystal
6 : the state of being vacant **:** VACUITY

va·cant \'vā-kənt\ *adjective* [Middle English, from Old French, from Latin *vacant-, vacans,* present participle of *vacare* to be empty, be free] (14th century)
1 : not occupied by an incumbent, possessor, or officer ⟨a *vacant* office⟩ ⟨*vacant* thrones⟩
2 : being without content or occupant ⟨a *vacant* seat in a bus⟩ ⟨a *vacant* room⟩
3 : free from activity or work **:** DISENGAGED ⟨*vacant* hours⟩
4 : devoid of thought, reflection, or expression ⟨a *vacant* smile⟩
5 : not lived in ⟨*vacant* houses⟩
6 a : not put to use ⟨*vacant* land⟩ **b :** having no heir or claimant **:** ABANDONED ⟨a *vacant* estate⟩
synonym see EMPTY
— **va·cant·ly** *adverb*
— **va·cant·ness** *noun*

va·cate \'vā-,kāt, vā-'\ *verb* **va·cat·ed; va·cat·ing** [Latin *vacatus,* past participle of *vacare*] (1643)
transitive verb
1 : to make legally void **:** ANNUL
2 a : to deprive of an incumbent or occupant **b :** to give up the incumbency or occupancy of
intransitive verb
: to vacate an office, post, or tenancy

¹va·ca·tion \vā-'kā-shən, və-\ *noun, often attributive* [Middle English *vacacioun,* from Middle French *vacation,* from Latin *vacation-, vacatio* freedom, exemption, from *vacare*] (14th century)
1 : a respite or a time of respite from something **:** INTERMISSION
2 a : a scheduled period during which activity (as of a court or school) is suspended **b :** a period of exemption from work granted to an employee for rest and relaxation
3 : a period spent away from home or business in travel or recreation ⟨had a restful *vacation* at the beach⟩
4 : an act or an instance of vacating

²vacation *intransitive verb* **-tioned; -tion·ing** \-sh(ə-)niŋ\ (1896)
: to take or spend a vacation
— **va·ca·tion·er** \-sh(ə-)nər\ *noun*

va·ca·tion·ist \-sh(ə-)nist\ *noun* (1885)
: a person taking a vacation

va·ca·tion·land \-shən-,land\ *noun* (1927)
: an area with recreational attractions and facilities for vacationists

vac·ci·nal \'vak-sə-nᵊl, vak-'sē-\ *adjective* (circa 1860)
: of or relating to vaccine or vaccination

vac·ci·nate \'vak-sə-,nāt\ *verb* **-nat·ed; -nat·ing** (1803)
transitive verb
1 : to inoculate (a person) with cowpox virus in order to produce immunity to smallpox
2 : to administer a vaccine to usually by injection
intransitive verb
: to perform or practice vaccination
— **vac·ci·na·tor** \-,nā-tər\ *noun*

vac·ci·na·tion \,vak-sə-'nā-shən\ *noun* (1800)
1 : the act of vaccinating
2 : the scar left by vaccinating

vac·cine \vak-'sēn, 'vak-,\ *noun* [Latin *vaccinus,* adjective, of or from cows, from *vacca* cow; akin to Sanskrit *vaśa* cow] (1803)
1 : matter or a preparation containing the virus of cowpox in a form used for vaccination
2 : a preparation of killed microorganisms, living attenuated organisms, or living fully virulent organisms that is administered to produce or artificially increase immunity to a particular disease ◆
— **vaccine** *adjective*

vac·ci·nee \,vak-sə-'nē\ *noun* (1889)
: a vaccinated individual

vac·cin·ia \vak-'si-nē-ə\ *noun* [New Latin, from *vaccinus*] (1803)
1 : COWPOX
2 : the virus that is the causative agent of cowpox
— **vac·cin·i·al** \-nē-əl\ *adjective*

vac·il·late \'va-sə-,lāt\ *intransitive verb* **-lat·ed; -lat·ing** [Latin *vacillatus,* past participle of *vacillare* to sway, waver — more at WINK] (1597)
1 a : to sway through lack of equilibrium **b :** FLUCTUATE, OSCILLATE
2 : to waver in mind, will, or feeling **:** hesitate in choice of opinions or courses
synonym see HESITATE
— **vac·il·lat·ing·ly** \-,lā-tiŋ-lē\ *adverb*
— **vac·il·la·tor** \-,lā-tər\ *noun*

vac·il·la·tion \,va-sə-'lā-shən\ *noun* (15th century)
1 : an act or instance of vacillating
2 : inability to take a stand **:** IRRESOLUTION, INDECISION

va·cu·i·ty \va-'kyü-ə-tē, və-\ *noun, plural* **-ties** [Latin *vacuitas,* from *vacuus* empty] (1541)
1 : an empty space
2 : the state, fact, or quality of being vacuous
3 : something (as an idea) that is vacuous or inane

vac·u·o·late \'va-kyə-(,)wō-,lāt\ *or* **vac·u·o·lat·ed** \-,lā-təd\ *adjective* (1859)
: containing one or more vacuoles ⟨highly *vacuolated* cells⟩

vac·u·o·la·tion \,va-kyə-(,)wō-'lā-shən\ *noun* (1858)
: the development or formation of vacuoles

vac·u·ole \'va-kyə-,wōl\ *noun* [French, literally, small vacuum, from Latin *vacuum*] (1853)
1 : a small cavity or space in the tissues of an organism containing air or fluid
2 : a cavity or vesicle in the cytoplasm of a cell usually containing fluid — see CELL illustration
— **vac·u·o·lar** \,va-kyə-'wō-lər, -,lär\ *adjective*

vac·u·ous \'va-kyə-wəs\ *adjective* [Latin *vacuus*] (1655)
1 : emptied of or lacking content
2 : marked by lack of ideas or intelligence **:** STUPID, INANE ⟨a *vacuous* mind⟩ ⟨a *vacuous* expression⟩
3 : devoid of serious occupation **:** IDLE
synonym see EMPTY
— **vac·u·ous·ly** *adverb*
— **vac·u·ous·ness** *noun*

¹vac·u·um \'va-(,)kyüm, -kyəm *also* -kyü-əm\ *noun, plural* **vac·u·ums** *or* **vac·ua** \-kyə-wə\ [Latin, from neuter of *vacuus* empty, from *vacare* to be empty] (1550)
1 : emptiness of space
2 a : a space absolutely devoid of matter **b :** a space partially exhausted (as to the highest degree possible) by artificial means (as an air pump) **c :** a degree of rarefaction below atmospheric pressure

◇ WORD HISTORY

vaccine Toward the end of the 18th century the English physician Edward Jenner set about investigating the truth of the folk wisdom that people such as dairymaids who contracted cowpox thereby gained immunity from smallpox, a much more serious disease. Working from this premise, he inoculated an eight-year-old boy with material taken from a milkmaid's cowpox sores. After the boy contracted and recovered from cowpox, Jenner proceeded to inoculate him with smallpox. Immunity had been provided, however, and the boy did not contract the disease. When Jenner published his documentation of 23 such cases in June 1798, he employed the medical Latin name for cowpox, *variolae vaccinae,* literally, "cow pustules." This led to the use of *vaccine matter* or *vaccine virus* for the cowpox inoculum (the virus-containing material used in inoculations), and *vaccination* as a name for the inoculation procedure. French authors writing about Jenner's work used the word *vaccine* alone in 1799 as a term for cowpox, and *vaccin* (a masculine derivative of *vaccine*) in 1801 as a term for the cowpox inoculum. English began to employ *vaccine* in the same sense not long after.

3 a : a state or condition resembling a vacuum **:** VOID ⟨the power *vacuum* in Indochina after the departure of the French —Norman Cousins⟩ **b :** a state of isolation from outside influences ⟨people who live in a *vacuum* . . . so that the world outside them is of no moment —W. S. Maugham⟩
4 : a device creating or utilizing a partial vacuum; *especially* **:** VACUUM CLEANER

[2]vacuum *adjective* (1825)
1 : of, containing, producing, or utilizing a partial vacuum ⟨separated by means of *vacuum* distillation⟩
2 : of or relating to a vacuum device or system

[3]vacuum (1922)
transitive verb
: to use a vacuum device (as a vacuum cleaner) on
intransitive verb
: to operate a vacuum device

vacuum bottle *noun* (1910)
: THERMOS

vacuum cleaner *noun* (1903)
: a household appliance for cleaning (as floors, carpets, or upholstery) by suction — called also *vacuum sweeper*

vacuum flask *noun* (1917)
: THERMOS

vacuum gauge *noun* (circa 1864)
: a gauge indicating degree of rarefaction below atmospheric pressure

vacuum–packed *adjective* (circa 1926)
: having much of the air removed before being hermetically sealed

vacuum pan *noun* (1833)
: a tank with a vacuum pump for rapid evaporation and condensation (as of sugar syrup) by boiling at a low temperature

vacuum pump *noun* (circa 1858)
: a pump for exhausting gas from an enclosed space

vacuum tube *noun* (1859)
: an electron tube evacuated to a high degree of vacuum

va·de me·cum \ˌvā-dē-ˈmē-kəm, ˌvä-dē-ˈmā-\ *noun, plural* **vade mecums** [Latin, go with me] (1629)
1 : a book for ready reference **:** MANUAL
2 : something regularly carried about by a person

va·dose \ˈvā-ˌdōs\ *adjective* [Latin *vadosus* shallow, from *vadum,* noun, shallow, ford; akin to Latin *vadere* to go — more at WADE] (1894)
: of, relating to, or being water or solutions in the earth's crust above the permanent groundwater level

vag- *or* **vago-** *combining form* [International Scientific Vocabulary, from New Latin *vagus*]
: vagus nerve ⟨vagal⟩ ⟨*vago*tomy⟩

[1]vag·a·bond \ˈva-gə-ˌbänd\ *adjective* [Middle English, from Middle French, from Late Latin *vagabundus,* from Latin *vagari* to wander] (15th century)
1 : moving from place to place without a fixed home **:** WANDERING
2 a : of, relating to, or characteristic of a wanderer **b :** leading an unsettled, irresponsible, or disreputable life
— vag·a·bond·ish \-ˌbän-dish\ *adjective*

[2]vagabond *noun* (15th century)
: one leading a vagabond life; *especially* **:** TRAMP
— vag·a·bond·age \-ˌbän-dij\ *noun*
— vag·a·bond·ism \-ˌbän-ˌdi-zəm\ *noun*

[3]vagabond *intransitive verb* (1586)
: to wander in the manner of a vagabond **:** roam about

va·gal \ˈvā-gəl\ *adjective* [International Scientific Vocabulary] (1854)
: of, relating to, mediated by, or being the vagus nerve
— va·gal·ly \-gə-lē\ *adverb*

va·gar·i·ous \vā-ˈger-ē-əs, və-, -ˈgar-\ *adjective* (1798)
: marked by vagaries **:** CAPRICIOUS, WHIMSICAL
— va·gar·i·ous·ly *adverb*

va·ga·ry \ˈvā-gə-rē; və-ˈger-ē, -ˈgar-, vā-; *also* ˈva-gə-rē\ *noun, plural* **-ries** [probably from Latin *vagari* to wander, from *vagus* wandering] (1573)
: an erratic, unpredictable, or extravagant manifestation, action, or notion
synonym see CAPRICE

vag·ile \ˈva-jəl, -ˌjīl\ *adjective* [International Scientific Vocabulary, from Latin *vagus* wandering] (circa 1890)
: free to move about ⟨*vagile* organisms⟩
— va·gil·i·ty \və-ˈji-lə-tē, va-\ *noun*

va·gi·na \və-ˈjī-nə\ *noun, plural* **-nae** \-(ˌ)nē\ *or* **-nas** [Latin, literally, sheath] (1682)
1 : a canal in a female mammal that leads from the uterus to the external orifice of the genital canal
2 : a canal that is similar in function or location to the vagina and occurs in various animals other than mammals

vag·i·nal \ˈva-jə-nᵊl\ *adjective* (1726)
1 : of or relating to a theca
2 : of, relating to, or affecting the genital vagina
— vag·i·nal·ly \-nᵊl-ē\ *adverb*

vag·i·nis·mus \ˌva-jə-ˈniz-məs\ *noun* [New Latin, from Latin *vagina*] (1866)
: a painful spasmodic contraction of the vagina

vag·i·ni·tis \ˌva-jə-ˈnī-təs\ *noun* [New Latin] (1846)
: inflammation of the vagina or of a sheath (as a tendon sheath)

va·got·o·my \vā-ˈgä-tə-mē\ *noun, plural* **-mies** [International Scientific Vocabulary] (circa 1903)
: surgical division of the vagus nerve

va·go·to·nia \ˌvā-gə-ˈtō-nē-ə\ *noun* [New Latin] (circa 1915)
: excessive excitability of the vagus nerve resulting typically in vasomotor instability, constipation, and sweating
— va·go·ton·ic \-ˈtä-nik\ *adjective*

va·gran·cy \ˈvā-grən(t)-sē\ *noun, plural* **-cies** (1641)
1 : VAGARY
2 : the state or action of being vagrant
3 : the offense of being a vagrant

[1]va·grant \ˈvā-grənt\ *noun* [Middle English *vagraunt,* probably modification of Middle French *waucrant, wacrant* wandering, from Old French, from present participle of *waucrer, wacrer* to roll, wander, of Germanic origin; akin to Old English *wealcan* to roll — more at WALK] (15th century)
1 a : one who has no established residence and wanders idly from place to place without lawful or visible means of support **b :** one (as a prostitute or drunkard) whose conduct constitutes statutory vagrancy
2 : WANDERER, ROVER

[2]vagrant *adjective* (15th century)
1 : wandering about from place to place usually with no means of support
2 a : having a fleeting, wayward, or inconstant quality **b :** having no fixed course **:** RANDOM
— va·grant·ly *adverb*

va·grom \ˈvā-grəm\ *adjective* [by alteration] (1599)
: VAGRANT

vague \ˈvāg\ *adjective* **vagu·er; vagu·est** [Middle French, from Latin *vagus,* literally, wandering] (1548)
1 a : not clearly expressed **:** stated in indefinite terms ⟨*vague* accusation⟩ **b :** not having a precise meaning ⟨*vague* term of abuse⟩
2 a : not clearly defined, grasped, or understood **:** INDISTINCT ⟨only a *vague* notion of what's needed⟩; *also* **:** SLIGHT ⟨a *vague* hint of a thickening waistline⟩ ⟨hasn't the *vaguest* idea⟩ **b :** not clearly felt or sensed **:** somewhat subconscious ⟨a *vague* longing⟩
3 : not thinking or expressing one's thoughts clearly or precisely ⟨*vague* about dates and places⟩
4 : lacking expression **:** VACANT
5 : not sharply outlined **:** HAZY
synonym see OBSCURE
— vague·ly *adverb*
— vague·ness *noun*

va·gus nerve \ˈvā-gəs-\ *noun* [New Latin *vagus nervus,* literally, wandering nerve] (1840)
: either of the 10th pair of cranial nerves that arise from the medulla and supply chiefly the viscera especially with autonomic sensory and motor fibers — called also *vagus*

vail \ˈvā(ə)l\ *transitive verb* [Middle English *valen,* partly from Middle French *valer* (short for *avaler* to let fall) & partly short for Middle English *avalen* to let fall, from Middle French *avaler,* from Old French, from *aval* downward, from *a* to (from Latin *ad*) + *val* valley — more at AT, VALE] (14th century)
: to lower often as a sign of respect or submission

vain \ˈvān\ *adjective* [Middle English, from Middle French, from Latin *vanus* empty, vain — more at WANE] (14th century)
1 : having no real value **:** IDLE, WORTHLESS
2 : marked by futility or ineffectualness **:** UNSUCCESSFUL, USELESS ⟨*vain* efforts to escape⟩
3 *archaic* **:** FOOLISH, SILLY
4 : having or showing undue or excessive pride in one's appearance or achievements **:** CONCEITED ☆
synonym see FUTILE
— vain·ly *adverb*
— vain·ness \ˈvān-nəs\ *noun*
— in vain 1 : to no end **:** without success or result **2 :** in an irreverent or blasphemous manner ⟨you shall not take the name of the Lord your God *in vain* —Deuteronomy 5:11 (Revised Standard Version)⟩

vain·glo·ri·ous \ˌvān-ˈglōr-ē-əs, -ˈglȯr-\ *adjective* (15th century)
: marked by vainglory **:** BOASTFUL
— vain·glo·ri·ous·ly *adverb*
— vain·glo·ri·ous·ness *noun*

vain·glo·ry \ˈvān-ˌglōr-ē, -ˌglȯr-, ˌvān-ˈ\ *noun* (14th century)
1 : excessive or ostentatious pride especially in one's achievements
2 : vain display or show **:** VANITY

vair \ˈvar, ˈver\ *noun* [Middle English *veir,* from Old French *vair,* from *vair,* adjective, variegated, from Latin *varius* variegated, various] (14th century)
: the bluish gray and white fur of a squirrel prized for ornamental use in medieval times

Vaish·na·va \ˈvīsh-nə-və\ *noun* [Sanskrit *vaiṣṇava* of Vishnu, from *Viṣṇu* Vishnu] (1815)
: a member of a major Hindu sect devoted to the cult of Vishnu
— Vaishnava *adjective*
— Vaish·na·vism \-ˌvi-zəm\ *noun*

☆ SYNONYMS
Vain, nugatory, otiose, idle, empty, hollow mean being without worth or significance. VAIN implies either absolute or relative absence of value ⟨*vain* promises⟩. NUGATORY suggests triviality or insignificance ⟨a monarch with *nugatory* powers⟩. OTIOSE suggests that something serves no purpose and is either an encumbrance or a superfluity ⟨a film without a single *otiose* scene⟩. IDLE suggests being incapable of worthwhile use or effect ⟨*idle* speculations⟩. EMPTY and HOLLOW suggest a deceiving lack of real substance or soundness or genuineness ⟨an *empty* attempt at reconciliation⟩ ⟨a *hollow* victory⟩.

Vais·ya \'vīsh-yə, 'vī-shə\ *noun* [Sanskrit *vaiśya,* from *viś* settlement; akin to Greek *oikos* house — more at VICINITY] (1665)
: a Hindu of an upper caste traditionally assigned to commercial and agricultural occupations

va·lance \'va-lən(t)s, 'vā-\ *noun* [Middle English *vallance,* perhaps from *Valence,* France] (15th century)
1 : a drapery hung along the edge of a bed, table, altar, canopy, or shelf
2 : a short drapery or wood or metal frame used as a decorative heading to conceal the top of curtains and fixtures
— va·lanced \-lən(t)st\ *adjective*

vale \'vā(ə)l\ *noun* [Middle English, from Middle French, from Old French *val,* from Latin *valles, vallis;* perhaps akin to Latin *volvere* to roll — more at VOLUBLE] (14th century)
1 : VALLEY, DALE
2 : WORLD ⟨this *vale* of tears⟩

val·e·dic·tion \,va-lə-'dik-shən\ *noun* [Latin *valedicere* to say farewell, from *vale* farewell + *dicere* to say — more at DICTION] (1614)
1 : an act of bidding farewell
2 : VALEDICTORY

val·e·dic·to·ri·an \-,dik-'tōr-ē-ən, -'tȯr-\ *noun* (1759)
: the student usually having the highest rank in a graduating class who delivers the valedictory address at the commencement exercises

¹val·e·dic·to·ry \-'dik-t(ə-)rē\ *adjective* [Latin *valedicere*] (1651)
: of or relating to a valediction **:** expressing or containing a farewell

²valedictory *noun, plural* **-ries** (1779)
: an address or statement of farewell or leave-taking

va·lence \'vā-lən(t)s\ *noun* [Late Latin *valentia* power, capacity, from Latin *valent-, valens,* present participle of *valēre* to be strong — more at WIELD] (1884)
1 : the degree of combining power of an element as shown by the number of atomic weights of a univalent element (as hydrogen) with which the atomic weight of the element will combine or for which it can be substituted or with which it can be compared
2 a : relative capacity to unite, react, or interact (as with antigens or a biological substrate) **b :** the degree of attractiveness an individual, activity, or object possesses as a behavioral goal

valence band *noun* (1953)
: the range of permissible energy values that are the highest energies an electron of an atom can have and still be associated with the atom and be used to form bonds — compare CONDUCTION BAND

valence electron *noun* (1923)
: a single electron or one of two or more electrons in the outer shell of an atom that is responsible for the chemical properties of the atom

Va·len·cia orange \və-'len-ch(ē-)ə, -len(t)-sē-ə\ *noun* [*Valencia,* Spain] (1858)
: a sweet orange of a juicy thin-skinned cultivar grown in the U.S. — called also *Valencia*

Va·len·ci·ennes \və-,len(t)-sē-'en(z), ,va-lən-sē-\ *noun* [*Valenciennes,* France] (1717)
: a fine bobbin lace

-valent *adjective combining form* [International Scientific Vocabulary, from Latin *valent-, valens*]
1 : having a (specified) valence or valences ⟨bi*valent*⟩ ⟨multi*valent*⟩
2 : having (so many) chromosomal strands or homologous chromosomes ⟨uni*valent*⟩

val·en·tine \'va-lən-,tīn\ *noun* (15th century)
1 : a sweetheart chosen or complimented on Saint Valentine's Day
2 a : a gift or greeting sent or given especially to a sweetheart on Saint Valentine's Day; *especially* **:** a greeting card sent on this day **b :** something (as a movie or piece of writing) expressing uncritical praise or affection **:** TRIBUTE

Valentine's Day *also* **Valentine Day** *noun* (1668)
: SAINT VALENTINE'S DAY

val·er·ate \'va-lə-,rāt\ *noun* (1852)
: a salt or ester of valeric acid

va·le·ri·an \və-'lir-ē-ən\ *noun* [Middle English, from Middle French or Medieval Latin; Middle French *valeriane,* from Medieval Latin *valeriana,* probably from feminine of *valerianus* of Valeria, from *Valeria,* Roman province formerly part of Pannonia] (14th century)
1 : any of a genus (*Valeriana* of the family Valerianaceae, the valerian family) of perennial herbs many of which possess medicinal properties
2 : a drug consisting of the dried rootstock and roots of the garden heliotrope (*Valeriana officinalis*) formerly used as a carminative and sedative

va·le·ric acid \və-'lir-ik-, -'ler-\ *noun* [*valerian;* from its occurrence in the root of valerian] (1857)
: any of four isomeric fatty acids $C_5H_{10}O_2$ or a mixture of these; *especially* **:** a liquid acid of disagreeable odor obtained from valerian or made synthetically and used especially in organic synthesis

¹va·let \'va-lət, 'va-(,)lā, va-'lā\ *noun* [Middle French *vaslet, varlet, valet* young nobleman, page, domestic servant, from (assumed) Medieval Latin *vassellittus,* diminutive of Medieval Latin *vassus* servant — more at VASSAL] (1567)
1 a : a man's male servant who performs personal services (as taking care of clothing) **b :** an employee (as of a hotel or a public facility) who performs personal services for customers
2 : a device (as a rack or tray) for holding clothing or personal effects

²valet *transitive verb* (1840)
: to serve as a valet

va·let de cham·bre \(,)va-,lā-də-'shäⁿbrᵊ\ *noun, plural* **va·lets de chambre** *same*\ [French, literally, chamber valet] (1646)
: VALET 1a

valet parking *noun* (1960)
: a service that provides parking of motor vehicles by an attendant

¹val·e·tu·di·nar·i·an \,va-lə-,tü-dᵊn-'er-ē-ən, -,tyü-\ *noun* [Latin *valetudinarius* sickly, infirm, from *valetudin-, valetudo* state of health, sickness, from *valēre* to be strong, be well — more at WIELD] (1703)
: a person of a weak or sickly constitution; *especially* **:** one whose chief concern is being or becoming a chronic invalid

²valetudinarian *adjective* (1713)
: of, relating to, or characteristic of a valetudinarian **:** SICKLY, WEAK

val·e·tu·di·nar·i·an·ism \-ē-ə-,ni-zəm\ *noun* (1839)
: the condition or state of mind of a valetudinarian

¹val·e·tu·di·nary \-'tü-dᵊn-,er-ē, -'tyü-\ *adjective* [Latin *valetudinarius*] (1581)
: VALETUDINARIAN

²valetudinary *noun, plural* **-nar·ies** (1665)
: VALETUDINARIAN

val·gus \'val-gəs\ *adjective* [New Latin, from Latin, bowlegged] (1884)
: of, relating to, or being a deformity in which an anatomical part is turned outward away from the midline of the body to an abnormal degree ⟨*valgus* deformity of the ankle⟩
— valgus *noun*

Val·hal·la \val-'ha-lə *also* väl-'hä-\ *noun* [German & Old Norse; German *Walhalla,* from Old Norse *Valhǫll,* literally, hall of the slain, from *valr* the slain (akin to Old English *wæl* slaughter, the slain) + *hǫll* hall; akin to Old English *heall* hall]
: the great hall in Norse mythology where the souls of heroes slain in battle are received

val·iance \'val-yən(t)s\ *noun* (15th century)
: VALOR

val·ian·cy \-yən(t)-sē\ *noun* (15th century)
: VALOR

¹val·iant \'val-yənt\ *adjective* [Middle English *valiaunt,* from Middle French *vaillant,* from Old French, from present participle of *valoir* to be of worth, from Latin *valēre* to be strong — more at WIELD] (14th century)
1 : possessing or acting with bravery or boldness **:** COURAGEOUS ⟨*valiant* soldiers⟩
2 : marked by, exhibiting, or carried out with courage or determination **:** HEROIC ⟨*valiant* feats⟩
— val·iant·ly *adverb*
— val·iant·ness *noun*

²valiant *noun* (1609)
: a valiant person

val·id \'va-ləd\ *adjective* [Middle French or Medieval Latin; Middle French *valide,* from Medieval Latin *validus,* from Latin, strong, from *valēre*] (1571)
1 : having legal efficacy or force; *especially* **:** executed with the proper legal authority and formalities ⟨a *valid* contract⟩
2 a : well-grounded or justifiable **:** being at once relevant and meaningful ⟨a *valid* theory⟩ **b :** logically correct ⟨a *valid* argument⟩ ⟨*valid* inference⟩
3 : appropriate to the end in view **:** EFFECTIVE ⟨every craft has its own *valid* methods⟩
4 *of a taxon* **:** conforming to accepted principles of sound biological classification ☆
— va·lid·i·ty \və-'li-də-tē, va-\ *noun*
— val·id·ly \'va-ləd-lē\ *adverb*

val·i·date \'va-lə-,dāt\ *transitive verb* **-dat·ed; -dat·ing** (1648)
1 a : to make legally valid **b :** to grant official sanction to by marking **c :** to confirm the validity of (an election); *also* **:** to declare (a person) elected
2 : to support or corroborate on a sound or authoritative basis ⟨experiments designed to *validate* the hypothesis⟩
synonym see CONFIRM

val·i·da·tion \,va-lə-'dā-shən\ *noun* (circa 1656)
: an act, process, or instance of validating; *especially* **:** the determination of the degree of validity of a measuring device

va·line \'va-,lēn, 'vā-\ *noun* [International Scientific Vocabulary, from *valeric (acid)*] (1907)
: a crystalline essential amino acid $C_5H_{11}NO_2$ that is one of the building blocks of plant and animal proteins

va·lise \və-'lēs\ *noun* [French, from Italian *valigia*] (1615)
: SUITCASE

Val·ium \'va-lē-əm, 'val-yəm\ *trademark*
— used for a preparation of diazepam

Val·ky·rie \val-'kir-ē *also* val-'kī-rē, 'val-kə-rē\ *noun* [German & Old Norse; German

☆ SYNONYMS
Valid, sound, cogent, convincing, telling mean having such force as to compel serious attention and usually acceptance. VALID implies being supported by objective truth or generally accepted authority ⟨a *valid* reason for being absent⟩ ⟨a *valid* marriage⟩. SOUND implies a basis of flawless reasoning or of solid grounds ⟨a *sound* proposal for reviving the economy⟩. COGENT may stress either weight of sound argument and evidence or lucidity of presentation ⟨the prosecutor's *cogent* summation won over the jury⟩. CONVINCING suggests a power to overcome doubt, opposition, or reluctance to accept ⟨a *convincing* argument for welfare reform⟩. TELLING stresses an immediate and crucial effect striking at the heart of a matter ⟨a *telling* example of bureaucratic waste⟩.

Walküre, from Old Norse *valkyrja,* literally, chooser of the slain; akin to Old English *wælcyrige* witch, Old Norse *valr* the slain, Old High German *kiosan* to choose — more at CHOOSE]
: any of the maidens of Odin who choose the heroes to be slain in battle and conduct them to Valhalla

val·late \'va-ˌlāt\ *adjective* [Latin *vallatus,* past participle of *vallare* to surround with a wall, from *vallum* wall, rampart — more at WALL] (1878)
: having a raised edge surrounding a depression ⟨*vallate* papillae of the tongue⟩

val·lec·u·la \va-'le-kyə-lə, və-\ *noun, plural* **-u·lae** \-kyə-ˌlē, -ˌlī\ [New Latin, from Late Latin, little valley, diminutive of Latin *valles* valley — more at VALE] (1859)
: an anatomical groove, channel, or depression; *especially* **:** one between the base of the tongue and the epiglottis
— val·lec·u·lar \-lər\ *adjective*

val·ley \'va-lē\ *noun, plural* **valleys** [Middle English *valey,* from Old French *valee,* from *val* valley — more at VALE] (14th century)
1 a : an elongate depression of the earth's surface usually between ranges of hills or mountains **b :** an area drained by a river and its tributaries
2 : a low point or condition
3 a : HOLLOW, DEPRESSION **b :** the place of meeting of two slopes of a roof that form on the plan a reentrant angle

valley fever *noun* [from its prevalence in the San Joaquin valley of California] (1938)
: COCCIDIOIDOMYCOSIS

Va·lois \'val-ˌwä, val-'\ *adjective* [Philippe de *Valois* (Philip VI of France)] (circa 1888)
: of or relating to the French royal house that ruled from 1328 to 1589

va·lo·nia \və-'lō-nē-ə, -nyə\ *noun* [Italian *vallonia,* from Middle Greek *balanidia,* plural of *balanidion,* diminutive of Greek *balanos* acorn — more at GLAND] (1722)
: dried acorn cups especially from a Eurasian evergreen oak (*Quercus macrolepis* synonym *Q. aegilops*) used in tanning or dressing leather

val·or \'va-lər\ *noun* [Middle English, from Middle French *valour,* from Medieval Latin *valor* value, valor, from Latin *valēre* to be strong — more at WIELD] (15th century)
: strength of mind or spirit that enables a person to encounter danger with firmness **:** personal bravery

val·o·rize \'va-lə-ˌrīz\ *transitive verb* **-rized; -riz·ing** [Portuguese *valorizar,* from *valor* value, price, from Medieval Latin] (circa 1906)
: to enhance or try to enhance the price, value, or status of by organized and usually governmental action ⟨using subsidies to *valorize* coffee⟩
— val·o·ri·za·tion \ˌva-lə-rə-'zā-shən\ *noun*

val·or·ous \'va-lə-rəs\ *adjective* (15th century)
: VALIANT
— val·or·ous·ly *adverb*

val·our \'va-lər\ *chiefly British variant of* VALOR

val·po·li·cel·la \ˌval-ˌpo-lə-'che-lə, ˌval-\ *noun, often capitalized* [*Valpolicella,* district in northern Italy] (1903)
: a dry red Italian table wine

Val·sal·va maneuver \val-'sal-və-\ *noun* [Antonio Maria *Valsalva* (died 1723) Italian anatomist] (1886)
: the process of making a forceful attempt at expiration while holding the nostrils closed and keeping the mouth shut for the purpose of testing the patency of the eustachian tubes or of adjusting middle ear pressure — called also *Valsalva*

valse \väls\ *noun* [French, from German *Walzer* — more at WALTZ] (1796)
: WALTZ; *specifically* **:** a concert waltz

[1]**valu·able** \'val-yə-bəl, -yə-wə-bəl\ *adjective* (1589)
1 a : having monetary value **b :** worth a good price
2 a : having desirable or esteemed characteristics or qualities ⟨*valuable* friendships⟩ **b :** of great use or service ⟨*valuable* advice⟩
— valu·able·ness *noun*
— valu·ably \-blē\ *adverb*

[2]**valuable** *noun* (circa 1775)
: a usually personal possession (as jewelry) of relatively great monetary value — usually used in plural

valuable consideration *noun* (1638)
: an equivalent or compensation having value that is given for something acquired or promised (as money or marriage) and that may consist either in a benefit accruing to one party or a loss falling upon the other

val·u·ate \'val-yə-ˌwāt\ *transitive verb* **-at·ed; -at·ing** (1873)
: to place a value on **:** APPRAISE

val·u·a·tion \ˌval-yə-'wā-shən\ *noun* [Middle French, from *valuer* to value, from *value*] (1529)
1 : the act or process of valuing; *specifically* **:** appraisal of property
2 : the estimated or determined market value of a thing
3 : judgment or appreciation of worth or character
— val·u·a·tion·al \-shnəl, -shə-nᵊl\ *adjective*
— val·u·a·tion·al·ly *adverb*

val·u·a·tor \'val-yə-ˌwā-tər\ *noun* (1731)
: one that valuates; *specifically* **:** one that appraises

[1]**val·ue** \'val-(ˌ)yü\ *noun* [Middle English, from Middle French, from (assumed) Vulgar Latin *valuta,* from feminine of *valutus,* past participle of Latin *valēre* to be worth, be strong — more at WIELD] (14th century)
1 : a fair return or equivalent in goods, services, or money for something exchanged
2 : the monetary worth of something **:** marketable price
3 : relative worth, utility, or importance ⟨a good *value* at the price⟩ ⟨the *value* of base stealing in baseball⟩ ⟨had nothing of *value* to say⟩
4 a : a numerical quantity that is assigned or is determined by calculation or measurement ⟨let *x* take on positive *values*⟩ ⟨a *value* for the age of the earth⟩ **b :** precise signification ⟨*value* of a word⟩
5 : the relative duration of a musical note
6 a : relative lightness or darkness of a color **:** LUMINOSITY **b :** the relation of one part in a picture to another with respect to lightness and darkness
7 : something (as a principle or quality) intrinsically valuable or desirable ⟨sought material *values* instead of human *values* —W. H. Jones⟩
8 : DENOMINATION 2
— val·ue·less \-(ˌ)yü-ləs, -yə-\ *adjective*
— val·ue·less·ness *noun*

[2]**value** *transitive verb* **val·ued; val·u·ing** (15th century)
1 a : to estimate or assign the monetary worth of **:** APPRAISE ⟨*value* a necklace⟩ **b :** to rate or scale in usefulness, importance, or general worth **:** EVALUATE
2 : to consider or rate highly **:** PRIZE, ESTEEM ⟨*values* your opinion⟩
synonym see ESTIMATE, APPRECIATE
— val·u·er \-yə-wər\ *noun*

value–added tax *noun* (1935)
: an incremental excise that is levied on the value added at each stage of the processing of a raw material or the production and distribution of a commodity and that typically has the impact of a sales tax on the ultimate consumer

val·ued \'val-(ˌ)yüd, -yəd\ *adjective* (1595)
: having a value or values especially of a specified kind or number — often used in combination ⟨real-*valued*⟩

val·ue–free \'val-(ˌ)yü-'frē\ *adjective* (1948)
: making or having no value judgments ⟨*value-free* distinctions⟩ ⟨*value-free* economics⟩

value judgment *noun* (1899)
: a judgment assigning a value (as good or bad) to something

va·lu·ta \və-'lü-tə, -(ˌ)tä\ *noun* [Italian, value, from (assumed) Vulgar Latin *valuta*] (1920)
1 : the agreed upon or exchange value of a currency
2 : FOREIGN EXCHANGE 2

val·vate \'val-ˌvāt\ *adjective* (1829)
: having valves or parts resembling a valve; *especially* **:** meeting at the edges without overlapping in the bud (as in the calyx of a mallow)

valve \'valv\ *noun* [Latin *valva;* akin to Latin *volvere* to roll — more at VOLUBLE] (14th century)
1 *archaic* **:** a leaf of a folding or double door
2 [New Latin *valva,* from Latin] **:** a structure especially in a vein or lymphatic that closes temporarily a passage or orifice or permits movement of fluid in one direction only
3 a : any of numerous mechanical devices by which the flow of liquid, gas, or loose material in bulk may be started, stopped, or regulated by a movable part that opens, shuts, or partially obstructs one or more ports or passageways; *also* **:** the movable part of such a device **b :** a device in a brass instrument for quickly channeling air flow through an added length of tube in order to change the fundamental tone by some definite interval **c** *chiefly British* **:** ELECTRON TUBE
4 [New Latin *valva,* from Latin] **:** one of the distinct and usually movably articulated pieces of which the shell of some shell-bearing animals (as lamellibranch mollusks, brachiopods, and barnacles) consists
5 [New Latin *valva,* from Latin] **a :** one of the segments or pieces into which a dehiscing capsule or legume separates **b :** the portion of various anthers (as of the barberry) resembling a lid **c :** one of the two encasing membranes of a diatom
— valved \'valvd\ *adjective*
— valve·less \'valv-ləs\ *adjective*

val·vu·la \'val-vyə-lə\ *noun, plural* **-lae** \-ˌlē, -ˌlī\ [New Latin, diminutive of Latin *valva*] (1615)
: a small valve or fold

val·vu·lar \'val-vyə-lər\ *adjective* (1797)
1 : resembling or functioning as a valve; *also* **:** opening by valves
2 : of, relating to, or affecting a valve especially of the heart ⟨*valvular* heart disease⟩

val·vu·li·tis \ˌval-vyə-'lī-təs\ *noun* [New Latin] (circa 1891)
: inflammation of a valve especially of the heart

va·moose \və-'müs, va-\ *intransitive verb* **va·moosed; va·moos·ing** [Spanish *vamos* let us go, suppletive 1st plural imperative (from Latin *vadere* to go) of *ir* to go, from Latin *ire* — more at WADE, ISSUE] (1810)
: to depart quickly

[1]**vamp** \'vamp\ *verb* [[2]*vamp*] (1599)
transitive verb
1 a : to provide (a shoe) with a new vamp **b :** to piece (something old) with a new part **:** PATCH ⟨*vamp* up old sermons⟩
2 : INVENT, FABRICATE ⟨*vamp* up an excuse⟩
intransitive verb
: to play a musical vamp
— vamp·er *noun*

\ə\ **abut** \ᵊ\ kitten \ər\ **further** \a\ **ash** \ā\ **ace** \ä\ mop, mar \au̇\ **out** \ch\ **chin** \e\ **bet** \ē\ **easy** \g\ **go** \i\ **hit** \ī\ **ice** \j\ **job** \ŋ\ **sing** \ō\ **go** \ȯ\ **law** \ȯi\ **boy** \th\ **thin** \t͟h\ **the** \ü\ **loot** \u̇\ **foot** \y\ **yet** \zh\ **vision** *see also* Guide to Pronunciation

[2]**vamp** *noun* [Middle English *vampe* sock, from Old French *avantpié,* from *avant-* fore- + *pié* foot, from Latin *ped-, pes* — more at VANGUARD, FOOT] (1654)
1 : the part of a shoe upper or boot upper covering especially the forepart of the foot and sometimes also extending forward over the toe or backward to the back seam of the upper
2 [[1]*vamp*] **:** a short introductory musical passage often repeated several times (as in vaudeville) before a solo or between verses

[3]**vamp** *noun* [short for *vampire*] (circa 1911)
: a woman who uses her charm or wiles to seduce and exploit men
— vamp·ish \ˈvam-pish\ *adjective*

[4]**vamp** *transitive verb* (circa 1915)
: to practice seductive wiles on

vam·pire \ˈvam-ˌpīr\ *noun* [French, from German *Vampir,* from Serbo-Croatian *vampir*] (1734)
1 : the reanimated body of a dead person believed to come from the grave at night and suck the blood of persons asleep
2 a : one who lives by preying on others **b :** a woman who exploits and ruins her lover
3 : VAMPIRE BAT
— vam·pir·ish \-ish\ *adjective*

vampire bat *noun* (1790)
: any of several Central and South American bats (*Desmodus rotundus, Diaemus youngi,* and *Diphylla ecaudata* of the subfamily Desmodontinae of the family Phyllostomidae) that feed on the blood of birds and mammals and especially domestic animals and that are sometimes vectors of equine trypanosomiasis and of rabies; *also* **:** any of several other bats (as of the families Megadermatidae and Phyllostomidae) that do not feed on blood but are sometimes reputed to do so

vampire bat

vam·pir·ism \-ˌpīr-ˌi-zəm\ *noun* (1794)
1 : belief in vampires
2 : the actions of a vampire

[1]**van** \ˈvan\ *noun* [Middle English, from Middle French, from Latin *vannus* — more at WINNOW] (14th century)
1 *dialect English* **:** a winnowing device (as a fan)
2 : WING 1a

[2]**van** *noun* [by shortening] (1610)
: VANGUARD

[3]**van** *noun* [short for *caravan*] (1829)
1 a : a usually enclosed wagon or motortruck used for transportation of goods or animals; *also* **:** CARAVAN 2a **b :** a multipurpose enclosed motor vehicle having a boxlike shape, rear or side doors, and side panels often with windows **c :** a detachable passenger cabin transportable by aircraft or truck
2 *chiefly British* **:** an enclosed railroad freight or baggage car

[4]**van** *transitive verb* **vanned; van·ning** (1840)
: to transport by van

van·a·date \ˈva-nə-ˌdāt\ *noun* [New Latin *vanadium* + English [1]*-ate*] (1835)
: a salt derived from vanadium pentoxide and containing pentavalent vanadium

va·na·di·um \və-ˈnā-dē-əm\ *noun* [New Latin, from Old Norse *Vanadīs* Freya] (1833)
: a grayish malleable ductile metallic element found combined in minerals and used especially to form alloys (as vanadium steel) — see ELEMENT table

vanadium pentoxide *noun* (1885)
: a yellowish red crystalline compound V_2O_5 used especially in glass manufacture and as a catalyst

Van Al·len belt \van-ˈa-lən-\ *noun* [James A. *Van Allen*] (1958)
: a belt of intense radiation in the magnetosphere composed of energetic charged particles trapped by the earth's magnetic field; *also* **:** a similar belt surrounding another planet

va·nas·pa·ti \və-ˈnəs-pə-tē, -ˈnäs-\ *noun* [Sanskrit, forest tree, soma plant, literally, lord of the forest, from *vana* forest + *pati* lord, master — more at POTENT] (1941)
: a hydrogenated vegetable fat used as a butter substitute in India

van·dal \ˈvan-dᵊl\ *noun* [Latin *Vandalii* (plural), of Germanic origin] (1555)
1 *capitalized* **:** a member of a Germanic people who lived in the area south of the Baltic between the Vistula and the Oder, overran Gaul, Spain, and northern Africa in the 4th and 5th centuries A.D., and in 455 sacked Rome
2 : one who willfully or ignorantly destroys, damages, or defaces property belonging to another or to the public ◆
— vandal *adjective, often capitalized*
— Van·dal·ic \van-ˈda-lik\ *adjective*

van·dal·ise *British variant of* VANDALIZE

van·dal·ism \ˈvan-dᵊl-ˌi-zəm\ *noun* (1798)
: willful or malicious destruction or defacement of public or private property

van·dal·is·tic \ˌvan-dᵊl-ˈis-tik\ *adjective* (1897)
: of or relating to vandalism

van·dal·ize \ˈvan-dᵊl-ˌīz\ *transitive verb* **-ized; -iz·ing** (1845)
: to subject to vandalism **:** DAMAGE
— van·dal·i·za·tion \ˌvan-dᵊl-ə-ˈzā-shən\ *noun*

van·da orchid \ˈvan-də-\ *noun* [New Latin, from Hindi *vandā* mistletoe, from Sanskrit, a parasitic plant] (1943)
: any of a large genus (*Vanda*) of Indo-Malayan epiphytic orchids often grown for their loose racemes of showy flowers — called also *vanda*

Van de Graaff generator \ˈvan-də-ˌgraf-\ *noun* [Robert J. *Van de Graaff* (died 1967) American physicist] (1937)
: an apparatus for the production of electrical discharges at high voltage commonly consisting of an insulated hollow conducting sphere that accumulates in its interior the charge continuously conveyed from a source of direct current by an endless belt of flexible nonconducting material

van der Waals forces \ˈvan-dər-ˌwȯlz-, ˈvän-dər-ˌvälz-\ *noun plural* [Johannes D. *van der Waals* (died 1923) Dutch physicist] (1926)
: the relatively weak attractive forces that act on neutral atoms and molecules and that arise because of the electric polarization induced in each of the particles by the presence of other particles

Van·dyke \van-ˈdīk, vən-\ *noun* [Sir Anthony *Vandyke*] (1755)
1 a : a wide collar with a deeply indented edge **b :** one of several V-shaped points forming a decorative edging **c :** a border of such points
2 : a trim pointed beard
— van·dyked \-ˈdīkt\ *adjective*

Vandyke 2

Vandyke brown *noun* [from its use by the painter Vandyke] (circa 1850)
: a natural brown-black pigment of organic matter obtained from bog earth or peat or lignite deposits; *also* **:** any of various synthetic brown pigments

vane \ˈvān\ *noun* [Middle English (southern dialect), from Old English *fana* banner; akin to Old High German *fano* cloth, Latin *pannus* cloth, rag] (15th century)
1 a : a movable device attached to an elevated object (as a spire) for showing the direction of the wind **b :** one that is changeable or inconstant
2 : a thin flat or curved object that is rotated about an axis by a flow of fluid or that rotates to cause a fluid to flow or that redirects a flow of fluid ⟨the *vanes* of a windmill⟩
3 : the web or flat expanded part of a feather — see FEATHER illustration
4 : a feather fastened to the shaft near the nock of an arrow
word history see VIXEN
— vaned \ˈvānd\ *adjective*

van·guard \ˈvan-ˌgärd *also* ˈvaŋ-\ *noun* [Middle English *vantgard,* from Middle French *avant-garde,* from Old French, from *avant-* fore- (from *avant* before, from Late Latin *abante*) + *garde* guard — more at ADVANCE] (15th century)
1 : the troops moving at the head of an army
2 : the forefront of an action or movement
— van·guard·ism \-ˌgär-ˌdi-zəm\ *noun*
— van·guard·ist \-dist\ *noun*

[1]**va·nil·la** \və-ˈni-lə, -ˈne-\ *noun* [New Latin, from Spanish *vainilla* vanilla (plant and fruit), diminutive of *vaina* sheath, from Latin *vagina* sheath, vagina] (1662)
1 a : VANILLA BEAN **b :** a commercially important extract of the vanilla bean that is used especially as a flavoring
2 : any of a genus (*Vanilla*) of tropical American climbing epiphytic orchids

[2]**vanilla** *adjective* (1846)
1 : flavored with vanilla
2 : lacking distinction **:** PLAIN, ORDINARY

vanilla bean *noun* (1874)
: the long capsular fruit of a vanilla (especially *Vanilla planifolia*) that is an important article of commerce

van·il·lin \ˈva-nᵊl-ən\ *noun* (circa 1868)
: a crystalline phenolic aldehyde $C_8H_8O_3$ that is extracted from vanilla beans or prepared synthetically and is used especially in flavoring and in perfumery

Va·nir \ˈvän-ˌir\ *noun plural* [Old Norse]
: a race of Norse gods who become united with the Aesir

van·ish \ˈva-nish\ *verb* [Middle English *vanisshen,* from Middle French *evaniss-,* stem of *evanir,* from (assumed) Vulgar Latin *exvanire,* alteration of Latin *evanescere* to dissipate like vapor, vanish, from *e-* + *vanescere* to vanish, from *vanus* empty] (14th century)
intransitive verb
1 a : to pass quickly from sight **:** DISAPPEAR **b :** to pass completely from existence

◇ WORD HISTORY

vandal Beginning in the 3rd century A.D. the Germanic tribes of northern Europe took advantage of the decline of the Roman Empire and began invading most of the Mediterranean world. The Vandals, a people from the shores of the Baltic, crossed the Rhine River, swept through Gaul, and by the early 5th century had penetrated Spain. They then sailed across the Strait of Gibraltar and seized the Roman province of Africa in 429. In 455 they organized a naval expedition against Italy and succeeded in plundering Rome itself for two weeks. When the Eastern Roman Empire reasserted its hegemony over Africa in the 6th century, the Vandals disappeared from history. Though in reality they were no worse than other peoples in a turbulent and violent era, 18th century authors such as Voltaire fixed on the Vandals as the prototype of the destructive barbarian who ignorantly destroys monuments of civilization. *Vandal* is now used in English generically for anyone who willfully defaces or destroys property belonging to another person or to the public, and the word is probably most closely associated with juvenile criminal behavior.

2 : to assume the value zero
transitive verb
: to cause to disappear
— van·ish·er *noun*
vanishing cream *noun* (1916)
: a cosmetic preparation that is less oily than cold cream and is used chiefly as a foundation for face powder
van·ish·ing·ly \ˈva-ni-shiŋ-lē\ *adverb* (1870)
: so as to be almost nonexistent or invisible ⟨the difference is *vanishingly* small⟩
vanishing point *noun* (1797)
1 : a point at which receding parallel lines seem to meet when represented in linear perspective
2 : a point at which something disappears or ceases to exist
van·i·ty \ˈva-nə-tē\ *noun, plural* **-ties** [Middle English *vanite,* from Old French *vanité,* from Latin *vanitat-, vanitas* quality of being empty or vain, from *vanus* empty, vain — more at WANE] (13th century)
1 : something that is vain, empty, or valueless
2 : the quality or fact of being vain
3 : inflated pride in oneself or one's appearance **:** CONCEIT
4 : a fashionable trifle or knicknack
5 a : ³COMPACT a **b :** a small case or handbag for toilet articles used by women
6 : DRESSING TABLE
vanity fair *noun, often V&F capitalized* [*Vanity-Fair,* a fair held in the frivolous town of Vanity in *Pilgrim's Progress* (1678) by John Bunyan] (1816)
: a scene or place characterized by frivolity and ostentation
vanity plate *noun* (1966)
: a license plate bearing letters or numbers designated by the owner of the vehicle
vanity press *noun* (1950)
: a publishing house that publishes books at the author's expense — called also *vanity publisher*
van·ner \ˈva-nər\ *noun* (1927)
: a person who owns a usually customized van
van·pool \ˈvan-ˌpül\ *noun* (1973)
: an arrangement by which a group of people commute to work in a van
— van·pool·ing *noun*
van·quish \ˈvaŋ-kwish, ˈvan-\ *transitive verb* [Middle English *venquissen,* from Middle French *venquis,* preterit of *veintre* to conquer, from Latin *vincere* — more at VICTOR] (14th century)
1 : to overcome in battle **:** subdue completely
2 : to defeat in a conflict or contest
3 : to gain mastery over (an emotion, passion, or temptation)
synonym see CONQUER
— van·quish·able \-kwi-shə-bəl\ *adjective*
— van·quish·er *noun*
van·tage \ˈvan-tij\ *noun* [Middle English, from Anglo-French, from Middle French *avantage* — more at ADVANTAGE] (14th century)
1 *archaic* **:** BENEFIT, GAIN
2 : superiority in a contest
3 : a position giving a strategic advantage, commanding perspective, or comprehensive view
4 : ADVANTAGE 4
— to the vantage *obsolete* **:** in addition
vantage point *noun* (1865)
: a position or standpoint from which something is viewed or considered; *especially* **:** POINT OF VIEW
van·ward \ˈvan-wərd\ *adjective* (1820)
: located in the vanguard **:** ADVANCED
— vanward *adverb*
vap·id \ˈva-pəd, ˈvā-\ *adjective* [Latin *vapidus* flat-tasting; akin to Latin *vappa* vapid wine and perhaps to Latin *vapor* steam] (circa 1656)
: lacking liveliness, tang, briskness, or force **:** FLAT, DULL ⟨a gossipy, *vapid* woman, obsessed by her own elegance —R. F. Delderfield⟩ ⟨London was not all *vapid* dissipation —V. S. Pritchett⟩
synonym see INSIPID
— va·pid·ly *adverb*
— va·pid·ness *noun*
va·pid·i·ty \va-ˈpi-də-tē, vā-, və-\ *noun, plural* **-ties** (circa 1721)
1 : the quality or state of being vapid
2 : something vapid
¹va·por \ˈvā-pər\ *noun* [Middle English *vapour,* from Middle French *vapeur,* from Latin *vapor* steam, vapor] (14th century)
1 : diffused matter (as smoke or fog) suspended floating in the air and impairing its transparency
2 a : a substance in the gaseous state as distinguished from the liquid or solid state **b :** a substance (as gasoline, alcohol, mercury, or benzoin) vaporized for industrial, therapeutic, or military uses; *also* **:** a mixture (as the explosive mixture in an internal combustion engine) of such a vapor with air
3 a : something unsubstantial or transitory **:** PHANTASM **b :** a foolish or fanciful idea
4 *plural* **a** *archaic* **:** exhalations of bodily organs (as the stomach) held to affect the physical or mental condition **b :** a depressed or hysterical nervous condition
²vapor *intransitive verb* **va·pored; va·por·ing** \-p(ə-)riŋ\ (15th century)
1 a : to rise or pass off in vapor **b :** to emit vapor
2 : to indulge in bragging, blustering, or idle talk
— va·por·er \-pər-ər\ *noun*
vapor barrier *noun* (circa 1941)
: a layer of material (as roofing paper or polyethylene film) used to retard or prevent the absorption of moisture into a construction (as a wall or floor)
va·po·ret·to \ˌvä-pə-ˈre-(ˌ)tō\ *noun, plural* **-ret·ti** \-ˈre-tē\ *also* **-ret·tos** [Italian, diminutive of *vapore* steamboat, from French *vapeur,* from *bateau à vapeur* steamboat] (1926)
: a motorboat serving as a canal bus in Venice, Italy
va·por·ing \ˈvā-p(ə-)riŋ\ *noun* (circa 1630)
: the act or speech of one that vapors; *specifically* **:** an idle, extravagant, or high-flown expression or speech — usually used in plural
va·por·ise *British variant of* VAPORIZE
va·por·ish \ˈvā-p(ə-)rish\ *adjective* (circa 1644)
1 : resembling or suggestive of vapor
2 : given to fits of the vapors
— va·por·ish·ness *noun*
va·por·ize \ˈvā-pə-ˌrīz\ *verb* **-ized; -iz·ing** (1803)
transitive verb
1 : to convert (as by the application of heat or by spraying) into vapor
2 : to cause to become dissipated
intransitive verb
1 : to become vaporized
2 : VAPOR 2
— va·por·iz·able \-ˌrī-zə-bəl\ *adjective*
— va·por·i·za·tion \ˌvā-pər-ə-ˈzā-shən\ *noun*
va·por·iz·er \ˈvā-pə-ˌrī-zər\ *noun* (circa 1846)
: one that vaporizes; as **a :** ATOMIZER **b :** a device for converting water or a medicated liquid into a vapor for inhalation
vapor lock *noun* (1926)
: partial or complete interruption of flow of a fluid (as fuel in an internal combustion engine) caused by the formation of bubbles of vapor in the feeding system
va·por·ous \ˈvā-p(ə-)rəs\ *adjective* (1527)
1 : consisting or characteristic of vapor
2 : producing vapors **:** VOLATILE
3 : containing or obscured by vapors **:** MISTY
4 a : ETHEREAL, UNSUBSTANTIAL **b :** consisting of or indulging in vaporings
— va·por·ous·ly *adverb*
— va·por·ous·ness *noun*
vapor pressure *noun* (1875)
: the pressure exerted by a vapor that is in equilibrium with its solid or liquid form — called also *vapor tension*
vapor trail *noun* (1941)
: CONTRAIL
va·por·ware \ˈvā-pər-ˌwar, -ˌwer\ *noun* (1984)
: a new computer-related product that has been widely advertised but is not yet available
va·pory \ˈvā-p(ə-)rē\ *adjective* (1598)
: VAPOROUS, MISTY
va·pour *chiefly British variant of* VAPOR
va·que·ro \vä-ˈker-(ˌ)ō\ *noun, plural* **-ros** [Spanish — more at BUCKAROO] (1826)
: HERDSMAN, COWBOY
va·ra \ˈvär-ə\ *noun* [American Spanish, from Spanish, pole, from Latin, forked pole, from feminine of *varus* bent, bow-legged] (1831)
: a Texas unit of length equal to 33.33 inches (84.66 centimeters)
vari- *or* **vario-** *combining form* [Latin *varius*]
1 : varied **:** diverse ⟨*vari*colored⟩
2 : variation **:** variability ⟨*vario*meter⟩
var·ia \ˈver-ē-ə, ˈvar-\ *noun plural* [New Latin, from Latin, neuter plural of *varius* various] (1926)
: MISCELLANY; *especially* **:** a literary miscellany
¹var·i·able \ˈver-ē-ə-bəl, ˈvar-\ *adjective* [Middle English, from Middle French, from Latin *variabilis,* from *variare* to vary] (14th century)
1 a : able or apt to vary **:** subject to variation or changes ⟨*variable* winds⟩ ⟨*variable* costs⟩ **b :** FICKLE, INCONSTANT
2 : characterized by variations
3 : having the characteristics of a variable
4 : not true to type **:** ABERRANT — used of a biological group or character
— var·i·abil·i·ty \ˌver-ē-ə-ˈbi-lə-tē, ˌvar-\ *noun*
— var·i·able·ness \ˈver-ē-ə-bəl-nəs, ˈvar-\ *noun*
— var·i·ably \-blē\ *adverb*
²variable *noun* (1816)
1 a : a quantity that may assume any one of a set of values **b :** a symbol representing a variable
2 : something that is variable
3 : VARIABLE STAR
variable rate mortgage *noun* (1975)
: ADJUSTABLE RATE MORTGAGE
variable star *noun* (1788)
: a star whose brightness changes usually in more or less regular periods
var·i·ance \ˈver-ē-ən(t)s, ˈvar-\ *noun* (14th century)
1 : the fact, quality, or state of being variable or variant **:** DIFFERENCE, VARIATION ⟨yearly *variance* in crops⟩
2 : the fact or state of being in disagreement **:** DISSENSION, DISPUTE
3 : a disagreement between two parts of the same legal proceeding that must be consonant
4 : a license to do some act contrary to the usual rule ⟨a zoning *variance*⟩
5 : the square of the standard deviation
synonym see DISCORD
— at variance : not in harmony or agreement
¹var·i·ant \ˈver-ē-ənt, ˈvar-\ *adjective* (14th century)
1 *obsolete* **:** VARIABLE
2 : manifesting variety, deviation, or disagreement
3 : varying usually slightly from the standard form ⟨*variant* readings⟩
²variant *noun* (1848)
: one of two or more persons or things exhibiting usually slight differences: as **a :** one that exhibits variation from a type or norm **b :** one of two or more different spellings (as *labor*

and *labour*) or pronunciations (as of *economics* \ek-, ēk-\) of the same word **c :** one of two or more words (as *geographic* and *geographical*) or word elements (as *mon-* and *mono-*) of essentially the same meaning differing only in the presence or absence of an affix

var·i·ate \'ver-ē-ˌāt, 'var-, -ət\ *noun* (1899)
: RANDOM VARIABLE

var·i·a·tion \ˌver-ē-'ā-shən, ˌvar-\ *noun* (14th century)
1 a : the act or process of varying **:** the state or fact of being varied **b :** an instance of varying **c :** the extent to which or the range in which a thing varies
2 : DECLINATION 6
3 a : a change of algebraic sign between successive terms of a sequence **b :** a measure of the change in data, a variable, or a function
4 : the repetition of a musical theme with modifications in rhythm, tune, harmony, or key
5 a : divergence in the characteristics of an organism from the species or population norm or average **b :** something (as an individual or group) that exhibits variation
6 a : a solo dance in classic ballet **b :** a repetition in modern ballet of a movement sequence with changes
— var·i·a·tion·al \-shnəl, -shə-nᵊl\ *adjective*
— var·i·a·tion·al·ly *adverb*

var·i·cel·la \ˌvar-ə-'se-lə\ *noun* [New Latin, irregular diminutive of *variola*] (circa 1771)
: CHICKEN POX

var·i·co·cele \'var-ə-kō-ˌsēl\ *noun* [New Latin, from Latin *varic-, varix* + New Latin *-o-* + *-cele*] (1736)
: a varicose enlargement of the veins of the spermatic cord

vari·col·ored \'ver-i-ˌkə-lərd, 'var-\ *adjective* (1665)
: having various colors **:** VARIEGATED ⟨the *varicolored* breeding plumage of a bird⟩; *also* **:** of various colors

var·i·cose \'var-ə-ˌkōs\ *also* **var·i·cosed** \-ˌkōst\ *adjective* [Latin *varicosus* full of dilated veins, from *varic-, varix* dilated vein] (circa 1730)
1 : abnormally swollen or dilated ⟨*varicose* veins⟩
2 : affected with varicose veins ⟨*varicose* legs⟩

var·i·cos·i·ty \ˌvar-ə-'kä-sə-tē\ *noun, plural* **-ties** (circa 1842)
1 : the quality or state of being abnormally or markedly swollen or dilated
2 : VARIX

var·ied \'ver-ēd, 'var-\ *adjective* (1588)
1 : VARIOUS, DIVERSE
2 : VARIEGATED 1
— var·ied·ly *adverb*

var·ie·gate \'ver-ē-ə-ˌgāt, 'ver-i-ˌgāt, 'var-\ *transitive verb* **-gat·ed; -gat·ing** [Latin *variegatus,* past participle of *variegare,* from *varius* various + *-egare* (akin to Latin *agere* to drive) — more at AGENT] (1653)
1 : to diversify in external appearance especially with different colors **:** DAPPLE
2 : to enliven or give interest to by means of variety
— var·ie·ga·tor \-ˌgā-tər\ *noun*

var·ie·gat·ed \-ˌgā-təd\ *adjective* (1661)
1 : having discrete markings of different colors ⟨*variegated* leaves⟩
2 : VARIED 1

variegated cutworm *noun* (1922)
: a widespread noctuid moth (*Peridroma saucia*) whose larva is destructive to crops

var·ie·ga·tion \ˌver-ē-ə-'gā-shən, ˌver-i-'gā-, ˌvar-\ *noun* (1646)
: the act of variegating **:** the state of being variegated; *especially* **:** diversity of colors

var·i·er \'ver-ē-ər, 'var-\ *noun* (1860)
: one that varies

¹va·ri·e·tal \və-'rī-ə-tᵊl\ *adjective* (1866)
1 : of, relating to, or characterizing a variety ⟨*varietal* name⟩; *also* **:** being a variety in distinction from an individual or species
2 : of, relating to, or producing a varietal

²varietal *noun* (1950)
: a wine bearing the name of the principal grape from which it is made

va·ri·e·ty \və-'rī-ə-tē\ *noun, plural* **-ties** [Middle French or Latin; Middle French *varieté,* from Latin *varietat-, varietas,* from *varius* various] (15th century)
1 : the quality or state of having different forms or types **:** MULTIFARIOUSNESS
2 : a number or collection of different things especially of a particular class **:** ASSORTMENT
3 a : something differing from others of the same general kind **:** SORT **b :** any of various groups of plants or animals ranking below a species **:** SUBSPECIES
4 : VARIETY SHOW

variety meat *noun* (circa 1946)
: an edible part (as the liver or tongue) of a slaughter animal other than skeletal muscle

variety show *noun* (1882)
: a theatrical entertainment of successive separate performances (as of songs, dances, skits, acrobatic feats, and trained animal acts)

variety store *noun* (circa 1768)
: a retail store that carries a wide variety of merchandise especially of low unit value

vario- — see VARI-

va·ri·o·la \və-'rī-ə-lə; ˌver-ē-'ō-lə, ˌvar-\ *noun* [New Latin, from Medieval Latin, pustule, pox, from Late Latin, pustule, probably from *varius* various] (1543)
: SMALLPOX

var·i·om·e·ter \ˌver-ē-'ä-mə-tər, ˌvar-\ *noun* (circa 1900)
1 : an instrument for measuring magnetic declination
2 : an aeronautical instrument for indicating rate of climb

¹var·i·o·rum \ˌver-ē-'ōr-əm, ˌvar-, -'ȯr-\ *noun* [Latin *variorum* of various persons (genitive plural masculine of *varius*), in the phrase *cum notis variorum* with the notes of various persons] (1728)
1 : an edition or text with notes by different persons
2 : an edition containing variant readings of the text

²variorum *adjective* (1763)
: relating to or being a variorum; *also* **:** VARIANT ⟨*variorum* readings⟩

¹var·i·ous \'ver-ē-əs, 'var-\ *adjective* [Latin *varius*] (1552)
1 *archaic* **:** VARIABLE, INCONSTANT
2 : VARICOLORED ⟨birds of *various* plumage⟩
3 a : of differing kinds **:** MULTIFARIOUS **b :** dissimilar in nature or form **:** UNLIKE ⟨animals as *various* as the jaguar and the sloth⟩
4 : having a number of different aspects or characteristics ⟨a *various* place⟩ ⟨a *various* talent⟩
5 : of an indefinite number greater than one ⟨stop at *various* towns⟩
6 : INDIVIDUAL, SEPARATE ⟨rate increases granted in the *various* states⟩
synonym see DIFFERENT
— var·i·ous·ness *noun*

²various *pronoun, plural in construction* (1877)
: an indefinite number of separate individuals greater than one ⟨conversations with people from *various* of the schools —Patricia Linden⟩

var·i·ous·ly *adverb* (1627)
1 : in various ways **:** at various times ⟨was *variously* occupied teaching, farming, and clerking⟩
2 : by various designations ⟨known *variously* as principal, headmaster, and rector⟩

vari·sized \'ver-i-ˌsīzd, 'var-\ *adjective* (1936)
: of various sizes

va·ris·tor \va-'ris-tər, ve-\ *noun* [*vari-* + *resistor*] (1937)
: an electrical resistor whose resistance depends on the applied voltage

var·ix \'var-iks\ *noun, plural* **var·i·ces** \'var-ə-ˌsēz\ [Middle English, from Latin *varic-, varix*] (14th century)
: an abnormally dilated and lengthened vein, artery, or lymph vessel; *especially* **:** a varicose vein

var·let \'vär-lət\ *noun* [Middle English, from Middle French *vaslet, varlet* young nobleman, page — more at VALET] (15th century)
1 a : ATTENDANT, MENIAL **b :** a knight's page
2 : a base unprincipled person **:** KNAVE

var·let·ry \-lə-trē\ *noun* (1606)
archaic **:** RABBLE

var·mint \'vär-mənt\ *noun* [alteration of *vermin*] (1539)
1 : an animal considered a pest; *specifically* **:** one classed as vermin and unprotected by game law
2 : a contemptible person **:** RASCAL; *broadly* **:** PERSON, FELLOW

¹var·nish \'vär-nish\ *noun* [Middle English *vernisch,* from Middle French *vernis,* from Old Italian or Medieval Latin; Old Italian *vernice,* from Medieval Latin *veronic-, veronix* sandarac] (14th century)
1 a : a liquid preparation that when spread and allowed to dry on a surface forms a hard lustrous typically transparent coating **b :** the covering or glaze given by the application of varnish **c** (1) **:** something that suggests varnish by its gloss (2) **:** a coating (as of deposits in an internal combustion engine) comparable to varnish
2 : outside show **:** ¹GLOSS
3 *chiefly British* **:** a liquid nail polish
— var·nishy \-ni-shē\ *adjective*

²varnish *transitive verb* (14th century)
1 : to apply varnish to
2 : to cover or conceal (as something unpleasant) with something that gives a fair appearance **:** ²GLOSS
3 : ADORN, EMBELLISH
— var·nish·er \'vär-ni-shər\ *noun*

varnish tree *noun* (1758)
: any of various trees yielding a milky juice from which in some cases varnish or lacquer is prepared; *especially* **:** a Japanese sumac (*Rhus verniciflua*)

var·si·ty \'vär-sə-tē, -stē\ *noun, plural* **-ties** [by shortening & alteration from *university*] (1646)
1 *British* **:** UNIVERSITY
2 a : the principal squad representing a university, college, school, or club especially in a sport **b :** REGULAR 1d

Var·so·vi·an \ˌvär-'sō-vē-ən\ *noun* [French *varsovien,* from *Varsovie* Warsaw] (1764)
: a native or resident of Warsaw, Poland

Var·u·na \'var-ə-nə\ *noun* [Sanskrit *Varuṇa*]
: a chief Vedic god responsible for natural and moral order in the cosmos

var·us \'var-əs, 'ver-\ *noun* [New Latin, from Latin, bent, knock-kneed] (1800)
: a deformed position of a bodily part characterized by bending or turning inward toward the midline of the body to an abnormal degree ⟨a moderate right metatarsus *varus* —*Journal American Medical Association*⟩
— varus *adjective*

varve \'värv\ *noun* [Swedish *varv* turn, layer; akin to Old Norse *hvarf* ring, Old English *hweorfan* to turn — more at WHARF] (1912)
: a pair of layers of alternately finer and coarser silt or clay believed to comprise an annual cycle of deposition in a body of still water
— varved \'värvd\ *adjective*

vary \'ver-ē, 'var-\ *verb* **var·ied; vary·ing** [Middle English *varien,* from Middle French or Latin; Middle French *varier,* from Latin *variare,* from *varius* various] (14th century)
transitive verb

1 a : to make a partial change in **:** make different in some attribute or characteristic **b :** to make differences between items in **:** DIVERSIFY
2 : to present under new aspects ⟨*vary* the rhythm and harmonic treatment⟩
intransitive verb
1 : to exhibit or undergo change ⟨the sky was constantly *varying*⟩
2 : DEVIATE, DEPART
3 : to take on successive values ⟨*y varies* inversely with *x*⟩
4 : to exhibit divergence in structural or physiological characters from the typical form
synonym see CHANGE
— vary·ing·ly \-iŋ-lē\ *adverb*

varying hare *noun* (1781)
: SNOWSHOE HARE

vas \ˈvas\ *noun, plural* **va·sa** \ˈvā-zə\ [New Latin, from Latin, vessel] (1651)
: an anatomical vessel **:** DUCT
— va·sal \-zəl\ *adjective*

vas- *or* **vaso-** *combining form* [New Latin, from Latin *vas*]
1 : vessel: as **a :** blood vessel ⟨*vaso*motor⟩ **b :** vas deferens ⟨*vas*ectomy⟩
2 : vascular and ⟨*vaso*vagal⟩
3 : vasomotor ⟨*vaso*active⟩

va·sa ef·fer·en·tia \ˈvā-zə-ˌe-fə-ˈren(t)-sh(ē-)ə\ *noun plural* [New Latin, literally, efferent vessels] (circa 1860)
: the 12 to 20 ductules that lead from the rete of the testis to the vas deferens and except near their commencement are greatly convoluted and form the compact head of the epididymis

vas·cu·lar \ˈvas-kyə-lər\ *adjective* [New Latin *vascularis,* from Latin *vasculum* small vessel, diminutive of *vas*] (1672)
1 : of or relating to a channel for the conveyance of a body fluid (as blood of an animal or sap of a plant) or to a system of such channels; *also* **:** supplied with or made up of such channels and especially blood vessels ⟨a *vascular* tumor⟩ ⟨a *vascular* system⟩
2 : marked by vigor and ardor **:** SPIRITED, PASSIONATE
— vas·cu·lar·i·ty \ˌvas-kyə-ˈlar-ə-tē\ *noun*

vascular bundle *noun* (circa 1884)
: a strand of specialized vascular tissue of higher plants consisting mostly of xylem and phloem

vascular cylinder *noun* (circa 1889)
: STELE

vas·cu·lar·i·za·tion \ˌvas-kyə-lə-rə-ˈzā-shən\ *noun* (1818)
: the process of becoming vascular; *also* **:** abnormal or excessive formation of blood vessels (as in the retina or on the cornea)

vascular plant *noun* (1861)
: a plant having a specialized conducting system that includes xylem and phloem **:** TRACHEOPHYTE

vascular ray *noun* (1672)
: a band of usually parenchymatous cells partly in the xylem and partly in the phloem of a plant root or stem that conducts fluids radially and appears in a cross section like a spoke of a wheel

vascular tissue *noun* (1815)
: plant tissue concerned mainly with conduction; *especially* **:** the specialized tissue of higher plants consisting essentially of phloem and xylem

vas·cu·la·ture \ˈvas-kyə-lə-ˌchu̇r, -ˌtyu̇r, -ˌtu̇r\ *noun* [Latin *vasculum* vessel + English *-ature* (as in *musculature*)] (circa 1927)
: the arrangement of blood vessels in an organ or part

vas·cu·li·tis \ˌvas-kyə-ˈlī-təs\ *noun, plural* **-lit·i·des** \-ˈli-tə-ˌdēz\ [New Latin, from Latin *vasculum* vessel] (circa 1900)
: inflammation of a blood or lymph vessel

vas·cu·lum \ˈvas-kyə-ləm\ *noun, plural* **-la** \-lə\ [New Latin, from Latin, small vessel] (1782)
: a usually metal and commonly cylindrical or flattened covered box used in collecting plants

vas def·er·ens \ˈvas-ˈde-fə-rənz, -ˌrenz\ *noun, plural* **va·sa def·er·en·tia** \ˈvā-zə-ˌde-fə-ˈren(t)-sh(ē-)ə\ [New Latin, literally, deferent vessel] (1578)
: a spermatic duct especially of a higher vertebrate that in the human male is a thick-walled tube about two feet (0.61 meters) long that begins at and is continuous with the tail of the epididymis and eventually joins the duct of the seminal vesicle to form the ejaculatory duct

vase *US oftenest* ˈvās; *Canadian usually & US also* ˈvāz; *British usually, Canadian also, & US sometimes* ˈväz\ *noun* [French, from Latin *vas* vessel] (1563)
: a usually round vessel of greater depth than width used chiefly as an ornament or for holding flowers ▢
— vase·like \-ˌlīk\ *adjective*

va·sec·to·mize \və-ˈsek-tə-ˌmīz, vā-ˈzek-\ *transitive verb* **-mized; -miz·ing** (1900)
: to perform a vasectomy on

va·sec·to·my \-tə-mē\ *noun, plural* **-mies** [International Scientific Vocabulary] (1899)
: surgical division or resection of all or part of the vas deferens usually to induce sterility

Vas·e·line \ˈva-sə-ˌlēn, ˌva-sə-ˈ\ *trademark*
— used for petroleum jelly

va·so·ac·tive \ˌvā-zō-ˈak-tiv\ *adjective* (circa 1921)
: affecting the blood vessels especially in respect to the degree of their relaxation or contraction
— va·so·ac·tiv·i·ty \-ak-ˈti-və-tē\ *noun*

va·so·con·stric·tion \-kən-ˈstrik-shən\ *noun* [International Scientific Vocabulary] (1899)
: narrowing of the lumen of blood vessels especially as a result of vasomotor action

va·so·con·stric·tive \-ˈstrik-tiv\ *adjective* (1890)
: inducing vasoconstriction

va·so·con·stric·tor \-tər\ *noun* (1877)
: an agent (as a sympathetic nerve fiber or a drug) that induces or initiates vasoconstriction

va·so·di·la·tion \-dī-ˈlā-shən, -də-\ *or* **va·so·di·la·ta·tion** \-ˌdi-lə-ˈtā-shən, -ˌdī-lə-\ *noun* [International Scientific Vocabulary] (1896)
: widening of the lumen of blood vessels

va·so·di·la·tor \-dī-ˈlā-tər, -ˈdī-ˌlā-\ *noun* (1881)
: an agent (as a parasympathetic nerve fiber or a drug) that induces or initiates vasodilation

va·so·mo·tor \ˌvā-zə-ˈmō-tər\ *adjective* [International Scientific Vocabulary] (1865)
: of, relating to, or being nerves or centers controlling the size of blood vessels

va·so·pres·sin \ˌvā-zō-ˈpre-sᵊn\ *noun* [from *Vasopressin,* a trademark] (1927)
: a polypeptide hormone secreted by the posterior lobe of the pituitary gland that increases blood pressure and decreases urine flow — called also *antidiuretic hormone*

va·so·pres·sor \-ˈpre-sər\ *adjective* (1928)
: causing a rise in blood pressure by exerting a vasoconstrictor effect
— vasopressor *noun*

va·so·spasm \ˈvā-zō-ˌspa-zəm\ *noun* [International Scientific Vocabulary] (circa 1909)
: sharp and often persistent contraction of a blood vessel reducing its caliber and blood flow
— va·so·spas·tic \ˌvā-zō-ˈspas-tik\ *adjective*

va·so·to·cin \ˌvā-zə-ˈtō-sᵊn\ *noun* [*vaso-* + oxy*tocin*] (circa 1963)
: a polypeptide pituitary hormone of most vertebrates below mammals that is probably the phylogenetic precursor of oxytocin and vasopressin

va·so·va·gal \ˌvā-zō-ˈvā-gəl\ *adjective* (circa 1923)
: of, relating to, or involving both vascular and vagal factors

vas·sal \ˈva-səl\ *noun* [Middle English, from Middle French, from Medieval Latin *vassallus,* from *vassus* servant, vassal, of Celtic origin; akin to Welsh *gwas* young man, servant] (14th century)
1 : a person under the protection of a feudal lord to whom he has vowed homage and fealty **:** a feudal tenant
2 : one in a subservient or subordinate position
— vassal *adjective*

vas·sal·age \-sə-lij\ *noun* (1594)
1 : the state of being a vassal
2 : the homage, fealty, or services due from a vassal
3 : a position of subordination or submission (as to a political power)

¹vast \ˈvast\ *adjective* [Latin *vastus;* akin to Old High German *wuosti* empty, desolate, Old Irish *fás*] (1575)
: very great in size, amount, degree, intensity, or especially in extent or range
synonym see ENORMOUS
— vast·ly *adverb*
— vast·ness \ˈvas(t)-nəs\ *noun*

²vast *noun* (1604)
: a boundless space ⟨the *vast* of heaven —John Milton⟩

vas·ti·tude \ˈvas-tə-ˌtüd, -ˌtyüd\ *noun* [Latin *vastitudo,* from *vastus*] (1623)
: IMMENSITY, VASTNESS

vasty \ˈvas-tē\ *adjective* (1596)
: VAST ⟨call spirits from the *vasty* deep —Shakespeare⟩

¹vat \ˈvat\ *noun* [Middle English *fat, vat,* from Old English *fæt;* akin to Old High German *vaz* vessel, Lithuanian *puodas* pot] (13th century)
1 : a large vessel (as a cistern, tub, or barrel) especially for holding liquors in an immature state or preparations for dyeing or tanning
2 : a liquor containing a dye converted into a soluble reduced colorless or weakly colored form that on textile material steeped in the liquor and exposed to the air is converted by oxidation to the original insoluble dye and precipitated in the fiber
word history see VIXEN

²vat *transitive verb* **vat·ted; vat·ting** (1784)
: to put into or treat in a vat

vat dye *noun* (circa 1903)
: a water-insoluble generally fast dye used in the form of a vat liquor — called also *vat color*

vat–dyed \ˈvat-ˈdīd\ *adjective* (circa 1947)
: dyed with one or more vat dyes

vat·ic \ˈva-tik\ *adjective* [Latin *vates* seer, prophet; akin to Old English *wōth* poetry, Old High German *wuot* madness, Old Irish *fáith* seer, poet] (1603)
: PROPHETIC, ORACULAR

Vat·i·can \ˈva-ti-kən\ *noun* [Latin *Vaticanus* Vatican Hill (in Rome)] (1555)

▢ USAGE
vase This little word of four letters, one of which is silent, provokes two long-standing pronunciation questions. Should the *a* have the sound of *a* in *are* or *ate,* and should the *s* have the sound of *s* in *case* or *rase*? Swift and Byron rhymed *vase* with *face* and *place* but all four possibilities have been advocated in the past, along with a fifth variant which rhymed with *gauze.* One also finds a variety of spellings over the centuries to match each pronunciation. To some ears the \ˈväz\ variant sounds more elegant, but the others are also established by long usage and are equally acceptable. Which is the dominant form depends on where you are in the English-speaking world, as noted above.

1 : the papal headquarters in Rome
2 : the papal government
— **Vatican** *adjective*

va·tic·i·nal \və-'ti-sᵊn-əl, va-\ *adjective* [Latin *vaticinus,* from *vaticinari* to foretell, prophesy] (1586)
: PROPHETIC

va·tic·i·nate \-sᵊn-ˌāt\ *verb* **-nat·ed; -nat·ing** [Latin *vaticinatus,* past participle of *vaticinari,* from *vates* + *-cinari* (akin to Latin *canere* to sing) — more at CHANT] (circa 1623)
: PROPHESY, PREDICT
— **va·tic·i·na·tor** \-ˌā-tər\ *noun*

va·tic·i·na·tion \-ˌti-sᵊn-'ā-shən\ *noun* (1603)
1 : PREDICTION
2 : the act of prophesying

va·tu \'vä-ˌtü\ *noun, plural* **vatu** [probably alteration of *Vanuatu*] (1981)
— see MONEY table

vaude·ville \'vȯd-vəl, 'väd-, 'vōd-, -ˌvil; 'vȯ-də-, 'vä-, 'vō-\ *noun* [French, from Middle French, popular satirical song, alteration of *vaudevire,* from *vau-de-Vire* valley of Vire, town in northwest France where such songs were composed] (1739)
1 : a light often comic theatrical piece frequently combining pantomime, dialogue, dancing, and song
2 : stage entertainment consisting of various acts (as performing animals, acrobats, comedians, dancers, or singers)
— **vaude·vil·lian** \ˌvȯd-'vil-yən, ˌväd-, ˌvōd-; ˌvȯ-də-, ˌvä-, ˌvō-\ *noun or adjective*

Vau·dois \vō-'dwä, 'vō-ˌ\ *noun plural* [Middle French, from Medieval Latin *Valdenses*] (1560)
: WALDENSES

[1]vault \'vȯlt\ *noun* [Middle English *voute,* from Middle French, from (assumed) Vulgar Latin *volvita* turn, vault, from feminine of *volvitus,* alteration of Latin *volutus,* past participle of *volvere* to roll — more at VOLUBLE] (14th century)
1 a : an arched structure of masonry usually forming a ceiling or roof **b :** something (as the sky) resembling a vault **c :** an arched or dome-shaped anatomical structure
2 a : a space covered by an arched structure; *especially* **:** an underground passage or room **b :** an underground storage compartment **c :** a room or compartment for the safekeeping of valuables
3 a : a burial chamber **b :** a prefabricated container usually of metal or concrete into which a casket is placed at burial
— **vaulty** \'vȯl-tē\ *adjective*

[2]vault *transitive verb* (14th century)
: to form or cover with or as if with a vault **:** ARCH

[3]vault *verb* [Middle French *volter,* from Old Italian *voltare,* from (assumed) Vulgar Latin *volvitare* to turn, leap, frequentative of Latin *volvere*] (1538)
intransitive verb
1 : to bound vigorously; *especially* **:** to execute a leap using the hands or a pole
2 : to do or achieve something as if by a leap
transitive verb
: to leap over; *especially* **:** to leap over by or as if by aid of the hands or a pole

[4]vault *noun* (1576)
: an act of vaulting **:** LEAP

vault·ed \'vȯl-təd\ *adjective* (1533)
1 : built in the form of a vault **:** ARCHED
2 : covered with a vault

vault·er \-tər\ *noun* (circa 1552)
: one that vaults; *especially* **:** an athlete who competes in the pole vault

[1]vault·ing \-tiŋ\ *noun* (1512)
: vaulted construction

[2]vaulting *adjective* (1593)
1 : reaching or stretching for the heights ⟨*vaulting* ambition⟩ ⟨a *vaulting* imagination⟩
2 : designed for use in vaulting or in gymnastic exercises ⟨a *vaulting* block⟩
— **vault·ing·ly** \-tiŋ-lē\ *adverb*

vaulting horse *noun* (circa 1875)
1 : a gymnastics apparatus used in vaulting that consists of a padded rectangular or cylindrical form supported in a horizontal position above the floor
2 : an event in which vaults are made over a vaulting horse

[1]vaunt \'vȯnt, 'vänt\ *verb* [Middle English, from Middle French *vanter,* from Late Latin *vanitare,* frequentative of (assumed) Latin *vanare,* from Latin *vanus* vain] (15th century)
intransitive verb
: to make a vain display of one's own worth or attainments **:** BRAG
transitive verb
: to call attention to pridefully and often boastfully ⟨people who *vaunt* their ingenuity⟩
synonym see BOAST
— **vaunt·er** *noun*
— **vaunt·ing·ly** \'vȯn-tiŋ-lē, 'vän-\ *adverb*

[2]vaunt *noun* (15th century)
1 : a vainglorious display of what one is or has or has done
2 : a bragging assertive statement

vaunt–cou·ri·er \vȯnt-'ku̇r-ē-ər, vänt-, -'kər-ē-, -'kə-rē-\ *noun* [Middle French *avant-courrier,* literally, advance courier] (1560)
archaic **:** FORERUNNER

vaunt·ed \'vȯn-təd, 'vän-\ *adjective* (1579)
: highly or widely praised or boasted about ⟨his own much *vaunted* ferocity —Calvin Tomkins⟩

vaunt·ful \'vȯnt-fəl, 'vänt-\ *adjective* (1590)
: VAINGLORIOUS, BOASTFUL

vaunty \'vȯn-tē, 'vän-\ *adjective* (1724)
Scottish **:** PROUD, BOASTFUL, VAIN

vav *variant of* WAW

vav·a·sor *or* **vav·a·sour** \'va-və-ˌsȯr, -ˌsōr, -ˌsu̇r\ *noun* [Middle English *vavasour,* from Middle French *vavassor,* probably from Medieval Latin *vassus vassorum* vassal of vassals] (14th century)
: a feudal tenant ranking directly below a baron

va·ward \'vau̇-ˌȯrd, -ˌwȯrd\ *noun* [Middle English *vauntwarde, vaward,* from Old North French *avantwarde,* from *avant* before (from Late Latin *abante*) + *warde* guard, from *warder* to guard — more at ADVANCE, REWARD] (1597)
archaic **:** the foremost part **:** FOREFRONT ⟨the *vaward* of our youth —Shakespeare⟩

VCR \ˌvē-(ˌ)sē-'är\ *noun* [*v*ideo*c*assette *r*ecorder] (1971)
: a videotape recorder that uses videocassettes

V–day \'vē-ˌdā\ *noun* [*v*ictory *day*] (1941)
: a day of victory

've \v, əv\ *verb* [by contraction]
: HAVE ⟨we'*ve* been there⟩

Ve·adar \'vā-ˌä-ˌdär, 'vā-ə-\ *noun* [Hebrew *wĕ-Adhār,* literally, and Adar (i.e., the second Adar)] (circa 1864)
: the intercalary month of the Jewish calendar following Adar in leap years — see MONTH table

[1]veal \'vē(ə)l\ *noun* [Middle English *veel,* from Middle French, from Latin *vitellus* small calf, diminutive of *vitulus* calf — more at WETHER] (14th century)
1 : the flesh of a young calf
2 : CALF; *especially* **:** VEALER

[2]veal *transitive verb* (1847)
: to kill and dress (a calf) for veal

veal·er \'vē-lər\ *noun* (circa 1895)
: a calf grown for or suitable for veal

vealy \'vē-lē\ *adjective* (1769)
1 : resembling or suggesting veal or a calf
2 : IMMATURE

[1]vec·tor \'vek-tər\ *noun* [New Latin, from Latin, carrier, from *vehere* to carry — more at WAY] (1846)
1 a : a quantity that has magnitude and direction and that is commonly represented by a directed line segment whose length represents the magnitude and whose orientation in space represents the direction; *broadly* **:** an element of a vector space **b :** a course or compass direction especially of an airplane
2 a : an organism (as an insect) that transmits a pathogen **b :** POLLINATOR a
3 : a sequence of genetic material (as a transposon or the genome of a bacteriophage) used to introduce specific genes into the genome of an organism
— **vector** *adjective*
— **vec·to·ri·al** \vek-'tōr-ē-əl, -'tȯr-\ *adjective*
— **vec·to·ri·al·ly** \-ə-lē\ *adverb*

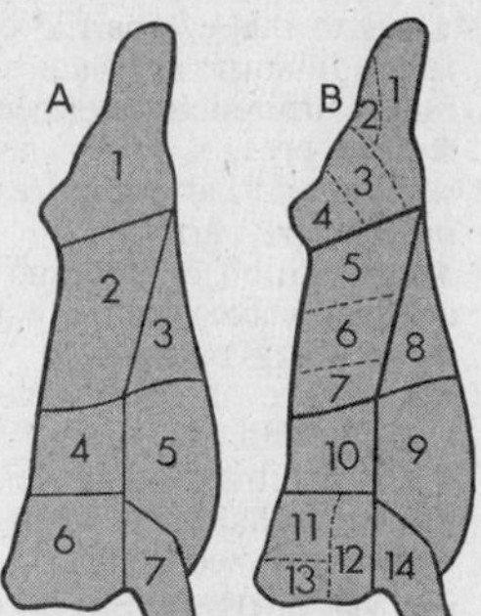

veal 1: ***A*** **wholesale cuts:** ***1*** **leg,** ***2*** **loin,** ***3*** **flank,** ***4*** **rib,** ***5*** **breast,** ***6*** **shoulder,** ***7*** **shank;** ***B*** **retail cuts:** ***1*** **hind shank,** ***2*** **heel of round,** ***3*** **round,** ***4*** **rump roast,** ***5*** **sirloin steak,** ***6*** **loin chops,** ***7*** **kidney chops,** ***8*** **flank,** ***9*** **breast,** ***10*** **rib roast,** ***11*** **blade steak,** ***12*** **arm steak,** ***13*** **shoulder roast,** ***14*** **foreshank**

[2]vector *transitive verb* **vec·tored; vec·tor·ing** \-t(ə-)riŋ\ (1941)
1 : to guide (as an airplane, its pilot, or a missile) in flight by means of a radioed vector
2 : to change the direction of (the thrust of a jet engine) for steering

vector product *noun* (1878)
: a vector *c* whose length is the product of the lengths of two vectors *a* and *b* and the sine of their included angle, whose direction is perpendicular to their plane, and whose direction is that in which a right-handed screw with axis *c* will move along *c* when *a* is rotated into *b* — called also *cross product*

vector space *noun* (1937)
: a set representing a generalization of a system of vectors and consisting of elements which comprise a commutative group under addition, each of which is left unchanged under multiplication by the multiplicative identity of a field, and for which multiplication by the multiplicative operation of the field is commutative, closed, distributive such that both $c(A + B) = cA + cB$ and $(c + d)A = cA + dA$, and associative such that $(cd)A = c(dA)$ where *A, B* are vectors and *c, d* are elements of the field

vector sum *noun* (circa 1890)
: the sum of a number of vectors that for the sum of two vectors is geometrically represented by the diagonal of a parallelogram whose sides represent the two vectors being added

Ve·da \'vā-də\ *noun* [Sanskrit, literally, knowledge; akin to Greek *eidenai* to know — more at WIT] (1734)
: any of four canonical collections of hymns, prayers, and liturgical formulas that comprise the earliest Hindu sacred writings

ve·da·lia \vi-'dāl-yə\ *noun* [New Latin, genus name] (1889)
: an Australian ladybug (*Rodolia cardinalis*) introduced to many countries to control scale insects

Ve·dan·ta \vā-'dän-tə, və-, -'dan-\ *noun* [Sanskrit *Vedānta,* literally, end of the Veda, from *Veda* + *anta* end; akin to Old English *ende* end] (1788)
: an orthodox system of Hindu philosophy developing especially in a qualified monism the speculations of the Upanishads on ultimate reality and the liberation of the soul
— **Ve·dan·tism** \-'dän-ˌti-zəm, -'dan-\ *noun*
— **Ve·dan·tist** \-'dän-tist, -'dan-\ *noun*

Ve·dan·tic \-'dän-tik, -'dan-\ *adjective* (1882)
1 : of or relating to the Vedanta philosophy
2 : VEDIC

Ved·da *or* **Ved·dah** \'ve-də\ *noun* [Sinhalese *vedda* hunter] (1681)
: a member of an aboriginal people of Sri Lanka

Ved·doid \'ve-ˌdȯid\ *noun* (1928)
: a member of an ancient race of southern Asia characterized by wavy to curly hair, chocolate-brown skin color, and slender body build
— **Veddoid** *adjective*

ve·dette \vi-'det\ *noun* [French, from Italian *vedetta,* alteration of *veletta,* probably from Spanish *vela* watch, from *velar* to keep watch, from Latin *vigilare* to wake, watch, from *vigil* awake — more at VIGIL] (circa 1702)
: a mounted sentinel stationed in advance of pickets

Ve·dic \'vā-dik\ *adjective* (1848)
: of or relating to the Vedas, the language in which they are written, or Hindu history and culture between 1500 B.C. and 500 B.C.

vee \'vē\ *noun* (circa 1883)
1 : something shaped like the letter V
2 : the letter *v*

vee·jay \'vē-ˌjā\ *noun* [video jockey] (1981)
: an announcer of a program (as on television) that features music videos

vee·na *variant of* VINA

veep \'vēp\ *noun* [from *v. p.* (abbreviation for *vice president*)] (1949)
: VICE PRESIDENT

1**veer** \'vir\ *transitive verb* [Middle English *veren,* of Low German or Dutch origin; akin to Middle Dutch *vieren* to slacken, Middle Low German *vīren*] (15th century)
: to let out (as a rope)

2**veer** *verb* [Middle French *virer,* from Old French, to throw with a twisting motion, perhaps modification of Latin *vibrare* to wave, propel suddenly — more at VIBRATE] (1582)
intransitive verb
1 *of the wind* **:** to shift in a clockwise direction — compare BACK
2 : to change direction or course
3 : to wear ship
transitive verb
: to direct to a different course; *specifically* **:** WEAR 7
synonym see SWERVE
— **veer·ing·ly** \-iŋ-lē\ *adverb*

3**veer** *noun* (1611)
: a change in course or direction ⟨a *veer* to the right⟩

vee·ry \'vir-ē\ *noun, plural* **veeries** [probably imitative] (1838)
: a thrush (*Catharus fuscescens*) common in the eastern U.S.

veg \'vej\ *noun, plural* **veg** (1918)
chiefly British **:** VEGETABLE

Ve·ga \'vē-gə, 'vā-\ *noun* [New Latin, from Arabic (*al-Nasr*) *al-Wāqiʻ*, literally, the falling (vulture)]
: a star of the first magnitude that is the brightest in the constellation Lyra

veg·an \'vē-gən *also* 'vā- *also* 've-jən *or* -ˌjan\ *noun* [by contraction from *vegetarian*] (1944)
: a strict vegetarian who consumes no animal food or dairy products; *also* **:** one who abstains from using animal products (as leather)
— **veg·an·ism** \'vē-gə-ˌni-zəm, 'vā-gə-, 've-jə-\ *noun*

1**veg·e·ta·ble** \'vej-tə-bəl, 've-jə-\ *adjective* [Middle English, from Medieval Latin *vegetabilis* vegetative, from *vegetare* to grow, from Latin, to animate, from *vegetus* lively, from *vegēre* to enliven — more at WAKE] (15th century)
1 a : of, relating to, constituting, or growing like plants **b :** consisting of plants **:** VEGETATIONAL
2 : made or obtained from plants or plant products
3 : resembling or suggesting a plant (as in inertness or passivity)

2**vegetable** *noun* (1582)
1 : PLANT 1b
2 : a usually herbaceous plant (as the cabbage, bean, or potato) grown for an edible part that is usually eaten as part of a meal; *also* **:** such edible part
3 : a human being whose mental and physical functioning is severely impaired (as from accident or disease)

vegetable ivory *noun* (1842)
1 : the hard white opaque endosperm of the ivory nut that takes a high polish and is used as a substitute for ivory
2 : IVORY NUT

vegetable marrow *noun* (1816)
: any of various smooth-skinned elongated summer squashes with creamy white to deep green skins

vegetable oil *noun* (1797)
: an oil of plant origin; *especially* **:** a fatty oil from seeds or fruits

vegetable oyster *noun* (circa 1818)
: SALSIFY

vegetable pear *noun* (1887)
: CHAYOTE

vegetable wax *noun* (1815)
: a wax of plant origin secreted commonly in thin flakes by the walls of epidermal cells

veg·e·ta·bly \'vej-tə-blē, 've-jə-\ *adverb* (1651)
: in the manner of or like a vegetable

veg·e·tal \'ve-jə-tᵊl\ *adjective* [Medieval Latin *vegetare* to grow] (15th century)
1 : VEGETABLE
2 : VEGETATIVE
3 : of or relating to the vegetal pole of an egg or to that part of an egg from which the endoderm normally develops ⟨*vegetal* blastomeres⟩

vegetal pole *noun* (1896)
: the point on the surface of an egg that is diametrically opposite to the animal pole and usually marks the center of the protoplasm containing more yolk, dividing more slowly and into larger blastomeres than that about the animal pole, and giving rise to the hypoblast of the embryo

1**veg·e·tar·i·an** \ˌve-jə-'ter-ē-ən\ *noun* [2*vegetable* + *-arian*] (1839)
1 : one who believes in or practices vegetarianism
2 : HERBIVORE

2**vegetarian** *adjective* (1849)
1 : of or relating to vegetarians
2 : consisting wholly of vegetables, fruits, and sometimes eggs or dairy products ⟨a *vegetarian* diet⟩

veg·e·tar·i·an·ism \-ē-ə-ˌni-zəm\ *noun* (circa 1851)
: the theory or practice of living on a diet made up of vegetables, fruits, grains, nuts, and sometimes eggs or dairy products

veg·e·tate \'ve-jə-ˌtāt\ *verb* **-tat·ed; -tat·ing** [Medieval Latin *vegetatus,* past participle of *vegetare* to grow] (1605)
intransitive verb
1 a : to grow in the manner of a plant; *also* **:** to grow exuberantly or with proliferation of fleshy or warty outgrowths **b :** to produce vegetation
2 : to lead a passive existence without exertion of body or mind
transitive verb
: to establish vegetation in or on

veg·e·ta·tion \ˌve-jə-'tā-shən\ *noun* (1564)
1 : the act or process of vegetating
2 : inert existence
3 : plant life or total plant cover (as of an area)
4 : an abnormal growth upon a body part
— **veg·e·ta·tion·al** \-shnəl, -shə-nᵊl\ *adjective*

veg·e·ta·tive \'ve-jə-ˌtā-tiv\ *adjective* (14th century)
1 a (1) **:** growing or having the power of growing (2) **:** of, relating to, or engaged in nutritive and growth functions as contrasted with reproductive functions ⟨a *vegetative* nucleus⟩ **b :** promoting plant growth ⟨the *vegetative* properties of soil⟩ **c :** of, relating to, or involving propagation by nonsexual processes or methods
2 : relating to, composed of, or suggesting vegetation ⟨*vegetative* cover⟩
3 : of or relating to the division of nature comprising the plant kingdom
4 : affecting, arising from, or relating to involuntary bodily functions
5 : VEGETABLE 3
— **veg·e·ta·tive·ly** *adverb*
— **veg·e·ta·tive·ness** *noun*

ve·gete \və-'jēt\ *adjective* [Latin *vegetus* — more at VEGETABLE] (1639)
archaic **:** LIVELY, HEALTHY

veg·gie *also* **veg·ie** \'ve-jē\ *noun* [by shortening & alteration] (1955)
1 : VEGETABLE
2 *slang* **:** VEGETARIAN

veg out \'vej-\ *intransitive verb* **vegged out; veg·ging out** [short for *vegetate*] (1980)
: to spend time idly or passively

ve·he·mence \'vē-ə-mən(t)s\ *noun* (15th century)
: the quality or state of being vehement **:** INTENSITY

ve·he·ment \-mənt\ *adjective* [Middle English, from Middle French, from Latin *vehement-, vehemens, vement-, vemens*] (15th century)
: marked by forceful energy **:** POWERFUL ⟨a *vehement* wind⟩: as **a :** intensely emotional **:** IMPASSIONED, FERVID ⟨*vehement* patriotism⟩ **b** (1) **:** deeply felt ⟨a *vehement* suspicion⟩ (2) **:** forcibly expressed ⟨*vehement* denunciations⟩ **c :** bitterly antagonistic ⟨a *vehement* debate⟩
— **ve·he·ment·ly** *adverb*

ve·hi·cle \'vē-ə-kəl *also* 'vē-ˌhi-kəl\ *noun* [French *véhicule,* from Latin *vehiculum* carriage, conveyance, from *vehere* to carry — more at WAY] (1612)
1 a : an inert medium (as a syrup) in which a medicinally active agent is administered **b :** any of various media acting usually as solvents, carriers, or binders for active ingredients or pigments
2 : an agent of transmission **:** CARRIER
3 : a medium through which something is expressed, achieved, or displayed
4 : a means of carrying or transporting something **:** CONVEYANCE: as **a :** MOTOR VEHICLE **b :** a piece of mechanized equipment

ve·hic·u·lar \vē-'hi-kyə-lər\ *adjective* (1616)
1 a : of, relating to, or designed for vehicles and especially motor vehicles **b :** transported by vehicle **c :** caused by or resulting from the operation of a vehicle ⟨*vehicular* homicide⟩
2 : serving as a vehicle

V–8 \'vē-'āt\ *noun* (1942)
: an internal combustion engine having two banks of four cylinders each with the banks at an angle to each other; *also* **:** an automobile having such an engine

1**veil** \'vā(ə)l\ *noun* [Middle English *veile,* from Old North French, from Latin *vela,* plural of *velum* sail, awning, curtain] (13th century)
1 a : a length of cloth worn by women as a covering for the head and shoulders and often especially in Eastern countries for the face; *specifically* **:** the outer covering of a nun's headdress **b :** a length of veiling or netting worn over the head or face or attached for protection or ornament to a hat or headdress **c :** any of various liturgical cloths; *especially* **:** a cloth used to cover the chalice
2 : the cloistered life of a nun

\ə\ **abut** \ᵊ\ kitten \ər\ **further** \a\ **ash** \ā\ **ace** \ä\ **mop, mar** \au̇\ **out** \ch\ **chin** \e\ **bet** \ē\ **easy** \g\ **go** \i\ **hit** \ī\ **ice** \j\ **job** \ŋ\ **sing** \ō\ **go** \ȯ\ **law** \ȯi\ **boy** \th\ **thin** \t͟h\ **the** \ü\ **loot** \u̇\ **foot** \y\ **yet** \zh\ **vision** *see also* Guide to Pronunciation

3 : a concealing curtain or cover of cloth
4 : something that hides or obscures like a veil
5 : a covering body part or membrane: as **a :** VELUM **b :** CAUL

²veil (14th century)
transitive verb
: to cover, provide, obscure, or conceal with or as if with a veil
intransitive verb
: to put on or wear a veil

veiled \ˈvā(ə)ld\ *adjective* (1593)
1 a : having or wearing a veil or a concealing cover ⟨a *veiled* hat⟩ **b :** characterized by a softening tonal distortion
2 : obscured as if by a veil **:** DISGUISED ⟨*veiled* threats⟩

veil·ing \ˈvā-liŋ\ *noun* (14th century)
1 : VEIL
2 : any of various light sheer fabrics

¹vein \ˈvān\ *noun* [Middle English *veine,* from Old French, from Latin *vena*] (14th century)
1 a : a narrow water channel in rock or earth or in ice **b** (1) **:** LODE 2, 3 (2) **:** a bed of useful mineral matter
2 : BLOOD VESSEL; *especially* **:** any of the tubular branching vessels that carry blood from the capillaries toward the heart
3 a : any of the vascular bundles forming the framework of a leaf **b :** any of the thickened cuticular ribs that serve to stiffen the wings of an insect
4 : something suggesting veins (as in reticulation); *specifically* **:** a wavy variegation (as in marble)
5 a : a distinctive mode of expression **:** STYLE **b :** a pervasive element or quality **:** STRAIN **c :** a line of thought or action
6 a : a special aptitude **:** TALENT **b :** a usually transitory and casually attained mood **c :** top form **:** FETTLE
— vein·al \ˈvā-nᵊl\ *adjective*

²vein *transitive verb* (1502)
: to pattern with or as if with veins

veined \ˈvānd\ *adjective* (circa 1529)
: patterned with or as if with veins **:** having venation **:** STREAKED ⟨a *veined* leaf⟩ ⟨*veined* marble⟩ ⟨*veined* cheese⟩

vein·er \ˈvā-nər\ *noun* (1895)
: a small V gouge used in wood carving

vein·ing \ˈvā-niŋ\ *noun* (1826)
: a pattern of veins **:** VENATION

vein·let \ˈvān-lət\ *noun* (1831)
: a small vein

veiny \ˈvā-nē\ *adjective* (1611)
: full of veins **:** noticeably veined

ve·la·men \və-ˈlā-mən\ *noun, plural* **ve·lam·i·na** \-ˈla-mə-nə\ [New Latin, from Latin, covering, from *velare* to cover, from *velum* curtain] (1882)
: the thick corky epidermis of aerial roots of an epiphytic orchid that absorbs water from the atmosphere

ve·lar \ˈvē-lər\ *adjective* [New Latin *velaris,* from *velum*] (1876)
1 : formed with the back of the tongue touching or near the soft palate ⟨the *velar* \k\ of \ˈkül\ *cool*⟩
2 : of, forming, or relating to a velum and especially the soft palate
— velar *noun*

ve·lar·i·um \vi-ˈlar-ē-əm, -ˈler-\ *noun, plural* **-ia** \-ē-ə\ [Latin, from *velum* curtain] (1834)
: an awning over an ancient Roman theater or amphitheater

ve·lar·i·za·tion \ˌvē-lə-rə-ˈzā-shən\ *noun* (1915)
1 : the quality or state of being velarized
2 : an act or instance of velarizing

ve·lar·ize \ˈvē-lə-ˌrīz\ *transitive verb* **-ized; -iz·ing** (1915)
: to modify (as the \l\ of \ˈpül\ *pool*) by a simultaneous velar articulation

Vel·cro \ˈvel-(ˌ)krō\ *trademark*
— used for a closure consisting of a piece of fabric of small hooks that sticks to a corresponding fabric of small loops

veld *or* **veldt** \ˈvelt, ˈfelt\ *noun* [Afrikaans *veld,* from Dutch, field; akin to Old English *feld* field] (1835)
: a grassland especially of southern Africa usually with scattered shrubs or trees

ve·li·ger \ˈvē-lə-jər, ˈve-\ *noun* [New Latin, from *velum* + *-ger* bearing, from *gerere* to bear] (1877)
: a larval mollusk in the stage when it has developed the velum

vel·le·i·ty \ve-ˈlē-ə-tē, və-\ *noun, plural* **-ties** [New Latin *velleitas,* from Latin *velle* to wish, will — more at WILL] (1618)
1 : the lowest degree of volition
2 : a slight wish or tendency **:** INCLINATION

¹vel·lum \ˈve-ləm\ *noun* [Middle English *velim,* from Middle French *veelin,* from *veelin,* adjective, of a calf, from *veel* calf — more at VEAL] (15th century)
1 : a fine-grained unsplit lambskin, kidskin, or calfskin prepared especially for writing on or for binding books
2 : a strong cream-colored paper

²vellum *adjective* (1565)
1 : of, resembling, or bound in vellum
2 : slightly rough ⟨paper with a *vellum* finish⟩

ve·lo·ce \vā-ˈlō-(ˌ)chā\ *adverb or adjective* [Italian, from Latin *veloc-, velox*] (circa 1823)
: in a rapid manner — used as a direction in music

ve·lo·cim·e·ter \ˌvē-lō-ˈsi-mə-tər, ˌve-\ *noun* [*velocity* + *-meter*] (1842)
: a device for measuring speed (as of fluid flow or sound)

ve·loc·i·pede \və-ˈlä-sə-ˌpēd\ *noun* [French *vélocipède,* from Latin *veloc-, velox* + *ped-, pes* foot — more at FOOT] (1818)
: a lightweight wheeled vehicle propelled by the rider: as **a** *archaic* **:** BICYCLE **b :** TRICYCLE **c :** a 3-wheeled railroad handcar

ve·loc·i·ty \və-ˈlä-sə-tē, -ˈläs-tē\ *noun, plural* **-ties** [Middle French *velocité,* from Latin *velocitat-, velocitas,* from *veloc-, velox* quick; probably akin to Latin *vegēre* to enliven — more at WAKE] (circa 1550)
1 a : quickness of motion **:** SPEED ⟨the *velocity* of sound⟩ **b :** rapidity of movement ⟨[my horse's] strong suit is grace & personal comeliness, rather than *velocity* —Mark Twain⟩ **c :** speed imparted to something ⟨the power pitcher relies on *velocity* —Tony Scherman⟩
2 : the rate of change of position along a straight line with respect to time **:** the derivative of position with respect to time
3 a : rate of occurrence or action **:** RAPIDITY ⟨the *velocity* of historical change —R. J. Lifton⟩ ⟨the narrative leaps from one frantic episode to another with impressive *velocity* —James Atlas⟩ **b :** rate of turnover ⟨the *velocity* of money⟩

ve·lo·drome \ˈvē-lə-ˌdrōm, ˈve-, ˈvā-\ *noun* [French *vélodrome,* from *vélo* cycle (short for *vélocipède*) + *-drome*] (1895)
: a track designed for cycling

ve·lour *or* **ve·lours** \və-ˈlu̇r\ *noun, plural* **ve·lours** \-ˈlu̇rz\ [French *velours* velvet, velour, from Middle French *velours, velour,* from Old French *velous,* from Old Provençal *velos,* from Latin *villosus* shaggy, from *villus* shaggy hair — more at VELVET] (circa 1706)
1 : any of various fabrics with a pile or napped surface resembling velvet used in heavy weights for upholstery and curtains and in lighter weights for clothing; *also* **:** the article of clothing itself
2 : a fur felt (as of rabbit or nutria) finished with a long velvety nap and used especially for hats

ve·lou·té \və-ˌlü-ˈtā\ *noun* [French, literally, velvetiness, from Middle French *velluté,* from Old Provençal *velut* velvety, from (assumed) Vulgar Latin *villutus*] (1830)
: a white sauce made of chicken, veal, or fish stock and cream and thickened with butter and flour

ve·lum \ˈvē-ləm\ *noun, plural* **ve·la** \-lə\ [New Latin, from Latin, curtain] (1753)
1 : a membrane or membranous part resembling a veil or curtain: as **a :** SOFT PALATE **b :** an annular membrane projecting inward from the margin of the umbrella in some jellyfishes (as the hydromedusans)
2 : a swimming organ that is especially well developed in the later larval stages of many marine gastropods

ve·lure \ve-ˈlu̇r, vel-ˈyu̇r, ˈvel-yər\ *noun* [modification of Middle French *velour*] (1587)
obsolete **:** VELVET; *also* **:** a fabric resembling velvet

¹vel·vet \ˈvel-vət\ *noun* [Middle English *veluet, velvet,* from Middle French *velu* shaggy, from (assumed) Vulgar Latin *villutus,* from Latin *villus* shaggy hair; akin to Latin *vellus* fleece — more at WOOL] (14th century)
1 : a clothing and upholstery fabric (as of silk, rayon, or wool) characterized by a short soft dense warp pile
2 a : something suggesting velvet **b :** a characteristic (as softness or smoothness) of velvet
3 : the soft vascular skin that envelops and nourishes the developing antlers of deer
4 a : the winnings of a player in a gambling game **b :** a profit or gain beyond ordinary expectation
— vel·vet·like \-ˌlīk\ *adjective*

²velvet *adjective* (14th century)
1 : made of or covered with velvet; *also* **:** clad in velvet
2 : resembling or suggesting velvet **:** VELVETY ⟨a *velvet* voice⟩

velvet ant *noun* (1748)
: any of various solitary usually brightly colored and hairy fossorial wasps (family Mutillidae) with the female wingless

velvet bean *noun* (1898)
: an annual legume (*Mucuna deeringiana* synonym *Stizolobium deeringianum*) grown especially in the southern U.S. for green manure and grazing; *also* **:** its seed often used as stock feed

vel·ve·teen \ˌvel-və-ˈtēn\ *noun* (1776)
1 : a clothing fabric usually of cotton in twill or plain weaves made with a short close weft pile in imitation of velvet
2 *plural* **:** clothes made of velveteen

vel·vety \ˈvel-və-tē\ *adjective* (1752)
1 : soft and smooth like velvet ⟨*velvety* hair⟩
2 : smooth to the taste **:** MILD ⟨a *velvety* wine⟩

ven- *or* **veni-** *or* **veno-** *combining form* [Latin *vena*]
: vein ⟨*veni*puncture⟩ ⟨*veno*graphy⟩

ve·na \ˈvē-nə\ *noun, plural* **ve·nae** \-(ˌ)nē\ [Middle English, from Latin] (14th century)
: VEIN

ve·na ca·va \ˌvē-nə-ˈkā-və\ *noun, plural* **ve·nae ca·vae** \-(ˌ)nē-ˈkā-(ˌ)vē\ [New Latin, literally, hollow vein] (1598)
: any of the large veins by which in air-breathing vertebrates the blood is returned to the right atrium of the heart
— vena ca·val \-vəl\ *adjective*

ve·nal \ˈvē-nᵊl\ *adjective* [Latin *venalis,* from *venum* (accusative) sale; akin to Greek *ōneisthai* to buy, Sanskrit *vasna* price] (1652)
1 : capable of being bought or obtained for money or other valuable consideration **:** PURCHASABLE; *especially* **:** open to corrupt influence and especially bribery **:** MERCENARY ⟨a *venal* legislator⟩
2 : originating in, characterized by, or associated with corrupt bribery ⟨a *venal* arrangement with the police⟩
— ve·nal·i·ty \vi-ˈna-lə-tē\ *noun*
— ve·nal·ly \ˈvē-nᵊl-ē\ *adverb*

ve·na·tion \ve-ˈnā-shən, vē-\ *noun* [Latin *vena* vein] (1646)
: an arrangement or system of veins (as in the tissue of a leaf or the wing of an insect)

1 2 3

venation a: *1* parallel-veined, *2* net-veined, *3* dichotomously veined

vend \ˈvend\ *verb* [Latin *vendere* to sell, transitive verb, contraction for *venum dare* to give for sale] (1622)
intransitive verb
: to dispose of something by sale **:** SELL; *also* **:** to engage in selling
transitive verb
1 a : to sell especially as a hawker or peddler **b :** to sell by means of vending machines
2 : to utter publicly

Ven·da \ˈven-də\ *noun* (1908)
: a Bantu language spoken by a people of Northern Transvaal; *also* **:** a member of this people

vend·ee \ven-ˈdē\ *noun* (1547)
: one to whom a thing is sold **:** BUYER

ven·det·ta \ven-ˈde-tə\ *noun* [Italian, literally, revenge, from Latin *vindicta* — more at VINDICTIVE] (1855)
1 : BLOOD FEUD
2 : an often prolonged series of retaliatory, vengeful, or hostile acts or exchange of such acts ⟨waged a personal *vendetta* against those who opposed his nomination⟩

ven·deuse \vän^{n}-ˈdə(r)z, vän-ˈdüz\ *noun* [French, feminine of *vendeur* salesman, from Middle French] (1913)
: a saleswoman especially in the fashion industry

vend·ible *or* **vend·able** \ˈven-də-bəl\ *adjective* (14th century)
: capable of being vended **:** SALABLE
— vend·ibil·i·ty \ˌven-də-ˈbi-lə-tē\ *noun*

vending machine *noun* (circa 1895)
: a coin-operated machine for selling merchandise

ven·dor \ˈven-dər, *for 1 also* ven-ˈdȯr\ *also* **vend·er** \-dər\ *noun* [Middle French *vendeur,* from *vendre* to sell, from Latin *vendere*] (1594)
1 : one that vends **:** SELLER
2 : VENDING MACHINE

ven·due \ˈven-ˌdü, ˈvän-, ˈfen-, -ˌdyü; ven-ˈ, vän-ˈ\ *noun* [obsolete French, from Middle French, from *vendre*] (1668)
: a public sale at auction

[1]**ve·neer** \və-ˈnir\ *noun* [German *Furnier,* from *furnieren* to veneer, from French *fournir* to furnish, equip — more at FURNISH] (1702)
1 : a thin sheet of a material: as **a :** a layer of wood of superior value or excellent grain to be glued to an inferior wood **b :** any of the thin layers bonded together to form plywood
2 : a protective or ornamental facing (as of brick or stone)
3 : a superficial or deceptively attractive appearance, display, or effect **:** FACADE, GLOSS

[2]**veneer** *transitive verb* (1720)
1 : to overlay or plate (as a common wood) with a thin layer of finer wood for outer finish or decoration; *broadly* **:** to face with a material giving a superior surface
2 : to cover over with a veneer; *especially* **:** to conceal (as a defect of character) under a superficial and deceptive attractiveness
— ve·neer·er *noun*

ve·neer·ing *noun* (circa 1706)
1 : a veneered surface
2 : material used as veneer

ven·er·a·ble \ˈve-nər(-ə)-bəl, ˈven-rə-bəl\ *adjective* (15th century)
1 : deserving to be venerated — used as a title for an Anglican archdeacon or for a Roman Catholic who has been accorded the lowest of three degrees of recognition for sanctity
2 : made sacred especially by religious or historical association
3 a : calling forth respect through age, character, and attainments; *broadly* **:** conveying an impression of aged goodness and benevolence **b :** impressive by reason of age ⟨under *venerable* pines⟩
synonym see OLD
— ven·er·a·bil·i·ty \ˌve-nə-rə-ˈbi-lə-tē, ˌven-rə-\ *noun*
— ven·er·a·ble·ness \ˈve-nər(-ə)-bəl-nəs, ˈven-rə-\ *noun*
— ven·er·a·bly \-blē\ *adverb*

ven·er·ate \ˈve-nə-ˌrāt\ *transitive verb* **-at·ed; -at·ing** [Latin *veneratus,* past participle of *venerari,* from *vener-, venus* love, charm — more at WIN] (circa 1623)
1 : to regard with reverential respect or with admiring deference
2 : to honor (as an icon or a relic) with a ritual act of devotion
synonym see REVERE
word history see VENOM
— ven·er·a·tor \-ˌrā-tər\ *noun*

ven·er·a·tion \ˌve-nə-ˈrā-shən\ *noun* (15th century)
1 : respect or awe inspired by the dignity, wisdom, dedication, or talent of a person
2 : the act of venerating
3 : the condition of one that is venerated

ve·ne·re·al \və-ˈnir-ē-əl\ *adjective* [Middle English *venereall,* from Latin *venereus,* from *vener-, venus* love, sexual desire] (15th century)
1 : of or relating to sexual pleasure or indulgence
2 a : resulting from or contracted during sexual intercourse ⟨*venereal* infections⟩ **b :** of, relating to, or affected with venereal disease ⟨a high *venereal* rate⟩ **c :** involving the genital organs ⟨*venereal* sarcoma⟩

venereal disease *noun* (1658)
: a contagious disease (as gonorrhea or syphilis) that is typically acquired in sexual intercourse — compare STD

[1]**ven·ery** \ˈve-nə-rē\ *noun* [Middle English *venerie,* from Middle French, from *vener* to hunt, from Latin *venari* — more at VENISON] (14th century)
1 : the art, act, or practice of hunting
2 : animals that are hunted **:** GAME

[2]**venery** *noun* [Middle English *venerie,* from Medieval Latin *veneria,* from Latin *vener-, venus* sexual desire] (15th century)
1 : the pursuit of or indulgence in sexual pleasure
2 : SEXUAL INTERCOURSE

ve·ne·sec·tion \ˈve-nə-ˌsek-shən, ˈvē-\ *noun* [New Latin *venae section-, venae sectio,* literally, cutting of a vein] (1661)
: the operation of opening a vein for letting blood **:** PHLEBOTOMY

Ven·e·ti \ˈve-nə-ˌtī\ *noun plural* [Latin *Veneti*] (1881)
1 : an ancient people in Gaul conquered by Caesar in 56 B.C.
2 : an ancient people in northeastern Italy allied politically to the Romans

ve·ne·tian blind \və-ˈnē-shən-\ *noun* [*Venetian* of Venice, Italy] (1770)
: a blind (as for a window) having numerous horizontal slats that may be set simultaneously at any of several angles so as to vary the amount of light admitted

venetian glass *noun, often V capitalized* (circa 1845)
: often colored glassware made at Murano near Venice of a soda-lime metal and typically elaborately decorated (as with gilt, enamel, or engraving)

Venetian red *noun* (circa 1753)
: an earthy hematite used as a pigment; *also* **:** a synthetic iron oxide pigment

Ve·net·ic \və-ˈne-tik\ *noun* [Latin *veneticus* of the Veneti, from *Veneti*] (1902)
: the language of the ancient Veneti of Italy — see INDO-EUROPEAN LANGUAGES table
— Venetic *adjective*

venge \ˈvenj\ *transitive verb* **venged; veng·ing** [Middle English, from Middle French *vengier,* from Old French] (14th century)
archaic **:** AVENGE

ven·geance \ˈven-jən(t)s\ *noun* [Middle English, from Old French, from *vengier* to avenge, from Latin *vindicare* to lay claim to, avenge — more at VINDICATE] (14th century)
: punishment inflicted in retaliation for an injury or offense **:** RETRIBUTION
— with a vengeance 1 : with great force or vehemence **2 :** to an extreme or excessive degree

venge·ful \ˈvenj-fəl\ *adjective* [obsolete English *venge* revenge] (1586)
: REVENGEFUL: as **a :** seeking to avenge **b :** serving to gain vengeance
— venge·ful·ly \-fə-lē\ *adverb*
— venge·ful·ness *noun*

V–en·gine \ˈvē-\ *noun* (circa 1922)
: an internal combustion engine whose cylinders are arranged in two banks forming an acute or right angle

veni- *or* **veno-** — see VEN-

ve·nial \ˈvē-nē-əl, -nyəl\ *adjective* [Middle English, from Middle French, from Late Latin *venialis,* from Latin *venia* favor, indulgence, pardon; akin to Latin *venus* love, charm — more at WIN] (14th century)
: of a kind that can be remitted **:** FORGIVABLE, PARDONABLE; *also* **:** meriting no particular censure or notice **:** EXCUSABLE ⟨*venial* faults⟩
— ve·nial·ly *adverb*
— ve·nial·ness *noun*

venial sin *noun* (14th century)
: a sin that is relatively slight or that is committed without full reflection or consent and so according to Thomist theology does not deprive the soul of sanctifying grace — compare MORTAL SIN

ve·ni·punc·ture \ˈvē-nə-ˌpəŋ(k)-chər, ˈve-\ *noun* (circa 1903)
: surgical puncture of a vein especially for the withdrawal of blood or for intravenous medication

ve·ni·re \və-ˈnī-rē\ *noun* [*venire facias*] (1665)
: an entire panel from which a jury is drawn

ve·ni·re fa·ci·as \-ˌnī-rē-ˈfā-shē-əs\ *noun* [Middle English, from Medieval Latin, you should cause to come] (15th century)
: a judicial writ directing the sheriff to summon a specified number of qualified persons to serve as jurors

ve·ni·re·man \və-ˈnī-rē-mən, -ˈnir-ē-\ *noun* (1776)
: a member of a venire

ven·i·son \ˈve-nə-sən *also* -zən, *British usually* ˈven-zən\ *noun, plural* **venisons** *also* **venison** [Middle English, from Old French *veneison* hunting, game, from Latin *venation-, venatio,* from *venari* to hunt, pursue, akin to Sanskrit *vanoti* he strives for — more at WIN] (14th century)
: the edible flesh of a game animal and especially a deer

Ve·ni·te \və-ˈnī-tē, -ˈnē-ˌtā\ *noun* [Middle English, from Latin, O come, from *venire* to come; from the opening word of Psalms 95:1 — more at COME] (13th century)
: a liturgical chant composed of parts of Psalms 95 and 96

Venn diagram \'ven-\ *noun* [John *Venn* (died 1923) English logician] (1918)
: a graph that employs closed curves and especially circles to represent logical relations between and operations on sets and the terms of propositions by the inclusion, exclusion, or intersection of the circles

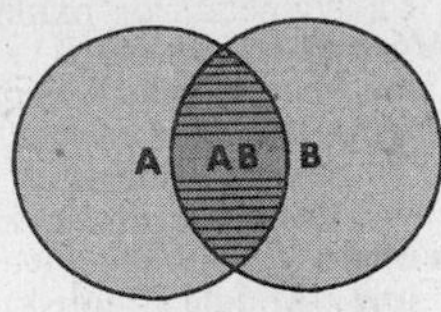

Venn diagram: *AB* represents the intersection of sets *A* and *B*

ve·nog·ra·phy \vi-'nä-grə-fē, vā-\ *noun* [International Scientific Vocabulary] (1935)
: roentgenography of a vein after injection of an opaque substance

¹ven·om \'ve-nəm\ *noun* [Middle English *venim, venom,* from Old French *venim,* from (assumed) Vulgar Latin *venimen,* alteration of Latin *venenum* magic charm, drug, poison; akin to Latin *venus* love, charm — more at WIN] (13th century)
1 : poisonous matter normally secreted by some animals (as snakes, scorpions, or bees) and transmitted to prey or an enemy chiefly by biting or stinging; *broadly* **:** material that is poisonous
2 : ILL WILL, MALEVOLENCE ◇

²venom *transitive verb* (14th century)
: ENVENOM

ven·om·ous \'ve-nə-məs\ *adjective* (14th century)
1 : full of venom: as **a :** POISONOUS, ENVENOMED **b :** NOXIOUS, PERNICIOUS ⟨expose a *venomous* dope ring —Don Porter⟩ **c :** SPITEFUL, MALEVOLENT ⟨*venomous* criticism⟩
2 : having a venom-producing gland and able to inflict a poisoned wound ⟨*venomous* snakes⟩
— ven·om·ous·ly *adverb*
— ven·om·ous·ness *noun*

ve·nous \'vē-nəs\ *adjective* [Latin *venosus,* from *vena* vein] (1626)
1 : of, relating to, or full of veins ⟨a *venous* rock⟩ ⟨a *venous* system⟩
2 *of blood* **:** having passed through the capillaries and given up oxygen for the tissues and become charged with carbon dioxide
— ve·nous·ly *adverb*

¹vent \'vent\ *transitive verb* [Middle English, probably from Middle French *esventer* to expose to the air, from *es-* ex- (from Latin *ex-*) + *vent* wind, from Latin *ventus* — more at WIND] (14th century)
1 : to provide with a vent
2 a : to serve as a vent for ⟨chimneys *vent* smoke⟩ **b :** DISCHARGE, EXPEL **c :** to give often vigorous or emotional expression to
3 : to relieve by means of a vent
synonym see EXPRESS

²vent *noun* (1508)
1 : an opportunity or means of escape, passage, or release **:** OUTLET ⟨finally gave *vent* to his pent-up hostility⟩
2 : an opening for the escape of a gas or liquid or for the relief of pressure: as **a :** the external opening of the rectum or cloaca **:** ANUS **b :** PIPE 3c, FUMAROLE **c :** an opening at the breech of a muzzle-loading gun through which fire is touched to the powder **d** *chiefly Scottish* **:** CHIMNEY, FLUE
— vent·less \-ləs\ *adjective*

³vent *noun* [Middle English *vente,* alteration of *fente,* from Middle French, slit, fissure, from *fendre* to split, from Latin *findere* — more at BITE] (15th century)
: a slit in a garment; *specifically* **:** an opening in the lower part of a seam (as of a jacket or skirt)
— vent·less *adjective*

vent·age \'ven-tij\ *noun* (1602)
: a small hole (as a flute stop)

ven·tail \'ven-,tāl\ *noun* [Middle English, from Middle French *ventaille,* from *vent* wind] (15th century)
: the lower movable front of a medieval helmet

ven·ter \'ven-tər\ *noun* [Anglo-French, from Latin, belly, womb; perhaps akin to Old High German *wanast* paunch, Latin *vesica* bladder, Sanskrit *vasti*] (1544)
1 : a wife or mother that is a source of offspring
2 : a protuberant and often hollow anatomical structure: as **a :** the undersurface of the abdomen of an arthropod **b :** the swollen basal portion of an archegonium in which an egg develops

ven·ti·fact \'ven-tə-,fakt\ *noun* [Latin *ventus* + English arti*fact*] (1911)
: a stone worn, polished, or faceted by windblown sand

ven·ti·late \'ven-tᵊl-,āt\ *transitive verb* **-lat·ed; -lat·ing** [Late Latin *ventilatus,* past participle of *ventilare,* from Latin, to fan, winnow, from *ventus* wind — more at WIND] (1527)
1 a : to examine, discuss, or investigate freely and openly **:** EXPOSE ⟨*ventilating* family quarrels in public⟩ **b :** to make public **:** UTTER ⟨*ventilated* their objections at length⟩
2 *archaic* **:** to free from chaff by winnowing
3 a : to expose to air and especially to a current of fresh air for purifying, curing, or refreshing ⟨*ventilate* stored grain⟩; *also* **:** OXYGENATE, AERATE ⟨*ventilate* blood in the lungs⟩ **b :** to subject the lungs to ventilation ⟨artificially *ventilate* a patient in respiratory distress⟩
4 a *of a current of air* **:** to pass or circulate through so as to freshen **b :** to cause fresh air to circulate through (as a room or mine)
5 : to provide an opening in (a burning structure) to permit escape of smoke and heat

ven·ti·la·tion \,ven-tᵊl-'ā-shən\ *noun* (1519)
1 : the act or process of ventilating
2 a : circulation of air ⟨a room with good *ventilation*⟩ **b :** the circulation and exchange of gases in the lungs or gills that is basic to respiration
3 : a system or means of providing fresh air

ven·ti·la·tor \'ven-tᵊl-,ā-tər\ *noun* (1743)
: one that ventilates: as **a :** a contrivance for introducing fresh air or expelling foul or stagnant air **b :** RESPIRATOR 2

ven·ti·la·to·ry \'ven-tᵊl-ə-,tōr-ē, -,tȯr-\ *adjective* (1850)
: of, relating to, or provided with ventilation ⟨*ventilatory* capacity of the lung⟩

ventr- *or* **ventro-** *combining form* [Latin *ventr-, venter* belly]
: ventral and ⟨*ventro*lateral⟩

¹ven·tral \'ven-trəl\ *adjective* [French, from Latin *ventralis,* from *ventr-, venter*] (1739)
1 a : of or relating to the belly **:** ABDOMINAL **b :** being or located near or on the anterior or lower surface of an animal opposite the back
2 : being or located on the lower surface of a dorsiventral plant structure
— ven·tral·ly \-trə-lē\ *adverb*

²ventral *noun* (1834)
: a ventral part (as a scale or fin)

ventral root *noun* (circa 1923)
: the one of the two roots of a spinal nerve that passes ventrally from the spinal cord and consists of motor fibers — compare DORSAL ROOT

ven·tri·cle \'ven-tri-kəl\ *noun* [Middle English, from Latin *ventriculus,* from diminutive of *ventr-, venter* belly] (14th century)
: a cavity of a bodily part or organ: as **a :** a chamber of the heart which receives blood from a corresponding atrium and from which blood is forced into the arteries — see HEART illustration **b :** any of the system of communicating cavities in the brain that are continuous with the central canal of the spinal cord — see BRAIN illustration

ven·tri·cose \-,kōs\ *adjective* [New Latin *ventricosus,* from Latin *ventr-, venter* + *-icosus* (as in *varicosus* varicose)] (1756)
: markedly swollen, distended, or inflated especially on one side ⟨*ventricose* corollas⟩

ven·tric·u·lar \ven-'tri-kyə-lər, vən-\ *adjective* (1822)
: of, relating to, or being a ventricle ⟨*ventricular* fibrillation⟩ ⟨*ventricular* pressure⟩ ⟨*ventricular* myocardium⟩

ven·tric·u·lus \ven-'tri-kyə-ləs, vən-\ *noun, plural* **-li** \-,lī, -,lē\ [New Latin, from Latin, diminutive of *venter*] (1693)
: a digestive cavity: as **a :** STOMACH **b :** GIZZARD 1a **c :** the digestive part of an insect's stomach

ven·tril·o·quism \ven-'tri-lə-,kwi-zəm\ *noun* [Late Latin *ventriloquus* ventriloquist, from Latin *ventr-, venter* + *loqui* to speak; from the belief that the voice is produced from the ventriloquist's stomach] (circa 1797)
: the production of the voice in such a way that the sound seems to come from a source other than the vocal organs of the speaker
— ven·tri·lo·qui·al \,ven-trə-'lō-kwē-əl\ *adjective*
— ven·tri·lo·qui·al·ly \-ə-lē\ *adverb*

ven·tril·o·quist \ven-'tri-lə-kwist\ *noun* (circa 1656)
: one who uses or is skilled in ventriloquism; *especially* **:** one who provides entertainment by using ventriloquism to carry on an apparent conversation with a hand-manipulated dummy
— ven·tril·o·quis·tic \(,)ven-,tri-lə-'kwis-tik\ *adjective*

ven·tril·o·quize \ven-'tri-lə-,kwīz\ *verb* **-quized; -quiz·ing** (1844)
intransitive verb
: to use ventriloquism
transitive verb
: to utter in the manner of a ventriloquist

ven·tril·o·quy \-kwē\ *noun* (1584)
: VENTRILOQUISM

ven·tro·lat·er·al \,ven-,trō-'la-tə-rəl, -'la-trəl\ *adjective* (1835)
: ventral and lateral

ven·tro·me·di·al \-'mē-dē-əl\ *adjective* (1908)
: ventral and medial

¹ven·ture \'ven(t)-shər\ *verb* **ven·tured; ven·tur·ing** \'ven(t)-sh(ə-)riŋ\ [Middle English *venteren,* by shortening & alteration

◇ WORD HISTORY
venom The Romans called their goddess of love *Venus,* a word that must first have meant simply "charm" or "epitome of qualities that arouse sexual desire" and was later applied to the goddess who personified sexual attractiveness. The unattested noun stem *venes-* that underlies the later attested Latin word *Venus* must also have meant "magical charm" or "magical power to propitiate the gods," to judge by its verbal derivative *venerari* "to solicit the favor of (a deity), worship." (From *veneratus,* the past participle of this verb, English has borrowed *venerate.*) Another derivative of *venes-* was, in its reconstructed shape, *venesnom,* which became Latin *venemum* "magical charm, potent drug." *Venemum* also developed the sense "deadly substance, poison," and in this meaning after some restructuring in spoken Latin became Old French *venim,* which Middle English borrowed as *venim* or *venom.* The word *poison* itself, the French outcome of Latin *potio,* first meant merely "drink" and then "drink with magical power" (hence English *potion*). The change in sense from the latter meaning to "deadly substance," which took place during the word's passage from Latin to French, is similar to the shift that produced *venom.*

from *aventuren,* from *aventure* adventure] (15th century)
transitive verb
1 : to expose to hazard **:** RISK, GAMBLE ⟨*ventured* a buck or two on the race⟩
2 : to undertake the risks and dangers of **:** BRAVE ⟨*ventured* the stormy sea⟩
3 : to offer at the risk of rebuff, rejection, or censure ⟨*venture* an opinion⟩
intransitive verb
: to proceed especially in the face of danger
— ven·tur·er \'ven(t)-sh(ə-)rər\ *noun*

²venture *noun* (15th century)
1 *obsolete* **:** DESTINY, FORTUNE, CHANCE
2 a : an undertaking involving chance, risk, or danger; *especially* **:** a speculative business enterprise **b :** a venturesome act
3 : something (as money or property) at hazard in a speculative venture
— at a venture : at random ⟨a certain man drew a bow *at a venture,* and smote the king —1 Kings 22:34 (Authorized Version)⟩

venture capital *noun* (1943)
: capital (as retained corporate earnings or individual savings) invested or available for investment in the ownership element of new or fresh enterprise — called also *risk capital*
— venture capitalist *noun*

ven·ture·some \'ven(t)-shər-səm\ *adjective* (1661)
1 : involving risk **:** HAZARDOUS ⟨a *venturesome* journey⟩
2 : inclined to court or incur risk or danger **:** DARING ⟨a *venturesome* hunter⟩
synonym see ADVENTUROUS
— ven·ture·some·ly *adverb*
— ven·ture·some·ness *noun*

ven·tu·ri \ven-'tu̇r-ē\ *noun* [G. B. *Venturi* (died 1822) Italian physicist] (1887)
: a short tube with a tapering constriction in the middle that causes an increase in the velocity of flow of a fluid and a corresponding decrease in fluid pressure and that is used especially in measuring fluid flow or for creating a suction (as for driving aircraft instruments or drawing fuel into the flow stream of a carburetor)

ven·tur·ous \'ven(t)-sh(ə-)rəs\ *adjective* (1565)
: VENTURESOME
— ven·tur·ous·ly *adverb*
— ven·tur·ous·ness *noun*

ven·ue \'ven-,yü\ *noun* [Middle English *venyw* action of coming, from Middle French *venue,* from *venir* to come, from Latin *venire* — more at COME] (1531)
1 a : the place or county in which take place the alleged events from which a legal action arises **b :** the place from which a jury is drawn and in which trial is held ⟨requested a change of *venue*⟩ **c :** a statement showing that a case is brought to the proper court or authority
2 : LOCALE 1

ve·nule \'vēn-(,)yü(ə)l, 'ven-\ *noun* [Latin *venula,* diminutive of *vena* vein] (circa 1850)
: a small vein; *especially* **:** any of the minute veins connecting the capillaries with the larger systemic veins

Ve·nus \'vē-nəs\ *noun* [Middle English, from Latin *Vener-, Venus*]
1 : the Roman goddess of love and beauty — compare APHRODITE
2 : the planet second in order from the sun — see PLANET table

Ve·nus·berg \-,bərg\ *noun*
: a mountain in central Germany containing a cavern where in medieval legend Venus held court

Ve·nus·hair \-,har, -,her\ *noun* (1548)
: a delicate maidenhair fern (*Adiantum capillus-veneris*) that grows chiefly on wet calcareous rocks

Ve·nu·sian \vi-'nü-zhən, -'nyü-\ *adjective* (1874)
: of or relating to the planet Venus

Ve·nus's–flow·er–bas·ket \'vē-nə-səz-, 'vē-nəs-\ *noun* (1872)
: any of several glass sponges (genus *Euplectella*) of the western Pacific and Indian oceans — called also *Venus flower basket*

Ve·nus's–fly·trap \-'flī-,trap\ *noun* (1770)
: an insectivorous plant (*Dionaea muscipula*) of the sundew family of the Carolina coast with the leaf apex modified into an insect trap — called also *Venus flytrap*

Venus's-flytrap

ve·ra·cious \və-'rā-shəs\ *adjective* [Latin *verac-, verax* — more at VERY] (1677)
1 : TRUTHFUL, HONEST
2 : marked by truth **:** ACCURATE
— ve·ra·cious·ly *adverb*
— ve·ra·cious·ness *noun*

ve·rac·i·ty \və-'ra-sə-tē\ *noun, plural* **-ties** (circa 1623)
1 : devotion to the truth **:** TRUTHFULNESS
2 : power of conveying or perceiving truth
3 : conformity with truth or fact **:** ACCURACY
4 : something true ⟨makes lies sound like *veracities*⟩

ve·ran·da *or* **ve·ran·dah** \və-'ran-də\ *noun* [Hindi *varaṇḍā*] (1711)
: a usually roofed open gallery or portico attached to the exterior of a building

ve·ran·daed *also* **ve·ran·dahed** \-dəd\ *adjective* (1818)
: having a veranda

ve·rap·a·mil \və-'ra-pə-,mil\ *noun* [International Scientific Vocabulary *val*eric (acid) + *-apam-* (probably alteration of *amino* + *propyl*) + nitr*il*e] (1967)
: a calcium channel blocker $C_{27}H_{38}N_2O_4$ used especially in the form of its hydrochloride

ve·rat·ri·dine \və-'ra-trə-,dēn\ *noun* [*veratrine* + *-idine*] (1907)
: a poisonous alkaloid $C_{36}H_{51}NO_{11}$ occurring especially in sabadilla seed

ve·ra·trine \'ver-ə-,trēn, və-'ra-trən\ *noun* [New Latin *veratrina,* from *Veratrum,* genus of herbs] (1822)
: a poisonous irritant mixture of alkaloids from sabadilla seed that has been used as a counterirritant and insecticide

ve·ra·trum \və-'rā-trəm\ *noun* [New Latin, genus name, from Latin, hellebore] (1577)
: HELLEBORE 2

verb \'vərb\ *noun* [Middle English *verbe,* from Middle French, from Latin *verbum* word, verb — more at WORD] (14th century)
: a word that characteristically is the grammatical center of a predicate and expresses an act, occurrence, or mode of being, that in various languages is inflected for agreement with the subject, for tense, for voice, for mood, or for aspect, and that typically has rather full descriptive meaning and characterizing quality but is sometimes nearly devoid of these especially when used as an auxiliary or linking verb
— verb·less \'vər-bləs\ *adjective*

¹ver·bal \'vər-bəl\ *adjective* [Middle French or Late Latin; Middle French, from Late Latin *verbalis,* from Latin *verbum* word] (15th century)
1 a : of, relating to, or consisting of words ⟨*verbal* instructions⟩ **b :** of, relating to, or involving words rather than meaning or substance ⟨a consistency that is merely *verbal* and scholastic —B. N. Cardozo⟩ **c :** consisting of or using words only and not involving action ⟨a *verbal* protest⟩
2 : of, relating to, or formed from a verb ⟨a *verbal* adjective⟩
3 : spoken rather than written ⟨a *verbal* contract⟩
4 : VERBATIM, WORD-FOR-WORD ⟨a *verbal* translation⟩
5 : of or relating to facility in the use and comprehension of words ⟨*verbal* aptitude⟩ ■
— ver·bal·ly \-bə-lē\ *adverb*

²verbal *noun* (1530)
: a word that combines characteristics of a verb with those of a noun or adjective — compare GERUND, INFINITIVE, PARTICIPLE

verbal auxiliary *noun* (circa 1958)
: an auxiliary verb

ver·bal·ism \'vər-bə-,li-zəm\ *noun* (1787)
1 a : a verbal expression **:** TERM **b :** PHRASING, WORDING
2 : words used as if they were more important than the realities they represent ⟨the emancipation of science from *verbalism* —G. A. L. Sarton⟩
3 a : a wordy expression of little meaning **b :** VERBOSITY

ver·bal·ist \-list\ *noun* (circa 1609)
1 : one who stresses words above substance or reality
2 : a person who uses words skillfully
— ver·bal·is·tic \,vər-bə-'lis-tik\ *adjective*

ver·bal·ize \'vər-bə-,līz\ *verb* **-ized; -iz·ing** (1609)
intransitive verb
1 : to speak or write verbosely
2 : to express something in words
transitive verb
1 : to convert into a verb
2 : to name or describe in words
— ver·bal·i·za·tion \,vər-bə-lə-'zā-shən\ *noun*
— ver·bal·iz·er \'vər-bə-,lī-zər\ *noun*

verbal noun *noun* (circa 1706)
: a noun derived directly from a verb or verb stem and in some uses having the sense and constructions of a verb

¹ver·ba·tim \(,)vər-'bā-təm\ *adverb* [Middle English, from Medieval Latin, from Latin *verbum* word] (15th century)
: in the exact words **:** word for word

²verbatim *adjective* (1737)
: being in or following the exact words **:** WORD-FOR-WORD

ver·be·na \(,)vər-'bē-nə\ *noun* [New Latin, genus of herbs or subshrubs, from Latin, leafy

☐ USAGE

verbal It is a cherished belief of numerous modern commentators that using *verbal* to mean "spoken, not written" is an abuse of the language, a confusion of *verbal* with *oral* that can lead to misunderstanding. But this sense of *verbal,* which has been in use since the end of the 16th century, seems not to cause misunderstanding when the writer takes care with the context ⟨from the written and *verbal* information of our envoys —Thomas Jefferson⟩ ⟨where the poor actors undergo the most merciless tortures of *verbal* criticism —Washington Irving⟩ ⟨he and Hayward and I had made a *verbal* contract —Ernest Hemingway⟩ ⟨professed to see no reason for a written document at all. Why not leave it with a *verbal* understanding? —George F. Kennan⟩. The adverb *verbally* is similarly used ⟨*verbally* assured me that he had constantly prayed in church for the President and Government —Abraham Lincoln⟩ ⟨expressed to me in a letter (as well as *verbally*) their desire —E. B. White⟩. A meaning that had been in reputable use for three centuries before anyone questioned it is no blunder. And *oral*—not quite as old in this sense as *verbal*—is always available if *verbal* might be misconstrued. *Oral* is regularly used in combinations such as *oral history* and *oral examination* where *verbal* is not.

branch used ceremonially or medicinally — more at VERVAIN] (1562)
: VERVAIN; *especially* **:** any of numerous garden plants of hybrid origin widely grown for their showy spikes of white, pink, red, or blue flowers which are borne in profusion over a long season

ver·biage \'vər-bē-ij *also* -bij\ *noun* [French, from Middle French *verbier* to chatter, from *verbe* speech, from Latin *verbum* word] (circa 1721)
1 : a profusion of words usually of little or obscure content ⟨such a tangled maze of evasive *verbiage* as a typical party platform —Marcia Davenport⟩
2 : manner of expressing oneself in words **:** DICTION ⟨sportswriters guarded their *verbiage* so jealously —Raymond Sokolov⟩

ver·bi·cide \'vər-bə-,sīd\ *noun* [Latin *verbum* word + English *-cide*] (1858)
1 : deliberate distortion of the sense of a word (as in punning)
2 : one who distorts the sense of a word

ver·bid \'vər-bəd\ *noun* (1914)
: VERBAL

ver·big·er·a·tion \(,)vər-,bi-jə-'rā-shən\ *noun* [International Scientific Vocabulary, from Latin *verbigerare* to talk, chat, from *verbum* word + *gerere* to carry, wield — more at WORD] (1886)
: continual repetition of stereotyped phrases (as in some forms of mental illness)

ver·bose \(,)vər-'bōs\ *adjective* [Latin *verbosus,* from *verbum*] (1672)
1 : containing more words than necessary **:** WORDY ⟨a *verbose* reply⟩; *also* **:** impaired by wordiness ⟨a *verbose* style⟩
2 : given to wordiness ⟨a *verbose* orator⟩
synonym see WORDY
— ver·bose·ly *adverb*
— ver·bose·ness *noun*
— ver·bos·i·ty \-'bä-sə-tē\ *noun*

ver·bo·ten \vər-'bō-tᵊn, fər-, ver-\ *adjective* [German, from Old High German *farboten,* past participle of *farbioten* to forbid (akin to Old English *forbēodan* to forbid), from *far-, fur-* for- + *biotan* to offer — more at BID] (1912)
: not allowed **:** FORBIDDEN; *especially* **:** prohibited by dictate

verb sap \'vərb-'sap\ (1841)
: VERBUM SAP

ver·bum sap \,vər-bəm-'sap\ [short for New Latin *verbum sapienti* (*sat est*) a word to the wise (is sufficient)] (1818)
: enough said — used to indicate that something left unsaid may or should be inferred

ver·dant \'vər-dᵊnt\ *adjective* [modification of Middle French *verdoyant,* from present participle of *verdoyer* to be green, from Old French *verdoier,* from *verd, vert* green, from Latin *viridis,* from *virēre* to be green] (1581)
1 a : green in tint or color ⟨*verdant* grass⟩ **b :** green with growing plants ⟨*verdant* fields⟩
2 : unripe in experience or judgment **:** GREEN 9a, b
— ver·dan·cy \-dᵊn(t)-sē\ *noun*
— ver·dant·ly \-dᵊnt-lē\ *adverb*

verd an·tique *or* **verde an·tique** \,vərd-,an-'tēk\ *noun* [Italian *verde antico,* literally, ancient green] (1745)
: a green mottled or veined serpentine marble or calcareous serpentine much used for indoor decoration especially by the ancient Romans

ver·der·er *also* **ver·der·or** \'vər-dər-ər\ *noun* [Anglo-French, from Old French *verdier,* from *verd* green] (circa 1538)
: a onetime English judicial officer in charge of the king's forest

ver·dict \'vər-(,)dikt\ *noun* [alteration of Middle English *verdit,* from Anglo-French, from Old French *ver* true (from Latin *verus*) + *dit* saying, dictum, from Latin *dictum* — more at VERY] (1533)
1 : the finding or decision of a jury on the matter submitted to it in trial
2 : OPINION, JUDGMENT

ver·di·gris \'vər-də-,grēs, -,gris, -grəs *also* -,grē\ *noun* [Middle English *vertegrez,* from Old French *vert de Grice,* literally, green of Greece] (14th century)
1 a : a green or greenish blue poisonous pigment resulting from the action of acetic acid on copper and consisting of one or more basic copper acetates **b :** normal copper acetate $Cu(C_2H_3O_2)_2 \cdot H_2O$
2 : a green or bluish deposit especially of copper carbonates formed on copper, brass, or bronze surfaces

ver·din \'vər-dᵊn\ *noun* [French, a green songbird of Indochina, alteration of *verdon, verdun* bunting, yellowhammer, from *vert* green, from Old French *verd, vert*] (1881)
: a very small yellow-headed titmouse (*Auriparus flaviceps*) found from Texas to California and southward

ver·dure \'vər-jər *also* -dyər\ *noun* [Middle English, from Middle French, from *verd* green] (14th century)
1 : the greenness of growing vegetation; *also* **:** such vegetation itself
2 : a condition of health and vigor
— ver·dur·ous \'verj-rəs; 'ver-jə-rəs, -dyə-\ *adjective*

ver·dured \'vər-jərd *also* -dyərd\ *adjective* (circa 1718)
: covered with verdure

[1]**verge** \'vərj\ *noun* [Middle English, from Middle French, from Latin *virga* rod, stripe] (15th century)
1 a (1) **:** a rod or staff carried as an emblem of authority or symbol of office (2) *obsolete* **:** a stick or wand held by a person being admitted to tenancy while he swears fealty **b :** the spindle of a watch balance; *especially* **:** a spindle with pallets in an old vertical escapement **c :** the male intromittent organ of any of various invertebrates
2 a : something that borders, limits, or bounds: as (1) **:** an outer margin of an object or structural part (2) **:** the edge of roof covering (as tiling) projecting over the gable of a roof (3) *British* **:** a paved or planted strip of land at the edge of a road **:** SHOULDER **b :** BRINK, THRESHOLD ⟨a country on the *verge* of destruction —Archibald MacLeish⟩

[2]**verge** *intransitive verb* **verged; verg·ing** (1787)
1 : to be contiguous
2 : to be on the verge or border

[3]**verge** *intransitive verb* **verged; verg·ing** [Latin *vergere* to bend, incline — more at WRENCH] (1610)
1 a *of the sun* **:** to incline toward the horizon **:** SINK **b :** to move or extend in some direction or toward some condition
2 : to be in transition or change

verg·er \'vər-jər\ *noun* (15th century)
1 *chiefly British* **:** an attendant that carries a verge (as before a bishop or justice)
2 : a church official who keeps order during services or serves as an usher or a sacristan

ve·rid·i·cal \və-'ri-di-kəl\ *adjective* [Latin *veridicus,* from *verus* true + *dicere* to say — more at VERY, DICTION] (1653)
1 : TRUTHFUL, VERACIOUS ⟨tried . . . to supply . . . a *veridical* background to the events and people portrayed —Laura Krey⟩
2 : not illusory **:** GENUINE ⟨it is assumed that . . . perception is *veridical* —George Lakoff⟩
— ve·rid·i·cal·i·ty \-,ri-də-'ka-lə-tē\ *noun*
— ve·rid·i·cal·ly \-'ri-di-k(ə-)lē\ *adverb*

ver·i·fi·able \,ver-ə-'fī-ə-bəl\ *adjective* (1593)
: capable of being verified
— ver·i·fi·abil·i·ty \,ver-ə-,fī-ə-'bi-lə-tē\ *noun*
— ver·i·fi·able·ness *noun*

ver·i·fi·ca·tion \,ver-ə-fə-'kā-shən\ *noun* (1523)
: the act or process of verifying **:** the state of being verified

ver·i·fy \'ver-ə-,fī\ *transitive verb* **-fied; -fy·ing** [Middle English *verifien,* from Middle French *verifier,* from Medieval Latin *verificare,* from Latin *verus* true] (14th century)
1 : to confirm or substantiate in law by oath
2 : to establish the truth, accuracy, or reality of
synonym see CONFIRM
— ver·i·fi·er \-,fī(-ə)r\ *noun*

ver·i·ly \'ver-ə-lē\ *adverb* [Middle English *verraily,* from *verray* very] (14th century)
1 : in truth **:** CERTAINLY
2 : TRULY, CONFIDENTLY

veri·sim·i·lar \,ver-ə-'si-mə-lər, -'sim-lər\ *adjective* [Latin *verisimilis*] (1681)
1 : having the appearance of truth **:** PROBABLE
2 : depicting realism (as in art or literature)
— veri·sim·i·lar·ly *adverb*

veri·si·mil·i·tude \-sə-'mi-lə-,tüd, -,tyüd\ *noun* [Latin *verisimilitudo,* from *verisimilis* verisimilar, from *veri similis* like the truth] (1603)
1 : the quality or state of being verisimilar
2 : something verisimilar
— veri·si·mil·i·tu·di·nous \-,mi-lə-'tüd-nəs, -'tyüd-; -'tü-dᵊn-əs, -'tyü-\ *adjective*

ve·rism \'vir-,i-zəm, 'ver-\ *noun* [Italian *verismo,* from *vero* true, from Latin *verus*] (1892)
: artistic use of contemporary everyday material in preference to the heroic or legendary especially in grand opera
— ve·rist \-ist\ *noun or adjective*
— ve·ris·tic \vir-'is-tik, ver-\ *adjective*

ve·ris·mo \vā-'rēz-(,)mō\ *noun* [Italian] (circa 1908)
: VERISM; *also* **:** REALISM 3

ver·i·ta·ble \'ver-ə-tə-bəl\ *adjective* [Middle English, from Middle French, from *verité*] (15th century)
: being in fact the thing named and not false, unreal, or imaginary — often used to stress the aptness of a metaphor ⟨a *veritable* mountain of references⟩
— ver·i·ta·ble·ness *noun*
— ver·i·ta·bly \-blē\ *adverb*

vé·ri·té \,ver-ə-'tā\ *noun* [*cinema verité*] (1966)
: the art or technique of filming (as a motion picture) so as to convey candid realism

ver·i·ty \'ver-ə-tē\ *noun, plural* **-ties** [Middle English *verite,* from Middle French *verité,* from Latin *veritat-, veritas,* from *verus* true] (14th century)
1 : the quality or state of being true or real
2 : something (as a statement) that is true; *especially* **:** a fundamental and inevitably true value ⟨such eternal *verities* as honor, love, and patriotism⟩
3 : the quality or state of being truthful or honest ⟨the king-becoming graces, as justice, *verity* —Shakespeare⟩

ver·juice \'vər-,jüs\ *noun* [Middle English *verjus,* from Middle French, from *vert jus,* literally, green juice] (14th century)
1 : the sour juice of crab apples or of unripe fruit (as grapes or apples); *also* **:** an acid liquor made from verjuice
2 : acidity of disposition or manner

ver·meil *noun* [Middle French, from *vermeil,* adjective — more at VERMILION] (1590)
1 \'vər-məl, -,māl\ **:** VERMILION
2 \ver-'mā\ **:** gilded silver
— vermeil *adjective*

vermi- *combining form* [New Latin, from Late Latin, from Latin *vermis* — more at WORM]
: worm ⟨*vermi*form⟩

ver·mi·cel·li \,vər-mə-'che-lē, -'se-\ *noun* [Italian, from plural of *vermicello,* diminutive of *verme* worm, from Latin *vermis*] (1669)
: pasta made in long solid strings smaller in diameter than spaghetti

ver·mi·cide \'vər-mə-,sīd\ *noun* (1849)
: an agent that destroys worms

ver·mic·u·lar \(,)vər-'mi-kyə-lər\ *adjective* [New Latin *vermicularis,* from Latin *vermiculus,* diminutive of *vermis*] (1672)

1 a : resembling a worm in form or motion **b :** VERMICULATE
2 : of, relating to, or caused by worms

ver·mic·u·late \-lət\ *or* **ver·mic·u·lat·ed** \-ˌlā-təd\ *adjective* [Latin *vermiculatus,* from *vermiculus*] (1605)
1 : TORTUOUS, INVOLUTE
2 : full of worms **:** WORM-EATEN
3 a : VERMIFORM **b :** marked with irregular fine lines or with wavy impressed lines ⟨a *vermiculate* nut⟩
— ver·mic·u·la·tion \-ˌmi-kyə-ˈlā-shən\ *noun*

ver·mic·u·lite \(ˌ)vər-ˈmi-kyə-ˌlīt\ *noun* [Latin *vermiculus* little worm] (1824)
: any of various micaceous minerals that are hydrous silicates resulting usually from expansion of the granules of mica at high temperatures to give a lightweight highly water-absorbent material

ver·mi·form \ˈvər-mə-ˌfȯrm\ *adjective* [New Latin *vermiformis,* from *vermi-* + *-formis* -form] (circa 1730)
: resembling a worm in shape

vermiform appendix *noun* (1778)
: a narrow blind tube usually about three to four inches (7.6 to 10.2 centimeters) long that extends from the cecum in the lower right-hand part of the abdomen

ver·mi·fuge \ˈvər-mə-ˌfyüj\ *adjective* (1697)
: serving to destroy or expel parasitic worms **:** ANTHELMINTIC
— vermifuge *noun*

ver·mil·ion *also* **ver·mil·lion** \vər-ˈmil-yən\ *noun* [Middle English *vermilioun,* from Old French *vermeillon,* from *vermeil,* adjective, bright red, vermilion, from Late Latin *vermiculus* kermes, from Latin, little worm] (13th century)
1 : a bright red pigment consisting of mercuric sulfide; *broadly* **:** any of various red pigments
2 : a vivid reddish orange

ver·min \ˈvər-mən\ *noun, plural* **vermin** [Middle English, from Middle French, from (assumed) Vulgar Latin *verminum,* from Latin *verminare* to be infested with maggots, have racking pains, from *vermina* racking pains] (14th century)
1 a : small common harmful or objectionable animals (as lice or fleas) that are difficult to control **b :** birds and mammals that prey on game **c :** animals that at a particular time and place compete (as for food) with humans or domestic animals
2 : an offensive person

ver·min·ous \ˈvər-mə-nəs\ *adjective* (circa 1616)
1 : consisting of or being vermin **:** NOXIOUS
2 : forming a breeding place for or infested by vermin **:** FILTHY ⟨*verminous* garbage⟩
3 : caused by vermin ⟨*verminous* disease⟩

ver·mouth \vər-ˈmüth\ *noun* [French *vermout,* from German *Wermut* wormwood, from Old High German *wermuota* — more at WORMWOOD] (1806)
: a dry or sweet aperitif wine flavored with aromatic herbs and often used in mixed drinks

[1]**ver·nac·u·lar** \və(r)-ˈna-kyə-lər\ *adjective* [Latin *vernaculus* native, from *verna* slave born in the master's house, native] (1601)
1 a : using a language or dialect native to a region or country rather than a literary, cultured, or foreign language **b :** of, relating to, or being a nonstandard language or dialect of a place, region, or country **c :** of, relating to, or being the normal spoken form of a language
2 : applied to a plant or animal in the common native speech as distinguished from the Latin nomenclature of scientific classification
3 : of, relating to, or characteristic of a period, place, or group; *especially* **:** of, relating to, or being the common building style of a period or place ◆
— ver·nac·u·lar·ly *adverb*

[2]**vernacular** *noun* (circa 1706)
1 : a vernacular language, expression, or mode of expression
2 : the mode of expression of a group or class
3 : a vernacular name of a plant or animal

ver·nac·u·lar·ism \və(r)-ˈna-kyə-lə-ˌri-zəm\ *noun* (circa 1841)
: a vernacular word or idiom

ver·nal \ˈvər-nᵊl\ *adjective* [Latin *vernalis,* alteration of *vernus,* from *ver* spring; akin to Greek *ear* spring, Sanskrit *vasanta*] (1534)
1 : of, relating to, or occurring in the spring ⟨*vernal* equinox⟩ ⟨*vernal* sunshine⟩
2 : fresh or new like the spring; *also* **:** YOUTHFUL
— ver·nal·ly \-nᵊl-ē\ *adverb*

ver·nal·i·za·tion \ˌvər-nᵊl-ə-ˈzā-shən\ *noun* (1933)
: the act or process of hastening the flowering and fruiting of plants by treating seeds, bulbs, or seedlings so as to induce a shortening of the vegetative period
— ver·nal·ize \ˈvər-nᵊl-ˌīz\ *transitive verb*

ver·na·tion \(ˌ)vər-ˈnā-shən\ *noun* [New Latin *vernation-, vernatio,* from Latin *vernare* to behave as in spring, from *vernus* vernal] (1793)
: the arrangement of foliage leaves within the bud

Ver·ner's law \ˈver-nərz-\ *noun* [Karl A. *Verner*] (1878)
: a statement in historical linguistics: in medial or final position in voiced environments and when the immediately preceding vowel did not bear the principal accent in Proto-Indo-European, the Proto-Germanic voiceless fricatives *f,* þ, and χ derived from the Proto-Indo-European voiceless stops *p, t,* and *k* and the Proto-Germanic voiceless fricative *s* derived from Proto-Indo-European *s* became the voiced fricatives ƀ, ð, ǥ, and *z* represented in various recorded Germanic languages by *b, d, g,* and *r*

ver·ni·cle *or* **ver·na·cle** \ˈvər-ni-kəl\ *noun* [Middle English *vernicle,* from Middle French *veronique, vernicle,* from Medieval Latin *veronica*] (14th century)
: [2]VERONICA

[1]**ver·ni·er** \ˈvər-nē-ər\ *noun* [Pierre *Vernier*] (1766)
1 : a short scale made to slide along the divisions of a graduated instrument for indicating parts of divisions
2 a : a small auxiliary device used with a main device to obtain fine adjustment **b :** any of two or more small supplementary rocket engines or gas nozzles on a missile or a rocket vehicle for making fine adjustments in the speed or course or controlling the attitude — called also *vernier engine*

[2]**vernier** *adjective* (1788)
: having or comprising a vernier

vernier caliper *noun* (circa 1876)
: a measuring device that consists of a main scale with a fixed jaw and a sliding jaw with an attached vernier

ver·nis·sage \ˌver-ni-ˈsäzh\ *noun* [French, day before an exhibition opens reserved for artists to varnish and put finishing touches to their paintings, literally, varnishing, from *vernis* varnish — more at VARNISH] (1912)
: a private showing or preview of an art exhibition

[1]**ve·ron·i·ca** \və-ˈrä-ni-kə\ *noun* [New Latin, genus of herbs] (1527)
: SPEEDWELL

[2]**veronica** *noun* [Medieval Latin, from *Veronica,* legendary saint of the 1st century A.D.] (circa 1700)
: an image of Christ's face said to have been impressed on the cloth that Saint Veronica gave him to wipe his face with on the way to his crucifixion; *also* **:** a cloth resembling the legendary one of Saint Veronica

[3]**veronica** *noun* [Spanish *verónica,* from Saint *Veronica*] (1926)
: a pase in bullfighting in which the cape is swung slowly away from the charging bull while the matador keeps his feet in the same position

Vé·ro·nique *also* **Ve·ro·nique** \ˌvā-rō-ˈnēk\ *adjective* [French *Véronique* Veronica] (1907)
: prepared or garnished with usually white seedless grapes ⟨sole *Véronique*⟩

ver·ru·ca \və-ˈrü-kə\ *noun, plural* **-cae** \-(ˌ)kē, -ˌkī, -ˌsī\ [Latin, wart, hillock; akin to Lithuanian *viršus* summit and probably to Old English *wearte* wart — more at WART] (1565)
1 : a wart or warty skin lesion
2 : a warty elevation on a plant or animal surface

verruca vul·ga·ris \-ˌvəl-ˈgar-əs, -ˈger-\ *noun* [New Latin, literally, common verruca] (1903)
: WART 1a

ver·ru·cose \və-ˈrü-ˌkōs\ *adjective* (1686)
: covered with warty elevations

ver·sal \ˈvər-səl, ˈvär-\ *adjective* [short for *universal*] (1592)
archaic **:** ENTIRE, WHOLE ⟨as pale as any clout in the *versal* world —Shakespeare⟩

ver·sant \ˈvər-sᵊnt\ *adjective* [Latin *versant-, versans,* present participle of *versare, versari* to turn, occupy oneself, meditate] (1645)
1 *archaic* **:** EXPERIENCED, PRACTICED
2 : CONVERSANT

ver·sa·tile \ˈvər-sə-tᵊl, *especially British* -ˌtīl\ *adjective* [French or Latin; French, from Latin *versatilis* turning easily, from *versare* to turn, frequentative of *vertere*] (1605)
1 : changing or fluctuating readily **:** VARIABLE ⟨a *versatile* disposition⟩
2 : embracing a variety of subjects, fields, or skills; *also* **:** turning with ease from one thing to another
3 a (1) **:** capable of turning forward or backward **:** REVERSIBLE ⟨a *versatile* toe of a bird⟩ (2) **:** capable of moving laterally and up and down ⟨*versatile* antennae⟩ **b** *of an anther* **:** having the filaments attached at or near the middle so as to swing freely
4 : having many uses or applications ⟨*versatile* building material⟩
— ver·sa·tile·ly \-tᵊl-(l)ē, -ˌtīl-lē\ *adverb*
— ver·sa·tile·ness \-tᵊl-nəs, -ˌtīl-nəs\ *noun*

ver·sa·til·i·ty \ˌvər-sə-ˈti-lə-tē\ *noun* (circa 1755)

◇ WORD HISTORY

vernacular In ancient Rome, where slaves constituted most of the labor force, a distinction was made between a slave born within a master's household, called in Latin a *verna,* and one who was acquired by other means. From *verna* the Romans derived the adjective *vernaculus,* meaning "domestic, homegrown" or "native, indigenous," presumably created by analogy with *masculus* "male" (noun and adjective), originally a diminutive of *mas* "male animal." *Vernaculus* was applied to language peculiar to the inhabitants of a particular area as well as to a wide variety of other things that were local or domestic. When Latin *vernaculus* was anglicized as *vernacular* in the early 17th century its scope was restricted to language, and it meant "using local rather than literary or foreign speech." A relatively modern application of *vernacular* is its use in distinguishing the common native names of plants and animals from the Latin nomenclature of scientific classification, the vernacular name for *Ursus arctos horribilis* being, for example, *grizzly bear.*

: the quality or state of being versatile ⟨a writer of great *versatility*⟩

vers de so·ci·é·té \ˌver-də-ˌsō-sē-ə-ˈtā\ *noun* [French, society verse] (1796)
: witty and typically ironic light verse

[1]verse \ˈvərs\ *noun* [Middle English *vers,* from Old French & Old English; both from Latin *versus,* literally, turning, from *vertere* to turn — more at WORTH] (before 12th century)
1 : a line of metrical writing
2 a (1) **:** metrical language (2) **:** metrical writing distinguished from poetry especially by its lower level of intensity (3) **:** POETRY 2 **b :** POEM **c :** a body of metrical writing (as of a period or country)
3 : STANZA
4 : one of the short divisions into which a chapter of the Bible is traditionally divided

[2]verse *verb* **versed; vers·ing** (before 12th century)
intransitive verb
: to make verse **:** VERSIFY
transitive verb
1 : to tell or celebrate in verse
2 : to turn into verse

[3]verse *transitive verb* **versed; vers·ing** [back-formation from *versed,* from Latin *versatus,* past participle of *versari* to be active, be occupied (in), passive of *versare* to turn] (1673)
: to familiarize by close association, study, or experience ⟨well *versed* in the theater⟩

vers·et \ˈvər-sət, -ˌset; ˌvər-ˈset\ *noun* [Middle English, from Old French, diminutive of *vers* verse] (13th century)
: a short verse especially from a sacred book (as the Koran)

ver·si·cle \ˈvər-si-kəl\ *noun* [Middle English, from Latin *versiculus,* diminutive of *versus* verse] (14th century)
1 : a short verse or sentence (as from a psalm) said or sung by a leader in public worship and followed by a response from the people
2 : a little verse

ver·sic·u·lar \ˌvər-ˈsi-kyə-lər\ *adjective* [Latin *versiculus* little verse] (1812)
: of or relating to verses or versicles

ver·si·fi·ca·tion \ˌvər-sə-fə-ˈkā-shən\ *noun* (1603)
1 : the making of verses
2 a : metrical structure **:** PROSODY **b :** a particular metrical structure or style
3 : a version in verse of something originally in prose

ver·si·fi·er \ˈvər-sə-ˌfī(-ə)r\ *noun* (14th century)
: one that versifies; *especially* **:** a writer of light or inferior verse

ver·si·fy \-ˌfī\ *verb* **-fied; -fy·ing** [Middle English *versifien,* from Middle French *versifier,* from Latin *versificare,* from *versus* verse, line] (14th century)
intransitive verb
: to compose verses
transitive verb
1 : to relate or describe in verse
2 : to turn into verse

ver·sion \ˈvər-zhən, -shən\ *noun* [Middle French, from Medieval Latin *version-, versio* act of turning, from Latin *vertere* to turn — more at WORTH] (1582)
1 : a translation from another language; *especially* **:** a translation of the Bible or a part of it
2 a : an account or description from a particular point of view especially as contrasted with another account **b :** an adaptation of a literary work ⟨the movie *version* of the novel⟩ **c :** an arrangement of a musical composition
3 : a form or variant of a type or original ⟨an experimental *version* of the plane⟩
4 a : a condition in which an organ and especially the uterus is turned from its normal position **b :** manual turning of a fetus in the uterus to aid delivery
— ver·sion·al \ˈvərzh-nəl, ˈvərsh-; ˈvər-zhə-nᵊl, -shə-\ *adjective*

vers li·bre \ver-ˈlēbrᵊ\ *noun, plural* **vers li·bres** *same*\ [French] (1902)
: FREE VERSE

vers–li·brist \-ˈlē-brist\ *noun* [French *verslibriste*] (1916)
: a writer of free verse

ver·so \ˈvər-(ˌ)sō\ *noun, plural* **versos** [New Latin *verso (folio)* the page being turned] (1839)
1 : the side of a leaf (as of a manuscript) that is to be read second
2 : a left-hand page — compare RECTO

verst \ˈvərst\ *noun* [French *verste* & German *Werst;* both from Russian *versta;* akin to Latin *vertere* to turn] (1555)
: a Russian unit of distance equal to 0.6629 mile (1.067 kilometers)

ver·sus \ˈvər-səs, -səz\ *preposition* [Medieval Latin, towards, against, from Latin, adverb, so as to face, from past participle of *vertere* to turn] (15th century)
1 : AGAINST
2 : in contrast to or as the alternative of ⟨free trade *versus* protection⟩

vert \ˈvərt\ *noun* [Middle English *verte,* from Middle French *vert,* from *vert* green — more at VERDANT] (15th century)
1 a : green forest vegetation especially when forming cover or providing food for deer **b :** the right or privilege (as in England) of cutting living wood or sometimes of pasturing animals in a forest
2 : the heraldic color green

ver·te·bra \ˈvər-tə-brə, -ˌbrā\ *noun, plural* **-brae** \-ˌbrā, -(ˌ)brē, -brə\ *or* **-bras** [Latin, joint, vertebra, from *vertere* to turn] (1578)
: one of the bony or cartilaginous segments composing the spinal column, consisting in some lower vertebrates of several distinct elements which never become united, and in higher vertebrates having a short more or less cylindrical body whose ends articulate by pads of elastic or cartilaginous tissue with those of adjacent vertebrae and a bony arch that encloses the spinal cord

ver·te·bral \(ˌ)vər-ˈtē-brəl, ˈvər-tə-\ *adjective* (circa 1681)
1 : of, relating to, or being vertebrae or the vertebral column **:** SPINAL
2 : composed of or having vertebrae

vertebral canal *noun* (1831)
: a canal that contains the spinal cord and is delimited by the arches on the dorsal side of the vertebrae — called also *spinal canal*

vertebral column *noun* (1822)
: SPINAL COLUMN

[1]ver·te·brate \ˈvər-tə-brət, -ˌbrāt\ *adjective* [New Latin *vertebratus,* from Latin, jointed, from *vertebra*] (1826)
1 a : having a spinal column **b :** of or relating to the vertebrates
2 : organized or constructed in orderly or developed form

[2]vertebrate *noun* [New Latin *Vertebrata,* from neuter plural of *vertebratus*] (1826)
: any of a subphylum (Vertebrata) of chordates that possess a spinal column including the mammals, birds, reptiles, amphibians, and fishes

ver·tex \ˈvər-ˌteks\ *noun, plural* **ver·ti·ces** \ˈvər-tə-ˌsēz\ *also* **ver·tex·es** [Latin *vertic-, vertex, vortic-, vortex* whirl, whirlpool, top of the head, summit, from *vertere* to turn] (1570)
1 a : the point opposite to and farthest from the base in a figure **b :** a point (as of an angle, polygon, polyhedron, graph, or network) that terminates a line or curve or comprises the intersection of two or more lines or curves **c :** a point where an axis of an ellipse, parabola, or hyperbola intersects the curve itself
2 : the top of the head
3 : a principal or highest point **:** SUMMIT ⟨the *vertex* of the hill⟩

ver·ti·cal \ˈvər-ti-kəl\ *adjective* [Middle French or Late Latin; Middle French, from Late Latin *verticalis,* from Latin *vertic-, vertex*] (1559)
1 a : situated at the highest point **:** directly overhead or in the zenith **b** *of an aerial photograph* **:** taken with the camera pointing straight down or nearly so
2 a : perpendicular to the plane of the horizon or to a primary axis **:** UPRIGHT **b** (1) **:** located at right angles to the plane of a supporting surface (2) **:** lying in the direction of an axis **:** LENGTHWISE
3 a : relating to, involving, or integrating economic activity from basic production to point of sale ⟨*vertical* monopoly⟩ **b :** of, relating to, or comprising persons of different status ⟨the *vertical* arrangement of society⟩ ☆
— vertical *noun*
— ver·ti·cal·i·ty \ˌvər-tə-ˈka-lə-tē\ *noun*
— ver·ti·cal·ly \ˈvər-ti-k(ə-)lē\ *adverb*
— ver·ti·cal·ness \-kəl-nəs\ *noun*

vertical angle *noun* (1571)
: either of two angles lying on opposite sides of two intersecting lines

vertical circle *noun* (1559)
: a great circle of the celestial sphere whose plane is perpendicular to that of the horizon

vertical file *noun* (1906)
: a collection of articles (as pamphlets and clippings) that is maintained (as in a library) to answer brief questions or to provide points of information not easily located

vertical union *noun* (1933)
: INDUSTRIAL UNION

ver·ti·cil \ˈvər-tə-ˌsil\ *noun* [New Latin *verticillus,* diminutive of Latin *vertex* whirl] (1793)
: a circle of similar parts (as flowers around a stem or sensory hairs around an antennal joint) about the same point on the axis **:** WHORL

ver·ti·cil·late \ˌvər-tə-ˈsi-lət\ *adjective* (circa 1793)
: arranged in verticils

ver·ti·cil·li·um wilt \ˌvər-tə-ˈsi-lē-əm-\ *noun* [New Latin *Verticillium,* from *verticillus*] (1916)
: a wilt disease of various plants that is caused by a soil-borne imperfect fungus (genus *Verticillium*)

ver·tig·i·nous \(ˌ)vər-ˈti-jə-nəs\ *adjective* [Latin *vertiginosus,* from *vertigin-, vertigo*] (1608)
1 a : characterized by or suffering from vertigo or dizziness **b :** inclined to frequent and often pointless change **:** INCONSTANT
2 : causing or tending to cause dizziness ⟨the *vertiginous* heights⟩
3 : marked by turning **:** ROTARY ⟨the *vertiginous* motion of the earth⟩
— ver·tig·i·nous·ly *adverb*

ver·ti·go \ˈvər-ti-ˌgō\ *noun, plural* **-goes** *or* **-gos** [Latin *vertigin-, vertigo,* from *vertere* to turn] (1528)
1 a : a disordered state in which the individual or the individual's surroundings seem to whirl dizzily **b :** a dizzy confused state of mind
2 : disordered vertiginous movement as a symptom of disease in lower animals; *also* **:** a disease (as gid) causing this

ver·tu \ˌvər-ˈtü, ver-\ *variant of* VIRTU

☆ SYNONYMS
Vertical, perpendicular, plumb mean being at right angles to a base line. VERTICAL suggests a line or direction rising straight upward toward a zenith ⟨the side of the cliff is almost *vertical*⟩. PERPENDICULAR may stress the straightness of a line making a right angle with any other line, not necessarily a horizontal one ⟨the parallel bars are *perpendicular* to the support posts⟩. PLUMB stresses an exact verticality determined (as with a plumb line) by earth's gravity ⟨make sure that the wall is *plumb*⟩.

ver•vain \ˈvər-ˌvān\ *noun* [Middle English *verveine,* from Middle French, from Latin *verbena* leafy branch; akin to Latin *verber* rod, Lithuanian *virbas,* and perhaps to Greek *rhabdos* rod] (14th century)
: any of a genus (*Verbena* of the family Verbenaceae, the vervain family) of chiefly American plants that have bracted spicate flowers, a regular corolla with a 5-lobed limb, and a fruit that separates into four nutlets

verve \ˈvərv\ *noun* [French, fantasy, caprice, animation, from Latin *verba,* plural of *verbum* word — more at WORD] (1697)
1 *archaic* **:** special ability or talent
2 a : the spirit and enthusiasm animating artistic composition or performance **:** VIVACITY **b :** ENERGY, VITALITY

ver•vet monkey \ˈvər-vət-\ *noun* [French *vervet*] (1893)
: a monkey of any of several African races of a guenon (*Cercopithecus aethiops*) having the face, chin, hands, and feet black — called also *vervet*

vervet monkey

[1]**very** \ˈver-ē\ *adjective* **veri•er; -est** [Middle English *verray, verry,* from Old French *verai,* from (assumed) Vulgar Latin *veracus,* alteration of Latin *verac-, verax* truthful, from *verus* true; akin to Old English *wǣr* true, Old High German *wāra* trust, care, Greek *ēra* (accusative) favor] (13th century)
1 a : properly entitled to the name or designation **:** TRUE ⟨the fierce hatred of a *very* woman —J. M. Barrie⟩ **b :** ACTUAL, REAL ⟨the *very* blood and bone of our grammar —H. L. Smith (died 1972)⟩ **c :** SIMPLE, PLAIN ⟨in *very* truth⟩
2 a : being exactly as stated ⟨the *very* heart of the city⟩ **b :** exactly suitable or necessary ⟨the *very* thing for the purpose⟩
3 a : ABSOLUTE, UTTER ⟨the *veriest* fool alive⟩ **b :** UNQUALIFIED, SHEER ⟨the *very* shame of it⟩
4 — used as an intensive especially to emphasize identity ⟨before my *very* eyes⟩
5 : MERE, BARE ⟨the *very* thought terrified him⟩
6 : being the same one **:** SELFSAME ⟨the *very* man I saw⟩
7 : SPECIAL, PARTICULAR ⟨the *very* essence of truth is plainness and brightness —John Milton⟩
synonym see SAME

[2]**very** *adverb* (14th century)
1 : in actual fact **:** TRULY ⟨the *very* best store in town⟩ ⟨told the *very* same story⟩
2 : to a high degree **:** EXCEEDINGLY ⟨*very* hot⟩ ⟨didn't hurt *very* much⟩ ■

very hard *adjective* (circa 1943)
of cheese **:** suitable chiefly for grating

very high frequency *noun* (1920)
: a radio frequency between ultrahigh frequency and high frequency — see RADIO FREQUENCY table

Ve•ry light \ˈver-ē-, ˈvir-ē-\ *noun* [Edward W. *Very* (died 1910) American naval officer] (1917)
: a pyrotechnic signal in a system of signaling using white or colored balls of fire projected from a special pistol

very low–density lipoprotein *noun* (1977)
: VLDL

very low frequency *noun* (1938)
: a radio frequency between low frequency and voice frequency — see RADIO FREQUENCY table

Ve•ry pistol \ˈver-ē-, ˈvir-ē-\ *noun* (1915)
: a pistol for firing Very lights

Very Reverend \ˈver-ē-\ (circa 1828)
— used as a title for various ecclesiastical officials (as cathedral deans and canons, rectors of Roman Catholic colleges and seminaries, and superiors of some religious houses)

ves•i•cal \ˈve-si-kəl\ *adjective* [Latin *vesica* bladder — more at VENTER] (1797)
: of or relating to the urinary bladder ⟨*vesical* burning⟩

ves•i•cant \-kənt\ *noun* [Latin *vesica* bladder, blister] (1661)
: an agent (as a drug or a chemical weapon) that induces blistering
— vesicant *adjective*

ves•i•cle \ˈve-si-kəl\ *noun* [Middle French *vesicule,* from Latin *vesicula* small bladder, blister, from diminutive of *vesica*] (1578)
1 a : a membranous and usually fluid-filled pouch (as a cyst, vacuole, or cell) in a plant or animal **b :** a small abnormal elevation of the outer layer of skin enclosing a watery liquid **:** BLISTER **c :** a pocket of embryonic tissue that is the beginning of an organ
2 : a small cavity in a mineral or rock

ve•sic•u•lar \və-ˈsi-kyə-lər, ve-\ *adjective* [New Latin *vesicula* vesicle, from Latin, small bladder] (1715)
1 : containing, composed of, or characterized by vesicles ⟨*vesicular* lava⟩
2 : having the form or structure of a vesicle
3 : of or relating to vesicles
— ve•sic•u•lar•i•ty \-ˌsi-kyə-ˈlar-ə-tē\ *noun*

vesicular stomatitis *noun* (circa 1903)
: an acute virus disease especially of various domesticated animals (as horses and cows) that is marked by erosive blisters in and about the mouth and that much resembles foot-and-mouth disease

ve•sic•u•late \və-ˈsi-kyə-ˌlāt, ve-\ *verb* **-lat•ed; -lat•ing** (1865)
transitive verb
: to make vesicular
intransitive verb
: to become vesicular
— ve•sic•u•la•tion \-ˌsi-kyə-ˈlā-shən\ *noun*

[1]**ves•per** \ˈves-pər\ *noun* [Middle English, from Latin, evening, evening star — more at WEST] (14th century)
1 *capitalized, archaic* **:** EVENING STAR
2 : a vesper bell
3 *archaic* **:** EVENING, EVENTIDE

[2]**vesper** *adjective* (1791)
: of or relating to vespers or the evening

ves•per•al \ˈves-p(ə-)rəl\ *adjective* (circa 1623)
: VESPER ⟨a *vesperal* breeze⟩

ves•pers \ˈves-pərz\ *noun plural but singular or plural in construction, often capitalized* [French *vespres,* from Medieval Latin *vesperae,* from Latin, plural of *vespera* evening; akin to Latin *vesper* evening star] (1611)
1 : the sixth of the canonical hours that is said or sung in the late afternoon
2 : a service of evening worship

ves•per•til•ian \ˌves-pər-ˈti-lē-ən, -ˈtil-yən\ *adjective* [Latin *vespertilio* bat, from *vesper*] (1874)
: of, relating to, or resembling a bat ⟨flaunts *vespertilian* wing and cloven hoof —Robert Graves⟩

ves•per•tine \ˈves-pər-ˌtīn\ *adjective* [Latin *vespertinus,* from *vesper*] (1502)
1 : of, relating to, or occurring in the evening ⟨*vespertine* shadows⟩
2 : active, flowering, or flourishing in the evening **:** CREPUSCULAR

ves•pid \ˈves-pəd\ *noun* [ultimately from Latin *vespa* wasp — more at WASP] (circa 1900)
: any of a cosmopolitan family (Vespidae) of chiefly social wasps that usually live in colonies like bees
— vespid *adjective*

ves•pine \ˈves-ˌpīn\ *adjective* [Latin *vespa* wasp] (1843)
: of, relating to, or resembling wasps and especially vespid wasps

ves•sel \ˈve-səl\ *noun* [Middle English, from Middle French *vaissel,* from Late Latin *vascellum,* diminutive of Latin *vas* vase, vessel] (14th century)
1 a : a container (as a hogshead, bottle, kettle, cup, or bowl) for holding something **b :** a person into whom some quality (as grace) is infused ⟨a child of light, a true *vessel* of the Lord —H. J. Laski⟩
2 : a watercraft bigger than a rowboat; *especially* **:** SHIP 1
3 a : a tube or canal (as an artery) in which a body fluid is contained and conveyed or circulated **b :** a conducting tube in the xylem of a vascular plant formed by the fusion and loss of end walls of a series of cells

[1]**vest** \ˈvest\ *verb* [Middle English, from Middle French *vestir* to clothe, invest, from Latin *vestire* to clothe, from *vestis* clothing, garment — more at WEAR] (15th century)
transitive verb
1 a : to place or give into the possession or discretion of some person or authority; *especially* **:** to give to a person a legally fixed immediate right of present or future enjoyment of (as an estate) **b :** to grant or endow with a particular authority, right, or property ⟨the retirement plan *vests* the workers absolutely with the company's contribution after 10 years of continuous employment⟩
2 : to clothe with or as if with a garment; *especially* **:** to robe in ecclesiastical vestments
intransitive verb
1 : to become legally vested
2 : to put on garments; *especially* **:** to put on ecclesiastical vestments

[2]**vest** *noun* [French *veste,* from Italian, from Latin *vestis* garment] (1613)
1 *archaic* **a :** a loose outer garment **:** ROBE **b :** CLOTHING, GARB
2 a : a man's sleeveless garment for the upper body usually worn under a suit coat; *also* **:** a similar garment for women **b :** a protective usually sleeveless garment (as a life preserver) that extends to the waist **c :** an insulated sleeveless waist-length garment often worn under or in place of a coat
3 a *chiefly British* **:** a man's sleeveless undershirt **b :** a knitted undershirt for women

□ USAGE

very The propriety of using *very* to modify a participial adjective, as in "very pleased" or "very flattering," has been the subject of considerable speculation since the 1870s. *Very,* an intensifier, does not modify verbs—you can "try hard" but you can't "try very"—and it was thus a matter of dispute whether *very* could modify participles. Many commentators believe that *very much* rather than *very* alone should be used. However, the question is not one of propriety, but of grammar. Participles have been moving into the class of adjectives since at least the 17th century. And one of the surest signs that a participle is established as an adjective is its being modified by *very.* The transition from verb form to adjective is a gradual one, and at any given time there will be participles that will sound right with *very* to some people and wrong to others. So writers have to trust their ear and use *very* where it sounds right and *very much* (or something else) where *very* alone sounds wrong. Dictionaries will not always be helpful. Participial adjectives are often omitted when their meaning is simply the meaning of the verb so that the space they would take up can be devoted to more words.

4 : a plain or decorative piece used to fill in the front neckline of a woman's outer garment (as a waist, coat, or gown)
— vest·like \-ˌlīk\ *adjective*

ves·ta \'ves-tə\ *noun* [Latin *Vesta*]
1 *capitalized* **:** the Roman goddess of the hearth — compare HESTIA
2 : a short match with a shank of wax-coated threads; *also* **:** a short wooden match

[1]ves·tal \'ves-tᵊl\ *adjective* (15th century)
1 : of or relating to the Roman goddess Vesta
2 a : of or relating to a vestal virgin **b :** CHASTE

[2]vestal *noun* (1549)
: VESTAL VIRGIN

vestal virgin *noun* (1600)
1 : a virgin consecrated to the Roman goddess Vesta and to the service of watching the sacred fire perpetually kept burning on her altar
2 : a chaste woman

vest·ed \'ves-təd\ *adjective* (1766)
1 : fully and unconditionally guaranteed as a legal right, benefit, or privilege ⟨the *vested* benefits of the pension plan⟩
2 : having a vest ⟨a *vested* suit⟩

vested interest *noun* (1818)
1 a : an interest (as a title to an estate) carrying a legal right of present or future enjoyment **b :** a right vested in an employee under a pension plan
2 : a special concern or stake in maintaining or influencing a condition, arrangement, or action especially for selfish ends
3 : one having a vested interest in something; *specifically* **:** a group enjoying benefits from an existing economic or political privilege

vest·ee \ve-'stē\ *noun* (1915)
1 : DICKEY; *especially* **:** one made to resemble a vest and worn under a coat
2 : VEST 4

ves·ti·ary \'ves-tē-ˌer-ē, 'vesh-chē-\ *noun* [Middle English *vestiarie,* from Middle French, vestry — more at VESTRY] (15th century)
1 : a room where clothing is kept
2 : CLOTHING, RAIMENT

ves·tib·u·lar \ve-'sti-byə-lər\ *adjective* (circa 1839)
: of, relating to, or functioning as a vestibule

ves·ti·bule \'ves-tə-ˌbyü(ə)l\ *noun* [Latin *vestibulum*] (1728)
1 : any of various bodily cavities especially when serving as or resembling an entrance to some other cavity or space: as **a :** the central cavity of the bony labyrinth of the ear or the parts of the membranous labyrinth that it contains **b :** the part of the left ventricle below the aortic orifice **c :** the space between the labia minora containing the orifice of the urethra **d :** the part of the mouth cavity outside the teeth and gums
2 a : a passage, hall, or room between the outer door and the interior of a building **:** LOBBY **b :** an enclosed entrance at the end of a railway passenger car
3 : a course that offers access (as to something new)
— ves·ti·buled \-ˌbyü(ə)ld\ *adjective*

vestibule school *noun* (1918)
: a school organized in an industrial plant to train new workers in specific skills

ves·tib·u·lo·co·chle·ar nerve \ve-ˌsti-byə-lō-'kō-klē-ər-, -'kä-\ *noun* (1962)
: AUDITORY NERVE

ves·tige \'ves-tij\ *noun* [French, from Latin *vestigium* footstep, footprint, track, vestige] (1602)
1 a (1) **:** a trace, mark, or visible sign left by something (as an ancient city or a condition or practice) vanished or lost (2) **:** the smallest quantity or trace **b :** FOOTPRINT 1
2 : a bodily part or organ that is small and degenerate or imperfectly developed in comparison to one more fully developed in an earlier stage of the individual, in a past generation, or in closely related forms
synonym see TRACE
— ves·ti·gial \ve-'sti-jē-əl, -jəl\ *adjective*
— ves·ti·gial·ly *adverb*

vest·ing \'ves-tiŋ\ *noun* (1944)
: the conveying to an employee of the inalienable right to share in a pension fund especially in the event of termination of employment prior to the normal retirement age; *also* **:** the right so conveyed

vest·ment \'ves(t)-mənt\ *noun* [Middle English *vestement,* from Middle French, from Latin *vestimentum,* from *vestire* to clothe] (14th century)
1 a : an outer garment; *especially* **:** a robe of ceremony or office **b** *plural* **:** CLOTHING, GARB
2 : a covering resembling a garment
3 : one of the articles of the ceremonial attire and insignia worn by ecclesiastical officiants and assistants as indicative of their rank and appropriate to the rite being celebrated
— vest·men·tal \ves(t)-'men-tᵊl\ *adjective*

vest–pocket *adjective* (1848)
1 : adapted to fit into the vest pocket ⟨a *vest-pocket* edition of a book⟩
2 : of very small size or scope

vest–pocket park *noun* (1966)
: a very small urban park

ves·try \'ves-trē\ *noun, plural* **vestries** [Middle English *vestrie,* probably modification of Middle French *vestiarie,* from Medieval Latin *vestiarium,* from Latin *vestire;* from its use as a robing room for the clergy] (14th century)
1 a : SACRISTY **b :** a room used for church meetings and classes
2 a : the business meeting of an English parish **b :** an elective body in an Episcopal parish composed of the rector and a group of elected parishioners administering the temporal affairs of the parish

ves·try·man \-trē-mən\ *noun* (1614)
: a member of a vestry

[1]ves·ture \'ves-chər, 'vesh-\ *noun* [Middle English, from Middle French, from *vestir* to clothe — more at VEST] (14th century)
1 a : a covering garment (as a robe or vestment) **b :** CLOTHING, APPAREL
2 : something that covers like a garment

[2]vesture *transitive verb* **ves·tured; ves·tur·ing** (1555)
: to cover with vesture **:** CLOTHE

ve·su·vi·an \və-'sü-vē-ən\ *noun* [*Vesuvian*] (1853)
: a match used especially formerly for lighting cigars

Ve·su·vi·an \və-'sü-vē-ən\ *adjective* (1673)
1 : of, relating to, or resembling the volcano Vesuvius
2 : marked by sudden outbursts ⟨has a *Vesuvian* temper, but quickly controls himself —Sidney Shalett⟩

ve·su·vi·an·ite \-vē-ə-ˌnīt\ *noun* (circa 1888)
: IDOCRASE

[1]vet \'vet\ *noun* (1862)
: VETERINARIAN, VETERINARY

[2]vet *transitive verb* **vet·ted; vet·ting** (1891)
1 a : to provide veterinary care for (an animal) or medical care for (a person) **b :** to subject (a person or animal) to a physical examination or checkup
2 : to subject to expert appraisal or correction **:** EVALUATE

[3]vet *adjective or noun* (1848)
: VETERAN

vetch \'vech\ *noun* [Middle English *vecche,* from Old North French *veche,* from Latin *vicia;* perhaps akin to Latin *vincire* to bind] (14th century)
: any of a genus (*Vicia*) of herbaceous twining leguminous plants including some grown for fodder and green manure

vetch·ling \-liŋ\ *noun* (1578)
: any of various leguminous herbs (genus *Lathyrus* and especially *L. pratensis*)

vet·er·an \'ve-tə-rən, 've-trən\ *noun* [Latin *veteranus,* from *veteranus,* adjective, old, of long experience, from *veter-, vetus* old — more at WETHER] (1509)
1 a : an old soldier of long service **b :** a former member of the armed forces
2 : a person of long experience in some occupation or skill (as politics or the arts)
— veteran *adjective*

Veterans Day *noun* (1952)
: November 11 set aside in commemoration of the end of hostilities in 1918 and 1945 and observed as a legal holiday in the U.S. to honor the veterans of the armed forces

veterans' preference *noun* (circa 1941)
: preferential treatment given qualified veterans of the U.S. armed forces under federal or state law; *specifically* **:** special consideration (as by allowance of points) on a civil service examination

vet·er·i·nar·i·an \ˌve-tə-rə-'ner-ē-ən, ˌve-trə-, ˌve-tᵊn-\ *noun* (1646)
: a person qualified and authorized to practice veterinary medicine

[1]vet·er·i·nary \'ve-tə-rə-ˌner-ē, 've-trə-, 've-tᵊn-\ *adjective* [Latin *veterinarius* of beasts of burden, from *veterinae* beasts of burden, from feminine plural of *veterinus* of beasts of burden; akin to Latin *veter-, vetus* old] (1790)
: of, relating to, or being the science and art of prevention, cure, or alleviation of disease and injury in animals and especially domestic animals

[2]veterinary *noun, plural* **-nar·ies** (1861)
: VETERINARIAN

veterinary surgeon *noun* (circa 1802)
British **:** VETERINARIAN

vet·i·ver \'ve-tə-vər\ *noun* [French *vétiver,* from Tamil *veṭṭivēr*] (circa 1858)
: an East Indian grass (*Vetiveria zizanioides*) cultivated in warm regions especially for its fragrant roots which are used especially in woven goods (as mats) and in perfumes; *also* **:** its root

[1]ve·to \'vē-(ˌ)tō\ *noun, plural* **vetoes** [Latin, I forbid, from *vetare* to forbid] (1629)
1 : an authoritative prohibition **:** INTERDICTION
2 a : a power of one department or branch of a government to forbid or prohibit finally or provisionally the carrying out of projects attempted by another department; *especially* **:** a power vested in a chief executive to prevent permanently or temporarily the enactment of measures passed by a legislature **b** (1) **:** the exercise of such authority (2) **:** a message communicating the reasons of an executive and especially the president of the U.S. for vetoing a proposed law

[2]veto *transitive verb* (1706)
: to refuse to admit or approve **:** PROHIBIT; *also* **:** to refuse assent to (a legislative bill) so as to prevent enactment or cause reconsideration
— ve·to·er \-ˌtō(-ə)r\ *noun*

ve·to–proof \-ˌprüf\ *adjective* (1972)
: having enough potential votes to be enacted over a veto or to override vetoes consistently ⟨a *veto-proof* bill⟩

vex \'veks\ *transitive verb* **vexed** *also* **vext; vex·ing** [Middle English, from Middle French *vexer,* from Latin *vexare* to agitate, harry; probably akin to Latin *vehere* to convey — more at WAY] (15th century)
1 a : to bring trouble, distress, or agitation to ⟨the restaurant is *vexed* by slow service⟩ **b :** to bring physical distress to ⟨a headache *vexed* him all morning⟩ **c :** to irritate or annoy by petty provocations **:** HARASS ⟨*vexed* by the children⟩ **d :** PUZZLE, BAFFLE ⟨a problem to *vex* the keenest wit⟩
2 : to shake or toss about
synonym see ANNOY

vex·a·tion \vek-'sā-shən\ *noun* (15th century)
1 : the act of harassing or vexing **:** TROUBLING
2 : the quality or state of being vexed **:** IRRITATION
3 : a cause of trouble **:** AFFLICTION

vex·a·tious \-shəs\ *adjective* (1534)

1 a : causing vexation **:** DISTRESSING **b :** intended to harass
2 : full of disorder or stress **:** TROUBLED
— vex·a·tious·ly *adverb*
— vex·a·tious·ness *noun*

vexed \'vekst\ *adjective* (1657)
: debated or discussed at length ⟨a *vexed* question⟩

vexed·ly \'vek-səd-lē, 'vekst-lē\ *adverb* (1748)
: with vexation

vex·il·lol·o·gy \,vek-sə-'lä-lə-jē\ *noun* [Latin *vexillum*] (1959)
: the study of flags
— vex·il·lo·log·ic \(,)vek-,si-lə-'lä-jik\ *or* **vex·il·lo·log·i·cal** \-'lä-ji-kəl\ *adjective*
— vex·il·lol·o·gist \,vek-sə-'lä-lə-jist\ *noun*

vex·il·lum \vek-'si-ləm\ *noun, plural* **-la** \-lə\ [Latin; akin to Latin *velum* curtain, awning] (1726)
1 : a square flag of the ancient Roman cavalry
2 : the web or vane of a feather

via \'vī-ə, 'vē-ə\ *preposition* [Latin, ablative of *via* way — more at WAY] (1779)
1 : by way of
2 : through the medium or agency of; *also* **:** by means of

vi·a·ble \'vī-ə-bəl\ *adjective* [French, from Middle French, from *vie* life, from Latin *vita* — more at VITAL] (circa 1832)
1 : capable of living; *especially* **:** capable of surviving outside the mother's womb without artificial support ⟨the normal human fetus is usually *viable* by the end of the seventh month⟩
2 : capable of growing or developing ⟨*viable* seeds⟩ ⟨*viable* eggs⟩
3 a : capable of working, functioning, or developing adequately ⟨*viable* alternatives⟩ **b :** capable of existence and development as an independent unit ⟨the colony is now a *viable* state⟩ **c** (1) **:** having a reasonable chance of succeeding ⟨a *viable* candidate⟩ (2) **:** financially sustainable ⟨a *viable* enterprise⟩
— vi·a·bil·i·ty \,vī-ə-'bi-lə-te\ *noun*
— vi·a·bly \'vī-ə-blē\ *adverb*

via·duct \'vī-ə-,dəkt\ *noun* [Latin *via* way, road + English aque*duct*] (1816)
: a long elevated roadway usually consisting of a series of short spans supported on arches, piers, or columns

viaduct

vi·al \'vī(-ə)l\ *noun* [Middle English *fiole, viole,* from Middle French *fiole,* from Old Provençal *fiola,* from Latin *phiala* — more at PHIAL] (14th century)
: a small closed or closable vessel especially for liquids

via me·dia \,vī-ə-'mē-dē-ə; ,vē-ə-'mā-dē-ə, -'me-\ *noun* [Latin] (1834)
: a middle way

vi·and \'vī-ənd\ *noun* [Middle English, from Middle French *viande,* from Medieval Latin *vivanda* food, alteration of Latin *vivenda,* neuter plural of *vivendus,* gerundive of *vivere* to live — more at QUICK] (15th century)
1 : an item of food; *especially* **:** a choice or tasty dish
2 *plural* **:** PROVISIONS, FOOD

vi·at·i·cum \vī-'a-ti-kəm, vē-\ *noun, plural* **-cums** *or* **-ca** \-kə\ [Latin — more at VOYAGE] (1562)
1 : the Christian Eucharist given to a person in danger of death
2 a : an allowance (as of transportation or supplies and money) for traveling expenses **b :** provisions for a journey

vibe \'vīb\ *noun* (1967)
: VIBRATION 4 ⟨seems to be in on every conversation, every deal, every *vibe* that is winging through the room —Albert Goldman⟩ — usually used in plural ⟨got bad *vibes* from him⟩

vibes \'vībz\ *noun plural* (1940)
: VIBRAPHONE
— vib·ist \'vī-bist\ *noun*

vi·bra·harp \'vī-brə-,härp\ *noun* [from *Vibra-Harp,* a trademark] (1930)
: VIBRAPHONE
— vi·bra·harp·ist \-,här-pist\ *noun*

vi·brance \'vī-brən(t)s\ *noun* (1921)
: VIBRANCY

vi·bran·cy \'vī-brən(t)-sē\ *noun* (circa 1890)
: the quality or state of being vibrant

vi·brant \-brənt\ *adjective* (1616)
1 a (1) **:** oscillating or pulsating rapidly (2) **:** pulsating with life, vigor, or activity ⟨a *vibrant* personality⟩ **b** (1) **:** readily set in vibration (2) **:** RESPONSIVE, SENSITIVE
2 : sounding as a result of vibration **:** RESONANT ⟨a *vibrant* voice⟩
3 : BRIGHT 4 ⟨a *vibrant* orange⟩
— vi·brant·ly *adverb*

vi·bra·phone \'vī-brə-,fōn\ *noun* [Latin *vibrare* + International Scientific Vocabulary *-phone*] (1926)
: a percussion instrument resembling the xylophone but having metal bars and motor-driven resonators for sustaining the tone and producing a vibrato
— vi·bra·phon·ist \-,fō-nist\ *noun*

vibraphone

vi·brate \'vī-,brāt, *especially British* vī-'\ *verb* **vi·brat·ed; vi·brat·ing** [Latin *vibratus,* past participle of *vibrare* to brandish, wave, rock — more at WIPE] (1616)
transitive verb
1 : to swing or move to and fro
2 : to emit with or as if with a vibratory motion
3 : to mark or measure by oscillation ⟨a pendulum *vibrating* seconds⟩
4 : to set in vibration
intransitive verb
1 a : to move to and fro or from side to side **:** OSCILLATE **b :** FLUCTUATE, VACILLATE ⟨*vibrate* between two choices⟩
2 : to have an effect as or as if of vibration ⟨music, when soft voices die, *vibrates* in the memory —P. B. Shelley⟩
3 : to be in a state of vibration **:** QUIVER
4 : to respond sympathetically **:** THRILL ⟨*vibrate* to the opportunity⟩
synonym see SWING

vi·bra·tile \'vī-brə-tᵊl, -,tīl\ *adjective* (circa 1826)
1 : characterized by vibration
2 : adapted to or used in vibratory motion ⟨the *vibratile* organs of insects⟩

vi·bra·tion \vī-'brā-shən\ *noun* (1655)
1 a : a periodic motion of the particles of an elastic body or medium in alternately opposite directions from the position of equilibrium when that equilibrium has been disturbed (as when a stretched cord produces musical tones or particles of air transmit sounds to the ear) **b :** the action of vibrating **:** the state of being vibrated or in vibratory motion: as (1) **:** OSCILLATION (2) **:** a quivering or trembling motion **:** QUIVER
2 : an instance of vibration
3 : vacillation in opinion or action **:** WAVERING
4 a : a characteristic emanation, aura, or spirit that infuses or vitalizes someone or something and that can be instinctively sensed or experienced — often used in plural **b :** a distinctive usually emotional atmosphere capable of being sensed — usually used in plural
— vi·bra·tion·al \-shnəl, -shə-nᵊl\ *adjective*
— vi·bra·tion·less \-shən-ləs\ *adjective*

vi·bra·to \vi-'brä-(,)tō, vī-\ *noun, plural* **-tos** [Italian, from past participle of *vibrare* to vibrate, from Latin] (circa 1876)
: a slightly tremulous effect imparted to vocal or instrumental tone for added warmth and expressiveness by slight and rapid variations in pitch
— vi·bra·to·less \-ləs\ *adjective*

vi·bra·tor \'vī-,brā-tər\ *noun* (1862)
: one that vibrates or causes vibration: as **a :** a vibrating electrical apparatus used in massage or for sexual stimulation **b :** a vibrating device (as in an electric bell or buzzer)

vi·bra·to·ry \'vī-brə-,tōr-ē, -,tȯr-\ *adjective* (1728)
1 : consisting in, capable of, or causing vibration or oscillation
2 : characterized by vibration

vib·rio \'vi-brē-,ō\ *noun, plural* **-rios** [New Latin, *Vibrion-, Vibrio,* from Latin *vibrare* to wave] (circa 1864)
: any of a genus (*Vibrio*) of short rigid motile bacteria typically shaped like a comma or an S
— vib·ri·on·ic \,vi-brē-'ä-nik\ *adjective*

vib·ri·on \'vi-brē-,än\ *noun* [New Latin *Vibrion-, Vibrio*] (1882)
: VIBRIO; *also* **:** a motile bacterium

vib·ri·o·sis \,vi-brē-'ō-səs\ *noun, plural* **-o·ses** \-,sēz\ [New Latin, from *Vibrio*] (1950)
: abortion in sheep and cattle caused by a bacterium (*Campylobacter fetus* synonym *Vibrio fetus*) that invades the uterine and placental capillaries, interferes with fetal nutrition, and causes the death of the developing fetus

vi·bris·sa \vī-'bri-sə, və-\ *noun, plural* **vi·bris·sae** \vī-'bri-(,)sē; və-'bri-(,)sē, -,sī\ [Medieval Latin, from Latin *vibrare*] (circa 1693)
1 : one of the stiff hairs that are located especially about the nostrils or on other parts of the face in many mammals and that often serve as tactile organs
2 : one of the bristly feathers near the mouth of many and especially insectivorous birds that may help to prevent the escape of insects

vi·bur·num \vī-'bər-nəm\ *noun* [New Latin, from Latin, a viburnum] (circa 1731)
: any of a genus (*Viburnum*) of widely distributed shrubs or trees of the honeysuckle family with simple leaves and white or rarely pink cymose flowers

vic·ar \'vi-kər\ *noun* [Middle English, from Latin *vicarius,* from *vicarius* vicarious] (14th century)
1 : one serving as a substitute or agent; *specifically* **:** an administrative deputy
2 : an ecclesiastical agent: as **a :** a Church of England incumbent receiving a stipend but not the tithes of a parish **b :** a member of the Episcopal clergy or laity who has charge of a mission or chapel **c :** a member of the clergy who exercises a broad pastoral responsibility as the representative of a prelate
— vic·ar·ship \-,ship\ *noun*

vic·ar·age \'vi-k(ə-)rij\ *noun* (15th century)
1 : the benefice of a vicar
2 : the house of a vicar
3 : VICARIATE 1

vicar apostolic *noun, plural* **vicars apostolic** (1766)
: a Roman Catholic titular bishop who administers a territory not organized as a diocese

vic·ar·ate \'vi-kə-rət, -,rāt\ *noun* (1883)
: VICARIATE

vicar–general *noun, plural* **vicars–general** (15th century)
: an administrative deputy of a Roman Catholic or Anglican bishop or of the head of a religious order

vi·car·i·al \vī-'ker-ē-əl, və-, -'kar-\ *adjective* [Latin *vicarius*] (1617)

1 : VICARIOUS 1
2 : of or relating to a vicar

vi·car·i·ance \-ē-ən(t)s\ *noun* (1957)
: fragmentation of the environment (as by splitting of a tectonic plate) in contrast to dispersal as a factor in promoting biological evolution by division of large populations into isolated subpopulations — called also *vicariance biogeography*

vi·car·i·ant \-ē-ənt\ *adjective* [translation of German *vikarirend,* present participle of *vikarieren* to act as a substitute, from *Vikar* representative, proxy, from Middle High German *vicar,* from Latin *vicarius* substitute] (1952)
: of, relating to, or being the process of vicariance or organisms that evolved through this process ⟨the possible *vicariant* origin of the Antillean arthropod fauna⟩
— vicariant *noun*

vi·car·i·ate \-ē-ət\ *noun* [Medieval Latin *vicariatus,* from Latin *vicarius* vicar] (1610)
1 : the office, jurisdiction, or tenure of a vicar
2 : the office or district of a governmental administrative deputy

vi·car·i·ous \vī-ˈker-ē-əs, və-, -ˈkar-\ *adjective* [Latin *vicarius,* from *vicis* change, alternation, stead — more at WEEK] (1637)
1 a : serving instead of someone or something else **b :** that has been delegated ⟨*vicarious* authority⟩
2 : performed or suffered by one person as a substitute for another or to the benefit or advantage of another **:** SUBSTITUTIONARY ⟨a *vicarious* sacrifice⟩
3 : experienced or realized through imaginative or sympathetic participation in the experience of another
4 : occurring in an unexpected or abnormal part of the body instead of the usual one ⟨bleeding from the gums sometimes occurs in the absence of the normal discharge from the uterus in *vicarious* menstruation⟩
— vi·car·i·ous·ly *adverb*
— vi·car·i·ous·ness *noun*

Vicar of Christ (1570)
: POPE 1

¹vice \ˈvīs\ *noun* [Middle English, from Old French, from Latin *vitium* fault, vice] (14th century)
1 a : moral depravity or corruption **:** WICKEDNESS **b :** a moral fault or failing **c :** a habitual and usually trivial defect or shortcoming **:** FOIBLE ⟨suffered from the *vice* of curiosity⟩
2 : BLEMISH, DEFECT
3 : a physical imperfection, deformity, or taint
4 a *often capitalized* **:** a character representing one of the vices in an English morality play **b :** BUFFOON, JESTER
5 : an abnormal behavior pattern in a domestic animal detrimental to its health or usefulness
6 : sexual immorality; *especially* **:** PROSTITUTION
synonym see FAULT, OFFENSE

²vice *chiefly British variant of* VISE

³vice \ˈvīs *also* ˈvī-sē\ *preposition* [Latin, ablative of *vicis* change, alternation, stead — more at WEEK] (1770)
: in the place of ⟨I will preside, *vice* the absent chairman⟩; *also* **:** rather than

vice- \ˈvīs, ˌvīs\ *prefix* [Middle English *vis-, vice-,* from Middle French, from Late Latin *vice-,* from Latin *vice,* ablative of *vicis*]
: one that takes the place of ⟨*vice*-chancellor⟩

vice admiral *noun* [Middle French *visamiral,* from *vis-* vice- + *amiral* admiral] (1520)
: a commissioned officer in the navy or coast guard who ranks above a rear admiral and whose insignia is three stars

vice–chan·cel·lor \ˈvīs-ˈchan(t)-s(ə-)lər, ˌvīs-\ *noun* [Middle English *vichauncellor,* from Middle French *vischancelier,* from *vis-* + *chancelier* chancellor] (15th century)
1 : an officer ranking next below a chancellor and serving as deputy to the chancellor
2 : chief administrative officer in a British university
3 : a judge appointed to act for or to assist a chancellor

vice–con·sul \-ˈkän(t)-səl\ *noun* (1559)
: a consular officer subordinate to a consul general or to a consul

vice·ge·ren·cy \-ˈjir-ən(t)-sē\ *noun, plural* **-cies** (1596)
: the office or jurisdiction of a vicegerent

vice·ge·rent \-ˈjir-ənt\ *noun* [Medieval Latin *vicegerent-, vicegerens,* from Late Latin *vice-* + Latin *gerent-, gerens,* present participle of *gerere* to carry, carry on] (1536)
: an administrative deputy of a king or magistrate

vi·cen·ni·al \vī-ˈse-nē-əl\ *adjective* [Late Latin *vicennium* period of 20 years, from Latin *vicies* 20 times + *annus* year; akin to Latin *viginti* twenty — more at VIGESIMAL, ANNUAL] (circa 1859)
: occurring once every 20 years

vice presidency *noun* (1804)
: the office of vice president

vice president *noun* (1574)
1 : an officer next in rank to a president and usually empowered to serve as president in that officer's absence or disability
2 : any of several officers serving as a president's deputies in charge of particular locations or functions
— vice presidential *adjective*

vice·re·gal \ˈvīs-ˈrē-gəl, ˌvīs-\ *adjective* (1836)
: of or relating to a viceroy or viceroyalty
— vice·re·gal·ly \-gə-lē\ *adverb*

vice–re·gent \-ˈrē-jənt\ *noun* (1556)
: a regent's deputy

vice·reine \ˈvīs-ˌrān\ *noun* [French, from *vice-* + *reine* queen, from Latin *regina,* feminine of *reg-, rex* king — more at ROYAL] (1823)
1 : the wife of a viceroy
2 : a woman who is a viceroy

vice·roy \ˈvīs-ˌrȯi\ *noun* [Middle French *viceroi,* from *vice-* + *roi* king, from Latin *reg-, rex*] (1524)
1 : the governor of a country or province who rules as the representative of a king or sovereign
2 : a showy American nymphalid butterfly (*Limenitis archippus*) closely mimicking the monarch in coloration but smaller

vice·roy·al·ty \ˈvīs-ˌrȯi(-ə)l-tē, ˌvīs-ˈ\ *noun* (1703)
: the office, jurisdiction, or term of service of a viceroy

vice·roy·ship \ˈvīs-ˌrȯi-ˌship\ *noun* (1609)
: VICEROYALTY

vice squad *noun* (1905)
: a police squad charged with enforcement of laws concerning gambling, pornography, prostitution, and the illegal use of liquor and narcotics

vice ver·sa \ˌvī-si-ˈvər-sə, ˈvīs-ˈvər-\ *adverb* [Latin] (1601)
: with the order changed **:** with the relations reversed **:** CONVERSELY

vi·chys·soise \ˌvi-shē-ˈswäz, ˌvē-\ *noun* [French, from feminine of *vichyssois* of Vichy, from *Vichy,* France] (1939)
: a soup made of pureed leeks or onions and potatoes, cream, and chicken stock and usually served cold

Vi·chy water \ˈvi-shē-\ *noun* (circa 1858)
: a natural sparkling mineral water from Vichy, France; *also* **:** an imitation of or substitute for this

vic·i·nage \ˈvi-sᵊn-ij, ˈvis-nij\ *noun* [Middle English *vesinage,* from Middle French, from *vesin* neighboring, from Latin *vicinus*] (14th century)
: a neighboring or surrounding district **:** VICINITY

vic·i·nal \ˈvi-sᵊn-əl, ˈvis-nəl\ *adjective* [Latin *vicinalis,* from *vicinus* neighbor, from *vicinus,* adjective, neighboring] (circa 1623)
1 : of or relating to a limited district **:** LOCAL
2 : of, relating to, or substituted in adjacent sites in a molecule ⟨a *vicinal* disulfide group⟩

vi·cin·i·ty \və-ˈsi-nə-tē\ *noun, plural* **-ties** [Middle French *vicinité,* from Latin *vicinitat-, vicinitas,* from *vicinus* neighboring, from *vicus* row of houses, village; akin to Gothic *weihs* village, Old Church Slavonic *vĭsĭ,* Greek *oikos, oikia* house] (1560)
1 : the quality or state of being near **:** PROXIMITY
2 : a surrounding area or district **:** NEIGHBORHOOD
3 : NEIGHBORHOOD 3b

vi·cious \ˈvi-shəs\ *adjective* [Middle English, from Middle French *vicieus,* from Latin *vitiosus* full of faults, corrupt, from *vitium* vice] (14th century)
1 : having the nature or quality of vice or immorality **:** DEPRAVED
2 : DEFECTIVE, FAULTY; *also* **:** INVALID
3 : IMPURE, NOXIOUS
4 a : dangerously aggressive **:** SAVAGE ⟨a *vicious* dog⟩ **b :** marked by violence or ferocity **:** FIERCE ⟨a *vicious* fight⟩
5 : MALICIOUS, SPITEFUL ⟨*vicious* gossip⟩
6 : worsened by internal causes that reciprocally augment each other ⟨a *vicious* wage-price spiral⟩ ☆
— vi·cious·ly *adverb*
— vi·cious·ness *noun*

vicious circle *noun* (circa 1792)
1 : a chain of events in which the response to one difficulty creates a new problem that aggravates the original difficulty
2 : an argument or definition that begs the question

vi·cis·si·tude \və-ˈsi-sə-ˌtüd, vī-, -ˌtyüd\ *noun* [Middle French, from Latin *vicissitudo,* from *vicissim* in turn, from *vicis* change, alternation — more at WEEK] (circa 1576)
1 a : the quality or state of being changeable **:** MUTABILITY **b :** natural change or mutation visible in nature or in human affairs
2 a : a favorable or unfavorable event or situation that occurs by chance **:** a fluctuation of state or condition ⟨the *vicissitudes* of daily life⟩ **b :** a difficulty or hardship attendant on a way of life, a career, or a course of action and usually beyond one's control **c :** alternating change **:** SUCCESSION

vi·cis·si·tu·di·nous \və-ˌsi-sə-ˈtüd-nəs, (ˌ)vī-, -ˈtyüd-; -ˈtü-dᵊn-əs, -ˈtyü-\ *adjective* [Latin *vicissitudin-, vicissitudo*] (circa 1846)
: marked by or filled with vicissitudes

vic·tim \ˈvik-təm\ *noun* [Middle English *vyctym,* from Latin *victima;* perhaps akin to Old High German *wīh* holy] (15th century)
1 : a living being sacrificed to a deity or in the performance of a religious rite

☆ SYNONYMS

Vicious, villainous, iniquitous, nefarious, corrupt, degenerate mean highly reprehensible or offensive in character, nature, or conduct. VICIOUS may directly oppose *virtuous* in implying moral depravity, or may connote malignancy, cruelty, or destructive violence ⟨a *vicious* gangster⟩. VILLAINOUS applies to any evil, depraved, or vile conduct or characteristic ⟨a *villainous* assault⟩. INIQUITOUS implies absence of all signs of justice or fairness ⟨an *iniquitous* system of taxation⟩. NEFARIOUS suggests flagrant breaching of time-honored laws and traditions of conduct ⟨the *nefarious* rackets of organized crime⟩. CORRUPT stresses a loss of moral integrity or probity causing betrayal of principle or sworn obligations ⟨city hall was rife with *corrupt* politicians⟩. DEGENERATE suggests having sunk to an especially vicious or enervated condition ⟨a *degenerate* regime propped up by foreign powers⟩.

2 : one that is acted on and usually adversely affected by a force or agent ⟨the schools are *victims* of the social system⟩: as **a** (1) **:** one that is injured, destroyed, or sacrificed under any of various conditions ⟨a *victim* of cancer⟩ ⟨a *victim* of the auto crash⟩ ⟨a murder *victim*⟩ (2) **:** one that is subjected to oppression, hardship, or mistreatment ⟨a frequent *victim* of political attacks⟩ **b :** one that is tricked or duped ⟨a con man's *victim*⟩
— **vic·tim·hood** \-ˌhu̇d\ *noun*
vic·tim·ise *British variant of* VICTIMIZE
vic·tim·ize \ˈvik-tə-ˌmīz\ *transitive verb* **-ized; -iz·ing** (1830)
1 : to make a victim of
2 : to subject to deception or fraud **:** CHEAT
— **vic·tim·i·za·tion** \ˌvik-tə-mə-ˈzā-shən\ *noun*
— **vic·tim·iz·er** \ˈvik-tə-ˌmī-zər\ *noun*
vic·tim·less \ˈvik-təm-ləs\ *adjective* (1965)
: having no victim **:** not of a nature that may produce a complainant ⟨a *victimless* crime⟩
vic·tim·ol·o·gy \ˌvik-tə-ˈmä-lə-jē\ *noun* (1958)
1 : the study of victims and victimization
2 : the claim that the problems of a person or group are the result of victimization
— **vic·tim·ol·o·gist** \-jist\ *noun*
vic·tor \ˈvik-tər\ *noun* [Middle English, from Latin, from *vincere* to conquer, win; akin to Old English *wīgan* to fight, Lithuanian *veikti* to be active] (14th century)
: one that defeats an enemy or opponent **:** WINNER
— **victor** *adjective*
Victor (1942)
— a communications code word for the letter *v*
vic·to·ria \vik-ˈtōr-ē-ə, -ˈtȯr-\ *noun* [Queen *Victoria*] (circa 1864)
: a low four-wheeled pleasure carriage for two with a folding top and a raised seat in front for the driver
Victoria Cross *noun* (1856)
: a bronze Maltese cross awarded to members of the British armed services for acts of remarkable valor
Victoria Day *noun* [Queen *Victoria*] (1901)
1 : formerly May 24 and now the Monday preceding May 25 observed in Canada as a legal holiday
2 : COMMONWEALTH DAY
[1]Vic·to·ri·an \vik-ˈtōr-ē-ən, -ˈtȯr-\ *adjective* (1839)
1 : of, relating to, or characteristic of the reign of Queen Victoria of England or the art, letters, or tastes of her time
2 : typical of the moral standards, attitudes, or conduct of the age of Victoria especially when considered stuffy or hypocritical
[2]Victorian *noun* (1876)
1 : a person living during Queen Victoria's reign; *especially* **:** a representative figure of that time
2 : a typically large and ornate house built during Queen Victoria's reign
Vic·to·ri·ana \(ˌ)vik-ˌtōr-ē-ˈa-nə, -ˌtȯr-, -ˈä-, -ˈā-\ *noun* [Queen *Victoria* + English *-ana*] (1940)
: materials concerning or characteristic of the Victorian age; *also* **:** a collection of such materials
Vic·to·ri·an·ism \vik-ˈtōr-ē-ə-ˌni-zəm, -ˈtȯr-\ *noun* (1905)
1 : a typical instance or product of Victorian expression, taste, or conduct
2 : the quality or state of being Victorian especially in taste or conduct
vic·to·ri·ous \vik-ˈtōr-ē-əs, -ˈtȯr-\ *adjective* (14th century)
1 a : having won a victory **b :** of, relating to, or characteristic of victory
2 : evincing moral harmony or a sense of fulfillment **:** FULFILLED
— **vic·to·ri·ous·ly** *adverb*
— **vic·to·ri·ous·ness** *noun*

vic·to·ry \ˈvik-t(ə-)rē\ *noun, plural* **-ries** [Middle English, from Middle French *victorie,* from Latin *victoria,* from *victor*] (14th century)
1 : the overcoming of an enemy or antagonist
2 : achievement of mastery or success in a struggle or endeavor against odds or difficulties
Vic·tro·la \vik-ˈtrō-lə\ *trademark*
— used for a phonograph
[1]vict·ual \ˈvi-tᵊl\ *noun* [alteration of Middle English *vitaille,* from Middle French, from Late Latin *victualia,* plural, provisions, victuals, from neuter plural of *victualis* of nourishment, from Latin *victus* nourishment, way of living, from *vivere* to live — more at QUICK] (1523)
1 : food usable by people
2 *plural* **:** supplies of food **:** PROVISIONS
[2]victual *verb* **-ualed** *or* **-ualled; -ual·ing** *or* **-ual·ling** (1558)
transitive verb
: to supply with food
intransitive verb
1 : EAT
2 : to lay in provisions
vict·ual·ler *or* **vict·ual·er** \ˈvi-tᵊl-ər\ *noun* (1568)
1 : the keeper of a restaurant or tavern
2 : one that provisions an army, a navy, or a ship with food
3 : an army or navy provision ship
vi·cu·ña *or* **vi·cu·na** \vi-ˈkün-yə, vī-; vī-ˈkü-nə, və-, -ˈkyü-\ *noun* [Spanish *vicuña,* from Quechua *wik'uña*] (1604)
1 : a wild ruminant (*Vicugna vicugna* synonym *Lama vicugna*) of the Andes from Peru to Argentina that is related to the llama and alpaca
2 a : the wool from the vicuña's fine lustrous undercoat **b :** a fabric made of vicuña wool; *also* **:** a sheep's wool imitation of this

vicuña 1

vi·de \ˈvī-dē, ˈvē-ˌdā\ *imperative verb* [Latin, from *vidēre* to see — more at WIT] (1565)
: SEE — used to direct a reader to another item
vi·de·li·cet \və-ˈde-lə-ˌset, vī-; vi-ˈdā-li-ˌket\ *adverb* [Middle English, from Latin, from *vidēre* to see + *licet* it is permitted, from *licēre* to be permitted] (15th century)
: that is to say **:** NAMELY
[1]vid·eo \ˈvi-dē-ˌō\ *noun* [Latin *vidēre* to see + *-o* (as in *audio*)] (1937)
1 : TELEVISION; *also* **:** the visual portion of television
2 : VIDEOTAPE: as **a :** a recording of a motion picture or television program for playing through a television set **b :** a videotaped performance of a song often featuring an interpretation of the lyrics through visual images
[2]video *adjective* (1938)
1 : being, relating to, or used in the transmission or reception of the television image ⟨*video* channel⟩ — compare AUDIO
2 : being, relating to, or involving images on a television screen or computer display ⟨*video* terminal⟩
vid·eo·cas·sette \ˌvi-dē-ō-kə-ˈset, -ka-\ *noun* (1970)
1 : a case containing videotape for use with a VCR
2 : a recording (as of a movie) on a videocassette
videocassette recorder *noun* (1976)
: VCR
vid·eo·con·fer·enc·ing \-ˈkän-f(ə-)rən(t)-siŋ, -fərn(t)-\ *noun* (1977)

: the holding of a conference among people at remote locations by means of transmitted audio and video signals
— **vid·eo·con·fer·ence** \-ˈkän-f(ə-)rən(t)s, -fərn(t)s\ *noun*
vid·eo·disc *or* **vid·eo·disk** \ˈvi-dē-ō-ˌdisk\ *noun* (1967)
1 : a disc similar in appearance and use to a phonograph record on which programs have been recorded for playback on a television set; *also* **:** OPTICAL DISK
2 : a recording (as of a movie) on a videodisc
video game *noun* (1973)
: an electronic game played by means of images on a video screen and often emphasizing fast action
vid·e·og·ra·phy \ˌvi-dē-ˈä-grə-fē\ *noun* (1972)
: the practice or art of recording images with a video camera
— **vid·e·og·ra·pher** \-fər\ *noun*
vid·eo·land \-ˌland\ *noun* (1967)
: television as a medium or industry
vid·eo·phile \ˈvi-dē-ə-ˌfīl\ *noun* (1966)
: a person fond of video; *especially* **:** one interested in video equipment or in producing videos
vid·eo·phone \ˈvi-dē-ə-ˌfōn\ *noun* (circa 1950)
: a telephone equipped for transmission of video as well as audio signals so that users can see each other
[1]vid·eo·tape \ˈvi-dē-ō-ˌtāp\ *noun* (1953)
: a recording of visual images and sound (as of a television production) made on magnetic tape; *also* **:** the magnetic tape used for such a recording
[2]videotape *transitive verb* (1958)
: to make a videotape of ⟨*videotape* a show⟩ ⟨*videotape* the president's speech⟩
videotape recorder *noun* (1953)
: a device for recording and playing back videotapes — called also *video recorder*
vid·eo·tex \-ˌteks\ *also* **vid·eo·text** \-ˌtekst\ *noun* [[2]*video* + *-tex* (alteration of *text*)] (1978)
: an electronic data retrieval system in which usually textual information is transmitted via telephone or cable-television lines and displayed on a television set or video display terminal; *especially* **:** such a system that is interactive — compare TELETEXT
video vé·ri·té \-ˌver-ə-ˈtā\ *noun* [*cinema verité*] (1969)
: the filming or videotaping of a television program (as a documentary) so as to convey candid realism
vi·dette *variant of* VEDETTE
vid·icon \ˈvi-di-ˌkän\ *noun, often capitalized* [[2]*video* + *iconoscope*] (1950)
: a camera tube using the principle of photoconductivity
vi·du·ity \vi-ˈdü-ə-tē, -ˈdyü-\ *noun* [Middle English (Scots) *viduite,* from Middle French *viduite,* from Latin *viduitat-, viduitas,* from *vidua* widow — more at WIDOW] (15th century)
: WIDOWHOOD
vie \ˈvī\ *verb* **vied; vy·ing** \ˈvī-iŋ\ [modification of Middle French *envier* to invite, challenge, wager, from Latin *invitare* to invite] (1577)
intransitive verb
: to strive for superiority **:** CONTEND, COMPETE
transitive verb
archaic **:** WAGER, HAZARD; *also* **:** to exchange in rivalry **:** MATCH
— **vi·er** \ˈvī(-ə)r\ *noun*
Vi·en·na sausage \vē-ˈe-nə-\ *noun* [*Vienna,* Austria] (circa 1902)
: a short slender frankfurter

\ə\ **abut** \ᵊ\ **kitten** \ər\ **further** \a\ **ash** \ā\ **ace** \ä\ **mop, mar** \au̇\ **out** \ch\ **chin** \e\ **bet** \ē\ **easy** \g\ **go** \i\ **hit** \ī\ **ice** \j\ **job** \ŋ\ **sing** \ō\ **go** \ȯ\ **law** \ȯi\ **boy** \th\ **thin** \t͟h\ **the** \ü\ **loot** \u̇\ **foot** \y\ **yet** \zh\ **vision** *see also* Guide to Pronunciation

Viet·cong \vē-ˈet-ˈkäŋ, vyet-, ˌvē-ət-, vēt-, -ˈkȯŋ\ *noun, plural* **Vietcong** [Vietnamese *Việt-cộng*] (1957)
: a guerrilla member of the Vietnamese communist movement

Viet·minh \-ˈmin\ *noun, plural* **Vietminh** [Vietnamese *Việt-Minh,* short for *Việt-Nam Độc-Lập Đồng-Minh* League for the Independence of Vietnam] (1945)
: an adherent of the Vietnamese communist movement from 1941 to 1951

Viet·nam·ese \vē-ˌet-nə-ˈmēz, ˌvyet-, ˌvē-ət-, ˌvēt-, -na-, -nä-, -ˈmēs\ *noun, plural* **Vietnamese** (1947)
1 : a native or inhabitant of Vietnam
2 : the language of the largest group in Vietnam and the official language of the country
— Vietnamese *adjective*

[1]**view** \ˈvyü\ *noun* [Middle English *vewe,* from Middle French *veue, vue,* from Old French, from feminine of *veu, vu,* past participle of *veeir, voir* to see, from Latin *vidēre* — more at WIT] (15th century)
1 : the act of seeing or examining **:** INSPECTION; *also* **:** SURVEY ⟨a *view* of English literature⟩
2 a : a mode or manner of looking at or regarding something **b :** an opinion or judgment colored by the feeling or bias of its holder ⟨in my *view* the conference has no chance of success⟩
3 : SCENE, PROSPECT ⟨the lovely *view* from the balcony⟩
4 : extent or range of vision **:** SIGHT ⟨tried to keep the ship in *view*⟩ ⟨sat high in the bleachers to get a good *view*⟩
5 : something that is looked toward or kept in sight **:** OBJECT ⟨studied hard with a *view* to getting an A⟩
6 : the foreseeable future ⟨no hope in *view*⟩
7 : a pictorial representation
synonym see OPINION
— in view of : in regard to **:** in consideration of
— on view : open to public inspection **:** on exhibition

[2]**view** *transitive verb* (1523)
1 : to look at attentively **:** SCRUTINIZE, OBSERVE ⟨*view* an exhibit⟩ ⟨*view* the landscape⟩
2 a : SEE, WATCH ⟨*view* a film⟩ **b :** to look on in a particular light **:** REGARD ⟨doesn't *view* himself as a rebel⟩
3 : to survey or examine mentally **:** CONSIDER ⟨*view* all sides of a question⟩
— view·able \-ə-bəl\ *adjective*

view·data \ˈvyü-ˌdā-tə *also* -ˌda-\ *noun* (1975)
: a videotex system usually employing telephone lines

view·er \ˈvyü-ər\ *noun* (15th century)
: one that views: as **a :** a person legally appointed to inspect and report on property **b :** an optical device used in viewing **c :** a person who watches television

view·er·ship \-ˌship\ *noun* (1954)
: a television audience especially with respect to size or makeup

view·find·er \ˈvyü-ˌfīn-dər\ *noun* (1889)
: a device on a camera for showing the area of the subject to be included in the picture

view hal·loo \ˌvyü-hə-ˈlü\ *interjection* (1761)
— used in fox hunting on seeing a fox break cover

view·ing *noun* (1535)
: an act of seeing, watching, or taking a look; *especially* **:** an instance or the practice of watching television

view·less \ˈvyü-ləs\ *adjective* (1603)
1 : not perceivable **:** INVISIBLE
2 : affording no view
3 : expressing no views or opinions
— view·less·ly *adverb*

view·point \-ˌpȯint\ *noun* (1855)
: POINT OF VIEW, STANDPOINT

viewy \ˈvyü-ē\ *adjective* (1848)
1 : possessing visionary, impractical, or fantastic views
2 : spectacular or arresting in appearance **:** SHOWY

vig \ˈvig\ *noun* (1968)
: VIGORISH

vi·ga \ˈvē-gə\ *noun* [Spanish, beam, rafter] (1844)
: one of the heavy rafters and especially a log supporting the roof in American Indian and Spanish architecture of the Southwest

vi·ges·i·mal \vī-ˈje-sə-məl\ *adjective* [Latin *vicesimus, vigesimus* twentieth; akin to Latin *viginti* twenty, Greek *eikosi*] (circa 1656)
: based on the number 20

vig·il \ˈvi-jəl\ *noun* [Middle English *vigile,* from Old French, from Late Latin & Latin; Late Latin *vigilia* watch on the eve of a feast, from Latin, wakefulness, watch, from *vigil* awake, watchful; akin to Latin *vigēre* to be vigorous, *vegēre* to enliven — more at WAKE] (13th century)
1 a : a watch formerly kept on the night before a religious feast with prayer or other devotions **b :** the day before a religious feast observed as a day of spiritual preparation **c :** evening or nocturnal devotions or prayers — usually used in plural
2 : the act of keeping awake at times when sleep is customary; *also* **:** a period of wakefulness
3 : an act or period of watching or surveillance **:** WATCH

vig·i·lance \ˈvi-jə-lən(t)s\ *noun* (1533)
: the quality or state of being vigilant

vigilance committee *noun* (1835)
: a committee of vigilantes

vig·i·lant \ˈvi-jə-lənt\ *adjective* [Middle English, from Middle French, from Latin *vigilant-, vigilans,* from present participle of *vigilare* to keep watch, stay awake, from *vigil* awake] (15th century)
: alertly watchful especially to avoid danger
synonym see WATCHFUL
— vig·i·lant·ly *adverb*

vig·i·lan·te \ˌvi-jə-ˈlan-tē\ *noun* [Spanish, watchman, guard, from *vigilante* vigilant, from Latin *vigilant-, vigilans*] (1865)
: a member of a volunteer committee organized to suppress and punish crime summarily (as when the processes of law appear inadequate); *broadly* **:** a self-appointed doer of justice
— vig·i·lan·tism \-ˈlan-ˌti-zəm\ *noun*

vigil light *noun* (circa 1931)
: a candle lighted devotionally (as in a Roman Catholic church) before a shrine or image — called also *vigil candle*

vi·gin·til·lion \ˌvī-ˌjin-ˈtil-yən\ *noun, often attributive* [Latin *viginti* twenty + English *-illion* (as in *million*) — more at VIGESIMAL] (circa 1903)
— see NUMBER table

vi·gne·ron \ˌvēn-yə-ˈrōn\ *noun* [Middle English, from Middle French, from Old French *vineron,* from *vine, vigne* vine, vineyard] (15th century)
: WINEGROWER

[1]**vi·gnette** \vin-ˈyet, vēn-\ *noun* [French, from Middle French *vignete,* from diminutive of *vigne* vine — more at VINE] (1751)
1 : a running ornament (as of vine leaves, tendrils, and grapes) put on or just before a title page or at the beginning or end of a chapter; *also* **:** a small decorative design or picture so placed
2 a : a picture (as an engraving or photograph) that shades off gradually into the surrounding paper **b :** the pictorial part of a postage stamp design as distinguished from the frame and lettering
3 a : a short descriptive literary sketch **b :** a brief incident or scene (as in a play or movie)
— vi·gnett·ist \-ˈye-tist\ *noun*

[2]**vignette** *transitive verb* **vi·gnett·ed; vi·gnett·ing** (1853)
1 : to finish (as a photograph) in the manner of a vignette
2 : to describe briefly
— vi·gnett·er *noun*

vig·or \ˈvi-gər\ *noun* [Middle English, from Middle French *vigor,* from Latin, from *vigēre* to be vigorous] (14th century)
1 : active bodily or mental strength or force
2 : active healthy well-balanced growth especially of plants
3 : intensity of action or effect **:** FORCE
4 : effective legal status

vig·o·rish \ˈvi-gə-rish\ *noun* [perhaps from Ukrainian *vygrash* or Russian *vyigrysh* winnings, profit] (1912)
1 : a charge taken (as by a bookie or a gambling house) on bets; *also* **:** the degree of such a charge ⟨a *vigorish* of five percent⟩
2 : interest paid to a moneylender

vi·go·ro·so \ˌvi-gə-ˈrō-(ˌ)sō, ˌvē-, -(ˌ)zō\ *adjective or adverb* [Italian, literally, vigorous, from Middle French *vigorous*] (circa 1724)
: energetic in style — used as a direction in music

vig·or·ous \ˈvi-g(ə-)rəs\ *adjective* [Middle English, from Middle French, from Old French, from *vigor*] (14th century)
1 : possessing vigor **:** full of physical or mental strength or active force **:** STRONG ⟨a *vigorous* youth⟩ ⟨a *vigorous* plant⟩
2 : done with vigor **:** carried out forcefully and energetically ⟨*vigorous* exercises⟩ ☆
— vig·or·ous·ly *adverb*
— vig·or·ous·ness *noun*

vig·our \ˈvi-gər\ *chiefly British variant of* VIGOR

Vi·king \ˈvī-kiŋ\ *noun* [Old Norse *vīkingr*] (1807)
1 a : one of the pirate Norsemen plundering the coasts of Europe in the 8th to 10th centuries **b** *not capitalized* **:** SEA ROVER
2 : SCANDINAVIAN

vile \ˈvī(ə)l\ *adjective* **vil·er** \ˈvī-lər\; **vil·est** \-ləst\ [Middle English, from Old French *vil,* from Latin *vilis*] (14th century)
1 a : morally despicable or abhorrent ⟨nothing is so *vile* as intellectual dishonesty⟩ **b :** physically repulsive **:** FOUL ⟨a *vile* slum⟩
2 : of little worth or account **:** COMMON; *also* **:** MEAN
3 : tending to degrade ⟨*vile* employments⟩
4 : disgustingly or utterly bad **:** OBNOXIOUS, CONTEMPTIBLE ⟨*vile* weather⟩ ⟨had a *vile* temper⟩
synonym see BASE
— vile·ly \ˈvī(ə)l-lē\ *adverb*
— vile·ness *noun*

vil·i·fi·ca·tion \ˌvi-lə-fə-ˈkā-shən\ *noun* (1630)
1 : the act of vilifying **:** ABUSE
2 : an instance of vilifying **:** a defamatory utterance

vil·i·fy \ˈvi-lə-ˌfī\ *transitive verb* **-fied; -fy·ing** [Middle English *vilifien,* from Late Latin *vilificare,* from Latin *vilis* cheap, vile] (15th century)
1 : to lower in estimation or importance

☆ SYNONYMS
Vigorous, energetic, strenuous, lusty, nervous mean having or showing great vitality and force. VIGOROUS further implies showing no signs of depletion or diminishing of freshness or robustness ⟨as *vigorous* as a youth half his age⟩. ENERGETIC suggests a capacity for intense activity ⟨an *energetic* campaigner⟩. STRENUOUS suggests a preference for coping with the arduous or the challenging ⟨the *strenuous* life on an oil rig⟩. LUSTY implies exuberant energy and capacity for enjoyment ⟨a *lusty* appetite for life⟩. NERVOUS suggests especially the forcibleness and sustained effectiveness resulting from mental vigor ⟨full of *nervous* energy⟩.

2 : to utter slanderous and abusive statements against **:** DEFAME
synonym see MALIGN
— **vil·i·fi·er** \-ˌfī(-ə)r\ *noun*

vil·i·pend \ˈvi-lə-ˌpend\ *transitive verb* [Middle English, from Middle French *vilipender,* from Medieval Latin *vilipendere,* from Latin *vilis* + *pendere* to weigh, estimate] (15th century)
1 : to hold or treat as of little worth or account **:** CONTEMN
2 : to express a low opinion of **:** DISPARAGE

vill \ˈvil\ *noun* [Anglo-French, from Old French *ville* village] (1596)
1 : a division of a hundred **:** TOWNSHIP
2 : VILLAGE

vil·la \ˈvi-lə\ *noun* [Italian, from Latin; akin to Latin *vicus* village — more at VICINITY] (1611)
1 : a country estate
2 : the rural or suburban residence of a wealthy person
3 *British* **:** a detached or semidetached urban residence with yard and garden space

vil·la·dom \ˈvi-lə-dəm\ *noun* (1880)
British **:** the world constituted by villas and their occupants

vil·lage \ˈvi-lij\ *noun, often attributive* [Middle English, from Middle French, from Old French, from *ville* farm, village, from Latin *villa*] (14th century)
1 a : a settlement usually larger than a hamlet and smaller than a town **b :** an incorporated minor municipality
2 : the residents of a village
3 : something (as an aggregation of burrows or nests) suggesting a village
4 : a territorial area having the status of a village especially as a unit of local government

vil·lag·er \ˈvi-li-jər\ *noun* (1570)
: an inhabitant of a village

vil·lage·ry \ˈvi-lij-rē, -li-jə-\ *noun* (1590)
: VILLAGES

vil·lain \ˈvi-lən\ *noun* [Middle English *vilain, vilein,* from Middle French, from Medieval Latin *villanus,* from Latin *villa*] (14th century)
1 : an uncouth person **:** BOOR
2 : VILLEIN
3 : a deliberate scoundrel or criminal
4 : a scoundrel in a story or play
5 : a person or thing blamed for a particular evil or difficulty ⟨automation as the *villain* in job . . . displacement —M. H. Goldberg⟩ ◆

vil·lain·ess \-lə-nəs\ *noun* (1586)
: a woman who is a villain

vil·lain·ous \-lə-nəs\ *adjective* (15th century)
1 a : befitting a villain (as in evil, depraved, or vile character) ⟨a *villainous* attack⟩ **b :** being or having the character of a villain **:** DEPRAVED ⟨the *villainous* foe⟩
2 : highly objectionable **:** WRETCHED
synonym see VICIOUS
— **vil·lain·ous·ly** *adverb*
— **vil·lain·ous·ness** *noun*

vil·lainy \-lə-nē\ *noun, plural* **-lain·ies** (13th century)
1 : villainous conduct; *also* **:** a villainous act
2 : the quality or state of being villainous **:** DEPRAVITY

vil·la·nel·la \ˌvi-lə-ˈne-lə\ *noun, plural* **-nel·le** \-ˈne-le\ [Italian, from *villano* villein, peasant, from Medieval Latin *villanus*] (1596)
1 : a 16th century Italian rustic part-song unaccompanied and in free form
2 : an instrumental piece in the style of a rustic dance

vil·la·nelle \ˌvi-lə-ˈnel\ *noun* [French, from Italian *villanella*] (1877)
: a chiefly French verse form running on two rhymes and consisting typically of five tercets and a quatrain in which the first and third lines of the opening tercet recur alternately at the end of the other tercets and together as the last two lines of the quatrain

vil·lat·ic \vi-ˈla-tik\ *adjective* [Latin *villaticus,* from *villa*] (1671)
: RURAL

-ville \ˌvil, *especially Southern* vəl\ *noun suffix* [*-ville,* suffix occurring in names of towns, from French, from Old French, from *ville* village]
: place, category, or quality of a specified nature ⟨dulls*ville*⟩

vil·lein \ˈvi-lən, ˈvi-ˌlān, vi-ˈlān\ *noun* [Middle English *vilain, vilein* — more at VILLAIN] (14th century)
1 : a free common villager or village peasant of any of the feudal classes lower in rank than the thane
2 : a free peasant of a feudal class lower than a sokeman and higher than a cotter
3 : an unfree peasant standing as the slave of a feudal lord but free in legal relations with respect to all others

vil·len·age \ˈvi-lə-nij\ *noun* [Middle English *vilenage,* from Middle French, from Old French, from *vilein, vilain*] (14th century)
1 : tenure at the will of a feudal lord by villein services
2 : the status of a villein

vil·li·form \ˈvi-lə-ˌfȯrm\ *adjective* [International Scientific Vocabulary] (1849)
: having the form or appearance of villi

vil·los·i·ty \vi-ˈlä-sə-tē\ *noun, plural* **-ties** (1777)
1 : the state of being villous
2 : a villous patch or area

vil·lous \ˈvi-ləs\ *adjective* [Middle English, from Latin *villosus* hairy, shaggy, from *villus*] (14th century)
1 : covered or furnished with villi
2 : having soft long hairs ⟨leaves *villous* underneath⟩ — compare PUBESCENT

vil·lus \ˈvi-ləs\ *noun, plural* **vil·li** \ˈvi-ˌlī, -(ˌ)lē\ [New Latin, from Latin, tuft of shaggy hair — more at VELVET] (circa 1704)
: a small slender often vascular process: as **a :** one of the minute finger-shaped processes of the mucous membrane of the small intestine that serve in the absorption of nutriment **b :** one of the branching processes of the surface of the chorion of the developing embryo of most mammals that help to form the placenta

vim \ˈvim\ *noun* [Latin, accusative of *vis* strength; akin to Greek *is* strength, Sanskrit *vaya* meal, strength] (1843)
: robust energy and enthusiasm

vi·na \ˈvē-nə\ *noun* [Sanskrit *vīṇā*] (1788)
: a stringed instrument of India having usually four strings on a long bamboo fingerboard with movable frets and a gourd resonator at each end

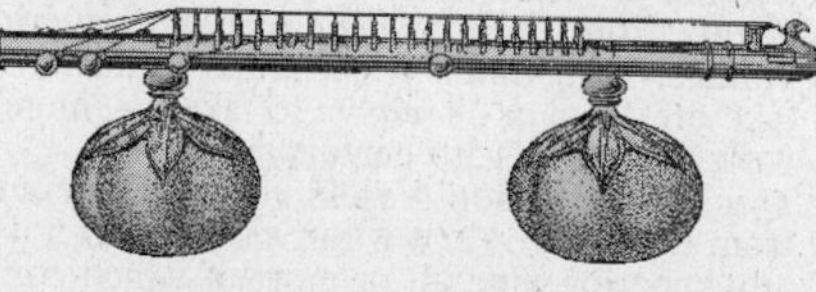

vina

vi·na·ceous \vī-ˈnā-shəs, vi-\ *adjective* [Latin *vinaceus* of wine, from *vinum* wine — more at WINE] (1688)
: of the color wine

vin·ai·grette \ˌvi-ni-ˈgret\ *noun* [French, from *vinaigre* vinegar] (1699)
1 : a sauce made typically of oil and vinegar, onions, parsley, and herbs and used especially on cold meats or fish — called also *vinaigrette dressing, vinaigrette sauce*
2 : a small ornamental box or bottle with perforated top used for holding an aromatic preparation (as smelling salts)

vi·nal \ˈvī-ˌnal\ *noun* [polyvinyl alcohol] (circa 1939)
: a synthetic textile fiber that is a long-chain polymer consisting largely of vinyl alcohol units

vin·blas·tine \(ˌ)vin-ˈblas-ˌtēn\ *noun* [contraction of *vincaleukoblastine,* from *vinca* + *leukoblast* developing leukocyte, from *leuk-* + *-blast*] (1962)
: an alkaloid $C_{46}H_{58}N_4O_9$ from Madagascar periwinkle used especially in the form of its sulfate to treat human neoplastic diseases

vin·ca \ˈviŋ-kə\ *noun* [New Latin, short for Latin *vincapervinca* periwinkle] (1868)
: [1]PERIWINKLE

Vin·cen·tian \vin-ˈsen(t)-shən\ *noun* (1854)
1 : a member of the Roman Catholic Congregation of the Mission founded by Saint Vincent de Paul in Paris, France, in 1625 and devoted to missions and seminaries
2 : a native or inhabitant of the island of Saint Vincent
— **Vincentian** *adjective*

Vin·cent's angina \ˈvin(t)-sən(t)s-, (ˌ)vaⁿ-ˈsäⁿz-\ *noun* [Jean Hyacinthe *Vincent* (died 1950) French bacteriologist] (circa 1903)
: Vincent's infection in which the ulceration has spread to surrounding tissues (as of the pharynx and tonsils) — called also *trench mouth*

Vincent's infection *noun* (circa 1922)
: a progressive painful disease of the mouth that is marked especially by dirty gray ulceration of the mucous membranes, spontaneous hemorrhaging of the gums, and a foul odor to the breath and that is associated with the presence of large numbers of a bacillus (*Fusobacterium nucleatum* synonym *F. fusiforme*) and a spirochete (*Treponema vincentii* synonym *Borrelia vincentii*) in the lesions — called also *trench mouth*

vin·ci·ble \ˈvin(t)-sə-bəl\ *adjective* [Latin *vincibilis,* from *vincere* to conquer — more at VICTOR] (1548)
: capable of being overcome or subdued

vin·cris·tine \(ˌ)vin-ˈkris-ˌtēn\ *noun* [*vin*circa + Latin *crista* crest + English [2]*-ine* — more at CREST] (circa 1962)
: an alkaloid $C_{46}H_{56}N_4O_{10}$ from Madagascar periwinkle used especially in the form of its sulfate to treat some human neoplastic diseases (as leukemias)

vin·cu·lum \ˈviŋ-kyə-ləm\ *noun, plural* **-lums** *or* **-la** \-lə\ [Latin, from *vincire* to bind] (1661)
1 : a unifying bond **:** LINK, TIE

◇ WORD HISTORY

villain In medieval France and England a *vilain* or *vilein* was an agricultural laborer, at some times and places free and at others bound through feudal obligation to the property of his lord. *Vilain* is the Old French outcome of a spoken Latin word that appears in early Medieval Latin as *villanus,* literally, "inhabitant of a *villa.*" *Villa* in classical Latin was a large country dwelling, though by the time of Late Latin (3d to 6th centuries) the word could also refer more generally to any aggregation of rural dwellings and was nearly equivalent in sense to "village," so that a *villanus* was in effect a "villager." Because the landed aristocracy that dominated medieval society had a preponderant influence on linguistic usage, Middle English *vilain* or *vilein* developed the depreciatory sense "a person of uncouth mind and manners." This pejorative tendency gained in strength and currency through the common equation of manners with morals, so that the modern *villain* is a deliberate scoundrel or criminal, or a person or thing blamed for a particular evil or difficulty.

2 : a straight horizontal mark that is placed over two or more members of a compound mathematical expression and is equivalent to parentheses or brackets placed around them (as in $a-\overline{b-c}=a-[b-c]$)

vin·da·loo \ˈvin-də-ˌlü\ *noun* [probably from Konkani *vindalu,* from Indo-Portuguese (Portuguese creole of India) *vinh d'alho,* literally, wine of garlic, from Portuguese *vinho de alho*] (1888)
: a curried dish of Indian origin made with meat or shellfish, garlic, and wine or vinegar

vin·di·ca·ble \ˈvin-di-kə-bəl\ *adjective* (1647)
: capable of being vindicated

vin·di·cate \ˈvin-də-ˌkāt\ *transitive verb* **-cat·ed; -cat·ing** [Latin *vindicatus,* past participle of *vindicare* to lay claim to, avenge, from *vindic-, vindex* claimant, avenger] (1570)
1 *obsolete* **:** to set free **:** DELIVER
2 : AVENGE
3 a : to free from allegation or blame **b** (1) **:** CONFIRM, SUBSTANTIATE (2) **:** to provide justification or defense for **:** JUSTIFY **c :** to protect from attack or encroachment **:** DEFEND
4 : to maintain a right to
synonym see EXCULPATE, MAINTAIN
— vin·di·ca·tor \-ˌkā-tər\ *noun*

vin·di·ca·tion \ˌvin-də-ˈkā-shən\ *noun* (1613)
: an act of vindicating **:** the state of being vindicated; *specifically* **:** justification against denial or censure **:** DEFENSE

vin·dic·a·tive \vin-ˈdi-kə-tiv\ *adjective* (1521)
1 *obsolete* **:** VINDICTIVE, VENGEFUL
2 *archaic* **:** PUNITIVE

vin·di·ca·to·ry *adjective* (1647)
1 \ˈvin-di-kə-ˌtōr-ē, -ˌtȯr-\ **:** providing vindication **:** JUSTIFICATORY
2 \vin-ˈdi-kə-\ **:** PUNITIVE, RETRIBUTIVE

vin·dic·tive \vin-ˈdik-tiv\ *adjective* [Latin *vindicta* revenge, vindication, from *vindicare*] (1616)
1 a : disposed to seek revenge **:** VENGEFUL **b :** intended for or involving revenge
2 : intended to cause anguish or hurt **:** SPITEFUL
— vin·dic·tive·ly *adverb*
— vin·dic·tive·ness *noun*

¹vine \ˈvīn\ *noun* [Middle English, from Middle French, from Old French *vigne,* from Latin *vinea* vine, vineyard, from feminine of *vineus* of wine, from *vinum* wine — more at WINE] (14th century)
1 : GRAPE 2
2 a : a plant whose stem requires support and which climbs by tendrils or twining or creeps along the ground; *also* **:** the stem of such a plant **b :** any of various sprawling herbaceous plants (as a tomato or potato) that lack specialized adaptations for climbing

²vine *intransitive verb* **vined; vin·ing** (1796)
: to form or grow in the manner of a vine

vine·dress·er \ˈvīn-ˌdre-sər\ *noun* (1560)
: a person who cultivates and prunes grapevines

vin·e·gar \ˈvi-ni-gər\ *noun* [Middle English *vinegre,* from Middle French *vinaigre,* from *vin* wine (from Latin *vinum*) + *aigre* keen, sour — more at EAGER] (14th century)
1 : a sour liquid obtained by fermentation of dilute alcoholic liquids and used as a condiment or preservative
2 : ill humor **:** SOURNESS
3 : VIM

vin·e·gared \-gərd\ *adjective* (1861)
: flavored or marinated with vinegar

vinegar eel *noun* (circa 1839)
: a minute nematode worm (*Turbatrix aceti*) often found in great numbers in vinegar or acid fermenting vegetable matter

vinegar fly *noun* [from its breeding in pickles] (1901)
: DROSOPHILA

vin·e·gar·ish \ˈvi-ni-g(ə-)rish\ *adjective* (1648)
: VINEGARY 2

vin·e·gary \ˈvi-ni-g(ə-)rē\ *adjective* (circa 1730)
1 a : resembling vinegar **:** SOUR **b :** flavored with vinegar
2 : disagreeable, bitter, or irascible in character or manner

vin·ery \ˈvīn-rē, ˈvī-nə-\ *noun, plural* **-er·ies** (15th century)
: an area or building in which vines are grown

vine·yard \ˈvin-yərd\ *noun* (14th century)
1 : a planting of grapevines
2 : a sphere of activity **:** field of endeavor ⟨toilers in the *vineyard* of diplomacy —Daniel Schorr⟩

vine·yard·ist \-yər-dist\ *noun* (1848)
: a person who owns or cultivates a vineyard

vingt–et–un \ˌvan-ˌtā-ˈən\ *noun* [French, literally, twenty-one] (1772)
: BLACKJACK 5

vi·ni·cul·ture \ˈvi-nə-ˌkəl-chər, ˈvī-\ *noun* [Latin *vinum* + International Scientific Vocabulary *-i-* + *culture*] (1871)
: VITICULTURE

vi·nif·era \vī-ˈni-f(ə-)rə\ *adjective* [New Latin, from Latin *vinifer* wine-producing, from *vinum* wine] (1900)
: of, relating to, being, or derived from a common European grape (*Vitis vinifera*) that is the chief source of Old World wine grapes and table grapes
— vinifera *noun*

vi·ni·fi·ca·tion \ˌvi-nə-fə-ˈkā-shən, ˌvī-\ *noun* [French, from *vin* wine + *-i-* + *-fication*] (1880)
: the conversion of fruit juices (as grape juice) into wine by fermentation

vin·i·fy \ˈvi-nə-ˌfī, ˈvī-\ *transitive verb* **-fied; -fy·ing** [probably back-formation from *vinification*] (1969)
1 : to make wine from (grapes often of a specified kind)
2 : to make (wine) from grapes

vi·no \ˈvē-(ˌ)nō\ *noun* [Italian & Spanish, from Latin *vinum*] (circa 1919)
: WINE

vi·nos·i·ty \vī-ˈnä-sə-tē\ *noun, plural* **-ties** (1658)
: the characteristic body, flavor, and color of a wine

vi·nous \ˈvī-nəs\ *adjective* [Latin *vinosus,* from *vinum* wine] (1664)
1 : of, relating to, or made with wine ⟨*vinous* medications⟩
2 : showing the effects of the use of wine
3 : VINACEOUS
— vi·nous·ly *adverb*

¹vin·tage \ˈvin-tij\ *noun* [Middle English, alteration of *vendage,* from Middle French *vendenge,* from Latin *vindemia* grape-gathering, vintage, from *vinum* wine, grapes + *demere* to take off, from *de-* + *emere* to take — more at WINE, REDEEM] (15th century)
1 a (1) **:** a season's yield of grapes or wine from a vineyard (2) **:** WINE; *especially* **:** a usually superior wine all or most of which comes from a single year **b :** a collection of contemporaneous and similar persons or things **:** CROP
2 : the act or time of harvesting grapes or making wine
3 a : a period of origin or manufacture ⟨a piano of 1845 *vintage*⟩ **b :** length of existence **:** AGE

²vintage *adjective* (1601)
1 *of wine* **:** of, relating to, or produced in a particular vintage
2 : of old, recognized, and enduring interest, importance, or quality **:** CLASSIC
3 a : dating from the past **:** OLD **b :** OUTMODED, OLD-FASHIONED
4 : of the best and most characteristic — used with a proper noun ⟨*vintage* Shaw: a wise and winning comedy —*Time*⟩

vin·tag·er \ˈvin-ti-jər\ *noun* (1589)
: a person concerned with the production of grapes and wine

vintage year *noun* (1933)
1 : a year of outstanding distinction or success
2 : a year in which a vintage wine is produced

vint·ner \ˈvint-nər\ *noun* [Middle English *vineter,* from Middle French *vinetier,* from Medieval Latin *vinetarius,* from Latin *vinetum* vineyard, from *vinum* wine] (15th century)
1 : a wine merchant
2 : a person who makes wine

viny \ˈvī-nē\ *adjective* **vin·i·er; -est** (1570)
1 : of, relating to, or resembling vines ⟨*viny* plants⟩
2 : covered with or abounding in vines

vi·nyl \ˈvī-nᵊl\ *noun* [International Scientific Vocabulary, from Latin *vinum* wine] (1863)
1 : a univalent group $CH_2{=}CH$ derived from ethylene by removal of one hydrogen atom
2 : a polymer of a vinyl compound or a product (as a resin or a textile fiber) made from such a polymer
— vi·nyl·ic \vī-ˈni-lik\ *adjective*

vinyl alcohol *noun* (1873)
: an unstable compound $CH_2{=}CHOH$ isolated only in the form of its polymers or derivatives

vinyl chloride *noun* (1872)
: a flammable gaseous carcinogenic compound C_2H_3Cl that is used especially to make vinyl resins

vi·nyl·i·dene \vī-ˈni-lə-ˌdēn\ *noun* [International Scientific Vocabulary *vinyl* + *-ide* + *-ene*] (1898)
: a bivalent group $CH_2{=}C$ derived from ethylene by removal of two hydrogen atoms from one carbon atom

vinyl resin *noun* (1934)
: any of various thermoplastic resinous materials that are essentially polymers of vinyl compounds

vi·ol \ˈvī(-ə)l, ˈvī-(ˌ)ōl\ *noun* [Middle English, from Middle French *viole* viol, viola, from Old Provençal *viola* viol] (15th century)
: a bowed stringed instrument chiefly of the 16th and 17th centuries made in treble, alto, tenor, and bass sizes and distinguished from members of the violin family especially in having a deep body, a flat back, sloping shoulders, usually six strings, a fretted fingerboard, and a low-arched bridge

¹vi·o·la \vī-ˈō-lə, vē-; ˈvī-ə-\ *noun* [Middle English, from Latin] (15th century)
: VIOLET 1a; *especially* **:** any of various garden hybrids with solitary white, yellow, or purple often variegated flowers resembling but smaller than typical pansies

²vi·o·la \vē-ˈō-lə\ *noun* [Italian & Spanish, viol, viola, from Old Provençal, viol] (circa 1724)
: a musical instrument of the violin family that is intermediate in size and compass between the violin and cello and is tuned a fifth below the violin
— vi·o·list \-list\ *noun*

vi·o·la·ble \ˈvī-ə-lə-bəl\ *adjective* (1552)
: capable of being or likely to be violated
— vi·o·la·bil·i·ty \ˌvī-ə-lə-ˈbi-lə-tē\ *noun*
— vi·o·la·ble·ness \ˈvī-ə-lə-bəl-nəs\ *noun*
— vi·o·la·bly \-blē\ *adverb*

vi·o·la·ceous \ˌvī-ə-ˈlā-shəs\ *adjective* [Latin *violaceus,* from *viola* violet] (1657)
: of the color violet

vi·o·la da gam·ba \vē-ˌō-lə-də-ˈgäm-bə, -ˈgam-\ *noun, plural* **vi·o·las da gamba** \-ləz-də-\ *or* **vi·o·le da gamba** \-(ˌ)lā-\ [Italian, leg viol] (1597)
: a bass member of the viol family having a range approximating the cello
— vi·o·list da gamba \-lis(t)-də-\ *noun*

viola d'a·mo·re \-də-ˈmōr-ē, -ˈmȯr-, -(ˌ)ā\ *noun, plural* **violas d'amore** *or* **viole d'amore** [Italian, viol of love] (circa 1700)
: a tenor viol having usually seven gut and seven wire strings

¹vi·o·late \ˈvī-ə-ˌlāt\ *transitive verb* **-lat·ed; -lat·ing** [Middle English, from Latin *violatus,* past participle of *violare,* probably from *violentus* violent] (15th century)
1 : BREAK, DISREGARD ⟨*violate* the law⟩

2 : to do harm to the person or especially the chastity of; *specifically* **:** RAPE
3 : to fail to show proper respect for **:** PROFANE ⟨*violate* a shrine⟩
4 : INTERRUPT, DISTURB ⟨*violate* the peace of a spring evening —Nancy Larter⟩
— vi·o·la·tive \-ˌlā-tiv\ *adjective*
— vi·o·la·tor \-ˌlā-tər\ *noun*

[2]vi·o·late \ˈvī-ə-lət\ *adjective* (1503)
archaic **:** subjected to violation

vi·o·la·tion \ˌvī-ə-ˈlā-shən\ *noun* (15th century)
: the act of violating **:** the state of being violated: as **a :** INFRINGEMENT, TRANSGRESSION; *specifically* **:** an infringement of the rules in sports that is less serious than a foul and usually involves technicalities of play **b :** an act of irreverence or desecration **:** PROFANATION **c :** DISTURBANCE, INTERRUPTION **d :** RAPE, RAVISHMENT

vi·o·lence \ˈvī-ə-lən(t)s\ *noun* (14th century)
1 a : exertion of physical force so as to injure or abuse (as in effecting illegal entry into a house) **b :** an instance of violent treatment or procedure
2 : injury by or as if by distortion, infringement, or profanation **:** OUTRAGE
3 a : intense, turbulent, or furious and often destructive action or force ⟨the *violence* of the storm⟩ **b :** vehement feeling or expression **:** FERVOR; *also* **:** an instance of such action or feeling **c :** a clashing or jarring quality **:** DISCORDANCE
4 : undue alteration (as of wording or sense in editing a text)

vi·o·lent \-lənt\ *adjective* [Middle English, from Middle French, from Latin *violentus;* akin to Latin *vis* strength — more at VIM] (14th century)
1 : marked by extreme force or sudden intense activity ⟨a *violent* attack⟩
2 a : notably furious or vehement ⟨a *violent* denunciation⟩ **b :** EXTREME, INTENSE ⟨*violent* pain⟩ ⟨*violent* colors⟩
3 : caused by force **:** not natural ⟨a *violent* death⟩
4 a : emotionally agitated to the point of loss of self-control ⟨a mental patient becoming *violent*⟩ **b :** prone to commit acts of violence ⟨*violent* prison inmates⟩
— vi·o·lent·ly *adverb*

violent storm *noun* (circa 1881)
: STORM 1c(1) — see BEAUFORT SCALE table

vi·o·let \ˈvī-(ə-)lət\ *noun* [Middle English, from Middle French *violete,* diminutive of *viole* violet, from Latin *viola*] (14th century)
1 a : any of a genus (*Viola* of the family Violaceae, the violet family) of chiefly herbs with alternate stipulate leaves and showy flowers in spring and cleistogamous flowers in summer; *especially* **:** one with smaller usually solid-colored flowers as distinguished from the usually larger-flowered violas and pansies **b :** any of several plants of genera other than that of the violet — compare DOGTOOTH VIOLET

violet 1a

2 : any of a group of colors of reddish blue hue, low lightness, and medium saturation

vi·o·lin \ˌvī-ə-ˈlin\ *noun* [Italian *violino,* diminutive of *viola*] (1579)
: a bowed stringed instrument having four strings tuned at intervals of a fifth and a usual range from G below middle C upward for more than 4½ octaves and having a shallow body, shoulders at right angles to the neck, a fingerboard without frets, and a curved bridge
— vi·o·lin·ist \-ist\ *noun*
— vi·o·lin·is·tic \-ə-lə-ˈnis-tik\ *adjective*

vi·o·lon·cel·lo \ˌvī-ə-lən-ˈche-(ˌ)lō, ˌvē-\ *noun* [Italian, diminutive of *violone,* augmentative of *viola*] (circa 1724)
: CELLO
— vi·o·lon·cel·list \-ˈche-list\ *noun*

violin: *1* bridge, *2* sound hole, *3* soundboard, *4* fingerboard, *5* pegs, *6* scroll, *7* tailpiece, *g* G-string, *d* D-string, *a* A-string, *e* E-string

vio·my·cin \ˌvī-ə-ˈmī-sᵊn\ *noun* [*vio*let + *-mycin;* from the color of the soil organism] (1950)
: a polypeptide antibiotic $C_{25}H_{43}N_{13}O_{10}$ that is produced by several soil actinomycetes (genus *Streptomyces*) and is administered in the form of its sulfate in the treatment of tuberculosis

VIP \ˌvē-ˌī-ˈpē\ *noun, plural* **VIPs** \-ˈpēz\ [*v*ery *i*mportant *p*erson] (1933)
: a person of great influence or prestige; *especially* **:** a high official with special privileges

vi·per \ˈvī-pər\ *noun* [Middle French *vipere,* from Latin *vipera*] (1526)
1 a : a common Eurasian venomous snake (*Vipera berus*) that attains a length of two feet (0.6 meter), varies in color from red, brown, or gray with dark markings to black, and is usually not fatal to humans; *broadly* **:** any of a family (Viperidae) of venomous snakes that includes Old World snakes (subfamily Viperinae) and the pit vipers **b :** a venomous or reputedly venomous snake
2 : a vicious or treacherous person

vi·per·ine \-pə-ˌrīn\ *adjective* (circa 1550)
: of, relating to, or resembling a viper **:** VENOMOUS

vi·per·ish \-p(ə-)rish\ *adjective* (1755)
: spitefully vituperative **:** VENOMOUS

vi·per·ous \-p(ə-)rəs\ *adjective* (1535)
1 : VIPERINE
2 : having the qualities attributed to a viper **:** MALIGNANT, VENOMOUS
— vi·per·ous·ly *adverb*

viper's bugloss *noun* (1597)
: a coarse Old World herb (*Echium vulgare*) of the borage family that is naturalized in North America and has showy blue tubular flowers with exserted stamens — called also *blueweed*

vi·ra·go \və-ˈrä-(ˌ)gō, -ˈrā-; ˈvir-ə-ˌgō\ *noun, plural* **-goes** *or* **-gos** [Middle English, from Latin *viragin-, virago,* from *vir* man — more at VIRILE] (14th century)
1 : a loud overbearing woman **:** TERMAGANT
2 : a woman of great stature, strength, and courage
— vi·rag·i·nous \və-ˈra-jə-nəs\ *adjective*

vi·ral \ˈvī-rəl\ *adjective* (1937)
: of, relating to, or caused by a virus
— vi·ral·ly \-rə-lē\ *adverb*

vir·e·lay \ˈvir-ə-ˌlā\ *noun* [Middle English, from Middle French *virelai*] (14th century)
: a chiefly French verse form consisting of stanzas of indeterminate length and number with alternating long and short lines and interlaced rhyme (as *abab bcbc cdcd dada*)

vi·re·mia \vī-ˈrē-mē-ə\ *noun* [New Latin, from *vir*us + *-emia*] (1946)
: the presence of virus in the blood of a host
— vi·re·mic \-mik\ *adjective*

vir·eo \ˈvir-ē-ˌō\ *noun, plural* **-e·os** [Latin, a small bird, from *virēre* to be green] (1834)
: any of various small insectivorous American oscine birds (family Vireonidae and especially genus *Vireo*) that are chiefly olivaceous and grayish in color

vires *plural of* VIS

vi·res·cence \və-ˈre-sᵊn(t)s, vī-\ *noun* (circa 1888)
: the state or condition of becoming green; *especially* **:** such a condition due to the development of chloroplasts in plant organs (as petals) normally white or colored

vi·res·cent \-sᵊnt\ *adjective* [Latin *virescent-, virescens,* present participle of *virescere* to become green, inchoative of *virēre* to be green] (1826)
1 : beginning to be green **:** GREENISH
2 : developing or displaying virescence

vir·ga \ˈvər-gə\ *noun* [New Latin, from Latin, branch, rod, streak in the sky suggesting rain] (1938)
: wisps of precipitation evaporating before reaching the ground

[1]vir·gate \ˈvər-ˌgāt\ *noun* [Medieval Latin *virgata,* from *virga,* a land measure, from Latin, rod] (1655)
: an old English unit of land area equal to one quarter of a hide or one quarter of an acre

[2]virgate *adjective* [New Latin *virgatus,* from Latin, made of twigs, from *virga*] (1821)
: shaped like a rod or wand ⟨a *virgate* one-flowered branch⟩

[1]vir·gin \ˈvər-jən\ *noun* [Middle English, from Old French *virgine,* from Latin *virgin-, virgo* young woman, virgin] (13th century)
1 a : an unmarried woman devoted to religion **b** *capitalized* **:** VIRGO
2 a : an absolutely chaste young woman **b :** an unmarried girl or woman
3 *capitalized* **:** VIRGIN MARY
4 a : a person who has not had sexual intercourse **b :** a person who is inexperienced in a usually specified sphere of activity ⟨a *virgin* in politics⟩
5 : a female animal that has never copulated

[2]virgin *adjective* (14th century)
1 : free of impurity or stain **:** UNSULLIED
2 : CHASTE
3 : characteristic of or befitting a virgin **:** MODEST
4 : FRESH, UNSPOILED; *specifically* **:** not altered by human activity ⟨a *virgin* forest⟩
5 a (1) **:** being used or worked for the first time (2) *of a metal* **:** produced directly from ore by primary smelting **b :** INITIAL, FIRST
6 *of a vegetable oil* **:** obtained from the first light pressing and without heating

[1]vir·gin·al \ˈvər-jə-nᵊl, ˈvərj-nəl\ *adjective* (15th century)
1 : of, relating to, or characteristic of a virgin or virginity; *especially* **:** PURE, CHASTE
2 : PRISTINE, UNSULLIED
— vir·gin·al·ly *adverb*

[2]virginal *noun* [probably from Latin *virginalis* of a virgin, from *virgin-, virgo*] (1530)
: a small rectangular spinet having no legs and only one wire to a note and popular in the 16th and 17th centuries — often used in plural; called also *pair of virginals*
— vir·gin·al·ist \ˈvər-jə-nᵊl-ist, ˈvərj-nə-list\ *noun*

virgin birth *noun* (1652)
1 : birth from a virgin
2 *often V&B capitalized* **:** the theological doctrine that Jesus was miraculously begotten of God and born of a virgin mother

Vir·gin·ia bluebells \vər-ˈjin-yə-, -ˈji-nē-ə-\ *noun plural* [*Virginia,* state of the U.S.] (circa 1922)
: BLUEBELL 2b

Virginia creeper *noun* (1704)
: a common North American tendril-climbing vine (*Parthenocissus quinquefolia*) of the grape family with palmately compound leaves and bluish black berries — called also *woodbine*

Virginia fence *noun* (1671)
: WORM FENCE — called also *Virginia rail fence*

Virginia ham *noun* (1824)

: a dry-cured, smoked, and aged ham especially from a peanut-fed hog
Virginia pine *noun* (1897)
: a pine (*Pinus virginiana*) of the eastern U.S. that has short needles occurring in pairs — called also *Jersey pine*
Virginia rail *noun* (1813)
: an American long-billed rail (*Rallus limicola*) that has gray cheeks
Virginia reel *noun* (1817)
: an American dance in which two lines of couples face each other and all couples in turn participate in a series of figures
Virginia snakeroot *noun* (1694)
: a birthwort (*Aristolochia serpentaria*) of the eastern U.S. with oblong leaves cordate at the base and a solitary basal very irregular flower
vir·gin·i·ty \(,)vər-'ji-nə-tē\ *noun, plural* **-ties** (14th century)
1 : the quality or state of being virgin; *especially* : MAIDENHOOD
2 : the unmarried life : CELIBACY, SPINSTERHOOD
Virgin Mary *noun*
: the mother of Jesus
virgin's bower *noun* (1597)
: any of several usually small-flowered and climbing clematises (especially *Clematis virginiana*)
virgin wool *noun* (1915)
: wool not used before in manufacture
Vir·go \'vər-(,)gō, 'vir-\ *noun* [Latin (genitive *Virginis*), literally, virgin]
1 : a zodiacal constellation on the celestial equator that lies due south of the handle of the Big Dipper and is pictured as a woman holding a spike of grain
2 **a** : the 6th sign of the zodiac in astrology — see ZODIAC table **b** : one born under the sign of Virgo
— **Vir·go·an** *noun*
vir·gule \'vər-(,)gyü(ə)l\ *noun* [French, from Latin *virgula* small stripe, obelus, from diminutive of *virga* rod] (1837)
: DIAGONAL 3
vi·ri·ci·dal \,vī-rə-'sī-dᵊl\ *adjective* [New Latin *virus* + English *-i-* + *-cide*] (1924)
: VIRUCIDAL
— **vi·ri·cide** \'vī-rə-,sīd\ *noun*
vir·id \'vir-əd\ *adjective* [Latin *viridis* green] (1600)
: vividly green : VERDANT
vir·i·des·cent \,vir-ə-'de-sᵊnt\ *adjective* [Latin *viridis* green] (circa 1847)
: slightly green : GREENISH
vi·rid·i·an \və-'ri-dē-ən\ *noun* [Latin *viridis*] (1882)
: a chrome green that is a hydrated oxide of chromium
vi·rid·i·ty \və-'ri-də-tē\ *noun* [Middle English *viridite*, from Middle French *viridité*, from Latin *viriditat-, viriditas*, from *viridis*] (15th century)
1 **a** : the quality or state of being green **b** : the color of grass or foliage
2 : naive innocence
vir·ile \'vir-əl, 'vir-,īl, *British also* 'vīr-,īl\ *adjective* [Middle French or Latin; Middle French *viril*, from Latin *virilis*, from *vir* man, male; akin to Old English & Old High German *wer* man, Sanskrit *vīra*] (15th century)
1 : having the nature, properties, or qualities of an adult male; *specifically* : capable of functioning as a male in copulation
2 : ENERGETIC, VIGOROUS
3 **a** : characteristic of or associated with men : MASCULINE **b** : having traditionally masculine traits especially to a marked degree
4 : MASTERFUL, FORCEFUL
— **vir·ile·ly** *adverb*
vir·il·ism \'vir-ə-,li-zəm\ *noun* (1922)
: the appearance of secondary male characteristics in the female
vi·ril·i·ty \və-'ri-lə-tē, *British also* vī-\ *noun* (1586)
: the quality or state of being virile: **a** : MANHOOD **b** : manly vigor : MASCULINITY
vi·ri·on \'vī-rē-,än, 'vir-ē-\ *noun* [French, from *virien* viral (from *virus* virus) + *-on* ²-on] (1959)
: a complete virus particle that consists of an RNA or DNA core with a protein coat sometimes with external envelopes and that is the extracellular infective form of a virus
virl \'vər(-ə)l\ *noun* [Middle English *virole* — more at FERRULE] (15th century)
Scottish : FERRULE 1
vi·roid \'vī-,ròid\ *noun* [New Latin *virus* + English *-oid*] (1971)
: any of several causative agents of plant disease that consist solely of a single-stranded RNA of low molecular weight arranged in a closed loop or a linear chain
vi·rol·o·gy \vī-'rä-lə-jē\ *noun* [New Latin *virus* + International Scientific Vocabulary *-logy*] (circa 1935)
: a branch of science that deals with viruses
— **vi·ro·log·i·cal** \,vī-rə-'lä-ji-kəl\ *or* **vi·ro·log·ic** \-jik\ *adjective*
— **vi·ro·log·i·cal·ly** \-ji-k(ə-)lē\ *adverb*
— **vi·rol·o·gist** \vī-'rä-lə-jist\ *noun*
vir·tu \,vər-'tü, vir-\ *noun* [Italian *virtù*, literally, virtue, from Latin *virtut-, virtus*] (1722)
1 : a love of or taste for curios or objets d'art
2 : productions of art especially of a curious or antique nature : OBJETS D'ART
vir·tu·al \'vər-chə-wəl, -chəl; 'vərch-wəl\ *adjective* [Middle English, possessed of certain physical virtues, from Medieval Latin *virtualis*, from Latin *virtus* strength, virtue] (1654)
1 : being such in essence or effect though not formally recognized or admitted ⟨a *virtual* dictator⟩
2 : of, relating to, or using virtual memory
3 : of, relating to, or being a hypothetical particle whose existence is inferred from indirect evidence ⟨*virtual* photons⟩ — compare REAL 3
virtual image *noun* (1859)
: an image (as seen in a plane mirror) formed of points from which divergent rays (as of light) seem to emanate without actually doing so
vir·tu·al·i·ty \,vər-chə-'wa-lə-tē\ *noun, plural* **-ties** (1646)
1 : ESSENCE
2 : potential existence : POTENTIALITY
vir·tu·al·ly \'vər-chə-wə-lē, -chə-lē; 'vərch-wə-lē\ *adverb* (15th century)
1 : almost entirely : NEARLY
2 : for all practical purposes ⟨*virtually* unknown⟩
virtual memory *noun* (1959)
: external memory (as magnetic disks) for a computer that can be used as if it were an extension of the computer's internal memory — called also *virtual storage*
virtual reality *noun* (1989)
: an artificial environment which is experienced through sensory stimuli (as sights and sounds) provided by a computer and in which one's actions partially determine what happens in the environment
vir·tue \'vər-(,)chü\ *noun* [Middle English *virtu*, from Old French, from Latin *virtut-, virtus* strength, manliness, virtue, from *vir* man — more at VIRILE] (13th century)
1 **a** : conformity to a standard of right : MORALITY **b** : a particular moral excellence
2 *plural* : an order of angels — see CELESTIAL HIERARCHY
3 : a beneficial quality or power of a thing
4 : manly strength or courage : VALOR
5 : a commendable quality or trait : MERIT
6 : a capacity to act : POTENCY
7 : chastity especially in a woman
— **vir·tue·less** \-(,)chü-ləs\ *adjective*
— **by virtue of** *or* **in virtue of** : through the force of : by authority of
vir·tu·o·sa \,vər-chü-'ō-sə, -zə\ *noun* [Italian, feminine of *virtuoso*] (1668)
: a girl or woman who is a virtuoso
vir·tu·os·i·ty \-'ä-sə-tē\ *noun, plural* **-ties** (1673)
1 : a taste for or interest in virtu
2 : great technical skill (as in the practice of a fine art)
vir·tu·o·so \-'ō-(,)sō, -(,)zō\ *noun, plural* **-sos** *or* **-si** \-(,)sē, -(,)zē\ [Italian, from *virtuoso*, adjective, virtuous, skilled, from Late Latin *virtuosus* virtuous, from Latin *virtus*] (1651)
1 : an experimenter or investigator especially in the arts and sciences : SAVANT
2 : one skilled in or having a taste for the fine arts
3 : one who excels in the technique of an art; *especially* : a highly skilled musical performer (as on the violin)
4 : a person who has great skill at some endeavor ⟨a computer *virtuoso*⟩ ⟨a *virtuoso* at public relations⟩
— **vir·tu·o·sic** \-'ō-sik, -zik\ *adjective*
— **virtuoso** *adjective*
vir·tu·ous \'vər-chə-wəs, 'vərch-wəs\ *adjective* (14th century)
1 : POTENT, EFFICACIOUS
2 **a** : having or exhibiting virtue **b** : morally excellent : RIGHTEOUS
3 : CHASTE
synonym see MORAL
— **vir·tu·ous·ly** *adverb*
— **vir·tu·ous·ness** *noun*
vi·ru·cid·al \,vī-rə-'sī-dᵊl\ *adjective* [New Latin, *virus* + English *-cide*] (1925)
: having the capacity to or tending to destroy or inactivate viruses ⟨*virucidal* agents⟩ ⟨*virucidal* activity⟩
— **vi·ru·cide** \'vī-rə-,sīd\ *noun*
vir·u·lence \'vir-ə-lən(t)s, 'vir-yə-\ *noun* (1663)
: the quality or state of being virulent: as **a** : extreme bitterness or malignity of temper : RANCOR **b** : MALIGNANCY, VENOMOUSNESS ⟨ameliorate the *virulence* of a disease⟩ **c** : the relative capacity of a pathogen to overcome body defenses
vir·u·len·cy \-lən(t)-sē\ *noun* (circa 1617)
: VIRULENCE
vir·u·lent \-lənt\ *adjective* [Middle English, from Latin *virulentus*, from *virus* poison] (14th century)
1 **a** : marked by a rapid, severe, and malignant course ⟨a *virulent* infection⟩ **b** : able to overcome bodily defensive mechanisms ⟨a *virulent* pathogen⟩
2 : extremely poisonous or venomous
3 : full of malice : MALIGNANT ⟨*virulent* racists⟩
4 : objectionably harsh or strong
— **vir·u·lent·ly** *adverb*
vir·u·lif·er·ous \,vir-ə-'li-f(ə-)rəs, ,vir-yə-\ *adjective* [*virulence* + *-iferous*] (circa 1899)
: containing, producing, or conveying an agent of infection and especially a virus ⟨*viruliferous* insects⟩
vi·rus \'vī-rəs\ *noun* [Latin, venom, poisonous emanation; akin to Greek *ios* poison, Sanskrit *viṣa*; in senses 2 & 4, from New Latin, from Latin] (1599)
1 *archaic* : VENOM 1
2 **a** : the causative agent of an infectious disease **b** : any of a large group of submicroscopic infective agents that are regarded either as extremely simple microorganisms or as extremely complex molecules, that typically contain a protein coat surrounding an RNA or DNA core of genetic material but no semipermeable membrane, that are capable of growth and multiplication only in living cells, and that cause various important diseases in humans, lower animals, or plants; *also* : FILTERABLE VIRUS **c** : a disease caused by a virus
3 : something that poisons the mind or soul ⟨the force of this *virus* of prejudice —V. S. Waters⟩
4 : a computer program usually hidden within another seemingly innocuous program that

produces copies of itself and inserts them into other programs and that usually performs a malicious action (as destroying data)

vis \'vis\ *noun, plural* **vi·res** \'vī-ˌrēz\ [Latin — more at VIM] (1601)
: FORCE, POWER

¹vi·sa \'vē-zə *also* -sə\ *noun* [French, from Latin, neuter plural of *visus,* past participle] (1831)
1 : an endorsement made on a passport by the proper authorities denoting that it has been examined and that the bearer may proceed
2 : a signature of formal approval by a superior upon a document

²visa *transitive verb* **vi·saed** \-zəd, -səd\; **vi·sa·ing** \-zə-iŋ, -sə-\ (circa 1847)
: to give a visa to (a passport)

vis·age \'vi-zij\ *noun* [Middle English, from Old French, from *vis* face, from Latin *visus* sight, from *vidēre* to see — more at WIT] (14th century)
1 : the face, countenance, or appearance of a person or sometimes an animal
2 : ASPECT, APPEARANCE ⟨grimy *visage* of a mining town⟩
— **vis·aged** \-zijd\ *adjective*

¹vis-à-vis \'vēz-ə-'vē, ˌvēs- *also* -ä-'vē\ *preposition* [French, literally, face-to-face] (1755)
1 : face-to-face with
2 : in relation to
3 : as compared with

²vis-à-vis *noun, plural* **vis-à-vis** \-ə-'vē(z), -ä-\ (circa 1757)
1 : one that is face-to-face with another
2 a : ESCORT, DATE **b :** COUNTERPART
3 : TÊTE-À-TÊTE 1

³vis-à-vis *adverb* (1870)
: in company : TOGETHER

Vi·sa·yan \və-'sī-ən\ *variant of* BISAYAN

vis·ca·cha \vis-'kä-chə\ *noun* [Spanish *vizcacha,* from Quechua *wisk'acha*] (1604)
: any of several South American burrowing rodents (genera *Lagostomus* and *Lagidium*) closely related to the chinchilla

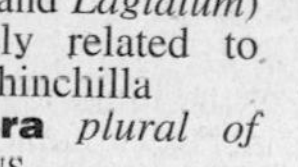
viscacha

viscera *plural of* VISCUS

vis·cer·al \'vi-sə-rəl, 'vis-rəl\ *adjective* (1575)
1 : felt in or as if in the viscera : DEEP ⟨*visceral* conviction⟩
2 : not intellectual : INSTINCTIVE, UNREASONING ⟨*visceral* drives⟩
3 : dealing with crude or elemental emotions : EARTHY ⟨a *visceral* novel⟩
4 : of, relating to, or located on or among the viscera : SPLANCHNIC
— **vis·cer·al·ly** \-rə-lē\ *adverb*

vis·cid \'vi-səd\ *adjective* [Late Latin *viscidus,* from Latin *viscum* birdlime — more at VISCOUS] (1635)
1 a : having an adhesive quality : STICKY **b :** having a glutinous consistency : VISCOUS
2 : covered with a sticky layer
— **vis·cid·i·ty** \vi-'si-də-tē\ *noun*
— **vis·cid·ly** \'vi-səd-lē\ *adverb*

vis·co·elas·tic \ˌvis-kō-ə-'las-tik\ *adjective* [*viscous* + *-o-* + *elastic*] (1935)
: having appreciable and conjoint viscous and elastic properties ⟨such *viscoelastic* materials as asphalt⟩; *also* : constituting or relating to the state of viscoelastic materials ⟨*viscoelastic* data⟩ ⟨*viscoelastic* properties⟩
— **vis·co·elas·tic·i·ty** \-ˌlas-'ti-sə-tē, -'tis-tē\ *noun*

vis·com·e·ter \vis-'kä-mə-tər\ *noun* [*viscosity* + *-meter*] (circa 1883)
: an instrument with which to measure viscosity
— **vis·co·met·ric** \ˌvis-kə-'me-trik\ *adjective*
— **vis·com·e·try** \-mə-trē\ *noun*

¹vis·cose \'vis-ˌkōs, -ˌkōz\ *noun* [obsolete *viscose,* adjective, viscous] (1896)
1 : a viscous golden-brown solution made by treating cellulose with caustic alkali solution and carbon disulfide and used in making rayon and films of regenerated cellulose
2 : viscose rayon

²viscose *adjective* (1900)
: of, relating to, or made from viscose

vis·co·sim·e·ter \ˌvis-kə-'si-mə-tər\ *noun* [International Scientific Vocabulary *viscosity* + *-meter*] (circa 1868)
: VISCOMETER
— **vis·cosi·met·ric** \(ˌ)vis-ˌkä-sə-'me-trik\ *adjective*

vis·cos·i·ty \vis-'kä-sə-tē\ *noun, plural* **-ties** [Middle English *viscosite,* from Middle French *viscosité,* from Medieval Latin *viscositat-, viscositas,* from Late Latin *viscosus* viscous] (15th century)
1 : the quality or state of being viscous
2 : the property of resistance to flow in a fluid or semifluid
3 : the ratio of the tangential frictional force per unit area to the velocity gradient perpendicular to the direction of flow of a liquid — called also *coefficient of viscosity*

viscosity index *noun* (1929)
: an arbitrary number assigned as a measure of the constancy of the viscosity of a lubricating oil with change of temperature with higher numbers indicating viscosities that change little with temperature

vis·count \'vī-ˌkau̇nt\ *noun* [Middle English *viscounte,* from Middle French *viscomte,* from Medieval Latin *vicecomit-, vicecomes,* from Late Latin *vice-* vice- + *comit-, comes* count — more at COUNT] (15th century)
: a member of the peerage in Great Britain ranking below an earl and above a baron
— **vis·count·cy** \-ˌkau̇n(t)-sē\ *noun*
— **vis·county** \-ˌkau̇n-tē\ *noun*

vis·count·ess \-ˌkau̇n-təs\ *noun* (15th century)
1 : the wife or widow of a viscount
2 : a woman who holds the rank of viscount in her own right

vis·cous \'vis-kəs\ *adjective* [Middle English *viscouse,* from Late Latin *viscosus* full of birdlime, viscous, from Latin *viscum* mistletoe, birdlime; akin to Old High German *wīhsila* cherry, Greek *ixos* mistletoe] (14th century)
1 : VISCID
2 : having or characterized by viscosity ⟨*viscous* flow⟩
— **vis·cous·ly** *adverb*
— **vis·cous·ness** *noun*

vis·cus \'vis-kəs\ *noun, plural* **vis·cera** \'vi-sə-rə\ [Latin (plural *viscera*)] (1651)
1 : an internal organ of the body; *especially* : one (as the heart, liver, or intestine) located in the great cavity of the trunk proper
2 *plural* **:** HEART 4

¹vise \'vīs\ *noun* [Middle French *vis* screw, something winding, from Latin *vitis* vine — more at WITHY] (1500)
1 : any of various tools with two jaws for holding work that close usually by a screw, lever, or cam
2 : something likened to a vise ⟨economic *vise* of slow growth and rampant price increases —David Milne⟩
— **vise·like** \-ˌlīk\ *adjective*

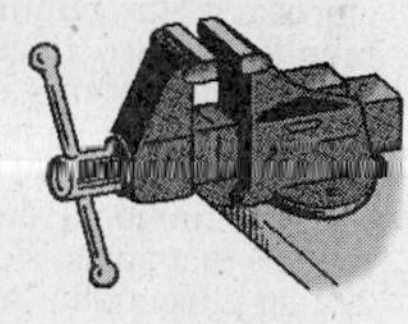
vise 1

²vise *transitive verb* **vised; vis·ing** (1602)
: to hold, force, or squeeze with or as if with a vise

¹vi·sé \'vē-ˌzā, vē-'\ *transitive verb* **vi·séd** *or* **vi·séed; vi·sé·ing** [French, past participle of *viser* to visa, from *visa*] (1810)
: VISA

²visé *noun* (1842)
: VISA

Vish·nu \'vish-(ˌ)nü\ *noun* [Sanskrit *Viṣṇu*] (1638)
: the preserver god of the Hindu sacred triad — compare BRAHMA, SIVA

vis·i·bil·i·ty \ˌvi-zə-'bi-lə-tē\ *noun, plural* **-ties** (1581)
1 : the quality or state of being visible
2 a : the degree of clearness of the atmosphere; *specifically* : the greatest distance toward the horizon at which prominent objects can be identified with the naked eye **b :** capability of being readily noticed **c :** capability of affording an unobstructed view **d :** PUBLICITY 2d
3 : a measure of the ability of radiant energy to evoke visual sensation

vis·i·ble \'vi-zə-bəl\ *adjective* [Middle English, from Middle French or Latin; Middle French, from Latin *visibilis,* from *visus,* past participle of *vidēre* to see] (14th century)
1 a : capable of being seen ⟨stars *visible* to the naked eye⟩ **b :** situated in the region of the electromagnetic spectrum perceptible to human vision ⟨*visible* light⟩ — used of radiation having a wavelength between about 400 nanometers and 700 nanometers
2 a : exposed to view ⟨the *visible* horizon⟩ **b :** CONSPICUOUS
3 : capable of being discovered or perceived : RECOGNIZABLE ⟨no *visible* means of support⟩
4 : ACCESSIBLE 4
5 : devised to keep a particular part or item always in full view or readily seen or referred to ⟨a *visible* index⟩
— **vis·i·ble·ness** *noun*
— **vis·i·bly** \-blē\ *adverb*

visible speech *noun* (1865)
1 : a set of phonetic symbols based on symbols for articulatory position
2 : speech reproduced spectrographically

Vis·i·goth \'vi-zə-ˌgäth\ *noun* [Late Latin *Visigothi,* plural] (1611)
: a member of the western division of the Goths
— **Vis·i·goth·ic** \ˌvi-zə-'gä-thik\ *adjective*

¹vi·sion \'vi-zhən\ *noun* [Middle English, from Old French, from Latin *vision-, visio,* from *vidēre* to see — more at WIT] (14th century)
1 a : something seen in a dream, trance, or ecstasy; *especially* : a supernatural appearance that conveys a revelation **b :** an object of imagination **c :** a manifestation to the senses of something immaterial ⟨look, not at *visions,* but at realities —Edith Wharton⟩
2 a : the act or power of imagination **b** (1) : mode of seeing or conceiving (2) : unusual discernment or foresight ⟨a man of *vision*⟩ **c :** direct mystical awareness of the supernatural usually in visible form
3 a : the act or power of seeing : SIGHT **b :** the special sense by which the qualities of an object (as color, luminosity, shape and size) constituting its appearance are perceived and which is mediated by the eye
4 a : something seen **b :** a lovely or charming sight
— **vi·sion·al** \'vizh-nəl, 'vi-zhə-nᵊl\ *adjective*
— **vi·sion·al·ly** *adverb*

²vision *transitive verb* **vi·sioned; vi·sion·ing** \'vi-zhə-niŋ, 'vizh-niŋ\ (1795)
: ENVISION

¹vi·sion·ary \'vi-zhə-ˌner-ē\ *adjective* (1648)
1 a : of the nature of a vision : ILLUSORY **b :** incapable of being realized or achieved : UTOPIAN ⟨a *visionary* scheme⟩ **c :** existing only in imagination : UNREAL
2 a : able or likely to see visions **b :** disposed to reverie or imagining : DREAMY

3 : of, relating to, or characterized by visions or the power of vision
4 : having or marked by foresight and imagination ⟨a *visionary* leader⟩ ⟨a *visionary* invention⟩
synonym see IMAGINARY
— vi·sion·ar·i·ness \-ē-nəs\ *noun*

²visionary *noun, plural* **-ar·ies** (1702)
1 : one whose ideas or projects are impractical **:** DREAMER
2 : one who sees visions **:** SEER

vi·sioned \'vi-zhənd\ *adjective* (1510)
1 : seen in a vision ⟨a *visioned* face⟩
2 : produced by or experienced in a vision ⟨*visioned* agony⟩
3 : endowed with vision **:** INSPIRED

vi·sion·less \'vi-zhən-ləs\ *adjective* (1820)
1 : SIGHTLESS, BLIND ⟨*visionless* eyes⟩
2 : lacking vision or inspiration ⟨a *visionless* leader⟩

¹vis·it \'vi-zət\ *verb* **vis·it·ed** \'vi-zə-təd, 'viz-təd\; **vis·it·ing** \'vi-zə-tiŋ, 'viz-tiŋ\ [Middle English, from Old French *visiter,* from Latin *visitare,* frequentative of *visere* to go to see, frequentative of *vidēre* to see] (13th century)
transitive verb
1 a *archaic* **:** COMFORT — used of the Deity ⟨*visit* us with Thy salvation —Charles Wesley⟩ **b** (1) **:** AFFLICT ⟨*visited* his people with distempers —Tobias Smollett⟩ (2) **:** INFLICT, IMPOSE ⟨*visited* his wrath upon them⟩ **c :** AVENGE ⟨*visited* the sins of the fathers upon the children⟩ **d :** to present itself to or come over momentarily ⟨was *visited* by a strange notion⟩
2 : to go to see in order to comfort or help
3 a : to pay a call on as an act of friendship or courtesy **b :** to reside with temporarily as a guest **c :** to go to see or stay at (a place) for a particular purpose (as business or sightseeing) **d :** to go or come officially to inspect or oversee ⟨a bishop *visiting* his parishes⟩
intransitive verb
1 : to make a visit; *also* **:** to make frequent or regular visits
2 : CHAT, CONVERSE

²visit *noun* (1621)
1 a : a short stay **:** CALL **b :** a brief residence as a guest **c :** an extended stay **:** SOJOURN
2 : a journey to and stay or short sojourn at a place
3 : an official or professional call or tour **:** VISITATION
4 : the act of a naval officer in boarding a merchant ship on the high seas in exercise of the right of search

vis·it·able \'vi-zə-tə-bəl, 'viz-tə-\ *adjective* (1605)
1 : subject to or allowing visitation or inspection
2 : socially eligible to receive visits

vis·i·tant \'vi-zə-tənt, 'viz-tənt\ *noun* (1599)
1 : VISITOR; *especially* **:** one thought to come from a spirit world
2 : a migratory bird that appears at intervals for a limited period
— visitant *adjective*

vis·i·ta·tion \,vi-zə-'tā-shən\ *noun* (14th century)
1 : an instance of visiting: as **a :** an official visit (as for inspection) **b :** ²WAKE 3 **c :** temporary custody of a child granted to a noncustodial parent ⟨*visitation* rights⟩
2 a : a special dispensation of divine favor or wrath **b :** a severe trial **:** AFFLICTION
3 *capitalized* **:** the visit of the Virgin Mary to Elizabeth recounted in Luke and celebrated July 2 by a Christian feast

vis·i·ta·to·ri·al \,vi-zə-tə-'tōr-ē-əl, ,viz-tə-, -'tȯr-\ *adjective* (1688)
: of or relating to visitation or to a judicial visitor or superintendent

visiting *adjective* (1949)
: invited to join or attend an institution (as a university) for a limited time ⟨a *visiting* professor⟩ ⟨a *visiting* fellow⟩

visiting card *noun* (1782)
: a small card presented when visiting that bears the name and sometimes the address of the visitor

visiting fireman *noun* (1926)
: a usually important or influential visitor whom it is desirable or expedient to entertain impressively

visiting nurse *noun* (circa 1924)
: a nurse employed by a hospital or social-service agency to perform public health services and especially to visit sick persons in a community

vis·i·tor \'vi-zə-tər, 'viz-tər\ *noun* (15th century)
: one that visits; *especially* **:** one that makes formal visits of inspection

vi·sive \'vi-ziv, 'vī-siv\ *adjective* [Medieval Latin *visivus,* from Latin *visus,* past participle of *vidēre* to see — more at WIT] (1543)
archaic **:** of, relating to, or serving for vision

vi·sor \'vī-zər\ *noun* [Middle English *viser,* from Anglo-French, from Old French *visiere,* from *vis* face — more at VISAGE] (14th century)
1 : the front piece of a helmet; *especially* **:** a movable upper piece
2 a : a face mask **b :** DISGUISE
3 a : a projecting front on a cap for shading the eyes **b :** a usually movable flat sunshade attached at the top of an automobile windshield
— vi·sored \-zərd\ *adjective*
— vi·sor·less \-zər-ləs\ *adjective*

vis·ta \'vis-tə\ *noun* [Italian, sight, from *visto,* past participle of *vedere* to see, from Latin *vidēre* — more at WIT] (1644)
1 : a distant view through or along an avenue or opening **:** PROSPECT
2 : an extensive mental view (as over a stretch of time or a series of events)

vis·taed \'vis-təd\ *adjective* (1835)
1 : affording or made to form a vista
2 : seen in or as if in a vista

¹vi·su·al \'vi-zhə-wəl, -zhəl; 'vizh-wəl\ *adjective* [Middle English, from Late Latin *visualis,* from Latin *visus* sight, from *vidēre* to see] (1603)
1 : of, relating to, or used in vision ⟨*visual* organs⟩
2 : attained or maintained by sight ⟨*visual* impressions⟩
3 : VISIBLE
4 : producing mental images **:** VIVID
5 : done or executed by sight only ⟨*visual* navigation⟩
6 : of, relating to, or employing visual aids
— vi·su·al·ly \'vi-zhə-wə-lē, -zhə-lē; 'vizh-wə-lē\ *adverb*

²visual *noun* (1938)
: something (as a picture, chart, or film) that appeals to the sight and is variously used (as for illustration, demonstration, or promotion) — usually used in plural

visual acuity *noun* (1889)
: the relative ability of the visual organ to resolve detail that is usually expressed as the reciprocal of the minimum angular separation in minutes of two lines just resolvable as separate and that forms in the average human eye an angle of one minute

visual aid *noun* (1911)
: an instructional device (as a chart, map, or model) that appeals chiefly to vision; *especially* **:** an educational motion picture or filmstrip

visual field *noun* (1880)
: the entire expanse of space visible at a given instant without moving the eyes — called also *field of vision*

vi·su·al·ise *British variant of* VISUALIZE

vi·su·al·i·za·tion \,vi-zhə-wə-lə-'zā-shən, ,vi-zhə-lə-, ,vizh-wə-lə-\ *noun* (1883)
1 : formation of mental visual images
2 : the act or process of interpreting in visual terms or of putting into visible form
3 : the process of making an internal organ visible by the introduction (as by swallowing, by an injection, or by an enema) of a radiopaque substance followed by roentgenography

vi·su·al·ize \'vi-zhə-wə-,līz, 'vi-zhə-,līz, 'vizh-wə-,līz\ *verb* **-ized; -iz·ing** (1863)
transitive verb
: to make visible: as **a :** to see or form a mental image of **:** ENVISAGE **b :** to make (an organ) visible by roentgenographic visualization
intransitive verb
: to form a mental visual image

vi·su·al·iz·er \-,lī-zər\ *noun* (1886)
: one that visualizes; *especially* **:** a person whose mental imagery is prevailingly visual

visual literacy *noun* (1971)
: the ability to recognize and understand ideas conveyed through visible actions or images (as pictures)

visual purple *noun* (1878)
: a photosensitive red or purple pigment in the retinal rods of various vertebrates; *especially* **:** RHODOPSIN

vi·ta \'vē-tə, 'vī-tə\ *noun, plural* **vi·tae** \'vē-,tī, -tē\ [Latin, literally, life] (1939)
1 : a brief biographical sketch
2 : CURRICULUM VITAE

vi·tal \'vī-tᵊl\ *adjective* [Middle English, from Middle French, from Latin *vitalis* of life, from *vita* life; akin to Latin *vivere* to live — more at QUICK] (14th century)
1 a : existing as a manifestation of life **b :** concerned with or necessary to the maintenance of life ⟨*vital* organs⟩ ⟨blood and other *vital* fluids⟩
2 : full of life and vigor **:** ANIMATED
3 : characteristic of life or living beings
4 a : fundamentally concerned with or affecting life or living beings: as (1) **:** tending to renew or refresh the living **:** INVIGORATING (2) **:** destructive to life **:** MORTAL **b :** of the utmost importance
5 : recording data relating to lives
6 : of, relating to, or constituting the staining of living tissues
synonym see ESSENTIAL
— vi·tal·ly \-tᵊl-ē\ *adverb*

vital capacity *noun* (1852)
: the breathing capacity of the lungs expressed as the number of cubic inches or cubic centimeters of air that can be forcibly exhaled after a full inspiration

vi·tal·ism \'vī-tᵊl-,i-zəm\ *noun* (1822)
1 : a doctrine that the functions of a living organism are due to a vital principle distinct from physicochemical forces
2 : a doctrine that the processes of life are not explicable by the laws of physics and chemistry alone and that life is in some part self-determining
— vi·tal·ist \-tᵊl-ist\ *noun or adjective*
— vi·tal·is·tic \,vī-tᵊl-'is-tik\ *adjective*

vi·tal·i·ty \vī-'ta-lə-tē\ *noun, plural* **-ties** (1592)
1 a : the peculiarity distinguishing the living from the nonliving **b :** capacity to live and develop; *also* **:** physical or mental vigor especially when highly developed
2 a : power of enduring **b :** lively and animated character

vi·tal·ize \'vī-tᵊl-,īz\ *transitive verb* **-ized; -iz·ing** (1678)
: to endow with vitality **:** ANIMATE
— vi·tal·i·za·tion \,vī-tᵊl-ə-'zā-shən\ *noun*

vi·tals \'vī-tᵊlz\ *noun plural* (circa 1610)
1 : vital organs (as the heart, liver, lungs, and brain)
2 : essential parts

vital signs *noun plural* (circa 1919)
: signs of life; *specifically* **:** the pulse rate, respiratory rate, body temperature, and often blood pressure of a person

vital statistics *noun plural* (1837)
1 : statistics relating to births, deaths, marriages, health, and disease

2 : facts (as physical dimensions or quantities) considered to be interesting or important; *especially* **:** a woman's bust, waist, and hip measurements

vi·ta·min \'vī-tə-mən, *British usually* 'vi-\ *noun* [Latin *vita* life + International Scientific Vocabulary *amine*] (1912)
: any of various organic substances that are essential in minute quantities to the nutrition of most animals and some plants, act especially as coenzymes and precursors of coenzymes in the regulation of metabolic processes but do not provide energy or serve as building units, and are present in natural foodstuffs or sometimes produced within the body

vitamin A *noun* (1920)
: any of several fat-soluble vitamins (as retinol) found especially in animal products (as egg yolk, milk, or fish-liver oils) or a mixture of them whose lack in the animal body causes epithelial tissues to become keratinous (as in the eye with resulting visual defects)

vitamin B *noun* (1920)
1 : VITAMIN B COMPLEX
2 *or* **vitamin B_1** \-'bē-'wən\ **:** THIAMINE

vitamin B complex *noun* (1928)
: a group of water-soluble vitamins found especially in yeast, seed germs, eggs, liver and flesh, and vegetables that have varied metabolic functions and include coenzymes and growth factors — called also *B complex*; compare BIOTIN, CHOLINE, NICOTINIC ACID, PANTOTHENIC ACID

vitamin B_6 \-'bē-'siks\ *noun* (1934)
: pyridoxine or a closely related compound found widely in combined form and considered essential to vertebrate nutrition

vitamin B_{12} \-'bē-'twelv\ *noun* (1948)
1 : a complex cobalt-containing compound $C_{63}H_{88}CoN_{14}O_{14}P$ that occurs especially in liver, is essential to normal blood formation, neural function, and growth, and is used especially in treating pernicious and related anemias and in animal feed as a growth factor — called also *cyanocobalamin*
2 : any of several compounds similar to vitamin B_{12} in action but having different chemistry

vitamin B_2 \-'bē-'tü\ *noun* (1928)
: RIBOFLAVIN

vitamin C *noun* (1920)
: a water-soluble vitamin $C_6H_8O_6$ found in plants and especially in fruits and leafy vegetables or made synthetically and used in the prevention and treatment of scurvy and as an antioxidant for foods — called also *ascorbic acid*

vitamin D *noun* (1921)
: any or all of several fat-soluble vitamins chemically related to steroids, essential for normal bone and tooth structure, and found especially in fish-liver oils, egg yolk, and milk or produced by activation (as by ultraviolet irradiation) of sterols: as **a** *or* **vitamin D_2** \-'dē-'tü\ **:** CALCIFEROL **b** *or* **vitamin D_3** \-'dē-'thrē\ **:** CHOLECALCIFEROL

vitamin E *noun* (1925)
: any of several fat-soluble vitamins that are chemically tocopherols, are essential in the nutrition of various vertebrates in which their absence is associated with infertility, degenerative changes in muscle, or vascular abnormalities, are found especially in leaves and in seed germ oils, and are used chiefly in animal feeds and as antioxidants

vitamin G *noun* (1929)
: RIBOFLAVIN

vitamin H *noun* (circa 1935)
: BIOTIN

vitamin K *noun* [Danish *koagulation* coagulation] (1935)
1 : either of two naturally occurring fat-soluble vitamins $C_{31}H_{46}O_2$ and $C_{41}H_{56}O_2$ essential for the clotting of blood because of their role in the production of prothrombin — called also respectively *vitamin K_1*, *vitamin K_2*
2 : any of several synthetic compounds closely related chemically to natural vitamins K_1 and K_2 and of similar biological activity

vi·tel·line \vī-'te-lən, -,lēn, -,līn\ *adjective* [Middle English, from Middle French *vitellin*, from Medieval Latin *vitellinus*, from Latin *vitellus*] (15th century)
1 : resembling the yolk of an egg especially in yellow color
2 : of, relating to, or producing yolk

vitelline membrane *noun* (1845)
: a membrane that encloses the egg proper, corresponds to the plasmalemma of an ordinary cell, and in many animals forms the fertilization membrane by separating from the plasma membrane immediately after fertilization

vi·tel·lo·gen·e·sis \vī-,te-lō-'je-nə-səs, və-\ *noun* [New Latin, from Latin *vitellus* + New Latin *-o-* + *genesis*] (1947)
: yolk formation

vi·tel·lus \-'te-ləs\ *noun* [Latin] (1728)
: the egg cell proper including the yolk but excluding any albuminous or membranous envelopes; *also* **:** YOLK 1c

vi·ti·ate \'vi-shē-,āt\ *transitive verb* **-at·ed; -at·ing** [Latin *vitiatus*, past participle of *vitiare*, from *vitium* fault, vice] (1534)
1 : to make faulty or defective **:** IMPAIR ⟨the comic impact is *vitiated* by obvious haste —William Styron⟩
2 : to debase in moral or aesthetic status ⟨a mind *vitiated* by prejudice⟩
3 : to make ineffective ⟨fraud *vitiates* a contract⟩
synonym see DEBASE
— vi·ti·a·tion \,vi-shē-'ā-shən\ *noun*
— vi·ti·a·tor \'vi-shē-,ā-tər\ *noun*

vi·ti·cul·ture \'vi-tə-,kəl-chər, 'vī-\ *noun* [Latin *vitis* vine + English *culture* — more at WITHY] (1872)
: the cultivation or culture of grapes especially for wine making
— vi·ti·cul·tur·al \,vi-tə-'kəl-chə-rəl, ,vī-, -'kəlch-rəl\ *adjective*
— vi·ti·cul·tur·al·ly *adverb*
— vi·ti·cul·tur·ist \-rist\ *noun*

vit·i·li·go \,vi-tᵊl-'ī-(,)gō, -'ē-\ *noun* [New Latin, from Latin, tetter] (1842)
: a skin disorder manifested by smooth white spots on various parts of the body

vit·rec·to·my \və-'trek-tə-mē\ *noun, plural* **-mies** [*vitr*eous humor + *-ectomy*] (1968)
: surgical removal of all or part of the vitreous humor

¹vit·re·ous \'vi-trē-əs\ *adjective* [Latin *vitreus*, from *vitrum* glass] (1646)
1 : of, relating to, derived from, or consisting of glass
2 a : resembling glass (as in color, composition, brittleness, or luster) **:** GLASSY ⟨*vitreous* rocks⟩ **b :** characterized by low porosity and usually translucence due to the presence of a glassy phase ⟨*vitreous* china⟩
3 : of, relating to, or constituting the vitreous humor

²vitreous *noun* (1869)
: VITREOUS HUMOR

vitreous humor *noun* (1663)
: the clear colorless transparent jelly that fills the eyeball posterior to the lens — see EYE illustration

vitreous silica *noun* (1925)
: a chemically stable and refractory glass made from silica alone — compare QUARTZ GLASS

vit·ri·fy \'vi-trə-,fī\ *verb* **-fied; -fy·ing** [Middle French *vitrifier*, from Latin *vitrum* glass] (1594)
transitive verb
: to convert into glass or a glassy substance by heat and fusion
intransitive verb
: to become vitrified
— vit·ri·fi·able \,vi-trə-'fī-ə-bəl\ *adjective*
— vit·ri·fi·ca·tion \,vi-trə-fə-'kā-shən\ *noun*

vi·trine \və-'trēn\ *noun* [French, from *vitre* pane of glass, from Old French, from Latin *vitrum*] (1880)
: a glass showcase or cabinet especially for displaying fine wares or specimens

vit·ri·ol \'vi-trē-əl\ *noun* [Middle English, from Middle French, from Medieval Latin *vitriolum*, alteration of Late Latin *vitreolum*, neuter of *vitreolus* glassy, from Latin *vitreus* vitreous] (14th century)
1 a : a sulfate of any of various metals (as copper, iron, or zinc); *especially* **:** a glassy hydrate of such a sulfate **b :** OIL OF VITRIOL
2 : something felt to resemble vitriol especially in caustic quality; *especially* **:** virulence of feeling or of speech
— vit·ri·ol·ic \,vi-trē-'ä-lik\ *adjective*

vit·ta \'vi-tə\ *noun, plural* **vit·tae** \'vi-tē, 'vi-,tē, 'vi-,tī\ [New Latin, from Latin, fillet; akin to Latin *viēre* to plait — more at WIRE] (1819)
1 : STRIPE, STREAK
2 : one of the oil tubes in the fruits of plants of the carrot family

vit·tles \'vi-tᵊlz\ *noun plural* (14th century)
: VICTUALS

vi·tu·per·ate \vī-'tü-pə-,rāt, və-, -'tyü-\ *verb* **-at·ed; -at·ing** [Latin *vituperatus*, past participle of *vituperare*, from *vitium* fault + *parare* to make, prepare — more at PARE] (1542)
transitive verb
: to abuse or censure severely or abusively **:** BERATE
intransitive verb
: to use harsh condemnatory language
synonym see SCOLD
— vi·tu·per·a·tor \-,rā-tər\ *noun*

vi·tu·per·a·tion \(,)vī-,tü-pə-'rā-shən, və-, -,tyü-\ *noun* (15th century)
1 : sustained and bitter railing and condemnation **:** vituperative utterance
2 : an act or instance of vituperating
synonym see ABUSE

vi·tu·per·a·tive \vī-'tü-p(ə-)rə-tiv, -pə-,rā-\ *adjective* (1727)
: uttering or given to censure **:** containing or characterized by verbal abuse
— vi·tu·per·a·tive·ly *adverb*

vi·tu·per·a·to·ry \-p(ə-)rə-,tōr-ē, -,tȯr-\ *adjective* (1586)
: VITUPERATIVE

vi·va \'vē-və, -,vä\ *interjection* [Italian & Spanish, long live, from 3d person singular present subjunctive of *vivere* to live, from Latin — more at QUICK] (circa 1700)
— used to express goodwill or approval

¹vi·va·ce \vē-'vä-(,)chā, -chē\ *adverb or adjective* [Italian, vivacious, from Latin *vivac-*, *vivax*] (1683)
: in a brisk spirited manner — used as a direction in music

²vivace *noun* (circa 1683)
: a musical composition or movement in vivace tempo

vi·va·cious \və-'vā-shəs *also* vī-\ *adjective* [Latin *vivac-*, *vivax*, literally, long-lived, from *vivere* to live] (circa 1645)
: lively in temper, conduct, or spirit **:** SPRIGHTLY
synonym see LIVELY
— vi·va·cious·ly *adverb*
— vi·va·cious·ness *noun*

vi·vac·i·ty \-'va-sə-tē\ *noun* (15th century)
: the quality or state of being vivacious

vi·van·dière \,vē-,väⁿ-'dyer\ *noun* [French, feminine of Middle French *vivandier*, from Medieval Latin *vivanda* food — more at VIAND] (1848)

: a female sutler

vi·var·i·um \vī-ˈvar-ē-əm, -ˈver-\ *noun, plural* **-ia** \-ē-ə\ *or* **-i·ums** [Latin, park, preserve, from *vivus* alive — more at QUICK] (1853)
: a terrarium used especially for small animals

[1]**vi·va vo·ce** \ˌvī-və-ˈvō-(ˌ)sē *or (as if from Italian)* ˌvē-və-ˈvō-(ˌ)chā\ *adverb* [Medieval Latin, with the living voice] (1563)
: by word of mouth **:** ORALLY

[2]**viva voce** *adjective* (1654)
: expressed or conducted by word of mouth **:** ORAL

[3]**viva voce** *noun* (1842)
: an examination conducted viva voce — called also *viva*

vi·vax malaria \ˈvī-ˌvaks-\ *noun* [New Latin *vivax,* specific epithet of *Plasmodium vivax,* parasite causing tertian] (circa 1941)
: malaria caused by a plasmodium (*Plasmodium vivax*) that induces paroxysms at 48-hour intervals

vi·ver·rid \vī-ˈver-əd\ *noun* [New Latin *Viverridae,* from *Viverra,* type genus, from Latin *viverra* ferret; akin to Old English *ācweorna* squirrel, Lithuanian *voverė̃*] (1902)
: any of a family (*Viverridae*) of carnivorous mammals (as a civet, a genet, or a mongoose) that are rarely larger than a domestic cat but are long, slender, and like a weasel in build with short more or less retractile claws and rounded feet
— viverrid *adjective*

vi·vers \ˈvē-vərz, ˈvī-\ *noun plural* [Middle French *vivres,* plural of *vivre* food, from *vivre* to live, from Latin *vivere*] (1536)
chiefly Scottish **:** VICTUALS, FOOD

Viv·i·an *or* **Viv·i·en** \ˈvi-vē-ən\ *noun*
: the mistress of Merlin in Arthurian legend — called also *Lady of the Lake*

viv·id \ˈvi-vəd\ *adjective* [Latin *vividus,* from *vivere* to live — more at QUICK] (1638)
1 : having the appearance of vigorous life or freshness **:** LIVELY ⟨a *vivid* sketch⟩
2 *of a color* **:** very strong **:** very high in chroma
3 : producing a strong or clear impression on the senses **:** SHARP, INTENSE; *specifically* **:** producing distinct mental images ⟨a *vivid* description⟩
4 : acting clearly and vigorously ⟨a *vivid* imagination⟩
synonym see GRAPHIC
— viv·id·ly *adverb*
— viv·id·ness *noun*

vi·vif·ic \vī-ˈvi-fik\ *adjective* [Latin *vivificus*] (1551)
: imparting spirit or vivacity

viv·i·fy \ˈvi-və-ˌfī\ *transitive verb* **-fied; -fy·ing** [Middle French *vivifier,* from Late Latin *vivificare,* from Latin *vivificus* enlivening, from *vivus* alive — more at QUICK] (1545)
1 : to endue with life or renewed life **:** ANIMATE ⟨rains that *vivify* the barren hills⟩
2 : to impart vitality or vividness to ⟨concentrating this union of quality and meaning in a way which *vivifies* both —John Dewey⟩
synonym see QUICKEN
— viv·i·fi·ca·tion \ˌvi-və-fə-ˈkā-shən\ *noun*
— viv·i·fi·er \ˈvi-və-ˌfī(-ə)r\ *noun*

vi·vi·par·i·ty \ˌvī-və-ˈpar-ət-ē, ˌvi-\ *noun* (1864)
: the quality or state of being viviparous

vi·vip·a·rous \vī-ˈvi-p(ə-)rəs, və-\ *adjective* [Latin *viviparus,* from *vivus* alive + *-parus* -parous] (1646)
1 : producing living young instead of eggs from within the body in the manner of nearly all mammals, many reptiles, and a few fishes
2 : germinating while still attached to the parent plant ⟨the *viviparous* seed of the mangrove⟩
— vi·vip·a·rous·ly *adverb*

viv·i·sect \ˈvi-və-ˌsekt\ *verb* [back-formation from *vivisection*] (1864)
transitive verb
: to perform vivisection on **:** subject to vivisection
intransitive verb
: to practice vivisection
— viv·i·sec·tor \-ˌsek-tər\ *noun*

viv·i·sec·tion \ˌvi-və-ˈsek-shən, ˈvi-və-ˌ\ *noun* [Latin *vivus* + English *section*] (1707)
1 : the cutting of or operation on a living animal usually for physiological or pathological investigation; *broadly* **:** animal experimentation especially if considered to cause distress to the subject
2 : minute or pitiless examination or criticism
— viv·i·sec·tion·al \ˌvi-və-ˈsek-shnəl, -shə-nᵊl\ *adjective*
— viv·i·sec·tion·ist \-ˈsek-sh(ə-)nist\ *noun*

vix·en \ˈvik-sən\ *noun* [(assumed) Middle English (southern dialect) *vixen,* alteration of Middle English *fixen,* from Old English *fyxe,* feminine of *fox*] (1590)
1 : a shrewish ill-tempered woman
2 : a female fox ◆
— vix·en·ish \-s(ə-)nish\ *adjective*

viz·ard \ˈvi-zərd, -ˌzärd\ *noun* [alteration of Middle English *viser* mask, visor] (circa 1555)
1 : a mask for disguise or protection
2 : DISGUISE, GUISE

viz·ca·cha *variant of* VISCACHA

vi·zier \və-ˈzir\ *noun* [Turkish *vezir,* from Arabic *wazīr*] (1599)
: a high executive officer of various Muslim countries and especially of the Ottoman Empire
— vi·zier·ate \-ˈzir-ət, -ˈzir-ˌāt\ *noun*
— vi·zier·ial \-ˈzir-ē-əl\ *adjective*
— vi·zier·ship \-ˈzir-ˌship\ *noun*

vi·zor *variant of* VISOR

vizs·la \ˈvēz-lə, ˈvēs-, ˈvizh-\ *noun* [Hungarian] (1945)
: any of a breed of hunting dogs of Hungarian origin that resemble the weimaraner but have a rich deep red coat and brown eyes

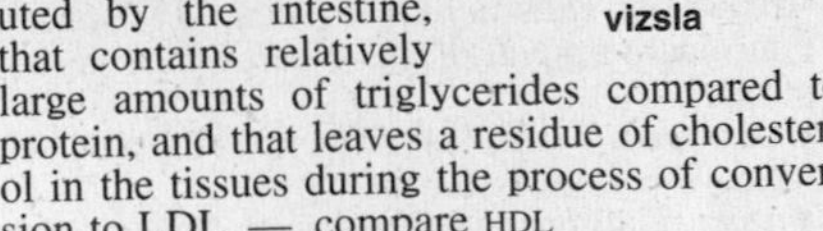
vizsla

VLDL \ˌvē-(ˌ)el-(ˌ)dē-ˈel\ *noun* [*v*ery *l*ow-*d*ensity *l*ipoprotein] (1977)
: a plasma lipoprotein that is produced primarily by the liver with lesser amounts contributed by the intestine, that contains relatively large amounts of triglycerides compared to protein, and that leaves a residue of cholesterol in the tissues during the process of conversion to LDL — compare HDL

V neck *noun* (1905)
: a V-shaped neck of a garment; *also* **:** a garment (as a sweater) with a V-shaped neck
— V–necked *adjective*

vo·ca·ble \ˈvō-kə-bəl\ *noun* [Middle French, from Latin *vocabulum,* from *vocare* to call, from *vox* voice — more at VOICE] (1530)
: TERM; *specifically* **:** a word composed of various sounds or letters without regard to its meaning

vo·cab·u·lar \vō-ˈka-byə-lər, və-\ *adjective* [back-formation from *vocabulary*] (1608)
: of or relating to words or phraseology **:** VERBAL

vo·cab·u·lary \vō-ˈka-byə-ˌler-ē, və-\ *noun, plural* **-lar·ies** [Middle French *vocabulaire,* probably from Medieval Latin *vocabularium,* from neuter of *vocabularius* verbal, from Latin *vocabulum*] (1532)
1 : a list or collection of words or of words and phrases usually alphabetically arranged and explained or defined **:** LEXICON
2 a : a sum or stock of words employed by a language, group, individual, or work or in a field of knowledge **b :** a list or collection of terms or codes available for use (as in an indexing system)
3 : a supply of expressive techniques or devices (as of an art form)

vocabulary entry *noun* (circa 1934)
: a word (as the noun *book*), hyphenated or open compound (as the verb *book-match* or the noun *book review*), word element (as the affix *pro-*), abbreviation (as *agt*), verbalized symbol (as *Na*), or term (as *man in the street*) entered alphabetically in a dictionary for the purpose of definition or identification or expressly included as an inflected form (as the noun *mice* or the verb *saw*) or as a derived form (as the noun *godlessness* or the adverb *globally*) or related phrase (as *one for the book*) run on at its base word and usually set in a type (as boldface) readily distinguishable from that of the lightface running text which defines, explains, or identifies the entry

[1]**vo·cal** \ˈvō-kəl\ *adjective* [Middle English, from Latin *vocalis,* from *voc-, vox* voice — more at VOICE] (14th century)
1 a : uttered by the voice **:** ORAL **b :** produced in the larynx **:** uttered with voice
2 : relating to, composed or arranged for, or sung by the human voice ⟨*vocal* music⟩
3 : VOCALIC
4 a : having or exercising the power of producing voice, speech, or sound **b :** EXPRESSIVE **c :** full of voices **:** RESOUNDING **d :** given to expressing oneself freely or insistently **:** OUTSPOKEN **e :** expressed in words
5 : of, relating to, or resembling the voice ⟨*vocal* impairment⟩
— vo·cal·i·ty \vō-ˈka-lə-tē\ *noun*
— vo·cal·ly \ˈvō-kə-lē\ *adverb*

[2]**vocal** *noun* (1582)
1 : a vocal sound
2 : a usually accompanied musical composition for the human voice **:** SONG; *also* **:** a performance of such a composition

vocal cords *noun plural* (1852)
: either of two pairs of folds of mucous membranes that project into the cavity of the larynx and have free edges extending dorsoventrally toward the middle line

vocal folds *noun plural* (1924)
: the lower pair of vocal cords each of which when drawn taut, approximated to the contralateral member of the pair, and subjected to a flow of breath produces the voice

[1]**vo·cal·ic** \vō-ˈka-lik, və-\ *adjective* [Latin *vocalis* vowel, from *vocalis* vocal] (1814)
1 : marked by or consisting of vowels

◇ WORD HISTORY

vixen Even though *very, voice,* and *visit* are now in everyday use, English originally had no words beginning with a \v\ sound. Almost all words beginning with *v* have been borrowed from other languages, such as French or Latin. The sound \v\ did exist in Old English, but it could only appear in the middle of words, not at the beginning or end. Because the \v\ sound was in a complementary relation with the sound \f\, which could appear at the beginning and end of words but not the middle, the single letter *f* sufficed in Old English to spell both \f\ and \v\. In parts of southern England, however, medieval speakers of English began to pronounce all instances of initial \f\ as \v\ and initial \s\ as \z\; this tendency continues today among rural dialect speakers in southwestern England, so that *finger* is pronounced as if it were spelled *vinger* and *seven* as if it were spelled *zeven.* A handful of such dialectal pronunciations with initial \v\ for \f\ became standard in early Modern English, and the words affected have consequently been respelled with *v.* These include *vat* (from Old English *fæt*), *vane* (from Old English *fana* "banner"), and *vixen* (from Old English *fyxe*). The change of consonant in *vixen* obscures the original relation of the word to *fox.*

2 a : being or functioning as a vowel **b :** of, relating to, or associated with a vowel
— **vo·cal·i·cal·ly** \-li-k(ə-)lē\ *adverb*

²vocalic *noun* (1924)
: a vowel sound or sequence in its function as the most sonorous part of a syllable

vo·cal·ise *British variant of* VOCALIZE

vo·cal·ism \ˈvō-kə-ˌli-zəm\ *noun* (circa 1859)
1 : VOCALIZATION
2 : vocal art or technique **:** SINGING
3 a : the vowel system of a language or dialect **b :** the pattern of vowels in a word or paradigm

vo·cal·ist \-kə-list\ *noun* (1834)
: ¹SINGER

vo·cal·ize \ˈvō-kə-ˌlīz\ *verb* **-ized; -iz·ing** (1669)
transitive verb
1 : to give voice to **:** UTTER; *specifically* **:** SING
2 a : to make voiced rather than voiceless **:** VOICE **b :** to convert to a vowel
3 : to furnish (as a consonantal Hebrew or Arabic text) with vowels or vowel points
intransitive verb
1 : to utter vocal sounds
2 : SING; *specifically* **:** to sing without words
— **vo·cal·i·za·tion** \ˌvō-kə-lə-ˈzā-shən\ *noun*
— **vo·cal·iz·er** \ˈvō-kə-ˌlī-zər\ *noun*

vo·ca·tion \vō-ˈkā-shən\ *noun* [Middle English *vocacioun,* from Latin *vocation-, vocatio* summons, from *vocare* to call, from *vox* voice — more at VOICE] (15th century)
1 a : a summons or strong inclination to a particular state or course of action; *especially* **:** a divine call to the religious life **b :** an entry into the priesthood or a religious order
2 a : the work in which a person is regularly employed **:** OCCUPATION **b :** the persons engaged in a particular occupation
3 : the special function of an individual or group

vo·ca·tion·al \-shnəl, -shə-n^{ə}l\ *adjective* (1652)
1 : of, relating to, or concerned with a vocation
2 : of, relating to, or being in training in a skill or trade to be pursued as a career
— **vo·ca·tion·al·ly** *adverb*

vo·ca·tion·al·ism \-shnə-ˌli-zəm, -shə-n^{ə}l-ˌi-zəm\ *noun* (1924)
: emphasis on vocational training in education
— **vo·ca·tion·al·ist** \-list, -ist\ *noun*

¹voc·a·tive \ˈvä-kə-tiv\ *adjective* [Middle English *vocatif,* from Middle French, from Latin *vocativus,* from *vocatus,* past participle of *vocare*] (15th century)
1 : of, relating to, or being a grammatical case marking the one addressed (as Latin *Domine* in *miserere, Domine* "have mercy, O Lord")
2 *of a word or word group* **:** marking the one addressed (as *mother* in "mother, come here")
— **voc·a·tive·ly** *adverb*

²vocative *noun* (circa 1522)
1 : the vocative case of a language
2 : a form in the vocative case

vo·cif·er·ant \vō-ˈsi-fə-rənt\ *adjective* (1609)
: CLAMOROUS, VOCIFEROUS

vo·cif·er·ate \-ˌrāt\ *verb* **-at·ed; -at·ing** [Latin *vociferatus,* past participle of *vociferari,* from *voc-, vox* voice + *ferre* to bear — more at VOICE, BEAR] (1599)
transitive verb
: to utter loudly **:** SHOUT
intransitive verb
: to cry out loudly **:** CLAMOR
— **vo·cif·er·a·tion** \-ˌsi-fə-ˈrā-shən\ *noun*
— **vo·cif·er·a·tor** \-ˈsi-fə-ˌrā-tər\ *noun*

vo·cif·er·ous \vō-ˈsi-f(ə-)rəs\ *adjective* (circa 1611)
: marked by or given to vehement insistent outcry ☆
— **vo·cif·er·ous·ly** *adverb*
— **vo·cif·er·ous·ness** *noun*

vo·cod·er \ˈvō-ˌkō-dər\ *noun* [voice coder] (1939)
: an electronic mechanism that reduces speech signals to slowly varying signals transmittable over communication systems of limited frequency bandwidth

vod·ka \ˈväd-kə\ *noun* [Russian, from *voda* water; akin to Old English *wæter* water] (circa 1803)
: a colorless liquor of neutral spirits distilled from a mash (as of rye or wheat)

vodka martini *noun* (1948)
: a martini made with vodka instead of gin

vo·dun *also* **vo·doun** \vō-ˈdü^n\ *noun* [Haitian Creole] (1920)
: VOODOO 1

vo·gie \ˈvō-gē\ *adjective* [origin unknown] (1712)
Scottish **:** PROUD, VAIN

¹vogue \ˈvōg\ *noun* [Middle French, action of rowing, course, fashion, from Old Italian *voga,* from *vogare* to row] (1571)
1 *archaic* **:** the leading place in popularity or acceptance
2 a : popular acceptation or favor **:** POPULARITY **b :** a period of popularity
3 : one that is in fashion at a particular time
synonym see FASHION
— **vogue** *adjective*

²vogue *intransitive verb* [from *Vogue,* a fashion magazine] **vogued; vogu·ing** *or* **vogue·ing** (1989)
: to strike poses in campy imitation of fashion models especially as a kind of dance
— **vogu·er** \ˈvō-gər\ *noun*

vogu·ish \ˈvō-gish\ *adjective* (1926)
1 : FASHIONABLE, SMART
2 : suddenly or temporarily popular
— **vogu·ish·ness** *noun*

¹voice \ˈvȯis\ *noun* [Middle English, from Old French *vois,* from Latin *voc-, vox;* akin to Old High German gi*wahanen* to mention, Greek *epos* word, speech, Sanskrit *vāk* voice] (14th century)
1 a : sound produced by vertebrates by means of lungs, larynx, or syrinx; *especially* **:** sound so produced by human beings **b** (1) **:** musical sound produced by the vocal folds and resonated by the cavities of head and throat (2) **:** the power or ability to produce musical tones (3) **:** SINGER (4) **:** one of the melodic parts in a vocal or instrumental composition (5) **:** condition of the vocal organs with respect to production of musical tones (6) **:** the use of the voice (as in singing or acting) ⟨studying *voice*⟩ **c :** expiration of air with the vocal cords drawn close so as to vibrate audibly (as in uttering vowels and consonant sounds as \v\ or \z\) **d :** the faculty of utterance ⟨lost my *voice*⟩
2 : a sound resembling or suggesting vocal utterance
3 : an instrument or medium of expression ⟨the party became the *voice* of the workers⟩
4 a : wish, choice, or opinion openly or formally expressed ⟨the *voice* of the people⟩ **b :** right of expression; *also* **:** influential power
5 : distinction of form or a system of inflections of a verb to indicate the relation of the subject of the verb to the action which the verb expresses ⟨active and passive *voices*⟩
— **with one voice :** without dissent **:** UNANIMOUSLY

²voice *transitive verb* **voiced; voic·ing** (15th century)
1 : to express in words **:** UTTER ⟨*voice* a complaint⟩
2 : to adjust for producing the proper musical sounds
3 : to pronounce (as a consonant) with voice
synonym see EXPRESS

voice box *noun* (1912)
: LARYNX

voiced \ˈvȯist\ *adjective* (1593)
1 : having or furnished with a voice especially of a specified kind — often used in combination ⟨soft-*voiced*⟩
2 : uttered with vocal cord vibration ⟨a *voiced* consonant⟩

voice·ful \ˈvȯis-fəl\ *adjective* (circa 1611)
: having a voice or vocal quality; *also* **:** having a loud voice or many voices
— **voice·ful·ness** *noun*

voice·less \-ləs\ *adjective* (1535)
1 : having no voice **:** MUTE
2 : not voiced **:** SURD ⟨a *voiceless* consonant⟩
— **voice·less·ly** *adverb*
— **voice·less·ness** *noun*

voice mail *noun* (1981)
: an electronic communication system in which spoken messages are recorded or digitized for later playback to the intended recipient

voice–over \ˈvȯis-ˌō-vər\ *noun* (circa 1947)
1 a : the voice of an unseen narrator speaking (as in a motion picture or television commercial) **b :** the voice of a visible character (as in a motion picture) expressing unspoken thoughts
2 : a recording of a voice-over

voice part *noun* (1776)
: VOICE 1b(4)

voice·print \ˈvȯis-print\ *noun* [*voice* + *fingerprint*] (1962)
: an individually distinctive pattern of certain voice characteristics that is spectrographically produced

voic·er \ˈvȯi-sər\ *noun* (1879)
: one that voices; *specifically* **:** one that voices organ pipes

voice vote *noun* (1924)
: a parliamentary vote taken by calling for ayes and noes and estimating which response is stronger

¹void \ˈvȯid\ *adjective* [Middle English *voide,* from Old French, from (assumed) Vulgar Latin *vocitus,* alteration of Latin *vocivus, vacivus* empty, from *vacare* to be empty] (14th century)
1 a : not occupied **:** VACANT ⟨a *void* bishopric⟩ **b :** not inhabited **:** DESERTED
2 : containing nothing ⟨*void* space⟩
3 : IDLE, LEISURE
4 a : being without **:** DEVOID ⟨a nature *void* of all malice⟩ **b :** having no members or examples; *specifically, of a suit* **:** having no cards represented in a particular hand
5 : VAIN, USELESS
6 a : of no legal force or effect **:** NULL ⟨a *void* contract⟩ **b :** VOIDABLE
synonym see EMPTY
— **void·ness** *noun*

²void *noun* (1616)
1 a : OPENING, GAP **b :** empty space **:** EMPTINESS, VACUUM

☆ SYNONYMS
Vociferous, clamorous, blatant, strident, boisterous, obstreperous mean so loud or insistent as to compel attention. VOCIFEROUS implies a vehement shouting or calling out ⟨*vociferous* cries of protest and outrage⟩. CLAMOROUS may imply insistency as well as vociferousness in demanding or protesting ⟨*clamorous* demands for prison reforms⟩. BLATANT implies an offensive bellowing or insensitive loudness ⟨*blatant* rock music⟩ ⟨a *blatant* clamor for impeachment⟩. STRIDENT suggests harsh and discordant noise ⟨heard the *strident* cry of the crow⟩. BOISTEROUS suggests a noisiness and turbulence due to high spirits ⟨a *boisterous* crowd of party goers⟩. OBSTREPEROUS suggests unruly and aggressive noisiness and resistance to restraint ⟨the *obstreperous* demonstrators were removed from the hall⟩.

2 **:** the quality or state of being without something **:** LACK, ABSENCE
3 **:** a feeling of want or hollowness
4 **:** absence of cards of a particular suit in a hand originally dealt to a player

³void *verb* [Middle English, from Middle French *vuidier,* from (assumed) Vulgar Latin *vocitare,* from *vocitus*] (14th century)
transitive verb
1 a : to make empty or vacant **:** CLEAR **b** *archaic* **:** VACATE, LEAVE
2 : DISCHARGE, EMIT ⟨*void* excrement⟩
3 : NULLIFY, ANNUL ⟨*void* a contract⟩
intransitive verb
: to eliminate solid or liquid waste from the body
— **void·er** *noun*

void·able \'vȯi-də-bəl\ *adjective* (15th century)
: capable of being voided; *specifically* **:** capable of being adjudged void
— **void·able·ness** *noun*

void·ance \'vȯi-dᵊn(t)s\ *noun* (14th century)
1 : the act of voiding
2 *of a benefice* **:** the state of being without an incumbent

void·ed \'vȯi-dəd\ *adjective* (circa 1539)
: having the inner part cut away or left vacant with a narrow border left at the sides — used of a heraldic charge

voi·là *or* **voi·la** \vwä-'lä\ *interjection* [French, literally, see there] (1739)
— used to call attention, to express satisfaction or approval, or to suggest an appearance as if by magic

voile \'vȯi(ə)l\ *noun* [French, veil, from Latin *vela,* neuter plural of *velum*] (1889)
: a fine soft sheer fabric used especially for women's summer clothing or curtains

voir dire \'vwär-'dir, 'wär-\ *noun* [Anglo-French, from Old French, to speak the truth] (1676)
: a preliminary examination to determine the competency of a witness or juror

vo·lant \'vō-lənt\ *adjective* [Middle French, from Latin *volant-, volans,* present participle of *volare* to fly] (1572)
1 : having the wings extended as if in flight — used of a heraldic bird
2 : flying or capable of flying
3 : QUICK, NIMBLE

vo·lan·te \vō-'län-(ˌ)tā\ *adjective* [Italian, literally, flying, from Latin *volant-, volans,* present participle] (circa 1854)
: moving with light rapidity — used as a direction in music

Vo·la·pük \'vō-lə-ˌpu̇k, 'vä-\ *noun* [Volapük, literally, world's speech, from *vola* of the world (genitive of *vol* world, modification of English *world*) + *pük* speech, modification of English *speak*] (1885)
: an artificial international language based largely on English but with some root words from German, French, and Latin

vo·lar \'vō-lər, -ˌlär\ *adjective* [Latin *vola* hollow in the palm of the hand or sole of the foot] (1814)
: relating to the palm of the hand or the sole of the foot; *specifically* **:** located on the same side as the palm of the hand ⟨the *volar* part of the forearm⟩

¹vol·a·tile \'vä-lə-tᵊl, *especially British* -ˌtīl\ *adjective* [French, from Latin *volatilis,* from *volare* to fly] (1605)
1 : readily vaporizable at a relatively low temperature
2 : flying or having the power to fly
3 a : LIGHTHEARTED, LIVELY **b :** easily aroused ⟨*volatile* suspicions⟩ **c :** tending to erupt into violence **:** EXPLOSIVE
4 a : unable to hold the attention fixed because of an inherent lightness or fickleness of disposition **b :** characterized by or subject to rapid or unexpected change
5 : difficult to capture or hold permanently **:** EVANESCENT, TRANSITORY
— **vol·a·tile·ness** *noun*
— **vol·a·til·i·ty** \ˌvä-lə-'ti-lə-tē\ *noun*

²volatile *noun* (1686)
: a volatile substance

volatile oil *noun* (1800)
: an oil that vaporizes readily; *especially* **:** ESSENTIAL OIL

vol·a·til·ise *British variant of* VOLATILIZE

vol·a·til·ize \'vä-lə-tᵊl-ˌīz, *British also* və-'la-\ *verb* **-ized; -iz·ing** (1657)
transitive verb
: to make volatile; *especially* **:** to cause to pass off in vapor
intransitive verb
: to pass off in vapor
— **vol·a·til·iz·able** \-ˌī-zə-bəl\ *adjective*
— **vol·a·til·i·za·tion** \ˌvä-lə-tᵊl-ə-'zā-shən, *British also* və-ˌla-\ *noun*

vol–au–vent \ˌvȯ-lō-'vä^n\ *noun* [French, literally, flight in the wind] (1828)
: a baked patty shell filled with meat, fowl, game, or seafood in sauce

¹vol·ca·nic \väl-'ka-nik, vȯl- *also* -'kā-\ *adjective* (1774)
1 a : of, relating to, or produced by a volcano **b :** characterized by volcanoes ⟨a *volcanic* range⟩ **c :** made of materials from volcanoes
2 : explosively violent **:** VOLATILE ⟨*volcanic* emotions⟩
— **vol·ca·ni·cal·ly** \-ni-k(ə-)lē\ *adverb*

²volcanic *noun* (1894)
: a volcanic rock

volcanic glass *noun* (circa 1840)
: natural glass produced by the cooling of molten lava too rapidly to permit crystallization

vol·ca·nic·i·ty \ˌväl-kə-'ni-sə-tē, ˌvȯl-\ *noun* (1836)
: VOLCANISM

vol·ca·nism \'väl-kə-ˌni-zəm, 'vȯl-\ *noun* (circa 1864)
: volcanic action or activity

vol·ca·no \väl-'kā-(ˌ)nō, vȯl-\ *noun, plural* **-noes** *or* **-nos** [Italian or Spanish; Italian *vulcano,* from Spanish *volcán,* ultimately from Latin *Volcanus* Vulcan] (1613)
1 : a vent in the crust of the earth or another planet from which usually molten or hot rock and steam issue; *also* **:** a hill or mountain composed wholly or in part of the ejected material
2 : something of explosively violent potential
◆

vol·ca·no·log·i·cal \ˌväl-kə-nᵊl-'ä-ji-kəl, ˌvȯl-\ *also* **vol·ca·no·log·ic** \-jik\ *adjective* (circa 1891)
: of, relating to, or involving volcanology or volcanic phenomena ⟨*volcanological* processes that shape the planets⟩

vol·ca·nol·o·gy \ˌväl-kə-'nä-lə-jē, ˌvȯl-\ *noun* (1886)
: a branch of science that deals with volcanic phenomena
— **vol·ca·nol·o·gist** \-jist\ *noun*

¹vole \'vōl\ *noun* [French, probably from *voler* to fly — more at VOLLEY] (1679)
: GRAND SLAM 1

²vole *noun* [earlier *vole-mouse,* from *vole-* (of Scandinavian origin; akin to Old Norse *vǫllr* field) + *mouse;* akin to Old English *weald* forest — more at WOLD] (1805)
: any of various small rodents (*Microtus* and related genera) that typically have a stout body, rather blunt nose, and short ears, inhabit both moist meadows and dry uplands and do much damage to crops, and are closely related to muskrats and lemmings

vo·li·tion \vō-'li-shən, və-\ *noun* [French, from Medieval Latin *volition-, volitio,* from Latin *vol-* (stem of *velle* to will, wish) + *-ition-, -itio* (as in Latin *position-, positio* position) — more at WILL] (1615)
1 : an act of making a choice or decision; *also* **:** a choice or decision made
2 : the power of choosing or determining **:** WILL
— **vo·li·tion·al** \-'lish-nəl, -'li-shə-nᵊl\ *adjective*

vol·i·tive \'vä-lə-tiv\ *adjective* (1660)
1 : of or relating to the will
2 : expressing a wish or permission

volks·lied \'fōks-ˌlēt, 'fōlks-\ *noun, plural* **volks·lie·der** \-ˌlē-dər\ [German, from *Volk* people + *Lied* song] (circa 1854)
: a folk song

¹vol·ley \'vä-lē\ *noun, plural* **volleys** [Middle French *volee* flight, from *voler* to fly, from Latin *volare*] (1573)
1 a : a flight of missiles (as arrows) **b :** simultaneous discharge of a number of missile weapons **c :** one round per gun in a battery fired as soon as a gun is ready without regard to order **d** (1) **:** the flight of the ball (as in volleyball or tennis) or its course before striking the ground; *also* **:** a return of the ball before it touches the ground (2) **:** a kick of the ball in soccer before it rebounds (3) **:** the exchange of the shuttlecock in badminton following the serve
2 a : a burst or emission of many things or a large amount at once ⟨received a *volley* of angry letters⟩ ⟨a *volley* of praise⟩ **b :** a burst of simultaneous or immediately sequential nerve impulses passing to an end organ, synapse, or center

²volley *verb* **vol·leyed; vol·ley·ing** (1591)
transitive verb
1 : to discharge in or as if in a volley
2 : to propel (an object) while in the air and before touching the ground; *especially* **:** to hit (a tennis ball) on the volley
intransitive verb
1 : to become discharged in or as if in a volley
2 : to make a volley; *specifically* **:** to volley an object of play (as in tennis)
— **vol·ley·er** *noun*

vol·ley·ball \'vä-lē-ˌbȯl\ *noun* (1896)
: a game played by volleying an inflated ball over a net; *also* **:** the ball used in this game

vol·plane \'väl-ˌplān, 'vȯl-\ *intransitive verb* **vol·planed; vol·plan·ing** [French *vol plané* gliding flight] (1909)
1 : to glide in or as if in an airplane
2 : GLIDE 3

Vol·sci \'vȯl-ˌskē, 'väl-ˌsī\ *noun plural* [Latin] (circa 1909)
: a people of ancient Italy dwelling between the Latins and Samnites

◇ WORD HISTORY

volcano Before the Age of Discoveries medieval Europe knew only the handful of active volcanoes long familiar to mariners in the Mediterranean—Vesuvius in Italy, Etna on Sicily, and the peaks of Stromboli and Vulcano among the Lipari Islands off the Sicilian coast. Apparently no need was felt for a generic word to describe a mountain that emitted fire. This situation changed in the 16th century, when the Spanish conquistadores came upon the great volcanic peaks of Mexico, Central America, and the Andes. The chroniclers of the conquistadores applied to these mountains the word *volcán,* whose roots lie in classical antiquity. *Vulcanus,* the ancient Roman god of fire, was particularly associated with the Lipari Islands—the *insulae Vulcaniae* "islands of Vulcan" to Cicero—and this association is perpetuated today in *Vulcano,* the modern Italian name of the southernmost island. Medieval Islamic authors applied *burkān,* an arabicized form of *Vulcanus,* to either Etna or the Lipari Island volcanoes, and this word appears retransformed into the latinized *vulcan* in a 13th century Spanish translation from Arabic. It resurfaces as *volcán* in a 1524 description of Guatemalan volcanoes, and from the Spanish chroniclers the word soon found its way into Italian, French, and English.

Vol·scian \'väl-shən, 'vȯl-skē-ən\ *noun, plural* **Volscians** (1513)
1 : a member of the Volsci
2 : the Italic language of the Volsci
— Volscian *adjective*

[1]volt \'vōlt, 'vȯlt\ *noun* [French *volte,* from Italian *volta* turn, from *voltare* to turn, from (assumed) Vulgar Latin *volvitare,* frequentative of Latin *volvere* to roll — more at VOLUBLE] (1688)
1 : a leaping movement in fencing to avoid a thrust
2 a : a tread or gait in which a horse going sideways makes a turn around a center **b :** a circle traced by a horse in this movement

[2]volt \'vōlt\ *noun* [Alessandro *Volta*] (1873)
: the practical meter-kilogram-second unit of electrical potential difference and electromotive force equal to the difference of potential between two points in a conducting wire carrying a constant current of one ampere when the power dissipated between these two points is equal to one watt and equivalent to the potential difference across a resistance of one ohm when one ampere is flowing through it

volt·age \'vōl-tij\ *noun* (1890)
1 : electric potential or potential difference expressed in volts
2 : intensity of feeling

voltage divider *noun* (1922)
: a resistor or series of resistors provided with taps at certain points and used to provide various potential differences from a single power source

vol·ta·ic \väl-'tā-ik, vōl-, vȯl-\ *adjective* [Alessandro *Volta*] (1812)
: of, relating to, or producing direct electric current by chemical action (as in a battery) **:** GALVANIC ⟨*voltaic* cell⟩

volt–am·pere \'vōlt-'am-ˌpir *also* -ˌper\ *noun* (1896)
: a unit of electric measurement equal to the product of a volt and an ampere that for direct current constitutes a measure of power equivalent to a watt

volte–face \ˌvȯlt-'fäs, ˌvȯl-tə-\ *noun* [French, from Italian *voltafaccia,* from *voltare* to turn + *faccia* face, from (assumed) Vulgar Latin *facia* — more at VOLT, FACE] (1819)
: a reversal in policy **:** ABOUT-FACE

volt·me·ter \'vōlt-ˌmē-tər\ *noun* [International Scientific Vocabulary] (1882)
: an instrument (as a galvanometer) for measuring in volts the differences in potential between different points of an electrical circuit

vol·u·ble \'väl-yə-bəl\ *adjective* [Middle French or Latin; Middle French, from Latin *volubilis,* from *volvere* to roll; akin to Old English *wealwian* to roll, Greek *eilyein* to roll, wrap] (15th century)
1 : easily rolling or turning **:** ROTATING
2 : characterized by ready or rapid speech **:** GLIB, FLUENT
synonym see TALKATIVE
— vol·u·bil·i·ty \ˌväl-yə-'bi-lə-tē\ *noun*
— vol·u·ble·ness \'väl-yə-bəl-nəs\ *noun*
— vol·u·bly \-blē\ *adverb*

[1]vol·ume \'väl-yəm, -(ˌ)yüm\ *noun* [Middle English, from Middle French, from Latin *volumen* roll, scroll, from *volvere* to roll] (14th century)
1 a : a series of printed sheets bound typically in book form **:** BOOK **b :** a series of issues of a periodical **c :** ALBUM 1c
2 : SCROLL 1a
3 : the amount of space occupied by a three-dimensional object as measured in cubic units (as quarts or liters) **:** cubic capacity — see METRIC SYSTEM table, WEIGHT table
4 a (1) **:** AMOUNT; *also* **:** BULK, MASS (2) **:** a considerable quantity **b :** the amount of a substance occupying a particular volume **c :** mass or the representation of mass in art or architecture
5 : the degree of loudness or the intensity of a sound; *also* **:** LOUDNESS
synonym see BULK
— vol·umed \-yəmd, -(ˌ)yümd\ *adjective*

VOLUME FORMULAS

FIGURE	FORMULA	MEANING OF LETTERS
cube	$V = a^3$	a=length of one edge
rectangular solid	$V = lwh$	*l=length of base; w=width of base; h=height*
pyramid	$V = \frac{Ah}{3}$	A=area of base; h=height
cylinder	$V = \pi r^2 h$	$\pi = 3.1416$; r=radius of the base; h=height
cone	$V = \frac{\pi r^2 h}{3}$	$\pi = 3.1416$; r=radius of the base; h=height
sphere	$V = \frac{4\pi r^3}{3}$	$\pi = 3.1416$; r=radius

[2]volume *verb* **vol·umed; vol·um·ing** (1815)
transitive verb
: to send or give out in volume
intransitive verb
: to roll or rise in volume

[3]volume *adjective* (circa 1945)
: involving large quantities ⟨offered *volume* discounts⟩

vol·u·me·ter \'väl-yü-ˌmē-tər\ *noun* [International Scientific Vocabulary, blend of *volume* and *-meter*] (1829)
: an instrument for measuring volumes (as of gases or liquids) directly or (as of solids) by displacement of a liquid

vol·u·met·ric \ˌväl-yü-'me-trik\ *adjective* (1857)
: of, relating to, or involving the measurement of volume
— vol·u·met·ri·cal·ly \-tri-k(ə-)lē\ *adverb*

vo·lu·mi·nos·i·ty \və-ˌlü-mə-'nä-sə-tē\ *noun* (1782)
: the quality or state of being voluminous

vo·lu·mi·nous \və-'lü-mə-nəs\ *adjective* [Late Latin *voluminosus,* from Latin *volumin-, volumen*] (1611)
1 : consisting of many folds, coils, or convolutions **:** WINDING
2 a : having or marked by great volume or bulk **:** LARGE ⟨long *voluminous* tresses⟩; *also* **:** FULL ⟨a *voluminous* skirt⟩ **b :** NUMEROUS ⟨trying to keep track of *voluminous* slips of paper⟩
3 a : filling or capable of filling a large volume or several volumes ⟨a *voluminous* literature on the subject⟩ **b :** writing or speaking much or at great length
— vo·lu·mi·nous·ly *adverb*
— vo·lu·mi·nous·ness *noun*

vol·un·ta·rism \'vä-lən-tə-ˌri-zəm\ *noun* (1838)
1 : the principle or system of doing something by or relying on voluntary action or volunteers
2 : a theory that conceives will to be the dominant factor in experience or in the world
— vol·un·ta·rist \-rist\ *noun*
— vol·un·ta·ris·tic \ˌvä-lən-tə-'ris-tik\ *adjective*

[1]vol·un·tary \'vä-lən-ˌter-ē\ *adjective* [Middle English, from Latin *voluntarius,* from *voluntas* will, from *velle* to will, wish — more at WILL] (14th century)
1 : proceeding from the will or from one's own choice or consent
2 : unconstrained by interference **:** SELF-DETERMINING
3 : done by design or intention **:** INTENTIONAL ⟨*voluntary* manslaughter⟩
4 : of, relating to, subject to, or regulated by the will ⟨*voluntary* behavior⟩
5 : having power of free choice
6 : provided or supported by voluntary action ⟨a *voluntary* organization⟩
7 : acting or done of one's own free will without valuable consideration or legal obligation ☆
— vol·un·tar·i·ly *adverb*
— vol·un·tar·i·ness *noun*

[2]voluntary *noun, plural* **-tar·ies** (1598)
1 a : a prefatory often extemporized musical piece **b :** an improvisatory organ piece played before, during, or after a religious service
2 : one who participates voluntarily **:** VOLUNTEER

vol·un·tary·ism \'vä-lən-ˌter-ē-ˌi-zəm\ *noun* (1835)
: VOLUNTARISM
— vol·un·tary·ist \-ē-ist\ *noun*

voluntary muscle *noun* (1788)
: muscle (as most striated muscle) under voluntary control

[1]vol·un·teer \ˌvä-lən-'tir\ *noun* [obsolete French *voluntaire* (now *volontaire*), from *voluntaire,* adjective, voluntary, from Latin *voluntarius*] (circa 1618)
1 : a person who voluntarily undertakes or expresses a willingness to undertake a service: as **a :** one who enters into military service voluntarily **b** (1) **:** one who renders a service or takes part in a transaction while having no legal concern or interest (2) **:** one who receives a conveyance or transfer of property without giving valuable consideration
2 : a volunteer plant
3 *capitalized* [*Volunteers of America*] **:** a member of a quasi-military religious and philanthropic organization founded in 1896 by Commander and Mrs. Ballington Booth

[2]volunteer *adjective* (1649)
1 : being, consisting of, or engaged in by volunteers ⟨a *volunteer* army⟩ ⟨*volunteer* activities to help the mentally handicapped⟩
2 : growing spontaneously without direct human control or supervision especially from seeds lost from a previous crop ⟨*volunteer* corn plants⟩

[3]volunteer (circa 1755)
intransitive verb
: to offer oneself as a volunteer
transitive verb
: to offer or bestow voluntarily ⟨*volunteer* one's services⟩

vol·un·teer·ism \ˌvä-lən-'tir-ˌi-zəm\ *noun* (1844)
1 : VOLUNTARISM 1
2 : the act or practice of doing volunteer work in community service

☆ SYNONYMS
Voluntary, intentional, deliberate, willing mean done or brought about of one's own will. VOLUNTARY implies freedom and spontaneity of choice or action without external compulsion ⟨a *voluntary* confession⟩. INTENTIONAL stresses an awareness of an end to be achieved ⟨the *intentional* concealment of vital information⟩. DELIBERATE implies full consciousness of the nature of one's act and its consequences ⟨*deliberate* acts of sabotage⟩. WILLING implies a readiness and eagerness to accede to or anticipate the wishes of another ⟨*willing* obedience⟩.

vo·lup·tu·ary \və-ˈləp(t)-shə-ˌwer-ē\ *noun, plural* **-ar·ies** (circa 1610)
: a person whose chief interests are luxury and the gratification of sensual appetites
— **voluptuary** *adjective*

vo·lup·tuous \-shə-wəs, -shəs\ *adjective* [Middle English, from Latin *voluptuosus,* irregular from *voluptas* pleasure, from *volup* pleasurable; akin to Greek *elpesthai* to hope, Latin *velle* to wish — more at WILL] (14th century)
1 a : full of delight or pleasure to the senses **:** conducive to or arising from sensuous or sensual gratification **:** LUXURIOUS ⟨a *voluptuous* dance⟩ ⟨*voluptuous* ornamentation⟩ **b :** suggesting sensual pleasure by fullness and beauty of form ⟨*voluptuous* nudes⟩
2 : given to or spent in enjoyments of luxury, pleasure, or sensual gratifications ⟨a long and *voluptuous* holiday —Edmund Wilson⟩
synonym see SENSUOUS
— **vo·lup·tuous·ly** *adverb*
— **vo·lup·tuous·ness** *noun*

vo·lute \və-ˈlüt\ *noun* [Latin *voluta,* from feminine of *volutus,* past participle of *volvere* to roll — more at VOLUBLE] (circa 1696)
1 : a spiral or scroll-shaped form
2 : a spiral scroll-shaped ornament forming the chief feature of the Ionic capital
3 a : any of various marine gastropod mollusks (family Volutidae) with a thick short-spired shell **b :** the shell of a volute
— **volute** *or* **vo·lut·ed** \-ˈlü-təd\ *adjective*

vo·lu·tin \ˈväl-yə-ˌtin, və-ˈlü-tᵊn\ *noun* [German, from New Latin *volutans,* specific epithet of the bacterium *Spirillum volutans* in which it was first found] (1908)
: a granular basophilic substance that is thought to be a nucleic acid compound and that is common in microorganisms

vol·va \ˈväl-və, ˈvȯl-\ *noun* [New Latin, from Latin *volva, vulva* integument — more at VULVA] (circa 1753)
: a membranous sac or cup about the base of the stipe in many gill fungi

vol·vox \-ˌväks\ *noun* [New Latin, from Latin *volvere* to roll] (1798)
: any of a genus (*Volvox*) of flagellated unicellular green algae that form spherical colonies

vol·vu·lus \ˈväl-vyə-ləs, ˈvȯl-\ *noun* [New Latin, from Latin *volvere*] (1679)
: a twisting of the intestine upon itself that causes obstruction

volva

vo·mer \ˈvō-mər\ *noun* [New Latin, from Latin, plowshare] (circa 1704)
: a bone of the skull of most vertebrates that is situated below the ethmoid region and in the human skull forms part of the nasal septum
— **vo·mer·ine** \ˈvō-mə-ˌrīn\ *adjective*

[1]**vom·it** \ˈvä-mət\ *noun* [Middle English, from Middle French, from Latin *vomitus,* from *vomere* to vomit; akin to Old Norse *vāma* seasickness, Greek *emein* to vomit] (14th century)
1 : an act or instance of disgorging the contents of the stomach through the mouth; *also* **:** the disgorged matter
2 : EMETIC

[2]**vomit** (15th century)
intransitive verb
1 : to disgorge the stomach contents
2 : to spew forth **:** BELCH, GUSH
transitive verb
1 : to disgorge (the contents of the stomach) through the mouth
2 : to eject violently or abundantly **:** SPEW
3 : to cause to vomit
— **vom·it·er** *noun*

vom·i·to·ry \ˈvä-mə-ˌtōr-ē, -ˌtȯr-\ *noun, plural* **-ries** [Late Latin *vomitorium,* from Latin *vomere;* from its disgorging the spectators] (1730)
: an entrance piercing the banks of seats of a theater, amphitheater, or stadium

vom·i·tus \ˈvä-mə-təs\ *noun* [Latin] (circa 1899)
: material discharged by vomiting

V–1 \ˈvē-ˈwən\ *noun* [German, abbreviation for *Vergeltungswaffe 1,* literally, reprisal weapon 1] (1944)
: BUZZ BOMB

[1]**voo·doo** \ˈvü-(ˌ)dü\ *noun, plural* **voodoos** [Louisiana Creole *voudou,* probably from Ewe *vódũ* tutelary deity, demon] (1850)
1 : a religion that is derived from African polytheism and ancestor worship and is practiced chiefly in Haiti
2 a : a person who deals in spells and necromancy **b** (1) **:** a sorcerer's spell **:** HEX (2) **:** a hexed object **:** CHARM
— **voodoo** *adjective*

[2]**voodoo** *transitive verb* (1880)
: to bewitch by or as if by means of voodoo **:** HEX

voo·doo·ism \ˈvü-(ˌ)dü-ˌi-zəm\ *noun* (1865)
1 : VOODOO 1
2 : the practice of witchcraft
— **voo·doo·ist** \-ist\ *noun*
— **voo·doo·is·tic** \ˌvü-(ˌ)dü-ˈis-tik\ *adjective*

vo·ra·cious \vȯ-ˈrā-shəs, və-\ *adjective* [Latin *vorac-, vorax,* from *vorare* to devour; akin to Old English *ācweorran* to guzzle, Latin *gurges* whirlpool, Greek *bibrōskein* to devour] (1635)
1 : having a huge appetite **:** RAVENOUS
2 : excessively eager **:** INSATIABLE ⟨a *voracious* reader⟩ ☆
— **vo·ra·cious·ly** *adverb*
— **vo·ra·cious·ness** *noun*

vo·rac·i·ty \vȯ-ˈra-sə-tē, və-\ *noun* (1526)
: the quality or state of being voracious

vor·la·ge \ˈfōr-ˌlä-gə, ˈfȯr-\ *noun* [German, literally, forward position, from *vor* fore + *Lage* position] (1936)
: the position of a skier leaning forward from the ankles usually without lifting the heels from the skis

-vorous *adjective combining form* [Latin *-vorus,* from *vorare* to devour]
: eating **:** feeding on ⟨frugi*vorous*⟩

vor·tex \ˈvȯr-ˌteks\ *noun, plural* **vor·ti·ces** \ˈvȯr-tə-ˌsēz\ *also* **vor·tex·es** \ˈvȯr-ˌtek-səz\ [New Latin *vortic-, vortex,* from Latin *vertex, vortex* whirlpool — more at VERTEX] (1652)
1 a : a mass of fluid (as a liquid) with a whirling or circular motion that tends to form a cavity or vacuum in the center of the circle and to draw toward this cavity or vacuum bodies subject to its action; *especially* **:** WHIRLPOOL, EDDY **b :** a region within a body of fluid in which the fluid elements have an angular velocity
2 : something that resembles a whirlpool ⟨the hellish *vortex* of battle —*Time*⟩

vor·ti·cal \ˈvȯr-ti-kəl\ *adjective* (1653)
: of, relating to, or resembling a vortex **:** SWIRLING
— **vor·ti·cal·ly** \-k(ə-)lē\ *adverb*

vor·ti·cel·la \ˌvȯr-tə-ˈse-lə\ *noun, plural* **-cel·lae** \-ˈse-(ˌ)lē\ *or* **-cellas** [New Latin, from Latin *vortic-, vortex*] (1787)
: any of a genus (*Vorticella*) of stalked bell-shaped ciliates

vor·ti·cism \ˈvȯr-tə-ˌsi-zəm\ *noun* [Latin *vortic-, vortex*] (1914)
: an English abstract art movement from about 1912–15 embracing cubist and futurist concepts
— **vor·ti·cist** \-sist\ *noun*

vor·tic·i·ty \vȯr-ˈti-sə-tē\ *noun* (1888)
1 : the state of a fluid in vortical motion; *broadly* **:** vortical motion
2 : a measure of vortical motion; *especially* **:** a vector measure of local rotation in a fluid flow

vor·ti·cose \ˈvȯr-ti-ˌkōs\ *adjective* (1783)
archaic **:** VORTICAL

vo·ta·ress \ˈvō-tə-rəs\ *noun* (1589)
: a female votary

vo·ta·rist \-rist\ *noun* (1603)
: VOTARY

vo·ta·ry \ˈvō-tə-rē\ *noun, plural* **-ries** [Latin *votum* vow] (1546)
1 *archaic* **:** a sworn adherent
2 a : ENTHUSIAST, DEVOTEE **b :** a devoted admirer
3 a : a devout or zealous worshiper **b :** a staunch believer or advocate

[1]**vote** \ˈvōt\ *noun* [Middle English (Scots), from Latin *votum* vow, wish — more at VOW] (15th century)
1 a : a usually formal expression of opinion or will in response to a proposed decision; *especially* **:** one given as an indication of approval or disapproval of a proposal, motion, or candidate for office **b :** the total number of such expressions of opinion made known at a single time (as at an election) **c :** an expression of opinion or preference that resembles a vote **d :** BALLOT 1
2 : the collective opinion or verdict of a body of persons expressed by voting
3 : the right to cast a vote; *specifically* **:** the right of suffrage **:** FRANCHISE
4 a : the act or process of voting ⟨brought the question to a *vote*⟩ **b :** a method of voting
5 : a formal expression of a wish, will, or choice voted by a meeting
6 a : VOTER **b :** a group of voters with some common and identifying characteristics ⟨the labor *vote*⟩
7 *chiefly British* **a :** a proposition to be voted on; *especially* **:** a legislative money item **b :** APPROPRIATION

[2]**vote** *verb* **vot·ed; vot·ing** (1552)
intransitive verb
1 : to express one's views in response to a poll; *especially* **:** to exercise a political franchise
2 : to express an opinion ⟨consumers . . . *vote* with their dollars —Lucia Mouat⟩
transitive verb
1 : to choose, endorse, decide the disposition of, defeat, or authorize by vote
2 a : to adjudge by general agreement **:** DECLARE **b :** to offer as a suggestion **:** PROPOSE ⟨I *vote* we all go home⟩
3 a : to cause to vote in a given way **b :** to cause to be cast for or against a proposal
4 : to vote in accordance with or in the interest of ⟨*vote* your conscience⟩ ⟨*voted* their pocketbooks⟩

vote·less \ˈvōt-ləs\ *adjective* (1672)
: having no vote; *especially* **:** denied the political franchise

vot·er \ˈvō-tər\ *noun* (circa 1578)
: one that votes or has the legal right to vote

voting machine *noun* (1900)
: a mechanical device for recording and counting votes cast in an election

☆ SYNONYMS
Voracious, gluttonous, ravenous, rapacious mean excessively greedy. VORACIOUS applies especially to habitual gorging with food or drink ⟨teenagers are often *voracious* eaters⟩. GLUTTONOUS applies to one who delights in eating or acquiring things especially beyond the point of necessity or satiety ⟨an admiral who was *gluttonous* for glory⟩. RAVENOUS implies excessive hunger and suggests violent or grasping methods of dealing with food or with whatever satisfies an appetite ⟨a nation with a *ravenous* lust for territorial expansion⟩. RAPACIOUS often suggests excessive and utterly selfish acquisitiveness or avarice ⟨*rapacious* developers indifferent to environmental concerns⟩.

vo·tive \ˈvō-tiv\ *adjective* [Latin *votivus,* from *votum* vow] (1597)
1 : consisting of or expressing a vow, wish, or desire ⟨a *votive* prayer⟩
2 : offered or performed in fulfillment of a vow or in gratitude or devotion
— **vo·tive·ly** *adverb*
— **vo·tive·ness** *noun*

votive candle *noun* (1824)
1 : a candle lit in devotion or gratitude
2 : a small squat candle

votive mass *noun* (1738)
: a mass celebrated for a special intention (as for a wedding or funeral) in place of the mass of the day

vo·tress \ˈvō-trəs\ *noun* [by alteration] (1590)
archaic **:** VOTARESS

[1]vouch \ˈvau̇ch\ *verb* [Middle English *vochen, vouchen,* from Middle French *vocher,* from Latin *vocare* to call, summon, from *vox* voice — more at VOICE] (14th century)
transitive verb
1 : to summon into court to warrant or defend a title
2 *archaic* **a :** ASSERT, AFFIRM **b :** ATTEST
3 *archaic* **:** to cite or refer to as authority or supporting evidence
4 a : PROVE, SUBSTANTIATE **b :** to verify (a business transaction) by examining documentary evidence
intransitive verb
1 : to give a guarantee **:** become surety
2 a : to supply supporting evidence or testimony **b :** to give personal assurance
synonym see CERTIFY

[2]vouch *noun* (1603)
obsolete **:** ALLEGATION, DECLARATION

vouch·ee \vau̇-ˈchē\ *noun* (15th century)
: a person for whom another vouches

[1]vouch·er \ˈvau̇-chər\ *noun* [Middle French *vocher, voucher* to vouch] (circa 1523)
1 : an act of vouching
2 a : a piece of supporting evidence **:** PROOF **b :** a documentary record of a business transaction **c :** a written affidavit or authorization **:** CERTIFICATE **d :** a form or check indicating a credit against future purchases or expenditures

[2]voucher *transitive verb* (1609)
1 : to establish the authenticity of
2 : to prepare a voucher for

[3]voucher *noun* [[1]*vouch* + [2]*-er*] (1612)
archaic **:** one that guarantees **:** SURETY

vouch·safe \vau̇ch-ˈsāf, ˈvau̇ch-ˌ\ *transitive verb* **vouch·safed; vouch·saf·ing** (1587)
1 a : to grant or furnish often in a gracious or condescending manner **b :** to give by way of reply ⟨refused to *vouchsafe* an explanation⟩
2 : to grant as a privilege or special favor
synonym see GRANT
— **vouch·safe·ment** \vau̇ch-ˈsāf-mənt\ *noun*

vous·soir \vü-ˈswär, ˈvü-ˌ\ *noun* [French, from (assumed) Vulgar Latin *volsorium,* from Latin *volvere* to roll — more at VOLUBLE] (1728)
: one of the wedge-shaped pieces forming an arch or vault — see ARCH illustration

Vou·vray \vü-ˈvrā\ *noun* [French, from *Vouvray,* village in France] (1885)
: a semidry to semisweet white wine from the Loire Valley of France that is often produced as a sparkling wine

[1]vow \ˈvau̇\ *noun* [Middle English *vowe,* from Old French *vou,* from Latin *votum,* from neuter of *votus,* past participle of *vovēre* to vow; akin to Greek *euchesthai* to pray, vow, Sanskrit *vāghat* sacrificer] (14th century)
: a solemn promise or assertion; *specifically* **:** one by which a person is bound to an act, service, or condition

[2]vow (14th century)
transitive verb
1 : to promise solemnly **:** SWEAR
2 : to bind or consecrate by a vow
intransitive verb
: to make a vow
— **vow·er** \ˈvau̇(-ə)r\ *noun*

[3]vow *transitive verb* [Middle English, short for *avowen*] (14th century)
: AVOW, DECLARE

vow·el \ˈvau̇(-ə)l\ *noun* [Middle English, from Middle French *vouel,* from Latin *vocalis* — more at VOCALIC] (14th century)
1 : one of a class of speech sounds in the articulation of which the oral part of the breath channel is not blocked and is not constricted enough to cause audible friction; *broadly* **:** the one most prominent sound in a syllable
2 : a letter or other symbol representing a vowel — usually used in English of *a, e, i, o, u,* and sometimes *y*

vow·el·ize \ˈvau̇-(ə-)ˌlīz\ *transitive verb* **-ized; -iz·ing** (1883)
: to furnish with vowel points

vowel point *noun* (1764)
: a mark placed below or otherwise near a consonant in some languages (as Hebrew) and representing the vowel sound that precedes or follows the consonant sound

vowel rhyme *noun* (1838)
: ASSONANCE 2b

vox po·pu·li \ˈväks-ˈpä-pyü-ˌlī, -pyə-(ˌ)lē, -pə-(ˌ)lē\ *noun* [Latin, voice of the people] (circa 1550)
: popular sentiment

[1]voy·age \ˈvȯi-ij, ˈvȯ(-)ij\ *noun* [Middle English, from Old French *voiage,* from Late Latin *viaticum,* from Latin, traveling money, from neuter of *viaticus* of a journey, from *via* way — more at WAY] (14th century)
1 : an act or instance of traveling **:** JOURNEY
2 : a course or period of traveling by other than land routes
3 : an account of a journey especially by sea

[2]voyage *verb* **voy·aged; voy·ag·ing** (15th century)
intransitive verb
: to take a trip **:** TRAVEL
transitive verb
: SAIL, TRAVERSE
— **voy·ag·er** *noun*

voya·geur \ˌvȯi-ə-ˈzhər, ˌvwä yä-\ *noun* [Canadian French, from French, traveler, from *voyager* to travel, from *voyage* voyage, from Old French *voiage*] (1793)
: a man employed by a fur company to transport goods to and from remote stations especially in the Canadian Northwest

voy·eur \vwä-ˈyər, vȯi-ˈər\ *noun* [French, literally, one who sees, from Middle French, from *voir* to see, from Latin *vidēre* — more at WIT] (1900)
1 : one obtaining sexual gratification from seeing sex organs and sexual acts; *broadly* **:** one who habitually seeks sexual stimulation by visual means
2 : a prying observer who is usually seeking the sordid or the scandalous
— **voy·eur·ism** \-ˌi-zəm\ *noun*
— **voy·eur·is·tic** \ˌvwä-(ˌ)yər-ˈis-tik, ˌvȯi-ər-\ *adjective*
— **voy·eur·is·ti·cal·ly** \-ti-k(ə-)lē\ *adverb*

vroom \ˈvrüm, və-ˈrüm\ *intransitive verb* [imitative of the noise of an engine] (1965)
: to operate a motor vehicle at high speed or so as to create a great deal of engine noise

V sign *noun* (1942)
: a sign made by raising the index and middle fingers in a V and used as a victory salute or a gesture of approval

V–2 \ˈvē-ˈtü\ *noun* [German, abbreviation for *Vergeltungswaffe 2,* literally, reprisal weapon 2] (1944)
: a rocket-propelled bomb of German invention

vug \ˈvəg\ *noun* [Cornish dialect *vooga* underground chamber, from Latin *fovea* small pit] (1818)
: a small unfilled cavity in a lode or in rock
— **vug·gy** \ˈvə-gē\ *adjective*

Vul·can \ˈvəl-kən\ *noun* [Latin *Volcanus, Vulcanus*]
: the Roman god of fire and metalworking — compare HEPHAESTUS

vul·ca·ni·an \ˌvəl-ˈkā-nē-ən\ *adjective* (1602)
1 *capitalized* **:** of or relating to Vulcan or to working in metals (as iron)
2 : of or relating to a volcanic eruption in which highly viscous or solid lava is blown into fragments and dust

vul·ca·nic·i·ty \ˌvəl-kə-ˈni-sə-tē\ *noun* (1873)
: VOLCANISM

vul·ca·ni·sate, vul·ca·ni·sa·tion, vul·ca·nise *British variant of* VULCANIZATE, VULCANIZATION, VULCANIZE

vul·ca·nism \ˈvəl-kə-ˌni-zəm\ *noun* (1877)
: VOLCANISM

vul·ca·ni·zate \ˈvəl-kə-nə-ˌzāt, ˌvəl-kə-ˈnī-\ *noun* [back-formation from *vulcanization*] (1926)
: a vulcanized product

vul·ca·ni·za·tion \ˌvəl-kə-nə-ˈzā-shən\ *noun* (1846)
: the process of treating crude or synthetic rubber or similar plastic material chemically to give it useful properties (as elasticity, strength, and stability)

vul·ca·nize \ˈvəl-kə-ˌnīz\ *verb* **-nized; -niz·ing** [International Scientific Vocabulary, from Latin *Vulcanus* Vulcan, fire] (1846)
transitive verb
: to subject to vulcanization
intransitive verb
: to undergo vulcanization
— **vul·ca·niz·er** *noun*

vulcanized fiber *noun* [from *Vulcanized Fibre,* a trademark] (circa 1884)
: a tough substance made by treatment of cellulose and used for luggage and electrical insulation and in packaging

vul·ca·nol·o·gy \ˌvəl-kə-ˈnä-lə-jē\ *noun* [International Scientific Vocabulary] (1858)
: VOLCANOLOGY
— **vul·ca·nol·o·gist** \-jist\ *noun*

vul·gar \ˈvəl-gər\ *adjective* [Middle English, from Latin *vulgaris* of the mob, vulgar, from *volgus, vulgus* mob, common people] (14th century)
1 a : generally used, applied, or accepted **b :** understood in or having the ordinary sense ⟨they reject the *vulgar* conception of miracle —W. R. Inge⟩
2 : VERNACULAR ⟨the *vulgar* name of a plant⟩
3 a : of or relating to the common people **:** PLEBEIAN **b :** generally current **:** PUBLIC ⟨the *vulgar* opinion of that time⟩ **c :** of the usual, typical, or ordinary kind
4 a : lacking in cultivation, perception, or taste **:** COARSE **b :** morally crude, undeveloped, or unregenerate **:** GROSS **c :** ostentatious or excessive in expenditure or display **:** PRETENTIOUS
5 a : offensive in language **:** EARTHY **b :** lewdly or profanely indecent
synonym see COMMON, COARSE
— **vul·gar·ly** *adverb*

vulgar era *noun* (1716)
: CHRISTIAN ERA

vul·gar·i·an \ˌvəl-ˈgar-ē-ən, -ˈger-\ *noun* (1804)
: a vulgar person

vul·gar·ise *British variant of* VULGARIZE

vul·gar·ism \ˈvəl-gə-ˌri-zəm\ *noun* (circa 1676)
1 : VULGARITY
2 a : a word or expression originated or used chiefly by illiterate persons **b :** a coarse word or phrase **:** OBSCENITY

vul·gar·i·ty \ˌvəl-ˈgar-ə-tē\ *noun, plural* **-ties** (1579)
1 : something vulgar

\ə\ **abut** \ᵊ\ **kitten** \ər\ **further** \a\ **ash** \ā\ **ace** \ä\ **mop, mar** \au̇\ **out** \ch\ **chin** \e\ **bet** \ē\ **easy** \g\ **go** \i\ **hit** \ī\ **ice** \j\ **job** \ŋ\ **sing** \ō\ **go** \ȯ\ **law** \ȯi\ **boy** \th\ **thin** \t͟h\ **the** \ü\ **loot** \u̇\ **foot** \y\ **yet** \zh\ **vision** *see also* Guide to Pronunciation

2 : the quality or state of being vulgar

vul·gar·ize \'vəl-gə-ˌrīz\ *transitive verb* **-ized; -iz·ing** (1709)
1 : to diffuse generally **:** POPULARIZE
2 : to make vulgar **:** COARSEN
— **vul·gar·i·za·tion** \ˌvəl-gə-rə-'zā-shən\ *noun*
— **vul·gar·iz·er** \'vəl-gə-ˌrī-zər\ *noun*

Vulgar Latin *noun* (1818)
: the nonclassical Latin of ancient Rome including the speech of plebeians and the informal speech of the educated established by comparative evidence as the chief source of the Romance languages

vul·gate \'vəl-ˌgāt, -gət\ *noun* [Medieval Latin *vulgata,* from Late Latin *vulgata editio* edition in general circulation] (1728)
1 *capitalized* **:** a Latin version of the Bible authorized and used by the Roman Catholic Church
2 : a commonly accepted text or reading
3 : the speech of the common people and especially of uneducated people

vul·gus \'vəl-gəs\ *noun* [probably alteration of obsolete *vulgars* English sentences to be translated into Latin] (1856)
: a short composition in Latin verse formerly common as an exercise in some English public schools

vul·ner·a·ble \'vəl-n(ə-)rə-bəl, 'vəl-nər-bəl\ *adjective* [Late Latin *vulnerabilis,* from Latin *vulnerare* to wound, from *vulner-, vulnus* wound; probably akin to Latin *vellere* to pluck, Greek *oulē* wound] (1605)
1 : capable of being physically wounded
2 : open to attack or damage **:** ASSAILABLE
3 : liable to increased penalties but entitled to increased bonuses after winning a game in contract bridge
— **vul·ner·a·bil·i·ty** \ˌvəl-n(ə-)rə-'bi-lə-tē\ *noun*
— **vul·ner·a·ble·ness** \'vəl-n(ə-)rə-bəl-nəs, 'vəl-nər-bəl-\ *noun*
— **vul·ner·a·bly** \-blē\ *adverb*

[1]vul·ner·ary \'vəl-nə-ˌrer-ē\ *adjective* [Latin *vulnerarius,* from *vulner-, vulnus*] (1599)
: used for or useful in healing wounds ⟨*vulnerary* plants⟩

[2]vulnerary *noun, plural* **-ar·ies** (1601)
: a vulnerary remedy

vul·pine \'vəl-ˌpīn\ *adjective* [Latin *vulpinus,* from *vulpes* fox; perhaps akin to Greek *alōpēx* fox — more at ALOPECIA] (1628)
1 : of, relating to, or resembling a fox
2 : FOXY, CRAFTY

vul·ture \'vəl-chər\ *noun* [Middle English, from Latin *vultur*] (14th century)
1 : any of various large raptorial birds (families Accipitridae and Cathartidae) that are related to the hawks, eagles, and falcons but have weaker claws and the head usually naked and that subsist chiefly or entirely on carrion
2 : a rapacious or predatory person
— **vul·tur·ish** \-chə-rish\ *adjective*

vul·tur·ine \-chə-ˌrīn\ *adjective* (1647)
1 : of, relating to, or characteristic of vultures
2 : RAPACIOUS, PREDATORY ⟨*vulturine* legislators⟩

vul·tur·ous \'vəl-chə-rəs, 'vəlch-rəs\ *adjective* (1623)
: resembling a vulture especially in rapacity or scavenging habits

vul·va \'vəl-və\ *noun, plural* **vul·vae** \-ˌvē, -ˌvī\ [New Latin, from Latin *volva, vulva* womb, female genitals; akin to Sanskrit *ulva* womb and perhaps to Latin *volvere* to roll — more at VOLUBLE] (circa 1577)
: the external parts of the female genital organs
— **vul·val** \'vəl-vəl\ *or* **vul·var** \-vər, -ˌvär\ *adjective*

vul·vo·vag·i·ni·tis \ˌvəl-(ˌ)vō-ˌva-jə-'nī-təs\ *noun* [New Latin] (1897)
: coincident inflammation of the vulva and vagina

vying *present participle of* VIE

W *is the twenty-third letter of the English alphabet. In form and origin,* W *is a ligature,* VV *or* UU*, called "double* u,*" which was introduced into English by French scribes in the 11th century to replace an Old English runic character called* wynn. *The sound of* w *is usually a consonant, as in* we. *In English, with minor exceptions,* w *represents a vowel only in diphthongs, as in* few *and* how. W *appears also in the frequent consonantal group* wh, *as in* what *and* white. *Written* W *is silent in some words, as* wrist *and* answer. *Words with* W *are almost wholly of Germanic origin, either by inheritance through Old English or by borrowing from Old Norse or from Old North French. In the latter, Germanic loanwords appear with* w *instead of with the* gu *or* g *that prevails in central Old French and in Modern French (as seen in Old North French* wage *"pledge" contrasting with Old French* guage *"pledge"). Small* w *was formed by a simple reduction in the size of the capital.*

w \'də-bəl-(ˌ)yü, 'də-bə-; 'dəb-(ˌ)yü, -yə; 'dəb-yē\ *noun, plural* **w's** *or* **ws** \-(ˌ)yüz, -yəz, -yēz\ *often capitalized, often attributive* (15th century)
1 a : the 23d letter of the English alphabet **b :** a graphic representation of this letter **c :** a speech counterpart of orthographic *w*
2 : a graphic device for reproducing the letter *w*
3 : one designated *w* especially as the 23d in order or class
4 : something shaped like the letter W

W *noun* (1960)
: W PARTICLE

wab·ble \'wä-bəl\ *variant of* WOBBLE

Wac \'wak\ *noun* [Women's Army Corps] (1943)
: a member of the Women's Army Corps

wacked–out *variant of* WHACKED-OUT

wacko \'wa-(ˌ)kō\ *adjective* [by alteration] (1975)
: WACKY
— wacko *noun*

wacky \'wa-kē\ *adjective* **wack·i·er; -est** [perhaps from English dialect *whacky* fool] (circa 1935)
: absurdly or amusingly eccentric or irrational **:** CRAZY
— wack·i·ly \'wa-kə-lē\ *adverb*
— wack·i·ness \'wa-kē-nəs\ *noun*

[1]wad \'wäd\ *noun* [Medieval Latin *wadda*] (1573)
1 : a small mass, bundle, or tuft: as **a :** a soft mass especially of a loose fibrous material variously used (as to stop an aperture, pad a garment, or hold grease around an axle) **b** (1) **:** a soft plug used to retain a powder charge or to avoid windage especially in a muzzle-loading gun (2) **:** a felt or paper disk used to separate the components of a shotgun cartridge **c :** a small mass of a chewing substance ⟨a *wad* of gum⟩
2 : a considerable amount (as of money)
3 a : a roll of paper money **b :** MONEY

[2]wad *transitive verb* **wad·ded; wad·ding** (1579)
1 a : to insert a wad into ⟨*wad* a gun⟩ **b :** to hold in by a wad ⟨*wad* a bullet in a gun⟩
2 : to form into a wad or wadding; *especially* **:** to roll or crush into a tight wad
3 : to stuff or line with some soft substance
— wad·der *noun*

wad·ding \'wä-diŋ\ *noun* (1627)
1 : wads or material for making wads
2 : a soft mass or sheet of short loose fibers used for stuffing or padding

[1]wad·dle \'wä-dᵊl\ *intransitive verb* **wad·dled; wad·dling** \'wäd-liŋ, 'wä-dᵊl-iŋ\ [frequentative of *wade*] (1592)
1 : to walk with short steps swinging the forepart of the body from side to side
2 : to move clumsily in a manner suggesting a waddle
— wad·dler \'wäd-lər, 'wä-dᵊl-ər\ *noun*

[2]waddle *noun* (1691)
: an awkward clumsy swaying gait

[1]wad·dy \'wä-dē\ *noun, plural* **waddies** [Dharuk (Australian aboriginal language of the Port Jackson area) *wadi* stick, wooden weapon] (circa 1790)
Australian **:** CLUB 1a

[2]waddy *transitive verb* **wad·died; wad·dy·ing** (1833)
Australian **:** to attack or beat with a waddy

[3]wad·dy *or* **wad·die** \'wä-dē\ *noun, plural* **waddies** [origin unknown] (1897)
West **:** COWBOY

[1]wade \'wād\ *verb* **wad·ed; wad·ing** [Middle English, from Old English *wadan;* akin to Old High German *watan* to go, wade, Latin *vadere* to go] (13th century)
intransitive verb
1 : to step in or through a medium (as water) offering more resistance than air
2 : to move or proceed with difficulty or labor
3 : to set to work or attack with determination or vigor — used with *in* or *into* ⟨*wade* into a task⟩
transitive verb
: to pass or cross by wading
— wad·able *or* **wade·able** \'wā-də-bəl\ *adjective*

[2]wade *noun* (1665)
: an act of wading ⟨a *wade* in the brook⟩

Wade–Giles \'wād-'gīlz\ *noun* [Thomas F. *Wade* (died 1895) British diplomat & Herbert A. *Giles* (died 1935) British sinologist] (1943)
: a system for romanizing Chinese ideograms in which tones are indicated by superscript numbers and consonantal aspiration is indicated by an apostrophe — compare PINYIN

wad·er \'wā-dər\ *noun* (1673)
1 : one that wades
2 : SHOREBIRD; *also* **:** WADING BIRD
3 *plural* **:** high waterproof boots or a one-piece waterproof garment usually consisting of pants with attached boots that are used for wading (as when fishing)

wa·di \'wä-dē\ *noun* [Arabic *wādiy*] (1839)
1 : the bed or valley of a stream in regions of southwestern Asia and northern Africa that is usually dry except during the rainy season and that often forms an oasis **:** GULLY, WASH
2 : a shallow usually sharply defined depression in a desert region

waders

wading bird *noun* (circa 1843)
: any of an order (Ciconiiformes) of long-legged birds (as herons, bitterns, storks, and flamingos) that wade in water in search of food

wading pool *noun* (1921)
: a shallow pool of portable or permanent construction used by children for wading

wad·mal *or* **wad·mol** *or* **wad·mel** \'wäd-məl\ *noun* [Middle English *wadmale,* from Old Norse *vathmāl,* literally, standard cloth, from *vāth* cloth, clothing + *māl* measure; akin to Latin *metiri* to measure — more at WEED, MEASURE] (14th century)
: a coarse rough woolen fabric formerly used in the British Isles and Scandinavia for protective coverings and warm clothing

wae·sucks \'wā-ˌsəks\ *interjection* [Scots *wae* woe (from Middle English *wa*) + *sucks,* alteration of English *sakes* — more at WOE] (circa 1774)
Scottish — used to express pity

Waf \'waf\ *noun* [Women in the Air Force] (1948)
: a member of the women's component of the air force formed after World War II

[1]wa·fer \'wā-fər\ *noun* [Middle English, from Old North French *waufre,* of Germanic origin; akin to Middle Dutch *wafel, wafer* waffle] (14th century)
1 a : a thin crisp cake, candy, or cracker **b :** a round thin piece of unleavened bread used in the celebration of the Eucharist
2 : an adhesive disk of dried paste with added coloring matter used as a seal
3 a : a thin disk or ring resembling a wafer and variously used (as for a valve or diaphragm) **b :** a thin slice of semiconductor (as silicon) used as a base for an electronic component or circuit

[2]wafer *transitive verb* **wa·fered; wa·fer·ing** \-f(ə-)riŋ\ (1748)
1 : to seal, close, or fasten with a wafer
2 : to divide (as a silicon rod) into wafers

waff \'waf\ *noun* [English dialect *waff* to wave] (1600)
1 *chiefly Scottish* **:** a waving motion
2 *chiefly Scottish* **:** PUFF, GUST

[1]waf·fle \'wä-fəl, 'wȯ-\ *noun* [Dutch *wafel,* from Middle Dutch *wafel, wafer;* akin to Old English *wefan* to weave] (1744)
: a crisp cake of batter baked in a waffle iron

[2]waffle *intransitive verb* **waf·fled; waf·fling** \-f(ə-)liŋ\ [frequentative of obsolete *woff* to yelp, of imitative origin] (1868)
1 : EQUIVOCATE, VACILLATE; *also* **:** YO-YO, FLIP-FLOP
2 : to talk or write foolishly **:** BLATHER ⟨can *waffle* . . . tiresomely off the point —*Times Literary Supplement*⟩
— waf·fler \-f(ə-)lər\ *noun*

[3]waffle *noun* (circa 1888)
: empty or pretentious words **:** TRIPE

waffle iron *noun* (1794)
: a cooking utensil having two hinged metal parts that shut upon each other and impress surface projections on waffles that are being cooked

waf·fle·stomp·er \'wä-fəl-ˌstäm-pər, 'wȯ-, -ˌstȯm-\ *noun* [from the pattern left by the soles] (1972)

\ə\ abut \ᵊ\ kitten \ər\ further \a\ ash \ā\ ace \ä\ mop, mar \au̇\ out \ch\ chin \e\ bet \ē\ easy \g\ go \i\ hit \ī\ ice \j\ job \ŋ\ sing \ō\ go \ȯ\ law \ȯi\ boy \th\ thin \th\ the \ü\ loot \u̇\ foot \y\ yet \zh\ vision *see also* Guide to Pronunciation

: a hiking boot with a lug sole

[1]waft \'wäft, 'waft\ *verb* [(assumed) Middle English *waughten* to guard, convoy, from Middle Dutch or Middle Low German *wachten* to watch, guard; akin to Old English *wæccan* to watch — more at WAKE] (circa 1562)
intransitive verb
: to move or go lightly on or as if on a buoyant medium
transitive verb
: to cause to move or go lightly by or as if by the impulse of wind or waves
— **waft·er** *noun*

[2]waft *noun* (1607)
1 : something (as an odor) that is wafted **:** WHIFF
2 : a slight breeze **:** PUFF
3 : the act of waving
4 : a pennant or flag used to signal or to show wind direction

waft·age \'wäf-tij, 'waf-\ *noun* (1558)
: the act of wafting or state of being wafted; *broadly* **:** CONVEYANCE

waf·ture \'wäf(t)-shər, 'waf(t)-\ *noun* (1601)
: the act of waving or a wavelike motion

[1]wag \'wag\ *verb* **wagged; wag·ging** [Middle English *waggen;* akin to Middle High German *wacken* to totter, Old English *wegan* to move — more at WAY] (13th century)
intransitive verb
1 : to be in motion **:** STIR
2 : to move to and fro or up and down especially with quick jerky motions
3 : to move in chatter or gossip ⟨scandal caused tongues to *wag*⟩
4 *archaic* **:** DEPART
5 : WADDLE
transitive verb
1 : to swing to and fro or up and down especially with quick jerky motions **:** SWITCH; *specifically* **:** to nod (the head) or shake (a finger) at (as in assent or mild reproof)
2 : to move (as the tongue) animatedly in conversation
— **wag·ger** *noun*

[2]wag *noun* (1589)
: an act of wagging **:** SHAKE

[3]wag *noun* [probably short for obsolete English *waghalter* gallows bird, from English [1]*wag* + *halter*] (circa 1553)
1 : WIT, JOKER
2 *obsolete* **:** a young man **:** CHAP

[1]wage \'wāj\ *noun* [Middle English, pledge, wage, from Old North French, of Germanic origin; akin to Gothic *wadi* pledge — more at WED] (14th century)
1 a : a payment usually of money for labor or services usually according to contract and on an hourly, daily, or piecework basis — often used in plural **b** *plural* **:** the share of the national product attributable to labor as a factor in production
2 : RECOMPENSE, REWARD — usually used in plural but singular or plural in construction ⟨the *wages* of sin is death —Romans 6:23 (Revised Standard Version)⟩
— **wage·less** \'wāj-ləs\ *adjective*

[2]wage *verb* **waged; wag·ing** [Middle English, to pledge, give as security, from Old North French *wagier,* from *wage*] (15th century)
transitive verb
: to engage in or carry on ⟨*wage* war⟩ ⟨*wage* a campaign⟩
intransitive verb
: to be in process of occurring ⟨the riot *waged* for several hours —*American Guide Series: Maryland*⟩

wage earner *noun* (1885)
: a person who works for wages or salary

[1]wa·ger \'wā-jər\ *noun* [Middle English, pledge, bet, from Anglo-French *wageure,* from Old North French *wagier* to pledge] (14th century)
1 a : something (as a sum of money) risked on an uncertain event **:** STAKE **b :** something on which bets are laid **:** GAMBLE ⟨do a stunt as a *wager*⟩
2 *archaic* **:** an act of giving a pledge to take and abide by the result of some action

[2]wager *verb* **wa·gered; wa·ger·ing** \'wāj-riŋ, 'wā-jə-\ (1602)
intransitive verb
: to make a bet
transitive verb
: to risk or venture on a final outcome; *specifically* **:** to lay as a gamble **:** BET ⟨*wager* $5 on a horse⟩
— **wa·ger·er** \'wā-jər-ər\ *noun*

wage scale *noun* (1902)
: a schedule of wage rates for related tasks; *broadly* **:** the general level of wages in an industry or region

wage slave *noun* (1886)
: a person dependent on wages or a salary for a livelihood

wage·work·er \'wāj-,wər-kər\ *noun* (1876)
: WAGE EARNER

wag·gery \'wa-gə-rē\ *noun, plural* **-ger·ies** (1594)
1 : mischievous merriment **:** PLEASANTRY
2 : JEST; *especially* **:** PRACTICAL JOKE

wag·gish \'wa-gish\ *adjective* (1589)
1 : resembling or characteristic of a wag ⟨a *waggish* friend⟩ ⟨a *waggish* prose style⟩
2 : done or made in waggery or for sport **:** HUMOROUS ⟨*waggish* spoofs of popular songs⟩
— **wag·gish·ly** *adverb*
— **wag·gish·ness** *noun*

[1]wag·gle \'wa-gəl\ *verb* **wag·gled; wag·gling** \-g(ə-)liŋ\ [frequentative of [1]*wag*] (1594)
transitive verb
: to move frequently one way and the other **:** WAG
intransitive verb
: to reel, sway, or move from side to side **:** WAG
— **wag·gly** \-g(ə-)lē\ *adjective*

[2]waggle *noun* (circa 1866)
1 : an instance of waggling **:** a jerky motion back and forth or up and down
2 : a preliminary swinging of a golf club head back and forth over the ball before the swing

wag·gon *chiefly British variant of* WAGON

[1]Wag·ne·ri·an \väg-'nir-ē-ən, -'ner-\ *adjective* [Richard *Wagner*] (1873)
: of, relating to, characteristic, or suggestive of Wagner or his music or theories

[2]Wagnerian *noun* (1882)
: an admirer of the musical theories and style of Wagner

Wag·ner·ite \'väg-nə-,rīt\ *noun* (1855)
: WAGNERIAN

[1]wag·on \'wa-gən\ *noun* [Dutch *wagen,* from Middle Dutch — more at WAIN] (15th century)
1 a : a usually four-wheel vehicle for transporting bulky commodities and drawn originally by animals **b :** a lighter typically horse-drawn vehicle for transporting goods or passengers **c :** PADDY WAGON
2 *British* **:** a railway freight car
3 : a low four-wheel vehicle with an open rectangular body and a retroflex tongue made for the play or use (as for carrying newspapers) of a child
4 : a small wheeled table used for the service of a dining room
5 : a delivery truck ⟨milk *wagon*⟩
6 : STATION WAGON
— **off the wagon :** no longer abstaining from alcoholic beverages
— **on the wagon :** abstaining from alcoholic beverages

[2]wagon (1606)
intransitive verb
: to travel or transport goods by wagon
transitive verb **:** to transport (goods) by wagon

wag·on·er \'wa-gə-nər\ *noun* (1544)
1 : a person who drives a wagon or transports goods by wagon
2 *capitalized, obsolete* **:** BOÖTES

wag·on·ette \,wa-gə-'net\ *noun* (circa 1858)
: a light wagon with two facing seats along the sides behind a transverse front seat

wa·gon–lit \vȧ-gō^{n}-lē\ *noun, plural* **wagons–lits** *or* **wagon–lits** \-gō^{n}-lē(z)\ [French, from *wagon* railroad car + *lit* bed] (1884)
: a railroad sleeping car

wagon master *noun* (1645)
: a person in charge of one or more wagons

wagon train *noun* (1810)
: a column of wagons (as of supplies for a group of settlers) traveling overland

wag·tail \'wag-,tāl\ *noun* (1510)
: any of various chiefly Old World oscine birds (family Motacillidae) related to the pipits and having a trim slender body and a very long tail that they habitually jerk up and down

Wah·habi *also* **Wa·habi** \wə-'hä-bē, wä-\ *noun* [Arabic *wahhābīy,* from Muḥammad born ʻAbd al-*Wahhāb* (Abdul-Wahhab) (died 1787) Arabian religious reformer] (1807)
: a member of a puritanical Muslim sect founded in Arabia in the 18th century by Muhammad ibn-Abdul Wahhab and revived by ibn-Saud in the 20th century
— **Wah·hab·ism** \-'hä-,bi-zəm\ *noun*
— **Wah·hab·ite** \-,bīt\ *adjective or noun*

wa·hi·ne \wä-'hē-nē, -(,)nā\ *noun* [Maori & Hawaiian, woman] (1773)
1 : a Polynesian woman
2 : a female surfer

[1]wa·hoo \'wä-,hü, 'wȯ-\ *noun, plural* **wahoos** [origin unknown] (1770)
: WINGED ELM

[2]wahoo *noun, plural* **wahoos** [Dakota *wāhu,* from *wā-* arrow + *hu* wood] (1857)
: a shrubby North American spindle tree (*Euonymus atropurpureus*) having purple capsules which in dehiscence expose the scarlet-ariled seeds — called also *burning bush*

[3]wahoo *noun, plural* **wahoos** [origin unknown] (circa 1900)
: a large vigorous mackerel (*Acanthocybium solandri*) that is common in warm seas and esteemed as a food and sport fish

[4]wa·hoo \'wä-'hü\ *interjection* (circa 1924)
chiefly West — used to express exuberance or enthusiasm or to attract attention

wah-wah pedal *variant of* WA-WA PEDAL

[1]waif \'wāf\ *noun* [Middle English, from Old North French, adjective, lost, unclaimed, probably of Scandinavian origin; akin to Old Norse *veif* something flapping, *veifa* to be in movement — more at WIPE] (14th century)
1 a : a piece of property found (as washed up by the sea) but unclaimed **b** *plural* **:** stolen goods thrown away by a thief in flight
2 a : something found without an owner and especially by chance **b :** a stray person or animal; *especially* **:** a homeless child
— **waif·like** \-,līk\ *adjective*

[2]waif *noun* (1530)
: WAFT 4

[1]wail \'wā(ə)l\ *verb* [Middle English, of Scandinavian origin; akin to Old Norse *væla, vāla* to wail; akin to Old Norse *vei* woe — more at WOE] (14th century)
intransitive verb
1 : to express sorrow audibly **:** LAMENT
2 : to make a sound suggestive of a mournful cry
3 : to express dissatisfaction plaintively **:** COMPLAIN
transitive verb
archaic **:** BEWAIL
— **wail·er** \'wā-lər\ *noun*

²wail *noun* (15th century)
1 : the act or practice of wailing **:** loud lamentation
2 a : a usually prolonged cry or sound expressing grief or pain **b :** a sound suggestive of wailing ⟨the *wail* of an air-raid siren⟩ **c :** a querulous expression of grievance **:** COMPLAINT

wail·ful \'wā(ə)l-fəl\ *adjective* (1544)
1 : uttering a sound suggestive of wailing
2 : expressing grief or pain **:** SORROWFUL, MOURNFUL
— wail·ful·ly \-fə-lē\ *adverb*

wailing wall *noun* (1919)
1 *capitalized* **:** a surviving section of the wall which in ancient times formed a part of the enclosure of Herod's temple near the Holy of Holies and at which Jews traditionally gather for prayer and religious lament
2 : a source of comfort and consolation in misfortune

wain \'wān\ *noun* [Middle English, wagon, chariot, from Old English *wægn;* akin to Middle Dutch *wagen* wagon, Old English *wegan* to move — more at WAY] (before 12th century)
1 : a usually large and heavy vehicle for farm use
2 *capitalized* [short for *Charles's Wain*] **:** BIG DIPPER

¹wain·scot \'wān-skət, -ˌskōt, -ˌskät\ *noun* [Middle English, from Middle Dutch *wagenschot,* probably from *wagen* wagon + *schot* shot, crossbar] (14th century)
1 *British* **:** a fine grade of oak imported for woodwork
2 a (1) **:** a usually paneled wooden lining of an interior wall (2) **:** a lining of an interior wall irrespective of material **b :** the lower three or four feet (about one meter) of an interior wall when finished differently from the remainder of the wall

²wainscot *transitive verb* **-scot·ed** *or* **-scot·ted; -scot·ing** *or* **-scot·ting** (circa 1570)
: to line with or as if with boards or paneling

wain·scot·ing *or* **wain·scot·ting** \-ˌskō-tiŋ, -ˌskä-, -skə-\ *noun* (1580)
1 : WAINSCOT 2
2 : material used to wainscot a surface

wain·wright \'wān-ˌrīt\ *noun* (before 12th century)
: a maker and repairer of wagons

waist \'wāst\ *noun* [Middle English *wast;* akin to Old English *wæstm* growth, *weaxan* to grow — more at WAX] (14th century)
1 a : the narrowed part of the body between the thorax and hips **b :** the greatly constricted basal part of the abdomen of some insects (as wasps and flies)
2 : the part of something corresponding to or resembling the human waist: as **a** (1) **:** the part of a ship's deck between the poop and forecastle (2) **:** the middle part of a sailing ship between foremast and mainmast **b :** the middle section of the fuselage of an airplane
3 : a garment or the part of a garment covering the body from the neck to the waistline or just below: **a :** BODICE 1 **b :** BLOUSE
4 : WAISTLINE 1b
— waist·ed \'wā-stəd\ *adjective*

waist·band \'wās(t)-ˌband\ *noun* (1584)
: a band (as of trousers or a skirt) fitting around the waist

waist·coat \'wes-kət, 'wās(t)-ˌkōt\ *noun* (1519)
1 : an ornamental garment worn under a doublet
2 *chiefly British* **:** VEST 2a
— waist·coat·ed \-ˌkō-təd\ *adjective*

waist·line \'wāst-ˌlīn\ *noun* (1896)
1 a : an arbitrary line encircling the narrowest part of the waist **b :** the part of a garment that covers the waistline or may be above or below it as fashion dictates
2 : body circumference at the waist

¹wait \'wāt\ *verb* [Middle English, from Old North French *waitier* to watch, of Germanic origin; akin to Old High German *wahta* watch, Old English *wæccan* to watch — more at WAKE] (14th century)
transitive verb
1 : to stay in place in expectation of **:** AWAIT
2 : to delay serving (a meal)
3 : to serve as waiter for ⟨*wait* tables⟩
intransitive verb
1 a : to remain stationary in readiness or expectation ⟨*wait* for a train⟩ **b :** to pause for another to catch up — usually used with *up*
2 a : to look forward expectantly ⟨just *waiting* to see his rival lose⟩ **b :** to hold back expectantly ⟨*waiting* for a chance to strike⟩
3 : to serve at meals — usually used in such phrases as *wait on tables* or *wait on table*
4 a : to be ready and available ⟨slippers *waiting* by the bed⟩ **b :** to remain temporarily neglected or unrealized ■
— wait on *also* **wait upon 1 a :** to attend as a servant **b :** to supply the wants of **:** SERVE **2 :** to make a formal call on **3 :** to wait for
— wait up : to delay going to bed **:** stay up

²wait *noun* [Middle English *waite* watchman, public musician, wait, from Old North French, watchman, watch, of Germanic origin; akin to Old High German *wahta* watch] (14th century)
1 a : a hidden or concealed position — used chiefly in the expression *lie in wait* **b :** a state or attitude of watchfulness and expectancy ⟨anchored in *wait* for early morning fishing —Fred Zimmer⟩
2 a : one of a band of public musicians in England employed to play for processions or public entertainments **b** (1) **:** one of a group who serenade for gratuities especially at the Christmas season (2) **:** a piece of music by such a group
3 : an act or period of waiting ⟨a long *wait* in line⟩

wait·er \'wā-tər\ *noun* (15th century)
1 : one that waits on another; *especially* **:** a person who waits tables (as in a restaurant)
2 : a tray on which something (as a tea service) is carried **:** SALVER

waiting game *noun* (1890)
: a strategy in which one or more participants withhold action temporarily in the hope of having a favorable opportunity for more effective action later

waiting list *noun* (1897)
: a list or roster of those waiting (as for election to a club or admission to an educational institution)

waiting room *noun* (1683)
: a room (as in a doctor's office) for the use of persons (as patients) who are waiting

wait out *transitive verb* (1941)
: to await an end to ⟨*wait* the storm *out*⟩

wait·per·son \'wāt-ˌpər-sᵊn\ *noun* (1976)
: a waiter or waitress

wait·ress \'wā-trəs\ *noun* (1834)
: a woman who waits tables (as in a restaurant)
usage see -ESS
— waitress *intransitive verb*

waive \'wāv\ *transitive verb* **waived; waiv·ing** [Middle English *weiven,* from Old North French *weyver,* from *waif* lost, unclaimed — more at WAIF] (14th century)
1 *archaic* **:** GIVE UP, FORSAKE
2 : to throw away (stolen goods)
3 *archaic* **:** to shunt aside (as a danger or duty) **:** EVADE
4 a : to relinquish voluntarily (as a legal right) ⟨*waive* a jury trial⟩ **b :** to refrain from pressing or enforcing (as a claim or rule) **:** FORGO
5 : to put off from immediate consideration **:** POSTPONE
6 : to dismiss with or as if with a wave of the hand ⟨*waived* the problem aside⟩
7 : to place (a ball player) on waivers; *also* **:** to release after placing on waivers
synonym see RELINQUISH

waiv·er \'wā-vər\ *noun* [Anglo-French *weyver,* from Old North French *weyver* to abandon, waive] (1628)
1 : the act of intentionally relinquishing or abandoning a known right, claim, or privilege; *also* **:** the legal instrument evidencing such an act
2 : the act of a club's waiving the right to claim a professional ball player who is being removed from another club's roster — often used in the phrase *on waivers* denoting the process by which a player to be removed from a roster is made available to other clubs ⟨was placed on *waivers* before being released⟩

Wa·kash·an \wo̦-'ka-shən, 'wo̦-ˌ\ *noun* (circa 1895)
: a family of American Indian languages spoken in coastal areas of British Columbia and northwest Washington

¹wake \'wāk\ *verb* **woke** \'wōk\ *also* **waked** \wākt\; **wo·ken** \'wō-kən\ *or* **waked** *also* **woke; wak·ing** [partly from Middle English *waken* (past *wook,* past participle *waken*), from Old English *wacan* to awake (past *wōc,* past participle *wacen*); partly from Middle English *wakien, waken* (past & past participle *waked*), from Old English *wacian* to be awake (past *wacode,* past participle *wacod*); akin to Old English *wæccan* to watch, Latin *vegēre* to enliven] (before 12th century)
intransitive verb
1 a : to be or remain awake **b** *archaic* **:** to remain awake on watch especially over a corpse **c** *obsolete* **:** to stay up late in revelry
2 : AWAKE — often used with *up*
transitive verb
1 : to stand watch over (as a dead body); *especially* **:** to hold a wake over
2 a : to rouse from or as if from sleep **:** AWAKE — often used with *up* **b :** STIR, EXCITE ⟨*woke* up latent possibilities —Norman Douglas⟩ **c :** to arouse conscious interest in **:** ALERT — usually used with *to* ⟨*woke* the publishers to

☐ USAGE
wait American dialectologists have evidence showing *wait on* (sense 3) to be more a Southern than a Northern form in speech. Handbook writers universally denigrate *wait on* and prescribe *wait for* in writing. Our evidence from printed sources does not show a regional preference; it does show that the handbooks' advice is not based on current usage ⟨settlement of the big problems still *waited on* Russia —*Time*⟩ ⟨I couldn't make out . . . whether Harper was *waiting on* me for approval —E. B. White⟩ ⟨the staggering bill that *waited on* them at the white commissary downtown —Maya Angelou⟩. One reason for the continuing use of *wait on* may lie in its being able to suggest protracted or irritating waits better than *wait for* ⟨for two days I've been *waiting on* weather —Charles A. Lindbergh⟩ ⟨the boredom of black Africans sitting there, *waiting on* the whims of a colonial bureaucracy —Vincent Canby⟩ ⟨doesn't care to sit around *waiting on* a House that's virtually paralyzed —Glenn A. Briere⟩. *Wait on* is less common than *wait for,* but if it seems natural, there is no reason to avoid it.

the fact that there was an enormous . . . audience —Harrison Smith⟩ ■
— **wak·er** *noun*

[2]**wake** *noun* (13th century)
1 : the state of being awake
2 a (1) **:** an annual English parish festival formerly held in commemoration of the church's patron saint (2) **:** VIGIL 1a **b :** the festivities originally connected with the wake of an English parish church — usually used in plural but singular or plural in construction **c** *British* **:** an annual holiday or vacation — usually used in plural but singular or plural in construction
3 : a watch held over the body of a dead person prior to burial and sometimes accompanied by festivity

[3]**wake** *noun* [of Scandinavian origin; akin to Old Norse *vǫk* hole in ice] (circa 1547)
1 : the track left by a moving body (as a ship) in a fluid (as water); *broadly* **:** a track or path left
2 : AFTERMATH 3
— **in the wake of 1 :** close behind and in the same path of travel ⟨*in the wake of* trappers and . . . riflemen came . . . settlers —*American Guide Series: Indiana*⟩ **2 :** as a result of **:** as a consequence of ⟨power vacuums left *in the wake of* the second world war —A. M. Schlesinger (born 1917)⟩

wake·ful \ˈwāk-fəl\ *adjective* (1549)
: not sleeping or able to sleep **:** SLEEPLESS
— **wake·ful·ly** \-fə-lē\ *adverb*
— **wake·ful·ness** *noun*

wake·less \ˈwā-kləs\ *adjective* (1824)
: SOUND, UNBROKEN ⟨*wakeless* sleep⟩

wak·en \ˈwā-kən\ *verb* **wak·ened; wak·en·ing** \ˈwāk-niŋ, ˈwā-kə-\ [Middle English *waknen*, from Old English *wæcnian;* akin to Old Norse *vakna* to awaken, Old English *wæccan* to watch] (before 12th century)
intransitive verb
: AWAKE — often used with *up*
transitive verb
: to rouse especially out of sleep **:** WAKE

wak·en·er \ˈwāk-nər, ˈwā-kə-\ *noun* (1573)
archaic **:** one that causes to waken

wake·rife \ˈwāk-ˌrīf\ *adjective* [Middle English (Scots) *walkryfe,* from *walk* awake (from *waken, walken* to wake) + *ryfe* rife] (15th century)
Scottish **:** WAKEFUL, ALERT

wake–rob·in \ˈwāk-ˌrä-bən\ *noun* (circa 1530)
1 : TRILLIUM
2 : JACK-IN-THE-PULPIT

wake–up \ˈwāk-ˌəp\ *adjective* (1946)
: serving to wake up ⟨a *wake-up* call⟩ ⟨a *wake-up* cup of coffee⟩

wak·ing \ˈwā-kiŋ\ *adjective* (1567)
: passed in a conscious or alert state ⟨every *waking* hour⟩

Wal·den·ses \wȯl-ˈden(t)-(ˌ)sēz, wäl-\ *noun plural* [Middle English *Waldensis,* from Medieval Latin *Waldenses, Valdenses,* from Peter *Waldo* (or *Valdo*)] (1537)
: a Christian sect arising in southern France in the 12th century, adopting Calvinist doctrines in the 16th century, and later living chiefly in Piedmont
— **Wal·den·sian** \-ˈden(t)-shən, -ˈden(t)-sē-ən\ *adjective or noun*

Wal·dorf salad \ˈwȯl-ˌdȯrf-\ *noun* [*Waldorf-*Astoria Hotel, New York City] (1902)
: a salad made typically of diced apples, celery, nuts, and mayonnaise

[1]**wale** \ˈwā(ə)l\ *noun* [Middle English, from Old English *walu;* akin to Old Norse *vǫlr* staff and perhaps to Old Norse *valr* round, Latin *volvere* to roll — more at VOLUBLE] (before 12th century)
1 a : a streak or ridge made on the skin especially by the stroke of a whip **:** WEAL **b :** a narrow raised surface **:** RIDGE
2 : any of a number of strakes usually of extra thick and strong planks in the sides of a wooden ship — usually used in plural
3 a : one of a series of even ribs in a fabric **b :** the texture especially of a fabric
4 : a horizontal constructional member (as of timber or steel) used for bracing vertical members

[2]**wale** *transitive verb* **waled; wal·ing** (15th century)
: to mark (as the skin) with welts

[3]**wale** *noun* [Middle English (Scots & northern dialect) *wal,* from Old Norse *val;* akin to Old High German *wala* choice, Old English *wyllan* to wish — more at WILL] (14th century)
1 *dialect British* **:** CHOICE
2 *dialect British* **:** the best part **:** PICK

[4]**wale** *verb* (14th century)
dialect British **:** CHOOSE

wal·er \ˈwā-lər\ *noun, often capitalized* [New South *Wales,* Australia] (1849)
: a horse from New South Wales; *especially* **:** a rather large rugged saddle horse of mixed ancestry formerly exported in quantity from Australia to British India for military use

Wal·hal·la \väl-ˈhä-lə\ *noun* [German]
: VALHALLA

[1]**walk** \ˈwȯk\ *verb* [partly from Middle English *walken* (past *welk,* past participle *walken*), from Old English *wealcan* to roll, toss, journey about (past *weolc,* past participle *wealcen*) and partly from Middle English *walkien* (past *walked,* past participle *walked*), from Old English *wealcian* to roll up, muffle up; akin to Middle Dutch *walken* to knead, press, full] (before 12th century)
intransitive verb
1 a *obsolete* **:** ROAM, WANDER **b** *of a spirit* **:** to move about in visible form **:** APPEAR **c** *of a ship* **:** to make headway
2 a : to move along on foot **:** advance by steps **b :** to come or go easily or readily **c :** to go on foot for exercise or pleasure **d :** to go at a walk
3 a : to pursue a course of action or way of life **:** conduct oneself **:** BEHAVE ⟨*walk* warily⟩ **b :** to be or act in association **:** continue in union ⟨the British and American peoples will . . . *walk* together side by side . . . in peace —Sir Winston Churchill⟩ **c :** WALK OUT ⟨the workers didn't like the new contract so they *walked*⟩
4 : to go to first base as a result of a base on balls
5 *of an inanimate object* **a :** to move in a manner that is suggestive of walking **b :** to stand with an appearance suggestive of strides ⟨pylons *walking* across the valley⟩
6 *of an astronaut* **:** to move about in space outside a spacecraft
transitive verb
1 a : to pass on foot or as if on foot through, along, over, or upon **:** TRAVERSE, PERAMBULATE ⟨*walk* the streets⟩ ⟨*walk* a tightrope⟩ **b :** to perform or accomplish by going on foot ⟨*walk* guard⟩
2 a : to cause (an animal) to go at a walk **:** take for a walk ⟨*walking* a dog⟩ **b** (1) **:** to cause to move by walking ⟨*walked* her bicycle up the hill⟩ (2) **:** to haul (as an anchor) by walking round the capstan
3 : to follow on foot for the purpose of measuring, surveying, or inspecting ⟨*walk* a boundary⟩
4 a : to accompany on foot **:** walk with ⟨*walked* her home⟩ **b :** to compel to walk (as by a command) **c :** to bring to a specified condition by walking ⟨*walked* us off our feet⟩
5 : to move (an object) in a manner suggestive of walking
6 : to perform (a dance) at a walking pace ⟨*walk* a quadrille⟩
7 : to give a base on balls to
— **walk away from 1 :** to outrun or get the better of without difficulty **2 :** to survive (an accident) with little or no injury
— **walk off with 1 a :** to steal and take away **b :** to take over unexpectedly from someone else **:** STEAL 1d ⟨*walked off with* the show⟩ **2 :** to win or gain especially by outdoing one's competitors without difficulty
— **walk on :** to take advantage of **:** ABUSE
— **walk over :** to treat contemptuously
— **walk the plank 1 :** to walk under compulsion over the side of a ship into the sea **2 :** to resign an office or position under compulsion
— **walk through 1 :** to go through (as a theatrical role or familiar activity) perfunctorily (as in an early stage of rehearsal) **2 :** to guide (as a novice) through an unfamiliar or complex procedure step-by-step **3 :** to deal with or carry out perfunctorily

[2]**walk** *noun* (14th century)
1 a : an act or instance of going on foot especially for exercise or pleasure ⟨go for a *walk*⟩ **b :** SPACE WALK
2 : an accustomed place of walking **:** HAUNT
3 : a place designed for walking: **a :** a railed platform above the roof of a dwelling house **b** (1) **:** a path specially arranged or paved for walking (2) **:** SIDEWALK **c :** a public avenue for promenading **:** PROMENADE **d :** ROPEWALK
4 : a place or area of land in which animals feed and exercise with minimal restraint
5 : distance to be walked ⟨a quarter mile *walk* from here⟩
6 *British* **:** a ceremonial procession
7 : manner of living **:** CONDUCT, BEHAVIOR
8 a : the gait of a biped in which the feet are lifted alternately with one foot not clear of the ground before the other touches **b :** the gait of a quadruped in which there are always at least two feet on the ground; *specifically* **:** a four-beat gait of a horse in which the feet strike the ground in the sequence near hind, near fore, off hind, off fore **c :** a low rate of speed ⟨the shortage of raw materials slowed production to a *walk*⟩
9 : a route regularly traversed by a person in the performance of a particular activity (as patrolling, begging, or vending)
10 : characteristic manner of walking ⟨his *walk* is just like his father's⟩
11 a : social or economic status ⟨all *walks* of life⟩ **b** (1) **:** range or sphere of action **:** FIELD, PROVINCE (2) **:** VOCATION
12 : BASE ON BALLS

walk·about \ˈwȯ-kə-ˌbau̇t\ *noun* (1908)
1 : a short period of wandering bush life engaged in by an Australian aborigine as an occasional interruption of regular work

□ USAGE
wake *Wake* has such a large selection of forms for the past and past participle because it is the modern result of the coalescence of two old verbs, one with regular parts, one with irregular. What makes *wake* interesting is that preference for these forms has changed over time and is still changing. The regular *waked* was the prevalent form in the 16th, 17th, and 18th centuries. During the 19th century *woke* came into wide use for the past tense while *waked* continued as the past participle. In the 20th century *woken* has emerged as a strong contender to be the dominant past participle. These changes have not taken place uniformly, and they seem to have become prevalent in British English earlier than they did in American English. For instance *woken* seems to be the usual past participle for British writers ⟨in the mornings he was *woken* by his butler —Julian Huxley⟩ ⟨one night he was *woken* by someone coming into the bedroom —Graham Greene⟩ while in American English *waked* and *woken* are both used ⟨during the night he has *waked* up sweating —Mary McCarthy⟩ ⟨he had *woken* them, clearing his throat —John Updike⟩. *Woke* is much less frequent as a past participle than *waked* and *woken.*

2 : a walking tour **:** walking trip

walk·a·thon \'wó-kə-,thän\ *noun* [*walk* + *-athon*] (1932)
: a walk covering a considerable distance organized especially to raise money for a cause

walk·away \'wó-kə-,wā\ *noun* (1888)
: an easily won contest

walk·er \'wó-kər\ *noun* (14th century)
1 : one that walks: as **a :** a competitor in a walking race **b :** a peddler going on foot **c :** a temporary male escort of socially prominent women attending usually public events
2 : something used in walking: as **a :** a framework designed to support a baby learning to walk or an infirm or handicapped person **b :** a walking shoe

walk·ie–talk·ie \,wó-kē-'tó-kē, 'wó-kē-,\ *noun* (1939)
: a compact easily transportable battery-operated radio transmitting and receiving set

¹walk–in \'wók-,in\ *adjective* (1926)
1 : large enough to be walked into ⟨a *walk-in* closet⟩
2 : arranged so as to be entered directly rather than through a lobby ⟨a *walk-in* apartment⟩
3 a : being a person who walks in without an appointment ⟨a *walk-in* blood donor⟩ **b :** of or relating to such persons ⟨*walk-in* clinic⟩

²walk–in \'wók-,in\ *noun* (1944)
1 : a walk-in refrigerator or cold storage room
2 : an easy election victory
3 : one who walks in without an appointment

¹walk·ing \'wó-kiŋ\ *noun* (15th century)
1 : the action of one that walks ⟨*walking* is good exercise⟩
2 : the condition of a surface for one going on foot ⟨the *walking* is slippery⟩

²walking *adjective* (15th century)
1 a : able to walk **:** AMBULATORY ⟨the *walking* wounded⟩ **b :** being the personification of a nonhuman quality or thing ⟨a *walking* encyclopedia⟩
2 a : used for or in walking ⟨*walking* shoes⟩ **b :** characterized by or consisting of the action of walking ⟨a *walking* tour⟩
3 : that moves or appears to move in a manner suggestive of walking; *especially* **:** that swings or rocks back and forth ⟨*walking* beam⟩
4 : guided or operated by a person on foot ⟨a *walking* plow⟩

walking catfish *noun* (1968)
: an Asian freshwater catfish (*Clarias batrachus* of the family Clariidae) that is able to move about on land and has become established in Florida waters

walking delegate *noun* (1889)
: a labor union representative appointed to visit members and their places of employment, to secure enforcement of union rules and agreements, and at times to represent the union in dealing with employers

walking leaf *noun* (1826)
: any of a family (Phylliidae) of phasmid insects with wings and legs resembling leaves

walking papers *noun plural* (1825)
: DISMISSAL, DISCHARGE — called also *walking ticket*

walking pneumonia *noun* (1964)
: a usually mild pneumonia caused by a mycoplasma (*Mycoplasma pneumoniae*) and characterized by malaise, cough, and often fever

walking stick *noun* (1580)
1 : a stick used in walking
2 *usually* **walk·ing·stick :** STICK INSECT; *especially* **:** a phasmid insect (*Diapheromera femorata*) common in parts of the U.S.

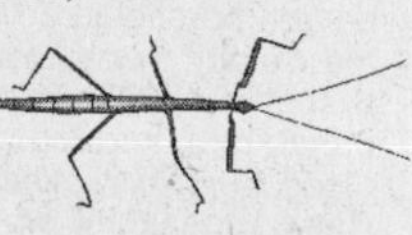
walkingstick

Walk·man \'wók-mən, -,man\ *trademark*
— used for a small portable radio or cassette player listened to by means of headphones or earphones

walk–on \'wók-,ón, -,än\ *noun* (1902)
1 : a minor part (as in a dramatic production); *also* **:** an actor having such a part
2 : a college athlete who tries out for an athletic team without having been recruited or offered a scholarship

walk·out \'wók-,au̇t\ *noun* (1888)
1 : STRIKE 3a
2 : the action of leaving a meeting or organization as an expression of disapproval

walk out *intransitive verb* (1840)
1 : to leave suddenly often as an expression of disapproval
2 : to go on strike
— walk out on : to leave in the lurch **:** ABANDON, DESERT

walk·over \'wók-,ō-vər\ *noun* (1838)
1 : a one-sided contest **:** an easy or uncontested victory
2 : a horse race with only one starter

walk–through \'wók-,thrü\ *noun* (1940)
1 : a perfunctory performance of a play or acting part (as in an early stage of rehearsal)
2 : a television rehearsal without cameras

¹walk–up \'wók-,əp\ *adjective* (1919)
1 : located above the ground floor in a building with no elevator ⟨a *walk-up* apartment⟩
2 : consisting of several stories and having no elevator ⟨a *walk-up* tenement⟩
3 : designed to allow pedestrians to be served without entering a building ⟨the *walk-up* window of a bank⟩

²walk–up *noun* (1924)
: a building or apartment house of several stories that has no elevator; *also* **:** an apartment or office in such a building

walk·way \'wók-,wā\ *noun* (1792)
: a passage for walking **:** WALK

Wal·ky·rie \väl-'kir-ē *also* väl-'kī-rē *or* 'väl-kə-rē\ *noun* [German *Walküre* & Old Norse *valkyrja*]
: VALKYRIE

¹wall \'wól\ *noun* [Middle English, from Old English *weall;* akin to Middle High German *wall;* both from Latin *vallum* rampart, from *vallus* stake, palisade; perhaps akin to Old Norse *vǫlr* staff — more at WALE] (before 12th century)
1 a : a high thick masonry structure forming a long rampart or an enclosure chiefly for defense — often used in plural **b :** a masonry fence around a garden, park, or estate **c :** a structure that serves to hold back pressure (as of water or sliding earth)
2 : one of the sides of a room or building connecting floor and ceiling or foundation and roof
3 : the side of a footpath next to buildings
4 : an extreme or desperate position or a state of defeat, failure, or ruin — usually used in the phrase *to the wall*
5 : a material layer enclosing space ⟨the *wall* of a container⟩ ⟨heart *walls*⟩
6 : something resembling a wall (as in appearance, function, or effect); *especially* **:** something that acts as a barrier or defense ⟨a *wall* of reserve⟩ ⟨tariff *wall*⟩
— walled \'wóld\ *adjective*
— wall–like \'wól-,līk\ *adjective*
— up the wall *slang* **:** into a state of intense agitation, annoyance, or frustration ⟨the noise drove me *up the wall*⟩

²wall *transitive verb* (13th century)
1 a : to provide, cover with, or surround with or as if with a wall ⟨*wall* in the garden⟩ **b :** to separate by or as if by a wall ⟨*walled* off half the house⟩
2 a : IMMURE **b :** to close (an opening) with or as if with a wall

³wall *verb* [Middle English (Scots) *wawlen,* probably from Middle English *wawil-* (in *wawil-eghed* walleyed)] (15th century)
intransitive verb
of the eyes **:** to roll in a dramatic manner
transitive verb
: to roll (one's eyes) in a dramatic manner

wal·la·by \'wä-lə-bē\ *noun, plural* **wallabies** *also* **wallaby** [Dharuk (Australian aboriginal language of the Port Jackson area) *walaba*] (circa 1798)
: any of various small or medium-sized kangaroos (especially genus *Macropus*) — compare ROCK WALLABY

wallaby

Wal·lace's line \'wä-lə-səz-\ *noun* [Alfred Russel *Wallace*] (1868)
: a hypothetical boundary that separates the highly distinctive floras and faunas of the Oriental and Australian biogeographic regions and passes between the islands of Bali and Lombok in Indonesia, between Borneo and Sulawesi, and between the Philippines and the Moluccas

wal·lah \'wä-lə, *in combination usually* ,wä-lə\ *noun* [Hindi *-vālā* one in charge, from Sanskrit *pāla* protector, from *pālayati, pārayati* he guards; akin to Sanskrit *piparti* he brings over, saves, Old English *faran* to go — more at FARE] (1782)
: a person who is associated with a particular work or who performs a specific duty or service — usually used in combination ⟨the book *wallah* was an itinerant peddler —George Orwell⟩

wal·la·roo \,wä-lə-'rü\ *noun, plural* **-roos** [Dharuk (Australian aboriginal language of the Port Jackson area) *walaru*] (1827)
1 : a large reddish gray kangaroo (*Macropus robustus*) — called also *euro*
2 : either of two kangaroos (*Macropus antelopinus* and *M. bernardus*) related to the wallaroo

wall·board \'wól-,bōrd, -,bórd\ *noun* (1906)
: a structural boarding of any of various materials (as wood pulp, gypsum, or plastic) made in large rigid sheets and used especially for sheathing interior walls and ceilings

wal·let \'wä-lət\ *noun* [Middle English *walet*] (14th century)
1 : a bag for carrying miscellaneous articles while traveling
2 a : a folding pocketbook with compartments for personal papers and usually unfolded paper money; *also* **:** BILLFOLD **b :** a container that resembles a money wallet: as (1) **:** a usually flexible folding case fitted for carrying specific items (as tools or fishing flies) (2) **:** FOLDER 3

wall·eye \'wó-,lī\ *noun* [back-formation from *walleyed*] (1523)
1 a : an eye with a whitish or bluish white iris **b :** an eye with an opaque white cornea
2 a : strabismus in which the eye turns outward away from the nose **b :** an eye affected with divergent strabismus
3 : a large vigorous North American freshwater food and sport fish (*Stizostedion vitreum*) that has prominent eyes and is related to the perches but resembles the true pike — called also *walleyed pike*

wall·eyed \-'līd\ *adjective* [by folk etymology from Middle English *wawil-eghed,* part translation of Old Norse *vagl-eygr* walleyed, from *vagl* beam (akin to Greek *ochleus* bar, Old English *wegan* to move, carry) + *eygr* eyed — more at WAY] (15th century)
1 : having walleyes or affected with walleye
2 : marked by a wild irrational staring of the eyes

walleye pollack *noun* (1907)

: POLLACK 2

wall·flow·er \ˈwȯl-ˌflau̇(-ə)r\ *noun* (1578)
1 a : any of several Old World herbaceous or somewhat woody perennial plants (genus *Cheiranthus*) of the mustard family; *especially* **:** a hardy erect herb (*C. cheiri*) widely cultivated for its showy fragrant flowers **b :** any of a related genus (*Erysimum*) of plants with showy flowers
2 : a person who from shyness or unpopularity remains on the sidelines of a social activity (as a dance)

Wal·loon \wä-ˈlün\ *noun* [Middle French *Wallon,* adjective & noun, of Germanic origin; probably akin to Old High German *Walah* Celtic, Roman, Old English *Wealh* Celt, Welshman — more at WELSH] (1567)
1 : a member of a people of southern and southeastern Belgium and adjacent parts of France
2 : a French dialect of the Walloons
— **Walloon** *adjective*

[1]**wal·lop** \ˈwä-ləp\ *verb* [Middle English *walopen* to gallop, from Old North French *waloper*] (1579)
intransitive verb
1 : to boil noisily
2 a : to move with reckless or disorganized haste **:** advance in a headlong rush **b :** WALLOW, FLOUNDER
transitive verb
1 a : to thrash soundly **:** LAMBASTE **b :** to beat by a wide margin **:** TROUNCE
2 : to hit with force **:** SOCK
— **wal·lop·er** *noun*

[2]**wallop** *noun* [Middle English, gallop, from Old North French *walop,* from *waloper* to gallop] (circa 1823)
1 a : a powerful blow **:** PUNCH **b :** something resembling a wallop especially in suddenness of force **c :** the ability (as of a boxer) to hit hard
2 a : emotional, sensory, or psychological force or influence **:** IMPACT ⟨a novel that packs a *wallop*⟩ **b :** an exciting emotional response **:** THRILL
3 *British* **:** BEER

wal·lop·ing *adjective* (circa 1847)
1 : LARGE, WHOPPING
2 : exceptionally fine or impressive **:** SMASHING

[1]**wal·low** \ˈwä-(ˌ)lō\ *intransitive verb* [Middle English *walwen,* from Old English *wealwian* to roll — more at VOLUBLE] (before 12th century)
1 : to roll oneself about in an indolent or ungainly manner
2 : to billow forth **:** SURGE
3 : to devote oneself entirely; *especially* **:** to take unrestrained pleasure **:** DELIGHT
4 a : to become abundantly supplied **:** LUXURIATE ⟨a family that *wallows* in money⟩ **b :** to indulge oneself immoderately ⟨*wallowing* in self-pity⟩
5 : to become or remain helpless ⟨allowed them to *wallow* in their ignorance⟩
— **wal·low·er** \ˈwä-lə-wər\ *noun*

[2]**wallow** *noun* (15th century)
1 : an act or instance of wallowing
2 a : a muddy area or one filled with dust used by animals for wallowing **b :** a depression formed by or as if by the wallowing of animals
3 : a state of degradation or degeneracy

wall painting *noun* (1688)
: FRESCO

[1]**wall·pa·per** \ˈwȯl-ˌpā-pər\ *noun* (1827)
: decorative paper for the walls of a room

[2]**wallpaper** (1924)
transitive verb
: to provide the walls of (a room) with wallpaper
intransitive verb
: to put wallpaper on a wall

wall plate *noun* (14th century)
: PLATE 5

wall plug *noun* (1888)
: an electric receptacle in a wall

wall rock *noun* (1876)
: a rock through which a fault or vein runs

wall rocket *noun* (1611)
: any of several Old World herbs (genus *Diplotaxis*) of the mustard family; *especially* **:** a yellow-flowered European herb (*D. tenuifolia*) adventive in North America

Wall Street \ˈwȯl-\ *noun* [*Wall Street,* New York City, site of the New York Stock Exchange] (1836)
: the influential financial interests of the U.S. economy

Wall Street·er \-ˌstrē-tər\ *noun* (1885)
: a person who is involved in Wall Street

wall system *noun* (1968)
: a set of shelves often with cabinets or bureaus that can be variously arranged along a wall

wall–to–wall *adjective* (1946)
1 : covering the entire floor ⟨*wall-to-wall* carpeting⟩
2 a : covering or filling one entire space or time ⟨a party crammed with *wall-to-wall* bodies⟩ **b :** occurring or found everywhere **:** UBIQUITOUS

wal·ly \ˈwā-lē\ *adjective* [probably from [3]*wale*] (circa 1520)
Scottish **:** FINE, STURDY

wal·ly·drai·gle \ˈwā-lē-ˌdrā-gəl, ˈwä-lē-\ *noun* [origin unknown] (1508)
chiefly Scottish **:** a feeble, imperfectly developed, or slovenly creature

wal·nut \ˈwȯl-(ˌ)nət\ *noun* [Middle English *walnot,* from Old English *wealhhnutu,* literally, foreign nut, from *Wealh* Welshman, foreigner + *hnutu* nut — more at WELSH, NUT] (before 12th century)
1 a : a nut of any of a genus (*Juglans* of the family Juglandaceae, the walnut family) of trees; *especially* **:** the large edible nut of a Eurasian tree (*J. regia*) with a hard richly figured wood **b :** a tree that bears walnuts **c :** the wood of a walnut that is often used for cabinetmaking and veneers
2 : a moderate reddish brown ◆

Wal·pur·gis·nacht \väl-ˈpu̇r-gəs-ˌnäkt\ *noun* [German] (1822)
: WALPURGIS NIGHT

Wal·pur·gis Night \väl-ˈpu̇r-gəs-\ *noun* [part translation of German *Walpurgisnacht,* from *Walpurgis* Saint Walburga (died A.D. 779) English saint whose feast day falls on May Day + German *Nacht* night] (1823)
1 : the eve of May Day on which witches are held to ride to an appointed rendezvous
2 : something (as an event or situation) having a nightmarish quality

wal·rus \ˈwȯl-rəs, ˈwäl-\ *noun, plural* **walrus** *or* **wal·rus·es** [Dutch, of Scandinavian origin; akin to Danish & Norwegian *hvalros* walrus, Old Norse *rosmhvalr*] (1728)
: a large gregarious marine mammal (*Odobenus rosmarus* of the family Odobenidae) related to the seal that is found in Arctic seas, has long ivory tusks, a tough wrinkled hide, and stiff whiskers, and feeds mostly on bivalve mollusks

Wal·ter Mit·ty \ˌwȯl-tər-ˈmi-tē\ *noun* [*Walter Mitty,* daydreaming hero of a story by James Thurber] (1949)
: a commonplace unadventurous person who seeks escape from reality through daydreaming
— **Walter Mit·ty·ish** \-ish\ *adjective*

[1]**waltz** \ˈwȯl(t)s\ *noun* [German *Walzer,* from *walzen* to roll, dance, from Old High German *walzan* to turn, roll — more at WELTER] (1781)
1 : a ballroom dance in ¾ time with strong accent on the first beat and a basic pattern of step-step-close
2 : music for a waltz or a concert composition in ¾ time

[2]**waltz** (circa 1794)
intransitive verb
1 : to dance a waltz
2 : to move or advance in a lively or conspicuous manner **:** FLOUNCE
3 a : to advance easily and successfully **:** BREEZE — often used with *through* **b :** to approach boldly — used with *up* ⟨can't just *waltz* up and introduce ourselves⟩
transitive verb
1 : to dance a waltz with
2 : to grab and lead (as a person) unceremoniously **:** MARCH
— **waltz·er** *noun*

[1]**wam·ble** \ˈwäm-bəl\ *intransitive verb* **wam·bled; wam·bling** \-b(ə-)liŋ\ [Middle English *wamlen;* akin to Danish *vamle* to become nauseated, Latin *vomere* to vomit — more at VOMIT] (14th century)
1 a : to feel nausea **b** *of a stomach* **:** RUMBLE 1
2 : to move unsteadily or with a weaving or rolling motion

[2]**wamble** *noun* (1552)
1 : a wambling especially of the stomach
2 : a reeling or staggering gait or movement

wame \ˈwām\ *noun* [Middle English, alteration of *wamb* — more at WOMB] (15th century)
chiefly Scottish **:** BELLY

Wam·pa·no·ag \ˌwäm-pə-ˈnō-(ˌ)ag, ˌwȯm-\ *noun, plural* **Wampanoag** *or* **Wampanoags** [Narraganset, literally, easterners] (1676)
: a member of an American Indian people of Rhode Island east of Narragansett Bay and neighboring parts of Massachusetts

wam·pum \ˈwäm-pəm\ *noun* [short for *wampumpeag*] (1636)
1 : beads of polished shells strung in strands, belts, or sashes and used by North American Indians as money, ceremonial pledges, and ornaments
2 : MONEY

wam·pum·peag \-ˌpēg\ *noun* [Massachuset *wampompeag,* from *wampan* white + *api* string + *-ag,* plural suffix] (1627)
: WAMPUM

[1]**wan** \ˈwän\ *adjective* **wan·ner; wan·nest** [Middle English, from Old English *wann* dark, livid] (14th century)
1 a : suggestive of poor health **:** SICKLY, PALLID **b :** lacking vitality **:** FEEBLE
2 : DIM, FAINT
3 : LANGUID ⟨a *wan* smile⟩
— **wan·ly** *adverb*
— **wan·ness** \ˈwän-nəs\ *noun*

◇ WORD HISTORY

walnut The English walnut (*Juglans regia*) is not native to England in spite of its name. This tree, important for both its wood and fruit, was known in southern Europe long before it was introduced to northern Europeans, to whom the only readily available nut was the hazelnut. *Wealhhnutu,* the Old English ancestor of *walnut,* reflects the foreign origin of the walnut: it is a compound of *wealh* "stranger, foreigner" and *hnutu* "nut." Other Germanic words of later date, such as Middle Low German *walnut* (from which ultimately Modern German *Walnuss* comes), Middle High German *walhisch nuz,* and Old Norse *valhnot,* contain the same elements. They have a parallel in a Medieval Latin name for the walnut, *nux gallica,* literally, "Gallic or French nut." Old English *wealh* "foreigner" is buried in other Modern English words in addition to *walnut. Wealh* was applied to the Celtic speakers who opposed the Anglo-Saxons' invasion of Britain—despite being the indigenous inhabitants they were "foreigners" to the Germanic tribes—and *Wealas,* the plural of *wealh,* developed into Modern English *Wales,* now designating the land where Celtic Britons took refuge and maintained their ancient speech.

[2]**wan** *intransitive verb* **wanned; wan·ning** (1582)
: to grow or become pale or sickly

wand \'wänd\ *noun* [Middle English, slender stick, from Old Norse *vǫndr;* probably akin to Old English *windan* to wind, twist — more at WIND] (15th century)
1 : a slender staff carried in a procession **:** VERGE
2 : a slender rod used by conjurers and magicians
3 : a slat 6 feet by 2 inches used as a target in archery; *also* **:** a narrow strip of paper pasted vertically on a target face
4 : any of various pipelike devices; *especially* **:** the rigid tube between the hose and the nozzle of a vacuum cleaner
5 : a handheld device used to enter information (as from a bar code) into a computer

wan·der \'wän-dər\ *verb* **wan·dered; wan·der·ing** \-d(ə-)riŋ\ [Middle English *wandren,* from Old English *wandrian;* akin to Middle High German *wandern* to wander, Old English *windan* to wind, twist] (before 12th century)
intransitive verb
1 a : to move about without a fixed course, aim, or goal **b :** to go idly about **:** RAMBLE
2 : to follow a winding course **:** MEANDER
3 a : to deviate (as from a course) **:** STRAY **b :** to go astray morally **:** ERR **c :** to lose normal mental contact **:** stray in thought
transitive verb
: to roam over ☆
— wander *noun*
— wan·der·er \-dər-ər\ *noun*

[1]**wandering** *adjective* (before 12th century)
: characterized by aimless, slow, or pointless movement: as **a :** that winds or meanders ⟨a *wandering* course⟩ **b :** not keeping a rational or sensible course **:** VAGRANT **c :** NOMADIC ⟨*wandering* tribes⟩ **d** *of a plant* **:** having long runners or tendrils

[2]**wandering** *noun* (14th century)
1 : a going about from place to place — often used in plural
2 : movement away from the proper, normal, or usual course or place — often used in plural

Wandering Jew *noun*
1 : a Jew of medieval legend condemned by Christ to wander the earth till Christ's second coming
2 *W not capitalized* **:** any of several plants (genera *Zebrina* and *Tradescantia*) of the spiderwort family; *especially* **:** either of two trailing or creeping plants (*Z. pendula* and *T. fluminensis*) cultivated for their showy and often white-striped foliage

wan·der·lust \'wän-dər-ˌləst\ *noun* [German, from *wandern* to wander + *Lust* desire, pleasure] (1902)
: strong longing for or impulse toward wandering

[1]**wane** \'wān\ *intransitive verb* **waned; wan·ing** [Middle English, from Old English *wanian;* akin to Old High German *wanōn* to wane, Old English *wan* wanting, deficient, Latin *vanus* empty, vain] (before 12th century)
1 : to decrease in size, extent, or degree **:** DWINDLE: as **a :** to diminish in phase or intensity — used chiefly of the moon **b :** to become less brilliant or powerful **:** DIM **c :** to flow out **:** EBB
2 : to fall gradually from power, prosperity, or influence
synonym see ABATE

[2]**wane** *noun* (14th century)
1 a : the act or process of waning ⟨strength on the *wane*⟩ **b :** a period or time of waning; *specifically* **:** the period from full phase of the moon to the new moon
2 [Middle English, defect, from Old English *wana;* akin to Old English *wan* deficient] **:** a defect in lumber characterized by bark or a lack of wood at a corner or edge

wan·gle \'waŋ-gəl\ *verb* **wan·gled; wan·gling** \-g(ə-)liŋ\ [perhaps alteration of *waggle*] (circa 1820)
intransitive verb
1 : to extricate oneself (as from difficulty) **:** WIGGLE
2 : to resort to trickery or devious methods
transitive verb
1 : SHAKE, WIGGLE
2 : to adjust or manipulate for personal or fraudulent ends
3 : to make or get by devious means **:** FINAGLE ⟨*wangle* an invitation⟩
— wan·gler \-g(ə-)lər\ *noun*

wan·i·gan *or* **wan·ni·gan** \'wä-ni-gən\ *noun* [Ojibwa *wa·nikka·n* pit] (circa 1848)
: a shelter (as for sleeping, eating, or storage) often mounted on wheels or tracks and towed by tractor or mounted on a raft or boat

wan·ion \'wän-yən\ *noun* [from the obsolete phrase *in the waniand* unluckily, literally, in the waning (moon), from Middle English, from *waniand,* northern present participle of *wanien, wanen* to wane] (1549)
archaic **:** PLAGUE, VENGEANCE — used in the phrase *with a wanion*

Wan·kel engine \'väŋ-kəl-, 'waŋ-\ *noun* [Felix *Wankel* (died 1988) German engineer] (1961)
: an internal combustion rotary engine that has a rounded triangular rotor functioning as a piston and rotating in a space in the engine and that has only two major moving parts

wan·na–be \'wä-nə-ˌbē\ *noun* [from the phrase *want to be*] (1981)
: a person who wants or aspires to be someone or something else or who tries to look or act like someone else

[1]**want** \'wȯnt *also* 'wänt & 'wənt\ *verb* [Middle English, from Old Norse *vanta;* akin to Old English *wan* deficient] (13th century)
intransitive verb
1 : to be needy or destitute
2 : to have or feel need ⟨never *wants* for friends⟩
3 : to be necessary or needed
4 : to desire to come, go, or be ⟨the cat *wants* in⟩ ⟨*wants* out of the deal⟩
transitive verb
1 : to fail to possess especially in customary or required amount **:** LACK ⟨the answer *wanted* courtesy⟩
2 a : to have a strong desire for ⟨*wanted* a chance to rest⟩ **b :** to have an inclination to **:** LIKE ⟨say what you *want,* he is efficient⟩
3 a : to have need of **:** REQUIRE ⟨the motor *wants* a tune-up⟩ **b :** to suffer from the lack of ⟨thousands still *want* food and shelter⟩
4 : OUGHT — used with the infinitive ⟨you *want* to be very careful what you say —Claudia Cassidy⟩
5 : to wish or demand the presence of
6 : to hunt or seek in order to apprehend ⟨*wanted* for murder⟩
synonym see DESIRE

[2]**want** *noun* (14th century)
1 a : DEFICIENCY, LACK ⟨suffers from a *want* of good sense⟩ **b :** grave and extreme poverty that deprives one of the necessities of life
2 : something wanted **:** NEED, DESIRE
3 : personal defect **:** FAULT
synonym see POVERTY

want ad *noun* (1897)
: a newspaper advertisement stating that something (as an employee, employment, or a specified item) is wanted

[1]**want·ing** *adjective* (15th century)
1 : not present or in evidence **:** ABSENT
2 a : not being up to standards or expectations **b :** lacking in ability or capacity **:** DEFICIENT

[2]**wanting** *preposition* (1693)
1 : LESS, MINUS ⟨a month *wanting* two days⟩
2 : WITHOUT ⟨a book *wanting* a cover⟩

[1]**wan·ton** \'wȯn-tᵊn, 'wän-\ *adjective* [Middle English, from *wan-* deficient, wrong, mis- (from Old English, from *wan* deficient) + *towen,* past participle of *teen* to draw, train, discipline, from Old English *tēon* — more at TOW] (14th century)
1 a *archaic* **:** hard to control **:** UNDISCIPLINED, UNRULY **b :** playfully mean or cruel **:** MISCHIEVOUS
2 a : LEWD, BAWDY **b :** causing sexual excitement **:** LUSTFUL, SENSUAL
3 a : MERCILESS, INHUMANE ⟨*wanton* cruelty⟩ **b :** having no just foundation or provocation **:** MALICIOUS ⟨a *wanton* attack⟩
4 : being without check or limitation: as **a :** luxuriantly rank ⟨*wanton* vegetation⟩ **b :** unduly lavish **:** EXTRAVAGANT
— wan·ton·ly *adverb*
— wan·ton·ness \-tᵊn-nəs\ *noun*

[2]**wanton** *noun* (1526)
1 : a pampered person or animal **:** PET; *especially* **:** a spoiled child
2 : a frolicsome child or animal
3 a : one given to self-indulgent flirtation or trifling — used especially in the phrase *play the wanton* **b :** a lewd or lascivious person

[3]**wanton** (1582)
intransitive verb
: to be wanton or act wantonly
transitive verb
: to pass or waste wantonly or in wantonness
— wan·ton·er *noun*

wa·pen·take \'wa-pən-ˌtāk, 'wä-\ *noun* [Middle English, from Old English *wǣpentæc,* from Old Norse *vāpnatak* act of grasping weapons, from *vāpn* weapon + *tak* act of grasping, from *taka* to take; probably from the brandishing of weapons as an expression of approval when the chief of the wapentake entered upon his office — more at WEAPON, TAKE] (before 12th century)
: a subdivision of some English shires corresponding to a hundred

wa·pi·ti \'wä-pə-tē\ *noun, plural* **wapiti** *or* **wapitis** [Shawnee *wa·piti,* literally, white rump] (1806)
: ELK 1b

wap·pen·schaw·ing \'wa-pən-ˌshȯ(-)iŋ, 'wä-\ *noun* [Middle English (northern dialect) *wapynschawing,* from *wapen* weapon (from Old Norse *vāpn*) + *schawing,* gerund of *schawen* to show, from Old English *scēawian* to look, look at — more at WEAPON, SHOW] (15th century)
: an inspection or muster of soldiers formerly held at various times in each district of Scotland

☆ SYNONYMS
Wander, roam, ramble, rove, traipse, meander mean to go about from place to place usually without a plan or definite purpose. WANDER implies an absence of or an indifference to a fixed course ⟨fond of *wandering* about the square just watching the people⟩. ROAM suggests wandering about freely and often far afield ⟨liked to *roam* through the woods⟩. RAMBLE stresses carelessness and indifference to one's course or objective ⟨the speaker *rambled* on without ever coming to the point⟩. ROVE suggests vigorous and sometimes purposeful roaming ⟨armed brigands *roved* over the countryside⟩. TRAIPSE implies a course that is erratic but may sometimes be purposeful ⟨*traipsed* all over town looking for the right dress⟩. MEANDER implies a winding or intricate course suggestive of aimless or listless wandering ⟨the river *meanders* for miles through rich farmland⟩.

[1]**war** \'wȯr\ *noun, often attributive* [Middle English *werre,* from Old North French, of Germanic origin; akin to Old High German *werra* strife; akin to Old High German *werran* to confuse] (12th century)
1 a (1) **:** a state of usually open and declared armed hostile conflict between states or nations (2) **:** a period of such armed conflict (3) **:** STATE OF WAR **b :** the art or science of warfare **c** (1) *obsolete* **:** weapons and equipment for war (2) *archaic* **:** soldiers armed and equipped for war
2 a : a state of hostility, conflict, or antagonism **b :** a struggle or competition between opposing forces or for a particular end ⟨a class *war*⟩ ⟨a *war* against disease⟩ **c :** VARIANCE, ODDS 3
— war·less \-ləs\ *adjective*
[2]**war** *intransitive verb* **warred; war·ring** (13th century)
1 : to be in active or vigorous conflict
2 : to engage in warfare
[3]**war** \'wär\ *adverb or adjective* [Middle English *werre,* from Old Norse *verri,* adjective, *verr,* adverb; akin to Old English *wiersa* worse — more at WORSE] (13th century)
chiefly Scottish **:** WORSE
[4]**war** \'wär\ *transitive verb* **warred; war·ring** (15th century)
Scottish **:** WORST, OVERCOME
war baby *noun* (1901)
: a person born or conceived during a war
[1]**war·ble** \'wȯr-bəl\ *noun* [Middle English *werble* tune, from Old North French, of Germanic origin; akin to Middle High German *wirbel* whirl, tuning peg, Old High German *wirbil* whirlwind — more at WHIRL] (14th century)
1 : a melodious succession of low pleasing sounds
2 : a musical trill
3 : the action of warbling
[2]**warble** *verb* **war·bled; war·bling** \-b(ə-)liŋ\ (15th century)
intransitive verb
1 : to sing in a trilling manner or with many turns and variations
2 : to become sounded with trills, quavers, and rapid modulations in pitch
3 : SING
transitive verb
: to render with turns, runs, or rapid modulations **:** TRILL
[3]**warble** *noun* [perhaps of Scandinavian origin; akin to obsolete Swedish *varbulde* boil, from *var* pus + *bulde* swelling] (circa 1585)
1 : a swelling under the hide especially of the back of cattle, horses, and wild mammals caused by the maggot of a botfly or warble fly
2 : the maggot of a warble fly
— war·bled \-bəld\ *adjective*
warble fly *noun* (1877)
: any of various dipteran flies (family Oestridae) whose larvae live under the skin of various mammals and cause warbles
war·bler \'wȯr-blər\ *noun* (1611)
1 : one that warbles **:** SINGER, SONGSTER
2 a : any of numerous small Old World oscine birds (family Sylviidae) many of which are noted songsters and are closely related to the thrushes **b :** any of numerous small brightly colored American oscine birds (family Parulidae) with a usually weak and unmusical song — called also *wood warbler*
war·bon·net \'wȯr-ˌbä-nət\ *noun* (1845)
: an American Indian ceremonial headdress with a feathered extension down the back
war bride *noun* (1916)
1 : a woman who marries a serviceman ordered into active service in time of war
2 : a woman who marries a serviceman especially of a foreign nation met during a time of war
war chest *noun* (1901)
: a fund accumulated to finance a war; *broadly* **:** a fund earmarked for a specific purpose, action, or campaign
war club *noun* (1776)
: a club-shaped implement used as a weapon especially by American Indians
war crime *noun* (1906)
: a crime (as genocide or maltreatment of prisoners) committed during or in connection with war — usually used in plural
— war criminal *noun*
war cry *noun* (1748)
1 : a cry used by a body of fighters in war
2 : a slogan used especially to rally people to a cause
[1]**ward** \'wȯrd\ *noun* [Middle English, from Old English *weard;* akin to Old High German *warta* act of watching, Old English *warian* to beware of, guard, *wær* careful — more at WARY] (before 12th century)

warbonnet

1 a : the action or process of guarding **b :** a body of guards
2 : the state of being under guard; *especially* **:** CUSTODY
3 a : the inner court of a castle or fortress **b :** a division (as a cell or block) of a prison **c :** a division in a hospital; *especially* **:** a large room in a hospital where a number of patients often requiring similar treatment are accommodated
4 a : a division of a city for representative, electoral, or administrative purposes **b :** a division of some English and Scottish counties corresponding to a hundred **c :** the Mormon local congregation having auxiliary organizations (as Sunday schools and relief societies) and one or more quorums of each office of the Aaronic priesthood
5 : a projecting ridge of metal in a lock casing or keyhole permitting only the insertion of a key with a corresponding notch; *also* **:** a corresponding notch in a bit of a key
6 : a person or thing under guard, protection, or surveillance: as **a :** a minor subject to wardship **b :** a person who by reason of incapacity (as minority or lunacy) is under the protection of a court either directly or through a guardian appointed by the court — called also *ward of court* **c :** a person or body of persons under the protection or tutelage of a government
7 : a means of defense **:** PROTECTION
— ward·ed \'wȯr-dəd\ *adjective*
[2]**ward** *transitive verb* [Middle English, from Old English *weardian;* akin to Old High German *wartēn* to watch, Old Norse *vartha* to guard, Old English *weard* ward] (before 12th century)
1 : to keep watch over **:** GUARD
2 : to turn aside (something threatening) **:** DEFLECT — usually used with *off*
[1]**-ward** *also* **-wards** *adjective suffix* [*-ward* from Middle English, from Old English *-weard;* akin to Old High German *-wart, -wert* -ward, Latin *vertere* to turn; *-wards* from *-wards,* adverb suffix — more at WORTH]
1 : that moves, tends, faces, or is directed toward ⟨river*ward*⟩
2 : that occurs or is situated in the direction of ⟨left*ward*⟩
[2]**-ward** *or* **-wards** *adverb suffix* [*-ward* from Middle English, from Old English *-weard,* from *-weard,* adjective suffix; *-wards* from Middle English, from Old English *-weardes,* genitive singular neuter of *-weard,* adjective suffix]
1 : in a (specified) spatial or temporal direction ⟨up*ward*⟩ ⟨after*ward*⟩
2 : toward a (specified) point, position, or area ⟨earth*ward*⟩
war dance *noun* (1711)
: a dance performed (as by American Indians) in preparation for battle or in celebration of victory
ward·ed \'wȯr-dəd\ *adjective* (1572)
: provided with a ward ⟨a *warded* lock⟩
war·den \'wȯr-dᵊn\ *noun* [Middle English *wardein,* from Old North French, from *warder* to guard, of Germanic origin; akin to Old High German *wartēn* to watch] (13th century)
1 : one having care or charge of something **:** GUARDIAN, KEEPER
2 a : REGENT 2 **b :** the governor of a town, district, or fortress **c :** a member of the governing body of a guild
3 a : an official charged with special supervisory duties or with the enforcement of specified laws or regulations ⟨game *warden*⟩ ⟨air raid *warden*⟩ **b :** an official in charge of the operation of a prison **c :** any of various British officials having designated administrative functions ⟨*warden* of the mint⟩
4 a : one of two ranking lay officers of an Episcopal parish **b :** any of various British college officials whose duties range from the administration of academic matters to the supervision of student discipline
war·den·ship \-ˌship\ *noun* (15th century)
: the office, jurisdiction, or powers of a warden
[1]**ward·er** \'wȯr-dər\ *noun* [Middle English, from Anglo-French *wardere,* from *warde* act of guarding, of Germanic origin; akin to Old High German *warta* act of watching] (15th century)
1 : WATCHMAN, PORTER
2 *British* **a :** WARDEN **b :** a prison guard
[2]**war·der** *noun* [Middle English, staff, perhaps from *warden* to ward] (circa 1548)
: a truncheon used by a king or commander in chief to signal orders
ward heeler *noun* (1888)
: a worker for a political boss in a ward or other local area
ward·ress \'wȯr-drəs\ *noun* (1878)
: a woman supervising female prisoners (as in a prison)
ward·robe \'wȯr-ˌdrōb\ *noun* [Middle English *warderobe,* from Old North French, from *warder* to guard + *robe* robe] (14th century)
1 a : a room or closet where clothes are kept **b :** CLOTHESPRESS **c :** a large trunk in which clothes may be hung upright
2 a : a collection of wearing apparel (as of one person or for one activity) ⟨a summer *wardrobe*⟩ **b :** a collection of stage costumes and accessories
3 : the department of a royal or noble household entrusted with the care of wearing apparel, jewels, and personal articles
ward·room \'wȯr-ˌdrüm, -ˌdrům\ *noun* (1758)
: the space in a warship allotted for living quarters to the commissioned officers excepting the captain; *specifically* **:** the mess assigned to these officers
ward·ship \'wȯrd-ˌship\ *noun* (15th century)
1 a : care and protection of a ward **b :** the right to the custody of an infant heir of a feudal tenant and of the heir's property
2 : the state of being under a guardian
[1]**ware** \'war, 'wer\ *adjective* [Middle English *war, ware* careful, aware, from Old English *wær* — more at WARY] (before 12th century)
1 : AWARE, CONSCIOUS ⟨was *ware* of black looks cast at me —Mary Webb⟩
2 *archaic* **:** WARY, VIGILANT
[2]**ware** *transitive verb* **wared; war·ing** [Middle English, from Old English *warian;* akin to Old High German *biwarōn* to protect, Old English *wær* aware] (before 12th century)
: to beware of **:** AVOID — used chiefly as a command to hunting animals

[3]**ware** *noun* [Middle English, from Old English *waru;* akin to Middle High German *ware* ware and probably to Sanskrit *vasna* price — more at VENAL] (before 12th century)
1 a : manufactured articles, products of art or craft, or farm produce **:** GOODS — often used in combination ⟨tin*ware*⟩ **b :** an article of merchandise
2 : articles (as pottery or dishes) of fired clay ⟨earthen*ware*⟩
3 : an intangible item (as a service or ability) that is a marketable commodity

[4]**ware** *transitive verb* **wared; war·ing** [Middle English, from Old Norse *verja* to clothe, invest, spend — more at WEAR] (14th century)
Scottish **:** SPEND, EXPEND

[1]**ware·house** \ˈwar-ˌhau̇s, ˈwer-\ *noun* (14th century)
: a structure or room for the storage of merchandise or commodities

[2]**ware·house** \-ˌhau̇z, -ˌhau̇s\ *transitive verb* (1799)
1 : to deposit, store, or stock in or as if in a warehouse
2 : to confine or house (a person) in conditions suggestive of a warehouse

ware·house·man \-ˌhau̇s-mən\ *noun* (1635)
: a person who manages or works in a warehouse

ware·hous·er \-ˌhau̇-zər, -sər\ *noun* (circa 1927)
: WAREHOUSEMAN

ware·room \ˈwar-ˌrüm, ˈwer-, -ˌru̇m\ *noun* (1811)
: a room in which goods are exhibited for sale

war·fare \ˈwȯr-ˌfar, -ˌfer\ *noun* [Middle English, from *werre, warre* war + *fare* journey, passage — more at FARE] (15th century)
1 : military operations between enemies **:** HOSTILITIES, WAR; *also* **:** an activity undertaken by a political unit (as a nation) to weaken or destroy another ⟨economic *warfare*⟩
2 : struggle between competing entities **:** CONFLICT

war·fa·rin \ˈwȯr-fə-rən\ *noun* [*W*isconsin *A*lumni *R*esearch *F*oundation (its patentee) + coum*arin*] (1950)
: a crystalline anticoagulant compound $C_{19}H_{16}O_4$ used as a rodent poison and in medicine

war footing *noun* (1847)
: the condition of being prepared to undertake or maintain war

war–game (1942)
transitive verb
: to plan or conduct in the manner of a war game ⟨*war-gamed* an invasion —*Newsweek*⟩
intransitive verb
: to conduct a war game

war game *noun* (1828)
1 : a simulated battle or campaign to test military concepts and usually conducted in conferences by officers acting as the opposing staffs
2 : a two-sided umpired training maneuver with actual elements of the armed forces participating

war hawk *noun* (1798)
: one who clamors for war; *especially* **:** an American jingo favoring war with Britain around 1812

war·head \ˈwȯr-ˌhed\ *noun* (1898)
: the section of a missile containing the explosive, chemical, or incendiary charge

war·horse \-ˌhȯrs\ *noun* (1653)
1 : a horse used in war **:** CHARGER
2 : a person with long experience in a field; *especially* **:** a veteran soldier or public person (as a politician)
3 : something (as a work of art or musical composition) that has become overly familiar or hackneyed due to much repetition in the standard repertoire

war·i·son \ˈwar-ə-sən\ *noun* [probably a misunderstanding by Sir Walter Scott in the *Lay of the Last Minstrel* (1805) of Middle English *waryson* reward, from Old North French *warison* defense, possessions, from *warir* to protect, provide, of Germanic origin; akin to Old High German *werien* to defend — more at WEIR] (1805)
: a bugle call to attack

war·like \ˈwȯr-ˌlīk\ *adjective* (15th century)
1 *obsolete* **:** ready for war **:** equipped to fight
2 : fit for, disposed to, or fond of war **:** BELLICOSE
3 : of, relating to, or useful in war
4 : befitting or characteristic of war or a soldier

war·lock \-ˌläk\ *noun* [Middle English *warloghe,* from Old English *wǣrloga* one that breaks faith, the Devil, from *wǣr* faith, troth + *-loga* (from *lēogan* to lie); akin to Old English *wǣr* true — more at VERY, LIE] (14th century)
1 : a man practicing the black arts **:** SORCERER — compare WITCH
2 : CONJURER

war·lord \-ˌlȯrd\ *noun* (1856)
1 : a supreme military leader
2 : a military commander exercising civil power by force usually in a limited area
— war·lord·ism \-ˌlȯr-ˌdi-zəm\ *noun*

[1]**warm** \ˈwȯrm\ *adjective* [Middle English, from Old English *wearm;* akin to Old High German *warm* warm and probably to Lithuanian *virti* to cook, boil] (before 12th century)
1 a : having or giving out heat to a moderate or adequate degree **b :** serving to maintain or preserve heat especially to a satisfactory degree ⟨a *warm* sweater⟩ **c :** feeling or causing sensations of heat brought about by strenuous exertion
2 : comfortably established **:** SECURE
3 a : marked by strong feeling **:** ARDENT **b :** marked by excitement, disagreement, or anger ⟨the argument grew *warm*⟩
4 : marked by or readily showing affection, gratitude, cordiality, or sympathy ⟨a *warm* welcome⟩ ⟨*warm* regards⟩
5 : emphasizing or exploiting sexual imagery or incidents
6 : accompanied or marked by extreme danger or duress
7 : newly made **:** FRESH ⟨a *warm* scent⟩
8 : having the color or tone of something that imparts heat; *specifically* **:** of a hue in the range yellow through orange to red
9 : near to a goal, object, or solution sought
— warm·ish \ˈwȯr-mish\ *adjective*
— warm·ness \ˈwȯrm-nəs\ *noun*

[2]**warm** (before 12th century)
transitive verb
1 : to make warm
2 a : to infuse with a feeling of love, friendship, well-being, or pleasure **b :** to fill with anger, zeal, or passion
3 : to reheat (cooked food) for eating — often used with *over*
4 : to make ready for operation or performance by preliminary exercise or operation — often used with *up*
intransitive verb
1 : to become warm
2 a : to become ardent, interested, or receptive — usually used with *to* or *toward* ⟨*warmed* to the idea⟩ **b :** to become filled with affection or love — used with *to* or *toward*
3 : to experience feelings of pleasure **:** BASK
4 : to become ready for operation or performance by preliminary activity — often used with *up*

[3]**warm** *adverb* (before 12th century)
: WARMLY — usually used in combination ⟨*warm*-clad⟩

warm–blood·ed \ˈwȯrm-ˈblə-dəd\ *adjective* (1793)
1 : having warm blood; *specifically* **:** having a relatively high and constant body temperature relatively independent of the surroundings
2 : fervent or ardent in spirit
— warm–blood·ed·ness *noun*

warmed–over \ˈwȯrmd-ˈō-vər\ *adjective* (1887)
1 : not fresh or new **:** STALE ⟨*warmed-over* ideas⟩
2 : heated again ⟨*warmed-over* beans⟩

warm·er \ˈwȯr-mər\ *noun* (circa 1595)
: one that warms; *especially* **:** a device for keeping something warm ⟨a hand *warmer*⟩

warm front *noun* (circa 1921)
: an advancing edge of a warm air mass

warm·heart·ed \ˈwȯrm-ˈhär-təd\ *adjective* (circa 1520)
: marked by ready affection, cordiality, generosity, or sympathy
— warm·heart·ed·ness *noun*

warming pan *noun* (15th century)
: a long-handled covered pan filled with live coals that is used to warm a bed

warm·ly \ˈwȯrm-lē\ *adverb* (1529)
1 : in a manner characterized or accompanied by warmth of emotion
2 : in a manner that causes or maintains warmth

war·mon·ger \ˈwȯr-ˌməŋ-gər, -ˌmäŋ-\ *noun* (1590)
: one who urges or attempts to stir up war **:** JINGO
— war·mon·ger·ing \-g(ə-)riŋ\ *noun*

war·mouth \ˈwȯr-ˌmau̇th\ *noun* [origin unknown] (circa 1883)
: a freshwater sunfish (*Lepomis gulosus*) chiefly of the eastern U.S. — called also *warmouth bass*

warm spot *noun* (1951)
: a lasting affection for a particular person or object

warmth \ˈwȯrm(p)th\ *noun* (13th century)
1 : the quality or state of being warm in temperature
2 : the quality or state of being warm in feeling ⟨a child needing human *warmth* and family life⟩
3 : a glowing effect produced by the use of warm colors

warm–up \ˈwȯrm-ˌəp\ *noun* (1915)
1 : the act or an instance of warming up; *also* **:** a preparatory activity or procedure
2 : a suit for exercise or casual wear consisting of a jacket or sweatshirt and pants — often used in plural; called also *warm-up suit*

warm up *intransitive verb* (1846)
: to engage in exercise or practice especially before entering a game or contest; *broadly* **:** to get ready

warn \ˈwȯrn\ *verb* [Middle English, from Old English *warnian;* akin to Old High German *warnōn* to take heed, Old English *wær* careful, aware — more at WARY] (before 12th century)
transitive verb
1 a : to give notice to beforehand especially of danger or evil **b :** to give admonishing advice to **:** COUNSEL **c :** to call to one's attention **:** INFORM
2 : to order to go or stay away — often used with *off*
intransitive verb
: to give a warning
— warn·er *noun*

[1]**warn·ing** \ˈwȯr-niŋ\ *noun* (before 12th century)
1 : the act of warning **:** the state of being warned ⟨he had *warning* of his illness⟩
2 : something that warns or serves to warn; *especially* **:** a notice or bulletin that alerts the public that a tornado has been reported in the immediate vicinity or that the approach of a severe storm is imminent

²warning *adjective* (circa 1552)
: serving as an alarm, signal, summons, or admonition ⟨*warning* bell⟩ ⟨*warning* shot⟩
— warn·ing·ly \ˈwȯr-niŋ-lē\ *adverb*

warning coloration *noun* (circa 1928)
: conspicuous coloration possessed by an animal (as an insect) otherwise effectively but not obviously defended that serves to warn off potential enemies

warning track *noun* (1966)
: a usually dirt or cinder strip around the outside edge of a baseball outfield to warn a fielder when running to make a catch that the fence is near — called also *warning path*

war of nerves (1939)
: a conflict characterized by psychological tactics (as bluff, threats, and intimidation) designed primarily to create confusion, indecision, or breakdown of morale

¹warp \ˈwȯrp\ *noun* [Middle English, from Old English *wearp;* akin to Old High German *warf* warp, Old English *weorpan* to throw, Old Norse *verpa*] (before 12th century)
1 a : a series of yarns extended lengthwise in a loom and crossed by the weft **b :** FOUNDATION, BASE ⟨the *warp* of the economic structure is agriculture —*American Guide Series: North Carolina*⟩
2 : a rope for warping or mooring a ship or boat
3 [²*warp*] **a :** a twist or curve that has developed in something originally flat or straight ⟨a *warp* in a door panel⟩ **b :** a mental twist or aberration
— warp·age \ˈwȯr-pij\ *noun*

²warp (14th century)
transitive verb
1 : to arrange (yarns) so as to form a warp
2 a : to turn or twist out of or as if out of shape; *especially* **:** to twist or bend out of a plane **b :** to cause to judge, choose, or act wrongly or abnormally **:** PERVERT **c :** DISTORT 1 ⟨intellect and learning . . . *warped* by prejudices —Irving Wallace⟩ **d :** to deflect from a course
3 : to move (as a ship) by hauling on a line attached to a fixed object
intransitive verb
1 : to become warped
2 : to move a ship by warping
synonym see DEFORM
— warp·er *noun*

war paint *noun* (1826)
1 : paint put on parts of the body (as the face) by American Indians as a sign of going to war
2 : MAKEUP 3a

warp and woof *noun* (1842)
: FOUNDATION, BASE ⟨the vigorous Anglo-Saxon base had become the *warp and woof* of English speech —H. R. Warfel⟩

war party *noun* (1755)
1 : a group of American Indians on the warpath
2 : a usually jingoistic political party advocating or upholding a war

war·path \ˈwȯr-ˌpath, -ˌpȧth\ *noun* (1755)
1 : the route taken by a party of American Indians going on a warlike expedition or to a war
2 : a hostile course of action or frame of mind

warp beam *noun* (circa 1833)
: a roll on which warp is wound for a loom

warp knit *noun* (1946)
: a knit fabric produced by machine with the yarns running in a lengthwise direction — compare WEFT KNIT
— warp–knit·ted \-ˈni-təd\ *adjective*
— warp knitting *noun*

war·plane \ˈwȯr-ˌplān\ *noun* (circa 1911)
: a military airplane; *specifically* **:** one armed for combat

war power *noun* (1766)
: the power to make war; *specifically* **:** an extraordinary power exercised usually by the executive branch of a government in the prosecution of a war

¹war·rant \ˈwȯr-ənt, ˈwär-\ *noun* [Middle English, protector, warrant, from Old North French *warant,* modification of a Germanic noun represented by Old High German *werēnto* guarantor, from present participle of *werēn* to warrant; akin to Old High German *wāra* trust, care — more at VERY] (14th century)
1 a (1) **:** SANCTION, AUTHORIZATION; *also* **:** evidence for or token of authorization (2) **:** GUARANTEE, SECURITY **b** (1) **:** GROUND, JUSTIFICATION (2) **:** CONFIRMATION, PROOF
2 a : a commission or document giving authority to do something; *especially* **:** a writing that authorizes a person to pay or deliver to another and the other to receive money or other consideration **b :** a precept or writ issued by a competent magistrate authorizing an officer to make an arrest, a seizure, or a search or to do other acts incident to the administration of justice **c :** an official certificate of appointment issued to an officer of lower rank than a commissioned officer **d** (1) **:** a short-term obligation of a governmental body (as a municipality) issued in anticipation of revenue (2) **:** an instrument issued by a corporation giving to the holder the right to purchase the capital stock of the corporation at a stated price either prior to a stipulated date or at any future time
— war·rant·less \-ləs\ *adjective*

²warrant *transitive verb* [Middle English, from Old North French *warantir,* from *warant*] (14th century)
1 a : to declare or maintain with certainty **:** be sure that ⟨I'll *warrant* he'll be here by noon⟩ **b :** to assure (a person) of the truth of what is said
2 a : to guarantee to a person good title to and undisturbed possession of (as an estate) **b :** to provide a guarantee of the security of (as title to property sold) usually by an express covenant in the deed of conveyance **c :** to guarantee to be as represented **d :** to guarantee (as goods sold) especially in respect of the quality or quantity specified
3 : to guarantee security or immunity to **:** SECURE ⟨I'll *warrant* him from drowning —Shakespeare⟩
4 : to give warrant or sanction to **:** AUTHORIZE ⟨the law *warrants* this procedure⟩
5 a : to give proof of the authenticity or truth of **b :** to give assurance of the nature of or for the undertaking of **:** GUARANTEE
6 : to serve as or give adequate ground or reason for ⟨promising enough to *warrant* further consideration⟩

war·rant·able \ˈwȯr-ən-tə-bəl, ˈwär-\ *adjective* (1581)
: capable of being warranted **:** JUSTIFIABLE
— war·rant·able·ness *noun*
— war·rant·ably \-blē\ *adverb*

war·ran·tee \ˌwȯr-ən-ˈtē, ˌwär-\ *noun* (1706)
: the person to whom a warranty is made

warrant officer *noun* (1693)
1 : an officer in the armed forces holding rank by virtue of a warrant and ranking above a noncommissioned officer and below a commissioned officer
2 : a commissioned officer ranking below an ensign in the navy or coast guard and below a second lieutenant in the marine corps

war·ran·tor \ˌwȯr-ən-ˈtȯr, ˌwär-; ˈwȯr-ən-tər, ˈwär-\ *also* **war·rant·er** \ˈwȯr-ən-tər, ˈwär-\ *noun* (1583)
: one that warrants or gives a warranty

war·ran·ty \ˈwȯr-ən-tē, ˈwär-\ *noun, plural* **-ties** [Middle English *warantie,* from Old North French, from *warantir* to warrant] (14th century)
1 a : a real covenant binding the grantor of an estate and his heirs to warrant and defend the title **b :** a collateral undertaking that a fact regarding the subject of a contract is or will be as it is expressly or by implication declared or promised to be
2 : something that authorizes, sanctions, supports, or justifies **:** WARRANT
3 : a usually written guarantee of the integrity of a product and of the maker's responsibility for the repair or replacement of defective parts

warranty deed *noun* (1779)
: a deed warranting that the grantor has a good title free and clear of all liens and encumbrances and will defend the grantee against all claims

war·ren \ˈwȯr-ən, ˈwär-\ *noun* [Middle English *warenne,* from Old North French, probably of Germanic origin; akin to Old High German *werien* to defend, protect — more at WEIR] (15th century)
1 *chiefly British* **a :** a place legally authorized for keeping small game (as hare or pheasant) **b :** the privilege of hunting game in such a warren
2 a (1) **:** an area (as of uncultivated ground) where rabbits breed (2) **:** a structure where rabbits are kept or bred **b :** the rabbits of a warren
3 a : a crowded tenement or district **b :** a maze of passageways or cubbies

war·ren·er \-ə-nər\ *noun* (13th century)
1 : GAMEKEEPER
2 : a person who maintains a rabbit warren

war·rior \ˈwȯr-yər, ˈwȯr-ē-ər, ˈwär-ē- *also* ˈwär-yər\ *noun, often attributive* [Middle English *werriour,* from Old North French *werreieur,* from *werreier* to make war, from *werre* war — more at WAR] (14th century)
: a man engaged or experienced in warfare; *broadly* **:** a person engaged in some struggle or conflict ⟨poverty *warriors*⟩

war room *noun* (1914)
1 : a room at a military headquarters where maps showing the current status of troops in battle are maintained
2 : a room (as at a business headquarters) used for conferences and planning that is often specially equipped (as with computers, maps, or charts)

war·saw grouper \ˈwȯr-(ˌ)sȯ-\ *noun* [*warsaw* modification of American Spanish *guasa*] (1949)
: any of several large groupers (especially *Epinephelus nigritus*) — called also *warsaw*

war·ship \ˈwȯr-ˌship\ *noun* (1533)
: a naval vessel

war·sle *or* **wars·tle** \ˈwä(r)-səl\ *verb* [Middle English *werstelen, warstelen,* alteration of *wrestlen, wrastlen*] (14th century)
Scottish **:** WRESTLE, STRUGGLE
— warsle *noun, Scottish*

war story *noun* (1981)
: a story of a memorable personal experience typically involving an element of danger, hardship, or adventure ⟨politicians swapping *war stories* from past campaigns⟩

wart \ˈwȯrt\ *noun* [Middle English, from Old English *wearte;* akin to Old High German *warza* wart, Old Church Slavonic *vrĕdŭ* injury] (before 12th century)
1 a : a horny projection on the skin usually of the extremities that is caused by a virus — called also *verruca vulgaris* **b :** any of numerous similar skin lesions
2 : an excrescence or protuberance resembling a true wart; *especially* **:** a glandular excrescence or hardened protuberance on a plant
3 a : one that suggests a wart especially in smallness, unpleasantness, or unattractiveness **b :** DEFECT, IMPERFECTION — often used in the phrase *warts and all*
— wart·ed \ˈwȯr-təd\ *adjective*
— wart·less \ˈwȯrt-ləs\ *adjective*
— warty \ˈwȯr-tē\ *adjective*

wart·hog \ˈwȯrt-ˌhȯg, -ˌhäg\ *noun* (1840)
: a wild African hog (*Phacochoerus aethiopicus*) which has large protruding tusks and the males of which have two pairs of rough warty excrescences on the face

warthog

war·time \ˈwȯr-ˌtīm\ *noun, often attributive* (14th century)
: a period during which a war is in progress

warts–and–all *adjective* (1930)
: showing defects or imperfections frankly **:** not idealized ⟨a *warts-and-all* biography⟩

war whoop *noun* (1739)
: a war cry especially of American Indians

wary \ˈwar-ē, ˈwer-\ *adjective* **war·i·er; -est** [[1]*ware,* from Middle English *war, ware,* from Old English *wær* careful, aware, wary; akin to Old High German gi*war* aware, attentive, Latin *vereri* to fear, Greek *horan* to see] (15th century)
: marked by keen caution, cunning, and watchful prudence especially in detecting and escaping danger
synonym see CAUTIOUS
— war·i·ly \ˈwar-ə-lē, ˈwer-\ *adverb*
— war·i·ness \ˈwar-ē-nəs, ˈwer-\ *noun*

war zone *noun* (1914)
1 : a zone in which belligerents are waging war
2 : a designated area especially on the high seas within which rights of neutrals are not respected by a belligerent nation in time of war

was [Middle English, from Old English, 1st & 3d singular past indicative of *wesan* to be; akin to Old Norse *vera* to be, *var* was, Sanskrit *vasati* he lives, dwells] *past 1st & 3d singular of* BE

wa·sa·bi \ˈwä-sə-bē\ *noun* [Japanese] (1903)
1 : a condiment that is prepared from the thick pungent greenish root of an Asian herb (*Eutrema wasabi*) of the mustard family and is similar in flavor and use to horseradish; *also* **:** the root
2 : the herb that yields wasabi

[1]**wash** \ˈwȯsh, ˈwäsh, *chiefly Midland also* ˈwȯrsh *or* ˈwärsh\ *verb* [Middle English, from Old English *wascan;* akin to Old High German *waskan* to wash and perhaps to Old English *wæter* water] (before 12th century)
transitive verb
1 a : to cleanse by or as if by the action of liquid (as water) **b :** to remove (as dirt) by rubbing or drenching with liquid
2 : to cleanse (fur) by licking or by rubbing with a paw moistened with saliva
3 a : to flush or moisten (a bodily part or injury) with a liquid **b** (1) **:** to wet thoroughly **:** DRENCH (2) **:** to overspread with light **:** SUFFUSE **c :** to pass a liquid (as water) over or through especially so as to carry off material from the surface or interior
4 : to flow along or dash or overflow against ⟨waves *washing* the shore⟩
5 : to move, carry, or deposit by or as if by the force of water in motion ⟨houses *washed* away by the flood⟩
6 a : to subject (as crushed ore) to the action of water to separate valuable material **b :** to separate (particles) from a substance (as ore) by agitation with or in water **c** (1) **:** to pass through a bath to carry off impurities or soluble components (2) **:** to pass (a gas or gaseous mixture) through or over a liquid to purify it especially by removing soluble components
7 a : to cover or daub lightly with or as if with an application of a thin liquid (as whitewash or varnish) **b :** to depict or paint by a broad sweep of thin color with a brush
8 : to cause to swirl ⟨*washing* coffee around in his cup⟩
9 : LAUNDER 3 ⟨how the mob *washes* its money through corrupt bankers —Vincent Teresa⟩
intransitive verb
1 : to wash oneself or a part of one's body
2 : to become worn away by the action of water
3 : to clean something by rubbing or dipping in water
4 a : to become carried along on water **:** DRIFT ⟨cakes of ice *washing* along⟩ **b :** to pour, sweep, or flow in a stream or current ⟨waves of pioneers *washing* westward —Green Peyton⟩
5 : to serve as a cleansing agent ⟨this soap *washes* thoroughly⟩
6 a : to undergo laundering ⟨this dress doesn't *wash* well⟩ **b** (1) **:** to undergo testing successfully **:** WORK 4 ⟨an interesting theory, but it just won't *wash*⟩ (2) **:** to gain acceptance **:** inspire belief ⟨the story didn't *wash* with me⟩
— wash one's hands of : to disclaim interest in, responsibility for, or further connection with

[2]**wash** *noun* (15th century)
1 a : a piece of ground washed by the sea or river **b :** BOG, MARSH **c** (1) **:** a shallow body of water (2) **:** a shallow creek **d** *West* **:** the dry bed of a stream — called also *dry wash*
2 a : the act or process or an instance of washing or being washed **b :** articles to be washed, being washed, or having been washed
3 : the surging action of waves
4 a : worthless especially liquid waste **:** REFUSE **b :** an insipid beverage **c :** vapid writing or speech
5 a : a sweep or splash especially of color made by or as if by a long stroke of a brush **b :** a thin coat of paint (as watercolor) **c :** a thin liquid used for coating a surface (as a wall)
6 : LOTION
7 : loose or eroded surface material of the earth (as rock debris) transported and deposited by running water
8 a : BACKWASH 1 **b :** a disturbance in the air produced by the passage of an airfoil or propeller
9 : a situation in which losses and gains or advantages and disadvantages balance each other

[3]**wash** *adjective* (1634)
1 : WASHABLE ⟨*wash* fabric⟩
2 : involving essentially simultaneous purchase and sale of the same security ⟨spurious market activity resulting from *wash* trading⟩

wash·able \ˈwȯ-shə-bəl, ˈwä-\ *adjective* (1821)
: capable of being washed without damage
— wash·abil·i·ty \ˌwȯ-shə-ˈbi-lət-ē, ˌwä-\ *noun*

wash–and–wear *adjective* (1956)
: of, relating to, or constituting a fabric or garment that needs little or no ironing after washing

wash·a·te·ria *also* **wash·e·te·ria** \ˌwä-shə-ˈtir-ē-ə, ˌwȯ-\ *noun* [[2]*wash* + *-ateria* or *-eteria* (as in *cafeteria*)] (1937)
chiefly Southern **:** a self-service laundry

wash·ba·sin \ˈwȯsh-ˌbā-sᵊn, ˈwäsh-\ *noun* (1812)
: WASHBOWL

wash·board \ˈwȯsh-ˌbōrd, ˈwäsh-, -ˌbȯrd\ *noun* (1742)
1 : a broad thin plank along a gunwale or on the sill of a lower deck port to keep out the sea
2 : BASEBOARD
3 a : a corrugated rectangular surface that is used for scrubbing clothes or as a percussion instrument **b :** a road or pavement so worn by traffic as to be corrugated transversely

wash·bowl \-ˌbōl\ *noun* (1816)
: a large bowl for water that is used to wash one's hands and face

wash·cloth \-ˌklȯth\ *noun* (circa 1900)
: a cloth that is used for washing one's face and body — called also *facecloth, washrag*

wash down *transitive verb* (1600)
1 : to move or carry downward by action of a liquid; *especially* **:** to facilitate the passage of (food) down the gullet with accompanying swallows of liquid ⟨pizza *washed down* with beer⟩
2 : to wash the whole length or extent of ⟨*washed down* and scrubbed the front porch⟩

wash drawing *noun* (1889)
: watercolor painting in or chiefly in washes especially in black, white, and gray tones only

washed–out \ˈwȯsht-ˈau̇t, ˈwäsht-\ *adjective* (1796)
1 : faded in color
2 : depleted in vigor or animation **:** EXHAUSTED

washed–up \ˈwȯsht-ˈəp, ˈwäsht-\ *adjective* (1928)
: no longer successful, skillful, popular, or needed

wash·er \ˈwȯ-shər, ˈwä-\ *noun* (13th century)
1 : a flat thin ring or a perforated plate used in joints or assemblies to ensure tightness, prevent leakage, or relieve friction
2 : one that washes; *especially* **:** WASHING MACHINE

wash·er·man \-mən\ *noun* (1715)
: LAUNDRYMAN; *also* **:** a man operating any of various industrial washing machines

wash·er·wom·an \-ˌwu̇-mən\ *noun* (1632)
: a woman whose occupation is washing clothes **:** LAUNDRESS

wash·house \ˈwȯsh-ˌhau̇s, ˈwäsh-\ *noun* (1577)
: a building used or equipped for washing; *especially* **:** one for washing clothes

wash·ing \ˈwȯ-shiŋ, ˈwä-\ *noun* (13th century)
1 : the act or action of one that cleanses with water
2 : material obtained by washing
3 : articles washed or to be washed **:** WASH

washing machine *noun* (circa 1754)
: a machine for washing; *especially* **:** one for washing clothes and household linen

washing soda *noun* (1865)
: a transparent crystalline hydrated sodium carbonate — called also *sal soda*

Wash·ing·ton pie \ˈwȯ-shiŋ-tən-, ˈwä-, *chiefly Midland also* ˈwȯr-shiŋ- *or* ˈwär-shiŋ-\ *noun* [George *Washington*] (1905)
: cake layers put together with a jam or jelly filling

Washington's Birthday *noun* [George *Washington*] (1829)
1 : February 22 formerly observed as a legal holiday in most of the states of the U.S.
2 : the third Monday in February observed as a legal holiday in most of the states of the U.S. — called also *Presidents' Day*

wash·out \ˈwȯsh-ˌau̇t, ˈwäsh-\ *noun* (1873)
1 a : the washing out or away of something and especially of earth in a roadbed by a freshet **b :** a place where earth is washed away
2 : one that fails to measure up **:** FAILURE: as **a :** one who fails in a course of training or study **b :** an unsuccessful enterprise or undertaking

wash out (1555)
transitive verb
1 : to wash free of an extraneous substance (as dirt)
2 a : to cause to fade by or as if by laundering **b :** to deplete the strength or vitality of **c :** to eliminate as useless or unsatisfactory **:** REJECT
3 a : to destroy or make useless by the force or action of water ⟨the storm *washed out* the bridge⟩ **b :** RAIN OUT ⟨the game was *washed out*⟩
intransitive verb
1 : to become depleted of color or vitality **:** FADE
2 : to fail to meet requirements or measure up to a standard

wash·rag \'wȯsh-ˌrag, 'wäsh-\ *noun* (1890)
: WASHCLOTH

wash·room \-ˌrüm, -ˌru̇m\ *noun* (1806)
: a room that is equipped with washing and toilet facilities : LAVATORY

wash·stand \-ˌstand\ *noun* (1789)
1 : a stand holding articles needed for washing one's face and hands
2 : a washbowl permanently set in place and attached to water and drainpipes

wash·tub \-ˌtəb\ *noun* (1602)
: a tub in which clothes are washed or soaked

wash·up \-ˌəp\ *noun* (1884)
: the act or process of washing clean

wash up (1751)
transitive verb
1 : to get rid of by washing ⟨*wash up* the spilled milk⟩
2 : EXHAUST, FINISH
intransitive verb
1 : to wash one's face and hands
2 *British* **:** to wash the dishes after a meal
3 : to be deposited by or as if by a swell of waves ⟨seaweed *washed up* on the shore⟩

wash·wom·an \'wȯsh-ˌwu̇-mən, 'wäsh-\ *noun* (1590)
: WASHERWOMAN

washy \'wȯ-shē, 'wä-\ *adjective* **wash·i·er; -est** (1615)
1 a : WEAK, WATERY ⟨*washy* tea⟩ **b :** deficient in color : PALLID **c :** lacking in vigor, individuality, or definiteness
2 : lacking in condition and in firmness of flesh

wasn't \'wə-zᵊnt, 'wä-, *dialect also* 'wə-tᵊn(t)\ (circa 1849)
: was not

wasp \'wäsp, 'wȯsp\ *noun* [Middle English *waspe,* from Old English *wæps, wæsp;* akin to Old High German *wafsa* wasp, Latin *vespa* wasp] (before 12th century)
1 : any of numerous social or solitary winged hymenopterous insects (especially families Sphecidae and Vespidae) that usually have a slender smooth body with the abdomen attached by a narrow stalk, well-developed wings, biting mouthparts, and in the females and workers an often formidable sting, and that are largely carnivorous and often provision their nests with caterpillars, insects, or spiders killed or paralyzed by stinging for their larvae to feed on — compare BEE

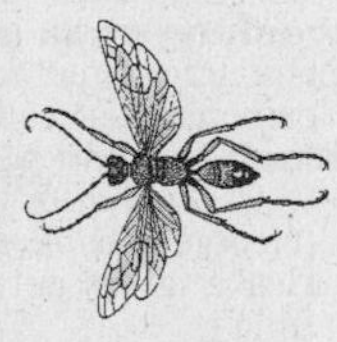
wasp 1

2 : any of various hymenopterous insects (as a chalcid or ichneumon fly) other than wasps with larvae that are parasitic on other arthropods
— wasp·like \-ˌlīk\ *adjective*

WASP *or* **Wasp** \'wäsp, 'wȯsp\ *noun, often attributive* [white Anglo-Saxon Protestant] (1957)
: an American of Northern European and especially British ancestry and of Protestant background; *especially* **:** a member of the dominant and the most privileged class of people in the U.S. — sometimes used disparagingly
— Wasp·dom \-dəm\ *noun*
— Wasp·ish \'wäs-pish, 'wȯs-\ *adjective*
— Wasp·ish·ness *noun*
— Waspy \-pē\ *adjective*

wasp·ish \'wäs-pish, 'wȯs-\ *adjective* (1566)
1 : resembling a wasp in behavior; *especially* : SNAPPISH, PETULANT
2 : resembling a wasp in form; *especially* : slightly built
— wasp·ish·ly *adverb*
— wasp·ish·ness *noun*

wasp waist *noun* (1870)
: a very slender waist
— wasp–waist·ed \'wäsp-'wā-stəd, 'wȯsp-\ *adjective*

¹was·sail \'wä-səl *also* wä-'sā(ə)l\ *noun* [Middle English *wæs hæil,* from Old Norse *ves heill* be well, from *ves* (imperative singular of *vera* to be) + *heill* healthy — more at WAS, WHOLE] (13th century)
1 : an early English toast to someone's health
2 : a hot drink that is made with wine, beer, or cider, spices, sugar, and usually baked apples and is traditionally served in a large bowl especially at Christmastime
3 : riotous drinking : REVELRY

²wassail (14th century)
intransitive verb
1 : to indulge in wassail : CAROUSE
2 *dialect English* **:** to sing carols from house to house at Christmas
transitive verb
: to drink to the health or thriving of

was·sail bowl \'wä-səl-\ *noun* (1606)
1 : a bowl that is used for the serving of wassail
2 : WASSAIL 2

was·sail·er \'wä-sə-lər *also* wä-'sā-lər\ *noun* (1634)
1 : one that carouses : REVELER
2 *archaic* **:** one who goes about singing carols

Was·ser·mann reaction \'wä-sər-mən-, 'vä-\ *noun* [August von *Wassermann*] (1911)
: the complement-fixing reaction that occurs in a positive complement-fixation test for syphilis using the serum of an infected individual

Wassermann test *noun* (1909)
: a test for the detection of syphilitic infection using the Wassermann reaction — called also *Wasserman*

wast \'wəst, 'wäst\ *archaic past 2d singular of* BE

wast·age \'wā-stij\ *noun* (1756)
: loss, decrease, or destruction of something (as by use, decay, erosion, or leakage); *especially* **:** wasteful or avoidable loss of something valuable

¹waste \'wāst\ *noun* [Middle English *waste, wast;* in sense 1, from Old North French *wast,* from *wast,* adjective, desolate, waste, from Latin *vastus;* in other senses, from Middle English *wasten* to waste — more at VAST] (13th century)
1 a : a sparsely settled or barren region : DESERT **b :** uncultivated land **c :** a broad and empty expanse (as of water)
2 : the act or an instance of wasting **:** the state of being wasted
3 a : loss through breaking down of bodily tissue **b :** gradual loss or decrease by use, wear, or decay
4 a : damaged, defective, or superfluous material produced by a manufacturing process: as (1) **:** material rejected during a textile manufacturing process and used usually for wiping away dirt and oil ⟨cotton *waste*⟩ (2) **:** SCRAP (3) **:** an unwanted by-product of a manufacturing process, chemical laboratory, or nuclear reactor ⟨toxic *waste*⟩ ⟨hazardous *waste*⟩ ⟨nuclear *waste*⟩ **b :** refuse from places of human or animal habitation: as (1) **:** GARBAGE, RUBBISH (2) **:** EXCREMENT — often used in plural (3) **:** SEWAGE **c :** material derived by mechanical and chemical weathering of the land and moved down sloping surfaces or carried by streams to the sea

²waste *verb* **wast·ed; wast·ing** [Middle English, from Old North French *waster,* from Latin *vastare,* from *vastus* desolate, waste] (13th century)
transitive verb
1 : to lay waste; *especially* **:** to damage or destroy gradually and progressively ⟨reclaiming land *wasted* by strip-mining⟩
2 : to cause to shrink in physical bulk or strength : EMACIATE, ENFEEBLE
3 : to wear away or diminish gradually : CONSUME
4 a : to spend or use carelessly : SQUANDER **b :** to allow to be used inefficiently or become dissipated
5 : KILL; *also* **:** to injure severely
intransitive verb
1 : to lose weight, strength, or vitality — often used with *away*
2 a : to become diminished in bulk or substance **b :** to become consumed
3 : to spend money or consume property extravagantly or improvidently
synonym see RAVAGE
— waste one's breath : to accomplish nothing by speaking

³waste *adjective* [Middle English *waste, wast,* from Old North French *wast*] (14th century)
1 a (1) **:** being wild and uninhabited : DESOLATE (2) **:** ARID, EMPTY **b :** not cultivated : not productive
2 : being in a ruined or devastated condition
3 [¹*waste*] **a :** discarded as worthless, defective, or of no use : REFUSE ⟨*waste* material⟩ **b :** excreted from or stored in inert form in a living body as a byproduct of vital activity ⟨*waste* disposal in birds⟩
4 [¹*waste*] **:** serving to conduct or hold refuse material; *specifically* **:** carrying off superfluous water
5 : WASTED 4

waste·bas·ket \'wās(t)-ˌbas-kət\ *noun* (1850)
: a receptacle for refuse and especially for wastepaper — called also *wastepaper basket*

wasted *adjective* (15th century)
1 : laid waste : RAVAGED
2 : impaired in strength or health : EMACIATED
3 *archaic* **:** gone by : ELAPSED ⟨the chronicle of *wasted* time —Shakespeare⟩
4 : unprofitably used, made, or expended ⟨*wasted* effort⟩
5 *slang* **:** intoxicated from drugs or alcohol

waste·ful \'wāst-fəl\ *adjective* (14th century)
: given to or marked by waste : LAVISH, PRODIGAL
— waste·ful·ly \-fə-lē\ *adverb*
— waste·ful·ness *noun*

waste·land \'wāst-ˌland *also* -lənd\ *noun* (14th century)
1 : barren or uncultivated land ⟨a desert *wasteland*⟩
2 : an ugly often devastated or barely inhabitable place or area
3 : something (as a way of life) that is spiritually and emotionally arid and unsatisfying

waste·pa·per \'wās(t)-'pā-pər\ *noun* (1585)
: paper discarded as used, superfluous, or not fit for use

waste pipe *noun* (circa 1512)
: a pipe for carrying off waste fluid

wast·er \'wā-stər\ *noun* (14th century)
1 a (1) **:** one that spends or consumes extravagantly and without thought for the future (2) **:** a dissolute person **b :** one that uses wastefully or causes or permits waste ⟨a procedure that is a *waster* of time⟩ **c :** one that lays waste : DESTROYER
2 : an imperfect or inferior manufactured article or object

waste·wa·ter \'wāst-ˌwȯ-tər, -ˌwä-\ *noun* (15th century)
: water that has been used (as in a manufacturing process) : SEWAGE

wast·ing \'wā-stiŋ\ *adjective* (13th century)
1 : laying waste : DEVASTATING
2 : undergoing or causing decay or loss of strength ⟨*wasting* diseases such as tuberculosis⟩

wast·rel \'wās-trəl *also* 'wäs-\ *noun* [irregular from ²*waste*] (circa 1841)
1 : VAGABOND, WAIF
2 : one who dissipates resources foolishly and self-indulgently : PROFLIGATE

¹watch \'wäch, 'wȯch\ *verb* [Middle English *wacchen,* from Old English *wæccan* — more at WAKE] (before 12th century)
intransitive verb

1 a : to keep vigil as a devotional exercise **b :** to be awake during the night
2 a : to be attentive or vigilant **b :** to keep guard
3 a : to keep someone or something under close observation **b :** to observe as a spectator ⟨the country *watched* as stocks fell sharply⟩
4 : to be expectant **:** WAIT ⟨*watch* for the signal⟩
transitive verb
1 : to keep under guard
2 a : to observe closely in order to check on action or change ⟨being *watched* by the police⟩ **b :** to look at **:** OBSERVE ⟨sat and *watched* the crowd⟩ **c :** to look on at ⟨*watch* television⟩ ⟨*watch* a ball game⟩
3 a : to take care of **:** TEND **b :** to be careful of ⟨*watches* his diet⟩
4 : to be on the alert for **:** BIDE ⟨*watched* her opportunity⟩
— watch it : look out **:** be careful ⟨*watch it* when you handle the glassware⟩
— watch one's step : to proceed with extreme care **:** act or talk warily
— watch over : to have charge of **:** SUPERINTEND

²watch *noun* (before 12th century)
1 a : the act of keeping awake to guard, protect, or attend **b** *obsolete* **:** the state of being wakeful **c :** a wake over a dead body **d :** a state of alert and continuous attention **e :** close observation **:** SURVEILLANCE **f :** a notice or bulletin that alerts the public to the possibility of severe weather conditions occurring in the near future ⟨winter storm *watch*⟩
2 a : any of the definite divisions of the night made by ancient peoples **b :** one of the indeterminate intervals marking the passage of night — usually used in plural ⟨the silent *watches* of the night⟩
3 a : LOOKOUT, WATCHMAN **b** *archaic* **:** the office or function of a sentinel or guard
4 a : a body of soldiers or sentinels making up a guard **b :** a watchman or body of watchmen formerly assigned to patrol the streets of a town at night, announce the hours, and act as police
5 a (1) **:** a portion of time during which a part of a ship's company is on duty (2) **:** the part of a ship's company required to be on duty during a particular watch (3) **:** a sailor's assigned duty period **b :** a period of duty **:** SHIFT **c :** a term as holder especially of an overseeing or managerial office ⟨the business grew on her *watch*⟩
6 : a portable timepiece designed to be worn (as on the wrist) or carried in the pocket — compare CLOCK

watch·able \'wä-chə-bəl, 'wȯ-\ *adjective* (1954)
: worth watching
— watchable *noun*

watch and ward *noun* (14th century)
1 : continuous unbroken vigilance and guard
2 : service as a watchman or sentinel required from a feudal tenant

watch·band \'wäch-ˌband, 'wȯch-\ *noun* (1924)
: the bracelet or strap of a wristwatch

watch cap *noun* (1886)
: a knitted close-fitting usually navy-blue cap worn especially by enlisted men in the U.S. navy in cold or stormy weather

watch·case \'wäch-ˌkās, 'wȯch-\ *noun* (1671)
: the outside metal covering of a watch

¹watch·dog \-ˌdȯg\ *noun* (1610)
1 : a dog kept to guard property
2 : one that guards against loss, waste, theft, or undesirable practices

²watchdog *transitive verb* (1902)
: to act as a watchdog for

watch·er \'wä-chər, 'wȯ-\ *noun* (1509)
: one that watches: as **a :** one that sits up or continues awake at night **b :** WATCHMAN **c** (1) **:** one that keeps watch beside a dead person (2) **:** one that attends a sick person at night **d :** a person who closely follows or observes someone or something ⟨a Supreme Court *watcher*⟩ — often used in combination ⟨celebrity-*watchers*⟩ **e :** a representative of a party or candidate who is stationed at the polls on an election day to watch the conduct of officials and voters

watch fire *noun* (1801)
: a fire lighted as a signal or for the use of a guard

watch·ful \'wäch-fəl, 'wȯch-\ *adjective* (1548)
1 *archaic* **a :** not able or accustomed to sleep or rest **:** WAKEFUL **b :** causing sleeplessness **c :** spent in wakefulness **:** SLEEPLESS
2 : carefully observant or attentive **:** being on the watch ☆
— watch·ful·ly \-fə-lē\ *adverb*
— watch·ful·ness *noun*

watch·mak·er \-ˌmā-kər\ *noun* (1630)
: one that makes or repairs watches or clocks
— watch·mak·ing \-ˌmā-kiŋ\ *noun*

watch·man \-mən\ *noun* (15th century)
: a person who keeps watch **:** GUARD

watch night *noun* (1742)
: a devotional service lasting until after midnight especially on New Year's Eve

watch out *intransitive verb* (1845)
: to be vigilant or alert **:** be on the lookout ⟨you'd better *watch out*⟩ ⟨*watch out* for the tree!⟩

watch pocket *noun* (1831)
: a small pocket just below the front waistband of men's trousers

watch·tow·er \'wäch-ˌtau̇r, 'wȯch-\ *noun* (1544)
: a tower for a lookout

watch·word \-ˌwərd\ *noun* (15th century)
1 : a word or phrase used as a sign of recognition among members of the same society, class, or group
2 a : a word or motto that embodies a principle or guide to action of an individual or group **:** SLOGAN ⟨"safety" is our *watchword*⟩ **b :** a guiding principle ⟨change is the *watchword* for both parties⟩

¹wa·ter \'wȯ-tər, 'wä-\ *noun, often attributive* [Middle English, from Old English *wæter;* akin to Old High German *wazzar* water, Greek *hydōr,* Latin *unda* wave] (before 12th century)
1 a : the liquid that descends from the clouds as rain, forms streams, lakes, and seas, and is a major constituent of all living matter and that when pure is an odorless, tasteless, very slightly compressible liquid oxide of hydrogen H_2O which appears bluish in thick layers, freezes at 0° C and boils at 100° C, has a maximum density at 4° C and a high specific heat, is feebly ionized to hydrogen and hydroxyl ions, and is a poor conductor of electricity and a good solvent **b :** a natural mineral water — usually used in plural
2 : a particular quantity or body of water: as **a** (1) *plural* **:** the water occupying or flowing in a particular bed (2) *chiefly British* **:** LAKE, POND **b :** a quantity or depth of water adequate for some purpose (as navigation) **c** *plural* (1) **:** a band of seawater abutting on the land of a particular sovereignty and under the control of that sovereignty (2) **:** the sea of a particular part of the earth **d :** WATER SUPPLY ⟨threatened to turn off the *water*⟩
3 : travel or transportation on water ⟨we went by *water*⟩
4 : the level of water at a particular state of the tide **:** TIDE
5 : liquid containing or resembling water: as **a** (1) **:** a pharmaceutical or cosmetic preparation made with water (2) **:** a watery solution of a gaseous or readily volatile substance — compare AMMONIA WATER **b** *archaic* **:** a distilled fluid (as an essence); *especially* **:** a distilled alcoholic liquor **c :** a watery fluid (as tears, urine, or sap) formed or circulating in a living body **d :** AMNIOTIC FLUID; *also* **:** BAG OF WATERS
6 a : the degree of clarity and luster of a precious stone **b :** degree of excellence ⟨a scholar of the first *water*⟩
7 : WATERCOLOR
8 a : capital stock not representing assets of the issuing company and not backed by earning power **b :** fictitious or exaggerated asset entries that give a stock an unrealistic book value
— above water : out of difficulty

²water (before 12th century)
transitive verb
1 : to moisten, sprinkle, or soak with water
2 : to supply with water for drink
3 : to supply water to
4 : to treat with or as if with water; *specifically* **:** to impart a lustrous appearance and wavy pattern to (cloth) by calendering
5 a : to dilute by the addition of water — often used with *down* ⟨*water* down the punch⟩ **b :** to add to the aggregate par value of (securities) without a corresponding addition to the assets represented by the securities
intransitive verb
1 : to form or secrete water or watery matter (as tears or saliva)
2 : to get or take water: as **a :** to take on a supply of water **b :** to drink water

water bag *noun* (1638)
1 : a bag for holding water; *especially* **:** one designed to keep water cool for drinking by evaporation through a slightly porous surface
2 : the fetal membranes enclosing the amniotic fluid — used especially of domestic animals

water balance *noun* (1911)
: the ratio between the water assimilated into the body and that lost from the body; *also* **:** the condition of the body when this ratio approximates unity

water ballet *noun* (1926)
: a synchronized sequence of movements performed by a group of swimmers

water bear *noun* (1852)
: TARDIGRADE

Water Bearer *noun*
: AQUARIUS 1, 2a

water bed *noun* (1844)
: a bed whose mattress is a watertight bag filled with water

water beetle *noun* (circa 1668)
: any of numerous oval flattened aquatic beetles (especially family Dytiscidae) that swim by means of their fringed hind legs which act together as oars

wa·ter·bird \'wȯ-tər-ˌbərd, 'wä-\ *noun* (15th century)
: a swimming or wading bird — compare WATERFOWL

water biscuit *noun* (circa 1790)
: a cracker of flour and water and sometimes fat

☆ SYNONYMS
Watchful, vigilant, wide-awake, alert mean being on the lookout especially for danger or opportunity. WATCHFUL is the least explicit term ⟨the *watchful* eye of the department supervisor⟩. VIGILANT suggests intense, unremitting, wary watchfulness ⟨eternally *vigilant* in the safeguarding of democracy⟩. WIDE-AWAKE applies to watchfulness for opportunities and developments more often than dangers ⟨*wide-awake* companies latched onto the new technology⟩. ALERT stresses readiness or promptness in meeting danger or in seizing opportunity ⟨*alert* traders anticipated the stock market's slide⟩.

water blister *noun* (1895)
: a blister with a clear watery content that is not purulent or sanguineous

water bloom *noun* (1903)
: an accumulation of algae and especially of blue-green algae at or near the surface of a body of water

water boatman *noun* (1815)
1 : BACK SWIMMER
2 : any of various aquatic bugs (family Corixidae) with one pair of legs modified into paddles

wa•ter•borne \'wȯ-tər-ˌbōrn, 'wä-, -ˌbȯrn\ *adjective* (circa 1559)
: supported or carried by water ⟨*waterborne* commerce⟩ ⟨*waterborne* infection⟩

water boy *noun* (1859)
: one who keeps a group (as of football players) supplied with drinking water

wa•ter•buck \'wȯ-tər-ˌbək, 'wä-\ *noun, plural* **waterbuck** *or* **waterbucks** (1850)
: an antelope (*Kobus ellipsiprymnus*) of sub-Saharan Africa that commonly frequent streams or wet areas

water buffalo *noun* (circa 1890)
: an often domesticated Asian buffalo (*Bubalus bubalis* synonym *B. arnee*)

water buffalo

water bug *noun* (1750)
: any of various small arthropods and especially insects that frequent damp or wet places: as **a :** GERMAN COCKROACH **b :** WATER BOATMAN **c :** any of various large aquatic bugs (family Belostomatidae) with the hind legs flattened and used for swimming

water cannon *noun* (1964)
: a large truck-mounted nozzle for directing a high-pressure stream of water (as at a crowd of rioters or demonstrators)

water chestnut *noun* (1854)
1 : any of a genus (*Trapa* and especially *T. natans* of the family Trapaceae, the water-chestnut family) of Old World aquatic herbs sometimes grown as ornamentals; *also* **:** its edible nutlike spiny fruit
2 : a whitish crunchy vegetable used especially in Chinese cooking that is the peeled and often sliced tuber of a sedge (*Eleocharis dulcis* synonym *E. tuberosa*) native to Asia but widely cultivated elsewhere; *also* **:** the tuber or the sedge itself

water clock *noun* (1601)
: an instrument designed to measure time by the fall or flow of a quantity of water — called also *clepsydra*

water closet *noun* (1755)
1 : a compartment or room for defecation and excretion into a toilet bowl
2 : a toilet bowl and its accessories

wa•ter•col•or \'wȯ-tər-ˌkə-lər, 'wä-\ *noun* (1596)
1 : a paint of which the liquid is a water dispersion of the binding material (as glue, casein, or gum)
2 : the art or method of painting with watercolors
3 : a picture or design executed in watercolors
— watercolor *adjective*
— wa•ter•col•or•ist \-ˌkə-lə-rist\ *noun*

wa•ter•cool•er \'wȯ-tər-ˌkü-lər, 'wä-\ *noun* (1846)
: a device for dispensing refrigerated drinking water

wa•ter•course \'wȯ-tər-ˌkōrs, -ˌkȯrs\ *noun* (1510)
1 : a natural or artificial channel through which water flows
2 : a stream of water (as a river, brook, or underground stream)

wa•ter•craft \-ˌkraft\ *noun* (1566)
1 : skill in aquatic activities (as managing boats)
2 a : SHIP, BOAT **b :** craft for water transport

wa•ter•cress \-ˌkres\ *noun* (14th century)
: any of several water-loving cresses; *especially* **:** a perennial European cress (*Nasturtium officinale*) naturalized in the U.S. and used especially in salads or as a potherb

water cycle *noun* (1928)
: HYDROLOGIC CYCLE

water dog *noun* (14th century)
1 : any of several large American salamanders; *especially* **:** any of a genus (*Necturus* of the family Proteidae) with external gills
2 : a person (as a skilled sailor) who is quite at ease in or on water

water down *transitive verb* (1850)
: to reduce or temper the force or effectiveness of
— wa•tered–down *adjective*

wa•ter•er \'wȯ-tər-ər, 'wä-\ *noun* (1549)
: one that waters: as **a :** a person who obtains or supplies drinking water **b :** a device used for supplying water to livestock and poultry — called also *drinker*

wa•ter•fall \-ˌfȯl\ *noun* (before 12th century)
1 a : a perpendicular or very steep descent of the water of a stream **b :** an artificial waterfall (as in a hotel lobby or a nightclub)
2 : something resembling a waterfall

water flea *noun* (circa 1585)
: any of various small active dark or brightly colored aquatic crustaceans (as a daphnia or cyclops)

¹wa•ter•flood \-ˌfləd\ *noun* (1928)
: the process of waterflooding an oil well

²waterflood *intransitive verb* (1928)
: to pump water into the ground around an oil well nearing depletion in order to loosen and force out additional oil

wa•ter•fowl \'wȯ-tər-ˌfau̇l, 'wä-\ *noun* (14th century)
1 : a bird that frequents water; *especially* **:** a swimming bird
2 waterfowl *plural* **:** swimming game birds as distinguished from upland game birds and shorebirds

wa•ter•fowl•er \-ˌfau̇-lər\ *noun* (1968)
: a hunter of waterfowl
— wa•ter•fowl•ing \-liŋ\ *noun*

wa•ter•front \-ˌfrənt\ *noun* (1766)
: land, land with buildings, or a section of a town fronting or abutting on a body of water

water gap *noun* (1756)
: a pass in a mountain ridge through which a stream runs — compare WIND GAP

water garden *noun* (1891)
1 : a garden in which aquatic plants predominate
2 : a garden built about a stream or pool as a central feature

water gas *noun* (1851)
: a poisonous flammable gaseous mixture that consists chiefly of carbon monoxide and hydrogen with small amounts of methane, carbon dioxide, and nitrogen, is usually made by blowing air and then steam over red-hot coke or coal, and is used as a fuel or after carbureting as an illuminant

Wa•ter•gate \-ˌgāt\ *noun* [*Watergate,* apartment and office complex in Washington, D.C.; from the scandal following the break-in at the Democratic National Committee headquarters there in 1972] (1973)
: a scandal usually involving abuses of office, skulduggery, and a cover-up

water gate *noun* (14th century)
1 : a gate (as of a building) giving access to a body of water
2 : FLOODGATE

water gauge *noun* (circa 1706)
: an instrument to measure or find the depth or quantity of water or to indicate the height of its surface especially in a steam boiler

water glass *noun* (1612)
1 : a glass vessel (as a drinking glass) for holding water
2 : an instrument consisting of an open box or tube with a glass bottom used for examining objects in or under water
3 : a substance that consists usually of the silicate of sodium, is found in commerce as a glassy mass, a stony powder, or dissolved in water as a viscous syrupy liquid, and is used especially as a cement, a protective coating, and as a fireproofing agent

water gun *noun* (1951)
: WATER PISTOL

water hammer *noun* (circa 1890)
: a concussion or sound of concussion of moving water against the sides of a containing pipe or vessel (as a steam pipe)

water haul *noun* [from the figure of a fishing net that catches nothing but water] (1823)
: a fruitless effort

water heater *noun* (circa 1876)
: an apparatus for heating and usually storing hot water (as for domestic use)

water hemlock *noun* (1764)
: a tall poisonous Eurasian perennial herb (*Cicuta virosa*) of the carrot family; *also* **:** any of several poisonous North American plants (especially *Cicuta maculata* and *C. douglasii*) of the same genus

water hen *noun* (circa 1529)
: any of various birds (as a coot or gallinule) of the rail family

water hole *noun* (circa 1653)
1 : a natural hole or hollow containing water
2 : a hole in a surface of ice

water hyacinth *noun* (circa 1890)
: a showy floating aquatic plant (*Eichhornia crassipes* of the family Pontederiaceae) of tropical America that often clogs waterways (as in the southern U.S.)

water ice *noun* (1818)
: a frozen dessert of water, sugar, and flavoring

watering can *noun* (1692)
: a vessel usually with a spout used to sprinkle or pour water especially on plants — called also *watering pot*

watering hole *noun* (1955)
: a place where people gather socially; *especially* **:** WATERING PLACE 3

watering place *noun* (15th century)
1 : a place where water may be obtained; *especially* **:** one where animals and especially livestock come to drink
2 : a health or recreational resort featuring mineral springs or bathing
3 : a place (as a nightclub, bar, or lounge) where drink is available

wa•ter•ish \'wȯ-tər-ish, 'wä-\ *adjective* (1542)
: somewhat watery
— wa•ter•ish•ness *noun*

water jacket *noun* (1869)
: an outer casing which holds water or through which water circulates to cool the interior; *specifically* **:** the enclosed space surrounding the cylinder block of an internal combustion engine and containing the cooling liquid

water jump *noun* (1875)
: an obstacle (as in a steeplechase) consisting of a pool, stream, or ditch of water

wa•ter•leaf \'wȯ-tər-ˌlēf, 'wä-\ *noun, plural* **-leafs** \-ˌlēfs\ (1760)
: any of a genus (*Hydrophyllum* of the family Hydrophyllaceae, the waterleaf family) of perennial North American woodland herbs with lobed or pinnate toothed leaves and cymes of bell-shaped flowers

wa•ter•less \-ləs\ *adjective* (before 12th century)
1 : lacking or destitute of water **:** DRY
2 : not requiring water (as for cooling or cooking)
— wa•ter•less•ness *noun*

water level *noun* (1563)

1 : an instrument to show the level by means of the surface of water in a trough or in a U-shaped tube
2 : the surface of still water: as **a :** the level assumed by the surface of a particular body or column of water **b :** the waterline of a vessel

water lily *noun* (15th century)
: any of various aquatic plants (especially genera *Nymphaea* and *Nuphar* of the family Nymphaeaceae, the water-lily family) with floating leaves and usually showy flowers; *broadly* **:** an aquatic plant (as a water hyacinth) with showy flowers

wa·ter·line \ˈwȯ-tər-ˌlīn, ˈwä-\ *noun* (circa 1625)
: a line that marks the level of the surface of water on something: as **a** (1) **:** the point on the hull of a ship or boat to which the water rises (2) **:** a line marked on the outside of a ship that corresponds with the water's surface when the ship is afloat on an even keel under specified conditions of loading **b :** SHORELINE 1

wa·ter·log \-ˌlȯg, -ˌläg\ *transitive verb* [back-formation from *waterlogged*] (1779)
: to make waterlogged

wa·ter·logged \-ˌlȯgd, -ˌlägd\ *adjective* [[1]*water* + *log* to accumulate in the hold] (circa 1776)
1 : so filled or soaked with water as to be heavy or hard to manage ⟨*waterlogged* boats⟩
2 : saturated with water ⟨*waterlogged* soil⟩

wa·ter·loo \ˌwȯ-tər-ˈlü, ˌwä-\ *noun, plural* **-loos** [*Waterloo,* Belgium, scene of Napoleon's defeat in 1815] (1816)
: a decisive or final defeat or setback

water main *noun* (1803)
: a pipe or conduit for conveying water

wa·ter·man \ˈwȯ-tər-mən, ˈwä-\ *noun* (15th century)
: one who works or lives on the water: as **a :** a man who makes his living from the water (as by fishing) **b :** a boatman who plies for hire usually on inland waters or harbors

wa·ter·man·ship \-ˌship\ *noun* (1882)
: the business, skill, or art of a waterman: as **a :** technique or expertness in rowing **b :** technique or expertness in swimming

[1]**wa·ter·mark** \ˈwȯ-tər-ˌmärk, ˈwä-\ *noun* (1678)
1 : a mark indicating the height to which water has risen
2 : a marking in paper resulting from differences in thickness usually produced by pressure of a projecting design in the mold or on a processing roll and visible when the paper is held up to the light; *also* **:** the design of or the metal pattern producing the marking

[2]**watermark** *transitive verb* (1866)
1 : to mark (paper) with a watermark
2 : to impress (a given design) as a watermark

wa·ter·mel·on \-ˌme-lən\ *noun* (1615)
1 : a large oblong or roundish fruit with a hard green or white rind often striped or variegated, a sweet watery pink, yellowish, or red pulp, and usually many seeds
2 : a widely grown African vine (*Citrullus lanatus* synonym *C. vulgaris*) of the gourd family that bears watermelons

water meter *noun* (circa 1858)
: an instrument for recording the quantity of water passing through a particular outlet

water milfoil *noun* (1578)
: any of a genus (*Myriophyllum* of the family Haloragaceae) of aquatic plants with finely pinnate submersed leaves

water mill *noun* (15th century)
: a mill whose machinery is moved by water

water moccasin *noun* (1821)
1 : a venomous semiaquatic pit viper (*Agkistrodon piscivorus*) chiefly of the southeastern U.S. that is closely related to the copperhead — called also *cottonmouth, cottonmouth moccasin*
2 : an American water snake (genus *Nerodia*)

water mold *noun* (1899)
: an aquatic fungus (as of the genus *Saprolegnia*)

water moccasin 1

water nymph *noun* (14th century)
: a nymph (as a naiad, Nereid, or Oceanid) associated with a body of water

water oak *noun* (1687)
: any of several American oaks (especially *Quercus nigra*) that thrive in wet soils

water of crystallization (1791)
: water of hydration present in many crystallized substances that is usually essential for maintenance of a particular crystal structure

water of hydration (1889)
: water that is chemically combined with a substance to form a hydrate and can be expelled (as by heating) without essentially altering the composition of the substance

water on the knee (circa 1934)
: an accumulation of inflammatory exudate in the knee joint often following an injury

water ouzel *noun* (1622)
: DIPPER 2

water parting *noun* (1859)
: DIVIDE 2a

water pepper *noun* (circa 1538)
: an annual smartweed (*Polygonum hydropiper*) of moist soils with extremely acrid peppery juice

water pimpernel *noun* (circa 1760)
: either of two small white-flowered herbs (*Samolus valerandi* of Eurasia and *S. floribundus* synonym *S. parviflorus* of America) of the primrose family that grow in wet places

water pipe *noun* (15th century)
1 : a pipe for conveying water
2 : a smoking device that consists of a bowl mounted on a vessel of water which is provided with a long tube and arranged so that smoke is drawn through the water where it is cooled and up the tube to the mouth

water pistol *noun* (1905)
: a toy pistol designed to squirt a jet of liquid — called also *water gun, squirt gun*

water plantain *noun* (circa 1538)
: any of a genus (*Alisma* of the family Alismataceae, the water-plantain family) of marsh or aquatic herbs with acrid sap and scapose 3-petaled flowers

water polo *noun* (1884)
: a goal game similar to soccer that is played in water by teams of swimmers using a ball resembling a soccer ball

wa·ter·pow·er \ˈwȯ-tər-ˌpau̇(-ə)r, ˈwä-\ *noun* (1817)
1 : the power of water employed to move machinery
2 : a fall of water suitable for being used to move machinery

water privilege *noun* (1804)
: the right to use water especially as a source of mechanical power

[1]**wa·ter·proof** \ˈwȯ-tər-ˌprüf, ˈwä-\ *adjective* (1736)
: impervious to water; *especially* **:** covered or treated with a material (as a solution of rubber) to prevent permeation by water
— wa·ter·proof·ness *noun*

[2]**waterproof** *noun* (1799)
1 : a waterproof fabric
2 *chiefly British* **:** RAINCOAT

[3]**waterproof** *transitive verb* (1841)
: to make waterproof
— wa·ter·proof·er *noun*

wa·ter·proof·ing \-ˌprü-fiŋ\ *noun* (1845)
1 a : the act or process of making something waterproof **b :** the condition of being made waterproof
2 : something (as a coating) capable of imparting waterproofness

water rat *noun* (1552)
1 : a rodent that frequents water
2 : a waterfront loafer or petty thief

wa·ter-re·pel·lent \ˈwȯ-tə(r)-ri-ˈpe-lənt, ˌwä-\ *adjective* (1896)
: treated with a finish that is resistant but not impervious to penetration by water

wa·ter-re·sis·tant \-ri-ˈzis-tənt\ *adjective* (1921)
: WATER-REPELLENT

water right *noun* (1793)
: a right to the use of water (as for irrigation); *especially* **:** RIPARIAN RIGHT

water sapphire *noun* (circa 1741)
: a deep blue cordierite sometimes used as a gem

wa·ter·scape \ˈwȯ-tər-ˌskāp, ˈwä-\ *noun* (circa 1854)
: a water or sea view **:** SEASCAPE 1

water scorpion *noun* (1681)
: any of numerous aquatic bugs (family Nepidae) with the end of the abdomen prolonged by a long breathing tube

wa·ter·shed \ˈwȯ-tər-ˌshed, ˈwä-\ *noun* (1803)
1 a : DIVIDE 2a **b :** a region or area bounded peripherally by a divide and draining ultimately to a particular watercourse or body of water
2 : a crucial dividing point, line, or factor **:** TURNING POINT
— watershed *adjective*

water shield *noun* (circa 1818)
: an aquatic plant (*Brasenia schreberi*) of the water-lily family having floating oval leaves with a gelatinous coating and small dull purple flowers; *also* **:** any of a related genus (*Cabomba*)

[1]**wa·ter·side** \ˈwȯ-tər-ˌsīd, ˈwä-\ *noun* (14th century)
: the margin of a body of water **:** WATERFRONT

[2]**waterside** *adjective* (1663)
1 : employed along the waterside ⟨*waterside* workers⟩; *also* **:** of or relating to the workers along the waterside ⟨a *waterside* strike⟩
2 : of, relating to, or located on the waterside ⟨a *waterside* café⟩

water ski *noun* (1931)
: a ski used in planing over water while being towed by a speedboat
— wa·ter-ski *intransitive verb*

wa·ter-ski·er \ˈwȯ-tər-ˌskē-ər, ˈwä-\ *noun* (1931)
: one that water-skis

wa·ter-ski·ing \-ˌskē-iŋ\ *noun* (1931)
: the sport of planing and jumping on water skis

water snake *noun* (1601)
: any of various snakes (especially genus *Nerodia* formerly included in the genus *Natrix*) that frequent or inhabit fresh waters and feed largely on aquatic animals

wa·ter-soak \ˈwȯ-tər-ˌsōk, ˈwä-\ *transitive verb* (1791)
: to soak in water

water spaniel *noun* (1566)
: a spaniel of either of two breeds: **a :** AMERICAN WATER SPANIEL **b :** IRISH WATER SPANIEL

water spot *noun* (1939)
: a physiological disorder of citrus fruits in the rainy season in which the epidermal air spaces of the rind become filled with liquid

wa·ter·spout \ˈwȯ-tər-ˌspau̇t, ˈwä-\ *noun* (14th century)
1 : a pipe, duct, or orifice from which water is spouted or through which it is carried
2 : a funnel-shaped or tubular column of rotating cloud-filled wind usually extending from the underside of a cumulus or cumulonimbus cloud down to a cloud of spray torn up by the whirling winds from the surface of an ocean or lake

water sprite *noun* (1798)

: a sprite believed to inhabit or haunt water : WATER NYMPH

water sprout *noun* (circa 1892)
: an extremely vigorous shoot from an adventitious or latent bud on a tree — compare SUCKER 2

water strider *noun* (1849)
: any of various long-legged bugs (family Gerridae) that move about on the surface of the water

water supply *noun* (circa 1882)
: a source, means, or process of supplying water (as for a community) usually including reservoirs, tunnels, and pipelines

water table *noun* (15th century)
1 : a stringcourse or similar member when projecting so as to throw off water
2 : the upper limit of the portion of the ground wholly saturated with water

water taxi *noun* (1928)
: a boat functioning (as within a harbor) as a taxi

wa·ter·thrush \'wȯ-tər-ˌthrəsh, 'wä-\ *noun* (circa 1813)
: either of two North American warblers (*Seiurus noveboracensis* and *S. motacilla*) found near freshwater (as a stream)

wa·ter·tight \ˌwȯ-tər-'tīt, ˌwä-\ *adjective* (14th century)
1 : of such tight construction or fit as to be impermeable to water except when under sufficient pressure to produce structural discontinuity
2 : leaving no possibility of misconstruction or evasion ⟨a *watertight* lease⟩
— wa·ter·tight·ness *noun*

water tower *noun* (circa 1883)
: a tower or standpipe serving as a reservoir to deliver water at a required head

water turkey *noun* (1836)
: a New World anhinga (*Anhinga anhinga*)

water vapor *noun* (1880)
: water in a vaporous form especially when below boiling temperature and diffused (as in the atmosphere)

water–vascular system *noun* (1870)
: a system of canals in echinoderms containing a circulating watery fluid that is used for the movement of the tentacles and tube feet

water wagon *noun* (1904)
: a wagon or motortruck equipped with a tank or barrels for hauling water or for sprinkling
— on the water wagon : abstaining from alcoholic beverages : on the wagon

water wave *noun* (1882)
: a method or style of setting hair by dampening with water and forming into waves
— wa·ter–waved \'wȯ-tər-ˌwāvd, 'wä-\ *adjective*

wa·ter·way \'wȯ-tər-ˌwā, 'wä-\ *noun* (15th century)
1 : a way or channel for water
2 : a navigable body of water

wa·ter·weed \-ˌwēd\ *noun* (1842)
: any of various aquatic plants (as a pondweed) with inconspicuous flowers — compare WATER LILY

wa·ter·wheel \-ˌhwēl, -ˌwēl\ *noun* (15th century)
1 : a wheel made to rotate by direct action of water
2 : a wheel for raising water

waterwheel 1

water wings *noun plural* (1907)
: a pneumatic device that buoys the body of a swimmer and is used especially in learning to swim

water witch *noun* (1817)
: one that dowses for water
— water witch·ing \-ˌwi-chiŋ\ *noun*

water witch·er \-ˌwi-chər\ *noun* (1917)
: WATER WITCH

wa·ter·works \'wȯ-tər-ˌwərks, 'wä-\ *noun plural* (circa 1586)
1 : an ornamental fountain or cascade
2 : the system of reservoirs, channels, mains, and pumping and purifying equipment by which a water supply is obtained and distributed (as to a city)
3 : the shedding of tears : TEARS

wa·ter·worn \-ˌwōrn, -ˌwȯrn\ *adjective* (1815)
: worn, smoothed, or polished by the action of water

wa·tery \'wȯ-tə-rē, 'wä-\ *adjective* (before 12th century)
1 a : consisting of, filled with, or surrounded by water **b :** containing, sodden with, or yielding water or a thin liquid ⟨a *watery* solution⟩ ⟨*watery* vesicles⟩
2 a : resembling water or watery matter especially in thin fluidity, soggy texture, paleness, or lack of savor ⟨*watery* blood⟩ ⟨*watery* sunlight⟩ ⟨a *watery* soup⟩ **b :** exhibiting weakness and vapidity : WISHY-WASHY ⟨a *watery* writing style⟩
— wa·ter·i·ly \-tə-rə-lē\ *adverb*
— wa·ter·i·ness \-tə-rē-nəs\ *noun*

wa·ter·zooi \'vä-tər-ˌzoi\ *noun* [Flemish, from *water* water + *zooi* quantity of cooked food] (1949)
: a stew of fish or chicken and vegetables in a seasoned stock thickened with cream and egg yolks

Wat·son–Crick \ˌwät-sən-'krik\ *adjective* (1964)
: of or relating to the Watson-Crick model ⟨*Watson-Crick* helix⟩ ⟨*Watson-Crick* structure⟩ ⟨*Watson-Crick* base pairs⟩

Watson–Crick model *noun* [J. D. *Watson* & F. H. C. *Crick*] (1958)
: a model of DNA structure in which the molecule is a cross-linked double-stranded helix, each strand is composed of alternating links of phosphate and deoxyribose, and the strands are cross-linked by pairs of purine and pyrimidine bases projecting inward from the deoxyribose sugars and joined by hydrogen bonds with adenine paired with thymine and with cytosine paired with guanine — compare DOUBLE HELIX

watt \'wät\ *noun* [James *Watt* (died 1819)] (1882)
: the absolute meter-kilogram-second unit of power equal to the work done at the rate of one joule per second or to the power produced by a current of one ampere across a potential difference of one volt : 1/746 horsepower

watt·age \'wä-tij\ *noun* (1903)
: amount of power expressed in watts

Wat·teau \(ˌ)wä-'tō\ *adjective* [Antoine *Watteau*] (1873)
1 *of women's dress* **:** having back pleats falling loosely from neckline to hem
2 *of a hat* **:** shallow-crowned and having a wide brim turned up at the back to hold flower trimmings

watt–hour \'wät-'au̇(-ə)r\ *noun* (1888)
: a unit of work or energy equivalent to the power of one watt operating for one hour

¹wat·tle \'wä-tᵊl\ *noun* [Middle English *wattel,* from Old English *watel;* akin to Old High German *wadal* bandage] (before 12th century)
1 a : a fabrication of poles interwoven with slender branches, withes, or reeds and used especially formerly in building **b :** material for such construction **c** *plural* **:** poles laid on a roof to support thatch
2 : a fleshy dependent process usually about the head or neck (as of a bird)
3 *Australian* **:** ACACIA 2
— wat·tled \-tᵊld\ *adjective*

²wattle *transitive verb* **wat·tled; wat·tling** \'wät-liŋ, 'wä-tᵊl-iŋ\ (14th century)
1 : to form or build of or with wattle
2 a : to form into wattle **:** interlace to form wattle **b :** to unite or make solid by interweaving light flexible material

wattle and daub *noun* (1808)
: a framework of woven rods and twigs covered and plastered with clay and used in building construction

wat·tle·bird \'wä-tᵊl-ˌbərd\ *noun* (1819)
: any of several Australasian honeyeaters (genus *Anthochaera*) having fleshy pendulous ear wattles

watt·me·ter \'wät-ˌmē-tər\ *noun* [International Scientific Vocabulary] (1887)
: an instrument for measuring electric power in watts

Wa·tu·si \wä-'tü-sē\ *noun, plural* **Watusi** *also* **Watusis** (1899)
: TUTSI

¹wave \'wāv\ *verb* **waved; wav·ing** [Middle English, from Old English *wafian* to wave with the hands; akin to *wæfan* to clothe and perhaps to Old English *wefan* to weave] (before 12th century)
intransitive verb
1 : to motion with the hands or with something held in them in signal or salute
2 : to float, play, or shake in an air current **:** move loosely to and fro **:** FLUTTER
3 *of water* **:** to move in waves **:** HEAVE
4 : to become moved or brandished to and fro
5 : to move before the wind with a wavelike motion ⟨field of *waving* grain⟩
6 : to follow a curving line or take a wavy form **:** UNDULATE
transitive verb
1 : to swing (something) back and forth or up and down
2 : to impart a curving or undulating shape to ⟨*waved* her hair⟩
3 a : to motion to (someone) to go in an indicated direction or to stop **:** FLAG, SIGNAL ⟨checked his name and *waved* him on⟩ ⟨*waved* down a passing car⟩ **b :** to gesture with (the hand or an object) in greeting or farewell or in homage **c :** to dismiss or put out of mind **:** DISREGARD — usually used with *aside* or *off* **d :** to convey by waving ⟨*waved* farewell⟩
4 : BRANDISH, FLOURISH ⟨*waved* a pistol menacingly⟩
synonym see SWING

²wave *noun* (1526)
1 a : a moving ridge or swell on the surface of a liquid (as of the sea) **b :** open water
2 a : a shape or outline having successive curves **b :** a waviness of the hair **c :** an undulating line or streak or a pattern formed by such lines
3 : something that swells and dies away: as **a :** a surge of sensation or emotion ⟨a *wave* of anger swept over her⟩ **b :** a movement sweeping large numbers in a common direction ⟨*waves* of protest⟩ **c :** a peak or climax of activity ⟨a *wave* of buying⟩
4 : a sweep of hand or arm or of some object held in the hand used as a signal or greeting
5 : a rolling or undulatory movement or one of a series of such movements passing along a surface or through the air
6 : a movement like that of an ocean wave: as **a :** a surging movement of a group ⟨a big new *wave* of women politicians⟩ **b :** one of a succession of influxes of people migrating into a region **c** (1) **:** a moving group of animals of one kind (2) **:** a sudden rapid increase in a population **d :** a line of attacking or advancing troops or airplanes **e :** a display of people in a large crowd (as at a sports event) successively rising, lifting their arms overhead, and quickly sitting so as to form a swell moving through the crowd
7 a : a disturbance or variation that transfers energy progressively from point to point in a medium and that may take the form of an elastic deformation or of a variation of pressure,

electric or magnetic intensity, electric potential, or temperature **b :** one complete cycle of such a disturbance
8 : a marked change in temperature **:** a period of hot or cold weather
9 : an undulating or jagged line constituting a graphic representation of an action
— **wave·less** \ˈwāv-ləs\ *adjective*
— **wave·less·ly** *adverb*
— **wave·like** \-ˌlīk\ *adjective*

Wave \ˈwāv\ *noun* [*W*omen *A*ccepted for *V*olunteer *E*mergency Service] (1942)
: a woman serving in the navy

wave band *noun* (1923)
: a band of radio-wave frequencies

waved \ˈwāvd\ *adjective* (1599)
: having a wavelike form or outline: as **a :** having wavy lines of color **:** WATERED ⟨*waved* cloth⟩ **b :** marked by undulations **:** CURVING ⟨the *waved* cutting edge of a bread knife⟩

wave equation *noun* (1926)
: a partial differential equation of the second order whose solutions describe wave phenomena

wave·form \ˈwāv-ˌfȯrm\ *noun* (1845)
: a usually graphic representation of the shape of a wave that indicates its characteristics (as frequency and amplitude) — called also *waveshape*

wave front *noun* (circa 1864)
: a surface composed at any instant of all the points just reached by a vibrational disturbance in its propagation through a medium

wave·guide \ˈwāv-ˌgīd\ *noun* (1932)
: a device (as a duct, coaxial cable, or glass fiber) designed to confine and direct the propagation of electromagnetic waves (as light); *especially* **:** a metal tube for channeling ultrahigh-frequency waves

wave·length \-ˌleŋ(k)th\ *noun* (1850)
1 : the distance in the line of advance of a wave from any one point to the next point of corresponding phase
2 : a particular course or line of thought especially as related to mutual understanding ⟨two people on different *wavelengths*⟩

wave·let \-lət\ *noun* (circa 1810)
: a little wave **:** RIPPLE

wave mechanics *noun plural but singular or plural in construction* (1926)
1 : the mathematical description of atomic and subatomic particles in terms of their wave characteristics
2 : QUANTUM MECHANICS

wave number *noun* (1873)
: the number of waves per unit distance of radiant energy of a given wavelength **:** the reciprocal of the wavelength

wave of the future (1940)
: an idea, product, or movement that is viewed as representing forces or a trend that will inevitably prevail

wave packet *noun* (1928)
: a pulse of radiant energy that is the resultant of a number of wave trains of differing wavelengths

¹wa·ver \ˈwā-vər\ *intransitive verb* **wa·vered; wa·ver·ing** \ˈwāv-riŋ, ˈwā-və-riŋ\ [Middle English; akin to Old English *wǣfre* restless, *wafian* to wave with the hands — more at WAVE] (14th century)
1 : to vacillate irresolutely between choices **:** fluctuate in opinion, allegiance, or direction
2 a : to weave or sway unsteadily to and fro **:** REEL, TOTTER **b :** QUIVER, FLICKER ⟨*wavering* flames⟩ **c :** to hesitate as if about to give way **:** FALTER
3 : to give an unsteady sound **:** QUAVER
synonym see SWING, HESITATE
— **wa·ver·er** \ˈwā-və-rər\ *noun*
— **wa·ver·ing·ly** \ˈwāv-riŋ-lē, ˈwā-və-\ *adverb*

²waver *noun* (1519)
: an act of wavering, quivering, or fluttering

³wav·er \ˈwā-vər\ *noun* (1835)
: one that waves

wa·very \ˈwāv-rē, ˈwā-və-rē\ *adjective* (1820)
: that waves **:** WAVERING

wave·shape \ˈwāv-ˌshāp\ *noun* (1907)
: WAVEFORM

wave theory *noun* (1833)
: a theory in physics: light is transmitted from luminous bodies to the eye and other objects by an undulatory movement — called also *undulatory theory*

wave train *noun* (1897)
: a succession of similar waves at equal intervals

wavy \ˈwā-vē\ *adjective* **wav·i·er; -est** (circa 1586)
1 : rising or swelling in waves; *also* **:** abounding in waves ⟨*wavy* hair⟩
2 : moving with an undulating motion **:** FLUCTUATING; *also* **:** marked by wavering
3 : marked by undulation **:** ROLLING
— **wav·i·ly** \ˈwā-və-lē\ *adverb*
— **wav·i·ness** \-vē-nəs\ *noun*

waw \ˈväv, ˈvȯv\ *noun* [Hebrew *wāw*] (14th century)
: the 6th letter of the Hebrew alphabet — see ALPHABET table

wa–wa pedal \ˈwä-ˌwä-\ *noun* [imitative] (1968)
: an electronic device that is connected to an amplifier and operated by a foot pedal and that is used (as with an electric guitar) to produce a fluctuating muted effect

¹wax \ˈwaks\ *noun* [Middle English, from Old English *weax;* akin to Old High German *wahs* wax, Lithuanian *vaškas*] (before 12th century)
1 : a substance that is secreted by bees and is used by them for constructing the honeycomb, that is a dull yellow solid plastic when warm, and that is composed of a mixture of esters, cerotic acid, and hydrocarbons — called also *beeswax*
2 : any of various substances resembling beeswax: as **a :** any of numerous substances of plant or animal origin that differ from fats in being less greasy, harder, and more brittle and in containing principally compounds of high molecular weight (as fatty acids, alcohols, and saturated hydrocarbons) **b :** a solid substance (as ozokerite or paraffin wax) of mineral origin consisting usually of hydrocarbons of high molecular weight **c :** a pliable or liquid composition used especially in uniting surfaces, excluding air, making patterns or impressions, or producing a polished surface
3 : something likened to wax as soft, impressionable, or readily molded
4 : a waxy secretion; *especially* **:** CERUMEN
5 : a phonograph recording
— **wax·like** \ˈwaks-ˌlīk\ *adjective*

²wax *transitive verb* (14th century)
1 : to treat or rub with wax usually for polishing or stiffening
2 : to record on phonograph records

³wax *intransitive verb* [Middle English, from Old English *weaxan;* akin to Old High German *wahsan* to increase, Greek *auxanein,* Latin *augēre* — more at EKE] (before 12th century)
1 a : to increase in size, numbers, strength, prosperity, or intensity **b :** to grow in volume or duration **c :** to grow toward full development
2 : to increase in phase or intensity — used chiefly of the moon, other satellites, and inferior planets
3 : to assume a (specified) characteristic, quality, or state **:** BECOME ⟨*wax* indignant⟩

⁴wax *noun* (14th century)
: INCREASE, GROWTH — usually used in the phrase *on the wax*

⁵wax *noun* [perhaps from ³*wax*] (1854)
: a fit of temper **:** RAGE

wax bean *noun* (1897)
: a kidney bean with pods that turn creamy yellow to bright yellow when mature enough for use as snap beans

wax·bill \ˈwaks-ˌbil\ *noun* (1757)
: any of numerous Old World oscine birds (family Estrildidae and especially genus *Estrilda*) having white, pink, or reddish bills of a waxy appearance

waxed paper *noun* (1853)
: paper coated or treated with wax to make it resistant to water and grease and used especially as a wrapping

wax·en \ˈwak-sən\ *adjective* (before 12th century)
1 : made of or covered with wax
2 : resembling wax: as **a :** easily molded **:** PLIABLE **b :** seeming to lack vitality or animation **:** PALLID **c :** lustrously smooth

wax·er \-sər\ *noun* (circa 1875)
1 : a device for applying wax
2 : one whose work is applying or polishing with wax

wax·ing *noun* (15th century)
1 : the act of applying wax (as in polishing)
2 : the process of removing body hair with a depilatory wax

wax light *noun* (1599)
: a wax candle **:** TAPER

wax moth *noun* (1766)
: a dull brownish or ashen pyralid moth (*Galleria mellonella*) with a larva that feeds on the comb wax of the honeybee

wax museum *noun* (1953)
: a place where wax effigies (as of famous historical persons) are exhibited

wax myrtle *noun* (1806)
: any of a genus (*Myrica* of the family Myricaceae, the wax-myrtle family) of trees or shrubs with aromatic foliage; *especially* **:** a shrub (*M. cerifera*) of the eastern U.S. having small hard berries with a thick coating of white wax used for candles

wax palm *noun* (circa 1828)
: any of several palms that yield wax: as **a :** an Andean pinnate-leaved palm (*Ceroxylon alpinum* synonym *C. andicola*) whose stem yields a resinous wax used in candles **b :** CARNAUBA

wax paper *noun* (circa 1844)
: WAXED PAPER

wax·wing \ˈwaks-ˌwiŋ\ *noun* (1817)
: any of a genus (*Bombycilla*) of American and Eurasian chiefly brown to gray oscine birds (as a cedar waxwing) having a showy crest, red waxy material on the tips of the secondaries, and a yellow band on the tip of the tail

waxwing

wax·work \-ˌwərk\ *noun* (1697)
1 : an effigy in wax usually of a person
2 *plural but singular or plural in construction* **:** WAX MUSEUM

waxy \ˈwak-sē\ *adjective* **wax·i·er; -est** (1552)
1 : made of, abounding in, or covered with wax **:** WAXEN ⟨a *waxy* surface⟩ ⟨*waxy* berries⟩
2 : resembling wax: as **a :** readily shaped or molded **b :** marked by smooth or lustrous whiteness ⟨a *waxy* complexion⟩
— **wax·i·ness** *noun*

¹way \ˈwā\ *noun* [Middle English, from Old English *weg;* akin to Old High German *weg* way, Old English *wegan* to move, Latin *vehere* to carry, *via* way] (before 12th century)
1 a : a thoroughfare for travel or transportation from place to place **b :** an opening for passage ⟨this door is the only *way* out of the room⟩
2 : the course traveled from one place to another **:** ROUTE
3 a : a course (as a series of actions or se-

\ə\ **abut** \ᵊ\ **kitten** \ər\ **further** \a\ **ash** \ā\ **ace** \ä\ **mop, mar** \au̇\ **out** \ch\ **chin** \e\ **bet** \ē\ **easy** \g\ **go** \i\ **hit** \ī\ **ice** \j\ **job** \ŋ\ **sing** \ō\ **go** \ȯ\ **law** \ȯi\ **boy** \th\ **thin** \t͟h\ **the** \ü\ **loot** \u̇\ **foot** \y\ **yet** \zh\ **vision** *see also* Guide to Pronunciation

quence of events) leading in a direction or toward an objective ⟨led the *way* to eventual open heart operations —*Current Biography*⟩ **b** (1) **:** a course of action ⟨took the easy *way* out⟩ (2) **:** opportunity, capability, or fact of doing as one pleases ⟨always manages to get her own *way*⟩ **c :** a possible decision, action, or outcome **:** POSSIBILITY ⟨they were rude—no two *ways* about it⟩
4 a : manner or method of doing or happening; *also* **:** method of accomplishing **:** MEANS **b :** FEATURE, RESPECT ⟨in no *way* resembles her mother⟩ **c :** a usually specified degree of participation in an activity or enterprise ⟨active in real estate in a small *way*⟩
5 a : characteristic, regular, or habitual manner or mode of being, behaving, or happening ⟨knows nothing of the *ways* of women⟩ **b :** ability to get along well or perform well ⟨she has a *way* with kids⟩ ⟨a *way* with words⟩
6 : the length of a course **:** DISTANCE ⟨has come a long *way* in her studies⟩ ⟨still have a *way* to go⟩
7 : movement or progress along a course ⟨worked her *way* up the corporate ladder⟩
8 a : DIRECTION ⟨is coming this *way*⟩ **b :** PARTICIPANT — usually used in combination ⟨three-*way* discussion⟩
9 : state of affairs **:** CONDITION, STATE ⟨that's the *way* things are⟩
10 a *plural but sometimes singular in construction* **:** an inclined structure upon which a ship is built or supported in launching **b** *plural* **:** the guiding surfaces on the bed of a machine along which a table or carriage moves
11 : CATEGORY, KIND — usually used in the phrase *in the way of* ⟨doesn't require much in the *way* of expensive equipment —*Forbes*⟩
12 : motion or speed of a ship or boat through the water
synonym see METHOD
— all the way : to the full or entire extent **:** as far as possible ⟨ran *all the way* home⟩ ⟨seated *all the way* in the back⟩
— by the way : by way of interjection or digression **:** INCIDENTALLY
— by way of 1 : for the purpose of **2 :** by the route through **:** VIA
— in a way 1 : within limits **:** with reservations **2 :** from one point of view
— in one's way *also* **in the way 1 :** in a position to be encountered by one **:** in or along one's course ⟨an opportunity had been put *in my way* —Ellen Glasgow⟩ **2 :** in a position to hinder or obstruct
— on the way *or* **on one's way :** moving along in one's course **:** in progress
— out of the way 1 : WRONG, IMPROPER ⟨didn't know I'd said anything *out of the way*⟩ **2 a :** in or to a secluded place **b :** UNUSUAL, REMARKABLE ⟨there's nothing *out of the way* about the plan⟩ **3 :** DONE, COMPLETED ⟨got his homework *out of the way*⟩
— the way 1 : in view of the manner in which ⟨you'd think she was rich, *the way* she spends money⟩ **2 :** LIKE, AS ⟨we have cats *the way* other people have mice —James Thurber⟩
[2]way *adjective* (1799)
: of, connected with, or constituting an intermediate point on a route ⟨visited five major countries plus *way* points⟩
[3]way *adverb* (1849)
1 a : AWAY 7 ⟨is *way* ahead of the class⟩ **b :** by far **:** MUCH ⟨ate *way* too much⟩
2 : all the way ⟨pull the switch *way* back⟩
— from way back : of long standing ⟨friends *from way back*⟩
way·bill \'wā-ˌbil\ *noun* (1821)
: a document prepared by the carrier of a shipment of goods that contains details of the shipment, route, and charges
way·far·er \'wā-ˌfar-ər, -ˌfer-\ *noun* [Middle English *weyfarere,* from *wey, way* way + *-farere* traveler, from *faren* to go — more at FARE] (15th century)
: a traveler especially on foot
— way·far·ing \-ˌfar-iŋ, -ˌfer-\ *adjective*
wayfaring tree *noun* (1597)
: a Eurasian shrub (*Viburnum lantana*) that has large ovate leaves and dense cymes of small white flowers and is common along waysides
way·go·ing \'wā-ˌgō-ən, -iŋ\ *noun* (1633)
chiefly Scottish **:** the act of leaving **:** DEPARTURE
Way·land \'wā-lən(d)\ *noun* [Old English *Wēland*]
: an heroic smith of Germanic legend
way·lay \'wā-ˌlā\ *transitive verb* **-laid** \-ˌlād\; **-lay·ing** (1513)
: to lie in wait for or attack from ambush
way·less \-ləs\ *adjective* (12th century)
: having no road or path
Way of the Cross (1868)
: STATIONS OF THE CROSS
way–out \'wā-'au̇t\ *adjective* (1954)
: FAR-OUT
ways \'wāz\ *noun plural but singular in construction* [Middle English *wayes,* from genitive of [1]*way*] (1588)
: WAY 6 ⟨a long *ways* from home⟩
-ways *adverb suffix* [Middle English, from *ways,* genitive of *way*]
: in (such) a way, course, direction, or manner ⟨side*ways*⟩
ways and means *noun plural* (15th century)
1 : methods and resources for accomplishing something and especially for defraying expenses
2 *W&M often capitalized* **a :** methods and resources for raising the necessary revenues for the expenses of a nation or state **b :** a legislative committee concerned with this function
way·side \'wā-ˌsīd\ *noun* (15th century)
: the side of or land adjacent to a road or path
— wayside *adjective*
— by the wayside : out of consideration **:** into a condition of neglect or disuse — usually used with *fall*
way station *noun* (1850)
1 : an intermediate station between principal stations on a line of travel (as a railroad)
2 : an intermediate stopping place
way·ward \'wā-wərd\ *adjective* [Middle English, short for *awayward* turned away, from *away,* adverb + *-ward*] (14th century)
1 : following one's own capricious, wanton, or depraved inclinations **:** UNGOVERNABLE
2 : following no clear principle or law **:** UNPREDICTABLE
3 : opposite to what is desired or expected **:** UNTOWARD ⟨*wayward* fate⟩
synonym see CONTRARY
— way·ward·ly *adverb*
— way·ward·ness *noun*
way·worn \-ˌwōrn, -ˌwȯrn\ *adjective* (1788)
: wearied by traveling
we \'wē\ *pronoun, plural in construction* [Middle English, from Old English *wē;* akin to Old High German *wir* we, Sanskrit *vayam*] (before 12th century)
1 : I and the rest of a group that includes me **:** you and I **:** you and I and another or others **:** I and another or others not including you — used as pronoun of the first person plural; compare I, OUR, OURS, US
2 : [1]I — used by sovereigns; used by writers to keep an impersonal character
weak \'wēk\ *adjective* [Middle English *weike,* from Old Norse *veikr;* akin to Old English *wīcan* to yield, Greek *eikein* to give way, Sanskrit *vijate* he speeds, flees] (14th century)
1 : lacking strength: as **a :** deficient in physical vigor **:** FEEBLE, DEBILITATED **b :** not able to sustain or exert much weight, pressure, or strain **c :** not able to resist external force or withstand attack
2 a : mentally or intellectually deficient **b :** not firmly decided **:** VACILLATING **c :** resulting from or indicating lack of judgment or discernment **d :** not able to withstand temptation or persuasion ⟨the spirit is willing but the flesh is *weak*⟩
3 : not factually grounded or logically presented ⟨a *weak* argument⟩
4 a : not able to function properly **b** (1) **:** lacking skill or proficiency ⟨tutoring for *weaker* students⟩ (2) **:** indicative of a lack of skill or aptitude ⟨history was my *weakest* subject⟩ **c :** wanting in vigor of expression or effect
5 a : deficient in the usual or required ingredients **:** DILUTE ⟨*weak* coffee⟩ **b :** lacking normal intensity or potency ⟨*weak* strain of virus⟩
6 a : not having or exerting authority or political power ⟨*weak* government⟩ **b :** INEFFECTIVE, IMPOTENT
7 : of, relating to, or constituting a verb or verb conjugation that in English forms the past tense and past participle by adding the suffix *-ed* or *-d* or *-t*
8 a : bearing the minimal degree of stress occurring in the language ⟨*weak* syllable⟩ **b :** having little or no stress and obscured vowel sound ⟨*'d* is the *weak* form of *would*⟩
9 : tending toward a lower price ⟨a *weak* market⟩
10 : ionizing only slightly in solution ⟨*weak* acids and bases⟩ ☆
— weak·ly *adverb*
weak anthropic principle *noun* (1985)
: ANTHROPIC PRINCIPLE a
weak·en \'wē-kən\ *verb* **weak·ened; weak·en·ing** \'wēk-niŋ, 'wē-kə-\ (1530)
transitive verb
1 : to make weak **:** lessen the strength of
2 : to reduce in intensity or effectiveness
intransitive verb
: to become weak ☆
— weak·en·er \'wēk-nər, 'wē-kə-\ *noun*

☆ SYNONYMS
Weak, feeble, frail, fragile, infirm, decrepit mean not strong enough to endure strain, pressure, or strenuous effort. WEAK applies to deficiency or inferiority in strength or power of any sort ⟨students too *weak* to resist peer pressure⟩. FEEBLE suggests extreme weakness inviting pity or contempt ⟨a *feeble* attempt to get out of bed⟩. FRAIL implies delicacy and slightness of constitution or structure ⟨a *frail* teenager who was unable to enjoy contact sports⟩. FRAGILE suggests frailty and brittleness unable to resist rough usage ⟨a reclusive poet too *fragile* for the rigors of this world⟩. INFIRM suggests instability, unsoundness, and insecurity due to old age or crippling illness ⟨*infirm* residents requiring constant care⟩. DECREPIT implies being worn-out or broken-down from long use or old age ⟨the dowager's faithful, *decrepit* retainers⟩.

Weaken, enfeeble, debilitate, undermine, sap, cripple, disable mean to lose or cause to lose strength or vigor. WEAKEN may imply loss of physical strength, health, soundness, or stability or of quality, intensity, or effective power ⟨a disease that *weakens* the body's defenses against infection⟩. ENFEEBLE implies an obvious and pitiable condition of weakness and helplessness ⟨*enfeebled* by starvation⟩. DEBILITATE suggests a less marked or more temporary impairment of strength or vitality ⟨the *debilitating* effects of surgery⟩. UNDERMINE and SAP suggest a weakening by something working surreptitiously and insidiously ⟨a poor diet *undermines* your health⟩ ⟨drugs had *sapped* his ability to think⟩. CRIPPLE implies causing a serious loss of functioning power through damaging or removing an essential part or element ⟨*crippled* by arthritis⟩. DISABLE suggests a usually sudden crippling or enfeebling ⟨*disabled* soldiers received an immediate discharge⟩.

weak·fish \'wēk-ˌfish\ *noun* [obsolete Dutch *weekvis,* from Dutch *week* soft + *vis* fish; from its tender flesh] (1791)
1 : a common marine bony fish (*Cynoscion regalis* of the family Sciaenidae) that is a sport and food fish of the eastern coast of the U.S. — called also *sea trout*
2 : any of several fishes congeneric with the weakfish

weak force *noun* (1971)
: a fundamental physical force that governs interactions between hadrons and leptons (as in the emission and absorption of neutrinos) and is responsible for particle decay processes (as beta decay) in radioactivity, that is 10^5 times weaker than the strong force, and that acts over distances smaller than those between nucleons in an atomic nucleus — called also *weak interaction, weak nuclear force;* compare ELECTROMAGNETISM 2a, GRAVITY 3a(2), STRONG FORCE

weak·heart·ed \-'här-təd\ *adjective* (1549)
: lacking courage **:** FAINTHEARTED

weak·ish \'wē-kish\ *adjective* (1594)
: somewhat weak ⟨*weakish* tea⟩

weak–kneed \'wēk-'nēd\ *adjective* (1863)
: lacking willpower or resolution

weak·ling \'wē-kliŋ\ *noun* (1548)
: one that is weak in body, character, or mind
— weakling *adjective*

weak·ly \'wē-klē\ *adjective* (1577)
: FEEBLE, WEAK
— weak·li·ness *noun*

weak–mind·ed \'wēk-'mīn-dəd\ *adjective* (1716)
: having or indicating a weak mind: **a :** lacking in judgment or good sense **:** FOOLISH **b :** FEEBLEMINDED
— weak–mind·ed·ness *noun*

weak·ness \-nəs\ *noun* (14th century)
1 : the quality or state of being weak; *also* **:** an instance or period of being weak ⟨agreed in a moment of *weakness* to go along⟩
2 : FAULT, DEFECT
3 a : a special desire or fondness ⟨has a *weakness* for chocolates⟩ **b :** an object of special desire or fondness ⟨pizza is my *weakness*⟩

weak side *noun* (1940)
1 : the side of a football formation having the smaller number of players; *specifically* **:** the side of a formation away from the tight end
2 : the side of a court or field (as in basketball or soccer) away from the ball
— weak·side \'wēk-ˌsīd\ *adjective*

weak sister *noun* (1857)
: a member of a group who needs aid; *also* **:** something that is weak and ineffective as compared with others in the group

[1]weal \'wē(ə)l\ *noun* [Middle English *wele,* from Old English *wela;* akin to Old English *wel* well] (before 12th century)
1 : a sound, healthy, or prosperous state **:** WELL-BEING
2 *obsolete* **:** BODY POLITIC, COMMONWEAL

[2]weal *noun* [alteration of *wale*] (circa 1798)
: WELT

weald \'wē(ə)ld\ *noun* [the *Weald,* England, from Middle English *weeld,* from Old English *weald* forest — more at WOLD] (before 12th century)
1 : a heavily wooded area **:** FOREST ⟨the *Weald* of Kent⟩
2 : a wild or uncultivated usually upland region

wealth \'welth *also* 'weltth\ *noun* [Middle English *welthe,* from *wele* weal] (13th century)
1 *obsolete* **:** WEAL, WELFARE
2 : abundance of valuable material possessions or resources
3 : abundant supply **:** PROFUSION
4 a : all property that has a money value or an exchangeable value **b :** all material objects that have economic utility; *especially* **:** the stock of useful goods having economic value in existence at any one time ⟨national *wealth*⟩

wealthy \'wel-thē *also* 'welt-thē\ *adjective* **wealth·i·er; -est** (15th century)
1 : having wealth **:** extremely affluent
2 : characterized by abundance **:** AMPLE
synonym see RICH
— wealth·i·ly \-thə-lē\ *adverb*
— wealth·i·ness \-thē-nəs\ *noun*

wean \'wēn\ *transitive verb* [Middle English *wenen,* from Old English *wenian* to accustom, wean; akin to Old English *wunian* to be used to — more at WONT] (before 12th century)
1 : to accustom (as a child) to take food otherwise than by nursing
2 : to detach from a source of dependence ⟨being *weaned* off the medication⟩ ⟨*wean* the bears from human food —*Sports Illustrated*⟩; *also* **:** to free from a usually unwholesome habit or interest ⟨*wean* him off his excessive drinking⟩ ⟨settling his soldiers on the land . . ., *weaning* them from habits of violence —Geoffrey Carnall⟩
3 : to accustom to something from an early age — used in the passive especially with *on* ⟨students *weaned* on the microcomputer⟩ ⟨I was *weaned* on greasepaint —Helen Hayes⟩ ⟨the principles upon which he had been *weaned* —J. A. Michener⟩

wean·er \'wē-nər\ *noun* (1579)
1 : one that weans
2 : a young animal recently weaned from its mother

wean·ling \'wēn-liŋ\ *noun* (1532)
: a child or animal newly weaned
— wean·ling *adjective*

[1]weap·on \'we-pən\ *noun* [Middle English *wepen,* from Old English *wǣpen;* akin to Old High German *wāffan* weapon, Old Norse *vāpn*] (before 12th century)
1 : something (as a club, knife, or gun) used to injure, defeat, or destroy
2 : a means of contending against another

[2]weapon *transitive verb* (before 12th century)
: ARM

weap·on·less \-ləs\ *adjective* (before 12th century)
: lacking weapons **:** UNARMED

weap·on·ry \-rē\ *noun* (1844)
1 : WEAPONS
2 : the science of designing and making weapons

[1]wear \'war, 'wer\ *verb* **wore** \'wōr, 'wȯr\; **worn** \'wōrn, 'wȯrn\; **wear·ing** [Middle English *weren,* from Old English *werian;* akin to Old Norse *verja* to clothe, invest, spend, Latin *vestis* clothing, garment, Greek *hennynai* to clothe] (before 12th century)
transitive verb
1 : to bear or have on the person ⟨*wore* a coat⟩
2 a : to use habitually for clothing, adornment, or assistance ⟨*wears* a toupee⟩ ⟨*wear* glasses⟩ **b :** to carry on the person ⟨*wear* a sword⟩
3 a : to hold the rank or dignity or position signified by (an ornament) ⟨*wear* the royal crown⟩ **b :** EXHIBIT, PRESENT ⟨*wore* a happy smile⟩ ⟨commend the book for *wearing* its research so lightly —Brad Leithauser⟩ **c :** to show or fly (a flag or colors) on a ship
4 a : to cause to deteriorate by use **b :** to impair or diminish by use or attrition **:** consume or waste gradually ⟨letters on the stone *worn* away by weathering⟩
5 : to produce gradually by friction or attrition ⟨*wear* a hole in the rug⟩
6 : to exhaust or lessen the strength of **:** WEARY, FATIGUE
7 : to cause (a ship) to go about with the stern presented to the wind
8 *British* **:** to accept or tolerate without complaint **:** put up with — usually used in negative constructions ⟨your mates wouldn't *wear* it —Colin MacInnes⟩
intransitive verb
1 a : to endure use **:** last under use or the passage of time ⟨material that will *wear* for years⟩ **b :** to retain quality or vitality
2 a : to diminish or decay through use ⟨the heels of his shoes began to *wear*⟩ **b :** to diminish or fail with the passage of time ⟨the effect of the drug *wore* off⟩ ⟨the day *wore* on⟩ **c :** to grow or become by attrition or use
3 *of a ship* **:** to change to an opposite tack by turning the stern to the wind — compare TACK
— wear·er *noun*
— wear on : IRRITATE, FRAY
— wear the trousers *or* **wear the pants :** to have the controlling authority in a household
— wear thin 1 : to become weak or ready to give way ⟨my patience was *wearing thin*⟩ **2 :** to become trite, unconvincing, or out-of-date ⟨an argument that quickly *wore thin*⟩

[2]wear *noun* (15th century)
1 : the act of wearing **:** the state of being worn **:** USE ⟨clothes for everyday *wear*⟩
2 a : clothing or an article of clothing usually of a particular kind; *especially* **:** clothing worn for a special occasion or popular during a specific period **b :** FASHION, VOGUE
3 : wearing quality **:** durability under use
4 : the result of wearing or use **:** diminution or impairment due to use ⟨*wear*-resistant surface⟩

[1]wear·able \'war-ə-bəl, 'wer-\ *adjective* (1590)
: capable of being worn **:** suitable to be worn
— wear·abil·i·ty \ˌwar-ə-'bi-lə-tē, ˌwer-\ *noun*

[2]wearable *noun* (1711)
: GARMENT — usually used in plural

wear and tear *noun* (1666)
: the loss, injury, or stress to which something is subjected by or in the course of use; *especially* **:** normal depreciation

wear down *transitive verb* (1803)
: to weary and overcome by persistent resistance or pressure

wea·ri·ful \'wir-ē-fəl\ *adjective* (15th century)
1 : causing weariness; *especially* **:** TEDIOUS
2 : full of weariness **:** WEARIED
— wea·ri·ful·ly \-fə-lē\ *adverb*
— wea·ri·ful·ness *noun*

wea·ri·less \'wir-ē-ləs\ *adjective* (15th century)
: TIRELESS
— wea·ri·less·ly *adverb*

[1]wear·ing \'war-iŋ, 'wer-\ *adjective* (15th century)
: intended for wear ⟨*wearing* apparel⟩

[2]wearing *adjective* (1811)
: subjecting to or inflicting wear; *especially* **:** causing fatigue ⟨a *wearing* journey⟩
— wear·ing·ly \-iŋ-lē\ *adverb*

wea·ri·some \'wir-ē-səm\ *adjective* (15th century)
: causing weariness **:** TIRESOME
— wea·ri·some·ly *adverb*
— wea·ri·some·ness *noun*

wear out (14th century)
transitive verb
1 : TIRE, EXHAUST
2 : to make useless especially by long or hard usage
3 : ERASE, EFFACE
4 : to endure through **:** OUTLAST ⟨*wear out* a storm⟩
5 : to consume (as time) tediously ⟨*wear out* idle days⟩
intransitive verb
: to become useless from long or excessive wear or use

[1]wea·ry \'wir-ē\ *adjective* **wea·ri·er; -est** [Middle English *wery,* from Old English *wērig;* akin to Old High German *wuorag* intoxicated and perhaps to Greek *aōros* sleep] (before 12th century)

\ə\ **a**but \ᵊ\ **k**itten \ər\ **fur**ther \a\ **a**sh \ā\ **a**ce \ä\ **m**op, m**a**r \au̇\ **ou**t \ch\ **ch**in \e\ **b**et \ē\ **ea**sy \g\ **g**o \i\ **h**it \ī\ **i**ce \j\ **j**ob \ŋ\ si**ng** \ō\ g**o** \ȯ\ l**aw** \ȯi\ b**oy** \th\ **th**in \t͟h\ **th**e \ü\ l**oo**t \u̇\ f**oo**t \y\ **y**et \zh\ vi**si**on *see also* Guide to Pronunciation

1 : exhausted in strength, endurance, vigor, or freshness
2 : expressing or characteristic of weariness
3 : having one's patience, tolerance, or pleasure exhausted — used with *of*
4 : WEARISOME
— **wea·ri·ly** \'wir-ə-lē\ *adverb*
— **wea·ri·ness** \'wir-ē-nəs\ *noun*

[2]**weary** *verb* **wea·ried; wea·ry·ing** (before 12th century)
intransitive verb
: to become weary
transitive verb
: to make weary
synonym see TIRE

wea·sand \'wē-zᵊnd, 'wi-zᵊn(d)\ *noun* [Middle English *wesand,* from (assumed) Old English *wǣsend* gullet; akin to Old English *wāsend* gullet, Old High German *weisunt* windpipe] (before 12th century)
: THROAT, GULLET; *also* **:** TRACHEA

[1]**wea·sel** \'wē-zəl\ *noun, plural* **weasels** [Middle English *wesele,* from Old English *weosule;* akin to Old High German *wisula* weasel] (before 12th century)
1 *or plural* **weasel :** any of various small slender active carnivorous mammals (genus *Mustela* of the family Mustelidae, the weasel family) that are able to prey on animals (as rabbits) larger than themselves, are mostly reddish brown with white or yellowish underparts, and in northern forms turn white in winter

weasel 1

2 : a light self-propelled tracked vehicle built either for traveling over snow, ice, or sand or as an amphibious vehicle

[2]**weasel** *intransitive verb* **wea·seled; wea·sel·ing** \'wēz-liŋ, 'wē-zə-\ [*weasel word*] (1900)
1 : to use weasel words **:** EQUIVOCATE
2 : to escape from or evade a situation or obligation — often used with *out*

wea·sel·ly *also* **wea·sely** \'wēz-lē, 'wē-zə-\ *adjective* (1838)
: resembling or suggestive of a weasel

weasel word *noun* [from the weasel's reputed habit of sucking the contents out of an egg while leaving the shell superficially intact] (1900)
: a word used in order to evade or retreat from a direct or forthright statement or position

[1]**weath·er** \'we-t͟hər\ *noun* [Middle English *weder,* from Old English; akin to Old High German *wetar* weather, Old Church Slavonic *vetrŭ* wind] (before 12th century)
1 : the state of the atmosphere with respect to heat or cold, wetness or dryness, calm or storm, clearness or cloudiness
2 : state or vicissitude of life or fortune
3 : disagreeable atmospheric conditions: as **a :** RAIN, STORM **b :** cold air with dampness
4 : WEATHERING
— **under the weather 1 :** ILL **2 :** DRUNK 1a

[2]**weather** *verb* **weath·ered; weath·er·ing** \'weth-riŋ, 'we-t͟hə-\ (15th century)
transitive verb
1 : to expose to the open air **:** subject to the action of the elements
2 : to bear up against and come safely through ⟨*weather* a storm⟩ ⟨*weather* a crisis⟩
intransitive verb
: to undergo or endure the action of the elements

[3]**weather** *adjective* (circa 1625)
: of or relating to the side facing the wind — compare LEE

weath·er·abil·i·ty \ˌweth-rə-'bi-lə-tē, ˌwe-t͟hə-\ *noun* (1947)
: capability of withstanding the weathering process ⟨*weatherability* of a plastic⟩

weath·er–beat·en \'we-t͟hər-ˌbē-tᵊn\ *adjective* (1530)
1 : toughened, tanned, or bronzed by the weather ⟨a *weather-beaten* face⟩
2 : worn or damaged by exposure to weather

weath·er·board \-ˌbōrd, -ˌbȯrd\ *noun* (circa 1540)
1 : CLAPBOARD, SIDING
2 : the weather side of a ship
— **weath·er·board·ed** *adjective*

weath·er·board·ing \-ˌbōr-diŋ, -ˌbȯr-\ *noun* (1632)
: CLAPBOARDS, SIDING

weath·er–bound \-ˌbau̇nd\ *adjective* (1590)
: kept in port or at anchor or from travel or sport by bad weather

weather bureau *noun* (1871)
: a bureau engaged in the collection of weather reports as a basis for weather predictions, storm warnings, and the compiling of statistical records

weath·er–burned \'we-t͟hər-ˌbərnd\ *adjective* (1906)
: browned by sun and wind

weath·er·cast \-ˌkast\ *noun* [[1]*weather* + [2]*forecast*] (1866)
: a weather forecast especially on radio or television

weath·er·cast·er \-ˌkas-tər\ *noun* (1607)
: a weather forecaster especially on radio or television

weath·er·cock \-ˌkäk\ *noun* (13th century)
1 : a vane often in the figure of a cock mounted so as to turn freely with the wind and show its direction
2 : a person or thing that changes readily or often

weather deck *noun* (1850)
: a deck having no overhead protection from the weather

weath·ered \'we-t͟hərd\ *adjective* (1789)
1 : seasoned by exposure to the weather
2 : altered in color, texture, composition, or form by such exposure or by artificial means producing a similar effect ⟨*weathered* oak⟩

weather eye *noun* (1839)
1 : an eye quick to observe coming changes in the weather
2 : constant and shrewd watchfulness and alertness

weath·er·glass \'we-t͟hər-ˌglas\ *noun* (1626)
: a simple instrument for showing changes in atmospheric pressure by the changing level of liquid in a spout connected with a closed reservoir; *broadly* **:** BAROMETER

weathering *noun* (1548)
: the action of the elements in altering the color, texture, composition, or form of exposed objects; *specifically* **:** the physical disintegration and chemical decomposition of earth materials at or near the earth's surface

weath·er·ize \'we-t͟hə-ˌrīz\ *transitive verb* **-ized; -iz·ing** (1943)
: to make (as a house) better protected against winter weather (as by adding insulation)
— **weath·er·i·za·tion** \ˌweth-rə-'zā-shən, ˌwe-t͟hə-\ *noun*

weath·er·ly \'we-t͟hər-lē\ *adjective* (1729)
: able to sail close to the wind with little leeway

weath·er·man \-ˌman\ *noun* (1859)
: one who reports and forecasts the weather **:** METEOROLOGIST

weather map *noun* (1871)
: a map or chart showing the principal meteorological elements at a given hour and over an extended region

weath·er·per·son \'we-t͟hər-ˌpər-sᵊn\ *noun* (1974)
: a person who reports and forecasts the weather **:** METEOROLOGIST

weath·er·proof \'we-t͟hər-ˌprüf\ *adjective* (1620)
: able to withstand exposure to weather without damage or loss of function
— **weatherproof** *transitive verb*
— **weath·er·proof·ness** *noun*

weather ship *noun* (1946)
: a ship that makes observations for use by meteorologists

weather station *noun* (circa 1895)
: a station for taking, recording, and reporting meteorological observations

weather strip *noun* (1846)
: a strip of material to cover the joint of a door or window and the sill, casing, or threshold so as to exclude rain, snow, and cold air — called also *weather stripping*
— **weath·er–strip** *transitive verb*

weather vane *noun* (circa 1721)
: VANE 1a

weath·er–wise \'we-t͟hər-ˌwīz\ *adjective* (14th century)
1 : skillful in forecasting changes in the weather
2 : skillful in forecasting changes in opinion or feeling ⟨a *weather-wise* politician⟩

weath·er·worn \-ˌwōrn, -ˌwȯrn\ *adjective* (1609)
: worn by exposure to the weather

[1]**weave** \'wēv\ *verb* **wove** \'wōv\ *or* **weaved; wo·ven** \'wō-vən\ *or* **weaved; weav·ing** [Middle English *weven,* from Old English *wefan;* akin to Old High German *weban* to weave, Greek *hyphainein* to weave, *hyphos* web] (before 12th century)
transitive verb
1 a : to form (cloth) by interlacing strands (as of yarn); *specifically* **:** to make (cloth) on a loom by interlacing warp and filling threads **b :** to interlace (as threads) into cloth **c :** to make (as a basket) by intertwining
2 : SPIN 2 — used of spiders and insects
3 : to interlace especially to form a texture, fabric, or design
4 a : to produce by elaborately combining elements **:** CONTRIVE **b :** to unite in a coherent whole **c :** to introduce as an appropriate element **:** work in — usually used with *in* or *into*
5 : to direct (as the body) in a winding or zigzag course especially to avoid obstacles
intransitive verb
1 : to work at weaving **:** make cloth
2 : to move in a devious, winding, or zigzag course especially to avoid obstacles

[2]**weave** *noun* (1581)
1 : something woven; *especially* **:** woven cloth
2 : any of the patterns or methods for interlacing the threads of woven fabrics

[3]**weave** *intransitive verb* **weaved; weaving** [Middle English *weven* to move to and fro, wave; akin to Old Norse *veifa* to be in movement — more at WIPE] (1596)
: to move waveringly from side to side **:** SWAY

weav·er \'wē-vər\ *noun* (14th century)
1 : one that weaves especially as an occupation
2 : WEAVERBIRD

weav·er·bird \-ˌbərd\ *noun* (1826)
: any of numerous Old World passerine birds (family Ploceidae) that resemble finches and mostly construct elaborate nests of interlaced vegetation

weaver's knot *noun* (1532)
: SHEET BEND — called also *weaver's hitch*

[1]**web** \'web\ *noun* [Middle English, from Old English; akin to Old Norse *vefr* web, Old English *wefan* to weave] (before 12th century)
1 : a fabric on a loom or in process of being removed from a loom
2 a : COBWEB, SPIDERWEB **b :** a network of silken thread spun especially by the larvae of various insects (as a tent caterpillar) and usually serving as a nest or shelter
3 : a tissue or membrane of an animal or plant; *especially* **:** that uniting fingers or toes

either at their bases (as in humans) or for a greater part of their length (as in many water-birds)
4 a : a thin metal sheet, plate, or strip **b :** the plate connecting the upper and lower flanges of a girder or rail **c :** the arm of a crank
5 : something resembling a web: **a :** SNARE, ENTANGLEMENT ⟨a *web* of intrigue⟩ ⟨ensnarled in a *web* of folly —David A. Stockman⟩ **b :** an intricate pattern or structure suggestive of something woven **:** NETWORK ⟨a *web* of little roads⟩ ⟨a complex *web* of relationships⟩
6 : the series of barbs implanted on each side of the shaft of a feather **:** VANE
7 a : a continuous sheet of paper manufactured or undergoing manufacture on a paper machine **b :** a roll of paper for use in a rotary printing press
8 : the part of a ribbed vault between the ribs
— webbed \ˈwebd\ *adjective*
— web·like \ˈweb-ˌlīk\ *adjective*

[2]**web** *verb* **webbed; web·bing** (1604)
intransitive verb
: to construct or form a web
transitive verb
1 : to cover with a web or network
2 : ENSNARE, ENTANGLE
3 : to provide with a web

web·bing \ˈwe-biŋ\ *noun* (1796)
1 : a strong narrow closely woven tape designed for bearing weight and used especially for straps, harness, or upholstery
2 : TRAP 3c

web·by \ˈwe-bē\ *adjective* (1661)
: of, relating to, or consisting of a web

we·ber \ˈwe-bər, ˈvā-bər\ *noun* [Wilhelm E. *Weber* (died 1891) German physicist] (1891)
: the practical meter-kilogram-second unit of magnetic flux equal to that flux which in linking a circuit of one turn produces in it an electromotive force of one volt as the flux is reduced to zero at a uniform rate in one second **:** 10^8 maxwells

web·fed \ˈweb-ˌfed\ *adjective* (1947)
: of, relating to, or printed by a web press

web·foot *noun* (1765)
1 \ˈweb-ˈfu̇t\ **:** a foot having webbed toes
2 \-ˌfu̇t\ **:** an animal having web feet
— web–foot·ed \-ˈfu̇-təd\ *adjective*

web member *noun* (circa 1890)
: one of the several members joining the top and bottom chords of a truss or lattice girder

web–off·set \ˈweb-ˈȯf-ˌset\ *noun, often attributive* (1959)
: offset printing by web press

web press *noun* (1875)
: a press that prints a continuous roll of paper

web spinner *noun* (circa 1907)
: an insect that spins a web; *especially* **:** any of an order (Embioptera synonym Embiidina) of small slender insects with biting mouthparts that live in silken tunnels which they spin

web·ster \ˈweb-stər\ *noun* [Middle English, from Old English *webbestre* female weaver, from *webbian* to weave; akin to Old English *wefan* to weave] (12th century)
archaic **:** WEAVER 1

web·work \ˈweb-ˌwərk\ *noun* (1790)
: WEB 5b ⟨a vast *webwork* of real estate holdings⟩

web·worm \-ˌwərm\ *noun* (1797)
: any of various caterpillars that are more or less gregarious and spin large webs

wed \ˈwed\ *verb* **wed·ded** *also* **wed; wed·ding** [Middle English *wedden,* from Old English *weddian;* akin to Middle High German *wetten* to pledge, Old English *wedd* pledge, Old High German *wetti,* Gothic *wadi,* Latin *vad-, vas* bail, security] (before 12th century)
transitive verb
1 : to take for wife or husband by a formal ceremony **:** MARRY
2 : to join in marriage
3 : to unite as if by the bond of marriage
intransitive verb
: to enter into matrimony
— wed·der *noun*

we'd \ˈwēd\ (1603)
: we had **:** we would **:** we should

Wed·dell seal \wi-ˈdel-, ˈwe-dᵊl-\ *noun* [James *Weddell* (died 1834) English navigator] (1914)
: a vocal Antarctic hair seal (*Leptonychotes weddelli*) noted for its deep dives in search of food

wed·ding \ˈwe-diŋ\ *noun, often attributive* (before 12th century)
1 : a marriage ceremony usually with its accompanying festivities **:** NUPTIALS
2 : an act, process, or instance of joining in close association
3 : a wedding anniversary or its celebration — usually used in combination ⟨a golden *wedding*⟩

wedding cake *noun* (1648)
1 : a usually elaborately decorated and tiered cake made for the celebration of a wedding
2 : something (as a large building) likened to a wedding cake especially in elaborate ornamentation

wedding march *noun* (1850)
: a march of slow tempo and stately character composed or played to accompany the bridal procession

wedding ring *noun* (14th century)
: a ring often of plain gold or platinum given by the groom to the bride during the wedding service; *also* **:** a similar ring given by the bride to the groom in a double-ring service

we·del \ˈvā-dᵊl\ *intransitive verb* [German *Wedeln*] (circa 1963)
: to ski downhill by means of wedeln

we·deln \ˈvā-dᵊln\ *noun* [German, from *wedeln,* literally, to fan, wag the tail, from *Wedel* fan, tail, from Old High German *wadal;* akin to Old Norse *vēli* bird's tail] (circa 1957)
: a style of skiing in which a skier rhythmically swings the rear of the skis from side to side while following the fall line

[1]**wedge** \ˈwej\ *noun* [Middle English *wegge,* from Old English *wecg;* akin to Old High German *wecki* wedge, Lithuanian *vagis*] (before 12th century)
1 : a piece of a substance (as wood or iron) that tapers to a thin edge and is used for splitting wood and rocks, raising heavy bodies, or for tightening by being driven into something
2 a : something (as a policy) causing a breach or separation **b :** something used to initiate an action or development
3 : something wedge-shaped: as **a :** an array of troops or tanks in the form of a wedge **b :** the wedge-shaped stroke in cuneiform characters **c :** a shoe having a heel extending from the back of the shoe to the front of the shank and a tread formed by an extension of the sole **d :** an iron golf club with a broad low-angled face for maximum loft

[2]**wedge** *verb* **wedged; wedg·ing** (15th century)
transitive verb
1 : to fasten or tighten by driving in a wedge
2 a : to force or press (something) into a narrow space **:** CRAM **b :** to force (one's way) into or through ⟨*wedged* his way into the crowd⟩
3 : to separate or force apart with or as if with a wedge
intransitive verb
: to become wedged

wedged \ˈwejd, ˈwe-jəd\ *adjective* (1552)
: shaped like a wedge

wedg·ie \ˈwe-jē\ *noun* (1939)
1 : a shoe having a wedge-shaped piece serving as the heel and joining the half sole to form a continuous flat undersurface
2 : the condition of having one's clothing wedged between the buttocks especially from having one's pants or underpants yanked up from behind as a prank — often used with *get* or *give*

Wedg·wood \ˈwej-ˌwu̇d\ *trademark*
— used for ceramic wares (as bone china or jasper)

wedgy \ˈwe-jē\ *adjective* (1799)
: resembling a wedge in shape

wed·lock \ˈwed-ˌläk\ *noun* [Middle English *wedlok,* from Old English *wedlāc* marriage bond, from *wedd* pledge + *-lāc,* suffix denoting activity] (13th century)
: the state of being married **:** MARRIAGE, MATRIMONY
— out of wedlock : with the natural parents not legally married to each other

Wednes·day \ˈwenz-dē, -(ˌ)dā; *British also* ˈwe-dᵊnz-\ *noun* [Middle English, from Old English *wōdnesdæg* (akin to Old Norse *ōthinsdagr* Wednesday); akin to Old English *Wōden* Odin and *dæg* day] (before 12th century)
: the fourth day of the week
— Wednes·days \-dēz, -(ˌ)dāz\ *adverb*

wee \ˈwē\ *adjective* [Middle English *we,* from *we,* noun, little bit, from Old English *wǣge* weight; akin to Old English *wegan* to move, weigh — more at WAY] (15th century)
1 : very small **:** DIMINUTIVE
2 : very early ⟨*wee* hours of the morning⟩

[1]**weed** \ˈwēd\ *noun* [Middle English, from Old English *wēod* weed, herb; akin to Old Saxon *wiod* weed] (before 12th century)
1 a (1) **:** a plant that is not valued where it is growing and is usually of vigorous growth; *especially* **:** one that tends to overgrow or choke out more desirable plants (2) **:** a weedy growth of plants **b :** an aquatic plant; *especially* **:** SEAWEED **c** (1) **:** TOBACCO (2) **:** MARIJUANA
2 a : an obnoxious growth, thing, or person **b :** something like a weed in detrimental quality; *especially* **:** an animal unfit to breed from

[2]**weed** (before 12th century)
intransitive verb
: to remove weeds or something harmful
transitive verb
1 a : to clear of weeds ⟨*weed* a garden⟩ **b** (1) **:** to free from something hurtful or offensive (2) **:** to remove the less desirable portions of
2 : to get rid of (something harmful or superfluous) — often used with *out*

[3]**weed** *noun* [Middle English *wede,* from Old English *wǣd, gewǣde;* akin to Old Norse *vāth* cloth, clothing and perhaps to Lithuanian *austi* to weave] (before 12th century)
1 : GARMENT — often used in plural
2 a : dress worn as a sign of mourning (as by a widow) — usually used in plural **b :** a band of crape worn on a man's hat as a sign of mourning — usually used in plural

weed·er \ˈwē-dər\ *noun* (15th century)
: one that weeds; *specifically* **:** any of various devices for removing weeds from an area

weedy \ˈwē-dē\ *adjective* **weed·i·er; -est** (15th century)
1 : abounding with or consisting of weeds
2 : resembling a weed especially in rank growth or ready propagation
3 : noticeably lean and scrawny **:** LANKY
— weed·i·ness *noun*

week \ˈwēk\ *noun* [Middle English *weke,* from Old English *wicu, wucu;* akin to Old High German *wehha* week and perhaps to Latin *vicis* change, alternation, Old High German *wehsal* exchange] (before 12th century)
1 a : any of a series of 7-day cycles used in various calendars; *especially* **:** a 7-day cycle beginning on Sunday and ending on Saturday **b** (1) **:** a week beginning with a specified day or containing a specified holiday ⟨the *week* of the 18th⟩ ⟨Easter *week*⟩ (2) **:** a week appointed for public recognition of some cause ⟨Fire Prevention *Week*⟩

2 a : any seven consecutive days **b :** a series of regular working, business, or school days during each 7-day period
3 : a time seven days before or after a specified day ⟨last Sunday *week*⟩
week·day \ˈwēk-ˌdā\ *noun* (15th century)
: a day of the week except Sunday or sometimes except Saturday and Sunday
week·days \-ˌdāz\ *adverb* (1777)
: on weekdays repeatedly **:** on any weekday ⟨takes a bus *weekdays*⟩
1week·end \ˈwēk-ˌend\ *noun* (1638)
: the end of the week; *specifically* **:** the period between the close of one working or business or school week and the beginning of the next
2weekend *intransitive verb* (1901)
: to spend the weekend
3weekend *adjective* (1935)
: active in a specified role only on weekends or part-time ⟨a *weekend* father⟩ ⟨*weekend* athletes⟩
weekend bag *noun* (1921)
: a suitcase of a size to carry clothing and personal articles for a weekend trip — called also *weekend case*
week·end·er \ˈwēk-ˈen-dər\ *noun* (1880)
1 : one that vacations or visits for a weekend
2 : WEEKEND BAG
week·ends \ˈwēk-ˌen(d)z\ *adverb* (1892)
: on weekends repeatedly **:** on any weekend ⟨travels *weekends*⟩
week·long \ˈwēk-ˈlȯŋ\ *adjective* (1898)
: lasting a week
1week·ly \ˈwē-klē\ *adverb* (15th century)
: every week **:** once a week **:** by the week
2weekly *adjective* (15th century)
1 : occurring, appearing, or done weekly
2 : reckoned by the week
3weekly *noun, plural* **weeklies** (1833)
: a weekly newspaper or periodical
week·night \ˈwēk-ˌnīt\ *noun* (1859)
: a weekday night
week·nights \-ˌnīts\ *adverb* (1965)
: on weeknights repeatedly **:** on any weeknight
ween \ˈwēn\ *transitive verb* [Middle English *wenen,* from Old English *wēnan;* akin to Old Norse *væna* to hope and probably to Latin *venus* love, charm — more at WIN] (before 12th century)
archaic **:** IMAGINE
wee·nie \ˈwē-nē\ *noun* [alteration of *wienie*] (circa 1906)
1 : FRANKFURTER
2 *slang* **:** PENIS
3 : NERD ⟨computer *weenies*⟩
wee·ny \ˈwē-nē\ *also* **ween·sy** \ˈwēn(t)-sē\ *adjective* [alteration of *wee*] (circa 1781)
: exceptionally small **:** TINY
weep \ˈwēp\ *verb* **wept** \ˈwept\; **weep·ing** [Middle English *wepen,* from Old English *wēpan;* akin to Old High German *wuoffan* to weep, Old Church Slavonic *vabiti* to call to] (before 12th century)
transitive verb
1 : to express deep sorrow for usually by shedding tears **:** BEWAIL
2 : to pour forth (tears) from the eyes
3 : to exude (a fluid) slowly **:** OOZE
intransitive verb
1 : to express passion (as grief) by shedding tears
2 a : to give off or leak fluid slowly **:** OOZE **b** *of a fluid* **:** to flow sluggishly or in drops
3 : to droop over **:** BEND
weep·er \ˈwē-pər\ *noun* (14th century)
1 a : one that weeps **b :** a professional mourner
2 : a small statue of a figure in mourning on a funeral monument
3 : a badge of mourning worn especially in the 18th and 19th centuries
4 *plural* **:** long and flowing side-whiskers
5 : TEARJERKER
weep hole *noun* (1851)
: a hole (as in a wall or foundation) that is designed to drain off accumulated water
weep·ie \ˈwē-pē\ *noun* (1928)
: TEARJERKER
weep·ing \ˈwē-piŋ\ *adjective* (before 12th century)
1 : TEARFUL
2 *archaic* **:** RAINY
3 : having slender pendent branches
weeping willow *noun* (circa 1731)
: an Asian willow (*Salix babylonica*) introduced into North America that has slender pendent branches
weepy \ˈwē-pē\ *adjective* (1863)
: inclined to weep **:** TEARFUL
weet \ˈwēt\ *verb* [Middle English *weten,* alteration of *witen* — more at WIT] (14th century)
archaic **:** KNOW
wee·vil \ˈwē-vəl\ *noun* [Middle English *wevel,* from Old English *wifel;* akin to Old High German *wibil* beetle, Old English *wefan* to weave] (before 12th century)
: any of a superfamily (Curculionoidea) of beetles which have the head prolonged into a more or less distinct snout and which include many that are injurious especially as larvae to nuts, fruit, and grain or to living plants; *especially* **:** any of a family (Curculionidae) having a well-developed snout curved downward with the jaws at the tip and clubbed usually elbowed antennae

weevil

— wee·vily *or* **wee·vil·ly** \ˈwēv-lē, ˈwē-və-lē\ *adjective*
weft \ˈweft\ *noun* [Middle English, from Old English; akin to Old Norse *veptr* weft, Old English *wefan* to weave — more at WEAVE] (before 12th century)
1 a : a filling thread or yarn in weaving **b :** yarn used for the weft
2 : WEB, FABRIC; *also* **:** an article of woven fabric
weft knit *noun* (1943)
: a knit fabric produced in machine or hand knitting with the yarns running crosswise or in a circle — compare WARP KNIT
— weft–knit·ted \-ˌni-təd\ *adjective*
— weft knit·ting *noun*
wei·ge·la \wī-ˈjē-lə\ *noun* [New Latin, from Christian E. *Weigel* (died 1831) German physician] (1846)
: any of a genus (*Weigela*) of showy shrubs of the honeysuckle family; *especially* **:** one (*W. florida*) of China widely grown for its usually pink or red flowers
1weigh \ˈwā\ *verb* [Middle English *weyen,* from Old English *wegan* to move, carry, weigh — more at WAY] (before 12th century)
transitive verb
1 : to ascertain the heaviness of by or as if by a balance
2 a : OUTWEIGH **b :** COUNTERBALANCE **c :** to make heavy **:** WEIGHT — often used with *down*
3 : to consider carefully especially by balancing opposing factors or aspects in order to reach a choice or conclusion **:** EVALUATE
4 : to heave up (an anchor) preparatory to sailing
5 : to measure or apportion (a definite quantity) on or as if on a scales
intransitive verb
1 a : to have a certain heaviness **:** experience a specific force due to gravity **b :** to register a weight (as on a scales) — used with *in* or *out;* compare WEIGH IN
2 : to merit consideration as important **:** COUNT ⟨evidence will *weigh* heavily against him⟩
3 a : to press down with or as if with a heavy weight **b :** to have a saddening or disheartening effect
4 : to weigh anchor
synonym see CONSIDER
— weigh·able \ˈwā-ə-bəl\ *adjective*
— weigh·er *noun*
2weigh *noun* [alteration of *way*] (1777)
: WAY — used in the phrase *under weigh*
weigh down *transitive verb* (14th century)
1 : to cause to bend down **:** OVERBURDEN
2 : OPPRESS, DEPRESS
weigh–in \ˈwā-ˌin\ *noun* (1939)
: an act or instance of weighing in as a contestant especially in sport
weigh in *intransitive verb* (1868)
1 : to have oneself or one's possessions (as baggage) weighed; *especially* **:** to have oneself weighed in connection with an athletic contest
2 : to enter as a participant
1weight \ˈwāt\ *noun* [Middle English *wight, weght,* from Old English *wiht;* akin to Old Norse *vætt* weight, Old English *wegan* to weigh] (before 12th century)
1 a : the amount that a thing weighs **b** (1) **:** the standard or established amount that a thing should weigh (2) **:** one of the classes into which contestants in a sports event are divided according to body weight (3) **:** poundage required to be carried by a horse in a handicap race
2 a : a quantity or thing weighing a fixed and usually specified amount **b :** a heavy object (as a metal ball) thrown, put, or lifted as an athletic exercise or contest
3 a : a unit of weight or mass — see METRIC SYSTEM table **b :** a piece of material (as metal) of known specified weight for use in weighing articles **c :** a system of related units of weight
4 a : something heavy **:** LOAD **b :** a heavy object to hold or press something down or to counterbalance
5 a : BURDEN, PRESSURE **b :** the quality or state of being ponderous **c :** CORPULENCE
6 a : relative heaviness **:** MASS **b :** the force with which a body is attracted toward the earth or a celestial body by gravitation and which is equal to the product of the mass and the local gravitational acceleration
7 a : the relative importance or authority accorded something **b :** measurable influence especially on others
8 : overpowering force
9 : the quality (as lightness) that makes a fabric or garment suitable for a particular use or season — often used in combination ⟨summer-*weight*⟩
10 : a numerical coefficient assigned to an item to express its relative importance in a frequency distribution
11 : the degree of thickness of the strokes of a type character
synonym see IMPORTANCE, INFLUENCE
2weight *transitive verb* (1647)
1 : to oppress with a burden ⟨*weighted* down with cares⟩
2 a : to load or make heavy with or as if with a weight **b :** to increase in heaviness by adding an ingredient
3 a : WEIGH 1 **b :** to feel the weight of **:** HEFT
4 : to assign a statistical weight to
5 : to cause to incline in a particular direction by manipulation ⟨the tax structure . . . which was *weighted* so heavily in favor of the upper classes —A. S. Link⟩
6 : to shift the burden of weight upon ⟨*weight* the inside ski⟩
weight·ed *adjective* (circa 1732)
1 : made heavy **:** LOADED ⟨*weighted* silk⟩
2 a : having a statistical weight attached ⟨a *weighted* test score⟩ **b :** compiled or calculated from weighted data ⟨a *weighted* mean⟩
3 : INCLINED 1
weight·less \ˈwāt-ləs\ *adjective* (circa 1547)
: having little weight **:** lacking apparent gravitational pull
— weight·less·ly *adverb*
— weight·less·ness *noun*
weight lifter *noun* (1897)
: one that lifts barbells in competition or as an exercise
— weight lifting *noun*
weight man *noun* (circa 1949)

WEIGHTS AND MEASURES[1]

UNIT	ABBREVIATION OR SYMBOL	EQUIVALENTS IN OTHER UNITS OF SAME SYSTEM	METRIC EQUIVALENT
		WEIGHT	
		Avoirdupois[2]	
ton			
short ton		*20 short hundredweight, 2000 pounds*	*0.907 metric ton*
long ton		*20 long hundredweight, 2240 pounds*	*1.016 metric tons*
hundredweight	*cwt*		
short hundredweight		*100 pounds, 0.05 short ton*	*45.359 kilograms*
long hundredweight		*112 pounds, 0.05 long ton*	*50.802 kilograms*
pound	*lb or* lb avdp *also #*	16 ounces, 7000 grains	0.454 kilogram
ounce	oz *or* oz avdp	16 drams, 437.5 grains, 0.0625 pound	28.350 grams
dram	dr *or* dr avdp	27.344 grains, 0.0625 ounce	1.772 grams
grain	gr	0.037 dram, 0.002286 ounce	0.0648 gram
		Troy	
pound	lb t	12 ounces, 240 pennyweight, 5760 grains	0.373 kilogram
ounce	oz t	20 pennyweight, 480 grains, 0.083 pound	31.103 grams
pennyweight	dwt *also* pwt	24 grains, 0.05 ounce	1.555 grams
grain	gr	0.042 pennyweight, 0.002083 ounce	0.0648 gram
		Apothecaries'	
pound	*lb ap*	*12 ounces, 5760 grains*	*0.373 kilogram*
ounce	*oz ap or ℥*	8 drams, 480 grains, 0.083 pound	31.103 grams
dram	dr ap *or* ʒ	3 scruples, 60 grains	3.888 grams
scruple	s ap *or* ℈	20 grains, 0.333 dram	1.296 grams
grain	gr	0.05 scruple, 0.002083 ounce, 0.0166 dram	0.0648 gram
		CAPACITY	
		U.S. liquid measure	
gallon	*gal*	*4 quarts (231 cubic inches)*	*3.785 liters*
quart	*qt*	*2 pints (57.75 cubic inches)*	*0.946 liter*
pint	*pt*	*4 gills (28.875 cubic inches)*	*473.176 milliliters*
gill	*gi*	*4 fluid ounces (7.219 cubic inches)*	*118.294 milliliters*
fluid ounce	*fl oz or f℥*	*8 fluid drams (1.805 cubic inches)*	*29.573 milliliters*
fluid dram	*fl dr or fʒ*	60 minims (0.226 cubic inch)	3.697 milliliters
minim	min *or* ♍	*1/60 fluid dram (0.003760 cubic inch)*	*0.061610 milliliter*
		U.S. dry measure	
bushel	*bu*	*4 pecks (2150.42 cubic inches)*	*35.239 liters*
peck	*pk*	*8 quarts (537.605 cubic inches)*	*8.810 liters*
quart	*qt*	*2 pints (67.201 cubic inches)*	*1.101 liters*
pint	*pt*	*½ quart (33.600 cubic inches)*	*0.551 liter*
		British imperial liquid and dry measure	
bushel	*bu*	*4 pecks (2219.36 cubic inches)*	*36.369 liters*
peck	*pk*	*2 gallons (554.84 cubic inches)*	*9.092 liters*
gallon	*gal*	*4 quarts (277.420 cubic inches)*	*4.546 liters*
quart	*qt*	*2 pints (69.355 cubic inches)*	*1.136 liters*
pint	*pt*	*4 gills (34.678 cubic inches)*	*568.26 milliliters*
gill	*gi*	*5 fluid ounces (8.669 cubic inches)*	*142.066 milliliters*
fluid ounce	*fl oz or f℥*	*8 fluid drams (1.7339 cubic inches)*	*28.412 milliliters*
fluid dram	*fl dr or fʒ*	*60 minims (0.216734 cubic inch)*	*3.5516 milliliters*
minim	*min or ♍*	*1/60 fluid dram (0.003612 cubic inch)*	*0.059194 milliliter*
		LENGTH	
mile	mi	5280 feet, 1760 yards, 320 rods	1.609 kilometers
rod	rd	5.50 yards, 16.5 feet	5.029 meters
yard	yd	3 feet, 36 inches	0.9144 meter
foot	ft *or* ′	12 inches, 0.333 yard	30.48 centimeters
inch	in *or* ″	0.083 foot, 0.028 yard	2.54 centimeters
		AREA	
square mile	sq mi *or* mi^2	640 acres, 102,400 square rods	2.590 square kilometers
acre		4840 square yards, 43,560 square feet	0.405 hectare, 4047 square meters
square rod	sq rd *or* rd^2	30.25 square yards, 0.00625 acre	25.293 square meters
square yard	sq yd *or* yd^2	1296 square inches, 9 square feet	0.836 square meter
square foot	sq ft *or* ft^2	144 square inches, 0.111 square yard	0.093 square meter
square inch	sq in *or* in^2	0.0069 square foot, 0.00077 square yard	6.452 square centimeters
		VOLUME	
cubic yard	cu yd *or* yd^3	27 cubic feet, 46,656 cubic inches	0.765 cubic meter
cubic foot	cu ft *or* ft^3	1728 cubic inches, 0.0370 cubic yard	0.028 cubic meter
cubic inch	cu in *or* in^3	0.00058 cubic foot, 0.000021 cubic yard	16.387 cubic centimeters

[1]For U.S. equivalents of the metric units see Metric System table.

[2]The U.S. uses the avoirdupois units as the common system of measuring weight.

: an athlete who competes in any of the field events in which a weight is thrown or put

weight room *noun* (1973)
: a room containing equipment for weight training

weight training *noun* (1955)
: a system of conditioning involving lifting weights especially for strength and endurance

weighty \ˈwā-tē\ *adjective* **weight·i·er; -est** (15th century)
1 a : of much importance or consequence **:** MOMENTOUS **b :** SOLEMN
2 a : weighing a considerable amount **b :** heavy in proportion to its bulk ⟨*weighty* metal⟩
3 : POWERFUL, TELLING ⟨*weighty* arguments⟩
synonym see HEAVY
— weight·i·ly \ˈwā-tᵊl-ē\ *adverb*
— weight·i·ness \ˈwā-tē-nəs\ *noun*

wei·ma·ra·ner \ˌvī-mə-ˈrä-nər, ˌwī-; ˈvī-mə-ˌ, ˈwī-\ *noun* [German, from *Weimar,* Germany] (1943)
: any of a breed of large light gray usually short-haired pointers of German origin

wei·ner *variant of* WIENER

weir \ˈwar, ˈwer, ˈwir\ *noun* [Middle English *were,* from Old English *wer;* akin to Old Norse *ver* fishing place, Old High German *werien, werren* to defend] (before 12th century)
1 : a fence or enclosure set in a waterway for taking fish
2 : a dam in a stream or river to raise the water level or divert its flow

[1]weird \ˈwird\ *noun* [Middle English *wird, werd,* from Old English *wyrd;* akin to Old Norse *urthr* fate, Old English *weorthan* to become — more at WORTH] (before 12th century)
1 : FATE, DESTINY; *especially* **:** ill fortune
2 : SOOTHSAYER ◆

[2]weird *adjective* (15th century)
1 : of, relating to, or caused by witchcraft or the supernatural **:** MAGICAL
2 : of strange or extraordinary character **:** ODD, FANTASTIC ☆
— weird·ly *adverb*
— weird·ness *noun*

weird·ie \ˈwir-dē\ *or* **weirdy** *noun, plural* **weird·ies** (1894)
: WEIRDO

weirdo \ˈwir-(ˌ)dō\ *noun, plural* **weird·os** (circa 1955)
: a person who is extraordinarily strange or eccentric

Weird Sisters *noun plural*
: FATES

wei·sen·hei·mer *variant of* WISENHEIMER

weka \ˈwe-kə\ *noun* [Maori] (1845)
: a flightless New Zealand rail (*Gallirallus australis*)

welch \ˈwelch\ *variant of* WELSH

Welch \ˈwelch\ *variant of* WELSH

[1]wel·come \ˈwel-kəm\ *transitive verb* **wel·comed; wel·com·ing** [Middle English, from Old English *welcumian, wylcumian,* from *wilcuma,* noun] (before 12th century)
1 : to greet hospitably and with courtesy or cordiality
2 : to accept with pleasure the occurrence or presence of ⟨*welcomes* danger⟩
— wel·com·er *noun*

[2]welcome *interjection* [Middle English, alteration of *wilcume,* from Old English, from *wilcuma* desirable guest (akin to Old High German *willicomo* desirable guest); akin to Old English *willa, will* desire, *cuman* to come — more at WILL, COME] (12th century)
— used to express a greeting to a guest or newcomer upon arrival

[3]welcome *adjective* (12th century)
1 : received gladly into one's presence or companionship ⟨was always *welcome* in their home⟩
2 : giving pleasure **:** received with gladness or delight especially in response to a need ⟨a *welcome* relief⟩
3 : willingly permitted or admitted ⟨he was *welcome* to come and go —W. M. Thackeray⟩
4 — used in the phrase "You're welcome" as a reply to an expression of thanks
— wel·come·ly *adverb*
— wel·come·ness *noun*

[4]welcome *noun* (1525)
1 : a greeting or reception usually upon arrival
2 : the state of being welcome ⟨overstayed their *welcome*⟩

welcome mat *noun* (1946)
: something likened to a mat placed before an entrance as a sign of welcome

[1]weld \ˈweld\ *verb* [alteration of obsolete English *well* to weld, from Middle English *wellen* to boil, well, weld — more at WELL] (1599)
intransitive verb
: to become or be capable of being welded
transitive verb
1 a : to unite (metallic parts) by heating and allowing the metals to flow together or by hammering or compressing with or without previous heating **b :** to unite (plastics) in a similar manner by heating **c :** to repair (as an article) by this method **d :** to produce or create as if by such a process
2 : to unite or reunite closely or intimately
— weld·able \ˈwel-də-bəl\ *adjective*

[2]weld *noun* (1831)
1 : a welded joint
2 : union by welding **:** the state or condition of being welded

weld·er \ˈwel-dər\ *noun* (circa 1828)
: one that welds: as **a** *or* **wel·dor :** one whose work is welding **b :** a machine used in welding

weld·ment \ˈwel(d)-mənt\ *noun* (1941)
: a unit formed by welding together an assembly of pieces

[1]wel·fare \ˈwel-ˌfar, -ˌfer\ *noun* [Middle English, from the phrase *wel faren* to fare well] (14th century)
1 : the state of doing well especially in respect to good fortune, happiness, well-being, or prosperity
2 a : aid in the form of money or necessities for those in need **b :** an agency or program through which such aid is distributed

[2]welfare *adjective* (1904)
1 : of, relating to, or concerned with welfare and especially with improvement of the welfare of disadvantaged social groups ⟨*welfare* legislation⟩
2 : receiving public welfare benefits ⟨*welfare* families⟩

welfare state *noun* (1941)
1 : a social system based on the assumption by a political state of primary responsibility for the individual and social welfare of its citizens
2 : a nation or state characterized by the operation of the welfare state system

wel·far·ism \ˈwel-ˌfar-ˌi-zəm, -ˌfer-\ *noun* (1949)
: the complex of policies, attitudes, and beliefs associated with the welfare state
— wel·far·ist \-ist\ *noun or adjective*

wel·kin \ˈwel-kən\ *noun* [Middle English, literally, cloud, from Old English *wolcen;* akin to Old High German *wolkan* cloud] (12th century)
1 a : the vault of the sky **:** FIRMAMENT **b :** the celestial abode of God or the gods **:** HEAVEN
2 : the upper atmosphere

[1]well \ˈwel\ *noun* [Middle English *welle,* from Old English *welle;* akin to Old English *weallan* to bubble, boil, Old High German *wella* wave, Lithuanian *vilnis*] (before 12th century)
1 a : an issue of water from the earth **:** a pool fed by a spring **b :** SOURCE, ORIGIN
2 a : a pit or hole sunk into the earth to reach a supply of water **b :** a shaft or hole sunk to obtain oil, brine, or gas
3 a : an enclosure in the middle of a ship's hold to protect from damage and facilitate the inspection of the pumps **b :** a compartment in the hold of a fishing boat in which fish are kept alive
4 : an open space extending vertically through floors of a structure
5 : a space having a construction or shape suggesting a well for water
6 a : something resembling a well in being damp, cool, deep, or dark **b :** a deep vertical hole **c :** a source from which something may be drawn as needed
7 : a pronounced minimum of a variable in physics ⟨a potential *well*⟩

[2]well [Middle English, from Old English *wellan* to cause to well; akin to Old English *weallan* to bubble, boil] (before 12th century)
intransitive verb
1 : to rise to the surface and usually flow forth ⟨tears *welled* from her eyes⟩
2 : to rise like a flood of liquid ⟨longing *welled* up in his breast⟩
transitive verb
: to emit in a copious free flow

[3]well *adverb* **bet·ter** \ˈbe-tər\; **best** \ˈbest\ [Middle English *wel,* from Old English; akin to Old High German *wela* well, Old English *wyllan* to wish — more at WILL] (before 12th century)
1 a : in a good or proper manner **:** JUSTLY, RIGHTLY **b :** satisfactorily with respect to conduct or action ⟨did *well* in math⟩
2 : in a kindly or friendly manner ⟨spoke *well* of your idea⟩
3 a : with skill or aptitude **:** EXPERTLY ⟨paints *well*⟩ **b :** SATISFACTORILY **c :** with good appearance or effect **:** ELEGANTLY ⟨carried himself *well*⟩
4 : with careful or close attention **:** ATTENTIVELY
5 : to a high degree ⟨*well* deserved the honor⟩ ⟨a *well*-equipped kitchen⟩ — often used as an intensifier or qualifier ⟨there are . . . vacancies pretty *well* all the time —*Listener*⟩
6 : FULLY, QUITE ⟨*well* worth the price⟩

☆ SYNONYMS

Weird, eerie, uncanny mean mysteriously strange or fantastic. WEIRD may imply an unearthly or supernatural strangeness or it may stress queerness or oddness ⟨*weird* creatures from another world⟩. EERIE suggests an uneasy or fearful consciousness that mysterious and malign powers are at work ⟨an *eerie* calm preceded the bombing raid⟩. UNCANNY implies disquieting strangeness or mysteriousness ⟨an *uncanny* resemblance between total strangers⟩.

◇ WORD HISTORY

weird *Weird* is a northern English and Scots outcome of Old English *wyrd,* the Anglo-Saxons' name for the unpredictable and uncontrollable force determining the outcome of human events that in Modern English we call *fate.* As early as the late 8th century, *wyrde,* the plural of *wyrd,* served as an Old English gloss for *Parcae,* the Latin name for the Fates, three goddesses who spun, measured, and cut the thread of life. In the 15th and 16th centuries Scots authors employed *werd* or *weird* attributively in the phrase *weird sisters* to refer to the Fates of classical myth. Shakespeare adopted this usage in his tragedy *Macbeth,* in which the *weird sisters* are depicted not in their ancient guise but rather as three witches. Subsequent adjectival use of *weird* to mean "unearthly," "mysterious," or "odd" has grown out of a reinterpretation of *weird* in the Shakespearean phrase, most notably by the Romantic poets Shelley and Keats in the early 19th century.

7 a : in a way appropriate to the facts or circumstances **:** FITTINGLY, RIGHTLY **b :** in a prudent manner **:** SENSIBLY — used with *do*
8 : in accordance with the occasion or circumstances **:** with propriety or good reason ⟨cannot *well* refuse⟩
9 a : as one could wish **:** FAVORABLY **b :** with material success **:** ADVANTAGEOUSLY ⟨married *well*⟩
10 a : EASILY, READILY ⟨could *well* afford a new car⟩ **b :** in all likelihood **:** INDEED ⟨it may *well* be true⟩
11 : in a prosperous or affluent manner ⟨he lives *well*⟩
12 : to an extent approaching completeness **:** THOROUGHLY ⟨after being *well* dried with a sponge⟩
13 : without doubt or question **:** CLEARLY ⟨*well* knew the penalty⟩
14 : in a familiar manner ⟨knew her *well*⟩
15 : to a large extent or degree **:** CONSIDERABLY, FAR ⟨*well* over a million⟩
usage see GOOD
— as well 1 : in addition **:** ALSO ⟨there were other features *as well*⟩ **2 :** to the same extent or degree **:** as much ⟨open *as well* to the poor as to the rich⟩ **3 :** with equivalent, comparable, or more favorable effect ⟨might just *as well* have stayed home⟩

⁴well *interjection* (before 12th century)
1 — used to indicate resumption of discourse or to introduce a remark
2 — used to express surprise or expostulation

⁵well *adjective* (before 12th century)
1 a : PROSPEROUS, WELL-OFF **b :** being in satisfactory condition or circumstances
2 : being in good standing or favor
3 : SATISFACTORY, PLEASING ⟨all's *well* that ends well⟩
4 : ADVISABLE, DESIRABLE ⟨it might be *well* for you to leave⟩
5 a : free or recovered from infirmity or disease **:** HEALTHY ⟨a *well* man⟩ **b :** completely cured or healed ⟨the wound is nearly *well*⟩
6 : pleasing or satisfactory in appearance
7 : being a cause for thankfulness **:** FORTUNATE ⟨it is *well* that this has happened⟩
synonym see HEALTHY
usage see GOOD

we'll \'wē(ə)l, wil\ (1578)
: we will **:** we shall

well–ad·just·ed \ˌwel-ə-'jəs-təd\ *adjective* (1692)
: WELL-BALANCED 2

well–ad·vised \'wel-əd-'vīzd\ *adjective* (14th century)
1 : acting with wisdom, wise counsel, or proper deliberation **:** PRUDENT
2 : resulting from, based on, or showing careful deliberation or wise counsel ⟨*well-advised* plans⟩

well–ap·point·ed \'wel-ə-'pȯin-təd\ *adjective* (1530)
: having good and complete equipment **:** properly fitted out ⟨a *well-appointed* house⟩

wel·la·way \ˌwe-lə-'wā, 'we-lə-ˌ\ *interjection* [Middle English *welaway,* from Old English *weilāwei,* literally, woe! lo! woe!, alteration of *wālāwā,* from *wā* woe + *lā* lo + *wā* woe — more at WOE] (before 12th century)
archaic — used to express sorrow or lamentation

well–ba·lanced \'wel-'ba-lən(t)st\ *adjective* (1629)
1 : nicely or evenly balanced, arranged, or regulated ⟨a *well-balanced* diet⟩ ⟨a *well-balanced* attack in football⟩
2 : emotionally or psychologically untroubled

well–be·ing \'wel-'bē-iŋ\ *noun* (circa 1613)
: the state of being happy, healthy, or prosperous **:** WELFARE

well–be·loved \ˌwel-bi-'ləvd\ *adjective* (14th century)
1 : sincerely and deeply loved ⟨my *well-beloved* wife⟩
2 : sincerely respected — used in various ceremonial forms of address

well·born \'wel-'bȯrn\ *adjective* (before 12th century)
: born of noble or wealthy lineage

well–bred \-'bred\ *adjective* (1597)
1 : having or displaying good breeding **:** REFINED
2 : having a good pedigree ⟨*well-bred* swine⟩

well–con·di·tioned \ˌwel-kən-'di-shənd\ *adjective* (15th century)
1 : characterized by proper disposition, morals, or behavior
2 : having a good physical condition **:** SOUND

well deck *noun* (1888)
: a space on the weather deck of a ship lying at a lower level between a raised forecastle or poop and the bridge superstructure

well–de·fined \ˌwel-di-'fīnd\ *adjective* (1704)
1 : having clearly distinguishable limits, boundaries, or features ⟨a *well-defined* scar⟩
2 : clearly stated or described ⟨*well-defined* policies⟩

well–dis·posed \-dis-'pōzd\ *adjective* (14th century)
: having a good disposition; *especially* **:** disposed to be friendly, favorable, or sympathetic

well–done \'wel-'dən\ *adjective* (15th century)
1 : rightly or properly performed
2 : cooked thoroughly

Wel·ler·ism \'we-lə-ˌri-zəm\ *noun* [Sam *Weller,* witty servant of Mr. Pickwick in the story *Pickwick Papers* (1836–37) by Charles Dickens] (1839)
: an expression of comparison comprising a usually well-known quotation followed by a facetious sequel (as "'every one to his own taste,' said the old woman as she kissed the cow")

well–fa·vored \'wel-'fā-vərd\ *adjective* (15th century)
: GOOD-LOOKING, HANDSOME
— well–fa·vored·ness *noun*

well–fixed \-'fikst\ *adjective* (1822)
: having plenty of money or property

well–found \-'fau̇nd\ *adjective* (1793)
: fully furnished **:** properly equipped ⟨a *well-found* ship⟩

well–found·ed \-'fau̇n-dəd\ *adjective* (14th century)
: based on excellent reasoning, information, judgment, or grounds

well–groomed \-'grümd, -'gru̇md\ *adjective* (1886)
1 : well-dressed and scrupulously neat ⟨*well-groomed* men⟩
2 : made neat, tidy, and attractive down to the smallest details ⟨a *well-groomed* lawn⟩

well–ground·ed \-'grau̇n-dəd\ *adjective* (14th century)
1 : having a firm foundation ⟨*well-grounded* in Latin and Greek⟩
2 : WELL-FOUNDED

well–han·dled \-'han-dᵊld\ *adjective* (15th century)
1 : managed or administered efficiently
2 : having been handled a great deal ⟨*well-handled* goods for sale⟩

well·head \'wel-ˌhed\ *noun* (14th century)
1 : the source of a spring or a stream
2 : principal source **:** FOUNTAINHEAD
3 : the top of or a structure built over a well

wellhead price *noun* (1953)
: the price less transportation costs charged by the producer for petroleum or natural gas

well–heeled \-'hē(ə)ld\ *adjective* (1897)
: having plenty of money **:** WELL-FIXED

wel·lie *or* **wel·ly** \'we-lē\ *noun, plural* **wel·lies** [by shortening & alteration] (1961)
chiefly British **:** WELLINGTON — usually used in plural

well–in·formed \-in-'fȯrmd\ *adjective* (15th century)
1 : having extensive knowledge especially of current topics and events
2 : thoroughly knowledgeable in a particular subject

Wel·ling·ton \'we-liŋ-tən\ *noun* [Arthur Wellesley, 1st Duke of *Wellington*] (1817)
: a boot having a loose top with the front usually coming to or above the knee — usually used in plural

well–in·ten·tioned \ˌwel-in-'ten(t)-shənd\ *adjective* (1598)
: WELL-MEANING

well–knit \'wel-'nit\ *adjective* (15th century)
: firmly knit ⟨a *well-knit* group⟩; *especially* **:** firmly and strongly constructed, compacted, or framed ⟨a *well-knit* drama⟩

well–known \-'nōn\ *adjective* (15th century)
: fully or widely known

well–mean·ing \-'mē-niŋ\ *adjective* (1532)
1 : having good intentions ⟨*well-meaning* but misguided idealists⟩
2 : based on good intentions ⟨*well-meaning* advice⟩

well–meant \-'ment\ *adjective* (15th century)
: WELL-MEANING 2

well·ness \'wel-nəs\ *noun* (1654)
: the quality or state of being in good health especially as an actively sought goal ⟨*wellness* clinics⟩ ⟨lifestyles that promote *wellness*⟩

well–nigh \-'nī\ *adverb* (12th century)
: ALMOST, NEARLY ⟨*well-nigh* impossible⟩

well–off \-'ȯf\ *adjective* (1733)
1 : being in good condition or favorable circumstances ⟨doesn't know when he's *well-off*⟩
2 : having no lack — usually used with *for*
3 a : being in easy or affluent circumstances **:** WELL-TO-DO **b :** suggesting prosperity ⟨the house had a sleek *well-off* look⟩

well–oiled \-'ȯild\ *adjective* (1847)
: smoothly functioning ⟨a *well-oiled* political machine⟩

well–or·dered \-'ȯr-dərd\ *adjective* (1589)
1 : having an orderly procedure or arrangement ⟨a *well-ordered* household⟩
2 : partially ordered with every subset containing a first element and exactly one of the relationships "greater than", "less than", or "equal to" holding for any given pair of elements
— well–or·der·ing \-'ȯrd-riŋ, -'ȯr-də-\ *noun*

well–read \-'red\ *adjective* (1596)
: well informed or deeply versed through reading ⟨*well-read* in history⟩

well–round·ed \-'rau̇n-dəd\ *adjective* (1875)
: fully or broadly developed: as **a :** having a broad educational background ⟨schools that turn out *well-rounded* graduates⟩ **b :** COMPREHENSIVE ⟨a *well-rounded* program of activities⟩

well–set \-'set\ *adjective* (14th century)
1 : well or firmly established ⟨*well-set* in his own values —William Johnson⟩
2 : strongly built ⟨a *well-set* athlete⟩

well–spo·ken \'wel-'spō-kən\ *adjective* (15th century)
1 : speaking well, fitly, or courteously
2 : spoken with propriety ⟨*well-spoken* words⟩

well·spring \-ˌspriŋ\ *noun* (before 12th century)
: a source of continual supply ⟨a *wellspring* of information⟩

well–tak·en \-'tā-kən\ *adjective* (1761)
: WELL-GROUNDED, JUSTIFIABLE ⟨your point is *well-taken*⟩

well–thought–of \wel-'thȯt-ˌəv, -ˌäv\ *adjective* (1579)
: being of good repute

well–timed \'wel-'tīmd\ *adjective* (circa 1656)
: happening at an opportune moment **:** TIMELY ⟨a *well-timed* announcement⟩

well–to–do \ˌwel-tə-'dü\ *adjective* (1825)

: having more than adequate financial resources **:** PROSPEROUS ⟨a *well-to-do* family⟩

well–turned \ˈwel-ˈtərnd\ *adjective* (1616)
1 : symmetrically shaped or rounded **:** SHAPELY
2 : concisely and appropriately expressed ⟨a *well-turned* phrase⟩
3 : expertly rounded or turned ⟨a *well-turned* column⟩

well–wish•er \ˈwel-ˌwi-shər, -ˈwi-\ *noun* (1590)
: one that wishes well to another
— well–wish•ing \-shiŋ\ *adjective or noun*

well–worn \-ˈwōrn, -ˈwȯrn\ *adjective* (1621)
1 a : made trite by overuse **:** HACKNEYED ⟨a *well-worn* quotation⟩ **b :** having been much used or worn ⟨*well-worn* shoes⟩
2 *archaic* **:** worn well or properly

welsh \ˈwelsh, ˈwelch\ *intransitive verb* [probably from *Welsh,* adjective] (1905)
1 : to avoid payment — used with *on* ⟨*welshed* on his debts⟩
2 : to break one's word **:** RENEGE ⟨*welshed* on their promises⟩
— welsh•er *noun*

Welsh \ˈwelsh *also* ˈwelch\ *noun* [Middle English *Walsche, Welsse,* from *walisch, welisch,* adjective, Welsh, from Old English *wælisc, welisc* Celtic, Welsh, foreign, from Old English *Wealh* Celtic, Welshman, foreigner, of Celtic origin; akin to the source of Latin *Volcae,* a Celtic people of southeastern Gaul] (before 12th century)
1 : the Celtic language of the Welsh people
2 *plural in construction* **:** the natives or inhabitants of Wales
3 : WELSH PONY
— Welsh *adjective*

Welsh black *noun* (1919)
: any of a breed of hardy medium-sized thick-haired horned black cattle of Welsh origin that are raised for meat and milk

Welsh cob *noun* (1947)
: any of a breed of medium-sized cobby horses that were developed by interbreeding Welsh mountain ponies with larger horses

Welsh corgi *noun* (1926)
: a short-legged long-backed dog with foxy head of either of two breeds of Welsh origin: **a :** CARDIGAN WELSH CORGI **b :** PEMBROKE WELSH CORGI

Welsh corgi: *1* Pembroke, *2* Cardigan

Welsh•man \-mən\ *noun* (13th century)
: a native or inhabitant of Wales

Welsh mountain pony *noun* (1947)
: any of a breed of small sturdy ponies native to the mountains of Wales that do not exceed 12.2 hands (124 centimeters) in height

Welsh pony *noun* (1771)
: a pony of any of several breeds of Welsh origin; *especially* **:** any of a breed of riding and light draft ponies measuring 12.2 to 13.2 hands (124 to 134 centimeters) in height

Welsh rabbit *noun* (1725)
: melted often seasoned cheese poured over toast or crackers

Welsh rare•bit \-ˈrar-bət, -ˈrer-\ *noun* [by alteration] (1785)
: WELSH RABBIT

Welsh springer spaniel *noun* (circa 1929)
: any of a breed of red and white or orange and white small-eared springer spaniels of Welsh origin

Welsh terrier *noun* (1885)
: any of a breed of wiry-coated terriers resembling Airedales but smaller and developed in Wales for hunting

Welsh•wom•an \ˈwelsh-ˌwu̇-mən *also* ˈwelch-\ *noun* (15th century)
: a woman who is a native or inhabitant of Wales

[1]welt \ˈwelt\ *noun* [Middle English *welte*] (15th century)
1 : a strip between a shoe sole and upper through which they are stitched or stapled together
2 : a doubled edge, strip, insert, or seam (as on a garment) for ornament or reinforcement
3 a : a ridge or lump raised on the body usually by a blow **b :** a heavy blow

[2]welt *transitive verb* (15th century)
1 : to furnish with a welt
2 a : to raise a welt on the body of **b :** to hit hard

welt•an•schau•ung \ˈvelt-ˌän-ˌshau̇-əŋ\ *noun, plural* **weltanschauungs** \-əŋz\ *or* **welt•an•schau•ung•en** \-əŋ-ən\ *often capitalized* [German, from *Welt* world + *Anschauung* view] (1868)
: a comprehensive conception or apprehension of the world especially from a specific standpoint

[1]wel•ter \ˈwel-tər\ *intransitive verb* **wel•tered; wel•ter•ing** \-t(ə-)riŋ\ [Middle English; akin to Middle Dutch *welteren* to roll, Old High German *walzan,* Lithuanian *volioti,* Latin *volvere* — more at VOLUBLE] (14th century)
1 a : WRITHE, TOSS; *also* **:** WALLOW **b :** to rise and fall or toss about in or with waves
2 : to become deeply sunk, soaked, or involved
3 : to be in turmoil

[2]welter *noun* (1596)
1 : a state of wild disorder **:** TURMOIL
2 : a chaotic mass or jumble ⟨a bewildering *welter* of data⟩

[3]welter *noun* (1900)
: WELTERWEIGHT

wel•ter•weight \ˈwel-tər-ˌwāt\ *noun* [*welter* (probably from [1]*welt*) + *weight*] (circa 1892)
: a boxer in a weight division having a maximum limit of 147 pounds — compare LIGHTWEIGHT, MIDDLEWEIGHT

welt•schmerz \ˈvelt-ˌshmerts\ *noun, often capitalized* [German, from *Welt* world + *Schmerz* pain] (1875)
1 : mental depression or apathy caused by comparison of the actual state of the world with an ideal state
2 : a mood of sentimental sadness

[1]wen \ˈwen\ *noun* [Middle English *wenn,* from Old English; akin to Middle Low German *wene* wen] (before 12th century)
: an abnormal growth or a cyst protruding from a surface especially of the skin

[2]wen *variant of* WYNN

[1]wench \ˈwench\ *noun* [Middle English *wenche,* short for *wenchel* child, from Old English *wencel;* akin to Old High German *wankōn* to totter, waver and probably to Old High German *winchan* to stagger — more at WINK] (14th century)
1 a : a young woman **:** GIRL **b :** a female servant
2 : a lewd woman **:** PROSTITUTE

[2]wench *intransitive verb* (1599)
: to consort with lewd women; *especially* **:** to practice fornication
— wench•er *noun*

wend \ˈwend\ *verb* [Middle English, from Old English *wendan;* akin to Old High German *wenten* to turn, Old English *windan* to twist — more at WIND] (before 12th century)
intransitive verb
: to direct one's course **:** TRAVEL
transitive verb
: to proceed on (one's way) **:** DIRECT

Wend \ˈwend\ *noun* [German *Wende,* from Old High German *Winida;* akin to Old English *Winedas,* plural, Wends] (1786)
: a member of a Slavic people of eastern Germany

[1]Wend•ish \ˈwen-dish\ *adjective* (1614)
: of or relating to the Wends or their language

[2]Wendish *noun* (1617)
: the West Slavic language of the Wends

went [Middle English, past & past participle of *wenden* to wend] *past of* GO

wen•tle•trap \ˈwen-tᵊl-ˌtrap\ *noun* [Dutch *wenteltrap* winding stair, from Middle Dutch *wendeltrappe,* from *wendel* turning + *trappe* stairs] (1758)
: any of a family (Epitoniidae) of marine snails with usually white shells; *also* **:** one of the shells

wept *past and past participle of* WEEP

were [Middle English *were* (suppletive singular past subjunctive & 2d singular past indicative of *been* to be), *weren* (suppletive past plural of *been*), from Old English *wǣre* (singular past subjunctive & 2d singular past indicative of *wesan* to be), *wǣron* (past plural indicative of *wesan*), *wǣren* (past plural subjunctive of *wesan*) — more at WAS] *past 2d singular, past plural, or past subjunctive of* BE

we're \ˈwir, ˈwər, ˈwē-ər\ (1608)
: we are

weren't \(ˈ)wərnt, ˈwər-ənt\ (1865)
: were not

were•wolf \ˈwir-ˌwu̇lf, ˈwer-, ˈwər-\ *noun, plural* **were•wolves** \-ˌwu̇lvz\ [Middle English, from Old English *werwulf* (akin to Old High German *werwolf* werewolf), from *wer* man + *wulf* wolf — more at VIRILE, WOLF] (before 12th century)
: a person transformed into a wolf or capable of assuming a wolf's form

wer•gild \ˈwər-ˌgild\ *or* **wer•geld** \-ˌgeld\ *noun* [Middle English *wergeld,* from Old English, from *wer* man + *-geld,* alteration of *gield, geld* payment, tribute — more at GELD] (13th century)
: the value set in Anglo-Saxon and Germanic law upon human life in accordance with rank and paid as compensation to the kindred or lord of a slain person

wert \ˈwərt\ *archaic past 2d singular of* BE

wes•kit \ˈwes-kət\ *noun* [alteration of *waistcoat*] (1856)
: VEST 2a

Wes•ley•an•ism \ˈwes-lē-ə-ˌni-zəm *also* ˈwez-\ *noun* (1774)
: METHODISM 1; *specifically* **:** the system of Arminian Methodism taught by John Wesley
— Wes•ley•an \-lē-ən\ *adjective or noun*

[1]west \ˈwest\ *adverb* [Middle English, from Old English; akin to Old High German *westar* to the west and probably to Latin *vesper* evening, Greek *hesperos*] (before 12th century)
: to, toward, or in the west

[2]west *adjective* (before 12th century)
1 : situated toward or at the west ⟨the *west* exit⟩
2 : coming from the west ⟨a *west* wind⟩

[3]west *noun* (12th century)
1 a : the general direction of sunset **:** the direction to the left of one facing north **b :** the compass point directly opposite to east
2 *capitalized* **a :** regions or countries lying to the west of a specified or implied point of orientation **b :** the noncommunist countries of Europe and America
3 : the end of a church opposite the chancel
4 *often capitalized* **a :** the one of four positions at 90-degree intervals that lies to the west or at the left of a diagram **b :** a person (as a bridge player) occupying this position during a specified activity

west•bound \ˈwes(t)-ˌbau̇nd\ *adjective* (1881)
: traveling or heading west

west by north (1760)
: a compass point that is one point north of due west **:** N78°45′W

west by south (1577)
: a compass point that is one point south of due west **:** S78°45′W

west·er \'wes-tər\ *intransitive verb* **west·ered; west·er·ing** \-t(ə-)riŋ\ [Middle English *westren,* from [1]*west*] (14th century)
: to turn or move westward ⟨the half moon *westers* low —A. E. Housman⟩

[1]**west·er·ly** \'wes-tər-lē\ *adjective or adverb* [obsolete *wester* western] (15th century)
1 : situated toward or belonging to the west ⟨the *westerly* end of the farm⟩
2 : coming from the west ⟨a *westerly* breeze⟩

[2]**westerly** *noun, plural* **-lies** (1876)
: a wind from the west

[1]**west·ern** \'wes-tərn\ *adjective* [Middle English *westerne,* from Old English; akin to Old High German *westrōni* western, Old English *west*] (before 12th century)
1 a : coming from the west ⟨a *western* storm⟩ **b :** lying toward the west
2 *capitalized* **:** of, relating to, or characteristic of a region conventionally designated West: as **a :** steeped in or stemming from the Greco-Roman traditions **b :** of or relating to the noncommunist countries of Europe and America **c :** of or relating to the American West
3 *capitalized* **:** of or relating to the Roman Catholic or Protestant segment of Christianity ⟨*Western* liturgies⟩
— west·ern·most \-ˌmōst\ *adjective*

[2]**western** *noun* (1612)
1 : one that is produced in or characteristic of a western region and especially the western U.S.
2 *often capitalized* **:** a novel, story, motion picture, or broadcast dealing with life in the western U.S. especially during the latter half of the 19th century

Western blot *noun* [after *Southern blot*] (1983)
: a blot that consists of a nitrocellulose sheet containing spots of protein for identification by a suitable molecular probe and is used especially for the detection of antibodies
— Western blotting *noun*

West·ern·er \'wes-tə(r)-nər\ *noun* (1599)
1 : a native or inhabitant of the West; *especially* **:** a native or resident of the western part of the U.S.
2 : one advocating the adoption of western European culture especially in 19th century Russia

western hemisphere *noun, often W&H capitalized* (1624)
: the half of the earth comprising North and South America and surrounding waters

west·ern·i·sa·tion, west·ern·ise *British variant of* WESTERNIZATION, WESTERNIZE

west·ern·i·za·tion \ˌwes-tər-nə-'zā-shən\ *noun, often capitalized* (1904)
: conversion to or adoption of western traditions or techniques

west·ern·ize \'wes-tər-ˌnīz\ *verb* **-ized; -iz·ing** *often capitalized* (1837)
transitive verb
: to imbue with qualities native to or associated with a western region and especially the noncommunist countries of Europe and America
intransitive verb
: to become westernized

western omelet *noun* (1931)
: an omelet made usually with diced ham, green pepper, and onion — called also *Denver omelet*

western red cedar *noun* (1886)
: RED CEDAR 2; *also* **:** its wood

western saddle *noun, often W capitalized* (1897)
: a deep-seated saddle with a high pommel and broad skirts and fenders used originally by cattlemen — called also *stock saddle;* see SADDLE illustration

western sandwich *noun* (1926)
: a sandwich filled with a western omelet — called also *Denver sandwich*

western swing *noun* (1973)
: swing music played typically on country-music instruments (as guitar, fiddle, or steel guitar)

West Germanic *noun* (1894)
: a subdivision of the Germanic languages including English, Frisian, Dutch, and German — see INDO-EUROPEAN LANGUAGES table

West Highland white terrier *noun* (circa 1904)
: any of a breed of small white long-coated terriers developed in Scotland — called also *West·ie* \'wes-tē\

West Highland white terrier

west·ing \'wes-tiŋ\ *noun* (1628)
: westerly progress
: a going westward

west–northwest *noun* (15th century)
: a compass point that is two points north of due west **:** N67°30′W

West·pha·lian ham \wes(t)-'fāl-yən-, -'fā-lē-ən-\ *noun* [*Westphalia,* Germany] (1664)
: a ham of distinctive flavor produced by smoking with juniper brush

West Saxon *noun* (14th century)
1 : a member or a descendant of the Saxons who occupied England largely south of the Thames, west of the Downs, and east of Dartmoor
2 : a dialect of Old English used as the chief literary dialect in England before the Norman Conquest

west–southwest *noun* (1555)
: a compass point that is two points south of due west **:** S67°30′W

[1]**west·ward** \'wes-twərd\ *adverb or adjective* (before 12th century)
: toward the west
— west·wards \-twərdz\ *adverb*

[2]**westward** *noun* (1652)
: westward direction or part ⟨sail to the *westward*⟩

[1]**wet** \'wet\ *adjective* **wet·ter; wet·test** [Middle English, partly from past participle of *weten* to wet & partly from Old English *wǣt* wet; akin to Old Norse *vātr* wet, Old English *wæter* water] (before 12th century)
1 a : consisting of, containing, covered with, or soaked with liquid (as water) **b** *of natural gas* **:** containing appreciable quantities of readily condensable hydrocarbons
2 : RAINY
3 : still moist enough to smudge or smear ⟨*wet* paint⟩
4 a : DRUNK 1a ⟨a *wet* driver⟩ **b :** having or advocating a policy permitting the manufacture and sale of alcoholic beverages ⟨a *wet* county⟩ ⟨a *wet* candidate⟩
5 : preserved in liquid
6 : employing or done by means of or in the presence of water or other liquid ⟨*wet* extraction of copper⟩
7 : overly sentimental
8 *British* **a :** lacking strength of character **:** WEAK, SPINELESS ⟨thought him *wet* and violence petrified him —William Golding⟩ **b :** belonging to the moderate or liberal wing of the Conservative party ☆
— wet·ly *adverb*
— wet·ness *noun*
— all wet : completely wrong **:** in error
— wet behind the ears : IMMATURE, INEXPERIENCED

[2]**wet** *noun* (before 12th century)
1 : WATER; *also* **:** MOISTURE, WETNESS
2 : rainy weather **:** RAIN
3 : an advocate of a policy of permitting the sale of intoxicating liquors
4 *British* **:** one who is wet

[3]**wet** *verb* **wet** *or* **wet·ted; wet·ting** [Middle English, from Old English *wǣtan,* from *wǣt,* adjective] (before 12th century)
transitive verb
1 : to make wet
2 : to urinate in or on
intransitive verb
1 : to become wet
2 : URINATE
— wet one's whistle : to take a drink especially of liquor

wet·back \'wet-ˌbak\ *noun* [from the practice of wading or swimming the Rio Grande where it forms the U.S.-Mexico border] (1929)
: a Mexican who enters the U.S. illegally — sometimes taken to be offensive

wet bar *noun* (1967)
: a bar for mixing drinks (as in a home) that contains a sink with running water

wet blanket *noun* (1857)
: one that quenches or dampens enthusiasm or pleasure

wet down *transitive verb* (1840)
: to dampen by sprinkling with water

wet dream *noun* (1851)
: an erotic dream culminating in orgasm and in the male accompanied by seminal emission

weth·er \'we-t͟hər\ *noun* [Middle English, ram, from Old English; akin to Old High German *widar* ram, Latin *vitulus* calf, *vetus* old, Greek *etos* year] (before 12th century)
: a male sheep castrated before sexual maturity; *also* **:** a castrated male goat

wet·land \'wet-ˌland, -lənd\ *noun* (1778)
: land or areas (as tidal flats or swamps) containing much soil moisture — usually used in plural

wet–nurse *transitive verb* (1784)
1 : to act as wet nurse to
2 : to give constant and often excessive care to

wet nurse *noun* (1620)
: a woman who cares for and suckles children not her own

wet suit *noun* (1955)
: a close-fitting suit made of material (as sponge rubber) that traps a thin layer of water against the body to retain body heat and that is worn (as by a skin diver) especially in cold water

wet·ta·bil·i·ty \ˌwe-tə-'bi-lə-tē\ *noun* (1913)
: the quality or state of being wettable **:** the degree to which something can be wet

wet·ta·ble \'we-tə-bəl\ *adjective* (1885)
: capable of being wetted

wet·ter \'we-tər\ *noun* (1737)
: one that wets; *also* **:** WETTING AGENT

wetting agent *noun* (1927)

☆ SYNONYMS
Wet, damp, dank, moist, humid mean covered or more or less soaked with liquid. WET usually implies saturation but may suggest a covering of a surface with water or something (as paint) not yet dry ⟨slipped on the *wet* pavement⟩. DAMP implies a slight or moderate absorption and often connotes an unpleasant degree of moisture ⟨clothes will mildew if stored in a *damp* place⟩. DANK implies a more distinctly disagreeable or unwholesome dampness ⟨a prisoner in a cold, *dank* cell⟩. MOIST applies to what is slightly damp or not felt as dry ⟨treat the injury with *moist* heat⟩. HUMID applies to the presence of much water vapor in the air ⟨a hot, *humid* climate⟩.

: a substance that by becoming adsorbed prevents a surface from being repellent to a wetting liquid and is used especially in mixing solids with liquids or spreading liquids on surfaces

wet·tish \'we-tish\ *adjective* (1648)
: somewhat wet **:** MOIST

wet wash *noun* (1916)
: laundry returned damp and not ironed

we've \'wēv\ (1742)
: we have

[1]**whack** \'hwak, 'wak\ *verb* [probably imitative of the sound of a blow] (1719)
transitive verb
1 a : to strike with a smart or resounding blow **b :** to cut with or as if with a whack **:** CHOP
2 *chiefly British* **:** to get the better of **:** DEFEAT
intransitive verb
: to strike a smart or resounding blow
— whack·er *noun*

[2]**whack** *noun* (1736)
1 a : a smart or resounding blow; *also* **:** the sound of or as if of such a blow **b :** a critical attack
2 : PORTION, SHARE
3 : CONDITION, STATE
4 a : an opportunity or attempt to do something ⟨take a *whack* at it⟩ **b :** a single action or occasion ⟨borrowed $50 all at one *whack*⟩
— out of whack 1 : out of proper order or shape ⟨threw his knee *out of whack*⟩ **2 :** not in accord ⟨feeling *out of whack* with her contemporaries —S. E. Rubin⟩

whacked–out \'(h)wakt-,au̇t, ,(h)wakt-'\ *adjective* (1967)
1 : WORN-OUT, EXHAUSTED
2 : WACKY ⟨a *whacked-out* parody⟩
3 : STONED ⟨*whacked-out* on drugs⟩

[1]**whack·ing** \'hwa-kiŋ, 'wa-\ *adjective* (1806)
: very large **:** WHOPPING

[2]**whacking** *adverb* (1853)
: VERY ⟨a *whacking* good story⟩

whacko \'hwa-(,)kō, 'wa-\ *variant of* WACKO

whack off *verb* (1969)
: MASTURBATE — usually considered vulgar

whack up *transitive verb* (circa 1893)
: to divide into shares

whacky \'hwa-kē, 'wa-\ *variant of* WACKY

[1]**whale** \'hwā(ə)l, 'wā(ə)l\ *noun, plural* **whales** *often attributive* [Middle English, from Old English *hwæl;* akin to Old High German *hwal* whale and perhaps to Latin *squalus* sea fish] (before 12th century)
1 *or plural* **whale :** CETACEAN; *especially* **:** one (as a sperm whale or killer whale) of larger size
2 : one that is impressive especially in size ⟨a *whale* of a difference⟩ ⟨a *whale* of a good time⟩
— whale·like \-,līk\ *adjective*

[2]**whale** *intransitive verb* **whaled; whal·ing** (1700)
: to engage in whale fishing

[3]**whale** *transitive verb* **whaled; whal·ing** [origin unknown] (circa 1790)
1 : LASH, THRASH
2 : to strike or hit vigorously
3 : to defeat soundly

whale·back \'hwā(ə)l-,bak, 'wā(ə)l-\ *noun* (1886)
: something shaped like the back of a whale

whale·boat \-,bōt\ *noun* (1682)
1 : a long narrow rowboat made with both ends sharp and raking, often steered with an oar, and formerly used by whalers for hunting whales
2 : a long narrow rowboat or motorboat resembling the original whaleboats and often carried by warships and merchant ships

whale·bone \-,bōn\ *noun* (1601)
1 : BALEEN
2 : an article made of whalebone

whalebone whale *noun* (1725)
: BALEEN WHALE

whal·er \'hwā-lər, 'wā-\ *noun* (1684)
1 : a person or ship engaged in whale fishing
2 : WHALEBOAT 2

whale shark *noun* (circa 1885)
: a shark (*Rhincodon typus*) of warm waters that has small teeth, feeds chiefly on plankton strained by its gill rakers, may attain a length of up to 60 feet (18.3 meters), and is the largest known fish

whal·ing \'hwā-liŋ, 'wā-\ *noun* (1688)
: the occupation of catching and extracting commercial products from whales

[1]**wham** \'hwam, 'wam\ *noun* [imitative] (1739)
1 : a solid blow
2 : the loud sound of a hard impact

[2]**wham** *or* **wham·mo** \'hwa-(,)mō, 'wa-\ *adverb* (1924)
: with violent abruptness ⟨everything seemed to be going well, when *wham* the deal fell through⟩

[3]**wham** *verb* **whammed; wham·ming** (1925)
transitive verb
: to propel, strike, or beat so as to produce a loud impact
intransitive verb
: to hit or explode with a loud impact

wham·my \'hwa-mē, 'wa-\ *noun, plural* **whammies** [probably from [1]*wham*] (1940)
1 a : a supernatural power bringing bad luck **b :** a magic curse or spell
2 : a potent force or attack; *specifically* **:** a paralyzing or lethal blow

[1]**whang** \'hwaŋ, 'waŋ\ *noun* [alteration of Middle English *thong, thwang*] (1536)
1 *dialect* **a :** THONG **b :** RAWHIDE
2 *British* **:** a large piece **:** CHUNK
3 : PENIS — often considered vulgar

[2]**whang** (1684)
transitive verb
1 *dialect* **:** BEAT, THRASH
2 : to propel or strike with force
intransitive verb
: to beat or work with force or violence

[3]**whang** *noun* [imitative] (circa 1824)
: a loud sharp vibrant or resonant sound

[4]**whang** (1875)
intransitive verb
: to make a whang
transitive verb
: to strike with a whang

whan·gee \hwaŋ-'ē, waŋ-, -'gē\ *noun* [probably modification of Chinese (Beijing) *huáng* bamboo] (1790)
1 : any of several Chinese bamboos (genus *Phyllostachys*)
2 : a walking stick or riding crop of whangee

whap \'hwäp, 'wäp\ *variant of* WHOP

wharf \'hwȯrf, 'wȯrf\ *noun, plural* **wharves** \'hwȯrvz, 'wȯrvz\ *also* **wharfs** [Middle English, from Old English *hwearf* embankment, wharf; akin to Old English *hweorfan* to turn, Old High German *hwerban,* Greek *karpos* wrist] (before 12th century)
1 : a structure built along or at an angle from the shore of navigable waters so that ships may lie alongside to receive and discharge cargo and passengers
2 *obsolete* **:** the bank of a river or the shore of the sea

wharf·age \'hwȯr-fij, 'wȯr-\ *noun* (15th century)
1 a : the provision or the use of a wharf **b :** the handling or stowing of goods on a wharf
2 : the charge for the use of a wharf
3 : the wharf accommodations of a place **:** WHARVES

wharf·in·ger \-fən-jər\ *noun* [irregular from *wharfage*] (1552)
: the operator or manager of a commercial wharf

wharf·mas·ter \'hwȯrf-,mas-tər, 'wȯrf-\ *noun* (1618)
: the manager of a wharf **:** WHARFINGER

[1]**what** \'hwät, 'hwət, 'wät, 'wət\ *pronoun* [Middle English, from Old English *hwæt,* neuter of *hwā* who — more at WHO] (before 12th century)
1 a (1) — used as an interrogative expressing inquiry about the identity, nature, or value of an object or matter ⟨*what* is this⟩ ⟨*what* is wealth without friends⟩ ⟨*what* does he earn⟩ ⟨*what* hath God wrought⟩ (2) — often used to ask for repetition of an utterance or part of an utterance not properly heard or understood ⟨you said *what*⟩ **b** (1) *archaic* **:** WHO 1 — used as an interrogative expressing inquiry about the identity of a person (2) — used as an interrogative expressing inquiry about the character, nature, occupation, position, or role of a person ⟨*what* do you think I am, a fool⟩ ⟨*what* is she, that all our swains commend her —Shakespeare⟩ **c** — used as an exclamation expressing surprise or excitement and frequently introducing a question ⟨*what,* no breakfast⟩ **d** — used in expressions directing attention to a statement that the speaker is about to make ⟨you know *what*⟩ **e** — used at the end of a question to express inquiry about additional possibilities ⟨is it raining, or snowing, or *what*⟩ **f** *chiefly British* — used at the end of an utterance as a form of tag question ⟨a clever play, *what*⟩
2 *chiefly dialect* **:** [4]THAT 1, WHICH 3, WHO 3
3 : that which **:** the one or ones that ⟨no income but *what* he gets from his writings⟩ — sometimes used in reference to a clause or phrase that is yet to come or is not yet complete ⟨gave also, *what* is more valuable, understanding⟩
4 a : WHATEVER 1a ⟨say *what* you will⟩ **b** *obsolete* **:** WHOEVER
— what for 1 : for what purpose or reason **:** WHY — usually used with the other words of a question between *what* and *for* ⟨*what* did you do that *for*⟩ except when used alone **2 :** harsh treatment especially by blows or by a sharp reprimand ⟨gave him *what for* in violent Spanish —*New Yorker*⟩
— what have you : WHATNOT ⟨novels, plays, short stories, travelogues, and *what have you* —Haldeen Braddy⟩
— what if 1 : what will or would be the result if **2 :** what does it matter if
— what of 1 : what is the situation with respect to **2 :** what importance can be assigned to
— what's what : the true state of things ⟨knows *what's what* when it comes to fashion⟩
— what though : what does it matter if ⟨*what though* the rose have prickles, yet 'tis plucked —Shakespeare⟩

[2]**what** *adverb* (before 12th century)
1 *obsolete* **:** WHY
2 : in what respect **:** HOW ⟨*what* does he care⟩
3 — used to introduce prepositional phrases in parallel construction or a prepositional phrase that expresses cause and usually has more than one object; used principally before phrases beginning with *with* ⟨*what* with unemployment increasing⟩ ⟨*what* with the war, *what* with the sweat, *what* with the gallows, and *what* with poverty, I am custom-shrunk —Shakespeare⟩

[3]**what** *adjective* (13th century)
1 a — used as an interrogative expressing inquiry about the identity, nature, or value of a person, object, or matter ⟨*what* minerals do we export⟩ **b :** how remarkable or striking for good or bad qualities — used especially in exclamatory utterances and dependent clauses ⟨*what* mountains⟩ ⟨remember *what* fun we had⟩ ⟨*what* a suggestion⟩ ⟨*what* a charming girl⟩
2 a (1) **:** WHATEVER 1a (2) **:** ANY ⟨ornament of *what* description soever⟩ **b :** the . . . that **:** as much or as many . . . as ⟨rescued *what* survivors they found⟩

what all \'hwä-,dȯl, 'wä-, '(h)wə-\ *pronoun* (1702)
: WHATNOT

what·cha·ma·call·it \'hwä-chə-mə-,kȯ-lət, 'wä-, '(h)wə-\ *noun* [alteration of *what you may call it*] (1928)

: THINGAMAJIG

1what·ev·er \hwät-ˈe-vər, wät-, (ˌ)(h)wət-\ *pronoun* (14th century)
1 a : anything or everything that ⟨take *whatever* you want⟩ **b :** no matter what ⟨*whatever* he says, they won't believe him⟩ **c :** WHATNOT ⟨buffalo or rhinoceros or *whatever* —Alan Moorehead⟩
2 : WHAT 1a(1) — used to express astonishment or perplexity ⟨*whatever* do you mean by that⟩

2whatever *adjective* (14th century)
1 a : any . . . that **:** all . . . that ⟨buy peace . . . on *whatever* terms could be obtained —C. S. Forester⟩ **b :** no matter what ⟨money, in *whatever* hands, will confer power —Samuel Johnson⟩
2 : of any kind at all — used after the substantive it modifies with *any* or with an expressed or implied negative ⟨in any order *whatever* —W. G. Moulton⟩ ⟨no food *whatever*⟩

3whatever *adverb* (1870)
: in any case **:** whatever the case may be — sometimes used interjectionally to suggest the unimportance of a decision between alternatives

what–if \ˌhwät-ˈif, ˌwät-, ˌ(h)wət-\ *noun* (1970)
: a suppositional question

what·ness \ˈhwät-nəs, ˈwät-, ˈ(h)wət-\ *noun* (1611)
: QUIDDITY 1

1what·not \ˈhwät-ˌnät, ˈwät-, ˈ(h)wət-\ *pronoun* [*what not?*] (1540)
: any of various other things that might also be mentioned ⟨paper clips, pins, and *whatnot*⟩

2whatnot *noun* (1602)
1 : a nondescript person or thing
2 : a light open set of shelves for bric-a-brac

what·sit \ˈhwät-sət, ˈwät-, ˈwət-\ *or* **what·sis** \-səs\ *also* **what–is–it** \-iz-ət\ *noun* [*whatsit* & *whatsis* contraction of *what-is-it*] (circa 1882)
: THINGAMAJIG

what·so·ev·er \ˌhwät-sə-ˈwe-vər, ˌwät-, ˌ(h)wət-\ *pronoun or adjective* (13th century)
: WHATEVER

whaup \ˈhwȯp, ˈwȯp\ *noun, plural* **whaup** *also* **whaups** [imitative] (circa 1512)
chiefly Scottish **:** a European curlew (*Numenius arquata*)

wheal \ˈhwē(ə)l, ˈwē(ə)l\ *noun* [alteration of 1*wale*] (1808)
: a suddenly formed elevation of the skin surface: as **a :** WELT **b :** a flat burning or itching eminence on the skin

wheat \ˈhwēt, ˈwēt\ *noun, often attributive* [Middle English *whete,* from Old English *hwǣte;* akin to Old High German *weizzi* wheat, *hwīz, wīz* white — more at WHITE] (before 12th century)
1 : a cereal grain that yields a fine white flour, is the chief breadstuff of temperate climates, is used also in pastas (as macaroni or spaghetti), and is important in animal feeds
2 : any of various annual grasses (genus *Triticum* and especially *T. aestivum* and *T. turgidum*) of wide climatic adaptability that are cultivated in most temperate areas for the wheat they yield
3 : a light yellow

wheat bread *noun* (14th century)
: a bread made of a combination of white and whole wheat flours as distinguished from bread made entirely of white or whole wheat flour

wheat cake *noun* (1772)
: a pancake made of wheat flour

wheat·ear \ˈhwēt-ˌir, ˈwēt-\ *noun* [back-formation from earlier *wheatears* wheatear, probably by folk etymology or euphemism from *white + arse*] (1591)
: any of various small thrushes (genus *Oenanthe*); *especially* **:** a white-rumped one (*O. oenanthe*) of Eurasia and northern North America

1wheat·en \ˈhwē-tᵊn, ˈwē-\ *adjective* (before 12th century)
: of, relating to, or made of wheat

2wheaten *noun* (circa 1931)
: a pale yellowish to ruddy fawn color characteristic of the coat of some dogs

wheat germ *noun* (1897)
: the embryo of the wheat kernel separated in milling and used especially as a source of vitamins and protein

wheat rust *noun* (1884)
: a destructive disease of wheat caused by rust fungi; *also* **:** a fungus (as *Puccinia graminis*) causing a wheat rust

Wheat·stone bridge \ˈhwēt-ˌstōn-, ˈwēt-, *chiefly British* -stən-\ *noun* [Sir Charles *Wheatstone*] (1872)
: a bridge for measuring electrical resistances that consists of a conductor joining two branches of a circuit

whee \ˈhwē, ˈwē\ *interjection* (1898)
— used to express delight or exuberance

whee·dle \ˈhwē-dᵊl, ˈwē-\ *verb* **whee·dled; whee·dling** \ˈ(h)wēd-liŋ, ˈ(h)wē-dᵊl-iŋ\ [origin unknown] (circa 1661)
transitive verb
1 : to influence or entice by soft words or flattery
2 : to gain or get by wheedling ⟨*wheedle* one's way into favor⟩
intransitive verb
: to use soft words or flattery

1wheel \ˈhwē(ə)l, ˈwē(ə)l\ *noun, often attributive* [Middle English, from Old English *hweogol, hwēol;* akin to Old Norse *hvēl* wheel, Greek *kyklos* circle, wheel, Sanskrit *cakra,* Latin *colere* to cultivate, inhabit, Sanskrit *carati* he moves, wanders] (before 12th century)
1 : a circular frame of hard material that may be solid, partly solid, or spoked and that is capable of turning on an axle
2 : a contrivance or apparatus having as its principal part a wheel: as **a :** a chiefly medieval instrument of torture designed for mutilating a victim (as by stretching or disjointing) **b :** BICYCLE **c :** any of many revolving disks or drums used as gambling paraphernalia **d :** POTTER'S WHEEL **e :** STEERING WHEEL

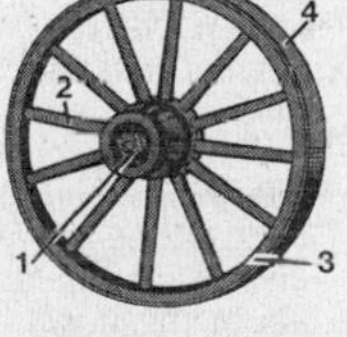

wheel 1: *1* **hub,** *2* **spoke,** *3* **felly,** *4* **tire**

3 a : an imaginary turning wheel symbolizing the inconstancy of fortune **b :** a recurring course, development, or action **:** CYCLE
4 : something resembling a wheel in shape or motion: as **a :** a round flat cheese **b :** a firework that rotates while burning **c :** a propeller on a boat
5 a : a curving or circular movement **b :** a rotation or turn usually about an axis or center; *specifically* **:** a turning movement of troops or ships in line in which the units preserve alignment and relative positions as they change direction
6 a : a moving or essential part of something compared to a machine ⟨the *wheels* of government⟩ **b :** a directing or controlling force **c :** a person of importance especially in an organization ⟨a big *wheel*⟩
7 : the refrain or burden of a song
8 a : a circuit of theaters or places of entertainment **b :** a sports league
9 *plural, slang* **:** a wheeled vehicle; *especially* **:** AUTOMOBILE
— wheel·less \ˈhwē(ə)l-ləs, ˈwē(ə)l-\ *adjective*

2wheel (13th century)
intransitive verb
1 : to turn on or as if on an axis **:** REVOLVE
2 : to change direction as if revolving on a pivot ⟨the battalion would have *wheeled* to the flank —Walter Bernstein⟩ ⟨her mind will *wheel* around to the other extreme —Liam O'Flaherty⟩
3 : to move or extend in a circle or curve ⟨birds in *wheeling* flight⟩ ⟨valleys where young cotton *wheeled* slowly in fanlike rows —William Faulkner⟩
4 : to drive or go on or as if on wheels or in a wheeled vehicle
transitive verb
1 : to cause to turn on or as if on an axis **:** ROTATE
2 : to convey or move on or as if on wheels or in a wheeled vehicle; *especially* **:** to drive (a vehicle) at high speed
3 : to cause to change direction as if revolving on a pivot
4 : to make or perform in a circle or curve
— wheel and deal : to make deals or do business especially shrewdly or briskly

wheel and axle *noun* (circa 1773)
: a mechanical device consisting of a grooved wheel turned by a cord or chain with a rigidly attached axle (as for winding up a weight) together with the supporting standards

wheel animal *noun* (1788)
: ROTIFER

wheel animalcule *noun* (1834)
: ROTIFER

1wheel·bar·row \ˈhwē(ə)l-ˌbar-(ˌ)ō, ˈwē(ə)l-\ *noun* (14th century)
: a small usually single-wheeled vehicle that is used for carrying small loads and is fitted with handles at the rear by which it can be pushed and guided

2wheelbarrow *transitive verb* (1721)
: to convey in a wheelbarrow

wheel·base \ˈhwē(ə)l-ˌbās, ˈwē(ə)l-\ *noun* (1886)
: the distance in inches between the front and rear axles of an automotive vehicle

wheel bug *noun* (1815)
: a large North American bug (*Arilus cristatus*) that has a high serrated crest on its prothorax and preys on other insects

wheel·chair \ˈhwē(ə)l-ˌcher, ˈwē(ə)l-, -ˌchar\ *noun* (circa 1700)
: a chair mounted on wheels especially for the use of disabled persons

wheeled \ˈhwē(ə)ld, ˈwē(ə)ld\ *adjective* (1606)
1 : equipped with wheels ⟨*wheeled* vehicles⟩
2 : moving or functioning by means of wheels ⟨*wheeled* traffic⟩

wheel·er \ˈhwē-lər, ˈwē-\ *noun* (1683)
1 : one that wheels
2 : a draft animal (as a horse) pulling in the position nearest the front wheels of a wagon
3 : something (as a vehicle or ship) that has wheels — used especially in combination ⟨side-*wheeler*⟩

wheeler and dealer *noun, plural* **wheelers and dealers** (1966)
: WHEELER-DEALER

wheel·er–deal·er \ˌhwē-lər-ˈdē-lər, ˌwē-\ *noun* (1954)
: a shrewd operator especially in business or politics

wheel·horse \ˈhwē(ə)l-ˌhȯrs, ˈwē(ə)l-\ *noun* (1708)
1 : a horse (as in a tandem) in a position nearest the wheels
2 : a steady and effective worker especially in a political body

wheel·house \-ˌhau̇s\ *noun* (1835)
: PILOTHOUSE

wheel·ie \ˈhwē-lē, ˈwē-\ *noun* (circa 1965)
: a maneuver in which a wheeled vehicle (as a bicycle) is momentarily balanced on its rear wheel or wheels

wheel·ing \'hwē-liŋ, 'wē-\ *noun* (15th century)
1 : the act or process of one that wheels
2 : the condition of a road relative to passage on wheels

wheel lock *noun* (1670)
: a gun's lock for a muzzle-loading firearm in which sparks are struck from a flint or a piece of iron pyrites by a revolving wheel

wheel·man \'hwē(ə)l-mən, 'wē(ə)l-\ *noun* (1865)
1 a : HELMSMAN **b :** the driver of an automobile
2 : CYCLIST

wheels·man \'hwē(ə)lz-mən, 'wē(ə)lz-\ *noun* (1866)
: one who steers with a wheel; *especially* **:** HELMSMAN

wheel–thrown \'hwēl-ˌthrōn, 'wēl-\ *adjective* (1975)
: made on a potter's wheel ⟨*wheel-thrown* pottery⟩

wheel·work \'hwē(ə)l-ˌwərk, 'wē(ə)l-\ *noun* (1670)
: wheels in gear and their connections in a machine or mechanism

wheel·wright \-ˌrīt\ *noun* (13th century)
: a maker and repairer of wheels and wheeled vehicles

[1]**ween** \'hwēn, 'wēn\ *adjective* [Middle English (Scots) *quheyne,* from Old English *hwǣne, hwēne,* adverb, somewhat, from instrumental of *hwōn* little, few] (14th century)
dialect British **:** FEW 2

[2]**ween** *noun* (1680)
dialect British **:** a considerable number or amount

[1]**wheeze** \'hwēz, 'wēz\ *intransitive verb* **wheezed; wheez·ing** [Middle English *whesen,* probably of Scandinavian origin; akin to Old Norse *hvæsa* to hiss; akin to Old English *hwǣst* action of blowing, Sanskrit *śvasiti* he blows, snorts] (15th century)
1 : to breathe with difficulty usually with a whistling sound
2 : to make a sound resembling that of wheezing

[2]**wheeze** *noun* (1834)
1 : a sound of wheezing
2 a : an often repeated and widely known joke used especially by entertainers **b :** a trite saying or proverb

wheezy \'hwē-zē, 'wē-\ *adjective* **wheez·i·er; -est** (1818)
1 : inclined to wheeze
2 : having a wheezing sound
— wheez·i·ly \-zə-lē\ *adverb*
— wheez·i·ness \-zē-nəs\ *noun*

[1]**whelk** \'hwelk, 'welk, 'wilk\ *noun* [Middle English *welke,* from Old English *weoloc;* akin to Middle Dutch *willoc* whelk and perhaps to Latin *volvere* to turn — more at VOLUBLE] (before 12th century)
: any of numerous large marine snails (as of the genus *Buccinum*); *especially* **:** one (*B. undatum*) much used as food in Europe

[2]**whelk** \'hwelk, 'welk\ *noun* [Middle English *whelke,* from Old English *hwylca,* from *hwelian* to suppurate] (before 12th century)
: PAPULE, PUSTULE

whelk

whelm \'hwelm, 'welm\ *verb* [Middle English] (14th century)
transitive verb
1 : to turn (as a dish or vessel) upside down usually to cover something **:** cover or engulf completely with usually disastrous effect
2 : to overcome in thought or feeling **:** OVERWHELM
intransitive verb
: to pass or go over something so as to bury or submerge it

[1]**whelp** \'hwelp, 'welp\ *noun* [Middle English, from Old English *hwelp;* akin to Old High German *hwelf* whelp] (before 12th century)
1 : any of the young of various carnivorous mammals and especially of the dog
2 : a young boy or girl

[2]**whelp** (13th century)
transitive verb
: to give birth to — used of various carnivores and especially the dog
intransitive verb
: to bring forth young

[1]**when** \'hwen, 'wen, (h)wən\ *adverb* [Middle English, from Old English *hwanne, hwenne;* akin to Old High German *hwanne* when, Old English *hwā* who — more at WHO] (before 12th century)
1 : at what time ⟨*when* will you return⟩
2 a : at or during which time **b :** and then
3 : at a former and usually less prosperous time ⟨brag fondly of having known him *when* —Vance Packard⟩

[2]**when** *conjunction* [Middle English, from Old English *hwanne, hwenne,* from *hwanne, hwenne,* adverb] (before 12th century)
1 a : at or during the time that **:** WHILE ⟨went fishing *when* he was a boy⟩ **b :** just at the moment that ⟨stop writing *when* the bell rings⟩ **c :** at any or every time that ⟨*when* he listens to music, he falls asleep⟩
2 : in the event that **:** IF ⟨a contestant is disqualified *when* he disobeys the rules⟩
3 a : considering that ⟨why use water at all *when* you can drown in it —Stuart Chase⟩ **b :** in spite of the fact that **:** ALTHOUGH ⟨quit politics *when* I might have had a great career in it⟩
4 : the time or occasion at or in which ⟨tomorrow is *when* we must decide⟩ ⟨humor is *when* you laugh —Earl Rovit⟩

[3]**when** \'hwen, 'wen\ *pronoun* (14th century)
: what or which time ⟨life-long homes for those . . . who have lived here since *when* —Kim Waller⟩

[4]**when** *noun* (1616)
: the time in which something is done or comes about ⟨troubled his head very little about the hows and *whens* of life —Laurence Sterne⟩

when·as \hwe-'naz, we-, (h)wə-\ *conjunction* [Middle English (Scots) *when as,* from Middle English *when* + *as*] (15th century)
archaic **:** WHEN

[1]**whence** \'hwen(t)s, 'wen(t)s\ *adverb* [Middle English *whennes,* from *whenne* whence (from Old English *hwanon*) + *-s,* adverb suffix, from *-s,* genitive singular ending; akin to Old High German *hwanān* whence, Old English *hwā* who] (13th century)
: from what place, source, or cause ⟨then *whence* comes this paradox —*Changing Times*⟩
— from whence : from what place, source, or cause ⟨no one could tell me *from whence* the gold had come —Graham Greene⟩

[2]**whence** *conjunction* (14th century)
1 : from what place, source, or cause ⟨inquired *whence* the water came —Maria Edgeworth⟩
2 a : from or out of which place, source, or cause ⟨the lawless society *whence* the ballads sprang —DeLancey Ferguson⟩ **b :** by reason of which fact **:** WHEREFORE ⟨nothing broke — *whence* I infer that my bones are not yet chalky —O. W. Holmes (died 1935)⟩

whence·so·ev·er \'hwen(t)(s)-sə-ˌwe-vər, 'wen(t)(s)-\ *conjunction* (1511)
: from whatever place or source

[1]**when·ev·er** \hwe-'ne-vər, we-, (h)wə-\ *conjunction* (14th century)
: at any or every time that

[2]**whenever** *adverb* (1667)
: at whatever time

[1]**when·so·ev·er** \'hwen(t)-sə-ˌwe-vər, 'wen(t)-\ *conjunction* (14th century)
: WHENEVER

[2]**whensoever** *adverb* (1526)
obsolete **:** at any time whatever

[1]**where** \'hwer, 'hwar, 'wer, 'war, (ˌ)(h)wər\ *adverb* [Middle English, from Old English *hwǣr;* akin to Old High German *hwār* where, Old English *hwā* who — more at WHO] (before 12th century)
1 a : at, in, or to what place ⟨*where* is the house⟩ ⟨*where* are we going⟩ **b :** at, in, or to what situation, position, direction, circumstances, or respect ⟨*where* does this plan lead⟩ ⟨*where* am I wrong⟩
2 *archaic* **:** HERE, THERE ⟨lo, *where* it comes again —Shakespeare⟩

[2]**where** *conjunction* (13th century)
1 a : at, in, or to what place ⟨knows *where* the house is⟩ **b :** at, in, or to what situation, position, direction, circumstances, or respect ⟨shows *where* the plan leads⟩ **c :** the place or point at, in, or to which ⟨couldn't see from *where* he was sitting⟩ ⟨kept that horse and gentled him to *where* I finally rode him —William Faulkner⟩
2 : WHEREVER ⟨goes *where* she likes⟩
3 a : at, in, or to which place ⟨the town *where* she lives⟩ **b :** at or in which ⟨has reached the size *where* traffic is a problem⟩ ⟨two fireplaces *where* you can bake bread in the ovens —Randall Jarrell⟩
4 a : at, in, or to the place at, in, or to which ⟨stay *where* you are⟩ ⟨send him away *where* he'll forget⟩ **b :** in a case, situation, or respect in which ⟨outstanding *where* endurance is called for⟩
5 : THAT ⟨I've read *where* they do it that way in some Middle Eastern countries —Andy Rooney⟩

[3]**where** \'hwer, 'hwar, 'wer, 'war\ *noun* (15th century)
1 : PLACE, LOCATION ⟨the *where* and the how of the accident⟩
2 : what place, source, or cause ⟨I know *where* that comes from⟩
— where it's at 1 a : a place of central interest or activity **b :** something (as a topic or field of interest) of primary concern or importance ⟨education is *where it's at*⟩ **2 :** the true nature of things
— where one is at : one's true position, state, or nature

[1]**where·abouts** \-ə-ˌbau̇ts\ *also* **where·about** \-ˌbau̇t\ *adverb* [Middle English *wherabouts* (from *wher aboute* + *-s,* adverb suffix) & *wher aboute,* from *where, wher* where + *about, aboute* about — more at WHENCE] (14th century)
: about where **:** near what place ⟨*whereabouts* is the house⟩

[2]**whereabouts** *also* **whereabout** *conjunction* (14th century)
1 *obsolete* **:** on what business or errand
2 : near what place **:** WHERE ⟨know *whereabouts* he lives⟩

[3]**whereabouts** *noun plural but singular or plural in construction, also* **whereabout** (1605)
: the place or general locality where a person or thing is ⟨their present *whereabouts* are a secret⟩

[1]**where·as** \hwer-'az, hwar-, wer-, war-, (ˌ)(h)wər-\ *conjunction* [Middle English *where as,* from *where* + *as*] (14th century)
1 a : while on the contrary **b :** ALTHOUGH
2 : in view of the fact that **:** SINCE — used especially to introduce a preamble

[2]**whereas** *noun* (1795)
1 : an introductory statement of a formal document **:** PREAMBLE
2 : a conditional or qualifying statement

where·at \-'at\ *conjunction* (15th century)
1 : at or toward which
2 : in consequence of which **:** WHEREUPON

[1]**where·by** \-'bī\ *conjunction* (13th century)
: by, through, or in accordance with which

[2]**whereby** *adverb* (14th century)
obsolete **:** by what **:** HOW

[1]**where·fore** \ˈhwer-ˌfōr, ˈhwar-, ˈwer-, ˈwar-, -ˌfȯr\ *adverb* [Middle English *wherfor, wherfore,* from *where, wher* + *for, fore* for] (13th century)
1 : for what reason or purpose **:** WHY
2 : THEREFORE
[2]**wherefore** *noun* (1590)
: an answer or statement giving an explanation **:** REASON ⟨wants to know the whys and *wherefores*⟩
where·from \-ˌfrəm, -ˌfräm\ *conjunction* (15th century)
: from which
[1]**where·in** \hwer-ˈin, hwar-, wer-, war-, (ˌ)(h)wər-\ *adverb* (13th century)
: in what **:** in what particular or respect ⟨*wherein* was I wrong⟩
[2]**wherein** *conjunction* (15th century)
1 a : in which **:** WHERE ⟨the city *wherein* he lives⟩ **b :** during which
2 : in what way **:** HOW ⟨showed me *wherein* I was wrong⟩
where·in·to \-ˈin-(ˌ)tü\ *conjunction* (1539)
: into which
[1]**where·of** \-ˈəv, -ˈäv\ *conjunction* (13th century)
1 *archaic* **:** with or by which
2 : of what ⟨knows *whereof* she speaks⟩
3 a : of which ⟨books *whereof* the best are lost⟩ **b :** of whom
[2]**whereof** *adverb* (15th century)
archaic **:** of what ⟨*whereof* are you made —Shakespeare⟩
[1]**where·on** \-ˈȯn, -ˈän\ *conjunction* (13th century)
1 *archaic* **:** on what ⟨tell me *whereon* the likelihood depends —Shakespeare⟩
2 : on which ⟨the base *whereon* it rests⟩
[2]**whereon** *adverb* (15th century)
archaic **:** on what ⟨*whereon* do you look —Shakespeare⟩
where·so·ev·er \ˈhwer-sə-ˌwe-vər, ˈhwar-, ˈwer-, ˈwar-\ *conjunction* (13th century)
: WHEREVER
where·through \ˈhwer-ˌthrü, ˈhwar-, ˈwer-, ˈwar-\ *conjunction* (13th century)
: through which
[1]**where·to** \-ˌtü\ *adverb* (13th century)
: to what place, purpose, or end ⟨*whereto* tends all this —Shakespeare⟩
[2]**whereto** *conjunction* (14th century)
: to which
where·un·to \hwer-ˈən-(ˌ)tü, hwar-, wer-, war-, (ˌ)(h)wər-\ *adverb or conjunction* (15th century)
: WHERETO
where·up·on \ˈhwer-ə-ˌpȯn, ˈhwar-, ˈwer-, ˈwar-, -ˌpän\ *conjunction* (14th century)
1 : on which
2 : closely following and in consequence of which
[1]**wher·ev·er** \hwer-ˈe-vər, hwar-, wer-, war-, (ˌ)(h)wər-\ *adverb* (13th century)
1 : where in the world ⟨*wherever* did you get that tie⟩
2 : anywhere at all ⟨explore northward or *wherever* —Bernard De Voto⟩
[2]**wherever** *conjunction* (14th century)
1 : at, in, or to any or all places that ⟨thrives *wherever* he goes⟩
2 : in any circumstance in which ⟨*wherever* it is possible, we try to help⟩
[1]**where·with** \ˈhwer-ˌwith, ˈhwar-, ˈwer-, ˈwar-, -ˌwith\ *pronoun* (13th century)
archaic **:** that with or by which — used with an infinitive ⟨so shall I have *wherewith* to answer him —Psalms 119:42 (Authorized Version)⟩
[2]**wherewith** *adverb* (13th century)
obsolete **:** with what ⟨*wherewith* shall it be salted —Matthew 5:13 (Authorized Version)⟩
[3]**wherewith** *conjunction* (14th century)
: with or by means of which ⟨metal tools *wherewith* to break ground —Russell Lord⟩
[1]**where·with·al** \ˈhwer-wi-ˌthȯl, ˈhwar-, ˈwer-, ˈwar-, -ˌthȯl\ *conjunction* [*where* + [2]*withal*] (1534)
: WHEREWITH
[2]**wherewithal** *pronoun* (1583)
: WHEREWITH
[3]**wherewithal** *noun* (1659)
: MEANS, RESOURCES; *specifically* **:** MONEY ⟨didn't have the *wherewithal* for an expensive dinner⟩
wher·ry \ˈhwer-ē, ˈwer-\ *noun, plural* **wherries** [Middle English *whery*] (15th century)
1 : any of various light boats: as **a :** a long light rowboat made sharp at both ends and used to transport passengers on rivers and about harbors **b :** a racing scull for one person
2 : a large light barge, lighter, or fishing boat varying in type in different parts of Great Britain
[1]**whet** \ˈhwet, ˈwet\ *transitive verb* **whet·ted; whet·ting** [Middle English *whetten,* from Old English *hwettan;* akin to Old High German *wezzen* to whet, *waz* sharp] (before 12th century)
1 : to sharpen by rubbing on or with something (as a stone) ⟨*whet* a knife⟩
2 : to make keen or more acute **:** EXCITE, STIMULATE ⟨*whet* the appetite⟩
— whet·ter *noun*
[2]**whet** *noun* (circa 1628)
1 *dialect* **a :** a spell of work done with a scythe between the time it is sharpened and the time it needs to be sharpened again **b :** TIME, WHILE
2 : something that sharpens or makes keen: **a :** GOAD, INCITEMENT **b :** APPETIZER; *also* **:** a drink of liquor
[1]**wheth·er** \ˈhwe-thər, ˈwe-, (ˌ)(h)wə-\ *pronoun* [Middle English, from Old English *hwæther, hwether;* akin to Old High German *hwedar* which of two, Latin *uter,* Greek *poteros,* Old English *hwā* who — more at WHO] (before 12th century)
1 *archaic* **:** which one of the two
2 *archaic* **:** whichever one of the two
[2]**whether** *conjunction* (before 12th century)
— used as a function word usually with correlative *or* or with *or whether* to indicate (1) until the early 19th century a direct question involving alternatives; (2) an indirect question involving stated or implied alternatives ⟨decide *whether* he should agree or raise objections⟩ ⟨wondered *whether* to stay⟩; (3) alternative conditions or possibilities ⟨see me no more, *whether* he be dead or no —Shakespeare⟩ ⟨seated him next to her *whether* by accident or design⟩
— whether or no *or* **whether or not :** in any case
whet·stone \ˈhwet-ˌstōn, ˈwet-\ *noun* (before 12th century)
: a stone for whetting edge tools
whew *often read as* ˈhwü, ˈwü, ˈhyü; *the interjection is a whistle concluded with a voiceless* ü\ *noun* [imitative] (1513)
: a whistling sound or a sound like a half-formed whistle uttered as an exclamation ⟨gave a long *whew* when he realized the size of the job⟩ — used interjectionally chiefly to express amazement, discomfort, or relief
whey \ˈhwā, ˈwā\ *noun* [Middle English, from Old English *hwæg;* akin to Middle Dutch *wey* whey] (before 12th century)
: the watery part of milk that is separated from the coagulable part or curd especially in the process of making cheese and that is rich in lactose, minerals, and vitamins and contains lactalbumin and traces of fat
— whey·like \-ˌlīk\ *adjective*
whey–face \ˈhwā-ˌfās, ˈwā-\ *noun* (1605)
: a person having a pale face (as from fear)
— whey–faced \-ˌfāst\ *adjective*
[1]**which** \ˈhwich, ˈwich\ *adjective* [Middle English, of what kind, which, from Old English *hwilc;* akin to Old High German *wilīh* of what kind, which, Old English *hwā* who, *gelīk* like — more at WHO, LIKE] (before 12th century)
1 : being what one or ones out of a group — used as an interrogative ⟨*which* tie should I wear⟩ ⟨kept a record of *which* employees took their vacations in July⟩
2 : WHICHEVER ⟨it will not fit, turn it *which* way you like⟩
3 — used as a function word to introduce a nonrestrictive relative clause and to modify a noun in that clause and to refer together with that noun to a word or word group in a preceding clause or to an entire preceding clause or sentence or longer unit of discourse ⟨in German, *which* language might . . . have been the medium of transmission —Thomas Pyles⟩ ⟨that this city is a rebellious city . . . : for *which* cause was this city destroyed —Ezra 4:15 (Authorized Version)⟩
[2]**which** *pronoun* (before 12th century)
1 : what one or ones out of a group — used as an interrogative ⟨*which* of those houses do you live in⟩ ⟨*which* of you want tea and *which* want lemonade⟩ ⟨he is swimming or canoeing, I don't know *which*⟩
2 : WHICHEVER ⟨take *which* you like⟩
3 — used as a function word to introduce a relative clause; used in any grammatical relation except that of a possessive; used especially in reference to animals, inanimate objects, groups, or ideas ⟨the bonds *which* represent the debt —G. B. Robinson⟩ ⟨the Samnite tribes, *which* settled south and southeast of Rome —Ernst Pulgram⟩; used freely in reference to persons as recently as the 17th century ⟨our Father *which* art in heaven —Matthew 6:9 (Authorized Version)⟩, and still occasionally so used but usually with some implication of emphasis on the function or role of the person rather than on the person as such ⟨chiefly they wanted husbands, *which* they got easily —Lynn White⟩; used by speakers on all educational levels and by many reputable writers, though disapproved by some grammarians, in reference to an idea expressed by a word or group of words that is not necessarily a noun or noun phrase ⟨he resigned that post, after *which* he engaged in ranching —*Current Biography*⟩
usage see [4]THAT, WHOSE
[1]**which·ev·er** \hwich-ˈe-vər, wich-\ *adjective* (14th century)
: being whatever one or ones out of a group **:** no matter which ⟨its soothing . . . effect will be the same *whichever* way you take it —*Punch*⟩
[2]**whichever** *pronoun* (15th century)
: whatever one or ones out of a group ⟨take two of the four elective subjects, *whichever* you prefer⟩
which·so·ev·er \ˌhwich-sə-ˈwe-vər, ˌwich-\ *pronoun or adjective* (15th century)
archaic **:** WHICHEVER
whick·er \ˈhwi-kər, ˈwi-\ *intransitive verb* **whick·ered; whick·er·ing** \-k(ə-)riŋ\ [imitative] (1753)
: NEIGH, WHINNY
— whicker *noun*
whid \ˈhwid, ˈwid\ *intransitive verb* **whid·ded; whid·ding** [Scots *whid* silent rapid motion] (1728)
Scottish **:** to move nimbly and silently
[1]**whiff** \ˈhwif, ˈwif\ *noun* [origin unknown] (1591)
1 a : a quick puff or slight gust especially of air, odor, gas, smoke, or spray **b :** an inhalation of odor, gas, or smoke **c :** a slight puffing or whistling sound
2 : a slight trace or indication
3 : STRIKEOUT

²whiff (1591)
intransitive verb
1 : to move with or as if with a puff of air
2 : to emit whiffs **:** PUFF
3 : to inhale an odor
4 : STRIKE OUT 3
transitive verb
1 a : to carry or convey by or as if by a whiff **:** BLOW **b :** to expel or puff out in a whiff **:** EXHALE **c :** SMOKE 3
2 : FAN 8

whif·fet \'hwi-fət, 'wi-\ *noun* [probably alteration of *whippet*] (1839)
: a small, young, or unimportant person

whif·fle \'hwi-fəl, 'wi-\ *verb* **whif·fled; whif·fling** \-f(ə-)liŋ\ [probably frequentative of *whiff*] (1568)
intransitive verb
1 a *of the wind* **:** to blow unsteadily or in gusts **b :** VACILLATE
2 : to emit or produce a light whistling or puffing sound
transitive verb
: to blow, disperse, emit, or expel with or as if with a whiff

¹whif·fler \'hwi-flər, 'wi-\ *noun* [alteration of earlier *wifler,* from obsolete *wifle* battle-ax] (1539)
British **:** one that clears the way for a procession

²whif·fler \'hwi-f(ə-)lər, 'wi-\ *noun* [*whiffle*] (1607)
1 : a person who frequently changes opinions or course
2 : a person who uses shifts and evasions in argument

whif·fle·tree \'hwi-fəl-(,)trē, 'wi-\ *noun* [alteration of *whippletree*] (circa 1806)
: the pivoted swinging bar to which the traces of a harness are fastened and by which a vehicle or implement is drawn

1 **whiffletree**

Whig \'hwig, 'wig\ *noun* [short for *Whiggamore,* member of a Scottish group that marched to Edinburgh in 1648 to oppose the court party] (circa 1680)
1 : a member or supporter of a major British political group of the late 17th through early 19th centuries seeking to limit the royal authority and increase parliamentary power — compare TORY
2 : an American favoring independence from Great Britain during the American Revolution
3 : a member or supporter of an American political party formed about 1834 in opposition to the Jacksonian Democrats, associated chiefly with manufacturing, commercial, and financial interests, and succeeded about 1854 by the Republican party
— Whig *adjective*
— Whig·gism \'hwi-,gi-zəm\ *noun*

Whig·gery \'hwi-gə-rē, 'wi-\ *noun* (1714)
: the principles or practices of Whigs

Whig·gish \'hwi-gish, 'wi-\ *adjective* (1684)
1 : characteristic of Whigs or Whiggery
2 : of, relating to, or characterized by a view which holds that history follows a path of inevitable progression and improvement and which judges the past in light of the present

whig·ma·lee·rie \,hwig-mə-'lir-ē, ,wig-\ *noun* [origin unknown] (1730)
1 *chiefly Scottish* **:** WHIM
2 *chiefly Scottish* **:** an odd or fanciful contrivance **:** GIMCRACK

¹while \'hwī(ə)l, 'wī(ə)l\ *noun* [Middle English, from Old English *hwīl;* akin to Old High German *hwīla* time, Latin *quies* rest, quiet] (before 12th century)
1 : a period of time especially when short and marked by the occurrence of an action or a condition **:** TIME ⟨stay here for a *while*⟩
2 : the time and effort used (as in the performance of an action) **:** TROUBLE ⟨worth your *while*⟩

²while *conjunction* (12th century)
1 a : during the time that ⟨take a nap *while* I'm out⟩ **b :** as long as ⟨*while* there's life there's hope⟩
2 a : when on the other hand **:** WHEREAS ⟨easy for an expert, *while* it is dangerous for a novice⟩ **b :** in spite of the fact that **:** ALTHOUGH ⟨*while* respected, he is not liked⟩
3 : similarly and at the same time that ⟨*while* the book will be welcomed by scholars, it will make an immediate appeal to the general reader —*British Book News*⟩

³while *preposition* (15th century)
dialect British **:** UNTIL

⁴while *transitive verb* **whiled; whil·ing** (1635)
: to cause to pass especially without boredom or in a pleasant manner — usually used with *away* ⟨*while* away the time⟩

¹whiles \'hwī(ə)lz, 'wī(ə)lz\ *conjunction* [Middle English, from *while* + *-s,* adverb suffix — more at WHENCE] (13th century)
archaic **:** WHILE

²whiles *adverb* (15th century)
chiefly Scottish **:** SOMETIMES

¹whi·lom \'hwī-ləm, 'wī-\ *adverb* [Middle English, literally, at times, from Old English *hwīlum,* dative plural of *hwīl* time, while] (13th century)
archaic **:** FORMERLY

²whilom *adjective* (1837)
: FORMER

whilst \'hwī(ə)lst, 'wī(ə)lst\ *conjunction* [Middle English *whilest,* alteration of *whiles*] (14th century)
chiefly British **:** WHILE

whim \'hwim, 'wim\ *noun* [short for *whim-wham*] (1697)
1 : a capricious or eccentric and often sudden idea or turn of the mind **:** FANCY
2 : a large capstan that is made with one or more radiating arms to which a horse may be yoked and that is used in mines for raising ore or water
synonym see CAPRICE

whim·brel \'hwim-brəl, 'wim-\ *noun* [origin unknown] (circa 1531)
: a small curlew (*Numenius phaeopus*) chiefly of the northern regions of North America and Eurasia; *broadly* **:** any small curlew

¹whim·per \'hwim-pər, 'wim-\ *intransitive verb* **whim·pered; whim·per·ing** \-p(ə-)riŋ\ [imitative] (1513)
1 : to make a low whining plaintive or broken sound
2 : to complain or protest with or as if with a whimper

²whimper *noun* (circa 1700)
1 : a whimpering cry or sound
2 : a petulant complaint or protest

whim·si·cal \'hwim-zi-kəl, 'wim-\ *adjective* [*whimsy*] (1653)
1 : full of, actuated by, or exhibiting whims
2 a : resulting from or characterized by whim or caprice; *especially* **:** lightly fanciful **b :** subject to erratic behavior or unpredictable change
— whim·si·cal·i·ty \,hwim-zə-'ka-lə-tē, ,wim-\ *noun*
— whim·si·cal·ly \'hwim-zi-k(ə-)lē, 'wim-\ *adverb*
— whim·si·cal·ness \-kəl-nəs\ *noun*

whim·sy *also* **whim·sey** \'hwim-zē, 'wim-\ *noun, plural* **whimsies** *also* **whimseys** [irregular from *whim-wham*] (1605)
1 : WHIM, CAPRICE
2 : the quality or state of being whimsical or fanciful ⟨the designer's new line showed a touch of *whimsy*⟩
3 : a fanciful or fantastic device, object, or creation especially in writing or art

whim–wham \'hwim-,hwam, 'wim-,wam\ *noun* [origin unknown] (1500)
1 : a whimsical object or device especially of ornament or dress
2 : FANCY, WHIM
3 *plural* **:** JITTERS

whin \'hwin, 'win\ *noun* [Middle English *whynne,* of Scandinavian origin; akin to Norwegian *kvein* bent grass] (15th century)
: GORSE

whin·chat \'hwin-,chat, 'win-\ *noun* [*whin*] (1678)
: a small brown and buff European singing bird (*Saxicola rubetra*) of grassy meadows

¹whine \'hwīn, 'wīn\ *verb* **whined; whin·ing** [Middle English, from Old English *hwīnan* to whiz; akin to Old Norse *hvīna* to whiz] (13th century)
intransitive verb
1 a : to utter a high-pitched plaintive or distressed cry **b :** to make a sound similar to such a cry ⟨the wind *whined* in the chimney⟩
2 : to complain with or as if with a whine
3 : to move or proceed with the sound of a whine ⟨the bullet *whined* . . . across the ice —Berton Roueché⟩
transitive verb
: to utter or express with or as if with a whine
— whin·er *noun*
— whin·ing·ly \'hwī-niŋ-lē, 'wī-\ *adverb*

²whine *noun* (1633)
1 a : a prolonged high-pitched cry usually expressive of distress or pain **b :** a sound resembling such a cry
2 : a complaint uttered with or as if with a whine
— whiny *or* **whin·ey** \'hwī-nē, 'wī-\ *adjective*

whing–ding \'wiŋ-,diŋ, 'hwiŋ-\ *noun* [by alteration] (circa 1945)
: WINGDING

whinge \'hwinj, 'winj\ *intransitive verb* **whinged; whing·ing** *or* **whinge·ing** [from (assumed) Middle English, from Old English *hwinsian;* akin to Old High German *winsōn* to moan] (12th century)
British **:** to complain fretfully **:** WHINE

¹whin·ny \'hwi-nē, 'wi-\ *verb* **whin·nied; whin·ny·ing** [probably imitative] (1530)
intransitive verb
: to neigh especially in a low or gentle way
transitive verb
: to utter with or as if with a whinny

²whinny *noun, plural* **whinnies** (circa 1823)
1 : the neigh of a horse especially when low or gentle
2 : a sound resembling a neigh

whin·stone \'hwin-,stōn, 'win-\ *noun* [*whin,* a hard rock] (1513)
: basaltic rock **:** TRAP; *also* **:** any of various other dark resistant rocks (as chert)

¹whip \'hwip, 'wip\ *verb* **whipped; whip·ping** [Middle English *wippen, whippen;* akin to Middle Dutch *wippen* to move up and down, sway, Old English *wīpian* to wipe] (14th century)
transitive verb
1 : to take, pull, snatch, jerk, or otherwise move very quickly and forcefully ⟨*whipped* out his gun —Green Peyton⟩
2 a (1) **:** to strike with a slender lithe implement (as a lash or rod) especially as a punishment (2) **:** SPANK **b :** to drive or urge on by or as if by using a whip **c :** to strike as a lash does ⟨rain *whipped* the pavement⟩
3 a : to bind or wrap (as a rope or fishing rod) with cord for protection and strength **b :** to wind or wrap around something
4 : to belabor with stinging words **:** ABUSE
5 : to seam or hem with shallow overcasting stitches
6 : to overcome decisively **:** DEFEAT

7 : to stir up **:** INCITE — usually used with *up* ⟨trying to *whip* up a new emotion —Ellen Glasgow⟩
8 : to produce in a hurry — usually used with *up* ⟨a sketch . . . an artist might *whip* up —*N.Y. Times*⟩
9 : to fish (water) with rod, line, and artificial lure
10 : to beat (as eggs or cream) into a froth with a utensil (as a whisk or fork)
11 : to gather together or hold together for united action in the manner of a party whip
intransitive verb
1 : to proceed nimbly or quickly ⟨*whipping* through the supper dishes —C. B. Davis⟩
2 : to thrash about flexibly in the manner of a whiplash ⟨a flag . . . *whipping* out from its staff —H. A. Calahan⟩
— whip·per *noun*
— whip into shape : to bring forcefully to a desired state or condition

²whip *noun* (14th century)
1 : an instrument consisting usually of a handle and lash forming a flexible rod that is used for whipping
2 : a stroke or cut with or as if with a whip
3 a : a dessert made by whipping a portion of the ingredients ⟨prune *whip*⟩ **b :** a kitchen utensil made of braided or coiled wire or perforated metal with a handle and used in whipping
4 : one that handles a whip: as **a :** a driver of horses **:** COACHMAN **b :** WHIPPER-IN 1
5 a : a member of a legislative body appointed by a political party to enforce party discipline and to secure the attendance of party members at important sessions **b** *often capitalized* **:** a notice of forthcoming business sent weekly to each member of a political party in the British House of Commons
6 : a whipping or thrashing motion
7 : the quality of resembling a whip especially in being flexible
8 : a flexible vertical rod radio antenna — called also *whip antenna*
— whip·like \ˈhwip-ˌlīk, ˈwip-\ *adjective*

whip·cord \ˈhwip-ˌkȯrd, ˈwip-\ *noun* [from its use in making whips] (14th century)
1 : a thin tough cord made of braided or twisted hemp or catgut
2 : a cloth that is made of hard-twisted yarns and has fine diagonal cords or ribs

whip hand *noun* (1680)
1 : positive control **:** ADVANTAGE
2 : the hand holding the whip in driving

whip in *transitive verb* (1742)
1 : to collect or keep together (members of a political party) for legislative action
2 : to keep (hounds in a pack) from scattering by use of a whip

whip·lash \ˈhwip-ˌlash, ˈwip-\ *noun* (circa 1580)
1 : the lash of a whip
2 : something resembling a blow from a whip ⟨the *whiplash* of fear —R. S. Banay⟩
3 : WHIPLASH INJURY

whiplash injury *noun* (circa 1953)
: injury resulting from a sudden sharp whipping movement of the neck and head (as of a person in a vehicle that is struck head-on or from the rear by another vehicle)

whip·per–in \ˌhwi-pər-ˈin, ˌwi-\ *noun, plural* **whip·pers–in** \-pərz-\ (1739)
1 : a huntsman's assistant who whips in the hounds
2 : WHIP 5a

whip·per·snap·per \ˈhwi-pər-ˌsna-pər, ˈwi-\ *noun* [alteration of *snippersnapper*] (1700)
: a diminutive, insignificant, or presumptuous person

whip·pet \ˈhwi-pət, ˈwi-\ *noun* [probably from ¹*whip*] (1610)
: any of a breed of small swift slender dogs that are widely used for racing

whipping *noun* (1540)
1 : the act of one that whips: as **a :** a severe beating or chastisement **b :** a stitching with small overcasting stitches
2 : material used to whip or bind

whipping boy *noun* (1647)
1 : a boy formerly educated with a prince and punished in his stead
2 : SCAPEGOAT 2

whipping cream *noun* (1921)
: a cream suitable for whipping that by law contains not less than 30 percent butterfat

whipping post *noun* (1600)
: a post to which offenders are tied to be legally whipped

whip·ple·tree \ˈhwi-pəl-(ˌ)trē, ˈwi-\ *noun* [perhaps irregular from *whip* + *tree*] (1733)
: WHIFFLETREE

whip·poor·will \ˈhwi-pər-ˌwil, ˌhwi-pər-ˈ, ˈwi-, ˌwi-\ *noun* [imitative] (1709)
: a nocturnal nightjar (*Caprimulgus vociferus*) of eastern North America with a loud repeated call suggestive of its name

whippoorwill

whip·py \ˈhwi-pē, ˈwi-\ *adjective* **whip·pi·er; -est** (1867)
1 : unusually resilient **:** SPRINGY ⟨a *whippy* fishing rod⟩
2 : of, relating to, or resembling a whip

whip–round \ˈhwip-ˌrau̇nd, ˈwip-\ *noun* (1887)
chiefly British **:** a collection of money made usually for a benevolent purpose ⟨had a *whip-round* to help the couple pay for a Paris honeymoon —*The People*⟩

¹whip·saw \ˈhwip-ˌsȯ, ˈwip-\ *noun* [²*whip*] (1538)
: a narrow pit saw averaging from 5 to 7½ feet (1.5 to 2.3 meters) in length

²whipsaw *transitive verb* (1842)
1 : to saw with a whipsaw
2 : to beset or victimize in two opposite ways at once, by a two-phase operation, or by the collusive action of two opponents ⟨wage earners were *whipsawed* by inflation and high taxes⟩

whip·sawed \-ˌsȯd\ *adjective* (1892)
: subjected to a double market loss through trying inopportunely to recoup a loss by a subsequent short sale of the same security

whip scorpion *noun* (circa 1890)
: any of an order (Uropygi) of arachnids somewhat resembling true scorpions but having a long slender caudal process and no sting

whip stall *noun* (1924)
: a stall during a vertical climb in which the nose of the airplane whips violently forward and then downward

¹whip·stitch \ˈhwip-ˌstich, ˈwip-\ *transitive verb* (1592)
: WHIP 5

²whipstitch *noun* (1640)
: a shallow overcasting stitch

whip·stock \-ˌstäk\ *noun* (circa 1530)
: the handle of a whip

whip·worm \-ˌwərm\ *noun* (1875)
: a parasitic nematode worm (genus *Trichuris*) with a body that is thickened posteriorly and that is very long and slender anteriorly; *especially* **:** one (*T. trichiura*) of the human intestine

¹whir *also* **whirr** \ˈhwər, ˈwər\ *verb* **whirred; whir·ring** [Middle English (Scots) *quirren,* probably of Scandinavian origin; akin to Danish *hvirre* to whirl, whir] (15th century)
intransitive verb
: to fly, revolve, or move rapidly with a whir
transitive verb
: to move or carry rapidly with a whir

²whir *also* **whirr** *noun* (1677)
: a continuous fluttering or vibratory sound made by something in rapid motion

¹whirl \ˈhwər(-ə)l, ˈwər(-ə)l\ *verb* [Middle English, probably of Scandinavian origin; akin to Old Norse *hvirfla* to whirl; akin to Old High German *wirbil* whirlwind, Old English *hweorfan* to turn — more at WHARF] (14th century)
intransitive verb
1 : to move in a circle or similar curve especially with force or speed
2 a : to turn on or around an axis like a wheel **:** ROTATE **b :** to turn abruptly around or aside **:** WHEEL
3 : to pass, move, or go quickly ⟨*whirled* down the hallway⟩
4 : to become giddy or dizzy **:** REEL ⟨my head is *whirling*⟩
transitive verb
1 : to drive, impel, or convey with or as if with a rotary motion
2 a : to cause to turn usually rapidly on or around an axis **:** ROTATE **b :** to cause to turn abruptly around or aside
3 *obsolete* **:** to throw or hurl violently with a revolving motion
— whirl·er \ˈhwər-lər, ˈwər-\ *noun*

²whirl *noun* (15th century)
1 a : a rapid rotating or circling movement **b :** something undergoing such a movement
2 a : a busy or fast-paced succession of events **:** BUSTLE ⟨a *whirl* of activity⟩ ⟨the social *whirl*⟩ **b :** a confused or disturbed mental state **:** TURMOIL ⟨a *whirl* of febrile excitement —Emily Skeel⟩
3 : an experimental or brief attempt **:** TRY ⟨gave it a *whirl*⟩

whirl·i·gig \ˈhwər-li-ˌgig, ˈwər-\ *noun* [Middle English *whirlegigg,* from *whirlen* to whirl + *gigg* top — more at GIG] (15th century)
1 : a child's toy having a whirling motion
2 : MERRY-GO-ROUND
3 a : one that continuously whirls, moves, or changes **b :** a whirling or circling course (as of events)

whirligig beetle *noun* (1855)
: any of a family (Gyrinidae) of beetles with two pairs of eyes and clubbed antennae that live mostly on the surface of water where they swim swiftly about in circles

whirl·pool \ˈhwər(-ə)l-ˌpül, ˈwər(-ə)l-\ *noun* (1529)
1 a : a confused tumult and bustle **:** WHIRL **b :** a magnetic or impelling force by which something may be engulfed ⟨refusing to be drawn into this *whirlpool* of intrigue —A. D. White⟩
2 a : water moving rapidly in a circle so as to produce a depression in the center into which floating objects may be drawn **:** EDDY, VORTEX **b :** WHIRLPOOL BATH

whirlpool bath *noun* (circa 1916)
: a therapeutic bath in which all or part of the body is exposed to forceful whirling currents of hot water

¹whirl·wind \-ˌwind\ *noun* (14th century)
1 : a small rotating windstorm of limited extent
2 a : a confused rush **:** WHIRL **b :** a violent or destructive force or agency

²whirlwind *adjective* (1614)
: resembling a whirlwind especially in speed or force ⟨a *whirlwind* campaign⟩ ⟨a *whirlwind* romance⟩

¹whirly \ˈhwər-lē, ˈwər-\ *adjective* (15th century)
: marked by or exhibiting a whirling motion

²whirly *noun, plural* **whirl·ies** (1914)
: a small whirlwind

whirly·bird \-ˌbərd\ *noun* (1951)
: HELICOPTER

whir·ry \'hwər-ē, 'wər-, '(h)wə-rē\ *verb* **whir·ried; whir·ry·ing** [perhaps blend of *whir* and *hurry*] (1582)
transitive verb
Scottish **:** to convey quickly
intransitive verb
Scottish **:** HURRY

¹whish \'hwish, 'wish\ *verb* [imitative] (1518)
transitive verb
: to urge on or cause to move with a whish
intransitive verb
1 : to make a sibilant sound
2 : to move with a whish especially at high speed ⟨an elevator . . . *whishes* down to the lower level —Natalie Cooper⟩

²whish *noun* (circa 1802)
: a rushing sound **:** SWISH

whisht \'hwisht, 'wisht\ *intransitive verb* [Middle English; imitative] (15th century)
chiefly Irish **:** HUSH — often used interjectionally to enjoin silence

¹whisk \'hwisk, 'wisk\ *noun* [Middle English *wisk,* probably of Scandinavian origin; akin to Old Norse *visk* wisp; akin to Old English *wiscian* to plait] (14th century)
1 : a quick light brushing or whipping motion
2 a : a usually wire kitchen utensil used for beating food by hand **b :** a flexible bunch (as of twigs, feathers, or straw) attached to a handle for use as a brush

²whisk (15th century)
intransitive verb
: to move nimbly and quickly
transitive verb
1 : to move or convey briskly ⟨*whisked* the children off to bed⟩
2 : to mix or fluff up by or as if by beating with a whisk ⟨*whisk* egg whites⟩
3 : to brush or wipe off lightly

whisk broom *noun* (1857)
: a small broom with a short handle used especially as a clothes brush

whis·ker \'hwis-kər, 'wis-\ *noun* [singular of *whiskers* mustache, from ²*whisk*] (circa 1600)
1 a : a hair of the beard **b** *plural* (1) *archaic* **:** MUSTACHE (2) **:** the part of the beard growing on the sides of the face or on the chin **c :** HAIRBREADTH ⟨lost the race by a *whisker*⟩
2 : one of the long projecting hairs or bristles growing near the mouth of an animal (as a cat or bird)
3 : an outrigger extending on each side of the bowsprit to spread the jib and flying jib guys — usually used in plural
4 a : a shred or filament resembling a whisker **b :** a thin hairlike crystal (as of sapphire or copper) of exceptional mechanical strength used especially to reinforce composite structural material
— whis·kered \-kərd\ *adjective*
— whis·kery \-k(ə-)rē\ *adjective*

whis·key *or* **whis·ky** \'hwis-kē, 'wis-\ *noun, plural* **whiskeys** *or* **whiskies** [Irish *uisce beathadh* & Scottish Gaelic *uisge beatha,* literally, water of life] (1715)
1 : a liquor distilled from the fermented mash of grain (as rye, corn, or barley)
2 : a drink of whiskey ◆

Whiskey (circa 1952)
— a communication code word for the letter *w*

whiskey sour *noun* (circa 1889)
: a cocktail usually consisting of whiskey, sugar, and lemon juice shaken with ice

¹whis·per \'hwis-pər, 'wis-\ *verb* **whis·pered; whis·per·ing** \-p(ə-)riŋ\ [Middle English, from Old English *hwisperian;* akin to Old High German *hwispalōn* to whisper, Old Norse *hvīsla* — more at WHISTLE] (before 12th century)
intransitive verb
1 : to speak softly with little or no vibration of the vocal cords especially to avoid being overheard
2 : to make a sibilant sound that resembles whispering
transitive verb
1 : to address in a whisper
2 : to utter or communicate in or as if in a whisper

²whisper *noun* (1595)
1 : something communicated by or as if by whispering; *especially* **:** RUMOR ⟨*whispers* of scandal⟩
2 a : an act or instance of whispering; *especially* **:** speech without vibration of the vocal cords **b :** a sibilant sound that resembles whispered speech
3 : HINT, TRACE

whis·per·er \-pər-ər\ *noun* (1547)
: one that whispers; *specifically* **:** RUMORMONGER

¹whispering *noun* (before 12th century)
1 a : whispered speech **b :** GOSSIP, RUMOR
2 : a sibilant sound **:** WHISPER

²whispering *adjective* (1547)
1 : making a sibilant sound
2 : spreading confidential and especially derogatory reports ⟨*whispering* tongues can poison truth —S. T. Coleridge⟩
— whis·per·ing·ly \-p(ə-)riŋ-lē\ *adverb*

whispering campaign *noun* (1920)
: the systematic dissemination by word of mouth of derogatory rumors or charges especially against a candidate for public office

whis·pery \'hwis-p(ə-)rē, 'wis-\ *adjective* (1834)
1 : resembling a whisper
2 : full of whispers

¹whist \'hwist, 'wist\ *intransitive verb* [Middle English; imitative] (14th century)
dialect British **:** to be silent **:** HUSH — often used interjectionally to enjoin silence

²whist *adjective* (15th century)
: QUIET, SILENT

³whist *noun* [alteration of earlier *whisk,* probably from ²*whisk;* from whisking up the tricks] (1663)
: a card game for four players in two partnerships that is played with a pack of 52 cards and that scores one point for each trick in excess of six

¹whis·tle \'hwi-səl, 'wi-\ *noun, often attributive* [Middle English, from Old English *hwistle;* akin to Old Norse *hvīsla* to whisper] (before 12th century)
1 a : a small wind instrument in which sound is produced by the forcible passage of breath through a slit in a short tube ⟨police *whistle*⟩ **b :** a device through which air or steam is forced into a cavity or against a thin edge to produce a loud sound ⟨a factory *whistle*⟩
2 a : a shrill clear sound produced by forcing breath out or air in through the puckered lips **b :** the sound produced by a whistle **c :** a signal given by or as if by whistling
3 : a sound that resembles a whistle; *especially* **:** a shrill clear note of or as if of a bird

²whistle *verb* **whis·tled; whis·tling** \-s(ə-)liŋ\ (before 12th century)
intransitive verb
1 a : to utter a shrill clear sound by blowing or drawing air through the puckered lips **b :** to utter a shrill note or call resembling a whistle **c :** to make a shrill clear sound especially by rapid movement ⟨the wind *whistled*⟩ **d :** to blow or sound a whistle
2 a : to give a signal or issue an order or summons by or as if by whistling **b :** to make a demand without result ⟨did a sloppy job so he can *whistle* for his money⟩
transitive verb
1 a : to send, bring, signal, or call by or as if by whistling **b :** to charge (as a basketball or hockey player) with an infraction
2 : to produce, utter, or express by whistling ⟨*whistle* a tune⟩
— whis·tle·able \-sə-lə-bəl\ *adjective*
— whistle in the dark : to keep up one's courage by or as if by whistling

whis·tle–blow·er \-ˌblō(-ə)r\ *noun* (1970)
: one who reveals something covert or who informs against another ⟨pledges to protect *whistle-blowers* who fear reprisals —*Wall Street Journal*⟩
— whis·tle–blow·ing \-ˌblō-iŋ\ *noun*

whis·tler \'hwi-s(ə-)lər, 'wi-\ *noun* (before 12th century)
: one that whistles: as **a :** any of various birds; *especially* **:** any of numerous Australian and Polynesian oscine birds (especially genus *Pachycephala*) that are related to the shrikes and have a whistling call **b :** a large mountain marmot (*Marmota caligata*) of northwestern North America **c :** a broken-winded horse **d :** a very-low-frequency radio signal that is generated by lightning discharge, travels along the earth's magnetic-field lines, and produces a sound resembling a whistle of descending pitch in radio receivers

¹whis·tle–stop \'hwi-səl-ˌstäp, 'wi-\ *noun, often attributive* (circa 1925)
1 a : a small station at which trains stop only on signal **:** FLAG STOP **b :** a small community
2 : a brief personal appearance especially by a political candidate usually on the rear platform of a train during the course of a tour

²whistle–stop *intransitive verb* (1952)
: to make a tour especially in a political campaign with many brief personal appearances in small communities

whistling *noun* (14th century)
: the act or sound of one that whistles **:** WHISTLE

whistling swan *noun* (1785)
: a native North American swan (*Cygnus columbianus* synonym *Olor columbianus*) that calls with a soft musical note, breeds in the Arctic tundra, and winters in shallow fresh or salt water especially along the eastern and western coasts of the U.S.

whit \'hwit, 'wit\ *noun* [Middle English, probably alteration of *wiht, wight* creature, thing — more at WIGHT] (15th century)
: the smallest part or particle imaginable **:** BIT ⟨what some people will do for a *whit* of publicity —Patrick Quinn⟩

¹white \'hwīt, 'wīt\ *adjective* **whit·er; whit·est** [Middle English, from Old English *hwīt;* akin to Old High German *hwīz* white and probably to Old Church Slavonic *světŭ* light, Sanskrit *śveta* white, bright] (before 12th century)
1 a : free from color **b :** of the color of new snow or milk; *specifically* **:** of the color white **c :** light or pallid in color ⟨*white* hair⟩ ⟨lips *white* with fear⟩ **d :** lustrous pale gray **:** SILVERY; *also* **:** made of silver
2 a : being a member of a group or race characterized by reduced pigmentation and usually

◇ WORD HISTORY

whiskey Of the relatively few English words that have come from the Celtic languages, certainly one of the most common is *whiskey. Whiskey* (or *whisky*) is a shortening from earlier *whiskeybae, usquebaugh,* and many other variants, all of which are borrowed from either Irish *uisce beathadh* or Scottish Gaelic *uisge beatha,* a drink traditionally distilled from malted barley. Both of these phrases mean literally "water of life" and are ultimately translations of Medieval Latin *aqua vitae,* a phrase that writers on alchemy used to describe distilled spirits. *Aqua vitae* turns up in other guises elsewhere in Europe. In Scandinavia it was directly borrowed into Danish, Norwegian, and Swedish as *akvavit,* usually rendered in English as *aquavit,* a clear liquor flavored with caraway seeds. French *eau-de-vie,* on the other hand, is, like the Gaelic expression, a translation from Latin; today it is applied to a clear brandy usually distilled from fermented fruit juice.

specifically distinguished from persons belonging to groups marked by black, brown, yellow, or red skin coloration **b :** of, relating to, characteristic of, or consisting of white people **c :** marked by upright fairness
3 : free from spot or blemish: as **a** (1) **:** free from moral impurity **:** INNOCENT (2) **:** marked by the wearing of white by the woman as a symbol of purity ⟨a *white* wedding⟩ **b :** unmarked by writing or printing **c :** not intended to cause harm ⟨a *white* lie⟩ ⟨*white* magic⟩ **d :** FAVORABLE, FORTUNATE ⟨one of the *white* days of his life —Sir Walter Scott⟩
4 a : wearing or habited in white **b :** marked by the presence of snow **:** SNOWY ⟨a *white* Christmas⟩
5 a : heated to the point of whiteness **b :** notably ardent **:** PASSIONATE ⟨*white* fury⟩
6 a : conservative or reactionary in political outlook and action **b :** instigated or carried out by reactionary forces as a counterrevolutionary measure ⟨a *white* terror⟩
7 : of, relating to, or constituting a musical tone quality characterized by a controlled pure sound, a lack of warmth and color, and a lack of resonance
8 : consisting of a wide range of frequencies — used of light, sound, and electromagnetic radiation
— **whit·ish** \'hwī-tish, 'wī-\ *adjective*

²white *noun* (before 12th century)
1 : the achromatic object color of greatest lightness characteristically perceived to belong to objects that reflect diffusely nearly all incident energy throughout the visible spectrum
2 a : a white or light-colored part of something: as (1) **:** a mass of albuminous material surrounding the yolk of an egg (2) **:** the white part of the ball of the eye (3) **:** the light-colored pieces in a two-handed board game; *also* **:** the player by whom these are played (4) **:** the area of a page unmarked by writing, printing, or illustration **b** (1) *archaic* **:** a white target (2) **:** the fifth or outermost circle of an archery target; *also* **:** a shot that hits it
3 : one that is or approaches white in color: as **a :** white clothing — often used in plural **b :** WHITE WINE **c :** a white mammal (as a horse or a hog) **d** (1) **:** a white-colored product (as flour, pins, or sugar) — usually used in plural (2) **:** any of numerous butterflies (subfamily Pierinae of the family Pieridae) that usually have the ground color of the wings white and are related to the sulphur butterflies
4 *plural* **:** LEUKORRHEA
5 : a person belonging to a light-skinned race
6 *often capitalized* **:** a member of a conservative or reactionary political group

³white *transitive verb* **whit·ed; whit·ing** [Middle English, from *white,* adjective] (before 12th century)
archaic **:** WHITEN

white amur \-ä-'mu̇r\ *noun* [*amur* from *Amur* River] (1968)
: GRASS CARP

white ant *noun* (1684)
: TERMITE

white ash *noun* (1683)
: an American ash (*Fraxinus americana*) having compound leaves with a pale green or oil very white underside; *also* **:** its hard brownish wood

white·bait \'hwīt-ˌbāt, 'wīt-\ *noun* (1758)
1 : the young of any of several European herrings and especially of the common herring (*Clupea harengus*) or of the sprat (*C. sprattus*)
2 : any of various small fishes likened to the European whitebait and used as food

white bass *noun* (1813)
: a North American freshwater bony fish (*Morone chrysops* of the family Percichthyidae) that is used for food

white bean *noun* (1969)
: a dried white kidney bean seed; *also* **:** a plant that is a source of white beans

white·beard \'hwīt-ˌbird, 'wīt-\ *noun* (15th century)
: an old man **:** GRAYBEARD

white birch *noun* (1789)
1 : PAPER BIRCH; *also* **:** GRAY BIRCH
2 : either of two European birches (*Betula pubescens* and *B. pendula*) with white or ash-colored bark that are often planted as ornamentals in the U.S.

white blood cell *noun* (1885)
: any of the blood cells that are colorless, lack hemoglobin, contain a nucleus, and include the lymphocytes, monocytes, neutrophils, eosinophils, and basophils — called also *white blood corpuscle*

white book *noun* (15th century)
: an official report of government affairs bound in white

white–bread \'hwīt-'bred, 'wīt-\ *adjective* (1979)
: being, typical of, or having qualities (as blandness) associated with the white middle class

white·cap \'hwīt-ˌkap, 'wīt-\ *noun* (1773)
: a wave crest breaking into white foam — usually used in plural

white cedar *noun* (1674)
1 : a strong-scented evergreen swamp tree (*Chamaecyparis thyoides*) of the cypress family that occurs along the eastern coast of the U.S. and that has smaller leaves than an arborvitae and globose cones with peltate scales; *also* **:** its wood
2 : NORTHERN WHITE CEDAR

white cell *noun* (1861)
: WHITE BLOOD CELL

white chip *noun* (1897)
1 : a white-colored poker chip usually of minimum value
2 : a thing or quantity of little worth — compare BLUE CHIP

white chocolate *noun* (1923)
: a confection of cocoa butter, sugar, milk solids, lecithin, and flavorings

white clover *noun* (before 12th century)
: a Eurasian clover (*Trifolium repens*) with round heads of white flowers that is widely used in lawn and pasture grass-seed mixtures and is an important source of nectar for bees — called also *white Dutch clover*

white–col·lar \'hwīt-'kä-lər, 'wīt-\ *adjective* (1920)
: of, relating to, or constituting the class of salaried employees whose duties do not call for the wearing of work clothes or protective clothing — compare BLUE-COLLAR

white crappie *noun* (circa 1926)
: a silvery North American sunfish (*Pomoxis annularis*) with 5 or 6 protruding spines on the dorsal fins that is used as a panfish and often for stocking small ponds

white–crowned sparrow \'hwīt-'kraund-, 'wīt-\ *noun* (1839)
: a migratory sparrow (*Zonotrichia leucophrys*) that breeds in northern and western North America and has a grayish breast, pink bill, and head striped with black and white

whit·ed \'hwī-təd, 'wī-\ *adjective* (14th century)
1 : covered with white or whiting and especially with whitewash
2 : made white **:** WHITENED

whited sepulcher *noun* [from the simile in Matthew 23:27 (Authorized Version)] (1582)
: a person inwardly corrupt or wicked but outwardly or professedly virtuous or holy **:** HYPOCRITE

white dwarf *noun, plural* **white dwarfs** (1924)
: a small whitish star of low intrinsic brightness usually with a mass approximately equal to that of the sun but with a density many times larger

white elephant *noun* (15th century)
1 : an Indian elephant of a pale color that is sometimes venerated in India, Sri Lanka, Thailand, and Myanmar
2 a : a property requiring much care and expense and yielding little profit **b :** an object no longer of value to its owner but of value to others **c :** something of little or no value

white·face \'hwīt-ˌfās, 'wīt-\ *noun* (1709)
1 : a white-faced animal; *specifically* **:** HEREFORD
2 : dead-white facial makeup ⟨a clown in *whiteface*⟩

white–faced \-'fāst\ *adjective* (1595)
1 : having a wan pale face
2 : having the face white in whole or in part — used especially of an animal otherwise dark in color

white feather *noun* [from the superstition that a white feather in the plumage of a gamecock is a mark of a poor fighter] (circa 1785)
: a mark or symbol of cowardice — used chiefly in the phrase *show the white feather*

white·fish \'hwīt-ˌfish, 'wīt-\ *noun* (15th century)
1 a : any of various freshwater salmonid food fishes (especially of genera *Coregonus* and *Prosopium*) **b :** any of various fishes resembling the true whitefishes **c** *British* **:** any of various market fishes with white flesh that is not oily
2 : the flesh of a whitefish especially as an article of food

white flag *noun* (1600)
1 : a flag of plain white used as a flag of truce or as a token of surrender
2 : a token of weakness or yielding

white flight *noun* (1967)
: the departure of white families usually from urban neighborhoods undergoing racial integration or from cities implementing school desegregation

white·fly \'hwīt-ˌflī, 'wīt-\ *noun* (circa 1889)
: any of numerous small homopterous insects (family Aleyrodidae) that are injurious plant pests

white–foot·ed mouse \'hwīt-ˌfu̇-təd-, 'wīt-\ *noun* (1869)
: any of various largely nocturnal mice (genus *Peromyscus*) of North and Central America typically having whitish feet and underparts; *especially* **:** a common woodland mouse (*P. leucopus*) of North America

white-footed mouse

white friar *noun, often W&F capitalized* [from his white habit] (15th century)
: CARMELITE

white–fringed beetle \'hwīt-ˌfrinj(d)-, 'wīt-\ *noun* (1939)
: any of a genus (*Graphognathus*) of South American flightless beetles of which one (*G. leucoloma*) has been accidentally introduced into the southeastern U.S. where it is a pest on cultivated plants

white gas *noun* (1926)
: unleaded gasoline — called also *white gasoline*

white gold *noun* (1666)
: a pale alloy of gold especially with nickel or palladium that resembles platinum in appearance

white goods *noun plural* (circa 1871)
1 a : white fabrics especially of cotton or linen **b :** articles (as sheets, towels, or curtains) originally or typically made of white cloth

2 : major household appliances (as stoves and refrigerators) that are typically finished in white enamel

white grub *noun* (circa 1817)
: a grub that is a destructive pest of grass roots and is the larva of various beetles and especially june bugs

White·hall \'hwīt-ˌhȯl, 'wīt-\ *noun* [*Whitehall,* thoroughfare of London in which are located the chief offices of British government] (1827)
: the British government

white·head \-ˌhed\ *noun* (circa 1931)
: MILIUM

white–head·ed \-'he-dəd\ *adjective* (1525)
1 : having the hair, fur, or plumage of the head white or very light
2 : specially favored **:** FORTUNATE — used especially in the phrase *white-headed boy*

white heat *noun* (circa 1710)
1 : a temperature (as for copper and iron from 1500° to 1600° C) which is higher than red heat and at which a body becomes brightly incandescent
2 : a state of intense mental or physical strain, emotion, or activity

white hole *noun* (1971)
: a hypothetical extremely dense celestial object that radiates enormous amounts of energy and matter — compare BLACK HOLE 1

white hope *noun* (1911)
1 : a white contender for a boxing championship held by a black; *also* **:** one who is felt to represent whites
2 : one from whom much is expected; *especially* **:** a person undertaking a difficult task

white–hot \'hwīt-'hät, 'wīt-\ *adjective* (1820)
1 : being at or radiating white heat
2 : ardently zealous **:** FERVID

White House \-ˌhau̇s\ *noun* [the *White House,* mansion in Washington, D.C., assigned to the use of the president of the U.S.] (1811)
1 : the executive department of the U.S. government
2 : a residence of the president of the U.S.

white hunter *noun* (1945)
: a white man serving as guide and professional hunter to an African safari

white knight *noun* (1951)
1 : one that comes to the rescue of another; *especially* **:** a corporation invited to buy out a second corporation in order to prevent an undesired takeover by a third
2 : one that champions a cause

white–knuck·le \'hwīt-'nə-kəl, 'wīt-\ *also* **white–knuck·led** \-kəld\ *adjective* (1974)
: showing or causing tense nervousness ⟨a *white-knuckle* ride on a roller coaster⟩

white lead *noun* (15th century)
: any of several white lead-containing pigments; *especially* **:** a heavy poisonous basic carbonate of lead of variable composition that is marketed as a powder or as a paste in linseed oil, has good hiding power, and is used chiefly in exterior paints

white lightning *noun* (1915)
: MOONSHINE 3

white line *noun* (15th century)
: a band or edge of something white; *especially* **:** a stripe painted on a road and used to guide traffic

white list \-ˌlist\ *noun* (1900)
: a list of approved or favored items — compare BLACKLIST
— white–list·ed \-ˌlis-təd\ *adjective*

white–liv·ered \-'li-vərd\ *adjective* [from the former belief that the choleric temperament depends on the body's producing large quantities of yellow bile] (1549)
: PUSILLANIMOUS, LILY-LIVERED

white·ly \'hwīt-lē, 'wīt-\ *adverb* (14th century)
: with an effect of whiteness **:** so as to show or appear white

white man's burden *noun* ["The White Man's Burden" (1899), poem by Rudyard Kipling] (1911)
: the alleged duty of the white peoples to manage the affairs of the less developed nonwhite peoples

white matter *noun* (circa 1847)
: neural tissue that consists largely of myelinated nerve fibers, has a whitish color, and underlies the gray matter of the brain and spinal cord or is gathered into nerves

white metal *noun* (1613)
1 : any of several light-colored alloys used especially as a base for plated silverware and ornaments and novelties
2 : any of several lead-base or tin-base alloys (as babbitt) used especially for bearings, fusible plugs, and type metal

white mustard *noun* (1731)
: a Eurasian mustard (*Brassica hirta*) grown for its seeds which yield mustard and mustard oil

whit·en \'hwī-tᵊn, 'wī-\ *verb* **whit·ened; whit·en·ing** \'hwīt-niŋ, 'wīt-, -ᵊn-iŋ\ (14th century)
transitive verb
: to make white or whiter ⟨snow *whitened* the hills⟩
intransitive verb
: to become white or whiter

whit·en·er \'hwīt-nər, 'wīt-; 'hwī-tᵊn-ər, 'wī-\ *noun* (1611)
: one that whitens; *specifically* **:** an agent (as a bleach) used to impart whiteness to something

white·ness \'hwīt-nəs, 'wīt-\ *noun* (before 12th century)
1 : the quality or state of being white: as **a :** white color **b :** PALLOR, PALENESS **c :** freedom from stain **:** CLEANNESS
2 : white substance

whitening *noun* (1601)
1 : the act or process of making or becoming white
2 : something that is used to make white **:** WHITING

white noise *noun* (1943)
: a heterogeneous mixture of sound waves extending over a wide frequency range

white oak *noun* (1634)
: any of various oaks (especially *Quercus alba* of eastern North America) with acorns that mature in one year and leaf veins that never extend beyond the margin of the leaf; *also* **:** its hard strong durable wood

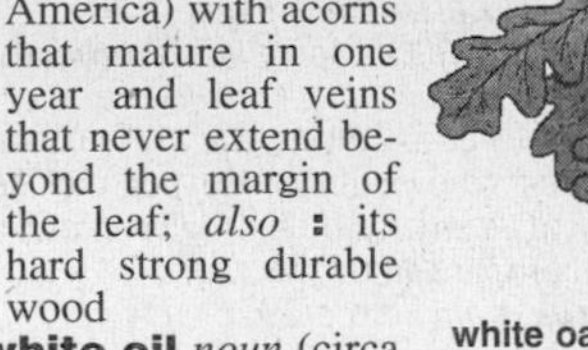
white oak: leaves and acorns

white oil *noun* (circa 1900)
: any of various colorless odorless tasteless mineral oils used especially in medicine and in pharmaceutical and cosmetic preparations

white·out \'hwīt-ˌau̇t, 'wīt-\ *noun* [*white* + black*out*] (1942)
: a surface weather condition in a snow-covered area (as a polar region) in which no object casts a shadow, the horizon cannot be seen, and only dark objects are discernible; *also* **:** a blizzard that severely reduces visibility

white pages *noun plural* (1952)
: the section of a telephone directory that lists individuals and businesses alphabetically

white paper *noun* (1899)
1 : a government report on any subject; *especially* **:** a British publication that is usually less extensive than a blue book
2 : a detailed or authoritative report

white pepper *noun* (14th century)
: a pungent condiment that consists of the fruit of an East Indian plant (*Piper nigrum*) ground after the black husk has been removed

white perch *noun* (1775)
1 : a silvery anadromous bass (*Morone americana*) chiefly of the coast and coastal streams of the eastern U.S.
2 : FRESHWATER DRUM
3 : WHITE CRAPPIE

white pine *noun* (1682)
1 a : a tall-growing pine (*Pinus strobus*) of eastern North America with long needles in clusters of five — called also *eastern white pine* **b :** any of several trees that resemble the white pine especially in having leaves in bundles of five
2 : the wood of a white pine and especially of the eastern white pine

white–pine blister rust *noun* (1911)
: a destructive disease of white pine caused by a rust fungus (*Cronartium ribicola*) that passes part of its complex life cycle on currant or gooseberry bushes; *also* **:** this fungus

white potato *noun* (circa 1890)
: POTATO 2b

white rice *noun* (1923)
: rice from which the hull and bran have been removed by milling

white room *noun* (1961)
: CLEAN ROOM

White Russian *noun* (1850)
1 : BELORUSSIAN
2 : a cocktail made of vodka, coffee liqueur, and cream or milk

white rust *noun* (circa 1848)
: any of various plant diseases caused by a fungus (genus *Albugo* of the order Peronosporales) and characterized by the presence of masses of white spores that escape through ruptures of the host tissue; *also* **:** a fungus causing a white rust

white sale *noun* (1914)
: a sale of white goods

white sauce *noun* (1723)
: a sauce consisting essentially of a roux with milk, cream, or stock and seasoning

white sea bass *noun* (1884)
: a large croaker (*Atractoscion nobilis*) of the Pacific coast that is an important sport and food fish

white shark *noun* (1674)
: GREAT WHITE SHARK

white slave *noun* (1882)
: a woman or girl held unwillingly for purposes of commercial prostitution

white slav·er \-'slā-vər\ *noun* (1912)
: one engaged in white-slave traffic

white slavery *noun* (1857)
: enforced prostitution

white·smith \'hwīt-ˌsmith, 'wīt-\ *noun* (14th century)
1 : TINSMITH
2 : a worker in iron who finishes or polishes the work

white space *noun* (1849)
: the areas of a page without print or pictures

white spruce *noun* (1770)
1 : any of several spruces; *especially* **:** a widely distributed spruce (*Picea glauca*) of coniferous forests of Canada and the northern U.S. that has short stiff blue-green needles and slender cones
2 : the wood of a white spruce; *especially* **:** the light pale tough straight-grained wood of the common white spruce (*Picea glauca*) used especially for construction and as a source of paper pulp

white sucker *noun* (circa 1902)
: a common and widespread edible sucker (*Catostomus commersoni*) of the U.S. and Canada

white supremacist *noun* (1945)
: an advocate of or believer in white supremacy

white supremacy *noun* (1867)
: a doctrine based on a belief in the inherent superiority of the white race over the black race and the correlative necessity for the subordination of blacks to whites in all relationships

white·tail \'hwīt-ˌtāl, 'wīt-\ *noun* (1872)
: WHITE-TAILED DEER

white–tailed deer \-ˌtāld-\ *noun* (1849)
: a North American deer (*Odocoileus virginianus*) with a rather long tail white on the undersurface and the males of which have forward-arching antlers

white-tailed deer

white·throat \'hwīt-ˌthrōt, 'wīt-\ *noun* (1676)
: any of several birds with white on the throat: as **a :** an Old World warbler (*Sylvia communis*) with rusty upper parts and largely pale buff underparts **b :** WHITE-THROATED SPARROW

white–throat·ed sparrow \-ˌthrō-təd-\ *noun* (1811)
: a common brown sparrow (*Zonotrichia albicollis*) chiefly of eastern North America with a black-and-white striped crown and a white patch on the throat

white–tie *adjective* (1936)
: characterized by or requiring the wearing of formal evening clothes consisting of white tie and tailcoat for men and a formal gown for women ⟨a *white-tie* dinner⟩ — compare BLACK-TIE

white trash *noun singular but plural in construction* (1831)
: POOR WHITE — usually used disparagingly

white·wall \'hwīt-ˌwȯl, 'wīt-\ *noun* (1953)
: an automobile tire having a white band on the sidewall

white walnut *noun* (1743)
1 : BUTTERNUT 1
2 : the light-colored wood of a butternut

¹white·wash \'hwīt-ˌwȯsh, 'wīt-, -ˌwäsh\ *transitive verb* (1591)
1 : to whiten with whitewash
2 a : to gloss over or cover up (as vices or crimes) **b :** to exonerate by means of a perfunctory investigation or through biased presentation of data
3 : to hold (an opponent) scoreless in a game or contest
— white·wash·er *noun*

²whitewash *noun* (1689)
1 : a liquid composition for whitening a surface: as **a :** a preparation for whitening the skin **b :** a composition (as of lime and water or whiting, size, and water) for whitening structural surfaces
2 : an act or instance of glossing over or of exonerating
3 : a defeat in a contest in which the loser fails to score

white·wash·ing *noun* (1663)
: an act or instance of applying whitewash; *also* **:** WHITEWASH 3

white water *noun* (1586)
: frothy water (as in breakers, rapids, or falls)
— white–water *adjective*

white way *noun* [the *Great White Way,* nickname for the theatrical section of Broadway, New York City] (1900)
: a brilliantly lighted street or avenue especially in a city's business or theater district

white whale *noun* (circa 1834)
: a cetacean (*Delphinapterus leucas*) that is about 10 feet (3.0 meters) long and white when adult — called also *beluga*

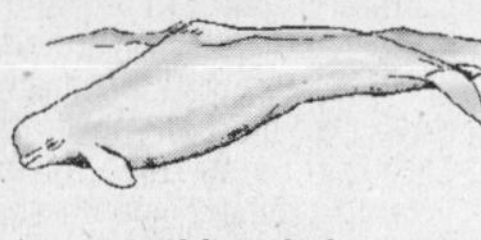
white whale

white wine *noun* (14th century)
: a wine ranging in color from faintly yellow to amber that is produced from the juice alone of dark- or light-colored grapes

white·wing \'hwīt-ˌwiŋ, 'wīt-\ *noun* (1898)
: a person and especially a street sweeper wearing a white uniform

white·wood \-ˌwu̇d\ *noun* (1663)
1 : any of various trees with pale or white wood: as **a :** TULIP TREE 1 **b :** an Australian tree (*Atalaya hemiglauca*) of the soapberry family
2 : the wood of a whitewood; *especially* **:** TULIPWOOD 1

whit·ey \'hwī-tē, 'wī-\ *noun, often capitalized* (1828)
: the white man **:** white society — usually used disparagingly

white zinfandel *noun* (1976)
: a blush wine made from zinfandel grapes

¹whith·er \'hwi-t͟hər, 'wi-\ *adverb* [Middle English, from Old English *hwider;* akin to Latin *quis* who and to Old English hi*der* hither — more at WHO, HITHER] (before 12th century)
1 : to what place ⟨*whither* will they go⟩
2 : to what situation, position, degree, or end ⟨*whither* will this abuse drive him⟩

²whither *conjunction* (before 12th century)
1 a : to what place ⟨knew *whither* to go —Daniel Defoe⟩ **b :** to what situation, position, degree, or end
2 a : to the place at, in, or to which **b :** to which place
3 : to whatever place

whith·er·so·ev·er \ˌhwi-t͟hər-sə-'we-vər, ˌwi-\ *conjunction* (14th century)
: to whatever place ⟨will go *whithersoever* you lead⟩

whith·er·ward \'hwi-t͟hər-wərd, 'wi-\ *adverb* (13th century)
archaic **:** toward what or which place

¹whit·ing \'hwī-tiŋ, 'wī-\ *noun* [Middle English, from Middle Dutch *witinc,* from *wit* white; akin to Old English *hwīt* white] (15th century)
: any of various marine food fishes: as **a :** a common European fish (*Merlangus merlangus*) of the cod family **b :** SILVER HAKE

²whiting *noun* [Middle English, from gerund of *whiten* to white] (15th century)
: calcium carbonate ground into fine powder, washed, and used especially as a pigment and extender, in putty, and in rubber compounding and paper coating

whit·low \'hwit-(ˌ)lō, 'wit-\ *noun* [Middle English *whitflawe, whitflowe, whitlowe*] (14th century)
: a deep usually suppurative inflammation of the finger or toe especially near the end or around the nail — called also *felon*

Whit·mon·day \'hwit-ˌmən-dē, 'wit-, -'mən-\ *noun* [*Whit*sunday + *Monday*] (1557)
: the day after Whitsunday observed as a legal holiday in England, Wales, and Ireland

Whit·sun \'hwit-sən, 'wit-\ *adjective* [Middle English *Whitson,* from *Whitsonday*] (14th century)
: of, relating to, or observed on Whitsunday or at Whitsuntide

Whit·sun·day \-'sən-dē, -sən-ˌdā\ *noun* [Middle English *Whitsonday,* from Old English *hwīta sunnandæg,* literally, white Sunday; probably from the custom of wearing white robes by those newly baptized at this season] (12th century)
: PENTECOST 2

Whit·sun·tide \-sən-ˌtīd\ *noun* (13th century)
: the week beginning with Whitsunday and especially the first three days of this week

¹whit·tle \'hwi-tᵊl, 'wi-\ *noun* [Middle English *whittel,* alteration of *thwitel,* from *thwiten* to whittle, from Old English *thwītan;* akin to Old Norse *thveita* to hew] (15th century)
archaic **:** a large knife

²whittle *verb* **whit·tled; whit·tling** \'hwit-liŋ, 'wit-; 'hwi-tᵊl-iŋ, 'wi-\ (1552)
transitive verb
1 a : to pare or cut off chips from the surface of (wood) with a knife **b :** to shape or form by so paring or cutting
2 : to reduce, remove, or destroy gradually as if by cutting off bits with a knife **:** PARE ⟨*whittle* down expenses⟩
intransitive verb
1 : to cut or shape something (as wood) by or as if by paring it with a knife
2 : to wear oneself or another out with fretting
— whit·tler \'hwit-lər, 'wit-; 'hwi-tᵊl-ər, 'wi-\ *noun*

whit·tling *noun* (1854)
1 : the act or art of whittling
2 : a piece cut away in whittling

whit·tret \'hwi-trət, 'wi-\ *noun* [Middle English *whitrat,* from *white, whit* white + *rat* rat] (15th century)
chiefly Scottish **:** WEASEL

whity *or* **whit·ey** \'hwī-tē, 'wī-\ *adjective* (1593)
: somewhat white **:** WHITISH — usually used in combination

¹whiz *or* **whizz** \'hwiz, 'wiz\ *verb* **whizzed; whiz·zing** [imitative] (1547)
intransitive verb
1 : to hum, whir, or hiss like a speeding object (as an arrow or ball) passing through air
2 : to fly or move swiftly especially with a whiz
transitive verb
: to cause to whiz; *especially* **:** to rotate very rapidly

²whiz *or* **whizz** *noun, plural* **whiz·zes** (1620)
1 : a hissing, buzzing, or whirring sound
2 : a movement or passage of something accompanied by a whizzing sound

³whiz *noun, plural* **whiz·zes** [probably by shortening & alteration] (1914)
: WIZARD 3 ⟨a *whiz* at math⟩

whiz·bang *also* **whizz·bang** \'hwiz-ˌbaŋ, 'wiz-, -'baŋ\ *noun* (1915)
: one that is conspicuous for noise, speed, excellence, or startling effect
— whiz–bang *adjective*

whiz kid *also* **whizz kid** *noun* [³*whiz*] (circa 1942)
: a person who is unusually intelligent, clever, or successful especially at an early age

whiz·zer \'hwi-zər, 'wi-\ *noun* (1881)
: one that whizzes; *especially* **:** a centrifugal machine for drying something (as grain, sugar, or nitrated cotton)

who \'hü, ü\ *pronoun* [Middle English, from Old English *hwā;* akin to Old High German *hwer,* interrogative pronoun, who, Latin *quis,* Greek *tis,* Latin *qui,* relative pronoun, who] (before 12th century)
1 : what or which person or persons — used as an interrogative ⟨*who* was elected?⟩ ⟨find out *who* they are⟩; used by speakers on all educational levels and by many reputable writers, though disapproved by some grammarians, as the object of a verb or a following preposition ⟨*who* did I see but a Spanish lady —Padraic Colum⟩ ⟨do not know *who* the message is from —G. K. Chesterton⟩
2 : the person or persons that **:** WHOEVER
3 — used as a function word to introduce a relative clause; used especially in reference to persons ⟨my father, *who* was a lawyer⟩ but also in reference to groups ⟨a generation *who* had known nothing but war —R. B. Woot⟩ or to animals ⟨dogs *who* . . . fawn all over tramps —Nigel Balchin⟩ or to inanimate objects especially with the implication that the reference is really to a person ⟨earlier sources *who* maintain a Davidic ancestry —F. M. Cross⟩; used by speakers on all educational levels and by many reputable writers, though disapproved by some grammarians, as the object of a verb or a following preposition ⟨a

character *who* we are meant to pity —*Times Literary Supplement*⟩
usage see WHOM, THAT
— as who *archaic* **:** as one that **:** as if someone
— as who should say *archaic* **:** so to speak
— who is who *or* **who's who** *or* **who was who :** the identity of or the noteworthy facts about each of a number of persons

whoa \'wō, 'hō, 'hwō\ *imperative verb* [Middle English *whoo, who*] (15th century)
— a command (as to a draft animal) to stand still

who'd \'hüd\ (1640)
: who had **:** who would

who·dun·it *also* **who·dun·nit** \hü-'də-nət\ *noun* [alteration of *who done it?*] (1930)
: a detective story or mystery story

who·ev·er \hü-'e-vər\ *pronoun* (13th century)
: whatever person **:** no matter who — used in any grammatical relation except that of a possessive ⟨sells to *whoever* has the money to buy⟩

¹whole \'hōl\ *adjective* [Middle English *hool* healthy, unhurt, entire, from Old English *hāl;* akin to Old High German *heil* healthy, unhurt, Old Norse *heill,* Old Church Slavonic *cělŭ*] (before 12th century)
1 a (1) **:** free of wound or injury **:** UNHURT (2) **:** recovered from a wound or injury **:** RESTORED (3) **:** being healed ⟨*whole* of an ancient evil, I sleep sound —A. E. Housman⟩ **b :** free of defect or impairment **:** INTACT **c :** physically sound and healthy **:** free of disease or deformity **d :** mentally or emotionally sound
2 : having all its proper parts or components **:** COMPLETE, UNMODIFIED ⟨*whole* milk⟩ ⟨a *whole* egg⟩
3 a : constituting the total sum or undiminished entirety **:** ENTIRE ⟨owns the *whole* island⟩ **b :** each or all of the ⟨took part in the *whole* series of athletic events⟩
4 a : constituting an undivided unit **:** UNBROKEN, UNCUT ⟨a *whole* roast suckling pig⟩ **b :** directed to one end **:** CONCENTRATED ⟨promised to give it his *whole* attention⟩
5 a : seemingly complete or total ⟨the *whole* idea is to help, not hinder⟩ **b :** very great in quantity, extent, or scope ⟨feels a *whole* lot better now⟩
6 : constituting the entirety of a person's nature or development ⟨educate the *whole* student⟩
7 : having the same father and mother ⟨*whole* brother⟩ ☆
synonym see PERFECT
— whole·ness *noun*

²whole *noun* (14th century)
1 : a complete amount or sum **:** a number, aggregate, or totality lacking no part, member, or element
2 : something constituting a complex unity **:** a coherent system or organization of parts fitting or working together as one
— in whole : to the full or entire extent **:** WHOLLY — usually used in the phrase *in whole or in part*
— on the whole 1 : in view of all the circumstances or conditions **:** all things considered **2 :** in general **:** in most instances **:** TYPICALLY

³whole *adverb* (14th century)
1 : WHOLLY, ENTIRELY ⟨a *whole* new age group —Henry Chauncey⟩
2 : as a complete entity

whole cloth *noun* (1840)
: pure fabrication — usually used in the phrase *out of whole cloth*

whole gale *noun* (circa 1805)
: wind having a speed of 55 to 63 miles (88 to 101 kilometers) per hour — see BEAUFORT SCALE table

whole·heart·ed \'hōl-'här-təd\ *adjective* (1840)
1 : completely and sincerely devoted, determined, or enthusiastic ⟨a *wholehearted* student of social problems⟩
2 : marked by complete earnest commitment **:** free from all reserve or hesitation ⟨gave the proposal *wholehearted* approval⟩
synonym see SINCERE
— whole·heart·ed·ly *adverb*

whole–hog *adjective* (1829)
: committed without reservation **:** THOROUGHGOING ⟨a *whole-hog* patriot⟩

¹whole hog *noun* (1829)
: the whole way or farthest limit — usually used adverbially in the phrase *go the whole hog*

²whole hog *adverb* (1844)
: to the fullest extent **:** without reservation **:** COMPLETELY ⟨accepting *whole hog* the standards . . . of the majority —R. B. Kaplan⟩

whole–life \'hōl-'līf\ *adjective* (1845)
: of, relating to, or being life insurance with a fixed premium for the life of the policyholder and a cash value that can be redeemed on sale of the policy or can be the basis of low-interest loans

whole note *noun* (1597)
: a musical note equal in time value to four quarter notes or two half notes — see NOTE illustration

whole number *noun* (1557)
: any of the set of nonnegative integers; *also* **:** INTEGER

whole rest *noun* (circa 1890)
: a musical rest corresponding in time value to a whole note — see REST illustration

¹whole·sale \'hōl-ˌsāl\ *noun* (15th century)
: the sale of commodities in quantity usually for resale (as by a retail merchant)

²wholesale *adjective* (1642)
1 : performed or existing on a large scale especially without discrimination ⟨*wholesale* slaughter⟩
2 : of, relating to, or engaged in the sale of commodities in quantity for resale ⟨a *wholesale* grocer⟩

³wholesale *adverb* (1759)
: in a wholesale manner

⁴wholesale *verb* **whole·saled; whole·sal·ing** (1800)
transitive verb
: to sell (something) in quantity usually for resale
intransitive verb
: to sell in quantity usually for resale

whole·sal·er \'hōl-ˌsā-lər\ *noun* (1857)
: a merchant middleman who sells chiefly to retailers, other merchants, or industrial, institutional, and commercial users mainly for resale or business use

whole·some \'hōl-səm\ *adjective* (13th century)
1 : promoting health or well-being of mind or spirit
2 : promoting health of body
3 a : sound in body, mind, or morals **b :** having the simple health or vigor of normal domesticity
4 a : based on well-grounded fear **:** PRUDENT ⟨a *wholesome* respect for the law⟩ **b :** SAFE ⟨it wouldn't be *wholesome* for you to go down there —Mark Twain⟩
synonym see HEALTHFUL, HEALTHY
— whole·some·ly *adverb*
— whole·some·ness *noun*

whole–souled \'hōl-'sōld\ *adjective* (1834)
: moved by ardent enthusiasm or single-minded devotion **:** WHOLEHEARTED

whole step *noun* (circa 1899)
: a musical interval (as C–D or G–A) comprising two half steps — called also *whole tone*

whole wheat *adjective* (1880)
: made of ground entire wheat kernels

who·lis·tic \hō-'lis-tik\ *variant of* HOLISTIC

whol·ly \'hō(l)-lē\ *adverb* [Middle English *hoolly,* from *hool* whole] (14th century)
1 : to the full or entire extent **:** COMPLETELY ⟨a *wholly* owned subsidiary⟩
2 : to the exclusion of other things **:** SOLELY ⟨a book dealing *wholly* with herbs⟩

whom \'hüm, üm\ *pronoun, objective case of* WHO [Middle English, from Old English *hwām,* dative of *hwā* who] (before 12th century)
— used as an interrogative or relative; used as object of a verb or a preceding preposition ⟨to know for *whom* the bell tolls —John Donne⟩ or less frequently as the object of a following preposition ⟨the man *whom* you wrote to⟩ though now often considered stilted especially as an interrogative and especially in oral use; occasionally used as predicate nominative with a copulative verb or as subject of a verb especially in the vicinity of a preposition or a verb of which it might mistakenly be considered the object ⟨*whom* say ye that I am —Matthew 16:15 (Authorized Version)⟩ ⟨people . . . *whom* you never thought would sympathize —Shea Murphy⟩ ■

whom·ev·er \hü-'me-vər\ *pronoun, objective case of* WHOEVER

¹whomp \'hwämp, 'hwȯmp, 'wämp, 'wȯmp\ *noun* [imitative] (1926)
: a loud slap, crash, or crunch

²whomp (1942)
intransitive verb
: to strike with a sharp noise or thump
transitive verb
1 : to hit or slap sharply
2 : to defeat decisively **:** TROUNCE
3 : to create or put together especially hastily — usually used with *up*

whomp up *transitive verb* (1949)

☆ SYNONYMS
Whole, entire, total, all mean including everything or everyone without exception. WHOLE implies that nothing has been omitted, ignored, abated, or taken away ⟨read the *whole* book⟩. ENTIRE may suggest a state of completeness or perfection to which nothing can be added ⟨the *entire* population was wiped out⟩. TOTAL implies that everything has been counted, weighed, measured, or considered ⟨the *total* number of people present⟩. ALL may equal WHOLE, ENTIRE, or TOTAL ⟨*all* proceeds go to charity⟩.

□ USAGE
whom Observers of the language have been predicting the demise of *whom* from about 1870 down to the present day ⟨one of the pronoun cases is visibly disappearing—the objective case *whom* —R. G. White (1870)⟩ ⟨*whom* is dying out in England, where "Whom did you see?" sounds affected —Anthony Burgess (1980)⟩. Our evidence shows that no one—English or not—should expect *whom* to disappear momentarily; it shows every indication of persisting quite a while yet. Actual usage of *who* and *whom*—accurately described at the entries in this dictionary—does not appear to be markedly different from the usage of Shakespeare's time. But the 18th century grammarians, propounding rules and analogies, rejecting other rules and analogies, and usually justifying both with appeals to Latin or Greek, have intervened between us and Shakespeare. It seems clear that the grammarians' rules have had little effect on the traditional uses. One thing they have accomplished is to encourage hypercorrect uses of *whom* ⟨*whom* shall I say is calling?⟩. Another is that they have made some people unsure of themselves ⟨said he was asked to step down, although it is not known exactly *who* or *whom* asked him —*Redding (Connecticut) Pilot*⟩.

: to stir up **:** AROUSE

whom·so \'hüm-(ˌ)sō\ *pronoun, objective case of* WHOSO

whom·so·ev·er \ˌhüm-sə-'we-vər\ *pronoun, objective case of* WHOSOEVER

1whoop \'hüp, 'hu̇p, 'hwüp, 'hwu̇p, 'wüp, 'wu̇p\ *verb* [Middle English *whopen,* from Middle French *houpper,* of imitative origin] (14th century)
intransitive verb
1 : to utter a whoop in expression of eagerness, enthusiasm, or enjoyment **:** SHOUT
2 : to utter the cry of an owl **:** HOOT
3 : to make the characteristic whoop of whooping cough
4 a : to go or pass with a loud noise **b :** to be rushed through by acclamation or with noisy support ⟨the bill *whooped* through both houses⟩
transitive verb
1 a : to utter or express with a whoop **b :** to urge, drive, or cheer on with a whoop
2 : to agitate in behalf of
3 : RAISE, BOOST ⟨*whoop* up the price⟩
— whoop it up 1 : to celebrate riotously **:** CAROUSE **2 :** to stir up enthusiasm

2whoop *noun* (1593)
1 a : a loud yell expressive of eagerness, exuberance, or jubilation — often used interjectionally **b :** a shout of hunters or of men in battle or pursuit
2 : the cry of an owl **:** HOOT
3 : the crowing intake of breath following a paroxysm in whooping cough
4 : a minimum amount or degree **:** the least bit ⟨not worth a *whoop*⟩

whoop–de–do *or* **whoop–de–doo** \ˌh(w)üp-dē-'dü, ˌh(w)u̇p-, -tē-\ *noun* [probably irregular from 2*whoop*] (1929)
1 : noisy and exuberant or attention-getting activity (as at a social affair or in a political campaign)
2 : a lively social affair
3 : agitated public discussion or debate

1whoop·ee \'(h)wu̇-(ˌ)pē, '(h)wü-; (h)wu̇-'pē, (h)wü-\ *interjection* [irregular from 2*whoop*] (1845)
— used to express exuberance

2whoop·ee \'(h)wu̇-(ˌ)pē, '(h)wü-\ *noun* (1924)
1 : boisterous convivial fun **:** MERRYMAKING — usually used with *make*
2 : sexual play — usually used with *make*

whoopee cushion *noun* (1953)
: a cushion that makes a sound like the breaking of wind when sat upon

whoop·er \'h(w)ü-pər, 'h(w)u̇-\ *noun* (1660)
: one that whoops; *specifically* **:** WHOOPING CRANE

whooper swan *noun* (1880)
: a chiefly Eurasian swan (*Cygnus cygnus*) with a yellow and black bill — compare TRUMPETER SWAN

whooping cough *noun* (circa 1670)
: an infectious disease especially of children caused by a bacterium (*Bordetella pertussis*) and marked by a convulsive spasmodic cough sometimes followed by a crowing intake of breath — called also *pertussis*

whooping crane *noun* (circa 1730)
: a large white nearly extinct North American crane (*Grus americana*) noted for its loud trumpeting call

whooping crane

whoop·la \'h(w)üp-ˌlä, 'h(w)u̇p-\ *noun* [alteration of *hoopla*] (1931)
1 : HOOPLA
2 : boisterous merrymaking

whoops \'(w)u̇(ə)ps\ *variant of* OOPS

1whoosh \'hwüsh, 'wüsh, '(h)wu̇sh\ *noun* [imitative] (1880)
: a swift or explosive rush; *also* **:** the sound created by such a rush — often used interjectionally

2whoosh (1909)
intransitive verb
: to rush past or move explosively ⟨cars *whooshing* along the expressway⟩
transitive verb
: to move (a person or thing) with or as if with a whoosh

1whop \'hwäp, 'wäp\ *transitive verb* **whopped; whop·ping** [Middle English *whappen,* alteration of *wappen* to throw violently] (15th century)
1 : to pull or whip out
2 a : BEAT, STRIKE **b :** to defeat totally

2whop *noun* (15th century)
: a heavy blow **:** THUMP

whop·per \'hwä-pər, 'wä-\ *noun* [1*whop*] (circa 1785)
1 : something unusually large or otherwise extreme of its kind
2 : an extravagant or monstrous lie

whop·ping \'hwä-piŋ, 'wä-\ *adjective* (circa 1625)
: extremely large; *also* **:** EXTRAORDINARY, INCREDIBLE

1whore \'hōr, 'hȯr, 'hu̇r\ *noun* [Middle English *hore,* from Old English *hōre;* akin to Old Norse *hōra* whore, *hōrr* adulterer, Latin *carus* dear — more at CHARITY] (before 12th century)
1 : a woman who engages in sexual acts for money **:** PROSTITUTE; *also* **:** a promiscuous or immoral woman
2 : a male who engages in sexual acts for money
3 : a venal or unscrupulous person

2whore *verb* **whored; whor·ing** (1583)
intransitive verb
1 : to have unlawful sexual intercourse as or with a whore
2 : to pursue a faithless, unworthy, or idolatrous desire
transitive verb
obsolete **:** to corrupt by lewd intercourse **:** DEBAUCH

whore·dom \'hōr-dəm, 'hȯr-, 'hu̇r-\ *noun* [Middle English *hordom* sexual immorality, idolatrous practices, from Old Norse *hōrdōmr* adultery, from *hōrr*] (12th century)
1 : the practice of whoring **:** PROSTITUTION
2 : faithless, unworthy, or idolatrous practices or pursuits

whore·house \'hōr-ˌhau̇s, 'hȯr-, 'hu̇r-\ *noun* (14th century)
: a building in which prostitutes are available **:** BORDELLO

whore·mas·ter \-ˌmas-tər\ *noun* (14th century)
: a man consorting with whores or given to lechery

whore·mon·ger \-ˌməŋ-gər, -ˌmäŋ-\ *noun* (1526)
: WHOREMASTER

whore·son \'hōr-sᵊn, 'hȯr-, 'hu̇r-\ *noun, often attributive* (14th century)
1 : BASTARD
2 : a coarse fellow — used as a generalized term of abuse

Whorf·ian hypothesis \'wȯr-fē-ən-, 'hwȯr-\ *noun* [Benjamin Lee *Whorf* (died 1941) American anthropologist] (1954)
: a theory in linguistics: one's language determines one's conception of the world

whor·ish \'hōr-ish, 'hȯr-, 'hu̇r-\ *adjective* (1535)
: of or befitting a whore

whorl \'hwȯr(-ə)l, 'wȯr(-ə)l, '(h)wər(-ə)l\ *noun* [Middle English *wharle, whorle,* probably alteration of *whirle,* from *whirlen* to whirl] (15th century)
1 : a drum-shaped section on the lower part of a spindle in spinning or weaving machinery serving as a pulley for the tape drive that rotates the spindle
2 : an arrangement of similar anatomical parts (as leaves) in a circle around a point on an axis
3 : something that whirls, coils, or spirals or whose form suggests such movement **:** SWIRL ⟨*whorls* of snow⟩
4 : one of the turns of a univalve shell
5 : a fingerprint in which the central papillary ridges turn through at least one complete circle

whorled \'hwȯr(-ə)ld, 'wȯr(-ə)ld, '(h)wər(-ə)ld\ *adjective* (circa 1776)
: having or arranged in whorls ⟨leaves *whorled* at the nodes of the stem⟩

whor·tle·ber·ry \'hwər-tᵊl-ˌber-ē, 'wər-\ *noun* [alteration of earlier *hurtleberry,* from Middle English *hurtilberye,* irregular from Old English *horte* whortleberry + Middle English *berye* berry] (1578)
1 : a European blueberry (*Vaccinium myrtillus*); *also* **:** its glaucous blackish edible berry
2 : BLUEBERRY

1whose \'hüz, üz\ *adjective* [Middle English *whos,* genitive of *who, what*] (before 12th century)
: of or relating to whom or which especially as possessor or possessors ⟨*whose* gorgeous vesture heaps the ground —Robert Browning⟩, agent or agents ⟨the law courts, *whose* decisions were important —F. L. Mott⟩, or object or objects of an action ⟨the first poem *whose* publication he ever sanctioned —J. W. Krutch⟩

2whose *pronoun, singular or plural in construction* (13th century)
: that which belongs to whom — used without a following noun as a pronoun equivalent in meaning to the adjective *whose* ⟨tell me *whose* it was —Shakespeare⟩ ◼

whose·so·ev·er \ˌhüz-sə-'we-vər\ *adjective* (1611)
: of or relating to whomsoever ⟨*whosesoever* sins ye remit —John 20:23 (Authorized Version)⟩

who·so \'hü-(ˌ)sō\ *pronoun* (12th century)
: WHOEVER

who·so·ev·er \ˌhü-sə-'we-vər\ *pronoun* (13th century)

◻ USAGE

whose Since neither *which* nor *that* has a possessive case, *whose* has long been pressed into duty as a substitute ⟨a land *whose* stones are iron, and out of *whose* hills thou mayest dig brass —Deuteronomy 8:9 (Authorized Version)⟩ ⟨that forbidden tree, *whose* mortal taste brought death into the world —John Milton⟩. A number of 18th century grammarians took to wondering about the propriety of such use, since *whose* is related to *who,* and stones, hills, and trees are not persons. But some of them also noticed that using *whose* made a more straightforward or even a more elegant sentence than the alternative construction with *of which* did. This characteristic probably explains why good writers have gone on using *whose* in spite of the grammarians ⟨a precaution *whose* necessity was demonstrated —Lewis Mumford⟩ ⟨I can see its lights through my window, *whose* sash rattles — John Updike⟩. Almost no modern handbook disparages this use of *whose,* but many people still believe that there must be something wrong with it. There is not. The belief that *whose* should not be used of inanimate things is a superstition.

: WHOEVER

who's who \ˈhüz-ˈhü\ *noun, often both Ws capitalized* (1917)
1 : a compilation of brief biographical sketches of prominent persons in a particular field ⟨a *who's who* of sports figures⟩
2 : the leaders of a group **:** ELITE; *also* **:** a listing of such figures

whump \ˈhwəmp, ˈwəmp\ *intransitive verb* [imitative] (1897)
: BANG, THUMP
— whump *noun*

1why \ˈhwī, ˈwī\ *adverb* [Middle English, from Old English *hwȳ*, instrumental case of *hwæt* what — more at WHAT] (before 12th century)
: for what cause, reason, or purpose ⟨*why* did you do it?⟩

2why *conjunction* (before 12th century)
1 : the cause, reason, or purpose for which ⟨know *why* you did it⟩ ⟨that is *why* you did it⟩
2 : for which **:** on account of which ⟨know the reason *why* you did it⟩

3why *noun, plural* **whys** (14th century)
1 : REASON, CAUSE ⟨wants to know the *whys* and wherefores⟩
2 : a baffling problem **:** ENIGMA

4why *interjection* (1519)
— used to express mild surprise, hesitation, approval, disapproval, or impatience ⟨*why*, here's what I was looking for⟩

whyd·ah \ˈhwi-də, ˈwi-\ *noun* [alteration of *widow (bird)*] (1783)
: any of various mostly black and white African weaverbirds (genera *Euplectes* and *Vidua*) often kept as cage birds and distinguished in the male by long drooping tail feathers during the breeding season

1wick \ˈwik\ *noun* [Middle English *weke, wicke*, from Old English *wēoce;* akin to Old High German *wiohha* wick, Middle Irish *figid* he weaves] (before 12th century)
: a bundle of fibers or a loosely twisted, braided, or woven cord, tape, or tube usually of soft spun cotton threads that by capillary attraction draws up to be burned a steady supply of the oil in lamps or the melted tallow or wax in candles

2wick *transitive verb* (1949)
: to carry (as moisture) by capillary action — often used with *away* ⟨a fabric that wicks away perspiration⟩

1wick·ed \ˈwi-kəd\ *adjective* [Middle English, alteration of *wicke* wicked] (13th century)
1 : morally very bad **:** EVIL
2 a : FIERCE, VICIOUS ⟨a *wicked* dog⟩ **b :** disposed to or marked by mischief **:** ROGUISH ⟨does *wicked* impersonations⟩
3 a : disgustingly unpleasant **:** VILE ⟨a *wicked* odor⟩ **b :** causing or likely to cause harm, distress, or trouble ⟨a *wicked* storm⟩
4 : going beyond reasonable or predictable limits **:** of exceptional quality or degree ⟨throws a *wicked* fastball⟩
— wick·ed·ly *adverb*

2wicked *adverb* (1980)
: VERY, EXTREMELY ⟨*wicked* fast⟩

wick·ed·ness *noun* (14th century)
1 : the quality or state of being wicked
2 : something wicked

wick·er \ˈwi-kər\ *noun* [Middle English *wiker*, of Scandinavian origin; akin to Swedish dialect *vikker* willow, Old Norse *veikr* weak — more at WEAK] (14th century)
1 : a small pliant twig or branch **:** OSIER, WITHE
2 a : WICKERWORK **b :** something made of wicker
— wicker *adjective*

wick·er·work \-ˌwərk\ *noun* (1719)
: work consisting of interlaced osiers, twigs, or rods ⟨a cage of *wickerwork*⟩

wick·et \ˈwi-kət\ *noun* [Middle English *wiket*, from Old North French, of Germanic origin; akin to Old Norse *vik* corner, *vīkja* to move, turn] (13th century)
1 : a small gate or door; *especially* **:** one forming part of or placed near a larger gate or door
2 : an opening like a window; *especially* **:** a grilled or grated window through which business is transacted
3 a : either of the two sets of three stumps topped by two crosspieces and set 66 feet apart at which the ball is bowled in cricket **b :** an area 10 feet wide bounded by these wickets **c :** one innings of a batsman; *specifically* **:** one that is not completed or never begun ⟨win by three *wickets*⟩
4 : an arch or hoop in croquet

wick·ing \ˈwi-kiŋ\ *noun* (1847)
: material for wicks

wick·i·up \ˈwi-kē-ˌəp\ *noun* [Fox (Algonquian language of the Fox, Sauk, and Kickapoo Indians) *wi·kiyaˑpi* house] (1852)
: a hut used by the nomadic Indians of the arid regions of the western and southwestern U.S. with a usually oval base and a rough frame covered with reed mats, grass, or brushwood; *also* **:** a rude temporary shelter or hut

wickiup

wid·der·shins \ˈwi-dər-shənz\ *adverb* [Middle Low German *weddersinnes*, from Middle High German *widersinnes*, from *widersinnen* to go against, from *wider* back against (from Old High German *widar*) + *sinnen* to travel, go; akin to Old High German *sendan* to send — more at WITH, SEND] (1513)
: in a left-handed, wrong, or contrary direction **:** COUNTERCLOCKWISE — compare DEASIL

wid·dy \ˈwi-dē\ *noun, plural* **widdies** [Middle English (Scots), from Middle English *withy*] (15th century)
1 *Scottish & dialect English* **:** a rope of osiers
2 *Scottish & dialect English* **:** a hangman's noose

1wide \ˈwīd\ *adjective* **wid·er; wid·est** [Middle English, from Old English *wīd;* akin to Old High German *wīt* wide] (before 12th century)
1 a : having great extent **:** VAST ⟨a *wide* area⟩ **b :** extending over a vast area **:** EXTENSIVE ⟨a *wide* reputation⟩ **c :** extending throughout a specified area or scope — usually used in combination ⟨nation*wide*⟩ ⟨industry-*wide*⟩ **d :** COMPREHENSIVE, INCLUSIVE ⟨a *wide* assortment⟩
2 a : having a specified extension from side to side ⟨3 feet *wide*⟩ **b :** having much extent between the sides **:** BROAD ⟨a *wide* doorway⟩ **c :** fully opened ⟨*wide*-eyed⟩ **d :** LAX 4
3 a : extending or fluctuating considerably between limits ⟨a *wide* variation⟩ **b :** straying or deviating from something specified — used with *of* ⟨the accusation was *wide* of the truth⟩
4 *of an animal ration* **:** relatively rich in carbohydrate as compared with protein
synonym see BROAD
— wide·ness *noun*

2wide *adverb* **wid·er; wid·est** (before 12th century)
1 a : over a great distance or extent **:** WIDELY ⟨searched far and *wide*⟩ **b :** over a specified distance, area, or extent — usually in combination ⟨expanded the business country-*wide*⟩
2 a : so as to leave much space or distance between **b :** so as to pass at or clear by a considerable distance ⟨ran *wide* around left end⟩
3 : to the fullest extent **:** COMPLETELY, FULLY ⟨*wide* open⟩

wide–an·gle \ˈwīd-ˈaŋ-gəl\ *adjective* (1878)
1 : having or covering an angle of view wider than the ordinary — used especially of lenses of shorter than normal focal length
2 : having, involving the use of, or relating to a wide-angle lens ⟨a *wide-angle* shot⟩

wide·awake \ˌwīd-ə-ˈwāk\ *noun* (1837)
1 : a soft felt hat with a low crown and a wide brim
2 : SOOTY TERN

wide–awake *adjective* (1818)
1 : fully awake
2 : alertly watchful especially for advantages or opportunities
synonym see WATCHFUL

wide·band \ˈwīd-ˌband\ *adjective* (1935)
: BROADBAND

wide–body \ˈwīd-ˌbä-dē\ *noun* (1968)
: a large jet aircraft

wide–eyed \ˈwīd-ˈīd\ *adjective* (1853)
1 : having the eyes wide open especially with wonder or astonishment
2 : having or marked by unsophisticated or uncritical acceptance or admiration **:** NAIVE ⟨*wide-eyed* innocence⟩

wide·ly *adverb* (1579)
1 : over or through a wide area ⟨has traveled *widely*⟩
2 : to a great extent ⟨departed *widely* from the previous edition⟩
3 : by or among a large well-dispersed group of people ⟨a *widely* known political figure⟩
4 : over a broad range ⟨persons with *widely* fluctuating incomes —*Current Biography*⟩

wide·mouthed \ˈwīd-ˈmau̇thd, -ˈmau̇tht\ *adjective* (1593)
1 : having one's mouth opened wide (as in awe)
2 : having a wide mouth ⟨*widemouthed* jars⟩

wid·en \ˈwī-dᵊn\ *verb* **wid·ened; wid·en·ing** \ˈwīd-niŋ, ˈwī-dᵊn-iŋ\ (1650)
transitive verb
: to increase the width, scope, or extent of
intransitive verb
: to become wide or wider
— wid·en·er \ˈwīd-nər, ˈwī-dᵊn-ər\ *noun*

wide–open \ˈwīd-ˈō-pən, -ˌō-\ *adjective* (1852)
: having virtually no limits or restrictions ⟨a *wide-open* town⟩

wide·out \ˈwīd-ˌau̇t\ *noun* (1979)
: WIDE RECEIVER

wide–rang·ing \ˈwīd-ˌrān-jiŋ\ *adjective* (1816)
: extensive in scope **:** COMPREHENSIVE ⟨*wide-ranging* interests⟩

wide receiver *noun* (1968)
: a football receiver who normally lines up several yards to the side of the offensive formation

wide–screen *adjective* (1931)
: of or relating to a projected picture whose aspect ratio is substantially greater than 1.33:1

wide·spread \ˈwīd-ˈspred\ *adjective* (1705)
1 : widely diffused or prevalent ⟨*widespread* public interest⟩
2 : widely extended or spread out ⟨low, *widespread* hood and fenders —*Time*⟩ ⟨a *widespread* erosion surface —C. B. Hitchcock⟩

wide–spread·ing \-ˌspre-diŋ\ *adjective* (1591)
: stretching or extending over a wide space or area ⟨*wide-spreading* thatch roofs —*Nat'l Geographic*⟩

wid·get \ˈwi-jət\ *noun* [alteration of *gadget*] (1926)
1 : GADGET
2 : an unnamed article considered for purposes of hypothetical example

wid·ish \ˈwī-dish\ *adjective* (1845)
: somewhat wide

1wid·ow \ˈwi-(ˌ)dō\ *noun* [Middle English *widewe*, from Old English *wuduwe;* akin to Old High German *wituwa* widow, Latin *vidua*, Sanskrit *vidhavā*, Latin *-videre* to separate] (before 12th century)
1 a : a woman who has lost her husband by death and usually has not remarried **b :** GRASS WIDOW 2 **c :** a woman whose husband leaves her alone frequently or for long periods to engage in a usually specified activity ⟨a golf *widow*⟩
2 : an extra hand or part of a hand of cards dealt face down and usually placed at the disposal of the highest bidder

3 : a single usually short last line (as of a paragraph) separated from its related text and appearing at the top of a printed page or column

²widow *transitive verb* (14th century)
1 : to cause to become a widow or widower
2 *obsolete* **:** to survive as the widow of
3 : to deprive of something greatly valued or needed

wid·ow·er \'wi-də-wər\ *noun* [Middle English *widewer,* alteration of *wedow* widow, widower, from Old English *wuduwa* widower; akin to Old English *wuduwe* widow] (14th century)
: a man who has lost his wife by death and usually has not remarried

wid·ow·er·hood \-ˌhu̇d\ *noun* (1796)
1 : the fact or state of being a widower
2 : the period during which a man remains a widower

wid·ow·hood \'wi-dō-ˌhu̇d, -də-\ *noun* (before 12th century)
1 : the fact or state of being a widow
2 : the period during which a woman remains a widow
3 : WIDOWERHOOD

widow's cruse *noun* [from the widow's cruse of oil that miraculously supplies Elijah during a famine (I Kings 17:8–16)] (1816)
: an inexhaustible supply

widow's peak *noun* (1849)
: a point formed by the hairline in front

widow's walk *noun* (1937)
: a railed observation platform atop a usually coastal house

width \'width, 'witth\ *noun* [¹*wide*] (1627)
1 : the horizontal measurement taken at right angles to the length **:** BREADTH
2 : largeness of extent or scope
3 : a measured and cut piece of material ⟨a *width* of calico⟩

wield \'wē(ə)ld\ *transitive verb* [Middle English *welden* to control, from Old English *wieldan;* akin to Old High German *waltan* to rule, Latin *valēre* to be strong, be worth] (before 12th century)
1 *chiefly dialect* **:** to deal successfully with **:** MANAGE
2 : to handle (as a tool) especially effectively ⟨*wield* a broom⟩
3 a : to exert one's authority by means of ⟨*wield* influence⟩ **b :** have at one's command or disposal ⟨did not *wield* appropriate credentials —G. W. Bonham⟩
— wield·er *noun*

wieldy \'wē(ə)l-dē\ *adjective* (14th century)
: capable of being wielded easily

wie·ner \'wē-nər, 'wē-nē *also* 'wi-nē\ *noun* [short for *wienerwurst*] (1900)
: FRANKFURTER

Wie·ner schnit·zel \'vē-nər-ˌshnit-səl, 'wē-nər-ˌsnit-\ *noun* [German, literally, Vienna cutlet] (1862)
: a thin breaded veal cutlet

wie·ner·wurst \'wē-nə(r)-ˌwərst *also* -ˌwu̇rst *sometimes* -ˌwu̇sht *or* -ˌwu̇st; *also with* v *for* w\ *noun* [German, from *Wiener* of Vienna + *Wurst* sausage] (1889)
1 : VIENNA SAUSAGE
2 : FRANKFURTER

wie·nie \'wē-nē *also* 'wi-nē\ *noun* [by shortening & alteration from *wienerwurst*] (1897)
: FRANKFURTER

wife \'wīf\ *noun, plural* **wives** \'wīvz\ [Middle English *wif,* from Old English *wīf;* akin to Old High German *wīb* wife] (before 12th century)
1 a *dialect* **:** WOMAN **b :** a woman acting in a specified capacity — used in combination ⟨fish*wife*⟩
2 : a female partner in a marriage
— wife·hood \'wīf-ˌhu̇d, 'wī-ˌfu̇d\ *noun*
— wife·less \'wī-fləs\ *adjective*

¹wife·like \'wī-ˌflīk\ *adverb* (1598)
: in a wifely manner

²wifelike *adjective* (1613)
: WIFELY

wife·ly \'wī-flē\ *adjective* (before 12th century)
: of, relating to, or befitting a wife
— wife·li·ness \-flē-nəs\ *noun*

Wif·fle \'wi-fᵊl\ *trademark*
— used for a hollow plastic ball with cutouts in one hemisphere

wif·ty \'wif-tē\ *adjective* [origin unknown] (1979)
: DITSY

¹wig \'wig\ *noun* [short for *periwig*] (1675)
1 a : a manufactured covering of natural or synthetic hair for the head **b :** TOUPEE 2
2 : an act of wigging **:** REBUKE

²wig *verb* **wigged; wig·ging** (1829)
transitive verb
: to scold severely **:** REBUKE
intransitive verb
slang **:** to lose one's composure or reason **:** FLIP — usually used with *out*

wig·an \'wi-gən\ *noun* [*Wigan,* England] (circa 1875)
: a stiff plain-weave cotton fabric used for interlining

wi·geon *or* **wid·geon** \'wi-jən\ *noun, plural* **wigeon** *or* **wigeons** *or* **widgeon** *or* **widgeons** [origin unknown] (1513)
: any of several freshwater ducks (genus *Anas*): as **a :** an Old World duck (*Anas penelope*) with a large white patch on each wing and in the male with a red brown head and a buff crown **b :** BALDPATE 2

wigged \'wigd\ *adjective* (1777)
: wearing a wig especially of a specified kind ⟨the mute, blond-*wigged* . . . member of the team —*Current Biography*⟩

wigged–out \'wigd-'au̇t\ *adjective* (1970)
slang **:** having lost touch with reality **:** CRAZY

¹wig·gle \'wi-gəl\ *verb* **wig·gled; wig·gling** \-g(ə-)liŋ\ [Middle English *wiglen,* from or akin to Middle Dutch or Middle Low German *wiggelen* to totter; akin to Old English *wegan* to move — more at WAY] (13th century)
intransitive verb
1 : to move to and fro with quick jerky or shaking motions **:** JIGGLE
2 : to proceed with or as if with twisting and turning movements **:** WRIGGLE
transitive verb
: to cause to wiggle

²wiggle *noun* (1816)
1 : the act of wiggling
2 : shellfish or fish in cream sauce with peas
— wig·gly \'wi-g(ə-)lē\ *adjective*

wig·gler \'wi-g(ə-)lər\ *noun* (1859)
1 : a larva or pupa of the mosquito — called also *wriggler*
2 : one that wiggles

¹wight \'wīt\ *noun* [Middle English, creature, thing, from Old English *wiht;* akin to Old High German *wiht* creature, thing, Old Church Slavonic *veštĭ* thing] (before 12th century)
: a living being **:** CREATURE; *especially* **:** a human being

²wight *adjective* [Middle English, of Scandinavian origin; akin to Old Norse *vīgr* skilled in fighting (neuter *vīgt*); akin to Old English *wīgan* to fight — more at VICTOR] (13th century)
archaic **:** VALIANT, STALWART

wig·let \'wi-glət\ *noun* (1831)
: a small wig used especially to enhance a hairstyle

¹wig·wag \'wig-ˌwag\ *verb* [English dialect *wig* to move + English *wag*] (1892)
transitive verb
1 : to signal by wigwagging
2 : to cause to wigwag
intransitive verb
1 : to send a signal by or as if by a flag or light waved according to a code
2 : to make a signal (as by waving the hand or arm)

²wigwag *noun* (1886)
: the art or practice of wigwagging

wig·wam \'wig-ˌwäm\ *noun* [Eastern Abenaki *wikəwαm* house] (1628)
: a hut of the American Indians of the Great Lakes region and eastward having typically an arched framework of poles overlaid with bark, rush mats, or hides; *also* **:** a rough hut

wigwam

wil·co \'wil-(ˌ)kō\ *interjection* [*will* comply] (circa 1938)
— used especially in radio and signaling to indicate that a message received will be complied with

¹wild \'wī(ə)ld\ *adjective* [Middle English *wilde,* from Old English; akin to Old High German *wildi* wild, Welsh *gwyllt*] (before 12th century)
1 a : living in a state of nature and not ordinarily tame or domesticated ⟨*wild* duck⟩ **b** (1) **:** growing or produced without human aid or care ⟨*wild* honey⟩ (2) **:** related to or resembling a corresponding cultivated or domesticated organism **c :** of or relating to wild organisms ⟨the *wild* state⟩
2 a : not inhabited or cultivated ⟨*wild* land⟩ **b :** not amenable to human habitation or cultivation; *also* **:** DESOLATE
3 a (1) **:** not subject to restraint or regulation **:** UNCONTROLLED; *also* **:** UNRULY (2) **:** emotionally overcome ⟨*wild* with grief⟩; *also* **:** passionately eager or enthusiastic ⟨was *wild* to own a toy train —J. C. Furnas⟩ **b :** marked by turbulent agitation **:** STORMY ⟨a *wild* night⟩ **c :** going beyond normal or conventional bounds **:** FANTASTIC ⟨*wild* ideas⟩; *also* **:** SENSATIONAL **d :** indicative of strong passion, desire, or emotion ⟨a *wild* gleam of delight in his eyes —*Irish Digest*⟩
4 : UNCIVILIZED, BARBARIC
5 : characteristic of, appropriate to, or expressive of wilderness, wildlife, or a simple or uncivilized society
6 a : deviating from the intended or expected course ⟨*wild* spelling —C. W. Cunnington⟩ ⟨the throw was *wild*⟩; *also* **:** tending to throw inaccurately **b :** having no basis in known or surmised fact ⟨a *wild* guess⟩
7 *of a playing card* **:** able to represent any card designated by the holder
— wild·ish \'wīl-dish\ *adjective*
— wild·ness \-nəs\ *noun*

²wild *adverb* (1549)
: in a wild manner: as **a :** without regulation or control **b :** off an intended or expected course

³wild *noun* (1596)
1 : a sparsely inhabited or uncultivated region or tract **:** WILDERNESS
2 : a wild, free, or natural state or existence

wild bergamot *noun* (1843)
: a fragrant North American herbaceous mint (*Monarda fistulosa*) having a terminal capitate cluster of rather large pink or purple flowers

wild boar *noun* (13th century)
: an Old World wild hog (*Sus scrofa*) from which most domestic swine have been derived

wild card *noun* [*wild card,* playing card with arbitrarily determined value] (1972)
1 : one picked to fill a leftover playoff or tournament berth after regularly qualifying competitors have all been determined
2 : an unknown or unpredictable factor

wild carrot *noun* (circa 1538)
: QUEEN ANNE'S LACE

[1]**wild·cat** \'wī(ə)l(d)-ˌkat\ *noun, plural* **wildcats** (14th century)
1 a : a cat (*Felis silvestris*) of Eurasia and Africa that resembles but is heavier in build than the domestic tabby cat and is usually held to be among the ancestors of the domestic cat **b** *or plural* **wildcat :** any of various small or medium-sized cats (as the lynx or ocelot) **c :** a feral domestic cat
2 : a savage quick-tempered person
3 a : wildcat money **b :** a wildcat oil or gas well **c :** a wildcat strike

[2]**wildcat** *adjective* (1838)
1 a (1) **:** financially irresponsible or unreliable ⟨*wildcat* banks⟩ (2) **:** issued by a financially irresponsible banking establishment ⟨*wildcat* currency⟩ **b :** operating, produced, or carried on outside the bounds of standard or legitimate business practices ⟨*wildcat* insurance schemes —H. H. Reichard⟩ **c :** of, relating to, or being an oil or gas well drilled in territory not known to be productive **d :** initiated by a group of workers without formal union approval or in violation of a contract ⟨a *wildcat* strike⟩ ⟨*wildcat* work stoppages⟩
2 a *of a cartridge* **:** having a bullet of standard caliber but using an expanded case or a case designed for a bullet of greater caliber necked down for the smaller bullet **b** *of a firearm* **:** using wildcat cartridges

[3]**wildcat** *intransitive verb* **wild·cat·ted; wild·cat·ting** (1883)
: to prospect and drill an experimental oil or gas well or sink a mine shaft in territory not known to be productive

wild·cat·ter \-ˌka-tər\ *noun* (1883)
1 : one that drills wells in the hope of finding oil in territory not known to be an oil field
2 : one that promotes unsafe and unreliable enterprises; *especially* **:** one that sells stocks in such enterprises
3 : one that designs, builds, or fires wildcat cartridges and firearms
4 : a worker who goes out on a wildcat strike

wild celery *noun* (1874)
: TAPE GRASS

wil·de·beest \'wil-də-ˌbēst\ *noun, plural* **wildebeests** *also* **wildebeest** [Afrikaans *wildebees,* from *wilde* wild + *bees* ox] (1824)
: GNU

wil·der \'wil-dər\ *verb* [probably irregular from *wilderness*] (1613)
transitive verb
1 *archaic* **:** to lead astray
2 *archaic* **:** BEWILDER, PERPLEX
intransitive verb
archaic **:** to move at random **:** WANDER
— wil·der·ment \-dər-mənt\ *noun, archaic*

wil·der·ness \'wil-dər-nəs\ *noun* [Middle English, from *wildern* wild, from Old English *wilddēoren* of wild beasts] (13th century)
1 a (1) **:** a tract or region uncultivated and uninhabited by human beings (2) **:** an area essentially undisturbed by human activity together with its naturally developed life community **b :** an empty or pathless area or region ⟨in remote *wildernesses* of space groups of nebulae are found —G. W. Gray (died 1960)⟩ **c :** a part of a garden devoted to wild growth
2 *obsolete* **:** wild or uncultivated state
3 a : a confusing multitude or mass **:** an indefinitely great number or quantity ⟨I would not have given it for a *wilderness* of monkeys —Shakespeare⟩ **b :** a bewildering situation ⟨those moral *wildernesses* of civilized life —Norman Mailer⟩

wilderness area *noun, often W&A capitalized* (1928)
: an often large tract of public land maintained essentially in its natural state and protected against introduction of intrusive artifacts (as roads and buildings)

wild–eyed \'wī(ə)ld-'īd\ *adjective* (1817)
1 : having a wild expression in the eyes
2 : consisting of or favoring extreme or visionary ideas ⟨*wild-eyed* schemes⟩

wild·fire \-ˌfīr\ *noun* (12th century)
1 : a sweeping and destructive conflagration especially in a wilderness or a rural area
2 : GREEK FIRE
3 : a phosphorescent glow (as ignis fatuus or fox fire)
4 : a destructive leaf-spot disease of tobacco caused by several strains of a bacterium (*Pseudomonas syringae*)
— like wildfire : very rapidly

wild·flow·er \-ˌflau̇(-ə)r\ *noun* (1797)
: the flower of a wild or uncultivated plant or the plant bearing it

wild·fowl \-ˌfau̇l\ *noun* (before 12th century)
: a game bird; *especially* **:** a game waterfowl (as a wild duck or goose)
— wild·fowl·er \-ˌfau̇-lər\ *noun*
— wild·fowl·ing \-liŋ\ *noun*

wild geranium *noun* (1840)
: a common geranium (*Geranium maculatum*) of eastern North America with deeply parted leaves and flowers of rosy purple; *also* **:** any of several related geraniums

wild ginger *noun* (1804)
: a North American perennial herb (*Asarum canadense*) of the birthwort family with a pungent creeping rhizome

wild–goose chase *noun* (1592)
: a complicated or lengthy and usually fruitless pursuit or search

wild hyacinth *noun* (1847)
: any of several plants with flowers suggestive of hyacinths: as **a :** a camas (*Camassia scilloides*) of the central U.S. and southern Ontario with white racemose flowers **b :** BLUEBELL 2a **c :** any of several western North American plants (genus *Brodiaea*) of the lily family with grasslike basal leaves and variously colored flowers

wild indigo *noun* (1744)
: BAPTISIA; *especially* **:** a tumbleweed (*Baptisia tinctoria*) with bright yellow flowers and trifoliolate leaves

[1]**wild·ing** \'wīl-diŋ\ *noun* [[1]*wild* + [2]*-ing*] (circa 1525)
1 a : a plant growing uncultivated in the wild either as a native or an escape; *especially* **:** a wild apple or crab apple **b :** the fruit of a wilding
2 : a wild animal

[2]**wilding** *adjective* (1697)
: not domesticated or cultivated **:** WILD

wild·land \'wī(ə)l(d)-ˌland\ *noun* (1813)
: land that is uncultivated or unfit for cultivation

wild·life \-ˌlīf\ *noun, often attributive* (1879)
: living things and especially mammals, birds, and fishes that are neither human nor domesticated

wild·ling \-liŋ\ *noun* (1840)
: WILDING

wild·ly \'wī(ə)l(d)-lē\ *adverb* (14th century)
1 : in a wild manner ⟨was talking *wildly*⟩
2 : EXTREMELY 2 ⟨*wildly* popular⟩

wild marjoram *noun* (1550)
: OREGANO 1

wild mustard *noun* (1597)
: CHARLOCK

wild oat *noun* (15th century)
1 : any of several wild grasses (genus *Avena*); *especially* **:** a European annual weed (*A. fatua*) common in meadows and pastures
2 *plural* **:** offenses and indiscretions ascribed to youthful exuberance — usually used in the phrase *sow one's wild oats*

wild pansy *noun* (circa 1900)
: JOHNNY-JUMP-UP

wild pink *noun* (1814)
: an American catchfly (*Silene caroliniana*) of the eastern U.S. with pink or whitish flowers

wild pitch *noun* (1867)
: a baseball pitch not hit by the batter that cannot be caught or controlled by the catcher with ordinary effort and that enables a base runner to advance — compare PASSED BALL

wild rice *noun* (1748)
: a tall aquatic North American perennial grass (*Zizania aquatica*) that yields an edible grain; *also* **:** the grain

wild rye *noun* (circa 1500)
: any of several grasses (genus *Elymus*)

wild sarsaparilla *noun* (1814)
: a common North American perennial herb (*Aralia nudicaulis*) of the ginseng family with long-stalked basal compound leaves, umbels of greenish flowers, and an aromatic root used as a substitute of true sarsaparilla

wild type *noun* (1914)
: a phenotype, genotype, or gene that predominates in a natural population of organisms or strain of organisms in contrast to that of natural or laboratory mutant forms; *also* **:** an organism or strain displaying the wild type
— wild–type *adjective*

Wild West *noun* (1849)
: the western U.S. in its frontier period characterized by roughness and lawlessness
— Wild West *adjective*

wild·wood \'wī(ə)l(d)-ˌwu̇d\ *noun* (12th century)
: a wood unaltered or unfrequented by humans

[1]**wile** \'wī(ə)l\ *noun* [Middle English *wil,* probably of Scandinavian origin; akin to Old Norse *vēl* deceit, artifice] (12th century)
1 : a trick or stratagem intended to ensnare or deceive; *also* **:** a beguiling or playful trick
2 : skill in outwitting **:** TRICKERY, GUILE
synonym see TRICK

[2]**wile** *transitive verb* **wiled; wil·ing** (14th century)
1 : to lure by or as if by a magic spell **:** ENTICE
2 [by alteration] **:** WHILE

[1]**will** \wəl, (ə)l, ᵊl, 'wil\ *verb, past* **would** \wəd, (ə)d, 'wu̇d\; *present singular & plural* **will** [Middle English (1st & 3d singular present indicative), from Old English *wille* (infinitive *wyllan*); akin to Old High German *wili* (3d singular present indicative) wills, Latin *velle* to wish, will] (before 12th century)
transitive verb
: DESIRE, WISH ⟨call it what you *will*⟩
verbal auxiliary
1 — used to express desire, choice, willingness, consent, or in negative constructions refusal ⟨no one *would* take the job⟩ ⟨if we *will* all do our best⟩ ⟨*will* you please stop that racket⟩
2 — used to express frequent, customary, or habitual action or natural tendency or disposition ⟨*will* get angry over nothing⟩ ⟨*will* work one day and loaf the next⟩
3 — used to express futurity ⟨tomorrow morning I *will* wake up in this first-class hotel suite —Tennessee Williams⟩
4 — used to express capability or sufficiency ⟨the back seat *will* hold three passengers⟩
5 — used to express probability and often equivalent to the simple verb ⟨that *will* be the milkman⟩
6 a — used to express determination, insistence, persistence, or willfulness ⟨I have made up my mind to go and go I *will*⟩ **b** — used to express inevitability ⟨accidents *will* happen⟩
7 — used to express a command, exhortation, or injunction ⟨you *will* do as I say, at once⟩
intransitive verb
: to have a wish or desire ⟨whether we *will* or no⟩
usage see SHALL
— if you will : if you wish to call it that ⟨a kind of preoccupation, or obsession *if you will* —Louis Auchincloss⟩

[2]**will** \'wil\ *noun* [Middle English, from Old English *willa* will, desire; akin to Old English *wille*] (before 12th century)
1 : DESIRE, WISH: as **a :** DISPOSITION, INCLINATION ⟨where there's a *will* there's a way⟩ **b :** APPETITE, PASSION **c :** CHOICE, DETERMINATION
2 a : something desired; *especially* **:** a choice or determination of one having authority or

power **b** (1) *archaic* **:** REQUEST, COMMAND (2) [from the phrase *our will is* which introduces it] **:** the part of a summons expressing a royal command
3 : the act, process, or experience of willing **:** VOLITION
4 a : mental powers manifested as wishing, choosing, desiring, or intending **b :** a disposition to act according to principles or ends **c :** the collective desire of a group ⟨the *will* of the people⟩
5 : the power of control over one's own actions or emotions ⟨a man of iron *will*⟩
6 : a legal declaration of a person's wishes regarding the disposal of his or her property or estate after death; *especially* **:** a written instrument legally executed by which a person makes disposition of his or her estate to take effect after death
— at will : as one wishes **:** as or when it pleases or suits oneself

[3]will \'wil\ (before 12th century)
transitive verb
1 a : to order or direct by a will **b :** to dispose of by or as if by a will **:** BEQUEATH
2 a : to determine by an act of choice **b :** DECREE, ORDAIN ⟨Providence *wills* it⟩ **c :** INTEND, PURPOSE **d :** to cause or change by an act of will; *also* **:** to try to do so
intransitive verb
1 : to exercise the will
2 : CHOOSE

willed \'wild\ *adjective* (14th century)
1 : having a will especially of a specified kind — usually used in combination ⟨strong-willed⟩
2 : DELIBERATE

wil·lem·ite \'wi-lə-,mīt\ *noun* [German *Willemit,* from *Willem* (William) I (died 1843) king of the Netherlands] (circa 1841)
: a mineral consisting of a silicate of zinc, occurring in hexagonal prisms and in massive or granular forms, and varying in color

wil·let \'wi-lət\ *noun, plural* **willet** [imitative] (1791)
: a large American shorebird (*Catoptrophorus semipalmatus*) that resembles the greater yellowlegs but has a thicker bill and gray legs and that displays a black and white wing pattern when in flight

will·ful *or* **wil·ful** \'wil-fəl\ *adjective* (13th century)
1 : obstinately and often perversely self-willed
2 : done deliberately **:** INTENTIONAL
synonym see UNRULY
— will·ful·ly \-fə-lē\ *adverb*
— will·ful·ness *noun*

Wil·liam Tell \,wil-yəm-'tel\ *noun*
: an heroic archer in Swiss legend who complies with an order to shoot an apple off his son's head

wil·lies \'wi-lēz\ *noun plural* [origin unknown] (circa 1896)
: a fit of nervousness **:** JITTERS — used with *the*

will·ing \'wi-liŋ\ *adjective* (14th century)
1 : inclined or favorably disposed in mind **:** READY
2 : prompt to act or respond
3 : done, borne, or accepted by choice or without reluctance
4 : of or relating to the will or power of choosing **:** VOLITIONAL
synonym see VOLUNTARY
— will·ing·ly \-liŋ-lē\ *adverb*
— will·ing·ness *noun*

wil·li·waw \'wi-li-,wȯ\ *noun* [origin unknown] (1842)
1 a : a sudden violent gust of cold land air common along mountainous coasts of high latitudes **b :** a sudden violent wind
2 : a violent commotion

will–less \'wil-ləs\ *adjective* (1747)
1 : involving no exercise of the will **:** INVOLUNTARY ⟨*will-less* obedience⟩
2 : not exercising the will ⟨*will-less* slaves⟩

will–o'–the–wisp \,wil-ə-thə-'wisp\ *noun* [*Will* (nickname for *William*) + *of* + *the* + *wisp*] (1661)
1 : IGNIS FATUUS 1
2 : a delusive or elusive goal
— will–o'–the–wisp *adjective*

wil·low \'wi-(,)lō\ *noun* [Middle English *wilghe, wilowe,* from Old English *welig;* akin to Middle High German *wilge* willow] (before 12th century)
1 : any of a genus (*Salix* of the family Salicaceae, the willow family) of trees and shrubs bearing catkins of apetalous flowers and including forms of value for wood, osiers, or tanbark and a few ornamentals
2 : an object made of willow wood; *especially* **:** a cricket bat
— wil·low·like \-lō-,līk\ *adjective*

willow herb *noun* (1578)
: any of a genus (*Epilobium*) of herbs of the evening-primrose family; *especially* **:** FIREWEED b

wil·low·ware \'wi-lə-,war, 'wi-lō-, -,wer\ *noun* (circa 1885)
: dinnerware that is usually blue-and-white and that is decorated with a story-telling design featuring a large willow tree by a little bridge

wil·lowy \'wi-lə-wē\ *adjective* (1766)
1 : abounding with willows
2 : resembling a willow: **a :** PLIANT **b :** gracefully tall and slender

will·pow·er \'wil-,pau̇(-ə)r\ *noun* (1874)
: energetic determination

will to power (1907)
1 : the drive of the superman in the philosophy of Nietzsche to perfect and transcend the self through the possession and exercise of creative power
2 : a conscious or unconscious desire to exercise authority over others

wil·ly–nil·ly \,wi-lē-'ni-lē\ *adverb or adjective* [alteration of *will I nill I* or *will ye nill ye* or *will he nill he*] (1608)
1 : by compulsion **:** without choice
2 : in a haphazard or spontaneous manner

Wilms' tumor \'vilm-zəz-, 'vilmz-\ *noun* [Max *Wilms* (died 1918) German surgeon] (circa 1910)
: a malignant tumor of the kidney that primarily affects children and is made up of embryonic elements

Wil·son's disease \'wil-sənz-\ *noun* [Samuel A. K. *Wilson* (died 1937) English neurologist] (circa 1915)
: a hereditary disease that is determined by an autosomal recessive gene and is marked especially by cirrhotic changes in the liver and severe mental disorder due to a ceruloplasmin deficiency and resulting inability to metabolize copper

[1]wilt \wəlt, 'wilt\ *archaic present 2d singular of* WILL

[2]wilt \'wilt\ *verb* [alteration of earlier *welk,* from Middle English *welken,* probably from Middle Dutch; akin to Old High German *erwelkēn* to wilt] (circa 1691)
intransitive verb
1 a : to lose turgor from lack of water ⟨the plants *wilted* in the heat⟩ **b :** to become limp
2 : to grow weak or faint **:** LANGUISH
transitive verb
: to cause to wilt

[3]wilt \'wilt\ *noun* (1855)
1 : an act or instance of wilting **:** the state of being wilted
2 a : a disorder (as a fungus disease) of plants marked by loss of turgidity in soft tissues with subsequent drooping and often shriveling — called also *wilt disease* **b :** a destructive virus disease of various caterpillars marked by visceral liquefaction and shriveling of the body

Wil·ton \'wil-tᵊn\ *noun* [*Wilton,* borough in England] (1774)
: a carpet woven with loops like the Brussels carpet but having a velvet cut pile and being generally of better materials

wily \'wī-lē\ *adjective* **wil·i·er; -est** (14th century)
: full of wiles **:** CRAFTY
synonym see SLY
— wil·i·ly \-lə-lē\ *adverb*
— wil·i·ness \-lē-nəs\ *noun*

[1]wim·ble \'wim-bəl\ *noun* [Middle English, from Anglo-French, from Middle Dutch *wimmel* auger; akin to Middle Low German *wimmel* auger] (13th century)
: any of various instruments for boring holes

[2]wimble *transitive verb* **wim·bled; wim·bling** \-b(ə-)liŋ\ (15th century)
archaic **:** to bore with or as if with a wimble

wimp \'wimp\ *noun* [origin unknown] (1920)
: a weak, cowardly, or ineffectual person
— wimp·i·ness \'wim-pē-nəs\ *noun*
— wimp·ish \'wim-pish\ *adjective*
— wimp·ish·ness \-nəs\ *noun*
— wimpy \'wim-pē\ *adjective*

[1]wim·ple \'wim-pəl\ *noun* [Middle English *wimpel,* from Old English; perhaps akin to Old English *wīpian* to wipe] (before 12th century)
1 : a cloth covering worn over the head and around the neck and chin especially by women in the late medieval period and by some nuns
2 *Scottish* **a :** a crafty turn **:** TWIST **b :** CURVE, BEND

wimple 1

[2]wimple *verb* **wim·pled; wim·pling** \-p(ə-)liŋ\ (13th century)
transitive verb
1 : to cover with or as if with a wimple **:** VEIL
2 : to cause to ripple
intransitive verb
1 *archaic* **:** to fall or lie in folds
2 *chiefly Scottish* **:** to follow a winding course **:** MEANDER
3 : RIPPLE

wimp out *intransitive verb* (1982)
: to behave like a wimp **:** chicken out; *especially* **:** to choose the easiest course of action

[1]win \'win\ *verb* **won** \'wən\; **win·ning** [Middle English *winnen,* from Old English *winnan* to struggle; akin to Old High German *winnan* to struggle and probably to Latin *venus* sexual desire, charm, Sanskrit *vanas* desire, *vanoti* he strives for] (before 12th century)
transitive verb
1 a : to get possession of by effort or fortune **b :** to obtain by work **:** EARN ⟨striving to *win* a living from the sterile soil⟩
2 a : to gain in or as if in battle or contest **b :** to be the victor in ⟨*won* the war⟩
3 a : to make friendly or favorable to oneself or to one's cause — often used with *over* ⟨*won* him over with persuasive arguments⟩ **b :** to induce to accept oneself in marriage
4 a : to obtain (as ore, coal, or clay) by mining **b :** to prepare (as a vein or bed) for regular mining **c :** to recover (as metal) from ore
5 : to reach by expenditure of effort
intransitive verb
1 : to gain the victory in a contest **:** SUCCEED
2 : to succeed in arriving at a place or a state
— win·less \'win-ləs\ *adjective*
— win·na·ble \'wi-nə-bəl\ *adjective*

[2]win *noun* (1862)

: VICTORY; *especially* : first place at the finish (as of a horse race)

wince \'win(t)s\ *intransitive verb* **winced; winc·ing** [Middle English *wynsen* to kick impatiently, from (assumed) Old North French *wincier* to turn aside, of Germanic origin; akin to Old High German *wankōn* to totter — more at WENCH] (circa 1748)
: to shrink back involuntarily (as from pain) : FLINCH
synonym see RECOIL
— **wince** *noun*

[1]**winch** \'winch\ *noun* [Middle English *winche* roller, reel, from Old English *wince;* akin to Old English *wincian* to wink] (before 12th century)
1 : any of various machines or instruments for hauling or pulling; *especially* **:** a powerful machine with one or more drums on which to coil a rope, cable, or chain for hauling or hoisting : WINDLASS
2 : a crank with a handle for giving motion to a machine (as a grindstone)

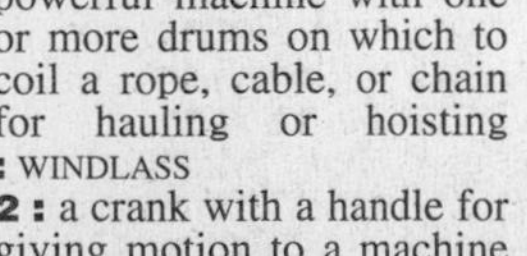
winch 1

[2]**winch** *transitive verb* (1529)
: to hoist or haul with or as if with a winch
— **winch·er** *noun*

Win·ches·ter \'win-ˌches-tər\ *adjective* [from the code name used by the original developer] (1973)
: relating to or being computer disk technology that permits high-density storage by sealing the rigid metal disks within the disk drive mechanism as protection against dust

[1]**wind** \'wind, *archaic or poetic* 'wīnd\ *noun, often attributive* [Middle English, from Old English; akin to Old High German *wint* wind, Latin *ventus,* Greek *aēnai* to blow, Sanskrit *vāti* it blows] (before 12th century)
1 a : a natural movement of air of any velocity; *especially* **:** the earth's air or the gas surrounding a planet in natural motion horizontally **b :** an artificially produced movement of air **c :** SOLAR WIND, STELLAR WIND
2 a : a destructive force or influence **b :** a force or agency that carries along or influences : TENDENCY, TREND ⟨withstood the *winds* of popular opinion —Felix Frankfurter⟩
3 a : BREATH 4a **b :** BREATH 2a **c :** the pit of the stomach : SOLAR PLEXUS
4 : gas generated in the stomach or the intestines
5 a : compressed air or gas **b** *archaic* : AIR
6 : something that is insubstantial: as **a :** mere talk : idle words **b :** NOTHING, NOTHINGNESS **c :** vain self-satisfaction
7 a : air carrying a scent (as of a hunter or game) **b :** slight information especially about something secret : INTIMATION ⟨got *wind* of the plan⟩
8 a : musical wind instruments especially as distinguished from strings and percussion **b** *plural* : players of wind instruments
9 a : a direction from which the wind may blow : a point of the compass; *especially* : one of the cardinal points **b :** the direction from which the wind is blowing
— **wind·less** \-ləs\ *adjective*
— **wind·less·ly** *adverb*
— **before the wind :** in the same direction as the main force of the wind
— **close to the wind :** as nearly as possible against the main force of the wind
— **have the wind of 1 :** to be to windward of **2 :** to be on the scent of **3 :** to have a superior position to
— **in the wind :** about to happen : ASTIR, AFOOT ⟨change is *in the wind*⟩
— **near the wind 1 :** close to the wind **2 :** close to a point of danger : near the permissible limit
— **off the wind :** away from the direction from which the wind is blowing
— **on the wind :** toward the direction from which the wind is blowing
— **to the wind** *or* **to the winds :** ASIDE, AWAY ⟨threw caution *to the wind*⟩
— **under the wind 1 :** to leeward **2 :** in a place protected from the wind : under the lee

[2]**wind** \'wind\ (15th century)
transitive verb
1 : to detect or follow by scent
2 : to expose to the air or wind : dry by exposing to air
3 : to make short of breath
4 : to regulate the wind supply of (an organ pipe)
5 : to rest (as a horse) in order to allow the breath to be recovered
intransitive verb
1 : to scent game
2 *dialect* : to pause for breath

[3]**wind** \'wīnd, 'wind\ *verb* **wind·ed** \'wīn-dəd, 'win-\ *or* **wound** \'waund\; **wind·ing** [[1]*wind*] (1586)
transitive verb
1 : to cause (as a horn) to sound by blowing : BLOW
2 : to sound (as a call or note) on a horn ⟨*wound* a rousing call —R. L. Stevenson⟩
intransitive verb
: to produce a sound on a horn

[4]**wind** \'wīnd\ *verb* **wound** \'waund\ *also* **wind·ed; wind·ing** [Middle English, from Old English *windan* to twist, move with speed or force, brandish; akin to Old High German *wintan* to wind, Umbrian *ohavendu* let him turn aside] (before 12th century)
transitive verb
1 a *obsolete* : WEAVE **b :** ENTANGLE, INVOLVE **c :** to introduce sinuously or stealthily : INSINUATE
2 a : to encircle or cover with something pliable : bind with loops or layers **b :** to turn completely or repeatedly about an object : COIL, TWINE **c** (1) **:** to hoist or haul by means of a rope or chain and a windlass (2) **:** to move (a ship) by hauling on a capstan **d** (1) **:** to tighten the spring of ⟨*wind* a clock⟩ (2) *obsolete* **:** to make tighter : TIGHTEN, TUNE (3) : CRANK **e :** to raise to a high level (as of excitement or tension) — usually used with *up*
3 a : to cause to move in a curving line or path **b** *archaic* **:** to turn the course of; *especially* **:** to lead (a person) as one wishes **c** (1) **:** to cause (as a ship) to change direction : TURN (2) **:** to turn (a ship) end for end **d :** to traverse on a curving course ⟨the river *winds* the valley⟩ **e :** to effect by or as if by curving
intransitive verb
1 : BEND, WARP
2 a : to have a curving course or shape : extend in curves **b :** to proceed as if by winding
3 : to move so as to encircle something
4 : to turn when lying at anchor

[5]**wind** \'wīnd\ *noun* (14th century)
1 : a mechanism (as a winch) for winding
2 : an act of winding : the state of being wound
3 : COIL, TURN
4 : a particular method of winding

wind·age \'win-dij\ *noun* [[1]*wind*] (circa 1710)
1 a : the space between the projectile of a smoothbore gun and the surface of the bore **b :** the difference between the diameter of the bore of a muzzle-loading rifled cannon and that of the projectile cylinder
2 a : the amount of sight deflection necessary to compensate for wind displacement in aiming a gun **b** (1) **:** the influence of the wind in deflecting the course of a projectile (2) **:** the amount of deflection due to the wind
3 : the surface exposed (as by a ship) to the wind

wind·bag \'win(d)-ˌbag\ *noun* (1827)
: an exhaustively talkative person

wind–bell \-ˌbel\ *noun* (1901)
1 : WIND CHIME — usually used in plural
2 : a bell that is light enough to be moved and sounded by the wind

wind·blast \-ˌblast\ *noun* (1942)
: the destructive effect of air friction on a pilot ejected from a high-speed airplane

wind·blown \-ˌblōn\ *adjective* (1600)
: blown by the wind; *especially* : having a permanent set or character of growth determined by the prevailing winds ⟨*windblown* trees⟩

wind–borne \-ˌbōrn, -ˌbȯrn\ *adjective* (1842)
: carried by the wind ⟨*wind-borne* pollen⟩ ⟨*wind-borne* soil deposits⟩

wind·break \-ˌbrāk\ *noun* (1861)
: a growth of trees or shrubs serving to break the force of wind; *broadly* : a shelter (as a fence) from the wind

wind·break·er \-ˌbrā-kər\ *noun* [from a trademark] (1918)
: an outer jacket made of wind-resistant material

wind–bro·ken \-ˌbrō-kən\ *adjective* (1603)
of a horse : affected with pulmonary emphysema or heaves

wind·burn \-ˌbərn\ *noun* (1925)
: irritation of the skin caused by wind
— **wind·burned** \-ˌbərnd\ *adjective*

Wind·cheat·er \'win(d)-ˌchē-tər\ *trademark*
— used for a windbreaker

wind·chill \'win(d)-ˌchil\ *noun* (1939)
: a still-air temperature that would have the same cooling effect on exposed human skin as a given combination of temperature and wind speed — called also *chill factor, windchill factor, windchill index*

wind chime *noun* (1927)
: a cluster of small often sculptured pieces (as of metal or glass) suspended so as to chime when blown by the wind — usually used in plural

wind down (1952)
intransitive verb
1 : to draw gradually toward an end
2 : RELAX, UNWIND
transitive verb
: to cause a gradual lessening of usually with the intention of bringing to an end

wind·er \'wīn-dər\ *noun* [[4]*wind*] (1552)
: one that winds: as **a :** a worker or machine that winds thread and yarn **b :** a key for winding a mechanism (as a clock) **c :** a step that is wider at one end than at the other (as in a spiral staircase)

wind·fall \'win(d)-ˌfȯl\ *noun* (15th century)
1 : something (as a tree or fruit) blown down by the wind
2 : an unexpected, unearned, or sudden gain or advantage

wind farm *noun* (1980)
: an area of land with a cluster of wind turbines for driving electrical generators

wind·flow·er \-ˌflau̇(-ə)r\ *noun* (1551)
: ANEMONE 1

wind·gall \-ˌgȯl\ *noun* (circa 1534)
: a soft tumor or synovial swelling on a horse's leg in the region of the fetlock joint

wind gap *noun* (1769)
: a notch in the crest of a mountain ridge : a pass not occupied by a stream — compare WATER GAP

wind harp *noun* (1813)
: AEOLIAN HARP

wind·hov·er \'wind-ˌhə-vər, -ˌhä-\ *noun* (1674)
British : KESTREL

[1]**wind·ing** \'wīn-diŋ\ *noun* (before 12th century)
1 : material (as wire) wound or coiled about an object (as an armature); *also* : a single turn of the wound material
2 a : the act of one that winds **b :** the manner of winding something
3 : a curved or sinuous course, line, or progress

[2]**winding** *adjective* (1530)

: marked by winding: as **a :** having a curved or spiral course or form ⟨a *winding* stairway⟩ **b :** having a course that winds ⟨a *winding* road⟩

wind·ing–sheet \'wīn-diŋ-ˌshēt\ *noun* (15th century)
: a sheet in which a corpse is wrapped

wind·ing–up \ˌwīn-diŋ-'əp\ *noun* (circa 1858)
British **:** the process of liquidating the assets of a partnership or corporation in order to pay creditors and make distributions to partners or shareholders upon dissolution

wind instrument *noun* (1582)
: a musical instrument (as a trumpet, clarinet, or organ) sounded by wind; *especially* **:** one sounded by the player's breath

wind·jam·mer \'win(d)-ˌja-mər\ *noun* (1880)
: a sailing ship; *also* **:** one of its crew
— wind·jam·ming \-miŋ\ *noun*

[1]**wind·lass** \'win(d)-ləs\ *noun* [Middle English *wyndlas,* alteration of *wyndas,* from Old Norse *vindass,* from *vinda* to wind (akin to Old High German *wintan* to wind) + *āss* pole; akin to Gothic *ans* beam] (15th century)
: any of various machines for hoisting or hauling: as **a :** a horizontal barrel supported on vertical posts and turned by a crank so that the hoisting rope is wound around the barrel **b :** a steam or electric winch with horizontal or vertical shaft and two drums used to raise a ship's anchor

[2]**windlass** *transitive verb* (1834)
: to hoist or haul with a windlass

win·dle·straw \'win-dᵊl-ˌstrȯ, 'wi-nᵊl-\ *noun* [(assumed) Middle English, from Old English *windelstrēaw,* from *windel* basket (from *windan* to wind) + *strēaw* straw] (before 12th century)
British **:** a dry thin stalk of grass

[1]**wind·mill** \'win(d)-ˌmil\ *noun* (14th century)
1 a : a mill or machine operated by the wind usually acting on oblique vanes or sails that radiate from a horizontal shaft; *especially* **:** a wind-driven water pump or electric generator **b :** the wind-driven wheel of a windmill
2 : something that resembles or suggests a windmill; *especially* **:** a calisthenic exercise that involves alternately lowering each outstretched hand to touch the toes of the opposite foot
3 [from the episode in *Don Quixote* by Cervantes in which the hero attacks windmills under the illusion that they are giants] **:** an imaginary wrong, evil, or opponent — usually used in the phrase *to tilt at windmills*

[2]**windmill** (1914)
transitive verb
: to cause to move like a windmill
intransitive verb
: to move like a windmill; *especially* **:** to spin from the force of wind

win·dow \'win-(ˌ)dō\ *noun, often attributive* [Middle English *windowe,* from Old Norse *vindauga,* from *vindr* wind (akin to Old English *wind*) + *auga* eye; akin to Old English *ēage* eye — more at EYE] (13th century)
1 a : an opening especially in the wall of a building for admission of light and air that is usually closed by casements or sashes containing transparent material (as glass) and capable of being opened and shut **b :** WINDOWPANE **c :** a space behind a window of a retail store containing displayed merchandise **d :** an opening in a partition or wall through which business is conducted ⟨a bank teller's *window*⟩
2 : a means of entrance or access; *especially* **:** a means of obtaining information ⟨a *window* on history⟩
3 : an opening (as a shutter, slot, or valve) that resembles or suggests a window
4 : the transparent panel or opening of a window envelope
5 : the framework (as a shutter or sash with its fittings) that closes a window opening
6 : CHAFF 4
7 : a range of wavelengths in the electromagnetic spectrum to which a planet's atmosphere is transparent
8 a : an interval of time within which a rocket or spacecraft must be launched to accomplish a particular mission **b :** an interval of time during which certain conditions or an opportunity exists ⟨a *window* of vulnerability⟩
9 : an area at the limits of the earth's sensible atmosphere through which a spacecraft must pass for successful reentry
10 : any of the areas into which a computer display may be divided and on which distinctly different types of information are displayed ◇
— win·dow·less \-dō-ləs, -də-\ *adjective*
— out the window : out of existence, use, or consideration

window box *noun* (circa 1885)
: a box designed to hold soil for growing plants at a windowsill

window dressing *noun* (1895)
1 : the display of merchandise in a retail store window
2 a : the act or an instance of making something appear deceptively attractive or favorable **b :** something used to create a deceptively favorable or attractive impression
— win·dow–dress \'win-dō-ˌdres\ *transitive verb*
— window dresser *noun*

win·dowed \'win-(ˌ)dōd, -dəd\ *adjective* (15th century)
: having windows especially of a specified kind — often used in combination

window envelope *noun* (1914)
: an envelope having an opening through which the address on the enclosure is visible

win·dow·pane \'win-dō-ˌpān, -də-\ *noun* (1819)
1 : a pane in a window
2 : TATTERSALL

window seat *noun* (circa 1745)
1 : a seat built into a window recess
2 : a seat next to a window (as in a bus or airplane)

window shade *noun* (1810)
: a shade or curtain for a window

win·dow–shop \'win-dō-ˌshäp, -də-\ *intransitive verb* (1922)
: to look at the displays in retail store windows without going inside the stores to make purchases
— win·dow–shop·per *noun*

win·dow·sill \-ˌsil\ *noun* (1703)
: the horizontal member at the bottom of a window opening

wind·pipe \'win(d)-ˌpīp\ *noun* (1530)
: TRACHEA 1

wind–pol·li·nat·ed \-'pä-lə-ˌnā-təd\ *adjective* (1884)
: pollinated by wind-borne pollen

wind·proof \-'prüf\ *adjective* (1616)
: impervious to wind ⟨a *windproof* jacket⟩

wind rose \'wind-ˌrōz\ *noun* [German *Windrose* compass card] (1846)
: a diagram showing for a given place the relative frequency or frequency and strength of winds from different directions

[1]**wind·row** \'win(d)-ˌrō\ *noun* (circa 1534)
1 a : a row of hay raked up to dry before being baled or stored **b :** a similar row of cut vegetation (as grain) for drying
2 : a row heaped up by or as if by the wind
3 a : a long low ridge of road-making material scraped to the side of a road **b :** BANK, RIDGE, HEAP

[2]**windrow** *transitive verb* (1729)
: to form (as hay) into a windrow

wind·screen \'win(d)-ˌskrēn\ *noun* (1858)
1 : a screen that protects against the wind
2 *chiefly British* **:** WINDSHIELD

wind shake *noun* (1545)
: shake in timber attributed to high winds

wind shear *noun* (1941)
: a radical shift in wind speed and direction that occurs over a very short distance

wind·shield \'win(d)-ˌshēld\ *noun* (1902)
: a transparent screen (as of glass) in front of the occupants of a vehicle

wind sock *noun* (1928)
: a truncated cloth cone open at both ends and mounted in an elevated position to indicate the direction of the wind

Wind·sor chair \'win-zər-\ *noun* [*Windsor,* England] (1740)
: a wooden chair with spindle back, raking legs, and usually a saddle seat — called also *Windsor*

Windsor chair

Windsor knot *noun* [probably after Edward, Duke of *Windsor*] (1947)
: a symmetrical necktie knot that is wider than the usual four-in-hand knot

Windsor tie *noun* (1895)
: a broad necktie usually tied in a loose bow

wind sprint *noun* (1948)
: a sprint performed as a training exercise to develop breathing capacity especially during exertion

wind·storm \'win(d)-ˌstȯrm\ *noun* (14th century)
: a storm marked by high wind with little or no precipitation

Wind·surf·er \-ˌsər-fər\ *trademark*
— used for a sailboard

wind·surf·ing \-ˌsər-fiŋ\ *noun* (1969)
: the sport or activity of riding a sailboard
— wind·surf \-ˌsərf\ *intransitive verb*

wind·swept \'win(d)-ˌswept\ *adjective* (1812)
: swept by or as if by wind

wind tee *noun* (1932)
: a large weather vane shaped like a horizontal letter T on or near a landing field

wind·throw \'win(d)-ˌthrō\ *noun* (1916)
: the uprooting and overthrowing of trees by the wind

wind tunnel *noun* (1911)
: a tunnellike passage through which air is blown at a known velocity to investigate air flow around an object (as an airplane part or model) placed in the passage

◇ WORD HISTORY

window To the ancient inhabitants of cold regions around the world, a window in a dwelling was a practical feature only when glass had become available to provide light and a view while sealing the opening from the weather. Consequently, in English and other tongues of northern Europe words for "window" developed relatively late, being either borrowings from languages of the Mediterranean region, where glass was known much earlier, or entirely new coinages. A metaphor common to several northern European languages equated the window in a dwelling with the human or animal eye, both being means of seeing out. In Old English "window" was *ēagduru,* literally, "eye-door," or *ēagthyrel* "eye-hole." The latter survived into early Middle English as *eie thurl,* but was soon supplanted by *windowe,* a loan from Old Norse *vindauga,* literally, "wind eye." *Windowe* in its turn competed for a while with *fenestre,* a borrowing from the Old French descendant of Latin *fenestra,* but the French loan died out in early Modern English, leaving us with *window* as our usual word.

wind turbine *noun* (1909)
: a wind-driven turbine for generating electricity

¹wind·up \'wīn-ˌdəp\ *noun* (1665)
1 a : the act of bringing to an end **b :** a concluding act or part **:** FINISH
2 a : a series of regular and distinctive motions (as swinging the arms) made by a pitcher preparatory to releasing a pitch **b :** an exaggerated backswing (as in tennis)

²windup *adjective* (1784)
: operated by a spring mechanism wound by hand

wind up (1583)
transitive verb
1 : to bring to a conclusion **:** END
2 a : to put in order for the purpose of bringing to an end ⟨*winds up* the meeting⟩ **b** *British* **:** to effectuate a winding up
intransitive verb
1 a : to come to a conclusion **b :** to arrive in a place, situation, or condition at the end or as a result of a course of action ⟨*wound up* as millionaires⟩
2 : to make a pitching windup

¹wind·ward \'win-(d)wərd\ *noun* (1549)
: the side or direction from which the wind is blowing
— to windward : into or in an advantageous position

²windward *adjective* (1627)
: being in or facing the direction from which the wind is blowing — compare LEEWARD

wind·way \'win(d)-ˌwā\ *noun* (circa 1875)
: a passage for air (as in an organ pipe)

windy \'win-dē\ *adjective* **wind·i·er; -est** (before 12th century)
1 a (1) **:** WINDSWEPT (2) **:** marked by strong wind or by more wind than usual **b :** VIOLENT, STORMY
2 : FLATULENT 1 ⟨a *windy* bellyache⟩
3 a : VERBOSE, BOMBASTIC **b :** lacking substance **:** EMPTY
— wind·i·ly \-də-lē\ *adverb*
— wind·i·ness \-dē-nəs\ *noun*

¹wine \'wīn\ *noun, often attributive* [Middle English *win,* from Old English *wīn;* akin to Old High German *wīn* wine; both from Latin *vinum* wine, probably of non-Indo-European origin; akin to the source of Greek *oinos* wine] (before 12th century)
1 a : the fermented juice of fresh grapes used as a beverage **b :** wine or a substitute used in Christian communion services
2 : the usually fermented juice of a plant product (as a fruit) used as a beverage
3 : something that invigorates or intoxicates
4 : a dark red

²wine *verb* **wined; win·ing** (1829)
intransitive verb
: to drink wine
transitive verb
: to give wine to ⟨*wined* and dined his friends⟩

wine cellar *noun* (14th century)
: a room for storing wines; *also* **:** a stock of wines

wine cooler *noun* (1815)
1 : a vessel or container in which wine is cooled
2 : a usually carbonated beverage that contains a mixture of wine and fruit juice

wine·glass \'wīn-ˌglas\ *noun* (1709)
: a stemware drinking glass for wine

wine·grow·er \-ˌgrō(-ə)r\ *noun* (1844)
: a person who cultivates a vineyard and makes wine

wine·press \'wīn-ˌpres\ *noun* (1526)
: a vat in which juice is expressed from grapes by treading or by means of a plunger

win·ery \'wī-nə-rē, 'wīn-rē\ *noun, plural* **-er·ies** (1882)
: a wine-making establishment

wine·shop \'wīn-ˌshäp\ *noun* (1848)
: a tavern that specializes in serving wine

wine·skin \-ˌskin\ *noun* (1821)
: a bag that is made from the skin of an animal (as a goat) and that is used for holding wine

wine taster *noun* (1632)
1 : one who tastes and evaluates wine especially professionally
2 : a small shallow vessel used to sample wine

win·ey *variant of* WINY

¹wing \'wiŋ\ *noun, often attributive* [Middle English *winge,* of Scandinavian origin; akin to Danish & Swedish *vinge* wing; akin to Sanskrit *vāti* it blows — more at WIND] (13th century)
1 a : one of the movable feathered or membranous paired appendages by means of which a bird, bat, or insect is able to fly; *also* **:** such an appendage even though rudimentary if possessed by an animal belonging to a group characterized by the power of flight **b :** any of various anatomical structures especially of a flying fish or flying lemur providing means of limited flight

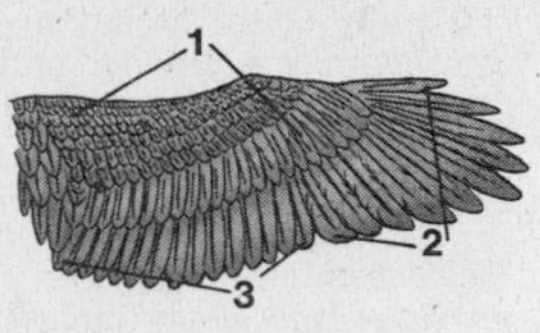

wing 1a: *1* coverts, *2* primaries, *3* secondaries

2 : an appendage or part resembling a wing in appearance, position, or function: as **a :** a device worn under the arms to aid a person in swimming or staying afloat **b :** ALA **c :** a turned-back or extended edge on an article of clothing **d :** a sidepiece at the top of an armchair **e** (1) **:** a foliaceous, membranous, or woody expansion of a plant especially along a stem or on a samara or capsule (2) **:** either of the two lateral petals of a papilionaceous flower **f :** a vane of a windmill or arrow **g :** SAIL **h :** an airfoil that develops a major part of the lift which supports a heavier-than-air aircraft **i** *chiefly British* **:** FENDER d
3 : a means of flight or rapid progress
4 : the act or manner of flying **:** FLIGHT
5 : a side or outlying region or district
6 : a part or feature usually projecting from and subordinate to the main or central part ⟨the servants' *wing* of the mansion⟩
7 a : one of the pieces of scenery at the side of a stage **b** *plural* **:** the area at the side of the stage out of sight
8 a : a left or right section of an army or fleet **:** FLANK **b :** one of the offensive positions or players on either side of a center position in certain team sports; *also* **:** FLANKER
9 a : either of two opposing groups within an organization or society **:** FACTION **b :** a section of an organized body (as a legislative chamber) representing a group or faction holding distinct opinions or policies — compare LEFT WING, RIGHT WING
10 a : a unit of the U.S. Air Force higher than a group and lower than a division **b :** two or more squadrons of naval airplanes
11 : a dance step marked by a quick outward and inward rolling glide of one foot
12 *plural* **:** insignia consisting of an outspread pair of stylized bird's wings which are awarded on completion of prescribed training to a qualified pilot, aircrew member, or military balloon pilot
— wingy \'wiŋ-ē\ *adjective*
— in the wings 1 : out of sight in the stage wings **2 :** close at hand in the background **:** readily available ⟨had a plan waiting *in the wings*⟩
— on the wing 1 : in flight **:** FLYING **2 :** in motion
— under one's wing : under one's protection **:** in one's care

²wing (1591)
transitive verb
1 a : to fit with wings **b :** to enable to fly or move swiftly
2 a : to traverse with or as if with wings **b :** to effect or achieve by flying
3 : to let fly **:** DISPATCH ⟨would start to *wing* punches —A. J. Liebling⟩
4 a : to wound in the wing **:** disable the wing of ⟨*winged* the duck⟩ **b :** to wound (as with a bullet) without killing ⟨*winged* by a sniper⟩
5 : to do or perform without preparation or guidelines **:** IMPROVISE ⟨*winging* it⟩
intransitive verb
: to go with or as if with wings **:** FLY

wing and wing *adverb* (1781)
: with sails extended on both sides

wing·back \'wiŋ-ˌbak\ *noun* (1933)
: an offensive back in football who lines up outside the tight end; *also* **:** the position of such a player

wing bar *noun* (1855)
: a line of contrasting color across the middle of a bird's wing made by markings on the wing coverts

wing case *noun* (1661)
: ELYTRON

wing chair *noun* (1904)
: an upholstered armchair with high solid back and sides that provide a rest for the head and protection from drafts

wing commander *noun* (1914)
: a commissioned officer in the British air force who ranks with a lieutenant colonel in the army

wing covert *noun* (1815)
: one of the feathers covering the bases of the wing quills

wing·ding \'wiŋ-ˌdiŋ\ *noun* [origin unknown] (1944)
: a wild, lively, or lavish party

winged \'wiŋd *also except for 1a*(2) 'wiŋ-əd\ *adjective* (14th century)
1 a (1) **:** having wings ⟨*winged* seeds⟩ (2) **:** having wings of a specified kind — used in combination ⟨strong-*winged*⟩ **b :** using wings in flight
2 a : soaring with or as if with wings **:** ELEVATED **b :** SWIFT, RAPID

winged bean \'wiŋd-\ *noun* (1910)
: an Asian twining legume (*Psophocarpus tetragonolobus*) cultivated in warm regions for its edible high-protein 4-winged pods; *also* **:** its pod

winged elm *noun* (1820)
: an elm (*Ulmus alata*) of the U.S. having twigs and young branches with prominent corky projections

wing·er \'wiŋ-ər\ *noun* (1896)
: a player (as in soccer or ice hockey) in a wing position

wing–foot·ed \'wiŋ-'fu̇-təd\ *adjective* (1591)
1 : having winged feet
2 : SWIFT

wing·less \'wiŋ-ləs\ *adjective* (1591)
: having no wings or very rudimentary wings
— wing·less·ness *noun*

wing·let \'wiŋ-lət\ *noun* (1816)
: a small wing; *also* **:** a small nearly vertical airfoil at an airplane's wingtip that reduces drag by inhibiting turbulence

wing·like \-ˌlīk\ *adjective* (circa 1804)
: resembling a wing in form or lateral position

wing·man \-mən\ *noun* (1942)
: a pilot who flies behind and outside the leader of a flying formation

wing nut *noun* (circa 1900)
: a nut with wings that provide a grip for the thumb and finger

wing·over \'wiŋ-ˌō-vər\ *noun* (1927)
: a flight maneuver in which a plane is put into a climbing turn until nearly stalled after which the nose is allowed to fall while the turn is continued until normal flight is attained in a direction opposite to that in which the maneuver was entered

wing shooting *noun* (1881)
: the act or practice of shooting at game birds in flight or at flying targets

wing·span \'wiŋ-ˌspan\ *noun* (circa 1917)
: the distance from the tip of one of a pair of wings to that of the other; *also* **:** SPAN 2c

wing·spread \-ˌspred\ *noun* (1897)
: the spread of the wings : WINGSPAN; *specifically* : the extreme measurement between the tips or outer margins of the wings (as of a bird or insect)

wing tip *noun* (circa 1908)
1 : a toe cap having a point that extends back toward the throat of the shoe and curving sides that extend toward the shank
2 : a shoe having a wing tip
3 a : the edge or outer margin of a bird's wing **b** *usually* **wingtip :** the outer end of an airplane wing

[1]wink \ˈwiŋk\ *verb* [Middle English, from Old English *wincian;* akin to Old High German *winchan* to stagger, wink and perhaps to Latin *vacillare* to sway, Sanskrit *vañcati* he goes crookedly] (before 12th century)
intransitive verb
1 : to shut one eye briefly as a signal or in teasing
2 : to close and open the eyelids quickly
3 : to avoid seeing or noting something — usually used with *at*
4 : to gleam or flash intermittently : TWINKLE
5 a : to come to an end — usually used with *out* **b :** to stop shining — usually used with *out*
6 : to signal a message with a light
transitive verb
1 : to cause to open and shut
2 : to affect or influence by or as if by blinking the eyes

[2]wink *noun* (14th century)
1 : a brief period of sleep : NAP
2 a : a hint or sign given by winking **b :** an act of winking
3 : the time of a wink : INSTANT ⟨quick as a *wink*⟩
4 : a flicker of the eyelids : BLINK

wink·er \ˈwiŋ-kər\ *noun* (1549)
1 : one that winks
2 : a horse's blinder

[1]win·kle \ˈwiŋ-kəl\ *noun* [by shortening] (1585)
: [2]PERIWINKLE

[2]winkle *intransitive verb* **win·kled; win·kling** \-k(ə-)liŋ\ [frequentative of *wink*] (1791)
: TWINKLE

[3]winkle *transitive verb* **win·kled; win·kling** \-k(ə-)liŋ\ [[1]*winkle;* from the process of extracting a winkle from its shell] (1918)
chiefly British **:** to displace, extract, or evict from a position — usually used with *out*

win·ner \ˈwi-nər\ *noun* (14th century)
: one that wins: as **a :** one that is successful especially through praiseworthy ability and hard work **b :** a victor especially in games and sports **c :** one that wins admiration **d :** a shot in a court game that is not returned and that scores for the player making it

winner's circle *noun* (1951)
: an enclosure near a racetrack where the winning horse and jockey are brought for photographs and awards

Win·nie \ˈwi-nē\ *noun* [*winner* + *-ie*] (circa 1944)
: an award presented annually by a professional organization for notable achievement in fashion design

[1]win·ning \ˈwi-niŋ\ *noun* (14th century)
1 : the act of one that wins : VICTORY
2 : something won: as **a :** a captured territory : CONQUEST **b :** money won by success in a game or competition — usually used in plural

[2]winning *adjective* (15th century)
1 a : of or relating to winning : that wins ⟨the *winning* ticket⟩ **b :** successful especially in competition ⟨a *winning* team⟩
2 : tending to please or delight ⟨a *winning* smile⟩
— win·ning·ly \-niŋ-lē\ *adverb*

win·nock \ˈwi-nək\ *noun* [Middle English (Scots) *windok, windowe*] (15th century)
Scottish **:** WINDOW

[1]win·now \ˈwi-(ˌ)nō\ *verb* [Middle English *winewen,* from Old English *windwian* to fan, winnow; akin to Old High German *wintōn* to fan, Latin *vannus* winnowing fan, *ventus* wind — more at WIND] (before 12th century)
transitive verb
1 a (1) **:** to remove (as chaff) by a current of air (2) **:** to get rid of (something undesirable or unwanted) : REMOVE — often used with *out* ⟨*winnow* out certain inaccuracies —Stanley Walker⟩ **b** (1) **:** SEPARATE, SIFT ⟨an old hand at *winnowing* what is true and significant —Oscar Lewis⟩ (2) **:** SELECT
2 a : to treat (as grain) by exposure to a current of air so that waste matter is eliminated **b :** to free of unwanted or inferior elements : PARE
3 : to blow on : FAN ⟨the wind *winnowing* his thin white hair —*Time*⟩
intransitive verb
1 : to separate chaff from grain by fanning
2 : to separate desirable and undesirable elements
— win·now·er \ˈwi-nə-wər\ *noun*

[2]winnow *noun* (1580)
1 : a device for winnowing
2 a : the action of winnowing **b :** a motion resembling that of winnowing

wino \ˈwī-(ˌ)nō\ *noun, plural* **win·os** (circa 1915)
: one who is chronically addicted to drinking wine

win·some \ˈwin(t)-səm\ *adjective* [Middle English *winsum,* from Old English *wynsum,* from *wynn* joy; akin to Old High German *wunna* joy, Latin *venus* desire — more at WIN] (before 12th century)
1 : generally pleasing and engaging often because of a childlike charm and innocence
2 : CHEERFUL, GAY
— win·some·ly *adverb*
— win·some·ness *noun*

[1]win·ter \ˈwin-tər\ *noun* [Middle English, from Old English; akin to Old High German *wintar* winter and perhaps to Lithuanian *vanduo* water, Old English *wæter* — more at WATER] (before 12th century)
1 : the season between autumn and spring comprising in the northern hemisphere usually the months of December, January, and February or as reckoned astronomically extending from the December solstice to the March equinox
2 : the colder half of the year
3 : YEAR ⟨happened many *winters* ago⟩
4 : a period of inactivity or decay

[2]winter *verb* **win·tered; win·ter·ing** \ˈwin-t(ə-)riŋ\ (14th century)
intransitive verb
1 : to pass the winter
2 : to feed or find food during the winter — used with *on*
transitive verb
: to keep, feed, or manage during the winter

[3]winter *adjective* (15th century)
1 : of, relating to, or suitable for winter ⟨a *winter* vacation⟩ ⟨*winter* clothes⟩
2 : sown in the autumn and harvested in the following spring or summer ⟨*winter* wheat⟩ ⟨*winter* rye⟩ — compare SUMMER

winter aconite *noun* (1794)
: a small European perennial herb (*Eranthis hyemalis*) of the buttercup family grown for its bright yellow flowers which often bloom through the snow

win·ter·ber·ry \ˈwin-tər-ˌber-ē\ *noun* (1759)
1 : an eastern North American shrub (*Ilex verticillata*) of the holly family with clusters of axillary flowers, usually bright red berries, and deciduous leaves that turn black in the fall — called also *black alder*
2 : a shrub (*Ilex laevigata*) of the holly family similar to the winterberry but of more restricted range

winter crookneck *noun* (circa 1909)
: any of several crooknecks that are winter squashes of the pumpkin group noted for their keeping qualities

win·ter·er \ˈwin-tər-ər\ *noun* (1783)
: one that winters; *specifically* : a winter resident or visitor

winter flounder *noun* (1814)
: a rusty brown flounder (*Pseudopleuronectes americanus* of the family Pleuronectidae) of the northwestern Atlantic important as a market fish especially in winter

win·ter·green \ˈwin-tər-ˌgrēn\ *noun* (1548)
1 : any of a genus (*Pyrola* of the family Pyrolaceae, the wintergreen family) of evergreen perennial herbs (as the shinleafs) that have basal leaves and racemose flowers
2 a : any of a genus (*Gaultheria*) of evergreen plants of the heath family; *especially* : a low creeping evergreen shrub (*G. procumbens*) of North America with white flowers and spicy red berries — compare CHECKERBERRY **b** (1) **:** an essential oil from this plant (2) **:** the flavor of this oil ⟨*wintergreen* lozenges⟩

win·ter·ize \ˈwin-tə-ˌrīz\ *transitive verb* **-ized; -iz·ing** (1934)
: to make ready for winter or winter use and especially resistant or proof against winter weather ⟨*winterize* a car⟩
— win·ter·i·za·tion \ˌwin-tə-rə-ˈzā-shən\ *noun*

win·ter–kill \ˈwin-tər-ˌkil\ (circa 1806)
transitive verb
: to kill (as a plant) by exposure to winter conditions
intransitive verb
: to die as a result of exposure to winter conditions
— winterkill *noun*

win·ter·ly \ˈwin-tər-lē\ *adjective* (1559)
: of, relating to, or occurring in winter : WINTRY

winter melon *noun* (circa 1900)
1 : any of several muskmelons (as a casaba or honeydew melon) that are fruits of a cultivated vine (*Cucumis melo indorus*)
2 : a large white fleshed melon that is the fruit of an Asian vine (*Benincasa hispida*) and is used especially in Chinese cooking

winter quarters *noun plural but singular or plural in construction* (1641)
: a winter residence or station (as of a military unit or a circus)

winter savory *noun* (1597)
: a perennial European mint (*Satureja montana*) with leaves used for seasoning — compare SUMMER SAVORY

winter squash *noun* (1775)
: any of various hard-shelled squashes that belong to cultivars derived from several species (especially *Cucurbita maxima, C. moschata,* and *C. pepo*) and that can be stored for several months

win·ter·tide \ˈwin-tər-ˌtīd\ *noun* (before 12th century)
: WINTERTIME

win·ter·time \-ˌtīm\ *noun* (14th century)
: the season of winter

win through *intransitive verb* (1644)
: to survive difficulties and reach a desired or satisfactory end

win·tle \ˈwi-n^{ə}l, ˈwin-t^{ə}l\ *intransitive verb* **win·tled; win·tling** \ˈwin(t)-liŋ; ˈwi-n^{ə}l-iŋ, ˈwin-t^{ə}l-\ [perhaps from Flemish *windtelen* to reel] (1786)
1 *Scottish* **:** STAGGER, REEL
2 *Scottish* **:** WRIGGLE

win·try \ˈwin-trē\ *also* **win·tery** \ˈwin-t(ə-)rē\ *adjective* **win·tri·er; -est** [Old English *wintrig,* from *winter*] (before 12th century)
1 : of, relating to, or characteristic of winter

\ə\ **abut** \$^{ə}$\ kitten \ər\ **further** \a\ **ash** \ā\ **ace** \ä\ **mop, mar** \au̇\ **out** \ch\ **chin** \e\ **bet** \ē\ **easy** \g\ **go** \i\ **hit** \ī\ **ice** \j\ **job** \ŋ\ **sing** \ō\ **go** \ȯ\ **law** \ȯi\ **boy** \th\ **thin** \t͟h\ **the** \ü\ **loot** \u̇\ **foot** \y\ **yet** \zh\ **vision** *see also* Guide to Pronunciation

2 a : weathered by or as if by winter **:** AGED, HOARY **b :** CHEERLESS, CHILLING ⟨a *wintry* greeting⟩
— win·tri·ness \'win-trē-nəs\ *noun*

winy \'wī-nē\ *adjective* **win·i·er; -est** (14th century)
1 : having the taste or qualities of wine **:** VINOUS
2 *of the air* **:** crisply fresh **:** EXHILARATING

[1]**winze** \'winz\ *noun* [alteration of earlier *winds,* probably from plural of [5]*wind*] (1757)
: a steeply inclined passageway in a mine

[2]**winze** *noun* [Flemish or Dutch *wensch* wish] (1785)
Scottish **:** CURSE

[1]**wipe** \'wīp\ *verb* **wiped; wip·ing** [Middle English *wipen,* from Old English *wīpian;* akin to Old High German *wīfan* to wind around, Latin *vibrare* to brandish, and probably to Old Norse *veipa* to be in movement, Sanskrit *vepate* it trembles] (before 12th century)
transitive verb
1 a : to rub with or as if with something soft for cleaning **b :** to clean or dry by rubbing **c :** to draw, pass, or move for or as if for rubbing or cleaning
2 a : to remove by or as if by rubbing **b :** to expunge completely ⟨*wipe* from memory the gruesome scenes —*American Guide Series: Delaware*⟩
3 : to spread by or as if by wiping
intransitive verb
: to make a motion of or as if of wiping something
— wipe one's boots on : to treat with indignity
— wipe the floor with *or* **wipe the ground with :** to defeat decisively

[2]**wipe** *noun* (1550)
1 a : BLOW, STRIKE **b :** JEER, GIBE
2 a : an act or instance of wiping **b :** a transition from one scene or picture to another (as in movies or television) made by a line moving across the screen
3 : something (as a towel) used for wiping

wiped out *adjective* (1965)
slang **:** INTOXICATED, HIGH

wipe·out \'wīp-,au̇t\ *noun* (1921)
1 : the act or an instance of wiping out **:** complete or utter destruction
2 : a fall or crash caused usually by losing control

wipe out (1535)
transitive verb
: to destroy completely **:** ANNIHILATE
intransitive verb
: to fall or crash usually as a result of losing control

wip·er \'wī-pər\ *noun* (1552)
1 : one that wipes
2 a : something (as a towel or sponge) used for wiping **b :** a moving contact for making connections with the terminals of an electrical device (as a rheostat) **c :** a usually motor-driven arm with a flexible blade for wiping a window (as the windshield of an automobile or airplane)

[1]**wire** \'wīr\ *noun, often attributive* [Middle English, from Old English *wīr;* akin to Old High German *wiara* fine gold work, Latin *viēre* to plait, and probably to Greek *iris* rainbow] (before 12th century)
1 a : metal in the form of a usually very flexible thread or slender rod **b :** a thread or rod of such material
2 a : WIREWORK **b :** the meshwork of parallel or woven wire on which the wet web of paper forms
3 : something (as a thin plant stem) that is wirelike
4 *plural* **a :** a system of wires used to operate the puppets in a puppet show **b :** hidden influences controlling the action of a person or organization
5 a : a line of wire for conducting electrical current — compare CORD 3b **b :** a telephone or telegraph wire or system; *especially* **:** WIRE SERVICE **c :** TELEGRAM, CABLEGRAM
6 : fencing or a fence of usually barbed wire
7 : the finish line of a race
8 : WIREHAIR
— wire·like \-,līk\ *adjective*
— under the wire 1 : at the finish line **2 :** at the last moment

[2]**wire** *verb* **wired; wir·ing** (15th century)
transitive verb
1 : to provide with wire **:** use wire on for a specific purpose
2 : to send or send word to by telegraph
3 : to connect by or as if by a wire
intransitive verb
: to send a telegraphic message
— wir·er \'wīr-ər\ *noun*

wire cloth *noun* (1798)
: a fabric of woven metallic wire (as for strainers)

wired \'wīrd\ *adjective* (15th century)
1 : reinforced by wire (as for strength)
2 : furnished with wires (as for electric connections)
3 : bound with wire ⟨a *wired* container⟩
4 : having a wirework netting or fence
5 : feverishly excited

wire·draw \'wīr-,dro̊\ *transitive verb* (1598)
1 : to draw or stretch forcibly **:** ELONGATE
2 : to draw or spin out to great length, tenuity, or overrefinement **:** ATTENUATE
— wire·draw·er \-dro̊(-ə)r\ *noun*

wire·drawn \-,dro̊n\ *adjective* (1603)
: excessively minute and subtle ⟨curious speculations, *wiredrawn* comparisons, obsolete erudition —Virginia Woolf⟩

wire fox terrier *noun* (1929)
: a wirehaired fox terrier

wire fraud *noun* (1976)
: fraud committed using a means of electronic communication (as a telephone)

wire gauge *noun* (1833)
1 : a gauge especially for measuring the diameter of wire or the thickness of sheet metal
2 : any of various systems consisting of a series of standard sizes used in describing the diameter of wire or the thickness of sheet metal

wire gauze *noun* (1816)
: a gauzelike wire cloth

wire grass *noun* (1751)
: any of various grasses or rushes having wiry culms or leaves: as **a :** a European slender-stemmed meadow grass (*Poa compressa*) widely naturalized in the U.S. and Canada **b :** any of several coarse grasses (genus *Aristida*) with a 3-awned lemma that grow extensively in open dry, sandy, or sterile areas especially of the southeastern and south-central U.S.

wire·hair \'wīr-,har, -,her\ *noun* (1884)
: a wirehaired dog or cat

wire·haired \-'hard, -'herd\ *adjective* (1801)
: having a stiff wiry outer coat of hair — used especially of a dog or cat; compare ROUGH, SMOOTH

wirehaired pointing griffon *noun* (1929)
: any of a breed of dogs of Dutch origin that both hunt and retrieve and have a long head and a harsh wiry often chestnut or white and chestnut colored coat

wirehaired terrier *noun* (circa 1885)
: WIRE FOX TERRIER

[1]**wire·less** \'wīr-ləs\ *adjective* (1894)
1 : having no wire or wires
2 *chiefly British* **:** of or relating to radiotelegraphy, radiotelephony, or radio

[2]**wireless** *noun* (1903)
1 : WIRELESS TELEGRAPHY
2 : two-way wireless transmission of sound using radio waves
3 *chiefly British* **:** RADIO

wireless telegraphy *noun* (1898)
: telegraphy carried on by radio waves and without connecting wires — called also *wireless telegraph*

wireless telephone *noun* (1894)
: RADIOTELEPHONE

wire·man \'wīr-mən\ *noun* (circa 1548)
1 : a maker of or worker with wire; *especially* **:** LINEMAN 1
2 : WIRETAPPER

wire netting *noun* (1801)
: a wire cloth coarser than wire gauze

wire·pho·to \'wīr-'fō-(,)tō\ *noun* [from *Wirephoto,* a trademark] (1935)
: a photograph transmitted by electrical signals over telephone wires

wire–pull·er \-,pu̇-lər\ *noun* (1832)
: one who uses secret or underhanded means to influence the acts of a person or organization
— wire–pull·ing \-,pu̇-liŋ\ *noun*

wire recorder *noun* (1943)
: a magnetic recorder using a thin wire as the recording medium
— wire–re·cord·ing *noun*

wire rope *noun* (1841)
: a rope formed wholly or chiefly of wires

wire service *noun* (1944)
: a news agency that sends out syndicated news copy to subscribers by wire or by satellite transmission

[1]**wire·tap** \'wīr-,tap\ (1904)
intransitive verb
: to tap a telephone or telegraph wire in order to get information
transitive verb
: to tap the telephone of

[2]**wiretap** *noun* (1948)
1 : the act or an instance of wiretapping
2 : an electrical connection for wiretapping

wire·tap·per \-,tap-ər\ *noun* (1893)
: one that taps telephone or telegraph wires

wire·work \-,wərk\ *noun* (1587)
1 : a work of wires; *especially* **:** meshwork, netting, or grillwork of wire ⟨plan the *wirework* for new circuitry⟩
2 : walking on wires especially by acrobats

wire·worm \-,wərm\ *noun* (1790)
: any of the slender hard-coated larvae of various click beetles that include some destructive especially to plant roots

wir·ing \'wīr-iŋ\ *noun* (1809)
1 : the act of providing or using wire
2 : a system of wires; *especially* **:** an arrangement of wires used for electric distribution

wir·ra \'wir-ə\ *interjection* [*oh wirra,* from Irish *a Mhuire,* literally, Mary!] (1829)
Irish — usually used to express lament, grief, or concern

wiry \'wīr-ē\ *adjective* **wir·i·er** \'wī-rē-ər\; **-est** (1588)
1 a *archaic* **:** made of wire **b :** resembling wire especially in form and flexibility **c** *of sound* **:** produced by or suggestive of the vibration of wire ⟨the violinist . . . often let his tone go nasal and *wiry* —D. J. Henahan⟩
2 : being lean, supple, and vigorous **:** SINEWY
— wir·i·ly \'wī-rə-lē\ *adverb*
— wir·i·ness \-rē-nəs\ *noun*

wis \'wis\ *verb* [by misdivision from *iwis* (understood as *I wis,* with *wis* taken to be an archaic present indicative of [1]*wit*)] (1508)
archaic **:** KNOW

wis·dom \'wiz-dəm\ *noun* [Middle English, from Old English *wīsdōm,* from *wīs* wise] (before 12th century)
1 a : accumulated philosophic or scientific learning **:** KNOWLEDGE **b :** ability to discern inner qualities and relationships **:** INSIGHT **c :** good sense **:** JUDGMENT **d :** generally accepted belief ⟨challenges what has become accepted *wisdom* among many historians —Robert Darnton⟩
2 : a wise attitude or course of action

3 : the teachings of the ancient wise men
synonym see SENSE

Wisdom *noun*
: a didactic book included in the Roman Catholic canon of the Old Testament and corresponding to the Wisdom of Solomon in the Protestant Apocrypha — see BIBLE table

Wisdom of Sol•o•mon \-'sä-lə-mən\
: a didactic book included in the Protestant Apocrypha — see BIBLE table

wisdom tooth *noun* [from being cut usually in the late teens] (1848)
: the third molar that is the last tooth to erupt on each side of the upper and lower jaws in humans

¹wise \'wīz\ *noun* [Middle English, from Old English *wīse;* akin to Old High German *wīsa* manner, Greek *eidos* form, *idein* to see — more at WIT] (before 12th century)
: MANNER, WAY ⟨in any *wise*⟩

²wise *adjective* **wis•er; wis•est** [Middle English *wis,* from Old English *wīs;* akin to Old High German *wīs* wise, Old English *witan* to know — more at WIT] (before 12th century)
1 a : characterized by wisdom **:** marked by deep understanding, keen discernment, and a capacity for sound judgment **b :** exercising sound judgment **:** PRUDENT
2 a : evidencing or hinting at the possession of inside information **:** KNOWING **b :** possessing inside information **c :** CRAFTY, SHREWD
3 *archaic* **:** skilled in magic or divination
4 : INSOLENT, SMART-ALECKY, FRESH ☆
— wise•ly *adverb*
— wise•ness *noun*

³wise *verb* **wised; wis•ing** (1905)
transitive verb
: to give instruction or information to **:** TEACH — usually used with *up*
intransitive verb
: to become informed or knowledgeable **:** LEARN — used with *up*

⁴wise *transitive verb* **wised; wis•ing** [Middle English, from Old English *wīsian;* akin to Old Norse *vīsa* to show the way, Old English *wīs* wise] (before 12th century)
1 *chiefly Scottish* **a :** DIRECT, GUIDE **b :** ADVISE, PERSUADE
2 *chiefly Scottish* **:** to divert or impel in a given direction **:** SEND

-wise *adverb combining form* [Middle English, from Old English *-wīsan,* from *wīse* manner]
1 a : in the manner of ⟨crab*wise*⟩ ⟨fan*wise*⟩ **b :** in the position or direction of ⟨slant*wise*⟩ ⟨clock*wise*⟩
2 : with regard to **:** in respect of ⟨dollar*wise*⟩

wise•acre \'wī-,zā-kər\ *noun* [Middle Dutch *wijssegger* soothsayer, modification of Old High German *wīzzago;* akin to Old English *wītega* soothsayer, *witan* to know] (1595)
: one who pretends to knowledge or cleverness; *especially* **:** SMART ALECK

wise•ass \'wīz-,as\ *noun* (1971)
: SMART ALECK

¹wise•crack \'wīz-,krak\ *noun* (1924)
: a clever or sarcastic remark

²wisecrack *intransitive verb* (1924)
: to make a wisecrack
— wise•crack•er *noun*

wised–up \'wīzd-'əp\ *adjective* (1926)
: KNOWING 1

wise guy \'wīz-,gī\ *noun* (1896)
: SMART ALECK

wise man *noun* (before 12th century)
1 : a man of unusual learning, judgment, or insight **:** SAGE
2 : a man versed in esoteric lore (as of magic or astrology); *especially* **:** MAGUS 1b

wi•sen•hei•mer \'wī-zᵊn-,hī-mər\ *noun* [²*wise* + *-enheimer* (as in family names such as *Guggenheimer, Oppenheimer*)] (1904)
: SMART ALECK

wi•sent \'vē-,zent\ *noun* [German, from Old High German *wisant* — more at BISON] (1866)
: a European bison (*Bison bonasus*) sometimes considered conspecific with the North American buffalo (*B. bison*) — called also *aurochs*

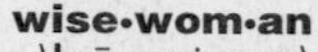
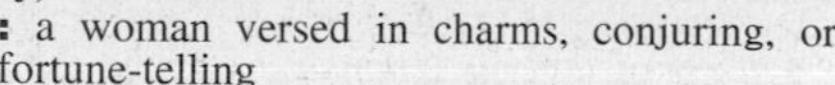
wisent

wise•wom•an \'wīz-,wu̇-mən\ *noun* (14th century)
: a woman versed in charms, conjuring, or fortune-telling

¹wish \'wish\ *verb* [Middle English *wisshen,* from Old English *wȳscan;* akin to Old High German *wunsken* to wish, Sanskrit *vāñchati* he wishes, *vanoti* he strives for — more at WIN] (before 12th century)
transitive verb
1 : to have a desire for (as something unattainable) ⟨*wished* he could live his life over⟩
2 : to give expression to as a wish **:** BID ⟨*wish* them good night⟩
3 a : to give form to (a wish) **b :** to express a wish for **c :** to request in the form of a wish **:** ORDER **d :** to desire (a person or thing) to be as specified ⟨cannot *wish* our problems away⟩
4 : to confer (something unwanted) on someone **:** FOIST
intransitive verb
1 : to have a desire **:** WANT
2 : to make a wish
synonym see DESIRE
— wish•er *noun*

²wish *noun* (14th century)
1 a : an act or instance of wishing or desire **:** WANT **b :** an object of desire **:** GOAL
2 a : an expressed will or desire **:** MANDATE **b :** a request or command couched as a wish
3 : an invocation of good or evil fortune on someone

wisha \'wi-shə\ *interjection* [Irish *mhuise*] (1826)
chiefly Irish — used as an intensive or to express surprise

wish•bone \'wish-,bōn\ *noun* [from the superstition that when two persons pull it apart the one getting the longer fragment will have a wish granted] (1853)
1 : a forked bone in front of the breastbone in a bird consisting chiefly of the two clavicles fused at their median or lower end
2 : a variation of the T formation in which the halfbacks line up farther from the line of scrimmage than the fullback does

wish•ful \'wish-fəl\ *adjective* (1593)
1 a : expressive of a wish **:** HOPEFUL **b :** having a wish **:** DESIROUS
2 : according with wishes rather than reality
— wish•ful•ly \-fə-lē\ *adverb*
— wish•ful•ness *noun*

wish fulfillment *noun* (1908)
: the gratification of a desire especially symbolically (as in dreams, daydreams, or neurotic symptoms)

wishful thinking *noun* (1932)
: the attribution of reality to what one wishes to be true or the tenuous justification of what one wants to believe

wish•ing *adjective* (circa 1530)
1 *archaic* **:** WISHFUL
2 : regarded as having the power to grant wishes ⟨threw a coin in the *wishing* well⟩

wish list *noun* (1972)
: a list of desired but often realistically unobtainable items

wish–wash \'wish-,wȯsh, -,wäsh\ *noun* [reduplication of ²*wash*] (1786)
1 : a weak drink
2 : insipid talk or writing

wishy–washy \'wi-shē-,wȯ-shē, -,wä-\ *adjective* [reduplication of *washy*] (circa 1693)
1 : lacking in character or determination **:** INEFFECTUAL
2 : lacking in strength or flavor **:** WEAK
— wishy–wash•i•ness *noun*

¹wisp \'wisp\ *noun* [Middle English] (14th century)
1 : a small handful (as of hay or straw)
2 a : a thin strip or fragment **b :** a thready streak ⟨a *wisp* of smoke⟩ **c :** something frail, slight, or fleeting ⟨a *wisp* of a girl⟩ ⟨a *wisp* of a smile⟩
3 *archaic* **:** WILL-O'-THE-WISP
— wisp•i•ly \'wis-pə-lē\ *adverb*
— wisp•i•ness \'wis-pē-nəs\ *noun*
— wispy \'wis-pē\ *adjective*

²wisp (1753)
transitive verb
1 : to roll into a wisp
2 : to make wisps of ⟨a cigarette *wisping* smoke at the corner of his mouth —Raymond Chandler⟩
intransitive verb
: to emerge or drift in wisps ⟨her hair began to *wisp* into her eyes —Mary Manning⟩

wisp•ish \'wis-pish\ *adjective* (1896)
: resembling a wisp **:** INSUBSTANTIAL

wist \'wist\ *transitive verb* [alteration of *wis*] (1508)
archaic **:** KNOW

wis•te•ria \wis-'tir-ē-ə\ *or* **wis•tar•ia** \-'tir-ē-ə *also* -'ter-\ *noun* [New Latin *Wisteria,* from Caspar *Wistar* (died 1818) American physician] (1842)
: any of a genus (*Wisteria*) of chiefly Asian mostly woody leguminous vines that have pinnately compound leaves and long racemes of showy blue, white, purple, or rose papilionaceous flowers and that include several grown as ornamentals

wisteria

wist•ful \'wist-fəl\ *adjective* [blend of *wishful* and obsolete English *wistly* intently] (1714)
1 : full of yearning or desire tinged with melancholy; *also* **:** inspiring such yearning ⟨a *wistful* memoir⟩

☆ SYNONYMS
Wise, sage, sapient, judicious, prudent, sensible, sane mean having or showing sound judgment. WISE suggests great understanding of people and of situations and unusual discernment and judgment in dealing with them ⟨*wise* beyond his tender years⟩. SAGE suggests wide experience, great learning, and wisdom ⟨the *sage* advice of my father⟩. SAPIENT suggests great sagacity and discernment ⟨the *sapient* musings of an old philosopher⟩. JUDICIOUS stresses a capacity for reaching wise decisions or just conclusions ⟨*judicious* parents using kindness and discipline in equal measure⟩. PRUDENT suggests exercise of the restraint of sound practical wisdom and discretion ⟨a *prudent* decision to wait out the storm⟩. SENSIBLE applies to action guided and restrained by good sense and rationality ⟨a *sensible* woman who was not fooled by flattery⟩. SANE stresses mental soundness, rationality, and levelheadedness ⟨remained *sane* even in times of crises⟩.

2 : musingly sad **:** PENSIVE
— **wist·ful·ly** \-fə-lē\ *adverb*
— **wist·ful·ness** *noun*

¹wit \'wit\ *verb* **wist** \'wist\; **wit·ting** *present 1st & 3d singular* **wot** \'wät\ [Middle English *witen* (1st & 3d singular present *wot,* past *wiste*), from Old English *witan* (1st & 3d singular present *wāt,* past *wisse, wiste*); akin to Old High German *wizzan* to know, Latin *vidēre* to see, Greek *eidenai* to know, *idein* to see] (before 12th century)
1 *archaic* **:** KNOW
2 *archaic* **:** to come to know **:** LEARN

²wit *noun* [Middle English, from Old English; akin to Old High German *wizzi* knowledge, Old English *witan* to know] (before 12th century)
1 a : MIND, MEMORY **b :** reasoning power **:** INTELLIGENCE
2 a : SENSE 2a — usually used in plural ⟨alone and warming his five *wits,* the white owl in the belfry sits —Alfred Tennyson⟩ **b** (1) **:** mental soundness **:** SANITY — usually used in plural (2) **:** mental capability and resourcefulness **:** INGENUITY
3 a : astuteness of perception or judgment **:** ACUMEN **b :** the ability to relate seemingly disparate things so as to illuminate or amuse **c** (1) **:** a talent for banter or persiflage (2) **:** a witty utterance or exchange **d :** clever or apt humor
4 a : a person of superior intellect **:** THINKER **b :** an imaginatively perceptive and articulate individual especially skilled in banter or persiflage ☆
— **at one's wit's end** *or* **at one's wits' end :** at a loss for a means of solving a problem

wi·tan \'wi-,tän\ *noun plural* [Old English, plural of *wita* sage, adviser; akin to Old High German *wizzo* sage, Old English *witan* to know] (1807)
: members of the witenagemot

¹witch \'wich\ *noun* [Middle English *wicche,* from Old English *wicca,* masculine, wizard & *wicce,* feminine, witch; akin to Middle High German *wicken* to bewitch, Old English *wigle* divination, and perhaps to Old High German *wīh* holy — more at VICTIM] (before 12th century)
1 : one that is credited with usually malignant supernatural powers; *especially* **:** a woman practicing usually black witchcraft often with the aid of a devil or familiar **:** SORCERESS — compare WARLOCK
2 : an ugly old woman **:** HAG
3 : a charming or alluring girl or woman
— **witch·like** \'wich-,līk\ *adjective*
— **witchy** \'wi-chē\ *adjective*

²witch (14th century)
transitive verb
1 : to affect injuriously with witchcraft
2 *archaic* **:** to influence or beguile with allure or charm
intransitive verb
: DOWSE

witch·craft \'wich-,kraft\ *noun* (before 12th century)
1 a : the use of sorcery or magic **b :** communication with the devil or with a familiar
2 : an irresistible influence or fascination

witch doctor *noun* (1718)
: a professional worker of magic usually in a primitive society who often works to cure sickness

witch·ery \'wi-chə-rē, 'wich-rē\ *noun, plural* **-er·ies** (1546)
1 a : the practice of witchcraft **:** SORCERY **b :** an act of witchcraft
2 : an irresistible fascination

witches' brew *noun* (1929)
: a potent or fearsome mixture ⟨a *witches' brew* of untamed sex and brutality —Harrison Smith⟩

witch·es'–broom \'wi-chəz-,brüm, -,bru̇m\ *noun* (1881)
: an abnormal tufted growth of small branches on a tree or shrub caused especially by fungi or viruses

witches' Sabbath *noun* (circa 1676)
: a midnight assembly of witches, devils, and sorcerers for the celebration of rites and orgies

witch·grass \'wich-,gras\ *noun* [probably alteration of *quitch (grass)*] (1790)
1 : QUACK GRASS
2 [¹*witch*] **:** a North American grass (*Panicum capillare*) with slender brushy panicles that is often a weed on cultivated land

witch ha·zel \'wich-,hā-zəl\ *noun* [*witch,* a tree with pliant branches, from Middle English *wyche,* from Old English *wice;* probably akin to Old English *wīcan* to yield — more at WEAK] (circa 1542)
1 : any of a genus (*Hamamelis* of the family Hamamelidaceae, the witch-hazel family) of shrubs or small trees with slender-petaled yellow flowers borne in late fall or early spring; *especially* **:** one (*H. virginiana*) of eastern North America that blooms in the fall
2 : an alcoholic solution of a distillate of the bark of a witch hazel (*H. virginiana*) used as a soothing and mildly astringent lotion

witch–hunt \'wich-,hənt\ *noun* (1885)
1 : a searching out for persecution of persons accused of witchcraft
2 : the searching out and deliberate harassment of those (as political opponents) with unpopular views
— **witch–hunt·er** *noun*
— **witch–hunt·ing** *noun or adjective*

¹witch·ing \'wi-chiŋ\ *noun* (before 12th century)
: the practice of witchcraft **:** SORCERY

²witching *adjective* (14th century)
: of, relating to, or suitable for sorcery or supernatural occurrences ⟨the very *witching* time of night —Shakespeare⟩

witch of Agne·si \-än-'yā-zē\ *noun* [Maria Gaetana *Agnesi* (died 1799) Italian mathematician; *witch,* translation of Italian *versiera* cubic curve (influenced by Italian *versiera* female demon)] (1875)
: a plane cubic curve that is symmetric about the y-axis and approaches the x-axis as an asymptote that is constructed by drawing lines from the origin intersecting an upright circle tangent to the x-axis at the origin and taking the locus of points of intersection of pairs of lines parallel to the x-axis and y-axis each pair of which consists of a line parallel to the x-axis through the point where a line through the origin intersects the circle and a line parallel to the y-axis through the point where the same line through the origin intersects the line parallel to the x-axis through the point of intersection of the circle and the y-axis and that has the equation $x^2y = 4a^2(2a - y)$ — called also *witch*

witch·weed \'wich-,wēd\ *noun* (1904)
: any of a genus (*Striga*) of yellow-flowered Old World plants of the snapdragon family that are damaging root parasites of grasses (as sorghum and maize) and that include one (*S. lutea*) which is an introduced pest in parts of the southeastern U.S.

¹wite \'wīt\ *transitive verb* **wit·ed; wit·ing** [Middle English, from Old English *wītan;* akin to Old High German *wīzan* to blame, Old English *witan* to know] (before 12th century)
chiefly Scottish **:** BLAME

²wite *noun* (13th century)
chiefly Scottish **:** BLAME, RESPONSIBILITY

wi·te·na·ge·mot *or* **wi·te·na·ge·mote** \'wi-tᵊn-ə-gə-,mōt, -yə-,mōt\ *noun* [Old English *witena gemōt,* from *witena* (genitive plural of *wita* sage, adviser) + *gemōt* gemot] (before 12th century)
: an Anglo-Saxon council made up of a varying number of nobles, prelates, and influential officials and convened from time to time to advise the king on administrative and judicial matters

with \'wit͟h, 'with, wət͟h, wəth\ *preposition* [Middle English, against, from, with, from Old English; akin to Old English *wither* against, Old High German *widar* against, back, Sanskrit *vi* apart] (before 12th century)
1 a : in opposition to **:** AGAINST ⟨had a fight *with* his brother⟩ **b :** so as to be separated or detached from ⟨broke *with* her family⟩
2 a — used as a function word to indicate a participant in an action, transaction, or arrangement ⟨works *with* his father⟩ ⟨a talk *with* a friend⟩ ⟨got into an accident *with* the car⟩ **b** — used as a function word to indicate the object of attention, behavior, or feeling ⟨get tough *with* him⟩ ⟨angry *with* her⟩ **c :** in respect to **:** so far as concerns ⟨on friendly terms *with* all nations⟩ **d** — used to indicate the object of an adverbial expression of imperative force ⟨off *with* his head⟩ **e :** OVER, ON ⟨no longer has any influence *with* them⟩ **f :** in the performance, operation, or use of ⟨the trouble *with* this machine⟩
3 a — used as a function word to indicate the object of a statement of comparison or equality ⟨a dress identical *with* her hostess's⟩ **b** — used as a function word to express agreement or sympathy ⟨must conclude, *with* you, that the painting is a forgery⟩ **c :** on the side of **:** FOR ⟨if he's for lower taxes, I'm *with* him⟩ **d :** as well as ⟨can pitch *with* the best of them⟩
4 a — used as a function word to indicate combination, accompaniment, presence, or addition ⟨heat milk *with* honey⟩ ⟨went there *with* her⟩ ⟨his money, *with* his wife's, comes to a million⟩ **b :** inclusive of ⟨costs $5 *with* the tax⟩
5 a : in the judgment or estimation of ⟨stood well *with* her classmates⟩ **b :** in or according to the experience or practice of ⟨*with* many of us, our ideas seem to fall by the wayside —W. J. Reilly⟩
6 a — used as a function word to indicate the means, cause, agent, or instrumentality ⟨hit him *with* a rock⟩ ⟨pale *with* anger⟩ ⟨threatened *with* tuberculosis⟩ ⟨he amused the crowd *with* his antics⟩ **b** *archaic* **:** by the direct act of
7 a — used as a function word to indicate manner of action ⟨ran *with* effort⟩ ⟨acknowledge your contribution *with* thanks⟩ **b** — used as a function word to indicate an attendant fact or circumstance ⟨stood there *with* his hat on⟩ **c** — used as a function word to indicate a result attendant on a specified action ⟨got off *with* a light sentence⟩
8 a (1) **:** in possession of **:** HAVING ⟨came *with* good news⟩ (2) **:** in the possession or care of ⟨left the money *with* her mother⟩ **b :** characterized or distinguished by ⟨a person *with* a sharp nose⟩

☆ SYNONYMS
Wit, humor, irony, sarcasm, satire, repartee mean a mode of expression intended to arouse amusement. WIT suggests the power to evoke laughter by remarks showing verbal felicity or ingenuity and swift perception especially of the incongruous ⟨a playful *wit*⟩. HUMOR implies an ability to perceive the ludicrous, the comical, and the absurd in human life and to express these usually without bitterness ⟨a sense of *humor*⟩. IRONY applies to a manner of expression in which the intended meaning is the opposite of what is seemingly expressed ⟨the *irony* of the title⟩. SARCASM applies to expression frequently in the form of irony that is intended to cut or wound ⟨given to heartless *sarcasm*⟩. SATIRE applies to writing that exposes or ridicules conduct, doctrines, or institutions either by direct criticism or more often through irony, parody, or caricature ⟨a *satire* on the Congress⟩. REPARTEE implies the power of answering quickly, pointedly, or wittily ⟨a dinner guest noted for *repartee*⟩.

9 a — used as a function word to indicate a close association in time ⟨*with* the outbreak of war they went home⟩ ⟨mellows *with* time⟩ **b** : in proportion to ⟨the pressure varies *with* the depth⟩

10 a : in spite of : NOTWITHSTANDING ⟨a really tip-top man, *with* all his wrongheadedness —H. J. Laski⟩ **b** : except for ⟨finds that, *with* one group of omissions and one important addition, they reflect that curriculum —Gilbert Highet⟩

11 : in the direction of ⟨*with* the wind⟩ ⟨*with* the grain⟩

[1]**with·al** \wi-ˈthȯl, -ˈthȯl\ *adverb* [Middle English, from *with* + *all, al* all] (13th century)
1 : together with this : BESIDES ⟨a supporter of all constructive work and *withal* an excellent businessman —A. W. Long⟩
2 *archaic* : THEREWITH 1
3 : on the other hand : NEVERTHELESS

[2]**withal** *preposition* (14th century)
archaic : WITH — used postpositively with a relative or interrogative pronoun as object

with·draw \with-ˈdrȯ, with-\ *verb* **-drew** \-ˈdrü\; **-drawn** \-ˈdrȯn\; **-draw·ing** \-ˈdrȯ(-)iŋ\ [Middle English, from *with* from + *drawen* to draw] (13th century)
transitive verb
1 a : to take back or away : REMOVE ⟨pressure upon educational administrators to *withdraw* academic credit —J. W. Scott⟩ **b** : to remove from use or cultivation **c** : to remove (money) from a place of deposit **d** : to turn away (as the eyes) from an object of attention ⟨*withdrew* her gaze⟩ **e** : to draw (as a curtain) back or aside
2 a : to remove from consideration or set outside a group ⟨*withdrew* his name from the list of nominees⟩ ⟨*withdrew* their child from the school⟩ **b** (1) : TAKE BACK, RETRACT (2) : to recall or remove (a motion) under parliamentary procedure
intransitive verb
1 a : to move back or away : RETIRE **b** : to draw back from a battlefield : RETREAT
2 a : to remove oneself from participation **b** : to become socially or emotionally detached ⟨had *withdrawn* farther and farther into herself —Ethel Wilson⟩
3 : to recall a motion under parliamentary procedure
— **with·draw·able** \-ˈdrȯ-ə-bəl\ *adjective*

with·draw·al \-ˈdrȯ(-ə)l\ *noun* (1749)
1 a : the act of taking back or away something that has been granted or possessed **b** : removal from a place of deposit or investment **c** (1) : the discontinuance of administration or use of a drug (2) : a period following the discontinuance of an addicting drug that is marked by often painful physical and psychological symptoms ⟨a heroin addict going through *withdrawal*⟩
2 a : retreat or retirement especially into a more secluded or less exposed place or position **b** : an operation by which a military force disengages from the enemy **c** (1) : social or emotional detachment (2) : a pathological retreat from objective reality (as in some schizophrenic states)
3 : RETRACTION, REVOCATION ⟨threatened us with *withdrawal* of consent⟩
4 a : the act of drawing someone or something back from or out of a place or position **b** : COITUS INTERRUPTUS

withdrawing room *noun* (1591)
: a room to retire to (as from a dining room); *especially* : DRAWING ROOM

with·drawn \with-ˈdrȯn, with-\ *adjective* (1615)
1 : removed from immediate contact or easy approach : ISOLATED
2 : socially detached and unresponsive : exhibiting withdrawal : INTROVERTED
— **with·drawn·ness** \-ˈdrȯn-nəs\ *noun*

withe \ˈwith, ˈwith, ˈwīth\ *noun* [Middle English, from Old English *withthe;* akin to Old English *wīthig* withy] (before 12th century)
: a slender flexible branch or twig; *especially* : one used as a band or line

[1]**with·er** \ˈwi-thər\ *verb* **with·ered; with·er·ing** \ˈwith-riŋ, ˈwi-thə-\ [Middle English *widren;* probably akin to Middle English *weder* weather] (14th century)
intransitive verb
1 : to become dry and sapless; *especially* : to shrivel from or as if from loss of bodily moisture
2 : to lose vitality, force, or freshness
transitive verb
1 : to cause to wither
2 : to make speechless or incapable of action : STUN ⟨*withered* him with a look —Dorothy Sayers⟩

[2]**wither** *noun* (1607)
chiefly British : WITHERS

withering *adjective* (1579)
: acting or serving to cut down or destroy : DEVASTATING ⟨a *withering* fire from the enemy⟩
— **with·er·ing·ly** \ˈwith-riŋ-lē, ˈwi-thə-\ *adverb*

with·er·ite \ˈwi-thə-ˌrīt\ *noun* [German *Witherit,* irregular from William *Withering* (died 1799) English physician] (1794)
: a mineral consisting of a carbonate of barium in the form of white or gray twin crystals or columnar or granular masses

withe rod *noun* (1846)
: a North American viburnum (*Viburnum cassinoides*) with tough slender shoots

with·ers \ˈwi-thərz\ *noun plural* [probably from obsolete English *wither-* against, from Middle English, from Old English, from *wither* against; from the withers being the parts which resist the pull in drawing a load — more at WITH] (1580)
1 : the ridge between the shoulder bones of a horse — see HORSE illustration
2 : a part corresponding to the withers in a quadruped other than a horse

with·er·shins \ˈwi-thər-shənz\ *variant of* WIDDERSHINS

with·hold \with-ˈhōld, with-\ *verb* **-held** \-ˈheld\; **-hold·ing** [Middle English, from *with* from + *holden* to hold — more at WITH] (13th century)
transitive verb
1 : to hold back from action : CHECK
2 *archaic* : to keep in custody
3 : to refrain from granting, giving, or allowing ⟨*withhold* permission⟩
4 : to deduct (withholding tax) from income
intransitive verb
: FORBEAR, REFRAIN
synonym see KEEP
— **with·hold·er** *noun*

withholding tax *noun* (1940)
: a deduction (as from wages, fees, or dividends) levied at a source of income as advance payment on income tax

[1]**with·in** \wi-ˈthin, -ˈthin\ *adverb* [Middle English *withinne,* from Old English *withinnan,* from *with* + *innan* inwardly, within, from *in*] (before 12th century)
1 : in or into the interior : INSIDE
2 : in one's inner thought, disposition, or character : INWARDLY ⟨search *within* for a creative impulse —Kingman Brewster, Jr.⟩

[2]**within** *preposition* (12th century)
1 — used as a function word to indicate enclosure or containment
2 — used as a function word to indicate situation or circumstance in the limits or compass of: as **a** : before the end of ⟨gone *within* a week⟩ **b** (1) : not beyond the quantity, degree, or limitations of ⟨live *within* your income⟩ (2) : in or into the scope or sphere of ⟨*within* the jurisdiction of the state⟩ (3) : in or into the range of ⟨*within* reach⟩ ⟨*within* sight⟩ (4) — used as a function word to indicate a specified difference or margin ⟨came *within* two points of a perfect mark⟩ ⟨*within* a mile of the town⟩
3 : to the inside of : INTO

[3]**within** *noun* (15th century)
: an inner place or area ⟨revolt from *within*⟩

[4]**within** *adjective* (1748)
: being inside : ENCLOSED ⟨the *within* indictment⟩

with·in·doors \wi-ˌthin-ˈdōrz, -ˌthin-, -ˈdȯrz\ *adverb* (1581)
: INDOORS

with–it \ˈwi-thət, -thət\ *adjective* (1959)
: socially or culturally up-to-date ⟨the intelligent, disaffected, *with-it* young —Eliot Fremont-Smith⟩

[1]**with·out** \wi-ˈthau̇t, -ˈthau̇t\ *preposition* [Middle English *withoute,* from Old English *withūtan,* from *with* + *ūtan* outside, from *ūt* out — more at OUT] (before 12th century)
1 : OUTSIDE
2 — used as a function word to indicate the absence or lack of something or someone ⟨fight *without* fear⟩ ⟨left *without* him⟩ ⟨looks *without* seeing⟩

[2]**without** *adverb* (before 12th century)
1 : on the outside : EXTERNALLY
2 : with something lacking or absent ⟨has learned to do *without*⟩

[3]**without** *conjunction* (14th century)
chiefly dialect : UNLESS ⟨you don't know about me *without* you have read a book —Mark Twain⟩

[4]**without** *noun* (15th century)
: an outer place or area ⟨came from *without*⟩

with·out·doors \wi-ˌthau̇t-ˈdōrz, -ˌthau̇t-\ *adverb* (1617)
: OUTDOORS

with·stand \with-ˈstand, with-\ *transitive verb* **-stood** \-ˈstu̇d\; **-stand·ing** [Middle English, from Old English *withstandan,* from *with* against + *standan* to stand] (before 12th century)
1 a : to stand up against : oppose with firm determination; *especially* : to resist successfully **b** : to be proof against : resist the effect of ⟨*withstand* the impact of a landing —*Current Biography*⟩
2 *archaic* : to stop or obstruct the course of
synonym see OPPOSE

[1]**withy** \ˈwi-thē\ *noun, plural* **with·ies** [Middle English, from Old English *wīthig;* akin to Old High German *wīda* willow, Latin *vitis* vine, *viēre* to plait — more at WIRE] (before 12th century)
1 : WILLOW; *especially* : OSIER 1
2 : a flexible slender twig or branch (as of osier) : WITHE

[2]**withy** \ˈwi-thē, ˈwi-thē, ˈwī-thē\ *adjective* [*withe*] (1598)
: flexibly tough

wit·less \ˈwit-ləs\ *adjective* (before 12th century)
1 : destitute of wit or understanding : FOOLISH
2 : mentally deranged : CRAZY ⟨drive one *witless* with anxiety —William Styron⟩
— **wit·less·ly** *adverb*
— **wit·less·ness** *noun*

wit·ling \-liŋ\ *noun* (1693)
1 : a would-be wit
2 : a person of little wit

wit·loof \ˈwit-ˌlōf, -ˌlüf\ *noun* [Dutch dialect *witloof* chicory, from Dutch *wit* white + *loof* foliage] (1885)
: CHICORY 1; *also* : ENDIVE 2

[1]**wit·ness** \ˈwit-nəs\ *noun* [Middle English *witnesse,* from Old English *witnes* knowledge, testimony, witness, from [2]*wit*] (before 12th century)
1 : attestation of a fact or event : TESTIMONY

2 : one that gives evidence; *specifically* **:** one who testifies in a cause or before a judicial tribunal
3 : one asked to be present at a transaction so as to be able to testify to its having taken place
4 : one who has personal knowledge of something
5 a : something serving as evidence or proof **:** SIGN **b :** public affirmation by word or example of usually religious faith or conviction ⟨the heroic *witness* to divine life —*Pilot*⟩
6 *capitalized* **:** a member of the Jehovah's Witnesses

²witness (14th century)
transitive verb
1 : to testify to **:** ATTEST
2 : to act as legal witness of
3 : to furnish proof of **:** BETOKEN
4 a : to have personal or direct cognizance of**:** see for oneself ⟨*witnessed* the historic event⟩ **b :** to take note of ⟨our grammar — *witness* our verb system—is a marvel of flexibility, variety, and exactitude —Charlton Laird⟩
5 : to constitute the scene or time of ⟨structures . . . which this striking Dorset hilltop once *witnessed* —*Times Literary Supplement*⟩
intransitive verb
1 : to bear witness **:** TESTIFY
2 : to bear witness to one's religious convictions ⟨opportunity to *witness* for Christ —Billy Graham⟩
synonym see CERTIFY

wit·ness–box \-ˌbäks\ *noun* (1806)
chiefly British **:** an enclosure in which a witness sits or stands while testifying in court

witness stand *noun* (1853)
: a stand or an enclosure from which a witness gives evidence in a court

wit·ted \ˈwi-təd\ *adjective* (14th century)
: having wit or understanding — usually used in combination ⟨dull-*witted*⟩ ⟨quick-*witted*⟩

wit·ti·cism \ˈwi-tə-ˌsi-zəm\ *noun* [*witty* + *-cism* (as in *criticism*)] (1651)
: a cleverly witty and often biting or ironic remark

¹wit·ting \ˈwi-tᵊn, -tiŋ\ *noun* (14th century)
1 *chiefly dialect* **:** knowledge or awareness of something **:** COGNIZANCE
2 *chiefly dialect* **:** information obtained or communicated **:** NEWS

²wit·ting \-tiŋ\ *adjective* (circa 1520)
1 : cognizant or aware of something **:** CONSCIOUS
2 : done deliberately **:** INTENTIONAL
— wit·ting·ly \-tiŋ-lē\ *adverb*

wit·tol \ˈwi-tᵊl\ *noun* [Middle English *wetewold,* from *weten, witen* to know + cokewold cuckold — more at WIT] (15th century)
1 *archaic* **:** a man who knows of his wife's infidelity and puts up with it
2 *archaic* **:** a witless person

wit·ty \ˈwi-tē\ *adjective* **wit·ti·er; -est** (before 12th century)
1 *archaic* **:** having good intellectual capacity **:** INTELLIGENT
2 : amusingly or ingeniously clever in conception or execution ⟨the costumes are sumptuous and *witty* —Virgil Thomson⟩ ⟨the musical background is . . . often *witty* —Wolcott Gibbs⟩
3 : marked by or full of wit **:** smartly facetious or jocular
4 : quick or ready to see or express illuminating or amusing relationships or insights ☆
— wit·ti·ly \ˈwi-tᵊl-ē\ *adverb*
— wit·ti·ness \ˈwi-tē-nəs\ *noun*

wive \ˈwīv\ *verb* **wived; wiv·ing** [Middle English, from Old English *wīfian,* from *wīf* woman, wife] (before 12th century)
intransitive verb
: to marry a woman
transitive verb
1 : to marry to a woman
2 : to take for a wife

wives *plural of* WIFE

wiz \ˈwiz\ *noun* (1902)
: WIZARD 3

¹wiz·ard \ˈwi-zərd\ *noun* [Middle English *wysard,* from *wis, wys* wise] (15th century)
1 *archaic* **:** a wise man **:** SAGE
2 : one skilled in magic **:** SORCERER
3 : a very clever or skillful person

²wizard *adjective* (1579)
1 *archaic* **:** having magical influence or power
2 *archaic* **:** of or relating to wizardry **:** ENCHANTED
3 *chiefly British* **:** worthy of the highest praise **:** EXCELLENT

wiz·ard·ly \ˈwi-zərd-lē\ *adjective* (1588)
1 : having characteristics of a wizard
2 : marvelous in construction or operation ⟨uses *wizardly* circuitry to distort images —*Time*⟩

wiz·ard·ry \ˈwi-zə(r)-drē\ *noun, plural* **-ries** (1583)
1 : the art or practices of a wizard **:** SORCERY
2 a : a seemingly magical transforming power or influence ⟨electronic *wizardry*⟩ **b :** great skill or cleverness in an activity

¹wiz·en \ˈwi-zᵊn *also* ˈwē-\ *verb* **wiz·ened; wiz·en·ing** \ˈwiz-niŋ *also* ˈwēz-; ˈwi-zᵊn-iŋ *also* ˈwē-\ [Middle English *wisenen,* from Old English *wisnian;* akin to Old High German *wesanēn* to wither, Lithuanian *vysti*] (before 12th century)
intransitive verb
: to become dry, shrunken, and wrinkled often as a result of aging or of failing vitality
transitive verb
: to cause to wizen

²wizen *adjective* [alteration of *wizened*] (1786)
: that is wizened

woad \ˈwōd\ *noun* [Middle English *wod,* from Old English *wād;* akin to Old High German *weit* woad, Latin *vitrum*] (before 12th century)
: a European herb (*Isatis tinctoria*) of the mustard family formerly grown for the blue dyestuff yielded by its leaves; *also* **:** this dyestuff

¹wob·ble \ˈwä-bəl\ *verb* **wob·bled; wob·bling** \-b(ə-)liŋ\ [probably from Low German *wabbeln;* akin to Old English *wǣfre* restless — more at WAVER] (1657)
intransitive verb
1 a : to move or proceed with an irregular rocking or staggering motion or unsteadily and clumsily from side to side **b :** TREMBLE, QUAVER
2 : WAVER, VACILLATE
transitive verb
: to cause to wobble
— wob·bler \-b(ə-)lər\ *noun*
— wob·bli·ness \ˈwä-blē-nəs\ *noun*
— wob·bly \ˈwä-b(ə-)lē\ *adjective*

²wobble *noun* (1699)
1 a : a hobbling or rocking unequal motion (as of a wheel unevenly mounted) **b :** an uncertainly directed movement
2 : an intermittent variation (as in volume of sound)

Wob·bly \ˈwä-blē\ *noun, plural* **Wobblies** [origin unknown] (1914)
: a member of the Industrial Workers of the World

Wo·den \ˈwō-dᵊn\ *noun* [Old English *Wōden*]
: ODIN

wodge \ˈwäj\ *noun* [probably alteration of *wedge*] (1860)
chiefly British **:** a bulky mass or chunk **:** LUMP, WAD

¹woe \ˈwō\ *interjection* [Middle English *wa, wo,* from Old English *wā;* akin to Old Norse *vei,* interjection, woe, Latin *vae*] (before 12th century)
— used to express grief, regret, or distress

²woe *noun* (13th century)
1 : a condition of deep suffering from misfortune, affliction, or grief
2 : ruinous trouble **:** CALAMITY, AFFLICTION ⟨economic *woes*⟩
synonym see SORROW

woe·be·gone \ˈwō-bi-ˌgȯn *also* -ˌgän\ *adjective* [Middle English *wo begon,* from *wo,* noun + *begon,* past participle of *begon* to go about, beset, from Old English *begān,* from *be-* + *gān* to go — more at GO] (14th century)
1 : strongly affected with woe **:** WOEFUL
2 a : exhibiting great woe, sorrow, or misery ⟨a *woebegone* expression⟩ **b :** being in a sorry state ⟨*woebegone* tattered clothes⟩
— woe·be·gone·ness *noun*

woe·ful *also* **wo·ful** \ˈwō-fəl\ *adjective* (14th century)
1 : full of woe **:** GRIEVOUS ⟨*woeful* prophecies⟩
2 : involving or bringing woe
3 : lamentably bad or serious **:** DEPLORABLE ⟨*woeful* ignorance⟩
— woe·ful·ly \-f(ə-)lē\ *adverb*
— woe·ful·ness \-fəl-nəs\ *noun*

wog \ˈwäg, ˈwȯg\ *noun* [perhaps short for *golliwog*] (circa 1929)
chiefly British **:** a dark-skinned foreigner; *especially* **:** one from the Middle East or Far East — usually used disparagingly

wok \ˈwäk\ *noun* [Chinese (Guangdong) *wohk*] (1952)
: a bowl-shaped cooking utensil used especially in stir-frying

woke *past and past participle of* WAKE

woken *past participle of* WAKE

wold \ˈwōld\ *noun* [Middle English *wald, wold,* from Old English *weald, wald* forest; akin to Old High German *wald* forest, Old Norse *vǫllr* field] (before 12th century)
1 : a usually upland area of open country
2 *capitalized* **:** a hilly or rolling region — used in names of various English geographical areas ⟨Yorkshire *Wolds*⟩

¹wolf \ˈwu̇lf\ *noun, plural* **wolves** \ˈwu̇lvz\ *often attributive* [Middle English, from Old English *wulf;* akin to Old High German *wolf* wolf, Latin *lupus,* Greek *lykos*] (before 12th century)
1 *plural also* **wolf a :** any of various large predatory canids (genus *Canis*) that live and hunt in packs and resemble the related dogs; *especially* **:** GRAY WOLF — compare COYOTE, JACKAL **b :** the fur of a wolf
2 a (1) **:** a fierce, rapacious, or destructive person (2) **:** a man forward, direct, and zealous in amatory attentions to women **b :** dire poverty **:** STARVATION ⟨keep the *wolf* from the door⟩ **c :** the maggot of a warble fly
3 [German; from the howling sound] **a** (1) **:** dissonance in some chords on organs, pianos, or other instruments with fixed tones tuned by unequal temperament (2) **:** an instance of such dissonance **b :** a harshness due to faulty vibration in various tones in a bowed instrument
— wolf·like \ˈwu̇lf-ˌlīk\ *adjective*
— wolf in sheep's clothing : one who cloaks a hostile intention with a friendly manner

²wolf *transitive verb* (1862)
: to eat greedily **:** DEVOUR

wolf·ber·ry \ˈwu̇lf-ˌber-ē\ *noun* (circa 1834)

☆ SYNONYMS
Witty, humorous, facetious, jocular, jocose mean provoking or intended to provoke laughter. WITTY suggests cleverness and quickness of mind ⟨a *witty* remark⟩. HUMOROUS applies broadly to anything that evokes usually genial laughter and may contrast with *witty* in suggesting whimsicality or eccentricity ⟨*humorous* anecdotes⟩. FACETIOUS stresses a desire to produce laughter and may be derogatory in implying dubious or ill-timed attempts at wit or humor ⟨*facetious* comments⟩. JOCULAR implies a usually habitual fondness for jesting and joking ⟨a *jocular* fellow⟩. JOCOSE is somewhat less derogatory than FACETIOUS in suggesting habitual waggishness or playfulness ⟨*jocose* proposals⟩.

: a white-berried North American shrub (*Symphoricarpos occidentalis*) of the honeysuckle family

wolf dog *noun* (1652)
1 : any of various large dogs formerly kept for hunting wolves
2 : the offspring of a wolf and a domestic dog

wolf·er \'wu̇l-fər\ *noun* (1872)
: a hunter of wolves

wolff·ian body \'wu̇l-fē-ən-\ *noun, often W capitalized* [Kaspar Friedrich *Wolff*] (circa 1844)
: MESONEPHROS

Wolffian duct *noun* (1879)
: the duct of the mesonephros that persists in the female chiefly as part of a vestigial organ and in the male as the duct system leaving the testis and including the epididymis, vas deferens, and ejaculatory duct

wolf·fish \'wu̇lf-,fish\ *noun* (1569)
: any of several large marine blennies (genus *Anarhichas* of the family Anarhicadidae) having strong canine teeth in the front of the jaws and molar teeth on the sides and that feed chiefly on shellfish, starfish, and sea urchins

wolf·hound \'wu̇lf-,hau̇nd\ *noun* (1786)
: any of several large dogs used especially formerly in hunting large animals (as wolves)

wolf·ish \'wu̇l-fish\ *adjective* (1570)
1 : of or relating to wolves
2 a : suggestive of a wolf ⟨*wolfish* mongrel dogs —Hoffman Birney⟩ ⟨a *wolfish* and withdrawn youth —Marshall Frady⟩ **b :** befitting or characteristic of a wolf ⟨a *wolfish* appetite⟩
— **wolf·ish·ly** *adverb*
— **wolf·ish·ness** *noun*

wolf pack *noun* (1941)
: a group of submarines that make a coordinated attack on shipping

wol·fram \'wu̇l-frəm\ *noun* [German] (1757)
1 : WOLFRAMITE
2 : TUNGSTEN

wol·fram·ite \'wu̇l-frə-,mīt\ *noun* [German *Wolframit,* from *Wolfram*] (circa 1868)
: a mineral that consists of a tungstate of iron and manganese usually of a brownish or grayish black color and slightly metallic luster, occurs in monoclinic crystals and in granular or columnar masses, and is used as a source of tungsten

wolfs·bane \'wu̇lfs-,bān\ *noun* (1548)
: MONKSHOOD; *especially* **:** a highly variable yellow-flowered poisonous European herb (*Aconitum vulparia*)

wolf spider *noun* (1608)
: any of various active wandering ground spiders (family Lycosidae)

wolf whistle *noun* (1946)
: a distinctive whistle sounded by a boy or man to express sexual admiration for a girl or woman in his vicinity

wol·las·ton·ite \'wu̇-lə-stə-,nīt, 'wä-\ *noun* [William H. *Wollaston*] (1823)
: a triclinic mineral consisting of a native calcium silicate occurring usually in cleavable masses

Wo·lof \'wō-,lōf\ *noun* (1823)
: a Niger-Congo language of Senegambia

wol·ver·ine \,wu̇l-və-'rēn\ *noun, plural* **wolverines** [probably irregular from *wolv-* (as in *wolves*)] (1574)
1 *plural also* **wolverine a :** a carnivorous usually solitary mammal (*Gulo gulo*) of the weasel family of northern forests and associated tundra that is blackish with a light brown band on each side of the body and is noted especially for its strength **b :** the fur of the wolverine
2 *capitalized* **:** a native or resident of Michigan — used as a nickname

wolverine 1a

wom·an \'wu̇-mən, *especially Southern* 'wō- *or* 'wə-\ *noun, plural* **wom·en** \'wi-mən\ [Middle English, from Old English *wīfman,* from *wīf* woman, wife + *man* human being, man] (before 12th century)
1 a : an adult female person **b :** a woman belonging to a particular category (as by birth, residence, membership, or occupation) — usually used in combination ⟨council*woman*⟩
2 : WOMENKIND
3 : distinctively feminine nature **:** WOMANLINESS
4 : a female servant or personal attendant
5 a *chiefly dialect* **:** WIFE **b :** MISTRESS **c :** GIRLFRIEND 2 ◆
— **woman** *adjective*
— **wom·an·less** \-ləs\ *adjective*

wom·an·hood \-,hu̇d\ *noun* (14th century)
1 a : the state of being a woman **b :** the distinguishing character or qualities of a woman or of womankind
2 : WOMEN, WOMENKIND

wom·an·ise *British variant of* WOMANIZE

wom·an·ish \'wu̇-mə-nish\ *adjective* (14th century)
1 : characteristic of or suitable for a woman
2 : unsuitable to a man or to a strong character of either sex **:** EFFEMINATE ⟨*womanish* fears⟩
— **wom·an·ish·ly** *adverb*
— **wom·an·ish·ness** *noun*

wom·an·ize \'wu̇-mə-,nīz\ *verb* **-ized; -iz·ing** (1593)
transitive verb
: to make effeminate
intransitive verb
: to pursue multiple casual relationships with women
— **wom·an·iz·er** *noun*

wom·an·kind \'wu̇-mən-,kīnd\ *noun singular but singular or plural in construction* (14th century)
: WOMENKIND

¹wom·an·like \-,līk\ *adjective* (15th century)
: WOMANLY

²womanlike *adverb* (15th century)
: in the manner of a woman

wom·an·ly \-lē\ *adjective* (13th century)
1 : having qualities generally associated with a woman
2 : appropriate in character to a woman
— **wom·an·li·ness** *noun*

woman of letters (1818)
1 : a woman who is a scholar
2 : a woman who is an author

woman of the street (1928)
: PROSTITUTE

wom·an·pow·er \'wu̇-mən-,pau̇(-ə)r\ *noun* (1938)
: women available and prepared for work (as in industry or a particular line of endeavor)

woman's rights *noun plural* (1840)
1 : legal, political, and social rights for women equal to those of men
2 : FEMINISM 2

woman suffrage *noun* (1867)
: possession and exercise of suffrage by women

womb \'wüm\ *noun* [Middle English *wamb, womb,* from Old English; akin to Old High German *wamba* belly] (before 12th century)
1 : UTERUS
2 a : a cavity or space that resembles a womb in containing and enveloping **b :** a place where something is generated
— **wombed** \'wümd\ *adjective*

wom·bat \'wäm-,bat\ *noun* [Dharuk (Australian aboriginal language of the Port Jackson area) *wambat*] (1798)
: any of several stocky Australian marsupials (genera *Vombatus* and *Lasiorhinus* of the family Vombatidae) resembling small bears

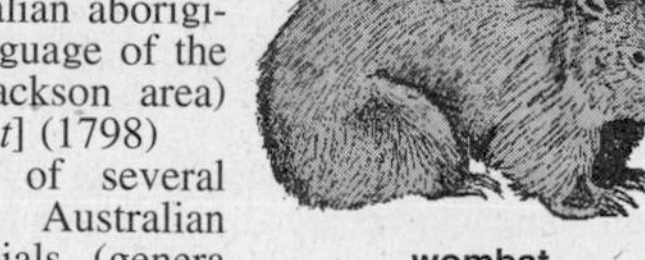
wombat

wom·en·folk \'wi-mən-,fōk\ *also* **wom·en·folks** \-,fōks\ *noun plural* (1833)
: WOMEN

wom·en·kind \-,kīnd\ *noun* (14th century)
: female human beings **:** women especially as distinguished from men

women's rights *noun plural* (1632)
: WOMAN'S RIGHTS

women's room *noun* (circa 1937)
: LADIES' ROOM

wom·mera \'wä-mə-rə\ *variant of* WOOMERA

¹won \'wən, 'wōn\ *intransitive verb* **wonned; won·ning** [Middle English, from Old English *wunian* — more at WONT] (before 12th century)
archaic **:** DWELL 2a, ABIDE 2

²won \'wən\ *past and past participle of* WIN

³won \'wȯn\ *noun, plural* **won** [Korean *wŏn*] (circa 1917)
— see MONEY table

¹won·der \'wən-dər\ *noun* [Middle English, from Old English *wundor;* akin to Old High German *wuntar* wonder] (before 12th century)
1 a : a cause of astonishment or admiration **:** MARVEL ⟨it's a *wonder* you weren't killed⟩ **b :** MIRACLE
2 : the quality of exciting amazed admiration
3 a : rapt attention or astonishment at something awesomely mysterious or new to one's experience **b :** a feeling of doubt or uncertainty

²wonder *verb* **won·dered; won·der·ing** \-d(ə-)riŋ\ (before 12th century)
intransitive verb
1 a : to be in a state of wonder **b :** to feel surprise
2 : to feel curiosity or doubt
transitive verb
: to be curious or in doubt about
— **won·der·er** \-dər-ər\ *noun*

³wonder *adjective* (12th century)
: WONDROUS, WONDERFUL: as **a :** exciting

◇ WORD HISTORY

woman *Woman* is descended from Old English *wīfman,* made up of *wīf* "woman" and *man* "human being." This seemingly redundant compound, with no counterpart in other Germanic languages, forms a pair with Old English *wæpnedman* or *wæpenman* "male human," literally, "weaponed man" or "weapon man," in which *wæpen* serves as a euphemism for the organ specific to males. Phrases such as *bitweone wepmon & wimmane* "between man and woman" are found in Middle English, but only up to about 1300, after which *wepman* is entirely supplanted by simple *man* in the sense "male human." *Wifman* lasted, however, becoming in Middle English *wimman,* with the \v\ sound (written *f*) of the Old English word blended into the following \m\ and the vowel of the first syllable shortened. The Modern English plural \'wi-mən\ perpetuates this vowel, while a further change, in which the initial *i* was altered to *u* under influence of the preceding \w\ sound, resulted in Middle English *wumman,* continued by the Modern English singular \'wu̇-mən\. Both the *i* and the *u* variants were also lengthened, ultimately producing early Modern English variants with *wee-* and *wo-* in the first syllable. The latter has provided our spelling—though not our pronunciation—for both singular *woman* and plural *women.*

amazement or admiration **b :** effective or efficient far beyond anything previously known or anticipated
wonder drug *noun* (1939)
: MIRACLE DRUG
won·der·ful \'wən-dər-fəl\ *adjective* (before 12th century)
1 : exciting wonder **:** MARVELOUS, ASTONISHING ⟨a sight *wonderful* to behold⟩
2 : unusually good **:** ADMIRABLE
— won·der·ful·ly \-f(ə-)lē\ *adverb*
— won·der·ful·ness \-fəl-nəs\ *noun*
won·der·land \'wən-dər-ˌland, -lənd\ *noun* (1790)
1 : an imaginary place of delicate beauty or magical charm
2 : a place that excites admiration or wonder
won·der·ment \-mənt\ *noun* (1535)
1 : ASTONISHMENT, SURPRISE
2 : a cause of or occasion for wonder
3 : curiosity about something
won·der·work \-də(r)-ˌwərk\ *noun* (before 12th century)
: a marvelous act, work, or accomplishment
won·der–work·er \-ˌwər-kər\ *noun* (1599)
: one that performs wonders
won·der–work·ing \-kiŋ\ *adjective* (1594)
: producing wonders
won·drous \'wən-drəs\ *adjective* [Middle English, alteration of *wonders,* from genitive of [1]*wonder*] (15th century)
: that is to be marveled at **:** EXTRAORDINARY
— wondrous *adverb, archaic*
— won·drous·ly *adverb*
— won·drous·ness *noun*
wonk \'wäŋk, 'wȯŋk\ *noun* [origin unknown] (1954)
: NERD
won·ky \'wäŋ-kē\ *adjective* [probably alteration of English dialect *wankle,* from Middle English *wankel,* from Old English *wancol;* akin to Old High German *wankōn* to totter — more at WENCH] (1919)
1 *British* **:** UNSTEADY, SHAKY
2 *British* **:** AWRY, WRONG
[1]**wont** \'wȯnt, 'wōnt *also* 'wənt, 'wänt\ *adjective* [Middle English *woned, wont,* from past participle of *wonen* to dwell, be used to, from Old English *wunian;* akin to Old High German *wonēn* to dwell, be used to, Sanskrit *vanoti* he strives for — more at WIN] (before 12th century)
1 : ACCUSTOMED, USED ⟨got up early as he is *wont* to do⟩
2 : INCLINED, APT ⟨revealing as letters are *wont* to be —Gladys M. Wrigley⟩
[2]**wont** *verb* **wont; wont** *or* **wont·ed; wont·ing** (15th century)
transitive verb
: ACCUSTOM, HABITUATE
intransitive verb
: to have the habit of doing something
[3]**wont** *noun* (1530)
: habitual way of doing **:** USE
synonym see HABIT
won't \'wōnt; *New England, upstate New York, northern Pennsylvania* ˌwənt, 'wənt; *greater New York City* 'wünt; *eastern South Carolina* 'wünt, 'wu̇nt\ (1652)
: will not
wont·ed \'wȯn-təd, 'wōn- *also* 'wən- *or* 'wän-\ *adjective* (15th century)
: usual or ordinary especially by reason of established habit ⟨flashed his *wonted* grin⟩
synonym see USUAL
— wont·ed·ly *adverb*
— wont·ed·ness *noun*
won·ton \'wän-ˌtän\ *noun* [Chinese (Guangdong) *wàhn-tān*] (1934)
: filled pockets of noodle dough served boiled in soup or fried
woo \'wü\ *verb* [Middle English *wowen,* from Old English *wōgian*] (before 12th century)
transitive verb
1 : to sue for the affection of and usually marriage with **:** COURT
2 : to solicit or entreat especially with importunity
3 : to seek to gain or bring about
intransitive verb
: to court a woman
— woo·er *noun*
[1]**wood** \'wu̇d, 'wōd, 'wu̇d\ *adjective* [Middle English, from Old English *wōd* insane; akin to Old High German *wuot* madness — more at VATIC] (before 12th century)
archaic **:** violently mad
[2]**wood** \'wu̇d\ *noun* [Middle English *wode,* from Old English *widu, wudu;* akin to Old High German *witu* wood, Old Irish *fid* tree] (before 12th century)
1 a : a dense growth of trees usually greater in extent than a grove and smaller than a forest — often used in plural but singular or plural in construction **b :** WOODLAND
2 a : the hard fibrous substance consisting basically of xylem that makes up the greater part of the stems, branches, and roots of trees or shrubs beneath the bark and is found to a limited extent in herbaceous plants **b :** wood suitable or prepared for some use (as burning or building)
3 a : something made of wood **b :** a golf club having a thick wooden head; *also* **:** a golf club having a similar head made of metal
— out of the woods : clear of danger or difficulty
[3]**wood** \'wu̇d\ *adjective* (1538)
1 : WOODEN
2 : suitable for cutting or working with wood ⟨a *wood* saw⟩
3 *or* **woods** \'wu̇dz\ **:** living, growing, or existing in woods ⟨*woods* trails⟩
[4]**wood** \'wu̇d\ (1630)
intransitive verb
: to gather or take on wood
transitive verb
: to cover with a growth of trees or plant with trees
wood alcohol *noun* (1861)
: METHANOL
wood anemone *noun* (1657)
: any of several anemones; *especially* **:** one (*Anemone quinquefolia*) of North America with solitary often pink-tinged flowers
wood bet·o·ny \-'be-tᵊn-ē\ *noun* [Middle English *betone,* from Old French *betoine,* from Latin *vettonica, betonica,* from *Vettones,* an ancient people inhabiting the Iberian peninsula] (1657)
: a lousewort (*Pedicularis canadensis*) of eastern North America with yellow or reddish flowers in bracted spikes
wood·bine \'wu̇d-ˌbīn\ *noun* [Middle English *wodebinde,* from Old English *wudubinde,* from *wudu* wood + *bindan* to tie, bind; from its winding around trees] (before 12th century)
1 : any of several honeysuckles; *especially* **:** a Eurasian twining shrub (*Lonicera periclymenum*)
2 : VIRGINIA CREEPER
wood·block \-ˌbläk\ *noun* (1837)
: WOODCUT
— wood–block *adjective*
wood–bor·ing \-ˌbōr-iŋ, -ˌbȯr-\ *adjective* (1815)
: excavating galleries in wood in feeding or in constructing a nest — used chiefly of an insect
wood carving *noun* (1847)
: the art of fashioning or ornamenting objects of wood by cutting with a sharp handheld implement; *also* **:** an object of wood so fashioned or ornamented
— wood–carv·er \-ˌkär-vər\ *noun*
wood·chat shrike \'wu̇d-ˌchat-\ *noun* (1781)
: a European shrike (*Lanius senator*) — called also *woodchat*
wood·chop·per \-ˌchä-pər\ *noun* (1779)
: one engaged especially in chopping down trees
wood·chuck \-ˌchək\ *noun* [by folk etymology from a word of Algonquian origin; akin to Narraganset *ockqutchaun* woodchuck] (1674)
: a grizzled thickset marmot (*Marmota monax*) chiefly of Alaska, Canada, and the northeastern U.S. — called also *groundhog*
wood·cock \'wu̇d-ˌkäk\ *noun, plural* **wood·cocks** (before 12th century)
1 *or plural* **woodcock :** a widespread Old World woodland bird (*Scolopax rusticola*) that is related to the sandpipers and snipes; *also* **:** a smaller related game bird (*Scolopax minor* synonym *Philohela minor*) of eastern North America
2 [from the ease with which the woodcock is snared] *archaic* **:** SIMPLETON
wood·craft \-ˌkraft\ *noun* (14th century)
1 : skill and practice in anything relating to the woods and especially in maintaining oneself and making one's way in the woods
2 : skill in shaping or constructing articles from wood
wood·cut \-ˌkət\ *noun* (1662)
1 : a relief printing surface consisting of a wooden block with a usually pictorial design cut with the grain
2 : a print from a woodcut
wood·cut·ter \-ˌkə-tər\ *noun* (1761)
: one that cuts wood
wood·cut·ting \-ˌkə-tiŋ\ *noun* (1683)
1 : the action or occupation of cutting wood or timber
2 : the producing of woodcuts
wood duck *noun* (1777)
: a showy American duck (*Aix sponsa*) which nests in tree cavities and the males of which have a large crest and iridescent plumage varied with green, purple, black, white, and chestnut

wood duck

wood ear *noun* (1917)
: TREE EAR
wood·ed \'wu̇-dəd\ *adjective* (1605)
: covered with growing trees
wood·en \'wu̇-dᵊn\ *adjective* (1538)
1 : made or consisting of wood
2 : lacking ease or flexibility **:** awkwardly stiff
— wood·en·ly *adverb*
— wood·en·ness \-dᵊn-(n)əs\ *noun*
wood engraving *noun* (1816)
1 : a relief printing surface consisting of a wooden block with a usually pictorial design cut in the end grain
2 : a print from a wood engraving
wood·en·head \'wu̇-dᵊn-ˌhed\ *noun* (1831)
: BLOCKHEAD
wood·en·head·ed \ˌwu̇-dᵊn-'he-dəd\ *adjective* (circa 1854)
: DENSE, STUPID
wooden Indian *noun* (1879)
: a wooden image of a standing American Indian brave used especially formerly as a sign for a cigar store
wood·en·ware \'wu̇-dᵊn-ˌwar, -ˌwer\ *noun* (1647)
: articles made of wood for domestic use
wood fiber *noun* (1875)
: any of various fibers in or associated with xylem
[1]**wood·land** \'wu̇d-lənd, -ˌland\ *noun* (before 12th century)
: land covered with woody vegetation **:** TIMBERLAND, FOREST
— wood·land·er \-lən-dər, -ˌlan-\ *noun*
[2]**woodland** *adjective* (14th century)
1 : growing, living, or existing in woodland
2 : of, relating to, or being woodland
wood·lore \'wu̇d-ˌlōr, -ˌlȯr\ *noun* (1918)
: knowledge of the woods
wood·lot \-ˌlät\ *noun* (1643)

: a restricted area of woodland usually privately maintained as a source of fuel, posts, and lumber

wood louse *noun* (1611)
: a terrestrial isopod crustacean (suborder Oniscoidea) with a flattened elliptical body often capable of being rolled into a ball — called also *pill bug, sow bug*

wood•man \ˈwu̇d mən\ *noun* (15th century)
1 : WOODSMAN
2 *capitalized* [Modern *Woodmen* of America & *Woodmen* of the World] **:** a member of either of two independent benevolent and fraternal societies

wood•note \-ˌnōt\ *noun* [from its likeness to the call of a bird in the woods] (1632)
: verbal expression that is natural and artless

wood nymph *noun* (1577)
: a nymph living in woods — called also *dryad*

wood•peck•er \ˈwu̇d-ˌpe-kər\ *noun* (circa 1530)
: any of numerous birds (family Picidae) with zygodactyl feet, stiff spiny tail feathers used in climbing or resting on tree trunks, a usually extensile tongue, a very hard bill used to drill the bark or wood of trees for insect food or to excavate nesting cavities, and generally showy parti-colored plumage

wood•pile \-ˌpīl\ *noun* (circa 1552)
: a pile of wood (as firewood)
— in the woodpile : doing or responsible for covert mischief

wood pulp *noun* (1866)
: pulp from wood used in making cellulose derivatives (as paper or rayon)

wood pussy *noun* (circa 1899)
: SKUNK

wood rat *noun* (1763)
: any of numerous cricetid rodents (especially genus *Neotoma*) of North and Central America with soft fur, well-furred tails, and large ears

wood ray *noun* (1925)
: XYLEM RAY

wood•ruff \ˈwu̇d-(ˌ)rəf\ *noun* [Middle English *woderove,* from Old English *wudurofe,* from *wudu* wood + *-rofe* (of unknown origin)] (before 12th century)
1 : any of a genus (*Asperula*) of Old World herbs of the madder family
2 : a small Eurasian and North African sweet-scented herb (*Galium odoratum* synonym *Asperula odorata*) used in perfumery and for flavoring wine

[1]wood•shed \-ˌshed\ *noun* (1844)
1 : a shed for storing wood and especially firewood
2 : a place, means, or session for administering discipline

[2]woodshed *intransitive verb* **-shed•ded; -shed•ding** [probably from the former use of woodsheds for private practicing] (1936)
: PRACTICE; *especially* **:** to practice on a musical instrument

wood shot *noun* (1927)
1 : a golf shot played with a wood
2 : a stroke in a racket game in which the ball or shuttlecock is hit with the frame of the racket rather than the strings

woods•man \ˈwu̇dz-mən\ *noun* (1688)
: a person who frequents or works in the woods; *especially* **:** one skilled in woodcraft

wood sorrel *noun* (1525)
: any of a genus (*Oxalis* of the family Oxalidaceae, the wood-sorrel family) of herbs with acid sap, compound leaves, and regular flowers; *especially* **:** either of two stemless herbs (*O. montana* of North America and *O. acetosella* of Eurasia) with trifoliolate leaves

wood spirit *noun* (1842)
: METHANOL

wood stork *noun* (1884)
: a white stork (*Mycteria americana*) with black wing flight feathers and tail that frequents wooded swamps from the southeastern U.S. to Argentina — called also *wood ibis*

wood•stove \ˈwu̇d-ˌstōv\ *noun* (1847)
: a stove that uses wood for fuel

wood sugar *noun* (circa 1900)
: XYLOSE

woodsy \ˈwu̇d-zē\ *adjective* (1860)
: characteristic or suggestive of woods

wood tar *noun* (1857)
: tar obtained by the destructive distillation of wood either as a deposit from pyroligneous acid or as a residue from the distillation of the acid or of wood turpentine

wood tick *noun* (1668)
: any of several ixodid ticks: as **a :** a widely distributed tick (*Dermacentor andersonii*) of western North America that is a vector of Rocky Mountain spotted fever **b :** AMERICAN DOG TICK

wood turning *noun* (circa 1876)
: the art or process of fashioning wooden pieces or blocks into various forms and shapes by means of a lathe

wood turpentine *noun* (circa 1909)
: TURPENTINE 2b

wood warbler *noun* (1817)
: WARBLER 2b

wood•wind \ˈwu̇d-ˌwind\ *noun* (1876)
1 : any of a group of wind instruments (as a clarinet, flute, oboe, or saxophone) that are characterized by a cylindrical or conical tube of wood or metal usually ending in a slightly flared bell, that produce tones by the vibration of one or two reeds in the mouthpiece or by the passing of air over a mouth hole, and that usually have finger holes or keys by which the player may produce all the tones within an instrument's range
2 *plural* **:** the woodwind section of a band or orchestra

wood•work \-ˌwərk\ *noun* (1650)
1 : work made of wood; *especially* **:** interior fittings (as moldings or stairways) of wood
2 : a place or state of concealment, seclusion, or anonymity ⟨witnesses came out of the *woodwork* when a reward was offered⟩

[1]wood•work•ing \-ˌwər-kiŋ\ *adjective* (1872)
: used for woodworking

[2]woodworking *noun* (1875)
: the act, process, or occupation of working wood into a useful or desired form
— wood•work•er \-kər\ *noun*

wood•worm \ˈwu̇d-ˌwərm\ *noun* (1725)
: an insect larva (as of a furniture beetle) that bores especially in dead wood; *also* **:** an infestation of woodworms

[1]woody \ˈwu̇-dē\ *adjective* **wood•i•er; -est** (14th century)
1 : abounding or overgrown with woods
2 a : of or containing wood or wood fibers **:** LIGNEOUS ⟨*woody* tissues⟩ **b :** having woody parts **:** rich in xylem and associated structures ⟨*woody* plants⟩
3 : characteristic of or suggestive of wood ⟨wine with a *woody* flavor⟩
— wood•i•ness *noun*

[2]woody *or* **wood•ie** \ˈwu̇-dē\ *noun, plural* **woodies** [alteration of [3]*wood*] (1961)
: a wood-paneled station wagon

[1]woof \ˈwu̇f, ˈwüf\ *noun* [alteration of Middle English *oof,* from Old English *ōwef,* from *ō-* (from *on*) + *wefan* to weave — more at WEAVE] (before 12th century)
1 a : WEFT 1a **b :** woven fabric; *also* **:** the texture of such a fabric
2 : a basic or essential element or material

[2]woof \ˈwu̇f\ *intransitive verb* [imitative] (1804)
1 : to make the low gruff sound typically produced by a dog
2 : to express oneself in a usually stylized boastful, aggressive, or exaggeratedly deceitful manner

[3]woof *noun* (1839)
1 : a low gruff sound typically produced by a dog
2 : a low note emitted by sound reproducing equipment

woof•er \ˈwu̇-fər\ *noun* (1935)
: a loudspeaker usually larger than a tweeter, responsive only to the lower acoustic frequencies, and used for reproducing sounds of low pitch — compare TWEETER

wool \ˈwu̇l\ *noun, often attributive* [Middle English *wolle,* from Old English *wull;* akin to Old High German *wolla* wool, Latin *vellus* fleece, *lana* wool] (before 12th century)
1 : the soft wavy or curly hypertrophied undercoat of various hairy mammals and especially the sheep made up of a matrix of keratin fibers and covered with minute scales
2 : a product of wool; *especially* **:** a woven fabric or garment of such fabric
3 a : a dense felted pubescence especially on a plant **:** TOMENTUM **b :** a filamentous mass — usually used in combination; compare MINERAL WOOL, STEEL WOOL

wooled *also* **woolled** \ˈwu̇ld\ *adjective* (circa 1859)
: having wool especially of a specified kind — used in combination ⟨long-*wooled*⟩

[1]wool•en *or* **wool•len** \ˈwu̇-lən\ *adjective* (before 12th century)
1 : made of wool
2 : of or relating to the manufacture or sale of woolen products ⟨*woolen* mills⟩ ⟨the *woolen* industry⟩

[2]woolen *or* **woollen** *noun* (14th century)
1 : a fabric made of wool and especially of woolen yarns having a fuzzy or napped face (as for use in clothing or blankets) — compare WORSTED
2 : garments of woolen fabric — usually used in plural

wool fat *noun* (1875)
: wool grease especially after refining **:** LANOLIN

wool–gath•er \ˈwu̇l-ˌga-t͟hər, -ˌge-t͟hər\ *intransitive verb* (1850)
: to indulge in woolgathering
— wool•gath•er•er \-t͟hər-ər\ *noun*

wool•gath•er•ing \-ˌga-t͟h(ə-)riŋ, -ˌge-t͟h(ə-)riŋ\ *noun* (1553)
: indulgence in idle daydreaming

wool grease *noun* (1875)
: a fatty slightly sticky wax coating the surface of the fibers of sheep's wool — compare WOOL FAT

[1]wool•ly *also* **wooly** \ˈwu̇-lē\ *adjective* **wool•li•er; -est** (1578)
1 a : of, relating to, or bearing wool **b :** resembling wool
2 a : lacking in clearness or sharpness of outline ⟨a *woolly* TV picture⟩ **b :** marked by mental confusion ⟨*woolly* thinking⟩
3 : marked by boisterous roughness or lack of order or restraint ⟨where the West is still *woolly* —Paul Schubert⟩ — used especially in the phrase *wild and woolly*
— wool•li•ly \-lə-lē\ *adverb*
— wool•li•ness *noun*

[2]wool•ly *also* **wool•ie** *or* **wooly** \ˈwu̇-lē\ *noun, plural* **wool•lies** *also* **wool•ies** (circa 1865)
1 : a garment made from wool; *especially* **:** underclothing of knitted wool — usually used in plural
2 *West & Australian* **:** SHEEP

woolly aphid *noun* (1812)
: any of several plant lice (especially genus *Eriosoma*) covered with a dense coat of white filaments — called also *woolly aphis*

woolly bear *noun* (circa 1841)
: any of various rather large very hairy moth caterpillars; *especially* **:** one of a tiger moth

wool•ly–head•ed \ˌwu̇-lē-ˈhe-dəd\ *adjective* (1650)
1 : having hair suggesting wool

2 : marked by vague or confused perception or thinking

woolly mammoth *noun* (1933)
: a heavy-coated mammoth (*Mammuthus primigenius*) of the colder parts of the northern hemisphere

wool·pack \'wu̇l-ˌpak\ *noun* (14th century)
1 : a wrapper of strong fabric into which fleeces are packed for shipment
2 : the complete package of wool and wrapper

wool·sack \-ˌsak\ *noun* (14th century)
1 *archaic* **:** WOOLPACK 2
2 : a cushion that is the official seat of the Lord Chancellor or his deputy in presiding over the House of Lords

wool·shed \-ˌshed\ *noun* (1850)
: a building or range of buildings (as on an Australian sheep station) in which sheep are sheared and wool is prepared for market

wool·skin \-ˌskin\ *noun* (15th century)
: a sheepskin having the wool still on it

wool·sort·er's disease \'wu̇l-ˌsȯr-tərz-\ *noun* (1880)
: pulmonary anthrax resulting especially from inhalation of bacterial spores (*Bacillus anthracis*) from contaminated wool or hair

wool stapler *noun* (1709)
: a dealer in wool

woom·era \'wu̇-mə-rə\ *noun* [Dharvk (Australian aboriginal language of the Port Jackson area) *wumera*] (1817)
: a wooden rod with a hooked end used by Australian aborigines for throwing a spear

woops \'(w)u̇(ə)ps\ *variant of* OOPS

woo·zy \'wü-zē, 'wu̇-\ *adjective* **woo·zi·er; -est** [origin unknown] (1897)
1 : mentally unclear or hazy ⟨seems a little *woozy*, not quite knowing what to say —J. A. Lukacs⟩
2 : affected with dizziness, mild nausea, or weakness
— woo·zi·ly *adverb*
— woo·zi·ness *noun*

wop \'wäp\ *noun, often capitalized* [Italian dialect *guappo* swaggerer, tough, from Spanish *guapo*, probably from Middle French dialect *vape, wape* weak, insipid, from Latin *vappa* wine gone flat] (1908)
: ITALIAN — usually used disparagingly

Worces·ter \'wu̇s-tər\ *noun* (1802)
: low-fired porcelain containing a frit and steatite produced at Worcester, England, from about 1751 — called also *Worcester china, Worcester porcelain*

Worces·ter·shire sauce \'wu̇s-tə(r)-ˌshir-, -shər- *also* -ˌshīr-\ *noun* [*Worcestershire*, England, where it was originally made] (1843)
: a pungent sauce whose ingredients include soy, vinegar, and garlic

¹word \'wərd\ *noun* [Middle English, from Old English; akin to Old High German *wort* word, Latin *verbum*, Greek *eirein* to say, speak, Hittite *weriya-* to call, name] (before 12th century)
1 a : something that is said **b** *plural* (1) **:** TALK, DISCOURSE ⟨putting one's feelings into *words*⟩ (2) **:** the text of a vocal musical composition **c :** a brief remark or conversation ⟨would like to have a *word* with you⟩
2 a (1) **:** a speech sound or series of speech sounds that symbolizes and communicates a meaning without being divisible into smaller units capable of independent use (2) **:** the entire set of linguistic forms produced by combining a single base with various inflectional elements without change in the part of speech elements **b** (1) **:** a written or printed character or combination of characters representing a spoken word ⟨the number of *words* to a line⟩ — sometimes used with the first letter of a real or pretended taboo word prefixed often as a humorous euphemism ⟨we were not afraid to use the d *word* and talk about death —Erma Bombeck⟩ ⟨the first man to utter the f *word* on British TV —*Time*⟩ (2) **:** any segment of written or printed discourse ordinarily appearing between spaces or between a space and a punctuation mark **c :** a number of bytes processed as a unit and conveying a quantum of information in communication and computer work
3 : ORDER, COMMAND ⟨don't move till I give the *word*⟩
4 *often capitalized* **a :** LOGOS **b :** GOSPEL 1a **c :** the expressed or manifested mind and will of God
5 a : NEWS, INFORMATION ⟨sent *word* that he would be late⟩ **b :** RUMOR
6 : the act of speaking or of making verbal communication
7 : SAYING, PROVERB
8 : PROMISE, DECLARATION ⟨kept her *word*⟩
9 : a quarrelsome utterance or conversation — usually used in plural ⟨they had *words* and parted⟩
10 : a verbal signal **:** PASSWORD
— good word 1 : a favorable statement ⟨put in a *good word* for me⟩ **2 :** good news ⟨what's the *good word*⟩
— in a word : in short
— in so many words 1 : in exactly those terms ⟨implied that such actions were criminal but did not say so *in so many words*⟩ **2 :** in plain forthright language ⟨*in so many words*, she wasn't fit to be seen —Jean Stafford⟩
— of few words : not inclined to say more than is necessary **:** LACONIC
— of one's word : that can be relied on to keep a promise — used only after *man* or *woman*
— upon my word : with my assurance **:** INDEED, ASSUREDLY ⟨*upon my word*, I've never heard of such a thing⟩

²word (13th century)
intransitive verb
archaic **:** SPEAK
transitive verb
: to express in words **:** PHRASE

word·age \'wər-dij\ *noun* (1829)
1 a : WORDS **b :** VERBIAGE 1
2 : the number or quantity of words
3 : WORDING

word–association test *noun* (1946)
: a test of personality and mental function in which the subject is required to respond to each of a series of words with the first one that comes to mind or with one of a specified class of words

word·book \'wərd-ˌbu̇k\ *noun* (1598)
: VOCABULARY, DICTIONARY

word class *noun* (1914)
: a linguistic form class whose members are words; *especially* **:** PART OF SPEECH

word–for–word *adjective* (1611)
: being in the exact words **:** VERBATIM

word for word *adverb* (14th century)
: in the exact words **:** VERBATIM

word–hoard \'wərd-ˌhōrd, -ˌhȯrd\ *noun* [translation of Old English *wordhord*] (circa 1869)
: a supply of words **:** VOCABULARY

word·ing \'wər-diŋ\ *noun* (1649)
: the act or manner of expressing in words **:** PHRASEOLOGY

word·less \'wərd-ləs\ *adjective* (15th century)
1 : not expressed in or accompanied by words
2 : SILENT, SPEECHLESS
— word·less·ly *adverb*
— word·less·ness *noun*

word·mon·ger \-ˌməŋ-gər, -ˌmäŋ-\ *noun* (1590)
: a writer who uses words for show or without particular regard for meaning

word–mon·ger·ing \-g(ə-)riŋ\ *noun* (1879)
: the use of empty or bombastic words

word–of–mouth \ˌwərd-ə(v)-'mau̇th\ *adjective* (1802)
: orally communicated; *also* **:** generated from or reliant on oral publicity ⟨*word-of-mouth* customers⟩ ⟨a *word-of-mouth* business⟩

word of mouth (1553)
: oral communication; *especially* **:** oral often inadvertent publicity

word order *noun* (1892)
: the order or arrangement of words in a phrase, clause, or sentence

word·play \'wərd-ˌplā\ *noun* (1855)
: verbal wit

word processing *noun* (1970)
: the production of typewritten documents (as business letters) with automated and usually computerized typing and text-editing equipment
— word process *verb*

word processor *noun* (1970)
: a keyboard-operated terminal usually with a video display and a magnetic storage device for use in word processing; *also* **:** software (as for a computer system) to perform word processing

word·smith \'wərd-ˌsmith\ *noun* (1896)
: a person who works with words; *especially* **:** a skillful writer
— word·smith·ery \-ˌsmi-thə-rē\ *noun*

word square *noun* (circa 1879)
: a series of words of equal length arranged in a square pattern to read the same horizontally and vertically

word stress *noun* (circa 1914)
: the manner in which stresses are distributed on the syllables of a word — called also *word accent*

wordy \'wər-dē\ *adjective* **word·i·er; -est** (12th century)
1 : using or containing many and usually too many words
2 : of or relating to words **:** VERBAL ☆
— word·i·ly \'wər-dᵊl-ē\ *adverb*
— word·i·ness \'wər-dē-nəs\ *noun*

wore *past of* WEAR

¹work \'wərk\ *noun* [Middle English *werk, work*, from Old English *werc, weorc;* akin to Old High German *werc*, Greek *ergon*, Avestan *varəzem* activity] (before 12th century)
1 : activity in which one exerts strength or faculties to do or perform something: **a :** sustained physical or mental effort to overcome obstacles and achieve an objective or result **b :** the labor, task, or duty that is one's accustomed means of livelihood **c :** a specific task, duty, function, or assignment often being a part or phase of some larger activity
2 a : energy expended by natural phenomena **b :** the result of such energy ⟨sand dunes are the *work* of sea and wind⟩ **c :** the transference of energy that is produced by the motion of the point of application of a force and is measured by multiplying the force and the displacement of its point of application in the line of action
3 a : something that results from a particular manner or method of working, operating, or devising ⟨careful police *work*⟩ ⟨clever camera *work*⟩ **b :** something that results from the use or fashioning of a particular material ⟨porcelain *work*⟩
4 a : a fortified structure (as a fort, earthen barricade, or trench) **b** *plural* **:** structures in engineering (as docks, bridges, or embankments) or mining (as shafts or tunnels)
5 *plural but singular or plural in construction* **:** a place where industrial labor is carried on **:** PLANT, FACTORY

☆ SYNONYMS
Wordy, verbose, prolix, diffuse mean using more words than necessary to express thought. WORDY may also imply loquaciousness or garrulity ⟨a *wordy* speech⟩. VERBOSE suggests a resulting dullness, obscurity, or lack of incisiveness or precision ⟨the *verbose* position papers⟩. PROLIX suggests unreasonable and tedious dwelling on details ⟨habitually transformed brief anecdotes into *prolix* sagas⟩. DIFFUSE stresses lack of compactness and pointedness of style ⟨*diffuse* memoirs that are so many shaggy-dog stories⟩.

6 *plural* **:** the working or moving parts of a mechanism ⟨*works* of a clock⟩
7 a : something produced or accomplished by effort, exertion, or exercise of skill ⟨this book is the *work* of many hands⟩ **b :** something produced by the exercise of creative talent or expenditure of creative effort **:** artistic production
8 *plural* **:** performance of moral or religious acts ⟨salvation by *works*⟩
9 a : effective operation **:** EFFECT, RESULT ⟨wait for time to do its healing *work*⟩ **b :** manner of working **:** WORKMANSHIP, EXECUTION
10 : the material or piece of material that is operated upon at any stage in the process of manufacture
11 *plural* **a :** everything possessed, available, or belonging ⟨the whole *works,* rod, reel, tackle box, went overboard⟩ ⟨ordered pizza with the *works*⟩ **b :** subjection to drastic treatment **:** all possible abuse — usually used with *get* ⟨get the *works*⟩ or *give* ⟨gave them the *works*⟩
☆ ☆
— at work 1 : engaged in working **:** BUSY; *especially* **:** engaged in one's regular occupation **2 :** having effect **:** OPERATING, FUNCTIONING
— in the works : in process of preparation, development, or completion
— in work 1 : in process of being done **2** *of a horse* **:** in training
— out of work : without regular employment **:** JOBLESS

[2]work *adjective* (14th century)
1 : used for work ⟨*work* elephant⟩
2 : suitable or styled for wear while working ⟨*work* clothes⟩
3 : involving or engaged in work ⟨*work* gang⟩ ⟨*work* hours⟩

[3]work *verb* **worked** \ˈwərkt\ *or* **wrought** \ˈrȯt\; **work·ing** [Middle English *werken, worken,* from Old English *wyrcan;* akin to Old English *weorc*] (before 12th century)
transitive verb
1 : to bring to pass **:** EFFECT ⟨*work* miracles⟩
2 a : to fashion or create a useful or desired product by expending labor or exertion on **:** FORGE, SHAPE ⟨*work* flint into tools⟩ **b :** to make or decorate with needlework; *especially* **:** EMBROIDER
3 a : to prepare for use by stirring or kneading **b :** to bring into a desired form by a gradual process of cutting, hammering, scraping, pressing, or stretching ⟨*work* cold steel⟩
4 : to set or keep in motion, operation, or activity **:** cause to operate or produce ⟨a pump *worked* by hand⟩ ⟨*work* farmland⟩
5 : to solve (a problem) by reasoning or calculation — often used with *out*
6 a : to cause to toil or labor ⟨*worked* their horses nearly to death⟩ **b :** to make use of **:** EXPLOIT **c :** to control or guide the operation of ⟨switches are *worked* from a central tower⟩
7 a : to carry on an operation or perform a job through, at, in, or along ⟨the salespeople *worked* both sides of the street⟩ ⟨a sportscaster hired to *work* the game⟩ **b :** to greet and talk with in a friendly way in order to ingratiate oneself or achieve a purpose ⟨politicians *working* the crowd⟩ ⟨*worked* the room⟩
8 : to pay for or achieve with labor or service ⟨*worked* my way through college⟩ ⟨*worked* my way up in the company⟩
9 a : to get (oneself or an object) into or out of a condition or position by gradual stages **b :** CONTRIVE, ARRANGE ⟨we can *work* it so that you can take your vacation⟩
10 a : to practice trickery or cajolery on for some end ⟨*worked* the management for a free ticket⟩ **b :** EXCITE, PROVOKE ⟨*worked* myself into a rage⟩
intransitive verb
1 a : to exert oneself physically or mentally especially in sustained effort for a purpose or under compulsion or necessity **b :** to perform or carry through a task requiring sustained effort or continuous repeated operations ⟨*worked* all day over a hot stove⟩ **c :** to perform work or fulfill duties regularly for wages or salary
2 : to function or operate according to plan or design ⟨hinges *work* better with oil⟩
3 : to exert an influence or tendency
4 : to produce a desired effect or result **:** SUCCEED
5 a : to make way slowly and with difficulty **:** move or progress laboriously ⟨*worked* up to the presidency⟩ **b :** to sail to windward
6 : to permit of being worked **:** react in a specified way to being worked ⟨this wood *works* easily⟩
7 a : to be in agitation or restless motion **b :** FERMENT 1 **c :** to move slightly in relation to another part **d :** to get into a specified condition by slow or imperceptible movements ⟨the knot *worked* loose⟩
usage see WRECK
— work on 1 : AFFECT ⟨*worked on* my sympathies⟩ **2 :** to strive to influence or persuade
— work upon : to have effect upon **:** operate on **:** INFLUENCE

work·able \ˈwər-kə-bəl\ *adjective* (1545)
1 : capable of being worked
2 : PRACTICABLE, FEASIBLE ⟨a *workable* system⟩
— work·abil·i·ty \ˌwər-kə-ˈbi-lə-tē\ *noun*
— work·able·ness \ˈwər-kə-bəl-nəs\ *noun*

work·a·day \ˈwər-kə-ˌdā\ *adjective* [alteration of earlier *workyday,* from obsolete *workyday,* noun workday] (1554)
1 : of, relating to, or suited for working days
2 : ORDINARY, PROSAIC

work·a·hol·ic \ˌwər-kə-ˈhȯ-lik, -ˈhä-\ *noun* [*work* + *-aholic,* alteration of *-oholic* (as in *alcoholic*)] (1968)
: a compulsive worker
— workaholic *adjective*
— work·a·hol·ism \ˈwər-kə-ˌhȯ-ˌli-zəm, -ˌhä-\ *noun*

work·bag \ˈwərk-ˌbag\ *noun* (1775)
: a bag for implements or materials for work; *especially* **:** a bag for needlework

work·bas·ket \-ˌbas-kət\ *noun* (1743)
: a basket for needlework

work·bench \-ˌbench\ *noun* (1781)
: a bench on which work especially of mechanics, machinists, and carpenters is performed

work·boat \-ˌbōt\ *noun* (1937)
: a boat used for work purposes (as commercial fishing and ferrying supplies) rather than for sport or for passenger or naval service

work·book \-ˌbu̇k\ *noun* (1910)
1 : a worker's manual
2 : a booklet outlining a course of study
3 : a record of work done
4 : a student's book of problems to be solved directly on the pages

work·box \-ˌbäks\ *noun* (1605)
: a box for work instruments and materials

work camp *noun* (1933)
: a camp for workers: as **a :** PRISON CAMP 1 **b :** a short-term group project in which individuals from one or more religious organizations volunteer their labor

work·day \ˈwərk-ˌdā\ *noun* (15th century)
1 : a day on which work is performed as distinguished from a day off
2 : the period of time in a day during which work is performed
— workday *adjective*

worked \ˈwərkt\ *adjective* (1740)
: that has been subjected to some process of development, treatment, or manufacture ⟨a newly *worked* field⟩

worked up *adjective* (1903)
: emotionally aroused **:** EXCITED

work·er \ˈwər-kər\ *noun* (14th century)
1 a : one that works especially at manual or industrial labor or with a particular material — often used in combination **b :** a member of the working class
2 : any of the sexually underdeveloped and usually sterile members of a colony of social ants, bees, wasps, or termites that perform most of the labor and protective duties of the colony

worker–priest *noun* (1949)
: a French Roman Catholic priest who for missionary purposes spends part of each weekday as a worker in a secular job

workers' compensation *noun* (1948)
: a system of insurance that reimburses an employer for damages that must be paid to an employee for injury occurring in the course of employment

work ethic *noun* (1951)
: a belief in work as a moral good

work·fare \ˈwərk-ˌfar, -ˌfer\ *noun* [*work* + welfare] (1968)
: a welfare program in which recipients are required to perform usually public-service work

work farm *noun* (1953)
: a farm on which persons guilty of minor law violations are confined

work·folk \ˈwərk-ˌfōk\ *or* **work·folks** \-ˌfōks\ *noun plural* (15th century)
: working people; *especially* **:** farm workers

work·force \ˈwərk-ˌfōrs, -ˌfȯrs\ *noun* (1943)
1 : the workers engaged in a specific activity or enterprise ⟨the factory's *workforce*⟩
2 : the number of workers potentially assignable for any purpose ⟨the nation's *workforce*⟩

work·horse \ˈwərk-ˌhȯrs\ *noun* (1543)
1 : a horse used chiefly for labor as distinguished from driving, riding, or racing

☆ SYNONYMS

Work, labor, travail, toil, drudgery, grind mean activity involving effort or exertion. WORK may imply activity of body, of mind, of a machine, or of a natural force ⟨too tired to do any *work*⟩. LABOR applies to physical or intellectual work involving great and often strenuous exertion ⟨farmers demanding fair compensation for their *labor*⟩. TRAVAIL is bookish for labor involving pain or suffering ⟨years of *travail* were lost when the house burned⟩. TOIL implies prolonged and fatiguing labor ⟨his lot would be years of backbreaking *toil*⟩. DRUDGERY suggests dull and irksome labor ⟨an editorial job with a good deal of *drudgery*⟩. GRIND implies labor exhausting to mind or body ⟨the *grind* of the assembly line⟩.

Work, employment, occupation, calling, pursuit, métier, business mean a specific sustained activity engaged in especially in earning one's living. WORK may apply to any purposeful activity whether remunerative or not ⟨her *work* as a hospital volunteer⟩. EMPLOYMENT implies work for which one has been engaged and is being paid by an employer ⟨your *employment* with this firm is hereby terminated⟩. OCCUPATION implies work in which one engages regularly especially as a result of training ⟨his *occupation* as a trained auto mechanic⟩. CALLING applies to an occupation viewed as a vocation or profession ⟨the ministry seemed my true *calling*⟩. PURSUIT suggests a trade, profession, or avocation followed with zeal or steady interest ⟨her family considered medicine the only proper *pursuit*⟩. MÉTIER implies a calling or pursuit for which one believes oneself to be especially fitted ⟨acting was my one and only *métier*⟩. BUSINESS suggests activity in commerce or the management of money and affairs ⟨the *business* of managing a hotel⟩.

2 a (1) **:** a person who performs most of the work of a group task (2) **:** a hardworking person **b :** a markedly useful or durable vehicle, craft, or machine **c :** HORSE 7
work·house \-ˌhau̇s\ *noun* (1652)
1 *British* **:** POORHOUSE
2 : a house of correction for persons guilty of minor law violations
work in *transitive verb* (1675)
1 : to insert or cause to penetrate by repeated or continued effort
2 : to interpose or insinuate gradually or unobtrusively ⟨*worked in* a few topical jokes⟩
¹work·ing *noun* (14th century)
1 : the manner of functioning or operating **:** OPERATION — usually used in plural
2 : an excavation or group of excavations made in mining, quarrying, or tunneling — usually used in plural
²working *adjective* (1613)
1 : engaged in work ⟨a *working* journalist⟩
2 : adequate to permit work to be done ⟨a *working* majority⟩
3 : assumed or adopted to permit or facilitate further work or activity ⟨*working* draft⟩
4 : spent at work ⟨*working* life⟩
5 : being in use or operation ⟨a *working* farm⟩
working asset *noun* (circa 1914)
: an asset other than a capital asset
working capital *noun* (circa 1901)
: capital actively turned over in or available for use in the course of business activity: **a :** the excess of current assets over current liabilities **b :** all capital of a business except that invested in capital assets
working–class *adjective* (1839)
: of, relating to, deriving from, or suitable to the class of wage earners ⟨*working-class* virtues⟩ ⟨*working-class* family⟩
working class *noun* (1789)
: the class of people who work for wages usually at manual labor
working day *noun* (15th century)
: WORKDAY
working dog *noun* (1891)
: a dog fitted by size, breeding, or training for useful work (as draft or herding) especially as distinguished from one fitted primarily for pet, show, or sporting use
working fluid *noun* (1903)
: a fluid working substance
work·ing·man \ˈwər-kiŋ-ˌman\ *noun* (1638)
: one who works for wages usually at manual labor
working papers *noun plural* (1928)
: official documents legalizing the employment of a minor
working substance *noun* (1897)
: a usually fluid substance that through changes of temperature, volume, and pressure is the means of carrying out thermodynamic processes or cycles (as in a heat engine)
work·ing·wom·an \ˈwər-kiŋ-ˌwu̇-mən\ *noun* (1853)
: WORKWOMAN
work·less \ˈwər-kləs\ *adjective* (15th century)
: being without work **:** UNEMPLOYED
— work·less·ness *noun*
work·load \ˈwərk-ˌlōd\ *noun* (1943)
1 : the amount of work or of working time expected or assigned
2 : the amount of work performed or capable of being performed (as by a mechanical device) usually within a specific period
work·man \ˈwərk-mən\ *noun* (before 12th century)
1 : WORKINGMAN
2 : ARTISAN
work·man·like \-ˌlīk\ *adjective* (1739)
: characterized by the skill and efficiency typical of a good workman
work·man·ly \-lē\ *adjective* (1545)
: WORKMANLIKE
work·man·ship \-ˌship\ *noun* (1523)
1 : something effected, made, or produced **:** WORK
2 : the art or skill of a workman **:** CRAFTSMANSHIP; *also* **:** the quality imparted to a thing in the process of making ⟨a vase of exquisite *workmanship*⟩
work·mate \ˈwərk-ˌmāt\ *noun* (1851)
chiefly British **:** a fellow worker
workmen's compensation insurance *noun* (circa 1917)
: WORKERS' COMPENSATION
work of art (1834)
1 : a product of one of the fine arts; *especially* **:** a painting or sculpture of high artistic quality
2 : something giving high aesthetic satisfaction to the viewer or listener
work off *transitive verb* (1678)
: to dispose of or get rid of by work or activity
work·out \ˈwər-ˌkau̇t\ *noun* (circa 1894)
1 : a practice or exercise to test or improve one's fitness for athletic competition, ability, or performance
2 : a test of one's ability, capacity, stamina, or suitability
work out (1534)
transitive verb
1 a : to bring about by labor and exertion ⟨*work out* your own salvation —Philippians 2:12 (Authorized Version)⟩ **b :** to solve (as a problem) by a process of reasoning or calculation **c :** to devise, arrange, or achieve by resolving difficulties ⟨after many years of wrangling, *worked out* a definite agreement —A. A. Butkus⟩ **d :** DEVELOP ⟨the final situation is not *worked out* with psychological profundity —Leslie Rees⟩
2 : to discharge (as a debt) by labor
3 : to exhaust (as a mine) by working
intransitive verb
1 a : to prove effective, practicable, or suitable ⟨how this will actually *work out* I don't know —Milton Kotler⟩ **b :** to amount to a total or calculated figure — used with *at* or *to*
2 : to engage in a workout ⟨*works out* in gymnasiums . . . to keep in shape —*Current Biography*⟩
work over *transitive verb* (1874)
1 : to do over **:** REWORK ⟨saved the play by *working* the first act *over*⟩
2 : to subject to thorough examination, study, or treatment ⟨shelf stock *worked over* by shoppers⟩
3 : to beat up or manhandle with thoroughness ⟨the gang *worked* me *over*⟩
work·peo·ple \ˈwərk-ˌpē-pəl\ *noun plural* (1708)
chiefly British **:** WORKERS, EMPLOYEES
work·piece \-ˌpēs\ *noun* (1926)
: a piece of work in process of manufacture
work·place \-ˌplās\ *noun* (circa 1828)
: a place (as a shop or factory) where work is done
work print *noun* (1937)
: a completely edited motion-picture print used as a guide in cutting the original negative from which the final production prints will be made
work·room \ˈwərk-ˌrüm, -ˌru̇m\ *noun* (1828)
: a room used for work
work·shop \-ˌshäp\ *noun* (1562)
1 : a small establishment where manufacturing or handicrafts are carried on
2 : WORKROOM
3 : a usually brief intensive educational program for a relatively small group of people that focuses especially on techniques and skills in a particular field
work song *noun* (1911)
: a song sung in rhythm with work
work·sta·tion \-ˌstā-shən\ *noun* (1931)
1 : an area with equipment for the performance of a specialized task usually by a single individual
2 a : an intelligent terminal or personal computer usually connected to a computer network **b :** a powerful microcomputer used especially for scientific or engineering work
work stoppage *noun* (1945)
: concerted cessation of work by a group of employees usually more spontaneous and less serious than a strike
work–study program *noun* (1946)
: a program planned to give high school or college students work experience
work·ta·ble \ˈwərk-ˌtā-bəl\ *noun* (1790)
: a table for holding working materials and implements; *especially* **:** a small table with drawers and other conveniences for needlework
work–to–rule *noun* (1950)
: the practice of working to the strictest interpretation of the rules as a job action
work·up \ˈwərk-ˌəp\ *noun* (1939)
: an intensive diagnostic study
work–up \ˈwərk-ˌəp\ *noun* (1903)
: an unintended mark on a printed sheet caused by the rising of spacing material
work up (15th century)
transitive verb
1 : to stir up **:** ROUSE
2 : to produce by mental or physical work ⟨*worked up* a comedy act⟩ ⟨*worked up* a sweat in the gymnasium⟩
intransitive verb
: to rise gradually in intensity or emotional tone
work·week \ˈwərk-ˌwēk\ *noun* (1921)
: the hours or days of work in a calendar week ⟨40-hour *workweek*⟩ ⟨a 5-day *workweek*⟩ ⟨a shortened *workweek*⟩
work·wom·an \-ˌwu̇-mən\ *noun* (circa 1530)
: a woman who works
¹world \ˈwər(-ə)ld\ *noun* [Middle English, from Old English *woruld* human existence, this world, age (akin to Old High German *weralt* age, world); akin to Old English *wer* man, *eald* old — more at VIRILE, OLD] (before 12th century)
1 a : the earthly state of human existence **b :** life after death — used with a qualifier ⟨the next *world*⟩
2 : the earth with its inhabitants and all things upon it
3 : individual course of life **:** CAREER
4 : the inhabitants of the earth **:** the human race
5 a : the concerns of the earth and its affairs as distinguished from heaven and the life to come **b :** secular affairs
6 : the system of created things **:** UNIVERSE
7 a : a division or generation of the inhabitants of the earth distinguished by living together at the same place or at the same time ⟨the medieval *world*⟩ **b :** a distinctive class of persons or their sphere of interest ⟨the academic *world*⟩ ⟨the sports *world*⟩
8 : human society ⟨withdraw from the *world*⟩
9 : a part or section of the earth that is a separate independent unit
10 : the sphere or scene of one's life and action ⟨living in your own little *world*⟩
11 : an indefinite multitude or a great quantity or distance ⟨makes a *world* of difference⟩ ⟨a *world* away⟩
12 : the whole body of living persons **:** PUBLIC ⟨announced their discovery to the *world*⟩
13 : KINGDOM 5 ⟨the animal *world*⟩
14 : a celestial body (as a planet)
— for all the world : in every way **:** EXACTLY ⟨copies which look *for all the world* like the original⟩
— in the world : among innumerable possibilities **:** EVER — used as an intensive ⟨what *in the world* is it⟩
— out of this world : of extraordinary excellence **:** SUPERB
²world *adjective* (13th century)
1 : of or relating to the world ⟨a *world* championship⟩
2 a : extending or found throughout the world **:** WORLDWIDE ⟨brought about *world* peace⟩ **b :** involving or applying to part of or the whole world ⟨a *world* tour⟩ ⟨a *world* state⟩

world–beat·er \'wərl(d)-ˌbē-tər\ *noun* (circa 1888)
: one that excels all others of its kind : CHAMPION

world–class *adjective* (1950)
: being of the highest caliber in the world ⟨a *world-class* polo player⟩

world federalism *noun* (1950)
1 : federalism on a worldwide basis
2 *W&F capitalized* **a :** the principles and policies of the World Federalists **b :** the body or movement composed of World Federalists

world federalist *noun* (1951)
1 : an adherent or advocate of world federalism
2 *W&F capitalized* **:** a member of a movement arising after World War II advocating the formation of a federal union of the nations of the world with limited but positive governmental powers

world·ling \'wər(-ə)ld-liŋ, 'wərl-liŋ\ *noun* (1549)
: a person engrossed in the concerns of this present world

world·ly \'wər(-ə)ld-lē, 'wərl-lē\ *adjective* (before 12th century)
1 : of, relating to, or devoted to this world and its pursuits rather than to religion or spiritual affairs
2 : WORLDLY-WISE
synonym see EARTHLY
— world·li·ness *noun*

world·ly–mind·ed \ˌwərl(d)-lē-'mīn-dəd\ *adjective* (1601)
: devoted to or engrossed in worldly interests
— world·ly–mind·ed·ness *noun*

world·ly–wise \'wərl(d)-lē-ˌwīz\ *adjective* (15th century)
: possessing a practical and often shrewd understanding of human affairs
synonym see SOPHISTICATED

world power *noun* (1860)
: a political unit (as a nation or state) powerful enough to affect the entire world by its influence or actions

world premiere *noun* (1925)
: the first regular performance (as of a theatrical production) anywhere in the world

World Series *noun* [from *World Series,* annual championship of U.S. major-league baseball] (1951)
: a contest or event that is the most important or prestigious of its kind ⟨the *World Series* of the equestrian world⟩

world's fair *noun* (1850)
: an international exposition featuring exhibits and participants from all over the world

world–shak·ing \'wərl(d)-ˌshā-kiŋ\ *adjective* (1598)
: EARTHSHAKING

world soul *noun* (1848)
: an animating spirit or creative principle related to the world as the soul is to the individual being

world·view \-ˌvyü\ *noun* (1858)
: WELTANSCHAUUNG

world war *noun* (1909)
: a war engaged in by all or most of the principal nations of the world; *especially, both Ws capitalized* **:** either of two such wars of the first half of the 20th century

world–wea·ry \'wərld-ˌwir-ē\ *adjective* (1768)
: feeling or showing fatigue from or boredom with the life of the world and especially material pleasures
— world–wea·ri·ness *noun*

¹world·wide \-'wīd\ *adjective* (1632)
: extended throughout or involving the entire world

²worldwide *adverb* (1892)
: throughout the world

¹worm \'wərm\ *noun, often attributive* [Middle English, from Old English *wyrm* serpent, worm; akin to Old High German *wurm* serpent, worm, Latin *vermis* worm] (before 12th century)
1 a : EARTHWORM; *broadly* **:** an annelid worm **b :** any of numerous relatively small elongated usually naked and soft-bodied animals: as (1) **:** an insect larva; *especially* **:** one that is a destructive grub, caterpillar, or maggot (2) **:** SHIPWORM (3) **:** BLINDWORM
2 a : a human being who is an object of contempt, loathing, or pity **:** WRETCH **b :** something that torments or devours from within
3 *archaic* **:** SNAKE, SERPENT
4 : HELMINTHIASIS — usually used in plural
5 : something (as a mechanical device) spiral or vermiculate in form or appearance: as **a :** the thread of a screw **b :** a short revolving screw whose threads gear with the teeth of a worm wheel or a rack **c :** a spiral condensing tube used in distilling **d :** ARCHIMEDES' SCREW; *also* **:** a conveyor working on the principle of such a screw
6 : a usually small self-contained computer program that invades computers on a network and usually performs a malicious action
— worm·like \-ˌlīk\ *adjective*

²worm (1610)
intransitive verb
: to move or proceed sinuously or insidiously
transitive verb
1 a : to proceed or make (one's way) insidiously or deviously ⟨*worm* their way into positions of power —Bill Franzen⟩ **b :** to insinuate or introduce (oneself) by devious or subtle means **c :** to cause to move or proceed in or as if in the manner of a worm
2 : to wind rope or yarn spirally round and between the strands of (a cable or rope) before serving
3 : to obtain or extract by artful or insidious questioning or by pleading, asking, or persuading — usually used with *out of*
4 : to treat (an animal) with a drug to destroy or expel parasitic worms

worm–eat·en \'wərm-ˌē-tᵊn\ *adjective* (14th century)
1 a : eaten or burrowed by worms ⟨*worm-eaten* timber⟩ **b :** PITTED
2 : WORN-OUT, ANTIQUATED

worm·er \'wər-mər\ *noun* (circa 1934)
: a drug used in veterinary medicine to worm an animal

worm fence *noun* (1652)
: a zigzag fence consisting of interlocking rails supported by crossed poles — called also *snake fence, Virginia fence*

worm fence

worm gear *noun* (circa 1876)
1 : WORM WHEEL
2 : a gear of a worm and a worm wheel working together

worm·hole \'wərm-ˌhōl\ *noun* (1593)
1 : a hole or passage burrowed by a worm
2 : a hypothetical structure of space-time envisioned as a long thin tunnel connecting points that are separated in space and time

worm·seed \-ˌsēd\ *noun* (15th century)
: any of various plants whose seeds possess anthelmintic properties: as **a :** any of several artemisias **b :** a goosefoot (*Chenopodium ambrosioides*)

worm's–eye \'wərm-ˌzī\ *adjective* (1908)
: seen from ground level or from the lowest levels of a hierarchy ⟨the bird's-eye view of the executive and the *worm's-eye* view of the employee —*Current Biography*⟩

worm snake *noun* (1885)
: a reddish terrestrial colubrid snake (*Carphophis amoenus*) of the eastern U.S.

worm wheel *noun* (1677)
: a toothed wheel gearing with the thread of a worm

worm·wood \'wərm-ˌwu̇d\ *noun* [Middle English *wormwode,* alteration of *wermode,* from Old English *wermōd;* akin to Old High German *wermuota* wormwood] (15th century)
1 : ARTEMISIA; *especially* **:** a European plant (*Artemisia absinthium*) that has silvery silky-haired leaves and drooping yellow nearly globular flower heads and yields a bitter dark green oil used in absinthe
2 : something bitter or grievous **:** BITTERNESS

wormy \'wər-mē\ *adjective* **worm·i·er; -est** (15th century)
1 : containing, abounding in, or infested with or as if with worms ⟨*wormy* flour⟩ ⟨a *wormy* dog⟩; *also* **:** damaged by worms **:** WORM-EATEN ⟨*wormy* timbers⟩
2 : resembling or suggestive of a worm

worn *past participle of* WEAR

worn–out \'wōrn-'au̇t, 'wȯrn-\ *adjective* (1589)
: exhausted or used up by or as if by wear

wor·ri·ment \'wər-ē-mənt, 'wə-rē-\ *noun* (1833)
: an act or instance of worrying; *also* **:** TROUBLE, WORRY

wor·ri·some \-səm\ *adjective* (1845)
1 : causing distress or worry
2 : inclined to worry or fret
— wor·ri·some·ly *adverb*
— wor·ri·some·ness *noun*

¹wor·ry \'wər-ē, 'wə-rē\ *verb* **wor·ried; wor·ry·ing** [Middle English *worien,* from Old English *wyrgan;* akin to Old High German *wurgen* to strangle, Lithuanian *veržti* to constrict] (before 12th century)
transitive verb
1 *dialect British* **:** CHOKE, STRANGLE
2 a : to harass by tearing, biting, or snapping especially at the throat **b :** to shake or pull at with the teeth ⟨a terrier *worrying* a rat⟩ **c :** to touch or disturb something repeatedly **d :** to change the position of or adjust by repeated pushing or hauling
3 a : to assail with rough or aggressive attack or treatment **:** TORMENT **b :** to subject to persistent or nagging attention or effort
4 : to afflict with mental distress or agitation **:** make anxious
intransitive verb
1 *dialect British* **:** STRANGLE, CHOKE
2 : to move, proceed, or progress by unceasing or difficult effort **:** STRUGGLE
3 : to feel or experience concern or anxiety **:** FRET ☆
— wor·ried·ly \-(r)ēd-lē, -(r)əd-\ *adverb*

☆ SYNONYMS
Worry, annoy, harass, harry, plague, pester, tease mean to disturb or irritate by persistent acts. WORRY implies an incessant goading or attacking that drives one to desperation ⟨pursued a policy of *worrying* the enemy⟩. ANNOY implies disturbing one's composure or peace of mind by intrusion, interference, or petty attacks ⟨you're doing that just to *annoy* me⟩. HARASS implies petty persecutions or burdensome demands that exhaust one's nervous or mental power ⟨*harassed* on all sides by creditors⟩. HARRY may imply heavy oppression or maltreatment ⟨the strikers had been *harried* by thugs⟩. PLAGUE implies a painful and persistent affliction ⟨*plagued* all her life by poverty⟩. PESTER stresses the repetition of petty attacks ⟨constantly *pestered* with trivial complaints⟩. TEASE suggests an attempt to break down one's resistance or rouse to wrath ⟨children *teased* the dog⟩.

— **wor•ri•er** \-(r)ē-ər\ *noun*

[2]**worry** *noun, plural* **worries** (1804)
1 a : mental distress or agitation resulting from concern usually for something impending or anticipated **:** ANXIETY **b :** an instance or occurrence of such distress or agitation
2 : a cause of worry **:** TROUBLE, DIFFICULTY
synonym see CARE

worry beads *noun plural* (1956)
: a string of beads that can be fingered to keep one's hands occupied

wor•ry•wart \ˈwər-ē-ˌwȯrt, ˈwə-rē-\ *noun* (1936)
: one who is inclined to worry unduly

[1]**worse** \ˈwərs\ *adjective, comparative of* BAD *or of* ILL [Middle English *werse, worse,* from Old English *wiersa, wyrsa;* akin to Old High German *wirsiro* worse] (before 12th century)
1 : of more inferior quality, value, or condition
2 a : more unfavorable, difficult, unpleasant, or painful **b :** more faulty, unsuitable, or incorrect **c :** less skillful or efficient
3 : bad, evil, or corrupt in a greater degree **:** more reprehensible
4 : being in poorer health **:** SICKER

[2]**worse** *noun* (before 12th century)
: one that is worse ⟨thought he was an atheist and *worse* —Van Wyck Brooks⟩

[3]**worse** *adverb, comparative of* BAD *or of* ILL (before 12th century)
1 : in a worse manner **:** to a worse extent or degree
2 : what is worse

wors•en \ˈwər-sᵊn\ *verb* **wors•ened; wors•en•ing** \ˈwərs-niŋ, ˈwər-sᵊn-iŋ\ (13th century)
transitive verb
: to make worse
intransitive verb
: to become worse

wors•er \ˈwər-sər\ *adjective or adverb* (15th century)
archaic **:** WORSE ⟨I cannot hate thee *worser* than I do —Shakespeare⟩

[1]**wor•ship** \ˈwər-shəp\ *noun* [Middle English *worshipe* worthiness, respect, reverence paid to a divine being, from Old English *weorthscipe* worthiness, respect, from *weorth* worthy, worth + *-scipe* -ship] (before 12th century)
1 *chiefly British* **:** a person of importance — used as a title for various officials (as magistrates and some mayors)
2 : reverence offered a divine being or supernatural power; *also* **:** an act of expressing such reverence
3 : a form of religious practice with its creed and ritual
4 : extravagant respect or admiration for or devotion to an object of esteem ⟨*worship* of the dollar⟩

[2]**worship** *verb* **-shiped** *or* **-shipped; -ship•ing** *or* **-ship•ping** (13th century)
transitive verb
1 : to honor or reverence as a divine being or supernatural power
2 : to regard with great or extravagant respect, honor, or devotion
intransitive verb
: to perform or take part in worship or an act of worship
synonym see REVERE
— **wor•ship•er** *or* **wor•ship•per** *noun*

wor•ship•ful \ˈwər-shəp-fəl\ *adjective* (14th century)
1 a *archaic* **:** NOTABLE, DISTINGUISHED **b** *chiefly British* — used as a title for various persons or groups of rank or distinction
2 : giving or expressing worship or veneration
— **wor•ship•ful•ly** \-fə-lē\ *adverb*
— **wor•ship•ful•ness** *noun*

wor•ship•less \-shə-pləs\ *adjective* (1765)
: lacking worship or worshipers

[1]**worst** \ˈwərst\ *adjective, superlative of* BAD *or of* ILL [Middle English *werste, worste,* from Old English *wierresta, wyrsta,* superlative of the root of Old English *wiersa* worse] (before 12th century)
1 : most corrupt, bad, evil, or ill
2 a : most unfavorable, difficult, unpleasant, or painful **b :** most unsuitable, faulty, unattractive, or ill-conceived **c :** least skillful or efficient
3 : most wanting in quality, value, or condition
— **the worst way :** very much ⟨such men . . . need indoctrination *the worst way* —J. G. Cozzens⟩ — often used with *in* ⟨wanted a new bicycle in *the worst way*⟩

[2]**worst** *adverb, superlative of* ILL *or* ILLY *or of* BAD *or* BADLY (before 12th century)
1 : to the extreme degree of badness or inferiority
2 : to the greatest or highest degree ⟨groups who need the subsidies *worst* lose out —T. W. Arnold⟩

[3]**worst** *noun, plural* **worst** (14th century)
: one that is worst
— **at worst :** under the worst circumstances

[4]**worst** *transitive verb* (1636)
: to get the better of **:** DEFEAT

worst–case *adjective* (1964)
: involving, projecting, or providing for the worst possible circumstances or outcome of a given situation

wor•sted \ˈwu̇s-təd, ˈwər-stəd\ *noun* [Middle English, from *Worsted* (now *Worstead*), England] (13th century)
: a smooth compact yarn from long wool fibers used especially for firm napless fabrics, carpeting, or knitting; *also* **:** a fabric made from worsted yarns
— **worsted** *adjective*

[1]**wort** \ˈwərt, ˈwȯrt\ *noun* [Middle English, from Old English *wyrt* root, herb, plant — more at ROOT] (before 12th century)
: PLANT; *especially* **:** an herbaceous plant — usually used in combination ⟨louse*wort*⟩

[2]**wort** *noun* [Middle English, from Old English *wyrt;* akin to Middle High German *würze* brewer's wort, Old English *wyrt* root, herb] (before 12th century)
: a liquid formed by soaking mash in hot water and then fermented to make beer

[1]**worth** \ˈwərth\ *intransitive verb* [Middle English, from Old English *weorthan;* akin to Old High German *werdan* to become, Latin *vertere* to turn, Lithuanian *versti* to overturn, Sanskrit *vartate* he turns] (before 12th century)
archaic **:** BECOME — usually used in the phrase *woe worth*

[2]**worth** *adjective* [Middle English, from Old English *weorth* worthy, of (a specified) value; akin to Old High German *werd* worthy, worth] (before 12th century)
1 *archaic* **:** having monetary or material value
2 *archaic* **:** ESTIMABLE
— **for all one is worth :** to the fullest extent of one's value or ability

[3]**worth** *noun* (before 12th century)
1 a : monetary value ⟨farmhouse and lands of little *worth*⟩ **b :** the equivalent of a specified amount or figure ⟨a dollar's *worth* of gas⟩
2 : the value of something measured by its qualities or by the esteem in which it is held ⟨a literary heritage of great *worth*⟩
3 a : moral or personal value ⟨trying to teach human *worth*⟩ **b :** MERIT, EXCELLENCE ⟨a field in which we have proved our *worth*⟩
4 : WEALTH, RICHES

[4]**worth** *preposition* (13th century)
1 a : equal in value to **b :** having assets or income equal to
2 : deserving of ⟨well *worth* the effort⟩
— **worth one's salt :** of substantial or significant value or merit

worth•ful \ˈwərth-fəl\ *adjective* (before 12th century)
1 : full of merit ⟨a good and *worthful* person⟩
2 : having value ⟨the *worthful* aspects of their culture⟩

worth•less \ˈwərth-ləs\ *adjective* (1588)
1 a : lacking worth **:** VALUELESS ⟨*worthless* currency⟩ **b :** USELESS ⟨*worthless* to continue searching⟩
2 : CONTEMPTIBLE, DESPICABLE
— **worth•less•ly** *adverb*
— **worth•less•ness** *noun*

worth•while \-ˈhwī(ə)l, -ˈwī(ə)l\ *adjective* (1900)
1 : being worth the time or effort spent
2 : WORTHY 1
— **worth•while•ness** *noun*

[1]**wor•thy** \ˈwər-t͟hē\ *adjective* **wor•thi•er; -est** (13th century)
1 a : having worth or value **:** ESTIMABLE ⟨a *worthy* cause⟩ **b :** HONORABLE, MERITORIOUS ⟨*worthy* candidates⟩
2 : having sufficient worth or importance ⟨*worthy* to be remembered⟩
— **wor•thi•ly** \ˈwər-t͟hə-lē\ *adverb*
— **wor•thi•ness** \-t͟hē-nəs\ *noun*

[2]**worthy** *noun, plural* **worthies** (14th century)
: a worthy or prominent person

-wor•thy \ˌwər-t͟hē\ *adjective combining form*
1 : fit or safe for ⟨a sea*worthy* vessel⟩
2 : of sufficient worth for ⟨a news*worthy* event⟩

[1]**wot** *present 1st & 3d singular of* WIT

[2]**wot** \ˈwät\ *verb* **wot•ted; wot•ting** [Middle English, alteration of *witen* — more at WIT] (14th century)
chiefly British **:** KNOW — often used with *of*

would \wəd, əd, d, ˈwu̇d\ *past of* WILL [Middle English *wolde,* from Old English; akin to Old High German *wolta* wished, desired] (before 12th century)
1 a *archaic* **:** WISHED, DESIRED **b** *archaic* **:** wish for **:** WANT **c** (1) **:** strongly desire **:** WISH ⟨I *would* I were young again⟩ — often used without a subject and with *that* in a past or conditional construction ⟨*would* that I had heeded your advice⟩ (2) — used in auxiliary function with *rather* or *sooner* to express preference ⟨he *would* sooner die than face them⟩
2 a — used in auxiliary function to express wish, desire, or intent ⟨those who *would* forbid gambling⟩ **b** — used in auxiliary function to express willingness or preference ⟨as ye *would* that men should do to you —Luke 6:31 (Authorized Version)⟩ **c** — used in auxiliary function to express plan or intention ⟨said we *would* come⟩
3 — used in auxiliary function to express custom or habitual action ⟨we *would* meet often for lunch⟩
4 — used in auxiliary function to express consent or choice ⟨*would* put it off if he could⟩
5 a — used in auxiliary function in the conclusion of a conditional sentence to express a contingency or possibility ⟨if he were coming, he *would* be here now⟩ **b** — used in auxiliary function in a noun clause (as one completing a statement of desire, request, or advice) ⟨we wish that he *would* go⟩
6 — used in auxiliary function to express probability or presumption in past or present time ⟨*would* have won if I had not tripped⟩
7 : COULD ⟨the barrel *would* hold 20 gallons⟩
8 — used in auxiliary function to express a request with which voluntary compliance is expected ⟨*would* you please help us⟩
9 — used in auxiliary function to express doubt or uncertainty ⟨the explanation . . . *would* seem satisfactory⟩
10 : SHOULD ⟨knew I *would* enjoy the trip⟩ ⟨*would* be glad to know the answer⟩ ■

would–be \ˈwu̇d-ˈbē\ *adjective* (1647)
: desiring, professing, or having the potential to be

would•est \ˈwu̇-dəst\ *archaic past 2d singular of* WILL

wouldn't \ˈwu̇-dᵊnt, -dᵊn, *dialect also* ˈwu̇-tᵊn(t) *or* ˌwu̇nt\ (circa 1828)
: would not

wouldst \wədst, ˈwu̇dst, wətst\ *archaic past 2d singular of* WILL

¹wound \'wünd, *archaic or dialect* 'waund\ *noun* [Middle English, from Old English *wund;* akin to Old High German *wunta* wound] (before 12th century)
1 a : an injury to the body (as from violence, accident, or surgery) that involves laceration or breaking of a membrane (as the skin) and usually damage to underlying tissues **b :** a cut or breach in a plant due to external violence
2 : a mental or emotional hurt or blow
3 : something resembling a wound in appearance or effect; *especially* **:** a rift in or blow to a political body or social group
²wound (before 12th century)
transitive verb
: to cause a wound to or in
intransitive verb
: to inflict a wound
³wound \'waund\ *past and past participle of* WIND
¹wound·ed \'wün-dəd\ *noun plural* (before 12th century)
: wounded persons
²wounded *adjective* (14th century)
: injured, hurt by, or suffering from a wound
wound·less \'wün(d)-ləs\ *adjective* (1579)
1 : free from wounds **:** UNWOUNDED
2 *obsolete* **:** INVULNERABLE ⟨the *woundless* air —Shakespeare⟩
wove *past of* WEAVE
¹woven *past participle of* WEAVE
²wo·ven \'wō-vən\ *noun* (1930)
: a woven fabric
wove paper \'wōv-\ *noun* [*wove* (archaic past participle of *weave*)] (1815)
: paper made with a revolving roller covered with wires so woven as to produce no fine lines running across the grain — compare LAID PAPER
¹wow \'waů\ *interjection* (1513)
— used to express strong feeling (as pleasure or surprise)
²wow *noun* (1920)
: a striking success **:** HIT
³wow *transitive verb* (1924)
: to excite to enthusiastic admiration or approval
⁴wow *noun* [imitative] (1932)
: a distortion in reproduced sound consisting of a slow rise and fall of pitch caused by speed variation in the reproducing system
wow·ser \'waů-zər\ *noun* [origin unknown] (1899)
chiefly Australian **:** an obtrusively puritanical person
W particle *noun* [weak] (1963)
: either of two particles about 80 times heavier than a proton that along with the Z particle are transmitters of the weak force and that can have a positive or negative charge
¹wrack \'rak\ *noun* [Middle English *wrak,* from Middle Dutch or Middle Low German; akin to Old English *wræc* something driven by the sea] (14th century)
1 a : a wrecked ship **b :** WRECKAGE **c :** WRECK **d** *dialect* **:** the violent destruction of a structure, machine, or vehicle
2 a : marine vegetation; *especially* **:** KELP **b :** dried seaweeds
²wrack *noun* [Middle English, from Old English *wræc* misery, punishment, something driven by the sea; akin to Old English *wrecan* to drive, punish — more at WREAK] (15th century)
1 : RUIN, DESTRUCTION
2 : a remnant of something destroyed
³wrack *transitive verb* (1562)
: to utterly ruin **:** WRECK
⁴wrack *verb* [by alteration] (circa 1555)
: ⁴RACK
⁵wrack *noun* (1591)
: ³RACK 2
⁶wrack *noun* (1796)
: ¹RACK
wrack·ful \'rak-fəl\ *adjective* (1558)
: DESTRUCTIVE

wraith \'rāth\ *noun, plural* **wraiths** \'rāths *also* 'rāthz\ [origin unknown] (1513)
1 a : the exact likeness of a living person seen usually just before death as an apparition **b :** GHOST, SPECTER
2 : an insubstantial form or semblance **:** SHADOW
3 : a barely visible gaseous or vaporous column
— **wraith·like** \-,līk\ *adjective*
¹wran·gle \'raŋ-gəl\ *verb* **wran·gled; wran·gling** \-g(ə-)liŋ\ [Middle English; akin to Old High German *ringan* to struggle — more at WRING] (14th century)
intransitive verb
1 : to dispute angrily or peevishly **:** BICKER
2 : to engage in argument or controversy
transitive verb
1 : to obtain by persistent arguing or maneuvering **:** WANGLE
2 [back-formation from *wrangler*] **:** to herd and care for (livestock and especially horses) on the range
²wrangle *noun* (1547)
1 : an angry, noisy, or prolonged dispute or quarrel
2 : the action or process of wrangling
wran·gler \-g(ə-)lər\ *noun* (circa 1515)
1 : a bickering disputant
2 [short for *horse-wrangler,* probably part translation of Mexican Spanish *caballerango* groom] **:** a ranch hand who takes care of the saddle horses; *broadly* **:** COWBOY
¹wrap \'rap\ *verb* **wrapped; wrap·ping** [Middle English *wrappen*] (14th century)
transitive verb
1 a : to cover especially by winding or folding **b :** to envelop and secure for transportation or storage **:** BUNDLE **c :** ENFOLD, EMBRACE **d :** to coil, fold, draw, or twine (as string or cloth) around something
2 a : SURROUND, ENVELOP **b :** to suffuse or surround with an aura or state ⟨the affair was *wrapped* in scandal⟩ **c :** to involve completely **:** ENGROSS — usually used with *up*
3 : to conceal or obscure as if by enveloping
4 : to enclose as if with a protective covering
5 : to finish filming or videotaping ⟨*wrap* a movie⟩
intransitive verb
1 : to wind, coil, or twine so as to encircle or cover something
2 : to put on clothing **:** DRESS — usually used with *up*
3 : to be subject to covering, enclosing, or packaging — usually used with *up*
4 : to come to completion in filming or videotaping
²wrap *noun* (15th century)
1 a (1) **:** WRAPPER, WRAPPING (2) **:** material used for wrapping ⟨plastic *wrap*⟩ **b :** an article of clothing that may be wrapped round a person; *especially* **:** an outer garment (as a coat or shawl) **c :** BLANKET
2 : a single turn or convolution of something wound round an object
3 *plural* **a :** RESTRAINT **b :** a shroud of secrecy ⟨a plan kept under *wraps*⟩
4 : the completion of a schedule or session for filming or videotaping
³wrap *adjective* (1948)
: WRAPAROUND 1
¹wrap·around \'rap-ə-,raůnd\ *noun* (1924)
1 : a garment (as a dress) made with a full-length opening and adjusted to the figure by wrapping around
2 : an object that encircles or especially curves and laps over another
²wraparound *adjective* (1926)
1 : made to be wrapped around something and especially the body ⟨a *wraparound* skirt⟩
2 a : shaped to follow a contour; *especially* **:** made to curve from the front around to the side ⟨*wraparound* sunglasses⟩ ⟨*wraparound* terraces⟩ **b :** extending laterally to the outermost limits of the field of vision ⟨a *wraparound* movie screen⟩
wrap·per \'ra-pər\ *noun* (15th century)
1 : that in which something is wrapped: as **a :** a tobacco leaf used for the outside covering especially of cigars **b** (1) **:** JACKET 3c(1) (2) **:** the paper cover of a book not bound in boards **c :** a paper wrapped around a newspaper or magazine in the mail
2 : one that wraps
3 : an article of clothing worn wrapped around the body
wrap·ping \'ra-piŋ\ *noun* (14th century)
: something used to wrap an object **:** WRAPPER
wrap–up \'rap-,əp\ *noun* (1951)
1 : a summarizing report
2 : a concluding part **:** FINALE
wrap up *transitive verb* (circa 1568)
1 : SUMMARIZE, SUM UP
2 a : to bring to a usually successful conclusion **b :** CINCH, SEW UP ⟨has the nomination *wrapped up*⟩
wrasse \'ras\ *noun, plural* **wrasses** *also* **wrasse** [Cornish *gwragh, wragh* hag, wrasse] (circa 1672)
: any of a large family (Labridae) of elongate compressed usually brilliantly colored marine bony fishes that usually bury themselves in sand at night and include important food fishes as well as a number of popular aquarium fishes
¹wrath \'rath, *chiefly British* 'rōth\ *noun* [Middle English, from Old English *wrǣththo,* from *wrāth* wroth — more at WROTH] (before 12th century)
1 : strong vengeful anger or indignation
2 : retributory punishment for an offense or a crime **:** divine chastisement
synonym see ANGER
²wrath *adjective* [alteration of *wroth*] (1535)
archaic **:** WRATHFUL
wrath·ful \-fəl\ *adjective* (14th century)
1 : filled with wrath **:** IRATE

□ USAGE
would *Would have* is sometimes substituted for the past subjunctive *had* in the subordinate clause of a conditional sentence ⟨if we *would have* known Max was going to die [we] should have made sound movies of him —Ernest Hemingway (letter)⟩. Most of the evidence for this construction comes from speech, and it must be common in student writing for it is mentioned as something to avoid in student handbooks dating from 1920 to the present. It is occasionally found in edited writing ⟨if the spurious signal *would have* turned up during the brief period it took to punch the keys, it would have triggered an actual abort —*N. Y. Times*⟩ ⟨if this *would have* been a real . . . test of his critical powers, in another sense it could have proved a greater surety for his critical survival —Louis Kronenberger⟩. The origin and development of this construction are imperfectly known, but it has been found as far back as the second half of the 19th century ⟨these characteristics, fixed and unchangeable as bone in Mr. Casaubon, might have remained longer unfelt by Dorothea . . . if he *would have* held her hands between his and listened —George Eliot⟩. It is apparently well established in speech, where it tends not to be noticed. It is better avoided in writing, however, where it tends to attract unfavorable notice.

2 : arising from, marked by, or indicative of wrath
— **wrath·ful·ly** \-fə-lē\ *adverb*
— **wrath·ful·ness** *noun*

wrathy \'ra-thē, *chiefly British* 'rȯ-\ *adjective* (1828)
: WRATHFUL

wreak \'rēk *also* 'rek\ *transitive verb* [Middle English *wreken,* from Old English *wrecan* to drive, punish, avenge; akin to Old High German *rehhan* to avenge and perhaps to Latin *urgēre* to drive on, urge] (before 12th century)
1 a *archaic* **:** AVENGE **b :** to cause the infliction of (vengeance or punishment)
2 : to give free play or course to (malevolent feeling)
3 : BRING ABOUT, CAUSE ⟨*wreak* havoc⟩
usage see WRECK

wreath \'rēth\ *noun, plural* **wreaths** \'rēthz, 'rēths\ [Middle English *wrethe,* from Old English *writha;* akin to Old English *wrīthan* to twist — more at WRITHE] (before 12th century)
: something intertwined into a circular shape; *especially* **:** GARLAND, CHAPLET

wreathe \'rēth\ *verb* **wreathed; wreath·ing** [*wreath*] (1530)
transitive verb
1 a : to shape into a wreath **b :** INTERWEAVE **c :** to cause to coil about something
2 : to twist or contort so as to show folds or creases
3 : to encircle or adorn with or as if with a wreath
intransitive verb
1 : to twist in coils **:** WRITHE
2 a : to take on the shape of a wreath **b :** to move or extend in circles or spirals

wreathy \'rē-thē, -thē\ *adjective* (1644)
1 : having the form of a wreath
2 : constituting a wreath

¹wreck \'rek\ *noun* [Middle English *wrek,* from Anglo-French, of Scandinavian origin; akin to Old Norse *rek* wreck; akin to Old English *wrecan* to drive] (13th century)
1 : something cast up on the land by the sea especially after a shipwreck
2 a : SHIPWRECK **b :** the action of wrecking or fact or state of being wrecked **:** DESTRUCTION
3 a : a hulk or the ruins of a wrecked ship **b :** the broken remains of something wrecked or otherwise ruined **c :** something disabled or in a state of ruin or dilapidation; *also* **:** a person or animal of broken constitution, health, or spirits

²wreck (15th century)
transitive verb
1 : to cast ashore
2 a : to reduce to a ruinous state by or as if by violence **b :** SHIPWRECK **c :** to ruin, damage, or imperil by a wreck **d :** to involve in disaster or ruin
3 : BRING ABOUT, WREAK ⟨*wreck* havoc⟩
intransitive verb
1 : to become wrecked
2 : to rob, salvage, or repair wreckage or a wreck ■

wreck·age \'re-kij\ *noun* (1837)
1 : the act of wrecking **:** the state of being wrecked
2 a : something that has been wrecked **b :** broken and disordered parts or material from something wrecked

wreck·er \'re-kər\ *noun* (1802)
1 a : one that searches for or works on the wrecks of ships (as for rescue or for plunder) **b :** TOW TRUCK
2 : one that wrecks; *especially* **:** one whose work is the demolition of buildings

wrecker's ball *noun* (1967)
: a heavy iron or steel ball swung or dropped by a derrick to demolish old buildings — called also *wrecking ball*

wrecking bar *noun* (1924)
: a small crowbar with a claw for pulling nails at one end and a slight bend for prying at the other end

wren \'ren\ *noun* [Middle English *wrenne,* from Old English *wrenna;* akin to Old High German *rentilo* wren] (before 12th century)
1 : any of a family (Troglodytidae) of numerous small more or less brown oscine singing birds; *especially* **:** a very small widely distributed bird (*Troglodytes troglodytes*) that has a short erect tail and is noted for its song
2 : any of various small singing birds resembling the true wrens in size and habits

wren 1

¹wrench \'rench\ *verb* [Middle English, from Old English *wrencan;* akin to Old High German *renken* to twist and perhaps to Latin *vergere* to bend, incline] (before 12th century)
intransitive verb
1 : to move with a violent twist; *also* **:** to undergo twisting
2 : to pull or strain at something with violent twisting
transitive verb
1 : to twist violently
2 : to injure or disable by a violent twisting or straining
3 : CHANGE; *especially* **:** DISTORT, PERVERT
4 a : to pull or tighten by violent twisting or with violence **b :** to snatch forcibly **:** WREST
5 : to cause to suffer mental anguish **:** RACK
— **wrench·ing·ly** \'ren-chiŋ-lē\ *adverb*

²wrench *noun* (1530)
1 a : a violent twisting or a pull with or as if with twisting **b :** a sharp twist or sudden jerk straining muscles or ligaments; *also* **:** the resultant injury (as of a joint) **c :** a distorting or perverting alteration **d :** acute emotional distress **:** sudden violent mental change
2 : a hand or power tool for holding, twisting, or turning an object (as a bolt or nut)
3 : MONKEY WRENCH 2

¹wrest \'rest\ *transitive verb* [Middle English *wrasten, wresten,* from Old English *wrǣstan;* akin to Old Norse *reista* to bend and probably to Old English *wrigian* to turn — more at WRY] (before 12th century)
1 : to pull, force, or move by violent wringing or twisting movements
2 : to gain with difficulty by or as if by force, violence, or determined labor

²wrest *noun* (14th century)
1 : the action of wresting **:** WRENCH
2 *archaic* **:** a key or wrench used for turning pins in a stringed instrument (as a harp or piano)

¹wres·tle \'re-səl, 'ra-\ *verb* **wres·tled; wres·tling** \'res-liŋ, 'ras-; 're-sᵊl-iŋ, 'ra-\ [Middle English *wrastlen, wrestlen,* from Old English *wrǣstlian,* frequentative of *wrǣstan*] (before 12th century)
intransitive verb
1 : to contend by grappling with and striving to trip or throw an opponent down or off balance
2 : to combat an opposing tendency or force ⟨*wrestling* with his conscience⟩
3 : to engage in deep thought, consideration, or debate
4 : to engage in or as if in a violent or determined struggle ⟨*wrestling* with cumbersome luggage⟩
transitive verb
1 a : to engage in (a match, bout, or fall) in wrestling **b :** to wrestle with ⟨*wrestle* an alligator⟩
2 : to move, maneuver, or force with difficulty
— **wres·tler** \'res-lər, 'ras-; 're-sᵊl-ər, 'ra-\ *noun*

²wrestle *noun* (1593)
: the action or an instance of wrestling **:** STRUGGLE; *especially* **:** a wrestling bout

wres·tling \'res-liŋ, 'ras-; 're-sᵊl-iŋ, 'ra-\ *noun* (before 12th century)
: a sport or contest in which two unarmed individuals struggle hand to hand with each attempting to subdue or unbalance the other

wretch \'rech\ *noun* [Middle English *wrecche,* from Old English *wrecca* outcast, exile; akin to Old High German *hrechjo* fugitive, Old English *wrecan* to drive, drive out — more at WREAK] (before 12th century)
1 : a miserable person **:** one who is profoundly unhappy or in great misfortune
2 : a base, despicable, or vile person

wretch·ed \'re-chəd\ *adjective* [Middle English, irregular from *wretch*] (12th century)
1 : deeply afflicted, dejected, or distressed in body or mind
2 : extremely or deplorably bad or distressing ⟨was in *wretched* health⟩ ⟨a *wretched* accident⟩
3 a : being or appearing mean, miserable, or contemptible ⟨dressed in *wretched* old clothes⟩ **b :** very poor in quality or ability **:** INFERIOR ⟨*wretched* workmanship⟩
— **wretch·ed·ly** *adverb*
— **wretch·ed·ness** *noun*

¹wrig·gle \'ri-gəl\ *verb* **wrig·gled; wrig·gling** \-g(ə-)liŋ\ [Middle English, from or akin to Middle Low German *wriggeln* to wriggle; akin to Old English *wrigian* to turn — more at WRY] (15th century)
intransitive verb
1 : to move the body or a bodily part to and fro with short writhing motions like a worm **:** SQUIRM
2 : to move or advance by twisting and turning
3 : to extricate or insinuate oneself or reach a goal as if by wriggling
transitive verb
1 : to cause to move in short quick contortions
2 : to introduce, insinuate, or bring into a state or place by or as if by wriggling
— **wrig·gly** \-g(ə-)lē\ *adjective*

²wriggle *noun* (1709)
1 : a short or quick writhing motion or contortion
2 : a formation or marking of sinuous design

wrig·gler \'ri-g(ə-)lər\ *noun* (1631)
: one that wriggles; *especially* **:** WIGGLER 1

wright \'rīt\ *noun* [Middle English, from Old English *wyrhta, wryhta* worker, maker; akin to Old English *weorc* work — more at WORK] (before 12th century)

□ USAGE
wreck The phrase *wreak havoc* is often spoken as *wreck havoc* ⟨the havoc we've *wrecked* on this whole cuisine —Barbara Kafka (Television Food Network)⟩ and occasionally written that way. It is frequently termed an error, but it appears to be merely a relic. It has a curious history. In the 16th, 17th, and 18th centuries *wreck* was used as a variant of *wreak* in various senses. In 1770 an early usage critic who favored *wreak* complained of its use by several writers ⟨satires and epigrams in vain you write, there's a securer way to *wreck* your spite —Jonathan Swift⟩. Printed evidence of this use of *wreck* is rare after the 18th century, but it has survived in speech. *Wrought,* a variant past and past participle of *work,* is also found with *havoc* ⟨*wrought* havoc with the civilian population —Daniel Pipes⟩. This too has been termed an error, but it is not, as *work* is sometimes used with *havoc* ⟨one agency in particular which is *working* havoc —Thomas R. Lounsbury⟩ as are *create, play,* and *raise.* The usual form of the phrase is, however, *wreak havoc.*

: a worker skilled in the manufacture especially of wooden objects — usually used in combination ⟨ship*wright*⟩ ⟨wheel*wright*⟩

wring \'riŋ\ *verb* **wrung** \'rəŋ\; **wring·ing** \'riŋ-iŋ\ [Middle English, from Old English *wringan;* akin to Old High German *ringan* to struggle, Lithuanian *rengtis* to bend down, Old English *wyrgan* to strangle — more at WORRY] (before 12th century)
transitive verb
1 : to squeeze or twist especially so as to make dry or to extract moisture or liquid ⟨*wring* a towel dry⟩
2 : to extract or obtain by or as if by twisting and compressing ⟨*wring* water from a towel⟩ ⟨*wring* a confession from the suspect⟩
3 a : to twist so as to strain or sprain into a distorted shape ⟨I could *wring* your neck⟩ **b :** to twist together (clasped hands) as a sign of anguish
4 : to affect painfully as if by wringing **:** TORMENT ⟨a tragedy that *wrings* the heart⟩
intransitive verb
: SQUIRM, WRITHE
— wring *noun*

wring·er \'riŋ-ər\ *noun* (14th century)
: one that wrings: as **a :** a machine or device for pressing out liquid or moisture ⟨a clothes *wringer*⟩ **b :** something that causes pain, hardship, or exertion

¹wrin·kle \'riŋ-kəl\ *noun* [Middle English, back-formation from *wrinkled* twisted, winding, probably from Old English *gewrinclod,* past participle of *gewrinclian* to wind, from *ge-,* perfective prefix + *-wrinclian* (akin to *wrencan* to wrench) — more at CO-] (15th century)
1 : a small ridge or furrow especially when formed on a surface by the shrinking or contraction of a smooth substance **:** CREASE; *specifically* **:** one in the skin especially when due to age, worry, or fatigue
2 a : METHOD, TECHNIQUE **b :** a change in a customary procedure or method **c :** something new or different **:** INNOVATION
3 : IMPERFECTION, IRREGULARITY
— wrin·kly \-k(ə-)lē\ *adjective*

²wrinkle *verb* **wrin·kled; wrin·kling** \-k(ə-)liŋ\ (1523)
intransitive verb
: to become marked with or contracted into wrinkles
transitive verb
: to contract into wrinkles **:** PUCKER

wrist \'rist\ *noun* [Middle English, from Old English; akin to Middle High German *rist* wrist, ankle, Old English *wrǣstan* to twist — more at WREST] (before 12th century)
1 : the joint or the region of the joint between the human hand and the arm or a corresponding part on a lower animal
2 : the part of a garment or glove covering the wrist

wrist·band \'ris(t)-,band\ *noun* (1571)
1 : the part of a sleeve covering the wrist
2 : a band encircling the wrist

wrist·let \'ris(t)-lət\ *noun* (circa 1847)
: a band encircling the wrist; *especially* **:** a close-fitting knitted band attached to the top of a glove or the end of a sleeve

wrist·lock \'rist-,läk\ *noun* (1921)
: a wrestling hold in which one contestant is thrown or made helpless by a twisting grip on the wrist

wrist pin *noun* (circa 1875)
: a stud or pin that forms a journal (as in a crosshead) for a connecting rod

wrist shot *noun* (circa 1899)
: a quick usually short-range shot in ice hockey made while the puck is against the blade of the stick by snapping the blade quickly forward

wrist·watch \'rist-,wäch\ *noun* (1896)
: a small watch that is attached to a bracelet or strap and is worn around the wrist

wrist wrestling *noun* (1968)
: a form of arm wrestling in which opponents interlock thumbs instead of gripping hands

wristy \'ris-tē\ *adjective* **wrist·i·er; -est** (1867)
: involving or using a lot of wrist movement (as in stroking a ball)

writ \'rit\ *noun* [Middle English, from Old English; akin to Old English *wrītan* to write] (before 12th century)
1 : something written **:** WRITING ⟨Sacred *Writ*⟩
2 a : a formal written document; *specifically* **:** a legal instrument in epistolary form issued under seal in the name of the English monarch **b :** an order or mandatory process in writing issued in the name of the sovereign or of a court or judicial officer commanding the person to whom it is directed to perform or refrain from performing an act specified therein ⟨*writ* of detinue⟩ ⟨*writ* of entry⟩ ⟨*writ* of execution⟩ **c :** the power and authority of the issuer of such a written order — usually used with *run* ⟨outside the United States where . . . our *writ* does not run —Dean Acheson⟩

writ·able \'rī-tə-bəl\ *adjective* (1782)
: capable of being put in writing

write \'rīt\ *verb* **wrote** \'rōt\; **writ·ten** \'rit-ᵊn\ *also* **writ** \'rit\ *or dialect* **wrote; writ·ing** \'rī-tiŋ\ [Middle English, from Old English *wrītan* to scratch, draw, inscribe; akin to Old High German *rīzan* to tear and perhaps to Greek *rhinē* file, rasp] (before 12th century)
transitive verb
1 a : to form (as characters or symbols) on a surface with an instrument (as a pen) **b :** to form (as words) by inscribing the characters or symbols of on a surface **c :** to spell in writing ⟨words *written* alike but pronounced differently⟩ **d :** to cover, fill, or fill in by writing ⟨*wrote* ten pages⟩ ⟨*write* a check⟩
2 : to set down in writing: as **a :** DRAW UP, DRAFT ⟨*write* a will⟩ **b** (1) **:** to be the author of **:** COMPOSE ⟨*writes* poems and essays⟩ (2) **:** to compose in musical form ⟨*write* a string quartet⟩ **c :** to express in literary form ⟨if I could *write* the beauty of your eyes —Shakespeare⟩ **d :** to communicate by letter ⟨*writes* that they are coming⟩ **e :** to use or exhibit (a specific script, language, or literary form or style) in writing ⟨*write* Braille⟩ ⟨*writes* French with ease⟩ **f :** to write contracts or orders for; *especially* **:** UNDERWRITE ⟨*write* life insurance⟩
3 : to make a permanent impression of
4 : to communicate with in writing ⟨we'll *write* you when we get there⟩
5 : ORDAIN, FATE ⟨so be it, it is *written* —D. C. Peattie⟩
6 : to make evident or obvious ⟨guilt *written* on his face⟩
7 : to force, effect, introduce, or remove by writing ⟨*write* oneself into fame and fortune —Charles Lee⟩
8 : to take part in or bring about (something worth recording)
9 a : to introduce (information) into the storage device or medium of a computer **b :** to transfer (information) from the main memory of a computer to a storage or output device
10 : SELL ⟨*write* a stock option⟩
intransitive verb
1 a : to make significant characters or inscriptions; *also* **:** to permit or be adapted to writing **b :** to form or produce written letters, words, or sentences
2 : to compose, communicate by, or send a letter
3 a : to produce a written work **b :** to compose music ▪
— write one's own ticket : to select a course of action or position entirely according to one's wishes
— writ large : on a larger scale or in a more prominent manner ⟨the problems of modern totalitarianism are only our own problems *writ large* —*Times Literary Supplement*⟩
— writ small : on a smaller scale

write–down \'rīt-,dau̇n\ *noun* (1932)
: a deliberate reduction in the book value of an asset (as to reflect the effect of obsolescence)

write down (1588)
transitive verb
1 : to record in written form
2 a : to depreciate, disparage, or injure by writing **b :** to reduce in status, rank, or value; *especially* **:** to reduce the book value of
intransitive verb
: to write so as to appeal to a lower level of taste, comprehension, or intelligence

write–in \'rīt-,in\ *noun* (1932)
1 : a vote cast by writing in the name of a candidate
2 : a candidate whose name is written in

write in *transitive verb* (14th century)
1 : to insert in a document or text
2 a : to insert (a name not listed on a ballot or voting machine) in an appropriate space **b :** to cast (a vote) in this manner

write–off \'rīt-,ȯf\ *noun* (1905)
1 : an elimination of an item from the books of account
2 : a reduction in book value of an item (as by way of depreciation)

write off *transitive verb* (1682)
1 a : to reduce the estimated or book value of **:** DEPRECIATE **b :** to take off the books **:** CANCEL ⟨*write off* a bad debt⟩
2 : to regard or concede to be lost ⟨most were content to *write off* 1979 and look optimistically ahead —*Money*⟩; *also* **:** DISMISS ⟨was *written off* as an expatriate highbrow —Brendan Gill⟩

write out *transitive verb* (1548)
: to write especially in a full and complete form

writ·er \'rī-tər\ *noun* (before 12th century)
: one that writes: as **a :** AUTHOR **b :** one who writes stock options

writ·er·ly \'rī-tər-lē\ *adjective* (1957)
: of, relating to, or typical of a writer

writer's block *noun* (1950)
: a psychological inhibition preventing a writer from proceeding with a piece

writer's cramp *noun* (1853)
: a painful spasmodic cramp of muscles of the hand or fingers brought on by excessive writing

write–up \'rīt-,əp\ *noun* (1885)
1 : a written account; *especially* **:** a flattering article
2 : a deliberate increase in the book value of an asset (as to reflect the effect of inflation)

write up *transitive verb* (15th century)
1 : to make a write-up of
2 : to report (a person) especially for some violation of law or rules

☐ USAGE
write Back in the 18th century, the past participles *wrote* and *writ* were both in common use ⟨I have *wrote* him word —George Washington⟩ ⟨I ought to have *wrote* to you long since —Benjamin Franklin⟩ ⟨you will have thought me very inexcusable for not having *writ* to you —Richard Brinsley Sheridan⟩. We can still find *writ* in the Civil War era ⟨I have *writ* to George again —Walt Whitman⟩ and it has continued in the 20th century in much diminished use—chiefly in facetious text ⟨yes, Little Eva, the foregoing sentence was *writ* deliberate —James J. Kilpatrick⟩ and in the fixed phrase *writ large* ⟨museums are collections *writ large* —*Smithsonian*⟩. *Written* is the usual past participle now; *wrote* has dropped out of standard use.

writhe \'rīth\ *verb* **writhed; writh·ing** [Middle English, from Old English *wrīthan;* akin to Old Norse *rītha* to twist] (before 12th century)
transitive verb
1 a : to twist into coils or folds **b :** to twist so as to distort **:** WRENCH **c :** to twist (the body or a bodily part) in pain
2 : INTERTWINE
intransitive verb
1 : to move or proceed with twists and turns
2 : to twist from or as if from pain or struggling
3 : to suffer keenly
— writhe *noun*

writh·en \'ri-thən\ *adjective* [Middle English, from Old English, from past participle of *wrīthan*] (before 12th century)
: being twisted or contorted ⟨*writhen* trees⟩ ⟨a *writhen* smile⟩

writ·ing \'rī-tiŋ\ *noun* (13th century)
1 : the act or process of one who writes: as **a :** the act or art of forming visible letters or characters; *specifically* **:** HANDWRITING 1 **b :** the act or practice of literary or musical composition
2 : something written: as **a :** letters or characters that serve as visible signs of ideas, words, or symbols **b :** a letter, note, or notice used to communicate or record **c :** a written composition **d :** INSCRIPTION
3 : a style or form of composition
4 : the occupation of a writer; *especially* **:** the profession of authorship
— writing on the wall : HANDWRITING ON THE WALL

writing desk *noun* (1611)
: a desk that often has a sloping top for writing on; *also* **:** a portable case that contains writing materials and has a surface for writing

writing paper *noun* (1548)
: paper that is usually finished with a smooth surface and sized and that can be written on with ink

Writ·ings \'rī-tiŋz\ *noun plural* [translation of Late Hebrew *kĕthūbhīm*] (14th century)
: HAGIOGRAPHA

writ of assistance (1706)
1 : a writ issued to a law officer (as a sheriff or marshal) for the enforcement of a court order or decree; *especially* **:** one used to enforce an order for the possession of lands
2 : a writ used especially in colonial America authorizing a law officer to search in unspecified locations for unspecified illegal goods

writ of certiorari (circa 1532)
: CERTIORARI

writ of error (15th century)
: a common law writ directing an inferior court to remit the record of a legal action to the reviewing court in order that an error of law may be corrected if it exists

writ of extent (circa 1861)
: a writ formerly used to recover debts of record to the British crown and under which the lands, goods, and person of the debtor might all be seized to secure payment

writ of habeas corpus (1771)
: HABEAS CORPUS

writ of mandamus (circa 1861)
: MANDAMUS

writ of prohibition (circa 1876)
: a writ issued by a superior court to prevent an inferior court from acting beyond its jurisdiction

writ of right *noun* (15th century)
1 : a common law writ for restoring to its owner property held by another
2 : a writ granted as a matter of right

writ of summons (1660)
: a writ issued on behalf of the British monarch summoning a lord spiritual or a lord temporal to attend parliament

¹wrong \'rȯŋ\ *noun* [Middle English, from Old English *wrang,* from (assumed) *wrang,* adjective, wrong] (before 12th century)
1 a : an injurious, unfair, or unjust act **:** action or conduct inflicting harm without due provocation or just cause **b :** a violation or invasion of the legal rights of another; *especially* **:** TORT
2 : something wrong, immoral, or unethical; *especially* **:** principles, practices, or conduct contrary to justice, goodness, equity, or law
3 : the state, position, or fact of being or doing wrong: as **a :** the state of being mistaken or incorrect **b :** the state of being guilty
synonym see INJUSTICE

²wrong *adjective* **wrong·er** \'rȯŋ-ər\; **wrong·est** \'rȯŋ-əst\ [Middle English, from (assumed) Old English *wrang,* of Scandinavian origin; akin to Old Norse *rangr* awry, wrong; akin to Old English *wringan* to wring] (13th century)
1 : not according to the moral standard **:** SINFUL, IMMORAL ⟨thought that war was *wrong*⟩
2 : not right or proper according to a code, standard, or convention **:** IMPROPER ⟨it was *wrong* not to thank your host⟩
3 : not according to truth or facts **:** INCORRECT ⟨gave a *wrong* date⟩
4 : not satisfactory (as in condition, results, health, or temper)
5 : not in accordance with one's needs, intent, or expectations ⟨took the *wrong* bus⟩
6 : of, relating to, or constituting the side of something that is usually held to be opposite to the principal one, that is the one naturally or by design turned down, inward, or away, or that is the least finished or polished
— wrong·ly \'rȯŋ-lē\ *adverb*
— wrong·ness *noun*
— wrong side of the tracks : a rundown or unfashionable neighborhood

³wrong *adverb* (13th century)
1 : without accuracy **:** INCORRECTLY ⟨guessed *wrong*⟩
2 : without regard for what is proper or just
3 : in a wrong direction
4 a : in an unsuccessful or unfortunate way **b :** out of working order or condition
5 : in a false light ⟨don't get me *wrong*⟩

⁴wrong *transitive verb* **wronged; wrong·ing** \'rȯŋ-iŋ\ (14th century)
1 a : to do wrong to **:** INJURE, HARM **b :** to treat disrespectfully or dishonorably **:** VIOLATE
2 : DEFRAUD — usually used with *of*
3 : DISCREDIT, MALIGN ☆
— wrong·er \'rȯŋ-ər\ *noun*

wrong·do·er \'rȯŋ-ˌdü-ər\ *noun* (15th century)
: one that does wrong; *especially* **:** one who transgresses moral laws

wrong·do·ing \-ˌdü-iŋ\ *noun* (14th century)
1 : evil or improper behavior or action
2 : an instance of doing wrong

wronged *adjective* (circa 1547)
: being injured unjustly **:** suffering a wrong

wrong·ful \'rȯŋ-fəl\ *adjective* (14th century)
1 : WRONG, UNJUST
2 a : having no legal sanction **:** UNLAWFUL **b :** ILLEGITIMATE
— wrong·ful·ly \-fə-lē\ *adverb*
— wrong·ful·ness *noun*

wrong·head·ed \'rȯŋ-'he-dəd\ *adjective* (1732)
: stubborn in adherence to wrong opinion or principles ⟨a *wrongheaded* policy⟩
— wrong·head·ed·ly *adverb*
— wrong·head·ed·ness *noun*

wrote *past and dialect past participle of* WRITE

wroth \'rȯth *also* 'rōth\ *adjective* [Middle English, from Old English *wrāth;* akin to Old High German *reid* twisted, Old English *wrīthan* to writhe] (before 12th century)
: intensely angry **:** highly incensed **:** WRATHFUL

¹wrought \'rȯt\ *past and past participle of* WORK
usage see WRECK

²wrought *adjective* [Middle English, from past participle of *worken* to work] (14th century)
1 : worked into shape by artistry or effort ⟨carefully *wrought* essays⟩
2 : elaborately embellished **:** ORNAMENTED
3 : processed for use **:** MANUFACTURED ⟨*wrought* silk⟩
4 : beaten into shape by tools **:** HAMMERED — used of metals
5 : deeply stirred **:** EXCITED — often used with *up* ⟨gets easily *wrought* up over nothing⟩

wrought iron *noun* (1678)
: a commercial form of iron that is tough, malleable, and relatively soft, contains less than 0.3 percent and usually less than 0.1 percent carbon, and carries 1 or 2 percent of slag mechanically mixed with it

wrung *past and past participle of* WRING

¹wry \'rī\ *verb* **wried; wry·ing** [Middle English *wrien,* from Old English *wrigian* to turn; akin to Middle High German *rigel* kerchief wound around the head, Greek *rhiknos* shriveled, Avestan *urvisyeiti* he turns] (14th century)
intransitive verb
: TWIST, WRITHE
transitive verb
: to pull out of or as if out of proper shape **:** make awry

²wry *adjective* **wry·er** \'rī(-ə)r\; **wry·est** \'rī-əst\ (1523)
1 : having a bent or twisted shape or condition ⟨a *wry* smile⟩; *especially* **:** turned abnormally to one side ⟨a *wry* neck⟩
2 : WRONGHEADED
3 : cleverly and often ironically or grimly humorous
— wry·ly \'rī-lē\ *adverb*
— wry·ness *noun*

wry·neck \'rī-ˌnek\ *noun* (1585)
1 : either of two Old World woodpeckers (*Jynx torquilla* or *J. ruficollis*) that differ from the typical woodpeckers in having soft tail feathers and a peculiar manner of writhing the neck
2 : TORTICOLLIS

Wu \'wü\ *noun* [Chinese (Beijing) *Wú,* historical kingdom coextensive with the dialect area] (1908)
: a group of Chinese dialects spoken principally in Jiangsu and Zhejiang provinces

wud \'wüd\ *adjective* [alteration of ¹*wood*] (1772)
chiefly Scottish **:** INSANE, MAD

wul·fen·ite \'wu̇l-fə-ˌnīt\ *noun* [German *Wulfenit,* from F. X. von *Wulfen* (died 1805) Austrian mineralogist] (1849)
: a tetragonal mineral that is a complex oxide of lead and molybdenum and occurs especially in bright orange-yellow tabular crystals

wun·der·kind \'vu̇n-dər-ˌkint\ *noun, plural* **wun·der·kin·der** \-ˌkin-dər\ [German, from *Wunder* wonder + *Kind* child] (1891)
: a child prodigy; *also* **:** one who succeeds in a competitive or highly difficult field or profession at an early age

wurst \'wərst *also* 'wu̇rst *sometimes* 'wu̇sht *or*

☆ SYNONYMS
Wrong, oppress, persecute, aggrieve mean to injure unjustly or outrageously. WRONG implies inflicting injury either unmerited or out of proportion to what one deserves ⟨a penal system that had *wronged* him⟩. OPPRESS suggests inhumane imposing of burdens one cannot endure or exacting more than one can perform ⟨a people *oppressed* by a warmongering tyrant⟩. PERSECUTE implies a relentless and unremitting subjection to annoyance or suffering ⟨a child *persecuted* by constant criticism⟩. AGGRIEVE implies suffering caused by an infringement or denial of rights ⟨a legal aid society representing *aggrieved* minority groups⟩.

ˈwu̇st; *also with* v *for* w\ *noun* [German, from Old High German] (1855)

: SAUSAGE

wurzel *noun* [short for *mangel-wurzel*] (circa 1888)

: MANGEL

wuss \ˈwu̇s\ *also* **wus•sy** \ˈwu̇-sē\ *noun, plural* **wuss•es** *also* **wus•sies** [origin unknown] (1983)

: WIMP

— wussy *adjective*

wuth•er \ˈwə-t͟hər\ *intransitive verb* [alteration of *whither* to rush, bluster, hurl] (circa 1825) *dialect English* **:** to blow with a dull roaring sound

Wy•an•dot \ˈwī-ən-ˌdät *also* ˈwīn-\ *noun* (1749)

: a member of an American Indian group formed in the 17th century by Hurons and other Indians fleeing the Iroquois

wy•an•dotte \-ˌdät\ *noun* [probably from *Wyandotte* Wyandot] (1884)

: any of a U.S. breed of medium-sized domestic fowls raised for meat and eggs

Wyc•liff•ite \ˈwi-klə-ˌfīt\ *noun* [John *Wycliffe*] (1580)

: LOLLARD

— Wycliffite *adjective*

wye \ˈwī\ *noun* (1857)

1 : a Y-shaped part or object

2 : the letter *y*

wy•lie•coat \ˈwī-lē-ˌkōt, ˈwi-\ *noun* [Middle English (Scots) *wyle cot*] (15th century)

1 *chiefly Scottish* **:** a warm undergarment

2 *chiefly Scottish* **:** PETTICOAT

wynd \ˈwīnd\ *noun* [Middle English (Scots) *wynde,* probably from *wynden* to wind, proceed, go, from Old English *windan* to twist — more at WIND] (15th century)

chiefly Scottish **:** a very narrow street

wynn *or* **wyn** \ˈwin\ *noun* [Old English *wynn,* literally, joy — more at WINSOME] (before 12th century)

: a runic letter used in Old English and Middle English to represent the consonant \w\

WYS•I•WYG \ˈwi-zē-ˌwig, -zə-\ *adjective* [*w*hat *y*ou *s*ee *i*s *w*hat *y*ou *g*et] (1982)

: of, relating to, or being a display generated by word-processing or desktop-publishing software that exactly reflects the appearance of the printed document

wy•vern \ˈwī-vərn\ *noun* [alteration of Middle English *wyvere* viper, from Old North French *wivre,* modification of Latin *vipera*] (1610)

: a mythical animal usually represented as a 2-legged winged creature resembling a dragon

X is the twenty-fourth letter of the English alphabet. Its form and pronunciation are from Latin, as is its name. The letter came into Latin from a western Greek alphabet in which it had the value of ks. *In an eastern Greek alphabet and in classical Greek, the same sign stood for the letter* chi, *which represented the breathy, guttural sound heard in English at the end of Scottish* loch, *while that of* ks *was indicated by the letter* xi. *In English,* x *represents an unusually wide variety of sounds for a consonant letter. Most prominent are these five: the sound of* ks *as in* tax, *of* z *as in* xylophone, *of* gz *as in* exact, *of* ksh *as in one pronunciation of* anxious, *and of* gzh *as in one pronunciation of* luxurious. *Initial* X *occurs most often in words of Greek origin, where it is pronounced like* z *and stands for classical Greek* xi. *In the middle and at the end of words, however,* X *appears frequently in words of Latin origin, directly or through French, such as* axis *and* mix. *Small* x *developed through the miniaturizing of the capital form.*

---

^{1}**x** \ˈeks\ *noun, plural* **x's** *or* **xs** \ˈek-səz\ *often capitalized, often attributive* (before 12th century)
1 a : the 24th letter of the English alphabet **b :** a graphic representation of this letter **c :** a speech counterpart of orthographic *x*
2 : TEN — see NUMBER table
3 : a graphic device for reproducing the letter *x*
4 : one designated *x* especially as the 24th in order or class, or the first in an order or class that includes x, y, and sometimes z
5 : an unknown quantity
6 : something shaped like or marked with the letter X

^{2}**x** *transitive verb* **x-ed** *also* **x'd** *or* **xed** \ˈekst\; **x-ing** *or* **x'ing** \ˈek-siŋ\ (circa 1849)
1 : to mark with an *x*
2 : to cancel or obliterate with a series of *x*'s — usually used with *out*

X \ˈeks\ *adjective* (1950)
of a motion picture **:** of such a nature that admission is denied to persons under a specified age (as 17) — used before the adoption of *NC-17*

Xan·a·du \ˈza-nə-ˌdü, -ˌdyü\ *noun* [*Xanadu,* locality in *Kubla Khan* (1798), poem by Samuel T. Coleridge] (1919)
: an idyllic, exotic, or luxurious place

xanth- *or* **xantho-** *combining form* [New Latin, from Greek, from *xanthos*]
: yellow ⟨*xanth*ene⟩

xan·than gum \ˈzan-thən-\ *noun* [*xanth-* (from New Latin *Xanthomonas,* genus name) + ^{3}*-an*] (1964)
: a polysaccharide that is produced by fermentation of carbohydrates by a gram-negative bacterium (*Xanthomonas campestris* of the family Pseudomonadaceae) and is a thickening and suspending agent used especially in pharmaceuticals and prepared foods — called also *xanthan*

xan·thate \ˈzan-ˌthāt\ *noun* (1831)
: a salt or ester of any of various thio acids and especially $C_3H_6OS_2$

xan·thene \-ˌthēn\ *noun* (1898)
1 : a white crystalline heterocyclic compound $C_{13}H_{10}O$; *also* **:** an isomer of this that is the parent of the colored forms of the xanthene dyes
2 : any of various derivatives of xanthene

xanthene dye *noun* (1930)
: any of various brilliant fluorescent yellow to pink to bluish red dyes that are characterized by the presence of the xanthene nucleus

xan·thine \ˈzan-ˌthēn\ *noun* [International Scientific Vocabulary] (1857)
: a feebly basic compound $C_5H_4N_4O_2$ that occurs especially in animal or plant tissue, is derived from guanine and hypoxanthine, and yields uric acid on oxidation; *also* **:** any of various derivatives of this

Xan·thip·pe \zan-ˈthi-pē, -ˈti-\ *or* **Xan·tip·pe** \-ˈti-pē\ *noun* [Greek *Xanthippē,* shrewish wife of Socrates] (1691)
: an ill-tempered woman

xan·thone \ˈzan-ˌthōn\ *noun* [International Scientific Vocabulary] (circa 1894)
: a ketone $C_{13}H_8O_2$ that is the parent of several natural yellow pigments

xan·tho·phyll \ˈzan(t)-thə-ˌfil\ *noun* [French *xanthophylle,* from *xanth-* + *-phylle* -phyll] (1838)
: any of several neutral yellow to orange carotenoid pigments that are oxygen derivatives of carotenes; *especially* **:** LUTEIN

Xa·ve·ri·an Brother \zā-ˈvir-ē-ən-, za-\ *noun* [*Xaverian* of Saint Francis Xavier] (1882)
: a member of a Roman Catholic congregation of lay brothers founded by Theodore J. Ryken in Bruges, Belgium, in 1839 and dedicated to education

x–ax·is \ˈeks-ˌak-səs\ *noun* (1886)
1 : the axis in a plane Cartesian coordinate system parallel to which abscissas are measured
2 : one of the three axes in a three-dimensional rectangular coordinate system

X band *noun* (circa 1946)
: a segment of the superhigh-frequency radio spectrum that lies between 5.2 GHz and 10.9 GHz and is used especially for radars and for spacecraft communication

X chromosome *noun* (1911)
: a sex chromosome that usually occurs paired in each female cell and single in each male cell in species in which the male typically has two unlike sex chromosomes — compare Y CHROMOSOME

x–co·or·di·nate \ˌeks-kō-ˈȯrd-nət; -ˈȯr-d^{ə}n-ət, -də-ˌnāt\ *noun* (1927)
: a coordinate whose value is determined by measuring parallel to an x-axis; *specifically* **:** ABSCISSA

xe·bec \ˈzē-ˌbek, zi-ˈ\ *noun* [modification of French *chebec,* from Arabic *shabbāk*] (1756)
: a usually 3-masted Mediterranean sailing ship with long overhanging bow and stern

xen- *or* **xeno-** *combining form* [Late Latin, from Greek, from *xenos* stranger, guest, host]
1 : guest **:** foreigner ⟨*xeno*phobia⟩
2 : strange **:** foreign ⟨*xeno*lith⟩

xe·nia \ˈzē-nē-ə, -nyə\ *noun* [New Latin, from Greek, hospitality, from *xenos* host] (1899)
: the effect of genes introduced by pollen especially on endosperm and embryo development

xe·no·bi·ot·ic \ˌze-nō-bī-ˈä-tik, ˌzē-, -bē-\ *noun* (1965)
: a chemical compound (as a drug, pesticide, or carcinogen) that is foreign to a living organism
— **xenobiotic** *adjective*

xe·no·di·ag·no·sis \-ˌdī-ig-ˈnō-səs\ *noun* [New Latin] (circa 1929)
: the detection of a parasite (as of humans) by feeding a suitable intermediate host (as an insect) on supposedly infected material (as blood) and later examining the host for the parasite
— **xe·no·di·ag·nos·tic** \-ˈnäs-tik\ *adjective*

xe·no·ge·ne·ic \-jə-ˈnē-ik\ *adjective* [*xen-* + *-geneic* (as in *isogeneic*)] (1961)
: derived from, originating in, or being a member of another species

xe·no·graft \ˈze-nə-ˌgraft, ˈzē-\ *noun* (1961)
: HETEROGRAFT

xe·no·lith \ˈze-n^{ə}l-ˌith, ˈzē-\ *noun* (1894)
: a fragment of a rock included in another rock
— **xe·no·lith·ic** \ˌze-n^{ə}l-ˈi-thik, ˌzē-\ *adjective*

xe·non \ˈzē-ˌnän, ˈze-\ *noun* [Greek, neuter of *xenos* strange] (1898)
: a heavy, colorless, and relatively inert gaseous element that occurs in air as about one part in 20 million by volume and is used especially in thyratrons and specialized flashtubes — see ELEMENT table

xe·no·phile \ˈze-nə-ˌfīl, ˈzē-\ *noun* [International Scientific Vocabulary] (1948)
: one attracted to foreign things (as styles or people)

xe·no·phobe \ˈze-nə-ˌfōb, ˈzē-\ *noun* [International Scientific Vocabulary] (1922)
: one unduly fearful of what is foreign and especially of people of foreign origin
— **xe·no·pho·bic** \ˌze-nə-ˈfō-bik, ˌzē-\ *adjective*
— **xe·no·pho·bi·cal·ly** \-bi-k(ə-)lē\ *adverb*

xe·no·pho·bia \ˌze-nə-ˈfō-bē-ə, ˌzē-\ *noun* [New Latin] (1903)
: fear and hatred of strangers or foreigners or of anything that is strange or foreign

xe·no·tro·pic \-ˈträ-pik, -ˈtrō-\ *adjective* (1973)
: replicating or reproducing only in cells other than those of the host species ⟨*xenotropic* viruses⟩

xer- *or* **xero-** *combining form* [Late Latin, from Greek *xēr-, xēro-,* from *xēros*]
: dry ⟨*xer*ic⟩ ⟨*xero*phyte⟩

xe·ric \ˈzir-ik, ˈzer-\ *adjective* (1926)
: characterized by, relating to, or requiring only a small amount of moisture ⟨a *xeric* habitat⟩ ⟨a *xeric* plant⟩ — compare HYDRIC, MESIC

Xeri·scape \ˈzir-ə-ˌskāp, ˈzer-\ *trademark*
— used for a water-conserving method of landscaping in arid or semiarid climates

xe·ro·der·ma pig·men·to·sum \ˌzir-ə-ˈdər-mə-ˌpig-mən-ˈtō-səm, -ˌmen-\ *noun* [New Latin, literally, pigmented dryness of the skin] (1884)
: a genetic condition inherited as a recessive autosomal trait that is caused by a defect in mechanisms that repair DNA mutations (as those caused by ultraviolet light) and is characterized by the development of pigment abnormalities and multiple skin cancers in body areas exposed to the sun

xe·rog·ra·phy \zə-ˈrä-grə-fē\ *noun* [International Scientific Vocabulary] (1948)
1 : a process for copying graphic matter by the action of light on an electrically charged photoconductive insulating surface in which the latent image is developed with a resinous powder
2 : XERORADIOGRAPHY
— **xe·ro·graph·ic** \ˌzir-ə-ˈgra-fik\ *adjective*
— **xe·ro·graph·i·cal·ly** \-fi-k(ə-)lē\ *adverb*

xe·roph·i·lous \zə-ˈrä-fə-ləs\ *or* **xe·ro·phile** \ˈzir-ə-ˌfīl\ *adjective* (1863)

: thriving in or tolerant or characteristic of a xeric environment
— **xe·roph·i·ly** \zə-'rä-fə-lē\ *noun*

xe·roph·thal·mia \ˌzir-ˌäf-'thal-mē-ə, -ˌäp-'thal-\ *noun* [Late Latin, from Greek *xērophthalmia,* from *xēr-* xer- + *ophthalmia* ophthalmia] (circa 1656)
: a dry thickened lusterless condition of the eyeball resulting especially from a severe systemic deficiency of vitamin A
— **xe·roph·thal·mic** \-mik\ *adjective*

xe·ro·phyte \'zir-ə-ˌfīt\ *noun* (1897)
: a plant structurally adapted for life and growth with a limited water supply especially by means of mechanisms that limit transpiration or that provide for the storage of water
— **xe·ro·phyt·ic** \ˌzir-ə-'fi-tik\ *adjective*
— **xe·ro·phyt·ism** \'zir-ə-ˌfī-ˌti-zəm\ *noun*

xe·ro·ra·di·og·ra·phy \ˌzir-ō-ˌrā-dē-'ä-grə-fē\ *noun* (1949)
: radiography used especially in mammographic screening for breast cancer that produces an image using X rays in a manner similar to the way an image is produced by light in xerography

xe·ro·ther·mic \ˌzir-ə-'thər-mik\ *adjective* (1904)
1 : characterized by heat and dryness
2 : adapted to or thriving in a hot dry environment

xe·rox \'zir-ˌäks, 'zē-ˌräks\ *transitive verb* [from *Xerox*] (1965)
1 : to copy on a Xerox copier
2 : to make (a copy) on a Xerox copier

Xe·rox \'zir-ˌäks, 'zē-ˌräks\ *trademark*
— used for a xerographic copier

x–height \'eks-ˌhīt, -ˌhītth\ *noun* (circa 1945)
: the height of a lowercase x used to represent the height of the main body of a lowercase letter

Xho·sa \'kō-sə, 'hō-, 'kō- *or with* ò for ō\ *noun* (1801)
1 : a member of a Bantu-speaking people of Eastern Cape province
2 : a Bantu language of the Xhosas

xi \'zī, 'ksī\ *noun, plural* **xis** [Greek *xei*] (1823)
: the 14th letter of the Greek alphabet — see ALPHABET table

x–in·ter·cept \'eks-'in-tər-ˌsept\ *noun* (circa 1939)
: the x-coordinate of a point where a line, curve, or surface intersects the x-axis

xi·phi·ster·num \ˌzī-fə-'stər-nəm, ˌzi-\ *noun, plural* **-na** \-nə\ [New Latin, from Greek *xiphos* sword + New Latin *sternum*] (circa 1860)
: XIPHOID PROCESS

xi·phoid \'zī-ˌfȯid, 'zi-\ *noun* [New Latin *xiphoides,* from Greek *xiphoeidēs,* from *xiphos*] (circa 1860)
: XIPHOID PROCESS
— **xiphoid** *adjective*

xiphoid process *noun* (1873)
: the segment of the human sternum that is the third and closest to the feet

x–ir·ra·di·a·tion \ˌeks-i-ˌrā-dē-'ā-shən\ *noun, often capitalized* (1927)
: X-RADIATION 1

Xmas \'kris-məs *also* 'eks-məs\ *noun* [X (symbol for *Christ,* from the Greek letter chi (X), initial of *Christos* Christ) + *-mas* (in *Christmas*)] (1551)
: CHRISTMAS

x–ra·di·a·tion \ˌeks-ˌrā-dē-'ā-shən\ *noun, often capitalized* (1896)
1 : exposure to X rays
2 : radiation composed of X rays

X–rated \'eks-'rā-təd\ *adjective* (1970)
1 : having a rating of X; *broadly* **:** relating to or characterized by explicit sexual material or activity ⟨an *X-rated* book⟩
2 : OBSCENE, VULGAR ⟨an *X-rated* gesture⟩

Xray \'eks-ˌrā\ (1943)
— a communications code word for the letter x ◆

x–ray \'eks-ˌrā\ *transitive verb, often capitalized* (1899)
: to examine, treat, or photograph with X rays

X ray \'eks-ˌrā\ *noun* (1896)
1 : any of the electromagnetic radiations of the same nature as visible radiation but having an extremely short wavelength of less than 100 angstroms that is produced by bombarding a metallic target with fast electrons in vacuum or by transition of atoms to lower energy states and that has the properties of ionizing a gas upon passage through it, of penetrating various thicknesses of all solids, of producing secondary radiations by impinging on material bodies, of acting on photographic films and plates as light does, and of causing fluorescent screens to emit light
2 : a photograph obtained by use of X rays
— **X–ray** *adjective*

X–ray astronomy *noun* (1963)
: astronomy dealing with investigations of celestial bodies by means of the X rays they emit

X–ray diffraction *noun* (1924)
: a scattering of X rays by the atoms of a crystal that produces an interference effect so that the diffraction pattern gives information on the structure of the crystal or the identity of a crystalline substance

X–ray star *noun* (1964)
: a luminous celestial object emitting a major portion of its radiation in the form of X rays — called also *X-ray source*

X–ray therapy *noun* (1926)
: medical treatment (as of cancer) by controlled application of X rays

X–ray tube *noun* (1896)
: a vacuum tube in which a concentrated stream of electrons strikes a metal target and produces X rays

x–sec·tion \'krȯs-'sek-shən, -ˌsek-\ *noun* [*x,* rebus for *cross*] (1962)
: CROSS SECTION
— **x–sec·tion·al** \-shnəl, -shə-nᵊl\ *adjective*

xu \'sü\ *noun, plural* **xu** [Vietnamese, from French *sou* sou] (1948)
1 : a coin formerly minted by South Vietnam equivalent to the cent
2 — see *dong* at MONEY table

xyl- *or* **xylo-** *combining form* [Latin, from Greek, from *xylon*]
1 : wood ⟨*xylo*phone⟩
2 : xylene ⟨*xyl*idine⟩

xy·lan \'zī-ˌlan\ *noun* [International Scientific Vocabulary] (circa 1894)
: a yellow gummy pentosan that yields xylose on hydrolysis and is abundantly present in plant cell walls and woody tissue

xy·lem \'zī-ləm, -ˌlem\ *noun* [German, from Greek *xylon*] (1873)
: a complex tissue in the vascular system of higher plants that consists of vessels, tracheids, or both usually together with wood fibers and parenchyma cells, functions chiefly in conduction of water and dissolved minerals but also in support and food storage, and typically constitutes the woody element (as of a plant stem) — compare PHLOEM

xylem ray *noun* (1875)
: a vascular ray or portion of a vascular ray located in xylem — called also *wood ray;* compare PHLOEM RAY

xy·lene \'zī-ˌlēn\ *noun* [International Scientific Vocabulary] (1851)
: any of three toxic flammable oily isomeric aromatic hydrocarbons C_8H_{10} that are dimethyl homologues of benzene and are usually obtained from petroleum or natural gas distillates; *also* **:** a mixture of xylenes and ethyl benzene used chiefly as a solvent

xy·li·dine \'zī-lə-ˌdēn\ *noun* [International Scientific Vocabulary] (1850)
: any or a mixture of six toxic liquid or low-melting crystalline isomeric amino derivatives $C_8H_{11}N$ of the xylenes used chiefly as intermediates for azo dyes and in organic synthesis

xy·li·tol \'zī-lə-ˌtȯl, -ˌtōl\ *noun* (1891)
: a crystalline alcohol $C_5H_{12}O_5$ that is a derivative of xylose, is obtained especially from birch bark, and is used as a sweetener

xy·log·ra·phy \zī-'lä-grə-fē\ *noun* [French *xylographie,* from *xyl-* + *-graphie* -graphy] (1816)
: the art of making engravings on wood especially for printing
— **xy·lo·graph** \'zī-lə-ˌgraf\ *noun*
— **xy·log·ra·pher** \zī-'lä-grə-fər\ *noun*
— **xy·lo·graph·ic** \ˌzī-lə-'gra-fik\ *also* **xy·lo·graph·i·cal** \-fi-kəl\ *adjective*

xy·lol \'zī-ˌlȯl, -ˌlōl\ *noun* [International Scientific Vocabulary] (1851)
: XYLENE

xy·loph·a·gous \zī-'lä-fə-gəs\ *adjective* [Greek *xylophagos,* from *xyl-* + *-phagos* -phagous] (1739)
: feeding on or in wood

xy·lo·phone \'zī-lə-ˌfōn *also* 'zi-\ *noun* (1866)
: a percussion instrument consisting of a series of wooden bars graduated in length to produce the musical scale, supported on belts of straw or felt, and sounded by striking with two small wooden hammers

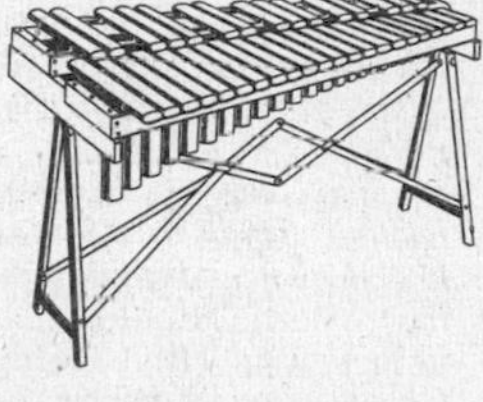
xylophone

— **xy·lo·phon·ist** \-ˌfō-nist\ *noun*

xy·lose \'zī-ˌlōs, -ˌlōz\ *noun* [International Scientific Vocabulary] (circa 1894)
: a crystalline aldose sugar $C_5H_{10}O_5$ that is not fermentable with ordinary yeasts and occurs especially as a constituent of xylans from which it is obtained by hydrolysis

◇ WORD HISTORY

Xray On November 8, 1895, Wilhelm Conrad Röntgen was conducting an experimental investigation in Würzburg, Germany, of the properties of cathode rays. He noticed that a fluorescent surface in the neighborhood of a cathode-ray tube would become luminous even if shielded from the direct light of the tube. A thick metal object placed between the tube and the affected surface impaired the fluorescence and cast a dark shadow. An object made of a less dense substance such as wood cast only a weak shadow. Röntgen's explanation for this phenomenon was that the tube produced some type of invisible radiation which could pass through substances that blocked ordinary visible light. Because he did not know the nature of this radiation he had discovered, he named it *X-Strahl,* based on the mathematical use of *x* to designate an unknown quantity. His German name was translated into English as *X ray.*

Y *is the twenty-fifth letter of the English alphabet. It comes from the Latin, which borrowed it from the* Greek upsilon *and used it as a vowel in the writing of Greek words. In Greek,* Y *was an eastern and classical form of the* V *which had been borrowed earlier from the western Greek alphabet by Latin. In English,* Y *is used as both a consonant and a vowel. At the beginning of a word or syllable, it is usually pronounced as a consonant with slight or no audible friction (as in* yes*); in the middle or at the end of a syllable, it is usually a vowel, with various values (as in* myth, happy, my, myrrh, *and* martyr*). In certain archaic forms such as the* ye *of "ye olde tavern,"* Y *represents an Old and Middle English character called* thorn *and is pronounced like* th. *This use caused confusion in later English speakers, who interpreted words like* ye *as though pronounced with the sound of consonant* y. *Actually, such words were originally pronounced with the* th *sound.*

y \'wī\ *noun, plural* **y's** *or* **ys** \'wīz\ *often capitalized, often attributive* (before 12th century)
1 a : the 25th letter of the English alphabet **b :** a graphic representation of this letter **c :** a speech counterpart of orthographic *y*
2 : a graphic device for reproducing the letter *y*
3 : one designated *y* especially as the 25th in order or class or the second in order or class when x is made the first
4 : something shaped like the letter Y

Y \'wī\ *noun* (circa 1915)
: YMCA, YWCA

[1]-y *also* **-ey** \ē; *in some dialects, especially British, Southern, & New England, often* i *but not shown at individual entries*\ *adjective suffix* [Middle English, from Old English *-ig;* akin to Old High German *-īg* -y, Latin *-icus,* Greek *-ikos,* Sanskrit *-ika*]
1 a : characterized by **:** full of ⟨blossom*y*⟩ ⟨dirt*y*⟩ ⟨mudd*y*⟩ ⟨clay*ey*⟩ **b :** having the character of **:** composed of ⟨ic*y*⟩ ⟨wax*y*⟩ **c :** like **:** like that of ⟨hom*ey*⟩ ⟨wintr*y*⟩ — often with a disparaging connotation ⟨stag*y*⟩
2 a : tending or inclined to ⟨sleep*y*⟩ ⟨chatt*y*⟩ **b :** giving occasion for (specified) action ⟨tear*y*⟩ **c :** performing (specified) action ⟨curl*y*⟩

[2]-y *same*\ *noun suffix, plural* **-ies** [Middle English *-ie,* from Old French, from Latin *-ia,* from Greek *-ia, -eia*]
1 : state **:** condition **:** quality ⟨beggar*y*⟩
2 : activity, place of business, or goods dealt with ⟨chandler*y*⟩ ⟨laundr*y*⟩
3 : whole body or group ⟨soldier*y*⟩

[3]-y *noun suffix, plural* **-ies** [Middle English *-ie,* from Anglo-French, from Latin *-ium*]
: instance of a (specified) action ⟨entreat*y*⟩ ⟨inquir*y*⟩

[4]-y — see -IE

yab·ber \'ya-bər\ *noun* [probably modification of Wuywurung (Australian aboriginal language of the Melbourne area) *yaba* speak] (1855)
Australian **:** TALK, JABBER ⟨all *yabber* and chatter ceased around the campfires —Francis Birtles⟩
— yabber *intransitive verb*

[1]yacht \'yät\ *noun* [obsolete Dutch *jaght,* from Middle Low German *jacht,* short for *jachtschip,* literally, hunting ship] (1557)
: any of various recreational watercraft: as **a :** a sailboat used for racing **b :** a large usually motor-driven craft used for pleasure cruising ◆

[2]yacht *intransitive verb* (1836)
: to race or cruise in a yacht

yacht club *noun* (1834)
: a club organized to promote and regulate yachting and boating

yacht·ing *noun* (1836)
: the action, fact, or pastime of racing or cruising in a yacht

yachts·man \'yäts-mən\ *noun* (1862)
: a person who owns or sails a yacht

YAG \'yag\ *noun* [*y*ttrium *a*luminum *g*arnet] (1964)
: a synthetic yttrium aluminum garnet of marked hardness and high refractive index that is used especially as a gemstone and in laser technology

ya·gi \'yä-gē, 'ya-\ *noun* [Hidetsugu *Yagi* (born 1886) Japanese engineer] (1943)
: a highly directional and selective shortwave antenna consisting of a horizontal conductor of one or two dipoles connected with the receiver or transmitter and of a set of nearly equal insulated dipoles parallel to and on a level with the horizontal conductor

ya·hoo \'yā-(ˌ)hü, 'yä-\ *noun, plural* **yahoos** (1726)
1 *capitalized* **:** a member of a race of brutes in Swift's *Gulliver's Travels* who have the form and all the vices of humans
2 : a boorish, crass, or stupid person
— ya·hoo·ism \-ˌi-zəm\ *noun*

Yah·weh \'yä-(ˌ)wā, -(ˌ)vā\ *also* **Yah·veh** \-(ˌ)vā\ *noun* [Hebrew *Yahweh*] (1869)
: GOD 1a — used especially by the Hebrews; compare TETRAGRAMMATON

Yah·wism \-ˌwi-zəm, -ˌvi-\ *noun* (1867)
: the worship of Yahweh among the ancient Hebrews

Yah·wis·tic \yä-'wis-tik, -'vis-\ *adjective* (1874)
1 : characterized by the use of *Yahweh* as the name of God
2 : of or relating to Yahwism

[1]yak \'yak\ *noun, plural* **yaks** *also* **yak** [Tibetan *gyagk*] (1795)
: a large long-haired wild or domesticated ox (*Bos grunniens* synonym *B. mutus*) of Tibet and adjacent elevated parts of central Asia

yak

[2]yak \'yäk, 'yak\ *noun* [imitative] (1948)
1 *slang* **:** LAUGH
2 *slang* **:** JOKE, GAG

[3]yak *also* **yack** *intransitive verb* **yakked** *also* **yacked; yak·king** *also* **yack·ing** [probably imitative] (1949)
: to talk persistently **:** CHATTER

[4]yak *also* **yack** \'yak\ *noun* (1950)
: persistent or voluble talk

Yak·i·ma \'ya-kə-ˌmȯ\ *noun, plural* **Yakima** *or* **Yakimas** (1838)
1 : a member of a group of Shahaptian peoples of the lower Yakima River valley, south central Washington
2 : the language of the Yakima people

ya·ki·to·ri \ˌyä-ki-'tȯr-ē\ *noun* [Japanese, grilled chicken, from *yaki* broil, roast + *tori* bird] (1962)
: bite-sized marinated pieces of beef, seafood, or chicken on skewers

y'all \'yȯl\ *variant of* YOU-ALL

yam \'yam\ *noun* [earlier *iname,* from Portuguese *inhame* & Spanish *ñame,* of African origin; akin to Fulani *nyami* to eat] (1657)
1 : the edible starchy tuberous root of various plants (genus *Dioscorea* of the family Dioscoreaceae) used as a staple food in tropical areas; *also* **:** a plant producing yams
2 : a moist-fleshed and usually orange-fleshed sweet potato

ya·men \'yä-mən\ *noun* [Chinese (Beijing) *yámen*] (1747)
: the headquarters or residence of a Chinese government official or department

yam·mer \'ya-mər\ *intransitive verb* **yammered; yam·mer·ing** \'ya-mə-riŋ, 'yam-riŋ\ [Middle English *yameren,* alteration of *yomeren* to murmur, be sad, from Old English *gēomrian;* akin to Old High German *jāmaron* to be sad] (15th century)
1 a : to utter repeated cries of distress or sorrow **b :** WHIMPER
2 : to utter persistent complaints **:** WHINE
3 : to talk persistently or volubly and often loudly ⟨caused the purists to *yammer* for censorship —D. W. Maurer⟩
— yammer *noun*

yang \'yäŋ, 'yaŋ\ *noun* [Chinese (Beijing) *yáng*] (1671)
: the masculine active principle in nature that in Chinese cosmology is exhibited in light, heat, or dryness and that combines with yin to produce all that comes to be

[1]yank \'yaŋk\ *noun* [origin unknown] (circa 1864)
: a strong sudden pull **:** JERK

[2]yank (1822)
intransitive verb
: to pull on something with a quick vigorous movement
transitive verb
1 : to pull or extract with a quick vigorous movement
2 : to remove in or as if in an abrupt manner ⟨*yanked* the story from the evening edition⟩

Yank \'yaŋk\ *noun* (1778)
: YANKEE

◇ WORD HISTORY

yacht In the 16th century the Dutch developed a kind of shallow-draft, fast-moving ship for use in coastal waters and river estuaries. These vessels were called by the name *jaght,* short for Middle Low German *jachtschip* or Middle Dutch *jageschip,* literally "hunting ship" or "pursuit ship," though *jacht* may simply imply that the boat was a speedy one. The name appears in 16th and 17th century English documents in a variety of forms, such as *yoath* and *yaught.* The form *jacht* or *yacht* became widespread after 1660, when King Charles II, on his return from exile on the Continent, was given such a craft, dubbed the *Mary,* by the Dutch East India Company. The king chose to use the boat for excursions and racing, and the word *yacht* became attached to other vessels used for the same purposes, the earliest being copied after the design of the *Mary.* Today the word *yacht* is applied to a wide variety of luxuriously outfitted craft—powerboats as well as sailboats—that are designed for pleasure cruising.

¹Yan·kee \'yaŋ-kē\ *noun* [origin unknown] (1758)
1 a : a native or inhabitant of New England **b :** a native or inhabitant of the northern U.S.
2 : a native or inhabitant of the U.S.
— Yan·kee·dom \-kē-dəm\ *noun*
— Yan·kee·ism \-kē-,i-zəm\ *noun*

²Yankee (1952)
— a communications code word for the letter *y*

Yan·kee–Doo·dle \,yaŋ-kē-'dü-dᵊl\ *noun* [*Yankee Doodle,* popular song during the American Revolution] (1770)
: YANKEE

yan·qui \'yäŋ-kē\ *noun, often capitalized* [Spanish, from English ¹*Yankee*] (1928)
: a citizen of the U.S. as distinguished from a Latin American

yan·tra \'yən-trə, 'yan-, 'yän-\ *noun* [Sanskrit] (1877)
: a geometrical diagram used like an icon usually in meditation

¹yap \'yap\ *intransitive verb* **yapped; yap·ping** [imitative] (1668)
1 : to bark snappishly **:** YELP
2 : to talk in a shrill insistent way **:** CHATTER
— yap·per *noun*

²yap *noun* (1826)
1 a : a quick sharp bark **:** YELP **b :** shrill insistent talk **:** CHATTER
2 : an unsophisticated, ignorant, or uncouth person **:** BUMPKIN
3 *slang* **:** MOUTH

Ya·qui \'yä-kē\ *noun* (1861)
1 : a member of an American Indian people of Sonora, Mexico
2 : the Uto-Aztecan language of the Yaqui people

Yar·bor·ough \'yär-,bər-ə, -,bə-rə, -b(ə-)rə\ *noun* [Charles Anderson Worsley, 2d Earl of *Yarborough* (died 1897) English nobleman said to have bet a thousand to one against the dealing of such a hand] (1900)
: a hand in bridge or whist containing no ace and no card higher than a nine

¹yard \'yärd\ *noun* [Middle English, from Old English *geard* enclosure, yard; akin to Old High German *gart* enclosure, Latin *hortus* garden] (before 12th century)
1 a : a small usually walled and often paved area open to the sky and adjacent to a building **:** COURT **b :** the grounds of a building or group of buildings
2 : the grounds immediately surrounding a house that are usually covered with grass
3 a : an enclosure for livestock (as poultry) **b** (1) **:** an area with its buildings and facilities set aside for a particular business or activity (2) **:** an assembly or storage area **c :** a system of tracks for storage and maintenance of cars and making up trains
4 : a locality in a forest where deer herd in winter

²yard *adjective* (15th century)
1 : of, relating to, or employed in the yard surrounding a building ⟨*yard* light⟩
2 : of, relating to, or employed in a railroad yard ⟨a *yard* engine⟩

³yard (1758)
transitive verb
1 : to drive into or confine in a restricted area **:** HERD, PEN
2 : to deliver to or store in a yard
intransitive verb
: to congregate in or as if in a yard

⁴yard *noun* [Middle English *yarde,* from Old English *gierd* twig, measure, yard; akin to Old High German *gart* stick, Latin *hasta* spear] (before 12th century)
1 : any of various units of measure: as **a :** a unit of length equal in the U.S. to 0.9144 meter — see WEIGHT table **b :** a unit of volume equal to a cubic yard
2 a : a great length or quantity ⟨remembered *yards* of facts and figures⟩ **b** *slang* **:** one hundred dollars
3 : a long spar tapered toward the ends to support and spread the head of a square sail, lateen, or lugsail
4 : a slender horn-shaped glass about three feet tall; *also* **:** the amount it contains ⟨a *yard* of ale⟩
— the whole nine yards : all of a related set of circumstances, conditions, or details ⟨who could learn the most about making records, about electronics and engineering, *the whole nine yards* —Stephen Stills⟩ — sometimes used adverbially with *go* to indicate an all-out effort

¹yard·age \'yär-dij\ *noun* [¹*yard*] (1867)
1 : the use of a livestock enclosure for animals in transit provided by a railroad at a station
2 : a charge made by a railroad for the use of a livestock enclosure

²yardage *noun* [⁴*yard*] (1900)
1 a : an aggregate number of yards **b :** the length, extent, or volume of something as measured in yards
2 : YARD GOODS

yard·arm \'yärd-,ärm\ *noun* (1553)
: either end of the yard of a square-rigged ship

yard·bird \-,bərd\ *noun* [¹*yard*] (circa 1941)
1 : a soldier assigned to a menial task or restricted to a limited area as a disciplinary measure
2 : an untrained or inept enlisted man

yard goods *noun plural* (1905)
: fabrics sold by the yard **:** PIECE GOODS

yard grass *noun* [¹*yard*] (1822)
: a coarse annual grass (*Eleusine indica*) with digitate spikes that is widely distributed as a weed — called also *goosegrass*

yard line *noun* (1949)
: any of a series of marked or imaginary lines one yard apart on a football field that are parallel to the goal lines and that indicate the distance to the nearest goal line

yard–long bean \'yärd-'lȯŋ-\ *noun* (1926)
: the edible 1- to 3-foot (0.3- to 0.9- meter) long thin stringless fruit of a plant of a subspecies (*Vigna unguiculata sesquipedalis*) of the cowpea; *also* **:** the plant

yard·man \'yärd-mən, -,man\ *noun* (circa 1825)
1 : a person employed to do outdoor work (as mowing lawns)
2 : a person who works in the yard of a commercial establishment; *especially* **:** one who supervises the handling of building materials in a lumberyard
3 : a railroad hand employed in yard service

yard·mas·ter \-,mas-tər\ *noun* (1864)
: the person in charge of operations in a railroad yard

yard sale *noun* (1972)
: GARAGE SALE

yard·stick \'yärd-,stik\ *noun* (1816)
1 a : a graduated measuring stick three feet (0.9144 meter) long **b :** a standard basis of calculation
2 : a standard for making a critical judgment **:** CRITERION
synonym see STANDARD

yare \'yar, 'yer, 'yär\ *adjective* [Middle English, from Old English *gearu;* akin to Old High German *garo* ready] (before 12th century)
1 *archaic* **:** set for action **:** READY
2 *or* **yar** \'yär\ **a :** characterized by speed and agility **:** NIMBLE, LIVELY **b :** HANDY 1c, MANEUVERABLE
— yare *adverb, archaic*
— yare·ly *adverb, archaic*

yar·mul·ke *also* **yar·mel·ke** \'yä-mə-kə, 'yär-mə(l)-kə\ *noun* [Yiddish *yarmlke,* from Polish *jarmułka* & Ukrainian *yarmulka* skullcap, of Turkic origin; akin to Turkish *yağmurluk* rainwear] (1903)
: a skullcap worn especially by Orthodox and Conservative Jewish males in the synagogue and the home

¹yarn \'yärn\ *noun* [Middle English, from Old English *gearn;* akin to Old High German *garn* yarn, Greek *chordē* string, Latin *hernia* rupture, Sanskrit *hira* band] (before 12th century)
1 a : a continuous often plied strand composed of either natural or man-made fibers or filaments and used in weaving and knitting to form cloth **b :** a similar strand of another material (as metal, glass, or plastic)
2 : a narrative of adventures; *especially* **:** a tall tale

²yarn *intransitive verb* (1812)
: to tell a yarn
— yarn·er *noun*

yarn–dye \'yärn-,dī\ *transitive verb* (1885)
: to dye before weaving or knitting

yar·row \'yar-(,)ō\ *noun* [Middle English *yarowe,* from Old English *gearwe;* akin to Old High German *garwa* yarrow] (before 12th century)
: a widely naturalized strong-scented Eurasian composite herb (*Achillea millefolium*) with finely dissected leaves and small usually white corymbose flowers; *also* **:** any of several congeneric plants

yash·mak *also* **yas·mak** \'yash-,mak, 'yas-\ *noun* [Turkish *yaşmak*] (1844)
: a veil worn by Muslim women that is wrapped around the upper and lower parts of the face so that only the eyes remain exposed to public view

yat·a·ghan \'ya-tə-,gan, 'ya-ti-gən\ *noun* [Turkish *yatağan*] (1819)
: a long knife or short saber that lacks a guard for the hand at the juncture of blade and hilt and that usually has a double curve to the edge and a nearly straight back

yauld \'yȯl(d)\ *adjective* [origin unknown] (1786)
chiefly Scottish **:** VIGOROUS, ENERGETIC

yau·pon \'yü-,pän *also* 'yō-, 'yȯ-\ *noun* [Catawba *yą́pą,* from *yą-* tree + *pą* leaf] (1709)
: a holly (*Ilex vomitoria*) of the southeastern U.S. that has smooth elliptical leaves with emetic and purgative properties

yau·tia \'yau̇-tē-ə\ *noun* [American Spanish *yautia,* from Taino] (1899)
: any of several aroid plants (genus *Xanthosoma* and especially *X. sagittifolium*) chiefly of tropical America with starchy edible shaggy brown tubers that are cooked and eaten like yams or potatoes; *especially* **:** one of these tubers

¹yaw \'yȯ\ *noun* [origin unknown] (1546)
1 : the action of yawing; *especially* **:** a side to side movement
2 : the extent of the movement in yawing

²yaw *intransitive verb* (1586)
1 a *of a ship* **:** to deviate erratically from a course (as when struck by a heavy sea); *especially* **:** to move from side to side **b** *of an airplane, spacecraft, or projectile* **:** to turn by angular motion about the vertical axis
2 : ALTERNATE ⟨restlessly *yawing* between apparent extremes —Martin Kasindorf⟩

yawl \'yȯl\ *noun* [Low German *jolle*] (1670)
1 : a ship's small boat **:** JOLLY BOAT
2 : a fore-and-aft rigged sailboat carrying a mainsail and one or more jibs with a mizzenmast far aft

yawl 2

\ə\ abut \ᵊ\ kitten \ər\ further \a\ ash \ā\ ace \ä\ mop, mar \au̇\ out \ch\ chin \e\ bet \ē\ easy \g\ go \i\ hit \ī\ ice \j\ job \ŋ\ sing \ō\ go \ȯ\ law \ȯi\ boy \th\ thin \t͟h\ the \ü\ loot \u̇\ foot \y\ yet \zh\ vision *see also* Guide to Pronunciation

¹yawn \'yȯn, 'yän\ *verb* [Middle English *yenen, yanen,* from Old English *ginian;* akin to Old High German *ginēn* to yawn, Latin *hiare,* Greek *chainein*] (before 12th century)
intransitive verb
1 : to open wide **:** GAPE
2 : to open the mouth wide usually as an involuntary reaction to fatigue or boredom
transitive verb
1 : to utter with a yawn
2 : to accomplish with or impel by yawns ⟨his grandchildren *yawned* him to bed —L. L. King⟩
²yawn *noun* (1602)
1 : GAP, CAVITY
2 : a deep usually involuntary intake of breath through the wide open mouth often as an involuntary reaction to fatigue or boredom; *also* **:** a reaction resembling a yawn ⟨a . . . success at the box office but drew only *yawns* from critics —*Current Biography*⟩
3 : ⁵BORE
yawn·er *noun* (1687)
1 : one that yawns
2 : something that causes boredom ⟨the show was a real *yawner*⟩
yawn·ing *adjective* (before 12th century)
1 : wide open **:** CAVERNOUS ⟨a *yawning* hole⟩ ⟨a dreadful, *yawning* gap of . . . weeks —*Times Educational Supplement*⟩
2 : showing fatigue or boredom by yawns ⟨a *yawning* audience⟩
— yawn·ing·ly \'yȯ-niŋ-lē\ *adverb*
¹yawp *or* **yaup** \'yȯp\ *intransitive verb* [Middle English *yolpen*] (14th century)
1 : to make a raucous noise **:** SQUAWK
2 : CLAMOR, COMPLAIN
— yawp·er *noun*
²yawp *also* **yaup** *noun* (1824)
1 : a raucous noise **:** SQUAWK
2 : something suggestive of a raucous noise; *specifically* **:** rough vigorous language
yawp·ing *noun* (1876)
: a strident utterance
yaws \'yȯz\ *noun plural but singular or plural in construction* [probably from an English-based creole of the Caribbean] (1679)
: an infectious contagious tropical disease caused by a spirochete (*Treponema pertenue*) closely resembling the causative agent of syphilis and marked by ulcerating lesions with later bone involvement — called also *frambesia*
y–ax·is \'wī-ˌak-səs\ *noun* (1875)
1 : the axis of a plane Cartesian coordinate system parallel to which ordinates are measured
2 : one of the three axes in a three-dimensional rectangular coordinate system
Y chromosome *noun* (1911)
: a sex chromosome that is characteristic of male cells in species in which the male typically has two unlike sex chromosomes — compare X CHROMOSOME
yclept *or* **ycleped** [Middle English, from Old English *geclipod,* past participle of *clipian* to cry out, name] *past participle of* CLEPE
y–co·or·di·nate \ˌwī-kō-'ȯrd-nət, -'ȯr-dᵊn-ət, -də-ˌnāt\ *noun* (1927)
: a coordinate whose value is determined by measuring parallel to a y-axis; *specifically* **:** ORDINATE
¹ye \'yē\ *pronoun* [Middle English, from Old English *gē;* akin to Old High German *ir* you — more at YOU] (before 12th century)
: YOU 1 — used originally only as a plural pronoun of the second person in the subjective case and now used especially in ecclesiastical or literary language and in various English dialects
word history see YOU
²ye \yē, yə; *originally same as* ¹THE\ *definite article* [alteration of Middle English *þe* the, from Old English *þē;* from the use of the letter y by printers and scribes of late Middle English to represent þ (*th*) of earlier manuscripts] (1551)
archaic **:** THE ⟨*Ye* Olde Gifte Shoppe⟩
¹yea \'yā\ *adverb* [Middle English *ye, ya,* from Old English *gēa;* akin to Old High German *jā* yes] (before 12th century)
1 : YES — used in oral voting
2 : more than this **:** not only so but — used as a function word to introduce a more explicit or emphatic phrase ⟨yet the impression, *yea* the evidence, is inescapable —J. G. Harrison⟩
²yea *noun* (13th century)
1 : AFFIRMATION, ASSENT
2 a : an affirmative vote **b :** a person casting a yea vote
yeah \'ye-ə, 'yeu̇, 'ya-ə\ *adverb* [by alteration] (1902)
: YES
yean \'yēn, ēn\ *intransitive verb* [Middle English *yenen,* from (assumed) Old English *geēanian,* from Old English *ge-,* perfective prefix + *ēanian* to yean; akin to Latin *agnus* lamb, Greek *amnos*] (1548)
: to bring forth young — used of a sheep or goat
yean·ling \-liŋ, -lən\ *noun* (1637)
: LAMB, KID 1a
year \'yir\ *noun* [Middle English *yere,* from Old English *gēar;* akin to Old High German *jār* year, Greek *hōros* year, *hōra* season, hour] (before 12th century)
1 a : the period of about 365¼ solar days required for one revolution of the earth around the sun **b :** the time required for the apparent sun to return to an arbitrary fixed or moving reference point in the sky **c :** the time in which a planet completes a revolution about the sun ⟨a *year* of Jupiter⟩
2 a : a cycle in the Gregorian calendar of 365 or 366 days divided into 12 months beginning with January and ending with December **b :** a period of time equal to one year of the Gregorian calendar but beginning at a different time
3 : a calendar year specified usually by a number ⟨died in the *year* 1900⟩
4 *plural* **:** a time or era having a special significance ⟨their glory *years*⟩
5 a : 12 months that constitute a measure of age or duration ⟨her 21st *year*⟩ — often used in combination ⟨a *year*-old child⟩ **b** *plural* **:** AGE ⟨an adult in *years* but a child in understanding⟩; *also* **:** the final stage of the normal life span
6 : a period of time (as the usually nine-month period in which a school is in session) other than a calendar year
year·book \-ˌbu̇k\ *noun* (1710)
1 : a book published yearly as a report or summary of statistics or facts **:** ANNUAL
2 : a school publication that is compiled usually by a graduating class and that serves as a record of the year's activities
¹year–end \'yir-'end\ *noun* (1872)
: the end of usually the fiscal year
²year–end *adjective* (1899)
: made, occurring, or existing at the year-end ⟨a *year-end* report⟩
year·ling \'yir-liŋ, 'yər-lən\ *noun* (15th century)
: one that is a year old: as **a :** an animal one year old or in the second year of its age **b :** a racehorse between January 1st of the year after the year in which it was foaled and the next January 1st
— yearling *adjective*
year·long \'yir-'lȯŋ\ *adjective* (1813)
: lasting through a year
¹year·ly \'yir-lē\ *adjective* (before 12th century)
1 : reckoned by the year
2 : occurring, appearing, made, done, or acted upon every year or once a year **:** ANNUAL
²yearly *adverb* (before 12th century)
: every year **:** ANNUALLY
Yearly Meeting *noun* (1688)
: an organization uniting several Quarterly Meetings of the Society of Friends
yearn \'yərn\ *intransitive verb* [Middle English *yernen,* from Old English *giernan;* akin to Old High German *gerōn* to desire, Latin *hortari* to urge, encourage, Greek *chairein* to rejoice] (before 12th century)
1 : to long persistently, wistfully, or sadly
2 : to feel tenderness or compassion
synonym see LONG
— yearn·er *noun*
— yearn·ing·ly \'yər-niŋ-lē\ *adverb*
yearn·ing *noun* (before 12th century)
: a tender or urgent longing
year of grace (14th century)
: a calendar year of the Christian era ⟨the *year of grace* 1993⟩
year–round \'yir-'rau̇nd\ *adjective* (1924)
: occurring, effective, employed, staying, or operating for the full year **:** not seasonal ⟨a *year-round* resort⟩
— year–round *adverb*
— year–round·er \-'rau̇n-dər\ *noun*
yea–say·er \'yā-ˌsā-ər, -ˌse(-ə)r\ *noun* (1920)
1 : one whose attitude is that of confident affirmation
2 : YES-MAN
¹yeast \'yēst, *especially Southern & Midland* 'ēst\ *noun* [Middle English *yest,* from Old English *gist;* akin to Old High German *jesen, gesen* to ferment, Greek *zein* to boil] (before 12th century)
1 a : a yellowish surface froth or sediment that occurs especially in saccharine liquids (as fruit juices) in which it promotes alcoholic fermentation, consists largely of cells of a fungus (family Saccharomycetaceae), and is used especially in the making of alcoholic liquors and as a leaven in baking **b :** a commercial product containing yeast plants in a moist or dry medium **c** (1) **:** a minute fungus (especially *Saccharomyces cerevisiae*) that is present and functionally active in yeast, usually has little or no mycelium, and reproduces by budding (2) **:** any of various similar fungi (especially orders Endomycetales and Moniliales)
2 *archaic* **:** the foam or spume of waves
3 : something that causes ferment or activity ⟨were all seething with the *yeast* of revolt —J. F. Dobie⟩
²yeast *intransitive verb* (1819)
: FERMENT, FROTH
yeasty \'yē-stē, 'ē-stē\ *adjective* **yeast·i·er; -est** (1598)
1 : of, relating to, or resembling yeast
2 a : IMMATURE, UNSETTLED **b :** marked by change **c :** full of vitality **d :** FRIVOLOUS 1a, 2
— yeast·i·ness \-stē-nəs\ *noun*
yech *or* **yecch** \'yək, 'yək̲, 'yek, 'yek̲\ *interjection* (1969)
— used to express rejection or disgust
yegg \'yeg, 'yāg\ *noun* [origin unknown] (1903)
: SAFECRACKER; *also* **:** ROBBER
¹yell \'yel\ *verb* [Middle English, from Old English *giellan;* akin to Old High German *gellan* to yell, Old English *galan* to sing] (before 12th century)
intransitive verb
1 : to utter a loud cry, scream, or shout
2 : to give a cheer usually in unison
transitive verb
: to utter or declare with or as if with a yell **:** SHOUT
— yell·er *noun*
²yell *noun* (14th century)
1 : SCREAM, SHOUT
2 : a usually rhythmic cheer used especially in schools or colleges to encourage athletic teams
¹yel·low \'ye-(ˌ)lō, *dialect* 'ye-lər *or* 'ya-\ *adjective* [Middle English *yelwe, yelow,* from Old English *geolu;* akin to Old High German *gelo* yellow, Latin *helvus* light bay, Greek *chlōros* greenish yellow, Sanskrit *hari* yellowish] (before 12th century)

1 a : of the color yellow **b :** become yellowish through age, disease, or discoloration **:** SALLOW **c :** having a yellow or light brown complexion or skin
2 a : featuring sensational or scandalous items or ordinary news sensationally distorted ⟨*yellow* journalism⟩ **b :** MEAN, COWARDLY
— yel·low·ish \'ye-lə-wish\ *adjective*

²yellow *noun* (before 12th century)
1 : something yellow or marked by a yellow color: as **a :** a person having yellow or light brown skin **b :** the yolk of an egg
2 a : a color whose hue resembles that of ripe lemons or sunflowers or is that of the portion of the spectrum lying between green and orange **b :** a pigment or dye that colors yellow
3 *plural* **:** JAUNDICE
4 *plural but singular in construction* **:** any of several plant diseases caused especially by mycoplasma-like organisms and marked by yellowing of the foliage and stunting

³yellow (1598)
transitive verb
: to make yellow **:** give a yellow tinge or color to ⟨*yellowed* by time⟩
intransitive verb
: to become or turn yellow

yellow bile *noun* (1881)
: a humor believed in medieval physiology to be secreted by the liver and to cause irascibility

yellow birch *noun* (1787)
: a North American birch (*Betula alleghaniensis* synonym *B. lutea*) with thin lustrous gray or yellow bark forming plates with ragged edges in older trees; *also* **:** its strong hard dark brown to reddish brown wood

yel·low–dog \,ye-lō-'dȯg, -lə-'dȯg\ *adjective* (1880)
1 : MEAN, CONTEMPTIBLE
2 : of or relating to opposition to trade unionism or a labor union

yellow–dog contract *noun* (1920)
: an employment contract in which a worker disavows membership in and agrees not to join a labor union in order to get a job

yellow dwarf *noun* (1928)
: any of several virus diseases especially of cereal grasses (as oats and barley) characterized by yellowing and stunting

yellow fever *noun* (1739)
: an acute destructive infectious disease of warm regions marked by sudden onset, prostration, fever, albuminuria, jaundice, and often hemorrhage and caused by a virus transmitted by the yellow-fever mosquito

yellow–fever mosquito *noun* (1905)
: a small dark-colored mosquito (*Aedes aegypti*) that is the usual vector of yellow fever

yel·low·fin tuna \'ye-lō-,fin-, 'ye-lə-\ *noun* (1922)
: a rather small and nearly cosmopolitan tuna (*Thunnus albacares*) with yellowish fins — called also *yellowfin*

yellow–green alga *noun* (1930)
: any of a class (Xanthophyceae of the division Chrysophyta) of algae with the chlorophyll masked by brown or yellow pigment

yel·low·ham·mer \'ye-lō-,ha-mər, 'ye-lə-\ *noun* [alteration of earlier *yelambre,* from (assumed) Middle English *yelwambre,* from Middle English *yelwe* yellow + (assumed) Middle English *ambre* yellowhammer, from Old English *amore;* akin to Old High German *amaro* yellowhammer, *amari* emmer] (1556)
1 : a common Palearctic finch (*Emberiza citrinella*) having the male largely bright yellow — called also *yellow bunting*
2 : YELLOW-SHAFTED FLICKER

yellow jack *noun* (1836)
1 : YELLOW FEVER
2 : a yellowish carangid marine food fish (*Caranx bartholomaei*) found from Massachusetts to Brazil

yellow jacket *noun* (1796)
1 : any of various small yellow-marked vespid wasps (especially genus *Vespula*) that commonly nest in the ground and can sting repeatedly and painfully
2 *slang* **:** pentobarbital especially in a yellow capsule

yellow jessamine *noun* (1709)
: a twining evergreen shrub (*Gelsemium sempervirens* of the family Loganiaceae) with fragrant yellow flowers — called also *yellow jasmine*

yel·low·legs \'ye-lō-,legz, 'ye-lə-, -,lāgz\ *noun plural but singular or plural in construction* (1772)
: either of two American shorebirds with yellow legs: **a :** GREATER YELLOWLEGS **b :** LESSER YELLOWLEGS

yellow ocher *noun* (15th century)
1 : a mixture of limonite usually with clay and silica used as a pigment
2 : a moderate orange yellow

yellow pages *noun plural, often Y&P capitalized* (1952)
: the section of a telephone book that lists business and professional firms alphabetically by category and that includes classified advertising; *also* **:** a listing of products or services that is independently published

yellow perch *noun* (1805)
: a common North American freshwater bony fish (*Perca flavescens*) of the perch family that is yellowish with dark green bands and is an excellent food and sport fish

yellow peril *noun, often Y&P capitalized* (1898)
1 : a danger to Western civilization held to arise from expansion of the power and influence of eastern Asian peoples
2 : a threat to Western living standards from the influx of eastern Asian laborers willing to work for very low wages

yellow pine *noun* (1709)
1 : any of several North American pines (as a Ponderosa pine or longleaf pine) with yellowish wood
2 : the wood of a yellow pine

yellow poplar *noun* (1774)
1 : TULIP TREE 1
2 : TULIPWOOD 1

yellow rain *noun* (1979)
: a yellow substance reported to occur as a mist or as spots on rocks and vegetation in Southeast Asia and variously held to be a chemical warfare agent used in the Vietnam War or a natural substance similar if not identical to pollen or the feces of bees

yel·low–shaft·ed flicker \'ye-lō-,shaf-təd-, 'ye-lə-\ *noun* (1888)
: a flicker of eastern North America that is golden yellow on the underside of the tail and wings, has a red mark on the nape, and in the male has a black streak on each cheek

yellow spot *noun* (1869)
: MACULA LUTEA

yel·low·tail \'ye-lō-,tāl, 'ye-lə-\ *noun, plural* **yellowtail** *or* **yellowtails** (1709)
: any of various fishes having a yellow or yellowish tail: as **a :** any of several carangid fishes (genus *Seriola*); *especially* **:** a sport fish (*S. lalandei*) of the California coast and southward that reaches a length of about three feet (one meter) **b :** SILVER PERCH a **c :** a common snapper (*Ocyurus chrysurus*) that is a sport and food fish of the tropical western Atlantic and West Indies and is olive above and broadly striped with yellow along the sides and on the tail — called also *yellowtail snapper*

yel·low·throat \-,thrōt\ *noun* (1702)
: any of several largely olive American warblers (genus *Geothlypis*); *especially* **:** one (*G. trichas*) with yellow breast and throat and a whitish belly

yel·low·ware \-,war, -,wer\ *noun* (1785)
: pottery made from buff clay and covered with a yellowish transparent clay

yel·low·wood \-,wu̇d\ *noun* (1666)
1 : any of various trees having yellowish wood or yielding a yellow extract; *especially* **:** a leguminous tree (*Cladrastis lutea*) of the southern U.S. having showy white fragrant flowers and yielding a yellow dye
2 : the wood of a yellowwood tree

¹yelp \'yelp\ *noun* [²*yelp*] (1501)
: a sharp shrill bark or cry (as of a dog); *also* **:** SQUEAL

²yelp *verb* [Middle English, to boast, cry out, from Old English *gielpan* to boast, exult; akin to Old High German *gelph* outcry] (1553)
intransitive verb
: to utter a sharp quick shrill cry ⟨dogs *yelp*⟩
transitive verb
: to utter with a yelp

yelp·er \'yel-pər\ *noun* (1673)
1 : one that yelps; *especially* **:** a yelping dog
2 : an instrument used by hunters to produce a call or whistle imitating the yelp of the wild turkey hen

¹yen \'yen\ *noun, plural* **yen** [Japanese *en*] (1875)
— see MONEY table

²yen *noun* [obsolete English argot *yen-yen* craving for opium, from Chinese (Guangdong) *yīn-yáhn,* from *yīn* opium + *yáhn* craving] (1906)
: a strong desire or propensity **:** LONGING; *also* **:** URGE

³yen *intransitive verb* **yenned; yen·ning** (1919)
: to have an intense desire **:** LONG, YEARN

yen–shee \'yen-'shē\ *noun* [Chinese (Guangdong) *yīn-sí,* from *yīn* opium + *sí* excrement, filth] (1882)
: the residue formed in the bowl of an opium pipe by smoking

yen·ta \'yen-tə\ *noun* [Yiddish *yente,* from the name *Yente*] (1923)
: one that meddles; *also* **:** BLABBERMOUTH, GOSSIP

yeo·man \'yō-mən\ *noun* [Middle English *yoman*] (14th century)
1 a : an attendant or officer in a royal or noble household **b :** a person attending or assisting another **:** RETAINER **c :** YEOMAN OF THE GUARD **d :** a naval petty officer who performs clerical duties
2 a : a person who owns and cultivates a small farm; *specifically* **:** one belonging to a class of English freeholders below the gentry **b :** a person of the social rank of yeoman
3 : one that performs great and loyal service ⟨did a *yeoman's* job in seeing the program through⟩

¹yeo·man·ly \-lē\ *adverb* (14th century)
archaic **:** in a manner befitting a yeoman **:** BRAVELY

²yeomanly *adjective* (1576)
1 : of, relating to, or having the rank of a yeoman
2 : becoming or suitable to a yeoman **:** STURDY, LOYAL

yeoman of the guard (circa 1520)
: a member of a military corps attached to the British royal household that serves as ceremonial attendants of the sovereign and as warders of the Tower of London

yeo·man·ry \'yō-mən-rē\ *noun* (14th century)
1 : the body of yeomen; *specifically* **:** the body of small landed proprietors of the middle class
2 : a British volunteer cavalry force created from yeomen in 1761 as a home defense force and reorganized in 1907 as part of the territorial force

yep \'yep, *or with glottal stop instead of* p\ *adverb* [by alteration] (1891)
: YES

-yer — see -ER

\ə\ **abut** \ᵊ\ **kitten** \ər\ **further** \a\ **ash** \ā\ **ace** \ä\ **mop, mar** \au̇\ **out** \ch\ **chin** \e\ **bet** \ē\ **easy** \g\ **go** \i\ **hit** \ī\ **ice** \j\ **job** \ŋ\ **sing** \ō\ **go** \ȯ\ **law** \ȯi\ **boy** \th\ **thin** \t͟h\ **the** \ü\ **loot** \u̇\ **foot** \y\ **yet** \zh\ **vision** *see also* Guide to Pronunciation

yer·ba ma·té \ˌyer-bə-ˈmä-ˌtā, ˌyər-\ *noun* [American Spanish *yerba mate,* from *yerba* herb + *mate* maté] (1839)
: MATÉ

[1]**yerk** \ˈyərk\ *transitive verb* [Middle English, to bind tightly] (circa 1520)
1 *dialect* **:** to beat vigorously **:** THRASH
2 *dialect* **:** to attack or excite vigorously **:** GOAD

[2]**yerk** *noun* (1581)
1 *Scottish* **:** a lashing out **:** KICK
2 *dialect* **:** JERK 1

[1]**yes** \ˈyes\ *adverb* [Middle English, from Old English *gēse*] (before 12th century)
1 — used as a function word to express assent or agreement ⟨are you ready? *Yes,* I am⟩
2 — used as a function word usually to introduce correction or contradiction of a negative assertion or direction ⟨don't say that! *Yes,* I will⟩
3 — used as a function word to introduce a more emphatic or explicit phrase
4 — used as a function word to indicate uncertainty or polite interest or attentiveness

[2]**yes** *noun* (1712)
: an affirmative reply **:** YEA

ye·shi·va *also* **ye·shi·vah** \yə-ˈshē-və\ *noun, plural* **yeshivas** *or* **ye·shi·vot** \-ˌshē-ˈvōt, -ˈvōth\ [Late Hebrew *yĕshībhāh*] (1851)
1 : a school for talmudic study
2 : an Orthodox Jewish rabbinical seminary
3 : a Jewish day school providing secular and religious instruction

yes–man \ˈyes-ˌman\ *noun* (1913)
: a person who agrees with everything that is said; *especially* **:** one who endorses or supports without criticism every opinion or proposal of an associate or superior

yes·ter \ˈyes-tər\ *adjective* (1577)
archaic **:** of or relating to yesterday

[1]**yes·ter·day** \ˈyes-tər-dē, -(ˌ)dā\ *adverb* [Middle English *yisterday,* from Old English *giestran dæg,* from *giestran* yesterday + *dæg* day; akin to Old High German *gestaron* yesterday, Latin *heri,* Greek *chthes*] (before 12th century)
1 : on the day last past **:** on the day preceding today
2 : at a time not long past **:** only a short time ago ⟨I wasn't born *yesterday*⟩
— yesterday *adjective*

[2]**yesterday** *noun* (before 12th century)
1 : the day last past **:** the day next before the present
2 : recent time **:** time not long past
3 : past time — usually used in plural

[1]**yes·ter·night** \ˈyes-tər-ˌnīt\ *adverb* [Middle English, from Old English *gystran niht,* from *giestran* yesterday + *niht* night] (before 12th century)
archaic **:** on the night last past

[2]**yesternight** *noun* (1513)
: the night last past

yes·ter·year \ˈyes-tər-ˌyir\ *noun* [*yester*day + *year*] (1870)
1 : last year
2 : time gone by; *especially* **:** the recent past
— yesteryear *adverb*

yes·treen \ye-ˈstrēn\ *noun* [Middle English (Scots) *yistrevin,* from *yisterday* + *evin* evening, alteration of Middle English *even*] (1785)
chiefly Scottish **:** last evening or night
— yestreen *adverb*

[1]**yet** \ˈyet\ *adverb* [Middle English, from Old English *gīet;* akin to Old Frisian *ieta* yet] (before 12th century)
1 a : in addition **:** BESIDES ⟨gives *yet* another reason⟩ **b :** EVEN 2c ⟨a *yet* higher speed⟩ **c :** on top of everything else **:** no less ⟨had wells going dry. Between two large lakes, *yet* —J. H. Buzard⟩
2 a (1) **:** up to now **:** so far ⟨hasn't done much *yet*⟩ — often used to imply the negative of a following infinitive ⟨have *yet* to win a game⟩ (2) **:** at this or that time **:** so soon as now ⟨not time to go *yet*⟩ **b :** continuously up to the present or a specified time **:** STILL ⟨is *yet* a new country⟩ **c :** at a future time **:** EVENTUALLY ⟨may *yet* see the light⟩
3 : NEVERTHELESS, HOWEVER
— as yet : up to the present or a specified time
— yet again : one more time ⟨arrived late *yet again*⟩

[2]**yet** *conjunction* (13th century)
: but nevertheless **:** BUT

ye·ti \ˈye-tē, ˈyā-\ *noun* [Tibetan] (1937)
: ABOMINABLE SNOWMAN

yeuk \ˈyük\ *intransitive verb* [Middle English (northern) *yukyn,* from Old English *giccan* — more at ITCH] (15th century)
chiefly Scottish **:** ITCH

yew \ˈyü\ *noun* [Middle English *ew,* from Old English *īw;* akin to Old High German *īwa* yew, Middle Irish *eó*] (before 12th century)
1 a : any of a genus (*Taxus* of the family Taxaceae, the yew family) of evergreen trees and shrubs with stiff linear leaves and fruits with a fleshy aril: as (1) **:** a long-lived Eurasian tree or shrub (*T. baccata*) — called also *English yew* (2) **:** a low straggling bush (*T. canadensis*) of the eastern U.S. and Canada **b :** the wood of a yew; *especially* **:** the heavy fine-grained wood of English yew
2 *archaic* **:** an archery bow made of yew

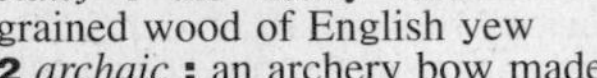
yew 1a

Ygerne \ē-ˈgern\ *noun*
: IGRAINE

Ygg·dra·sil \ˈig-drə-ˌsil\ *noun* [Old Norse]
: a huge ash tree in Norse mythology that overspreads the world and binds earth, hell, and heaven together

YHWH \ˈyä-(ˌ)wā, -(ˌ)vä\ *noun*
: YAHWEH — compare TETRAGRAMMATON

Yid \ˈyid\ *noun* [Yiddish, from Middle High German *Jüde,* from Old High German *Judeo,* from Latin *Judaeus* — more at JEW] (circa 1874)
: JEW — usually taken to be offensive

Yid·dish \ˈyi-dish\ *noun* [Yiddish *yidish,* short for *yidish daytsh,* literally, Jewish German, from Middle High German *jüdisch diutsch,* from *jüdisch* Jewish (from *Jude* Jew) + *diutsch* German] (1886)
: a High German language written in Hebrew characters that is spoken by Jews and descendants of Jews of central and eastern European origin
— Yiddish *adjective*

Yid·dish·ism \ˈyi-di-ˌshi-zəm\ *noun* (1904)
1 : a usage, word, phrase, or idiom peculiar to Yiddish
2 : a movement characterized by advocacy of the Yiddish language and culture
— Yid·dish·ist \-shist\ *noun or adjective*

[1]**yield** \ˈyē(ə)ld\ *verb* [Middle English, from Old English *gieldan;* akin to Old High German *geltan* to pay] (before 12th century)
transitive verb
1 *archaic* **:** RECOMPENSE, REWARD
2 : to give or render as fitting, rightfully owed, or required
3 : to give up possession of on claim or demand: as **a :** to give up (as one's breath) and so die **b :** to surrender or relinquish to the physical control of another **:** hand over possession of **c :** to surrender or submit (oneself) to another **d :** to give (oneself) up to an inclination, temptation, or habit **e :** to relinquish one's possession of (as a position of advantage or point of superiority) ⟨*yield* precedence⟩
4 a : to bear or bring forth as a natural product especially as a result of cultivation ⟨the tree always *yields* good fruit⟩ **b :** to produce or furnish as return ⟨this soil should *yield* good crops⟩ **c** (1) **:** to produce as return from an expenditure or investment **:** furnish as profit or interest ⟨a bond that *yields* 12 percent⟩ (2) **:** to produce as revenue **:** BRING IN ⟨the tax is expected to *yield* millions⟩
5 : to give up (as a hit or run) in baseball ⟨*yielded* two runs in the third inning⟩
intransitive verb
1 : to be fruitful or productive **:** BEAR, PRODUCE
2 : to give up and cease resistance or contention **:** SUBMIT, SUCCUMB
3 : to give way to pressure or influence **:** submit to urging, persuasion, or entreaty
4 : to give way under physical force (as bending, stretching, or breaking)
5 a : to give place or precedence **:** acknowledge the superiority of someone else **b :** to be inferior ⟨our dictionary *yields* to none⟩ **c :** to give way to or become succeeded by someone or something else
6 : to relinquish the floor of a legislative assembly ☆

[2]**yield** *noun* (15th century)
1 : something yielded **:** PRODUCT; *especially* **:** the amount or quantity produced or returned ⟨*yield* of wheat per acre⟩
2 : the capacity of yielding produce

yield·er \ˈyēl-dər\ *noun* (1590)
: one that yields: as **a :** a person who surrenders, concedes, or gives in **b :** something that yields produce or products

yield·ing \-diŋ\ *adjective* (1533)
1 : PRODUCTIVE ⟨a high-*yielding* wheat⟩
2 : lacking rigidity or stiffness **:** FLEXIBLE
3 : disposed to submit or comply

yikes \ˈyīks\ *interjection* [probably alteration of *yoicks*] (1957)
— used to express fear or astonishment

yin \ˈyin\ *noun* [Chinese (Beijing) *yīn*] (1671)
: the feminine passive principle in nature that in Chinese cosmology is exhibited in darkness, cold, or wetness and that combines with yang to produce all that comes to be

Yin·glish \ˈyiŋ-glish, -lish\ *noun* [blend of *Yiddish* and *English*] (1951)
: English marked by numerous borrowings from Yiddish

y–in·ter·cept \ˈwī-ˈin-tər-ˌsept\ *noun* (circa 1939)
: the y-coordinate of a point where a line, curve, or surface intersects the y-axis

yip \ˈyip\ *intransitive verb* **yipped; yip·ping** [imitative] (1907)
1 : to bark sharply, quickly, and often continuously

☆ SYNONYMS
Yield, submit, capitulate, succumb, relent, defer mean to give way to someone or something that one can no longer resist. YIELD may apply to any sort or degree of giving way before force, argument, persuasion, or entreaty ⟨*yields* too easily in any argument⟩. SUBMIT suggests full surrendering after resistance or conflict to the will or control of another ⟨a repentant sinner vowing to *submit* to the will of God⟩. CAPITULATE stresses the fact of ending all resistance and may imply either a coming to terms (as with an adversary) or hopelessness in the face of an irresistible opposing force ⟨officials *capitulated* to the protesters' demands⟩. SUCCUMB implies weakness and helplessness to the one that gives way or an overwhelming power to the opposing force ⟨a stage actor *succumbing* to the lure of Hollywood⟩. RELENT implies a yielding through pity or mercy by one who holds the upper hand ⟨finally *relented* and let the children stay up late⟩. DEFER implies a voluntary yielding or submitting out of respect or reverence for or deference and affection toward another ⟨I *defer* to your superior expertise in these matters⟩. See in addition RELINQUISH.

2 : to utter a short sharp cry
— **yip** *noun*

yip·pee \'yi-pē\ *interjection* (1914)
— used to express exuberant delight or triumph

yip·pie \'yi-pē\ *noun* [*Y*outh *I*nternational *P*arty + *-ie* (as in *hippie*)] (1968)
: a person belonging to or identified with a politically active group of hippies

yips \'yips\ *noun plural* [origin unknown] (1962)
: a state of nervous tension affecting an athlete (as a golfer) in the performance of a crucial action ⟨had a bad case of the *yips* on short putts⟩

-yl \əl, ᵊl, (ˌ)il, ˌēl, *chiefly British* ˌīl\ *noun combining form* [Greek *hylē* matter, material, literally, wood]
: chemical and usually univalent group ⟨eth*yl*⟩

ylang-ylang \ˌē-ˌläŋ-'ē-ˌläŋ\ *noun* [Tagalog] (1876)
1 : a tree (*Cananga odorata* synonym *Canangium odoratum*) of the custard-apple family of the Malay Archipelago, the Philippines, and adjacent areas that has very fragrant greenish yellow flowers
2 : a perfume distilled from the flowers of the ylang-ylang tree

YMCA \ˌwī-ˌem-(ˌ)sē-'ā\ *noun* [*Y*oung *M*en's *C*hristian *A*ssociation] (1881)
: an international organization that promotes the spiritual, intellectual, social, and physical welfare originally of young men

YMHA \ˌwī-ˌem-ˌā-'chā\ *noun* [*Y*oung *M*en's *H*ebrew *A*ssociation] (1918)
: an organization that promotes the religious, intellectual, social, and physical welfare of Jewish young men

Ymir \'ē-ˌmir\ *noun* [Old Norse]
: a giant from whose body the gods create the world in Norse mythology

yo \'yō\ *interjection* [Middle English *yo, io,* interjection] (15th century)
— used especially to call attention, to indicate attentiveness, or to express affirmation

yob \'yäb\ *noun* [backward spelling for *boy*] (1908)
British **:** YOBBO

yob·bo \'yä-bō\ *noun, plural* **yobbos** *or* **yobboes** [*yob* + [1]*-o*] (1922)
1 *British* **:** LOUT, YOKEL
2 *British* **:** HOODLUM

yock \'yäk\ *variant of* [2]YAK

yod \'yȯd, 'yu̇d\ *noun* [Hebrew *yōdh*] (1735)
: the 10th letter of the Hebrew alphabet — see ALPHABET table

[1]**yo·del** \'yō-dᵊl\ *verb* **-deled** *or* **-delled; -del·ing** *or* **-del·ling** \'yōd-liŋ, 'yō-dᵊl-iŋ\ [German *jodeln*] (1838)
intransitive verb
: to sing by suddenly changing from a natural voice to a falsetto and back; *also* **:** to shout or call in a similar manner
transitive verb
: to sing (a tune) by yodeling
— **yo·del·er** \'yōd-lər, 'yō-dᵊl-ər\ *noun*

[2]**yodel** *noun* (1849)
: a song or refrain sung by yodeling; *also* **:** a yodeled shout or cry

yo·ga \'yō-gə\ *noun* [Sanskrit, literally, yoking, from *yunakti* he yokes; akin to Latin *jungere* to join — more at YOKE] (1820)
1 *capitalized* **:** a Hindu theistic philosophy teaching the suppression of all activity of body, mind, and will in order that the self may realize its distinction from them and attain liberation
2 : a system of exercises for attaining bodily or mental control and well-being
— **yo·gic** \-gik\ *adjective, often capitalized*

yogh \'yōk, 'yōg, 'yōk̲\ *noun* [Middle English] (14th century)
: the letter ȝ used especially in Middle English chiefly to represent voiced and voiceless velar and palatal fricatives

yo·gi \'yō-gē\ *also* **yo·gin** \-gən, -ˌgin\ *noun* [Sanskrit *yogin,* from *yoga*] (1619)
1 : a person who practices yoga
2 *capitalized* **:** an adherent of Yoga philosophy
3 : a markedly reflective or mystical person

yo·gurt *also* **yo·ghurt** \'yō-gərt\ *noun* [Turkish *yoğurt*] (1625)
: a fermented slightly acid often flavored semisolid food made of whole or skimmed cow's milk and milk solids to which cultures of two bacteria (*Lactobacillus bulgaricus* and *Streptococcus thermophilus*) have been added

yo·him·bine \yō-'him-ˌbēn, -bən\ *noun* [International Scientific Vocabulary, from *yohimbé,* a tropical African tree (*Corynanthe yohimbe*) from which it is obtained] (1898)
: an alkaloid $C_{21}H_{26}N_2O_3$ that is a weak blocker of alpha-adrenergic receptors and has been used as an aphrodisiac

yoicks \'yȯiks\ *interjection* (1774)
— used as a cry of encouragement to foxhounds

[1]**yoke** \'yōk\ *noun, plural* **yokes** [Middle English *yok,* from Old English *geoc;* akin to Old High German *joh* yoke, Latin *jugum,* Greek *zygon,* Sanskrit *yuga,* Latin *jungere* to join] (before 12th century)
1 a : a wooden bar or frame by which two draft animals (as oxen) are joined at the heads or necks for working together **b :** an arched device formerly laid on the neck of a defeated person **c :** a frame fitted to a person's shoulders to carry a load in two equal portions **d :** a bar by which the end of the tongue of a wagon or carriage is suspended from the collars of the harness **e** (1) **:** a crosspiece on the head of a boat's rudder (2) **:** the control device for an airplane's ailerons that is mounted on a column which also serves to operate the elevator **f :** a frame from which a bell is hung **g :** a clamp or similar piece that embraces two parts to hold or unite them in position
2 *plural usually* **yoke :** two animals yoked or worked together
3 a (1) **:** an oppressive agency (2) **:** SERVITUDE, BONDAGE **b :** TIE, LINK; *especially* **:** MARRIAGE
4 : a fitted or shaped piece at the top of a skirt or at the shoulder of various garments

[2]**yoke** *verb* **yoked; yok·ing** (before 12th century)
transitive verb
1 a (1) **:** to put a yoke on (2) **:** to join in or with a yoke **b :** to attach a draft animal to; *also* **:** to attach (a draft animal) to something
2 : to join as if by a yoke
3 : to put to work
intransitive verb
: to become joined or linked

yoke·fel·low \'yōk-ˌfe-(ˌ)lō\ *noun* (1526)
: a close companion **:** MATE

yo·kel \'yō-kəl\ *noun* [perhaps from English dialect *yokel* green woodpecker, of imitative origin] (circa 1812)
: a naive or gullible inhabitant of a rural area or small town

yolk \'yōk, 'yelk (*in cultivated speech, especially Southern*), *also* 'yōlk, 'yȯlk, 'yälk, 'yəlk\ *also* **yoke** \'yōk\ *noun* [Middle English *yolke,* from Old English *geoloca,* from *geolu* yellow — more at YELLOW] (before 12th century)
1 a : the yellow spheroidal mass of stored food that forms the inner portion of the egg of a bird or reptile and is surrounded by the white — see EGG illustration **b** *archaic* **:** the whole contents of an ovum consisting of a protoplasmic formative portion and an inert nutritive portion **c :** material stored in an ovum that supplies food to the developing embryo and consists chiefly of proteins, lecithin, and cholesterol
2 [akin to Middle Dutch *ieke* yolk (of wool), Old English *ēowu* ewe] **:** oily material in unprocessed sheep wool consisting of wool fat, suint, and debris
— **yolked** *adjective*
— **yolky** *adjective*

yolk sac *noun* (1861)
: a membranous sac that is attached to an embryo and encloses food yolk, that is continuous in most forms through the yolk stalk with the intestinal cavity of the embryo, that being abundantly supplied with blood vessels is throughout embryonic life and in some forms later the chief organ of nutrition, and that in placental mammals is nearly vestigial and functions chiefly prior to the elaboration of the placenta

yolk stalk *noun* (1900)
: the narrow tubular stalk connecting the yolk sac with the embryo

Yom Kip·pur \ˌyōm-ki-'pu̇r, ˌyȯm-, ˌyäm-; -'ki-pər, -(ˌ)pu̇r\ *noun* [Hebrew *yōm kippūr,* literally, day of atonement] (1854)
: a Jewish holiday observed with fasting and prayer on the 10th day of Tishri in accordance with the rites described in Leviticus 16 — called also *Day of Atonement*

[1]**yon** \'yän\ *adjective* [Middle English, from Old English *geon;* akin to Old High German *ienēr,* adjective, that, Greek *enē* day after tomorrow] (before 12th century)
: YONDER

[2]**yon** *pronoun* (14th century)
dialect **:** that or those yonder

[3]**yon** *adverb* (15th century)
1 : YONDER
2 : THITHER ⟨ran hither and *yon*⟩

[1]**yond** \'yänd\ *adverb* [Middle English, from Old English *geond;* akin to Old English *geon*] (before 12th century)
archaic **:** YONDER

[2]**yond** *adjective* (13th century)
dialect **:** YONDER

[1]**yon·der** \'yän-dər\ *adverb* [Middle English, from *yond* + *-er* (as in *hither*)] (14th century)
: at or in that indicated more or less distant place usually within sight

[2]**yonder** *adjective* (14th century)
1 : farther removed **:** more distant
2 : being at a distance within view or at a place or in a direction known or indicated

[3]**yonder** *pronoun* (14th century)
: something that is or is in an indicated more or less distant place

yo·ni \'yō-nē\ *noun* [Sanskrit, vulva] (1799)
: a stylized representation of the female genitalia symbolizing the feminine principle in Hindu cosmology — compare LINGAM
— **yo·nic** \'yō-nik\ *adjective*

yoo-hoo \'yü-(ˌ)hü\ *interjection* (circa 1924)
— used to attract attention or as a call to persons

yore \'yōr, 'yȯr\ *noun* [Middle English, from *yore,* adverb, long ago, from Old English *gēara,* from *gēar* year — more at YEAR] (14th century)
: time past and especially long past — usually used in the phrase *of yore*

York·ie \'yȯr-kē\ *noun* [by shortening & alteration] (1946)
: YORKSHIRE TERRIER

York·ist \'yȯr-kist\ *adjective* [Edward, Duke of *York* (Edward IV of England)] (1823)
: of or relating to the English royal house that ruled from 1461 to 1485
— **Yorkist** *noun*

York rite \'yȯrk-\ *noun* [*York,* England] (circa 1878)
1 : a ceremonial observed by one of the Masonic systems
2 : a system or organization that observes the York rite and confers in the U.S. 13 degrees of which the last three are in commanderies of Knights Templar — compare SCOTTISH RITE

York·shire \'yȯrk-ˌshir, -shər\ *noun* (1902)

: a white swine of any of several breeds or strains originated in Yorkshire, England

Yorkshire pudding *noun* (1747)
: batter consisting of eggs, flour, and milk that is baked in meat drippings

Yorkshire terrier *noun* (1872)
: any of a breed of compact toy terriers with long straight silky hair mostly bluish gray but tan on the head and chest

Yor·u·ba \'yȯr-ə-bə\ *noun, plural* **Yoruba** *or* **Yorubas** (1841)
: a Niger-Congo language of southwestern Nigeria and parts of Benin and Togo; *also* : a member of any of the Yoruba-speaking peoples of this region
— **Yo·ru·ban** \'yȯr-ə-bən\ *noun or adjective*

you \'yü, yə *also* yē\ *pronoun* [Middle English, from Old English *ēow,* dative & accusative of *gē* you; akin to Old High German *iu,* dative of *ir* you, Sanskrit *yūyam* you] (before 12th century)
1 : the one or ones being addressed — used as the pronoun of the second person singular or plural in any grammatical relation except that of a possessive ⟨*you* may sit in that chair⟩ ⟨*you* are my friends⟩ ⟨can I pour *you* a cup of tea⟩; used formerly only as a plural pronoun of the second person in the dative or accusative case as direct or indirect object of a verb or as object of a preposition; compare THEE, THOU, YE, YOUR, YOURS
2 : ONE 2a ▢ ◇

you–all \yü-'ȯl, 'yü-,; 'yȯl\ *pronoun* (1824)
chiefly Southern : YOU — usually used in addressing two or more persons or sometimes one person as representing also another or others

you'd \'yüd, 'yu̇d, yəd\ (1602)
: you had : you would

you'll \'yü(ə)l, 'yu̇l, yəl\ (1592)
: you will : you shall

[1]**young** \'yəŋ\ *adjective* **youn·ger** \'yəŋ-gər\; **youn·gest** \'yəŋ-gəst\ [Middle English *yong,* from Old English *geong;* akin to Old High German *jung* young, Latin *juvenis*] (before 12th century)
1 a : being in the first or an early stage of life, growth, or development **b** : JUNIOR 1a **c** : of an early, tender, or desirable age for use as food or drink ⟨fresh *young* lamb⟩ ⟨a *young* wine⟩
2 : having little experience
3 a : recently come into being : NEW **b** : YOUTHFUL 5
4 : of, relating to, or having the characteristics of youth or a young person
5 *capitalized* : representing a new or rejuvenated especially political group or movement
— **young·ish** \'yəŋ-ish\ *adjective*
— **young·ness** \'yəŋ-nəs\ *noun*

[2]**young** *noun, plural* **young** (before 12th century)
1 *plural* **a** : young persons : YOUTH **b** : immature offspring especially of lower animals
2 : a single recently born or hatched animal
— **with young** : PREGNANT — used of a female animal

young·ber·ry \'yəŋ-,ber-ē\ *noun* [B. M. *Young* (flourished 1900) American fruit grower] (1927)
: the large sweet reddish black fruit of a cultivar of a hybrid bramble closely related to the boysenberry and loganberry and grown in the western and southern U.S.; *also* : a bramble bearing youngberries

youn·ger \'yəŋ-gər\ *noun* (before 12th century)
: an inferior in age : JUNIOR — usually used with a possessive pronoun ⟨is several years his *younger*⟩

youn·gest \'yəŋ-gəst\ *noun, plural* **youngest** (13th century)
: one that is the least old; *especially* : the youngest child or member of a family

young·ling \'yəŋ-liŋ\ *noun* (before 12th century)
: one that is young; *especially* : a young person or animal
— **youngling** *adjective*

young·ster \'yəŋ(k)-stər\ *noun* (1589)
1 a : a young person : YOUTH **b** : CHILD
2 : a young mammal, bird, or plant especially of a domesticated or cultivated breed or type

Young Turk *noun* [*Young Turks,* a 20th century revolutionary party in Turkey] (1908)
: an insurgent or a member of an insurgent group especially in a political party : RADICAL; *broadly* : one advocating changes within a usually established group

youn·ker \'yəŋ-kər\ *noun* [Dutch *jonker* young nobleman] (1513)
1 : a young man
2 : CHILD, YOUNGSTER

your \yər, 'yu̇r, 'yōr, 'yȯr\ *adjective* [Middle English, from Old English *ēower;* akin to Old English *ēow* you — more at YOU] (before 12th century)
1 : of or relating to you or yourself or yourselves especially as possessor or possessors ⟨*your* bodies⟩, agent or agents ⟨*your* contributions⟩, or object or objects of an action ⟨*your* discharge⟩
2 : of or relating to one or oneself ⟨when you face the north, east is at *your* right⟩
3 — used with little or no meaning almost as an equivalent to the definite article *the* ⟨a trait . . . that sets him apart from *your* average professor —James Breckenridge⟩

you're \yər, 'yu̇r, 'yōr, 'yȯr, ,yü-ər\ (1593)
: you are

yours \'yu̇rz, 'yōrz, 'yȯrz\ *pronoun, singular or plural in construction* [Middle English from *your* + *-s* -'s] (1526)
: that which belongs to you — used without a following noun as a pronoun equivalent in meaning to the adjective *your;* often used especially with an adverbial modifier in the complimentary close of a letter ⟨*yours* truly⟩
— **yours truly** : I, ME, MYSELF ⟨I can take care of *yours truly*⟩

your·self \yər-'self, *Southern also* -'sef\ *pronoun* (14th century)
1 a : that identical one that is you — used reflexively ⟨you might hurt *yourself*⟩, for emphasis ⟨carry them *yourself*⟩, or in absolute constructions **b** : your normal, healthy, or sane condition
2 : ONESELF ▢

your·selves \-'selvz, *Southern also* -'sevz\ *pronoun plural* (1523)
1 : those identical ones that are you — used reflexively ⟨get *yourselves* a treat⟩, for emphasis, or in absolute constructions
2 : your normal, healthy, or sane condition

youth \'yüth\ *noun, plural* **youths** \'yüt͟hz, 'yüths\ *often attributive* [Middle English *youthe,* from Old English *geoguth;* akin to Old English *geong* young — more at YOUNG] (before 12th century)
1 a : the time of life when one is young; *especially* : the period between childhood and maturity **b** : the early period of existence, growth, or development
2 a : a young person; *especially* : a young male between adolescence and maturity **b** : young persons or creatures — usually plural in construction
3 : the quality or state of being youthful : YOUTHFULNESS

youth·ful \'yüth-fəl\ *adjective* (1590)
1 : of, relating to, or characteristic of youth
2 : being young and not yet mature
3 : marked by or possessing youth
4 : having the vitality or freshness of youth : VIGOROUS
5 : having accomplished or undergone little erosion
— **youth·ful·ly** \-fə-lē\ *adverb*
— **youth·ful·ness** *noun*

youth hostel *noun* (1929)
: HOSTEL 2b

▢ USAGE

you Just about everybody agrees that the use of *you* to address the reader contributes to informality and directness, and helps avoid labored circumlocutions to boot. But when *you* is used indefinitely ⟨in seventh grade they were always assigning *you* to write about things like farm produce — Russell Baker⟩ some commentators are reluctant to consider it acceptable, preferring *one* instead. *One* sometimes presents problems of pronoun agreement in American English; in British English "one does what one can" is normal, but the second *one* sounds a bit odd to American ears. Should it be replaced with *he, she, he or she, they?* The question is neatly sidestepped by using *you. You* is appropriately used for *one* in all contexts except the most formal.

yourself *Yourself* and, less often, *yourselves* turn up in contexts in which *you* might be used ⟨get me some good left-handers like *yourself* and Robinson —Robert Frost (letter)⟩. Such use has been censured in handbooks by many of the same authors who have censured the similar use of *myself.* But the commentators have failed to note that such use is not random, but rather is confined to contexts in which the person addressed is the subject of the discourse rather than a participant in it. A letter written to Alfred Lord Tennyson makes this point: "How are you standing this tropical heat, and Mrs. Tennyson? Let us have a good account of yourselves." The *you* is in direct address—Tennyson is a participant here. But the *yourselves*—the plural showing more certainly than *you* alone would that Mrs. Tennyson is to be included—makes the Tennysons the subjects about whom the writer wants to hear. Substituting *you,* or even *you both,* for *yourselves* sounds awkward. This use of the reflexive is standard in English and is found especially in conversation, dialogue, and letters.

◇ WORD HISTORY

you Any English speaker who has studied another European language has discovered that most foreign tongues have more than one word corresponding to *you.* Typically there are separate forms to express distinctions of number and grammatical function. Many languages also use different forms depending on degree of familiarity with the person addressed; a common pattern, exemplified by French and Russian, is to address an elder, social superior, or stranger with the plural pronoun. The leveling of all these possibilities into one word, *you,* is in fact relatively recent in our language. Early Modern English inherited four second person pronouns: *thou* and *thee* as subject and object forms for singular usage, and *ye* and *you* for plural. However, in weakly stressed positions both *ye* and *you* were pronounced \yə\, which led to the blurring of the grammatical distinction between them. *You* emerged predominant around 1600. In the 14th century English had copied French and begun to use *ye* and *you* as a polite mode of addressing a single person. *You* gradually became the common neutral form, while *thou* and *thee* conveyed either intimacy or disrespect. By the late 17th century, use of *thou* and *thee* was confined to the church, Quaker speech, and rural dialects.

youth·quake \-ˌkwāk\ *noun* [*youth* + *earthquake*] (1966)
: the impact of the values, tastes, and mores of youth on the established norms of society

you've \ˈyüv, yəv\ (1691)
: you have

[1]yowl \ˈyau̇(ə)l\ *verb* [Middle English] (13th century)
intransitive verb
1 : to utter a loud long cry of grief, pain, or distress **:** WAIL
2 : to complain or protest with or as if with yowls
transitive verb
: to express with yowling

[2]yowl *noun* (15th century)
: a loud long mournful wail or howl (as of a cat)

[1]yo–yo \ˈyō-(ˌ)yō\ *noun, plural* **yo–yos** [perhaps from a language of the Philippines] (1915)
1 : a thick grooved double disk with a string attached to its center which is made to fall and rise to the hand by unwinding and rewinding on the string
2 : one that resembles a yo-yo especially in moving up and down unexpectedly or repeatedly
3 : a stupid or foolish person

[2]yo–yo *adjective* (1932)
: shifting back and forth or up and down uncertainly or unexpectedly

[3]yo–yo *intransitive verb* **yo–yoed; yo–yo·ing** (1967)
: to move from one position to another repeatedly **:** FLUCTUATE

yt·ter·bi·um \i-ˈtər-bē-əm\ *noun* [New Latin, from *Ytterby,* town in southern Sweden] (1879)
: a metallic element of the rare-earth group that resembles yttrium and occurs with it and related elements in several minerals — see ELEMENT table

yt·tri·um \ˈi-trē-əm\ *noun* [New Latin, from *yttria* yttrium oxide (Y_2O_3), irregular from *Ytterby,* town in southern Sweden] (1822)
: a metallic element usually included among the rare earth metals which it resembles chemically and with which it usually occurs in minerals — see ELEMENT table

yu·an \ˈyü-ən, yu̇-ˈän\ *noun, plural* **yuan** [Chinese (Beijing) *yuán*] (1914)
1 — see MONEY table
2 : the dollar of the Republic of China (Taiwan)

yu·ca \ˈyü-kə\ *noun* [New Latin *jucca,* from Taino *yuca*] (1555)
: CASSAVA

Yu·ca·tec \ˈyü-kə-ˌtek\ *noun* [Spanish *yucateco*] (1843)
1 : a member of an American Indian people of the Yucatán peninsula, Mexico
2 : the Mayan language of the Yucatecs
— **Yu·ca·tec·an** \ˌyü-kə-ˈte-kən\ *adjective or noun*

yuc·ca \ˈyə-kə\ *noun* [New Latin, from Spanish *yuca,* of unknown origin] (1664)
1 : any of a genus (*Yucca*) of sometimes arborescent plants of the agave family that occur in warm regions chiefly of western North America and have long often rigid fibrous-margined leaves on a woody base and bear a large panicle of white blossoms
2 : CASSAVA

yucca

[1]yuck \ˈyək\ *variant of* [2]YAK

[2]yuck *also* **yuk** *interjection* (1966)
— used to express rejection or disgust ⟨spending hours over some dish and getting, "*yuck,* I hate that" —Anne Dowie⟩

yucky \ˈyə-kē\ *adjective* [[2]*yuck*] (1970)
: OFFENSIVE, DISTASTEFUL

yu·ga \ˈyu̇-gə, ˈyü-\ *noun* [Sanskrit, yoke, age — more at YOKE] (1784)
: one of the four ages of a Hindu world cycle

yuk \ˈyək\ *variant of* [2]YAK

yule \ˈyü(ə)l\ *noun, often capitalized* [Middle English *yol,* from Old English *geōl;* akin to Old Norse *jōl,* a pagan midwinter festival] (before 12th century)
: the feast of the nativity of Jesus Christ **:** CHRISTMAS

Yule log *noun* (1725)
: a large log formerly put on the hearth on Christmas Eve as the foundation of the fire

yule·tide \ˈyü(ə)l-ˌtīd\ *noun, often capitalized* (15th century)
: CHRISTMASTIDE

Yu·man \ˈyü-mən\ *noun* (1891)
: an American Indian language family of southwestern U.S. and northern Mexico
— **Yuman** *adjective*

yum·my \ˈyə-mē\ *adjective* **yum·mi·er; -est** [*yum-yum*] (1899)
: highly attractive or pleasing; *especially* **:** DELICIOUS, DELECTABLE

yum–yum \ˈyəm-ˈyəm\ *interjection* (1878)
— used to express pleasurable satisfaction especially in the taste of food

[1]yup \ˈyəp\ *variant of* YEP

[2]yup *noun* (1983)
: YUPPIE

Yu·pik \ˈyü-pək\ *noun, plural* **Yupiks** *or* **Yupik** [Central Alaskan Yupik *yúppik,* literally, authentic person] (1951)
1 : the language of the Eskimo people of southwestern Alaska
2 : a member of the Yupik-speaking people

yup·pie \ˈyə-pē\ *noun, often capitalized* [probably from young urban professional + *-ie*] (1983)
: a young college-educated adult who is employed in a well-paying profession and who lives and works in or near a large city

yurt \ˈyu̇rt\ *noun* [Russian dialect *yurta,* of Turkic origin; akin to Turkish *yurt* home] (1784)
: a circular domed tent of skins or felt stretched over a collapsible lattice framework and used by pastoral peoples of inner Asia; *also* **:** a structure that resembles a yurt usually in size and design

yurt

YWCA \ˌwī-ˌdə-bəl-yu̇-(ˌ)sē ˈā, -ˌdə-bə-yu̇-\ *noun* [Young Women's Christian Association] (1887)
: an international organization that promotes the spiritual, intellectual, social, and physical welfare originally of young women

YWHA \-ˌāch-ˈā\ *noun* [Young Women's Hebrew Association] (1918)
: an organization that promotes the religious, intellectual, social, and physical welfare of Jewish young women

Z is the twenty-sixth and last letter of the English alphabet. It comes from Latin, which borrowed it, after the Roman conquest of the eastern Mediterranean, from the Greek zeta *and used it to render that letter in words from Greek. The Greek letter is from Phoenician* zayin. *In English, the letter is usually pronounced as in* maze, *but that sound is more often represented by* s *than by* z. *It sometimes has the palatalized sound of* z *in* azure. *The letter Z occurs for a Middle English character called* yogh, *which looked like cursive* z *but had the sound of* y, *in a few Scottish words such as* capercailzie *"a European grouse" and names such as* Mackenzie. *English Z occurs in some native words, such as* graze *and* ooze, *to represent the voiced sound of an earlier written* s *(that is, an* s *pronounced with vibration of the vocal cords), but for the most part it is restricted to loanwords, especially from Greek. Small* z *developed through a simple reduction in size of the capital form. The name of this letter is usually* zee *in American English but usually* zed *in other varieties.* Zed *comes by way of French and Latin from Greek* zeta.

z \'zē, *Canadian, British, & Australian* 'zed, *chiefly dialect* 'i-zərd\ *noun, plural* **z's** *or* **zs** *often capitalized, often attributive* (before 12th century)
1 a : the 26th and last letter of the English alphabet **b :** a graphic representation of this letter **c :** a speech counterpart of orthographic *z*
2 : a graphic device for reproducing the letter *z*
3 : one designated *z* especially as the 26th in order or class or the third in order or class when *x* is made the first
4 : something shaped like the letter Z
5 : WINK 1 — usually used in plural ⟨catch some *z's* before dinner⟩

Z *noun* (1967)
: Z PARTICLE

za·ba·glio·ne \ˌzä-bəl-'yō-nē\ *noun* [Italian] (1899)
: a whipped dessert consisting of a mixture of egg yolks, sugar, and usually Marsala wine that is often served on fruit

Zach·a·ri·as \ˌza-kə-'rī-əs\ *noun* [Late Latin, from Greek, from Hebrew *Zĕkharyāh*]
: ZECHARIAH

zad·dik \'tsä-dik\ *noun, plural* **zad·dik·im** \tsä-'di-kəm\ [Yiddish *tsadek*, from Hebrew *ṣaddīq* just, righteous] (1873)
1 : a righteous and saintly person by Jewish religious standards
2 : the spiritual leader of a modern Hasidic community

zaf·tig \'zäf-tig, 'zȯf-\ *adjective* [Yiddish *zaftik* juicy, succulent, from *zaft* juice, sap, from Middle High German *saf, saft*, from Old High German *saf* — more at SAP] (circa 1936)
of a woman **:** having a full rounded figure **:** pleasingly plump

¹zag \'zag\ *noun* [zig*zag*] (1793)
1 a : one of the sharp turns, angles, or alterations in a zigzag course **b :** one of the short straight lines or sections of a zigzag course at an angle to a zig
2 : ZIG 2

²zag *intransitive verb* **zagged; zag·ging** (1900)
: to execute a zag — usually contrasted with *zig*

zaire \'zīr, zä-'ir\ *noun, plural* **zaires** *or* **zaire** [French *zaïre*, from *Zaïre*, former country in central Africa] (1967)
— see MONEY table

za·mia \'zā-mē-ə\ *noun* [New Latin, from Latin *zamiae nuces*, false manuscript reading for *azaniae nuces* pine nuts] (1819)
: any of a genus (*Zamia* of the family Zamiaceae) of American cycads with a short thick woody base, a crown of palmlike leaves, and oblong cones

za·min·dar *or* **ze·min·dar** \'za-mən-ˌdär, 'ze-; zə-ˌmēn-'där\ *noun* [Hindi *zamīndār*, from Persian, from *zamīn* land + *-dār* holder] (1683)
1 : a collector of the land revenue of a district for the government during the period of Mogul rule in India
2 : a feudal landlord in British India paying the government a fixed revenue

za·min·dari *or* **ze·min·dary** \ˌza-mən-'där-ē, ˌze-; zə-ˌmēn-\ *noun, plural* **-dar·is** *or* **-dar·ies** [Hindi *zamīndārī*, from Persian, from *zamīndār*] (1757)
1 : the system of landholding and revenue collection by zamindars
2 : the land held or administered by a zamindar

zan·der \'zan-dər, 'tsän-\ *noun, plural* **zander** *or* **zanders** [German] (1854)
: a pike perch (*Stizostedion lucioperca*) of central Europe related to the walleye

¹za·ny \'zā-nē\ *noun, plural* **zanies** [Italian *zanni*, a traditional masked clown, from Italian dialect *Zanni*, nickname for Italian *Giovanni* John] (1588)
1 : a subordinate clown or acrobat in old comedies who mimics ludicrously the tricks of the principal **:** MERRY-ANDREW
2 : a slavish follower **:** TOADY
3 a : one who acts the buffoon to amuse others **b :** NUT, KOOK

²zany *adjective* **za·ni·er; -est** (1616)
1 : being or having the characteristics of a zany
2 : fantastically or absurdly ludicrous
— za·ni·ly \'zā-nə-lē, 'zā-nᵊl-ē\ *adverb*
— za·ni·ness \'zā-nē-nəs\ *noun*

¹zap \'zap\ *interjection* [imitative] (1929)
1 — used to express a sound made by or as if by a gun
2 — used to indicate a sudden or instantaneous occurrence

²zap *verb* **zapped; zap·ping** (1942)
transitive verb
1 a : to get rid of, destroy, or kill especially with or as if with sudden force **b :** to hit with or as if with a sudden concentrated application of force or energy **c :** to irradiate especially with microwaves
2 a : to propel suddenly or speedily **b :** to transport instantaneously
3 : to avoid watching (as a television commercial) by changing channels especially with a remote control or by fast-forwarding a videotape
intransitive verb
1 : to move with speed or force
2 : to change television channels using a remote control

³zap *noun* (1963)
: a pungent or zestful quality **:** ZIP; *also* **:** a sudden forceful blow

za·pa·te·ado \ˌzä-pə-tā-'ä-(ˌ)dō, ˌsä-pə-tā-'au̇\ *noun* [Spanish, from *zapatear* to strike or tap with the shoe, from *zapato* shoe] (1845)
: a Latin-American dance marked by rhythmic stamping or tapping of the feet

za·pa·teo \ˌzä-pə-'tā-(ˌ)ō, ˌsä-\ *noun* [Spanish, from *zapatear*] (1922)
: ZAPATEADO

Za·po·tec \ˌzä-pə-'tek, ˌsä-\ *noun* (1797)
: a member of an American Indian people of Oaxaca state, Mexico

zap·per \'za-pər\ *noun* (1969)
: one that zaps: as **a :** an electronic device designed to attract and kill insects **b :** a person who habitually changes channels (as to avoid commercials) **c :** a remote control device used for zapping

zap·py \'za-pē\ *adjective* (1969)
: ZIPPY

za·re·ba *or* **za·ri·ba** \zə-'rē-bə\ *noun* [Arabic *zarībah* enclosure] (1849)
: an improvised stockade constructed in parts of Africa especially of thorny bushes

zar·zue·la \ˌzär-zə-'wā-lə, ˌzär-'zwā-\ *noun* [Spanish, probably from *La Zarzuela*, royal residence near Madrid where it was first performed] (1888)
: a usually comic Spanish operetta

z–ax·is \'zē-ˌak-səs; *Canadian, British, & Australian* 'zed-\ *noun* (circa 1949)
: one of the axes in a three-dimensional rectangular coordinate system

za·yin \'zä-yən, 'zī(-ə)n\ *noun* [Hebrew] (1823)
: the 7th letter of the Hebrew alphabet — see ALPHABET table

z–co·or·di·nate \ˌzē-kō-'ȯrd-nət; -'ȯr-dᵊn-ət, -də-ˌnāt; *Canadian, British, & Australian* ˌzed-\ *noun* (circa 1956)
: a coordinate whose value is determined by measuring parallel to a z-axis

zeal \'zē(ə)l\ *noun* [Middle English *zele*, from Late Latin *zelus*, from Greek *zēlos*] (15th century)
: eagerness and ardent interest in pursuit of something **:** FERVOR
synonym see PASSION

zeal·ot \'ze-lət\ *noun* [Late Latin *zelotes*, from Greek *zēlōtēs*, from *zēlos*] (1537)
1 *capitalized* **:** a member of a fanatical sect arising in Judea during the first century A.D. and militantly opposing the Roman domination of Palestine
2 : a zealous person; *especially* **:** a fanatical partisan

zeal·ot·ry \'ze-lə-trē\ *noun, plural* **-ries** (1656)
: excess of zeal **:** fanatical devotion

zeal·ous \'ze-ləs\ *adjective* (1535)
: filled with or characterized by zeal ⟨*zealous* missionaries⟩
— zeal·ous·ly *adverb*
— zeal·ous·ness *noun*

ze·a·tin \'zē-ə-tən\ *noun* [New Latin *Zea*, genus of grasses including Indian corn + kine*tin* — more at ZEIN] (1963)
: a cytokinin first isolated from the endosperm of Indian corn

ze·bra \'zē-brə, *Canadian & British also* 'ze-\ *noun, plural* **zebras** [Portuguese *zebra, zebro* wild ass, perhaps from Latin *equiferus*, kind of wild horse, from *equus* horse + *ferus* wild — more at EQUINE, FIERCE] (1600)
1 *plural also* **zebra :** any of several fleet African mammals (*Equus burchelli, E. grevyi,*

and *E. zebra*) related to the horse but distinctively and conspicuously patterned in stripes of black or dark brown and white or buff
2 [from the shirts patterned in black and white stripes worn by football referees] **:** REFEREE 2
3 : ZEBRA CROSSING
— **ze·brine** \-ˌbrīn\ *adjective or noun*

ze·bra crossing \'ze-brə-, 'zē-\ *noun* (1950)
British **:** a crosswalk marked by a series of broad white stripes to indicate a crossing point at which pedestrians have the right of way

zebra finch *noun* (1889)
: a small largely gray-and-white Australian waxbill (*Poephila guttata*) that has black bars on the tail coverts and is often kept as a cage bird

zebra fish *noun* (1771)
: any of various barred fishes; *especially* **:** a very small blue-and-silver-striped Indian danio (*Brachydanio rerio* synonym *Danio rerio*) often kept in the tropical aquarium

zebra mussel *noun* (1883)
: a freshwater Eurasian lamellibranch mollusk (*Dreissena polymorpha*) that was accidentally introduced into the Great Lakes and is spreading to surrounding waterways where it colonizes and clogs water intake pipes and competes with native fish for food

ze·bra·wood \'zē-brə-ˌwu̇d, 'ze-\ *noun* (1783)
1 : any of several trees or shrubs having mottled or striped wood: as **a :** any of various leguminous African timber trees (genus *Brachystegia*) **b :** a tall South American timber tree (*Astronium fraxinifolium*) of the cashew family
2 : the wood of a zebrawood

ze·bu \'zē-(ˌ)bü, -(ˌ)byü\ *noun* [French *zébu*] (1774)
: any of various breeds of domestic oxen developed in India that are often considered conspecific with the common ox (*Bos taurus*) or sometimes as a separate species (*B. indicus*) and are characterized by a large fleshy hump over the shoulders, a dewlap, pendulous ears, and marked resistance to the injurious effects of heat and insect attack

zebu

Zeb·u·lun \'ze-byə-lən\ *noun* [Hebrew *Zĕbhūlūn*]
: a son of Jacob and the traditional eponymous ancestor of one of the tribes of Israel

zec·chi·no \ze-'kē-(ˌ)nō, tse-\ *noun, plural* **-ni** \-(ˌ)nē\ *or* **-nos** [Italian — more at SEQUIN] (1617)
: SEQUIN 1

Zech·a·ri·ah \ˌze-kə-'rī-ə\ *noun* [Hebrew *Zĕkharyāh*]
1 : a Hebrew prophet of the 6th century B.C.
2 : a prophetic book of canonical Jewish and Christian Scripture — see BIBLE table

ze·chin \'ze-kən, zē-'kēn\ *noun* [Italian *zecchino*] (1575)
: SEQUIN 1

zed \'zed\ *noun* [Middle English, from Middle French *zede,* from Late Latin *zeta* zeta, from Greek *zēta*] (15th century)
chiefly British **:** the letter *z*

zee \'zē\ *noun* (1677)
: the letter *z*

ze·in \'zē-ən\ *noun* [New Latin *Zea,* genus of grasses including Indian corn, from Greek *zeai,* plural, wheat; akin to Sanskrit *yava* barley] (1822)
: a protein from Indian corn that lacks lysine and tryptophan and is used especially in making textile fibers, plastics, printing inks, coatings, and adhesives and sizes

zeit·ge·ber \'tsīt-ˌgā-bər, 'zīt-\ *noun* [German, from *Zeit* time + *Geber* giver] (1964)
: an environmental agent or event (as the occurrence of light or dark) that provides the stimulus setting or resetting a biological clock of an organism

zeit·geist \'tsīt-ˌgīst, 'zīt-\ *noun, often capitalized* [German, from *Zeit* + *Geist* spirit] (1884)
: the general intellectual, moral, and cultural climate of an era

zel·ko·va \'zel-kə-və, zel-'kō-və\ *noun* [New Latin, from Russian *zel'kova, zel'kva,* from Georgian *dzelkva*] (circa 1890)
: a tall widely spreading Japanese tree (*Zelkova serrata*) of the elm family that is often used as an ornamental and shade tree in place of the American elm because of its resistance to Dutch elm disease

zemst·vo \'zem(p)st-(ˌ)vō, -və\ *noun, plural* **zemstvos** [Russian; akin to Russian *zemlya* earth, land, Latin *humus* — more at HUMBLE] (1865)
: one of the district and provincial assemblies established in Russia in 1864

Zen \'zen\ *noun* [Japanese, religious meditation] (1727)
: a Japanese sect of Mahayana Buddhism that aims at enlightenment by direct intuition through meditation

ze·na·na \zə-'nä-nə\ *noun* [Hindi *zanāna,* from Persian, from *zan* woman] (1760)
: HAREM 1a

Zend–Aves·ta \ˌzend-ə-'ves-tə\ *noun* [French, from Middle Persian *Avastāk va Zand* Avesta and commentary] (1630)
: AVESTA

ze·ner diode \'zē-nər-, 'ze-\ *noun, often Z capitalized* [Clarence M. Zener (born 1905) American physicist] (1957)
: a silicon semiconductor device used especially as a voltage regulator

ze·nith \'zē-nəth, *Canadian also & British usually* 'ze-nəth, -nith\ *noun* [Middle English *senith,* from Middle French *cenith,* from Medieval Latin, from Old Spanish *zenit,* modification of Arabic *samt* (*ar-ra's*) way (over one's head)] (14th century)
1 : the point of the celestial sphere that is directly opposite the nadir and vertically above the observer
2 : the highest point reached in the heavens by a celestial body
3 : culminating point **:** ACME ⟨at the *zenith* of his powers —John Buchan⟩ ◆

ze·nith·al \-nə-thəl\ *adjective* (1860)
1 : of, relating to, or located at or near the zenith
2 : showing correct directions from the center ⟨a *zenithal* map⟩

ze·o·lite \'zē-ə-ˌlīt\ *noun* [Swedish *zeolit,* from Greek *zein* to boil + *-o-* + Swedish *-lit* -lite, from French *-lite* — more at YEAST] (circa 1777)
: any of various hydrous silicates that are analogous in composition to the feldspars, occur as secondary minerals in cavities of lavas, and can act as ion-exchangers; *also* **:** any of various natural or synthesized silicates of similar structure used especially in water softening and as adsorbents and catalysts
— **ze·o·lit·ic** \ˌzē-ə-'li-tik\ *adjective*

Zeph·a·ni·ah \ˌze-fə-'nī-ə\ *noun* [Hebrew *Ṣĕphanyāh*]
1 : a Hebrew prophet of the 7th century B.C.
2 : an apocalyptic book of canonical Jewish and Christian Scripture — see BIBLE table

zeph·yr \'ze-fər\ *noun* [Middle English *Zephirus,* west wind (personified), from Latin *Zephyrus,* god of the west wind & *zephyrus* west wind, zephyr, from Greek *Zephyros* & *zephyros*] (1611)
1 a : a breeze from the west **b :** a gentle breeze
2 : any of various lightweight fabrics and articles of clothing

Zeph·y·rus \'ze-fə-rəs\ *noun* [Latin]
: the Greek god of the west wind

zep·pe·lin \'ze-p(ə-)lən\ *noun* [Count Ferdinand von *Zeppelin*] (1900)
: a rigid airship consisting of a cylindrical trussed and covered frame supported by internal gas cells; *broadly* **:** AIRSHIP

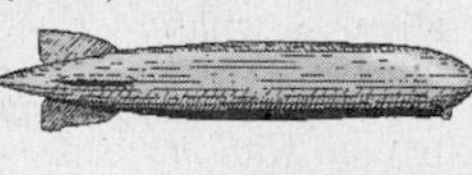
zeppelin

zerk \'zərk\ *noun* [Oscar U. *Zerk* (died 1968) American (Austrian-born) inventor] (1926)
: a grease fitting

¹ze·ro \'zē-(ˌ)rō, 'zir-(ˌ)ō\ *noun, plural* **zeros** *also* **zeroes** [French or Italian; French *zéro,* from Italian *zero,* from Medieval Latin *zephirum,* from Arabic *ṣifr*] (1604)
1 a : the arithmetical symbol 0 or ∅ denoting the absence of all magnitude or quantity **b :** ADDITIVE IDENTITY; *specifically* **:** the number between the set of all negative numbers and the set of all positive numbers **c :** a value of an independent variable that makes a function equal to zero ⟨+2 and −2 are *zeros* of $f(x)=x^2-4$⟩
2 — see NUMBER table
3 a (1) **:** the point of departure in reckoning; *specifically* **:** the point from which the graduation of a scale (as of a thermometer) begins (2) **:** the temperature represented by the zero mark on a thermometer **b :** the setting or adjustment of the sights of a firearm that causes it to shoot to point of aim at a desired range
4 : an insignificant person or thing **:** NONENTITY
5 a : a state of total absence or neutrality **b :** the lowest point **:** NADIR
6 : something arbitrarily or conveniently designated zero

²zero *adjective* (1810)
1 a : of, relating to, or being a zero **b :** having no magnitude or quantity **:** not any ⟨*zero* growth⟩ **c** (1) **:** having no phonetic manifestation ⟨the *zero* modification in the past tense of *cut*⟩ (2) **:** having no modified inflectional form ⟨a *zero* plural⟩
2 a *of a cloud ceiling* **:** limiting vision to 50 feet (15 meters) or less **b** *of horizontal visibility* **:** limited to 165 feet (50.3 meters) or less

³zero (1913)
transitive verb
1 : to determine or adjust the zero of (as a rifle)

◇ WORD HISTORY

zenith The zenith is the point of the celestial sphere vertically above the observer. The Arabs called this point *samt ar-ra's* "the way over one's head" (literally, "the way of the head"). When *samt* was borrowed into Medieval Latin, it was presumably first rendered *zemt,* in keeping with transcription conventions of the time. But most likely by a scribe's error in copying a text—*m* could easily have been misread as *ni*—it appears in manuscripts, initially in the 12th century, as *zenit,* later as *cenith* or *zenith.* The first vernacular appearance of the word was in Spanish astronomical texts of the 13th century; from Spanish or Medieval Latin the word eventually spread to most languages of Europe, appearing in English in the 14th century. By the 1600s *zenith* had acquired the additional senses "the upper region of the heavens" and then "the highest point reached in the heavens by a celestial body." The generalized meaning "culminating point, acme" soon followed.

2 a : to concentrate firepower on the exact range of — usually used with *in* **b :** to bring to bear on the exact range of a target — usually used with *in*
intransitive verb
1 : to adjust fire on a specific target — usually used with *in*
2 : to close in on or focus attention on an objective — usually used with *in*
zero–based *or* **zero–base** *adjective* (1970)
: having each item justified on the basis of cost or need ⟨*zero-based* budgeting⟩
zero coupon *adjective* (1979)
: of, relating to, or being an investment security that is sold at a deep discount, is redeemable at face value on maturity, and that pays no periodic interest ⟨*zero coupon* municipal bonds⟩
zero gravity *noun* (1951)
: the state or condition of lacking apparent gravitational pull **:** WEIGHTLESSNESS
zero hour *noun* [from its being marked by the count of zero in a countdown] (1917)
1 a : the hour at which a planned military operation is scheduled to start **b :** the time at which a usually significant or notable event is scheduled to take place
2 : a time when a vital decision or decisive change must be made
zero–sum *adjective* (1944)
: of, relating to, or being a situation (as a game or relationship) in which a gain for one side entails a corresponding loss for the other side
ze·roth \ˈzē-(ˌ)rōth, ˈzir-(ˌ)ōth\ *adjective* (1896)
: being numbered zero in a series; *also* **:** ZERO 1 ⟨the *zeroth* power of a number⟩
zero tillage *noun* (1963)
: NO-TILLAGE
zero vector *noun* (circa 1901)
: a vector which is of zero length and all of whose components are zero
zero–zero *adjective* (circa 1939)
1 : characterized by or being atmospheric conditions that reduce ceiling and visibility to zero
2 : limited to zero by atmospheric conditions
zest \ˈzest\ *noun* [obsolete French (now *zeste*), orange or lemon peel (used as flavoring)] (circa 1674)
1 : a piece of the peel or of the thin outer skin of an orange or lemon used as flavoring
2 : an enjoyably exciting quality **:** PIQUANCY
3 : keen enjoyment **:** RELISH, GUSTO ◆
— zest·ful \-fəl\ *adjective*
— zest·ful·ly \-fə-lē\ *adverb*
— zest·ful·ness *noun*
— zest·less \-ləs\ *adjective*
zest·er \ˈzes-tər\ *noun* (1973)
: a small utensil for peeling zest
zesty \ˈzes-tē\ *adjective* **zest·i·er; -est** (1930)
: having or characterized by zest
ze·ta \ˈzā-tə, ˈzē-\ *noun* [Greek *zēta*] (1823)
: the 6th letter of the Greek alphabet — see ALPHABET table
zeug·ma \ˈzüg-mə\ *noun* [Latin, from Greek, literally, joining, from *zeugnynai* to join; akin to Latin *jungere* to join — more at YOKE] (1523)
: the use of a word to modify or govern two or more words usually in such a manner that it applies to each in a different sense or makes sense with only one (as in "opened the door and her heart to the homeless boy")
Zeus \ˈzüs\ *noun* [Greek]
: the king of the gods and husband of Hera in Greek mythology — compare JUPITER
zib·e·line *or* **zib·el·line** \ˈzi-bə-ˌlēn, -ˌlīn\ *noun* [Middle French, sable, from Old Italian *zibellino,* of Slavic origin; akin to Russian *sobol'* sable] (1892)
: a soft lustrous wool fabric with mohair, alpaca, or camel's hair
zi·do·vu·dine \zi-ˈdō-vyü-ˌdēn\ *noun* [by shortening & alteration from *azidothymidine*] (1987)
: AZIDOTHYMIDINE
¹zig \ˈzig\ *noun* [*zigzag*] (1840)
1 a : one of the sharp turns, angles, or alterations in a zigzag course **b :** one of the short straight lines or sections of a zigzag course at an angle to a zag
2 : a sharp alteration or change of direction (as in a process or policy) ⟨the quick *zigs* and zags of his international maneuverings —*N.Y. Times*⟩
²zig *intransitive verb* **zigged; zig·ging** (1940)
: to execute a zig — usually contrasted with *zag*
zig·gu·rat \ˈzi-gə-ˌrat\ *noun* [Akkadian *ziqqurratu*] (1877)
: an ancient Mesopotamian temple tower consisting of a lofty pyramidal structure built in successive stages with outside staircases and a shrine at the top

ziggurat

¹zig·zag \ˈzig-ˌzag\ *noun* [French] (1712)
: one of a series of short sharp turns, angles, or alterations in a course; *also* **:** something having the form or character of such a series ⟨a blouse with green *zigzags*⟩ ⟨endured the *zigzags* of policy —Richard Bernstein⟩
²zigzag *adverb* (circa 1730)
: in or by a zigzag path or course
³zigzag *adjective* (1750)
: having short sharp turns or angles ⟨a *zigzag* trail⟩
⁴zigzag *verb* **zig·zagged; zig·zag·ging** (1777)
transitive verb
: to form into a zigzag or move along a zigzag course
intransitive verb
: to lie in, proceed along, or consist of a zigzag course
zilch \ˈzilch\ *adjective or noun* [origin unknown] (circa 1966)
: ZERO, NOTHING
zil·lion \ˈzil-yən\ *noun* [*z* + *-illion* (as in *million*)] (1934)
: an indeterminately large number ⟨*zillions* of mosquitoes⟩
— zil·lionth \-yən(t)th\ *adjective*
zil·lion·aire \ˌzil-yə-ˈnar, -ˈner\ *noun* [*zillion* + *-aire* (as in *millionaire*)] (1946)
: an immeasurably wealthy person
zin \ˈzin\ *noun, often capitalized* (1980)
: ZINFANDEL
¹zinc \ˈziŋk\ *noun, often attributive* [German *Zink*] (1651)
: a bluish white crystalline metallic element of low to intermediate hardness that is ductile when pure but in the commercial form is brittle at ordinary temperatures and becomes ductile on slight heating, occurs abundantly in minerals, is an essential micronutrient for both plants and animals, and is used especially as a protective coating for iron and steel — see ELEMENT table
²zinc *transitive verb* **zinced** *or* **zincked** \ˈziŋ(k)t\; **zinc·ing** *or* **zinck·ing** \ˈziŋ-kiŋ\ (1841)
: to treat or coat with zinc **:** GALVANIZE
zinc blende *noun* (1842)
: SPHALERITE
zinc chloride *noun* (1851)
: a poisonous caustic deliquescent salt $ZnCl_2$ used especially as a wood preservative, drying agent, and catalyst
zinc·ite \ˈziŋ-ˌkīt\ *noun* [German *Zinkit,* from *Zink*] (1854)
: a brittle deep-red to orange-yellow hexagonal mineral that consists essentially of zinc oxide and occurs in massive or granular form
zinc oxide *noun* (1849)
: an infusible white solid ZnO used especially as a pigment, in compounding rubber, and in pharmaceutical and cosmetic preparations
zinc oxide ointment *noun* (1936)
: an ointment that contains about 20 percent of zinc oxide and is used in treating skin disorders
zinc sulfate *noun* (1851)
: a crystalline salt $ZnSO_4$ used especially in making a white paint pigment, in printing and dyeing, in sprays and fertilizers, and in medicine as an astringent, emetic, and weak antiseptic
zinc sulfide *noun* (1851)
: a fluorescent white to yellowish compound ZnS used especially as a white pigment and a phosphor
zinc white *noun* (1847)
: a white pigment that consists of zinc oxide
zin·eb \ˈzi-ˌneb\ *noun* [*z*inc + *e*thylene + *b*is-] (1950)
: a zinc-containing agricultural fungicide $C_4H_6N_2S_4Zn$ used on fruits and vegetables
zin·fan·del \ˈzin-fən-ˌdel\ *noun* [origin unknown] (1896)
: a dry red table wine made from a small black grape that is grown chiefly in California; *also* **:** the grape
¹zing \ˈziŋ\ *noun* [imitative] (1911)
1 : a shrill humming noise
2 a : an enjoyably exciting or stimulating quality **:** ZEST ⟨really put some *zing* into this industry —Erwin Fine⟩ **b :** a sharply piquant flavor ⟨barbecue sauce with *zing*⟩
²zing (1920)
intransitive verb
1 : to make or move with a humming sound
2 : ZIP, SPEED
transitive verb
1 : to hit suddenly **:** ZAP
2 : to criticize in a pointed or witty manner
zing·er \ˈziŋ-ər\ *noun* (1955)
1 : something causing or meant to cause interest, surprise, or shock
2 : a pointed witty remark or retort
zingy \ˈziŋ-ē\ *adjective* **zing·i·er; -est** [¹*zing*] (1945)
1 : enjoyably exciting ⟨a *zingy* musical⟩
2 : strikingly attractive or appealing ⟨wore a *zingy* new outfit⟩
3 : sharply piquant ⟨a *zingy* salad⟩
zin·nia \ˈzi-nē-ə, ˈzē-; ˈzin-yə, ˈzēn-\ *noun* [New Latin, from Johann G. *Zinn* (died 1759) German botanist] (1767)
: any of a small genus (*Zinnia*) of tropical

◇ WORD HISTORY

zest *Zest* was borrowed into English in the 17th century from French *zest* (now spelled *zeste*), meaning "orange or lemon peel." In French the word appears to be related to the onomatopoeic interjection *zeste,* which expressed suddenness of occurrence or contemptuous dismissal (somewhat like English *poof*). As a noun, *zeste* could refer to something of no importance or value. Apparently because a rind, shell, or the like is usually worthless, *zeste* was applied more specifically to the membrane enclosing the sections of a nut and then the peel of a citrus fruit. Belying the logic of its name, however, citrus peel turned out to be quite useful as a flavoring in drinks or food. Upon borrowing the word, English extended the "flavoring" meaning of *zest* beyond the culinary domain, and in the 18th century the word began to denote any quality that adds enjoyment or piquancy to something. By the end of that century it could simply mean "keen enjoyment."

American composite herbs and low shrubs with showy flower heads and long-lasting ray flowers

Zi·on \'zī-ən\ *noun* [Middle English *Sion,* from Old English, citadel in Palestine which was the nucleus of Jerusalem, from Late Latin, from Greek *Seiōn,* from Hebrew *Ṣīyōn*] (14th century)
1 a : the Jewish people **:** ISRAEL **b :** the Jewish homeland that is symbolic of Judaism or of Jewish national aspiration **c :** the ideal nation or society envisaged by Judaism
2 : HEAVEN
3 : UTOPIA

Zi·on·ism \'zī-ə-ˌni-zəm\ *noun* (1896)
: an international movement originally for the establishment of a Jewish national or religious community in Palestine and later for the support of modern Israel
— **Zi·on·ist** \-nist\ *adjective or noun*
— **Zi·on·is·tic** \ˌzī-ə-'nis-tik\ *adjective*

[1]zip \'zip\ *verb* **zipped; zip·ping** [imitative of the sound of a speeding object] (1852)
intransitive verb
1 : to move, act, or function with speed and vigor
2 : to travel with a sharp hissing or humming sound
transitive verb
1 : to impart speed or force to
2 : to add zest, interest, or life to — often used with *up*
3 : to transport or propel with speed

[2]zip *noun* (1875)
1 : a sudden sharp hissing or sibilant sound
2 : ENERGY, VIM
— **zip·less** \-ləs\ *adjective*

[3]zip *noun* [origin unknown] (circa 1900)
: NOTHING, ZERO ⟨the final score was 27 to *zip*⟩

[4]zip *noun* (1920)
chiefly British **:** ZIPPER

[5]zip *verb* **zipped; zip·ping** [back-formation from *zipper*] (1932)
transitive verb
1 a : to close or open with or as if with a zipper **b :** to enclose or wrap by fastening a zipper
2 : to cause (a zipper) to open or shut
intransitive verb
: to become open, closed, or attached by means of a zipper

[6]zip *noun, often Z&I&P capitalized* (1965)
: ZIP CODE

zip–code *transitive verb* (1964)
: to furnish with a zip code

zip code *noun, often Z&I&P capitalized* [zone improvement plan] (1963)
: a number that identifies each postal delivery area in the U.S.

zip fastener *noun* (1927)
chiefly British **:** ZIPPER

zip gun *noun* (1950)
: a crudely homemade single-shot pistol

[1]zip·per \'zi-pər\ *noun* [from *Zipper,* a trademark] (1926)
: a fastener consisting of two rows of metal or plastic teeth on strips of tape and a sliding piece that closes an opening by drawing the teeth together

[2]zipper *transitive verb* (1930)
: [5]ZIP

zip·pered \-pərd\ *adjective* (1939)
: equipped with a zipper

zip·py \'zi-pē\ *adjective* **zip·pi·er; -est** (1904)
: full of zip **:** BRISK, SNAPPY

zi·ram \'zī-ˌram\ *noun* [*zi*nc + dithiocarbamate] (1949)
: an organic zinc salt $C_6H_{12}N_2S_4Zn$ used especially as an agricultural fungicide

zir·con \'zər-ˌkän, -kən\ *noun* [German, modification of French *jargon* jargoon, zircon, from Italian *giargone*] (1794)
: a tetragonal mineral consisting of a silicate of zirconium and occurring usually in brown or grayish square prisms of adamantine luster or sometimes in transparent forms which are used as gems

zir·co·nia \ˌzər-'kō-nē-ə\ *noun* [New Latin, from International Scientific Vocabulary *zircon*] (1797)
: a white crystalline compound ZrO_2 used especially in refractories, in thermal and electric insulation, in abrasives, and in enamels and glazes — called also *zirconium oxide*

zir·co·ni·um \ˌzər-'kō-nē-əm\ *noun* [New Latin, from International Scientific Vocabulary *zircon*] (1808)
: a steel-gray strong ductile metallic element with a high melting point that occurs widely in combined form (as in zircon), is highly resistant to corrosion, and is used especially in alloys and in refractories and ceramics — see ELEMENT table

zit \'zit\ *noun* [origin unknown] (circa 1966)
slang **:** PIMPLE 1

zith·er \'zi-t͟hər, -thər\ *noun* [German, from Latin *cithara* cithara — more at CITHER] (1850)
: a stringed instrument having usually 30 to 40 strings over a shallow horizontal soundboard and played with pick and fingers
— **zith·er·ist** \-t͟hə-rist, -thə-\ *noun*

zither

zi·ti \'zē-tē\ *noun, plural* **ziti** [Italian, plural of *zito,* alteration of *zita* piece of tubular pasta, probably short for *maccheroni di zita,* literally, bride's macaroni] (circa 1845)
: medium-sized tubular pasta

zi·zith \'tsit-səs, tsēt-'sēt\ *noun plural* [Hebrew *ṣīṣīth*] (1675)
: the fringes or tassels worn on traditional or ceremonial garments by Jewish males as reminders of the commandments of Deuteronomy 22:12 and Numbers 15:37–41

Z line *noun* (1916)
: any of the dark thin lines across a striated muscle fiber that mark the boundaries between adjacent sarcomeres

zlo·ty \'zlȯ-tē\ *noun, plural* **zlo·tys** \-tēz\ *or* **zloty** [Polish *złoty*] (1915)
— see MONEY table

zo- *or* **zoo-** *combining form* [Greek *zōi-, zōio-,* from *zōion;* akin to Greek *zōē* life — more at QUICK]
1 : animal **:** animal kingdom or kind ⟨*zo*oid⟩ ⟨*zoo*logy⟩
2 [Greek *zō-* alive, from *zōos;* akin to Greek *zōē*] **:** motile ⟨*zoo*spore⟩

-zoa \'zō-ə\ *noun plural combining form* [New Latin, from Greek *zōia,* plural of *zōion*]
: animals — in taxa ⟨Meta*zoa*⟩ ⟨Proto*zoa*⟩

zo·an·thar·i·an \ˌzō-ən-'ther-ē-ən, -'thar-\ *noun* [ultimately from *zo-* + Greek *anthos* flower — more at ANTHOLOGY] (1887)
: any of a subclass (Zoantharia) of anthozoans having a hexamerous arrangement of tentacles or septa or both and including most of the recent corals and sea anemones
— **zoantharian** *adjective*

zo·di·ac \'zō-dē-ˌak\ *noun* [Middle English, from Middle French *zodiaque,* from Latin *zodiacus,* from Greek *zōidiakos,* from *zōidiakos,* adjective, of carved figures, of the zodiac, from *zōidion* carved figure, sign of the zodiac, from diminutive of *zōion* living being, figure; akin to Greek *zōē* life] (14th century)
1 a : an imaginary band in the heavens centered on the ecliptic that encompasses the apparent paths of all the planets except Pluto and is divided into 12 constellations or signs each taken for astrological purposes to extend 30 degrees of longitude **b :** a figure representing the signs of the zodiac and their symbols
2 : a cyclic course ⟨a *zodiac* of feasts and fasts —R. W. Emerson⟩
— **zo·di·a·cal** \zō-'dī-ə-kəl, zə-\ *adjective*

SIGNS OF THE ZODIAC

NUMBER and NAME	SYMBOL	SUN ENTERS
1 Aries the Ram	♈	March 21
2 Taurus the Bull	♉	April 20
3 Gemini the Twins	♊	May 21
4 Cancer the Crab	♋	June 22
5 Leo the Lion	♌	July 23
6 Virgo the Virgin	♍	August 23
7 Libra the Balance	♎	September 23
8 Scorpio the Scorpion	♏	October 24
9 Sagittarius the Archer	♐	November 22
10 Capricorn the Goat	♑	December 22
11 Aquarius the Water Bearer	♒	January 20
12 Pisces the Fishes	♓	February 19

zodiacal light *noun* (1734)
: a diffuse glow seen in the west after twilight and in the east before dawn

zo·ea \zō-'ē-ə\ *noun, plural* **zo·eae** \-'ē-ˌē\ *or* **zo·eas** \-'ē-əz\ [New Latin, from Greek *zōē* life] (circa 1890)
: a free-swimming planktonic larval form of many decapod crustaceans and especially crabs that has a relatively large cephalothorax, conspicuous eyes, and fringed antennae and mouthparts

zof·tig \'zȯf-tig\ *variant of* ZAFTIG

[1]-zoic *adjective combining form* [Greek *zōikos* of animals, from *zōion* animal — more at ZO-]
: having a (specified) animal mode of existence ⟨holo*zoic*⟩

[2]-zoic *adjective combining form* [Greek *zōē* life]
: of, relating to, or being a (specified) geological era ⟨Archeo*zoic*⟩ ⟨Meso*zoic*⟩

zois·ite \'zȯi-ˌsīt\ *noun* [German *Zoisit,* from Baron Sigismund *Zois* von Edelstein (died 1819) Slovenian nobleman] (1805)
: an orthorhombic mineral that consists of a basic silicate of calcium and aluminum and is related to epidote

zom·bie *also* **zom·bi** \'zäm-bē\ *noun* [Louisiana Creole or Haitian Creole *zōbi,* of Bantu origin; akin to Kimbundu *nzúmbe* ghost] (circa 1871)
1 *usually zombi* **a :** the supernatural power that according to voodoo belief may enter into and reanimate a dead body **b :** a will-less and speechless human in the West Indies capable only of automatic movement who is held to have died and been supernaturally reanimated
2 a : a person held to resemble the so-called walking dead; *especially* **:** AUTOMATON **b :** a

\ə\ **abut** \ᵊ\ kitten \ər\ **further** \a\ **ash** \ā\ **ace** \ä\ **mop, mar** \au̇\ **out** \ch\ **chin** \e\ **bet** \ē\ **easy** \g\ **go** \i\ **hit** \ī\ **ice** \j\ **job** \ŋ\ **sing** \ō\ **go** \ȯ\ **law** \ȯi\ **boy** \th\ **thin** \t͟h\ **the** \ü\ **loot** \u̇\ **foot** \y\ **yet** \zh\ **vision** *see also* Guide to Pronunciation

person markedly strange in appearance or behavior
3 : a mixed drink made of several kinds of rum, liqueur, and fruit juice ◆
— **zom·bie·like** \-bē-ˌlīk\ *adjective*
zom·bi·fy \ˈzäm-bə-ˌfī\ *transitive verb* **-fied; -fy·ing** (1946)
: to turn (a human being) into a zombie
— **zom·bi·fi·ca·tion** \ˌzäm-bə-fə-ˈkā-shən\ *noun*
zom·bi·ism \-bē-ˌi-zəm\ *noun* (1932)
: the beliefs and practices of the cult of the zombi
zon·al \ˈzō-nᵊl\ *adjective* (1873)
1 : of, relating to, affecting, or having the form of a zone ⟨a *zonal* boundary⟩
2 : of, relating to, or being a soil or a major soil group marked by well-developed characteristics that are determined primarily by the action of climate and organisms especially vegetation — compare AZONAL, INTRAZONAL
— **zon·al·ly** \-nᵊl-ē\ *adverb*
zo·na pel·lu·ci·da \ˌzō-nə-pə-ˈlü-sə-də, -pel-ˈyü-\ *noun, plural* **zo·nae pel·lu·ci·dae** \-(ˌ)nē . . . -(ˌ)dē, -ˌnī . . . -ˌdī\ [New Latin, transparent zone] (1841)
: the transparent more or less elastic outer layer or envelope of a mammalian ovum often traversed by numerous radiating striae
zo·na·tion \zō-ˈnā-shən\ *noun* (1902)
1 : structure or arrangement in zones
2 : distribution of kinds of organisms in biogeographic zones
¹zone \ˈzōn\ *noun* [Middle English, from Latin *zona* belt, zone, from Greek *zōnē;* akin to Lithuanian *juosti* to gird] (15th century)
1 a : any of five great divisions of the earth's surface with respect to latitude and temperature — compare FRIGID ZONE, TEMPERATE ZONE, TORRID ZONE **b :** a portion of the surface of a sphere included between two parallel planes
2 *archaic* **:** GIRDLE, BELT
3 a : an encircling anatomical structure **b** (1) **:** a subdivision of a biogeographic region that supports a similar fauna and flora throughout its extent (2) **:** such a zone dominated by a particular life form **c :** a distinctive belt, layer, or series of layers of earth materials (as rock)
4 : a region or area set off as distinct from surrounding or adjoining parts
5 : one of the sections of an area or territory created for a particular purpose: as **a :** a zoned section of a city **b :** any of the eight concentric bands of territory centered on a given postal shipment point designated as a distance bracket for U.S. parcel post to which mail is charged at a single rate **c :** a distance within which the same fare is charged by a common carrier **d :** an area on a field of play **e :** a stretch of roadway or a space in which certain traffic regulations are in force
²zone *transitive verb* **zoned; zon·ing** (1795)
1 : to surround with a zone **:** ENCIRCLE
2 : to arrange in or mark off into zones; *specifically* **:** to partition (a city, borough, or township) by ordinance into sections reserved for different purposes (as residence or business)
— **zon·er** *noun*
³zone *adjective* (1795)
1 : ZONAL 1
2 : of, relating to, or being a system of defense (as in basketball or football) in which each player guards an assigned area rather than a specified opponent
zone out *intransitive verb* (1984)
: to become oblivious to one's surroundings especially in order to relax
zone refining *noun* (1952)
: a technique for the purification of a crystalline material and especially a metal in which a molten region travels through the material to be refined, picks up impurities at its advancing edge, and then allows the purified part to recrystallize at its opposite edge — called also *zone melting*
— **zone–refined** *adjective*
Zon·ian \ˈzō-nē-ən, -nyən\ *noun* (1910)
: a U.S. citizen who lives in the Panama Canal Zone
zonk \ˈzäŋk, ˈzȯŋk\ *verb* [probably imitative] (1950)
transitive verb
: STUN, STUPEFY; *also* **:** STRIKE, ZAP — often used with *out*
intransitive verb
: to pass out from or as if from alcohol or a drug — often used with *out*
zonked \ˈzäŋkt, ˈzȯŋ(k)t\ *adjective* (circa 1959)
: stupefied by or as if by alcohol or a drug
zonked–out \-ˈau̇t\ *adjective* (1967)
: ZONKED
Zon·ti·an \ˈzän-tē-ən\ *noun* [*Zonta International,* a service club] (1934)
: a member of a service club made up of executive women each of whom is a sole representative of one business or profession in a community
zoo \ˈzü\ *noun, plural* **zoos** [short for *zoological garden*] (circa 1847)
1 a : ZOOLOGICAL GARDEN **b :** a collection of living animals usually for public display
2 : a place, situation, or group marked by crowding, confusion, or unrestrained behavior ⟨the convention was a *zoo*⟩
zoo- — see ZO-
zo·oe·ci·um *or* **zo·e·ci·um** \zō-ˈē-shē-əm\ *noun, plural* **-cia** [New Latin, from *zo-* + Greek *oikia* house — more at VICINITY] (1880)
: a sac or chamber secreted and lived in by a bryozoan zooid
zoo·gen·ic \ˌzō-ə-ˈje-nik\ *adjective* [International Scientific Vocabulary] (circa 1864)
: caused by or associated with animals or their activities ⟨*zoogenic* humus⟩
zoo·ge·og·ra·phy \ˌzō-ə-jē-ˈä-grə-fē\ *noun* [International Scientific Vocabulary] (1868)
: a branch of biogeography concerned with the geographical distribution of animals and especially with the determination of the areas characterized by special groups of animals and the study of the causes and significance of such groups
— **zoo·ge·og·ra·pher** \-fər\ *noun*
— **zoo·geo·graph·ic** \-ˌjē-ə-ˈgra-fik\ *or* **zoo·geo·graph·i·cal** \-fi-kəl\ *adjective*
— **zoo·geo·graph·i·cal·ly** \-fi-k(ə-)lē\ *adverb*
zo·oid \ˈzō-ˌȯid\ *noun* (1851)
: one of the asexually produced individuals of a compound organism (as a bryozoan, hydroid, or coral colony)
zoo·keep·er \ˈzü-ˌkē-pər\ *noun* (1924)
: one who maintains or cares for animals in a zoo
zooks \ˈzu̇ks\ *interjection* [short for *gadzooks*] (1634)
archaic — used as a mild oath
zo·ol·a·try \zō-ˈä-lə-trē, zə-ˈwä-\ *noun* [New Latin *zoolatria,* from *zo-* + Late Latin *-latria* -latry] (1817)
: animal worship
zoo·log·i·cal \ˌzō-ə-ˈlä-ji-kəl\ *also* **zoo·log·ic** \-jik\ *adjective* (1807)
1 : of, relating to, or occupied with zoology
2 : of, relating to, or affecting lower animals often as distinguished from humans
— **zoo·log·i·cal·ly** \-ji-k(ə-)lē\ *adverb*
zoological garden *noun* (1829)
: a garden or park where wild animals are kept for exhibition
zo·ol·o·gy \zō-ˈä-lə-jē, zə-ˈwä-\ *noun* [New Latin *zoologia,* from *zo-* + *-logia* -logy] (1669)
1 : a branch of biology concerned with the classification and the properties and vital phenomena of animals
2 a : animal life (as of a region) **:** FAUNA **b :** the properties and vital phenomena exhibited by an animal, animal type, or group
— **zo·ol·o·gist** \-jist\ *noun*
¹zoom \ˈzüm\ *verb* [imitative] (1886)
intransitive verb
1 a : to move with a loud low hum or buzz **b :** to go speedily **:** ZIP
2 *of an airplane* **:** to climb for a short time at an angle greater than that which can be maintained in steady flight so that the machine is carried upward at the expense of stored kinetic energy
3 : to focus a camera or microscope on an object using a zoom lens so that the object's apparent distance from the observer changes — often used with *in* or *out*
4 : to increase sharply ⟨retail sales *zoomed*⟩
transitive verb
: to cause to zoom ■
²zoom *noun* (1918)
1 a : an act or process of zooming; *especially* **:** a sharp upward movement **b :** an image created by zooming
2 : a zooming sound
3 : ZOOM LENS
zoom lens *noun* (1936)
: a lens (as of a camera or projector) in which the image size can be varied continuously while the image remains in focus
zoo·mor·phic \ˌzō-ə-ˈmȯr-fik\ *adjective* [International Scientific Vocabulary] (1872)
1 : having the form of an animal
2 : of, relating to, or being a deity conceived of in animal form or with animal attributes
— **zoo·morph** \ˈzō-ə-ˌmȯrf\ *noun*

□ USAGE
zoom A recent commentator tells us that it is technically correct to zoom up (although the *up* is redundant), that informally *zoom* may apply to level movement, and that *zoom* is nonstandard in reference to downward movement. The distinction is tidy but bears no relation to the real use of *zoom.* A cursory examination of the entry above will show that there is a technical meaning for airplanes but that it is neither the earliest nor the only sense. One may zoom in any direction without transgressing the bounds of standard English: skiers are unlikely to zoom up the slopes.

◇ WORD HISTORY
zombie The word *zombie* or *zombi* is now inextricably associated with the walking dead of Haitian folk belief, though its first attestation in English has no connection with Haiti. In a collection of Americanisms published in 1871, the philologist Maximilian Schele de Vere claims that *zombi,* "a phantom or ghost," is "not infrequently heard in the Southern States in nurseries and among the servants." The source of the word in the American South is the French creole spoken by African slaves and their descendants in Louisiana—a language akin to Haitian Creole. *Zombi* as an English loan from Haitian Creole was first noted in an 1884 book by Spencer St. John, British consul in Port-au-Prince. St. John's book initiated a series of naive and sensational popular depictions of Haitian life. One of the more successful such works was W. H. Seabrook's *The Magic Island* (1929), which did much to increase the currency of *zombie* in American English. The ultimate source of French creole *zōbi* is most likely Kimbundu *nzúmbe* "departed spirit," or cognate words in other Bantu languages of west central Africa. The Bantu word also surfaces as *jumby* "ghost" in English creoles of the Caribbean.

-zoon *noun combining form, plural* **-zoa** [New Latin, from Greek *zōion*]
: animal **:** zooid ⟨spermato*zoon*⟩

zoo·no·sis \zō-ˈä-nə-səs, ˌzō-ə-ˈnō-səs\ *noun, plural* **-no·ses** \-ˌsēz\ [New Latin, from *zo-* + Greek *nosos* disease] (1876)
: a disease communicable from animals to humans under natural conditions
— zoo·not·ic \ˌzō-ə-ˈnä-tik\ *adjective*

zoo·phil·ic \ˌzō-ə-ˈfi-lik\ *or* **zo·oph·i·lous** \zō-ˈä-fə-ləs, zə-ˈwä-\ *adjective* (1886)
: having an attraction to or preference for animals; *especially, of an insect* **:** preferring animals to humans as a source of food

zoo·phyte \ˈzō-ə-ˌfīt\ *noun* [Greek *zōophyton,* from *zōi-, zō-* zo- + *phyton* plant — more at PHYT-] (1621)
: an invertebrate animal (as a coral or sponge) more or less resembling a plant in appearance or mode of growth

zoo·plank·ter \ˈzō-ə-ˌplaŋ(k)-tər\ *noun* [*zo-* + *plankter*] (1943)
: a planktonic animal

zoo·plank·ton \ˌzō-ə-ˈplaŋ(k)-tən, -ˌtän\ *noun* (1901)
: plankton composed of animals
— zoo·plank·ton·ic \-ˌplaŋ(k)-ˈtä-nik\ *adjective*

zoo·spo·ran·gi·um \ˌzō-ə-spə-ˈran-jē-əm\ *noun* [New Latin] (1874)
: a spore case or sporangium bearing zoospores

zoo·spore \ˈzō-ə-ˌspōr, -ˌspȯr\ *noun* [International Scientific Vocabulary] (1846)
: an independently motile spore; *especially* **:** a motile usually naked and flagellated asexual spore especially of an alga or lower fungus
— zoo·spor·ic \ˌzō-ə-ˈspōr-ik, -ˈspȯr-\ *adjective*

zo·os·ter·ol \zō ˈäs-tə-ˌrȯl, -ˌrōl\ *noun* (1926)
: a sterol (as cholesterol) of animal origin — compare PHYTOSTEROL

zoo·tech·ni·cal \ˌzō-ə-ˈtek-ni-kəl\ *adjective* (1926)
: of or relating to the technology of animal husbandry
— zoo·tech·nics \-ˈtek-niks\ *noun plural but singular or plural in construction*

zoot suit \ˈzüt-\ *noun* [coined by Harold C. Fox (born 1910) American clothier and bandleader] (1942)
: a suit of extreme cut typically consisting of a thigh-length jacket with wide padded shoulders and peg pants with narrow cuffs
— zoot–suit·er \-ˌsü-tər\ *noun*

zooty \ˈzü-tē\ *adjective* (1946)
: typical of a zoot-suiter **:** flashy in manner or style ⟨a *zooty* haircut⟩

zo·o·xan·thel·la \ˌzō-ə-zan-ˈthe-lə\ *noun, plural* **-lae** \-(ˌ)lē\ [New Latin, from *zo-* + *xanth-* + *-ella* (diminutive suffix)] (circa 1891)
: any of various symbiotic dinoflagellates that live within the cells of other organisms (as reef-building coral polyps)

zo·ri \ˈzōr-ē, ˈzȯr-\ *noun, plural* **zori** *also* **zoris** [Japanese *zōri*] (1823)
: a flat thonged sandal usually made of straw, cloth, leather, or rubber

Zorn's lemma \ˈzȯrnz-, ˈtsȯrnz-\ *noun* [Max August *Zorn* (died 1993) German mathematician] (circa 1950)
: a lemma in set theory: if a set S is partially ordered and if each subset for which every pair of elements is related by exactly one of the relationships "less than," "equal to," or "greater than" has an upper bound in S, then S contains at least one element for which there is no greater element in S — compare AXIOM OF CHOICE

Zo·ro·as·tri·an·ism \ˌzōr-ə-ˈwas-trē-ə-ˌni-zəm\ *noun* (1854)
: a Persian religion founded in the 6th century B.C. by the prophet Zoroaster, promulgated in the Avesta, and characterized by worship of a supreme god Ahura Mazda who requires good deeds for help in his cosmic struggle against the evil spirit Ahriman
— Zo·ro·as·tri·an \-trē-ən\ *adjective or noun*

zoster *noun* [Latin, from Greek *zōstēr* girdle; akin to Greek *zōnē* zone] (circa 1706)
: HERPES ZOSTER

Zou·ave \zü-ˈäv\ *noun* [French, from Berber *Zwāwa,* Berber tribe] (1830)
1 : a member of a French infantry unit originally composed of Algerians wearing a brilliant uniform and conducting a quick spirited drill
2 : a member of a military unit adopting the dress and drill of the Zouaves

zounds \ˈzau̇n(d)z, ˈzün(d)z, ˈzwau̇n(d)z, ˈzwün(d)z\ *interjection* [euphemism for *God's wounds*] (circa 1600)
— used as a mild oath

zow·ie \ˈzau̇-ē\ *interjection* [imitative of the sound of a speeding vehicle] (1902)
— used to express astonishment or admiration especially in response to something sudden or speedy

zoy·sia \ˈzȯi-shə, -zhə, -sē-ə, -zē-ə\ *noun* [New Latin, alteration of *Zoisia,* from Karl von *Zois* (died 1800) German botanist] (1924)
: any of a genus (*Zoysia*) of creeping perennial grasses of southeastern Asia and New Zealand having fine wiry leaves and including some suitable for lawn grasses especially in warm regions

Z particle *noun* (1979)
: a neutral elementary particle about 90 times heavier than a proton that along with the W particle is a transmitter of the weak force — called also Z^0 or *Z^0 particle*

zuc·chet·to \zü-ˈke-(ˌ)tō, tsü-\ *noun, plural* **-tos** [Italian, diminutive of *zucca* gourd, head] (1853)
: a small round skullcap worn by Roman Catholic ecclesiastics in colors that vary according to the rank of the wearer

zuc·chi·ni \zü-ˈkē-nē\ *noun, plural* **-ni** *or* **-nis** [Italian, plural of *zucchino,* diminutive of *zucca* gourd] (1929)
: a summer squash of bushy growth with smooth cylindrical usually dark green fruits; *also* **:** its fruit

[1]Zu·lu \ˈzü-(ˌ)lü\ *noun, plural* **Zulu** *or* **Zulus** (1824)
1 : a member of a Bantu-speaking people of Natal
2 : the Bantu language of the Zulus
— Zulu *adjective*

[2]Zulu (1952)
— a communications code word for the letter *z*

Zu·ni \ˈzü-nē\ *also* **Zu·ñi** \ˈzün-yē\ *noun, plural* **Zuni** *or* **Zunis** *also* **Zuñi** *or* **Zuñis** [American Spanish *Zuñi*] (1834)
1 : a member of an American Indian people of western New Mexico
2 : the language of the Zuni people

zup·pa in·gle·se \ˌtsü-pə-iŋ-ˈglā-(ˌ)zā, ˌzü-, -in-, -(ˌ)sā\ *noun, often I capitalized* [Italian, literally, English soup] (1941)
: a dessert consisting of sponge cake and custard or pudding that is flavored with rum, covered with cream, and garnished with fruit

zwie·back \ˈswē-ˌbak, ˈswī-, ˈzwē-, ˈzwī-, -ˌbäk\ *noun* [German, literally, twice baked, from *zwie-* twice (from Old High German *zwi-*) + *backen* to bake, from Old High German *bahhan* — more at TWI-, BAKE] (1894)
: a usually sweetened bread enriched with eggs that is baked and then sliced and toasted until dry and crisp

Zwing·li·an \ˈzwiŋ-glē-ən, ˈswiŋ-, -lē-; ˈtsfiŋ-lē-\ *adjective* (1532)
: of or relating to Ulrich Zwingli or his teachings and especially his doctrine that Christ's presence in the Eucharist is not corporeal but symbolic
— Zwinglian *noun*
— Zwing·li·an·ism \-ə-ˌni-zəm\ *noun*

zwit·ter·ion \ˈtsvi-tər-ˌī-ˌän *also* ˈzwi-\ *noun* [German, from *Zwitter* hybrid (from Old High German *zwitaran,* from *zwi-*) + *Ion* ion — more at TWI-] (1906)
: a dipolar ion
— zwit·ter·ion·ic \ˌzwi-tər-ī-ˈä-nik, ˌswi-\ *adjective*

zy·de·co \ˈzī-də-ˌkō\ *noun, often attributive* [perhaps modification of French *les haricots* beans, from the Cajun dance tune *Les Haricots Sont Pas Salés*] (1960)
: popular music of southern Louisiana that combines tunes of French origin with elements of Caribbean music and the blues and that features guitar, washboard, and accordion

zyg- *or* **zygo-** *combining form* [New Latin, from Greek, from *zygon* — more at YOKE]
1 : yoke ⟨*zygo*morphic⟩
2 : pair ⟨*zygo*dactyl⟩
3 : union ⟨*zygo*spore⟩

zyg·apoph·y·sis \ˌzī-gə-ˈpä-fə-səs\ *noun, plural* **-y·ses** \-ˌsēz\ [New Latin] (1854)
: any of the articular processes of the neural arch of a vertebra of which there are usually two anterior and two posterior

zy·go·dac·tyl \ˌzī-gə-ˈdak-tᵊl\ *adjective* [International Scientific Vocabulary *zyg-* + Greek *daktylos* toe] (1831)
: having the toes arranged two in front and two behind — used of a bird

zy·go·dac·ty·lous \-tə-ləs\ *adjective* (circa 1828)
: ZYGODACTYL

zy·go·ma \zī-ˈgō-mə\ *noun, plural* **-ma·ta** \-mə-tə\ *also* **-mas** [New Latin *zygomat-, zygoma,* from Greek *zygōma,* from *zygoun* to join, from *zygon* yoke] (circa 1684)
1 a : ZYGOMATIC ARCH **b :** a slender bony process of the zygomatic arch
2 : ZYGOMATIC BONE

zy·go·mat·ic \ˌzī-gə-ˈma-tik\ *adjective* (1709)
: of, relating to, constituting, or situated in the region of the zygoma and especially the zygomatic arch

zygomatic arch *noun* (1825)
: the arch of bone that extends along the front or side of the skull beneath the orbit

zygomatic bone *noun* (1709)
: a bone of the side of the face below the eye that in mammals forms part of the zygomatic arch and part of the orbit — called also *cheekbone*

zygomatic process *noun* (1741)
: any of several bony processes that enter into or strengthen the zygomatic arch

zy·go·mor·phic \ˌzī-gə-ˈmȯr-fik\ *adjective* (1875)
: having one or more similar parts unequal in size or form so that the whole structure is capable of division into essentially symmetrical halves by only one longitudinal plane passing through the axis ⟨the flower of the pea is *zygomorphic*⟩
— zy·go·mor·phy \ˈzī-gə-ˌmȯr-fē\ *noun*

zy·gos·i·ty \zī-ˈgä-sə-tē\ *noun* [probably from *-zygous*] (1946)
: the makeup or characteristics of a particular zygote

zy·go·spore \ˈzī-gə-ˌspōr, -ˌspȯr\ *noun* [International Scientific Vocabulary] (1864)
: a thick-walled spore of some algae and fungi that is formed by union of two similar sexual cells, usually serves as a resting spore, and produces the sporophytic phase of the plant — compare OOSPORE

zy·gote \ˈzī-ˌgōt\ *noun* [Greek *zygōtos* yoked,

\ə\ **abut** \ᵊ\ **kitten** \ər\ **further** \a\ **ash** \ā\ **ace** \ä\ **mop, mar** \au̇\ **out** \ch\ **chin** \e\ **bet** \ē\ **easy** \g\ **go** \i\ **hit** \ī\ **ice** \j\ **job** \ŋ\ **sing** \ō\ **go** \ȯ\ **law** \ȯi\ **boy** \th\ **thin** \t͟h\ **the** \ü\ **loot** \u̇\ **foot** \y\ **yet** \zh\ **vision** *see also* Guide to Pronunciation

from *zygoun* to join — more at ZYGOMA] (circa 1887)
: a cell formed by the union of two gametes; *broadly* **:** the developing individual produced from such a cell
— **zy·got·ic** \zī-ˈgä-tik\ *adjective*

zy·go·tene \ˈzī-gə-ˌtēn\ *noun* [International Scientific Vocabulary] (1911)
: the stage of meiotic prophase which immediately follows the leptotene and during which synapsis of homologous chromosomes occurs
— **zygotene** *adjective*

-zygous *adjective combining form* [Greek *-zygos* yoked, from *zygon* yoke — more at YOKE]
: having (such) a zygotic constitution ⟨heterozygous⟩

zym- *or* **zymo-** *combining form* [International Scientific Vocabulary, from Greek, leaven, from *zymē* — more at JUICE]
1 : yeast ⟨*zymo*san⟩
2 : enzyme ⟨*zymo*gen⟩

zy·mase \ˈzī-ˌmās, -ˌmāz\ *noun* [International Scientific Vocabulary] (1875)
: an enzyme or enzyme complex of yeast that promotes fermentation of sugar

-zyme *noun combining form* [Greek *zymē* leaven]
: enzyme ⟨lyso*zyme*⟩

zy·mo·gen \ˈzī-mə-jən\ *noun* [International Scientific Vocabulary] (1877)
: an inactive protein precursor of an enzyme secreted by living cells and activated by catalysis (as by a kinase or an acid) — called also *proenzyme*

zy·mo·gram \ˈzī-mə-ˌgram\ *noun* (1957)
: an electrophoretic strip (as of starch gel) or a representation of it exhibiting the pattern of separated enzymes and especially isoenzymes after electrophoresis

zy·mo·san \ˈzī-mə-ˌsan\ *noun* [*zym-* + *-osan* (as in *hexosan*)] (1943)
: an insoluble largely polysaccharide fraction of yeast cell walls

Abbreviations

AND SYMBOLS FOR CHEMICAL ELEMENTS

The following list of abbreviations is not intended to be all-inclusive; it does, however, contain many of those most commonly used. Symbols for chemical elements are also included. Most of these abbreviations have been normalized to one form. In practice, however, there is considerable variation in the use of periods and in capitalization (as *mph, m.p.h., Mph,* and *MPH*), and stylings others than those given in this dictionary are often acceptable. For a small number of abbreviations about which the question of punctuation is raised very frequently, a note is added at the entry.

For a list of abbreviations regulary used in this dictionary, see the section Abbreviations in This Work in the front matter. Many of these are also in general use, but as a rule an abbreviation is entered either in that list or this one, not both.

a absent, acceleration, acre, adult, alto, anode, answer, ante, anterior, are, area, atto-, author
A ace, adenine, ampere, argon
Å angstrom unit
aa ana
AA administrative assistant, Alcoholics Anonymous, antiaircraft, associate in arts, author's alterations
AAA Agricultural Adjustment Administration, American Automobile Association
AAAL American Academy of Arts and Letters
AAAS American Association for the Advancement of Science
AAFP American Academy of Family Physicians
AAMC American Association of Medical Colleges
A and M agricultural and mechanical, ancient and modern
A and R artists and repertory
AAR against all risks
AARP American Association of Retired Persons
AAS associate in applied science
AASCU American Association of State Colleges and Universities
AAU Amateur Athletic Union
AAUP American Association of University Professors
AAUW American Association of University Women
ab about
AB able-bodied seaman, airborne, airman basic, Alberta, [New Latin *artium baccalaureus*] bachelor of arts
ABA Amateur Boxing Association, American Bankers Association, American Bar Association, American Booksellers Association
abbr abbreviation
ABC American Bowling Congress American Broadcasting Companies, Australian Broadcasting Corporation
ABCD accelerated business collection and delivery
abd *or* **abdom** abdomen, abdominal
abl ablative
abn airborne
abp archbishop
abr abridged, abridgment
abs absolute, abstract
ABS American Bible Society, antilock braking system
abstr abstract
ac account, acre
Ac actinium, altocumulus
AC air-conditioning, alternating current, [Medieval Latin *ante Christum*] before Christ; [Latin *ante cibum*] before meals; area code, athletic club
acad academic, academy
acc accusative
accel accelerando
acct account, accountant
acous accusative
ACE American Council on Education
ack acknowledge, acknowledgment
ACLU American Civil Liberties Union
ACP African, Caribbean and Pacific (states), American College of Physicians
acpt acceptance
ACS American Chemical Society, American College of Surgeons
act active, actor, actual
ACT Action for Children's Television, American College Test, Association of Classroom Teachers, Australian Capital Territory
actg acting
ACV actual cash value, air-cushion vehicle
AD active duty, after date, anno Domini — often printed in small capitals and often punctuated; assembly district, assistant director, athletic director
A/D analog/digital
ADA American Dental Association, Americans for Democratic Action, average daily attendance
ADC aide-de-camp, Aid to Dependent Children, Air Defense Command, assistant division commander
ADD American Dialect Dictionary, attention deficit disorder
addn addition
addnl additional
ADF automatic direction finder
ADHD attention deficit/hyperactivity disorder
ad int ad interim
ADIZ air defense identification zone
adj adjective, adjunct, adjustment adjutant
ad loc [Latin *ad locum*] to or at the place
adm administration, administrative
ADM admiral
admin administration, administrative
adv adverb, [Latin *adversus*] against; advertisement, advertising, advisory
ad val ad valorem
advt advertisement
AEC Atomic Energy Commission
AEF American Expeditionary Force
aeq [Latin *aequalis*] equal
aero aeronautical, aeronautics
aet *or* **aetat** [Latin *aetatis*] of age, aged
af affix
AF air force, audio frequency
AFB air force base
AFC American Football Conference, automatic frequency control
A/1C airman first class
AFDC Aid to Families with Dependent Children
aff affirmative
afft affidavit
AFL American Football League
AFL–CIO American Federation of Labor and Congress of Industrial Organizations
AFP alpha-fetoprotein
Afr Africa, African
aft afternoon
AFT American Federation of Teachers, automatic fine tuning
AFTRA American Federation of Television and Radio Artists
Ag [Latin *argentum*] silver
AG adjutant general, attorney general
AGC advanced graduate certificate
agcy agency
agl above ground level
agr *or* **agric** agricultural, agriculture
agt agent
AH ampere-hour, anno hegirae, arts and humanities
AHL American Hockey League
AI ad interim, airborne intercept, air interception, artificial insemination, artificial intelligence
AIA American Institute of Architects
AID Agency for International Development, artificial insemination by donor
AIM American Indian Movement
AK Alaska
aka also known as
AKC American Kennel Club
Al aluminum
AL Alabama, American League, American Legion
Ala Alabama
ALA American Library Association
Alb Albania, Albanian
alc alcohol
ALCS American League Championship Series
ald alderman
alg algebra
alk alkaline
alky alkalinity
ALS amyotrophic lateral sclerosis, autographed letter signed
alt alternate, altitude, alto
Alta Alberta
alum aluminum
alw allowance
Am America, American, americium
AM airmail, Air Medal, [Latin *anno mundi*] in the year of the world — often printed in small capitals; ante meridiem — often not capitalized and often punctuated; [New Latin *artium magister*] master of arts
AMA American Medical Association
amb ambassador
amdt amendment
AmE American English
Amer America, American
AmerInd American Indian
AMG allied military government
Amn airman
amp hr ampere-hour
AMSLAN American Sign Language
amt amount
AMU atomic mass unit
AMVETS American Veterans (of World War II)
an annum
AN airman (Navy)
ANA American Nurses Association
anal analogy, analysis, analytic
anat anatomical, anatomy
anc ancient
ANC African National Congress
Angl Anglican
anhyd anhydrous
ann annals, annual
anon anonymous, anonymously
ANOVA analysis of variance
ans answer
ANSI American National Standards Institute
ant antenna, antonym
Ant Antarctica, Antrim
anthrop anthropological, anthropology
antiq antiquarian, antiquary
AO account of, and others
aor aorist
ap apostle, apothecaries'
AP additional premium, adjective phrase, airplane, American plan, antipersonnel, arithmetic progression, armor-piercing, Associated Press, author's proof
APB all points bulletin
APC armored personnel carrier
API air position indicator
APO army post office
Apoc Apocalypse, Apocrypha, apocryphal
app apparatus, appendix, appliance
appl applied
appro approval
approx approximate, approximately
appt appoint, appointed, appointment
apptd appointed
Apr April
APR annual percentage rate
apt apartment, aptitude
aq aqua, aqueous
ar arrival, arrive
Ar Arabic, argon

AR accounts receivable, acknowledgment of receipt, all rail, all risks, annual return, Arkansas, army regulation, autonomous republic
Arab Arabian, Arabic
ARC AIDS-related complex, American Red Cross
arch archaic, archery, architect, architectural, architecture
Arch Archbishop
archeol archeology
arg argent, argument
Arg Argentina
arith arithmetic, arithmetical
Ariz Arizona
Ark Arkansas
Arm Armagh, Armenian
ARM adjustable rate mortgage
ARP air-raid precautions
arr arranged, arrival, arrive
ARRT American Registry of Radiologic Technologists
art article, artificial, artillery
arty artillery
ARV American Revised Version
ARVN Army of the Republic of Vietnam (South Vietnam)
As altostratus, arsenic
AS after sight, airspeed, American Samoa, Anglo-Saxon, antisubmarine, associate in science
ASAP as soon as possible
asb asbestos
ASCAP American Society of Composers, Authors and Publishers
ASE American Stock Exchange
ASEAN Association of Southeast Asian Nations
asgd assigned
asgmt assignment
ASI airspeed indicator
ASL American Sign Language
ASPCA American Society for the Prevention of Cruelty to Animals
ASR airport surveillance radar, air-sea rescue
assn association
assoc associate, associated, association
ASSR Autonomous Soviet Socialist Republic
asst assistant, assorted
asstd assented, assorted
assy assembly
Assyr Assyrian
AST Alaska standard time
ASTM American Society for Testing and Materials
ASTP army specialized training program
astrol astrologer, astrology
astron astronomer, astronomy
ASV American Standard Version
ASW antisubmarine warfare
at airtight, atmosphere, atomic
At astatine
AT air temperature, ampere-turn, automatic transmission
ATC air traffic control
Atl Atlantic
atm atmosphere, atmospheric
ATM automated teller machine, automatic teller machine
at no atomic number
att attached, attention, attorney
attn attention
attrib attributive, attributively
atty attorney
atty gen attorney general
ATV all-terrain vehicle
at wt atomic weight
Au [Latin *aurum*] gold; author
AU angstrom unit, astronomical unit
AUC [Latin *ab urbe condita*] from the year of the founding of the city (of Rome)
aud audit, auditor
aug augmentative
Aug August
Aus Austria, Austrian
AUS Army of the United States
Austral Australia
auth authentic, author, authorized
auto automatic
aux auxiliary verb
av avenue, average, avoirdupois
AV ad valorem, audiovisual, Authorized Version
AVC automatic volume control
avdp avoirdupois
ave avenue
avg average
avn aviation
AW actual weight, aircraft warning, articles of war, automatic weapon
AWACS airborne warning and control system
ax axiom, axis
AYC American Youth Congress
AYD American Youth for Democracy
AYH American Youth Hostels
az azimuth, azure
AZ Arizona
b bachelor, bacillus, back, bag, bale, bass, basso, bat, Baumé, before, Bible, billion, bishop, black, blue, bolivar, book, born, bottom, brick, brightness, British, bulb, butut
B boron
Ba barium
BA bachelor of arts, batting average, Buenos Aires
BAA bachelor of applied arts
BAAE bachelor of aeronautical and astronautical engineering
bac [Medieval Latin *baccalaureus*] bachelor
bact bacterial, bacteriology, bacterium
BAE bachelor of aeronautical engineering, bachelor of agricultural engineering, bachelor of architectural engineering, bachelor of art education, bachelor of arts in education
BAEd bachelor of arts in education
BAeE bachelor of aeronautical engineering
BAEE bachelor of arts in elementary education
BAg bachelor of agriculture
bal balance
BAM bachelor of applied mathematics, bachelor of arts in music
B and B bed-and-breakfast
B and E breaking and entering
b and w black and white
Bap *or* **Bapt** Baptist
bar barometer, barometric, barrel
Bar Baruch
BAr bachelor of architecture
BAR Browning automatic rifle
Bart baronet
BAS bachelor of applied science, bachelor of arts and sciences
BAT bachelor of arts in teaching
Bav Bavaria, Bavarian
BB bachelor of business, ball bearing, base on balls, blue book, B'nai B'rith
BBA bachelor of business administration
BBB Better Business Bureau
BBC British Broadcasting Corporation
BBE bachelor of business education
bbl barrel, barrels
BBQ barbecue
BBS bulletin board system
BC before Christ — often printed in small capitals and often punctuated; British Columbia
BCD bad conduct discharge
BCE bachelor of chemical engineering, bachelor of civil engineering, before the Christian Era — often punctuated; before the Common Era — often punctuated
bcf billion cubic feet
BCh bachelor of chemistry
BChE bachelor of chemical engineering
BCL bachelor of canon law, bachelor of civil law
bcn beacon
BCS bachelor of chemical science, bachelor of commercial science
bd barrels per day, board, bound, boundary, bundle
BD bachelor of divinity, bank draft, bills discounted, bomb disposal, brought down
bd ft board foot
bdl *or* **bdle** bundle
bdrm bedroom
Be beryllium
BE bachelor of education, bachelor of engineering, bill of exchange, Black English
Bé Baumé
BEC Bureau of Employees' Compensation
BEd bachelor of education
Beds Bedfordshire
BEE bachelor of electrical engineering
bef before
BEF British Expeditionary Force
beg begin, beginning
Belg Belgian, Belgium
BEM bachelor of engineering of mines, British Empire Medal
BEngr bachelor of engineering
BEngS bachelor of engineering science
Berks Berkshire
bet between
BeV billion electron volts
BEV Black English vernacular
bf boldface
BF bachelor of forestry, board foot, brought forward
BFA bachelor of fine arts
bg background, bag, beige, being
BG *or* **B Gen** brigadier general
Bh bohrium
BH bill of health, Brinell hardness
bhd bulkhead
BHL bachelor of Hebrew letters, bachelor of Hebrew literature
BHN Brinell hardness number
bhp bishop
Bi bismuth
BIA bachelor of industrial administration, Bureau of Indian Affairs
bib Bible, biblical
bibliog bibliography
bid [Latin *bis in die*] twice a day
BID bachelor of industrial design
BIE bachelor of industrial engineering
bil billion
biog biographer, biographical, biography
biol biologic, biological, biologist, biology
BJ bachelor of journalism
bk bank, book, break, brook
Bk berkelium
bkg banking, bookkeeping, breakage
bkgd background
bkt basket, bracket
bl bale, barrel, black, block, blue
BL bachelor of law, bachelor of letters, baseline, bats left, bill of lading, breadth-length
bld blond, blood
bldg building
bldr builder
BLEVE boiling liquid expanding vapor explosion
BLitt *or* **BLit** [Medieval Latin *baccalaureus litterarum*] bachelor of letters, bachelor of literature
blk black, block, bulk
BLM Bureau of Land Management
BLS bachelor of liberal studies, bachelor of library science, Bureau of Labor Statistics
blvd boulevard
bm beam
BM bachelor of medicine, bachelor of music, basal metabolism, bill of material, board measure, bowel movement, bronze medal
BME bachelor of mechanical engineering, bachelor of mining engineering, bachelor of music education
BMOC big man on campus
BMR basal metabolic rate
BMS bachelor of marine science
BMT bachelor of medical technology
BMX bicycle motocross
bn baron, battalion, beacon, been
BN bachelor of nursing, bank note, Bureau of Narcotics
BNDD Bureau of Narcotics and Dangerous Drugs
BNS bachelor of naval sciences
BO back order, best offer, body odor, box office, branch office, buyer's option
BOD biochemical oxygen demand, biological oxygen demand
BOQ bachelor officers' quarters
bor borough
bot botanical, botanist, botany, bottle, bottom, bought
botan botanical
bp baptized, base pair, birthplace, bishop
BP batting practice, beautiful people, before the present, bills payable, blood pressure, blueprint, boiling point
bpd barrels per day
BPE bachelor of petroleum engineering, bachelor of physical education
BPh bachelor of philosophy
BPH benign prostatic hyperplasia; benign prostatic hypertrophy
BPharm bachelor of pharmacy
bpi bits per inch, bytes per inch
bpl birthplace
BPOE Benevolent and Protective Order of Elks
bps bits per second
BPW Board of Public Works, Business and Professional Women's Clubs
br branch, brass, brown
Br Britain, British, bromine
BR bats right, bedroom, bills receivable
Braz Brazil, Brazilian
BRE bachelor of religious education, business reply envelope
Breck Brecknockshire
brig brigade, brigadier
Brig Gen brigadier general
Brit Britain, British
brl barrel
bros brothers
BS bachelor of science, balance sheet, bill of sale, British standard, bullshit — sometimes considered vulgar
BSA bachelor of science in agriculture, Boy Scouts of America
BSAA bachelor of science in applied arts
BSAE bachelor of science in aeronautical engineering, bachelor of science in agricultural engineering, bachelor of science in architectural engineering
BSAg bachelor of science in agriculture
BSArch bachelor of science in architecture
BSB bachelor of science in business
BSc bachelor of science
BSCE bachelor of science in chemical engineering
BSCh bachelor of science in chemistry
BSE bovine spongiform encephalopathy
BSEc *or* **BSEcon** bachelor of science in economics
BSEd *or* **BSE** bachelor of science in education
BSEE bachelor of science in electrical engineering, bachelor of science in elementary education
BSFor bachelor of science in forestry
BSFS bachelor of science in foreign service
BSI British Standards Institution
bskt basket
BSL bachelor of sacred literature, bachelor of science in languages, bachelor of science in law, bachelor of science in linguistics

BSME bachelor of science in mechanical engineering
bsmt basement
BSN bachelor of science in nursing
BSW bachelor of social work
Bt baronet
btry battery
Btu British thermal unit
bu bureau, bushel
Bucks Buckinghamshire
buff buffalo
Bulg Bulgaria, Bulgarian
bull bulletin
bur bureau
bus business
BV Blessed Virgin
bvt brevet
BW bacteriological warfare, biological warfare, black and white
BWI British West Indies
bx box
BX base exchange
by billion years
BYO bring your own
BYOB bring your own beer, bring your own booze, bring your own bottle
byp bypass
c calm, calorie, Canadian, canceled, candle, carat, case, castle, catcher, Catholic, cedi, cent, centavo, center, centi-, centime, centimeter, centum, century, chairman, chapter, circa, circuit, circumference, clockwise, cloudy, cocaine, codex, coefficient, college, colon, color, colt, [Latin *congius*] gallon; congress, conservative, contralto, copyright, cost, cubic, cup, curie
C capacitance, carbon, Celsius, centigrade, Coulomb, cytosine
ca circa
Ca calcium
CA California, Central America, certified acupuncturist, chartered accountant, chief accountant, chronological age, commercial agent, controller of accounts, current account
CAB Civil Aeronautics Board
CAD computer-aided design
CAF cost and freight
CAFE corporate average fuel economy
CAGS Certificate of Advanced Graduate Study
CAI computer-aided instruction, computer-assisted instruction
cal calendar, caliber, calorie, small calorie
Cal California, large calorie
calc calculate, calculated
Calif California
cam camera
CAM computer-aided manufacturing
Cambs Cambridgeshire
can canceled, cancellation, cannon, canto
Can *or* **Canad** Canada, Canadian
canc canceled
C and F cost and freight
C and W country and western
Cant Canticle of Canticles, Cantonese
cap capacity, capital, capitalize, capitalized
CAP Civil Air Patrol
caps capitals, capsule
Capt captain
Car Carlow
CAR civil air regulations
card cardinal
CARE Cooperative for American Relief to Everywhere
CAS certificate of advanced study
cat catalog, catalyst
CAT clear-air turbulence, college ability test, computerized axial tomography
cath cathedral, cathode
CATV community antenna television
caus causative
cav cavalry, cavity
CAVU ceiling and visibility unlimited
Cb columbium, cumulonimbus
CBC Canadian Broadcasting Corporation
CBD cash before delivery, central business district
CBE commander of the Order of the British Empire, companion of the Order of the British Empire
CBI computer-based instruction, Cumulative Book Index
CBO Congressional Budget Office
CBS Columbia Broadcasting System
CBW chemical and biological warfare
cc cubic centimeter
Cc cirrocumulus
CC carbon copy, chief clerk, common carrier, community college, country club
CCC Civilian Conservation Corps
CCD Confraternity of Christian Doctrine
CCF Chinese communist forces
cckw counterclockwise
CCTV closed-circuit television
CCU cardiac care unit, coronary care unit, critical care unit
ccw counterclockwise
cd candela, candle, cord
Cd cadmium
CD carried down, certificate of deposit, civil defense, [French *corps diplomatique*] diplomatic corps
CDC Centers for Disease Control
CDD certificate of disability for discharge
cdg commanding
Cdn Canadian
CDP certificate in data processing
CDR commander
CDT central daylight time
Ce cerium
CE chemical engineer, civil engineer, Christian Era — often punctuated; Common Era — often punctuated; Corps of Engineers
CEA College English Association, Council of Economic Advisors
CED Committee for Economic Development
cem cement
CEMF counter electromotive force
cent centigrade, central, centum, century
Cent Central
CENTO Central Treaty Organization
CERN [*Conseil Européen pour la Recherche Nucléaire*] European Organization for Nuclear Research
cert certificate, certification, certified, certify
CETA Comprehensive Employment and Training Act
CEU continuing education credit
cf calf, [Latin *confer,* imperative of *conferre* to compare] compare
Cf californium
CF carried forward, centrifugal force, cost and freight, cystic fibrosis
CFC chlorofluorocarbon
CFI certified flight instructor, chief flying instructor; cost, freight, and insurance
CFL Canadian Football League
cfm cubic feet per minute
CFO chief financial officer
CFP Certified Financial Planner
cfs cubic feet per second
CFS chronic fatigue syndrome
cg centigram
CG center of gravity, coast guard, commanding general
cgs centimeter-gram-second
CGT [French *Confédération Germanénérale du Travail*] General Confederation of Labor
ch chain, champion, chaplain, chapter, chief, child, children, church
CH clearinghouse, courthouse, customhouse
chan channel
chap chapter
CHD coronary heart disease
chem chemical, chemist, chemistry
Ches Cheshire
chg change, charge
Chin Chinese
chm chairman, checkmate
Chmn chairman
chron chronicle, chronological, chronology
Chron Chronicles
Ci cirrus, curie
CI cast iron, certificate of insurance, Channel Islands
CIA Central Intelligence Agency, certified internal auditor
cía [Spanish *compañía*] company
CIAA Central Intercollegiate Athletic Association
CIC counterintelligence corps
CID Criminal Investigation Department, cubic inch displacement
cie [French *compagnie*] company
CIF central information file; cost, insurance, and freight
C in C commander in chief
CIP Cataloging in Publication
cir circle, circuit, circular, circumference
circ circular
cit citation, cited, citizen
civ civil, civilian, civilization
CIWS close-in weapons system
CJ chief justice
ck cask, check
cl centiliter, claiming, class, clause, close, closet, cloth
Cl chlorine
CL carload, center line, civil law, common law
cld called, cleared
CLEP College Level Examination Program
clin clinical
clk clerk
clo clothing
clr clear, clearance
CLU chartered life underwriter
cm centimeter, cumulative
Cm curium
CM center matched, circular mil, common meter, [Commonwealth of the Northern Mariana Islands] Northern Mariana Islands, Congregation of the Mission
CMA certified medical assistant
cmd command
cmdg commanding
cmdr commander
CMG Companion of the Order of St. Michael and St. George
cml commercial
CMSgt chief master sergeant
CMV cytomegalovirus
CN credit note
CNA certified nurse's aid
CNC computer numerical control
CNO chief of naval operations
CNS central nervous system
co company, county
Co cobalt
CO cash order, Colorado, commanding officer, conscientious objector
c/o care of
cod codex
COD carrier onboard delivery, cash on delivery, collect on delivery
coeff *or* **coef** coefficient
C of C Chamber of Commerce
C of S chief of staff
cog cognate
col colonial, colony, color, colored, column, counsel
col *or* **coll** collateral, collect, collected, collection, college, collegiate
Col colonel, Colorado, Colossians
COL colonel, cost of living
COLA cost-of-living allowance
collat collateral
colloq colloquial
Colo Colorado
colog cologarithm
com comedy, comic, comma
COM computer output microfilm, computer output microfilmer
comb combination, combined, combining, combustion
comd command
comdg commanding
comdr commander
comdt commandant
COMECON Council for Mutual Economic Assistance
coml commercial
comm command, commandant, commander, commanding, commentary, commerce, commercial, commission, commissioned, commissioner, committee, common, commoner, commonwealth, commune, communication, communist, community
commo commodore
comp comparative, compare, compensation, compiled, compiler, composition, compositor, compound, comprehensive, comptroller
compd compound
comr commissioner
con [Latin *conjunx*] consort; consolidated, consul, continued
conc concentrate, concentrated, concentration, concrete
concn concentration
cond condition, conductivity
conf conference, confidential
Confed Confederate
cong congress, congressional
conj conjunction, conjunctive
Conn Connecticut
cons consecrated, conservative, consigned, consignment, consol, consolidated, consonant, constable, constitution, construction, consul, consulting
consol consolidated
const constant, constitution, constitutional, construction
constr construction
cont containing, contents, continent, continental, continued, control
contd continued
contg containing
contr contract, contraction, contralto, contrary, control, controller
contrib contribution, contributor
conv convention, conventional, convertible, convocation
COO chief operating officer
cop copper, copulative, copy, copyright
cor corner, coroner, corpus, corresponding
Cor Corinthians
CORE Congress of Racial Equality
Corn Cornish, Cornwall
corp corporal, corporation
corr correct, corrected, correction, correspondence, correspondent, corresponding, corrupt, corruption
cos companies, consul, consulship, cosine, counties
COS cash on shipment, chief of staff
cosec cosecant
cot cotangent
cp compare, coupon
CP candlepower, Cape Province, center of pressure, cerebral palsy, charter party, chemically pure, command post, Communist party, Congregation of the Passion, custom of port
CPA certified public accountant
CPB Corporation for Public Broadcasting
CPCU chartered property casualty underwriter
cpd compound
CPFF cost plus fixed fee
cpi characters per inch
CPI consumer price index
cpl complete, compline
Cpl corporal
CPM cost per thousand
CPO chief petty officer
CPOM master chief petty officer
CPOS senior chief petty officer
CPR cardiopulmonary resuscitation

CPS cards per second, certified professional secretary, characters per second, Civilian Public Service, cycles per second
CPSC Consumer Product Safety Commission
CPT captain
cpu central processing unit
CQ call to quarters, charge of quarters, commercial quality
cr center, circular, commander, cream, creased, credit, creditor, creek, crescendo
Cr chromium
CR carrier's risk, cathode ray, class rate, conditioned reflex, conditioned response, consciousness-raising, Costa Rica, current rate
CRC Civil Rights Commission
cresc crescendo
crim criminal
crim con criminal conversation
criminol criminologist, criminology
crit critical, criticism, criticized
CRNA certified registered nurse anesthetist
CRT carrier route
cryst crystalline, crystallized
cs case, cases, census, consciousness, consul
Cs cesium, cirrostratus
CS capital stock, cesarean section, chief of staff, Christian Science practitioner, civil service, conditioned stimulus, county seat
CSA Confederate States of America
csc cosecant
CSC Civil Service Commission
CSF cerebrospinal fluid
CSM command sergeant major
CST central standard time
ct carat, cent, count, county, court
CT central time, certificated teacher, certified teacher, code telegram, computed tomography, computerized tomography, Connecticut
CTC centralized traffic control
ctf certificate
ctn carton, cotangent
c to c center to center
ctr center, counter
cu cubic, cumulative
Cu cumulus, [Latin *cuprum*] copper
CU close-up
cum cumulative
Cumb Cumbria
cur currency, current
cv *or* **cvt** convertible
CV cardiovascular, chief value, curriculum vitae
CVA Columbia Valley Authority
cw clockwise
CW chemical warfare, chief warrant officer, continuous wave
CWO cash with order, chief warrant officer
cwt hundredweight
CY calendar year
cyl cylinder
CYO Catholic Youth Organization
cytol cytological, cytology
CZ Canal Zone
d date, daughter, day, dead, deceased, deci-, degree, [Latin *denarius, denarii*] penny, pence; depart, departure, diameter, differential, dimensional, distance, dorsal, drive, driving
D Democrat, derivative, deuterium, Dutch
da deka-
DA days after acceptance, delayed action, deposit account, Dictionary of Americanisms, district attorney, doctor of arts, documents against acceptance, documents for acceptance, don't answer
DAB Dictionary of American Biography
DAE Dictionary of American English
dag dekagram
DAH Dictionary of American History
dal dekaliter
dam dekameter
Dan Daniel, Danish
D & C dilation and curettage
DAR Daughters of the American Revolution
DARE Dictionary of American Regional English
DARPA Defense Advanced Research Projects Agency
dat dative
DAT differential aptitude test, digital audiotape
dau daughter
DAV Disabled American Veterans
db debenture
dB decibel
Db dubnium
DB daybook
d/b/a doing business as
DBA doctor of business administration
DBE Dame Commander of the Order of the British Empire
DBH diameter at breast height
dbl *or* **dble** double
DBMS data base management system
DBS direct broadcast satellite
DC [Italian *da capo*] from the beginning; decimal classification, direct current, District of Columbia, doctor of chiropractic, double crochet
DChE doctor of chemical engineering
DCL doctor of canon law, doctor of civil law
dd dated, delivered
DD days after date, demand draft, dishonorable discharge, doctor of divinity, due date
DDC Dewey Decimal Classification
DDD direct distance dialing
DDS doctor of dental science, doctor of dental surgery
DE defensive end, Delaware, doctor of engineering
DEA Drug Enforcement Administration
deb debenture
dec deceased, declaration, declared, declination, decorated, decorative, decrease, decrescendo
Dec December
decd deceased
def defendant, defense, deferred, defined, definite, definition
deg degree
del delegate, delegation, delete
Del Delaware
dely delivery
dem demonstrative, demurrage
Dem Democrat, Democratic
Den Denmark
dent dental, dentist, dentistry
dep depart, department, departure, deponent, deposed, deposit, depot, deputy
depr depreciation, depression
dept department
der *or* **deriv** derivation, derivative
Derbys Derbyshire
derm dermatologist, dermatology
det detached, detachment, detail, determine
detd determined
detn detention, determination
Deut Deuteronomy
dev deviation
Devon Devonshire
DEW distant early warning
DF damage free, direction finder, direction finding
DFA doctor of fine arts
DFC Distinguished Flying Cross
dft defendant, draft
dg decigram
DG [Late Latin *Dei gratia*] by the grace of God; director general
DH doctor of humanities
DHL doctor of Hebrew letters, doctor of Hebrew literature
DHS Department of Homeland Security
DI drill instructor
dia diameter
diag diagonal, diagram
dial dialect, dialectical
diam diameter
dict dictionary
dif *or* **diff** difference
dig digest
dil dilute
dim dimension, diminished, diminuendo, diminutive
dimin diminuendo
din dinar
DIN [German *Deutsche Industrie-Normen*] German Industrial Standards
dip diploma
dir direction, director
dis discharge, discount, distance
disc discount
disp dispensary
diss dissertation
dist distance, district
distn distillation
distr distribute, distribution
div divided, dividend, division, divorced
DIY do it yourself
DJ district judge, doctor of jurisprudence, dust jacket
DJIA Dow-Jones Industrial Average
dk dark, deck, dock
dl deciliter
DL disabled list
DLitt *or* **DLit** [New Latin *doctor litterarum*] doctor of letters, doctor of literature
DLO dead letter office, dispatch loading only
DLS doctor of library science
dm decimeter
DM deutsche mark
DMA doctor of musical arts
DMD [New Latin *dentariae medicinae doctor*] doctor of dental medicine
DMin doctor of ministry
DML doctor of modern languages
DMZ demilitarized zone
dn down
DNB Dictionary of National Biography
DNF did not finish
DNR do not resuscitate
do ditto
DO defense order, doctor of osteopathy
DOA dead on arrival
DOB date of birth
doc document
DOD Department of Defense
DOE Department of Energy
dol dollar
dom domestic, dominant, dominion
Don Donegal
Dors Dorset
DOS disk operating system
DOT Department of Transportation
doz dozen
DP data processing, degree of polymerization, dew point, doctor of podiatry, double play
DPE doctor of physical education
DPh doctor of philosophy
DPH department of public health, doctor of public health
DPM doctor of podiatric medicine
dpt department, deponent
DPT diphtheria, pertussis, tetanus
DQ disqualification, disqualify
dr debtor, drachma, dram, drive, drum
Dr doctor
DR dead reckoning, dining room
dram dramatic, dramatist
Ds darmstadtium
DS [Italian *dal segno*] from the sign; days after sight, detached service, document signed
DSc doctor of science
DSC Distinguished Service Cross, doctor of surgical chiropody
DSL digital subscriber line
DSM Distinguished Service Medal
DSO Distinguished Service Order
DSP [Latin *decessit sine prole*] died without issue
DST daylight saving time, doctor of sacred theology
DSW doctor of social welfare, doctor of social work
DT daylight time, doctor of theology, double time
DTh doctor of theology
DTP desktop publishing; diphtheria, tetanus, pertussis
Du Dutch
Dub Dublin
DUI driving under the influence
Dumf Gal Dumfries and Galloway
dup duplex, duplicate
Dur Durham
DV [Latin *Deo volente*] God willing; Douay Version
DVM doctor of veterinary medicine
DW deadweight, delayed weather, distilled water, dust wrapper
DWI driving while intoxicated, Dutch West Indies
dwt deadweight ton, pennyweight
DX distance
dy delivery, deputy, duty
Dy dysprosium
dynam dynamics
dz dozen
e earth, east, easterly, eastern, edge, eldest, ell, empty, end, energy, erg, error, excellent
E electromotive force, energy, English, exponent
ea each
E and OE errors and omissions excepted
EB eastbound
EBC Educational Broadcasting Corporation
EBV Epstein-Barr virus
EC European Community
eccl ecclesiastic, ecclesiastical
Eccles Ecclesiastes
Ecclus Ecclesiasticus
ECG electrocardiogram
ECM electronic countermeasure, European Common Market
ecol ecological, ecology
econ economics, economist, economy
ECT electroconvulsive therapy
Ecua Ecuador
ed edited, edition, editor, education
EDB ethylene dibromide
EdD doctor of education
EDD English Dialect Dictionary
EDP electronic data processing
EdS specialist in education
EDT eastern daylight time
edu educational organization
educ education, educational
EE electrical engineer
EEC European Economic Community
EEG electroencephalogram, electroencephalograph
EENT eye, ear, nose, and throat
EEO equal employment opportunity
eff efficiency
EFT *or* **EFTS** electronic funds transfer (system)
e.g. [Latin *exempli gratia*] for example
Eg Egypt, Egyptian
EGF epidermal growth factor
EHF extremely high frequency
EHP effective horsepower, electric horsepower
EHV extra high voltage
EKG [German *Elektrokardiogramm*] electrocardiogram, electrocardiograph
el elevation
elec electric, electrical, electricity
elem elementary
elev elevation
ELF extremely low frequency
Eliz Elizabethan

EM electromagnetic, electron microscope, electron microscopy, end matched, engineer of mines, enlisted man

emer emeritus

EMG electromyogram, electromyograph, electromyography

emp emperor, empress

EMP electromagnetic pulse

emu electromagnetic unit

enc *or* **encl** enclosure

ency *or* **encyc** encyclopedia

ENE east-northeast

eng engine, engineer, engineering

Eng England, English

engr engineer, engraved, engraver, engraving

enl enlarged, enlisted

ENS ensign

ENT ear, nose, and throat

entom *or* **entomol** entomological, entomology

env envelope

EO executive order

EOE equal opportunity employer

EOM end of month

EP estimated position, European plan, extended play

EPA Environmental Protection Agency

Eph *or* **Ephes** Ephesians

Episc Episcopal

eq equal, equation

equip equipment

equiv equivalency, equivalent

Er erbium

ER earned run, emergency room

ERA earned run average, Equal Rights Amendment

ERISA Employee Retirement Income Security Act

ERT estrogen replacement therapy

Es einsteinium

ESB electrical stimulation of the brain

Esd Esdras

ESE east-southeast

Esk Eskimo

ESL English as a second language

ESOP employee stock ownership plan

esp especially

Esq *or* **Esqr** esquire

est established, estimate, estimated

EST eastern standard time

Esth Esther

Et ethyl

ET eastern time

ETA estimated time of arrival

et al [Latin *et alii* (masculine), *et aliae* (feminine), or *et alia* (neuter)] and others

etc et cetera

ETD estimated time of departure

ETO European theater of operations

et seq [Latin *et sequens*] and the following one; [Latin *et sequentes* (masculine & feminine plural), or *et sequentia* (neuter plural)] and the following ones

et ux [Latin *et uxor*] and wife

ETV educational television

Eu europium

Eur Europe, European

eV electron volt

EVA extravehicular activity

eval evaluation

evap evaporate

evg evening

EW enlisted woman

ex example, exchange, executive, express, extra

Ex Exodus

exc excellent, except

exch exchange, exchanged

excl exclude, excluded, excluding

exhbn exhibition

Exod Exodus

exor executor

exp expense, experience, experiment, experimental, exponent, export, express

expt experiment

exptl experimental

expwy expressway

exrx executrix

ext extension, exterior, external, externally, extra, extract

Ez *or* **Ezr** Ezra

Ezech Ezechiel

Ezek Ezekiel

f failure, false, family, faraday, feast, female, feminine, femto-, fermi, fine, finish, fluid, fluidness, focal length, [*f*ollowing] and the following one; force, forte, fragile, frequency, from, full

F Fahrenheit, farad, fluorine, French, Friday

FA field artillery, fielding average, football association

FAA Federal Aviation Administration, free of all average

fac facsimile, faculty

FADM fleet admiral

Fah *or* **Fahr** Fahrenheit

fam familiar, family

F and A fore and aft

FAO Food and Agriculture Organization of the United Nations

FAQ fair average quality, frequently asked question

far farthing

FAS firsts and seconds, free alongside (ship)

fasc fascicle

fath fathom

FB foreign body, freight bill

FBI Federal Bureau of Investigation

FC fire control, fire controlman, follow copy, food control, footcandle

FCA Farm Credit Administration

FCC Federal Communications Commission

fcp foolscap

fcy fancy

FD fire department, free dock

FDA Food and Drug Administration

FDIC Federal Deposit Insurance Corporation

Fe [Latin *ferrum*] iron

Feb February

fec [Latin *fecit*] he made it

fed federal, federation

fedn federation

fem female, feminine

FEMA Federal Emergency Management Agency

FEPA Fair Employment Practices Act

FEPC Fair Employment Practices Commission

Ferm Fermanagh

FET Federal excise tax, field-effect transistor

ff folios, [*f*ollowing] and the following ones; fortissimo

FHA Federal Housing Administration

FICA Federal Insurance Contributions Act

fict fiction, fictitious

fi fa fieri facias

FIFO first in, first out

fig figurative, figuratively, figure

fin finance, financial, finish

FIO free in and out

fl flanker, floor, florin, [Latin *floruit*] flourished; fluid

FL Florida, focal length, foreign language

Fla Florida

fl dr fluid dram

Flem Flemish

Flint *or* **Flints** Flintshire

FLIR forward-looking infrared

fl oz fluid ounce

FLSA Fair Labor Standards Act

fm fathom

Fm fermium

FM field manual

FMCS Federal Mediation and Conciliation Service

fn footnote

fo *or* **fol** folio

FO field officer, field order, finance officer, flight officer, foreign office, forward observer

FOB free on board

FOC free of charge

FOE Fraternal Order of Eagles

FOIA Freedom of Information Act

for foreign, forestry

FOR free on rail

forz forzando

FOS free on steamer

FOT free on truck

4WD four-wheel drive

fow first open water

fp freezing point

FPA Foreign Press Association, free of particular average

FPC fish protein concentrate

fpm feet per minute

FPO fleet post office

fps feet per second, foot-pound-second, frames per second

fr father, franc, friar, from

Fr France, francium, French, Friday

FRB Federal Reserve Board

freq frequency, frequent, frequentative, frequently

FRG Federal Republic of Germany

Fri Friday

FRM fixed rate mortgage

front frontispiece

FRS Federal Reserve System

frt freight

frwy freeway

fs femtosecond

FS filmstrip, Foreign Service

FSH follicle-stimulating hormone

FSLIC Federal Savings and Loan Insurance Corporation

FSO Foreign Service Officer

ft feet, foot, fort

FT Fourier transform, free throw, full time

FTC Federal Trade Commission

FTE full-time equivalent

fth fathom

ft lb foot-pound

FTP file transfer protocol

fund fundamental

fur furlong

fut future

FV [Latin *folio verso* the page being turned] on the back of the page

fwd foreword, forward

FWD front-wheel drive

FX foreign exchange

FY fiscal year

FYI for your information

g acceleration of gravity, game, gauge, gelding, gender, good, gram, grand, gravity

G German, giga-, guanine, Gulf

ga gauge

Ga gallium, Georgia

GA Gamblers Anonymous, general agent, general assembly, general average, general of the army, Georgia

GAAP generally accepted accounting principles

gal gallery, gallon

Gal Galatians

galv galvanized

GAO General Accounting Office

gar garage

GAR Grand Army of the Republic

GATT General Agreement on Tariffs and Trade

gaz gazette

GB Great Britain

GBF Great Books Foundation

GC gas chromatograph, gas chromatography

GCA ground-controlled approach

GCB Knight Grand Cross of the Bath

GCD greatest common divisor

GCF greatest common factor

gd good

Gd gadolinium

GDP gross domestic product

GDR German Democratic Republic

Ge germanium

GE gilt edges

GED General Educational Development (tests), general equivalency diploma

GEM ground-effect machine

gen general, genitive, genus

Gen Genesis

Gen AF general of the air force

genl general

geog geographic, geographical, geography

geol geologic, geological, geology

geom geometric, geometrical, geometry

ger gerund

Ger German, Germany

GeV giga-electron-volt

GHQ general headquarters

GHz gigahertz

gi gill

GI galvanized iron, gastrointestinal, general issue, government issue

Gib *or* **Gibr** Gibraltar

GIFT gamete intrafallopian transfer; gamete intrafallopian tube transfer

GIGO garbage in, garbage out

GIT Group Inclusive Tour

Gk Greek

Glos Gloucestershire

gm gram

GM general manager, grand master, guided missile

GMT Greenwich mean time

GMW gram-molecular weight

gn guinea

GNI gross national income

GNP gross national product

GO general order

GOP Grand Old Party (Republican)

Goth Gothic

gov government, governor

govt government

gp group

GP general practice, general practitioner, geometric progression

GPA grade point average

GPD gallons per day

GPH gallons per hour

GPM gallons per minute

GPO general post office, Government Printing Office

GPS gallons per second

GQ general quarters

gr grade, grain, gram, gravity, gross

Gr Greece, Greek

grad graduate, graduated

gram grammar, grammatical

Gramp Grampian

GRAS generally recognized as safe

GRE Graduate Record Examination

gro gross

GRP glass-reinforced plastic

GRU [Russian *Germanlavnoe razvedyvatel'noe upravlenie*] Chief Intelligence Directorate

gr wt gross weight

GS general staff, giant slalom, government service, ground speed

GSA General Services Administration, Girl Scouts of America

GSC general staff corps

GSL Guaranteed Student Loan

GSM [French *Groupe spéciale mobile,* team appointed by European telecommunications administrations in 1982 to develop standards for wireless networks] global system for mobile communications

GSO general staff officer

GSR galvanic skin response

GST Greenwich sidereal time

GSUSA Girl Scouts of the United States of America

gt gilt top, great

GT gross ton

Gt Brit Great Britain

gtd guaranteed

Gtr Man Greater Manchester

gtt [Latin *gutta,* plural *guttae*] drop

GU genitourinary, Guam

GUI graphical user interface

GUT grand unified theory, grand unification theory

GWRBI game winning run batted in

Gwyn Gwynedd

gyn *or* **gynecol** gynecology

Gy Sgt gunnery sergeant
h half, harbor, hard, hardness, hecto-, height, high, hit, horse, hour, humidity, hundred, Hungary, husband
H Hamiltonian, henry, heroin, hydrogen
ha hectare
HA hour angle
Hab Habacuc, Habakkuk
hab corp habeas corpus
Hag Haggai
Hants Hampshire
Hb hemoglobin
HBM Her Britannic Majesty, His Britannic Majesty
HC Holy Communion, House of Commons, hydrocarbon
HCF highest common factor
HCG human chorionic gonadotropin
HCL high cost of living
hd head
HD heavy-duty
hdbk handbook
hdkf handkerchief
HDTV high-definition television
hdwe hardware
He helium
HE Her Excellency, high explosive, His Eminence, His Excellency
Heb Hebrew, Hebrews
her heraldry
Heref/Worcs Hereford and Worcester
Herts Hertfordshire
HEW Department of Health, Education, and Welfare
hex hexagon
hf half
Hf hafnium
HF height finding, high frequency, home forces
hg hectogram, heliogram, hemoglobin
Hg [New Latin *hydrargyrum,* literally, water silver] mercury
HGH human growth hormone
hgt height
hgwy highway
HH Her Highness, His Highness, His Holiness
HHD [New Latin *humanitatum doctor*] doctor of humanities
HHS Department of Health and Human Services
HI Hawaii, high intensity, humidity index
hist historian, historical, history
Hitt Hittite
HJ [Latin *hic jacet*] here lies
HJR House joint resolution
hl hectoliter
HL House of Lords
hld hold
hlqn harlequin
HLS [Latin *hoc loco situs*] laid in this place; holograph letter signed
hlt halt
hm hectometer
HM Her Majesty, Her Majesty's, His Majesty, His Majesty's
HMAS Her Majesty's Australian ship, His Majesty's Australian ship
HMBS Her Majesty's British ship, His Majesty's British ship
HMC Her Majesty's Customs, His Majesty's Customs
HMCS Her Majesty's Canadian ship, His Majesty's Canadian ship
HMS Her Majesty's ship, His Majesty's ship
HN head nurse
Ho holmium
hom homiletics, homily
hon honor, honorable, honorary
Hon *or* **Hond** Honduras
HOPE Health Opportunity for People Everywhere
hor horizontal
hort horticultural, horticulture
Hos Hosea
hosp hospital
HOV high-occupancy vehicle
hp horsepower
HP half pay, high pressure
HPA high-power amplifier
HPF highest possible frequency, high power field
HPLC high-performance liquid chromatography
HPV human-powered vehicle
HQ headquarters
hr here, hour
HR home run, House of Representatives
H Res House resolution
HRH Her Royal Highness, His Royal Highness
hrzn horizon
Hs hassium
HS high school
HSL high-speed launch
HST Hawaiian standard time, hypersonic transport
ht height
HT half time, halftone, hardtop, Hawaii time, high-tension, high tide, [Latin *hoc tempore*] at this time; [Latin *hoc titulo*] under this title
http hypertext transfer protocol
HUAC House Un-American Activities Committee
HUD Department of Housing and Urban Development
Humber Humberside
Hung Hungarian, Hungary
hv have
HV high velocity, high voltage
HVAC heating, ventilating, and air-conditioning
hvy heavy
hw how
HW high water, highway, hot water
HWM high-water mark
hwy highway
hyd hydraulics, hydrostatics
hyp hypothesis, hypothetical
Hz hertz
i industrial, initial, intelligence, intensity, intransitive, island, isle
I electric current, Indian, interstate, iodine, Israeli
Ia *or* **IA** Iowa
IAA indoleacetic acid
IAAF International Amateur Athletic Federation
IABA International Amateur Boxing Association
IAEA International Atomic Energy Agency
IALC instrument approach and landing chart
IAM International Association of Machinists and Aerospace Workers
IAP international airport
IAS indicated airspeed
IATA International Air Transport Association
IAU International Association of Universities, International Astronomical Union
ib *or* **ibid** ibidem
IB in bond, incendiary bomb
IBRD International Bank for Reconstruction and Development
IBS irritable bowel syndrome
ICA International Cooperative Alliance
ICAO International Civil Aviation Organization
ICC International Chamber of Commerce, Interstate Commerce Commission
Ice Iceland
ICE internal combustion engine, International Cultural Exchange
Icel Icelandic
ICF intermediate care facility
ICFTU International Confederation of Free Trade Unions
ICJ International Court of Justice
ICRC International Committee of the Red Cross
ICU intensive care unit
id idem
ID Idaho, identification, independent distributor, industrial design, inner diameter, inside dimensions, intelligence department, internal diameter
IDA International Development Association
IDDM insulin-dependent diabetes mellitus
IDP international driving permit
i.e. [Latin *id est*] that is
IE industrial engineer
IEEE The Institute of Electrical and Electronics Engineers
IF intermediate frequency
IFC International Finance Corporation
IFF identification, friend or foe
IFO identified flying object
IFR instrument flight rules
Ig immunoglobulin
IG inspector general
IGY International Geophysical Year
ihp indicated horsepower
IIE Institute of Industrial Engineers
IL Illinois
ILA International Longshoremen's Association
ILGWU International Ladies' Garment Workers' Union
ill illustrated, illustration, illustrator
Ill Illinois
illust *or* **illus** illustrated, illustration
ILO International Labor Organization
ILS instrument landing system
IL-2 interleukin-2
IM individual medley, intramural
IMCO Inter-Governmental Maritime Consultative Organization
IMF International Monetary Fund
IMHO in my humble opinion
imit imitative
immed immediate, immediately
immun immunity, immunization
immunol immunology
imp imperative, imperfect, imperial, import, imported
IMP international match point
imperf imperfect, imperforate
in inch, inlet
In indium
IN Indiana
inc incomplete, incorporated, increase
incl include, included, including, inclusive
incog incognito
incr increase, increased
ind independent, index, industrial, industry
Ind Indian, Indiana
IND investigational new drug
IndE industrial engineer
indef indefinite
indic indicative
indiv individual
indn indication
Indon Indonesia, Indonesian
indus industrial, industry
inf infantry, infinitive
INF intermediate range nuclear forces
infl influenced
INH [*iso-nicotinic acid hydrazide*] isoniazid
inorg inorganic
INP International News Photo
inq inquire
INRI [Latin *Iesus Nazarenus Rex Iudaeorum*] Jesus of Nazareth, King of the Jews
ins inches, insurance
INS Immigration and Naturalization Service, inertial navigation system
insol insoluble
insp inspector
inst instant, institute, institution, institutional
instr instructor, instrument, instrumental
insur insurance
int intelligence, intercept, interest, interim, interior, interjection, interleaved, intermediate, internal, international, interpreter, intersection, interval, interview, intransitive
intel intelligence
interj interjection
Interpol International Criminal Police Organization
interrog interrogative
intl *or* **intnl** international
intrans intransitive
in trans [Latin *in transitu*] in transit
introd introduction
inv inventor, invoice
I/O input/output
IOC International Olympic Committee
IOM Isle of Man
Ion Ionic
IOOF Independent Order of Odd Fellows
IORM Improved Order of Red Men
IOW Isle of Wight
IP initial point, innings pitched, intermediate pressure
IPA individual practice association
ipm inches per minute
IPO initial public offering
ips inches per second
IPTS International Practical Temperature Scale
iq [Latin *idem quod*] the same as
Ir iridium, Irish
IR information retrieval, infrared, inland revenue, intelligence ratio, internal revenue
IRA Irish Republican Army
IRBM intermediate range ballistic missile
Ire Ireland
irid iridescent
irred irredeemable
irreg irregular
IRS Internal Revenue Service
is island, isle
Isa *or* **Is** Isaiah
ISBN International Standard Book Number
ISDN integrated services digital network
ISC interstate commerce
isl island
ISO International Organization for Standardization
ISP Internet service provider
isoln isolation
Isr Israel, Israeli
ISSN International Standard Serial Number
isth isthmus
ISV International Scientific Vocabulary
It Italian, Italy
IT information technology
ital italic, italicized
Ital Italian
ITO International Trade Organization
ITU International Telecommunication Union, International Typographical Union
ITV instructional television
IU international unit
IV intravenous, intravenously
IVF in vitro fertilization
IW inside width, isotopic weight
IWW Industrial Workers of the World
j jack, journal, judge, justice
J joule, jump shot
JA joint account, judge advocate
JAG judge advocate general
Jam Jamaica
Jan January
Jas James
Jav Javanese
JBS John Birch Society
JC junior college
JCAHO Joint Commission on Accreditation of Healthcare Organizations
JCB [New Latin *juris canonici baccalaureus*] bachelor of canon law
JCD [New Latin *juris canonici doctor*] doctor of canon law
JCL [New Latin *juris canonici licentiatus*] licentiate in canon law
JCS joint chiefs of staff

jct junction
JD [New Latin *juris doctor*] doctor of jurisprudence, doctor of law; [New Latin *jurum doctor*] doctor of laws; justice department, juvenile delinquent
Jer Jeremiah, Jeremias
jg junior grade
JIT job instruction training, just in time
Jn *or* **Jno** John
JND just noticeable difference
jnr *British* junior
Jo Joel
Jon Jonah, Jonas
Josh Joshua
jour journal, journeyman
JP jet propulsion, justice of the peace
Jpn Japan, Japanese
Jr junior
JSD [New Latin *juris scientiae doctor*] doctor of science of law
jt *or* **jnt** joint
Jud Judith
Judg Judges
Jul July
jun junior
Jun June
junc junction
juv juvenile
JV junior varsity
k karat, kindergarten, king, kitchen, knit, knot, koruna, kosher — often enclosed in a circle; kyat
K [New Latin *kalium*] potassium; Kelvin, kilometer
ka [German *Kathode*] cathode
Kan *or* **Kans** Kansas
kb kilobar, kilobase
KB kilobyte
kbar kilobar
kc kilocycle
KC Kansas City, King's Counsel, Knights of Columbus
kcal kilocaloric, kilogram calorie
KCB knight commander of the Order of the Bath
kc/s kilocycles per second
KD kiln-dried, knockdown, knocked down
Ker Kerry
keV kilo-electron volt
kg keg, kilogram, king
kG kilogauss
KG knight of the Order of the Garter
KGB [Russian *Komitet gosudarstvennoĭ bezopasnosti*] (Soviet) State Security Committee
kgps kilograms per second
kHz kilohertz
KIA killed in action
Kild Kildare
Kilk Kilkenny
kit kitchen
kJ kilojoule
KJV King James Version
KKK Ku Klux Klan
kl kiloliter
km kilometer
kmh *or* **kmph** kilometers per hour
kmps kilometers per second
kn knot
K of C Knights of Columbus
Kor Korea, Korean
kPa kilopascal
kpc kiloparsec
kph kilometers per hour
Kr krypton
KS Kansas, Kaposi's sarcoma
kt karat, knight, knot
KT kiloton
kV kilovolt
kW kilowatt
kWh kilowatt-hour
Ky *or* **KY** Kentucky
l lady, lake, lambert, land, large, late, left, [Latin *libra*] pound; line, liquid, lira, lire, liter, little, low
L Lagrangian, Latin, long
La lanthanum, Louisiana
LA law agent, legislative assistant, Los Angeles, Louisiana
Lab Labrador
lam laminated
Lam Lamentations
Lancs Lancashire
lang language
lat latitude
Lat Latin, Latvia
LAT local apparent time
lav lavatory
lb [Latin *libra*] pound
LB Labrador
LBO leveraged buyout
lc lowercase
LC landing craft, left center, letter of credit, Library of Congress
LCD least common denominator, lowest common denominator
LCDR lieutenant commander
LCL less-than-carload lot
LCM least common multiple, lowest common multiple, [New Latin *legis comparativae magister*] master of comparative law
LCpl lance corporal
LCS League Championship Series
LCT local civil time
ld load, lord
LD learning disabled, learning disability; lethal dose — often used with a numerical subscript to indicate the percent of a test group of organisms the dose is expected to kill ⟨LD_{50}⟩; line of departure
LDC less developed country
ldg landing, loading
LDH lactate dehydrogenase, lactic dehydrogenase
ldr leader
LDS Latter-day Saints
LE leading edge
lea leather
Leb Lebanese, Lebanon
lect lecture, lecturer
leg legal, legato, legislative, legislature
legis legislation, legislative, legislature
Leics Leicestershire
Leit Leitrim
LEM lunar excursion module
LEP limited English proficiency, limited English proficient
Lev *or* **Levit** Leviticus
lf lightface
LF ledger folio, low frequency
lg large, long
LH left hand, lower half, luteinizing hormone
LHD [New Latin *litterarum humaniorum doctor*] doctor of humane letters, doctor of humanities
li link
Li lithium
LI Long Island
lib liberal, librarian, library
lic license
lieut lieutenant
LIFO last in, first out
Lim Limerick
lin lineal, linear
Lincs Lincolnshire
ling linguistics
liq liquid, liquor
lit liter, literal, literally, literary, literature
lith lithographic, lithography
Litt B *or* **Lit B** [Medieval Latin *litterarum baccalaureus*] bachelor of letters, bachelor of literature
Litt D *or* **Lit D** [Medieval Latin *litterarum doctor*] doctor of letters, doctor of literature
Lk Luke
ll lines
LL lending library, limited liability, lower left
LLB [New Latin *legum baccalaureus*] bachelor of laws
LLC limited liability company
LLD [New Latin *legum doctor*] doctor of laws
LLM [New Latin *legum magister*] master of laws
lm lumen
LM Legion of Merit, long meter, lunar module
LMG light machine gun
LMT local mean time
ln lane, natural logarithm
lndg landing
LNG liquefied natural gas
LOA length overall
loc cit [Latin *loco citato*] in the place cited
log logic
Lond London, Londonderry
long longitude
Long Longford
loq [Latin *loquitur*] he speaks, she speaks
LOS line of scrimmage, line of sight
Loth Lothian
Lou Louth
LP low pressure
LPG liquefied petroleum gas
LPGA Ladies Professional Golf Association
Lr lawrencium
LR living room, log run, lower right
LRT light-rail transit
LRV light-rail vehicle
LS left side, letter signed, library science, [Latin *locus sigilli*] place of the seal; long shot
LSAT Law School Admission Test
LSI large-scale integrated circuit, large-scale integration
LSM letter-sorting machine
LSS lifesaving service, lifesaving station, life-support system
LST landing ship, tank; local sidereal time
lt light
Lt lieutenant
LT long ton, low-tension
LTC lieutenant colonel, long term care
Lt Col lieutenant colonel
Lt Comdr lieutenant commander
ltd limited
LTG *or* **Lt Gen** lieutenant general
lt gov lieutenant governor
LTh licentiate in theology
LTJG lieutenant, junior grade
LTL less than truckload
ltr letter, lighter
LTS launch telemetry station, launch tracking system
Lu lutetium
lub lubricant, lubricating
Luth Lutheran
lv leave
LVT landing vehicle, tracked
LW low water
LWM low-water mark
LWV League of Women Voters
lx lux
LZ landing zone
m male, manual, married, martyr, masculine, mass, measure, meridian, [Latin *meridies*] noon; meter, middle, mile, [Latin *mille*] thousand; milli-, minute, molal, molality, molar, molarity, mole, month, moon, morning, muscle
m- meta-
M Mach, March, May, medium, mega-, million, monsieur
mA milliampere
MA [Medieval Latin *magister artium*] master of arts; Massachusetts, mental age, Middle Ages
MAA master of applied arts
Mac Machabees
Mac *or* **Macc** Maccabees
MAC military airlift command
mach machine, machinery, machinist
MAD mutual assured destruction
MAE *or* **MA Ed** master of arts in education
mag magnesium, magnetism, magneto, magnitude
Maj major
Maj Gen major general
Mal Malachi
MALS master of arts in liberal studies, master of arts in library science
man manual
Man Manitoba
M&A mergers and acquisitions
manuf manufacture, manufacturing
MAO monoamine oxidase
MAP modified American plan
mar maritime
Mar March
MARC machine readable cataloging
MARV maneuverable reentry vehicle
masc masculine
MASH mobile army surgical hospital
Mass Massachusetts
MAT master of arts in teaching, Miller analogy test
math mathematical, mathematician
matric matriculated, matriculation
Matt Matthew
MATV master antenna television
max maximum
mb millibar
Mb megabit
MB bachelor of medicine, Manitoba, megabyte, municipal borough
MBA master of business administration
mbd million barrels per day
MBD minimal brain dysfunction
MBE member of the Order of the British Empire
MBO management by objective
Mbps megabits per second
MBS Mutual Broadcasting System
mc megacycle, millicurie
MC member of Congress
MCAT Medical College Admission Test
MCC mission control center
mcf thousand cubic feet
mcg microgram
MCL Marine Corps League, master of civil law, master of comparative law
MCP male chauvinist pig
MCPO master chief petty officer
Md Maryland, mendelevium
MD [New Latin *medicinae doctor*] doctor of medicine; [Italian *mano destra*] right hand; Maryland, medical department, months after date, muscular dystrophy
MDC more developed country
MDiv master of divinity
mdnt midnight
mdse merchandise
MDT mountain daylight time
Me Maine, methyl
ME Maine, managing editor, mechanical engineer, medical examiner
Mea Meath
meas measure
mech mechanical, mechanics
med medicine, medieval, medium
Med Mediterranean
MEd master of education
meg megohm
MEGO my eyes glaze over
MEK methyl ethyl ketone
mem member, memoir, memorial
MEng master of English
MEP member of the European Parliament
mer meridian
Mersey Merseyside
met meteorological, meteorology, metropolitan
metal *or* **metall** metallurgical, metallurgy
metaph metaphysics
METO Middle East Treaty Organization
MeV million electron volts
Mex Mexican, Mexico
mf mezzo forte
mF millifarad
MF medium frequency, microfiche
MFA master of fine arts
mfd manufactured
mfg manufacturing
MFH master of foxhounds

MFN most favored nation
mfr manufacture, manufacturer
mg milligram
Mg magnesium
MG machine gun, major general, military government
mgal milligal
MGB [Russian *Ministerstvo gosudarstvennoĭ bezopasnosti*] Ministry of State Security
mgd million gallons per day
mgr manager, monseigneur, monsignor
mgt *or* **mgmt** management
MGy Sgt master gunnery sergeant
MH medal of honor, mental health, mobile home
MHA master of hospital administration
MHC major histocompatibility complex
MHD magnetohydrodynamic, magnetohydrodynamics
mhg mahogany
MHW mean high water
MHz megahertz
mi mile, mileage, mill
MI Michigan, military intelligence
Mic Micah
MIC methyl isocyanate
Mich Michigan
MICR magnetic ink character recognition
mid middle
Middx Middlesex
Mid Glam Mid Glamorgan
midn midshipman
mil military, million
MIME multipurpose Internet mail extensions
min minim, minimum, mining, minister, minor, minute
Minn Minnesota
MIO minimum identifiable odor
MIPS *or* **mips** million instructions per second
MIS management information systems
misc miscellaneous
Miss Mississippi
mixt mixture
mk mark, markka
Mk Mark
mks meter-kilogram-second
mkt market
mktg marketing
ml milliliter
mL millilambert
MLA Member of the Legislative Assembly
MLD median lethal dose, minimum lethal dose
MLF multilateral force
Mlle [French] mademoiselle
Mlles [French] mesdemoiselles
MLS master of library science
MLW mean low water
mm measures, millimeter
MM [French] messieurs; mutatis mutandis
Mme [French] madame
Mmes [French] mesdames
mmf magnetomotive force
MMPI Minnesota Multiphasic Personality Inventory
Mn manganese
MN magnetic north, Minnesota
MNC multinational company, multinational corporation
mo month
Mo Missouri, molybdenum, Monday
MO mail order, medical officer, Missouri, modus operandi, money order
mod moderate, modification, modified, modulo, modulus
modif modification
mol mole, molecular, molecule
MOL manned orbiting laboratory
mol wt molecular weight
MOM middle of month
mon monastery, monetary
Mon Monaghan, Monday
Mont Montana
mor morocco
MOR middle of the road
morph morphology
mos months
MOS metal-oxide semiconductor, military occupational specialty
MOSFET metal-oxide-semiconductor field-effect transistor
MP melting point, member of parliament, metropolitan police, milepost, military police, military policeman
MPA master of public administration
MPAA Motion Picture Association of America
mpg miles per gallon
mph miles per hour
MPH master of public health
MPhil master of philosophy
MPM meters per minute
MPS meters per second
MPV multi-purpose vehicle
MPX multiplex
mR milliroentgen
MR map reference, mentally retarded
MRE meals ready to eat
MRI magnetic resonance imaging
mRNA messenger RNA
ms millisecond
MS [Italian *mano sinistra*] left hand; manuscript, master of science, military science, Mississippi, motor ship, multiple sclerosis
MSc master of science
msec millisecond
msg message
MSG master sergeant, monosodium glutamate
msgr monseigneur, monsignor
MSgt master sergeant
MSH melanocyte-stimulating hormone
MSL mean sea level
MSN master of science in nursing
MSS manuscripts
MST mountain standard time
MSW master of social welfare, master of social work
mt mount, mountain
Mt Matthew
MT machine translation, metric ton, Montana, mountain time
mtg meeting, mortgage
mtge mortgage
mth month
mtn mountain
MTO Mediterranean theater of operations
mun *or* **munic** municipal
mus museum, music, musical, musician
mV millivolt
MV main verb, mean variation, motor vessel
MVD [Russian *Ministerstvo vnutrennikh del*] Ministry of Internal Affairs
MVP most valuable player
mW milliwatt
MW megawatt
MWe megawatts electric
mxd mixed
my million years
myc *or* **mycol** mycology
Myr million years
n name, nano-, navy, net, neuter, *usually italic* neutron, noon, normal, north, northern, note, noun, number
N newton, nitrogen
Na [New Latin *natrium*] sodium
NA national association, no account, North America, not applicable, not available
NAACP National Association for the Advancement of Colored People
NAB New American Bible
NAD no appreciable disease
NAFTA North American Free Trade Agreement
Nah Nahum
NAIA National Association of Intercollegiate Athletes
NAS National Academy of Sciences, naval air station
NASA National Aeronautics and Space Administration
NASCAR National Association for Stock Car Auto Racing
NASD National Association of Securities Dealers
NASL North American Soccer League
NASW National Association of Social Workers
nat national, native, natural
natl national
NATO North Atlantic Treaty Organization
naut nautical
nav naval, navigable, navigation
Nb niobium
NB New Brunswick, northbound, nota bene
NBA National Basketball Association, National Boxing Association
NBC National Broadcasting Company
NBS National Bureau of Standards
NC no charge, no credit, North Carolina, nurse corps
NCAA National Collegiate Athletic Association
NCE New Catholic Edition
NCV no commercial value
nd no date
Nd neodymium
ND doctor of naturopathy, North Dakota
N Dak North Dakota
NDE near-death experience
NDP New Democratic Party (Canada)
Ne neon
NE Nebraska, New England, no effects, northeast
NEA National Education Association, National Endowment for the Arts
Neb *or* **Nebr** Nebraska
NEB New English Bible
NED New English Dictionary
neg negative, negotiable
Neh Nehemiah
NEH National Endowment for the Humanities
NEI not elsewhere included
nem con [New Latin *nemine contradicente*] no one contradicting
nem diss [New Latin *nemine dissentiente*] no one dissenting
NEP New Economic Policy
NES not elsewhere specified
Neth Netherlands
neurol neurological, neurology
neut neuter
Nev Nevada
New Eng New England
NF Newfoundland, no funds
NFC National Football Conference
NFL National Football League
Nfld Newfoundland
NFP natural family planning
NFS not for sale
ng nanogram
NG national guard, no good
NGF nerve growth factor
NGO nongovernmental organization
NGU nongonococcal urethritis
NH never hinged, New Hampshire
NHL National Hockey League
Ni nickel
NIC newly industrialized country, newly industrializing country
NIDDM non-insulin-dependent diabetes mellitus
NIH National Institutes of Health, not invented here
NIMBY not in my backyard
NIT National Invitational Tournament
NIU network interface unit
NJ New Jersey
NKVD [Russian *Narodnyĭ komissariat vnutrennikh del*] People's Commissariat of Internal Affairs
NL National League, new line, night letter, [Latin *non licet*] it is not permitted; north latitude
NLCS National League Championship Series
NLF National Liberation Front
NLRB National Labor Relations Board
NLT night letter
nm nanometer
NM nautical mile, New Mexico, no mark, not marked
N Mex New Mexico
NMHA National Mental Health Association
NMI no middle initial
NMR nuclear magnetic resonance
NNE north-northeast
NNW north-northwest
no north, northern, [Latin *numero*, ablative of *numerus*] number
No nobelium
NOAA National Oceanic and Atmospheric Administration
NOIBN not otherwise indexed by name
nom nominative
non obst *or* **non obs** non obstante
non seq non sequitur
NOP not otherwise provided for
Nor Norway, Norwegian
NORAD North American Air Defense Command
Norf Norfolk
norm normal
Northants Northamptonshire
Norw Norway, Norwegian
nos numbers
NOS not otherwise specified
Notts Nottinghamshire
nov novelist
Nov November
NOW National Organization for Women
NO_x nitrogen oxide
np no pagination, no place (of publication)
Np neptunium
NP neuropsychiatric, neuropsychiatry, no protest, notary public, noun phrase
NPF not provided for
NPN nonprotein nitrogen
NPR National Public Radio
NPS National Park Service
nr near
NR not rated
NRA National Recovery Administration, National Rifle Association
NRC National Research Council, Nuclear Regulatory Commission
ns *also* **nsec** nanosecond
Ns nimbostratus
NS new series, new style, not specified, not sufficient, Nova Scotia
NSA National Security Agency
NSC National Security Council
NSF National Science Foundation, not sufficient funds
NSW New South Wales
NT New Territories, New Testament, Northern Territory, Northwest Territories
Nthmb Northumberland
NTP normal temperature and pressure
NTSB National Transportation Safety Board
nt wt *or* **n wt** net weight
NU name unknown
num numeral
Num *or* **Numb** Numbers
numis numismatic, numismatical, numismatics
NV Nevada, nonvoting
NW northwest
NWT Northwest Territories
NY New York
NYC New York City
NYSE New York Stock Exchange
NZ New Zealand
o ocean, ohm, old, order, oriental, over
O Ohio, oxygen, [New Latin *octarius*] pint
o- ortho-
o/a on or about
OAS Organization of American States

OASDHI Old Age, Survivors, Disability, and Health Insurance
OAU Organization of African Unity
ob [Latin *obiit*] he died, she died; observation
Ob *or* **Obad** Obadiah
OB obstetric, obstetrician, obstetrics
OBE officer of the Order of the British Empire, out-of-body experience
obj object, objective
obl oblique, oblong
obstet obstetrical, obstetrics
obv obverse
oc ocean
OC off center, officer candidate, on center, on course, over-the-counter
occas occasionally
OCD obsessive-compulsive disorder
oceanog oceanography
OCR optical character reader, optical character recognition
OCS officer candidate school
oct octavo
Oct October
OD doctor of optometry, [Latin *oculus dexter*] right eye; officer of the day, olive drab, on demand, outside diameter, outside dimension, overdraft, overdrawn
Oe oersted
OECD Organization for Economic Cooperation and Development
OED Oxford English Dictionary
OEO Office of Economic Opportunity
OER officer efficiency report
OES Order of the Eastern Star
OF outfield
off office, officer, official
offic official
OG officer of the guard, original gum
OH Ohio
OHMS on Her Majesty's service, on His Majesty's service
OIT Office of International Trade
OJ orange juice
OJT on-the-job training
OK Oklahoma, outer keel
Okla Oklahoma
OM order of merit
OMB Office of Management and Budget
ON *or* **Ont** Ontario
OOD officer of the deck
op operation, operative, operator, opportunity, opus
OP observation post, out of print
op cit [Latin *opere citato*] in the work cited
OPEC Organization of Petroleum Exporting Countries
opp opposite
opt optical, optician, optics, option, optional
OR operating room, operational research, operations research, Oregon, owner's risk, own recognizance
orch orchestra
ord order, ordnance
Ore *or* **Oreg** Oregon
org organic, organization, organized
orig original, originally, originator
Ork Orkney
ornith ornithology
ORV off-road vehicle
Os osmium
OS [Latin *oculus sinister*] left eye; old series, old style, ordinary seaman, out of stock
OSHA Occupational Safety and Health Administration
OSS Office of Strategic Services
OT occupational therapy, Old Testament, overtime
OTA Office of Technology Assessment
OTB offtrack betting
OTC over-the-counter
OTR occupational therapist, registered
OTS officers' training school
OW one-way
Oxfam Oxford Committee for Famine Relief
Oxon [Medieval Latin *Oxonia*] Oxford; Oxfordshire, [Medieval Latin *Oxoniensis*] of Oxford
oz [obs. Italian *onza* (now *oncia*)] ounce, ounces
p page, pages, parental generation, part, participle, past, pater, pawn, pence, penny, per, peseta, peso, petite, piano, pico-, pint, pipe, pitch, pole, port, power, pro, proton, purl
P phosphorus, pressure, [French *poids*] weight
p- para-
Pa pascal, Pennsylvania, protactinium
PA particular average, passenger agent, Pennsylvania, per annum, personal appearance, personal assistant, power amplifier, power of attorney, press agent, private account, professional association, public address, purchasing agent
Pac Pacific
PAC political action committee
paleon paleontology
pam pamphlet
Pan Panama
p and h postage and handling
P&I principal and interest
P&L profit and loss
par paragraph, parallel, parish
para paragraph
part participial, participle, particular
PAS para-aminosalicylic acid
pass passenger, passive
pat patent
PAT point after touchdown
path *or* **pathol** pathological, pathology
PAYE *British* pay as you earn
payt payment
pb paperback
Pb [Latin *plumbum*] lead
PB personal best, power brakes
PB&J peanut butter and jelly
PBS Public Broadcasting Service
pc parsec
PC Peace Corps, percent, percentage, politically correct, postcard, [Latin *post cibum*] after meals; printed circuit, professional corporation
PCP pneumocystis carinii pneumonia, primary care physician
PCR polymerase chain reaction
pct percent, percentage
pd paid
Pd palladium
PD per diem, police department, postal district, potential difference, program director, public domain
PDA predicted drift angle, public display of affection
PDD past due date
PDT Pacific daylight time
PE physical education, printer's error, probable error, professional engineer
P/E price/earnings
PEI Prince Edward Island
pen peninsula
PEN International Association of Poets, Playwrights, Editors, Essayists and Novelists
Penn *or* **Penna** Pennsylvania
per period, person
perf perfect, perforated, performance
perh perhaps
perm permanent
perp perpendicular
pers person, personal, personnel
Pers Persia, Persian
pert pertaining
pet petroleum
Pet Peter
PET positron-emission tomography
pf personal foul, pfennig, picofarad, preferred
PF power factor, pianoforte, [Italian *più forte*] louder
PFC *or* **Pfc** private first class
pfd preferred
PFD personal flotation device
pg page, picogram
Pg Portugal, Portuguese
PG paying guest, postgraduate, prostaglandin
PGA Professional Golfers' Association
ph phase
PH pinch hit, public health, Purple Heart
phar pharmacopoeia, pharmacy
pharm pharmaceutical, pharmacist, pharmacy
PhB [New Latin *philosophiae baccalaureus*] bachelor of philosophy
PhD [New Latin *philosophiae doctor*] doctor of philosophy
phil *or* **philol** philological, philology
Phil Philippians
philos philosopher, philosophy
phon phonetics
photog photographic, photography
phr phrase
phys physical, physics
physiol physiologist, physiology
PI Philippine Islands, private investigator, programmed instruction
PID pelvic inflammatory disease
PIK payment in kind
PIN personal identification number
PINS persons in need of supervision
pinx [Latin *pinxit*] he painted it, she painted it
PIRG Public Interest Research Group
pizz pizzicato
pk park, peak, peck, pike
PK preacher's kid
pkg package
pkt packet, pocket
PKU phenylketonuria
pkwy parkway
pl place, plate, plural
PL partial loss, private line, Public Law
plat plateau, platoon
PLC *British* public limited company
plf plaintiff
PLO Palestine Liberation Organization
pls please
PLSS portable life-support system
pm phase modulation, premium
Pm promethium
PM paymaster, permanent magnet, police magistrate, postmaster, post meridiem — often not capitalized and often punctuated; postmortem, prime minister, provost marshal
PMB private mailbox
pmk postmark
PMS premenstrual syndrome
pmt payment
PN promissory note
pnxt [Latin *pinxit*] he painted it, she painted it
Po polonium
PO [Latin *per os*] by mouth, orally; petty officer, postal order, post office, purchase order
POB post office box
POC port of call
POD pay on delivery, post office department
POE port of embarkation, port of entry
Pol Poland, Polish
polit political, politician
poly polytechnic
pon pontoon
pop population
POP point of purchase
por portrait
POR pay on return, price on request
port portable, portrait
Port Portugal, Portuguese
pos position, positive
POS point of sale
poss possessive
pot potential, potentiometer
POV point of view
pp pages, [Latin *per procurationem*] by proxy; pianissimo
PP parcel post, past participle, postpaid, prepaid
ppa per power of attorney
ppb parts per billion
ppd postpaid, prepaid
PPI plan position indicator
ppm parts per million
PPS [New Latin *post postscriptum*] an additional postscript
ppt parts per thousand, parts per trillion, precipitate
pptn precipitation
PQ Province of Quebec
pr pair, price, printed
Pr praseodymium, propyl
PR payroll, proportional representation, public relations, Puerto Rico
PRC People's Republic of China
prec preceding
pred predicate
pref preface, preference, preferred, prefix
prem premium
prep preparatory, preposition
prepd prepared
prepg preparing
prepn preparation
pres present, president
Presb Presbyterian
prev previous, previously
prf proof
prim primary, primitive
prin principal, principle
priv private, privately, privative
PRN [Latin *pro re nata*] for the emergency, as needed
PRO public relations officer
prob probable, probably, probate, problem
proc proceedings
prod product, production
prof professional
prog program
proj project, projector
prom promontory
pron pronoun, pronounced, pronunciation
prop property, proposition, proprietor
pros prosody
Prot Protestant
prov province, provincial, provisional
Prov Proverbs
prox proximo
ps picosecond
Ps *or* **Psa** Psalms
PS [New Latin *postscriptum*] postscript; power steering, power supply, public school
PSA prostate-specific antigen, public service announcement
psec picosecond
pseud pseudonym, pseudonymous
psf pounds per square foot
PSG platoon sergeant
psi pounds per square inch
psig pounds per square inch gauge
PST Pacific standard time
psych psychology
psychol psychologist, psychology
pt part, payment, pint, point, port
Pt platinum
PT Pacific time, part-time, physical therapy, physical training
pta peseta
PTA Parent-Teacher Association
pte *British* private
ptg printing
PTO Parent-Teacher Organization, please turn over, power takeoff
PTSD post-traumatic stress disorder
PTV public television
Pty *British* proprietary
Pu plutonium
PU pickup
pub public, publication, publicity, published, publisher, publishing
publ publication, published, publisher
PUD pickup and delivery
pulv [Latin *pulvis*] powder
PUSH People United to Serve Humanity
PV photovoltaic, polyvinyl
PVA polyvinyl acetate
PVC polyvinyl chloride
PVO private voluntary organization
pvt private

PVT pressure, volume, temperature
PW prisoner of war
PWA person with AIDS
pwr power
PWR pressurized water reactor
pwt pennyweight
PX please exchange, post exchange
PYO pick your own
q quart, quartile, quarto, queen, query, question, quetzal, quire
QA quality assurance
QB quarterback, queen's bench
QC quality control, queen's counsel
qd [Latin *quaque die*] daily
QED quantum electrodynamics, [Latin *quod erat demonstrandum*] which was to be demonstrated
QEF [Latin *quod erat faciendum*] which was to be done
QEI [Latin *quod erat inveniendum*] which was to be found out
QF quick-firing
qid [Latin *quater in die*] four times a day
Qld Queensland
qm [Latin *quoque matutino*] every morning
QM quartermaster
QMC quartermaster corps
QMG quartermaster general
qp *or* **q pl** [Latin *quantum placet*] as much as you please
qq questions
qq v [Latin *quae vide*] which (*plural*) see
qr quarter, quire
qs [Latin *quantum sufficit*] as much as suffices
qt quantity, quart
qtd quartered
qto quarto
qty quantity
qu *or* **ques** question
quad quadrant
qual qualitative, quality
quant quantitative
quar quarterly
Que Quebec
quot quotation
qv [Latin *quod vide*] which see
qy query
r rabbi, radius, rain, range, Rankine, rare, real, Reaumur, recto, red, repeat, rerun, resistance, right, river, roentgen, rook, rough, run
R radial, radical — used especially of a univalent hydrocarbon radical; recipe; registered trademark — often enclosed in a circle; regular, Republican
ra range
Ra radium
RA regular army, right ascension, Royal Academician, Royal Academy
RAAF Royal Australian Air Force
rad radical, radian, radiator, radio, radius, radix
RADM rear admiral
RAF Royal Air Force
R & B rhythm and blues
R and R rest and recreation, rest and recuperation
rap rapid
Rb rubidium
RBC red blood cells, red blood count
RBE relative biological effectiveness
RC Red Cross, resistance-capacitance, Roman Catholic
RCAF Royal Canadian Air Force
RCMP Royal Canadian Mounted Police
RCN Royal Canadian Navy
rct recruit
rd road, rod, round
RD registered dietitian, rural delivery
RDA recommended daily allowance, recommended dietary allowance
RDF radio direction finder, radio direction finding, Rapid Deployment Force, refuse-derived fuel
Re rhenium
REA Rural Electrification Administration
reas reasonable
rec receipt, record, recording, recreation
recd received
recip reciprocal, reciprocity
rec sec recording secretary
rect receipt, rectangle, rectangular, rectified
red reduce, reduction
ref reference, referred, refining, reformed, refunding
refl reflex, reflexive
refr refraction
refrig refrigerating, refrigeration
reg region, register, registered, registration, regular
regd registered
regt regiment
REIT real estate investment trust
rel relating, relative, released, religion, religious
relig religion
rep repair, repeat, report, reporter, representative, republic
Rep Republican
repl replace, replacement
rept report
req request, require, required, requisition
reqd required
res research, reservation, reserve, reservoir, residence, resident, resolution
RES reticuloendothelial system
resp respective, respectively
ret retain, retired, return
retd retained, retired, returned
rev revenue, reverse, review, reviewed, revised, revision, revolution
Rev Revelation, reverend
Revd *British* reverend
rf refunding
Rf rutherfordium
RF radio frequency
RFD rural free delivery
RFP request for proposal
Rh rhodium
RH relative humidity, right hand
rhet rhetoric
RHIP rank has its privileges
RI refractive index, Rhode Island
RIA radioimmunoassay
RICE rest, ice, compression, elevation
RICO Racketeer Influenced and Corrupt Organizations (Act)
RIF reduction in force
RIO radar intercept officer
RIP [Latin *requiescat in pace*] may he rest in peace, may she rest in peace; [Latin *requiescant in pace*] may they rest in peace
RISC reduced instruction-set computer; reduced instruction-set computing
rit ritardando
riv river
RJ road junction
rm ream, room
rms root-mean-square
Rn radon
RN Royal Navy
rnd round
RNZAF Royal New Zealand Air Force
ROC Republic of China
ROG receipt of goods
ROI return on investment
ROK Republic of Korea (South Korea)
Rom Roman, Romance, Romania, Romanian, Romans
ROP record of production, run-of-paper
ROR release on own recognizance
Ros *or* **Rosc** Roscommon
rot rotating, rotation
ROTC Reserve Officers' Training Corps
RP Received Pronunciation, relief pitcher, reply paid, reprint, reprinting, Republic of the Philippines
RPh registered pharmacist
rpm revolutions per minute
rps revolutions per second
rpt repeat, report
RQ respiratory quotient
RR railroad, rural route
RRT registered record technician
RS Received Standard, recording secretary, revised statutes, right side, Royal Society
RSA Republic of South Africa
RSFSR [Russian *Rossiĭskaya Sovetskaya Frenchederativnaya Sotsialisticheskaya Respublika*] Russian Soviet Federated Socialist Republic
RSV Revised Standard Version
RSVP [French *répondez s'il vous plaît*] please reply
RSWC right side up with care
rt right, route
RT radiologic technologist, radiotelephone, respiratory therapy, room temperature, round-trip
rte route
rtw ready-to-wear
Ru ruthenium
Rum Rumania, Rumanian
Russ Russia, Russian
RV Revised Version
RW radiological warfare, right worshipful, right worthy
rwy *or* **ry** railway
s sabbath, saint, schilling, scruple, second, secondary, section, senate, series, shilling, [Latin *signa*] label; siemens, signor, sine, singular, small, smooth, snow, society, son, sou, south, southern, subject, symmetrical
S satisfactory, short, standard deviation of a sample, sulfur, svedberg
Sa Saturday
SA Salvation Army, seaman apprentice, sex appeal, [Latin *sine anno* without year] without date; South Africa, South America, South Australia, subject to approval
sac sacrifice
SAC special agent in charge, Strategic Air Command
SAD seasonal affective disorder
SAE self-addressed envelope, Society of Automotive Engineers, stamped addressed envelope
SAG Screen Actors Guild
sal salary
SALT Strategic Arms Limitation Talks
Sam *or* **Saml** Samuel
san sanatorium
S&H shipping and handling
S and M sadism and masochism, sadist and masochist
sanit sanitary, sanitation
SAR search and rescue
SASE self-addressed stamped envelope
Sask Saskatchewan
sat saturate, saturated, saturation
Sat Saturday
SAT Scholastic Assessment Test
satd saturated
S Aust South Australia
sb substantive
Sb [Latin *stibium*] antimony
SB [New Latin *scientiae baccalaureus*] bachelor of science; simultaneous broadcast, southbound
SBA Small Business Administration
SBN Standard Book Number
sc scale, scene, science, scilicet, screw, [Latin *sculpsit*] he carved it, she carved it, he engraved it, she engraved it
Sc scandium, Scots, stratocumulus
SC small capitals, South Carolina, supercalendered, supreme court
Scand Scandinavia, Scandinavian
SCAT School and College Ability Test, supersonic commercial air transport
ScD doctor of science
sch school
sci science, scientific
SCID severe combined immune deficiency, severe combined immunodeficiency
scil scilicet
SCLC Southern Christian Leadership Conference
Scot Scotland, Scottish
SCPO senior chief petty officer
script scripture
sct scout
sctd scattered
sd said, sewed
SD sea-damaged, sine die, South Dakota, special delivery, stage direction, standard deviation
S Dak South Dakota
SDI Strategic Defense Initiative
SDRs special drawing rights
SDS Students for a Democratic Society
Se selenium
SE self-explanatory, southeast, Standard English, stock exchange, straight edge
SEAL sea, air, land (team)
SEATO Southeast Asia Treaty Organization
sec secant, second, secondary, secretary, section, [Latin *secundum*] according to; security
SEC Securities and Exchange Commission
sect section, sectional
secy secretary
sed sediment, sedimentation
sel select, selected, selection
sem semicolon, seminar, seminary
Sem Semitic
sen senate, senator, senior
sep separate, separated
Sep September
SEP simplified employee pension
sepd separated
sepg separating
sepn separation
Sept September
seq [Latin *sequens, sequentes, sequentia*] the following
seqq [Latin *sequentia*] the following ones
ser serial, series, service
Serb Serbian
serg *or* **sergt** sergeant
serv service
SES socioeconomic status
sess session
sf *or* **sfz** sforzando
SF sacrifice fly, science fiction, sinking fund, square feet, square foot
SFC sergeant first class
Sg seaborgium
SG senior grade, sergeant, solicitor general, *often not capitalized* specific gravity; surgeon general
sgd signed
Sgt sergeant
Sgt Maj sergeant major
sh share
Shak Shakespeare
shd should
Shet Shetland
SHF superhigh frequency
shipt shipment
shp shaft horsepower
shpt shipment
sht sheet
shtg shortage
Si silicon
SI [French *Système International d'Unités*] International System of Units
SIDS sudden infant death syndrome
sig signal, signature, signor
Sig [Latin *signa*] label
SIG special interest group
sigill [Latin *sigillum*] seal
sin sine
sing singular
SIS Secret Intelligence Service (British)
SJ Society of Jesus
SK Saskatchewan
SKU stock-keeping unit
sl slightly, slip, slow
SL salvage loss, sea level, south latitude

SLAN [Latin *sine loco, anno, (vel) nomine*] without place, year, or name
SLBM submarine-launched ballistic missile
sld sailed, sealed, sold
SLE systemic lupus erythematosus
Slo Sligo
SLR single-lens reflex
sm small
Sm samarium
SM [New Latin *scientiae magister*] master of science; sergeant major, service mark, soldier's medal, stage manager, station master
S–M *or* **S/M** sadomasochism, sadomasochist
SMA sergeant major of the army
SMaj sergeant major
SMSA Standard Metropolitan Statistical Area
SMSgt senior master sergeant
SMV slow-moving vehicle
Sn [Late Latin *stannum*] tin
SN seaman
SNCC Student Nonviolent Coordinating Committee
SNF skilled nursing facility
SNG substitute natural gas, synthetic natural gas
Snr *British* senior
so south, southern
SO seller's option, strikeout
soc social, society, sociology
sociol sociologist, sociology
sol solicitor, soluble, solution
soln solution
Som Somersetshire
SOP standard operating procedure, standing operating procedure
soph sophomore
SO_x sulfur oxide
sp special, species, specific, specimen, spelling, spirit
Sp Spain, Spanish
SP self-propelled, shore patrol, shore patrolman, shore police, [Latin *sine prole*] without issue; single pole, specialist
Span Spanish
Spc specialist
SPCA Society for the Prevention of Cruelty to Animals
SPCC Society for the Prevention of Cruelty to Children
spd speed
spec special, specialist, specifically, specification
specif specific, specifically
SPF sun protection factor
sp gr specific gravity
sp ht specific heat
SPOT satellite positioning and tracking
spp species (*plural*)
SPQR [Latin *senatus populusque Romanus*] the senate and the people of Rome; small profits, quick returns
sps [Latin *sine prole superstite*] without surviving issue
sq squadron, square
Sr senior, senor, señor, sister, strontium
SR seaman recruit, sedimentation rate, shipping receipt
Sra senora, señora
SRO standing room only
Srta senorita, señorita
ss scilicet — used in legal documents, [Latin *semis*] one half
SS saints, same size, Social Security, steamship, sworn statement
SSA Social Security Administration
SSE south-southeast
SSG *or* **SSgt** staff sergeant
SSI supplemental security income
SSM staff sergeant major
SSN Social Security number
ssp subspecies
SSPE subacute sclerosing panencephalitis
SSR Soviet Socialist Republic
SSS Selective Service System
SSW south-southwest
st stanza, state, stitch, stone, street
St saint, stratus
ST short ton, single throw, standard time
sta station, stationary
Staffs Staffordshire
START strategic arms limitation talks
stat [Latin *statim*] immediately; statute
STB [New Latin *sacrae theologiae baccalaureus*] bachelor of sacred theology; [New Latin *scientiae theologicae baccalaureus*] bachelor of theology
stbd starboard
std standard
STD [New Latin *sacrae theologiae doctor*] doctor of sacred theology
Ste [French *sainte*] saint (female)
ster *or* **stg** sterling
stge storage
stk stock
STL [New Latin *sacrae theologiae licentiatus*] licentiate of sacred theology
STM [New Latin *sacrae theologiae magister*] master of sacred theology, scanning tunneling microscope; scanning tunneling microscopy
STOL short takeoff and landing
stor storage
STP standard temperature and pressure
str steamer, strophe
stud student
STV subscription television
Su Sunday
sub subaltern, subscription, subtract, suburb
subj subject, subjunctive
suff sufficient, suffix
Suff Suffolk
Sun Sunday
sup superior, supplement, supplementary, supply, supra
supp *or* **suppl** supplement, supplementary
supr supreme
supt superintendent
supvr supervisor
sur surface
surg surgeon, surgery, surgical
surv survey, surveying, surveyor
Suss Sussex
SUV sport-utility vehicle
sv sailing vessel, saves, [Latin *sub verbo* or *sub voce*] under the word
svc *or* **svce** service
svgs savings
sw switch
Sw Sweden
SW seawater, shipper's weight, shortwave, southwest
SWAK sealed with a kiss
SWAT Special Weapons and Tactics
Swed Sweden
SWG standard wire gauge
Switz Switzerland
syl *or* **syll** syllable
sym symbol, symmetrical
syn synonym, synonymous, synonymy
syst system
t metric ton, tablespoon, target, teaspoon, technical, temperature, [Latin *tempore*] in the time of; tense, tension, tertiary, time, ton, township, transitive, troy, true
T tera-, tesla, thymine, toddler, tritium, T-shirt
Ta tantalum
TA teaching assistant, transactional analysis
TAC Tactical Air Command
TAG the adjutant general
tan tangent
TAT thematic apperception test
taxon taxonomic, taxonomy
Tay Tayside
tb tablespoon, tablespoonful
Tb terbium
TB thoroughbred, trial balance, tubercle bacillus
TBA *often not capitalized* to be announced
TBD to be determined
tbs *or* **tbsp** tablespoon, tablespoonful
TBS talk between ships
tc tierce
Tc technetium
TC terra-cotta, till countermanded
TCE trichloroethylene
tchr teacher
TD tank destroyer, touchdown, Treasury Department
TDD telecommunications device for the deaf
TDN total digestible nutrients
TDY temporary duty
Te tellurium
tec technical, technician
tech technical, technically, technician, technological, technology
technol technological, technology
TEFL teaching English as a foreign language
tel telegram, telegraph, telephone
teleg telegraphy
temp temperance, template, temporal, temporary, [Latin *tempore*] in the time of
Tenn Tennessee
ter terrace, territory
terr territory
TESL teaching English as a second language
TESOL Teachers of English to Speakers of Other Languages
Test Testament
Tex Texas
TF task force, territorial force
tfr transfer
TFR total fertility rate
TG transformational grammar, type genus
TGIF thank God it's Friday
tgt target
TGV [French *train à grande vitesse*] high-speed train
Th thorium, Thursday
TH true heading
Thai Thailand
ThD [New Latin *theologiae doctor*] doctor of theology
theat theater, theatrical
theol theological, theology
therm thermometer
Thess Thessalonians
ThM [New Latin *theologiae magister*] master of theology
Thurs *or* **Thur** *or* **Thu** Thursday
Ti titanium
TIA transient ischemic attack
tid [Latin *ter in die*] three times a day
Tim Timothy
TIN taxpayer identification number
tinc tincture
Tip Tipperary
tit title
Tit Titus
tk tank, truck
tkt ticket
Tl thallium
TL total loss, truckload
TLC tender loving care, thin-layer chromatography
TLO total loss only
tlr tailor, trailer
Tm thulium
TM trademark, transcendental meditation
TMJ temporomandibular joint
TMO telegraph money order
tn ton, town, train
TN Tennessee, true north
TNF tumor necrosis factor
tng training
tnpk turnpike
TO table of organization, telegraph office, traditional orthography, turn over
Tob Tobit
TOEFL Test of English as a Foreign Language
tol tolerance
tonn tonnage
topo topographic, topographical
topog topography
tot total
TOT time on target
tp title page, township
TP triple play
TPA tissue plasminogen activator
tpk *or* **tpke** turnpike
tps townships, troops
tr translated, translation, translator, transpose, troop, trustee
trag tragedy, tragic
trans transaction, transitive, translated, translation, translator, transmission, transportation, transverse
transf transfer, transferred
transl translated, translation
transp transportation
trav travel, traveler, travels
treas treasurer, treasury
trib tributary
trit triturate
trop tropic, tropical
ts tensile strength
TSgt technical sergeant
TSH thyroid-stimulating hormone
tsp teaspoon, teaspoonful
TSS toxic shock syndrome
TT telegraphic transfer, teletypewriter, Trust Territories
TTY teletypewriter
Tu Tuesday
TU trade union, transmission unit
TUC Trades Union Congress
Tues *or* **Tue** Tuesday
Turk Turkey, Turkish
TV terminal velocity, transvestite
TVA Tennessee Valley Authority
2WD two-wheel drive
twp township
TWX teletypewriter exchange
TX Texas
Tyr Tyrone
u uncle, unit, unsymmetrical, upper
U [*Union of Orthodox Hebrew Congregations*] kosher certification — often enclosed in a circle; university, unsatisfactory, uracil, uranium
UAE United Arab Emirates
UAR United Arab Republic
UAW United Automobile Workers
UC undercharge, uppercase
ugt urgent
UHF ultrahigh frequency
UI unemployment insurance
UK United Kingdom
ult ultimate, ultimo
UMWA United Mine Workers of America
UN United Nations
unan unanimous
Unc uncirculated
UNCF United Negro College Fund
UNESCO United Nations Educational, Scientific, and Cultural Organization
Unh unnilhexium
uni uniform
UNICEF [*United Nations International Children's Emergency Fund,* its former name] United Nations Children's Fund
univ universal, university
unp unpaged
Unp unnilpentium
Unq unnilquadium
UNRWA United Nations Relief and Works Agency
uns unsymmetrical
UP underproof, Upper Peninsula (of Michigan)
UPC Universal Product Code
UPI United Press International
UPS uninterruptible power supply
urol urological, urology
US [Latin *ubi supra*] where above mentioned; United States, [Latin *ut supra*] as above
USA United States Army, United States of America

USAF United States Air Force
USCG United States Coast Guard
USDA United States Department of Agriculture
USES United States Employment Service
USG United States government
USGA United States Golf Association
USGS United States Geological Survey
USIA United States Information Agency
USMC United States Marine Corps
USN United States Navy
USNR United States Naval Reserve
USNS United States Naval Ship
USO United Service Organizations
USP United States Pharmacopeia
USPS United States Postal Service
USS United States ship
USSR Union of Soviet Socialist Republics
USTA United States Tennis Association
usu usual, usually
UT Universal time, Utah
UTC Coordinated Universal Time
ut dict [Latin *ut dictum*] as directed
util utility
UV ultraviolet
UW underwriter
ux [Latin *uxor*] wife
UXB unexploded bomb
UXO unexploded ordnance
v vector, velocity, verb, verse, verso, versus, very, vice, victory, vide, voice, voltage, volume, vowel
V vanadium, volt
Va Virginia
VA Veterans Administration, vicar apostolic, vice admiral, Virginia, visual aid, volt-ampere
vac vacuum
VADM vice admiral
vag vagrancy
val value, valued
var variable, variant, variation, variety, various
VAR visual-aural range, volt-ampere reactive
VAT value-added tax
vb verb, verbal
VC veterinary corps, vice-chancellor, vice-consul, Victoria Cross, Vietcong
VD vapor density, various dates, venereal disease
VDRL venereal disease research laboratory
VDT video display terminal
VDU visual display unit
veg vegetable
vel vellum, velocity
Ven venerable
ver verse
vert vertebrate, vertical
VF very fair, very fine, video frequency, visual field, voice frequency
VFD volunteer fire department
VFR visual flight rules
VFW Veterans of Foreign Wars
VG very good, vicar-general
VHF very high frequency
vi verb intransitive, [Latin *vide infra*] see below
VI Virgin Islands, viscosity index, volume indicator
vic vicinity
Vic Victoria
vil village
VIN vehicle identification number
vis visibility, visible, visual
VISTA Volunteers in Service to America
viz videlicet
VJ veejay
VLF very low frequency
VMD [New Latin *veterinariae medicinae doctor*] doctor of veterinary medicine
VNA Visiting Nurse Association
VOA Voice of America
voc vocational, vocative
vocab vocabulary
vol volcano, volume, volunteer
VOM volt-ohmmeter
VOR very-high-frequency omnidirectional radio range
vou voucher
VP variable pitch, various places, verb phrase, vice president
VRM variable rate mortgage
vs verse, versus
VS veterinary surgeon, [Latin *vide supra*] see above
vss verses, versions
V/STOL vertical or short takeoff and landing
vt verb transitive
Vt Vermont
VT vacuum tube, variable time, Vermont, voice tube
VTOL vertical takeoff and landing
VTR videotape recorder
VU volume unit
Vulg Vulgate
vv verses, vice versa
w warden, water, week, weight, Welsh, west, western, white, wicket, wide, width, wife, with, withdrawal, work
W energy, [German *Wolfram*] tungsten; watt
WA Washington, Western Australia
war warrant
Warks Warwickshire
Wash Washington
Wat Waterford
WATS Wide-Area Telecommunications Service
W Aust Western Australia
Wb weber
WB water ballast, waybill, weather bureau, westbound, wheelbase
WBC white blood cells
WBF wood-burning fireplace
WC water closet, without charge
WCTU Women's Christian Temperance Union
wd wood, word, would
WD War Department
We *or* **Wed** Wednesday
Westm Westmeath, Westmorland
Wex Wexford
wf wrong font
WFTU World Federation of Trade Unions
wh which, white
WH watt-hour, withholding
whf wharf
WHO World Health Organization
whr watt-hour
whs *or* **whse** warehouse
whsle wholesale
wi when issued
WI West Indies, Wisconsin, wrought iron
WIA wounded in action
Wick Wicklow
wid widow, widower
Wilts Wiltshire
Wis *or* **Wisc** Wisconsin
Wisd Wisdom
wk week, work
wkly weekly
WL waterline, wavelength
wm wattmeter
WMD weapons of mass destruction
wmk watermark
WNW west-northwest
WO warrant officer
w/o without
WOC without compensation
WP weather permitting, wettable powder, white phosphorus, without prejudice, word processing, word processor
WPA Works Progress Administration
wpc watts per candle, watts per channel
WPI Wholesale Price Index
WPM words per minute
wpn weapon
WR warehouse receipt, world record
WRAC Women's Royal Army Corps
WRAF Women's Royal Air Force
WRNS Women's Royal Naval Service
wrnt warrant
WRVS Women's Royal Voluntary Service
WSW west-southwest
wt weight
WT watertight, wireless telegraphy
wtd wanted
WV *or* **W Va** West Virginia
WVS Women's Voluntary Services
WW warehouse warrant, with warrants, world war
w/w wall-to-wall
WWW World Wide Web
WY *or* **Wyo** Wyoming
x cross, ex, experimental, extra
X–C cross-country
XD *or* **x div** ex dividend
Xe xenon
XL extra large, extra long
Xn Christian
Xnty Christianity
XO executive officer
XS extra small
XW ex warrants
y yard, year, yeoman
Y yttrium
YA young adult
Yb ytterbium
YB yearbook
YBP years before present
yd yard
yeo *or* **yeom** yeomanry
YO year old
YOB year of birth
Yorks Yorkshire
yr year, younger, your
yrbk yearbook
YT Yukon Territory
Yug Yugoslavia
z zero, zone
Z *or* **ZD** zenith distance
Zach Zacharias
Zech Zechariah
Zeph Zephaniah
ZI zone of interior
ZIFT zygote intrafallopian transfer
Zn [azimuth + north] azimuth; zinc
zool zoological, zoology
ZPG zero population growth
Zr zirconium

Biographical, Biblical, and Mythological Names

This section constitutes a pronouncing dictionary of the names of important figures from contemporary life, history, biblical tradition, legend, and myth. In cases where figures have alternate names, they are entered under the name by which they are best known. Names containing connectives like *d', de, di, van,* or *von* are alphabetized generally under the part of the name following the connective. Parts of names that are not commonly used are often given in parentheses. When two sets of dates are given, the first set indicates the dates of the person's birth and death, and the second pertains only to the particular office, honor, or achievement which it immediately follows. Italicized names within an entry refer to a person's nickname, original name, title, or other name.

Aar·on \'ar-ən, 'er-\ brother of Moses and high priest of the Hebrews in the Bible

Aaron Hank 1934– *Henry Louis Aaron* American baseball player

Abel \'ā-bəl\ son of Adam and Eve and brother of Cain in the Bible

Ab·er·nathy \'ab-ər-,nath-ē\ Ralph David 1926–1990 American clergyman and civil rights leader

Abra·ham \'ā-brə-,ham\ patriarch and founder of the Hebrew people in the Bible; also revered by Muslims

Achil·les \ə-'kil-ēz\ hero of the Trojan War in Greek mythology

Ad·am \'ad-əm\ the first man in biblical tradition

Ad·ams \'ad-əmz\ Abigail 1744–1818 née *Smith* American writer; wife of John Adams

Adams Ansel Easton 1902–1984 American photographer

Adams John 1735–1826 2nd president of the U.S. (1797–1801)

Adams John Quin·cy \'kwin-zē, -sē\ 1767–1848 6th president of the U.S. (1825–29); son of John and Abigail Adams

Adams Samuel 1722–1803 American Revolutionary patriot

Ad·dams \'ad-əmz\ Jane 1860–1935 American social worker; Nobel Prize winner (1931)

Ad·di·son \'ad-ə-sən\ Joseph 1672–1719 English essayist

Ado·nis \ə-'dän-əs, -'dō-nəs\ beautiful youth in Greek mythology who is loved by Aphrodite

Ae·ne·as \i-'nē-əs\ Trojan hero in classical mythology

Ae·o·lus \'ē-ə-ləs\ god of the winds in Greek mythology

Aes·chy·lus \'es-kə-ləs, 'ēs-\ 525–456 B.C. Greek playwright

Aes·cu·la·pi·us \'es-kyə-'lā-pē-əs\ god of medicine in Roman mythology — compare ASCLEPIUS

Ae·sop \'ē-,säp, -səp\ Greek writer of fables; probably legendary

Ag·a·mem·non \,ag-ə-'mem-,nän, -nən\ leader of the Greeks during the Trojan War in Greek mythology

Aggeus — see HAGGAI

Ag·nes \'ag-nəs\ Saint *died* 304 A.D. Christian martyr

Ag·rip·pi·na \,ag-rə-'pī-nə, -'pē-\ *about* 14 B.C.–33 A.D. mother of Caligula

Ahab \'ā-,hab\ king of Israel in the 9th century B.C. and husband of Jezebel

Ajax \'ā-,jaks\ hero in Greek mythology who kills himself during the Trojan War because the armor of Achilles is awarded to Odysseus

Alad·din \ə-'lad-n\ youth in the *Arabian Nights' Entertainments* who comes into possession of a magic lamp

Al·a·ric \'al-ə-rik\ *about* 370–410 A.D. king of the Visigoths; conqueror of Rome

Al·ber·tus Mag·nus \al-'bərt-ə-'smag-nəs\ Saint *about* 1200–1280 German philosopher and theologian

Al·bright \'ól-,brīt\ Madeleine 1937– née *Korbel* American (Czech-born) diplomat; U.S. secretary of state (1997–2001)

Al·ci·bi·a·des \,al-sə-'bī-ə-,dēz\ *about* 450–404 B.C. Athenian general and politician

Al·cott \'ól-kət\ Louisa May 1832–1888 American author

Al·ex·an·der \,al-ig-'zan-dər, ,el-\ name of 8 popes: especially **VI** (*Rodrigo Borgia*) 1431–1503 (pope 1492–1503)

Alexander name of 3 emperors of Russia: **I** 1777–1825 (reigned 1801–25); **II** 1818–1881 (reigned 1855–81); **III** 1845–1894 (reigned 1881–94)

Alexander the Great 356–323 B.C. *Alexander III* king of Macedonia (336–323)

Al·fred \'al-frəd, -fərd\ 849–899 *Alfred the Great* king of Wessex (871–899)

Ali \ä-'lē\ Muhammad 1942– originally *Cassius Marcellus Clay* American boxer

Al·len \'al-ən\ Ethan 1738–1789 American Revolutionary soldier

Am·brose \'am-,brōz\ Saint 339–397 A.D. bishop of Milan

Amerigo Vespucci — see VESPUCCI

Am·herst \'am-ərst, -,ərst\ Jeffery 1717–1797 Baron *Amherst* British general in America

Amos \'ā-məs\ Hebrew prophet of the 8th century B.C.

Amund·sen \'äm-ən-sən\ Roald 1872–1928 Norwegian explorer; discoverer of the South Pole (1911)

An·a·ni·as \,an-ə-'nī-əs\ early Christian who in the Bible is struck dead for lying

An·chi·ses \an-'kī-sēz, aŋ-\ father of Aeneas in classical mythology

An·der·sen \'an-dər-sən\ Hans Christian 1805–1875 Danish writer of fairy tales

An·der·son \'an-dər-sən\ Marian 1897–1993 American contralto

An·drea del Sar·to \än-,drā-ə-,del-'särt-ō\ 1486–1530 Florentine painter

An·dro·cles \'an-drə-,klēz\ legendary Roman slave spared in the arena by a lion from whose foot he had once taken a thorn

An·drom·a·che \an-'dräm-ə-kē\ wife of Hector in Greek mythology

An·drom·e·da \an-'dräm-ə-də\ Ethiopian princess rescued from a monster by Perseus in Greek mythology

An·ge·li·co \än-'jä-lē-kō\ Fra *about* 1400–1455 originally *Guido di Pietro* Florentine painter and Dominican friar

An·ge·lou \'an-jə-,lō\ Maya 1928– originally *Marguerite Johnson* American author

An·nan \ä-'nän\ Kofi 1938– Ghanaian United Nations official; secretary-general (1997–)

Anne \'an\ 1665–1714 queen of Great Britain (1702–14); daughter of James II

An·tho·ny \'an-thə-nē, *chiefly British* 'an-tə-nē\ Saint *about* 250–355 A.D. Egyptian monk

Anthony Mark — see ANTONY

Anthony Susan Brownell 1820–1906 American suffragist

Anthony of Padua Saint 1195–1231 Franciscan friar

An·tig·o·ne \an-'tig-ə-nē\ daughter of Oedipus and Jocasta in Greek mythology

An·to·ni·nus Marcus Au·re·lius — see MARCUS AURELIUS

An·to·ny \'an-tə-nē\ Mark *about* 82–30 B.C. *Marc Anthony; Marcus An·to·ni·us* \an-'tō-nē-əs\ Roman general and triumvir (43–30)

Aph·ro·di·te \,af-rə-'dīt-ē\ goddess of love and beauty in Greek mythology — compare VENUS

Apol·lo \ə-'päl-ō\ *or* **Phoe·bus** \'fē-bəs\ god of sunlight, prophecy, music, and poetry in classical mythology

Apol·lyon \ə-'päl-yən, -'päl-ē-ən\ angel of the bottomless pit in the Book of Revelation

Ap·ple·seed \'ap-əl-,sēd\ Johnny 1774–1845 *John Chapman* American pioneer

Aqui·nas \ə-'kwī-nəs\ Saint Thomas 1224 (or 1225)–1274 Italian theologian

Arach·ne \ə-'rak-nē\ girl in Greek mythology who is changed into a spider for challenging Athena to a contest in weaving

Ar·chi·me·des \,är-kə-'mēd-ēz\ *about* 287–212 B.C. Greek mathematician and inventor

Ares \'aər-ēz, 'eər-; 'ā-,rēz\ god of war in Greek mythology — compare MARS

Ar·gus \'är-gəs\ hundred-eyed monster in Greek mythology

Ar·i·ad·ne \,ar-ē-'ad-nē\ daughter of Minos who helps Theseus escape from a labyrinth in Greek mythology

Ar·is·ti·des \,ar-ə-'stīd-ēz\ *about* 530–*about* 468 B.C. *Aristides the Just* Athenian statesman and general

Ar·is·toph·a·nes \,ar-ə-'stäf-ə-,nēz\ *about* 450–*about* 388 B.C. Greek playwright

Ar·is·tot·le \'ar-ə-,stät-l\ 384–322 B.C. Greek philosopher

\ə\ **abut** \ᵊ\ **kitten** \ər\ **further** \a\ **ash** \ā\ **ace** \ä\ **mop, mar** \au̇\ **out** \ch\ **chin** \e\ **bet** \ē\ **easy** \g\ **go** \i\ **hit** \ī\ **ice** \j\ **job** \ŋ\ **sing** \ō\ **go** \ó\ **law** \ói\ **boy** \th\ **thin** \t͟h\ **the** \ü\ **loot** \u̇\ **foot** \y\ **yet** \zh\ **vision** *see also* Guide to Pronunciation

Ari•us \ə-ˈrī-əs; ˈar-ē-əs, ˈer-\ *about* 250–336 A.D. Greek theologian

Arm•strong \ˈärm-ˌstrȯŋ\ Lance 1971– American cyclist

Armstrong Louis 1901–1971 *Satch·mo* \ˈsach-ˌmō\ American jazz musician

Armstrong Neil Alden 1930– American astronaut; first man on the moon (1969)

Ar•nold \ˈärn-ld\ Benedict 1741–1801 American Revolutionary general and traitor

Arnold Matthew 1822–1888 English poet and critic

Ar•te•mis \ˈär-tə-məs\ goddess of the moon, wild animals, and hunting in Greek mythology — compare DIANA

Ar•thur \ˈär-thər\ legendary king of the Britons whose story is based on traditions of a 6th-century military leader — **Ar•thu•ri•an** \är-ˈthu̇r-ē-ən, -ˈthyu̇r-\ *adj*

Arthur Chester Alan 1829–1886 21st president of the U.S. (1881–85)

As•cle•pi•us \ə-ˈsklē-pē-əs\ god of medicine in Greek mythology — compare AESCULAPIUS

Ashe \ˈash\ Arthur Robert 1943–1993 American tennis player

As•tar•te \ə-ˈstärt-ē\ Phoenician goddess of love and fertility

As•tor \ˈas-tər\ John Jacob 1763–1848 American (German-born) fur trader and capitalist

At•a•lan•ta \ˌat-l-ˈant-ə\ beautiful fleet-footed heroine in Greek mythology who challenges her suitors to a race and is defeated when she stops to pick up three golden apples

Ath•el•stan \ˈath-əl-ˌstan\ *died* 939 Anglo-Saxon ruler

Athe•na \ə-ˈthē-nə\ *or* **Athe•ne** \-nē\ goddess of wisdom in Greek mythology — compare MINERVA

At•las \ˈat-ləs\ Titan in Greek mythology forced to bear the heavens on his shoulders

Atreus \ˈā-ˌtrüs, -trē-əs\ king of Mycenae and father of Agamemnon and Menelaus in Greek mythology

At•ti•la \ˈat-l-ə, ə-ˈtil-ə\ 406?–453 A.D. king of the Huns

At•tucks \ˈat-əks\ Crispus 1723?–1770 American patriot

Au•du•bon \ˈȯd-ə-bən, -ˌbän\ John James 1785–1851 American (Haitian-born) artist and ornithologist

Au•gus•tine \ˈȯ-gə-ˌstēn; ȯ-ˈgəs-tən, ə-\ Saint 354–430 A.D. church father; bishop of Hippo (396–430)

Augustine of Canterbury Saint *died about* 604 A.D. *Apostle of the English* 1st archbishop of Canterbury (601–604)

Au•gus•tus \ȯ-ˈgəs-təs, ə-\ *or* **Caesar Augustus** *or* **Oc•ta•vi•an** \äk-ˈtā-vē-ən\ 63 B.C.–14 A.D. originally *Gaius Octavius* 1st Roman emperor 27 B.C.–14 A.D.

Au•ro•ra \ə-ˈrȯr-ə, ȯ-, -ˈrȯr-\ goddess of the dawn in Roman mythology — compare EOS

Aus•ten \ˈȯs-tən, ˈäs-\ Jane 1775–1817 English author

Bab•bage \ˈbab-ij\ Charles 1791–1871 British mathematician and inventor

Bac•chus \ˈbak-əs\ — see DIONYSUS

Bach \ˈbäk, ˈbäk\ Johann Sebastian 1685–1750 German composer and organist

Ba•con \ˈbā-kən\ Francis 1561–1626 English philosopher and author

Bacon Roger *about* 1220–1292 English philosopher and scientist

Ba•den–Pow•ell \ˌbād-n-ˈpō-əl\ Robert Stephenson Smyth 1857–1941 Baron *Baden-Powell* British general and founder of Boy Scout movement

Bal•boa \bal-ˈbō-ə\ Vasco Núñez de 1475–1519 Spanish explorer and discoverer of the Pacific Ocean (1513)

Bald•win \ˈbȯld-wən\ James 1924–1987 American author

Baltimore Lord — see CALVERT

Bal•zac \ˈbȯl-ˌzak, ˈbal-, *French* bȧl-zȧk\ Honoré de 1799–1850 French author

Ba•rab•bas \bə-ˈrab-əs\ prisoner in the Bible released in preference to Jesus at the demand of the multitude

Barbarossa — see FREDERICK I

Bar•num \ˈbär-nəm\ P. T. 1810–1891 *Phineas Taylor Barnum* American showman

Bar•rie \ˈbar-ē\ Sir James Matthew 1860–1937 Scottish author

Bar•ry•more \ˈbar-i-ˌmōr, -ˌmȯr\ family of American actors: Maurice (originally *Herbert Blythe*) 1847–1905; his wife Georgiana Emma (née *Drew*) 1854–1893; their children Lionel 1878–1954, Ethel 1879–1959, and John Blythe 1882–1942

Bar•thol•di \bär-ˈtäl-dē, -ˈtȯl-, -ˈthäl-, -ˈthȯl-\ Frédéric-Auguste 1834–1904 French sculptor of the Statue of Liberty

Bar•tók \ˈbar-ˌtäk, -ˌtȯk\ Bé·la \ˈbā-lə\ 1881–1945 Hungarian composer

Bar•ton \ˈbärt-n\ Clara 1821–1912 founder of American Red Cross Society

Ba•rysh•ni•kov \bə-ˈrish-nə-ˌkȯf\ Mikhail 1948– American (Russian-born) dancer

Ba•sie \ˈbā-sē\ William 1904–1984 *Count Basie* American bandleader and pianist

Ba•sil \ˈbaz-əl, ˈbās-, ˈbas-, ˈbāz-\ Saint *about* 329–379 A.D. *Basil the Great* church father; bishop of Caesarea

Bau•de•laire \bōd-ˈlaər, -ˈleər\ Charles (-Pierre) 1821–1867 French poet

Beau•re•gard \ˈbōr-ə-ˌgärd, ˈbȯr-\ Pierre Gustave Toutant 1818–1893 American Confederate general

Beck•et \ˈbek-ət\ Saint Thomas *about* 1118–1170 *Thomas à Becket* archbishop of Canterbury (1162–70)

Bede \ˈbēd\ Saint *about* 672–735 A.D. *the Venerable Bede* Anglo-Saxon historian and theologian

Beel•ze•bub \bē-ˈel-zi-ˌbəb, ˈbēl-zi-, ˈbel-\ prince of the demons identified with Satan in the New Testament

Bee•tho•ven \ˈbā-ˌtō-vən\ Ludwig van 1770–1827 German composer

Bell \ˈbel\ Alexander Graham 1847–1922 American (Scottish-born) inventor of the telephone

Bel•ler•o•phon \bə-ˈler-ə-fən, -ˌfän\ hero in Greek mythology who slays the monster Chimera with the help of his horse Pegasus

Bel•li•ni \bə-ˈlē-nē\ Vincenzo 1801–1835 Italian composer

Bel•low \ˈbel-ō\ Saul 1915– American (Canadian-born) author; Nobel Prize winner (1976)

Ben•e•dict \ˈben-ə-dikt\ name of 16 popes: especially **XIV** (*Prospero Lambertini*) 1675–1758 (pope 1740–58); **XV** (*Giacomo della Chiesa*) 1854–1922 (pope 1914–22); **XVI** (*Joseph Ratzinger*) 1927– (pope 2005–)

Benedict of Nur•sia \ˈnər-shə, -shē-ə\ Saint *about* 480–*about* 547 A.D. Italian founder of Benedictine order

Be•nét \bə-ˈnā\ Stephen Vincent 1898–1943 American author

Ben•ja•min \ˈbenj-mən, -ə-mən\ Jacob's youngest son and ancestor of one of the 12 tribes of Israel in the Bible

Ben•tham \ˈben-thəm\ Jeremy 1748–1832 English jurist and philosopher

Ben•ton \ˈbent-n\ Thomas Hart 1889–1975 American painter

Be•o•wulf \ˈbā-ə-ˌwu̇lf\ legendary Scandinavian warrior and hero of the Old English poem *Beowulf*

Be•ring \ˈbiər-iŋ, ˈbeər-\ Vitus 1681–1741 Danish navigator and explorer for Russia

Ber•lin \bər-ˈlin, ˌbər-\ Irving 1888–1989 American (Russian-born) composer and songwriter

Ber•li•oz \ˈber-lē-ˌōz\ (Louis-) Hector 1803–1869 French composer

Bern•hardt \ˈbərn-ˌhärt, ber-ˈnär\ Sarah 1844–1923 originally *Henriette-Rosine Ber·nard* \ber-ˈnär\ French actress

Ber•ni•ni \bər-ˈnē-nē\ Gian Lorenzo 1598–1680 Italian sculptor, architect, and painter

Bern•stein \ˈbərn-ˌstīn *also* -ˌstēn\ Leonard 1918–1990 American conductor and composer

Bes•se•mer \ˈbes-ə-mər\ Sir Henry 1813–1898 English engineer and inventor

Be•thune \bə-ˈthün, -ˈthyün\ Mary 1875–1955 née *McLeod* American educator

Beyle Marie-Henri — see STENDHAL

Bierce \ˈbiərs\ Ambrose Gwinnett 1842–?1914 American author

Bi•ko \ˈbē-ˌkō\ (Bantu) Stephen 1946–1977 South African black nationalist

Bil•ly the Kid \ˈbil-ē-\ 1859?–1881 originally *William H. Bon·ney* \ˈbän-ē\, *Jr.* or *Henry McCarty?* American outlaw

Bird \ˈbərd\ Larry 1956– American basketball player

Bis•marck \ˈbiz-ˌmärk\ Prince Otto Eduard Leopold von 1815–1898 1st chancellor of German empire (1871–90)

Bi•zet \bē-ˈzā\ Georges 1838–1875 originally *Alexandre-César-Léopold Bizet* French composer

Black Hawk \ˈblak-ˌhȯk\ 1767–1838 American Indian chief

Black•stone \ˈblak-ˌstōn, *chiefly British* -stən\ Sir William 1723–1780 English jurist

Black•well \ˈblak-ˌwel, -wəl\ Elizabeth 1821–1910 American (English-born) physician

Blair \ˈblaər, ˈbleər\ Tony 1953– *Anthony Charles Lynton Blair* British prime minister (1997–)

Blake \ˈblāk\ William 1757–1827 English poet and artist

Bloom•er \ˈblü-mər\ Amelia 1818–1894 née *Jenks* American social reformer

Boc•cac•cio \bō-ˈkäch-ē-ˌō, -ˈkäch-ō\ Giovanni 1313–1375 Italian author

Bohr \ˈbōr, ˈbȯr\ Niels 1885–1962 Danish physicist; Nobel Prize winner (1922)

Bo•leyn \bu̇-ˈlin, ˈbu̇l-ən\ Anne 1507?–1536 2nd wife of Henry VIII of England; mother of Elizabeth I

Bo•lí•var Si·món \sē-ˌmōn-bə-ˈlē-ˌvär, ˌsī-mən-ˈbäl-ə-vər\ 1783–1830 South American liberator

Bo•na•parte \ˈbō-nə-ˌpärt\ *or Italian* **Buo•na•par•te** \ˌbwȯn-ə-ˈpärt-ē\ Corsican family: Jérôme 1784–1860 king of Westphalia; Joseph 1768–1844 king of Naples and Spain; Louis 1778–1846 king of Holland; Lucien 1775–1840 prince of Canino; all brothers of Napoléon I

Bon•i•face \ˈbän-ə-fəs, -ˌfās\ name of 9 popes: especially **VIII** (*Benedetto Caetani*) *about* 1235 (or 1240)–1303 (pope 1294–1303)

Boniface Saint *about* 675–754 A.D. originally *Wynfrid* or *Wynfrith* English missionary in Germany

Boone \ˈbün\ Daniel 1734–1820 American pioneer

Booth \ˈbüth\ John Wilkes 1838–1865 American actor; assassin of Abraham Lincoln

Booth \ˈbüth, *chiefly British* ˈbüth\ William 1829–1912 English founder of the Salvation Army

Bo•re•as \ˈbōr-ē-əs, ˈbȯr-\ god of the north wind in Greek mythology

Bor•gia \ˈbȯr-ˌjä, -jə, -zhə\ Cesare 1475 (or 1476)–1507 Italian cardinal and military leader; son of Rodrigo Borgia

Borgia Lucrezia 1480–1519 duchess of Ferrara; daughter of Rodrigo Borgia

Borgia Rodrigo — see Pope ALEXANDER VI

Bo•ro•din \ˌbȯr-ə-ˈdēn, ˌbär-\ Aleksandr Porfiryevich 1833–1887 Russian composer and chemist

Bosch \ˈbäsh, ˈbȯsh, *Dutch* ˈbäs, ˈbȯs\ Hieronymus *about* 1450–1516 Dutch painter

Bos•co \ˈbäs-kō, ˈbȯs-\ Saint Giovanni Melchior 1815–1888 Italian priest and founder of the Salesians

Bos•well \ˈbäz-ˌwel, -wəl\ James 1740–1795 Scottish biographer of Samuel Johnson

Bot•ti•cel•li \ˌbät-ə-ˈchel-ē\ Sandro 1445–1510 Italian painter

Bow•ie \ˈbü-ē, ˈbō-\ Jim 1796?–1836 *James Bowie* American popular hero of the Texas revolution

Boyle \ˈbȯil\ Robert 1627–1691 British physicist and chemist

Brad•bury \ˈbrad-ˌber-ē, -bə-rē, -brē\ Ray Douglas 1920– American author

Brad•dock \ˈbrad-ək\ Edward 1695–1755 British general in America

Brad•ford \ˈbrad-fərd\ William 1590–1657 Pilgrim leader; 2nd governor of Plymouth colony

Brad•street \ˈbrad-ˌstrēt\ Anne *about* 1612–1672 American poet

Bra•dy \ˈbrād-ē\ Mathew B. 1823?–1896 American photographer

Brahe \ˌbrä; ˈbrä-hē, -hə\ Tycho 1546–1601 Danish astronomer

Brah•ma \ˈbräm-ə\ creator god of the Hindu sacred triad — compare SHIVA, VISHNU

Brahms \ˈbrämz\ Johannes 1833–1897 German composer

Braille \ˈbrāl, ˈbrī\ Louis 1809–1852 French blind teacher of the blind

Bran•deis \ˈbran-ˌdīs, -ˌdīz\ Louis Dembitz 1856–1941 American jurist

Brant \ˈbrant\ Joseph 1742–1807 *Thayendanegea* Mohawk Indian chief

Brant Mary 1736?–1796 *Molly Brant* Mohawk Indian leader; sister of Joseph Brant

Braun \ˈbrau̇n\ Wernher von 1912–1977 American (German-born) engineer

Brezh•nev \ˈbrezh-ˌnef\ Leonid Ilyich 1906–1982 Russian politician; 1st secretary of Communist party (1964–82); president of the U.S.S.R. (1960–64; 1977–82)

Bri•an Bo•ru \ˌbrī-ən-bə-ˈrü\ 941–1014 king of Ireland (1002–14)

Brig•id \ˈbrij-əd, ˈbrē-əd\ Saint *died about* 524–528 A.D. a patron saint of Ireland

Brit·ten \'brit-ⁿn\ (Edward) Benjamin 1913–1976 English composer

Bron·të \'bränt-ē, 'brän-ˌtā\ family of English authors: Charlotte 1816–1855 and her sisters Emily 1818–1848 and Anne 1820–1849

Brooks \'brůks\ Gwendolyn Elizabeth 1917–2000 American poet

Brown \'braůn\ John 1800–1859 American abolitionist

Brow·ning \'braů-niŋ\ Elizabeth Barrett 1806–1861 English poet

Browning Robert 1812–1889 English poet; husband of Elizabeth

Broz Josip — see TITO

Bruce the — see ROBERT I

Bruck·ner \'brůk-nər\ Anton 1824–1896 Austrian composer

Brue·ghel *or* **Breu·ghel** \'brü-gəl, 'brȯi-\ Pieter *about* 1525–1569 *the Elder* Flemish painter

Brun·hild \'brün-ˌhilt\ legendary Germanic queen won by Siegfried for Gunther

Bru·tus \'brüt-əs\ Marcus Junius 85–42 B.C. Roman politician and one of Julius Caesar's assassins

Bry·an \'brī-ən\ William Jennings 1860–1925 American lawyer and politician

Bry·ant \'brī-ənt\ William Cullen 1794–1878 American poet

Bu·chan·an \byü-'kan-ən, bə-\ James 1791–1868 15th president of the U.S. (1857–61)

Buck \'bək\ Pearl S. 1892–1973 née *Sydenstricker* American author; Nobel Prize winner (1938)

Bud·dha \'bü-də, 'bů-\ *about* 563–*about* 483 B.C. originally *Sid·dhar·tha Gau·ta·ma* \si-'där-tə-'gaůt-ə-mə\ Indian founder of Buddhism

Buffalo Bill — see William Frederick CODY

Bunche \'bənch\ Ralph Johnson 1904–1971 American diplomat

Bun·yan \'bən-yən\ John 1628–1688 English preacher and author

Bunyan Paul — see PAUL BUNYAN

Bur·bank \'bər-ˌbaŋk\ Luther 1849–1926 American horticulturist

Bur·ger \'bər-gər\ Warren Earl 1907–1995 American jurist; chief justice U.S. Supreme Court (1969–86)

Bur·goyne \'bər-ˌgȯin, ˌbər-'\ John 1722–1792 British general in America

Burke \'bərk\ Edmund 1729–1797 British statesman and author

Burns \'bərnz\ Robert 1759–1796 Scottish poet

Burn·side \'bərn-ˌsīd\ Ambrose Everett 1824–1881 American general

Burr \'bər\ Aaron 1756–1836 American politician; vice president of the U.S. (1801–5)

Bur·roughs \'bər-ˌōz, 'bə-ˌrōz\ Edgar Rice 1875–1950 American author

Bush \'bůsh\ George (Herbert Walker) 1924– 41st president of the U.S. (1989–93)

Bush George W. 1946– *George Walker Bush* 43rd president of the U.S. (2001–); son of the preceding

But·ler \'bət-lər\ Samuel 1835–1902 English author

Byrd \'bərd\ Richard Evelyn 1888–1957 American admiral and polar explorer

By·ron \'bī-rən\ Lord 1788–1824 *George Gordon Byron,* 6th Baron *Byron* English poet

Cab·ot \'kab-ət\ John *about* 1450–*about* 1499 *Giovanni Caboto* Italian navigator and explorer of coast of North America for England

Ca·bri·ni \kə-'brē-nē\ Saint Frances Xavier 1850–1917 *Mother Cabrini* 1st American (Italian-born) saint (1946)

Cad·mus \'kad-məs\ founder of Thebes in Greek mythology

Caed·mon \'kad-mən\ *flourished* 658–680 A.D. English poet

Cae·sar \'sē-zər\ (Gaius) Julius 100?–44 B.C. Roman general, statesman, and writer

Cain \'kān\ son of Adam and Eve and brother of Abel in the Bible

Calamity Jane \-'jān\ 1852?–1903 *Martha Jane Burk* \'bərk\ née *Can·nary* \'kan-ə-rē\ American frontier figure

Cal·der \'kȯl-dər\ Alexander 1898–1976 American sculptor

Cal·houn \kal-'hün\ John Caldwell 1782–1850 American politician; vice president of the U.S. (1825–32)

Ca·lig·u·la \kə-'lig-yə-lə\ 12–41 A.D. *Gaius Caesar* Roman emperor (37–41)

Cal·li·o·pe \kə-'lī-ə-pē\ Muse of heroic poetry in Greek mythology

Cal·vert \'kal-vərt\ George 1580?–1632 Baron *Baltimore* English colonist in America

Cal·vin \'kal-vən\ John 1509–1564 *Jean Calvin* or *Cauvin* French theologian and reformer

Ca·lyp·so \kə-'lip-sō\ sea nymph in Homer's *Odyssey* who keeps Odysseus for seven years on an island

Ca·mus \kȧ-'mᵫ\ Albert 1913–1960 French author; Nobel Prize winner (1957)

Ca·nute \kə-'nüt, -'nyüt\ *died* 1035 *Canute the Great* Danish king of England (1016–35); of Denmark (1018–35); of Norway (1028–35)

Ča·pek \'chäp-ˌek\ Karel 1890–1938 Czech author

Capet Hugh — see HUGH CAPET

Ca·pone \kə-'pōn\ Al 1899–1947 *Alphonse Capone* American gangster

Ca·ra·vag·gio \ˌkar-ə-'vä-jō\ 1571?–1610 originally *Michelangelo Merisi* Italian painter

Car·lyle \kär-'līl, 'kär-ˌ\ Thomas 1795–1881 Scottish essayist and historian

Car·ne·gie \'kär-nə-gē, kär-'neg-ē\ Andrew 1835–1919 American (Scottish-born) industrialist and philanthropist

Car·roll \'kar-əl\ Lewis 1832–1898 pseudonym of *Charles Lut·widge* \'lət-wij\ *Dodg·son* \'däj-sən, 'däd-\ English author and mathematician

Car·son \'kärs-ⁿn\ Kit 1809–1868 *Christopher Carson* American frontiersman and scout

Carson Rachel Louise 1907–1964 American biologist and writer

Car·ter \'kärt-ər\ Jimmy 1924– originally *James Earl Carter, Jr.* 39th president of the U.S. (1977–81); Nobel Prize winner (2002)

Car·tier \kär-'tyā, 'kärt-ē-ˌā\ Jacques 1491–1557 French navigator and explorer of Saint Lawrence River

Ca·ru·so \kə-'rü-sō, -zō\ En·ri·co \en-'rē-kō\ 1873–1921 Italian tenor

Car·ver \'kär-vər\ George Washington 1861?–1943 American agricultural chemist and agronomist

Ca·sals \kə-'sälz, -'zälz\ Pablo 1876–1973 Spanish-born cellist and conductor

Ca·sa·no·va \ˌkaz-ə-'nō-və, ˌkas-\ Giovanni Giacomo 1725–1798 Italian adventurer

Cas·san·dra \kə-'san-drə\ daughter of Priam in Greek mythology who is endowed with the gift of prophecy but fated never to be believed

Cas·satt \kə-'sat\ Mary 1845–1926 American painter

Cas·tor \'kas-tər\ mortal twin of Pollux in classical mythology

Cas·tro (Ruz) \'kas trō(-'rüs), 'käs-\ Fi·del \fē-'del\ 1926– Cuban leader (1959–)

Cath·er \'kath-ər\ Willa 1873–1947 American author

Cath·er·ine \'kath-rən, -ə-rən\ name of 1st, 5th, and 6th wives of Henry VIII of England: Catherine of Aragon 1485–1536; Catherine Howard 1520?–1542; Catherine Parr 1512–1548

Catherine I 1684–1727 wife of Peter the Great; empress of Russia (1725–27)

Catherine II 1729–1796 *Catherine the Great* empress of Russia (1762–96)

Cath·er·ine de Mé·di·cis \'kath-ə-rən-də-'mā-də-ˌsēs, -'med-ə-ˌchē\ 1519–1589 Italian *Ca·te·ri·na de' Me·di·ci* \ˌkä-te-'rē-nä-dā-'med-ē-ˌchē\ queen consort of Henry II of France (1547–59); regent of France (1560–74)

Cat·i·line \'kat-l-ˌīn\ *about* 108–62 B.C. Roman politician and conspirator

Ca·to \'kāt-ō\ Marcus Porcius 234–149 B.C. *Cato the Elder; Cato the Censor* Roman statesman

Cato Marcus Porcius 95–46 B.C. *Cato the Younger* Roman statesman; great-grandson of the preceding

Ca·tul·lus \kə-'təl-əs\ Gaius Valerius *about* 84–*about* 54 B.C. Roman poet

Cav·en·dish \'kav-ən-ˌdish\ Henry 1731–1810 English scientist

Cax·ton \'kak-stən\ William *about* 1422–1491 1st English printer

Ce·ci·lia \sə-'sēl-yə, -'sil-\ Saint *flourished* 3rd century A.D. Christian martyr; patron saint of music

Cel·li·ni \chə-'lē-nē\ Benvenuto 1500–1571 Italian goldsmith and sculptor

Cer·ber·us \'sər-bə-rəs, -brəs\ 3-headed dog in classical mythology who guards the entrance to Hades

Ce·res \'siər-ˌēz\ goddess of agriculture in Roman mythology — compare DEMETER

Cer·van·tes (Saa·ve·dra) \sər-'van-ˌtēz(-ˌsä-ə-'vā-drə)\ Miguel de 1547–1616 Spanish author

Cé·zanne \sā-'zan\ Paul 1839–1906 French painter

Cha·gall \shə-'gäl, -'gal\ Marc 1887–1985 Russian painter in France

Cham·ber·lain \'chām-bər-lən\ (Arthur) Neville 1869–1940 British prime minister (1937–40)

Chamberlain Wilt 1936–1999 *Wilton Norman Chamberlain* American basketball player

Cham·plain \sham-'plān, shäⁿ-'plaⁿ\ Samuel de 1567–1635 French navigator, explorer, and founder of Quebec

Chap·lin \'chap-lən\ Charlie 1889–1977 *Sir Charles Spencer Chaplin* British actor, director, and producer

Char·le·magne \'shär-lə-ˌmān\ *about* 742–814 A.D. *Charles the Great* or *Charles I* Frankish king (768–814); emperor of the West (800–814)

Charles \'chärlz\ name of 10 kings of France: especially **II** 823–877 A.D. *Charles the Bald* (reigned 840–77), Holy Roman emperor (875–77); **IV** 1294–1328 *Charles the Fair* (reigned 1322–28); **V** 1337–1380 *Charles the Wise* (reigned 1364–80); **VI** 1368–1422 *Charles the Mad* or *the Beloved* (reigned 1380–1422); **VII** 1403–1461 *Charles the Well-Served* or *the Victorious* (reigned 1422–61); **IX** 1550–1574 (reigned 1560–74); **X** 1757–1836 (reigned 1824–30)

Charles name of 2 kings of Great Britain: **I** 1600–1649 (reigned 1625–49); **II** 1630–1685 (reigned 1660–85) son of Charles I

Charles 1948– prince of Wales; son of Elizabeth II

Charles I 1887–1922 *Charles Francis Joseph* emperor of Austria and (as *Charles IV*) king of Hungary (1916–18)

Charles V 1500–1558 Holy Roman emperor (1519–56); king of Spain as *Charles I* (1516–56)

Charles XII 1682–1718 king of Sweden (1697–1718)

Charles Edward — see STUART

Charles Mar·tel \mär-'tel\ *about* 688–741 A.D. grandfather of Charlemagne; Frankish ruler (719–41)

Char·on \'kar-ən, 'ker-\ boatman in Greek mythology who ferries the souls of the dead across the river Styx to Hades

Cha·teau·bri·and \sha-ˌtō-brē-'äⁿ\ (François-Auguste-) René 1768–1848 *Vicomte de Chateaubriand* French author

Chau·cer \'chȯ-sər\ Geoffrey *about* 1342–1400 English poet

Che·khov \'chek-ˌȯf, -ȯv\ Anton Pavlovich 1860–1904 Russian author

Che·ney \'chē-nē\ Richard Bruce 1941– vice president of the U.S. (2001–)

Cheops — see KHUFU

Ches·ter·field \'ches-tər-ˌfēld\ 4th Earl of 1694–1773 *Philip Dormer Stanhope* English statesman and author

Ches·ter·ton \'ches-tər-tən\ G. K. 1874–1936 *Gilbert Keith Chesterton* English author

Chiang Kai-shek \jē-'äŋ-'kī-'shek, 'chaŋ-\ 1887–1975 Chinese general and statesman; head of Chinese Nationalist government (1948–49; Taiwan, 1950–75)

Chi·ron \'kī-rən, 'kī-ˌrän\ wise centaur and tutor to many heroes in Greek mythology

Cho·pin \'shō-ˌpan, -ˌpaⁿ\ Frédéric François 1810–1849 Polish pianist and composer

Chou En-lai *or* **Zhou Enlai** \'jō-'en-'lī\ 1898–1976 Chinese Communist politician; premier (1949–76)

Chré·tien \krā-'tyeⁿ\ (Joseph Jacques) Jean 1934– prime minister of Canada (1993–2003)

Christ Jesus — see JESUS

Chris·tie \'kris-tē\ Dame Agatha 1890–1976 née *Miller* English author

Chry·sos·tom \'kris-əs-təm, kris-'äs-təm\ Saint John *about* 347–407 A.D. church father; patriarch of Constantinople

Chur·chill \'chər-ˌchil, 'chərch-ˌhil\ Randolph Henry Spencer 1849–1895 *Lord Randolph Churchill* British statesman

Churchill Sir Winston Leonard Spencer 1874–1965 British prime minister (1940–45; 1951–55); Nobel Prize winner (1953); son of the preceding

Cic·ero \'sis-ə-ˌrō\ Marcus Tullius 106–43 B.C. Roman statesman, orator, and author

Cid, El \el-'sid\ *about* 1043–1099 *Rodrigo Díaz de Vi·var* \bē-'vär\ Spanish soldier and hero

Cir·ce \'sər-sē\ enchantress in Greek mythology who turns her victims into swine

\ə\ abut \ᵊ\ kitten \ər\ **further** \a\ ash \ā\ ace \ä\ mop, mar \aů\ **out** \ch\ **chin** \e\ bet \ē\ **easy** \g\ go \i\ hit \ī\ ice \j\ job \ŋ\ sing \ō\ go \ȯ\ law \ȯi\ boy \th\ **thin** \t͟h\ **the** \ü\ **loot** \ů\ **foot** \y\ yet \zh\ **vision** *see also* Guide to Pronunciation

Clark \'klärk\ George Rogers 1752–1818 American soldier and frontiersman
Clark William 1770–1838 American explorer (with Meriwether Lewis)
Clay \'klā\ Henry 1777–1852 American statesman and orator
Cle·men·ceau \,klem-ən-'sō, klā-mäⁿ-'sō\ Georges 1841–1929 French statesman; premier (1906–9; 1917–20)
Clemens Samuel Langhorne — see TWAIN
Clem·ent \'klem-ənt\ name of 14 popes
Cle·o·pa·tra \,klē-ə-'pa-trə, -'pā-, -'pä-\ 69–30 B.C. queen of Egypt (51–30)
Cleve·land \'klēv-lənd\ (Stephen) Grover 1837–1908 22nd and 24th president of the U.S. (1885–89; 1893–97)
Clin·ton \'klint-n\ William Jefferson 1946– *Bill Clinton* 42nd president of the U.S. (1993–2001)
Clio \'klī-ō, 'klē-\ Muse of history in Greek mythology
Clo·vis I \'klō-vəs\ *about* 466–511 A.D. Frankish king (481–511)
Cly·tem·nes·tra \,klīt-əm-'nes-trə\ wife of Agamemnon in Greek mythology
Cobb \'käb\ Ty 1886–1961 *Tyrus Raymond Cobb* American baseball player
Co·chise \kō-'chēs\ 1812?–1874 Apache Indian chief
Coch·ran \'käk-rən\ Jacqueline 1910?–1980 American aviator
Co·dy \'kōd-ē\ William Frederick 1846–1917 *Buffalo Bill* American scout and showman
Co·han \'kō-,han\ George Michael 1878–1942 American actor, playwright, and songwriter
Cole·ridge \'kōl-rij, 'kō-lə-rij\ Samuel Taylor 1772–1834 English poet and critic
Co·lette \kò-'let\ 1873–1954 originally *Sidonie-Gabrielle Colette* French author
Co·lum·bus \kə-'ləm-bəs\ Christopher 1451–1506 Italian navigator and discoverer of America for Spain (1492)
Con·fu·cius \kən-'fyü-shəs\ 551–479 B.C. Chinese philosopher
Con·rad \'kän-,rad\ Joseph 1857–1924 British (Ukrainian-born of Polish parents) author
Con·stan·tine I \'kän-stən-,tēn, -,tīn\ *after* 280?–337 A.D. *Constantine the Great* Roman emperor (306–37)
Cook \'kůk\ Captain James 1728–1779 English navigator
Coo·lidge \'kü-lij\ (John) Calvin 1872–1933 30th president of the U.S. (1923–29)
Coo·per \'kü-pər, 'kůp-ər\ James Fen·i·more \'fen-ə-,mōr, -,mȯr\ 1789–1851 American author
Co·per·ni·cus \kō-'pər-ni-kəs\ Nicolaus 1473–1543 Polish astronomer
Cop·land \'kō-plənd\ Aaron 1900–1990 American composer
Cop·ley \'käp-lē\ John Singleton 1738–1815 American portrait painter
Corn·plan·ter \'kȯrn-,plant-ər\ *about* 1732–1836 *John O'Bail* Seneca Indian leader of partly European ancestry
Corn·wal·lis \kȯrn-'wäl-əs\ Charles 1738–1805 1st *Marquess Cornwallis* British general in America
Co·ro·na·do \,kȯr-ə-'näd-ō, ,kär-\ Francisco Vásquez de *about* 1510–1554 Spanish explorer and conquistador
Cor·tés *or* **Cor·tez** \kȯr-'tez, 'kȯr-,\ Hernán *or* Hernando 1485–1547 Spanish conqueror of Mexico
Cous·teau \kü-'stō\ Jacques (-Yves) 1910–1997 French marine explorer
Cow·per \'kü-pər, 'kůp-ər, 'kaů-pər\ William 1731–1800 English poet
Crane \'krān\ Stephen 1871–1900 American author
Crazy Horse \'krā-zē-,hȯrs\ 1842?–1877 *Ta-sunko-witko* Sioux Indian chief
Cres·si·da \'kres-əd-ə\ Trojan woman who in medieval legend is unfaithful to her lover Troilus
Crock·ett \'kräk-ət\ Davy 1786–1836 *David Crockett* American frontiersman and politician
Croe·sus \'krē-səs\ *died* 546 B.C. king of Lydia (560–546)
Crom·well \'kräm-,wel, 'krəm-, -wəl\ Oliver 1599–1658 English general and statesman; lord protector of England (1653–58)
Cro·nus \'krō-nəs\ Titan dethroned by his son Zeus in Greek mythology
Cum·mings \'kəm-ingz\ Edward Estliu 1894–1962 known as *e. e. cummings* American poet
Cu·pid \'kyü-pəd\ god of love in Roman mythology — compare EROS
Cu·rie \kyů-'rē, 'kyůr-ē\ Marie 1867–1934 née *Skłodowska* French (Polish-born) chemist; Nobel Prize winner (1903, 1911)
Curie Pierre 1859–1906 French chemist; husband of Marie Curie; Nobel Prize winner (1903)
Cus·ter \'kəs-tər\ George Armstrong 1839–1876 American general
Cyb·e·le \'sib-ə-lē\ a nature goddess of ancient Asia Minor
Cy·ra·no de Ber·ge·rac \,sir-ə-,nō-də-'ber-zhə-,rak\ Savinien 1619–1655 French satirist and playwright
Cyr·il \'sir-əl\ Saint *about* 827–869 A.D. apostle to the Slavs; brother of Methodius
Cy·rus II \'sī-rəs\ *about* 585–*about* 529 B.C. *Cyrus the Great* king of Persia (*about* 550–529)
Cyrus 424?–401 B.C. *the Younger* Persian prince and satrap
Dae·da·lus \'ded-l-əs, 'dēd-\ builder in Greek mythology of the Cretan labyrinth and inventor of wings by which he and his son Icarus escape imprisonment
Dahl \'däl\ Roald 1916–1990 British author
Da·lai La·ma \,däl-ī-'läm-ə\ 1935– *Tenzin Gyatso* Tibetan religious and political leader
Da·lí \'dä-lē, dä-'lē\ Salvador 1904–1989 Spanish painter
Dal·ton \'dȯlt-n\ John 1766–1844 English chemist and physicist
Dam·o·cles \'dam-ə-,klēz\ courtier of ancient Syracuse who according to legend was seated at a banquet beneath a sword hung by a single hair
Da·mon \'dā-mən\ legendary Sicilian who pledges his life for that of his condemned friend Pythias
Da·na \'dā-nə\ Richard Henry 1815–1882 American author
Dan·aë \'dan-ə-,ē\ imprisoned princess in Greek mythology who is visited by Zeus as a shower of gold; mother of Perseus
Dan·iel \'dan-yəl\ prophet in the Bible who is held captive in Babylon and divinely delivered from a den of lions
Dan·te \'dän-tā, 'dan-, -tē\ 1265–1321 *Dante Alighieri* Italian poet
Daph·ne \'daf-nē\ nymph in Greek mythology who is transformed into a laurel tree to escape the pursuing Apollo
Dare \'daər, 'deər\ Virginia 1587–? 1st child born in America of English parents
Da·ri·us I \də-'rī-əs\ 550–486 B.C. *Darius the Great* king of Persia (522–486)
Dar·row \'dar-ō\ Clarence Seward 1857–1938 American lawyer
Dar·win \'där-wən\ Charles Robert 1809–1882 English naturalist
Da·vid \'dā-vəd\ youth in the Bible who slays Goliath and succeeds Saul as king of Israel
David Saint *about* 520–600 patron saint of Wales
Da·vid \dä-'vēd\ Jacques-Louis 1748–1825 French painter
Da·vis \'dā-vəs\ Jefferson 1808–1889 president of the Confederate States of America (1861–65)
Davis Miles 1926–1991 American jazz musician
Da·vy \'dā-vē\ Sir Humphry 1778–1829 English chemist
Debs \'debz\ Eugene Victor 1855–1926 American socialist and labor organizer
De·bus·sy \,deb-yů-'sē, ,dāb-; də-'byü-sē\ (Achille-) Claude 1862–1918 French composer
De·ca·tur \di-'kāt-ər\ Stephen 1779–1820 American naval officer
De·foe \di-'fō\ Daniel 1660–1731 English author
De·gas \də-'gä\ (Hilaire-Germain-) Edgar 1834–1917 French painter
de Gaulle \di-'gōl, -'gȯl\ Charles (-André-Marie-Joseph) 1890–1970 French general; president of France's Fifth Republic (1958–69)
de Klerk \də-'klərk\ F. W. 1936– *Frederik Willem de Klerk* president of South Africa (1989–94); vice president (1994–96); Nobel Prize winner (1993)
De·la·croix \,del-ə-'krwä, -'kwä\ (Ferdinand-Victor-) Eugène 1798–1863 French painter
De·li·lah \di-'lī-lə\ mistress and betrayer of Samson in the Bible
De·me·ter \di-'mēt-ər\ goddess of agriculture in Greek mythology — compare CERES
de Mille \də-'mil\ Agnes George 1905–1993 American dancer and choreographer
de·Mille Cec·il \'ses-əl\ Blount \'blənt\ 1881–1959 American film director and producer
De·mos·the·nes \di-'mäs-thə-,nēz\ 384–322 B.C. Athenian orator and statesman
Demp·sey \'demp-sē\ Jack 1895–1983 originally *William Harrison Dempsey* American boxer
De·nis *or* **De·nys** \'den-əs, də-'nē\ Saint *died* 258? A.D. 1st bishop of Paris; patron saint of France
De Quin·cey \di-'kwin-sē, -'kwin-zē\ Thomas 1785–1859 English author
Des·cartes \dā-'kärt\ René 1596–1650 French mathematician and philosopher
de So·to \di-'sōt-ō\ Hernando *about* 1496–1542 Spanish explorer of Mississippi River (1540)
Dew·ey \'dü-ē, 'dyü-\ George 1837–1917 American admiral
Dewey John 1859–1952 American philosopher and educator
Di·ana \dī-'an-ə\ ancient Italian goddess of the forest and of childbirth who was identified with Artemis in Roman mythology
Di·as *or* **Di·az** \'dē-,äsh\ Bartholomeu *about* 1450–1500 Portuguese navigator
Dick·ens \'dik-ənz\ Charles (John Huffam) 1812–1870 *Boz* \'bäz, 'bōz\ English author
Dick·in·son \'dik-ən-sən\ Emily Elizabeth 1830–1886 American poet
Di·de·rot \dē-'drō, 'dēd-ə-,rō\ Denis 1713–1784 French philosopher and author
Di·do \'dīd-ō\ legendary queen of Carthage who falls in love with Aeneas and kills herself upon his departure
Di·Mag·gio \də-'mäzh-ē-ō, -'maj-\ Joe 1914–1999 *Joseph Paul DiMaggio* American baseball player
Di·o·cle·tian \,dī-ə-'klē-shən\ 245–316 A.D. Roman emperor (284–305)
Di·og·e·nes \dī-'äj-ə-,nēz\ *died about* 320 B.C. Greek Cynic philosopher
Di·o·ny·sus \,dī-ə-'nī-səs, -'nē-\ god of wine and ecstasy in classical mythology
Dis \'dis\ god of the underworld in Roman mythology — compare PLUTO
Dis·ney \'diz-nē\ Walt 1901–1966 *Walter Elias Disney* American film producer and cartoonist
Dis·rae·li \diz-'rā-lē\ Benjamin 1804–1881 Earl of *Bea·cons·field* \'bē-kənz-,fēld\ British politician and author; prime minister (1868; 1874–80)
Dix \'diks\ Dorothea Lynde 1802–1887 American social reformer
Dodgson Charles Lutwidge — see CARROLL
Dom·i·nic \'däm-ə-nik\ Saint *about* 1170–1221 Spanish-born founder of the Dominican order of friars
Do·mi·tian \də-'mish-ən\ 51–96 A.D. Roman emperor (81–96)
Don·i·zet·ti \,dän-əd-'zet-ē, ,dōn-, -ə-'zet-\ Gaetano 1797–1848 Italian composer
Donne \'dən\ John 1572–1631 English poet and clergyman
Dos·to·yev·sky \,däs-tə-'yef-skē, -'yev-\ Fyodor Mikhaylovich 1821–1881 Russian author
Doug·las \'dəg-ləs\ Stephen Arnold 1813–1861 American politician
Doug·lass \'dəg-ləs\ Frederick 1817–1895 American abolitionist
Doyle \'dȯil\ Sir Arthur Co·nan \'kō-nən, 'kō-\ 1859–1930 British author and physician
Dra·co \'drā-kō\ *flourished* 7th century B.C. Athenian lawgiver
Drake \'drāk\ Sir Francis 1540 (or 1543)–1596 English navigator, explorer, and admiral
Drei·ser \'drī-sər, -zər\ Theodore 1871–1945 American author
Drey·fus \'drī-fəs, 'drā-\ Alfred 1859–1935 French army officer
Dry·den \'drīd-n\ John 1631–1700 English author
DuBois \dü-'bȯis, dyü-\ William Edward Burghardt 1868–1963 American educator
Du·mas \dü-'mä, dyü-; 'dü-,mä, 'dyü-\ Alexandre 1802–1870 *Dumas père* \'peər\ French author
Dumas Alexandre 1824–1895 *Dumas fils* \'fēs\ French author
Dun·bar \'dən-,bär\ Paul Laurence 1872–1906 American poet
Dun·can \'dəng-kən\ Isadora 1877–1927 American dancer
Dü·rer \'důr-ər, 'dyůr-, 'dɯ̄r-\ Albrecht 1471–1528 German painter and engraver
Du·se \'dü-zā\ Eleonora 1858–1924 Italian actress
Dvo·řák \də-'vȯr-,zhäk, 'vȯr-,zhäk\ Antonín 1841–1904 Bohemian (Czech) composer

Ea•kins \'ā-kənz\ Thomas 1844–1916 American artist

Ear•hart \'er-,härt, 'ir-\ Amelia 1897–1937 American aviator

Earp \'ərp\ Wyatt 1848–1929 American frontiersman and lawman

Ed•dy \'ed-ē\ Mary Baker 1821–1910 American founder of the Christian Science religious faith

Ed•i•son \'ed-ə-sən\ Thomas Alva 1847–1931 American inventor

Ed•ward \'ed-wərd\ name of 8 post-Norman kings of England: **I** 1239–1307 *Edward Longshanks* (reigned 1272–1307); **II** 1284–1327 (reigned 1307–27); **III** 1312–1377 (reigned 1327–77); **IV** 1442–1483 (reigned 1461–70; 1471–83); **V** 1470–1483 (reigned 1483); **VI** 1537–1553 (reigned 1547–53) son of Henry VIII and Jane Seymour; **VII** 1841–1910 (reigned 1901–10) son of Queen Victoria; **VIII** 1894–1972 (reigned 1936; abdicated) *Duke of Windsor* son of George V

Edward 1330–1376 *the Black Prince* prince of Wales; son of Edward III

Edward 1003?–1066 *the Confessor* king of the English (1042–66)

Ed•wards \'ed-wərdz\ Jonathan 1703–1758 American theologian

Ein•stein \'īn-,stīn\ Albert 1879–1955 American (German-born) physicist; Nobel Prize winner (1921)

Ei•sen•how•er \'īz-n-,hau̇-ər, -,hau̇r\ Dwight David 1890–1969 American general; 34th president of the U.S. (1953–61)

Elec•tra \i-'lek-trə\ sister of Orestes in Greek mythology who aids him in avenging their father's murder

El•gar \'el-,gär, -gər\ Sir Edward 1857–1934 English composer

Eli \'ē-,lī\ early Hebrew judge and priest in the Bible

Eli•jah \i-'lī-jə\ Hebrew prophet of the 9th century B.C.

El•i•on \'el-ē-ən\ Gertrude Belle 1918–1999 American biochemist; Nobel Prize winner (1988)

El•iot \'el-ē-ət, 'el-yət\ George 1819–1880 pseudonym of *Mary Ann Evans* English author

Eliot T. S. 1888–1965 *Thomas Stearns Eliot* British (American-born) poet and critic; Nobel prize winner (1948)

Elis•a•beth \i-'liz-ə-bəth\ mother of John the Baptist in the Bible

Eli•sha \i-'lī-shə\ Hebrew prophet in the Bible who is disciple and successor of Elijah

Eliz•a•beth I \i-'liz-ə-bəth\ 1533–1603 queen of England (1558–1603); daughter of Henry VIII and Anne Boleyn

Elizabeth II 1926– queen of the United Kingdom (1952–); daughter of George VI

El•ling•ton \'el-ing-tən\ Duke 1899–1974 originally *Edward Kennedy Ellington* American bandleader and composer

El•li•son \'el-ə-sən\ Ralph (Waldo) 1914–1994 American author

Em•er•son \'em-ər-sən\ Ralph Waldo 1803–1882 American essayist and poet

En•dym•i•on \em-'dim-ē-ən\ beautiful youth loved by the moon goddess Selene in Greek mythology

Eos \'ē-,äs\ goddess of the dawn in Greek mythology — compare AURORA

Ep•i•cu•rus \,ep-i-'kyu̇r-əs\ 341–270 B.C. Greek philosopher

Eras•mus \i-'raz-məs\ Desiderius 1466?–1536 Dutch scholar

Er•a•to \'er-ə-,tō\ Muse of lyric and especially love poetry in Greek mythology

Erik the Red \'er-ik\ *flourished* 10th century originally *Erik Thor·vald·son* \'thȯr-vȧl-sən\ Norwegian explorer of Greenland coast (*about* 986); father of Leif Eriksson

Eriksson, Leif — see LEIF ERIKSSON

Eriny•es \i-'rin-ē-,ēz\ *or* **Eu•men•i•des** \yü-'men-ə-,dēz\ the Furies in Greek mythology

Ernst \'eərnst, 'ərnst\ Max 1891–1976 German painter

Eros \'eər-,äs, 'ir-\ god of love in Greek mythology — compare CUPID

Esau \'ē-,sȯ\ son of Isaac and Rebekah and elder twin brother of Jacob in the Bible

Es•ther \'es-tər\ Hebrew woman in the Bible who as the queen of Persia delivers her people from destruction

Eu•clid \'yü-kləd\ *flourished about* 300 B.C. Greek mathematician

Eumenides — see ERINYES

Eu•rip•i•des \yu̇-'rip-ə-,dēz\ *about* 484–406 B.C. Greek playwright

Eu•ro•pa \yu̇-'rō-pə\ Phoenician princess in Greek mythology who is abducted by Zeus disguised as a white bull

Eu•ryd•i•ce \yu̇-'rid-ə-sē\ wife of Orpheus in Greek mythology

Eu•ter•pe \yu̇-'tər-pē\ Muse of music in Greek mythology

Eve \'ēv\ the first woman in biblical tradition

Eze•kiel \i-'zē-kyəl, -kē-əl\ Hebrew prophet of the 6th century B.C.

Ez•ra \'ez-rə\ Hebrew priest, scribe, and reformer of the 5th century B.C.

Fa•bi•us Max•i•mus Cunc•ta•tor \'fā-bē-əs-'mak-si-məs-,kəngk-tā-tər\ Quintus *died* 203 B.C. Roman general against Hannibal

Fahd \'fäd\ 1923–2005 king of Saudi Arabia (1982–2005)

Fahr•en•heit \'far-ən-,hīt, 'fär-\ Daniel Gabriel 1686–1736 German physicist

Far•a•day \'far-ə-,dā, -əd-ē\ Michael 1791–1867 English chemist and physicist

Far•mer \'fär-mər\ James Leonard 1920–1999 American civil rights leader

Far•ra•gut \'far-ə-gət\ David Glasgow 1801–1870 American admiral

Faulk•ner \'fȯk-nər\ William 1897–1962 American author; Nobel Prize winner (1949)

Faust *or* **Fau•stus** \'fau̇-stəs, 'fȯ-\ magician in German legend and literature who sells his soul to the devil for knowledge and power

Fawkes \'fȯks\ Guy 1570–1606 English conspirator

Fer•ber \'fər-bər\ Edna 1887–1968 American writer

Fer•di•nand I \'fərd-n-,and\ 1016 (or 1018)–1065 *Ferdinand the Great* king of Castile (1035–65); of León (1037–65)

Ferdinand V of Castile *or* **II** of Aragon 1452–1516 *Ferdinand the Catholic* king of Castile (1474–1504); of Aragon (1479–1516); of Naples (1504–16); founder of the Spanish monarchy; husband of Isabella I

Fer•mi \'feər-mē\ Enrico 1901–1954 American (Italian-born) physicist; Nobel Prize winner (1938)

Feyn•man \'fīn-mən\ Richard Phillips 1918–1988 American physicist; Nobel Prize winner (1965)

Fiel•ding \'fēl-ding\ Henry 1707–1754 English author

Fill•more \'fil-,mōr, -,mȯr\ Millard 1800–1874 13th president of the U.S. (1850–53)

Fitz•ger•ald \fits-'jer-əld\ Ella 1917–1996 American singer

Fitzgerald F. Scott 1896–1940 *Francis Scott Key Fitzgerald* American author

Fitz•Ger•ald \fits-'jer-əld\ Edward 1809–1883 English poet

Flau•bert \flō-'beər\ Gustave 1821–1880 French author

Flem•ing \'flem-ing\ Sir Alexander 1881–1955 British bacteriologist; Nobel Prize winner (1945)

Flo•ra \'flōr-ə, 'flȯr-\ goddess of flowers in Roman mythology

Flying Dutchman legendary Dutch mariner condemned to sail the seas until Judgment Day

Foch \'fȯsh, 'fäsh\ Ferdinand 1851–1929 French general; marshal of France (1918)

Ford \'fōrd, 'fȯrd\ Gerald Rudolph 1913– 38th president of the U.S. (1974–77)

Ford Henry 1863–1947 American automobile manufacturer

Fos•sey \'fȯs-ē, 'fäs-\ Dian 1932–1985 American zoologist

Fos•ter \'fȯs-tər, 'fäs-\ Stephen Collins 1826–1864 American songwriter

Fox \'fäks\ George 1624–1691 English founder of Society of Friends (Quakers)

Fran•cis I \'fran-səs\ 1494–1547 king of France (1515–47)

Francis II 1768–1835 last Holy Roman emperor (1792–1806); emperor of Austria (as *Francis I*) (1804–35)

Francis Ferdinand 1863–1914 archduke of Austria

Francis Joseph 1830–1916 emperor of Austria (1848–1916); king of Hungary (1867–1916)

Francis of As•si•si \ə-'sē-sē, -zē, -'sis-ē\ Saint 1181 (or 1182)–1226 Italian friar; founder of Franciscan order

Franck \'fräŋk\ César Auguste 1822–1890 French (Belgian-born) composer

Fran•co \'fräng-kō, 'frang-\ Francisco 1892–1975 Spanish general, dictator, and head of Spanish state (1936–75)

Frank \'frangk\ Anne 1929–1945 Jewish (German-born) diarist during the Holocaust

Frank•lin \'frang-klən\ Benjamin 1706–1790 American statesman, philosopher, and inventor

Fred•er•ick I \'fred-rik, -ə-rik\ *about* 1123–1190 *Frederick Bar·ba·ros·sa* \,bär-bə-'räs-ə, -'rȯs-\ Holy Roman emperor (1152–90)

Frederick II 1194–1250 Holy Roman emperor (1215–50); king of Sicily (1198–1250)

Frederick I 1657–1713 king of Prussia (1701–13)

Frederick II 1712–1786 *Frederick the Great* king of Prussia (1740–86)

Frederick IX 1899–1972 king of Denmark (1947–72)

Fré•mont \'frē-,mänt\ John Charles 1813–1890 American general and explorer

French \'french\ Daniel Chester 1850–1931 American sculptor

Freud \'frȯid\ Sigmund 1856–1939 Austrian neurologist; founder of psychoanalysis

Frig•ga \'frig-ə\ *or* **Frigg** \'frig\ wife of Odin and goddess of married love and the hearth in Norse mythology

Fron•te•nac (et Pal•lu•au) \'fränt-n-,ak(-,ā-pà-'lwʸō)\ Comte de 1622–1698 French colonial administrator in New France (Canada)

Frost \'frȯst\ Robert Lee 1874–1963 American poet

Ful•ler \'fu̇l-ər\ (Richard) Buckminster 1895–1963 American engineer and architect of the geodesic dome

Ful•ton \'fu̇lt-n\ Robert 1765–1815 American inventor

Ga•bri•el \'gā-bre-əl\ one of the four archangels named in Hebrew tradition — compare MICHAEL, RAPHAEL, URIEL

Ga•ga•rin \gə-'gär-ən\ Yu·ry \'yu̇r-ē\ Alekseyevich 1934–1968 Russian cosmonaut; 1st man in space (1961)

Gage \'gāj\ Thomas 1721–1787 British general in America

Gains•bor•ough \'gānz-,bər-ə, -bə-rə, -brə\ Thomas 1727–1788 English painter

Gal•a•had \'gal-ə-,had\ knight of the Round Table in medieval legend who finds the Holy Grail

Gal•a•tea \,gal-ə-'tē-ə\ female figure carved by Pygmalion in Greek mythology and given life by Aphrodite in response to the sculptor's prayer

Ga•len \'gā-lən\ 129–*about* 216 A.D. Greek physician and writer

Ga•li•leo \,gal-ə-'lē-ō, -'lā-\ 1564–1642 *Galileo Ga·li·lei* \,gal-ə-'lā-,ē\ Italian astronomer and physicist

Gall \'gȯl\ 1840?–1894 Sioux Indian leader

Ga•lois \gal-'wä\ Évariste 1811–1832 French mathematician

Gals•wor•thy \'gȯlz-,wər-thē\ John 1867–1933 English author; Nobel Prize winner (1932)

Ga•ma \'gam-ə, 'gäm-\ Vasco da *about* 1460–1524 Portuguese navigator and explorer

Gan•dhi \'gän-dē, 'gan-\ Indira 1917–1984 Indian prime minister (1966–77; 1980–84); daughter of Jawaharlal Nehru

Gandhi Mohandas Karamchand 1869–1948 *Ma·hat·ma* \mə-'hät-mə, -'hat-\ *Gandhi* Indian nationalist leader

Gan•y•mede \'gan-i-,mēd\ cupbearer of the gods in Greek mythology

Gar•field \'gär-,fēld\ James Abram 1831–1881 20th president of the U.S. (1881)

Gar•i•bal•di \,gar-ə-'bȯl-dē\ Giuseppe 1807–1882 Italian patriot

Gar•ri•son \'gar-ə-sən\ William Lloyd 1805–1879 American abolitionist

Gar•vey \'gär-vē\ Marcus Moziah 1887–1940 Jamaican black leader in America (1919–26)

Gates \'gāts\ Horatio *about* 1728–1806 American (British-born) Revolutionary general

Gates Bill 1955– *William Henry Gates III* American computer software manufacturer

Gau•guin \gō-'gaⁿ\ (Eugène-Henri-) Paul 1848–1903 French painter

Gauss \'gau̇s\ Carl Friedrich 1777–1855 German mathematician and astronomer

Gautama Buddha — see BUDDHA
Ga·wain \gə-ˈwān, ˈgä-ˌwān, ˈgau̇-ən\ nephew of King Arthur and knight of the Round Table in medieval legend
Gay \ˈgā\ John 1685–1732 English author
Geh·rig \ˈger-ig\ Lou 1903–1941 *Henry Louis Gehrig* American baseball player
Gei·sel \ˈgī-zəl\ Theodor Seuss 1904–1991 pseudonym *Dr. Seuss* \ˈsüs\ American author and illustrator
Gen·ghis Khan \ˌjeng-gə-ˈskän, ˌgeng-\ *about* 1162–1227 Mongol conqueror
George \ˈjȯrj\ Saint *about* 3rd century A.D. Christian martyr; patron saint of England
George name of 6 kings of Great Britain: **I** 1660–1727 (reigned 1714–27); **II** 1683–1760 (reigned 1727–60); **III** 1738–1820 (reigned 1760–1820); **IV** 1762–1830 (reigned 1820–30); **V** 1865–1936 (reigned 1910–36); **VI** 1895–1952 (reigned 1936–52) father of Elizabeth II
George I 1845–1913 king of Greece (1863–1913)
George II 1890–1947 king of Greece (1922–23; 1935–47)
George David Lloyd — see LLOYD GEORGE
Ge·ron·i·mo \jə-ˈrän-ə-ˌmō\ 1829–1909 Apache Indian leader
Gersh·win \ˈgərsh-wən\ George 1898–1937 American composer
Gib·bon \ˈgib-ən\ Edward 1737–1794 English historian
Gide \ˈzhēd\ André 1869–1951 French author; Nobel Prize winner (1947)
Gid·e·on \ˈgid-ē-ən\ Hebrew hero in the Bible
Gil·bert \ˈgil-bərt\ Sir William Schwenck 1836–1911 English librettist and poet; collaborated with Sir Arthur Sullivan
Gil·les·pie \gə-ˈles-pē\ Dizzy 1917–1993 *John Birks Gillespie* American jazz musician
Gins·burg \ˈginz-bərg\ Ruth Bader 1933– American jurist
Giot·to \ˈjȯt-tō, ˈjȯ-tō, jē-ˈät-ō\ 1266?–1337 *Giotto di Bondone* Florentine painter
Glad·stone \ˈglad-ˌstōn, *chiefly British* -stən\ William Ewart 1809–1898 British prime minister (1868–74; 1880–85; 1886; 1892–94)
Glenn \ˈglen\ John Herschel 1921– American astronaut; 1st American to orbit the earth (1962)
God·dard \ˈgäd-ərd\ Robert Hutchings 1882–1945 American physicist and inventor
Go·di·va \gə-ˈdī-və\ English earl's wife noted in legend for riding naked through Coventry to save its citizens from a tax levied by her husband
Goeb·bels \ˈgərb-əlz, ˈgœb-əls\ (Paul) Joseph 1897–1945 German Nazi propagandist
Goe·thals \ˈgō-thəlz\ George Washington 1858–1928 American engineer who directed the building of the Panama Canal
Goe·the \ˈgər-tə, ˈgœ̄-tə\ Johann Wolfgang von 1749–1832 German author
Gogh, van \van-ˈgō, -ˈgäk, -ˈkȯk\ Vincent Willem 1853–1890 Dutch painter
Go·gol \ˈgȯ-gəl, ˈgō-ˌgȯl\ Nikolay Vasilyevich 1809–1852 Russian author
Gol·ding \ˈgōld-ing\ William Gerald 1911–1993 English author; Nobel Prize winner (1983)
Gold·smith \ˈgōld-ˌsmith, ˈgōl-\ Oliver 1730–1774 British author
Go·li·ath \gə-ˈlī-əth\ Philistine giant in the Bible who is killed by David with a sling
Gom·pers \ˈgäm-pərz\ Samuel 1850–1924 American (British-born) labor leader
Goo·dall \ˈgu̇d-ˌȯl\ Jane 1934– British zoologist
Good·year \ˈgu̇d-ˌyiər, ˈgu̇j-ˌiər\ Charles 1800–1860 American inventor
Gor·ba·chev \ˌgȯr-bə-ˈchȯf\ Mikhail Sergeyevich 1931– Soviet politician; 1st secretary of Communist party (1985–91); president of the U.S.S.R. (1990–91); Nobel Prize winner (1990)
Gore \ˈgōr, ˈgȯr\ Albert, Jr. 1948– vice president of the U.S. (1993–2001)
Gor·gas \ˈgȯr-gəs\ William Crawford 1854–1920 American army surgeon
Gor·ky \ˈgȯr-kē\ Maksim 1868–1936 pseudonym of *Aleksey Maksimovich Pesh·kov* \ˈpesh-ˌkȯf, -ˌkȯv\ Russian author
Gou·nod \ˈgü-ˌnō\ Charles (-François) 1818–1893 French composer
Go·ya (y Lu·cien·tes) \ˈgȯi-ə(-ˌē-ˌlü-sē-ˌen-ˌtās)\ Francisco José de 1746–1828 Spanish painter
Grac·chus \ˈgrak-əs\ Gaius Sempronius 153?–121 B.C. and his brother Tiberius Sempronius 163?–133 B.C. *the Grac·chi* \ˈgrak-ˌī\ Roman statesmen
Gra·ham \ˈgrā-əm, ˈgra-əm, ˈgram\ Martha 1893–1991 American dancer and choreographer
Grant \ˈgrant\ Ulysses S. 1822–1885 originally *Hiram Ulysses Grant* American general; 18th president of the U.S. (1869–77)
Gray \ˈgrā\ Thomas 1716–1771 English poet
Gre·co, El \el-ˈgrek-ō, -ˈgräk-, -ˈgrēk-\ 1541–1614 *Doménikos Theotokópoulos* Spanish (Cretan-born) painter
Gree·ley \ˈgrē-lē\ Horace 1811–1872 American journalist and politician
Greene \ˈgrēn\ (Henry) Graham 1904–1991 British author
Greene Nathanael 1742–1786 American Revolutionary general
Greg·o·ry \ˈgreg-rē, -ə-rē\ name of 16 popes: especially **I** Saint *about* 540–604 A.D. *Gregory the Great* (pope 590–604); **VII** Saint (*Hil·de·brand* \ˈhil-də-ˌbrand\) *about* 1020–1085 (pope 1073–85); **XIII** 1502–1585 (pope 1572–85)
Grey \ˈgrā\ Lady Jane 1537–1554 queen of England for nine days (1553)
Grey (Pearl) Zane 1872–1939 American author
Grieg \ˈgrēg, ˈgrig\ Edvard Hagerup 1843–1907 Norwegian composer
Grimm \ˈgrim\ Jacob 1785–1863 and his brother Wilhelm 1786–1859 German philologists and folklorists
Guin·e·vere \ˈgwin-ə-ˌviər, ˈgwen-\ wife of King Arthur and mistress of Lancelot in Arthurian legend
Gun·ther \ˈgu̇nt-ər\ Burgundian king and husband of Brunhild in Germanic legend
Gu·ten·berg \ˈgüt-n-ˌbərg\ Johannes *about* 1400–1468 German inventor of method of printing from movable type
Ha·dri·an \ˈhā-drē-ən\ 76–138 A.D. Roman emperor (117–138)
Ha·gar \ˈhā-ˌgär, -gər\ mistress of Abraham and mother of Ishmael in the Bible
Hag·gai \ˈhag-ē-ˌī, ˈhag-ˌī\ *or* **Ag·ge·us** \a-ˈgē-əs\ Hebrew prophet of the 6th century B.C.
Hai·le Se·las·sie \ˌhī-lē-sə-ˈlas-ē, -ˈläs-\ 1892–1975 emperor of Ethiopia (1930–36; 1941–74)
Hale \ˈhāl\ Edward Everett 1822–1909 American clergyman and author
Hale Nathan 1755–1776 American Revolutionary hero
Hal·ley \ˈhal-ē\ Edmond *or* Edmund 1656–1742 English astronomer and mathematician
Hal·sey \ˈhȯl-sē, -zē\ William Frederick 1882–1959 American admiral
Ham \ˈham\ son of Noah and ancestor of the Hamitic peoples in biblical tradition
Ha·man \ˈhā-mən\ Old Testament enemy of the Jews hanged for plotting their destruction
Ha·mil·car Bar·ca \hə-ˈmil-ˌkär-ˈbär-kə, ˈham-əl-\ 270?–229 (or 228) B.C. Carthaginian general; father of Hannibal
Ham·il·ton \ˈham-əl-tən, -əlt-n\ Alexander 1755–1804 American statesman
Hamilton Edith 1867–1963 American classicist
Ham·mar·skjöld \ˈham-ər-ˌshəld\ Dag 1905–1961 Swedish U.N. secretary-general (1953–61); Nobel Prize winner (1961, posthumously)
Ham·mu·ra·bi \ˌham-ə-ˈräb-ē\ *died about* 1750 B.C. king of Babylon (*about* 1792–50)
Ham·sun \ˈhäm-sən\ Knut 1859–1952 pseudonym of *Knut Pedersen* Norwegian author; Nobel Prize winner (1920)
Han·cock \ˈhan-ˌkäk\ John 1737–1793 American Revolutionary leader
Han·del \ˈhan-dl\ George Frideric 1685–1759 British (German-born) composer
Han·dy \ˈhan-dē\ W. C. 1873–1958 *William Christopher Handy* American blues musician and composer
Han·ni·bal \ˈhan-ə-bəl\ 247–183? B.C. Carthaginian general
Har·de·ca·nute \ˌhärd-i-kə-ˈnüt, -ˈnyüt\ *about* 1019–1042 king of the English (1040–42); king of Denmark (1028–42)
Har·ding \ˈhärd-ing\ Warren Gamaliel 1865–1923 29th president of the U.S. (1921–23)
Har·dy \ˈhärd-ē\ Thomas 1840–1928 English author
Har·old I \ˈhar-əld\ *died* 1040 *Harold Hare·foot* \ˈhaər-ˌfu̇t, ˈheər-\ king of the English (1035–40)
Harold II *about* 1022–1066 king of the English (1066)
Har·per \ˈhär-pər\ Stephen 1959– prime minister of Canada (2006–)
Har·ris \ˈhar-əs\ Joel Chandler 1848–1908 American author
Har·ri·son \ˈhar-ə-sən\ Benjamin 1833–1901 23rd president of the U.S. (1889–93); grandson of W. H. Harrison
Harrison William Henry 1773–1841 American general; 9th president of the U.S. (1841)
Harte \ˈhärt\ Bret 1836–1902 originally *Francis Brett Harte* American author
Har·vey \ˈhär-vē\ William 1578–1657 English physician and anatomist
Ha·vel \ˈhav-el\ Vá·clav\ˈvät-ˌsläf\ 1936– Czech playwright and politician; president of Czech Republic (1993–2003)
Haw·king \ˈhȯ-king\ Stephen William 1942– British physicist
Haw·thorne \ˈhȯ-ˌthȯrn\ Nathaniel 1804–1864 American author
Haydn \ˈhīd-n\ Franz Joseph 1732–1809 Austrian composer
Hayes \ˈhāz\ Rutherford Birchard 1822–1893 19th president of the U.S. (1877–81)
Haz·litt \ˈhaz-lət, ˈhāz-\ William 1778–1830 English essayist
Hearst \ˈhərst\ William Randolph 1863–1951 American newspaper publisher
Hec·ate \ˈhek-ət-ē\ goddess associated especially with the underworld, night, and witchcraft in Greek mythology
Hec·tor \ˈhek-tər\ son of Priam and Hecuba; Trojan hero slain by Achilles in Greek mythology
Hec·u·ba \ˈhek-yə-bə\ wife of Priam and mother of Hector and Paris in Greek mythology
He·gel \ˈhā-gəl\ Georg Wilhelm Friedrich 1770–1831 German philosopher
Hei·deg·ger \ˈhī-ˌdeg-ər, ˈhīd-i-gər\ Martin 1889–1976 German philosopher
Hei·ne \ˈhī-nə *also* -nē\ Heinrich 1797–1856 German author
Hei·sen·berg \ˈhī-zn-bərg\ Werner Karl 1901–1976 German physicist; Nobel Prize winner (1932)
Hel·en of Troy \ˌhel-ə-nəv-ˈtrȯi\ wife of Menelaus whose abduction by Paris causes the Trojan War in Greek mythology
He·li·os \ˈhē-lē-ˌōs, -əs\ god of the sun in Greek mythology — compare SOL
Hell·man \ˈhel-mən\ Lillian 1905–1984 American playwright
Hem·ing·way \ˈhem-ing-ˌwā\ Ernest Miller 1899–1961 American author; Nobel Prize winner (1954)
Hen·ry \ˈhen-rē\ name of 8 kings of England: **I** 1068–1135 (reigned 1100–1135); **II** 1133–1189 (reigned 1154–89); **III** 1207–1272 (reigned 1216–72); **IV** 1366–1413 (reigned 1399–1413); **V** 1387–1422 (reigned 1413–22); **VI** 1421–1471 (reigned 1422–61; 1470–71); **VII** 1457–1509 (reigned 1485–1509); **VIII** 1491–1547 (reigned 1509–47)
Henry name of 4 kings of France: **I** *about* 1008–1060 (reigned 1031–60); **II** 1519–1559 (reigned 1547–59); **III** 1551–1589 (reigned 1574–89); **IV** 1553–1610 *Henry of Navarre* (reigned 1589–1610)
Henry O. 1862–1910 pseudonym of *William Sydney Porter* American author
Henry Patrick 1736–1799 American statesman and orator
Hen·son \ˈhen-sən\ Matthew Alexander 1866–1955 American arctic explorer
He·phaes·tus \hi-ˈfes-təs, -ˈfēs-\ god of fire and of metalworking in Greek mythology — compare VULCAN
He·ra \ˈhir-ə, ˈhē-rə\ sister and wife of Zeus and goddess of women and marriage in Greek mythology — compare JUNO
Her·bert \ˈhər-bərt\ Victor 1859–1924 American (Irish-born) composer and conductor
Her·cu·les \ˈhər-kyə-ˌlēz\ *or* **Her·a·cles** \ˈher-ə-ˌklēz\ hero in classical mythology noted for his strength and for performing 12 labors imposed on him by Hera
Her·maph·ro·di·tus \hər-ˌmaf-rə-ˈdīt-əs\ son of Hermes and Aphrodite who in Greek mythology is bodily joined with a nymph
Her·mes \ˈhər-ˌmēz, -mēz\ god of commerce, eloquence, invention, travel, and theft who serves as herald and messenger for the other gods in Greek mythology — compare MERCURY
He·ro \ˈhē-rō, ˈhiər-ō\ priestess of Aphrodite loved by Leander in Greek mythology
Her·od \ˈher-əd\ 73–4 B.C. *Herod the Great* Roman king of Judea (37–4)
Herod An·ti·pas \ˈant-ə-ˌpas, -pəs\ 21 B.C.–39 A.D. Roman tetrarch of Galilee (4 B.C.–39 A.D.); son of Herod the Great

He·rod·o·tus \hi-'räd-ə-təs\ *about* 484–*between* 430 *and* 420 B.C. Greek historian
Her·rick \'her-ik\ Robert 1591–1674 English poet
He·si·od \'hē-sē-əd, 'hes-ē-\ *flourished about* 800 B.C. Greek poet
Hes·se \'hes-ə\ Hermann 1877–1962 German author; Nobel Prize winner (1946)
Hes·tia \'hes-tē-ə; 'hes-chə, 'hesh-\ goddess of the hearth and domestic activity in Greek mythology — compare VESTA
Hey·er·dahl \'hā-ər-ˌdäl, 'hī-\ Thor 1914–2002 Norwegian explorer and author
Hi·a·wa·tha \ˌhī-ə-'wȯ-thə, ˌhē-ə-, -'wäth-ə\ legendary Iroquois Indian chief
Hick·ok \'hik-ˌäk\ Wild Bill 1837–1876 originally *James Butler Hickok* American frontiersman and U.S. marshal
Hildebrand — see GREGORY VII
Hil·la·ry \'hil-ə-rē\ Sir Edmund Percival 1919– New Zealand mountaineer and explorer
Hil·ton \'hilt-ᵊn\ James 1900–1954 English novelist
Hin·den·burg \'hin-dən-ˌbərg, -ˌbu̇rg\ Paul von 1847–1934 German field marshal; president of Germany (1925–34)
Hip·poc·ra·tes \hip-'äk-rə-ˌtēz\ *about* 460–*about* 377 B.C. *father of medicine* Greek physician
Hi·ro·hi·to \ˌhir-ō-'hē-tō\ 1901–1989 emperor of Japan (1926–89)
Hit·ler \'hit-lər\ Adolf 1889–1945 German (Austrian-born) chancellor and dictator (1933–45)
Hobbes \'häbz\ Thomas 1588–1679 English philosopher
Ho Chi Minh \'hō-'chē-'min\ 1890–1969 originally *Nguyen Sinh Cung* Vietnamese nationalist; president of North Vietnam (1945–69)
Hodg·kin \'häj-kin\ Dorothy Mary 1910–1994 née *Crowfoot* British physicist; Nobel Prize winner (1964)
Ho·gan \'hō-gən\ Ben 1912–1997 *William Benjamin Hogan* American golfer
Ho·garth \'hō-ˌgärth\ William 1697–1764 English painter and engraver
Hol·bein \'hōl-ˌbīn, 'hȯl-\ Hans *the Elder* 1465?–1524 and his son Hans *the Younger* 1497?–1543 German painters
Hol·i·day \'häl-ə-ˌdā\ Billie 1915–1959 originally *Eleanora Fagan* American jazz singer
Holmes \'hōmz, 'hōlmz\ Oliver Wendell 1809–1894 American physician and author
Holmes Oliver Wendell, Jr. 1841–1935 American jurist; son of the preceding
Ho·mer \'hō-mər\ *flourished* 9th *or* 8th century B.C. Greek epic poet
Homer Winslow 1836–1910 American painter
Hooke \'hu̇k\ Robert 1635–1703 English scientist
Hook·er \'hu̇k-ər\ Thomas 1586?–1647 English Puritan clergyman and a founder of Connecticut
Hoo·ver \'hü-vər\ Herbert Clark 1874–1964 31st president of the U.S. (1929–33)
Hoover John Edgar 1895–1972 American criminologist; director of the Federal Bureau of Investigation (1924–72)
Hop·kins \'häp-kənz\ Gerard Manley 1844–1889 English poet
Hop·per \'häp-ər\ Grace 1906–1992 née *Murray* American admiral, mathematician, and computer scientist
Hor·ace \'hȯr-əs, 'här-\ 65–8 B.C. Roman poet
Ho·ra·tius \hə-'rā-shē-əs, -shəs\ hero in Roman legend noted for his defense of a bridge over the Tiber against the Etruscans
Ho·sea \hō-'zē-ə, -'zā-\ Hebrew prophet of the 8th century B.C.
Hou·di·ni \hü-'dē-nē\ Harry 1874–1926 originally *Erik Weisz* American magician
Hous·man \'hau̇-smən\ Alfred Edward 1859–1936 English classical scholar and poet
Hous·ton \'hyü-stən, 'yü-\ Sam 1793–1863 *Samuel Houston* American politician; president of the Republic of Texas (1836–38; 1841–44)
Howe \'hau̇\ Elias 1819–1867 American inventor
Howe Julia 1819–1910 née *Ward* American suffragist and reformer
How·ells \'hau̇-əlz\ William Dean 1837–1920 American author
Hud·son \'həd-sən\ Henry *about* 1565–1611 English navigator and explorer
Hugh Ca·pet \'kā-pət, 'kap-ət, ka-'pā\ *about* 938–996 A.D. king of France (987–996)
Hughes \'hyüz *also* 'yüz\ Charles Evans 1862–1948 chief justice of the U.S. Supreme Court (1930–41)
Hughes (James) Langston 1902–1967 American author
Hu·go \'hyü-gō, yü-\ Victor (-Marie) 1802–1885 French author
Hu Jin·tao \'hü-'jin-'tau̇\ 1942– Chinese Communist party leader (2002–); president of China (2003–)
Hume \'hyüm *also* 'yüm\ David 1711–1776 Scottish philosopher
Hus *or* **Huss** \'həs, 'hu̇s\ Jan *about* 1370–1415 Czech religious reformer
Hus·sein I \hü-'sān\ 1935–1999 king of Jordan (1952–99)
Hussein Saddam al-Tikriti 1937– leader of Iraq (1979–2003)
Hux·ley \'hək-slē\ Aldous Leonard 1894–1963 English author
Hy·ge·ia \hī-'jē-ə, -yə\ goddess of health in Greek mythology
Hy·men \'hī-mən\ god of marriage in Greek mythology
Hy·pe·ri·on \hī-'pir-ē-ən\ Titan and the father of Eos, Selene, and Helios in Greek mythology
Ib·sen \'ib-sən, 'ip-\ Henrik 1828–1906 Norwegian playwright
Ic·a·rus \'ik-ə-rəs\ son of Daedalus who falls into the sea when the wax of his artificial wings melts as he flies too near the sun
Ig·na·tius \ig-'nā-shē-əs, -shəs\ Saint 1491–1556 *Ignatius of Loy·o·la* \lȯi-'ō-lə\ Spanish founder of the Society of Jesus (Jesuits)
In·no·cent \'in-ə-sənt\ name of 13 popes: especially **II** *died* 1143 (pope 1130–43); **III** 1160 (or 1161)–1216 (pope 1198–1216); **IV** *died* 1254 (pope 1243–54); **XI** 1611–1689 (pope 1676–89)
Iph·i·ge·nia \ˌif-ə-jə-'nī-ə\ daughter of Agamemnon offered by him as a sacrifice but saved and made a priestess of Artemis in Greek mythology
Iris \'ī-rəs\ goddess of the rainbow and a messenger of the gods in Greek mythology
Ir·ving \'ər-viŋ\ Washington 1783–1859 American author
Isaac \'ī-zik, -zək\ son of Abraham and father of Jacob in the Bible
Is·a·bel·la I \ˌiz-ə-'bel-ə\ 1451–1504 queen of Castile (1474–1504) and of Aragon (1479–1504); wife of Ferdinand V of Castile
Isa·iah \ī-'zā-ə\ Hebrew prophet of the 8th century B.C.
Ish·ma·el \'ish-mē-əl, -mā-\ outcast son of Abraham and Hagar in the Bible
Isis \'ī-səs\ ancient Egyptian nature goddess
Isol·de \i-'zōl-də\ legendary Irish princess married to King Mark of Cornwall and loved by Tristram
Ivan III \ē-'vän, 'ī-vən\ 1440–1505 *Ivan the Great* grand prince of Moscow (1462–1505)
Ivan IV 1530–1584 *Ivan the Terrible* ruler of Russia (1533–84) and 1st czar (1547–84)
Ives \'īvz\ Charles Edward 1874–1954 American composer
Jack·son \'jak-sən\ Andrew 1767–1845 American general; 7th president of the U.S. (1829–37)
Jackson Jesse Louis 1941– American clergyman and civil rights leader
Jackson Mahalia 1911–1972 American gospel singer
Jackson Thomas Jonathan 1824–1863 *Stonewall Jackson* American Confederate general
Ja·cob \'jā-kəb\ son of Isaac and Rebekah and younger twin brother of Esau in the Bible
James \'jāmz\ one of the 12 apostles and brother of the apostle John in the Bible
James *the Less* one of the 12 apostles in the Bible
James name of 2 kings of Great Britain: **I** 1566–1625 (reigned 1603–25); king of Scotland as *James VI* (reigned 1567–1625); **II** 1633–1701 (reigned 1685–88)
James Henry 1843–1916 British (American-born) author
James William 1842–1910 American psychologist and philosopher; brother of Henry James
James Edward — see STUART
Ja·nus \'jā-nəs\ god of gates and doors and of all beginnings in Roman mythology and that is portrayed with two opposite faces
Ja·pheth \'jā-fəth\ son of Noah and ancestor of the Medes and Greeks in biblical tradition
Ja·son \'jās-ᵊn\ hero in Greek mythology noted for his successful quest of the Golden Fleece
Jay \'jā\ John 1745–1829 American jurist and statesman; 1st chief justice of the U.S. Supreme Court (1789–95)
Jef·fer·son \'jef-ər-sən\ Thomas 1743–1826 3rd president of the U.S. (1801–09)
Jen·ner \'jen-ər\ Edward 1749–1823 English physician
Jer·e·mi·ah \ˌjer-ə-'mī-ə\ Hebrew prophet of the 7th–6th century B.C.
Je·rome \jə-'rōm\ Saint *about* 347–419 (or 420) A.D. church father and biblical translator
Je·sus \'jē-zəs, -zəz\ *or* **Jesus Christ** *about* 6 B.C.–*about* 30 A.D. source of the Christian religion and Savior in the Christian faith
Jez·e·bel \'jez-ə-ˌbel\ queen of Israel and wife of Ahab who is notable for her wickedness in the Bible
Ji·ang Ze·min \jē-'äŋ-zə-'min\ 1926– Chinese Communist party leader (1989–2002) and president of China (1993–2003)
Joan of Arc \ˌjō-nə-'värk\ Saint *about* 1412–1431 *the Maid of Orléans* French national heroine
Job \'jōb\ man in the Bible who endures afflictions with fortitude and faith
Jo·cas·ta \jō-'kas-tə\ queen of Thebes in Greek mythology who unknowingly marries her son Oedipus
Jo·el \'jō-əl\ Hebrew prophet in the Bible
John \'jän\ one of the 12 apostles and the traditional author of the 4th Gospel, three Epistles, and the Book of Revelation
John name of 21 popes: especially **XXIII** (*Angelo Giuseppe Roncalli*) 1881–1963 (pope 1958–63)
John 1167–1216 *John Lack·land* \'lak-ˌland\ king of England (1199–1216)
John of Gaunt \-'gȯnt, -'gänt\ 1340–1399 Duke of Lancaster; son of Edward III of England
John Paul name of two popes: **I** (*Albino Luciani*) 1912–1978 (pope 1978); **II** (*Karol Józef Wojtyła*) 1920–2005 (pope 1978–2005)
John·son \'jän-sən\ Andrew 1808–1875 17th president of the U.S. (1865–69)
Johnson Jack 1878–1946 *John Arthur Johnson* American boxer
Johnson Lyndon Baines 1908–1973 36th president of the U.S. (1963–69)
Johnson Magic 1959– *Earvin Johnson, Jr.* American basketball player
Johnson Philip Cortelyou 1906– American architect
Johnson Samuel 1709–1784 *Dr. Johnson* English lexicographer and author
John the Baptist Saint, 1st century A.D. prophet who in the Bible foretells Jesus' ministry and baptizes him
Jol·liet *or* **Jo·liet** \zhȯl-'yā\ Louis 1645–1700 French-Canadian explorer
Jo·nah \'jō-nə\ Hebrew prophet who in the Bible spends three days in the belly of a great fish
Jon·a·than \'jän-ə-thən\ son of Saul and friend of David in the Bible
Jones \'jōnz\ John Paul 1747–1792 originally *John Paul* American (Scottish-born) naval officer
Jon·son \'jän-sən\ Ben 1572–1637 *Benjamin Johnson* English author
Jop·lin \'jäp-lən\ Scott 1868–1917 American pianist and composer
Jor·dan \'jȯrd-ᵊn\ Michael Jeffrey 1963– American basketball player
Jo·seph \'jō-zəf *also* -səf\ son of Jacob who in the Bible rises to high office in Egypt after being sold into slavery by his brothers
Joseph Chief *about* 1840–1904 Nez Percé Indian chief
Joseph Saint, husband of Mary, the mother of Jesus, in the Bible
Jo·se·phine \'jō-zə-ˌfēn *also* -sə-\ 1763–1814 1st wife of Napoléon I; empress of France (1804–09)
Joseph of Ar·i·ma·thea \-ˌar-ə-mə-'thē-ə\ member of the Sanhedrin (supreme council) who in the Bible places the body of Jesus in his own tomb
Jo·se·phus \jō-'sē-fəs\ Flavius *about* 37–*about* 100 A.D. originally *Joseph Ben Matthias* Jewish historian
Josh·ua \'jäsh-wə, -ə-wə\ Hebrew leader in the Bible who succeeds Moses during the settlement of the Israelites in Canaan
Joyce \'jȯis\ James Augustine 1882–1941 Irish author
Juan Car·los \'hwän-'kär-ˌlōs, 'wän-\ 1938– king of Spain (1975–)

\ə\ **abut** \ᵊ\ **kitten** \ər\ **further** \a\ **ash** \ā\ **ace** \ä\ **mop, mar** \au̇\ **out** \ch\ **chin** \e\ **bet** \ē\ **easy** \g\ **go** \i\ **hit** \ī\ **ice** \j\ **job** \ŋ\ **sing** \ō\ **go** \ȯ\ **law** \ȯi\ **boy** \th\ **thin** \t͟h\ **the** \ü\ **loot** \u̇\ **foot** \y\ **yet** \zh\ **vision** *see also* Guide to Pronunciation

Ju·dah \'jüd-ə\ son of Jacob and ancestor of one of the 12 tribes of Israel in the Bible
Ju·das \'jüd-əs\ *or* **Judas Is·car·i·ot** \-is-'kar-ē-ət\ one of the 12 apostles and betrayer of Jesus in the Bible
Ju·das Mac·ca·bae·us \'jud-ə-,smak-ə-'bē-əs\ *died about* 161 B.C. Jewish patriot
Ju·lian \'jül-yən\ *about* 331–363 A.D. *Julian the Apostate* Roman emperor (361–63)
Jung \'yůŋ\ Carl Gustav 1875–1961 Swiss psychologist
Ju·no \'jü-nō\ queen of heaven, wife of Jupiter, and goddess of light, birth, women, and marriage in Roman mythology — compare HERA
Ju·pi·ter \'jü-pət-ər\ chief god, husband of Juno, and god of light, of the sky and weather, and of the state in Roman mythology — compare ZEUS
Jus·tin·i·an I \jə-'stin-ē-ən\ 483–565 A.D. *Justinian the Great* Byzantine emperor (527–565)
Ju·ve·nal \'jü-vən-l\ 55 to 60–*about* 127 A.D. Roman satirist
Kaf·ka \'käf-kə, 'kaf-\ Franz 1883–1924 Czech-born author who wrote in German
Kalb \'kälp, 'kalb\ Johann 1721–1780 Baron *de Kalb* \di-'kalb\ German general in American Revolutionary army
Ka·me·ha·me·ha I \kə-,mā-ə-'mā-hä\ 1758?–1819 *Kamehameha the Great* originally *Pai·ea* \pī-'ā-ə\ Hawaiian king (1795–1819)
Kant \'kant, 'känt\ Immanuel 1724–1804 German philosopher
Keats \'kēts\ John 1795–1821 English poet
Kel·ler \'kel-ər\ Helen Adams 1880–1968 American deaf and blind lecturer and author
Kel·vin \'kel-vən\ 1st Baron 1824–1907 *William Thomson* British mathematician and physicist
Kempis Thomas à — see THOMAS A KEMPIS
Ken·ne·dy \'ken-əd-ē\ John Fitzgerald 1917–1963 35th president of the U.S. (1961–63); brother of R. F. Kennedy
Kennedy Robert Francis 1925–1968 American politician
Ke·o·kuk \'kē-ə-,kək\ 1780?–1848 American Indian chief
Kep·ler \'kep-lər\ Johannes 1571–1630 German astronomer
Ke·ren·sky \'ker-ən-skē\ Aleksandr Fyodorovich 1881–1970 Russian revolutionary and head of provisional government (1917)
Key \'kē\ Francis Scott 1779–1843 American lawyer and author of "The Star-Spangled Banner"
Keynes \'kānz\ John Maynard 1883–1946 Baron *Keynes of Tilton* English economist — **Keynes·ian** \'kān-zē-ən\ *adj or n*
Khayyám Omar — see OMAR KHAYYAM
Khru·shchev \krůsh-'chóf, -'óf, -'chóv, -'óv, -'chef, -'ef, 'krüsh-\ Ni·ki·ta \nə-'kēt-ə\ Sergeyevich 1894–1971 premier of U.S.S.R. (1958–64)
Khu·fu \'kü-,fü\ *or Greek* **Che·ops** \'kē-,äps\ *flourished* 25th century B.C. king of Egypt and pyramid builder
Kidd \'kid\ William *about* 1645–1701 *Captain Kidd* Scottish pirate
Kier·ke·gaard \'kir-kə-,gärd, -,gär, -,gór\ Søren Aabye 1813–1855 Danish philosopher and theologian
King \'king\ Billie Jean 1943– née *Moffitt* American tennis player
King Ernest Joseph 1878–1956 American admiral
King Martin Luther, Jr. 1929–1968 American clergyman and civil rights leader; Nobel Prize winner (1964)
Kip·ling \'kip-ling\ (Joseph) Rud·yard \'rəd-yərd, 'rəj-ərd\ 1865–1936 English author; Nobel Prize winner (1907)
Kis·sin·ger \'kis-n-jər\ Henry Alfred 1923– American (German-born) government official; U.S. secretary of state (1973–77); Nobel Prize winner (1973)
Klee \'klā\ Paul 1879–1940 Swiss painter
Knox \'näks\ John *about* 1514–1572 Scottish religious reformer
Koch \'kók, 'kók, 'kōk, 'kōk, 'käk, 'käk\ Robert 1843–1910 German bacteriologist; Nobel Prize winner (1905)
Kohl \'kōl\ Helmut 1930– chancellor of Germany (1990–98)
Koś·ciusz·ko \,käs-ē-'əs-,kō, kósh-'chůsh-kō\ Tadeusz 1746–1817 Polish patriot and general in American Revolutionary army
Krish·na \'krish-nə\ deity or deified hero of later Hinduism worshiped as an incarnation of Vishnu
Ku·blai Khan \,kü-blə-'kän, -,blī-\ 1215–1294 founder of Mongol dynasty in China; grandson of Genghis Khan
La·fa·yette \,läf-ē-'et, ,laf-\ Marquis de 1757–1834 French general in American Revolutionary army
La Fon·taine \lə-,fän-'tān, -,fōn-'ten\ Jean de 1621–1695 French writer of fables
La·ius \'lā-əs, 'lī-\ king of Thebes who in Greek mythology is slain by his son Oedipus
La·marck \lə-'märk\ Chevalier de 1744–1829 *Jean-Baptiste de Monet* French naturalist
Lamb \'lam\ Charles 1775–1834 English author
Lan·ce·lot \'lan-sə-,lät\ knight of the Round Table and lover of Queen Guinevere in Arthurian legend
Lange \'lang\ Dorothea 1895–1965 American photographer
Lang·land \'lang-lənd\ William *about* 1330–*about* 1400 English poet
Lang·ley \'lang-lē\ Samuel Pierpont 1834–1906 American astronomer and airplane pioneer
La·oc·o·ön \lā-'äk-ə-,wän\ Trojan priest in Greek mythology killed by sea serpents after warning against the wooden horse
Lao–tzu \'laůd-'zə\ *flourished* 6th century B.C. Chinese philosopher
La·place \lə-'pläs\ Pierre-Simon 1749–1827 Marquis *de Laplace* French astronomer and mathematician
La Roche·fou·cauld \lä-,rōsh-,fü-'kō, -,rōsh-\ François 1613–1680 Duc *de la Rochefoucauld* French author and moralist
La Salle \lə-'sal\ Sieur de 1643–1687 *René-Robert Cavelier* French explorer in North America
La·voi·sier \ləv-'wäz-ē-,ā\ Antoine-Laurent 1743–1794 French chemist
Law·rence \'lór-əns, 'lär-\ D.H. 1885–1930 *David Herbert Lawrence* English author
Lawrence Sir Thomas 1769–1830 English painter
Lawrence Thomas Edward 1888–1935 *Lawrence of Arabia* British archaeologist, soldier, and author
Laz·a·rus \'laz-rəs, -ə-rəs\ brother of Mary and Martha who in the Bible is raised by Jesus from the dead
Lazarus beggar in the biblical parable of the rich man and the beggar
Lea·key \'lē-kē\ family of British anthropologists and paleontologists: Louis 1903–1972; his wife Mary 1913–1996 née *Nicol;* their son Richard 1944–
Le·an·der \lē-'an-dər\ youth in Greek mythology who swims the Hellespont nightly to visit his lover Hero
Le·da \'lēd-ə\ Spartan princess in Greek mythology who is visited by Zeus in the form of a swan
Lee \'lē\ Ann 1736–1784 English mystic and founder of Shaker society in the U.S.
Lee Henry 1756–1818 *Light-Horse Harry* American general
Lee Robert Edward 1807–1870 American Confederate general; son of the preceding
Leeu·wen·hoek *or* **Leu·wen·hoek** \'lā-vən-,hůk\ Antonie van 1632–1723 Dutch naturalist
Leib·niz \'līb-nəts, 'līp-nits\ Gottfried Wilhelm 1646–1716 German philosopher and mathematician
Leif Er·iks·son \'lā-'ver-ik-sən, 'lē-'fer-\ *flourished* 1000 Norwegian explorer; son of Erik the Red
Le·nin \'len-ən\ 1870–1924 originally *Vladimir Ilyich Ul·ya·nov* \ül-'yän-əf, -,óf, -,óv\ Russian Communist leader
Len·non \'len-ən\ John Winston 1940–1980 British popular singer and songwriter
Leo \'lē-ō\ name of 13 popes: especially **I** Saint *died* 461 A.D. *Leo the Great* (pope 440–61); **III** Saint *died* 816 A.D. (pope 795–816); **XIII** 1810–1903 (pope 1878–1903)
Le·o·nar·do da Vin·ci \,lē-ə-'när-dō-də-'vin-chē, ,lā-\ 1452–1519 Italian painter, sculptor, architect, and engineer
Le·on·ca·val·lo \,lā-,ōn-kə-'väl-ō\ Ruggero 1858–1919 Italian composer and librettist
Le·on·i·das \lē-'än-əd-əs\ *died* 480 B.C. Greek hero; king of Sparta (490?–480)
Lep·i·dus \'lep-əd-əs\ Marcus Aemilius *died* 13 (or 12) B.C. Roman triumvir (43–36)
Le·vi \'lē-,vī\ son of Jacob and ancestor of one of the 12 tribes of Israel in the Bible
Lew·is \'lü-əs\ C. S. 1898–1963 *Clive Staples Lewis* British author
Lewis John Llewellyn 1880–1969 American labor leader
Lewis Meriwether 1774–1809 American explorer (with William Clark)
Lewis (Harry) Sinclair 1885–1951 American author; Nobel Prize winner (1930)
Lin·coln \'ling-kən\ Abraham 1809–1865 16th president of the U.S. (1861–65)
Lind·bergh \'lind-,bərg, 'lin-\ Charles Augustus 1902–1974 American aviator
Lin·nae·us \lə-'nē-əs, -'nā-\ Carolus 1707–1778 *Carl von Lin·né* \lə-'nā\ Swedish botanist
Lip·pi \'lip-ē\ Fra Fi·lip·po \fə-'lip-ō\ *about* 1406–1469 Florentine painter
Lis·ter \'lis-tər\ Joseph 1827–1912 English surgeon and medical scientist
Liszt \'list\ Franz 1811–1886 Hungarian pianist and composer
Liv·ing·stone \'liv-ing-stən\ David 1813–1873 Scottish explorer and missionary in Africa
Livy \'liv-ē\ 59 B.C.–17 A.D. *Titus Livius* Roman historian
Lloyd George \'lòid-'jórj\ David 1863–1945 Earl of *Dwy·for* \'dü-ē-,vór\ British prime minister (1916–22)
Locke \'läk\ John 1632–1704 English philosopher
Lo·hen·grin \'lō-ən-,grin\ son of Parsifal and knight of the Holy Grail in German legend
Lon·don \'lən-dən\ Jack 1876–1916 *John Griffith London* American author
Long·fel·low \'lóng-,fel-ō\ Henry Wads·worth \'wädz-wərth, -,wərth\ 1807–1882 American poet
Lo·re·lei \'lōr-ə-,lī, 'lór-\ siren in German legend whose singing lures boatmen to destruction on a reef in the Rhine River
Lot \'lät\ nephew of Abraham in biblical tradition whose wife is turned into a pillar of salt for looking back on the doomed city of Sodom
Lou·is \'lü-ē, 'lü-əs\ name of 18 kings of France: especially **IX** Saint 1214–1270 (reigned 1226–70); **XI** 1423–1483 (reigned 1461–83); **XII** 1462–1515 (reigned 1498–1515); **XIII** 1601–1643 (reigned 1610–43); **XIV** 1638–1715 (reigned 1643–1715); **XV** 1710–1774 (reigned 1715–74); **XVI** 1754–1793 (reigned 1774–92; guillotined); **XVII** 1785–1795 (nominally reigned 1793–95); **XVIII** 1755–1824 (reigned 1814–15; 1815–24)
Louis \'lü-əs\ Joe 1914–1981 originally *Joseph Louis Barrow* American boxer
Louis Napoléon — see NAPOLEON III
Louis Phi·lippe \fi-'lēp\ 1773–1850 *the Citizen King* king of the French (1830–48)
Low \'lō\ Juliette 1860–1927 née *Gordon* American founder of the Girl Scouts
Low·ell \'lō-əl\ Amy 1874–1925 American poet and critic
Lowell James Russell 1819–1891 American author
Loyola — see IGNATIUS
Lu·cre·tius \lü-'krē-shē-əs, -shəs\ *about* 96–*about* 55 B.C. *Titus Lucretius Carus* Roman poet and philosopher
Luke \'lük\ physician and companion of the apostle Paul and the traditional author of the 3rd Gospel and the Book of Acts
Lu·ther \'lü-thər\ Martin 1483–1546 German Reformation leader
Ly·on \'lī-ən\ Mary 1797–1849 American educator
Mac·Ar·thur \mə-'kär-thər\ Douglas 1880–1964 American general
Ma·cau·lay \mə-'kó-lē\ Thomas Babington 1800–1859 1st Baron *Macaulay* English historian, author, and statesman
Ma·chi·a·vel·li \,mak-ē-ə-'vel-ē\ Niccolò 1469–1527 Italian political philosopher
Mad·i·son \'mad-ə-sən\ James 1751–1836 4th president of the U.S. (1809–17)
Mae·ter·linck \'māt-ər-,lingk *also* 'met-, 'mat-\ Count Maurice 1862–1949 Belgian author; Nobel Prize winner (1911)
Ma·gel·lan \mə-'jel-ən\ Ferdinand *about* 1480–1521 Portuguese navigator and explorer
Mah·ler \'mäl-ər\ Gustav 1860–1911 Austrian composer
Ma·jor \'mā-jər\ John 1943– prime minister of Great Britain (1990–97)
Malcolm X \,mal-kə-'meks\ 1925–1965 originally *Malcolm Little* American civil rights leader
Mal·o·ry \'mal-rē, -ə-rē\ Sir Thomas *flourished about* 1470 English author
Mal·thus \'mal-thəs\ Thomas Robert 1766–1834 English economist and demographer

Man·dela \man-ˈdel-ə\ Nelson Rolihlahla 1918– South African black political leader; president of South Africa (1994–99); Nobel Peace Prize winner (1993)
Ma·net \ma-ˈnā, mä-\ Édouard 1832–1883 French painter
Mann \ˈman\ Horace 1796–1859 American educator
Mann \ˈmän, ˈman\ Thomas 1875–1955 American (German-born) author; Nobel Prize winner (1929)
Mao Zedong *or* **Mao Tse–tung** \ˌmau̇d-zə-ˈdu̇ŋ, ˌmau̇-zə-, ˌmau̇t-sə-\ 1893–1976 Chinese Communist leader of the People's Republic of China (1949–76)
Ma·rat \mə-ˈrä\ Jean-Paul 1743–1793 French (Swiss-born) revolutionary
Mar·co·ni \mär-ˈkō-nē\ Guglielmo 1874–1937 Italian physicist and inventor; Nobel Prize winner (1909)
Marco Polo — see POLO
Mar·cus Au·re·lius \ˈmär-kəs-ȯ-ˈrēl-yəs\ 121–180 A.D. *Marcus Aurelius An·to·ni·nus* \ˌan-tə-ˈnī-nəs\ Roman emperor (161–80) and philosopher
Ma·ria The·re·sa \mə-ˌrē-ə-tə-ˈrā-sə, -ˈrā-zə\ 1717–1780 wife of Holy Roman Emperor Francis I; queen of Hungary and Bohemia (1740–80)
Ma·rie An·toi·nette \mə-ˈrē-ˌan-twə-ˈnet, -tə-ˈnet\ 1755–1793 wife of Louis XVI of France; daughter of the preceding
Mark \ˈmärk\ evangelist believed to be the author of the 2nd Gospel
Mark Antony — see ANTONY
Mar·lowe \ˈmär-ˌlō\ Christopher 1564–1593 English playwright
Mar·quette \mär-ˈket\ Jacques 1637–1675 *Père* \ˌpiər, ˌpeər\ *Marquette* Jesuit missionary and French explorer in America
Mars \ˈmärz\ god of war in Roman mythology — compare ARES
Mar·shall \ˈmär-shəl\ George Catlett 1880–1959 American general and diplomat; Nobel Prize winner (1953)
Marshall John 1755–1835 American jurist; chief justice of the U.S. Supreme Court (1801–35)
Marshall Thurgood 1908–1993 American jurist
Martel Charles — see CHARLES MARTEL
Mar·tha \ˈmär-thə\ sister of Lazarus and Mary and friend of Jesus in the Bible
Mar·tial \ˈmär-shəl\ *about* 40–*about* 103 A.D. Roman poet and epigrammatist
Mar·tin \ˈmärt-n, mär-ˈtaⁿ\ Saint 316–397 A.D. *Martin of Tours* \-ˈtu̇r\ patron saint of France
Martin Paul 1938– prime minister of Canada (2003–2006)
Marx \ˈmärks\ Karl Heinrich 1818–1883 German political philosopher and socialist
Mary \ˈmeər-ē, ˈmaər-ē, ˈmā-rē\ *Saint Mary; Virgin Mary* mother of Jesus
Mary sister of Lazarus and Martha in the Bible
Mary I 1516–1558 *Mary Tudor; Bloody Mary* queen of England (1553–58)
Mary II 1662–1694 joint British sovereign with William III (1689–94)
Mary Mag·da·lene \-ˈmag-də-ˌlēn, -ˌmag-də-ˈlē-nē\ woman in the Bible who is healed of evil spirits by Jesus and who later sees the risen Christ
Mary, Queen of Scots 1542–1587 *Mary Stuart* queen of Scotland (1542–67)
Ma·sca·gni \mä-ˈskän-yē, ma-\ Pietro 1863–1945 Italian composer
Mase·field \ˈmās-ˌfēld\ John 1878–1967 English author
Mason \ˈmās-n\ George 1725–1792 American Revolutionary statesman
Mas·sa·soit \ˌmas-ə-ˈsȯit\ *died* 1661 American Indian chief
Mas·se·net \ˌmas-n-ˈā, ma-ˈsnā\ Jules (-Émile-Frédéric) 1842–1912 French composer
Math·er \ˈmath-ər, ˈmath-\ Cotton 1663–1728 American clergyman and author
Mather Increase 1639–1723 American clergyman and author; father of Cotton Mather
Ma·tisse \ma-ˈtēs, mə-\ Henri 1869–1954 French painter
Mat·thew \ˈmath-yü\ one of the 12 apostles and the traditional author of the 1st Gospel
Maugham \ˈmȯm\ (William) Somerset 1874–1965 English author
Mau·pas·sant \ˌmō-pə-ˈsäⁿ\ (Henri-René-Albert-) Guy de 1850–1893 French author
Max·i·mil·ian \ˌmak-sə-ˈmil-yən\ 1832–1867 emperor of Mexico (1864–67); brother of Francis Joseph I of Austria
Maximilian I 1459–1519 Holy Roman emperor (1493–1519)
Maximilian II 1527–1576 Holy Roman emperor (1564–76)
Max·well \ˈmak-ˌswel, -swəl\ James Clerk \ˈklärk\ 1831–1879 Scottish physicist
Mayo \ˈmā-ō\ William James 1861–1939 and his brother Charles Horace 1865–1939 American surgeons
Mays \ˈmāz\ Willie Howard 1931– American baseball player
Ma·za·rin \ˌmaz-ə-ˈraⁿ\ Jules 1602–1661 French cardinal and statesman
Maz·zi·ni \mät-ˈsē-nē, mäd-ˈzē-\ Giuseppe 1805–1872 Italian patriot and revolutionary
Mc·Car·thy \mə-ˈkär-thē\ Joseph Raymond 1908–1957 American politician
Mc·Cart·ney \mə-ˈkärt-nē\ (James) Paul 1942– *Sir Paul McCartney* British singer and songwriter
Mc·Clel·lan \mə-ˈklel-ən\ George Brinton 1826–1885 American general
Mc·Clin·tock \mə-ˈklin-tək\ Barbara 1902–1992 American botanist; Nobel Prize winner (1983)
Mc·Cor·mick \mə-ˈkȯr-mik\ Cyrus Hall 1809–1884 American inventor of a mechanical reaper
Mc·Cul·lers \mə-ˈkəl-ərz\ Carson 1917–1967 née *Smith* American writer
Mc·Kin·ley \mə-ˈkin-lē\ William 1843–1901 25th president of the U.S. (1897–1901)
Mead \ˈmēd\ Margaret 1901–1978 American anthropologist
Meade \ˈmēd\ George Gordon 1815–1872 American Civil War general
Mea·ny \ˈmē-nē\ George 1894–1980 American labor leader
Me·dea \mə-ˈdē-ə\ enchantress in Greek mythology who helps Jason to win the Golden Fleece and kills her children when he deserts her
Medici Catherine de — see CATHERINE DE MEDICIS
Medici Lorenzo de' 1449–1492 *Lorenzo the Magnificent* Florentine ruler, statesman, and patron of the arts
Me·du·sa \mi-ˈdü-sə, -ˈdyü-, -zə\ Gorgon in Greek mythology slain by Perseus
Mel·pom·e·ne \mel-ˈpäm-ə-nē\ Muse of tragedy in Greek mythology
Mel·ville \ˈmel-ˌvil\ Herman 1819–1891 American author
Men·del \ˈmen-dl\ Gregor Johann 1822–1884 Austrian botanist
Men·de·le·yev \ˌmen-də-ˈlā-əf\ Dmitry Ivanovich 1834–1907 Russian chemist
Men·dels·sohn (–Bar·thol·dy) \ˈmen-dl-sən(-bär-ˈtȯl-dē, -ˈthȯl-)\ (Jakob Ludwig) Felix 1809–1847 German composer
Men·e·la·us \ˌmen-l-ˈā-əs\ king of Sparta, brother of Agamemnon, and husband of Helen of Troy in Greek mythology
Meph·is·toph·e·les \ˌmef-ə-ˈstäf-ə-ˌlēz\ chief devil in the Faust legend
Mer·ca·tor \ˌmər-ˈkāt-ər\ Gerhardus 1512–1594 originally *Gerhard Kremer* Flemish cartographer
Mer·cu·ry \ˈmər-kyə-rē, -kə-rē, -krē\ god of commerce, eloquence, travel, and theft who serves as herald and messenger of the other gods in Roman mythology — compare HERMES
Mer·e·dith \ˈmer-əd-əth\ George 1828–1909 English author
Mer·lin \ˈmər-lən\ prophet and magician in Arthurian legend
Met·a·com \ˈmet-ə-ˌkäm\ *or* **King Philip** *about* 1638–1676 *Met·a·com·et* \ˌmet-ə-ˈkäm-ət\ American Indian chief; son of Massasoit
Me·tho·di·us \mə-ˈthōd-ē-əs\ Saint *about* 825–884 A.D. Apostle to the Slavs; brother of Cyril
Me·thu·se·lah \mə-ˈthüz-lə, -ˈthyüz-, -ə-lə\ ancestor of Noah who lived 969 years according to biblical tradition
Met·ter·nich \ˈmet-ər-nik, -nik\ Klemens (Wenzel Nepomuk Lothar) 1773–1859 Fürst (Prince) *von Metternich* Austrian statesman
Mey·er·beer \ˈmī-ər-ˌbiər, -ˌbeər\ Giacomo 1791–1864 originally *Jakob Liebmann Meyer Beer* German composer
Mfume \əm-ˈfü-mā\ Kweisi 1948– originally *Frizell Gray* American civil rights leader, politician, and head of the NAACP (1996–)
Mi·cah \ˈmī-kə\ Hebrew prophet of the 8th century B.C.
Mi·chael \ˈmī-kəl\ one of the four archangels named in Hebrew tradition — compare GABRIEL, RAPHAEL, URIEL
Mi·chel·an·ge·lo \ˌmī-kə-ˈlan-jə-ˌlō, ˌmik-ə-ˈlan-, ˌmē-kə-ˈlän-\ 1475–1564 *Michelangelo di Lodovico Buonarroti Simoni* Italian sculptor, painter, architect, and poet
Mich·e·ner \ˈmich-nər, -ə-nər\ James Albert 1907–1997 American author
Mi·das \ˈmīd-əs\ legendary king of Phrygia having the power to turn everything he touched into gold
Mill \ˈmil\ John Stuart 1806–1873 British philosopher and economist
Mil·lay \mil-ˈā\ Edna St. Vincent 1892–1950 American poet
Mil·ler \ˈmil-ər\ Arthur 1915–2005 American playwright
Mil·let \mē-ˈyā, mi-ˈlā\ Jean-François 1814–1875 French painter
Milne \ˈmiln, ˈmil\ A. A. 1882–1956 *Alan Alexander Milne* English author
Mil·ti·a·des \mil-ˈtī-ə-ˌdēz\ *about* 544–489? B.C. *Miltiades the Younger* Athenian general
Mil·ton \ˈmilt-n\ John 1608–1674 English poet
Mi·ner·va \mə-ˈnər-və\ goddess of wisdom in Roman mythology — compare ATHENE
Mi·nos \ˈmī-nəs\ just king of Crete in Greek mythology who upon his death is made supreme judge of the underworld
Min·o·taur \ˈmin-ə-ˌtȯr, ˈmī-nə-\ monster in Greek mythology shaped half like a man and half like a bull
Min·u·it \ˈmin-yə-wət\ Peter *about* 1580–1638 Dutch colonial administrator in America
Mi·ró \mē-ˈrō\ Joan \zhü-ˈän\ 1893–1983 Spanish painter
Mitch·ell \ˈmich-əl\ Margaret 1900–1949 American author
Mitchell Maria 1818–1889 American astronomer
Mith·ras \ˈmith-rəs\ ancient Persian god of light who was the savior hero of an Iranian mystery cult for men flourishing in the late Roman Empire
Mit·ter·rand \ˌmē-ter-ˈäⁿ\ François (-Maurice) 1916–1996 president of France (1981–95)
Mne·mos·y·ne \ni-ˈmäs-n-ē\ goddess of memory and mother of the Muses by Zeus in Greek mythology
Mo·dred \ˈmō-drəd, ˈmäd-rəd\ knight of the Round Table and rebellious nephew of King Arthur in Arthurian legend
Mohammed — see MUHAMMAD
Mo·lière \mōl-ˈyeər, ˈmōl-ˌ\ 1622–1673 pseudonym of *Jean-Baptiste Poque·lin* \pō-ˈklaⁿ, -kə-ˈlaⁿ\ French actor and playwright
Mol·och \ˈmäl-ək, ˈmō-ˌläk\ *or* **Mol·ech** \ˈmäl-ək, ˈmō-ˌlek\ Semitic deity to whom children were sacrificed
Mo·lo·tov \ˈmäl-ə-ˌtȯf, ˈmȯl-, ˈmōl-, -ˌtȯv\ Vyacheslav Mikhaylovich 1890–1986 Soviet statesman
Mo·net \mō-ˈnā\ Claude 1840–1926 French painter
Mon·roe \mən-ˈrō\ James 1758–1831 5th president of the U.S. (1817–25)
Mon·taigne \män-ˈtān, mōⁿ-ˈtenʸ\ Michel (Eyquem) de 1533–1592 French essayist
Mont·calm (de Saint–Véran) \mänt-ˈkäm(-də-ˌsaⁿ-vā-ˈräⁿ), -ˈkälm-\ Marquis de 1712–1759 *Louis-Joseph de Montcalm-Grozon* French field marshal in Canada
Mon·tes·quieu \ˌmänt-əs-ˈkyü, -ˈkyər, -ˈkyœ\ Baron *de La Brède et de* 1689–1755 *Charles-Louis de Secondat* French political philosopher
Mon·tes·so·ri \ˌmänt-ə-ˈsōr-ē, -ˈsȯr-\ Maria 1870–1952 Italian physician and educator
Mon·te·ver·di \ˌmänt-ə-ˈveərd-ē, -ˈvərd-\ Claudio Giovanni Antonio 1567–1643 Italian composer
Mon·te·zu·ma II \ˌmänt-ə-ˈzü-mə\ 1466–1520 last Aztec emperor of Mexico (1502–20)
Moore \ˈmōr, ˈmȯr, ˈmu̇r\ Marianne 1887–1972 American poet
Moore Thomas 1779–1852 Irish poet
Mor·de·cai \ˈmȯrd-i-ˌkī\ cousin of Esther in the Bible who saves the Jews from the destruction planned by Haman
More \ˈmōr, ˈmȯr\ Sir Thomas 1478–1535 *Saint Thomas More* English statesman and author

\ə\ **abut** \ᵊ\ **kitten** \ər\ **further** \a\ **ash** \ā\ **ace** \ä\ **mop, mar** \au̇\ **out** \ch\ **chin** \e\ **bet** \ē\ **easy** \g\ **go** \i\ **hit** \ī\ **ice** \j\ **job** \ŋ\ **sing** \ō\ **go** \ȯ\ **law** \ȯi\ **boy** \th\ **thin** \th\ **the** \ü\ **loot** \u̇\ **foot** \y\ **yet** \zh\ **vision** *see also* Guide to Pronunciation

Mor·gan \'mȯr-gən\ Sir Henry 1635–1688 English buccaneer

Morgan J. P. 1837–1913 *John Pierpont Morgan* American financier

Mor·pheus \'mȯr-fē-əs, -ˌfyüs, -ˌfüs\ god of dreams in Greek mythology

Mor·ris \'mȯr-əs, 'mär-\ William 1834–1896 English poet, artist, and socialist

Mor·ri·son \'mȯr-ə-sən\ Toni 1931– originally *Chloe Anthony Wofford* American author; Nobel Prize winner (1993)

Morse \'mȯrs\ Samuel Finley Breese 1791–1872 American artist and inventor of the electrical telegraph

Mo·ses \'mō-zəz *also* -zəs\ Hebrew prophet and lawgiver who in the Bible is the liberator of the Israelites from Egypt

Moses Grandma 1860–1961 *Anna Mary Moses* née *Robertson* American painter

Mo·zart \'mōt-ˌsärt\ Wolfgang Amadeus 1756–1791 Austrian composer

Mu·bar·ak \mu̇-'bär-ək\ (Muhammad) Hosni 1929– president of Egypt (1981–)

Mu·ham·mad \mō-'ham-əd, -häm- *also* mü-\ *or* **Mo·ham·med** \mō-'häm-əd\ *about* 570–632 A.D. Arab prophet and founder of Islam

Mu·ham·mad \mu̇-'häm-əd, -'ham-\ Elijah 1897–1975 originally *Elijah Poole* American religious leader

Muir \'myu̇r\ John 1838–1914 American (Scottish-born) naturalist

Mul·ro·ney \məl-'rü-nē\ (Martin) Brian 1939– prime minister of Canada (1984–93)

Mus·so·li·ni \ˌmü-sə-'lē-nē, ˌmu̇s-ə-\ Be·ni·to \bə-'nēt-ō\ 1883–1945 *Il Du·ce* \ēl-'dü-chā\ Italian Fascist premier (1922–43)

My·ron \'mī-rən\ *flourished about* 480–440 B.C. Greek sculptor

Na·bo·kov \nə-'bȯ-kəf, -ˌkȯf\ Vladimir Vladimirovich 1899–1977 American (Russian-born) author

Na·hum \'nā-əm, -həm\ Hebrew prophet of the 7th century B.C.

Na·o·mi \nā-'ō-mē\ mother-in-law of the biblical heroine Ruth

Na·pier \'nā-pē-ər, -ˌpiər; nə-'piər\ John 1550–1617 Scottish mathematician

Na·po·léon I \nə-'pōl-yən, -'pō-lē-ən\ *or* **Napo·léon Bo·na·parte** \'bō-nə-ˌpärt\ 1769–1821 French general and emperor of the French (1804–15) — **Na·po·le·on·ic** \nə-ˌpō-lē-'än-ik\ *adj*

Napoléon III 1808–1873 *Louis-Napoléon* emperor of the French (1852–70); son of Louis Bonaparte and nephew of Napoléon I

Nar·cis·sus \när-'sis-əs\ beautiful youth in Greek mythology who pines away for love of his own reflection and is then transformed into the narcissus flower

Nash \'nash\ Ogden 1902–1971 American poet

Nas·ser \'näs-ər, 'nas-\ Ga·mal \gə-'mäl\ Ab·del \'ab-dl\ 1918–1970 Egyptian politician; president of Egypt (1956–70)

Na·tion \'nā-shən\ Car·ry \'kar-ē\ Amelia 1846–1911 née *Moore* American temperance agitator

Nav·ra·ti·lo·va \ˌnav-rət-ə-'lō-və\ Martina 1956– American (Czech-born) tennis player

Neb·u·cha·drez·zar II \ˌneb-yə-kə-'drez-ər, ˌneb-ə-kə-\ *or* **Neb·u·chad·nez·zar** \-kəd-'nez-\ *about* 630–*about* 561 B.C. Chaldean king of Babylon (605–562)

Ne·he·mi·ah \ˌnē-ə-'mī-ə, ˌnē-hə-\ Hebrew leader of the 5th century B.C.

Neh·ru \'neər-ˌü, 'nā-ˌrü\ Ja·wa·har·lal \jə-'wä-hər-ˌläl\ 1889–1964 Indian nationalist; 1st prime minister of the Republic of India (1947–64)

Nel·son \'nel-sən\ Horatio 1758–1805 Viscount *Nelson* British admiral

Nem·e·sis \'nem-ə-səs\ goddess of reward and punishment in Greek mythology

Nep·tune \'nep-ˌtün, -ˌtyün\ god of the sea in Roman mythology — compare POSEIDON

Ne·ro \'nē-ˌrō, 'niər-ˌō\ 37–68 A.D. Roman emperor (54–68)

Nes·tor \'nes-tər\ wise old counselor of the Greeks during the Trojan War in Greek mythology

Nev·el·son \'nev-əl-sən\ Louise *about* 1900–1988 originally *Leah Berliavsky* American (Russian-born) sculptor

New·man \'nü-mən, 'nyü-\ John Henry 1801–1890 English cardinal and author

New·ton \'nüt-n, 'nyüt-\ Sir Isaac 1642–1727 English mathematician and physicist

Nich·o·las \'nik-ləs, -ə-ləs\ Saint *flourished* 4th century A.D. Christian bishop

Nicholas I 1796–1855 czar of Russia (1825–55)

Nicholas II 1868–1918 czar of Russia (1894–1917)

Nick·laus \'nik-ləs\ Jack William 1940– American golfer

Nietz·sche \'nē-chə, -chē\ Friedrich Wilhelm 1844–1900 German philosopher

Night·in·gale \'nīt-n-ˌgāl, -ing-\ Florence 1820–1910 *Lady of the Lamp* English nurse and philanthropist

Ni·jin·sky \nə-'zhin-skē, -'jin-\ Vas·lav \'vät-släf\ Fomich 1890–1950 Russian dancer

Ni·ke \'nī-kē\ goddess of victory in Greek mythology

Nim·itz \'nim-əts\ Chester William 1885–1966 American admiral

Nim·rod \'nim-ˌräd\ ruler in the Bible renowned as a mighty hunter

Ni·o·be \'nī-ə-bē, nī-'ō-bē\ bereaved mother in Greek mythology who while weeping for her slain children is turned into a stone from which her tears continue to flow

Nix·on \'nik-sən\ Richard Mil·hous \'mil-ˌhau̇s\ 1913–1994 37th president of the U.S. (1969–74)

No·ah \'nō-ə\ biblical builder of the ark in which he, his family, and living creatures of every kind survive the Flood

No·bel \nō-'bel\ Alfred Bernhard 1833–1896 Swedish manufacturer, inventor, and philanthropist

Nor·man \'nȯr-mən\ Jessye 1945– American soprano

Nos·tra·da·mus \ˌnäs-trə-'dā-məs, ˌnōs-trə-'däm-əs\ 1503–1566 *Michel de Notredame* French physician, astrologer, and seer

Nu·re·yev \nu̇-'rā-yəf\ Rudolf Hametovich 1938–1993 Russian dancer

Oak·ley \'ōk-lē\ Annie 1860–1926 originally *Phoebe Anne Oakley Moses* American sharpshooter

Oba·di·ah \ˌō-bə-'dī-ə\ Hebrew prophet of Old Testament times

Ober·on \'ō-bə-ˌrän, -rən\ king of the fairies in medieval folklore and in Shakespeare's *A Midsummer Night's Dream*

O'·Ca·sey \ō-'kā-sē\ Sean 1880–1964 Irish playwright

Oce·anus \ō-'sē-ə-nəs\ god of the great outer sea that in Greek mythology encircles the earth

O'·Con·nor \ō-'kän-ər\ (Mary) Flannery 1925–1964 American author

O'Connor Sandra Day 1930– American jurist

Octavian — see AUGUSTUS

Odin \'ōd-n\ *or* **Wo·den** \'wōd-n\ chief god, god of war, and patron of heroes in Norse mythology

Odys·seus \ō-'dish-ˌüs, -'dis-ˌyüs, -'dis-ē-əs\ *or* **Ulys·ses** \yü-'lis-ēz\ king of Ithaca in Greek mythology who after the Greek victory in the Trojan War wanders for 10 years before reaching home

Oe·di·pus \'ed-ə-pəs, 'ēd-\ son of the king and queen of Thebes who in Greek mythology unknowingly kills his father and marries his mother as foretold by an oracle

Of·fen·bach \'ȯf-ən-ˌbäk, -ˌbäk\ Jacques 1819–1880 French composer

Ogle·thorpe \'ō-gəl-ˌthȯrp\ James Edward 1696–1785 English general and founder of Georgia

O'·Keeffe \ō-'kēf\ Georgia 1887–1986 American painter

Olaf I Trygg·va·son \'ō-ləf . . .'trig-və-sən\ *about* 964–*about* 1000 king of Norway (995–*about* 1000)

Olaf II Har·alds·son \'har-əld-sən\ *about* 995–1030 *Saint Olaf* king of Norway (1016–28)

Olav V \'ō-ləf\ 1903–1991 king of Norway (1957–91)

Oliv·i·er \ō-'liv-ē-ˌā\ Laurence Kerr 1907–1989 Baron *Olivier of Brighton* British actor and director

Omar Khay·yám \ˌō-ˌmär-ˌkī-'äm, -'yäm, -'am, -'yam\ 1048–?1131 Persian poet and astronomer

O'·Neill \ō-'nēl\ Eugene Gladstone 1888–1953 American playwright; Nobel Prize winner (1936)

Op·pen·hei·mer \'äp-ən-ˌhī-mər\ (Julius) Robert 1904–1967 American physicist

Ores·tes \ə-'res-tēz, ȯ-\ son of Agamemnon and Clytemnestra who in Greek mythology avenges his father's murder by slaying his mother and her lover

Or·pheus \'ȯr-ˌfyüs, -fē-əs\ poet and musician in Greek mythology who almost rescues his wife Eurydice from Hades by charming Pluto and Persephone with his lyre

Or·well \'ȯr-ˌwel, -wəl\ George 1903–1950 pseudonym of *Eric Arthur Blair* English author — **Or·well·ian** \ȯr-'wel-ē-ən\ *adj*

Osce·o·la \ˌäs-ē-'ō-lə, ˌō-sē-\ *about* 1804–1838 Seminole Indian chief

Osi·ris \ō-'sī-rəs\ god of the underworld in ancient Egyptian mythology

Otis \'ōt-əs\ James 1725–1783 American Revolutionary statesman

Ot·to I \'ät-ō\ 912–973 A.D. *Otto the Great* German king (936–73) and Holy Roman emperor (962–73)

Ov·id \'äv-əd\ 43 B.C.–17 A.D. Roman poet

Ow·en \'ō-ən\ Robert 1771–1858 Welsh social reformer

Ow·ens \'ō-ənz\ Jesse 1913–1980 originally *James Cleveland Owens* American track-and-field athlete

Paine \'pān\ Thomas 1737–1809 American (English-born) political philosopher and author

Pa·le·stri·na \ˌpal-ə-'strē-nə\ Giovanni Pierluigi da *about* 1525–1594 Italian composer

Pan \'pan\ god of pastures, flocks, and shepherds in Greek mythology who is usually represented as having the legs, ears, and horns of a goat

Pan·da·rus \'pan-də-rəs\ procurer of Cressida for Troilus during the Trojan War according to medieval legend

Pan·do·ra \pan-'dōr-ə, -'dȯr-\ curious woman in Greek mythology who opens a box and lets loose a swarm of evils upon mankind

Pank·hurst \'pangk-ˌhərst\ Emmeline 1858–1928 née *Goulden* English suffragist

Par·a·cel·sus \ˌpar-ə-'sel-səs\ 1493–1541 *Philippus Aureolus Theophrastus Bombastus von Hohenheim* Swiss-born alchemist and physician

Par·is \'par-əs\ son of Priam whose abduction of Helen of Troy leads to the Trojan War in Greek mythology

Park \'pärk\ Mungo 1771–1806 Scottish explorer

Park·man \'pärk-mən\ Francis 1823–1893 American historian

Parks \'pärks\ Rosa 1913–2005 née *McCauley* American civil rights activist

Par·nell \pär-'nel\ Charles Stewart 1846–1891 Irish nationalist

Pas·cal \pas-'kal, pás-kál\ Blaise 1623–1662 French mathematician and philosopher

Pas·ter·nak \'pas-tər-ˌnak\ Boris Leonidovich 1890–1960 Russian author; Nobel Prize winner (1958)

Pas·teur \pas-'tər\ Louis 1822–1895 French chemist and microbiologist

Pat·rick \'pa-trik\ Saint *flourished* 5th century A.D. apostle and patron saint of Ireland

Pa·tro·clus \pə-'trō-kləs, -'träk-ləs\ warrior in Greek mythology who is slain during the Trojan War by Hector and avenged by his friend Achilles

Pat·ton \'pat-n\ George Smith 1885–1945 American general

Paul \'pȯl\ Saint *died about* 67 A.D. Christian missionary and author of several New Testament epistles — **Paul·ine** \'pȯ-ˌlīn\ *adj*

Paul name of 6 popes: especially **III** 1468–1549 (pope 1534–49); **V** 1552–1621 (pope 1605–21); **VI** 1897–1978 (pope 1963–78)

Paul Bun·yan \'pȯl-'bən-yən\ giant lumberjack in American folklore

Pau·ling \'pȯ-ling\ Linus Carl 1901–1994 American chemist; Nobel Prize winner (1954, 1962)

Pav·lov \'päv-ˌlȯf, 'pav-, -ˌlȯv\ Ivan Petrovich 1849–1936 Russian physiologist; Nobel Prize winner (1904)

Pav·lo·va \'pav-lə-və, pav-'lō-və\ Anna 1881–1931 Russian ballerina

Peale \'pēl\ Charles Wilson 1741–1827 and his son Rembrandt 1778–1860 American painters

Pea·ry \'piər-ē\ Robert Edwin 1856–1920 American arctic explorer

Peg·a·sus \'peg-ə-səs\ winged horse in Greek mythology

Pei·sis·tra·tus *or* **Pi·sis·tra·tus** \pī-'sis-trət-əs, pə-\ *died* 527 B.C. tyrant of Athens

Pe·nel·o·pe \pə-'nel-ə-pē\ wife of Odysseus who in Greek mythology waits faithfully for him during his 20 years' absence

Penn \'pen\ William 1644–1718 English Quaker and founder of Pennsylvania

Pepys \'pēps\ Samuel 1633–1703 English diarist

Per·ce·val \'pər-sə-vəl\ knight in Arthurian legend who wins a sight of the Holy Grail

Per·i·cles \'per-ə-ˌklēz\ *about* 495–429 B.C. Athenian statesman

Per·ry \'per-ē\ Matthew Calbraith 1794–1858 American commodore

Perry Oliver Hazard 1785–1819 American naval officer; brother of the preceding

Per·seph·o·ne \pər-'sef-ə-nē\ daughter of Zeus and Demeter who in Greek mythology is abducted by Pluto and made his wife and queen

Per·seus \'pər-,süs, -sē-əs\ son of Zeus and Danaë and slayer of Medusa in Greek mythology

Per·shing \'pər-shing, -zhing\ John Joseph 1860–1948 American general

Pé·tain \pā-'ta^{n}\ (Henri-) Philippe 1856–1951 French general; marshal of France; premier of Vichy France (1940–44)

Pe·ter \'pēt-ər\ Saint *died about* 64 A.D. *Si·mon Peter* \'sī-mən-\ originally *Simon* one of the 12 apostles in the Bible

Peter I 1672–1725 *Peter the Great* czar of Russia (1682–1725)

Peter the Hermit *about* 1050–1115 French preacher of the First Crusade

Pe·trarch \'pē-,trärk, 'pe-\ 1304–1374 Italian *Francesco Pe·trar·ca* \pā-'trär-kə\ Italian poet — **Pe·trarch·an** \pē-'trär-kən, pe-\ *adj*

Phae·dra \'fē-drə\ wife of Theseus in Greek mythology who falls in love with her stepson Hippolytus

Pha·ë·thon \'fā-ət-n; 'fā-ə-tən, -,thän\ youth in Greek mythology who is allowed by his father Helios to drive the sun-chariot across the sky but loses control and is struck down with a thunderbolt by Zeus

Phid·i·as \'fid-ē-əs\ *about* 490–430 B.C. Greek sculptor

Phil·ip \'fil-əp\ Saint one of the 12 apostles in the Bible

Philip King — see METACOM

Philip name of 6 kings of France: especially **II** *or* **Philip Augustus** 1165–1223 (reigned 1179–1223); **IV** 1268–1314 (reigned 1285–1314) *Philip the Fair;* **VI** 1293–1350 (reigned 1328–50)

Philip name of 5 kings of Spain: especially **II** 1527–1598 (reigned 1556–98); **V** 1683–1746 (reigned 1700–24; 1724–46)

Philip II 382–336 B.C. king of Macedonia (359–336); father of Alexander the Great

Philip Prince 1921– Duke of *Edinburgh* consort of Elizabeth II of the United Kingdom

Phoebus — see APOLLO

Pi·cas·so \pi-'käs-ō, -'kas-\ Pablo 1881–1973 Spanish painter and sculptor in France

Pick·ett \'pik-ət\ George Edward 1825–1875 American Confederate general

Pierce \'piərs\ Franklin 1804–1869 14th president of the U.S. (1853–57)

Pilate Pontius *died after* 36 A.D. Roman prefect of Judea (26–36)

Pin·dar \'pin-dər, -,där\ *about* 522–*about* 438 B.C. Greek poet

Pi·ran·del·lo \,pir-ən-'del-ō\ Luigi 1867–1936 Italian author; Nobel Prize winner (1934)

Pitt \'pit\ William 1708–1778 Earl of *Chatham; the Elder Pitt* British statesman

Pitt William 1759–1806 *the Younger Pitt* British prime minister (1783–1801; 1804–6); son of the preceding

Pi·us \'pī-əs\ name of 12 popes: especially **VII** 1742–1823 (pope 1800–23); **IX** 1792–1878 (pope 1846–78); **X** Saint 1835–1914 (pope 1903–14); **XI** 1857–1939 (pope 1922–39); **XII** 1876–1958 (pope 1939–58)

Pi·zar·ro \pə-'zär-ō\ Francisco *about* 1475–1541 Spanish conqueror of Peru

Planck \'plängk\ Max (Karl Ernst Ludwig) 1858–1947 German physicist; Nobel Prize winner (1918)

Pla·to \'plāt-ō\ *about* 428–348 (or 347) B.C. Greek philosopher

Plau·tus \'plȯt-əs\ *about* 254–184 B.C. Roman playwright

Pliny the Elder \'plin-ē\ 23–79 A.D. Roman scholar

Pliny the Younger 61 (or 62)–*about* 113 A.D. Roman author; nephew of the preceding

Plu·tarch \'plü-,tärk\ *about* 46–*after* 119 A.D. Greek biographer

Plu·to \'plüt-ō\ god of the underworld in Greek mythology — compare DIS

Po·ca·hon·tas \,pō-kə-'hänt-əs\ *about* 1595–1617 American Indian friend to the colonists at Jamestown; daughter of Powhatan

Poe \'pō\ Edgar Allan 1809–1849 American author

Polk \'pōk\ James Knox 1795–1849 11th president of the U.S. (1845–49)

Pol·lock \'päl-ək\ Jackson 1912–1956 American painter

Pol·lux \'päl-əks\ immortal twin of Castor in classical mythology

Po·lo \'pō-lō\ Mar·co \'mär-kō\ *about* 1254–1324 Venetian merchant and traveler

Pol·y·hym·nia \,päl-i-'him-nē-ə\ Muse of sacred song in Greek mythology

Poly·phe·mus \,päl-ə-'fē-məs\ Cyclops in Greek mythology whom Odysseus blinds in order to escape from his cave

Pom·pa·dour \'päm-pə-,dōr, -,dȯr, -,du̇r\ Madame de 1721–1764 *Jeanne-Antoinette Poisson* mistress of Louis XV of France

Pom·pey the Great \'päm-pē\ 106–48 B.C. Roman general and statesman

Ponce de Le·ón \,päns-də-'lē-ən, ,pän-sə-,dā-lē-'ōn \ Juan 1460–1521 Spanish explorer of Florida (1513)

Pon·ti·ac \'pänt-ē-,ak\ *about* 1720–1769 Ottawa Indian chief

Pon·tius Pi·late — see PILATE

Pope \'pōp\ Alexander 1688–1744 English poet

Por·ter \'pōrt-ər, 'pȯrt-\ Cole Albert 1891–1964 American composer and songwriter

Porter Katherine Anne 1890–1980 American author

Porter William Sydney — see O. HENRY

Po·sei·don \pə-'sīd-n\ god of the sea in Greek mythology — compare NEPTUNE

Pot·ter \'pät-ər\ (Helen) Beatrix 1866–1943 British author and illustrator

Pound \'pau̇nd\ Ezra Loomis 1885–1972 American poet

Pound·mak·er \'pau̇nd-,mā-kər\ 1826–1886 Cree Indian chief

Pow·ell \'pau̇-əl\ Adam Clayton, Jr. 1908–1972 American clergyman and politician

Powell Colin Luther 1937– American general; U.S. secretary of state (2001–2004)

Powell John Wesley 1834–1902 American geologist and explorer

Pow·ha·tan \,pau̇-ə-'tan, pau̇-'hat-n\ 1550?–1618 chief of a confederacy of Algonquian-speaking Indians; father of Pocahontas

Prax·it·e·les \prak-'sit-l-,ēz\ *flourished* 370–330 B.C. Athenian sculptor

Pres·ley \'pres-lē, 'prez-\ Elvis Aaron 1935–1977 American popular singer

Pri·am \'prī-əm, -,am\ king of Troy during the Trojan War and father of Hector and Paris in Greek mythology

Price \'prīs\ (Mary) Leontyne 1927– American soprano

Priest·ley \'prēst-lē\ Joseph 1733–1804 English clergyman and chemist

Pro·crus·tes \prə-'krəs-tēz, pə-, prō-\ robber in Greek mythology who forces travelers to fit one of two unequally long beds by stretching their bodies or cutting off their legs

Pro·kof·iev \prə-'kȯf-yəf, -,yef, -,yev\ Sergey Sergeyevich 1891–1953 Russian composer

Pro·me·theus \prə-'me-thyüs, -thüs, -thē-əs\ Titan in Greek mythology whom Zeus tortures for giving fire to humans

Pro·tag·o·ras \prō-'tag-ə-rəs\ *about* 485–410 B.C. Greek philosopher and teacher

Pro·teus \'prō-,tyüs, -,tüs; 'prōt-ē-əs\ sea god in Greek mythology capable of assuming different forms

Proust \'prüst\ Marcel 1871–1922 French novelist

Psy·che \'sī-kē\ beautiful princess in classical mythology loved by Cupid

Ptol·e·my \'täl-ə-mē\ name of 15 kings of Egypt 323–30 B.C.

Ptolemy *flourished* 2nd century A.D. Greco-Egyptian astronomer, geographer, and mathematician in Alexandria

Puc·ci·ni \pü-'chē-nē\ Giacomo 1858–1924 Italian composer

Puck — see ROBIN GOODFELLOW

Pu·łas·ki \pə-'las-kē, pyü-\ Kazimierz 1747–1779 Polish soldier in American Revolutionary army

Pu·litz·er \'pu̇l-ət-sər, 'pyü-lət-sər\ Joseph 1847–1911 American (Hungarian-born) journalist

Push·kin \'pu̇sh-kən\ Aleksandr Sergeyevich 1799–1837 Russian author

Pu·tin \'püt-in\ Vladimir Vladimirovich 1952– president of Russia (2000–)

Pyg·ma·lion \pig-'māl-yən, -'mā-lē-ən\ sculptor in Greek mythology who creates Galatea

Pyr·a·mus \'pir-ə-məs\ legendary Babylonian youth who dies for the love of Thisbe

Py·thag·o·ras \pə-'thag-ə-rəs, pī-\ *about* 580–*about* 500 B.C. Greek philosopher and mathematician

Pyth·i·as \'pith-ē-əs\ condemned Sicilian in Greek legend whose life is spared when his devoted friend Damon offers to take his place

Quin·til·ian \kwin-'til-yən\ *about* 35–*about* 100 A.D. Roman rhetorician

Ra \'rä, 'rȯ\ god of the sun and chief deity of ancient Egypt

Ra·be·lais \'rab-ə-,lā, ,rab-ə-'lā\ François *about* 1483–1553 French author

Ra·bin \rä-'bēn\ Yitzhak 1922–1995 prime minister of Israel (1974–77; 1992–95); Nobel Prize winner (1994)

Ra·chel \'rā-chəl\ one of the wives of Jacob in the Bible

Rach·ma·ni·noff \räk-'män-ə-,nȯf, rak-'man-, -,nȯv\ Sergey Vasilyevich 1873–1943 Russian composer and pianist

Ra·cine \ra-'sēn, rə-\ Jean (-Baptiste) 1639–1699 French playwright

Ra·leigh *or* **Ra·legh** \'rȯl-ē, 'räl- *also* 'ral-\ Sir Walter 1554?–1618 English navigator, courtier, and writer

Ra·ma \'räm-ə\ deity or deified hero of later Hinduism worshiped as an incarnation of Vishnu

Ram·say \'ram-zē\ Sir William 1852–1916 British chemist; Nobel Prize winner (1904)

Ram·ses \'ram-,sēz\ *or* **Ram·e·ses** \'ram-ə-,sēz\ name of 11 kings of Egypt: especially **II** (reigned 1279–1213 B.C.); **III** (reigned 1187–1156 B.C.)

Ran·dolph \'ran-,dȯlf\ Asa Philip 1889–1979 American labor and civil rights leader

Ra·pha·el \'raf-ē-əl, 'rā-fē-\ one of the four archangels named in Hebrew tradition — compare GABRIEL, MICHAEL, URIEL

Ra·pha·el \'raf-ē-əl, 'rā-fē-, 'räf-ē-\ 1483–1520 originally *Raffaello Sanzio* or *Santi* Italian painter

Ras·pu·tin \ra-'spyüt-n, 'spüt-, -'spu̇t-\ Grigory Yefimovich 1872–1916 Russian mystic

Ra·vel \rə-'vel, ra-\ (Joseph) Mau·rice \mȯ-'rēs\ 1875–1937 French composer

Rea·gan \'rā-gən, 'rē-\ Ronald Wilson 1911–2004 40th president of the U.S. (1981–89)

Re·bek·ah \ri-'bek-ə\ wife of Isaac and mother of Jacob in the Bible

Red Cloud \'red-,klau̇d\ 1822–1909 Sioux Indian chief

Red Jack·et \'red-,jak-ət\ 1758?–1830 *Sagoyewatha* Seneca Indian chief

Reed \'rēd\ Walter 1851–1902 American army surgeon

Rehn·quist \'ren-,kwist\ William Hubbs 1924–2005 American jurist; chief justice U.S. Supreme Court (1986–2005)

Re·marque \rə-'märk\ Erich Maria 1898–1970 American (German-born) author

Rem·brandt \'rem-,brant\ 1606–1669 *Rembrandt (Harmenszoon) van Rijn* Dutch painter

Rem·ing·ton \'rem-ing-tən\ Frederic 1861–1909 American painter and sculptor

Re·mus \'rē-məs\ legendary founder of Rome killed by his twin brother Romulus

Re·noir \ren-'wär, 'ren-,wär\ (Pierre-) Auguste 1841–1919 French painter

Re·vere \ri-'viər\ Paul 1735–1818 American patriot and silversmith

Reyn·olds \'ren-ldz, -lz\ Sir Joshua 1723–1792 English painter

Rhodes \'rōdz\ Cecil John 1853–1902 British administrator and financier in South Africa

Rice \'rīs\ Condoleezza 1954– U.S. Secretary of State (2005–)

Rich·ard \'rich-ərd\ name of 3 kings of England: **I** 1157–1199 *Richard the Lion-Hearted* (reigned 1189–99); **II** 1367–1400 (reigned 1377–99); **III** 1452–1485 (reigned 1483–85)

Rich·ard·son \'rich-ərd-sən\ Samuel 1689–1761 English author

Ri·che·lieu \'rish-əl-,ü, -,yü; rē-shə-'lyœ\ Duc de 1585–1642 originally *Armand-Jean du Plessis* French cardinal and statesman

Ride \'rīd\ Sally Kristen 1951– American astronaut; first American woman in space

Rim·sky–Kor·sa·kov \,rim-skē-'kȯr-sə-,kȯf, -,kȯv, -,kȯr-sə-'\ Nikolay Andreyevich 1844–1908 Russian composer

Ri·ve·ra \ri-'ver-ə\ Diego 1886–1957 Mexican painter

\ə\ abut \ᵊ\ kitten \ər\ further \a\ ash \ā\ ace
\ä\ mop, mar \au̇\ out \ch\ chin \e\ bet \ē\ easy
\g\ go \i\ hit \ī\ ice \j\ job \ŋ\ sing \ō\ go
\ȯ\ law \ȯi\ boy \th\ thin \th\ the \ü\ loot \u̇\ foot
\y\ yet \zh\ vision *see also* Guide to Pronunciation

Rob·ert I \'räb-ərt\ 1274–1329 *Robert the Bruce* \'brüs\ king of Scotland (1306–29)
Roberts \'räb-ərts\ John Glover, Jr. 1955– American jurist; chief justice U.S. Supreme Court (2005–)
Robe·son \'rōb-sən\ Paul Bustill 1898–1976 American actor and singer
Robes·pierre \'rōbz-,piər, -,pyeər; ,rō-,bes-'pyeər\ Maximilien (-François-Marie-Isidore) de 1758–1794 French revolutionary
Rob·in Good·fel·low \,räb-ən-'gu̇d-,fel-ō\ *or* **Puck** \'pək\ mischievous sprite in English folklore
Rob·in Hood \,räb-ən-'hu̇d\ legendary English outlaw noted for his skill in archery and for his robbing the rich to help the poor
Rob·in·son \'räb-ən-sən\ Edwin Arlington 1869–1935 American poet
Robinson Jackie 1919–1972 *Jack Roosevelt Robinson* American baseball player; 1st black player in the major leagues (1947–56)
Ro·cham·beau \,rō-,sham-'bō\ Comte de 1725–1807 originally *Jean-Baptiste-Donatien de Vimeur* French general in American Revolution
Rocke·fel·ler \'räk-i-,fel-ər, 'räk-,fel-\ John Davison 1839–1937 and his son John Davison, Jr. 1874–1960 American oil magnates and philanthropists
Rod·gers \'räj-ərz\ Richard 1902–1979 American musical theater composer
Ro·din \'rō-,daⁿ, -,daⁿn\ (François-) Auguste (-René) 1840–1917 French sculptor
Roeb·ling \'rōbling\ John Augustus 1806–1869 and his son Washington 1837–1926 American civil engineers and designers of the Brooklyn Bridge
Rog·ers \'räj-ərz\ Will 1879–1935 *William Penn Adair Rogers* American actor and humorist
Ro·land \'rō-lənd\ stalwart defender of the Christians against the Saracens in the Charlemagne legends
Röl·vaag \'rōl-,väg\ Ole \'ō-lə\ Ed·vart \'ed-,värt\ 1876–1931 American (Norwegian-born) educator and author
Ro·ma·nov \rō-'män-əf, 'rō-mə-,näf\ Michael 1596–1645 1st czar (1613–45) of Russian Romanov dynasty (1613–1917)
Rom·mel \'räm-əl\ Erwin 1891–1944 German field marshal
Rom·u·lus \'räm-yə-ləs\ legendary founder of Rome who killed his twin brother Remus
Rönt·gen *or* **Roent·gen** \'rent-gən, 'rənt-, -jən\ Wilhelm Conrad 1845–1923 German physicist; Nobel Prize winner (1901)
Roo·se·velt \'rō-zə-vəlt *(Roosevelts' usual pronunciation),* -,velt *also* 'rü-\ (Anna) Eleanor 1884–1962 American humanitarian and writer; wife of Franklin Delano Roosevelt
Roosevelt Franklin Del·a·no \'del-ə-,nō\ 1882–1945 32nd president of the U.S. (1933–45)
Roosevelt Theodore 1858–1919 26th president of the U.S. (1901–09); Nobel Prize winner (1906)
Root \'rüt, 'ru̇t\ Elihu 1845–1937 American lawyer and statesman; Nobel Prize winner (1912)
Ross \'rȯs\ Betsy 1752–1836 née *Griscom* reputed maker of 1st American flag
Ros·set·ti \rō-'zet-ē, -'set-\ Christina Georgina 1830–1894 English poet; sister of Dante Gabriel Rossetti
Rossetti Dante Gabriel 1828–1882 English painter and poet
Ros·si·ni \rō-'sē-nē, rə-\ Gioacchino Antonio 1792–1868 Italian composer
Ros·tand \rȯ-'stäⁿ, 'räs-,tand\ Edmond 1868–1918 French playwright
Roth·schild \'rȯths-,chīld, 'rȯth-, 'rȯs-; *German* 'rōt-,shilt\ Mayer Amschel 1744–1812 German financier
Rothschild Nathan Mayer 1777–1836 German financier in London; son of the preceding
Rous·seau \ru̇-'sō, 'rü-,\ Jean-Jacques 1712–1778 French (Swiss-born) philosopher and author
Row·ling \'rō-ling\ J. K. 1965– *Joanne Kathleen Rowling* British author
Ru·bens \'rü-bənz\ Peter Paul 1577–1640 Flemish painter
Ru·bin·stein \'rü-bən-,stīn\ An·ton \än-'tȯn\ Grigoryevich 1829–1894 Russian pianist and composer
Ru·dolph \'rü-,dälf\ Wilma Glodean 1940–1994 American athlete
Ru·pert \'rü-pərt\ Prince 1619–1682 English (German-born) royalist general and admiral
Rus·kin \'rəs-kən\ John 1819–1900 English art critic
Rus·sell \'rəs-əl\ Bertrand Arthur William 1872–1970 3rd Earl *Russell* English mathematician and philosopher; Nobel Prize winner (1950)
Russell Bill 1934– *William Felton Russell* American basketball player
Ruth \'rüth\ Moabite woman in the Bible who is an ancestor of King David
Ruth Babe 1895–1948 *George Herman Ruth* American baseball player
Ruth·er·ford \'rəth-ər-fərd, 'rəth-ə-, 'rəth-\ Ernest 1871–1937 Baron *Rutherford* British physicist; Nobel Prize winner (1908)
Sa·bin \'sā-bin\ Albert Bruce 1906–1993 American (Polish-born) physician and microbiologist
Sac·a·ga·wea \,sak-ə-jə-'wē-ə\ 1786?–1812 Shoshone Indian guide to Lewis and Clark
Sā·dāt, \sə-'dat, -'dät\ (Muhammad) Anwar el- 1918–1981 president of Egypt (1970–81); Nobel Prize winner (1978)
Sa·gan \'sā-gən\ Carl Edward 1934–1996 American astronomer and science writer
Saint–Gau·dens \sānt-'gȯd-nz, sənt-\ Augustus 1848–1907 American (Irish-born) sculptor
Saint–Saëns \saⁿ-'säⁿs\ (Charles-) Camille 1835–1921 French composer
Sal·a·din \'sal-əd-ən, ,sal-ə-'dēn\ 1137 (or 1138)–1193 sultan of Egypt and Syria
Sal·in·ger \'sal-ən-jər\ J. D. 1919– *Jerome David Salinger* American author
Salk \'sȯk, 'sȯlk\ Jonas Edward 1914–1995 American physician and medical researcher
Sa·lo·me \sə-'lō-mē\ niece of Herod Antipas who in the Bible is given the head of John the Baptist as a reward for her dancing
Sa·mo·set \'sam-ə-,set, sə-'mäs-ət\ *died about* 1653 American Indian leader and friend of the Pilgrims
Sam·son \'sam-sən, 'samp-\ powerful, long-haired Hebrew hero in the Bible who wreaks havoc among the Philistines but is betrayed by Delilah
Sam·u·el \'sam-yəl, -yə-wəl\ Hebrew judge in the Bible who anoints Saul and then David king
Sand \'sand, 'säⁿd, 'säⁿnd, 'säⁿ\ George 1804–1876 pseudonym of *Amandin-Aurore-Lucie Dudevant* née *Dupin* French author
Sand·burg \'sand-,bərg, 'san-\ Carl 1878–1967 American author
Sang·er \'sang-ər\ Margaret 1883–1966 née *Higgins* American birth-control leader
San·ta Claus \'sant-ə-,klȯz, -,klȯs\ plump white-bearded and red-suited old man of modern folklore who delivers presents to good children at Christmastime
Sap·pho \'saf-ō\ *flourished about* 610–*about* 580 B.C. Greek poet
Sa·rah \'ser-ə, 'sar-ə, 'sā-rə\ wife of Abraham and mother of Isaac in the Bible
Sar·gent \'sär-jənt\ John Singer 1856–1925 American painter
Sar·tre \'särtr\ Jean-Paul 1905–1980 French philosopher and author
Sat·urn \'sat-ərn\ god of agriculture in Roman mythology
Saul \'sȯl\ 1st king of Israel in the Bible
Saul *or* **Saul of Tarsus** the apostle Paul
Sa·vo·na·ro·la \,sav-ə-nə-'rō-lə, sə-,vän-ə-'rō-\ Gi·ro·la·mo \ji-'rȯl-ə-,mō\ 1452–1498 Italian friar and reformer
Scar·lat·ti \skär-'lät-ē\ (Pietro) Alessandro 1660–1725 and his son (Giuseppe) Domenico 1685–1757 Italian composers
Sche·her·a·zade \shə-,her-ə-'zäd, -'zäd-ə, -'zäd-ē\ fictional wife of a sultan and narrator of the tales in the *Arabian Nights' Entertainments*
Schil·ler \'shil-ər\ (Johann Christoph) Friedrich von 1759–1805 German poet and playwright
Schin·dler \'shind-lər\ Oskar 1908–1974 German humanitarian during the Holocaust
Schoen·berg \'shərn-,bərg, 'shœn-,berk\ Arnold Franz Walter 1874–1951 American (Austrian-born) composer
Scho·pen·hau·er \'shō-pən-,hau̇-ər, -,hau̇r\ Arthur 1788–1860 German philosopher
Schu·bert \'shü-bərt, -,bert\ Franz Peter 1797–1828 Austrian composer
Schu·mann \'shü-,män, -mən\ Robert Alexander 1810–1856 German composer
Schweit·zer \'shwīt-sər, 'swīt-, 'shvīt-\ Albert 1875–1965 French theologian, philosopher, physician, and music scholar; Nobel Prize winner (1952)
Scip·io Ae·mil·i·a·nus Af·ri·ca·nus Nu·man·ti·nus \'sip-ē-,ō-i-,mil-ē-'ā-nəs-,af-rə-'kan-əs-,nü-mən-'tē-nəs, -,nyü-\ Publius Cornelius 185 (or 184)–129 B.C. *Scipio Africanus the Younger* Roman general
Scipio Africanus Publius Cornelius 236–184 (or 183) B.C. *Scipio Africanus the Elder* Roman general; adopted grandson of the preceding
Scott \'skät\ Dred \'dred\ 1795?–1858 American slave
Scott Robert Falcon 1868–1912 British polar explorer
Scott Sir Walter 1771–1832 Scottish author
Scott Winfield 1786–1866 American general
Seaborg \'sē-,bȯrg\ Glenn Theodore 1912–1999 American chemist; Nobel Prize winner (1951)
Se·at·tle *or* **Se·atlh** \sē-'at-l\ 1786?–1866 American Indian chief
Se·le·ne \sə-'lē-nē, -nə\ goddess of the moon in classical mythology
Se·leu·cus I Ni·ca·tor \sə-'lü-kəs-nī-'kāt-ər\ 358 (*or* 354)–281 B.C. Macedonian general and ruler (306–281) of an empire centering on Syria and Iran; founder of the Seleucid dynasty
Sen·e·ca \'sen-i-kə\ Lucius Annaeus 4 B.C.?–65 A.D. Roman philosopher, statesman, and playwright
Sen·nach·er·ib \sə-'nak-ə-rəb\ *died* 681 B.C. king of Assyria (704–681)
Se·quoyah *or* **Se·quoia** \si-'kwȯi-ə\ *about* 1760–1843 *George Guess* Cherokee Indian scholar
Ser·ra \'ser-ə\ Junipero 1713–1784 Spanish missionary in Mexico and California
Se·ton \'sēt-n\ Saint Elizabeth Ann 1774–1821 *Mother Seton* née *Bayley* American religious leader
Seu·rat \sə-'rä\ Georges 1859–1891 French painter
Sew·ard \'sü-ərd, 'su̇-ərd, 'su̇rd\ William Henry 1801–1872 American statesman; U.S. secretary of state (1861–69)
Shack·le·ton \'shak-əl-tən\ Sir Ernest Henry 1874–1922 British polar explorer
Shake·speare \'shāk-,spiər\ William 1564–1616 English playwright and poet
Shaw \'shȯ\ George Bernard 1856–1950 British playwright; Nobel Prize winner (1925)
Shaw Robert Gould 1837–1863 American soldier
Shel·ley \'shel-ē\ Mary Woll·stone·craft \'wu̇l-stən-,kraft\ 1797–1851 née *Godwin* English novelist; wife of Percy Bysshe Shelley
Shelley Percy Bysshe \'bish\ 1792–1822 English poet
Shem \'shem\ eldest son of Noah and ancestor of the Semitic peoples in biblical tradition
Shep·ard \'shep-ərd\ Alan Bartlett, Jr. 1923–1998 American astronaut; 1st American in space (1961)
Sher·i·dan \'sher-əd-n\ Philip Henry 1831–1888 American general
Sheridan Richard Brins·ley \'brinz-lē\ 1751–1816 British playwright
Sher·man \'shər-mən\ John 1823–1900 American statesman; brother of William Tecumseh Sherman
Sherman William Tecumseh 1820–1891 American general
Shi·va \'shiv-ə, 'shē-və\ *or* **Si·va** \'shiv-ə, 'siv-; 'shē-və, 'sē-\ god of destruction and regeneration in the Hindu sacred triad — compare BRAHMA, VISHNU
Sho·sta·ko·vich \,shäs-tə-'kō-vich\ Dmitri Dmitrievich 1906–1975 Russian composer
Si·be·lius \sə-'bāl-yəs, -'bā-lē-əs\ Jean 1865–1957 Finnish composer
Sid·ney \'sid-nē\ Sir Philip 1554–1586 English poet
Sieg·fried \'sig-,frēd, 'sēg-\ *or* **Sig·urd** \'sig-u̇rd, -ərd\ hero in Germanic legend who slays a dragon guarding a gold hoard
Si·kor·sky \sə-'kȯr-skē\ Igor Ivan 1889–1972 American (Russian-born) aeronautical engineer
Si·mon \'sī-mən\ — see PETER
Simon *or* **Simon the Zealot** one of the 12 apostles in the Bible
Si·na·tra \sə-'nät-rə\ Frank 1915–1998 *Francis Albert Sinatra* American singer and actor
Sind·bad the Sailor \'sin-,bad-\ citizen of Baghdad whose adventures are narrated in the *Arabian Nights' Entertainments*
Sing·er \'siŋ-ər\ Isaac Bashevis 1904–1991 American (Polish-born) author
Sis·y·phus \'sis-i-fəs\ king of Corinth in Greek mythology condemned to roll a heavy stone up a hill in Hades only to have it roll down again as it nears the top — **Sis·y·phe·an** \,sis-i-'fē-ən\ *adj*
Sit·ting Bull \,sit-ing-'bu̇l\ *about* 1831–1890 Sioux Indian chief

Smith \'smith\ Adam 1723–1790 Scottish economist

Smith Bessie 1894?–1937 American blues singer

Smith John *about* 1580–1631 English colonist in America

Smith Joseph 1805–1844 American founder of the Mormon Church

Smol·lett \'smäl-ət\ Tobias George 1721–1771 British author

Snead \'snēd\ Sam 1912–2002 *Samuel Jackson Snead* American golfer

Soc·ra·tes \'säk-rə-ˌtēz\ *about* 470–399 B.C. Greek philosopher

Sol \'säl\ god of the sun in Roman mythology — compare HELIOS

Sol·o·mon \'säl-ə-mən\ son of David and 10th century B.C. king of Israel noted for his wisdom

Sol·zhe·ni·tsyn \ˌsōl-zhə-'nēt-sən\ Aleksandr Isayevich 1918– Russian author; Nobel Prize winner (1970)

Soph·o·cles \'säf-ə-ˌklēz\ *about* 496–406 B.C. Greek playwright

Sou·sa \'sü-zə, 'sü-sə\ John Philip 1854–1932 American bandmaster and composer

Spar·ta·cus \'spärt-ə-kəs\ *died* 71 B.C. Roman slave and gladiator; leader of a slave rebellion

Spen·ser \'spen-sər\ Edmund 1552 (*or* 1553)–1599 English poet

Spiel·berg \'spēl-bərg\ Steven 1947– American filmmaker

Spi·no·za \spin-'ō-zə\ Benedict de 1632–1677 *Baruch Spinoza* Dutch philosopher of Portuguese-Jewish ancestry

Squan·to \'skwän-tō, 'skwȯn-\ *died* 1622 American Indian friend of the Pilgrims

Sta·lin \'stäl-ən, 'stal-, -ˌēn\ Joseph 1879–1953 originally *Iosif Vissarionovich Dzhu·gash·vi·li* \ˌjü-gəsh-'vē-lē\ Soviet Communist Party leader (1922–53), premier (1941–53), and dictator

Stan·dish \'stan-dish\ Myles *or* Miles 1584?–1656 English colonist in America

Stan·ley \'stan-lē\ Sir Henry Morton 1841–1904 British explorer

Stan·ton \'stant-ᵊn\ Elizabeth Cady 1815–1902 née *Cady* American suffragist

Steele \'stēl\ Sir Richard 1672–1729 English author

Stein \'stīn\ Gertrude 1874–1946 American author

Stein·beck \'stīn-ˌbek\ John Ernst 1902–1968 American author; Nobel Prize winner (1962)

Sten·dhal \sten-'däl, stan-, *French* sta^{n}-'dȧl\ 1783–1842 pseudonym of *Marie-Henri Beyle* \'bel\ French author

Ste·phen \'stē-vən\ Saint *died about* 36 A.D. Christian martyr

Stephen *about* 1097–1154 *Stephen of Blois* king of England (1135–54)

Sterne \'stərn\ Laurence 1713–1768 English author

Steu·ben \'stü-bən, 'styü-, 'shtȯi-\ Friedrich Wilhelm 1730–1794 Prussian-born general in American Revolutionary army

Ste·ven·son \'stē-vən-sən\ Adlai Ewing 1900–1965 American politician

Stevenson Robert Louis Balfour 1850–1894 Scottish author

Sto·ker \'stō-kər\ Bram 1847–1912 *Abraham Stoker* Irish author

Stowe \'stō\ Harriet Beecher 1811–1896 née *Beecher* American author

Stra·di·va·ri \ˌstrad-ə-'vär-ē, -'var-, -'ver-\ Antonio 1644?–1737 Italian violin maker

Strauss \'straủs, 'shtraủs\ Johann 1804–1849 and his sons Johann, Jr. 1825–1899 and Josef 1827–1870 Austrian composers

Strauss Ri·chard \'rik-ˌärt, 'rik-\ 1864–1949 German composer

Stra·vin·sky \strə-'vin-skē\ Igor \'ē-ˌgȯr\ Fyodorovich 1882–1971 American (Russian-born) composer

Stu·art \'stü-ərt, 'styü-, 'styủ-, 'styủrt\ Charles Edward 1720–1788 *the Young Pretender; Bonnie Prince Charlie* claimant to the British throne; son of James Edward Stuart

Stuart Gilbert Charles 1755–1828 American painter

Stuart James Edward 1688–1766 *the Old Pretender* claimant to the British throne; son of James II

Stuart Jeb 1833–1864 *James Ewell Brown Stuart* American Confederate general

Stuy·ve·sant \'stī-və-sənt\ Peter *about* 1610–1672 Dutch colonial administrator in America

Sue·to·ni·us \swē-'tō-nē-əs, sü-ə-'tō-\ *about* 69–*after* 122 A.D. Roman biographer and historian

Sü·ley·man \'sü-lā-ˌmän, -li-\ 1494 (or 1495)–1566 *Süleyman the Magnificent* sultan of the Ottoman Empire (1520–66)

Sul·la \'səl-ə\ 138–78 B.C. Roman general and statesman

Sul·li·van \'səl-ə-vən\ Sir Arthur Seymour 1842–1900 English composer and collaborator with Sir William S. Gilbert

Sullivan Louis Henri 1856–1924 American architect

Sum·ner \'səm-nər\ Charles 1811–1874 American statesman

Sun Yat–sen \'sủn-'yät-'sen\ 1866–1925 Chinese statesman

Swift \'swift\ Jonathan 1667–1745 English (Irish-born) author

Swin·burne \'swin-ˌbərn, -bərn\ Algernon Charles 1837–1909 English poet

Tac·i·tus \'tas-ət-əs\ Cornelius *about* 56–*about* 120 A.D. Roman historian

Taft \'taft\ William Howard 1857–1930 27th president of the U.S. (1909–13); chief justice of the U.S. Supreme Court (1921–30)

Ta·gore \tə-'gȯr\ Ra·bin·dra·nath \rə-'bin-drə-ˌnät\ 1861–1941 Indian poet; Nobel Prize winner (1913)

Tall·chief \'tȯl-ˌchēf\ Maria 1925– American dancer

Tal·ley·rand (–Pé·ri·gord) \'tal-ē-ˌrand(-ˌper-ə-'gȯr), -ˌran-, *French* tȧl-e-'rä^{n}-, tȧl-'rä^{n}-\ Charles-Maurice de 1754–1838 French statesman

Tam·er·lane — see TIMUR

Tan \'tan\ Amy 1952– American author

Tan·cred \'tang-krəd\ 1078?–1112 Norman crusader

Ta·ney \'tȯ-nē\ Roger Brooke 1777–1864 American jurist; chief justice of the U.S. Supreme Court (1836–64)

Tan·ta·lus \'tant-ᵊl-əs\ king in Greek mythology condemned to stand up to his chin in a pool of water in Hades and beneath fruit-laden boughs only to have the water or fruit recede at each attempt to eat or drink

Tay·lor \'tā-lər\ Zachary 1784–1850 American general; 12th president of the U.S. (1849–50)

Tchai·kov·sky \chī-'kȯf-skē, chə-, -'kȯv-\ Pyotr Ilich 1840–1893 Russian composer

Te·cum·seh \tə-'kəm-sə, -'kəmp-, -sē\ 1768–1813 Shawnee Indian chief

Tek·a·kwitha \ˌtek-ə-'kwith-ə\ Ka·teri \'kät-ə-rē\ 1656–1680 *Lily of the Mohawks* beatified Mohawk Indian religious

Te·lem·a·chus \tə-'lem-ə-kəs\ son of Odysseus and Penelope who aids his father in the slaying of his mother's suitors

Ten·ny·son \'ten-ə-sən\ Alfred 1809–1892 Baron *Tennyson* known as *Alfred, Lord Tennyson* English poet

Ter·ence \'ter-əns\ *about* 195–159? B.C. Roman playwright

Te·re·sa \tə-'rē-sə, -'rā-\ Mother 1910–1997 beatified Albanian religious in India; Nobel Prize winner (1979)

Te·re·sa of Avi·la \tə-'rē-sə, -'rā-sə, -'rā-zə . . .'äv-ē-ˌlä\ Saint 1515–1582 Spanish Carmelite nun and mystic

Terp·sich·o·re \ˌtərp-'sik-ə-rē\ Muse of dancing and choral song in Greek mythology

Tes·la \'tes-lə\ Nikola 1856–1943 American (Croatian-born) electrical engineer and inventor

Thack·er·ay \'thak-rē, -ə-rē\ William Makepeace 1811–1863 English author

Tha·les \'thā-ˌlēz\ *flourished* 6th century B.C. Greek philosopher

Tha·lia \thə-'lī-ə\ Muse of comedy in Greek mythology

Thatch·er \'thach-ər\ Margaret Hilda 1925– Baroness *Roberts of Kesteven* née *Roberts* British prime minister (1979–90)

The·mis·to·cles \thə-'mis-tə-ˌklēz\ *about* 524–*about* 460 B.C. Athenian general and statesman

The·oc·ri·tus \thē-'äk-rət-əs\ *about* 310–250 B.C. Greek poet

The·od·o·ric \thē-'äd-ə-rik\ 454–526 A.D. *Theodoric the Great* king of the Ostrogoths (471–526) and of Italy (493–526)

The·o·do·sius I \ˌthē-ə-'dō-shəs, -shē-əs\ 347–395 A.D. *Theodosius the Great* Roman general and emperor (379–395)

The·seus \'thē-ˌsüs, -sē-əs\ hero in Greek mythology who slays Procrustes and the Minotaur and conquers the Amazons

Thes·pis \'thes-pəs\ *flourished* 6th century B.C. Greek poet

The·tis \'thēt-əs\ sea goddess and mother of Achilles in Greek mythology

This·be \'thiz-bē\ legendary Babylonian maiden who dies for the love of Pyramus

Thom·as \'täm-əs\ apostle in the Bible who demands proof of Jesus' resurrection

Thomas Clarence 1948– American jurist

Thomas Dyl·an \'dil-ən\ 1914–1953 Welsh poet

Thomas à Becket — see BECKET

Thomas à Kem·pis \ə-'kem-pəs, ä-'kem-\ 1379 (or 1380)–1471 Dutch ecclesiastic and author

Thomas Aquinas Saint — see AQUINAS

Thomp·son \'täm-sən, 'tämp-\ Benjamin 1753–1814 Count *Rum·ford* \'rəm-fərd, 'rəmp-\ British (American-born) physicist and statesman

Thor \'thȯr\ god of thunder, weather, and crops in Norse mythology

Tho·reau \thə-'rō, thȯ-; 'thȯr-ō\ Henry David 1817–1862 American author

Thorpe \'thȯrp\ Jim 1888–1953 *James Francis Thorpe* American athlete

Thu·cyd·i·des \thü-'sid-ə-ˌdēz, thyü-\ *died about* 401 B.C. Greek historian

Thur·ber \'thər-bər\ James Grover 1894–1961 American author

Ti·be·ri·us \tī-'bir-ē-əs\ 42 B.C.–37 A.D. Roman emperor (14–37)

Tim·o·thy \'tim-ə-thē\ a disciple of the apostle Paul

Tim·ur \'tim-ủr\ *or* **Tam·er·lane** \'tam-ər-ˌlān\ *or* **Tam·bur·laine** \'tam-bər-ˌlān\ 1336–1405 *Timur Lenk* Turkic conqueror

Tin·to·ret·to \ˌtin-tə-'ret-ō\ *about* 1518–1594 *Jacopo Robusti* Italian painter

Ti·ta·nia \tə-'tān-yə, -'tän-, tī-'tān-\ queen of the fairies and wife of Oberon in Shakespeare's *A Midsummer Night's Dream*

Ti·tian \'tish-ən\ *about* 1488–1576 *Tiziano Vecellio* Italian painter

Ti·to \'tēt-ō\ 1892–1980 originally *Josip Broz* known as *Marshal Tito* Yugoslavian leader (1943–80)

Ti·tus \'tīt-əs\ disciple of the apostle Paul

Titus 39–81 A.D. Roman emperor (79–81)

Tocque·ville \'tōk-ˌvil, 'tȯk-, 'täk-, -ˌvēl, -vəl\ Alexis (-Charles-Henri Clérel) de 1805–1859 French statesman and author

Tol·kien \'tȯl-ˌkēn, 'tōl-, 'täl-\ J. R. R. 1892–1973 *John Ronald Reuel Tolkien* British author

Tol·stoy \tōl-'stȯi, tōl-', täl-', 'tōl-ˌ, 'tōl-ˌ, 'täl-ˌ\ Leo 1828–1910 *Count Lev Nikolayevich Tolstoy* Russian author

Tor·que·ma·da \ˌtȯr-kə-'mäd-ə\ Tomás de 1420–1498 Spanish grand inquisitor

Tou·louse–Lau·trec (–Mon·fa) \tủ-ˌlüz-lō-'trek(-mō^{n}-'fä)\ Henri (-Marie-Raymond) de 1864–1901 French painter

Tous·saint–Lou·ver·ture \tü-ˌse^{n}-ˌlü-ver-'tuẽr\ *about* 1743–1803 Haitian general and liberator

Toyn·bee \'tȯin-bē\ Arnold Joseph 1889–1975 British historian

Tra·jan \'trā-jən\ 53–117 A.D. Roman emperor (98–117)

Tris·tram \'tris-trəm\ *or* **Tris·tan** \'tris-tən, -ˌtän, -ˌtan\ hero in medieval romance who drinks a love potion and falls in love with the Irish princess Isolde

Tri·ton \'trīt-ᵊn\ sea god in Greek mythology who is half man and half fish

Troi·lus \'trȯi-ləs, 'trō-ə-ləs\ son of Priam who in medieval legend loves Cressida but loses her to Diomedes

Trol·lope \'träl-əp\ Anthony 1815–1882 English author

Trots·ky \'trät-skē, 'trȯt-\ Leon 1879–1940 originally *Lev Davidovich Bronshtein* Russian Communist leader

Tru·deau \'trü-dō, trü-'\ Pierre Elliott 1919–2000 prime minister of Canada (1968–79; 1980–84)

Tru·man \'trü-mən\ Harry S. 1884–1972 33rd president of the U.S. (1945–53)

Truth \'trüth\ Sojourner *about* 1797–1883 American abolitionist

Tub·man \'təb-mən\ Harriet *about* 1820–1913 American abolitionist

Tur·ge·nev \tủr-'gān-yəf\ Ivan Sergeyevich 1818–1883 Russian author

\ə\ **a**but \$^{ə}$\ kitten \ər\ **fur**ther \a\ **a**sh \ā\ **a**ce \ä\ m**o**p, m**ar** \aủ\ **ou**t \ch\ **ch**in \e\ b**e**t \ē\ **ea**sy \g\ **g**o \i\ h**i**t \ī\ **i**ce \j\ **j**ob \ŋ\ si**ng** \ō\ g**o** \ȯ\ l**aw** \ȯi\ b**oy** \th\ **th**in \th\ **th**e \ü\ l**oo**t \ủ\ f**oo**t \y\ **y**et \zh\ vi**si**on *see also* Guide to Pronunciation

Tur·ner \'tər-nər\ J. M. W. 1775–1851 *Joseph Mallord William Turner* English painter
Turner Nat 1800–1831 American leader of a slave rebellion
Tut·ankh·a·men \,tü-,tang-'käm-ən, -,täng-\ originally *Tut·ankh·a·ten* \-,teng-'kät-ən\ *about* 1370–1352 B.C. king of Egypt (1361–52)
Tu·tu \'tü-,tü\ Desmond Mpilo 1931– South African clergyman and political activist; Nobel Prize winner (1984)
Twain \'twān\ Mark 1835–1910 pseudonym of *Samuel Langhorne Clem·ens* \'klem-ənz\ American author
Tweed \'twēd\ William Marcy 1823–1878 *Boss Tweed* American politician
Ty·ler \'tī-lər\ John 1790–1862 10th president of the U.S. (1841–45)
Ulysses — see ODYSSEUS
Up·dike \'əp-,dīk\ John Hoyer 1932– American author
Ura·nia \yu̇-'rā-nē-ə, -nyə\ Muse of astronomy in Greek mythology
Ura·nus \'yu̇r-ə-nəs, yu̇-'rā-nəs\ the sky personified as a god and the father of the Titans in Greek mythology
Ur·ban \'ər-bən\ name of 8 popes: especially **II** 1035–1099 (pope 1088–99)
Uri·el \'yu̇r-ē-əl\ one of the four archangels named in Hebrew tradition — compare GABRIEL, MICHAEL, RAPHAEL
Va·le·ri·an \və-'lir-ē-ən\ *died* 260 A.D. Roman emperor (253–260)
Van Bu·ren \van-'byu̇r-ən, vən-\ Martin 1782–1862 8th president of the U.S. (1837–41)
Van·der·bilt \'van-dər-,bilt\ Cornelius 1794–1877 American shipping and railroad magnate
Van Dyck *or* **Van·dyke** \van-'dīk, vən-\ Sir Anthony 1599–1641 Flemish painter
van Gogh Vincent — see GOGH, VAN
Ve·ga \'vā-gə\ Lo·pe \'lō-pā\ de 1562–1635 Spanish playwright
Ve·láz·quez \və-'las-kəs\ Diego Rodríguez de Silva 1599–1660 Spanish painter
Ve·nus \'vē-nəs\ goddess of love and beauty in Roman mythology — compare APHRODITE
Ver·di \'veərd-ē\ Giuseppe 1813–1901 Italian composer
Vergil — see VIRGIL
Ver·meer \vər-'mer, -'mir\ Jan *or* Johannes 1632–1675 Dutch painter
Verne \'vərn, 'vern\ Jules \'jülz, 'zhu̅e̅l\ 1828–1905 French author
Ve·ro·ne·se \,ver-ə-'nā-sē, -'nā-zē\ Paolo 1528–1588 Italian painter
Ves·puc·ci \ve-'spü-chē\ Ame·ri·go \,äm-ə-'rē-gō\ 1454–1512 *Amer·i·cus Ves·pu·cius* \ə-'mer-ə-kəs-,ves-'pyü-shəs, -shē-əs\ Italian navigator for Spain and namesake of America
Ves·ta \'ves-tə\ goddess of the hearth in Roman mythology — compare HESTIA
Vic·tor Em·man·u·el I \'vik-tər-i-'man-yə-wəl, -'man-yəl\ 1759–1824 king of Sardinia (1802–21)
Victor Emmanuel II 1820–1878 king of Sardinia (1849–61); 1st king of Italy (1861–78)
Victor Emmanuel III 1869–1947 king of Italy (1900–46)
Vic·to·ria \vik-'tōr-ē-ə, -'tȯr-\ 1819–1901 *Alexandrina Victoria* queen of the United Kingdom (1837–1901)
Vil·lon \vē-'ōⁿ, -'yōⁿ\ François 1431–*after* 1463 French poet
Vin·cent de Paul \,vin-sənt-də-'pȯl\ Saint 1581–1660 French priest and founder of the Vincentians
Vinci, da Leonardo — see LEONARDO DA VINCI
Vir·gil *or* **Ver·gil** \'vər-jəl\ 70–19 B.C. Roman poet — **Vir·gil·ian** *also* **Ver·gil·ian** \,vər-'jil-ē-ən\ *adj*
Vish·nu \'vish-nü\ god of preservation in the Hindu sacred triad — compare BRAHMA SHIVA
Vi·val·di \vi-'väl-dē, -'vȯl-\ Antonio 1678–1741 Italian composer
Vol·ta \'vōl-tə, 'väl-, 'vȯl-\ Alessandro 1745–1827 Italian physicist
Vol·taire \vōl-'taər, vȯl-, väl-, -'teər\ 1694–1778 originally *François-Marie Arouet* French author
Vul·can \'vəl-kən\ god of fire and metalworking in Roman mythology — compare HEPHAESTUS

Wag·ner \'väg-nər\ (Wilhelm) Ri·chard \'rik-,ärt, 'rik-\ 1813–1883 German composer
Wal·cott \'wȯl-kət, -,kät\ Derek Alton 1930– West Indian author; Nobel Prize winner (1992)
Wa·le·sa \vä-'len-sə\ Lech 1943– Polish labor leader and president of Poland (1990–95); Nobel Prize winner (1983)
Walk·er \'wȯk-ər\ Alice Malsenior 1944– American author
Wal·len·berg \'wäl-ən-,bərg\ Raoul 1912–?1947 Swedish diplomat and hero of the Holocaust
Wal·pole \'wȯl-,pōl, 'wäl-\ Horace 1717–1797 4th Earl of *Or·ford* \'ȯr-fərd\ English author
Wal·ton \'wȯlt-n\ Izaak \'ī-zik, -zək\ 1593–1683 English author
War·ren \'wȯr-ən, 'wär-\ Earl 1891–1974 American jurist; chief justice of the U.S. Supreme Court (1953–69)
War·ren \'wȯr-ən, 'wär-\ Robert Penn 1905–1989 American author
Wash·ing·ton \'wȯsh-ing-tən, 'wäsh-\ Book·er \'bu̇k-ər\ Tal·ia·ferro \'täl-ə-vər\ 1856–1915 American educator
Washington George 1732–1799 American general; 1st president of the U.S. (1789–97)
Wat·son \'wät-sən\ James Dewey 1928– American geneticist; Nobel Prize winner (1962)
Watt \'wät\ James 1736–1819 Scottish inventor
Wayne \'wān\ Anthony 1745–1796 *Mad Anthony* American Revolutionary general
We·ber \'vā-bər\ Carl Maria von 1786–1826 German composer
Web·ster \'web-stər\ Daniel 1782–1852 American statesman
Webster Noah 1758–1843 American lexicographer
Welles \'welz\ (George) Orson 1915–1985 American film director and producer
Wel·ling·ton \'wel-ing-tən\ Duke of 1769–1852 *Arthur Wellesley; the Iron Duke* British general and statesman
Wel·ty \'wel-tē\ Eudora 1909–2001 American author
Wells \'welz\ H. G. 1866–1946 *Herbert Gordon Wells* English author
Wes·ley \'wes-lē, 'wez-\ John 1703–1791 English founder of Methodism
West \'west\ Benjamin 1738–1820 American painter in England
Wes·ting·house \'wes-ting-,hau̇s\ George 1846–1914 American inventor and industrialist
Whar·ton \'hwȯrt-n, 'wȯrt-\ Edith 1862–1937 née *Jones* American author
Wheat·ley \'wēt-lē, 'hwēt-\ Phillis *about* 1753–1784 American (African-born) poet
Whis·tler \'hwis-lər, 'wis-\ James (Abbott) McNeill 1834–1903 American artist
Whit·man \'hwit-mən, 'wit-\ Walt 1819–1892 American poet
Whit·ney \'hwit-nē, 'wit-\ Eli 1765–1825 American inventor
Whit·ti·er \'hwit-ē-ər, 'wit-\ John Greenleaf 1807–1892 American poet
Wie·sel \vē-'zel, wē-\ Elie 1928– American (Romanian-born) author; Nobel Prize winner (1986)
Wilde \'wīld\ Oscar (Fingal O'Flahertie Wills) 1854–1900 Irish author
Wil·der \'wīl-dər\ Thornton Niven 1897–1975 American author
Wil·hel·mi·na \,wil-,hel-'mē-nə, ,wil-ə-'mē-\ 1880–1962 queen of the Netherlands (1890–1948)
Wil·kins \'wil-kənz\ Roy 1901–1981 American civil rights leader
Wil·lard \'wil-ərd\ Emma 1787–1870 née *Hart* American educator
Wil·liam \'wil-yəm\ name of 4 kings of England: **I** *about* 1028–1087 *William the Conqueror* (reigned 1066–87); **II** *about* 1056–1100 *William Ru·fus* \'rü-fəs\ (reigned 1087–1100); **III** 1650–1702 (reigned 1689–1702); **IV** 1765–1837 (reigned 1830–37)
William I 1533–1584 *William the Silent* prince of Orange and founder of the Dutch Republic
William I 1797–1888 king of Prussia (1861–88) and emperor of Germany (1871–88)
William II 1859–1941 *Kaiser Wilhelm* emperor of Germany and king of Prussia (1888–1918; abdicated)
Wil·liam Tell \,wil-yəm-'tel\ legendary Swiss patriot sentenced to shoot an apple off his son's head

Wil·liams \'wil-yəmz\ Roger 1603?–1683 American (English-born) clergyman and founder of Rhode Island
Williams Ted 1918–2002 *Theodore Samuel Williams* American baseball player
Williams Tennessee 1911–1983 *Thomas Lanier Williams* American playwright
Williams Venus 1980– and her sister Serena 1981– American tennis players
Williams William Carlos 1883–1963 American poet and physician
Wil·son \'wil-sən\ August 1945– American playwright
Wilson (Thomas) Wood·row \'wu̇d-rō\ 1856–1924 28th president of the U.S. (1913–21); Nobel Prize winner (1919)
Windsor Duke of — see EDWARD VIII
Win·throp \'win-thrəp, 'wint-\ John 1588–1649 English colonist in America; 1st governor of Massachusetts Bay Colony
Wo·den \'wōd-n\ — see ODIN
Wolfe \'wu̇lf\ James 1727–1759 British general
Wolfe Thomas Clayton 1900–1938 American author
Woll·stone·craft \'wu̇l-stən-,kraft\ Mary 1759–1797 English feminist and author; mother of Mary Wollstonecraft Shelley
Wol·sey \'wu̇l-zē\ Thomas *about* 1475–1530 English cardinal and statesman
Woods \'wu̇dz\ Tiger 1975– *Eldrick Woods* American golfer
Woolf \'wu̇lf\ Virginia 1882–1941 née *Stephen* English author
Words·worth \'wərdz-wərth, -,wərth\ William 1770–1850 English poet
Wo·vo·ka \wō-'vō-kə\ 1858?–1932 *Jack Wilson* Paiute Indian mystic
Wren \'ren\ Sir Christopher 1632–1723 English architect
Wright \'rīt\ Frank Lloyd 1867–1959 American architect
Wright Or·ville \'ȯr-vəl\ 1871–1948 and his brother Wilbur 1867–1912 American pioneers in aviation
Wright Richard 1908–1960 American author
Wyc·liffe \'wik-,lif, -ləf\ John *about* 1330–1384 English reformer and Bible translator
Wy·eth \'wī-əth\ Andrew Newell 1917– American painter
Xa·vi·er \'zāv-yər, 'zā-vē-ər, ig-'zā-\ Saint Francis 1506–1552 Spanish *Francisco Javier* Spanish Jesuit missionary
Xen·o·phon \'zen-ə-fən\ 431–*about* 350 B.C. Greek historian
Xer·xes I \'zərk-,sēz\ *about* 519–465 B.C. *Xerxes the Great* king of Persia (486–465); son of Darius I
Yeats \'yāts\ William Butler 1865–1939 Irish author
Yelt·sin \'yelt-sən, 'yel-sin\ Boris Nikolayevich 1931– president of Russia (1990–99)
Young \'yəng\ Brig·ham \'brig-əm\ 1801–1877 American Mormon leader
Young Whitney Moore 1921–1971 American civil rights leader
Za·har·i·as \zə-'har-ē-əs\ Babe Didrikson 1914–1956 *Mildred Ella Zaharias* née *Didrikson* American athlete
Zech·a·ri·ah \,zek-ə-'rī-ə\ Hebrew prophet of the 6th century B.C.
Zeng·er \'zeng-ər, 'zeng-gər\ John Peter 1697–1746 American (German-born) journalist and printer
Ze·no of Ci·ti·um \'zē-nō-əv-'sish-ē-əm\ *about* 335–*about* 263 B.C. Greek philosopher and founder of Stoic school
Zeph·a·ni·ah \,zef-ə-'nī-ə\ Hebrew prophet of the 7th century B.C.
Zeph·y·rus \'zef-ə-rəs\ god of the west wind in Greek mythology
Zeus \'züs\ chief god, ruler of the elements, and husband of Hera in Greek mythology — compare JUPITER
Zo·la \'zō-lə, 'zō-,lä, zō-'lä\ Émile 1840–1902 French author
Zo·ro·as·ter \'zōr-ə-,was-tər, 'zȯr-\ *or* **Zar·a·thu·shtra** \,zar-ə-'thüsh-trə, -'thəsh-\ *about* 628–*about* 551 B.C. founder of a Persian religion
Zwing·li \'zwing-lē, 'swing-, -glē; 'tsfing-lē\ Huldrych 1484–1531 Swiss Reformation leader

Geographical Names

This section constitutes a pronouncing dictionary of names of current, historical, mythological, and legendary places. It complements the general vocabulary by entering many adjectives and nouns derived from these names, such as **Florentine** at **Florence** and **Libyan** at **Libya.**

In the entries the letters N, E, S, and W, singly or in combination indicate direction and are not part of the name. They may represent either the direction (as *north*) or the adjective derived from it (as *northern*); thus, west-northwest of Santiago appears as WNW of Santiago, and southern California appears as S California. The only other special abbreviations used in this section are U.S. for United States, and U.S.S.R. for Union of Soviet Socialist Republics.

Aa·chen \'äk-ən\ *or French* **Aix–la–Cha·pelle** \,äk-,slä-shə-'pel, ,ek-\ city W Germany WSW of Cologne

Aarhus — see ARHUS

Aba·dan \,ab-ə-'dän, ,ab-ə-'dan\ town W Iran on Abadan Island in delta of Shatt al Arab

Ab·er·deen \,ab-ər-'dēn\ **1** *or* **Ab·er·deen·shire** \-,shir, -shər\ administrative area NE Scotland **2** city NE Scotland constituting an administrative area (**Aberdeen City**) — **Ab·er·do·ni·an** \-'dō-nē-ən\ *adj or n*

Ab·i·djan \,ab-i-'jän\ city, seat of government of Ivory Coast

Ab·i·lene \'ab-ə-,lēn\ city NW central Texas

Abruz·zi \ä-'brüt-sē\ region central Italy on the Adriatic E of Latium; capital, L'Aquila

Abu Dha·bi \,äb-ü-'däb-ē\ city, capital of United Arab Emirates

Abu·ja \ä-'bü-jä\ city, capital of Nigeria

Abyssinia — see ETHIOPIA

Aca·dia \ə-'kād-ē-ə\ *or French* **Aca·die** \á-ká-'dē\ NOVA SCOTIA — an early name

Acadia National Park section of coast of Maine including areas on Mount Desert Island & Isle au Haut

Aca·pul·co \,äk-ə-'pül-kō, ,ak-\ city S Mexico on the North Pacific

Ac·ar·na·nia \,ak-ər-'nā-nē-ə, -'nā-nyə\ region W Greece on Ionian Sea

Accad — see AKKAD

Ac·cra \ə-'krä\ city, capital of Ghana

Achaea \ə-'kē-ə\ *or* **Acha·ia** \ə-'kī-ə, -'kā-ə, -'kā-yə\ region S Greece in N Peloponnese

Ach·er·on \'ak-ə-,rän, -rən\ a river of Hades in Greek mythology

Acon·ca·gua \,ak-ən-'käg-wə, ,äk- [illegible]-\ mountain 22,834 feet (6960 meters) W Argentina; highest in the Andes & Western Hemisphere

Açores — see AZORES

Ac·ti·um \'ak-shē-əm, 'ak-tē-\ promontory & ancient town W Greece in NW Acarnania

Ada·na \'äd-ə-nə, -,nä; ə-'dän-ə\ city S Turkey

Ad·dis Aba·ba \,ad-ə-'sab-ə-bə\ city, capital of Ethiopia

Ad·e·laide \'ad-l-,ād\ city, capital of South Australia

Aden \'äd-n, 'ād-, 'ad-\ **1** former British protectorate S Arabia E of Yemen **2** former British colony in SW Aden Protectorate **3** city & port S Yemen; formerly capital of People's Democratic Republic of Yemen

Aden, Gulf of arm of Indian Ocean between Yemen (Arabia) & Somalia (Africa)

Adi·ge \'äd-ə-,jā\ river 255 miles (410 kilometers) long N Italy flowing SE into the Adriatic

Ad·i·ron·dack \,ad-ə-'rän-,dak\ mountains NE New York; highest Mount Marcy 5344 feet (1629 meters)

Ad·mi·ral·ty \'ad-mrəl-tē, -mə-rəl-\ **1** island SE Alaska in N Alexander Archipelago **2** islands W Pacific N of New Guinea in Bismarck Archipelago; belong to Papua New Guinea

Adri·at·ic Sea \,ā-drē-'at-ik, ,ad-rē-\ arm of Mediterranean between Italy & Balkan Peninsula

Ae·ge·an Sea \i-'jē-ən\ arm of Mediterranean between Asia Minor & Greece

Ae·gi·na \i-'jī-nə\ *or Greek* **Aí·yi·na** \'ā-yē-,nä\ island & ancient state SE Greece in Saronic Gulf

Ae·o·lis \'ē-ə-ləs\ *or* **Ae·o·lia** \ē-'ō-lē-ə, -'ōl-yə\ ancient country NW Asia Minor

Afars and the Issas, French Territory of the — see DJIBOUTI 1

Af·ghan·i·stan \af-'gan-ə-,stan\ country W Asia E of Iran; capital, Kabul

Af·ri·ca \'af-ri-kə\ continent of Eastern Hemisphere S of the Mediterranean

Aga·na \ə-'gän-yə\ town, capital of Guam

Agra \'äg-rə\ city N India in W Uttar Pradesh

Agri Dagi — see ARARAT

Aguas·ca·lien·tes \,äg-wə-,skäl-yen-,tās\ **1** state central Mexico **2** city, its capital, NE of Guadalajara

Agul·has, Cape \ə-'gəl-əs\ headland Republic of South Africa; most southerly point of Africa, at 34°50' S latitude

Agulhas Current warm current of the Indian Ocean flowing SW along SE coast of Africa

Ahag·gar \ə-'häg-ər, ,ä-hə-'gär\ *or* **Hog·gar** \'häg-ər, hə-'gär\ mountains S Algeria in W central Sahara; highest peak 9842 feet (3000 meters)

Ah·mad·abad \'äm-əd-ə-,bäd\ city W India in Gujarat

Ah·waz \ä-'wäz\ city SW Iran

Aisne \'ān\ river 165 miles (265 kilometers) long N France flowing from Argonne Forest into the Oise

Aix–la–Chapelle — see AACHEN

Aj·mer \,əj-'miər, -'meər\ city NW India in central Rajasthan SW of Delhi

Aki·ta \ä-'kēt-ə, 'äk-i-,tä\ city Japan in N Honshu on Sea of Japan

Ak·kad *or* **Ac·cad** \'ak-,ad, 'äk-,äd\ **1** N division of ancient Babylonia **2** *or* **Aga·de** \ə-'gäd-ə\ ancient city, its capital

Ak·ron \'ak-rən\ city NE Ohio

Al·a·bama \,al-ə-'bam-ə\ state SE U.S.; capital, Montgomery — **Al·a·bam·i·an** \-'bam-ē-ən\ *or* **Al·a·bam·an** \-'bam-ən\ *adj or n*

Åland \'ō-,länd\ *or* **Ah·ven·an·maa** \'äk-və-,nän-,mä, 'ä-və-\ archipelago SW Finland in Baltic Sea

Alas·ka \ə-'las-kə\ **1** peninsula SW Alaska (state) SW of Cook Inlet **2** state of U.S. in NW North America; capital, Juneau **3** mountain range S Alaska (state) extending from Alaska Peninsula to Yukon boundary — **Alas·kan** \-kən\ *adj or n*

Alaska, Gulf of inlet of North Pacific off S Alaska between Alaska Peninsula on W & Alexander Archipelago on E

Al·ba Lon·ga \,al-bə-'lȯŋ-gə\ ancient city central Italy in Latium SE of Rome

Al·ba·nia \al-'bā-nē-ə, -nyə\ country S Europe in Balkan Peninsula on Adriatic; capital, Tirane

Al·ba·ny \'ȯl-bə-nē\ city, capital of New York

Al·be·marle Sound \'al-bə-,märl\ inlet of North Atlantic in NE North Carolina

Al·bert, Lake \'al-bərt\ lake E Africa between Uganda & Democratic Republic of the Congo in course of the Nile

Al·ber·ta \al-'bərt-ə\ province W Canada; capital, Edmonton — **Al·ber·tan** \-'bərt-n\ *adj or n*

Albert Nile — see NILE

Al·bi·on \'al-bē-ən\ **1** the island of Great Britain **2** ENGLAND

Al·bu·quer·que \'al-bə-,kər-kē, -byə-\ city central New Mexico

Al·da·bra \'al-də-brə\ island (atoll) Seychelles, in NW Indian Ocean N of Madagascar

Al·der·ney \'ȯl-dər-nē\ island in English Channel — see CHANNEL

Alep·po \ə-'lep-ō\ city N Syria

Aleu·tian \ə-'lü-shən\ islands SW Alaska extending in an arc 1700 miles (2735 kilometers) W from Alaska Peninsula

Al·ex·an·der \,al-ig-'zan-dər, ,el-\ archipelago SE Alaska

Al·ex·an·dria \,al-ig-'zan-drē-ə, ,el-\ **1** city N Virginia on the Potomac **2** city N Egypt on the Mediterranean — **Al·ex·an·dri·an** \-drē-ən\ *adj or n*

Al·ge·ria \al-'jir-ē-ə\ country NW Africa on the Mediterranean; capital, Algiers — **Al·ge·ri·an** \-ē-ən\ *adj or n*

Al·giers \al-'jiərz\ **1** Algeria especially as one of former Barbary States **2** city, capital of Algeria — **Al·ge·rine** \,al-jə-'rēn\ *adj or n*

Al·i·garh \,al-i-'gär\ city N India in NW Uttar Pradesh N of Agra

Al Jīzah — see GIZA

Al·lah·abad \'al-ə-hə-,bad, -,bäd\ city N India in S Uttar Pradesh W of Varanasi

Al·le·ghe·ny \,al-ə-'gā-nē\ mountains of Applachian system E U.S. in Pennsylvania, Maryland, Virginia, & West Virginia

Al·len·town \'al-ən-,taůn\ city E Pennsylvania

\ə\ abut \ᵊ\ kitten \ər\ **further** \a\ **ash** \ā\ **ace** \ä\ mop, mar \aů\ **out** \ch\ **chin** \e\ **bet** \ē\ **easy** \g\ go \i\ **hit** \ī\ **ice** \j\ **job** \ŋ\ **sing** \ō\ **go** \ȯ\ **law** \ȯi\ **boy** \th\ **thin** \t͟h\ **the** \ü\ **loot** \ů\ **foot** \y\ yet \zh\ vision *see also* Guide to Pronunciation

Al·maty \əl-ˈmät-ē\ *or* **Al·ma–Ata** \əl-ˈmä-ə-ˈtä\ city, former capital of Kazakhstan

Alps \ˈalps\ mountain system central Europe — see MONT BLANC

Al·sace \al-ˈsas, -sās, ˈal-ˌ\ *or German* **El·sass** \ˈel-ˌzäs\ *or ancient* **Al·sa·tia** \al-ˈsā-shē-ə, -shə\ region & former province NE France between Rhine River & Vosges Mountains — **Al·sa·tian** \al-ˈsā-shən\ *adj or n*

Al·sace–Lor·raine \-lə-ˈrān, -lȯ-ˈrān\ region N France W of the Rhine including Alsace & part of Lorraine

Al·tai *or* **Al·tay** \ˌal-ˈtī\ mountain system central Asia between Outer Mongolia & W China & between Kazakhstan & Russia in Asia; highest peak about 15,000 feet (4570 meters)

Al·ta·mi·ra \ˌal-tə-ˈmir-ə\ caverns N Spain WSW of Sandander

Al·to Adi·ge \ˌäl-tō-ˈäd-i-ˌjā\ *or* **South Ti·rol** \-tə-ˈrōl, -ˈtī-ˌrōl, -ˌtī-ˈ; -ˈtir-əl\ district N Italy in S Tirol in N Trentino-Alto Adige region

Al·to Pa·ra·ná \ˌal-tō-ˌpar-ə-ˈnä\ upper course of the Paraná

Ama·ga·sa·ki \ˌam-ə-gə-ˈsäk-ē\ city Japan in W central Honshu on Osaka Bay

Am·a·ril·lo \ˌam-ə-ˈril-ō, -ˈril-ə\ city NW Texas

Am·a·zon \ˈam-ə-ˌzän, -zən\ *or Portuguese and Spanish* **Ama·zo·nas** \ˌam-ə-ˈzō-nəs\ river about 3900 miles (6276 kilometers) long N South America flowing from Peruvian Andes into Atlantic in N Brazil

Am·a·zo·nia \ˌam-ə-ˈzō-nē-ə\ region N South America; basin of the Amazon

Amer·i·ca \ə-ˈmer-ə-kə\ **1** either continent (North America or South America) of Western Hemisphere **2** *or* **the Amer·i·cas** \-kəz\ lands of Western Hemisphere including North, Central & South America, & West Indies **3** UNITED STATES OF AMERICA

American Samoa *or* **Eastern Samoa** islands SW central Pacific; capital, Pago Pago (on Tutuila Island)

American Samoa National Park reservation at three locations in American Samoa

Am·man \ä-ˈmän, -ˈman\ city, capital of Jordan

Amoy — see XIAMEN

Am·rit·sar \ˌəm-ˈrit-sər\ city N India in NW Punjab

Am·ster·dam \ˈam-stər-ˌdam, ˈamp-\ city, official capital of the Netherlands

Amu Dar'·ya \ˌäm-ü-ˈdär-yə\ *or ancient* **Ox·us** \ˈäk-səs\ river over 1500 miles (2400 kilometers) long in central & W Asia flowing from the Pamirs into Aral Sea

Amur \ä-ˈmu̇r\ river about 1780 miles (2865 kilometers) long E Asia formed by junction of Shilka & Argun rivers flowing into the North Pacific at N end of Tatar Strait & forming part of boundary between China & Russia in Asia

An·a·heim \ˈan-ə-ˌhīm\ city SW California E of Long Beach

Aná·huac \ə-ˈnä-ˌwäk\ the central plateau of Mexico

An·a·to·lia \ˌan-ə-ˈtō-lē-ə, -ˈtōl-yə\ — see ASIA MINOR — **An·a·to·li·an** \-ən, -yən\ *adj or n*

An·chor·age \ˈaŋ-kə-rij, -krij\ city S central Alaska; largest in state

An·co·hu·ma \ˌaŋ-kə-ˈhü-mə, -ˈhyü-\ mountain peak 20,958 feet (6388 meters) W Bolivia; highest of Illampu

An·co·na \aŋ-ˈkō-nə, an-\ city central Italy, capital of Marche

An·da·lu·sia \ˌan-də-ˈlü-zhē-ə, -zhə\ *or Spanish* **An·da·lu·cía** \ˌan-də-ˈlü-ˈsē-ə\ region S Spain including Sierra Nevada & valley of the Guadalquivir — **An·da·lu·sian** \ˌan-də-ˈlü-zhən\ *adj or n*

An·da·man \ˈan-də-mən, -ˌman\ **1** islands India in Bay of Bengal S of Myanmar & N of Nicobar Islands; in **Andaman and Nic·o·bar** \ˈnik-ə-ˌbär\ territory **2** sea arm of Bay of Bengal S of Myanmar — **An·da·man·ese** \ˌan-də-mə-ˈnēz, -ˈnēs\ *adj or n*

An·des \ˈan-dēz, -ˌdēz\ mountain system W South America extending from Panama to Tierra del Fuego — see ACONCAGUA — **An·de·an** \ˈan-dē-ən, an-ˈ\ *adj* — **An·dine** \ˈan-ˌdēn, -ˌdīn\ *adj*

An·dhra Pra·desh \ˌän-drə-prə-ˈdāsh, -ˈdesh\ state S India N of Tamil Nadu bordering on Bay of Bengal; capital, Hyderabad

An·dor·ra \an-ˈdȯr-ə, -ˈdär-ə\ country SW Europe in E Pyrenees between France & Spain; capital, Andorra la Vella — **An·dor·ran** \-ən\ *adj or n*

An·dros **1** \ˈan-drəs\ island, largest of Bahamas **2** \ˈan-drəs, -ˌdräs\ island Greece in N Cyclades SE of Euboea

An·gel Falls \ˌän-jəl-\ waterfall 3212 feet (979 meters) SE Venezuela on Auyán-tepuí Mountain

An·gers \äⁿ-ˈzhā\ city W France ENE of Nantes

Ang·kor \ˈaŋ-ˌkȯr\ ruins of ancient city NW Cambodia

An·gle·sey \ˈaŋ-gəl-sē\ island NW Wales

An·glo–Egyp·tian Sudan \ˌaŋ-glō-i-ˌjip-shən-\ former territory NE Africa under joint British & Egyptian rule; since 1956 has formed republic of Sudan

An·go·la \aŋ-ˈgō-lə, an-\ *or formerly* **Portuguese West Africa** country SW Africa S of mouth of Congo River; until 1975 a dependency of Portugal; capital, Luanda — **An·go·lan** \-lən\ *adj or n*

An·gus \ˈaŋ-gəs\ administrative area E Scotland

An·hui *or* **An·hwei** \ˈän-ˈhwā, -ˈwā\ province E China W of Jiangsu; capital, Hefei

An·i·ak·chak Crater \ˌan-ē-ˈak-ˌchak\ volcanic crater SW Alaska on Alaska Peninsula; crater 6 miles (10 kilometers) in diameter

An·jou \ˈan-ˌjü, äⁿ-ˈzhü\ region & former province NW France in valley of the Loire SE of Brittany; chief city, Angers

An·ka·ra \ˈaŋ-kə-rə, ˈäŋ-\ *or formerly* **An·go·ra** \aŋ-ˈgōr-ə, an-, -ˈgȯr-\ city, capital of Turkey in N central Anatolia

An·na·ba \ə-ˈnäb-ə\ *or formerly* **Bône** \ˈbōn\ city NE Algeria

An·nam \a-ˈnam, ə-; ˈan-ˌam\ region & former kingdom E Indochina in central Vietnam; capital, Hue

An·nap·o·lis \ə-ˈnap-ləs, -ə-ləs\ city, capital of Maryland

An·shan \ˈän-ˈshän\ city NE China in E central Liaoning

An·ta·nan·a·ri·vo \ˌan-tə-ˌnan-ə-ˈrē-vō\ *or formerly* **Ta·nan·a·rive** \tə-ˈnan-ə-ˌrēv\ city, capital of Madagascar

Ant·arc·ti·ca \ant-ˈärk-ti-kə, ˈant-, -ˈärt-i-\ body of land around the South Pole; plateau covered by great ice cap

Antarctic Peninsula *or formerly* **Palm·er Peninsula** \ˌpäm-ər-, ˌpäl-mər-\ peninsula about 700 miles (1126 kilometers) long W Antarctica S of S end of South America

An·ti·gua \an-ˈtē-gə\ island British West Indies in the Leewards; with Barbuda constitutes independent country of **Antigua and Barbuda**; capital, Saint John's

Anti–Leb·a·non \ˈant-i-ˈleb-ə-nən, -nän\ mountains SW Asia on Lebanon–Syria border — see HERMON (Mount)

An·til·les \an-ˈtil-ēz\ the West Indies excluding Bahamas — see GREATER ANTILLES, LESSER ANTILLES — **An·til·le·an** \-ˈtil-ē-ən\ *adj*

An·ti·och \ˈant-ē-ˌäk\ city of ancient Syria on the Orontes; site at modern Antakya, Turkey

An·trim \ˈan-trəm\ traditional county E Northern Ireland

Antung — see DANDONG

Ant·werp \ˈant-wərp\ city N Belgium on the Scheldt

An·yang \ˈän-ˈyäŋ\ city E China in N Henan

An·zio \ˈan-zē-ō, ˈän-\ *or ancient* **An·ti·um** \ˈan-shē-əm\ Mediterranean seaport Italy in Latium SSE of Rome

Aomen — see MACAO

Ao·mo·ri \ˈau̇-mə-rē\ city N Japan in NE Honshu

Aorangi — see COOK (Mount)

Aos·ta \ä-ˈȯs-tə\ city NW Italy

Ap·en·nines \ˈap-ə-ˌnīnz\ mountain chain Italy extending length of the peninsula; highest peak Monte Corno (NE of Rome) 9560 feet (2897 meters) — **Ap·en·nine** \-ˌnīn\ *adj*

Apia \ə-ˈpē-ə\ town, capital of independent Samoa on Upolu Island

Apo, Mount \ˈäp-ō\ volcano Philippines in SE Mindanao 9692 feet (2954 meters); highest peak in the Philippines

Ap·pa·la·chia \ˌap-ə-ˈlā-chə, -ˈlach-ə, -ˈlā-shə\ region E U.S. including Appalachian Mountains from S central New York to central Alabama

Ap·pa·la·chian \ˌap-ə-ˈlā-chən, -ˈlach-ən, -ˈlā-shən\ mountain system E North America extending from S Quebec to central Alabama — see MITCHELL (Mount)

Apu·lia \ə-ˈpyül-yə, -ˈpyü-lē-ə\ — see PUGLIA — **Apu·lian** \ə-ˈpyül-yən, -ˈpyü-lē-ən\ *adj or n*

Aqa·ba, Gulf of \ˈäk-ə-bə, ˈak-\ arm of Red Sea E of Sinai Peninsula

Aquid·neck Island \ə-ˈkwid-ˌnek\ *or* **Rhode Island** \ˈrōd\ island SE Rhode Island in Narragansett Bay

Aq·ui·taine \ˈak-wə-ˌtān\ region of SW France

Aq·ui·ta·nia \ˌak-wə-ˈtā-nyə, -nē-ə\ a Roman division of SW Gaul

Ara·bia \ə-ˈrā-bē-ə\ peninsula of SW Asia including Saudi Arabia, Yemen, Oman, & Persian Gulf States

Ara·bi·an \ə-ˈrā-bē-ən\ **1** desert E Egypt between Red Sea & the Nile **2** sea NW section of Indian Ocean between Arabia & India

Ara·ca·ju \ˌar-ə-kə-ˈzhü\ city E Brazil NE of Salvador

Arad \ä-ˈräd\ city W Romania

Ar·a·fu·ra \ˌar-ə-ˈfu̇r-ə\ sea between N Australia & W New Guinea

Ar·a·gon \ˈar-ə-ˌgän, -gən\ region NE Spain bordering on France — **Ar·a·go·nese** \ˌar-ə-gə-ˈnēz, -ˈnēs\ *adj or n*

Arak \är-ˈäk, ə-ˈrak\ city W Iran SW of Tehran

Ar·al Sea \ˈar-əl\ *or Russian* **Aral·sko·ye Mo·re** \ə-ˌral-skə-yə-ˈmȯr-ə, -yə\ *or formerly* **Lake Aral** lake W Asia between Kazakhstan & Uzbekistan

Ar·a·rat \ˈar-ə-ˌrat\ *or Turkish* **Ağ·ri Da·gi** \ˌä-rē-dä-ˈē, ˌäg-rē-däg-ˈē\ mountain 16,946 feet (5165 meters) E Turkey near border of Iran

Ar·bil *or* **Ir·bīl** *or* **Er·bil** \ər-ˈbēl\ city N Iraq

Ar·ca·dia \är-ˈkād-ē-ə\ mountain region S Greece in central Peloponnese

Arch·es National Park \ˈär-chəz\ reservation E Utah

Arc·tic \ˈärk-tik, ˈärt-ik\ **1** ocean N of Arctic Circle **2** Arctic regions **3** archipelago N Canada in Nunavut & Northwest Territories

Ar·da·bil *or* **Ar·de·bil** \ˌär-də-ˈbēl\ city NW Iran

Ar·dennes \är-ˈden\ wooded plateau NE France, W Luxembourg, & SE Belgium E of the Meuse

Are·ci·bo \ˌä-rā-ˈsē-bō\ city & port N Puerto Rico

Are·qui·pa \ˌar-ə-ˈkē-pə\ city S Peru

Ar·gen·ti·na \ˌär-jən-ˈtē-nə\ country S South America between the Andes & the South Atlantic S of the Pilcomayo; capital, Buenos Aires — **Ar·gen·tine** \ˈär-jən-ˌtēn\ *adj or n* — **Ar·gen·tin·ean** *or* **Ar·gen·tin·i·an** \ˌär-jən-ˈtin-ē-ən\ *adj or n*

Ar·go·lis \ˌär-gə-lis\ district S Greece in E Peloponnese

Ar·gonne \är-ˈgän, ˈär-ˌ\ *or* **Argonne Forest** wooded plateau NE France S of the Ardennes between Meuse & Aisne rivers

Ar·gos \ˌär-ˌgäs, -gəs\ ancient Greek city & state S Greece in Argolis; site at present town of Argos

Ar·gyll and Bute \är-ˈgīl-\ administrative area W Scotland

År·hus *or* **Aar·hus** \ˈȯr-ˌhüs\ city & port Denmark in E Jutland

Ar·i·zo·na \ˌar-ə-ˈzō-nə\ state SW U.S.; capital, Phoenix — **Ar·i·zo·nan** \-nən\ *or* **Ar·i·zo·nian** \-nē-ən, -nyən\ *adj or n*

Ar·kan·sas \ˈär-kən-ˌsȯ; *1 is also* är-ˈkan-zəs\ **1** river 1450 miles (2334 kilometers) long SW central U.S. flowing SE into the Mississippi **2** state S central U.S.; capital, Little Rock — **Ar·kan·san** \är-ˈkan-zən\ *adj or n*

Ar·khan·gelsk \är-ˈkan-ˌgelsk\ *or* **Arch·an·gel** \ˈär-ˌkān-jəl\ city N Russia in Europe, on the Northern Dvina

Ar·ling·ton \ˈär-liŋ-tən\ city N Texas E of Fort Worth

Ar·magh \är-ˈmä, ˈär-ˌ\ traditional county S Northern Ireland

Ar·me·nia \är-ˈmē-nē-ə, -nyə\ **1** region W Asia in mountainous area SE of Black Sea & SW of Caspian Sea divided between Iran, Turkey, Armenia (country), & Azerbaijan **2** country E Europe, capital, Yerevan; a constituent republic of U.S.S.R. 1936–91 — see LESSER ARMENIA — **Ar·me·ni·an** \-nē-ən, -nyən\ *adj or n*

Arn·hem Land \ˈärn-ˌhem, ˈär-nəm\ region N Australia on N coast of Northern Territory

Ar·no \ˈär-nō\ river 150 miles (241 kilometers) long central Italy flowing through Florence into Ligurian Sea

Aru·ba \ə-ˈrü-bə\ island Netherlands Antilles off coast of NW Venezuela NW of Curaçao; chief town, Oranjestad

Arun·a·chal Pra·desh \ˌär-ə-ˌnäch-əl-prə-ˈdāsh, -ˈdesh\ *or formerly* **North East Frontier Agency** state NE India N of Assam; capital, Itanagar

Ar·vada \är-ˈvad-ə\ city N central Colorado

Asa·hi·ka·wa \ˌäs-ə-he-ˈkä-wə\ *or* **Asa·hi·ga·wa** \-ˈgä-wə\ city Japan in central Hokkaido

Ashkh•a•bad \'ash-kə-ˌbad, -ˌbäd\ city, capital of Turkmenistan

Asia \'ā-zhə, -shə\ continent Eastern Hemisphere N of Equator — see EURASIA

Asia Mi•nor \-'mī-nər\ *or* **An•a•to•lia** \ˌan-ə-'tō-lē-ə, -'tōl-yə\ peninsula in modern Turkey between Black Sea on N & the Mediterranean on S

As•ma•ra \az-'mär-ə, -'mar-ə\ city, capital of Eritrea

As•sam \ə-'sam, a-; 'as-ˌam\ state NE India on edge of the Himalaya NW of Myanmar; capital, Dispur — **As•sam•ese** \ˌas-ə-'mēz, -'mēs\ *adj or n*

As•syr•ia \ə-'sir-ē-ə\ ancient empire W Asia extending along the middle Tigris & over foothills to the E; early capital Calah, later capital Nineveh — **As•syr•i•an** \-ən\ *adj or n*

As•ta•na \ä-stä-'nä\ city, capital of Kazakhstan

As•tra•khan \'as-trə-ˌkan, -kən\ city Russia in Europe, on the Volga at head of its delta

As•tu•ri•as \ə-'stu̇r-ē-əs\ region NW Spain on Bay of Biscay

Asun•ción \ə-ˌsün-sē-'ōn, ä-\ city, capital of Paraguay

As•wân \a-'swän, ä-\ city S Egypt on the Nile near site of **Aswân High Dam** which forms Lake Nasser

As•yût \ˌas-ē-'üt, ˌäs-\ city central Egypt on the Nile

Ata•ca•ma \ˌat-ə-'käm-ə\ **1** desert N Chile between Copiapó & Peru border **2** — see PUNA DE ATACAMA

Atchaf•a•laya \ə-ˌchaf-ə-'lī-ə, ˌchaf-\ river 225 miles (362 kilometers) long S Louisiana flowing S into Gulf of Mexico; receives waters of Red & Mississippi rivers

Ath•a•bas•ca \ˌath-ə-'bas-kə\ river 765 miles (1231 kilometers) long NE Alberta flowing into Lake Athabasca

Athabasca, Lake lake W central Canada on Alberta–Saskatchewan border

Ath•ens \'ath-ənz\ city, capital of Greece — **Athe•nian** \ə-'thē-nē-ən, -nyən\ *adj or n*

At•lan•ta \ət-'lant-ə, at-\ city, capital of Georgia

At•lan•tic \ət-'lant-ik, at-\ ocean separating North America & South America from Europe & Africa; often divided into **North Atlantic Ocean** & **South Atlantic Ocean**

At•lan•tis \ət-'lant-əs, at-\ fabled island that was traditionally placed W of Strait of Gibraltar and that sank into the sea

At•las \'at-ləs\ mountains NW Africa extending from SW Morocco to N Tunisia

At•ti•ca \'at-i-kə\ region E Greece; chief city, Athens

Auck•land \'ȯ-klənd\ city N New Zeland on NW North Island

Augs•burg \'ȯgz-ˌbərg, 'au̇gz-ˌbu̇rg\ city S Germany in S Bavaria NW of Munich

Au•gus•ta \ȯ-'gəst-ə, ə-\ city, capital of Maine

Au•la•vik National Park \'au̇-lə-ˌvik\ reservation Northwest Territories on N Banks Island

Au•ro•ra \ə-'rȯr-ə\ **1** city N central Colorado **2** city NE Illinois

Auschwitz — see OSWIECIM

Aus•tin \'ȯs-tən, 'äs-\ city, capital of Texas

Austral — see TUBUAI

Aus•tral•asia \ˌȯs-trə-'lā-shə, ˌäs-, -'lā-shə\ Australia, Tasmania, New Zealand, & Melanesia — **Aus•tral•asian** \-zhən, -shən\ *adj or n*

Aus•tra•lia \ȯ-'strāl-yə, ä-, ə-\ **1** continent of Eastern Hemisphere SE of Asia **2** *or in full* **Commonwealth of Australia** independent country including continent of Australia & island of Tasmania; capital, Canberra — **Aus•tra•lian** \-yən\ *adj or n*

Australian Alps mountain range SE Australia in E Victoria & SE New South Wales; part of Great Dividing Range

Australian Capital Territory district SE Australia including two areas, one containing Canberra (capital of Australia) & the other on Jervis Bay; surrounded by New South Wales

Aus•tria \'ȯs-trē-ə, 'äs-\ country central Europe; capital, Vienna — **Aus•tri•an** \-ən\ *adj or n*

Aus•tria–Hun•ga•ry \-'həŋ-gə-rē\ dual monarchy 1867–1918, central Europe including what is now Austria, Hungary, the Czech Republic, Bukovina & Transylvania in Romania, Slovenia, Croatia, Galicia in Poland, & part of NE Italy — **Aus•tro–Hun•gar•i•an** \'ȯs-trō-ˌhəŋ-'gar-ē-ən, 'äs-, -'ger-\ *adj or n*

Aus•tro•ne•sia \ˌȯs-trə-'nē-zhə, ˌäs-, -'nē-shə\ **1** the islands of the South Pacific **2** area extending from Madagascar through Malay Peninsula & Malay Archipelago to Hawaii & Easter Island — **Aus•tro•ne•sian** \-zhən, -shən\ *adj or n*

Au•vergne \ō-'veərn, -'veərn-yə, -'vərn\ **1** region & former province S central France **2** mountains S central France in Massif Central; highest peak 6188 feet (1886 meters)

Au•yuit•tuq National Park \au̇-'yü-ə-tək\ reservation Nunavut on Baffin Island

Ave•lla•ne•da \ˌav-ə-zhə-'nā-də\ city E Argentina on Río de la Plata E of Buenos Aires

Avon \'ā-vən, 'av-ən, *in the U.S. also* 'ā-ˌvän\ river 96 miles (154 kilometers) long central England flowing WSW past Stratford-upon-Avon into the Severn

Ayers Rock \'aərz-, 'eərz-\ outcrop central Australia in SW Northern Territory SW of Alice Springs

Azer•bai•jan \ˌaz-ər-ˌbī-'jän, ˌäz-\ country SE Europe, capital, Baku; a constituent republic of U.S.S.R. 1936–91

Azores \'ā-ˌzōrz, -ˌzȯrz, ə-'\ *or Portuguese* **Aço•res** \ə-'sōr-ēsh\ islands North Atlantic belonging to Portugal & lying 800 miles (1287 kilometers) W of Portuguese coast — **Azor•e•an** *or* **Azor•i•an** \ā-'zōr-ē-ən, ə-, -'zȯr-\ *adj or n*

Az•ov, Sea of \'az-ˌȯf, 'äz-, -ˌäv\ gulf of Black Sea between Ukraine & Russia

Baalbek — see HELIOPOLIS

Ba•bel•thu•ap \ˌbäb-əl-'tü-ˌäp\ island W Pacific; chief island of Palau

Bab•y•lon \'bab-ə-lən, -ˌlän\ ancient city, capital of Babylonia; site about 55 miles (89 kilometers) S of Baghdad near the Euphrates — **Bab•y•lo•nian** \ˌbab-ə-'lō-nyən, -nē-ən\ *adj or n*

Bab•y•lo•nia \ˌbab-ə-'lō-nyə, -nē-ə\ ancient country W Asia in valley of lower Euphrates and Tigris rivers; capital, Babylon — **Bab•y•lo•nian** \-nyən, -nē-ən\ *adj or n*

Ba•co•lod \bäk-'ō-ˌlȯd\ city Philippines on Negros

Bac•tria \'bak-trē-ə\ ancient country W Asia between the Hindu Kush & upper Oxus in present NE Afghanistan — **Bac•tri•an** \-ən\ *adj or n*

Ba•den–Würt•tem•berg \ˌbäd-n-'wərt-əm-ˌbərg, -'wu̇rt-; -'vuert-əm-ˌberk\ state SW Germany W of Bavaria; capital, Stuttgart

Bad•lands National Park \'bad-ˌlandz-, -ˌlanz-\ reservation SW South Dakota E of Black Hills

Baf•fin \'baf-ən\ island NE Canada in Arctic Archipelago N of Hudson Strait

Baffin Bay inlet of the North Atlantic between W Greenland & E Baffin Island

Bagh•dad \'bag-ˌdad, ˌbäg-'däd\ city, capital of Iraq on the middle Tigris

Ba•guio \ˌbäg-ē-'ō\ city, former summer capital of the Philippines in NW central Luzon

Ba•ha•ma \bə-'häm-ə, *by outsiders also* -'hā-mə\ islands in North Atlantic SE of Florida; an independent country; capital, Nassau — **Ba•ha•mi•an** \-'hā-mē-ən, -'häm-ē-ən\ *or* **Ba•ha•man** \-'hā-mən, -'häm-ən\ *adj or n*

Bahia — see SALVADOR

Bah•rain \bä-'rān\ islands in Persian Gulf off coast of Arabia forming an independent country; capital, Manama

Bai•kal *or* **Bay•kal** \bī-'kȯl, -'käl\ lake Russia in Asia, in mountains N of Mongolia

Baile Atha Cliath — see DUBLIN

Ba•ja California \ˌbä-hä-\ **1** peninsula NW Mexico W of Gulf of California **2** state NW Mexico in N Baja California Peninsula; capital, Mexicali

Baja California Sur \'su̇r\ state NW Mexico in S Baja California Peninsula; capital, La Paz

Bakh•ta•ran \ˌbäk-tə-'rän\ city W Iran

Ba•ku \bä-'kü\ city, capital of Azerbaijan on W coast of Caspian Sea

Bakwanga — see MBUJI-MAYI

Bal•a•ton \'bal-ə-ˌtän, 'bȯl-ə-ˌtōn\ lake W Hungary

Bal•boa Heights \bal-ˌbō-ə-\ town Panama; former administrative center for Canal Zone

Bâle — see BASEL

Bal•e•ar•ic \ˌbal-ē-'ar-ik\ islands E Spain in the W Mediterranean — see MAJORCA, MINORCA, IBIZA

Ba•li \'bäl-ē\ island Indonesia off E end of Java — **Ba•li•nese** \ˌbäl-i-'nēz, ˌbal-, -'nēs\ *adj or n*

Bal•kan \'bȯl-kən\ **1** mountains N Bulgaria extending from Serbia border to Black Sea; highest 7793 feet (2375 meters) **2** peninsula SE Europe between Adriatic & Ionian seas on the W & Aegean & Black seas on the E

Bal•kans \-kənz\ *or* **Balkan States** countries occupying the Balkan Peninsula: Slovenia, Croatia, Bosnia and Herzegovina, independent Macedonia, Serbia and Montenegro, Romania, Bulgaria, Albania, Greece, & Turkey (in Europe)

Bal•khash \bal-'kash, bäl-'käsh\ lake E Kazakhstan

Bal•tic Sea \'bȯl-tik\ arm of North Atlantic N Europe E of Scandinavian Peninsula

Bal•ti•more \'bȯl-tə-ˌmōr, -ˌmȯr; 'bȯl-tə-mər, 'bȯl-mər\ city N central Maryland

Ba•lu•chi•stan \bə-ˌlü-chə-'stan\ **1** arid region S Asia bordering on Arabian Sea in SW Pakistan & SE Iran **2** province SW Pakistan

Ba•ma•ko \ˌbäm-ə-'kō\ city, capital of Mali on the Niger

Ban•dar — see MACHILIPATNAM

Ban•dar Lam•pung \ˌbən-dər-'läm-ˌpu̇ŋ\ city & port Indonesia in S Sumatra

Ban•dar Se•ri Be•ga•wan \ˌbən-dər-ˌser-ē-bə-'gä-wən\ town, capital of Brunei

Ban•dung \'bän-ˌdu̇ŋ\ city Indonesia in W Java SE of Jakarta

Banff National Park \'bamf\ reservation SW Alberta on E slope of Rocky Mountains

Ban•ga•lore \'baŋ-gə-ˌlōr, -ˌlȯr\ city S India W of Madras, capital of Karnataka

Bang•kok \'baŋ-ˌkäk, baŋ-'\ city, capital of Thailand

Ban•gla•desh \ˌbäŋ-glə-'desh, ˌbaŋ-, -'dāsh\ country S Asia E of India; formerly part of Pakistan; an independent republic since 1971; capital, Dhaka — see EAST PAKISTAN

Ban•gui \bäⁿ-'gē\ city, capital of Central African Republic

Ban•jul \'bän-ˌjül\ *or formerly* **Bath•urst** \'bath-ˌərst, -ərst\ city, capital of Gambia

Bao•ding *or* **Pao•ting** \'bau̇-'diŋ\ city NE China in Hebei SW of Beijing

Bao•tou *or* **Pao–t'ou** \'bau̇-'tō\ city N China in SW Inner Mongolia

Bar•ba•dos \bär-'bād-əs, -ōz, äs, -ōs\ island British West Indies in Lesser Antilles E of Windward Islands; an independent country since 1966; capital, Bridgetown — **Bar•ba•di•an** \-'bād-ē-ən\ *adj or n*

Bar•ba•ry States \'bär-bə-rē, -brē\ the states of Morocco, Algeria, Tunisia, & Tripolitania while under Turkish rule

Bar•bu•da \bär-'büd-ə\ island British West Indies in the Leewards — see ANTIGUA

Bar•ce•lo•na \ˌbär-sə-'lō-nə\ city NE Spain on the Mediterranean; chief city of Catalonia

Ba•reil•ly *or* **Ba•re•li** \bə-'rā-lē\ city N India in NW central Uttar Pradesh

Ba•rents \'bar-ənts, 'bär-\ sea comprising part of Arctic Ocean between Spitsbergen & Novaya Zemlya

Ba•ri \'bär-ē\ city SE Italy, capital of Puglia on the Adriatic

Bar•king \'bär-kiŋ\ borough of E Greater London, England

Bar•na•ul \ˌbär-nə-'ül\ city S Russia in Asia, on the Ob

Bar•net \'bär-nət\ borough of N Greater London, England

Ba•ro•da \bə-'rōd-ə\ — see VADODARA

Bar•qui•si•me•to \ˌbär-kə-sə-'māt-ō\ city NW Venezuela

Bar•ran•qui•lla \ˌbar-ən-'kē-ə, -'kē-yə\ city N Colombia on the Magdalena

Barren Grounds treeless plains N Canada W of Hudson Bay

Bar•row, Point \'bar-ō\ most northerly point of Alaska & of U.S. at about 71°25' N latitude

Ba•sel \'bäz-əl\ *or French* **Bâle** \'bäl\ city NW Switzerland

Ba•si•lan \bä-'sē-ˌlän\ **1** island S Philippines **2** city on the island

Bas•il•don \'baz-əl-dən\ town SE England in Essex

Ba•si•li•ca•ta \bə-ˌzil-ə-'kät-ə, -ˌsil-\ region S Italy on Gulf of Taranto; capital, Potenza

Basque Country \'bask\ region N Spain on Bay of Biscay

Bas•ra \'bäs-rə, 'bəs-, 'bas-\ city S Iraq on Shatt al Arab

Bass \'bas\ strait separating Tasmania & continent of Australia

Bas•sein \bə-'sān\ city S Myanmar

Basse–Nor•man•die \ˌbäs-ˌnȯr-mäⁿ-'dē\ region N France on English Channel

Basse•terre \bäs-'ter\ town, capital of Saint Kitts and Nevis

Bas·tille \ba-'stēl\ medieval fortress, Paris; used as prison until destroyed by mobs on July 14, 1789
Basutoland — see LESOTHO
Ba·taan \bə-'tan, -'tän\ peninsula Philippines in W Luzon on W side of Manila Bay
Batavia — see JAKARTA
Bathurst — see BANJUL
Bat·on Rouge \,bat-n-'rüzh\ city, capital of Louisiana
Ba·var·ia \bə-'ver-ē-ə, -'var-\ *or German* **Bay·ern** \'bī-ərn\ state SE Germany bordering on the Czech Republic & Austria; capital, Munich — **Ba·var·i·an** \bə-'ver-ē-ən, -'var-\ *adj or n*
Ba·ya·mon \,bī-ə-'mōn\ city NE central Puerto Rico
Baykal — see BAIKAL
Beard·more \'biərd-,mōr, -,mȯr\ glacier Antarctica, one of world's largest
Beau·fort \'bō-fərt\ sea comprising part of Arctic Ocean NE of Alaska & NW of Canada
Beau·mont \'bō-,mänt, bō-'\ city SE Texas
Bech·u·a·na·land \,bech-'wän-ə-,land, -ə-'wän-\ **1** region S Africa N of Orange River **2** — see BOTSWANA — **Bech·u·a·na** \,bech-'wän-ə, -ə-'wän-\ *adj or n*
Bed·ford·shire \'bed-fərd-,shiər, -shər\ *or* **Bedford** county SE England
Bedloe's — see LIBERTY
Beer·she·ba \bir-'shē-bə, ber-\ town S Israel
Bei·jing \'bā-'jiŋ\ *or* **Pe·king** \'pē-'kiŋ, 'pā-\ city, capital of China
Bei·rut \bā-'rüt\ *or ancient* **Be·ry·tus** \bə-'rīt-əs\ city, capital of Lebanon
Be·la·rus \,bē-lə-'rüs, ,byel-ə\ country central Europe; capital, Minsk — see BELORUSSIA — **Be·la·ru·si·an** \-'rü-sē-ən, -'rəsh-ən\ *or* **Be·la·rus·sian** \-'rəsh-ən\ *adj or n*
Belau — see PALAU
Be·lém \bə-'lem\ *or* **Pa·rá** \pə-'rä\ city N Brazil on Pará River
Bel·fast \'bel-,fast, bel-'\ city, capital of Northern Ireland
Belgian Congo — see CONGO 2
Bel·gium \'bel-jəm\ *or French* **Bel·gique** \bel-'zhēk\ *or Flemish* **Bel·gië** \'bel-gē-ə\ country W Europe; capital, Brussels — **Bel·gian** \'bel-jən\ *adj or n*
Bel·grade \'bel-,grād, -,gräd, -,grad, bel-'\ *or* **Be·o·grad** \'beu̇-,gräd\ city, capital of Serbia and Montenegro on the Danube
Be·lize \bə-'lēz\ *or Spanish* **Be·li·ce** \bā-'lē-sā\ **1** *or formerly* **British Hon·du·ras** \hän-'du̇r-əs, -'dyu̇r-\ country Central America on the Caribbean; capital, Belmopan **2** city E Belize on the Caribbean
Belle·vue \'bel-,vyü\ city W Washington
Bel·mo·pan \,bel-mō-'pan\ city, capital of Belize
Be·lo Ho·ri·zon·te \'bā-lō-,hȯr-ə-'zänt-ē, 'bel-ō-, -,här-\ city E Brazil N of Rio de Janeiro
Be·lo·rus·sia \,bel-ō-'rəsh-ə, ,byel-\ *or* **Bye·lo·rus·sia** \bē-,el-ō-, ,byel-ō-\ former constituent republic of U.S.S.R.; became independent Belarus 1991 — **Be·lo·rus·sian** \,bel-ō-'rəsh-ən, ,byel-\ *adj or n*
Beloye More — see WHITE SEA
Be·ne·lux \'ben-l-,əks\ economic union comprising Belgium, Luxembourg, & the Netherlands; formed 1947
Ben·gal \ben-'gȯl, beŋ-\ region S Asia including delta of Ganges & Brahmaputra rivers; divided between West Bengal, India & Bangladesh — see EAST BENGAL, WEST BENGAL — **Ben·gal·ese** \,beŋ-gə-'lēz, ,ben-, -'lēs\ *adj or n*
Bengal, Bay of arm of Indian Ocean between India & Myanmar
Ben·gha·zi \ben-'gäz-ē, beŋ-, -'gaz-\ city NE Libya; formerly a capital of Libya
Ben·guela Current \ben-'gwel-ə, 'beŋ-, -'gel-\ cold current of the Atlantic Ocean flowing N along SW coast of Africa
Be·nin \bə-'nin, -'nēn; 'ben-ən\ **1** *or formerly* **Da·ho·mey** \də-'hō-mē\ country W Africa on Gulf of Guinea; capital, Porto-Novo **2** city SW Nigeria — **Ben·i·nese** \bə-,nin-'ēz, -,nēn-; ,ben-i-'nēz, -'nēs\ *adj or n*
Benin, Bight of the N section of Gulf of Guinea
Ben Nev·is \ben-'nev-əs\ mountain 4406 feet (1343 meters) W Scotland in the Grampians; highest in Great Britain
Be·no·ni \bə-'nō-nē\ city NE Republic of South Africa
Ber·ga·mo \'beər-gə-,mō, 'bər-\ city N Italy in Lombardy NE of Milan
Ber·gen \'bər-gən, 'beər-\ city & port SW Norway
Be·ring \'biər-iŋ, 'beər-\ **1** sea arm of the North Pacific between Alaska & NE Siberia **2** strait 53 miles (85 kilometers) wide between North America (Alaska) & Asia (Russia)
Berke·ley \'bər-klē\ city W California on San Francisco Bay N of Oakland
Berk·shire \'bərk-,shiər, -shər\ hills W Massachusetts; highest peak Mount Greylock 3491 feet (1064 meters)
Ber·lin \bər-'lin, ,bər-\ city, capital of Germany; divided 1945–90 into East Berlin & West Berlin; comprises a state of present-day Germany — **Ber·lin·er** \-'lin-ər\ *n*
Berlin, East former city, capital of East Germany 1945–90
Berlin, West former city, West Germany; an enclave lying wholly within East Germany
Ber·mu·da \bər-'myüd-ə, ,bər-\ islands W North Atlantic ESE of Cape Hatteras; a British colony; capital, Hamilton — **Ber·mu·dan** \-'myüd-n\ *or* **Ber·mu·di·an** \-'myüd-ē-ən\ *adj or n*
Bern \'bərn, 'beərn\ city, capital of Switzerland — **Ber·nese** \bər-'nēz, ,bər-, -'nēs\ *adj or n*
Berytus — see BEIRUT
Bes·kids \'bes-,kidz, be-'skēdz\ mountain ranges central Europe in the W Carpathians including **West Beskids** (in Poland, NW Slovakia, & E Czech Republic W of Tatra Mountains) & **East Beskids** (in NE Slovakia)
Bes·sa·ra·bia \,bes-ə-'rā-bē-ə\ region SE Europe between Dniester & Prut rivers now chiefly in Moldova — **Bes·sa·ra·bi·an** \-bē-ən\ *adj or n*
Beth·le·hem \'beth-li-,hem, -lē-həm, -lē-əm\ town of ancient Palestine in Judea SW of Jerusalem; now in West Bank
Bex·ley \'bek-slē\ borough of E Greater London, England
Bezwada — see VIJAYAWADA
Bhav·na·gar \bau̇-'nəg-ər\ city W India in S Gujarat
Bho·pal \bō-'päl\ city N central India NW of Nagpur, capital of Madhya Pradesh
Bhu·tan \bü-'tan, -'tän\ country S Asia in the Himalaya on NE border of India; a monarchy; capital, Thimphu — **Bhu·ta·nese** \,büt-n-'ēz, -'ēs\ *adj or n*
Bi·af·ra, Bight of \bē-'af-rə, bī-, -'äf-\ the E section of Gulf of Guinea in W Africa
Bia·ly·stok \bē-'äl-i-,stȯk\ city NE Poland
Bie·le·feld \'bē-lə-,felt\ city NW central Germany E of Münster
Big Bend National Park reservation SW Texas on Rio Grande
Big Thicket wilderness area E Texas NE of Houston
Bi·har \bi-'här\ state E India bordering on Nepal; capital, Patna
Bi·ki·ni \bə-'kē-nē\ island (atoll) W Pacific in Marshall Islands
Bil·bao \bil-'bä-,ō, -'bau̇, -'bā-ō\ city N Spain
Bil·lings \'bil-iŋz\ city S central Montana; largest in state
Bio·ko \bē-'ō-(,)kō\ *or formerly* **Fer·nan·do Póo** \fər-,nan-(,)dō-'pō\ island Equatorial Guinea in Bight of Biafra
Bir·ken·head \'bər-kən-,hed, ,bər-kən-'\ borough NW England
Bir·ming·ham \'bər-miŋ-,ham, *British usually* -miŋ-əm\ **1** city N central Alabama **2** city W central England
Bisayas — see VISAYAN
Bis·cay, Bay of \'bis-,kā, -kē\ inlet of North Atlantic between W coast of France & N coast of Spain
Bis·cayne National Park \bis-'kān-, 'bis-,\ reservation S Florida
Bish·kek \bish-'kek\ *or 1926–91* **Frun·ze** \'frün-zə\ city, capital of Kyrgyzstan
Bis·marck \'biz-,märk\ **1** city, capital of North Dakota **2** archipelago W Pacific N of E end of New Guinea
Bis·sau \bis-'au̇\ city, capital of Guinea-Bissau
Bi·thyn·ia \bə-'thin-ē-ə\ ancient country NW Asia Minor bordering on Sea of Marmara and Black Sea — **Bi·thyn·i·an** \-ē-ən\ *adj or n*
Black·burn \'blak-bərn, -,bərn\ borough NW England in Lancashire
Black Canyon of the Gun·ni·son National Park \'gən-ə-sən\ reservation SW central Colorado along the Gunnison River (150 miles or 241 kilometers flowing W & NW into the Colorado River)
Black Forest *or German* **Schwarz·wald** \'shfärts-,vält, 'shwȯrt-,swȯld\ forested mountain region SW Germany along E bank of the upper Rhine
Black Hills mountains W South Dakota & NE Wyoming; highest Harney Peak 7242 feet (2207 meters)
Black·pool \'blak-,pül\ town NW England in Lancashire
Black Sea *or ancient* **Pon·tus Eux·i·nus** \,pän-təs-yük-'sī-nəs\ *or* **Pon·tus** \'pän-təs\ sea between Europe & Asia connected with Aegean Sea through the Bosporus, Sea of Marmara, & Dardanelles
Blae·nau Gwent \'blī-,nī-'gwent\ administrative area SE Wales
Blan·tyre \'blan-,tīr\ city S Malawi
Bloem·fon·tein \'blüm-fən-,tān, -,fän-\ city, judicial capital of the Republic of South Africa
Blue Nile river 850 miles (1368 kilometers) long Ethiopia & Sudan flowing NNW into the Nile at Khartoum
Blue Ridge E range of the Appalachians E U.S. extending from S Pennsylvania to N Georgia
Bo·bo–Diou·las·so \'bō-,bō-dyü-,las-ō\ town W Burkina Faso
Bo·chum \'bō-kəm\ city W Germany in valley of the Ruhr
Bodh Ga·ya \'bȯd-'gī-ä\ village NE India in central Bihar
Boe·o·tia \bē-'ō-shē-ə, -shə\ district E central Greece NW of Attica; chief ancient city, Thebes — **Boe·o·tian** \-shē-ən, -shən\ *adj or n*
Bo·go·tá \,bō-gə-'tó, -'tä\ city, capital of Colombia
Bo Hai *or* **Po Hai** \'bō-'hī\ *or* **Gulf of Chih·li** \'chē-lē, 'jiər-'lē\ arm of Yellow Sea NE China N of Shandong Peninsula
Bo·he·mia \bō-'hē-mē-ə\ region W Czech Republic; once a kingdom & later a province; chief city, Prague
Boi·se \'bȯi-sē, -zē\ city, capital of Idaho
Bokhara — see BUKHARA
Boks·burg \'bäks-,bərg\ city NE Republic of South Africa
Bo·liv·ia \bə-'liv-ē-ə\ country W central South America; administrative capital, La Paz; constitutional capital, Sucre — **Bo·liv·ian** \-ē-ən\ *adj or n*
Bo·lo·gna \bə-'lōn-yə, -'lōn-ə\ city N Italy N of Florence, capital of Emilia-Romagna
Bol·ton \'bōlt-n\ town NW England in Greater Manchester
Bom·bay \bäm-'bā\ *or* **Mum·bai** \'məm-,bī\ city, capital of Maharashtra, India
Bône — see ANNABA
Bo·nin \'bō-nən\ *or* **Oga·sa·wa·ra** \ō-,gäs-ə-'wär-ə\ islands Japan in W Pacific SE of Honshu
Bonn \'bän, 'bȯn\ city W Germany on the Rhine SSE of Cologne; formerly capital of West Germany
Bon·ne·ville Salt Flats \'bän-ə-,vil\ broad level area of desert NW Utah
Boo·thia \'bü-thē-ə\ peninsula N Nunavut, Canada W of Baffin Island; its N tip is most northerly point on North American mainland
Bor·deaux \bȯr-'dō\ city SW France on the Garonne
Bor·neo \'bȯr-nē-,ō\ island Malay Archipelago SW of the Philippines
Bos·nia \'bäz-nē-ə\ region S Europe; with Herzegovina constitutes country of **Bosnia and Herzegovina**; capital, Sarajevo (in Bosnia) — **Bos·ni·an** \-nē-ən\ *adj or n*
Bos·po·rus \'bäs-pə-rəs, -prəs\ *or ancient* **Bosporus Thra·ci·us** \-'thrā-shē-əs, -shəs\ strait about 18 miles (29 kilometers) long between Turkey in Europe & Turkey in Asia connecting Sea of Marmara & Black Sea
Bos·ton \'bȯ-stən\ city, capital of Massachusetts — **Bos·to·nian** \bȯ-'stō-nē-ən, -nyən\ *adj or n*
Bot·a·ny Bay \'bät-n-ē, 'bät-nē\ inlet of South Pacific SE Australia in New South Wales S of Sydney
Both·nia, Gulf of \'bäth-nē-ə\ arm of Baltic Sea between Sweden & Finland
Bo·tswa·na \bät-'swän-ə\ country S Africa N of Molopo River; formerly British protectorate of Bechuanaland; now an independent republic; capital, Gaborone
Boulder Dam — see HOOVER DAM
Bourgogne — see BURGUNDY
Bourne·mouth \'bȯrn-məth, 'bȯrn-, 'bu̇rn-\ town S England in Dorset on English Channel
Brad·ford \'brad-fərd\ city N England
Brah·ma·pu·tra \,bräm-ə-'pü trə, -'pyü-\ river 1800 miles (2900 kilometers) long S Asia flowing from the Himalaya in Tibet to delta of the Ganges

Bra·ila \brə-ˈē-lə\ city E Romania
Bran·den·burg \ˈbran-dən-ˌbərg\ state E Germany bordering on Poland; capital, Potsdam
Bra·sí·lia \brə-ˈzil-yə\ city, capital of Brazil in Federal District
Bra·sov \bräsh-ˈóv\ city central Romania
Bra·ti·sla·va \ˌbrat-ə-ˈsläv-ə, ˌbrät-\ city, capital of Slovakia
Bratsk \ˈbrätsk\ city S central Russia in Asia, NNE of Irkutsk
Braunschweig — see BRUNSWICK
Bra·zil \brə-ˈzil\ country E & central South America; capital, Brasília — **Bra·zil·ian** \brə-ˈzil-yən\ *adj or n*
Brazil Current warm current of the Atlantic Ocean flowing S along coast of Brazil
Braz·za·ville \ˈbraz-ə-ˌvil, ˈbräz-ə-ˌvēl\ city, capital of Republic of the Congo on W bank of Pool Malebo in lower Congo River
Bre·men \ˈbrem-ən, ˈbrā-mən\ **1** state NW Germany **2** city, its capital
Bren·ner \ˈbren-ər\ pass 4495 feet (1370 meters) in the Alps between Austria & Italy
Brent \ˈbrent\ borough of W Greater London, England
Bre·scia \ˈbresh-ə, ˈbrā-shə\ city N Italy in Lombardy ENE of Milan
Breslau — see WROCLAW
Brest \ˈbrest\ **1** city SW Belarus **2** city NW France in Brittany
Bret·on, Cape \kāp-ˈbret-n, kə-ˈbret-, -ˈbrit-\ headland Canada; most easterly point of Cape Breton Island & of Nova Scotia
Bridg·end \ˌbrij-ˈend\ administrative area S Wales
Bridge·port \ˈbrij-ˌpōrt, -ˌpȯrt\ city SW Connecticut on Long Island Sound
Bridge·town \ˈbrij-ˌtau̇n\ city, capital of Barbados
Brigh·ton \ˈbrīt-n\ town S England in East Sussex on English Channel
Bris·bane \ˈbriz-bən, -ˌbān\ city E Australia, capital of Queensland
Bris·tol \ˈbris-tl\ **1** city SW England **2** channel between S Wales & SW England
Brit·ain \ˈbrit-n\ **1** the island of Great Britain **2** UNITED KINGDOM
British Columbia province W Canada on North Pacific coast; capital, Victoria
British Commonwealth — see COMMONWEALTH (the)
British Guiana — see GUYANA
British Honduras — see BELIZE 1
British Indian Ocean Territory British colony in Indian Ocean comprising Chagos Archipelago & formerly Aldabra, Farquhar, & Desroches islands (returned to Seychelles 1976)
British Isles island group W Europe comprising Great Britain, Ireland, & adjacent islands
British Solomon Islands former British protectorate comprising the Solomon Islands (except Bougainville, Buka, & adjacent small islands) & Santa Cruz Islands; capital, Honiara
British Somaliland former British protectorate E Africa bordering on Gulf of Aden; capital, Hargeisa; since 1960 part of Somalia
British Virgin Islands E islands of Virgin Islands group; a British dependency; capital, Road Town (on Tortola Island)
British West Indies islands of the West Indies belonging to the Commonwealth & including Jamaica, Trinidad and Tobago, & the Bahama, Cayman, Windward, Leeward, & British Virgin islands
Brit·ta·ny \ˈbrit-n-ē\ *or French* **Bre·tagne** \brə-ˈtänʸ\ region & former province NW France SW of Normandy
Brno \ˈbər-nō\ city SE Czech Republic, chief city of Moravia
Brom·ley \ˈbräm-lē\ borough of SE Greater London, England
Bronx \ˈbräŋs, ˈbräŋks\ *or* **The Bronx** borough of New York City on mainland NE of Manhattan Island
Brook·lyn \ˈbru̇k-lən\ borough of New York City at SW end of Long Island
Brooks Range \ˈbru̇ks\ mountains N Alaska
Bruce Peninsula National Park \ˈbrüs\ reservation SE Ontario
Bruges \ˈbrüzh, ˈbrüezh\ *or Flemish* **Brug·ge** \ˈbrüeg-ə\ city NW Belgium
Bru·nei \bru̇n-ˈī, ˈbrü-ˌnī\ sultanate NE Borneo; formerly a British protectorate; capital, Bandar Seri Begawan — **Bru·nei·an** \bru̇n-ˈī-ən\ *adj or n*
Bruns·wick \ˈbrənz-wik\ *or German* **Braun·schweig** \ˈbrau̇n-ˌshwīg, -ˌshfīk\ city central Germany W of Berlin
Brus·sels \ˈbrəs-əlz\ city, capital of Belgium
Bryansk \brē-ˈänsk\ city W Russia in Europe, SW of Moscow
Bryce Canyon National Park \ˈbrīs\ reservation S Utah NE of Zion National Park
Bu·ca·ra·man·ga \ˌbü-kə-rə-ˈmäŋ-gə\ city N Colombia NNE of Bogotá
Bu·cha·rest \ˈbü-kə-ˌrest, ˈbyü-\ city, capital of Romania
Bucheon — see PUCHON
Buck·ing·ham·shire \ˈbək-iŋ-əm-ˌshiər, -shər, *in the U.S. also* -iŋ-ˌham-\ *or* **Buckingham** county SE central England
Bu·da·pest \ˈbüd-ə-ˌpest *also* ˈbyüd-, ˈbu̇d-, -ˌpesht\ city, capital of Hungary
Bue·nos Ai·res \ˌbwā-nə-ˈsaər-ēz, ˌbō-nə-, -ˈseər-, -ˈsīr-\ city, capital of Argentina on Rio de la Plata
Buf·fa·lo \ˈbəf-ə-ˌlō\ city W New York on Lake Erie
Bu·jum·bu·ra \ˌbü-jəm-ˈbu̇r-ə\ *or formerly* **Usum·bu·ra** \ˌü-səm-ˈbu̇r-ə\ city, capital of Burundi
Bu·ka·vu \bü-ˈkäv-ü\ city E Democratic Republic of the Congo
Bu·kha·ra \bü-ˈkär-ə, -ˈkar-, -ˈhär-, -ˈhar-\ *or* **Bo·khara** \bō-\ city Uzbekistan E of the Amu Dar'ya
Bu·la·wayo \ˌbu̇l-ə-ˈwā-ō, -ˈwī-\ city SW Zimbabwe
Bul·gar·ia \ˌbəl-ˈgar-ē-ə, bu̇l-, -ˈger-\ country SE Europe on Black Sea; capital, Sofia
Bur·bank \ˈbər-ˌbaŋk\ city SW California
Bur·gun·dy \ˈbər-gən-dē\ *or French* **Bour·gogne** \bür-ˈgȯnʸ\ region E France; a former kingdom, duchy, & province — **Bur·gun·di·an** \bər-ˈgən-dē-ən, ˌbər-\ *adj or n*
Bur·ki·na Fa·so \bu̇r-ˈkē-nə-ˈfäs-ō, bər-\ *or formerly* **Upper Vol·ta** \ˈvōl-tə, ˈvȯl-, ˈväl-\ country W Africa N of Ivory Coast, Ghana, & Togo; capital, Ouagadougou
Bur·ling·ton \ˈbər-liŋ-tən\ city NW Vermont; largest in state
Burma — see MYANMAR
Bur·sa \bu̇r-ˈsä, ˈbər-sə\ city NW Turkey in Asia near Sea of Marmara
Bu·run·di \bu̇-ˈrün-dē\ *or formerly* **Urun·di** \u̇-ˈrün-dē\ country E central Africa; capital, Bujumbura — see RUANDA-URUNDI
Busan — see PUSAN
Bute \ˈbyüt\ island SW Scotland in Firth of Clyde
Byd·goszcz \ˈbid-ˌgȯshch, -ˌgȯsh\ *or German* **Brom·berg** \ˈbräm-ˌbərg, ˈbrȯm ˌberk\ city NW central Poland
Byelorussia — see BELORUSSIA
Byzantium — see ISTANBUL
Caen \ˈkäⁿ\ city NW France
Caer·phil·ly \kər-ˈfil-ē\ administrative area SE Wales
Cae·sa·rea \ˌsē-zə-ˈrē-ə; ˌses-ə-, ˌsez-ə-\ city of ancient Palestine in Samaria on the Mediterranean; Roman capital of Palestine
Ca·glia·ri \ˈkäl-yə-rē\ city Italy, capital of Sardinia
Ca·guas \ˈkäg-ˌwäs\ town E central Puerto Rico
Cai·ro \ˈkī-rō\ city, capital of Egypt — **Cai·rene** \kī-ˈrēn\ *adj or n*
Ca·la·bria \kə-ˈlā-brē-ə, -ˈläb-rē-\ **1** district of ancient Italy comprising area forming heel of peninsula of Italy; now S part of Puglia **2** *or ancient* **Brut·ti·um** \ˈbru̇t-ē-əm, ˈbrət-\ region S Italy occupying toe of peninsula of Italy; capital, Catanzaro — **Ca·la·bri·an** \kə-ˈlā-brē-ən, -ˈläb-rē-\ *adj or n*
Cal·cut·ta \kal-ˈkət-ə\ *or* **Kol·ka·ta** \kōl-ˈkä-tä\ city E India on Hugli River capital of West Bengal — **Cal·cut·tan** \-ˈkət-n\ *adj or n*
Cal·e·do·nia \ˌkal-ə-ˈdō-nyə, -nē-ə\ — see SCOTLAND — **Cal·e·do·nian** \-nyən, -nē-ən\ *adj or n*
Cal·ga·ry \ˈkal-gə-rē\ city SW Alberta
Ca·li \ˈkäl-ē\ city W Colombia
Cal·i·cut \ˈkal-i-ˌkət\ *or* **Ko·zhi·kode** \ˈkō-zhə-ˌkōd\ city & port SW India
Cal·i·for·nia \ˌkal-ə-ˈfȯr-nyə\ state SW U.S.; capital, Sacramento — **Cal·i·for·nian** \-nyən\ *adj or n*
California, Gulf of arm of the North Pacific NW Mexico
California Current cold current of the North Pacific flowing SE along W coast of North America
Ca·llao \kə-ˈyä-ō, kə-ˈyau̇\ city W Peru W of Lima
Cal·va·ry \ˈkalv-rē, -ə-rē\ *or Hebrew* **Gol·go·tha** \ˈgäl-gə-thə, gäl-ˈgäth-ə\ place outside ancient Jerusalem where Jesus was crucified
Ca·ma·güey \ˌkam-ə-ˈgwā\ city E central Cuba
Cam·bay, Gulf of \kam-ˈbā\ inlet of Arabian Sea India N of Bombay
Cam·ber·well \ˈkam-bər-ˌwel, -wəl\ city SE Australia in S Victoria E of Melbourne
Cam·bo·dia \kam-ˈbōd-ē-ə\ *or* **Kam·pu·chea** \ˌkam-pu̇-ˈchē-ə\ country SE Asia bordering on Gulf of Thailand; capital, Phnom Penh — **Cam·bo·di·an** \-ē-ən\ *adj or n*
Cam·bria \ˈkam-brē-ə\ — see WALES
Cam·bridge \ˈkām-brij\ **1** city E Massachusetts W of Boston **2** city E England in Cambridgeshire
Cam·bridge·shire \ˈkām-brij-ˌshiər, -shər\ *or* **Cambridge** county E England
Cam·den \ˈkam-dən\ **1** city SW New Jersey **2** borough of N Greater London, England
Cam·er·oon *or French* **Cam·er·oun** \ˌkam-ə-ˈrün\ country W equatorial Africa; capital, Yaoundé — **Cam·er·oo·nian** \-ˈrü-nē-ən, -ˈrü-nyən\ *adj or n*
Cam·er·oons \ˌkam-ə-ˈrünz\ region W Africa on NE Gulf of Guinea formerly belonging to the British and French but now divided between Nigeria & Cameroon
Cam·pa·nia \kam-ˈpā-nyə, -nē-ə\ region S Italy bordering on Tyrrhenian Sea; capital, Naples — **Cam·pa·nian** \-nyən, -nē-ən\ *adj or n*
Cam·pe·che \kam-ˈpē-chē, käm-ˈpā-chā\ state SE Mexico in W Yucatán Peninsula; capital, Campeche
Cam·pi·nas \kam-ˈpē-nəs\ city SE Brazil N of São Paulo
Cam·po·bas·so \ˌkäm-pō-ˈbäs-ˌō\ city central Italy
Cam·po Gran·de \ˌkam-pü-ˈgran-də, -dē\ city SW Brazil
Cam·pos \ˈkam-pəs\ city SE Brazil NE of Rio de Janeiro
Cam Ranh Bay \ˈkäm-ˈrän\ inlet of South China Sea SE Vietnam
Ca·naan \ˈkā-nən\ ancient region corresponding approximately to later Palestine — **Ca·naan·ite** \ˈkā-nə-ˌnīt\ *adj or n*
Can·a·da \ˈkan-ə-də\ country N North America; capital, Ottawa
Canadian Shield *or* **Lau·ren·tian Plateau** \lȯ-ˈren-chən-\ plateau region E Canada & NE U.S. extending from Mackenzie Basin E to Davis Strait & S to S Quebec, S central Ontario, NE Minnesota, N Wisconsin, NW Michigan, & NE New York including the Adirondacks
Canal Zone *or* **Panama Canal Zone** strip of territory Panama; under U.S. control through 1999 for administration of the Panama Canal
Ca·nary \kə-ˈner-ē\ islands in the North Atlantic off NW coast of Africa belonging to Spain; capital, Las Palmas
Ca·nav·er·al, Cape \kə-ˈnav-rəl, -ə-rəl\ *or 1963–73 officially* **Cape Ken·ne·dy** \-ˈken-ə-dē\ headland E Florida in the North Atlantic on Canaveral Peninsula E of Indian River
Can·ber·ra \ˈkan-bə-rə, -brə, -ˌber-ə\ city, capital of Australia in Australian Capital Territory
Cannes \ˈkan, ˈkän\ town & port SE France
Can·ta·bria \kän-ˈtäb-rē-ə\ region N Spain on Bay of Biscay
Can·ter·bury \ˈkant-ər-ˌber-ē, ˈkant-ə-, -bə-rē, -brē\ **1** city SE Australia in E New South Wales **2** city SE England in Kent
Can·ton \ˈkant-n\ city NE Ohio
Canton — see GUANGZHOU
Can·yon·lands National Park \ˈkan-yən-ˌlandz, -ˌlanz\ reservation SE Utah
Cape Bret·on Highlands National Park \kāp-ˌbret-n, kə-ˌbret-, -ˌbrit-\ reservation NE Nova Scotia near N end of Cape Breton Island
Cape Breton Island island NE Nova Scotia
Cape Coral city SW Florida
Cape of Good Hope **1** — see GOOD HOPE (Cape of) **2** *or* **Cape Province** *or before 1910* **Cape Colony** former province S Republic of South Africa; capital, Cape Town
Ca·per·na·um \kə-ˈpər-nē-əm\ city of ancient Palestine on NW shore of Sea of Galilee
Cape Town \ˈkāp-ˌtau̇n\ city Republic of South Africa, its legislative capital
Cape Verde \ˈvərd\ **1** islands in the North Atlantic off W Africa; a republic; capital, Praia; until 1975 belonged to Portugal **2** — see VERDE (Cape)

Cape York Peninsula \'yörk\ peninsula NE Australia in N Queensland

Capitol Reef National Park reservation S central Utah

Cap·pa·do·cia \ˌkap-ə-'dō-shə, -shē-ə\ ancient country & Roman province E Asia Minor; capital, Caesarea Mazaca

Ca·pri \ka-'prē, kə-; 'käp-rē, 'kap-\ island Italy S of Bay of Naples

Ca·ra·cas \kə-'rak-əs, -'räk-\ city, capital of Venezuela

Car·diff \'kärd-əf\ **1** administrative area S Wales **2** city, capital of Wales

Ca·rib·be·an Sea \ˌkar-ə-'bē-ən, kə-'rib-ē-ən\ arm of the North Atlantic bounded on N & E by West Indies, on S by South America, & on W by Central America

Ca·rin·thia \kə-'rin-thē-ə, -'rint-\ region central Europe in E Alps in S Austria & Slovenia

Car·low \'kär-ˌlō\ county SE Ireland in Leinster

Carls·bad Caverns \'kärlz-ˌbad\ limestone caves SE New Mexico in **Carlsbad Caverns National Park**

Car·mar·then·shire \kär-'mär-thən-ˌshiər, kər-, kə-, -shər\ *or* **Carmarthen** administrative area S Wales

Carmel, Mount \'kär-məl\ mountain ridge N Israel; highest point 1791 feet (546 meters)

Car·o·li·na \ˌkar-ə-'lī-nə\ English colony on E coast of North America founded 1663 & divided 1729 into North Carolina & South Carolina (the **Carolinas**) — **Car·o·lin·i·an** \-'lin-ē-ən\ *adj or n*

Ca·ro·li·na \ˌkär-ə-'lē-nə\ city NE central Puerto Rico

Car·o·line \'kar-ə-ˌlīn, -lən\ islands W Pacific E of S Philippines comprising Palau and the Federated States of Micronesia; formerly part of Trust Territory of the Pacific Islands

Car·pa·thi·an \kär-'pā-thē-ən\ mountains E central Europe along boundary between Slovakia & Poland & in N & central Romania; highest Gerlachovsky 8711 feet (2655 meters)

Carpathian Ruthenia — see RUTHENIA

Car·pen·tar·ia, Gulf of \ˌkär-pən-'ter-ē-ə, -'tar-\ inlet of Arafura Sea N of Australia

Car·roll·ton \'kar-əl-tən\ city N Texas

Car·son City \'kärs-ⁿn\ city, capital of Nevada

Car·ta·ge·na \ˌkärt-ə-'gā-nə, -'hā-\ **1** city NW Colombia **2** city SE Spain

Car·thage \'kär-thij\ ancient city N Africa NE of modern Tunis; capital of an empire that included at greatest extent much of NW Africa, E Spain, & Sicily — **Car·tha·gin·ian** \ˌkär-thə-'jin-yən, -'jin-ē-ən\ *adj or n*

Ca·sa·blan·ca \ˌkas-ə-'blaŋ-kə, ˌkaz-\ *or Arabic* **Dar el Bei·da** \ˌdär-ˌel-bā-'dä\ city W Morocco on the North Atlantic

Cas·cade Range \kas-'kād, 'kas-'kād\ mountains NW U.S. in Washington, Oregon, & N California — see RAINIER (Mount)

Cas·pi·an Sea \'kas-pē-ən\ salt lake between Europe & Asia about 90 feet (27 meters) below sea level

Cas·tile \kas-'tēl\ *or Spanish* **Cas·ti·lla** \kä-'stē-lʸä, -'stē-yä\ region & ancient kingdom central & N Spain

Cas·tries \ka-'strē, 'kas-ˌtrēz\ city, capital of St. Lucia

Cat·a·lo·nia \ˌkat-ᵊl-'ō-nyə, -nē-ə\ *or Spanish* **Ca·ta·lu·ña** \ˌkät-ᵊl-'ü-nyə\ region NE Spain bordering on France & the Mediterranean; chief city, Barcelona — **Cat·a·lo·nian** \-'ō-nyən, -nē-ən\ *adj or n*

Ca·ta·nia \kə-'tän-yə, -'tän-\ city Italy in E Sicily at foot of Mount Etna

Ca·tan·za·ro \ˌkät-ˌän-'zär-ˌō, -ˌänd-\ city S Italy

Ca·thay \ka-'thā, ka-\ CHINA — an old name

Cats·kill \'kat-ˌskil\ mountains in Appalachian system SE New York W of the Hudson

Cau·ca·sia \kò-'kā-zhə, -shə\ *or* **Caucasus** region SE Europe between Black & Caspian seas

Cau·ca·sus \'kò-kə-səs\ mountain system in Caucasia — see ELBRUS (Mount)

Cav·an \'kav-ən\ county NE Ireland (republic) in Ulster

Cay·enne \kī-'en, kā-\ city, capital of French Guiana

Cay·man \kā-'man, 'kā-, *attributively* 'kā-mən\ islands British West Indies NW of Jamaica; a British colony; capital, Georgetown (on Grand Cayman Island)

Ce·bu \sā-'bü\ **1** island E central Philippines in Visayan Islands **2** city on E coast of Cebu Island

Ce·dar Rapids \'sēd-ər\ city E Iowa

Celebes — see SULAWESI

Celestial Empire the former Chinese Empire

Cel·le \'tsel-ə, 'sel-ə\ city N central Germany NE of Hannover

Cel·tic \'kel-tik, 'sel-\ sea inlet of the North Atlantic in British Isles SE of Ireland, SW of Wales, & W of Cornwall

Central African Republic country N central Africa; formerly the French territory of **Uban·gi-Sha·ri** \ü-'baŋ-gē-'shär-ē, yü-, -'baŋ-ē-\; capital, Bangui

Central America narrow portion of North America from S border of Mexico to South America — **Central American** *adj or n*

Centre \'säⁿtrᵊ\ region central France

Ce·ram *or* **Se·ram** \'sā-ˌräm\ island E Indonesia in central Moluccas

Ce·re·dig·i·on \ˌker-ə-'dig-ē-ˌän\ administrative area SW Wales

Cé·vennes \sā-'ven\ mountain range S France W of the Rhone in SE Massif Central; highest peak Mount Mézenc 5755 feet (1754 meters)

Cey·lon \si-'län, sā-\ **1** island in Indian Ocean off S India **2** — see SRI LANKA — **Cey·lon·ese** \ˌsā-lə-'nēz, ˌsē-lə-, ˌsel-ə-, -'nēs\ *adj or n*

Chad *or French* **Tchad** \'chad\ country N central Africa; capital, N'Djamana — **Chad·ian** \'chad-ē-ən\ *adj or n*

Chad, Lake shallow lake N central Africa at junction of boundaries of Chad, Niger, & Nigeria

Cha·gos \'chä-gəs\ archipelago central Indian Ocean; comprises British Indian Ocean Territory — see DIEGO GARCIA

Chal·cid·i·ce \kal-'sid-ə-sē\ peninsula NE Greece in E Macedonia

Chal·dea \kal-'dē-ə\ ancient region SW Asia on Euphrates River & Persian Gulf — **Chal·de·an** \-'dē-ən\ *adj or n* — **Chal·dee** \'kal-ˌdē\ *n*

Cham·pagne–Ar·denne \sham-'pān-är-'den\ region N France bordering on Belgium

Cham·plain, Lake \sham-'plān\ lake between New York & Vermont extending N into Quebec

Chan·di·garh \'chən-dē-gər\ city N India N of Delhi, constitutes a territory administered by the national government

Chan·dler \'chand-lər\ city SW central Arizona

Chang \'chäŋ\ *or* **Yang·tze** \'yaŋ-'sē; 'yaŋt-sē, 'yaŋkt-\ river 3434 miles (5525 kilometers) long central China flowing into East China Sea

Changan — see XI'AN

Ch'ang–chia–k'ou — see ZHANGJIAKOU

Ch'ang–chou — see ZHANGZHOU

Chang–chun \'chäŋ-'chün\ city NE China, capital of Jilin

Chang·de *or* **Chang·te** \'chäŋ-'də\ city SE central China in N Hunan

Chang·sha \'chäŋ-'shä\ city SE central China, capital of Hunan

Channel **1** — see SANTA BARBARA 2 **2** islands in English Channel including Jersey, Guernsey, & Alderney & belonging to United Kingdom

Channel Islands National Park reservation California off SW coast

Cha·pa·la \chə-'päl-ə\ lake W central Mexico SE of Guadalajara

Charles \'chärlz\ river 47 miles (76 kilometers) long E Massachusetts

Charles, Cape cape E Virginia N of entrance to Chesapeake Bay

Charles·ton \'chärl-stən\ city, capital of West Virginia

Char·lotte \'shär-lət\ city S North Carolina

Charlotte Ama·lie \ə-'mäl-yə\ city, capital of Virgin Islands of the U.S.; on Saint Thomas Island

Char·lotte·town \'shär-lət-ˌtaun\ city, capital of Prince Edward Island, Canada

Chat·ta·noo·ga \ˌchat-ə-'nü-gə, ˌchat-n-'ü-\ city SE Tennessee

Che·bok·sa·ry \ˌcheb-ˌäk-'sär-ē\ city central Russia in Europe on the Volga

Chech·nya \chech-'nyä, 'chech-nyə\ area of SE Russia in Europe; capital, Grozny

Chekiang — see ZHEJIANG

Chelsea — see KENSINGTON AND CHELSEA

Che·lya·binsk \chel-'yä-bənsk\ city W Russia in Asia S of Sverdlovsk

Chemnitz \'kem-ˌnits, -nəts\ *or 1953–90* **Karl-Marx-Stadt** \ˌkärl-'märk-ˌshtät, -ˌstät\ city E Germany SE of Leipzig

Chen–chiang — see ZHENJIANG

Cheng–chou — see ZHENGZHOU

Cheng·du *or* **Ch'eng–tu** \'chəŋ-'dü\ city SW central China, capital of Sichuan

Chennai — see MADRAS

Cher·no·byl \chər-'nō-bəl, cher-\ site N Ukraine of town abandoned after 1986 nuclear accident

Ches·a·peake Bay \'ches-ˌpēk, -ə-ˌpēk\ inlet of the North Atlantic in Virginia & Maryland

Chesh·ire \'chesh-ər, 'chesh-ˌiər\ *or* **Ches·ter** \'ches-tər\ county W England bordering on Wales

Chev·i·ot \'chev-ē-ət, 'chē-vē-ət\ hills along English–Scottish border

Chey·enne \shī-'an, -'en\ city, capital of Wyoming

Chhat·tis·garh \chə-'tēz-gər\ state central India; capital, Raipur

Chia–mu–ssu — see JIAMUSI

Chia·pas \chē-'äp-əs\ state SE Mexico; capital, Tuxtla Gutiérrez

Chi·ba \'chē-bə\ city E Japan in Honshu on Tokyo Bay E of Tokyo

Chi·ca·go \shə-'käg-ō, -'kòg-\ city NE Illinois — **Chi·ca·go·an** \-'käg-ə-wən, -'kòg-\ *n*

Chi·chén It·zá \chə-ˌchen-ət-'sä\ ruined Mayan city SE Mexico in Yucatán ESE of Mérida

Ch'i–ch'i–ha–erh — see QIQIHAR

Chihli, Gulf of — see BO HAI

Chi·hua·hua \chə-'wä-wä, shə-, -wə\ **1** state N Mexico bordering on U.S. **2** city, its capital

Chile \'chil-ē\ country SW South America; capital, Santiago — **Chil·ean** \'chil-ē-ən, chə-'lā-ən\ *adj or n*

Chi–lung *or* **Keelung** city & port N Taiwan

Chim·bo·ra·zo \ˌchim-bə-'räz-ō, ˌshim-\ mountain 20,561 feet (6267 meters) W central Ecuador

Chim·kent \chim-'kent\ city S Kazakhstan N of Tashkent

Chi·na, People's Republic of \'chī-nə\ country E Asia; capital, Beijing

China, Republic of — see TAIWAN

China, Sea of sea section of the W Pacific E & SE of China; divided at Taiwan Strait into **East China** & **South China** seas

Chinan — see JINAN

Chin–chou *or* **Chinchow** — see JINZHOU

Chi·os \'kī-ˌäs\ island Greece in the Aegean off W coast of Turkey

Chi·si·nau \ˌkē-shē-'naù\ *or 1940–91* **Ki·shi·nev** \'kish-i-ˌnef\ city, capital of Moldova

Chi·ta \chit-'ä\ city S Russia in Asia, E of Lake Baikal

Chit·ta·gong \'chit-ə-ˌgäŋ, -ˌgòŋ\ city SE Bangladesh on Bay of Bengal

Chkalov — see ORENBURG

Chong·jin \'chòŋ-ˌjin\ city & port NE North Korea on Sea of Japan (East Sea)

Chong·ju *or* **Cheong·ju** \'chəŋ-ˌjü\ city central South Korea

Chong·qing *or* **Ch'ung–ch'ing** \'chùŋ-'chiŋ\ *or* **Chung·king** \'chùŋ-'kiŋ\ city SW central China in SE Sichuan

Chon·ju *or* **Jeon·ju** \'jən-ˌjü\ city W South Korea

Chosen — see KOREA

Christ·church \'kris-ˌchərch, 'krīst-\ city New Zealand on E coast of South Island

Christ·mas \'kris-məs\ **1** island E Indian Ocean SW of Java; administered by Australia **2** — see KIRITIMATI

Ch'üan–chou *or* **Chuanchow** — see QUANZHOU

Chu–chou *or* **Chuchow** — see ZHUZHOU

Chu·la Vis·ta \ˌchü-lə-'vis-tə\ city SW California S of San Diego

Chuuk \'chùk\ *or* **Truk** \'trək, 'trùk\ islands central Carolines W Pacific; part of the Federated States of Micronesia

Ci·li·cia \sə-'lish-ə, -'lish-ē-ə\ ancient country SE Asia Minor on coast S of Taurus Mountains

Cin·cin·na·ti \ˌsin-sə-'nat-ē, -'nat-ə\ city SW Ohio on Ohio River

Cis·al·pine Gaul \sis-ˌal-ˌpīn-\ the part of Gaul lying S & E of the Alps

Citlaltepetl — see ORIZABA

Città del Vaticano — see VATICAN CITY

Ci·u·dad Gua·ya·na \ˌsē-ü-ˌthäth-gwə-'yän-ə\ city E Venezuela

Ciudad Juá·rez *or* **Juárez** \'hwär-əs, 'wär-\ city N Mexico in Chihuahua on Rio Grande opposite El Paso, Texas

Ciudad Trujillo — see SANTO DOMINGO

Clack·man·nan·shire \klak-'man-ən-ˌshir, -shər\ administrative area central Scotland

Clare \'klaər, 'kleər\ county W Ireland in Munster

Clarks·ville \ˈklärks-ˌvil\ city N Tennessee
Clear·wa·ter \ˈklir-ˌwȯt-ər, -ˌwät-\ city W Florida on Gulf of Mexico
Cler·mont–Fer·rand \ˌkler-ˌmō^{n}-fə-ˈrä^{n}\ city S central France
Cleve·land \ˈklēv-lənd\ city & port NE Ohio on Lake Erie
Cluj–Na·po·ca \ˈklüzh-ˈnäp-ō-kə\ city NW central Romania
Clyde \ˈklīd\ river 106 miles (171 kilometers) long SW Scotland flowing into **Firth of Clyde** (estuary)
Clydes·dale \ˈklīdz-ˌdāl\ valley of upper Clyde River, Scotland
Cnossus — see KNOSSOS
Coa·hui·la \ˌkō-ə-ˈwē-lə, kwä-ˈwē-\ state N Mexico bordering on U.S.; capital, Saltillo
Coast Mountains mountain range W British Columbia; the N continuation of Cascade Range
Coast Ranges chain of mountain ranges W North America extending along Pacific coast W of Sierra Nevada & Cascade Range & through Vancouver Island into S Alaska to Kenai Peninsula & Kodiak Island
Co·cha·bam·ba \ˌkō-chə-ˈbäm-bə\ city W central Bolivia
Co·chin China \ˌkō-chən-\ region S Vietnam
Cod, Cape \ˈkäd\ peninsula SE Massachusetts
Coim·ba·tore \ˌkȯim-bə-ˈtōr, -ˈtȯr\ city S India in W Tamil Nadu
Col·chis \ˈkäl-kəs\ ancient country bordering on Black Sea S of Caucasus Mountains; district now in W Republic of Georgia
Co·li·ma \kə-ˈlē-mə\ **1** state SW Mexico **2** city, its capital
Co·logne \kə-ˈlōn\ *or German* **Köln** \ˈkœln\ city W Germany on the Rhine
Co·lom·bia \kə-ˈləm-bē-ə\ country NW South America; capital, Bogotá — **Co·lom·bi·an** \-bē-ən\ *adj or n*
Co·lom·bo \kə-ˈləm-bō\ city, capital of Sri Lanka
Co·lón \kə-ˈlōn\ city Panama on the Caribbean
Colón Archipelago — see GALAPAGOS ISLANDS
Col·o·ra·do \ˌkäl-ə-ˈrad-ō, -ˈräd-\ **1** river 1450 miles (2334 kilometers) long SW U.S. & NW Mexico flowing from N Colorado into Gulf of California **2** desert SE California **3** plateau region SW U.S. W of Rocky Mountains **4** state W U.S.; capital, Denver — **Col·o·rad·an** \-ˈrad-n, -ˈräd-n\ *or* **Co·lo·ra·do·an** \-ˈrad-ə-wən, -ˈräd-\ *adj or n*
Colorado Springs city central Colorado E of Pikes Peak
Co·lum·bia \kə-ˈləm-bē-ə\ **1** river 1214 miles (1953 kilometers) long SW Canada & NW U.S. rising in SE British Columbia & flowing S & W into North Pacific **2** plateau in basin of Columbia River in E Washington, E Oregon, & SW Idaho **3** city, capital of South Carolina
Co·lum·bus \kə-ˈləm-bəs\ **1** city W Georgia **2** city, capital of Ohio
Commonwealth of Independent States association of the former constituent republics of the U.S.S.R. except for Lithuania, Latvia, & Estonia; formed 1991
Commonwealth, the *or* **Commonwealth of Nations** *or formerly* **British Commonwealth** political organization consisting of the United Kingdom and a number of its former dependencies
Co·mo, Lake \ˈkō-mō\ lake N Italy in Lombardy
Com·o·ros \ˈkäm-ə-ˌrōz\ islands off SE Africa NW of Madagascar; formerly a French possession; a republic (except for Mayotte Island remaining French) since 1975; capital, Moroni
Com·stock Lode \ˈkäm-ˌstäk\ gold & silver deposit at Virginia City, Nevada, discovered 1859
Con·a·kry \ˈkän-ə-krē\ city, capital of Guinea
Con·cep·ción \kən-ˌsep-sē-ˈōn, -ˈsep-shən\ city S central Chile
Con·cord \ˈkäŋ-kərd\ **1** city, capital of New Hampshire **2** town E Massachusetts NW of Boston
Con·ga·ree National Park reservation central South Carolina along the **Congaree River** (60 miles or 95 kilometers)
Con·go \ˈkäŋ-gō\ **1** *or* **Zaire** river more than 2700 miles (4344 kilometers) long W equatorial Africa flowing into the South Atlantic **2** *or* **Democratic Republic of the Congo** *or 1971–97* **Zaire** *or 1908–60* **Belgian Congo** country central Africa comprising most of basin of Congo River E of lower Congo River; capital, Kinshasa **3** *or* **Republic of the Congo** *or formerly* **Middle Congo** country W central Africa W of lower Congo River; capital, Brazzaville — **Con·go·lese** \ˌkäŋ-gə-ˈlēz, -ˈlēs\ *adj or n*
Con·nacht \ˈkän-ˌȯt\ province W Ireland
Con·nect·i·cut \kə-ˈnet-i-kət\ **1** river 407 miles (655 kilometers) long NE U.S. flowing S from N New Hampshire into Long Island Sound **2** state NE U.S.; capital, Hartford
Con·stan·tine \ˈkän-stən-ˌtēn\ city NE Algeria
Constantinople — see ISTANBUL
Con·stan·tsa \kən-ˈstän-sə, -ˈstänt-\ city SE Romania
Con·wy \ˈkän-wē\ administrative area N Wales
Cook \ˈku̇k\ **1** inlet of the North Pacific S Alaska W of Kenai Peninsula **2** islands South Pacific SW of Society Islands belonging to New Zealand; capital, Avarua (on Rarotonga Island) **3** strait New Zealand between North Island & South Island
Cook, Mount *or formerly* **Ao·rangi** \au̇-ˈräŋ-ē\ mountain 12,316 feet (3754 meters) New Zealand in W central South Island in Southern Alps; highest in New Zealand
Co·pen·ha·gen \ˌkō-pən-ˈhā-gən, -ˈhäg-ən\ city, capital of Denmark
Coquilhatville — see MBANDAKA
Coral Sea arm of the W Pacific NE of Australia
Coral Springs city SE Florida
Cór·do·ba \ˈkȯrd-ə-bə, -ə-və\ **1** *or* **Cor·do·va** \ˈkȯrd-ə-və\ *or ancient* **Cor·du·ba** \ˈkȯrd-yə-bə, ˈkȯrd-u̇-bə\ city S Spain on the Guadalquivir **2** city N central Argentina
Cor·fu \kȯr-ˈfü; ˈkȯr-ˌfü, -ˌfyü\ island NW Greece in Ionian Islands
Cor·inth \ˈkȯr-ənth, ˈkär-, -ənth\ **1** region of ancient Greece occupying most of Isthmus of Corinth & part of NE Peloponnese **2** ancient city, its capital; site SW of present city of Corinth — **Co·rin·thi·an** \kə-ˈrin-thē-ən, -ˈrint-\ *adj or n*
Corinth, Gulf of inlet of Ionian Sea central Greece N of the Peloponnese
Corinth, Isthmus of neck of land connecting the Peloponnese with rest of Greece
Cork \ˈkȯrk\ **1** county S Ireland in Munster **2** city S Ireland in County Cork
Corn·wall \ˈkȯrn-ˌwȯl, -wəl\ *or since 1974* **Cornwall and Isles of Scil·ly** \ˈsil-ē\ county SW England
Cor·o·man·del \ˌkȯr-ə-ˈman-dl, ˌkär-\ coast region SE India on Bay of Bengal
Co·ro·na \kə-ˈrō-nə\ city SW California E of Los Angeles
Cor·pus Chris·ti \ˌkȯr-pə-ˈskris-tē\ city & port S Texas
Cor·reg·i·dor \kə-ˈreg-ə-ˌdȯr\ island Philippines at entrance to Manila Bay
Cor·si·ca \ˈkȯr-si-kə\ *or French* **Corse** \kȯrs\ island France in the Mediterranean N of Sardinia — **Cor·si·can** \ˈkȯr-si-kən\ *adj or n*
Cos·ta Bra·va \ˌkäs-tə-ˈbräv-ə, ˌkȯs-, ˌkōs-\ coast region NE Spain on the Mediterranean extending NE from Barcelona
Costa del Sol \-del-ˈsȯl, -ˈsōl\ coast region S Spain on the Mediterranean extending E from Gibraltar
Cos·ta Me·sa \ˈkōs-tə-ˈmā-sə\ city SW California on Pacific coast
Cos·ta Ri·ca \ˌkäs-tə-ˈrē-kə, ˌkȯs-, ˌkōs-\ country Central America between Nicaragua & Panama; capital, San José — **Cos·ta Ri·can** \-ˈrē-kən\ *adj or n*
Côte d'A·zur \ˌkōt-də-ˈzu̇r\ region SE France on Mediterranean coast; part of the Riviera
Côte d'Ivoire — see IVORY COAST
Co·to·nou \ˌkōt-ə-ˈnü\ city & port S Benin
Cots·wold \ˈkät-ˌswōld\ hills SW central England
Cov·en·try \ˈkäv-ən-trē, ˈkəv-\ city central England
Co·zu·mel \ˌkō-zə-ˈmel\ island SE Mexico off NE coast of Quintana Roo
Cracow — see KRAKOW
Cra·io·va \krə-ˈyō-və\ city S Romania
Cra·ter \ˈkrāt-ər\ lake SW Oregon in Cascade Range; main feature of **Crater Lake National Park** — see MAZAMA (Mount)
Crete \ˈkrēt\ island Greece in E Mediterranean; capital, Iráklion — **Cre·tan** \ˈkrēt-n\ *adj or n*
Cri·mea \krī-ˈmē-ə, krə-\ peninsula E Europe, extending into Black Sea — **Cri·me·an** \krī-ˈmē-ən, krə-\ *adj*
Cro·atia \krō-ˈā-shə, -shē-ə\ country S Europe; capital, Zagreb; formerly a constituent republic of Yugoslavia
Croy·don \ˈkrȯid-n\ borough of S Greater London, England
Cu·ba \ˈkyü-bə\ island in the West Indies; a republic; capital, Havana — **Cu·ban** \-bən\ *adj or n*
Cú·cu·ta \ˈkü-kət-ə\ city N Colombia
Cu·lia·cán \ˌkül-yə-ˈkän\ city NW Mexico, capital of Sinaloa
Cu·ma·ná \ˌkü-mə-ˈnä\ city NE Venezuela
Cum·ber·land \ˈkəm-bər-lənd\ former county NW England — see CUMBRIA
Cumberland Plateau mountain region E U.S.; part of S Appalachian Mountains W of Tennessee River extending from S West Virginia to NE Alabama
Cum·bria \ˈkəm-brē-ə\ county NW England including former counties of Cumberland & Westmorland — **Cum·bri·an** \-ən\ *adj or n*
Cumbrian Mountains range NW England chiefly in Cumbria
Cu·ra·çao \ˈku̇r-ə-ˌsō, ˈkyu̇r-, -ˌsau̇\ island Netherlands Antilles in the S Caribbean; chief town, Willemstad
Cu·ri·ti·ba \ˌku̇r-ə-ˈtē-bə\ city S Brazil SW of São Paulo
Cush \ˈkəsh, ˈku̇sh\ ancient country NE Africa in upper valley of the Nile S of Egypt — **Cush·ite** \-ˌīt\ *n* — **Cush·it·ic** \ˌkəsh-ˈit-ik, ku̇sh-\ *adj*
Cut·tack \ˈkət-ək\ city E India in Orissa
Cuy·a·hoga Valley National Park \ˌkī-ə-ˈhō-gə\ reservation NE Ohio along the **Cuyahoga River** (100 miles or 161 kilometers flowing into Lake Erie)
Cyc·la·des \ˈsik-lə-ˌdēz\ islands Greece in S Aegean
Cymru — see WALES
Cy·prus \ˈsī-prəs\ island E Mediterranean S of Turkey; a republic; capital, Nicosia — **Cyp·ri·ot** \ˈsip-rē-ət, -rē-ˌät\ *or* **Cyp·ri·ote** \-ˌōt, -ət\ *adj or n*
Cy·re·na·ica \ˌsir-ə-ˈnā-ə-kə, ˌsī-rə-\ **1** ancient region N Africa on coast W of Egypt; capital, Cyrene **2** region E Libya; formerly a province — **Cy·re·na·i·can** \-kən\ *adj or n*
Czecho·slo·va·kia \ˌchek-ə-slō-ˈväk-ē-ə, -ˈvak-\ former country central Europe; capital, Prague; divided 1993 into Czech Republic & Slovakia — **Czecho·slo·vak** \-ˈslō-ˌväk, -ˌvak\ *adj or n* — **Czecho·slo·va·ki·an** \-slō-ˈväk-ē-ən, -ˈvak-\ *adj or n*
Czech Republic country central Europe; capital, Prague
Cze·sto·cho·wa \ˌchen-stə-ˈkō-və\ city S Poland
Dacca — see DHAKA
Da·cia \ˈdā-shə, -shē-ə\ ancient country & Roman province SE Europe roughly equivalent to Romania & Bessarabia
Da·dra and Na·gar Ha·ve·li \də-ˈdrä-ən-ˌnəg-ər-ə-ˌvel-ē\ territory India bordering on Gujarat & Maharashtra
Daegu — see TAEGU
Daejeon — see TAEJON
Da·ho·mey \də-ˈhō-mē\ — see BENIN — **Da·ho·man** \-mən\ *or* **Da·ho·me·an** *or* **Da·ho·mey·an** \-mē-ən\ *adj or n*
Da·kar \ˈdak-ˌär, də-ˈkär\ city, capital of Senegal
Da·ko·ta \də-ˈkōt-ə\ — see JAMES
Dakota Territory territory 1861–89 NW U.S. divided 1889 into states of North Dakota & South Dakota (the **Da·ko·tas**)
Da·lian *or* **Ta·lien** \ˈdäl-ˈyen\ *or* **Lü·da** *or* **Lü·ta** \ˈlü̅-ˈdä\ *or* **Dai·ren** \ˈdī-ˈren\ city NE China in S Liaoning
Dal·las \ˈdal-əs, ˈda-lis\ city NE Texas
Dal·ma·tia \dal-ˈmā-shə, -shē-ə\ region W Balkan Peninsula on the Adriatic — **Dal·ma·tian** \-shən\ *adj or n*
Da·ly City \ˈdā-lē\ city W California S of San Francisco
Da·man and Diu \də-ˈmän-ən-ˈdē-ˌü, -ˈman-\ territory W India on Gulf of Cambay
Da·mas·cus \də-ˈmas-kəs\ city, capital of Syria
Da·ma·vand \ˈdam-ə-ˌvand\ *or* **Dem·a·vend** \ˈdem-ə-ˌvend\ mountain 18,934 feet (5771 meters) N Iran NE of Tehran
Da Nang \dä-ˈnäŋ, ˈdä-\ *or formerly* **Tou·rane** \tü-ˈrän\ city S Vietnam in Annam SE of Hue

Dan·dong \'dän-'du̇ŋ\ *or* **An·tung** \'än-'du̇ŋ\ *or* **Tan–tung** \'dän-'du̇ŋ\ city NE China in SE Liaoning at mouth of the Yalu

Dan·ube \'dan-yüb\ *or German* **Do·nau** \'dō-ˌnau̇\ river 1771 miles (2850 kilometers) long S Europe flowing from S Germany into Black Sea — **Da·nu·bi·an** \da-'nyü-bē-ən\ *adj*

Danzig — see GDANSK

Dar·da·nelles \ˌdärd-n-'elz\ *or* **Hel·les·pont** \'hel-ə-ˌspänt\ strait NW Turkey connecting Sea of Marmara & the Aegean

Dar el Beida — see CASABLANCA

Dar es Sa·laam \ˌdär-ˌes-sə-'läm\ city, capital of Tanzania

Darien, Isthmus of — see PANAMA (Isthmus of)

Dar·ling \'där-liŋ\ river about 1700 miles (2735 kilometers) long SE Australia in Queensland & New South Wales flowing SW into the Murray

Dar·win \'där-wən\ city Australia, capital of Northern Territory

Da·tong *or* **Ta·tung** \'dä-'tu̇ŋ\ city NE China in N Shanxi

Da·vao \'däv-ˌau̇, dä-'vau̇\ city S Philippines in E Mindanao on **Davao Gulf** (inlet of the North Pacific)

Da·vis \'dā-vəs\ strait between SW Greenland & E Baffin Island connecting Baffin Bay & the North Atlantic

Day·ton \'dāt-n\ city SW Ohio

Dead Sea \'ded\ salt lake between Israel & Jordan; 1312 feet (400 meters) below the level of the Mediterranean

Dear·born \'diər-ˌbȯrn, -bərn\ city SE Michigan SW of Detroit

Death Valley \'deth\ arid valley E California & S Nevada containing lowest point in U.S. (282 feet or 86 meters below sea level); most of area included in **Death Valley National Park**

De·bre·cen \'deb-rət-ˌsen\ city E Hungary

Dec·can \'dek-ən, -ˌan\ plateau region S India

Del·a·ware \'del-ə-ˌwaər, -ˌweər, -wər\ **1** river 296 miles (476 kilometers) long E U.S. flowing S from S New York into Delaware Bay **2** state E U.S.; capital, Dover — **Del·a·war·ean** *or* **Del·a·war·ian** \ˌdel-ə-'war-ē-ən, -'wer-\ *adj or n*

Delaware Bay inlet of the North Atlantic between SW New Jersey & E Delaware

Del·hi \'del-ē\ **1** territory N India W of Uttar Pradesh **2** city, its capital — see NEW DELHI

De·los \'dē-ˌläs\ island Greece in central Cyclades — **De·lian** \'dē-lē-ən, 'dēl-yən\ *adj or n*

Del·phi \'del-ˌfī\ ancient town central Greece in Phocis on S slope of Mount Parnassus

Democratic Republic of the Congo — see CONGO 2

Denali, Denali National Park — see MCKINLEY (Mount)

Den·bigh·shire \'den-bē-ˌshiər, -shər\ administrative area N Wales

Den·mark \'den-ˌmärk\ country N Europe occupying most of Jutland & adjacent islands; capital, Copenhagen

Den·ver \'den-vər\ city, capital of Colorado

Der·by \'där-bē, *chiefly in the U.S.* 'dər-bē\ borough N central England in Derbyshire

Der·by·shire \'där-bē-ˌshiər, -shər, *U.S. also* 'dər-\ *or* **Derby** county N central England

Der·ry \'der-ē\ *or* **Lon·don·der·ry** \ˌlən-dən-'der-ē, 'lən-dən-ˌ\ **1** traditional county NW Northern Ireland **2** seaport NW Northern Ireland

Des Moines \di-'mȯin\ city, capital of Iowa

De·troit \di-'trȯit\ **1** river 31 miles (50 kilometers) long between Michigan & Ontario connecting Lake Saint Clair & Lake Erie **2** city SE Michigan

Dev·on \'dev-ən\ *or* **De·von·shire** \-ˌshiər, -shər\ county SW England

Dha·ka *or* **Dac·ca** \'dak-ə, 'däk-\ city, capital of Bangladesh

Dhau·la·gi·ri, Mount \ˌdau̇-lə-'giər-ē\ mountain 26,810 feet (8172 meters) W central Nepal in the Himalaya

Di·a·mond Bar \'dī-mənd-ˌbär, 'dī-ə-\ city S California E of Los Angeles

Di·e·go Gar·cia \dē-ˌā-gō-ˌgär-'sē-ə\ island in Indian Ocean; chief island of Chagos Archipelago

Di·jon \dē-'zhōn\ city E France N of Lyon

Di·li \'dil-ē\ city & port, capital of East Timor

Di·nar·ic Alps \də-ˌnar-ik\ range of the E Alps in W Balkan Peninsula

Diospolis — see THEBES 1

District of Co·lum·bia \kə-'ləm-bē-ə\ federal district E U.S. coextensive with city of Washington

Dix·ie \'dik-sē\ the states of the SE U.S. and especially those which constituted the Confederacy

Djakarta — see JAKARTA

Dji·bou·ti *or* **Ji·bu·ti** \jə-'büt-ē\ **1** *or formerly* **French Somaliland** *or later* **French Territory of the Afars and the Is·sas** \ä-'fär, -'färz . . .ē-'sä, -'säz\ republic E Africa on Gulf of Aden **2** city, its capital

Dnie·per \'nē-pər\ river 1420 miles (2285 kilometers) long flowing from Valdai Hills, Russia through Belarus & Ukraine into Black Sea

Dnies·ter \'nēs-tər\ river 877 miles (1411 kilometers) long W Ukraine & E Moldova flowing SE from the Carpathians into Black Sea

Dni·pro·pe·trovs'k *or* **Dne·pro·pe·trovsk** \də-ˌnyep-rə-pē-'trȯfsk\ city E Ukraine

Do·dec·a·nese \dō-'dek-ə-ˌnēz, 'dō-di-kə-, -ˌnēs\ islands Greece in the SE Aegean — see RHODES

Do·ha \'dō-hä\ city & port, capital of Qatar on Persian Gulf

Do·lo·mites \'dō-lə-ˌmīts, 'däl-ə-\ *or Italian* **Do·lo·mi·ti** \ˌdȯ-lə-'mēt-ē\ range of the E Alps in NE Italy

Dom·i·ni·ca \ˌdäm-ə-'nē-kə\ island British West Indies in the Leewards; an independent republic; capital, Roseau

Do·min·i·can Republic \də-ˌmin-i-kən\ *or formerly* **San·to Do·min·go** \ˌsant-əd-ə-'miŋ-gō\ country West Indies in E Hispaniola; capital, Santo Domingo — **Do·min·i·can** \də-'min-i-kən\ *adj or n*

Don \'dän\ river 1224 miles (1969 kilometers) long S Russia in Europe flowing into Sea of Azov

Donau — see DANUBE

Don·e·gal \ˌdän-i-'gȯl, ˌdən-\ county NW Ireland (republic) in Ulster

Do·nets Basin \də-ˌnets\ *or* **Don·bass** *or* **Don·bas** \'dän-ˌbas\ region E Ukraine SW of **Donets River** (over 630 miles or 1014 kilometers flowing SE into Don River)

Do·netsk \də-'netsk\ city E Ukraine in Donets Basin

Dor·set \'dȯr-sət\ *or* **Dor·set·shire** \-ˌshiər, -shər\ county S England on English Channel

Dort·mund \'dȯrt-ˌmu̇nt, -mənd\ city W Germany in the Ruhr

Dou·a·la \du̇-'äl-ə\ city SW Cameroon

Doug·las \'dəg-ləs\ town British Isles, capital of Isle of Man

Dou·ro \'dōr-ü, 'dȯr-\ *or Spanish* **Due·ro** \'dweər-ō\ *or ancient* **Du·ri·us** \'du̇r-ē-əs, 'dyu̇r-\ river 556 miles (895 kilometers) long N Spain & N Portugal flowing into the North Atlantic

Do·ver \'dō-vər\ **1** city, capital of Delaware **2** borough SE England in Kent on Strait of Dover

Dover, Strait of channel between SE England & N France; the most easterly section of English Channel

Down \'dau̇n\ traditional county SE Northern Ireland

Dow·ney \'dau̇-nē\ city SW California SE of Los Angeles

Dra·kens·berg \'dräk-ənz-ˌbərg\ mountain range E Republic of South Africa & Lesotho; highest peak Thabana Ntlenyana 11,425 feet (3482 meters)

Dres·den \'drez-dən\ city E Germany in Saxony

Dry Tor·tu·gas \tȯr-'tü-gəz\ island group S Florida; site of **Dry Tortugas National Park**

Du·bayy *or* **Du·bai** \dü-'bī\ city United Arab Emirates on Persian Gulf

Dub·lin \'dəb-lən\ *or Gaelic* **Bai·le Atha Cli·ath** \blä-'klē-ə\ **1** county E Ireland in Leinster **2** city, capital of Ireland (republic) in County Dublin

Dud·ley \'dəd-lē\ borough W central England

Duis·burg \'dü-əs-ˌbərg; 'düz-ˌbərg, 'dyüz-; German 'dües-ˌbu̇rk\ city W Germany at junction of Rhine & Ruhr rivers

Du·luth \də-'lüth\ city & port NE Minnesota at W end of Lake Superior

Dum·fries and Gal·lo·way \ˌdəm-'frēs-ənd-'gal-ə-ˌwā\ administrative area S Scotland

Dun·dee \ˌdən-'dē\ city E Scotland constituting an administrative area (**Dundee City**)

Dun·e·din \ˌdə-'nēd-n\ city New Zealand in SE South Island

Du·que de Ca·xi·as \ˌdü-kə-də-kə-'shē-əs\ city SE Brazil NW of Rio de Janeiro

Du·ran·go \du̇-'raŋ-gō, dyu̇-\ **1** state NW central Mexico **2** city, its capital

Dur·ban \'dər-bən\ city E Republic of South Africa

Dur·ham \'dər-əm, 'də-rəm, 'du̇r-əm\ **1** city N central North Carolina **2** county N England on North Sea

Du·shan·be \dü-'sham-bē, dyü-, -'shäm-\ city, capital of Tajikistan

Düs·sel·dorf \'düs-əl-ˌdȯrf, 'dyüs-, 'dues-\ city W Germany, capital of North Rhine-Westphalia

Dutch Borneo — see KALIMANTAN

Dutch Guiana — see SURINAME

Dzaudzhikau — see VLADIKAVKAZ

Dzer·zhinsk \dər-'zhinsk\ city central Russia in Europe, ENE of Moscow

Ea·ling \'ē-liŋ\ borough of W Greater London, England

East Africa region E Africa — usually considered to include Tanzania, Kenya, Uganda, Rwanda, Burundi, & Somalia

East An·glia \'aŋ-glē-ə\ region E England including Norfolk & Suffolk

East Ayrshire administrative area W Scotland

East Bengal the part of Bengal now in Bangladesh

East China Sea — see CHINA

East Dunbartonshire administrative area W Scotland

Eas·ter \'ē-stər\ island SE Pacific about 2000 miles (5200 kilometers) W of Chilean coast; belongs to Chile

Eastern Cape province SE Republic of South Africa

Eastern Ghats \'gȯts\ chain of low mountains SE India along coast

Eastern Samoa — see AMERICAN SAMOA

East Germany the former German Democratic Republic — see GERMANY

East Indies collective name for India, Indochina, and the Malay Archipelago — **East Indian** *adj or n*

East London city S Republic of South Africa

East Lo·thi·an \'lō-t͟hē-ən\ *or* **Had·ding·ton** \'ha-diŋ-tən\ *or* **Had·ding·ton·shire** \-ˌshir, -shər\ administrative area SE Scotland

East Pakistan the former E division of Pakistan comprising E portion of Bengal; now the independent republic of Bangladesh

East Prussia region N Europe on the Baltic; formerly a part of Germany; divided 1945 between Poland & U.S.S.R. (Russia & Lithuania)

East Renfrewshire administrative area W Scotland

East River strait SE New York connecting upper New York Bay & Long Island Sound & separating Manhattan Island and Long Island

East Sus·sex \'səs-iks, *U.S. also* -ˌeks\ county SE England

East Timor country SE Asia; capital, Dili

Ebro \'ā-brō\ river 565 miles (909 kilometers) long NE Spain flowing into the Mediterranean

Ec·ua·dor \'ek-wə-ˌdȯr\ country W South America; capital, Quito — **Ec·ua·dor·an** \ˌek-wə-'dȯr-ən, -'dōr-\ *or* **Ec·ua·dor·ean** *or* **Ec·ua·dor·ian** \-ē-ən\ *adj or n*

Ede \'ā-ˌdā\ city SW Nigeria

Ed·in·burgh \'ed-n-ˌbər-ə, -ˌbə-rə, -bə-rə, -brə\ city, capital of Scotland constituting an administrative area (**City of Edinburgh**)

Ed·mon·ton \'ed-mən-tən\ city, capital of Alberta

Edo — see TOKYO

Edom \'ēd-əm\ *or* **Id·u·maea** *or* **Id·u·mea** \ˌij-ə-'mē-ə\ ancient country SW Asia S of Judea & Dead Sea — **Edom·ite** \'ēd-ə-ˌmīt\ *n*

Egypt \'ē-jəpt\ country NE Africa & Sinai Peninsula of SW Asia bordering on Mediterranean & Red seas; capital, Cairo

Eilean Siar — see WESTERN ISLES

Eire — see IRELAND 2

Elam \'ē-ləm\ ancient country SW Asia at head of Persian Gulf E of Babylonia; capital, Susa (Shushan) — **Elam·ite** \'ē-lə-ˌmīt\ *n*

El·ba \'el-bə\ island Italy E of N Corsica off coast of Tuscany; chief town, Portoferraio

Elbe \'el-bə, 'elb\ *or Czech* **La·be** \'lä-be\ *or ancient* **Al·bis** \'al-bəs\ river 720 miles (1159 kilometers) long N Czech Republic & N Germany flowing NW into North Sea

El·bert, Mount \'el-bərt\ mountain 14,433 feet (4399 meters) W central Colorado; highest in Colorado & the Rocky Mountains

El·brus, Mount \el-'bru̇z\ mountain 18,510 feet (5642 meters) S Russia in Europe in NW Caucasus Mountains

El·burz \el-'bu̇rz\ mountains N Iran

El Gîza — see GIZA

Elis \'ē-ləs\ ancient country in NW Peloponnese, Greece

Elisabethville — see LUBUMBASHI

Eliz·a·beth \i-'liz-ə-bəth\ city NE New Jersey on Newark Bay

Elk Island National Park reservation E central Alberta

Ellás — see GREECE

Elles·mere \'elz-,miər\ island N Canada in Nunavut

Ellice — see TUVALU

El·lis \'el-əs\ island between New Jersey & New York just SW of mouth of the Hudson; served as immigration station 1892–1954

El Mon·te \el-'mänt-ē\ city SW California E of Los Angeles

El Paso \el-'pas-ō\ city W Texas on Rio Grande

El Sal·va·dor \el-'sal-və-,dȯr, -,sal-və-'\ country Central America bordering on the North Pacific; capital, San Salvador

Elsass — see ALSACE

Ely, Isle of \'ē-lē\ district E England — see CAMBRIDGESHIRE

Emi·lia–Ro·ma·gna \ā-,mēl-yə-rō-'män-yə\ region N Italy on the Adriatic S of the Po; capital, Bologna

En·field \'en-,fēld\ borough of N Greater London, England

En·gland \'iŋ-glənd *also* iŋ-lənd\ country S Great Britain; a division of United Kingdom; capital, London

English Channel arm of the North Atlantic between S England & N France

En·se·na·da \,en-sə-'näd-ə\ city NW Mexico in Baja California

Ephra·im \'ē-frē-əm\ **1** hilly region N Jordan E of River Jordan **2** — see ISRAEL — **Ephra·im·ite** \'ē-frē-ə-,mīt\ *n*

Epi·rus \i-'pī-rəs\ region NW Greece on Ionian Sea

Equatorial Guinea *or formerly* **Spanish Guinea** country W Africa on Bight of Biafra including Mbini & Bioko; capital, Malabo

Erbil — see ARBIL

Ere·bus, Mount \'er-ə-bəs\ volcano 12,450 feet (3795 meters) Antarctica on Ross Island in SW Ross Sea

Er·furt \'eər-fərt, -,fu̇rt\ city central Germany WSW of Leipzig

Erie \'iər-ē\ **1** city NW Pennsylvania **2** canal New York between Hudson River at Albany & Lake Erie at Buffalo; built 1817–25; now superseded by New York State Barge Canal

Erie, Lake lake E central North America in U.S. & Canada; one of the Great Lakes

Er·in \'er-ən\ IRELAND — a poetic name

Er·i·trea \,er-ə-'trē-ə, -'trā-\ country NE Africa on Red Sea; capital, Asmara

Er Rif *or* **Er Riff** \er-'rif\ mountain region N Morocco on Mediterranean coast E of Strait of Gibraltar

Erz·ge·bir·ge \'erts-gə-,bir-gə\ mountains E central Germany & NW Czech Republic

Escaut — see SCHELDT

Es·con·di·do \,es-kən-'dēd-ō\ city SW California N of San Diego

Es·fa·han \,es-fä-'hän\ *or* **Is·fa·han** \,is-\ city W central Iran

Es·ki·se·hir \,es-ki-shə-'hiər\ city W central Turkey

España — see SPAIN

Española — see HISPANIOLA

Es·sen \'es-n\ city W Germany in the Ruhr

Es·sex \'es-iks\ county SE England on North Sea

Es·to·nia \e-'stō-nē-ə, -nyə\ country E Europe on Baltic Sea; capital, Tallinn; a constituent republic of U.S.S.R. 1940–91

Ethi·o·pia \,ē-thē-'ō-pē-ə\ **1** ancient country NE Africa S of Egypt **2** *or formerly* **Ab·ys·sin·ia** \,ab-ə-'sin-yə, -'sin-ē-ə\ country E Africa; a republic since 1975; capital, Addis Ababa

Et·na \'et-nə\ volcano 10,902 feet (3323 meters) Italy in NE Sicily

Eto·bi·coke \et-'ō-bik-,ō\ former city Canada in SE Ontario; now part of Toronto

Etru·ria \i-'tru̇r-ē-ə\ ancient country central Italy coextensive with modern Tuscany & part of Umbria

Eu·boea \yu̇-'bē-ə\ island E Greece NE of Attica & Boeotia

Eu·phra·tes \yu̇-'frāt-ēz\ river 1700 miles (2736 kilometers) long SW Asia flowing from E Turkey & uniting with the Tigris to form the Shatt al Arab

Eur·asia \yu̇-'rā-zhə, -shə\ landmass comprising Europe & Asia — **Eur·asian** \-zhən, -shən\ *adj or n*

Eu·rope \'yu̇r-əp\ continent of the Eastern Hemisphere between Asia & the North Atlantic

European Union *or formerly* **European Community** economic, scientific, & political organization consisting of Belgium, France, Italy, Luxembourg, Netherlands, Germany, Denmark, Greece, Ireland, United Kingdom, Spain, Portugal, Austria, Finland, Sweden, Cyprus, Czech Republic, Estonia, Hungary, Latvia, Lithuania, Malta, Poland, Slovakia & Slovenia

Ev·ans·ville \'ev-ənz-,vil\ city SW Indiana

Ev·er·est, Mount \'ev-rəst, -ə-rəst\ mountain 29,035 feet (8850 meters) S Asia in the Himalaya on border between Nepal & Tibet; highest in the world

Ev·er·glades \'ev-ər-,glādz\ swamp region S Florida now partly drained; SW part forms **Everglades National Park**

Ex·tre·ma·du·ra \,ek-strə-mə-'du̇r-ə\ region W Spain bordering on Portugal

Eyre, Lake \'aər, 'eər\ intermittent lake central Australia in N South Australia

Faer·oe *or* **Far·oe** \'faər-ō, 'feər-\ islands NE Atlantic NW of the Shetlands belonging to Denmark; capital, Tórshavn — **Faero·ese** \,far-ə-'wēz, ,fer-, -'wēs\ *adj or n*

Fair·banks \'faər-,baŋks, 'feər-\ city E central Alaska

Fai·sa·la·bad \,fī-,säl-ə-'bäd, -,sal-ə-'bad\ *or formerly* **Ly·all·pur** \lē-,äl-'pu̇r\ city NE Pakistan W of Lahore

Fal·kirk \'fȯl-kərk\ administrative area central Scotland

Falk·land Islands \'fȯ-klənd-, 'fȯl-\ *or Spanish* **Islas Mal·vi·nas** \,ēz-läz-mäl-'vē-näs\ islands SW Atlantic E of S end of Argentina; a British colony; capital, Stanley

Far East the countries of E Asia & the Malay Archipelago — usually considered as comprising the Asian countries bordering on the Pacific but sometimes as including also India, Sri Lanka, Bangladesh, Tibet, & Myanmar — **Far Eastern** *adj*

Far·go \'fär-gō\ city E North Dakota; largest in state

Fay·ette·ville \'fā-ət-,vil, 'fed-vəl\ city SE central North Carolina

Fear, Cape \'fiər\ cape SE North Carolina at mouth of **Cape Fear River** (202 miles or 325 kilometers flowing from central North Carolina into North Atlantic)

Federated Malay States former British protectorate (1895–1945) comprising states of Negri Sembilan, Pahang, Perak, & Selangor; now part of Malaysia

Fengtien — see SHENYANG

Fer·man·agh \fər-'man-ə\ traditional county SW Northern Ireland

Fernando Póo — see BIOKO

Fer·ra·ra \fə-'rär-ə\ city N Italy NE of Bologna

Fez \'fez\ *or* **Fès** \'fes\ city N central Morocco

Fife \'fīf\ *or* **Fife·shire** \-,shiər, -shər\ administrative area E Scotland

Fi·ji \'fē-jē\ islands SW Pacific; an independent country; capital, Suva — **Fi·ji·an** \-jē-ən\ *adj or n*

Fin·is·terre, Cape \,fin-ə-'steər, -'ster-ē\ cape NW Spain

Fin·land \'fin-lənd\ *or Finnish* **Suo·mi** \'swȯ-mē\ country NE Europe on Gulf of Bothnia and Gulf of Finland; capital, Helsinki — **Fin·land·er** \'fin-lənd-ər\ *n*

Finland, Gulf of arm of the Baltic between Finland & Estonia

Fiume — see RIJEKA

Flan·ders \'flan-dərz\ **1** region W Belgium & N France on North Sea **2** semiautonomous region W Belgium

Flat·tery, Cape \'flat-ə-rē\ cape NW Washington at entrance to Strait of Juan de Fuca

Flint \'flint\ city SE Michigan NNW of Detroit

Flint·shire \'flint-,shiər, -shər\ administrative area NE Wales

Flor·ence \'flȯr-əns, 'flär-\ *or Italian* **Fi·ren·ze** \fē-'rent-sā\ *or ancient* **Flo·ren·tia** \flə-'ren-chə, -chē-ə\ city central Italy, capital of Tuscany — **Flor·en·tine** \'flȯr-ən-,tēn, 'flär-, -,tīn\ *adj or n*

Flo·res \'flōr-əs, 'flȯr-\ island Indonesia in Lesser Sunda Islands

Flo·ri·a·nó·po·lis \,flōr-ē-ə-'näp-ə-ləs, ,flȯr-\ city S Brazil on an island NE of Porto Alegre

Flor·i·da \'flȯr-əd-ə, 'flär-\ state SE U.S.; capital, Tallahassee — **Flo·rid·i·an** \flə-'rid-ē-ən\ *or* **Flor·i·dan** \'flȯr-əd-n, 'flär-\ *adj or n*

Florida, Straits of channel between Florida Keys on NW & Cuba & Bahamas on S & E connecting Gulf of Mexico & North Atlantic

Florida Keys chain of islands off S tip of Florida

Fog·gia \'fȯ-jə, -jä\ city SE Italy in Puglia

Foochow — see FUZHOU

For·a·ker, Mount \'fȯr-i-kər, 'fär-\ mountain 17,400 feet (5304 meters) S central Alaska in Alaska Range

Fo·ril·lon National Park \,fȯr-ē-'yōⁿ\ reservation Quebec in Gaspé Peninsula

For·mo·sa \fȯr-'mō-sə, fər-, -zə\ — see TAIWAN — **For·mo·san** \-sən, -zən\ *adj or n*

For·ta·le·za \,fȯrt-l-'ā-zə\ city NE Brazil NW of Recife

Fort Col·lins \'käl-ənz\ city N Colorado

Fort–de–France \,fȯrd-ə-'fräⁿs\ city French West Indies, capital of Martinique on W coast

Forth \'fōrth, 'fȯrth\ river 116 miles (187 kilometers) long S central Scotland flowing E into North Sea through **Firth of Forth** (estuary)

Fort Knox \'näks\ military reservation N central Kentucky SSW of Louisville; location of U.S. Gold Bullion Depository

Fort–Lamy — see N'DJAMENA

Fort Lau·der·dale \'lȯd-ər-,dāl\ city SE Florida on Atlantic

Fort Wayne \'wān\ city NE Indiana

Fort Worth \'wərth\ city NE Texas

Foxe Basin \'fäks\ inlet of North Atlantic N Canada W of Baffin Island

France \'frans\ country W Europe between the English Channel & the Mediterranean; capital, Paris

Franche–Com·té \,fräⁿsh-kōⁿ-'tā\ region E France bordering on Switzerland

Frank·fort \'fraŋk-fərt\ city, capital of Kentucky

Frank·furt \'fraŋk-fərt, 'fräŋk-,fu̇rt\ *or in full* **Frankfurt am Main** \-äm-'mīn\ city SW central Germany on Main River

Frank·lin \'fraŋ-klən\ former district N Canada in Northwest Territories including Arctic Islands & Boothia & Melville peninsulas

Fra·ser \'frā-zər, -zhər\ river 850 miles (1368 kilometers) long Canada in S central British Columbia flowing into North Pacific

Fred·er·ic·ton \'fred-rik-tən, -ə-rik-\ city, capital of New Brunswick

Free State *or formerly* **Or·ange Free State** \'ȯr-inj, 'är-, -ənj\ province E central Republic of South Africa

Free·town \'frē-,tau̇n\ city, capital of Sierra Leone

Fre·mont \'frē-,mänt\ city W California SE of Oakland

French Equatorial Africa former country W central Africa N of Congo River comprising a federation of Chad, Gabon, Middle Congo, & Ubangi-Shari territories

French Guiana country N South America on North Atlantic; a dependency of France; capital, Cayenne

French Guinea — see GUINEA

French Indochina — see INDOCHINA

French Morocco — see MOROCCO

French Polynesia islands in South Pacific belonging to France and including Society, Marquesas, Tuamotu, Gambier, & Tubuai groups; capital, Papeete

French Somaliland — see DJIBOUTI

French Sudan — see MALI

French Territory of the Afars and the Issas — see DJIBOUTI

French Togo — see TOGO

French West Indies islands of the West Indies belonging to France and including Guadeloupe, Martinique, Désirade, Les Saintes, Marie Galante, Saint Barthélemy, & part of Saint Martin

Fres·no \'frez-nō\ city S central California

Fri·sian \'frizh-ən, 'frē-zhən\ islands N Europe in North Sea including **West Frisian** islands off N Netherlands, **East Frisian** islands off NW Germany, & **North Frisian** islands off N Germany and W Denmark

Fri·u·li–Ve·ne·zia Giu·lia \frē-'ü-lē-və-,net-sē-ə-'jü-lyə\ region N Italy; capital, Trieste

Frunze — see BISHKEK

Fu·ji, Mount \'fü-jē, 'fyü-\ *or* **Fu·ji·ya·ma** \,fü-jē-'äm-ə, ,fyü-, -'yäm-\ *or* **Fu·ji·san** \-jē-'sän\ mountain 12,388 feet (3776 meters) Japan in S central Honshu; highest in Japan

Fu·jian \'fü-'jän, -jē-'än\ *or* **Fu·kien** \'fü-'kyen, -kē-'en\ province SE China on Formosa Strait; capital, Fuzhou

Fu·ji·sa·wa \,fü-jē-'sä-wə\ city Japan in SE Honshu

Fu·ku·o·ka \,fü-kə-'wō-kə\ city Japan in N Kyushu

\ə\ abut \ᵊ\ kitten \ər\ further \a\ ash \ā\ ace \ä\ mop, mar \au̇\ out \ch\ chin \e\ bet \ē\ easy \g\ go \i\ hit \ī\ ice \j\ job \ŋ\ sing \ō\ go \ȯ\ law \ȯi\ boy \th\ thin \t͟h\ the \ü\ loot \u̇\ foot \y\ yet \zh\ vision *see also* Guide to Pronunciation

Fu·ku·ya·ma \ˌfü-kə-ˈyäm-ə\ city Japan in SW Honshu
Fu·na·bashi \ˌfü-nə-ˈbäsh-ē\ city Japan in SE Honshu on Tokyo Bay
Fu·na·fu·ti \ˌfü-nə-ˈfüt-ē, ˌfyü-, -ˈfyüt-\ city, capital of Tuvalu
Fun·dy, Bay of \ˈfən-dē\ inlet of North Atlantic SE Canada between New Brunswick & Nova Scotia
Fundy National Park reservation New Brunswick on Bay of Fundy
Fu·shun \ˈfü-ˈshün\ city NE China in NE Liaoning
Fu·zhou \ˈfü-ˈjō\ *or* **Foo·chow** \ˈfü-ˈjō, -ˈchau̇\ *or formerly* **Min·how** \ˈmin-ˈhō\ city SE China, capital of Fujian
Ga·bon \ga-ˈbōⁿ\ country W equatorial Africa; capital, Libreville — **Gab·o·nese** \ˌgab-ə-ˈnēz, -ˈnēs\ *adj or n*
Ga·bo·rone \ˌgäb-ə-ˈrōn\ city, capital of Botswana
Gads·den Purchase \ˈgadz-dən\ tract of land S of Gila River in present Arizona & New Mexico purchased 1853 by the U.S. from Mexico
Gaeseong — see KAESONG
Ga·lá·pa·gos Islands \gə-ˈläp-ə-gəs, -ˈlap-\ *or* **Co·lón Archipelago** \kə-ˈlōn\ island group Ecuador in the Pacific about 600 miles (965 kilometers) W of mainland
Ga·lati \gə-ˈläts, -ˈlät-sē\ city E Romania on the Danube
Ga·la·tia \gə-ˈlā-shə, -shē-ə\ ancient country & Roman province central Asia Minor in region centering on modern Ankara, Turkey — **Ga·la·tian** \-shən\ *adj or n*
Ga·li·cia \gə-ˈlish-ə, -ˈlish-ē-ə\ **1** region E central Europe now divided between Poland & Ukraine **2** region NW Spain on North Atlantic — **Ga·li·cian** \-shən\ *adj or n*
Gal·i·lee \ˈgal-ə-ˌlē\ hilly region N Israel — **Gal·i·le·an** \ˌgal-ə-ˈlē-ən\ *adj or n*
Galilee, Sea of *or modern* **Lake Ti·be·ri·as** \tī-ˈbir-ē-əs\ *or biblical* **Lake of Gen·nes·a·ret** \gə-ˈnes-ə-ˌret, -rət\ *or* **Sea of Tiberias** lake N Israel on Syrian border traversed by Jordan River
Gal·lip·o·li \gə-ˈlip-ə-lē\ *or Turkish* **Ge·li·bo·lu** \ge-ˈlē-bó-ˌlü\ peninsula Turkey in Europe between the Dardanelles & Saros Gulf
Gal·way \ˈgȯl-ˌwā\ county W Ireland in Connacht
Gam·bia \ˈgam-bē-ə\ **1** river 700 miles (1126 kilometers) long W Africa **2** country W Africa; capital, Banjul — **Gam·bi·an** \-ən\ *adj or n*
Gand — see GHENT
Gan·ges \ˈgan-ˌjēz\ river 1550 miles (2494 kilometers) long N India flowing from the Himalaya SE & E to unite with the Brahmaputra & empty into Bay of Bengal through the vast **Ganges Delta** — see HUGLI — **Gan·get·ic** \gan-ˈjet-ik\ *adj*
Gan·su *or* **Kan·su** \ˈgän-ˈsü\ province NW China; capital, Lanzhou
Gar·da, Lake \ˈgärd-ə\ lake N Italy NW of Verona
Garden Grove city SW California SW of Los Angeles
Ga·ronne \gə-ˈrän, -ˈrōn\ river about 355 miles (571 kilometers) long SE France flowing into Gironde Estuary
Gary \ˈgaər-ē, ˈgeər-ē\ city NW Indiana on Lake Michigan
Gas·co·ny \ˈgas-kə-nē\ *or French* **Gas·cogne** \gȧ-ˈskōnʸ\ region and former province SW France
Gas·pé \gas-ˈpā, ˈgas-ˌ\ peninsula SE Quebec E of mouth of the Saint Lawrence — **Gas·pe·sian** \ga-ˈspē-zhən\ *adj or n*
Gasteiz — see VITORIA
Gates·head \ˈgāts-ˌhed\ borough N England
Gates of the Arctic National Park reservation N central Alaska in Brooks Range
Gaul \ˈgȯl\ *or Latin* **Gal·lia** \ˈgal-ē-ə\ ancient country W Europe chiefly comprising region occupied by modern France & Belgium but at one time including also Po valley in N Italy — see CISALPINE GAUL, TRANSALPINE GAUL
Gau·teng \ˈgau̇-ˌteng\ province central NE Republic of South Africa
Ga·za Strip \ˈgäz-ə, ˈgaz-, ˈgāz-\ strip of land on SE Mediterranean Sea adjoining Sinai Peninsula; chief city, Gaza
Ga·zi·an·tep \ˌgäz-ē-ˌän-ˈtep, -än-\ city S Turkey
Gdansk \gə-ˈdänsk, -ˈdansk\ *or German* **Dan·zig** \ˈdan-sig, ˈdän-\ city N Poland on Gulf of Danzig
Gdyn·ia \gə-ˈdin-ē-ə\ city N Poland
Gebel Musa — see MUSA (Gebel)
Gee·long \jə-ˈlȯŋ\ city SE Australia in S Victoria
Gel·sen·kir·chen \ˌgel-zən-ˈkir-kən\ city W Germany in the Ruhr W of Dortmund
Ge·ne·ral San Mar·tín \ˌhā-nā-ˌräl-ˌsan-mär-ˈtēn\ city E Argentina NW of Buenos Aires
Ge·ne·va \jə-ˈnē-və\ city SW Switzerland on Lake Geneva — **Ge·ne·van** \-vən\ *adj or n*
Geneva, Lake of *or* **Lake Le·man** \ˈlē-mən, ˈlem-ən, lə-ˈman\ *ancient* **Le·man·nus** \li-ˈman-əs\ *or* **Le·ma·nus** \li-ˈmān-əs\ lake on border between SW Switzerland & E France traversed by the Rhone
Gen·oa \ˈjen-ə-wə\ *or Italian* **Ge·no·va** \ˈje-nō-vä\ city NW Italy, capital of Liguria — **Gen·o·ese** \ˌjen-ə-ˈwēz, -ˈwēs\ *or* **Gen·o·vese** \-ə-ˈvēz, -ˈvēs\ *adj or n*
George·town \ˈjȯrj-ˌtau̇n\ **1** a W section of Washington, District of Columbia **2** city, capital of Guyana
George Town *or* **Pe·nang** \pə-ˈnaŋ\ city Malaysia, on an island in Peninsular Malaysia
Geor·gia \ˈjȯr-jə\ **1** state SE U.S.; capital, Atlanta **2** country SE Europe on Black Sea S of Caucasus Mountains; capital, Tbilisi; a constituent republic of U.S.S.R. 1936–91 — **Geor·gian** \ˈjȯr-jən\ *adj or n*
Georgia, Strait of channel Canada & U.S. between Vancouver Island & mainland NW of Puget Sound
Georgian Bay inlet of Lake Huron in S Ontario
Georgian Bay Islands National Park reservation Ontario in Georgian Bay
Ger·man·town \ˈjər-mən-ˌtau̇n\ a NW section of Philadelphia, Pennsylvania
Ger·ma·ny \ˈjərm-nē, -ə-nē\ country central Europe bordering on North & Baltic seas; capital, Berlin; divided into two republics 1940–90: the **Federal Republic of Germany** (capital, Bonn) & the **German Democratic Republic** (capital, East Berlin)
Ger·mis·ton \ˈjər-mə-stən\ city NE Republic of South Africa E of Johannesburg
Gha·na \ˈgän-ə, ˈgan-ə\ *or formerly* **Gold Coast** country W Africa on Gulf of Guinea; a republic; capital, Accra — **Gha·na·ian** \gä-ˈnā-ən, ga-, -yən; -ˈnī-ən\ *or* **Gha·ni·an** \ˈgän-ē-ən, ˈgän-yən, ˈgan-\ *adj or n*
Ghats \ˈgȯts\ two mountain chains S India — see EASTERN GHATS, WESTERN GHATS
Ghent \ˈgent\ *or Flemish* **Gent** \ˈkent\ *or French* **Gand** \ˈgäⁿ\ city NW central Belgium
Gi·bral·tar \jə-ˈbrȯl-tər\ British colony & fortress on S coast of Spain including Rock of Gibraltar
Gibraltar, Rock of headland on S coast of Spain in Gibraltar colony at E end of Strait of Gibraltar; highest point 1396 feet (426 meters) — see PILLARS OF HERCULES
Gibraltar, Strait of passage between Spain & Africa connecting North Atlantic & the Mediterranean
Gi·fu \ˈgē-ˌfü\ city Japan in central Honshu
Gi·jón \hē-ˈhōn\ city & port NW Spain on Bay of Biscay
Gi·la \ˈhē-lə\ river 630 miles (1014 kilometers) long SW New Mexico and S Arizona flowing W into the Colorado
Gil·bert \ˈgil-bərt\ **1** city SW central Arizona **2** islands Kiribati in W Pacific; until 1975 formed with Ellice Islands the British colony of **Gilbert and El·lice Islands** \ˈel-əs\ — see KIRIBATI, TUVALU
Gil·e·ad \ˈgil-ē-əd\ mountain region of ancient Palestine E of Jordan River; now in NW Jordan — **Gil·e·ad·ite** \-ē-ə-ˌdīt\ *n*
Gi·ronde \jə-ˈränd, zhə-; zhē-ˈrōⁿd\ estuary W France formed by junction of Garonne & Dordogne rivers
Gi·za \ˈgē-zə\ *or* **El Gî·za** \el-\ *or* **Al Jī·zah** \al-jē-zə\ city N Egypt on the Nile SW of Cairo
Gla·cier Bay \ˌglā-shər-\ inlet SE Alaska at S end of Saint Elias Range in **Glacier Bay National Park**
Glacier National Park — see WATERTON-GLACIER INTERNATIONAL PEACE PARK
Glades \ˈglādz\ EVERGLADES
Glas·gow \ˈglas-kō, ˈglas-gō, ˈglaz-gō\ city S central Scotland on the Clyde constituting an administrative area (**Glasgow City**) — **Glas·we·gian** \glas-ˈwē-jən\ *adj or n*
Glen·dale \ˈglen-ˌdāl\ **1** city central Arizona NW of Phoenix **2** city S California NE of Los Angeles
Glouces·ter \ˈgläs-tər, ˈglȯs-\ city SW central England
Glouces·ter·shire \ˈgläs-tər-ˌshiər, ˈglȯs-, -shər\ *or* **Gloucester** county SW central England
Gnossus — see KNOSSOS
Goa \ˈgō-ə\ state India on W coast belonging before 1962 to Portugal; capital, Panaji
Go·bi \ˈgō-bē\ desert E central Asia in Mongolia & N China
Godthåb — see NUUK
Godwin Austen — see K2
Goi·â·nia \gȯi-ˈan-ē-ə\ city SE central Brazil SW of Brasília
Go·lan Heights \ˌgō-ˌlän, -lən\ hilly region between NE Israel & SW Syria
Gol·con·da \gäl-ˈkän-də\ ruined city central India W of Hyderabad
Gold Coast **1** coast region W Africa on N shore of Gulf of Guinea E of Ivory Coast **2** — see GHANA
Golden Gate strait W California connecting San Francisco Bay and North Pacific
Golden Horn inlet of the Bosporus, Turkey in Europe; harbor of Istanbul
Golgotha — see CALVARY
Gomel — see HOMYEL'
Go·mor·rah \gə-ˈmär-ə, -ˈmȯr-\ ancient city thought to be in the area now covered by the SW part of Dead Sea
Good Hope, Cape of \ˌgu̇d-ˈhōp\ cape S Republic of South Africa extending into South Atlantic
Go·rakh·pur \ˈgōr-ək-ˌpu̇r, ˈgȯr-\ city NE India in E Uttar Pradesh N of Varanasi
Gorki — see NIZHNIY NOVGOROD
Gor·lov·ka \gȯr-ˈlóf-kə, -ˈlȯv-\ city E Ukraine in Donets Basin
Go·shen \ˈgō-shən\ district of ancient Egypt E of delta of the Nile
Gö·te·borg \ˌyərt-ə-ˈbȯr-ē, *Swedish* ˌyœ-tə-ˈbȯry\ *or* **Goth·en·burg** \ˈgäth-ən-ˌbərg\ city & port SW Sweden
Got·land \ˈgät-ˌland, -lənd\ island Sweden in the Baltic; capital, Visby
Göt·ting·en \ˈgərt-iŋ-ən, ˈget-, ˈgœt-\ city central Germany SSW of Brunswick
Gram·pi·an \ˈgram-pē-ən\ hills N central Scotland — see BEN NEVIS
Gra·na·da \grə-ˈnäd-ə\ city S Spain in Andalusia
Grand Banks shoal area in the W Atlantic SE of Newfoundland
Grand Canyon gorge of Colorado River NW Arizona; area largely in **Grand Canyon National Park**
Grand Canyon of the Snake — see HELLS CANYON
Grande, Rio — see RIO GRANDE
Grand Prairie city NE central Texas
Grand Rapids city SW Michigan
Grand Te·ton National Park \ˈtē-ˌtän\ reservation NW Wyoming S of Yellowstone National Park
Grass·lands National Park \ˈgras-ˌlandz\ reservation SW Saskatchewan
Gravenhage, 's — see HAGUE (The)
Graz \ˈgräts\ city S Austria
Great Australian Bight wide bay on S coast of Australia
Great Barrier Reef coral reef Australia off NE coast of Queensland
Great Basin region W U.S. between Sierra Nevada & Wasatch ranges including most of Nevada & parts of California, Idaho, Utah, Wyoming, & Oregon; has no drainage to ocean
Great Basin National Park reservation E Nevada
Great Bear lake Canada in Northwest Territories draining through Great Bear River into Mackenzie River
Great Brit·ain \ˈbrit-ⁿn\ **1** island W Europe NW of France comprising England, Scotland, & Wales **2** UNITED KINGDOM
Great Dividing Range mountain system E Australia & Tasmania extending S from Cape York Peninsula — see KOSCIUSKO (Mount)
Greater An·til·les \an-ˈtil-ēz\ group of islands of the West Indies including Cuba, Hispaniola, Jamaica, & Puerto Rico — see LESSER ANTILLES
Greater London the City of London & 32 surrounding boroughs
Greater Manchester metropolitan county NW England including city of Manchester
Greater Sunda — see SUNDA
Great Lakes **1** chain of five lakes (Superior, Michigan, Huron, Erie, & Ontario) central North America in U.S. & Canada **2** group of lakes E central Africa including Turkana, Albert, Victoria, Tanganyika, & Malawi

Great Plains elevated plains region W central U.S. & W Canada E of the Rockies; chiefly W of the 100th meridian extending from W Texas to NE British Columbia & NW Alberta

Great Rift Valley \'rift\ depression SW Asia & E Africa extending with several breaks from valley of the Jordan S to central Mozambique

Great Salt lake N Utah having strongly saline waters & no outlet

Great Sand Dunes National Park reservation S central Colorado

Great Slave lake NW Canada in SE Northwest Territories drained by Mackenzie River

Great Smoky mountains between W North Carolina & E Tennessee partly in **Great Smoky Mountains National Park**; highest Clingmans Dome 6643 feet (2025 meters)

Greece \'grēs\ *or ancient* **Hel·las** \'hel-əs\ *or Greek* **El·lás** \e-'läs\ country S Europe at S end of Balkan Peninsula; capital, Athens

Green \'grēn\ **1** mountains E North America in the Appalachians extending from S Quebec S through Vermont into W Massachusetts **2** river 730 miles (1175 kilometers) long W U.S. flowing from Wind River Range in W Wyoming S into the Colorado in SE Utah

Green Bay **1** inlet of NW Lake Michigan about 120 miles (193 kilometers) long in NW Michigan & NE Wisconsin **2** city NE Wisconsin

Green·land \'grēn-lənd, -ˌland\ island in the North Atlantic off NE North America belonging to Denmark; capital, Nuuk

Greens·boro \'grēnz-ˌbər-ə, -ˌbə-rə\ city N central North Carolina

Green·wich \'grin-ij, 'gren-, -ich\ SE borough of Greater London, England

Green·wich Village \ˌgren-ich-, ˌgrin-, -ij-\ section of New York City in Manhattan on lower W side

Gre·na·da \grə-'nād-ə\ island British West Indies in S Windwards; an independent country since 1974; capital, Saint George's

Gren·a·dines \ˌgren-ə-'dēnz\ islands West Indies in central Windwards; divided between Grenada & Saint Vincent and the Grenadines

Gre·no·ble \grə-'nō-bəl\ city SE France

Grisons — see GRAUBÜNDEN

Gro·ning·en \'grō-ning-ən\ city NE Netherlands

Gros Morne National Park \grō-'mȯrn\ reservation Newfoundland along W coast

Groz·ny \'grȯz-nē, 'gräz-\ city S Russia in Europe N of Caucasus Mountains, capital of Chechnya

Gua·da·la·ja·ra \ˌgwäd-ə-lə-'här-ə\ city W central Mexico, capital of Jalisco

Gua·dal·ca·nal \ˌgwäd-l-kə-'nal, ˌgwäd-ə-kə-\ island W Pacific in the SE Solomons

Gua·dal·qui·vir \ˌgwäd-l-'kwiv-ər, -ki-'viər\ river 408 miles (656 kilometers) long S Spain flowing into North Atlantic

Gua·da·lupe Mountains National Park \'gwäd-l-ˌüp\ reservation W Texas

Gua·de·loupe \'gwäd-l-ˌüp\ two islands, Basse-Terre (or Guadeloupe proper) & Grande-Terre, separated by a narrow channel in French West Indies in central Leewards; capital, Basse-Terre (on Basse-Terre Island)

Gua·di·a·na \ˌgwäd-ē-'än-ə, -'an-\ river 515 miles (829 kilometers) long S Spain & SE Portugal flowing into North Atlantic

Guaíra *or* **Guayra** — see SETE QUEDAS

Guam \'gwäm\ island W Pacific in S Mariana Islands belonging to U.S.; capital, Agana — **Gua·ma·ni·an** \gwä-'mā-nē-ən\ *adj or n*

Gua·na·ba·coa \ˌgwän-ə-bə-'kō-ə\ city W Cuba

Gua·na·ba·ra Bay \ˌgwän-ə-'bar-ə, -'bär-\ inlet of South Atlantic SE Brazil on which city of Rio de Janeiro is situated

Gua·na·jua·to \ˌgwän-ə-'hwät-ō, -'wät-\ **1** state central Mexico **2** city, its capital

Guang·dong *or* **Kwang·tung** \'gwäng-'du̇ng\ province SE China bordering on South China Sea & Gulf of Tonkin; capital Guangzhou

Guang·xi Zhuang·zu \'gwäng-'shē-jə-'wäng-'zü\ *or* **Kwang·si–Chuang** \'gwäng-'shē-jə-'wäng\ region & former province S China; capital, Nanning

Guang·zhou *or* **Kuang–chou** \'gwäng-'jō\ *or* **Can·ton** \'kan-ˌtän, kan-'\ city SE China, capital of Guangdong

Guan·ta·na·mo Bay \gwän-'tän-ə-ˌmō\ inlet of the Caribbean in SE Cuba; site of U.S. naval station

Gua·te·ma·la \ˌgwät-ə-'mä-lə\ **1** country Central America; a republic **2** *or* **Guatemala City** city, its capital — **Gua·te·ma·lan** \-lən\ *adj or n*

Gua·ya·quil \ˌgwī-ə-'kēl, -'kil\ city W Ecuador

Guay·na·bo \gwī-'näb-ō\ city NE central Puerto Rico

Guern·sey \'gərn-zē\ island in English Channel — see CHANNEL

Guer·re·ro \gə-'reər-ō\ state S Mexico on North Pacific; capital, Chilpancingo

Gui·a·na \gē-'an-ə, -'än-ə; gī-'an-ə\ region N South America on the Atlantic bounded on W & S by Orinoco, Negro, & Amazon rivers; includes Guyana, French Guiana, Suriname, & adjacent parts of Brazil & Venezuela — **Gui·a·nan** \-ən\ *adj or n*

Gui·lin \'gwē-'lin\ *or* **Kwei·lin** *or* **Kuei–lin** \'gwā-'lin\ city S China in NE Guangxi Zhuangzu

Guin·ea \'gin-ē\ **1** region W Africa on the Atlantic extending along coast from Gambia to Angola **2** *or formerly* **French Guinea** country W Africa N of Sierra Leone & Liberia; a republic; capital, Conakry — **Guin·ean** \'gin-ē-ən\ *adj or n*

Guinea, Gulf of arm of the Atlantic W central Africa

Guin·ea–Bis·sau \ˌgin-ē-bis-'au̇\ *or formerly* **Portuguese Guinea** country W Africa on North Atlantic; capital, Bissau

Gui·yang \'gwē-'yäng\ *or* **Kuei–yang** \'gwā-'yäng\ city S China, capital of Guizhou

Gui·zhou \'gwē-'jō\ *or* **Kuei–chou** \'gwā-'jō\ *or* **Kwei·chow** \'gwā-'chau̇\ province S China S of Sichuan; capital, Guiyang

Gu·ja·rat *or* **Gu·je·rat** \ˌgü-jə-'rät, ˌgu̇j-ə-\ state W India N & E of Gulf of Cambay; capital, Gandhinagar

Guj·ran·wa·la \ˌgu̇j-rən-'wäl-ə, ˌgu̇j-\ city NE Pakistan

Gulf States states of U.S. bordering on Gulf of Mexico: Florida, Alabama, Mississippi, Louisiana, and Texas

Gulf Stream warm current of North Atlantic flowing from Gulf of Mexico NE along coast of U.S. to Nantucket Island and thence eastward

Gun·tur \gu̇n-'tu̇r\ city E India in central Andhra Pradesh

Gus·ta·vo A. Ma·de·ro \gu̇s-'täv-ˌō-ˌä-mə-'der-ˌō\ city central Mexico N of Mexico City

Guy·ana \gī-'an-ə\ *or formerly* **British Guiana** country N South America on the Atlantic; a republic since 1970, capital, Georgetown

Gwa·li·or \'gwäl-ē-ˌȯr\ city N central India in NW Andhra Pradesh SSE of Agra

Gwyn·edd \'gwin-eth\ administrative area NW Wales

Habana, La — see HAVANA

Ha·chi·ō·ji \ˌhäch-ē-'ō-jē\ city Japan in SE central Honshu W of Tokyo

Hack·ney \'hak-nē\ borough of N Greater London, England

Ha·des \'hād-ēz, -ˌēz\ underground abode of the dead in Greek mythology

Hague, The \thə-'hāg\ *or Dutch* **'s Gra·ven·ha·ge** \ˌskräv-ən-'häg-ə, ˌskräv-\ city SW Netherlands; seat of government of the Netherlands

Haidarabad — see HYDERABAD

Hai·fa \'hī-fə\ city & port NW Israel

Hai·kou \'hī-'kau̇, -'kō\ city SE China, capital of Hainan

Hai·nan \'hī-'nän\ island SE China in South China Sea; a province; capital, Haikou

Hai·phong \'hī-'fȯng\ city N Vietnam

Hai·ti \'hāt-ē\ **1** — see HISPANIOLA **2** country West Indies in W Hispaniola; capital, Port-au-Prince — **Hai·tian** \'hā-shən\ *adj or n*

Ha·ko·da·te \ˌhäk-ə-'dät-ē\ city & port Japan in SW Hokkaido

Ha·le·a·ka·la Crater \ˌhäl-ē-ˌäk-ə-'lä\ crater more than 2500 feet (762 meters) deep Hawaii in E Maui Island in **Haleakala National Park**

Hal·i·car·nas·sus \ˌhal-ə-kär-'nas-əs\ ancient city SW Asia Minor in SW Caria on Aegean Sea

Hal·i·fax \'hal-ə-ˌfaks\ city, capital of Nova Scotia

Hal·le \'häl-ə\ city E central Germany NW of Leipzig

Hal·ma·hera \ˌhal-mə-'her-ə, ˌhäl-\ island E Indonesia; largest of the Moluccas

Ha·ma \'ham-ˌä\ city W Syria

Ha·ma·dan \ˌham-ə-'dan, -'dän\ city W Iran

Ha·ma·ma·tsu \ˌhäm-ə-'mät-sü\ city Japan in S Honshu

Ham·burg \'ham-ˌbərg, 'häm-ˌbu̇rg\ city N Germany on the Elbe; comprises a state of Germany — **Ham·burg·er** \-ˌbər-gər, -ˌbu̇r-\ *n*

Ham·hung \'häm-ˌhu̇ng\ city E central North Korea

Ham·il·ton \'ham-əl-tən, -əlt-n\ **1** city S Ontario **2** town, capital of Bermuda

Ham·mer·smith and Ful·ham \'ham-ər-ˌsmith-ənd-'fu̇l-əm\ borough of SW Greater London, England

Ham·mond \'ham-ənd\ city NW Indiana

Hamp·shire \'hamp-ˌshiər, 'ham-, -shər\ county S England on English Channel

Hamp·ton \'hamp-tən, 'ham-\ city SE Virginia on Hampton Roads

Hampton Roads channel SE Virginia through which James & Elizabeth rivers flow into Chesapeake Bay

Hang·zhou \'häng-'jō\ *or* **Hang–chou** \-'jō\ *or* **Hang·chow** \'hang-'chau̇, 'häng-jō\ city E China, capital of Zhejiang

Han·kow \'hang-'kau̇, -'kō; 'häng-'kō\ former city E central China — see WUHAN

Han·no·ver *or* **Han·o·ver** \'han-ˌō-vər, 'han-ə-vər; *German* hä-'nō-vər\ city N central Germany, capital of Lower Saxony

Ha·noi \ha-'nȯi, hə-, hä-\ city, capital of Vietnam

Han·yang \'hän-'yäng\ former city E central China — see WUHAN

Ha·ra·re \hə-'rä-ˌrā\ *or formerly* **Salis·bury** \'sȯlz-ˌber-ē, 'salz-, -bə-rē, -brē\ city, capital of Zimbabwe

Har·bin \'här-bən, här-'bin\ *or* **Ha–erh–pin** \'hä-'er-'bin\ *or formerly* **Pin·kiang** \'bin-jē-'äng\ city NE China, capital of Heilongjiang

Har·in·gey \'har-ing-ˌgā\ borough of N Greater London, England

Har·lem \'här-ləm\ section of New York City in N Manhattan

Har·ris·burg \'har-əs-ˌbərg\ city, capital of Pennsylvania

Har·row \'har-ō\ borough of NW Greater London, England

Hart·ford \'härt-fərd\ city, capital of Connecticut on Connecticut River

Hart·le·pool \'härt-lē-ˌpül\ borough N England

Ha·ry·a·na \ˌhə-rē-'än-ə\ state N India; capital, Chandigarh

Harz \'härts\ mountains central Germany between Elbe & Leine rivers

Hat·ter·as, Cape \'hat-ə-rəs, 'ha-trəs\ cape, North Carolina on **Hatteras Island** (barrier island between Pamlico Sound and North Atlantic)

Haute–Nor·man·die \ˌōt-ˌnȯr-mäⁿ-'dē\ region N France on English Channel

Ha·vana \hə-'van-ə\ *or Spanish* **La Ha·ba·na** \ˌlä-ä-'vän-ə, ˌlä-'vän-ə\ city, capital of Cuba — **Ha·van·an** \hə-'van-ən\ *adj or n*

Hav·ant \'hav-ənt\ town S England

Ha·ver·ing \'hāv-ring, -ə-ring\ borough of NE Greater London, England

Ha·waii \hə-'wä-ē, -'wī-, -'wȯ-, -yē\ **1** *or formerly* **Sand·wich Islands** \ˌsan-ˌwich-, ˌsand-\ group of islands central Pacific belonging to U.S. **2** island, largest of the group **3** state of U.S. comprising Hawaiian Islands except Midway; capital, Honolulu

Hawaii Volcanoes National Park reservation Hawaii on Hawaii Island including Mauna Loa & Kilauea

Hay·ward \'hā-wərd\ city W California SE of Oakland

He·bei \'həb-'ā\ *or* **Ho·peh** *or* **Ho·pei** \'hō-'bā\ province NE China; capital, Shijiazhuang

Heb·ri·des \'heb-rə-ˌdēz\ islands W Scotland in North Atlantic comprising **Outer Hebrides** (to W) and **Inner Hebrides** (to E) — see WESTERN ISLES — **Heb·ri·de·an** \ˌheb-rə-'dē-ən\ *adj or n*

He·fei *or* **Ho·fei** \'həf-'ā\ *or formerly* **Lu·chow** \'lü-'jō\ city E China, capital of Anhui W of Nanjing

Hei·long·jiang *or* **Hei·lung·kiang** \'hā-'lu̇ng-'jyäng\ province NE China in N Manchuria; capital, Harbin

He·jaz \hej-'az, hij-\ region W Saudi Arabia on Red Sea

Hel·e·na \'hel-ə-nə\ city, capital of Montana

He·li·op·o·lis \ˌhē-lē-'äp-ə-ləs\ **1** either of two cities of ancient Egypt near modern Cairo **2** city of ancient Syria; site at modern town of **Baal·bek** \'bā-əl-ˌbek, 'bäl-ˌbek\ in E Lebanon N of Damascus

Hellas — see GREECE

Hellespont — see DARDANELLES

\ə\ abut \ᵊ\ kitten \ər\ further \a\ ash \ā\ ace \ä\ mop, mar \au̇\ out \ch\ chin \e\ bet \ē\ easy \g\ go \i\ hit \ī\ ice \j\ job \ŋ\ sing \ō\ go \ȯ\ law \ȯi\ boy \th\ thin \th\ the \ü\ loot \u̇\ foot \y\ yet \zh\ vision *see also* Guide to Pronunciation

Hells Canyon \'helz\ *or* **Grand Canyon of the Snake** canyon of Snake River on Idaho–Oregon boundary

Hel·sin·ki \'hel-ˌsiŋ-kē, hel-'\ *or Swedish* **Hel·sing·fors** \'hel-siŋ-ˌfȯrz\ city, capital of Finland

Helvetia — see SWITZERLAND

He·nan \'hən-'än\ *or* **Ho·nan** \'hō-'nän\ province E central China; capital, Zhengzhou

Hen·der·son \'hen-dər-sən\ city S Nevada

Heng·yang \'həŋ-'yäŋ\ city SE central China in SE Hunan

Henry, Cape \'hen-rē\ headland E Virginia S of entrance to Chesapeake Bay

Her·mon, Mount \'hər-mən\ mountain 9232 feet (2814 meters) on Lebanon–Syria border; highest in Anti-Lebanon Mountains

Her·mo·si·llo \ˌer-mə-'sē-ō, -yō\ city NW Mexico, capital of Sonora

Hert·ford·shire \'här-fərd-ˌshiər, *also* 'härt-, *in the U.S. also* 'hərt-\ *or* **Hertford** county SE England

Her·ze·go·vi·na \ˌhert-sə-gō-'vē-nə, ˌhərt-\ *or* **Her·ce·go·vi·na** \'hert-sä-ˌgō-vē-nä\ region S Europe S of Bosnia; now part of Bosnia and Herzegovina

Hesse \'hes, 'hes-ē\ *or German* **Hes·sen** \'hes-n\ state central Germany E of the Rhine & N of the Main; capital, Wiesbaden

Hi·a·le·ah \ˌhī-ə-'lē-ə\ city SE Florida N of Miami

Hibernia — see IRELAND

Hi·dal·go \hid-'al-gō\ state central Mexico; capital, Pachuca

Hi·ga·shi·ōsa·ka \hē-ˌgä-shē-ō-'säk-ə\ city Japan in S Honshu E of Osaka

High·land \'hī-lənd\ administrative area NW Scotland

High·lands \'hī-ləndz, -lənz\ the mountainous N part of Scotland lying N & W of the Lowlands

High Plains the Great Plains especially from Nebraska southward

Hil·ling·don \'hil-iŋ-dən\ borough of W Greater London, England

Hi·ma·chal Pra·desh \hi-ˌmäch-əl-prə-'desh, -'dāsh\ state NW India comprising two areas NW of Uttar Pradesh; capital, Simla

Hi·ma·la·ya, the \ˌhim-ə-'lā-ə; hə-'mäl-yə, -'mäl-ə-yə\ *or* **the Hi·ma·la·yas** \-əz, -yəz\ mountain system S Asia on border between India & Tibet & in Kashmir, Nepal, & Bhutan — see EVEREST (Mount) — **Hi·ma·la·yan** \ˌhim-ə-'lā-ən; hə-'mäl-yən, -'mäl-ə-yən\ *adj*

Hi·me·ji \hi-'mej-ē\ city Japan in W Honshu WNW of Kobe

Hin·du Kush \ˌhin-dü-'ku̇sh, -'kəsh\ mountain range central Asia SW of the Pamirs on border of Kashmir & in Afghanistan

Hin·du·stan \ˌhin-dü-'stan, -də-; -'stän\ **1** region N India N of the Deccan **2** historical name for India (country)

Hip·po \'hip-ō\ ancient city N Africa; chief town of Numidia

Hi·ra·ka·ta \ˌhir-ə-'kät-ə\ city Japan on Honshu

Hi·ro·shi·ma \ˌhir-ə-'shē-mə, hə-'rō-shə-mə\ city Japan in SW Honshu on Inland Sea

Hispalis — see SEVILLE

Hispania — see SPAIN

His·pan·io·la \ˌhis-pən-'yō-lə\ *or Spanish* **Es·pa·ño·la** \ˌes-ˌpän-'yō-lə\ *or formerly* **Hai·ti** \'hāt-ē\ *or* **San·to Do·min·go** \ˌsant-əd-ə-'miŋ-gō\ island West Indies in Greater Antilles divided between Haiti on W & Dominican Republic on E

Ho·bart \'hō-ˌbärt\ city Australia, capital of Tasmania

Ho Chi Minh City \ˌhō-ˌchē-ˌmin-, -ˌshē-\ *or formerly* **Sai·gon** \sī-'gän, 'sī-ˌ\ city S Vietnam

Hofei — see HEFEI

Hoggar — see AHAGGAR

Hoh·hot \'hō-'hōt\ *or* **Hu·he·hot** \'hü-ˌhā-'hōt\ *or* **Hu–ho–hao–t'e** \'hü-'hō-'hau̇-'tə\ city N China, capital of Inner Mongolia

Hok·kai·do \hä-'kīd-ō\ *or formerly* **Ye·zo** \'yez-ō\ island N Japan N of Honshu

Hol·land \'häl-ənd\ **1** medieval county of Holy Roman Empire bordering on North Sea & comprising area now forming North Holland & South Holland provinces of the Netherlands **2** — see NETHERLANDS — **Hol·land·er** \-ən-dər\ *n*

Hol·ly·wood \'häl-ē-ˌwu̇d\ **1** section of Los Angeles, California, NW of downtown district **2** city SE Florida

Hol·stein \'hōl-ˌstīn, -ˌstēn\ region NW Germany S of Jutland Peninsula adjoining Schleswig — see SCHLESWIG-HOLSTEIN

Holy Land the lands comprising ancient Palestine and including the holy land of the Jewish, Christian, & Islamic religions

Homs \'hȯmz, 'hu̇ms\ city W Syria

Ho·myel' \hȯ-'myel, kȯ-\ *or* **Go·mel** \gȯ-'mel, -'myel\ *or* **Ho·mel** \hȯ-\ city SE Belarus

Honan — see HENAN

Hon·du·ras \hän-'du̇r-əs, -'dyu̇r-\ country Central America; capital, Tegucigalpa — **Hon·du·ran** \-ən\ *or* **Hon·du·ra·ne·an** *or* **Hon·du·ra·ni·an** \ˌhän-du̇-'rā-nē-ən, -dyu̇-\ *adj or n*

Hong Kong \'häŋ-ˌkäŋ, -'käŋ; 'hȯŋ-ˌkȯŋ, -'kȯŋ\ *or Chinese* **Xiang·gang** *or* **Hsiang Kang** \'shyäŋ-ˌgäŋ\ special administrative region China including Hong Kong Island & Jiulong Peninsula; formerly a British colony with Victoria as capital

Ho·ni·a·ra \ˌhō-nē-'är-ə\ town, capital of Solomon Islands

Ho·no·lu·lu \ˌhän-l-'ü-lü, ˌhōn-l-\ city, capital of Hawaii on Oahu Island

Hon·shu \'hän-shü\ *or* **Hon·do** \'hän-dō\ island Japan; largest of the four chief islands

Hood, Mount \'hu̇d\ mountain 11,235 feet (3424 meters) NW Oregon in Cascade Range

Hoo·ver Dam \ˌhü-vər\ *or* **Boul·der Dam** \ˌbōl-dər\ dam 726 feet (221 meters) high in Colorado River between Arizona & Nevada — see MEAD (Lake)

Hopeh *or* **Hopei** — see HEBEI

Ho·reb \'hōr-ˌeb, 'hȯr-\ *or* **Si·nai** \'sī-ˌnī *also* -nē-ˌī\ mountain where according to the Bible the Law was given to Moses; generally thought to be on Sinai Peninsula

Hor·muz, Strait of \'hȯr-ˌməz, hȯr-'müz\ strait SW Asia connecting Persian Gulf & Gulf of Oman

Horn, Cape \'hȯrn\ headland S Chile on Horn Island in Tierra del Fuego; the most southerly point of South America at 56° S latitude

Horn of Africa the easternmost projection of Africa -- variously used to refer to Somalia, SE or all of Ethiopia, & sometimes Djibouti

Hos·pi·ta·let \ˌäs-ˌpit-l-'et, ˌhäs-\ city NE Spain

Hot Springs National Park reservation SW central Arkansas adjoining city of Hot Springs

Houns·low \'hau̇nz-ˌlō\ borough of SW Greater London, England

Hous·ton \'hyü-stən, 'yü-\ city & port SE Texas

How·rah \'hau̇-rə\ city E India in West Bengal on Hugli River opposite Calcutta

Hsia–men — see XI'AMEN

Hsiang–t'an — see XIANGTAN

Huai·nan \hü-ī-'nän, 'hwī-\ city E China in N central Anhui

Huang *or* **Hwang** \'hwäŋ\ *or* **Yellow** river about 3000 miles (4828 kilometers) long N China flowing into Bo Hai

Huas·ca·rán \ˌwäs-kə-'rän\ mountain 22,205 feet (6768 meters) W Peru

Hu·bei \'hü-'bā\ *or* **Hu·peh** \'hü-'bē\ *or* **Hu·pei** \'hü-'bā, -'pā\ province E central China; capital, Wuhan

Hu·bli–Dhar·war \ˌhu̇b-lē-ˌdär-'wär\ city SW India in W Karnataka

Hud·ders·field \'həd-ərz-ˌfēld\ town N England NE of Manchester

Hud·son \'həd-sən\ **1** river 306 miles (492 kilometers) long E New York flowing S into New York Bay **2** bay of North Atlantic in N Canada **3** strait NE Canada connecting Hudson Bay & North Atlantic

Hue \'hwā, 'wā, hü-'ā, hyü-'ā\ city central Vietnam in Annam

Hu·gli *or* **Hoo·ghly** \'hü-glē\ river 120 miles (193 kilometers) long E India flowing S into Bay of Bengal; most westerly channel of the Ganges in its delta

Huhehot *or* **Hu–ho–hao–t'e** — see HOHHOT

Hull \'həl\ *or* **Kings·ton upon Hull** \'kiŋ-stən, 'kiŋk-\ city N England

Hum·ber \'həm-bər\ estuary E England formed by the Ouse & the Trent and flowing into North Sea

Hum·boldt \'həm-ˌbōlt\ glacier NW Greenland

Humboldt Current — see PERU CURRENT

Hu·nan \'hü-'nän\ province SE central China; capital, Changsha

Hun·ga·ry \'həŋ-grē, -gə-rē\ country central Europe; capital, Budapest

Hunt·ing·ton Beach \'hən-tiŋ-tən\ city SW California SE of Los Angeles

Hunts·ville \'hənts-ˌvil, -vəl\ city N Alabama

Hu·ron, Lake \'hyu̇r-ən, 'yu̇r-, -ˌän\ lake E central North America in U.S. & Canada; one of the Great Lakes

Hy·der·abad \'hīd-rə-ˌbad, -ə-rə-, -ˌbäd\ **1** *or* **Hai·dar·abad** *same*\ city S central India; capital of Andhra Pradesh **2** city SE Pakistan on the Indus

Hy·met·tus \hī-'met-əs\ mountain ridge about 3370 feet (1027 meters) central Greece E & SE of Athens

Ia·si \'yäsh, 'yäsh-ē\ city NE Romania

Iba·dan \i-'bäd-n, -'bad-\ city SW Nigeria

Ibe·ri·an \ī-'bir-ē-ən\ peninsula SW Europe occupied by Spain & Portugal

Ibi·za \ē-'vē-thə, -'bē-\ island Spain in Balearic Islands SW of Majorca

Ice·land \'ī-slənd, -ˌsland\ island SE of Greenland between Arctic & North Atlantic oceans; capital, Reykjavik — **Ice·land·er** \-slən-dər, -ˌslan-dər\ *n*

Ichi·ka·wa \i-'chē-ˌkä-wə\ city Japan in SE Honshu E of Tokyo

Ida, Mount — see KAZ DAGI

Ida·ho \'īd-ə-ˌhō\ state NW U.S.; capital, Boise — **Ida·ho·an** \ˌīd-ə-'hō-ən\ *adj or n*

Idumaea *or* **Idumea** — see EDOM

If·ni \'if-nē\ former territory SW Morocco on North Atlantic; administered by Spain 1934–69; capital, Sidi Ifni

Igua·zú *or* **Igua·çu** \ˌē-gwä-'sü\ river 745 miles (1199 kilometers) long S Brazil flowing W into the Alto Paraná; contains **Iguazú Falls** (waterfall over 2 miles or 3.2 kilometers wide composed of numerous cataracts averaging 200 feet or 61 meters in height)

IJs·sel *or* **Ijs·sel** *or* **Ys·sel** \'ī-səl\ river 70 miles (113 kilometers) long E Netherlands flowing out of Rhine N into IJsselmeer

IJs·sel·meer \ˌī-səl-'meər\ *or* **Lake Ijs·sel** \'ī-səl\ freshwater lake N Netherlands separated from North Sea by a dike; part of former Zuider Zee (inlet of North Sea)

Ika·ria \ˌē-kə-'rē-ə\ *or ancient* **Icar·ia** \ī-'ker-ē-ə, -'kar-; ik-'er-, -'ar-\ island Greece in central Aegean W of Samos

Île–de–France \ˌēl-də-'fräⁿs\ region N France containing Paris

Ilium *or* **Ilion** — see TROY

Illam·pu \ē-'äm-pü, -'yäm-\ *or* **So·ra·ta** \sə-'rät-ə\ mountain W Bolivia in the Andes E of Lake Titicaca — see ANCOHUMA

Il·li·nois \ˌil-ə-'nȯi *also* -'nȯiz\ state N central U.S.; capital, Springfield — **Il·li·nois·an** \-'nȯi-ən, -'nȯiz-\ *adj or n*

Il·lyr·ia \il-'ir-ē-ə\ ancient country S Europe in Balkan Peninsula on the Adriatic — **Il·lyr·i·an** \-ē-ən\ *adj or n*

Ilo·ilo \ˌē-lə-'wē-lō\ city central Philippines on S coast of Panay Island

Imperial Valley valley SE corner of California partly in Baja California, Mexico

In·chon *or* **In·cheon** \'in-ˌchən\ city South Korea on Yellow Sea

In·de·pen·dence \ˌin-də-'pen-dəns\ city W Missouri E of Kansas City

In·dia \'in-dē-ə\ **1** peninsula region (often called a subcontinent) S Asia S of the Himalaya between Bay of Bengal & Arabian Sea **2** country comprising major portion of the peninsula; capital, New Delhi **3** *or* **Indian Empire** before 1947 those parts of the Indian subcontinent under British rule or protection

In·di·an \'in-dē-ən\ **1** ocean E of Africa, S of Asia, W of Australia, & N of Antarctica **2** — see THAR

In·di·ana \ˌin-dē-'an-ə\ state E central U.S.; capital, Indianapolis — **In·di·an·an** \-'an-ən\ *or* **In·di·an·i·an** \-'an-ē-ən\ *adj or n*

In·di·a·nap·o·lis \ˌin-dē-ə-'nap-ləs, -ə-ləs\ city, capital of Indiana

Indian River lagoon 165 miles (266 kilometers) long E Florida between mainland & coastal islands

Indian Territory former territory S central U.S. in present state of Oklahoma

In·dies \'in-dēz\ **1** EAST INDIES **2** WEST INDIES

In·do·chi·na \'in-dō-'chī-nə\ **1** peninsula SE Asia including Myanmar, Malay Peninsula, Thailand, Cambodia, Laos, & Vietnam **2** *or* **French Indochina** former country SE Asia comprising area now forming Cambodia, Laos, & Vietnam — **In·do–Chi·nese** \-chī-'nēz, -'nēs\ *adj or n*

In·do·ne·sia \ˌin-də-'nē-zhə, -shə\ country SE Asia in Malay Archipelago comprising Sumatra, Java, S

& E Borneo, Sulawesi, W New Guinea, W Timor, & many smaller islands; capital, Jakarta — see NETHERLANDS EAST INDIES — **In·do·ne·sian** \-zhən, -shən\ *adj or n*

In·dore \in-'dōr, -'dȯr\ city W central India in W Madhya Pradesh

In·dus \'in-dəs\ river 1800 miles (2897 kilometers) long S Asia flowing from Tibet NW & SSW through Pakistan into Arabian Sea

In·land \'in-,land, -lənd\ sea inlet of North Pacific in SW Japan between Honshu Island on N & Shikoku Island & Kyushu Island on S

Inner Hebrides — see HEBRIDES

Inner Mongolia region N China in SE Mongolia & W Manchuria; capital, Hohhot

Inns·bruck \'inz-,bru̇k, 'ins-\ city W Austria in Tirol

Inside Passage protected shipping route between Puget Sound, Washington, & Skagway, Alaska

In·ver·clyde \,in-vər-'klīd\ administrative area W Scotland

Io·ni·an \ī-'ō-nē-ən\ **1** sea arm of the Mediterranean between SE Italy & W Greece **2** islands W Greece in Ionian Sea

Io·wa \'ī-ə-wə\ state N central U.S.; capital, Des Moines — **Io·wan** \-wən\ *adj or n*

I–pin — see YIBIN

Ipoh \'ē-pō\ city Malaysia NNW of Kuala Lumpur

Ips·wich \'ip-swich\ town SE England

Iqa·lu·it \e-'kal-ü-ət\ town Canada on Baffin Island; capital of Nunavut

Iran \i-'ran, -'rän; ī-'ran\ *or formerly* **Per·sia** \'pər-zhə\ country SW Asia S of Caspian Sea; capital, Tehran — **Irani** \i-'ran-ē, -'rän-\ *adj or n* — **Ira·nian** \i-'ran-ē-ən, -'rän-, -'rān-\ *adj or n*

Iraq \i-'räk, -'rak\ country SW Asia in Mesopotamia; capital, Baghdad — **Iraqi** \-'räk-ē, -'rak-\ *adj or n*

Irbīl — see ARBIL

Ire·land \'īr-lənd\ **1** *or Latin* **Hi·ber·nia** \hī-'bər-nē-ə\ island W Europe in North Atlantic; one of the British Isles **2** *or* **Ei·re** \'er-ə\ country occupying major portion of the island; capital, Dublin

Irian — see NEW GUINEA

Irian Jaya — see WEST PAPUA

Irish \'ī-rish\ sea arm of North Atlantic between Great Britain & Ireland

Ir·kutsk \iər-'kütsk, ,ər-\ city S Russia in Asia, near Lake Baikal

Ir·ra·wad·dy \,ir-ə-'wäd-ē\ river 1300 miles (2092 kilometers) long Myanmar flowing S into Bay of Bengal

Ir·tysh \iər-'tish, ,ər-\ river over 2600 miles (4180 kilometers) long central Asia flowing NW & N from Altai Mountains in China, through Kazakhstan, & into the Ob in Russia

Ir·vine \'ər-,vīn\ city SW California

Isfahan — see ESFAHAN

Is·lam·abad \is-'läm-ə-,bäd, iz-'lam-ə-,bad\ city, capital of Pakistan

Isle of Anglesey administrative area NW Wales

Isle of Man — see MAN (Isle of)

Isle Roy·ale \'īl-'rȯi-əl, -'rȯil\ island Michigan in Lake Superior in **Isle Royale National Park**

Is·ling·ton \'iz-ling-tən\ borough of N Greater London, England

Is·ma·ilia \,iz-mā-ə-'lē-ə\ city NE Egypt on Suez Canal

Is·ra·el \'iz-rē-əl\ **1** kingdom in ancient Palestine comprising lands occupied by the Hebrew people **2** *or* **Northern Kingdom** *or* **Ephra·im** \'ē-frē-əm\ the N part of the Hebrew kingdom after about 933 B.C. — see JUDAH **3** country SW Asia bordering on the Mediterranean; capital, Jerusalem; established 1948 — **Is·rae·li** \iz-'rā-lē\ *adj or n*

Is·tan·bul \,is-təm-'bül, -,täm-, -,tam-, -,tän-\ *or formerly* **Con·stan·ti·no·ple** \,kän-,stant-n-'ō-pəl\ *or ancient* **By·zan·tium** \bə-'zan-shəm, -shē-əm; -'zant-ē-əm\ city NW Turkey on Sea of Marmara & both sides of the Bosporus; former capital of Turkey

Is·tria \'is-trē-ə\ peninsula S central Europe extending into the N Adriatic; belongs to Croatia & Slovenia except for Trieste (to Italy) — **Is·tri·an** \-trē-ən\ *adj or n*

Italian Somaliland former country E Africa now part of Somalia

It·a·ly \'it-l-ē\ **1** peninsula about 760 miles (1223 kilometers) long S Europe extending into the Mediterranean between Adriatic & Tyrrhenian seas **2** country including the peninsula of Italy, Sicily, & Sardinia; capital, Rome

Itas·ca, Lake \ī-'tas-kə\ lake NW central Minnesota; source of the Mississippi

Ith·a·ca \'ith-i-kə\ island W Greece in Ionian Islands

Iva·no·vo \i-'vän-ə-və\ city central Russia in Europe WNW of Nizhniy Novgorod

Ivory Coast *or French* **Côte d'Ivoire** \,kōt-dēv-'wär\ country W Africa on Gulf of Guinea; official capital, Yamoussoukro; seat of government, Abidjan

Iv·vav·ik National Park \'iv-ə-,vik\ reservation extreme NW Yukon Territory

Iwo \'ē-wō\ city SW Nigeria NE of Ibadan

Iwo Ji·ma \,ē-wō-'jē-mə\ island Japan in W Pacific in Volcano Islands about 759 miles (1221 kilometers) S of Tokyo

Izhevsk \'ē-,zhefsk\ city E central Russia in Europe NE of Kazan

Iz·mir \iz-'miər\ *or formerly* **Smyr·na** \'smər-nə\ city W Turkey

Ja·bal·pur \'jəb-əl-,pu̇r\ city central India in central Madhya Pradesh

Jack·son \'jak-sən\ city, capital of Mississippi

Jack·son·ville \'jak sən-,vil\ city NE Florida

Jadotville — see LIKASI

Jaf·fa \'jaf-ə, 'yaf-ə\ *or ancient* **Jop·pa** \'jäp-ə\ former city W Israel; now part of Tel Aviv

Jai·pur \'jī-,pu̇r\ city NW India, capital of Rajasthan

Ja·kar·ta *or* **Dja·kar·ta** \jə-'kär-tə\ *or formerly* **Ba·ta·via** \bə-'tā-vē-ə, -'tä-\ city, capital of Indonesia in NW Java

Ja·lan·dhar \'jəl-ən-dər\ city N India in Punjab SE of Amritsar

Ja·la·pa \hə-'läp-ə\ city E Mexico, capital of Veracruz

Ja·lis·co \hə-'lis-kō\ state W central Mexico; capital, Guadalajara

Ja·mai·ca \jə-'mā-kə\ island West Indies in Greater Antilles; an independent country; capital, Kingston — **Ja·mai·can** \-kən\ *adj or n*

James \'jāmz\ **1** river in North Dakota & South Dakota — see DAKOTA **2** river 340 miles (547 kilometers) long Virginia flowing E into Chesapeake Bay

James Bay the S extension of Hudson Bay between NE Ontario & W Quebec

James·town \'jām-,stau̇n\ ruined village E Virginia on James River; first permanent English settlement in America (1607)

Jam·mu and Kashmir \'jəm-,ü\ *or* **Kashmir** disputed territory N Indian subcontinent; claimed as a state (capital, Srinagar; winter capital, Jammu) & partly administered by India, but also claimed & partly controlled by Pakistan

Jam·na·gar \jäm-'nəg-ər\ city W India in W Gujarat

Jam·shed·pur \'jäm-,shed-,pu̇r\ city E India in Jharkhand

Ja·pan \jə-'pan, ji-, ja-\ **1** country E Asia comprising Honshu, Hokkaido, Kyushu, Shikoku, & other islands in the W Pacific; a constitutional monarchy; capital, Tokyo **2** warm current of the North Pacific flowing from E coast of the Philippines N along E coast of Japan & thence eastward

Japan, Sea of arm of North Pacific between Japan & Asian mainland

Jas·per National Park \'jas-pər\ reservation W Alberta on E slope of Rocky Mountains

Ja·va \'jäv-ə, 'jav-ə\ island Indonesia SW of Borneo; chief city, Jakarta — **Ja·van** \-ən\ *adj or n*

Jebel Musa — see MUSA (Jebel)

Jef·fer·son City \'jef-ər-sən\ city, capital of Missouri

Je·rez \hə-'rās\ *or in full* **Je·rez de la Fron·te·ra** \hə-'rez-də-lə-,frən-'ter-ə\ city SW Spain

Jer·i·cho \'jer-i-,kō\ city of ancient Palestine N of Dead Sea

Jer·sey \'jər-zē\ **1** island in English Channel — see CHANNEL **2** NEW JERSEY — **Jer·sey·ite** \-zē-,īt\ *n*

Jersey City city NE New Jersey on Hudson River

Je·ru·sa·lem \jə-'rü-sə-ləm, -sləm; -'rüz-ə-ləm, -'rüz-ləm\ city NW of Dead Sea divided 1948–67 between Israel & Jordan; capital of Israel since 1950 & formerly of ancient kingdoms of Israel & Judah

Jhan·si \'jän-sē\ city N India in S Uttar Pradesh SW of Kanpur

Jhar·khand \'jär-kənd\ state NE India; capital, Ranchi

Jia·mu·si \jē-'ä-'mü-'sē\ *or* **Chia–mu–ssu** \jē-'ä-'mü-'sü\ *or* **Kia·mu·sze** \jē-'ä-'mü-'sə\ city NE China in E Heilongjiang

Jiang·su *or* **Kiang·su** \jē-'äng-'sü\ province E China; capital, Nanjing

Jiang·xi *or* **Kiang·si** \jē-'äng-shē\ province SE China; capital, Nanchang

Jibuti — see DJIBOUTI

Jid·da \'jid-ə\ *or* **Jed·da** \'jed-ə\ city W Saudi Arabia in Hejaz on Red Sea; port for Mecca

Ji·lin \'jē-'lin\ *or* **Ki·rin** \'kē-'rin\ **1** province NE China; capital, Changchun **2** city in Jilin province

Ji·nan *or* **Chi·nan** *or* **Tsi·nan** \'jē-'nän\ city E China, capital of Shandong

Jin·zhou *or* **Chin–chou** *or* **Chin·chow** \'jin-'jō\ city NE China in SW Liaoning

Jiu·long \'jü-'lȯŋ\ *or* **Kow·loon** \'kau̇-'lün\ **1** peninsula SE China in Hong Kong opposite Hong Kong Island **2** city on Jiulong Peninsula

João Pes·soa \,zhau̇ⁿ-pə-'sō-ə, ,zhau̇ⁿm-\ city NE Brazil N of Recife

Jodh·pur \'jäd-pər, -,pu̇r\ city NW India in central Rajasthan

Jo·han·nes·burg \jō-'han-əs-,bərg, -'hän-\ city NE Republic of South Africa

Jo·hore Bah·ru \jə-'hōr-'bä-rü, -'hȯr-\ city Malaysia in S Peninsular Malaysia opposite Singapore Island

Jo·li·et \,jō-lē-'et\ city NE Illinois

Jor·dan \'jord-n\ **1** river 200 miles (322 kilometers) long SW Asia rising in Syria & flowing S from Anti-Lebanon Mountains in Lebanon into Dead Sea in Israel **2** *or formerly* **Trans·jor·dan** \trans-, tranz-, 'trans-, 'tranz-\ country SW Asia in NW Arabia; capital, Amman — **Jor·da·ni·an** \jȯr-'dā-nē-ən\ *adj or n*

Josh·ua Tree National Park \'jäsh-ə-wə\ reservation S California

Juan de Fu·ca \,hwän-də-'fyü-kə, ,wän-\ strait 100 miles (161 kilometers) long between Vancouver Island, British Columbia, & Olympic Peninsula, Washington

Juan Fer·nán·dez \,hwän-fər-'nan-dəs, ,wän-\ group of three islands SE Pacific about 400 miles (645 kilometers) W of Chile; belongs to Chile

Juárez — see CIUDAD JUAREZ

Ju·dah \'jüd-ə\ kingdom S ancient Palestine; capital, Jerusalem — see ISRAEL

Ju·dea *or* **Ju·daea** \jü-'dē-ə, -'dā-\ region of ancient Palestine constituting the S division (Judah) of the country under Persian, Greek, & Roman rule — **Ju·de·an** \-ən\ *adj or n*

Jugoslavia — see YUGOSLAVIA

Juiz de Fo·ra \zhwēzh-də-'fōr-ə, -'fȯr-\ city E Brazil N of Rio de Janeiro

Ju·neau \'jü-nō, jü-'\ city, capital of Alaska

Jung·frau \'yu̇ŋ-,frau̇\ mountain 13,642 feet (4158 meters) SW central Switzerland in Bernese Alps

Ju·ra \'ju̇r-ə\ mountain range extending along boundary between France & Switzerland N of Lake Geneva

Jut·land \'jət-lənd\ **1** peninsula N Europe extending into North Sea & comprising mainland of Denmark & N portion of Schleswig-Holstein, Germany **2** the mainland of Denmark

Ka·bul \'käb-əl, kä-'bül\ city, capital of Afghanistan

Kadiyevka — see STAKHANOV

Kae·song \'kā-,sȯŋ\ *or* **Gae·seong** \'gā-,səŋ\ city North Korea SE of Pyongyang

Ka·go·shi·ma \,käg-ə-'shē-mə, kä-'gō-shə-mə\ city Japan in S Kyushu

Kai·feng \'kī-'fəŋ\ city E central China in NE Henan

Ka Lae \kä-'lä-ā\ *or* **South Cape** *or* **South Point** most southerly point of Hawaii & of U.S.

Kal·a·ha·ri \,kal-ə-'här-ē\ desert region S Africa N of Orange River in S Botswana & NW Republic of South Africa

Kalgan — see ZHANGJIAKOU

Ka·li·man·tan \,kal-ə-'man-,tan, ,käl-ə-'män-,tän\ **1** BORNEO — its Indonesian name **2** the S & E portion of Borneo belonging to Indonesia; formerly (as **Dutch Borneo**) part of Netherlands East Indies

Kalinin — see TVER

Ka·li·nin·grad \kə-'lē-nən-,grad, -nyən-\ *or German* **Kö·nigs·berg** \'kā-nigz-,bərg, 'kərn-igz-, -,beərg, *German* 'kœ-niks-,berk\ city W Russia; formerly capital of East Prussia

Ka·lu·ga \kə-'lü-gə\ city Russia in Europe WNW of Oka

\ə\ abut \ᵊ\ kitten \ər\ further \a\ ash \ā\ ace \ä\ mop, mar \au̇\ out \ch\ chin \e\ bet \ē\ easy \g\ go \i\ hit \ī\ ice \j\ job \ŋ\ sing \ō\ go \ȯ\ law \ȯi\ boy \th\ thin \t͟h\ the \ü\ loot \u̇\ foot \y\ yet \zh\ vision *see also* Guide to Pronunciation

Kam·chat·ka \kam-'chat-kə\ peninsula 750 miles (1207 kilometers) long NE Russia in Asia between Sea of Okhotsk & Bering Sea
Kam·pa·la \käm-'päl-ə\ city, capital of Uganda
Kampuchea — see CAMBODIA
Ka·nan·ga \kə-'näng-gə\ *or formerly* **Lu·lua·bourg** \lü-'lü-ə-,bu̇rg, -,bu̇r\ city S central Democratic Republic of the Congo
Ka·na·za·wa \kə-'näz-ə-wə, ,kan-ə-'zä-wə\ city Japan in W Honshu N of Nagoya near Sea of Japan
Kan·chen·jun·ga \,kan-chən-'jəng-gə, -'ju̇ng-\ mountain 28,169 feet (8586 meters) Nepal & Sikkim (India) in the Himalaya; third highest in the world
Kan·da·har \'kan-də-,här\ city SE Afghanistan
Ka·no \'kän-ō\ city N central Nigeria
Kan·pur \'kän-,pu̇r\ city N India in S Uttar Pradesh on the Ganges
Kan·sas \'kan-zəs\ state W central U.S.; capital, Topeka — **Kan·san** \-zən\ *adj or n*
Kansas City **1** city NE Kansas adjoining Kansas City, Missouri **2** city W Missouri
Kansu — see GANSU
Kao·hsiung \'kau̇-shē-'u̇ng, 'gau̇-\ city SW Taiwan
Ka·ra·chi \kə-'räch-ē\ city S Pakistan on Arabian Sea
Karafuto — see SAKHALIN
Ka·ra·gan·da \,kar-ə-gən-'dä\ city central Kazakhstan
Ka·raj \kə-'räj\ city N Iran NW of Tehran
Kar·a·ko·ram \,kar-ə-'kōr-əm, -'kȯr-\ mountain system S central Asia in N Kashmir & NW Tibet connecting the Himalaya & the Pamirs
Karakoram Pass mountain pass through Karakoram Range
Ka·ra Sea \'kär-ə\ arm of Arctic Ocean off N coast of Russia E of Novaya Zemlya
Ka·re·lia \kə-'rē-lē-ə, -'rēl-yə\ region NE Europe between Gulf of Finland & White Sea; now chiefly in Russia — **Ka·re·lian** \-'rē-lē-ən, -'rēl-yən\ *adj or n*
Karl–Marx–Stadt — see CHEMNITZ
Karls·ru·he \'kärlz-,rü-ə\ city SW Germany
Kar·na·ta·ka \kär-'nät-ə-kə\ *or formerly* **My·sore** \mī-'sōr, -'sȯr\ state S India; capital, Bangalore
Kar·roo \kə-'rü\ plateau region W Republic of South Africa W of Drakensberg Mountains; divided into **Little** (*or* **Southern**) **Karroo**, **Great** (*or* **Central**) **Karroo**, & **Northern Karroo**
Kashi \'kash-ē, 'käsh-\ *or* **Kash·gar** \'kash-,gär, 'käsh-\ city W China in SW Xinjiang Uygur
Ka·shi·wa \'kä-shē-,wä\ city Japan on Honshu
Kash·mir \'kash-,mir, 'kazh-, kash-', kazh-'\ — see JAMMU AND KASHMIR — **Kash·miri** \kash-'miər-ē, kazh-\ *adj or n*
Kas·sel \'kas-əl, 'käs-\ city central Germany WNW of Erfurt
Ka·thi·a·war \,kät-ē-ə-'wär\ peninsula W India in Gujarat N of Gulf of Cambay
Kath·man·du *or* **Kat·man·du** \,kat-,man-'dü, ,kät-,män-\ city, capital of Nepal
Kat·mai, Mount \'kat-,mī\ volcano 6715 feet (2047 meters) S Alaska in **Katmai National Park**
Ka·to·wi·ce \,kät-ə-'vēt-sə\ city S Poland WNW of Krakow
Kat·te·gat \'kat-i-,gat\ arm of North Sea between Sweden & E coast of Jutland Peninsula of Denmark
Kau·ai \'kau̇-,ī\ island Hawaii NW of Oahu
Kau·nas \'kau̇-nəs, -,näs\ *or Russian* **Kov·no** \'kȯv-nō\ city central Lithuania
Ka·wa·goe \kə-'wäg-,ȯi\ city Japan on SE central Honshu
Ka·wa·gu·chi \,kä-wə-'gü-chē, kä-'wäg-ü-chē\ city Japan in E Honshu N of Tokyo
Ka·wa·sa·ki \,kä-wə-'säk-ē\ city Japan in E Honshu S of Tokyo
Kay·se·ri \,kī-zə-'rē\ city central Turkey
Ka·zakh·stan \kə-,zak-'stan; kə-,zäk-'stän, ,kä-\ country NW central Asia; capital, Astana; a constituent republic of U.S.S.R. 1936–91
Ka·zan \kə-'zan, -'zän, -'zän-yə\ city E central Russia in Europe
Kazan Retto — see VOLCANO ISLANDS
Kaz Da·gi \,käz-'dī\ *or* **Mount Ida** \'īd-ə\ mountain 5797 feet (1767 meters) NW Turkey in Asia SE of ancient Troy
Kee·lung \'kē-'lu̇ŋ\ — see CHI-LUNG
Kee·wa·tin \kē-'wāt-n\ former district N Canada in E Northwest Territories NW of Hudson Bay
Kej·im·ku·jik National Park \,kej-mə-'kü-jik, -ə-mə-\ reservation E Canada in SW Nova Scotia
Ke·me·ro·vo \'kem-ə-rə-və, -,rō-və, -rə-,vō\ city S Russia in Asia in Kuznetsk Basin
Ke·nai \'kē-,nī\ peninsula S Alaska E of Cook Inlet; site of **Kenai Fjords National Park**
Ke·ni·tra \kə-'nē-trə\ *or formerly* **Port Lyau·tey** \,pȯr-lē-,ō-'tā, -'ō-,\ city N Morocco
Kennedy, Cape — see CANAVERAL (Cape)
Ken·sing·ton and Chel·sea \'ken-zing-tən-ən-'chel-sē, 'ken-sing-\ borough of W Greater London, England
Kent \'kent\ county SE England — **Kent·ish** \'kent-ish\ *adj*
Ken·tucky \kən-'tək-ē\ state E central U.S.; capital, Frankfort — **Ken·tuck·i·an** \-ē-ən\ *adj or n*
Ken·ya \'ken-yə, 'kēn-\ **1** mountain 17,058 feet (5199 meters) central Kenya **2** country E Africa S of Ethiopia; capital, Nairobi — **Ken·yan** \-yən\ *adj or n*
Ker·a·la \'ker-ə-lə\ state SW India bordering on Arabian Sea; capital, Trivandrum
Ker·gue·len \'kər-gə-lən, ,kər-gə-'len\ **1** archipelago S Indian Ocean belonging to France **2** chief island of the archipelago
Ker·man \kər-'män, ker-\ city SE central Iran
Ker·ry \'ker-ē\ county SW Ireland in Munster
Kes·te·ven, Parts of \ke-'stē-vən\ district E England
Kha·ba·rovsk \kə-'bär-əfsk\ city SE Russia in Asia on the Amur
Khan·ka \'kang-kə\ lake E Asia between Russia & China
Khar·kiv \'kär-kəf, 'kär-\ *or* **Khar·kov** \'kär-,kȯf, -kȯv, -kəf\ city NE Ukraine
Khar·toum \kär-'tüm\ city, capital of Sudan
Kher·son \keər-'sȯn\ city S Ukraine
Khy·ber \'kī-bər\ pass 33 miles (53 kilometers) long on border between Afghanistan & Pakistan WNW of Peshawar
Kiamusze — see JIAMUSI
Kiangsi — see JIANGXI
Kiangsu — see JIANGSU
Ki·bo \'kē-bō\ mountain peak 19,340 feet (5895 meters) NE Tanzania; highest peak of Kilimanjaro & highest point in Africa
Kiel \'kēl\ city N Germany, capital of Schleswig-Holstein
Kiel — see NORD-OSTSEE
Kiel·ce \kē-'elt-sä\ city S Poland S of Warsaw
Ki·ev \'kē-,ef, -,ev, -if\ *or Ukrainian* **Kyiv** \'kyē-ü\ city, capital of Ukraine
Ki·ga·li \ki-'gäl-ē\ city, capital of Rwanda
Ki·lau·ea \,kē-,lau̇-'ā-ə\ volcanic crater Hawaii on Hawaii Island on E slope of Mauna Loa in Hawaii Volcanoes National Park
Kil·dare \kil-'daər, -'deər\ county E Ireland in Leinster
Kil·i·man·ja·ro \,kil-ə-mən-'jär-ō, -'jar-\ mountain NE Tanzania; highest in Africa — see KIBO
Kil·ken·ny \kil-'ken-ē\ county SE Ireland in Leinster
Kil·lar·ney, Lakes of \kil-'är-nē\ three lakes SW Ireland in Kerry
Kim·ber·ley \'kim-bər-lē\ city central Republic of South Africa
Kings Canyon National Park \'kingz-\ reservation SE central California in Sierra Nevada N of Sequoia National Park
Kings·ton \'king-stən\ city, capital of Jamaica
Kingston upon Hull — see HULL
Kingston upon Thames \'temz\ borough of SW Greater London, England
Kin·sha·sa \kin-'shäs-ə\ *or formerly* **Lé·o·pold·ville** \'lē-ə-,pōld-,vil, 'lā-\ city, capital of Democratic Republic of the Congo
Kirghiz Soviet Socialist Republic *or* **Kirgiz Soviet Socialist Republic** \kir-'gēz\ former constituent republic of U.S.S.R.; became independent Kyrgyzstan 1991
Kir·i·bati \'kir-ə-,bas — *sic*\ islands W Pacific including the Gilberts; an independent country; capital, Tarawa
Kirin — see JILIN
Ki·riti·mati \kə-'ris-məs — *sic*\ *or formerly* **Christ·mas** \'kris-məs\ island in Line Islands; largest atoll in the Pacific
Kir·kuk \kiər-'kük\ city NE Iraq
Ki·rov \'kē-,rȯf, -,rȯv, -rəf\ *or* **Vyat·ka** \vē-'at-kə, -'ät-\ city central Russia in Europe N of Kazan
Ki·ro·vo·hrad \,kē-rə-və-'hrät\ *or* **Ki·ro·vo·grad** \ki-'rō-və-,grad\ city S central Ukraine
Ki·san·ga·ni \,kē-sən-'gän-ē\ *or formerly* **Stan·ley·ville** \'stan-lē-,vil\ city NE Democratic Republic of the Congo
Kishinev — see CHISINAU
Ki·ta·kyu·shu \kē-,tä-kē-'ü-shü\ city Japan in N Kyushu
Kitch·e·ner \'kich-nər, -ə-nər\ city Canada in SE Ontario
Klai·pe·da \'klī-pəd-ə\ *or* **Me·mel** \'mā-məl\ city & port W Lithuania
Klon·dike \'klän-,dīk\ region NW Canada in central Yukon Territory in valley of **Klondike River** (90 miles or 145 kilometers flowing W into the Yukon River)
Klu·ane National Park \klü-'ò-nē-, -'än-ē-\ reservation SW Yukon Territory on Alaska border
Knos·sos *or* **Cnos·sus** \kə-'näs-əs, 'näs-əs\ *or* **Gnos·sus** \gə-'näs-əs, 'näs-əs\ ruined city, capital of ancient Crete near N coast
Knox·ville \'näks-,vil, -vəl\ city E Tennessee
Ko·be \'kō-bē, -,bā\ city Japan in S Honshu
Ko·buk Valley National Park \kō-'bu̇k\ reservation NW Alaska N of Arctic Circle
Ko·chi \'kō-chē\ city Japan on S coast of Shikoku
Ko·di·ak \'kōd-ē-,ak\ island S Alaska E of Alaska Peninsula
Kokand — see QUQON
Ko·la \'kō-lə\ peninsula NW Russia in Europe between Barents & White seas
Ko·lar Gold Fields \kō-'lär\ region S India in SE Karnataka NE of Bangalore
Kol·ha·pur \'kō-lə-,pu̇r\ city W India in SW Maharashtra SSE of Bombay
Kolkata — see CALCUTTA
Köln — see COLOGNE
Ko·mo·do \kə-'mōd-ō\ island Indonesia in the Lesser Sundas W of Flores Island
Königsberg — see KALININGRAD
Kon·ya \kȯn-'yä\ city SW central Turkey
Koo·te·nay National Park \'küt-n-,ā, -n-ē\ reservation SE British Columbia
Ko·rea \kə-'rē-ə, *especially South* kō-\ *or Japanese* **Cho·sen** \'chō-'sen\ former kingdom E Asia between Yellow Sea & Sea of Japan (East Sea); capital, Seoul; divided after World War II at 38th parallel into republics of **North Korea** (capital, Pyongyang) & **South Korea** (capital, Seoul)
Korea Bay arm of Yellow Sea between China & North Korea
Kos·ci·us·ko, Mount \,käz-ē-'əs-kō\ mountain 7310 feet (2228 meters) SE Australia in SE New South Wales; highest in Great Dividing Range & in Australia
Ko·shi·ga·ya \kō-'shē-gä-yə; ,kō-shig-'ä-yə\ city Japan on Honshu
Ko·so·vo \'kȯ-sȯ-,vō, 'käs-ō-\ province SW Serbia
Kos·tro·ma \,käs-trə-'mä\ city central Russia in Europe on the Volga
Ko·ta Bha·ru \,kōt-ə-'bär-,ü, -ü\ city Malaysia in N Peninsular Malaysia
Kou·chi·bou·guac National Park \kü-,shē-bü-'gwäk\ reservation E New Brunswick along the coast
Kovno — see KAUNAS
Kowloon — see JIULONG
Kozhikode — see CALICUT
Krak·a·toa \,krak-ə-'tō-ə\ *or* **Krak·a·tau** \-'tau̇\ island & volcano Indonesia between Sumatra & Java
Kra·kow *or* **Cra·cow** \'kräk-,au̇, 'krak-, 'kräk-, -ō, *Polish* 'kräk-,üf\ city S Poland
Kras·no·dar \'kras-nə-,där\ city S Russia in Europe in N Caucasus
Kras·no·yarsk \,kras-nə-'yärsk\ city S central Russia in Asia on the upper Yenisey
Kre·feld \'krā-,felt\ city W Germany on the Rhine WSW of Essen
Kru·ger National Park \'krü-gər\ game reserve NE Republic of South Africa on Mozambique border
Kru·gers·dorp \'krü-gərz-,dȯrp\ city NE Republic of South Africa
Kry·vyy Rih \kri-,vē-'rik\ *or* **Kri·voy Rog** \,kri-,vȯi-'rōg, -'rȯk\ city SE central Ukraine
K2 \'kā-'tü\ *or* **God·win Aus·ten** \'gȯd-wən-'ȯs-tən, 'gäd-, -'äs-\ mountain 28,250 feet (8611 meters) N Kashmir in Karakoram Range; second highest in the world
Kua·la Lum·pur \,kwäl-ə-'lu̇m-,pu̇r, -'ləm-\ city, capital of Malaysia in Peninsular Malaysia
Kuang–chou — see GUANGZHOU
Kuei–chou — see GUIZHOU
Kuei–yang — see GUIYANG

Ku·ma·mo·to \ˌküm-ə-ˈmōt-ō\ city Japan in W Kyushu
Ku·ma·si \kü-ˈmäs-ē, -ˈmas-\ city S central Ghana
Kun·lun \ˈkün-ˈlün\ mountain system W China extending E from the Pamirs to SE Qinghai; highest peak 25,340 feet (7724 meters)
Kun·ming \ˈku̇n-ˈmiŋ\ *or formerly* **Yun·nan** \yü-ˈnän\ *or* **Yun·nan·fu** \-ˈfü\ city S China, capital of Yunnan
Ku·ra·shi·ki \ku̇-ˈrä-shē-kē, ˌku̇r-ə-ˈshē-kē\ city Japan in W Honshu WSW of Okayama
Kur·di·stan \ˌku̇rd-ə-ˈstan, ˌkərd-\ region SW Asia chiefly in E Turkey, NW Iran, & N Iraq
Ku·re \ˈku̇r-ē, ˈkyu̇r-ē, ˈkü-rä\ city Japan in SW Honshu on Inland Sea SE of Hiroshima
Kur·gan \ku̇r-ˈgan, -ˈgän\ city W Russia in Asia, SE of Yekaterinburg
Ku·ril *or* **Ku·rile** \ˈkyu̇r-ˌēl, kyu̇-ˈrēl\ islands Russia in North Pacific between S Kamchatka Peninsula & NE Hokkaido Island
Kur·nool \kər-ˈnül\ city S India in W Andhra Pradesh SSW of Hyderabad
Kursk \ˈku̇rsk\ city SW Russia in Europe, N of Kharkiv, Ukraine
Ku·wait \ku̇-ˈwāt\ **1** country SW Asia in Arabia at head of Persian Gulf **2** city, its capital — **Ku·waiti** \-ˈwāt-ē\ *adj or n*
Kuybyshev — see SAMARA
Kuz·netsk Basin \ku̇z-ˈnetsk\ *or* **Kuz·bass** *or* **Kuz·bas** \ˈku̇z-ˌbas\ basin of Tom River S Russia in Asia, extending from Tomsk to Novokuznetsk
Kwa·ja·lein \ˈkwäj-ə-lən, -ˌlān\ island (atoll) W Pacific in Ralik Chain of Marshall Islands
Kwang·ju *or* **Gwang·ju** \ˈgwäŋ-ˈjü\ city SW South Korea
Kwangsi–Chuang — see GUANGXI ZHUANGZU
Kwangtung — see GUANGDONG
Kwa·Zu·lu–Na·tal \kwä-ˈzü-lü-nä-ˈtäl\ province E Republic of South Africa between Drakensberg Mountains & Indian Ocean
Kweichow — see GUIZHOU
Kweilin *or* **Kuei–lin** — see GUILIN
Kyiv — see KIEV
Kyo·to \kē-ˈōt-ō\ city Japan in W central Honshu; formerly capital of Japan
Kyr·gyz·stan \ˌkir-gi-ˈstan, -ˈstän; ˈkir-gi-ˌ\ country W central Asia; capital, Bishkek; a constituent republic of U.S.S.R. 1936–91
Kyu·shu \kē-ˈü-shü\ island Japan S of W end of Honshu
Labe — see ELBE
Lab·ra·dor \ˈlab-rə-ˌdȯr\ **1** peninsula E Canada between Hudson Bay & North Atlantic divided between the provinces of Quebec & Newfoundland and Labrador **2** the part of the peninsula belonging to Newfoundland and Labrador — **Lab·ra·dor·ean** *or* **Lab·ra·dor·ian** \ˌlab-rə-ˈdȯr-ē-ən, -ˈdōr-\ *adj or n*
Labrador Current cold current flowing S from Baffin Bay through Davis Strait past Labrador & Newfoundland
Lac·ca·dive \ˈlak-ə-ˌdēv, -ˌdīv\ islands India in Arabian Sea N of Maldive Islands
Lacedaemon — see SPARTA
La·co·nia \lə-ˈkō-nē-ə, -nyə\ ancient country S Greece in SE Peloponnese; capital, Sparta — **La·co·nian** \-nē-ən, -nyən\ *adj or n*
Lad·o·ga \ˈlad-ə-gə, ˈläd-\ lake W Russia in Europe, near Finland border
La·fay·ette \ˌlaf-ē-ˈet, ˌläf-\ city S Louisiana
La·gos \ˈlā-ˌgäs\ city & port, former capital of Nigeria
La Habana — see HAVANA
La·hon·tan, Lake \lə-ˈhänt-n\ prehistoric lake NW Nevada & NE California
La·hore \lə-ˈhōr, -ˈhȯr\ city Pakistan in E Punjab province
Lake Clark National Park \ˈklärk\ reservation S central Alaska WSW of Anchorage
Lake District region NW England containing many lakes & mountains
Lak·shad·weep \lək-ˈshäd-ˌwēp, ˌlək-shəd-ˈ\ territory India comprising the Laccadive Islands
La Mau·ri·cie National Park \lä-ˌmȯr-ē-ˈsē\ reservation S Quebec
Lam·beth \ˈlam-bəth, -ˌbeth\ borough of S Greater London, England
La·nai \lə-ˈnī\ island Hawaii W of Maui
Lan·ca·shire \ˈlaŋ-kə-ˌshiər, -shər\ *or* **Lan·cas·ter** \ˈlaŋ-kə-stər\ county NW England — **Lan·cas·tri·an** \laŋ-ˈkas-trē-ən, lan-\ *adj or n*
Lan·cas·ter \ˈlan-ˌkas-tər, ˈlaŋ-kəs-tər\ **1** city SW California NE of Los Angeles **2** city NW England
Land's End \ˈland-ˈzend, ˈlan-\ *or ancient* **Bo·le·ri·um** \bə-ˈlir-ē-əm\ cape SW England containing England's westernmost point
Lan·gue·doc \ˌlaŋ-gə-ˈdäk; ˌläⁿ-gə-ˈdȯk, ˌläⁿg-\ region & former province S France on the Mediterranean W of Provence
Lan·sing \ˈlan-siŋ\ city, capital of Michigan
La·nús \lə-ˈnüs\ city E Argentina S of Buenos Aires
Lan·zhou *or* **Lan–chou** \ˈlän-ˈjō\ city W central China, capital of Gansu
Laoighis \ˈlāsh, ˈlēsh\ *or* **Leix** \ˈlāsh, ˈlēsh\ *or formerly* **Queen's** \ˈkwēnz\ county central Ireland in Leinster
Laos \ˈlau̇s, ˈlā-ˌäs, ˈlä-ōs\ country SE Asia in Indochina NE of Thailand; capital, Vientiane
La Paz \lə-ˈpaz, -ˈpäz, -ˈpäs\ city, administrative capital of Bolivia
Lap·land \ˈlap-ˌland, -lənd\ region N Europe above the Arctic Circle in N Norway, N Sweden, N Finland, & Kola Peninsula of Russia — **Lap·land·er** \-ˌlan-dər, -lən-\ *n*
La Pla·ta \lə-ˈplät-ə\ city E Argentina SE of Buenos Aires
L'Aqui·la \ˈläk-wi-lə, ˈlak-\ city central Italy NE of Rome
Las Pal·mas \lä-ˈspäl-məs\ city Spain in the Canary Islands on Grand Canary Island
La Spe·zia \lä-ˈspet-sē-ə\ city NW Italy in Liguria SE of Genoa
Las·sen Peak \ˈlas-n\ volcano 10,457 feet (3187 meters) N California at S end of Cascade Range in **Lassen Volcanic National Park**
Las Ve·gas \läs-ˈvā-gəs\ city SE corner of Nevada
Lat·a·kia \ˌlat-ə-ˈkē-ə\ city NW Syria
Latin America **1** Spanish America and Brazil **2** all of the Americas S of the U.S. — **Latin-American** *adj* — **Latin American** *n*
La·tium \ˈlā-shē-əm, -shəm\ *or Italian* **La·zio** \ˈlät-sē-ō\ region central Italy on Tyrrhenian Sea; capital, Rome
Lat·via \ˈlat-vē-ə\ country E Europe on Baltic Sea; capital, Riga; a constituent republic of U.S.S.R. 1940–91
Lau·ren·tian Mountains \lȯ-ˈren-chən-\ range E Canada in S Quebec N of the Saint Lawrence on S edge of Canadian Shield
Laurentian Plateau — see CANADIAN SHIELD
La·val \lə-ˈval\ city S Quebec NW of Montreal
League of Nations political organization established by the Allied powers at end of World War I; replaced by United Nations 1946
Leb·a·non \ˈleb-ə-nən, -ˌnän\ **1** *or ancient* **Lib·a·nus** \ˈlib-ə-nəs\ mountains Lebanon running parallel to coast; highest 10,131 feet (3088 meters) **2** country SW Asia on the Mediterranean; capital, Beirut — **Leb·a·nese** \ˌleb-ə-ˈnēz, -ˈnēs\ *adj or n*
Leeds \ˈlēdz\ city N England
Lee·ward \ˈlē-wərd\ **1** islands Hawaii extending WNW from main islands of the group **2** islands South Pacific in W Society Islands **3** islands West Indies in N Lesser Antilles
Le Ha·vre \lə-ˈhävr\ city N France on English Channel
Leices·ter \ˈles-tər\ city central England ENE of Birmingham
Leices·ter·shire \ˈles-tər-ˌshiər, -shər\ *or* **Leicester** county central England
Lein·ster \ˈlen-stər\ province E Ireland
Leip·zig \ˈlīp-sig, -sik\ city E central Germany in Saxony
Lei·trim \ˈlē-trəm\ county NW Ireland in Connacht
Leix — see LAOIGHIS
Leman, Lake *or* **Lemannus** *or* **Lemanus** — see GENEVA (Lake)
Lemberg — see L'VIV
Lem·nos \ˈlem-ˌnäs, -nəs\ *or Greek* **Lím·nos** \ˈlēm-ˌnȯs\ island Greece in the N Aegean
Le·na \ˈlē-nə, ˈlā-\ river about 2700 miles (4345 kilometers) long E central Russia in Asia, flowing NE & N into Arctic Ocean
Leningrad — see SAINT PETERSBURG 2
Le·ón \lā-ˈōn\ **1** city central Mexico in Guanajuato **2** region & ancient kingdom NW Spain
Léopoldville — see KINSHASA
Le Puglie — see PUGLIA
Les·bos \ˈlez-ˌbäs, -bəs\ *or* **Myt·i·le·ne** \ˌmit-l-ˈē-nē\ island Greece in the Aegean off NW coast of Asia Minor
Le·so·tho \lə-ˈsō-tō, ˈsü-ˌtü\ country S Africa surrounded by Republic of South Africa; formerly British territory of **Ba·su·to·land** \bə-ˈsüt-ə-ˌland\, now an independent monarchy; capital, Maseru
Lesser An·til·les \an-ˈtil-ēz\ islands in the West Indies including Virgin, Leeward, & Windward islands, Barbados, Trinidad, Tobago, & islands in the S Caribbean N of Venezuela — see GREATER ANTILLES
Lesser Armenia region S Turkey corresponding to ancient Cilicia
Lesser Sunda — see SUNDA
Le·vant \lə-ˈvant\ the countries bordering on the E Mediterranean — **Lev·an·tine** \ˈlev-ən-ˌtīn, -ˌtēn, lə-ˈvan-\ *adj or n*
Lew·i·sham \ˈlü-ə-shəm\ borough of SE Greater London, England
Lew·is with Har·ris \ˌlü-ə-swəth-ˈhar-əs, -swəth-\ island NW Scotland in Outer Hebrides
Lex·ing·ton \ˈlek-siŋ-tən\ city N central Kentucky
Ley·te \ˈlāt-ē\ island Philippines in Visayan Islands S of Samar
Lha·sa \ˈläs-ə, ˈlas-\ city SW China, capital of Tibet
Liao·ning \lē-ˈau̇-ˈniŋ\ province NE China in S Manchuria; capital, Shenyang
Liao·yang \lē-ˈau̇-ˈyäŋ\ city NE China in central Liaoning NE of Anshan
Libanus — see LEBANON
Li·be·ria \lī-ˈbir-ē-ə\ country W Africa on North Atlantic; capital, Monrovia — **Li·be·ri·an** \-ē-ən\ *adj or n*
Lib·er·ty \ˈlib-ərt-ē\ *or formerly* **Bed·loe's** \ˈbed-ˌlōz\ island SE New York in Upper New York Bay; the Statue of Liberty is on it
Li·bre·ville \ˈlē-brə-ˌvil, -ˌvēl\ city, capital of Gabon
Lib·ya \ˈlib-ē-ə\ **1** the part of Africa N of the Sahara between Egypt & Gulf of Sidra — an ancient name **2** N Africa W of Egypt — an ancient name **3** country N Africa on the Mediterranean W of Egypt; capital, Tripoli — **Lib·y·an** \ˈlib-ē-ən\ *adj or n*
Libyan desert N Africa W of the Nile in Libya, Egypt, & Sudan
Li·do \ˈlēd-ō\ island Italy in Adriatic Sea
Liech·ten·stein \ˈlik-tən-ˌstīn, -ˌshtīn\ country W Europe between Austria & Switzerland; a principality; capital, Vaduz — **Liech·ten·stein·er** \-ˌstī-nər, -ˌshtī-\ *n*
Li·ège \lē-ˈezh, -ˈāzh\ *or Flemish* **Luik** \ˈlīk\ city E Belgium
Lif·fey \ˈlif-ē\ river 50 miles (80 kilometers) long E Ireland flowing into Dublin Bay
Li·gu·ria \lə-ˈgyu̇r-ē-ə\ region NW Italy; capital, Genoa — **Li·gu·ri·an** \-ē-ən\ *adj or n*
Ligurian Sea arm of the Mediterranean N of Corsica
Li·ka·si \li-ˈkäs-ē\ *or formerly* **Ja·dot·ville** \ˌzhad-ō-ˈvēl\ city SE Democratic Republic of the Congo
Lille \ˈlēl\ city N France
Li·long·we \li-ˈlȯŋ-wā\ city, capital of Malawi
Li·ma \ˈlē-mə\ city, capital of Peru
Lim·burg \ˈlim-ˌbərg\ *or French* **Lim·bourg** \ˈlim-ˌbərg, laⁿ-ˈbu̇r\ region W Europe E of the Meuse in Belgium & Netherlands
Lim·er·ick \ˈlim-rik, -ə-rik\ county SW Ireland in Munster
Límnos — see LEMNOS
Li·mou·sin \ˌlē-mü-ˈzaⁿ\ region S central France
Lim·po·po \lim-ˈpō-pō\ **1** river 1000 miles (1609 kilometers) long Africa flowing from NE Republic of South Africa into Indian Ocean in Mozambique **2** *or formerly* **Northern** province NE Republic of South Africa
Lin·coln \ˈliŋ-kən\ **1** city, capital of Nebraska **2** city E England
Lin·coln·shire \ˈliŋ-kən-ˌshiər, -shər\ *or* **Lincoln** county E England
Line \ˈlīn\ islands Kiribati S of Hawaii, formerly divided between U.S. & United Kingdom
Lip·a·ri \ˈlip-ə-rē\ islands Italy off NE Sicily
Li·petsk \ˈlē-ˌpetsk\ city S central Russia in Europe, N of Voronezh
Lis·bon \ˈliz-bən\ *or Portuguese* **Lis·boa** \lēzh-ˈvō-ə\ city, capital of Portugal

\ə\ abut \ᵊ\ kitten \ər\ further \a\ ash \ā\ ace \ä\ mop, mar \au̇\ out \ch\ chin \e\ bet \ē\ easy \g\ go \i\ hit \ī\ ice \j\ job \ŋ\ sing \ō\ go \ȯ\ law \ȯi\ boy \th\ thin \th\ the \ü\ loot \u̇\ foot \y\ yet \zh\ vision *see also* Guide to Pronunciation

Lith·u·a·nia \,lith-ə-'wā-nē-ə, ,lith-yə-, -nyə\ country E Europe; capital, Vilnius; a constituent republic of U.S.S.R. 1940–91

Lit·tle Rock \'lit-l-,räk\ city, capital of Arkansas

Liv·er·pool \'liv-ər-,pül\ city NW England on Mersey Estuary

Li·vo·nia \lə-'vō-nē-ə, -nyə\ **1** city SE Michigan W of Detroit **2** region E Europe on Baltic Sea in Latvia & Estonia

Lju·blja·na \lē-,ü-blē-'än-ə\ city central Slovenia on Sava River

Lla·no Es·ta·ca·do \'lan-ō-,es-tə-'käd-ō, 'län-\ *or* **Staked Plain** \'stākt-, 'stāk-\ plateau region SE New Mexico & NW Texas

Lo·bam·ba \lō-'bäm-bə\ town, legislative capital of Swaziland

Lodz \'lüj, 'lädz\ city central Poland WSW of Warsaw

Lo·fo·ten \'lō-,fōt-n\ islands NW Norway

Lo·gan, Mount \'lō-gən\ mountain 19,524 feet (5951 meters) NW Canada in Saint Elias Range; highest in Canada & second highest in North America

Loire \lə-'wär\ river 634 miles (1020 kilometers) long central France flowing NW & W into Bay of Biscay

Lo·mas de Za·mo·ra \'lō-,mäz-də-zə-'mōr-ə, -'mȯr-\ city E Argentina SW of Buenos Aires

Lom·bar·dy \'läm-,bärd-ē, -bərd-\ *or Italian* **Lom·bar·dia** \,läm-bər-'dē-ə, ,lōm-\ region N Italy N of Po River; capital, Milan

Lo·mé \lō-'mā\ city, capital of Togo

Lo·mond, Loch \'lō-mənd\ lake S central Scotland

Lon·don \'lən-dən\ **1** city S Ontario, Canada **2** city, capital of England & of United Kingdom on the Thames; comprises **City of London** & 12 inner boroughs of Greater London — **Lon·don·er** \-də-nər\ *n*

Londonderry — see DERRY

Long Beach city & port SW California S of Los Angeles

Long·ford \'lȯŋ-fərd\ county E central Ireland in Leinster

Long Island island 118 miles (190 kilometers) long SE New York S of Connecticut

Long Island Sound inlet of North Atlantic between Connecticut & Long Island, New York

Lon·gueuil \lȯŋ-'gāl\ city Canada in S Quebec E of Montreal

Lor·raine \lə-'rān, lȯ-\ region NE France around upper Moselle & Meuse rivers — see ALSACE-LORRAINE

Los An·ge·les \lȯ-'san-jə-ləs *also* -'saŋ-gə-ləs\ city SW California

Lou·ise, Lake \lü-'ēz\ lake SW Alberta in Banff National Park

Lou·i·si·ana \lü-,ē-zē-'an-ə, ,lü-ə-zē-, ,lü-zē-\ state S U.S.; capital, Baton Rouge — **Lou·i·si·an·ian** \-'an-ē-ən, -'an-yən\ *or* **Lou·i·si·an·an** \-'an-ən\ *adj or n*

Louisiana Purchase area W central U.S. between Rocky Mountains & the Mississippi purchased 1803 from France

Lou·is·ville \'lü-i-,vil, -vəl\ city N Kentucky on the Ohio River

Lourenço Marques — see MAPUTO

Louth \'lauth\ county E Ireland in Leinster

Low Countries region W Europe comprising modern Belgium, Luxembourg, & the Netherlands

Low·ell \'lō-əl\ city NE Massachusetts

Lower California — see BAJA CALIFORNIA

Lower Canada former province, Canada in S & E parts of present-day Quebec

Lower 48 the continental states of the U.S. excluding Alaska

Lower Saxony *or German* **Nie·der·sach·sen** \,nēd-ər-'zäk-sən\ state NW Germany; capital, Hannover

Low·lands \'lō-ləndz, -lənz, -,landz, -,lanz\ the central & E part of Scotland lying between the Highlands & the Southern Uplands

Lu·an·da \lü-'an-də\ city, capital of Angola

Lub·bock \'ləb-ək\ city NW Texas

Lü·beck \'lü-,bek, 'lue-\ city N Germany NE of Hamburg

Lu·blin \'lü-blən, -,blēn\ city E Poland SE of Warsaw

Lu·bum·ba·shi \,lü-büm-'bäsh-ē\ *or formerly* **Elis·a·beth·ville** \i-'liz-ə-bəth-,vil\ city SE Democratic Republic of the Congo

Lu·cerne, Lake of \lü-'sərn\ lake central Switzerland

Luchow — see HEFEI

Luck·now \'lək-,nau\ city N India, capital of Uttar Pradesh

Lüda — see DALIAN

Lu·dhi·a·na \,lüd-ē-'än-ə\ city NW India in Punjab SE of Amritsar

Luik — see LIEGE

Luluabourg — see KANANGA

Lu·sa·ka \lü-'säk-ə\ city, capital of Zambia

Lü·shun \'lü-'shün\ *or* **Port Ar·thur** \'är-thər\ city NE China in S Liaoning

Lusitania — see PORTUGAL

Lü–ta — see DALIAN

Lu·ton \'lüt-n\ town SE central England

Lux·em·bourg *or* **Lux·em·burg** \'lək-səm-,bərg, 'lük-səm-,bürg\ **1** country W Europe bordered by Belgium, France, & Germany; a grand duchy **2** city, its capital — **Lux·em·bourg·er** \-,bər-gər, -,bür-\ *n* — **Lux·em·bourg·ian** \,lək-səm-'bər-gē-ən, ,lük-səm-'bür-\ *adj*

Lu·zon \lü-'zän\ island N Philippines

L'viv \lə-'vē-ü, -'vēf\ *or* **L'vov** \lə-'vȯf, -'vȯv\ *or Polish* **Lwów** \lə-'vüf, -'vüv\ *or German* **Lem·berg** \'lem-,bərg, -,berg\ city W Ukraine

Lyallpur — see FAISALABAD

Ly·cia \'lish-ə, 'lish-ē-ə\ ancient district & Roman province SW Asia Minor

Lyd·ia \'lid-ē-ə\ ancient country W Asia Minor on the Aegean; capital, Sardis — **Lyd·i·an** \-ē-ən\ *adj or n*

Lyon \'lyō^{n}\ *or* **Ly·ons** \lē-'ō^{n}, 'lī-ənz\ *or ancient* **Lug·du·num** \lüg-'dü-nəm, ,ləg-\ city SE central France

Maas — see MEUSE

Ma·cao *or Portuguese* **Ma·cau** \mə-'kau\ *or Chinese* **Ao·men** \'au-'mən\ **1** special administrative region on coast of SE China W of Hong Kong; formerly a Portuguese overseas territory **2** city & port Macao — **Mac·a·nese** \,mak-ə-'nēz, -'nēs\ *n*

Macassar — see UJUNG PANDANG

Mac·e·do·nia \,mas-ə-'dō-nyə, -nē-ə\ **1** region S Europe in Balkan Peninsula in NE Greece, independent country of Macedonia, & SW Bulgaria including territory of ancient kingdom of Macedonia (**Mac·e·don** \'mas-əd-ən, -ə-,dän\) **2** country S central Balkan Peninsula; formerly a constituent republic of Yugoslavia; capital, Skopje — **Mac·e·do·nian** \,mas-ə-'dō-nyən, -nē-ən\ *adj or n*

Ma·ceió \,mas-ā-'ō\ city NE Brazil

Mac·gil·li·cud·dy's Reeks \mə-,gil-ə-,kəd-ēz-'rēks\ mountains SW Ireland in Kerry; highest Carrantuohill 3414 feet (1041 meters)

Ma·chi·da \mə-'chē-də, 'mä-chi-,dä\ city Japan on Honshu

Ma·chi·li·pat·nam \,məch-ə-lə-'pət-nəm\ *or* **Ban·dar** \'bənd-ər\ city SE India in E Andhra Pradesh

Ma·chu Pic·chu \,mäch-ü-'pēk-chü\ site SE Peru of ancient Inca city NW of Cuzco

Mac·ken·zie \mə-'ken-zē\ **1** river 1120 miles (1802 kilometers) long NW Canada flowing from Great Slave Lake NW into Beaufort Sea **2** former district NW Canada in W Northwest Territories in basin of Mackenzie River; area now split between Northwest Territories & Nunavut

Mack·i·nac, Straits of \'mak-ə-,nȯ\ channel N Michigan connecting Lake Huron & Lake Michigan

Ma·con \'mā-kən\ city central Georgia

Mad·a·gas·car \,mad-ə-'gas-kər\ *or formerly* **Mal·a·gasy Republic** \,mal-ə-,gas-ē\ island W Indian Ocean off SE Africa; capital, Antananarivo — **Mad·a·gas·can** \,mad-ə-'gas-kən\ *adj or n*

Ma·dei·ra \mə-'dir-ə, -'der-\ **1** river 2013 miles (3239 kilometers) long W Brazil flowing NE into the Amazon **2** islands in the North Atlantic N of the Canaries belonging to Portugal; capital, Funchal **3** island; chief of the Madeira group — **Ma·dei·ran** \-ən\ *adj or n*

Ma·dhya Pra·desh \,mäd-yə-prə-'desh, -'dāsh\ state central India; capital, Bhopal

Mad·i·son \'mad-ə-sən\ city, capital of Wisconsin

Ma·dras \mə-'dras, -'dräs\ **1** — see TAMIL NADU **2** *or* **Chen·nai** \'chen-,ī\ city SE India, capital of Tamil Nadu

Ma·drid \mə-'drid\ city, capital of Spain

Ma·du·ra \mə-'dür-ə\ island Indonesia NE of Java

Ma·du·rai \,mäd-ə-'rī\ *or* **Ma·du·ra** \'maj-ə-rə\ city S India in S Tamil Nadu

Mag·da·len \'mag-də-lən\ *or French* **Ma·de·leine** \mäd-'len, mä-də-\ islands Quebec in Gulf of Saint Lawrence

Mag·de·burg \'mäg-də-,bürg, 'mag-də-,bərg\ city central Germany WSW of Berlin; capital of Saxony-Anhalt

Ma·gel·lan, Strait of \mə-'jel-ən\ strait at S end of South America between mainland & Tierra del Fuego Archipelago

Mageröy — see NORTH CAPE

Mag·gio·re, Lake \mə-'jōr-ē, -'jȯr-\ lake N Italy & S Switzerland

Ma·ghreb \'mäg-rəb\ NW Africa & at one time Spain — now considered to include Morocco, Algeria, Tunisia, & sometimes Libya

Mag·ni·to·gorsk \mag-'nēt-ə-,gȯrsk\ city SW Russia in Asia, on Ural River

Ma·hal·la el Ku·bra \mə-,hal-ə-el-'kü-brə\ city N Egypt in Nile Delta

Ma·ha·rash·tra \,mä-hə-'räsh-trə\ state W India on Arabian Sea; capital, Bombay

Ma·hi·lyow *or* **Mo·gi·lev** \mə-gil-'yȯf\ city E Belarus

Main \'mīn, 'mān\ river 325 miles (523 kilometers) long S central Germany flowing W into the Rhine

Maine \'mān\ state NE U.S.; capital, Augusta

Mainz \'mīns\ city W Germany on the Rhine, capital of Rhineland-Palatinate

Ma·jor·ca \mə-'jȯr-kə, -'yȯr-\ *or Spanish* **Ma·llor·ca** \mə-'yȯr-kə\ island Spain; largest of the Balearic Islands — **Ma·jor·can** \-'jȯr-kən, -'yȯr-\ *adj or n*

Ma·ju·ro \mə-'jür-ō\ atoll W Pacific; contains capital of Marshall Islands

Ma·ka·lu \'mək-ə-,lü\ mountain 27,824 miles (8481 meters) NE Nepal in the Himalaya

Ma·kas·sar \mə-'kas-ər\ **1** strait Indonesia between Borneo & Sulawesi **2** — see UJUNG PANDANG

Ma·ke·yev·ka *or* **Ma·ki·yiv·ka** \mə-'kā-əf-kə, -yəf-\ city E Ukraine in Donets Basin

Ma·khach·ka·la \mə-,käch-kə-'lä\ city S Russia in Europe, on the Caspian

Mal·a·bar Coast \'mal-ə-,bär\ region SW India on Arabian Sea in Karnataka & Kerala states

Ma·la·bo \mä-'läb-ō\ *or formerly* **San·ta Isa·bel** \,san-tə-'iz-ə-bel\ city, capital of Equatorial Guinea

Ma·lac·ca, Strait of \mə-'lak-ə, -'läk-\ channel between S Malay Peninsula & island of Sumatra

Má·la·ga \'mal-ə-gə\ city S Spain in Andalusia

Ma·lang \mə-'läŋ\ city Indonesia in E Java

Ma·la·wi \mə-'lä-wē, -'lau-ē\ *or formerly* **Ny·asa·land** \nī-'as-ə-,land, nē-\ country SE Africa on Lake Malawi; a former British protectorate; capital, Lilongwe

Malawi, Lake *or* **Lake Nya·sa** \'nyä-sä, nī-'as-ə\ lake SE Africa in Malawi, Mozambique, & Tanzania

Ma·lay \mə-'lā, 'mā-lā\ **1** archipelago SE Asia including Sumatra, Java, Borneo, Sulawesi, Moluccas, & Timor; usually considered to include the Philippines & sometimes New Guinea **2** peninsula SE Asia divided between Thailand & Malaysia

Ma·laya, Federation of \mə-'lā-ə, mā-\ former country SE Asia on Malay Peninsula; since 1963 part of Malaysia

Ma·lay·sia \mə-'lā-zhə, -shə, -zhē-ə, -shē-ə\ **1** — see MALAY 1 **2** country SE Asia; a limited constitutional monarchy; capital, Kuala Lumpur — **Ma·lay·sian** \mə-'lā-zhən, -shən\ *adj or n*

Mal·dives \'mȯl-,dēvz, -,dīvz\ islands in Indian Ocean S of the Laccadives; formerly a sultanate under British protection; independent since 1965; capital, **Ma·le** \'mäl-ē\ — **Mal·div·i·an** \mȯl-'div-ē-ən\ *adj or n*

Ma·le·bo, Pool \mä-'lā-,bō\ expansion of Congo River between Democratic Republic of the Congo & Republic of the Congo

Ma·li \'mäl-ē, 'mal-ē\ *or formerly* **French Sudan** country W Africa; capital, Bamako — **Ma·li·an** \-ē-ən\ *adj or n*

Malmö \'mal-,mər, 'mal-,mœ\ city & port SW Sweden

Mal·ta \'mȯl-tə\ **1** islands in the Mediterranean S of Sicily; a former British colony; independent since 1964; capital, Valletta **2** island, chief of the group

Maluku — see MOLUCCAS

Malvinas, Islas — see FALKLAND ISLANDS

Mam·moth Cave \,mam-əth\ limestone caverns SW central Kentucky in **Mammoth Cave National Park**

Man, Isle of \'man\ island British Isles in Irish Sea; capital, Douglas; has own legislature & laws

Ma·na·do \mə-'näd-,ō\ city & port Indonesia on NE Sulawesi

Ma·na·gua \mə-'näg-wə\ city, capital of Nicaragua

Ma·na·ma \mə-'nam-ə\ city, capital of Bahrain

Ma·naus \mə-'nau̇s\ city W Brazil on Rio Negro 12 miles (19 kilometers) above its junction with the Amazon

Man·ches·ter \'man-,ches-tər, -chə-stər\ city NW England ENE of Liverpool — see GREATER MANCHESTER

Man·chu·kuo \'man-'chü-'kwō, man-'chü-,\ former country (1931–45) E Asia in Manchuria & E Inner Mongolia; capital, Changchun

Man·chu·ria \man-'chu̇r-ē-ə\ region NE China S of the Amur — **Man·chu·ri·an** \man-'chu̇r-ē-ən\ *adj or n*

Man·da·lay \,man-də-'lā\ city central Myanmar

Man·hat·tan \man-'hat-n, mən-\ **1** island SE New York in New York City **2** borough of New York City comprising chiefly Manhattan Island

Manihiki — see NORTHERN COOK

Ma·nila \mə-'nil-ə\ city, capital of Philippines

Ma·ni·pur \,man-ə-'pu̇r, ,mən-\ state NE India between Assam & Myanmar; capital, Imphal

Man·i·to·ba \,man-ə-'tō-bə\ province central Canada; capital, Winnipeg — **Man·i·to·ban** \-'tō-bən\ *adj or n*

Man·i·tou·lin \,man-ə-'tü-lən\ island 80 miles (129 kilometers) long S Ontario in Lake Huron

Ma·ni·za·les \,man-ə-'zäl-əs, -'zal-\ city W central Colombia

Man·nar, Gulf of \mə-'när\ inlet of Indian Ocean between Sri Lanka & S tip of India

Mann·heim \'man-,hīm, 'män-\ city SW Germany on the Rhine NW of Stuttgart

Man·za·nil·lo \,man-zə-'nē-ō, -yō\ city SW Mexico in Colima

Ma·pu·to \mä-'pü-tō\ *or formerly* **Lou·ren·ço Mar·quez** \lə-,ren-sō-,mär-'kes, -'märks, -'märk\ city, capital of Mozambique

Mar·a·cai·bo \,mar-ə-'kī-bō\ city NW Venezuela

Maracaibo, Lake the S extension of Gulf of Venezuela in NW Venezuela

Ma·ra·cay \,mär-ə-'kī\ city N Venezuela

Mar·a·thon \'mar-ə-,thän, -thən\ plain E Greece in Attica NE of Athens

Marche \'märsh\ region central Italy on the Adriatic; capital, Ancona

Mar del Pla·ta \,mär-del-'plät-ə\ city & port E Argentina

Mar·i·ana \,mar-ē-'an-ə, ,mer-\ islands W Pacific N of Caroline Islands including the Northern Mariana Islands & Guam

Ma·ri·a·nao \,mär-ē-ə-'nau̇\ city W Cuba W of Havana

Mariana Trench ocean trench W Pacific extending from SE of Guam to NW of Mariana Islands; deepest in world

Maritime Alps section of the W Alps SE France & NW Italy extending N from Mediterranean coast

Maritime Provinces the Canadian provinces of New Brunswick, Nova Scotia, & Prince Edward Island & sometimes thought to include Newfoundland and Labrador

Ma·ri·u·pol' \,mar-ē-'ü-,pōl\ *or 1949–89* **Zhda·nov** \zhe-'dä-nəf, 'shtä-\ city E Ukraine on Sea of Azov

Mark·ham, Mount \'mär-kəm\ mountain 14,275 feet (4351 meters) Antarctica E of Ross Ice Shelf

Mar·ma·ra, Sea of *or* **Sea of Mar·mo·ra** \'mär-mə-rə\ *or ancient* **Pro·pon·tis** \prə-'pänt-əs\ sea NW Turkey connected with Black Sea by the Bosporus & with Aegean Sea by the Dardanelles

Marne \'märn\ river 325 miles (523 kilometers) long NE France flowing W into the Seine

Mar·que·sas \mär-'kā-zəz, -zəs, -səz, -səs\ islands South Pacific N of Tuamotu Archipelago in French Polynesia — **Mar·que·san** \-zən, -sən\ *adj or n*

Mar·ra·kech \mə-'räk-ish, ,mar-ə-'kesh\ *or formerly* **Mo·roc·co** \mə-'räk-ō\ city central Morocco

Mar·seille \mär-'sā\ *or* **Mar·seilles** \mär-'sā, -'sālz\ *or ancient* **Mas·sil·ia** \mə-'sil-ē-ə\ city SE France

Mar·shall Islands \'mär-shəl\ islands W Pacific E of the Carolines; since 1986 an independent republic in association with the U.S.; capital, Majuro

Mar·tha's Vineyard \,mär-thəz\ island SE Massachusetts off SW coast of Cape Cod WNW of Nantucket

Mar·ti·nique \,märt-n-'ēk\ island West Indies in the Windwards; an overseas department of France; capital, Fort-de-France

Mary·land \'mer-ə-lənd\ state E U.S.; capital, Annapolis — **Mary·land·er** \-lən-dər, -,lan-\ *n*

Ma·san \'mäs-,än\ city South Korea W of Pusan

Mas·e·ru \'maz-ə-,rü\ city, capital of Lesotho

Mash·had \mə-'shad\ city NE Iran

Ma·son–Dix·on Line \,mās-n-'dik-sən\ boundary between Maryland & Pennsylvania; was in part boundary between free & slave states

Mas·qat \'məs-,kät\ *or* **Mus·cat** \-,kät, -kət\ city E Arabia, capital of Oman

Mas·sa·chu·setts \,mas-ə-'chü-səts, ,mas-'chü-, -zəts\ state NE U.S.; capital, Boston

Mas·sif Cen·tral \ma-,sēf-,sen-'träl, -,säⁿ-'träl\ plateau central France W of the Rhone-Saône valley — see AUVERGNE CEVENNES

Mat·a·be·le·land \,mat-ə-'bē-lē-,land\ region SW Zimbabwe; chief town, Bulawayo

Ma·to Gros·so \,mat-ə-'grō-sō\ plateau region SW Brazil in E central Mato Grosso state

Mat·su·do \mät-'sü-dō\ city Japan in SE Honshu NE of Tokyo

Ma·tsu·shi·ma \,mät-sü-'shē-mə, mät-'sü-shi-mə\ group of islets Japan off N Honshu NE of Sendai

Ma·tsu·ya·ma \,mät-sə-'yäm-ə\ city Japan in W Shikoku

Mat·ter·horn \'mat-ər-,hȯrn, 'mät-\ mountain 14,691 feet (4478 meters) in Pennine Alps on border between Switzerland & Italy

Maui \'mau̇-ē\ island Hawaii NW of Hawaii Island

Mau·na Kea \,mau̇-nə-'kā-ə\ extinct volcano 13,796 feet (4205 meters) Hawaii in N central Hawaii Island

Mau·na Loa \,mau̇-nə-'lō-ə\ volcano 13,680 feet (4170 meters) Hawaii in S central Hawaii Island in Hawaii Volcanoes National Park

Mau·re·ta·nia *or* **Mau·ri·ta·nia** \,mȯr-ə-'tā-nē-ə, ,mär-, -nyə\ ancient country NW Africa in modern Morocco & W Algeria — **Mau·re·ta·ni·an** *or* **Mau·ri·ta·ni·an** \-nē-ən, -nyən\ *adj or n*

Mauritania country NW Africa on North Atlantic N of Senegal River; capital, Nouakchott — **Maurita·nian** *adj or n*

Mau·ri·tius \mȯ-'rish-əs, -'rish-ē-əs\ island in Indian Ocean E of Madagascar; an independent country; capital, Port Louis — **Mau·ri·tian** \-'rish-ən\ *adj or n*

May, Cape \'mā\ cape S New Jersey at entrance to Delaware Bay

Mayo \'mā-ō\ county NW Ireland in Connacht

Ma·yon \mä-'yōn\ volcano 8077 feet (2462 meters) Philippines in SE Luzon

Ma·yotte \mä-'yät, -'yȯt\ island Comoros group; a French dependency

Ma·za·ma, Mount \mə-'zäm-ə\ prehistoric mountain SW Oregon the collapse of whose summit formed Crater Lake

Ma·za·tlán \,mäz-ə-'tlän, ,mäs-\ city W Mexico in Sinaloa on the Pacific

Mba·bane \,em-bə-'bän-ā\ city, capital of Swaziland

Mban·da·ka \,em-,bän-'däk-ə\ *or formerly* **Co·qui·lhat·ville** \,kō-kē-'at-,vil\ city W Democratic Republic of the Congo

Mbi·ni \em-'bē-nē\ *or formerly* **Río Mu·ni** \,rē-ō-'mü-nē\ mainland portion of Equatorial Guinea on Gulf of Guinea

Mbu·ji–Ma·yi \em-,bü-jē-'mī-,ē\ *or formerly* **Ba·kwan·ga** \bə-'kwäng-gə\ city S Democratic Republic of the Congo

Mc·Al·len \mə-'kal-ən\ city S Texas

Mc·Kin·ley, Mount \mə-'kin-lē\ *or* **De·na·li** \də-'näl-ē\ mountain 20,320 feet (6194 meters) S central Alaska in Alaska Range; highest in U.S. & in North America; in **Denali National Park**

Mead, Lake \'mēd\ reservoir NW Arizona & SE Nevada formed by Hoover Dam in Colorado River

Meath \'mēth, 'mēth\ county E Ireland in Leinster

Mec·ca \'mek-ə\ city W Saudi Arabia, capital of Hejaz

Meck·len·burg–West Pomerania \'mek-lən-,bərg-\ state NE Germany on the Baltic; capital, Schwerin

Me·dan \mā-'dän\ city Indonesia in NE Sumatra

Me·de·llín \,med-l-'ēn, ,mā-thə-'yēn\ city NW Colombia NW of Bogotá

Me·dia \'mēd-ē-ə\ ancient country & province of Persian Empire

Me·di·na \mə-'dē-nə\ city W Saudi Arabia

Mediolanum — see MILAN

Med·i·ter·ra·nean \,med-ə-tə-'rā-nē-ən, -nyən\ sea 2300 miles (3700 kilometers) long between Europe & Africa connecting with North Atlantic through Strait of Gibraltar

Mee·rut \'mā-rət, 'mir-ət\ city N India in NW Uttar Pradesh

Me·gha·la·ya \,mā-gə-'lā-ə\ state NE India; capital, Shillong

Méjico — see MEXICO

Mek·nes \mek-'nes\ city N Morocco; former capital of the country

Me·kong \'mā-'kȯng, -'käng\ river about 2600 miles (4184 kilometers) long SE Asia flowing from E Tibet S & SE into South China Sea in S Vietnam

Mel·a·ne·sia \,mel-ə-'nē-zhə, -shə\ islands of South Pacific NE of Australia & S of Micronesia including Bismarck Archipelago, the Solomons, Vanuatu, New Caledonia, & the Fijis

Mel·bourne \'mel-bərn\ city SE Australia, capital of Victoria

Me·los *or Greek* **Mí·los** \'mē-,läs\ island Greece in SW Cyclades — **Me·li·an** \'mē-lē-ən\ *adj or n*

Mel·ville \'mel-,vil\ **1** island N Canada in Parry Islands; split between Northwest Territories & Nunavut **2** peninsula in Nunavut

Me·mel \'mā-məl\ — see KLAIPEDA

Mem·phis \'mem-fəs, 'memp-\ **1** city SW Tennessee **2** ancient city N Egypt S of modern Cairo

Mem·phre·ma·gog, Lake \,mem-fri-'mā-,gäg\ lake 30 miles (48 kilometers) long on border between Quebec & Vermont

Men·do·ci·no, Cape \,men-də-'sē-nō\ headland NW California

Menorca — see MINORCA

Mer·cia \'mər-shə, -shē-ə\ ancient Anglo-Saxon kingdom central England — **Mer·cian** \'mər-shən\ *adj or n*

Mé·ri·da \'mer-əd-ə\ city SE Mexico, capital of Yucatán

Mer·sey \'mər-zē\ river 70 miles (113 kilometers) long NW England flowing NW & W into Irish Sea through a large estuary

Mer·sey·side \'mər-zē-,sīd\ metropolitan county NW England; includes Liverpool

Mer·thyr Tyd·fil \'mər-thər-'tid-vil\ administrative area S Wales

Mer·ton \'mərt-n\ borough of SW Greater London, England

Me·sa \'mā-sə\ city SW central Arizona E of Phoenix

Me·sa·bi Range \mə-'säb-ē\ region NE Minnesota that contains iron ore

Me·sa Verde National Park \,mā-sə-'vərd-ē, -'vərd\ reservation SW Colorado containing prehistoric cliff dwellings

Mes·o·po·ta·mia \,mes-ə-pə-'tā-mē-ə, -myə\ **1** region SW Asia between Euphrates & Tigris rivers **2** the entire Tigris-Euphrates valley — **Mes·o·po·ta·mian** \-mē-ən, -myən\ *adj or n*

Mes·quite \mə-'skēt\ city NE Texas E of Dallas

Mes·si·na \mə-'sē-nə\ city Italy in NE Sicily

Messina, Strait of channel between NE Sicily & SW tip of peninsula of Italy

Meuse \'myüz, 'myərz, *French* mœz\ *or Dutch* **Maas** \'mäs\ river about 580 miles (933 kilometers) long W Europe flowing from NE France into North Sea in the Netherlands

Mex·i·cali \,mek-si-'kal-ē\ city NW Mexico, capital of Baja California

Mex·i·co \'mek-si-,kō\ *or Spanish* **Mé·ji·co** \'me-hē-kō\ **1** country S North America; capital Mexico City **2** *or* **Mexico City** city, its capital, in Federal District **3** state S central Mexico; capital, Toluca

Mexico, Gulf of inlet of North Atlantic SE North America

Mez·zo·gior·no \,met-sō-'jȯr-nō, ,med-zō-\ the peninsula of Italy S of roughly the latitude of Rome

Mi·ami \mī-'am-ē, -'am-ə\ city SE Florida

Mich·i·gan \'mish-i-gən\ state N central U.S.; capital, Lansing — **Mich·i·gan·der** \,mish-i-'gan-dər\ *n* — **Mich·i·ga·ni·an** \,mish-ə-'gā-nē-ən, -'gan-ē-\ *n* — **Mich·i·gan·ite** \'mish-i-gə-,nīt\ *n*

Michigan, Lake lake N central U.S.; one of the Great Lakes

Mi·cho·a·cán \,mē-chə-wä-'kän\ state SW Mexico on North Pacific; capital, Morelia

Mi·cro·ne·sia \,mī-krə-'nē-zhə, -shə\ islands of the W Pacific E of the Philippines & N of Melanesia including Caroline, Kiribati, Mariana, & Marshall groups — **Mi·cro·ne·sian** \-zhən, -shən\ *adj or n*

Micronesia, Federated States of islands W Pacific in the Carolines comprising Kosrae, Pohnpei,

\ə\ abut \ᵊ\ kitten \ər\ further \a\ ash \ā\ ace \ä\ mop, mar \au̇\ out \ch\ chin \e\ bet \ē\ easy \g\ go \i\ hit \ī\ ice \j\ job \ŋ\ sing \ō\ go \ȯ\ law \ȯi\ boy \th\ thin \th\ the \ü\ loot \u̇\ foot \y\ yet \zh\ vision *see also* Guide to Pronunciation

Chuuk, & Yap; part of former Trust Territory of the Pacific Islands; republic in association with U.S.; capital, Palikir

Middle Congo — see CONGO 3

Middle East the countries of SW Asia & N Africa — usually considered as including the countries extending from Libya on the W to Afghanistan on the E — **Middle Eastern** *or* **Mid·east·ern** \'mid-'ē-stərn\ *adj*

Mid·dles·brough \'mid-lz-brə\ town N England

Mi·di \mē-'dē\ the south of France

Mid·i·an \'mid-ē-ən\ ancient region NW Arabia E of Gulf of Aqaba — **Mid·i·an·ite** \-ē-ə-ˌnīt\ *n*

Mid·lands \'mid-ləndz, -lənz\ the central counties of England — see WEST MIDLANDS

Mid·lo·thi·an \mid-'lō-t͟hē-ən\ administrative area SE Scotland; chief city, Edinburgh

Mid·way \'mid-ˌwā\ islands (atoll) central Pacific in Hawaiian group 1300 miles (2092 kilometers) WNW of Honolulu belonging to U.S.; not included in state of Hawaii

Mid·west *or* **Middle West** \'mid-'west\ region N central U.S. including area around Great Lakes & in upper Mississippi valley from Ohio on the E to North Dakota, South Dakota, Nebraska, & Kansas on the W — **Mid·west·ern** \'mid-'wes-tərn\ *or* **Middle Western** *adj* — **Mid·west·ern·er** \'mid-'wes-tər-nər, -tə-nər\ *or* **Middle Westerner** *n*

Mi·lan \mə-'lan, -'län\ *or Italian* **Mi·la·no** \mi-'län-ō\ *or ancient* **Me·di·o·la·num** \ˌmed-ē-ō-'lä-nəm\ city NW Italy, capital of Lombardy — **Mil·a·nese** \ˌmil-ə-'nēz, -'nēs\ *adj or n*

Mílos — see MELOS

Mil·wau·kee \mil-'wȯ-kē\ city SE Wisconsin

Mi·nas Basin \ˌmī-nəs\ landlocked bay central Nova Scotia; NE extension of Bay of Fundy

Min·da·nao \ˌmin-də-'nä-ō, -'naù\ island S Philippines

Min·do·ro \min-'dōr-ō, -'dȯr-\ island central Philippines

Minhow — see FUZHOU

Min·ne·ap·o·lis \ˌmin-ē-'ap-ləs, -ə-ləs\ city SE Minnesota

Min·ne·so·ta \ˌmin-ə-'sōt-ə\ state N central U.S.; capital, Saint Paul — **Min·ne·so·tan** \-'sōt-n\ *adj or n*

Mi·nor·ca \mə-'nȯr-kə\ *or Spanish* **Me·nor·ca** \mā-\ island Spain in Balearic Islands — **Mi·nor·can** \mə-'nȯr-kən\ *adj or n*

Minsk \'minsk\ city, capital of Belarus

Mi·que·lon \'mik-ə-ˌlän, *French* mēk-'lō^n, mēk-ə-\ island off S coast of Newfoundland belonging to France — see SAINT PIERRE

Mission Vie·jo \-vē-'ā-hō\ city SW California

Mis·sis·sau·ga \ˌmis-ə-'sȯg-ə\ city Canada in S Ontario

Mis·sis·sip·pi \ˌmis-ə-'sip-ē, mis-'sip-ē\ **1** river 2340 miles (3765 kilometers) long central U.S. flowing into Gulf of Mexico — see ITASCA (Lake) **2** state S U.S.; capital, Jackson

Mis·sou·la \mə-'zü-lə\ city W Montana

Mis·sou·ri \mə-'zùr-ē, -'zùr-ə\ **1** river 2466 miles (3968 kilometers) long W U.S. flowing from SW Montana to the Mississippi in E Missouri **2** state central U.S.; capital, Jefferson City — **Mis·sou·ri·an** \-'zùr-ē-ən\ *adj or n*

Mitch·ell, Mount \'mich-əl\ mountain 6684 feet (2037 meters) W North Carolina in Black Mountains of the Appalachians; highest in U.S. E of the Mississippi

Mi·ya·za·ki \mē-ˌäz-'äk-ē, -ˌyäz-; mē-'äz-ə-kē, -'yäz-\ city Japan in SE Kyushu

Mi·zo·ram \mi-'zȯr-əm\ state NE India; capital, Aizawl

Mo·ab \'mō-ˌab\ region Jordan E of Dead Sea; in biblical times a kingdom

Mo·bile \mō-'bēl, 'mō-ˌbēl\ city SW Alabama on **Mobile Bay** (inlet of Gulf of Mexico)

Moçambique — see MOZAMBIQUE

Mo·de·na \'mȯd-n-ə, -n-ˌä\ city N Italy SW of Venice

Moe·sia \'mē-shə, -shē-ə\ ancient country & Roman province S of the Danube in modern Bulgaria & Serbia

Mog·a·di·shu \ˌmäg-ə-'dish-ü, -'dēsh-\ *or* **Mog·a·di·scio** \-ō\ city, capital of Somalia

Mogilev — see MAHILYOW

Mo·hen·jo–Da·ro \mō-ˌhen-jō-'där-ō\ prehistoric city Pakistan in valley of the Indus NE of modern Karachi

Mo·ja·ve *or* **Mo·ha·ve** \mə-'häv-ē\ desert S California SE of S end of Sierra Nevada

Mo·ji \'mō-jē\ city Japan in N Kyushu

Mol·da·via \mäl-'dā-vē-ə, -vyə\ **1** region E Europe in NE Romania & Moldova W of the Dniester **2** — see MOLDOVA — **Mol·da·vian** \-vē-ən, -vyən\ *adj or n*

Mol·do·va \mäl-'dō-və, mȯl-\ country E Moldavia region; capital Chisinau; formerly (as **Moldavia**) a constituent republic of the U.S.S.R.

Mo·li·se \'mȯ-li-ˌzā\ region central Italy on the Adriatic; capital, Campobasso

Mol·o·kai \ˌmäl-ə-'kī, ˌmō-lə-\ island Hawaii ESE of Oahu

Molotov — see PERM

Mo·luc·cas \mə-'lək-əz\ *or* **Spice Islands** \'spīs\ *or Indonesian* **Ma·lu·ku** \mə-'lü-kü\ islands Indonesia E of Sulawesi — **Mo·luc·ca** \mə-'lək-ə\ *adj* — **Mo·luc·can** \-ən\ *adj or n*

Mom·ba·sa \mäm-'bäs-ə\ city S Kenya on Mombasa Island

Mo·na·co \'män-ə-ˌkō *also* mə-'näk-ō\ country W Europe on Mediterranean coast of France; a principality; capital, Monaco — **Mo·na·can** \'män-ə-kən, mə-'näk-ən\ *adj or n* — **Mon·e·gasque** \ˌmän-i-'gask\ *adj or n*

Mon·a·ghan \'män-ə-hən, -ˌhan\ county NE Ireland (republic) in Ulster

Mön·chen·glad·bach \ˌmœn-kən-'glät-ˌbäk\ city W Germany

Mon·go·lia \män-'gōl-yə, mäng-, -'gō-lē-ə\ **1** region E Asia E of Altai Mountains; includes Gobi Desert **2** INNER MONGOLIA **3** *or* **Outer Mongolia** country E Asia comprising major portion of Mongolia; capital, Ulaanbaatar

Mon·mouth·shire \'män-məth-ˌshiər, 'mən-, -shər\ *or* **Monmouth** administrative area SE Wales bordering on England

Mon·ro·via \mən-'rō-vē-ə, ˌmən-\ city, capital of Liberia

Mon·tana \män-'tan-ə\ state NW U.S.; capital, Helena — **Mon·tan·an** \-ən\ *adj or n*

Mont Blanc \mō^n-'blä^n\ mountain 15,771 feet (4807 meters) SE France on Italian border; highest in the Alps

Mon·te·go Bay \män-'tē-gō\ city & port NW Jamaica on Montego Bay (inlet of the Caribbean)

Mon·te·ne·gro \ˌmänt-ə-'nē-grō, -'nā-\ federated republic S Serbia and Montenegro on the Adriatic; capital, Cetinje — **Mon·te·ne·grin** \-grən\ *adj or n*

Mon·ter·rey \ˌmänt-ə-'rā\ city NE Mexico, capital of Nuevo León

Mon·te·vi·deo \ˌmänt-ə-və-'dā-ō, -'vid-ē-ˌō\ city, capital of Uruguay

Mont·gom·ery \mənt-'gəm-rē, mänt-, mən-, män-, -'gäm-, -ə-rē\ city, capital of Alabama

Mont·pe·lier \mänt-'pēl-yər, -'pil-\ city, capital of Vermont

Mon·tre·al \ˌmän-trē-'ȯl, ˌmən-\ city S Quebec on Montreal Island in the Saint Lawrence

Mont–Saint–Mi·chel \mō^n-sa^n-mē-'shel\ islet NW France off coast of Brittany in Gulf of Saint-Malo

Mont·ser·rat \ˌmän-sə-'rat\ island British West Indies in the Leewards; capital, Plymouth

Mo·ra·via \mə-'rā-vē-ə\ region E Czech Republic; chief city, Brno

Mor·ay \'mər-ē, 'mə-rē\ *or* **Mor·ay·shire** \-ˌshiər, -shər\ administrative area NE Scotland

Mo·re·los \mə-'rā-ləs\ state S central Mexico; capital, Cuernavaca

Mo·re·no Valley \mə-'rē-nō\ city S California

Mo·roc·co \mə-'räk-ō\ **1** country NW Africa; a kingdom; capital, Rabat; formerly divided into **French Morocco** (capital, Rabat), **Spanish Morocco** (capital, Tetuán) & **International Zone** of Tangier **2** — see MARRAKECH — **Mo·roc·can** \-'räk-ən\ *adj or n*

Mo·ro·ni \mȯ-'rō-nē\ city, capital of Comoros

Morris Jes·up, Cape \ˌmȯr-əs-'jes-əp, ˌmär-\ headland N Greenland in Arctic Ocean

Mos·cow \'mäs-ˌkaù, -kō\ *or Russian* **Mos·kva** \mäsk-'vä\ city, capital of Russia, on Moskva River

Mo·selle \mō-'zel\ *or German* **Mo·sel** \'mō-zəl\ river about 340 miles (545 kilometers) long E France & W Germany flowing from Vosges Mountains into the Rhine at Koblenz

Mosquito Coast region Central America bordering on the Caribbean in E Honduras & E Nicaragua

Mo·sul \mō-'sül, 'mō-səl\ city N Iraq on the Tigris

Moul·mein \mül-'mān, mōl-, -'mīn\ city S Myanmar at mouth of the Salween

Mount Rainier National Park — see RAINIER (Mount)

Mount Rev·el·stoke National Park \'rev-əl-ˌstōk\ reservation SE British Columbia including Mount Revelstoke (over 7000 feet or 2130 meters)

Mo·zam·bique \ˌmō-zəm-'bēk\ *or Portuguese* **Mo·çam·bi·que** \ˌmü-səm-'bē-kə\ **1** channel SE Africa between Mozambique & Madagascar **2** *or formerly* **Portuguese East Africa** country SE Africa; formerly a dependency of Portugal; capital, Maputo

Mpu·ma·lan·ga \əm-ˌpü-mä-'läng-gä\ province NE Republic of South Africa

Mukden — see SHENYANG

Mul·tan \mùl-'tän\ city NE Pakistan SW of Lahore

Mumbai — see BOMBAY

Mu·nich \'myü-nik\ *or German* **Mün·chen** \'mœn-kən\ city S Germany, capital of Bavaria

Mun·ster \'mən-stər\ province S Ireland

Mün·ster \'mən-stər, 'mün-, 'myün-, 'mœn-\ city W Germany NNE of Dortmund

Mur·cia \'mər-shə, -shē-ə\ **1** region, province, & ancient kingdom SE Spain **2** city, its capital — **Mur·cian** \-shən\ *adj or n*

Mur·mansk \mùr-'mansk, -'mänsk\ city NW Russia in Europe on Barents Sea

Mur·ray \'mər-ē, 'mə-rē\ river 1609 miles (2589 kilometers) long SE Australia flowing W from E Victoria into Indian Ocean in South Australia — see DARLING

Mur·rum·bidg·ee \ˌmər-əm-'bij-ē, ˌmə-rəm-\ river almost 1000 miles (1609 kilometers) long SE Australia in New South Wales flowing W into the Murray

Mu·sa, Ge·bel \ˌjeb-əl-'mü-sə\ mountain group NE Egypt in S Sinai Peninsula; highest Gebel Katherina 8652 feet (2637 meters) — see HOREB

Mu·sa, Je·bel \ˌjeb-əl-'mü-sə\ mountain 2775 feet (846 meters) N Morocco opposite Rock of Gibraltar — see PILLARS OF HERCULES

Muscat — see MASQAT

Muscat and Oman — see OMAN

Mus·co·vy \mə-'skō-vē; 'məs-kə-vē, -ˌkō-\ **1** the principality of Moscow (founded 1295) which in 15th century came to dominate Russia **2** RUSSIA — an old name

Myan·mar \'myän-ˌmär\ *or formerly* **Bur·ma** \'bər-mə\ country SE Asia; capital, Yangon

My·ce·nae \mī-'sē-nē\ ancient city S Greece in NE Peloponnese N of Argos

My·ko·la·yiv \ˌmē-kə-'lä-yif\ *or* **Ni·ko·la·yev** \ˌnē-kə-'lä-yif\ city S Ukraine

Myk·o·nos \'mik-ə-ˌnōs\ island Greece in the Aegean in NE Cyclades

Myr·tle Beach \'mərt-l\ city E South Carolina on the Atlantic

My·sia \'mish-ə, 'mish-ē-ə\ ancient country NW Asia Minor bordering on the Propontis

My·sore \mī-'sōr, -'sȯr\ **1** — see KARNATAKA **2** city in S Karnataka

Myt·i·le·ne \ˌmit-l-'ē-nē\ **1** ancient city Greece on E coast of Lesbos Island; site at modern town of Mytilene **2** — see LESBOS

Nab·a·taea *or* **Nab·a·tea** \ˌnab-ə-'tē-ə\ ancient Arab kingdom SE of Palestine — **Nab·a·tae·an** *or* **Nab·a·te·an** \-'tē-ən\ *adj or n*

Na·be·rezh·nye Chel·ny \ˌnäb-ə-'rezh-nə-'chel-nē, -nyə-\ city E Russia in Europe

Na·ga·land \'näg-ə-ˌland\ state E India N of Manipur in Naga Hills; capital, Kohima

Na·ga·no \nä-'gän-ō\ city Japan in SE Honshu NW of Tokyo

Na·ga·sa·ki \ˌnäg-ə-'säk-ē, ˌnag-ə-'sak-ē\ city & port Japan in W Kyushu

Na·goya \nə-'gȯi-ə, 'näg-ə-ˌyä\ city Japan in S central Honshu

Nag·pur \'näg-ˌpùr\ city E central India in NE Maharashtra

Na·ha \'nä-hä\ city Japan in Ryukyu Islands; capital of Okinawa

Na·huel Hua·pí \nä-ˌwel-wä-'pē\ lake SW Argentina in the Andes

Nai·ro·bi \nī-'rō-bē\ city, capital of Kenya

Najd — see NEJD

Na·mib·ia \nə-'mib-ē-ə\ *or formerly* **South–West Africa** country SW Africa, on South Atlantic; administered by South Africa 1919–90; capital, Windhoek

Nan·chang \'nän-'chäng\ city SE China, capital of Jiangxi

Nan·cy \'nan-sē, nä^{n}-sē\ city NE France
Nan·jing *or* **Nan·king** \'nan-'jiŋ, 'nän-\ city E China on the Chang, capital of Jiangsu
Nan·ning \'nän-'niŋ\ city S China, capital of Guangxi Zhuangzu
Nantes \'nans\ city NW France on the Loire
Nan·tuck·et \nan-'tək-ət\ island SE Massachusetts S of Cape Cod
Na·per·ville \'nā-pər-,vil\ city NE Illinois W of Chicago
Na·ples \'nā-pəlz\ *or Italian* **Na·po·li** \'näp-ə-lē\ *or ancient* **Ne·ap·o·lis** \nē-'ap-ə-ləs\ city S Italy on Bay of Naples — **Ne·a·pol·i·tan** \,nē-ə-'päl-ət-n\ *adj or n*
Na·ra \'när-ə\ city Japan in W central Honshu E of Osaka
Nar·ra·gan·sett Bay \,nar-ə-'gan-sət\ inlet of the North Atlantic SE Rhode Island
Nash·ville \'nash-,vil, -vəl\ city, capital of Tennessee
Nas·sau \'nas-,ȯ\ city, capital of Bahamas on New Providence Island
Na·tal \nə-'tal, -'täl\ **1** city & port NE Brazil **2** former province E Republic of South Africa; now part of KwaZulu-Natal
Natch·ez Trace \'nach-əz\ pioneer road between Nashville & **Natchez** (city SW Mississippi on Mississippi River); constructed in early 19th century
Na·u·ru \nä-'ü-rü\ island (atoll) W Pacific; formerly a joint British, New Zealand, & Australian trust territory; an independent republic since 1968; capital, Yaren
Na·varre \nə-'vär\ *or Spanish* **Na·var·ra** \nə-'vär-ə\ region & former kingdom N Spain & SW France in W Pyrenees
Nax·ci·van \,nək-chi-'vän\ exclave of Azerbaijan separated from the rest of the country by Armenia
Nax·os \'nak-səs, -,säs\ island Greece in the Aegean; largest of the Cyclades
Na·ya·rit \,nī-ə-'rēt\ state W Mexico on North Pacific; capital, Tepic
Naz·a·reth \'naz-rəth, -ə-rəth\ town of ancient Palestine in central Galilee; now a city of N Israel
N'Dja·me·na \ən jä-mā-nä\ *or formerly* **Fort–Lamy** \,fȯr-lə-'mē\ city, capital of Chad
Neagh, Lough \'nā\ lake Northern Ireland; largest in British Isles
Near East the countries of NE Africa & SW Asia — **Near Eastern** *adj*
Neath and Port Tal·bot \'nēth-ənd-,pȯrt-'tȯl-bət\ administrative area S Wales
Ne·bras·ka \nə-'bras-kə\ state central U.S.; capital, Lincoln — **Ne·bras·kan** \-kən\ *adj or n*
Neg·ev \'neg-,ev\ *or* **Neg·eb** \-,eb\ desert region S Israel
Ne·gro \'nā-grō\ river 1400 miles (2253 kilometers) long in E Colombia & N Brazil flowing into the Amazon
Ne·gros \'nā-grōs\ island S central Philippines in Visayan Islands
Neis·se \'nī-sə\ river 150 miles (256 kilometers) long N Europe flowing from N Czech Republic N into the Oder; forms part of boundary between Poland & Germany
Nejd \'nejd, 'nezhd\ *or* **Najd** \'najd, 'nazhd\ region central & E Saudi Arabia; capital, Riyadh
Ne·pal \nə-'pȯl, -'päl, -'pal\ country Asia on NE border of India in the Himalaya; a kingdom; capital, Kathmandu — **Nep·a·lese** \,nep-ə-'lēz, -'lēs\ *adj or n* — **Ne·pali** \nə-'pȯl-ē, -'päl-, -'pal-\ *adj or n*
Neth·er·lands \'neth-ər-lənz, -ləndz\ **1** *or Dutch* **Ne·der·land** \'nād-ər-,länt\ *also* **Holland** country NW Europe on North Sea; a kingdom; capital, Amsterdam; seat of government, The Hague **2** LOW COUNTRIES — an historical usage — **Neth·er·land** \'neth-ər-lənd\ *adj* — **Neth·er·land·er** \-,lan-dər, -lən-\ *n* — **Neth·er·land·ish** \-dish\ *adj*
Netherlands An·til·les \an-'til-ēz\ islands of the West Indies belonging to the Netherlands; Bonaire, Curaçao, and formerly Aruba, Saba, Saint Eustatius, & S part of Saint Martin; capital, Willemstad (on Curaçao)
Netherlands East Indies *or* **Netherlands India** *or* **Dutch East Indies** former Dutch possessions in the East Indies including Indonesia
Netherlands New Guinea — see WEST PAPUA
Ne·va \'nē-və, 'nā-\ river 40 miles (64 kilometers) long W Russia in Europe, flowing from Lake Ladoga into Gulf of Finland at Saint Petersburg
Ne·vada \nə-'vad-ə, -'väd-ə\ state W U.S.; capital, Carson City — **Ne·vad·an** \-'vad-n, -'väd-n\ *or* **Ne·vad·i·an** \-'vad-ē-ən, -'väd-\ *adj or n*
Ne·vis \'nē-vəs\ island British West Indies in the Leewards — see SAINT KITTS
Nevis, Ben — see BEN NEVIS
New Am·ster·dam \'am-stər-'dam, 'amp-\ town founded 1625 on Manhattan Island by the Dutch; renamed New York 1664 by the British
New·ark \'nü-ərk, 'nü-, 'nyü-, 'nyu̇-\ city & port NE New Jersey
New Bed·ford \'bed-fərd\ city & port SE Massachusetts
New Brit·ain \'brit-n\ island W Pacific, largest in Bismarck Archipelago
New Bruns·wick \'brənz-wik\ province SE Canada; capital, Fredericton
New Cal·e·do·nia \,kal-ə-'dō-nyə, -nē-ə\ island SW Pacific SW of Vanuatu; an overseas department of France; capital, Nouméa
New·cas·tle \'nü-,kas-əl, 'nyü-\ city SE Australia in E New South Wales
New·cas·tle up·on Tyne \nü-'kas-əl-ə-,pȯn-'tīn, nyü-, -,pän-, -pən-, 'nü-,, 'nyü-,\ city N England
New Del·hi \-'del-ē\ city, capital of India S of city of Delhi
New England section of NE U.S. comprising states of Maine, New Hampshire, Vermont, Massachusetts, Rhode Island, & Connecticut — **New En·gland·er** \-'iŋ-glən-dər *also* -'iŋ-lən-\ *n*
New·found·land \'nü-fən-dlənd, 'nyü-, -lənd, -,dland, -,land; ,nü-fən-'dland, ,nyü-, -'land\ island Canada in North Atlantic E of Gulf of Saint Lawrence — **New·found·land·er** \-ər\ *n*
Newfoundland and Labrador *or 1949–2001* **Newfoundland** province E Canada; capital, Saint John's
New France the possessions of France in North America before 1763
New Guin·ea \'gin-ē\ *or Indonesian* **Iri·an** \,ir-ē-'än\ **1** island W Pacific N of E Australia divided between Indonesia & Papua New Guinea **2** the NE portion of the island of New Guinea together with Bismarck Archipelago, Bougainville, Buka, & adjacent small islands; now part of Papua New Guinea — **New Guin·ean** \'gin-ē-ən\ *adj or n*
New·ham \'nü-əm, 'nyü-\ borough of E Greater London, England
New Hamp·shire \'ham shər, 'hamp-, -,shiər\ state NE U.S.; capital, Concord — **New Hamp·shire·man** \-mən\ *n* — **New Hamp·shir·ite** \-,īt\ *n*
New Ha·ven \'hā-vən\ city & port S Connecticut
New Hebrides — see VANUATU
New Jer·sey \'jər-zē\ state E U.S.; capital, Trenton — **New Jer·sey·an** \-ən\ *n* — **New Jer·sey·ite** \-,īt\ *n*
New Mex·i·co \'mek-si-,kō\ state SW U.S.; capital, Santa Fe — **New Mex·i·can** \-si-kən\ *adj or n*
New Neth·er·land \'neth-ər-lənd\ former Dutch colony (1613–64) North America along Hudson & lower Delaware rivers; capital, New Amsterdam
New Or·leans \'ȯr-lē-ənz; 'ȯrl-ənz, -yənz; ȯr-'lēnz\ city & port SE Louisiana between Lake Pontchartrain & Mississippi River
New·port \'nü-,pōrt, 'nyü-, -,pȯrt\ **1** administrative area SE Wales **2** city SE Wales in Newport administrative area
New·port News \,nü-,pōrt-'nüz, -,pȯrt-, -pərt-, ,nyü...'nyüz\ city & port SE Virginia
New Prov·i·dence \'präv-əd-əns, -ə-,dens\ island NW central Bahamas; chief town, Nassau
New South Wales state SE Australia; capital, Sydney
New Spain former Spanish possessions in North America, Central America, West Indies, & the Philippines; capital, Mexico City
New Sweden former Swedish colony (1638–55) North America on W bank of Delaware River
New York \'yȯrk\ **1** state NE U.S.; capital, Albany **2** *or* **New York City** city SE New York; includes Bronx, Brooklyn, Manhattan, Queens, & Staten Island — **New York·er** \'yȯr-kər\ *n*
New Zea·land \'zē-lənd\ country SW Pacific ESE of Australia; capital, Wellington — **New Zea·land·er** \-lən-dər\ *n*
Ni·ag·a·ra Falls \nī-'ag-rə, -ə-rə\ falls New York & Ontario in **Niagara River** (flowing N from Lake Erie into Lake Ontario); divided by Goat Island into Horseshoe, or Canadian, Falls (158 feet or 48 meters high) & American Falls (167 feet or 51 meters high)
Nia·mey \nē-'äm-ā, nyä-'mā\ city, capital of Niger
Ni·caea \nī-'sē-ə\ *or* **Nice** \'nīs\ ancient city W Bithynia; site at modern village of Iznik in NW Turkey — **Ni·cae·an** \nī-'sē-ən\ *adj or n* — **Ni·cene** \'nī-,sēn, nī-'sēn\ *adj*
Ni·ca·ra·gua \,nik-ə-'räg-wə\ **1** lake 102 miles (164 kilometers) long S Nicaragua **2** country Central America; capital, Managua — **Ni·ca·ra·guan** \-wən\ *adj or n*
Nice \'nēs\ *or ancient* **Ni·caea** \nī-'sē-ə\ city SE France on the Mediterranean
Nic·o·bar \'nik-ə-,bär\ islands India in Bay of Bengal S of the Andamans
Nic·o·sia \,nik-ə-'sē-ə\ city, capital of Cyprus
Niedersachsen — see LOWER SAXONY
Ni·ger \'nī-jər\ **1** river 2600 miles (4183 kilometers) long W Africa flowing into Gulf of Guinea **2** country W Africa N of Nigeria; capital, Niamey
Ni·ge·ria \nī-'jir-ē-ə\ country W Africa on Gulf of Guinea; capital, Abuja — **Ni·ge·ri·an** \-ē-ən\ *adj or n*
Nii·ga·ta \nē-'gät-ə, 'nē-gə-,tä\ city Japan in N Honshu on Sea of Japan
Nii·hau \'nē-,hau̇\ island Hawaii WSW of Kauai
Nikolayev — see MYKOLAYIV
Nile \'nīl\ river about 4160 miles (6693 kilometers) long E Africa flowing from Lake Victoria in Uganda N into the Mediterranean in Egypt; in various sections called specifically **Vic·to·ria** \vik-'tȯr-ē-ə, -'tȯr-\ *or* **Som·er·set** \'səm-ər-sət, -,set\, **Nile** between Lake Victoria & Lake Albert; **Al·bert** \'al-bərt\ **Nile** between Lake Albert & Lake No; & **White Nile** from Lake No to Khartoum — see BLUE NILE
Nil·gi·ri \'nil-gə-rē\ hills S India in W Tamil Nadu
Nîmes \'nēm\ city S France NW of Marseille
Nin·e·veh \'nin-ə-və\ ancient city, capital of Assyria; ruins in N Iraq
Ning·xia Hui·zu \'niŋ-shē-'ä-'hwēd-'zü\ *or* **Ning·sia Hui** \'niŋ-shē-'ä-'hwē\ region N China; before 1954 part of former province of **Ningsia**
Nip·i·gon, Lake \'nip-ə-,gän\ lake Canada in W Ontario N of Lake Superior
Niš *or* **Nish** \'nish\ city E Serbia
Ni·shi·no·mi·ya \,nish-ə-'nō-mē-,ä, -,yä\ city Japan in central Honshu E of Kobe
Ni·te·rói \,nēt-ə-'rȯi\ city SE Brazil on Guanabara Bay opposite Rio de Janeiro
Ni·ue \nē-'ü-ā\ island S central Pacific; a self-governing territory of New Zealand
Nizh·niy Nov·go·rod *or* **Nizh·ni Novgorod** *or* \,nizh-ne-'näv-gə-,räd, -'nȯv-gə-rət\ *or formerly* **Gor·ki** \'gȯr-kē\ city central Russia in Europe, on the Volga E of Moscow
Nizh·ny Ta·gil *or* **Nizh·ni Tagil** \,nizh-nē-tə-'gil\ city W Russia in Asia, in the Urals
Nord–Ost·see \'nȯrt-,ȯst-'zā\ *or* **Kiel** \'kēl\ ship canal 61 miles (98 kilometers) long N Germany connecting Baltic & North seas
Nord–Pas–de–Ca·lais \,nȯr-,päd-kȧ-'lā\ region northernmost France
Nor·folk \'nȯr-fək, *in the U.S. also* -,fȯk\ **1** city & port SE Virginia **2** county E England on North Sea
Nor·i·cum \'nȯr-i-kəm, 'när-\ ancient country & Roman province S central Europe S of the Danube in modern Austria & Germany
Nor·man·dy \'nȯr-mən-dē\ *or French* **Nor·man·die** \nȯr-mä^{n}-'dē\ region & former province NW France NE of Brittany; capital, Rouen
North **1** river estuary of the Hudson between NE New Jersey & SE New York **2** sea arm of North Atlantic E of Great Britain **3** island N New Zealand
North America continent of Western Hemisphere NW of South America & N of the Equator — **North American** *adj or n*
North·amp·ton \nȯrth-'am-tən, -'ham-, -'amp-, -'hamp-\ city central England in Northamptonshire
North·amp·ton·shire \-,shiər, -shər\ *or* **Northampton** county central England
North Ayrshire administrative area W Scotland
North Cape **1** headland New Zealand at N end of North Island **2** headland NE Norway on **Ma·ger·öy** \,mäg-ə-'rəi\ Island

\ə\ abut \ᵊ\ kitten \ər\ further \a\ ash \ā\ ace \ä\ mop, mar \au̇\ out \ch\ chin \e\ bet \ē\ easy \g\ go \i\ hit \ī\ ice \j\ job \ŋ\ sing \ō\ go \ȯ\ law \ȯi\ boy \th\ thin \th\ the \ü\ loot \u̇\ foot \y\ yet \zh\ vision *see also* Guide to Pronunciation

North Car·o·li·na \ˌkar-ə-ˈlī-nə\ state E U.S.; capital, Raleigh — **North Car·o·lin·ian** \-ˈlin-ē-ən, -ˈlin-yən\ *adj or n*

North Cas·cades National Park \kas-ˈkādz, ˈkas-ˌ\ reservation N Washington

North Da·ko·ta \də-ˈkō-tə\ state N U.S.; capital, Bismarck — **North Da·ko·tan** \-ˈkō-tᵊn\ *adj or n*

North East Frontier Agency — see ARUNACHAL PRADESH

North–East New Guinea the NE section of Papua New Guinea on New Guinea mainland

Northern — see LIMPOPO 2

Northern Cape province W Republic of South Africa

Northern Cook \ˈku̇k\ *or* **Ma·ni·hi·ki** \ˌmän-ə-ˈhē-kē\ islands S central Pacific N of Cook Islands

Northern Ireland region N Ireland; a division of United Kingdom; capital, Belfast

Northern Kingdom — see ISRAEL

Northern Mariana Islands islands W Pacific; commonwealth in association with U.S.; capital, Saipan

Northern Rhodesia — see ZAMBIA

Northern Territory territory N & central Australia; capital, Darwin

North Korea — see KOREA

North Lan·ark·shire \ˈlan-ərk-ˌshir, -shər\ administrative area W Scotland

North Las Vegas city SE Nevada

North Rhine–Westphalia *or German* **Nord·rhein–West·fa·len** \ˈnȯrt-ˌrīn-ˌvest-ˈfä-lən\ state W Germany; capital, Düsseldorf

North Slope region N Alaska between Brooks Range & Arctic Ocean

North·um·ber·land \nȯr-ˈthəm-bər-lənd\ county N England — **North·um·bri·an** \-ˈthəm-brē-ən\ *adj or n*

North·um·bria \nȯr-ˈthəm-brē-ə\ ancient country Great Britain between the Humber & Firth of Forth — **North·um·bri·an** \-brē-ən\ *adj or n*

North Vietnam — see VIETNAM

North West province N Republic of South Africa

North–West Frontier province Pakistan on Afghanistan border

Northwest Passage navigable sea route between North Atlantic & North Pacific along N coast of North America

Northwest Territories territory N Canada comprising the mainland N of 60° between Yukon Territory & Nunavut, & the westernmost part of the Arctic Archipelago; capital, Yellowknife

North York \ˈyȯrk\ former city Canada in SE Ontario; now part of Toronto

North York·shire \ˈyȯrk-ˌshiər, -shər\ county N England

Nor·walk \ˈnȯr-ˌwȯk\ city SW California

Nor·way \ˈnȯr-ˌwā\ country N Europe in Scandinavia; a kingdom; capital, Oslo

Nor·we·gian \nȯr-ˈwē-jən\ sea between North Atlantic & Arctic oceans W of Norway

Nor·wich \ˈnȯr-wich; ˈnȯr-ich, ˈnär-\ city E England in Norfolk

Not·ting·ham \ˈnät-iŋ-əm, *in the U.S. also* -ˌham\ city N central England in Nottinghamshire

Not·ting·ham·shire \-ˌshiər, -shər\ *or* **Nottingham** county N central England

Nouak·chott \nu̇-ˈäk-ˌshät\ city, capital of Mauritania

Nou·méa \nü-ˈmā-ə\ city, capital of New Caledonia

No·va Igua·çu \ˌnȯ-və-ˌē-gwə-ˈsü\ city SE Brazil NW of Rio de Janeiro

No·va Sco·tia \ˌnō-və-ˈskō-shə\ province SE Canada; capital, Halifax — **No·va Sco·tian** \-ˈskō-shən\ *adj or n*

No·va·ya Zem·lya \ˌnō-və-yə-ˌzem-lē-ˈä\ two islands N of Russia in Europe, in Arctic Ocean between Barents & Kara seas

Nov·go·rod \ˈnäv-gə-ˌräd, ˈnȯv-gə-rət\ city W Russia in Europe SSE of Saint Petersburg

No·vi Sad \ˌnō-vē-ˈsäd\ city N Serbia on the Danube

No·vo·kuz·netsk \ˌnō-vō-ku̇z-ˈnetsk\ city S Russia in Asia at S end of Kuznetsk Basin

No·vo·si·birsk \ˌnō-vō-sə-ˈbiərsk\ city S Russia in Asia, on the Ob

Nu·bia \ˈnü-bē-ə, ˈnyü-\ region NE Africa in valley of the Nile in S Egypt & N Sudan — **Nu·bi·an** \-bē-ən\ *adj or n*

Nubian desert NE Sudan E of the Nile

Nue·vo Le·ón \nu̇-ˌā-vō-lā-ˈōn\ state N Mexico; capital, Monterrey

Nu·ku·a·lo·fa \ˌnü-kə-wə-ˈlō-fə\ seaport, capital of Tonga

Nul·lar·bor Plain \ˈnəl-ə-ˌbȯr\ treeless area SW Australia in Western Australia & South Australia

Nu·mid·ia \nu̇-ˈmid-ē-ə, nyu̇-\ ancient country N Africa E of Mauretania in modern Algeria; chief city, Hippo — **Nu·mid·i·an** \-ē-ən\ *adj or n*

Nu·na·vut \ˈnü-nə-ˌvüt\ territory N Canada comprising the mainland N of 60° between Northwest Territories & Hudson Bay & the easternmost & northernmost part of the Arctic Archipelago; capital, Iqaluit

Nu·rem·berg \ˈnu̇r-əm-ˌbərg, ˈnyu̇r-\ *or German* **Nürn·berg** \ˈnʊrn-ˌberk\ city S central Germany in N Bavaria

Nu·ri·stan \ˌnu̇r-i-ˈstan\ mountainous area E Afghanistan

Nuuk \ˈnük\ *or* **Godt·håb** \ˈgȯt-ˌhȯp\ town, capital of Greenland on SW coast

Nyasa, Lake — see MALAWI (Lake)

Nyasaland — see MALAWI

Oa·hu \ə-ˈwä-hü\ island Hawaii; site of Honolulu

Oak·land \ˈō-klənd\ city W California on San Francisco Bay E of San Francisco

Oa·xa·ca \wə-ˈhä-kə\ **1** state SE Mexico **2** city, its capital

Ob' \ˈäb, ˈȯb\ river over 2250 miles (3620 kilometers) long W central Russia in Asia flowing NW & N into **Gulf of Ob'** (inlet of Arctic Ocean)

Oce·a·nia \ˌō-shē-ˈan-ē-ə, -ˈä-nē-ə\ lands of the central & S Pacific: Micronesia, Melanesia, Polynesia including New Zealand, & sometimes Australia & Malay Archipelago — **Oce·a·ni·an** \-ˈan-ē-ən, -ˈä-nē-\ *adj or n*

Ocean·side \ˈō-shən-ˌsīd\ city SW California NNW of San Diego

Oden·se \ˈōd-ᵊn-sə, ˈü-ən-zə\ city central Denmark on Fyn Island

Oder \ˈōd-ər\ *or* **Odra** \ˈō-drə\ river about 565 miles (909 kilometers) long central Europe flowing from E Czech Republic NW into Baltic Sea; forms part of boundary between Poland & Germany — see NEISSE

Odes·sa \ō-ˈdes-ə\ city S Ukraine on Black Sea

Of·fa·ly \ˈȯf-ə-lē, ˈäf-\ county central Ireland in Leinster

Ogasawara — see BONIN

Og·bo·mo·sho \ˌäg-bə-ˈmō-shō\ city W Nigeria

Ohio \ō-ˈhī-ō\ **1** river 981 miles (1578 kilometers) long E U.S. flowing from W Pennsylvania into the Mississippi **2** state E central U.S.; capital, Columbus — **Ohio·an** \ō-ˈhī-ə-wən\ *adj or n*

Oi·ta \ˈȯi-ˌtä, ō-ˈēt-ə\ city Japan in NE Kyushu

Oka·ya·ma \ˌō-kə-ˈyäm-ə\ city Japan in W Honshu on Inland Sea

Oka·za·ki \ˌō-kə-ˈzäk-ē\ city Japan in S central Honshu

Okee·cho·bee, Lake \ˌō-kē-ˈchō-bē\ lake S central Florida

Oke·fe·no·kee \ˌō-kē-fə-ˈnō-kē\ swamp SE Georgia & NE Florida

Okhotsk, Sea of \ō-ˈkätsk\ inlet of North Pacific E Russia in Asia, W of Kamchatka Peninsula & Kuril Islands

Oki·na·wa \ˌō-kə-ˈnä-wə, -ˈnau̇-ə\ **1** islands Japan in central Ryukyus; capital, Naha **2** island, chief of group — **Oki·na·wan** \-ˈnä-wən, -ˈnau̇-ən\ *adj or n*

Okla·ho·ma \ˌō-klə-ˈhō-mə\ state S U.S.; capital, Oklahoma City — **Okla·ho·man** \-mən\ *adj or n*

Oklahoma City city, capital of Oklahoma

Old Faithful geyser Yellowstone National Park NW Wyoming

Old·ham \ˈōl-dəm\ city NW England in Greater Manchester

Old Point Comfort cape SE Virginia N of entrance to Hampton Roads

Ol·du·vai Gorge \ˈōl-də-ˌvī\ canyon N Tanzania SE of Serengeti Plain; site of fossil beds

Olives, Mount of *or* **Ol·i·vet** \ˈäl-ə-ˌvet, ˌäl-ə-ˈ\ mountain ridge W Jordan on E side of Jerusalem

Olym·pia \ə-ˈlim-pē-ə, ō-\ **1** city, capital of Washington **2** plain S Greece in NW Peloponnese

Olym·pic \ə-ˈlim-pik, ō-\ mountains NW Washington on Olympic Peninsula, partly in **Olympic National Park**; highest Mt. Olympus 7965 feet (2428 meters)

Olym·pus \ə-ˈlim-pəs, ō-\ mountains NE Greece in Thessaly; home of the gods in Greek mythology

Oma·ha \ˈō-mə-ˌhȯ, -ˌhä\ **1** city E Nebraska **2** beach NW France; in World War II landing place of American army June 6, 1944

Oman \ō-ˈmän, -ˈman\ *or formerly* **Mus·cat and Oman** \ˈməs-ˌkat, -kət\ country SW Asia in SE Arabia; a sultanate; capital, Masqat — **Omani** \ō-ˈmän-ē\ *adj or n*

Oman, Gulf of arm of Arabian Sea between Oman & SE Iran

Om·dur·man \ˌäm-dər-ˈman, -ˈmän\ city central Sudan on left bank of the Nile opposite Khartoum

Omi·ya \ō-ˈmē-ə, ˈō-mē-ˌä\ city Japan in SE Honshu NW of Tokyo

Omsk \ˈȯmsk, ˈȯmpsk, ˈämsk, ˈämpsk\ city S Russia in Asia, at confluence of Irtysh & Om rivers

On·ta·ke \ȯn-ˈtäk-ā\ mountain 10,049 feet (3063 meters) Japan in central Honshu

On·tar·io \än-ˈter-ē-ˌō\ **1** city SW California **2** province E Canada; capital, Toronto — **On·tar·i·an** \-ē-ən\ *adj or n*

Ontario, Lake lake E central North America in U.S. & Canada; one of the Great Lakes

Ophir \ˈō-fər\ a biblical land rich in gold; probably in Arabia

Oporto — see PORTO

Ora·dea \ȯ-ˈräd-ē-ə\ city NW Romania

Oral \ȯ-ˈräl\ *or* **Uralsk** \u̇-ˈrälsk, yu̇-ˈralsk\ city W Kazakhstan on Ural River

Oran \ȯ-ˈrän\ city & port NW Algeria

Or·ange \ˈȯr-inj, ˈär-, -ənj\ **1** city SW California **2** river 1300 miles (2092 kilometers) long S Africa flowing W from Drakensberg Mountains into South Atlantic

Orange Free State — see FREE STATE

Ordzhonikidze — see VLADIKAVKAZ

Or·e·gon \ˈȯr-i-gən, ˈär-, -ˌgän\ state NW U.S.; capital, Salem — **Or·e·go·nian** \ˌȯr-i-ˈgō-nē-ən, ˌär-, -nyən\ *adj or n*

Oregon Trail pioneer route to the Pacific Northwest over 2000 miles (3225 kilometers) long from vicinity of Independence, Missouri, to Vancouver, Washington

Orel \ȯ-ˈrel, ȯr-ˈyȯl\ city SW Russia in Europe, S of Moscow

Oren·burg \ˈōr-ən-ˌbərg, ˈȯr-, -ˌbu̇rg\ *or formerly* **Chka·lov** \chə-ˈkäl-əf\ city E Russia in Europe, on Ural River

Ori·no·co \ˌōr-ə-ˈnō-kō, ˌȯr-\ river 1336 miles (2150 kilometers) long Venezuela flowing into North Atlantic

Oris·sa \ō-ˈris-ə\ state E India; capital, Bhubaneswar

Ori·za·ba \ˌōr-ə-ˈzäb-ə, ˌȯr-\ **1** *or* **Ci·tlal·te·petl** \sēt-ˌläl-ˈtā-ˌpet-ᵊl\ inactive volcano 18,700 feet (5700 meters) SE Mexico on Puebla–Veracruz border; highest point in Mexico & third highest in North America **2** city E Mexico in Veracruz state

Ork·ney Islands \ˈȯrk-nē\ islands N Scotland; an administrative area

Or·lan·do \ȯr-ˈlan-dō\ city E central Florida

Or·léans \ȯr-lā-ˈäⁿ\ city N central France

Oru·mi·yeh \ˌu̇r-ü-ˈmē-yə\ city NW Iran

Osa·ka \ō-ˈsäk-ə\ city Japan in S Honshu

Osh·a·wa \ˈäsh-ə-ˌwä\ city Canada in SE Ontario on Lake Ontario ENE of Toronto

Os·lo \ˈäz-lō, ˈäs-\ city, capital of Norway

Os·sa, Mount \ˈäs-ə\ mountain 6490 feet (1967 meters) NE Greece in E Thessaly near Mount Pelion

Ostra·va \ˈō-strə-və\ city E Czech Republic in Moravia

Os·wie·cim \ˌȯsh-ˈfyen-chēm\ *or German* **Ausch·witz** \ˈau̇sh-ˌvits\ town S Poland W of Krakow; site of Nazi concentration camp during World War II

Otran·to, Strait of \ō-ˈtran-tō, ˈō-trən-ˌtō\ strait between SE Italy & Albania

Ot·ta·wa \ˈät-ə-ˌwä, -wə, -ˌwȯ\ city, capital of Canada in SE Ontario on Ottawa River

Oua·ga·dou·gou \ˌwäg-ə-ˈdü-ˌgü\ city, capital of Burkina Faso

Ouj·da \u̇zh-ˈdä\ city NE Morocco near Algerian border

Outer Hebrides — see HEBRIDES

Outer Mongolia — see MONGOLIA

Overland Park city NE Kansas S of Kansas City

Ovie·do \ˌō-vē-ˈā-thō\ city NW Spain

Ox·ford \ˈäks-fərd\ city central England in Oxfordshire

Ox·ford·shire \ˈäks-fərd-ˌshiər, -shər\ *or* **Oxford** county central England

Oxus — see AMU DAR'YA

Ozark Plateau \ˈō-ˌzärk\ eroded tableland N Arkansas, S Missouri, & NE Oklahoma

Pa·cif·ic \pə-ˈsif-ik\ ocean extending from Arctic Circle to Antarctica & from W North America & W

South America to E Asia & Australia; often divided into **North Pacific Ocean** & **South Pacific Ocean**

Pacific Islands, Trust Territory of the former U.S. trust territory W Pacific comprising the Northern Mariana Islands (until 1978), the Federated States of Micronesia (until 1991), the Marshall Islands (until 1991), & Palau (until 1994)

Pacific Rim the countries bordering on or located in the Pacific Ocean -- used especially of Asian countries on the Pacific

Pacific Rim National Park reservation Vancouver Island, British Columbia

Pa·dang \'pä-ˌdäng\ city & port Indonesia in W Sumatra

Pa·dre \'päd-rē, 'pad-\ island about 113 miles (182 kilometers) long S Texas in Gulf of Mexico

Padus — see PO

Pa·go Pa·go \ˌpäng-gō-'päng-gō, ˌpäng-ō-'päng-ō, ˌpäg-ō-'päg-ō\ town, capital of American Samoa on Tutuila Island

Pa·kan·ba·ru \ˌpäk-ən-'bär-ü\ city Indonesia in central Sumatra

Pak·i·stan \'pak-i-ˌstan, ˌpäk-i-'stän\ country S Asia in Indian subcontinent NW of India; until 1971 included also an E division E of India; capital, Islamabad — see EAST PAKISTAN — **Pak·i·stani** \-'stanē, -'stän-ē\ *adj or n*

Pal·at·i·nate \pə-'lat-n-ət\ *or German* **Pfalz** \'pfälts, 'fälts\ either of two districts SW Germany once ruled by counts of the Holy Roman Empire: **Rhenish Palatinate** *or German* **Rhein·pfalz** \'rīn-ˌpfälts, -ˌfälts\ (on the Rhine E of Saarland) & **Upper Palatinate** (on the Danube around Regensburg) — see RHINELAND-PALATINATE

Pa·lau \pə-'lau̇\ *or* **Be·lau** \bə-\ islands W Pacific; since 1994 an independent republic in association with the U.S.; capital, Koror

Pa·la·wan \pə-'lä-wən, -ˌwän\ island W Philippines between South China & Sulu seas

Pa·lem·bang \ˌpäl-əm-'bäng\ city Indonesia in SE Sumatra

Pa·ler·mo \pə-'lər-mō, -'leər-\ city Italy, capital of Sicily

Pal·es·tine \'pal-ə-ˌstīn, -ˌstēn\ ancient region SW Asia bordering on E coast of the Mediterranean and extending E of Jordan River; now approximately coextensive with Israel & the West Bank — **Pal·es·tin·ian** \ˌpal-ə-'stin-ē-ən, -'stin-yən\ *adj or n*

Pa·li·kir \ˌpäl-ē-'kir\ town, capital of Federated States of Micronesia

Pal·ma \'päl-mə\ *or* **Palma de Ma·llor·ca** \-ˌdä-mə-'yȯr-kə, -məl-\ city Spain on Majorca

Palm·dale \'päm-ˌdāl\ city SW California NE of Los Angeles

Palmer Peninsula — see ANTARCTIC PENINSULA

Pa·mirs \pə-'miərz\ elevated mountainous region central Asia in E Tajikistan & on borders of Xinjiang Uygur, Kashmir, & Afghanistan; many peaks over 20,000 feet (6096 meters)

Pam·li·co Sound \'pam-li-ˌkō\ inlet of the North Atlantic E North Carolina between mainland & offshore islands

Pam·plo·na \pam-'plō-nə\ city N Spain in Navarre

Pan·a·ma \'pan-ə-ˌmä, -ˌmȯ, ˌpan-ə-'\ **1** country S Central America **2** *or* **Panama City** city, its capital on North Pacific **3** canal 51 miles (82 kilometers) long Panama connecting North Atlantic & North Pacific oceans — **Pan·a·ma·ni·an** \ˌpan-ə-'mā-nē-ən\ *adj or n*

Panama, Isthmus of *or formerly* **Isthmus of Dar·i·en** \-ˌdar-ē-'en\ strip of land central Panama connecting North America & South America

Panama Canal Zone — see CANAL ZONE

Pa·nay \pə-'nī\ island Philippines in Visayan Islands; chief city, Iloilo

Pan·gaea \pan-'jē-ə\ hypothetical land area believed to have once connected the landmasses of the Southern Hemisphere with those of the Northern Hemisphere

Panjab — see PUNJAB

Pan·mun·jom *or* **Pan·mun·jeom** \ˌpän-ˌmün-'jəm\ village on border between North Korea & South Korea

Paoting — see BAODING

Pao–t'ou — see BAOTOU

Papal States *or* **States of the Church** temporal domain of the popes in central Italy 755–1870

Pa·pee·te \ˌpäp-ē-'āt-ē; pə-'pāt-ē, -'pēt-\ city Society Islands on Tahiti, capital of French Polynesia

Pap·ua, Territory of \'pap-yə-wə, 'päp-ə-wə\ former British territory comprising SE New Guinea & offshore islands; now part of Papua New Guinea

Papua New Guinea country SW Pacific combining former territories of Papua & New Guinea; formerly a United Nations trust territory administered by Australia; independent since 1975; capital, Port Moresby

Pará — see BELEM

Par·a·guay \'par-ə-ˌgwī, -ˌgwā\ **1** river 1584 miles (2549 kilometers) long central South America flowing from Brazil S into the Paraná in Paraguay **2** country central South America; capital, Asunción — **Par·a·guay·an** \ˌpar-ə-'gwī-ən, -'gwā-ən\ *adj or n*

Par·a·mar·i·bo \ˌpar-ə-'mar-ə-ˌbō\ city, capital of Suriname

Pa·ra·ná \ˌpar-ə-'nä\ **1** river about 2500 miles (4022 kilometers) long central South America flowing S from Brazil into Río de la Plata in Argentina **2** city NE Argentina on Paraná River

Par·is \'par-əs\ city, capital of France — **Pa·ri·sian** \pə-'rizh-ən, -'rēzh-\ *adj or n*

Par·ma \'pär-mə\ **1** city NE Ohio S of Cleveland **2** city N Italy in Emilia-Romagna SE of Milan

Par·nas·sus, Mount \pär-'nas-əs\ massif central Greece N of Gulf of Corinth; highest point 8061 feet (2457 meters)

Par·os \'par-ˌäs, 'per-\ island Greece in central Cyclades — **Par·i·an** \'par-ē-ən, 'per-\ *adj*

Par·ra·mat·ta \ˌpar-ə-'mat-ə\ city SE Australia in New South Wales NW of Sydney

Par·thia \'pär-thē-ə\ ancient country SW Asia in NE modern Iran — **Par·thi·an** \-thē-ən\ *adj or n*

Pas·a·de·na \ˌpas-ə-'dē-nə\ city SW California E of Glendale

Pat·a·go·nia \ˌpat-ə-'gō-nyə, -nē-ə\ barren region South America S of about 40° S latitude in S Argentina & S tip of Chile; sometimes considered to include Tierra del Fuego — **Pat·a·go·nian** \-nyən, -nē-ən\ *adj or n*

Pat·er·son \'pat-ər-sən\ city NE New Jersey N of Newark

Pat·mos \'pat-məs\ island Greece in the Dodecanese SSW of Samos

Pat·na \'pət-nə\ city NE India on the Ganges, capital of Bihar

Pat·ras \pə-'tras, 'pa-trəs\ city W Greece in N Peloponnese on **Gulf of Patras** (inlet of Ionian Sea)

Pays de la Loire \ˌpād-lə-'lwär\ region W France containing the point where Loire River empties into the Atlantic

Pearl Harbor inlet of North Pacific in Hawaii on S coast of Oahu W of Honolulu

Peking — see BEIJING

Pe·li·on, Mount \'pē-lē-ən\ mountain 5089 feet (1551 meters) NE Greece in E Thessaly near Mount Ossa

Pel·o·pon·nese \'pel-ə-pə-ˌnēz, -ˌnēs\ *or* **Pel·o·pon·ne·sus** \ˌpel-ə-pə-'nē-səs\ *or* **Pel·o·pon·ne·sos** \-'ne-səs\ peninsula forming S part of mainland of Greece — **Pel·o·pon·ne·sian** \ˌpel-ə-pə-'nē-zhən, -shən\ *adj or n*

Pe·lo·tas \pə-'lōt-əs\ city S Brazil SW of Porto Alegre

Pem·broke Pines \'pem-ˌbrōk\ city SE Florida

Pem·broke·shire \'pem-bru̇k-ˌshiər, -shər\ administrative area SW Wales

Penang — see GEORGE TOWN

Pen·de·li·kón \ˌpen-ˌdel-ē-'kōn\ *or* **Pen·te·li·kon** \ˌpen-ˌtel-ē-'kän\ mountain 3638 feet (1109 meters) E Greece NE of Athens

Peninsular Malaysia territory W Malaysia comprising that part of Malaysia contained on Malay Peninsula

Pen·nine Chain \'pen-ˌīn\ mountains N England; highest Cross Fell 2930 feet (893 meters)

Penn·syl·va·nia \ˌpen-səl-'vā-nyə, -nē-ə\ state E U.S.; capital, Harrisburg

Pen·za \'pen-zə\ city S central Russia in Europe

People's Democratic Republic of Yemen — see YEMEN

People's Republic of China — see CHINA (People's Republic of)

Pe·o·ria \pē-'ōr-ē-ə, -'ȯr-\ city N central Illinois

Per·ga·mum \'pər-gə-məm\ *or* **Per·ga·mus** \-məs\ ancient Greek kingdom including most of Asia Minor; at its height 263–133 B.C.; capital, Pergamum (in what is now W Turkey)

Perm \'pərm, 'peərm\ *or formerly* **Mo·lo·tov** \'mäl-ə-ˌtȯf, 'mȯl-, 'mōl-, -ˌtȯv\ city E Russia in Europe

Pernambuco — see RECIFE

Per·pi·gnan \per-pē-'nyäⁿ\ city S France SE of Toulouse

Per·sep·o·lis \pər-'sep-ə-ləs\ city of ancient Persia; site in SW Iran NE of Shiraz

Persia — see IRAN

Per·sian Gulf \'pər-zhən\ arm of Arabian Sea between Iran & Arabia

Persian Gulf States Kuwait, Bahrain, Qatar, & United Arab Emirates

Perth \'pərth\ city, capital of Western Australia

Perth and Kin·ross \-kin-'rȯs\ administrative area E central Scotland

Pe·ru \pə-'rü\ country W South America; capital, Lima — **Pe·ru·vi·an** \pə-'rü-vē-ən\ *adj or n*

Peru Current *or* **Hum·boldt Current** \'həm-ˌbōlt\ cold current of the South Pacific flowing N & NW along coast of N Chile, Peru, & Ecuador

Pe·ru·gia \pə-'rü-jə, -jē-ə\ city central Italy SE of Florence

Pe·sha·war \pə-'shä-wər, -'shau̇-ər\ city N Pakistan ESE of Khyber Pass

Pe·ta·re \pet-'är-ˌā\ city N Venezuela; a suburb of Caracas

Pe·tra \'pē-trə, 'pe-trə\ ancient city NW Arabia; site in SW Jordan

Petrified Forest National Park reservation E Arizona

Petrograd — see SAINT PETERSBURG

Pe·tro·pav·lovsk \ˌpe-trə-'pav-ˌlȯfsk\ city N Kazakhstan

Pe·tro·za·vodsk \ˌpe-trə-zə-'vätsk\ city NW Russia in Europe, on Lake Onega

Pfalz — see PALATINATE

Phil·a·del·phia \ˌfil-ə-'del-fyə, -fē-ə\ city SE Pennsylvania on Delaware River — **Phil·a·del·phian** \-fyən, -fē-ən\ *adj or n*

Phi·lae \'fī-lē\ former island S Egypt in the Nile above Aswân; now submerged in Lake Nasser

Phil·ip·pines \ˌfil-ə-'pēnz, 'fil-ə-ˌ\ country, an archipelago approximately 500 miles (805 kilometers) off SE coast of Asia; capital, Manila — **Phil·ip·pine** \-'pēn, -ˌpēn\ *adj*

Phi·lis·tia \fə-'lis-tē-ə\ country SW ancient Palestine on the coast; the land of the Philistines

Phnom Penh \pə-'nȯm-'pen, 'nȯm-, pə-'näm-, 'näm-\ city, capital of Cambodia

Phoe·ni·cia \fi-'nish-ə, -'nēsh-, -ē-ə\ ancient country SW Asia on the Mediterranean in modern Syria & Lebanon

Phoe·nix \'fē-niks\ city, capital of Arizona

Phry·gia \'frij-ə, 'frij-e-ə\ ancient country W central Asia Minor

Pia·cen·za \pyä-'chen-sə, ˌpē-ə-'chen-\ city N Italy SE of Milan

Pic·ar·dy \'pik-ərd-ē\ *or French* **Pi·car·die** \pē-kȧr-dē\ region & former province N France N of Normandy

Pied·mont \'pēd-ˌmänt\ **1** plateau region E U.S. E of the Appalachians between SE New York & NE Alabama **2** *or Italian* **Pie·mon·te** \pyā-'mōn-tā\ region NW Italy W of Lombardy; capital, Turin — **Pied·mon·tese** \ˌpēd-mən-'tēz, -män-, -'tēs\ *adj or n*

Pierre \'piər\ city, capital of South Dakota

Pie·ter·mar·itz·burg \ˌpēt-ər-'mar-əts-ˌbərg\ city E Republic of South Africa in KwaZulu-Natal

Pigs, Bay of bay W Cuba on S coast

Pikes Peak \'pīks\ mountain 14,110 feet (4301 meters) E central Colorado in a range of the Rockies

Pillars of Her·cu·les \'hər-kyə-ˌlēz\ two promontories at E end of Strait of Gibraltar: Rock of Gibraltar (in Europe) & Jebel Musa (in Africa)

Pin·dus \'pin-dəs\ mountains W Greece W of Thessaly; highest point 8136 feet (2480 meters)

Pinkiang — see HARBIN

Pi·rae·us \pī-'rē-əs\ *or Greek* **Pi·rai·évs** \ˌpē-rē-'efs\ city E Greece on Saronic Gulf; port for Athens

Pi·sa \'pē-zə, *Italian* 'pē-sä\ city W central Italy W of Florence

Pit·cairn \'pit-ˌkaərn, -ˌkeərn\ island South Pacific SE of Tuamotu Archipelago; a British colony

Pitts·burgh \'pits-ˌbərg\ city SW Pennsylvania

\ə\ abut \ᵊ\ kitten \ər\ further \a\ ash \ā\ ace \ä\ mop, mar \au̇\ out \ch\ chin \e\ bet \ē\ easy \g\ go \i\ hit \ī\ ice \j\ job \ŋ\ sing \ō\ go \ȯ\ law \ȯi\ boy \th\ thin \th\ the \ü\ loot \u̇\ foot \y\ yet \zh\ vision *see also* Guide to Pronunciation

Pla·no \'plän-ō\ city NE Texas N of Dallas
Pla·ta, Río de la \ˌrē-ō-ˌdel-ə-'plät-ə\ estuary of Paraná & Uruguay rivers between Uruguay & Argentina
Plov·div \'plòv-ˌdif, -ˌdiv\ city S Bulgaria
Plym·outh \'plim-əth\ city & port SW England
Po \'pō\ *or ancient* **Pa·dus** \'pād-əs\ river 405 miles (652 kilometers) N Italy flowing into the Adriatic
Po Hai — see BO HAI
Point Pe·lee National Park \ˌpòint-'pē-lē\ reservation SE Ontario on Point Pelee (cape projecting into Lake Erie)
Poi·tou–Cha·rentes \pwä-'tü-shä-'rä^{n}t\ region W France on the Atlantic
Po·land \'pō-lənd\ country central Europe on Baltic Sea; capital, Warsaw
Pol·y·ne·sia \ˌpäl-ə-'nē-zhə, -shə\ islands of the central & S Pacific including Hawaii, the Line, Tuvalu, Phoenix, Tonga, Cook, & Samoa Islands, French Polynesia, & often New Zealand
Pom·er·a·nia \ˌpäm-ə-'rā-nē-ə, -'rā-nyə\ region N Europe on Baltic Sea; formerly in Germany, now mostly in Poland
Po·mo·na \pə-'mō-nə\ city SW California E of Los Angeles
Pom·peii \päm-'pā, -'pā-ˌē\ ancient city S Italy SE of Naples destroyed 79 A.D. by eruption of Vesuvius — **Pom·pe·ian** \-'pā-ən\ *adj or n*
Po·na·pe \'pō-nə-ˌpā\ island W Pacific in the E Carolines
Pon·ce \'pòn-sā\ city & port S Puerto Rico
Pon·di·cher·ry \ˌpän-də-'cher-ē, -'sher-\ territory SE India SSW of Madras; a settlement of French India before 1954
Pon·ta Del·ga·da \ˌpänt-ə-del-'gäd-ə, -'gad-\ city Portugal in the Azores on São Miguel Island
Pont·char·train, Lake \'pän-chər-ˌtrān, ˌpän-chər-'\ lake SE Louisiana E of the Mississippi & N of New Orleans
Pon·ti·a·nak \ˌpän-tē-'ä-ˌnäk\ city Indonesia on SW coast of Borneo
Pon·tine Marshes \'pän-ˌtīn, -ˌtēn\ district central Italy in SW Latium; marshes now reclaimed
Pon·tus \'pänt-əs\ **1** ancient country & Roman province NE Asia Minor **2** *or* **Pontus Euxinus** — see BLACK SEA — **Pon·tic** \'pänt-ik\ *adj or n*
Poole \'pül\ town S England on English Channel
Pool Malebo — see MALEBO (Pool)
Poona — see PUNE
Po·po·ca·te·petl \ˌpō-pə-ˌkat-ə-'pet-l\ volcano 17,887 feet (5452 meters) SE central Mexico in Puebla
Port Arthur — see LUSHUN
Port–au–Prince \ˌpōrt-ō-'prins, ˌport-, -'prans, -'pra^{n}s\ city, capital of Haiti
Port Eliz·a·beth \i-'liz-ə-bəth\ city S Republic of South Africa in Eastern Cape
Port Jack·son \'jak-sən\ inlet of South Pacific SE Australia in New South Wales; harbor of Sydney
Port·land \'pōrt-lənd, 'pòrt-\ city NW Oregon
Port Lou·is \'lü-əs, 'lü-ē, lü-'ē\ city, capital of Mauritius
Port Lyautey — see KENITRA
Port Mores·by \'mōrz-bē, 'mòrz-\ city, capital of Papua New Guinea
Por·to \'pòr-tü\ *or* **Opor·to** \ō-'pòr-tü\ city & port NW Portugal
Por·to Ale·gre \ˌpōrt-ō-ə-'leg-rə, ˌpòrt-\ city S Brazil
Port of Spain city NW Trinidad, capital of Trinidad and Tobago
Por·to–No·vo \ˌpōrt-ə-'nō-vō, ˌpòrt-\ city, capital of Benin
Porto Rico — see PUERTO RICO
Port Phil·lip Bay \'fil-əp\ inlet of South Pacific SE Australia in Victoria; harbor of Melbourne
Port Said \sä-'ēd, 'sīd\ city NE Egypt on the Mediterranean at N end of Suez Canal
Ports·mouth \'pōrt-sməth, 'pòrt-\ **1** city SE Virginia **2** city S England on an island in English Channel
Port Su·dan \sü-'dan, -'dän\ city NE Sudan on Red Sea
Por·tu·gal \'pōr-chi-gəl, 'pòr-\ *or ancient* **Lu·si·ta·nia** \ˌlü-sə-'tā-nē-ə, -nyə\ country SW Europe; capital, Lisbon
Portuguese East Africa — see MOZAMBIQUE
Portuguese Guinea — see GUINEA-BISSAU
Portuguese India former Portuguese possession on W coast of India including Goa, Daman, & Diu; annexed to India 1962
Portuguese West Africa — see ANGOLA
Port–Vi·la \ˌpòr-'vē-lä, -vē-'lä\ *or* **Vi·la** \'vē-lä, vē-'lä\ seaport, capital of Vanuatu
Po·ten·za \pə-'tent-sə; -'ten-sə, -zə\ city S Italy
Po·to·mac \pə-'tō-mək, -mik\ river 287 miles (462 kilometers) long flowing from West Virginia into Chesapeake Bay & forming boundary between Maryland & Virginia
Po·wys \'pō-əs\ administrative area E central Wales
Poz·nan \'pōz-ˌnan-yə, 'pòz-, -ˌnän-yə, -ˌnan, -ˌnän\ city W central Poland
Prague \'präg\ *or Czech* **Pra·ha** \'prä-hä\ city, capital of Czech Republic and formerly of Czechoslovakia
Praia \'prī-ə\ town, capital of Cape Verde
Prairie Provinces the Canadian provinces of Alberta, Manitoba, & Saskatchewan
Pres·ton \'pres-tən\ city NW England
Pre·to·ria \pri-'tōr-ē-ə, -'tòr-\ city Republic of South Africa, administrative capital of the country
Prib·i·lof \'prib-ə-ˌlòf\ islands Alaska in Bering Sea
Prince Al·bert National Park \'al-bərt\ reservation Canada in central Saskatchewan
Prince Ed·ward Island \ˌed-wərd\ island SE Canada in Gulf of Saint Lawrence; a province; capital, Charlottetown
Prince Edward Island National Park reservation on N coast of Prince Edward Island
Prince Ru·pert's Land \'rü-pərts\ historical region N & W Canada comprising drainage basin of Hudson Bay granted 1670 by King Charles II to Hudson's Bay Company
Prín·ci·pe \'prin-sə-pə\ island W Africa in Gulf of Guinea — see SAO TOME
Pro·ko·pyevsk \prə-'kòp-yəfsk\ city S Russia in Asia, NW of Novokuznetsk
Propontis — see MARMARA (Sea of)
Pro·vence \prə-'vä^{n}s\ region & former province SE France on the Mediterranean
Prov·i·dence \'präv-əd-əns, -ə-ˌdens\ city, capital of Rhode Island
Pro·vo \'prō-vō\ city N central Utah
Prud·hoe Bay \'prüd-ō, 'prəd-\ inlet of Beaufort Sea N Alaska
Prus·sia \'prəsh-ə\ former kingdom &, later, state Germany; capital, Berlin — **Prus·sian** \-ən\ *adj or n*
Pu·chon \'pü-ˌchòn\ *or* **Bu·cheon** \'bü-ˌchən\ city NW South Korea
Pu·eb·la \pü-'eb-lə, 'pweb-, pyü-'eb-\ **1** state SE central Mexico **2** city, its capital
Puer·to Ri·co \ˌpōrt-ə-'rē-kō, ˌpòrt-, ˌpwert-\ *or formerly* **Por·to Ri·co** \ˌpōrt-, ˌpòrt-\ island West Indies E of Hispaniola; a self-governing commonwealth associated with U.S.; capital, San Juan — **Puer·to Ri·can** \-'rē-kən\ *adj or n*
Pu·get Sound \ˌpyü-jət-\ arm of North Pacific W Washington
Pu·glia \'pül-yä\ *or* **Apu·lia** \ä-'pül-yä\ *or* **Le Pu·glie** \lə-pül-'yā\ region SE Italy bordering on the Adriatic Sea & Gulf of Taranto; capital, Bari
Pu·kas·kwa National Park \pü-'käs-kwə\ reservation Ontario bordering on Lake Superior
Pu·na de Ata·ca·ma \'pü-nə-ˌdā-ˌat-ə-'käm-ə, -ˌät-\ plateau region NW Argentina NW of San Miguel de Tucumán
Pu·ne *or* **Poo·na** \'pü-nə\ city W India in Maharashtra ESE of Bombay
Pun·jab *or* **Pan·jab** \ˌpən-'jäb, -'jab, 'pən-ˌ\ **1** region NW Indian subcontinent in Pakistan & NW India in valley of the Indus **2** *or* **Pun·jabi Su·ba** \ˌpən-ˌjäb-ē-'sü-bə, -ˌjab-\ state NW India in E Punjab region; capital, Chandigarh — see HARYANA **3** *or formerly* **West Punjab** province NE Pakistan
Pu·rus \pə-'rüs\ river about 2000 miles (3219 kilometers) long NW central South America in SE Peru & NW Brazil flowing into the Amazon
Pu·san \'pü-ˌsän\ *or* **Bu·san** \'bü-\ city South Korea on Korea Strait
Pyong·yang \pē-'òng-ˌyäng, pē-'əng-, -ˌyang\ city, capital of North Korea
Pyr·e·nees \'pir-ə-ˌnēz\ mountains on French–Spanish border extending from Bay of Biscay to the Mediterranean; highest Pico de Aneto (Pic de Néthou) 11,168 feet (3404 meters)
Qa·tar \'kät-ər, 'gät-, 'gət-\ country E Arabia on peninsula extending into Persian Gulf; an independent emirate; capital, Doha
Qing·dao \'ching-'daù\ *or* **Tsing·tao** \'ching-'daù, 'tsing-, 'sing-\ city E China in E Shandong
Qing·hai *or* **Tsing·hai** \'ching-'hī\ province W China W of Gansu; capital, Xining
Qi·qi·har \'chē-'chē-'här\ *or* **Ch'i–ch'i–ha–erh** \'chē-'chē-'hä-'ər\ city NE China in W Heilongjiang
Qom \'kùm\ city NW central Iran
Quan·zhou *or* **Ch'üan–chou** *or* **Chuan·chow** \chə-'wän-'jō\ city SE China in Fujian on Taiwan Strait
Que·bec \kwi-'bek, ki-\ *or French* **Qué·bec** \kā-bek\ **1** province E Canada **2** city, its capital, on the Saint Lawrence — **Que·bec·er** *or* **Que·beck·er** \kwi-'bek-ər, ki-\ *n*
Queens \'kwēnz\ borough of New York City on Long Island E of Brooklyn
Queen's — see LAOIGHIS
Queens·land \'kwēnz-ˌland, -lənd\ state NE Australia; capital, Brisbane — **Queens·land·er** \-ər\ *n*
Que·moy \kwi-'mòi, ki-, 'kwē-ˌ\ island E China in Taiwan Strait
Que·ré·ta·ro \kə-'ret-ə-ˌrō\ **1** state central Mexico **2** city, its capital
Quet·ta \'kwet-ə\ city Pakistan in N Baluchistan
Que·zon City \'kā-ˌsòn\ city Philippines in Luzon NE of Manila; former (1948–76) official capital of the Philippines
Quil·mes \'kēl-ˌmās, -ˌmes\ city E Argentina SE of Buenos Aires
Quin·ta·na Roo \kēn-ˌtän-ə-'rō\ state SE Mexico in E Yucatán; capital, Chetumal
Qui·to \'kē-tō\ city, capital of Ecuador
Qŭ·qon \kə-'kän\ *or* **Ko·kand** \kə-'känt\ city E Uzbekistan SE of Tashkent
Ra·bat \rə-'bät\ city, capital of Morocco
Ra·dom \'räd-ˌòm\ city E central Poland
Rai·nier, Mount \rə-'niər, rā-\ mountain 14,410 feet (4392 meters) W central Washington in **Mount Rainier National Park**; highest in Cascade Mountains
Ra·ja·sthan \'räj-ə-ˌstän\ **1** state NW India bordering on Pakistan; capital, Jaipur **2** RAJPUTANA
Raj·kot \'räj-ˌkōt\ city W India in Gujarat
Raj·pu·ta·na \ˌräj-pə-'tän-ə\ region NW India S of Punjab now largely included in Rajasthan state
Ra·leigh \'rò-lē, 'räl-ē\ city, capital of North Carolina
Ran·chi \'rän-chē\ city E India NW of Calcutta
Ran·cho Cu·ca·mon·ga \'ran-chō-ˌkü-kə-'məng-gə\ city SW California
Rand — see WITWATERSRAND
Range·ley Lakes \'rānj-lē\ chain of lakes W Maine & N New Hampshire
Rangoon — see YANGON
Rasht \'rasht\ city NW Iran
Rat \'rat\ islands SW Alaska in W Aleutians
Ra·wal·pin·di \ˌrä-wəl-'pin-dē, raùl-'pin-, ròl-'pin-\ city NE Pakistan NNW of Lahore
Read·ing \'red-ing\ town S England
Re·ci·fe \rə-'sē-fə\ *or formerly* **Per·nam·bu·co** \ˌpər-nəm-'bü-kō, -'byü-, ˌper-nəm-'bü-\ city NE Brazil
Red \'red\ **1** river 1018 miles (1638 kilometers) long flowing E on Oklahoma–Texas boundary & into the Atchafalaya & the Mississippi in Louisiana **2** sea between Arabia & NE Africa
Red·bridge \'red-brij\ borough of NE Greater London, England
Red·wood National Park \'red-ˌwùd\ reservation NW California
Reg·gio \'rej-ō, 'rej-ē-ˌō\ **1** *or* **Reggio di Ca·la·bria** \-ˌdē-kə-'läb-rē-ə\ city S Italy on Strait of Messina **2** *or* **Reggio nel·l'Emi·lia** \-ˌnel-ə-'mēl-yə\ city N Italy NW of Bologna
Re·gi·na \ri-'jī-nə\ city, capital of Saskatchewan
Reims *or* **Rheims** \'rēmz, *French* 'ra^{n}s\ city NE France ENE of Paris
Ren·frew·shire \-ˌshiər, -shər\ administrative area SW Scotland
Rennes \'ren\ city NW France
Ré·union \rē-'yün-yən\ island W Indian Ocean E of Madagascar; an overseas department of France; capital, Saint-Denis
Revel — see TALLINN
Rey·kja·vik \'rāk-yə-ˌvik, -ˌvēk\ city, capital of Iceland
Rey·no·sa \rā-'nōs-ə\ city NE Mexico in Tamaulipas
Rheinpfalz — see PALATINATE
Rhine *or German* **Rhein** \'rīn\ *or French* **Rhin** \'ra^n\ *or Dutch* **Rijn** \'rīn\ *or ancient* **Rhe·nus** \'rē-nəs\ river 820 miles (1320 kilometers) long W Europe

flowing from SE Switzerland to North Sea in the Netherlands — **Rhen·ish** \ˈren-ish, ˈrē-nish\ *adj or n*

Rhine·land \ˈrīn-ˌland, -lənd\ *or German* **Rhein·land** \ˈrīn-ˌlänt\ the part of Germany W of the Rhine — **Rhine·land·er** \ˈrīn-ˈlan-dər, -lən-\ *n*

Rhineland–Palatinate *or German* **Rhein·land–Pfalz** \-ˈpfälts, -ˈfälts\ state W Germany chiefly W of the Rhine; capital, Mainz

Rhode Is·land \rō-ˈdī-lənd\ **1** *or officially* **Rhode Island and Providence Plantations** state NE U.S.; capital, Providence **2** — see AQUIDNECK ISLAND — **Rhode Is·land·er** \-lən-dər\ *n*

Rhodes \ˈrōdz\ **1** island Greece in the SE Aegean; chief island of the Dodecanese **2** city, its capital

Rhodesia — see ZIMBABWE

Rhon·dda \ˈrän-də, ˈrän-thə, ˈhrän-thə\ town SE Wales NW of Cardiff

Rhondda Cy·non Taff \ˈkən-ən-ˈtaf\ administrative area S Wales

Rhone *or French* **Rhône** \ˈrōn\ *or ancient* **Rhod·a·nus** \ˈräd-n-əs\ river 505 miles (813 kilometers) long Switzerland & SE France

Rhône–Alpes \-ˈälp\ region E France bordering on Switzerland & Italy

Ri·bei·rão Prê·to \ˌrē-və-ˈrau̇^{n}-ˈprā-tü\ city SE Brazil NNW of São Paulo

Rich·mond \ˈrich-mənd\ **1** — see STATEN ISLAND 2 **2** city, capital of Virginia **3** *or* **Rich·mond upon Thames** borough of SW Greater London, England

Rid·ing Mountain National Park \ˌrīd-iŋ\ reservation Canada in SW Manitoba

Rift Valley GREAT RIFT VALLEY

Ri·ga \ˈrē-gə\ city, capital of Latvia

Ri·je·ka *or* **Ri·e·ka** \rē-ˈek-ə, -ˈyek-\ *or Italian* **Fiu·me** \ˈfyü-ˌmā, fē-ˈü-\ city W Croatia

Rio \ˈrē-ō\ RIO DE JANEIRO

Rio de Ja·nei·ro \ˈrē-ō-ˌdā-zhə-ˈner-ō, -ˌdē-, -də-, -jə ˈner-\ city SE Brazil on Guanabara Bay

Río de Oro \ˌrē-ōd-ē-ˈōr-o, -ˈȯr-\ territory NW Africa comprising S zone of Western Sahara

Rio Grande \ˌrē-ō-ˈgrand, -ˈgrand-ē\ *or Mexican* **Río Bra·vo** \-ˈbräv-ō\ river 1885 miles (3034 kilometers) long SW U.S. forming part of U.S.–Mexico boundary & flowing into Gulf of Mexico

Río Muni — see MBINI

Riv·er·side \ˈriv-ər-ˌsīd\ city S California

Riv·i·era \ˌriv-ē-ˈer-ə\ coast region SE France & NW Italy

Ri·yadh \rē-ˈyäd\ city, capital of Saudi Arabia

Ro·a·noke \ˈrō-ə-ˌnōk, ˈrō-ˌnōk\ island North Carolina S of entrance to Albemarle Sound

Rob·son, Mount \ˈräb-sən\ mountain 12,972 feet (3954 meters) W Canada in E British Columbia; highest in the Canadian Rockies

Roch·es·ter \ˈräch-ə-stər, ˈräch-ˌes-tər\ city W New York

Rock·ford \ˈräk-fərd\ city N Illinois NW of Chicago

Rocky Mountains \ˈräk-ē\ *or* **Rock·ies** \ˈräk-ēz\ mountains W North America extending SE from N Alaska to central New Mexico — see ELBERT (Mount), ROBSON (Mount)

Rocky Mountain National Park reservation N Colorado

Ro·ma·nia \ru̇-ˈmā-nē-ə, rō-\ *or* **Ru·ma·nia** \ru̇-ˈmā-nē-ə\ country SE Europe on Black Sea; capital, Bucharest

Rome \ˈrōm\ **1** *or Italian* **Ro·ma** \ˈrō-mä\ city, capital of Italy **2** the Roman Empire

Ron·ces·va·lles \ˌrȯn-səs-ˈvī-əs\ commune N Spain in the Pyrenees

Roo·de·poort \ˈrōd-ə-ˌpȯrt\ city NE Republic of South Africa in Gauteng

Ro·sa·rio \rō-ˈzär-ē-ˌō, -ˈsär-\ city E central Argentina on the Paraná

Ros·com·mon \rä-ˈskäm-ən\ county central Ireland in Connacht

Ro·seau \rō-ˈzō\ seaport, capital of Dominica

Ross Sea \ˈrȯs\ arm of South Pacific extending into Antarctica E of Victoria Land

Ros·tock \ˈräs-ˌtäk, ˈrȯ-ˌstȯk\ city NE Germany near Baltic coast

Ros·tov \rə-ˈstȯf, -ˈstȯv\ city S Russia in Europe on the Don

Rot·ter·dam \ˈrät-ər-ˌdam\ city & port SW Netherlands

Rou·baix \rü-ˈbā\ city N France NE of Lille

Rou·en \rü-ˈä^{n}, rü-ˈä^{n}n\ city N France on the Seine

Ru·an·da–Urun·di \ru̇-ˌän-də-ü-ˈrün-dē\ former trust territory E central Africa bordering on Lake Tanganyika and administered by Belgium; divided 1962 into independent nations of Burundi (formerly Urundi) & Rwanda (formerly Ruanda)

Rudolf, Lake — see TURKANA (Lake)

Ruhr \ˈru̇r\ industrial district W Germany E of the Rhine in valley of **Ruhr River** (146 miles or 235 kilometers long flowing NW & W into the Rhine)

Ru·me·lia \rü-ˈmēl-yə, -ˈmē-lē-ə\ a division of the old Ottoman Empire including Albania, Macedonia, & Thrace

Run·ny·mede \ˈrən-ē-ˌmēd\ meadow S England in Surrey on S bank of the Thames where Magna Carta was signed 1215

Rupert's Land PRINCE RUPERT'S LAND

Ru·se \ˈrü-sā\ city NE Bulgaria on the Danube

Rush·more, Mount \ˈrəsh-ˌmōr, -ˌmȯr\ mountain 5600 feet (1707 meters) W South Dakota in Black Hills SW of Rapid City

Rus·sia \ˈrəsh-ə\ **1** former empire largely coextensive with later U.S.S.R.; capital, Petrograd (Saint Petersburg) **2** UNION OF SOVIET SOCIALIST REPUBLICS **3** country N Asia (**Russia in Asia**) & E Europe (**Russia in Europe**) bordering on Arctic & North Pacific oceans & Baltic & Black seas; capital, Moscow; a constituent republic (**Russian Soviet Federated Socialist Republic** *or* **Soviet Russia**) of U.S.S.R. 1922–91

Ru·the·nia \rü-ˈthē-nyə, -nē-ə\ *or* **Car·pa·thi·an Ruthenia** \kär-ˈpā-thē-ən\ region W Ukraine W of the N Carpathians — **Ru·the·nian** \rü-ˈthē-nyən, -nē-ən\ *adj or n*

Ru·wen·zo·ri \ˌrü-ən-ˈzōr-ē, -ˈzȯr-\ mountain group E central Africa between Uganda & Democratic Republic of the Congo; highest Margherita Peak (highest peak of Mount Stanley) 16,763 feet (5109 meters)

Rwan·da *or formerly* **Ru·an·da** \rü-ˈän-də\ country E central Africa, until 1962 part of Ruanda–Urundi trust territory; capital, Kigali — **Rwan·dan** \rü-ˈän-dən\ *adj or n*

Rya·zan \ˌrē-ə-ˈzan-yə, -ˈzan\ city W Russia in Europe SE of Moscow

Ry·binsk \ˈrib-ənsk\ *or formerly* **Shcher·ba·kov** \ˌshcher-bə-ˈkȯf, ˌsher-, -ˈkȯv\ city central Russia in Europe NNE of Moscow

Ryu·kyu \rē-ˈü-kyü, -ˈyü-, -kü\ islands Japan extending in an arc from Kyushu, Japan, to Taiwan, China — **Ryu·kyu·an** \-kyü-ən, -kü-ən\ *adj or n*

Saar \ˈsär, ˈzär\ **1** river about 150 miles (241 kilometers) long Europe flowing from Vosges Mountains in E France into the Moselle in Germany **2** *or* **Saar·land** \ˈsär-ˌland, ˈzär-\ district W Europe in valley of Saar River; a state of W Germany; capital, Saarbrücken

Sa·ba \ˈsäb-ə\ island West Indies in Netherlands Antilles; capital, The Bottom

Sachsen — see SAXONY

Sac·ra·men·to \ˌsak-rə-ˈment-ō\ **1** river 382 miles (615 kilometers) long N California flowing S into Suisun Bay **2** city, capital of California

Sa·ga·mi Sea \sə-ˈgäm-ē\ inlet of North Pacific in central Honshu, Japan

Sa·ga·mi·ha·ra \sə-ˌgäm-ē-ˈhär-ə\ city Japan on Honshu

Saguaro National Park reservation SE Arizona E of Tucson

Sag·ue·nay \ˈsag-ə-ˌnā, ˌsag-ə-ˈ\ river 105 miles (169 kilometers) long Canada in S Quebec flowing from Lake Saint John E into the Saint Lawrence

Sa·hara \sə-ˈhar-ə, -ˈher-, -ˈhär-\ desert region N Africa N of Sudan region extending from North Atlantic coast to Red Sea or, as sometimes considered, to the Nile — **Sa·har·an** \-ən\ *adj*

Saigon — see HO CHI MINH CITY

Saint Ber·nard \ˌsānt-bər-ˈnärd, -bə-\ either of two mountain passes in the Alps: the **Great Saint Bernard** (8090 feet or 2468 meters between Italy & Switzerland E of Mont Blanc) & the **Little Saint Bernard** (7178 feet or 2188 meters between France & Italy S of Mont Blanc)

Saint Cath·a·rines \ˈkath-rənz, -ə-rənz\ city Canada in SE Ontario

Saint Christopher — see SAINT KITTS

Saint Clair, Lake \ˈklaər, ˈkleər\ lake SE Michigan & SE Ontario connected by **Saint Clair River** (about 40 miles or 64 kilometers long) with Lake Huron and draining by Detroit River into Lake Erie

Saint Croix \sānt-ˈkrȯi, sənt-\ **1** river 129 miles (208 kilometers) long Canada & U.S. on border between New Brunswick & Maine **2** island West Indies; largest of Virgin Islands of the U.S.

Saint Eli·as, Mount \ˌsānt-l-ˈī-əs\ mountain 18,008 feet (5489 meters) on Alaska–Yukon boundary in **Saint Elias Mountains** (range of the Coast Ranges)

Saint George's \ˈjȯr-jəz\ **1** channel British Isles between SW Wales & Ireland **2** town, capital of Grenada

Saint Gott·hard \sānt-ˈgät-ərd, -ˈgäth-, sənt-, ˌsa^{n}-gə-ˈtär\ **1** pass S central Switzerland in Saint Gotthard Range of the Alps **2** tunnel 3.5 miles (5.6 kilometers) long near the pass

Saint He·le·na \ˌsānt-l-ˈē-nə, ˌsānt-hə-ˈlē-\ island South Atlantic; a British colony; capital, Jamestown

Saint Hel·ens \sānt-ˈhel-ənz, sənt-\ town NW England ENE of Liverpool

Saint Helens, Mount volcano about 8366 feet (2550 meters) SW Washington

Saint John \sānt-ˈjän, sənt-\ city Canada in S New Brunswick

Saint John's \sānt-ˈjänz, sənt-\ **1** city, capital of Antigua and Barbuda **2** city & port Canada, capital of Newfoundland and Labrador on SE Newfoundland

Saint Kitts \ˈkits\ *or* **Saint Chris·to·pher** \ˈkris-tə-fər\ island British West Indies in the Leewards; with Nevis constitutes country of **Saint Kitts and Nevis**; capital, Basseterre (on Saint Kitts)

Saint Law·rence \sānt-ˈlȯr-əns, sənt-, -ˈlär-\ **1** river 760 miles (1223 kilometers) long E Canada in Ontario & Quebec bordering on U.S. in New York and flowing from Lake Ontario NE into the **Gulf of Saint Lawrence** (inlet of North Atlantic) **2** seaway Canada & U.S. in & along the Saint Lawrence between Lake Ontario & Montreal

Saint Lawrence Islands National Park reservation SE Ontario

Saint Lou·is \sānt-ˈlü-əs, sənt-\ city E Missouri on the Mississippi

Saint Lu·cia \sānt-ˈlü-shə, sənt-\ island British West Indies in the Windwards S of Martinique; an independent country; capital, Castries

Saint Mo·ritz \ˌsānt-mə-ˈrits, ˌsa^{n}-\ town E Switzerland

Saint Paul \ˈpȯl\ city, capital of Minnesota

Saint Pe·ters·burg \ˈpēt-ərz-ˌbərg\ **1** city W Florida **2** *or 1914–24* **Pet·ro·grad** \ˈpe-trə-ˌgrad\ *or 1924–91* **Le·nin·grad** \ˈlen-ən-ˌgrad\ city W Russia, on Gulf of Finland

Saint Pierre \sānt-ˈpiər, sənt-, -pē-ˈeər, *French* sa^{n}-ˈpyer\ **1** island in North Atlantic off S Newfoundland; with nearby island of Miquelon constitutes French territory of **Saint Pierre and Miquelon** **2** town, capital of Saint Pierre and Miquelon

Saint Thom·as \ˈtäm-əs\ **1** island West Indies, one of Virgin Islands of the U.S.; chief town, Charlotte Amalie **2** — see SAO TOME

Saint–Tro·pez \ˌsa^{n}-trō-ˈpā\ town SE France on the Mediterranean

Saint Vin·cent \sānt-ˈvin-sənt, sənt-\ island British West Indies in the central Windwards; with N Grenadines constitutes independent country of **Saint Vincent and the Grenadines**; capital, Kingstown (on Saint Vincent)

Sai·pan \sī-ˈpan, -ˈpän, ˈsī-ˌ\ island W Pacific in S central Mariana Islands; contains capital of Northern Mariana Islands

Sa·kai \sä-ˈkī, ˈsä-\ city Japan in S Honshu on Osaka Bay

Sa·kha·lin \ˈsak-ə-ˌlēn, -lən; ˌsak-ə-ˈlēn\ *or formerly* **Sa·ghal·ien** \ˈsag-ə-ˌlēn, ˌsag-ə-ˈ\ *or Japanese* **Ka·ra·fu·to** \kə-ˈräf-ə-ˌtō\ island Russia in W Pacific N of Hokkaido, Japan; until 1945 divided between Japan & U.S.S.R.

Sa·la·do \sə-ˈläd-ō\ river 1120 miles (1802 kilometers) N Argentina flowing from the Andes SE into the Paraná

Sal·a·man·ca \ˌsal-ə-ˈmaŋ-kə, ˌsäl-ə-ˈmäŋ-\ city W Spain WNW of Madrid

Sal·a·mis \ˈsal-ə-məs\ **1** ancient city Cyprus on E coast **2** island Greece in Saronic Gulf off Attica

Sa·lé \sal-ˈā\ city & port NW Morocco

Sa·lem \ˈsā-ləm\ **1** city, capital of Oregon **2** city S India in N Tamil Nadu SW of Madras

\ə\ abut \ᵊ\ kitten \ər\ **further** \a\ **ash** \ā\ **ace** \ä\ **mop, mar** \au̇\ **out** \ch\ **chin** \e\ **bet** \ē\ **easy** \g\ **go** \i\ **hit** \ī\ **ice** \j\ **job** \ŋ\ **sing** \ō\ **go** \ȯ\ **law** \ȯi\ **boy** \th\ **thin** \t͟h\ **the** \ü\ **loot** \u̇\ **foot** \y\ **yet** \zh\ **vision** *see also* Guide to Pronunciation

Sa•ler•no \sə-'lər-nō, -'leər-\ city S Italy on Gulf of Salerno
Sal•ford \'sȯl-fərd\ city NW England adjacent to Manchester
Sa•li•nas \sə-'lē-nəs\ city W California
Salisbury — see HARARE
Salonika — see THESSALONIKI
Salop — see SHROPSHIRE
Sal•ta \'säl-tə\ city NW Argentina
Salt Lake City city, capital of Utah
Sal•ton Sea \'sȯlt-n\ saline lake SE California
Sal•va•dor \'sal-və-,dȯr, ,sal-və-'\ *or formerly* **São Salvador** \saůⁿ-\ *or* **Ba•hia** \bä-'ē-ə\ city NE Brazil on South Atlantic — **Sal•va•dor•an** \,sal-və-'dōr-ən, -'dȯr-\ *or* **Sal•va•do•re•an** *or* **Sal•va•do•ri•an** \-ē-ən\ *adj or n*
Sal•ween \'sal-,wēn\ river about 1500 miles (2415 kilometers) long SE Asia flowing from Tibet S into Bay of Bengal in Myanmar
Salz•burg \'sȯlz-,bərg, 'sälz-, 'salz-, 'sȯlts-, -,bůrg, *German* 'zälts-,bůrk\ city W Austria ESE of Munich, Germany
Sa•mar \'säm-,är\ island central Philippines in Visayan Islands
Sa•ma•ra \sə-'mär-ə\ *or 1935–91* **Kuy•by•shev** \'kwē-bə-,shef, 'kü-ē-, -,shev\ city E Russia in Europe, on the Volga
Sa•mar•ia \sə-'mer-ē-ə, -'mar-\ **1** district of ancient Palestine W of the Jordan between Galilee & Judea **2** ancient city, its capital & capital of the Northern Kingdom (Israel)
Sam•a•rin•da \,sam-ə-'rin-də\ city Indonesia in E Borneo
Sam•ar•qand *or* **Sam•ar•kand** \'sam-ər-,kand\ city E Uzbekistan
Sam•ni•um \'sam-nē-əm\ ancient country S central peninsula of Italy SE of Latium — **Sam•nite** \'sam-,nīt\ *adj or n*
Sa•moa \sə-'mō-ə\ **1** islands SW central Pacific N of Tonga Islands; divided at longitude 171° W into American, or Eastern, Samoa & independent Samoa **2** *or formerly* **Western Samoa** islands W of American Samoa; an independent country; capital, Apia — **Sa•mo•an** \-ən\ *adj or n*
Sa•mos \'sā-,mäs\ island Greece in the Aegean off coast of Turkey N of the Dodecanese — **Sa•mi•an** \-mē-ən\ *adj or n*
Sam•o•thrace \'sam-ə-,thrās\ island Greece in the NE Aegean
San•aa *or* **San•'a** \'san-,ä, sän-'ä\ city SW Arabia, capital of Yemen & formerly of Yemen Arab Republic
San An•dre•as Fault \,san-an-'drā-əs\ zone of faults California extending from N coast SE toward head of Gulf of California
San An•to•nio \,san-ən-'tō-nē-,ō\ city S Texas
San Ber•nar•di•no \,san-,bər-nə-'dē-nō, -nər-'dē-\ city SW California E of Los Angeles
San Cris•tó•bal \,san-kris-'tō-bəl\ city W Venezuela SSW of Lake Maracaibo
Sanc•ti Spi•ri•tus \,säng-tē-'spir-ə-,tüs, ,sängk-\ city W central Cuba
San Di•ego \,san-dē-'ā-gō\ city & port SW California
Sandwich Islands — see HAWAII
Sandy Hook peninsula E New Jersey extending N
San Fran•cis•co \,san-frən-'sis-kō\ city W California on **San Francisco Bay** & North Pacific
San•i•bel \'san-ə-bəl\ island SW Florida
San Isi•dro \,san-ə-'sē-drō\ city E Argentina NW of Buenos Aires
San Joa•quin \,san-wä-'kēn, -wȯ-\ river 350 miles (563 kilometers) long central California flowing NW into the Sacramento
San Jo•se \,san-ə-'zā\ city W California SE of San Francisco
San Jo•sé \,san-ə-'zā, -ō-'zā, -hō-'zā\ city, capital of Costa Rica
San Juan \san-'hwän, -'wän\ city, capital of Puerto Rico
San Lu•is Po•to•sí \,sän-lü-,ē-,spōt-ə-'sē\ **1** state central Mexico **2** city, its capital
San Ma•ri•no \,san-mə-'rē-nō\ **1** country S Europe on peninsula of Italy ENE of Florence near Adriatic Sea **2** town, its capital
San Mi•guel de Tu•cu•mán \,san-mig-,el-də-,tü-kə-'män\ *or* **Tu•cu•mán** \,tü-kə-'män\ city NW Argentina
San Pe•dro Su•la \san-,pā-,drō-'sü-lə\ city NW Honduras
San Sal•va•dor \san-'sal-və-,dȯr\ **1** *or formerly* **Wat•lings** \'wät-lingz\ island central Bahama Islands **2** city, capital of El Salvador
San•ta Ana \,sant-ə-'an-ə\ **1** city SW California ESE of Long Beach **2** city NW El Salvador NW of San Salvador
San•ta Bar•ba•ra \-'bär-brə, -bə-rə\ **1** city S California **2** *or* **Channel** islands California off SW coast
San•ta Clara \-'klar-ə, -'kler-ə\ **1** city W California NW of San Jose **2** city W central Cuba
San•ta Cla•ri•ta \-klə-'rēt-ə\ city S California
San•ta Cruz \-'krüz\ city E Bolivia
San•ta Cruz de Te•ne•rife \-də-,ten-ə-'rēf-ā, -'rēf, -'rif\ city Spain in W Canary Islands on Tenerife Island
San•ta Fe \,sant-ə-'fā\ **1** city, capital of New Mexico **2** city central Argentina
Santa Fe Trail pioneer route to the Southwest about 1200 miles (1930 kilometers) long used especially 1821–80 from vicinity of Kansas City, Missouri, to Santa Fe, New Mexico
Santa Isabel — see MALABO
San•ta Mar•ta \,sant-ə-'märt-ə\ city N Colombia on the Caribbean
San•tan•der \,sän-,tän-'deər, ,san-,tan-\ city N Spain WNW of Bilbao
San•ta Ro•sa \,sant-ə-'rō-zə\ city W California N of San Francisco
San•ti•a•go \,sant-ē-'äg-ō, ,sänt-\ **1** city, capital of Chile **2** *or* **Santiago de los Ca•ba•lle•ros** \-də-,lōs-,käb-ə-'yeər-ōs\ city N central Dominican Republic
Santiago de Cu•ba \-də-'kyü-bə\ city SE Cuba
San•to Do•min•go \,sant-əd-ə-'ming-gō\ **1** — see HISPANIOLA **2** — see DOMINICAN REPUBLIC **3** *or formerly* **Ci•u•dad Tru•ji•llo** \,sē-ü-,thä-trü-'hē-ō, ,sē-ü-,dad-\ city, capital of Dominican Republic
San•tos \'sant-əs\ city SE Brazil
São Lu•ís \,saůⁿ-lü-'ēs\ city NE Brazil on Maranhão Island
Saône \'sōn\ river E France flowing into the Rhone
São Pau•lo \saůⁿ-'paů-lü, saůⁿm-, -lō\ city SE Brazil
São Salvador — see SALVADOR
São To•mé *or* **São Tho•mé** \,saůⁿt-ə-'mā, ,saůⁿnt-\ *or* **Saint Thom•as** \sānt-'täm-əs\ island W Africa in Gulf of Guinea; with Príncipe Island, forms country of **São Tomé and Príncipe**; capital, São Tomé; until 1975 a Portuguese colony
Sap•po•ro \'säp-ə-,rō; sə-'pōr-ō, -'pȯr-\ city Japan on W Hokkaido
Saragossa — see ZARAGOZA
Sa•ra•je•vo \'sär-ə-ye-,vō\ city, capital of Bosnia and Herzegovina
Sa•ransk \sə-'ränsk, -'ransk\ city central Russia in Europe
Sa•ra•tov \sə-'rät-əf\ city S central Russia in Europe, on the Volga
Sar•din•ia \sär-'din-ē-ə, -'din-yə\ *or Italian* **Sar•de•gna** \sär-'dā-nyä\ island Italy in the Mediterranean S of Corsica; a region; capital, Cagliari — **Sar•din•ian** \-'din-ē-ən, -'din-yən\ *adj or n*
Sar•dis \'särd-əs\ ancient city W Asia Minor, capital of Lydia
Sar•gas•so Sea \sär-,gas-ō\ area of comparatively still water in North Atlantic lying chiefly between 25° & 35° N latitude & 40° & 70° W longitude
Sa•ron•ic Gulf \sə-,rän-ik\ inlet of the Aegean SE Greece between Attica & Peloponnese
Sas•katch•e•wan \sə-'skach-ə-wən, sa-, -,wän\ province W Canada; capital, Regina
Sas•ka•toon \,sas-kə-'tün\ city Canada in central Saskatchewan
Sas•sa•ri \'säs-ə-rē\ city Italy in NW Sardinia
Sau•di Arabia \,saůd-ē-ə-'rā-bē-ə, ,sȯd-ē-, sä-,üd-ē-\ country SW Asia occupying largest part of Arabian Peninsula; a kingdom; capital, Riyadh — **Saudi** *adj or n* — **Saudi Arabian** *adj or n*
Sault Sainte Ma•rie Canals \,sü-,sānt-mə-'rē\ *or* **Soo Canals** \,sü-\ three ship canals, two in U.S. (Michigan) & one in Canada (Ontario), at rapids in Saint Marys River (70 miles or 115 kilometers long connecting Lake Superior & Lake Huron)
Sa•vaii \sə-'vī-,ē\ island, largest in independent Samoa
Sa•van•nah \sə-'van-ə\ city E Georgia
Sa•voy \sə-'vȯi\ *or French* **Sa•voie** \sá-'vwá\ region SE France SW of Switzerland bordering on Italy — **Sa•voy•ard** \sə-'vȯi-,ärd, ,sav-,ȯi-'ärd; ,sav-,wä-'yär, -'yärd\ *adj or n*
Sax•o•ny \'sak-sə-nē, 'sak-snē\ *or German* **Sach•sen** \'zäk-sən\ region E Germany N of the Erzgebirge — see LOWER SAXONY
Saxony–An•halt \-'än-'hält\ state NE central Germany; capital, Magdeburg
Sca•fell Pike \,skȯ-'fel\ mountain 3210 feet (978 meters) NW England; highest in Cumbrian Mountains & in England
Scan•di•na•via \,skan-də-'nā-vē-ə, -vyə\ **1** peninsula N Europe occupied by Norway & Sweden **2** Denmark, Norway, Sweden, & sometimes also Iceland & Finland
Scar•bor•ough \'skär-,bər-ō\ former city Canada in SE Ontario; now part of Toronto
Scheldt \'skelt\ *or* **Schel•de** \'skel-də\ *or French* **Es•caut** \es-kō\ river 270 miles (434 kilometers) long W Europe flowing from N France through Belgium into North Sea in Netherlands
Schles•wig–Hol•stein \'shles-wig-'hōl-,stīn, 'sles-, -vik-'hōl-\ state N Germany consisting of Holstein & part of Schleswig; capital, Kiel
Schuyl•kill \'skü-kl, 'skül-,kil\ river 131 miles (211 kilometers) long SE Pennsylvania flowing SE into the Delaware River at Philadelphia
Schwarzwald — see BLACK FOREST
Schwe•rin \shvā-'rēn\ city N Germany, capital of Mecklenburg-West Pomerania
Schweiz — see SWITZERLAND
Scil•ly \'sil-ē\ islands SW England off Land's End in county of Cornwall and Isles of Scilly
Sco•tia \'skō-shə\ SCOTLAND — the Medieval Latin name
Scot•land \'skät-lənd\ *or Latin* **Cal•e•do•nia** \,kal-ə-'dō-nyə, -nē-ə\ country N Great Britain; a division of United Kingdom of Great Britain and Northern Ireland; capital, Edinburgh
Scottish Borders administrative area S Scotland
Scotts•dale \'skäts-,dāl\ city SW central Arizona E of Phoenix
Scran•ton \'skrant-n\ city NE Pennsylvania
Scyth•ia \'sith-ē-ə, 'sith-\ ancient country comprising parts of Europe & Asia in regions N & NE of Black Sea & E of Aral Sea — **Scyth•i•an** \-ē-ən\ *adj or n*
Se•at•tle \sē-'at-l\ city & port W Washington
Seine \'sān, 'sen\ river 480 miles (772 kilometers) long N France flowing NW into English Channel
Sel•kirk \'sel-,kərk\ range of the Rocky Mountains SE British Columbia; highest peak, Mount Sir Sandford 11,555 feet (3522 meters)
Se•ma•rang \sə-'mär-,äng\ city Indonesia in central Java
Se•mey \'sem-ā\ *or* **Sem•i•pa•la•tinsk** \,sem-i-pə-'lä-,tinsk\ city NE Kazakhstan on the Irtysh
Sen•dai \sen-'dī, 'sen-\ city Japan in NE Honshu
Sen•e•gal \,sen-i-'gȯl\ **1** river 1015 miles (1633 kilometers) long W Africa flowing W into North Atlantic **2** country W Africa; capital, Dakar — **Sen•e•ga•lese** \,sen-i-gə-'lēz, -'lēs\ *adj or n*
Sen•e•gam•bia \,sen-ə-'gam-bē-ə\ region W Africa around Senegal & Gambia rivers
Seongnam — see SONGNAM
Seoul \'sōl\ city, capital of South Korea
Se•quoia National Park \si-'kwȯi-ə\ reservation SE central California; includes Mount Whitney
Seram — see CERAM
Ser•bia \'sər-bē-ə\ region Balkan Peninsula comprising a federated republic of Serbia and Montenegro
Serbia and Montenegro *or formerly* **Yugosla•via** country S Europe on Balkan Peninsula; capital, Belgrade
Ser•en•ge•ti Plain \,ser-ən-'get-ē\ area N Tanzania including **Serengeti National Park**
Se•te Que•das \,sāt-ə-'kā-thəsh\ *or formerly* **Guaí•ra** *or* **Guay•ra** \gwī-'rä\ former cataract in Alto Paraná on Brazil–Paraguay boundary; now submerged in dam-created lake
Se•vas•to•pol \sə-'vas-tə-,pōl, -,pȯl, -pəl; ,sev-ə-'stō-pəl, -'stō-\ city & port SW Crimea
Sev•ern \'sev-ərn\ river 210 miles (338 kilometers) long Wales & England flowing from E central Wales into Bristol Channel
Se•ville \sə-'vil\ *or Spanish* **Se•vi•lla** \sā-'vē-ä, -yä\ *or ancient* **His•pa•lis** \'his-pə-ləs\ city SW Spain
Sew•ard \'sü-ərd\ peninsula W Alaska projecting into Bering Sea

Sey·chelles \sā-'shel, -'shelz\ islands W Indian Ocean NE of Madagascar; formerly a British colony; became independent 1976; capital, Victoria (on Mahé Island)

's Gravenhage — see HAGUE (The)

Shaan·xi \'shän-'shē\ *or* **Shen·si** \'shen-'sē, 'shən-'shē\ province N central China; capital, Xi'an

Sha·ba \'shäb-ə\ region SE Democratic Republic of the Congo; rich in mineral deposits

Shan·dong \'shän-'dùŋ\ *or* **Shan·tung** \'shan-'təŋ\ **1** peninsula E China extending into Yellow Sea **2** province E China including Shandong Peninsula; capital, Jinan

Shang·hai \shaŋ-'hī\ city E China in SE Jiangsu

Shan·non \'shan-ən\ river 230 miles (370 kilometers) long W Ireland flowing S & W into North Atlantic

Shan·tou \'shän-'tō\ *or* **Swa·tow** \'swä-'taù\ city SE China in E Guangdong

Shan·xi \'shän-'shē\ *or* **Shan·si** \'shän-'sē, -'shē\ province N China bordering on Yellow River; capital, Taiyuan

Shar·on, Plain of \shar-ən\ region Israel on coast between Mount Carmel & Jaffa

Shas·ta, Mount \'shas-tə\ mountain 14,162 feet (4316 meters) N California in Cascade Range

Shatt al Ar·ab \,shat-,al-'ar-əb\ river 120 miles (193 kilometers) long SE Iraq formed by confluence of Euphrates & Tigris rivers & flowing SE into Persian Gulf

Shcherbakov — see RYBINSK

She·ba \'shē-bə\ ancient country S Arabia

She·chem \'shē-kəm, -,kem\ city of ancient Palestine in Samaria; site in present West Bank

Shef·field \'shef-,ēld\ city N England

Shen·an·do·ah National Park \,shen-ən-'dō-ə, ,shan-ə-'dō-ə\ reservation N Virginia in Blue Ridge Mountains

Shen·yang \'shən-'yäŋ\ *or* **Muk·den** \'mùk-dən, 'mək-; 'mùk-'den\ *or formerly* **Feng·tien** \'fəŋ-tē-'en\ city NE China, capital of Liaoning

Sher·wood Forest \,shər-,wùd\ ancient royal forest central England chiefly in Nottinghamshire

Shet·land Islands \'shet-lənd\ island group N Scotland NE of the Orkneys constituting an administrative area

Shi·jia·zhuang *or* **Shih–chia chuang** \'shiər-jē-'äj-'wäŋ, 'shē-jē-\ city NE China, capital of Hebei

Shi·ko·ku \shi-'kō-kü\ island S Japan E of Kyushu

Shim·la \'shim-lə\ *or* **Sim·la** \'sim \ town N India, capital of Himachal Pradesh

Shi·raz \shi-'räz\ city SW Iran

Shi·zu·o·ka \,shiz-ə-'wō-kə, ,shē-zə-'ō-kə\ city Japan in central Honshu SW of Tokyo

Sho·la·pur \'shō-lə-,pùr\ city W India in SE Maharashtra SE of Bombay

Shreve·port \'shrēv-,pōrt, -,pȯrt\ city NW Louisiana on Red River

Shrop·shire \'shräp-shər, -,shir\ *or officially 1974–80* **Sal·op** \'sal-əp, -,äp\ county W England bordering on Wales

Shushan — see SUSA

Si·al·kot \sē-'äl-,kōt\ city NE Pakistan NNE of Lahore

Siam — see THAILAND

Siam, Gulf of — see THAILAND (Gulf of)

Sian — see XI'AN

Siangtan — see XIANGTAN

Si·be·ria \sī-'bir-ē-ə\ region N Asia in Russia between the Urals & North Pacific — **Si·be·ri·an** \-ē-ən\ *adj or n*

Si·chuan \'sē-'chwän\ *or* **Sze·chwan** \'sech-'wän, 'sesh-\ province SW China; capital, Chengdu

Sic·i·ly \'sis-ə-lē, 'sis-lē\ *or Italian* **Si·ci·lia** \sē-'chēl-yä\ island S Italy SW of toe of peninsula of Italy; a region; capital, Palermo — **Si·cil·ian** \sə-'sil-yən\ *adj or n*

Sid·ra, Gulf of \'sid-rə\ inlet of the Mediterranean on coast of Libya

Si·er·ra Le·one \sē-,er-ə-lē-'ōn, ,sir-ə-\ country W Africa on North Atlantic; capital, Freetown — **Sierra Le·on·ean** \-'ō-nē-ən\ *adj or n*

Si·er·ra Ma·dre \sē-,er-ə-'mäd-rē\ mountain system Mexico including **Sierra Madre Oc·ci·den·tal** \-,äk-sə-,den-'täl\ range W of the central plateau, **Sierra Madre Ori·en·tal** \-,ōr-ē-,en-'täl, -,ȯr-\ range E of the plateau, & **Sierra del Sur** \sē-,er-ə-,del-'sùr\ range to the S

Sierra Ne·vada \-nə-'vad-ə, -'väd-\ **1** mountain range E California & W Nevada — see WHITNEY (Mount) **2** mountain range S Spain; highest peak Mulhacén 11,410 feet (3478 meters), highest in Spain

Sik·kim \'sik-əm, -,im\ former country SE Asia on S slope of the Himalaya between Nepal & Bhutan; a state of Republic of India since 1975; capital, Gangtok

Si·le·sia \sī-'lē-zhə, sə-, -zhē-ə, -shə, -shē-ə\ region E central Europe in valley of the upper Oder bordering on Sudeten Mountains; formerly chiefly in Germany now chiefly in NE Czech Republic & SW Poland — **Si·le·sian** \-zhən, -shən\ *adj or n*

Silk Road ancient trade route that extended from China to the Mediterranean Sea

Sim·birsk \sim-'birsk\ *or 1924–91* **Ul·ya·novsk** \ül-'yän-əfsk\ city central Russia in Europe

Sim·coe, Lake \'sim-kō\ lake Canada in SE Ontario

Sim·fe·ro·pol \,sim-fə-'rò-pəl, ,simp-, -'rō-\ city S Ukraine in central Crimea Peninsula

Si·mi Valley \sē-'mē\ city SW California W of Los Angeles

Sim·plon \'sim-,plän\ **1** pass between Italy & Switzerland in Lepontine Alps **2** tunnel 12.5 miles (20 kilometers) long near the pass

Si·nai \'sī-,nī\ **1** — see HOREB **2** peninsula extension of continent of Asia NE Egypt between Red Sea & the Mediterranean

Si·na·loa \,sē-nə-'lō-ə, ,sin-ə-\ state W Mexico on Gulf of California; capital, Culiacán

Sind \'sind\ province S Pakistan in lower Indus River valley; chief city Karachi

Sin·ga·pore \'siŋ-ə-,pōr, -gə-, -,pȯr\ **1** island off S end of Malay Peninsula; an independent republic **2** city, its capital — **Sin·ga·por·ean** \,siŋ-ə-'pōr-ē-ən, -gə-, -'pȯr-\ *adj or n*

Sining — see XINING

Sinkiang–Uighur — see XINJIANG UYGUR

Sion — see ZION

Siracusa — see SYRACUSE

Sjæl·land \'shel-,än\ *or* **Zea·land** \'zē-lənd\ island, largest of islands of Denmark; site of Copenhagen

Skag·ge·rak \'skag-ə-,rak\ arm of North Sea between S Norway & N Denmark

Skop·je \'skȯp-,yā\ city, capital of independent Macedonia

Skye \'skī\ island Scotland; one of the Inner Hebrides

Sky·ros \'skī-rəs, -,räs\ *or Greek* **Skí·ros** \'skē-,rȯs\ island Greece in Northern Sporades E of Euboea

Sla·vo·nia \slə-'vō-nē-ə, -nyə\ region E Croatia between Sava, Drava, & Danube rivers — **Sla·vo·nian** \-nē-ən, -nyən\ *adj or n*

Sli·go \'slī-gō\ county N Ireland (republic) in Connacht

Slo·va·kia \slō-'väk-ē-ə, -'vak-\ country central Europe; capital, Bratislava

Slo·ve·nia \slō-'vē-nē-ə, -nyə\ country S Europe N & W of Croatia; formerly a constituent republic of Yugoslavia; capital, Ljubljana — **Slo·ve·nian** \-nē-ən, -nyən\ *adj or n*

Smo·lensk \smō-'lensk\ city W Russia in Europe, WSW of Moscow

Smyrna — see IZMIR

Snow·do·nia \snō-'dō-nē-ə, -nyə\ mountainous district NW Wales centering around **Snow·don** \'snōd-n\ (massif 3560 feet or 1085 meters; highest point in Wales)

So·chi \'sō-chē\ city S Russia in Europe, on Black Sea

So·ci·e·ty \sə-'sī-ət-ē\ islands South Pacific in French Polynesia; capital, Papeete (on Tahiti)

So·co·tra \sə-'kō-trə\ island Yemen in Indian Ocean E of Gulf of Aden; capital, Tamridah

Sod·om \'säd-əm\ ancient city thought to have been in the area now covered by the SW part of the Dead Sea

So·fia \'sō-fē-ə, 'sȯ-, sō-'\ city, capital of Bulgaria

So·ho \'sō-,hō\ district of central London, England, in Westminster

So·li·hull \,sō-li-'həl\ town central England SE of Birmingham

Sol·o·mon \'säl-ə-mən\ **1** islands W Pacific E of New Guinea divided between Papua New Guinea & independent **Solomon Islands** (capital, Honiara) **2** sea arm of Coral Sea W of the Solomons

So·ma·lia \sō-'mäl-ē-ə, sə-, -'mäl-yə\ *or* **So·ma·li Republic** \-'mäl-ē\ country E Africa on Gulf of Aden & Indian Ocean; capital, Mogadishu — **So·ma·li·an** \-'mäl-ē-ən, -'mäl-yən\ *adj or n*

So·ma·li·land \sō-'mäl-ē-,land, sə-\ region E Africa comprising Somalia, Djibouti, & part of E Ethiopia

Som·er·set \'səm-ər-,set, -sət\ *or* **Som·er·set·shire** \-,shiər, -shər\ county SW England

Somerset Nile — see NILE

Song·nam *or* **Seong·nam** \'səŋ-'näm\ city NW South Korea

So·no·ra \sə-'nōr-ə, -'nȯr-\ state NW Mexico bordering on U.S.; capital, Hermosillo

So·nor·an \sə-'nōr-ən, -'nȯr-\ *or* **Sonora** desert SW U.S. & NW Mexico in S Arizona, SE California, & N Sonora

Soo Canals — see SAULT SAINTE MARIE CANALS

Soochow — see SUZHOU

Sorata — see ILLAMPU

So·ro·ca·ba \,sōr-ə-'kab-ə, ,sȯr-\ city SE Brazil W of São Paulo

Sos·no·wiec \säs-'nō-,vyets\ city SW Poland

South \'saùth\ island S New Zealand

South Africa, Republic of country S Africa; an independent republic; until 1961 (as **Union of South Africa**) a British dominion; administrative capital, Pretoria; legislative capital, Cape Town; judicial capital, Bloemfontein

South America continent of Western Hemisphere SE of North America & chiefly S of the Equator — **South American** *adj or n*

South·amp·ton \saùth-'am-tən, -'ham-, -'amp-, -'hamp-\ city S England

South Australia state S Australia; capital, Adelaide — **South Australian** *adj or n*

South Ayrshire administrative area W Scotland

South Bend \'bend\ city N Indiana

South Cape *or* **South Point** — see KA LAE

South Car·o·li·na \,kar-ə-'lī-nə\ state SE U.S.; capital, Columbia — **South Car·o·lin·i·an** \-'lin-ē-ən, -'lin-yən\ *adj or n*

South China Sea — see CHINA

South Da·ko·ta \də-'kōt-ə\ state NW central U.S.; capital, Pierre — **South Da·ko·tan** \-'kōt-n\ *adj or n*

South·end–on–Sea \,saù-,thend-än-'sē, -ȯn-\ resort SE England E of London

Southern Alps mountain range New Zealand in W South Island extending almost the length of the island

Southern Ocean the waters surrounding Antarctica including the S parts of the South Atlantic, South Pacific, & Indian oceans

Southern Rhodesia — see ZIMBABWE

Southern Yemen — see YEMEN

South Georgia island South Atlantic E of Tierra del Fuego; administered by Britain

South Korea — see KOREA

South Lanarkshire administrative area W Scotland

South Seas the areas of the Atlantic, Indian, & Pacific oceans in the Southern Hemisphere; especially, the South Pacific

South Shields \'shēldz, 'shēlz\ city N England

South Tirol — see ALTO ADIGE

South Vietnam — see VIETNAM

South·wark \'sə<u>th</u>-ərk, 'saùth-wərk\ borough of S Greater London, England

South–West Africa — see NAMIBIA

South York·shire \'yȯrk-,shiər, -shər\ metropolitan county N England

Soviet Central Asia formerly used name for the portion of central & SW Asia belonging to U.S.S.R. & comprising the Kirghiz, Tadzhik, Turkmen, & Uzbek republics & sometimes thought to also include all or part of the Kazakh republic

Soviet Russia **1** — see RUSSIA **2** — see UNION OF SOVIET SOCIALIST REPUBLICS

Soviet Union — see UNION OF SOVIET SOCIALIST REPUBLICS

So·we·to \sō-'wāt-ō\ residential area NE Republic of South Africa adjoining SW Johannesburg

Spain \'spān\ *or Spanish* **Es·pa·ña** \ā-'spän-yä\ *or ancient* **His·pa·nia** \his-'pān-ē-ə, -'pān-yə, -'pan-\ country SW Europe in Iberian Peninsula; a kingdom; capital, Madrid

Spanish America **1** the Spanish-speaking countries of America **2** the parts of America settled & formerly governed by the Spanish

Spanish Guinea — see EQUATORIAL GUINEA

\ə\ abut \ᵊ\ kitten \ər\ further \a\ ash \ā\ ace \ä\ mop, mar \aù\ out \ch\ chin \e\ bet \ē\ easy \g\ go \i\ hit \ī\ ice \j\ job \ŋ\ sing \ō\ go \ȯ\ law \ȯi\ boy \th\ thin \<u>th</u>\ the \ü\ loot \ù\ foot \y\ yet \zh\ vision *see also* Guide to Pronunciation

Spanish Main \'mān\ **1** the mainland of Spanish America especially along N coast of South America **2** the Caribbean Sea & adjacent waters especially when region was infested with pirates
Spanish Morocco — see MOROCCO
Spanish Sahara former Spanish territory NW Africa SW of Morocco comprising Río de Oro & Saguia el Hamra — see WESTERN SAHARA
Spar·ta \'spärt-ə\ *or* **Lac·e·dae·mon** \ˌlas-ə-'dē-mən\ ancient city S Greece in Peloponnese, capital of Laconia
Spey·er \'shpī-ər, 'spī-; 'shpīr, 'spīr\ *or* **Spires** \'spīrz\ city SW Germany on the Rhine
Spice Islands — see MOLUCCAS
Spits·ber·gen \'spits-ˌbər-gən\ islands in Arctic Ocean N of Norway; chief island, West Spitsbergen — see SVALBARD
Split \'split\ *or* **Spljet** \'splʸet, splē-'et\ city S Croatia
Spo·kane \spō-'kan\ city E Washington
Spor·a·des \'spȯr-ə-ˌdēz, 'spär-\ two island groups Greece in the Aegean: the **Northern Sporades** (chief island, Skyros, E of Euboea) & **Southern Sporades** (including Samos, Icaria, & the Dodecanese, off SW Turkey)
Sprat·ly \'sprat-lē\ islands central South China Sea; claimed by several countries
Spring·field \'spring-ˌfēld\ **1** city, capital of Illinois **2** city SW Massachusetts on Connecticut River **3** city SW Missouri
Springs \'springz\ city NE Republic of South Africa in Gauteng
Sri Lan·ka \srē-'läng-kə, 'srē-\ *or formerly* **Cey·lon** \si-'län, sā-\ country coextensive with island of Ceylon; an independent republic; capital, Colombo
Sri·na·gar \sri-'nəg-ər\ city, summer capital of Jammu and Kashmir, in W Kashmir
Staf·ford·shire \'staf-ərd-ˌshiər, -shər\ *or* **Stafford** county W central England
Staked Plain — see LLANO ESTACADO
Sta·kha·nov \stə-'kän-əf\ *or formerly* **Ka·di·yev·ka** \kə-'dē-yəf-kə\ city E Ukraine in Donets Basin
Stalingrad — see VOLGOGRAD
Stam·ford \'stam-fərd, 'stamp-\ city SW Connecticut
Stan·ley \'stan-lē\ town, capital of Falkland Islands
Stanley, Mount — see RUWENZORI
Stanleyville — see KISANGANI
Stat·en Island \'stat-ⁿn\ **1** island SE New York SW of mouth of the Hudson **2** *or formerly* **Rich·mond** \'rich-mənd\ borough of New York City including Staten Island
States of the Church — see PAPAL STATES
Stavropol — see TOL'YATTI
Stir·ling \'stər-ling\ administrative area central Scotland
Stock·holm \'stäk-ˌhōlm, -ˌhōm\ city, capital of Sweden
Stock·port \'stäk-ˌpōrt, -ˌpȯrt\ town NW England S of Manchester
Stock·ton \'stäk-tən\ city central California
Stoke on Trent \ˌstō-ˌkȯn-'trent, -ˌkän-\ city central England
Stone·henge \'stōn-ˌhenj, stōn-'henj\ assemblage of megaliths S England on Salisbury Plain; erected by a prehistoric people
Stone Mountain mountain 1686 feet (514 meters) NW Georgia E of Atlanta
Straits Settlements former British crown colony SE Asia on Strait of Malacca comprising Singapore Island & George Town & Malacca settlements on Malay Peninsula
Stras·bourg \'sträs-ˌbu̇rg, 'sträz-, -ˌbərg\ city NE France
Strat·ford–upon–Avon \'strat-fərd\ town central England
Strom·bo·li \'sträm-bə-lē\ volcano 3038 feet (926 meters) Italy in Lipari Islands on Stromboli Island
Stutt·gart \'shtu̇t-ˌgärt, 'stu̇t-, 'stət-\ city SW Germany, capital of Baden-Württemberg
Styx \'stiks\ chief river of Hades in Greek mythology
Su·bic Bay \'sü-bik\ inlet of South China Sea in W Luzon, Philippines
Süchow **1** — see XUZHOU **2** — see YIBIN
Su·cre \'sü-krā\ city, constitutional capital of Bolivia
Su·dan \sü-'dan, -'dän\ **1** region N Africa S of the Sahara between the Atlantic & the upper Nile **2** country NE Africa S of Egypt; capital, Khartoum — see ANGLO-EGYPTIAN SUDAN — **Su·da·nese** \ˌsüd-ⁿn-'ēz, -'ēs\ *adj or n*
Su·de·ten \sü-'dāt-ⁿn\ *or* **Su·de·ten·land** \sü-'dāt-ⁿn-ˌland\ region NE Czech Republic in Sudety Mountains
Su·de·ty \'süd-et-ē, sü-'det-\ mountains central Europe between Czech Republic & Poland
Su·ez \sü-'ez, 'sü-ˌez\ **1** city NE Egypt at S end of Suez Canal on **Gulf of Suez** (arm of Red Sea) **2** canal over 100 miles (161 kilometers) long NE Egypt across Isthmus of Suez
Suez, Isthmus of neck of land NE Egypt between Mediterranean & Red seas connecting Africa & Asia
Suf·folk \'səf-ək\ county E England on North Sea
Su·i·ta \sü-'ēt-ə\ city Japan in S Honshu N of Osaka
Su·la·we·si \ˌsü-lə-'wā-sē\ *or* **Ce·le·bes** \'sel-ə-ˌbēz, sə-'lē-bēz\ island Indonesia E of Borneo
Su·lu \'sü-lü\ archipelago SW Philippines SW of Mindanao — see BASILAN
Su·ma·tra \su̇-'mä-trə\ island W Indonesia S of Malay Peninsula — **Su·ma·tran** \-trən\ *adj or n*
Su·mer \'sü-mər\ the S division of ancient Babylonia — **Su·me·ri·an** \sü-'mer-ē-ən, -'mir-\ *adj or n*
Sun·belt \'sən-ˌbelt\ region S & SW U.S.
Sun·da \'sün-də\ **1** islands Malay Archipelago comprising the **Greater Sunda Islands** (Sumatra, Borneo, Java, Sulawesi, & adjacent islands) & the **Lesser Sunda Islands** (extending from Bali to Timor); with exception of N Borneo & East Timor belong to Indonesia **2** strait between Java & Sumatra
Sun·der·land \'sən-dər-lənd\ seaport N England on North Sea
Suomi — see FINLAND
Su·pe·ri·or, Lake \su̇-'pir-ē-ər\ lake E central North America in U.S. & Canada; largest of the Great Lakes
Su·ra·ba·ya \ˌsu̇r-ə-'bī-ə\ city Indonesia in NE Java
Su·ra·kar·ta \ˌsu̇r-ə-'kärt-ə\ city Indonesia in central Java
Su·rat \'su̇r-ət, sə-'rat\ city W India in SE Gujarat
Su·ri·na·me \ˌsu̇r-ə-'näm-ə\ *or* **Su·ri·nam** \'su̇r-ə-ˌnam, ˌsu̇r-ə-'näm\ *or formerly* **Dutch Guiana** country N South America between Guyana & French Guiana; capital, Paramaribo
Sur·rey \'sər-ē, 'sə-rē\ **1** county SE England SW of London **2** city Canada in SW British Columbia
Surts·ey \'sərt-ˌsā\ island Iceland off S coast
Su·sa \'sü-zə\ *or biblical* **Shu·shan** \'shü-shən, -ˌshan\ ancient city, capital of Elam; ruins in SW Iran
Sut·ton \'sət-ⁿn\ borough of S Greater London, England
Su·va \'sü-və\ city, capital of Fiji on Viti Levu Island
Su·wan·nee \sə-'wän-ē\ river 250 miles (400 kilometers) SE Georgia & N Florida flowing SW into Gulf of Mexico
Su·won \'sü-ˌwän\ city NW South Korea S of Seoul
Su·zhou \'sü-'jō\ *or* **Soo·chow** \'sü-'jō, -'chau̇\ *or formerly* **Wu·hsien** \'wü-shē-'en\ city E China in Jiangsu W of Shanghai
Sval·bard \'sfäl-ˌbär\ islands in Arctic Ocean including Spitsbergen under Norwegian administration
Sverdlovsk — see YEKATERINBURG
Swa·bia \'swāb-ē-ə\ region & medieval county SW Germany
Swan·sea \'swän-zē\ **1** administrative area S Wales **2** city & port S Wales in Swansea administrative area
Swatow — see SHANTOU
Swa·zi·land \'swäz-ē-ˌland\ country SE Africa between Republic of South Africa & Mozambique; an independent kingdom; capital, Mbabane — **Swa·zi** \'swäz-ē\ *adj or n*
Swe·den \'swēd-ⁿn\ country N Europe on Scandinavian Peninsula bordering on Baltic Sea; a kingdom; capital, Stockholm
Swit·zer·land \'swit-sər-lənd\ *or Latin* **Hel·ve·tia** \hel-'vē-shə, -shē-ə\ *or French* **Suisse** \'swʸēs\ *or German* **die Schweiz** \dē-'shfīts\ *or Italian* **Sviz·ze·ra** \'zvēt-sä-rä\ country W Europe in the Alps; capital, Bern
Syd·ney \'sid-nē\ city SE Australia, capital of New South Wales
Syr·a·cuse \'sir-ə-ˌkyüs, -kyüz\ **1** city central New York **2** ancient city Italy in SE Sicily; site at modern city of **Si·ra·cu·sa** \ˌsē-rə-'kü-zə\
Syr Dar'·ya \sir-'där-yə\ river about 1370 miles (2204 kilometers) Tajikistan & S Kazakhstan flowing from Tian Shan W & NW into Aral Sea
Syr·ia \'sir-ē-ə\ **1** ancient region SW Asia bordering on the Mediterranean **2** former French mandate (1920–44) including present Syria & Lebanon **3** country S of Turkey; capital, Damascus — **Syr·i·an** \'sir-ē-ən\ *adj or n*
Syrian Desert desert region N Saudi Arabia, SE Syria, W Iraq, & NE Jordan
Szcze·cin \'shchet-ˌsēn\ city NW Poland on the Oder
Szechwan — see SICHUAN
Ta·bas·co \tə-'bas-kō\ state SE Mexico SW of Yucatán Peninsula; capital, Villahermosa
Ta·ble Bay \ˌtā-bəl\ harbor of Cape Town, Republic of South Africa
Ta·briz \tə-'brēz\ city NW Iran
Ta·co·ma \tə-'kō-mə\ city W Washington on Puget Sound
Tae·gu \'tā-gü, 'dā-\ *or* **Dae·gu** \'dā-\ city South Korea NNW of Pusan
Tae·jon \'tā-'jən, 'dā-\ *or* **Dae·jeon** \'dā-\ city South Korea NW of Taegu
Ta·gan·rog \'tag-ən-ˌräg\ city SW Russia in Europe, W of Rostov
Ta·gus \'tā-gəs\ *or Spanish* **Ta·jo** \'tä-hō\ *or Portuguese* **Te·jo** \'tā-zhü\ river 626 miles (1007 kilometers) long Spain & Portugal flowing W into North Atlantic
Ta·hi·ti \tə-'hēt-ē\ island South Pacific in French Polynesia in Society Islands; chief town, Papeete — **Ta·hi·tian** \-'hē-shən\ *adj or n*
Tai·chung \'tī-'chu̇ng\ city W Taiwan
T'ai·nan \'tī-'nän\ city SW Taiwan
Tai·pei \'tī-'pā, -'bā\ *or formerly* **Dai·ho·ku** \'dī-'hō-ˌkü\ city, capital of (Nationalist) Republic of China in N Taiwan
Tai·wan \'tī-'wän\ *or formerly* **For·mo·sa** \fȯr-'mō-sə, fər-, -zə\ **1** island off SE coast of mainland Asia; since 1949 seat of government of (Nationalist) Republic of China; capital, Taipei **2** strait between Taiwan & Fujian, China connecting East China & South China seas — **Tai·wan·ese** \ˌtī-wə-'nēz, -'nēs\ *adj or n*
Tai·yuan \'tī-yü-'än\ *or formerly* **Yang·ku** \'yäng-'kü\ city N China, capital of Shanxi
Tai·zhou *or* **T'ai–chou** \'tī-'jō\ city E China in central Jiangsu NW of Shanghai
Ta·jik·i·stan \tä-'jik-i-ˌstan, -'jēk-\ country W central Asia; capital, Dushanbe; a constituent republic (**Ta·dzhik·i·stan** *same*\ *or* **Ta·dzhik Soviet Socialist Republic** \tä-'jik, -'jēk\) of U.S.S.R. 1929–91
Ta·ka·ma·tsu \ˌtäk-ə-'mät-sü, tä-'käm-ət-ˌsü\ city Japan in NE Shikoku
Ta·kat·su·ki \tə-'kät-su̇-kē\ city Japan in S Honshu NNE of Osaka
Ta·kli·ma·kan *or* **Ta·kla Ma·kan** \ˌtäk-lə-mə-'kän\ desert W China in Xinjiang Uygur
Ta–lien — see DALIAN
Tal·la·has·see \ˌtal-ə-'has-ē\ city, capital of Florida
Tal·la·hatch·ie \ˌtal-ə-'hach-ē\ river 230 miles (370 kilometers) long N Mississippi
Tal·linn \'tal-ən, 'täl-\ *or formerly* **Re·vel** \'rā-vəl\ city, capital of Estonia
Ta·mau·li·pas \ˌtäm-ˌau̇-'lē-pəs, təm-\ state NE Mexico; capital, Ciudad Victoria
Tam·bov \täm-'bȯf, -'bȯv\ city S central Russia in Europe, SE of Moscow
Tam·il Na·du \ˌtam-əl-'näd-ü\ *or formerly* **Ma·dras** \mə-'dras, -'dräs\ state S India on Bay of Bengal; capital, Madras
Tam·pa \'tam-pə\ city W Florida on Tampa Bay
Tam·pe·re \'tam-pə-ˌrā, 'täm-\ city SW Finland
Tam·pi·co \tam-'pē-kō\ city E Mexico in S Tamaulipas
Tananarive — see ANTANANARIVO
Tan·gan·yi·ka \ˌtan-gən-'yē-kə, ˌtang-gən-, -gə-'nē-\ former country E Africa S of Kenya; became part of Tanzania 1964
Tanganyika, Lake lake E Africa between Tanzania & Democratic Republic of the Congo
Tang·shan \'täng-'shäng\ city NE China in E Hebei
Tan·ta \'tänt-ə\ city N Egypt in central delta of the Nile
Tan–tung — see DANDONG
Tan·za·nia \ˌtan-zə-'nē-ə, ˌtän-\ country E Africa on Indian Ocean; a republic formed 1964 by union of Tanganyika & Zanzibar; capital, Dar es Salaam — **Tan·za·ni·an** \-'nē-ən\ *adj or n*
Taor·mi·na \tau̇r-'mē-nə\ city Italy in NE Sicily
Ta·ran·to \'tär-ən-ˌtō, tə-'rant-ō\ *or ancient* **Ta·ren·tum** \tə-'rent-əm\ city SE Italy on Gulf of Taranto
Ta·ra·wa \tə-'rä-wə\ island central Pacific containing capital of Kiribati

Ta·rim \'dä-'rēm, 'tä-\ river 1250 miles (2012 kilometers) long W China in Xinjiang Uygur flowing into a marshy depression

Tar·lac \'tär-,läk\ city Philippines in central Luzon

Tar·shish \'tär-shish\ ancient maritime country referred to in the Bible & often identified with Tartessus

Tar·sus \'tär-səs\ ancient city of S Asia Minor, capital of Cilicia; now a city of S Turkey

Tar·ta·ry \'tärt-ə-rē\ vast historical region in Asia & E Europe roughly extending from Sea of Japan (East Sea) to the Dnieper

Tar·tes·sus *or* **Tar·tes·sos** \tär-'tes-əs\ ancient kingdom on SW coast of Spain near mouth of the Guadalquivir — see TARSHISH

Tar·tu \'tär-,tü\ city E Estonia

Tash·kent \tash-'kent\ city, capital of Uzbekistan

Tas·man Sea \'taz-mən\ the part of the South Pacific between SE Australia & New Zealand

Tas·ma·nia \taz-'mā-nē-ə, -nyə\ *or earlier* **Van Die·men's Land** \van-'dē-mənz\ island SE Australia S of Victoria; a state; capital, Hobart — **Tas·ma·nian** \-nē-ən, -nyən\ *adj or n*

Ta·try \'tä-trē\ *or* **Ta·tra** \'tä-trə\ mountains N Slovakia & S Poland in central Carpathian Mountains

Tatung — see DATONG

Tau·rus \'tȯr-əs\ mountains S Turkey parallel to Mediterranean coast; highest more than 12,000 feet (3660 meters)

Tbi·li·si \tə-'bē-lə-sē, tə-bə-'lē-sē\ *or* **Tif·lis** \'tif-ləs, tə-'flēs\ city, capital of Republic of Georgia

Tchad — see CHAD

Te·gu·ci·gal·pa \tə-,gü-si-'gal-pə\ city, capital of Honduras

Teh·ran *or* **Te·he·ran** \,tā-ə-'ran, -'rän\ city, capital of Iran

Tel·a·nai·pura \,tel-ə-'nī-,pu̇r-ə\ city & port Indonesia in Sumatra

Tel Aviv \,tel-ə-'vēv\ city W Israel on the Mediterranean

Te·ne·ri·fe \,ten-ə-'rē-'rē-fā, -'rēf, -'rif\ island Spain, largest of the Canary Islands

Ten·nes·see \,ten-ə-'sē, 'ten-ə-,\ state E central U.S.; capital, Nashville — **Ten·nes·se·an** *or* **Ten·nes·see·an** \,ten-ə-'sē-ən\ *adj or n*

Te·re·si·na \,ter-ə-'zē-nə\ city NE Brazil

Ter·ra–No·va National Park \,ter-ə-'nō-və\ reservation E Newfoundland

Té·tou·an \tā-'twäⁿ\ *or* **Te·tuán** \te-'twän, ,tet-ə-'wän\ city N Morocco on the Mediterranean

Tex·as \'tek-səs, -siz\ state S U.S.; capital, Austin — **Tex·an** \-sən\ *adj or n*

Thai·land \'tī-,land, -lənd\ *or formerly* **Si·am** \sī-'am\ country SE Asia on Gulf of Thailand; capital, Bangkok — **Thai·land·er** \'tī-,lan-dər, -lən-dər\ *n*

Thailand, Gulf of *or formerly* **Gulf of Siam** arm of South China Sea between Indochina & Malay Peninsula

Thames \'temz\ river over 200 miles (322 kilometers) long S England flowing E from the Cotswolds into the North Sea

Thar \'tär\ *or* **Indian** desert E Pakistan & NW Republic of India E of Indus River

Thebes \'thēbz\ **1** *or ancient* **The·bae** \'thē-bē\ *or later* **Di·os·po·lis** \dī-'äs-pə-ləs\ ancient city S Egypt, capital of Upper Egypt on the Nile on site including modern towns of Karnak & Luxor **2** ancient city E Greece NNW of Athens on site of modern village of Thivai — **The·ban** \'thē-bən\ *adj or n*

Theodore Roosevelt National Park reservation W North Dakota

Thes·sa·lo·niki \,thes-ə-lō-'nē-kē\ *or formerly* **Sa·lon·i·ka** \sə-'län-i-kə\ *or ancient* **Thes·sa·lo·ni·ca** \,thes-ə-lə-'nī-kə, -'län-i-kə\ city N Greece in Macedonia — **Thes·sa·lo·nian** \-'lō-nē-ən, -'lō-nyən\ *adj or n*

Thes·sa·ly \'thes-ə-lē\ region central Greece between Pindus Mountains & the Aegean — **Thes·sa·lian** \thə-'sā-lē-ən, -'sāl-yən\ *adj or n*

The Vale of Gla·mor·gan \glə-'mȯr-gən\ administrative area S Wales

Thim·phu \'thim-pü\ city, capital of Bhutan

Thousand islands Canada & U.S. in the Saint Lawrence in Ontario & New York

Thousand Oaks city SW California W of Los Angeles

Thrace \'thrās\ *or ancient* **Thra·cia** \'thrā-shə, -shē-ə\ region SE Europe in Balkan Peninsula N of the Aegean now divided between Greece & Turkey; in ancient times extended N to the Danube — **Thra·cian** \'thrā-shən\ *adj or n*

Thu·le \'thü-lē, 'thyü-\ *or* **Ul·ti·ma Thule** \,əl-tə-mə-\ northernmost part of the habitable ancient world

Thunder Bay city Canada in SW Ontario on Lake Superior

Thu·rin·gia \thu̇-'rin-jē-ə\ region central Germany

Tian·jin \tē-'än-'jin\ *or* **Tien·tsin** \tē-'ent-'sin, 'tint-\ city NE China in Hebei

Tian Shan \tē-'än-'shän\ *or* **Tien Shan** \tē-'en-'shän\ mountain system central Asia extending NE from Pamirs into Xinjiang Uygur; highest Pobeda Peak (in Kyrgyzstan) 24,406 feet (7439 meters)

Ti·ber \'tī-bər\ *or Italian* **Te·ve·re** \'tā-vā-rā\ *or ancient* **Ti·ber·is** \'tī-bə-rəs\ river 252 miles (405 kilometers) long central Italy flowing through Rome into Tyrrhenian Sea

Tiberias, Lake — see GALILEE (Sea of)

Ti·bes·ti \tə-'bes-tē\ mountains N central Africa in central Sahara in NW Chad; highest 11,204 feet (3415 meters)

Ti·bet \tə-'bet\ *or* **Xi·zang** \'shēd-'zäng\ autonomous region SW China on high plateau N of the Himalaya; capital, Lhasa

Tier·ra del Fue·go \tē-'er-ə-,del-fü-'ā-gō, -fyu̇-\ **1** archipelago off S South America S of Strait of Magellan **2** chief island of the group; divided between Argentina & Chile

Tiflis — see TBILISI

Ti·gris \'tī-grəs\ river 1180 miles (1899 kilometers) long Turkey & Iraq flowing SSE & uniting with the Euphrates to form the Shatt al Arab

Ti·jua·na \,tē-ə-'wän-ə, tē-'wän-\ city NW Mexico on U.S. border in Baja California

Til·burg \'til-,bərg\ city S Netherlands SE of Rotterdam

Timbuktu — see TOMBOUCTOU

Ti·mi·soa·ra \,tē-mish-ə-'wär-ə, -mish-'war-\ city W Romania

Ti·mor \'tē-,mȯr, tē-'\ island SE Asia SE of Sulawesi; W half formerly belonged to Netherlands and is now part of Indonesia, E half formerly belonged to Portugal and is now independent East Timor

Tip·pe·rary \,tip-ə-'reər-ē\ county S Ireland in Munster

Ti·ra·ne *or* **Ti·ra·na** \ti-'rän-ə\ city, capital of Albania

Ti·rol *or* **Ty·rol** \tə-'rōl; 'tī-,rōl, tī-'; 'tir-əl\ *or Italian* **Ti·ro·lo** \tē-'rō-lō\ region in E Alps in W Austria & NE Italy — **Ti·ro·le·an** \tə-'rō-lē-ən, tī-; ,tir-ə-', ,tī-rə-'\ *or* **Tir·o·lese** \,tir-ə-'lēz, ,tī-rə-, -'lēs\ *adj or n*

Ti·ruch·chi·rap·pal·li \,tir-ə-chə-'räp-ə-lē\ city S India in Tamil Nadu

Ti·ti·ca·ca, Lake \,tit-i-'käk-ə\ lake on Bolivia–Peru boundary at altitude of 12,500 feet (3810 meters)

Tlax·ca·la \tlä-'skäl-ə\ state SE central Mexico; capital, Tlaxcala

To·ba·go \tə-'bā-gō\ island West Indies NE of Trinidad; part of independent Trinidad and Tobago — **To·ba·go·ni·an** \,tō-bə-'gō-nē-ən\ *n*

To·go \'tō-gō\ *or* **To·go·land** \-,land\ region W Africa on Gulf of Guinea between Benin & Ghana; until 1919 a German protectorate, then divided into two trust territories: British Togoland (in W, since 1957 part of Ghana) & French Togo (in E, since 1958 independent Togo; capital, Lomé) — **To·go·land·er** \-,lan-dər\ *n* — **To·go·lese** \,tō-gō-'lēz, -'lēs\ *adj or n*

To·ko·ro·za·wa \,tō-kə-'rō-zə-,wä\ city Japan on Honshu; a suburb of Tokyo

To·ku·shi·ma \,tō-kə-'shē-mə\ city Japan in E Shikoku

To·kyo \'tō-kē-,ō\ *or formerly* **Edo** \'ed-ō\ city, capital of Japan in SE Honshu on **Tokyo Bay** (inlet of North Pacific) — **To·kyo·ite** \'tō-kē-,ō-,īt\ *n*

To·le·do \tə-'lēd-ō, -'lēd-ə\ **1** city NW Ohio **2** city central Spain SW of Madrid

To·lu·ca \tə-'lü-kə\ city central Mexico, capital of Mexico state

Tol'·yat·ti *or* **To·gliat·ti** \tȯl-'yät-ē\ *or formerly* **Stav·ro·pol** \stav-'rȯ-pəl, -'rō-\ city Russia in Europe, NW of Samara

Tom·bouc·tou \tōⁿ-bük-'tü\ *or* **Tim·buk·tu** \,tim-,bək-'tü, tim-'bək-tü\ town W Africa in Mali near Niger River

Tombstone city SE corner of Arizona

Tomsk \'tämsk, 'tämpsk, 'tȯmsk, 'tȯmpsk\ city S central Russia in Asia

Ton·ga \'täng-gə, 'täng-ə\ islands SW Pacific E of Fiji Islands; a kingdom; capital, Nukualofa — **Ton·gan** \-gən, -ən\ *adj or n*

Tong·hua *or* **T'ung–hua** \'tȯng-'hwä, -'wä\ city NE China in SW Jilin

Ton·kin, Gulf of \'täng-kən\ arm of South China Sea E of N Vietnam

To·pe·ka \tə-'pē-kə\ city, capital of Kansas

Tor·bay \tȯr-'bā, 'tȯr-\ town SW England on **Tor Bay** (inlet of English Channel)

Tor·faen \'tȯr-,vīn\ administrative area SE Wales

Torino — see TURIN

To·ron·to \tə-'ränt-ō, -'ränt-ə\ city, capital of Ontario

Tor·rance \'tȯr-əns, 'tär-\ city SW California SSW of Los Angeles

Tor·re·ón \,tȯr-ē-'ōn\ city N Mexico in SW Coahuila

Tor·res \'tȯr-əs\ strait between New Guinea & Cape York Peninsula, Australia

Tor·tu·ga \tȯr-'tü-gə\ island Haiti off N coast; a stronghold of pirates in 17th century

To·ruń \'tȯr-,ün-yə, -,ün\ city N Poland on the Vistula

Toscana — see TUSCANY

Tou·lon \tü-'lōⁿ\ city SE France ESE of Marseille

Tou·louse \tu̇-'lüz\ city S France on the Garonne

Tou·raine \tu̇-'rān, -'ren\ region NW central France; chief city **Tours** \'tu̇r\

Tourane — see DA NANG

Tow·er Hamlets \'tau̇-ər-, 'tau̇r-\ borough of E Greater London, England

To·ya·ma \tō-'yäm-ə\ city Japan in W central Honshu

To·yo·ha·shi \,tȯi-ə-'häsh-ē\ city Japan in S Honshu

To·yo·na·ka \,tȯi-ə-'nä-kə\ city Japan on S Honshu

To·yo·ta \tȯi-'ōt-ə\ city Japan on S Honshu

Tra·fal·gar, Cape \trə-'fal-gər, *Spanish* ,trä-fäl-'gär\ headland SW Spain at W end of Strait of Gibraltar

Trans·al·pine Gaul \trans-,al-,pīn-, tranz-\ the part of Gaul included in modern France & Belgium

Transjordan — see JORDAN

Trans·vaal \trans-'väl, tranz-\ former province NE Republic of South Africa between Vaal & Limpopo rivers; capital, Pretoria

Tran·syl·va·nia \,trans-əl-'vā-nyə, -nē-ə\ region W Romania — **Tran·syl·va·nian** \-nyən, -nē-ən\ *adj or n*

Transylvanian Alps a S extension of Carpathian Mountains in central Romania

Treb·i·zond \'treb-ə-,zänd\ Greek empire 1204–1461, an offshoot of Byzantine Empire; at greatest extent included Crimea, Georgia, & N coast of Black Sea E of Sakarya River; capital, Trebizond (modern Trabon, in Turkey)

Tren·ti·no–Al·to Adi·ge \tren 'tē-,nō-,äl-,tō-'äd-i-,jā\ region N Italy; capital, **Tren·to** \'tren-,tō\

Tren·ton \'trent-n\ city, capital of New Jersey

Trier \'triər\ *or* **Treves** \'trēvz\ city SW Germany on the Moselle

Tri·este \trē-'est, trē-'es-tē\ city NE Italy on the Adriatic

Trin·i·dad \'trin-ə-,dad\ island West Indies off NE coast of Venezuela; with Tobago forms (since 1962) the country of **Trinidad and Tobago**; capital, Port of Spain — **Trin·i·da·di·an** \,trin-ə-'dād-ē-ən, -'dad-\ *adj or n*

Trip·o·li \'trip-ə-lē\ **1** city, capital of Libya **2** city NW Lebanon NNE of Beirut **3** Tripolitania when it was one of the Barbary States

Tri·pol·i·ta·nia \trip-,äl-ə-'tān-yə, ,trip-ə-lə-\ region NW Libya; chief city, Tripoli

Tri·pu·ra \'trip-ə-rə\ state E India between Bangladesh & Assam; capital, Agartala

Tris·tan da Cu·nha \,tris-tən-də-'kü-nə\ island South Atlantic, chief of the Tristan da Cunha Islands (a dependency of the British colony of Saint Helena)

Tri·van·drum \triv-'an-drəm\ city S India NW of Cape Comorin, capital of Kerala

Tro·as \'trō-,as\ *or* **Tro·ad** \-,ad\ territory surrounding ancient city of Troy in NW Mysia

Tro·bri·and \'trō-brē-,änd\ islands SW Pacific in Solomon Sea belonging to Papua New Guinea

Trond·heim \'trän-,hām\ city & port central Norway

Troy \'trȯi\ *or* **Il·i·um** \'il-ē-əm\ *or ancient* **Troja** \'trō-jə, -yə\ ancient city NW Asia Minor SW of the Dardanelles

Trucial States *or* **Trucial Oman** — see UNITED ARAB EMIRATES

Tru·ji·llo \trü-'hē-ō, -yō\ city NW Peru NW of Lima

Truk — see CHUUK

Tsaritsyn — see VOLGOGRAD

Tsinan — see JINAN
Tsinghai — see QINGHAI
Tsingtao — see QINGDAO
Tu·a·mo·tu \ˌtü-ə-ˈmō-tü\ archipelago South Pacific in French Polynesia E of Society Islands
Tu·buai \tüb-ˈwä-ē\ *or* **Aus·tral** \ˈȯs-trəl, ˈäs-\ islands South Pacific in French Polynesia S of Tahiti
Tuc·son \tü-ˈsän, ˈtü-ˌ\ city SE Arizona
Tucumán — see SAN MIGUEL DE TUCUMAN
Tuk·tut No·gait National Park \ˈtu̇k-ˌtət-ˈnäg-ˌgīd\ reservation N Northwest Territories on Nunavut border
Tu·la \ˈtü-lə\ city W Russia in Europe, S of Moscow
Tul·sa \ˈtəl-sə\ city NE Oklahoma on Arkansas River
T'ung–hua — see TONGHUA
Tu·nis \ˈtü-nəs, ˈtyü-\ **1** city, capital of Tunisia **2** Tunisia especially as one of the former Barbary States — **Tu·ni·sian** \tü-ˈnē-zhən, tyü-, -zhē-ən; -ˈnizh-ən, -ē-ən\ *adj or n*
Tu·ni·sia \tü-ˈnē-zhə, tyü-, -zhē-ə; -ˈnizh-ə, -ē-ə\ country N Africa on the Mediterranean E of Algeria; capital, Tunis — **Tu·ni·sian** \-zhən, -zhē-ən; -ən, -ē-ən\ *adj or n*
Tu·rin \ˈtu̇r-ən, ˈtyu̇r-; tu̇-ˈrin, tyu̇-\ *or Italian* **To·ri·no** \tō-ˈrē-nō\ city NW Italy on the Po, capital of Piedmont
Tur·kana, Lake \tər-ˈkan-ə\ *or* **Lake Ru·dolf** \ˈrü-ˌdȯlf, -ˌdälf\ lake N Kenya in Great Rift Valley
Tur·key \ˈtər-kē\ country W Asia (**Turkey in Asia**) & SE Europe (**Turkey in Europe**) between Mediterranean & Black seas; capital, Ankara
Turk·men·i·stan \ˌtərk-ˈmen-ə-ˌstan\ country central Asia bordering on Afghanistan, Iran, & Caspian Sea; capital, Ashkhabad; a constituent republic (**Turk·men Soviet Socialist Republic** \ˌtərk-mən\) of U.S.S.R. 1925–91 — **Turk·me·ni·an** \ˌtərk-ˈmē-nē-ən\ *adj*
Turks and Cai·cos \ˌtərk-sən-ˈkā-kəs\ two groups of islands (Turks Islands & Caicos Islands) British West Indies at SE end of Bahamas; a British colony
Tur·ku \ˈtu̇r-kü\ city & port SW Finland
Tus·ca·ny \ˈtəs-kə-nē\ *or Italian* **To·sca·na** \tō-ˈskän-ə\ region NW central Italy; capital, Florence
Tu·tu·ila \ˌtüt-ə-ˈwē-lə\ island South Pacific, chief of American Samoa group
Tu·va·lu \tü-ˈväl-ü, -ˈvär-\ *or formerly* **El·lice** \ˈel-əs\ islands W Pacific N of Fiji; an independent country; capital, Funafuti — see GILBERT
Tver' \ˈtver\ *or 1932–90* **Ka·li·nin** \kä-ˈlē-nin\ city Russia in Europe on the Volga
Twin Cities the cities of Minneapolis & Saint Paul, Minnesota
Tyne and Wear \ˈtī-nən-ˈdwiər, -ˈwiər\ metropolitan county N England; includes Newcastle upon Tyne
Tyre \ˈtīər\ ancient city, capital of Phoenicia; now a town of S Lebanon — **Tyr·i·an** \ˈtir-ē-ən\ *adj or n*
Tyrol — see TIROL
Ty·rone \tir-ˈōn\ traditional county W central Northern Ireland
Tyr·rhe·ni·an Sea \tə-ˈrē-nē-ən\ part of the Mediterranean SW of Italy, N of Sicily, & E of Sardinia & Corsica
Tyu·men \tyü-ˈmen\ city W Russia in Asia, ENE of Yekaterinburg
Tzu–kung — see ZIGONG
Tzu–po — see ZIBO
Ubangi–Shari — see CENTRAL AFRICAN REPUBLIC
Uca·ya·li \ˌü-kä-ˈyäl-ē\ river about 1000 miles (1609 kilometers) long central & N Peru
Udi·ne \ˈüd-i-ˌnā\ city NE Italy NE of Venice
Ufa \ü-ˈfä\ city E Russia in Europe, NE of Samara
Ugan·da \yü-ˈgan-də, -ˈgän-, -ˈgän-\ country E Africa N of Lake Victoria; capital, Kampala — **Ugan·dan** \-dən\ *adj or n*
Ujung Pan·dang \ü-ˌju̇ng-pän-ˈdäng\ *or formerly* **Ma·kas·sar** \mə-ˈkas-ər\ city Indonesia in SW Sulawesi
Ukraine \yü-ˈkrān, ˈyü-ˌ\ country E Europe on N coast of Black Sea; capital, Kiev; a constituent republic of U.S.S.R. 1923–91
Ulaan·baa·tar *or* **Ulan Ba·tor** \ˌü-ˌlän-ˈbä-ˌtȯr\ *or formerly* **Ur·ga** \ˈu̇r-gə\ city, capital of Mongolia
Ulan–Ude \ˌü-ˌlän-u̇-ˈdā\ city S Russia in Asia, E of Lake Baikal
Ul·san \ˈül-ˈsän\ city SE South Korea
Ul·ster \ˈəl-stər\ **1** region N Ireland (island) comprising Northern Ireland & N Ireland (republic); a province until 1921 **2** province N Ireland (republic) comprising counties Donegal, Cavan, & Monaghan **3** NORTHERN IRELAND
Ultima Thule — see THULE
Ulyanovsk — see SIMBIRSK
Um·bria \ˈəm-brē-ə\ region central Italy in the Apennines; capital, Perugia — **Um·bri·an** \-brē-ən\ *adj or n*
Un·ga·va \ˌən-ˈgav-ə\ **1** bay inlet of Hudson Strait NE Canada **2** peninsula region NE Canada in N Quebec
Union of South Africa — see SOUTH AFRICA (Republic of)
Union of Soviet Socialist Republics *or* **Soviet Union** *or* **Soviet Russia** country (1922–91) E Europe & N Asia; a union of 15 constituent republics; capital, Moscow
United Arab Emirates *or formerly* **Tru·cial States** \ˈtrü-shəl\ *or* **Trucial Oman** country E Arabia on Persian Gulf; a federation of seven emirates; capital, Abu Dhabi
United Arab Republic former name (1961–71) of Egypt & previously (1958–61) of union of Egypt & Syria
United Kingdom *or in full* **United Kingdom of Great Britain and Northern Ireland** country W Europe in British Isles comprising England, Scotland, Wales, Northern Ireland, Channel Islands, & Isle of Man; capital, London
United Nations international territory; a small area in New York City in E central Manhattan; seat of permanent headquarters of the United Nations
United States of America *or* **United States** country North America bordering on North Atlantic, North Pacific, & Arctic oceans & including Hawaii; a federal republic; capital, Washington
Upper Canada former province, Canada in S part of present-day Ontario
Upper Volta — see BURKINA FASO
Ural \ˈyu̇r-əl\ **1** mountains W central Russia extending about 1640 miles (2640 kilometers) S from point near Kara Sea; usually considered to be dividing line between Europe & Asia; highest 6214 feet (1894 meters) **2** river over 1500 feet (2414 kilometers) long Russia flowing from S end of Ural Mountains into Caspian Sea
Uralsk \yu̇-ˈralsk\ — see ORAL
Ura·wa \u̇-ˈrä-wə\ city Japan in central Honshu N of Tokyo
Uru·guay \ˈu̇r-ə-ˌgwī, ˈyu̇r-; ˈyu̇r-ə-ˌgwā\ **1** river about 1000 miles (1609 kilometers) long SE South America rising in Brazil & flowing into Río de la Plata **2** country SE South America; capital, Montevideo — **Uru·guay·an** \ˌu̇r-ə-ˈgwī-ən, ˌyu̇r-; ˌyu̇r-ə-ˈgwā-\ *adj or n*
Ürüm·qi \ˈtē-ˈrēm-ˈchē\ *or* **Urum·chi** \ü-ˈrüm-chē, ˌu̇r-əm-ˈchē\ *or* **Wu–lu–mu–ch'i** \ˈwü-ˈlü-ˈmü-ˈchē\ city NW China, capital of Xinjiang Uygur
Urundi — see BURUNDI
Us·pa·lla·ta \ˌü-spə-ˈyät-ə, -ˈzhät-\ mountain pass 12,572 feet (3832 meters) S South America in the Andes between Argentina & Chile
Usumbura — see BUJUMBURA
Utah \ˈyü-ˌtȯ, -ˌtä\ state W U.S.; capital, Salt Lake City — **Utah·an** \ˈyü-ˌtȯ-ən, -ˌtȯn, -ˌtä-ən, -ˌtän\ *adj or n* — **Utahn** \-ˌtȯ-ən, -ˌtȯn, -ˌtä-ən, -ˌtän\ *n*
Utrecht \ˈyü-ˌtrekt\ city central Netherlands
Utsu·no·mi·ya \ˌüt-sə-ˈnō-mē-ˌä, -ˌyä\ city Japan in central Honshu N of Tokyo
Ut·ta·ran·chal \ˈu̇t-ə-ˌrän-chəl\ state N India; capital Dehra Dun
Ut·tar Pra·desh \ˌu̇t-ər-prə-ˈdesh, -ˈdāsh\ state N India bordering on Tibet & Nepal; capital, Lucknow
Uz·bek·i·stan \u̇z-ˌbek-i-ˈstan, ˌəz-, -ˈstän\ country W central Asia between Aral Sea & Afghanistan; capital, Tashkent; a constituent republic (**Uz·bek Soviet Socialist Republic** \ˈu̇z-ˌbek, ˈəz-, u̇z-ˈ\) of U.S.S.R. 1924–91
Va·do·da·ra \və-ˈdō-də-ˌrä\ *or* **Ba·ro·da** \bə-ˈrō-də\ city W India in Gujarat
Va·duz \vä-ˈdüts\ town, capital of Liechtenstein
Va·len·cia \və-ˈlen-chə, -chē-ə, -ˈlen-sē-ə\ **1** region & ancient kingdom E Spain between Andalusia & Catalonia **2** city, its capital, on the Mediterranean **3** city N Venezuela WSW of Caracas
Val·la·do·lid \ˌval-əd-ə-ˈlid, -ˈlē\ city NW central Spain NNW of Madrid
Val·le d'Ao·sta \ˌväl-ā-dä-ˈȯs-tə\ region NW Italy bordering on France & Switzerland; capital, Aosta
Val·le·jo \və-ˈlā-ō\ city W California
Val·let·ta \və-ˈlet-ə\ city, capital of Malta
Val·pa·rai·so \ˌval-pə-ˈrī-zō, -ˈrā-\ *or Spanish* **Val·pa·ra·í·so** \ˌväl-pä-rä-ˈē-sō\ city central Chile on the South Pacific WNW of Santiago
Van·cou·ver \van-ˈkü-vər\ **1** city SW Washington on Columbia River opposite Portland, Oregon **2** island W Canada in SW British Columbia **3** city SW British Columbia
Van Diemen's Land — see TASMANIA
Va·nu·a·tu \ˌvä-nü-ˈä-ˌtü\ *or formerly* **New Heb·ri·des** \-ˈheb-rə-ˌdēz\ islands SW Pacific W of Fiji; an independent republic; capital, Port-Vila
Va·ra·na·si \və-ˈrän-ə-sē\ city N India in SE Uttar Pradesh
Var·na \ˈvär-nə\ city & port E Bulgaria on Black Sea
Vat·i·can City \ˌvat-i-kən\ *or Italian* **Cit·tà del Va·ti·ca·no** \chēt-ˈtä-del-ˌvä-tē-ˈkä-nō\ independent papal state within city of Rome, Italy; created 1929
Ve·ne·to \ˈven-ə-ˌtō, ˈvā-nə-\ region NE Italy; capital, Venice
Ven·e·zu·e·la \ˌven-əz-ə-ˈwā-lə, -əz-ˈwā-, -ˈwē-\ country N South America; capital, Caracas — **Ven·e·zu·e·lan** \-lən\ *adj or n*
Ven·ice \ˈven-əs\ *or Italian* **Ve·ne·zia** \və-ˈnet-sē-ə\ city N Italy on islands in Lagoon of Venice — **Ve·ne·tian** \və-ˈnē-shən\ *adj or n*
Ven·tu·ra \ven-ˈtu̇r-ə, -ˈtyu̇r-\ *or officially* **San Buen·a·ven·tu·ra** \ˌsan-ˌbwen-ə-ven-\ city & port SW California
Ve·ra·cruz \ˌver-ə-ˈkrüz, -ˈkrüs\ **1** state E Mexico; capital, Jalapa **2** city E Mexico in Veracruz state on Gulf of Mexico
Verde, Cape \ˈvərd\ *or* **Cape Vert** \ˈvərt\ promontory W Africa in Senegal; most westerly point of Africa
Ver·ee·ni·ging \fə-ˈrā-nə-ging, -nə-kəng\ city NE Republic of South Africa S of Johannesburg
Ver·mont \vər-ˈmänt\ state NE U.S.; capital, Montpelier — **Ver·mont·er** \-ər\ *n*
Ve·ro·na \və-ˈrō-nə\ city N Italy W of Venice
Ver·sailles \ˌvər-ˈsī, ver-\ city N France; suburb of Paris
Ve·su·vi·us \və-ˈsü-vē-əs\ volcano 4190 feet (1277 meters) S Italy near Bay of Naples
Vi·cen·te Ló·pez \və-ˌsent-ə-ˈlō-ˌpez\ city E Argentina N of Buenos Aires
Vi·cen·za \vi-ˈchen-sə\ city NE Italy W of Venice
Vic·to·ria \vik-ˈtōr-ē-ə, -ˈtȯr-\ **1** city, capital of British Columbia on Vancouver Island **2** island N Canada in Arctic Archipelago S of Melville Sound; split between Northwest Territories & Nunavut **3** state SE Australia; capital, Melbourne **4** city & port Hong Kong — **Vic·to·ri·an** \-ē-ən\ *adj or n*
Victoria, Lake lake E Africa in Tanzania, Kenya, & Uganda
Victoria Falls waterfall 355 feet (108 meters) high S Africa in the Zambezi on border between Zambia & Zimbabwe
Victoria Nile — see NILE
Vi·en·na \vē-ˈen-ə\ *or German* **Wien** \ˈvēn\ city, capital of Austria on the Danube — **Vi·en·nese** \ˌvē-ə-ˈnēz, -ˈnēs\ *adj or n*
Vien·tiane \vyen-ˈtyän\ city, capital of Laos
Vie·ques \vē-ˈā-kās\ island Puerto Rico off E coast of main island
Viet·nam \vē-ˈet-ˈnäm, vyet-, ˌvē-ət-, vēt-, -ˈnam\ country SE Asia in Indochina; capital, Hanoi; established 1945–46 & divided 1954–75 at 17th parallel into republics of **North Vietnam** (capital, Hanoi) & **South Vietnam** (capital, Saigon)
Vi·go \ˈvē-ˌgō\ city & port NW Spain on **Vigo Bay** (inlet of North Atlantic)
Vi·ja·ya·wa·da \ˌvij-ə-yə-ˈwäd-ə\ *or formerly* **Bez·wa·da** \bez-ˈwäd-ə\ city SE India in E Andhra Pradesh
Vila — see PORT-VILA
Vil·ni·us \ˈvil-nē-əs\ *or Polish* **Wil·no** \ˈvil-nō\ *or Russian* **Vil·na** \ˈvil-nə\ *or* **Vil·no** \-nō\ city, capital of Lithuania
Vin·land \ˈvin-lənd\ a portion of E coast of North America visited & so called by Norse voyagers about 1000 A.D.; perhaps Newfoundland
Vin·ny·tsya *or* **Vin·ni·tsa** \ˈvin-ət-syə\ city W central Ukraine
Vin·son Massif \ˈvin-sən\ mountain 16,066 feet (4897 meters) W Antarctica in Ellsworth Mountains; highest in Antarctica
Vir·gin·ia \vər-ˈjin-yə, -ˈjin-ē-ə\ state E U.S.; capital, Richmond — **Vir·gin·ian** \-yən, -ē-ən\ *adj or n*
Virginia Beach city SE Virginia on North Atlantic
Vir·gin Islands \ˌvər-jən\ island group West Indies E of Puerto Rico — see BRITISH VIRGIN ISLANDS, VIRGIN ISLANDS OF THE UNITED STATES
Virgin Islands National Park reservation Saint John, Virgin Islands of the United States
Virgin Islands of the United States the W islands of the Virgin Islands group including Saint

Croix, Saint John, & Saint Thomas; capital, Charlotte Amalie (on Saint Thomas)

Vi·sa·yan \və-'sī-ən\ *or* **Bi·sa·yas** \bə-'sī-əz\ islands central Philippines including Bohol, Cebu, Leyte, Masbate, Negros, Panay, & Samar

Vish·a·kha·pat·nam \vish-,äk-ə-'pət-nəm\ city & port E India in Andhra Pradesh

Vis·tu·la \'vis-chə-lə, 'vish-; 'vis-tə-lə\ *or Polish* **Wis·la** \'vē-slä\ river over 660 miles (1062 kilometers) long Poland flowing N from the Carpathians into Gulf of Danzig

Vi·ti Le·vu \,vēt-ē-'lev-ü\ island SW Pacific; largest of the Fiji group

Vi·to·ria \vi-'tōr-ē-ə, -'tȯr-\ *or* **Gas·teiz** \'gäsh-,tās\ city N Spain

Vi·tó·ria \vi-'tōr-ē-ə, -'tȯr-\ city E Brazil NE of Rio de Janeiro

Vit·syebsk \'vēts-yipsk\ *or* **Vi·tebsk** \'vē-,tipsk\ city NE Belarus

Vlad·i·kav·kaz \,vlad-ə-,käf-'käz, -'kaz\ *or 1932–43 & 1955–91* **Or·dzho·ni·kid·ze** \,ȯr-,jän-ə-'kid-zə\ *or 1944–54* **Dzau·dzhi·kau** \dzaů-'jē-,kaů, zaů-\ city S Russia in Europe, in the Caucasus

Vlad·i·vos·tok \,vlad-ə-və-'stäk, -'väs-,täk\ city SE Russia in Asia, on Sea of Japan

Voj·vo·dina \'voi-vȯ-,dē-nä\ province N Serbia

Volcano Islands *or* **Ka·zan Ret·to** \,käz-,än-'ret-ō\ island chain Japan in W Pacific S of Bonin Islands — see IWO JIMA

Vol·ga \'väl-gə, 'vȯl-, 'vōl-\ river 2293 miles (3689 kilometers) long W Russia flowing into Caspian Sea

Vol·go·grad \'väl-gə-,grad, 'vȯl-, 'vōl-\ *or formerly* **Sta·lin·grad** \'stäl-ən-,grad, 'stal-\ *or earlier* **Tsa·ri·tsyn** \tsə-'rēt-sən, sə-\ city S Russia in Europe, on the Volga

Vol·ta \'väl-tə, 'vōl-, 'vȯl-\ river Ghana flowing into Bight of Benin and including **Lake Volta** (reservoir)

Vo·ro·nezh \və-'rȯ-nish\ city SW Russia in Europe, near Don River

Vo·ro·shi·lov·grad \,vȯr-ə-'shē-ləf-,grad, ,vär-, -ləv-\ city E Ukraine in Donets Basin

Vosges \'vōzh\ mountains NE France on W side of valley of the Rhine; highest 4672 feet (1424 meters)

Voy·a·geurs National Park \,vȯi-ə-'zhərz\ reservation N Minnesota on Canadian border

Vun·tut National Park \'vůn-,tůt\ reservation NW Yukon Territory on Alaska border

Vyatka — see KIROV

Wad·den·zee \,väd-n-'zā\ inlet of North Sea N Netherlands

Wai·ki·ki \,wī-ki-'kē\ resort section of Honolulu, Hawaii

Wa·ka·ya·ma \,wäk-ə-'yäm-ə\ city Japan in SW Honshu SW of Osaka

Wake \'wāk\ island North Pacific N of Marshall Islands; U.S. territory

Wa·la·chia *or* **Wal·la·chia** \wä-'lā-kē-ə\ region S Romania between Transylvanian Alps & the Danube

Wales \'wālz\ *or Welsh* **Cym·ru** \'kəm-rē\ *or Latin* **Cambria** principality SW Great Britain; a division of United Kingdom; capital, Cardiff

Wal·la·sey \'wäl-ə-sē\ town NW England on Irish Sea

Wal·lis \'wäl-əs\ islands SW Pacific NE of Fiji; with Futuna Islands constitute a French overseas territory (**Wallis and Futuna Islands**)

Wal·lo·nia \wä-'lō-nē-ə\ semiautonomous region S Belgium

Wal·sall \'wȯl-,sȯl, -səl\ town W central England NNW of Birmingham

Wal·tham Forest \,wȯl-thəm-\ borough of NE Greater London, England

Wands·worth \'wändz-wərth, 'wänz-\ borough of SW Greater London, England

Wan·ne–Eick·el \,vän-ə-'ī-kəl\ city W Germany in the Ruhr

Wa·pusk National Park \'wä-,půsk\ reservation NE Manitoba bordering Hudson Bay

War·ley \'wȯr-lē\ town W central England; a suburb of Birmingham

War·ren \'wȯr-ən, 'wär-\ city SE Michigan N of Detroit

War·saw \'wȯr-,sȯ\ *or Polish* **War·sza·wa** \vär-'shäv-ə\ city, capital of Poland

War·wick·shire \'wär-ik-,shiər, -shər\ *or* **Warwick** county central England

Wa·satch \'wȯ-,sach\ range of the Rockies SE Idaho & N central Utah; highest Mount Timpanogos 12,008 feet or 3660 meters (in Utah)

Wash·ing·ton \'wȯsh-ing-tən, 'wäsh-\ **1** state NW U.S.; capital, Olympia **2** city, capital of U.S.; coextensive with District of Columbia — **Wash·ing·to·nian** \,wȯsh-ing-'tō-nē-ən, ,wäsh-, -nyən\ *adj or n*

Washington, Mount mountain 6288 feet (1916 meters) N New Hampshire; highest in White Mountains

Wa·ter·bury \'wȯt-ər-,ber-ē, 'wȯt-ə-, 'wät-\ city W central Connecticut

Wa·ter·ford \'wȯt-ər-fərd, 'wät-\ county S Ireland in Munster

Wa·ter·ton–Glacier International Peace Park \'wȯt-ər-tən-, 'wät-\ international park, comprising **Waterton Lakes National Park** in S Alberta, and **Glacier National Park** in NW Montana

Watlings — see SAN SALVADOR

Wed·dell Sea \wə-'del, 'wed-l\ arm of the South Atlantic E of Antarctic Peninsula

Wei·mar Republic \'vī-,mär, 'wī-\ the German republic 1919–33

Wel·land \'wel-ənd\ canal 27 miles (44 kilometers) long SE Ontario connecting Lake Erie & Lake Ontario

Wel·ling·ton \'wel-ing-tən\ city, capital of New Zealand

Wes·sex \'wes-iks\ ancient Anglian kingdom S England; capital, Winchester

West Australian Current warm current flowing N off W coast of Australia

West Bank area Middle East W of Jordan River

West Bengal state E India; capital, Calcutta

West Brom·wich \'brəm-ij, 'bräm-, -ich\ town W central England NW of Birmingham

West Co·vi·na \kō-'vē-nə\ city SW California

West Dunbartonshire administrative area W Scotland

Western Australia state W Australia; capital, Perth

Western Cape province SW Republic of South Africa

Western Ghats \'gȯts\ chain of low mountains SW India

Western Isles *or* **Ei·lean Siar** \'el-ən 'shēər\ the Outer Hebrides constituting an administrative area of W Scotland

Western Sahara *or formerly* **Spanish Sahara** region NW Africa divided 1975 between Mauritania which gave up its claim in 1979 & Morocco which thereafter occupied the entire territory

Western Samoa — see SAMOA

West Germany the former Federal Republic of Germany — see GERMANY

West Indies islands lying between SE North America & N South America & comprising the Greater Antilles, Lesser Antilles, & Bahamas — **West Indian** *adj or n*

West Lothian administrative area S Scotland

West·meath \west-'mēth, wes-, -'mēth\ county E central Ireland in Leinster

West Midlands metropolitan county W central England

West·min·ster \'west-,min-stər\ **1** city N central Colorado NW of Denver **2** *or* **City of Westminster** borough of W central Greater London, England

West Pakistan the former W division of Pakistan now coextensive with Pakistan

West Papua *or formerly* **Iri·an Ja·ya** \'ir-ē-,än-'jī-ä\ *or* **West Irian** *or earlier* **Netherlands New Guinea** territory of Indonesia comprising W half of New Guinea

West·pha·lia \west-'fāl-yə, -'fā-lē-ə, wes-\ region NW Germany E of the Rhine; now part of North Rhine-Westphalia — **West·pha·lian** \-'fāl-yən, -'fā-lē-ən\ *adj or n*

West Punjab — see PUNJAB 3

West Quod·dy Head \,kwäd-ē\ cape; most easterly point of Maine & of the Lower 48

West Sus·sex \'səs-iks\ county SE England

West Valley City city N Utah S of Salt Lake City

West Virginia state E U.S.; capital, Charleston — **West Virginian** *adj or n*

West York·shire \'yȯrk-,shiər, -shər\ metropolitan county NW England

Wex·ford \'weks-fərd\ county SE Ireland in Leinster

White mountains N New Hampshire in the Appalachians — see WASHINGTON (Mount)

White·horse \'hwīt-,hȯrs, 'wīt-\ city NW Canada, capital of Yukon Territory

White Nile — see NILE

White Sea *or* **Be·lo·ye Mo·re** \,bel-ə-yə-'mȯr-yə\ inlet of Barents Sea NW Russia

Whit·ney, Mount \'hwit-nē, 'wit-\ mountain 14,495 feet (4418 meters) SE central California in Sierra Nevada in Sequoia National Park; highest in U.S. outside of Alaska

Wich·i·ta \'wich-ə-,tȯ\ city S central Kansas on Arkansas River

Wichita Falls city N Texas

Wick·low \'wik-lō\ county E Ireland in Leinster

Wien — see VIENNA

Wies·ba·den \'vēs-,bäd-n, 'vis-\ city W Germany on the Rhine W of Frankfurt, capital of Hesse

Wight, Isle of \'wīt\ island S England in English Channel

Wil·helms·ha·ven \,vil-,helmz-'häf-ən, 'vil-əmz-,\ city NW Germany NW of Bremen

Wil·lem·stad \'vil-əm-,stät\ city, capital of Netherlands Antilles on Curaçao Island

Wil·ming·ton \'wil-ming-tən\ city N Delaware; largest in state

Wilno — see VILNIUS

Wilt·shire \'wilt-,shiər, 'wil-chər, 'wilt-shər\ county S England

Wim·ble·don \'wim-bəl-dən\ section of Merton in Greater London, England

Wind Cave National Park reservation SW South Dakota in Black Hills

Win·der·mere \'win-dər-,miər, -də-\ lake NW England in Lake District

Wind·hoek \'vint-,hůk\ city, capital of Namibia

Wind·sor \'win-zər\ city S Ontario on Detroit River

Wind·ward \'win-dwərd\ islands West Indies in the S Lesser Antilles extending S from Martinique but not including Barbados, Tobago, or Trinidad

Win·ni·peg \'win-ə-,peg\ city, capital of Manitoba

Winnipeg, Lake lake S central Manitoba

Win·ni·pe·sau·kee, Lake \,win-ə-pə-'sȯ-kē\ lake central New Hampshire

Win·ston–Sa·lem \,win-stən-'sā-ləm\ city N central North Carolina

Wis·con·sin \wis-'kän-sən\ state N central U.S.; capital, Madison — **Wis·con·sin·ite** \-sə-,nīt\ *n*

Wisla — see VISTULA

Wit·wa·ters·rand \'wit-,wȯt-ərz-,rand, -,wät-, -,ränd, -,ränt\ *or* **Rand** \'rand, 'ränd, 'ränt\ ridge of gold-bearing rock NE Republic of South Africa

Wol·lon·gong \'wůl-ən-,gäng, -,gong\ city SE Australia in E New South Wales S of Sydney

Wol·ver·hamp·ton \'wůl-vər-,ham-tən, ,hamp-\ town W central England NW of Birmingham

Won·san \'wən-,sän\ city North Korea on E coast

Wood Buffalo National Park reservation N Alberta & SE Northwest Territories

Worces·ter \'wůs-tər\ **1** city E central Massachusetts W of Boston **2** *or* **Worces·ter·shire** \-,shir, -shər\ county W central England

Wounded Knee locality SW South Dakota

Wran·gell, Mount \'rang-gəl\ volcano 14,163 feet (4317 meters) S Alaska in Wrangell Range; highest volcano in U.S.

Wrangell–Saint Eli·as National Park \-sānt-i-'lī-əs\ reservation S central Alaska E of Anchorage

Wrex·ham \'rek-səm\ administrative area NE Wales

Wro·claw \'vrȯt-,släf, -,släv\ *or German* **Bres·lau** \'bres-,laů\ city SW Poland in Silesia

Wu·chang \'wü-'chäng\ former city E central China — see WUHAN

Wu·han \'wü-'hän\ city S China, capital of Hubei; formed from former separate cities of Hankow, Hanyang, & Wuchang

Wuhsien — see SUZHOU

Wu–lu–mu–ch'i — see ÜRÜMQI

Wup·per·tal \'vůp-ər-,täl\ city W Germany in valley of the Ruhr ENE of Düsseldorf

Würt·tem·berg \'wərt-əm-,bərg, 'wůrt-; 'vœrt-əm-,berk\ region SW Germany between Baden & Bavaria; chief city, Stuttgart; now part of Baden-Württemberg

Wy·o·ming \wī-'ō-ming\ state NW U.S.; capital, Cheyenne — **Wy·o·ming·ite** \-ming-,īt\ *n*

Xia·men *or* **Hsia–men** \shē-'ä-'mən\ *or* **Amoy** \ä-'mȯi\ city SE China in S Fujian on two islands

\ə\ **abut** \ᵊ\ **kitten** \ər\ **further** \a\ **ash** \ā\ **ace** \ä\ **mop, mar** \aů\ **out** \ch\ **chin** \e\ **bet** \ē\ **easy** \g\ **go** \i\ **hit** \ī\ **ice** \j\ **job** \ŋ\ **sing** \ō\ **go** \ȯ\ **law** \ȯi\ **boy** \th\ **thin** \t͟h\ **the** \ü\ **loot** \ů\ **foot** \y\ **yet** \zh\ **vision** *see also* Guide to Pronunciation

Xi'·an *or* **Si·an** \'shē-'än\ *or formerly* **Chang·an** \'chäng-'än\ city E central China, capital of Shaanxi
Xianggang — see HONG KONG
Xiang·tan *or* **Hsiang–t'an** *or* **Siang·tan** \shē-'äng-'tän\ city SE China in Hunan
Xi·ning *or* **Si·ning** \'shē-'ning\ city NW China, capital of Qinghai
Xin·jiang Uy·gur *or* **Sin·kiang–Ui·ghur** \'shin-jē-'äng-'wē-gər\ region & former province W China; capital, Ürümqi
Xizang — see TIBET
Xu·zhou \'shü-'jō\ *or* **Hsü–chou** *or* **Süchow** \'shü-'jō, 'sü-; 'sü-'chaü\ city E China in NW Jiangsu
Ya·kutsk \yə-'kütsk\ city E central Russia in Asia
Yal·ta \'yȯl-tə\ city & port Ukraine on S coast of Crimea
Ya·lu \'yäl-ü\ *or* **Am·nok** \'am-,näk\ river about 500 miles (804 kilometers) long SE Manchuria, China & North Korea flowing into Korea Bay
Ya·mous·sou·kro \,yä-mü-'sü-krō\ town central Ivory Coast, official capital of the country
Yangku — see TAIYUAN
Yan·gon \,yän-'gōn\ *or formerly* **Ran·goon** \ran-'gün, rang-\ city, capital of Myanmar
Yangtze — see CHANG
Yao \'yaü\ city Japan in S Honshu E of Osaka
Yaoun·dé \yaün-'dā\ city, capital of Cameroon
Yap \'yap, 'yäp\ island W Pacific in the W Carolines
Ya·ren \'yä-,ren\ town, capital of Nauru
Ya·ro·slavl \,yär-ə-'släv-əl\ city central Russia in Europe, NE of Moscow
Ye·ka·te·rin·burg \yi-'kat-ə-rən-,bərg, yi-,kät-ə-rən-'bürk\ *or 1924–91* **Sverd·lovsk** \sverd-'lȯfsk\ city W Russia in Asia, in central Ural Mountains
Yellow **1** — see HUANG **2** sea section of East China Sea between N China & Korea
Yel·low·knife \'yel-ə-,nīf\ town Canada, capital of Northwest Territories
Yel·low·stone National Park \'yel-ə-,stōn\ reservation NW Wyoming, E Idaho, & S Montana
Ye·men \'yem-ən\ country S Arabia bordering on Red Sea & Gulf of Aden; formed 1990 by merger of **Yemen Arab Republic** (capital, Sanaa) with **People's Democratic Republic of Yemen** *or* **Southern Yemen** (capital, Aden); capital, Sanaa — **Ye·me·ni** \'yem-ə-nē\ *adj or n* — **Ye·men·ite** \-ə-,nīt\ *n*
Yen·i·sey *or* **Yen·i·sei** \,yen-ə-'sā\ river over 2500 miles (4022 kilometers) long central Russia, flowing N into Arctic Ocean
Ye·re·van \,yer-ə-'vän\ city, capital of Armenia
Yezo — see HOKKAIDO
Yi·bin \'yē-'bēn\ *or* **I–pin** \'ē-'bēn, -'pin\ *or formerly* **Sü·chow** \'shü-'jō, 'sü-; 'sü-'chaü\ city central China in S Sichuan
Yog·ya·kar·ta \,yōg-yə-'kär-tə\ city Indonesia in S Java
Yo·ho National Park \'yō-hō\ reservation W Canada in SE British Columbia on Alberta boundary
Yo·ko·ha·ma \,yō-kə-'häm-ə\ city Japan in SE Honshu on Tokyo Bay S of Tokyo
Yo·ko·su·ka \yō-'kȯ-sə-kə, -'kȯ-skə\ city Japan in E Honshu W of entrance to Tokyo Bay
Yon·kers \'yäng-kərz\ city SE New York N of New York City
York \'yȯrk\ city N England
York, Cape cape NE Australia in Queensland at N tip of Cape York Peninsula
Yo·sem·i·te Falls \yō-'sem-ət-ē\ waterfall E California in **Yosemite National Park** (reservation in the Sierra Nevada); includes two falls, the upper 1430 feet (436 meters) & the lower 320 feet (98 meters), connected by a series of cascades; total drop 2425 feet (739 meters)
Youngs·town \'yəngz-,staün\ city NE Ohio
Youth, Isle of island W Cuba in the Caribbean
Yssel — see IJSSEL
Yu·ca·tán \,yü-kə-'tan, -'tän\ **1** peninsula SE Mexico & N Central America including Belize & N Guatemala **2** state SE Mexico; capital, Mérida
Yu·go·sla·via *or formerly* **Ju·go·sla·via** \,yü-gō-'släv-ē-ə\ **1** former country S Europe on the Adriatic consisting of Serbia, Montenegro, Slovenia, Croatia, Bosnia and Herzegovina, & Macedonia; capital, Belgrade **2** — see SERBIA AND MONTENEGRO — **Yu·go·slav** \,yü-gō-'släv, -'slav\ *or* **Yu·go·sla·vi·an** \-'släv-ē-ən\ *adj or n*
Yu·kon \'yü-,kän\ **1** river 1979 miles (3185 kilometers) long NW Canada & Alaska flowing into Bering Sea **2** *or* **Yukon Territory** territory NW Canada; capital, Whitehorse
Yun·nan \yü-'nän\ **1** province SW China bordering on Myanmar & Indochina; capital, Kunming **2** — see KUNMING
Yunnanfu — see KUNMING
Zab·rze \'zäb-zhā\ city SW Poland in Silesia
Za·ca·te·cas \,zak-ə-'tā-kəs, -'tek-əs\ **1** state N central Mexico **2** city; its capital
Zag·a·zig \'zag-ə-,zig\ city N Egypt NNE of Cairo
Za·greb \'zäg-,reb\ city, capital of Croatia
Za·he·dan \,zä-hi-'dän\ city E Iran
Zaire \zä-'iər\ **1** river in Africa — see CONGO 1 **2** country in Africa — see CONGO 2
Zam·be·zi *or* **Zam·be·si** \zam-'bē-zē\ river about 1700 miles (2735 kilometers) long SE Africa flowing from NW Zambia into Mozambique Channel
Zam·bia \'zam-bē-ə\ country S Africa N of the Zambezi; formerly the British protectorate of **Northern Rhodesia**; an independent republic since 1964; capital, Lusaka
Zam·bo·an·ga \,zam-bə-'wäng-gə\ city S Philippines in SW Mindanao
Zan·zi·bar \'zan-zə-,bär\ island Tanzania off NE Tanganyika coast; formerly a sultanate & British protectorate including also Pemba & other islands; became independent 1963; united 1964 with Tanganyika forming Tanzania
Za·po·rizh·zhya *or* **Za·po·ro·zh'ye** \,zäp-ə-'rȯ-zhə\ city SE Ukraine
Za·ra·go·za \,zar-ə-'gō-zə\ *or* **Sar·a·gos·sa** \,sar-ə-'gäs-ə\ city NE Spain in W Aragon
Zealand — see SJÆLLAND
Zhang·jia·kou \'jäng-jē-'ä-'kō\ *or* **Ch'ang–chia–k'ou** \'chäng-jē-'ä-'kō\ *or* **Kal·gan** \'kal-'gan\ city NE China in NW Hebei NW of Beijing
Zhang·zhou *or* **Chang–chou** \'jäng-'jō\ city SE China in S Fujian
Zhdanov — see MARIUPOL'
Zhe·jiang *or* **Che·kiang** \'jəj-ē-'äng\ province E China bordering on East China Sea; capital, Hangzhou
Zheng·zhou *or* **Cheng–chou** \'jəng-'jō\ city NE central China, capital of Henan
Zhen·jiang *or* **Chen–chiang** \'jən-jē-'äng\ city E China in NW central Jiangsu
Zhu·zhou *or* **Chu–chou** *or* **Chu·chow** \'jü-'jō\ city SE China in E Hunan
Zhy·to·myr *or* **Zhi·to·mir** \zhi-'tȯ-,miər\ city W Ukraine
Zi·bo *or* **Tzu–po** \'dzə-'bō, 'zə-\ city E China in central Shandong
Zi·gong *or* **Tzu–kung** *or* **Tze·kung** \'dzə-'güng, 'zə-\ city S central China in S Sichuan
Zim·ba·bwe \zim-'bäb-wā, -wē\ *or before 1979* **Rho·de·sia** \rō-'dē-zhə, -zhē-ə\ *or* **Southern Rhodesia** country S Africa S of Zambezi River; capital, Harare — **Zim·ba·bwe·an** \-ən\ *adj or n*
Zi·on \'zī-ən\ *or* **Si·on** \'sī-ən\ the stronghold of Jerusalem conquered by David; occupied in ancient times by the Jewish Temple
Zion National Park reservation SW Utah
Zla·to·ust \,zlät-ə-'üst\ city W Russia in Asia in the S Urals
Zom·ba \'zäm-bə\ city S Malawi S of Lake Malawi
Zui·der Zee \,zīd-ər-'zā, -'zē\ former inlet of North Sea N Netherlands; now (as IJsselmeer) partly reclaimed
Zu·lu·land \'zü-lü-,land\ territory E Republic of South Africa on Indian Ocean
Zu·rich \'zu̇r-ik\ city N Switzerland on **Lake of Zurich** (25 miles or 40 kilometers long)
Zwick·au \'tsfik-,aü, 'zwik-\ city E Germany S of Leipzig

A Handbook of Style

Punctuation

The English writing system uses punctuation marks to separate groups of words for meaning and emphasis; to convey an idea of the variations of pitch, volume, pauses, and intonations of speech; and to help avoid contextual ambiguity. The use of the standard English punctuation marks is discussed in the following pages; examples are provided to illustrate the general rules.

’ Apostrophe

1. Indicates the possessive case of nouns and indefinite pronouns. The possessive case of almost all singular nouns may be formed by adding *’s*. Traditionally, however, only the apostrophe is added when the *s* would not be pronounced in normal speech. The possessive case of plural nouns ending in *s* or in an \s\ or \z\ sound is generally formed by adding an apostrophe only; the possessive of irregular plurals is formed by adding *’s*.

    her mother-in-law’s car

    anyone’s guess

    the boy’s mother

    the boys’ mothers

    Degas’s drawings

    Knox’s products

    Aristophanes’ play

    for righteousness’ sake

    the Stephenses’ house

    children’s laughter

2. Marks omission of letters in contracted words.

    didn’t

    o’clock

    hang ’em up

3. Marks omission of digits in numbers.

    class of ’83

4. Is often used to form plurals of letters, figures, punctuated abbreviations, symbols, and words referred to as words.

    Dot your *i*’s and cross your *t*’s.

    His *1*’s and his *7*’s looked alike.

    Two of the junior faculty have Ph.D’s.

    She has trouble pronouncing her *the*’s.

[ ] Brackets

1. Set off interpolated editorial matter within quoted material.

    He wrote, “I ain’t [sic] going.”

    Vaulting ambition, which o’erleaps itself
    And falls on the other [side].
    —Shakespeare

2. Function as parentheses within parentheses.

    Bowman Act (22 Stat., ch. 4, § [or sec.] 4, p. 50)

3. Set off phonetic symbols and transcriptions.

    [t] in British *duty*

    the word is pronounced [ˈek-sə-jənt]

: Colon

1. Introduces a clause or phrase that explains, illustrates, amplifies, or restates what has gone before.

    The sentence was poorly constructed: it lacked both unity and coherence.

2. Directs attention to an appositive.

    He had only one pleasure: eating.

3. Introduces a series.

    Three abstained: England, France, and Belgium.

4. Introduces lengthy quoted material set off from the rest of a text by indentation but not by quotation marks.

    I quote from the text of Chapter One:

5. Separates elements in page references, in bibliographical and biblical citations, and in set formulas used to express ratios and time.

    Journal of the American Medical Association 48:356

Stendhal, *Love* (New York: Penguin, 1975)

John 4:10   a ratio of 3:5   8:30 a.m.

6. Separates titles and subtitles (as of books).

Battle Cry of Freedom: The Era of the Civil War

7. Follows the salutation in formal correspondence.

Dear Sir or Madam:

Ladies and Gentlemen:

8. Punctuates headings in memorandums and formal correspondence.

TO: VIA:

SUBJECT: REFERENCE:

Comma ,

1. Separates main clauses joined by a coordinating conjunction (such as *and, but, or, nor,* or *for*), and sometimes short parallel clauses not joined by conjunctions.

She knew very little about him, and he volunteered nothing.

I came, I saw, I conquered.

2. Sets off an adverbial clause or a long adverbial phrase that precedes or interrupts the main clause.

When she discovered the answer, she reported it to us.

The report, after being read aloud, was put up for consideration.

3. Sets off transitional words and expressions (such as *on the contrary, on the other hand*), conjunctive adverbs (such as *consequently, furthermore, however*), and expressions that introduce an illustration or example (such as *namely, for example*).

My partner, on the other hand, remains unconvinced.

The regent's whim, however, threw the negotiations into chaos.

She responded as completely as she could; that is, she answered each of the individual questions specifically.

4. Sets off contrasting and opposing expressions within sentences.

The cost is not $65.00, but $56.65.

He changed his style, not his ethics.

5. Separates words, phrases, or clauses in series. (Many omit the comma before the conjunction introducing the last item in a series when no ambiguity results.)

He was young, eager, and restless.

It requires one to travel constantly, to have no private life, and to live on almost nothing.

Be sure to pack a flashlight, a sweater and an extra pair of socks.

6. Separates coordinate adjectives modifying a noun. However, a comma is not used between two adjectives when the first modifies the combination of the second adjective and the word or phrase it modifies.

The harsh, damp, piercing wind cut through his jacket.

a low common denominator

7. Sets off parenthetical elements such as nonrestrictive clauses and phrases.

Our guide, who wore a blue beret, was an experienced traveler.

We visited Gettysburg, site of the famous battle.

The book's author, Marie Jones, was an accomplished athlete.

8. Introduces a direct quotation, terminates a direct quotation that is neither a question nor an exclamation, and sets off split quotations. The comma is not used with quotations that are tightly integrated into the sentences in which they appear (e.g., as subject or predicate nominatives) or those that do not represent actual dialogue.

Mary said, "I am leaving."

"I am leaving," Mary said.

"I am leaving," Mary said with determination, "even if you want me to stay."

"The computer is down" was the reply she feared.

The fact that he said he was about to "faint from hunger" doesn't mean he actually fainted.

9. Sets off words in direct address, absolute phrases, and mild interjections.

You may go, John, if you wish.

I fear their encounter, his temper being what it is.

Ah, that's my idea of an excellent dinner.

10. Separates a tag question from the rest of the sentence.

It's a fine day, isn't it?

11. Indicates the omission of a word or words used in a parallel construction earlier in the sentence. When the meaning of the sentence is quite clear without the comma, the comma is omitted.

Common stocks are preferred by some investors; bonds, by others.

He was in love with her and she with him.

12. Is used to avoid ambiguity that might arise from adjacent words.

To Mary, Jane was someone special.

13. Is used to divide digits in numbers into groups of three; however, it is generally not used in pagination, in dates, or in street numbers, and sometimes not used in numbers with four digits.

Smithville, pop. 100,000

4,550 cars

but

page 1411   4507 Main St.

3600 rpm   the year 1983

14. Punctuates an inverted name.

Morton, William A.

15. Separates a surname from a following title or degree and often from the words "Junior" and "Senior" and their abbreviations.

Sandra H. Cobb, Vice President

Jesse Ginsburg, D.D.M.

16. Sets off geographical names (such as state or country from city), elements of dates, and addresses. When just the month and the year are given in a date, the comma is usually omitted.

Shreveport, Louisiana, is the site of a large air base.

On Sunday, June 23, 1940, he was wounded.

Number 10 Downing Street, London, is a famous address.

She began her career in April 1993 at a modest salary.

17. Follows the salutation in informal correspondence, and follows the complimentary close of a letter.

Dear Mark,

Affectionately,

Very truly yours,

— Dash

1. Usually marks an abrupt change or break in the continuity of a sentence.

When in 1960 the stockpile was sold off—indeed, dumped as surplus—natural rubber sales were hard hit.—Barry Commoner

2. Is sometimes used in place of commas or parentheses when special emphasis is required.

The presentations—and especially the one by Ms. Dow—impressed the audience.

3. Introduces a statement that explains, summarizes, or expands on what precedes it.

Oil, steel, and wheat—these are the sinews of industrialization.

The motion was then tabled—that is, removed indefinitely from consideration.

4. Often precedes the attribution of a quotation.

My foot is on my native heath. . . .—Sir Walter Scott

5. Sets off an interrupting clause or phrase. The dash takes the place of a comma that would ordinarily set off the clause, but an exclamation point or question mark is retained.

If we don't succeed—and the critics say we won't—then the whole project is in jeopardy.

They are demanding that everything—even the marshland!—be transferred to the new trust.

Your question—it was *your* question, wasn't it, Mr. Jones?—just can't be answered.

. . . . Ellipsis (or Suspension Points)

1. Indicates the omission of one or more words within a quoted passage. When four dots are used, the ellipsis indicates the omission of one or more sentences within the passage or the omission of words at the end of a sentence. The first or the last of the four dots is a period.

In the little world in which children have their existence, . . . there is nothing so finely perceived and so finely felt as injustice.—Charles Dickens

Security is mostly a superstition. . . . Avoiding danger is no safer in the long run than outright exposure. . . . Life is either a daring adventure or nothing.—Helen Keller

2. Usually indicates omission of one or more lines of poetry when ellipsis is extended the length of the line.

I think that I shall never see
A poem lovely as a tree
. . . .
Poems are made by fools like me,
But only God can make a tree.
—Joyce Kilmer

3. Indicates halting speech or an unfinished sentence in dialogue.

"I'd like to . . . that is . . . if you don't mind. . . ."

! Exclamation Point

1. Ends an emphatic phrase or sentence.

Get out of here!

Her notorious ostentation—she flew her friends to Bangkok for her birthday parties!—was feasted on by the popular press.

2. Ends an emphatic interjection.

Encore!

All of this proves—at long last!—that we were right from the start.

- Hyphen

The hyphen is often used between parts of a compound. The styling of such words varies; when in doubt, see the entry in the dictionary at its own place or in a list of undefined words at an individual prefix. For unentered compounds, advice will be found in *Merriam-Webster's Manual for Writers and Editors* or a comparable guide.

1. Is often used between a prefix and root, especially whenever the root is capitalized, when two identical vowels come together, or when the resulting word could be confused with another identically spelled word.

pre-Renaissance

co-opted

anti-inflationary

re-cover a sofa

but

recover from an illness

2. Is used in some compounds, especially those containing prepositions.

president-elect	sister-in-law
good-for-nothing	over-the-counter
falling-out	write-off

3. Is often used in compound modifiers in attributive position.

traveling in a fast-moving van

She has gray-green eyes.

a come-as-you-are party

4. Suspends the first element of a hyphenated compound or a prefix (hyphenated or not) when the second element or base word is part of a following hyphenated compound or derived form.

a six- or eight-cylinder engine

pre- and postadolescent trauma

5. Marks division of a word at the end of a line.

The ruling pas-
sion of his life

6. Is used in writing out compound numbers between 21 and 99.

thirty-four

one hundred thirty-eight

7. Is often used between the numerator and the denominator in writing out fractions, especially when they are used as modifiers. However, fractions used as nouns are often written as open compounds, especially when either the numerator or the denominator already contains a hyphen.

a two-thirds majority of the vote

three fifths of her paycheck

one seventy-second of an inch

8. Serves as an equivalent of *through* or (*up*) *to and including* when used between indicators of range such as numbers and dates. (In typeset material the longer en dash is used.)

pages 40–98

the years 1980–89

9. Serves as the equivalent of *to, and,* or *versus* in indicating linkage or opposition. (In typeset material the longer en dash is used.)

the New York–Paris flight

the Hardy–Weinberg law

the Lincoln–Douglas Debates

The final score was 7–2.

Hyphen, Double ⸗

Is used at the end-of-line division of a hyphenated compound to indicate that the compound is hyphenated and not closed.

self ⸗ [end of line] seeker

but

self- [end of line] same

Parentheses ( )

1. Enclose words, numbers, phrases, or clauses that provide examples, explanations, or supplementary material that does not essentially alter the meaning of the sentence.

Three old destroyers (all now out of commission) will be scrapped.

He has followed the fortunes of the modern renaissance (*al-Nahdad*) in the Arabic-speaking world.

2. Enclose numerals that confirm a written number in a text.

Delivery will be made in thirty (30) days.

3. Enclose numbers or letters in a series.

We must set forth (1) our long-term goals, (2) our immediate objectives, and (3) the means at our disposal.

4. Enclose abbreviations that follow their spelled-out forms or spelled-out forms that follow their abbreviations.

a ruling by the Federal Communications Commission (FCC)

the manufacture and disposal of PVC (polyvinyl chloride)

5. Indicate alternative terms.

Please indicate the lecture(s) you would like to attend.

6. Enclose publication data in footnotes and endnotes.

Marguerite Yourcenar, *The Dark Brain of Piranesi and Other Essays* (New York: Farrar, Straus and Giroux, 1985), p. 9.

7. Are used with other punctuation marks in the following ways:

If the parenthetic expression is an independent sentence standing alone, its first word is capitalized and a period is included *inside* the last parenthesis. However, if the parenthetic expression, even if it could stand alone as a sentence, occurs within a sentence, it is uncapitalized and has no sentence period but may have an exclamation point, a question mark, a period for an abbreviation, or quotation marks within the closing parenthesis.

The discussion was held in the boardroom. (The results are still confidential.)

Although we liked the restaurant (their Italian food was the best), we seldom went there.

After waiting in line for an hour (why do we do these things?), we finally left.

Years ago, someone (I wish I could remember who!) told me about it.

What was once informally known as A.B.D. status is now often recognized by the degree of Master of Philosophy (M. Phil.).

He was depressed ("I must resign") and refused to do anything.

No punctuation mark should be placed directly before parenthetical material in a sentence; if a break is required, punctuation should be placed *after* the final parenthesis.

I'll get back to you tomorrow (Friday), when I have more details.

Period .

1. Ends sentences or sentence fragments that are neither interrogatory nor exclamatory.

Not bad.

Give it your best.

I gave it my best.

He asked if she had given it her best.

2. Follows some abbreviations and contractions.

Dr. A.D. ibid. i.e.

Jr. etc. cont.

3. Is normally used with an individual's initials.

F. Scott Fitzgerald

T. S. Eliot

4. Is used after numerals and letters in vertical enumerations and outlines.

Required skills are:
1. Shorthand
2. Typing
3. Transcription

I. Objectives
A. Economy
1. low initial cost
2. low maintenance cost
B. Ease of operation

Question Mark ?

1. Ends a direct question.

How did she do it?

"How did she do it?" he asked.

2. Ends a question that is part of a larger sentence, but not an indirect question.

How did she do it? was the question on each person's mind.

He wondered, Will it work?

He wondered whether it would work.

3. Indicates the writer's ignorance or uncertainty.

Geoffrey Chaucer, English poet (1342?–1400)

Quotation Marks, Double " "

1. Enclose direct quotations but not indirect quotations.

She said, "I am leaving."

She said that she was leaving.

2. Enclose words or phrases borrowed from others, words used in a special way, and words of marked informality when introduced into formal writing.

Much of the population in the hellish future he envisions is addicted to "derms," patches that deliver potent drug doses instantaneously through the skin.

He called himself "emperor," but he was really just a dictator.

He was arrested for smuggling "smack."

3. Enclose titles of poems, short stories, articles, lectures, chapters of books, short musical compositions, and radio and TV programs.

Robert Frost's "After Apple-Picking"

Cynthia Ozick's "Rosa"

The third chapter of *Treasure Island* is entitled "The Black Spot."

"All the Things You Are"

Debussy's "Clair de lune"

NBC's "Today Show"

4. Are used with other punctuation marks in the following ways:

The period and the comma fall *within* the quotation marks.

"I am leaving," she said.

It was unclear how she maintained such an estate on "a small annuity."

The colon and semicolon fall *outside* the quotation marks.

There was only one thing to do when he said "I may not run": promise him a large campaign contribution.

He spoke of his "little cottage in the country"; he might better have called it a mansion.

The dash, the question mark, and the exclamation point fall *within* the quotation marks when they refer to the quoted matter only; they fall *outside* when they refer to the whole sentence.

"I can't see how—" he started to say.

He asked, "When did she leave?"

What is the meaning of "the open door"?

The sergeant shouted "Halt!"

Save us from his "mercy"!

5. Are not used with *yes* or *no* except in direct discourse.

She said yes to all our requests.

6. Are not used with lengthy quotations set off from the text.

He took the title for his biography of Thoreau from a passage in *Walden*:

> I long ago lost a hound, a bay horse, and a turtle-dove, and am still on their trail. . . . I have met one or two who had heard the hound, and the tramp of the horse, and even seen the dove disappear behind a cloud, and they seemed as anxious to recover them as if they had lost them themselves.

However, the title *A Hound, a Bay Horse, and a Turtle-Dove* probably puzzled some readers.

' ' Quotation Marks, Single

1. Enclose a quotation within a quotation in American usage. When both single and double quotation marks occur at the end of a sentence, the period typically falls within *both* sets of marks.

The witness said, "I distinctly heard him say, 'Don't be late,' and then heard the door close."

The witness said, "I distinctly heard him say, 'Don't be late.' "

2. Are sometimes used in place of double quotation marks, especially in British usage. In this case a quotation within a quotation is set off by double quotation marks.

The witness said, 'I distinctly heard him say, "Don't be late," and then heard the door close.'

; Semicolon

1. Links independent clauses not joined by a coordinating conjunction.

Some people have the ability to write well; others do not.

2. Links clauses joined by a conjunctive adverb (such as *consequently, furthermore, however*).

Speeding is illegal; furthermore, it is very dangerous.

3. Often occurs before expressions that introduce expansions or series (such as *for example, for instance, that is, e.g.,* or *i.e.*).

As a manager she tried to do the best job she could; that is, to keep her project on schedule and under budget.

4. Separates phrases that contain commas.

The country's resources consist of large ore deposits; lumber, waterpower, and fertile soils; and a strong, rugged people.

Send copies to our offices in Portland, Maine; Springfield, Illinois; and Savannah, Georgia.

5. Is placed outside quotation marks and parentheses.

They again demanded "complete autonomy"; the demand was again rejected.

/ Virgule (or Slash)

1. Separates alternatives.

high-heat and/or high-speed applications

. . . sit hour after hour . . . and finally year after year in a catatonic/frenzied trance rewriting the Bible
—William Saroyan

2. Replaces the word *to* or *and* between related terms that are compounded.

the fiscal year 1983/1984

in the May/June issue

3. Divides run-in lines of poetry.

Say, sages, what's the charm on earth/Can turn death's dart aside?—Robert Burns

4. Divides elements in dates and divides numerators and denominators in fractions.

offer expires 5/19/2007

Fifteen and 44/100 dollars

5. Often represents *per* or *to* when used with units of measure or to indicate the terms of a ratio.

9 ft/sec

29 mi/gal

risk/reward trade-off

6. Sets off phonemes and phonemic transcription.

/b/ as in *but*

Capitalization

Capitals are used for two broad purposes in English: they mark a beginning (as of a sentence) and they signal a proper noun, pronoun, or adjective. The following principles, each with examples, describe the most common uses of capital letters.

Beginnings

1. The first word of a sentence or sentence fragment is capitalized.

The play lasted nearly three hours.

How are you feeling?

Bravo!

2. The first word of a sentence contained within parentheses is capitalized if it does not occur within another sentence. The first word of a parenthetical sentence within another sentence is not capitalized.

The discussion was held in the boardroom. (The results are still confidential.)

Although we liked the restaurant (their Italian food was the best), we seldom ate there.

After waiting in line for an hour (why do we do these things?), we finally left.

3. The first word of a direct quotation is capitalized. However, if the quotation is interrupted in the middle of a sentence, the second part does not begin with a capital. When a quotation, whether a sentence fragment or a complete sentence, is syntactically dependent on the sentence in which it occurs, the quotation does not begin with a capital.

The President said, "We have rejected this report entirely."

"We have rejected this report entirely," the President said, "and we will not comment on it further."

The President made it clear that "there is no room for compromise."

4. The first word of a sentence within a sentence is usually capitalized when it represents a direct question, a motto or aphorism, or spoken or unspoken dialogue. The first word following a colon may be either lowercased or capitalized if it introduces a complete sentence. While the former is more usual, the latter is common when the sentence is fairly lengthy and distinctly separate from the preceding clause.

The question, as Disraeli said, is this: Is man an ape or an angel?

My first thought was, How can I avoid this assignment?

The advantage of this particular system is clear: it's inexpensive.

The situation is critical: This company cannot hope to recoup the fourth-quarter losses that were sustained in five operating divisions.

5. The first word of a line of poetry is traditionally capitalized; however, in much twentieth-century poetry the line beginnings are lowercased.

The best lack all conviction, while the worst
Are full of passionate intensity.

—W. B. Yeats

6. The first words of run-in enumerations that form complete sentences are capitalized, as are usually the first words of vertical lists and enumerations. However, enumerations of words or phrases run in with the introductory text are generally lowercased.

Do the following tasks at the end of the day:
1. Clear your desktop of papers. 2. Cover office machines. 3. Straighten the contents of your desk drawers, cabinets, and bookcases.

This is the agenda:
Call to order
Roll call
Minutes of the previous meeting
Treasurer's report

On the agenda will be (1) call to order, (2) roll call, (3) minutes of the previous meeting, (4) treasurer's report. . . .

7. The first word in an outline heading is capitalized.

I. Editorial tasks
II. Production responsibilities
A. Cost estimates
B. Bids

8. The first word of the salutation of a letter and the first word of a complimentary close are capitalized.

Dear Mary,

Ladies and Gentlemen:

Sincerely yours,

Proper Nouns, Pronouns, and Adjectives

Capitals are used with almost all proper nouns—that is, nouns that name particular persons, places, or things (including abstract entities), distinguishing them from others of the same class—and proper adjectives—that is, adjectives that take their meaning from what is named by the proper noun. The essential distinction in the use of capitals and lowercase letters at the beginnings of words lies in this individualizing significance of capitals as against the generalizing significance of lowercase. The following subject headings are in alphabetical order.

ARMED FORCES

1. Branches and units of the armed forces are capitalized, as are easily recognized short forms of full branch and unit designations. However, the words *army, navy,* etc., are lowercased when used in their plural forms or when they are not part of an official title.

United States Army

a contract with the Army

Corps of Engineers

a bridge built by the Engineers

allied armies

AWARDS

2. Names of awards and prizes are capitalized.

the Nobel Prize in Chemistry

Distinguished Service Cross

Academy Award

DERIVATIVES OF PROPER NAMES

3. Derivatives of proper names are capitalized when used in their primary sense. However, if the derived term has taken on a specialized meaning, it is usually not capitalized.

Roman customs

Shakespearean comedies

Edwardian era

but

quixotic

herculean

bohemian tastes

GEOGRAPHICAL REFERENCES

4. Divisions of the earth's surface and names of distinct areas, regions, places, or districts are capitalized, as are most derivative adjectives and some derivative nouns and verbs.

The Eastern Hemisphere

Midwest

Tropic of Cancer

Springfield, Massachusetts

the Middle Eastern situation

an Americanism

but

french fries

a japan finish

manila envelope

5. Popular names of localities are capitalized.

the Corn Belt	the Loop
The Big Apple	the Gold Coast
the Pacific Rim	

6. Words designating global, national, regional, or local political divisions are capitalized when they are essential elements of specific names. However, they are usually lowercased when they precede a proper name or stand alone. (In legal documents, these words are often capitalized regardless of position.)

the British Empire	Washington State
New York City	Ward 1
but	
the fall of the empire	the state of Washington
the city of New York	fires in three wards

7. Generic geographical terms (such as *lake, mountain, river, valley*) are capitalized if they are part of a specific proper name.

Hudson Bay	Long Island
Niagara Falls	Crater Lake
the Shenandoah Valley	

8. Generic terms preceding names are usually capitalized.

Lakes Michigan and Superior

Mounts Whitney and Rainier

9. Generic terms following names are usually lowercased, as are singular or plural generic terms that are used descriptively or alone.

the Himalaya and Andes mountains

the Atlantic coast of Labrador

the Hudson valley

the river valley

the valley

10. Compass points are capitalized when they refer to a geographical region or when they are part of a street name, but they are lowercased when they refer to simple direction.

up North

back East

the Northwest

West Columbus Avenue

Park Avenue South

but

west of the Rockies

the east coast of Florida

11. Adjectives derived from compass points and nouns designating the inhabitants of some geographical regions are capitalized. When in doubt, see the entry in the dictionary.

a Southern accent

Northerners

12. Terms designating public places are capitalized if they are part of a proper name.

Brooklyn Bridge

Lincoln Park

the St. Regis Hotel

Independence Hall

but

Wisconsin and Connecticut avenues

the Plaza and St. Regis hotels

GOVERNMENTAL AND JUDICIAL BODIES

13. Full names of legislative, deliberative, executive, and administrative bodies are capitalized, as are short forms of these names. However, nonspecific noun and adjective references to them are usually lowercased.

the U.S. House of Representatives

the House

the Federal Bureau of Investigation

but

both houses of Congress

a federal agency

14. Names of international courts, the U.S. Supreme Court, and other higher courts are capitalized. However, names of city and county courts are usually lowercased.

The International Court of Arbitration

the Supreme Court of the United States

the Supreme Court

the United States Court of Appeals for the Second Circuit

the Michigan Court of Appeals

Lawton municipal court

Newark night court

HISTORICAL PERIODS AND EVENTS

15. Names of congresses, councils, and expositions are capitalized.

the Yalta Conference

the Republican National Convention

16. Names of historical events, some historical periods, and some cultural periods and movements are capitalized. When in doubt, consult the entry in the dictionary, especially for periods.

the Boston Tea Party

the Renaissance

Prohibition

the Augustan Age

the Enlightenment

but

the space age

neoclassicism

17. Numerical designations of historical time periods are capitalized when they are part of a proper name; otherwise they are lowercased.

the Third Reich

the Roaring Twenties

but

the eighteenth century

the eighties

18. Names of treaties, laws, and acts are capitalized.

Treaty of Versailles

The Clean Air Act of 1990

ORGANIZATIONS

19. Names of firms, corporations, schools, and organizations and their members are capitalized. However, common nouns occurring after the names of two or more organizations are lowercased. The word *the* at the beginning of such names is only capitalized when the full legal name is used.

Thunder's Mouth Press

University of Wisconsin

European Community

Rotary International

Kiwanians

American and United airlines

20. Words such as *group, division, department, office,* or *agency* that designate a corporate and organizational unit are capitalized only when used with its specific name.

in the Editorial Department of Merriam-Webster

but

a notice to all department heads

PEOPLE

21. Names of persons are capitalized. However, the capitalization of particles such as *de, della, der, du, l', la, ten,* and *van* varies widely, especially in names of people in English-speaking countries.

Noah Webster

W.E.B. Du Bois

Daphne du Maurier

Wernher von Braun

Anthony Van Dyck

22. Titles preceding the name of a person and epithets used instead of a name are capitalized. However, titles following a name or used alone are usually lowercased.

President Roosevelt

Professor Kaiser

Queen Elizabeth

Old Hickory

the Iron Chancellor

but

Henry VIII, king of England

23. Corporate titles are capitalized when used with an individual's name; otherwise, they are lowercased.

Lisa Dominguez, Vice President

The sales manager called me.

24. Words of family relationship preceding or used in place of a person's name are capitalized; however, these words are lowercased if they are part of a noun phrase used in place of a name.

Cousin Julia

I know when Mother's birthday is.

but

I know when my mother's birthday is.

25. Words designating peoples, nationalities, religious groups, tribes, races, and languages are capitalized. Other terms used to refer to groups of people are often lowercased. Designations based on color are usually lowercased.

Canadians	Iroquois
Ibo	African-American
Latin	Indo-European

highlander (an inhabitant of a highland)

Highlander (an inhabitant of the Highlands of Scotland)

black	white

PERSONIFICATIONS

26. Personifications are capitalized.

She dwells with Beauty—Beauty that must die;
And Joy, whose hand is ever at his lips
Bidding adieu.
—John Keats

obey the commands of Nature

PRONOUNS

27. The pronoun I is capitalized. For pronouns referring to the Deity, see rule 29 below.

. . . no one but I myself had yet printed any of my work.—Paul Bowles

RELIGIOUS TERMS

28. Words designating the Deity are capitalized.

An anthropomorphic, vengeful Jehovah became a spiritual, benevolent Supreme Being.—A.R. Katz

29. Personal pronouns referring to the Deity are usually capitalized, even when they closely follow their antecedent. However, many writers never capitalize such pronouns.

All Thy works, O Lord, shall bless Thee.
—*Oxford American Hymnal*

God's in his heaven—
All's right with the world!
—Robert Browning

30. Traditional designations of revered persons, such as prophets, apostles, and saints, are often capitalized.

our Lady

the Prophet

the Lawgiver

31. Names of religions, creeds and confessions, denominations, and religious orders are capitalized, as is the word *Church* when used as part of a proper name.

Judaism

Apostles' Creed

the Thirty-nine Articles of the Church of England

Society of Jesus

Hunt Memorial Church

but

the local Baptist church

32. Names for the Bible or parts, versions, or editions of it and names of other sacred books are capitalized but not italicized. Adjectives derived from the names of sacred books are irregularly capitalized or lowercased; when in doubt, see the entry in the dictionary.

Authorized Version	New English Bible
Old Testament	Pentateuch
Apocrypha	Gospel of Saint Mark
Talmud	Koran
biblical	Koranic

SCIENTIFIC TERMS

33. Names of planets and their satellites, asteroids, stars, constellations and groups of stars, and other unique celestial objects are capitalized. However, the words *sun, earth,* and *moon* are usually lowercased unless they occur with other astronomical names.

Venus	Ganymede
Sirius	the Pleiades

the Milky Way

enjoying the beauty of the moon

probes heading for the Moon and Mars

34. New Latin genus names in zoology and botany are capitalized; the second term in binomial scientific names, identifying the species, is not.

a cabbage butterfly (*Pieris rapae*)

a common buttercup (*Ranunculus acris*)

35. New Latin names of all groups above genus in zoology and botany (such as class or family) are capitalized; however, their derivative adjectives and nouns are not.

Gastropoda *but* gastropod

Mantidae *but* mantid

36. Names of geological eras, periods, epochs, and strata and names of prehistoric divisions are capitalized.

Silurian period	Pleistocene epoch
Age of Reptiles	Neolithic age

SEASONS, MONTHS, DAYS

37. Names of months, days of the week, and holidays and holy days are capitalized.

January	Ramadan
Tuesday	Thanksgiving
Yom Kippur	Easter

38. Names of seasons are not capitalized except when personified.

last spring

the sweet breath of Spring

TITLES OF PRINTED MATTER AND WORKS OF ART

39. Words in titles are capitalized, with the exception of internal conjunctions, prepositions, and articles. In some publications, prepositions of five or more letters are capitalized also.

Of Mice and Men

"The Man Who Would Be King"

"To His Coy Mistress"

Slouching Towards Bethlehem

40. Capitalization of the titles of movies, plays, paintings, sculpture, and musical compositions follow similar conventions. For more details, see the Italicization section below.

41. Major sections of books, long articles, or reports are capitalized when they are referred to within the same material.

See the Appendix for further information.

The Introduction explains the scope of this book.

discussed later in Chapter 4

42. Nouns used with numbers or letters to designate major reference headings are capitalized. Nouns designating minor elements are typically lowercased.

Volume V	Table 3
page 101	note 10

TRADEMARKS

43. Registered trademarks and service marks are capitalized.

Express Mail	Orlon
Kleenex	Walkman

VEHICLES

44. Names of ships, aircraft, and spacecraft are capitalized.

Titanic

Lindbergh's *Spirit of St. Louis*

Apollo 13

Italicization

The following are usually italicized in print and underlined in manuscript and typescript.

1. Words and passages that are to be emphasized.

   This was their fatal error: there *was* no cache of supplies in the now-abandoned depot.

2. Titles of books, magazines, newspapers, plays, long poems, movies, paintings, sculpture, and long musical compositions (but not musical compositions identified by the name of their genre).

   Dickens's *Bleak House*

   National Geographic

   Christian Science Monitor

   Shakespeare's *Othello*

   Eliot's *The Waste Land*

   the movie *Back to the Future*

   Gainsborough's *Blue Boy*

   Mozart's *Don Giovanni*

   but

   Schubert's Sonata in B-flat Major, D. 960

NOTE: In the plurals of such italicized titles, the *s* or *es* endings are usually in roman type.

   hidden under a stack of *New Yorker*s

3. Names of ships, aircraft, and spacecraft.

   Titanic

   Lindbergh's *Spirit of St. Louis*

   Apollo 13

4. Words, letters, and figures when referred to as such.

   The *g* in *align* is silent.

   The first *2* and the last *0* are barely legible.

5. Unfamiliar words when first introduced and defined in a text.

   Heart failure is often accompanied by *edema,* an accumulation of fluid which tends to produce swelling of the lower extremities.

6. Foreign words and phrases that have not been naturalized in English. In general, any word entered in the main A–Z vocabulary of this dictionary need not be italicized.

   c'est la vie

   aere perennius

   che sarà, sarà

   sans peur et sans reproche

   but

   pasta ad hoc ex officio

7. New Latin scientific names of genera, species, subspecies, races, and varieties (but not groups of higher rank, such as phyla, classes, or orders) in botanical or zoological names.

   a thick-shelled American clam *(Mercenaria mercenaria)*

   a mallard *(Anas platyrhynchos)*

   but

   the family Hominidae

8. Case titles in legal citations, both in full and shortened form; "v." for "versus" is set in either roman or italic.

   Jones v. *Ohio*

   Smith et al v. Jones

   the *Jones* case

   Jones

Documentation of Sources

Writers and editors use various methods to indicate the source of a quotation or piece of information borrowed from another work. In works published for the general public and traditionally in scholarly works in the humanities, footnotes or endnotes have been preferred. In this system, sequential numbers within the text refer the reader to notes at the bottom of the page or at the end of the article, chapter, or book; these notes contain full bibliographical information on the works cited. In scholarly works in the social and natural sciences, and increasingly in the humanities as well, parenthetical references within the text refer the reader to an alphabetically arranged list of sources at the end of the work. The system of footnotes or endnotes is the more flexible, in that it allows for commentary on the work or subject and can also be used for brief peripheral discussions not tied to any specific work. However, style manuals tend to encourage the use of parenthetical references in addition to or instead of footnotes or endnotes, since for most kinds of material they are efficient and convenient for both writer and reader. In a carefully documented work, a bibliography or list of sources normally follows the entire text (including any endnotes) regardless of which system is used.

Though different publishers and journals have adopted slightly varying styles, the following examples illustrate standard styles for references, notes, and bibliographic entries. For more extensive treatment than can be provided here, *Merriam-Webster's Manual for Writers and Editors, The Chicago Manual of Style, The MLA Style Manual,* or the *Publication Manual of the American Psychological Association* may be consulted.

Footnotes and Endnotes

Footnotes and endnotes are indicated by superscript Arabic numerals placed immediately after the material to be documented. The numbering is consecutive throughout an article or monograph; in a book, it usually starts over with each new chapter or section. Footnotes appear at the bottom of the page; endnotes, which take the same form as footnotes, are gathered at the end of the article, chapter, or book. Endnotes are generally preferred over footnotes by writers and publishers because they are easier to handle when preparing both manuscript and printed pages, though they can be less convenient for the reader. All of the examples shown reflect humanities citation style. All of the cited works appear again in the Lists of Sources section below.

Books

One author

[1]Elizabeth Bishop, *The Complete Poems: 1927–1979* (New York: Farrar, Straus & Giroux, 1983), 46.

Two or more authors

[2]Bert Holldobler and Edward O. Wilson, *The Ants* (Cambridge, Mass.: Belknap–Harvard Univ. Press, 1990), 119.

[3]Randolph Quirk et al., *A Comprehensive Grammar of the English Language* (London: Longman, 1985), 135.

Edition and/or translation

[4]Arthur S. Banks, ed., *Political Handbook of the World: 1992* (Binghamton, N.Y.: CSA Publications, 1992), 293–95.

[5]Simone de Beauvoir, *The Second Sex,* trans. and ed. H. M. Parshley (New York: Knopf, 1953; Random House, 1974), 446.

Second or later edition

[6]Albert C. Baugh and Thomas Cable, *A History of the English Language,* 3d ed. (Englewood Cliffs, N.J.: Prentice Hall, 1978), 14.

Article in a collection or festschrift

[7]Ernst Mayr, "Processes of Speciation in Animals," in *Mechanisms of Speciation,* ed. C. Barigozzi (New York: Alan R. Liss, 1982), 1–3.

Work in two or more volumes

[8]Ronald M. Nowak, *Walker's Mammals of the World,* 5th ed. (Baltimore: Johns Hopkins Univ. Press, 1991), 2:661.

Corporate author

[9]Commission on the Humanities, *The Humanities in American Life* (Berkeley: Univ. of California Press, 1980), 46.

Book lacking publication data

[10]*Photographic View Album of Cambridge* [England], n.p., n.d., n.pag.

Subsequent reference

[11]Baugh and Cable, 18–19.

Articles

Journal paginated consecutively throughout annual volume

[12]Stephen Jay Gould and Niles Eldredge, "Punctuated Equilibria: The Tempo and Mode of Evolution Reconsidered," *Paleobiology* 3 (1977): 121.

Journal paginated consecutively only within each issue

[13]Roseann Duenas Gonzalez, "Teaching Mexican American Students to Write: Capitalizing on the Culture," *English Journal* 71.7 (Nov. 1982): 22–24.

Monthly magazine

[14]John Lukacs, "The End of the Twentieth Century," *Harper's,* Jan. 1993: 40.

Weekly magazine

[15]Richard Preston, "A Reporter at Large: Crisis in the Hot Zone," *New Yorker,* 26 Oct. 1992: 58.

Newspaper

[16]William J. Broad, "Big Science Squeezes Small-Scale Researchers," *New York Times,* 29 Dec. 1992: C1.

Signed review

[17]George Steiner, review of *Oeuvres en Prose Complètes, Tome 3,* by Charles Péguy, *Times Literary Supplement,* 25 Dec. 1992: 3.

Parenthetical References

Parenthetical references are highly abbreviated bibliographical citations that appear within the text itself, enclosed in parentheses. Such references direct the reader to a detailed bibliography or list of sources at the end of the work, often removing the need for footnotes or endnotes. The parenthetical references usually include only the author's last name and a page reference. (In the social and natural sciences, the year of publication is included after the author's name, and the page number is often omitted.) Any element of the reference that is clear from the context may be omitted. To distinguish among cited works published by the same author, the author's name may be followed by the specific work's title, which is usually shortened. (If the author-date system is being used, a lowercase letter can be added after the year—e.g., 1992a, 1992b—to distinguish between works published in the same year.) Each of the following references is keyed to an entry in the Lists of Sources section below.

Humanities style

(Quirk et al., 135)
(Baugh and Cable, *History,* 14)
(Commission on the Humanities, 46)

Sciences style

(Mayr 1982, 1–3)
(Nowak 1991, 2:661)
(Gould and Eldredge 1977)

Lists of Sources

A bibliography or list of sources in alphabetical order usually appears at the end of the work. The following lists of cited works illustrate standard styles employed in, respectively, the humanities and the social and natural sciences. The principal differences between the two styles are these. In the sciences, (1) an initial is generally used instead of the author's first name, (2) the date is placed directly after the author's name, (3) all words in titles are lowercased except the first word and the first word of any subtitle as well as proper nouns and adjectives, and (4) article titles are not set off by quotation marks. (In some scientific publications, book and journal titles are not italicized.)

Humanities style

Baugh, Albert C., and Thomas Cable. *A History of the English Language.* 3d ed. Englewood Cliffs, N.J.: Prentice Hall, 1978.

Beauvoir, Simone de. *The Second Sex.* Trans. and ed. H. M. Parshley. New York: Alfred A. Knopf, 1953. Reprint. New York: Random House, 1974.

Bishop, Elizabeth. *The Complete Poems: 1927–1979.* New York: Farrar, Straus & Giroux, 1983.

Commission on the Humanities. *The Humanities in American Life.* Berkeley: University of California Press, 1980.

Gonzalez, Roseann Duenas. "Teaching Mexican American Students to Write: Capitalizing on the Culture." *English Journal* 71.7 (November 1982): 22–24.

Lukacs, John. "The End of the Twentieth Century." *Harper's,* January 1993: 39–58.

Photographic View Album of Cambridge [England]. N.d., n.p., n. pag.

Quirk, Randolph, Sidney Greenbaum, Geoffrey Leech, and Jan Svartvik. *A Comprehensive Grammar of the English Language.* London: Longman, 1985.

Steiner, George. Review of *Oeuvres en Prose Complètes, Tome 3,* by Charles Péguy. *Times Literary Supplement,* 25 December 1992: 3–4.

Sciences style

Banks, A. S., ed. 1992. *Political handbook of the world: 1992.* Binghamton, N.Y.: CSA Publications.

Broad, W. J. 1992. Big science squeezes small-scale researchers. *New York Times,* 29 Dec.: C1+.

Gould, S. J., and N. Eldredge. 1977. Punctuated equilibria: The tempo and mode of evolution reconsidered. *Paleobiology* 3: 115–151.

Holldobler, B., and E. O. Wilson. 1990. *The ants.* Cambridge, Mass.: Belknap–Harvard Univ. Press.

Mayr, E. 1982. Processes of speciation in animals. In C. Barigozzi, ed., *Mechanisms of speciation.* New York: Alan R. Liss: 1–19.

Nowak, R.M. 1991. *Walker's mammals of the world.* 5th ed. 2 vols. Baltimore: Johns Hopkins Univ. Press.

Preston, R. 1992. A reporter at large: Crisis in the hot zone. *New Yorker,* 26 Oct.: 58–81.